FORMULAS/EQUATIONS

Distance Formula

If $P_1 = (x_1, y_1)$ and $P_2 = (x_2, y_2)$, the distance from P_1 to P_2 is

$$d(P_1, P_2) = \sqrt{(x_2 - x_1)^2 + (y_2 - y_1)^2}$$

Standard Equation of a Circle

The standard equation of a circle of radius r with center at (h, k) is

$$(x - h)^2 + (y - k)^2 = r^2$$

Slope Formula

The slope m of the line containing the points $P_1 = (x_1, y_1)$ and $P_2 = (x_2, y_2)$ is

$$m = \frac{y_2 - y_1}{x_2 - x_1} \quad \text{if } x_1 \neq x_2$$

$$m \text{ is undefined} \quad \text{if } x_1 = x_2$$

Point-Slope Equation of a Line

The equation of a line with slope m containing the point (x_1, y_1) is

$$y - y_1 = m(x - x_1)$$

Slope-Intercept Equation of a Line

The equation of a line with slope m and y-intercept b is

$$y = mx + b$$

Quadratic Formula

The solutions of the equation $ax^2 + bx + c = 0, a \neq 0$, are

$$x = \frac{-b \pm \sqrt{b^2 - 4ac}}{2a}$$

If $b^2 - 4ac > 0$, there are two unequal real solutions.
If $b^2 - 4ac = 0$, there is a repeated real solution.
If $b^2 - 4ac < 0$, there are two complex solutions that are not real.

GEOMETRY FORMULAS

Circle

r = Radius, A = Area, C = Circumference
$A = \pi r^2 \quad C = 2\pi r$

Triangle

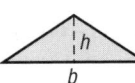

b = Base, h = Altitude (Height), A = area
$A = \frac{1}{2}bh$

Rectangle

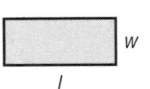

l = Length, w = Width, A = area, P = perimeter
$A = lw \quad P = 2l + 2w$

Rectangular Box

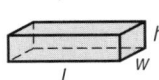

l = Length, w = Width, h = Height, V = Volume
$V = lwh$

Sphere

r = Radius, V = Volume, S = Surface area
$V = \frac{4}{3}\pi r^3 \quad S = 4\pi r^2$

Right Circular Cylinder

r = Radius, h = Height, V = Volume
$V = \pi r^2 h$

TRIGONOMETRY

Enhanced with Graphing Utilities

Third Edition

TRIGONOMETRY

THIRD
EDITION

Enhanced with Graphing Utilities

Michael Sullivan
Chicago State University

Michael Sullivan, III
Joliet Junior College

Prentice Hall
Upper Saddle River, New Jersey 07458

Library of Congress Cataloging-in-Publication Data

Sullivan, Michael
   Trigonometry: enhanced with graphing utilities / Michael Sullivan, Michael Sullivan III.—3rd ed.
      p. cm.
   Includes index.
   ISBN 0-13-065914-2
   1. Trigonometry. 2. Graphic calculators. I. Sullivan, Michael II. Title.

QA531 .S86 2003
516.24—dc21   2002021835

Editor-in-Chief: Sally Yagan
Acquisitions Editor: Eric Frank
Associate Editor: Dawn Murrin
Vice President/Director of Production and Manufacturing: David W. Riccardi
Executive Managing Editor: Kathleen Schiaparelli
Senior Managing Editor: Linda Mihatov Behrens
Production Editor: Bob Walters
Manufacturing Buyer: Alan Fischer
Manufacturing Manager: Trudy Pisciotti
Executive Marketing Manager: Patrice Lumumba Jones
Marketing Assistant: Rachel Beckman
Assistant Managing Editor, Math Media Production: John Matthews
Editorial Assistant/Supplements Editor: Aja Shevelew
Art Director: Kenny Beck
Interior Designer: Lee Goldstein
Cover Designer: Tom Nery
Creative Director: Carole Anson
Art Editor: Thomas Benfatti
Director of Creative Services: Paul Belfanti
Photo Editor: Beth Boyd
Cover Photo: Galen Rowell/CORBIS
Art Studio: Artworks:
          Senior Manager: Patricia Burns
          Production Manager: Ronda Whitson
          Manager, Production Technologies: Matt Haas
          Project Coordinator: Jessica Einsig
          Illustrators: Kathryn Anderson, Mark Landis
          Art Quality Assurance: Timothy Nguyen, Stacy Smith, Pamela Taylor

© 2003, 2000, 1996 by Prentice-Hall, Inc.
Upper Saddle River, New Jersey 07458

Printed in the United States of America

10 9 8 7 6 5 4 3 2 1

ISBN: 0-13-065914-2

Pearson Education Ltd., *London*
Pearson Education Australia Pty. Ltd., *Sydney*
Pearson Education Singapore, Pte. Ltd.
Pearson Education North Asia Ltd., *Hong Kong*
Pearson Education Canada, Inc., *Toronto*
Pearson Educacíon de Mexico, S.A. de C.V.
Pearson Education—Japan, *Tokyo*
Pearson Education Malaysia, Pte. Ltd

For Our Students...
Past and Present

CONTENTS

As professors at both an urban public university and a community college, Michael Sullivan and Michael Sullivan III are aware of the varied needs of trigonometry students. As a teacher, and as an author of engineering calculus, finite mathematics, and business calculus texts, Michael understands what students must know if they are to be focused and successful in upper level mathematics courses. As a father of four, including the co-author, he also understands the realities of college life. His co-author and son, Michael III, believes passionately in the value of technology as a tool for learning that enhances understanding without sacrificing important skills.

Together, Michael and Michael III have taken great pains to ensure that this text contains solid, student-friendly examples and exercises, as well as a clear, seamless writing style. Please share with them your experiences teaching from this text.

The Third Edition

The Third Edition builds upon a strong foundation by integrating new features and techniques that further enhance student interest and involvement. The elements of previous editions that have proved successful remain, while many changes, some obvious, others subtle, have been made. One important benefit of authoring a successful series is the broad-based feedback upon which improvements and additions are ultimately based. Virtually every change to this edition is the result of thoughtful comments and suggestions from colleagues and students who used previous editions. We are sincerely grateful for this feedback and have tried to make changes that improve the usefulness of the text for both instructors and students.

New to the Third Edition

Preparing for This Section

Most sections now open with a referenced list (by section and page number) of key items to review in preparation for the section ahead. This provides a just-in-time review for students.

Objectives

Each section also contains a numbered list of learning objectives. As the learning objective is addressed in the text, its number will appear.

Concepts and Vocabulary

At the end of every section, there is a short list of Fill-in-the-Blank and True/False items that test concepts and vocabulary in a short answer format. Several quick-answer questions are also included.

Cumulative Reviews

At the end of Chapters 2–6, exercises are provided that require skills learned in the earlier chapters. These cumulative reviews serve to continually reinforce the important concepts of trigonometry. They also make it easier for the student to prepare for a comprehensive final examination.

Content

- The formula for the area of a sector and related exercises are now part of Section 2.1 Angles and Their Measure.
- Combining Waves is a new subsection in Section 4.5 Simple Harmonic Motion; Damped Motion; Combining Waves.
- The Cross Product is a new section in Chapter 5 Polar Coordinates; Vectors.
- Section 7.1, Exponential Functions, now contains a subsection on exponential equations; Section 7.2, Logarithmic Functions, now contains a subsection on logarithmic equations.
- The Appendix has been expanded to include more review material appropriate to trigonometry.
- New Chapter Projects have been added that discuss topics of current interest.

Organization

- Scatter Diagrams, formerly found in Section 1.1, has been relocated to the Appendix as part of Section A.8. This change positions the content to where it is being used.
- Graphs of the Trigonometric Functions and Sinusoidal Graphs; Sinusoidal Curve Fitting, formerly Sections 2.6 and 2.7, now is covered in three sections, 2.6, 2.7, and 2.8. This change makes it possible to teach the sections in one period each.
- The Inverse Trigonometric Functions, formerly Section 3.5, now is covered in two sections at the beginning of the chapter. This change makes it possible to teach the sections in one period each. It also places the content closer to the discussion of the trigonometric functions and their graphs.

Features in the 3rd Edition

- Section **OBJECTIVES** appear in a numbered list to begin each section.
- NOW WORK PROBLEM XX. appears after a concept has been introduced. This directs the student to a problem in the exercises that tests the concept, insuring that the concept has been mastered before moving on. The Now Work problems are identified in the exercises using orange numbers and a pencil icon .
- References to Calculus are identified by a calculus icon.
- Discussion, Writing, and Research problems appear in most exercise sets, identified by an icon and red numbers. These problems provide a basis for class discussion, writing projects, and library projects.
- Historical Perspectives, sometimes with exercises, are presented in context, and provide interesting anecdotal information.
- Varied applications and real-world, sourced data are abundant in Examples and Exercises.
- Concepts and Vocabulary, a short list of Fill-in-the-Blank, True/False, and open-ended questions that test concepts and vocabulary in a quick-answer format, are given at the end of every section.
- An extensive Chapter Review provides a list of important formulas and key definitions and theorems. The objectives of the chapter are listed by section, with page references and review exercises that relate to the

objective. The authors' suggestions for a practice test are indicated by a blue number in the review exercise set.

- Chapter Projects that are relevant and current, many based on newspaper articles, appear at the end of each chapter. These can serve as the basis for collaborative learning experiences.
- Cumulative Reviews appear at the end of Chapters 2–6. These problem sets serve to continually reinforce skills from earlier chapters.

Using the 3ʳᵈ Edition Effectively and Efficiently with Your Syllabus.

To meet the varied needs of diverse syllabi, this book contains more content than expected in a trigonometry course. The illustration shows the dependencies of chapters on each other.

As the chart indicates, this book has been organized with flexibility of use in mind. Even within a given chapter, certain sections are optional and can be skipped without fear of future problems.

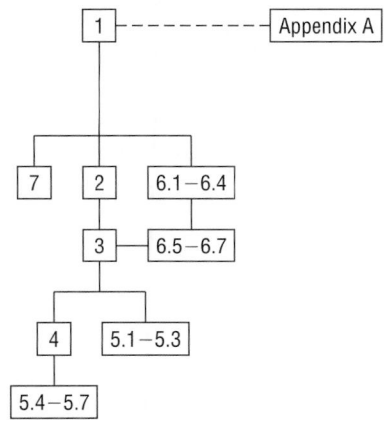

Chapter 1 Functions and Their Graphs

A quick coverage of this chapter, which is mainly review material, will enable you to get to Chapter 2 Trigonometric Functions earlier.

Chapter 2 Trigonometric Functions

The sections follow in sequence. Section 2.8 is optional.

Chapter 3 Analytic Trigonometry

The sections follow in sequence. Sections 3.2, 3.6, and 3.8 may be skipped in a brief course.

Chapter 4 Applications of Trigonometry

The sections follow in sequence. Sections 4.4 and 4.5 are optional.

Chapter 5 Polar Coordinates; Vectors

Sections 5.1–5.3 and Sections 5.4–5.7 are independent and may be covered separately.

Chapter 6 Analytic Geometry

Sections 6.1–6.4 follow in sequence. Sections 6.5, 6.6, and 6.7 are independent of each other, but do depend on sections 6.1–6.4.

Chapter 7 Exponential and Logarithmic Functions

Sections 7.1–7.4 follow in sequence. Sections 7.5, 7.6, and 7.7 each require Section 7.3.

Appendix Review

This consists of review material, which can be used as the first part of a course in trigonometry or as a just-in-time review. Specific references to this material occur throughout the text to assist in the review process.

Acknowledgments

Textbooks are written by authors, but evolve from an idea into final form through the efforts of many people. Special thanks to Don Dellen, who first suggested this book and the other books in this series. Don's extensive contributions to publishing and mathematics are well known; we all miss him dearly.

There are many colleagues we would like to thank for their input, encouragement, patience, and support. They have our deepest thanks and appreciation. We apologize for any omissions.

James Africh, *College of DuPage*
Steve Agronsky, *Cal Poly State University*
Grant Alexander, *Joliet Junior College*
Dave Anderson, *South Suburban College*
Joby Milo Anthony, *University of Central Florida*
James E. Arnold, *University of Wisconsin-Milwaukee*
Carolyn Autray, *University of West Georgia*
Agnes Azzolino, *Middlesex County College*
Wilson P Banks, *Illinois State University*
Sudeshna Basu, *Howard University*
Dale R. Bedgood, *East Texas State University*
Beth Beno, *South Suburban College*
Carolyn Bernath, *Tallahassee Community College*
William H. Beyer, *University of Akron*
Annette Blackwelder, *Florida State University*
Richelle Blair, *Lakeland Community College*
Trudy Bratten, *Grossmont College*
Joanne Brunner, *Joliet Junior College*
Warren Burch, *Brevard Community College*
Mary Butler, *Lincoln Public Schools*
William J. Cable, *University of Wisconsin-Stevens Point*
Lois Calamia, *Brookdale Community College*
Jim Campbell, *Lincoln Public Schools*
Roger Carlsen, *Moraine Valley Community College*
Elena Catoiu, *Joliet Junior College*
John Collado, *South Suburban College*
Nelson Collins, *Joliet Junior College*
Jim Cooper, *Joliet Junior College*
Denise Corbett, *East Carolina University*
Theodore C. Coskey, *South Seattle Community College*
John Davenport, *East Texas State University*
Faye Dang, *Joliet Junior College*
Antonio David, *Del Mar College*
Duane E. Deal, *Ball State University*
Timothy Deis, *University of Wisconsin-Platteville*
Vivian Dennis, *Eastfield College*
Guesna Dohrman, *Tallahassee Community College*
Karen R. Dougan, *University of Florida*
Louise Dyson, *Clark College*
Paul D. East, *Lexington Community College*
Don Edmondson, *University of Texas-Austin*
Erica Egizio, *Joliet Junior College*
Christopher Ennis, *University of Minnesota*
Ralph Esparza, Jr., *Richland College*
Garret J. Etgen, *University of Houston*
Pete Falzone, *Pensacola Junior College*
W.A. Ferguson, *University of Illinois-Urbana/Champaign*
Iris B. Fetta, *Clemson University*
Mason Flake, *student at Edison Community College*

Timothy W. Flood, *Pittsburg State University*
Merle Friel, *Humboldt State University*
Richard A. Fritz, *Moraine Valley Community College*
Carolyn Funk, *South Suburban College*
Dewey Furness, *Ricke College*
Dawit Getachew, *Chicago State University*
Wayne Gibson, *Rancho Santiago College*
Robert Gill, *University of Minnesota Duluth*
Sudhir Kumar Goel, *Valdosta State University*
Joan Goliday, *Sante Fe Community College*
Frederic Gooding, *Goucher College*
Sue Graupner, *Lincoln Public Schools*
Jennifer L. Grimsley, *University of Charleston*
Ken Gurganus, *University of North Carolina*
James E. Hall, *University of Wisconsin-Madison*
Judy Hall, *West Virginia University*
Edward R. Hancock, *DeVry Institute of Technology*
Julia Hassett, *DeVry Institute-Dupage*
Michah Heibel, *Lincoln Public Schools*
LaRae Helliwell, *San Jose City College*
Brother Herron, *Brother Rice High School*
Robert Hoburg, *Western Connecticut State University*
Lee Hruby, *Naperville North High School*
Kim Hughes, *California State College-San Bernardino*
Ron Jamison, *Brigham Young University*
Richard A. Jensen, *Manatee Community College*
Sandra G. Johnson, *St. Cloud State University*
Tuesday Johnson, *New Mexico State University*
Moana H. Karsteter, *Tallahassee Community College*
Arthur Kaufman, *College of Staten Island*
Thomas Kearns, *North Kentucky University*
Shelia Kellenbarger, *Lincoln Public Schools*
Keith Kuchar, *Manatee Community College*
Tor Kwembe, *Chicago State University*
Linda J. Kyle, *Tarrant Country Jr. College*
H.E. Lacey, *Texas A & M University*
Harriet Lamm, *Coastal Bend College*
Matt Larson, *Lincoln Public Schools*
Christopher Lattin, *Oakton Community College*
Adele LeGere, *Oakton Community College*
Kevin Leith, *University of Houston*
Jeff Lewis, *Johnson County Community College*
Stanley Lukawecki, *Clemson University*
Janice C. Lyon, *Tallahassee Community College*
Virginia McCarthy, *Iowa State University*
Jean McArthur, *Joliet Junior College*
Tom McCollow, *DeVry Institute of Technology*

Laurence Maher, *North Texas State University*
Jay A. Malmstrom, *Oklahoma City Community College*
Sherry Martina, *Naperville North High School*
Alec Matheson, *Lamar University*
James Maxwell, *Oklahoma State University-Stillwater*
Judy Meckley, *Joliet Junior College*
David Meel, *Bowling Green State University*
Carolyn Meitler, *Concordia University*
Sarmia Metwali, *Erie Community College*
Rich Meyers, *Joliet Junior College*
Eldon Miller, *University of Mississippi*
James Miller, *West Virginia University*
Michael Miller, *Iowa State University*
Kathleen Miranda, *SUNY at Old Westbury*
Thomas Monaghan, *Naperville North High School*
Craig Morse, *Naperville North High School*
Samad Mortabit, *Metropolitan State University*
A. Muhundan, *Manatee Community College*
Jane Murphy, *Middlesex Community College*
Richard Nadel, *Florida International University*
Gabriel Nagy, *Kansas State University*
Bill Naegele, *South Suburban College*
Lawrence E. Newman, *Holyoke Community College*
James Nymann, *University of Texas-El Paso*
Sharon O'Donnell, *Chicago State University*
Seth F. Oppenheimer, *Mississippi State University*
Linda Padilla, *Joliet Junior College*
E. James Peake, *Iowa State University*
Kelly Pearson, *Murray State University*
Thomas Radin, *San Joaquin Delta College*
Ken A. Rager, *Metropolitan State College*
Kenneth D. Reeves, *San Antonio College*
Elsi Reinhardt, *Truckee Meadows Community College*
Jane Ringwald, *Iowa State University*
Stephen Rodi, *Austin Community College*
Bill Rogge, *Lincoln Public Schools*
Howard L. Rolf, *Baylor University*
Phoebe Rouse, *Lousiana State University*
Edward Rozema, *University of Tennessee at Chattanooga*
Dennis C. Runde, *Manatee Community College*
John Sanders, *Chicago State University*
Susan Sandmeyer, *Jamestown Community College*
A.K. Shamma, *University of West Florida*
Martin Sherry, *Lower Columbia College*
Tatrana Shubin, *San Jose State University*

Anita Sikes, *Delgado Community College*
Timothy Sipka, *Alma College*
Lori Smellegar, *Manatee Community College*
John Spellman, *Southwest Texas State University*
Rajalakshmi Sriram, *Okaloosa-Walton Community College*
Becky Stamper, *Western Kentucky University*
Judy Staver, *Florida Community College-South*
Neil Stephens, *Hinsdale South High School*

Christopher Terry, *Augusta State University*
Diane Tesar, *South Suburban College*
Tommy Thompson, *Brookhaven College*
Richard J. Tondra, *Iowa State University*
Marvel Townsend, *University of Florida*
Jim Trudnowski, *Carroll College*
Robert Tuskey, *Joliet Junior College*
Richard G. Vinson, *University of South Alabama*
Mary Voxman, *University of Idaho*
Jennifer Walsh, *Daytona Beach Community College*

Donna Wandke, *Naperville North High School*
Darlene Whitkenack, *Northern Illinois University*
Christine Wilson, *West Virginia University*
Brad Wind, *Florida International University*
Canton Woods, *Auburn University*
Tamara S. Worner, *Wayne State College*
Terri Wright, *New Hampshire Community Technical College, Manchester*
George Zazi, *Chicago State University*

Recognition and thanks are due particularly to the following individuals for their valuable assistance in the preparation of this edition:
- Sally Yagan, for her continued support and genuine concern;
- Patrice Jones, for his innovative marketing efforts;
- Eric Frank, for jumping right in and contributing as the new editor;
- Dawn Murrin, for her skill at getting all the supplements out on time;
- Aja Shevelew, for her attention to detail, particularly in getting timely reviews;
- Bob Walters, for his organizational skills as production manager;
- Teri Lovelace and Cindy Trimble of Laurel technical Services, for their dedication to accuracy in proofreading and checking our answers;
- And, most importantly, the entire Prentice-Hall sales staff, for their continuing confidence in this book.

Michael Sullivan

Michael Sullivan, III

As you begin your study of Trigonometry, you may feel overwhelmed by the number of theorems, definitions, procedures, and equations that confront you. You may even wonder whether or not you can learn all of this material in the time allotted. These concerns are normal. Keep in mind that the elements of Trigonometry are all around us as we go through our daily routines. Many of the concepts you will learn to express mathematically, you already know intuitively. For many of you, this may be your last math course, while for others, just the first in a series of many. Either way, this text was written with you in mind. We have spent countless hours teaching Trigonometry courses. We know what you're going through. You'll find that we have written a text that doesn't overwhelm, or unnecessarily complicate Trigonometry, but at the same time gives you the skills and practice you need to be successful.

This text is designed to help you, the student, master the terminology and basic concepts of Trigonometry. These aims have helped to shape every aspect of the book. Many learning aids are built into the format of the text to make your study of the material easier and more rewarding. This book is meant to be a "machine for learning," that can help you focus your efforts, ensuring that you get the most from the time and energy you invest.

How to Use This Book Effectively and Efficiently

First, and most important, this book is meant to be read—so please, begin by reading the material assigned. You will find that the text has additional explanation and examples that will help you. Also, it is best to read the section before the lecture, so you can ask questions right away about anything you didn't understand.

Many sections begin with "Preparing for This Section," a list of concepts that will be used in the section. Take the short amount of time required to refresh your memory. This will make the section easier to understand and will actually save you time and effort.

A list of **OBJECTIVES** is provided at the beginning of each section. Read them. They will help you recognize the important ideas and skills developed in the section.

After a concept has been introduced and an example given, you will see **NOW WORK PROBLEM XX**. Go to the exercises at the end of the section, work the problem cited, and check your answer in the back of the book. If you get it right, you can be confident in continuing on in the section. If you don't get it right, go back over the explanations and examples to see what you might have missed. Then rework the problem. Ask for help if you miss it again.

If you follow these practices throughout the section, you will find that you have probably done many of your homework problems. In the exercises, every "Now Work Problem" number is in orange with a pencil icon. All the odd-numbered problems have answers in the back of the book and worked-out solutions in the Student Solutions Manual supplement. Be sure you have made an honest effort before looking at a worked-out solution.

Spend the few minutes necessary to answer the "Concepts and Vocabulary" items. These are quick and valuable questions to answer.

At the end of each chapter is a Chapter Review. Use it to be sure you are completely familiar with the equations and formulas listed under "Things to Know." If you are unsure of an item here, use the page reference to go back and review it. Go through the Objectives and be sure you can answer "Yes" to the question "I should be able to . . . ." If you are uncertain, a page reference to the objective is provided.

Do the problems identified with blue numbers in the Review Exercises. These are our suggestions for a Practice Test. Do some of the other problems in the review for more practice to prepare for your exam.

Finally, beginning with Chapter 2, we have included Cumulative Review exercises. These exercises help to reinforce skills developed from previous chapters. By doing these exercises, you will keep the material presented in the course fresh. This will make studying for the final exam much easier.

Please do not hesitate to contact us, through Prentice Hall, with any suggestions or comments that would improve this text. We look forward to hearing from you.

Best Wishes!

Michael Sullivan

Michael Sullivan, III

OVERVIEW

CHAPTER 2

TRIGONOMETRIC FUNCTIONS

Tidal Coastline and Pots of Water

In Florida, they post times of tides coming in and going out very precisely, like 11:23 A.M. How can they be so precise?

There is more to tides, the rise and fall of ocean waters, than the gravitational pull of the moon and sun.

These are the chief factors, of course. And because the movement of the Earth, sun and moon in relationship to each other are known precisely, the rhythm of tides rising and falling along coastlines is easy to predict.

Yet the time and heights of high and low tide may vary along different stretches of the same coast though they're reacting to similar driving forces and pressures.

Historic observation makes possible the exact timing of high tides and low tides along a particular section of coast for a month, a year, or far into the future.

The reason for the difference is oscillation. Think of pots and pans filled with varying levels of water on a table, says Charles O'Reilly, Chief of Tidal Analysis for the Geological Survey of Canada's hydrographic service in Dartmouth, N.S. Then kick the table.

"You'll notice the water in the pots and pans will slosh differently. That's their natural oscillation," he said. "If you kick the table rhythmically, you'll find each pot continues to slosh differently because it has its own rhythm.

"Now, if you join those pots and pans together, that's sort of like a coastal ocean. They're all feeling the same 'kick,' but they are all responding differently. In order to predict a tide, you have to measure them for some period of time."

SOURCE: *Toronto Star*, June 13, 2001, p. GT02

SEE CHAPTER PROJECT 1.

OUTLINE

2.1 Angles and Their Measure
2.2 Right Triangle Trigonometry
2.3 Computing the Values of Trigonometric Functions of Given Angles
2.4 Trigonometric Functions of General Angles
2.5 ~~Properties of the Trigonometric Functions; Unit Circle Approach~~
~~...t, and Secant Functions~~

 For additional study help, go to
www.prenhall.com/sullivanegu3e

Materials include:
- Graphing Calculator Help
- Chapter Quiz
- Chapter Test
- PowerPoint Downloads
- Chapter Projects
- Student Tips

97

Page 97

Chapter Projects

1. Tides A partial tide table for September, 2001, for Sabine Pass along the Texas Gulf Coast is given in the table on page 195:

(a) On September 15, when was the tide high? This is called "high tide". On September 19, when was the tide low? This is called "low tide". Most days will have two low tides and two high tides.

(b) Why do you think there is a negative height for the low tide on September 14? What is the tide height measured against?

(c) On your graphing utility, draw a scatter diagram for the data in the table. Let T (time) be the independent variable, with $T = 0$ being 12:00 A.M. on September 1, $T = 24$ being 12:00 A.M. on September 2, etc. Remember that there are 60 minutes in an hour. Let H be the height in feet when converting the times. Also, make sure your graphing utility is in radian mode.

(d) What shape does the data take? What is the period of the data? What is the amplitude? Is the amplitude constant? Explain.

| Sept | High Tide | | High Tide | | Low Tide | | Low Tide | | Sun/Moon phase |
	Time	Ht(ft)	Time	Ht(ft)	Time	Ht(ft)	Time	Ht(ft)	Rise/Set
F 14	03:08a	2.4	11:12a	2.2	08:14a	2.0	07:19p	−0.1	7:00a/7:23p
S 15	03:33a	2.4	12:56p	2.2	08:15a	1.9	08:13p	0.0	7:00a/7:22p
S 16	03:57a	2.3	02:17p	2.3	08:45a	1.6	09:05p	0.3	7:01a/7:20p
M17	04:20a	2.2	03:33p	2.3	09:24a	1.4	09:54p	0.5	7:01a/7:19p
T18	04:41a	2.2	04:47p	2.3	10:08a	1.0	10:43p	1.0	7:02a/7:08p
W19	05:01a	2.0	06:04p	2.3	10:54a	0.7	11:32p	1.4	7:02a/7:17p
T20	05:20a	2.0	07:27p	2.3	11:44a	0.4			7:03a/7:15p

Date from: *www.harbortides.com*

Page 194-195

CLEAR WRITING STYLE

The Sullivans' accessible writing style is apparent throughout, often utilizing various approaches to the same concept. This clear writing style makes potentially difficult concepts intuitive and class time more productive.

Sometimes it is helpful to think of a function f as a machine that receives as input a number from the domain, manipulates it, and outputs the value. See Figure 46.

Figure 46

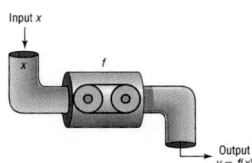

The restrictions on this input/output machine are

1. It only accepts numbers from the domain of the function.
2. For each input, there is exactly one output (which may be repeated for different inputs).

For a function $y = f(x)$, the variable x is called the **independent variable**, because it can be assigned any of the numbers from the domain. The variable y is called the **dependent variable**, because its value depends on x.

Any symbol can be used to represent the independent and dependent variables. For example, if f is the *cube function*, then f can be defined by $f(x) = x^3$ or $f(t) = t^3$ or $f(z) = z^3$. All three functions are the same: Each tells us to cube the independent variable. In practice, the symbols used for the independent and dependent variables are based on common usage, such as using C for cost in business.

Page 37

PREPARING FOR THIS SECTION

Before getting started, review the following:

✓ Unit Circle (Section 1.3, p. 27)   ✓ Functions (Section 1.4, pp. 33–41)

✓ Symmetry (Section 1.3, pp. 20–22)   ✓ Even and Odd Functions (Section 1.5, pp. 55–57)

2.5 PROPERTIES OF THE TRIGONOMETRIC FUNCTIONS; UNIT CIRCLE APPROACH

OBJECTIVES
1. Find the Exact Value of the Trigonometric Functions Using the Unit Circle
2. Know the Domain and Range of the Trigonometric Functions
3. Use the Periodic Properties to Find the Exact Value of the Trigonometric Functions
4. Use Even-Odd Properties to Find the Exact Value of the Trigonometric Functions

Page 142

PREPARING FOR THIS SECTION

The Preparing for this Section feature provides you and your instructor with a list of skills and concepts needed to approach the section, along with page references. You can use the feature to determine what you should know before tackling each section.

STEP-BY-STEP EXAMPLES

Step-by-step examples ensure that you follow the entire solution process and give you an opportunity to check your understanding of each step.

EXAMPLE 5   Finding a Sinusoidal Function for Hours of Daylight

According to the *Old Farmer's Almanac*, the number of hours of sunlight in Boston on the summer solstice is 15.283 and the number of hours of sunlight on the winter solstice is 9.067.

(a) Find a sinusoidal function of the form $y = A\sin(\omega x - \phi) + B$ that fits the data.*
(b) Use the function found in part (a) to predict the number of hours of sunlight on April 1, the 91st day of the year.
(c) Draw a graph of the function found in part (a).
(d) Look up the number of hours of sunlight for April 1 in the *Old Farmer's Almanac* and compare the actual hours of daylight to the results found in part (b).

Solution   (a) STEP 1: Amplitude $= \dfrac{\text{largest data value} - \text{smallest data value}}{2}$

$$= \frac{15.283 - 9.067}{2} = 3.108$$

STEP 2: Vertical shift $= \dfrac{\text{largest data value} + \text{smallest data value}}{2}$

$$= \frac{15.283 + 9.067}{2} = 12.175$$

STEP 3: The data repeat every 365 days. Since $T = \dfrac{2\pi}{\omega} = 365$, we find

$$\omega = \frac{2\pi}{365}$$

So far, we have $y = 3.108 \sin\left(\dfrac{2\pi}{365} x - \phi\right) + 12.175$.

Page 184

xix

REAL-WORLD DATA

Real-world data is incorporated into examples and exercise sets to emphasize that mathematics is a tool used to understand the world around us. As you use these problems and examples, you will see the relevance and utility of the skills being covered.

Finding Sinusoidal Functions from Data

3   Scatter diagrams of data sometimes take the form of a sinusoidal function. Let's look at an example.

The data given in Table 12 represent the average monthly temperatures in Denver, Colorado. Since the data represent average monthly temperatures collected over many years, the data will not vary much from year to year and so will essentially repeat each year. In other words, the data are periodic. Figure 112 shows the scatter diagram of these data repeated over two years, where $x = 1$ represents January, $x = 2$ represents February, and so on.

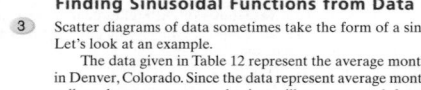

TABLE 12

Month, x	Average Monthly Temperature, °F
January, 1	29.7
February, 2	33.4
March, 3	39.0
April, 4	48.2
May, 5	57.2
June, 6	66.9
July, 7	73.5
August, 8	71.4
September, 9	62.3
October, 10	51.4
November, 11	39.0
December, 12	31.0

Source: U.S. National Oceanic and Atmospheric Administration

Figure 112

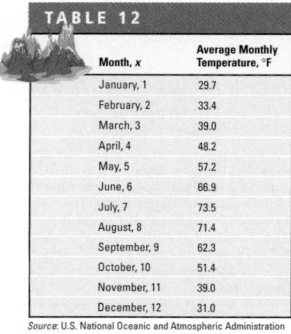

Page 181

Solution   Figure 18(a) shows the satellite dish. We draw the parabola used to form the dish on a rectangular coordinate system so that the vertex of the parabola is at the origin and its focus is on the positive y-axis. See Figure 18(b).

Figure 18

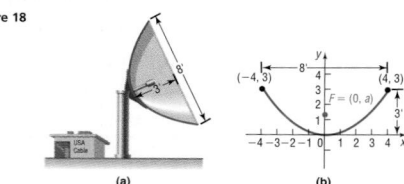

(a)   (b)

The form of the equation of the parabola is

$$x^2 = 4ay$$

and its focus is at $(0, a)$. Since $(4, 3)$ is a point on the graph, we have

$$4^2 = 4a(3)$$

$$a = \frac{4}{3}$$

The receiver should be located $1\frac{1}{3}$ feet from the base of the dish, along its axis of symmetry. ∎

━━━ NOW WORK PROBLEM 57.

Page 407

"NOW WORK" PROBLEMS

Many examples end with **"Now Work" Problems.** The problems suggested here are similar to the corresponding examples and provide a great way to check your understanding as you work through the chapter. The solutions to all "Now Work" problems can be found in the back of the text as well as in the *Student Solutions Manual.*

57. **Satellite Dish**   A satellite dish is shaped like a paraboloid of revolution. The signals that emanate from a satellite strike the surface of the dish and are reflected to a single point, where the receiver is located. If the dish is 10 feet across at its opening and is 4 feet deep at its center, at what position should the receiver be placed?

Page 409

PROCEDURES

Alternative **procedures** are clearly expressed throughout the text.

Finding the Values of the Trigonometric Functions When One Is Known

Given the value of one trigonometric function of an acute angle θ, the exact value of each of the remaining five trigonometric functions of θ can be found in either of two ways.

Method 1 Using the Definition

STEP 1: Draw a right triangle showing the acute angle θ.

STEP 2: Two of the sides can then be assigned values based on the given trigonometric function.

STEP 3: Find the length of the third side by using the Pythagorean Theorem.

STEP 4: Use the definitions in equation (1) to find the value of each of the remaining trigonometric functions.

Method 2 Using Identities

Use appropriately selected identities to find the value of each of the remaining trigonometric functions.

Page 117

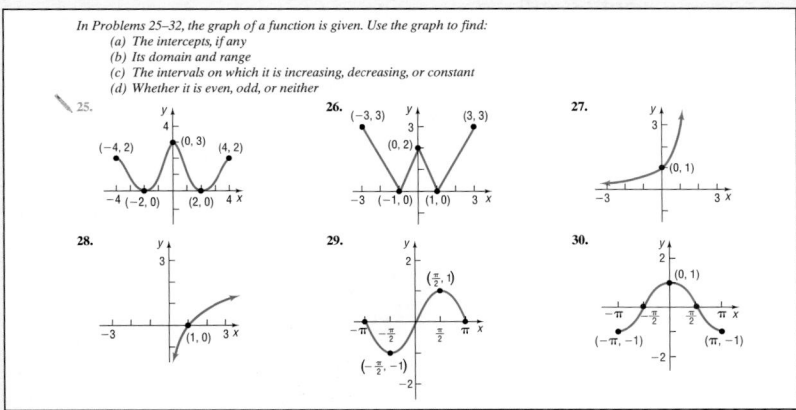

In Problems 25–32, the graph of a function is given. Use the graph to find:
(a) The intercepts, if any
(b) Its domain and range
(c) The intervals on which it is increasing, decreasing, or constant
(d) Whether it is even, odd, or neither

Page 62

END-OF-SECTION EXERCISES

The Sullivans' exercises are unparalleled in terms of thorough coverage and accuracy. Each **end-of-section exercise** set begins with visual and concept based problems, starting you out with the basics of the section. Well-thought-out exercises better prepare you for exams.

ALGEBRAIC AND GRAPHING SOLUTIONS

When appropriate, examples are solved using both an **algebraic and graphing approach**. The graphing solution will appear after the algebraic solution to reinforce the concept that graphing utilities often provide approximate, rather than exact, solutions. In this way, the utility is used to support the algebra.

---

④ EXAMPLE 6   Rate of Interest Required to Double an Investment

What annual rate of interest compounded annually should you seek if you want to double your investment in 5 years?

Algebraic Solution   If P is the principal and we want P to double, the amount A will be $2P$. We use the compound interest formula with $n = 1$ and $t = 5$ to find r.

$$2P = P(1 + r)^5$$
$$2 = (1 + r)^5$$
$$1 + r = \sqrt[5]{2}$$
$$r = \sqrt[5]{2} - 1 = 1.148698 - 1 = 0.148698$$

The annual rate of interest needed to double the principal in 5 years is 14.87%.

Graphing Solution   We solve the equation

$$2 = (1 + r)^5$$

for r by graphing the two functions $Y_1 = 2$ and $Y_2 = (1 + x)^5$. The x-coordinate of their point of intersection is the rate r that we seek. See Figure 28.

Figure 28

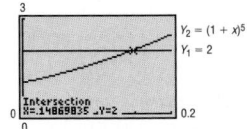

Pages 523

xxi

1. **Biology** A strain of E-coli Beu 397-recA441 is placed into a petri dish at 30° Celsius and allowed to grow. The following data are collected. Theory states that the number of bacteria in the petri dish will initially grow according to the law of uninhibited growth. The population is measured using an optical device in which the amount of light that passes through the petri dish is measured.

Time (hours), x	Population, y
0	0.09
2.5	0.18
3.5	0.26
4.5	0.35
6	0.50

Source: Dr. Polly Lavery, Joliet Junior College

(a) Draw a scatter diagram treating time as the predictor variable.
(b) Using a graphing utility, fit an exponential function to the data.
(c) Express the function found in part (b) in the form $N(t) = N_0 e^{kt}$.
(d) Graph the exponential function found in part (b) or (c) on the scatter diagram.
(e) Use the exponential function from part (b) or (c) to predict the population at $x = 7$ hours.
(f) Use the exponential function from part (b) or (c) to predict when the population will reach 0.75.

Page 544

MODELING

Many examples and exercises connect real-world situations to mathematical concepts. Learning to work with **models** is a skill that transfers to many disciplines.

GRAPHING UTILITIES AND TECHNIQUES

Increase your understanding, visualize, discover, explore, and solve problems using a **graphing utility**. Sullivan uses the graphing utility to further your understanding of concepts, not to circumvent essential math skills.

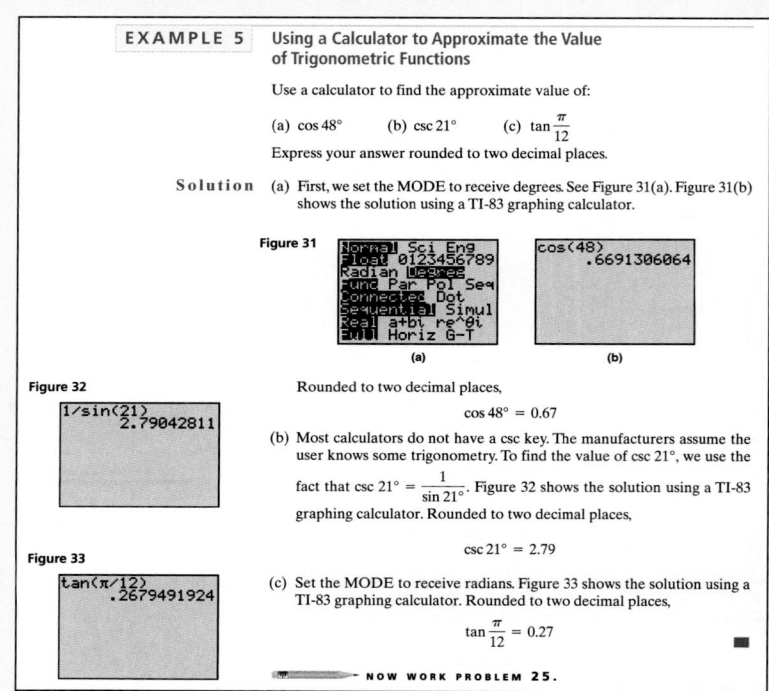

EXAMPLE 5 Using a Calculator to Approximate the Value of Trigonometric Functions

Use a calculator to find the approximate value of:

(a) $\cos 48°$ (b) $\csc 21°$ (c) $\tan \dfrac{\pi}{12}$

Express your answer rounded to two decimal places.

Solution (a) First, we set the MODE to receive degrees. See Figure 31(a). Figure 31(b) shows the solution using a TI-83 graphing calculator.

Rounded to two decimal places,
$$\cos 48° = 0.67$$

(b) Most calculators do not have a csc key. The manufacturers assume the user knows some trigonometry. To find the value of $\csc 21°$, we use the fact that $\csc 21° = \dfrac{1}{\sin 21°}$. Figure 32 shows the solution using a TI-83 graphing calculator. Rounded to two decimal places,
$$\csc 21° = 2.79$$

(c) Set the MODE to receive radians. Figure 33 shows the solution using a TI-83 graphing calculator. Rounded to two decimal places,
$$\tan \dfrac{\pi}{12} = 0.27$$

— NOW WORK PROBLEM 25.

Page 127

DISCUSSION WRITING AND READING PROBLEMS

These **problems** are designed to get you to "think outside the box," therefore fostering an intuitive understanding of key mathematical concepts. In this example, matching the graph to the functions ensures that you understand functions at a fundamental level.

79. Match each of the following functions with the graphs that best describe the situation.
(a) The cost of building a house as a function of its square footage
(b) The height of an egg dropped from a 300-foot building as a function of time
(c) The height of a human as a function of time
(d) The demand for Big Macs as a function of price
(e) The height of a child on a swing as a function of time

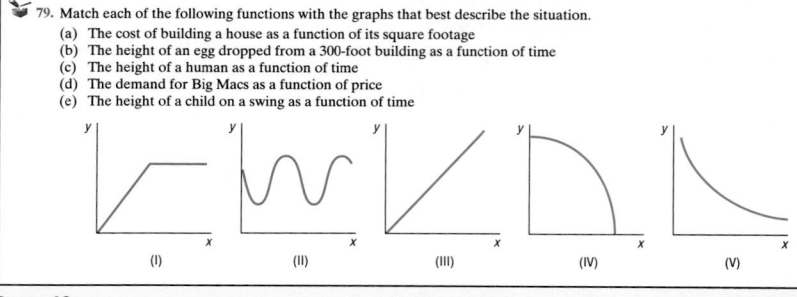

Page 49

Library of Functions

Linear function (p. 57)
$f(x) = mx + b$
Graph is a line with slope m and y-intercept b (see Figure 61)

Constant function (p. 58)
$f(x) = b$
Graph is a horizontal line with y-intercept b (see Figure 62)

Identity function (p. 58)
$f(x) = x$
Graph is a line with slope 1 and y-intercept 0 (see Figure 63)

Square function (p. 58)
$f(x) = x^2$
Graph is a parabola with vertex at $(0,0)$ (see Figure 64)

Cube function (p. 59)
$f(x) = x^3$ See Figure 65.

Square root function (p. 59)
$f(x) = \sqrt{x}$ See Figure 66.

Reciprocal function (p. 59)
$f(x) = \dfrac{1}{x}$ See Figure 67.

Absolute value function (p. 60)
$f(x) = |x|$ See Figure 68.

Things to Know

Distance formula (p. 5)
$$d = \sqrt{(x_2 - x_1)^2 + (y_2 - y_1)^2}$$

Midpoint formula (p. 8)
$$M = (x, y) = \left(\frac{x_1 + x_2}{2}, \frac{y_1 + y_2}{2} \right)$$

Standard form of the equation of a circle (p. 26)
$(x - h)^2 + (y - k)^2 = r^2$
(h, k) is the center; r is the radius

Function (p. 34)
A relation between two sets of real numbers so that each number x in the first set, the domain, has corresponding to it exactly one number y in the second set. The range is the set of y values of the function for the x values in the domain.
A function can also be characterized as a set of order pairs (x, y) or $(x, f(x))$ in which no two ordered pairs have the same first element, but different second elements.
A function $f(x)$ may be defined implicitly by an equation involving x and y or explicitly by writing $y =$

Function notation (p. 36)
$y = f(x)$
f is a symbol for the functi
x is the argument, or indep

Pages 90-92

CHAPTER REVIEW

The **Chapter Review** helps check your understanding of the chapter materials in several ways. **"Things to Know"** gives a general overview of review topics. The **"Objectives"** section lists each skill you are expected to have mastered along with page references and review exercises to test your skills. The **"Review Exercises"** then serve as a chance to practice the concepts presented within the chapter. The review materials are designed to make you, the student, confident in knowing the chapter material.

Objectives

Section		You should be able to:	Review Exercises
1.1	1	Use the distance formula (p. 5)	1, 2, 89, 90, 91, 92
	2	Use the midpoint formula (p. 7)	1, 2, 91
1.2	1	Graph equations by plotting points (p. 12)	3–8
	2	Graph equations using a graphing utility (p. 14)	3–8
	3	Find intercepts (p. 17)	3–8

Review Exercises

Blue problem numbers indicate the authors' suggestions for use in a Practice Test.

1. What is the distance between the points $(7, 4)$ and $(-3, 2)$? What is their midpoint?

2. What is the distance between the points $(2, 5)$ and $(6, -3)$? What is their midpoint?

In Problems 3–8, graph each equation by plotting points. Be sure to label any intercepts. Verify your results using a graphing utility.

3. $2x - 3y = 6$

4. $2x + 5y = 10$

5. $y = x^2 - 9$

6. $y = x^2 + 4$

7. $x^2 + 2y = 16$

8. $2x^2 - 4y = 24$

1. Find all the solutions of the equation $\sin(2\theta) = 0.5$.
2. Find a polar equation for the line containing the origin that makes an angle of $30°$ with the positive x-axis.
3. Find a polar equation for the circle with center at the point $(0, 4)$ and radius 4. Graph this circle.
4. What is the domain of the function $f(x) = \dfrac{3}{\sin x + \cos x}$?
5. Find the exact value of $\sin\left(\cos^{-1}\dfrac{3}{5} + \tan^{-1}1 \right)$
6. Find the exact value of $\sin\dfrac{\pi}{12}$ using a half-angle-formula.
   Find the exact value using a difference formlua
7. Graph $y = -2\sin(\pi x)$.
8. Graph $y = \cos(2x + 1) - 3$.
9. Solve the equation $\cot(2\theta) = 1$, where $0° < \theta < 90°$.
10. Find an equation for each of the following graphs:

(a) Line:
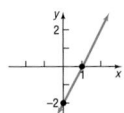

(b) Circle:
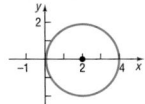

(c) Ellipse:
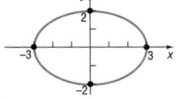

(d) Parabola:
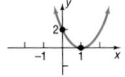

(e) Hyperbola:
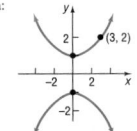

CUMULATIVE REVIEW

Chapters 2 to 6 each end with a **cumulative review**. These reviews serve to continually reinforce the important concepts of trigonometry.

Page 474

Sullivan M@thP@k
An Integrated Learning Environment

Today's textbooks offer a wide variety of ancillary materials to students, from solutions manuals to tutorial software to text-specific Websites. Making the most of all of these resources can be difficult. Sullivan **M@thP@k** helps students get it together. **M@thP@k** for Algebra and Trigonometry can be customized to fit Trigonometry courses.

MathPro 5

MathPro 5 is online, customizable tutorial software integrated with the text at the Learning Objective level for anytime, anywhere learning. MathPro5's "watch" feature integrates lecture videos into the algorithmic tutorial environment. The easy-to-use course management system enables instructors to track and assess student performance on tutorial work, quizzes and tests. A robust reports wizard provides a grade book, individual student reports and class summaries. The customizable syllabus allows instructors to remove and reorganize chapters, sections and objectives. MathPro 5's messaging system enhances communication between students and instructors.

The combination of MathPro5's richly integrated tutorial, testing and robust course management tools provides an unparalleled tutorial experience for students, and new assessment and time-saving tools for instructors.

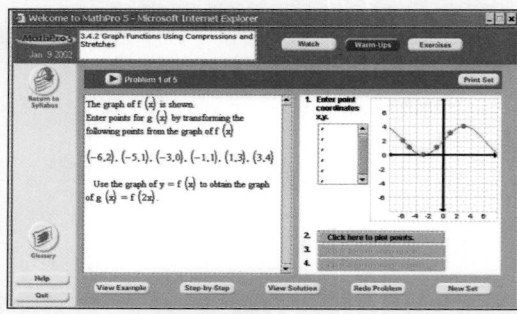

The Sullivan M@thP@k Website

This robust pass-code protected site features quizzes, homework starters, live animated examples, graphing calculator manuals, and much more. It offers the student many, many ways to test and reinforce their understanding of the course material.

Student Solutions Manual

Written by Mark McCombs, University of North Carolina, Chapel Hill and a long term user of the Sullivan series, the *Student Solutions Manual* offers thorough, accurate solutions that are consistent with the precise mathematics found in the text.

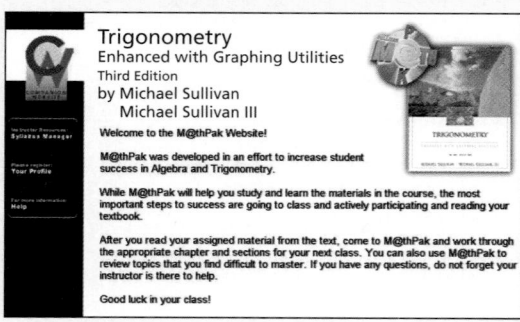

Sullivan M@thP@k.
Helping students *Get it Together.*

Additional Media

Sullivan Companion Website

www.prenhall.com/sullivanegu3e
This text-specific website beautifully complements the text. Here
students can find chapter tests, section-specific links, and
PowerPoint downloads in addition to other helpful features.

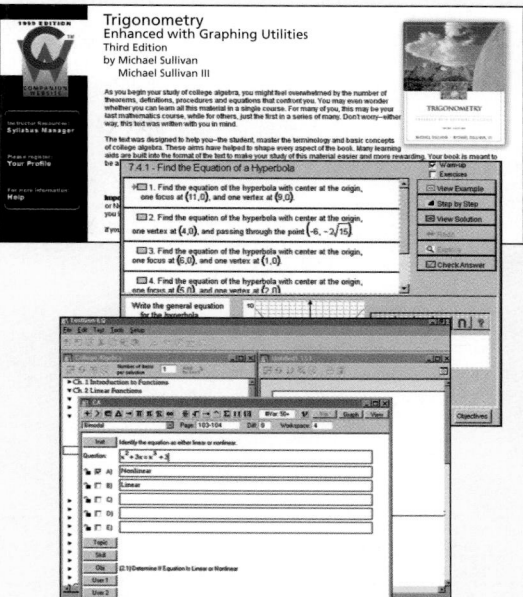

TestGen-EQ

CD-ROM (Windows/Macintosh)

- Algorithmically driven, text-specific testing program
- Networkable for administering tests and capturing grades
- Edit existing test items or add your own questions to create a
  nearly unlimited number of tests and drill worksheets

ISBN: 0-13-044964-4

S U P P L E M E N T S

Student Supplements

Student Solutions Manual
Worked solutions to all odd-numbered exercises from
the text and complete solutions for chapter review
problems and chapter tests. ISBN: 0-13-044958-X

New York Times Themes of the Times
A *free* newspaper from Prentice Hall and
The New York Times. Interesting and current articles on
mathematics which invite discussion and writing about
mathematics.

Mathematics on the Internet
Free guide providing a brief history of the Internet,
discussing the use of the World Wide Web, and
describing how to find your way within the Internet
and how to find others on it.

Prentice Hall Math Tutor Center
The PH Math Tutor Center provides tutoring for students
enrolled in developmental and precalculus mathematics
using selected Prentice Hall titles. Registration is
required; once registered, students can receive help via
toll free phone, fax, and email, only during the hours of
operation. (Sunday through Thursday, 5:00 PM – 12:00
AM EST) ISBN: 0-13-064604-0

Instructor Supplements

Instructor's Resource Manual
Contains complete step-by-step worked-out
solutions to all exercises in the textbook.
ISBN: 0-13-044961-X

Test Item File
Hard copy of the algorithmic computerized testing
materials. ISBN: 0-13-044963-6

PHOTO AND ILLUSTRATION CREDITS

CHAPTER 1	Pages 1 and 95, Jajlah Feany/Stock Boston
CHAPTER 2	Pages 97 and 194, Ned Maines/Photo Researchers, Inc.
CHAPTER 3	Pages 197 and 264, Steve Starr/Stock Boston
CHAPTER 4	Pages 267 and 314, Photo Researchers, Inc.
CHAPTER 5	Pages 317 and 394, Art Matrix/Visuals Unlimited; Page 343, CORBIS; Page 352, CORBIS; Page 363, Library of Congress
CHAPTER 6	Pages 397 and 473, CORBIS
CHAPTER 7	Pages 475 and 553, Amy C. Etra/PhotoEdit; Page 509, The Granger Collection

TRIGONOMETRY

Enhanced with Graphing Utilities

Third Edition

FUNCTIONS AND THEIR GRAPHS

The Use of Statistics in Making Investment Decisions: Using Beta

One of the most common statistics used by investors is beta, which measures a security or portfolio's movement relative to the S&P 500. When determining beta graphically, returns are plotted on a graph, with the horizontal axis representing the S&P returns and the vertical axis representing the security or portfolio returns; the points that are plotted represent the return of the security versus the return of the S&P 500 at the same time period. A regression line is then drawn, which is a straight line that most closely approximates the points in the graph. Beta represents the slope of that line; a steep slope (45 degrees or greater) indicates that when the stock market moved up or down, the security or portfolio moved up or down on average to the same degree or greater, and a flatter slope indicates that when the stock market moved up or down, the security or portfolio moved to a lesser degree. Since beta measures variability, it is used as a measure of risk; the greater the beta, the greater the risk.

SOURCE: Paul E. Hoffman, *Journal of the American Association of Individual Investors*, September, 1991; http://www.aaii.com

SEE CHAPTER PROJECT 1.

 For additional
study help, go to
www.prenhall.com/sullivanegu3e

Materials include:

- Graphing Calculator Help
- Chapter Quiz
- Chapter Test
- PowerPoint Downloads
- Chapter Projects
- Student Tips

A Look Back, A Look Forward

You have already had courses in algebra and geometry. In this chapter, we make the connection between algebra and geometry through the rectangular coordinate system. The idea of using a system of rectangular coordinates dates back to ancient times, when such a system was used for surveying and city planning. Sporadic use of rectangular coordinates continued until the 1600s. By that time, algebra had developed sufficiently so that René Descartes (1596–1650) and Pierre de Fermat (1601–1665) could take the crucial step, which was the use of rectangular coordinates to translate geometry problems into algebra problems, and vice versa. With the advent of technology, in particular graphing utilities, we are now able to solve many problems that required advanced methods before this technology. We will also develop techniques for graphing equations involving two variables. Then we will look at a special type of equation involving two variables, called a *function*. We will discuss what a function is, properties of functions, and techniques for graphing functions.

PREPARING FOR THIS BOOK

Before getting started, read the Preface to the Student (p. xvi).

PREPARING FOR THIS SECTION

Before getting started, review the following:

✓ Algebra Review (Section A.1, pp. 555–563)   ✓ Geometry Review (Section A.2, pp. 565–568)

1.1 RECTANGULAR COORDINATES; GRAPHING UTILITIES

OBJECTIVES   **1**   Use the Distance Formula

         2   Use the Midpoint Formula

Figure 1

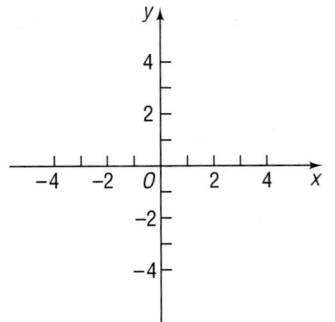

We locate a point on the real number line by assigning it a single real number, called the *coordinate of the point*. For work in a two-dimensional plane, we locate points by using two numbers.

     We begin with two real number lines located in the same plane: one horizontal and the other vertical. We call the horizontal line the **x-axis**, the vertical line the **y-axis**, and the point of intersection the **origin O**. We assign coordinates to every point on these number lines as shown in Figure 1, using a convenient scale. In mathematics, we usually use the same scale on each axis; in applications, a different scale is often used on each axis.

     The origin O has a value of 0 on both the x-axis and the y-axis. We follow the usual convention that points on the x-axis to the right of O are associated with positive real numbers, and those to the left of O are associated with negative real numbers. Points on the y-axis above O are associated with positive real numbers, and those below O are associated with negative real numbers. In Figure 1, the x-axis and y-axis are labeled as x and y, respectively, and we have used an arrow at the end of each axis to denote the positive direction.

Figure 2

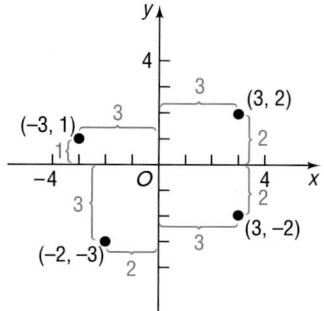

Figure 3

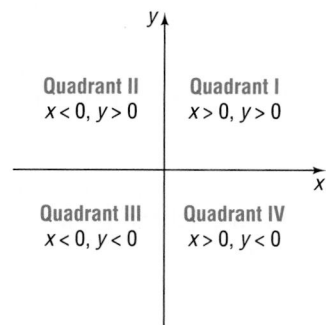

Figure 4
$y = 2x$

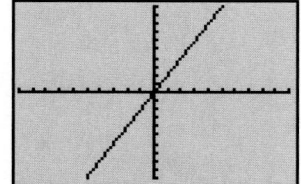

Figure 5

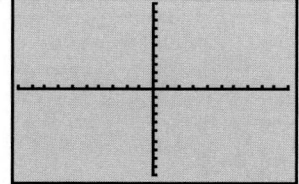

The coordinate system described here is called a **rectangular** or **Cartesian*** coordinate system. The plane formed by the x-axis and y-axis is sometimes called the **xy-plane**, and the x-axis and y-axis are referred to as the **coordinate axes**.

Any point P in the xy-plane can then be located by using an **ordered pair** (x, y) of real numbers. Let x denote the signed distance of P from the y-axis (*signed* in the sense that, if P is to the right of the y-axis, then $x > 0$, and if P is to the left of the y-axis, then $x < 0$); and let y denote the signed distance of P from the x-axis. The ordered pair (x, y), also called the **coordinates** of P, then gives us enough information to locate the point P in the plane.

For example, to locate the point whose coordinates are $(-3, 1)$, go 3 units along the x-axis to the left of O and then go straight up 1 unit. We **plot** this point by placing a dot at this location. See Figure 2, in which the points with coordinates $(-3, 1)$, $(-2, -3)$, $(3, -2)$, and $(3, 2)$ are plotted.

The origin has coordinates $(0, 0)$. Any point on the x-axis has coordinates of the form $(x, 0)$, and any point on the y-axis has coordinates of the form $(0, y)$.

If (x, y) are the coordinates of a point P, then x is called the **x-coordinate**, or **abscissa**, of P and y is the **y-coordinate**, or **ordinate**, of P. We identify the point P by its coordinates (x, y) by writing $P = (x, y)$. Usually, we will simply say "the point (x, y)" rather than "the point whose coordinates are (x, y)."

The coordinate axes divide the xy-plane into four sections called **quadrants**, as shown in Figure 3. In quadrant I, both the x-coordinate and the y-coordinate of all points are positive; in quadrant II, x is negative and y is positive; in quadrant III, both x and y are negative; and in quadrant IV, x is positive and y is negative. Points on the coordinate axes belong to no quadrant.

┈┈┈┈ **NOW WORK PROBLEM 1.**

Graphing Utilities

All graphing utilities, that is, all graphing calculators and all computer software graphing packages, graph equations by plotting points on a screen. The screen itself actually consists of small rectangles, called **pixels**. The more pixels the screen has, the better the resolution. Most graphing calculators have 48 pixels per square inch; most computer screens have 32 to 108 pixels per square inch. When a point to be plotted lies inside a pixel, the pixel is turned on (lights up). The graph of an equation is a collection of lighted pixels. Figure 4 shows how the graph of $y = 2x$ looks on a TI-83 graphing calculator.

The screen of a graphing utility will display the coordinate axes of a rectangular coordinate system. However, you must set the scale on each axis. You must also include the smallest and largest values of x and y that you want included in the graph. This is called **setting the viewing rectangle** or **viewing window**. Figure 5 illustrates a typical viewing window.

To select the viewing window, we must give values to the following expressions:

Xmin:	the smallest value of x
Xmax:	the largest value of x
Xscl:	the number of units per tick mark on the x-axis
Ymin:	the smallest value of y
Ymax:	the largest value of y
Yscl:	the number of units per tick mark on the y-axis

*Named after René Descartes (1596–1650), a French mathematician, philosopher, and theologian.

Figure 6 illustrates these settings and their relation to the Cartesian coordinate system.

Figure 6

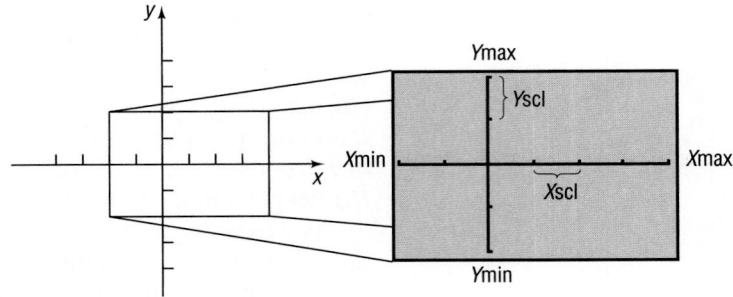

If the scale used on each axis is known, we can determine the minimum and maximum values of x and y shown on the screen by counting the tick marks. Look again at Figure 5. For a scale of 1 on each axis, the minimum and maximum values of x are -10 and 10, respectively; the minimum and maximum values of y are also -10 and 10. If the scale is 2 on each axis, then the minimum and maximum values of x are -20 and 20, respectively; and the minimum and maximum values of y are -20 and 20, respectively.

Conversely, if we know the minimum and maximum values of x and y, we can determine the scales being used by counting the tick marks displayed. We shall follow the practice of showing the minimum and maximum values of x and y in our illustrations so that you will know how the window was set. See Figure 7.

Figure 7

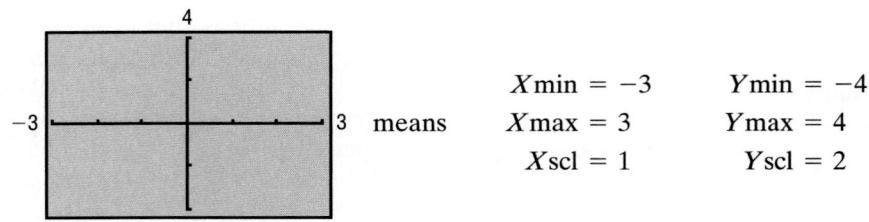

means

$X\text{min} = -3$ $Y\text{min} = -4$
$X\text{max} = 3$ $Y\text{max} = 4$
$X\text{scl} = 1$ $Y\text{scl} = 2$

EXAMPLE 1   **Finding the Coordinates of a Point Shown on a Graphing Utility Screen**

Find the coordinates of the point shown in Figure 8. Assume that the coordinates are integers.

Figure 8

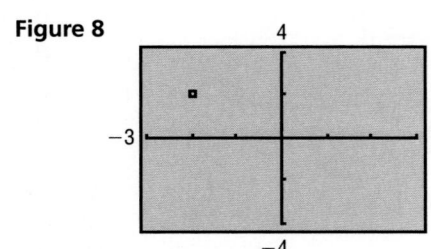

Solution First we note that the viewing window used in Figure 8 is

$$X\text{min} = -3 \qquad\qquad Y\text{min} = -4$$
$$X\text{max} = 3 \qquad\qquad Y\text{max} = 4$$
$$X\text{scl} = 1 \qquad\qquad Y\text{scl} = 2$$

The point shown is 2 tick units to the left on the horizontal axis (scale = 1) and 1 tick up on the vertical scale (scale = 2). Thus, the coordinates of the point shown are $(-2, 2)$. ■

 **NOW WORK PROBLEMS 5 AND 15.**

Distance between Points

① If the same units of measurement, such as inches, centimeters, and so on, are used for both the x-axis and the y-axis, then all distances in the xy-plane can be measured using this unit of measurement.

EXAMPLE 2 | **Finding the Distance between Two Points**

Find the distance d between the points $(1, 3)$ and $(5, 6)$.

Solution First we plot the points $(1, 3)$ and $(5, 6)$ and connect them with a straight line. See Figure 9(a). We are looking for the length d. We begin by drawing a horizontal line from $(1, 3)$ to $(5, 3)$ and a vertical line from $(5, 3)$ to $(5, 6)$, forming a right triangle, as in Figure 9(b). One leg of the triangle is of length 4 (since $|5 - 1| = 4$) and the other is of length 3 (since $|6 - 3| = 3$). By the Pythagorean Theorem, the square of the distance d that we seek is

$$d^2 = 4^2 + 3^2 = 16 + 9 = 25$$
$$d = \sqrt{25} = 5$$

Figure 9

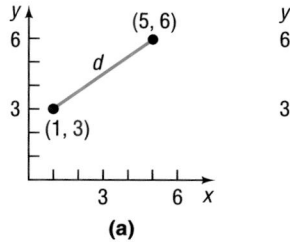
(a)

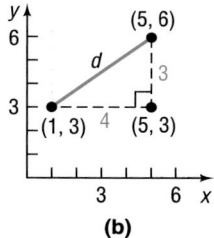
(b)

■

The **distance formula** provides a straightforward method for computing the distance between two points.

Figure 10

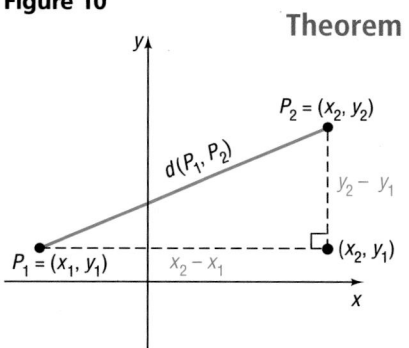

Theorem | **Distance Formula**

The distance between two points $P_1 = (x_1, y_1)$ and $P_2 = (x_2, y_2)$, denoted by $d(P_1, P_2)$, is

$$d(P_1, P_2) = \sqrt{(x_2 - x_1)^2 + (y_2 - y_1)^2} \qquad (1)$$

■

That is, to compute the distance between two points, find the difference of the x-coordinates, square it, and add this to the square of the difference of the y-coordinates. The square root of this sum is the distance. See Figure 10.

Proof of the Distance Formula   Let (x_1, y_1) denote the coordinates of point P_1, and let (x_2, y_2) denote the coordinates of point P_2. Assume that the line joining P_1 and P_2 is neither horizontal nor vertical. Refer to Figure 11(a). The coordinates of P_3 are (x_2, y_1). The horizontal distance from P_1 to P_3 is the absolute value of the difference of the x-coordinates, $|x_2 - x_1|$. The vertical distance from P_3 to P_2 is the absolute value of the difference of the y-coordinates, $|y_2 - y_1|$. See Figure 11(b). The distance $d(P_1, P_2)$ that we seek is the length of the hypotenuse of the right triangle, so, by the Pythagorean Theorem, it follows that

$$[d(P_1, P_2)]^2 = |x_2 - x_1|^2 + |y_2 - y_1|^2$$
$$= (x_2 - x_1)^2 + (y_2 - y_1)^2$$
$$d(P_1, P_2) = \sqrt{(x_2 - x_1)^2 + (y_2 - y_1)^2}$$

Figure 11

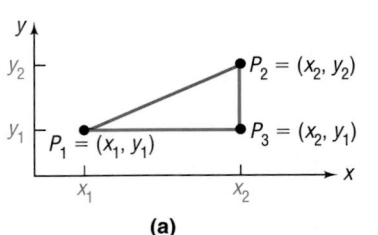
(a)

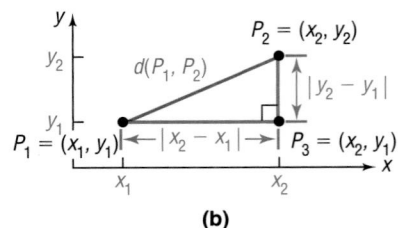
(b)

Now, if the line joining P_1 and P_2 is horizontal, then the y-coordinate of P_1 equals the y-coordinate of P_2; that is, $y_1 = y_2$. Refer to Figure 12(a). In this case, the distance formula (1) still works, because, for $y_1 = y_2$, it reduces to

$$d(P_1, P_2) = \sqrt{(x_2 - x_1)^2 + 0^2} = \sqrt{(x_2 - x_1)^2} = |x_2 - x_1|$$

Figure 12

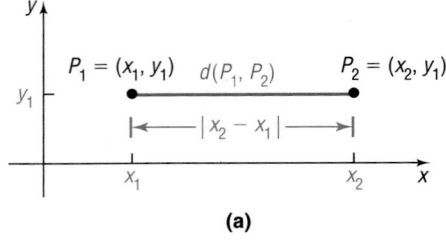
(a)

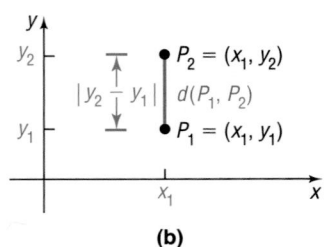
(b)

A similar argument holds if the line joining P_1 and P_2 is vertical. See Figure 12(b).

The distance formula is valid in all cases.   ■

EXAMPLE 3   **Finding the Length of a Line Segment**

Figure 13

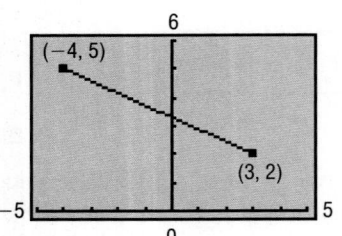

Find the length of the line segment shown in Figure 13.

Solution   The length of the line segment is the distance between the points $(-4, 5)$ and $(3, 2)$. Using the distance formula (1), the length d is

$$d = \sqrt{[3 - (-4)]^2 + (2 - 5)^2} = \sqrt{7^2 + (-3)^2}$$
$$= \sqrt{49 + 9} = \sqrt{58} \approx 7.62$$ ■

✎ ━ **NOW WORK PROBLEM 21.**

The distance between two points $P_1 = (x_1, y_1)$ and $P_2 = (x_2, y_2)$ is never a negative number. Furthermore, the distance between two points is 0 only when the points are identical, that is, when $x_1 = x_2$ and $y_1 = y_2$. Also, because $(x_2 - x_1)^2 = (x_1 - x_2)^2$ and $(y_2 - y_1)^2 = (y_1 - y_2)^2$, it makes no difference whether the distance is computed from P_1 to P_2 or from P_2 to P_1; that is, $d(P_1, P_2) = d(P_2, P_1)$.

The introduction to this chapter mentioned that rectangular coordinates enable us to translate geometry problems into algebra problems, and vice versa. The next example shows how algebra (the distance formula) can be used to solve geometry problems.

EXAMPLE 4   Using Algebra to Solve Geometry Problems

Consider the three points $A = (-2, 1)$, $B = (2, 3)$, and $C = (3, 1)$.

(a) Plot each point and form the triangle ABC.

(b) Find the length of each side of the triangle.

(c) Verify that the triangle is a right triangle.

(d) Find the area of the triangle.

Solution   (a) Points A, B, C, and triangle ABC are plotted in Figure 14.

Figure 14

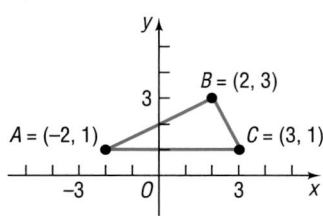

(b) $d(A, B) = \sqrt{[2 - (-2)]^2 + (3 - 1)^2} = \sqrt{16 + 4} = \sqrt{20} = 2\sqrt{5}$

$d(B, C) = \sqrt{(3 - 2)^2 + (1 - 3)^2} = \sqrt{1 + 4} = \sqrt{5}$

$d(A, C) = \sqrt{[3 - (-2)]^2 + (1 - 1)^2} = \sqrt{25 + 0} = 5$

(c) To show that the triangle is a right triangle, we need to show that the sum of the squares of the lengths of two of the sides equals the square of the length of the third side. (Why is this sufficient?) Looking at Figure 14 it seems reasonable to conjecture that the right angle is at vertex B. We shall check to see whether

$$[d(A, B)]^2 + [d(B, C)]^2 = [d(A, C)]^2$$

We find that

$$[d(A, B)]^2 + [d(B, C)]^2 = (2\sqrt{5})^2 + (\sqrt{5})^2$$
$$= 20 + 5 = 25 = [d(A, C)]^2$$

so it follows from the converse of the Pythagorean Theorem that triangle ABC is a right triangle.

(d) Because the right angle is at vertex B, the sides AB and BC form the base and altitude of the triangle. Its area is

$$\text{Area} = \frac{1}{2}(\text{Base})(\text{Altitude}) = \frac{1}{2}(2\sqrt{5})(\sqrt{5}) = 5 \text{ square units} \quad \blacksquare$$

 NOW WORK PROBLEM 39.

Midpoint Formula

(2)   We now derive a formula for the coordinates of the **midpoint of a line segment**. Let $P_1 = (x_1, y_1)$ and $P_2 = (x_2, y_2)$ be the endpoints of a line segment, and let $M = (x, y)$ be the point on the line segment that is the same

Figure 15

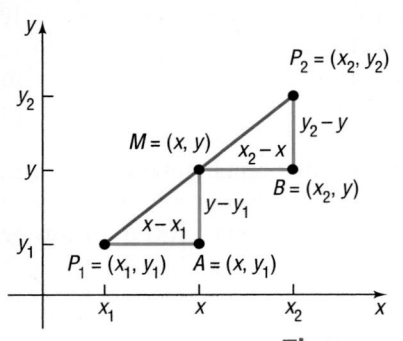

distance from P_1 as it is from P_2. See Figure 15. The triangles P_1AM and MBP_2 are congruent.* [Do you see why? Angle AP_1M = angle BMP_2,[†] angle P_1MA = angle MP_2B, and $d(P_1, M) = d(M, P_2)$ is given. So, we have angle–side–angle.] Hence, corresponding sides are equal in length. That is,

$$x - x_1 = x_2 - x \quad \text{and} \quad y - y_1 = y_2 - y$$
$$2x = x_1 + x_2 \qquad\qquad 2y = y_1 + y_2$$
$$x = \frac{x_1 + x_2}{2} \qquad\qquad y = \frac{y_1 + y_2}{2}$$

Theorem

Midpoint Formula

The midpoint $M = (x, y)$ of the line segment from $P_1 = (x_1, y_1)$ to $P_2 = (x_2, y_2)$ is

$$M = (x, y) = \left(\frac{x_1 + x_2}{2}, \frac{y_1 + y_2}{2} \right) \qquad (2)$$

To find the midpoint of a line segment, we average the x-coordinates and the y-coordinates of the endpoints.

EXAMPLE 5 **Finding the Midpoint of a Line Segment**

Find the midpoint of a line segment from $P_1 = (-5, 5)$ to $P_2 = (3, 1)$. Plot the points P_1 and P_2 and their midpoint. Check your answer.

Solution We apply the midpoint formula (2) using $x_1 = -5$, $y_1 = 5$, $x_2 = 3$, and $y_2 = 1$. Then the coordinates (x, y) of the midpoint M are

$$x = \frac{x_1 + x_2}{2} = \frac{-5 + 3}{2} = -1 \quad \text{and} \quad y = \frac{y_1 + y_2}{2} = \frac{5 + 1}{2} = 3$$

Figure 16

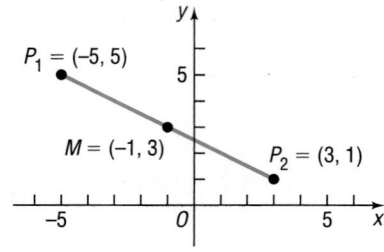

That is, $M = (-1, 3)$. See Figure 16.

✔ CHECK: Because M is the midpoint, we check the answer by verifying that $d(P_1, M) = d(M, P_2)$:

$$d(P_1, M) = \sqrt{[-1 - (-5)]^2 + (3 - 5)^2} = \sqrt{16 + 4} = \sqrt{20}$$
$$d(M, P_2) = \sqrt{[3 - (-1)]^2 + (1 - 3)^2} = \sqrt{16 + 4} = \sqrt{20}$$

NOW WORK PROBLEM **49.**

---

*The following statement is a postulate from geometry. Two triangles are congruent if their sides are the same length (SSS), or if two sides and the included angle are the same (SAS), or if two angles and the included side are the same (ASA).

[†]Another postulate from geometry states that the transversal $\overline{P_1P_2}$ forms equal corresponding angles with the parallel line segments $\overline{P_1A}$ and $\overline{MB}$.

1.1 Concepts and Vocabulary

In Problems 1–3, fill in the blanks.

1. If (x, y) are the coordinates of a point P in the xy-plane, then x is called the _____ of P and y is the _____ of P.

2. The coordinate axes divide the xy-plane into four sections called _____.

3. If three distinct points P, Q, and R all lie on a line and if $d(P, Q) = d(Q, R)$ then Q is called the _____ of the line segment from P to R.

In Problems 4–6, answer True or False to each statement.

4. The distance between two points is sometimes a negative number.

5. The point $(-1, 4)$ lies in quadrant IV of the Cartesian plane.

6. The midpoint of a line segment is found by averaging the x-coordinates and averaging the y-coordinates of the endpoints.

7. Explain what is meant by setting the viewing window.

1.1 Exercises

In Problems 1 and 2, plot each point in the xy-plane. Tell in which quadrant or on what coordinate axis each point lies.

1. (a) $A = (-3, 2)$
(b) $B = (6, 0)$
(c) $C = (-2, -2)$
(d) $D = (6, 5)$
(e) $E = (0, -3)$
(f) $F = (6, -3)$

2. (a) $A = (1, 4)$
(b) $B = (-3, -4)$
(c) $C = (-3, 4)$
(d) $D = (4, 1)$
(e) $E = (0, 1)$
(f) $F = (-3, 0)$

3. Plot the points $(2, 0)$, $(2, -3)$, $(2, 4)$, $(2, 1)$, and $(2, -1)$. Describe the set of all points of the form $(2, y)$, where y is a real number.

4. Plot the points $(0, 3)$, $(1, 3)$, $(-2, 3)$, $(5, 3)$ and $(-4, 3)$. Describe the set of all points of the form $(x, 3)$, where x is a real number.

In Problems 5–8, determine the coordinates of the points shown. Tell in which quadrant each point lies. Assume the coordinates are integers.

5.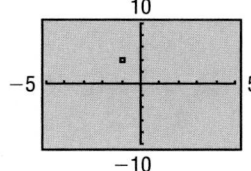

6.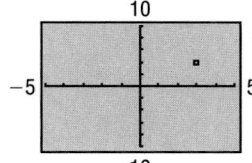

7.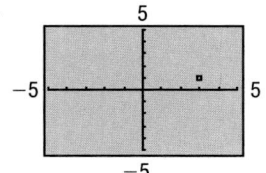

8.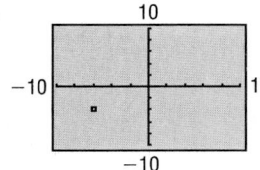

In Problems 9–14, select a setting so that each of the given points will lie within the viewing window.

9. $(-10, 5)$, $(3, -2)$, $(4, -1)$

10. $(5, 0)$, $(6, 8)$, $(-2, -3)$

11. $(40, 20)$, $(-20, -80)$, $(10, 40)$

12. $(-80, 60)$, $(20, -30)$, $(-20, -40)$

13. $(0, 0)$, $(100, 5)$, $(5, 150)$

14. $(0, -1)$, $(100, 50)$, $(-10, 30)$

In Problems 15–20, determine the viewing window used.

15.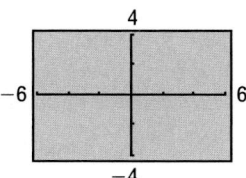

16.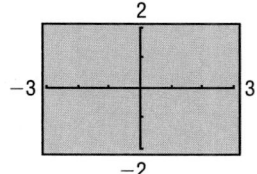

17.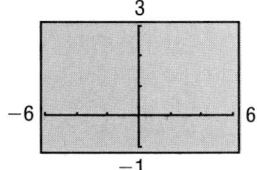

18.
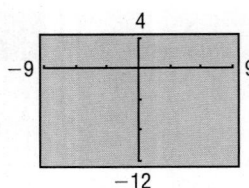

19.
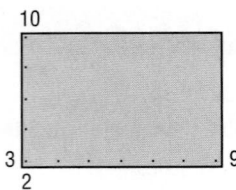

20.
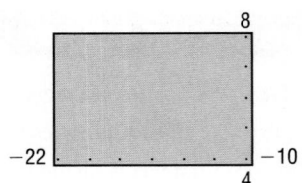

In Problems 21–34, find the distance $d(P_1, P_2)$ between the points P_1 and P_2.

21.
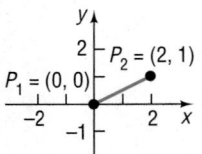

22.
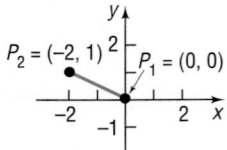

23.
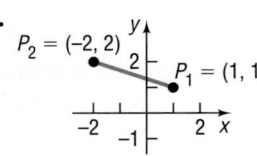

24.
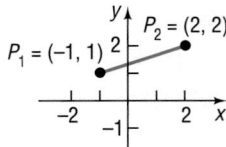

25. $P_1 = (3, 8);$ $P_2 = (5, 4)$

26. $P_1 = (6, 0);$ $P_2 = (2, 4)$

27. $P_1 = (-3, 2);$ $P_2 = (6, 0)$

28. $P_1 = (2, -3);$ $P_2 = (4, 2)$

29. $P_1 = (4, -3);$ $P_2 = (6, 4)$

30. $P_1 = (-4, -3);$ $P_2 = (6, 2)$

31. $P_1 = (-0.2, 0.3);$ $P_2 = (2.3, 1.1)$

32. $P_1 = (1.2, 2.3);$ $P_2 = (-0.3, 1.1)$

33. $P_1 = (a, b);$ $P_2 = (0, 0)$

34. $P_1 = (a, a);$ $P_2 = (0, 0)$

In Problems 35–38, find the length of the line segment. Assume that the end points of each line segment have integer coordinates.

35.
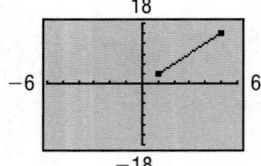

36.
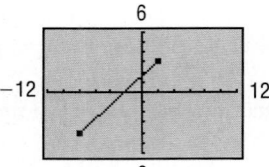

37.
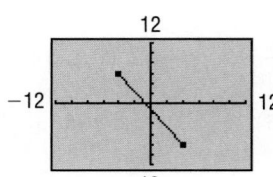

38.
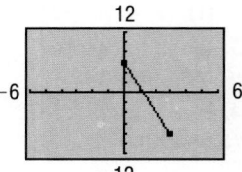

In Problems 39–44, plot each point and form the triangle ABC. Verify that the triangle is a right triangle. Find its area.

39. $A = (-2, 5);$ $B = (1, 3);$ $C = (-1, 0)$

40. $A = (-2, 5);$ $B = (12, 3);$ $C = (10, -11)$

41. $A = (-5, 3);$ $B = (6, 0);$ $C = (5, 5)$

42. $A = (-6, 3);$ $B = (3, -5);$ $C = (-1, 5)$

43. $A = (4, -3);$ $B = (0, -3);$ $C = (4, 2)$

44. $A = (4, -3);$ $B = (4, 1);$ $C = (2, 1)$

45. Find all points having an x-coordinate of 2 whose distance from the point $(-2, -1)$ is 5.

46. Find all points having a y-coordinate of -3 whose distance from the point $(1, 2)$ is 13.

47. Find all points on the x-axis that are 5 units from the point $(4, -3)$.

48. Find all points on the y-axis that are 5 units from the point $(4, 4)$.

In Problems 49–58, find the midpoint of the line segment joining the points P_1 and P_2.

49. $P_1 = (5, 4);$ $P_2 = (3, 2)$

50. $P_1 = (6, 0);$ $P_2 = (2, 4)$

51. $P_1 = (-3, 2);$ $P_2 = (6, 0)$

52. $P_1 = (2, -3);$ $P_2 = (4, 2)$

53. $P_1 = (4, -3);$ $P_2 = (6, 1)$

54. $P_1 = (-4, -3);$ $P_2 = (2, 2)$

55. $P_1 = (-0.2, 0.3);$ $P_2 = (2.3, 1.1)$

56. $P_1 = (1.2, 2.3);$ $P_2 = (-0.3, 1.1)$

57. $P_1 = (a, b);$ $P_2 = (0, 0)$

58. $P_1 = (a, a);$ $P_2 = (0, 0)$

59. The **medians** of a triangle are the line segments from each vertex to the midpoint of the opposite side (see the figure). Find the lengths of the medians of the triangle with vertices at $A = (0, 0)$, $B = (6, 0)$, and $C = (4, 4)$.

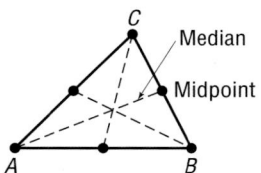

60. An **equilateral triangle** is one in which all three sides are of equal length. If two vertices of an equilateral triangle are $(0, 4)$ and $(0, 0)$, find the third vertex. How many of these triangles are possible?

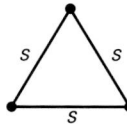

*In Problems 61–64, find the length of each side of the triangle determined by the three points P_1, P_2, and P_3. State whether the triangle is an isosceles triangle, a right triangle, neither of these, or both. (An **isosceles triangle** is one in which at least two of the sides are of equal length.)*

61. $P_1 = (2, 1)$; $P_2 = (-4, 1)$; $P_3 = (-4, -3)$

62. $P_1 = (-1, 4)$; $P_2 = (6, 2)$; $P_3 = (4, -5)$

63. $P_1 = (-2, -1)$; $P_2 = (0, 7)$; $P_3 = (3, 2)$

64. $P_1 = (7, 2)$; $P_2 = (-4, 0)$; $P_3 = (4, 6)$

---

65. Baseball   A major league baseball "diamond" is actually a square, 90 feet on a side (see the figure). What is the distance directly from home plate to second base (the diagonal of the square)?

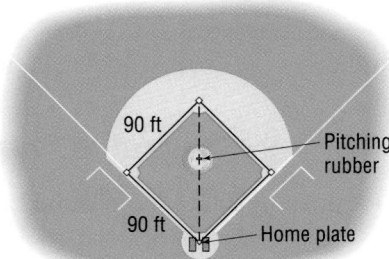

66. Little League Baseball   The layout of a Little League playing field is a square, 60 feet on a side.* How far is it directly from home plate to second base (the diagonal of the square)?

67. Baseball   Refer to Problem 65. Overlay a rectangular coordinate system on a major league baseball diamond so that the origin is at home plate, the positive x-axis lies in the direction from home plate to first base, and the positive y-axis lies in the direction from home plate to third base.

(a)   What are the coordinates of first base, second base, and third base? Use feet as the unit of measurement.

(b)   If the right fielder is located at $(310, 15)$, how far is it from there to second base?

(c)   If the center fielder is located at $(300, 300)$, how far is it from there to third base?

68. Little League Baseball   Refer to Problem 66. Overlay a rectangular coordinate system on a Little League baseball diamond so that the origin is at home plate, the positive x-axis lies in the direction from home plate to first base, and the positive y-axis lies in the direction from home plate to third base.

(a)   What are the coordinates of first base, second base, and third base? Use feet as the unit of measurement.

(b)   If the right fielder is located at $(180, 20)$, how far is it from there to second base?

(c)   If the center fielder is located at $(220, 220)$, how far is it from there to third base?

69. A Dodge Intrepid and a Mack truck leave an intersection at the same time. The Intrepid heads east at an average speed of 30 miles per hour, while the truck heads south at an average speed of 40 miles per hour. Find an expression for their distance apart d (in miles) at the end of t hours.

70. A hot-air balloon, headed due east at an average speed of 15 miles per hour and at a constant altitude of 100 feet, passes over an intersection (see the figure). Find an expression for its distance d (measured in feet) from the intersection t seconds later.

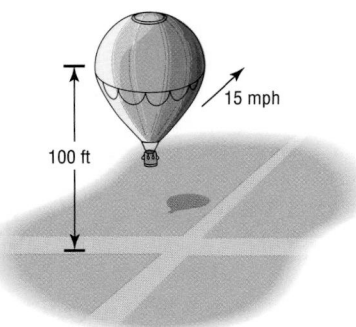

*Source: Little League Baseball, Official Regulations and Playing Rules, 2001.

PREPARING FOR THIS SECTION

Before getting started, review the following:

✓ Zero-Product Property (Section A.1, p. 557)   ✓ Equations (Section A.3, pp. 571–583)

1.2 INTRODUCTION TO GRAPHING EQUATIONS

OBJECTIVES ① Graph Equations by Plotting Points
② Graph Equations Using a Graphing Utility
③ Find Intercepts

An **equation in two variables**, say x and y, is a statement in which two expressions involving x and y are equal. The expressions are called the **sides** of the equation. Since an equation is a statement, it may be true or false, depending on the value of the variables. Any values of x and y that result in a true statement are said to **satisfy** the equation.

For example, the following are all equations in two variables x and y:

$$x^2 + y^2 = 5 \qquad 2x - y = 6 \qquad y = 2x + 5 \qquad x^2 = y$$

The first of these, $x^2 + y^2 = 5$, is satisfied for $x = 1$, $y = 2$, since $1^2 + 2^2 = 1 + 4 = 5$. Other choices of x and y also satisfy this equation. It is not satisfied for $x = 2$ and $y = 3$, since $2^2 + 3^2 = 4 + 9 = 13 \neq 5$.

 The **graph of an equation in two variables** x and y consists of the set of points in the xy-plane whose coordinates (x, y) satisfy the equation.

EXAMPLE 1 **Determining Whether a Point Is on the Graph of an Equation**

Determine if the following points are on the graph of the equation $2x - y = 6$.

(a) $(2, 3)$ (b) $(2, -2)$

Solution (a) For the point $(2, 3)$, we check to see if $x = 2$, $y = 3$ satisfies the equation $2x - y = 6$.

$$2x - y = 2(2) - 3 = 4 - 3 = 1 \neq 6$$

The equation is not satisfied, so the point $(2, 3)$ is not on the graph.

(b) For the point $(2, -2)$, we have

$$2x - y = 2(2) - (-2) = 4 + 2 = 6$$

The equation is satisfied, so the point $(2, -2)$ is on the graph. ■

 NOW WORK PROBLEM 1.

EXAMPLE 2 **Graphing an Equation by Plotting Points**

Graph the equation: $y = 2x + 5$

Solution We want to find all points (x, y) that satisfy the equation. To locate some of these points (and thus get an idea of the pattern of the graph), we assign some numbers to x and find corresponding values for y.

Figure 17
$y = 2x + 5$

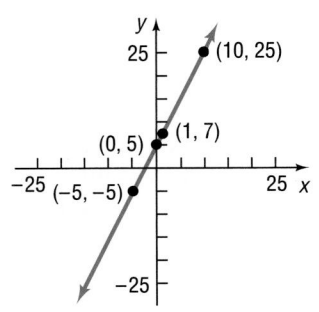

If	Then	Point on Graph
$x = 0$	$y = 2(0) + 5 = 5$	$(0, 5)$
$x = 1$	$y = 2(1) + 5 = 7$	$(1, 7)$
$x = -5$	$y = 2(-5) + 5 = -5$	$(-5, -5)$
$x = 10$	$y = 2(10) + 5 = 25$	$(10, 25)$

By plotting these points and then connecting them, we obtain the graph of the equation (a *line*), as shown in Figure 17. ■

EXAMPLE 3   **Graphing an Equation by Plotting Points**

Graph the equation: $y = x^2$

Solution Table 1 provides several points on the graph. In Figure 18 we plot these points and connect them with a smooth curve to obtain the graph (a *parabola*).

TABLE 1

x	$y = x^2$	(x, y)
-4	16	$(-4, 16)$
-3	9	$(-3, 9)$
-2	4	$(-2, 4)$
-1	1	$(-1, 1)$
0	0	$(0, 0)$
1	1	$(1, 1)$
2	4	$(2, 4)$
3	9	$(3, 9)$
4	16	$(4, 16)$

Figure 18
$y = x^2$

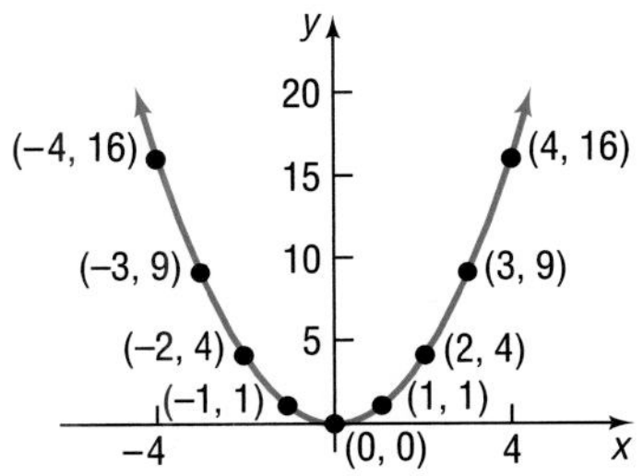

The graphs of the equations shown in Figures 17 and 18 do not show all points. For example, in Figure 17, the point $(20, 45)$ is a part of the graph of $y = 2x + 5$, but it is not shown. Since the graph of $y = 2x + 5$ could be extended out as far as we please, we use arrows to indicate that the pattern shown continues. It is important when illustrating a graph to present enough of the graph so that any viewer of the illustration will "see" the rest of it as an obvious continuation of what is actually there. This is referred to as a **complete graph**.

One way to obtain a complete graph of an equation is to plot a sufficient number of points on the graph until a pattern becomes evident. Then these points are connected with a smooth curve following the suggested pattern. But how many points are sufficient? Sometimes knowledge about the equation tells us. For example, if an equation is of the form $y = mx + b$, then its graph is a line*. In this case, two points would suffice to obtain the graph.

One of the purposes of this book is to investigate the properties of equations in order to decide whether a graph is complete. Sometimes we shall graph equations by plotting a sufficient number of points on the graph until a pattern becomes evident; then we connect these points with a smooth curve following the suggested pattern. (Shortly, we shall investigate various techniques that will enable us to graph an equation without plotting so many points.) Other times we shall graph equations using a graphing utility.

* Lines are discussed in the Appendix, Section A.7.

Using a Graphing Utility to Graph Equations

（2）From Examples 2 and 3, we see that a graph can be obtained by plotting points in a rectangular coordinate system and connecting them. Graphing utilities perform these same steps when graphing an equation. For example, the TI-83 determines 95 evenly spaced input values,* uses the equation to determine the output values, plots these points on the screen, and finally (if in the connected mode) draws a line between consecutive points.

To graph an equation in two variables x and y using a graphing utility requires that the equation be written in the form $y = \{\text{expression in } x\}$. If the original equation is not in this form, replace it by equivalent equations until the form $y = \{\text{expression in } x\}$ is obtained. In general, there are four ways to obtain equivalent equations.

Procedures That Result in Equivalent Equations

1. Interchange the two sides of the equation:

   Replace $3x + 5 = y$ by $y = 3x + 5$

2. Simplify the sides of the equation by combining like terms, eliminating parentheses, and so on:

   Replace $2y + 2 + 6 = 2x + 5(x + 1)$
   by $2y + 8 = 7x + 5$

3. Add or subtract the same expression on both sides of the equation:

   Replace $y + 3x - 5 = 4$
   by $y + 3x - 5 + 5 = 4 + 5$

4. Multiply or divide both sides of the equation by the same nonzero expression:

   Replace $3y = 6 - 2x$

   by $\dfrac{1}{3} \cdot 3y = \dfrac{1}{3}(6 - 2x)$

EXAMPLE 4   **Expressing an Equation in the Form $y = \{\text{expression in } x\}$**

Solve for y: $2y + 3x - 5 = 4$

Solution   We replace the original equation by a succession of equivalent equations.

$$2y + 3x - 5 = 4$$
$$2y + 3x - 5 + 5 = 4 + 5 \qquad \text{Add 5 to both sides.}$$
$$2y + 3x = 9 \qquad \text{Simplify.}$$
$$2y + 3x - 3x = 9 - 3x \qquad \text{Subtract 3x from both sides.}$$
$$2y = 9 - 3x \qquad \text{Simplify.}$$
$$\frac{2y}{2} = \frac{9 - 3x}{2} \qquad \text{Divide both sides by 2.}$$
$$y = \frac{9 - 3x}{2} \qquad \text{Simplify.}$$ ∎

*These input values depend on the values of Xmin and Xmax. For example, if Xmin $= -10$ and Xmax $= 10$, then the first input value will be -10 and the next input value will be $-10 + (10 - (-10))/94 = -9.7872$, and so on.

Now we are ready to graph equations using a graphing utility. Most graphing utilities require the following steps.

Steps for Graphing an Equation Using a Graphing Utility

STEP 1: Solve the equation for y in terms of x.

STEP 2: Get into the graphing mode of your graphing utility. The screen will usually display $Y =$ ____, prompting you to enter the expression involving x that you found in Step 1. (Consult your manual for the correct way to enter the expression; for example, $y = x^2$ might be entered as x^2 or as $x*x$ or as $x\ x^Y 2$).

STEP 3: Select the viewing window. Without prior knowledge about the behavior of the graph of the equation, it is common to select the **standard viewing window*** initially. The viewing window is then adjusted based on the graph that appears. In this text, the standard viewing window will be

$$X\text{min} = -10 \qquad\qquad Y\text{min} = -10$$
$$X\text{max} = 10 \qquad\qquad Y\text{max} = 10$$
$$X\text{scl} = 1 \qquad\qquad Y\text{scl} = 1$$

STEP 4: Graph.

STEP 5: Adjust the viewing window until a complete graph is obtained.

EXAMPLE 5 **Graphing an Equation on a Graphing Utility**

Graph the equation: $6x^2 + 3y = 36$.

Solution **STEP 1:** We solve for y in terms of x.

$$6x^2 + 3y = 36$$
$$3y = -6x^2 + 36 \qquad \text{Subtract } 6x^2 \text{ from both sides of the equation.}$$
$$y = -2x^2 + 12 \qquad \text{Divide both sides of the equation by 3 and simplify.}$$

STEP 2: From the graphing mode, enter the expression $-2x^2 + 12$ after the prompt $Y =$ ____. Figure 19 shows the expression entered in a TI-83.

STEP 3: Set the viewing window to the standard viewing window.

STEP 4: Graph. The screen should look like Figure 20.

STEP 5: The graph of $y = -2x^2 + 12$ is not complete. The value of Ymax must be increased so that the top portion of the graph is visible.

Figure 19

Figure 20

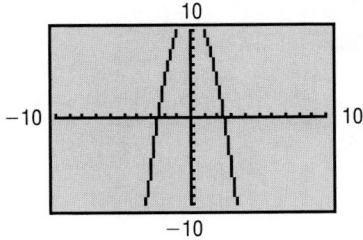

---

*Some graphing utilities have a ZOOM-STANDARD feature that automatically sets the viewing window to the standard viewing window and graphs the equation.

After increasing the value of Ymax to 12, we obtain the graph in Figure 21.* The graph is now complete.

Figure 21

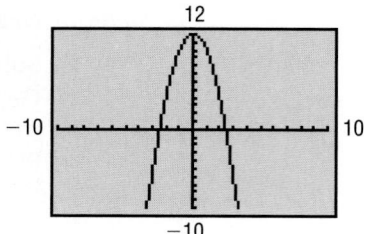

Using a Graphing Utility to Create Tables

In addition to graphing equations, graphing utilities can also be used to create a table of values that satisfy the equation. This feature is especially useful in determining an appropriate viewing window when graphing an equation. Many graphing utilities require the following steps to create a table.

Steps for Creating a Table of Values Using a Graphing Utility

STEP 1: Solve the equation for y in terms of x.

STEP 2: Enter the expression in x following the $Y =$ prompt of the graphing utility.

STEP 3: Set up the table. Graphing utilities typically have two modes for creating tables. In the AUTO mode, the user determines a starting point for the table (TblStart) and ΔTbl (pronounced "delta-table"). The ΔTbl feature determines the increment for x. The ASK mode requires the user to enter values of x and then the utility determines the corresponding value of y.

STEP 4: Create the table. The user can scroll within the table if the table was created in AUTO mode.

EXAMPLE 6   **Creating a Table Using a Graphing Utility**

Create a table that displays the points on the graph of $6x^2 + 3y = 36$ for $x = -3, -2, -1, 0, 1, 2,$ and 3.

Solution   **STEP 1:** We solved the equation for y in Example 5 and obtained $y = -2x^2 + 12$.

STEP 2: Enter the expression in x following the $Y =$ prompt.

STEP 3: We set up the table in the AUTO mode with TblStart $= -3$ and ΔTbl $= 1$.

STEP 4: Create the table. The screen should look like Table 2.

In looking at Table 2, we notice that $y = 12$ when $x = 0$. This information could have been used to help to create the initial viewing window by letting us know that Ymax needs to be at least 12 in order to get a complete graph.

TABLE 2

X	Y1
-3	-6
-2	4
-1	10
0	12
1	10
2	4
3	-6

Y1 ▆ -2X²+12

---

*Some graphing utilities have a ZOOM-FIT feature that determines the appropriate Ymin and Ymax for a given Xmin and Xmax. Consult your owner's manual for the appropriate keystrokes.

Intercepts

③ The points, if any, at which a graph crosses or touches the coordinate axes are called the **intercepts**. See Figure 22. The *x*-coordinate of a point at which the graph crosses or touches the *x*-axis is an ***x*-intercept**, and the *y*-coordinate of a point at which the graph crosses or touches the *y*-axis is a ***y*-intercept**. In order for a graph to be complete, all of its intercepts must be displayed.

Figure 22

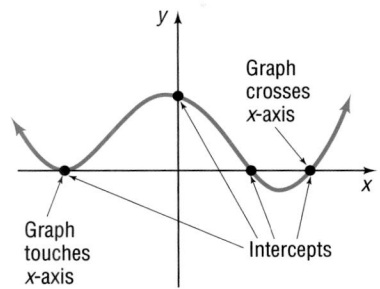

Graph crosses *x*-axis

Graph touches *x*-axis

Intercepts

EXAMPLE 7   **Finding Intercepts from a Graph**

Find the intercepts of the graph in Figure 23. What are its *x*-intercepts? What are its *y*-intercepts?

Solution   The intercepts of the graph are the points

Figure 23

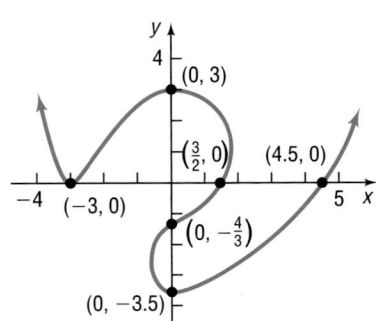

$$(-3, 0), \quad (0, 3), \quad \left(\frac{3}{2}, 0\right), \quad \left(0, -\frac{4}{3}\right), \quad (0, -3.5), \quad (4.5, 0)$$

The *x*-intercepts are -3, $\frac{3}{2}$, and 4.5; the *y*-intercepts are -3.5, $-\frac{4}{3}$, and 3. ■

✎ **NOW WORK PROBLEM 7.**

The intercepts of the graph of an equation can be found from the equation by using the fact that points on the *x*-axis have *y*-coordinates equal to 0 and points on the *y*-axis have *x*-coordinates equal to 0.

> ### Procedure for Finding Intercepts
> 1. To find the *x*-intercept(s), if any, of the graph of an equation, let $y = 0$ in the equation and solve for *x*.
> 2. To find the *y*-intercept(s), if any, of the graph of an equation, let $x = 0$ in the equation and solve for *y*.

Because the *x*-intercepts of the graph of an equation are those *x*-values for which $y = 0$, they are also called the **zeros** (or **roots**) of the equation.

EXAMPLE 8   **Finding Intercepts from an Equation**

Find the *x*-intercept(s) and the *y*-intercept(s) of the graph of $y = x^2 - 4$.

Solution   To find the x-intercept(s), we let $y = 0$ and obtain the equation

$$x^2 - 4 = 0$$
$$(x + 2)(x - 2) = 0 \qquad \text{Factor.}$$
$$x + 2 = 0 \qquad \text{or} \qquad x - 2 = 0 \qquad \text{Zero-Product Property}$$
$$x = -2 \qquad \text{or} \qquad x = 2$$

The equation has two solutions, -2 and 2. Thus, the x-intercepts (or zeros) are -2 and 2.

To find the y-intercept(s), we let $x = 0$ in the equation.

$$y = x^2 - 4$$
$$= 0^2 - 4 = -4$$

The y-intercept is -4.   ■

Sometimes the TABLE feature of a graphing utility will reveal the intercepts of an equation. Table 3 shows a table for the equation $y = x^2 - 4$. Can you find the intercepts in the table?

Although a table can be used to find the intercepts of the graph of an equation, it is usually easier to graph an equation and use the graph to find the intercepts.

TABLE 3

X	Y1
-3	5
-2	0
-1	-3
0	-4
1	-3
2	0
3	5

$Y_1 \boxminus X^2 - 4$

EXAMPLE 9   **Finding Intercepts from the Graph Using a Graphing Utility**

Use a graphing utility to find the intercepts of the equation $y = x^3 - 16$.

Solution   Figure 24(a) shows the graph of $y = x^3 - 16$.

The eVALUEate feature of a TI-83 graphing calculator* accepts as input a value of x and determines the value of y. If we let $x = 0$, we find that the y-intercept is -16. See Figure 24(b).

The ZERO feature of a TI-83 is used to find the x-intercept(s). See Figure 24(c). Rounded to two decimal places, the x-intercept is 2.52.

Figure 24

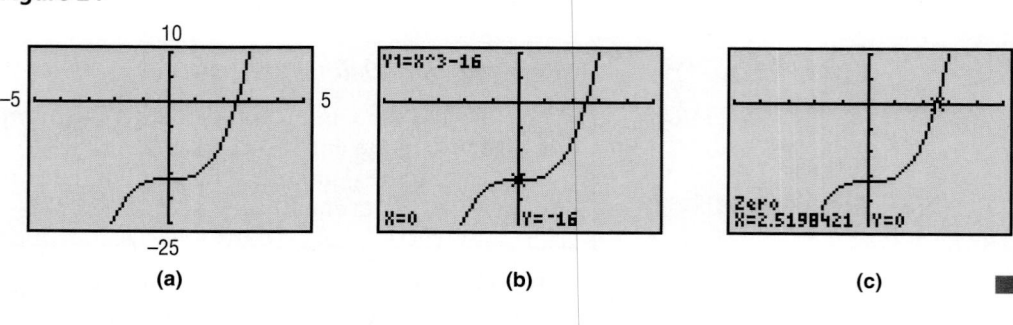

(a)   (b)   (c)   ■

NOW WORK PROBLEMS **23** AND **33**.

---

*Consult your manual to determine the appropriate keystrokes for these features.

1.2   Concepts and Vocabulary

In Problems 1 and 2, fill in the blanks.

1. The points, if any, at which a graph crosses or touches the coordinate axes are called _____.

2. Because the x-intercepts of the graph of an equation are those x-values for which $y = 0$, they are also called _____ or _____.

In Problems 3 and 4, answer True or False to each statement.

3. To find the x-intercepts of the graph of an equation, let $y = 0$ and solve for x.

4. To graph an equation using a graphing utility, we must first solve the equation for x in terms of y.

5. Explain what is meant by a complete graph.

6. What is the standard viewing window?

7. Draw a graph of an equation that contains two x-intercepts. At one the graph crosses the x-axis, and at the other the graph touches the x-axis.

1.2   Exercises

In Problems 1–6, tell whether the given points are on the graph of the equation.

1 Equation: $y = x^4 - \sqrt{x}$
Points: $(0,0); (1,1); (-1,0)$

2. Equation: $y = x^3 - 2\sqrt{x}$
Points: $(0,0); (1,1); (1,-1)$

3. Equation: $y^2 = x^2 + 9$
Points: $(0,3); (3,0); (-3,0)$

4. Equation: $y^3 = x + 1$
Points: $(1,2); (0,1); (-1,0)$

5. Equation: $x^2 + y^2 = 4$
Points: $(0,2); (-2,2); (\sqrt{2}, \sqrt{2})$

6. Equation: $x^2 + 4y^2 = 4$
Points: $(0,1); (2,0); \left(2, \dfrac{1}{2}\right)$

In Problems 7–14, the graph of an equation is given. List the intercepts of the graph.

7.
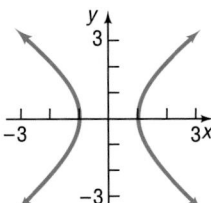

8.
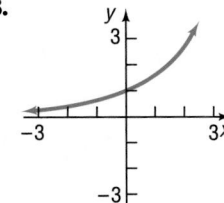

9.
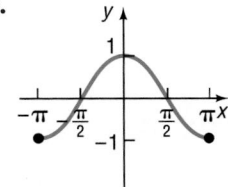

10.
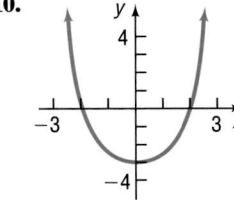

11.
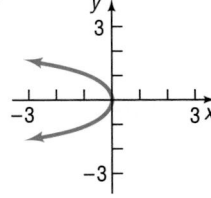

12.
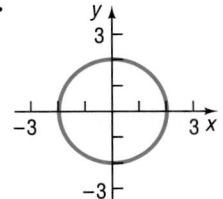

13.
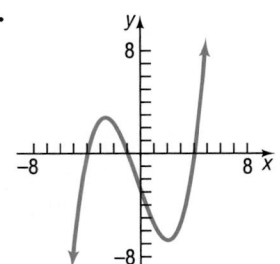

14.
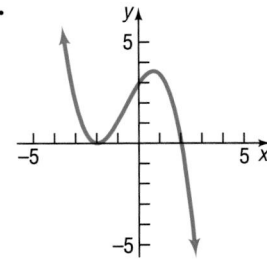

15. If $(a, 2)$ is a point on the graph of $y = 5x + 4$, what is a?

16. If $(2, b)$ is a point on the graph of $y = x^2 + 3x$, what is b?

17. If (a, b) is a point on the graph of $2x + 3y = 6$, write an equation that relates a to b.

18. If $(2, 0)$ and $(0, 5)$ are points on the graph of $y = mx + b$, what are m and b?

In Problems 19–30, determine the intercepts algebraically and graph each equation by plotting points. Be sure to label the intercepts. Verify your results using a graphing utility.

19. $y = x + 2$

20. $y = x - 6$

21. $y = 2x + 8$

22. $y = 3x - 9$

23. $y = x^2 - 1$

24. $y = x^2 - 9$

25. $y = -x^2 + 4$

26. $y = -x^2 + 1$

27. $2x + 3y = 6$

28. $5x + 2y = 10$

29. $9x^2 + 4y = 36$

30. $4x^2 + y = 4$

In Problems 31–38, graph each equation using a graphing utility. Use a graphing utility to approximate the intercepts rounded to two decimal places. Use the TABLE feature to help establish the viewing window.

31. $y = 2x - 13$

32. $y = -3x + 14$

33. $y = 2x^2 - 15$

34. $y = -3x^2 + 19$

35. $3x - 2y = 43$

36. $4x + 5y = 82$

37. $5x^2 + 3y = 37$

38. $2x^2 - 3y = 35$

In Problem 39, you may use a graphing utility, but it is not required.

39. (a) Graph $y = \sqrt{x^2}$, $y = x$, $y = |x|$, and $y = (\sqrt{x})^2$, noting which graphs are the same.
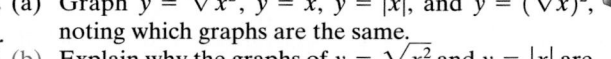
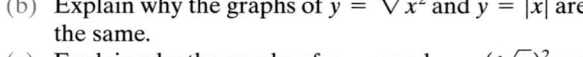
   (b) Explain why the graphs of $y = \sqrt{x^2}$ and $y = |x|$ are the same.
   (c) Explain why the graphs of $y = x$ and $y = (\sqrt{x})^2$ are not the same.
   (d) Explain why the graphs of $y = \sqrt{x^2}$ and $y = x$ are not the same.

40. Make up an equation with the intercepts $(2, 0)$, $(4, 0)$, and $(0, 1)$. Compare your equation with a friend's equation. Comment on any similarities.

41. Draw a graph that contains the points $(-2, -1)$, $(0, 1)$, $(1, 3)$, and $(3, 5)$. Compare your graph with those of other students. Are most of the graphs almost straight lines? How many are "curved"? Discuss the various ways that these points might be connected.

PREPARING FOR THIS SECTION

Before getting started, review the following:

✓ Completing the Square (Section A.3, pp. 579–580)   ✓ The Square Root Method (Section A.3, p. 578)

1.3 SYMMETRY; GRAPHING KEY EQUATIONS; CIRCLES

OBJECTIVES
1. Test an Equation for Symmetry with Respect to (a) the *x*-Axis, (b) the *y*-Axis, and (c) the Origin
2. Know How to Graph Key Equations
3. Write the Standard Form of the Equation of a Circle
4. Graph a Circle by Hand and by Using a Graphing Utility
5. Find the Center and Radius of a Circle from an Equation in General Form and Graph It

Symmetry

1. In Section 1.2 we saw the role that intercepts play in obtaining key points on the graph of an equation. Another helpful tool for graphing equations by hand involves *symmetry*, particularly symmetry with respect to the *x*-axis, the *y*-axis, and the origin.

A graph is said to be **symmetric with respect to the x-axis** if, for every point (x, y) on the graph, the point $(x, -y)$ is also on the graph.

A graph is said to be **symmetric with respect to the y-axis** if, for every point (x, y) on the graph, the point $(-x, y)$ is also on the graph.

A graph is said to be **symmetric with respect to the origin** if, for every point (x, y) on the graph, the point $(-x, -y)$ is also on the graph.

Figure 25 illustrates the definition. Notice that, when a graph is symmetric with respect to the x-axis, the part of the graph above the x-axis is a reflection or mirror image of the part below it, and vice versa. And when a graph is symmetric with respect to the y-axis, the part of the graph to the right of the y-axis is a reflection of the part to the left of it, and vice versa. Symmetry with respect to the origin may be viewed in two ways:

1. As a reflection about the y-axis, followed by a reflection about the x-axis

2. As a projection along a line through the origin so that the distances from the origin are equal

Figure 25

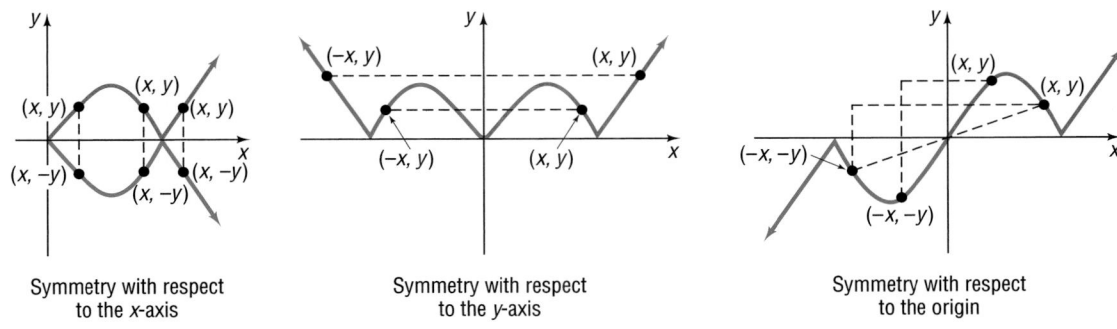

Symmetry with respect
to the x-axis

Symmetry with respect
to the y-axis

Symmetry with respect
to the origin

EXAMPLE 1   **Symmetric Points**

(a) If a graph is symmetric with respect to the x-axis and the point $(4, 2)$ is on the graph, then the point $(4, -2)$ is also on the graph.

(b) If a graph is symmetric with respect to the y-axis and the point $(4, 2)$ is on the graph, then the point $(-4, 2)$ is also on the graph.

(c) If a graph is symmetric with respect to the origin and the point $(4, 2)$ is on the graph, then the point $(-4, -2)$ is also on the graph. ■

NOW WORK PROBLEM 1.

When the graph of an equation is symmetric with respect to a coordinate axis or the origin, the number of points that you need to plot in order to see the pattern is reduced. For example, if the graph of an equation is symmetric with respect to the y-axis, then, once points to the right of the y-axis are plotted, an equal number of points on the graph can be obtained by reflecting them about the y-axis. Because of this, before we graph an equation, we first want to determine whether it has any symmetry. The following tests are used for this purpose.

Tests for Symmetry

To test the graph of an equation for symmetry with respect to the

x-Axis   Replace y by $-y$ in the equation. If an equivalent equation results, the graph of the equation is symmetric with respect to the x-axis.

y-Axis   Replace x by $-x$ in the equation. If an equivalent equation results, the graph of the equation is symmetric with respect to the y-axis.

Origin   Replace x by $-x$ and y by $-y$ in the equation. If an equivalent equation results, the graph of the equation is symmetric with respect to the origin.

EXAMPLE 2   **Testing an Equation for Symmetry**

Test $y = \dfrac{4x^2}{x^2 + 1}$ for symmetry.

Solution   *x-Axis:*   To test for symmetry with respect to the x-axis, replace y by $-y$. Since the result, $-y = \dfrac{4x^2}{x^2 + 1}$, is not equivalent to $y = \dfrac{4x^2}{x^2 + 1}$, we conclude that the graph of the equation is not symmetric with respect to the x-axis.

y-Axis: To test for symmetry with respect to the y-axis, replace x by $-x$. Since the result, $y = \dfrac{4(-x)^2}{(-x)^2 + 1} = \dfrac{4x^2}{x^2 + 1}$, is equivalent to $y = \dfrac{4x^2}{x^2 + 1}$, we conclude that the graph of the equation is symmetric with respect to the y-axis.

Origin: To test for symmetry with respect to the origin, replace x by $-x$ and y by $-y$.

$$-y = \frac{4(-x)^2}{(-x)^2 + 1} \qquad \text{Replace x by } -x \text{ and y by } -y.$$

$$-y = \frac{4x^2}{x^2 + 1} \qquad \text{Simplify.}$$

$$y = -\frac{4x^2}{x^2 + 1} \qquad \text{Multiply both sides by } -1.$$

Figure 26

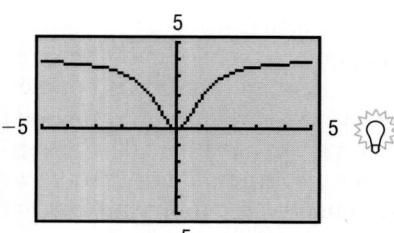

Since the result is not equivalent to the original equation, the graph of the equation $y = \dfrac{4x^2}{x^2 + 1}$ is not symmetric with respect to the origin. ∎

SEEING THE CONCEPT   Figure 26 shows the graph of $y = \dfrac{4x^2}{x^2 + 1}$ using a graphing utility. Do you see the symmetry with respect to the y-axis?

━━ **NOW WORK PROBLEM 27.**

Graphs of Key Equations

② The next three examples use intercepts, symmetry, and point plotting to obtain the graphs of key equations. The first of these is $y = x^3$.

EXAMPLE 3   **Graphing the Equation $y = x^3$ by Finding Intercepts and Checking for Symmetry**

Graph the equation $y = x^3$ by plotting points. Find any intercepts and check for symmetry first.

Solution   First, we seek the intercepts. When $x = 0$, then $y = 0$; and when $y = 0$, then $x = 0$. The origin $(0, 0)$ is the only intercept. Now we test for symmetry.

x-Axis: Replace y by $-y$. Since the result, $-y = x^3$, is not equivalent to $y = x^3$, the graph is not symmetric with respect to the x-axis.

y-Axis: Replace x by $-x$. Since the result, $y = (-x)^3 = -x^3$, is not equivalent to $y = x^3$, the graph is not symmetric with respect to the y-axis.

Origin: Replace x by $-x$ and y by $-y$. Since the result, $-y = (-x)^3 = -x^3$, is equivalent to $y = x^3$ (multiply both sides by -1), the graph is symmetric with respect to the origin.

To graph by hand, we use the equation to obtain several points on the graph. Because of the symmetry, we only need to locate points on the graph for which $x \geq 0$. See Table 4. Points on the graph could also be obtained using the TABLE feature on a graphing utility. See Table 5. Do you see the symmetry with respect to the origin from the table? Figure 27 shows the graph.

Figure 27
$y = x^3$

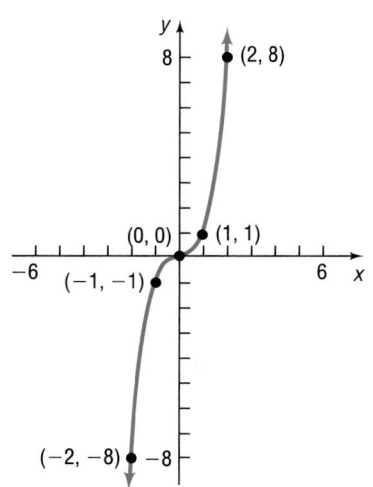

TABLE 4		
x	**y = x³**	**(x, y)**
0	0	(0, 0)
1	1	(1, 1)
2	8	(2, 8)
3	27	(3, 27)

TABLE 5	
X	Y1
-3	-27
-2	-8
-1	-1
0	0
1	1
2	8
3	27
Y1▉X^3	

EXAMPLE 4   **Graphing the Equation $x = y^2$**

Graph the equation $x = y^2$. Find any intercepts and check for symmetry first.

Solution   The lone intercept is $(0, 0)$. The graph is symmetric with respect to the x-axis since $x = (-y)^2$ is equivalent to $x = y^2$. The graph is not symmetric with respect to the y-axis or the origin.

To graph $x = y^2$ by hand, we use the equation to obtain several points on the graph. Because the equation is solved for x, it is easier to assign values to y and use the equation to determine the corresponding values of x. See Table 6. Because of the symmetry, we can restrict ourselves to points whose y-coordinates are positive. We then use the symmetry to find additional points on the graph. For example, since $(1, 1)$ is on the graph, so is $(1, -1)$. Since $(4, 2)$ is on the graph, so is $(4, -2)$, and so on. We plot these points and connect them with a smooth curve to obtain Figure 28.

TABLE 6		
y	**x = y²**	**(x, y)**
0	0	(0, 0)
1	1	(1, 1)
2	4	(4, 2)
3	9	(9, 3)

Figure 28
$x = y^2$

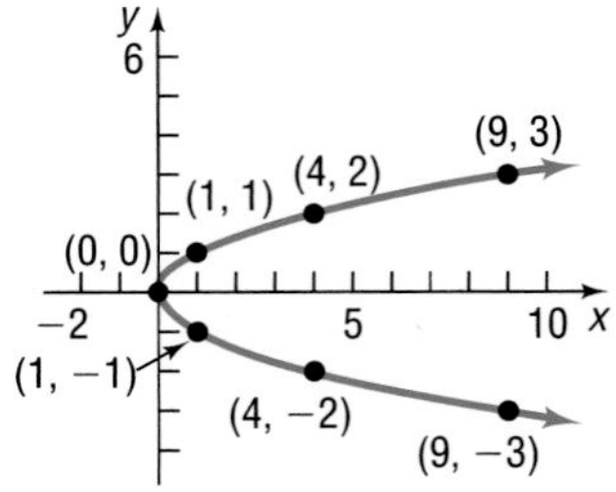

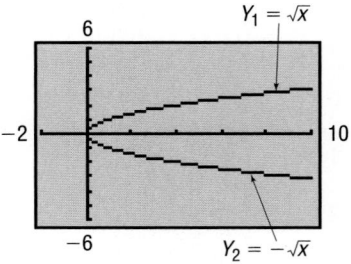

To graph the equation $x = y^2$ using a graphing utility, we must write the equation in the form $y = \{$expression in $x\}$. We proceed to solve for y.

$$x = y^2$$
$$y^2 = x$$
$$y = \pm\sqrt{x} \qquad \text{Square Root Method}$$

To graph $x = y^2$, we need to graph both $Y_1 = \sqrt{x}$ and $Y_2 = -\sqrt{x}$ on the same screen. Figure 29 shows the result. Table 7 shows various values of y for a given value of x when $Y_1 = \sqrt{x}$ and $Y_2 = -\sqrt{x}$. Notice that when $x < 0$ we get an error. Can you explain why?

Figure 29

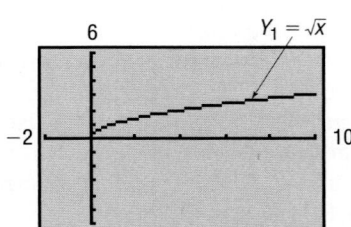

TABLE 7		
X	**Y1**	Y2
-1	ERROR	ERROR
0	0	0
1	1	-1
2	1.4142	-1.414
3	1.7321	-1.732
4	2	-2
5	2.2361	-2.236

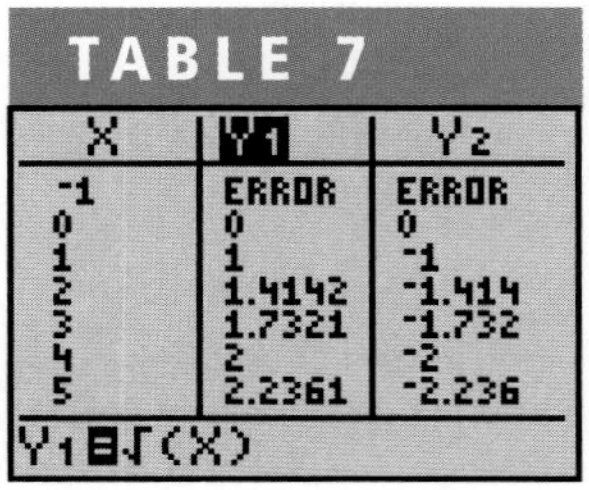

Y1=√(X)

If we restrict y so that $y \geq 0$, the equation $x = y^2$, $y \geq 0$, may be written as $y = \sqrt{x}$. The portion of the graph of $x = y^2$ in quadrant I is the graph of $y = \sqrt{x}$. See Figure 30.

Figure 30
$y = \sqrt{x}$

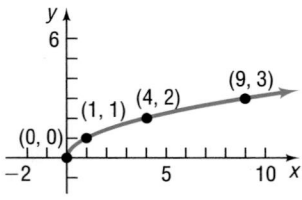

EXAMPLE 5

Graphing the Equation $y = \dfrac{1}{x}$

Consider the equation: $y = \dfrac{1}{x}$

(a) Graph this equation using a graphing utility. Set the viewing window as

$$X\text{min} = -3 \qquad Y\text{min} = -4$$
$$X\text{max} = 3 \qquad Y\text{max} = 4$$
$$X\text{scl} = 1 \qquad Y\text{scl} = 1$$

(b) Use algebra to find any intercepts and test for symmetry.
(c) Draw the graph by hand.

Solution

Figure 31

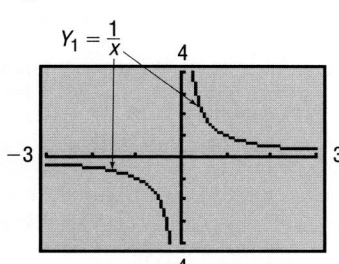

(a) Figure 31 illustrates the graph. We infer from the graph that there are no intercepts; we may also infer that symmetry with respect to the origin is a possibility. The TRACE feature on a graphing utility can provide further evidence of symmetry with respect to the origin. Using TRACE, we observe that for any ordered pair (x, y) the ordered pair $(-x, -y)$ is also a point on the graph. For example, the points $(0.95744681, 1.0444444)$ and $(-0.95744681, -1.0444444)$ both lie on the graph.

(b) We check for intercepts first. If we let $x = 0$, we obtain a 0 in the denominator, which is not defined. We conclude that there is no y-intercept. If we let $y = 0$, we get the equation $\dfrac{1}{x} = 0$, which has no solution. We conclude that there is no x-intercept. The graph of $y = \dfrac{1}{x}$ does not cross the coordinate axes.

Next we check for symmetry.

x-Axis: Replacing y by $-y$ yields $-y = \dfrac{1}{x}$, which is not equivalent to $y = \dfrac{1}{x}$.

y-Axis: Replacing x by $-x$ yields $y = \dfrac{1}{-x} = -\dfrac{1}{x}$, which is not equivalent to $y = \dfrac{1}{x}$.

Origin: Replacing x by $-x$ and y by $-y$ yields $-y = -\dfrac{1}{x}$, which is equivalent to $y = \dfrac{1}{x}$.

The graph is symmetric only with respect to the origin.
The inferences drawn in part (a) of the solution are now confirmed.

(c) We can use the equation to form Table 8 (page 26) and obtain some points on the graph. Because of symmetry, we only find points (x, y) for which x is positive. From Table 8 we infer that, if x is a large and positive number, then $y = \dfrac{1}{x}$ is a positive number close to 0. We also infer that if x is a positive number close to 0 then $y = \dfrac{1}{x}$ is a large and positive number.

Armed with this information, we can graph the equation. Figure 32 illustrates some of these points and graph of $y = \dfrac{1}{x}$.

TABLE 8

x	$y = \dfrac{1}{x}$	(x, y)
$\dfrac{1}{10}$	10	$\left(\dfrac{1}{10}, 10\right)$
$\dfrac{1}{3}$	3	$\left(\dfrac{1}{3}, 3\right)$
$\dfrac{1}{2}$	2	$\left(\dfrac{1}{2}, 2\right)$
1	1	$(1, 1)$
2	$\dfrac{1}{2}$	$\left(2, \dfrac{1}{2}\right)$
3	$\dfrac{1}{3}$	$\left(3, \dfrac{1}{3}\right)$
10	$\dfrac{1}{10}$	$\left(10, \dfrac{1}{10}\right)$

Figure 32

$y = \dfrac{1}{x}$

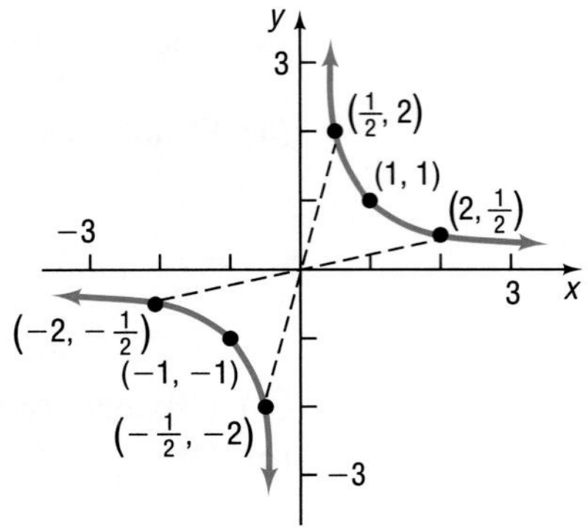

Observe how the absence of intercepts and the existence of symmetry with respect to the origin were utilized. ■

Circles

3

One advantage of a coordinate system is that it enables us to translate a geometric statement into an algebraic statement, and vice versa. Consider, for example, the following geometric statement that defines a circle.

> A **circle** is a set of points in the xy-plane that are a fixed distance r from a fixed point (h, k). The fixed distance r is called the **radius**, and the fixed point (h, k) is called the **center** of the circle.

Figure 33 shows the graph of a circle. To find an equation that has this graph, we let (x, y) represent the coordinates of any point on a circle with radius r and center (h, k). Then the distance between the points (x, y) and (h, k) must always equal r. That is, by the distance formula

$$\sqrt{(x - h)^2 + (y - k)^2} = r$$

or, equivalently,

$$(x - h)^2 + (y - k)^2 = r^2$$

Figure 33

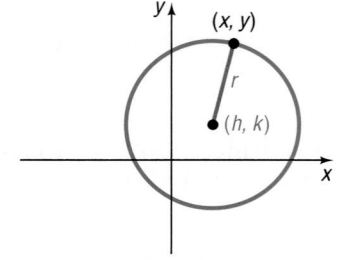

> The **standard form of an equation of a circle** with radius r and center (h, k) is
>
> $$(x - h)^2 + (y - k)^2 = r^2 \qquad (1)$$

The standard form of an equation of a circle of radius r with center at the origin $(0, 0)$ is

$$x^2 + y^2 = r^2$$

If the radius $r = 1$, the circle whose center is at the origin is called the **unit circle** and has the equation

$$x^2 + y^2 = 1$$

Notice that any circle whose center is at the origin has symmetry with respect to the x-axis, the y-axis, and the origin.

See Figure 34 for the graph of the unit circle.

Figure 34
Unit circle $x^2 + y^2 = 1$

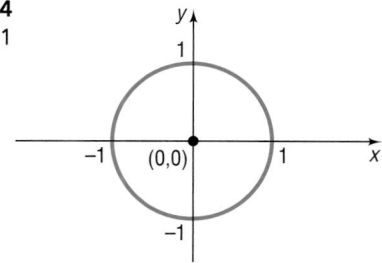

EXAMPLE 6   **Writing the Standard Form of the Equation of a Circle**

Write the standard form of the equation of the circle with radius 5 and center $(-3, 6)$.

Solution   Using the form of equation (1) and substituting the values $r = 5$, $h = -3$, and $k = 6$, we have

$$(x - h)^2 + (y - k)^2 = r^2$$

$$(x + 3)^2 + (y - 6)^2 = 25$$

━━━ NOW WORK PROBLEM 39.

Figure 35

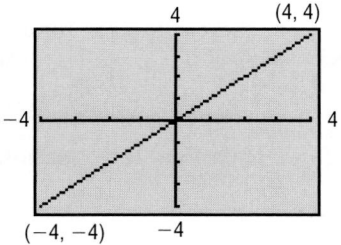

Figure 36

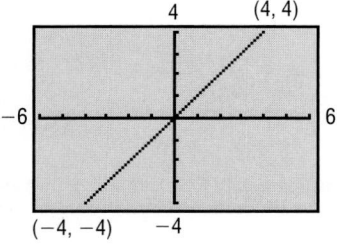

Square Screens

To get an undistorted view of a graph on a graphing utility, the same scale must be used on each axis. However, most graphing utilities have a rectangular screen. Because of this, using the same interval for both x and y will result in a distorted view. For example, Figure 35 shows the graph of the line $y = x$ connecting the points $(-4, -4)$ and $(4, 4)$.

We expect the line to bisect the first and third quadrants, but it doesn't. We need to adjust the selections for Xmin, Xmax, Ymin, and Ymax so that a **square screen** results. On many graphing utilities, this is accomplished by setting the ratio of x to y at $3:2$.*

Figure 36 shows the graph of the line $y = x$ on a square screen using a TI-83. Notice that the line now bisects the first and third quadrants. Compare this illustration to Figure 35.

*Most graphing utilities have a feature that automatically squares the viewing window. Consult your owner's manual for the appropriate keystrokes.

④ The graph of any equation of the form $(x - h)^2 + (y - k)^2 = r^2$ is that of a circle with radius r and center (h, k).

EXAMPLE 7   Graphing a Circle by Hand and by Using a Graphing Utility

Graph the equation: $(x + 3)^2 + (y - 2)^2 = 16$

Solution   The graph of the equation is a circle. To graph the equation by hand, we first compare the given equation to the standard form of the equation of a circle. The comparison yields information about the circle.

$$(x + 3)^2 + (y - 2)^2 = 16$$
$$(x - (-3))^2 + (y - 2)^2 = 4^2$$
$$\qquad \uparrow \qquad\qquad \uparrow \qquad \uparrow$$
$$(x - h)^2 + (y - k)^2 = r^2$$

We see that $h = -3, k = 2$, and $r = 4$. The circle has center $(-3, 2)$ and a radius of 4 units. To graph this circle, we first plot the center $(-3, 2)$. Since the radius is 4, we can locate four points on the circle by plotting points 4 units to the left, to the right, up, and down from the center. These four points can then be used as guides to obtain the graph. See Figure 37.

To graph a circle on a graphing utility, we must write the equation in the form $y = \{\text{expression involving } x\}$.* We solve for y in the equation

$$(x + 3)^2 + (y - 2)^2 = 16$$
$$(y - 2)^2 = 16 - (x + 3)^2 \qquad \text{Subtract } (x + 3)^2 \text{ from both sides.}$$
$$y - 2 = \pm\sqrt{16 - (x + 3)^2} \qquad \text{Use the Square Root method.}$$
$$y = 2 \pm \sqrt{16 - (x + 3)^2} \qquad \text{Add 2 to both sides.}$$

Figure 37

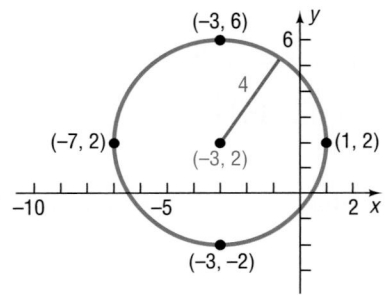

To graph the circle, we graph the top half

$$Y_1 = 2 + \sqrt{16 - (x + 3)^2}$$

and the bottom half

$$Y_2 = 2 - \sqrt{16 - (x + 3)^2}$$

Also, be sure to use a square screen. Otherwise, the circle will appear distorted. Figure 38 shows the graph on a TI-83.   ∎

Figure 38

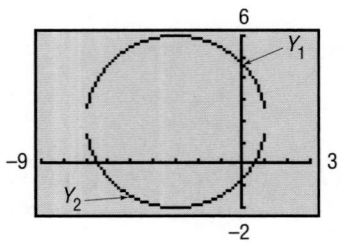

- NOW WORK PROBLEM 55.

If we eliminate the parentheses from the standard form of the equation of the circle given in Example 7, we get

$$(x + 3)^2 + (y - 2)^2 = 16$$
$$x^2 + 6x + 9 + y^2 - 4y + 4 = 16$$

which we find, upon simplifying, is equivalent to

$$x^2 + y^2 + 6x - 4y - 3 = 0$$

It can be shown that any equation of the form

$$x^2 + y^2 + ax + by + c = 0$$

---

*Some graphing utilities (e.g., TI-82, TI-83, TI-85, and TI-86) have a CIRCLE function that allows the user to enter only the coordinates of the center of the circle and its radius to graph the circle.

has a graph that is a circle, or a point, or has no graph at all. For example, the graph of the equation $x^2 + y^2 = 0$ is the single point $(0, 0)$. The equation $x^2 + y^2 + 5 = 0$, or $x^2 + y^2 = -5$, has no graph, because sums of squares of real numbers are never negative. When its graph is a circle, the equation

$$x^2 + y^2 + ax + by + c = 0$$

is referred to as the **general form of the equation of a circle**.

⑤   If an equation of a circle is in the general form, we use the method of completing the square to put the equation in standard form so we can identify its center and radius.

EXAMPLE 8   ## Graphing a Circle Whose Equation Is in General Form

Graph the equation: $x^2 + y^2 + 4x - 6y + 12 = 0$

Solution   We complete the square in both x and y to put the equation in standard form. Group the terms involving x, group the terms involving y, and put the constant on the right side of the equation. The result is

$$(x^2 + 4x) + (y^2 - 6y) = -12$$

Next, complete the square of each expression in parentheses. Remember that any number added on the left side of the equation must be added on the right.

$$(x^2 + 4x + 4) + (y^2 - 6y + 9) = -12 + 4 + 9$$

$$\left(\frac{4}{2}\right)^2 = 4 \qquad \left(\frac{-6}{2}\right)^2 = 9$$

$$(x + 2)^2 + (y - 3)^2 = 1 \qquad \text{Factor.}$$

We recognize this equation as the standard form of the equation of a circle with radius 1 and center $(-2, 3)$. To graph the equation by hand, use the center $(-2, 3)$ and the radius 1. See Figure 39(a).

To graph the equation using a graphing utility, we need to solve for y.

$$(y - 3)^2 = 1 - (x + 2)^2$$

$$y - 3 = \pm\sqrt{1 - (x + 2)^2} \qquad \text{Use the Square Root Method.}$$

$$y = 3 \pm \sqrt{1 - (x + 2)^2} \qquad \text{Add 3 to both sides.}$$

Figure 39(b) illustrates the graph.

Figure 39
$x^2 + y^2 + 4x - 6y + 12 = 0$

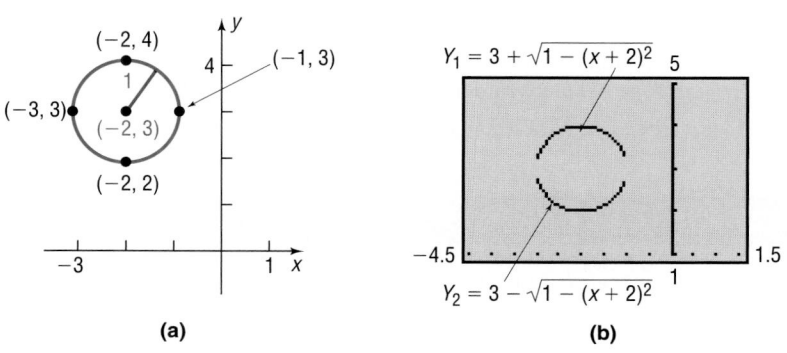

(a)   (b)

NOW WORK PROBLEM **57.**

EXAMPLE 9   Finding the General Form of the Equation of a Circle

Find the general form of the equation of the circle whose center is $(1, -2)$ and whose graph contains the point $(4, -2)$.

Solution   To find the equation of a circle, we need to know its center and its radius. Here, we know that the center is $(1, -2)$. Since the point $(4, -2)$ is on the graph, the radius r will equal the distance from $(4, -2)$ to the center $(1, -2)$. See Figure 40. Thus,

$$r = \sqrt{(4 - 1)^2 + [-2 - (-2)]^2}$$
$$= \sqrt{9} = 3$$

The standard form of the equation of the circle is

$$(x - 1)^2 + (y + 2)^2 = 9$$

Eliminating the parentheses and rearranging terms, we get the general equation

$$x^2 + y^2 - 2x + 4y - 4 = 0 \qquad \blacksquare$$

Figure 40
$x^2 + y^2 - 2x + 4y - 4 = 0$

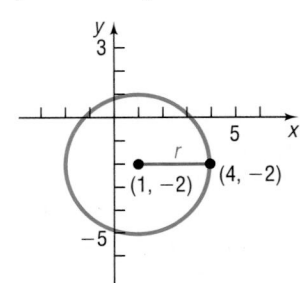

1.3 Concepts and Vocabulary

In Problems 1–3, fill in the blanks.

1. If for every point (x, y) on the graph of an equation, the point $(-x, y)$ is also on the graph, then the graph is symmetric with respect to the _____.

2. The graph of the equation $y = x^3 - 5x$ is symmetric with respect to the _____.

3. For a circle, the _____ is the distance from the center to any point on the circle.

In Problems 4–6, answer True or False to each statement.

4. The graph of the equation $y = x^4 + x^2 + 1$ is symmetric with respect to the y-axis.

5. The graph of a circle whose center is at the origin will have symmetry with respect to the x-axis, the y-axis, and the origin.

6. The center of the circle $(x + 3)^2 + (y - 2)^2 = 13$ is $(3, -2)$.

7. If the graph of an equation is symmetric with respect to the y-axis and 6 is an x-intercept of this graph, name another x-intercept.

8. An equation is being tested for symmetry with respect to the x-axis, the y-axis, and the origin. Explain why, if two of these symmetries are present, the remaining one must also be present.

9. Explain how the center and radius of a circle can be used to graph a circle by hand.

1.3 Exercises

In Problems 1–10, plot each point. Then plot the point that is symmetric to it with respect to: (a) the x-axis; (b) the y-axis; (c) the origin.

1. $(3, 4)$ **2.** $(5, 3)$ **3.** $(-2, 1)$ **4.** $(4, -2)$ **5.** $(1, 1)$

6. $(-1, -1)$ **7.** $(-3, -4)$ **8.** $(4, 0)$ **9.** $(0, -3)$ **10.** $(-3, 0)$

In Problems 11–18, the graph of an equation is given. Indicate whether the graph is symmetric with respect to the x-axis, the y-axis or the origin.

11.
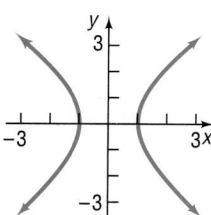

12.
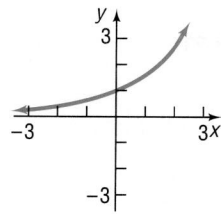

13.
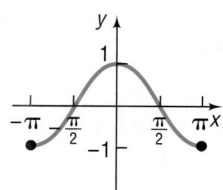

14.
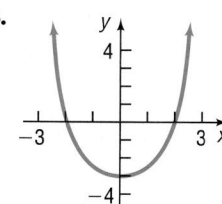

15.
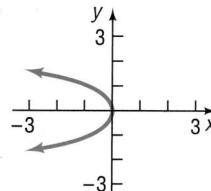

16.
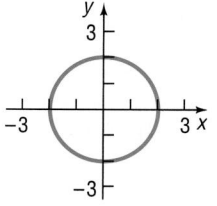

17.
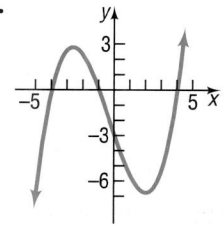

18.
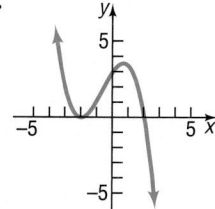

In Problems 19–22, draw a complete graph so that it has the type of symmetry indicated.

19. *y*-axis
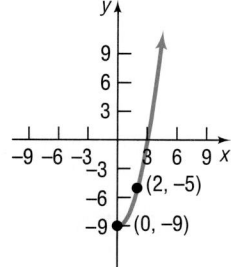

20. *x*-axis
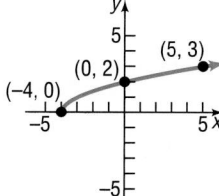

21. Origin
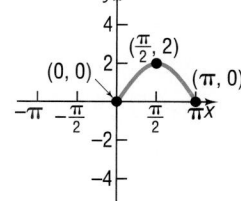

22. *y*-axis
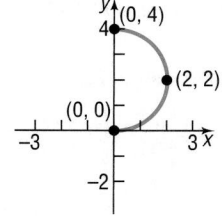

In Problems 23–34, test each equation for symmetry with respect to the x-axis, the y-axis, and the origin. Graph each equation using a graphing utility to verify your conclusions.

23. $x^2 = y + 5$ **24.** $y^2 = x + 3$ **25.** $y = 3x$ **26.** $y = -5x$

27. $x^2 + y - 9 = 0$ **28.** $y^2 - x - 4 = 0$ **29.** $y = x^3 - 27$ **30.** $y = x^4 - 1$

31. $y = x^2 - 3x - 4$ **32.** $y = x^2 + 4$ **33.** $y = \dfrac{x}{x^2 + 9}$ **34.** $y = \dfrac{x^2 - 4}{x}$

In Problems 35–38, draw a quick sketch of the equation by hand. Be sure to label all intercepts.

35. $y = x^3$ **36.** $x = y^2$ **37.** $y = \sqrt{x}$ **38.** $y = \dfrac{1}{x}$

In Problems 39–42, find the center and radius of each circle. Write the standard form of the equation.

39.
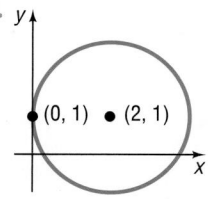

40.
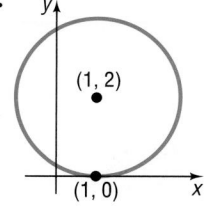

41.
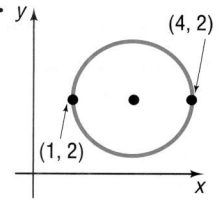

42.
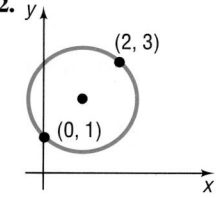

In Problems 43–52, write the standard form of the equation and the general form of the equation of each circle of radius r and center (h, k). By hand, graph each circle.

43. $r = 2$; $(h, k) = (0, 0)$ **44.** $r = 3$; $(h, k) = (0, 0)$ **45.** $r = 2$; $(h, k) = (0, 2)$ **46.** $r = 3$; $(h, k) = (1, 0)$

47. $r = 5$; $(h, k) = (4, -3)$

48. $r = 4$; $(h, k) = (2, -3)$

49. $r = 4$; $(h, k) = (-2, 1)$

50. $r = 7$; $(h, k) = (-5, -2)$

51. $r = \dfrac{1}{2}$; $(h, k) = \left(\dfrac{1}{2}, 0\right)$

52. $r = \dfrac{1}{2}$; $(h, k) = \left(0, -\dfrac{1}{2}\right)$

In Problems 53–62, find the center (h, k) and radius r of each circle. By hand, graph each circle.

53. $x^2 + y^2 = 25$

54. $x^2 + (y - 1)^2 = 9$

55. $(x - 2)^2 + y^2 = 4$

56. $(x + 3)^2 + (y - 1)^2 = 18$

57. $x^2 + y^2 + 4x - 4y - 1 = 0$

58. $x^2 + y^2 - 6x + 2y + 9 = 0$

59. $x^2 + y^2 - x + 2y + 1 = 0$

60. $x^2 + y^2 + x + y - \dfrac{1}{2} = 0$

61. $2x^2 + 2y^2 - 12x + 8y - 24 = 0$

62. $2x^2 + 2y^2 + 8x + 7 = 0$

In Problems 63–68, find the general form of the equation of each circle.

63. Center at the origin and containing the point $(-2, 3)$

64. Center $(1, 0)$ and containing the point $(-3, 2)$

65. Center $(2, 3)$ and tangent to the x-axis

66. Center $(-3, 1)$ and tangent to the y-axis

67. With end points of a diameter at $(1, 4)$ and $(-3, 2)$

68. With end points of a diameter at $(4, 3)$ and $(0, 1)$

In Problems 69–72, match each graph with the correct equation.

(a) $(x - 3)^2 + (y + 3)^2 = 9$ (c) $(x - 1)^2 + (y + 2)^2 = 4$

(b) $(x + 1)^2 + (y - 2)^2 = 4$ (d) $(x + 3)^2 + (y - 3)^2 = 9$

69.

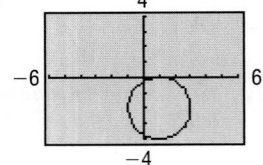

70.

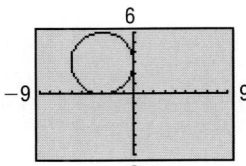

71.

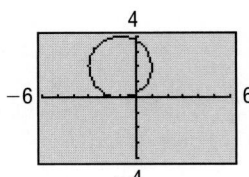

72.

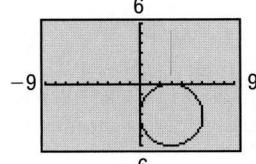

In Problems 73–76, find the standard form of the equation of each circle. Assume that the center has integer coordinates and that the radius is an integer.

73.

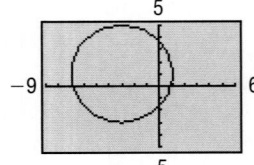

74.

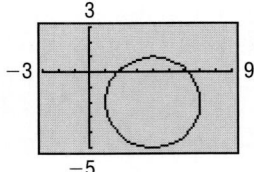

75.

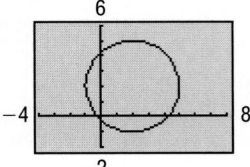

76.

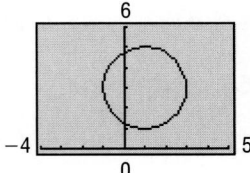

 77. Which of the following equations might have the graph shown? (More than one answer is possible.)

(a) $(x - 2)^2 + (y + 3)^2 = 13$
(b) $(x - 2)^2 + (y - 2)^2 = 8$
(c) $(x - 2)^2 + (y - 3)^2 = 13$
(d) $(x + 2)^2 + (y - 2)^2 = 8$
(e) $x^2 + y^2 - 4x - 9y = 0$
(f) $x^2 + y^2 + 4x - 2y = 0$
(g) $x^2 + y^2 - 9x - 4y = 0$
(h) $x^2 + y^2 - 4x - 4y = 4$

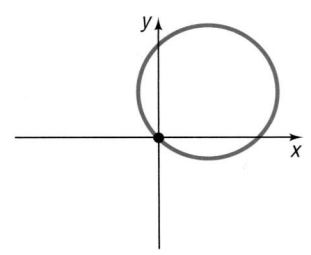

78. Which of the following equations might have the graph shown? (More than one answer is possible.)

(a) $(x - 2)^2 + y^2 = 3$ (e) $x^2 + y^2 + 10x + 16 = 0$

(b) $(x + 2)^2 + y^2 = 3$ (f) $x^2 + y^2 + 10x - 2y = 1$

(c) $x^2 + (y - 2)^2 = 3$ (g) $x^2 + y^2 + 9x + 10 = 0$

(d) $(x + 2)^2 + y^2 = 4$ (h) $x^2 + y^2 - 9x - 10 = 0$

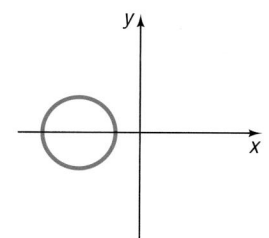

79. Weather Satellites   Earth is represented on a map of a portion of the solar system so that its surface is the circle with equation $x^2 + y^2 + 2x + 4y - 4091 = 0$. A weather satellite circles 0.6 unit above Earth with the center of its circular orbit at the center of Earth. Find the equation for the orbit of the satellite on this map.

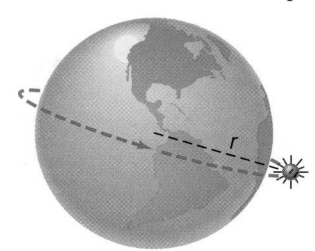

PREPARING FOR THIS SECTION

Before getting started, review the following:

✓ Intervals (Section A.5, pp. 594–595)

✓ Evaluating Algebraic Expressions, Domain of a Variable (Section A.1, pp. 557–558)

✓ Linear Equations (Section A.3, p. 576)

✓ Quadratic Equations (Section A.3, pp. 576–583)

✓ Solving Inequalities (Section A.5, pp. 598–599)

✓ Intercepts (Section 1.2, pp. 17–18)

1.4 FUNCTIONS

OBJECTIVES

1 Determine Whether a Relation Represents a Function

2 Find the Value of a Function

3 Find the Domain of a Function

4 Identify the Graph of a Function

5 Obtain Information from or about the Graph of a Function

A **relation** is a correspondence between two sets. If x and y are two elements in these sets and if a relation exists between x and y, then we say that x **corresponds** to y or that y **depends on** x, and we write $x \rightarrow y$. We may also write $x \rightarrow y$ as the ordered pair (x, y).

EXAMPLE 1   **An Example of a Relation**

Figure 41 depicts a relation between four individuals and their birthdays. The relation might be named "was born on." Then Katy corresponds to June 20, Dan corresponds to September 4, and so on. Using ordered pairs, this relation would be expressed as

$\{(\text{Katy}, \text{June 20}), (\text{Dan}, \text{September 4}), (\text{Patrick}, \text{December 31}), (\text{Phoebe}, \text{December 31})\}$

Figure 41

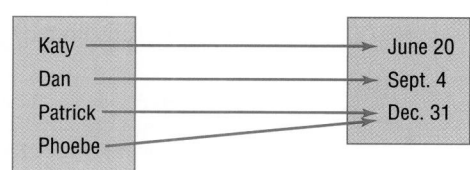

Often, we are interested in specifying the type of relation (such as an equation) that might exist between two variables. For example, the relation between the revenue R resulting from the sale of x items selling for \$10 each may be expressed by the equation $R = 10x$. If we know how many items have been sold, then we can calculate the revenue by using the equation $R = 10x$. This equation is an example of a *function*.

As another example, suppose that an icicle falls off a building from a height of 64 feet above the ground. According to a law of physics, the distance s (in feet) of the icicle from the ground after t seconds is given (approximately) by the formula $s = 64 - 16t^2$. When $t = 0$ seconds, the icicle is $s = 64$ feet above the ground. After 1 second, the icicle is $s = 64 - 16(1)^2 = 48$ feet above the ground. After 2 seconds, the icicle strikes the ground. The formula $s = 64 - 16t^2$ provides a way of finding the distance s for any time t ($0 \le t \le 2$). There is a correspondence between each time t in the interval $0 \le t \le 2$ and the distance s. We say that the distance s is a function of the time t because:

1. There is a correspondence between the set of times and the set of distances.
2. There is exactly one distance s obtained for any time t in the interval $0 \le t \le 2$.

Let's now look at the definition of a function.

Definition of Function

> Let X and Y be two nonempty sets.* A **function** from X into Y is a relation that associates with each element of X exactly one element of Y.

Figure 42

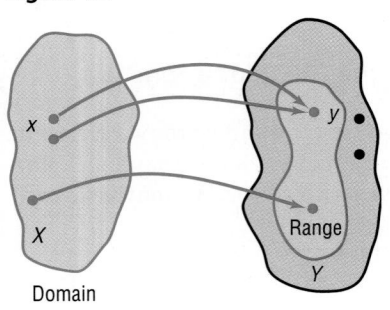

X
Domain

y
Range
Y

The set X is called the **domain** of the function. For each element x in X, the corresponding element y in Y is called the **value** of the function at x, or the **image** of x. The set of all images of the elements in the domain is called the **range** of the function. See Figure 42.

Since there may be some elements in Y that are not the image of some x in X, it follows that the range of a function may be a subset of Y, as shown in Figure 42.

Not all relations between two sets are functions. The next example shows how to determine whether a relation is a function or not.

① **EXAMPLE 2** **Determining Whether a Relation Represents a Function**

Determine whether the following relations represent functions.

(a) See Figure 43. For this relation, the domain represents four individuals and the range represents their birthdays.

---

*The sets X and Y will usually be sets of real numbers, in which case a (real) function results. The two sets can also be sets of complex numbers (discussed in Section A.4), and then we have defined a complex function. In the broad definition (due to Lejeune Dirichlet), X and Y can be any two sets.

Figure 43

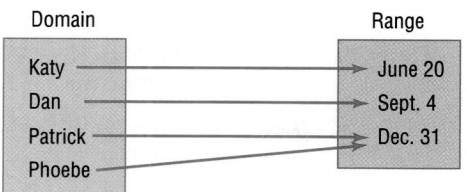

(b) See Figure 44. For this relation, the domain represents the employees of Sara's Pre-Owned Car Mart and the range represents their phone number(s).

Figure 44

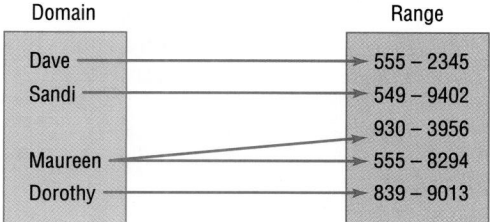

Solution (a) The relation in Figure 43 is a function because each element in the domain corresponds to exactly one element in the range. Notice that more than one element in the domain can correspond to the same element in the range. (Phoebe and Patrick were born on the same day of the year).

(b) The relation in Figure 44 is not a function because each element in the domain does not correspond to exactly one element in the range. Maureen has two telephone numbers; therefore, if Maureen is chosen from the domain, a single telephone number cannot be assigned to her. ∎

 NOW WORK PROBLEM 1.

We may think of a function as a set of ordered pairs (x, y) in which no two ordered pairs have the same first element, but different second elements. The set of all first elements x is the domain of the function, and the set of all second elements y is its range. Each element x in the domain corresponds to exactly one element y in the range.

EXAMPLE 3 **Determining Whether a Relation Represents a Function**

Determine whether each relation represents a function. If it is a function, state the domain and range.

(a) $\{(1, 4), (2, 5), (3, 6), (4, 7)\}$

(b) $\{(1, 4), (2, 4), (3, 5), (6, 10)\}$

(c) $\{(-3, 9), (-2, 4), (0, 0), (1, 1), (-3, 8)\}$

Solution (a) This relation is a function because there are no ordered pairs with the same first element and different second elements. The domain of this function is $\{1, 2, 3, 4\}$, and its range is $\{4, 5, 6, 7\}$.

(b) This relation is a function because there are no ordered pairs with the same first element and different second elements. The domain of this function is $\{1, 2, 3, 6\}$, and its range is $\{4, 5, 10\}$.

(c) This relation is not a function because there are two ordered pairs $(-3, 9)$ and $(-3, 8)$ that have the same first element, but different second elements. ■

In Example 3(b), notice that 1 and 2 in the domain each correspond to 4. This does not violate the definition of a function; two different first elements can have the same second element. A violation of the definition occurs when two ordered pairs have the same first element and different second elements, as in Example 3(c).

NOW WORK PROBLEM 5.

Example 2(a) demonstrates that a function may be defined by a correspondence between two sets. Examples 3(a) and 3(b) demonstrate that a function may be defined by a set of ordered pairs. A function may also be defined by an equation in two variables, usually denoted x and y.

EXAMPLE 4 **Example of a Function**

Consider the equation

$$y = 2x - 5 \qquad 1 \le x \le 6$$

Notice that for each input x there corresponds exactly one output y. For example, if $x = 1$, then $y = 2(1) - 5 = -3$. If $x = 3$, then $y = 2(3) - 5 = 1$. For this reason, the equation is a function. Since we restrict the inputs to the real numbers between 1 and 6, inclusive, the domain of the function is $\{x \mid 1 \le x \le 6\}$. The function specifies that in order to get the image of x we multiply x by 2 and then subtract 5 from this product. ■

Function Notation

Functions are often denoted by letters such as f, F, g, G, and so on. If f is a function, then for each number x in its domain the corresponding image in the range is designated by the symbol $f(x)$, read as "f of x" or as "f at x." We refer to $f(x)$ as the **value of f at the number x**; $f(x)$ is the number that results when x is given and the function f is applied; $f(x)$ does *not* mean "f times x." For example, the function given in Example 4 may be written as $y = f(x) = 2x - 5$, $1 \le x \le 6$. Then $f(1) = -3$.

Figure 45 illustrates some other functions. Note that, in every function illustrated, for each x in the domain there is one value in the range.

Figure 45

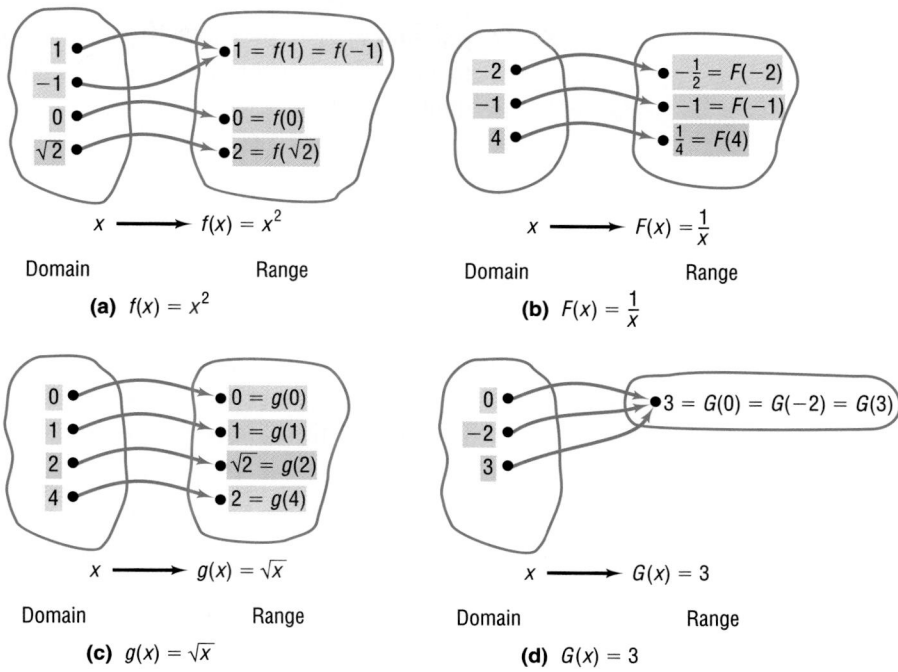

(a) $f(x) = x^2$

(b) $F(x) = \dfrac{1}{x}$

(c) $g(x) = \sqrt{x}$

(d) $G(x) = 3$

Sometimes it is helpful to think of a function f as a machine that receives as input a number from the domain, manipulates it, and outputs the value. See Figure 46.

Figure 46

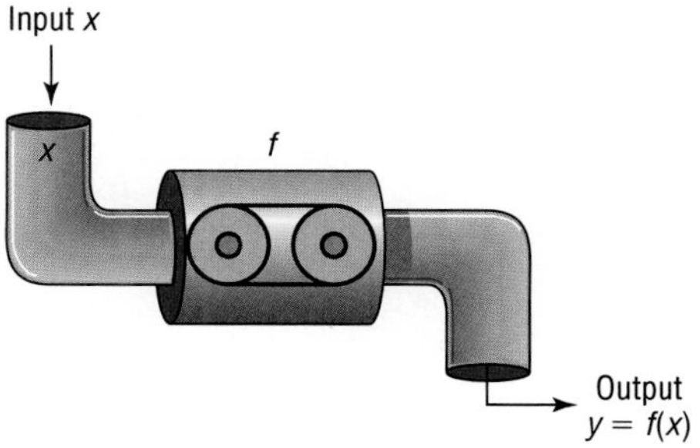

The restrictions on this input/output machine are

1. It only accepts numbers from the domain of the function.
2. For each input, there is exactly one output (which may be repeated for different inputs).

For a function $y = f(x)$, the variable x is called the **independent variable**, because it can be assigned any of the numbers from the domain. The variable y is called the **dependent variable**, because its value depends on x.

Any symbol can be used to represent the independent and dependent variables. For example, if f is the *cube function*, then f can be defined by $f(x) = x^3$ or $f(t) = t^3$ or $f(z) = z^3$. All three functions are the same: Each tells us to cube the independent variable. In practice, the symbols used for the independent and dependent variables are based on common usage, such as using C for cost in business.

2   The independent variable is also called the **argument** of the function. Thinking of the independent variable as an argument can sometimes make it easier to find the value of a function. For example, if f is the function defined by $f(x) = x^3$, then f tells us to cube the argument. Thus, $f(2)$ means to cube 2, $f(a)$ means to cube the number a, and $f(x + h)$ means to cube the quantity $x + h$.

EXAMPLE 5   **Finding Values of a Function**

For the function f defined by $f(x) = 2x^2 - 3x$, evaluate:

(a) $f(3)$ (b) $f(x) + f(3)$ (c) $f(-x)$

(d) $-f(x)$ (e) $f(x + 3)$ (f) $\dfrac{f(x + h) - f(x)}{h}$

Solution   (a) We substitute 3 for x in the equation for f to get

$$f(3) = 2(3)^2 - 3(3) = 18 - 9 = 9$$

(b) $f(x) + f(3) = (2x^2 - 3x) + (9) = 2x^2 - 3x + 9$

(c) We substitute $-x$ for x in the equation for f:

$$f(-x) = 2(-x)^2 - 3(-x) = 2x^2 + 3x$$

(d) $-f(x) = -(2x^2 - 3x) = -2x^2 + 3x$

(e) $f(x + 3) = 2(x + 3)^2 - 3(x + 3)$ Notice the use of parentheses here.

$= 2(x^2 + 6x + 9) - 3x - 9$

$= 2x^2 + 12x + 18 - 3x - 9$

$= 2x^2 + 9x + 9$

(f) $\dfrac{f(x + h) - f(x)}{h} = \dfrac{[2(x + h)^2 - 3(x + h)] - [2x^2 - 3x]}{h}$

↑
$f(x + h) = 2(x + h)^2 - 3(x + h)$

$= \dfrac{2(x^2 + 2xh + h^2) - 3x - 3h - 2x^2 + 3x}{h}$ Simplify.

$= \dfrac{2x^2 + 4xh + 2h^2 - 3h - 2x^2}{h}$

$= \dfrac{4xh + 2h^2 - 3h}{h}$

$= \dfrac{h(4x + 2h - 3)}{h}$ Factor out h.

$= 4x + 2h - 3$ Cancel h.   ■

 Notice in this example that $f(x + 3) \neq f(x) + f(3)$ and $f(-x) \neq -f(x)$. Also, the expression in part (f) is called the **difference quotient** of f, an important expression in calculus.

 NOW WORK PROBLEMS **15** AND **73.**

Graphing calculators have special keys that enable you to find the value of certain commonly used functions. For example, you should be able to find the square function $f(x) = x^2$, the square root function $f(x) = \sqrt{x}$, the reciprocal function $f(x) = \dfrac{1}{x} = x^{-1}$, and many others that will be discussed later in this book (such as $\ln x$, $\log x$, and so on). Verify the results of Example 6, which follows, on your calculator.

EXAMPLE 6   **Finding Values of a Function on a Calculator**

(a) $f(x) = x^2$; $f(1.23) = 1.5129$

(b) $F(x) = \dfrac{1}{x}$; $F(1.6) = 0.625$

(c) $g(x) = \sqrt{x}$; $g(1.234) \approx 1.110855526$ ■

Graphing calculators can also be used to evaluate any function that you wish. Figure 47 shows the result obtained in Example 5(a) on a TI-83 graphing calculator with the function to be evaluated, $f(x) = 2x^2 - 3x$, in Y_1.*

Figure 47

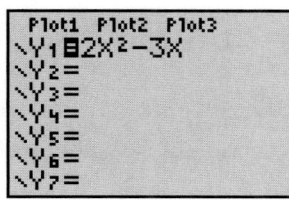 

Implicit Form of a Function

In general, when a function f is defined by an equation in x and y, we say that the function f is given **implicitly**. If it is possible to solve the equation for y in terms of x, then we write $y = f(x)$ and say that the function is given **explicitly**. For example,

Implicit Form	**Explicit Form**
$3x + y = 5$	$y = f(x) = -3x + 5$
$x^2 - y = 6$	$y = f(x) = x^2 - 6$
$xy = 4$	$y = f(x) = \dfrac{4}{x}$

Not all equations in x and y define a function $y = f(x)$. If an equation is solved for y and two or more values of y can be obtained for a given x, then the equation does not define a function.

EXAMPLE 7   **Determining Whether an Equation Is a Function**

Determine if the equation $x^2 + y^2 = 1$ is a function.

---

*Consult your owner's manual for the required keystrokes.

Solution   To determine whether the equation $x^2 + y^2 = 1$, which defines the unit circle, is a function, we need to solve the equation for y.

$$x^2 + y^2 = 1$$
$$y^2 = 1 - x^2$$
$$y = \pm\sqrt{1 - x^2}$$

For values of x between -1 and 1, two values of y result. For example, if $x = 0$, then $y = \pm\sqrt{1} = \pm1$. This means that the equation $x^2 + y^2 = 1$ is not a function. ∎

NOW WORK PROBLEM 27.

COMMENT   The explicit form of a function is the form required by a graphing calculator. Now do you see why it is necessary to graph a circle in two "pieces"? ∎

We list next a summary of some important facts to remember about a function f.

SUMMARY   **Important Facts About Functions**

1. For each x in the domain of a function f, there is one and only one image $f(x)$ in the range.
2. f is the symbol that we use to denote the function. It is symbolic of the equation that we use to get from an x in the domain to $f(x)$ in the range.
3. If $y = f(x)$, then x is called the independent variable or argument of f, and y is called the dependent variable or the value of f at x.

Domain of a Function

③   Often the domain of a function f is not specified; instead, only the equation defining the function is given. In such cases, we agree that the domain of f is the largest set of real numbers for which the value $f(x)$ is a real number. The domain of f is the same as the domain of the variable x in the expression $f(x)$.

EXAMPLE 8   **Finding the Domain of a Function**

Find the domain of each of the following functions:

(a) $f(x) = x^2 + 5x$ (b) $g(x) = \dfrac{3x}{x^2 - 4}$ (c) $h(t) = \sqrt{4 - 3t}$

Solution   (a) The function f tells us to square a number and then add five times the number. Since these operations can be performed on any real number, we conclude that the domain of f is all real numbers.

(b) The function g tells us to divide $3x$ by $x^2 - 4$. Since division by 0 is not defined, the denominator $x^2 - 4$ can never be 0, so x can never equal -2 or 2. The domain of the function g is $\{x | x \neq -2, x \neq 2\}$.

(c) The function h tells us to take the square root of $4 - 3t$. But only non-negative numbers have real square roots, so the expression under the square root must be greater than or equal to 0. This requires that

$$4 - 3t \geq 0$$
$$-3t \geq -4$$
$$t \leq \frac{4}{3}$$

The domain of h is $\left\{ t \mid t \leq \frac{4}{3} \right\}$ or the interval $\left(-\infty, \frac{4}{3} \right]$. ∎

NOW WORK PROBLEM **37**.

If x is in the domain of a function f, we say that **f is defined at x**, or **$f(x)$ exists**. If x is not in the domain of f, we say that **f is not defined at x**, or **$f(x)$ does not exist**. For example, if $f(x) = \dfrac{x}{x^2 - 1}$, then $f(0)$ exists, but $f(1)$ and $f(-1)$ do not exist. (Do you see why?)

We have not said much about finding the range of a function. The reason is that when a function is defined by an equation it is often difficult to find the range.* Therefore, we shall usually be content to find just the domain of a function when only the rule for the function is given. We shall express the domain of a function using inequalities, interval notation, set notation, or words, whichever is most convenient.

The Graph of a Function

In applications, a graph often demonstrates more clearly the relationship between two variables than an equation or table would. For example, Table 9 shows the price per share of Microsoft stock at the end of each month from June 2000 through June 2001.

If we plot the data in Table 9, using the date as the x-coordinate and the price as the y-coordinate, and then connect the points, we obtain Figure 48.

Figure 48

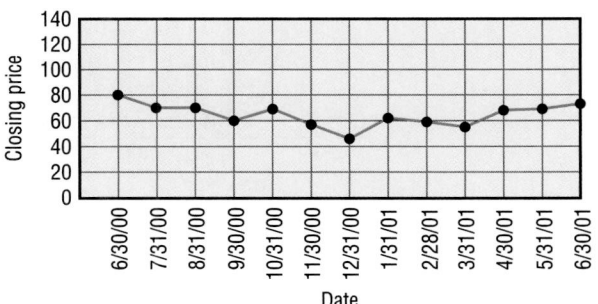

We can see from the graph that the price of the stock was falling during November and December and was rising in the month of January. The graph also shows that the lowest price occurred at the end of December, while the highest occurred at the end of June, 2000. Equations and tables, on the other hand, usually require some calculations and interpretation before this kind of information can be "seen."

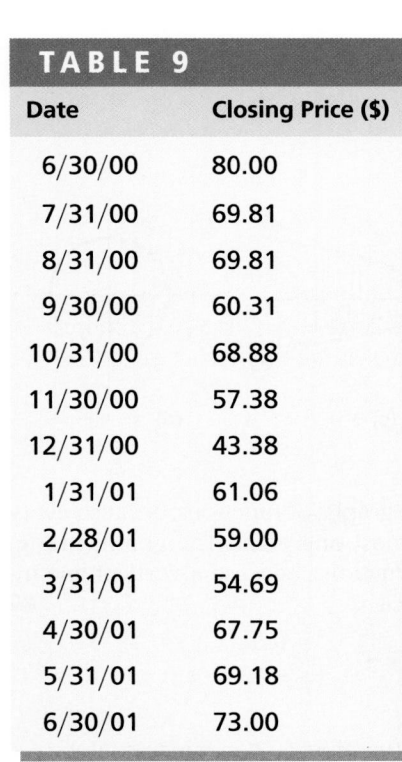

TABLE 9

Date	Closing Price ($)
6/30/00	80.00
7/31/00	69.81
8/31/00	69.81
9/30/00	60.31
10/31/00	68.88
11/30/00	57.38
12/31/00	43.38
1/31/01	61.06
2/28/01	59.00
3/31/01	54.69
4/30/01	67.75
5/31/01	69.18
6/30/01	73.00

Source: Courtesy of A.G. Edwards & Sons, Inc.

* In Section 1.7, we discuss a way to find the range of a certain class of functions.

Look again at Figure 48. The graph shows that for each date on the horizontal axis there is only one price on the vertical axis. Thus, the graph represents a function, although the exact rule for getting from date to price is not given.

When a function is defined by an equation in x and y, the **graph of the function** is the graph of the equation, that is, the set of points (x, y) in the xy- plane that satisfies the equation.

④   Not every collection of points in the xy-plane represents the graph of a function. Remember, for a function, each number x in the domain has one and only one image y. This means that the graph of a function cannot contain two points with the same x-coordinate and different y-coordinates. Therefore, the graph of a function must satisfy the following **vertical-line test**.

Theorem   **Vertical-line Test**

A set of points in the xy-plane is the graph of a function if and only if every vertical line intersects the graph in at most one point.

Another way of stating this result is as follows: If any vertical line intersects a graph at more than one point, the graph is not the graph of a function.

EXAMPLE 9   **Identifying the Graph of a Function**

Which of the graphs in Figure 49 are graphs of functions?

Figure 49

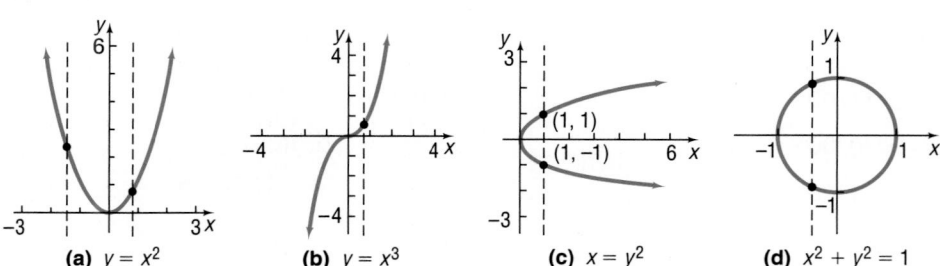

(a) $y = x^2$ (b) $y = x^3$ (c) $x = y^2$ (d) $x^2 + y^2 = 1$

Solution   The graphs in Figures 49(a) and 49(b) are graphs of functions, because every vertical line intersects each graph in at most one point. The graphs in Figures 49(c) and 49(d) are not graphs of functions, because a vertical line intersects each graph in more than one point.

━ **NOW WORK PROBLEM 53.**

⑤   If (x, y) is a point on the graph of a function f, then y is the value of f at x, that is, $y = f(x)$. The next example illustrates how to obtain information about a function if its graph is given.

EXAMPLE 10   Obtaining Information from the Graph of a Function

Figure 50

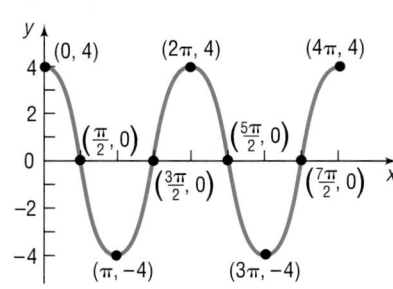

Let f be the function whose graph is given in Figure 50. (The graph of f might represent the distance that the bob of a pendulum is from its at-rest position. Negative values of y mean that the pendulum is to the left of the at-rest position, and positive values of y mean that the pendulum is to the right of the at-rest position.)

(a) What are $f(0), f\left(\dfrac{3\pi}{2}\right)$, and $f(3\pi)$?

(b) What is the domain of f?

(c) What is the range of f?

(d) List the intercepts. (Recall that these are the points, if any, where the graph crosses or touches the coordinate axes.)

(e) How often does the line $y = 2$ intersect the graph?

(f) For what values of x does $f(x) = -4$?

(g) For what values of x is $f(x) > 0$?

Solution

(a) Since $(0, 4)$ is on the graph of f, the y-coordinate 4 is the value of f at the x-coordinate 0; that is, $f(0) = 4$. In a similar way, we find that when $x = \dfrac{3\pi}{2}$, then $y = 0$, so $f\left(\dfrac{3\pi}{2}\right) = 0$. When $x = 3\pi$, then $y = -4$, so $f(3\pi) = -4$.

(b) To determine the domain of f, we notice that the points on the graph of f have x-coordinates between 0 and 4π, inclusive; and, for each number x between 0 and 4π there is a point $(x, f(x))$ on the graph. The domain of f is $\{x | 0 \le x \le 4\pi\}$ or the interval $[0, 4\pi]$.

(c) The points on the graph all have y-coordinates between -4 and 4, inclusive; and, for each such number y, there is at least one number x in the domain. The range of f is $\{y | -4 \le y \le 4\}$ or the interval $[-4, 4]$.

(d) The intercepts are $(0, 4)$, $\left(\dfrac{\pi}{2}, 0\right)$, $\left(\dfrac{3\pi}{2}, 0\right)$, $\left(\dfrac{5\pi}{2}, 0\right)$, and $\left(\dfrac{7\pi}{2}, 0\right)$.

(e) Draw the horizontal line $y = 2$ on the graph in Figure 50. We find that it intersects the graph four times.

(f) Since $(\pi, -4)$ and $(3\pi, -4)$ are the only points on the graph for which $y = f(x) = -4$, we have $f(x) = -4$ when $x = \pi$ and $x = 3\pi$.

(g) To determine where $f(x) > 0$, we look at Figure 50 and determine the x-values for which the y-coordinate is positive. This occurs on the intervals $\left[0, \dfrac{\pi}{2}\right)$, $\left(\dfrac{3\pi}{2}, \dfrac{5\pi}{2}\right)$, and $\left(\dfrac{7\pi}{2}, 4\pi\right]$. Using inequality notation $f(x) > 0$ for $0 \le x < \dfrac{\pi}{2}$, $\dfrac{3\pi}{2} < x < \dfrac{5\pi}{2}$, and $\dfrac{7\pi}{2} < x \le 4\pi$.

When the graph of a function is given, its domain may be viewed as the shadow created by the graph on the x-axis by vertical beams of light. Its range can be viewed as the shadow created by the graph on the y-axis by horizontal beams of light. Try this technique with the graph given in Figure 50.

NOW WORK PROBLEMS 47 AND 51.

EXAMPLE 11 | Obtaining Information about the Graph of a Function

Consider the function: $f(x) = \dfrac{x}{x+2}$

(a) Is the point $\left(1, \dfrac{1}{2}\right)$ on the graph of f?

(b) If $x = -1$, what is $f(x)$? What point is on the graph of f?
(c) If $f(x) = 2$, what is x? What point is on the graph of f?

Solution (a) When $x = 1$, then

$$f(x) = \frac{x}{x+2}$$

$$f(1) = \frac{1}{1+2} = \frac{1}{3}$$

The point $\left(1, \dfrac{1}{3}\right)$ is on the graph of f; the point $\left(1, \dfrac{1}{2}\right)$ is not.

(b) If $x = -1$, then

$$f(x) = \frac{x}{x+2}$$

$$f(-1) = \frac{-1}{-1+2} = -1$$

The point $(-1, -1)$ is on the graph of f.

(c) If $f(x) = 2$, then

$$f(x) = 2$$

$$\frac{x}{x+2} = 2$$

$$x = 2(x+2) \qquad \text{Multiply both sides by x + 2.}$$

$$x = 2x + 4 \qquad \text{Remove parentheses.}$$

$$x = -4 \qquad \text{Solve for x.}$$

If $f(x) = 2$, then $x = -4$. The point $(-4, 2)$ is on the graph of f. ∎

NOW WORK PROBLEM 63.

We can use the TABLE feature on a graphing calculator to verify the results of Example 11. We enter $Y_1 = \dfrac{x}{x+2}$ into a graphing utility and create Table 10. For example, when $x = -4$, $Y_1 = 2$, so $f(-4) = 2$. Why does the graphing utility display ERROR when $x = -2$?

TABLE 10

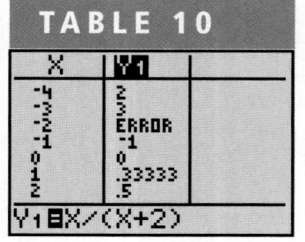

X	Y1
-4	2
-3	3
-2	ERROR
-1	-1
0	0
1	.33333
2	.5

Y₁■X/(X+2)

Application

When we use functions in applications, the domain may be restricted by physical or geometric considerations. For example, the domain of the function f defined by $f(x) = x^2$ is the set of all real numbers. However, if f is used to obtain the area of a square when the length x of a side is known, then we must restrict the domain of f to the positive real numbers, since the length of a side can never be 0 or negative.

EXAMPLE 12   **Area of a Circle**

Express the area of a circle as a function of its radius.

Figure 51

Solution   See Figure 51. We know that the formula for the area A of a circle of radius r is $A = \pi r^2$. If we use r to represent the independent variable and A to represent the dependent variable, the function expressing this relationship is

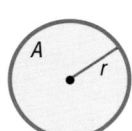

$$A(r) = \pi r^2$$

In this setting, the domain is $\{r \mid r > 0\}$. (Do you see why?)   ■

➤ **NOW WORK PROBLEM 87.**

SUMMARY

We list here some of the important vocabulary introduced in this section, with a brief description of each term.

Function	A relation between two sets of real numbers so that each number x in the first set, the domain, has corresponding to it exactly one number y in the second set. A set of ordered pairs (x, y) or $(x, f(x))$ in which no two ordered pairs have the same first element, but different second elements. The range is the set of y values of the function for the x values in the domain. A function f may be defined implicitly by an equation involving x and y or explicitly by writing $y = f(x)$.
Unspecified domain	If a function f is defined by an equation and no domain is specified, then the domain will be taken to be the largest set of real numbers for which the equation defines a real number.
Function notation	$y = f(x)$ f is a symbol for the function. x is the independent variable or argument. y is the dependent variable. $f(x)$ is the value of the function at x, or the image of x.
Graph of a function	The collection of points (x, y) that satisfies the equation $y = f(x)$. A collection of points is the graph of a function provided that every vertical line intersects the graph in at most one point (vertical-line test).

1.4 Concepts and Vocabulary

In Problems 1–3, fill in the blanks.

1. If f is a function defined by the equation $y = f(x)$, then x is called the _____ variable and y is the _____ variable.

2. A set of points in the xy-plane is the graph of a function if and only if every _____ line intersects the graph in at most one point.

3. The set of all images of the elements in the domain of a function is called the _____ .

In Problems 4–6, answer True or False to each statement.

4. Every relation is a function.

5. The y-intercept of the graph of the function $y = f(x)$, whose domain is all real numbers, is $f(0)$.

6. The independent variable is sometimes referred to as the argument of the function.

7. Describe how you would proceed to find the domain and range of a function if you were given its graph. How would your strategy change if you were given the equation defining the function instead of its graph?

8. How many x-intercepts can the graph of a function have? How many y-intercepts can it have?

9. Is a graph that consists of a single point the graph of a function? Can you write the equation of such a function?

1.4 Exercises

In Problems 1–12, determine whether each relation represents a function. For each function, state the domain and range.

1.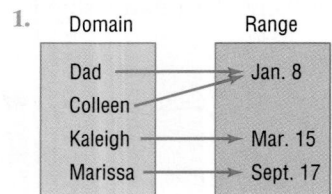

2.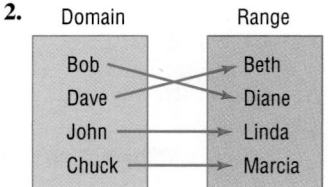

3.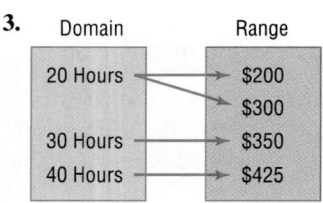

4.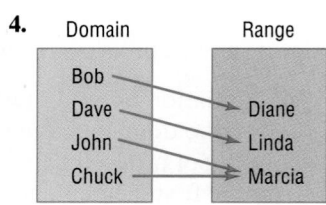

5. $\{(2,6), (-3,6), (4,9), (2,10)\}$

6. $\{(-2,5), (-1,3), (3,7), (4,12)\}$

7. $\{(1,3), (2,3), (3,3), (4,3)\}$

8. $\{(0,-2), (1,3), (2,3), (3,7)\}$

9. $\{(-2,4), (-2,6), (0,3), (3,7)\}$

10. $\{(-4,4), (-3,3), (-2,2), (-1,1), (-4,0)\}$

11. $\{(-2,4), (-1,1), (0,0), (1,1)\}$

12. $\{(-2,16), (-1,4), (0,3), (1,4)\}$

In Problems 13–20, find the following values for each function:

(a) $f(0)$ (b) $f(1)$ (c) $f(-1)$ (d) $f(-x)$ (e) $-f(x)$ (f) $f(x+1)$ (g) $f(2x)$ (h) $f(x+h)$

13. $f(x) = 2x + 5$

14. $f(x) = -3x + 1$

15. $f(x) = 3x^2 + 2x - 4$

16. $f(x) = -2x^2 + x - 1$

17. $f(x) = \dfrac{x}{x^2 + 1}$

18. $f(x) = \dfrac{x^2 - 1}{x + 4}$

19. $f(x) = |x| + 4$

20. $f(x) = \sqrt{x^2 + x}$

In Problems 21–32, determine whether the equation is a function.

21. $y = x^2$

22. $y = x^3$

23. $y = \dfrac{1}{x}$

24. $y = |x|$

25. $y^2 = 4 - x^2$

26. $y = \pm\sqrt{1 - 2x}$

27. $x = y^2$

28. $x + y^2 = 1$

29. $y = 2x^2 - 3x + 4$

30. $y = \dfrac{3x - 1}{x + 2}$

31. $2x^2 + 3y^2 = 1$

32. $x^2 - 4y^2 = 1$

In Problems 33–46, find the domain of each function.

33. $f(x) = -5x + 4$

34. $f(x) = x^2 + 2$

35. $f(x) = \dfrac{x}{x^2 + 1}$

36. $f(x) = \dfrac{x^2}{x^2 + 1}$

37. $g(x) = \dfrac{x}{x^2 - 16}$

38. $h(x) = \dfrac{2x}{x^2 - 4}$

39. $F(x) = \dfrac{x - 2}{x^3 + x}$

40. $G(x) = \dfrac{x + 4}{x^3 - 4x}$

41. $h(x) = \sqrt{3x - 12}$ **42.** $G(x) = \sqrt{1 - x}$ **43.** $f(x) = \dfrac{4}{\sqrt{x - 9}}$ **44.** $f(x) = \dfrac{x}{\sqrt{x - 4}}$

45. $p(x) = \sqrt{\dfrac{2}{x - 1}}$ **46.** $q(x) = \sqrt{-x - 2}$

47. Use the graph of the function f given below to answer parts (a)–(n).

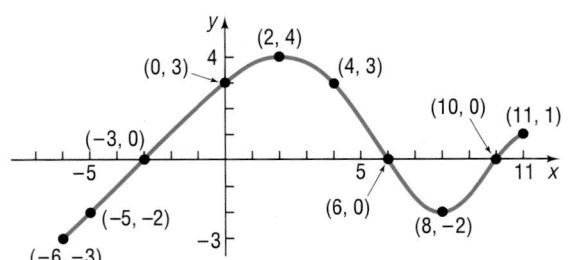
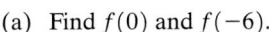

48. Use the graph of the function f given below to answer parts (a)–(n).

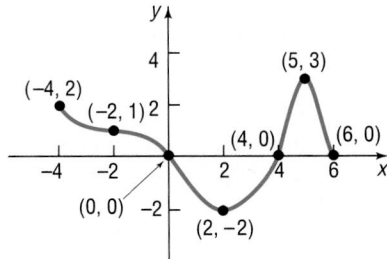

(a) Find $f(0)$ and $f(-6)$.
(b) Find $f(6)$ and $f(11)$.
(c) Is $f(3)$ positive or negative?
(d) Is $f(-4)$ positive or negative?
(e) For what numbers x is $f(x) = 0$?
(f) For what numbers x is $f(x) > 0$?
(g) What is the domain of f?
(h) What is the range of f?
(i) What are the x-intercepts?
(j) What is the y-intercept?

(k) How often does the line $y = \dfrac{1}{2}$ intersect the graph?

(l) How often does the line $x = 5$ intersect the graph?
(m) For what values of x does $f(x) = 3$?
(n) For what values of x does $f(x) = -2$?

(a) Find $f(0)$ and $f(6)$.
(b) Find $f(2)$ and $f(-2)$.
(c) Is $f(3)$ positive or negative?
(d) Is $f(-1)$ positive or negative?
(e) For what numbers x is $f(x) = 0$?
(f) For what numbers x is $f(x) < 0$?
(g) What is the domain of f?
(h) What is the range of f?
(i) What are the x-intercepts?
(j) What is the y-intercept?

(k) How often does the line $y = -1$ intersect the graph?
(l) How often does the line $x = 1$ intersect the graph?
(m) For what value of x does $f(x) = 3$?
(n) For what value of x does $f(x) = -2$?

In Problems 49–60, determine whether the graph is that of a function by using the vertical-line test. If it is, use the graph to find:
  (a) Its domain and range   (b) The intercepts, if any

49.
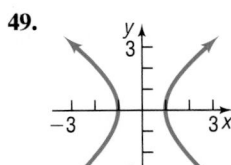

50.
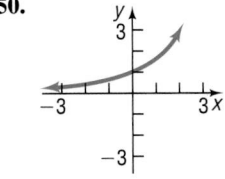

51.
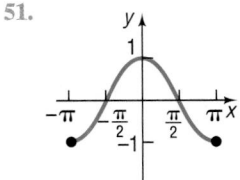

52.
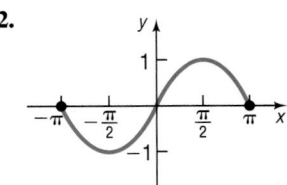

53.
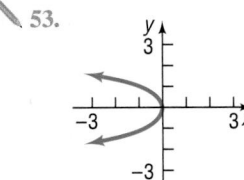

54.
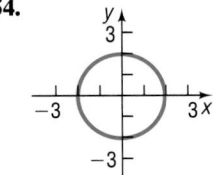

55.
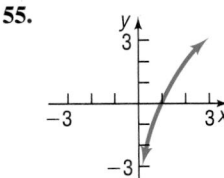

56.
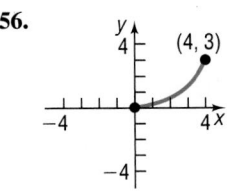

57.

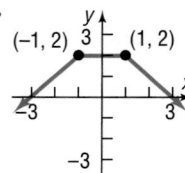

58.

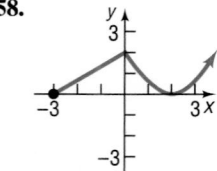

59.

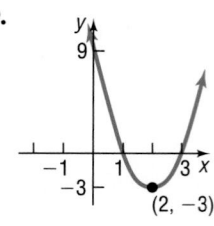

60.

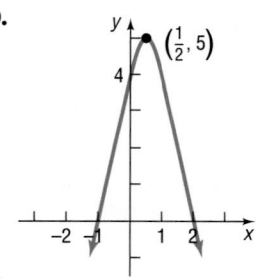

In Problems 61–66, answer the questions about the given function.

61. $f(x) = 2x^2 - x - 1$

   (a) Is the point $(-1, 2)$ on the graph of f?

   (b) If $x = -2$, what is $f(x)$? What point is on the graph of f?

   (c) If $f(x) = -1$, what is x? What point(s) are on the graph of f?

   (d) What is the domain of f?

   (e) List the x-intercepts, if any, of the graph of f.

   (f) List the y-intercept, if there is one, of the graph of f.

62. $f(x) = -3x^2 + 5x$

   (a) Is the point $(-1, 2)$ on the graph of f?

   (b) If $x = -2$, what is $f(x)$? What point is on the graph of f?

   (c) If $f(x) = -2$, what is x? What point(s) are on the graph of f?

   (d) What is the domain of f?

   (e) List the x-intercepts, if any, of the graph of f.

   (f) List the y-intercept, if there is one, of the graph of f.

63. $f(x) = \dfrac{x + 2}{x - 6}$

   (a) Is the point $(3, 14)$ on the graph of f?

   (b) If $x = 4$, what is $f(x)$? What point is on the graph of f?

   (c) If $f(x) = 2$, what is x? What point(s) are on the graph of f?

   (d) What is the domain of f?

   (e) List the x-intercepts, if any, of the graph of f.

   (f) List the y-intercept, if there is one, of the graph of f.

64. $f(x) = \dfrac{x^2 + 2}{x + 4}$

   (a) Is the point $\left(1, \dfrac{3}{5}\right)$ on the graph of f?

   (b) If $x = 0$, what is $f(x)$? What point is on the graph of f?

   (c) If $f(x) = \dfrac{1}{2}$, what is x? What point(s) are on the graph of f?

   (d) What is the domain of f?

   (e) List the x-intercepts, if any, of the graph of f.

   (f) List the y-intercept, if there is one, of the graph of f.

65. $f(x) = \dfrac{2x^2}{x^4 + 1}$

   (a) Is the point $(-1, 1)$ on the graph of f?

   (b) If $x = 2$, what is $f(x)$? What point is on the graph of f?

   (c) If $f(x) = 1$, what is x? What point(s) are on the graph of f?

   (d) What is the domain of f?

   (e) List the x-intercepts, if any, of the graph of f.

   (f) List the y-intercept, if there is one, of the graph of f.

66. $f(x) = \dfrac{2x}{x - 2}$

   (a) Is the point $\left(\dfrac{1}{2}, -\dfrac{2}{3}\right)$ on the graph of f?

   (b) If $x = 4$, what is $f(x)$? What point is on the graph of f?

   (c) If $f(x) = 1$, what is x? What point(s) are on the graph of f?

   (d) What is the domain of f?

   (e) List the x-intercepts, if any, of the graph of f.

   (f) List the y-intercept, if there is one, of the graph of f.

67. If $f(x) = 2x^3 - 4x^2 + 4x + C$ and $f(2) = 5$, what is the value of C?

68. If $f(x) = 3x^2 - 5x + C$ and $f(-1) = 12$, what is the value of C?

69. If $f(x) = \dfrac{3x + 8}{2x - A}$ and $f(0) = 2$, what is the value of A?

70. If $f(x) = \dfrac{2x - B}{3x + 4}$ and $f(2) = \dfrac{1}{2}$, what is the value of B?

71. If $f(x) = \dfrac{2x - A}{x - 3}$ and $f(4) = 0$, what is the value of A? Where is f not defined?

72. If $f(x) = \dfrac{x - B}{x - A}$, $f(2) = 0$, and $f(1)$ is undefined, what are the values of A and B?

In Problems 73–78, find the value of $\dfrac{f(x + h) - f(x)}{h}, h \neq 0,$ *for each function. Be sure to simplify.*

73. $f(x) = 4x + 3$ **74.** $f(x) = -3x + 1$ **75.** $f(x) = x^2 - x + 4$

76. $f(x) = x^2 + 5x - 1$ **77.** $f(x) = x^3 - 2$ **78.** $f(x) = \dfrac{1}{x + 3}$

79. Match each of the following functions with the graph that best describes the situation.
  (a) The cost of building a house as a function of its square footage
  (b) The height of an egg dropped from a 300-foot building as a function of time
  (c) The height of a human as a function of time
  (d) The demand for Big Macs as a function of price
  (e) The height of a child on a swing as a function of time

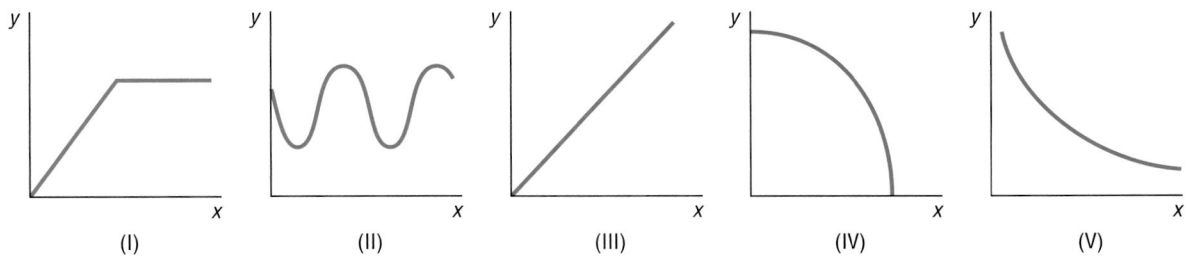

     (I)   (II)   (III)   (IV)   (V)

80. Match each of the following functions with the graph that best describes the situation.
  (a) The temperature of a bowl of soup as a function of time
  (b) The number of hours of daylight per day over a two year period
  (c) The population of Florida as a function of time
  (d) The distance of a car traveling at a constant velocity as a function of time
  (e) The height of a golf ball hit with a 7-iron as a function of time

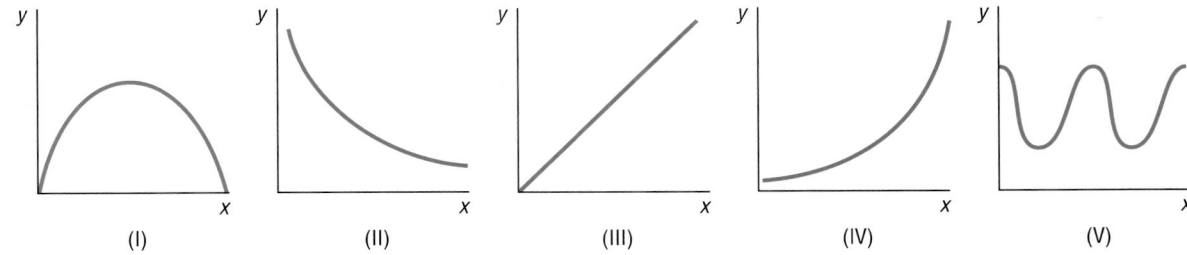

     (I)   (II)   (III)   (IV)   (V)

81. Consider the following scenario: Barbara decides to take a walk. She leaves home, walks 2 blocks in 5 minutes at a constant speed, and realizes that she forgot to lock the door. So Barbara runs home in 1 minute. While at her doorstep, it takes her 1 minute to find her keys and lock the door. Barbara walks 5 blocks in 15 minutes and then decides to jog home. It takes her 7 minutes to get home. Draw a graph of Barbara's distance from home (in blocks) as a function of time.

82. Consider the following scenario: Jayne enjoys riding her bicycle through the woods. At the forest preserve, she gets on her bicycle and rides up a 2,000-foot incline in 10 minutes. She then travels down the incline in 3 minutes. The next 5,000 feet is level terrain and she covers the distance in 20 minutes. She rests for 15 minutes. Jayne then travels 10,000 feet in 30 minutes. Draw a graph of Jayne's distance traveled (in feet) as a function of time.

83. The following sketch represents the distance d (in miles) that Kevin is from home as a function of time t (in hours). Answer the questions based on the graph. In parts (a)-(g), how many hours elapsed and how far was Kevin from home during this time?

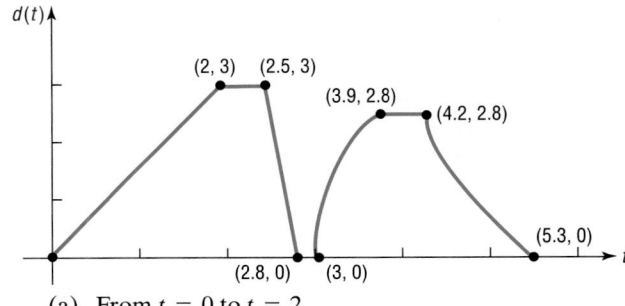

  (a) From $t = 0$ to $t = 2$
  (b) From $t = 2$ to $t = 2.5$

(c) From $t = 2.5$ to $t = 2.8$
(d) From $t = 2.8$ to $t = 3$
(e) From $t = 3$ to $t = 3.9$
(f) From $t = 3.9$ to $t = 4.2$
(g) From $t = 4.2$ to $t = 5.3$
(h) What is the farthest distance that Kevin is from home?
(i) How many times did Kevin return home?

84. The following sketch represents the speed v (in miles per hour) of Michael's car as a function of time t (in minutes).

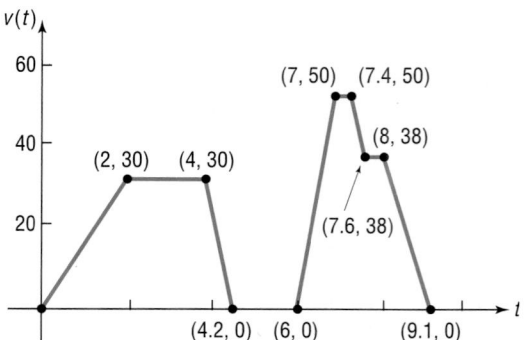

(a) Over what interval of time is Michael traveling fastest?
(b) Over what interval(s) of time is Michael's speed zero?
(c) What is Michael's speed between 2 and 4 minutes?
(d) What is Michael's speed between 4.2 and 6 minutes?
(e) What is Michael's speed between 7 and 7.4 minutes?
(f) When is Michael's speed constant?

85. Effect of Gravity on Earth If a rock falls from a height of 20 meters on Earth, the height H (in meters) after x seconds is approximately
$$H(x) = 20 - 4.9x^2$$

(a) Using a graphing utility, graph $H(x)$.
(b) What is the height of the rock when $x = 1$ second? $x = 1.1$ seconds? $x = 1.2$ seconds? $x = 1.3$ seconds?
(c) When is the height of the rock 15 meters? When is it 10 meters? When is it 5 meters?
(d) When does the rock strike the ground?

86. Effect of Gravity on Jupiter If a rock falls from a height of 20 meters on the planet Jupiter, its height H (in meters) after x seconds is approximately
$$H(x) = 20 - 13x^2$$

(a) Using a graphing utility, graph $H(x)$.
(b) What is the height of the rock when $x = 1$ second? $x = 1.1$ seconds? $x = 1.2$ seconds?
(c) When is the height of the rock 15 meters? When is it 10 meters? When is it 5 meters?
(d) When does the rock strike the ground?

87. Geometry Express the area A of a rectangle as a function of the length x if the length is twice the width of the rectangle.

88. Geometry Express the area A of an isosceles right triangle as a function of the length x of one of the two equal sides.

89. Express the gross salary G of a person who earns $10 per hour as a function of the number x of hours worked.

90. Tiffany, a commissioned salesperson, earns $100 base pay plus $10 per item sold. Express her gross salary G as a function of the number x of items sold.

91. Motion of a Golf Ball A golf ball is hit with an initial velocity of 130 feet per second at an inclination of 45° to the horizontal. In physics, it is established that the height h of the golf ball is given by the function
$$h(x) = \frac{-32x^2}{130^2} + x$$

where x is the horizontal distance that the golf ball has traveled.

(a) Determine the height of the golf ball after it has traveled 100 feet.
(b) What is the height after it has traveled 300 feet?
(c) What is the height after it has traveled 500 feet?
(d) Graph the function $h = h(x)$.
(e) Algebraically, determine the distance that the ball has traveled when the height of the ball is 90 feet. Verify your results graphically.
(f) Create a TABLE with TblStart $= 0$ and ΔTbl $= 25$. To the nearest 25 feet, how far does the ball travel before it reaches a maximum height? What is the maximum height?
(g) Adjust the value of ΔTbl until you determine the distance, to within 1 foot, that the ball travels before it reaches a maximum height.
(h) Find the domain of h.
Hint: For what values of x is $h(x) \geq 0$?

92. Cross-sectional Area The cross-sectional area of a beam cut from a log with radius 1 foot is given by the function $A(x) = 4x\sqrt{1 - x^2}$, where x represents the length of half the base of the beam. See the figure.

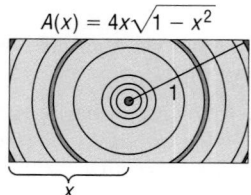

$$A(x) = 4x\sqrt{1 - x^2}$$

(a) Find the domain of A.
Determine the cross-sectional area of the beam if the length of half the base of the beam is as follows:
(b) One-third foot

(c)   One-half of a foot
(d)   Two-thirds of a foot
(e)   Graph the function $A = A(x)$.
(f)   Create a TABLE with TblStart = 0 and ΔTbl = 0.1 for $0 \le x \le 1$. Which value of x maximizes the cross-sectional area? What should be the length of the base of the beam to maximize the cross-sectional area?

93. **Cost of Trans-Atlantic Travel**   A Boeing 747 crosses the Atlantic Ocean (3000 miles) with an airspeed of 500 miles per hour. The cost C (in dollars) per passenger is given by

$$C(x) = 100 + \frac{x}{10} + \frac{36{,}000}{x}$$

where x is the ground speed (airspeed $\pm$ wind).

(a)   What is the cost per passenger for quiescent (no wind) conditions?
(b)   What is the cost per passenger with a head wind of 50 miles per hour?
(c)   What is the cost per passenger with a tail wind of 100 miles per hour?
(d)   What is the cost per passenger with a head wind of 100 miles per hour?
(e)   Graph the function $C = C(x)$.
(f)   Create a TABLE with TblStart = 0 and ΔTbl = 50. To the nearest 50 miles per hour, what ground speed minimizes the cost per passenger?

94. **Effect of Elevation on Weight**   If an object weighs m pounds at sea level, then its weight W (in pounds) at a height of h miles above sea level is given approximately by

$$W(h) = m\left(\frac{4000}{4000 + h}\right)^2$$

(a)   If Amy weighs 120 pounds at sea level, how much will she weigh on Pike's Peak, which is 14,110 feet above sea level?
(b)   Use a graphing utility to graph the function $W = W(h)$. Use $m = 120$ pounds.
(c)   Create the Table with TblStart = 0 and ΔTbl = 0.5 to see how the weight W varies as h changes from 0 to 5 miles.
(d)   At what height will Amy weigh 119.95 pounds?
(e)   Does your answer to part (d) seem reasonable?

95. Some functions f have the property that $f(a + b) = f(a) + f(b)$ for all real numbers a and b. Which of the following functions have this property?
(a) $h(x) = 2x$
(b) $g(x) = x^2$
(c) $F(x) = 5x - 2$
(d) $G(x) = \dfrac{1}{x}$

96. Draw the graph of a function whose domain is $\{x \mid -3 \le x \le 8,\ x \ne 5\}$ and whose range is $\{y \mid -1 \le y \le 2,\ y \ne 0\}$. What point(s) in the rectangle $-3 \le x \le 8$, $-1 \le y \le 2$ cannot be on the graph? Compare your graph with those of other students. What differences do you see?

97. Are the functions $f(x) = x - 1$ and $g(x) = \dfrac{x^2 - 1}{x + 1}$ the same? Explain.

98. Investigate when, historically, the use of function notation $y = f(x)$ first appeared.

PREPARING FOR THIS SECTION

Before getting started, review the following:

✓ Intervals (Section A.5, pp. 594–595)

✓ Symmetry (Section 1.3, pp. 20–22)

✓ Graphs of Certain Equations (Section 1.2: Example 3, p. 13; Section 1.3: Example 3, p. 23; Example 4, p. 23; Example 5, p. 25)

1.5   PROPERTIES OF FUNCTIONS; LIBRARY OF FUNCTIONS

OBJECTIVES

1   Use a Graph to Determine Where a Function Is Increasing, Is Decreasing, or Is Constant

2   Use a Graph to Locate Local Maxima and Minima

3   Use a Graphing Utility to Approximate Local Maxima and Minima and to Determine Where a Function Is Increasing or Decreasing

4   Determine Even and Odd Functions from a Graph

5   Identify Even and Odd Functions from the Equation

6   Graph the Functions Listed in the Library of Functions

Increasing and Decreasing Functions

 Consider the graph given in Figure 52. If you look from left to right along the graph of the function, you will notice that parts of the graph are rising, parts are falling, and parts are horizontal. In such cases, the function is described as *increasing, decreasing*, or *constant*, respectively.

Figure 52

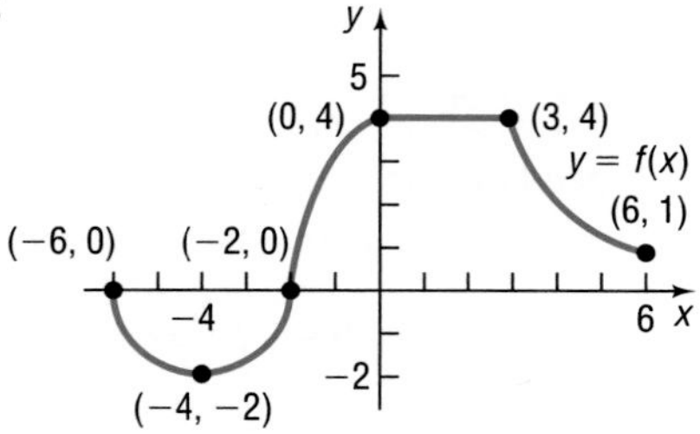

EXAMPLE 1 | **Determining Where a Function Is Increasing, Decreasing, or Constant from Its Graph**

Where is the function in Figure 52 increasing? Where is it decreasing? Where is it constant?

Solution | To answer the question of where a function is increasing, where it is decreasing, and where it is constant, we use strict inequalities involving the independent variable x, or we use open intervals* of x-coordinates. The graph in Figure 52 is rising (increasing) from the point $(-4, -2)$ to the point $(0, 4)$, so we conclude that it is increasing on the open interval $(-4, 0)$ $[-4 < x < 0]$. The graph is falling (decreasing) from the point $(-6, 0)$ to the point $(-4, -2)$ and from the point $(3, 4)$ to the point $(6, 1)$. We conclude that the graph is decreasing on the open intervals $(-6, -4)$ and $(3, 6)$ $[-6 < x < -4$ and $3 < x < 6]$. The graph is constant on the open interval $(0, 3)$ $[0 < x < 3]$. ∎

More precise definitions follow:

A function f is **increasing** on an open interval I if, for any choice of x_1 and x_2 in I, with $x_1 < x_2$, we have $f(x_1) < f(x_2)$.

A function f is **decreasing** on an open interval I if, for any choice of x_1 and x_2 in I, with $x_1 < x_2$, we have $f(x_1) > f(x_2)$.

A function f is **constant** on an open interval I if, for all choices of x in I, the values $f(x)$ are equal.

*The open interval (a, b) consists of all real numbers x for which $a < x < b$.

Figure 53 illustrates the definitions. The graph of an increasing function goes up from left to right, the graph of a decreasing function goes down from left to right, and the graph of a constant function remains at a fixed height.

Figure 53

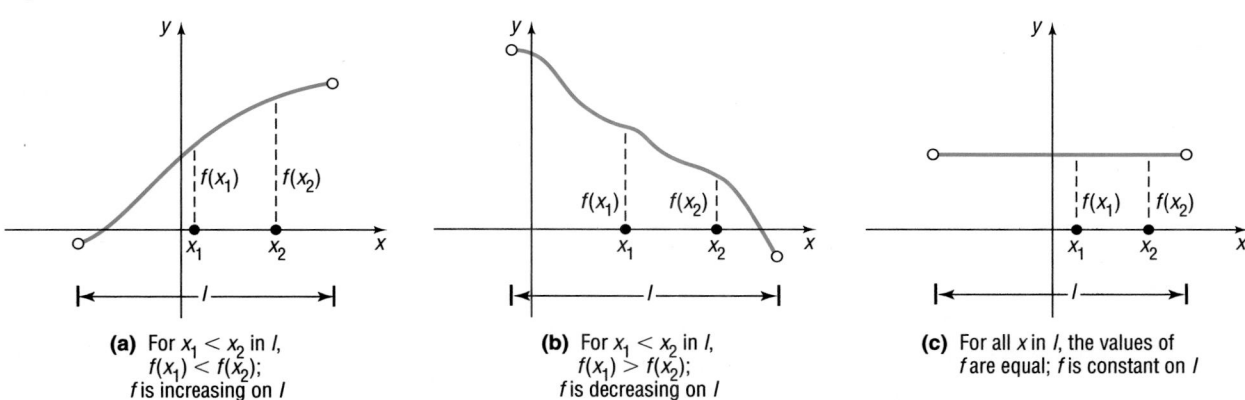

(a) For $x_1 < x_2$ in I, $f(x_1) < f(x_2)$; f is increasing on I

(b) For $x_1 < x_2$ in I, $f(x_1) > f(x_2)$; f is decreasing on I

(c) For all x in I, the values of f are equal; f is constant on I

 NOW WORK PROBLEMS 15, 17, AND 19.

Local Maximum; Local Minimum

2 When the graph of a function is increasing to the left of $x = c$ and decreasing to the right of $x = c$, then at c the value of f is largest. This value is called a *local maximum* of f. See Figure 54(a).

When the graph of a function is decreasing to the left of $x = c$ and is increasing to the right of $x = c$, then at c the value of f is the smallest. This value is called a *local minimum* of f. See Figure 54(b).

Figure 54

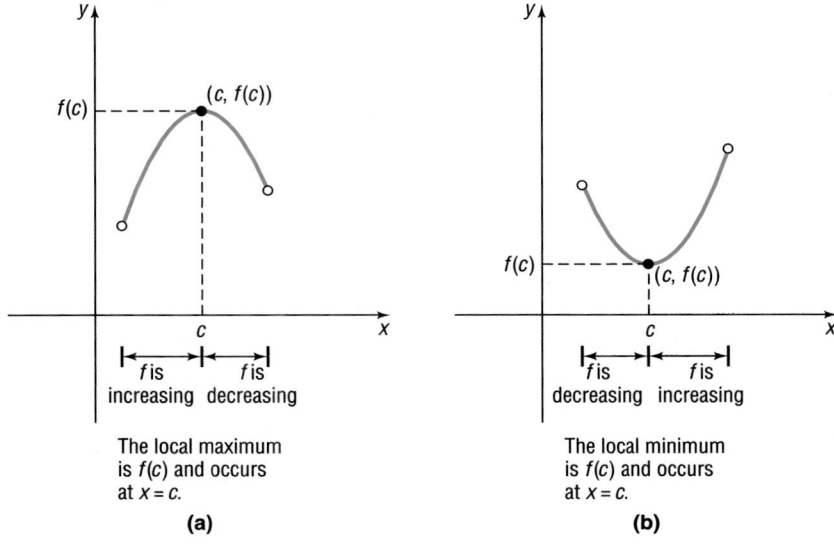

(a) The local maximum is $f(c)$ and occurs at $x = c$.

(b) The local minimum is $f(c)$ and occurs at $x = c$.

A function f has a **local maximum at c** if there is an open interval I containing c so that, for all $x \neq c$ in I, $f(x) < f(c)$. We call $f(c)$ a **local maximum of f**.

A function f has a **local minimum at c** if there is an open interval I containing c so that, for all $x \neq c$ in I, $f(x) > f(c)$. We call $f(c)$ a **local minimum of f.**

If f has a local maximum at c then the value of f at c is greater than the values of f near c. If f has a local minimum at c, then the value of f at c is less than the values of f near c. The word *local* is used to suggest that it is only near c that the value $f(c)$ is largest or smallest.

EXAMPLE 2   **Finding Local Maxima and Local Minima from the Graph of a Function and Determining Where the Function Is Increasing, Decreasing, or Constant**

Figure 55

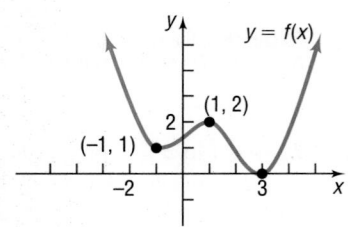

Figure 55 shows the graph of a function f.

(a) At what number(s), if any, does f have a local maximum?
(b) What are the local maxima?
(c) At what number(s), if any, does f have a local minimum?
(d) What are the local minima?
(e) List the intervals on which f is increasing. List the intervals on which f is decreasing.

Solution   The domain of f is the set of real numbers.

(a) f has a local maximum at 1, since for all x close to 1, $x \neq 1$, we have $f(x) < f(1)$.
(b) The local maximum is $f(1) = 2$.
(c) f has a local minimum at -1 and at 3.
(d) The local minima are $f(-1) = 1$ and $f(3) = 0$.
(e) The function whose graph is given in Figure 55 is increasing on the interval $(-1, 1)$. The function is also increasing for all values of x greater than 3. That is, the function is increasing on the intervals $(-1, 1)$ and $(3, \infty)$ $[-1 < x < 1$ and $x > 3]$. The function is decreasing for all values of x less than -1. The function is also decreasing on the interval $(1, 3)$. That is, the function is decreasing on the intervals $(-\infty, -1)$ and $(1, 3)$ $[x < -1$ and $1 < x < 3]$.   ∎

NOW WORK PROBLEMS **21** AND **23.**

To locate the exact value at which a function f has a local maximum or a local minimum usually requires calculus. However, a graphing utility may be used to approximate these values by using the MAXIMUM and MINIMUM features.*

EXAMPLE 3   **Using a Graphing Utility to Approximate Local Maxima and Minima and to Determine Where a Function Is Increasing or Decreasing**

(a) Use a graphing utility to graph $f(x) = 6x^3 - 12x + 5$ for $-2 < x < 2$. Approximate where f has a local maximum and where f has a local minimum.
(b) Determine where f is increasing and where it is decreasing.

*Consult your owner's manual for the appropriate keystrokes.

Solution   (a) Graphing utilities have a feature that finds the maximum or minimum point of a graph within a given interval. Graph the function f for $-2 < x < 2$. Using MAXIMUM, we find that the local maximum is 11.53 and it occurs at $x = -0.82$, rounded to two decimal places. See Figure 56(a). Using MINIMUM, we find that the local minimum is -1.53 and it occurs at $x = 0.82$, rounded to two decimal places. See Figure 56(b).

Figure 56

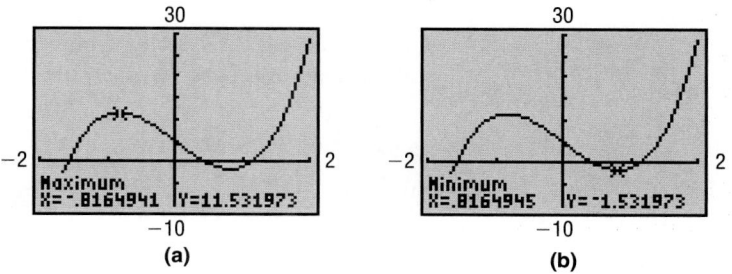

(a)   (b)

(b) Looking at Figures 56(a) and (b), we see that the graph of f is rising (increasing) from $x = -2$ to $x = -0.82$ and from $x = 0.82$ to $x = 2$, so f is increasing on the intervals $(-2, -0.82)$ and $(0.82, 2)$ $[-2 < x < -0.82$ and $0.82 < x < 2]$. The graph is falling (decreasing) from $x = -0.82$ to $x = 0.82$, so f is decreasing on the interval $(-0.82, 0.82)$ $[-0.82 < x < 0.82]$. ∎

NOW WORK PROBLEM 51.

Even and Odd Functions

4 A function f is even if and only if whenever the point (x, y) is on the graph of f then the point $(-x, y)$ is also on the graph. Algebraically, we define an even function as follows:

A function f is **even** if for every number x in its domain the number $-x$ is also in the domain and

$$f(-x) = f(x)$$

A function f is odd if and only if whenever the point (x, y) is on the graph of f then the point $(-x, -y)$ is also on the graph. Algebraically, we define an odd function as follows:

A function f is **odd** if for every number x in its domain the number $-x$ is also in the domain and

$$f(-x) = -f(x)$$

Refer to Section 1.3, where the tests for symmetry are listed. The following results are then evident.

Theorem	A function is even if and only if its graph is symmetric with respect to the y-axis. A function is odd if and only if its graph is symmetric with respect to the origin.

EXAMPLE 4   **Determining Even and Odd Functions from the Graph**

Determine whether each graph given in Figure 57 is the graph of an even function, an odd function, or a function that is neither even nor odd.

Figure 57

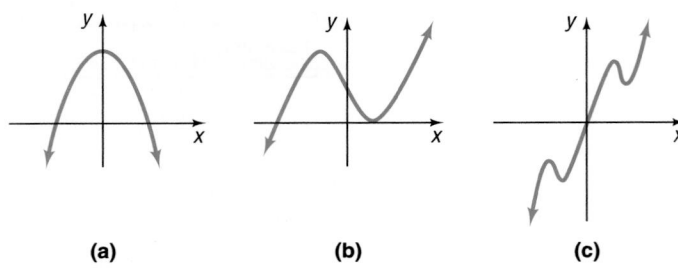

(a)   (b)   (c)

Solution   (a) The graph in Figure 57(a) is that of an even function, because the graph is symmetric with respect to the y-axis.

(b) The function whose graph is given in Figure 57(b) is neither even nor odd, because the graph is neither symmetric with respect to the y-axis nor symmetric with respect to the origin.

(c) The function whose graph is given in Figure 57(c) is odd, because its graph is symmetric with respect to the origin. ∎

 **NOW WORK PROBLEM 25.**

A graphing utility can be used to conjecture whether a function is even, odd, or neither. As stated, when the graph of an even function contains the point (x, y), it must also contain the point $(-x, y)$. Therefore, if the graph indicates evidence of symmetry with respect to the y-axis, then we would conjecture that the function is even.

In addition, if the graph shows evidence of symmetry with respect to the origin, then we would conjecture that the function is odd.

⑤   In the next example, we use a graphing utility to conjecture whether a function is even, odd, or neither. Then we verify our conjecture algebraically.

EXAMPLE 5   **Identifying Even and Odd Functions**

Use a graphing utility to conjecture whether each of the following functions is even, odd, or neither. Verify the conjecture algebraically. Then state whether the graph is symmetric with respect to the y-axis or with respect to the origin.

(a) $f(x) = x^2 - 5$ (b) $g(x) = x^3 - 1$ (c) $h(x) = 5x^3 - x$

Figure 58

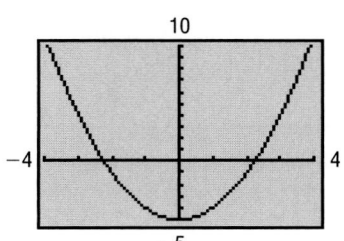

Solution

(a) Graph the function. See Figure 58. It appears the graph is symmetric with respect to the y-axis. We conjecture that the function is even.

To algebraically verify the conjecture, we replace x by $-x$ in $f(x) = x^2 - 5$. Then

$$f(-x) = (-x)^2 - 5 = x^2 - 5 = f(x)$$

Since $f(-x) = f(x)$, we conclude that f is an even function, and the graph is symmetric with respect to the y-axis.

Figure 59

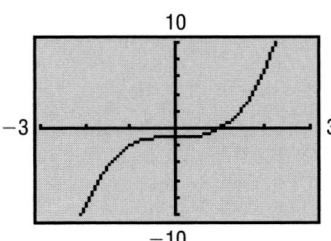

(b) Graph the function. See Figure 59. It appears there is no symmetry.

To algebraically verify that the function is not even, we find $g(-x)$.

$$g(-x) = (-x)^3 - 1 = -x^3 - 1; \quad g(x) = x^3 - 1$$

Since $g(-x) \neq g(x)$, the function is not even.

To algebraically verify that the function is not odd, we find $-g(x)$.

$$-g(x) = -(x^3 - 1) = -x^3 + 1; \quad g(-x) = -x^3 - 1$$

Since $g(-x) \neq -g(x)$, the function is not odd.

The graph is not symmetric with respect to the y-axis nor is it symmetric with respect to the origin.

Figure 60

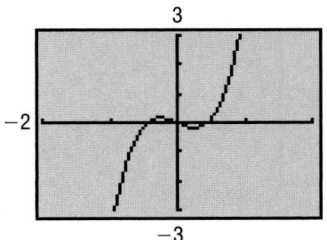

(c) Graph the function. See Figure 60. It appears there is symmetry with respect to the origin. We conjecture that the function is odd.

To algebraically verify the conjecture, we replace x by $-x$ in $h(x) = 5x^3 - x$. Then

$$h(-x) = 5(-x)^3 - (-x) = -5x^3 + x = -(5x^3 - x) = -h(x)$$

Since $h(-x) = -h(x)$, h is an odd function, and the graph of h is symmetric with respect to the origin. ■

— NOW WORK PROBLEM 37.

Library of Functions

6

We now give names to some of the functions that we have encountered. In going through this list, pay special attention to the properties of each function, particularly to the shape of each graph. Knowing these graphs will lay the foundation for later graphing techniques.

Figure 61

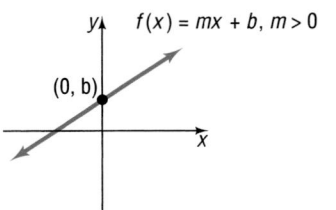

Linear Function

$$f(x) = mx + b \qquad m \text{ and } b \text{ are real numbers}$$

See Figure 61.

The domain of a linear function f consists of all real numbers. The graph of this function is a nonvertical line* with slope m and y-intercept b. A linear function is increasing if $m > 0$, decreasing if $m < 0$, and constant if $m = 0$.

Constant Function

$$f(x) = b \qquad b \text{ is a real number}$$

Figure 62

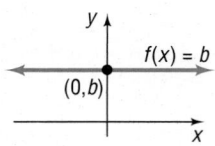

See Figure 62.

A **constant function** is a special linear function ($m = 0$). Its domain is the set of all real numbers; its range is the set consisting of a single number b. Its graph is a horizontal line whose y-intercept is b. The constant function is an even function whose graph is constant over its domain.

Identity Function

$$f(x) = x$$

See Figure 63.

Figure 63

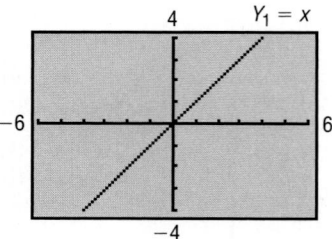

The **identity function** is also a special linear function. Its domain and range are the set of all real numbers. Its graph is a line whose slope is $m = 1$ and whose y-intercept is 0. The line consists of all points for which the x-coordinate equals the y-coordinate. The identity function is an odd function that is increasing over its domain. Note that the graph bisects quadrants I and III.

Square Function

$$f(x) = x^2$$

See Figure 64.

Figure 64

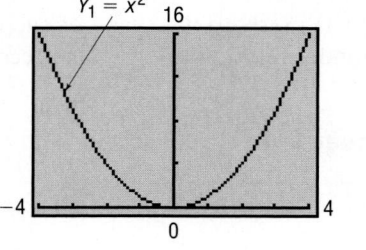
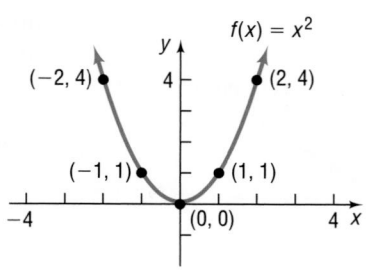

* Lines are discussed in the Appendix, Section A.7.

The domain of the **square function** f is the set of all real numbers; its range is the set of nonnegative real numbers. The graph of this function is a parabola whose intercept is at $(0, 0)$. The square function is an even function that is decreasing on the interval $(-\infty, 0)$ and increasing on the interval $(0, \infty)$.

Cube Function

$$f(x) = x^3$$

See Figure 65.

Figure 65

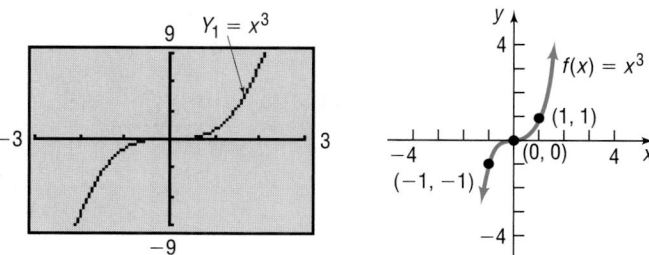

The domain and range of the **cube function** are the set of all real numbers. The intercept of the graph is at $(0, 0)$. The cube function is an odd function that is increasing on the interval $(-\infty, \infty)$.

Square Root Function

$$f(x) = \sqrt{x}$$

See Figure 66.

Figure 66

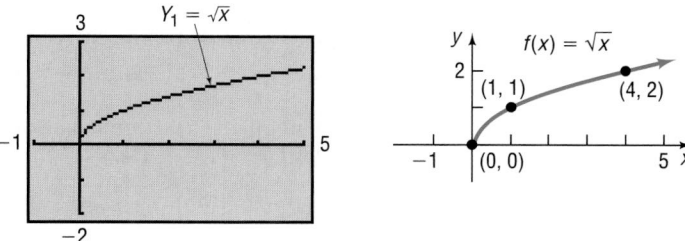

The domain and range of the **square root function** are the set of nonnegative real numbers. The intercept of the graph is at $(0, 0)$. The square root function is neither even nor odd and is increasing on the interval $(0, \infty)$.

Reciprocal Function

$$f(x) = \frac{1}{x}$$

Refer to Example 5, page 25, for a discussion of the equation $y = \dfrac{1}{x}$. See Figure 67.

Figure 67

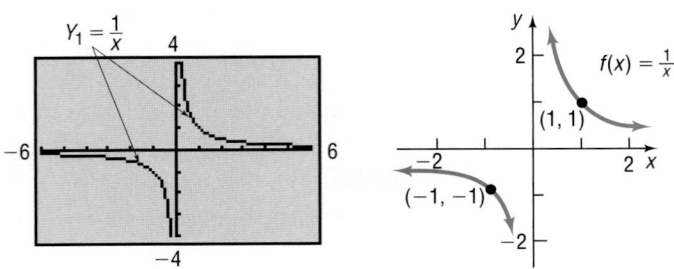

The domain and range of the **reciprocal function** are the set of all non-zero real numbers. The graph has no intercepts. The reciprocal function is an odd function and is decreasing on the intervals $(-\infty, 0)$ and $(0, \infty)$.

Absolute Value Function

$$f(x) = |x|$$

See Figure 68.

Figure 68

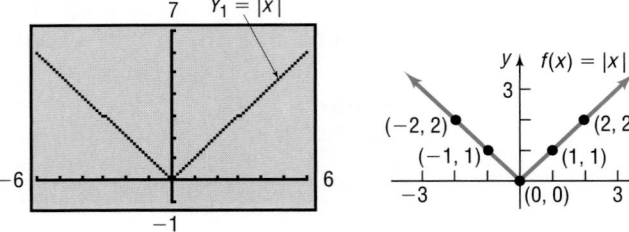

The domain of the **absolute value function** is the set of all real numbers; its range is the set of nonnegative real numbers. The intercept of the graph is at $(0, 0)$. If $x \geq 0$, then $f(x) = x$, and the graph of f is part of the line $y = x$; if $x < 0$, then $f(x) = -x$, and the graph of f is part of the line $y = -x$. The absolute value function is an even function; it is decreasing on the interval $(-\infty, 0)$ and increasing on the interval $(0, \infty)$.

NOW WORK PROBLEMS 1–7.

1.5   Concepts and Vocabulary

In Problems 1–3, fill in the blanks.

1. For the linear function $f(x) = mx + b$, the number m is called the _____ .

2. A function f is _____ on an open interval I if for any choice of x_1 and x_2 in I, with $x_1 < x_2$, we have $f(x_1) < f(x_2)$.

3. An _____ function f is one for which $f(-x) = f(x)$ for every x in the domain of f; an _____ function f is one for which $f(-x) = -f(x)$ for every x in the domain of f.

In Problems 4–6, answer True or False for each statement.

4. A function f is decreasing on an open interval I if, for any choice of x_1 and x_2 in I, with $x_1 < x_2$, we have $f(x_1) > f(x_2)$.

5. A function f has a local maximum at c if there is an open interval I containing c so that, for all $x \neq c$ in I, $f(x) < f(c)$.

6. Even functions have graphs that are symmetric with respect to the origin.

7. Describe the square function. Include a graph.

8. Suppose that a friend of yours does not understand the idea of increasing and decreasing functions. Provide an explanation complete with graphs that clarifies the idea.

9. Can a function be both even and odd? Explain.

1.5 Exercises

In Problems 1–7, match each graph to the function listed whose graph most resembles the one given.

A. *Constant function*   B. *Linear function*   C. *Square function*
D. *Cube function*   E. *Square root function*   F. *Reciprocal function*
G. *Absolute value function*

In Problems 8–14, sketch the graph of each function. Label at least three points.

8. $f(x) = 2$ **9.** $f(x) = x$ **10.** $f(x) = x^2$ **11.** $f(x) = x^3$

12. $f(x) = \sqrt{x}$ **13.** $f(x) = \dfrac{1}{x}$ **14.** $f(x) = |x|$

In Problems 15–24, use the graph of the function f given below.

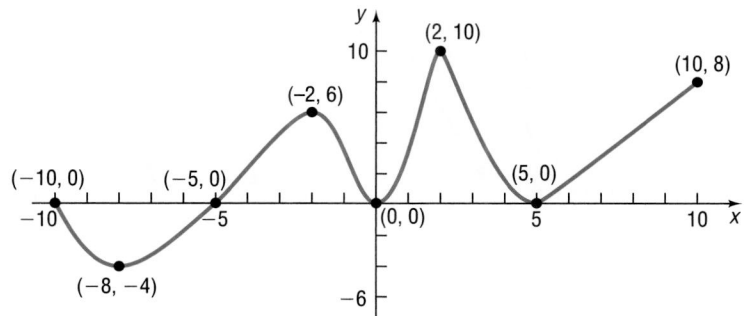

15. Is f increasing on the interval $(-8, -2)$?

16. Is f decreasing on the interval $(-8, -4)$?

17. Is f increasing on the interval $(2, 10)$?

18. Is f decreasing on the interval $(2, 5)$?

19. List the interval(s) on which f is increasing.

20. List the interval(s) on which f is decreasing.

21. Is there a local maximum at 2? If yes, what is it?

22. Is there a local maximum at 5? If yes, what is it?

23. List the numbers at which f has a local maximum. What are these local maxima?

24. List the numbers at which f has a local minimum. What are these local minima?

In Problems 25–32, the graph of a function is given. Use the graph to find:
  (a) *The intercepts, if any*
  (b) *Its domain and range*
  (c) *The intervals on which it is increasing, decreasing, or constant*
  (d) *Whether it is even, odd, or neither*

25.
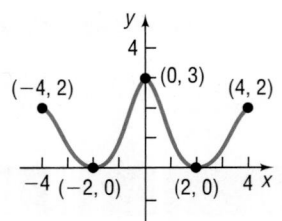

26.
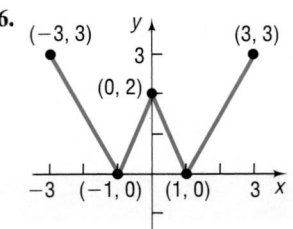

27.
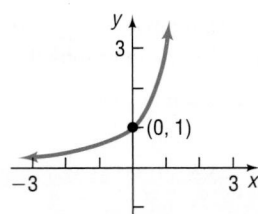

28.
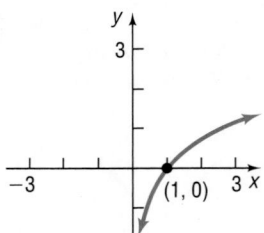

29.
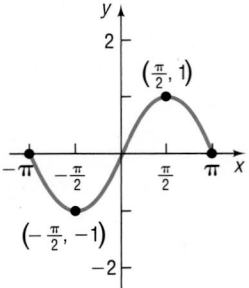

30.
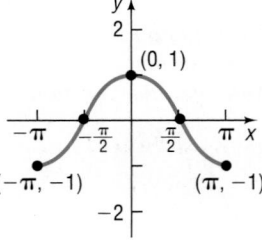

31.
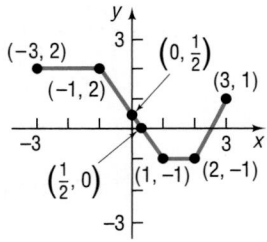

32.
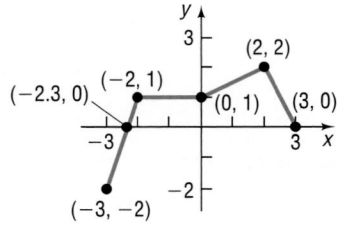

In Problems 33–36, the graph of a function f is given. Use the graph to find
  (a) *The numbers, if any, at which f has a local maximum. What are these local maxima?*
  (b) *The numbers, if any, at which f has a local minimum. What are these local minima?*

33.
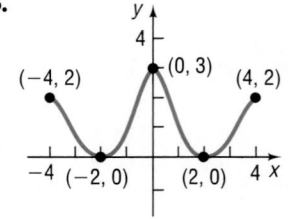

34.
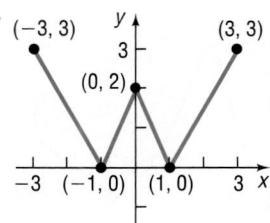

35.
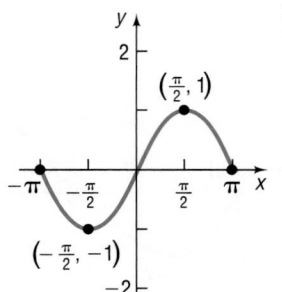

36.
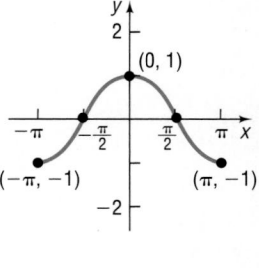

In Problems 37–48, use a graphing utility to graph each function; from the graph, conjecture whether the function is even, odd, or neither. Finally, algebraically verify the conjecture.

37. $f(x) = 4x^3$

38. $f(x) = 2x^4 - x^2$

39. $g(x) = -3x^2 - 5$

40. $h(x) = 3x^3 + 5$

41. $F(x) = \sqrt[3]{x}$

42. $G(x) = \sqrt{x}$

43. $f(x) = x + |x|$

44. $f(x) = \sqrt[3]{2x^2 + 1}$

45. $g(x) = \dfrac{1}{x^2}$

46. $h(x) = \dfrac{x}{x^2 - 1}$

47. $h(x) = \dfrac{-x^3}{3x^2 - 9}$

48. $F(x) = \dfrac{2x}{|x|}$

49. How many *x*-intercepts can a function defined on an interval have if it is increasing on that interval? Explain.

50. How many *y*-intercepts can a function have? Explain.

In Problems 51–58, use a graphing utility to graph each function over the indicated interval and approximate any local maxima and local minima. Determine where the function is increasing and where it is decreasing. Round answers to two decimal places.

51. $f(x) = x^3 - 3x + 2$ $(-2, 2)$

52. $f(x) = x^3 - 3x^2 + 5$ $(-1, 3)$

53. $f(x) = x^5 - x^3$ $(-2, 2)$

54. $f(x) = x^4 - x^2$ $(-2, 2)$

55. $f(x) = -0.2x^3 - 0.6x^2 + 4x - 6$ $(-6, 4)$

56. $f(x) = -0.4x^3 + 0.6x^2 + 3x - 2$ $(-4, 5)$

57. $f(x) = 0.25x^4 + 0.3x^3 - 0.9x^2 + 3$ $(-3, 2)$

58. $f(x) = -0.4x^4 - 0.5x^3 + 0.8x^2 - 2$ $(-3, 2)$

Problems 59–66 require the following definition of a secant line.

The slope of the line containing the two points $(x, f(x))$ and $(x + h, f(x + h))$ on the graph of a function $y = f(x)$ may be given as

$$\frac{\Delta y}{\Delta x} = \frac{f(x + h) - f(x)}{(x + h) - x} = \frac{f(x + h) - f(x)}{h}, \qquad h \neq 0$$

*In calculus, this expression is called the **difference quotient** of f.*

(a) *Express the difference quotient of each function in terms of x and h. Be sure to simplify your answer.*

(b) *Find* $\dfrac{\Delta y}{\Delta x}$ *for h = 0.5, 0.1, and 0.01 at x = 1. What value does* $\dfrac{\Delta y}{\Delta x}$ *approach as h approaches 0?*

(c) *Find the equation for the line at x = 1 with h = 0.01.*

(d) *Graph f and the line found in part (c) on the same viewing window.*

59. $f(x) = 2x + 5$

60. $f(x) = -3x + 2$

61. $f(x) = x^2 + 2x$

62. $f(x) = 2x^2 + x$

63. $f(x) = 2x^2 - 3x + 1$

64. $f(x) = -x^2 + 3x - 2$

65. $f(x) = \dfrac{1}{x}$

66. $f(x) = \dfrac{1}{x^2}$

---

67. Maximizing the Volume of a Box   An open box with a square base is to be made from a square piece of cardboard 24 inches on a side by cutting out a square from each corner and turning up the sides. (See the illustration.) The volume *V* of the box as a function of the length *x* of the side of the square cut from each corner is

$$V(x) = x(24 - 2x)^2$$

Graph *V* and determine where *V* is largest.

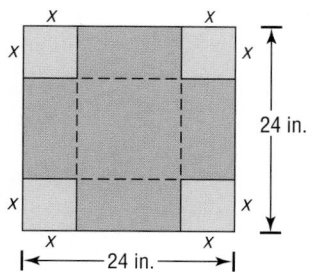

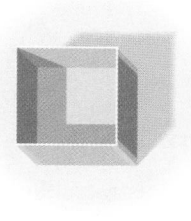

68. Minimizing the Material Needed to Make a Box   An open box with a square base is required to have a volume

of 10 cubic feet. The amount *A* of material used to make such a box as a function of the length *x* of a side of the square base is

$$A(x) = x^2 + \frac{40}{x}$$

Graph *A* and determine where *A* is smallest.

69. Maximum Height of a Ball   The height *s* of a ball (in feet) thrown with an initial velocity of 80 feet per second from an initial height of 6 feet is given as a function of the time *t* (in seconds) by

$$s(t) = -16t^2 + 80t + 6$$

(a) Graph *s*.

(b) Determine the time at which height is maximum.

(c) What is the maximum height?

70. Minimum Average Cost   The average cost of producing *x* riding lawn mowers per hour is given by

$$C(x) = 0.3x^2 + 21x - 251 + \frac{2500}{x}$$

(a) Graph C.

(b) Determine the number of riding lawn mowers to produce in order to minimize average cost.

(c) What is the minimum average cost?

71. Exploration   Graph $y = x^2$. Then on the same screen graph $y = x^2 + 2$, followed by $y = x^2 + 4$, followed by $y = x^2 - 2$. What pattern do you observe? Can you predict the graph of $y = x^2 - 4$? Of $y = x^2 + 5$.

72. Exploration   Graph $y = x^2$. Then on the same screen graph $y = (x - 2)^2$, followed by $y = (x - 4)^2$, followed by $y = (x + 2)^2$. What pattern do you observe? Can you predict the graph of $y = (x + 4)^2$? Of $y = (x - 5)^2$?

73. Exploration   Graph $y = |x|$. Then on the same screen graph $y = 2|x|$, followed by $y = 4|x|$, followed by $y = \frac{1}{2}|x|$. What pattern do you observe? Can you predict the graph of $y = \frac{1}{4}|x|$? Of $y = 5|x|$?

74. Exploration   Graph $y = x^2$. Then on the same screen graph $y = -x^2$. What do you observe? Now try $y = |x|$ and $y = -|x|$. What do you conclude?

75. Exploration   Graph $y = \sqrt{x}$. Then on the same screen graph $y = \sqrt{-x}$. What do you observe? Now try $y = 2x + 1$ and $y = 2(-x) + 1$. What do you conclude?

76. Exploration   Graph $y = x^3$. Then on the same screen graph $y = (x - 1)^3 + 2$. Could you have predicted the result?

77. Exploration   Graph $y = x^2$, $y = x^4$, and $y = x^6$ on the same screen. What do you notice is the same about each graph? What do you notice that is different?

78. Exploration   Graph $y = x^3$, $y = x^5$, and $y = x^7$ on the same screen. What do you notice is the same about each graph? What do you notice that is different?

79. Define some functions that pass through $(0, 0)$ and $(1, 1)$ and are increasing for $x \geq 0$. Begin your list with $y = \sqrt{x}$, $y = x$, and $y = x^2$. Can you propose a general result about such functions?

1.6   GRAPHING TECHNIQUES: TRANSFORMATIONS

OBJECTIVES

1. Graph Functions Using Horizontal and Vertical Shifts
2. Graph Functions Using Reflections about the x-Axis or y-Axis
3. Graph Functions Using Compressions and Stretches

At this stage, if you were asked to graph any of the functions defined by $y = x$, $y = x^2$, $y = x^3$, $y = \sqrt{x}$, $y = |x|$, or $y = \dfrac{1}{x}$, your response should be, "Yes, I recognize these functions and know the general shapes of their graphs." (If this is not your answer, review the previous section and Figures 63 through 68).

Sometimes we are asked to graph a function that is "almost" like one that we already know how to graph. In this section, we look at some of these functions and develop techniques for graphing them. Collectively, these techniques are referred to as **transformations**.

1   Vertical Shifts

EXPLORATION   On the same screen, graph each of the following functions:

$$Y_1 = x^2$$
$$Y_2 = x^2 + 2$$
$$Y_3 = x^2 - 2$$

What do you observe?

RESULT   Figure 69 illustrates the graphs. You should have observed a general pattern. With $Y_1 = x^2$ on the screen, the graph of $Y_2 = x^2 + 2$ is identical to that of $Y_1 = x^2$, except that it is shifted vertically up 2 units. The graph of $Y_3 = x^2 - 2$ is identical to that of $Y_1 = x^2$, except that it is shifted vertically down 2 units.

Figure 69

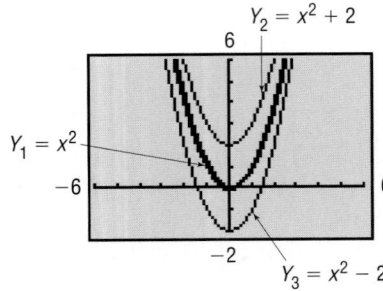

We are led to the following conclusion:

> If a real number k is added to the right side of a function $y = f(x)$, the graph of the new function $y = f(x) + k$ is the graph of f **shifted vertically** up (if $k > 0$) or down (if $k < 0$).

Let's look at an example.

EXAMPLE 1   **Vertical Shift Down**

Use the graph of $f(x) = x^2$ to obtain the graph of $h(x) = x^2 - 4$.

Solution   Table 11 lists some points on the graphs of $f = Y_1$ and $h = Y_2$. Notice that each y-coordinate of h is 4 units less than the corresponding y-coordinate of f.

The graph of h is identical to that of f, except that it is shifted down 4 units. See Figure 70.

TABLE 11

X	Y₁	Y₂
-3	9	5
-2	4	0
-1	1	-3
0	0	-4
1	1	-3
2	4	0
3	9	5

Y₂ ▣ X²−4

Figure 70

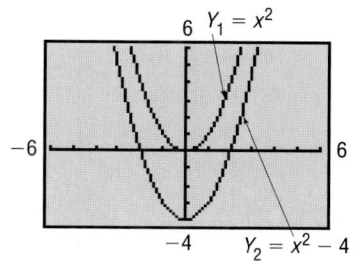

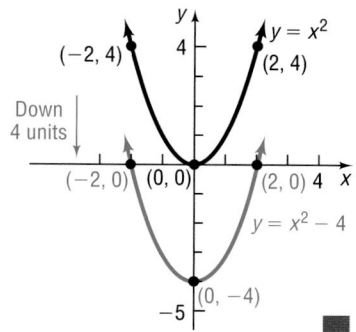

✏ **NOW WORK PROBLEM 33.**

Horizontal Shifts

EXPLORATION   On the same screen, graph each of the following functions:

$$Y_1 = x^2$$
$$Y_2 = (x - 3)^2$$
$$Y_3 = (x + 2)^2$$

What do you observe?

Figure 71

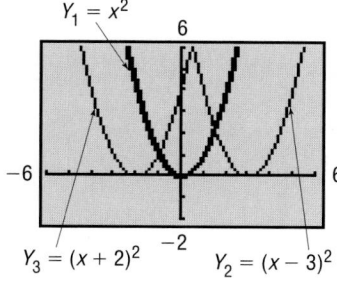

RESULT   Figure 71 illustrates the graphs.
You should have observed the following pattern. With the graph of $Y_1 = x^2$ on the screen, the graph of $Y_2 = (x - 3)^2$ is identical to that of $Y_1 = x^2$, except it is shifted horizontally to the right 3 units. The graph of $Y_3 = (x + 2)^2$ is identical to that of $Y_1 = x^2$, except it is shifted horizontally to the left 2 units.

We are led to the following conclusion.

> If the argument x of a function f is replaced by $x - h$, h a real number, the graph of the new function $y = f(x - h)$ is the graph of f **shifted horizontally** left (if $h < 0$) or right (if $h > 0$).

NOW WORK PROBLEM 37.

EXAMPLE 2 **Graphing a Function Using Transformations**

Graph the function $f(x) = (x + 3)^2 - 5$.

Solution We graph f in stages. First, we note that the rule for f is basically a square function, so we begin with the graph of $y = x^2$ as shown in Figure 72(a). To get the graph of $y = (x + 3)^2$, we shift the graph of $y = x^2$ horizontally 3 units to the left. See Figure 72(b). Finally, to get the graph of $y = (x + 3)^2 - 5$, we shift the graph of $y = (x + 3)^2$ vertically down 5 units. See Figure 72(c). Note the points plotted on each graph. Using key points can be helpful in keeping track of the transformation that has taken place.

Figure 72

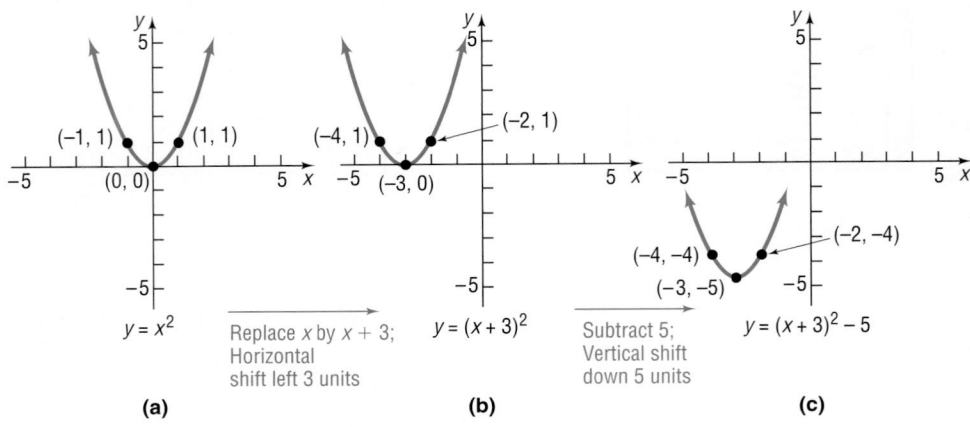

(a) $y = x^2$ Replace x by $x + 3$; Horizontal shift left 3 units   (b) $y = (x + 3)^2$ Subtract 5; Vertical shift down 5 units   (c) $y = (x + 3)^2 - 5$

✔ CHECK: Graph $Y_1 = f(x) = (x + 3)^2 - 5$ and compare the graph to Figure 72(c). ■ ■

In Example 2, if the vertical shift had been done first, followed by the horizontal shift, the final graph would have been the same. (Try it for yourself.)

NOW WORK PROBLEMS 39 AND 75.

2 **Reflections About the *x*-Axis and the *y*-Axis**

EXPLORATION **Reflection about the *x*-Axis**

(a) Graph $Y_1 = x^2$, followed by $Y_2 = -x^2$.
(b) Graph $Y_1 = |x|$, followed by $Y_2 = -|x|$.
(c) Graph $Y_1 = x^2 - 4$, followed by $Y_2 = -(x^2 - 4) = -x^2 + 4$.

RESULT See Tables 12(a), (b), and (c) and Figures 73(a), (b), and (c). In each instance, Y_2 is the reflection about the x-axis of Y_1.

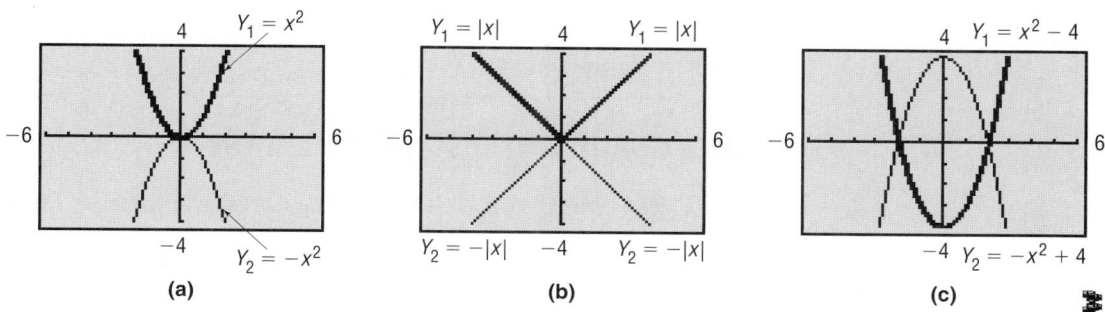

TABLE 12

(a) (b) (c)

Figure 73

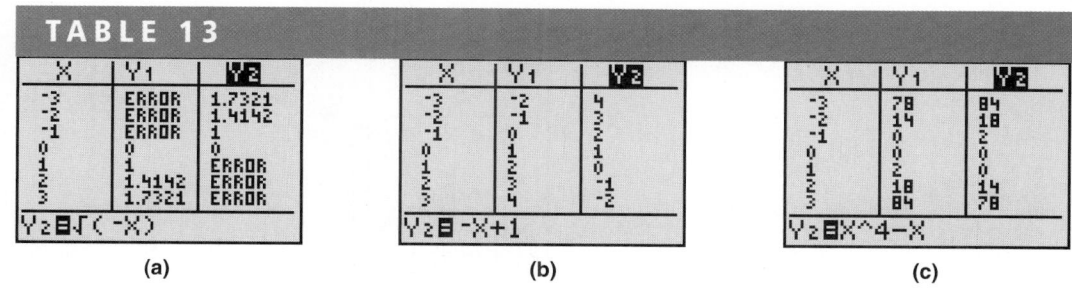

(a) (b) (c)

When the right side of a function $y = f(x)$ is multiplied by -1, the graph of the new function $y = -f(x)$ is the **reflection about the x-axis** of the graph of the function $y = f(x)$.

NOW WORK PROBLEM 45.

EXPLORATION   **Reflection about the y-axis**

(a) Graph $Y_1 = \sqrt{x}$, followed by $Y_2 = \sqrt{-x}$.

(b) Graph $Y_1 = x + 1$, followed by $Y_2 = -x + 1$.

(c) Graph $Y_1 = x^4 + x$, followed by $Y_2 = (-x)^4 + (-x) = x^4 - x$.

RESULT See Tables 13(a), (b), and (c) and Figures 74(a), (b), and (c). In each instance, the Y_2 is the reflection about the y-axis of the Y_1.

TABLE 13

(a) (b) (c)

Figure 74

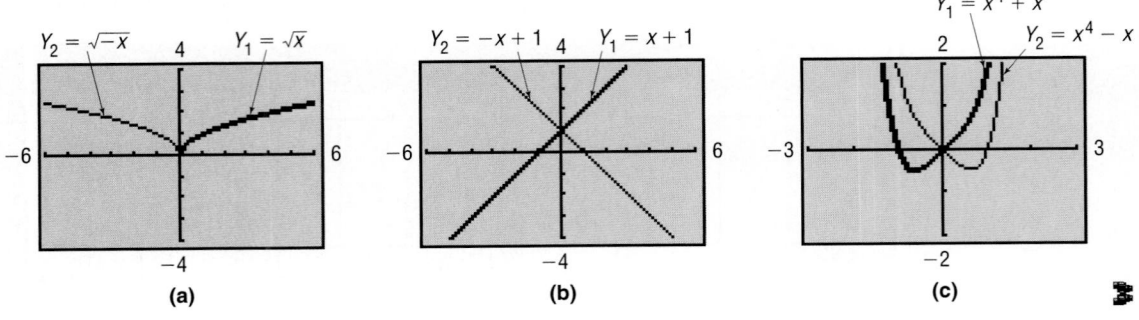

(a)   (b)   (c)

When the graph of the function $y = f(x)$ is known, the graph of the new function $y = f(-x)$ is the **reflection about the y-axis** of the graph of the function $y = f(x)$.

③ Compressions and Stretches

EXPLORATION On the same screen, graph each of the following functions:

$$Y_1 = |x|$$
$$Y_2 = 2|x|$$
$$Y_3 = \frac{1}{2}|x|$$

Figure 75

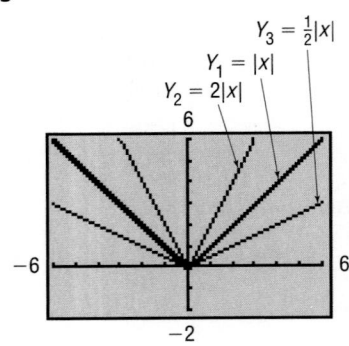

RESULT Figure 75 illustrates the graphs. You should have observed the following pattern. The graph of $Y_2 = 2|x|$ can be obtained from the graph of $Y_1 = |x|$ by multiplying each y-coordinate of $Y_1 = |x|$ by 2. This is sometimes referred to as a vertical *stretch* using a factor of 2.

The graph of $Y_3 = \frac{1}{2}|x|$ can be obtained from the graph of $Y_1 = |x|$ by multiplying each y-coordinate by $\frac{1}{2}$. This is sometimes referred to as a vertical *compression* using a factor of $\frac{1}{2}$.

Look at Tables 14 and 15, where $Y_1 = |x|$, $Y_2 = 2|x|$, and $Y_3 = \frac{1}{2}|x|$.

Notice that the values for Y_2 in Table 14 are two times the values of Y_1 for each x-value. Therefore, the graph of Y_2 will be vertically *stretched* by a factor of 2. Likewise, the values of Y_3 in Table 15 are half the values of Y_1 for each x-value.

Therefore, the graph of Y_3 will be vertically *compressed* by a factor of $\frac{1}{2}$.

TABLE 14

X	Y1	Y2
-2	2	4
-1	1	2
0	0	0
1	1	2
2	2	4
3	3	6
4	4	8

Y2■2abs(X)

TABLE 15

X	Y1	Y3
-2	2	1
-1	1	.5
0	0	0
1	1	.5
2	2	1
3	3	1.5
4	4	2

Y3■.5abs(X)

> When the right side of a function $y = f(x)$ is multiplied by a positive number a, the graph of the new function $y = af(x)$ is obtained by multiplying each y-coordinate on the graph of $y = f(x)$ by a. The new graph is a **vertically compressed** (if $0 < a < 1$) or a **vertically stretched** (if $a > 1$) version of the graph of $y = f(x)$.

✏ **NOW WORK PROBLEM 41.**

What happens if the argument x of a function $y = f(x)$ is multiplied by a positive number a, creating a new function $y = f(ax)$? To find the answer, we look at the following Exploration.

🖳 **EXPLORATION**   On the same screen, graph each of the following functions:

$$Y_1 = f(x) = x^2$$
$$Y_2 = f(4x) = (4x)^2$$
$$Y_3 = f\left(\frac{1}{2}x\right) = \left(\frac{1}{2}x\right)^2$$

RESULT   You should have obtained the graphs shown in Figure 76. Look at Table 16(a).

Figure 76

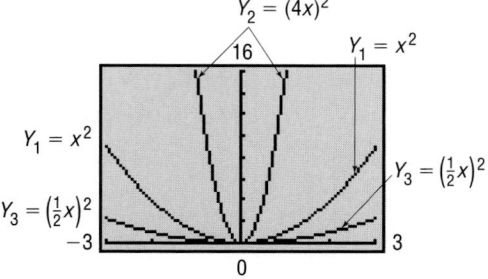

TABLE 16

X	Y1	Y2
0	0	0
.25	.0625	1
1	1	16
4	16	256
16	256	4096

Y2⊟(4X)²

X	Y1	Y3
0	0	0
.5	.25	.0625
1	1	.25
2	4	1
4	16	4
8	64	16
16	256	64

Y3⊟(X/2)²

(a)   (b)

Notice that $(1, 1)$, $(4, 16)$, and $(16, 256)$ are points on the graph of $Y_1 = x^2$. Also, $(0.25, 1)$, $(1, 16)$, and $(4, 256)$ are points on the graph of $Y_2 = (4x)^2$. For each y-coordinate, the x-coordinate on the graph of of Y_2 is $\frac{1}{4}$ the x-coordinate on Y_1. The graph of $Y_2 = (4x)^2$ is obtained by multiplying the x-coordinate of each point on the graph of $Y_1 = x^2$ by $\frac{1}{4}$.

The graph of $Y_2 = (4x)^2$ is the graph of $Y_1 = x^2$ compressed horizontally.
   Look at Table 16(b). Notice that $(0.5, 0.25)$, $(1, 1)$, $(2, 4)$, and $(4, 16)$ are points on the graph of $Y_1 = x^2$. Also, $(1, 0.25)$, $(2, 1)$, $(4, 4)$, and $(8, 16)$ are points on the graph of $Y_3 = \left(\frac{1}{2}x\right)^2$. For each y-coordinate, the x-coordinate on the graph of Y_3 is 2 times the x-coordinate on Y_1.

The graph of $Y_3 = \left(\frac{1}{2}x\right)^2$ is obtained by multiplying the x-coordinate of each point on the graph of $Y_1 = x^2$ by a factor of 2. The graph of $Y_3 = \left(\frac{1}{2}x\right)^2$ is the graph of $Y_1 = x^2$ stretched horizontally. ⬛

If the argument x of a function $y = f(x)$ is multiplied by a positive number a, the graph of the new function $y = f(ax)$ is obtained by multiplying each x-coordinate of $y = f(x)$ by $\dfrac{1}{a}$. A **horizontal compression** results if $a > 1$, and a **horizontal stretch** occurs if $0 < a < 1$.

Let's look at an example.

EXAMPLE 3   **Graphing Using Stretches and Compressions**

The graph of $y = f(x)$ is given in Figure 77. Use this graph to find the graphs of

(a) $y = 2f(x)$ (b) $y = f(3x)$

Figure 77

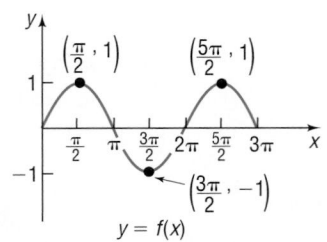

$y = f(x)$

Solution   (a) The graph of $y = 2f(x)$ is obtained by multiplying each y-coordinate of $y = f(x)$ by a factor of 2. See Figure 78(a).

(b) The graph of $y = f(3x)$ is obtained from the graph of $y = f(x)$ by multiplying each x-coordinate of $y = f(x)$ by a factor of $\dfrac{1}{3}$. See Figure 78(b).

Figure 78

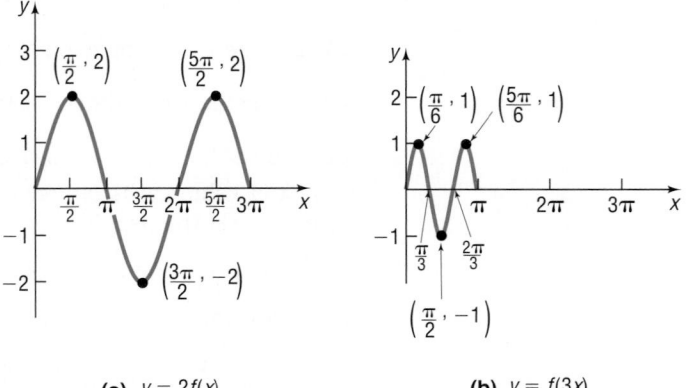

(a) $y = 2f(x)$ (b) $y = f(3x)$

━ **NOW WORK PROBLEMS 63(e) AND (g).**

SUMMARY	Summary of Graphing Techniques

Table 17 summarizes the graphing procedures that we have just discussed.

TABLE 17

To Graph:	Draw the Graph of *f* and:	Functional Change to *f(x)*
Vertical shifts		
$y = f(x) + k, \quad k > 0$	Add k to each y-coordinate of $y = f(x)$ and raise the graph of f by k units.	Add k to $f(x)$.
$y = f(x) - k, \quad k > 0$	Subtract k from each y-coordinate of $y = f(x)$ and lower the graph of f by k units.	Subtract k from $f(x)$.
Horizontal shifts		
$y = f(x + h), \quad h > 0$	Subtract h from each x-coordinate of $y = f(x)$ and shift the graph of f to the left h units.	Replace x by $x + h$.
$y = f(x - h), \quad h > 0$	Add h to each x-coordinate of $y = f(x)$ and shift the graph of f to the right h units.	Replace x by $x - h$.
Reflection about the x-axis		
$y = -f(x)$	Multiply each y-coordinate of $y = f(x)$ by -1. Reflect the graph of f about the x-axis.	Multiply $f(x)$ by -1.
Reflection about the y-axis		
$y = f(-x)$	Multiply each x-coordinate of $y = f(x)$ by -1. Reflect the graph of f about the y-axis.	Replace x by $-x$.
Compressing or stretching		
$y = af(x), \quad a > 0$	Multiply each y-coordinate of $y = f(x)$ by a. Stretch the graph of f vertically if $a > 1$. Compress the graph of f vertically if $0 < a < 1$.	Multiply $f(x)$ by a.
$y = f(ax), \quad a > 0$	Multiply each x-coordinate of $y = f(x)$ by $\dfrac{1}{a}$. Stretch the graph of f horizontally if $0 < a < 1$. Compress the graph of f horizontally if $a > 1$.	Replace x by ax.

The examples that follow combine some of the procedures outlined in this section to get the required graph.

EXAMPLE 4	**Determining the Function Obtained from a Series of Transformations**

Find the function that is finally graphed after the following three transformations are applied to the graph of $y = |x|$.

1. Shift left 2 units.   2. Shift up 3 units.   3. Reflect about the y-axis.

Solution   1. Shift left 2 units: Replace x by $x + 2$. $y = |x + 2|$
2. Shift up 3 units: Add 3. $y = |x + 2| + 3$
3. Reflect about the y-axis: Replace x by $-x$. $y = |-x + 2| + 3$ ■

▰▱▱▱ **NOW WORK PROBLEM 25.**

EXAMPLE 5   **Combining Graphing Procedures**

Graph the function: $f(x) = \dfrac{3}{x - 2} + 1$

Solution   We use the following steps to obtain the graph of f:

STEP 1: $y = \dfrac{1}{x}$ Reciprocal function.

STEP 2: $y = \dfrac{3}{x}$ Multiply by 3; vertical stretch of the graph of $y = \dfrac{1}{x}$ by a factor of 3.

STEP 3: $y = \dfrac{3}{x - 2}$ Replace x by x − 2; horizontal shift to the right 2 units.

STEP 4: $y = \dfrac{3}{x - 2} + 1$ Add 1; vertical shift up 1 unit.

See Figure 79.

Figure 79

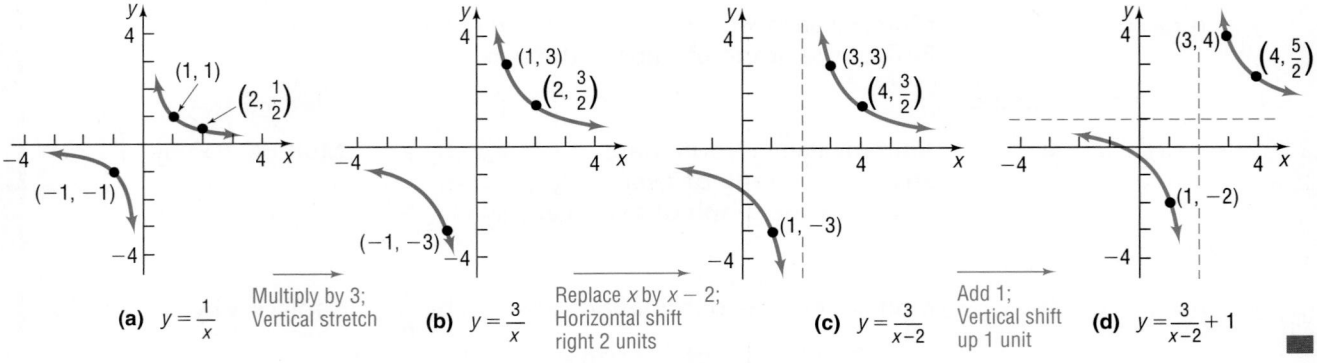

(a) $y = \dfrac{1}{x}$ Multiply by 3; Vertical stretch   (b) $y = \dfrac{3}{x}$ Replace x by x − 2; Horizontal shift right 2 units   (c) $y = \dfrac{3}{x-2}$ Add 1; Vertical shift up 1 unit   (d) $y = \dfrac{3}{x-2} + 1$ ■

Other orderings of the steps shown in Example 5 would also result in the graph of f. For example, try this one:

STEP 1: $y = \dfrac{1}{x}$ Reciprocal function.

STEP 2: $y = \dfrac{1}{x - 2}$ Replace x by x − 2; horizontal shift to the right 2 units.

STEP 3: $y = \dfrac{3}{x - 2}$ Multiply by 3; vertical stretch of the graph of $y = \dfrac{1}{x - 2}$ by a factor of 3.

STEP 4: $y = \dfrac{3}{x - 2} + 1$ Add 1; vertical shift up 1 unit.

EXAMPLE 6   **Combining Graphing Procedures**

Graph the function: $f(x) = \sqrt{1 - x} + 2$

Solution   We use the following steps to obtain the graph of $y = \sqrt{1 - x} + 2$:

STEP 1: $y = \sqrt{x}$ Square root function.

STEP 2: $y = \sqrt{x + 1}$ Replace x by x + 1; horizontal shift left 1 unit.

STEP 3: $y = \sqrt{-x + 1} = \sqrt{1 - x}$ Replace x by −x; reflect about y-axis.

STEP 4: $y = \sqrt{1 - x} + 2$ Add 2; vertical shift up 2 units.

See Figure 80.

Figure 80

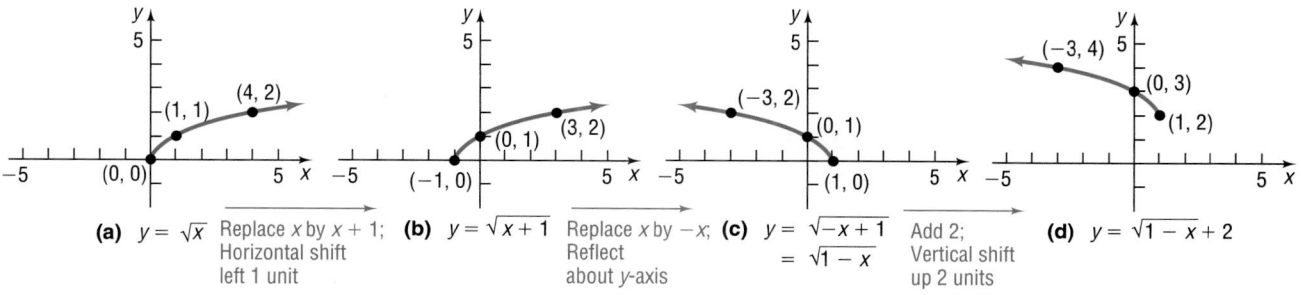

(a) $y = \sqrt{x}$ Replace x by x + 1; **(b)** $y = \sqrt{x + 1}$ Replace x by −x; **(c)** $y = \sqrt{-x + 1}$ Add 2; **(d)** $y = \sqrt{1 - x} + 2$
         Horizontal shift   Reflect $= \sqrt{1 - x}$ Vertical shift
         left 1 unit   about y-axis   up 2 units

✔ CHECK: Graph $Y_1 = f(x) = \sqrt{1 - x} + 2$ and compare the graph to Figure 80(d).

NOW WORK PROBLEM **55.**

1.6 Concepts and Vocabulary

In Problems 1–3, fill in the blanks.

1. Suppose that the graph of a function f is known. Then the graph of $y = f(x - 2)$ may be obtained by a(n) _____ shift of the graph of f to the _____ a distance of 2 units.

2. Suppose that the graph of a function f is known. Then the graph of $y = f(-x)$ may be obtained by a reflection about the _____-axis of the graph of the function $y = f(x)$.

3. Suppose that the graph of a function f is known. Then the graph of $y = f(3x)$ may be obtained by a horizontal _____ of the graph of f by a factor of _____.

In Problems 4–6, answer True or False to each statement.

4. The graph of $y = -f(x)$ is the reflection about the x-axis of the graph of $y = f(x)$.

5. To obtain the graph of $y = f(x + 2) - 3$, shift the graph of $y = f(x)$ horizontally to the right 2 units and vertically down 3 units.

6. To obtain the graph of $y = 4f(x)$, vertically stretch the graph of $y = f(x)$ by a factor of 4. That is, multiply each y-coordinate on the graph of $y = f(x)$ by 4.

7. Suppose that the graph of a function f is known. Explain how the graph of $y = 4f(x)$ differs from the graph of $y = f(4x)$.

1.6 Exercises

In Problems 1–12, match each graph to one of the following functions without using a graphing utility.

A. $y = x^2 + 2$

B. $y = -x^2 + 2$

C. $y = |x| + 2$

D. $y = -|x| + 2$

E. $y = (x - 2)^2$

F. $y = -(x + 2)^2$

G. $y = |x - 2|$

H. $y = -|x + 2|$

I. $y = 2x^2$

J. $y = -2x^2$

K. $y = 2|x|$

L. $y = -2|x|$

1.
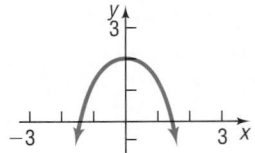

2.
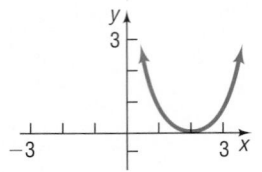

3.
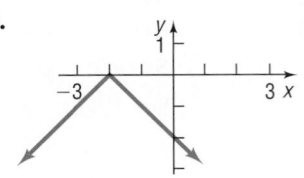

4.
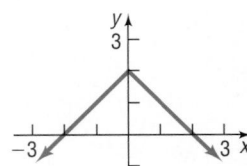

5.
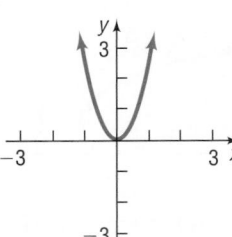

6.
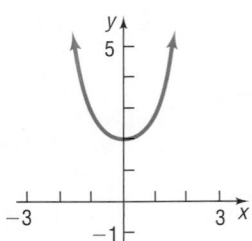

7.
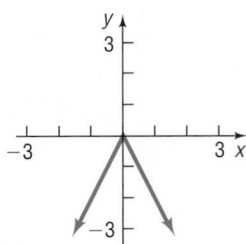

8.
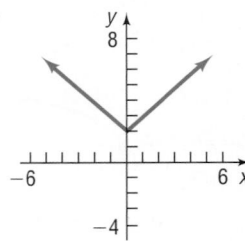

9.
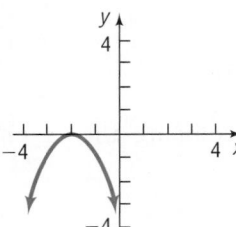

10.
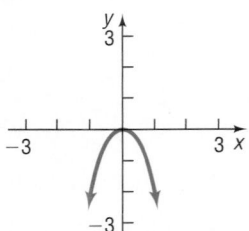

11.
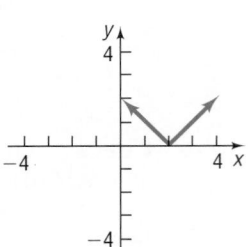

12.
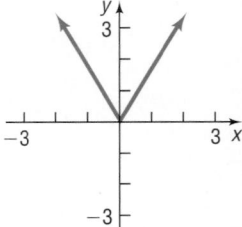

In Problems 13–16, match each graph to one of the following functions without using a graphing utility.

A. $y = x^3$

B. $y = (x + 2)^3$

C. $y = -2x^3$

D. $y = x^3 + 2$

13.
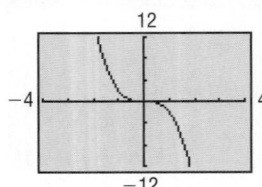

14.
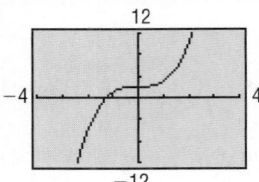

15.
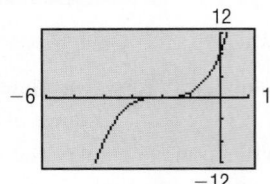

16.
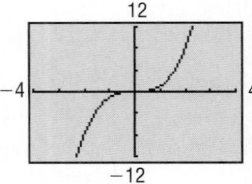

In Problems 17–24, write the function whose graph is the graph of $y = x^3$, but is:

17. Shifted to the right 4 units

18. Shifted to the left 4 units

19. Shifted up 4 units

20. Shifted down 4 units

21. Reflected about the *y*-axis

22. Reflected about the *x*-axis

23. Vertically stretched by a factor of 4

24. Horizontally stretched by a factor of 4

In Problems 25–28, find the function that is finally graphed after the following transformations are applied to the graph of $y = \sqrt{x}$.

25. (1) Shift up 2 units
   (2) Reflect about the *x*-axis
   (3) Reflect about the *y*-axis

26. (1) Reflect about the *x*-axis
   (2) Shift right 3 units
   (3) Shift down 2 units

27. (1) Reflect about the x-axis
(2) Shift up 2 units
(3) Shift left 3 units

28. (1) Shift up 2 units
(2) Reflect about the y-axis
(3) Shift left 3 units

29. If $(3, 0)$ is a point on the graph of $y = f(x)$, which of the following must be on the graph of $y = -f(x)$?
(a) $(0, 3)$ (b) $(0, -3)$
(c) $(3, 0)$ (d) $(-3, 0)$

30. If $(3, 0)$ is a point on the graph of $y = f(x)$, which of the following must be on the graph of $y = f(-x)$?
(a) $(0, 3)$ (b) $(0, -3)$
(c) $(3, 0)$ (d) $(-3, 0)$

31. If $(0, 3)$ is a point on the graph of $y = f(x)$, which of the following must be on the graph of $y = 2f(x)$?
(a) $(0, 3)$ (b) $(0, 2)$
(c) $(0, 6)$ (d) $(6, 0)$

32. If $(3, 0)$ is a point on the graph of $y = f(x)$, which of the following must be on the graph of $y = \dfrac{1}{2}f(x)$?
(a) $(3, 0)$ (b) $\left(\dfrac{3}{2}, 0\right)$
(c) $\left(0, \dfrac{3}{2}\right)$ (d) $\left(\dfrac{1}{2}, 0\right)$

In Problems 33–62, graph each function using transformations (shifting, compressing, stretching, and/or reflecting). Start with the graph of the basic function (for example, $y = x^2$) and show all stages. Verify your results using a graphing utility.

33. $f(x) = x^2 - 1$

34. $f(x) = x^2 + 4$

35. $g(x) = x^3 + 1$

36. $g(x) = x^3 - 1$

37. $h(x) = \sqrt{x} - 2$

38. $h(x) = \sqrt{x + 1}$

39. $f(x) = (x - 1)^3 + 2$

40. $f(x) = (x + 2)^3 - 3$

41. $g(x) = 4\sqrt{x}$

42. $g(x) = \dfrac{1}{2}\sqrt{x}$

43. $h(x) = \dfrac{1}{2x}$

44. $h(x) = \dfrac{4}{x}$

45. $f(x) = -\dfrac{1}{x}$

46. $f(x) = -\sqrt{x}$

47. $g(x) = |-x|$

48. $g(x) = (-x)^3$

49. $h(x) = -x^3 + 2$

50. $h(x) = \dfrac{1}{-x} + 2$

51. $f(x) = 2(x + 1)^2 - 3$

52. $f(x) = 3(x - 2)^2 + 1$

53. $g(x) = \sqrt{x - 2} + 1$

54. $g(x) = |x + 1| - 3$

55. $h(x) = \sqrt{-x} - 2$

56. $h(x) = \dfrac{4}{x} + 2$

57. $f(x) = -(x + 1)^3 - 1$

58. $f(x) = -4\sqrt{x - 1}$

59. $g(x) = 2|1 - x|$

60. $g(x) = 4\sqrt{2 - x}$

61. $h(x) = 2\,\text{int}(x - 1)$

62. $h(x) = \text{int}(-x)$

In Problems 63–68, the graph of a function f is illustrated. Use the graph of f as the first step toward graphing each of the following functions:

(a) $F(x) = f(x) + 3$ (b) $G(x) = f(x + 2)$ (c) $P(x) = -f(x)$ (d) $H(x) = f(x + 1) - 2$

(e) $Q(x) = \dfrac{1}{2}f(x)$ (f) $g(x) = f(-x)$ (g) $h(x) = f(2x)$

63.

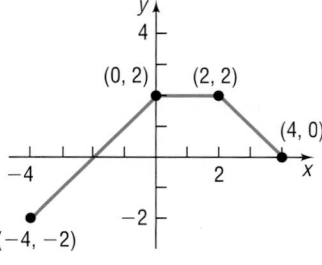

64.

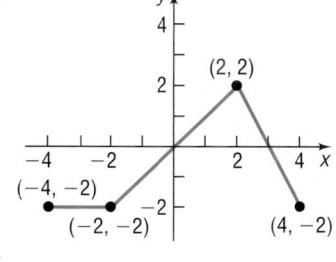

65.

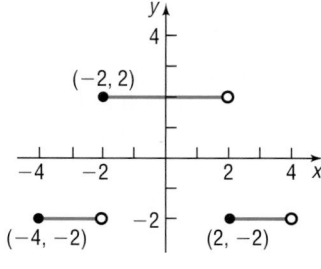

66.

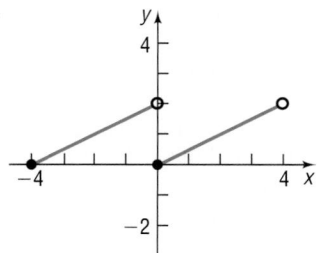

67.

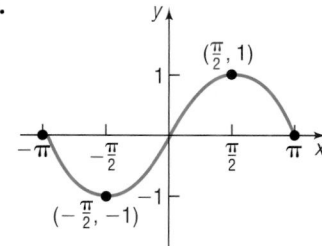

68.

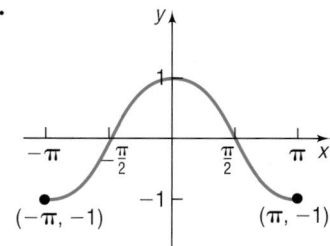

69. Exploration
(a) Use a graphing utility to graph $y = x + 1$ and $y = |x + 1|$.
(b) Graph $y = 4 - x^2$ and $y = |4 - x^2|$.
(c) Graph $y = x^3 + x$ and $y = |x^3 + x|$.
(d) What do you conclude about the relationship between the graphs of $y = f(x)$ and $y = |f(x)|$?

70. Exploration
(a) Use a graphing utility to graph $y = x + 1$ and $y = |x| + 1$.
(b) Graph $y = 4 - x^2$ and $y = 4 - |x|^2$.
(c) Graph $y = x^3 + x$ and $y = |x|^3 + |x|$.
(d) What do you conclude about the relationship between the graphs of $y = f(x)$ and $y = f(|x|)$?

71. The graph of a function f is illustrated in the figure.
(a) Draw the graph of $y = |f(x)|$.
(b) Draw the graph of $y = f(|x|)$.

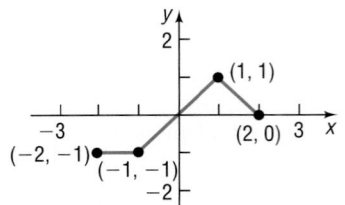

72. The graph of a function f is illustrated in the figure.
(a) Draw the graph of $y = |f(x)|$.
(b) Draw the graph of $y = f(|x|)$.

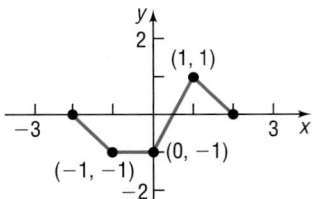

In Problems 73–78, complete the square of each quadratic expression. Then graph each function by hand using the technique of shifting. Verify your results using a graphing utility. (If necessary, refer to Section A.3 to review completing the square.)

73. $f(x) = x^2 + 2x$
74. $f(x) = x^2 - 6x$
75. $f(x) = x^2 - 8x + 1$
76. $f(x) = x^2 + 4x + 2$
77. $f(x) = x^2 + x + 1$
78. $f(x) = x^2 - x + 1$

79. The equation $y = (x - c)^2$ defines a *family of parabolas*, one parabola for each value of c. On one set of coordinate axes, graph the members of the family for $c = 0$, $c = 3$, and $c = -2$.

80. Repeat Problem 79 for the family of parabolas $y = x^2 + c$.

81. Temperature Measurements The relationship between the Celsius (°C) and Fahrenheit (°F) scales for measuring temperature is given by the equation

$$F = \frac{9}{5}C + 32$$

The relationship between the Celsius (°C) and Kelvin (K) scales is $K = C + 273$. Graph the equation $F = \frac{9}{5}C + 32$ using degrees Fahrenheit on the y-axis and degrees Celsius on the x-axis. Use the techniques introduced in this section to obtain the graph showing the relationship between Kelvin and Fahrenheit temperatures.

82. Period of a Pendulum The period T (in seconds) of a simple pendulum is a function of its length l (in feet) defined by the equation

$$T = 2\pi\sqrt{\frac{l}{g}}$$

where $g \approx 32.2$ feet per second per second is the acceleration of gravity.
(a) Use a graphing utility to graph the function $T = T(l)$.
(b) Now graph the functions $T = T(l + 1), T = T(l + 2)$, and $T = T(l + 3)$.
(c) Discuss how adding to the length l changes the period T.
(d) Now graph the functions $T = T(2l), T = T(3l)$, and $T = T(4l)$.

(e) Discuss how multiplying the length l by factors of 2, 3, and 4 changes the period T.

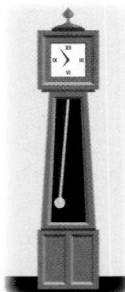

83. Cigar Company Profits The daily profits of a cigar company from selling x cigars are given by

$$p(x) = -0.05x^2 + 100x - 2000$$

The government wishes to impose a tax on cigars (sometimes called a *sin tax*) that gives the company the option of either paying a flat tax of $10,000 per day or a tax of 10% on profits. As chief financial officer (CFO) of the company, you need to decide which tax is the better option for the company.
(a) On the same screen, graph $Y_1 = p(x) - 10,000$ and $Y_2 = (1 - 0.10)p(x)$.
(b) Based on the graph, which option would you select? Why?
(c) Using the terminology learned in this section, describe each graph in terms of the graph of $p(x)$.
(d) Suppose that the government offered the options of a flat tax of $4800 or a tax of 10% on profits. Which would you select? Why?

PREPARING FOR THIS SECTION

Before getting started, review the following:

✓ Functions (Section 1.4, pp. 33–34)   ✓ Increasing/Decreasing Functions (Section 1.5, pp. 52–53)

1.7 ONE-TO-ONE FUNCTIONS; INVERSE FUNCTIONS

OBJECTIVES   ① Determine the Inverse of a Function
                  ② Obtain the Graph of the Inverse Function from the Graph of
                     the Function
                  ③ Find the Inverse Function f^{-1}

① In Section 1.4 we said that a function f can be thought of as a machine that
    receives as input a number, say x, from the domain, manipulates it, and out-
    puts the value $f(x)$. The **inverse of f** receives as input a number $f(x)$, ma-
    nipulates it, and outputs the value x.

EXAMPLE 1   **Finding the Inverse of a Function**

Find the inverse of the following functions.

(a) Let the domain of the function represent the employees of Yolanda's
    Preowned Car Mart and let the range represent their base salaries.

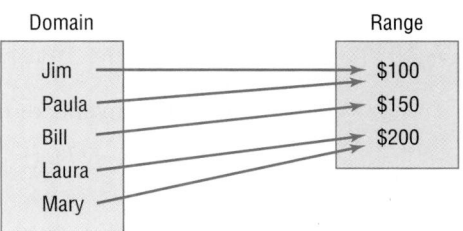

(b) Let the domain of the function represent the employees of Yolanda's
    Preowned Car Mart and let the range represent their spouse's names.

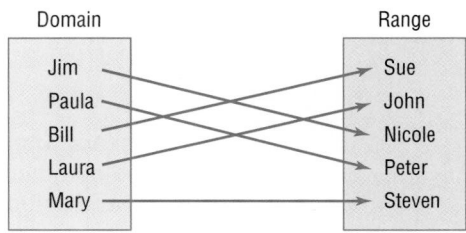

Solution   (a) The elements in the domain represent inputs to the function, and the el-
                ements in the range represent the outputs. To find the inverse, inter-
                change the elements in the domain with the elements in the range. For
                example, the function receives as input Bill and outputs $150. So the

inverse receives as input $150 and outputs Bill. The inverse of the given function takes the form

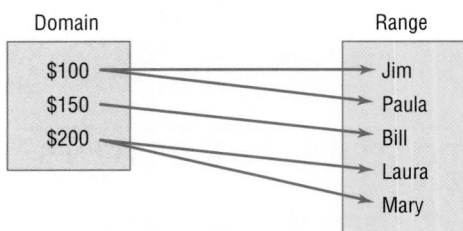

(b) The inverse of the given function is

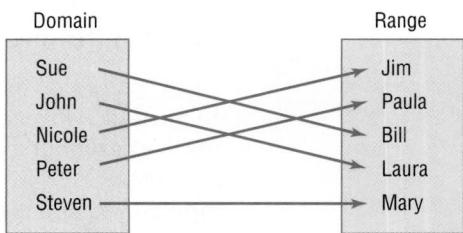

Notice that the inverse found in Example 1(b) is a function, since each element in the domain corresponds to exactly one element in the range. The inverse found in Example 1(a) is not a function, since each element in the domain does not correspond to exactly one element in the range.

If the function f is a set of ordered pairs (x, y), then the inverse of f is the set of ordered pairs (y, x).

EXAMPLE 2   **Finding the Inverse of a Function**

Find the inverse of the following functions:

(a) $\{(-3, -27), (-2, -8), (-1, -1), (0, 0), (1, 1), (2, 8), (3, 27)\}$
(b) $\{(-3, 9), (-2, 4), (-1, 1), (0, 0), (1, 1), (2, 4), (3, 9)\}$

Solution   (a) The inverse of the given function is found by interchanging the entries in each ordered pair and so is given by

$$\{(-27, -3), (-8, -2), (-1, -1), (0, 0), (1, 1), (8, 2), (27, 3)\}$$

(b) The inverse of the given function is

$$\{(9, -3), (4, -2), (1, -1), (0, 0), (1, 1), (4, 2), (9, 3)\}$$

The inverse obtained in the solution to Example 2(a) is a function, but the inverse obtained in Example 2(b) is not a function. So, sometimes the inverse of a function is a function, and sometimes it is not.

When the inverse of a function f is itself a function, then f is said to be a **one-to-one function**. That is, f is **one-to-one** if, for any choice of elements x_1 and x_2 in the domain of f, with $x_1 \neq x_2$, the corresponding values $f(x_1)$ and $f(x_2)$ are unequal, $f(x_1) \neq f(x_2)$.

In other words, a function f is one-to-one if no y in the range is the image of more than one x in the domain. A function is not one-to-one if two different elements in the domain correspond to the same element in the range. In Example 2(b), the elements -3 and 3 both correspond to 9, so the function is not one-to-one. Figure 81 illustrates the definition.

Figure 81

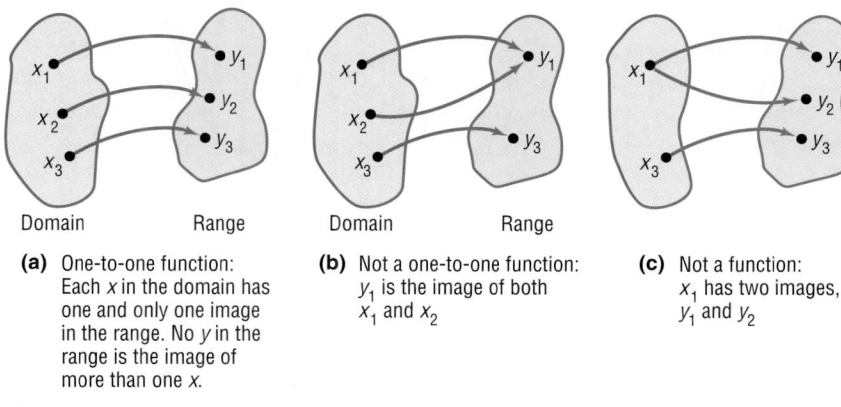

Domain	Range	Domain	Range		
(a) One-to-one function: Each x in the domain has one and only one image in the range. No y in the range is the image of more than one x.		**(b)** Not a one-to-one function: y_1 is the image of both x_1 and x_2		**(c)** Not a function: x_1 has two images, y_1 and y_2	

NOW WORK PROBLEMS 1 AND 5.

If the graph of a function f is known, there is a simple test, called the **horizontal-line test**, to determine whether f is one-to-one.

Theorem | **Horizontal-line Test**

If every horizontal line intersects the graph of a function f in at most one point, then f is one-to-one.

The reason that this test works can be seen in Figure 82, where the horizontal line $y = h$ intersects the graph at two distinct points, (x_1, h) and (x_2, h). Since h is the image of both x_1 and x_2, $x_1 \neq x_2$, f is not one-to-one. Based on Figure 82, we can state the horizontal-line test in another way: If the graph of any horizontal line intersects the graph of a function f in more than one point, then f is not one-to-one.

Figure 82
$f(x_1) = f(x_2) = h$,
and $x_1 \neq x_2$;
f is not a
one-to-one function.

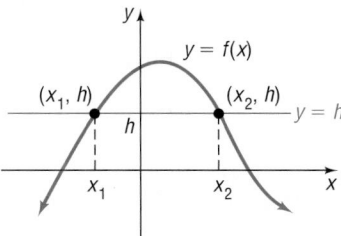

EXAMPLE 3 | **Using the Horizontal-line Test**

For each function, use the graph to determine whether the function is one-to-one.

(a) $f(x) = x^2$ (b) $g(x) = x^3$

Solution   (a) Figure 83(a) illustrates the horizontal-line test for $f(x) = x^2$. The horizontal line $y = 1$ intersects the graph of f twice, at $(1, 1)$ and at $(-1, 1)$, so f is not one-to-one.

(b) Figure 83(b) illustrates the horizontal-line test for $g(x) = x^3$. Because every horizontal line will intersect the graph of g exactly once, it follows that g is one-to-one.

Figure 83

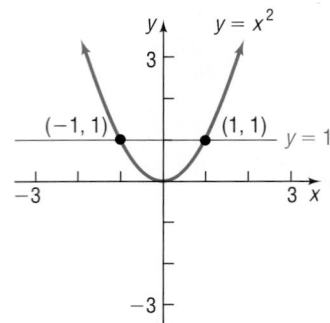

(a) A horizontal line intersects the graph twice; f is not one-to-one

(b) Every horizontal line intersects the graph exactly once; g is one-to-one

NOW WORK PROBLEM 9.

Let's look more closely at the one-to-one function $g(x) = x^3$. This function is an increasing function. Because an increasing (or decreasing) function will always have different y values for unequal x values, it follows that a function that is increasing (or decreasing) over its domain is also a one-to-one function.

Theorem

A function that is increasing over its domain is a one-to-one function.
A function that is decreasing over its domain is a one-to-one function.

Inverse Function of $y = f(x)$

If f is a one-to-one function, its inverse is a function. Then, to each x in the domain of f, there is exactly one y in the range (because f is a function); and to each y in the range of f, there is exactly one x in the domain (because f is one-to-one). The correspondence from the range of f back to the domain of f is called the **inverse function of f** and is denoted by the symbol f^{-1}. Figure 84 illustrates this definition.

Figure 84

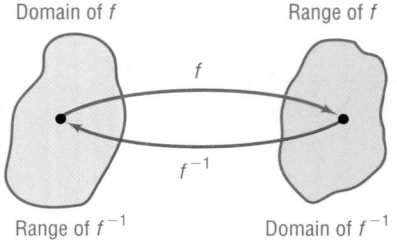

Domain of f Range of f

Range of f^{-1} Domain of f^{-1}

WARNING: Be careful! f^{-1} is a symbol for the inverse function of f. The -1 used in f^{-1} is not an exponent. That is, f^{-1} does *not* mean the reciprocal of f; $f^{-1}(x)$ is not equal to $\dfrac{1}{f(x)}$.

Two facts are now apparent about a function f and its inverse f^{-1}.

$$\text{Domain of } f = \text{Range of } f^{-1} \qquad \text{Range of } f = \text{Domain of } f^{-1}$$

Look again at Figure 84 to visualize the relationship. If we start with x, apply f, and then apply f^{-1}, we get x back again. If we start with x, apply f^{-1}, and then apply f, we get the number x back again. To put it simply, what f does f^{-1} undoes, and vice versa.

$$\boxed{\text{Input } x} \xrightarrow{\text{Apply } f} \boxed{f(x)} \xrightarrow{\text{Apply } f^{-1}} \boxed{f^{-1}(f(x)) = x}$$

$$\boxed{\text{Input } x} \xrightarrow{\text{Apply } f^{-1}} \boxed{f^{-1}(x)} \xrightarrow{\text{Apply } f} \boxed{f(f^{-1}(x)) = x}$$

In other words,

$$f^{-1}(f(x)) = x \quad \text{and} \quad f(f^{-1}(x)) = x$$

Consider the function $f(x) = 2x$, which multiplies the argument x by 2. The inverse function f^{-1} undoes whatever f does. So the inverse function of f is $f^{-1}(x) = \dfrac{1}{2}x$, which divides the argument by 2. For example, $f(3) = 2(3) = 6$ and $f^{-1}(6) = \dfrac{1}{2}(6) = 3$, so f^{-1} undoes what f did. We can verify this by showing that

$$f^{-1}(f(x)) = f^{-1}(2x) = \frac{1}{2}(2x) = x \quad \text{and} \quad f(f^{-1}(x)) = f\left(\frac{1}{2}x\right) = 2\left(\frac{1}{2}x\right) = x$$

See Figure 85.

Figure 85

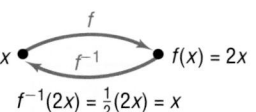

$f^{-1}(2x) = \frac{1}{2}(2x) = x$

EXAMPLE 4 **Verifying Inverse Functions**

(a) We verify that the inverse of $g(x) = x^3$ is $g^{-1}(x) = \sqrt[3]{x}$ by showing that

$$g^{-1}(g(x)) = g^{-1}(x^3) = \sqrt[3]{x^3} = x$$

and

$$g(g^{-1}(x)) = g(\sqrt[3]{x}) = (\sqrt[3]{x})^3 = x$$

(b) We verify that the inverse of $h(x) = 3x$ is $h^{-1}(x) = \dfrac{1}{3}x$ by showing that

$$h^{-1}(h(x)) = h^{-1}(3x) = \frac{1}{3}(3x) = x$$

and

$$h(h^{-1}(x)) = h\left(\frac{1}{3}x\right) = 3\left(\frac{1}{3}x\right) = x$$

(c) We verify that the inverse of $f(x) = 2x + 3$ is $f^{-1}(x) = \dfrac{1}{2}(x - 3)$ by showing that

$$f^{-1}(f(x)) = f^{-1}(2x + 3) = \frac{1}{2}[(2x + 3) - 3] = \frac{1}{2}(2x) = x$$

and

$$f(f^{-1}(x)) = f\left(\frac{1}{2}(x - 3)\right) = 2\left[\frac{1}{2}(x - 3)\right] + 3 = (x - 3) + 3 = x \quad \blacksquare$$

EXPLORATION Simultaneously graph $Y_1 = x$, $Y_2 = x^3$, and $Y_3 = \sqrt[3]{x}$ on a square screen with $-3 \le x \le 3$. What do you observe about the graphs of $Y_2 = x^3$, its inverse $Y_3 = \sqrt[3]{x}$, and the line $Y_1 = x$?

Repeat this experiment by simultaneously graphing $Y_1 = x, Y_2 = 2x + 3$, and $Y_3 = \dfrac{1}{2}(x - 3)$ on a square screen with $-6 \le x \le 3$. Do you see the symmetry of the graph of Y_2 and its inverse Y_3 with respect to the line $Y_1 = x$?

═══ NOW WORK PROBLEM **21.**

② **Geometric Interpretation**

For the functions in Example 4(c), we list points on the graph of $f = Y_1$ and on the graph of $f^{-1} = Y_2$ in Table 18.

We notice that whenever (a, b) is on the graph of f then (b, a) is on the graph of f^{-1}. Figure 86 shows these points plotted. Also shown is the graph of $y = x$, which you should observe is a line of symmetry of the points.

Figure 86

TABLE 18

X	Y1
-5	-7
-4	-5
-3	-3
-2	-1
-1	1
0	3
1	5

Y1目2X+3

X	Y2
-7	-5
-5	-4
-3	-3
-1	-2
1	-1
3	0
5	1

Y2目(1/2)(X-3)

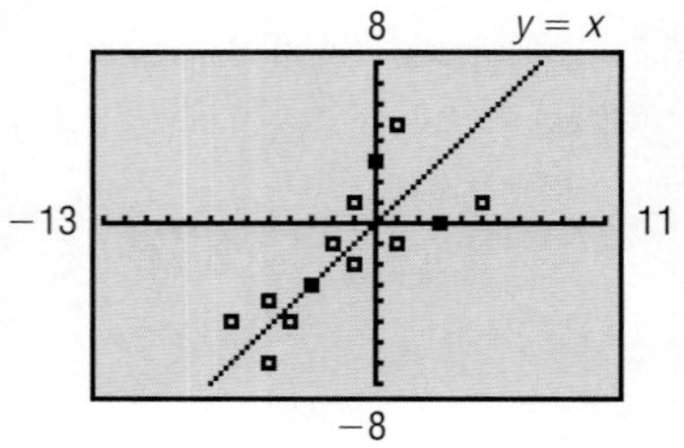

Figure 87

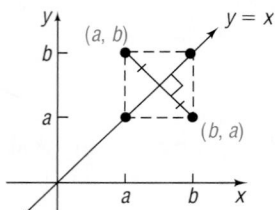

Suppose that (a, b) is a point on the graph of a one-to-one function f defined by $y = f(x)$. Then $b = f(a)$. This means that $a = f^{-1}(b)$, so (b, a) is a point on the graph of the inverse function f^{-1}. The relationship between the point (a, b) on f and the point (b, a) on f^{-1} is shown in Figure 87. The line segment containing (a, b) and (b, a) is perpendicular to the line $y = x$ and is bisected by the line $y = x$. (Do you see why?) It follows that the point (b, a) on f^{-1} is the reflection about the line $y = x$ of the point (a, b) on f.

Theorem   The graph of a function f and the graph of its inverse f^{-1} are symmetric with respect to the line $y = x$.

Figure 88 illustrates this result. Notice that, once the graph of f is known, the graph of f^{-1} may be obtained by reflecting the graph of f about the line $y = x$.

Figure 88

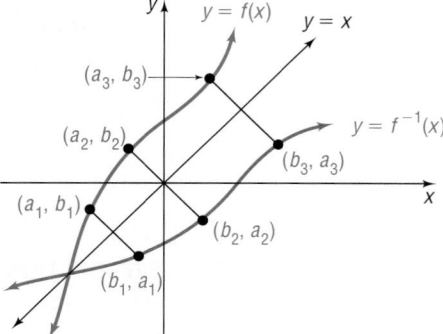

EXAMPLE 5   **Graphing the Inverse Function**

The graph in Figure 89(a) is that of a one-to-one function $y = f(x)$. Draw the graph of its inverse.

Solution   We begin by adding the graph of $y = x$ to Figure 89(a). Since the points $(-2, -1), (-1, 0)$, and $(2, 1)$ are on the graph of f, we know that the points $(-1, -2), (0, -1)$, and $(1, 2)$ must be on the graph of f^{-1}. Keeping in mind that the graph of f^{-1} is the reflection about the line $y = x$ of the graph of f, we can draw f^{-1}. See Figure 89(b).

Figure 89

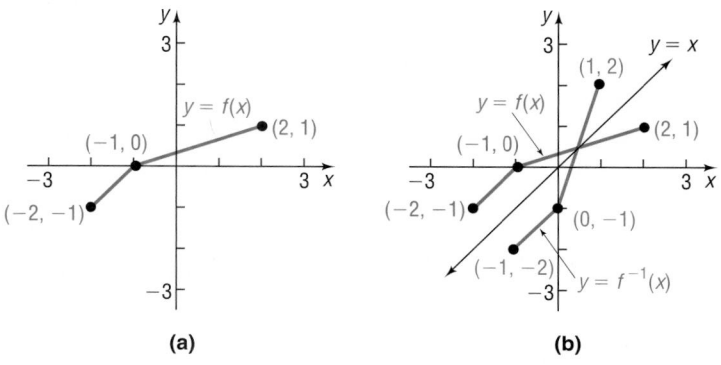

(a)   (b)

▰

NOW WORK PROBLEM 15.

③ Finding the Inverse Function

The fact that the graph of a one-to-one function f and its inverse function f^{-1} are symmetric with respect to the line $y = x$ tells us more. It says that we can obtain f^{-1} by interchanging the roles of x and y in f. Look again at Figure 88. If f is defined by the equation

$$y = f(x)$$

then f^{-1} is defined by the equation

$$x = f(y)$$

The equation $x = f(y)$ defines f^{-1} *implicitly*. If we can solve this equation for y, we will have the *explicit* form of f^{-1}, that is,

$$y = f^{-1}(x)$$

Let's use this procedure to find the inverse of $f(x) = 2x + 3$. (Since f is a linear function and is increasing, we know that f is one-to-one and so has an inverse function.)

EXAMPLE 6   **Finding the Inverse Function**

Find the inverse of $f(x) = 2x + 3$. Also find the domain and range of f and f^{-1}. Graph f and f^{-1} on the same coordinate axes.

Solution   In the equation $y = 2x + 3$, interchange the variables x and y. The result,

$$x = 2y + 3$$

is an equation that defines the inverse f^{-1} implicitly. To find the explicit form, we solve for y.

$$2y + 3 = x$$
$$2y = x - 3$$
$$y = \frac{1}{2}(x - 3)$$

The explicit form of the inverse f^{-1} is therefore

$$f^{-1}(x) = \frac{1}{2}(x - 3)$$

which we verified in Example 4(c).

Next we find

$$\text{Domain of } f = \text{Range of } f^{-1} = (-\infty, \infty)$$
$$\text{Range of } f = \text{Domain of } f^{-1} = (-\infty, \infty)$$

The graphs of $Y_1 = f(x) = 2x + 3$ and its inverse $Y_2 = f^{-1}(x) = \frac{1}{2}(x - 3)$ are shown in Figure 90. Note the symmetry of the graphs with respect to the line $Y_3 = x$.

Figure 90

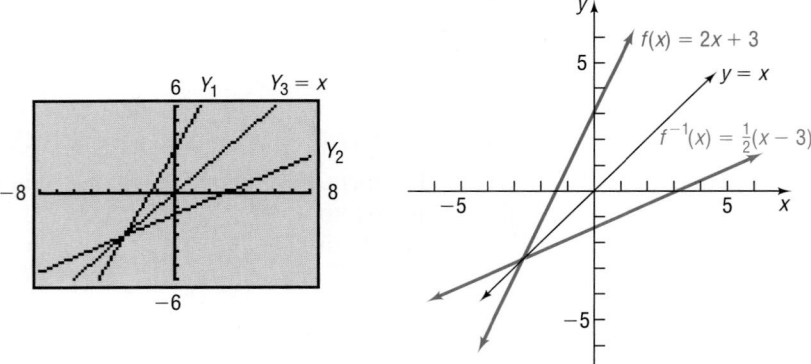

We outline next the steps to follow for finding the inverse of a one-to-one function.

Procedure for Finding the Inverse of a One-to-One Function

STEP 1:   In $y = f(x)$, interchange the variables x and y to obtain

$$x = f(y)$$

This equation defines the inverse function f^{-1} implicitly.

STEP 2:   If possible, solve the implicit equation for y in terms of x to obtain the explicit form of f^{-1}.

$$y = f^{-1}(x)$$

STEP 3:   Check the result by showing that

$$f^{-1}(f(x)) = x \quad \text{and} \quad f(f^{-1}(x)) = x$$

EXAMPLE 7   **Finding the Inverse Function**

The function

$$f(x) = \frac{2x + 1}{x - 1}, \qquad x \neq 1$$

is one-to-one. Find its inverse and check the result.

Solution   **STEP 1:**   Interchange the variables x and y in

$$y = \frac{2x + 1}{x - 1}$$

to obtain

$$x = \frac{2y + 1}{y - 1}$$

STEP 2:   Solve for y.

$$x = \frac{2y + 1}{y - 1}$$

$$x(y - 1) = 2y + 1 \qquad \text{Multiply both sides by } y - 1.$$

$$xy - x = 2y + 1 \qquad \text{Apply the distributive property.}$$

$$xy - 2y = x + 1 \qquad \text{Subtract } 2y \text{ from both sides; add } x \text{ to both sides.}$$

$$(x - 2)y = x + 1 \qquad \text{Factor.}$$

$$y = \frac{x + 1}{x - 2} \qquad \text{Divide by } x - 2.$$

The inverse is

$$f^{-1}(x) = \frac{x + 1}{x - 2}, \qquad x \neq 2 \qquad \text{Replace } y \text{ by } f^{-1}(x).$$

STEP 3:   ✔ CHECK:

$$f^{-1}(f(x)) = f^{-1}\left(\frac{2x + 1}{x - 1}\right) = \frac{\dfrac{2x + 1}{x - 1} + 1}{\dfrac{2x + 1}{x - 1} - 2} = \frac{2x + 1 + x - 1}{2x + 1 - 2(x - 1)} = \frac{3x}{3} = x$$

$$f(f^{-1}(x)) = f\left(\frac{x + 1}{x - 2}\right) = \frac{2\left(\dfrac{x + 1}{x - 2}\right) + 1}{\dfrac{x + 1}{x - 2} - 1} = \frac{2(x + 1) + x - 2}{x + 1 - (x - 2)} = \frac{3x}{3} = x$$

   ■   ■

EXPLORATION   In Example 7, we found that, if $f(x) = \dfrac{2x + 1}{x - 1}$, then $f^{-1}(x) = \dfrac{x + 1}{x - 2}$. Compare the vertical and horizontal asymptotes of f and f^{-1}. What did you find? Are you surprised?

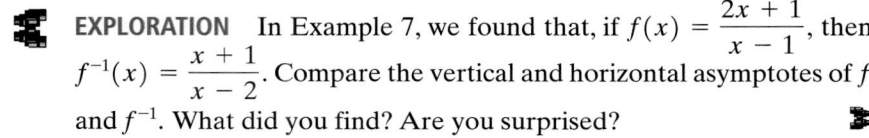

━ **NOW WORK PROBLEM 33.**

We said in Section 1.4 that finding the range of a function f is not easy. However, if f is one-to-one, we can find its range by finding the domain of the inverse function f^{-1}.

EXAMPLE 8   Finding the Range of a Function

Find the domain and range of

$$f(x) = \frac{2x + 1}{x - 1}$$

Solution   The domain of f is $\{x \mid x \neq 1\}$. To find the range of f, we first find the inverse f^{-1}. Based on Example 7, we have

$$f^{-1}(x) = \frac{x + 1}{x - 2}$$

The domain of f^{-1} is $\{x \mid x \neq 2\}$, so the range of f is $\{y \mid y \neq 2\}$.   ■

NOW WORK PROBLEM 47.

If a function is not one-to-one, then its inverse is not a function. Sometimes, though, an appropriate restriction on the domain of such a function will yield a new function that is one-to-one. Let's look at an example of this common practice.

EXAMPLE 9   Finding the Inverse of a Domain-restricted Function

Find the inverse of $y = f(x) = x^2$ if $x \geq 0$.

Solution   The function $y = x^2$ is not one-to-one. [Refer to Example 3(a).] However, if we restrict this function to only that part of its domain for which $x \geq 0$, as indicated, we have a new function that is increasing and therefore is one-to-one. As a result, the function defined by $y = f(x) = x^2, x \geq 0$, has an inverse function, f^{-1}.

We follow the steps given previously to find f^{-1}.

STEP 1:   In the equation $y = x^2, x \geq 0$, interchange the variables x and y. The result is

$$x = y^2, \quad y \geq 0$$

This equation defines (implicitly) the inverse function.

STEP 2:   We solve for y to get the explicit form of the inverse. Since $y \geq 0$, only one solution for y is obtained.

$$y = \sqrt{x}$$

So $f^{-1}(x) = \sqrt{x}$.

STEP 3:   ✔ CHECK: $f^{-1}(f(x)) = f^{-1}(x^2) = \sqrt{x^2} = |x| = x$, since $x \geq 0$

$$f(f^{-1}(x)) = f(\sqrt{x}) = (\sqrt{x})^2 = x$$ ■

Figure 91 illustrates the graphs of $Y_1 = f(x) = x^2, x \geq 0$, and $Y_2 = f^{-1}(x) = \sqrt{x}$.

Figure 91

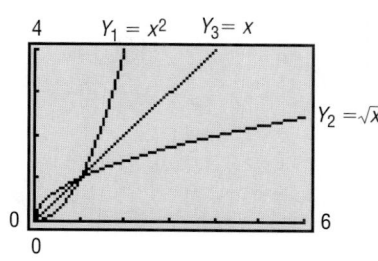

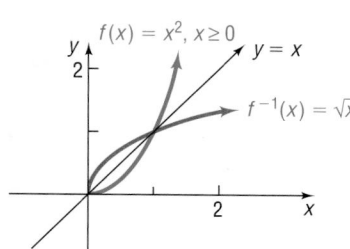

SUMMARY

1. If a function f is one-to-one, then it has an inverse function f^{-1}.
2. Domain of f = Range of f^{-1}; Range of f = Domain of f^{-1}.
3. To verify that f^{-1} is the inverse of f, show that $f^{-1}(f(x)) = x$ and $f(f^{-1}(x)) = x$.
4. The graphs of f and f^{-1} are symmetric with respect to the line $y = x$.
5. To find the range of a one-to-one function f, find the domain of the inverse function f^{-1}.

1.7 Concepts and Vocabulary

In Problems 1–3, fill in the blanks.

1. If every horizontal line intersects the graph of a function f at no more than one point, then f is a(n) _____ function.
2. If f^{-1} denotes the inverse of a function f, then the graphs of f and f^{-1} are symmetric with respect to the line _____.
3. If the domain of a one-to-one function f is $[4, \infty)$, then the range of its inverse, f^{-1}, is _____.

In Problems 4 and 5, answer True or False to each statement.

4. If f and g are inverse functions, then the domain of f is the same as the domain of g.
5. If f and g are inverse functions, then their graphs are symmetric with respect to the line $y = x$.

6. If a function f is even, can it be one-to-one? Explain.
7. Is every odd function one-to-one? Explain.
8. If the graph of a function and its inverse intersect, where must this necessarily occur? Can they intersect anywhere else? Must they intersect?

1.7 Exercises

In Problems 1–8, (a) find the inverse and (b) determine whether the inverse represents a function.

1.

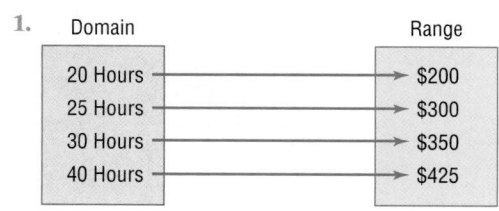

2.

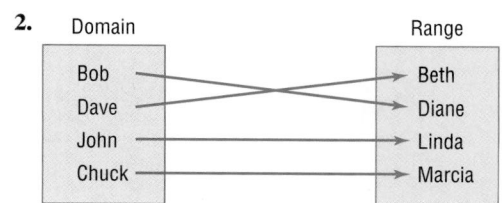

3.   Domain   Range

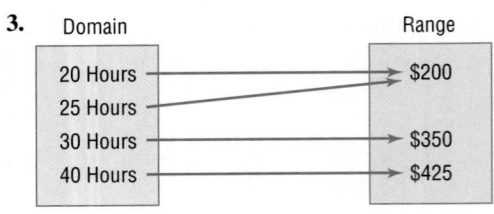

4.   Domain   Range

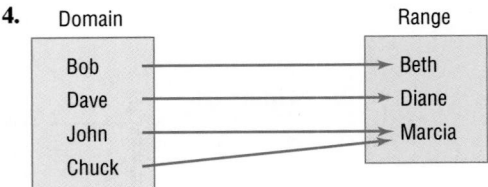

5. $\{(2,6),(-3,6),(4,9),(1,10)\}$

6. $\{(-2,5),(-1,3),(3,7),(4,12)\}$

7. $\{(0,0),(1,1),(2,16),(3,81)\}$

8. $\{(1,2),(2,8),(3,18),(4,32)\}$

In Problems 9–14, the graph of a function f is given. Use the horizontal line test to determine whether f is one-to-one.

9.

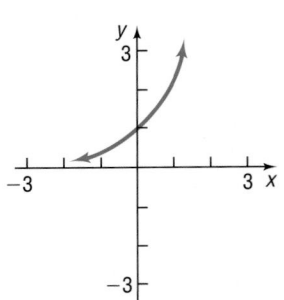

10.

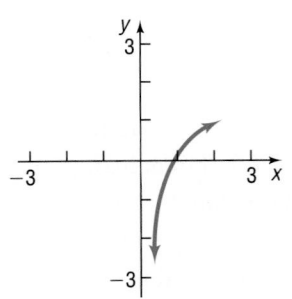

11.

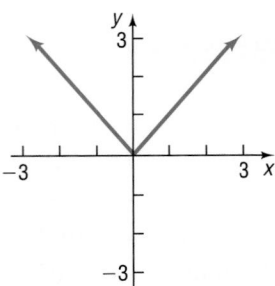

12.

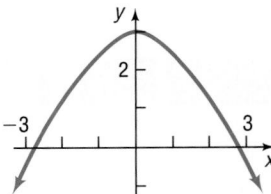

13.

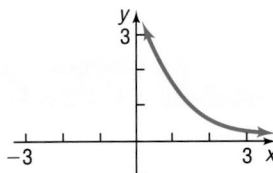

14.

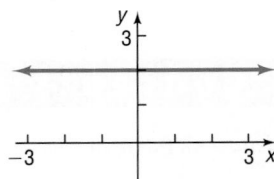

In Problems 15–20, the graph of a one-to-one function f is given. Draw the graph of the inverse function f^{-1}. For convenience (and as a hint), the graph of $y = x$ is also given.

15.

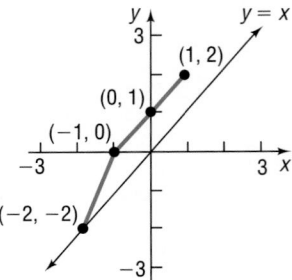

16.

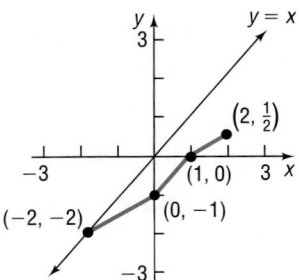

17.

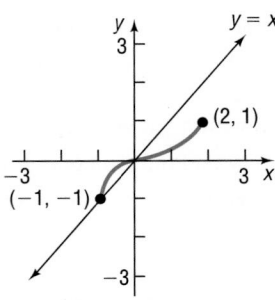

18.

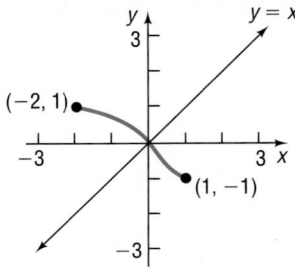

19.

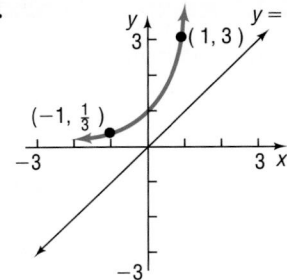

20.

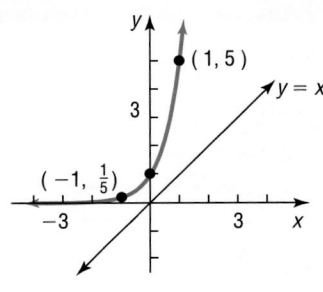

In Problems 21–30, verify that the functions f and g are inverses of each other by showing that $f(g(x)) = x$ and $g(f(x)) = x$. Using a graphing utility, simultaneously graph f, g, and $y = x$ on the same square screen.

21. $f(x) = 3x + 4;\quad g(x) = \dfrac{1}{3}(x - 4)$

22. $f(x) = 3 - 2x;\quad g(x) = -\dfrac{1}{2}(x - 3)$

23. $f(x) = 4x - 8;\quad g(x) = \dfrac{x}{4} + 2$

24. $f(x) = 2x + 6;\quad g(x) = \dfrac{1}{2}x - 3$

25. $f(x) = x^3 - 8;\quad g(x) = \sqrt[3]{x + 8}$

26. $f(x) = (x - 2)^2,\quad x \geq 2;\quad g(x) = \sqrt{x} + 2,\quad x \geq 0$

27. $f(x) = \dfrac{1}{x};\quad g(x) = \dfrac{1}{x}$

28. $f(x) = x;\quad g(x) = x$

29. $f(x) = \dfrac{2x + 3}{x + 4};\quad g(x) = \dfrac{4x - 3}{2 - x}$

30. $f(x) = \dfrac{x - 5}{2x + 3};\quad g(x) = \dfrac{3x + 5}{1 - 2x}$

In Problems 31–42, the function f is one-to-one. Find its inverse and check your answer. State the domain and range of f and f^{-1}. By hand, graph f, f^{-1}, and $y = x$ on the same coordinate axes. Check your results using a graphing utility.

31. $f(x) = 3x$

32. $f(x) = -4x$

33. $f(x) = 4x + 2$

34. $f(x) = 1 - 3x$

35. $f(x) = x^3 - 1$

36. $f(x) = x^3 + 1$

37. $f(x) = x^2 + 4,\quad x \geq 0$

38. $f(x) = x^2 + 9,\quad x \geq 0$

39. $f(x) = \dfrac{4}{x}$

40. $f(x) = -\dfrac{3}{x}$

41. $f(x) = \dfrac{1}{x - 2}$

42. $f(x) = \dfrac{4}{x + 2}$

In Problems 43–54, the function f is one-to-one. Find its inverse and check your answer. State the domain of f and find its range using f^{-1}. Using a graphing utility, simultaneously graph f, f^{-1}, and $y = x$ on the same square screen.

43. $f(x) = \dfrac{2}{3 + x}$

44. $f(x) = \dfrac{4}{2 - x}$

45. $f(x) = (x + 2)^2,\quad x \geq -2$

46. $f(x) = (x - 1)^2,\quad x \geq 1$

47. $f(x) = \dfrac{2x}{x - 1}$

48. $f(x) = \dfrac{3x + 1}{x}$

49. $f(x) = \dfrac{3x + 4}{2x - 3}$

50. $f(x) = \dfrac{2x - 3}{x + 4}$

51. $f(x) = \dfrac{2x + 3}{x + 2}$

52. $f(x) = \dfrac{-3x - 4}{x - 2}$

53. $f(x) = 2\sqrt[3]{x}$

54. $f(x) = \dfrac{4}{\sqrt{x}}$

---

55. Find the inverse of the linear function
$$f(x) = mx + b, \quad m \neq 0$$

56. Find the inverse of the function
$$f(x) = \sqrt{r^2 - x^2}, \quad 0 \leq x \leq r$$

57. A function f has an inverse function. If the graph of f lies in quadrant I, in which quadrant does the graph of f^{-1} lie?

58. A function f has an inverse function. If the graph of f lies in quadrant II, in which quadrant does the graph of f^{-1} lie?

59. The function $f(x) = |x|$ is not one-to-one. Find a suitable restriction on the domain of f so that the new function that results is one-to-one. Then find the inverse of f.

60. The function $f(x) = x^4$ is not one-to-one. Find a suitable restriction on the domain of f so that the new function that results is one-to-one. Then find the inverse of f.

61. **Temperature Conversion**   To convert from x degrees Celsius to y degrees Fahrenheit, we use the formula $y = f(x) = \dfrac{9}{5}x + 32$. To convert from x degrees Fahrenheit to y degrees Celsius, we use the formula $y = g(x) = \dfrac{5}{9}(x - 32)$. Show that f and g are inverse functions.

62. **Demand for Corn**   The demand for corn obeys the equation $p(x) = 300 - 50x$, where p is the price per bushel (in dollars) and x is the number of bushels produced, in millions. Express the production amount x as a function of the price p.

63. **Period of a Pendulum**   The period T (in seconds) of a simple pendulum is a function of its length l (in feet), given by $T(l) = 2\pi\sqrt{\dfrac{l}{g}}$, where $g \approx 32.2$ feet per second per second is the acceleration of gravity. Express the length l as a function of the period T.

64. Given

$$f(x) = \frac{ax + b}{cx + d}$$

find $f^{-1}(x)$. If $c \neq 0$, under what conditions on a, b, c, and d is $f = f^{-1}$?

65. Can a one-to-one function and its inverse be equal? What must be true about the graph of f for this to happen? Give some examples to support your conclusion.

66. Draw the graph of a one-to-one function that contains the points $(-2, -3)$, $(0, 0)$, and $(1, 5)$. Now draw the graph of its inverse. Compare your graph to those of other students. Discuss any similarities. What differences do you see?

Chapter Review

Library of Functions

Linear function (p. 57)

$f(x) = mx + b$
Graph is a line with slope m and y-intercept b (see Figure 61)

Constant function (p. 58)

$f(x) = b$
Graph is a horizontal line with y-intercept b (see Figure 62)

Identity function (p. 58)

$f(x) = x$
Graph is a line with slope 1 and y-intercept 0 (see Figure 63)

Square function (p. 58)

$f(x) = x^2$
Graph is a parabola with vertex at $(0, 0)$ (see Figure 64)

Cube function (p. 59)

$f(x) = x^3$ See Figure 65.

Square root function (p. 59)

$f(x) = \sqrt{x}$ See Figure 66.

Reciprocal function (p. 59)

$f(x) = \dfrac{1}{x}$ See Figure 67.

Absolute value function (p. 60)

$f(x) = |x|$ See Figure 68.

Things to Know

Distance formula (p. 5)

$$d = \sqrt{(x_2 - x_1)^2 + (y_2 - y_1)^2}$$

Midpoint formula (p. 8)

$$M = (x, y) = \left(\frac{x_1 + x_2}{2}, \frac{y_1 + y_2}{2} \right)$$

Standard form of the equation of a circle (p. 26)

$(x - h)^2 + (y - k)^2 = r^2$
(h, k) is the center; r is the radius

Function (p. 34)

A relation between two sets of real numbers so that each number x in the first set, the domain, has corresponding to it exactly one number y in the second set. The range is the set of y values of the function for the x values in the domain.

A function can also be characterized as a set of order pairs (x, y) or $(x, f(x))$ in which no two ordered pairs have the same first element, but different second elements.

A function $f(x)$ may be defined implicitly by an equation involving x and y or explicitily by writing $y = f(x)$.

Function notation (p. 36)

$y = f(x)$
f is a symbol for the function.
x is the argument, or independent variable.
y is the dependent variable
$f(x)$ is the value of the function at x, or the image of x.

Domain (p. 40)

If unspecified, the domain of a function f is the largest set of real numbers for which $f(x)$ is a real number.

Vertical-line test (p. 42)

A set of points in the plane is the graph of a function if and only if every vertical line intersects the graph in at most one point.

Increasing function (p. 52) A function f is increasing on an open interval I if, for any choice of x_1 and x_2 in I, with $x_1 < x_2$, we have $f(x_1) < f(x_2)$.

Decreasing function (p. 52) A function f is decreasing on an open interval I if, for any choice of x_1 and x_2 in I, with $x_1 < x_2$, we have $f(x_1) > f(x_2)$.

Constant function (p. 52) A function f is constant on an open interval I if, for all choices of x in I, the values of $f(x)$ are equal.

Local maximum (p. 53) A function f has a local maximum at c if there is an open interval I containing c so that, for all $x \neq c$ in I, $f(x) < f(c)$.

Local minimum (p. 54) A function f has a local minimum at c if there is an open interval I containing c so that, for all $x \neq c$ in I, $f(x) > f(c)$.

Even function f (p. 55) $f(-x) = f(x)$ for every x in the domain ($-x$ must also be in the domain).

Odd function f (p. 55) $f(-x) = -f(x)$ for every x in the domain ($-x$ must also be in the domain).

Difference Quotient of f (p. 63) $\dfrac{f(x + h) - f(x)}{h}, \quad h \neq 0$

One-to-one function f (p. 78) A function whose inverse is also a function. For any choice of elements x_1, x_2 in the domain of f, if $x_1 \neq x_2$, then $f(x_1) \neq f(x_2)$.

Horizontal-line test (p. 79) If every horizontal line intersects the graph of a function f in at most one point, then f is one-to-one.

Inverse function f^{-1} of f (pp. 80–83) Domain of f = Range of f^{-1}; Range of f = Domain of f^{-1}. $f^{-1}(f(x)) = x$ and $f(f^{-1}(x)) = x$. Graphs of f and f^{-1} are symmetric with respect to the line $y = x$.

Objectives

Section		You should be able to:	Review Exercises
1.1	1	Use the distance formula (p. 5)	1, 2, 89, 90, 91, 92
	2	Use the midpoint formula (p. 7)	1, 2, 91
1.2	1	Graph equations by plotting points (p. 12)	3–8
	2	Graph equations using a graphing utility (p. 14)	3–8
	3	Find intercepts (p. 17)	3–8
1.3	1	Test an equation for symmetry with respect to (a) the x-axis, (b) the y-axis, and (c) the origin (p. 20)	9–16
	2	Know how to graph key equations (p. 23)	17, 18
	3	Write the standard form of the equation of a circle (p. 26)	19–22
	4	Graph a circle by hand and by using a graphing utility (p. 28)	23–26
	5	Find the center and radius of a circle from an equation in general form and graph it (p. 29)	23–26
1.4	1	Determine whether a relation represents a function (p. 34)	27, 28
	2	Find the value of function (p. 38)	29–32, 35–40
	3	Find the domain of a function (p. 40)	41–48
	4	Identify the graph of a function (p. 42)	33
	5	Obtain information from or about the graph of a function (p. 42)	34, 49(a), 50(a)
1.5	1	Use a graph to determine where a function is increasing, is decreasing, or is constant (p. 52)	49(b), 50(b)
	2	Use a graph to locate local maxima and minima (p. 53)	49(c), 50(c)
	3	Use a graphing utility to approximate local maxima and minima and to determine where a function is increasing or decreasing (p. 54)	73–76

Review Exercises

Blue problem numbers indicate the authors' suggestions for use in a Practice Test.

1. What is the distance between the points $(7, 4)$ and $(-3, 2)$? What is their midpoint?

2. What is the distance between the points $(2, 5)$ and $(6, -3)$? What is their midpoint?

In Problems 3–8, graph each equation by plotting points. Be sure to label any intercepts. Verify your results using a graphing utility.

3. $2x - 3y = 6$

4. $2x + 5y = 10$

5. $y = x^2 - 9$

6. $y = x^2 + 4$

7. $x^2 + 2y = 16$

8. $2x^2 - 4y = 24$

In Problems 9–16, test each equation for symmetry with respect to the x-axis, the y-axis, and the origin.

9. $2x = 3y^2$

10. $y = 5x$

11. $x^2 + 4y^2 = 16$

12. $9x^2 - y^2 = 9$

13. $y = x^4 + 2x^2 + 1$

14. $y = x^3 - x$

15. $x^2 + x + y^2 + 2y = 0$

16. $x^2 + 4x + y^2 - 2y = 0$

17. Sketch a graph of $y = x^3$

18. Sketch a graph of $y = \sqrt{x}$.

In Problems 19–22, write the equation of each circle in standard form.

19. Center $(1, 3)$
radius 2

20. Center $(-1, 1)$
radius 4

21. Center $(-3, 2)$
radius 3

22. Center $(-2, -1)$
radius 1

In Problems 23–26, find the center and radius of each circle. Graph each circle by hand.

23. $x^2 + y^2 - 2x + 4y - 4 = 0$

24. $x^2 + y^2 + 4x - 4y - 1 = 0$

25. $3x^2 + 3y^2 - 6x + 12y = 0$

26. $2x^2 + 2y^2 - 4x = 0$

27. Is the relation $\{(-2, 2), (0, 2), (2, 2)\}$ a function?

28. Is the relation $\{(-2, 2), (0, 2), (2, 0)\}$ a function?

29. Given that f is a linear function, $f(4) = -5$, and $f(0) = 3$, write the equation that defines f.

30. Given that g is a linear function with slope $= -4$ and $g(-2) = 2$, write the equation that defines g.

31. A function f is defined by

$$f(x) = \frac{Ax + 5}{6x - 2}$$

If $f(1) = 4$, find A.

32. A function g is defined by

$$g(x) = \frac{A}{x} + \frac{8}{x^2}$$

If $g(-1) = 0$, find A.

33. Tell which of the following graphs are graphs of functions.

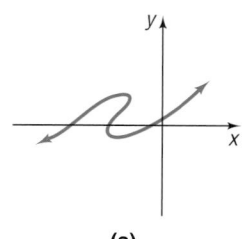

(a)

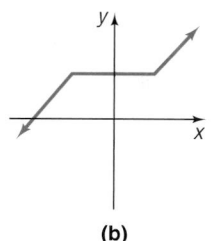

(b)

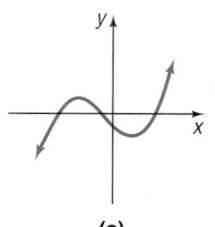

(c)

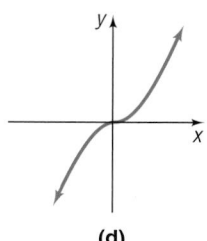

(d)

34. Use the graph of the function f shown to find
   (a) The domain and range of f
   (b) $f(-1)$
   (c) The intercepts of f
   (d) For what value of x does $f(x) = -3$?
   (e) For what values of x is $f(x) < 0$?

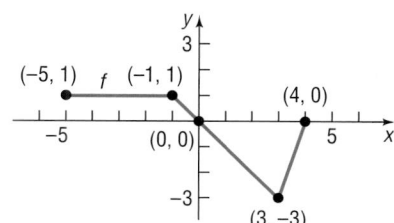

In Problems 35–40, find the following for each function:

 (a) $f(-x)$ (b) $-f(x)$ (c) $f(x + 2)$ (d) $f(x - 2)$

35. $f(x) = \dfrac{3x}{x^2 - 4}$

36. $f(x) = \dfrac{x^2}{x + 2}$

37. $f(x) = \sqrt{x^2 - 4}$

38. $f(x) = |x^2 - 4|$

39. $f(x) = \dfrac{x^2 - 4}{x^2}$

40. $f(x) = \dfrac{x^3}{x^2 - 4}$

In Problems 41–48, find the domain of each function.

41. $f(x) = \dfrac{x}{x^2 - 9}$

42. $f(x) = \dfrac{3x^2}{x - 2}$

43. $f(x) = \sqrt{2 - x}$

44. $f(x) = \sqrt{x + 2}$

45. $h(x) = \dfrac{\sqrt{x}}{|x|}$

46. $g(x) = \dfrac{|x|}{x}$

47. $f(x) = \dfrac{x}{x^2 + 2x - 3}$

48. $F(x) = \dfrac{1}{x^2 - 3x - 4}$

In Problems 49 and 50, use the graph of the function f to find

   (a) The domain and range of f
   (b) The intervals on which f is increasing, decreasing, or constant
   (c) The local minima and local maxima
   (d) Whether the graph is symmetric with respect to the x-axis, the y-axis, or the origin
   (e) Whether the function is even, odd, or neither
   (f) The intercepts, if any

49.

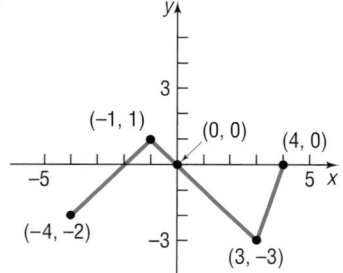

50.

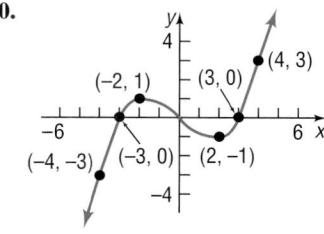

51. Sketch a graph of $f(x) = x^2$.

52. Sketch a graph of $f(x) = |x|$.

In Problems 53–60, determine (algebraically) whether the given function is even, odd, or neither.

53. $f(x) = x^3 - 4x$

54. $g(x) = \dfrac{4 + x^2}{1 + x^4}$

55. $h(x) = \dfrac{1}{x^4} + \dfrac{1}{x^2} + 1$

56. $F(x) = \sqrt{1 - x^3}$

57. $G(x) = 1 - x + x^3$

58. $H(x) = 1 + x + x^2$

59. $f(x) = \dfrac{x}{1 + x^2}$

60. $g(x) = \dfrac{1 + x^2}{x^3}$

In Problems 61–72, graph each function using transformations (shifting, compressing or stretching, and reflections). Identify any intercepts on the graph. State the domain and, based on the graph, find the range.

61. $F(x) = |x| - 4$

62. $f(x) = |x| + 4$

63. $g(x) = -2|x|$

64. $g(x) = \dfrac{1}{2}|x|$

65. $h(x) = \sqrt{x - 1}$

66. $h(x) = \sqrt{x} - 1$

67. $f(x) = \sqrt{1 - x}$

68. $f(x) = -\sqrt{x + 3}$

69. $h(x) = (x - 1)^2 + 2$

70. $h(x) = (x + 2)^2 - 3$

71. $g(x) = 3(x - 1)^3 + 1$

72. $g(x) = -2(x + 2)^3 - 8$

In Problems 73–76, use a graphing utility to graph each function over the indicated interval. Approximate any local maxima and local minima. Determine where the function is increasing and where it is decreasing.

73. $f(x) = 2x^3 - 5x + 1 \quad (-3, 3)$

74. $f(x) = -x^3 + 3x - 5 \quad (-3, 3)$

75. $f(x) = 2x^4 - 5x^3 + 2x + 1 \quad (-2, 3)$

76. $f(x) = -x^4 + 3x^3 - 4x + 3 \quad (-2, 3)$

In Problems 77 and 78, (a) find the inverse of the function, and (b) determine whether the inverse represents a function.

77. $\{(1, 2), (3, 5), (5, 8), (6, 10)\}$

78. $\{(-1, 4), (0, 2), (1, 4), (3, 7)\}$

In Problems 79 and 80, the graph of a one-to-one function is given. Draw the graph of the inverse function f^{-1}. For convenience (and as a hint), the graph of $y = x$ is also given.

79.

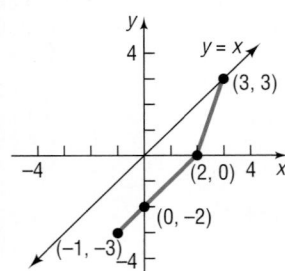

80.

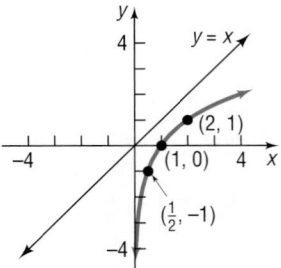

In Problems 81–86, the function f is one-to-one. Find the inverse of each function and check your answer. Find the domain and range of f and f^{-1}. Use a graphing utility to simultaneously graph f, f^{-1}, and $y = x$ on the same square screen.

81. $f(x) = \dfrac{2x + 3}{5x - 2}$

82. $f(x) = \dfrac{2 - x}{3 + x}$

83. $f(x) = \dfrac{1}{x - 1}$

84. $f(x) = \sqrt{x - 2}$

85. $f(x) = \dfrac{3}{x^{1/3}}$

86. $f(x) = x^{1/3} + 1$

87. For the following graph of the function f draw the graph of:
  (a) $y = f(-x)$ (b) $y = -f(x)$ (c) $y = f(x + 2)$
  (d) $y = f(x) + 2$ (e) $y = 2f(x)$ (f) $y = f(3x)$

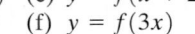

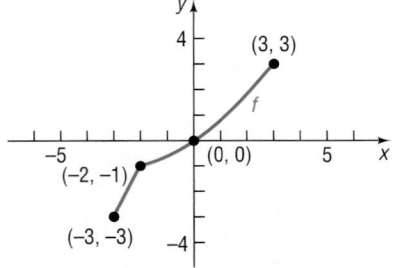

88. Repeat Problem 87 for the following graph of the function g.

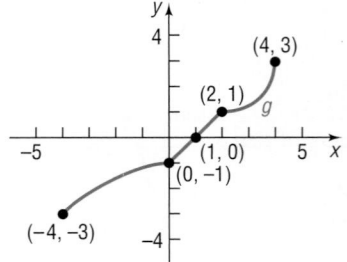

89. Show that the points $A = (3, 4)$, $B = (1, 1)$, and $C = (-2, 3)$ are the vertices of an isosceles triangle.

90. Show that the points $A = (1, 5)$, $B = (2, 4)$, and $C = (-3, 5)$ lie on a circle with center $(-1, 2)$. What is the radius of this circle?

91. The endpoints of the diameter of a circle are $(-3, 2)$ and $(5, -6)$. Find the center and radius of the circle. Write the general equation of this circle.

92. Find two numbers y such that the distance from $(-3, 2)$ to $(5, y)$ is 10.

Chapter Projects

1. Analyzing a Stock   The beta, β, of a stock represents the relative risk of a stock compared with a market basket of stocks, such as Standard and Poor's 500 Index of stocks. Beta is computed by finding the slope of the line of best fit between the rate of return of the stock and the rate of return of the S&P 500. The rates of return are computed on a weekly basis.

(a) Find the weekly closing price of your favorite stock and the S&P 500 for 20 weeks. One good source is on the Internet at *http://finance.yahoo.com.*

(b) Compute the rate of return by computing the weekly percentage change in the closing price of your stock

and the weekly percentage change in the S&P 500 using the following formula:

$$\text{Weekly \% change} = \frac{P_2 - P_1}{P_1}$$

where P_1 is last week's price and P_2 is this week's price.

(c) Using a graphing utility, find the line of best fit treating the weekly rate of return of the S&P 500 as the independent variable and the weekly rate of return of your stock as dependent variable.

(d) What is the beta of your stock?

(e) Compare your result with that of the Value Line Investment Survey found in your library. What might account for any differences?

2. Financing a Car   Advertisements for cars are bewildering. Suppose you have decided to purchase a car whose list price is $12,400. You have saved $2000 to use as a down payment. After shopping around at several dealerships, the offers given in the table are available to you.

(a) The first concern that you have is your monthly payment. The formula

$$M = A \left[\frac{1 - (1 + i)^{-n}}{i} \right]^{-1}$$

can be used to compute your monthly payment M. In this formula, A is the amount borrowed, i is the

	List Price	Dealer Discount	Down Payment	Amount Borrowed	Length of Loan	APR Annual Rate of Interest
Dealer 1	$12,400	1000	2000	9400	3 years	2.9%
Dealer 1	$12,400	1000	2000	9400	5 years	5.9%
Dealer 1	$12,400	1500	2000	8900	4 years	10%
Dealer 2	$12,400	1500	2000	8900	3 years	9%
Dealer 2	$12,400	1800	2000	8600	5 years	12%
Dealer 3	$12,400	400	2000	10,000	2 years	0.5%

monthly interest rate so i = APR/12 (expressed as a decimal), and n is the length of the loan, in months. Compute the monthly payment for each of the six possibilities.

(b) In each case, how much money do you actually pay for the car?

(c) In each case, how much interest is paid?

(d) Which of the six deals is "best"? Be sure to have reasons!

3. **Economics** An **isocost line** represents the various combinations of two inputs that can be used in a production process while maintaining a constant level of expenditures. The equation for an isocost line is

$$P_L \cdot L + P_K \cdot K = E$$

where

P_L represents the price of labor

L represents the amount of labor used in the production process

P_K represents the price of capital (machinery, etc.)

K represents the amount of capital used in the production process

E represents the total expenditures for the two inputs, labor and capital

(a) What values of L and K make sense as they are defined above?

(b) Suppose that P_L = $400 per hour and P_K = $500 per hour. If L = 100 hours, how many hours of capital, K, may be purchased if expenditures are to be $100,000?

(c) Suppose that P_L = $400 per hour and P_K = $500 per hour; graph the isocost line if E = $100,000 with the number of hours of labor, L, on the x-axis and the number of hours of capital, K, on the y-axis. In which quadrant is your graph located? Why?

(d) Graph the isocost line if E increases to $120,000. What effect does this have on the graph? What do you think would happen if expenditures were to decrease?

(e) Determine the slope of the isocost line. Interpret its value.

(f) Graph the isocost line with P_L = $500 per hour, P_K = $500 per hour, and E = $100,000. Compare this graph with the graph obtained in part (c). What effect does the increase in the price of labor have on the graph?

TRIGONOMETRIC FUNCTIONS

Tidal Coastline and Pots of Water

In Florida, they post times of tides coming in and going out very precisely, like 11:23 A.M. How can they be so precise?

There is more to tides, the rise and fall of ocean waters, than the gravitational pull of the moon and sun.

These are the chief factors, of course. And because the movement of the Earth, sun and moon in relationship to each other are known precisely, the rhythm of tides rising and falling along coastlines is easy to predict.

Yet the time and heights of high and low tide may vary along different stretches of the same coast though they're reacting to similar driving forces and pressures.

Historic observation makes possible the exact timing of high tides and low tides along a particular section of coast for a month, a year, or far into the future.

The reason for the difference is oscillation. Think of pots and pans filled with varying levels of water on a table, says Charles O'Reilly, Chief of Tidal Analysis for the Geological Survey of Canada's hydrographic service in Dartmouth, N.S. Then kick the table.

"You'll notice the water in the pots and pans will slosh differently. That's their natural oscillation," he said. "If you kick the table rhythmically, you'll find each pot continues to slosh differently because it has its own rhythm.

"Now, if you join those pots and pans together, that's sort of like a coastal ocean. They're all feeling the same 'kick,' but they are all responding differently. In order to predict a tide, you have to measure them for some period of time."

SOURCE: *Toronto Star*, June 13, 2001, p. GT02

SEE CHAPTER PROJECT 1.

OUTLINE

 For additional study help, go to

www.prenhall.com/sullivanegu3e

Materials include:

- Graphing Calculator Help
- Chapter Quiz
- Chapter Test
- PowerPoint Downloads
- Chapter Projects
- Student Tips

A Look Back, A Look Forward

In Chapter 1, we began our discussion of functions. We defined domain and range and independent and dependent variables; we found the value of a function and graphed functions. We continued our study of functions by listing properties that a function might have, like being even or odd, and we created a library of functions, naming key functions and listing their properties, including the graph.

In this chapter we define the trigonometric functions, six functions that have a wide application. We shall talk about their domain and range, how to find values, graph them, and develop a list of their properties.

There are two widely accepted approaches to the development of the trigonometric functions: one uses right triangles; the other uses circles, especially the unit circle. In this book, we develop the trigonometric functions using right triangles. In Section 2.5, we introduce trigonometric functions using the unit circle and show that this approach leads to the definition using right triangles.

PREPARING FOR THIS SECTION

Before getting started, review the following:

✓ Circumference and Area of a Circle
(Section A.2, p. 568)

✓ Rectangular Coordinates
(Section 1.1, pp. 2–3)

2.1   ANGLES AND THEIR MEASURE

OBJECTIVES

1. Convert between Degrees, Minutes, Seconds, and Decimal Forms for Angles
2. Find the Arc Length of a Circle
3. Convert from Degrees to Radians
4. Convert from Radians to Degrees
5. Find the Area of a Sector of a Circle
6. Find the Linear Speed of an Object Traveling in Circular Motion

A **ray**, or **half-line**, is that portion of a line that starts at a point V on the line and extends indefinitely in one direction. The starting point V of a ray is called its **vertex**. See Figure 1.

If two rays are drawn with a common vertex, they form an **angle**. We call one of the rays of an angle the **initial side** and the other the **terminal side**. The angle that is formed is identified by showing the direction and amount of rotation from the initial side to the terminal side. If the rotation is in the counterclockwise direction, the angle is **positive**; if the rotation is clockwise, the angle is **negative**. See Figure 2. Lowercase Greek letters, such as α (alpha), β (beta), γ (gamma), and θ (theta), will be used to denote angles. Notice in Figure 2(a) that the angle α is positive because the direction of the rotation from the initial side to the terminal side is counterclockwise. The angle β in Figure 2(b) is negative because the rotation is clockwise. The angle γ in Figure 2(c) is positive. Notice that the angle α in Figure 2(a) and the angle γ in Figure 2(c) have the same initial side and the same terminal side. However, α and γ are unequal, because the amount of rotation required to go from the initial side to the terminal side is greater for angle γ than for angle α.

Figure 1

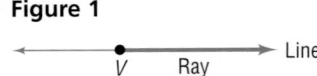

Figure 2

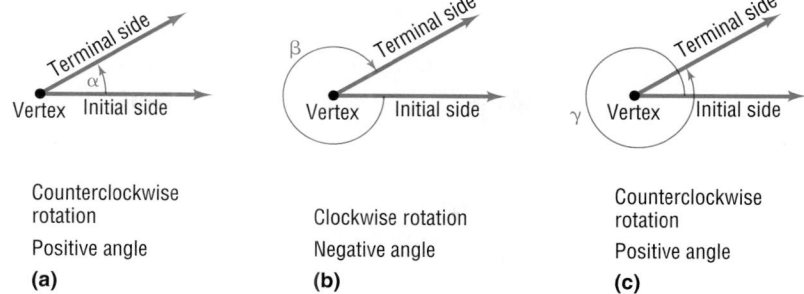

Counterclockwise rotation

Positive angle

(a)

Clockwise rotation

Negative angle

(b)

Counterclockwise rotation

Positive angle

(c)

An angle θ is said to be in **standard position** if its vertex is at the origin of a rectangular coordinate system and its initial side coincides with the positive *x*-axis. See Figure 3.

Figure 3

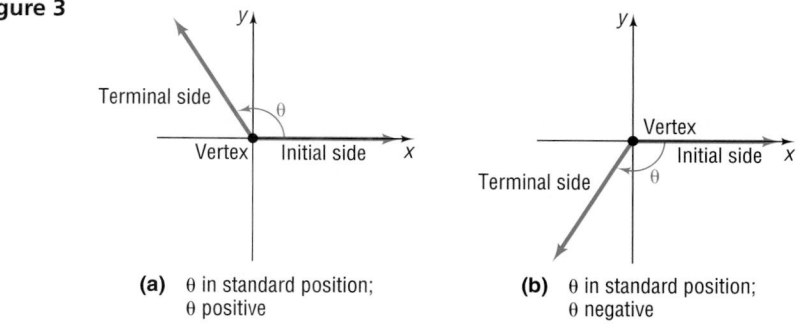

(a) θ in standard position; θ positive

(b) θ in standard position; θ negative

When an angle θ is in standard position, the terminal side will lie either in a quadrant, in which case we say that θ **lies in that quadrant**, or on the *x*-axis or the *y*-axis, in which case we say that θ is a **quadrantal angle**. For example, the angle θ in Figure 4(a) lies in quadrant II, the angle θ in Figure 4(b) lies in quadrant IV, and the angle θ in Figure 4(c) is a quadrantal angle.

Figure 4

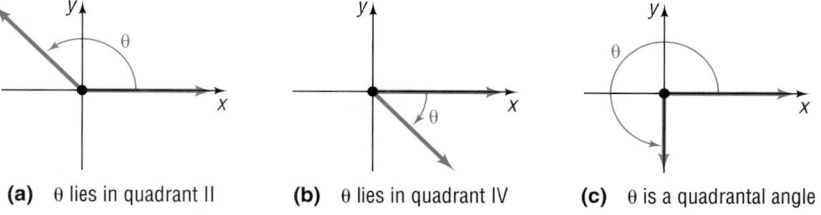

(a) θ lies in quadrant II

(b) θ lies in quadrant IV

(c) θ is a quadrantal angle

We measure angles by determining the amount of rotation needed for the initial side to become coincident with the terminal side. The two commonly used measures for angles are *degrees* and *radians*.

Degrees

The angle formed by rotating the initial side exactly once in the counterclockwise direction until it coincides with itself (1 revolution) is said to measure 360 degrees, abbreviated 360°. **One degree, 1°,** is $\dfrac{1}{360}$ revolution. A **right angle** is an angle that measures 90°, or $\dfrac{1}{4}$ revolution; a **straight angle** is an angle that measures $180°$, or $\dfrac{1}{2}$ revolution. See Figure 5. As Figure 5(b) shows, it is customary to indicate a right angle by using the symbol ⌐.

Figure 5

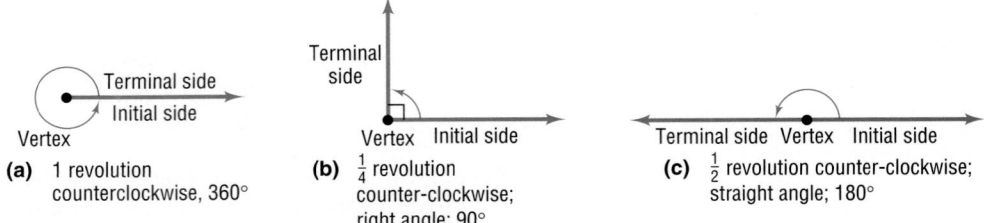

(a) 1 revolution counterclockwise, 360°

(b) $\frac{1}{4}$ revolution counter-clockwise; right angle; 90°

(c) $\frac{1}{2}$ revolution counter-clockwise; straight angle; 180°

It is also customary to refer to an angle that measures θ degrees as an angle of θ degrees.

EXAMPLE 1   **Drawing an Angle**

Draw each angle.

(a) 45°   (b) −90°   (c) 225°   (d) 405°

Solution   (a) An angle of $45°$ is $\dfrac{1}{2}$ of a right angle. See Figure 6.

(b) An angle of $−90°$ is $\dfrac{1}{4}$ revolution in the clockwise direction. See Figure 7.

Figure 6

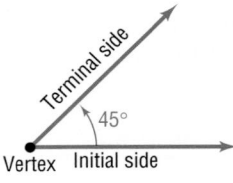

Figure 7

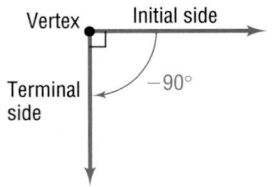

(c) An angle of $225°$ consists of a rotation through $180°$ followed by a rotation through $45°$. See Figure 8.

(d) An angle of $405°$ consists of 1 revolution ($360°$) followed by a rotation through $45°$. See Figure 9.

Figure 8

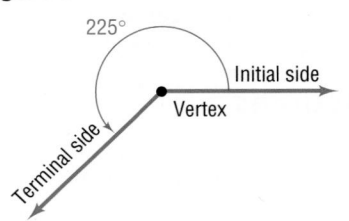

Figure 9

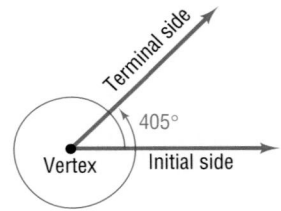

NOW WORK PROBLEM **1.**

1   Although subdivisions of a degree may be obtained by using decimals, we also may use the notion of *minutes* and *seconds*. **One minute,** denoted by **1′,** is defined as $\dfrac{1}{60}$ degree. **One second,** denoted by **1″,** is defined as $\dfrac{1}{60}$ minute or, equivalently, $\dfrac{1}{3600}$ degree. An angle of, say, 30 degrees, 40 minutes, 10 seconds is written compactly as 30°40′10″. To summarize:

$$\boxed{\begin{array}{c} 1 \text{ counterclockwise revolution} = 360° \\[4pt] 1° = 60' \qquad 1' = 60'' \end{array}} \tag{1}$$

It is sometimes necessary to convert from the degree, minute, second notation (D°M′S″) to a decimal form, and vice versa. Check your calculator; it should be capable of doing the conversion for you.

Before getting started you must set the mode to degrees, because there are two common ways to measure angles: degree mode and radian mode. (We will define radians shortly.) Usually, a menu is used to change from one mode to another. Check your owner's manual to find out how your particular calculator works.

Now let's see how to convert by hand from the degree, minute, second notation (D°M′S″) to a decimal form, and vice versa, by looking at some examples: $15°30' = 15.5°$ because $30' = \dfrac{1}{2}^{\circ} = 0.5°$

$$32.25° = 32°15' \text{ because } 0.25° = \frac{1}{4}^{\circ} = \frac{1}{4}(60') = 15'.$$

EXAMPLE 2   **Converting between Degrees, Minutes, Seconds, and Decimal Forms**

(a) Convert 50°6′21″ to a decimal in degrees.
(b) Convert 21.256° to the D°M′S″ form.

Algebraic Solution   (a) Because $1' = \dfrac{1}{60}^{\circ}$ and $1'' = \dfrac{1}{60}' = \left(\dfrac{1}{60} \cdot \dfrac{1}{60}\right)^{\circ}$, we convert as follows:

$$50°6'21'' = 50° + 6' + 21'' = 50° + 6 \cdot \frac{1}{60}^{\circ} + 21 \cdot \frac{1}{60} \cdot \frac{1}{60}^{\circ}$$

$$\approx 50° + 0.1° + 0.005833°$$

$$= 50.105833°$$

(b) We start with the decimal part of 21.256°, that is, 0.256°.

$$0.256° = (0.256)(1°) = (0.256)(60') = 15.36'$$

$$\uparrow$$
$$1° = 60'$$

Now we work with the decimal part of 15.36′, that is, 0.36′.

$$0.36' = (0.36)(1') = (0.36)(60'') = 21.6'' \approx 22''$$

$$\uparrow$$
$$1' = 60''$$

Thus,

$$21.256° = 21° + 0.256° = 21° + 15.36' = 21° + 15' + 0.36'$$

$$= 21° + 15' + 21.6'' \approx 21°15'22''$$

Graphing Solution
(a) Figure 10 shows the solution using a TI-83 graphing calculator.

Figure 10

(b) Figure 11 shows the solution using a TI-83 graphing calculator.

Figure 11

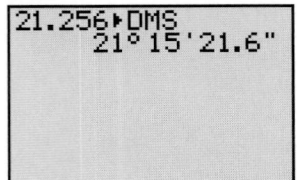

 NOW WORK PROBLEMS **69** AND **75**.

 In many applications, such as describing the exact location of a star or the precise position of a boat at sea, angles measured in degrees, minutes, and even seconds are used. For calculation purposes, these are transformed to decimal form. In other applications, especially those in calculus, angles are measured using *radians*.

Radians

A **central angle** is an angle whose vertex is at the center of a circle. The rays of a central angle subtend (intersect) an arc on the circle. If the radius of the circle is r and the length of the arc subtended by the central angle is also r, then the measure of the angle is **1 radian.** See Figure 12(a).

For a circle of radius 1, the rays of a central angle with measure 1 radian would subtend an arc of length 1. For a circle of radius 3, the rays of a central angle with measure 1 radian would subtend an arc of length 3. See Figure 12(b).

Figure 12

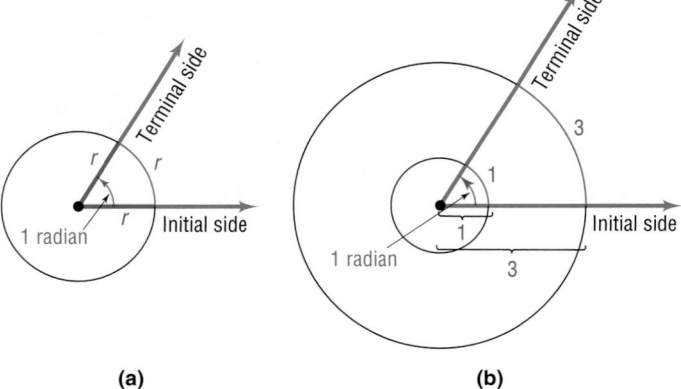

(a)   (b)

Figure 13

$$\frac{\theta}{\theta_1} = \frac{s}{s_1}$$

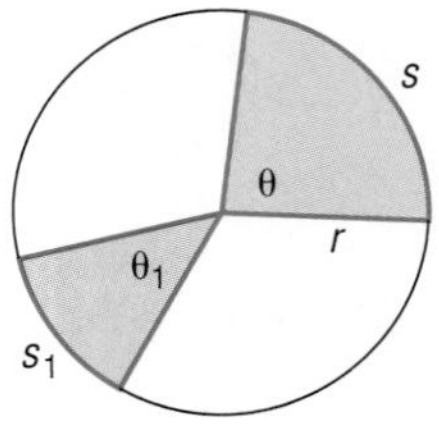

2   Now consider a circle of radius r and two central angles, θ and θ_1, measured in radians. Suppose that these central angles subtend arcs of lengths s and s_1, respectively, as shown in Figure 13. From geometry, we know that the ratio of the measures of the angles equals the ratio of the corresponding lengths of the arcs subtended by these angles; that is,

$$\frac{\theta}{\theta_1} = \frac{s}{s_1} \qquad (2)$$

Suppose that $\theta_1 = 1$ radian. Refer again to Figure 12(a). The amount of arc s_1 subtended by the central angle $\theta_1 = 1$ radian equals the radius r of the circle. Then $s_1 = r$, so formula (2) reduces to

$$\frac{\theta}{1} = \frac{s}{r} \quad \text{or} \quad s = r\theta \qquad (3)$$

Theorem   **Arc Length**

For a circle of radius r, a central angle of θ radians subtends an arc whose length s is

$$s = r\theta \qquad (4)$$

Note: Formulas must be consistent with regard to the units used. In equation (4), we write

$$s = r\theta$$

To see the units, however, we must go back to equation (3) and write

$$\frac{\theta \text{ radians}}{1 \text{ radian}} = \frac{s \text{ length units}}{r \text{ length units}}$$

$$s \text{ length units} = r \text{ length units} \frac{\theta \text{ radians}}{1 \text{ radian}}$$

Since the radians cancel, we are left with

$$s \text{ length units} = (r \text{ length units})\theta \qquad s = r\theta$$

where θ appears to be "dimensionless" but, in fact, is measured in radians. So, in using the formula $s = r\theta$, the dimension for θ is radians, and any convenient unit of length (such as inches or meters) may be used for s and r. ■

EXAMPLE 3   **Finding the Length of Arc of a Circle**

Find the length of the arc of a circle of radius 2 meters subtended by a central angle of 0.25 radian.

Solution   We use equation (4) with $r = 2$ meters and $\theta = 0.25$. The length s of the arc is

$$s = r\theta = 2(0.25) = 0.5 \text{ meter}$$ ■

 **NOW WORK PROBLEM 37.**

Figure 14
1 revolution = 2π radians

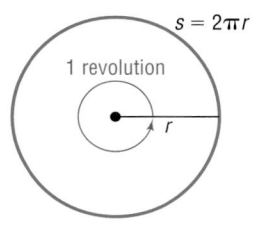

$s = 2\pi r$

1 revolution

r

Relationship Between Degrees And Radians

Consider a circle of radius r. A central angle of 1 revolution will subtend an arc equal to the circumference of the circle (Figure 14). Because the circumference of a circle equals $2\pi r$, we use $s = 2\pi r$ in equation (4) to find that, for an angle θ of 1 revolution,

$$s = r\theta$$
$$2\pi r = r\theta \qquad \theta = 1 \text{ revolution}; s = 2\pi r$$
$$\theta = 2\pi \text{ radians} \qquad \text{Solve for } \theta$$

From this we have,

$$1\ \text{revolution} = 2\pi\ \text{radians} \qquad (5)$$

so

$$360° = 2\pi\ \text{radians}$$

or

$$180° = \pi\ \text{radians} \qquad (6)$$

Divide both sides of equation (6) by 180. Then

$$1\ \text{degree} = \frac{\pi}{180}\text{radian}$$

Divide both sides of (6) by π. Then

$$\frac{180}{\pi}\text{degrees} = 1\ \text{radian}$$

We have the following two conversion formulas:

$$1\ \text{degree} = \frac{\pi}{180}\text{radian} \qquad 1\ \text{radian} = \frac{180}{\pi}\text{degrees} \qquad (7)$$

3 **EXAMPLE 4** **Converting from Degrees to Radians**

Convert each angle in degrees to radians.

(a) 60°   (b) 150°   (c) −45°   (d) 90°

Solution (a) $60° = 60 \cdot 1\ \text{degree} = 60 \cdot \frac{\pi}{180}\text{radian} = \frac{\pi}{3}\text{radians}$

(b) $150° = 150 \cdot \frac{\pi}{180}\text{radian} = \frac{5\pi}{6}\text{radians}$

(c) $-45° = -45 \cdot \frac{\pi}{180}\text{radian} = -\frac{\pi}{4}\text{radian}$

(d) $90° = 90 \cdot \frac{\pi}{180}\text{radian} = \frac{\pi}{2}\text{radians}$

Example 4 illustrates that angles that are fractions of a revolution are expressed in radian measure as fractional multiples of π, rather than as decimals. For example, a right angle, as in Example 4(d), is left in the form $\frac{\pi}{2}$ radians, which is exact, rather than using the approximation $\frac{\pi}{2} \approx \frac{3.1416}{2} = 1.5708$ radians.

NOW WORK PROBLEM 13.

(4) **EXAMPLE 5**   **Converting Radians to Degrees**

Convert each angle in radians to degrees.

(a) $\dfrac{\pi}{6}$ radian   (b) $\dfrac{3\pi}{2}$ radians   (c) $-\dfrac{3\pi}{4}$ radians   (d) $\dfrac{7\pi}{3}$ radians

Solution   (a) $\dfrac{\pi}{6}$ radian $= \dfrac{\pi}{6} \cdot 1$ radian $= \dfrac{\pi}{6} \cdot \dfrac{180}{\pi}$ degrees $= 30°$

(b) $\dfrac{3\pi}{2}$ radians $= \dfrac{3\pi}{2} \cdot \dfrac{180}{\pi}$ degrees $= 270°$

(c) $-\dfrac{3\pi}{4}$ radians $= -\dfrac{3\pi}{4} \cdot \dfrac{180}{\pi}$ degrees $= -135°$

(d) $\dfrac{7\pi}{3}$ radians $= \dfrac{7\pi}{3} \cdot \dfrac{180}{\pi}$ degrees $= 420°$

 **NOW WORK PROBLEM 25.**

Table 1 lists the degree and radian measures of some commonly encountered angles. You should learn to feel equally comfortable using degree or radian measure for these angles.

TABLE 1

Degrees	0°	30°	45°	60°	90°	120°	135°	150°	180°
Radians	0	$\dfrac{\pi}{6}$	$\dfrac{\pi}{4}$	$\dfrac{\pi}{3}$	$\dfrac{\pi}{2}$	$\dfrac{2\pi}{3}$	$\dfrac{3\pi}{4}$	$\dfrac{5\pi}{6}$	π
Degrees		210°	225°	240°	270°	300°	315°	330°	360°
Radians		$\dfrac{7\pi}{6}$	$\dfrac{5\pi}{4}$	$\dfrac{4\pi}{3}$	$\dfrac{3\pi}{2}$	$\dfrac{5\pi}{3}$	$\dfrac{7\pi}{4}$	$\dfrac{11\pi}{6}$	2π

EXAMPLE 6   **Finding the Distance between Two Cities**

See Figure 15(a). The latitude of a location L is the angle formed by a ray drawn from the center of Earth to the Equator and a ray drawn from the center of Earth to L. See Figure 15(b). Glasgow, Montana, is due north of Albuquerque, New Mexico. Find the distance between Glasgow (48°9′ north latitude) and Albuquerque (35°5′ north latitude). Assume that the radius of Earth is 3960 miles.

Figure 15

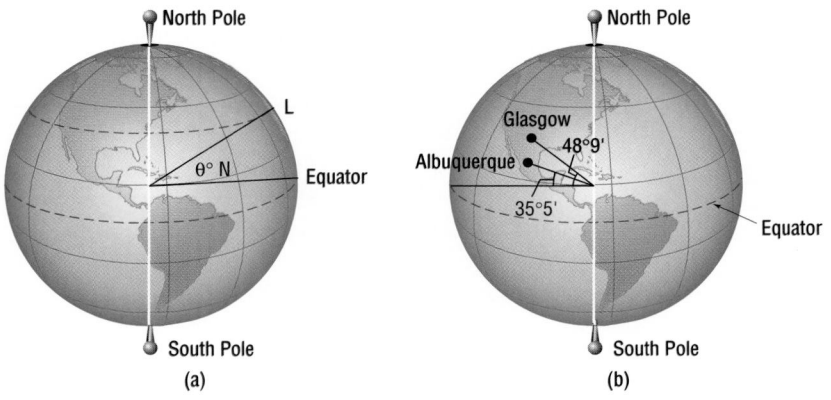

(a)   (b)

Solution The measure of the central angle between the two cities is $48°9' - 35°5' = 13°4'$. We use equation (4), $s = r\theta$, but first we must convert the angle of $13°4'$ to radians.

$$\theta = 13°4' \approx 13.0667° = 13.0667 \cdot \frac{\pi}{180}\text{radian} \approx 0.228 \text{ radian}$$

We use $\theta = 0.228$ radian and $r = 3960$ miles in equation (4). The distance between the two cities is

$$s = r\theta = 3960 \cdot 0.228 \approx 903 \text{ miles}$$ ■

When an angle is measured in degrees, the degree symbol will always be shown. However, when an angle is measured in radians, we will follow the usual practice and omit the word *radians*. So, if the measure of an angle is given as $\frac{\pi}{6}$ it is understood to mean $\frac{\pi}{6}$ radian.

✏️ ━ **NOW WORK PROBLEM 91.**

Area of a Sector

5 Consider a circle of radius r. Suppose that θ, measured in radians, is a central angle of this circle. See Figure 16. We seek a formula for the area A of the sector formed by the angle θ (shown in blue).

Figure 16

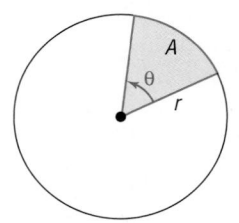

Now consider a circle of radius r and two central angles θ and θ_1, both measured in radians. See Figure 17. From geometry, we know the ratio of the measures of the angles equals the ratio of the corresponding areas of the sectors formed by these angles. That is,

$$\frac{\theta}{\theta_1} = \frac{A}{A_1}$$

Figure 17

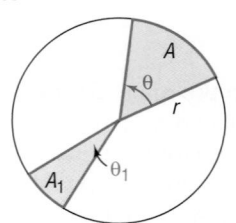

Suppose that $\theta_1 = 2\pi$ radians. Then $A_1 = $ area of the circle $= \pi r^2$. Solving for A, we find

$$A = A_1\frac{\theta}{\theta_1} = \pi r^2\frac{\theta}{2\pi} = \frac{1}{2}r^2\theta$$

Theorem   **Area of a Sector**

The area A of the sector of a circle of radius r formed by a central angle of θ radians is

$$A = \frac{1}{2}r^2\theta \qquad (8)$$

■

EXAMPLE 7   **Finding the Area of a Sector of a Circle**

Find the area of the sector of a circle of radius 2 feet formed by an angle of $30°$. Round the answer to two decimal places.

Solution   We use equation (8) with $r = 2$ feet and $\theta = 30° = \dfrac{\pi}{6}$ radian. [Remember, in equation (8), θ must be in radians.] The area A of the sector is

$$A = \frac{1}{2}r^2\theta = \frac{1}{2}(2)^2\frac{\pi}{6} = \frac{\pi}{3}\text{ square feet} \approx 1.05\text{ square feet}$$

rounded to two decimal places.

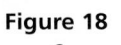 **NOW WORK PROBLEM 45.**

Circular Motion

⑥ We have already defined the average speed of an object as the distance traveled divided by the elapsed time. Suppose that an object moves around a circle of radius r at a constant speed. If s is the distance traveled in time t around this circle, then the **linear speed** v of the object is defined as

$$v = \frac{s}{t} \tag{9}$$

Figure 18

$$v = \frac{s}{t}; \omega = \frac{\theta}{t}$$

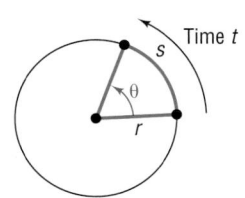

As this object travels around the circle, suppose that θ (measured in radians) is the central angle swept out in time t. See Figure 18. Then the **angular speed** ω (the Greek letter omega) of this object is the angle (measured in radians) swept out divided by the elapsed time; that is,

$$\omega = \frac{\theta}{t} \tag{10}$$

Angular speed is the way the turning rate of an engine is described. For example, an engine idling at 900 rpm (revolutions per minute) is one that rotates at an angular speed of

$$900\,\frac{\text{revolutions}}{\text{minute}} = 900\,\frac{\cancel{\text{revolutions}}}{\text{minute}} \cdot 2\pi\,\frac{\text{radians}}{\cancel{\text{revolution}}} = 1800\pi\,\frac{\text{radians}}{\text{minute}}$$

There is an important relationship between linear speed and angular speed:

$$\text{linear speed} = v = \underset{\underset{(9)}{\uparrow}}{\frac{s}{t}} = \underset{\underset{s\,=\,r\theta}{\uparrow}}{\frac{r\theta}{t}} = r\left(\frac{\theta}{t}\right)$$

Then, using equation (10), we obtain

$$v = r\omega \tag{11}$$

where ω is measured in radians per unit time.

When using equation (11), remember that $v = \dfrac{s}{t}$ (the linear speed) has the dimensions of length per unit of time (such as feet per second or miles per hour), r (the radius of the circular motion) has the same length dimension as s, and ω (the angular speed) has the dimensions of radians per unit of time. If the angular speed is given in terms of *revolutions* per unit of time (as is often the case), be sure to convert it to *radians* per unit of time before attempting to use equation (11).

EXAMPLE 8   Finding Linear Speed

A child is spinning a rock at the end of a 2-foot rope at the rate of 180 revolutions per minute (rpm). Find the linear speed of the rock when it is released.

Figure 19

Solution   Look at Figure 19. The rock is moving around a circle of radius $r = 2$ feet. The angular speed ω of the rock is

$$\omega = 180\frac{\text{revolutions}}{\text{minute}} = 180\frac{\text{revolutions}}{\text{minute}} \cdot 2\pi\frac{\text{radians}}{\text{revolution}} = 360\pi\frac{\text{radians}}{\text{minute}}$$

From equation (11), the linear speed v of the rock is

$$v = r\omega = 2 \text{ feet} \cdot 360\pi\frac{\text{radians}}{\text{minute}} = 720\pi\frac{\text{feet}}{\text{minute}} \approx 2262\frac{\text{feet}}{\text{minute}}$$

The linear speed of the rock when it is released is $2262 \text{ ft/min} \approx 25.7 \text{ mi/hr}$.
■

─── NOW WORK PROBLEM **87.**

HISTORICAL FEATURE   Trigonometry was developed by Greek astronomers, who regarded the sky as the inside of a sphere, so it was natural that triangles on a sphere were investigated early (by Menelaus of Alexandria about AD 100) and that triangles in the plane were studied much later. The first book containing a systematic treatment of plane and spherical trigonometry was written by the Persian astronomer Nasîr Eddîn (about AD 1250).

Regiomontanus (1436–1476) is the person most responsible for moving trigonometry from astronomy into mathematics. His work was improved by Copernicus (1473–1543) and Copernicus's student Rhaeticus (1514–1576). Rhaeticus's book was the first to define the six trigonometric functions as ratios of sides of triangles, although he did not give the functions their present names. Credit for this is due to Thomas Finck (1583), but Finck's notation was by no means universally accepted at the time. The notation was finally stabilized by the textbooks of Leonhard Euler (1707–1783).

Trigonometry has since evolved from its use by surveyors, navigators, and engineers to present applications involving ocean tides, the rise and fall of food supplies in certain ecologies, brain wave patterns, and many other phenomena.

2.1   Concepts and Vocabulary

In Problems 1–3, fill in the blanks.

1. An angle θ is in _____ _____ if its vertex is at the origin of a rectangular coordinate system and its initial side coincides with the positive x-axis.

2. On a circle of radius r, a central angle of θ radians subtends an arc of length $s =$ _____ ; the area of the sector formed by this angle θ is $A =$ _____ .

3. An object travels around a circle of radius r with constant speed. If s is the distance traveled in time t around the circle and θ is the central angle (in radians) swept out in time t, then the linear speed of the object is $v =$ _____ and the angular speed of the object is $\omega =$ _____ .

In Problems 4–6, answer True or False to each statement.

4. $\pi = 180$.

5. $180° = \pi$ radians

6. On the unit circle, if s is the length of the arc subtended by a central angle θ, measured in radians, then $s = \theta$.

7. What is 1 radian?

8. Which angle has the larger measure: 1 degree or 1 radian? Or are they equal?

9. Explain the difference between linear speed and angular speed.

2.1 Exercises

In Problems 1–12, draw each angle.

1. $30°$ 2. $60°$ 3. $135°$ 4. $-120°$ 5. $450°$ 6. $540°$

7. $\dfrac{3\pi}{4}$ 8. $\dfrac{4\pi}{3}$ 9. $-\dfrac{\pi}{6}$ 10. $-\dfrac{2\pi}{3}$ 11. $\dfrac{16\pi}{3}$ 12. $\dfrac{21\pi}{4}$

In Problems 13–24, convert each angle in degrees to radians. Express your answer as a multiple of π.

13. $30°$ 14. $120°$ 15. $240°$ 16. $330°$ 17. $-60°$ 18. $-30°$
19. $180°$ 20. $270°$ 21. $-135°$ 22. $-225°$ 23. $-90°$ 24. $-180°$

In Problems 25–36, convert each angle in radians to degrees.

25. $\dfrac{\pi}{3}$ 26. $\dfrac{5\pi}{6}$ 27. $-\dfrac{5\pi}{4}$ 28. $-\dfrac{2\pi}{3}$ 29. $\dfrac{\pi}{2}$ 30. 4π

31. $\dfrac{\pi}{12}$ 32. $\dfrac{5\pi}{12}$ 33. $-\dfrac{\pi}{2}$ 34. $-\pi$ 35. $-\dfrac{\pi}{6}$ 36. $-\dfrac{3\pi}{4}$

In Problems 37–44, s denotes the length of the arc of a circle of radius r subtended by the central angle θ. Find the missing quantity. Round answers to three decimal places, if necessary.

37. $r = 10$ meters, $\theta = \dfrac{1}{2}$ radian, $s = ?$

38. $r = 6$ feet, $\theta = 2$ radians, $s = ?$

39. $\theta = \dfrac{1}{3}$ radian, $s = 2$ feet, $r = ?$

40. $\theta = \dfrac{1}{4}$ radian, $s = 6$ centimeters, $r = ?$

41. $r = 5$ miles, $s = 3$ miles, $\theta = ?$

42. $r = 6$ meters, $s = 8$ meters, $\theta = ?$

43. $r = 2$ inches, $\theta = 30°$, $s = ?$

44. $r = 3$ meters, $\theta = 120°$, $s = ?$

In Problems 45–52, A denotes the area of the sector of a circle of radius r formed by the central angle θ. Find the missing quantity. Round answers to three decimal places, if necessary.

45. $r = 10$ meters, $\theta = \dfrac{1}{2}$ radian, $A = ?$

46. $r = 6$ feet, $\theta = 2$ radians, $A = ?$

47. $\theta = \dfrac{1}{3}$ radian, $A = 2$ square feet, $r = ?$

48. $\theta = \dfrac{1}{4}$ radian, $A = 6$ square centimeters, $r = ?$

49. $r = 5$ miles, $A = 3$ square miles, $\theta = ?$

50. $r = 6$ meters, $A = 8$ square meters, $\theta = ?$

51. $r = 2$ inches, $\theta = 30°$, $A = ?$

52. $r = 3$ meters, $\theta = 120°$, $A = ?$

In Problems 53–56, find the length s and area A. Round answers to three decimal places.

53.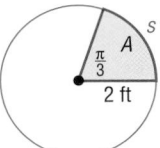
54.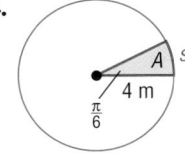
55.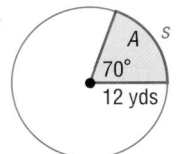
56.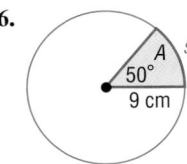

In Problems 57–62, convert each angle in degrees to radians. Express your answer in decimal form, rounded to two decimal places.

57. $17°$ 58. $73°$ 59. $-40°$ 60. $-51°$ 61. $125°$ 62. $350°$

In Problems 63–68, convert each angle in radians to degrees. Express your answer in decimal form, rounded to two decimal places.

63. 3.14 64. 0.75 65. 2 66. 3 67. 6.32 68. $\sqrt{2}$

In Problems 69–74, convert each angle to a decimal in degrees. Round your answer to two decimal places.

69. $40°10'25''$ 70. $61°42'21''$ 71. $1°2'3''$ 72. $73°40'40''$ 73. $9°9'9''$ 74. $98°22'45''$

In Problems 75–80, convert each angle to $D°M'S''$ form. Round your answer to the nearest second.

75. $40.32°$ 76. $61.24°$ 77. $18.255°$ 78. $29.411°$ 79. $19.99°$ 80. $44.01°$

81. **Minute Hand of a Clock** The minute hand of a clock is 6 inches long. How far does the tip of the minute hand move in 15 minutes? How far does it move in 25 minutes?

82. **Movement of a Pendulum** A pendulum swings through an angle of 20° each second. If the pendulum is 40 inches long, how far does its tip move each second?

83. **Area of a Sector** Find the area of the sector of a circle of radius 4 meters formed by an angle of 45°. Round the answer to two decimal places.

84. **Area of a Sector** Find the area of the sector of a circle of radius 3 centimeters formed by an angle of 60°. Round the answer to two decimal places.

85. **Watering a Lawn** A water sprinkler sprays water over a distance of 30 feet while rotating through an angle of 135°. What area of lawn receives water?

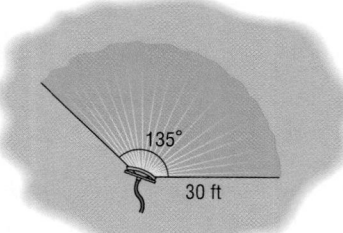

86. **Designing a Water Sprinkler** An engineer is asked to design a water sprinkler that will cover a field of 100 square yards that is in the shape of a sector of a circle of radius 50 yards. Through what angle should the sprinkler rotate?

87. **Motion on a Circle** An object is traveling around a circle with a radius of 5 centimeters. If in 20 seconds a central angle of $\frac{1}{3}$ radian is swept out, what is the angular speed of the object? What is its linear speed?

88. **Motion on a Circle** An object is traveling around a circle with a radius of 2 meters. If in 20 seconds the object travels 5 meters, what is its angular speed? What is its linear speed?

89. **Bicycle Wheels** The diameter of each wheel of a bicycle is 26 inches. If you are traveling at a speed of 35 miles per hour on this bicycle, through how many revolutions per minute are the wheels turning?

90. **Car Wheels** The radius of each wheel of a car is 15 inches. If the wheels are turning at the rate of 3 revolutions per second, how fast is the car moving? Express your answer in inches per second and in miles per hour.

In Problems 91–94, the latitude of a location L is the angle formed by a ray drawn from the center of Earth to the Equator and a ray drawn from the center of Earth to L. See the figure.

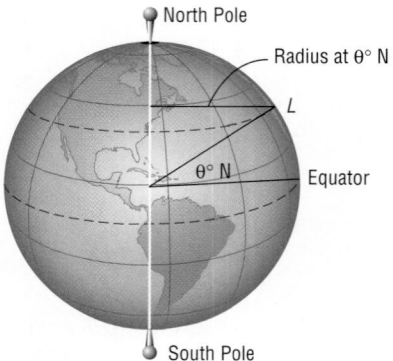

91. **Distance between Cities** Memphis, Tennessee, is due north of New Orleans, Louisiana. Find the distance between Memphis (35°9′ north latitude) and New Orleans (29°57′ north latitude). Assume that the radius of Earth is 3960 miles.

92. **Distance between Cities** Charleston, West Virginia, is due north of Jacksonville, Florida. Find the distance between Charleston (38°21′ north latitude) and Jacksonville (30°20′ north latitude). Assume that the radius of Earth is 3960 miles.

93. **Linear Speed on Earth** Earth rotates on an axis through its poles. The distance from the axis to a location on Earth 30° north latitude is about 3429.5 miles. Therefore, a location on Earth at 30° north latitude is spinning on a circle of radius 3429.5 miles. Compute the linear speed on the surface of Earth at 30° north latitude.

94. **Linear Speed on Earth** Earth rotates on an axis through its poles. The distance from the axis to a location on Earth 40° north latitude is about 3033.5 miles. Therefore, a location on Earth at 40° north latitude is spinning on a circle of radius 3033.5 miles. Compute the linear speed on the surface of Earth at 40° north latitude.

95. **Speed of the Moon** The mean distance of the Moon from Earth is 2.39×10^5 miles. Assuming that the orbit of the Moon around Earth is circular and that 1 revolution takes 27.3 days, find the linear speed of the Moon. Express your answer in miles per hour.

96. **Speed of Earth** The mean distance of Earth from the Sun is 9.29×10^7 miles. Assuming that the orbit of Earth around the Sun is circular and that 1 revolution takes 365 days, find the linear speed of Earth. Express your answer in miles per hour.

97. **Pulleys** Two pulleys, one with radius 2 inches and the other with radius 8 inches, are connected by a belt. (See the figure on page 111.) If the 2-inch pulley is caused to rotate at 3 revolutions per minute, determine the revolutions per minute of the 8-inch pulley.

[**Hint:** The linear speeds of the pulleys, that is, the speed of the belt, are the same.]

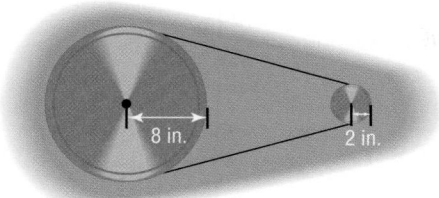

98. Ferris Wheels A neighborhood carnival has a Ferris wheel whose radius is 30 feet. You measure the time it takes for one revolution to be 70 seconds. What is the linear speed (in feet per second) of this Ferris wheel? What is the angular speed in radians per second?

99. Computing the Speed of a River Current To approximate the speed of the current of a river, a circular paddle wheel with radius 4 feet is lowered into the water. If the current causes the wheel to rotate at a speed of 10 revolutions per minute, what is the speed of the current? Express your answer in miles per hour.

100. Spin Balancing Tires A spin balancer rotates the wheel of a car at 480 revolutions per minute. If the diameter of the wheel is 26 inches, what road speed is being tested? Express your answer in miles per hour. At how many revolutions per minute should the balancer be set to test a road speed of 80 miles per hour?

101. The Cable Cars of San Francisco At the Cable Car Museum you can see the four cable lines that are used to pull cable cars up and down the hills of San Francisco. Each cable travels at a speed of 9.55 miles per hour, caused by a rotating wheel whose diameter is 8.5 feet. How fast is the wheel rotating? Express your answer in revolutions per minute.

102. Difference in Time of Sunrise Naples, Florida, is approximately 90 miles due west of Ft. Lauderdale. How much sooner would a person in Ft. Lauderdale first see the rising Sun than a person in Naples?

[**Hint:** Consult the figure. When a person at Q sees the first rays of the Sun, a person at P is still in the dark. The person at P sees the first rays after Earth has rotated so that P is at the location Q. Now use the fact that at the

latitude of Ft. Lauderdale in 24 hours a length of arc of $2\pi(3559)$ miles is subtended.]

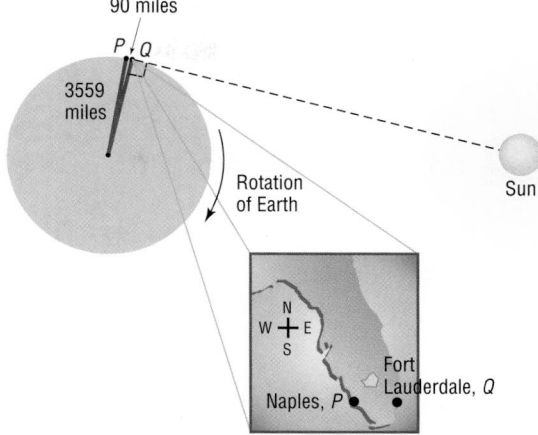

103. Keeping Up with the Sun How fast would you have to travel on the surface of Earth at the equator to keep up with the Sun (that is, so that the Sun would appear to remain in the same position in the sky)?

104. Nautical Miles A **nautical mile** equals the length of arc subtended by a central angle of 1 minute on a great circle* on the surface of Earth. (See the figure.) If the radius of Earth is taken as 3960 miles, express 1 nautical mile in terms of ordinary, or **statute,** miles.

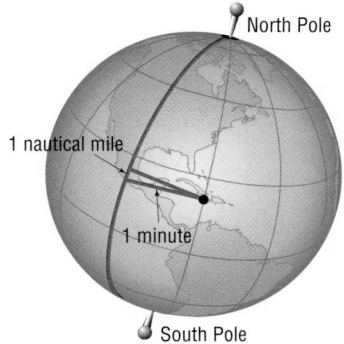

105. Pulleys Two pulleys, one with radius r_1 and the other with radius r_2, are connected by a belt. The pulley with radius r_1 rotates at ω_1 revolutions per minute, whereas the pulley with radius r_2 rotates at ω_2 revolutions per minute. Show that $\dfrac{r_1}{r_2} = \dfrac{\omega_2}{\omega_1}$.

106. Do you prefer to measure angles using degrees or radians? Provide justification and a rationale for your choice.

107. Discuss why ships and airplanes use nautical miles to measure distance. Explain the difference between a nautical mile and a statute mile.

108. Investigate the way that speed bicycles work. In particular, explain the differences and similarities between 5-speed and 9-speed derailleurs. Be sure to include a discussion of linear speed and angular speed.

*Any circle drawn on the surface of Earth that divides Earth into two equal hemispheres.

PREPARING FOR THIS SECTION

Before getting started, review the following:

✓ Pythagorean Theorem (Section A.2, pp. 565–567)   ✓ Functions (Section 1.4, pp. 33–44)

2.2 RIGHT TRIANGLE TRIGONOMETRY

OBJECTIVES   1 Find the Value of Trigonometric Functions of Acute Angles
2 Use the Fundamental Identities
3 Find the Remaining Trigonometric Functions Given the Value of One of Them
4 Use the Complementary Angle Theorem

A triangle in which one angle is a right angle ($90°$) is called a **right triangle.** Recall that the side opposite the right angle is called the **hypotenuse,** and the remaining two sides are called the **legs** of the triangle. In Figure 20 we have labeled the hypotenuse as c to indicate that its length is c units, and, in a like manner, we have labeled the legs as a and b. Because the triangle is a right triangle, the Pythagorean Theorem tells us that

$$c^2 = a^2 + b^2$$

Figure 20

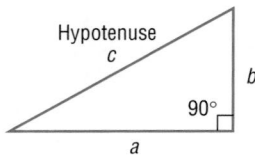

Now, suppose that θ is an **acute angle;** that is, $0° < \theta < 90°$ (if θ is measured in degrees) and $0 < \theta < \dfrac{\pi}{2}$ (if θ is measured in radians). See Figure 21(a). Using this acute angle θ, we can form a right triangle, like the one illustrated in Figure 21(b), with hypotenuse of length c and legs of lengths a and b. Using the three sides of this triangle, we can form exactly six ratios:

$$\frac{b}{c}, \quad \frac{a}{c}, \quad \frac{b}{a}, \quad \frac{c}{b}, \quad \frac{c}{a}, \quad \frac{a}{b}$$

Figure 21

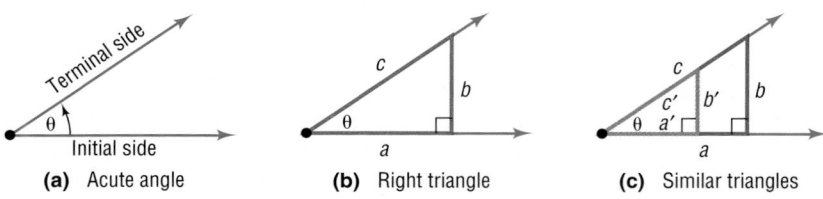

(a) Acute angle   (b) Right triangle   (c) Similar triangles

In fact, these ratios depend only on the size of the angle θ and not on the triangle formed. To see why, look at Figure 21(c). Any two right triangles formed using the angle θ will be similar and, hence, corresponding ratios will be equal. As a result,

$$\frac{b}{c} = \frac{b'}{c'} \qquad \frac{a}{c} = \frac{a'}{c'} \qquad \frac{b}{a} = \frac{b'}{a'} \qquad \frac{c}{b} = \frac{c'}{b'} \qquad \frac{c}{a} = \frac{c'}{a'} \qquad \frac{a}{b} = \frac{a'}{b'}$$

Because the ratios depend only on the angle θ and not on the triangle itself, we give each ratio a name that involves θ: sine of θ, cosine of θ, tangent of θ, cosecant of θ, secant of θ, and cotangent of θ.

The six ratios of a right triangle are called **trigonometric functions of acute angles** and are defined as follows:

Function Name	Abbreviation	Value
sine of θ	$\sin\theta$	$\dfrac{b}{c}$
cosine of θ	$\cos\theta$	$\dfrac{a}{c}$
tangent of θ	$\tan\theta$	$\dfrac{b}{a}$
cosecant of θ	$\csc\theta$	$\dfrac{c}{b}$
secant of θ	$\sec\theta$	$\dfrac{c}{a}$
cotangent of θ	$\cot\theta$	$\dfrac{a}{b}$

Figure 22

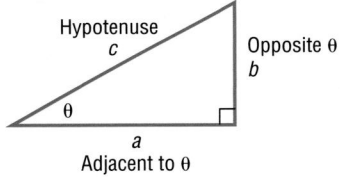

Hypotenuse
c

Opposite θ
b

θ

a
Adjacent to θ

As an aid to remembering these definitions, it may be helpful to refer to the lengths of the sides of the triangle by the names *hypotenuse (c)*, *opposite (b)*, and *adjacent (a)*. See Figure 22. In terms of these names, we have the following ratios:

$$
\sin\theta = \frac{\text{opposite}}{\text{hypotenuse}} = \frac{b}{c} \qquad \cos\theta = \frac{\text{adjacent}}{\text{hypotenuse}} = \frac{a}{c} \qquad \tan\theta = \frac{\text{opposite}}{\text{adjacent}} = \frac{b}{a}
$$

$$
\csc\theta = \frac{\text{hypotenuse}}{\text{opposite}} = \frac{c}{b} \qquad \sec\theta = \frac{\text{hypotenuse}}{\text{adjacent}} = \frac{c}{a} \qquad \cot\theta = \frac{\text{adjacent}}{\text{opposite}} = \frac{a}{b}
$$

(1)

Notice that each of the trigonometric functions of the acute angle θ is positive.

① **EXAMPLE 1** **Finding the Value of Trigonometric Functions**

Figure 23

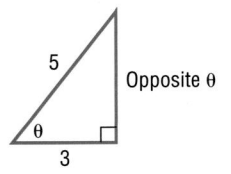

5

Opposite θ

θ

3

Find the value of each of the six trigonometric functions of the angle θ in Figure 23.

Solution We see in Figure 23 that the two given sides of the triangle are

$$c = \text{hypotenuse} = 5 \qquad a = \text{adjacent} = 3$$

To find the length of the opposite side, we use the Pythagorean Theorem.

$$(\text{adjacent})^2 + (\text{opposite})^2 = (\text{hypotenuse})^2$$
$$3^2 + (\text{opposite})^2 = 5^2$$
$$(\text{opposite})^2 = 25 - 9 = 16$$
$$\text{opposite} = 4$$

Now that we know the lengths of the three sides, we use the ratios in (1) to find the value of each of the six trigonometric functions:

$$\sin\theta = \frac{\text{opposite}}{\text{hypotenuse}} = \frac{4}{5} \qquad \cos\theta = \frac{\text{adjacent}}{\text{hypotenuse}} = \frac{3}{5} \qquad \tan\theta = \frac{\text{opposite}}{\text{adjacent}} = \frac{4}{3}$$

$$\csc\theta = \frac{\text{hypotenuse}}{\text{opposite}} = \frac{5}{4} \qquad \sec\theta = \frac{\text{hypotenuse}}{\text{adjacent}} = \frac{5}{3} \qquad \cot\theta = \frac{\text{adjacent}}{\text{opposite}} = \frac{3}{4}$$ ■

NOW WORK PROBLEM 1.

Fundamental Identities

2 You may have observed some relationships that exist among the six trigonometric functions of acute angles. For example, the **reciprocal identities** are

Reciprocal Identities

$$\csc\theta = \frac{1}{\sin\theta} \qquad \sec\theta = \frac{1}{\cos\theta} \qquad \cot\theta = \frac{1}{\tan\theta} \qquad (2)$$

Two other fundamental identities that are easy to see are the **quotient identities.**

Quotient Identities

$$\tan\theta = \frac{\sin\theta}{\cos\theta} \qquad \cot\theta = \frac{\cos\theta}{\sin\theta} \qquad (3)$$

If $\sin\theta$ and $\cos\theta$ are known, formulas (2) and (3) make it easy to find the values of the remaining trigonometric functions.

EXAMPLE 2 **Finding the Value of the Remaining Trigonometric Functions, Given $\sin\theta$ and $\cos\theta$**

Given $\sin\theta = \dfrac{\sqrt{5}}{5}$ and $\cos\theta = \dfrac{2\sqrt{5}}{5}$, find the value of each of the four remaining trigonometric functions of θ.

Solution Based on formula (3), we have

$$\tan\theta = \frac{\sin\theta}{\cos\theta} = \frac{\frac{\sqrt{5}}{5}}{\frac{2\sqrt{5}}{5}} = \frac{1}{2}$$

Then we use the reciprocal identities from formula (2) to get

$$\csc\theta = \frac{1}{\sin\theta} = \frac{1}{\frac{\sqrt{5}}{5}} = \frac{5}{\sqrt{5}} = \sqrt{5} \qquad \sec\theta = \frac{1}{\cos\theta} = \frac{1}{\frac{2\sqrt{5}}{5}} = \frac{5}{2\sqrt{5}} = \frac{\sqrt{5}}{2} \qquad \cot\theta = \frac{1}{\tan\theta} = \frac{1}{\frac{1}{2}} = 2$$ ■

NOW WORK PROBLEM 11.

Figure 24
$a^2 + b^2 = c^2$

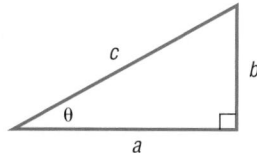

Refer now to the triangle in Figure 24. The Pythagorean Theorem states that $a^2 + b^2 = c^2$, which we write as

$$b^2 + a^2 = c^2$$

Dividing each side by c^2, we get

$$\frac{b^2}{c^2} + \frac{a^2}{c^2} = 1 \quad \text{or} \quad \left(\frac{b}{c}\right)^2 + \left(\frac{a}{c}\right)^2 = 1$$

In terms of trigonometric functions of the angle θ, this equation states that

$$(\sin\theta)^2 + (\cos\theta)^2 = 1 \tag{4}$$

Equation (4) is, in fact, an identity, since the equation is true for any acute angle θ.

It is customary to write $\sin^2\theta$ instead of $(\sin\theta)^2$, $\cos^2\theta$ instead of $(\cos\theta)^2$, and so on. With this notation, we can rewrite equation (4) as

$$\sin^2\theta + \cos^2\theta = 1 \tag{5}$$

Another identity can be obtained from equation (5) by dividing each side by $\cos^2\theta$.

$$\frac{\sin^2\theta}{\cos^2\theta} + 1 = \frac{1}{\cos^2\theta}$$

Now use formulas (2) and (3) to get

$$\tan^2\theta + 1 = \sec^2\theta \tag{6}$$

Similarly, by dividing each side of equation (5) by $\sin^2\theta$, we get

$$1 + \cot^2\theta = \csc^2\theta \tag{7}$$

Collectively, the identities in equations (5), (6), and (7) are referred to as the **Pythagorean identities.**

Let's pause here to summarize the fundamental identities.

Fundamental Identities

$$\tan\theta = \frac{\sin\theta}{\cos\theta} \qquad \cot\theta = \frac{\cos\theta}{\sin\theta}$$

$$\csc\theta = \frac{1}{\sin\theta} \qquad \sec\theta = \frac{1}{\cos\theta} \qquad \cot\theta = \frac{1}{\tan\theta}$$

$$\sin^2\theta + \cos^2\theta = 1 \qquad \tan^2\theta + 1 = \sec^2\theta \qquad 1 + \cot^2\theta = \csc^2\theta$$

EXAMPLE 3   **Finding the Exact Value of a Trigonometric Expression Using Identities**

Find the exact value of each expression. Do not use a calculator.

(a) $\tan 20° - \dfrac{\sin 20°}{\cos 20°}$ (b) $\sin^2\dfrac{\pi}{12} + \dfrac{1}{\sec^2\dfrac{\pi}{12}}$

Solution   (a) $\tan 20° - \underset{\substack{\big\uparrow \\ \frac{\sin\theta}{\cos\theta} = \tan\theta}}{\dfrac{\sin 20°}{\cos 20°}} = \tan 20° - \tan 20° = 0$

(b) $\sin^2\dfrac{\pi}{12} + \dfrac{1}{\sec^2\dfrac{\pi}{12}} = \sin^2\dfrac{\pi}{12} + \cos^2\dfrac{\pi}{12} = 1$

$\cos\theta = \dfrac{1}{\sec\theta}$ $\sin^2\theta + \cos^2\theta = 1$

NOW WORK PROBLEM 29.

③   Once the value of one trigonometric function is known, it is possible to find the value of each of the remaining five trigonometric functions.

EXAMPLE 4   **Finding the Value of the Remaining Trigonometric Functions, Given $\sin\theta$, θ Acute**

Given that $\sin\theta = \dfrac{1}{3}$ and θ is an acute angle, find the exact value of each of the remaining five trigonometric functions of θ.

Solution   We solve this problem in two ways: The first way uses the definition of the trigonometric functions; the second method uses the fundamental identities.

Solution 1
Using the Definition   We draw a right triangle with acute angle θ, opposite side of length $b = 1$, and hypotenuse of length $c = 3$ $\left(\text{because } \sin\theta = \dfrac{1}{3} = \dfrac{b}{c}\right)$. See Figure 25. The adjacent side a can be found by using the Pythagorean Theorem.

Figure 25

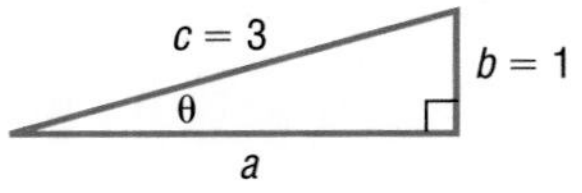

$$a^2 + 1^2 = 3^2$$
$$a^2 + 1 = 9$$
$$a^2 = 8$$
$$a = 2\sqrt{2}$$

Now the definitions given in equation (1) can be used to find the value of each of the remaining five trigonometric functions. (Refer back to the method used in Example 1.) Using $a = 2\sqrt{2}$, $b = 1$, and $c = 3$, we have

$$\cos\theta = \frac{a}{c} = \frac{2\sqrt{2}}{3} \qquad\qquad \tan\theta = \frac{b}{a} = \frac{1}{2\sqrt{2}} = \frac{\sqrt{2}}{4}$$

$$\csc\theta = \frac{c}{b} = \frac{3}{1} = 3 \qquad \sec\theta = \frac{c}{a} = \frac{3}{2\sqrt{2}} = \frac{3\sqrt{2}}{4} \qquad \cot\theta = \frac{a}{b} = \frac{2\sqrt{2}}{1} = 2\sqrt{2}$$

Solution 2
Using Identities   We begin by seeking $\cos\theta$, which can be found by using the Pythagorean identity from equation (5).

$$\sin^2\theta + \cos^2\theta = 1$$

$$\frac{1}{9} + \cos^2\theta = 1 \qquad\qquad \sin\theta = \frac{1}{3}$$

$$\cos^2\theta = 1 - \frac{1}{9} = \frac{8}{9}$$

Because $\cos \theta > 0$ for an acute angle θ, we have

$$\cos \theta = \sqrt{\frac{8}{9}} = \frac{2\sqrt{2}}{3}$$

Now we know that $\sin \theta = \frac{1}{3}$ and $\cos \theta = \frac{2\sqrt{2}}{3}$, so we can proceed as we did in Example 2.

$$\tan \theta = \frac{\sin \theta}{\cos \theta} = \frac{\frac{1}{3}}{\frac{2\sqrt{2}}{3}} = \frac{1}{2\sqrt{2}} = \frac{\sqrt{2}}{4} \qquad \cot \theta = \frac{1}{\tan \theta} = \frac{1}{\frac{\sqrt{2}}{4}} = \frac{4}{\sqrt{2}} = 2\sqrt{2}$$

$$\sec \theta = \frac{1}{\cos \theta} = \frac{1}{\frac{2\sqrt{2}}{3}} = \frac{3}{2\sqrt{2}} = \frac{3\sqrt{2}}{4} \qquad \csc \theta = \frac{1}{\sin \theta} = \frac{1}{\frac{1}{3}} = 3$$

Finding the Values of the Trigonometric Functions When One Is Known

Given the value of one trigonometric function of an acute angle θ, the exact value of each of the remaining five trigonometric functions of θ can be found in either of two ways.

Method 1 Using the Definition

STEP 1: Draw a right triangle showing the acute angle θ.

STEP 2: Two of the sides can then be assigned values based on the given trigonometric function.

STEP 3: Find the length of the third side by using the Pythagorean Theorem.

STEP 4: Use the definitions in equation (1) to find the value of each of the remaining trigonometric functions.

Method 2 Using Identities

Use appropriately selected identities to find the value of each of the remaining trigonometric functions.

EXAMPLE 5   **Given One Value of a Trigonometric Function, Find the Remaining Ones**

Given $\tan \theta = \frac{1}{2}$, θ an acute angle, find the exact value of each of the remaining five trigonometric functions of θ.

Solution 1 Using the Definition

Figure 26 shows a right triangle with acute angle θ, where

$$\tan \theta = \frac{1}{2} = \frac{\text{opposite}}{\text{adjacent}} = \frac{b}{a}$$

Choose $b = 1$ and $a = 2$. The hypotenuse c can be found by using the Pythagorean Theorem.

$$c^2 = a^2 + b^2 = 2^2 + 1^2 = 5$$
$$c = \sqrt{5}$$

Figure 26

$\tan \theta = \frac{1}{2}$

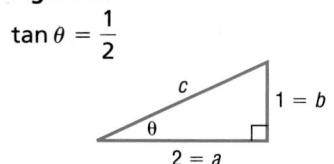

Now apply the definitions using $a = 2$, $b = 1$, and $c = \sqrt{5}$.

$$\sin\theta = \frac{b}{c} = \frac{1}{\sqrt{5}} = \frac{\sqrt{5}}{5} \qquad \cos\theta = \frac{a}{c} = \frac{2}{\sqrt{5}} = \frac{2\sqrt{5}}{5}$$

$$\csc\theta = \frac{c}{b} = \frac{\sqrt{5}}{1} = \sqrt{5} \qquad \sec\theta = \frac{c}{a} = \frac{\sqrt{5}}{2} \qquad \cot\theta = \frac{a}{b} = \frac{2}{1} = 2$$

Solution 2
Using Identities

We use the Pythagorean identity that involves $\tan\theta$:

$$\tan^2\theta + 1 = \sec^2\theta$$

$$\left(\frac{1}{2}\right)^2 + 1 = \sec^2\theta \qquad\qquad \tan\theta = \frac{1}{2}$$

$$\sec^2\theta = \frac{1}{4} + 1 = \frac{5}{4} \qquad\qquad \textit{Proceed to solve for sec }\theta.$$

$$\sec\theta = \frac{\sqrt{5}}{2} \qquad\qquad \textit{sec }\theta > 0, \theta \textit{ acute}$$

Now

$$\cos\theta = \frac{1}{\sec\theta} = \frac{1}{\frac{\sqrt{5}}{2}} = \frac{2}{\sqrt{5}} = \frac{2\sqrt{5}}{5}$$

$$\tan\theta = \frac{\sin\theta}{\cos\theta}, \quad \text{so} \quad \sin\theta = (\tan\theta)(\cos\theta) = \frac{1}{2}\cdot\frac{2\sqrt{5}}{5} = \frac{\sqrt{5}}{5}$$

$$\csc\theta = \frac{1}{\sin\theta} = \frac{1}{\frac{\sqrt{5}}{5}} = \sqrt{5}$$

$$\cot\theta = \frac{1}{\tan\theta} = \frac{1}{\frac{1}{2}} = 2$$

NOW WORK PROBLEM 15.

Complementary Angles; Cofunctions

④ Two acute angles are called **complementary** if their sum is a right angle. Because the sum of the angles of any triangle is 180°, it follows that, for a right triangle, the two acute angles are complementary.

Refer now to Figure 27; we have labeled the angle opposite side b as β and the angle opposite side a as α. Notice that side a is adjacent to angle β and is opposite angle α. Similarly, side b is opposite angle β and is adjacent to angle α. As a result,

Figure 27

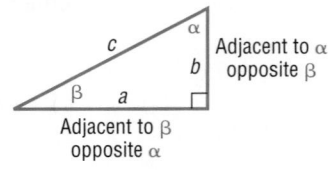

$$\sin\beta = \frac{b}{c} = \cos\alpha \qquad \cos\beta = \frac{a}{c} = \sin\alpha \qquad \tan\beta = \frac{b}{a} = \cot\alpha$$

$$\csc\beta = \frac{c}{b} = \sec\alpha \qquad \sec\beta = \frac{c}{a} = \csc\alpha \qquad \cot\beta = \frac{a}{b} = \tan\alpha$$

(8)

Because of these relationships, the functions sine and cosine, tangent and cotangent, and secant and cosecant are called **cofunctions** of each other. The identities (8) may be expressed in words as follows:

Theorem

Complementary Angle Theorem

Cofunctions of complementary angles are equal.

Here are examples of this theorem.

Complementary angles $\quad$ Complementary angles $\quad$ Complementary angles

$$\sin 30° = \cos 60° \qquad \tan 40° = \cot 50° \qquad \sec 80° = \csc 10°$$

Cofunctions $\qquad\qquad$ Cofunctions $\qquad\qquad$ Cofunctions

If an angle θ is measured in degrees, we will use the degree symbol when writing a trigonometric function of θ, as, for example, in $\sin 30°$ and $\tan 45°$. If an angle θ is measured in radians, then no symbol is used when writing a trigonometric function of θ, as, for example, in $\cos \pi$ and $\sec \dfrac{\pi}{3}$.

If θ is an acute angle measured in degrees, the angle $90° - \theta$ (or $\dfrac{\pi}{2} - \theta$, if θ is in radians) is the angle complementary to θ. Table 2 restates the preceding theorem on cofunctions.

TABLE 2

θ (Degrees)	θ (Radians)
$\sin \theta = \cos(90° - \theta)$	$\sin \theta = \cos\left(\dfrac{\pi}{2} - \theta\right)$
$\cos \theta = \sin(90° - \theta)$	$\cos \theta = \sin\left(\dfrac{\pi}{2} - \theta\right)$
$\tan \theta = \cot(90° - \theta)$	$\tan \theta = \cot\left(\dfrac{\pi}{2} - \theta\right)$
$\csc \theta = \sec(90° - \theta)$	$\csc \theta = \sec\left(\dfrac{\pi}{2} - \theta\right)$
$\sec \theta = \csc(90° - \theta)$	$\sec \theta = \csc\left(\dfrac{\pi}{2} - \theta\right)$
$\cot \theta = \tan(90° - \theta)$	$\cot \theta = \tan\left(\dfrac{\pi}{2} - \theta\right)$

Although the angle θ in Table 2 is acute, we will see later (Section 3.4) that these results are valid for any angle θ.

EXAMPLE 6 **Using the Complementary Angle Theorem**

(a) $\sin 62° = \cos(90° - 62°) = \cos 28°$

(b) $\tan \dfrac{\pi}{12} = \cot\left(\dfrac{\pi}{2} - \dfrac{\pi}{12}\right) = \cot \dfrac{5\pi}{12}$

(c) $\cos \dfrac{\pi}{4} = \sin\left(\dfrac{\pi}{2} - \dfrac{\pi}{4}\right) = \sin \dfrac{\pi}{4}$

(d) $\csc \dfrac{\pi}{6} = \sec\left(\dfrac{\pi}{2} - \dfrac{\pi}{6}\right) = \sec \dfrac{\pi}{3}$

EXAMPLE 7 **Using the Complementary Angle Theorem**

Find the exact value of each expression. Do not use a calculator.

(a) $\sec 28° - \csc 62°$

(b) $\dfrac{\sin 35°}{\cos 55°}$

Solution (a) $\sec 28° = \csc(90° - 28°) - \csc 62°$
$$= \csc 62° - \csc 62° = 0$$

(b) $\dfrac{\sin 35°}{\cos 55°} = \dfrac{\cos(90° - 35°)}{\cos 55°} = \dfrac{\cos 55°}{\cos 55°} = 1$

NOW WORK PROBLEMS 33 AND 47.

HISTORICAL FEATURE

The name *sine* for the sine function is due to a medieval confusion. The name comes from the Sanskrit word *jīva* (meaning chord), first used in India by Araybhata the Elder (AD 510). He really meant half-chord, but abbreviated it. This was brought into Arabic as *jība*, which was meaningless. Because the proper Arabic word *jaib* would be written the same way (short vowels are not written out in Arabic), *jība* was pronounced as *jaib*, which meant bosom or hollow, and *jaib* remains as the Arabic word for sine to this day. Scholars translating the Arabic works into Latin found that the word *sinus* also meant bosom or hollow, and from *sinus* we get the word *sine*.

The name *tangent*, due to Thomas Finck (1583), can be understood by looking at Figure 28. The line segment $\overline{DC}$ is tangent to the circle at C. If $d(O, B) = d(O, C) = 1$, then the length of the line segment $\overline{DC}$ is

$$d(D, C) = \dfrac{d(D, C)}{1} = \dfrac{d(D, C)}{d(O, C)} = \tan \alpha$$

The old name for the tangent is *umbra versa* (meaning turned shadow), referring to the use of the tangent in solving height problems with shadows.

Figure 28

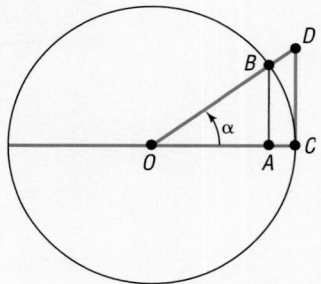

The names of the cofunctions came about as follows. If α and β are complementary angles, then $\cos \alpha = \sin \beta$. Because β is the complement of α, it was natural to write the cosine of α as *sin co* α. Probably for reasons involving ease of pronunciation, the *co* migrated to the front, and then cosine received a three-letter abbreviation to match sin, sec, and tan. The two other cofunctions were similarly treated, except that the long forms *cotan* and *cosec* survive to this day in some countries.

2.2 Concepts and Vocabulary

In Problems 1–3, fill in the blanks.

1. Two acute angles whose sum is a right angle are called _____.

2. The sine and _____ functions are cofunctions.

3. $\tan 28° = \cot$ _____.

In Problems 4–6, answer True or False to each statement.

4. $1 + \tan^2 \theta = \csc^2 \theta$.

5. If θ is an acute angle and $\sec \theta = 3$, then $\cos \theta = \dfrac{1}{3}$.

6. $\tan \dfrac{\pi}{5} = \cot \dfrac{4\pi}{5}$.

7. Write down the Reciprocal Identities.

8. Explain the Complementary Angle Theorem. Give an example that uses it.

9. If you know the value of $\tan \theta$, θ an acute angle, how would you find $\sec \theta$?

10. Name the two functions that are reciprocals and cofunctions of each other.

2.2   Exercises

In Problems 1–10, find the value of the six trigonometric functions of the angle θ in each figure.

1.
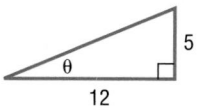

2.
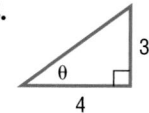

3.
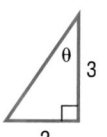

4.
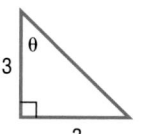

5.
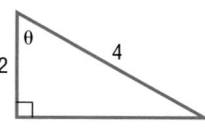

6.
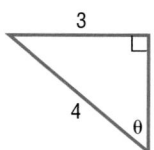

7.
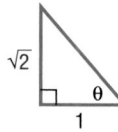

8.
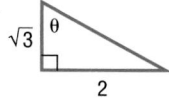

9.

10.
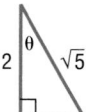

In Problems 11–14, use identities to find the exact value of each of the four remaining trigonometric functions of the acute angle θ.

11. $\sin \theta = \dfrac{1}{2}$, $\cos \theta = \dfrac{\sqrt{3}}{2}$

12. $\sin \theta = \dfrac{\sqrt{3}}{2}$, $\cos \theta = \dfrac{1}{2}$

13. $\sin \theta = \dfrac{2}{3}$, $\cos \theta = \dfrac{\sqrt{5}}{3}$

14. $\sin \theta = \dfrac{1}{3}$, $\cos \theta = \dfrac{2\sqrt{2}}{3}$

In Problems 15–26, use the definition or identities to find the exact value of each of the remaining five trigonometric functions of the acute angle θ.

15. $\sin \theta = \dfrac{\sqrt{2}}{2}$

16. $\cos \theta = \dfrac{\sqrt{2}}{2}$

17. $\cos \theta = \dfrac{1}{3}$

18. $\sin \theta = \dfrac{\sqrt{3}}{4}$

19. $\tan \theta = \dfrac{1}{2}$

20. $\cot \theta = \dfrac{1}{2}$

21. $\sec \theta = 3$

22. $\csc \theta = 5$

23. $\tan \theta = \sqrt{2}$

24. $\sec \theta = \dfrac{5}{3}$

25. $\csc \theta = 2$

26. $\cot \theta = 2$

In Problems 27–44, use Fundamental Identities and/or the Complementary Angle Theorem to find the exact value of each expression. Do not use a calculator.

27. $\sin^2 20° + \cos^2 20°$

28. $\sec^2 28° - \tan^2 28°$

29. $\sin 80°\csc 80°$

30. $\tan 10° \cot 10°$

31. $\tan 50° - \dfrac{\sin 50°}{\cos 50°}$

32. $\cot 25° - \dfrac{\cos 25°}{\sin 25°}$

33. $\sin 38° - \cos 52°$

34. $\tan 12° - \cot 78°$

35. $\dfrac{\cos 10°}{\sin 80°}$

36. $\dfrac{\cos 40°}{\sin 50°}$

37. $1 - \cos^2 20° - \cos^2 70°$

38. $1 + \tan^2 5° - \csc^2 85°$

39. $\tan 20° - \dfrac{\cos 70°}{\cos 20°}$

40. $\cot 40° - \dfrac{\sin 50°}{\sin 40°}$

41. $\tan 35° \cdot \sec 55° \cdot \cos 35°$

42. $\cot 25° \cdot \csc 65° \cdot \sin 25°$

43. $\cos 35°\sin 55° + \cos 55°\sin 35°$

44. $\sec 35° \csc 55° - \tan 35°\cot 55°$

45. Given $\sin 30° = \dfrac{1}{2}$, use trigonometric identities to find the exact value of

(a) $\cos 60°$ (b) $\cos^2 30°$

(c) $\csc \dfrac{\pi}{6}$ (d) $\sec \dfrac{\pi}{3}$

46. Given $\sin 60° = \dfrac{\sqrt{3}}{2}$, use trigonometric identities to find the exact value of

(a) $\cos 30°$ (b) $\cos^2 60°$

(c) $\sec \dfrac{\pi}{6}$ (d) $\csc \dfrac{\pi}{3}$

47. Given $\tan \theta = 4$, use trigonometric identities to find the exact value of

(a) $\sec^2 \theta$ (b) $\cot \theta$

(c) $\cot\left(\dfrac{\pi}{2} - \theta\right)$ (d) $\csc^2 \theta$

48. Given $\sec \theta = 3$, use trigonometric identities to find the exact value of

(a) $\cos \theta$ (b) $\tan^2 \theta$
(c) $\csc(90° - \theta)$ (d) $\sin^2 \theta$

49. Given $\csc \theta = 4$, use trigonometric identities to find the exact value of

(a) $\sin \theta$ (b) $\cot^2 \theta$
(c) $\sec(90° - \theta)$ (d) $\sec^2 \theta$

50. Given $\cot \theta = 2$, use trigonometric identities to find the exact value of

(a) $\tan \theta$ (b) $\csc^2 \theta$

(c) $\tan\left(\dfrac{\pi}{2} - \theta\right)$ (d) $\sec^2 \theta$

51. Given the approximation $\sin 38° \approx 0.62$, use trigonometric identities to find the approximate value of

(a) $\cos 38°$ (b) $\tan 38°$
(c) $\cot 38°$ (d) $\sec 38°$
(e) $\csc 38°$ (f) $\sin 52°$
(g) $\cos 52°$ (h) $\tan 52°$

52. Given the approximation $\cos 21° \approx 0.93$, use trigonometric identities to find the approximate value of

(a) $\sin 21°$ (b) $\tan 21°$
(c) $\cot 21°$ (d) $\sec 21°$
(e) $\csc 21°$ (f) $\sin 69°$
(g) $\cos 69°$ (h) $\tan 69°$

53. If $\sin \theta = 0.3$, find the exact value of $\sin \theta + \cos\left(\dfrac{\pi}{2} - \theta\right)$.

54. If $\tan \theta = 4$, find the exact value of $\tan \theta + \tan\left(\dfrac{\pi}{2} - \theta\right)$.

55. Find an acute angle θ that satisfies the equation $\sin \theta = \cos(2\theta + 30°)$.

56. Find an acute angle θ that satisfies the equation $\tan \theta = \cot(\theta + 45°)$.

57. Calculating the Time of a Trip   From a parking lot you want to walk to a house on the ocean. The house is located 1500 feet down a paved path that parallels the beach, which is 500 feet wide. Along the path you can walk 300

feet per minute, but in the sand on the beach you can only walk 100 feet per minute. See the illustration.

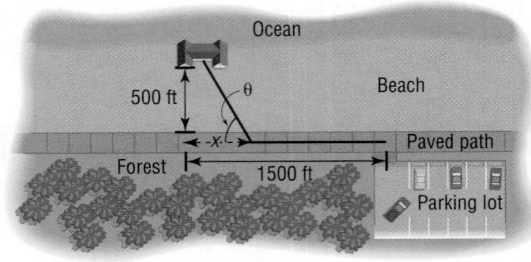

(a) Calculate the time T if you walk 1500 feet along the paved path and then 500 feet in the sand to the house.
(b) Calculate the time T if you walk in the sand directly toward the ocean for 500 feet and then turn left and walk along the beach for 1500 feet to the house.
(c) Express the time T to get from the parking lot to the beachhouse as a function of the angle θ shown in the illustration.
(d) Calculate the time T if you walk directly from the parking lot to the house.
   [**Hint:** $\tan \theta = 500/1500$]
(e) Calculate the time T if you walk 1000 feet along the paved path and then walk directly to the house.
(f) Use a graphing utility to graph $T = T(\theta)$. For what angle θ is T least? What is x for this angle? What is the minimum time?
(g) Explain why $\tan \theta = \dfrac{1}{3}$ gives the smallest angle θ that is possible.

58. Carrying a Ladder around a Corner   A ladder of length L is carried horizontally around a corner from a hall 3 feet wide into a hall 4 feet wide. See the illustration. Find the length L of the ladder as a function of the angle θ shown in the illustration.

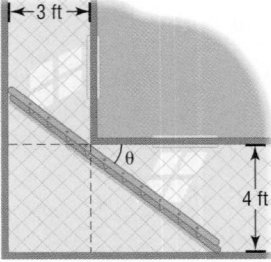

59. Suppose that the angle θ is a central angle of a circle of radius 1 (see the figure). Show that

(a) Angle $OAC = \dfrac{\theta}{2}$

(b) $|CD| = \sin \theta$ and $|OD| = \cos \theta$

(c) $\tan \dfrac{\theta}{2} = \dfrac{\sin \theta}{1 + \cos \theta}$

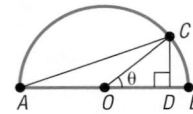

60. Show that the area A of an isosceles triangle is $A = a^2 \sin \theta \cos \theta$, where a is the length of one of the two equal sides and θ is the measure of one of the two equal angles (see the figure).

61. Let $n \geq 1$ be any real number and let θ be any angle for which $0 < n\theta < \dfrac{\pi}{2}$. Then we can draw a triangle with the angles θ and $n\theta$ and included side of length 1 (do you see why?) and place it on the unit circle as illustrated. Now, drop the perpendicular from C to $D = (x, 0)$ and show that

$$x = \frac{\tan(n\theta)}{\tan \theta + \tan(n\theta)}$$

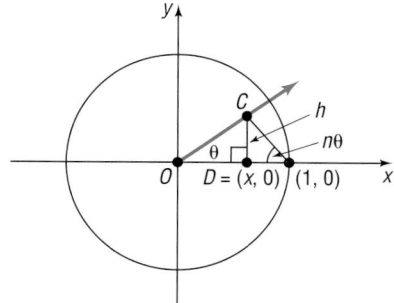

62. Refer to the figure. The smaller circle, whose radius is a, is tangent to the larger circle, whose radius is b. The ray $\overrightarrow{OA}$ contains a diameter of each circle, and the ray $\overrightarrow{OB}$ is tangent to each circle. Show that

$$\cos \theta = \frac{\sqrt{ab}}{\dfrac{a + b}{2}}$$

(This shows that $\cos \theta$ equals the ratio of the geometric mean of a and b to the arithmetic mean of a and b.)

[**Hint:** First show that $\sin \theta = \dfrac{b - a}{b + a}$.]

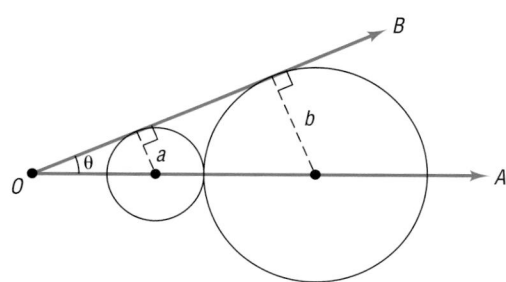

63. Refer to the figure. If $|OA| = 1$, show that

(a) Area $\triangle OAC = \dfrac{1}{2} \sin \alpha \cos \alpha$

(b) Area $\triangle OCB = \dfrac{1}{2} |OB|^2 \sin \beta \cos \beta$

(c) Area $\triangle OAB = \dfrac{1}{2} |OB| \sin(\alpha + \beta)$

(d) $|OB| = \dfrac{\cos \alpha}{\cos \beta}$

(e) $\sin(\alpha + \beta) = \sin \alpha \cos \beta + \cos \alpha \sin \beta$

[**Hint:** Area $\triangle OAB$ = Area $\triangle OAC$ + Area $\triangle OCB$]

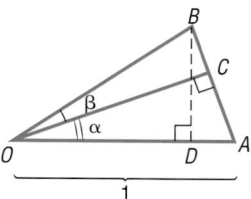

64. Refer to the figure, where a unit circle is drawn. The line DB is tangent to the circle.

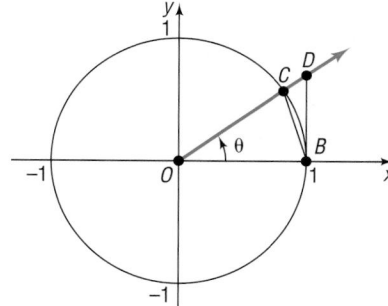

(a) Express the area of $\triangle OBC$ in terms of $\sin \theta$ and $\cos \theta$.
    [**Hint:** Use the altitude from C to the base $\overline{OB} = 1$.]
(b) Express the area of $\triangle OBD$ in terms of $\sin \theta$ and $\cos \theta$.
(c) The area of the sector of the circle OBC is $\dfrac{1}{2}\theta$, where θ is measured in radians. Use the results of parts (a) and (b) and the fact that

$$\text{Area } \triangle OBC < \text{Area } \overset{\frown}{OBC} < \text{Area } \triangle OBD$$

to show that

$$1 < \frac{\theta}{\sin \theta} < \frac{1}{\cos \theta}$$

65. If $\cos \alpha = \tan \beta$ and $\cos \beta = \tan \alpha$, where α and β are acute angles, show that

$$\sin \alpha = \sin \beta = \sqrt{\frac{3 - \sqrt{5}}{2}}$$

 66. If θ is an acute angle, explain why $\sec \theta > 1$.

67. If θ is an acute angle, explain why $0 < \sin \theta < 1$.

68. How would you explain the meaning of the sine function to a fellow student who has just completed college algebra?

2.3 COMPUTING THE VALUES OF TRIGONOMETRIC FUNCTIONS OF GIVEN ANGLES

OBJECTIVES
1. Find the Exact Value of the Trigonometric Functions of $\frac{\pi}{4} = 45°$
2. Find the Exact Value of the Trigonometric Functions of $\frac{\pi}{6} = 30°$ and $\frac{\pi}{3} = 60°$
3. Use a Calculator to Approximate the Value of the Trigonometric Functions of Acute Angles

In the previous section, we developed ways to find the value of each trigonometric function of an acute angle when one of the functions is known. In this section, we discuss the problem of finding the value of each trigonometric function of an acute angle when the angle is given.

For three special acute angles, we can use some results from plane geometry to find the exact value of each of the six trigonometric functions.

1. **Trigonometric Functions of $\frac{\pi}{4} = 45°$**

EXAMPLE 1   **Finding the Exact Value of the Trigonometric Functions of $\frac{\pi}{4} = 45°$**

Find the exact value of the six trigonometric functions of $\frac{\pi}{4} = 45°$.

Figure 29

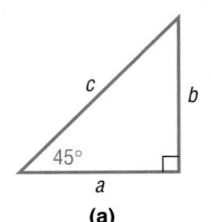

(a)

(b)

Solution   Form the right triangle in Figure 29(a), in which one of the angles is $\frac{\pi}{4} = 45°$. It follows that the other acute angle is also $\frac{\pi}{4} = 45°$, and hence the triangle is isosceles. As a result, side a and side b are equal in length. Because the values of the trigonometric functions of an angle depend only on the angle and not on the size of the triangle, we may choose to use the triangle for which

$$a = b = 1$$

Then, by the Pythagorean Theorem,

$$c^2 = a^2 + b^2 = 1 + 1 = 2$$
$$c = \sqrt{2}$$

As a result, we have the triangle in Figure 29(b), from which we find

$$\sin\frac{\pi}{4} = \sin 45° = \frac{b}{c} = \frac{1}{\sqrt{2}} = \frac{\sqrt{2}}{2} \qquad \cos\frac{\pi}{4} = \cos 45° = \frac{a}{c} = \frac{1}{\sqrt{2}} = \frac{\sqrt{2}}{2}$$

Using Quotient and Reciprocal Identities, we find

$$\tan\frac{\pi}{4} = \tan 45° = \frac{\sin 45°}{\cos 45°} = \frac{\frac{\sqrt{2}}{2}}{\frac{\sqrt{2}}{2}} = 1 \qquad \cot\frac{\pi}{4} = \cot 45° = \frac{1}{\tan 45°} = \frac{1}{1} = 1$$

$$\sec\frac{\pi}{4} = \sec 45° = \frac{1}{\cos 45°} = \frac{1}{\frac{1}{\sqrt{2}}} = \sqrt{2} \qquad \csc\frac{\pi}{4} = \csc 45° = \frac{1}{\sin 45°} = \frac{1}{\frac{1}{\sqrt{2}}} = \sqrt{2}$$

EXAMPLE 2 **Finding the Exact Value of a Trigonometric Expression**

Find the exact value of each expression.

(a) $(\sin 45°)(\tan 45°)$ (b) $\left(\sec \dfrac{\pi}{4}\right)\left(\cot \dfrac{\pi}{4}\right)$

Solution We use the results obtained in Example 1.

(a) $(\sin 45°)(\tan 45°) = \dfrac{\sqrt{2}}{2} \cdot 1 = \dfrac{\sqrt{2}}{2}$

(b) $\left(\sec \dfrac{\pi}{4}\right)\left(\cot \dfrac{\pi}{4}\right) = \sqrt{2} \cdot 1 = \sqrt{2}$

NOW WORK PROBLEMS 1 AND 13.

2 **Trigonometric Functions of** $\dfrac{\pi}{6} = 30°$ **and** $\dfrac{\pi}{3} = 60°$

EXAMPLE 3 **Finding the Exact Value of the Trigonometric Functions**

of $\dfrac{\pi}{6} = 30°$ **and** $\dfrac{\pi}{3} = 60°$

Find the exact value of the six trigonometric functions of $\dfrac{\pi}{6} = 30°$ and $\dfrac{\pi}{3} = 60°$.

Figure 30

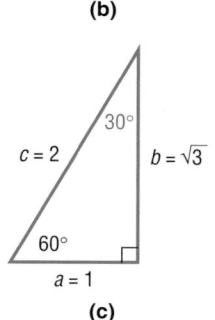

(a)

(b)

(c)

Solution Form a right triangle in which one of the angles is $\dfrac{\pi}{6} = 30°$. It then follows that the other angle is $\dfrac{\pi}{3} = 60°$. Figure 30(a) illustrates such a triangle with hypotenuse of length 2. Our problem is to determine a and b.

We begin by placing next to the triangle in Figure 30(a) another triangle congruent to the first, as shown in Figure 30(b). Notice that we now have a triangle whose angles are each 60°. This triangle is therefore equilateral, so each side is of length 2. In particular, the base is $2a = 2$, so $a = 1$. By the Pythagorean Theorem, b satisfies the equation $a^2 + b^2 = c^2$, so we have

$$a^2 + b^2 = c^2$$
$$1^2 + b^2 = 2^2 \qquad a = 1, c = 2$$
$$b^2 = 4 - 1 = 3$$
$$b = \sqrt{3}$$

Using the triangle in Figure 30(c) and the fact that $\dfrac{\pi}{6} = 30°$ and $\dfrac{\pi}{3} = 60°$ are complementary angles, we find

$$\sin \dfrac{\pi}{6} = \sin 30° = \dfrac{\text{opposite}}{\text{hypotenuse}} = \dfrac{1}{2} \qquad\qquad \cos \dfrac{\pi}{3} = \cos 60° = \dfrac{1}{2}$$

$$\cos \dfrac{\pi}{6} = \cos 30° = \dfrac{\text{adjacent}}{\text{hypotenuse}} = \dfrac{\sqrt{3}}{2} \qquad\qquad \sin \dfrac{\pi}{3} = \sin 60° = \dfrac{\sqrt{3}}{2}$$

$$\tan \dfrac{\pi}{6} = \tan 30° = \dfrac{\sin 30°}{\cos 30°} = \dfrac{\frac{1}{2}}{\frac{\sqrt{3}}{2}} = \dfrac{1}{\sqrt{3}} = \dfrac{\sqrt{3}}{3} \qquad\qquad \cot \dfrac{\pi}{3} = \cot 60° = \dfrac{\sqrt{3}}{3}$$

$$\csc\frac{\pi}{6} = \csc 30° = \frac{1}{\sin 30°} = \frac{1}{\frac{1}{2}} = 2 \qquad\qquad \sec\frac{\pi}{3} = \sec 60° = 2$$

$$\sec\frac{\pi}{6} = \sec 30° = \frac{1}{\cos 30°} = \frac{1}{\frac{\sqrt{3}}{2}} = \frac{2}{\sqrt{3}} = \frac{2\sqrt{3}}{3} \qquad \csc\frac{\pi}{3} = \csc 60° = \frac{2\sqrt{3}}{3}$$

$$\cot\frac{\pi}{6} = \cot 30° = \frac{1}{\tan 30°} = \frac{1}{\frac{\sqrt{3}}{3}} = \frac{3}{\sqrt{3}} = \sqrt{3} \qquad \tan\frac{\pi}{3} = \tan 60° = \sqrt{3}$$

Table 3 summarizes the information just derived for the angles $\frac{\pi}{6} = 30°$, $\frac{\pi}{4} = 45°$, and $\frac{\pi}{3} = 60°$. Until you memorize the entries in Table 3, you should draw the appropriate triangle to determine the values given in the table.

TABLE 3

θ (Radians)	θ (Degrees)	$\sin\theta$	$\cos\theta$	$\tan\theta$	$\csc\theta$	$\sec\theta$	$\cot\theta$
$\frac{\pi}{6}$	30°	$\frac{1}{2}$	$\frac{\sqrt{3}}{2}$	$\frac{\sqrt{3}}{3}$	2	$\frac{2\sqrt{3}}{3}$	$\sqrt{3}$
$\frac{\pi}{4}$	45°	$\frac{\sqrt{2}}{2}$	$\frac{\sqrt{2}}{2}$	1	$\sqrt{2}$	$\sqrt{2}$	1
$\frac{\pi}{3}$	60°	$\frac{\sqrt{3}}{2}$	$\frac{1}{2}$	$\sqrt{3}$	$\frac{2\sqrt{3}}{3}$	2	$\frac{\sqrt{3}}{3}$

EXAMPLE 4   **Finding the Exact Value of a Trigonometric Expression**

Find the exact value of each expression.

(a) $\sin 45°\cos 30°$ (b) $\tan\frac{\pi}{4} - \sin\frac{\pi}{3}$ (c) $\tan^2\frac{\pi}{6} + \sin^2\frac{\pi}{4}$

Solution (a) $\sin 45° \cos 30° = \dfrac{\sqrt{2}}{2}\cdot\dfrac{\sqrt{3}}{2} = \dfrac{\sqrt{6}}{4}$

(b) $\tan\dfrac{\pi}{4} - \sin\dfrac{\pi}{3} = 1 - \dfrac{\sqrt{3}}{2} = \dfrac{2 - \sqrt{3}}{2}$

(c) $\tan^2\dfrac{\pi}{6} + \sin^2\dfrac{\pi}{4} = \left(\dfrac{\sqrt{3}}{3}\right)^2 + \left(\dfrac{\sqrt{2}}{2}\right)^2 = \dfrac{1}{3} + \dfrac{1}{2} = \dfrac{5}{6}$

═ NOW WORK PROBLEMS **5** AND **15.**

The exact values of the trigonometric functions for the angles $\frac{\pi}{6} = 30°$, $\frac{\pi}{4} = 45°$, and $\frac{\pi}{3} = 60°$ are relatively easy to calculate, because the triangles that contain such angles have "nice" geometric features. For most other angles, we can only approximate the value of each trigonometric function. To do this, we will need a calculator.

Using a Calculator to Find Values of Trigonometric Functions

3   Before getting started, you must first decide whether to enter the angle in the calculator using radians or degrees and then set the calculator to the correct MODE. (Check your instruction manual to find out how your calculator handles degrees and radians.) Your calculator has the keys marked $\boxed{\sin}$, $\boxed{\cos}$, and $\boxed{\tan}$. To find the values of the remaining three trigonometric functions (secant, cosecant, and cotangent), we use the reciprocal identities.

$$\sec\theta = \frac{1}{\cos\theta} \qquad \csc\theta = \frac{1}{\sin\theta} \qquad \cot\theta = \frac{1}{\tan\theta}$$

EXAMPLE 5   **Using a Calculator to Approximate the Value of Trigonometric Functions**

Use a calculator to find the approximate value of:

(a) $\cos 48°$ (b) $\csc 21°$ (c) $\tan\dfrac{\pi}{12}$

Express your answer rounded to two decimal places.

Solution   (a) First, we set the MODE to receive degrees. See Figure 31(a). Figure 31(b) shows the solution using a TI-83 graphing calculator.

Figure 31

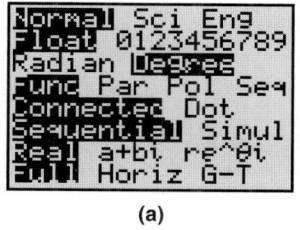

(a)

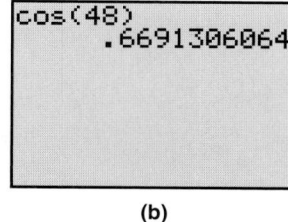

(b)

Figure 32

Rounded to two decimal places,

$$\cos 48° = 0.67$$

(b) Most calculators do not have a csc key. The manufacturers assume the user knows some trigonometry. To find the value of csc 21°, we use the fact that $\csc 21° = \dfrac{1}{\sin 21°}$. Figure 32 shows the solution using a TI-83 graphing calculator. Rounded to two decimal places,

$$\csc 21° = 2.79$$

Figure 33

(c) Set the MODE to receive radians. Figure 33 shows the solution using a TI-83 graphing calculator. Rounded to two decimal places,

$$\tan\frac{\pi}{12} = 0.27$$

■

NOW WORK PROBLEM **25.**

EXAMPLE 6   **Constructing a Rain Gutter**

Figure 34

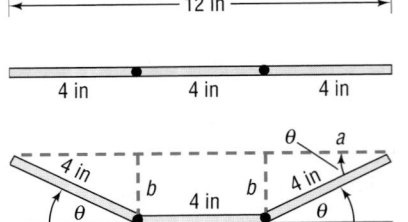

A rain gutter is to be constructed of aluminum sheets 12 inches wide. After marking off a length of 4 inches from each edge, this length is bent up at an angle θ. See Figure 34.

(a) Express the area A of the opening as a function of θ.

   [**Hint:** Let b denote the vertical height of the bend.]

(b) Find the area A of the opening for $\theta = 30°, \theta = 45°, \theta = 60°,$ and $\theta = 75°$.

(c) Graph $A = A(\theta)$. Find the angle θ that makes A largest. (This bend will allow the most water to flow through the gutter.)

Solution   (a) Look again at Figure 34. The area A of the opening is the sum of the areas of two congruent right triangles and one rectangle. Look at Figure 35, showing the triangle in Figure 34 redrawn. We see that

Figure 35

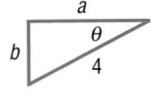

$$\cos \theta = \frac{a}{4} \quad \text{so} \quad a = 4 \cos \theta \qquad \sin \theta = \frac{b}{4} \quad \text{so} \quad b = 4 \sin \theta$$

The area of the triangle is

$$\text{area} = \frac{1}{2}(\text{base})(\text{height}) = \frac{1}{2}ab = \frac{1}{2}(4 \cos \theta)(4 \sin \theta) = 8 \sin \theta \cos \theta$$

So the area of the two triangles is $16 \sin \theta \cos \theta$.

The rectangle has length 4 and height b, so its area is

$$4b = 4(4 \sin \theta) = 16 \sin \theta$$

The area A of the opening is

$$A = \text{area of the two triangles} + \text{area of the rectangle}$$
$$A(\theta) = 16 \sin \theta \cos \theta + 16 \sin \theta = 16 \sin \theta(\cos \theta + 1)$$

(b) For $\theta = 30°$: $A(30°) = 16 \sin 30°(\cos 30° + 1)$

$$= 16\left(\frac{1}{2}\right)\left(\frac{\sqrt{3}}{2} + 1\right) = 4\sqrt{3} + 8$$

The area of the opening for $\theta = 30°$ is about 14.9 square inches.

For $\theta = 45°$: $A(45°) = 16 \sin 45°(\cos 45° + 1)$

$$= 16\left(\frac{\sqrt{2}}{2}\right)\left(\frac{\sqrt{2}}{2} + 1\right) = 8 + 8\sqrt{2}$$

The area of the opening for $\theta = 45°$ is about 19.3 square inches.

For $\theta = 60°$: $A(60°) = 16 \sin 60°(\cos 60° + 1)$

$$= 16\left(\frac{\sqrt{3}}{2}\right)\left(\frac{1}{2} + 1\right) = 12\sqrt{3}$$

The area of the opening for $\theta = 60°$ is about 20.8 square inches.

For $\theta = 75°$: $A(75°) = 16 \sin 75°(\cos 75° + 1) \approx 19.5$

The area of the opening for $\theta = 75°$ is about 19.5 square inches.

Figure 36

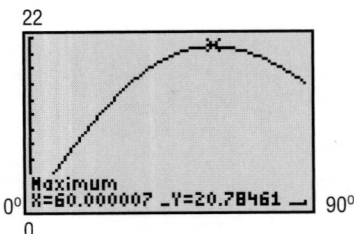

(c) Figure 36 shows the graph of $A = A(\theta)$. Using MAXIMUM, the angle θ that makes A largest is $60°$.

2.3   Concepts and Vocabulary

In Problems 1 and 2, fill in the blanks.

1. $\tan\dfrac{\pi}{4} + \sin 30° =$ _____.

2. Using a calculator, $\sin 2 =$ _____, rounded to two decimal places.

In Problems 3 and 4, answer True or False to each statement.

3. Exact values can be found for the trigonometric functions of 60°.

4. Exact values can be found for the sine of any angle.

5. Explain how you remember that $\sin 45° = \dfrac{\sqrt{2}}{2}$.

6. Explain how you remember that $\sin\dfrac{\pi}{6} = \dfrac{1}{2}$.

2.3   Exercises

1. Write down the exact value of each of the six trigonometric functions of 45°.

2. Write down the exact value of each of the six trigonometric functions of 30° and of 60°.

In Problems 3–12, find the exact value of each expression if $\theta = 60°$. Do not use a calculator.

3. $\sin\theta$
4. $\cos\theta$
5. $\sin\dfrac{\theta}{2}$
6. $\cos\dfrac{\theta}{2}$
7. $(\sin\theta)^2$

8. $(\cos\theta)^2$
9. $2\sin\theta$
10. $2\cos\theta$
11. $\dfrac{\sin\theta}{2}$
12. $\dfrac{\cos\theta}{2}$

In Problems 13–24, find the exact value of each expression. Do not use a calculator.

13. $4\cos 45° - 2\sin 45°$
14. $2\sin 45° + 4\cos 30°$
15. $6\tan 45° - 8\cos 60°$

16. $\sin 30° \cdot \tan 60°$
17. $\sec\dfrac{\pi}{4} + 2\csc\dfrac{\pi}{3}$
18. $\tan\dfrac{\pi}{4} + \cot\dfrac{\pi}{4}$

19. $\sec^2\dfrac{\pi}{6} - 4$
20. $4 + \tan^2\dfrac{\pi}{3}$
21. $\sin^2 30° + \cos^2 60°$

22. $\sec^2 60° - \tan^2 45°$
23. $1 - \cos^2 30° - \cos^2 60°$
24. $1 + \tan^2 30° - \csc^2 45°$

In Problems 25–42, use a calculator to find the approximate value of each expression. Round the answer to two decimal places.

25. $\sin 28°$
26. $\cos 14°$
27. $\tan 21°$
28. $\cot 70°$
29. $\sec 41°$
30. $\csc 55°$

31. $\sin\dfrac{\pi}{10}$
32. $\cos\dfrac{\pi}{8}$
33. $\tan\dfrac{5\pi}{12}$
34. $\cot\dfrac{\pi}{18}$
35. $\sec\dfrac{\pi}{12}$
36. $\csc\dfrac{5\pi}{13}$

37. $\sin 1$
38. $\tan 1$
39. $\sin 1°$
40. $\tan 1°$
41. $\tan 0.3$
42. $\tan 0.1$

Projectile Motion   The path of a projectile fired at an inclination θ to the horizontal with initial speed v_0 is a parabola (see the figure). The range R of the projectile, that is, the horizontal distance that the projectile travels, is found by using the formula

$$R = \frac{2v_0^2 \sin\theta \cos\theta}{g}$$

where $g \approx 32.2$ feet per second per second ≈ 9.8 meters per second per second is the acceleration due to gravity.

The maximum height H of the projectile is

$$H = \frac{v_0^2 \sin^2\theta}{2g}$$

v_0 = Initial speed

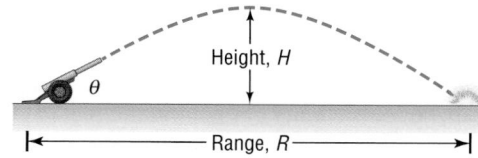

In Problems 43–46, find the range R and maximum height H. Round answers to two decimal places.

43. The projectile is fired at an angle of 45° to the horizontal with an initial speed of 100 feet per second.

44. The projectile is fired at an angle of 30° to the horizontal with an initial speed of 150 meters per second.

45. The projectile is fired at an angle of 25° to the horizontal with an initial speed of 500 meters per second.

46. The projectile is fired at an angle of 50° to the horizontal with an initial speed of 200 feet per second.

47. Inclined Plane   If friction is ignored, the time t (in seconds) required for a block to slide down an inclined plane (see the figure) is given by the formula

$$t = \sqrt{\frac{2a \sin \theta}{g \cos \theta}}$$

where a is the length (in feet) of the base and $g \approx 32$ feet per second per second is the acceleration of gravity. How long does it take a block to slide down an inclined plane with base $a = 10$ feet when

(a) $\theta = 30°$?   (b) $\theta = 45°$?   (c) $\theta = 60°$?

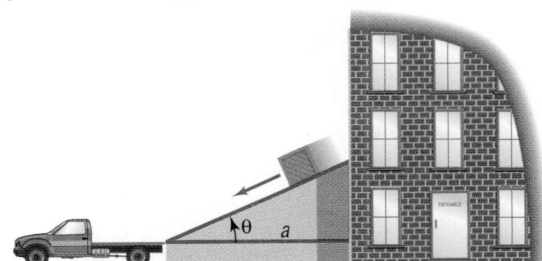

48. Piston Engines   In a certain piston engine, the distance x (in meters) from the center of the drive shaft to the head of the piston is given by

$$x = \cos \theta + \sqrt{16 + 0.5(2 \cos^2 \theta - 1)}$$

where θ is the angle between the crank and the path of the piston head (see the figure). Find x when $\theta = 30°$ and when $\theta = 45°$.

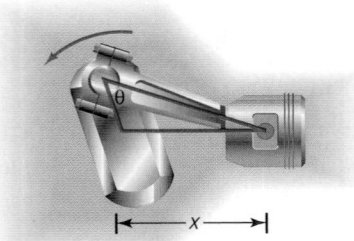

49. Calculating the Time of a Trip   Two oceanfront homes are located 8 miles apart on a straight stretch of beach, each a distance of 1 mile from a paved road that parallels the ocean. Sally can jog 8 miles per hour along the paved road, but only 3 miles per hour in the sand on the beach. Because of a river between the two houses, it is necessary

to jog on the sand to the road, continue on the road, and then jog on the sand to get from one house to the other. See the illustration.

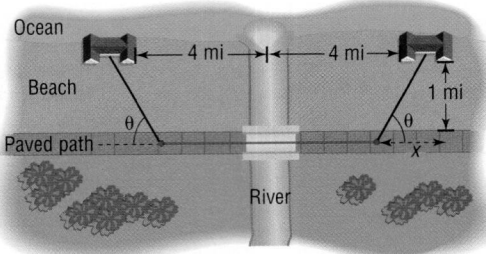

(a) Express the time T to get from one house to the other as a function of the angle θ shown in the illustration.

(b) Calculate the time T for $\theta = 30°$. How long is Sally on the paved road?

(c) Calculate the time T for $\theta = 45°$. How long is Sally on the paved road?

(d) Calculate the time T for $\theta = 60°$. How long is Sally on the paved road?

(e) Calculate the time T for $\theta = 90°$. Describe the path taken.

(f) Calculate the time T for $\tan \theta = \dfrac{1}{4}$. Describe the path taken. Explain why θ must be larger than 14°.

(g) Use a graphing utility to graph $T = T(\theta)$. What angle θ results in the least time? What is the least time? How long is Sally on the paved road?

50. Designing Fine Decorative Pieces   A designer of decorative art plans to market solid gold spheres encased in clear crystal cones. Each sphere is of fixed radius R and will be enclosed in a cone of height h and radius r. See the illustration. Many cones can be used to enclose the sphere, each having a different slant angle θ.

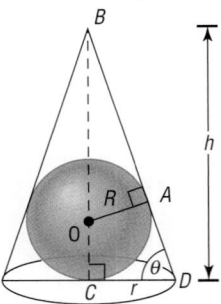

(a) Express the volume V of the cone as a function of the slant angle θ of the cone.

[**Hint:** The volume V of a cone of height h and radius r is $V = \dfrac{1}{3} \pi r^2 h$.]

(b) What volume V is required to enclose a sphere of radius 2 centimeters in a cone whose slant angle θ is 30°? 45°? 60°?

(c) What slant angle θ should be used for the volume V of the cone to be a minimum? (This choice minimizes the amount of crystal required and gives maximum emphasis to the gold sphere.)

51. Use a calculator set in radian mode to complete the following table. What can you conclude about the ratio $\dfrac{\sin\theta}{\theta}$ as θ approaches 0?

θ	0.5	0.4	0.2	0.1	0.01	0.001	0.0001	0.00001
$\sin\theta$								
$\dfrac{\sin\theta}{\theta}$								

52. Use a calculator set in radian mode to complete the following table. What can you conclude about the ratio $\dfrac{\cos\theta-1}{\theta}$ as θ approaches 0?

θ	0.5	0.4	0.2	0.1	0.01	0.001	0.0001	0.00001
$\cos\theta-1$								
$\dfrac{\cos\theta-1}{\theta}$								

53. Find the exact value of $\tan 1° \cdot \tan 2° \cdot \tan 3° \cdot \ldots \cdot \tan 89°$.

54. Find the exact value of $\cot 1° \cdot \cot 2° \cdot \cot 3° \cdot \ldots \cdot \cot 89°$.

55. Find the exact value of $\cos 1° \cdot \cos 2° \cdot \ldots \cdot \cos 45° \cdot \csc 46° \cdot \ldots \cdot \csc 89°$.

56. Find the exact value of $\sin 1° \cdot \sin 2° \cdot \ldots \cdot \sin 45° \cdot \sec 46° \cdot \ldots \cdot \sec 89°$.

57. Write a brief paragraph that explains how to quickly compute the trigonometric functions of $30°, 45°$, and $60°$.

2.4 TRIGONOMETRIC FUNCTIONS OF GENERAL ANGLES

OBJECTIVES
1. Find the Exact Value of the Trigonometric Functions for General Angles
2. Use Coterminal Angles to Find the Exact Value of a Trigonometric Function
3. Determine the Sign of the Trigonometric Functions of an Angle in a Given Quadrant
4. Find the Reference Angle of a General Angle
5. Use the Theorem on Reference Angles
6. Find the Exact Value of Trigonometric Functions of an Angle Given One of them and the Quadrant of the Angle

Figure 37

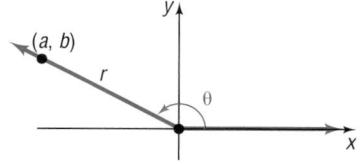

To extend the definitions of the trigonometric functions to include angles that are not acute, we employ a rectangular coordinate system and place the angle in the standard position so that its vertex is at the origin and its initial side is along the positive x-axis. See Figure 37.

Let θ be any angle in standard position, and let (a, b) denote the coordinates of any point, except the origin $(0, 0)$, on the terminal side of θ. If $r = \sqrt{a^2 + b^2}$ denotes the distance from $(0, 0)$ to (a, b), then the **six trigonometric functions of θ** are defined as the ratios

$$\sin \theta = \frac{b}{r} \qquad \cos \theta = \frac{a}{r} \qquad \tan \theta = \frac{b}{a}$$

$$\csc \theta = \frac{r}{b} \qquad \sec \theta = \frac{r}{a} \qquad \cot \theta = \frac{a}{b}$$

provided no denominator equals 0. If a denominator equals 0, that trigonometric function of the angle θ is not defined.

Figure 38

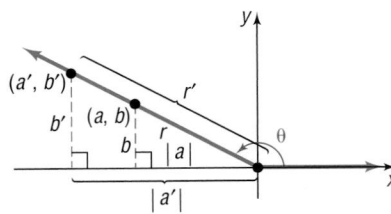

Notice in the preceding definitions that if $a = 0$, that is, the point $(0, b)$ is on the y-axis, then the tangent function and the secant function are undefined. Also, if $b = 0$, that is, the point $(a, 0)$ is on the x-axis, then the cosecant function and the cotangent function are undefined.

By constructing similar triangles, you should be convinced that the values of the six trigonometric functions of an angle θ do not depend on the selection of the point (a, b) on the terminal side of θ, but rather depend only on the angle θ itself. See Figure 38 for an illustration when θ lies in quadrant II.

Figure 39

$$\sin \theta = \frac{b}{r} = \frac{\text{opposite}}{\text{hypotenuse}}$$

$$\cos \theta = \frac{a}{r} = \frac{\text{adjacent}}{\text{hypotenuse}} \quad \text{and so on.}$$

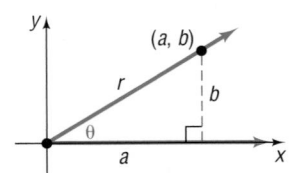

Since the triangles are similar, the ratio $\dfrac{b}{r}$ equals the ratio $\dfrac{b'}{r'}$, the common value being $\sin \theta$. Also, the ratio $\dfrac{|a|}{r}$ equals the ratio $\dfrac{|a'|}{r'}$, so $\dfrac{a}{r} = \dfrac{a'}{r'}$, the common value being $\cos \theta$. And so on.

Also, observe that if θ is acute these definitions reduce to the right triangle definitions given in Section 2.2, as illustrated in Figure 39.

Finally, from the definition of the six trigonometric functions of a general angle, we see that the Quotient and Reciprocal Identities hold. Also, using $r^2 = a^2 + b^2$ and dividing each side by r^2, we can derive the Pythagorean Identities for general angles.

① **EXAMPLE 1** **Finding the Exact Value of the Six Trigonometric Functions of θ Given a Point on the Terminal Side**

Find the exact value of each of the six trigonometric functions of a positive angle θ if $(4, -3)$ is a point on its terminal side.

Solution Figure 40 illustrates the situation. For the point $(a, b) = (4, -3)$, we have

Figure 40 $a = 4$ and $b = -3$. Then $r = \sqrt{a^2 + b^2} = \sqrt{16 + 9} = 5$, so that

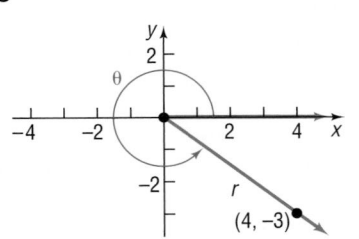

$$\sin \theta = \frac{b}{r} = -\frac{3}{5} \qquad \cos \theta = \frac{a}{r} = \frac{4}{5} \qquad \tan \theta = \frac{b}{a} = -\frac{3}{4}$$

$$\csc \theta = \frac{r}{b} = -\frac{5}{3} \qquad \sec \theta = \frac{r}{a} = \frac{5}{4} \qquad \cot \theta = \frac{a}{b} = -\frac{4}{3}$$

— NOW WORK PROBLEM **1.**

In the next example, we find the exact value of each of the six trigonometric functions at the quadrantal angles $0, \dfrac{\pi}{2}, \pi,$ and $\dfrac{3\pi}{2}$.

EXAMPLE 2 | **Finding the Exact Value of the Six Trigonometric Functions of Quadrantal Angles**

Find the exact value of each of the six trigonometric functions of

(a) $\theta = 0 = 0°$

(b) $\theta = \dfrac{\pi}{2} = 90°$

(c) $\theta = \pi = 180°$

(d) $\theta = \dfrac{3\pi}{2} = 270°$

Solution (a) The point $P = (1, 0)$ is on the terminal side of $\theta = 0 = 0°$ and is a distance of $r = 1$ unit from the origin. See Figure 41. Then,

Figure 41
$\theta = 0 = 0°$

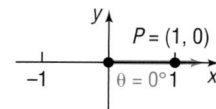

$$\sin 0 = \sin 0° = \frac{0}{1} = 0 \qquad \cos 0 = \cos 0° = \frac{1}{1} = 1$$

$$\tan 0 = \tan 0° = \frac{0}{1} = 0 \qquad \sec 0 = \sec 0° = \frac{1}{1} = 1$$

Since the y-coordinate of P is 0, $\csc 0$ and $\cot 0$ are not defined.

(b) The point $P = (0, 1)$ is on the terminal side of $\theta = \dfrac{\pi}{2} = 90°$ and is a distance of $r = 1$ unit from the origin. See Figure 42. Then

Figure 42
$\theta = \dfrac{\pi}{2} = 90°$

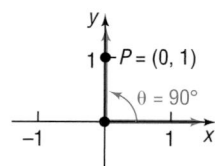

$$\sin \frac{\pi}{2} = \sin 90° = \frac{1}{1} = 1 \qquad \cos \frac{\pi}{2} = \cos 90° = \frac{0}{1} = 0$$

$$\csc \frac{\pi}{2} = \csc 90° = \frac{1}{1} = 1 \qquad \cot \frac{\pi}{2} = \cot 90° = \frac{0}{1} = 0$$

Since the x-coordinate of P is 0, $\tan \dfrac{\pi}{2}$ and $\sec \dfrac{\pi}{2}$ are not defined.

(c) The point $P = (-1, 0)$ is on the terminal side of $\theta = \pi = 180°$ and is a distance of $r = 1$ unit from the origin. See Figure 43. Then,

Figure 43
$\theta = \pi = 180°$

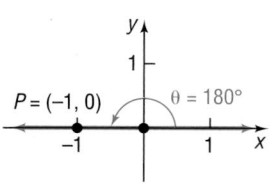

$$\sin \pi = \sin 180° = \frac{0}{1} = 0 \qquad \cos \pi = \cos 180° = \frac{-1}{1} = -1$$

$$\tan \pi = \tan 180° = \frac{0}{-1} = 0 \qquad \sec \pi = \sec 180° = \frac{1}{-1} = -1$$

Since the y-coordinate of P is 0, $\csc \pi$ and $\cot \pi$ are not defined.

(d) The point $P = (0, -1)$ is on the terminal side of $\theta = \dfrac{3\pi}{2} = 270°$ and is a distance of $r = 1$ unit from the origin. See Figure 44. Then,

Figure 44
$\theta = \dfrac{3\pi}{2} = 270°$

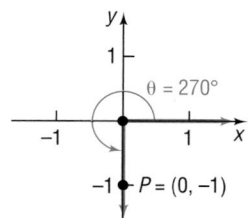

$$\sin \frac{3\pi}{2} = \sin 270° = \frac{-1}{1} = -1 \qquad \cos \frac{3\pi}{2} = \cos 270° = \frac{0}{1} = 0$$

$$\csc \frac{3\pi}{2} = \csc 270° = \frac{1}{-1} = -1 \qquad \cot \frac{3\pi}{2} = \cot 270° = \frac{0}{-1} = 0$$

Since the x-coordinate of P is 0, $\tan \dfrac{3\pi}{2}$ and $\sec \dfrac{3\pi}{2}$ are not defined. ∎

Table 4 summarizes the values of the trigonometric functions found in Example 2.

TABLE 4

θ (Radians)	θ (Degrees)	sin θ	cos θ	tan θ	csc θ	sec θ	cot θ
0	0°	0	1	0	Not defined	1	Not defined
$\dfrac{\pi}{2}$	90°	1	0	Not defined	1	Not defined	0
π	180°	0	−1	0	Not defined	−1	Not defined
$\dfrac{3\pi}{2}$	270°	−1	0	Not defined	−1	Not defined	0

There is no need to memorize Table 4. To find the value of a trigonometric function of a quadrantal angle, draw the angle and apply the definition as we did in Example 2.

Coterminal Angles

Two angles in standard position are said to be **coterminal** if they have the same terminal side.

See Figure 45.

Figure 45

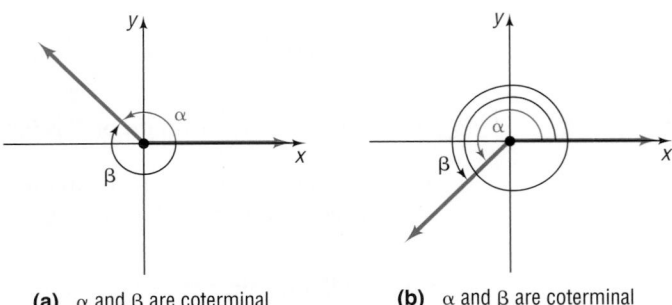

(a) α and β are coterminal   **(b)** α and β are coterminal

In general, if θ is an angle measured in degrees, then θ ± 360°k, where k is any integer, is an angle coterminal with θ. If θ is measured in radians, then θ ± 2πk, where k is any integer, is an angle coterminal with θ.

② Because coterminal angles have the same terminal side, it follows that the values of the trigonometric functions of coterminal angles are equal. We use this fact in the next example.

EXAMPLE 3   Using the Coterminal Angle to Find the Exact Value of a Trigonometric Function

Find the exact value of each of the following:

(a) $\sin 390°$ (b) $\cos 420°$ (c) $\tan \dfrac{9\pi}{4}$ (d) $\sec\left(-\dfrac{7\pi}{4}\right)$ (e) $\csc(-270°)$

Solution   (a) It is best to sketch the angle first. See Figure 46. The angle 390° is coterminal with 30°.

$$\sin 390° = \sin(360° + 30°)$$
$$= \sin 30° = \frac{1}{2}$$

(b) See Figure 47. The angle 420° is coterminal with 60°.

$$\cos 420° = \cos(360° + 60°)$$
$$= \cos 60° = \frac{1}{2}$$

Figure 46

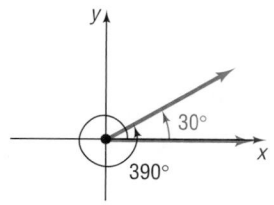

Figure 47

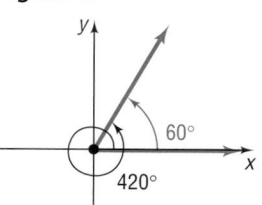

Figure 48

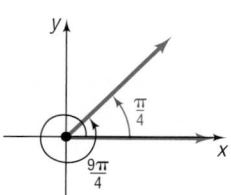

(c) The angle $\dfrac{9\pi}{4}$ is coterminal with $\dfrac{\pi}{4} \left(\dfrac{9\pi}{4} = \dfrac{8\pi}{4} + \dfrac{\pi}{4} = 2\pi + \dfrac{\pi}{4} \right)$. See Figure 48.

$$\tan \frac{9\pi}{4} = \tan\left(2\pi + \frac{\pi}{4}\right)$$
$$= \tan \frac{\pi}{4} = 1$$

Figure 49

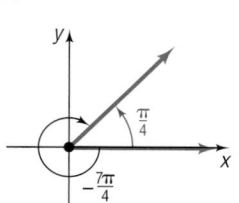

(d) See Figure 49. The angle $-\dfrac{7\pi}{4}$ is coterminal with $\dfrac{\pi}{4}$.

$$\sec\left(-\frac{7\pi}{4}\right) = \sec\left(-2\pi + \frac{\pi}{4}\right) = \sec\frac{\pi}{4} = \sqrt{2}$$

(e) See Figure 50. The angle −270° is coterminal with 90°.

$$\csc(-270°) = \csc(-360° + 90°)$$
$$= \csc 90° = 1$$

Figure 50

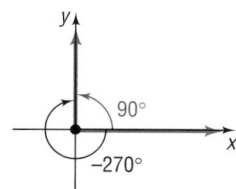

As Example 3 illustrates, the value of a trigonometric function of any angle is equal to the value of the same trigonometric function of an angle θ coterminal to it where $0° \le \theta < 360°$ (or $0 \le \theta < 2\pi$). Because the angles θ and $\theta \pm 360°k$ (or $\theta \pm 2\pi k$), where k is any integer, are coterminal, and because the values of the trigonometric functions are equal for coterminal angles, it follows that

$$\begin{array}{ll}
\sin(\theta \pm 360°k) = \sin\theta & \sin(\theta \pm 2\pi k) = \sin\theta \\
\cos(\theta \pm 360°k) = \cos\theta & \cos(\theta \pm 2\pi k) = \cos\theta \\
\tan(\theta \pm 360°k) = \tan\theta & \tan(\theta \pm 2\pi k) = \tan\theta \\
\csc(\theta \pm 360°k) = \csc\theta & \csc(\theta \pm 2\pi k) = \csc\theta \\
\sec(\theta \pm 360°k) = \sec\theta & \sec(\theta \pm 2\pi k) = \sec\theta \\
\cot(\theta \pm 360°k) = \cot\theta & \cot(\theta \pm 2\pi k) = \cot\theta
\end{array} \qquad (1)$$

These formulas show that the values of the trigonometric functions repeat themselves every 360° (or 2π radians).

NOW WORK PROBLEM **37**.

The Signs of the Trigonometric Functions

3 If θ is not a quadrantal angle, then it will lie in a particular quadrant. In such a case, the signs of the x-coordinate and the y-coordinate of a point (a, b) on the terminal side of θ are known. Because $r = \sqrt{a^2 + b^2} > 0$, it follows that the signs of the trigonometric functions of an angle θ can be found if we know in which quadrant θ lies.

For example, if θ lies in quadrant II, as shown in Figure 51, then a point (a, b) on the terminal side of θ has a negative x-coordinate and a positive y-coordinate. Thus,

$$\sin\theta = \frac{b}{r} > 0 \qquad \cos\theta = \frac{a}{r} < 0 \qquad \tan\theta = \frac{b}{a} < 0$$
$$\csc\theta = \frac{r}{b} > 0 \qquad \sec\theta = \frac{r}{a} < 0 \qquad \cot\theta = \frac{a}{b} < 0$$

Figure 51
θ in quadrant II; $a < 0$, $b > 0$, $r > 0$

y
(a, b)
r
θ
x

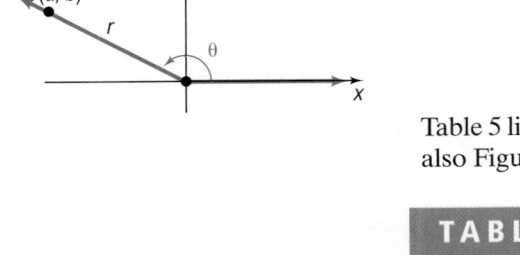

Table 5 lists the signs of the six trigonometric functions for each quadrant. See also Figure 52.

TABLE 5

Quadrant of θ	$\sin\theta$, $\csc\theta$	$\cos\theta$, $\sec\theta$	$\tan\theta$, $\cot\theta$
I	Positive	Positive	Positive
II	Positive	Negative	Negative
III	Negative	Negative	Positive
IV	Negative	Positive	Negative

Figure 52

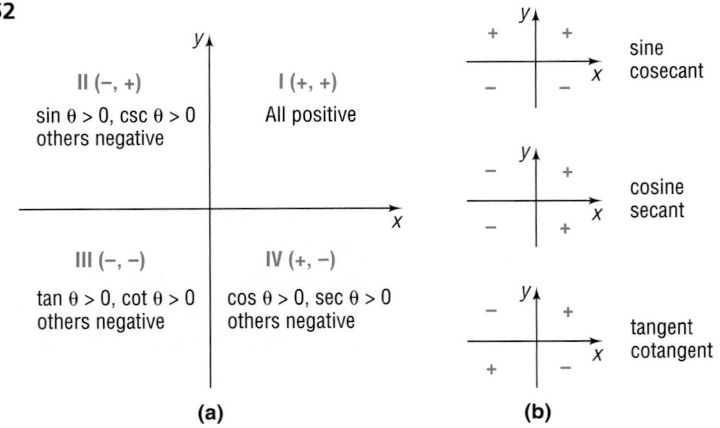

y
II (−, +)
sin θ > 0, csc θ > 0
others negative
I (+, +)
All positive
x
III (−, −)
tan θ > 0, cot θ > 0
others negative
IV (+, −)
cos θ > 0, sec θ > 0
others negative
(a)
sine
cosecant
cosine
secant
tangent
cotangent
(b)

EXAMPLE 4 **Finding the Quadrant in Which an Angle Lies**

If $\sin\theta < 0$ and $\cos\theta < 0$, name the quadrant in which the angle θ lies.

Solution If $\sin\theta < 0$, then θ lies in quadrant III or IV. If $\cos\theta < 0$, then θ lies in quadrant II or III. Therefore, θ lies in quadrant III. ■

NOW WORK PROBLEM 11.

Reference Angle

4 Once we know in which quadrant an angle lies, we know the sign of each trigonometric function of this angle. The use of a certain reference angle may help us to evaluate the trigonometric functions of such an angle.

> Let θ denote a nonacute angle that lies in a quadrant. The acute angle formed by the terminal side of θ and either the positive x-axis or the negative x-axis is called the **reference angle** for θ.

Figure 53 illustrates the reference angle for some general angles θ. Note that a reference angle is always an acute angle, that is, an angle whose measure is between $0°$ and $90°$.

Figure 53

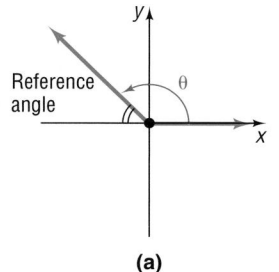

(a)

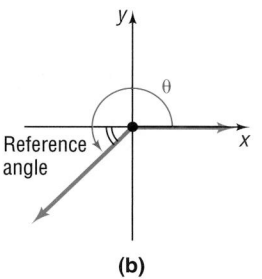

(b)

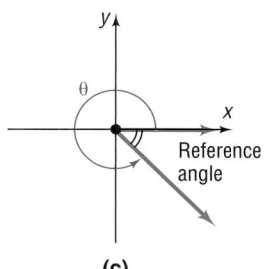

(c)

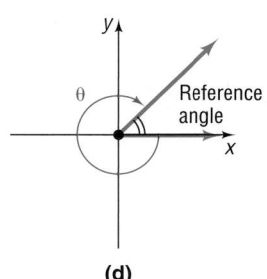

(d)

Although formulas can be given for calculating reference angles, usually it is easier to find the reference angle for a given angle by making a quick sketch of the angle.

EXAMPLE 5 **Finding Reference Angles**

Find the reference angle for each of the following angles:

(a) $150°$ (b) $-45°$ (c) $\dfrac{9\pi}{4}$ (d) $-\dfrac{5\pi}{6}$

Solution (a) Refer to Figure 54. The reference angle for $150°$ is $30°$.

Figure 54

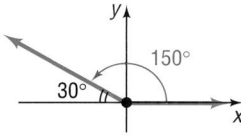

(b) Refer to Figure 55. The reference angle for $-45°$ is $45°$.

Figure 55

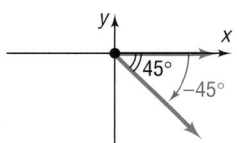

(c) Refer to Figure 56. The reference angle for $\dfrac{9\pi}{4}$ is $\dfrac{\pi}{4}$.

Figure 56

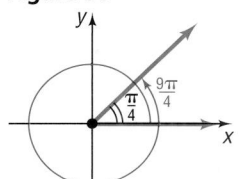

(d) Refer to Figure 57. The reference angle for $-\dfrac{5\pi}{6}$ is $\dfrac{\pi}{6}$.

Figure 57

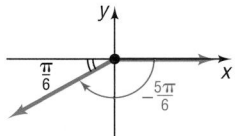

══ NOW WORK PROBLEM **19.**

⑤   The advantage of using reference angles is that, except for the correct sign, the values of the trigonometric functions of a general angle θ equal the values of the trigonometric functions of its reference angle.

Theorem

Reference Angles

If θ is an angle that lies in a quadrant and if α is its reference angle, then

$$\sin \theta = \pm\sin \alpha \quad \cos \theta = \pm\cos \alpha \quad \tan \theta = \pm\tan \alpha$$
$$\csc \theta = \pm\csc \alpha \quad \sec \theta = \pm\sec \alpha \quad \cot \theta = \pm\cot \alpha$$

(2)

where the $+$ or $-$ sign depends on the quadrant in which θ lies.

Figure 58
$\sin \theta = b/r, \sin \alpha = b/r,$
$\cos \theta = a/r, \cos \alpha = |a|/r$

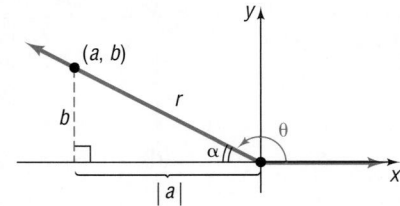

For example, suppose that θ lies in quadrant II and α is its reference angle. See Figure 58. If (a, b) is a point on the terminal side of θ and if $r = \sqrt{a^2 + b^2}$, we have

$$\sin \theta = \frac{b}{r} = \sin \alpha \qquad \cos \theta = \frac{a}{r} = \frac{-|a|}{r} = -\cos \alpha$$

and so on.

The next example illustrates how the theorem on reference angles is used.

EXAMPLE 6   **Using the Reference Angle to Find the Exact Value of a Trigonometric Function**

Find the exact value of each of the following trigonometric functions using reference angles.

(a) $\sin 135°$ (b) $\cos 240°$ (c) $\cos \dfrac{5\pi}{6}$ (d) $\tan\left(-\dfrac{\pi}{3}\right)$

Solution   (a) Refer to Figure 59. The angle $135°$ is in quadrant II, where the sine function is positive. The reference angle for $135°$ is $45°$.

$$\sin 135° = \sin 45° = \frac{\sqrt{2}}{2}$$

Figure 59

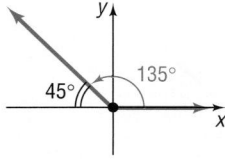

(b) Refer to Figure 60. The angle $240°$ is in quadrant III, where the cosine function is negative. The reference angle for $240°$ is $60°$.

$$\cos 240° = -\cos 60° = -\frac{1}{2}$$

Figure 60

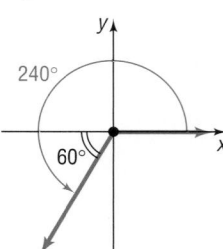

(c) Refer to Figure 61. The angle $\dfrac{5\pi}{6}$ is in quadrant II, where the cosine function is negative. The reference angle for $\dfrac{5\pi}{6}$ is $\dfrac{\pi}{6}$.

$$\cos\frac{5\pi}{6} = -\cos\frac{\pi}{6} = -\frac{\sqrt{3}}{2}$$

Figure 61

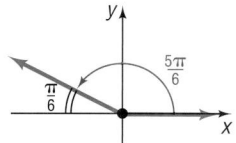

(d) Refer to Figure 62. The angle $-\dfrac{\pi}{3}$ is in quadrant IV, where the tangent function is negative. The reference angle for $-\dfrac{\pi}{3}$ is $\dfrac{\pi}{3}$.

$$\tan\left(-\frac{\pi}{3}\right) = -\tan\frac{\pi}{3} = -\sqrt{3}$$

Figure 62

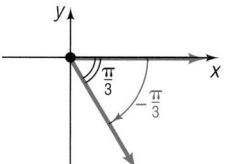

 **NOW WORK PROBLEM 53.**

⑥ **EXAMPLE 7** **Finding the Exact Value of Trigonometric Functions**

Given that $\cos\theta = -\dfrac{2}{3}$, $\dfrac{\pi}{2} < \theta < \pi$, find the exact value of each of the remaining trigonometric functions.

Solution The angle θ lies in quadrant II, so we know that $\sin\theta$ and $\csc\theta$ are positive, whereas the other trigonometric functions are negative. If α is the reference angle for θ, then $\cos\alpha = \dfrac{2}{3}$. The values of the remaining trigonometric functions of the angle α can be found by drawing the appropriate triangle and using the Pythagorean Theorem. We use Figure 63 to obtain

Figure 63

$\cos\alpha = \dfrac{2}{3}$

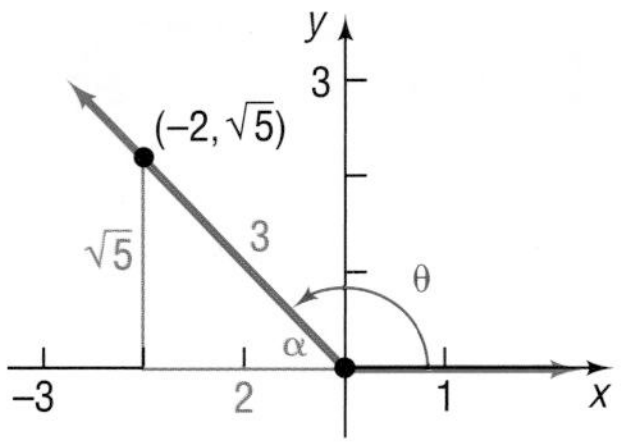

$$\sin\alpha = \frac{\sqrt{5}}{3} \qquad \cos\alpha = \frac{2}{3} \qquad \tan\alpha = \frac{\sqrt{5}}{2}$$

$$\csc\alpha = \frac{3}{\sqrt{5}} = \frac{3\sqrt{5}}{5} \qquad \sec\alpha = \frac{3}{2} \qquad \cot\alpha = \frac{2}{\sqrt{5}} = \frac{2\sqrt{5}}{5}$$

Now we assign the appropriate signs to each of these values to find the values of the trigonometric functions of θ.

$$\sin\theta = \frac{\sqrt{5}}{3} \qquad \cos\theta = -\frac{2}{3} \qquad \tan\theta = -\frac{\sqrt{5}}{2}$$

$$\csc\theta = \frac{3\sqrt{5}}{5} \qquad \sec\theta = -\frac{3}{2} \qquad \cot\theta = -\frac{2\sqrt{5}}{5}$$

NOW WORK PROBLEM 79.

EXAMPLE 8 **Finding the Exact Value of Trigonometric Functions**

If $\tan\theta = -4$ and $\sin\theta < 0$, find the exact value of each of the remaining trigonometric functions of θ.

Solution   Since $\tan \theta = -4 < 0$ and $\sin \theta < 0$, it follows that θ lies in quadrant IV. If α is the reference angle for θ, then $\tan \alpha = 4$. See Figure 64. Then

Figure 64
$\tan \alpha = 4$

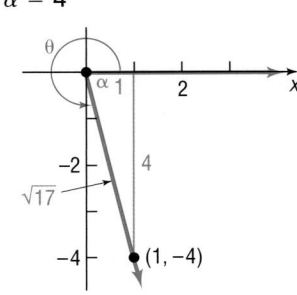

$$\sin \alpha = \frac{4}{\sqrt{17}} = \frac{4\sqrt{17}}{17} \qquad \cos \alpha = \frac{1}{\sqrt{17}} = \frac{\sqrt{17}}{17} \qquad \tan \alpha = \frac{4}{1} = 4$$

$$\csc \alpha = \frac{\sqrt{17}}{4} \qquad \sec \alpha = \frac{\sqrt{17}}{1} = \sqrt{17} \qquad \cot \alpha = \frac{1}{4}$$

We assign the appropriate sign to each of these to obtain the values of the trigonometric functions of θ.

$$\sin \theta = -\frac{4\sqrt{17}}{17} \qquad \cos \theta = \frac{\sqrt{17}}{17} \qquad \tan \theta = -4$$

$$\csc \theta = -\frac{\sqrt{17}}{4} \qquad \sec \theta = \sqrt{17} \qquad \cot \theta = -\frac{1}{4}$$

NOW WORK PROBLEM 89.

SUMMARY

To find the values of the trigonometric functions of a general angle:

STEP 1: If the angle θ is a quadrantal angle, draw the angle, pick a point on its terminal side, and apply the definitions of the trigonometric functions.

STEP 2: If the angle θ lies in a quadrant, determine the correct signs of the trigonometric functions in that quadrant and determine the reference angle α for θ. Now express each trigonometric function of θ in terms of the same value (except for the sign) of the trigonometric function of α, an acute angle. Finally, apply the correct sign to each function.

2.4 Concepts and Vocabulary

In Problems 1–3, fill in the blanks.

1. For an angle θ that lies in quadrant III, the trigonometric functions _____ and _____ are positive.

2. Two angles in standard position that have the same terminal side are _____.

3. The reference angle of $240°$ is _____.

In Problems 4–6, answer True or False to each statement.

4. $\sin 182° = \cos 2°$.

5. $\tan \dfrac{\pi}{2}$ is not defined.

6. The reference angle is always an acute angle.

7. Explain how you would find the reference angle of $240°$.

8. In which quadrants is the cosine function positive?

9. If $0 \le \theta < 2\pi$, for what angles θ, if any, is $\tan \theta$ undefined?

10. Find an acute angle that is coterminal with $-130°$.

2.4 Exercises

In Problems 1–10, a point on the terminal side of an angle θ is given. Find the exact value of each of the six trigonometric functions of θ.

1. $(-3, 4)$

2. $(5, -12)$

3. $(2, -3)$

4. $(-1, -2)$

5. $(-3, -3)$

6. $(2, -2)$

7. $\left(\dfrac{\sqrt{3}}{2}, \dfrac{1}{2}\right)$

8. $\left(-\dfrac{1}{2}, \dfrac{\sqrt{3}}{2}\right)$

9. $\left(\dfrac{\sqrt{2}}{2}, -\dfrac{\sqrt{2}}{2}\right)$

10. $\left(-\dfrac{\sqrt{2}}{2}, -\dfrac{\sqrt{2}}{2}\right)$

In Problems 11–18, name the quadrant in which the angle θ lies.

11. $\sin\theta > 0, \quad \cos\theta < 0$

12. $\sin\theta < 0, \quad \cos\theta > 0$

13. $\sin\theta < 0, \quad \tan\theta < 0$

14. $\cos\theta > 0, \quad \tan\theta > 0$

15. $\cos\theta > 0, \quad \cot\theta < 0$

16. $\sin\theta < 0, \quad \cot\theta > 0$

17. $\sec\theta < 0, \quad \tan\theta > 0$

18. $\csc\theta > 0, \quad \cot\theta < 0$

In Problems 19–36, find the reference angle of each angle.

19. $-30°$

20. $60°$

21. $120°$

22. $300°$

23. $210°$

24. $330°$

25. $\dfrac{5\pi}{4}$

26. $\dfrac{5\pi}{6}$

27. $\dfrac{8\pi}{3}$

28. $\dfrac{7\pi}{4}$

29. $-135°$

30. $-240°$

31. $-\dfrac{2\pi}{3}$

32. $-\dfrac{7\pi}{6}$

33. $440°$

34. $490°$

35. $-\dfrac{3\pi}{4}$

36. $-\dfrac{11\pi}{6}$

In Problems 37–78, find the exact value of each expression. Do not use a calculator.

37. $\sin 405°$

38. $\cos 420°$

39. $\tan 405°$

40. $\sin 390°$

41. $\csc 450°$

42. $\sec 540°$

43. $\cot 390°$

44. $\sec 420°$

45. $\cos \dfrac{33\pi}{4}$

46. $\sin \dfrac{9\pi}{4}$

47. $\tan(21\pi)$

48. $\csc \dfrac{9\pi}{2}$

49. $\sec \dfrac{17\pi}{4}$

50. $\cot \dfrac{17\pi}{4}$

51. $\tan \dfrac{19\pi}{6}$

52. $\sec \dfrac{25\pi}{6}$

53. $\sin 150°$

54. $\cos 210°$

55. $\cos 315°$

56. $\sin 120°$

57. $\sec 240°$

58. $\csc 300°$

59. $\cot 330°$

60. $\tan 225°$

61. $\sin \dfrac{3\pi}{4}$

62. $\cos \dfrac{2\pi}{3}$

63. $\cot \dfrac{7\pi}{6}$

64. $\csc \dfrac{7\pi}{4}$

65. $\cos(-60°)$

66. $\tan(-120°)$

67. $\sin\left(-\dfrac{2\pi}{3}\right)$

68. $\cot\left(-\dfrac{\pi}{6}\right)$

69. $\tan \dfrac{14\pi}{3}$

70. $\sec \dfrac{11\pi}{4}$

71. $\csc(-315°)$

72. $\sec(-225°)$

73. $\sin(8\pi)$

74. $\cos(-2\pi)$

75. $\tan(7\pi)$

76. $\cot(5\pi)$

77. $\sec(-3\pi)$

78. $\csc\left(-\dfrac{5\pi}{2}\right)$

In Problems 79–96, find the exact value of each of the remaining trigonometric functions of θ.

79. $\sin\theta = \dfrac{12}{13}, \quad \theta$ in quadrant II

80. $\cos\theta = \dfrac{3}{5}, \quad \theta$ in quadrant IV

81. $\cos\theta = -\dfrac{4}{5}, \quad \theta$ in quadrant III

82. $\sin\theta = -\dfrac{5}{13}, \quad \theta$ in quadrant III

83. $\sin\theta = \dfrac{5}{13}, \quad 90° < \theta < 180°$

84. $\cos\theta = \dfrac{4}{5}, \quad 270° < \theta < 360°$

85. $\cos\theta = -\dfrac{1}{3}, \quad 180° < \theta < 270°$

86. $\sin\theta = -\dfrac{2}{3}, \quad 180° < \theta < 270°$

87. $\sin\theta = \dfrac{2}{3}, \quad \tan\theta < 0$

88. $\cos\theta = -\dfrac{1}{4}, \quad \tan\theta > 0$

89. $\sec\theta = 2, \quad \sin\theta < 0$

90. $\csc\theta = 3, \quad \cot\theta < 0$

91. $\tan\theta = \dfrac{3}{4}, \quad \sin\theta < 0$

92. $\cot\theta = \dfrac{4}{3}, \quad \cos\theta < 0$

93. $\tan\theta = -\dfrac{1}{3}, \quad \sin\theta > 0$

94. $\sec\theta = -2, \quad \tan\theta > 0$

95. $\csc\theta = -2, \quad \tan\theta > 0$

96. $\cot\theta = -2, \quad \sec\theta > 0$

97. Find the exact value of
$\sin 45° + \sin 135° + \sin 225° + \sin 315°$.

98. Find the exact value of
$\tan 60° + \tan 150°$.

99. If $\sin \theta = 0.2$, find $\sin(\theta + \pi)$.

100. If $\cos \theta = 0.4$, find $\cos(\theta + \pi)$.

101. If $\tan \theta = 3$, find $\tan(\theta + \pi)$.

102. If $\cot \theta = -2$, find $\cot(\theta + \pi)$.

103. If $\sin \theta = \dfrac{1}{5}$, find $\csc \theta$.

104. If $\cos \theta = \dfrac{2}{3}$, find $\sec \theta$.

105. Find the exact value of
$\sin 1° + \sin 2° + \sin 3° + \cdots + \sin 358° + \sin 359°$

106. Find the exact value of
$\cos 1° + \cos 2° + \cos 3° + \cdots + \cos 358° + \cos 359°$

---

107. Projectile Motion An object is propelled upward at an angle θ, $45° < \theta < 90°$, to the horizontal with an initial velocity of v_0 feet per second from the base of a plane that makes an angle of $45°$ with the horizontal. See the illustration. If air resistance is ignored, the distance R that it travels up the inclined plane is given by

$$R = \frac{v_0^2 \sqrt{2}}{32}[\sin (2\theta) - \cos (2\theta) - 1]$$

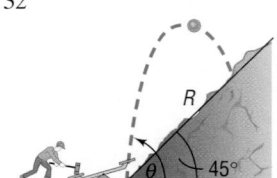

(a) Find the distance R that the object travels along the inclined plane if the initial velocity is 32 feet per second and $\theta = 60°$.

(b) Use a graphing utility to graph $R = R(\theta)$ if the initial velocity is 32 feet per second.

(c) What value of θ makes R largest?

108. Give three examples that demonstrate how to use the theorem on reference angles.

109. Write a brief paragraph that explains how to quickly compute the value of the trigonometric functions of $0°$, $90°$, $180°$, and $270°$.

PREPARING FOR THIS SECTION

Before getting started, review the following:

✓ Unit Circle (Section 1.3, p. 27)   ✓ Functions (Section 1.4, pp. 33–41)

✓ Symmetry (Section 1.3, pp. 20–22)   ✓ Even and Odd Functions (Section 1.5, pp. 55–57)

2.5 PROPERTIES OF THE TRIGONOMETRIC FUNCTIONS; UNIT CIRCLE APPROACH

OBJECTIVES

1 Find the Exact Value of the Trigonometric Functions Using the Unit Circle

2 Know the Domain and Range of the Trigonometric Functions

3 Use the Periodic Properties to Find the Exact Value of the Trigonometric Functions

4 Use Even-Odd Properties to Find the Exact Value of the Trigonometric Functions

In this section, we develop important properties of the trigonometric functions. We begin by introducing the trigonometric functions using the unit circle. This approach will lead to the definition given earlier of the trigonometric functions of a general angle.

The Unit Circle

1 Recall that the unit circle is a circle whose radius is 1 and whose center is at the origin of a rectangular coordinate system. Also recall that any circle of radius r has circumference of length $2\pi r$. Therefore, the unit circle (radius $= 1$) has a circumference of length 2π. In other words, for 1 revolution around the unit circle the length of arc is 2π units.

The following discussion sets the stage for defining the trigonometric functions.

Let $t \geq 0$ be any real number and let s be the distance from the origin to t on the real number line. See the red portion of Figure 65(a). Now look at the unit circle in Figure 65(a). Beginning at the the point $(1, 0)$ on the unit circle, travel $s = t$ units in the counterclockwise direction along the circle to arrive at the point $P = (a, b)$. In this sense, the length $s = t$ units is being **wrapped** around the unit circle.

Figure 65

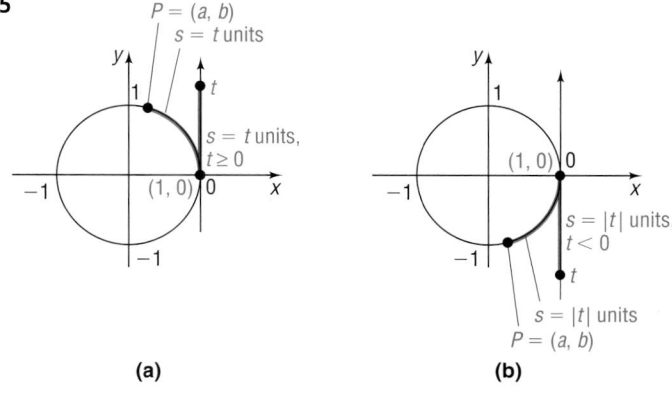

(a)   (b)

If $t < 0$, we begin at the point $(1, 0)$ on the unit circle and travel $s = |t|$ units in the clockwise direction to arrive at the point $P = (a, b)$. See Figure 65(b).

If $t > 2\pi$ or if $t < -2\pi$, it will be necessary to travel around the unit circle more than once before arriving at point P. Do you see why?

Let's describe this process another way. Picture a string of length $s = |t|$ units being wrapped around a circle of radius 1 unit. We start wrapping the string around the circle at the point $(1, 0)$. If $t \geq 0$, we wrap the string in the counterclockwise direction; if $t < 0$, we wrap the string in the clockwise direction. The point $P = (a, b)$ is the point where the string ends.

This discussion tells us that, for any real number t, we can locate a unique point $P = (a, b)$ on the unit circle. This point is called **the point P on the unit circle that corresponds to t.** This is the important idea here. No matter what real number t is chosen, there is a unique point P on the unit circle corresponding to it. We use the coordinates of the point $P = (a, b)$ on the unit circle corresponding to the real number t to define the **six trigonometric functions of t.**

Let t be a real number and let $P = (a, b)$ be the point on the unit circle that corresponds to t.

The **sine function** associates with t the y-coordinate of P and is denoted by

$$\sin t = b$$

The **cosine function** associates with t the x-coordinate of P and is denoted by

$$\cos t = a$$

If $a \neq 0$, the **tangent function** is defined as

$$\tan t = \frac{b}{a}$$

If $b \neq 0$, the **cosecant function** is defined as

$$\csc t = \frac{1}{b}$$

If $a \neq 0$, the **secant function** is defined as

$$\sec t = \frac{1}{a}$$

If $b \neq 0$, the **cotangent function** is defined as

$$\cot t = \frac{a}{b}$$

Once again, notice in these definitions that if $a = 0$ [that is, the point $P = (0, b)$ is on the y-axis] the tangent function and the secant function are undefined. Also, if $b = 0$ [that is, the point $P = (a, 0)$ is on the x-axis] the cosecant function and the cotangent function are undefined.

Because we use the unit circle in these definitions of the trigonometric functions, they are also sometimes referred to as **circular functions**.

EXAMPLE 1

Finding the Value of the Trigonometric Functions Using a Point on the Unit Circle

Figure 66

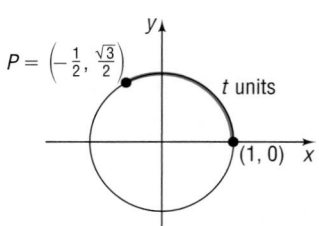

Find the value of $\sin t$, $\cos t$, $\tan t$, $\csc t$, $\sec t$, and $\cot t$ if $P = \left(-\frac{1}{2}, \frac{\sqrt{3}}{2}\right)$ is the point on the unit circle that corresponds to the real number t.

Solution   See Figure 66. We follow the definition of the six trigono-metric functions using $P = \left(-\frac{1}{2}, \frac{\sqrt{3}}{2}\right) = (a, b)$. Then, with $a = -\frac{1}{2}$, $b = \frac{\sqrt{3}}{2}$, we have

$$\sin t = b = \frac{\sqrt{3}}{2} \qquad \cos t = a = -\frac{1}{2} \qquad \tan t = \frac{b}{a} = \frac{\frac{\sqrt{3}}{2}}{-\frac{1}{2}} = -\sqrt{3}$$

$$\csc t = \frac{1}{b} = \frac{1}{\frac{\sqrt{3}}{2}} = \frac{2}{\sqrt{3}} = \frac{2\sqrt{3}}{3} \qquad \sec t = \frac{1}{a} = \frac{1}{-\frac{1}{2}} = -2 \qquad \cot t = \frac{a}{b} = \frac{-\frac{1}{2}}{\frac{\sqrt{3}}{2}} = -\frac{1}{\sqrt{3}} = -\frac{\sqrt{3}}{3}$$ ■

NOW WORK PROBLEM 1.

Trigonometric Functions of Angles

Let $P = (a, b)$ be the point on the unit circle corresponding to the real number t. See Figure 67(a). Let θ be the angle in standard position, measured in radians, whose terminal side is the ray from the origin through P. See Figure 67(b). Since the unit circle has radius 1 unit, from the formula for arc length, $s = r\theta$, we find that

$$s = r\theta = \theta$$
$$\uparrow$$
$$r = 1$$

So, if $s = |t|$ units, then $\theta = t$ radians. See Figures 67(c) and (d).

Figure 67

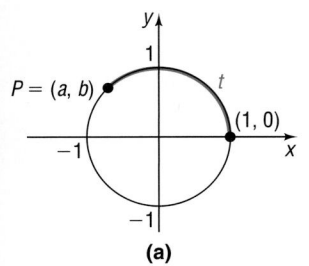
(a)

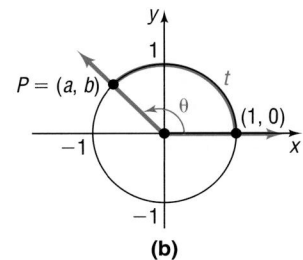
(b)

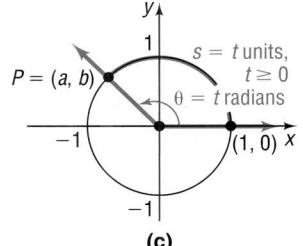
(c)

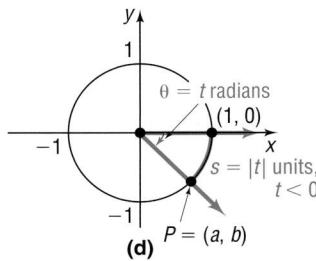
(d)

The point $P = (a, b)$ on the unit circle that corresponds to the real number t is the point P on the terminal side of the angle $\theta = t$ radians. As a result, we can say that

$$\sin t = \sin \theta$$
$$\uparrow \qquad \uparrow$$
Real number $\qquad \theta = t$ radians

and so on. We can now define the trigonometric functions of the angle θ.

If $\theta = t$ radians, the six **trigonometric functions of the angle θ** are defined as

$\sin \theta = \sin t$	$\cos \theta = \cos t$	$\tan \theta = \tan t$
$\csc \theta = \csc t$	$\sec \theta = \sec t$	$\cot \theta = \cot t$

Even though the distinction between trigonometric functions of real numbers and trigonometric functions of angles is important, it is customary to refer to trigonometric functions of real numbers and trigonometric functions of angles collectively as *the trigonometric functions*. We will follow this practice from now on.

Since the values of the trigonometric functions of an angle θ are determined by the coordinates of the point $P = (a, b)$ on the unit circle corresponding to θ, the units used to measure the angle θ are irrelevant. For example, it does not matter whether we write $\theta = \dfrac{\pi}{2}$ radians or $\theta = 90°$. The point on the unit circle corresponding to this angle is $P = (0, 1)$. Hence,

$$\sin \frac{\pi}{2} = \sin 90° = 1 \quad \text{and} \quad \cos \frac{\pi}{2} = \cos 90° = 0$$

To find the exact value of a trigonometric function of an angle θ requires that we locate the corresponding point $P^* = (a^*, b^*)$ on the unit circle. In fact, though, any circle whose center is at the origin can be used.

Let θ be any nonquadrantal angle placed in standard position. Let $P = (a, b)$ be the point on the circle $x^2 + y^2 = r^2$ that corresponds to θ. See Figure 68.

Notice that the triangles OA^*P^* and OAP are similar so that the ratios of the corresponding sides are equal.

Figure 68

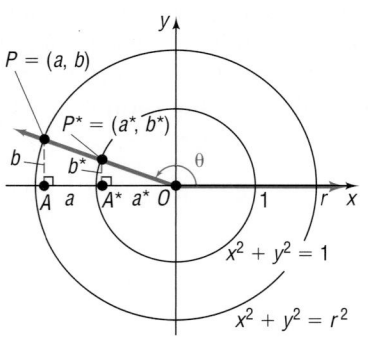

$$\frac{b^*}{1} = \frac{b}{r} \qquad \frac{a^*}{1} = \frac{a}{r} \qquad \frac{b^*}{a^*} = \frac{b}{a}$$

$$\frac{1}{b^*} = \frac{r}{b} \qquad \frac{1}{a^*} = \frac{r}{a} \qquad \frac{a^*}{b^*} = \frac{a}{b}$$

These results lead us to formulate the following theorem:

Theorem

For an angle θ in standard position, let $P = (a, b)$ be any point on the terminal side of θ that is also on the circle $x^2 + y^2 = r^2$. Then

$\sin \theta = \dfrac{b}{r}$	$\cos \theta = \dfrac{a}{r}$	$\tan \theta = \dfrac{b}{a}, \quad a \neq 0$
$\csc \theta = \dfrac{r}{b}, \quad b \neq 0$	$\sec \theta = \dfrac{r}{a}, \quad a \neq 0$	$\cot \theta = \dfrac{a}{b}, \quad b \neq 0$

This result coincides with the definition given in Section 2.4 for the six trigonometric functions of a general angle θ.

Figure 69

═══ ➤ **NOW WORK PROBLEM 7.**

Domain and Range of the Trigonometric Functions

② Let θ be an angle in standard position, and let $P = (a, b)$ be the point on the unit circle that corresponds to θ. See Figure 69. Then, by definition,

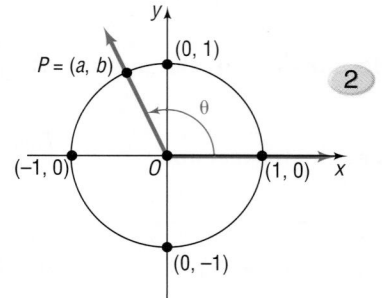

$\sin \theta = b$	$\cos \theta = a$	$\tan \theta = \dfrac{b}{a}, \quad a \neq 0$
$\csc \theta = \dfrac{1}{b}, \quad b \neq 0$	$\sec \theta = \dfrac{1}{a}, \quad a \neq 0$	$\cot \theta = \dfrac{a}{b}, \quad b \neq 0$

For $\sin\theta$ and $\cos\theta$, θ can be any angle, so it follows that the domain of the sine function and cosine function is the set of all real numbers.

> The domain of the sine function is the set of all real numbers.
>
> The domain of the cosine function is the set of all real numbers.

If $a = 0$, then the tangent function and the secant function are not defined. That is, for the tangent function and secant function, the x-coordinate of $P = (a, b)$ cannot be 0. On the unit circle, there are two such points $(0, 1)$ and $(0, -1)$. These two points correspond to the angles $\dfrac{\pi}{2}$ (90°) and $\dfrac{3\pi}{2}$ (270°) or, more generally, to any angle that is an odd multiple of $\dfrac{\pi}{2}$ (90°), such as $\dfrac{\pi}{2}$ (90°), $\dfrac{3\pi}{2}$ (270°), $\dfrac{5\pi}{2}$ (450°), $-\dfrac{\pi}{2}$ (−90°), and $-\dfrac{3\pi}{2}$ (−270°). Such angles must therefore be excluded from the domain of the tangent function and secant function.

> The domain of the tangent function is the set of all real numbers, except odd multiples of $\dfrac{\pi}{2}$ (90°).
>
> The domain of the secant function is the set of all real numbers, except odd multiples of $\dfrac{\pi}{2}$ (90°).

If $b = 0$, then the cotangent function and the cosecant function are not defined. For the cotangent function and cosecant function, the y-coordinate of $P = (a, b)$ cannot be 0. On the unit circle, there are two such points, $(1, 0)$ and $(-1, 0)$. These two points correspond to the angles 0 (0°) and π (180°) or, more generally, to any angle that is an integral multiple of π (180°), such as 0 (0°), π (180°), 2π (360°), 3π (540°), and $-\pi$ (−180°). Such angles must therefore be excluded from the domain of the cotangent function and cosecant function.

> The domain of the cotangent function is the set of all real numbers, except integral multiples of π (180°).
>
> The domain of the cosecant function is the set of all real numbers, except integral multiples of π (180°).

Next, we determine the range of each of the six trigonometric functions. Refer again to Figure 69. Let $P = (a, b)$ be the point on the unit circle that corresponds to the angle θ. It follows that $-1 \leq a \leq 1$ and $-1 \leq b \leq 1$. Since $\sin\theta = b$ and $\cos\theta = a$, we have

> $$-1 \leq \sin\theta \leq 1 \quad \text{and} \quad -1 \leq \cos\theta \leq 1$$

The range of both the sine function and the cosine function consists of all real numbers between −1 and 1, inclusive. Using absolute value notation, we have $|\sin\theta| \leq 1$ and $|\cos\theta| \leq 1$.

If θ is not a multiple of π $(180°)$, then $\csc\theta = \dfrac{1}{b}$. Since $b = \sin\theta$ and $|b| = |\sin\theta| \leq 1$, it follows that $|\csc\theta| = \dfrac{1}{|\sin\theta|} = \dfrac{1}{|b|} \geq 1$. The range of the cosecant function consists of all real numbers less than or equal to -1 or greater than or equal to 1. That is,

$$\csc\theta \leq -1 \quad \text{or} \quad \csc\theta \geq 1$$

Using absolute value notation, we have $|\csc\theta| \geq 1$.

If θ is not an odd multiple of $\dfrac{\pi}{2}$ $(90°)$, then, by definition, $\sec\theta = \dfrac{1}{a}$. Since $a = \cos\theta$ and $|a| = |\cos\theta| \leq 1$, it follows that $|\sec\theta| = \dfrac{1}{|\cos\theta|} = \dfrac{1}{|a|} \geq 1$. The range of the secant function consists of all real numbers less than or equal to -1 or greater than or equal to 1. That is,

$$\sec\theta \leq -1 \quad \text{or} \quad \sec\theta \geq 1$$

Using absolute value notation, we have $|\sec\theta| \geq 1$.

The range of both the tangent function and the cotangent function consists of all real numbers. That is,

$$-\infty < \tan\theta < \infty \quad \text{and} \quad -\infty < \cot\theta < \infty$$

You are asked to prove this in Problems 81 and 82.
Table 6 summarizes these results.

TABLE 6

Function	Symbol	Domain	Range
sine	$f(\theta) = \sin\theta$	All real numbers	All real numbers from -1 to 1, inclusive
cosine	$f(\theta) = \cos\theta$	All real numbers	All real numbers from -1 to 1, inclusive
tangent	$f(\theta) = \tan\theta$	All real numbers, except odd multiples of $\dfrac{\pi}{2}$ $(90°)$	All real numbers
cosecant	$f(\theta) = \csc\theta$	All real numbers, except integral multiples of π $(180°)$	All real numbers greater than or equal to 1 or less than or equal to -1
secant	$f(\theta) = \sec\theta$	All real numbers, except odd multiples of $\dfrac{\pi}{2}$ $(90°)$	All real numbers greater than or equal to 1 or less than or equal to -1
cotangent	$f(\theta) = \cot\theta$	All real numbers, except integral multiples of $\pi(180°)$	All real numbers

NOW WORK PROBLEMS 53 AND 57.

Period of the Trigonometric Functions

Figure 70

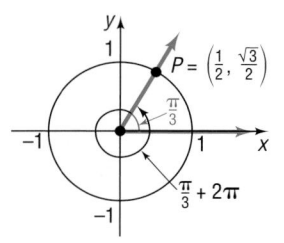

③ Look at Figure 70. This figure shows that for an angle of $\dfrac{\pi}{3}$ radians the corresponding point P on the unit circle is $\left(\dfrac{1}{2}, \dfrac{\sqrt{3}}{2}\right)$. Notice that for an angle of $\dfrac{\pi}{3} + 2\pi$ radians the corresponding point P on the unit circle is also $\left(\dfrac{1}{2}, \dfrac{\sqrt{3}}{2}\right)$. Thus,

$$\sin\frac{\pi}{3} = \frac{\sqrt{3}}{2} \quad \text{and} \quad \sin\left(\frac{\pi}{3} + 2\pi\right) = \frac{\sqrt{3}}{2}$$

$$\cos\frac{\pi}{3} = \frac{1}{2} \quad \text{and} \quad \cos\left(\frac{\pi}{3} + 2\pi\right) = \frac{1}{2}$$

This example illustrates a more general situation. For a given angle θ, measured in radians, suppose that we know the corresponding point $P = (a, b)$ on the unit circle. Now add 2π to θ. The point on the unit circle corresponding to $\theta + 2\pi$ is identical to the point P on the unit circle corresponding to θ. See Figure 71. The values of the trigonometric functions of $\theta + 2\pi$ are equal to the values of the corresponding trigonometric functions of θ.

Figure 71

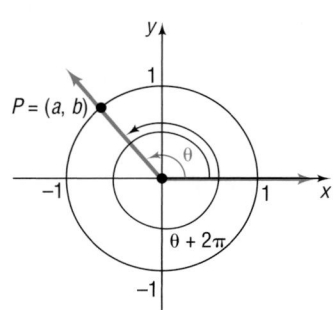

If we add (or subtract) integral multiples of 2π to θ, the trigonometric values remain unchanged. That is, for all θ,

$$\sin(\theta + 2\pi k) = \sin\theta \qquad \cos(\theta + 2\pi k) = \cos\theta \qquad (1)$$
$$\text{where } k \text{ is any integer}$$

Functions that exhibit this kind of behavior are called *periodic functions*.

> A function f is called **periodic** if there is a positive number p such that, whenever θ is in the domain of f, so is $\theta + p$, and
>
> $$f(\theta + p) = f(\theta)$$

If there is a smallest such number p, this smallest value is called the **(fundamental) period** of f.

Based on equation (1), the sine and cosine functions are periodic. In fact, the sine and cosine functions have period 2π. You are asked to prove this fact in Problems 83 and 84. The secant and cosecant functions are also periodic with period 2π; and the tangent and cotangent functions are periodic with period π. You are asked to prove these statements in Problems 85 through 88. These facts are summarized as follows:

Periodic Properties

$$\sin(\theta + 2\pi) = \sin\theta \quad \cos(\theta + 2\pi) = \cos\theta \quad \tan(\theta + \pi) = \tan\theta$$
$$\csc(\theta + 2\pi) = \csc\theta \quad \sec(\theta + 2\pi) = \sec\theta \quad \cot(\theta + \pi) = \cot\theta$$

Because the sine, cosine, secant, and cosecant functions have period 2π, once we know their values for $0 \le \theta < 2\pi$, we know all their values; similarly, since the tangent and cotangent functions have period π, once we know their values for $0 \le \theta < \pi$, we know all their values.

EXAMPLE 2   **Using Periodic Properties to Find Exact Values**

Find the exact value of: (a) $\sin 420°$ (b) $\tan \dfrac{5\pi}{4}$ (c) $\cos \dfrac{11\pi}{4}$

Solution   (a) $\sin 420° = \sin(360° + 60°) = \sin 60° = \dfrac{\sqrt{3}}{2}$

(b) $\tan \dfrac{5\pi}{4} = \tan\left(\pi + \dfrac{\pi}{4}\right) = \tan \dfrac{\pi}{4} = 1$

(c) $\cos \dfrac{11\pi}{4} = \cos\left(\dfrac{3\pi}{4} + \dfrac{8\pi}{4}\right) = \cos\left(\dfrac{3\pi}{4} + 2\pi\right) = \cos \dfrac{3\pi}{4} = -\dfrac{\sqrt{2}}{2}$ ∎

NOW WORK PROBLEMS 13 AND 71.

The periodic properties of the trigonometric functions will be very helpful to us when we study their graphs in the next section.

Even–Odd Properties

④   Recall that a function f is even if $f(-\theta) = f(\theta)$ for all θ in the domain of f; a function f is odd if $f(-\theta) = -f(\theta)$ for all θ in the domain of f. We will now show that the trigonometric functions sine, tangent, cotangent, and cosecant are odd functions, whereas the functions cosine and secant are even functions.

Theorem   **Even–Odd Properties**

$$\sin(-\theta) = -\sin \theta \qquad \cos(-\theta) = \cos \theta \qquad \tan(-\theta) = -\tan \theta$$
$$\csc(-\theta) = -\csc \theta \qquad \sec(-\theta) = \sec \theta \qquad \cot(-\theta) = -\cot \theta$$

∎

Figure 72

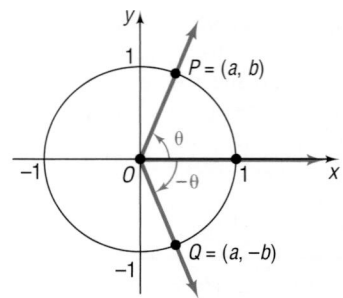

Proof   Let $P = (a, b)$ be the point on the unit circle that corresponds to the angle θ. See Figure 72. The point Q on the unit circle that corresponds to the angle $-\theta$ will have coordinates $(a, -b)$. Using the definition for the trigonometric functions, we have

$$\sin \theta = b \qquad \cos \theta = a \qquad \sin(-\theta) = -b \qquad \cos(-\theta) = a$$

so

$$\sin(-\theta) = -b = -\sin \theta \qquad \cos(-\theta) = a = \cos \theta$$

Now, using these results and some of the Fundamental Identities, we have

$$\tan(-\theta) = \frac{\sin(-\theta)}{\cos(-\theta)} = \frac{-\sin \theta}{\cos \theta} = -\tan \theta$$

$$\cot(-\theta) = \frac{1}{\tan(-\theta)} = \frac{1}{-\tan \theta} = -\cot \theta$$

$$\sec(-\theta) = \frac{1}{\cos(-\theta)} = \frac{1}{\cos\theta} = \sec\theta$$

$$\csc(-\theta) = \frac{1}{\sin(-\theta)} = \frac{1}{-\sin\theta} = -\csc\theta$$

EXAMPLE 3   **Finding Exact Values Using Even–Odd Properties**

Find the exact value of:

(a) $\sin(-45°)$ (b) $\cos(-\pi)$ (c) $\cot\left(-\dfrac{3\pi}{2}\right)$ (d) $\tan\left(-\dfrac{37\pi}{4}\right)$

Solution   (a) $\sin(-45°) \underset{\uparrow}{=} -\sin 45° = -\dfrac{\sqrt{2}}{2}$ (b) $\cos(-\pi) \underset{\uparrow}{=} \cos\pi = -1$

Odd function   *Even function*

(c) $\cot\left(-\dfrac{3\pi}{2}\right) \underset{\uparrow}{=} -\cot\dfrac{3\pi}{2} = 0$

Odd function

(d) $\tan\left(-\dfrac{37\pi}{4}\right) \underset{\uparrow}{=} -\tan\dfrac{37\pi}{4} = -\tan\left(\dfrac{\pi}{4} + 9\pi\right) \underset{\uparrow}{=} -\tan\dfrac{\pi}{4} = -1$

Odd function   *Period is π*

SEEING THE CONCEPT   To see that the cosine function is even, use a graphing utility to graph $Y_1 = \cos x$. Do you see the symmetry? Why would you conjecture that the cosine function is even? Clear the screen. Now graph $Y_1 = \sin x$. Do you see the symmetry? Do you conjecture that the sine function is odd?

NOW WORK PROBLEMS 29 AND 65.

2.5 Concepts and Vocabulary

In Problems 1–3, fill in the blanks.

1. The sine, cosine, cosecant, and secant functions have period _____; the tangent and cotangent functions have period _____.

2. The domain of the tangent function is _____.

3. The range of the sine function is _____.

In Problems 4–6, answer True or False to each statement.

4. The only even trigonometric functions are the cosine and secant functions.

5. All the trigonometric functions are periodic, with period 2π.

6. The range of the secant function is the set of positive real numbers.

7. Explain how you would find the value of $\sin 390°$ using periodic properties.

8. Explain how you would find the value of $\cos(-45°)$ using even–odd properties.

2.5 Exercises

In Problems 1–6, the point P on the unit circle that corresponds to a real number t is given. Find $\sin t$, $\cos t$, $\tan t$, $\csc t$, $\sec t$, *and* $\cot t$.

1. $\left(\dfrac{\sqrt{3}}{2}, -\dfrac{1}{2}\right)$ **2.** $\left(-\dfrac{\sqrt{3}}{2}, -\dfrac{1}{2}\right)$ **3.** $\left(-\dfrac{\sqrt{2}}{2}, -\dfrac{\sqrt{2}}{2}\right)$

4. $\left(\dfrac{\sqrt{2}}{2}, -\dfrac{\sqrt{2}}{2}\right)$ **5.** $\left(\dfrac{\sqrt{5}}{3}, \dfrac{2}{3}\right)$ **6.** $\left(-\dfrac{\sqrt{5}}{5}, \dfrac{2\sqrt{5}}{5}\right)$

In Problems 7–12, the point P on the circle $x^2 + y^2 = r^2$ *that is also on the terminal side of an angle* θ *in standard position is given. Find* $\sin\theta$, $\cos\theta$, $\tan\theta$, $\csc\theta$, $\sec\theta$, *and* $\cot\theta$.

7. $(3, -4)$ **8.** $(4, -3)$ **9.** $(-2, 3)$ **10.** $(2, -4)$ **11.** $(-1, -1)$ **12.** $(-3, 1)$

In Problems 13–28, use the fact that the trigonometric functions are periodic to find the exact value of each expression. Do not use a calculator.

13. $\sin 405°$ **14.** $\cos 420°$ **15.** $\tan 405°$ **16.** $\sin 390°$

17. $\csc 450°$ **18.** $\sec 540°$ **19.** $\cot 390°$ **20.** $\sec 420°$

21. $\cos\dfrac{33\pi}{4}$ **22.** $\sin\dfrac{9\pi}{4}$ **23.** $\tan(21\pi)$ **24.** $\csc\dfrac{9\pi}{2}$

25. $\sec\dfrac{17\pi}{4}$ **26.** $\cot\dfrac{17\pi}{4}$ **27.** $\tan\dfrac{19\pi}{6}$ **28.** $\sec\dfrac{25\pi}{6}$

In Problems 29–46, use the even-odd properties to find the exact value of each expression. Do not use a calculator.

29. $\sin(-60°)$ **30.** $\cos(-30°)$ **31.** $\tan(-30°)$ **32.** $\sin(-135°)$

33. $\sec(-60°)$ **34.** $\csc(-30°)$ **35.** $\sin(-90°)$ **36.** $\cos(-270°)$

37. $\tan\left(-\dfrac{\pi}{4}\right)$ **38.** $\sin(-\pi)$ **39.** $\cos\left(-\dfrac{\pi}{4}\right)$ **40.** $\sin\left(-\dfrac{\pi}{3}\right)$

41. $\tan(-\pi)$ **42.** $\sin\left(-\dfrac{3\pi}{2}\right)$ **43.** $\csc\left(-\dfrac{\pi}{4}\right)$ **44.** $\sec(-\pi)$

45. $\sec\left(-\dfrac{\pi}{6}\right)$ **46.** $\csc\left(-\dfrac{\pi}{3}\right)$

In Problems 47–52, find the exact value of each expression. Do not use a calculator.

47. $\sin(-\pi) + \cos(5\pi)$ **48.** $\tan\left(-\dfrac{5\pi}{6}\right) - \cot\dfrac{7\pi}{2}$ **49.** $\sec(-\pi) + \csc\left(-\dfrac{\pi}{2}\right)$

50. $\tan(-6\pi) + \cos\dfrac{9\pi}{4}$ **51.** $\sin\left(-\dfrac{9\pi}{4}\right) - \tan\left(-\dfrac{9\pi}{4}\right)$ **52.** $\cos\left(-\dfrac{17\pi}{4}\right) - \sin\left(-\dfrac{3\pi}{2}\right)$

53. What is the domain of the sine function?   **54.** What is the domain of the cosine function?

55. For what numbers θ is $f(\theta) = \tan\theta$ not defined?   **56.** For what numbers θ is $f(\theta) = \cot\theta$ not defined?

57. For what numbers θ is $f(\theta) = \sec\theta$ not defined?   **58.** For what numbers θ is $f(\theta) = \csc\theta$ not defined?

59. What is the range of the sine function?   **60.** What is the range of the cosine function?

61. What is the range of the tangent function?   **62.** What is the range of the cotangent function?

63. What is the range of the secant function?   **64.** What is the range of the cosecant function?

65. Is the sine function even, odd, or neither? Is its graph symmetric? With respect to what?

66. Is the cosine function even, odd, or neither? Is its graph symmetric? With respect to what?

67. Is the tangent function even, odd, or neither? Is its graph symmetric? With respect to what?

68. Is the cotangent function even, odd, or neither? Is its graph symmetric? With respect to what?

69. Is the secant function even, odd, or neither? Is its graph symmetric? With respect to what?

70. Is the cosecant function even, odd, or neither? Is its graph symmetric? With respect to what?

71. If $\sin \theta = 0.3$, find the value of

$$\sin \theta + \sin(\theta + 2\pi) + \sin(\theta + 4\pi)$$

72. If $\cos \theta = 0.2$, find the value of

$$\cos \theta + \cos(\theta + 2\pi) + \cos(\theta + 4\pi)$$

73. If $\tan \theta = 3$, find the value of

$$\tan \theta + \tan(\theta + \pi) + \tan(\theta + 2\pi).$$

74. If $\cot \theta = -2$, find the value of

$$\cot \theta + \cot(\theta - \pi) + \cot(\theta - 2\pi).$$

In Problems 75–80, use the periodic and even-odd properties.

75. If $f(x) = \sin x$ and $f(a) = \dfrac{1}{3}$, find the exact value of:
 (a) $f(-a)$ (b) $f(a) + f(a + 2\pi) + f(a + 4\pi)$

76. If $f(x) = \cos x$ and $f(a) = \dfrac{1}{4}$, find the exact value of:
 (a) $f(-a)$ (b) $f(a) + f(a + 2\pi) + f(a - 2\pi)$

77. If $f(x) = \tan x$ and $f(a) = 2$, find the exact value of:
 (a) $f(-a)$ (b) $f(a) + f(a + \pi) + f(a + 2\pi)$

78. If $f(x) = \cot x$ and $f(a) = -3$, find the exact value of:
 (a) $f(-a)$ (b) $f(a) + f(a + \pi) + f(a + 4\pi)$

79. If $f(x) = \sec x$ and $f(a) = -4$, find the exact value of:
 (a) $f(-a)$ (b) $f(a) + f(a + 2\pi) + f(a + 4\pi)$

80. If $f(x) = \csc x$ and $f(a) = 2$, find the exact value of:
 (a) $f(-a)$ (b) $f(a) + f(a + 2\pi) + f(a + 4\pi)$

81. Show that the range of the tangent function is the set of all real numbers.

82. Show that the range of the cotangent function is the set of all real numbers.

83. Show that the period of $f(\theta) = \sin \theta$ is 2π.

 [**Hint:** Assume that $0 < p < 2\pi$ exists so that $\sin(\theta + p) = \sin \theta$ for all θ. Let $\theta = 0$ to find p. Then let $\theta = \dfrac{\pi}{2}$ to obtain a contradiction.]

84. Show that the period of $f(\theta) = \cos \theta$ is 2π.

85. Show that the period of $f(\theta) = \sec \theta$ is 2π.

86. Show that the period of $f(\theta) = \csc \theta$ is 2π.

87. Show that the period of $f(\theta) = \tan \theta$ is π.

88. Show that the period of $f(\theta) = \cot \theta$ is π.

89. If $\theta\,(0 < \theta < \pi)$ is the angle between a horizontal ray directed to the right (say, the positive x-axis) and a non-horizontal, nonvertical line L, show that the slope m of L equals $\tan \theta$. The angle θ is called the **inclination** of L.
 [**Hint:** See the illustration, where we have drawn the line L^* parallel to L and passing through the origin. Use the fact that L^* intersects the unit circle at the point $(\cos \theta, \sin \theta)$.]

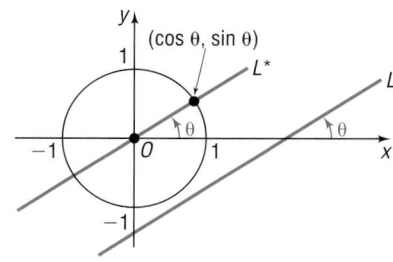

90. Write down five properties of the tangent function. Explain the meaning of each.

91. Describe your understanding of the meaning of a periodic function.

PREPARING FOR THIS SECTION

Before getting started, review the following:

✓ Functions (Section 1.4, pp. 41–44)

✓ Graphing Techniques: Transformations
(Section 1.6, pp. 64–73)

2.6 GRAPHS OF THE SINE AND COSINE FUNCTIONS

OBJECTIVES

1. Graph Transformations of the Sine Function
2. Graph Transformations of the Cosine Function
3. Determine the Amplitude and Period of Sinusoidal Functions
4. Graph Sinusoidal Functions: $y = A \sin(\omega x)$
5. Find an Equation for a Sinusoidal Graph

Since we want to graph the trigonometric functions in the xy-plane, we shall use the traditional symbols x for the independent variable (or argument) and y for the dependent variable (or value at x) for each function. So we write the six trigonometric functions as

$$
\begin{array}{lll}
y = f(x) = \sin x & y = f(x) = \cos x & y = f(x) = \tan x \\
y = f(x) = \csc x & y = f(x) = \sec x & y = f(x) = \cot x
\end{array}
$$

 Here the independent variable x represents an angle, measured in radians. In calculus, x will usually be treated as a real number. As we said earlier, these are equivalent ways of viewing x.

The Graph of $y = \sin x$

Since the sine function has period 2π, we need to graph $y = \sin x$ only on the interval $[0, 2\pi]$. The remainder of the graph will consist of repetitions of this portion of the graph.

We begin by constructing Table 7, which lists some points on the graph of $y = \sin x$, $0 \le x \le 2\pi$. As the table shows, the graph of $y = \sin x$, $0 \le x \le 2\pi$, begins at the origin. As x increases from 0 to $\dfrac{\pi}{2}$, the value of $y = \sin x$ increases from 0 to 1; as x increases from $\dfrac{\pi}{2}$ to π to $\dfrac{3\pi}{2}$, the value of y decreases from 1 to 0 to -1; as x increases from $\dfrac{3\pi}{2}$ to 2π, the value of y increases from -1 to 0. If we plot the points listed in Table 7 and connect them with a smooth curve, we obtain the graph shown in Figure 73.

TABLE 7

x	$y = \sin x$	(x, y)
0	0	$(0, 0)$
$\dfrac{\pi}{6}$	$\dfrac{1}{2}$	$\left(\dfrac{\pi}{6}, \dfrac{1}{2}\right)$
$\dfrac{\pi}{2}$	1	$\left(\dfrac{\pi}{2}, 1\right)$
$\dfrac{5\pi}{6}$	$\dfrac{1}{2}$	$\left(\dfrac{5\pi}{6}, \dfrac{1}{2}\right)$
π	0	$(\pi, 0)$
$\dfrac{7\pi}{6}$	$-\dfrac{1}{2}$	$\left(\dfrac{7\pi}{6}, -\dfrac{1}{2}\right)$
$\dfrac{3\pi}{2}$	-1	$\left(\dfrac{3\pi}{2}, -1\right)$
$\dfrac{11\pi}{6}$	$-\dfrac{1}{2}$	$\left(\dfrac{11\pi}{6}, -\dfrac{1}{2}\right)$
2π	0	$(2\pi, 0)$

Figure 73

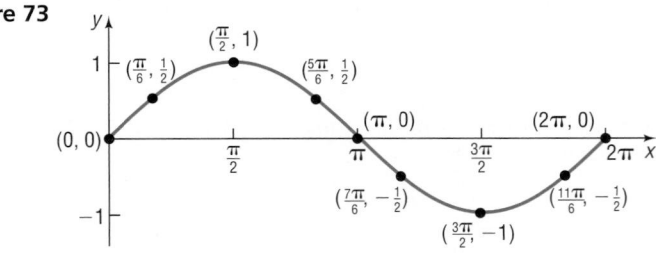

The graph in Figure 73 is one period, or **cycle**, of the graph of $y = \sin x$. To obtain a more complete graph of $y = \sin x$, we repeat this period in each direction, as shown in Figure 74(a). Figure 74(b) shows the graph on a TI-83 graphing calculator.

Figure 74
$y = \sin x,\ -\infty < x < \infty$

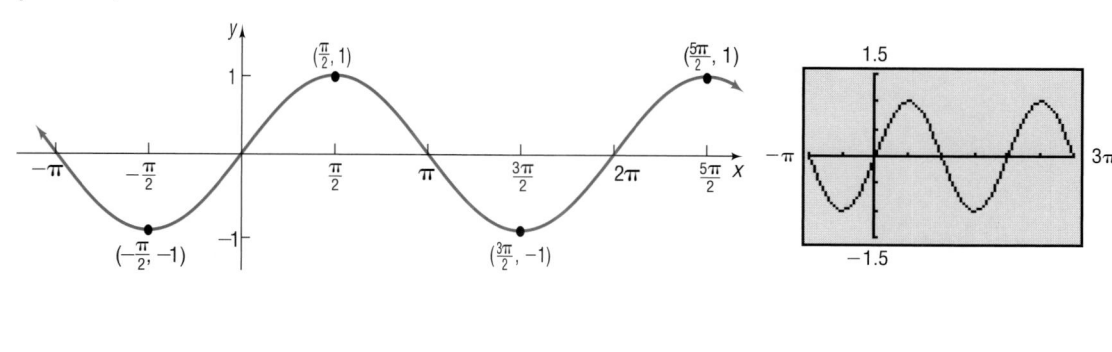

(a)   (b)

The graph of $y = \sin x$ illustrates some of the facts that we already know about the sine function.

Properties of the Sine Function

1. The domain is the set of all real numbers.
2. The range consists of all real numbers from -1 to 1, inclusive.
3. The sine function is an odd function, as the symmetry of the graph with respect to the origin indicates.
4. The sine function is periodic, with period 2π.
5. The x-intercepts are $\ldots, -2\pi, -\pi, 0, \pi, 2\pi, 3\pi, \ldots$; the y-intercept is 0.
6. The maximum value is 1 and occurs at $x = \ldots, -\dfrac{3\pi}{2}, \dfrac{\pi}{2}, \dfrac{5\pi}{2},$ $\dfrac{9\pi}{2}, \ldots$; the minimum value is -1 and occurs at $x = \ldots, -\dfrac{\pi}{2},$ $\dfrac{3\pi}{2}, \dfrac{7\pi}{2}, \dfrac{11\pi}{2}, \ldots.$

NOW WORK PROBLEMS 1, 3, AND 5.

1   The graphing techniques introduced in Chapter 1 may be used to graph functions that are transformations of the sine function (refer to Section 1.6).

EXAMPLE 1   **Graphing Variations of $y = \sin x$ Using Transformations**

Use the graph of $y = \sin x$ to graph $y = \sin\left(x - \dfrac{\pi}{4}\right)$.

Solution   Figure 75 illustrates the steps.

Figure 75

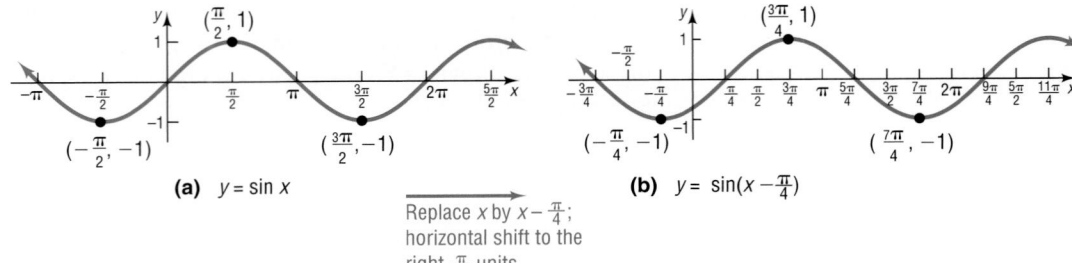

(a) $y = \sin x$

Replace x by $x - \frac{\pi}{4}$;
horizontal shift to the
right $\frac{\pi}{4}$ units.

(b) $y = \sin(x - \frac{\pi}{4})$

✔ CHECK: Figure 76 shows the graph of $Y_1 = \sin\left(x - \dfrac{\pi}{4}\right)$.

Figure 76

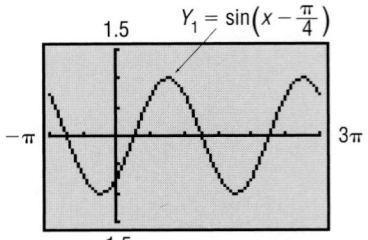

EXAMPLE 2   **Graphing Variations of $y = \sin x$ Using Transformations**

Use the graph of $y = \sin x$ to graph $y = -\sin x + 2$.

Solution   Figure 77 illustrates the steps.

Figure 77

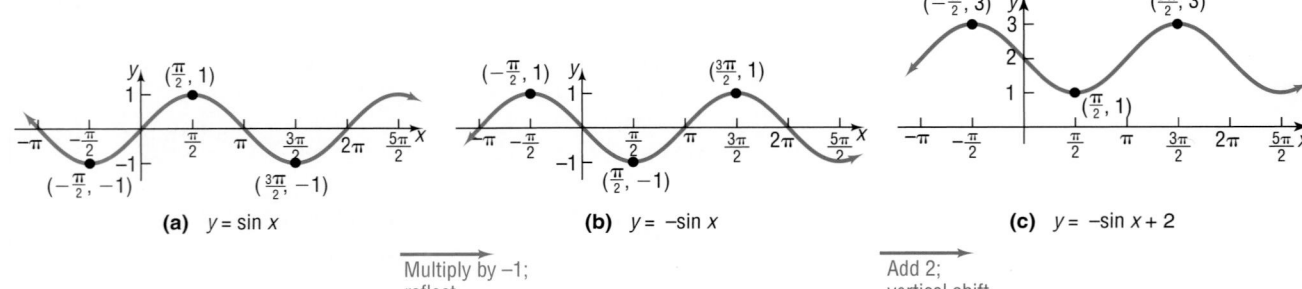

(a) $y = \sin x$

Multiply by -1;
reflect
about x-axis.

(b) $y = -\sin x$

Add 2;
vertical shift.

(c) $y = -\sin x + 2$

✔ CHECK: Figure 78 shows the graph of $Y_1 = -\sin x + 2$.

Figure 78

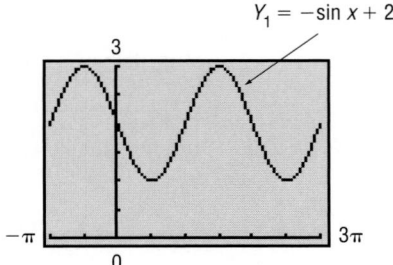

NOW WORK PROBLEM **17.**

TABLE 8

x	$y = \cos x$	(x, y)
0	1	$(0, 1)$
$\dfrac{\pi}{3}$	$\dfrac{1}{2}$	$\left(\dfrac{\pi}{3}, \dfrac{1}{2}\right)$
$\dfrac{\pi}{2}$	0	$\left(\dfrac{\pi}{2}, 0\right)$
$\dfrac{2\pi}{3}$	$-\dfrac{1}{2}$	$\left(\dfrac{2\pi}{3}, -\dfrac{1}{2}\right)$
π	-1	$(\pi, -1)$
$\dfrac{4\pi}{3}$	$-\dfrac{1}{2}$	$\left(\dfrac{4\pi}{3}, -\dfrac{1}{2}\right)$
$\dfrac{3\pi}{2}$	0	$\left(\dfrac{3\pi}{2}, 0\right)$
$\dfrac{5\pi}{3}$	$\dfrac{1}{2}$	$\left(\dfrac{5\pi}{3}, \dfrac{1}{2}\right)$
2π	1	$(2\pi, 1)$

The Graph of $y = \cos x$

The cosine function also has period 2π. We proceed as we did with the sine function by constructing Table 8, which lists some points on the graph of $y = \cos x$, $0 \le x \le 2\pi$. As the table shows, the graph of $y = \cos x$, $0 \le x \le 2\pi$, begins at the point $(0, 1)$. As x increases from 0 to $\dfrac{\pi}{2}$ to π, the value of y decreases from 1 to 0 to -1; as x increases from π to $\dfrac{3\pi}{2}$ to 2π, the value of y increases from -1 to 0 to 1. As before, we plot the points in Table 8 to get one period or cycle of the graph. See Figure 79.

Figure 79
$y = \cos x$, $0 \le x \le 2\pi$

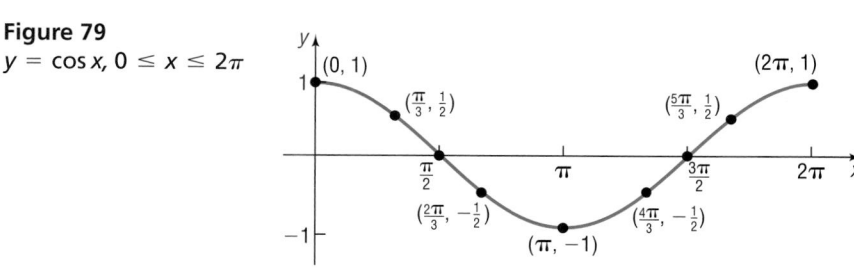

A more complete graph of $y = \cos x$ is obtained by repeating this period in each direction, as shown in Figure 80(a). Figure 80(b) shows the graph on a TI-83 graphing calculator.

Figure 80
$y = \cos x$, $-\infty < x < \infty$

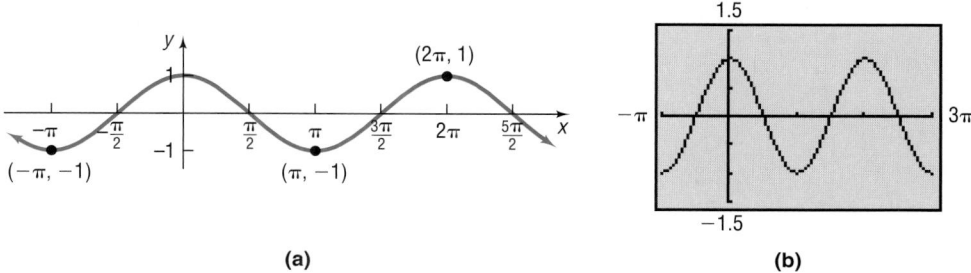

(a)   (b)

The graph of $y = \cos x$ illustrates some of the facts that we already know about the cosine function.

Properties of the Cosine Function

1. The domain is the set of all real numbers.
2. The range consists of all real numbers from -1 to 1, inclusive.
3. The cosine function is an even function, as the symmetry of the graph with respect to the y-axis indicates.
4. The cosine function is periodic, with period 2π.
5. The x-intercepts are $\ldots, -\dfrac{3\pi}{2}, -\dfrac{\pi}{2}, \dfrac{\pi}{2}, \dfrac{3\pi}{2}, \dfrac{5\pi}{2}, \ldots$; the y-intercept is 1.
6. The maximum value is 1 and occurs at $x = \ldots, -2\pi, 0, 2\pi, 4\pi, 6\pi, \ldots$; the minimum value is -1 and occurs at $x = \ldots, -\pi, \pi, 3\pi, 5\pi, \ldots$.

②   Again, the graphing techniques from Chapter 1 may be used to graph transformations of the cosine function.

EXAMPLE 3	**Graphing Variations of $y = \cos x$ Using Transformations**

Use the graph of $y = \cos x$ to graph $y = 2 \cos x$.

Solution   Figure 81 illustrates the steps.

Figure 81

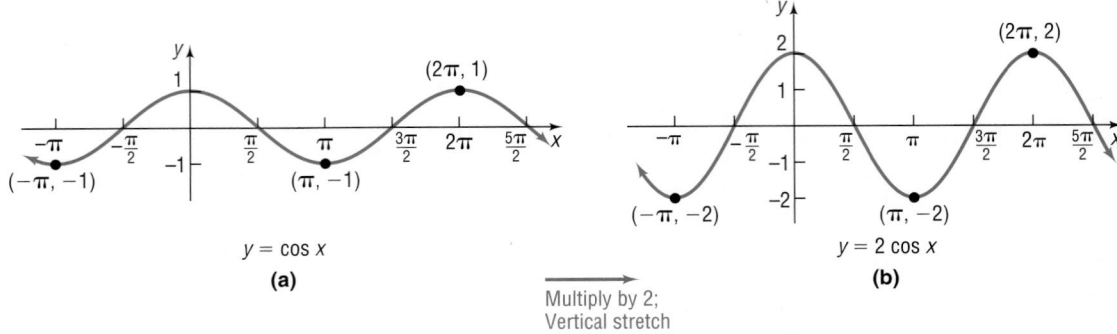

$y = \cos x$
(a)

Multiply by 2;
Vertical stretch
by a factor of 2.

$y = 2 \cos x$
(b)

✔ CHECK: Figure 82 shows the graph of $Y_1 = 2 \cos x$.

Figure 82

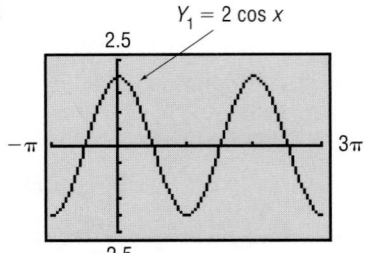

■ ■

EXAMPLE 4	**Graphing Variations of $y = \cos x$ Using Transformations**

Use the graph of $y = \cos x$ to graph $y = \cos(3x)$.

Solution   Figure 83 illustrates the graph, which is a horizontal compression of the graph of $y = \cos x$. (Multiply each x-coordinate by $\frac{1}{3}$.) Notice that, due to this compression, the period of $y = \cos(3x)$ is $\frac{2\pi}{3}$, whereas the period of $y = \cos x$ is 2π.

Figure 83

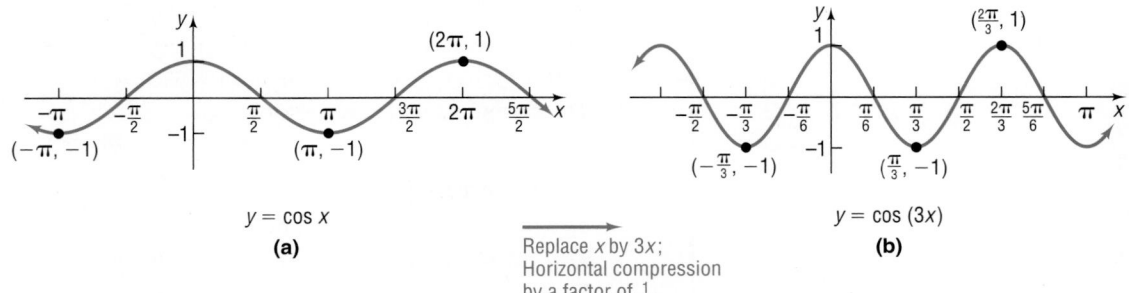

$y = \cos x$
(a)

Replace x by $3x$;
Horizontal compression
by a factor of $\frac{1}{3}$.

$y = \cos(3x)$
(b)

✔ CHECK: Figure 84 shows the graph of $Y_1 = \cos(3x)$.

Figure 84

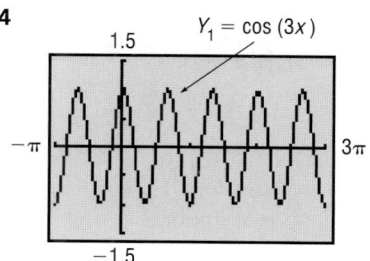

$Y_1 = \cos(3x)$

SEEING THE CONCEPT   Graph $Y_1 = \cos(3x)$ with $X\text{min} = 0$, $X\text{max} = \dfrac{2\pi}{3}$ and $X\text{scl} = \dfrac{\pi}{6}$ to verify that the period is $\dfrac{2\pi}{3}$.

▸ NOW WORK PROBLEM 25.

Sinusoidal Graphs

The graph of $y = \cos x$, when compared to the graph of $y = \sin x$, suggests that the graph of $y = \sin x$ is the same as the graph of $y = \cos\left(x - \dfrac{\pi}{2}\right)$. See Figure 85.

Figure 85

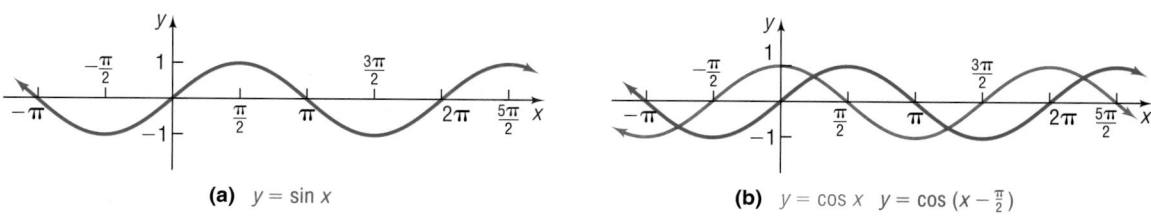

(a) $y = \sin x$ **(b)** $y = \cos x$ $y = \cos\left(x - \frac{\pi}{2}\right)$

Based on Figure 85, we conjecture that

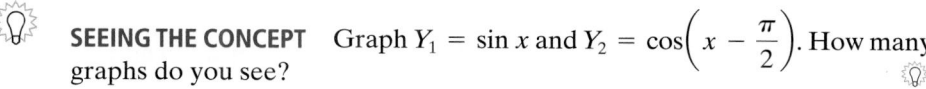

$$\sin x = \cos\left(x - \frac{\pi}{2}\right)$$

(We shall prove this fact in Chapter 3.) Because of this similarity, the graphs of sine functions and cosine functions are referred to as **sinusoidal graphs**.

SEEING THE CONCEPT   Graph $Y_1 = \sin x$ and $Y_2 = \cos\left(x - \dfrac{\pi}{2}\right)$. How many graphs do you see?

Let's look at some general properties of sinusoidal graphs.

③   In Example 3 we obtained the graph of $y = 2\cos x$, which we reproduce in Figure 86. Notice that the values of $y = 2\cos x$ lie between -2 and 2, inclusive.

Figure 86
$y = 2\cos x$

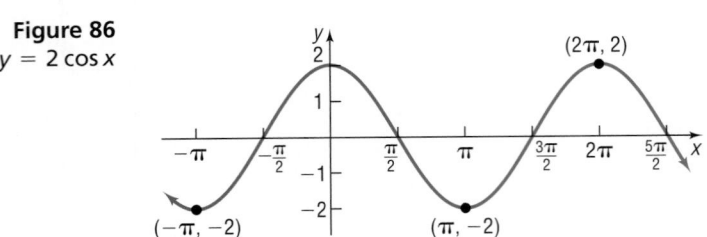

In general, the values of the functions $y = A\sin x$ and $y = A\cos x$, where $A \neq 0$, will always satisfy the inequalities

$$-|A| \leq A\sin x \leq |A| \quad \text{and} \quad -|A| \leq A\cos x \leq |A|$$

respectively. The number $|A|$ is called the **amplitude** of $y = A\sin x$ or $y = A\cos x$. See Figure 87.

Figure 87

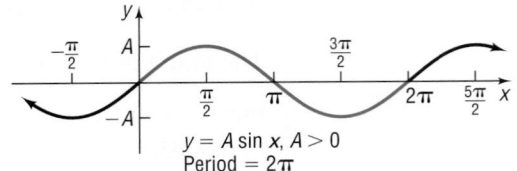

$y = A\sin x,\ A > 0$
Period $= 2\pi$

In Example 4, we obtained the graph of $y = \cos(3x)$, which we reproduce in Figure 88. Notice that the period of this function is $\dfrac{2\pi}{3}$.

Figure 88
$y = \cos(3x)$

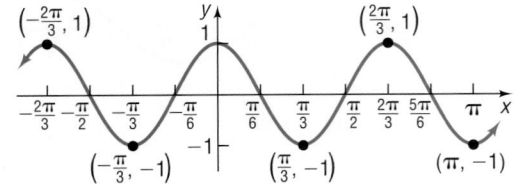

In general, if $\omega > 0$, the functions $y = \sin(\omega x)$ and $y = \cos(\omega x)$ will have period $T = \dfrac{2\pi}{\omega}$. To see why, recall that the graph of $y = \sin(\omega x)$ is obtained from the graph of $y = \sin x$ by performing a horizontal compression or stretch by a factor $\dfrac{1}{\omega}$. This horizontal compression replaces the interval $[0, 2\pi]$, which contains one period of the graph of $y = \sin x$, by the interval $\left[0, \dfrac{2\pi}{\omega}\right]$, which contains one period of the graph of $y = \sin(\omega x)$. The period of the functions $y = \sin(\omega x)$ and $y = \cos(\omega x)$, $\omega > 0$, is $\dfrac{2\pi}{\omega}$.

For example, for the function $y = \cos(3x)$, graphed in Figure 88, $\omega = 3$, so the period is $\dfrac{2\pi}{\omega} = \dfrac{2\pi}{3}$.

One period of the graph of $y = \sin(\omega x)$ or $y = \cos(\omega x)$ is called a **cycle**. Figure 89 illustrates the general situation. The blue portion of the graph is one cycle.

Figure 89

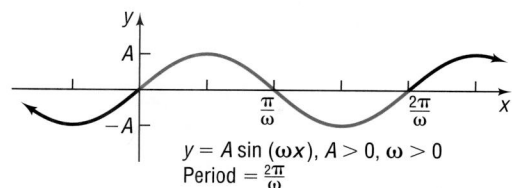

$y = A \sin(\omega x),\ A > 0,\ \omega > 0$
Period $= \frac{2\pi}{\omega}$

If $\omega < 0$ in $y = \sin(\omega x)$ or $y = \cos(\omega x)$, we use the even-odd properties of the sine and cosine functions as follows:

$$\sin(-\omega x) = -\sin(\omega x) \quad \text{and} \quad \cos(-\omega x) = \cos(\omega x)$$

This gives us an equivalent form in which the coefficient of x is positive. For example,

$$\sin(-2x) = -\sin(2x) \quad \text{and} \quad \cos(-\pi x) = \cos(\pi x)$$

Theorem

If $\omega > 0$, the amplitude and period of $y = A \sin(\omega x)$ and $y = A \cos(\omega x)$ are given by

$$\text{Amplitude} = |A| \qquad \text{Period} = T = \frac{2\pi}{\omega} \qquad (1)$$

EXAMPLE 5   **Finding the Amplitude and Period of a Sinusoidal Function**

Determine the amplitude and period of $y = 3\sin(4x)$.

Solution   Comparing $y = 3\sin(4x)$ to $y = A\sin(\omega x)$, we find that $A = 3$ and $\omega = 4$. From equation (1),

$$\text{Amplitude} = |A| = 3 \qquad \text{Period} = T = \frac{2\pi}{\omega} = \frac{2\pi}{4} = \frac{\pi}{2}$$

NOW WORK PROBLEM **33.**

④   Earlier, we graphed sine and cosine functions using transformations. We now introduce another method that can be used to graph these functions.

Figure 90 shows one cycle of the graphs of $y = \sin x$ and $y = \cos x$ on the interval $[0, 2\pi]$. Notice that each graph consists of four parts corresponding to the four subintervals:

$$\left[0, \frac{\pi}{2}\right], \quad \left[\frac{\pi}{2}, \pi\right], \quad \left[\pi, \frac{3\pi}{2}\right], \quad \left[\frac{3\pi}{2}, 2\pi\right]$$

Each of these subintervals is of length $\dfrac{\pi}{2}$ (the period 2π divided by 4) and the endpoints of these intervals give rise to five key points, as shown in Figure 90.

Figure 90

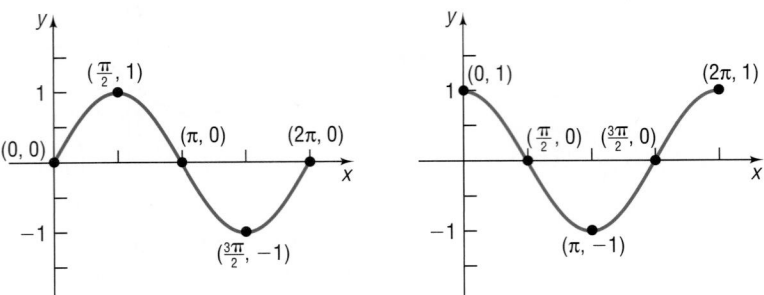

When graphing a sinusoidal function of the form $y = A\sin(\omega x)$ or $y = A\cos(\omega x)$ by hand, we use the amplitude to determine the maximum and minimum values of the function. The period is used to divide the x-axis into four subintervals. The endpoints of the subintervals give rise to five key points on the graph, which are used to sketch one cycle. Finally, extend the graph in either direction to make it complete.

To graph a sinusoidal function using a graphing utility, we use the amplitude to set Ymin and Ymax and use the period to set Xmin and Xmax.

Let's look at an example.

EXAMPLE 6   **Graphing a Sinusoidal Function**

Graph: $y = 3\sin(4x)$

Solution   From Example 5, the amplitude is 3 and the period is $\dfrac{\pi}{2}$. The graph of $y = 3\sin(4x)$ will lie between -3 and 3 on the y-axis. One cycle will begin at $x = 0$ and end at $x = \dfrac{\pi}{2}$.

We divide the interval $\left[0, \dfrac{\pi}{2}\right]$ into four subintervals, each of length $\dfrac{\pi}{2} \div 4 = \dfrac{\pi}{8}$:

$$\left[0, \frac{\pi}{8}\right], \quad \left[\frac{\pi}{8}, \frac{\pi}{4}\right], \quad \left[\frac{\pi}{4}, \frac{3\pi}{8}\right], \quad \left[\frac{3\pi}{8}, \frac{\pi}{2}\right]$$

The endpoints of these intervals give rise to five key points on the graph:

$$(0,0), \quad \left(\frac{\pi}{8}, 3\right), \quad \left(\frac{\pi}{4}, 0\right), \quad \left(\frac{3\pi}{8}, -3\right), \quad \left(\frac{\pi}{2}, 0\right)$$

We plot these five points and fill in the graph of the sine curve as shown in Figure 91(a). If we extend the graph in either direction, we obtain the complete graph shown in Figure 91(b).

Figure 91

Figure 91(c) shows the graph using a graphing utility.

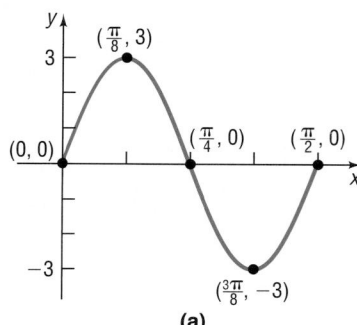

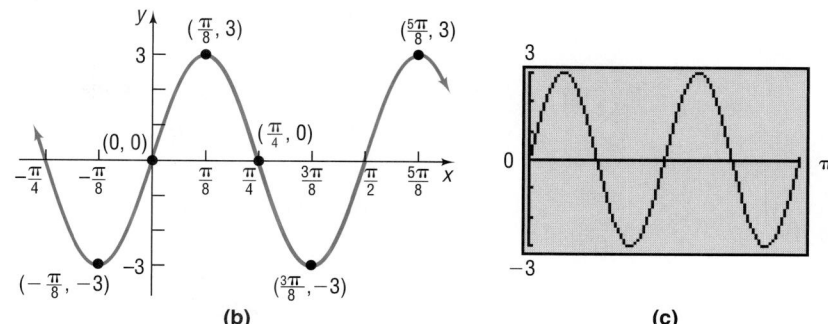

(a)   (b)   (c)

✔ CHECK: Graph $y = 3\sin(4x)$ by hand using transformations. Which graphing method do you prefer? ▪ ▪

━━ NOW WORK PROBLEM 39.

EXAMPLE 7 **Finding the Amplitude and Period of a Sinusoidal Function and Graphing It**

Determine the amplitude and period of $y = -4\cos(\pi x)$, and graph the function.

Solution Comparing $y = -4\cos(\pi x)$ with $y = A\cos(\omega x)$, we find that $A = -4$ and $\omega = \pi$. The amplitude is $|A| = |-4| = 4$, and the period is $T = \dfrac{2\pi}{\omega} = \dfrac{2\pi}{\pi} = 2$.

The graph of $y = -4\cos(\pi x)$ will lie between -4 and 4 on the y-axis. One cycle will begin at $x = 0$ and end at $x = 2$. We divide the interval $[0, 2]$ into four subintervals, each of length $2 \div 4 = \dfrac{1}{2}$:

$$\left[0, \frac{1}{2}\right], \quad \left[\frac{1}{2}, 1\right], \quad \left[1, \frac{3}{2}\right], \quad \left[\frac{3}{2}, 2\right].$$

The five key points on the graph are

$$(0, -4), \quad \left(\frac{1}{2}, 0\right), \quad (1, 4), \quad \left(\frac{3}{2}, 0\right), \quad (2, -4).$$

We plot these five points and fill in the graph of the cosine function as shown in Figure 92(a). Extending the graph in either direction, we obtain Figure 92(b). Figure 92(c) shows the graph using a graphing utility.

Figure 92
$y = -4\cos(\pi x)$

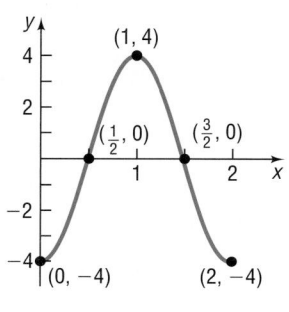

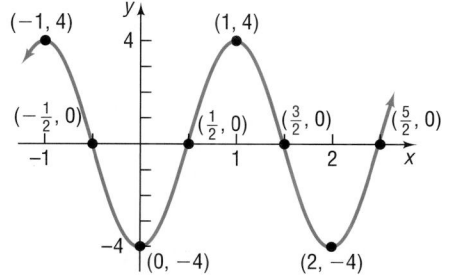

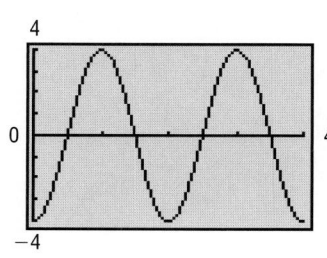

(a)   (b)   (c)

✔ CHECK: Graph $y = -4\cos(\pi x)$ by hand using transformations. Which graphing method do you prefer?   ■ ■

EXAMPLE 8   Finding the Amplitude and Period of a Sinusoidal Function and Graphing It

Determine the amplitude and period of $y = 2\sin\left(-\dfrac{\pi}{2}x\right)$, and graph the function.

Solution   Since the sine function is odd, we use the equivalent form:

$$y = -2\sin\left(\dfrac{\pi}{2}x\right)$$

Comparing $y = -2\sin\left(\dfrac{\pi}{2}x\right)$ to $y = A\sin(\omega x)$, we find that $A = -2$ and $\omega = \dfrac{\pi}{2}$. The amplitude is $|A| = 2$, and the period is $T = \dfrac{2\pi}{\omega} = \dfrac{2\pi}{\dfrac{\pi}{2}} = 4$.

The graph of $y = -2\sin\left(\dfrac{\pi}{2}x\right)$ will lie between -2 and 2 on the y-axis. One cycle will begin at $x = 0$ and end at $x = 4$. We divide the interval $[0,4]$ into four subintervals, each of length $4 \div 4 = 1$:

$$[0,1], \quad [1,2], \quad [2,3], \quad [3,4]$$

The five key points on the graph are

$$(0,0), \quad (1,-2), \quad (2,0), \quad (3,2), \quad (4,0)$$

We plot these five points and fill in the graph of the sine function as shown in Figure 93(a). Extending the graph in either direction, we obtain Figure 93(b).

Figure 93(c) shows the graph using a graphing utility.

Figure 93

$y = 2\sin\left(-\dfrac{\pi}{2}x\right)$

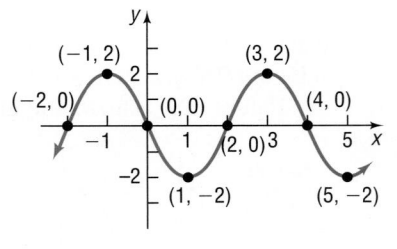

(a)   (b)

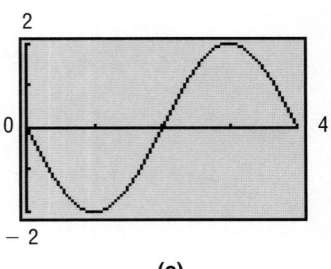

(c)

✔ CHECK: Graph $y = 2\sin\left(-\dfrac{\pi}{2}x\right)$ by hand using transformations. Which graphing method do you prefer?   ■ ■

—— NOW WORK PROBLEM **53.**

⑤   We can also use the ideas of amplitude and period to identify a sinusoidal function when its graph is given.

EXAMPLE 9   **Finding an Equation for a Sinusoidal Graph**

Find an equation for the graph shown in Figure 94.

Figure 94

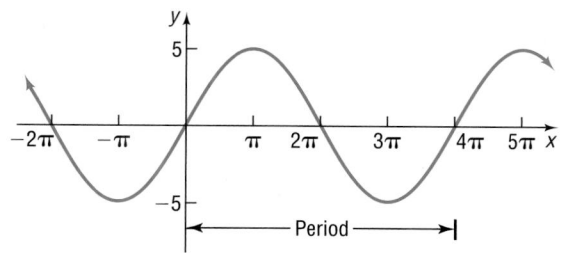

Solution   This graph can be viewed as the graph of a sine function* with amplitude $A = 5$. The period T is observed to be 4π. By equation (1),

$$T = \frac{2\pi}{\omega}$$

$$4\pi = \frac{2\pi}{\omega}$$

$$\omega = \frac{2\pi}{4\pi} = \frac{1}{2}$$

The sine function whose graph is given in Figure 94 is

$$y = A\sin(\omega x) = 5\sin\left(\frac{1}{2}x\right)$$

✔ CHECK: Graph $Y_1 = 5\sin\left(\frac{1}{2}x\right)$ and compare the result with Figure 94.

EXAMPLE 10   **Finding an Equation for a Sinusoidal Graph**

Find an equation for the graph shown in Figure 95.

Figure 95

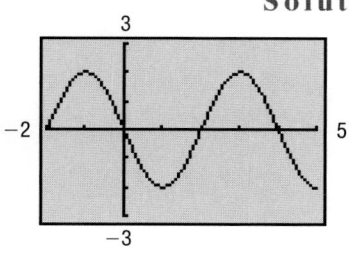

Solution   The graph is sinusoidal, with amplitude $|A| = 2$. The period is 4, so $\dfrac{2\pi}{\omega} = 4$ or $\omega = \dfrac{\pi}{2}$. Since the graph passes through the origin, it is easiest to view the equation as a sine function, but notice that the graph is actually the reflection of a sine function about the x-axis (since the graph is decreasing near the origin). As a result, $A = -2$ and we have

$$y = A\sin(\omega x) = -2\sin\left(\frac{\pi}{2}x\right)$$

*The equation could also be viewed as a cosine function with a horizontal shift, but viewing it as a sine function is easier.

✔ CHECK: Graph $Y_1 = -2 \sin\left(\dfrac{\pi}{2}x\right)$ and compare the result with Figure 95.

■ ■

■ NOW WORK PROBLEMS **63** AND **67**.

2.6 Concepts and Vocabulary

In Problems 1–3, fill in the blanks.

1. The maximum value of $y = \sin x$ is _____ and occurs at $x =$ _____.

2. The function $y = A \sin(\omega x)$, $A > 0$, has amplitude 3 and period 2; then $A =$ _____ and $\omega =$ _____.

3. The function $y = 3 \cos(6x)$ has amplitude _____ and period _____.

In Problems 4–6, answer True or False to each statement.

4. The graphs of $y = \sin x$ and $y = \cos x$ are identical except for a horizontal shift.

5. For $y = 2 \sin(\pi x)$, the amplitude is 2 and the period is $\dfrac{\pi}{2}$.

6. The graph of the sine function has infinitely many x-intercepts.

7. Draw a quick sketch of $y = \sin x$. Be sure to label at least five points.

8. Explain how you would scale the x-axis and the y-axis before graphing $y = 3 \cos(\pi x)$.

9. Explain the term *amplitude* as it relates to the graph of a sinusoidal function.

2.6 Exercises

In Problems 1–10, if necessary, refer to the graphs to answer each question.

1. What is the y-intercept of $y = \sin x$?

2. What is the y-intercept of $y = \cos x$?

3. For what numbers x, $-\pi \le x \le \pi$, is the graph of $y = \sin x$ increasing?

4. For what numbers x, $-\pi \le x \le \pi$, is the graph of $y = \cos x$ decreasing?

5. What is the largest value of $y = \sin x$?

6. What is the smallest value of $y = \cos x$?

7. For what numbers x, $0 \le x \le 2\pi$, does $\sin x = 0$?

8. For what numbers x, $0 \le x \le 2\pi$, does $\cos x = 0$?

9. For what numbers x, $-2\pi \le x \le 2\pi$, does $\sin x = 1$? What about $\sin x = -1$?

10. For what numbers x, $-2\pi \le x \le 2\pi$, does $\cos x = 1$? What about $\cos x = -1$?

In Problems 11 and 12, match the graph to a function. Three answers are possible.

A. $y = -\sin x$ **B.** $y = -\cos x$ **C.** $y = \sin\left(x - \dfrac{\pi}{2}\right)$

D. $y = -\cos\left(x - \dfrac{\pi}{2}\right)$ **E.** $y = \sin(x + \pi)$ **F.** $y = \cos(x + \pi)$

11.

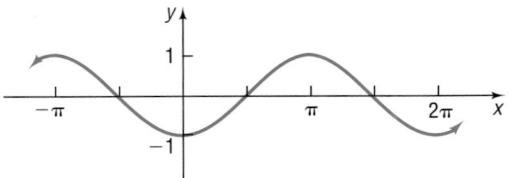

12.

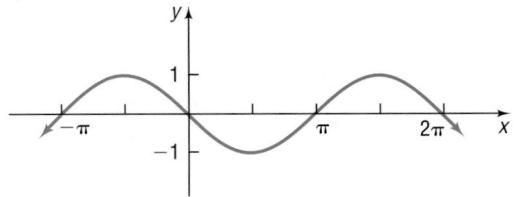

In Problems 13–28, use transformations to graph each function.

13. $y = 3 \sin x$

14. $y = 4 \cos x$

15. $y = \cos\left(x + \dfrac{\pi}{4}\right)$

16. $y = \sin(x - \pi)$

17. $y = \sin x - 1$

18. $y = \cos x + 1$

19. $y = -2 \sin x$

20. $y = -3 \cos x$

21. $y = \sin(\pi x)$

22. $y = \cos\left(\dfrac{\pi}{2} x\right)$

23. $y = 2 \sin x + 2$

24. $y = 3 \cos x + 3$

25. $y = -2 \cos\left(x - \dfrac{\pi}{2}\right)$

26. $y = -3 \sin\left(x + \dfrac{\pi}{2}\right)$

27. $y = 3 \sin(\pi - x)$

28. $y = 2 \cos(\pi - x)$

In Problems 29–38, determine the amplitude and period of each function without graphing.

29. $y = 2 \sin x$

30. $y = 3 \cos x$

31. $y = -4 \cos(2x)$

32. $y = -\sin\left(\dfrac{1}{2} x\right)$

33. $y = 6 \sin(\pi x)$

34. $y = -3 \cos(3x)$

35. $y = -\dfrac{1}{2} \cos\left(\dfrac{3}{2} x\right)$

36. $y = \dfrac{4}{3} \sin\left(\dfrac{2}{3} x\right)$

37. $y = \dfrac{5}{3} \sin\left(-\dfrac{2\pi}{3} x\right)$

38. $y = \dfrac{9}{5} \cos\left(-\dfrac{3\pi}{2} x\right)$

In Problems 39–48, match the given function to one of the graphs (A)–(J).

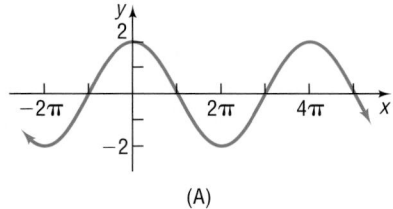

(A)

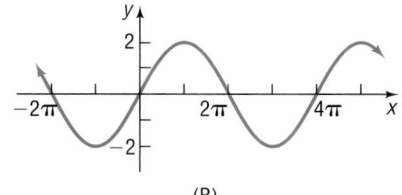

(B)

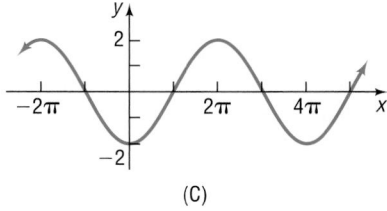

(C)

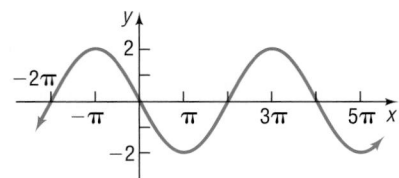

(D)

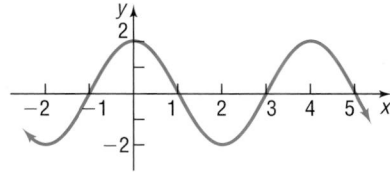

(E)

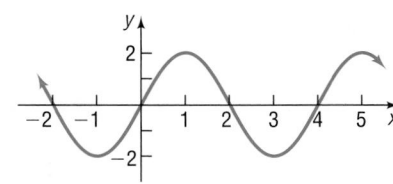

(F)

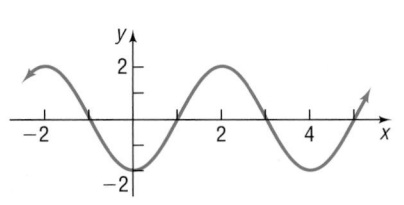

(G)

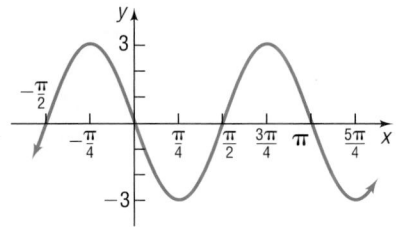

(H)

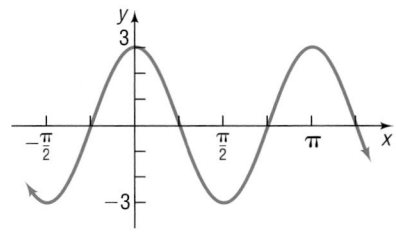

(I)

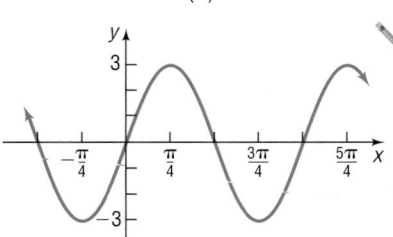

(J)

39. $y = 2 \sin\left(\dfrac{\pi}{2} x\right)$

40. $y = 2 \cos\left(\dfrac{\pi}{2} x\right)$

41. $y = 2 \cos\left(\dfrac{1}{2} x\right)$

42. $y = 3 \cos(2x)$

43. $y = -3 \sin(2x)$

44. $y = 2 \sin\left(\dfrac{1}{2} x\right)$

45. $y = -2 \cos\left(\dfrac{1}{2} x\right)$

46. $y = -2 \cos\left(\dfrac{\pi}{2} x\right)$

47. $y = 3 \sin(2x)$

48. $y = -2 \sin\left(\dfrac{1}{2} x\right)$

In Problems 49–52, match the given function to one of the graphs (A)–(D).

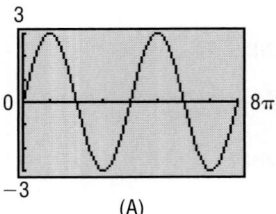
(A)

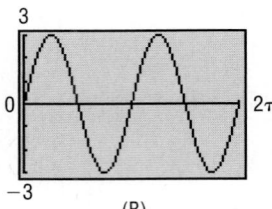
(B)

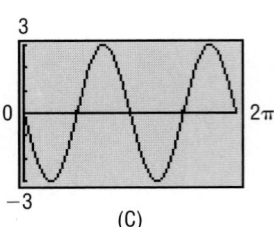
(C)

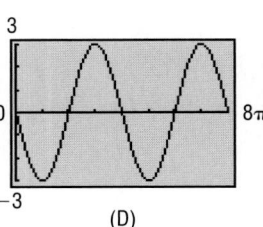
(D)

49. $y = 3\sin\left(\dfrac{1}{2}x\right)$

50. $y = -3\sin(2x)$

51. $y = 3\sin(2x)$

52. $y = -3\sin\left(\dfrac{1}{2}x\right)$

In Problems 53–62, graph each sinusoidal function.

53. $y = 5\sin(4x)$

54. $y = 4\cos(6x)$

55. $y = 5\cos(\pi x)$

56. $y = 2\sin(\pi x)$

57. $y = -2\cos(2\pi x)$

58. $y = -5\cos(2\pi x)$

59. $y = -4\sin\left(\dfrac{1}{2}x\right)$

60. $y = -2\cos\left(\dfrac{1}{2}x\right)$

61. $y = \dfrac{3}{2}\sin\left(-\dfrac{2}{3}x\right)$

62. $y = \dfrac{4}{3}\cos\left(-\dfrac{1}{3}x\right)$

In Problems 63–66, write the equation of a sine function $y = A\sin(\omega x)$, $A > 0$, that has the given characteristics.

63. Amplitude: 3
   Period: π

64. Amplitude: 2
   Period: 4π

65. Amplitude: 3
   Period: 2

66. Amplitude: 4
   Period: 1

In Problems 67–78, find an equation for each graph.

67.

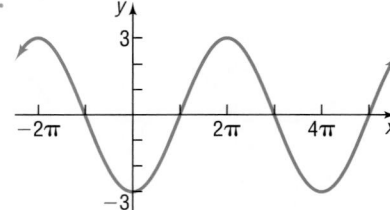

68.

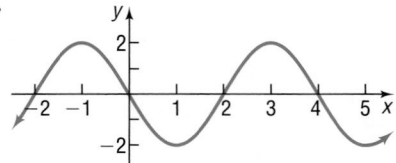

69.

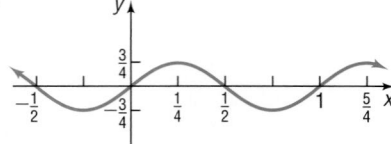

70.

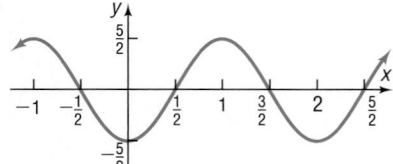

71.

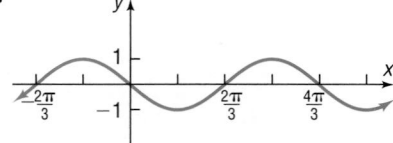

72.

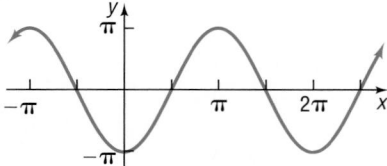

73.

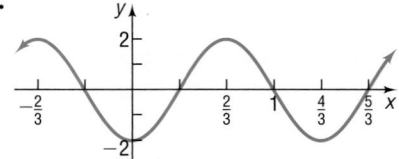

74.

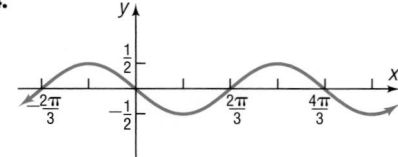

75.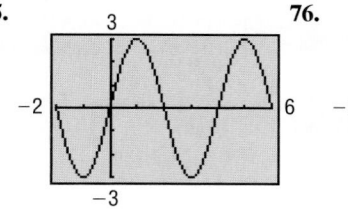

76.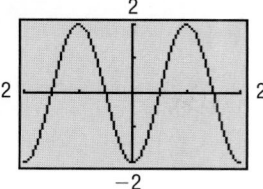

77.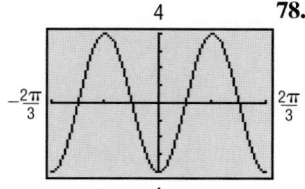

78.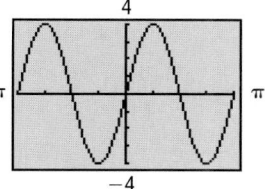

79. Alternating Current (ac) Circuits The current I, in amperes, flowing through an ac (alternating current) circuit at time t is

$$I = 220 \sin(60\pi t), \qquad t \geq 0$$

What is the period? What is the amplitude? Graph this function over two periods.

80. Alternating Current (ac) Circuits The current I, in amperes, flowing through an ac (alternating current) circuit at time t is

$$I = 120 \sin(30\pi t), \qquad t \geq 0$$

What is the period? What is the amplitude? Graph this function over two periods.

81. Alternating Current (ac) Generators The voltage V produced by an ac generator is

$$V = 220 \sin(120\pi t)$$

(a) What is the amplitude? What is the period?
(b) Graph V over two periods, beginning at $t = 0$.
(c) If a resistance of $R = 10$ ohms is present, what is the current I?
   [**Hint:** Use Ohm's Law, $V = IR$.]
(d) What is the amplitude and period of the current I?
(e) Graph I over two periods, beginning at $t = 0$.

82. Alternating Current (ac) Generators The voltage V produced by an ac generator is

$$V = 120 \sin(120\pi t)$$

(a) What is the amplitude? What is the period?
(b) Graph V over two periods, beginning at $t = 0$.
(c) If a resistance of $R = 20$ ohms is present, what is the current I?
   [**Hint:** Use Ohm's Law, $V = IR$.]
(d) What is the amplitude and period of the current I?
(e) Graph I over two periods, beginning at $t = 0$.

83. Alternating Current (ac) Generators The voltage V produced by an ac generator is sinusoidal. As a function of time, the voltage V is

$$V = V_0 \sin(2\pi f t)$$

where f is the **frequency**, the number of complete oscillations (cycles) per second. [In the United States and

Canada, f is 60 hertz (Hz).] The **power** P delivered to a resistance R at any time t is defined as

$$P = \frac{V^2}{R}$$

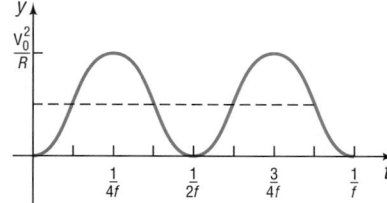

Power in an ac generator

(a) Show that $P = \dfrac{V_0^2}{R} \sin^2(2\pi f t)$.
(b) The graph of P is shown in the figure. Express P as a sinusoidal function.
(c) Deduce that

$$\sin^2(2\pi f t) = \frac{1}{2}[1 - \cos(4\pi f t)]$$

84. Biorhythms In the theory of biorhythms, a sine function of the form

$$P = 50 \sin(\omega t) + 50$$

is used to measure the percent P of a person's potential at time t, where t is measured in days and $t = 0$ is the person's birthday. Three characteristics are commonly measured:

  Physical potential: period of 23 days
  Emotional potential: period of 28 days
  Intellectual potential: period of 33 days

(a) Find ω for each characteristic.
(b) Use a graphing utility to graph all three functions.
(c) Is there a time t when all three characteristics have 100% potential? When is it?
(d) Suppose that you are 20 years old today ($t = 7305$ days). Describe your physical, emotional, and intellectual potential for the next 30 days.

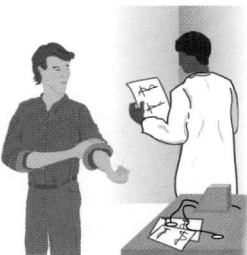

85. Graph $y = |\cos x|, -2\pi \leq x \leq 2\pi$.
86. Graph $y = |\sin x|, -2\pi \leq x \leq 2\pi$.
87. Explain how the amplitude and period of a sinusoidal graph are used to establish the scale on each coordinate axis.
88. Find an application in your major field that leads to a sinusoidal graph. Write a paper about your findings.

PREPARING FOR THIS SECTION

Before getting started, review the following:

✓ Functions (Section 1.4, pp. 41–44)

✓ Graphing Techniques: Transformations
(Section 1.6, pp. 64–73)

2.7 GRAPHS OF THE TANGENT, COTANGENT, COSECANT, AND SECANT FUNCTIONS

OBJECTIVES   ① Graph Transformations of the Tangent Function and Cotangent Function

② Graph Transformations of the Cosecant Function and Secant Function

The Graphs of $y = \tan x$ and $y = \cot x$

① Because the tangent function has period π, we only need to determine the graph over some interval of length π. The rest of the graph will consist of repetitions of that graph. Because the tangent function is not defined at $\dots, -\dfrac{3\pi}{2}, -\dfrac{\pi}{2}, \dfrac{\pi}{2}, \dfrac{3\pi}{2}, \dots$, we will concentrate on the interval $\left(-\dfrac{\pi}{2}, \dfrac{\pi}{2}\right)$, of length π, and construct Table 9, which lists some points on the graph of $y = \tan x$, $-\dfrac{\pi}{2} < x < \dfrac{\pi}{2}$. We plot the points in the table and connect them with a smooth curve. See Figure 96 for a partial graph of $y = \tan x$, where $-\dfrac{\pi}{3} \le x \le \dfrac{\pi}{3}$.

TABLE 9

x	$y = \tan x$	(x, y)
$-\dfrac{\pi}{3}$	$-\sqrt{3}$	$\left(-\dfrac{\pi}{3}, -\sqrt{3}\right)$
$-\dfrac{\pi}{4}$	-1	$\left(-\dfrac{\pi}{4}, -1\right)$
$-\dfrac{\pi}{6}$	$-\dfrac{\sqrt{3}}{3}$	$\left(-\dfrac{\pi}{6}, -\dfrac{\sqrt{3}}{3}\right)$
0	0	$(0, 0)$
$\dfrac{\pi}{6}$	$\dfrac{\sqrt{3}}{3}$	$\left(\dfrac{\pi}{6}, \dfrac{\sqrt{3}}{3}\right)$
$\dfrac{\pi}{4}$	1	$\left(\dfrac{\pi}{4}, 1\right)$
$\dfrac{\pi}{3}$	$\sqrt{3}$	$\left(\dfrac{\pi}{3}, \sqrt{3}\right)$

Figure 96

$y = \tan x, \ -\dfrac{\pi}{3} \le x \le \dfrac{\pi}{3}$

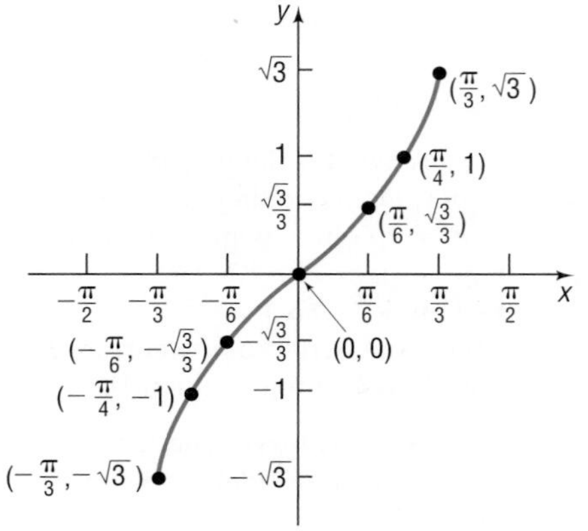

To complete one period of the graph of $y = \tan x$, we need to investigate the behavior of the function as x approaches $-\dfrac{\pi}{2}$ and $\dfrac{\pi}{2}$. We must be careful, though, because $y = \tan x$ is not defined at these numbers. To determine this behavior, we use the identity

$$\tan x = \frac{\sin x}{\cos x}$$

If x is close to $\dfrac{\pi}{2} \approx 1.5708$, but remains less than $\dfrac{\pi}{2}$, then $\sin x$ will be close to 1 and $\cos x$ will be positive and close to 0. (Refer back to the graphs of the sine function and the cosine function.) Hence, the ratio $\dfrac{\sin x}{\cos x}$ will be positive and large. In fact, the closer x gets to $\dfrac{\pi}{2}$, the closer $\sin x$ gets to 1 and $\cos x$ gets to 0, so $\tan x$ approaches $\infty \left(\lim\limits_{x \to \frac{\pi}{2}^-} \tan x = \infty \right)$. In other words, the vertical line $x = \dfrac{\pi}{2}$ is a vertical asymptote to the graph of $y = \tan x$. See Table 10.

If x is close to $-\dfrac{\pi}{2}$, but remains greater than $-\dfrac{\pi}{2}$, then $\sin x$ will be close to -1 and $\cos x$ will be positive and close to 0. Hence, the ratio $\dfrac{\sin x}{\cos x}$ approaches $-\infty \left(\lim\limits_{x \to -\frac{\pi}{2}^+} \tan x = -\infty \right)$. In other words, the vertical line $x = -\dfrac{\pi}{2}$ is also a vertical asymptote to the graph.

With these observations, we can complete one period of the graph. We obtain the complete graph of $y = \tan x$ by repeating this period, as shown in Figure 97(a).

Figure 97(b) shows the graph of $y = \tan x$, $-\infty < x < \infty$ using a graphing utility. Notice we used dot mode when graphing $y = \tan x$. Do you know why?

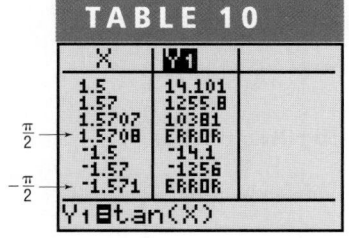

TABLE 10

X	Y1
1.5	14.101
1.57	1255.8
1.5707	10381
1.5708	ERROR
-1.5	-14.1
-1.57	-1256
-1.571	ERROR

Y1⊟tan(X)

Figure 97
$y = \tan x$, $-\infty < x < \infty$, x
not equal to odd multiples of $\dfrac{\pi}{2}$

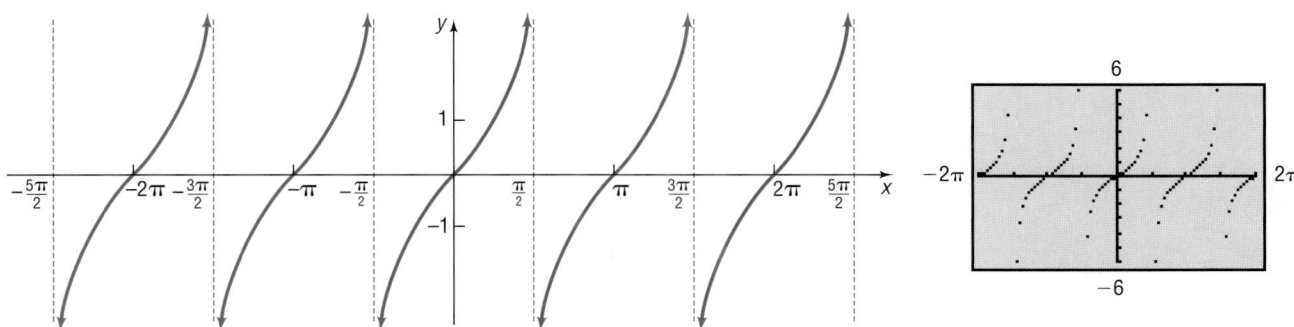

(a)   (b)

The graph of $y = \tan x$ illustrates some facts that we already know about the tangent function.

Properties of the Tangent Function

1. The domain is the set of all real numbers, except odd multiples of $\dfrac{\pi}{2}$.
2. The range consists of all real numbers.
3. The tangent function is an odd function, as the symmetry of the graph with respect to the origin indicates.
4. The tangent function is periodic, with period π.
5. The x-intercepts are $\ldots, -2\pi, -\pi, 0, \pi, 2\pi, 3\pi, \ldots$; the y-intercept is 0.
6. Vertical asymptotes occur at $x = \ldots, -\dfrac{3\pi}{2}, -\dfrac{\pi}{2}, \dfrac{\pi}{2}, \dfrac{3\pi}{2}, \ldots.$

✏ — **NOW WORK PROBLEMS 1 AND 9.**

EXAMPLE 1   **Graphing Variations of $y = \tan x$ Using Transformations**

Graph: $y = 2 \tan x$

Solution   We start with the graph of $y = \tan x$ and vertically stretch it by a factor of 2. See Figure 98.

Figure 98

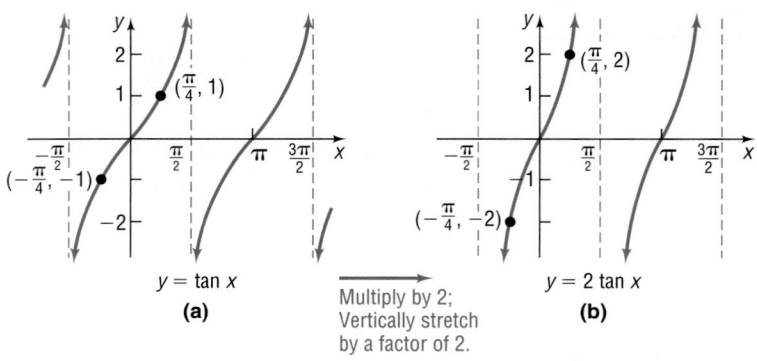

$y = \tan x$
(a)

Multiply by 2;
Vertically stretch
by a factor of 2.

$y = 2 \tan x$
(b)

Figure 99 shows the graph using a graphing utility.

Figure 99

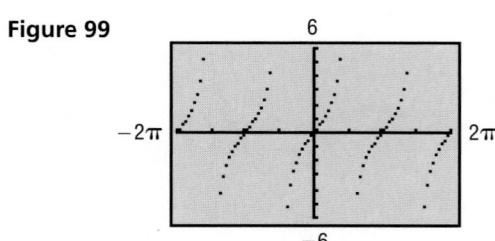

EXAMPLE 2   **Graphing Variations of $y = \tan x$ Using Transformations**

Graph: $y = -\tan\left(x + \dfrac{\pi}{4}\right)$

Solution We start with the graph of $y = \tan x$. See Figure 100.

Figure 100

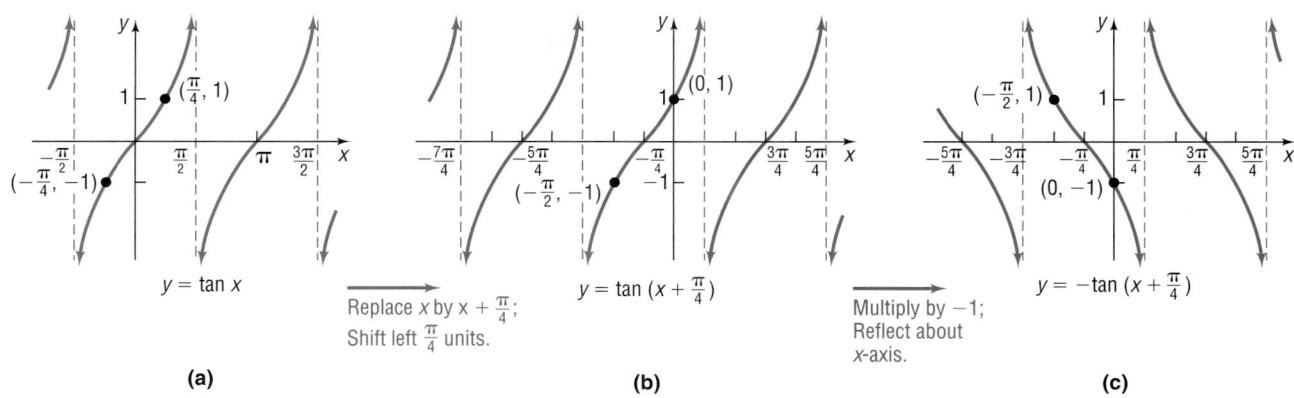

$y = \tan x$

Replace x by $x + \frac{\pi}{4}$;
Shift left $\frac{\pi}{4}$ units.

$y = \tan\left(x + \frac{\pi}{4}\right)$

Multiply by -1;
Reflect about
x-axis.

$y = -\tan\left(x + \frac{\pi}{4}\right)$

(a)   (b)   (c)

✔ CHECK: Graph $Y_1 = -\tan\left(x + \dfrac{\pi}{4}\right)$ and compare the result to Figure 100(c).

◼ ◼

NOW WORK PROBLEM 19.

We obtain the graph of $y = \cot x$ as we did the graph of $y = \tan x$. The period of $y = \cot x$ is π. Because the cotangent function is not defined for integral multiples of π, we will concentrate on the interval $(0, \pi)$. Table 11 lists some points on the graph of $y = \cot x, 0 < x < \pi$. As x approaches 0, but remains greater than 0, the value of $\cos x$ will be close to 1 and the value of $\sin x$ will be positive and close to 0. Hence, the ratio $\dfrac{\cos x}{\sin x} = \cot x$ will be positive and large; so as x approaches 0, with $x > 0$, $\cot x$ approaches ∞ $\left(\lim\limits_{x \to 0^+} \cot x = \infty\right)$. Similarly, as x approaches π, but remains less than π, the value of $\cos x$ will be close to -1, and the value of $\sin x$ will be positive and close to 0. Hence, the ratio $\dfrac{\cos x}{\sin x} = \cot x$ will be negative and will approach $-\infty$ as x approaches π $\left(\lim\limits_{x \to \pi^-} \cot x = -\infty\right)$. Figure 101 shows the graph.

TABLE 11

x	$y = \cot x$	(x, y)
$\dfrac{\pi}{6}$	$\sqrt{3}$	$\left(\dfrac{\pi}{6}, \sqrt{3}\right)$
$\dfrac{\pi}{4}$	1	$\left(\dfrac{\pi}{4}, 1\right)$
$\dfrac{\pi}{3}$	$\dfrac{\sqrt{3}}{3}$	$\left(\dfrac{\pi}{3}, \dfrac{\sqrt{3}}{3}\right)$
$\dfrac{\pi}{2}$	0	$\left(\dfrac{\pi}{2}, 0\right)$
$\dfrac{2\pi}{3}$	$-\dfrac{\sqrt{3}}{3}$	$\left(\dfrac{2\pi}{3}, -\dfrac{\sqrt{3}}{3}\right)$
$\dfrac{3\pi}{4}$	-1	$\left(\dfrac{3\pi}{4}, -1\right)$
$\dfrac{5\pi}{6}$	$-\sqrt{3}$	$\left(\dfrac{5\pi}{6}, -\sqrt{3}\right)$

Figure 101
$y = \cot x, -\infty < x < \infty,$
x not equal to integral
multiples of $\pi, -\infty < y < \infty$

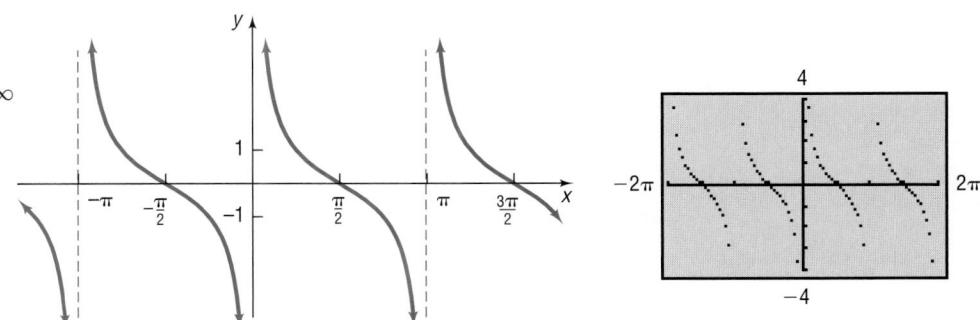

NOW WORK PROBLEM 25.

The Graphs of $y = \csc x$ and $y = \sec x$

2 The cosecant and secant functions, sometimes referred to as **reciprocal functions**, are graphed by making use of the reciprocal identities

$$\csc x = \frac{1}{\sin x} \quad \text{and} \quad \sec x = \frac{1}{\cos x}$$

For example, the value of the cosecant function $y = \csc x$ at a given number x equals the reciprocal of the corresponding value of the sine function, provided that the value of the sine function is not 0. If the value of $\sin x$ is 0, then, at such numbers x, (integral multiples of π) the cosecant function is not defined. In fact, the graph of the cosecant function has vertical asymptotes at integral multiples of π. Figure 102 shows the graph.

Figure 102
$y = \csc x$, $-\infty < x < \infty$, x not equal to integral multiples of π, $|y| \geq 1$

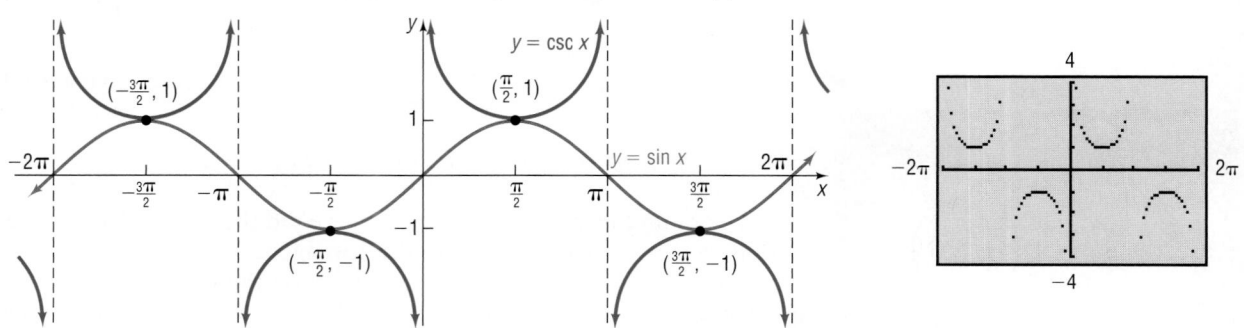

EXAMPLE 3   Graphing Variations of $y = \csc x$ Using Transformations

Graph: $y = 2 \csc\left(x - \frac{\pi}{2}\right)$

Solution   Figure 103 shows the required steps.

Figure 103

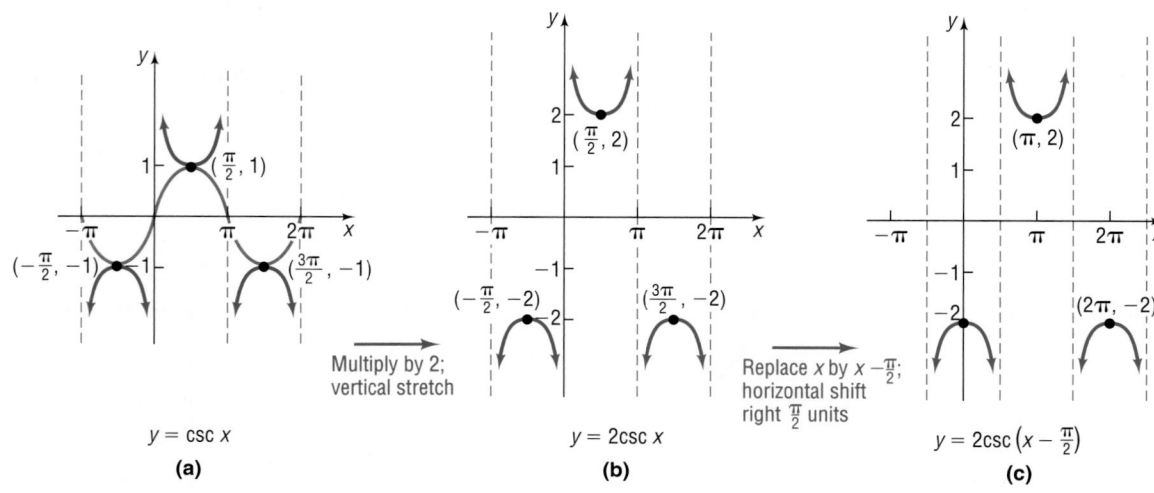

$y = \csc x$
(a)

Multiply by 2; vertical stretch

$y = 2\csc x$
(b)

Replace x by $x - \frac{\pi}{2}$; horizontal shift right $\frac{\pi}{2}$ units

$y = 2\csc\left(x - \frac{\pi}{2}\right)$
(c)

✔ CHECK: Graph $Y_1 = 2 \csc\left(x - \frac{\pi}{2}\right)$ and compare the result with Figure 103(c). ■ ■

NOW WORK PROBLEM **31.**

Using the idea of reciprocals, we can similarly obtain the graph of $y = \sec x$. See Figure 104.

Figure 104

$y = \sec x$, $-\infty < x < \infty$, x not equal to odd multiples of $\dfrac{\pi}{2}$, $|y| \geq 1$

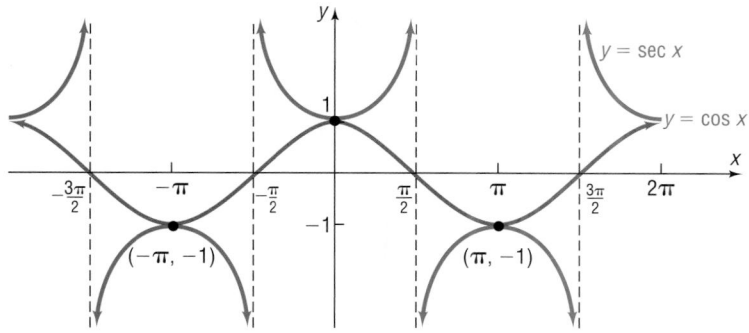

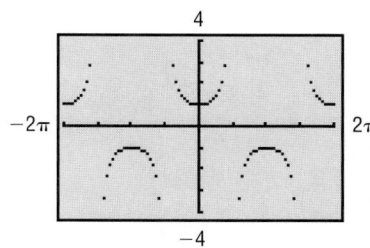

2.7 Concepts and Vocabulary

In Problems 1–3, fill in the blanks.

1. The graph of $y = \tan x$ is symmetric with respect to the _____ and has vertical asymptotes at _____.

2. The graph of $y = \sec x$ is symmetric with respect to the _____ and has vertical asymptotes at _____.

3. It is easiest to graph $y = \sec x$ by first sketching the graph of _____.

In Problem 4, answer True of False.

4. The graphs of $y = \tan x$, $y = \cot x$, $y = \sec x$, and $y = \csc x$ each have infinitely many vertical asymptotes.

5. Explain how you would graph $y = \csc x$ using the idea of a reciprocal function.

6. Sketch the graph of $y = \tan x$.

2.7 Exercises

In Problems 1–10, if necessary, refer to the graphs to answer each question.

1. What is the y-intercept of $y = \tan x$?

2. What is the y-intercept of $y = \cot x$?

3. What is the y-intercept of $y = \sec x$?

4. What is the y-intercept of $y = \csc x$?

5. For what numbers x, $-2\pi \leq x \leq 2\pi$, does $\sec x = 1$? What about $\sec x = -1$?

6. For what numbers x, $-2\pi \leq x \leq 2\pi$, does $\csc x = 1$? What about $\csc x = -1$?

7. For what numbers x, $-2\pi \leq x \leq 2\pi$, does the graph of $y = \sec x$ have vertical asymptotes?

8. For what numbers x, $-2\pi \leq x \leq 2\pi$, does the graph of $y = \csc x$ have vertical asymptotes?

9. For what numbers x, $-2\pi \leq x \leq 2\pi$, does the graph of $y = \tan x$ have vertical asymptotes?

10. For what numbers x, $-2\pi \leq x \leq 2\pi$, does the graph of $y = \cot x$ have vertical asymptotes?

In Problems 11–14, match each function to its graph.

A. $y = -\tan x$ B. $y = \tan\left(x + \dfrac{\pi}{2}\right)$ C. $y = \tan(x + \pi)$ D. $y = -\tan\left(x - \dfrac{\pi}{2}\right)$

11.

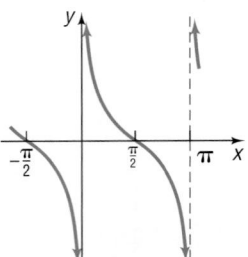

12.

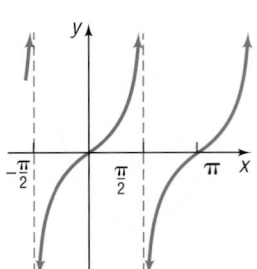

13.

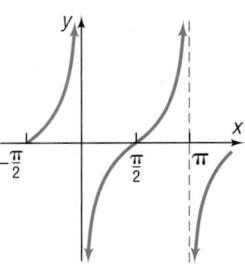

14.

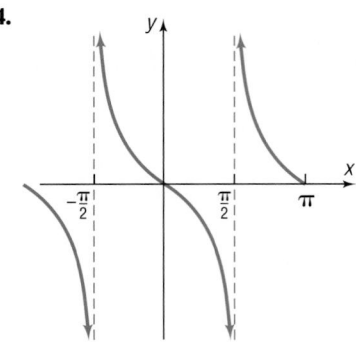

In Problems 15–34, use tranformations to graph each function. Verify your result using a graphing utility.

15. $y = -\sec x$

16. $y = -\cot x$

17. $y = \sec\left(x - \dfrac{\pi}{2}\right)$

18. $y = \csc(x - \pi)$

19. $y = \tan(x - \pi)$

20. $y = \cot(x - \pi)$

21. $y = 3\tan(2x)$

22. $y = 4\tan\left(\dfrac{1}{2}x\right)$

23. $y = \sec(2x)$

24. $y = \csc\left(\dfrac{1}{2}x\right)$

25. $y = \cot(\pi x)$

26. $y = \cot(2x)$

27. $y = -3\tan(4x)$

28. $y = -3\tan(2x)$

29. $y = 2\sec\left(\dfrac{1}{2}x\right)$

30. $y = 2\sec(3x)$

31. $y = -3\csc\left(x + \dfrac{\pi}{4}\right)$

32. $y = -2\tan\left(x + \dfrac{\pi}{4}\right)$

33. $y = \dfrac{1}{2}\cot\left(x - \dfrac{\pi}{4}\right)$

34. $y = 3\sec\left(x + \dfrac{\pi}{2}\right)$

---

35. Carrying a Ladder around a Corner A ladder of length L is carried horizontally around a corner from a hall 3 feet wide into a hall 4 feet wide. See the illustration below.

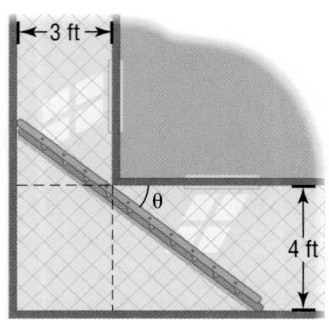

(a) Show that the length L of the ladder as a function of the angle θ is
$$L(\theta) = 3\sec\theta + 4\csc\theta$$

(b) Use a graphing utility to graph $L, 0 < \theta < \dfrac{\pi}{2}$.

(c) For what value of θ is L the least?

(d) What is the length of the longest ladder that can be carried around the corner? Why is this also the least value of L?

36. Exploration Graph
$$y = \tan x \quad \text{and} \quad y = -\cot\left(x + \dfrac{\pi}{2}\right)$$

Do you think that $\tan x = -\cot\left(x + \dfrac{\pi}{2}\right)$?

2.8 PHASE SHIFT; SINUSOIDAL CURVE FITTING

OBJECTIVES
1. Determine the Phase Shift of a Sinusoidal Function
2. Graph Sinusoidal Functions: $y = A \sin(\omega x - \phi)$
3. Find a Sinusoidal Function from Data

Phase Shift

1. We have seen that the graph of $y = A \sin(\omega x)$, $\omega > 0$, has amplitude $|A|$ and period $T = \dfrac{2\pi}{\omega}$. One cycle can be drawn as x varies from 0 to $\dfrac{2\pi}{\omega}$ or, equivalently, as ωx varies from 0 to 2π. See Figure 105.

We now want to discuss the graph of

$$y = A \sin(\omega x - \phi) = A \sin\left[\omega\left(x - \frac{\phi}{\omega}\right)\right]$$

Figure 105
One cycle
$y = A \sin(\omega x)$, $A > 0$, $\omega > 0$

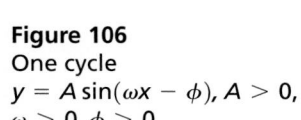

where $\omega > 0$ and ϕ (the Greek letter phi) are real numbers. The graph will be a sine curve of amplitude $|A|$. As $\omega x - \phi$ varies from 0 to 2π, one period will be traced out. This period will begin when

$$\omega x - \phi = 0 \quad \text{or} \quad x = \frac{\phi}{\omega}$$

and will end when

$$\omega x - \phi = 2\pi \quad \text{or} \quad x = \frac{2\pi}{\omega} + \frac{\phi}{\omega}$$

Figure 106
One cycle
$y = A \sin(\omega x - \phi)$, $A > 0$,
$\omega > 0$, $\phi > 0$

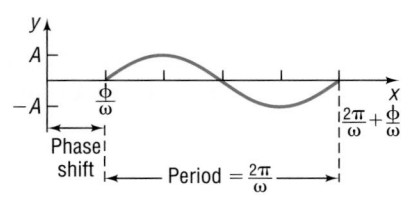

See Figure 106.

We see that the graph of $y = A \sin(\omega x - \phi) = A \sin\left[\omega\left(x - \dfrac{\phi}{\omega}\right)\right]$ is the same as the graph of $y = A \sin(\omega x)$, except that it has been shifted $\dfrac{\phi}{\omega}$ units (to the right if $\phi > 0$ and to the left if $\phi < 0$). This number $\dfrac{\phi}{\omega}$ is called the **phase shift** of the graph of $y = A \sin(\omega x - \phi)$.

For the graphs of $y = A \sin(\omega x - \phi)$ or $y = A \cos(\omega x - \phi)$, $\omega > 0$,

$$\text{Amplitude} = |A| \qquad \text{Period} = T = \frac{2\pi}{\omega} \qquad \text{Phase shift} = \frac{\phi}{\omega}$$

The phase shift is to the left if $\phi < 0$ and to the right if $\phi > 0$.

---

2. **EXAMPLE 1**   **Finding the Amplitude, Period, and Phase Shift of a Sinusoidal Function and Graphing It**

Find the amplitude, period, and phase shift of $y = 3 \sin(2x - \pi)$, and graph the function.

Solution   Comparing

$$y = 3\sin(2x - \pi) = 3\sin\left[2\left(x - \frac{\pi}{2}\right)\right]$$

to

$$y = A\sin(\omega x - \phi) = A\sin\left[\omega\left(x - \frac{\phi}{\omega}\right)\right]$$

we find that $A = 3, \omega = 2$, and $\phi = \pi$. The graph is a sine curve with amplitude $|A| = 3$, period $T = \dfrac{2\pi}{\omega} = \dfrac{2\pi}{2} = \pi$, and phase shift $= \dfrac{\phi}{\omega} = \dfrac{\pi}{2}$.

The graph of $y = 3\sin(2x - \pi)$ will lie between -3 and 3 on the y-axis. One cycle will begin at $x = \dfrac{\phi}{\omega} = \dfrac{\pi}{2}$ and end at $x = \dfrac{2\pi}{\omega} + \dfrac{\phi}{\omega} = \pi + \dfrac{\pi}{2} = \dfrac{3\pi}{2}$. We divide the interval $\left[\dfrac{\pi}{2}, \dfrac{3\pi}{2}\right]$ into four subintervals, each of length $\pi \div 4 = \dfrac{\pi}{4}$:

$$\left[\frac{\pi}{2}, \frac{3\pi}{4}\right], \quad \left[\frac{3\pi}{4}, \pi\right], \quad \left[\pi, \frac{5\pi}{4}\right], \quad \left[\frac{5\pi}{4}, \frac{3\pi}{2}\right]$$

The five key points on the graph are

$$\left(\frac{\pi}{2}, 0\right), \quad \left(\frac{3\pi}{4}, 3\right), \quad (\pi, 0), \quad \left(\frac{5\pi}{4}, -3\right), \quad \left(\frac{3\pi}{2}, 0\right)$$

We plot these five points and fill in the graph of the sine function as shown in Figure 107(a). Extending the graph in either direction, we obtain Figure 107(b).

Figure 107

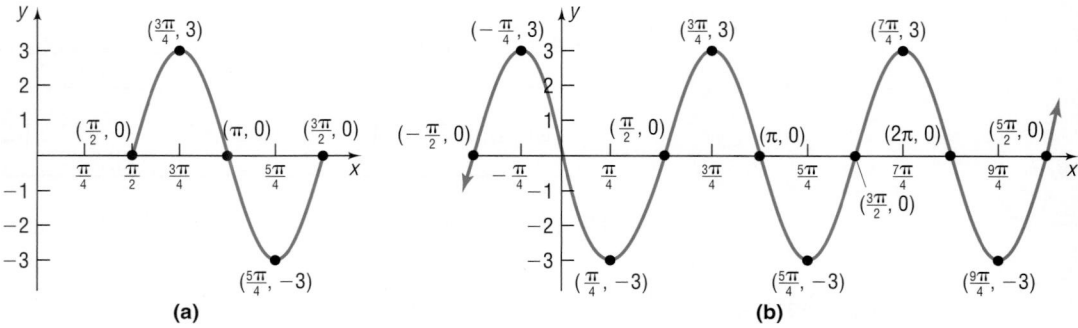

(a)

(b)

The graph of $y = 3\sin(2x - \pi) = 3\sin\left[2\left(x - \dfrac{\pi}{2}\right)\right]$ may also be obtained using transformations. See Figure 108.

Figure 108

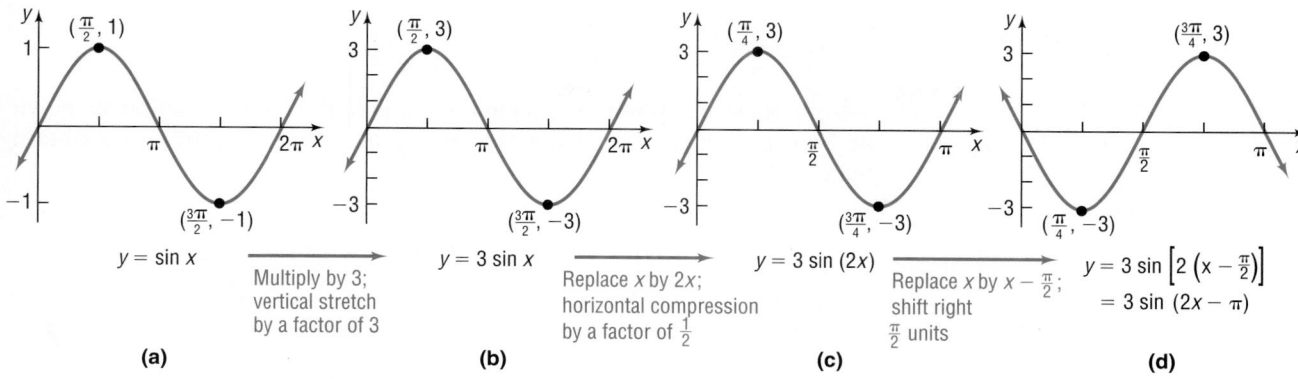

(a) $y = \sin x$ — Multiply by 3; vertical stretch by a factor of 3

(b) $y = 3 \sin x$ — Replace x by $2x$; horizontal compression by a factor of $\frac{1}{2}$

(c) $y = 3 \sin (2x)$ — Replace x by $x - \frac{\pi}{2}$; shift right $\frac{\pi}{2}$ units

(d) $y = 3 \sin\left[2\left(x - \frac{\pi}{2}\right)\right] = 3 \sin(2x - \pi)$

✔ CHECK: Figure 109 shows the graph of $Y_1 = 3 \sin(2x - \pi)$ using a graphing utility.

Figure 109

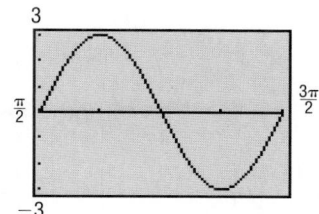

EXAMPLE 2   **Finding the Amplitude, Period, and Phase Shift of a Sinusoidal Function and Graphing It**

Find the amplitude, period, and phase shift of $y = 2 \cos(4x + 3\pi)$, and graph the function.

Solution   Comparing

$$y = 2 \cos(4x + 3\pi) = 2 \cos\left[4\left(x + \frac{3\pi}{4}\right)\right]$$

to

$$y = A \cos(\omega x - \phi) = A \cos\left[\omega\left(x - \frac{\phi}{\omega}\right)\right]$$

we see that $A = 2$, $\omega = 4$, and $\phi = -3\pi$. The graph is a cosine curve with amplitude $|A| = 2$, period $T = \dfrac{2\pi}{\omega} = \dfrac{2\pi}{4} = \dfrac{\pi}{2}$, and phase shift $= \dfrac{\phi}{\omega} = -\dfrac{3\pi}{4}$.

The graph of $y = 2 \cos(4x + 3\pi)$ will lie between -2 and 2 on the y-axis. One cycle will begin at $x = \dfrac{\phi}{\omega} = -\dfrac{3\pi}{4}$ and end at $x = \dfrac{2\pi}{\omega} + \dfrac{\phi}{\omega} = \dfrac{\pi}{2} + \left(-\dfrac{3\pi}{4}\right) = -\dfrac{\pi}{4}$. We divide the interval $\left[-\dfrac{3\pi}{4}, -\dfrac{\pi}{4}\right]$ into four subintervals, each of the length $\dfrac{\pi}{2} \div 4 = \dfrac{\pi}{8}$:

$$\left[-\frac{3\pi}{4}, -\frac{5\pi}{8}\right], \quad \left[-\frac{5\pi}{8}, -\frac{\pi}{2}\right], \quad \left[-\frac{\pi}{2}, -\frac{3\pi}{8}\right], \quad \left[-\frac{3\pi}{8}, -\frac{\pi}{4}\right]$$

The five key points on the graph are

$$\left(-\frac{3\pi}{4}, 2\right), \quad \left(-\frac{5\pi}{8}, 0\right), \quad \left(-\frac{\pi}{2}, -2\right), \quad \left(-\frac{3\pi}{8}, 0\right), \quad \left(-\frac{\pi}{4}, 2\right)$$

We plot these five points and fill in the graph of the cosine function as shown in Figure 110(a). Extending the graph in either direction, we obtain Figure 110(b).

Figure 110

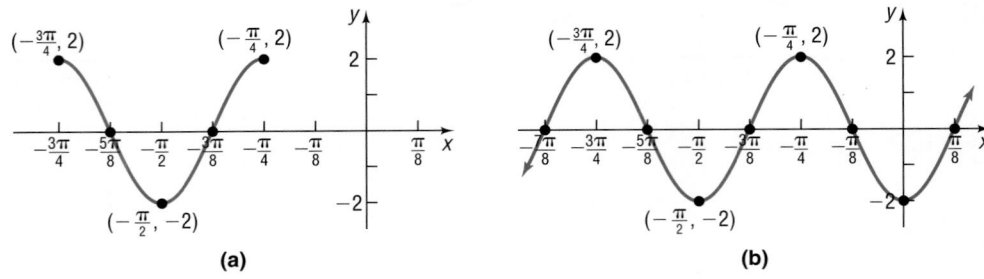

(a)   (b)

The graph of $y = 2\cos(4x + 3\pi) = 2\cos\left[4\left(x + \frac{3\pi}{4}\right)\right]$ may also be

Figure 111   obtained using transformations. See Figure 111.

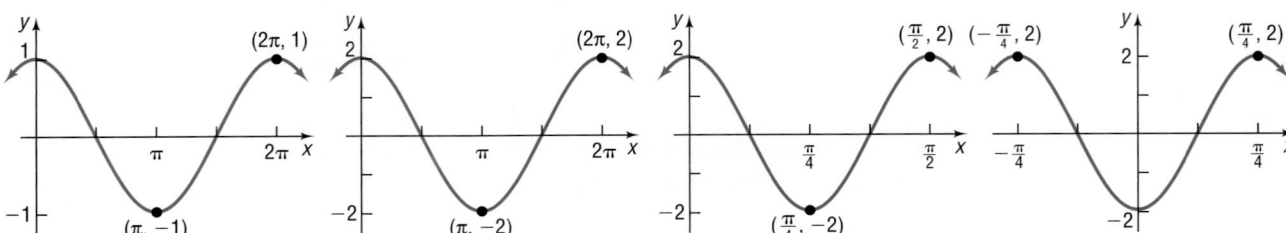

$y = \cos x$ $y = 2\cos x$ $y = 2\cos(4x)$ $y = 2\cos\left[4\left(x + \frac{3\pi}{4}\right)\right]$
                   $= 2\cos(4x + 3\pi)$

Multiply by 2;   Replace x by $4x$;   Replace x by $x + \frac{3\pi}{4}$;
vertical stretch   horizontal compression   shift left $\frac{3\pi}{4}$ units
by a factor of 2   by a factor of $\frac{1}{4}$

(a)   (b)   (c)   (d)

✔ CHECK: Graph $Y_1 = 2\cos(4x + 3\pi)$ using a graphing utility.   ■ ■

━ **NOW WORK PROBLEM 1.**

SUMMARY   **Steps for Graphing Sinusoidal Functions**

To graph sinusoidal functions of the form $y = A\sin(\omega x - \phi)$ or $y = A\cos(\omega x - \phi)$:

STEP 1: Determine the amplitude $|A|$ and period $T = \dfrac{2\pi}{\omega}$.

STEP 2: Determine the starting point of one cycle of the graph, $\dfrac{\phi}{\omega}$.

STEP 3: Determine the ending point of one cycle of the graph, $\dfrac{2\pi}{\omega} + \dfrac{\phi}{\omega}$.

STEP 4: Divide the interval $\left[\dfrac{\phi}{\omega}, \dfrac{2\pi}{\omega} + \dfrac{\phi}{\omega}\right]$ into four subintervals, each of length $\dfrac{2\pi}{\omega} \div 4$.

STEP 5: Use the endpoints of the subintervals to find the five key points on the graph.

STEP 6: Fill in one cycle of the graph.

STEP 7: Extend the graph in each direction to make it complete.

Finding Sinusoidal Functions from Data

3 Scatter diagrams of data sometimes take the form of a sinusoidal function. Let's look at an example.

The data given in Table 12 represent the average monthly temperatures in Denver, Colorado. Since the data represent average monthly temperatures collected over many years, the data will not vary much from year to year and so will essentially repeat each year. In other words, the data are periodic. Figure 112 shows the scatter diagram of these data repeated over two years, where $x = 1$ represents January, $x = 2$ represents February, and so on.

Figure 112

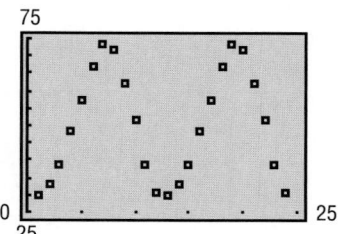

TABLE 12	
Month, x	**Average Monthly Temperature, °F**
January, 1	29.7
February, 2	33.4
March, 3	39.0
April, 4	48.2
May, 5	57.2
June, 6	66.9
July, 7	73.5
August, 8	71.4
September, 9	62.3
October, 10	51.4
November, 11	39.0
December, 12	31.0

Source: U.S. National Oceanic and Atmospheric Administration

Notice that the scatter diagram looks like the graph of a sinusoidal function. We choose to fit the data to a sine function of the form

$$y = A \sin(\omega x - \phi) + B$$

where A, B, ω, and ϕ are constants.

EXAMPLE 3   **Finding a Sinusoidal Function from Temperature Data**

Fit a sine function to the data in Table 12.

Figure 113

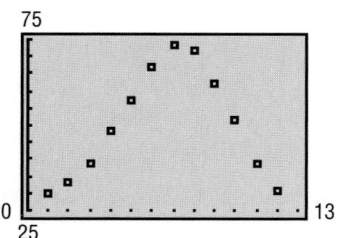

Solution   We begin with a scatter diagram of the data for one year. See Figure 113. The data will be fitted to a sine function of the form

$$y = A \sin(\omega x - \phi) + B$$

STEP 1:   To find the amplitude A, we compute

$$\text{Amplitude} = \frac{\text{largest data value} - \text{smallest data value}}{2}$$

$$= \frac{73.5 - 29.7}{2} = 21.9$$

To see the remaining steps in this process, we superimpose the graph of the function $y = 21.9 \sin x$, where x represents months, on the scatter diagram. Figure 114 shows the two graphs.

To fit the data, the graph needs to be shifted vertically, shifted horizontally, and stretched horizontally.

Figure 114

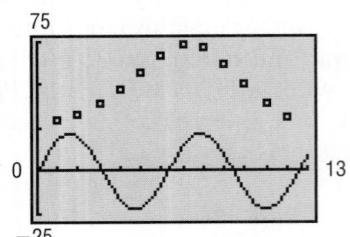

STEP 2:   We determine the vertical shift by finding the average of the highest and lowest data value.

$$\text{Vertical shift} = \frac{73.5 + 29.7}{2} = 51.6$$

Now we superimpose the graph of $y = 21.9 \sin x + 51.6$ on the scatter diagram. See Figure 115.

We see that the graph needs to be shifted horizontally and stretched horizontally.

Figure 115

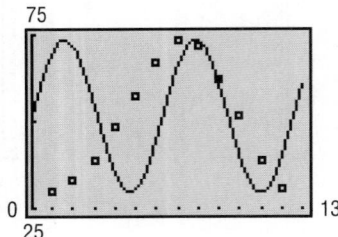

STEP 3:   It is easier to find the horizontal stretch factor first. Since the temperatures repeat every 12 months, the period of the function is $T = 12$. Since $T = \dfrac{2\pi}{\omega} = 12$,

$$\omega = \frac{2\pi}{12} = \frac{\pi}{6}$$

Now we superimpose the graph of $y = 21.9 \sin\left(\dfrac{\pi}{6} x\right) + 51.6$ on the scatter diagram. See Figure 116.

We see that the graph still needs to be shifted horizontally.

Figure 116

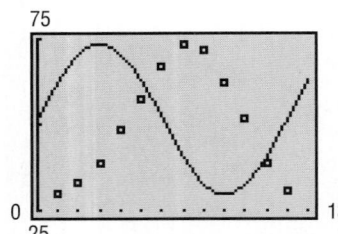

STEP 4:   To determine the horizontal shift, we solve the equation

$$y = 21.9 \sin\left(\frac{\pi}{6} x - \phi\right) + 51.6$$

for ϕ by letting $y = 29.7$ and $x = 1$ (the average temperature in Denver in January).*

$$29.7 = 21.9 \sin\left(\frac{\pi}{6} \cdot 1 - \phi\right) + 51.6$$

$$-21.9 = 21.9 \sin\left(\frac{\pi}{6} - \phi\right) \qquad \text{Subtract 51.6 from both sides of the equation.}$$

$$-1 = \sin\left(\frac{\pi}{6} - \phi\right) \qquad \text{Divide both sides of the equation by 21.9.}$$

$$\frac{\pi}{6} - \phi = -\frac{\pi}{2} \qquad \sin\theta = -1 \text{ when } \theta = -\frac{\pi}{2}.$$

$$\phi = \frac{2\pi}{3} \qquad \text{Solve for } \phi.$$

The sine function that fits the data is

$$y = 21.9 \sin\left(\frac{\pi}{6} x - \frac{2\pi}{3}\right) + 51.6$$

---

*The data point selected here to find ϕ is arbitrary. Selecting a different data point will usually result in a different value for ϕ. To maintain consistency, we will always choose the data point for which y is smallest (in this case, January gives the lowest temperature).

Figure 117

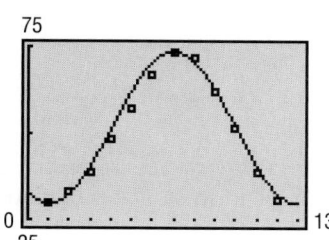

The graph of $y = 21.9 \sin\left(\dfrac{\pi}{6}x - \dfrac{2\pi}{3}\right) + 51.6$ and the scatter diagram of the data are shown in Figure 117.

The steps to fit a sine function

$$y = A \sin(\omega x - \phi) + B$$

to sinusoidal data follow:

Steps for Fitting Data to a Sine Function $y = A \sin(\omega x - \phi) + B$

STEP 1:   Determine A, the amplitude of the function.

$$\text{Amplitude} = \frac{\text{largest data value} - \text{smallest data value}}{2}$$

STEP 2:   Determine B, the vertical shift of the function.

$$\text{Vertical shift} = \frac{\text{largest data value} + \text{smallest data value}}{2}$$

STEP 3:   Determine ω. Since the period T, the time it takes for the data to repeat, is $T = \dfrac{2\pi}{\omega}$, we have

$$\omega = \frac{2\pi}{T}$$

STEP 4:   Determine the horizontal shift of the function by solving the equation

$$y = A \sin(\omega x - \phi) + B$$

for ϕ by choosing an ordered pair (x, y) from the data. Since answers will vary depending on the ordered pair selected, we will always choose the ordered pair for which y is smallest in order to maintain consistency.

NOW WORK PROBLEM 19(a)–(c).

Certain graphing utilities (such as a TI-83 and TI-86) have the capability of finding the sine function of best fit for sinusoidal data. At least four data points are required for this process.

EXAMPLE 4   **Finding the Sine Function of Best Fit**

Use a graphing utility to find the sine function of best fit for the data in Table 12. Graph this function with the scatter diagram of the data.

Figure 118

Solution   Enter the data from Table 12 and execute the SINe REGression program. The result is shown in Figure 118.

The output that the utility provides shows us the equation

$$y = a \sin(bx + c) + d$$

The sinusoidal function of best fit is

$$y = 21.15 \sin(0.55x - 2.35) + 51.19$$

Figure 119

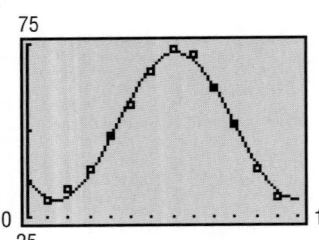

where *x* represents the month and *y* represents the average temperature.

Figure 119 shows the graph of the sinusoidal function of best fit on the scatter diagram.

━━━━━━ **NOW WORK PROBLEM 19(d) AND (e).**

Since the number of hours of sunlight in a day cycles annually, the number of hours of sunlight in a day for a given location can be modeled by a sinusoidal function.

The longest day of the year (in terms of hours of sunlight) occurs on the day of the summer solstice. The summer solstice is the time when the sun is farthest north (for locations in the northern hemisphere). In 2001, the summer solstice occurred on June 21 (the 172nd day of the year) at 3.38 AM EDT. The shortest day of the year occurs on the day of the winter solstice. The winter solstice is the time when the sun is farthest south (again, for locations in the northern hemisphere). In 2001, the winter solstice occurred on December 21 (the 355th day of the year) at 2:22 PM (EST).

EXAMPLE 5 **Finding a Sinusoidal Function for Hours of Daylight**

According to the *Old Farmer's Almanac*, the number of hours of sunlight in Boston on the summer solstice is 15.283 and the number of hours of sunlight on the winter solstice is 9.067.

(a) Find a sinusoidal function of the form $y = A\sin(\omega x - \phi) + B$ that fits the data.*

(b) Use the function found in part (a) to predict the number of hours of sunlight on April 1, the 91st day of the year.

(c) Draw a graph of the function found in part (a).

(d) Look up the number of hours of sunlight for April 1 in the *Old Farmer's Almanac* and compare the actual hours of daylight to the results found in part (b).

Solution (a) **STEP 1:** $\text{Amplitude} = \dfrac{\text{largest data value} - \text{smallest data value}}{2}$

$$= \frac{15.283 - 9.067}{2} = 3.108$$

STEP 2: $\text{Vertical shift} = \dfrac{\text{largest data value} + \text{smallest data value}}{2}$

$$= \frac{15.283 + 9.067}{2} = 12.175$$

STEP 3: The data repeat every 365 days. Since $T = \dfrac{2\pi}{\omega} = 365$, we find

$$\omega = \frac{2\pi}{365}$$

So far, we have $y = 3.108\sin\left(\dfrac{2\pi}{365}x - \phi\right) + 12.175$.

---

*Notice that only two data points are given, so a graphing utility cannot be used to find the sine function of best fit.

STEP 4: To determine the horizontal shift, we solve the equation

$$y = 3.108 \sin\left(\frac{2\pi}{365}x - \phi\right) + 12.175$$

for ϕ by letting $y = 9.067$ and $x = 355$ (the number of hours of daylight in Boston on December 21).

$$9.067 = 3.108 \sin\left(\frac{2\pi}{365} \cdot 355 - \phi\right) + 12.175$$

$$-3.108 = 3.108 \sin\left(\frac{2\pi}{365} \cdot 355 - \phi\right) \qquad \text{Subtract 12.175 from both sides of the equation.}$$

$$-1 = \sin\left(\frac{2\pi}{365} \cdot 355 - \phi\right) \qquad \text{Divide both sides of the equation by 3.108.}$$

$$\frac{2\pi}{365} \cdot 355 - \phi = -\frac{\pi}{2} \qquad \sin\theta = -1 \text{ when } \theta = -\frac{\pi}{2}.$$

$$\phi = \frac{357}{146}\pi \qquad \text{Solve for } \phi.$$

The function that provides the number of hours of daylight in Boston for any day, x, is given by

$$y = 3.108 \sin\left(\frac{2\pi}{365}x - \frac{357}{146}\pi\right) + 12.175$$

(b) To predict the number of hours of daylight on April 1, we let $x = 91$ in the function found in part (a) and obtain

$$y = 3.108 \sin\left(\frac{2\pi}{365} \cdot 91 - \frac{357}{146}\pi\right) + 12.175$$

$$\approx 12.69$$

So we predict that there will be about 12.69 hours of sunlight on April 1 in Boston.

(c) The graph of the function found in part (a) is given in Figure 120.

(d) According to the *Old Farmer's Almanac*, there will be 12 hours 43 minutes of sunlight on April 1 in Boston. Our prediction of 12.69 hours converts to 12 hours 41 minutes. ∎

Figure 120

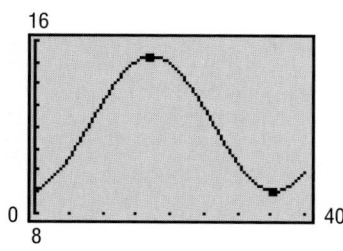

NOW WORK PROBLEM 25.

2.8 Concepts and Vocabulary

In Problem 1, fill in the blank.

1. For the graph of $y = A \sin(\omega x - \phi)$, the number $\frac{\phi}{\omega}$ is called the _____ .

In Problem 2, answer True or False.

2. Only two data points are required by a graphing utility to find the sine function of best fit.

2.8 Exercises

In Problems 1–12, find the amplitude, period, and phase shift of each function. Graph each function by hand. Show at least one period. Verify the result using a graphing utility.

1. $y = 4\sin(2x - \pi)$

2. $y = 3\sin(3x - \pi)$

3. $y = 2\cos\left(3x + \dfrac{\pi}{2}\right)$

4. $y = 3\cos(2x + \pi)$

5. $y = -3\sin\left(2x + \dfrac{\pi}{2}\right)$

6. $y = -2\cos\left(2x - \dfrac{\pi}{2}\right)$

7. $y = 4\sin(\pi x + 2)$

8. $y = 2\cos(2\pi x + 4)$

9. $y = 3\cos(\pi x - 2)$

10. $y = 2\cos(2\pi x - 4)$

11. $y = 3\sin\left(-2x + \dfrac{\pi}{2}\right)$

12. $y = 3\cos\left(-2x + \dfrac{\pi}{2}\right)$

In Problems 13–16, write the equation of a sine function, $y = A\sin(\omega x - \phi)$, $A > 0$, that has the given characteristics.

13. Amplitude: 2

Period: π

Phase shift: $\dfrac{1}{2}$

14. Amplitude: 3

Period: $\dfrac{\pi}{2}$

Phase shift: 2

15. Amplitude: 3

Period: 3π

Phase shift: $-\dfrac{1}{3}$

16. Amplitude: 2

Period: π

Phase shift: -2

---

17. Alternating Current (ac) Circuits The current I, in amperes, flowing through an ac (alternating current) circuit at time t is

$$I = 120\sin\left(30\pi t - \dfrac{\pi}{3}\right), \qquad t \geq 0$$

What is the period? What is the amplitude? What is the phase shift? Graph this function over two periods.

18. Alternating Current (ac) Circuits The current I, in amperes, flowing through an ac (alternating current) circuit at time t is

$$I = 220\sin\left(60\pi t - \dfrac{\pi}{6}\right), \qquad t \geq 0$$

What is the period? What is the amplitude? What is the phase shift? Graph this function over two periods.

19. Monthly Temperature The following data represent the average monthly temperatures for Juneau, Alaska.

Month, x	Average Monthly Temperature, °F
January, 1	24.2
February, 2	28.4
March, 3	32.7
April, 4	39.7
May, 5	47.0
June, 6	53.0
July, 7	56.0
August, 8	55.0
September, 9	49.4
October, 10	42.2
November, 11	32.0
December, 12	27.1

Source: U.S. National Oceanic and Atmospheric Administration.

(a) Use a graphing utility to draw a scatter diagram of the data for one period.

(b) By hand, find a sinusoidal function of the form $y = A\sin(\omega x - \phi) + B$ that fits the data.

(c) Draw the sinusoidal function found in part (b) on the scatter diagram.

(d) Use a graphing utility to find the sinusoidal function of best fit.

(e) Draw the sinusoidal function of best fit on the scatter diagram.

20. Monthly Temperature The following data represent the average monthly temperatures for Washington, D.C.

Month, x	Average Monthly Temperature, °F
January, 1	34.6
February, 2	37.5
March, 3	47.2
April, 4	56.5
May, 5	66.4
June, 6	75.6
July, 7	80.0
August, 8	78.5
September, 9	71.3
October, 10	59.7
November, 11	49.8
December, 12	39.4

Source: U.S. National Oceanic and Atmospheric Administration.

(a) Use a graphing utility to draw a scatter diagram of the data for one period.

(b) By hand, find a sinusoidal function of the form $y = A \sin(\omega x - \phi) + B$ that fits the data.

(c) Draw the sinusoidal function found in part (b) on the scatter diagram.

(d) Use a graphing utility to find the sinusoidal function of best fit.

(e) Graph the sinusoidal function of best fit on the scatter diagram.

21. Monthly Temperature The following data represent the average monthly temperatures for Indianapolis, Indiana.

Month, x	Average Monthly Temperature, °F
January, 1	25.5
February, 2	29.6
March, 3	41.4
April, 4	52.4
May, 5	62.8
June, 6	71.9
July, 7	75.4
August, 8	73.2
September, 9	66.6
October, 10	54.7
November, 11	43.0
December, 12	30.9

Source: U.S. National Oceanic and Atmospheric Administration.

(a) Use a graphing utility to draw a scatter diagram of the data for one period.

(b) By hand, find a sinusoidal function of the form $y = A \sin(\omega x - \phi) + B$ that fits the data.

(c) Draw the sinusoidal function found in part (b) on the scatter diagram.

(d) Use a graphing utility to find the sinusoidal function of best fit.

(e) Graph the sinusoidal function of best fit on the scatter diagram.

22. Monthly Temperature The data at the top of the right column represent the average monthly temperatures for Baltimore, Maryland.

(a) Use a graphing utility to draw a scatter diagram of the data for one period.

(b) By hand, find a sinusoidal function of the form $y = A \sin(\omega x - \phi) + B$ that fits the data.

(c) Draw the sinusoidal function found in part (b) on the scatter diagram.

(d) Use a graphing utility to find the sinusoidal function of best fit.

(e) Graph the sinusoidal function of best fit on the scatter diagram.

Month, x	Average Monthly Temperature, °F
January, 1	31.8
February, 2	34.8
March, 3	44.1
April, 4	53.4
May, 5	63.4
June, 6	72.5
July, 7	77.0
August, 8	75.6
September, 9	68.5
October, 10	56.6
November, 11	46.8
December, 12	36.7

Source: U.S. National Oceanic and Atmospheric Administration.

23. Tides Suppose that the length of time between consecutive high tides is approximately 12.5 hours. According to the National Oceanic and Atmospheric Administration, on Saturday, June 28, 1997, in Savannah, Georgia, high tide occurred at 3:38 AM (3.6333 hours) and low tide occurred at 10:08 AM (10.1333 hours). Water heights are measured as the amounts above or below the mean lower low water. The height of the water at high tide was 8.2 feet and the height of the water at low tide was −0.6 foot.

(a) Approximately when will the next high tide occur?

(b) Find a sinusoidal function of the form $y = A \sin(\omega x - \phi) + B$ that fits the data.

(c) Draw a graph of the function found in part (b).

(d) Use the function found in part (b) to predict the height of the water at the next high tide.

24. Tides Suppose that the length of time between consecutive high tides is approximately 12.5 hours. According to the National Oceanic and Atmospheric Administration, on Saturday, June 28, 1997, in Juneau, Alaska, high tide occurred at 8:11 AM (8.1833 hours) and low tide occurred at 2:14 PM (14.2333 hours). Water heights are measured as the amounts above or below the mean lower low water. The height of the water at high tide was 13.2 feet and the height of the water at low tide was 2.2 feet.

(a) Approximately when will the next high tide occur?

(b) Find a sinusoidal function of the form $y = A \sin(\omega x - \phi) + B$ that fits the data.

(c) Draw a graph of the function found in part (b).

(d) Use the function found in part (b) to predict the height of the water at the next high tide.

25. Hours of Daylight According to the *Old Farmer's Almanac*, in Miami, Florida, the number of hours of sunlight on the summer solstice is 12.75 and the number of hours of sunlight on the winter solstice is 10.583.

(a) Find a sinusoidal function of the form $y = A \sin(\omega x - \phi) + B$ that fits the data.

(b) Draw a graph of the function found in part (a).
(c) Use the function found in part (a) to predict the number of hours of sunlight on April 1, the 91st day of the year.

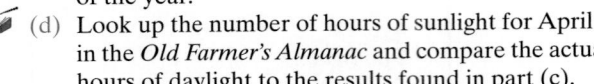

(d) Look up the number of hours of sunlight for April 1 in the *Old Farmer's Almanac* and compare the actual hours of daylight to the results found in part (c).

26. Hours of Daylight   According to the *Old Farmer's Almanac*, in Detroit, Michigan, the number of hours of sunlight on the summer solstice is 13.65 and the number of hours of sunlight on the winter solstice is 9.067.
(a) Find a sinusoidal function of the form
  $y = A \sin(\omega x - \phi) + B$ that fits the data.
(b) Draw a graph of the function found in part (a).
(c) Use the function found in part (a) to predict the number of hours of sunlight on April 1, the 91st day of the year.
(d) Look up the number of hours of sunlight for April 1 in the *Old Farmer's Almanac* and compare the actual hours of daylight to the results found in part (c).

27. Hours of Daylight   According to the *Old Farmer's Almanac*, in Anchorage, Alaska, the number of hours of sunlight on the summer solstice is 16.233 and the number of hours of sunlight on the winter solstice is 5.45.

(a) Find a sinusoidal function of the form
  $y = A \sin(\omega x - \phi) + B$ that fits the data.
(b) Draw a graph of the function found in part (a).
(c) Use the function found in part (a) to predict the number of hours of sunlight on April 1, the 91st day of the year.
(d) Look up the number of hours of sunlight for April 1 in the *Old Farmer's Almanac* and compare the actual hours of daylight to the results found in part (c).

28. Hours of Daylight   According to the *Old Farmer's Almanac*, in Honolulu, Hawaii, the number of hours of sunlight on the summer solstice is 12.767 and the number of hours of sunlight on the winter solstice is 10.783.
(a) Find a sinusoidal function of the form
  $y = A \sin(\omega x - \phi) + B$ that fits the data.
(b) Draw a graph of the function found in part (a).
(c) Use the function found in part (a) to predict the number of hours of sunlight on April 1, the 91st day of the year.
(d) Look up the number of hours of sunlight for April 1 in the *Old Farmer's Almanac* and compare the actual hours of daylight to the results found in part (c).

Chapter Review

Things To Know

Definitions

Angle in standard position (p. 99)	Vertex is at the origin; initial side is along the positive x-axis
Degree (1°) (p. 99)	$1° = \dfrac{1}{360}$ revolution
Radian (1) (p. 102)	The measure of a central angle of a circle whose rays subtend an arc whose length is the radius of the circle
Acute angle (p. 112)	An angle θ whose measure is $0° < \theta < 90°$ $\left(\text{or } 0 < \theta < \dfrac{\pi}{2}\right)$
Trigonometric functions (p. 113)	$P = (a, b)$ is the point on the terminal side of θ a distance r from the origin:

$$\sin\theta = \frac{b}{r} \qquad \cos\theta = \frac{a}{r} \qquad \tan\theta = \frac{b}{a}, \quad a \neq 0$$

$$\csc\theta = \frac{r}{b}, \quad b \neq 0 \qquad \sec\theta = \frac{r}{a}, \quad a \neq 0 \qquad \cot\theta = \frac{a}{b}, \quad b \neq 0$$

Complementary angles (p. 118)	Two acute angles whose sum is $90°\left(\dfrac{\pi}{2}\right)$
Cofunction (p. 118)	The following pairs of functions are cofunctions of each other: sine and cosine; tangent and cotangent; secant and cosecant
Complementary Angle Theorem (p. 119)	Cofunctions of complementary angles are equal
Reference angle of θ (p. 137)	The acute angle formed by the terminal side of θ and either the positive or negative x-axis
Periodic function (p. 149)	$f(\theta + p) = f(\theta)$, for all θ, $p > 0$, where the smallest such p is the fundamental period

Formulas

1 revolution counterclockwise $= 360° = 2\pi$ (p. 101)

counterclockwise $= 2\pi$ radians (p. 104)

$s = r\theta$ (p. 103)

θ is measured in radians; s is the length of arc subtended by the central angle θ of the circle of radius r; A is the area of the sector

$A = \dfrac{1}{2}r^2\theta$ (p. 106)

$v = r\omega$ (p. 107)

v is the linear speed along the circle of radius r; ω is the angular speed (measured in radians per unit time)

TABLE OF VALUES

θ (Radians)	θ (Degrees)	$\sin\theta$	$\cos\theta$	$\tan\theta$	$\csc\theta$	$\sec\theta$	$\cot\theta$
0	0°	0	1	0	Not defined	1	Not defined
$\dfrac{\pi}{6}$	30°	$\dfrac{1}{2}$	$\dfrac{\sqrt{3}}{2}$	$\dfrac{\sqrt{3}}{3}$	2	$\dfrac{2\sqrt{3}}{3}$	$\sqrt{3}$
$\dfrac{\pi}{4}$	45°	$\dfrac{\sqrt{2}}{2}$	$\dfrac{\sqrt{2}}{2}$	1	$\sqrt{2}$	$\sqrt{2}$	1
$\dfrac{\pi}{3}$	60°	$\dfrac{\sqrt{3}}{2}$	$\dfrac{1}{2}$	$\sqrt{3}$	$\dfrac{2\sqrt{3}}{3}$	2	$\dfrac{\sqrt{3}}{3}$
$\dfrac{\pi}{2}$	90°	1	0	Not defined	1	Not defined	0
π	180°	0	−1	0	Not defined	−1	Not defined
$\dfrac{3\pi}{2}$	270°	−1	0	Not defined	−1	Not defined	0

Fundamental Identities (p. 115)

$$\tan\theta = \frac{\sin\theta}{\cos\theta}, \quad \cot\theta = \frac{\cos\theta}{\sin\theta}$$

$$\cot\theta = \frac{1}{\tan\theta}, \quad \sec\theta = \frac{1}{\cos\theta}, \quad \csc\theta = \frac{1}{\sin\theta}$$

$$\sin^2\theta + \cos^2\theta = 1, \quad \tan^2\theta + 1 = \sec^2\theta, \quad 1 + \cot^2\theta = \csc^2\theta$$

Properties of the Trigonometric Functions

$y = \sin x$ Domain: $-\infty < x < \infty$

(p. 155)   Range: $-1 \le y \le 1$

       Periodic: period $= 2\pi(360°)$

       Odd function

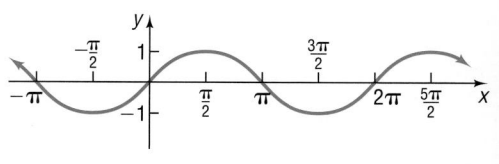

$y = \cos x$ Domain: $-\infty < x < \infty$

(p. 157)   Range: $-1 \le y \le 1$

       Periodic: period $= 2\pi(360°)$

       Even function

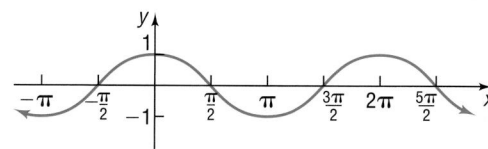

$y = \tan x$
(p. 172)

Domain: $-\infty < x < \infty$, except odd multiples of $\dfrac{\pi}{2}$ $(90°)$

Range: $-\infty < y < \infty$

Periodic: period $= \pi(180°)$

Odd function

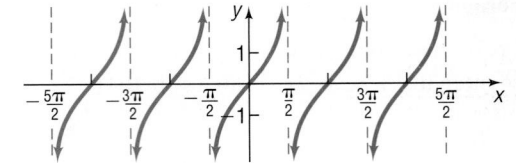

$y = \cot x$
(p. 173)

Domain: $-\infty < x < \infty$, except integral multiples of $\pi(180°)$

Range: $-\infty < y < \infty$

Periodic: period $= \pi(180°)$

Odd function

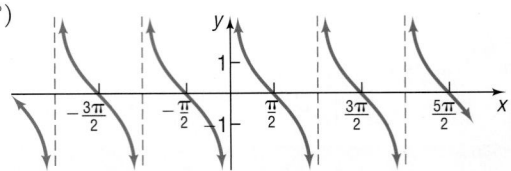

$y = \csc x$
(p. 174)

Domain: $-\infty < x < \infty$, except integral multiples of $\pi(180°)$

Range: $|y| \geq 1$

Periodic: period $= 2\pi(360°)$

Odd function

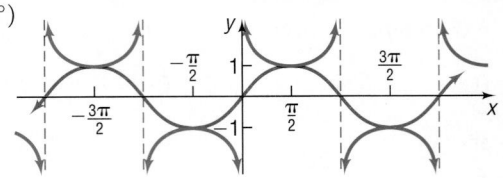

$y = \sec x$
(p. 175)

Domain: $-\infty < x < \infty$, except odd multiples of $\dfrac{\pi}{2}$ $(90°)$

Range: $|y| \geq 1$

Periodic: period $= 2\pi(360°)$

Even function

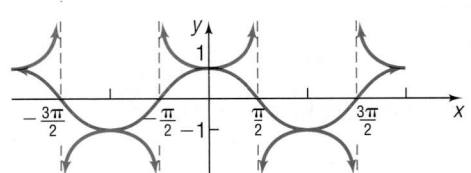

Sinusoidal graphs (pp. 161 and 177)

$y = A \sin(\omega x), \quad \omega > 0$

$y = A \cos(\omega x), \quad \omega > 0$

$y = A \sin(\omega x - \phi) = A \sin\left[\omega\left(x - \dfrac{\phi}{\omega}\right)\right]$

$y = A \cos(\omega x - \phi) = A \cos\left[\omega\left(x - \dfrac{\phi}{\omega}\right)\right]$

Period $= \dfrac{2\pi}{\omega}$

Amplitude $= |A|$

Phase shift $= \dfrac{\phi}{\omega}$

▉ Objectives

Section	You should be able to:	Review Exercises
2.1	**1** Convert between degrees, minutes, seconds, and decimal forms for angles (p. 101)	82
	2 Find the arc length of a circle (p. 102)	83–84
	3 Convert from degrees to radians (p. 104)	1–4
	4 Convert from radians to degrees (p. 105)	5–8
	5 Find the area of a sector of a circle (p. 106)	83
	6 Find the linear speed of an object traveling in circular motion (p. 107)	85–88
2.2	**1** Find the value of trigonometric functions of acute angles (p. 113)	75
	2 Use the fundamental identities (p. 114)	21–24
	3 Find the remaining trigonometric functions given the value of one of them (p. 116)	31–32
	4 Use the complementary angle theorem (p. 118)	25–26

2.3	①	Find the exact value of the trigonometric functions of $\frac{\pi}{4} = 45°$ (p. 124)	9–12
	②	Find the exact value of the trigonometric functions of $\frac{\pi}{6} = 30°$ and $\frac{\pi}{3} = 60°$ (p. 125)	9–12
	③	Use a calculator to approximate the value of the trigonometric functions of acute angles (p. 127)	76
2.4	①	Find the exact value of the trigonometric functions for general angles (p. 132)	77
	②	Use coterminal angles to find the exact value of a trigonometric function (p. 134)	19–20
	③	Determine the sign of the trigonometric functions of an angle in a given quadrant (p. 136)	78
	④	Find the reference angle of a general angle (p. 137)	79
	⑤	Use the theorem on reference angles (p. 138)	13–16, 19–20
	⑥	Find the exact value of the trigonometric functions of an angle given one of them and the quadrant of the angle (p. 139)	31–46
2.5	①	Find the exact value of the trigonometric functions using the unit circle (p. 142)	80
	②	Know the domain and range of the trigonometric functions (p. 146)	81
	③	Use the periodic properties to find the exact values of the trigonometric functions (p. 149)	19–20
	④	Use even-odd properties to find the exact value of the trigonometric functions (p. 150)	13, 15–16, 18–20, 27–30
2.6	①	Graph transformations of the sine function (p. 155)	47, 50
	②	Graph transformations of the cosine function (p. 158)	48, 49
	③	Determine the amplitude and period of sinusoidal functions (p. 160)	59–64, 89
	④	Graph sinusoidal functions: $y = A\sin(\omega x)$ (p. 161)	47, 48, 63, 64, 89
	⑤	Find an equation for a sinusoidal graph (p. 165)	71–74
2.7	①	Graph transformations of the tangent function and cotangent function (p. 170)	51–56
	②	Graph transformations of the cosecant function and secant function (p. 174)	57–58
2.8	①	Determine the phase shift of a sinusoidal function (p. 177)	65–70, 90
	②	Graph sinusoidal functions: $y = A\sin(\omega x - \phi)$ (p. 177)	65–70, 90
	③	Find a sinusoidal function from data (p. 181)	91–94

Review Exercises

Blue problem numbers indicate the authors' suggestions for use in a Practice Test.

In Problems 1–4, convert each angle in degrees to radians. Express your answer as a multiple of π.
1. 135°   2. 210°   3. 18°   4. 15°

In Problems 5–8, convert each angle in radians to degrees.
5. $\frac{3\pi}{4}$ 6. $\frac{2\pi}{3}$ 7. $-\frac{5\pi}{2}$ 8. $-\frac{3\pi}{2}$

In Problems 9–30, find the exact value of each expression. Do not use a calculator.

9. $\tan\dfrac{\pi}{4} - \sin\dfrac{\pi}{6}$

10. $\cos\dfrac{\pi}{3} + \sin\dfrac{\pi}{4}$

11. $3\sin 45° - 4\tan\dfrac{\pi}{6}$

12. $4\cos 60° + 3\tan\dfrac{\pi}{3}$

13. $6\cos\dfrac{3\pi}{4} + 2\tan\left(-\dfrac{\pi}{3}\right)$

14. $3\sin\dfrac{2\pi}{3} - 4\cos\dfrac{5\pi}{2}$

15. $\sec\left(-\dfrac{\pi}{3}\right) - \cot\left(-\dfrac{5\pi}{4}\right)$

16. $4\csc\dfrac{3\pi}{4} - \cot\left(-\dfrac{\pi}{4}\right)$

17. $\tan\pi + \sin\pi$

18. $\cos\dfrac{\pi}{2} - \csc\left(-\dfrac{\pi}{2}\right)$

19. $\cos 540° - \tan(-45°)$

20. $\sin 630° + \cos(-180°)$

21. $\sin^2 20° + \dfrac{1}{\sec^2 20°}$

22. $\dfrac{1}{\cos^2 40°} - \dfrac{1}{\cot^2 40°}$

23. $\sec 50° \cos 50°$

24. $\tan 10° \cot 10°$

25. $\dfrac{\sin 50°}{\cos 40°}$

26. $\dfrac{\tan 20°}{\cot 70°}$

27. $\dfrac{\sin(-40°)}{\cos 50°}$

28. $\tan(-20°)\cot 20°$

29. $\sin 400° \sec(-50°)$

30. $\cot 200° \cot(-70°)$

In Problems 31–46, find the exact value of each of the remaining trigonometric functions.

31. $\sin\theta = \dfrac{4}{5}$, θ acute

32. $\cos\theta = \dfrac{3}{5}$, θ acute

33. $\tan\theta = \dfrac{12}{5}$, $\sin\theta < 0$

34. $\cot\theta = \dfrac{12}{5}$, $\cos\theta < 0$

35. $\sec\theta = -\dfrac{5}{4}$, $\tan\theta < 0$

36. $\csc\theta = -\dfrac{5}{3}$, $\cot\theta < 0$

37. $\sin\theta = \dfrac{12}{13}$, θ in quadrant II

38. $\cos\theta = -\dfrac{3}{5}$, θ in quadrant III

39. $\sin\theta = -\dfrac{5}{13}$, $\dfrac{3\pi}{2} < \theta < 2\pi$

40. $\cos\theta = \dfrac{12}{13}$, $\dfrac{3\pi}{2} < \theta < 2\pi$

41. $\tan\theta = \dfrac{1}{3}$, $180° < \theta < 270°$

42. $\tan\theta = -\dfrac{2}{3}$, $90° < \theta < 180°$

43. $\sec\theta = 3$, $\dfrac{3\pi}{2} < \theta < 2\pi$

44. $\csc\theta = -4$, $\pi < \theta < \dfrac{3\pi}{2}$

45. $\cot\theta = -2$, $\dfrac{\pi}{2} < \theta < \pi$

46. $\tan\theta = -2$, $\dfrac{3\pi}{2} < \theta < 2\pi$

In Problems 47–58, graph each function. Each graph should contain at least one period.

47. $y = 2\sin(4x)$

48. $y = -3\cos(2x)$

49. $y = -2\cos\left(x + \dfrac{\pi}{2}\right)$:

50. $y = 3\sin(x - \pi)$

51. $y = \tan(x + \pi)$

52. $y = -\tan\left(x - \dfrac{\pi}{2}\right)$

53. $y = -2\tan(3x)$

54. $y = 4\tan(2x)$

55. $y = \cot\left(x + \dfrac{\pi}{4}\right)$

56. $y = -4\cot(2x)$

57. $y = \sec\left(x - \dfrac{\pi}{4}\right)$

58. $y = \csc\left(x + \dfrac{\pi}{4}\right)$

In Problems 59–62, determine the amplitude and period of each function without graphing.

59. $y = 4\cos x$

60. $y = \sin(2x)$

61. $y = -8\sin\left(\dfrac{\pi}{2}x\right)$

62. $y = -2\cos(3\pi x)$

In Problems 63–70, find the amplitude, period, and phase shift of each function. Graph each function. Show at least one period.

63. $y = 4\sin(3x)$

64. $y = 2\cos\left(\dfrac{1}{3}x\right)$

65. $y = 2\sin(2x - \pi)$

66. $y = -\cos\left(\dfrac{1}{2}x + \dfrac{\pi}{2}\right)$

67. $y = \dfrac{1}{2}\sin\left(\dfrac{3}{2}x - \pi\right)$

68. $y = \dfrac{3}{2}\cos(6x + 3\pi)$

69. $y = -\dfrac{2}{3}\cos(\pi x - 6)$

70. $y = -7\sin\left(\dfrac{\pi}{3}x + \dfrac{4}{3}\right)$

In Problems 71–74, find a function whose graph is given.

71.

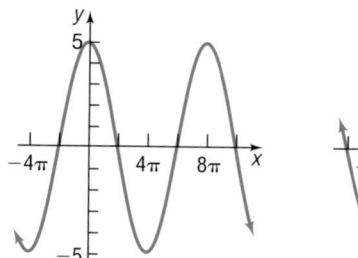

72.

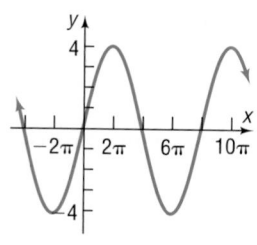

73.

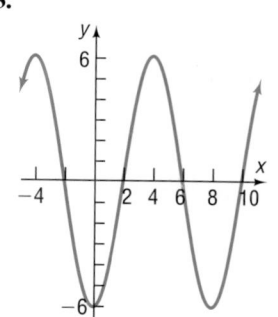

74.

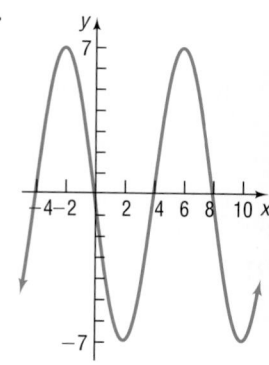

75. Find the value of each of the six trigonometric functions of the angle θ in the illustration.

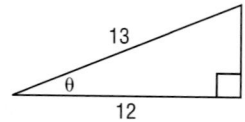

76. Use a calculator to approximate $\sec 10°$. Round the answer to two decimal places.

77. Find the exact value of each of the six trigonometric functions of an angle θ if $(3, -4)$ is a point on the terminal side of θ.

78. Name the quadrant θ lies in if $\cos \theta > 0$ and $\tan \theta < 0$.

79. Find the reference angle of $-\dfrac{4\pi}{5}$.

80. Find the exact value of $\sin t$, $\cos t$, and $\tan t$ if $P = \left(-\dfrac{3}{5}, \dfrac{4}{5}\right)$ is the point on the unit circle that corresponds to t.

81. Find the domain and range of the secant function.

82. (a) Convert the angle $32°20'35''$ to a decimal in degrees. Round the answer to two decimal places.
 (b) Convert the angle $63.18°$ to $D°M'S''$ form. Express the answer to the nearest second.

83. Find the length of arc subtended by a central angle of $30°$ on a circle of radius 2 feet. What is the area of the sector?

84. The minute hand of a clock is 8 inches long. How far does the tip of the minute hand move in 30 minutes? How far does it move in 20 minutes?

85. Angular Speed of a Race Car   A race car is driven around a circular track at a constant speed of 180 miles per hour. If the diameter of the track is $\dfrac{1}{2}$ mile, what is the angular speed of the car? Express your answer in revolutions per hour (which is equivalent to laps per hour).

86. Merry-Go-Rounds   A neighborhood carnival has a merry-go-round whose radius is 25 feet. If the time for one revolution is 30 seconds, how fast is the merry-go-round going?

87. Lighthouse Beacons   The Montauk Point Lighthouse on Long Island has dual beams (two light sources opposite each other). Ships at sea observe a blinking light every 5 seconds. What rotation speed is required to do this?

88. Spin Balancing Tires   The radius of each wheel of a car is 16 inches. At how many revolutions per minute should a spin balancer be set to balance the tires at a speed of 90 miles per hour? Is the setting different for a wheel of radius 14 inches? If so, what is this setting?

89. Alternating Voltage   The electromotive force E, in volts, in a certain ac circuit obeys the equation
$$E = 120 \sin(120\pi t), \quad t \geq 0$$
where t is measured in seconds.
 (a) What is the maximum value of E?
 (b) What is the period?
 (c) Graph this function over two periods.

90. Alternating Current   The current I, in amperes, flowing through an ac (alternating current) circuit at time t is
$$I = 220 \sin\left(30\pi t + \dfrac{\pi}{6}\right), \quad t \geq 0$$
 (a) What is the period?
 (b) What is the amplitude?
 (c) What is the phase shift?
 (d) Graph this function over two periods.

91. Monthly Temperature   The following data represent the average monthly temperatures for Phoenix, Arizona.

Month, m	Average Monthly Temperature, T
January, 1	51
February, 2	55
March, 3	63
April, 4	67
May, 5	77
June, 6	86
July, 7	90
August, 8	90
September, 9	84
October, 10	71
November, 11	59
December, 12	52

Source: U.S. National Oceanic and Atmospheric Administration.

(a) Use a graphing utility to draw a scatter diagram of the data for one period.
(b) By hand, find a sinusoidal function of the form $y = A \sin(\omega x - \phi) + B$ that fits the data.
(c) Draw the sinusoidal function found in part (b) on the scatter diagram.
(d) Use a graphing utility to find the sinusoidal function of best fit.
(e) Graph the sinusoidal function of best fit on the scatter diagram.

92. Monthly Temperature   The following data represent the average monthly temperatures for Chicago, Illinois.

Month, m	Average Monthly Temperature, T
January, 1	25
February, 2	28
March, 3	36
April, 4	48
May, 5	61
June, 6	72
July, 7	74
August, 8	75
September, 9	66
October, 10	55
November, 11	39
December, 12	28

Source: U.S. National Oceanic and Atmospheric Administration.

(a) Use a graphing utility to draw a scatter diagram of the data for one period.

(b) By hand, find a sinusoidal function of the form $y = A \sin(\omega x - \phi) + B$ that fits the data.
(c) Draw the sinusoidal function found in part (b) on the scatter diagram.
(d) Use a graphing utility to find the sinusoidal function of best fit.
(e) Graph the sinusoidal function of best fit on the scatter diagram.

93. Hours of Daylight   According to the *Old Farmer's Almanac*, in Las Vegas, Nevada, the number of hours of sunlight on the summer solstice is 13.367 and the number of hours of sunlight on the winter solstice is 9.667.
(a) Find a sinusoidal function of the form
$$y = A \sin(\omega x - \phi) + B$$
that fits the data.
(b) Draw a graph of the function found in part (a).
(c) Use the function found in part (a) to predict the number of hours of sunlight on April 1, the 91st day of the year.
(d) Look up the number of hours of sunlight for April 1 in the *Old Farmer's Almanac* and compare the actual hours of daylight to the results found in part (c).

94. Hours of Daylight   According to the *Old Farmer's Almanac*, in Seattle, Washington, the number of hours of sunlight on the summer solstice is 13.967 and the number of hours of sunlight on the winter solstice is 8.417.
(a) Find a sinusoidal function of the form
$$y = A \sin(\omega x - \phi) + B$$
that fits the data.
(b) Draw a graph of the function found in part (a).
(c) Use the function found in part (a) to predict the number of hours of sunlight on April 1, the 91st day of the year.
(d) Look up the number of hours of sunlight for April 1 in the *Old Farmer's Almanac* and compare the actual hours of daylight to the results found in part (c).

Chapter Projects

1. **Tides**   A partial tide table for September, 2001, for Sabine Pass along the Texas Gulf Coast is given in the table on page 195:

(a) On September 15, when was the tide high? This is called "high tide". On September 19, when was the tide low? This is called "low tide". Most days will have two low tides and two high tides.

(b) Why do you think there is a negative height for the low tide on September 14? What is the tide height measured against?

(c) On your graphing utility, draw a scatter diagram for the data in the table. Let T (time) be the independent variable, with $T = 0$ being 12:00 A.M. on September 1, $T = 24$ being 12:00 A.M. on September 2, etc. Remember that there are 60 minutes in an hour. Let H be the height in feet when converting the times. Also, make sure your graphing utility is in radian mode.

(d) What shape does the data take? What is the period of the data? What is the amplitude? Is the amplitude constant? Explain.

Sept	High Tide Time	Ht(ft)	High Tide Time	Ht(ft)	Low Tide Time	Ht(ft)	Low Tide Time	Ht(ft)	Sun/Moon phase Rise/Set
F 14	03:08a	2.4	11:12a	2.2	08:14a	2.0	07:19p	−0.1	7:00a/7:23p
S 15	03:33a	2.4	12:56p	2.2	08:15a	1.9	08:13p	0.0	7:00a/7:22p
S 16	03:57a	2.3	02:17p	2.3	08:45a	1.6	09:05p	0.3	7:01a/7:20p
M17	04:20a	2.2	03:33p	2.3	09:24a	1.4	09:54p	0.5	7:01a/7:19p
T18	04:41a	2.2	04:47p	2.3	10:08a	1.0	10:43p	1.0	7:02a/7:08p
W19	05:01a	2.0	06:04p	2.3	10:54a	0.7	11:32p	1.4	7:02a/7:17p
T20	05:20a	2.0	07:27p	2.3	11:44a	0.4			7:03a/7:15p

Date from: *www.harbortides.com*

(e) Using Steps 1–4, given on page 183, fit a sine curve to the data. Let the amplitude be the average of the amplitudes that you found in part (c), unless the amplitude was constant. Is there a vertical shift? Is there a phase shift?

(f) Using your graphing utility, find the sinusoidal function of best fit. How does it compare to your equation?

(g) Using the equation found in part (e) and using the sinusoidal equation of best fit found in part (f), predict the high tides and the low tides on September 21.

(h) Looking at the times of day that the low tides occur, what do you think causes the low tides to vary so much each day? Explain. Does this seem to have the same type of effect on the high tides? Explain.

2. **Identifying Mountain Peaks in Hawaii** Suppose that you are standing on the southeastern shore of Oahu and you see three mountain peaks on the horizon. You want to determine which mountains are visible from Oahu. The possible mountain peaks that can be seen from Oahu and the height (above sea level) of their peaks are:

Island	Distance (miles)	Mountain	Height (feet)
Lanai	65	Lanaihale	3,370
Maui	110	Haleakala	10,023
Hawaii	190	Mauna Kea	13,796
Molokai	40	Kamakou	4,961

(a) To determine which of these mountain peaks would be visible from Oahu, consider that you are standing on the shore and looking "straight out" so that your line of sight is tangent to the surface of Earth at the point you're standing. Make a sketch of the right triangle formed by your sight line, the radius from the center of Earth to the point you are standing, and the line from the center of Earth through Lanai.

(b) Assuming that the radius of Earth is 3960 miles, determine the angle formed at the center of Earth.

(c) Determine the length of the hypotenuse of the triangle. Is Lanaihale visible from Oahu?

(d) Repeat parts (a)–(d) for the other three islands.

(e) Which three mountains are visible from Oahu?

3. **CBL Experiment** Using a CBL, the microphone probe and a tuning fork, record the amplitude, frequency, and period of the sound from the graph of the sound created by the tuning fork over time. Repeat the experiment for different tuning forks.

Cumulative Review

1. Find the real solutions, if any, of the equation
   $2x^2 + x - 1 = 0$.

2. Find an equation for the line with slope -3, containing the point $(-2, 5)$.

3. Find an equation for the circle of radius 4 and center at the point $(0, -2)$.

4. Graph the equation $2x - 3y = 12$.

5. Graph the equation $x^2 + y^2 - 2x + 4y - 4 = 0$.

6. Use transformations to graph the function
   $y = (x - 3)^2 + 2$.

7. Sketch a graph of each of the following functions. Label at least three points on each graph.
   (a) $y = x^2$
   (b) $y = x^3$
   (c) $y = \sin x$
   (d) $y = \tan x$

8. Find the inverse function of $f(x) = 3x - 2$.

9. Find the exact value of $(\sin 14°)^2 + (\sin 76°)^2 - 3$.

10. Graph $y = 3\sin(2x)$.

11. Find the exact value of $\tan \dfrac{\pi}{4} - 3\cos \dfrac{\pi}{6} + \csc \dfrac{\pi}{6}$.

12. Find a sinusoidal function for the following graph.

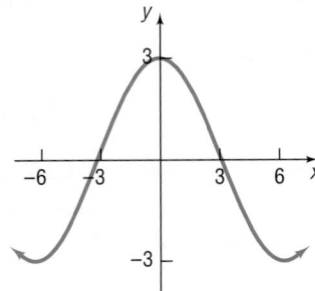

ANALYTIC TRIGONOMETRY

Tremor Brought First Hint of Doom

Underwater earthquakes that triggered the devastating tsunami in PNG would have been felt by villagers about 30 minutes before the waves struck, scientists said yesterday. "The tremor was felt by coastal residents who may not have realised its significance or did not have time to retreat," said associate Professor Ted Bryant, a geoscientist at the University of Wollongong.

The tsunami would have sounded like a fleet of bombers as it crashed into a 30-kilometre stretch of coast. "The tsunami would have been caused by a rapid uplift or drop of the sea floor," said an applied mathematician and cosmologist from Monash University, Professor Joe Monaghan, who is one of Australia's leading experts on tsunamis . . . .

Tsunamis, ridges of water hundreds of kilometres long and stretching from front to back for several kilometres, line up parallel to the beach. "We are talking about a huge volume of water moving very fast-300 kilometres per hour would be typical," Professor Monaghan said.

(SOURCE: Peter Spinks, The Age, Tuesday, July 21, 1998.)

SEE CHAPTER PROJECT 1.

OUTLINE

For additional study help, go to

www.prenhall.com/sullivanegu3e

Materials include:

- Graphing Calculator Help
- Chapter Quiz
- Chapter Test
- PowerPoint Downloads
- Chapter Projects
- Student Tips

197

A Look Back, A Look Forward

In Chapter 1, we defined inverse functions and developed their properties, particularly the relationship between the domain and range of a function and its inverse. We learned that the graph of a function and its inverse are symmetric with respect to the line $y = x$. In the first two sections of this chapter, we define the six inverse trigonometric functions and investigate their properties.

In Chapter 2, we derived several identities involving the trigonometric functions. In Sections 3.3 through 3.6 of this chapter, we continue the derivation of identities. These identities play an important role in calculus, the physical and life sciences, and economics, where they are used to simplify complicated expressions. The last two sections of this chapter deal with equations that contain trigonometric functions.

PREPARING FOR THIS SECTION

Before getting started, review the following:

✓ Inverse Functions (Section 1.7, pp. 77–87)

✓ Definition of the Trigonometric Functions (Section 2.2, p. 113)

✓ Graphs of the Sine, Cosine, and Tangent Functions (Section 2.6, pp. 154–159, and Section 2.8, pp. 177–180)

✓ Values of the Trigonometric Functions of Certain Angles (Section 2.3, p. 126, and Section 2.4, p. 134)

✓ Domain and Range of the Sine, Cosine, and Tangent Functions (Section 2.5, pp. 146–147)

3.1 THE INVERSE SINE, COSINE, AND TANGENT FUNCTIONS

OBJECTIVES

1. Find the Exact Value of the Inverse Sine, Cosine, and Tangent Functions
2. Find an Approximate Value of the Inverse Sine, Cosine, and Tangent Functions

In Section 1.7 we discussed inverse functions, and we noted that if a function is one-to-one it will have an inverse function. We also observed that if a function is not one-to-one it may be possible to restrict its domain in some suitable manner so that the restricted function is one-to-one.

Next, we review some properties of a function f and its inverse function f^{-1}.

1. $f^{-1}(f(x)) = x$ for every x in the domain of f and $f(f^{-1}(x)) = x$ for every x in the domain of f^{-1}.

2. Domain of f = range of f^{-1} and range of f = domain of f^{-1}.

3. The graph of f and the graph of f^{-1} are symmetric with respect to the line $y = x$.

4. If a function $y = f(x)$ has an inverse function, the equation of the inverse function is $x = f(y)$. The solution of this equation is $y = f^{-1}(x)$.

The Inverse Sine Function

In Figure 1, we reproduce the graph of $y = \sin x$. Because every horizontal line $y = b$, where b is between -1 and 1, intersects the graph of $y = \sin x$ infinitely many times, it follows from the horizontal-line test that the function $y = \sin x$ is not one-to-one.

Figure 1
$y = \sin x$,
$-\infty < x < \infty, -1 \leq y \leq 1$

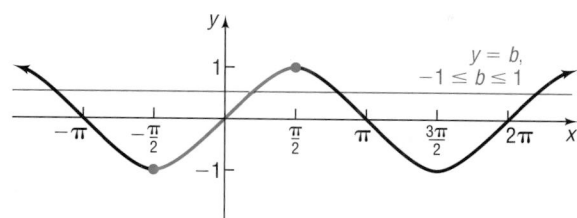

However, if we restrict the domain of $y = \sin x$ to the interval $\left[-\dfrac{\pi}{2}, \dfrac{\pi}{2}\right]$, the restricted function

$$y = \sin x, \qquad -\frac{\pi}{2} \leq x \leq \frac{\pi}{2}$$

Figure 2
$y = \sin x$,

$-\dfrac{\pi}{2} \leq x \leq \dfrac{\pi}{2}, -1 \leq y \leq 1$

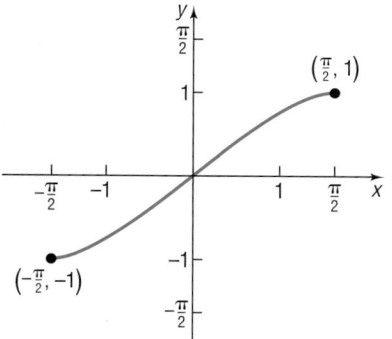

is one-to-one and, hence, will have an inverse function.* See Figure 2.

An equation for the inverse of $y = f(x) = \sin x$ is obtained by interchanging x and y. The implicit form of the inverse function is $x = \sin y$, $-\dfrac{\pi}{2} \leq y \leq \dfrac{\pi}{2}$. The explicit form is called the **inverse sine** of x and is symbolized by $y = f^{-1}(x) = \sin^{-1} x$.

> $$y = \sin^{-1} x \quad \text{means} \quad x = \sin y \tag{1}$$
>
> $$\text{where} \quad -1 \leq x \leq 1 \quad \text{and} \quad -\frac{\pi}{2} \leq y \leq \frac{\pi}{2}$$

Because $y = \sin^{-1} x$ means $x = \sin y$, we read $y = \sin^{-1} x$ as "y is the angle or real number whose sine equals x." Alternatively, we can say that "y is the inverse sine of x." Be careful about the notation used. The superscript -1 that appears in $y = \sin^{-1} x$ is not an exponent, but is reminiscent of the symbolism f^{-1} used to denote the inverse function of f. [To avoid this notation, some books use the notation $y = \arcsin x$ instead of $y = \sin^{-1} x$.]

The inverse of a function f receives as input an element from the range of f and returns as output an element in the domain of f. The restricted sine function, $y = f(x) = \sin x$, receives as input an angle or real number x in the interval $\left[-\dfrac{\pi}{2}, \dfrac{\pi}{2}\right]$ and outputs a real number in the interval $[-1, 1]$. Therefore, the inverse sine function receives as input a real number in the

---

*Although there are many other ways to restrict the domain and obtain a one-to-one function, mathematicians have agreed on a consistent use of the interval $\left[-\dfrac{\pi}{2}, \dfrac{\pi}{2}\right]$ in order to define the inverse of $y = \sin x$.

interval $[-1, 1]$ and outputs an angle or real number in the interval $\left[-\dfrac{\pi}{2}, \dfrac{\pi}{2}\right]$.

Since the domain of f = range of f^{-1} and the range of f = domain of f^{-1}, the domain of the inverse sine function, $y = f^{-1}(x) = \sin^{-1} x$, is $[-1, 1]$ or $-1 \le x \le 1$, and the range of the inverse sine function is $\left[-\dfrac{\pi}{2}, \dfrac{\pi}{2}\right]$ or $-\dfrac{\pi}{2} \le y \le \dfrac{\pi}{2}$. The graph of the inverse sine function can be obtained by reflecting the restricted portion of the graph of $y = f(x) = \sin x$ about the line $y = x$, as shown in Figure 3(a). Figure 3(b) shows the graph using a graphing utility.

Figure 3
$y = \sin^{-1} x$,
$-1 \le x \le 1, -\dfrac{\pi}{2} \le y \le \dfrac{\pi}{2}$

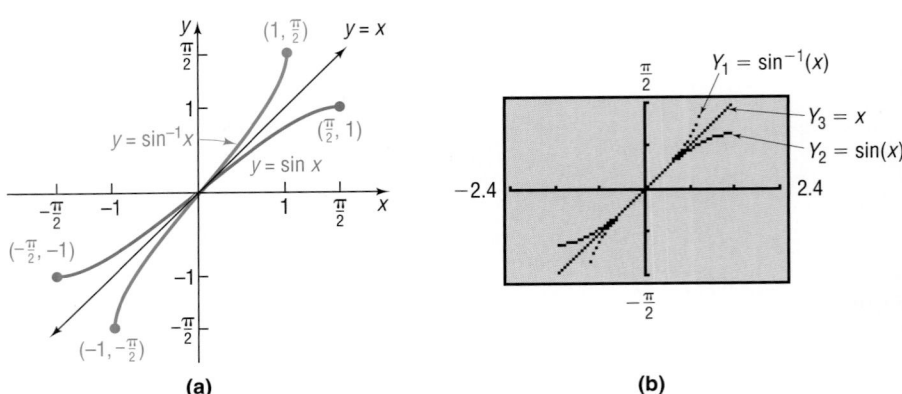

(a)   (b)

When we discussed functions and their inverses in Section 1.7, we found that $f^{-1}(f(x)) = x$ and $f(f^{-1}(x)) = x$. In terms of the sine function and its inverse, these properties are of the form

$$f^{-1}(f(x)) = \sin^{-1}(\sin x) = x, \quad \text{where } -\dfrac{\pi}{2} \le x \le \dfrac{\pi}{2} \qquad \text{(2a)}$$

$$f(f^{-1}(x)) = \sin(\sin^{-1} x) = x, \quad \text{where } -1 \le x \le 1 \qquad \text{(2b)}$$

For example, because $\dfrac{\pi}{8}$ lies in the interval $\left[-\dfrac{\pi}{2}, \dfrac{\pi}{2}\right]$, the restricted domain of the sine function, we have

$$\sin^{-1}\left[\sin\left(\dfrac{\pi}{8}\right)\right] = \dfrac{\pi}{8}$$

Also, because 0.8 lies in the interval $[-1, 1]$, the domain of the inverse sine function, we have

$$\sin[\sin^{-1}(0.8)] = 0.8$$

Figure 4

See Figure 4 for these calculations on a graphing calculator.

However, because $\dfrac{5\pi}{8}$ is not in the interval $\left[-\dfrac{\pi}{2}, \dfrac{\pi}{2}\right]$

$$\sin^{-1}\left[\sin\left(\dfrac{5\pi}{8}\right)\right] \neq \dfrac{5\pi}{8}$$

See Figure 5.

Also, because 1.8 is not in the interval $[-1, 1]$,

$$\sin[\sin^{-1}(1.8)] \neq 1.8$$

See Figure 6. Can you explain why the error appears?

Figure 5

Figure 6

① For some numbers x it is possible to find the exact value of $y = \sin^{-1} x$.

| EXAMPLE 1 | **Finding the Exact Value of an Inverse Sine Function** |

Find the exact value of: $\sin^{-1} 1$

Solution   Let $\theta = \sin^{-1} 1$. We seek the angle θ, $-\dfrac{\pi}{2} \leq \theta \leq \dfrac{\pi}{2}$, whose sine equals 1.

$$\theta = \sin^{-1} 1, \qquad -\frac{\pi}{2} \leq \theta \leq \frac{\pi}{2}$$

$$\sin \theta = 1, \qquad -\frac{\pi}{2} \leq \theta \leq \frac{\pi}{2} \qquad \text{By definition of } y = \sin^{-1} x.$$

Now look at Table 1 and Figure 7.

Figure 7

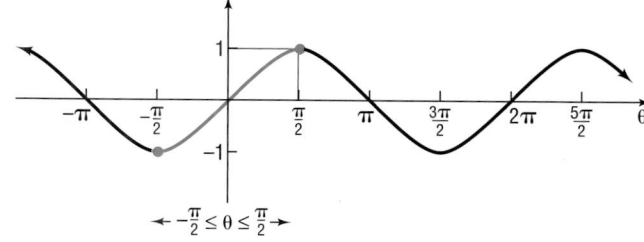

$$\longleftarrow -\tfrac{\pi}{2} \leq \theta \leq \tfrac{\pi}{2} \longrightarrow$$

TABLE 1

θ	$\sin \theta$
$-\dfrac{\pi}{2}$	-1
$-\dfrac{\pi}{3}$	$-\dfrac{\sqrt{3}}{2}$
$-\dfrac{\pi}{4}$	$-\dfrac{\sqrt{2}}{2}$
$-\dfrac{\pi}{6}$	$-\dfrac{1}{2}$
0	0
$\dfrac{\pi}{6}$	$\dfrac{1}{2}$
$\dfrac{\pi}{4}$	$\dfrac{\sqrt{2}}{2}$
$\dfrac{\pi}{3}$	$\dfrac{\sqrt{3}}{2}$
$\dfrac{\pi}{2}$	1

We see that the only angle θ within the interval $\left[-\dfrac{\pi}{2}, \dfrac{\pi}{2} \right]$ whose sine is 1 is $\dfrac{\pi}{2}$. [Note that $\sin \dfrac{5\pi}{2}$ also equals 1, but $\dfrac{5\pi}{2}$ lies outside the interval $\left[-\dfrac{\pi}{2}, \dfrac{\pi}{2} \right]$ and hence is not admissible.] Since $\sin \dfrac{\pi}{2} = 1$ and $\dfrac{\pi}{2}$ is in the interval $\left[-\dfrac{\pi}{2}, \dfrac{\pi}{2} \right]$, we conclude that

$$\sin^{-1} 1 = \frac{\pi}{2}$$

Figure 8

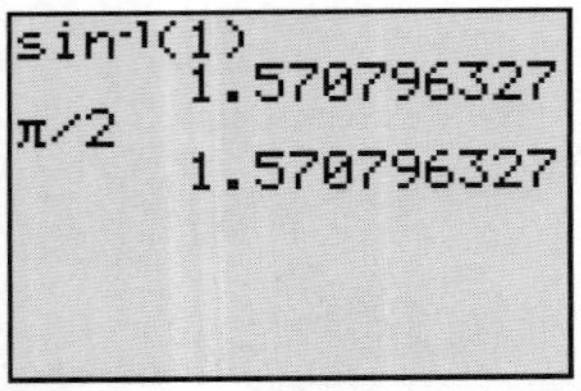

```
sin⁻¹(1)
        1.570796327
π/2
        1.570796327
```

✔ CHECK: We can verify the solution by evaluating $\sin^{-1} 1$ with our graphing calculator in radian mode. See Figure 8.   ■ ■

For the remainder of the section, the reader is encouraged to verify the solutions obtained using a graphing utility.

◗━━━━ NOW WORK PROBLEM **1**.

EXAMPLE 2   **Finding the Exact Value of an Inverse Sine Function**

Find the exact value of: $\sin^{-1}\left(-\dfrac{1}{2}\right)$

Solution   Let $\theta = \sin^{-1}\left(-\dfrac{1}{2}\right)$. We seek the angle θ, $-\dfrac{\pi}{2} \le \theta \le \dfrac{\pi}{2}$, whose sine equals $-\dfrac{1}{2}$.

$$\theta = \sin^{-1}\left(-\frac{1}{2}\right), \qquad -\frac{\pi}{2} \le \theta \le \frac{\pi}{2}$$

$$\sin\theta = -\frac{1}{2}, \qquad -\frac{\pi}{2} \le \theta \le \frac{\pi}{2}$$

(Refer to Table 1 and Figure 7, if necessary.) The only angle within the interval $\left[-\dfrac{\pi}{2}, \dfrac{\pi}{2}\right]$ whose sine is $-\dfrac{1}{2}$ is $-\dfrac{\pi}{6}$. So, since $\sin\left(-\dfrac{\pi}{6}\right) = -\dfrac{1}{2}$ and $-\dfrac{\pi}{6}$ is in the interval $\left[-\dfrac{\pi}{2}, \dfrac{\pi}{2}\right]$, we conclude that

$$\sin^{-1}\left(-\frac{1}{2}\right) = -\frac{\pi}{6}$$

■

◗━━━━ NOW WORK PROBLEM **7**.

② For most numbers x, the value $y = \sin^{-1} x$ must be approximated.

EXAMPLE 3   **Finding an Approximate Value of an Inverse Sine Function**

Find an approximate value of:

(a) $\sin^{-1}\dfrac{1}{3}$ (b) $\sin^{-1}\left(-\dfrac{1}{4}\right)$

Express the answer in radians rounded to two decimal places.

Solution   Because we want the angle measured in radians, we first set the mode to radians.

(a) Figure 9(a) shows the solution using a TI-83 graphing calculator.

(b) Figure 9(b) shows the solution using a TI-83 graphing calculator.

Figure 9(a)

Figure 9(b)

We have $\sin^{-1}\dfrac{1}{3} = 0.34$, rounded to two decimal places.

We have $\sin^{-1}\left(-\dfrac{1}{4}\right) = -0.25$, rounded to two decimal places.

 **NOW WORK PROBLEM 13.**

The Inverse Cosine Function

In Figure 10 we reproduce the graph of $y = \cos x$. Because every horizontal line $y = b$, where b is between -1 and 1, intersects the graph of $y = \cos x$ infinitely many times, it follows that the cosine function is not one-to-one.

Figure 10
$y = \cos x$,
$-\infty < x < \infty, -1 \le y \le 1$

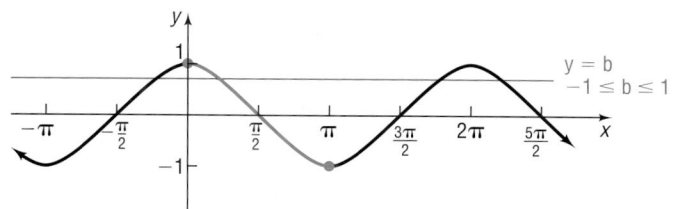

However, if we restrict the domain of $y = \cos x$ to the interval $[0, \pi]$, the restricted function

$$y = \cos x, \qquad 0 \le x \le \pi$$

is one-to-one and hence will have an inverse function.* See Figure 11.

Figure 11
$y = \cos x, 0 \le x \le \pi, -1 \le y \le 1$

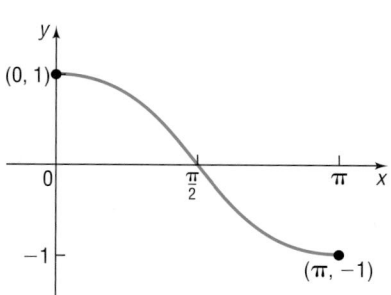

An equation for the inverse of $y = f(x) = \cos x$ is obtained by interchanging x and y. The implicit form of the inverse function is $x = \cos y$, $0 \le y \le \pi$. The explicit form is called the **inverse cosine** of x and is symbolized by $y = f^{-1}(x) = \cos^{-1} x$ (or by $y = \arccos x$).

$$y = \cos^{-1} x \quad \text{means} \quad x = \cos y \qquad (3)$$
$$\text{where} \quad -1 \le x \le 1 \quad \text{and} \quad 0 \le y \le \pi$$

Here, y is the angle whose cosine is x. The domain of the function $y = \cos^{-1} x$ is $-1 \le x \le 1$, and its range is $0 \le y \le \pi$. (Do you know why?) The graph of $y = \cos^{-1} x$ can be obtained by reflecting the restricted portion of the graph of $y = \cos x$ about the line $y = x$, as shown in Figure 12(a). Figure 12(b) shows the graph using a graphing utility.

*This is the generally accepted restriction.

Figure 12
$y = \cos^{-1} x,$
$-1 \le x \le 1, 0 \le y \le \pi$

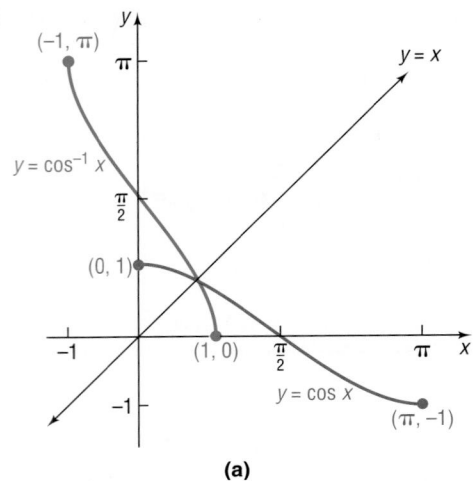

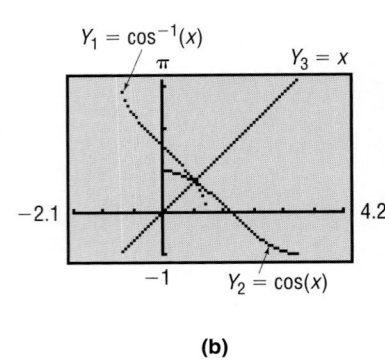

(a)

(b)

EXAMPLE 4 Finding the Exact Value of an Inverse Cosine Function

Find the exact value of: $\cos^{-1} 0$

Solution Let $\theta = \cos^{-1} 0$. We seek the angle $\theta, 0 \le \theta \le \pi$, whose cosine equals 0.

$$\theta = \cos^{-1} 0, \qquad 0 \le \theta \le \pi$$
$$\cos \theta = 0, \qquad 0 \le \theta \le \pi$$

Look at Table 2 and Figure 13.

TABLE 2

θ	$\cos \theta$
0	1
$\dfrac{\pi}{6}$	$\dfrac{\sqrt{3}}{2}$
$\dfrac{\pi}{4}$	$\dfrac{\sqrt{2}}{2}$
$\dfrac{\pi}{3}$	$\dfrac{1}{2}$
$\dfrac{\pi}{2}$	0
$\dfrac{2\pi}{3}$	$-\dfrac{1}{2}$
$\dfrac{3\pi}{4}$	$-\dfrac{\sqrt{2}}{2}$
$\dfrac{5\pi}{6}$	$-\dfrac{\sqrt{3}}{2}$
π	-1

Figure 13

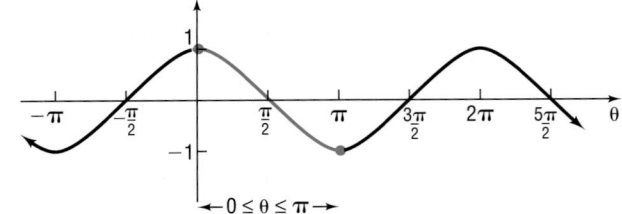

We see that the only angle θ within the interval $[0, \pi]$ whose cosine is 0 is $\dfrac{\pi}{2}$. [Note that $\cos \dfrac{3\pi}{2}$ also equals 0, but $\dfrac{3\pi}{2}$ lies outside the interval $[0, \pi]$ and hence is not admissible.] Since $\cos \dfrac{\pi}{2} = 0$ and $\dfrac{\pi}{2}$ is in the interval $[0, \pi]$, we conclude that

$$\cos^{-1} 0 = \frac{\pi}{2}$$

EXAMPLE 5 Finding the Exact Value of an Inverse Cosine Function

Find the exact value of: $\cos^{-1} \dfrac{\sqrt{2}}{2}$

Solution   Let $\theta = \cos^{-1}\dfrac{\sqrt{2}}{2}$. We seek the angle θ, $0 \le \theta \le \pi$, whose cosine equals $\dfrac{\sqrt{2}}{2}$.

$$\theta = \cos^{-1}\frac{\sqrt{2}}{2}, \qquad 0 \le \theta \le \pi$$

$$\cos\theta = \frac{\sqrt{2}}{2}, \qquad 0 \le \theta \le \pi$$

Look at Table 2 and Figure 14.

Figure 14

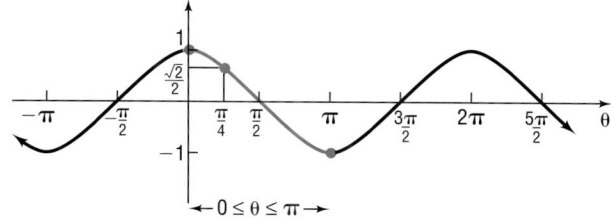

We see that the only angle θ within the interval $[0, \pi]$, whose cosine is $\dfrac{\sqrt{2}}{2}$ is $\dfrac{\pi}{4}$. Since $\cos\dfrac{\pi}{4} = \dfrac{\sqrt{2}}{2}$ and $\dfrac{\pi}{4}$ is in the interval $[0, \pi]$, we conclude that

$$\cos^{-1}\frac{\sqrt{2}}{2} = \frac{\pi}{4}$$

◢ **NOW WORK PROBLEM 11.**

For the cosine function $f(x) = \cos x$ and its inverse $f^{-1}(x) = \cos^{-1}x$, the following properties hold:

$$f^{-1}(f(x)) = \cos^{-1}(\cos x) = x, \qquad \text{where } 0 \le x \le \pi \qquad (4a)$$
$$f(f^{-1}(x)) = \cos(\cos^{-1}x) = x, \qquad \text{where } -1 \le x \le 1 \qquad (4b)$$

EXAMPLE 6   **Finding the Exact Value of a Composite Function**

Find the exact value of: (a) $\cos^{-1}\left[\cos\left(\dfrac{\pi}{12}\right)\right]$ (b) $\cos[\cos^{-1}(-0.4)]$

Solution   (a) $\cos^{-1}\left[\cos\left(\dfrac{\pi}{12}\right)\right] = \dfrac{\pi}{12}$ By Property (4a).

(b) $\cos[\cos^{-1}(-0.4)] = -0.4$ By Property (4b).

◢ **NOW WORK PROBLEM 27.**

The Inverse Tangent Function

In Figure 15 we reproduce the graph of $y = \tan x$. Because every horizontal line intersects the graph infinitely many times, it follows that the tangent function is not one-to-one.

Figure 15
$y = \tan x,\ -\infty < x < \infty,$
x not equal to odd multiples
of $\dfrac{\pi}{2},\ -\infty < y < \infty$

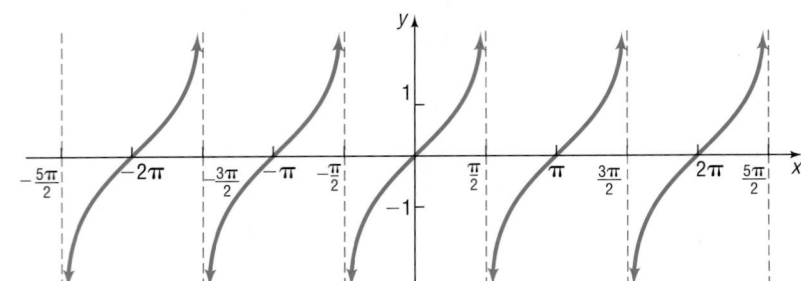

However, if we restrict the domain of $y = \tan x$ to the interval $\left(-\dfrac{\pi}{2}, \dfrac{\pi}{2}\right)$, the restricted function

$$y = \tan x, \qquad -\frac{\pi}{2} < x < \frac{\pi}{2}$$

is one-to-one and hence has an inverse function.* See Figure 16.

Figure 16
$y = \tan x,$
$-\dfrac{\pi}{2} < x < \dfrac{\pi}{2},\ -\infty < y < \infty$

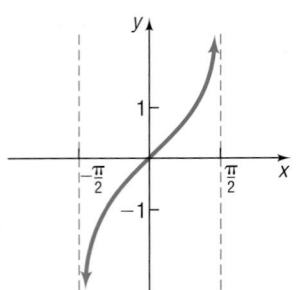

An equation for the inverse of $y = f(x) = \tan x$ is obtained by interchanging x and y. The implicit form of the inverse function is $x = \tan y$, $-\dfrac{\pi}{2} < y < \dfrac{\pi}{2}$. The explicit form is called the **inverse tangent** of x and is symbolized by $y = f^{-1}(x) = \tan^{-1} x$ (or by $y = \arctan x$).

$$y = \tan^{-1} x \quad \text{means} \quad x = \tan y \tag{5}$$
$$\text{where} \quad -\infty < x < \infty \quad \text{and} \quad -\frac{\pi}{2} < y < \frac{\pi}{2}$$

Here, y is the angle whose tangent is x. The domain of the function $y = \tan^{-1} x$ is $-\infty < x < \infty$, and its range is $-\dfrac{\pi}{2} < y < \dfrac{\pi}{2}$. The graph of $y = \tan^{-1} x$ can be obtained by reflecting the restricted portion of the graph of $y = \tan x$ about the line $y = x$, as shown in Figure 17(a). Figure 17(b) shows the graph using a graphing utitity.

Figure 17
$y = \tan^{-1} x,$
$-\infty < x < \infty,\ -\dfrac{\pi}{2} < y < \dfrac{\pi}{2}$

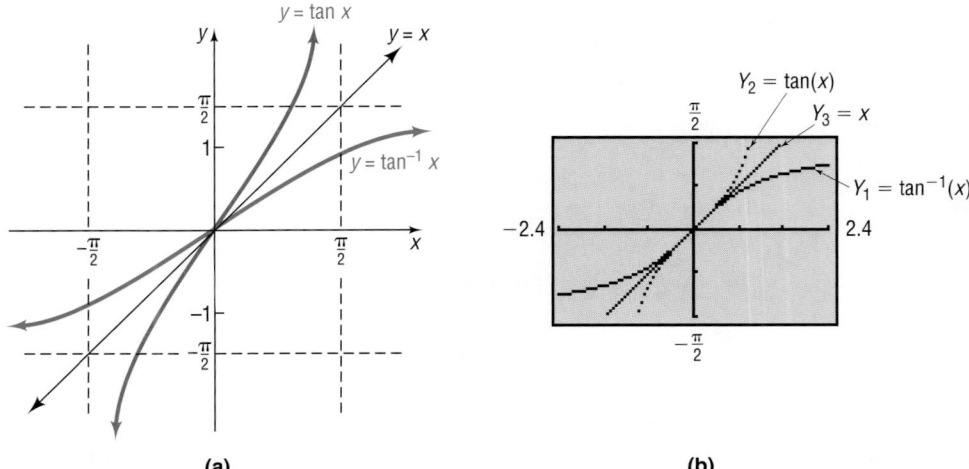

(a)   (b)

*This is the generally accepted restriction.

For the tangent function and its inverse, the following properties hold:

$$f^{-1}(f(x)) = \tan^{-1}(\tan x) = x, \quad \text{where } -\frac{\pi}{2} < x < \frac{\pi}{2}$$

$$f(f^{-1}(x)) = \tan(\tan^{-1} x) = x, \quad \text{where } -\infty < x < \infty$$

EXAMPLE 7   **Finding the Exact Value of an Inverse Tangent Function**

Find the exact value of: $\tan^{-1} 1$

TABLE 3

θ	$\tan \theta$
$-\dfrac{\pi}{2}$	Undefined
$-\dfrac{\pi}{3}$	$-\sqrt{3}$
$-\dfrac{\pi}{4}$	-1
$-\dfrac{\pi}{6}$	$-\dfrac{\sqrt{3}}{3}$
0	0
$\dfrac{\pi}{6}$	$\dfrac{\sqrt{3}}{3}$
$\dfrac{\pi}{4}$	1
$\dfrac{\pi}{3}$	$\sqrt{3}$
$\dfrac{\pi}{2}$	Undefined

Solution   Let $\theta = \tan^{-1} 1$. We seek the angle θ, $-\dfrac{\pi}{2} < \theta < \dfrac{\pi}{2}$, whose tangent equals 1.

$$\theta = \tan^{-1} 1, \quad -\frac{\pi}{2} < \theta < \frac{\pi}{2}$$

$$\tan \theta = 1, \quad -\frac{\pi}{2} < \theta < \frac{\pi}{2}$$

Look at Table 3 or Figure 16. The only angle θ within the interval $\left(-\dfrac{\pi}{2}, \dfrac{\pi}{2}\right)$ whose tangent is 1 is $\dfrac{\pi}{4}$. Since $\tan \dfrac{\pi}{4} = 1$ and $\dfrac{\pi}{4}$ is in the interval $\left(-\dfrac{\pi}{2}, \dfrac{\pi}{2}\right)$, we conclude that

$$\tan^{-1} 1 = \frac{\pi}{4}$$

■

EXAMPLE 8   **Finding the Exact Value of an Inverse Tangent Function**

Find the exact value of: $\tan^{-1}(-\sqrt{3})$

Solution   Let $\theta = \tan^{-1}(-\sqrt{3})$. We seek the angle θ, $-\dfrac{\pi}{2} < \theta < \dfrac{\pi}{2}$, whose tangent equals $-\sqrt{3}$.

$$\theta = \tan^{-1}(-\sqrt{3}), \quad -\frac{\pi}{2} < \theta < \frac{\pi}{2}$$

$$\tan \theta = -\sqrt{3}, \quad -\frac{\pi}{2} < \theta < \frac{\pi}{2}$$

Look at Table 3 or Figure 16 if necessary. The only angle θ within the interval $\left(-\dfrac{\pi}{2}, \dfrac{\pi}{2}\right)$ whose tangent is $-\sqrt{3}$ is $-\dfrac{\pi}{3}$. Since $\tan\left(-\dfrac{\pi}{3}\right) = -\sqrt{3}$ and $-\dfrac{\pi}{3}$ is in the interval $\left(-\dfrac{\pi}{2}, \dfrac{\pi}{2}\right)$, we conclude that

$$\tan^{-1}(-\sqrt{3}) = -\frac{\pi}{3}$$

■

NOW WORK PROBLEM **5**.

3.1 Concepts and Vocabulary

In Problems 1–3, fill in the blanks.

1. $y = \sin^{-1} x$ means _____ , where $-1 \leq x \leq 1$ and $-\dfrac{\pi}{2} \leq y \leq \dfrac{\pi}{2}$.

2. The value of $\sin^{-1}\left[\cos \dfrac{\pi}{2}\right]$ is _____ .

3. $\cos^{-1}\left[\cos \dfrac{\pi}{5}\right] =$ _____ .

In Problems 4–6, answer True or False to each statement.

4. The domain of $y = \sin^{-1} x$ is $-\dfrac{\pi}{2} \leq x \leq \dfrac{\pi}{2}$.

5. $\cos(\sin^{-1} 0) = 1$ and $\sin(\cos^{-1} 0) = 1$.

6. $y = \tan^{-1} x$ means $x = \tan y$, where $-\infty < x < \infty$ and $-\dfrac{\pi}{2} < y < \dfrac{\pi}{2}$.

7. How would you quickly sketch the graph of $y = \cos^{-1} x$?

8. What restrictions are on x if $\sin^{-1}(\sin x) = x$?

9. What restrictions are on x if $\sin(\sin^{-1} x) = x$?

10. Explain how you would find the exact value of $\tan^{-1}(-1)$.

3.1 Exercises

In Problems 1–12, find the exact value of each expression. Verify your results using a graphing utility.

1. $\sin^{-1} 0$ **2.** $\cos^{-1} 1$ **3.** $\sin^{-1}(-1)$ **4.** $\cos^{-1}(-1)$

5. $\tan^{-1} 0$ **6.** $\tan^{-1}(-1)$ **7.** $\sin^{-1} \dfrac{\sqrt{2}}{2}$ **8.** $\tan^{-1} \dfrac{\sqrt{3}}{3}$

9. $\tan^{-1} \sqrt{3}$ **10.** $\sin^{-1}\left(-\dfrac{\sqrt{3}}{2}\right)$ **11.** $\cos^{-1}\left(-\dfrac{\sqrt{3}}{2}\right)$ **12.** $\sin^{-1}\left(-\dfrac{\sqrt{2}}{2}\right)$

In Problems 13–24, use a calculator to find the value of each expression rounded to two decimal places.

13. $\sin^{-1} 0.1$ **14.** $\cos^{-1} 0.6$ **15.** $\tan^{-1} 5$ **16.** $\tan^{-1} 0.2$

17. $\cos^{-1} \dfrac{7}{8}$ **18.** $\sin^{-1} \dfrac{1}{8}$ **19.** $\tan^{-1}(-0.4)$ **20.** $\tan^{-1}(-3)$

21. $\sin^{-1}(-0.12)$ **22.** $\cos^{-1}(-0.44)$ **23.** $\cos^{-1} \dfrac{\sqrt{2}}{3}$ **24.** $\sin^{-1} \dfrac{\sqrt{3}}{5}$

In Problems 25–32, find the exact value of the expression. Do not use a calculator.

25. $\sin[\sin^{-1}(0.54)]$ **26.** $\tan[\tan^{-1}(7.4)]$ **27.** $\cos^{-1}\left[\cos\left(\dfrac{4\pi}{5}\right)\right]$ **28.** $\sin^{-1}\left[\sin\left(-\dfrac{\pi}{10}\right)\right]$

29. $\tan[\tan^{-1}(-3.5)]$ **30.** $\cos[\cos^{-1}(-0.05)]$ **31.** $\sin^{-1}\left[\sin\left(-\dfrac{3\pi}{7}\right)\right]$ **32.** $\tan^{-1}\left[\tan\left(\dfrac{2\pi}{5}\right)\right]$

In Problems 33–44, do not use a calculator.

33. Does $\sin^{-1}\left[\sin\left(-\dfrac{\pi}{6}\right)\right] = -\dfrac{\pi}{6}$? Why or why not?

34. Does $\sin^{-1}\left[\sin\left(\dfrac{2\pi}{3}\right)\right] = \dfrac{2\pi}{3}$? Why or why not?

35. Does $\sin[\sin^{-1}(2)] = 2$? Why or why not?

36. Does $\sin\left[\sin^{-1}\left(-\dfrac{1}{2}\right)\right] = -\dfrac{1}{2}$? Why or why not?

37. Does $\cos^{-1}\left[\cos\left(-\dfrac{\pi}{6}\right)\right] = -\dfrac{\pi}{6}$? Why or why not?

38. Does $\cos^{-1}\left[\cos\left(\dfrac{2\pi}{3}\right)\right] = \dfrac{2\pi}{3}$? Why or why not?

39. Does $\cos\left[\cos^{-1}\left(-\dfrac{1}{2}\right)\right] = -\dfrac{1}{2}$? Why or why not?

40. Does $\cos[\cos^{-1}(2)] = 2$? Why or why not?

41. Does $\tan^{-1}\left[\tan\left(-\dfrac{\pi}{3}\right)\right] = -\dfrac{\pi}{3}$? Why or why not?

42. Does $\tan^{-1}\left[\tan\left(\dfrac{2\pi}{3}\right)\right] = \dfrac{2\pi}{3}$? Why or why not?

43. Does $\tan[\tan^{-1}(2)] = 2$? Why or why not?

44. Does $\tan\left[\tan^{-1}\left(-\dfrac{1}{2}\right)\right] = -\dfrac{1}{2}$? Why or why not?

In Problems 45–50, use the following: The formula

$$D = 24\left[1 - \frac{\cos^{-1}(\tan i \tan \theta)}{\pi}\right]$$

can be used to approximate the number of hours of daylight when the declination of the Sun is $i°$ at a location $\theta°$ north latitude for any date between the vernal equinox and autumnal equinox. The declination of the Sun is defined as the angle i between the equatorial plane and any ray of light from the Sun. The latitude of a location is the angle θ formed by a ray from the center of Earth to the Equator and a ray drawn from the center of Earth to the location. See the figure. To use the formula, $\cos^{-1}(\tan i \tan \theta)$ must be expressed in radians.

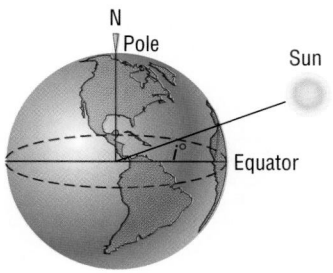
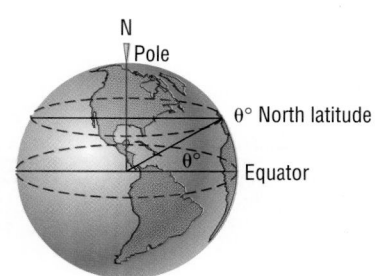

45. Approximate the number of hours of daylight in Houston, Texas (29°45′ north latitude), for the following dates:
(a) Summer solstice ($i = 23.5°$)
(b) Vernal equinox ($i = 0°$)
(c) July 4 ($i = 22°48′$)

46. Approximate the number of hours of daylight in New York, New York (40°45′ north latitude), for the following dates:
(a) Summer solstice ($i = 23.5°$)
(b) Vernal equinox ($i = 0°$)
(c) July 4 ($i = 22°48′$)

47. Approximate the number of hours of daylight in Honolulu, Hawaii (21°18′ north latitude), for the following dates:
(a) Summer solstice ($i = 23.5°$)
(b) Vernal equinox ($i = 0°$)
(c) July 4 ($i = 22°48′$)

48. Approximate the number of hours of daylight in Anchorage, Alaska (61°10′ north latitude), for the following dates:
(a) Summer solstice ($i = 23.5°$)
(b) Vernal equinox ($i = 0°$)
(c) July 4 ($i = 22°48′$)

49. Approximate the number of hours of daylight at the Equator ($0°$ north latitude) for the following dates:

   (a) Summer solstice ($i = 23.5°$)
   (b) Vernal equinox ($i = 0°$)
   (c) July 4 ($i = 22°48'$)
   (d) What do you conclude about the number of hours of daylight throughout the year for a location at the Equator?

50. Approximate the number of hours of daylight for any location that is $66°30'$ north latitude for the following dates:

   (a) Summer solstice ($i = 23.5°$)
   (b) Vernal equinox ($i = 0°$)
   (c) July 4 ($i = 22°48'$)
   (d) The number of hours of daylight on the winter solstice may be found by computing the number of hours of daylight on the summer solstice and subtracting this result from 24 hours, due to the symmetry of the orbital path of Earth around the Sun. Compute the number of hours of daylight for this location on the winter solstice. What do you conclude about daylight for a location at $66°30'$ north latitude?

51. Being the First to See the Rising Sun   Cadillac Mountain, elevation 1530 feet, is located in Acadia National Park, Maine, and is the highest peak on the east coast of the United States. It is said that a person standing on the summit will be the first person in the United States to see the rays of the rising Sun. How much sooner would a person atop Cadillac Mountain see the first rays than a person standing below, at sea level?

[**Hint:** Consult the figure. When the person at D sees the first rays of the Sun, the person at P does not. The person at P sees the first rays of the Sun only after Earth has rotated so that P is at location Q. Compute the length of the arc subtended by the central angle θ. Then use the fact that, at the latitude of Cadillac Mountain, in 24 hours a length of $2\pi(2710)$ miles is subtended, and find the time it takes to subtend this length.]

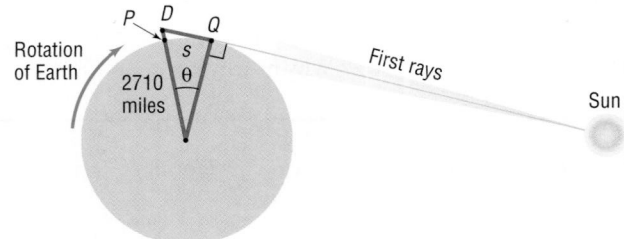

PREPARING FOR THIS SECTION

Before getting started, review the following:

✓ Finding Exact Values Given the Value of a Trigonometric Function and the Quadrant of the Angle (Section 2.4, pp. 139–140)

✓ Graphs of the Secant, Cosecant, and Cotangent Functions (Section 2.7, pp. 173–175)

✓ Domain and Range of the Secant, Cosecant, and Cotangent Functions (Section 2.5, pp. 146–148)

3.2   **THE INVERSE TRIGONOMETRIC FUNCTIONS [CONTINUED]**

OBJECTIVES   ① Find the Exact Value of Expressions Involving the Inverse Sine, Cosine, and Tangent Functions
② Find the Exact Value of the Inverse Secant, Cosecant, and Cotangent Functions
③ Use a Calculator to Evaluate $\sec^{-1} x$, $\csc^{-1} x$, and $\cot^{-1} x$

①   In this section we continue our discussion of the inverse trigonometric functions.

EXAMPLE 1   **Finding the Exact Value of Expressions Involving Inverse Trigonometric Functions**

Find the exact value of: $\sin^{-1}\left(\sin\dfrac{5\pi}{4}\right)$

Solution $\sin^{-1}\left(\sin\dfrac{5\pi}{4}\right) = \sin^{-1}\left(-\dfrac{\sqrt{2}}{2}\right) = -\dfrac{\pi}{4}$

Figure 18

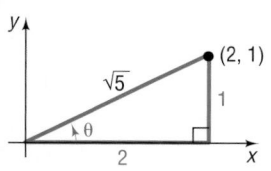

```
sin⁻¹(sin(5π/4))
          -.7853981634
-π/4
          -.7853981634
```

✔ CHECK: We can verify the solution by evaluating $\sin^{-1}\left(\sin\dfrac{5\pi}{4}\right)$ with our graphing calculator in radian mode. See Figure 18. ■ ■

Notice in the solution to Example 1 that we did not use Property (2a), page 200. This is because the argument of the sine function is not in the interval $\left[-\dfrac{\pi}{2}, \dfrac{\pi}{2}\right]$, as required. If we use the fact that

$$\sin\frac{5\pi}{4} = -\sin\frac{\pi}{4} \underset{\underset{y = \sin x \text{ is odd}}{\uparrow}}{=} \sin\left(-\frac{\pi}{4}\right)$$

then we can use Property (2a):

$$\sin^{-1}\left(\sin\frac{5\pi}{4}\right) = \sin^{-1}\left[\sin\left(-\frac{\pi}{4}\right)\right] \underset{\underset{\text{Property (2a)}}{\uparrow}}{=} -\frac{\pi}{4}$$

For the remainder of the section, the reader is encouraged to verify the solutions obtained using a graphing utility.

NOW WORK PROBLEM **13.**

EXAMPLE 2 **Finding the Exact Value of Expressions Involving Inverse Trigonometric Functions**

Find the exact value of: $\sin\left(\tan^{-1}\dfrac{1}{2}\right)$

Figure 19

$\tan\theta = \dfrac{1}{2}$

Solution Let $\theta = \tan^{-1}\dfrac{1}{2}$. Then $\tan\theta = \dfrac{1}{2}$, where $-\dfrac{\pi}{2} < \theta < \dfrac{\pi}{2}$. Because $\tan\theta > 0$, it follows that $0 < \theta < \dfrac{\pi}{2}$, so θ lies in quadrant I. Now, in Figure 19 we draw a triangle in quadrant I depicting $\tan\theta = \dfrac{1}{2}$. The hypotenuse of this triangle is found to be $\sqrt{5}$. Then $\sin\theta = \dfrac{1}{\sqrt{5}}$, and

$$\sin\left(\tan^{-1}\frac{1}{2}\right) = \sin\theta = \frac{1}{\sqrt{5}} = \frac{\sqrt{5}}{5}$$ ■

EXAMPLE 3 **Finding the Exact Value of Expressions Involving Inverse Trigonometric Functions**

Find the exact value of: $\cos\left[\sin^{-1}\left(-\dfrac{1}{3}\right)\right]$

Figure 20

$\sin \theta = -\dfrac{1}{3}$

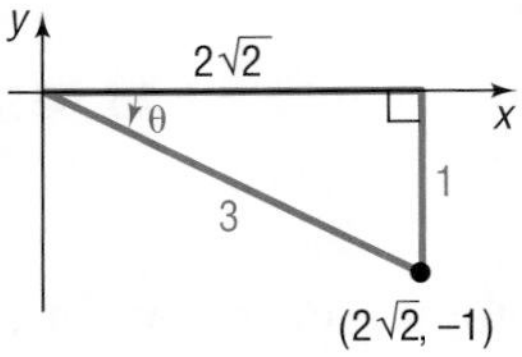

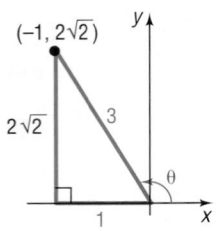

$(2\sqrt{2}, -1)$

Solution Let $\theta = \sin^{-1}\left(-\dfrac{1}{3}\right)$. Then $\sin \theta = -\dfrac{1}{3}$ and $-\dfrac{\pi}{2} \le \theta \le \dfrac{\pi}{2}$. Because $\sin \theta < 0$, it follows that $-\dfrac{\pi}{2} \le \theta < 0$, so θ lies in quadrant IV. Figure 20 illustrates $\sin \theta = -\dfrac{1}{3}$ for θ in quadrant IV. Then

$$\cos\left[\sin^{-1}\left(-\dfrac{1}{3}\right)\right] = \cos \theta = \dfrac{2\sqrt{2}}{3}$$ ■

EXAMPLE 4   Finding the Exact Value of Expressions Involving Inverse Trigonometric Functions

Find the exact value of: $\tan\left[\cos^{-1}\left(-\dfrac{1}{3}\right)\right]$

Figure 21

$\cos \theta = -\dfrac{1}{3}$

$(-1, 2\sqrt{2})$

$2\sqrt{2}$ 3

θ

1 x

Solution Let $\theta = \cos^{-1}\left(-\dfrac{1}{3}\right)$. Then $\cos \theta = -\dfrac{1}{3}$ and $0 \le \theta \le \pi$. Because $\cos \theta < 0$, it follows that $\dfrac{\pi}{2} < \theta \le \pi$, so θ lies in quadrant II. Figure 21 illustrates $\cos \theta = -\dfrac{1}{3}$ for θ in quadrant II. Then

$$\tan\left[\cos^{-1}\left(-\dfrac{1}{3}\right)\right] = \tan \theta = \dfrac{2\sqrt{2}}{-1} = -2\sqrt{2}$$ ■

✎— NOW WORK PROBLEMS 1 AND 19.

The Remaining Inverse Trigonometric Functions

② The inverse secant, inverse cosecant, and inverse cotangent functions are defined as follows:

$$y = \sec^{-1} x \quad \text{means} \quad x = \sec y \tag{1}$$

$$\text{where} \quad |x| \ge 1 \quad \text{and} \quad 0 \le y \le \pi, \quad y \ne \dfrac{\pi}{2} \ *$$

$$y = \csc^{-1} x \quad \text{means} \quad x = \csc y \tag{2}$$

$$\text{where} \quad |x| \ge 1 \quad \text{and} \quad -\dfrac{\pi}{2} \le y \le \dfrac{\pi}{2}, \quad y \ne 0 \ \dagger$$

$$y = \cot^{-1} x \quad \text{means} \quad x = \cot y \tag{3}$$

$$\text{where} \quad -\infty < x < \infty \quad \text{and} \quad 0 < y < \pi$$

You are encouraged to review the graphs of the cotangent, cosecant, and secant functions in Figures 101 (p. 173), 102 (p. 174), and 104 (p. 175) in Section 2.7 to help you to see the basis for these definitions.

*Most books use this definition. A few use the restriction $0 \le y < \dfrac{\pi}{2}, \pi \le y < \dfrac{3\pi}{2}$.

†Most books use this definition. A few use the restriction $-\pi < y \le -\dfrac{\pi}{2}, 0 < y \le \dfrac{\pi}{2}$.

EXAMPLE 5   **Finding the Exact Value of an Inverse Cosecant Function**

Find the exact value of: $\csc^{-1} 2$

Solution   Let $\theta = \csc^{-1} 2$. We seek the angle θ, $-\dfrac{\pi}{2} \le \theta \le \dfrac{\pi}{2}, \theta \ne 0$, whose cosecant equals 2.

$$\theta = \csc^{-1} 2, \qquad -\frac{\pi}{2} \le \theta \le \frac{\pi}{2}, \qquad \theta \ne 0$$

$$\csc \theta = 2, \qquad -\frac{\pi}{2} \le \theta \le \frac{\pi}{2}, \qquad \theta \ne 0$$

The only angle θ in the interval $-\dfrac{\pi}{2} \le \theta \le \dfrac{\pi}{2}, \theta \ne 0$, whose cosecant is 2 is $\dfrac{\pi}{6}$, so $\csc^{-1} 2 = \dfrac{\pi}{6}$.   ■

━ **NOW WORK PROBLEM 31.**

③   Most calculators do not have keys for evaluating the inverse cotangent, cosecant, and secant functions. The easiest way to evaluate them is to convert to an inverse trigonometric function whose range is the same as the one to be evaluated. In this regard, notice that $y = \cot^{-1} x$ and $y = \sec^{-1} x$ (except where undefined) each have the same range as $y = \cos^{-1} x$; also, $y = \csc^{-1} x$, except where undefined, has the same range as $y = \sin^{-1} x$.

EXAMPLE 6   **Approximating the Value of Inverse Trigonometric Functions**

Use a calculator to approximate each expression in radians rounded to two decimal places.

(a) $\sec^{-1} 3$ (b) $\csc^{-1}(-4)$ (c) $\cot^{-1} \dfrac{1}{2}$ (d) $\cot^{-1}(-2)$

Solution   First, set your calculator to radian mode.

(a) Let $\theta = \sec^{-1} 3$. Then $\sec \theta = 3$ and $0 \le \theta \le \pi, \theta \ne \dfrac{\pi}{2}$. Since $\sec \theta = \dfrac{1}{\cos \theta} = 3$, we have $\cos \theta = \dfrac{1}{3}$, so that $\theta = \cos^{-1}\left(\dfrac{1}{3}\right)$. Therefore,

$$\sec^{-1} 3 = \theta = \cos^{-1} \frac{1}{3} \underset{\underset{\text{Use a calculator.}}{\uparrow}}{\approx} 1.23$$

(b) Let $\theta = \csc^{-1}(-4)$. Then $\csc \theta = -4, -\dfrac{\pi}{2} \le \theta \le \dfrac{\pi}{2}, \theta \ne 0$. Since $\sin \theta = \dfrac{1}{\csc \theta} = -\dfrac{1}{4}$, then $\theta = \sin^{-1}\left(-\dfrac{1}{4}\right)$, and

$$\csc^{-1}(-4) = \theta = \sin^{-1}\left(-\frac{1}{4}\right) \approx -0.25$$

Figure 22

$\cot \theta = \dfrac{1}{2}, 0 < \theta < \pi$

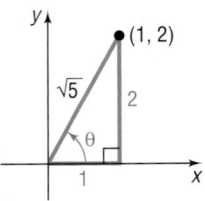

Figure 23

$\cot \theta = -2, 0 < \theta < \pi$

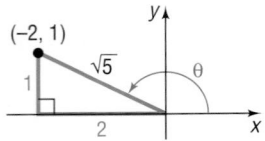

(c) Let $\theta = \cot^{-1}\dfrac{1}{2}$. Then $\cot \theta = \dfrac{1}{2}, 0 < \theta < \pi$. From these facts we know that θ lies in quadrant I. We draw Figure 22 to help us to find $\cos \theta$. Then

$$\cos \theta = \dfrac{1}{\sqrt{5}}, \text{ so } \theta = \cos^{-1}\left(\dfrac{1}{\sqrt{5}}\right), \text{ and}$$

$$\cot^{-1}\dfrac{1}{2} = \theta = \cos^{-1}\left(\dfrac{1}{\sqrt{5}}\right) \approx 1.11$$

(d) Let $\theta = \cot^{-1}(-2)$. Then $\cot \theta = -2, 0 < \theta < \pi$. From these facts we know that θ lies in quadrant II. We draw Figure 23 to help us to find $\cos \theta$. Then $\cos \theta = -\dfrac{2}{\sqrt{5}}$, so $\theta = \cos^{-1}\left(-\dfrac{2}{\sqrt{5}}\right)$, and

$$\cot^{-1}(-2) = \theta = \cos^{-1}\left(-\dfrac{2}{\sqrt{5}}\right) \approx 2.68$$

━ **NOW WORK PROBLEM 37.**

3.2   Concepts and Vocabulary

In Problems 1 and 2, fill in the blanks.

1. $y = \sec^{-1} x$ means _____ , where $|x|$ _____ and _____ $\le y \le$ _____ , $y \ne \dfrac{\pi}{2}$.

2. $\cos(\tan^{-1} 1) =$ _____ .

In Problems 3–5, answer True or False to each statement.

3. It is impossible to obtain exact values for the inverse secant function.

4. $\csc^{-1} 0.5$ is not defined.

5. The domain of the inverse cotangent function is the set of real numbers.

6. Explain how you would evaluate $\sec^{-1} 2$ using a calculator.

7. Explain how you would evaluate $\cot^{-1}(-4)$ using a calculator.

3.2   Exercises

In Problems 1–28, find the exact value of each expression. Verify your results using a graphing utility.

1. $\cos\left(\sin^{-1}\dfrac{\sqrt{2}}{2}\right)$

2. $\sin\left(\cos^{-1}\dfrac{1}{2}\right)$

3. $\tan\left[\cos^{-1}\left(-\dfrac{\sqrt{3}}{2}\right)\right]$

4. $\tan\left[\sin^{-1}\left(-\dfrac{1}{2}\right)\right]$

5. $\sec\left(\cos^{-1}\dfrac{1}{2}\right)$

6. $\cot\left[\sin^{-1}\left(-\dfrac{1}{2}\right)\right]$

7. $\csc(\tan^{-1} 1)$

8. $\sec(\tan^{-1} \sqrt{3})$

9. $\sin[\tan^{-1}(-1)]$

10. $\cos\left[\sin^{-1}\left(-\dfrac{\sqrt{3}}{2}\right)\right]$

11. $\sec\left[\sin^{-1}\left(-\dfrac{1}{2}\right)\right]$

12. $\csc\left[\cos^{-1}\left(-\dfrac{\sqrt{3}}{2}\right)\right]$

13. $\cos^{-1}\left(\cos\dfrac{5\pi}{4}\right)$

14. $\tan^{-1}\left(\tan\dfrac{2\pi}{3}\right)$

15. $\sin^{-1}\left[\sin\left(-\dfrac{7\pi}{6}\right)\right]$

16. $\cos^{-1}\left[\cos\left(-\dfrac{\pi}{3}\right)\right]$

17. $\tan\left(\sin^{-1}\dfrac{1}{3}\right)$

18. $\tan\left(\cos^{-1}\dfrac{1}{3}\right)$

19. $\sec\left(\tan^{-1}\dfrac{1}{2}\right)$

20. $\cos\left(\sin^{-1}\dfrac{\sqrt{2}}{3}\right)$

21. $\cot\left[\sin^{-1}\left(-\dfrac{\sqrt{2}}{3}\right)\right]$ **22.** $\csc[\tan^{-1}(-2)]$ **23.** $\sin[\tan^{-1}(-3)]$ **24.** $\cot\left[\cos^{-1}\left(-\dfrac{\sqrt{3}}{3}\right)\right]$

25. $\sec\left(\sin^{-1}\dfrac{2\sqrt{5}}{5}\right)$ **26.** $\csc\left(\tan^{-1}\dfrac{1}{2}\right)$ **27.** $\sin^{-1}\left(\cos\dfrac{3\pi}{4}\right)$ **28.** $\cos^{-1}\left(\sin\dfrac{7\pi}{6}\right)$

In Problems 29–36, find the exact value of each expression. Verify your results using a graphing utility.

29. $\cot^{-1}\sqrt{3}$ **30.** $\cot^{-1}1$ **31.** $\csc^{-1}(-1)$ **32.** $\csc^{-1}\sqrt{2}$

33. $\sec^{-1}\dfrac{2\sqrt{3}}{3}$ **34.** $\sec^{-1}(-2)$ **35.** $\cot^{-1}\left(-\dfrac{\sqrt{3}}{3}\right)$ **36.** $\csc^{-1}\left(-\dfrac{2\sqrt{3}}{3}\right)$

In Problems 37–48, use a calculator to find the value of each expression rounded to two decimal places.

37. $\sec^{-1}4$ **38.** $\csc^{-1}5$ **39.** $\cot^{-1}2$ **40.** $\sec^{-1}(-3)$

41. $\csc^{-1}(-3)$ **42.** $\cot^{-1}\left(-\dfrac{1}{2}\right)$ **43.** $\cot^{-1}\sqrt{5}$ **44.** $\cot^{-1}(-8.1)$

45. $\csc^{-1}\left(-\dfrac{3}{2}\right)$ **46.** $\sec^{-1}\left(-\dfrac{4}{3}\right)$ **47.** $\cot^{-1}\left(-\dfrac{3}{2}\right)$ **48.** $\cot^{-1}(-\sqrt{10})$

---

49. Using a graphing utility, graph $y = \cot^{-1}x$.
50. Using a graphing utility, graph $y = \sec^{-1}x$.
51. Using a graphing utility, graph $y = \csc^{-1}x$.

52. Explain in your own words how you would use your calculator to find the value of $\cot^{-1}10$.
53. Consult three books on calculus and write down the definition in each of $y = \sec^{-1}x$ and $y = \csc^{-1}x$. Compare these with the definition given in this book.

PREPARING FOR THIS SECTION

Before getting started, review the following:

✓ Fundamental Identities (Section 2.2, p. 113)

3.3   TRIGONOMETRIC IDENTITIES

OBJECTIVES   ① Establish Identities

We saw in the previous chapter that the trigonometric functions lend themselves to a wide variety of identities. Before establishing some additional identities, let's review the definition of an *identity*.

Two functions f and g are said to be **identically equal** if

$$f(x) = g(x)$$

for every value of x for which both functions are defined. Such an equation is referred to as an **identity**. An equation that is not an identity is called a **conditional equation**.

For example, the following are identities:

$$(x + 1)^2 = x^2 + 2x + 1 \qquad \sin^2 x + \cos^2 x = 1 \qquad \csc x = \dfrac{1}{\sin x}$$

The following are conditional equations:

$$2x + 5 = 0 \qquad \text{True only if } x = -\frac{5}{2}.$$

$$\sin x = 0 \qquad \text{True only if } x = k\pi, k \text{ an integer.}$$

$$\sin x = \cos x \qquad \text{True only if } x = \frac{\pi}{4} + k\pi, k \text{ an integer.}$$

The following boxes summarize the trigonometric identities that we have established thus far.

Quotient Identities

$$\tan \theta = \frac{\sin \theta}{\cos \theta} \qquad \cot \theta = \frac{\cos \theta}{\sin \theta}$$

Reciprocal Identities

$$\csc \theta = \frac{1}{\sin \theta} \qquad \sec \theta = \frac{1}{\cos \theta} \qquad \cot \theta = \frac{1}{\tan \theta}$$

Pythagorean Identities

$$\sin^2 \theta + \cos^2 \theta = 1 \qquad \tan^2 \theta + 1 = \sec^2 \theta$$
$$1 + \cot^2 \theta = \csc^2 \theta$$

Even–Odd Identities

$$\sin(-\theta) = -\sin \theta \qquad \cos(-\theta) = \cos \theta \qquad \tan(-\theta) = -\tan \theta$$
$$\csc(-\theta) = -\csc \theta \qquad \sec(-\theta) = \sec \theta \qquad \cot(-\theta) = -\cot \theta$$

This list of identities comprises what we shall refer to as the **basic trigonometric identities**. These identities should not merely be memorized, but should be *known* (just as you know your name rather than have it memorized). In fact, minor variations of a basic identity are often used. For example, we might want to use $\sin^2 \theta = 1 - \cos^2 \theta$ or $\cos^2 \theta = 1 - \sin^2 \theta$ instead of $\sin^2 \theta + \cos^2 \theta = 1$. For this reason, among others, you need to know these relationships and be quite comfortable with variations of them.

In the examples that follow, the directions will read "Establish the identity . . . ." As you will see, this is accomplished by starting with one side of the given equation (usually the one containing the more complicated expression) and, using appropriate basic identities and algebraic manipulations, arriving at the other side. The selection of appropriate basic identities to obtain the desired result is learned only through experience and lots of practice.

EXAMPLE 1   **Establishing an Identity**

Establish the identity: $\csc \theta \cdot \tan \theta = \sec \theta$

Solution   We start with the left side, because it contains the more complicated expression, and apply a reciprocal identity and a quotient identity.

$$\csc \theta \cdot \tan \theta = \frac{1}{\sin \theta} \cdot \frac{\sin \theta}{\cos \theta} = \frac{1}{\cos \theta} = \sec \theta$$

Having arrived at the right side, the identity is established. ■

COMMENT: A graphing utility can be used to provide evidence of an identity. For example, if we graph $Y_1 = \csc \theta \cdot \tan \theta$ and $Y_2 = \sec \theta$, the graphs appear to be the same. This provides evidence that $Y_1 = Y_2$. However, it does not prove their equality. A graphing utility *cannot be used to establish an identity*—identities must be established algebraically. ■

✏ **NOW WORK PROBLEM 1.**

EXAMPLE 2   **Establishing an Identity**

Establish the identity: $\sin^2(-\theta) + \cos^2(-\theta) = 1$

Solution   We begin with the left side and apply even–odd identities.

$$
\begin{aligned}
\sin^2(-\theta) + \cos^2(-\theta) &= [\sin(-\theta)]^2 + [\cos(-\theta)]^2 \\
&= (-\sin \theta)^2 + (\cos \theta)^2 &&\text{Even–odd identities.} \\
&= (\sin \theta)^2 + (\cos \theta)^2 \\
&= 1 &&\text{Pythagorean Identity.} \quad ■
\end{aligned}
$$

EXAMPLE 3   **Establishing an Identity**

Establish the identity: $\dfrac{\sin^2(-\theta) - \cos^2(-\theta)}{\sin(-\theta) - \cos(-\theta)} = \cos \theta - \sin \theta$

Solution   We begin with two observations: The left side appears to contain the more complicated expression. Also, the left side contains expressions with the argument $-\theta$, whereas the right side contains expressions with the argument θ. We decide, therefore, to start with the left side and apply even–odd identities.

$$
\begin{aligned}
\frac{\sin^2(-\theta) - \cos^2(-\theta)}{\sin(-\theta) - \cos(-\theta)} &= \frac{[\sin(-\theta)]^2 - [\cos(-\theta)]^2}{\sin(-\theta) - \cos(-\theta)} \\
&= \frac{(-\sin \theta)^2 - (\cos \theta)^2}{-\sin \theta - \cos \theta} &&\text{Even–odd identities.} \\
&= \frac{(\sin \theta)^2 - (\cos \theta)^2}{-\sin \theta - \cos \theta} &&\text{Simplify.} \\
&= \frac{(\sin \theta - \cos \theta)(\sin \theta + \cos \theta)}{-(\sin \theta + \cos \theta)} &&\text{Factor.} \\
&= \cos \theta - \sin \theta &&\text{Cancel and simplify.}
\end{aligned}
$$

■

EXAMPLE 4   Establishing an Identity

Establish the identity: $\dfrac{1 + \tan\theta}{1 + \cot\theta} = \tan\theta$

Solution $\dfrac{1 + \tan\theta}{1 + \cot\theta} = \dfrac{1 + \tan\theta}{1 + \dfrac{1}{\tan\theta}} = \dfrac{1 + \tan\theta}{\dfrac{\tan\theta + 1}{\tan\theta}} = \dfrac{\tan\theta(1 + \tan\theta)}{\tan\theta + 1} = \tan\theta$

■

━ **NOW WORK PROBLEM 9.**

When sums or differences of quotients appear, it is usually best to rewrite them as a single quotient, especially if the other side of the identity consists of only one term.

EXAMPLE 5   Establishing an Identity

Establish the identity: $\dfrac{\sin\theta}{1 + \cos\theta} + \dfrac{1 + \cos\theta}{\sin\theta} = 2\csc\theta$

Solution The left side is more complicated, so we start with it and proceed to add.

$$\dfrac{\sin\theta}{1 + \cos\theta} + \dfrac{1 + \cos\theta}{\sin\theta} = \dfrac{\sin^2\theta + (1 + \cos\theta)^2}{(1 + \cos\theta)(\sin\theta)} \qquad \text{Add the quotients.}$$

$$= \dfrac{\sin^2\theta + 1 + 2\cos\theta + \cos^2\theta}{(1 + \cos\theta)(\sin\theta)} \qquad \begin{array}{l}\text{Remove parentheses}\\\text{in numerator.}\end{array}$$

$$= \dfrac{(\sin^2\theta + \cos^2\theta) + 1 + 2\cos\theta}{(1 + \cos\theta)(\sin\theta)} \qquad \text{Regroup.}$$

$$= \dfrac{2 + 2\cos\theta}{(1 + \cos\theta)(\sin\theta)} \qquad \text{Pythagorean Identity.}$$

$$= \dfrac{2(1 + \cos\theta)}{(1 + \cos\theta)(\sin\theta)} \qquad \text{Factor and cancel.}$$

$$= \dfrac{2}{\sin\theta}$$

$$= 2\csc\theta \qquad \text{Reciprocal Identity.}$$

■

━ **NOW WORK PROBLEM 31.**

Sometimes it helps to write one side in terms of sines and cosines only.

EXAMPLE 6   Establishing an Identity

Establish the identity: $\dfrac{\tan\theta + \cot\theta}{\sec\theta\csc\theta} = 1$

Solution

$$\frac{\tan\theta + \cot\theta}{\sec\theta\csc\theta} \underset{\uparrow}{=} \frac{\dfrac{\sin\theta}{\cos\theta} + \dfrac{\cos\theta}{\sin\theta}}{\dfrac{1}{\cos\theta} \cdot \dfrac{1}{\sin\theta}} \underset{\uparrow}{=} \frac{\dfrac{\sin^2\theta + \cos^2\theta}{\cos\theta\sin\theta}}{\dfrac{1}{\cos\theta\sin\theta}}$$

Change to sines
and cosines.   Add the quotients
in the numerator.

$$\underset{\uparrow}{=} \frac{1}{\cos\theta\sin\theta} \cdot \frac{\cos\theta\sin\theta}{1} = 1$$

Divide quotient;
$\sin^2\theta + \cos^2\theta = 1$

- **NOW WORK PROBLEM 51.**

Sometimes, multiplying the numerator and denominator by an appropriate factor will result in a simplification.

EXAMPLE 7   Establishing an Identity

Establish the identity: $\quad \dfrac{1 - \sin\theta}{\cos\theta} = \dfrac{\cos\theta}{1 + \sin\theta}$

Solution We start with the left side and multiply the numerator and the denominator by $1 + \sin\theta$. (Alternatively, we could multiply the numerator and denominator of the right side by $1 - \sin\theta$.)

$$\frac{1 - \sin\theta}{\cos\theta} = \frac{1 - \sin\theta}{\cos\theta} \cdot \frac{1 + \sin\theta}{1 + \sin\theta} \qquad \text{\small Multiply numerator and denominator by } 1 + \sin\theta.$$

$$= \frac{1 - \sin^2\theta}{\cos\theta(1 + \sin\theta)}$$

$$= \frac{\cos^2\theta}{\cos\theta(1 + \sin\theta)} \qquad \text{\small $1 - \sin^2\theta = \cos^2\theta$.}$$

$$= \frac{\cos\theta}{1 + \sin\theta} \qquad \text{\small Cancel.}$$

- **NOW WORK PROBLEM 35.**

EXAMPLE 8   Establishing an Identity Involving Inverse Trigonometric Functions

Show that $\sin(\tan^{-1} v) = \dfrac{v}{\sqrt{1 + v^2}}$.

Solution Let $\theta = \tan^{-1} v$ so that $\tan\theta = v$, $-\dfrac{\pi}{2} < \theta < \dfrac{\pi}{2}$. As a result, we know that $\sec\theta > 0$.

$$\sin(\tan^{-1} v) = \sin\theta = \sin\theta \cdot \frac{\cos\theta}{\cos\theta} \underset{\uparrow}{=} \tan\theta\cos\theta = \frac{\tan\theta}{\sec\theta} \underset{\uparrow}{=} \frac{\tan\theta}{\sqrt{1 + \tan^2\theta}} = \frac{v}{\sqrt{1 + v^2}}$$

$\dfrac{\sin\theta}{\cos\theta} = \tan\theta$ $\sec^2\theta = 1 + \tan^2\theta$
$\sec\theta > 0$

- **NOW WORK PROBLEM 81.**

Although a lot of practice is the only real way to learn how to establish identities, the following guidelines should prove helpful.

> **Guidelines for Establishing Identities**
>
> 1. It is almost always preferable to start with the side containing the more complicated expression.
> 2. Rewrite sums or differences of quotients as a single quotient.
> 3. Sometimes, rewriting one side in terms of sines and cosines only will help.
> 4. Always keep your goal in mind. As you manipulate one side of the expression, you must keep in mind the form of the expression on the other side.

WARNING: Be careful not to handle identities to be established as if they were conditional equations. You *cannot* establish an identity by such methods as adding the same expression to each side and obtaining a true statement. This practice is not allowed, because the original statement is precisely the one that you are trying to establish. You do not know until it has been established that it is, in fact, true.   ■

3.3   Concepts and Vocabulary

In Problems 1–3, fill in the blanks.

1. Suppose that f and g are two functions with the same domain. If $f(x) = g(x)$ for every x in the domain, the equation is called a(n) _____ . Otherwise, it is called a(n) _____ equation.

2. $\tan^2 \theta - \sec^2 \theta =$ _____ .

3. $\cos(-\theta) - \cos \theta =$ _____ .

In Problems 4–6, answer True or False to each statement.

4. $\sin(-\theta) + \sin \theta = 0$ for any value of θ.

5. In establishing an identity, it is often easiest to just multiply both sides by a well-chosen nonzero expression involving the variable.

6. $[\tan \theta][\cos \theta] = \sin \theta$.

7. Write down the three Pythagorean Identities.

8. Why do you think it is usually preferable to start with the side containing the more complicated expression when establishing an identity?

9. Make up an identity that is not a Fundamental Identity.

3.3   Exercises

In Problems 1–80, establish each identity.

1. $\csc \theta \cdot \cos \theta = \cot \theta$

2. $\sec \theta \cdot \sin \theta = \tan \theta$

3. $1 + \tan^2(-\theta) = \sec^2 \theta$

4. $1 + \cot^2(-\theta) = \csc^2 \theta$

5. $\cos \theta(\tan \theta + \cot \theta) = \csc \theta$

6. $\sin \theta(\cot \theta + \tan \theta) = \sec \theta$

7. $\tan \theta \cot \theta - \cos^2 \theta = \sin^2 \theta$

8. $\sin \theta \csc \theta - \cos^2 \theta = \sin^2 \theta$

9. $(\sec \theta - 1)(\sec \theta + 1) = \tan^2 \theta$

10. $(\csc \theta - 1)(\csc \theta + 1) = \cot^2 \theta$

11. $(\sec \theta + \tan \theta)(\sec \theta - \tan \theta) = 1$

12. $(\csc \theta + \cot \theta)(\csc \theta - \cot \theta) = 1$

13. $\cos^2 \theta(1 + \tan^2 \theta) = 1$

14. $(1 - \cos^2 \theta)(1 + \cot^2 \theta) = 1$

15. $(\sin \theta + \cos \theta)^2 + (\sin \theta - \cos \theta)^2 = 2$

16. $\tan^2 \theta \cos^2 \theta + \cot^2 \theta \sin^2 \theta = 1$

17. $\sec^4 \theta - \sec^2 \theta = \tan^4 \theta + \tan^2 \theta$

18. $\csc^4 \theta - \csc^2 \theta = \cot^4 \theta + \cot^2 \theta$

19. $\sec\theta - \tan\theta = \dfrac{\cos\theta}{1 + \sin\theta}$

20. $\csc\theta - \cot\theta = \dfrac{\sin\theta}{1 + \cos\theta}$

21. $3\sin^2\theta + 4\cos^2\theta = 3 + \cos^2\theta$

22. $9\sec^2\theta - 5\tan^2\theta = 5 + 4\sec^2\theta$

23. $1 - \dfrac{\cos^2\theta}{1 + \sin\theta} = \sin\theta$

24. $1 - \dfrac{\sin^2\theta}{1 - \cos\theta} = -\cos\theta$

25. $\dfrac{1 + \tan\theta}{1 - \tan\theta} = \dfrac{\cot\theta + 1}{\cot\theta - 1}$

26. $\dfrac{\csc\theta - 1}{\csc\theta + 1} = \dfrac{1 - \sin\theta}{1 + \sin\theta}$

27. $\dfrac{\sec\theta}{\csc\theta} + \dfrac{\sin\theta}{\cos\theta} = 2\tan\theta$

28. $\dfrac{\csc\theta - 1}{\cot\theta} = \dfrac{\cot\theta}{\csc\theta + 1}$

29. $\dfrac{1 + \sin\theta}{1 - \sin\theta} = \dfrac{\csc\theta + 1}{\csc\theta - 1}$

30. $\dfrac{\cos\theta + 1}{\cos\theta - 1} = \dfrac{1 + \sec\theta}{1 - \sec\theta}$

31. $\dfrac{1 - \sin\theta}{\cos\theta} + \dfrac{\cos\theta}{1 - \sin\theta} = 2\sec\theta$

32. $\dfrac{\cos\theta}{1 + \sin\theta} + \dfrac{1 + \sin\theta}{\cos\theta} = 2\sec\theta$

33. $\dfrac{\sin\theta}{\sin\theta - \cos\theta} = \dfrac{1}{1 - \cot\theta}$

34. $1 - \dfrac{\sin^2\theta}{1 + \cos\theta} = \cos\theta$

35. $\dfrac{1 - \sin\theta}{1 + \sin\theta} = (\sec\theta - \tan\theta)^2$

36. $\dfrac{1 - \cos\theta}{1 + \cos\theta} = (\csc\theta - \cot\theta)^2$

37. $\dfrac{\cos\theta}{1 - \tan\theta} + \dfrac{\sin\theta}{1 - \cot\theta} = \sin\theta + \cos\theta$

38. $\dfrac{\cot\theta}{1 - \tan\theta} + \dfrac{\tan\theta}{1 - \cot\theta} = 1 + \tan\theta + \cot\theta$

39. $\tan\theta + \dfrac{\cos\theta}{1 + \sin\theta} = \sec\theta$

40. $\dfrac{\sin\theta\cos\theta}{\cos^2\theta - \sin^2\theta} = \dfrac{\tan\theta}{1 - \tan^2\theta}$

41. $\dfrac{\tan\theta + \sec\theta - 1}{\tan\theta - \sec\theta + 1} = \tan\theta + \sec\theta$

42. $\dfrac{\sin\theta - \cos\theta + 1}{\sin\theta + \cos\theta - 1} = \dfrac{\sin\theta + 1}{\cos\theta}$

43. $\dfrac{\tan\theta - \cot\theta}{\tan\theta + \cot\theta} = \sin^2\theta - \cos^2\theta$

44. $\dfrac{\sec\theta - \cos\theta}{\sec\theta + \cos\theta} = \dfrac{\sin^2\theta}{1 + \cos^2\theta}$

45. $\dfrac{\tan\theta - \cot\theta}{\tan\theta + \cot\theta} + 1 = 2\sin^2\theta$

46. $\dfrac{\tan\theta - \cot\theta}{\tan\theta + \cot\theta} + 2\cos^2\theta = 1$

47. $\dfrac{\sec\theta + \tan\theta}{\cot\theta + \cos\theta} = \tan\theta\sec\theta$

48. $\dfrac{\sec\theta}{1 + \sec\theta} = \dfrac{1 - \cos\theta}{\sin^2\theta}$

49. $\dfrac{1 - \tan^2\theta}{1 + \tan^2\theta} + 1 = 2\cos^2\theta$

50. $\dfrac{1 - \cot^2\theta}{1 + \cot^2\theta} + 2\cos^2\theta = 1$

51. $\dfrac{\sec\theta - \csc\theta}{\sec\theta\csc\theta} = \sin\theta - \cos\theta$

52. $\dfrac{\sin^2\theta - \tan\theta}{\cos^2\theta - \cot\theta} = \tan^2\theta$

53. $\sec\theta - \cos\theta - \sin\theta\tan\theta = 0$

54. $\tan\theta + \cot\theta - \sec\theta\csc\theta = 0$

55. $\dfrac{1}{1 - \sin\theta} + \dfrac{1}{1 + \sin\theta} = 2\sec^2\theta$

56. $\dfrac{1 + \sin\theta}{1 - \sin\theta} - \dfrac{1 - \sin\theta}{1 + \sin\theta} = 4\tan\theta\sec\theta$

57. $\dfrac{\sec\theta}{1 - \sin\theta} = \dfrac{1 + \sin\theta}{\cos^3\theta}$

58. $\dfrac{1 - \sin\theta}{1 + \sin\theta} = (\sec\theta - \tan\theta)^2$

59. $\dfrac{(\sec\theta - \tan\theta)^2 + 1}{\csc\theta(\sec\theta - \tan\theta)} = 2\tan\theta$

60. $\dfrac{\sec^2\theta - \tan^2\theta + \tan\theta}{\sec\theta} = \sin\theta + \cos\theta$

61. $\dfrac{\sin\theta + \cos\theta}{\cos\theta} - \dfrac{\sin\theta - \cos\theta}{\sin\theta} = \sec\theta\csc\theta$

62. $\dfrac{\sin\theta + \cos\theta}{\sin\theta} - \dfrac{\cos\theta - \sin\theta}{\cos\theta} = \sec\theta\csc\theta$

63. $\dfrac{\sin^3\theta + \cos^3\theta}{\sin\theta + \cos\theta} = 1 - \sin\theta\cos\theta$

64. $\dfrac{\sin^3\theta + \cos^3\theta}{1 - 2\cos^2\theta} = \dfrac{\sec\theta - \sin\theta}{\tan\theta - 1}$

65. $\dfrac{\cos^2\theta - \sin^2\theta}{1 - \tan^2\theta} = \cos^2\theta$

66. $\dfrac{\cos\theta + \sin\theta - \sin^3\theta}{\sin\theta} = \cot\theta + \cos^2\theta$

67. $\dfrac{(2\cos^2\theta - 1)^2}{\cos^4\theta - \sin^4\theta} = 1 - 2\sin^2\theta$

68. $\dfrac{1 - 2\cos^2\theta}{\sin\theta\cos\theta} = \tan\theta - \cot\theta$

69. $\dfrac{1 + \sin\theta + \cos\theta}{1 + \sin\theta - \cos\theta} = \dfrac{1 + \cos\theta}{\sin\theta}$

70. $\dfrac{1 + \cos\theta + \sin\theta}{1 + \cos\theta - \sin\theta} = \sec\theta + \tan\theta$

71. $(a\sin\theta + b\cos\theta)^2 + (a\cos\theta - b\sin\theta)^2 = a^2 + b^2$

72. $(2a\sin\theta\cos\theta)^2 + a^2(\cos^2\theta - \sin^2\theta)^2 = a^2$

73. $\dfrac{\tan\alpha + \tan\beta}{\cot\alpha + \cot\beta} = \tan\alpha\tan\beta$

74. $(\tan\alpha + \tan\beta)(1 - \cot\alpha\cot\beta) + (\cot\alpha + \cot\beta)(1 - \tan\alpha\tan\beta) = 0$

75. $(\sin\alpha + \cos\beta)^2 + (\cos\beta + \sin\alpha)(\cos\beta - \sin\alpha) = 2\cos\beta(\sin\alpha + \cos\beta)$

76. $(\sin\alpha - \cos\beta)^2 + (\cos\beta + \sin\alpha)(\cos\beta - \sin\alpha) = -2\cos\beta(\sin\alpha - \cos\beta)$

77. $\ln|\sec \theta| = -\ln|\cos \theta|$

78. $\ln|\tan \theta| = \ln|\sin \theta| - \ln|\cos \theta|$

79. $\ln|1 + \cos \theta| + \ln|1 - \cos \theta| = 2\ln|\sin \theta|$

80. $\ln|\sec \theta + \tan \theta| + \ln|\sec \theta - \tan \theta| = 0$

81. Show that $\sec(\tan^{-1}v) = \sqrt{1 + v^2}$.

82. Show that $\tan(\sin^{-1}v) = \dfrac{v}{\sqrt{1 - v^2}}$.

83. Show that $\tan(\cos^{-1}v) = \dfrac{\sqrt{1 - v^2}}{v}$.

84. Show that $\sin(\cos^{-1}v) = \sqrt{1 - v^2}$.

85. Show that $\cos(\sin^{-1}v) = \sqrt{1 - v^2}$.

86. Show that $\cos(\tan^{-1}v) = \dfrac{1}{\sqrt{1 + v^2}}$.

87. Write a few paragraphs outlining your strategy for establishing identities.

PREPARING FOR THIS SECTION

Before getting started, review the following:

✓ Distance Formula (Section 1.1, pp. 5–7)

✓ Values of the Trigonometric Functions of Certain Angles (Section 2.3, p. 126, and Section 2.4, p. 134)

3.4 SUM AND DIFFERENCE FORMULAS

OBJECTIVES

1. Use Sum and Difference Formulas to Find Exact Values
2. Use Sum and Difference Formulas to Establish Identities
3. Use Sum and Difference Formulas Involving Inverse Trigonometric Functions

In this section, we continue our derivation of trigonometric identities by obtaining formulas that involve the sum or difference of two angles, such as $\cos(\alpha + \beta)$, $\cos(\alpha - \beta)$, or $\sin(\alpha + \beta)$. These formulas are referred to as the **sum and difference formulas**. We begin with the formulas for $\cos(\alpha + \beta)$ and $\cos(\alpha - \beta)$.

Theorem

Sum and Difference Formulas for Cosines

$$\cos(\alpha + \beta) = \cos \alpha \cos \beta - \sin \alpha \sin \beta \qquad (1)$$
$$\cos(\alpha - \beta) = \cos \alpha \cos \beta + \sin \alpha \sin \beta \qquad (2)$$

In words, formula (1) states that the cosine of the sum of two angles equals the cosine of the first angle times the cosine of the second angle minus the sine of the first angle times the sine of the second angle.

Proof We will prove formula (2) first. Although this formula is true for all numbers α and β, we shall assume in our proof that $0 < \beta < \alpha < 2\pi$. We begin with the unit circle and place the angles α and β in standard position, as shown in Figure 24(a). The point P_1 lies on the terminal side of β, so its coordinates are $(\cos \beta, \sin \beta)$; and the point P_2 lies on the terminal side of α so its coordinates are $(\cos \alpha, \sin \alpha)$.

Figure 24

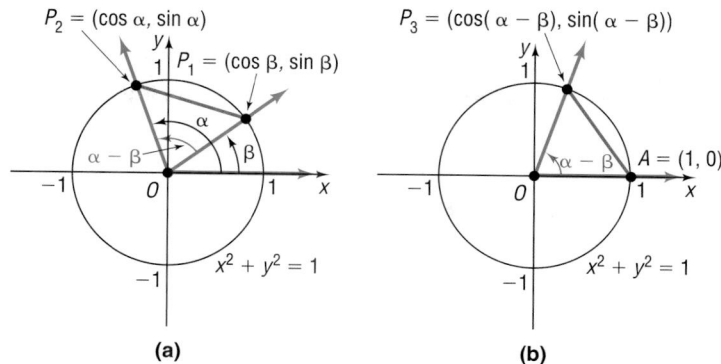

$P_2 = (\cos\alpha, \sin\alpha)$

$P_1 = (\cos\beta, \sin\beta)$

$P_3 = (\cos(\alpha - \beta), \sin(\alpha - \beta))$

$A = (1, 0)$

$x^2 + y^2 = 1$

$x^2 + y^2 = 1$

(a)   **(b)**

Now, place the angle $\alpha - \beta$ in standard position, as shown in Figure 24(b). The point A has coordinates $(1,0)$, and the point P_3 is on the terminal side of the angle $\alpha - \beta$, so its coordinates are $(\cos(\alpha - \beta), \sin(\alpha - \beta))$.

Looking at triangle OP_1P_2 in Figure 24(a) and triangle OAP_3 in Figure 24(b), we see that these triangles are congruent. (Do you see why? Two sides and the included angle, $\alpha - \beta$, are equal.) As a result, the unknown side of each triangle must be equal; that is,

$$d(A, P_3) = d(P_1, P_2)$$

Using the distance formula, we find that

$$\sqrt{[\cos(\alpha - \beta) - 1]^2 + [\sin(\alpha - \beta) - 0]^2} = \sqrt{(\cos\alpha - \cos\beta)^2 + (\sin\alpha - \sin\beta)^2} \qquad d(A, P_3) = d(P_1, P_2).$$

$$[\cos(\alpha - \beta) - 1]^2 + \sin^2(\alpha - \beta) = (\cos\alpha - \cos\beta)^2 + (\sin\alpha - \sin\beta)^2 \qquad \text{Square both sides.}$$

$$\cos^2(\alpha - \beta) - 2\cos(\alpha - \beta) + 1 + \sin^2(\alpha - \beta) = \cos^2\alpha - 2\cos\alpha\cos\beta + \cos^2\beta \qquad \text{Multiply out the squared terms.}$$
$$+ \sin^2\alpha - 2\sin\alpha\sin\beta + \sin^2\beta$$

$$2 - 2\cos(\alpha - \beta) = 2 - 2\cos\alpha\cos\beta - 2\sin\alpha\sin\beta \qquad \text{Apply a Pythagorean Identity (3 times).}$$

$$-2\cos(\alpha - \beta) = -2\cos\alpha\cos\beta - 2\sin\alpha\sin\beta \qquad \text{Subtract 2 from each side.}$$

$$\cos(\alpha - \beta) = \cos\alpha\cos\beta + \sin\alpha\sin\beta \qquad \text{Divide each side by } -2.$$

which is formula (2).

The proof of formula (1) follows from formula (2) and the Even–Odd Identities. We use the fact that $\alpha + \beta = \alpha - (-\beta)$. Then

$$\cos(\alpha + \beta) = \cos[\alpha - (-\beta)]$$
$$= \cos\alpha\cos(-\beta) + \sin\alpha\sin(-\beta) \qquad \text{Use formula (2).}$$
$$= \cos\alpha\cos\beta - \sin\alpha\sin\beta \qquad \text{Even–odd Identities} \qquad \blacksquare$$

1   One use of formulas (1) and (2) is to obtain the exact value of the cosine of an angle that can be expressed as the sum or difference of angles whose sine and cosine are known exactly.

EXAMPLE 1   Using the Sum Formula to Find Exact Values

Find the exact value of $\cos 75°$.

Solution Since $75° = 45° + 30°$, we use formula (1) to obtain

$$\cos 75° = \cos(45° + 30°) = \cos 45° \cos 30° - \sin 45° \sin 30°$$

Formula (1)

$$= \frac{\sqrt{2}}{2} \cdot \frac{\sqrt{3}}{2} - \frac{\sqrt{2}}{2} \cdot \frac{1}{2} = \frac{1}{4}(\sqrt{6} - \sqrt{2})$$ ∎

EXAMPLE 2   Using the Difference Formula to Find Exact Values

Find the exact value of $\cos \dfrac{\pi}{12}$.

Solution
$$\cos \frac{\pi}{12} = \cos\left(\frac{3\pi}{12} - \frac{2\pi}{12}\right) = \cos\left(\frac{\pi}{4} - \frac{\pi}{6}\right)$$

$$= \cos \frac{\pi}{4} \cos \frac{\pi}{6} + \sin \frac{\pi}{4} \sin \frac{\pi}{6}$$ Use formula (2).

$$= \frac{\sqrt{2}}{2} \cdot \frac{\sqrt{3}}{2} + \frac{\sqrt{2}}{2} \cdot \frac{1}{2} = \frac{1}{4}(\sqrt{6} + \sqrt{2})$$ ∎

NOW WORK PROBLEM 3.

2 Another use of formulas (1) and (2) is to establish other identities. One important pair of identities is given next.

$$\cos\left(\frac{\pi}{2} - \theta\right) = \sin \theta \qquad \text{(3a)}$$

$$\sin\left(\frac{\pi}{2} - \theta\right) = \cos \theta \qquad \text{(3b)}$$

SEEING THE CONCEPT Graph $Y_1 = \cos\left(\frac{\pi}{2} - \theta\right)$ and $Y_2 = \sin \theta$ in radian mode on the same screen. Does this demonstrate the result 3(a)? How would you demonstrate the result 3(b)?

Proof To prove formula (3a), we use the formula for $\cos(\alpha - \beta)$ with $\alpha = \dfrac{\pi}{2}$ and $\beta = \theta$.

$$\cos\left(\frac{\pi}{2} - \theta\right) = \cos \frac{\pi}{2} \cos \theta + \sin \frac{\pi}{2} \sin \theta$$

$$= 0 \cdot \cos \theta + 1 \cdot \sin \theta$$

$$= \sin \theta$$

To prove formula (3b), we make use of the identity (3a) just established.

$$\sin\left(\frac{\pi}{2} - \theta\right) = \cos\left[\frac{\pi}{2} - \left(\frac{\pi}{2} - \theta\right)\right] = \cos\theta$$

Use (3a).

Formulas (3a) and (3b) should look familiar. They are the basis for the theorem stated in Chapter 2: Cofunctions of complementary angles are equal. Also, since

$$\cos\left(\frac{\pi}{2} - \theta\right) = \cos\left[-\left(\theta - \frac{\pi}{2}\right)\right] = \cos\left(\theta - \frac{\pi}{2}\right)$$

Even Property of Cosine

and

$$\cos\left(\frac{\pi}{2} - \theta\right) = \sin\theta$$

3(a)

it follows that $\cos\left(\theta - \frac{\pi}{2}\right) = \sin\theta$. The graphs of $y = \cos\left(\theta - \frac{\pi}{2}\right)$ and $y = \sin\theta$ are identical, a fact that we conjectured earlier in Section 2.6.

Formulas For $\sin(\alpha + \beta)$ and $\sin(\alpha - \beta)$

Having established the identities in formulas (3a) and (3b), we now can derive the sum and difference formulas for $\sin(\alpha + \beta)$ and $\sin(\alpha - \beta)$.

Proof

$$\sin(\alpha + \beta) = \cos\left[\frac{\pi}{2} - (\alpha + \beta)\right] \qquad \text{Formula (3a)}$$

$$= \cos\left[\left(\frac{\pi}{2} - \alpha\right) - \beta\right]$$

$$= \cos\left(\frac{\pi}{2} - \alpha\right)\cos\beta + \sin\left(\frac{\pi}{2} - \alpha\right)\sin\beta \quad \text{Formula (2)}$$

$$= \sin\alpha\cos\beta + \cos\alpha\sin\beta \qquad \text{Formulas (3a) and (3b)}$$

$$\sin(\alpha - \beta) = \sin[\alpha + (-\beta)]$$

$$= \sin\alpha\cos(-\beta) + \cos\alpha\sin(-\beta) \qquad \begin{array}{l}\text{Use the sum formula for}\\ \text{sine just obtained.}\end{array}$$

$$= \sin\alpha\cos\beta + \cos\alpha(-\sin\beta) \qquad \text{Even–odd Identities}$$

$$= \sin\alpha\cos\beta - \cos\alpha\sin\beta$$

Theorem   **Sum and Difference Formulas for Sines**

$$\sin(\alpha + \beta) = \sin\alpha\cos\beta + \cos\alpha\sin\beta \qquad (4)$$

$$\sin(\alpha - \beta) = \sin\alpha\cos\beta - \cos\alpha\sin\beta \qquad (5)$$

In words, formula (4) states that the sine of the sum of two angles equals the sine of the first angle times the cosine of the second angle plus the cosine of the first angle times the sine of the second angle.

EXAMPLE 3 Using the Sum Formula to Find Exact Values

Find the exact value of $\sin\dfrac{7\pi}{12}$.

Solution
$$\sin\frac{7\pi}{12} = \sin\left(\frac{3\pi}{12} + \frac{4\pi}{12}\right) = \sin\left(\frac{\pi}{4} + \frac{\pi}{3}\right)$$

$$= \sin\frac{\pi}{4}\cos\frac{\pi}{3} + \cos\frac{\pi}{4}\sin\frac{\pi}{3} \qquad \text{Formula (4)}$$

$$= \frac{\sqrt{2}}{2}\cdot\frac{1}{2} + \frac{\sqrt{2}}{2}\cdot\frac{\sqrt{3}}{2} = \frac{1}{4}(\sqrt{2} + \sqrt{6}) \qquad ■$$

 NOW WORK PROBLEM 9.

EXAMPLE 4 Using the Difference Formula to Find Exact Values

Find the exact value of $\sin 80°\cos 20° - \cos 80°\sin 20°$.

Solution The form of the expression $\sin 80°\cos 20° - \cos 80°\sin 20°$ is that of the right side of the formula (5) for $\sin(\alpha - \beta)$ with $\alpha = 80°$ and $\beta = 20°$. Thus,

$$\sin 80°\cos 20° - \cos 80°\sin 20° = \sin(80° - 20°) = \sin 60° = \frac{\sqrt{3}}{2} \qquad ■$$

NOW WORK PROBLEMS 15 AND 19.

EXAMPLE 5 Finding Exact Values

If it is known that $\sin\alpha = \dfrac{4}{5}, \dfrac{\pi}{2} < \alpha < \pi$, and that

$\sin\beta = -\dfrac{2}{\sqrt{5}} = -\dfrac{2\sqrt{5}}{5}, \pi < \beta < \dfrac{3\pi}{2}$, find the exact value of

(a) $\cos\alpha$ (b) $\cos\beta$ (c) $\cos(\alpha + \beta)$ (d) $\sin(\alpha + \beta)$

Solution (a) Since $\sin\alpha = \dfrac{4}{5} = \dfrac{b}{r}$ and $\dfrac{\pi}{2} < \alpha < \pi$, we let $b = 4$ and $r = 5$ and place α in quadrant II. See Figure 25. Since $(a, 4)$ is in quadrant II, we have $a < 0$. The distance from $(a, 4)$ to $(0, 0)$ is 5, so

$$a^2 + 4^2 = 5^2,$$
$$a^2 + 16 = 25$$
$$a^2 = 25 - 16 = 9$$
$$a = -3 \qquad a < 0$$

Then,

$$\cos\alpha = \frac{a}{r} = -\frac{3}{5}$$

Figure 25

Given $\sin\alpha = \dfrac{4}{5}, \dfrac{\pi}{2} < \alpha < \pi$

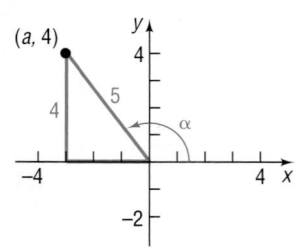

Alternatively, we can find $\cos \alpha$ using identities, as follows:

$$\cos \alpha = -\sqrt{1 - \sin^2 \alpha} = -\sqrt{1 - \frac{16}{25}} = -\sqrt{\frac{9}{25}} = -\frac{3}{5}$$

$\uparrow$
α in quadrant II,
$\cos \alpha < 0$

(b) Since $\sin \beta = \dfrac{-2}{\sqrt{5}} = \dfrac{b}{r}$ and $\pi < \beta < \dfrac{3\pi}{2}$, we let $b = -2$ and $r = \sqrt{5}$

Figure 26

Given $\sin \beta = \dfrac{-2}{\sqrt{5}}, \pi < \beta < \dfrac{3\pi}{2}$

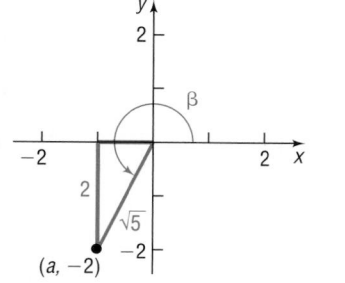

and place β in quadrant III. See Figure 26. Since $(a, -2)$ is in quadrant III, we have $a < 0$. The distance from $(a, -2)$ to $(0, 0)$ is $\sqrt{5}$, so

$$a^2 + 4 = (\sqrt{5})^2,$$
$$a^2 = 5 - 4 = 1$$
$$a = -1 \quad a < 0$$

Then,

$$\cos \beta = \frac{a}{r} = -\frac{1}{\sqrt{5}} = -\frac{\sqrt{5}}{5}$$

Alternatively, we can find $\cos \beta$ using identities, as follows:

$$\cos \beta = -\sqrt{1 - \sin^2 \beta} = -\sqrt{1 - \frac{4}{5}} = -\sqrt{\frac{1}{5}} = -\frac{\sqrt{5}}{5}$$

(c) Using the results found in parts (a) and (b) and formula (1), we have

$$\cos(\alpha + \beta) = \cos \alpha \cos \beta - \sin \alpha \sin \beta$$
$$= -\frac{3}{5}\left(-\frac{\sqrt{5}}{5}\right) - \frac{4}{5}\left(-\frac{2\sqrt{5}}{5}\right) = \frac{11\sqrt{5}}{25}$$

(d) $\sin(\alpha + \beta) = \sin \alpha \cos \beta + \cos \alpha \sin \beta$
$$= \frac{4}{5}\left(-\frac{\sqrt{5}}{5}\right) + \left(-\frac{3}{5}\right)\left(-\frac{2\sqrt{5}}{5}\right) = \frac{2\sqrt{5}}{25}$$

NOW WORK PROBLEM **23 (a), (b),** AND **(c).**

EXAMPLE 6   **Establishing an Identity**

Establish the identity: $\dfrac{\cos(\alpha - \beta)}{\sin \alpha \sin \beta} = \cot \alpha \cot \beta + 1$

Solution $\dfrac{\cos(\alpha - \beta)}{\sin \alpha \sin \beta} = \dfrac{\cos \alpha \cos \beta + \sin \alpha \sin \beta}{\sin \alpha \sin \beta}$

$$= \frac{\cos \alpha \cos \beta}{\sin \alpha \sin \beta} + \frac{\sin \alpha \sin \beta}{\sin \alpha \sin \beta}$$

$$= \frac{\cos \alpha}{\sin \alpha} \frac{\cos \beta}{\sin \beta} + 1$$

$$= \cot \alpha \cot \beta + 1$$

NOW WORK PROBLEMS **31** AND **43.**

Formulas for $\tan(\alpha + \beta)$ and $\tan(\alpha - \beta)$

We use the identity $\tan\theta = \dfrac{\sin\theta}{\cos\theta}$ and the sum formulas for $\sin(\alpha + \beta)$ and $\cos(\alpha + \beta)$ to derive a formula for $\tan(\alpha + \beta)$.

Proof

$$\tan(\alpha + \beta) = \frac{\sin(\alpha + \beta)}{\cos(\alpha + \beta)} = \frac{\sin\alpha\cos\beta + \cos\alpha\sin\beta}{\cos\alpha\cos\beta - \sin\alpha\sin\beta}$$

Now we divide the numerator and denominator by $\cos\alpha\cos\beta$.

$$\tan(\alpha + \beta) = \frac{\dfrac{\sin\alpha\cos\beta + \cos\alpha\sin\beta}{\cos\alpha\cos\beta}}{\dfrac{\cos\alpha\cos\beta - \sin\alpha\sin\beta}{\cos\alpha\cos\beta}} = \frac{\dfrac{\sin\alpha\cancel{\cos\beta}}{\cos\alpha\cancel{\cos\beta}} + \dfrac{\cancel{\cos\alpha}\sin\beta}{\cancel{\cos\alpha}\cos\beta}}{\dfrac{\cancel{\cos\alpha}\cancel{\cos\beta}}{\cancel{\cos\alpha}\cancel{\cos\beta}} - \dfrac{\sin\alpha\sin\beta}{\cos\alpha\cos\beta}}$$

$$= \frac{\dfrac{\sin\alpha}{\cos\alpha} + \dfrac{\sin\beta}{\cos\beta}}{1 - \dfrac{\sin\alpha\sin\beta}{\cos\alpha\cos\beta}} = \frac{\tan\alpha + \tan\beta}{1 - \tan\alpha\tan\beta}$$

We use the sum formula for $\tan(\alpha + \beta)$ and even–odd properties to get the difference formula.

$$\tan(\alpha - \beta) = \tan[\alpha + (-\beta)] = \frac{\tan\alpha + \tan(-\beta)}{1 - \tan\alpha\tan(-\beta)} = \frac{\tan\alpha - \tan\beta}{1 + \tan\alpha\tan\beta}$$

∎

We have proved the following results:

Theorem

Sum and Difference Formulas for Tangents

$$\tan(\alpha + \beta) = \frac{\tan\alpha + \tan\beta}{1 - \tan\alpha\tan\beta} \qquad (6)$$

$$\tan(\alpha - \beta) = \frac{\tan\alpha - \tan\beta}{1 + \tan\alpha\tan\beta} \qquad (7)$$

In words, formula (6) states that the tangent of the sum of two angles equals the tangent of the first angle plus the tangent of the second angle, all divided by 1 minus their product.

NOW WORK PROBLEM 23(d).

EXAMPLE 7   Establishing an Identity

Prove the identity: $\tan(\theta + \pi) = \tan \theta$

Solution $\tan(\theta + \pi) = \dfrac{\tan \theta + \tan \pi}{1 - \tan \theta \tan \pi} = \dfrac{\tan \theta + 0}{1 - \tan \theta \cdot 0} = \tan \theta$ ■

The result obtained in Example 7 verifies that the tangent function is periodic with period π, a fact that we mentioned earlier.

WARNING: Be careful when using formulas (6) and (7). These formulas can be used only for angles α and β for which $\tan \alpha$ and $\tan \beta$ are defined, that is, all angles except odd multiples of $\dfrac{\pi}{2}$. ■

EXAMPLE 8   Establishing an Identity

Prove the identity: $\tan\left(\theta + \dfrac{\pi}{2} \right) = -\cot \theta$

Solution   We cannot use formula (6), since $\tan \dfrac{\pi}{2}$ is not defined. Instead, we proceed as follows:

$$\tan\left(\theta + \dfrac{\pi}{2} \right) = \dfrac{\sin\left(\theta + \dfrac{\pi}{2} \right)}{\cos\left(\theta + \dfrac{\pi}{2} \right)} = \dfrac{\sin \theta \cos \dfrac{\pi}{2} + \cos \theta \sin \dfrac{\pi}{2}}{\cos \theta \cos \dfrac{\pi}{2} - \sin \theta \sin \dfrac{\pi}{2}}$$

$$= \dfrac{(\sin \theta)(0) + (\cos \theta)(1)}{(\cos \theta)(0) - (\sin \theta)(1)} = \dfrac{\cos \theta}{-\sin \theta} = -\cot \theta$$ ■

③ **EXAMPLE 9**   Finding the Exact Value of Expressions Involving Inverse Trigonometric Functions

Find the exact value of: $\sin\left(\cos^{-1} \dfrac{1}{2} + \sin^{-1} \dfrac{3}{5} \right)$

Solution   We seek the sine of the sum of two angles, $\alpha = \cos^{-1} \dfrac{1}{2}$ and $\beta = \sin^{-1} \dfrac{3}{5}$. Then

$$\cos \alpha = \dfrac{1}{2}, \quad 0 \le \alpha \le \pi, \quad \text{and} \quad \sin \beta = \dfrac{3}{5}, \quad -\dfrac{\pi}{2} \le \beta \le \dfrac{\pi}{2}$$

We use Pythagorean Identities to obtain $\sin \alpha$ and $\cos \beta$. Since $\sin \alpha > 0$ and $\cos \beta > 0$ (do you know why?), we find

$$\sin \alpha = \sqrt{1 - \cos^2 \alpha} = \sqrt{1 - \dfrac{1}{4}} = \sqrt{\dfrac{3}{4}} = \dfrac{\sqrt{3}}{2}$$

$$\cos \beta = \sqrt{1 - \sin^2 \beta} = \sqrt{1 - \dfrac{9}{25}} = \sqrt{\dfrac{16}{25}} = \dfrac{4}{5}$$

As a result,

$$\sin\left(\cos^{-1}\frac{1}{2} + \sin^{-1}\frac{3}{5}\right) = \sin(\alpha + \beta) = \sin\alpha\cos\beta + \cos\alpha\sin\beta$$

$$= \frac{\sqrt{3}}{2}\cdot\frac{4}{5} + \frac{1}{2}\cdot\frac{3}{5} = \frac{4\sqrt{3}+3}{10} \qquad \blacksquare$$

━━━ **NOW WORK PROBLEM 59.**

EXAMPLE 10   **Writing a Trigonometric Expression as an Algebraic Expression**

Write $\sin(\sin^{-1}u + \cos^{-1}v)$ as an algebraic expression containing u and v (that is, without any trigonometric functions).

Solution   Let $\alpha = \sin^{-1}u$ and $\beta = \cos^{-1}v$. Then

$$\sin\alpha = u, \quad -\frac{\pi}{2} \le \alpha \le \frac{\pi}{2} \quad \text{and} \quad \cos\beta = v, \quad 0 \le \beta \le \pi$$

Since $-\dfrac{\pi}{2} \le \alpha \le \dfrac{\pi}{2}$, we know that $\cos\alpha \ge 0$. As a result,

$$\cos\alpha = \sqrt{1 - \sin^2\alpha} = \sqrt{1 - u^2}$$

Similarly, since $0 \le \beta \le \pi$, we know that $\sin\beta \ge 0$. As a result,

$$\sin\beta = \sqrt{1 - \cos^2\beta} = \sqrt{1 - v^2}$$

Now

$$\sin(\sin^{-1}u + \cos^{-1}v) = \sin(\alpha + \beta) = \sin\alpha\cos\beta + \cos\alpha\sin\beta$$
$$= uv + \sqrt{1 - u^2}\sqrt{1 - v^2} \qquad \blacksquare$$

━━━ **NOW WORK PROBLEM 69.**

SUMMARY

The following box summarizes the sum and difference formulas.

Sum and Difference Formulas

$$\cos(\alpha + \beta) = \cos\alpha\cos\beta - \sin\alpha\sin\beta \qquad \cos(\alpha - \beta) = \cos\alpha\cos\beta + \sin\alpha\sin\beta$$

$$\sin(\alpha + \beta) = \sin\alpha\cos\beta + \cos\alpha\sin\beta \qquad \sin(\alpha - \beta) = \sin\alpha\cos\beta - \cos\alpha\sin\beta$$

$$\tan(\alpha + \beta) = \frac{\tan\alpha + \tan\beta}{1 - \tan\alpha\tan\beta} \qquad \tan(\alpha - \beta) = \frac{\tan\alpha - \tan\beta}{1 + \tan\alpha\tan\beta}$$

3.4   Concepts and Vocabulary

In Problems 1 and 2, fill in the blanks.

1. $\cos(\alpha + \beta) = \cos \alpha \cos \beta \underline{\hspace{2cm}} \sin \alpha \sin \beta$.

2. $\sin(\alpha - \beta) = \sin \alpha \cos \beta \underline{\hspace{2cm}} \cos \alpha \sin \beta$.

In Problems 3–5, answer True or False to each statement.

3. $\sin(\alpha + \beta) = \sin \alpha + \sin \beta + 2 \sin \alpha \sin \beta$.

4. $\tan 75° = \tan 30° + \tan 45°$.

5. $\cos\left(\dfrac{\pi}{2} - \theta\right) = \cos \theta$.

6. Use a sum formula to find the exact value of $\sin 75°$.

7. Use a difference formula to find the exact value of $\sin 75°$.

8. Find the exact value of $\cos 65° \cos 25° - \sin 65° \sin 25°$.

3.4   Exercises

In Problems 1–12, find the exact value of each trigonometric function.

1. $\sin \dfrac{5\pi}{12}$ **2.** $\sin \dfrac{\pi}{12}$ **3.** $\cos \dfrac{7\pi}{12}$ **4.** $\tan \dfrac{7\pi}{12}$ **5.** $\cos 165°$ **6.** $\sin 105°$

7. $\tan 15°$ **8.** $\tan 195°$ **9.** $\sin \dfrac{17\pi}{12}$ **10.** $\tan \dfrac{19\pi}{12}$ **11.** $\sec\left(-\dfrac{\pi}{12}\right)$ **12.** $\cot\left(-\dfrac{5\pi}{12}\right)$

In Problems 13–22, find the exact value of each expression.

13. $\sin 20° \cos 10° + \cos 20° \sin 10°$

14. $\sin 20° \cos 80° - \cos 20° \sin 80°$

15. $\cos 70° \cos 20° - \sin 70° \sin 20°$

16. $\cos 40° \cos 10° + \sin 40° \sin 10°$

17. $\dfrac{\tan 20° + \tan 25°}{1 - \tan 20° \tan 25°}$

18. $\dfrac{\tan 40° - \tan 10°}{1 + \tan 40° \tan 10°}$

19. $\sin \dfrac{\pi}{12} \cos \dfrac{7\pi}{12} - \cos \dfrac{\pi}{12} \sin \dfrac{7\pi}{12}$

20. $\cos \dfrac{5\pi}{12} \cos \dfrac{7\pi}{12} - \sin \dfrac{5\pi}{12} \sin \dfrac{7\pi}{12}$

21. $\cos \dfrac{\pi}{12} \cos \dfrac{5\pi}{12} + \sin \dfrac{5\pi}{12} \sin \dfrac{\pi}{12}$

22. $\sin \dfrac{\pi}{18} \cos \dfrac{5\pi}{18} + \cos \dfrac{\pi}{18} \sin \dfrac{5\pi}{18}$

In Problems 23–28, find the exact value of each of the following under the given conditions:

(a) $\sin(\alpha + \beta)$ (b) $\cos(\alpha + \beta)$ (c) $\sin(\alpha - \beta)$ (d) $\tan(\alpha - \beta)$

23. $\sin \alpha = \dfrac{3}{5}$, $0 < \alpha < \dfrac{\pi}{2}$; $\cos \beta = \dfrac{2\sqrt{5}}{5}$, $-\dfrac{\pi}{2} < \beta < 0$

24. $\cos \alpha = \dfrac{\sqrt{5}}{5}$, $0 < \alpha < \dfrac{\pi}{2}$; $\sin \beta = -\dfrac{4}{5}$, $-\dfrac{\pi}{2} < \beta < 0$

25. $\tan \alpha = -\dfrac{4}{3}$, $\dfrac{\pi}{2} < \alpha < \pi$; $\cos \beta = \dfrac{1}{2}$, $0 < \beta < \dfrac{\pi}{2}$

26. $\tan \alpha = \dfrac{5}{12}$, $\pi < \alpha < \dfrac{3\pi}{2}$; $\sin \beta = -\dfrac{1}{2}$, $\pi < \beta < \dfrac{3\pi}{2}$

27. $\sin \alpha = \dfrac{5}{13}$, $-\dfrac{3\pi}{2} < \alpha < -\pi$; $\tan \beta = -\sqrt{3}$, $\dfrac{\pi}{2} < \beta < \pi$

28. $\cos \alpha = \dfrac{1}{2}$, $-\dfrac{\pi}{2} < \alpha < 0$; $\sin \beta = \dfrac{1}{3}$, $0 < \beta < \dfrac{\pi}{2}$

29. If $\sin \theta = \dfrac{1}{3}, \theta$ in quadrant II, find the exact value of:

(a) $\cos \theta$

(b) $\sin\left(\theta + \dfrac{\pi}{6}\right)$

(c) $\cos\left(\theta - \dfrac{\pi}{3}\right)$

(d) $\tan\left(\theta + \dfrac{\pi}{4}\right)$

30. If $\cos \theta = \dfrac{1}{4}, \theta$ in quadrant IV, find the exact value of:

(a) $\sin \theta$

(b) $\sin\left(\theta - \dfrac{\pi}{6}\right)$

(c) $\cos\left(\theta + \dfrac{\pi}{3}\right)$

(d) $\tan\left(\theta - \dfrac{\pi}{4}\right)$

In Problems 31–56, establish each identity.

31. $\sin\left(\dfrac{\pi}{2} + \theta\right) = \cos \theta$

32. $\cos\left(\dfrac{\pi}{2} + \theta\right) = -\sin \theta$

33. $\sin(\pi - \theta) = \sin \theta$

34. $\cos(\pi - \theta) = -\cos \theta$

35. $\sin(\pi + \theta) = -\sin \theta$

36. $\cos(\pi + \theta) = -\cos \theta$

37. $\tan(\pi - \theta) = -\tan \theta$

38. $\tan(2\pi - \theta) = -\tan \theta$

39. $\sin\left(\dfrac{3\pi}{2} + \theta\right) = -\cos \theta$

40. $\cos\left(\dfrac{3\pi}{2} + \theta\right) = \sin \theta$

41. $\sin(\alpha + \beta) + \sin(\alpha - \beta) = 2 \sin \alpha \cos \beta$

42. $\cos(\alpha + \beta) + \cos(\alpha - \beta) = 2 \cos \alpha \cos \beta$

43. $\dfrac{\sin(\alpha + \beta)}{\sin \alpha \cos \beta} = 1 + \cot \alpha \tan \beta$

44. $\dfrac{\sin(\alpha + \beta)}{\cos \alpha \cos \beta} = \tan \alpha + \tan \beta$

45. $\dfrac{\cos(\alpha + \beta)}{\cos \alpha \cos \beta} = 1 - \tan \alpha \tan \beta$

46. $\dfrac{\cos(\alpha - \beta)}{\sin \alpha \cos \beta} = \cot \alpha + \tan \beta$

47. $\dfrac{\sin(\alpha + \beta)}{\sin(\alpha - \beta)} = \dfrac{\tan \alpha + \tan \beta}{\tan \alpha - \tan \beta}$

48. $\dfrac{\cos(\alpha + \beta)}{\cos(\alpha - \beta)} = \dfrac{1 - \tan \alpha \tan \beta}{1 + \tan \alpha \tan \beta}$

49. $\cot(\alpha + \beta) = \dfrac{\cot \alpha \cot \beta - 1}{\cot \beta + \cot \alpha}$

50. $\cot(\alpha - \beta) = \dfrac{\cot \alpha \cot \beta + 1}{\cot \beta - \cot \alpha}$

51. $\sec(\alpha + \beta) = \dfrac{\csc \alpha \csc \beta}{\cot \alpha \cot \beta - 1}$

52. $\sec(\alpha - \beta) = \dfrac{\sec \alpha \sec \beta}{1 + \tan \alpha \tan \beta}$

53. $\sin(\alpha - \beta) \sin(\alpha + \beta) = \sin^2 \alpha - \sin^2 \beta$

54. $\cos(\alpha - \beta) \cos(\alpha + \beta) = \cos^2 \alpha - \sin^2 \beta$

55. $\sin(\theta + k\pi) = (-1)^k \sin \theta, \quad k$ any integer

56. $\cos(\theta + k\pi) = (-1)^k \cos \theta, \quad k$ any integer

In Problems 57–68, find the exact value of each expression.

57. $\sin\left(\sin^{-1}\dfrac{1}{2} + \cos^{-1}0\right)$

58. $\sin\left(\sin^{-1}\dfrac{\sqrt{3}}{2} + \cos^{-1}1\right)$

59. $\sin\left[\sin^{-1}\dfrac{3}{5} - \cos^{-1}\left(-\dfrac{4}{5}\right)\right]$

60. $\sin\left[\sin^{-1}\left(-\dfrac{4}{5}\right) - \tan^{-1}\dfrac{3}{4}\right]$

61. $\cos\left(\tan^{-1}\dfrac{4}{3} + \cos^{-1}\dfrac{5}{13}\right)$

62. $\cos\left[\tan^{-1}\dfrac{5}{12} - \sin^{-1}\left(-\dfrac{3}{5}\right)\right]$

63. $\cos\left(\sin^{-1}\dfrac{5}{13} - \tan^{-1}\dfrac{3}{4}\right)$

64. $\cos\left(\tan^{-1}\dfrac{4}{3} + \cos^{-1}\dfrac{12}{13}\right)$

65. $\tan\left(\sin^{-1}\dfrac{3}{5} + \dfrac{\pi}{6}\right)$

66. $\tan\left(\dfrac{\pi}{4} - \cos^{-1}\dfrac{3}{5}\right)$

67. $\tan\left(\sin^{-1}\dfrac{4}{5} + \cos^{-1}1\right)$

68. $\tan\left(\cos^{-1}\dfrac{4}{5} + \sin^{-1}1\right)$

In Problems 69–74, write each trigonometric expression as an algebraic expression containing u and v.

69. $\cos(\cos^{-1}u + \sin^{-1}v)$

70. $\sin(\sin^{-1}u - \cos^{-1}v)$

71. $\sin(\tan^{-1}u - \sin^{-1}v)$

72. $\cos(\tan^{-1}u + \tan^{-1}v)$

73. $\tan(\sin^{-1}u - \cos^{-1}v)$

74. $\sec(\tan^{-1}u + \cos^{-1}v)$

75. Show that $\sin^{-1}v + \cos^{-1}v = \dfrac{\pi}{2}$.

76. Show that $\tan^{-1}v + \cot^{-1}v = \dfrac{\pi}{2}$.

77. Show that $\tan^{-1}\left(\dfrac{1}{v}\right) = \dfrac{\pi}{2} - \tan^{-1}v$, if $v > 0$.

78. Show that $\cot^{-1}e^v = \tan^{-1}e^{-v}$.

79. Show that $\sin(\sin^{-1}v + \cos^{-1}v) = 1$.

80. Show that $\cos(\sin^{-1}v + \cos^{-1}v) = 0$.

81. Calculus Show that the difference quotient for $f(x) = \sin x$ is given by

$$\frac{f(x+h) - f(x)}{h} = \frac{\sin(x+h) - \sin x}{h}$$

$$= \cos x \cdot \frac{\sin h}{h} - \sin x \cdot \frac{1 - \cos h}{h}$$

82. Calculus Show that the difference quotient for $f(x) = \cos x$ is given by

$$\frac{f(x+h) - f(x)}{h} = \frac{\cos(x+h) - \cos x}{h}$$

$$= -\sin x \cdot \frac{\sin h}{h} - \cos x \cdot \frac{1 - \cos h}{h}$$

83. Explain why formula (7) cannot be used to show that

$$\tan\left(\frac{\pi}{2} - \theta\right) = \cot\theta$$

Establish this identity by using formulas (3a) and (3b).

84. If $\tan\alpha = x + 1$ and $\tan\beta = x - 1$, show that $2\cot(\alpha - \beta) = x^2$.

85. Geometry: Angle between Two Lines Let L_1 and L_2 denote two nonvertical intersecting lines, and let θ denote the acute angle between L_1 and L_2 (see the figure). Show that

$$\tan\theta = \frac{m_2 - m_1}{1 + m_1 m_2}$$

where m_1 and m_2 are the slopes of L_1 and L_2, respectively.

[**Hint:** Use the facts that $\tan\theta_1 = m_1$ and $\tan\theta_2 = m_2$.]

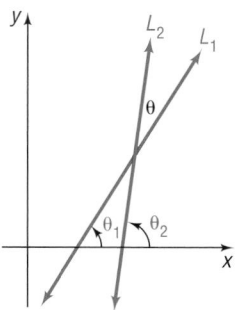

86. If $\alpha + \beta + \gamma = 180°$ and

$$\cot\theta = \cot\alpha + \cot\beta + \cot\gamma, \quad 0 < \theta < 90°$$

show that

$$\sin^3\theta = \sin(\alpha - \theta)\sin(\beta - \theta)\sin(\gamma - \theta)$$

87. Discuss the following derivation:

$$\tan\left(\theta + \frac{\pi}{2}\right) = \frac{\tan\theta + \tan\frac{\pi}{2}}{1 - \tan\theta\tan\frac{\pi}{2}}$$

$$= \frac{\dfrac{\tan\theta}{\tan\frac{\pi}{2}} + 1}{\dfrac{1}{\tan\frac{\pi}{2}} - \tan\theta} = \frac{0+1}{0 - \tan\theta}$$

$$= \frac{1}{-\tan\theta} = -\cot\theta$$

Can you justify each step?

3.5 # DOUBLE-ANGLE AND HALF-ANGLE FORMULAS

OBJECTIVES   1 Use Double-Angle Formulas to Find Exact Values
2 Use Double-Angle and Half-Angle Formulas to Establish Identities
3 Use Half-Angle Formulas to Find Exact Values

In this section we derive formulas for $\sin(2\theta)$, $\cos(2\theta)$, $\sin\left(\frac{1}{2}\theta\right)$, and $\cos\left(\frac{1}{2}\theta\right)$ in terms of $\sin\theta$ and $\cos\theta$. They are easily derived using the sum formulas.

Double-Angle Formulas
In the sum formulas for $\sin(\alpha + \beta)$ and $\cos(\alpha + \beta)$, let $\alpha = \beta = \theta$. Then

$$\sin(\alpha + \beta) = \sin\alpha\cos\beta + \cos\alpha\sin\beta$$
$$\sin(\theta + \theta) = \sin\theta\cos\theta + \cos\theta\sin\theta$$
$$\sin(2\theta) = 2\sin\theta\cos\theta \tag{1}$$

and

$$\cos(\alpha + \beta) = \cos \alpha \cos \beta - \sin \alpha \sin \beta$$
$$\cos(\theta + \theta) = \cos \theta \cos \theta - \sin \theta \sin \theta$$
$$\cos(2\theta) = \cos^2 \theta - \sin^2 \theta \qquad (2)$$

An application of the Pythagorean Identity $\sin^2 \theta + \cos^2 \theta = 1$ results in two other ways to write formula (2) for $\cos(2\theta)$.

$$\cos(2\theta) = \cos^2 \theta - \sin^2 \theta = (1 - \sin^2 \theta) - \sin^2 \theta = 1 - 2\sin^2 \theta$$

and

$$\cos(2\theta) = \cos^2 \theta - \sin^2 \theta = \cos^2 \theta - (1 - \cos^2 \theta) = 2\cos^2 \theta - 1$$

We have established the following **double-angle formulas:**

Theorem   **Double-Angle Formulas**

$\sin(2\theta) = 2 \sin \theta \cos \theta$	(1)
$\cos(2\theta) = \cos^2 \theta - \sin^2 \theta$	(2)
$\cos(2\theta) = 1 - 2\sin^2 \theta$	(3)
$\cos(2\theta) = 2\cos^2 \theta - 1$	(4)

 **E X A M P L E 1**   **Finding Exact Values Using the Double-Angle Formulas**

If $\sin \theta = \dfrac{3}{5}, \dfrac{\pi}{2} < \theta < \pi$, find the exact value of:

(a) $\sin(2\theta)$ (b) $\cos(2\theta)$

Solution   (a) Because $\sin(2\theta) = 2 \sin \theta \cos \theta$ and we already know that $\sin \theta = \dfrac{3}{5}$, we only need to find $\cos \theta$. Since $\sin \theta = \dfrac{3}{5} = \dfrac{b}{r}, \dfrac{\pi}{2} < \theta < \pi$, we let $b = 3$ and $r = 5$, and place θ in quadrant II. See Figure 27. The point $(a, 3)$ is in quadrant II, so $a < 0$. The distance from $(a, 3)$ to $(0, 0)$ is 5, so

Figure 27

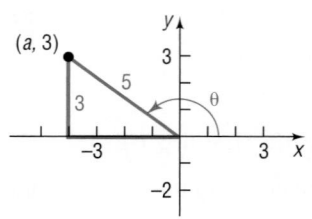

$$a^2 + 3^2 = 5^2,$$
$$a^2 + 9 = 25$$
$$a^2 = 25 - 9 = 16$$
$$a = -4 \qquad a < 0$$

We find that $\cos \theta = \dfrac{a}{r} = -\dfrac{4}{5}$. Now we use formula (1) to obtain

$$\sin(2\theta) = 2 \sin \theta \cos \theta = 2\left(\dfrac{3}{5}\right)\left(-\dfrac{4}{5}\right) = -\dfrac{24}{25}$$

(b) Because we are given $\sin\theta = \dfrac{3}{5}$, it is easiest to use formula (3) to get $\cos(2\theta)$.

$$\cos(2\theta) = 1 - 2\sin^2\theta = 1 - 2\left(\frac{9}{25}\right) = 1 - \frac{18}{25} = \frac{7}{25}$$ ∎

WARNING: In finding $\cos(2\theta)$ in Example 1(b), we chose to use a version of the double-angle formula, formula (3). Note that we are unable to use the Pythagorean Identity $\cos(2\theta) = \pm\sqrt{1 - \sin^2(2\theta)}$, with $\sin(2\theta) = -\dfrac{24}{25}$, because we have no way of knowing which sign to choose. ∎

- **NOW WORK PROBLEMS 1(a) AND (b).**

② **EXAMPLE 2**   **Establishing Identities**

(a) Develop a formula for $\tan(2\theta)$ in terms of $\tan\theta$.
(b) Develop a formula for $\sin(3\theta)$ in terms of $\sin\theta$ and $\cos\theta$.

Solution (a) In the sum formula for $\tan(\alpha + \beta)$, let $\alpha = \beta = \theta$. Then

$$\tan(\alpha + \beta) = \frac{\tan\alpha + \tan\beta}{1 - \tan\alpha\tan\beta}$$

$$\tan(\theta + \theta) = \frac{\tan\theta + \tan\theta}{1 - \tan\theta\tan\theta}$$

$$\tan(2\theta) = \frac{2\tan\theta}{1 - \tan^2\theta} \tag{5}$$

(b) To get a formula for $\sin(3\theta)$, we use the sum formula and write 3θ as $2\theta + \theta$.

$$\sin(3\theta) = \sin(2\theta + \theta) = \sin(2\theta)\cos\theta + \cos(2\theta)\sin\theta$$

Now use the double-angle formulas to get

$$\sin(3\theta) = (2\sin\theta\cos\theta)(\cos\theta) + (\cos^2\theta - \sin^2\theta)(\sin\theta)$$
$$= 2\sin\theta\cos^2\theta + \sin\theta\cos^2\theta - \sin^3\theta$$
$$= 3\sin\theta\cos^2\theta - \sin^3\theta$$ ∎

The formula obtained in Example 2(b) can also be written as

$$\sin(3\theta) = 3\sin\theta\cos^2\theta - \sin^3\theta = 3\sin\theta(1 - \sin^2\theta) - \sin^3\theta$$
$$= 3\sin\theta - 4\sin^3\theta$$

That is, $\sin(3\theta)$ is a third-degree polynomial in the variable $\sin\theta$. In fact, $\sin(n\theta)$, n a positive odd integer, can always be written as a polynomial of degree n in the variable $\sin\theta$.*

- **NOW WORK PROBLEM 47.**

*Due to the work done by P.L. Chebyshëv, these polynomials are sometimes called *Chebyshëv polynomials*.

Other Variations of the Double-Angle Formulas

By rearranging the double-angle formulas (3) and (4), we obtain other formulas that we will use later in this section.

We begin with formula (3) and proceed to solve for $\sin^2\theta$.

$$\cos(2\theta) = 1 - 2\sin^2\theta$$
$$2\sin^2\theta = 1 - \cos(2\theta)$$

$$\sin^2\theta = \frac{1 - \cos(2\theta)}{2} \tag{6}$$

Similarly, using formula (4), we proceed to solve for $\cos^2\theta$.

$$\cos(2\theta) = 2\cos^2\theta - 1$$
$$2\cos^2\theta = 1 + \cos(2\theta)$$

$$\cos^2\theta = \frac{1 + \cos(2\theta)}{2} \tag{7}$$

Formulas (6) and (7) can be used to develop a formula for $\tan^2\theta$.

$$\tan^2\theta = \frac{\sin^2\theta}{\cos^2\theta} = \frac{\dfrac{1 - \cos(2\theta)}{2}}{\dfrac{1 + \cos(2\theta)}{2}}$$

$$\tan^2\theta = \frac{1 - \cos(2\theta)}{1 + \cos(2\theta)} \tag{8}$$

Formulas (6) through (8) do not have to be memorized since their derivations are so straightforward.

 Formulas (6) and (7) are important in calculus. The next example illustrates a problem that arises in calculus requiring the use of formula (7).

EXAMPLE 3   **Establishing an Identity**

Write an equivalent expression for $\cos^4\theta$ that does not involve any powers of sine or cosine greater than 1.

Solution   The idea here is to apply formula (7) twice.

$$\cos^4\theta = (\cos^2\theta)^2 = \left(\frac{1 + \cos(2\theta)}{2}\right)^2 \qquad \text{Formula (7)}$$

$$= \frac{1}{4}[1 + 2\cos(2\theta) + \cos^2(2\theta)]$$

$$= \frac{1}{4} + \frac{1}{2}\cos(2\theta) + \frac{1}{4}\cos^2(2\theta)$$

$$= \frac{1}{4} + \frac{1}{2}\cos(2\theta) + \frac{1}{4}\left\{\frac{1 + \cos[2(2\theta)]}{2}\right\} \quad \text{Formula (7)}$$

$$= \frac{1}{4} + \frac{1}{2}\cos(2\theta) + \frac{1}{8}[1 + \cos(4\theta)]$$

$$= \frac{3}{8} + \frac{1}{2}\cos(2\theta) + \frac{1}{8}\cos(4\theta)$$

■

NOW WORK PROBLEM 23.

Identities, such as the double-angle formulas, can sometimes be used to rewrite expressions in a more suitable form. Let's look at an example.

EXAMPLE 4

Projectile Motion

An object is propelled upward at an angle θ to the horizontal with an initial velocity of v_0 feet per second. See Figure 28. If air resistance is ignored, the **range** R, the horizontal distance that the object travels, is given by

Figure 28

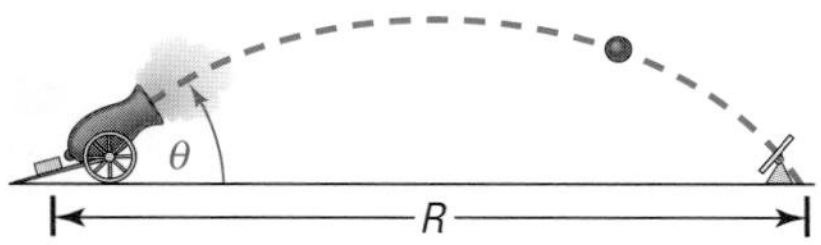

$$R = \frac{1}{16}v_0^2 \sin\theta\cos\theta$$

(a) Show that $R = \dfrac{1}{32}v_0^2 \sin(2\theta)$.

(b) Find the angle θ for which R is a maximum.

Solution (a) We rewrite the given expression for the range using the double-angle formula $\sin(2\theta) = 2\sin\theta\cos\theta$. Then

$$R = \frac{1}{16}v_0^2 \sin\theta\cos\theta = \frac{1}{16}v_0^2 \frac{2\sin\theta\cos\theta}{2} = \frac{1}{32}v_0^2 \sin(2\theta)$$

(b) In this form, the largest value for the range R can be found. For a fixed initial speed v_0, the angle θ of inclination to the horizontal determines the value of R. Since the largest value of a sine function is 1, occurring when the argument 2θ is $90°$, it follows that for maximum R we must have

$$2\theta = 90°$$
$$\theta = 45°$$

An inclination to the horizontal of $45°$ results in maximum range.

✔ CHECK: Graph $Y_1 = \dfrac{1}{16}v_0^2 \sin\theta\cos\theta$ with $v_0 = 1$, in degree mode. Use MAXIMUM to determine the angle θ that maximizes the range R. ■ ■

Half-Angle Formulas

Another important use of formulas (6) through (8) is to prove the **half-angle formulas**. In formulas (6) through (8), let $\theta = \dfrac{\alpha}{2}$. Then

$$\sin^2\frac{\alpha}{2} = \frac{1 - \cos\alpha}{2} \qquad \cos^2\frac{\alpha}{2} = \frac{1 + \cos\alpha}{2} \qquad \tan^2\frac{\alpha}{2} = \frac{1 - \cos\alpha}{1 + \cos\alpha} \qquad (9)$$

Note: The identities in box (9) will prove useful in integral calculus.

If we solve for the trigonometric functions on the left sides of equations (9), we obtain the half-angle formulas.

Theorem   **Half-Angle Formulas**

$$\sin \frac{\alpha}{2} = \pm \sqrt{\frac{1 - \cos \alpha}{2}} \qquad (10a)$$

$$\cos \frac{\alpha}{2} = \pm \sqrt{\frac{1 + \cos \alpha}{2}} \qquad (10b)$$

$$\tan \frac{\alpha}{2} = \pm \sqrt{\frac{1 - \cos \alpha}{1 + \cos \alpha}} \qquad (10c)$$

where the $+$ or $-$ sign is determined by the quadrant of the angle $\frac{\alpha}{2}$.

We use the half-angle formulas in the next example.

③   **EXAMPLE 5**   **Finding Exact Values Using Half-Angle Formulas**

Find the exact value of:

(a) $\cos 15°$ (b) $\sin(-15°)$

Solution   (a) Because $15° = \frac{30°}{2}$, we can use the half-angle formula for $\cos \frac{\alpha}{2}$ with $\alpha = 30°$. Also, because $15°$ is in quadrant I, $\cos 15° > 0$, so we choose the $+$ sign in using formula (10b):

$$\cos 15° = \cos \frac{30°}{2} = \sqrt{\frac{1 + \cos 30°}{2}}$$

$$= \sqrt{\frac{1 + \frac{\sqrt{3}}{2}}{2}} = \sqrt{\frac{2 + \sqrt{3}}{4}} = \frac{\sqrt{2 + \sqrt{3}}}{2}$$

(b) We use the fact that $\sin(-15°) = -\sin 15°$ and then apply formula (10a).

$$\sin(-15°) = -\sin \frac{30°}{2} = -\sqrt{\frac{1 - \cos 30°}{2}}$$

$$= -\sqrt{\frac{1 - \frac{\sqrt{3}}{2}}{2}} = -\sqrt{\frac{2 - \sqrt{3}}{4}} = -\frac{\sqrt{2 - \sqrt{3}}}{2}$$

It is interesting to compare the answer found in Example 5(a) with the answer to Example 2 of Section 3.4. There we calculated

$$\cos \frac{\pi}{12} = \cos 15° = \frac{1}{4}(\sqrt{6} + \sqrt{2})$$

Based on this and the result of Example 5(a), we conclude that

$$\frac{1}{4}(\sqrt{6} + \sqrt{2}) \quad \text{and} \quad \frac{\sqrt{2 + \sqrt{3}}}{2}$$

are equal. (Since each expression is positive, you can verify this equality by squaring each expression.) Two very different looking, yet correct, answers can be obtained, depending on the approach taken to solve a problem.

NOW WORK PROBLEM **13**.

EXAMPLE 6 **Finding Exact Values Using Half-Angle Formulas**

If $\cos \alpha = -\frac{3}{5}, \pi < \alpha < \frac{3\pi}{2}$, find the exact value of:

(a) $\sin \dfrac{\alpha}{2}$ (b) $\cos \dfrac{\alpha}{2}$ (c) $\tan \dfrac{\alpha}{2}$

Solution First, we observe that if $\pi < \alpha < \dfrac{3\pi}{2}$, then $\dfrac{\pi}{2} < \dfrac{\alpha}{2} < \dfrac{3\pi}{4}$. As a result, $\dfrac{\alpha}{2}$ lies in quadrant II.

(a) Because $\dfrac{\alpha}{2}$ lies in quadrant II, $\sin \dfrac{\alpha}{2} > 0$, so we use the $+$ sign in formula (10a) to get

$$\sin \frac{\alpha}{2} = \sqrt{\frac{1 - \cos \alpha}{2}} = \sqrt{\frac{1 - \left(-\dfrac{3}{5}\right)}{2}} = \sqrt{\frac{\dfrac{8}{5}}{2}}$$

$$= \sqrt{\frac{4}{5}} = \frac{2}{\sqrt{5}} = \frac{2\sqrt{5}}{5}$$

(b) Because $\dfrac{\alpha}{2}$ lies in quadrant II, $\cos \dfrac{\alpha}{2} < 0$, so we use the $-$ sign in formula (10b) to get

$$\cos \frac{\alpha}{2} = -\sqrt{\frac{1 + \cos \alpha}{2}} = -\sqrt{\frac{1 + \left(-\dfrac{3}{5}\right)}{2}} = -\sqrt{\frac{\dfrac{2}{5}}{2}}$$

$$= -\frac{1}{\sqrt{5}} = -\frac{\sqrt{5}}{5}$$

(c) Because $\dfrac{\alpha}{2}$ lies in quadrant II, $\tan \dfrac{\alpha}{2} < 0$, so we use the $-$ sign in formula (10c) to get

$$\tan \frac{\alpha}{2} = -\sqrt{\frac{1 - \cos \alpha}{1 + \cos \alpha}} = -\sqrt{\frac{1 - \left(-\dfrac{3}{5}\right)}{1 + \left(-\dfrac{3}{5}\right)}} = -\sqrt{\frac{\dfrac{8}{5}}{\dfrac{2}{5}}} = -2$$

Another way to solve Example 6(c) is to use the solutions found in parts (a) and (b).

$$\tan\frac{\alpha}{2} = \frac{\sin\dfrac{\alpha}{2}}{\cos\dfrac{\alpha}{2}} = \frac{\dfrac{2\sqrt5}{5}}{-\dfrac{\sqrt5}{5}} = -2$$

NOW WORK PROBLEMS 1(c) AND (d).

There is a formula for $\tan\dfrac{\alpha}{2}$ that does not contain $+$ and $-$ signs, making it more useful than formula 10(c). Because

$$1 - \cos\alpha = 2\sin^2\frac{\alpha}{2} \qquad \text{Formula (9)}$$

and

$$\sin\alpha = \sin\left[2\left(\frac{\alpha}{2}\right)\right] = 2\sin\frac{\alpha}{2}\cos\frac{\alpha}{2} \qquad \text{Double-angle formula}$$

we have

$$\frac{1 - \cos\alpha}{\sin\alpha} = \frac{2\sin^2\dfrac{\alpha}{2}}{2\sin\dfrac{\alpha}{2}\cos\dfrac{\alpha}{2}} = \frac{\sin\dfrac{\alpha}{2}}{\cos\dfrac{\alpha}{2}} = \tan\frac{\alpha}{2}$$

Since it also can be shown that

$$\frac{1 - \cos\alpha}{\sin\alpha} = \frac{\sin\alpha}{1 + \cos\alpha}$$

we have the following two half-angle formulas:

Half-Angle Formulas for $\tan\dfrac{\alpha}{2}$

$$\tan\frac{\alpha}{2} = \frac{1 - \cos\alpha}{\sin\alpha} = \frac{\sin\alpha}{1 + \cos\alpha} \qquad (11)$$

With this formula, the solution to Example 6(c) can be given as

$$\cos\alpha = -\frac{3}{5}$$

$$\sin\alpha = -\sqrt{1 - \cos^2\alpha} = -\sqrt{1 - \frac{9}{25}} = -\sqrt{\frac{16}{25}} = -\frac{4}{5}$$

Then, by equation (11),

$$\tan\frac{\alpha}{2} = \frac{1 - \cos\alpha}{\sin\alpha} = \frac{1 - \left(-\dfrac{3}{5}\right)}{-\dfrac{4}{5}} = \frac{\dfrac{8}{5}}{-\dfrac{4}{5}} = -2$$

3.5 Concepts and Vocabulary

In Problems 1–3, fill in the blanks.

1. $\cos(2\theta) = \cos^2\theta - \underline{\hspace{1.5cm}} = \underline{\hspace{1.5cm}} - 1 = 1 - \underline{\hspace{1.5cm}}$.

2. $\sin^2\dfrac{\theta}{2} = \dfrac{\underline{\hspace{0.8cm}}}{2}$.

3. $\tan\dfrac{\theta}{2} = \dfrac{1 - \cos\theta}{\underline{\hspace{0.8cm}}}$

In Problems 4 and 5, answer True or False to each statement.

4. $\cos(2\theta)$ has three equivalent forms: $\cos^2\theta - \sin^2\theta$; $1 - 2\sin^2\theta$; and $2\cos^2\theta - 1$.

5. $\sin(2\theta)$ has two equivalent forms: $2\sin\theta\cos\theta$ and $\sin^2\theta - \cos^2\theta$.

6. If you are given the value of $\cos\theta$ and want the exact value of $\cos(2\theta)$, what form of the double-angle formula for $\cos(2\theta)$ is most efficient to use?

7. Use a half-angle formula to find the exact value of $\sin 15°$.

8. Use a difference formula to find the exact value of $\sin 15°$.

9. Show that the answers found in Problems 7 and 8 are the same.

3.5 Exercises

In Problems 1–12, use the information given about the angle θ, $0 \le \theta < 2\pi$, to find the exact value of

(a) $\sin(2\theta)$ (b) $\cos(2\theta)$ (c) $\sin\dfrac{\theta}{2}$ (d) $\cos\dfrac{\theta}{2}$

1. $\sin\theta = \dfrac{3}{5}$, $0 < \theta < \dfrac{\pi}{2}$

2. $\cos\theta = \dfrac{3}{5}$, $0 < \theta < \dfrac{\pi}{2}$

3. $\tan\theta = \dfrac{4}{3}$, $\pi < \theta < \dfrac{3\pi}{2}$

4. $\tan\theta = \dfrac{1}{2}$, $\pi < \theta < \dfrac{3\pi}{2}$

5. $\cos\theta = -\dfrac{\sqrt{6}}{3}$, $\dfrac{\pi}{2} < \theta < \pi$

6. $\sin\theta = -\dfrac{\sqrt{3}}{3}$, $\dfrac{3\pi}{2} < \theta < 2\pi$

7. $\sec\theta = 3$, $\sin\theta > 0$

8. $\csc\theta = -\sqrt{5}$, $\cos\theta < 0$

9. $\cot\theta = -2$, $\sec\theta < 0$

10. $\sec\theta = 2$, $\csc\theta < 0$

11. $\tan\theta = -3$, $\sin\theta < 0$

12. $\cot\theta = 3$, $\cos\theta < 0$

In Problems 13–22, use the half-angle formulas to find the exact value of each trigonometric function.

13. $\sin 22.5°$

14. $\cos 22.5°$

15. $\tan\dfrac{7\pi}{8}$

16. $\tan\dfrac{9\pi}{8}$

17. $\cos 165°$

18. $\sin 195°$

19. $\sec\dfrac{15\pi}{8}$

20. $\csc\dfrac{7\pi}{8}$

21. $\sin\left(-\dfrac{\pi}{8}\right)$

22. $\cos\left(-\dfrac{3\pi}{8}\right)$

23. Show that $\sin^4\theta = \dfrac{3}{8} - \dfrac{1}{2}\cos(2\theta) + \dfrac{1}{8}\cos(4\theta)$.

24. Develop a formula for $\cos(3\theta)$ as a third-degree polynomial in the variable $\cos\theta$.

25. Show that $\sin(4\theta) = (\cos\theta)(4\sin\theta - 8\sin^3\theta)$.

26. Develop a formula for $\cos(4\theta)$ as a fourth-degree polynomial in the variable $\cos\theta$.

27. Find an expression for $\sin(5\theta)$ as a fifth-degree polynomial in the variable $\sin\theta$.

28. Find an expression for $\cos(5\theta)$ as a fifth-degree polynomial in the variable $\cos\theta$.

In Problems 29–50, establish each identity.

29. $\cos^4\theta - \sin^4\theta = \cos(2\theta)$

30. $\dfrac{\cot\theta - \tan\theta}{\cot\theta + \tan\theta} = \cos(2\theta)$

31. $\cot(2\theta) = \dfrac{\cot^2\theta - 1}{2\cot\theta}$

32. $\cot(2\theta) = \dfrac{1}{2}(\cot\theta - \tan\theta)$

33. $\sec(2\theta) = \dfrac{\sec^2\theta}{2 - \sec^2\theta}$

34. $\csc(2\theta) = \dfrac{1}{2}\sec\theta\csc\theta$

35. $\cos^2(2\theta) - \sin^2(2\theta) = \cos(4\theta)$

36. $(4\sin\theta\cos\theta)(1 - 2\sin^2\theta) = \sin(4\theta)$

37. $\dfrac{\cos(2\theta)}{1 + \sin(2\theta)} = \dfrac{\cot\theta - 1}{\cot\theta + 1}$

38. $\sin^2\theta\cos^2\theta = \dfrac{1}{8}[1 - \cos(4\theta)]$

39. $\sec^2\dfrac{\theta}{2} = \dfrac{2}{1 + \cos\theta}$

40. $\csc^2\dfrac{\theta}{2} = \dfrac{2}{1 - \cos\theta}$

41. $\cot^2\dfrac{\theta}{2} = \dfrac{\sec\theta + 1}{\sec\theta - 1}$

42. $\tan\dfrac{\theta}{2} = \csc\theta - \cot\theta$

43. $\cos\theta = \dfrac{1 - \tan^2\dfrac{\theta}{2}}{1 + \tan^2\dfrac{\theta}{2}}$

44. $1 - \dfrac{1}{2}\sin(2\theta) = \dfrac{\sin^3\theta + \cos^3\theta}{\sin\theta + \cos\theta}$

45. $\dfrac{\sin(3\theta)}{\sin\theta} - \dfrac{\cos(3\theta)}{\cos\theta} = 2$

46. $\dfrac{\cos\theta + \sin\theta}{\cos\theta - \sin\theta} - \dfrac{\cos\theta - \sin\theta}{\cos\theta + \sin\theta} = 2\tan(2\theta)$

47. $\tan(3\theta) = \dfrac{3\tan\theta - \tan^3\theta}{1 - 3\tan^2\theta}$

48. $\tan\theta + \tan(\theta + 120°) + \tan(\theta + 240°) = 3\tan(3\theta)$

49. $\ln|\sin\theta| = \dfrac{1}{2}(\ln|1 - \cos(2\theta)| - \ln 2)$

50. $\ln|\cos\theta| = \dfrac{1}{2}(\ln|1 + \cos(2\theta)| - \ln 2)$

In Problems 51–62, find the exact value of each expression.

51. $\sin\left(2\sin^{-1}\dfrac{1}{2}\right)$

52. $\sin\left[2\sin^{-1}\dfrac{\sqrt{3}}{2}\right]$

53. $\cos\left(2\sin^{-1}\dfrac{3}{5}\right)$

54. $\cos\left(2\cos^{-1}\dfrac{4}{5}\right)$

55. $\tan\left[2\cos^{-1}\left(-\dfrac{3}{5}\right)\right]$

56. $\tan\left(2\tan^{-1}\dfrac{3}{4}\right)$

57. $\sin\left(2\cos^{-1}\dfrac{4}{5}\right)$

58. $\cos\left[2\tan^{-1}\left(-\dfrac{4}{3}\right)\right]$

59. $\sin^2\left(\dfrac{1}{2}\cos^{-1}\dfrac{3}{5}\right)$

60. $\cos^2\left(\dfrac{1}{2}\sin^{-1}\dfrac{3}{5}\right)$

61. $\sec\left(2\tan^{-1}\dfrac{3}{4}\right)$

62. $\csc\left[2\sin^{-1}\left(-\dfrac{3}{5}\right)\right]$

63. If $x = 2\tan\theta$, express $\sin(2\theta)$ as a function of x.

64. If $x = 2\tan\theta$, express $\cos(2\theta)$ as a function of x.

65. Find the value of the number C:

$$\dfrac{1}{2}\sin^2 x + C = -\dfrac{1}{4}\cos(2x)$$

66. Find the value of the number C:

$$\dfrac{1}{2}\cos^2 x + C = \dfrac{1}{4}\cos(2x)$$

67. If $z = \tan\dfrac{\alpha}{2}$, show that $\sin\alpha = \dfrac{2z}{1 + z^2}$.

68. If $z = \tan\dfrac{\alpha}{2}$, show that $\cos\alpha = \dfrac{1 - z^2}{1 + z^2}$.

69. Area of an Isosceles Triangle Show that the area A of an isosceles triangle whose equal sides are of length s and θ is the angle between them is

$$\dfrac{1}{2}s^2\sin\theta$$

[**Hint:** See the illustration. The height h bisects the angle θ and is the perpendicular bisector of the base.]

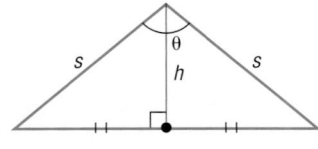

70. Geometry A rectangle is inscribed in a semicircle of radius 1. See the illustration.

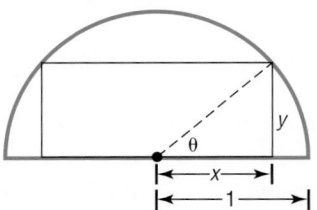

(a) Express the area A of the rectangle as a function of the angle θ shown in the illustration.
(b) Show that $A = \sin(2\theta)$.
(c) Find the angle θ that results in the largest area A.
(d) Find the dimensions of this largest rectangle.

71. Graph $f(x) = \sin^2 x = \dfrac{1 - \cos(2x)}{2}$ for $0 \le x \le 2\pi$ by using transformations.

72. Repeat Problem 71 for $g(x) = \cos^2 x$.

73. Use the fact that

$$\cos \frac{\pi}{12} = \frac{1}{4}(\sqrt{6} + \sqrt{2})$$

to find $\sin \dfrac{\pi}{24}$ and $\cos \dfrac{\pi}{24}$.

74. Show that

$$\cos \frac{\pi}{8} = \frac{\sqrt{2 + \sqrt{2}}}{2}$$

and use it to find $\sin \dfrac{\pi}{16}$ and $\cos \dfrac{\pi}{16}$.

75. Show that

$$\sin^3 \theta + \sin^3(\theta + 120°) + \sin^3(\theta + 240°) = -\frac{3}{4}\sin(3\theta)$$

76. If $\tan \theta = a \tan \dfrac{\theta}{3}$, express $\tan \dfrac{\theta}{3}$ in terms of a.

77. Projectile Motion   An object is propelled upward at an angle θ, $45° < \theta < 90°$, to the horizontal with an initial velocity of v_0 feet per second from the base of a plane that makes an angle of $45°$ with the horizontal. See the illustration. If air resistance is ignored, the distance R that it travels up the inclined plane is given by

$$R = \frac{v_0^2 \sqrt{2}}{16} \cos \theta(\sin \theta - \cos \theta)$$

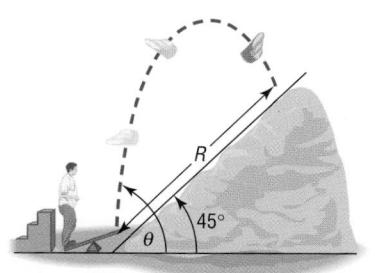

(a) Show that
$$R = \frac{v_0^2 \sqrt{2}}{32}[\sin(2\theta) - \cos(2\theta) - 1]$$

(b) Graph $R = R(\theta)$. (Use $v_0 = 32$ feet per second.)

(c) What value of θ makes R the largest? (Use $v_0 = 32$ feet per second.)

78. Sawtooth Curve   An oscilloscope often displays a sawtooth curve. This curve can be approximated by sinusoidal curves of varying periods and amplitudes. A first approximation to the sawtooth curve is given by

$$y = \frac{1}{2}\sin(2\pi x) + \frac{1}{4}\sin(4\pi x)$$

Show that $y = \sin(2\pi x)\cos^2(\pi x)$.

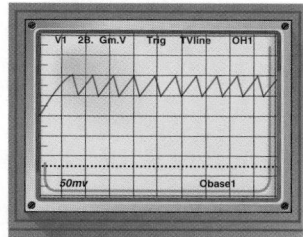

79. Go to the library and research Chebyshëv polynomials. Write a report on your findings.

3.6   **PRODUCT-TO-SUM AND SUM-TO-PRODUCT FORMULAS**

OBJECTIVES

1 Express Products as Sums

2 Express Sums as Products

1   The sum and difference formulas can be used to derive formulas for writing the products of sines and/or cosines as sums or differences. These identities are usually called the **Product-to-Sum Formulas.**

Theorem   **Product-to-Sum Formulas**

$$\sin \alpha \sin \beta = \frac{1}{2}[\cos(\alpha - \beta) - \cos(\alpha + \beta)] \qquad (1)$$

$$\cos \alpha \cos \beta = \frac{1}{2}[\cos(\alpha - \beta) + \cos(\alpha + \beta)] \qquad (2)$$

$$\sin \alpha \cos \beta = \frac{1}{2}[\sin(\alpha + \beta) + \sin(\alpha - \beta)] \qquad (3)$$

These formulas do not have to be memorized. Instead, you should remember how they are derived. Then, when you want to use them, either look them up or derive them, as needed.

To derive formulas (1) and (2), write down the sum and difference formulas for the cosine:

$$\cos(\alpha - \beta) = \cos \alpha \cos \beta + \sin \alpha \sin \beta \tag{4}$$
$$\cos(\alpha + \beta) = \cos \alpha \cos \beta - \sin \alpha \sin \beta \tag{5}$$

Subtract equation (5) from equation (4) to get

$$\cos(\alpha - \beta) - \cos(\alpha + \beta) = 2 \sin \alpha \sin \beta$$

from which

$$\sin \alpha \sin \beta = \frac{1}{2}[\cos(\alpha - \beta) - \cos(\alpha + \beta)]$$

Now, add equations (4) and (5) to get

$$\cos(\alpha - \beta) + \cos(\alpha + \beta) = 2 \cos \alpha \cos \beta$$

from which

$$\cos \alpha \cos \beta = \frac{1}{2}[\cos(\alpha - \beta) + \cos(\alpha + \beta)]$$

To derive Product-to-Sum Formula (3), use the Sum and Difference Formulas for sine in a similar way. (You are asked to do this in Problem 41.)

EXAMPLE 1   **Expressing Products as Sums**

Express each of the following products as a sum containing only sines or cosines.

(a) $\sin(6\theta) \sin(4\theta)$ (b) $\cos(3\theta) \cos \theta$ (c) $\sin(3\theta) \cos(5\theta)$

Solution   (a) We use formula (1) to get

$$\sin(6\theta) \sin(4\theta) = \frac{1}{2}[\cos(6\theta - 4\theta) - \cos(6\theta + 4\theta)]$$

$$= \frac{1}{2}[\cos(2\theta) - \cos(10\theta)]$$

(b) We use formula (2) to get

$$\cos(3\theta) \cos \theta = \frac{1}{2}[\cos(3\theta - \theta) + \cos(3\theta + \theta)]$$

$$= \frac{1}{2}[\cos(2\theta) + \cos(4\theta)]$$

(c) We use formula (3) to get

$$\sin(3\theta) \cos(5\theta) = \frac{1}{2}[\sin(3\theta + 5\theta) + \sin(3\theta - 5\theta)]$$

$$= \frac{1}{2}[\sin(8\theta) + \sin(-2\theta)] = \frac{1}{2}[\sin(8\theta) - \sin(2\theta)] ∎$$

NOW WORK PROBLEM 1.

② The **Sum-to-Product Formulas** are given next.

Theorem   **Sum-to-Product Formulas**

$$\sin \alpha + \sin \beta = 2 \sin \frac{\alpha + \beta}{2} \cos \frac{\alpha - \beta}{2} \tag{6}$$

$$\sin \alpha - \sin \beta = 2 \sin \frac{\alpha - \beta}{2} \cos \frac{\alpha + \beta}{2} \tag{7}$$

$$\cos \alpha + \cos \beta = 2 \cos \frac{\alpha + \beta}{2} \cos \frac{\alpha - \beta}{2} \tag{8}$$

$$\cos \alpha - \cos \beta = -2 \sin \frac{\alpha + \beta}{2} \sin \frac{\alpha - \beta}{2} \tag{9}$$

We will derive formula (6) and leave the derivations of formulas (7) through (9) as exercises (see Problems 42 through 44).

Proof

$$2 \sin \frac{\alpha + \beta}{2} \cos \frac{\alpha - \beta}{2} = 2 \cdot \frac{1}{2} \left[\sin \left(\frac{\alpha + \beta}{2} + \frac{\alpha - \beta}{2} \right) + \sin \left(\frac{\alpha + \beta}{2} - \frac{\alpha - \beta}{2} \right) \right]$$

Product-to-Sum Formula (3)

$$= \sin \frac{2\alpha}{2} + \sin \frac{2\beta}{2} = \sin \alpha + \sin \beta$$

EXAMPLE 2   **Expressing Sums (or Differences) as a Product**

Express each sum or difference as a product of sines and/or cosines.

(a) $\sin(5\theta) - \sin(3\theta)$ (b) $\cos(3\theta) + \cos(2\theta)$

Solution   (a) We use formula (7) to get

$$\sin(5\theta) - \sin(3\theta) = 2 \sin \frac{5\theta - 3\theta}{2} \cos \frac{5\theta + 3\theta}{2}$$

$$= 2 \sin \theta \cos(4\theta)$$

(b) $\cos(3\theta) + \cos(2\theta) = 2 \cos \dfrac{3\theta + 2\theta}{2} \cos \dfrac{3\theta - 2\theta}{2}$ Formula (8)

$$= 2 \cos \frac{5\theta}{2} \cos \frac{\theta}{2}$$

═ NOW WORK PROBLEM **11.**

3.6 Exercises

In Problems 1–10, express each product as a sum containing only sines or cosines.

1. $\sin(4\theta) \sin(2\theta)$ **2.** $\cos(4\theta) \cos(2\theta)$ **3.** $\sin(4\theta) \cos(2\theta)$ **4.** $\sin(3\theta) \sin(5\theta)$

5. $\cos(3\theta)\cos(5\theta)$ **6.** $\sin(4\theta)\cos(6\theta)$ **7.** $\sin\theta\sin(2\theta)$ **8.** $\cos(3\theta)\cos(4\theta)$

9. $\sin\dfrac{3\theta}{2}\cos\dfrac{\theta}{2}$ **10.** $\sin\dfrac{\theta}{2}\cos\dfrac{5\theta}{2}$

In Problems 11–18, express each sum or difference as a product of sines and/or cosines.

11. $\sin(4\theta) - \sin(2\theta)$ **12.** $\sin(4\theta) + \sin(2\theta)$ **13.** $\cos(2\theta) + \cos(4\theta)$ **14.** $\cos(5\theta) - \cos(3\theta)$

15. $\sin\theta + \sin(3\theta)$ **16.** $\cos\theta + \cos(3\theta)$ **17.** $\cos\dfrac{\theta}{2} - \cos\dfrac{3\theta}{2}$ **18.** $\sin\dfrac{\theta}{2} - \sin\dfrac{3\theta}{2}$

In Problems 19–36, establish each identity.

19. $\dfrac{\sin\theta + \sin(3\theta)}{2\sin(2\theta)} = \cos\theta$ **20.** $\dfrac{\cos\theta + \cos(3\theta)}{2\cos(2\theta)} = \cos\theta$ **21.** $\dfrac{\sin(4\theta) + \sin(2\theta)}{\cos(4\theta) + \cos(2\theta)} = \tan(3\theta)$

22. $\dfrac{\cos\theta - \cos(3\theta)}{\sin(3\theta) - \sin\theta} = \tan(2\theta)$ **23.** $\dfrac{\cos\theta - \cos(3\theta)}{\sin\theta + \sin(3\theta)} = \tan\theta$ **24.** $\dfrac{\cos\theta - \cos(5\theta)}{\sin\theta + \sin(5\theta)} = \tan(2\theta)$

25. $\sin\theta[\sin\theta + \sin(3\theta)] = \cos\theta[\cos\theta - \cos(3\theta)]$ **26.** $\sin\theta[\sin(3\theta) + \sin(5\theta)] = \cos\theta[\cos(3\theta) - \cos(5\theta)]$

27. $\dfrac{\sin(4\theta) + \sin(8\theta)}{\cos(4\theta) + \cos(8\theta)} = \tan(6\theta)$ **28.** $\dfrac{\sin(4\theta) - \sin(8\theta)}{\cos(4\theta) - \cos(8\theta)} = -\cot(6\theta)$

29. $\dfrac{\sin(4\theta) + \sin(8\theta)}{\sin(4\theta) - \sin(8\theta)} = -\dfrac{\tan(6\theta)}{\tan(2\theta)}$ **30.** $\dfrac{\cos(4\theta) - \cos(8\theta)}{\cos(4\theta) + \cos(8\theta)} = \tan(2\theta)\tan(6\theta)$

31. $\dfrac{\sin\alpha + \sin\beta}{\sin\alpha - \sin\beta} = \tan\dfrac{\alpha+\beta}{2}\cot\dfrac{\alpha-\beta}{2}$ **32.** $\dfrac{\cos\alpha + \cos\beta}{\cos\alpha - \cos\beta} = -\cot\dfrac{\alpha+\beta}{2}\cot\dfrac{\alpha-\beta}{2}$

33. $\dfrac{\sin\alpha + \sin\beta}{\cos\alpha + \cos\beta} = \tan\dfrac{\alpha+\beta}{2}$ **34.** $\dfrac{\sin\alpha - \sin\beta}{\cos\alpha - \cos\beta} = -\cot\dfrac{\alpha+\beta}{2}$

35. $1 + \cos(2\theta) + \cos(4\theta) + \cos(6\theta) = 4\cos\theta\cos(2\theta)\cos(3\theta)$

36. $1 - \cos(2\theta) + \cos(4\theta) - \cos(6\theta) = 4\sin\theta\cos(2\theta)\sin(3\theta)$

---

37. Touch-Tone Phones On a Touch-Tone phone, each button produces a unique sound. The sound produced is the sum of two tones, given by

$$y = \sin(2\pi lt) \quad \text{and} \quad y = \sin(2\pi ht)$$

where l and h are the low and high frequencies (cycles per second) shown on the illustration. For example, if you

Touch-Tone phone

touch 7, the low frequency is $l = 852$ cycles per second and the high frequency is $h = 1209$ cycles per second. The sound emitted by touching 7 is

$$y = \sin[2\pi(852)t] + \sin[2\pi(1209)t]$$

(a) Write this sound as a product of sines and/or cosines.
(b) Determine the maximum value of y.
(c) Graph the sound emitted by touching 7.

38. Touch-Tone Phones
(a) Write the sound emitted by touching the # key as a product of sines and/or cosines.
(b) Determine the maximum value of y.
(c) Graph the sound emitted by touching the # key.

39. If $\alpha + \beta + \gamma = \pi$, show that

$$\sin(2\alpha) + \sin(2\beta) + \sin(2\gamma) = 4\sin\alpha\sin\beta\sin\gamma$$

40. If $\alpha + \beta + \gamma = \pi$, show that

$$\tan\alpha + \tan\beta + \tan\gamma = \tan\alpha\tan\beta\tan\gamma$$

41. Derive formula (3).
42. Derive formula (7).
43. Derive formula (8).
44. Derive formula (9).

PREPARING FOR THIS SECTION

Before getting started, review the following:

✓ Solving Equations (Section A.3, pp. 571–576)

✓ Values of the Trigonometric Functions of Certain Angles (Section 2.3, p. 126, and Section 2.4, p. 134)

3.7 TRIGONOMETRIC EQUATIONS (I)

OBJECTIVES   ① Solve Equations Involving a Single Trigonometric Function

① The previous four sections of this chapter were devoted to trigonometric identities, that is, equations involving trigonometric functions that are satisfied by every value in the domain of the variable. In the remaining two sections, we discuss **trigonometric equations**, that is, equations involving trigonometric functions that are satisfied only by some values of the variable (or, possibly, are not satisfied by any values of the variable). The values that satisfy the equation are called **solutions** of the equation.

EXAMPLE 1   Checking Whether a Given Number Is a Solution of a Trigonometric Equation

Determine whether $\theta = \dfrac{\pi}{4}$ is a solution of the equation $\sin\theta = \dfrac{1}{2}$. Is $\theta = \dfrac{\pi}{6}$ a solution?

Solution   Replace θ by $\dfrac{\pi}{4}$ in the given equation. The result is

$$\sin\frac{\pi}{4} = \frac{\sqrt{2}}{2} \neq \frac{1}{2}$$

We conclude that $\dfrac{\pi}{4}$ is not a solution.

Next, replace θ by $\dfrac{\pi}{6}$ in the equation. The result is

$$\sin\frac{\pi}{6} = \frac{1}{2}$$

We conclude that $\dfrac{\pi}{6}$ is a solution of the given equation.   ■

Figure 29

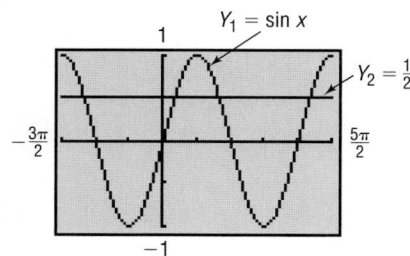

The equation given in Example 1 has other solutions besides $\theta = \dfrac{\pi}{6}$. For example, $\theta = \dfrac{5\pi}{6}$ is also a solution, as is $\theta = \dfrac{13\pi}{6}$. (You should check this for yourself.) In fact, the equation has an infinite number of solutions due to the periodicity of the sine function. See Figure 29.

As before, our practice will be to solve equations, whenever possible, by finding exact solutions. In such cases, we will also verify the solution obtained by using a graphing utility. When traditional methods cannot be used, approximate solutions will be obtained using a graphing utility. The reader is encouraged to pay particular attention to the form of equations for which exact solutions are possible.

Unless the domain of the variable is restricted, we need to find *all* the solutions of a trigonometric equation. As the next example illustrates, finding all the solutions can be accomplished by first finding solutions over an interval whose length equals the period of the function and then adding multiples of that period to the solutions found. Let's look at some examples.

EXAMPLE 2   **Finding All the Solutions of a Trigonometric Equation**

Solve the equation: $\cos \theta = \dfrac{1}{2}$

Give a general formula for all the solutions. List six solutions.

Figure 30
$\cos \theta = \dfrac{1}{2}$

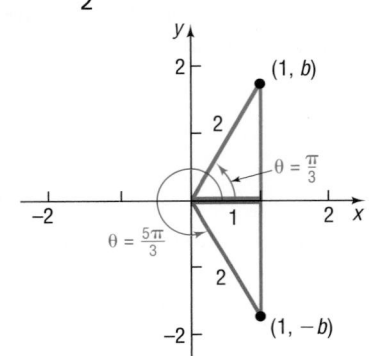

Solution   The period of the cosine function is 2π. In the interval $[0, 2\pi)$, the two angles θ for which $\cos \theta = \dfrac{1}{2}$ are $\theta = \dfrac{\pi}{3}$ and $\theta = \dfrac{5\pi}{3}$. See Figure 30. Because the cosine function has period 2π, all the solutions of $\cos \theta = \dfrac{1}{2}$ may be given by the general formula

$$\theta = \frac{\pi}{3} + 2k\pi \quad \text{or} \quad \theta = \frac{5\pi}{3} + 2k\pi \qquad \text{k any integer}$$

Some of the solutions are

$$\underbrace{\frac{\pi}{3}, \frac{5\pi}{3}}_{k = 0}, \quad \underbrace{\frac{7\pi}{3}, \frac{11\pi}{3}}_{k = 1}, \quad \underbrace{\frac{13\pi}{3}, \frac{17\pi}{3}}_{k = 2}, \quad \text{and so on}$$

Figure 31

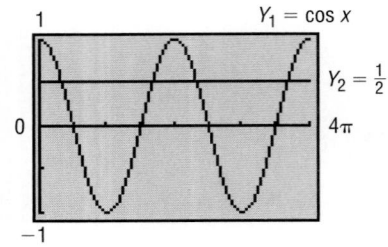

✔ CHECK: We can verify the solutions by graphing $Y_1 = \cos x$ and $Y_2 = \dfrac{1}{2}$ to determine where the graphs intersect. (Be sure to graph in radian mode.) See Figure 31. The graph of Y_1 intersects the graph of Y_2 at $x = 1.05 \left(\approx \dfrac{\pi}{3} \right), 5.24 \left(\approx \dfrac{5\pi}{3} \right), 7.33 \left(\approx \dfrac{7\pi}{3} \right)$, and $11.52 \left(\approx \dfrac{11\pi}{3} \right)$, rounded to two decimal places.   ■ ■

✏— **NOW WORK PROBLEM 1.**

In most of our work, we shall be interested only in finding solutions of trigonometric equations for $0 \le \theta < 2\pi$.

EXAMPLE 3   **Solving a Linear Trigonometric Equation**

Solve the equation: $2 \sin \theta + \sqrt{3} = 0, \quad 0 \le \theta < 2\pi$

Solution   We solve the equation for $\sin \theta$.

$$2 \sin \theta + \sqrt{3} = 0$$
$$2 \sin \theta = -\sqrt{3} \qquad \text{Subtract } \sqrt{3} \text{ from both sides.}$$
$$\sin \theta = -\frac{\sqrt{3}}{2} \qquad \text{Divide both sides by 2.}$$

The period of the sine function is 2π. In the interval $[0, 2\pi)$, the two angles θ for which $\sin \theta = -\dfrac{\sqrt{3}}{2}$ are $\theta = \dfrac{4\pi}{3}$ and $\theta = \dfrac{5\pi}{3}$. ■

NOW WORK PROBLEM **11.**

EXAMPLE 4   **Solving a Trigonometric Equation**

Solve the equation: $\sin (2\theta) = \dfrac{1}{2}, \quad 0 \le \theta < 2\pi$

Solution   The period of the sine function is 2π. In the interval $[0, 2\pi)$, the sine function has a value $\dfrac{1}{2}$ at $\dfrac{\pi}{6}$ and $\dfrac{5\pi}{6}$. See Figure 32. Because the argument is 2θ in the equation $\sin (2\theta) = \dfrac{1}{2}$, we have

Figure 32

$\sin (2\theta) = \dfrac{1}{2}, 0 \le \theta < 2\pi$

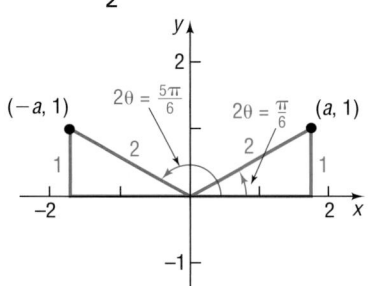

$$2\theta = \dfrac{\pi}{6} + 2k\pi \quad \text{or} \quad 2\theta = \dfrac{5\pi}{6} + 2k\pi \qquad \text{\textit{k} any integer}$$

$$\theta = \dfrac{\pi}{12} + k\pi \qquad\qquad \theta = \dfrac{5\pi}{12} + k\pi \qquad \text{Divide by 2.}$$

Then

$$\theta = \dfrac{\pi}{12} + (-1)\pi = -\dfrac{11\pi}{12} \quad {\scriptstyle k = -1} \qquad \theta = \dfrac{5\pi}{12} + (-1)\pi = -\dfrac{7\pi}{12}$$

$$\theta = \dfrac{\pi}{12} + (0)\pi = \dfrac{\pi}{12} \quad {\scriptstyle k = 0} \qquad \theta = \dfrac{5\pi}{12} + (0)\pi = \dfrac{5\pi}{12}$$

$$\theta = \dfrac{\pi}{12} + (1)\pi = \dfrac{13\pi}{12} \quad {\scriptstyle k = 1} \qquad \theta = \dfrac{5\pi}{12} + (1)\pi = \dfrac{17\pi}{12}$$

$$\theta = \dfrac{\pi}{12} + (2)\pi = \dfrac{25\pi}{12} \quad {\scriptstyle k = 2} \qquad \theta = \dfrac{5\pi}{12} + (2)\pi = \dfrac{29\pi}{12}$$

In the interval $[0, 2\pi)$, the solutions of $\sin (2\theta) = \dfrac{1}{2}$ are $\theta = \dfrac{\pi}{12}$, $\theta = \dfrac{13\pi}{12}, \theta = \dfrac{5\pi}{12}$, and $\theta = \dfrac{17\pi}{12}$.

✔ CHECK: Verify these solutions by graphing $Y_1 = \sin (2x)$ and $Y_2 = \dfrac{1}{2}$ for $0 \le x \le 2\pi$. ■ ■

WARNING: In solving a trigonometric equation for $\theta, 0 \le \theta < 2\pi$, in which the argument is not θ (as in Example 4), you must write down all the solutions first and then list those that are in the interval $[0, 2\pi)$. Otherwise, solutions may be lost. For example, in solving $\sin (2\theta) = \dfrac{1}{2}$, if you merely write the solutions $2\theta = \dfrac{\pi}{6}$ and $2\theta = \dfrac{5\pi}{6}$, you will find only $\theta = \dfrac{\pi}{12}$ and $\theta = \dfrac{5\pi}{12}$ and miss the other solutions. ■

NOW WORK PROBLEM **17.**

EXAMPLE 5   **Solving a Trigonometric Equation**

Solve the equation: $\tan\left(\theta - \dfrac{\pi}{2}\right) = 1, \quad 0 \le \theta < 2\pi$

Solution   The period of the tangent function is π. In the interval $[0, \pi)$, the tangent function has the value 1 when the argument is $\dfrac{\pi}{4}$. Because the argument is $\theta - \dfrac{\pi}{2}$ in the given equation, we have

$$\theta - \frac{\pi}{2} = \frac{\pi}{4} + k\pi \qquad \text{k any integer}$$

$$\theta = \frac{3\pi}{4} + k\pi$$

In the interval $[0, 2\pi)$, $\theta = \dfrac{3\pi}{4}$ and $\theta = \dfrac{3\pi}{4} + \pi = \dfrac{7\pi}{4}$ are the only solutions.

✔ CHECK: Verify these solutions using a graphing utility.   ■ ■

The next example illustrates how to solve trigonometric equations using a calculator. Remember that the function keys on a calculator will only give values consistent with the definition of the function.

EXAMPLE 6   **Solving a Trigonometric Equation with a Calculator**

Use a calculator to solve the equation: $\sin\theta = 0.3, \quad 0 \le \theta < 2\pi$ Express any solutions in radians, rounded to two decimal places.

Solution   To solve $\sin\theta = 0.3$ on a calculator, first set the mode to radians. Then use the $\boxed{\sin^{-1}}$ key to obtain

$$\theta = \sin^{-1}(0.3) \approx 0.304692654$$

Figure 33
$\sin\theta = 0.3$

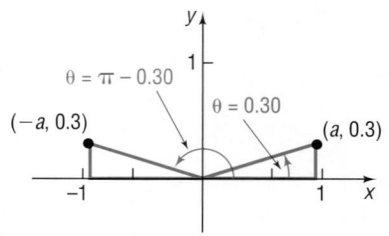

Rounded to two decimal places, $\theta = \sin^{-1}(0.3) = 0.30$ radian. Because of the definition of $y = \sin^{-1}x$, the angle θ that we obtain is the angle $-\dfrac{\pi}{2} \le \theta \le \dfrac{\pi}{2}$ for which $\sin\theta = 0.3$. Another angle for which $\sin\theta = 0.3$ is $\pi - 0.30$. See Figure 33. The angle $\pi - 0.30$ is the angle in quadrant II, where $\sin\theta = 0.3$. The solutions for $\sin\theta = 0.3, 0 \le \theta < 2\pi$, are

$$\theta = 0.30 \text{ radian} \quad \text{and} \quad \theta = \pi - 0.30 \approx 2.84 \text{ radians} \quad ■$$

A second method for solving $\sin\theta = 0.3, 0 \le \theta < 2\pi$ would be to graph $Y_1 = \sin x$ and $Y_2 = 0.3$ for $0 \le x < 2\pi$ and find their point(s) of intersection. Try this method for yourself to verify the results obtained in Example 6.

WARNING: Example 6 illustrates that caution must be exercised when solving trigonometric equations on a calculator. Remember that the calculator supplies an angle only within the restrictions of the definition of the inverse trigonometric function. To find the remaining solutions, you must identify other quadrants, if any, in which the angle may be located.   ■

NOW WORK PROBLEM **35.**

3.7 Concepts and Vocabulary

In Problems 1 and 2, fill in the blanks.

1. Two solutions of the equation $\sin \theta = \dfrac{1}{2}$ are _____ and _____.

2. All the solutions of the equation $\sin \theta = \dfrac{1}{2}$ are _____.

In Problems 3–5, answer True or False to each statement.

3. Most trigonometric equations have unique solutions.

4. The equation $\tan^{-1} \theta = \dfrac{\pi}{2}$ has no solution.

5. The equation $\sin \theta = 2$ has a solution that can be found using a graphing calculator.

3.7 Exercises

In Problems 1–10, solve each equation. Give a general formula for all the solutions. List six solutions.

1. $\sin \theta = \dfrac{1}{2}$
    2. $\tan \theta = 1$
    3. $\tan \theta = -\dfrac{\sqrt{3}}{3}$
    4. $\cos \theta = -\dfrac{\sqrt{3}}{2}$
    5. $\cos \theta = 0$

6. $\sin \theta = \dfrac{\sqrt{2}}{2}$
    7. $\cos (2\theta) = -\dfrac{1}{2}$
    8. $\sin (2\theta) = -1$
    9. $\sin \dfrac{\theta}{2} = -\dfrac{\sqrt{3}}{2}$
    10. $\tan \dfrac{\theta}{2} = -1$

In Problems 11–34, solve each equation on the interval $0 \le \theta < 2\pi$.

11. $2 \sin \theta + 3 = 2$
    12. $1 - \cos \theta = \dfrac{1}{2}$
    13. $4 \cos^2 \theta = 1$
    14. $\tan^2 \theta = \dfrac{1}{3}$

15. $2 \sin^2 \theta - 1 = 0$
    16. $4 \cos^2 \theta - 3 = 0$
    17. $\sin (3\theta) = -1$
    18. $\tan \dfrac{\theta}{2} = \sqrt{3}$

19. $\cos (2\theta) = -\dfrac{1}{2}$
    20. $\tan (2\theta) = -1$
    21. $\sec \dfrac{3\theta}{2} = -2$
    22. $\cot \dfrac{2\theta}{3} = -\sqrt{3}$

23. $\cos \left(2\theta - \dfrac{\pi}{2} \right) = -1$
    24. $\sin \left(3\theta + \dfrac{\pi}{18} \right) = 1$
    25. $\tan \left(\dfrac{\theta}{2} + \dfrac{\pi}{3} \right) = 1$
    26. $\cos \left(\dfrac{\theta}{3} - \dfrac{\pi}{4} \right) = \dfrac{1}{2}$

27. $2 \sin \theta + 1 = 0$
    28. $\cos \theta + 1 = 0$
    29. $\tan \theta + 1 = 0$
    30. $\sqrt{3} \cot \theta + 1 = 0$

31. $4 \sec \theta + 6 = -2$
    32. $5 \csc \theta - 3 = 2$
    33. $3\sqrt{2} \cos \theta + 2 = -1$
    34. $4 \sin \theta + 3\sqrt{3} = \sqrt{3}$

In Problems 35–42, use a calculator to solve each equation on the interval $0 \le \theta < 2\pi$. Round answers to two decimal places.

35. $\sin \theta = 0.4$
    36. $\cos \theta = 0.6$
    37. $\tan \theta = 5$
    38. $\cot \theta = 2$

39. $\cos \theta = -0.9$
    40. $\sin \theta = -0.2$
    41. $\sec \theta = -4$
    42. $\csc \theta = -3$

The following discussion of **Snell's Law of Refraction** (named after Willebrord Snell, 1580–1626) is needed for Problems 43–49. Light, sound, and other waves travel at different speeds, depending on the media (air, water, wood, and so on) through which they pass. Suppose that light travels from a point A in one medium, where its speed is v_1, to a point B in another medium, where its speed is v_2. Refer to the figure (page 252), where the angle θ_1 is called the **angle of incidence** and the angle θ_2 is the **angle of refraction**. Snell's Law,* which can be proved using calculus, states that

$$\frac{\sin \theta_1}{\sin \theta_2} = \frac{v_1}{v_2}$$

*Because this law was also deduced by René Descartes in France, it is also known as Descartes' Law.

*The ratio $\dfrac{v_1}{v_2}$ is called the **index of refraction**. Some values are given in the following table.*

SOME INDEXES OF REFRACTION	
Medium	**Index of Refraction***
Water	1.33
Ethyl alcohol	1.36
Carbon bisulfide	1.63
Air (1 atm and 20°C)	1.0003
Methylene iodide	1.74
Fused quartz	1.46
Glass, crown	1.52
Glass, dense flint	1.66
Sodium chloride	1.53

*For light of wavelength 589 nanometers, measured with respect to a vacuum. The index with respect to air is negligibly different in most cases.

43. The index of refraction of light in passing from a vacuum into water is 1.33. If the angle of incidence is 40°, determine the angle of refraction.

44. The index of refraction of light in passing from a vacuum into dense glass is 1.66. If the angle of incidence is 50°, determine the angle of refraction.

45. Ptolemy, who lived in the city of Alexandria in Egypt during the second century AD, gave the measured values in the table below for the angle of incidence θ_1 and the angle of refraction θ_2 for a light beam passing from air into water. Do these values agree with Snell's Law? If so, what index of refraction results? (These data are interesting as the oldest recorded physical measurements.)*

θ_1	θ_2	θ_1	θ_2
10°	7°45'	50°	35°0'
20°	15°30'	60°	40°30'
30°	22°30'	70°	45°30'
40°	29°0'	80°	50°0'

46. The speed of yellow sodium light (wavelength of 589 nanometers) in a certain liquid is measured to be 1.92×10^8 meters per second. What is the index of refraction of this liquid, with respect to air, for sodium light?[†]

[**Hint:** The speed of light in air is approximately 2.99×10^8 meters per second.]

47. A beam of light with a wavelength of 589 nanometers traveling in air makes an angle of incidence of 40° on a slab of transparent material, and the refracted beam makes an angle of refraction of 26°. Find the index of refraction of the material.[†]

48. A light ray with a wavelength of 589 nanometers (produced by a sodium lamp) traveling through air makes an angle of incidence of 30° on a smooth, flat slab of crown glass. Find the angle of refraction.[†]

49. A light beam passes through a thick slab of material whose index of refraction is n_2. Show that the emerging beam is parallel to the incident beam.[†]

50. Explain in your own words how you would use your calculator to solve the equation $\sin x = 0.3, 0 \le x < 2\pi$. How would you modify your approach in order to solve the equation $\cot x = 5, 0 < x < 2\pi$?

*Adapted from Halliday and Resnick, *Physics, Parts 1 & 2*, 3rd ed. New York: Wiley, 1978, p. 953.
†Adapted from Serway, *Physics*, 3rd ed. Philadelphia: W. B. Saunders, p. 805.

PREPARING FOR THIS SECTION

Before getting started, review the following:

✓ Solving Quadratic Equations by Factoring
  (Section A.3, pp. 576–577)

✓ Solving Equations Using a Graphing Utility
  (Section A.3, pp. 572–573)

✓ The Quadratic Formula (Section A.3, p. 581)

3.8 TRIGONOMETRIC EQUATIONS (II)

OBJECTIVES
1. Solve Trigonometric Equations Quadratic in Form
2. Solve Trigonometric Equations Using Identities
3. Solve Trigonometric Equations Linear in Sine and Cosine
4. Solve Trigonometric Equations Using a Graphing Utility

1. In this section we continue our study of trigonometric equations. Many trigonometric equations can be solved by applying techniques that we already know, such as applying the quadratic formula (if the equation is a second-degree polynomial) or factoring.

EXAMPLE 1   Solving a Trigonometric Equation Quadratic in Form

Solve the equation: $2\sin^2\theta - 3\sin\theta + 1 = 0,\ \ 0 \le \theta < 2\pi$

Solution   The equation that we wish to solve is a quadratic equation (in $\sin\theta$) that can be factored.

$$2\sin^2\theta - 3\sin\theta + 1 = 0 \qquad\qquad 2x^2 - 3x + 1 = 0, \quad x = \sin\theta$$
$$(2\sin\theta - 1)(\sin\theta - 1) = 0 \qquad\qquad (2x - 1)(x - 1) = 0$$
$$2\sin\theta - 1 = 0 \quad \text{or} \quad \sin\theta - 1 = 0$$
$$\sin\theta = \frac{1}{2} \qquad\qquad \sin\theta = 1$$

Solving each equation in the interval $[0, 2\pi)$, we obtain

$$\theta = \frac{\pi}{6}, \qquad \theta = \frac{5\pi}{6}, \qquad \theta = \frac{\pi}{2}$$

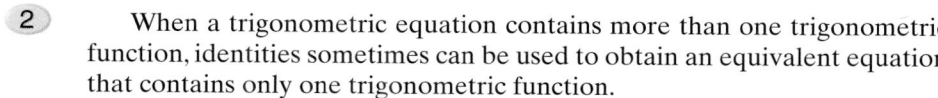 NOW WORK PROBLEM 3.

2. When a trigonometric equation contains more than one trigonometric function, identities sometimes can be used to obtain an equivalent equation that contains only one trigonometric function.

EXAMPLE 2   Solving a Trigonometric Equation Using Identities

Solve the equation: $3\cos\theta + 3 = 2\sin^2\theta,\ \ 0 \le \theta < 2\pi$

Solution   The equation in its present form contains sines and cosines. However, a form of the Pythagorean Identity can be used to transform the equation into an equivalent expression containing only cosines.

$$3 \cos \theta + 3 = 2 \sin^2 \theta$$
$$3 \cos \theta + 3 = 2(1 - \cos^2 \theta) \qquad \text{\small $\sin^2 \theta = 1 - \cos^2 \theta$}$$
$$3 \cos \theta + 3 = 2 - 2 \cos^2 \theta$$
$$2 \cos^2 \theta + 3 \cos \theta + 1 = 0 \qquad \text{\small Quadratic in $\cos \theta$}$$
$$(2 \cos \theta + 1)(\cos \theta + 1) = 0 \qquad \text{\small Factor.}$$
$$2 \cos \theta + 1 = 0 \quad \text{or} \quad \cos \theta + 1 = 0$$
$$\cos \theta = -\frac{1}{2} \qquad\qquad \cos \theta = -1$$

Solving each equation in the interval $[0, 2\pi)$, we obtain

$$\theta = \frac{2\pi}{3}, \qquad \theta = \frac{4\pi}{3}, \qquad \theta = \pi$$

✔ CHECK: Graph $Y_1 = 3 \cos x + 3$ and $Y_2 = 2 \sin^2 x$, $0 \le x \le 2\pi$, and find the points of intersection. How close are your approximate solutions to the exact ones found in this example?   ■ ■

EXAMPLE 3   Solving a Trigonometric Equation Using Identities

Solve the equation: $\cos (2\theta) + 3 = 5 \cos \theta, \quad 0 \le \theta < 2\pi$

Solution   First, we observe that the given equation contains two cosine functions, but with different arguments, θ and 2θ. We use the Double-Angle Formula $\cos (2\theta) = 2 \cos^2 \theta - 1$ to obtain an equivalent equation containing only $\cos \theta$.

$$\cos (2\theta) + 3 = 5 \cos \theta$$
$$(2 \cos^2 \theta - 1) + 3 = 5 \cos \theta$$
$$2 \cos^2 \theta - 5 \cos \theta + 2 = 0$$
$$(\cos \theta - 2)(2 \cos \theta - 1) = 0$$
$$\cos \theta = 2 \quad \text{or} \quad \cos \theta = \frac{1}{2}$$

For any angle θ, $-1 \le \cos \theta \le 1$; therefore, the equation $\cos \theta = 2$ has no solution. The solutions of $\cos \theta = \frac{1}{2}$, $0 \le \theta < 2\pi$, are

$$\theta = \frac{\pi}{3}, \qquad \theta = \frac{5\pi}{3}$$

✔ CHECK: Graph $Y_1 = \cos (2x) + 3$ and $Y_2 = 5 \cos x$, $0 \le x \le 2\pi$, and find the points of intersection.   ■ ■

✏ ─ **NOW WORK PROBLEM 19.**

EXAMPLE 4   Solving a Trigonometric Equation Using Identities

Solve the equation: $\cos^2 \theta + \sin \theta = 2, \quad 0 \le \theta < 2\pi$

Solution   This equation involves two trigonometric functions, sine and cosine. Since it is easier to work with only one, we use a form of the Pythagorean Identity, $\sin^2 \theta + \cos^2 \theta = 1$ to rewrite the equation.

$$\cos^2 \theta + \sin \theta = 2$$

$$(1 - \sin^2 \theta) + \sin \theta = 2 \qquad \cos^2 \theta = 1 - \sin^2 \theta$$

$$\sin^2 \theta - \sin \theta + 1 = 0$$

Figure 34

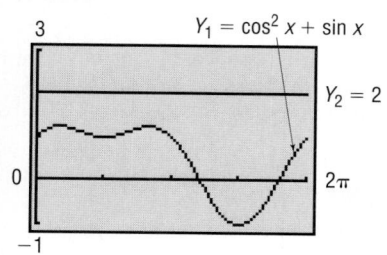

This is a quadratic equation in $\sin \theta$. The discriminant is $b^2 - 4ac = 1 - 4 = -3 < 0$. Therefore, the equation has no real solution.

✔ CHECK: Graph $Y_1 = \cos^2 x + \sin x$ and $Y_2 = 2$. See Figure 34. The two graphs do not intersect, so the equation $Y_1 = Y_2$ has no real solution. ■ ■

EXAMPLE 5   **Solving a Trigonometric Equation Using Identities**

Solve the equation: $\sin \theta \cos \theta = -\dfrac{1}{2}, \quad 0 \le \theta < 2\pi$

Solution   The left side of the given equation is in the form of the Double-Angle Formula $2 \sin \theta \cos \theta = \sin (2\theta)$, except for a factor of 2. We multiply each side by 2.

$$\sin \theta \cos \theta = -\frac{1}{2}$$

$$2 \sin \theta \cos \theta = -1 \qquad \text{Multiply each side by 2.}$$

$$\sin (2\theta) = -1 \qquad \text{Double-Angle Formula}$$

The argument here is 2θ. So we need to write all the solutions of this equation and then list those that are in the interval $[0, 2\pi)$.

$$2\theta = \frac{3\pi}{2} + 2k\pi \qquad k \text{ any integer}$$

$$\theta = \frac{3\pi}{4} + k\pi$$

$$\theta = \underset{\substack{\uparrow \\ k = -1}}{\frac{3\pi}{4}} + (-1)\pi = -\frac{\pi}{4}, \quad \theta = \underset{\substack{\uparrow \\ k = 0}}{\frac{3\pi}{4}} + (0)\pi = \frac{3\pi}{4}, \quad \theta = \underset{\substack{\uparrow \\ k = 1}}{\frac{3\pi}{4}} + (1)\pi = \frac{7\pi}{4}, \quad \theta = \underset{\substack{\uparrow \\ k = 2}}{\frac{3\pi}{4}} + (2)\pi = \frac{11\pi}{4}$$

The solutions in the interval $[0, 2\pi)$ are

$$\theta = \frac{3\pi}{4}, \qquad \theta = \frac{7\pi}{4} \qquad\qquad ■$$

③   Sometimes it is necessary to square both sides of an equation in order to obtain expressions that allow the use of identities. Remember, however, that when squaring both sides, extraneous solutions may be introduced. As a result, apparent solutions must be checked.

| EXAMPLE 6 | Other Methods for Solving a Trigonometric Equation |

Solve the equation: $\sin \theta + \cos \theta = 1, \quad 0 \le \theta < 2\pi$

Solution A   Attempts to use available identities do not lead to equations that are easy to solve. (Try it yourself.) Given the form of this equation, we decide to square each side.

$$\sin \theta + \cos \theta = 1$$
$$(\sin \theta + \cos \theta)^2 = 1 \qquad \text{Square each side.}$$
$$\sin^2 \theta + 2 \sin \theta \cos \theta + \cos^2 \theta = 1 \qquad \text{Remove parentheses.}$$
$$2 \sin \theta \cos \theta = 0 \qquad \sin^2 \theta + \cos^2 \theta = 1$$
$$\sin \theta \cos \theta = 0$$

Setting each factor equal to zero, we obtain

$$\sin \theta = 0 \quad \text{or} \quad \cos \theta = 0$$

The apparent solutions are

$$\theta = 0, \qquad \theta = \pi, \qquad \theta = \frac{\pi}{2}, \qquad \theta = \frac{3\pi}{2}$$

Because we squared both sides of the original equation, we must check these apparent solutions to see if any are extraneous.

$\theta = 0$: $\sin 0 + \cos 0 = 0 + 1 = 1$ A solution

$\theta = \pi$: $\sin \pi + \cos \pi = 0 + (-1) = -1$ Not a solution

$\theta = \frac{\pi}{2}$: $\sin \frac{\pi}{2} + \cos \frac{\pi}{2} = 1 + 0 = 1$ A solution

$\theta = \frac{3\pi}{2}$: $\sin \frac{3\pi}{2} + \cos \frac{3\pi}{2} = -1 + 0 = -1$ Not a solution

Therefore, $\theta = \frac{3\pi}{2}$ and $\theta = \pi$ are extraneous. The only solutions are $\theta = 0$ and $\theta = \frac{\pi}{2}$.   ∎

We can solve the equation given in Example 6 in another way.

Solution B   We start with the equation

$$\sin \theta + \cos \theta = 1$$

and divide each side by $\sqrt{2}$. (The reason for this choice will become apparent shortly.) Then

$$\frac{1}{\sqrt{2}} \sin \theta + \frac{1}{\sqrt{2}} \cos \theta = \frac{1}{\sqrt{2}}$$

The left side now resembles the formula for the sine of the sum of two angles, one of which is θ. The other angle is unknown (call it ϕ.) Then

$$\sin(\theta + \phi) = \sin \theta \cos \phi + \cos \theta \sin\phi = \frac{1}{\sqrt{2}} = \frac{\sqrt{2}}{2} \qquad (1)$$

where

$$\cos \phi = \frac{1}{\sqrt{2}} = \frac{\sqrt{2}}{2}, \qquad \sin \phi = \frac{1}{\sqrt{2}} = \frac{\sqrt{2}}{2}, \qquad 0 \le \phi < 2\pi$$

The angle ϕ is therefore $\dfrac{\pi}{4}$. As a result, equation (1) becomes

$$\sin\left(\theta + \frac{\pi}{4}\right) = \frac{\sqrt{2}}{2}$$

Figure 35

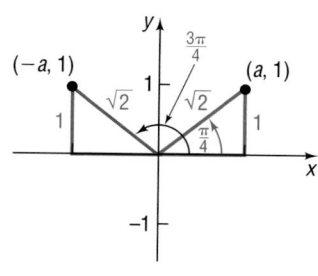

There are two angles whose sine is $\dfrac{\sqrt{2}}{2}$: $\dfrac{\pi}{4}$ and $\dfrac{3\pi}{4}$. See Figure 35. As a result,

$$\theta + \frac{\pi}{4} = \frac{\pi}{4} \quad \text{or} \quad \theta + \frac{\pi}{4} = \frac{3\pi}{4}$$
$$\theta = 0 \qquad\qquad \theta = \frac{\pi}{2}$$

These solutions agree with the solutions found earlier.   ■

This second method of solution can be used to solve any linear equation in the variables $\sin\theta$ and $\cos\theta$.

EXAMPLE 7   **Solving a Trigonometric Equation Linear in $\sin\theta$ and $\cos\theta$**

Solve:
$$a\sin\theta + b\cos\theta = c, \qquad 0 \le \theta < 2\pi \qquad\qquad (2)$$

where a, b, and c are constants and either $a \ne 0$ or $b \ne 0$.

Solution   We divide each side of equation (2) by $\sqrt{a^2 + b^2}$. Then

$$\frac{a}{\sqrt{a^2 + b^2}}\sin\theta + \frac{b}{\sqrt{a^2 + b^2}}\cos\theta = \frac{c}{\sqrt{a^2 + b^2}} \qquad\qquad (3)$$

There is a unique angle ϕ, $0 \le \phi < 2\pi$, for which

$$\cos\phi = \frac{a}{\sqrt{a^2 + b^2}} \quad \text{and} \quad \sin\phi = \frac{b}{\sqrt{a^2 + b^2}} \qquad\qquad (4)$$

Figure 36

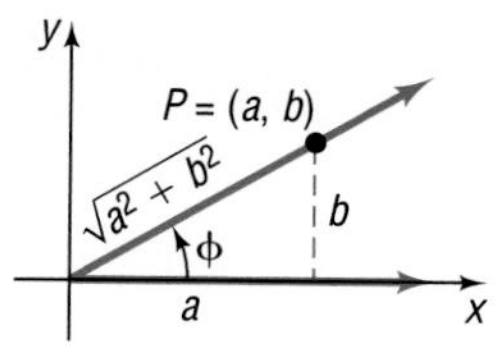

(see Figure 36). Equation (3) may be written as

$$\sin\theta\cos\phi + \cos\theta\sin\phi = \frac{c}{\sqrt{a^2 + b^2}}$$

or, equivalently,

$$\sin(\theta + \phi) = \frac{c}{\sqrt{a^2 + b^2}} \qquad\qquad (5)$$

where ϕ satisfies equations (4).

If $|c| > \sqrt{a^2 + b^2}$, then $\sin(\theta + \phi) > 1$ or $\sin(\theta + \phi) < -1$, and equation (5) has no solution.

If $|c| \le \sqrt{a^2 + b^2}$, then the solutions of equation (5) are

$$\theta + \phi = \sin^{-1}\frac{c}{\sqrt{a^2 + b^2}} \quad \text{or} \quad \theta + \phi = \pi - \sin^{-1}\frac{c}{\sqrt{a^2 + b^2}}$$

Because the angle ϕ is determined by equations (4), these are the solutions to equation (2).   ■

✎ ━━ **NOW WORK PROBLEM 33.**

4   ## Graphing Utility Solutions

The techniques introduced in this section apply only to certain types of trigonometric equations. Solutions for other types are usually studied in calculus, using numerical methods. In the next example, we show how a graphing utility may be used to obtain solutions.

EXAMPLE 8   ### Solving Trigonometric Equations Using a Graphing Utility

Solve: $5 \sin x + x = 3$
Express the solution(s) rounded to two decimal places.

Solution   This type of trigonometric equation cannot be solved by previous methods. A graphing utility, though, can be used here. The solutions of this equation are the same as the points of intersection of the graphs of $Y_1 = 5 \sin x + x$ and $Y_2 = 3$. See Figure 37.

Figure 37

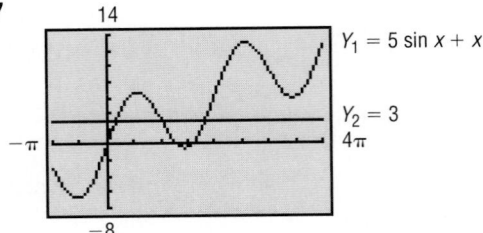

There are three points of intersection; the x-coordinates are the solutions that we seek. Using INTERSECT, we find

$$x = 0.52, \qquad x = 3.18, \qquad x = 5.71$$

rounded to two decimal places.

◢━ **NOW WORK PROBLEM 45.**

3.8 Exercises

In Problems 1–38, solve each equation on the interval $0 \leq \theta < 2\pi$.

1. $2 \cos^2 \theta + \cos \theta = 0$

2. $\sin^2 \theta - 1 = 0$

3. $2 \sin^2 \theta - \sin \theta - 1 = 0$

4. $2 \cos^2 \theta + \cos \theta - 1 = 0$

5. $(\tan \theta - 1)(\sec \theta - 1) = 0$

6. $(\cot \theta + 1)\left(\csc \theta - \dfrac{1}{2}\right) = 0$

7. $\sin^2 \theta - \cos^2 \theta = 1 + \cos \theta$

8. $\cos^2 \theta - \sin^2 \theta + \sin \theta = 0$

9. $\sin^2 \theta = 6(\cos \theta + 1)$

10. $2 \sin^2 \theta = 3(1 - \cos \theta)$

11. $\cos(2\theta) + 6 \sin^2 \theta = 4$

12. $\cos(2\theta) = 2 - 2 \sin^2 \theta$

13. $\cos \theta = \sin \theta$

14. $\cos \theta + \sin \theta = 0$

15. $\tan \theta = 2 \sin \theta$

16. $\sin(2\theta) = \cos \theta$

17. $\sin \theta = \csc \theta$

18. $\tan \theta = \cot \theta$

19. $\cos(2\theta) = \cos \theta$

20. $\sin(2\theta) \sin \theta = \cos \theta$

21. $\sin(2\theta) + \sin(4\theta) = 0$

22. $\cos(2\theta) + \cos(4\theta) = 0$

23. $\cos(4\theta) - \cos(6\theta) = 0$

24. $\sin(4\theta) - \sin(6\theta) = 0$

25. $1 + \sin \theta = 2 \cos^2 \theta$

26. $\sin^2 \theta = 2 \cos \theta + 2$

27. $\tan^2 \theta = \dfrac{3}{2} \sec \theta$

28. $\csc^2 \theta = \cot \theta + 1$

29. $3 - \sin \theta = \cos(2\theta)$

30. $\cos(2\theta) + 5 \cos \theta + 3 = 0$

31. $\sec^2 \theta + \tan \theta = 0$

32. $\sec \theta = \tan \theta + \cot \theta$

33. $\sin \theta - \sqrt{3} \cos \theta = 1$

34. $\sqrt{3}\sin\theta + \cos\theta = 1$

35. $\tan(2\theta) + 2\sin\theta = 0$

36. $\tan(2\theta) + 2\cos\theta = 0$

37. $\sin\theta + \cos\theta = \sqrt{2}$

38. $\sin\theta + \cos\theta = -\sqrt{2}$

In Problems 39–44, solve each equation for x, $-\pi \le x \le \pi$. Express the solution(s) rounded to two decimal places.

39. Solve the equation $\cos x = e^x$ by graphing $Y_1 = \cos x$ and $Y_2 = e^x$ and finding their point(s) of intersection.

40. Solve the equation $\cos x = e^x$ by graphing $Y_1 = \cos x - e^x$ and finding the x-intercept(s).

41. Solve the equation $2\sin x = 0.7x$ by graphing $Y_1 = 2\sin x$ and $Y_2 = 0.7x$ and finding their point(s) of intersection.

42. Solve the equation $2\sin x = 0.7x$ by graphing $Y_1 = 2\sin x - 0.7x$ and finding the x-intercept(s).

43. Solve the equation $\cos x = x^2$ by graphing $Y_1 = \cos x$ and $Y_2 = x^2$ and finding their point(s) of intersection.

44. Solve the equation $\cos x = x^2$ by graphing $Y_1 = \cos x - x^2$ and finding the x-intercept(s).

In Problems 45–56, use a graphing utility to solve each equation. Express the solution(s) rounded to two decimal places.

45. $x + 5\cos x = 0$

46. $x - 4\sin x = 0$

47. $22x - 17\sin x = 3$

48. $19x + 8\cos x = 2$

49. $\sin x + \cos x = x$

50. $\sin x - \cos x = x$

51. $x^2 - 2\cos x = 0$

52. $x^2 + 3\sin x = 0$

53. $x^2 - 2\sin(2x) = 3x$

54. $x^2 = x + 3\cos(2x)$

55. $6\sin x - e^x = 2, \quad x > 0$

56. $4\cos(3x) - e^x = 1, \quad x > 0$

---

57. Constructing a Rain Gutter   A rain gutter is to be constructed of aluminum sheets 12 inches wide. After marking off a length of 4 inches from each edge, this length is bent up at an angle θ. See the illustration. The area A of the opening as a function of θ is given by

$$A(\theta) = 16\sin\theta(\cos\theta + 1), \quad 0° < \theta < 90°$$

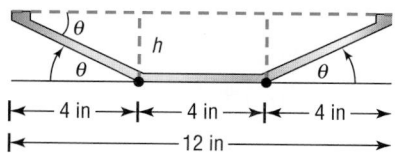

(a) In calculus, you will be asked to find the angle θ that maximizes A by solving the equation

$$\cos(2\theta) + \cos\theta = 0, \quad 0° < \theta < 90°$$

Solve this equation for θ by using the Double-Angle Formula.

(b) Solve the equation for θ by writing the sum of the two cosines as a product.

(c) What is the maximum area A of the opening?

(d) Graph A, $0° \le \theta \le 90°$, and find the angle θ that maximizes the area A. Also find the maximum area. Compare the results to the answers found earlier.

58. Projectile Motion   An object is propelled upward at an angle θ, $45° < \theta < 90°$, to the horizontal with an initial velocity of v_0 feet per second from the base of a plane that makes an angle of $45°$ with the horizontal. See the illustration. If air resistance is ignored, the distance R that it travels up the inclined plane is given by

$$R = \frac{v_0^2\sqrt{2}}{32}[\sin(2\theta) - \cos(2\theta) - 1]$$

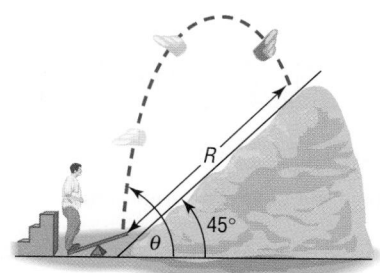

(a) In calculus, you will be asked to find the angle θ that maximizes R by solving the equation

$$\sin(2\theta) + \cos(2\theta) = 0$$

Solve this equation for θ by squaring both sides of the equation.

(b) Solve this equation for θ by dividing each side by $\cos(2\theta)$.

(c) What is the maximum distance R if $v_0 = 32$ feet per second?

(d) Graph R, $45° \le \theta \le 90°$, and find the angle θ that maximizes the distance R. Also find the maximum distance. Use $v_0 = 32$ feet per second. Compare the results with the answers found earlier.

59. Heat Transfer   In the study of heat transfer, the equation $x + \tan x = 0$ occurs. Graph $Y_1 = -x$ and $Y_2 = \tan x$ for $x \ge 0$. Conclude that there are an infinite number of points of intersection of these two graphs. Now find the first two positive solutions of $x + \tan x = 0$ rounded to two decimal places.

60. Carrying a Ladder around a Corner   A ladder of length L is carried horizontally around a corner from a hall 3 feet wide into a hall 4 feet wide. See the illustration.

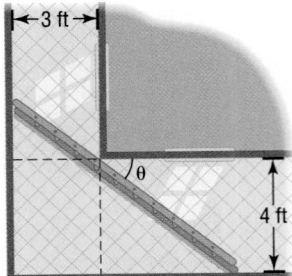

(a) Express L as a function of θ.

✍ (b) In calculus, you will be asked to find the length of the longest ladder that can turn the corner by solving the equation

$$3 \sec \theta \tan \theta - 4 \csc \theta \cot \theta = 0, \quad 0° < \theta < 90°$$

Solve this equation for θ.

(c) What is the length of the longest ladder that can be carried around the corner?

(d) Graph L, $0° \le \theta \le 90°$, and find the angle θ that minimizes the length L.

(e) Compare the result with the one found in part (b). Explain why the two answers are the same.

61. Projectile Motion   The horizontal distance that a projectile will travel in the air is given by the equation

$$R = \frac{v_0^2 \sin(2\theta)}{g}$$

where v_0 is the initial velocity of the projectile, θ is the angle of elevation, and g is acceleration due to gravity (9.8 meters per second squared).

(a) If you can throw a baseball with an initial speed of 34.8 meters per second, at what angle of elevation θ should you direct the throw so that the ball travels a distance of 107 meters before striking the ground?

(b) Determine the maximum distance that you can throw the ball.

(c) Graph R, with $v_0 = 34.8$ meters per second.

(d) Verify the results obtained in parts (a) and (b) using ZERO or ROOT.

62. Projectile Motion   Refer to Problem 61.

(a) If you can throw a baseball with an initial speed of 40 meters per second, at what angle of elevation θ should you direct the throw so that the ball travels a distance of 110 meters before striking the ground?

(b) Determine the maximum distance that you can throw the ball.

(c) Graph R, with $v_0 = 40$ meters per second.

(d) Verify the results obtained in parts (a) and (b) using ZERO or ROOT.

Chapter Review

Things to Know

Definitions of the six inverse trigonometric functions

$y = \sin^{-1} x$	means	$x = \sin y$	where $-1 \le x \le 1$, $\quad -\dfrac{\pi}{2} \le y \le \dfrac{\pi}{2}$	(p. 199)
$y = \cos^{-1} x$	means	$x = \cos y$	where $-1 \le x \le 1$, $\quad 0 \le y \le \pi$	(p. 203)
$y = \tan^{-1} x$	means	$x = \tan y$	where $-\infty < x < \infty$, $\quad -\dfrac{\pi}{2} < y < \dfrac{\pi}{2}$	(p. 206)
$y = \sec^{-1} x$	means	$x = \sec y$	where $\lvert x \rvert \ge 1$, $\quad 0 \le y \le \pi$, $\quad y \ne \dfrac{\pi}{2}$	(p. 212)
$y = \csc^{-1} x$	means	$x = \csc y$	where $\lvert x \rvert \ge 1$, $\quad -\dfrac{\pi}{2} \le y \le \dfrac{\pi}{2}$, $\quad y \ne 0$	(p. 212)
$y = \cot^{-1} x$	means	$x = \cot y$	where $-\infty < x < \infty$, $\quad 0 < y < \pi$	(p. 212)

Sum and Difference Formulas (pp. 222, 225, and 228)

$$\cos(\alpha + \beta) = \cos \alpha \cos \beta - \sin \alpha \sin \beta \qquad \cos(\alpha - \beta) = \cos \alpha \cos \beta + \sin \alpha \sin \beta$$

$$\sin(\alpha + \beta) = \sin \alpha \cos \beta + \cos \alpha \sin \beta \qquad \sin(\alpha - \beta) = \sin \alpha \cos \beta - \cos \alpha \sin \beta$$

$$\tan(\alpha + \beta) = \frac{\tan \alpha + \tan \beta}{1 - \tan \alpha \tan \beta} \qquad \tan(\alpha - \beta) = \frac{\tan \alpha - \tan \beta}{1 + \tan \alpha \tan \beta}$$

Double-Angle Formulas (pp. 234 and 235)

$$\sin(2\theta) = 2 \sin \theta \cos \theta \qquad \cos(2\theta) = \cos^2 \theta - \sin^2 \theta \qquad \cos(2\theta) = 1 - 2 \sin^2 \theta$$

$$\cos(2\theta) = 2 \cos^2 \theta - 1 \qquad \tan(2\theta) = \frac{2 \tan \theta}{1 - \tan^2 \theta}$$

Half-Angle Formulas (pp. 237, 238, and 240)

$$\sin^2 \frac{\alpha}{2} = \frac{1 - \cos \alpha}{2} \qquad \cos^2 \frac{\alpha}{2} = \frac{1 + \cos \alpha}{2} \qquad \tan^2 \frac{\alpha}{2} = \frac{1 - \cos \alpha}{1 + \cos \alpha}$$

$$\sin \frac{\alpha}{2} = \pm \sqrt{\frac{1 - \cos \alpha}{2}} \qquad \cos \frac{\alpha}{2} = \pm \sqrt{\frac{1 + \cos \alpha}{2}} \qquad \tan \frac{\alpha}{2} = \pm \sqrt{\frac{1 - \cos \alpha}{1 + \cos \alpha}} = \frac{1 - \cos \alpha}{\sin \alpha} = \frac{\sin \alpha}{1 + \cos \alpha}$$

where the $+$ or $-$ sign is determined by the quadrant of $\frac{\alpha}{2}$

Product-to-Sum Formulas (p. 243)

$$\sin \alpha \sin \beta = \frac{1}{2} [\cos(\alpha - \beta) - \cos(\alpha + \beta)]$$

$$\cos \alpha \cos \beta = \frac{1}{2} [\cos(\alpha - \beta) + \cos(\alpha + \beta)]$$

$$\sin \alpha \cos \beta = \frac{1}{2} [\sin(\alpha + \beta) + \sin(\alpha - \beta)]$$

Sum-to-Product Formulas (p. 245)

$$\sin \alpha + \sin \beta = 2 \sin \frac{\alpha + \beta}{2} \cos \frac{\alpha - \beta}{2} \qquad \sin \alpha - \sin \beta = 2 \sin \frac{\alpha - \beta}{2} \cos \frac{\alpha + \beta}{2}$$

$$\cos \alpha + \cos \beta = 2 \cos \frac{\alpha + \beta}{2} \cos \frac{\alpha - \beta}{2} \qquad \cos \alpha - \cos \beta = -2 \sin \frac{\alpha + \beta}{2} \sin \frac{\alpha - \beta}{2}$$

Objectives

Section		You should be able to:	Review Exercises
3.1	1	Find the exact value of the inverse sine, cosine, and tangent functions (p. 201)	1–6
	2	Find an approximate value of the inverse sine, cosine, and tangent functions (p. 202)	101–104
3.2	1	Find the exact value of expressions involving the inverse sine, cosine and tangent functions (p. 210)	9–20
	2	Find the exact value of the inverse secant, cosecant, and cotangent functions (p. 212)	7–8
	3	Use a calculator to evaluate $\sec^{-1} x$, $\csc^{-1} x$, and $\cot^{-1} x$ (p. 213)	105–106
3.3	1	Establish identities (p. 216)	21–38
3.4	1	Use sum and difference formulas to find exact values (p. 223)	53–56, 59–60, 61–70(a)–(d)
	2	Use sum and difference formulas to establish identities (p. 224)	39–42
	3	Use sum and difference formulas involving inverse trigonometric functions (p. 229)	71–74
3.5	1	Use double-angle formulas to find exact values (p. 234)	61–70(e)–(f)
	2	Use double-angle and half-angle formulas to establish identities (p. 235)	43–48
	3	Use half-angle formulas to find exact values (p. 238)	61–70(g)–(h)
3.6	1	Express products as sums (p. 243)	49–50
	2	Express sums as products (p. 245)	51–52
3.7	1	Solve equations involving a single trigonometric function (p. 247)	77–86
3.8	1	Solve trigonometric equations quadratic in form (p. 253)	93–94
	2	Solve trigonometric equations using identities (p. 253)	87–92, 95–98
	3	Solve trigonometric equations linear in sine and cosine (p. 255)	99–100
	4	Solve trigonometric equations using a graphing utility (p. 258)	107–112

Review Exercises

Blue problem numbers indicate the authors' suggestions for use in a Practice Test.

In Problems 1–20, find the exact value of each expression. Do not use a calculator.

1. $\sin^{-1} 1$

2. $\cos^{-1} 0$

3. $\tan^{-1} 1$

4. $\sin^{-1}\left(-\dfrac{1}{2}\right)$

5. $\cos^{-1}\left(-\dfrac{\sqrt{3}}{2}\right)$

6. $\tan^{-1}(-\sqrt{3})$

7. $\sec^{-1}\sqrt{2}$

8. $\cot^{-1}(-1)$

9. $\tan\left[\sin^{-1}\left(-\dfrac{\sqrt{3}}{2}\right)\right]$

10. $\tan\left[\cos^{-1}\left(-\dfrac{1}{2}\right)\right]$

11. $\sec\left(\tan^{-1}\dfrac{\sqrt{3}}{3}\right)$

12. $\csc\left(\sin^{-1}\dfrac{\sqrt{3}}{2}\right)$

13. $\sin\left(\tan^{-1}\dfrac{3}{4}\right)$

14. $\cos\left(\sin^{-1}\dfrac{3}{5}\right)$

15. $\tan\left[\sin^{-1}\left(-\dfrac{4}{5}\right)\right]$

16. $\tan\left[\cos^{-1}\left(-\dfrac{3}{5}\right)\right]$

17. $\sin^{-1}\left(\cos\dfrac{2\pi}{3}\right)$

18. $\cos^{-1}\left(\tan\dfrac{3\pi}{4}\right)$

19. $\tan^{-1}\left(\tan\dfrac{7\pi}{4}\right)$

20. $\cos^{-1}\left(\cos\dfrac{7\pi}{6}\right)$

In Problems 21–52, establish each identity.

21. $\tan\theta\cot\theta - \sin^2\theta = \cos^2\theta$

22. $\sin\theta\csc\theta - \sin^2\theta = \cos^2\theta$

23. $\cos^2\theta(1 + \tan^2\theta) = 1$

24. $(1 - \cos^2\theta)(1 + \cot^2\theta) = 1$

25. $4\cos^2\theta + 3\sin^2\theta = 3 + \cos^2\theta$

26. $4\sin^2\theta + 2\cos^2\theta = 4 - 2\cos^2\theta$

27. $\dfrac{1 - \cos\theta}{\sin\theta} + \dfrac{\sin\theta}{1 - \cos\theta} = 2\csc\theta$

28. $\dfrac{\sin\theta}{1 + \cos\theta} + \dfrac{1 + \cos\theta}{\sin\theta} = 2\csc\theta$

29. $\dfrac{\cos\theta}{\cos\theta - \sin\theta} = \dfrac{1}{1 - \tan\theta}$

30. $1 - \dfrac{\cos^2\theta}{1 + \sin\theta} = \sin\theta$

31. $\dfrac{\csc\theta}{1 + \csc\theta} = \dfrac{1 - \sin\theta}{\cos^2\theta}$

32. $\dfrac{1 + \sec\theta}{\sec\theta} = \dfrac{\sin^2\theta}{1 - \cos\theta}$

33. $\csc\theta - \sin\theta = \cos\theta\cot\theta$

34. $\dfrac{\csc\theta}{1 - \cos\theta} = \dfrac{1 + \cos\theta}{\sin^3\theta}$

35. $\dfrac{1 - \sin\theta}{\sec\theta} = \dfrac{\cos^3\theta}{1 + \sin\theta}$

36. $\dfrac{1 - \cos\theta}{1 + \cos\theta} = (\csc\theta - \cot\theta)^2$

37. $\dfrac{1 - 2\sin^2\theta}{\sin\theta\cos\theta} = \cot\theta - \tan\theta$

38. $\dfrac{(2\sin^2\theta - 1)^2}{\sin^4\theta - \cos^4\theta} = 1 - 2\cos^2\theta$

39. $\dfrac{\cos(\alpha + \beta)}{\cos\alpha\sin\beta} = \cot\beta - \tan\alpha$

40. $\dfrac{\sin(\alpha - \beta)}{\sin\alpha\cos\beta} = 1 - \cot\alpha\tan\beta$

41. $\dfrac{\cos(\alpha - \beta)}{\cos\alpha\cos\beta} = 1 + \tan\alpha\tan\beta$

42. $\dfrac{\cos(\alpha + \beta)}{\sin\alpha\cos\beta} = \cot\alpha - \tan\beta$

43. $(1 + \cos\theta)\left(\tan\dfrac{\theta}{2}\right) = \sin\theta$

44. $\sin\theta\tan\dfrac{\theta}{2} = 1 - \cos\theta$

45. $2\cot\theta\cot(2\theta) = \cot^2\theta - 1$

46. $2\sin(2\theta)(1 - 2\sin^2\theta) = \sin(4\theta)$

47. $1 - 8\sin^2\theta\cos^2\theta = \cos(4\theta)$

48. $\dfrac{\sin(3\theta)\cos\theta - \sin\theta\cos(3\theta)}{\sin(2\theta)} = 1$

49. $\dfrac{\sin(2\theta) + \sin(4\theta)}{\cos(2\theta) + \cos(4\theta)} = \tan(3\theta)$

50. $\dfrac{\sin(2\theta) + \sin(4\theta)}{\sin(2\theta) - \sin(4\theta)} + \dfrac{\tan(3\theta)}{\tan\theta} = 0$

51. $\dfrac{\cos(2\theta) - \cos(4\theta)}{\cos(2\theta) + \cos(4\theta)} - \tan\theta\tan(3\theta) = 0$

52. $\cos(2\theta) - \cos(10\theta) = \tan(4\theta)[\sin(2\theta) + \sin(10\theta)]$

In Problems 53–60, find the exact value of each expression.

53. $\sin 165°$

54. $\tan 105°$

55. $\cos\dfrac{5\pi}{12}$

56. $\sin\left(-\dfrac{\pi}{12}\right)$

57. $\cos 80°\cos 20° + \sin 80°\sin 20°$

58. $\sin 70°\cos 40° - \cos 70°\sin 40°$

59. $\tan\dfrac{\pi}{8}$

60. $\sin\dfrac{5\pi}{8}$

In Problems 61–70, use the information given about the angles α and β to find the exact value of:

(a) $\sin(\alpha + \beta)$

(b) $\cos(\alpha + \beta)$

(c) $\sin(\alpha - \beta)$

(d) $\tan(\alpha + \beta)$

(e) $\sin(2\alpha)$

(f) $\cos(2\beta)$

(g) $\sin\dfrac{\beta}{2}$

(h) $\cos\dfrac{\alpha}{2}$

61. $\sin\alpha = \dfrac{4}{5},\quad 0 < \alpha < \dfrac{\pi}{2};\quad \sin\beta = \dfrac{5}{13},\quad \dfrac{\pi}{2} < \beta < \pi$

62. $\cos\alpha = \dfrac{4}{5},\quad 0 < \alpha < \dfrac{\pi}{2};\quad \cos\beta = \dfrac{5}{13},\quad -\dfrac{\pi}{2} < \beta < 0$

63. $\sin \alpha = -\dfrac{3}{5}, \quad \pi < \alpha < \dfrac{3\pi}{2}; \quad \cos \beta = \dfrac{12}{13}, \quad \dfrac{3\pi}{2} < \beta < 2\pi$

64. $\sin \alpha = -\dfrac{4}{5}, \quad -\dfrac{\pi}{2} < \alpha < 0; \quad \cos \beta = -\dfrac{5}{13}, \quad \dfrac{\pi}{2} < \beta < \pi$

65. $\tan \alpha = \dfrac{3}{4}, \quad \pi < \alpha < \dfrac{3\pi}{2}; \quad \tan \beta = \dfrac{12}{5}, \quad 0 < \beta < \dfrac{\pi}{2}$

66. $\tan \alpha = -\dfrac{4}{3}, \quad \dfrac{\pi}{2} < \alpha < \pi; \quad \cot \beta = \dfrac{12}{5}, \quad \pi < \beta < \dfrac{3\pi}{2}$

67. $\sec \alpha = 2, \quad -\dfrac{\pi}{2} < \alpha < 0; \quad \sec \beta = 3, \quad \dfrac{3\pi}{2} < \beta < 2\pi$

68. $\csc \alpha = 2, \quad \dfrac{\pi}{2} < \alpha < \pi; \quad \sec \beta = -3, \quad \dfrac{\pi}{2} < \beta < \pi$

69. $\sin \alpha = -\dfrac{2}{3}, \quad \pi < \alpha < \dfrac{3\pi}{2}; \quad \cos \beta = -\dfrac{2}{3}, \quad \pi < \beta < \dfrac{3\pi}{2}$

70. $\tan \alpha = -2, \quad \dfrac{\pi}{2} < \alpha < \pi; \quad \cot \beta = -2, \quad \dfrac{\pi}{2} < \beta < \pi$

In Problems 71–76, find the exact value of each expression.

71. $\cos\left(\sin^{-1}\dfrac{3}{5} - \cos^{-1}\dfrac{1}{2} \right)$

72. $\sin\left(\cos^{-1}\dfrac{5}{13} - \cos^{-1}\dfrac{4}{5} \right)$

73. $\tan\left[\sin^{-1}\left(-\dfrac{1}{2}\right) - \tan^{-1}\dfrac{3}{4} \right]$

74. $\cos\left[\tan^{-1}(-1) + \cos^{-1}\left(-\dfrac{4}{5}\right) \right]$

75. $\sin\left[2\cos^{-1}\left(-\dfrac{3}{5}\right) \right]$

76. $\cos\left(2\tan^{-1}\dfrac{4}{3} \right)$

In Problems 77–100, solve each equation on the interval $0 \le \theta < 2\pi$.

77. $\cos \theta = \dfrac{1}{2}$

78. $\sin \theta = -\dfrac{\sqrt{3}}{2}$

79. $2\cos \theta + \sqrt{2} = 0$

80. $\tan \theta + \sqrt{3} = 0$

81. $\sin(2\theta) + 1 = 0$

82. $\cos(2\theta) = 0$

83. $\tan(2\theta) = 0$

84. $\sin(3\theta) = 1$

85. $\sec^2 \theta = 4$

86. $\csc^2 \theta = 1$

87. $\sin \theta = \tan \theta$

88. $\cos \theta = \sec \theta$

89. $\sin \theta + \sin(2\theta) = 0$

90. $\cos(2\theta) = \sin \theta$

91. $\sin(2\theta) - \cos \theta - 2\sin \theta + 1 = 0$

92. $\sin(2\theta) - \sin \theta - 2\cos \theta + 1 = 0$

93. $2\sin^2 \theta - 3\sin \theta + 1 = 0$

94. $2\cos^2 \theta + \cos \theta - 1 = 0$

95. $4\sin^2 \theta = 1 + 4\cos \theta$

96. $8 - 12\sin^2 \theta = 4\cos^2 \theta$

97. $\sin(2\theta) = \sqrt{2}\cos \theta$

98. $1 + \sqrt{3}\cos \theta + \cos(2\theta) = 0$

99. $\sin \theta - \cos \theta = 1$

100. $\sin \theta - \sqrt{3}\cos \theta = 2$

In Problems 101–106, use a calculator to find an approximate value for each expression, rounded to two decimal places.

101. $\sin^{-1} 0.7$

102. $\cos^{-1} \dfrac{4}{5}$

103. $\tan^{-1}(-2)$

104. $\cos^{-1}(-0.2)$

105. $\sec^{-1} 3$

106. $\cot^{-1}(-4)$

In Problems 107–112, use a graphing utility to solve each equation on the interval $0 \le x \le 2\pi$. Approximate any solutions rounded to two decimal places.

107. $2x = 5\cos x$

108. $2x = 5\sin x$

109. $2\sin x + 3\cos x = 4x$

110. $3\cos x + x = \sin x$

111. $\sin x = \ln x$

112. $\sin x = e^{-x}$

Chapter Projects

1. **Waves**   A stretched string that is attached at both ends, pulled in a direction perpendicular to the string, and released, has motion that is described as wave motion. If we assume no friction and a length such that there are no "echoes" (that is, the wave doesn't bounce back), the transverse motion (motion perpendicular to the string) can be described by the equation

$$y = y_m \sin(kx - \omega t)$$

where y_m is the amplitude measured in meters

k and ω are constants

The height of the sound wave depends on the distance x from one endpoint of the string and time t. Thus, a typical wave has horizontal and vertical motion over time.

(a) What is the amplitude of the wave
     $y = 0.00421 \sin(68.3x - 2.68t)$?

(b) The value of ω is the angular frequency measured in radians per second. What is the angular frequency of the wave given in part (a)?

(c) The frequency f is the number of vibrations per second (hertz) made by the wave as it passes a certain point. Its value is found using the formula $f = \dfrac{\omega}{2\pi}$.
     What is the frequency of the wave given in part (a)?

(d) The wavelength, λ, of a wave is the shortest distance at which the wave pattern repeats itself for a constant t. Thus, $\lambda = \dfrac{2\pi}{k}$. What is the wavelength of the wave given in part (a)?

(e) Graph the height of the string a distance $x = 1$ meter from an endpoint.

(f) If two waves travel simultaneously along the same stretched string, the vertical displacement of the string when both waves act is $y = y_1 + y_2$, where y_1 is the vertical displacement of the first wave and y_2 is the vertical displacement of the second wave. This result is called the Principle of Superposition and was analyzed by the French mathematician Jean Baptiste Fourier (1768–1830). When two waves travel along

the same string, one wave will differ from the other wave by a phase constant ϕ.

$$y_1 = y_m \sin(kx - \omega t)$$
$$y_2 = y_m \sin(kx - \omega t + \phi)$$

assuming that each wave has the same amplitude. Write $y_1 + y_2$ as a product using the Sum-to-Product formulas.

(g) Suppose that two waves are moving in the same direction along a stretched string. The amplitude of each string is 0.0045 meter and the phase difference between them is 2.5 radians. The wavelength, λ, of each wave is 0.09 meter and the frequency, f, is 2.3 hertz. Find y_1, y_2 and $y_1 + y_2$.

(h) Using a graphing utility, graph y_1, y_2 and $y_1 + y_2$ on the same viewing window.

(i) Redo parts (g) and (h) with the phase difference between the waves being 0.4 radian.

(j) What effect does the phase difference have on the amplitude of $y_1 + y_2$?

2. **Jacob's Field**   In Jacob's Field, home of the Cleveland Indians, the distance from home plate to the outfield fence varies from 325 feet (first and third base lines) to 410 feet (dead center field). We wish to estimate the minimum distance from home plate to the highest point of Jacob's Field and from the base of the outfield fence to the highest point of the stadium so that we can estimate the minimum distance that a ball must travel to be hit out of Jacob's Field.

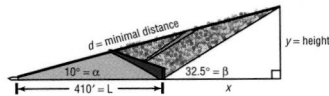

(a) Sketch a picture showing home plate, the fence, and the highest point of the stadium. Draw lines from home plate to the highest point and from the base of the fence to the highest point.

(b) Using a transit, the angle of elevation from home plate to the highest point of the stadium in dead center field is 10° and the angle of elevation from the base of the fence to the highest point of the stadium in dead center field is 32.5°. Use the distance from home plate to the base of the wall in dead center field to compute the minimum distance from home plate to the highest point of the stadium and the distance from the base of the outfield wall to the highest point of the stadium.

(c) Develop a general formula for computing the distance from home plate to the highest point of the stadium and the distance from the base of the outfield wall to the highest point of the stadium. Let α and β denote the angles of elevation to the top of the roof from home plate and from the base of the outfield fence, respectively. Let L be the distance from home plate to the outfield fence.

(d) Write each formula found in part (c) in the form

$$\text{Height} = \frac{L \sin \alpha \sin \beta}{\sin(\beta - \alpha)} \qquad \text{Distance} = \frac{L \sin \beta}{\sin(\beta - \alpha)}$$

3. Suppose $f(x) = \sin x$.

(a) Build a table of values for $f(x)$ where $x = 0, \dfrac{\pi}{6}, \dfrac{\pi}{4},$ $\dfrac{\pi}{3}, \dfrac{\pi}{2}, \dfrac{2\pi}{3}, \dfrac{3\pi}{4}, \dfrac{5\pi}{6}, \pi, \dfrac{7\pi}{6}, \dfrac{5\pi}{4}, \dfrac{4\pi}{3}, \dfrac{3\pi}{2}, \dfrac{5\pi}{3}, \dfrac{7\pi}{4},$ $\dfrac{11\pi}{6}, 2\pi.$ Use exact values.

(b) Find the **first differences** for each consecutive pair of values in part (a). That is, evaluate $g(x_i) = \dfrac{\Delta f(x_i)}{\Delta x_i} = \dfrac{f(x_{i+1}) - f(x_i)}{x_{i+1} - x_i}$, where $x_1 = 0$, $x_2 = \dfrac{\pi}{6}, \dots,$ $x_{17} = 2\pi$. Use your calculator to approximate each value rounded to three decimal places.

(c) Plot the points $(x_i, g(x_i))$ for $i = 1, \dots, 16$ on a scatter diagram. What shape does the set of points give? What function does this resemble? Fit a sine curve of best fit to the points. How does that relate to your guess?

(d) Find the first differences for each consecutive pair of values in part (b). That is, evaluate $h(x_i) = \dfrac{\Delta g(x_i)}{\Delta x_i} = \dfrac{g(x_{i+1}) - g(x_i)}{x_{i+1} - x_i}$ where $x_1 = 0$, $x_2 = \dfrac{\pi}{6}, \dots, x_{16} = \dfrac{11\pi}{6}$. This is the set of **second differences** of $f(x)$. Use your calculator to approximate each value rounded to three decimal places. Plot the points $(x_i, h(x_i))$ for $i = 1, \dots, 15$ on a scatter diagram. What shape does the set of points give? What function does this resemble? Fit a sine curve of best fit to the points. How does that relate to your guess?

(e) Find the first differences for each consecutive pair of values in part (d). That is, evaluate $k(x_i) = \dfrac{\Delta h(x_i)}{\Delta x_i}$ $= \dfrac{h(x_{i+1}) - h(x_i)}{x_{i+1} - x_i}$, where $x_1 = 0$, $x_2 = \dfrac{\pi}{6}, \dots, x_{15}$ $= \dfrac{7\pi}{4}$. This is the set of **third differences** of $f(x)$. Use your calculator to approximate each value rounded to three decimal places. Plot the points $(x_i, k(x_i))$ for $i = 1, \dots, 14$ on a scatter diagram. What shape does the set of points give? What function does this resemble? Fit a sine curve of best fit to the points. How does that relate to your guess?

(f) Find the first differences for each consecutive pair of values in part (e). That is, evaluate $m(x_i) = \dfrac{\Delta k(x_i)}{\Delta x_i}$ $= \dfrac{k(x_{i+1}) - k(x_i)}{x_{i+1} - x_i}$, where $x_1 = 0$, $x_2 = \dfrac{\pi}{6}, \dots,$ $x_{14} = \dfrac{5\pi}{3}$. This is the set of **fourth differences** of $f(x)$.

Use your calculator to approximate each value rounded to three decimal places. Plot the points $(x_i, m(x_i))$ for $i = 1, \dots, 13$ on a scatter diagram. What shape does the set of points give? What function does this resemble? Fit a sine curve of best fit to the points. How does that relate to your guess?

(g) What pattern do you notice about the curves that you found? What happened in part (f)? Can you make a generalization about what happened as you computed the differences? Explain your answers.

Cumulative Review

1. Find the real solutions, if any, of the equation $3x^2 + x - 1 = 0$.

2. Find an equation for the line containing the points $(-2, 5)$ and $(4, -1)$. What is the distance between these points? What is their midpoint?

3. Test the equation $3x + y^2 = 9$ for symmetry with respect to the x-axis, the y-axis, and the origin. List the intercepts.

4. Use transformations to graph the equation $y = |x - 3| + 2$.

5. Use transformations to graph the equation $y = -2 \sin x$.

6. Use transformations to graph the equation $y = \cos\left(x - \dfrac{\pi}{2}\right) - 1$.

7. Sketch a graph of each of the following functions. Label at least three points on each graph. Name the inverse function of each one and show its graph.

(a) $y = x^3$

(b) $y = \sin x$, $-\dfrac{\pi}{2} \le x \le \dfrac{\pi}{2}$

(c) $y = \cos x$, $0 \le x \le \pi$

8. If $\sin \theta = -\dfrac{1}{3}$ and $\pi < \theta < \dfrac{3\pi}{2}$, find the exact value of:

(a) $\cos \theta$ (b) $\tan \theta$ (c) $\sin(2\theta)$

(d) $\cos(2\theta)$ (e) $\sin\left(\dfrac{1}{2}\theta\right)$ (f) $\cos\left(\dfrac{1}{2}\theta\right)$

9. Find the exact value of $\cos(\tan^{-1} 2)$.

10. If $\sin\alpha = \dfrac{1}{3}, \dfrac{\pi}{2} < \alpha < \pi$, and $\cos\beta = -\dfrac{1}{3}, \pi < \beta < \dfrac{3\pi}{2}$, find the exact value of

(a) $\cos\alpha$ (b) $\sin\beta$ (c) $\cos(2\alpha)$

(d) $\cos(\alpha + \beta)$ (e) $\sin\dfrac{\beta}{2}$

11. For the triangle shown, find h if $\theta = 20°$.

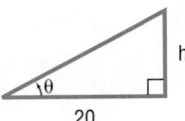

APPLICATIONS OF TRIGONOMETRIC FUNCTIONS

New maps pinpoint Lewis and Clark's journey through Missouri

KANSAS CITY, Mo.—Nearly two centuries ago, Congress commissioned Meriwether Lewis and William Clark to explore trade routes to the West.

Their long-documented journey took them through Missouri, but their exact route has been up for debate.

Now, modern computer technology combined with 19th century land surveys may provide a precise picture. The latest computer-generated maps of Lewis and Clark's expedition were unveiled here this week by Missouri Secretary of State Matt Blunt.

The maps combine the terrain of the early 19th century with contemporary geographical markers. They are the most precise to date of Lewis and Clark's journeys through Missouri in 1804 and 1806, said the project's lead researcher, Jim Harlan.

"We've known what they've done, but not with this much certainty," said Harlan, geographic resource project director at the University of Missouri-Columbia. "I think we need more information and less speculation. That's what this is all about."

SOURCE: Sophia Maines, "New Maps Pinpoint Lewis and Clark's Journey Through Missouri," *The Kansas City Star*, August 1, 2001 Distributed by Knight Ridder/Tribune Information Services..

SEE CHAPTER PROJECT 1.

 For additional study help, go to

www.prenhall.com/sullivanegu3e

Materials include:

- Graphing Calculator Help
- Chapter Quiz
- Chapter Test
- PowerPoint Downloads
- Chapter Projects
- Student Tips

A Look Back, A Look Forward

In Chapter 2, we defined the six trigonometric functions using right triangles and then extended the definition to include general angles. In particular, we learned to evaluate the trigonometric functions. We also learned how to graph sinusoidal functions.

In this chapter, we use the trigonometric functions to solve applied problems. The first four sections deal with applications involving right triangles and *oblique*

triangles, triangles that do not have a right angle. To solve problems involving oblique triangles, we will develop the Law of Sines and the Law of Cosines. We will also develop formulas for finding the area of a triangle.

The final section deals with applications of sinusoidal functions involving simple harmonic motion and damped motion.

PREPARING FOR THIS SECTION

Before getting started, review the following:

✓ Pythagorean Theorem (Section A.2, pp. 565–567)

✓ Complementary Angle Theorem (Section 2.2, pp. 118–120)

✓ Trigonometric Equations (I) (Section 3.7, pp. 247–250)

4.1 APPLICATIONS INVOLVING RIGHT TRIANGLES

OBJECTIVES
1. Solve Right Triangles
2. Solve Applied Problems

Solving Right Triangles

Figure 1

1. In the discussion that follows, we will always label a right triangle so that side a is opposite angle α, side b is opposite angle β, and side c is the hypotenuse, as shown in Figure 1. **To solve a right triangle** means to find the missing lengths of its sides and the measurements of its angles. We shall follow the practice of expressing the lengths of the sides rounded to two decimal places and of expressing angles in degrees rounded to one decimal place. (Be sure that your calculator is in degree mode.)

To solve a right triangle, we need to know one of the acute angles α or β and a side, or else two sides. Then we make use of the Pythagorean Theorem and the fact that the sum of the angles of a triangle is 180°. The sum of the angles α and β in a right triangle is therefore 90°.

For the right triangle shown in Figure 1, we have

$$c^2 = a^2 + b^2, \qquad \alpha + \beta = 90°$$

Figure 2

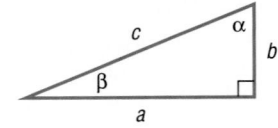

EXAMPLE 1 Solving a Right Triangle

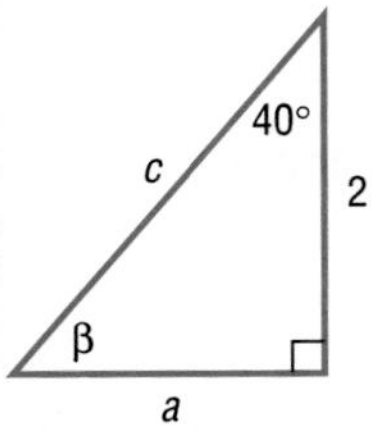

Use Figure 2. If $b = 2$ and $\alpha = 40°$, find a, c, and β.

Solution Since $\alpha = 40°$ and $\alpha + \beta = 90°$, we find that $\beta = 50°$. To find the sides a and c, we use the facts that

$$\tan 40° = \frac{a}{2} \quad \text{and} \quad \cos 40° = \frac{2}{c}$$

Now solve for a and c.

$$a = 2 \tan 40° \approx 1.68 \quad \text{and} \quad c = \frac{2}{\cos 40°} \approx 2.61$$

◾

✏️ NOW WORK PROBLEM **1**.

EXAMPLE 2   Solving a Right Triangle

Figure 3

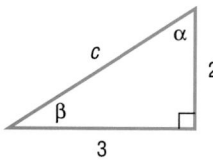

Use Figure 3. If $a = 3$ and $b = 2$, find c, α, and β.

Solution   Since $a = 3$ and $b = 2$, then, by the Pythagorean Theorem, we have

$$c^2 = a^2 + b^2 = 3^2 + 2^2 = 9 + 4 = 13$$
$$c = \sqrt{13} \approx 3.61$$

To find angle α, we use the fact that

$$\tan \alpha = \frac{3}{2} \quad \text{so} \quad \alpha = \tan^{-1} \frac{3}{2}$$

Set the mode on your calculator to degrees. Then, rounded to one decimal place, we find that $\alpha = 56.3°$. Since $\alpha + \beta = 90°$, we find that $\beta = 33.7°$.

◾

Note: To avoid round-off errors when using a calculator, we will store unrounded values in memory for use in subsequent calculations.

✏️ NOW WORK PROBLEM **11**.

Applications

② One common use for trigonometry is to measure heights and distances that are either awkward or impossible to measure by ordinary means.

EXAMPLE 3   Finding the Width of a River

A surveyor can measure the width of a river by setting up a transit* at a point C on one side of the river and taking a sighting of a point A on the other side. Refer to Figure 4. After turning through an angle of 90° at C, the surveyor walks a distance of 200 meters to point B. Using the transit at B, the angle β is measured and found to be 20°. What is the width of the river rounded to the nearest meter?

Figure 4

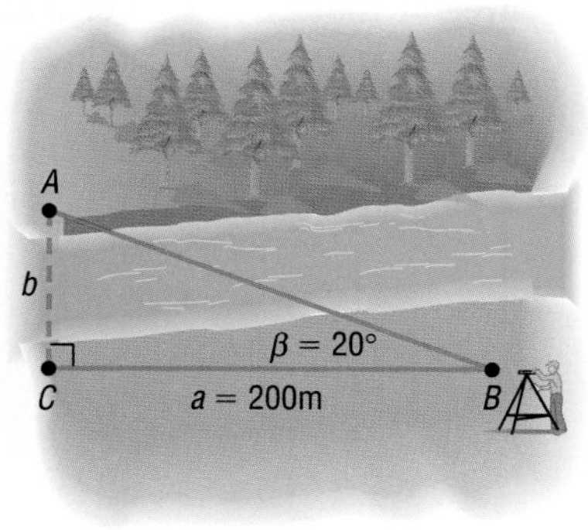

Solution   We seek the length of side b. We know a and β, so we use the fact that

$$\tan \beta = \frac{b}{a}$$

to get

$$\tan 20° = \frac{b}{200}$$
$$b = 200 \tan 20° \approx 72.79 \text{ meters}$$

The width of the river is 73 meters, rounded to the nearest meter.

◾

✏️ NOW WORK PROBLEM **21**.

*An instrument used in surveying to measure angles.

Figure 5

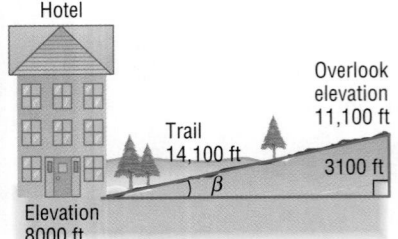

Hotel

Overlook
elevation
11,100 ft

Trail
14,100 ft

3100 ft

β

Elevation
8000 ft

EXAMPLE 4   Finding the Inclination of a Mountain Trail

A straight trail leads from the Alpine Hotel, elevation 8000 feet, to a scenic overlook, elevation 11,100 feet. The length of the trail is 14,100 feet. What is the inclination of the trail? That is, what is the angle β in Figure 5?

Solution   As Figure 5 illustrates, the angle β obeys the equation

$$\sin \beta = \frac{3100}{14,100}$$

Using a calculator,

$$\beta = \sin^{-1} \frac{3100}{14,100} \approx 12.7°$$

The inclination of the trail is approximately 12.7°.   ◼

Vertical heights can sometimes be measured using either the angle of elevation or the angle of depression. If a person is looking up at an object, the acute angle measured from the horizontal to a line-of-sight observation of the object is called the **angle of elevation**. See Figure 6(a).

If a person is standing on a cliff looking down at an object, the acute angle made by the line-of-sight observation of the object and the horizontal is called the **angle of depression**. See Figure 6(b).

Figure 6

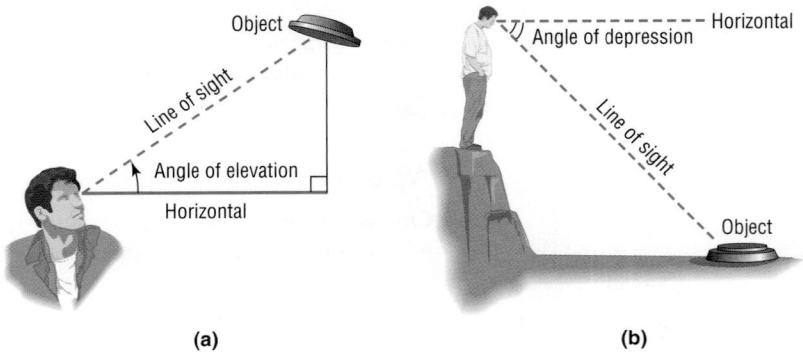

Object

Line of sight

Angle of elevation

Horizontal

Horizontal

Angle of depression

Line of sight

Object

(a)

(b)

EXAMPLE 5   Finding the Height of a Cloud

Meteorologists find the height of a cloud using an instrument called a **ceilometer**. A ceilometer consists of a **light projector** that directs a vertical light beam up to the cloud base and a **light detector** that scans the cloud to detect the light beam. See Figure 7(a). On December 1, 2001, at Midway Airport in Chicago, a ceilometer with a base of 300 feet was employed to find the height of the cloud cover. If the angle of elevation of the light detector is 75°, what is the height of the cloud cover?

Figure 7

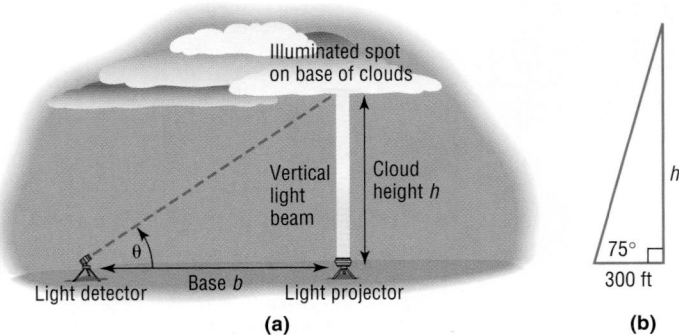

Illuminated spot
on base of clouds

Vertical
light
beam

Cloud
height h

θ

Base b

Light detector

Light projector

(a)

h

75°

300 ft

(b)

Solution   Figure 7(b) illustrates the situation. To find the height h, we use the fact that $\tan 75° = \dfrac{h}{300}$, so

$$h = 300 \tan 75° \approx 1120 \text{ feet}$$

The ceiling (height to the base of the cloud cover) is approximately 1120 feet.

 **NOW WORK PROBLEM 23.**

The idea behind Example 5 can also be used to find the height of an object with a base that is not accessible to the horizontal.

EXAMPLE 6   Finding the Height of a Statue on a Building

Adorning the top of the Board of Trade building in Chicago is a statue of Ceres, the Greek goddess of wheat. From street level, two observations are taken 400 feet from the center of the building. The angle of elevation to the base of the statue is found to be 45°; the angle of elevation to the top of the statue is 47.2°. See Figure 8(a). What is the height of the statue?

Figure 8

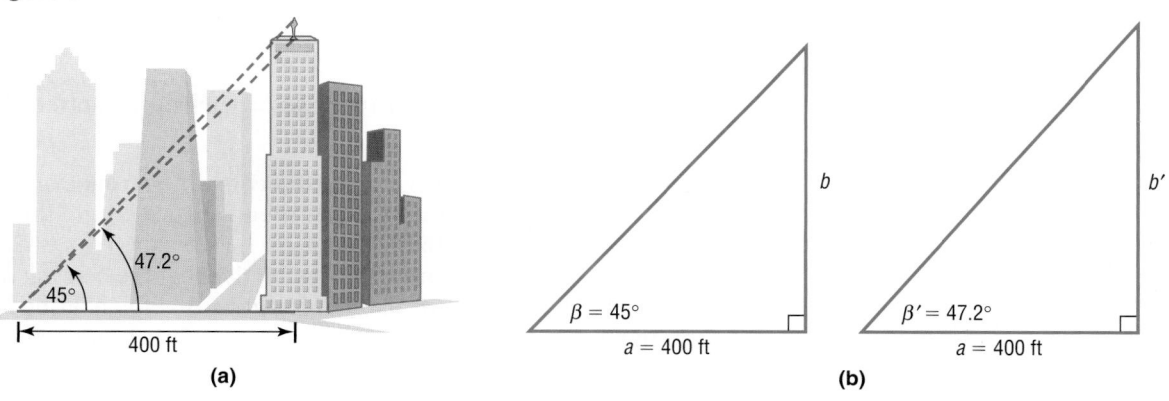

(a)   (b)

Solution   Figure 8(b) shows two triangles that replicate Figure 8(a). The height of the statue of Ceres will be $b' - b$. To find b and b', we refer to Figure 8(b).

$$\tan 45° = \frac{b}{400} \qquad\qquad \tan 47.2° = \frac{b'}{400}$$

$$b = 400 \tan 45° = 400 \qquad\qquad b' = 400 \tan 47.2° \approx 432$$

The height of the statue is approximately $432 - 400 = 32$ feet.

NOW WORK PROBLEM 31.

EXAMPLE 7   The Gibb's Hill Lighthouse, Southampton, Bermuda

In operation since 1846, the Gibb's Hill Lighthouse stands 117 feet high on a hill 245 feet high, so its beam of light is 362 feet above sea level. A brochure states that the light can be seen on the horizon about 26 miles distant. Verify the accuracy of this statement.

Solution Figure 9 illustrates the situation. The central angle θ, positioned at the center of Earth, radius 3960 miles, obeys the equation

Figure 9

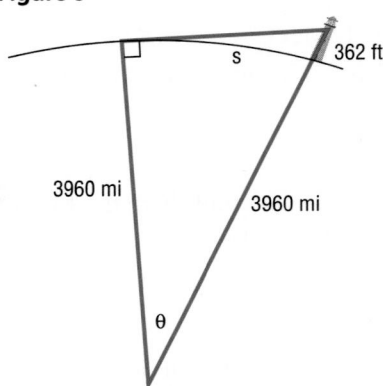

$$\cos\theta = \frac{3960}{3960 + \dfrac{362}{5280}} \approx 0.999982687 \qquad \text{1 mile = 5280 feet}$$

Solving for θ, we find

$$\theta \approx 0.33715° \approx 20.23'$$

The brochure does not indicate whether the distance is measured in nautical miles or statute miles. Let's calculate both distances.

The distance s in nautical miles (refer to Problem 104, p. 111) is the measure of angle θ in minutes, so $s = 20.23$ nautical miles.

The distance s in statute miles is given by the formula $s = r\theta$, where θ is measured in radians. Then, since

$$\theta = 20.23' = 0.33715° = 0.00588 \text{ radian}$$

$$\underset{1' = \frac{1}{60}°}{\uparrow} \qquad \underset{1° = \frac{\pi}{180} \text{radian}}{\uparrow}$$

we find that

$$s = r\theta = (3960)(0.00588) = 23.3 \text{ miles}$$

In either case, it would seem that the brochure overstated the distance somewhat. ∎

In navigation and surveying, the **direction** or **bearing** from a point O to a point P equals the acute angle θ between the ray OP and the vertical line through O, the north–south line.

Figure 10 illustrates some bearings. Notice that the bearing from O to P_1 is denoted by the symbolism N30°E, indicating that the bearing is 30° east of north. In writing the bearing from O to P, the direction north or south always appears first, followed by an acute angle, followed by east or west. In Figure 10, the bearing from O to P_2 is S50°W, and from O to P_3 it is N70°W.

Figure 10

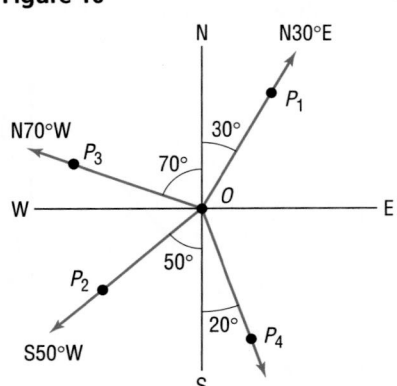

EXAMPLE 8 Finding the Bearing of an Object

In Figure 10, what is the bearing from O to an object at P_4?

Solution The acute angle between the ray OP_4 and the north–south line through O is given as 20°. The bearing from O to P_4 is S20°E. ∎

EXAMPLE 9 Finding the Bearing of an Airplane

A Boeing 777 aircraft takes off from O'Hare Airport on runway 2 LEFT which has a bearing of N20°E.* After flying for 1 mile, the pilot of the air craft requests permission to turn 90° and head toward the northwest. The request is granted. After the plane goes 2 miles in this direction, what bearing should the control tower use to locate the aircraft?

---

*In air navigation, the term **azimuth** is employed to denote the positive angle measured clockwise from the north (N) to a ray OP. In Figure 10, the azimuth from O to P_1 is 30°; the azimuth from O to P_2 is 230°; the azimuth from O to P_3 is 290°. In naming runways, the units digit is left off the azimuth. Runway 2 LEFT means the left runway with a direction of azimuth 20° (bearing N20°E). Runway 23 is the runway with azimuth 230° and bearing S50°W.

Figure 11

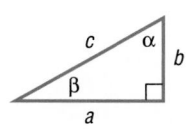

Solution Figure 11 illustrates the situation. After flying 1 mile from the airport O (the control tower), the aircraft is at P. After turning $90°$ toward the northwest and flying 2 miles, the aircraft is at the point Q. In triangle OPQ, the angle θ obeys the equation

$$\tan\theta = \frac{2}{1} = 2 \quad \text{so} \quad \theta = \tan^{-1}2 \approx 63.4°$$

The acute angle between the north-south line and the ray OQ is $63.4° - 20° = 43.4°$. The bearing of the aircraft from O to Q is N43.4°W. ■

NOW WORK PROBLEM 39.

4.1 Concepts and Vocabulary

In Problems 1 and 2, fill in the blanks.

1. When you look up at an object, the acute angle measured from the horizontal to a line-of-sight observation of the object is called the _____ _____ _____.

2. When you look down at an object, the acute angle described in Problem 1 is called the _____ _____ _____.

In Problems 3 and 4, answer True or False to each statement.

3. In a right triangle, if two sides are known, we can solve the triangle.

4. In a right triangle, if we know the two acute angles, we can solve the triangle.

5. Explain how you would measure the width of the Grand Canyon from a point on its ridge.

6. Explain how you would measure the height of a TV tower that is on the roof of a tall building.

4.1 Exercises

In Problems 1–14, use the right triangle shown in the margin. Then, using the given information, solve the triangle.

1. $b = 5$, $\beta = 20°$; find a, c, and α
2. $b = 4$, $\beta = 10°$; find a, c, and α
3. $a = 6$, $\beta = 40°$; find b, c, and α
4. $a = 7$, $\beta = 50°$; find b, c, and α
5. $b = 4$, $\alpha = 10°$; find a, c, and β
6. $b = 6$, $\alpha = 20°$; find a, c, and β
7. $a = 5$, $\alpha = 25°$; find b, c, and β
8. $a = 6$, $\alpha = 40°$; find b, c, and β
9. $c = 9$, $\beta = 20°$; find b, a, and α
10. $c = 10$, $\alpha = 40°$; find b, a, and β
11. $a = 5$, $b = 3$; find c, α, and β
12. $a = 2$, $b = 8$; find c, α, and β
13. $a = 2$, $c = 5$; find b, α, and β
14. $b = 4$, $c = 6$; find a, α, and β

15. **Geometry** A right triangle has a hypotenuse of length 8 inches. If one angle is 35°, find the length of each leg.

16. **Geometry** A right triangle has a hypotenuse of length 10 centimeters. If one angle is 40°, find the length of each leg.

17. **Geometry** A right triangle contains a 25° angle. If one leg is of length 5 inches, what is the length of the hypotenuse?

[**Hint:** Two answers are possible.]

18. Geometry   A right triangle contains an angle of $\pi/8$ radian. If one leg is of length 3 meters, what is the length of the hypotenuse?

[**Hint:** Two answers are possible.]

19. Geometry   The hypotenuse of a right triangle is 5 inches. If one leg is 2 inches, find the degree measure of each angle.

20. Geometry   The hypotenuse of a right triangle is 3 feet. If one leg is 1 foot, find the degree measure of each angle.

21. Finding the Width of a Gorge   Find the distance from A to C across the gorge illustrated in the figure.

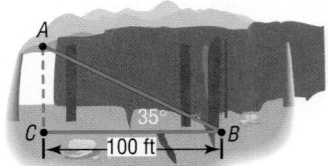

22. Finding the Distance across a Pond   Find the distance from A to C across the pond illustrated in the figure.

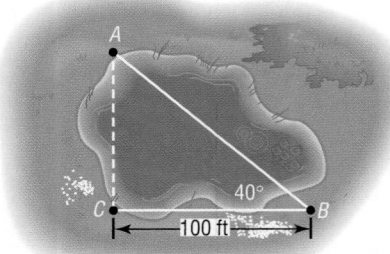

23. The Eiffel Tower   The tallest tower built before the era of television masts, the Eiffel Tower was completed on March 31, 1889. Find the height of the Eiffel Tower (before a television mast was added to the top) using the information given in the illustration.

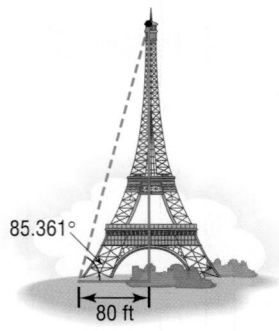

24. Finding the Distance of a Ship from Shore   A ship, offshore from a vertical cliff known to be 100 feet in height, takes a sighting of the top of the cliff. If the angle of elevation is found to be 25°, how far offshore is the ship?

25. Finding the Distance to a Plateau   Suppose that you are headed toward a plateau 50 meters high. If the angle of elevation to the top of the plateau is 20°, how far are you from the base of the plateau?

26. Statue of Liberty   A ship is just offshore of New York City. A sighting is taken of the Statue of Liberty, which is about 305 feet tall. If the angle of elevation to the top of the statue is 20°, how far is the ship from the base of the statue?

27. Finding the Reach of a Ladder   A 22-foot extension ladder leaning against a building makes a 70° angle with the ground. How far up the building does the ladder touch?

28. Finding the Height of a Building   To measure the height of a building, two sightings are taken a distance of 50 feet apart. If the first angle of elevation is 40° and the second is 32°, what is the height of the building?

29. Directing a Laser Beam   A laser beam is to be directed through a small hole in the center of a circle of radius 10 feet. The origin of the beam is 35 feet from the circle (see the figure). At what angle of elevation should the beam be aimed to ensure that it goes through the hole?

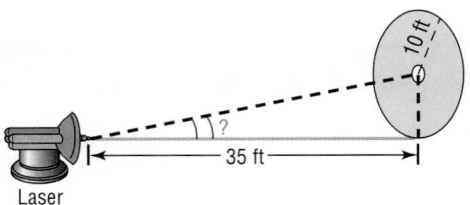

30. Finding the Angle of Elevation of the Sun   At 10 AM on April 26, 2000, a building 300 feet high casts a shadow 50 feet long. What is the angle of elevation of the Sun?

31. Mt. Rushmore   To measure the height of Lincoln's caricature on Mt. Rushmore, two sightings 800 feet from the base of the mountain are taken. If the angle of elevation to the bottom of Lincoln's face is 32° and the angle of elevation to the top is 35°, what is the height of Lincoln's face?

32. Finding the Distance between Two Objects   A blimp, suspended in the air at a height of 500 feet, lies directly over a line from Soldier Field to the Adler Planetarium on Lake Michigan (see the figure). If the angle of depression from the blimp to the stadium is 32° and from the blimp to the planetarium is 23°, find the distance between Soldier Field and the Adler Planetarium.

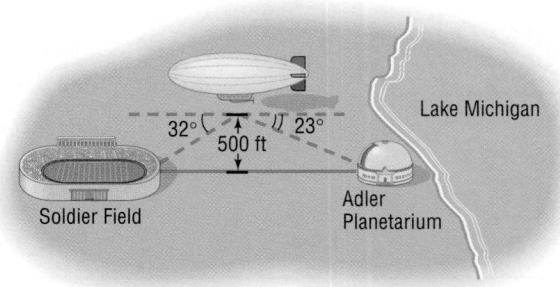

33. Finding the Length of a Guy Wire   A radio transmission tower is 200 feet high. How long should a guy wire be if it is to be attached to the tower 10 feet from the top and is to make an angle of 21° with the ground?

34. Finding the Height of a Tower A guy wire 80 feet long is attached to the top of a radio transmission tower, making an angle of 25° with the ground. How high is the tower?

35. Washington Monument The angle of elevation to the top of the Washington Monument is 35.1° at the instant it casts a shadow 789 feet long. Use this information to calculate the height of the monument.

36. Finding the Length of a Mountain Trail A straight trail with an inclination of 17° leads from a hotel at an elevation of 9000 feet to a mountain lake at an elevation of 11,200 feet. What is the length of the trail?

37. Finding the Speed of a Truck A state trooper is hidden 30 feet from a highway. One second after a truck passes, the angle θ between the highway and the line of observation from the patrol car to the truck is measured. See the illustration.

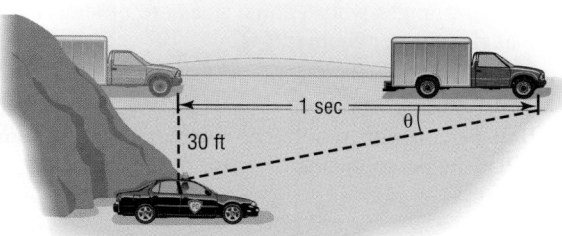

(a) If the angle measures 15°, how fast is the truck traveling? Express the answer in feet per second and in miles per hour.
(b) If the angle measures 20°, how fast is the truck traveling? Express the answer in feet per second and in miles per hour.
(c) If the speed limit is 55 miles per hour and a speeding ticket is issued for speeds of 5 miles per hour or more over the limit, for what angles should the trooper issue a ticket?

38. Security A security camera in a neighborhood bank is mounted on a wall 9 feet above the floor. What angle of depression should be used if the camera is to be directed to a spot 6 feet above the floor and 12 feet from the wall?

39. Finding the Bearing of an Aircraft A DC-9 aircraft leaves Midway Airport from runway 4 RIGHT, whose bearing is N40°E. After flying for 1/2 mile, the pilot requests permission to turn 90° and head toward the southeast. The permission is granted. After the airplane goes 1 mile in this direction, what bearing should the control tower use to locate the aircraft?

40. Finding the Bearing of a Ship A ship leaves the port of Miami with a bearing of S80°E and a speed of 15 knots. After 1 hour, the ship turns 90° toward the south. After 2 hours, maintaining the same speed, what is the bearing to the ship from port?

41. Shooting Free Throws in Basketball The eyes of a basketball player are 6 feet above the floor. The player is at the free-throw line, which is 15 feet from the center of the basket rim (see the figure). What is the angle of elevation from the player's eyes to the center of the rim? [**Hint:** The rim is 10 feet above the floor.]

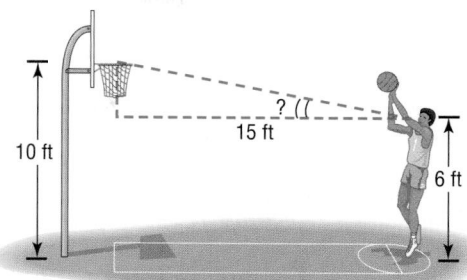

42. Finding the Pitch of a Roof A carpenter is preparing to put a roof on a garage that is 20 feet by 40 feet by 20 feet. A steel support beam 46 feet in length is positioned in the center of the garage. To support the roof, another beam will be attached to the top of the center beam (see the figure). At what angle of elevation is the new beam? In other words, what is the pitch of the roof?

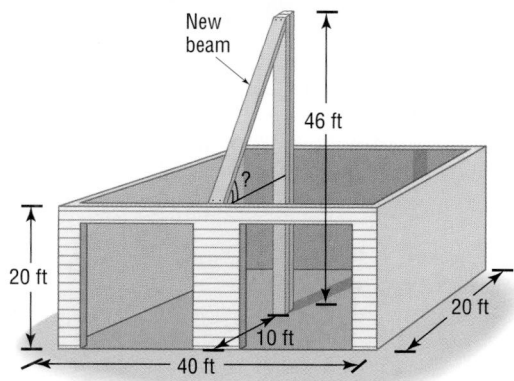

43. Constructing a Highway A highway whose primary directions are north–south is being constructed along the west coast of Florida. Near Naples, a bay obstructs the straight path of the road. Since the cost of a bridge is prohibitive, engineers decide to go around the bay. The illustration shows the path that they decide on and the measurements taken. What is the length of highway needed to go around the bay?

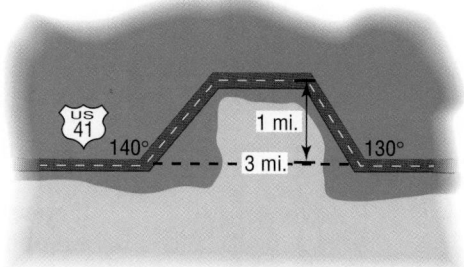

44. Surveillance Satellites   A surveillance satellite circles Earth at a height of h miles above the surface. Suppose that d is the distance, in miles, on the surface of Earth that can be observed from the satellite. See the illustration.
   (a) Find an equation that relates the central angle θ to the height h.
   (b) Find an equation that relates the observable distance d and θ.
   (c) Find an equation that relates d and h.
   (d) If d is to be 2500 miles, how high must the satellite orbit above Earth?
   (e) If the satellite orbits at a height of 300 miles, what distance d on the surface can be observed?

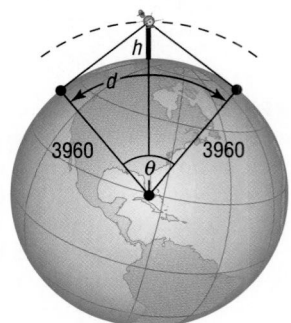

45. Photography   A camera is mounted on a tripod 4 feet high at a distance of 10 feet from George, who is 6 feet tall. See the illustration. If the camera lens has angles of depression and elevation of 20°, will George's feet and head be seen by the lens? If not, how far back will the camera need to be moved to include George's feet and head?

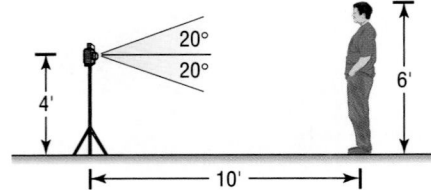

46. Construction   A ramp for wheelchair accessibility is to be constructed with an angle of elevation of 15° and a final height of 5 feet. How long is the ramp?

47. Geometry   A rectangle is inscribed in a semicircle of radius 1. See the illustration.

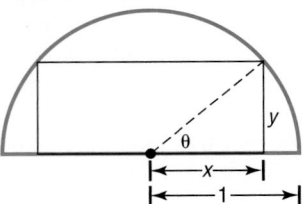

   (a) Express the area A of the rectangle as a function of the angle θ shown in the illustration.
   (b) Show that $A = \sin(2\theta)$.
   (c) Find the angle θ that results in the largest area A.
   (d) Find the dimensions of this largest rectangle.

48. Area of an Isosceles Triangle   Show that the area A of an isosceles triangle, whose equal sides are of length s and the angle between them is θ, is

$$A = \frac{1}{2}s^2 \sin \theta$$

[**Hint:** See the illustration. The height h bisects the angle θ and is the perpendicular bisector of the base.]

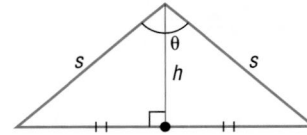

49. The Gibb's Hill Lighthouse, Southampton, Bermuda   In operation since 1846, the Gibb's Hill Lighthouse stands 117 feet high on a hill 245 feet high, so its beam of light is 362 feet above sea level. A brochure states that ships 40 miles away can see the light and planes flying at 10,000 feet can see it 120 miles away. Verify the accuracy of these statements. What assumption did the brochure make about the height of the ship?

PREPARING FOR THIS SECTION

Before getting started, review the following:

✓ Trigonometric Equations (I) (Section 3.7, pp. 247–250)   ✓ Difference Formulas for Sine (Section 3.4, p. 225)

4.2   THE LAW OF SINES

OBJECTIVES   ① Solve SAA or ASA Triangles
   ② Solve SSA Triangles
   ③ Solve Applied Problems Using the Law of Sines

If none of the angles of a triangle is a right angle, the triangle is called **oblique**. An oblique triangle will have either three acute angles or two acute angles and one obtuse angle (an angle between 90° and 180°). See Figure 12.

Figure 12

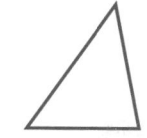 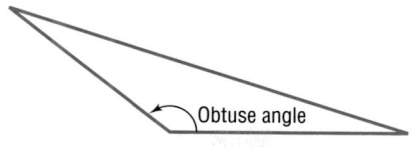

(a) All angles are acute   **(b)** Two acute angles and one obtuse angle

Figure 13

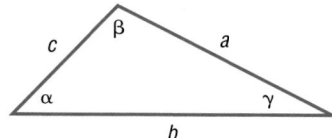

In the discussion that follows, we will always label an oblique triangle so that side a is opposite angle α, side b is opposite angle β, and side c is opposite angle γ, as shown in Figure 13.

To **solve an oblique triangle** means to find the lengths of its sides and the measurements of its angles. To do this, we need to know the length of one side along with: (1) two angles or (2) the other two sides or (3) one angle and one other side. Knowing three angles of a triangle determines a family of *similar triangles,* that is, triangles that have the same shape but different sizes. There are four possibilities to consider:

> **CASE 1:** One side and two angles are known (ASA or SAA).
> **CASE 2:** Two sides and the angle opposite one of them are known (SSA).
> **CASE 3:** Two sides and the included angle are known (SAS).
> **CASE 4:** Three sides are known (SSS).

Figure 14 illustrates the four cases.

Figure 14

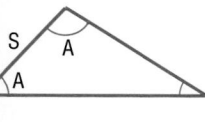 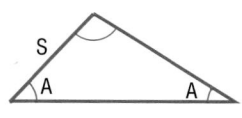 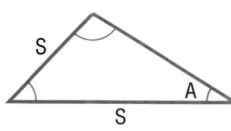 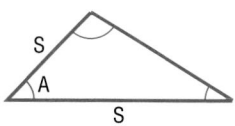 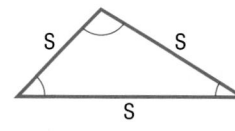

Case 1: ASA   Case 1: SAA   Case 2: SSA   Case 3: SAS   Case 4: SSS

The **Law of Sines** is used to solve triangles for which Case 1 or 2 holds. Cases 3 and 4 are considered when we study the Law of Cosines in Section 4.3.

Theorem

Law of Sines

For a triangle with sides a, b, c and opposite angles α, β, γ, respectively,

$$\frac{\sin \alpha}{a} = \frac{\sin \beta}{b} = \frac{\sin \gamma}{c} \tag{1}$$

Proof   To prove the Law of Sines, we construct an altitude of length h from one of the vertices of a triangle. Figure 15(a) shows h for a triangle with three acute angles, and Figure 15(b) shows h for a triangle with an obtuse angle. In each case, the altitude is drawn from the vertex at β. Using either illustration, we have

$$\sin \gamma = \frac{h}{a}$$

Figure 15

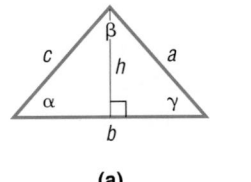

(a)

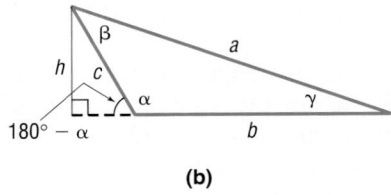

(b)

from which

$$h = a \sin \gamma \qquad (2)$$

From Figure 15(a), it also follows that

$$\sin \alpha = \frac{h}{c}$$

from which

$$h = c \sin \alpha \qquad (3)$$

From Figure 15(b), it follows that

$$\sin(180° - \alpha) = \sin \alpha = \frac{h}{c}$$

Difference formula

which again gives

$$h = c \sin \alpha$$

So, whether the triangle has three acute angles or has two acute angles and one obtuse angle, equations (2) and (3) hold. As a result, we may equate the expressions for h in equations (2) and (3) to get

$$a \sin \gamma = c \sin \alpha$$

from which

$$\frac{\sin \alpha}{a} = \frac{\sin \gamma}{c} \qquad (4)$$

In a similar manner, by constructing the altitude h' from the vertex of angle α as shown in Figure 16, we can show that

$$\sin \beta = \frac{h'}{c} \quad \text{and} \quad \sin \gamma = \frac{h'}{b}$$

Figure 16

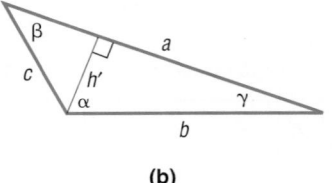

(a)

(b)

Equating the expressions for h', we find that

$$h' = c \sin \beta = b \sin \gamma$$

from which

$$\frac{\sin \beta}{b} = \frac{\sin \gamma}{c} \qquad (5)$$

When equations (4) and (5) are combined, we have equation (1), the Law of Sines. ∎

In applying the Law of Sines to solve triangles, we use the fact that the sum of the angles of any triangle equals 180°; that is,

$$\boxed{\alpha + \beta + \gamma = 180° \qquad (6)}$$

1   Our first two examples show how to solve a triangle when one side and two angles are known (Case 1: SAA or ASA).

EXAMPLE 1   **Using the Law of Sines to Solve a SAA Triangle**

Solve the triangle: $\alpha = 40°, \beta = 60°, a = 4$

Solution   Figure 17 shows the triangle that we want to solve. The third angle γ is found using equation (6).

Figure 17

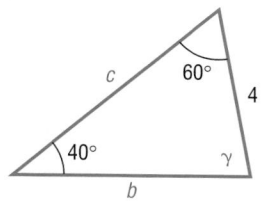

$$\alpha + \beta + \gamma = 180°$$
$$40° + 60° + \gamma = 180°$$
$$\gamma = 80°$$

Now we use the Law of Sines (twice) to find the unknown sides b and c.

$$\frac{\sin \alpha}{a} = \frac{\sin \beta}{b} \qquad \frac{\sin \alpha}{a} = \frac{\sin \gamma}{c}$$

Because $a = 4, \alpha = 40°, \beta = 60°$, and $\gamma = 80°$, we have

$$\frac{\sin 40°}{4} = \frac{\sin 60°}{b} \qquad \frac{\sin 40°}{4} = \frac{\sin 80°}{c}$$

Solving for b and c, we find that

$$b = \frac{4 \sin 60°}{\sin 40°} \approx 5.39 \qquad c = \frac{4 \sin 80°}{\sin 40°} \approx 6.13$$

Notice in Example 1 that we found b and c by working with the given side a. This is better than finding b first and working with a rounded value of b to find c.

━ NOW WORK PROBLEM **1.**

EXAMPLE 2   **Using the Law of Sines to Solve an ASA Triangle**

Solve the triangle: $\alpha = 35°, \beta = 15°, c = 5$

Solution   Figure 18 illustrates the triangle that we want to solve. Because we know two angles ($\alpha = 35°$ and $\beta = 15°$), we find the third angle using equation (6).

Figure 18

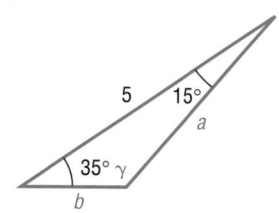

$$\alpha + \beta + \gamma = 180°$$
$$35° + 15° + \gamma = 180°$$
$$\gamma = 130°$$

Now we know the three angles and one side ($c = 5$) of the triangle. To find the remaining two sides a and b, we use the Law of Sines (twice).

$$\frac{\sin \alpha}{a} = \frac{\sin \gamma}{c} \qquad \frac{\sin \beta}{b} = \frac{\sin \gamma}{c}$$

$$\frac{\sin 35°}{a} = \frac{\sin 130°}{5} \qquad \frac{\sin 15°}{b} = \frac{\sin 130°}{5}$$

$$a = \frac{5 \sin 35°}{\sin 130°} \approx 3.74 \qquad b = \frac{5 \sin 15°}{\sin 130°} \approx 1.69$$

━ NOW WORK PROBLEM **15.**

The Ambiguous Case

2 Case 2 (SSA), which applies to triangles for which two sides and the angle opposite one of them are known, is referred to as the **ambiguous case**, because the known information may result in one triangle, two triangles, or no triangle at all. Suppose that we are given sides a and b and angle α, as illustrated in Figure 19. The key to determining the possible triangles, if any, that may be formed from the given information lies primarily with the height h and the fact that $h = b \sin \alpha$.

Figure 19

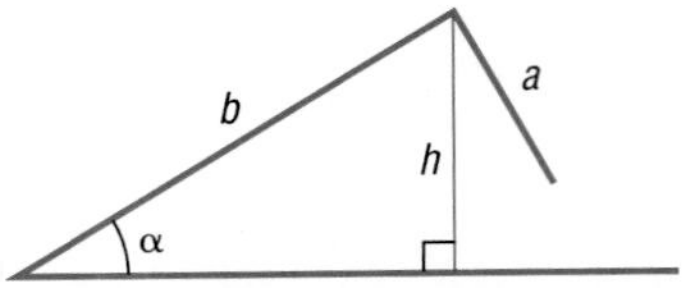

No Triangle If $a < h = b \sin \alpha$, then side a is not sufficiently long to form a triangle. See Figure 20.

One Right Triangle If $a = h = b \sin \alpha$, then side a is just long enough to form a right triangle. See Figure 21.

Figure 20
$a < h = b \sin \alpha$

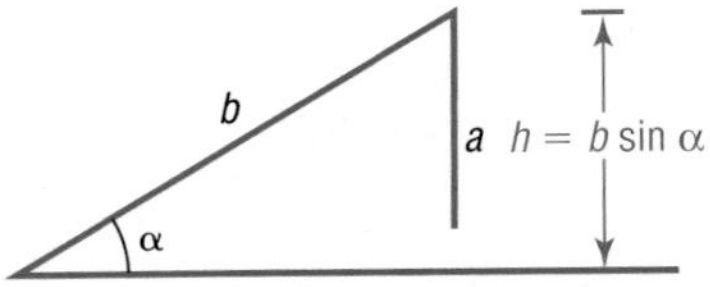

Figure 21
$a = b \sin \alpha$

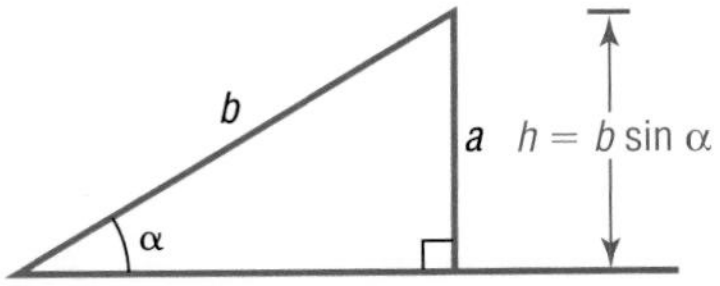

Two Triangles If $a < b$ and $h = b \sin \alpha < a$, then two distinct triangles can be formed from the given information. See Figure 22.

One Triangle If $a \geq b$, then only one triangle can be formed. See Figure 23.

Figure 22
$b \sin \alpha < a$ and $a < b$

Figure 23
$a \geq b$

Fortunately, we do not have to rely on an illustration to draw the correct conclusion in the ambiguous case. The Law of Sines will lead us to the correct determination. Let's see how.

EXAMPLE 3 **Using the Law of Sines to Solve a SSA Triangle (One Solution)**

Solve the triangle: $a = 3, b = 2, \alpha = 40°$

Solution See Figure 24(a). Because $a = 3, b = 2$, and $\alpha = 40°$ are known, we use the Law of Sines to find the angle β.

Figure 24(a)

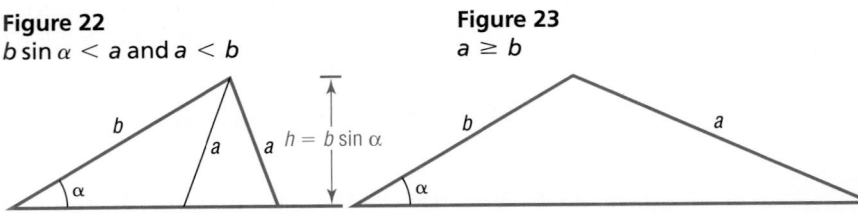

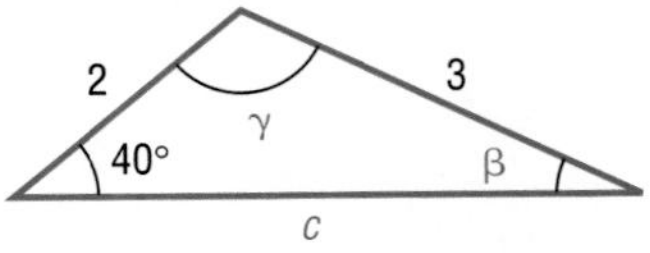

$$\frac{\sin \alpha}{a} = \frac{\sin \beta}{b}$$

Then

$$\frac{\sin 40°}{3} = \frac{\sin \beta}{2}$$

$$\sin \beta = \frac{2 \sin 40°}{3} \approx 0.43$$

There are two angles $\beta, 0° < \beta < 180°$, for which $\sin \beta \approx 0.43$.

$$\beta_1 \approx 25.4° \quad \text{and} \quad \beta_2 \approx 154.6°$$

Note: Here we computed β using the stored value of $\sin \beta$. If you use the rounded value, $\sin \beta \approx 0.43$, you will obtain slightly different results.

The second possibility, $\beta_2 \approx 154.6°$, is ruled out, because $\alpha = 40°$, making $\alpha + \beta_2 \approx 194.6° > 180°$. Now, using $\beta_1 \approx 25.4°$, we find that

$$\gamma = 180° - \alpha - \beta_1 \approx 180° - 40° - 25.4° = 114.6°$$

The third side c may now be determined using the Law of Sines.

$$\frac{\sin \alpha}{a} = \frac{\sin \gamma}{c}$$

$$\frac{\sin 40°}{3} = \frac{\sin 114.6°}{c}$$

$$c = \frac{3 \sin 114.6°}{\sin 40°} \approx 4.24$$

Figure 24(b)

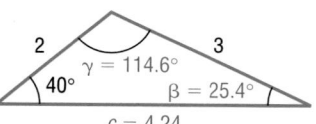

Figure 24(b) illustrates the solved triangle.   ■

EXAMPLE 4   Using the Law of Sines to Solve a SSA Triangle (Two Solutions)

Solve the triangle: $a = 6, b = 8, \alpha = 35°$

Solution See Figure 25(a). Because $a = 6$, $b = 8$, and $\alpha = 35°$ are known, we use the Law of Sines to find the angle β.

Figure 25(a)

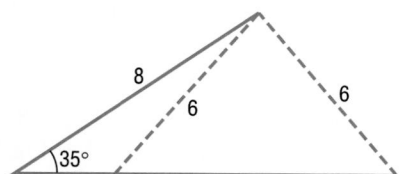

$$\frac{\sin \alpha}{a} = \frac{\sin \beta}{b}$$

Then

$$\frac{\sin 35°}{6} = \frac{\sin \beta}{8}$$

$$\sin \beta = \frac{8 \sin 35°}{6} \approx 0.76$$

$$\beta_1 \approx 49.9° \quad \text{or} \quad \beta_2 \approx 130.1°$$

For both choices of β, we have $\alpha + \beta < 180°$. There are two triangles one containing the angle $\beta_1 \approx 49.9°$ and the other containing the angle $\beta_2 \approx 130.1°$. The third angle γ is either

$$\gamma_1 = 180° - \alpha - \beta_1 \approx 95.1° \quad \text{or} \quad \gamma_2 = 180° - \alpha - \beta_2 \approx 14.9°$$

$$\uparrow \qquad\qquad\qquad\qquad\qquad\qquad \uparrow$$

$$\alpha = 35° \qquad\qquad\qquad\qquad\qquad \alpha = 35°$$

$$\beta_1 = 49.9° \qquad\qquad\qquad\qquad \beta_2 = 130.1°$$

Figure 25(b)

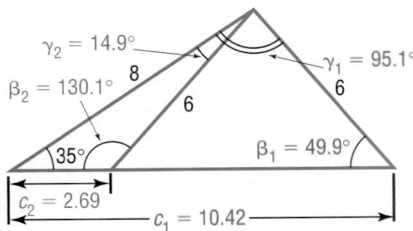

The third side c obeys the Law of Sines, so we have

$$\frac{\sin \alpha}{a} = \frac{\sin \gamma_1}{c_1} \qquad\qquad \frac{\sin \alpha}{a} = \frac{\sin \gamma_2}{c_2}$$

$$\frac{\sin 35°}{6} = \frac{\sin 95.1°}{c_1} \qquad\qquad \frac{\sin 35°}{6} = \frac{\sin 14.9°}{c_2}$$

$$c_1 = \frac{6 \sin 95.1°}{\sin 35°} \approx 10.42 \qquad\qquad c_2 = \frac{6 \sin 14.9°}{\sin 35°} \approx 2.69$$

The two solved triangles are illustrated in Figure 25(b).   ◼

EXAMPLE 5   **Using the Law of Sines to Solve a SSA Triangle (No Solution)**

Solve the triangle: $a = 2, c = 1, \gamma = 50°$

Solution   Because $a = 2, c = 1$, and $\gamma = 50°$ are known, we use the Law of Sines to find the angle α.

$$\frac{\sin \alpha}{a} = \frac{\sin \gamma}{c}$$

$$\frac{\sin \alpha}{2} = \frac{\sin 50°}{1}$$

$$\sin \alpha = 2 \sin 50° \approx 1.53$$

Figure 26

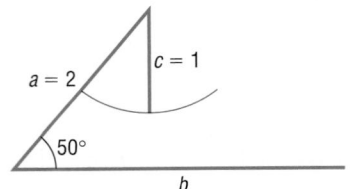

Since there is no angle α for which $\sin \alpha > 1$, there can be no triangle with the given measurements. Figure 26 illustrates the measurements given. Notice that, no matter how we attempt to position side c, it will never touch side b to form a triangle.   ◼

✎→ NOW WORK PROBLEMS **17** AND **23**.

Applications

3   The Law of Sines is particularly useful for solving certain applied problems.

EXAMPLE 6   **Finding the Height of a Mountain**

To measure the height of a mountain, a surveyor takes two sightings of the peak at a distance 900 meters apart on a direct line to the mountain.* See Figure 27(a). The first observation results in an angle of elevation of 47°, whereas the second results in an angle of elevation of 35°. If the transit is 2 meters high, what is the height h of the mountain?

Figure 27

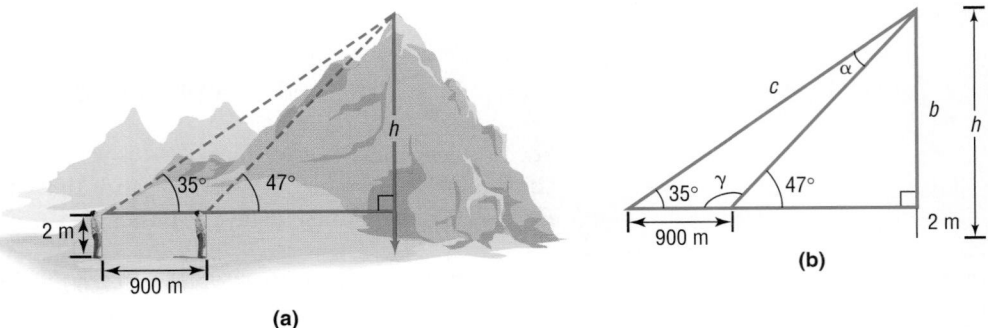

(a)

*For simplicity, we assume that these sightings are at the same level.

Solution   Figure 27(b) shows the triangles that replicate the illustration in Figure 27(a). Since $\gamma + 47° = 180°$, we find that $\gamma = 133°$. Also, since $\alpha + \beta + \gamma = 180°$, we find that $\alpha = 180° - 35° - 133° = 12°$. We use the Law of Sines to find c.

$$\frac{\sin \alpha}{a} = \frac{\sin \gamma}{c}$$

$\alpha = 12°, \gamma = 133°, a = 900$

$$c = \frac{900 \sin 133°}{\sin 12°} = 3165.86$$

Using the larger right triangle, we have

$$\sin 35° = \frac{b}{c} \qquad c = 3165.86$$

$$b = 3165.86 \sin 35° = 1815.86 \approx 1816 \text{ meters}$$

The height of the peak from ground level is approximately $1816 + 2 = 1818$ meters. ■

▬▬▬ **NOW WORK PROBLEM 31.**

EXAMPLE 7   **Rescue at Sea**

Coast Guard Station Zulu is located 120 miles due west of Station X-ray. A ship at sea sends an SOS call that is received by each station. The call to Station Zulu indicates that the bearing of the ship from Zulu is N40°E (40° east of north). The call to Station X-ray indicates that the bearing of the ship from X-ray is N30°W (30° west of north).

(a) How far is each station from the ship?
(b) If a helicopter capable of flying 200 miles per hour is dispatched from the nearest station to the ship, how long will it take to reach the ship?

Solution   (a) Figure 28 illustrates the situation. The angle γ is found to be

$$\gamma = 180° - 50° - 60° = 70°$$

Figure 28

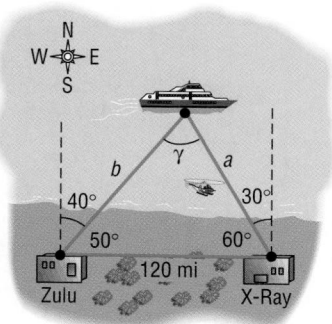

The Law of Sines can now be used to find the two distances a and b that we seek.

$$\frac{\sin 50°}{a} = \frac{\sin 70°}{120}$$

$$a = \frac{120 \sin 50°}{\sin 70°} \approx 97.82 \text{ miles}$$

$$\frac{\sin 60°}{b} = \frac{\sin 70°}{120}$$

$$b = \frac{120 \sin 60°}{\sin 70°} \approx 110.59 \text{ miles}$$

Station Zulu is about 111 miles from the ship, and Station X-ray is about 98 miles from the ship.

(b) The time t needed for the helicopter to reach the ship from Station X-ray is found by using the formula

$$(\text{Velocity}, v)(\text{Time}, t) = \text{Distance}, a$$

Then

$$t = \frac{a}{v} = \frac{97.82}{200} \approx 0.49 \text{ hour} \approx 29 \text{ minutes}$$

It will take about 29 minutes for the helicopter to reach the ship. ∎

NOW WORK PROBLEM **29.**

4.2 Concepts and Vocabulary

In Problems 1 and 2, fill in the blanks.

1. If none of the angles of a triangle is a right angle, the triangle is called _____.

2. For a triangle with sides a, b, c and opposite angles α, β, γ, the Law of Sines states that _____.

In Problems 3–5, answer True or False to each statement.

3. An oblique triangle in which two sides and an angle are given always results in at least one triangle.

4. The sum of the angles of any triangle equals $180°$.

5. The ambiguous case refers to the fact that, when two sides and the angle opposite one of them is given, sometimes the Law of Sines cannot be used.

6. What do you do first if you are asked to solve a triangle and are given one side and two angles?

7. What do you do first if you are asked to solve a triangle and are given two sides and the angle opposite one of them?

4.2 Exercises

In Problems 1–8 solve each triangle.

1.
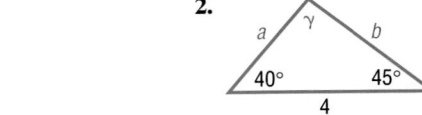

2.

3.
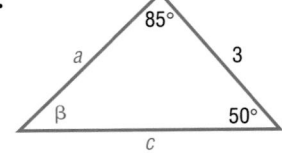

4.

5.

6.

7.
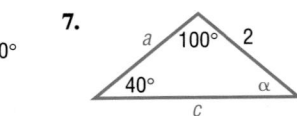

8.
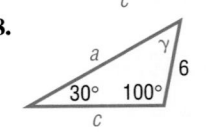

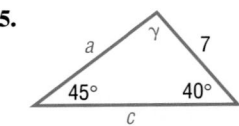

In Problems 9–14 solve each triangle.

9. $\alpha = 40°$, $\beta = 20°$, $a = 2$

10. $\alpha = 50°$, $\gamma = 20°$, $a = 3$

11. $\beta = 70°$, $\gamma = 10°$, $b = 5$

12. $\alpha = 70°$, $\beta = 60°$, $c = 4$

13. $\alpha = 110°$, $\gamma = 30°$, $c = 3$

14. $\beta = 10°$, $\gamma = 100°$, $b = 2$

15. $\alpha = 40°$, $\beta = 40°$, $c = 2$

16. $\beta = 20°$, $\gamma = 70°$, $a = 1$

In Problems 17–28, two sides and an angle are given. Determine whether the given information results in one triangle, two triangles, or no triangle at all. Solve any resulting triangle(s).

17. $a = 3$, $b = 2$, $\alpha = 50°$

18. $b = 4$, $c = 3$, $\beta = 40°$

19. $b = 5$, $c = 3$, $\beta = 100°$

20. $a = 2$, $c = 1$, $\alpha = 120°$

21. $a = 4$, $b = 5$, $\alpha = 60°$

22. $b = 2$, $c = 3$, $\beta = 40°$

23. $b = 4$, $c = 6$, $\beta = 20°$

24. $a = 3$, $b = 7$, $\alpha = 70°$

25. $a = 2$, $c = 1$, $\gamma = 100°$

26. $b = 4$, $c = 5$, $\beta = 95°$

27. $a = 2$, $c = 1$, $\gamma = 25°$

28. $b = 4$, $c = 5$, $\beta = 40°$

29. Rescue at Sea   Coast Guard Station Able is located 150 miles due south of Station Baker. A ship at sea sends an SOS call that is received by each station. The call to Station Able indicates that the ship is located N55°E; the call to Station Baker indicates that the ship is located S60°E.
(a) How far is each station from the ship?
(b) If a helicopter capable of flying 200 miles per hour is dispatched from the nearest station to the ship, how long will it take to reach the ship?

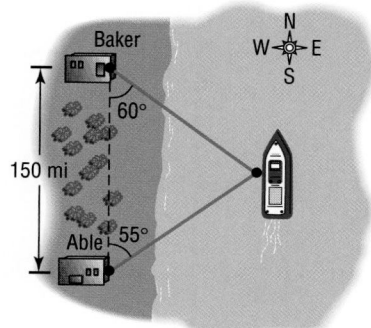

30. Surveying   Consult the figure. To find the distance from the house at A to the house at B, a surveyor measures the angle BAC to be 40° and then walks off a distance of 100 feet to C and measures the angle ACB to be 50°. What is the distance from A to B?

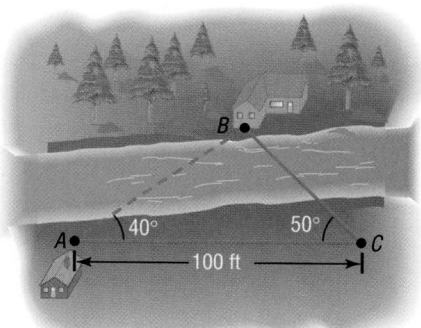

31. Finding the Length of a Ski Lift   Consult the figure. To find the length of the span of a proposed ski lift from A to B, a surveyor measures the angle DAB to be 25° and then walks off a distance of 1000 feet to C and measures the angle ACB to be 15°. What is the distance from A to B?

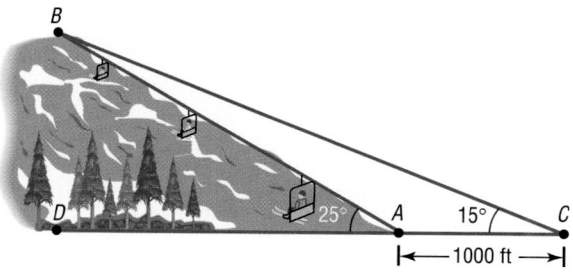

32. Finding the Height of a Mountain   Use the illustration in Problem 31 to find the height BD of the mountain at B.

33. Finding the Height of an Airplane   An aircraft is spotted by two observers who are 1000 feet apart. As the airplane passes over the line joining them, each observer takes a sighting of the angle of elevation to the plane, as indicated in the figure. How high is the airplane?

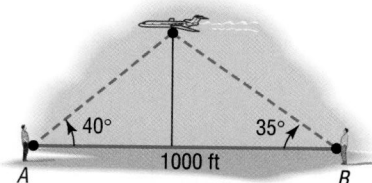

34. Finding the Height of the Bridge over the Royal Gorge   The highest bridge in the world is the bridge over the Royal Gorge of the Arkansas River in Colorado.* Sightings to the same point at water level directly under the bridge are taken from each side of the 880-foot-long bridge, as indicated in the figure. How high is the bridge?

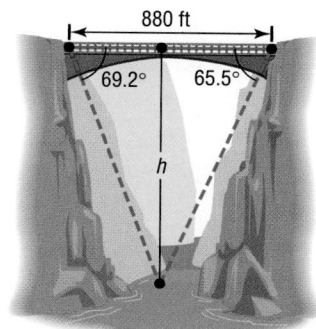

35. Navigation   An airplane flies from city A to city B, a distance of 150 miles, and then turns through an angle of 40° and heads toward city C, as shown in the figure.
(a) If the distance between cities A and C is 300 miles, how far is it from city B to city C?
(b) Through what angle should the pilot turn at city C to return to city A?

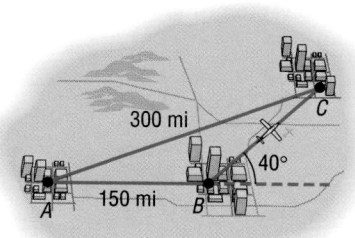

Source: Guinness Book of World Records.

36. Time Lost due to a Navigation Error In attempting to fly from city A to city B, an aircraft followed a course that was $10°$ in error, as indicated in the figure. After flying a distance of 50 miles, the pilot corrected the course by turning at point C and flying 70 miles farther. If the constant speed of the aircraft was 250 miles per hour, how much time was lost due to the error?

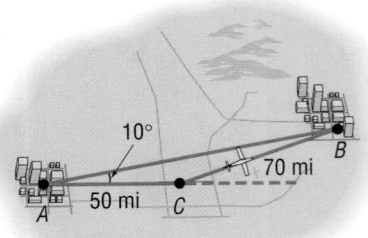

37. Finding the Lean of the Leaning Tower of Pisa The famous Leaning Tower of Pisa was originally 184.5 feet high. At a distance of 123 feet from the base of the tower, the angle of elevation to the top of the tower is found to be $60°$. Find the angle CAB indicated in the figure. Also, find the perpendicular distance from C to AB.

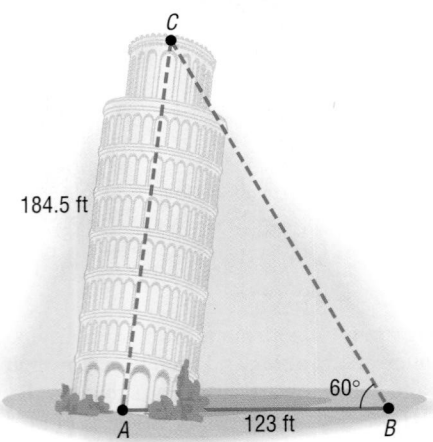

38. Crankshafts on Cars On a certain automobile, the crankshaft is 3 inches long and the connecting rod is 9 inches long (see the figure). At the time when the angle OPA is $15°$, how far is the piston (P) from the center (O) of the crankshaft?

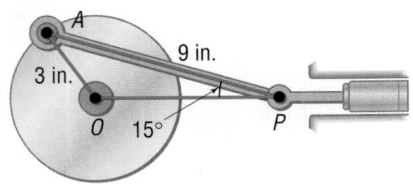

39. Constructing a Highway U.S. 41, a highway whose primary directions are north–south, is being constructed along the west coast of Florida. Near Naples, a bay obstructs the straight path of the road. Since the cost of a bridge is prohibitive, engineers decide to go around the bay. The illustration shows the path that they decide on and the measurements taken. What is the length of highway needed to go around the bay?

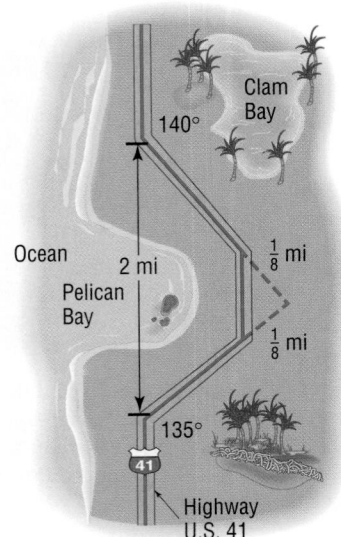

40. Determining Distances at Sea The navigator of a ship at sea spots two lighthouses that she knows to be 3 miles apart along a straight seashore. She determines that the angles formed between two line-of-sight observations of the lighthouses and the line from the ship directly to shore are $15°$ and $35°$. See the illustration.
(a) How far is the ship from lighthouse A?
(b) How far is the ship from lighthouse B?
(c) How far is the ship from shore?

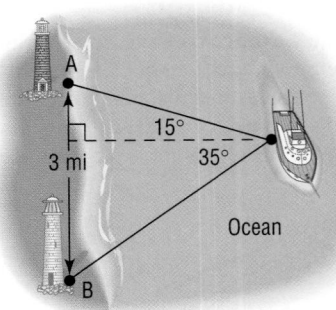

41. Calculating Distance at Sea The navigator of a ship at sea has the harbor in sight at which the ship is to dock. She spots a lighthouse that she knows is 1 mile up the coast from the mouth of the harbor, and she measures the angle between the line-of-sight observations of the harbor and lighthouse to be 20°. With the ship heading directly toward the harbor, she repeats this measurement after 5 minutes of traveling at 12 miles per hour. If the new angle is 30°, how far is the ship from the harbor?

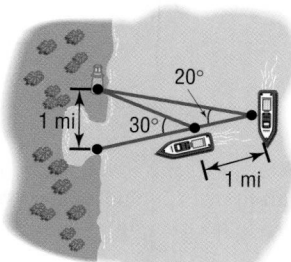

42. Finding Distances A forest ranger is walking on a path inclined at 5° to the horizontal directly toward a 100-foot-tall fire observation tower. The angle of elevation from the path to the top of the tower is 40°. How far is the ranger from the tower at this time?

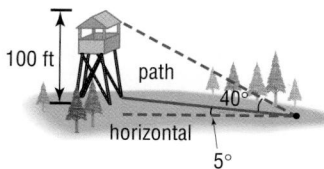

43. Great Pyramid of Cheops One of the original Seven Wonders of the World, the Great Pyramid of Cheops was built about 2580 BC. Its original height was 480 feet 11 inches, but due to the loss of its topmost stones, it is now shorter.* Find the current height of the Great Pyramid, using the information given in the illustration.

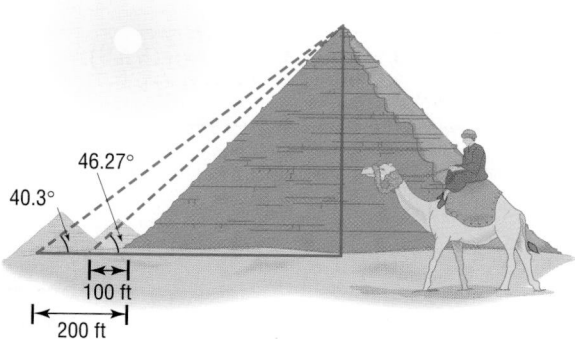

44. Determining the Height of an Aircraft Two sensors are spaced 700 feet apart along the approach to a small airport. When an aircraft is nearing the airport, the angle of elevation from the first sensor to the aircraft is 20°, and from the second sensor to the aircraft it is 15°. Determine how high the aircraft is at this time.

Source: Guinness Book of World Records.

45. Landscaping Pat needs to determine the height of a tree before cutting it down to be sure that it will not fall on a nearby fence. The angle of elevation of the tree from one position on a flat path from the tree is 30°, and from a second position 40 feet farther along this path it is 20°. What is the height of the tree?

46. Construction A loading ramp 10 feet long that makes an angle of 18° with the horizontal is to be replaced by one that makes an angle of 12° with the horizontal. How long is the new ramp?

47. Finding the Height of a Helicopter Two observers simultaneously measure the angle of elevation of a helicopter. One angle is measured as 25°, the other as 40° (see the figure). If the observers are 100 feet apart and the helicopter lies over the line joining them, how high is the helicopter?

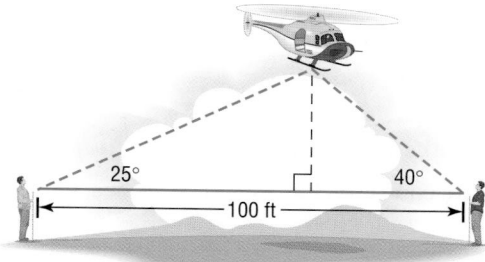

48. Mollweide's Formula For any triangle, Mollweide's Formula (named after Karl Mollweide, 1774–1825) states that

$$\frac{a+b}{c} = \frac{\cos\left[\frac{1}{2}(\alpha - \beta)\right]}{\sin\left(\frac{1}{2}\gamma\right)}$$

Derive it.

[**Hint:** Use the Law of Sines and then a sum-to-product formula. Notice that this formula involves all six parts of a triangle. As a result, it is sometimes used to check the solution of a triangle.]

49. Mollweide's Formula Another form of Mollweide's Formula is

$$\frac{a-b}{c} = \frac{\sin\left[\frac{1}{2}(\alpha - \beta)\right]}{\cos\left(\frac{1}{2}\gamma\right)}$$

Derive it.

50. For any triangle, derive the formula

$$a = b\cos\gamma + c\cos\beta$$

[**Hint:** Use the fact that $\sin\alpha = \sin(180° - \beta - \gamma)$.]

51. Law of Tangents   For any triangle, derive the Law of Tangents.

$$\frac{a - b}{a + b} = \frac{\tan\left[\frac{1}{2}(\alpha - \beta)\right]}{\tan\left[\frac{1}{2}(\alpha + \beta)\right]}$$

[**Hint:** Use Mollweide's Formula.]

52. Circumscribing a Triangle   Show that

$$\frac{\sin \alpha}{a} = \frac{\sin \beta}{b} = \frac{\sin \gamma}{c} = \frac{1}{2r}$$

where r is the radius of the circle circumscribing the triangle ABC whose sides are a, b, and c, as shown in the figure. [**Hint:** Draw the diameter AB'. Then β = angle ABC = angle $AB'C$, and angle $ACB' = 90°$.]

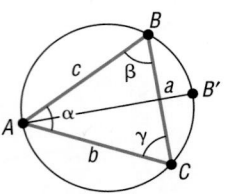

53. Make up three problems involving oblique triangles. One should result in one triangle, the second in two triangles, and the third in no triangle.

PREPARING FOR THIS SECTION

Before getting started, review the following:

✓ Trigonometric Equations (I) (Section 3.7, pp. 247–250)   ✓ The Distance Formula (Section 1.1, pp. 5–7)

4.3   THE LAW OF COSINES

OBJECTIVES   1 Solve SAS Triangles
2 Solve SSS Triangles
3 Solve Applied Problems Using the Law of Cosines

In Section 4.2, we used the Law of Sines to solve Case 1 (SAA or ASA) and Case 2 (SSA) of an oblique triangle. In this section, we derive the Law of Cosines and use it to solve the remaining cases, 3 and 4.

> **CASE 3:**   Two sides and the included angle are known (SAS).
> **CASE 4:**   Three sides are known (SSS).

Theorem   **Law of Cosines**

For a triangle with sides a, b, c and opposite angles α, β, γ, respectively,

$$c^2 = a^2 + b^2 - 2ab \cos \gamma \qquad (1)$$
$$b^2 = a^2 + c^2 - 2ac \cos \beta \qquad (2)$$
$$a^2 = b^2 + c^2 - 2bc \cos \alpha \qquad (3)$$

Proof   We will prove only formula (1) here. Formulas (2) and (3) may be proved using the same argument.

Figure 29

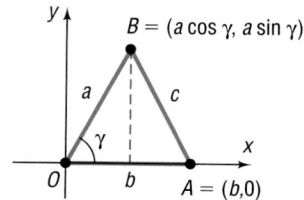

(a) Angle γ is acute

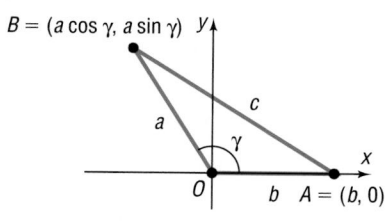

(b) Angle γ is obtuse

We begin by strategically placing a triangle on a rectangular coordinate system so that the vertex of angle γ is at the origin and side b lies along the positive x-axis. Regardless of whether γ is acute, as in Figure 29(a), or obtuse, as in Figure 29(b), the vertex B has coordinates $(a \cos \gamma, a \sin \gamma)$. Vertex A has coordinates $(b, 0)$.

We can now use the distance formula to compute c^2.

$$c^2 = (b - a \cos \gamma)^2 + (0 - a \sin \gamma)^2$$
$$= b^2 - 2ab \cos \gamma + a^2 \cos^2 \gamma + a^2 \sin^2 \gamma$$
$$= b^2 - 2ab \cos \gamma + a^2(\cos^2 \gamma + \sin^2 \gamma)$$
$$= a^2 + b^2 - 2ab \cos \gamma$$

Each of formulas (1), (2), and (3) may be stated in words as follows:

Theorem

Law of Cosines

The square of one side of a triangle equals the sum of the squares of the other two sides minus twice their product times the cosine of their included angle.

Observe that if the triangle is a right triangle (so that, say, $\gamma = 90°$) then formula (1) becomes the familiar Pythagorean Theorem: $c^2 = a^2 + b^2$. Thus, the Pythagorean Theorem is a special case of the Law of Cosines.

Let's see how to use the Law of Cosines to solve Case 3 (SAS), which applies to triangles for which two sides and the included angle are known.

1

EXAMPLE 1   Using the Law of Cosines to Solve a SAS Triangle

Solve the triangle: $a = 2, b = 3, \gamma = 60°$.

Figure 30

Solution   See Figure 30. The Law of Cosines makes it easy to find the third side, c.

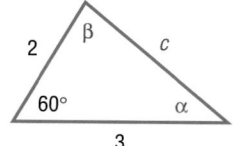

$$c^2 = a^2 + b^2 - 2ab \cos \gamma$$
$$= 4 + 9 - 2 \cdot 2 \cdot 3 \cdot \cos 60°$$
$$= 13 - \left(12 \cdot \frac{1}{2}\right) = 7$$
$$c = \sqrt{7}$$

Side c is of length $\sqrt{7}$. To find the angles α and β, we may use either the Law of Sines or the Law of Cosines. It is preferable to use the Law of Cosines, since it will lead to an equation with one solution. Using the Law of Sines would lead to an equation with two solutions that would need to be checked to determine which solution fits the given data. We choose to use formulas (2) and (3) of the Law of Cosines to find α and β.

For α:

$$a^2 = b^2 + c^2 - 2bc \cos \alpha$$
$$2bc \cos \alpha = b^2 + c^2 - a^2$$
$$\cos \alpha = \frac{b^2 + c^2 - a^2}{2bc} = \frac{9 + 7 - 4}{2 \cdot 3\sqrt{7}} = \frac{12}{6\sqrt{7}} = \frac{2\sqrt{7}}{7}$$
$$\alpha \approx 40.9°$$

For β:

$$b^2 = a^2 + c^2 - 2ac \cos \beta$$
$$\cos \beta = \frac{a^2 + c^2 - b^2}{2ac} = \frac{4 + 7 - 9}{4\sqrt{7}} = \frac{1}{2\sqrt{7}} = \frac{\sqrt{7}}{14}$$
$$\beta \approx 79.1°$$

Notice that $\alpha + \beta + \gamma = 40.9° + 79.1° + 60° = 180°$, as required. ■

NOW WORK PROBLEM **1**.

2   The next example illustrates how the Law of Cosines is used when three sides of a triangle are known, Case 4 (SSS).

EXAMPLE 2   **Using the Law of Cosines to Solve a SSS Triangle**

Solve the triangle: $a = 4, b = 3, c = 6$.

Solution   See Figure 31. To find the angles α, β, and γ, we proceed as we did in the latter part of the solution to Example 1.

For α:

$$\cos \alpha = \frac{b^2 + c^2 - a^2}{2bc} = \frac{9 + 36 - 16}{2 \cdot 3 \cdot 6} = \frac{29}{36}$$
$$\alpha \approx 36.3°$$

For β:

$$\cos \beta = \frac{a^2 + c^2 - b^2}{2ac} = \frac{16 + 36 - 9}{2 \cdot 4 \cdot 6} = \frac{43}{48}$$
$$\beta \approx 26.4°$$

Since we know α and β,

$$\gamma = 180° - \alpha - \beta \approx 180° - 36.3° - 26.4° = 117.3° \qquad ■$$

Figure 31

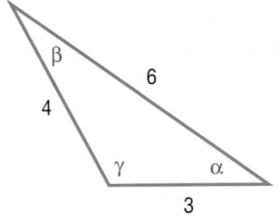

NOW WORK PROBLEM **7**.

3   **EXAMPLE 3**   **Correcting a Navigational Error**

A motorized sailboat leaves Naples, Florida, bound for Key West, 150 miles away. Maintaining a constant speed of 15 miles per hour, but encountering heavy crosswinds and strong currents, the crew finds, after 4 hours, that the sailboat is off course by 20°.

(a) How far is the sailboat from Key West at this time?

(b) Through what angle should the sailboat turn to correct its course?

(c) How much time has been added to the trip because of this? (Assume that the speed remains at 15 miles per hour.)

Figure 32

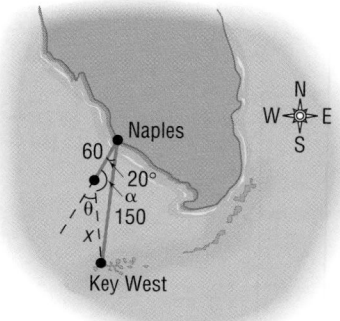

Solution See Figure 32. With a speed of 15 miles per hour, the sailboat has gone 60 miles after 4 hours. We seek the distance x of the sailboat from Key West. We also seek the angle θ that the sailboat should turn through to correct its course.

(a) To find x, we use the Law of Cosines, since we know two sides and the included angle.

$$x^2 = 150^2 + 60^2 - 2(150)(60) \cos 20° \approx 9186$$
$$x \approx 95.8$$

The sailboat is about 96 miles from Key West.

(b) We now know three sides of the triangle, so we can use the Law of Cosines again to find the angle α opposite the side of length 150 miles.

$$150^2 = 96^2 + 60^2 - 2(96)(60) \cos \alpha$$
$$9684 = -11,520 \cos \alpha$$
$$\cos \alpha \approx -0.8406$$
$$\alpha \approx 147.2°$$

The sailboat should turn through an angle of

$$\theta = 180° - \alpha \approx 180° - 147.2° = 32.8°$$

The sailboat should turn through an angle of about 33° to correct its course.

(c) The total length of the trip is now $60 + 96 = 156$ miles. The extra 6 miles will only require about 0.4 hour or 24 minutes more if the speed of 15 miles per hour is maintained. ∎

━━ **NOW WORK PROBLEM 27.**

HISTORICAL FEATURE

The Law of Sines was known vaguely long before it was explicitly stated by Nasîr Eddîn (about AD 1250). Ptolemy (about AD 150) was aware of it in a form using a chord function instead of the sine function. But it was first clearly stated in Europe by Regiomontanus, writing in 1464.

The Law of Cosines appears first in Euclid's *Elements* (Book II), but in a well-disguised form in which squares built on the sides of triangles are added and a rectangle representing the cosine term is subtracted. It was thus known to all mathematicians because of their familiarity with Euclid's work. An early modern form of the Law of Cosines, that for finding the angle when the sides are known, was stated by François Viète (in 1593).

The Law of Tangents (see Problem 51 of Exercise 4.2) has become obsolete. In the past it was used in place of the Law of Cosines, because the Law of Cosines was very inconvenient for calculation with logarithms or slide rules. Mixing of addition and multiplication is now very easy on a calculator, however, and the Law of Tangents has been shelved along with the slide rule.

4.3 Concepts and Vocabulary

In Problems 1–3, fill in the blanks.

1. If three sides of a triangle are given, the Law of _____ is used to solve the triangle.

2. If one side and two angles of a triangle are given, the Law of _____ is used to solve the triangle.

3. If two sides and the included angle of a triangle are given, the Law of _____ is used to solve the triangle.

In Problems 4–6, answer True or False to each statement.

4. Given only the three sides of a triangle, there is insufficient information to solve the triangle.

5. Given two sides and the included angle, the first thing to do to solve the triangle is to use the Law of Sines.

6. A special case of the Law of Cosines is the Pythagorean Theorem.

7. What do you do first if you are asked to solve a triangle and are given two sides and the included angle?

8. What do you do first if you are asked to solve a triangle and are given three sides?

9. Make up an applied problem that requires using the Law of Cosines.

4.3 Exercises

In Problems 1–8, solve each triangle.

1.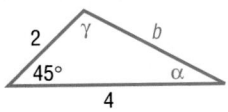

2.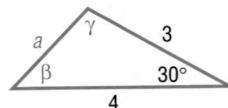

3.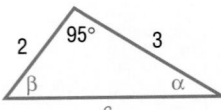

4.

5.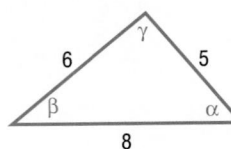

6.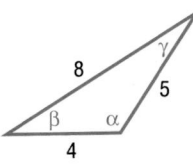

7.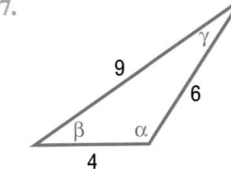

8.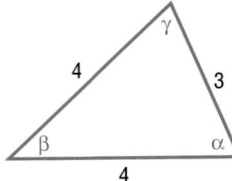

In Problems 9–24, solve each triangle.

9. $a = 3$, $b = 4$, $\gamma = 40°$

10. $a = 2$, $c = 1$, $\beta = 10°$

11. $b = 1$, $c = 3$, $\alpha = 80°$

12. $a = 6$, $b = 4$, $\gamma = 60°$

13. $a = 3$, $c = 2$, $\beta = 110°$

14. $b = 4$, $c = 1$, $\alpha = 120°$

15. $a = 2$, $b = 2$, $\gamma = 50°$

16. $a = 3$, $c = 2$, $\beta = 90°$

17. $a = 12$, $b = 13$, $c = 5$

18. $a = 4$, $b = 5$, $c = 3$

19. $a = 2$, $b = 2$, $c = 2$

20. $a = 3$, $b = 3$, $c = 2$

21. $a = 5$, $b = 8$, $c = 9$

22. $a = 4$, $b = 3$, $c = 6$

23. $a = 10$, $b = 8$, $c = 5$

24. $a = 9$, $b = 7$, $c = 10$

---

25. **Surveying** Consult the figure. To find the distance from the house at *A* to the house at *B*, a surveyor measures the angle *ACB*, which is found to be 70°, and then walks off the distance to each house, 50 feet and 70 feet, respectively. How far apart are the houses?

(a) How far is it from Ft. Myers to Orlando?

(b) Through what angle should the pilot turn at Orlando to return to Ft. Myers?

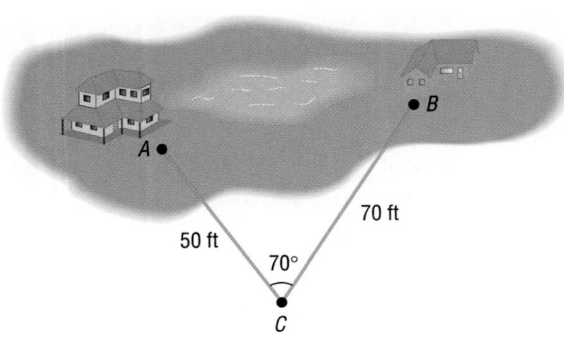

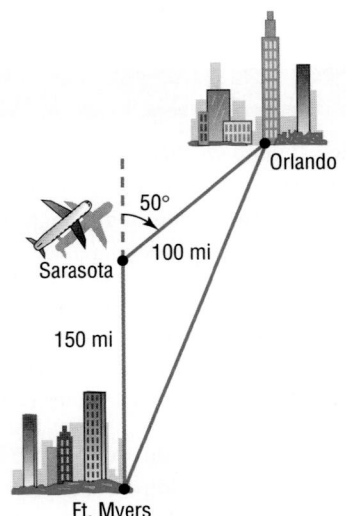

26. **Navigation** An airplane flies from Ft. Myers to Sarasota, a distance of 150 miles, and then turns through an angle of 50° and flies to Orlando, a distance of 100 miles (see the figure).

27. **Revising a Flight Plan** In attempting to fly from Chicago to Louisville, a distance of 330 miles, a pilot inadvertently took a course that was 10° in error, as indicated in the figure.
    (a) If the aircraft maintains an average speed of 220 miles per hour and if the error in direction is discovered after 15 minutes, through what angle should the pilot turn to head toward Louisville?
    (b) What new average speed should the pilot maintain so that the total time of the trip is 90 minutes?

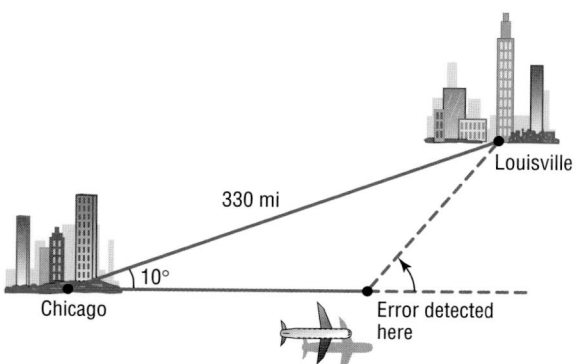

28. **Avoiding a Tropical Storm** A cruise ship maintains an average speed of 15 knots in going from San Juan, Puerto Rico, to Barbados, West Indies, a distance of 600 nautical miles. To avoid a tropical storm, the captain heads out of San Juan in a direction of 20° off a direct heading to Barbados. The captain maintains the 15-knot speed for 10 hours, after which time the path to Barbados becomes clear of storms.
    (a) Through what angle should the captain turn to head directly to Barbados?
    (b) Once the turn is made, how long will it be before the ship reaches Barbados if the same 15-knot speed is maintained?

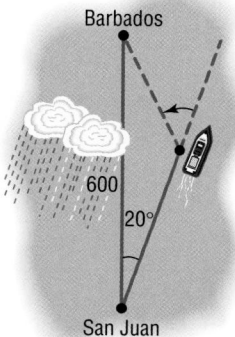

29. **Major League Baseball Field** A Major League baseball diamond is actually a square 90 feet on a side. The pitching rubber is located 60.5 feet from home plate on a line joining home plate and second base.
    (a) How far is it from the pitching rubber to first base?
    (b) How far is it from the pitching rubber to second base?
    (c) If a pitcher faces home plate, through what angle does he need to turn to face first base?

30. **Little League Baseball Field** According to Little League baseball official regulations, the diamond is a square 60 feet on a side. The pitching rubber is located 46 feet from home plate on a line joining home plate and second base.
    (a) How far is it from the pitching rubber to first base?
    (b) How far is it from the pitching rubber to second base?
    (c) If a pitcher faces home plate, through what angle does he need to turn to face first base?

31. **Finding the Length of a Guy Wire** The height of a radio tower is 500 feet, and the ground on one side of the tower slopes upward at an angle of 10° (see the figure).
    (a) How long should a guy wire be if it is to connect to the top of the tower and be secured at a point on the sloped side 100 feet from the base of the tower?
    (b) How long should a second guy wire be if it is to connect to the middle of the tower and be secured at a point 100 feet from the base on the flat side?

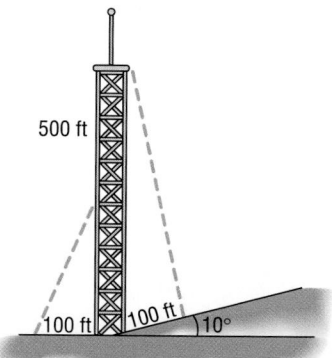

32. **Finding the Length of a Guy Wire** A radio tower 500 feet high is located on the side of a hill with an inclination to the horizontal of 5° (see the figure). How long should two guy wires be if they are to connect to the top of the tower and be secured at two points 100 feet directly above and directly below the base of the tower?

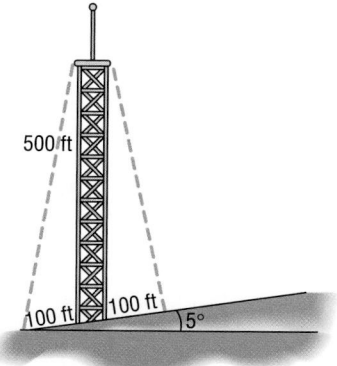

33. **Wrigley Field, Home of the Chicago Cubs** The distance from home plate to the fence in dead center in Wrigley Field is 400 feet (see the figure). How far is it from the fence in dead center to third base?

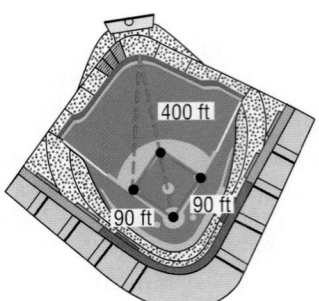

400 ft

90 ft   90 ft

34. **Little League Baseball** The distance from home plate to the fence in dead center at the Oak Lawn Little League field is 280 feet. How far is it from the fence in dead center to third base?

[**Hint:** The distance between the bases in Little League in 60 feet.]

35. **Rods and Pistons** Rod OA (see the figure) rotates about the fixed point O so that point A travels on a circle of radius r. Connected to point A is another rod AB of length $L > 2r$, and point B is connected to a piston. Show that the distance x between point O and point B is given by

$$x = r \cos \theta + \sqrt{r^2 \cos^2 \theta + L^2 - r^2}$$

where θ is the angle of rotation of rod OA.

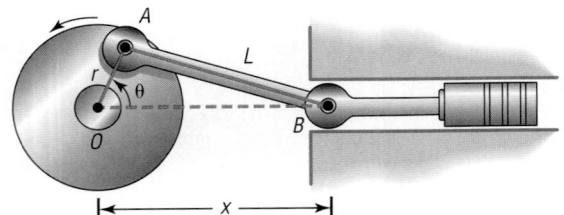

A

L

r

θ

O

B

x

36. **Geometry** Show that the length d of a chord of a circle of radius r is given by the formula

$$d = 2r \sin \frac{\theta}{2}$$

where θ is the central angle formed by the radii to the ends of the chord (see the figure). Use this result to derive the fact that $\sin \theta < \theta$, where $\theta > 0$ is measure in radians.

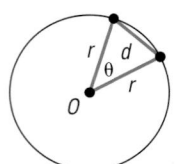

r   d

θ

O   r

37. For any triangle, show that

$$\cos \frac{\gamma}{2} = \sqrt{\frac{s(s-c)}{ab}}$$

where $s = \frac{1}{2}(a + b + c)$.

[**Hint:** Use a Half-angle Formula and the Law of Cosines.]

38. For any triangle show that

$$\sin \frac{\gamma}{2} = \sqrt{\frac{(s-a)(s-b)}{ab}}$$

where $s = \frac{1}{2}(a + b + c)$.

39. Use the Law of Cosines to prove the identity

$$\frac{\cos \alpha}{a} + \frac{\cos \beta}{b} + \frac{\cos \gamma}{c} = \frac{a^2 + b^2 + c^2}{2abc}$$

40. Write down your strategy for solving an oblique triangle.

PREPARING FOR THIS SECTION

Before getting started, review the following:

✓ Geometry Review (Section A.2, pp. 565–568)

4.4 AREA OF A TRIANGLE

OBJECTIVES   ① Find the Area of SAS Triangles
② Find the Area of SSS Triangles

In this section, we will derive several formulas for calculating the area A of a triangle. The most familiar of these is the following:

Theorem The area A of a triangle is

$$A = \frac{1}{2}bh \tag{1}$$

where b is the base and h is an altitude drawn to that base.

■

Figure 33
$A = \frac{1}{2}bh$

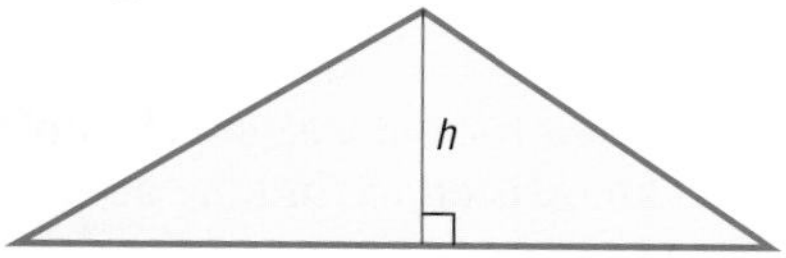

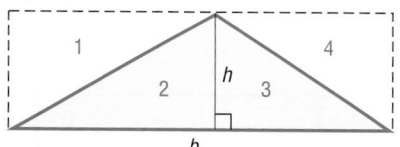

Proof The derivation of this formula is rather easy once a rectangle of base b and height h is constructed around the triangle. See Figures 33 and 34.

Triangles 1 and 2 in Figure 34 are equal in area, as are triangles 3 and 4. Consequently, the area of the triangle with base b and altitude h is exactly half the area of the rectangle, which is bh.

■

Figure 34

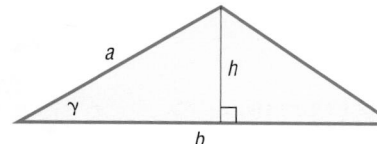

If the base b and altitude h to that base are known, then we can find the area of such a triangle using formula (1). Usually, though, the information required to use formula (1) is not given. Suppose, for example, that we know two sides a and b and the included angle γ (see Figure 35). Then the altitude h can be found by noting that

$$\frac{h}{a} = \sin \gamma$$

Figure 35
$h = a \sin \gamma$

so that

$$h = a \sin \gamma$$

Using this fact in formula (1) produces

$$A = \frac{1}{2}bh = \frac{1}{2}b(a \sin \gamma) = \frac{1}{2}ab \sin \gamma$$

We now have the formula

$$A = \frac{1}{2}ab \sin \gamma \tag{2}$$

By dropping altitudes from the other two vertices of the triangle, we obtain the following corresponding formulas:

$$A = \frac{1}{2}bc \sin \alpha \tag{3}$$

$$A = \frac{1}{2}ac \sin \beta \tag{4}$$

It is easiest to remember these formulas using the following wording:

Theorem The area A of a triangle equals one-half the product of two of its sides and the sine of their included angle.

■

① **EXAMPLE 1** **Finding the Area of a SAS Triangle**

Find the area A of the triangle for which $a = 8$, $b = 6$, and $\gamma = 30°$.

Figure 36

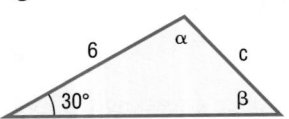

Solution See Figure 36. We use formula (2) to get

$$A = \frac{1}{2}ab \sin \gamma = \frac{1}{2} \cdot 8 \cdot 6 \sin 30° = 12$$

∎

➤ **NOW WORK PROBLEM 1.**

② If the three sides of a triangle are known, another formula, called **Heron's Formula** (named after Heron of Alexandria), can be used to find the area of a triangle.

Theorem **Heron's Formula**

The area A of a triangle with sides a, b, and c is

$$A = \sqrt{s(s-a)(s-b)(s-c)} \qquad (5)$$

where $s = \frac{1}{2}(a + b + c)$.

∎

Proof The proof we shall give uses formula (2) on page 295.

$$A = \frac{1}{2}ab \sin \gamma$$

$$A = \frac{1}{2}ab\sqrt{1 - \cos^2 \gamma} \qquad \qquad \sin \gamma = \sqrt{1 - \cos^2 \gamma}, 0° < \gamma < 180°$$

$$A = \frac{1}{2}ab\sqrt{1 - \left(\frac{a^2 + b^2 - c^2}{2ab}\right)^2} \qquad \qquad \text{Use the Law of Cosines}$$

$$A = \frac{1}{2}ab\sqrt{\frac{4a^2b^2 - (a^2 + b^2 - c^2)^2}{4a^2b^2}} \qquad \qquad \text{Write the expression under the radical as a single quotient}$$

$$A = \frac{1}{4}\sqrt{[2ab + (a^2 + b^2 - c^2)][2ab - (a^2 + b^2 - c^2)]} \qquad \qquad \text{Simply; factor}$$

$$A = \frac{1}{4}\sqrt{(a^2 + 2ab + b^2 - c^2)(c^2 - (a^2 - 2ab + b^2))} \qquad \qquad \text{Rearrange terms}$$

$$A = \frac{1}{4}\sqrt{[(a + b)^2 - c^2][c^2 - (a - b)^2]} \qquad \qquad \text{Factor}$$

$$A = \sqrt{\frac{1}{16}[(a+b)+c][(a+b)-c][c+(a-b)][c-(a-b)]}$$ Difference of two squares; $\frac{1}{4} = \sqrt{\frac{1}{16}}$

$$A = \sqrt{\frac{1}{2}(a+b+c)\cdot\frac{1}{2}(b+c-a)\cdot\frac{1}{2}(a+c-b)\cdot\frac{1}{2}(a+b-c)}$$ Simplify; Rearrange terms

$$A = \sqrt{\frac{1}{2}(a+b+c)\cdot\left[\frac{1}{2}(a+b+c)-a\right]\cdot\left[\frac{1}{2}(a+b+c)-b\right]\cdot\left[\frac{1}{2}(a+b+c)-c\right]}$$

If we let $s = \frac{1}{2}(a+b+c)$, then we have $A = \sqrt{s\cdot(s-a)\cdot(s-b)\cdot(s-c)}$. ■

EXAMPLE 2   **Finding the Area of a SSS Triangle**

Find the area of a triangle whose sides are 4, 5, and 7.

Solution   We let $a = 4$, $b = 5$, and $c = 7$. Then

$$s = \frac{1}{2}(a+b+c) = \frac{1}{2}(4+5+7) = 8$$

Heron's Formula then gives the area A as

$$A = \sqrt{s(s-a)(s-b)(s-c)} = \sqrt{8\cdot4\cdot3\cdot1} = \sqrt{96} = 4\sqrt{6}$$ ■

✏️ — **NOW WORK PROBLEM 7.**

HISTORICAL FEATURE   Heron's formula (also known as *Hero's Formula*) is due to Heron of Alexandria (first century AD), who had, besides his mathematical talents, a good deal of engineering skills. In various temples his mechanical devices produced effects that seemed supernatural, and visitors presumably were thus influenced to generosity. Heron's book *Metrica*, on making such devices, has survived and was discovered in 1896 in the city of Constantinople.

Heron's Formulas for the area of a triangle caused some mild discomfort in Greek mathematics, because a product with two factors was an area, while one with three factors was a volume, but four factors seemed contradictory in Heron's time.

4.4 Concepts and Vocabulary

In Problem 1, fill in the blank.

1. If three sides of a triangle are given, _____ Formula is used to find the area of the triangle.

In Problems 2 and 3, answer True or False to each statement.

2. No formula exists for finding the area of a triangle when only three sides are given.

3. Given two sides and the included angle, there is a formula that can be used to find the area of the triangle.

4. What do you do first if you are asked to find the area of a triangle and are given two sides and the included angle?

5. What do you do first if you are asked to find the area of a triangle and are given three sides?

4.4 Exercises

In Problems 1–8, find the area of each triangle. Round answers to two decimal places.

1.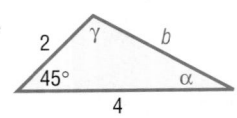

2.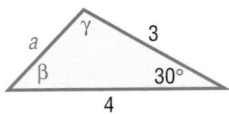

3.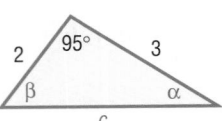

4.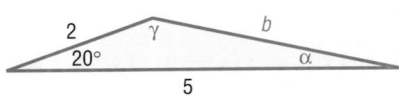

5.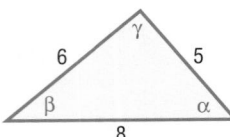

6.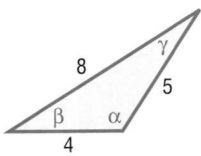

7.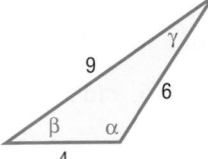

8.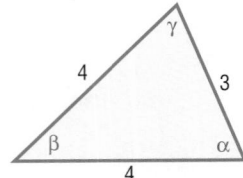

In Problems 9–24, find the area of each triangle. Round answers to two decimal places.

9. $a = 3$, $b = 4$, $\gamma = 40°$

10. $a = 2$, $c = 1$, $\beta = 10°$

11. $b = 1$, $c = 3$, $\alpha = 80°$

12. $a = 6$, $b = 4$, $\gamma = 60°$

13. $a = 3$, $c = 2$, $\beta = 110°$

14. $b = 4$, $c = 1$, $\alpha = 120°$

15. $a = 2$, $b = 2$, $\gamma = 50°$

16. $a = 3$, $c = 2$, $\beta = 90°$

17. $a = 12$, $b = 13$, $c = 5$

18. $a = 4$, $b = 5$, $c = 3$

19. $a = 2$, $b = 2$, $c = 2$

20. $a = 3$, $b = 3$, $c = 2$

21. $a = 5$, $b = 8$, $c = 9$

22. $a = 4$, $b = 3$, $c = 6$

23. $a = 10$, $b = 8$, $c = 5$

24. $a = 9$, $b = 7$, $c = 10$

---

25. Area of a Segment   Find the area of the segment (shaded in blue in the figure) of a circle whose radius is 8 feet, formed by a central angle of 70°.

[**Hint:** Subtract the area of the triangle from the area of the sector to obtain the area of the segment.]

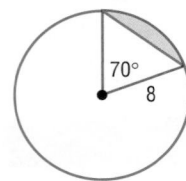

26. Area of a Segment   Find the area of the segment of a circle whose radius is 5 inches, formed by a central angle of 40°.

27. Cost of a Triangular Lot   The dimensions of a triangular lot are 100 feet by 50 feet by 75 feet. If the price of such land is $3 per square foot, how much does the lot cost?

28. Amount of Materials to Make a Tent   A cone-shaped tent is made from a circular piece of canvas 24 feet in diameter by removing a sector with central angle 100° and connecting the ends. What is the surface area of the tent?

29. Computing Areas   Find the area of the shaded region enclosed in a semicircle of diameter 8 centimeters. The length of the chord AB is 6 centimeters.

[**Hint:** Triangle ABC is a right triangle.]

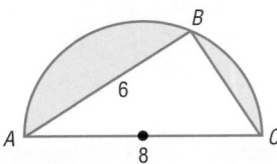

30. Computing Areas   Find the area of the shaded region enclosed in a semicircle of diameter 10 inches. The length of the chord AB is 8 inches.

[**Hint:** Triangle ABC is a right triangle.]

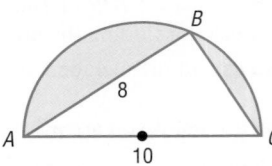

31. Area of a Triangle Prove that the area A of a triangle is given by the formula

$$A = \frac{a^2 \sin \beta \sin \gamma}{2 \sin \alpha}$$

32. Area of a Triangle Prove the two other forms of the formula given in Problem 31.

$$A = \frac{b^2 \sin \alpha \sin \gamma}{2 \sin \beta} \quad \text{and} \quad A = \frac{c^2 \sin \alpha \sin \beta}{2 \sin \gamma}$$

In Problems 33–38, use the results of Problem 31 or 32 to find the area of each triangle. Round answers to two decimal places.

33. $\alpha = 40°, \quad \beta = 20°, \quad a = 2$

34. $\alpha = 50°, \quad \gamma = 20°, \quad a = 3$

35. $\beta = 70°, \quad \gamma = 10°, \quad b = 5$

36. $\alpha = 70°, \quad \beta = 60°, \quad c = 4$

37. $\alpha = 110°, \quad \gamma = 30°, \quad c = 3$

38. $\beta = 10°, \quad \gamma = 100°, \quad b = 2$

39. Geometry Consult the figure, which shows a circle of radius r with center at O. Find the area A of the shaded region as a function of the central angle θ.

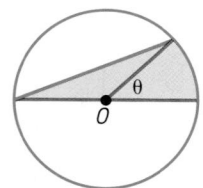

40. Approximating the Area of a Lake To approximate the area of a lake, a surveyor walks around the perimeter of the lake, taking the measurements shown in the illustration. Using this technique, what is the approximate area of the lake?

[**Hint:** Use the Law of Cosines on the three triangles shown and then find the sum of their areas.]

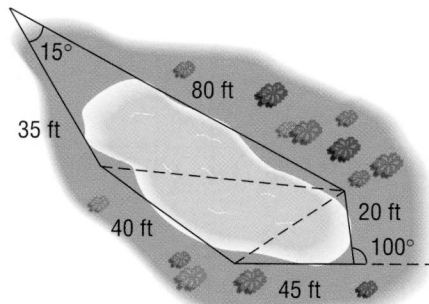

41. The Cow Problem* A cow is tethered to one corner of a square barn, 10 feet by 10 feet, with a rope 100 feet long. What is the maximum grazing area for the cow?

[**Hint:** See the illustration.]

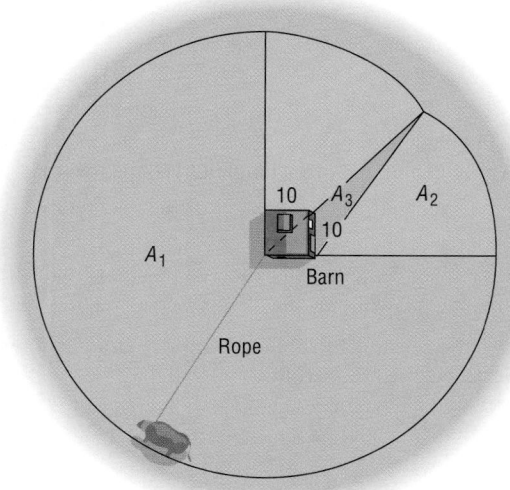

42. Another Cow Problem If the barn in Problem 41 is rectangular, 10 feet by 20 feet, what is the maximum grazing area for the cow?

43. If h_1, h_2, and h_3 are the altitudes dropped from A, B, and C, respectively, in a triangle (see the figure), show that

$$\frac{1}{h_1} + \frac{1}{h_2} + \frac{1}{h_3} = \frac{s}{K}$$

where K is the area of the triangle and $s = \frac{1}{2}(a + b + c)$.

[**Hint:** $h_1 = \frac{2K}{a}$.]

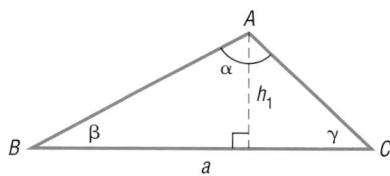

*Suggested by Professor Teddy Koukounas of SUNY at Old Westbury, who learned of it from an old farmer in Virginia. Solution provided by Professor Kathleen Miranda of SUNY at Old Westbury.

44. Show that a formula for the altitude h from a vertex to the opposite side a of a triangle is

$$h = \frac{a \sin \beta \sin \gamma}{\sin \alpha}$$

Inscribed Circle   *For Problems 45–48, the lines that bisect each angle of a triangle meet in a single point O, and the perpendicular distance r from O to each side of the triangle is the same. The circle with center at O and radius r is called the **inscribed circle** of the triangle (see the figure).*

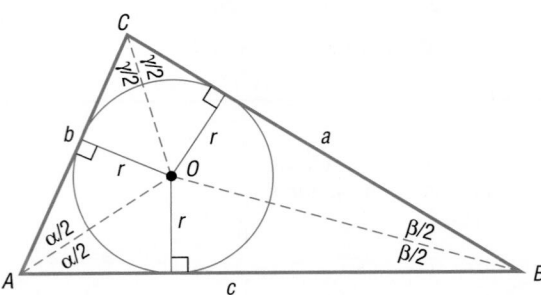

45. Apply Problem 44 to triangle OAB to show that

$$r = \frac{c \sin \dfrac{\alpha}{2} \sin \dfrac{\beta}{2}}{\cos \dfrac{\gamma}{2}}$$

46. Use the results of Problem 45 (above) and Problem 38 in Section 4.3 to show that

$$\cot \frac{\gamma}{2} = \frac{s - c}{r}$$

47. Show that

$$\cot \frac{\alpha}{2} + \cot \frac{\beta}{2} + \cot \frac{\gamma}{2} = \frac{s}{r}$$

48. Show that the area K of triangle ABC is $K = rs$. Then show that

$$r = \sqrt{\frac{(s - a)(s - b)(s - c)}{s}}$$

where $s = \dfrac{1}{2}(a + b + c)$.

PREPARING FOR THIS SECTION

Before getting started, review the following:

✓ Sinusoidal Graphs (Section 2.6, pp. 159–166)

4.5 SIMPLE HARMONIC MOTION; DAMPED MOTION; COMBINING WAVES

OBJECTIVES   1 Find an Equation for an Object in Simple Harmonic Motion
2 Analyze Simple Harmonic Motion
3 Analyze an Object in Damped Motion
4 Graph the Sum of Two Functions

Simple Harmonic Motion

Many physical phenomena can be described as simple harmonic motion. Radio and television waves, light waves, sound waves, and water waves exhibit motion that is simple harmonic.

The swinging of a pendulum, the vibrations of a tuning fork, and the bobbing of a weight attached to a coiled spring are examples of vibrational motion. In this type of motion, an object swings back and forth over the same path. In each illustration in Figure 37, the point B is the **equilibrium (rest) position** of the vibrating object. The **amplitude** of vibration is the distance from the object's rest position to its point of greatest displacement (either point A or point C in Figure 37). The **period** of a vibrating object is the time required to complete one vibration, that is, the time it takes to go from, say, point A through B to C and back to A.

Figure 37

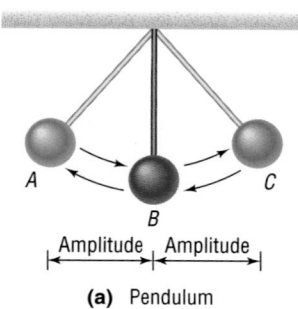

(a) Pendulum

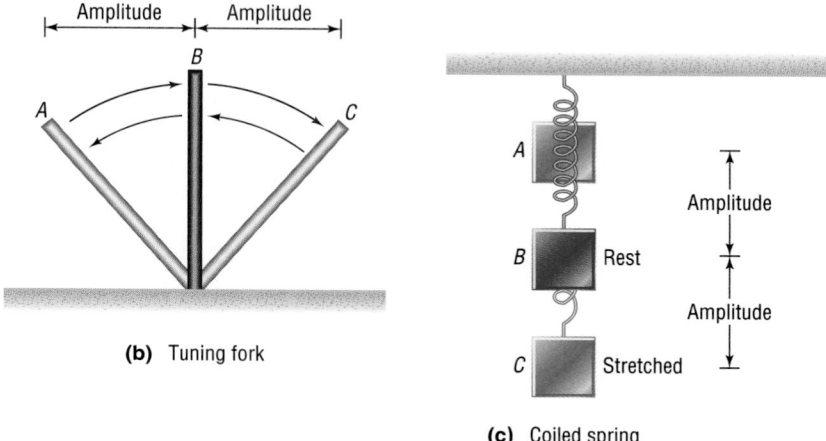

(b) Tuning fork

(c) Coiled spring

Simple harmonic motion is a special kind of vibrational motion in which the acceleration a of the object is directly proportional to the negative of its displacement d from its rest position. That is, $a = -kd, k > 0$.

For example, when the mass hanging from the spring in Figure 37(c) is pulled down from its rest position B to the point C, the force of the spring tries to restore the mass to its rest position. Assuming that there is no frictional force* to retard the motion, the amplitude will remain constant. The force increases in direct proportion to the distance that the mass is pulled from its rest position. Since the force increases directly, the acceleration of the mass of the object must do likewise, because (by Newton's Second Law of Motion) force is directly proportional to acceleration. Thus, the acceleration of the object varies directly with its displacement, and the motion is an example of simple harmonic motion.

Simple harmonic motion is related to circular motion. To see this relationship, consider a circle of radius a, with center at $(0,0)$. See Figure 38. Suppose that an object initially placed at $(a,0)$ moves counterclockwise around the circle at constant angular speed ω. Suppose further that after time t has elapsed the object is at the point $P = (x, y)$ on the circle. The angle θ, in radians, swept out by the ray $\overrightarrow{OP}$ in this time t is

$$\theta = \omega t$$

The coordinates of the point P at time t are

$$x = a \cos \theta = a \cos(\omega t)$$
$$y = a \sin \theta = a \sin(\omega t)$$

Figure 38

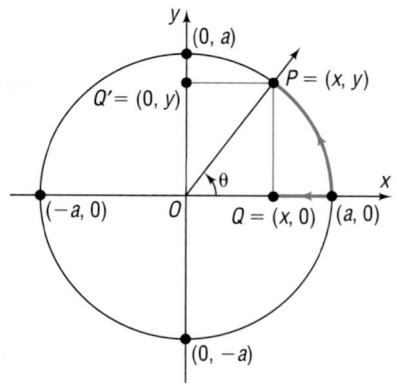

*If friction is present, the amplitude will decrease with time to 0. This type of motion is an example of **damped motion**, which is discussed later in this section.

Corresponding to each position $P = (x, y)$ of the object moving about the circle, there is the point $Q = (x, 0)$, called the **projection of P on the x-axis.** As P moves around the circle at a constant rate, the point Q moves back and forth between the points $(a, 0)$ and $(-a, 0)$ along the x-axis with a motion that is simple harmonic. Similarly, for each point P there is a point $Q' = (0, y)$, called the **projection of P on the y-axis.** As P moves around the circle, the point Q' moves back and forth between the points $(0, a)$ and $(0, -a)$ on the y-axis with a motion that is simple harmonic. Simple harmonic motion can be described as the projection of constant circular motion on a coordinate axis.

To put it another way, again consider a mass hanging from a spring where the mass is pulled down from its rest position to the point C and then released. See Figure 39(a). The graph shown in Figure 39(b) describes the displacement d of the object from its rest position as a function of time t, assuming that no frictional force is present.

Figure 39

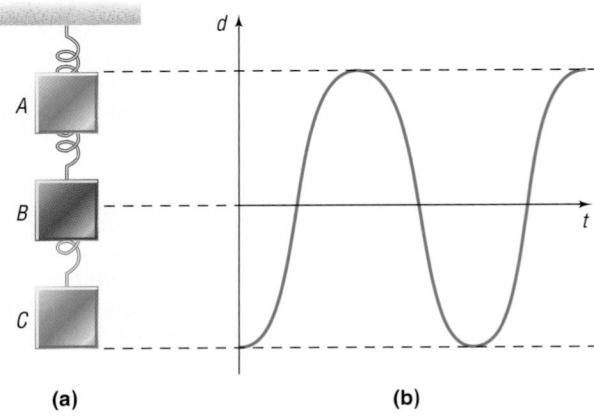

(a)   (b)

Theorem

Simple Harmonic Motion

An object that moves on a coordinate axis so that its distance d from the origin at time t is given by either

$$d = a\cos(\omega t) \quad \text{or} \quad d = a\sin(\omega t)$$

where a and $\omega > 0$ are constants, moves with simple harmonic motion. The motion has amplitude $|a|$ and period $T = \dfrac{2\pi}{\omega}$.

The **frequency** f of an object in simple harmonic motion is the number of oscillations per unit time. Since the period is the time required for one oscillation, it follows that frequency is the reciprocal of the period; that is,

$$f = \frac{\omega}{2\pi}, \quad \omega > 0$$

① | **EXAMPLE 1**   Finding an Equation for an Object in Harmonic Motion

Suppose that an object attached to a coiled spring is pulled down a distance of 5 inches from its rest position and then released. If the time for one oscillation is 3 seconds, write an equation that relates the displacement d of the object from its rest position after time t (in seconds). Assume no friction.

Figure 40

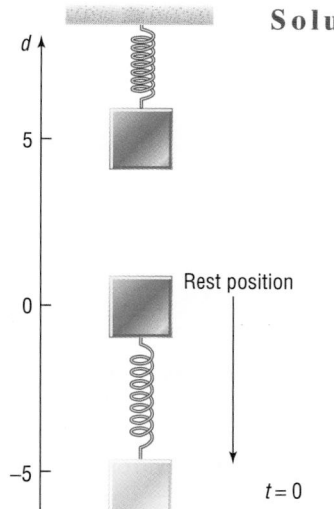

Solution   The motion of the object is simple harmonic. See Figure 40. When the object is released ($t = 0$), the displacement of the object from the rest position is -5 units (since the object was pulled down). Because $d = -5$ when $t = 0$, it is easier to use the cosine function*

$$d = a \cos(\omega t)$$

to describe the motion. Now the amplitude is $|-5| = 5$ and the period is 3, so

$$a = -5 \quad \text{and} \quad \frac{2\pi}{\omega} = \text{period} = 3, \quad \omega = \frac{2\pi}{3}$$

An equation of the motion of the object is

$$d = -5 \cos\left(\frac{2\pi}{3}t\right)$$

Note: In the solution to Example 1, we let $a = -5$, since the initial motion is down. If the initial direction were up, we would let $a = 5$.

NOW WORK PROBLEM **1.**

② | **EXAMPLE 2**   Analyzing the Motion of an Object

Suppose that the displacement d (in meters) of an object at time t (in seconds) satisfies the equation

$$d = 10 \sin(5t)$$

(a) Describe the motion of the object.
(b) What is the maximum displacement from its resting position?
(c) What is the time required for one oscillation?
(d) What is the frequency?

Solution   We observe that the given equation is of the form

$$d = a \sin(\omega t) \qquad d = 10 \sin(5t)$$

where $a = 10$ and $\omega = 5$.

(a) The motion is simple harmonic.
(b) The maximum displacement of the object from its resting position is the amplitude: $|a| = 10$ meters.
(c) The time required for one oscillation is the period:

$$\text{Period} = \frac{2\pi}{\omega} = \frac{2\pi}{5} \text{ seconds}$$

(d) The frequency is the reciprocal of the period. Thus,

$$\text{Frequency} = f = \frac{5}{2\pi} \text{ oscillations per second}$$

NOW WORK PROBLEM **9.**

*No phase shift is required if a cosine function is used.

Damped Motion*

③ Most physical phenomena are affected by friction or other resistive forces. These forces remove energy from a moving system and thereby damp its motion. For example, when a mass hanging from a spring is pulled down a distance a and released, the friction in the spring causes the distance that the mass moves from its at-rest position to decrease over time. Thus, the amplitude of any real oscillating spring or swinging pendulum decreases with time due to air resistance, friction, and so forth. See Figure 41.

Figure 41

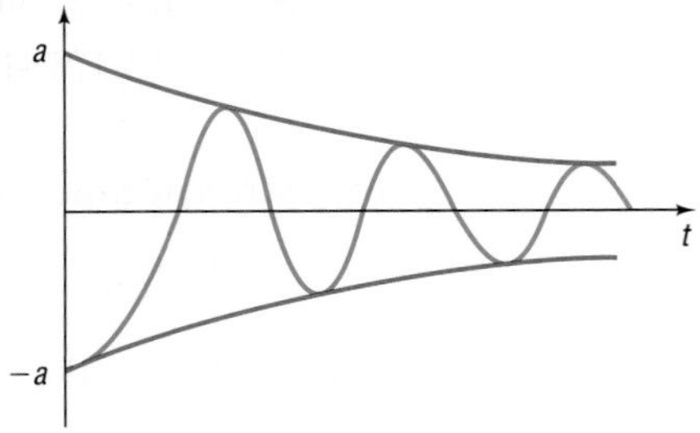

A function that describes this phenomenon maintains a sinusoidal component, but the amplitude of this component will decrease with time in order to account for the damping effect. In addition, the period of the oscillating component will be affected by the damping. The next result, from physics, describes damped motion.

Theorem

Damped Motion

The displacement d of an oscillating object from its at-rest position at time t is given by

$$d(t) = ae^{-bt/2m} \cos\left(\sqrt{\omega^2 - \frac{b^2}{4m^2}}\, t\right)$$

where b is a **damping factor** (most physics texts call this a **damping coefficient**) and m is the mass of the oscillating object.

    ■

Notice for $b = 0$ (zero damping) that we have the formula for simple harmonic motion with amplitude $|a|$ and period $\dfrac{2\pi}{\omega}$.

Figure 42

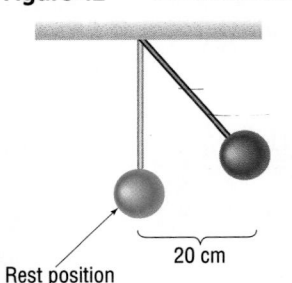

Rest position

20 cm

EXAMPLE 3

Analyzing the Motion of a Pendulum with Damped Motion

Suppose that a simple pendulum with a bob of mass 10 grams and a damping factor of 0.8 gram/second is pulled 20 centimeters from its at-rest position and released, see Figure 42. The period of the pendulum without the damping effect is 4 seconds.

(a) Find an equation that describes the position of the pendulum bob.
(b) Using a graphing utility, graph the function found in part (a).
(c) Determine the maximum displacement of the bob after the first oscillation.

* This subsection uses exponential functions, discussed in Chapter 7.

(d) What happens to the displacement of the bob as time increases without bound?

Solution (a) We have $m = 10$, $a = 20$, and $b = 0.8$. Since the period of the pendulum under simple harmonic motion is 4 seconds, we have

$$4 = \frac{2\pi}{\omega}$$

$$\omega = \frac{2\pi}{4} = \frac{\pi}{2}$$

Substituting these values into the equation for damped motion, we obtain

$$d = 20e^{-0.8t/2(10)} \cos\left(\sqrt{\left(\frac{\pi}{2}\right)^2 - \frac{0.8^2}{4(10)^2}}\, t\right)$$

$$d = 20e^{-0.8t/20} \cos\left(\sqrt{\frac{\pi^2}{4} - \frac{0.64}{400}}\, t\right)$$

(b) See Figure 43 for the graph of d.

(c) See Figure 44. After the first oscillation, the maximum displacement is approximately 17.05 centimeters.

Figure 43

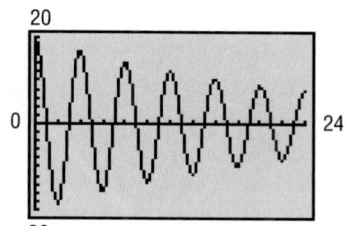

Figure 44

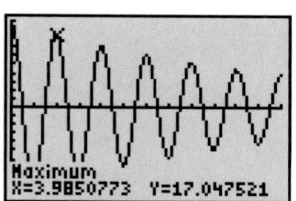

Maximum
X=3.9850773 Y=17.047521

(d) As t increases without bound $e^{-0.8t/20} \to 0$, so the displacement of the bob approaches zero. As a result, the pendulum will eventually come to rest. ∎

 **NOW WORK PROBLEM 17.**

Combining Waves

④ Many physical and biological applications require the graph of the sum of two functions, such as

$$f(x) = x + \sin x \quad \text{or} \quad g(x) = \sin x + \cos(2x)$$

For example, if two tones are emitted, the sound produced is the sum of the waves produced by the two tones. See Problem 39 for an explanation of Touch-Tone phones.

To graph the sum of two (or more) functions, we can use the method of adding y-coordinates described next.

EXAMPLE 4 | **Graphing the Sum of Two Functions**

Use the method of adding y-coordinates to graph $f(x) = x + \sin x$.

Solution First, we graph the component functions,

$$y = f_1(x) = x \qquad y = f_2(x) = \sin x$$

in the same coordinate system. See Figure 45(a). Now, select several values of x, say, $x = 0$, $x = \dfrac{\pi}{2}$, $x = \pi$, $x = \dfrac{3\pi}{2}$, and $x = 2\pi$, at which we compute $f(x) = f_1(x) + f_2(x)$. Table 1 shows the computation. We plot these points and connect them to get the graph, as shown in Figure 45(b).

TABLE 1

x	0	$\dfrac{\pi}{2}$	π	$\dfrac{3\pi}{2}$	2π
$y = f_1(x) = x$	0	$\dfrac{\pi}{2}$	π	$\dfrac{3\pi}{2}$	2π
$y = f_2(x) = \sin x$	0	1	0	-1	0
$f(x) = x + \sin x$	0	$\dfrac{\pi}{2} + 1 \approx 2.57$	π	$\dfrac{3\pi}{2} - 1 \approx 3.71$	2π
Point on Graph of f	$(0, 0)$	$\left(\dfrac{\pi}{2}, 2.57\right)$	(π, π)	$\left(\dfrac{3\pi}{2}, 3.71\right)$	$(2\pi, 2\pi)$

Figure 45

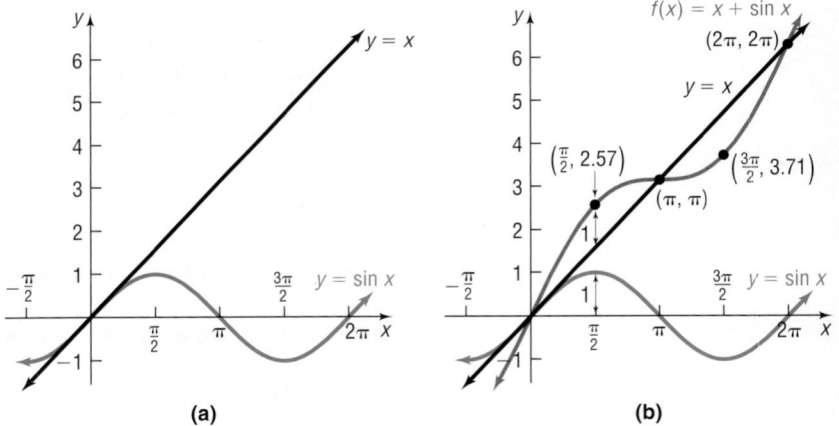

(a)　　　　(b)

Note that the graph of $f(x) = x + \sin x$ intersects the line $y = x$ whenever $\sin x = 0$. Also, notice that the graph of f is not periodic.

✔ CHECK:　Graph $Y_1 = x + \sin x$ and compare the result with Figure 45(b). Use INTERSECT to verify that the graphs intersect when $\sin x = 0$. ■ ■

The next example shows a periodic graph of the sum of two functions.

EXAMPLE 5　Graphing the Sum of two Sinusoidal Functions

Use the method of adding y-coordinates to graph

$$f(x) = \sin x + \cos(2x)$$

Solution Table 2 shows the steps for computing several points on the graph of f. Figure 46 illustrates the graphs of the component functions, $y = f_1(x) = \sin x$ and $y = f_2(x) = \cos(2x)$, and the graph of $f(x) = \sin x + \cos(2x)$, which is shown in red.

TABLE 2

x	$-\dfrac{\pi}{2}$	0	$\dfrac{\pi}{2}$	π	$\dfrac{3\pi}{2}$	2π
$y = f_1(x) = \sin x$	-1	0	1	0	-1	0
$y = f_2(x) = \cos(2x)$	-1	1	-1	1	-1	1
$f(x) = \sin x + \cos(2x)$	-2	1	0	1	-2	1
Point on Graph of f	$\left(-\dfrac{\pi}{2}, -2\right)$	$(0, 1)$	$\left(\dfrac{\pi}{2}, 0\right)$	$(\pi, 1)$	$\left(\dfrac{3\pi}{2}, -2\right)$	$(2\pi, 1)$

Figure 46
$f(x) = \sin x + \cos(2x)$

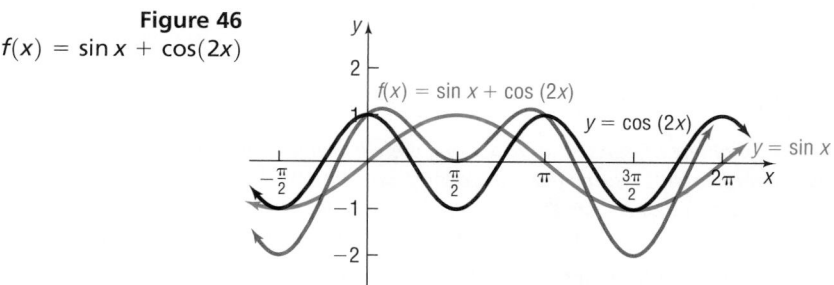

✔ CHECK: Graph $Y_1 = \sin x + \cos(2x)$ and compare the result with Figure 46.

■ ■

➤ **NOW WORK PROBLEM 29.**

4.5 Concepts and Vocabulary

In Problems 1 and 2, fill in the blanks.

1. The motion of an object obeys the equation $d = 4\cos(6t)$. Such motion is described as _____ _____. The number 4 is called the _____.

2. When a mass hanging from a spring is pulled down and then released, the motion is called _____ _____ if there is no frictional force to retard the motion, and the motion is called _____ _____ if there is friction.

In Problem 3, answer True or False to the statement.

3. If the distance d of an object from the origin at time t is given by a sinusoidal graph, the motion of the object is simple harmonic motion.

4.5 Exercises

In Problems 1–4, an object attached to a coiled spring is pulled down a distance a from its rest position and then released. Assuming that the motion is simple harmonic with period T, write an equation that relates the displacement d of the object from its rest position after t seconds. Also assume that the positive direction of the motion is up.

1. $a = 5$; $T = 2$ seconds

2. $a = 10$; $T = 3$ seconds

3. $a = 6$; $T = \pi$ seconds

4. $a = 4$; $T = \dfrac{\pi}{2}$ seconds

5. Rework Problem 1 under the same conditions except that, at time $t = 0$, the object is at its resting position and moving down.

6. Rework Problem 2 under the same conditions except that, at time $t = 0$, the object is at its resting position and moving down.

7. Rework Problem 3 under the same conditions except that, at time $t = 0$, the object is at its resting position and moving down.

8. Rework Problem 4 under the same conditions except that, at time $t = 0$, the object is at its resting position and moving down.

In Problems 9–16, the displacement d (in meters) of an object at time t (in seconds) is given.
  (a) Describe the motion of the object.
  (b) What is the maximum displacement from its resting position?
  (c) What is the time required for one oscillation?
  (d) What is the frequency?

9. $d = 5\sin(3t)$

10. $d = 4\sin(2t)$

11. $d = 6\cos(\pi t)$

12. $d = 5\cos\left(\dfrac{\pi}{2}t\right)$

13. $d = -3\sin\left(\dfrac{1}{2}t\right)$

14. $d = -2\cos(2t)$

15. $d = 6 + 2\cos(2\pi t)$

16. $d = 4 + 3\sin(\pi t)$

In Problems 17–22, an object of mass m attached to a coiled spring with damping factor b is pulled down a distance a from its rest position and then released. Assume that the positive direction of the motion is up and the period of the first oscillation is T.
  (a) Write an equation that relates the distance d of the object from its rest position after t seconds.
  (b) Graph the equation found in part (a) for 5 oscillations using a graphing utility.

17. $m = 25$ grams; $a = 10$ centimeters; $b = 0.7$ gram/second; $T = 5$ seconds
18. $m = 20$ grams; $a = 15$ centimeters; $b = 0.75$ gram/second; $T = 6$ seconds
19. $m = 30$ grams; $a = 18$ centimeters; $b = 0.6$ gram/second; $T = 4$ seconds
20. $m = 15$ grams; $a = 16$ centimeters; $b = 0.65$ gram/second; $T = 5$ seconds
21. $m = 10$ grams; $a = 5$ centimeters; $b = 0.8$ gram/second; $T = 3$ seconds
22. $m = 10$ grams; $a = 5$ centimeters; $b = 0.7$ gram/second; $T = 3$ seconds

In Problems 23–28, the distance d (in meters) of the bob of a pendulum of mass m (in kilograms) from its rest position at time t (in seconds) is given.
  (a) Describe the motion of the object. Be sure to give the mass and damping factor.
  (b) What is the initial displacement of the bob? That is, what is the displacement at $t = 0$?
  (c) Graph the motion using a graphing utility.
  (d) What is the maximum displacement after the first oscillation?
  (e) What happens to the displacement of the bob as time increases without bound?

23. $d = -20e^{-0.7t/40}\cos\left(\sqrt{\left(\dfrac{2\pi}{5}\right)^2 - \dfrac{0.49}{1600}}\,t\right)$

24. $d = -20e^{-0.8t/40}\cos\left(\sqrt{\left(\dfrac{2\pi}{5}\right)^2 - \dfrac{0.64}{1600}}\,t\right)$

25. $d = -30e^{-0.6t/80}\cos\left(\sqrt{\left(\dfrac{2\pi}{7}\right)^2 - \dfrac{0.36}{6400}}\,t\right)$

26. $d = -30e^{-0.5t/70}\cos\left(\sqrt{\left(\dfrac{\pi}{2}\right)^2 - \dfrac{0.25}{4900}}\,t\right)$

27. $d = -15e^{-0.9t/30}\cos\left(\sqrt{\left(\dfrac{\pi}{3}\right)^2 - \dfrac{0.81}{900}}\,t\right)$

28. $d = -10e^{-0.8t/50}\cos\left(\sqrt{\left(\dfrac{2\pi}{3}\right)^2 - \dfrac{0.64}{2500}}\,t\right)$

In Problem 29–36, use the method of adding y-coordinates to graph each function. Verify your result using a graphing utility.

29. $f(x) = x + \cos x$

30. $f(x) = x + \cos(2x)$

31. $f(x) = x - \sin x$

32. $f(x) = x - \cos x$

33. $f(x) = \sin x + \cos x$

34. $f(x) = \sin(2x) + \cos x$

35. $g(x) = \sin x + \sin(2x)$

36. $g(x) = \cos(2x) + \cos x$

---

37. Charging a Capacitor If a charged capacitor is connected to a coil by closing a switch (see the figure), energy is transferred to the coil and then back to the capacitor in an oscillatory motion. The voltage V (in volts) across the capacitor will gradually diminish to 0 with time t (in seconds).

(a) By hand, graph the equation relating V and t:
$$V(t) = e^{-t/3}\cos(\pi t), \quad 0 \le t \le 3$$

(b) At what times t will the graph of V touch the graph of $y = e^{-t/3}$? When does V touch the graph of $y = -e^{-t/3}$?

(c) When will the voltage V be between -0.4 and 0.4 volt?

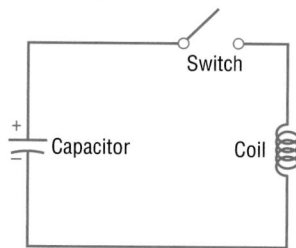
Switch

Capacitor   Coil

38. The Sawtooth Curve An oscilloscope often displays a *sawtooth curve*. This curve can be approximated by sinusoidal curves of varying periods and amplitudes.

(a) Use a graphing utility to graph the following function, which can be used to approximate the sawtooth curve.
$$f(x) = \frac{1}{2}\sin(2\pi x) + \frac{1}{4}\sin(4\pi x), \quad 0 \le x \le 2$$

(b) A better approximation to the sawtooth curve is given by
$$f(x) = \frac{1}{2}\sin(2\pi x) + \frac{1}{4}\sin(4\pi x) + \frac{1}{8}\sin(8\pi x)$$
Use a graphing utility to graph this function for $0 \le x \le 4$ and compare the result to the graph obtained in part (a).

(c) A third and even better approximation to the sawtooth curve is given by
$$f(x) = \frac{1}{2}\sin(2\pi x) + \frac{1}{4}\sin(4\pi x) + \frac{1}{8}\sin(8\pi x) + \frac{1}{16}\sin(16\pi x)$$
Use a graphing utility to graph this function for $0 \le x \le 4$ and compare the result to the graphs obtained in parts (a) and (b).

(d) What do you think the next approximation to the sawtooth curve is?

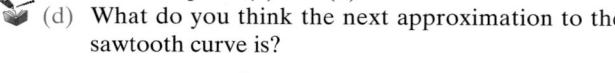

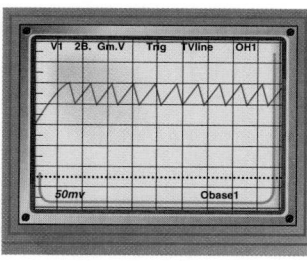

39. Touch-Tone Phones On a Touch-Tone phone, each button produces a unique sound. The sound produced is the sum of two tones, given by
$$y = \sin(2\pi l t) \quad \text{and} \quad y = \sin(2\pi h t)$$
where l and h are the low and high frequencies (cycles per second) shown on the illustration. For example, if you touch 7, the low frequency is $l = 852$ cycles per second and the high frequency is $h = 1209$ cycles per second. The sound emitted by touching 7 is
$$y = \sin[2\pi(852)t] + \sin[2\pi(1209)t]$$
Use a graphing utility to graph the sound emitted by touching 7.

Touch-Tone phone

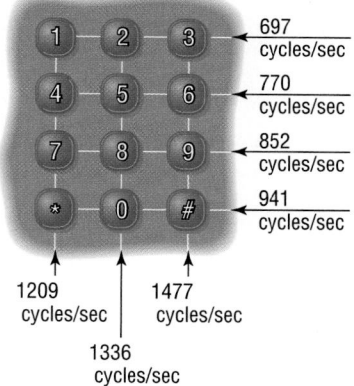

40. Use a graphing utility to graph the sound emitted by the * key on a Touch-Tone phone. See Problem 39.

41. Use a graphing utility to graph the function
$$f(x) = \frac{\sin x}{x}, x > 0.$$
Based on the graph, what do you conjecture about the value of $\frac{\sin x}{x}$ for x close to 0?

42. Use a graphing utility to graph $y = x\sin x$, $y = x^2 \sin x$, and $y = x^3 \sin x$ for $x > 0$. What patterns do you observe?

43. Use a graphing utility to graph $y = \frac{1}{x}\sin x$, $y = \frac{1}{x^2}\sin x$, and $y = \frac{1}{x^3}\sin x$ for $x > 0$. What patterns do you observe?

44. CBL Experiment Pendulum motion is analyzed to estimate simple harmonic motion. A plot is generated with the position of the pendulum over time. The graph is used to find a sinusoidal curve of the form $y = A\cos B(x - C) + D$. Determine the amplitude, period and frequency. (Activity 16, Real-World Math with the CBL System.)

45. CBL Experiment   The sound from a tuning fork is collected over time. The amplitude, frequency, and period of the graph are determined. A model of the form $y = A \cos B(x - C)$ is fitted to the data. (Activity 23, Real-World Math with the CBL System.)

46. How would you explain to a friend what simple harmonic motion is? How would you explain damped motion?

Chapter Review

Things to Know

Formulas

Law of Sines (p. 277)

$$\frac{\sin \alpha}{a} = \frac{\sin \beta}{b} = \frac{\sin \gamma}{c}$$

Law of Cosines (p. 288)

$$c^2 = a^2 + b^2 - 2ab \cos \gamma$$
$$b^2 = a^2 + c^2 - 2ac \cos \beta$$
$$a^2 = b^2 + c^2 - 2bc \cos \alpha$$

Area of a triangle (pp. 295–296)

$$A = \frac{1}{2}bh$$

$$A = \frac{1}{2}ab \sin \gamma$$

$$A = \frac{1}{2}bc \sin \alpha$$

$$A = \frac{1}{2}ac \sin \beta$$

$$A = \sqrt{s(s - a)(s - b)(s - c)}, \quad \text{where} \quad s = \frac{1}{2}(a + b + c)$$

Objectives

Section	You should be able to:	Review Exercises
4.1	1 Solve right triangles (p. 268)	1–4
	2 Solve applied problems using right triangle trigonometry (p. 269)	35–40
4.2	1 Solve SAA or ASA triangles (p. 278)	5–6, 22
	2 Solve SSA triangles (p. 280)	7–10, 12, 17–18, 21
	3 Solve applied problems using the Law of Sines (p. 282)	41, 43, 44
4.3	1 Solve SAS triangles (p. 289)	11, 15–16, 23–24
	2 Solve SSS triangles (p. 290)	13–14, 19–20
	3 Solve applied problems using the Law of Cosines (p. 290)	42, 45, 46, 47
4.4	1 Find the area of SAS triangles (p. 296)	25–28, 47–49
	2 Find the area of SSS triangles (p. 296)	29–32
4.5	1 Find an equation for an object in simple harmonic motion (p. 303)	57–58
	2 Analyze simple harmonic motion (p. 303)	53–58
	3 Analyze an object in damped motion (p. 304)	59–62
	4 Graph the sum of two functions (p. 305)	63, 64

Review Exercises

Blue problem numbers indicate the authors' suggestions for use in a Practice Test.

In Problems 1–4, solve each triangle.

1.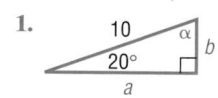

2.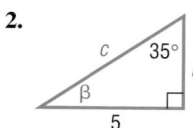

3.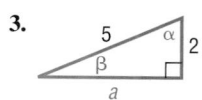

4.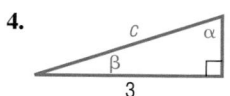

In Problems 5–24, find the remaining angle(s) and side(s) of each triangle, if it (they) exists. If no triangle exists, say "No triangle."

5. $\alpha = 50°$, $\beta = 30°$, $a = 1$
6. $\alpha = 10°$, $\gamma = 40°$, $c = 2$
7. $\alpha = 100°$, $a = 5$, $c = 2$

8. $a = 2$, $c = 5$, $\alpha = 60°$
9. $a = 3$, $c = 1$, $\gamma = 110°$
10. $a = 3$, $c = 1$, $\gamma = 20°$

11. $a = 3$, $c = 1$, $\beta = 100°$
12. $a = 3$, $b = 5$, $\beta = 80°$
13. $a = 2$, $b = 3$, $c = 1$

14. $a = 10$, $b = 7$, $c = 8$
15. $a = 1$, $b = 3$, $\gamma = 40°$
16. $a = 4$, $b = 1$, $\gamma = 100°$

17. $a = 5$, $b = 3$, $\alpha = 80°$
18. $a = 2$, $b = 3$, $\alpha = 20°$
19. $a = 1$, $b = \dfrac{1}{2}$, $c = \dfrac{4}{3}$

20. $a = 3$, $b = 2$, $c = 2$
21. $a = 3$, $\alpha = 10°$, $b = 4$
22. $a = 4$, $\alpha = 20°$, $\beta = 100°$

23. $c = 5$, $b = 4$, $\alpha = 70°$
24. $a = 1$, $b = 2$, $\gamma = 60°$

In Problems 25–34, find the area of each triangle.

25. $a = 2$, $b = 3$, $\gamma = 40°$
26. $b = 5$, $c = 5$, $\alpha = 20°$
27. $b = 4$, $c = 10$, $\alpha = 70°$

28. $a = 2$, $b = 1$, $\gamma = 100°$
29. $a = 4$, $b = 3$, $c = 5$
30. $a = 10$, $b = 7$, $c = 8$

31. $a = 4$, $b = 2$, $c = 5$
32. $a = 3$, $b = 2$, $c = 2$
33. $\alpha = 50°$, $\beta = 30°$, $a = 1$

34. $\alpha = 10°$, $\gamma = 40°$, $c = 3$

---

35. **Measuring the Length of a Lake**   From a stationary hot-air balloon 500 feet above the ground, two sightings of a lake are made (see the figure). How long is the lake?

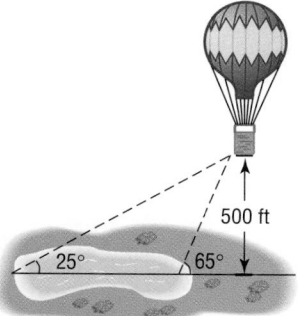

500 ft
25°   65°

36. **Finding the Speed of a Glider**   From a glider 200 feet above the ground, two sightings of a stationary object directly in front are taken 1 minute apart (see the figure). What is the speed of the glider?

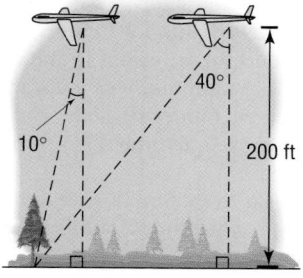

40°
10°
200 ft

37. **Finding the Width of a River**   Find the distance from A to C across the river illustrated in the figure.

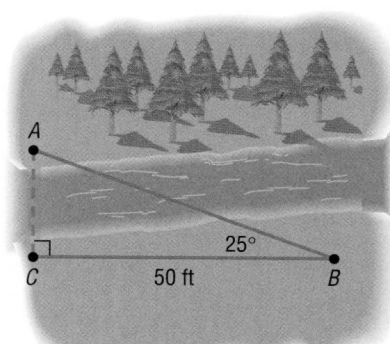

A
25°
C   50 ft   B

38. **Finding the Height of a Building**   Find the height of the building shown in the figure.

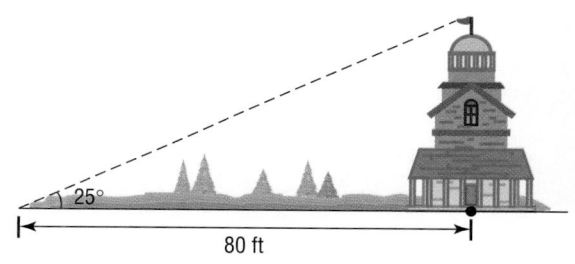

25°
80 ft

39. Finding the Distance to Shore   The Sears Tower in Chicago is 1454 feet tall and is situated about 1 mile inland from the shore of Lake Michigan, as indicated in the figure. An observer in a pleasure boat on the lake directly in front of the Sears Tower looks at the top of the tower and measures the angle of elevation as 5°. How far offshore is the boat?

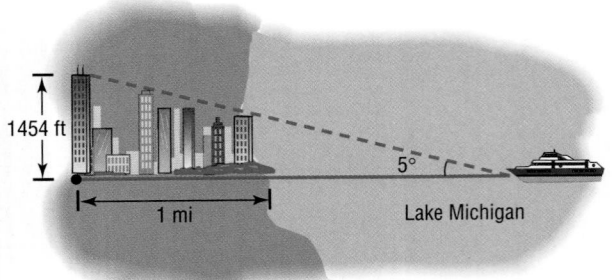

40. Finding the Grade of a Mountain Trail   A straight trail with a uniform inclination leads from a hotel, elevation 5000 feet, to a lake in a valley, elevation 4100 feet. The length of the trail is 4100 feet. What is the inclination of the trail?

41. Navigation   An airplane flies from city A to city B, a distance of 100 miles, and then turns through an angle of 20° and heads toward city C, as indicated in the figure. If the distance from A to C is 300 miles, how far is it from city B to city C?

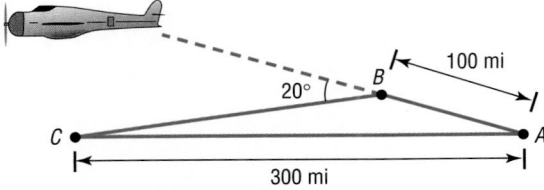

42. Correcting a Navigation Error   Two cities A and B are 300 miles apart. In flying from city A to city B, a pilot inadvertently took a course that was 5° in error.
(a) If the error was discovered after flying 10 minutes at a constant speed of 420 miles per hour, through what angle should the pilot turn to correct the course? (Consult the figure.)
(b) What new constant speed should be maintained so that no time is lost due to the error? (Assume that the speed would have been a constant 420 miles per hour if no error had occurred.)

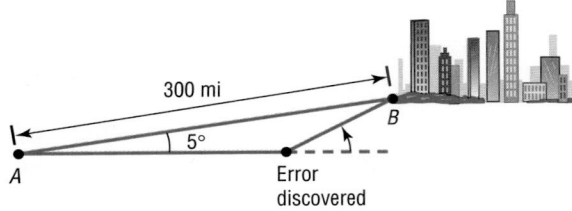

43. Determining Distances at Sea   Rebecca, the navigator of a ship at sea, spots two lighthouses that she knows to be 2 miles apart along a straight shoreline. She determines that the angles formed between two line-of-sight observations of the lighthouses and the line from the ship directly to shore are 12° and 30°. See the illustration.
(a) How far is the ship from lighthouse A?
(b) How far is the ship from lighthouse B?
(c) How far is the ship from shore?

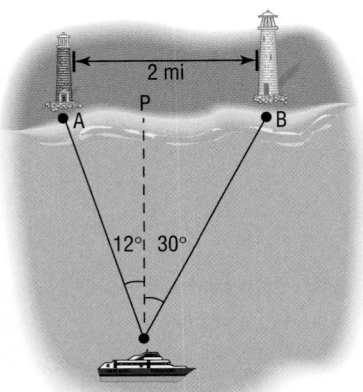

44. Constructing a Highway   A highway whose primary directions are north–south is being constructed along the west coast of Florida. Near Naples, a bay obstructs the straight path of the road. Since the cost of a bridge is prohibitive, engineers decide to go around the bay. The illustration shows the path that they decide on and the measurements taken. What is the length of highway needed to go around the bay?

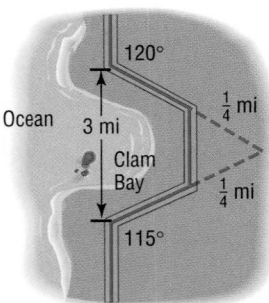

45. Correcting a Navigational Error   A sailboat leaves St. Thomas bound for an island in the British West Indies, 200 miles away. Maintaining a constant speed of 18 miles per hour, but encountering heavy crosswinds and strong currents, the crew finds after 4 hours that the sailboat is off course by 15°.
(a) How far is the sailboat from the island at this time?
(b) Through what angle should the sailboat turn to correct its course?

(c) How much time has been added to the trip because of this? (Assume that the speed remains at 18 miles per hour.)

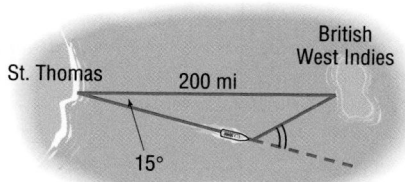

46. Surveying Two homes are located on opposite sides of a small hill. See the illustration. To measure the distance between them, a surveyor walks a distance of 50 feet from house *A* to point *C*, uses a transit to measure the angle *ACB*, which is found to be 80°, and then walks to house *B*, a distance of 60 feet. How far apart are the houses?

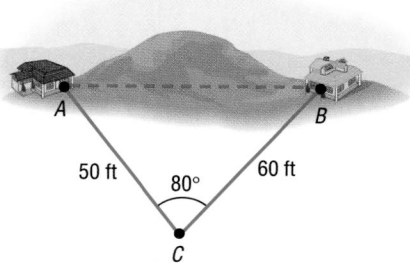

47. Approximating the Area of a Lake To approximate the area of a lake, Cindy walks around the perimeter of the lake, taking the measurements shown in the illustration. Using this technique, what is the approximate area of the lake?

[**Hint:** Use the Law of Cosines on the three triangles shown and then find the sum of their areas.]

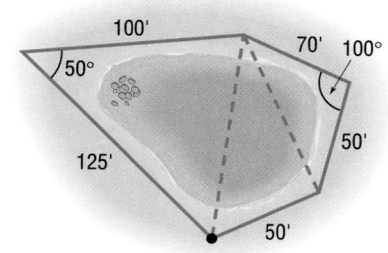

48. Calculating the Cost of Land The irregular parcel of land shown in the Figure is being sold for $100 per square foot. What is the cost of this parcel?

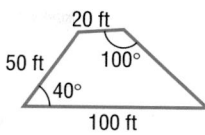

49. Area of a Segment Find the area of the segment of a circle whose radius is 6 inches formed by a central angle of 50°.

50. Finding the Bearing of a Ship The *Majesty* leaves the Port at Boston for Bermuda with a bearing of S80°E at an average speed of 10 knots. After 1 hour, the ship turns 90° toward the southwest. After 2 hours at an average speed of 20 knots, what is the bearing of the ship from Boston?

51. The drive wheel of an engine is 13 inches in diameter, and the pulley on the rotary pump is 5 inches in diameter. If the shafts of the drive wheel and the pulley are 2 feet apart, what length of belt is required to join them as shown in the figure?

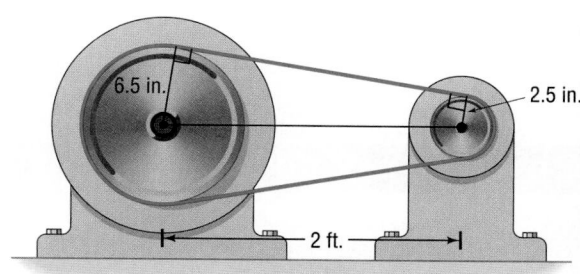

52. Rework Problem 51 if the belt is crossed, as shown in the figure.

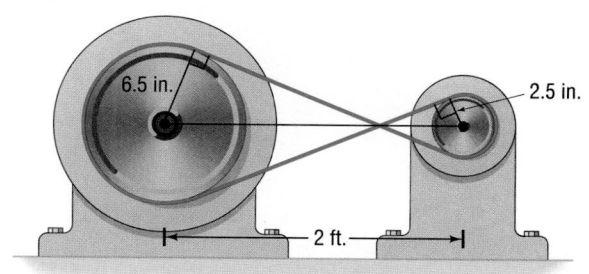

In Problems 53–56, the distance d (in feet) that an object travels in time t (in seconds) is given.
    (a) *Describe the motion of the object.*
    (b) *What is the maximum displacement from its resting position?*
    (c) *What is the time required for one oscillation?*
    (d) *What is the frequency?*

53. $d = 6\sin(2t)$ **54.** $d = 2\cos(4t)$ **55.** $d = -2\cos(\pi t)$ **56.** $d = -3\sin\left[\dfrac{\pi}{2}t\right]$

In Problems 57 and 58, an object attached to a coiled spring is pulled down a distance a from its rest position and then released. Assuming that the motion is simple harmonic with period T, write an equation that relates the displacement d of the object from its rest position after t seconds. Also assume that the positive direction of the motion is up.

57. $a = 4$, $T = 3$ seconds

58. $a = 5$, $T = 2$ seconds

In Problems 59 and 60, an object of mass m attached to a coiled spring with damping factor b is pulled down a distance a from its rest position and then released. Assume that the positive direction of the motion is up and the period of the first oscillation is T.

    (a)   *Write an equation that relates the distance d of the object from its rest position after t seconds.*
    (b)   *Graph the equation found in part (a) for 5 oscillations.*

59. $m = 40$ grams; $a = 15$ centimeters; $b = 0.75$ gram/second; $T = 5$ seconds

60. $m = 25$ grams; $a = 13$ centimeters; $b = 0.65$ gram/second; $T = 4$ seconds

In Problems 61 and 62, the distance d (in meters) of the bob of a pendulum of mass m (in kilograms) from its rest position at time t (in seconds) is given.

    (a)   *Describe the motion of the object.*
    (b)   *What is the initial displacement of the bob? That is, what is the displacement at $t = 0$?*
    (c)   *Graph the motion using a graphing utility.*
    (d)   *What is the maximum displacement after the first oscillation?*
    (e)   *What happens to the displacement of the bob as time increases without bound?*

61. $d = -15e^{-0.6t/40} \cos\left(\sqrt{\left(\dfrac{2\pi}{5}\right)^2 - \dfrac{0.36}{1600}}\, t\right)$

62. $d = -20e^{-0.5t/60} \cos\left(\sqrt{\left(\dfrac{2\pi}{3}\right)^2 - \dfrac{0.25}{3600}}\, t\right)$

In Problems 63 and 64, use the method of adding y-coordinates to graph each function. Verify your result using a graphing utility.

63. $y = 2 \sin x + \cos(2x)$

64. $y = 2 \cos(2x) + \sin \dfrac{x}{2}$

Chapter Projects

(a) Draw a spherical triangle and label each vertex by A, B, and C. Then connect each vertex by a radius to the center O of the sphere. Now, draw tangent lines to the sides a and b of the triangle that go through C. Extend the lines OA and OB to intersect the tangent lines at P and Q, respectively. See the diagram. List the plane right triangles. Determine the measures of the central angles.

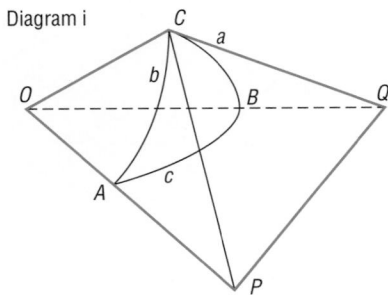

Diagram i

1. **A.** When the distance between two locations on the surface of the earth is small we can compute the distance in statutory miles. Using this assumption, we can use the Law of Sines and the Law of Cosines to approximate distances and angles. However, if you look at a globe, you notice that the Earth is a sphere, so as the distance between two points on its surface increases, the linear distance is less accurate because of curvature. Under this circumstance, we need to take into account the curvature of the Earth when using the Law of Sines and the Law of Cosines.

(b) Apply the Law of Cosines to triangles OPQ and CPQ to find two expressions for the length of PQ.
(c) Subtract the expressions in part (b) from each other. Solve for the term containing $\cos C$.
(d) Use the Pythagorean Theorem to find another value for $OQ^2 - CQ^2$ and $OP^2 - CP^2$. Now solve for $\cos C$.

(e) Replacing the ratios in part (d) by the cosines of the sides of the spherical triangle, you should now have the Law of Cosines for spherical triangles:

$$\cos C = \cos A \cos B + \sin A \sin B \cos C.$$

B. Lewis and Clark followed several rivers in their trek from what is now Great Falls, MT, to the Pacific coast. First, they went down the Missouri and Jefferson Rivers from Great Falls to Lemhi, ID. Because the two cities are on different longitudes and different latitudes, we must account for the curvature of the earth when computing the distance that they traveled. Assume that the radius of the earth is 3960 miles.

(a) Great Falls is at approximately 47.5°N and 111.3°W. Lemhi is at approximately 45.0°N and 113.5°W. (We will assume that the rivers flow straight from Great Falls to Lemhi on the surface of the earth.) This line is called a geodesic line. Apply the Law of Cosines for a spherical triangle to find the angle between Great Falls and Lemhi. (The central angles are found by using the differences in the latitudes and longitudes of the towns. See the diagram.) Then find the length of the arc joining the two towns. (Recall $s = r\theta$)

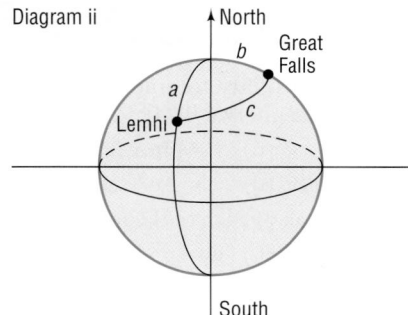

Diagram ii

(b) From Lemhi, they went up the Bitteroot River and the Snake River to what is now Lewiston and Clarkston on the border of Idaho and Washington. Although this is not really a side to a triangle, we will make a side that goes from Lemhi to Lewiston/Clarkston. If Lewiston and Clarkston is at about 46.5°N 117.0°W, find the distance from Lemhi using the Law of Cosines for a spherical triangle and the arc length.

(c) How far did the explorers travel just to get that far?

(d) Draw a plane triangle connecting the three towns. If the distance from Lewiston to Great Falls is 282 miles and the angle at Great Falls is 42° and the angle at Lewiston is 48.5°, find the distance from Great Falls to Lemhi and from Lemhi to Lewiston. How do these distances compare with the ones computed in parts (a) and (b)?

For spherical Law of Cosines: *Mathematics from the Birth of Numbers* by Jan Gullberg. WW Norton & Co Publishers. 1996. Pg. 491–494.

For Lewis and Clark Expedition: *American Journey: The Quest for Liberty to 1877, Texas Edition.* Prentice Hall. 1992. Pg. 345

For map coordinates: *National Geographic Atlas of the World*, published by National Geographic Society, 1981. Pg. 74–75

2. **Leaning Tower of Pisa**   PISA, Italy—Workers began removing two sets of steel suspenders attached to the leaning Tower of Pisa on Tuesday, in one of the final phases of a bold plan to partially straighten the famously tilted monument.

The 340 foot-long cables had been recurred to the tower in 1998 as a precaution in case it needed to be yanked back up while the soil under its foundation was being excavated. Anchored to giant winches dug into the ground about 100 yards from the tower, the suspenders haven't been needed.

Removal of the suspenders will be completed in time for an inauguration ceremony of the newly straightened tower scheduled for June 16, said Paolo Heiniger, who overseas the project.

The tower was closed to the public more than a decade ago, when officials feared it was beginning to lean so much that it might topple over. Work to stop the tower's increasing tilt has taken far longer than planned, but officials expect it to be open to tourists in the fall.

When work began the tower leaned 6 degrees, or 13 feet, off the perpendicular on its south side. By removing a small amount of soil, the tower has settled better and now leans about 16 inches less—nearly the tilt it had 300 years ago.

The decrease in lean isn't enough for the naked eye to detect but sufficient to stabilize the monument, experts have said.

Use the fact that the tower was 184.5 feet tall when it stood upright to answer the following questions:

(a) The article states that when work began, the tower leaned 6°, or 13 feet, off the perpendicular. Draw a sketch and label the two measurements.

(b) How high is the tower with a lean of 6°?

(c) The article goes on to say that the tower now leans 16 inches less. Draw a new sketch that shows this measurement.

(d) What is the degree measure of this lean?

(e) What is the height of the tower with this lean?

(f) Comment on the fact that $\sin 4° = \dfrac{13}{184.5} = 0.07046$, while $\sin 6° = 0.10453$. Can you reconcile this seeming discrepancy?

(g) Investigate further and write a report on your findings.

Source: Naples Daily News, Wednesday, May 16, 2001.

3. **Locating Lost Treasure**   While Scuba diving off Wreck Hill in Bermuda, a group of five entrepreneurs discovered a treasure map in a small watertight cask on a pirate schooner that had sunk in 1747. The map directed them to an area of Bermuda now known as The Flatts.

The directions on the map read as follows:

(1) From the tallest palm tree, sight the highest hill. Drop your eyes vertically until you sight the base of the hill.
(2) Turn 40° clockwise from that line and walk 70 paces to the big red rock.
(3) From the red rock walk 50 paces back to the sight line between the palm tree and the hill. Dig there.

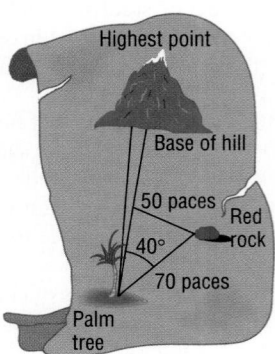

Upon reaching The Flatts, the five entrepreneurs believed that they had found the red rock and the highest hill in the vicinity, but the "tallest palm tree" had long since fallen and disintegrated. It occurred to them that the treasure must be located on a circle with radius 50 "paces" centered around the red rock, but they decided against digging a trench 942 feet in circumference, especially since they had no assurance that the treasure was still there. (They had decided that a "pace" must be about a yard.)

(a) Determine a plan to locate the position of the lost palm tree.
(b) One solution follows: From the location of the palm tree, turn 40° counter-clockwise from the rock to the hill, then go about 50 yards to the circle traced about the rock. Verify this solution.
(c) This location did not yield any treasure. Find the other solution and the treasure. Where is the treasure? How far is it from the palm tree?

C u m u l a t i v e   R e v i e w

1. Find the real solutions, if any, of the equation $3x^2 + 1 = 4x$.

2. Find an equation for the circle with center at the point $(-5, 1)$ and radius 3. Graph this circle.

3. What is the domain of the function $f(x) = \sqrt{x^2 - 3x - 4}$?

4. Graph the function $y = 3\sin(\pi x)$.

5. Graph the function $y = -2\cos(2x - \pi)$.

6. If $\tan \theta = -2$ and $\dfrac{3\pi}{2} < \theta < 2\pi$, find the exact value of:

   (a) $\sin \theta$ (b) $\cos \theta$ (c) $\sin(2\theta)$

   (d) $\cos(2\theta)$ (e) $\sin\left(\dfrac{1}{2}\theta\right)$ (f) $\cos\left(\dfrac{1}{2}\theta\right)$

7. Use a graphing utility to graph each of the following functions on the interval $[0, 4]$:

   (a) $y = e^x$ (b) $y = \sin x$
   (c) $y = e^x \sin x$ (d) $y = 2x + \sin x$

8. Sketch the graph of each of the following functions:

   (a) $y = x$ (b) $y = x^2$ (c) $y = \sqrt{x}$
   (d) $y = x^3$ (e) $y = \sin x$ (f) $y = \cos x$
   (g) $y = \tan x$

9. Solve the triangle and find its area.

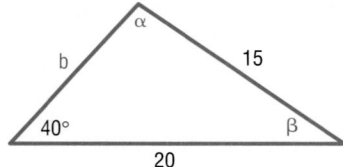

POLAR COORDINATES; VECTORS

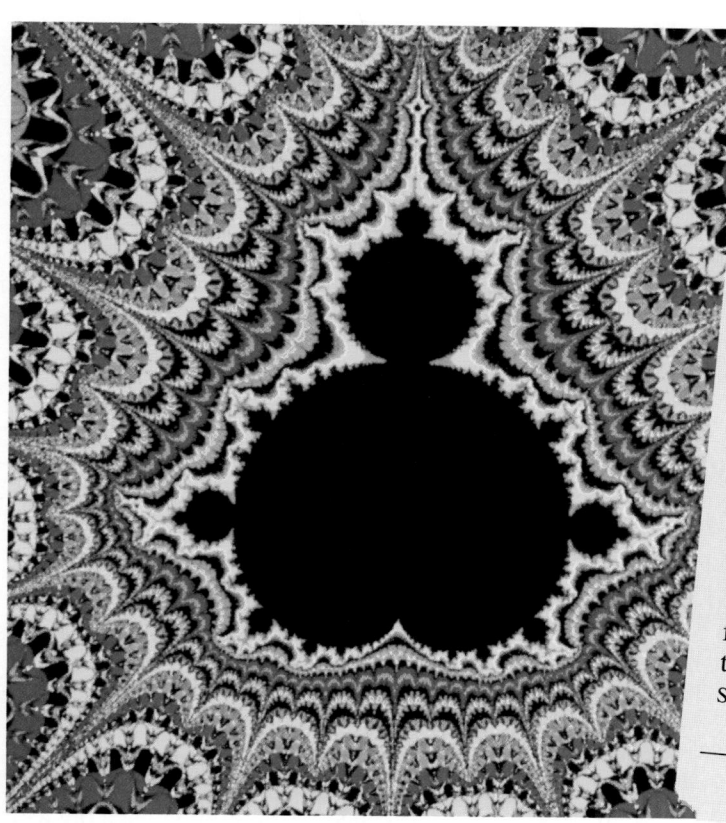

Multifractals and the Market

An extensive mathematical basis already exists for fractals and multifractals. Fractal patterns appear not just in the price changes of securities but in the distribution of galaxies throughout the cosmos, in the shape of coastlines and in the decorative designs generated by innumerable computer programs.

A fractal is a geometric shape that can be separated into parts, each of which is a reduced scale version of the whole. In finance, this concept is not a rootless abstraction but a theoretical reformulation of a down-to-earth bit of market folklore—namely, that movements of stock or currency all look alike when a market chart is enlarged or reduced so that it fits the same time and price scale. An observer then cannot tell which of the data concern prices that change from week to week, day to day or hour to hour. This quality defines charts as fractal curves and makes available many powerful tools of mathematical and computer analysis.

SOURCE: Benoit Mandlelbrot, Scientific American, February, 1999.

SEE CHAPTER PROJECT 1.

OUTLINE

For additional study help, go to

www.prenhall.com/sullivanegu3e

Materials include:

- Graphing Calculator Help
- Chapter Quiz
- Chapter Test
- PowerPoint Downloads
- Chapter Projects
- Student Tips

317

A Look Back, A Look Forward

This chapter is in two parts: Polar Coordinates, Sections 5.1–5.3, and Vectors, Sections 5.4–5.7. They are independent of each other and may be covered in any order.

Sections 5.1–5.3: In Chapter 1 we introduced rectangular coordinates (x, y) and discussed the graph of an equation in two variables involving x and y. In Sections 5.1 and 5.2, we introduce an alternative to rectangular coordinates, polar coordinates, and discuss graphing equations that involve polar coordinates. In

earlier courses, raising a real number to a real power was discussed. In Section 5.3 we extend this idea by raising a complex number to a real power. As it turns out, polar coordinates are useful for the discussion.

Sections 5.4–5.7: Often we are required to solve an equation to obtain a solution to applied problems. In the last four sections of this chapter, we develop the notion of a vector, which is the basis for solving certain types of applied problems, particularly in physics and engineering.

PREPARING FOR THIS SECTION

Before getting started, review the following:

✓ Rectangular Coordinates (Section 1.1, pp. 2–3)

✓ Definitions of the Sine and Cosine Functions (Section 2.4, p. 132)

✓ Inverse Tangent Function (Section 3.1, pp. 205–207)

✓ Completing the Square (Section A.3, pp. 579–580)

5.1   POLAR COORDINATES

OBJECTIVES
1. Plot Points Using Polar Coordinates
2. Convert from Polar Coordinates to Rectangular Coordinates
3. Convert from Rectangular Coordinates to Polar Coordinates

So far, we have always used a system of rectangular coordinates to plot points in the plane. Now we are ready to describe another system called *polar coordinates*. As we shall soon see, in many instances polar coordinates offer certain advantages over rectangular coordinates.

In a rectangular coordinate system, you will recall, a point in the plane is represented by an ordered pair of numbers (x, y), where x and y equal the signed distance of the point from the y-axis and x-axis, respectively. In a polar coordinate system, we select a point, called the **pole,** and then a ray with vertex at the pole, called the **polar axis.** Comparing the rectangular and polar coordinate systems, we see (in Figure 1) that the origin in rectangular coordinates coincides with the pole in polar coordinates, and the positive x-axis in rectangular coordinates coincides with the polar axis in polar coordinates.

A point P in a polar coordinate system is represented by an ordered pair of numbers (r, θ). If $r > 0$, then r is the distance of the point from the pole, while θ is the angle (in degrees or radians) formed by the polar axis and a ray from the pole through the point. We call the ordered pair (r, θ) the **polar coordinates** of the point. See Figure 2.

As an example, suppose that the polar coordinates of a point P are $\left(2, \dfrac{\pi}{4}\right)$. We locate P by first drawing an angle of $\dfrac{\pi}{4}$ radian, placing its vertex at the pole and its initial side along the polar axis. Then we go out a distance of 2 units along the terminal side of the angle to reach the point P. See Figure 3.

Figure 1

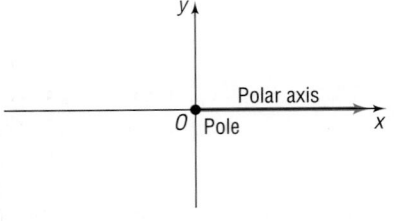

Figure 2

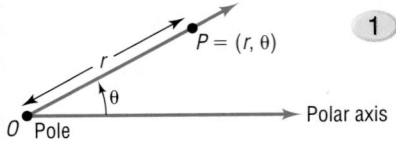

Figure 3

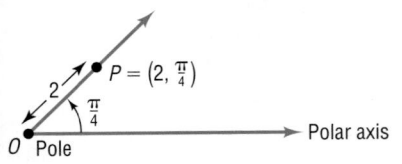

 NOW WORK PROBLEM **9.**

Recall that an angle measured counterclockwise is positive, whereas an angle measured clockwise is negative. This convention has some interesting consequences relating to polar coordinates. Let's see what these consequences are.

EXAMPLE 1   Finding Several Polar Coordinates of a Single Point

Consider again the point P with polar coordinates $\left(2, \dfrac{\pi}{4}\right)$, as shown in Figure 4(a). Because $\dfrac{\pi}{4}$, $\dfrac{9\pi}{4}$, and $-\dfrac{7\pi}{4}$ all have the same terminal side, we also could have located this point P by using the polar coordinates $\left(2, \dfrac{9\pi}{4}\right)$ or $\left(2, -\dfrac{7\pi}{4}\right)$, as shown in Figures 4(b) and (c).

Figure 4

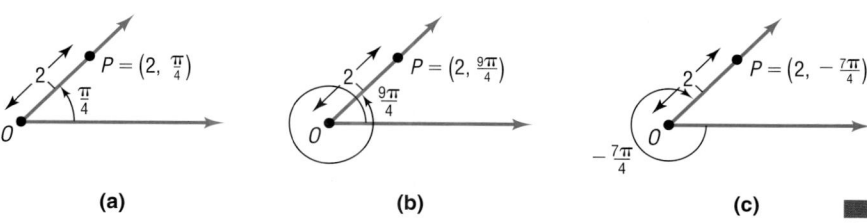

(a)   (b)   (c)   ■

Figure 5

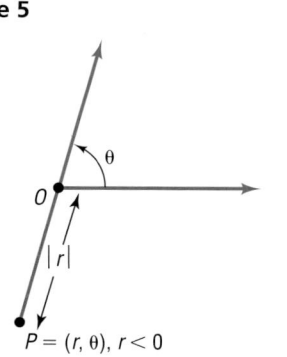

$P = (r, \theta),\ r < 0$

In using polar coordinates (r, θ), it is possible for the first entry r to be negative. When this happens, instead of the point being on the terminal side of θ, it is on the ray from the pole extending in the direction *opposite* the terminal side of θ at a distance $|r|$ from the pole. See Figure 5 for an illustration.

EXAMPLE 2   Polar Coordinates $(r, \theta),\ r < 0$

Consider again the point P with polar coordinates $\left(2, \dfrac{\pi}{4}\right)$, as shown in Figure 6(a). This same point P can be assigned the polar coordinates $\left(-2, \dfrac{5\pi}{4}\right)$, as indicated in Figure 6(b). To locate the point $\left(-2, \dfrac{5\pi}{4}\right)$, we use the ray in the opposite direction of $\dfrac{5\pi}{4}$ and go out 2 units along that ray to find the point P.

Figure 6

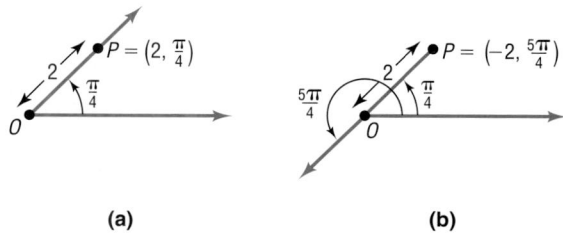

(a)   (b)   ■

These examples show a major difference between rectangular coordinates and polar coordinates. In the former, each point has exactly one pair of rectangular coordinates; in the latter, a point can have infinitely many pairs of polar coordinates.

EXAMPLE 3   Plotting Points Using Polar Coordinates

Plot the points with the following polar coordinates:

(a) $\left(3, \dfrac{5\pi}{3}\right)$ (b) $\left(2, -\dfrac{\pi}{4}\right)$ (c) $(3, 0)$ (d) $\left(-2, \dfrac{\pi}{4}\right)$

Solution   Figure 7 shows the points.

Figure 7

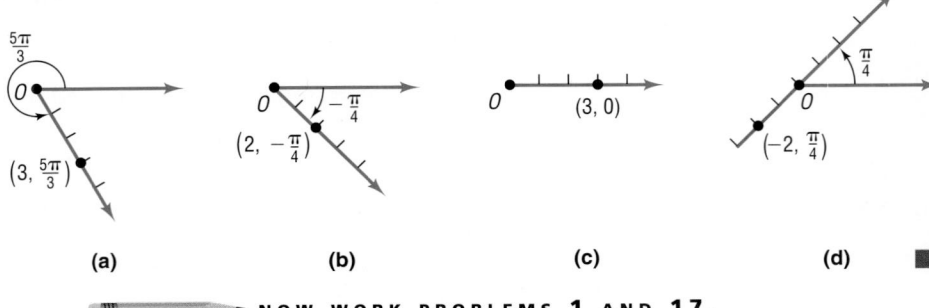

(a)   (b)   (c)   (d)

NOW WORK PROBLEMS **1** AND **17**.

EXAMPLE 4   Finding Other Polar Coordinates of a Given Point

Plot the point P with polar coordinates $\left(3, \dfrac{\pi}{6}\right)$, and find other polar coordinates (r, θ) of this same point for which:

(a) $r > 0$, $2\pi \le \theta < 4\pi$ (b) $r < 0$, $0 \le \theta < 2\pi$
(c) $r > 0$, $-2\pi \le \theta < 0$

Solution   The point $\left(3, \dfrac{\pi}{6}\right)$ is plotted in Figure 8.

Figure 8

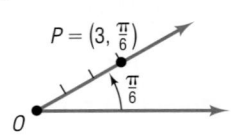

$P = \left(3, \dfrac{\pi}{6}\right)$

(a) We add 1 revolution (2π radians) to the angle $\dfrac{\pi}{6}$ to get
$P = \left(3, \dfrac{\pi}{6} + 2\pi\right) = \left(3, \dfrac{13\pi}{6}\right)$. See Figure 9(a).

(b) We add $\dfrac{1}{2}$ revolution (π radians) to the angle $\dfrac{\pi}{6}$ and replace 3 by -3 to get $P = \left(-3, \dfrac{\pi}{6} + \pi\right) = \left(-3, \dfrac{7\pi}{6}\right)$. See Figure 9(b).

(c) We subtract 2π from the angle $\dfrac{\pi}{6}$ to get $P = \left(3, \dfrac{\pi}{6} - 2\pi\right) = \left(3, -\dfrac{11\pi}{6}\right)$. See Figure 9(c).

Figure 9

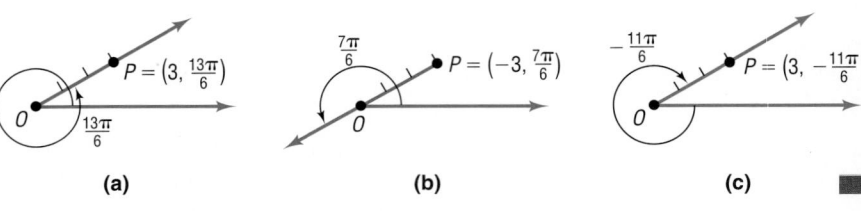

(a)   (b)   (c)

NOW WORK PROBLEM **21**.

SUMMARY

A point with polar coordinates (r, θ) also can be represented by either of the following:

$$(r, \theta + 2k\pi) \quad \text{or} \quad (-r, \theta + \pi + 2k\pi), \quad k \text{ any integer}$$

The polar coordinates of the pole are $(0, \theta)$, where θ can be any angle.

Conversion from Polar Coordinates to Rectangular Coordinates, and Vice Versa

2 It is sometimes convenient and, indeed, necessary to be able to convert co-ordinates or equations in rectangular form to polar form, and vice versa. To do this, we recall that the origin in rectangular coordinates is the pole in polar coordinates and that the positive x-axis in rectangular coordinates is the polar axis in polar coordinates.

Theorem

Conversion from Polar Coordinates to Rectangular Coordinates

If P is a point with polar coordinates (r, θ), the rectangular coordinates (x, y) of P are given by

$$x = r \cos \theta \qquad y = r \sin \theta \qquad (1)$$

Figure 10

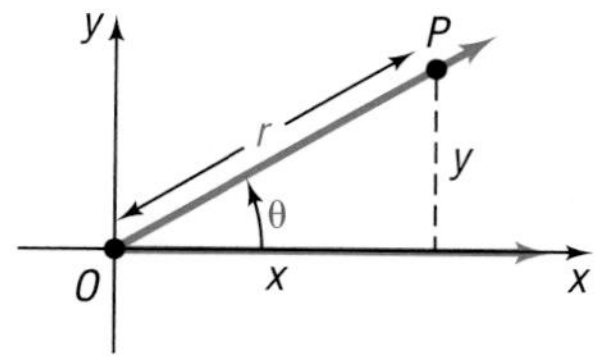

Proof Suppose that P has the polar coordinates (r, θ). We seek the rectangular coordinates (x, y) of P. Refer to Figure 10.

If $r = 0$, then, regardless of θ, the point P is the pole, for which the rectangular coordinates are $(0, 0)$. Formula (1) is valid for $r = 0$.

If $r > 0$, the point P is on the terminal side of θ, and $r = d(O, P) = \sqrt{x^2 + y^2}$. Since

$$\cos \theta = \frac{x}{r} \qquad \sin \theta = \frac{y}{r}$$

we have

$$x = r \cos \theta \qquad y = r \sin \theta$$

If $r < 0$, then the point $P = (r, \theta)$ can be represented as $(-r, \pi + \theta)$, where $-r > 0$. Since

$$\cos(\pi + \theta) = -\cos \theta = \frac{x}{-r} \qquad \sin(\pi + \theta) = -\sin \theta = \frac{y}{-r}$$

we have

$$x = r \cos \theta \qquad y = r \sin \theta$$

EXAMPLE 5 **Converting from Polar Coordinates to Rectangular Coordinates**

Find the rectangular coordinates of the points with the following polar coordinates:

(a) $\left(6, \dfrac{\pi}{6}\right)$ \qquad (b) $\left(-4, -\dfrac{\pi}{4}\right)$

Solution   We use formula (1): $x = r\cos\theta$ and $y = r\sin\theta$.

Figure 11

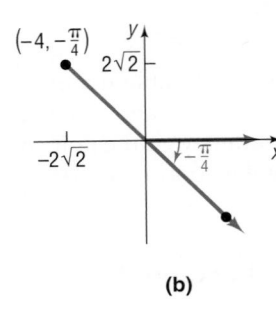

(a)

(b)

(a) Figure 11(a) shows $\left(6, \dfrac{\pi}{6}\right)$ plotted. With $r = 6$ and $\theta = \dfrac{\pi}{6}$, we have

$$x = r\cos\theta = 6\cos\frac{\pi}{6} = 6\cdot\frac{\sqrt{3}}{2} = 3\sqrt{3}$$

$$y = r\sin\theta = 6\sin\frac{\pi}{6} = 6\cdot\frac{1}{2} = 3$$

The rectangular coordinates of the point $\left(6, \dfrac{\pi}{6}\right)$ are $(3\sqrt{3}, 3)$.

(b) Figure 11(b) shows $\left(-4, -\dfrac{\pi}{4}\right)$ plotted. With $r = -4$ and $\theta = -\dfrac{\pi}{4}$, we have

$$x = r\cos\theta = -4\cos\left(-\frac{\pi}{4}\right) = -4\cdot\frac{\sqrt{2}}{2} = -2\sqrt{2}$$

$$y = r\sin\theta = -4\sin\left(-\frac{\pi}{4}\right) = -4\left(-\frac{\sqrt{2}}{2}\right) = 2\sqrt{2}$$

The rectangular coordinates of the point $\left(-4, -\dfrac{\pi}{4}\right)$ are $(-2\sqrt{2}, 2\sqrt{2})$. ∎

Figure 12

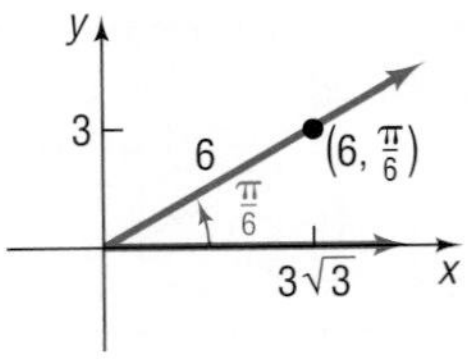

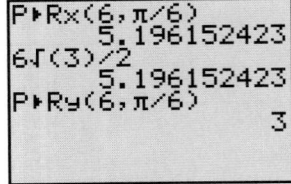

Most graphing calculators have the capability of converting from polar coordinates to rectangular coordinates. Consult you owner's manual for the proper keystrokes. Figure 12 verifies the results obtained in Example 5(a) using a TI-83. Note that the calculator is in radian mode.

━━━ **NOW WORK PROBLEMS 29 AND 41.**

③   Converting from rectangular coordinates (x, y) to polar coordinates (r, θ) is a little more complicated. Notice that we begin each example by plotting the given rectangular coordinates.

EXAMPLE 6   **Converting from Rectangular Coordinates to Polar Coordinates**

Find polar coordinates of a point whose rectangular coordinates are $(0, 3)$.

Figure 13

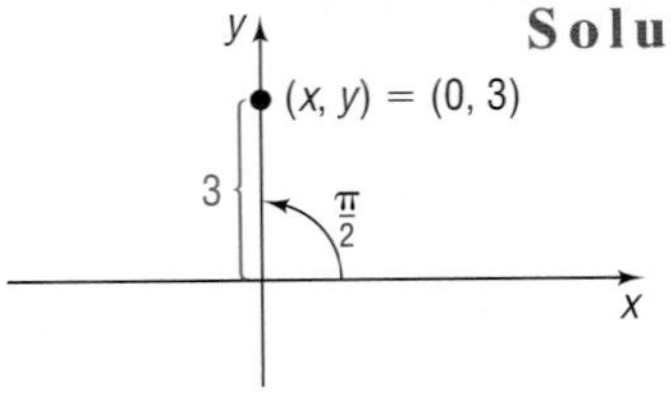

Solution   See Figure 13. The point $(0, 3)$ lies on the y-axis a distance of 3 units from the origin (pole), so $r = 3$. A ray with vertex at the pole through $(0, 3)$ forms an angle $\theta = \dfrac{\pi}{2}$ with the polar axis. Polar coordinates for this point can be given by $\left(3, \dfrac{\pi}{2}\right)$. ∎

Figure 14

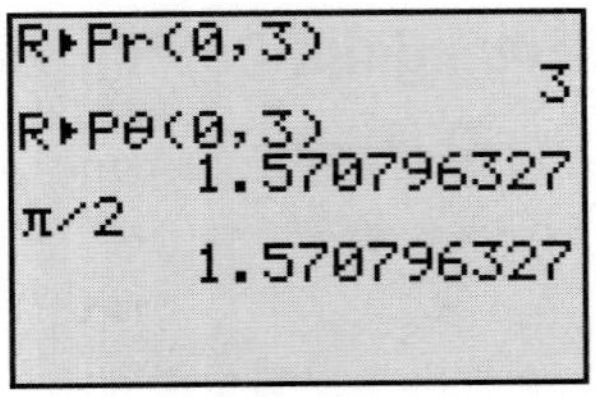

```
R▸Pr(0,3)
              3
R▸Pθ(0,3)
      1.570796327
π/2
      1.570796327
```

Most graphing calculators have the capability of converting from rectangular coordinates to polar coordinates. Consult you owner's manual for the proper keystrokes. Figure 14 verifies the results obtained in Example 6 using a TI-83. Note that the calculator is in radian mode.

Figure 15 shows polar coordinates of points that lie on either the x-axis or the y-axis. In each illustration, $a > 0$.

Figure 15

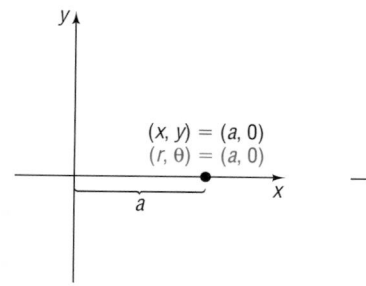

(a) $(x, y) = (a, 0), a > 0$

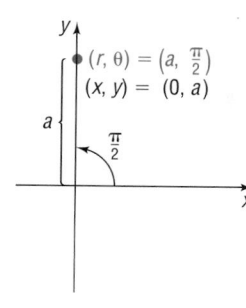

(b) $(x, y) = (0, a), a > 0$

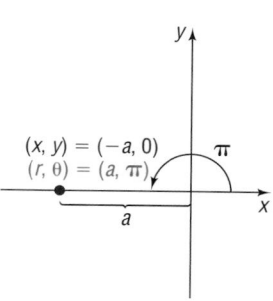

(c) $(x, y) = (-a, 0), a > 0$

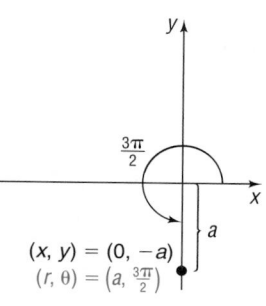

(d) $(x, y) = (0, -a), a > 0$

NOW WORK PROBLEM **45.**

EXAMPLE 7   **Converting from Rectangular Coordinates to Polar Coordinates**

Find polar coordinates of a point whose rectangular coordinates are:

(a) $(2, -2)$ (b) $(-1, -\sqrt{3})$

Solution   (a) See Figure 16(a). The distance r from the origin to the point $(2, -2)$ is

$$r = \sqrt{x^2 + y^2} = \sqrt{(2)^2 + (-2)^2} = \sqrt{8} = 2\sqrt{2}$$

We find θ by recalling that $\tan \theta = \dfrac{y}{x}$, so $\theta = \tan^{-1}\dfrac{y}{x}, \; -\dfrac{\pi}{2} < \theta < \dfrac{\pi}{2}$.
Since $(2, -2)$ lies in quadrant IV, we know that $-\dfrac{\pi}{2} < \theta < 0$. As a result,

$$\theta = \tan^{-1}\frac{y}{x} = \tan^{-1}\left(\frac{-2}{2}\right) = \tan^{-1}(-1) = -\frac{\pi}{4}$$

Figure 16

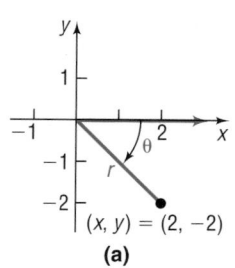

(a)

A set of polar coordinates for this point is $\left(2\sqrt{2}, -\dfrac{\pi}{4}\right)$. Other possible

representations include $\left(2\sqrt{2}, \dfrac{7\pi}{4}\right)$ and $\left(-2\sqrt{2}, \dfrac{3\pi}{4}\right)$.

(b) See Figure 16(b). The distance r from the origin to the point $(-1, -\sqrt{3})$ is

$$r = \sqrt{(-1)^2 + (-\sqrt{3})^2} = \sqrt{4} = 2$$

To find θ, we use $\theta = \tan^{-1}\dfrac{y}{x}, \; -\dfrac{\pi}{2} < \theta < \dfrac{\pi}{2}$. Since the point $(-1, -\sqrt{3})$

lies in quadrant III and the inverse tangent function gives an angle in
quadrant I, we add π to the result to obtain an angle in quadrant III.

$$\theta = \pi + \tan^{-1}\left(\frac{-\sqrt{3}}{-1}\right) = \pi + \tan^{-1}\sqrt{3} = \pi + \frac{\pi}{3} = \frac{4\pi}{3}$$

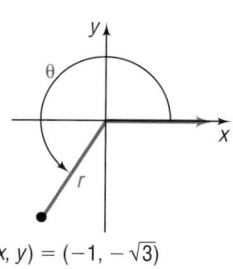

$(x, y) = (-1, -\sqrt{3})$

(b)

A set of polar coordinates is $\left(2, \dfrac{4\pi}{3}\right)$. Other possible representations

include $\left(-2, \dfrac{\pi}{3}\right)$ and $\left(2, -\dfrac{2\pi}{3}\right)$.

Figure 17 shows how to find polar coordinates of a point that lies in a quadrant when its rectangular coordinates (x, y) are given.

Figure 17

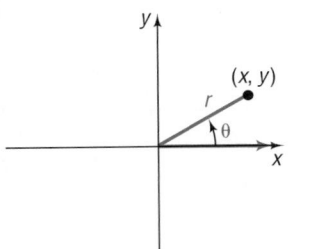
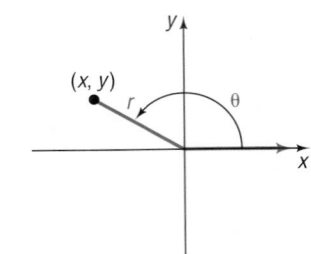
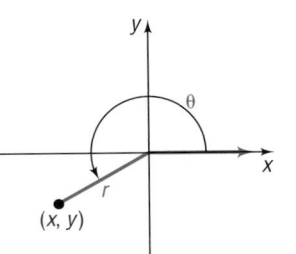
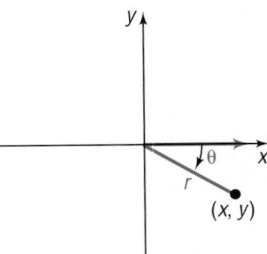

(a) $r = \sqrt{x^2 + y^2}$
$\theta = \tan^{-1}\dfrac{y}{x}$

(b) $r = \sqrt{x^2 + y^2}$
$\theta = \pi + \tan^{-1}\dfrac{y}{x}$

(c) $r = \sqrt{x^2 + y^2}$
$\theta = \pi + \tan^{-1}\dfrac{y}{x}$

(d) $r = \sqrt{x^2 + y^2}$
$\theta = \tan^{-1}\dfrac{y}{x}$

Based on the preceding discussion, we have the formulas

$$r^2 = x^2 + y^2 \qquad \tan\theta = \frac{y}{x} \qquad \text{if } x \neq 0 \qquad (2)$$

To use formula (2) effectively, follow these steps:

> **Steps for Converting from Rectangular to Polar Coordinates**
>
> **STEP 1:** Always plot the point (x, y) first, as we did in Examples 6 and 7.
> **STEP 2:** To find r, compute the distance from the origin to (x, y).
> **STEP 3:** To find θ, first determine the quadrant that the point lies in.
>
> Quadrant I: $\theta = \tan^{-1}\dfrac{y}{x}$ Quadrant II: $\theta = \pi + \tan^{-1}\dfrac{y}{x}$
>
> Quadrant III: $\theta = \pi + \tan^{-1}\dfrac{y}{x}$ Quadrant IV: $\theta = \tan^{-1}\dfrac{y}{x}$

 **NOW WORK PROBLEM 49.**

Formulas (1) and (2) may also be used to transform equations.

EXAMPLE 8   **Transforming an Equation from Polar to Rectangular Form**

Transform the equation $r = 4\sin\theta$ from polar coordinates to rectangular coordinates, and identify the graph.

Solution   If we multiply each side by r, it will be easier to apply formulas (1) and (2).

$$r = 4\sin\theta$$
$$r^2 = 4r\sin\theta \qquad \text{Multiply each side by } r.$$
$$x^2 + y^2 = 4y \qquad r^2 = x^2 + y^2; \quad y = r\sin\theta$$

This is the equation of a circle; we proceed to complete the square to obtain the standard form of the equation.

$$x^2 + (y^2 - 4y) = 0 \qquad \text{\small General form}$$

$$x^2 + (y^2 - 4y + 4) = 4 \qquad \text{\small Complete the square in } y.$$

$$x^2 + (y - 2)^2 = 4 \qquad \text{\small Standard form}$$

The center of the circle is at $(0, 2)$, and its radius is 2. ■

NOW WORK PROBLEM 65.

EXAMPLE 9 **Transforming an Equation from Rectangular to Polar Form**

Transform the equation $4xy = 9$ from rectangular coordinates to polar coordinates.

Solution We use formula (1).

$$4xy = 9$$

$$4(r \cos \theta)(r \sin \theta) = 9 \qquad \text{\small } x = r\cos\theta, y = r\sin\theta$$

$$4r^2 \cos \theta \sin \theta = 9$$

$$2r^2(2 \sin \theta \cos \theta) = 9$$

$$2r^2 \sin(2\theta) = 9 \qquad \text{\small Double-angle formula} \quad ■$$

5.1 Concepts and Vocabulary

In Problems 1–3, fill in the blanks.

1. In polar coordinates, the origin is called the _____ and the positive x-axis is referred to as the _____ _____ _____.

2. Another representation in polar coordinates for the point $\left(2, \dfrac{\pi}{3}\right)$ is $\left(\underline{\hspace{1cm}}, \dfrac{4\pi}{3}\right)$.

3. The polar coordinates $\left(-2, \dfrac{\pi}{6}\right)$ are represented in rectangular coordinates by $(\underline{\hspace{1cm}}, \underline{\hspace{1cm}})$.

In Problems 4–6, answer True or False to each statement.

4. The polar coordinates of a point are unique.
5. The rectangular coordinates of a point are unique.
6. In (r, θ), the number r can be negative.

7. In converting from polar coordinates to rectangular coordinates, what formulas will you use?
8. Explain how you proceed to convert from rectangular coordinates to polar coordinates.
9. Is the street system in your town based on a rectangular coordinate system, a polar coordinate system, or some other system? Explain.

5.1 Exercises

In Problems 1–8, match each point in polar coordinates with either A, B, C, or D on the graph.

1. $\left(2, -\dfrac{11\pi}{6}\right)$ **2.** $\left(-2, -\dfrac{\pi}{6}\right)$ **3.** $\left(-2, \dfrac{\pi}{6}\right)$ **4.** $\left(2, \dfrac{7\pi}{6}\right)$

5. $\left(2, \dfrac{5\pi}{6}\right)$ **6.** $\left(-2, \dfrac{5\pi}{6}\right)$ **7.** $\left(-2, \dfrac{7\pi}{6}\right)$ **8.** $\left(2, \dfrac{11\pi}{6}\right)$

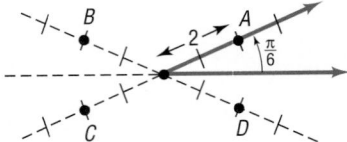

In Problems 9–20, plot each point given in polar coordinates.

9. $(3, 90°)$ **10.** $(4, 270°)$ **11.** $(-2, 0)$ **12.** $(-3, \pi)$

13. $\left(6, \dfrac{\pi}{6}\right)$ **14.** $\left(5, \dfrac{5\pi}{3}\right)$ **15.** $(-2, 135°)$ **16.** $(-3, 120°)$

17. $\left(-1, -\dfrac{\pi}{3}\right)$ **18.** $\left(-3, -\dfrac{3\pi}{4}\right)$ **19.** $(-2, -\pi)$ **20.** $\left(-3, -\dfrac{\pi}{2}\right)$

In Problems 21–28, plot each point given in polar coordinates, and find other polar coordinates (r, θ) of the point for which:

(a) $r > 0,\quad -2\pi \le \theta < 0$ (b) $r < 0,\quad 0 \le \theta < 2\pi$ (c) $r > 0,\quad 2\pi \le \theta < 4\pi$

21. $\left(5, \dfrac{2\pi}{3}\right)$ **22.** $\left(4, \dfrac{3\pi}{4}\right)$ **23.** $(-2, 3\pi)$ **24.** $(-3, 4\pi)$

25. $\left(1, \dfrac{\pi}{2}\right)$ **26.** $(2, \pi)$ **27.** $\left(-3, -\dfrac{\pi}{4}\right)$ **28.** $\left(-2, -\dfrac{2\pi}{3}\right)$

In Problems 29–44, polar coordinates of a point are given. Find the rectangular coordinates of each point. Verify your results using a graphing utility.

29. $\left(3, \dfrac{\pi}{2}\right)$ **30.** $\left(4, \dfrac{3\pi}{2}\right)$ **31.** $(-2, 0)$ **32.** $(-3, \pi)$

33. $(6, 150°)$ **34.** $(5, 300°)$ **35.** $\left(-2, \dfrac{3\pi}{4}\right)$ **36.** $\left(-2, \dfrac{2\pi}{3}\right)$

37. $\left(-1, -\dfrac{\pi}{3}\right)$ **38.** $\left(-3, -\dfrac{3\pi}{4}\right)$ **39.** $(-2, -180°)$ **40.** $(-3, -90°)$

41. $(7.5, 110°)$ **42.** $(-3.1, 182°)$ **43.** $(6.3, 3.8)$ **44.** $(8.1, 5.2)$

In Problems 45–56, the rectangular coordinates of a point are given. Find polar coordinates for each point. Verify your results using a graphing utility.

45. $(3, 0)$ **46.** $(0, 2)$ **47.** $(-1, 0)$ **48.** $(0, -2)$

49. $(1, -1)$ **50.** $(-3, 3)$ **51.** $(\sqrt{3}, 1)$ **52.** $(-2, -2\sqrt{3})$

53. $(1.3, -2.1)$ **54.** $(-0.8, -2.1)$ **55.** $(8.3, 4.2)$ **56.** $(-2.3, 0.2)$

In Problems 57–64, the letters x and y represent rectangular coordinates. Write each equation using polar coordinates (r, θ).

57. $2x^2 + 2y^2 = 3$ **58.** $x^2 + y^2 = x$ **59.** $x^2 = 4y$ **60.** $y^2 = 2x$

61. $2xy = 1$ **62.** $4x^2 y = 1$ **63.** $x = 4$ **64.** $y = -3$

In Problems 65–72, the letters r and θ represent polar coordinates. Write each equation using rectangular coordinates (x, y).

65. $r = \cos\theta$ **66.** $r = \sin\theta + 1$ **67.** $r^2 = \cos\theta$ **68.** $r = \sin\theta - \cos\theta$

69. $r = 2$ **70.** $r = 4$ **71.** $r = \dfrac{4}{1 - \cos\theta}$ **72.** $r = \dfrac{3}{3 - \cos\theta}$

73. Show that the formula for the distance d between two points $P_1 = (r_1, \theta_1)$ and $P_2 = (r_2, \theta_2)$ is

$$d = \sqrt{r_1^2 + r_2^2 - 2r_1 r_2 \cos(\theta_2 - \theta_1)}$$

PREPARING FOR THIS SECTION

Before getting started, review the following:

✓ Introduction to Graphing Equations
(Section 1.2, pp. 12–16)

✓ Even-Odd Properties of Trigonometric Functions
(Section 2.5, p. 150)

✓ Circles (Section 1.3, pp. 26–30)

✓ Difference Formulas (Section 3.4, pp. 222 and 225)

✓ Value of the Sine and Cosine Functions at Certain Angles
(Section 2.3, p. 126 and Section 2.4, p. 134)

5.2 POLAR EQUATIONS AND GRAPHS

OBJECTIVES

1. Graph and Identify Polar Equations by Converting to Rectangular Equations
2. Graph Polar Equations Using a Graphing Utility
3. Test Polar Equations for Symmetry
4. Graph Polar Equations by Plotting Points

Just as a rectangular grid may be used to plot points given by rectangular coordinates, as in Figure 18(a), we can use a grid consisting of concentric circles (with centers at the pole) and rays (with vertices at the pole) to plot points given by polar coordinates, as shown in Figure 18(b). We shall use such **polar grids** to graph *polar equations*.

Figure 18

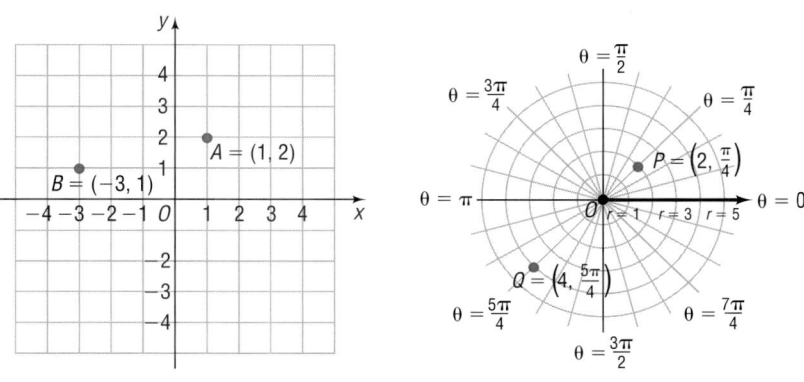

(a) Rectangular grid   **(b)** Polar grid

An equation whose variables are polar coordinates is called a **polar equation**. The **graph of a polar equation** consists of all points whose polar coordinates satisfy the equation.

1

One method that we can use to graph a polar equation is to convert the equation to rectangular coordinates. In the discussion that follows, (x, y) represent the rectangular coordinates of a point P, and (r, θ) represent polar coordinates of the point P.

EXAMPLE 1   **Identifying and Graphing a Polar Equation by Hand (Circle)**

Identify and graph the equation: $r = 3$

Solution   We convert the polar equation to a rectangular equation.

$$r = 3$$

$$r^2 = 9 \qquad \text{Square both sides.}$$

$$x^2 + y^2 = 9 \qquad r^2 = x^2 + y^2$$

The graph of $r = 3$ is a circle, with center at the pole and radius 3. See Figure 19.

Figure 19
$r = 3$ or $x^2 + y^2 = 9$

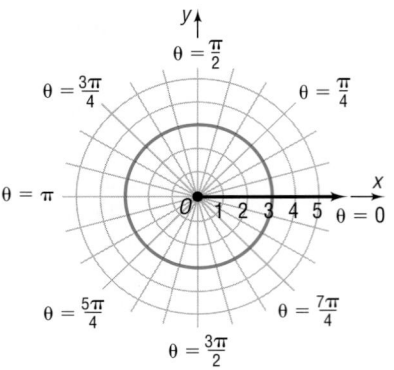

NOW WORK PROBLEM 1.

EXAMPLE 2   **Identifying and Graphing a Polar Equation by Hand (Line)**

Identify and graph the equation: $\theta = \dfrac{\pi}{4}$

Figure 20

$\theta = \dfrac{\pi}{4}$ or $y = x$

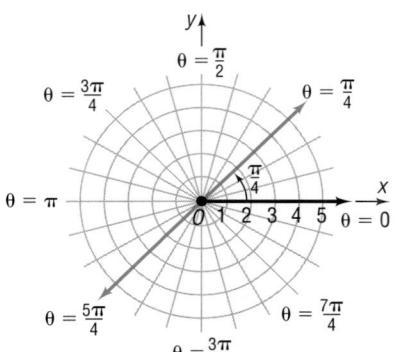

Solution   We convert the polar equation to a rectangular equation.

$$\theta = \frac{\pi}{4}$$

$$\tan \theta = \tan \frac{\pi}{4} = 1$$

$$\frac{y}{x} = 1 \qquad \tan \theta = \frac{y}{x}$$

$$y = x$$

The graph of $\theta = \dfrac{\pi}{4}$ is a line passing through the pole making an angle of $\dfrac{\pi}{4}$ with the polar axis. See Figure 20.

NOW WORK PROBLEM 3.

EXAMPLE 3   **Identifying and Graphing a Polar Equation by Hand (Horizontal Line)**

Identify and graph the equation: $r \sin \theta = 2$

Solution   Since $y = r \sin \theta$, we can write the equation as

$$y = 2$$

We conclude that the graph of $r \sin \theta = 2$ is a horizontal line 2 units above the pole. See Figure 21.

Figure 21
$r \sin \theta = 2$ or $y = 2$

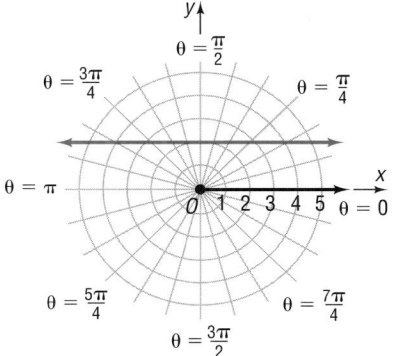

2   A second method we can use to graph a polar equation is to graph the equation using a graphing utility.

Most graphing utilities require the following steps in order to obtain the graph of an equation:

> ### *Graphing a Polar Equation Using a Graphing Utility*
>
> **STEP 1:**   Solve the equation for r in terms of θ.
>
> **STEP 2:**   Select the viewing window in POLar mode. In addition to setting Xmin, Xmax, Xscl, and so forth, the viewing window in polar mode requires setting minimum and maximum values for θ and an increment setting for θ (θstep). Finally, a square screen and radian measure should be used.
>
> **STEP 3:**   Enter the expression involving θ that you found in Step 1. (Consult your manual for the correct way to enter the expression.)
>
> **STEP 4:**   Graph.

EXAMPLE 4   **Graphing a Polar Equation Using a Graphing Utility**

Use a graphing utility to graph the polar equation $r \sin \theta = 2$.

Solution   **STEP 1:** We solve the equation for r in terms of θ.

$$r \sin \theta = 2$$

$$r = \frac{2}{\sin \theta}$$

STEP 2: From the polar mode, select a square viewing window. We will use the one given next.

$$\theta\text{min} = 0 \qquad X\text{min} = -9 \qquad Y\text{min} = -6$$
$$\theta\text{max} = 2\pi \qquad X\text{max} = 9 \qquad Y\text{max} = 6$$
$$\theta\text{step} = \frac{\pi}{24} \qquad X\text{scl} = 1 \qquad Y\text{scl} = 1$$

θstep determines the number of points the graphing utility will plot. For example, if θstep is $\dfrac{\pi}{24}$, then the graphing utility will evaluate r at $\theta = 0(\theta\text{min}), \dfrac{\pi}{24}, \dfrac{2\pi}{24}, \dfrac{3\pi}{24}$, and so forth, up to $2\pi(\theta\text{max})$. The smaller θstep, the more points the graphing utility will plot. The student is encouraged to experiment with different values for θmin, θmax, and θstep to see how the graph is affected.

Figure 22

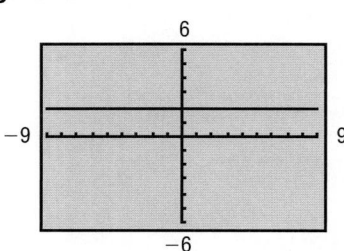

STEP 3: Enter the expression $\dfrac{2}{\sin\theta}$ after the prompt $r = \quad$.

STEP 4: Graph.

The graph is shown in Figure 22. ■

EXAMPLE 5   Identifying and Graphing a Polar Equation (Vertical Line)

Identify and graph the equation: $r\cos\theta = -3$

Solution   Since $x = r\cos\theta$, we can write the equation as

$$x = -3$$

We conclude that the graph of $r\cos\theta = -3$ is a vertical line 3 units to the left of the pole. Figure 23(a) shows the graph drawn by hand. Figure 23(b) shows the graph using a graphing utility with $\theta\text{min} = 0, \theta\text{max} = 2\pi$, and $\theta\text{step} = \dfrac{\pi}{24}$.

Figure 23
$r\cos\theta = -3$ or $x = -3$

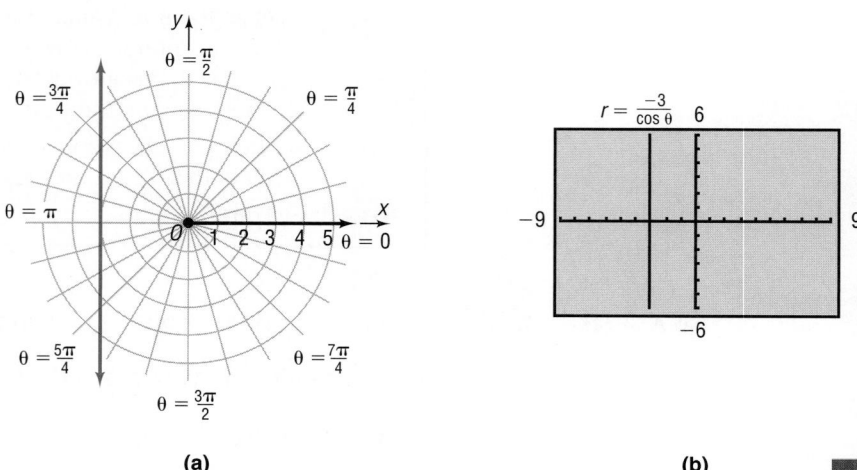

(a)   (b) ■

Based on Examples 3, 4, and 5, we are led to the following results. (The proofs are left as exercises.)

Theorem   Let a be a nonzero real number. Then the graph of the equation

$$r \sin \theta = a$$

is a horizontal line a units above the pole if $a > 0$ and $|a|$ units below the pole if $a < 0$.
   The graph of the equation

$$r \cos \theta = a$$

is a vertical line a units to the right of the pole if $a > 0$ and $|a|$ units to the left of the pole if $a < 0$.

───── **NOW WORK PROBLEM 7.**

EXAMPLE 6   **Identifying and Graphing a Polar Equation (Circle)**

Identify and graph the equation: $r = 4 \sin \theta$

Solution   To transform the equation to rectangular coordinates, we multiply each side by r.

$$r^2 = 4r \sin \theta$$

Now we use the facts that $r^2 = x^2 + y^2$ and $y = r \sin \theta$. Then

$$x^2 + y^2 = 4y$$
$$x^2 + (y^2 - 4y) = 0$$
$$x^2 + (y^2 - 4y + 4) = 4 \qquad \text{Complete the square in } y.$$
$$x^2 + (y - 2)^2 = 4 \qquad \text{Standard equation of a circle}$$

This is the equation of a circle with center at $(0, 2)$ in rectangular coordinates and radius 2. Figure 24(a) shows the graph drawn by hand. Figure 24(b) shows the graph using a graphing utility with $\theta\text{min} = 0, \theta\text{max} = 2\pi$, and $\theta\text{step} = \dfrac{\pi}{24}$.

Figure 24
$r = 4 \sin \theta$ or $x^2 + (y - 2)^2 = 4$

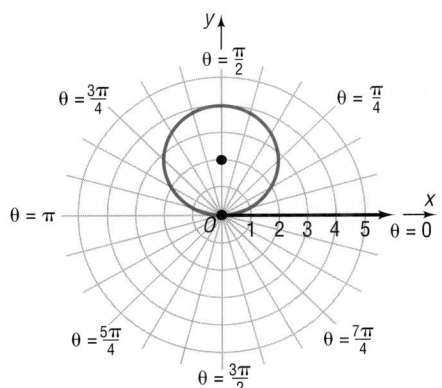

(a)

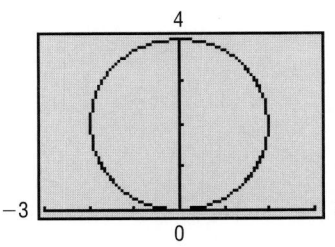

(b)

| EXAMPLE 7 | Identifying and Graphing a Polar Equation (Circle) |

Identify and graph the equation: $r = -2\cos\theta$

Solution We proceed as in Example 6.

$$r^2 = -2r\cos\theta \qquad \text{Multiply both sides by } r.$$

$$x^2 + y^2 = -2x \qquad r^2 = x^2 + y^2; \quad x = r\cos\theta$$

$$x^2 + 2x + y^2 = 0$$

$$(x^2 + 2x + 1) + y^2 = 1 \qquad \text{Complete the square in } x.$$

$$(x + 1)^2 + y^2 = 1 \qquad \text{Standard equation of a circle}$$

This is the equation of a circle with center at $(-1, 0)$ in rectangular coordinates and radius 1. Figure 25(a) shows the graph drawn by hand. Figure 25(b) shows the graph using a graphing utility with $\theta\text{min} = 0$, $\theta\text{max} = 2\pi$, and $\theta\text{step} = \dfrac{\pi}{24}$.

Figure 25
$r = -2\cos\theta$ or $(x + 1)^2 + y^2 = 1$

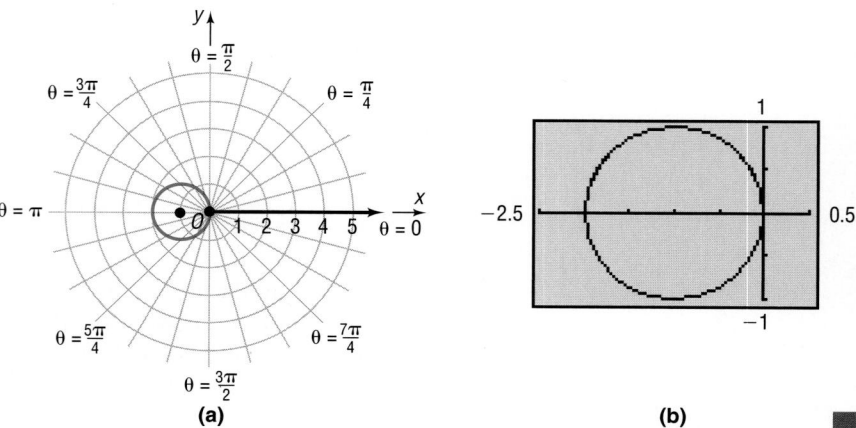

(a)   (b)

 EXPLORATION Using a square screen, graph $r_1 = \sin\theta$, $r_2 = 2\sin\theta$, and $r_3 = 3\sin\theta$. Do you see the pattern? Clear the screen and graph $r_1 = -\sin\theta$, $r_2 = -2\sin\theta$, and $r_3 = -3\sin\theta$. Do you see the pattern? Clear the screen and graph $r_1 = \cos\theta$, $r_2 = 2\cos\theta$, and $r_3 = 3\cos\theta$. Do you see the pattern? Clear the screen and graph $r_1 = -\cos\theta$, $r_2 = -2\cos\theta$, and $r_3 = -3\cos\theta$. Do you see the pattern?

Based on Examples 6 and 7 and the preceding Exploration, we are led to the following results. (The proofs are left as exercises.)

Theorem	Let a be a positive real number. Then,	

Equation	Description
(a) $r = 2a \sin \theta$	Circle: radius a; center at $(0, a)$ in rectangular coordinates
(b) $r = -2a \sin \theta$	Circle: radius a; center at $(0, -a)$ in rectangular coordinates
(c) $r = 2a \cos \theta$	Circle: radius a; center at $(a, 0)$ in rectangular coordinates
(d) $r = -2a \cos \theta$	Circle: radius a; center at $(-a, 0)$ in rectangular coordinates

Each circle passes through the pole.

━━━ **NOW WORK PROBLEM 9.**

The method of converting a polar equation to an identifiable rectangular equation in order to obtain the graph is not always helpful, nor is it always necessary. Usually, we set up a table that lists several points on the graph. By checking for symmetry, it may be possible to reduce the number of points needed to draw the graph.

Symmetry

③ In polar coordinates, the points (r, θ) and $(r, -\theta)$ are symmetric with respect to the polar axis (and to the x-axis). See Figure 26(a). The points (r, θ) and $(r, \pi - \theta)$ are symmetric with respect to the line $\theta = \dfrac{\pi}{2}$ (the y-axis). See Figure 26(b). The points (r, θ) and $(-r, \theta)$ are symmetric with respect to the pole (the origin). See Figure 26(c).

Figure 26

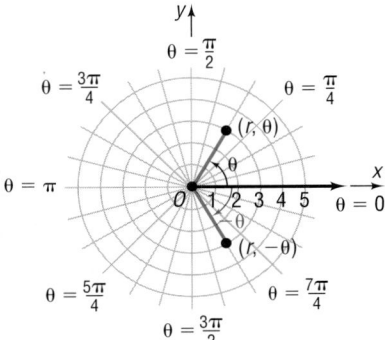

(a) Points symmetric with respect to the polar axis

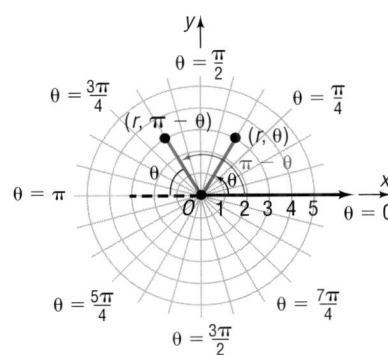

(b) Points symmetric with respect to the line $\theta = \dfrac{\pi}{2}$

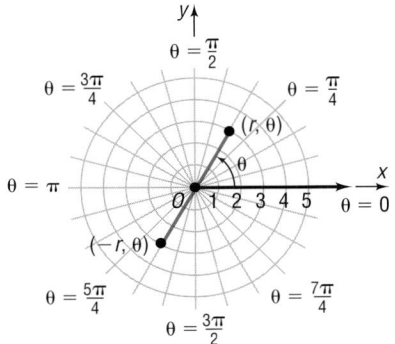

(c) Points symmetric with respect to the pole

The following tests are a consequence of these observations.

Theorem

Tests for Symmetry

Symmetry with Respect to the Polar Axis (*x*-Axis)

In a polar equation, replace θ by $-\theta$. If an equivalent equation results, the graph is symmetric with respect to the polar axis.

Symmetry with Respect to the Line $\theta = \dfrac{\pi}{2}$ (*y*-Axis)

In a polar equation, replace θ by $\pi - \theta$. If an equivalent equation results, the graph is symmetric with respect to the line $\theta = \dfrac{\pi}{2}$.

Symmetry with Respect to the Pole (Origin)

In a polar equation, replace r by $-r$. If an equivalent equation results, the graph is symmetric with respect to the pole.

The three tests for symmetry given here are *sufficient* conditions for symmetry, but they are not *necessary* conditions. That is, an equation may fail these tests and still have a graph that is symmetric with respect to the polar axis, the line $\theta = \dfrac{\pi}{2}$, or the pole. For example, the graph of $r = \sin(2\theta)$ turns out to be symmetric with respect to the polar axis, the line $\theta = \dfrac{\pi}{2}$, and the pole, but all three tests given here fail. See also Problems 71, 72, and 73.

---

④   **EXAMPLE 8**   **Graphing a Polar Equation (Cardioid)**

Graph the equation: $r = 1 - \sin\theta$

Solution   We check for symmetry first.

Polar Axis: Replace θ by $-\theta$. The result is

$$r = 1 - \sin(-\theta) = 1 + \sin\theta$$

The test fails, so the graph may or may not be symmetric with respect to the polar axis.

The Line $\theta = \dfrac{\pi}{2}$:   Replace θ by $\pi - \theta$. The result is

$$r = 1 - \sin(\pi - \theta) = 1 - (\sin\pi\cos\theta - \cos\pi\sin\theta)$$
$$= 1 - [0 \cdot \cos\theta - (-1)\sin\theta] = 1 - \sin\theta$$

The test is satisfied, so the graph is symmetric with respect to the line $\theta = \dfrac{\pi}{2}$.

The Pole: Replace r by $-r$. Then the result is $-r = 1 - \sin\theta$, so $r = -1 + \sin\theta$. The test fails, so the graph may or may not be symmetric with respect to the pole.

Next, we identify points on the graph by assigning values to the angle θ and calculating the corresponding values of r. Due to the symmetry with respect to the line $\theta = \dfrac{\pi}{2}$, we only need to assign values to θ from $-\dfrac{\pi}{2}$ to $\dfrac{\pi}{2}$, as given in Table 1.

TABLE 1

θ	$r = 1 - \sin\theta$
$-\dfrac{\pi}{2}$	$1 - (-1) = 2$
$-\dfrac{\pi}{3}$	$1 - \left(-\dfrac{\sqrt{3}}{2}\right) \approx 1.87$
$-\dfrac{\pi}{6}$	$1 - \left(-\dfrac{1}{2}\right) = \dfrac{3}{2}$
0	$1 - 0 = 1$
$\dfrac{\pi}{6}$	$1 - \dfrac{1}{2} = \dfrac{1}{2}$
$\dfrac{\pi}{3}$	$1 - \dfrac{\sqrt{3}}{2} \approx 0.13$
$\dfrac{\pi}{2}$	$1 - 1 = 0$

Now we plot the points (r, θ) from Table 1 and trace out the graph, beginning at the point $\left(2, -\dfrac{\pi}{2}\right)$ and ending at the point $\left(0, \dfrac{\pi}{2}\right)$. Then we reflect this portion of the graph about the line $\theta = \dfrac{\pi}{2}$ (the y-axis) to obtain the complete graph. Figure 27(a) shows the graph drawn by hand. Figure 27(b) shows the graph using a graphing utility with $\theta\text{min} = 0$, $\theta\text{max} = 2\pi$, and $\theta\text{step} = \dfrac{\pi}{24}$.

Figure 27
$r = 1 - \sin\theta$

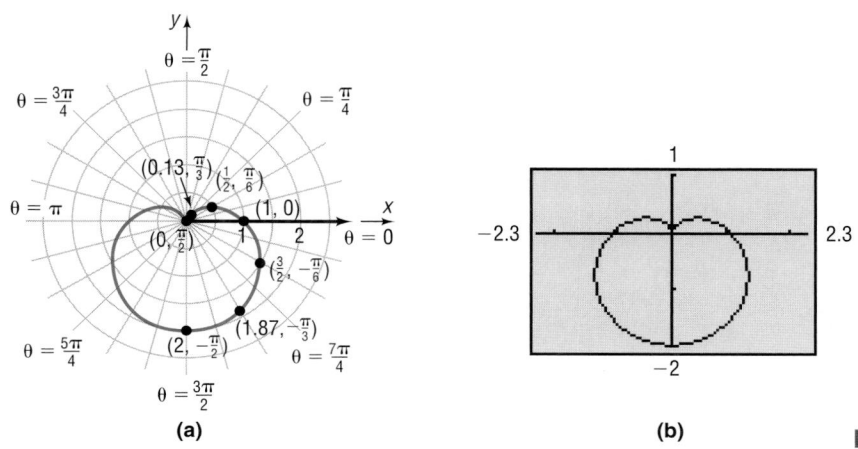

(a)   (b)

EXPLORATION Graph $r_1 = 1 + \sin\theta$. Clear the screen and graph $r_1 = 1 - \cos\theta$. Clear the screen and graph $r_1 = 1 + \cos\theta$. Do you see a pattern?

The curve in Figure 27 is an example of a *cardioid* (a heart-shaped curve).

Cardioids are characterized by equations of the form

$$r = a(1 + \cos\theta) \qquad r = a(1 + \sin\theta)$$
$$r = a(1 - \cos\theta) \qquad r = a(1 - \sin\theta)$$

where $a > 0$. The graph of a cardioid passes through the pole.

 **NOW WORK PROBLEM 31.**

EXAMPLE 9 | **Graphing a Polar Equation (Limaçon without Inner Loop)**

Graph the equation: $r = 3 + 2\cos\theta$

Solution We check for symmetry first.

Polar Axis: Replace θ by $-\theta$. The result is

$$r = 3 + 2\cos(-\theta) = 3 + 2\cos\theta$$

The test is satisfied, so the graph is symmetric with respect to the polar axis.

TABLE 2	
θ	$r = 3 + 2\cos\theta$
0	$3 + 2(1) = 5$
$\dfrac{\pi}{6}$	$3 + 2\left(\dfrac{\sqrt{3}}{2}\right) \approx 4.73$
$\dfrac{\pi}{3}$	$3 + 2\left(\dfrac{1}{2}\right) = 4$
$\dfrac{\pi}{2}$	$3 + 2(0) = 3$
$\dfrac{2\pi}{3}$	$3 + 2\left(-\dfrac{1}{2}\right) = 2$
$\dfrac{5\pi}{6}$	$3 + 2\left(-\dfrac{\sqrt{3}}{2}\right) \approx 1.27$
π	$3 + 2(-1) = 1$

The Line $\theta = \dfrac{\pi}{2}$: Replace θ by $\pi - \theta$. The result is

$$r = 3 + 2\cos(\pi - \theta) = 3 + 2(\cos\pi\cos\theta + \sin\pi\sin\theta)$$
$$= 3 - 2\cos\theta$$

The test fails, so the graph may or may not be symmetric with respect to the line $\theta = \dfrac{\pi}{2}$.

The Pole: Replace r by $-r$. The test fails, so the graph may or may not be symmetric with respect to the pole.

Next, we identify points on the graph by assigning values to the angle θ and calculating the corresponding values of r. Due to the symmetry with respect to the polar axis, we only need to assign values to θ from 0 to π, as given in Table 2.

Now we plot the points (r, θ) from Table 2 and trace out the graph, beginning at the point $(5, 0)$ and ending at the point $(1, \pi)$. Then we reflect this portion of the graph about the polar axis (the x-axis) to obtain the complete graph. Figure 28(a) shows the graph drawn by hand. Figure 28(b) shows the graph using a graphing utility with $\theta\text{min} = 0$, $\theta\text{max} = 2\pi$, and $\theta\text{step} = \dfrac{\pi}{24}$.

Figure 28
$r = 3 + 2\cos\theta$

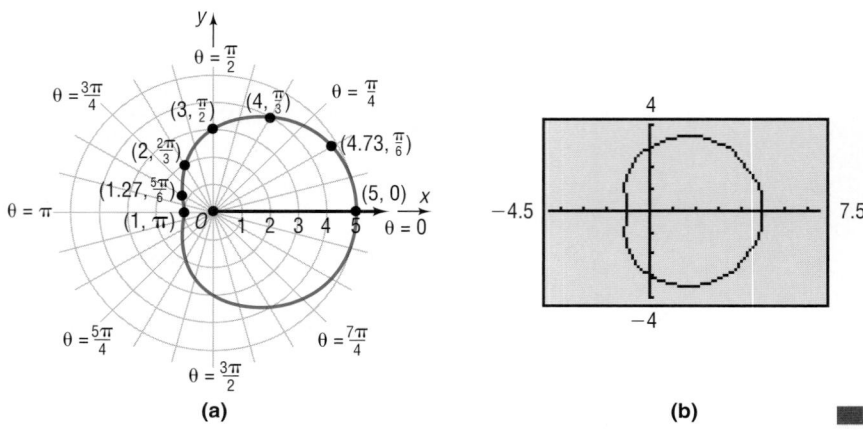

(a)   (b)

 EXPLORATION Graph $r_1 = 3 - 2\cos\theta$. Clear the screen and graph $r_1 = 3 + 2\sin\theta$. Clear the screen and graph $r_1 = 3 - 2\sin\theta$. Do you see a pattern?

The curve in Figure 28 is an example of a limaçon (the French word for *snail*) without an inner loop.

Limaçons without an inner loop are characterized by equations of the form

$r = a + b\cos\theta$	$r = a + b\sin\theta$
$r = a - b\cos\theta$	$r = a - b\sin\theta$

where $a > 0$, $b > 0$, and $a > b$. The graph of a limaçon without an inner loop does not pass through the pole.

━━ **NOW WORK PROBLEM 37.**

EXAMPLE 10   **Graphing a Polar Equation (Limaçon with Inner Loop)**

Graph the equation: $r = 1 + 2\cos\theta$

Solution   First, we check for symmetry.

Polar Axis: Replace θ by $-\theta$. The result is

$$r = 1 + 2\cos(-\theta) = 1 + 2\cos\theta$$

The test is satisfied, so the graph is symmetric with respect to the polar axis.

The Line $\theta = \dfrac{\pi}{2}$:   Replace θ by $\pi - \theta$. The result is

$$r = 1 + 2\cos(\pi - \theta) = 1 + 2(\cos\pi\cos\theta + \sin\pi\sin\theta)$$
$$= 1 - 2\cos\theta$$

The test fails, so the graph may or may not be symmetric with respect to the line $\theta = \dfrac{\pi}{2}$.

The Pole: Replace r by $-r$. The test fails, so the graph may or may not be symmetric with respect to the pole.

Next, we identify points on the graph of $r = 1 + 2\cos\theta$ by assigning values to the angle θ and calculating the corresponding values of r. Due to the symmetry with respect to the polar axis, we only need to assign values to θ from 0 to π, as given in Table 3.

Now we plot the points (r, θ) from Table 3, beginning at $(3, 0)$ and ending at $(-1, \pi)$. See Figure 29(a). Finally, we reflect this portion of the graph about the polar axis (the x-axis) to obtain the complete graph. See Figure 29(b). Figure 29(c) shows the graph using a graphing utility with $\theta\text{min} = 0$, $\theta\text{max} = 2\pi$, and $\theta\text{step} = \dfrac{\pi}{24}$.

TABLE 3

θ	$r = 1 + 2\cos\theta$
0	$1 + 2(1) = 3$
$\dfrac{\pi}{6}$	$1 + 2\left(\dfrac{\sqrt{3}}{2}\right) \approx 2.73$
$\dfrac{\pi}{3}$	$1 + 2\left(\dfrac{1}{2}\right) = 2$
$\dfrac{\pi}{2}$	$1 + 2(0) = 1$
$\dfrac{2\pi}{3}$	$1 + 2\left(-\dfrac{1}{2}\right) = 0$
$\dfrac{5\pi}{6}$	$1 + 2\left(-\dfrac{\sqrt{3}}{2}\right) \approx -0.73$
π	$1 + 2(-1) = -1$

Figure 29

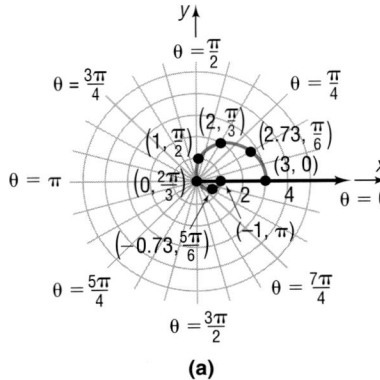

(a)

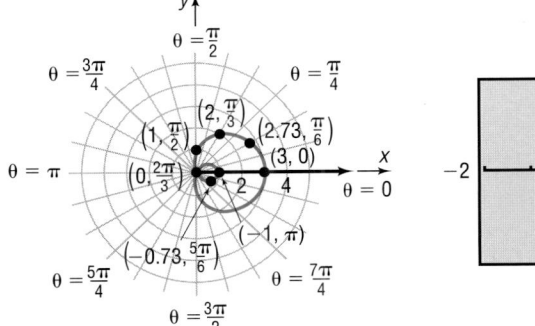

(b) $r = 1 + 2\cos\theta$

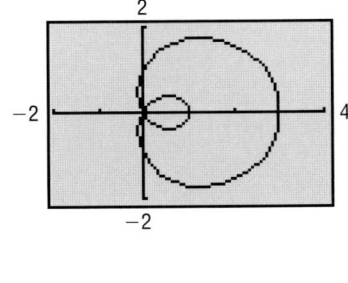

(c)

EXPLORATION   Graph $r_1 = 1 - 2\cos\theta$. Clear the screen and graph $r_1 = 1 + 2\sin\theta$. Clear the screen and graph $r_1 = 1 - 2\sin\theta$. Do you see a pattern?

The curve in Figure 29(b) or 29(c) is an example of a limaçon with an inner loop.

Limaçons with an inner loop are characterized by equations of the form

$r = a + b \cos \theta$	$r = a + b \sin \theta$
$r = a - b \cos \theta$	$r = a - b \sin \theta$

where $a > 0, b > 0$, and $a < b$. The graph of a limaçon with an inner loop will pass through the pole twice.

⟶ **NOW WORK PROBLEM 39.**

EXAMPLE 11   **Graphing a Polar Equation (Rose)**

Graph the equation: $r = 2 \cos(2\theta)$

Solution   We check for symmetry.

Polar Axis: If we replace θ by $-\theta$, the result is

$$r = 2 \cos[2(-\theta)] = 2 \cos(2\theta)$$

The test is satisfied, so the graph is symmetric with respect to the polar axis.

The Line $\theta = \dfrac{\pi}{2}$:   If we replace θ by $\pi - \theta$, we obtain

$$r = 2 \cos[2(\pi - \theta)] = 2 \cos(2\pi - 2\theta) = 2 \cos(2\theta)$$

The test is satisfied, so the graph is symmetric with respect to the line $\theta = \dfrac{\pi}{2}$.

The Pole: Since the graph is symmetric with respect to both the polar axis and the line $\theta = \dfrac{\pi}{2}$, it must be symmetric with respect to the pole.

Next, we construct Table 4. Due to the symmetry with respect to the polar axis, the line $\theta = \dfrac{\pi}{2}$, and the pole, we consider only values of θ from 0 to $\dfrac{\pi}{2}$.

We plot and connect these points in Figure 30(a). Finally, because of symmetry, we reflect this portion of the graph first about the polar axis (the x-axis) and then about the line $\theta = \dfrac{\pi}{2}$ (the y-axis) to obtain the complete

TABLE 4

θ	$r = 2 \cos(2\theta)$
0	$2(1) = 2$
$\dfrac{\pi}{6}$	$2\left(\dfrac{1}{2}\right) = 1$
$\dfrac{\pi}{4}$	$2(0) = 0$
$\dfrac{\pi}{3}$	$2\left(-\dfrac{1}{2}\right) = -1$
$\dfrac{\pi}{2}$	$2(-1) = -2$

Figure 30

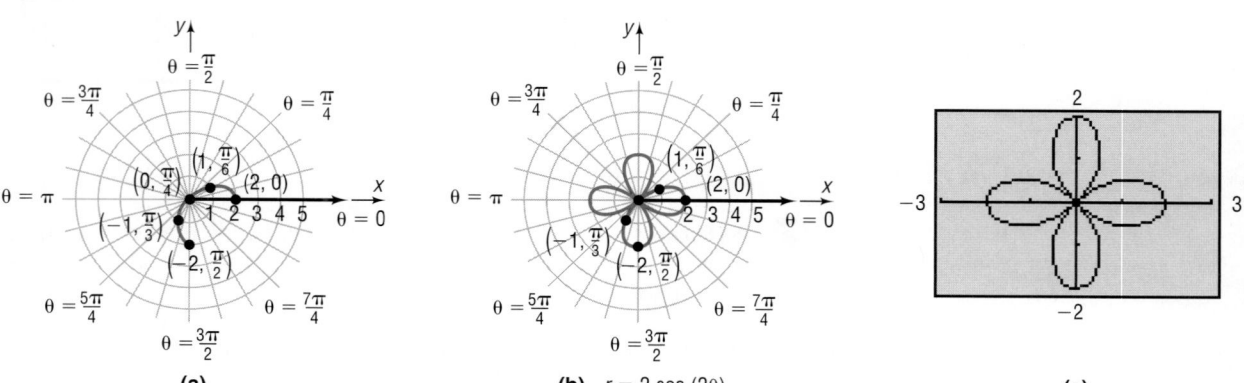

(a)   **(b)** $r = 2 \cos(2\theta)$ **(c)**

graph. See Figure 30(b). Figure 30(c) shows the graph using a graphing utility with θmin = 0, θmax = 2π, and θstep = $\dfrac{\pi}{24}$.

   EXPLORATION   Graph $r = 2\cos(4\theta)$; clear the screen and graph $r = 2\cos(6\theta)$. How many petals did each of these graphs have?
Clear the screen and graph, in order, each on a clear screen, $r = 2\cos(3\theta), r = 2\cos(5\theta)$, and $r = 2\cos(7\theta)$. What do you notice about the number of petals?

The curve in Figure 30(b) or (c) is called a *rose* with four petals.

Rose curves are characterized by equations of the form

$$r = a\cos(n\theta), \qquad r = a\sin(n\theta), \qquad a \neq 0$$

and have graphs that are rose shaped. If $n \neq 0$ is even, the rose has $2n$ petals; if $n \neq \pm1$ is odd, the rose has n petals.

 **NOW WORK PROBLEM 43.**

EXAMPLE 12   **Graphing a Polar Equation (Lemniscate)**

Graph the equation: $r^2 = 4\sin(2\theta)$

TABLE 5

θ	$r^2 = 4\sin(2\theta)$	r
0	$4(0) = 0$	0
$\dfrac{\pi}{6}$	$4\left(\dfrac{\sqrt{3}}{2}\right) = 2\sqrt{3}$	±1.9
$\dfrac{\pi}{4}$	$4(1) = 4$	±2
$\dfrac{\pi}{3}$	$4\left(\dfrac{\sqrt{3}}{2}\right) = 2\sqrt{3}$	±1.9
$\dfrac{\pi}{2}$	$4(0) = 0$	0

Solution   We leave it to you to verify that the graph is symmetric with respect to the pole. Table 5 lists points on the graph for values of $\theta = 0$ through $\theta = \dfrac{\pi}{2}$. Note that there are no points on the graph for $\dfrac{\pi}{2} < \theta < \pi$ (quadrant II), since $\sin(2\theta) < 0$ for such values. The points from Table 5 where $r \geq 0$ are plotted in Figure 31(a). The remaining points on the graph may be obtained by using symmetry. Figure 31(b) shows the final graph drawn by hand. Figure 31(c) shows the graph using a graphing utility with θmin = 0, θmax = 2π, and θstep = $\dfrac{\pi}{24}$.

Figure 31

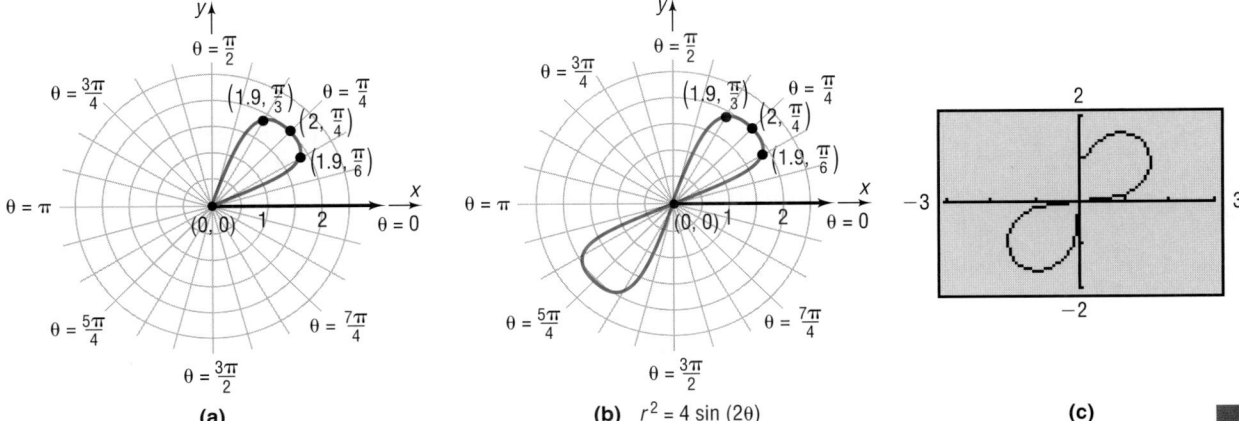

(a)   (b) $r^2 = 4\sin(2\theta)$ (c)

The curve in Figure 31(b) or (c) is an example of a *lemniscate*.

Lemniscates are characterized by equations of the form

$$r^2 = a^2 \sin(2\theta) \qquad r^2 = a^2 \cos(2\theta)$$

where $a \neq 0$, and have graphs that are propeller shaped.

NOW WORK PROBLEM **47.**

EXAMPLE 13   **Graphing a Polar Equation (Spiral)**

Graph the equation: $r = e^{\theta/5}$

Solution   The tests for symmetry with respect to the pole, the polar axis, and the line $\theta = \dfrac{\pi}{2}$ fail. Furthermore, there is no number θ for which $r = 0$, so the graph does not pass through the pole. We observe that r is positive for all θ, r increases as θ increases, $r \to 0$ as $\theta \to -\infty$, and $r \to \infty$ as $\theta \to \infty$. With the help of a calculator, we obtain the values in Table 6. See Figure 32(a) for the graph drawn by hand. Figure 32(b) shows the graph using a graphing utility with $\theta\min = -4\pi$, $\theta\max = 3\pi$, and $\theta\text{step} = \dfrac{\pi}{24}$.

TABLE 6

θ	$r = e^{\theta/5}$
$-\dfrac{3\pi}{2}$	0.39
$-\pi$	0.53
$-\dfrac{\pi}{2}$	0.73
$-\dfrac{\pi}{4}$	0.85
0	1
$\dfrac{\pi}{4}$	1.17
$\dfrac{\pi}{2}$	1.37
π	1.87
$\dfrac{3\pi}{2}$	2.57
2π	3.51

Figure 32
$r = e^{\theta/5}$

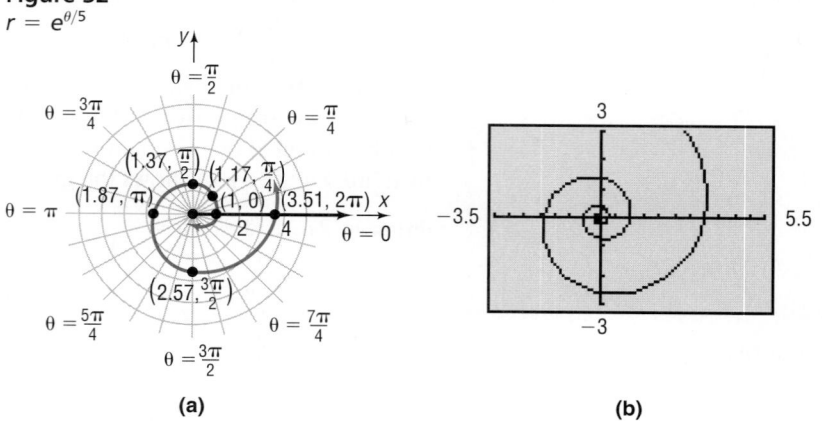

(a)   (b)

The curve in Figure 32 is called a **logarithmic spiral**, since its equation may be written as $\theta = 5 \ln r$ and it spirals infinitely both toward the pole and away from it.

Classification of Polar Equations

The equations of some lines and circles in polar coordinates and their corresponding equations in rectangular coordinates are given in Table 7 on pages 341–342. Also included are the names and the graphs of a few of the more frequently encountered polar equations.

TABLE 7

Lines

Description	Line passing through the pole making an angle α with the polar axis	Vertical line	Horizontal line
Rectangular equation	$y = (\tan \alpha)x$	$x = a$	$y = b$
Polar equation	$\theta = \alpha$	$r \cos \theta = a$	$r \sin \theta = b$
Typical graph	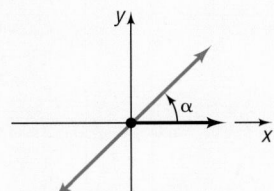	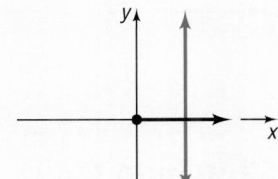	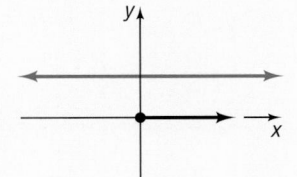

Circles

Description	Center at the pole, radius a	Passing through the pole, tangent to the y-axis, center on the x-axis, radius a	Passing through the pole, tangent to the x-axis, center on the y-axis, radius a
Rectangular equation	$x^2 + y^2 = a^2, \quad a > 0$	$x^2 + y^2 = \pm 2ax, \quad a > 0$	$x^2 + y^2 = \pm 2ay, \quad a > 0$
Polar equation	$r = a, \quad a > 0$	$r = \pm 2a \cos \theta, \quad a > 0$	$r = \pm 2a \sin \theta, \quad a > 0$
Typical graph	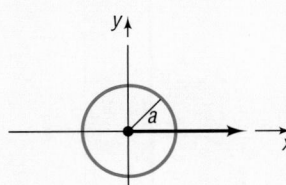	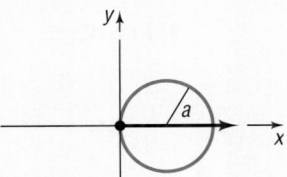	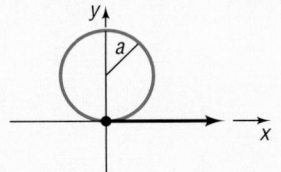

Other Equations

Name	Cardioid	Limaçon without inner loop	Limaçon with inner loop
Polar equations	$r = a \pm a \cos \theta, \quad a > 0$ $r = a \pm a \sin \theta, \quad a > 0$	$r = a \pm b \cos \theta, \quad 0 < b < a$ $r = a \pm b \sin \theta, \quad 0 < b < a$	$r = a \pm b \cos \theta, \quad 0 < a < b$ $r = a \pm b \sin \theta, \quad 0 < a < b$
Typical graph	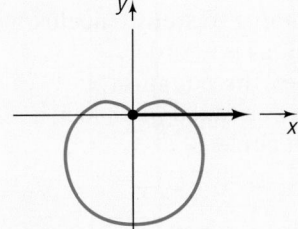	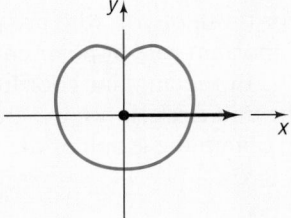	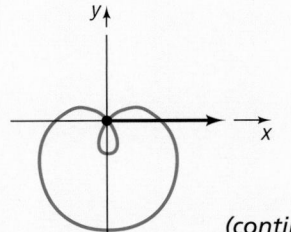

(continued)

TABLE 7 (continued)

Other Equations

Name	Lemniscate	Rose with three petals	Rose with four petals
Polar equations	$r^2 = a^2 \cos(2\theta), \quad a > 0$ $r^2 = a^2 \sin(2\theta), \quad a > 0$	$r = a \sin(3\theta), \quad a > 0$ $r = a \cos(3\theta), \quad a > 0$	$r = a \sin(2\theta), \quad a > 0$ $r = a \cos(2\theta), \quad a > 0$
Typical graph	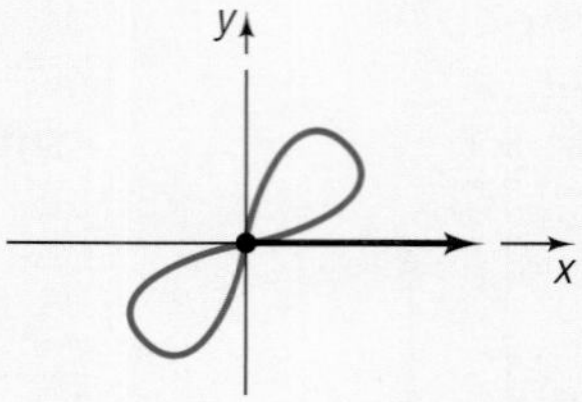	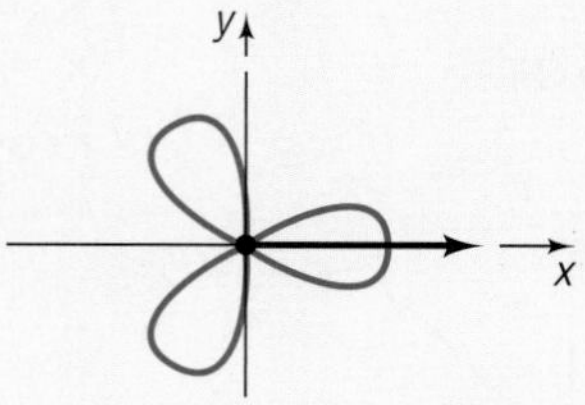	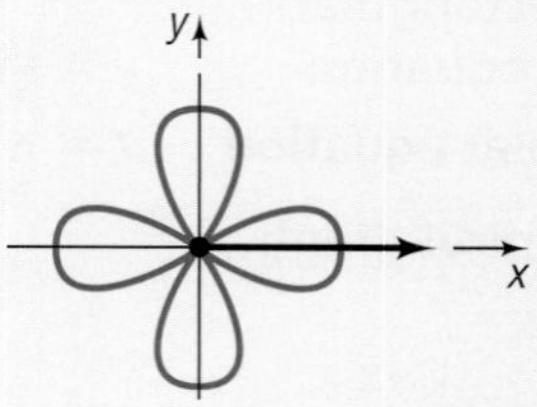

Sketching Quickly

If a polar equation only involves a sine (or cosine) function, you can quickly obtain a sketch of its graph by making use of Table 7, periodicity, and a short table.

EXAMPLE 14   **Sketching the Graph of a Polar Equation Quickly by Hand**

Graph the equation: $r = 2 + 2\sin\theta$

Solution   We recognize the polar equation: Its graph is a cardioid. The period of $\sin\theta$ is 2π, so we form a table using $0 \le \theta \le 2\pi$, compute r, plot the points (r, θ), and sketch the graph of a cardioid as θ varies from 0 to 2π. See Table 8 and Figure 33.

TABLE 8

θ	$r = 2 + 2\sin\theta$
0	$2 + 2(0) = 2$
$\dfrac{\pi}{2}$	$2 + 2(1) = 4$
π	$2 + 2(0) = 2$
$\dfrac{3\pi}{2}$	$2 + 2(-1) = 0$
2π	$2 + 2(0) = 2$

Figure 33

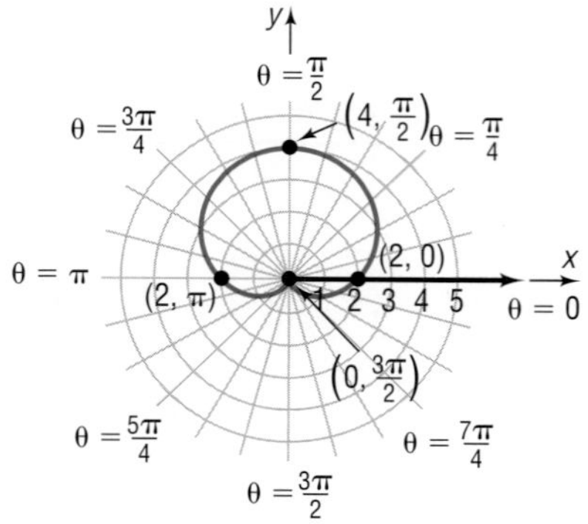

Calculus Comment

For those of you who are planning to study calculus, a comment about one important role of polar equations is in order.

In rectangular coordinates, the equation $x^2 + y^2 = 1$, whose graph is the unit circle, is not the graph of a function. In fact, it requires two functions to obtain the graph of the unit circle:

$$y_1 = \sqrt{1 - x^2} \qquad \text{Upper semicircle}$$

$$y_2 = -\sqrt{1 - x^2} \qquad \text{Lower semicircle}$$

In polar coordinates, the equation $r = 1$, whose graph is also the unit circle, does define a function. That is, for each choice of θ there is only one corresponding value of r, namely, $r = 1$. Since many uses of calculus require that functions be used, the opportunity to express nonfunctions in rectangular coordinates as functions in polar coordinates becomes extremely useful.

Note also that the vertical-line test for functions is valid only for equations in rectangular coordinates.

HISTORICAL FEATURE

Jakob Bernoulli
1654–1705

Polar coordinates seem to have been invented by Jakob Bernoulli (1654–1705) in about 1691, although, as with most such ideas, earlier traces of the notion exist. Early users of calculus remained committed to rectangular coordinates, and polar coordinates did not become widely used until the early 1800s. Even then, it was mostly geometers who used them for describing odd curves. Finally, about the mid-1800s, applied mathematicians realized the tremendous simplification that polar coordinates make possible in the description of objects with circular or cylindrical symmetry. From then on their use became widespread.

5.2   Concepts and Vocabulary

In Problems 1–3, fill in the blanks.

1. An equation whose variables are polar coordinates is called a _____ _____.

2. Using polar coordinates (r, θ), the circle $x^2 + y^2 = 2x$ takes the form _____.

3. A polar equation is symmetric with respect to the pole if an equivalent equation results when r is replaced by _____.

In Problems 4–6, answer True or False to each statement.

4. The tests for symmetry in polar coordinates are necessary, but not sufficient.

5. The graph of a cardioid never passes through the pole.

6. All polar equations have a symmetric feature.

7. What is the graph of $r = 4$?

8. What is the graph of $\theta = \dfrac{\pi}{6}$?

9. What is the shape of a cardioid?

10. What is the shape of a lemniscate?

5.2   Exercises

In Problems 1–16, transform each polar equation to an equation in rectangular coordinates. Then identify and graph the equation. Verify your graph using a graphing utility.

1. $r = 4$
2. $r = 2$
3. $\theta = \dfrac{\pi}{3}$
4. $\theta = -\dfrac{\pi}{4}$

5. $r \sin \theta = 4$
6. $r \cos \theta = 4$
7. $r \cos \theta = -2$
8. $r \sin \theta = -2$

9. $r = 2 \cos \theta$
10. $r = 2 \sin \theta$
11. $r = -4 \sin \theta$
12. $r = -4 \cos \theta$

13. $r \sec \theta = 4$
14. $r \csc \theta = 8$
15. $r \csc \theta = -2$
16. $r \sec \theta = -4$

In Problems 17–24, match each of the graphs (A) through (H) to one of the following polar equations.

17. $r = 2$

18. $\theta = \dfrac{\pi}{4}$

19. $r = 2\cos\theta$

20. $r\cos\theta = 2$

21. $r = 1 + \cos\theta$

22. $r = 2\sin\theta$

23. $\theta = \dfrac{3\pi}{4}$

24. $r\sin\theta = 2$

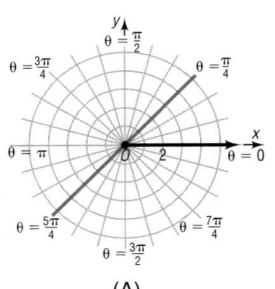

(A)

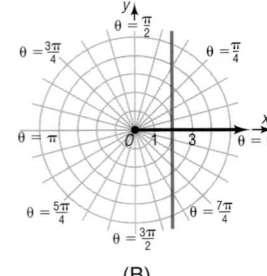

(B)

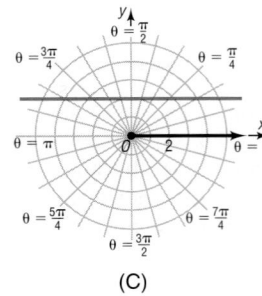

(C)

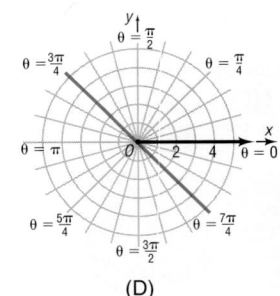

(D)

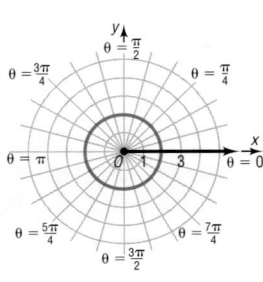

(E)

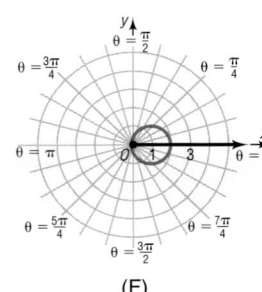

(F)

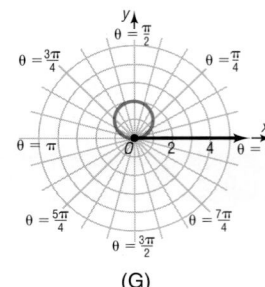

(G)

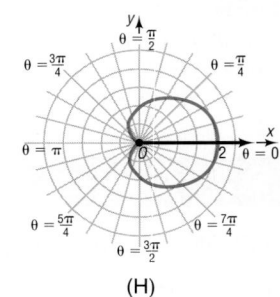

(H)

In Problems 25–30, match each of the graphs (A) through (F) to one of the following polar equations.

25. $r = 4$

26. $r = 3\cos\theta$

27. $r = 3\sin\theta$

28. $r\sin\theta = 3$

29. $r\cos\theta = 3$

30. $r = 2 + \sin\theta$

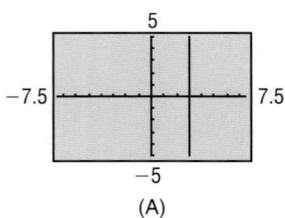

(A)

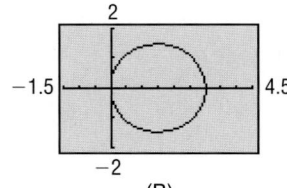

(B)

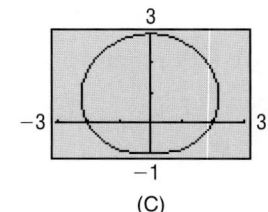

(C)

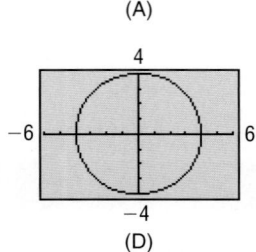

(D)

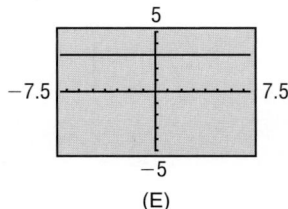

(E)

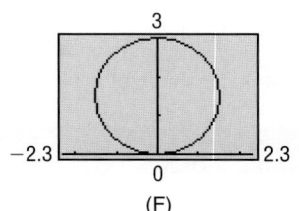

(F)

In Problems 31–54, identify and graph each polar equation. Verify your graph using a graphing utility.

31. $r = 2 + 2\cos\theta$

32. $r = 1 + \sin\theta$

33. $r = 3 - 3\sin\theta$

34. $r = 2 - 2\cos\theta$

35. $r = 2 + \sin\theta$

36. $r = 2 - \cos\theta$

37. $r = 4 - 2\cos\theta$

38. $r = 4 + 2\sin\theta$

39. $r = 1 + 2\sin\theta$

40. $r = 1 - 2\sin\theta$

41. $r = 2 - 3\cos\theta$

42. $r = 2 + 4\cos\theta$

43. $r = 3\cos(2\theta)$

44. $r = 2\sin(3\theta)$

45. $r = 4\sin(5\theta)$

46. $r = 3\cos(4\theta)$

47. $r^2 = 9\cos(2\theta)$ **48.** $r^2 = \sin(2\theta)$ **49.** $r = 2^\theta$ **50.** $r = 3^\theta$

51. $r = 1 - \cos\theta$ **52.** $r = 3 + \cos\theta$ **53.** $r = 1 - 3\cos\theta$ **54.** $r = 4\cos(3\theta)$

In Problems 55–64, graph each polar equation. Verify your graph using a graphing utility.

55. $r = \dfrac{2}{1 - \cos\theta}$ (parabola)

56. $r = \dfrac{2}{1 - 2\cos\theta}$ (hyperbola)

57. $r = \dfrac{1}{3 - 2\cos\theta}$ (ellipse)

58. $r = \dfrac{1}{1 - \cos\theta}$ (parabola)

59. $r = \theta, \quad \theta \geq 0$ (spiral of Archimedes)

60. $r = \dfrac{3}{\theta}$ (reciprocal spiral)

61. $r = \csc\theta - 2, \quad 0 < \theta < \pi$ (conchoid)

62. $r = \sin\theta\tan\theta$ (cissoid)

63. $r = \tan\theta, \quad -\dfrac{\pi}{2} < \theta < \dfrac{\pi}{2},$ (kappa curve)

64. $r = \cos\dfrac{\theta}{2}$

---

65. Show that the graph of the equation $r\sin\theta = a$ is a horizontal line a units above the pole if $a > 0$ and $|a|$ units below the pole if $a < 0$.

66. Show that the graph of the equation $r\cos\theta = a$ is a vertical line a units to the right of the pole if $a > 0$ and $|a|$ units to the left of the pole if $a < 0$.

67. Show that the graph of the equation $r = 2a\sin\theta, a > 0$, is a circle of radius a with center at $(0, a)$ in rectangular coordinates.

68. Show that the graph of the equation $r = -2a\sin\theta, a > 0$, is a circle of radius a with center at $(0, -a)$ in rectangular coordinates.

69. Show that the graph of the equation $r = 2a\cos\theta, a > 0$, is a circle of radius a with center at $(a, 0)$ in rectangular coordinates.

70. Show that the graph of the equation $r = -2a\cos\theta$, $a > 0$, is a circle of radius a with center at $(-a, 0)$ in rectangular coordinates.

71. Explain why the following test for symmetry is valid: Replace r by $-r$ and θ by $-\theta$ in a polar equation. If an equivalent equation results, the graph is symmetric with respect to the line $\theta = \dfrac{\pi}{2}$ (y-axis).

   (a) Show that the test on page 334 fails for $r^2 = \cos\theta$, but this new test works.

   (b) Show that the test on page 334 works for $r^2 = \sin\theta$, yet this new test fails.

72. Develop a new test for symmetry with respect to the pole.

   (a) Find a polar equation for which this new test fails, yet the test on page 334 works.

   (b) Find a polar equation for which the test on page 334 fails yet the new test works.

73. Write down two different tests for symmetry with respect to the polar axis. Find examples in which one test works and the other fails. Which test do you prefer to use? Justify your answer.

PREPARING FOR THIS SECTION

Before getting started, review the following:

✓ Complex Numbers (Section A.4, pp. 585–592)

✓ Definitions of the Sine and Cosine Functions (Section 2.4, p. 132)

✓ Value of the Sine and Cosine Functions at Certain Angles (Section 2.3, p. 126 and Section 2.4, p. 134)

✓ Sum and Difference Formulas for Sine and Cosine (Section 3.4, pp. 222 and 225)

5.3   THE COMPLEX PLANE; DE MOIVRE'S THEOREM

OBJECTIVES

  1   Convert a Complex Number from Rectangular Form to Polar Form

  2   Plot Points in the Complex Plane

  3   Find Products and Quotients of Complex Numbers in Polar Form

  4   Use De Moivre's Theorem

  5   Find Complex Roots

When we first introduced complex numbers, we were not prepared to give a geometric interpretation of a complex number. Now we are ready. Although we could give several interpretations, the one that follows is the easiest to understand.

A complex number $z = x + yi$ can be interpreted geometrically as the point (x, y) in the xy-plane. Each point in the plane corresponds to a complex number and, conversely, each complex number corresponds to a point in the plane. We shall refer to the collection of such points as the **complex plane**. The x-axis will be referred to as the **real axis**, because any point that lies on the real axis is of the form $z = x + 0i = x$, a real number. The y-axis is called the **imaginary axis**, because any point that lies on it is of the form $z = 0 + yi = yi$, a pure imaginary number. See Figure 34.

Figure 34
Complex plane

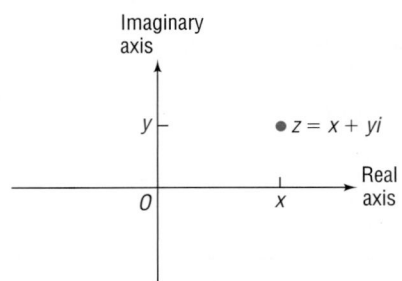

> Let $z = x + yi$ be a complex number. The **magnitude** or **modulus** of z, denoted by $|z|$, is defined as the distance from the origin to the point (x, y). That is,
> $$|z| = \sqrt{x^2 + y^2} \qquad (1)$$

Figure 35

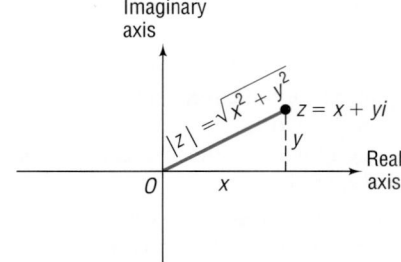

See Figure 35 for an illustration.

This definition for $|z|$ is consistent with the definition for the absolute value of a real number: If $z = x + yi$ is real, then $z = x + 0i$ and
$$|z| = \sqrt{x^2 + 0^2} = \sqrt{x^2} = |x|$$

For this reason, the magnitude of z is sometimes called the absolute value of z.

Recall (Section A.7) that if $z = x + yi$ then its **conjugate**, denoted by $\bar{z}$, is $\bar{z} = x - yi$. Because $z\bar{z} = x^2 + y^2$, it follows from equation (1) that the magnitude of z can be written as
$$|z| = \sqrt{z\bar{z}} \qquad (2)$$

Polar Form of a Complex Number

When a complex number is written in the standard form $z = x + yi$, we say that it is in **rectangular**, or **Cartesian, form** because (x, y) are the rectangular coordinates of the corresponding point in the complex plane. Suppose that (r, θ) are the polar coordinates of this point. Then
$$x = r \cos \theta \qquad y = r \sin \theta \qquad (3)$$

Figure 36

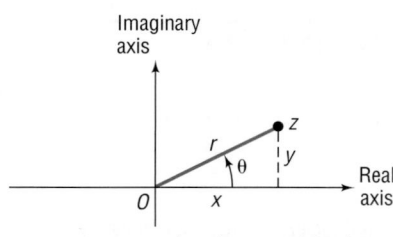

$z = x + yi = r(\cos \theta + i \sin \theta)$,
$r \ge 0, 0 \le \theta < 2\pi$

If $r \ge 0$ and $0 \le \theta < 2\pi$, the complex number $z = x + yi$ may be written in **polar form** as
$$z = x + yi = (r \cos \theta) + (r \sin \theta)i = r(\cos \theta + i \sin \theta) \qquad (4)$$

See Figure 36.

If $z = r(\cos \theta + i \sin \theta)$ is the polar form of a complex number, the angle $\theta, 0 \le \theta < 2\pi$, is called the **argument of** z.

Also, because $r \geq 0$, we have $r = \sqrt{x^2 + y^2}$. From equation (1) it follows that the magnitude of $z = r(\cos\theta + i\sin\theta)$ is

$$|z| = r$$

---

② **EXAMPLE 1**   **Plotting a Point in the Complex Plane and Writing a Complex Number in Polar Form**

Plot the point corresponding to $z = \sqrt{3} - i$ in the complex plane, and write an expression for z in polar form.

Figure 37

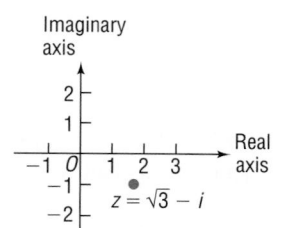

Solution   The point corresponding to $z = \sqrt{3} - i$ has the rectangular coordinates $(\sqrt{3}, -1)$. The point, located in quadrant IV, is plotted in Figure 37. Because $x = \sqrt{3}$ and $y = -1$, it follows that

$$r = \sqrt{x^2 + y^2} = \sqrt{(\sqrt{3})^2 + (-1)^2} = \sqrt{4} = 2$$

and

$$\sin\theta = \frac{y}{r} = \frac{-1}{2}, \qquad \cos\theta = \frac{x}{r} = \frac{\sqrt{3}}{2}, \qquad 0 \leq \theta < 2\pi$$

Then $\theta = \dfrac{11\pi}{6}$ and $r = 2$, so the polar form of $z = \sqrt{3} - i$ is

$$z = r(\cos\theta + i\sin\theta) = 2\left(\cos\frac{11\pi}{6} + i\sin\frac{11\pi}{6}\right)$$

✏ **NOW WORK PROBLEM 1.**

EXAMPLE 2   **Plotting a Point in the Complex Plane and Converting from Polar to Rectangular Form**

Plot the point corresponding to $z = 2(\cos 30° + i\sin 30°)$ in the complex plane, and write an expression for z in rectangular form.

Figure 38

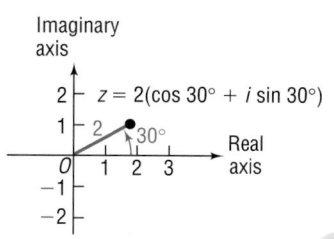

Solution   To plot the complex number $z = 2(\cos 30° + i\sin 30°)$, we plot the point whose polar coordinates are $(r, \theta) = (2, 30°)$, as shown in Figure 38. In rectangular form,

$$z = 2(\cos 30° + i\sin 30°) = 2\left(\frac{\sqrt{3}}{2} + \frac{1}{2}i\right) = \sqrt{3} + i$$

✏ **NOW WORK PROBLEM 13.**

③   The polar form of a complex number provides an alternative method for finding products and quotients of complex numbers.

Theorem   Let $z_1 = r_1(\cos\theta_1 + i\sin\theta_1)$ and $z_2 = r_2(\cos\theta_2 + i\sin\theta_2)$ be two complex numbers. Then

$$z_1 z_2 = r_1 r_2[\cos(\theta_1 + \theta_2) + i\sin(\theta_1 + \theta_2)] \qquad (5)$$

If $z_2 \neq 0$, then

$$\frac{z_1}{z_2} = \frac{r_1}{r_2}[\cos(\theta_1 - \theta_2) + i\sin(\theta_1 - \theta_2)] \qquad (6)$$

■

Proof   We will prove formula (5). The proof of formula (6) is left as an exercise (see Problem 56).

$$\begin{aligned}
z_1 z_2 &= [r_1(\cos\theta_1 + i\sin\theta_1)][r_2(\cos\theta_2 + i\sin\theta_2)] \\
&= r_1 r_2[(\cos\theta_1 + i\sin\theta_1)(\cos\theta_2 + i\sin\theta_2)] \\
&= r_1 r_2[(\cos\theta_1\cos\theta_2 - \sin\theta_1\sin\theta_2) + i(\sin\theta_1\cos\theta_2 + \cos\theta_1\sin\theta_2)] \\
&= r_1 r_2[\cos(\theta_1 + \theta_2) + i\sin(\theta_1 + \theta_2)] \qquad ■
\end{aligned}$$

Because the magnitude of a complex number z is r and its argument is θ, when $z = r(\cos\theta + i\sin\theta)$, we can restate this theorem as follows:

Theorem   The magnitude of the product (quotient) of two complex numbers equals the product (quotient) of their magnitudes; the argument of the product (quotient) of two complex numbers is determined by the sum (difference) of their arguments.

■

Let's look at an example of how this theorem can be used.

EXAMPLE 3   **Finding Products and Quotients of Complex Numbers in Polar Form**

If $z = 3(\cos 20° + i\sin 20°)$ and $w = 5(\cos 100° + i\sin 100°)$, find the following (leave your answers in polar form):

(a) zw \qquad\qquad\qquad (b) $\dfrac{z}{w}$

Solution   (a) $\begin{aligned}[t] zw &= [3(\cos 20° + i\sin 20°)][5(\cos 100° + i\sin 100°)] \\
&= (3\cdot 5)[\cos(20° + 100°) + i\sin(20° + 100°)] \\
&= 15(\cos 120° + i\sin 120°) \end{aligned}$

(b) $\begin{aligned}[t] \frac{z}{w} &= \frac{3(\cos 20° + i\sin 20°)}{5(\cos 100° + i\sin 100°)} \\[6pt]
&= \frac{3}{5}[\cos(20° - 100°) + i\sin(20° - 100°)] \\[6pt]
&= \frac{3}{5}[\cos(-80°) + i\sin(-80°)] \\[6pt]
&= \frac{3}{5}(\cos 280° + i\sin 280°) \quad \text{\small Argument must lie between 0° and 360°.} \end{aligned}$ ■

━ **NOW WORK PROBLEM 23.**

De Moivre's Theorem

④ De Moivre's Theorem, stated by Abraham De Moivre (1667–1754) in 1730, but already known to many people by 1710, is important for the following reason: The fundamental processes of algebra are the four operations of addition, subtraction, multiplication, and division, together with powers and the extraction of roots. De Moivre's Theorem allows these latter fundamental algebraic operations to be applied to complex numbers.

De Moivre's Theorem, in its most basic form, is a formula for raising a complex number z to the power n, where $n \geq 1$ is a positive integer. Let's see if we can guess the form of the result.

Let $z = r(\cos \theta + i \sin \theta)$ be a complex number. Then, based on equation (5), we have

$$n = 2: \quad z^2 = r^2[\cos(2\theta) + i \sin(2\theta)] \qquad \text{Equation (5)}$$

$$n = 3: \quad z^3 = z^2 \cdot z$$
$$= \{r^2[\cos(2\theta) + i \sin(2\theta)]\}[r(\cos \theta + i \sin \theta)]$$
$$= r^3[\cos(3\theta) + i \sin(3\theta)] \qquad \text{Equation (5)}$$

$$n = 4: \quad z^4 = z^3 \cdot z$$
$$= \{r^3[\cos(3\theta) + i \sin(3\theta)]\}[r(\cos \theta + i \sin \theta)]$$
$$= r^4[\cos(4\theta) + i \sin(4\theta)] \qquad \text{Equation (5)}$$

The pattern should now be clear.

Theorem

De Moivre's Theorem

If $z = r(\cos \theta + i \sin \theta)$ is a complex number, then

$$z^n = r^n[\cos(n\theta) + i \sin(n\theta)] \tag{7}$$

where $n \geq 1$ is a positive integer.

We will not prove De Moivre's Theorem because the proof requires mathematical induction, which is not discussed in this book.

Let's look at some examples.

EXAMPLE 4   **Using De Moivre's Theorem**

Write $[2(\cos 20° + i \sin 20°)]^3$ in the standard form $a + bi$.

Solution
$$[2(\cos 20° + i \sin 20°)]^3 = 2^3[\cos(3 \cdot 20°) + i \sin(3 \cdot 20°)]$$
$$= 8(\cos 60° + i \sin 60°)$$
$$= 8\left(\frac{1}{2} + \frac{\sqrt{3}}{2}i\right) = 4 + 4\sqrt{3}i$$

✎─ **NOW WORK PROBLEM 31.**

EXAMPLE 5   Using De Moivre's Theorem

Write $(1 + i)^5$ in the standard form $a + bi$.

Algebraic Solution   To apply De Moivre's Theorem, we must first write the complex number in polar form. Since the magnitude of $1 + i$ is $\sqrt{1^2 + 1^2} = \sqrt{2}$, we begin by writing

$$1 + i = \sqrt{2}\left(\frac{1}{\sqrt{2}} + \frac{1}{\sqrt{2}}i\right) = \sqrt{2}\left(\cos\frac{\pi}{4} + i\sin\frac{\pi}{4}\right)$$

Now

$$
\begin{aligned}
(1 + i)^5 &= \left[\sqrt{2}\left(\cos\frac{\pi}{4} + i\sin\frac{\pi}{4}\right)\right]^5 \\
&= (\sqrt{2})^5\left[\cos\left(5\cdot\frac{\pi}{4}\right) + i\sin\left(5\cdot\frac{\pi}{4}\right)\right] \\
&= 4\sqrt{2}\left(\cos\frac{5\pi}{4} + i\sin\frac{5\pi}{4}\right) \\
&= 4\sqrt{2}\left[-\frac{1}{\sqrt{2}} + \left(-\frac{1}{\sqrt{2}}\right)i\right] = -4 - 4i
\end{aligned}
$$

Graphing Solution   Using a TI-83 graphing calculator, we obtain the solution shown in Figure 39.

Figure 39

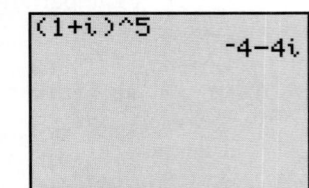

```
(1+i)^5
              -4-4i
```

Complex Roots

5   Let w be a given complex number, and let $n \geq 2$ denote a positive integer. Any complex number z that satisfies the equation

$$z^n = w$$

is called a **complex nth root** of w. In keeping with previous usage, if $n = 2$, the solutions of the equation $z^2 = w$ are called **complex square roots** of w, and if $n = 3$, the solutions of the equation $z^3 = w$ are called **complex cube roots** of w.

Theorem

Finding Complex Roots

Let $w = r(\cos\theta_0 + i\sin\theta_0)$ be a complex number and let $n \geq 2$ be an integer. If $w \neq 0$, there are n distinct complex roots of w, given by the formula

$$z_k = \sqrt[n]{r}\left[\cos\left(\frac{\theta_0}{n} + \frac{2k\pi}{n}\right) + i\sin\left(\frac{\theta_0}{n} + \frac{2k\pi}{n}\right)\right] \qquad (8)$$

where $k = 0, 1, 2, \ldots, n - 1$.

Proof (Outline)   We will not prove this result in its entirety. Instead, we shall show only that each z_k in equation (8) satisfies the equation $z_k^n = w$, proving that each z_k is a complex nth root of w.

$$z_k^n = \left\{ \sqrt[n]{r} \left[\cos\left(\frac{\theta_0}{n} + \frac{2k\pi}{n}\right) + i \sin\left(\frac{\theta_0}{n} + \frac{2k\pi}{n}\right) \right] \right\}^n$$

$$= (\sqrt[n]{r})^n \left\{ \cos\left[n\left(\frac{\theta_0}{n} + \frac{2k\pi}{n}\right) \right] + i \sin\left[n\left(\frac{\theta_0}{n} + \frac{2k\pi}{n}\right) \right] \right\} \qquad \text{De Moivre's Theorem}$$

$$= r\left[\cos(\theta_0 + 2k\pi) + i \sin(\theta_0 + 2k\pi) \right]$$

$$= r(\cos\theta_0 + i \sin\theta_0) = w \qquad\qquad \text{Periodic Property}$$

So, each z_k, $k = 0, 1, \ldots, n-1$, is a complex nth root of w. To complete the proof, we would need to show that each z_k, $k = 0, 1, \ldots, n-1$, is, in fact, distinct and that there are no complex nth roots of w other than those given by equation (8). ∎

EXAMPLE 6   **Finding Complex Cube Roots**

Find the complex cube roots of $-1 + \sqrt{3}i$. Leave your answers in polar form, with the argument in degrees.

Solution   First, we express $-1 + \sqrt{3}i$ in polar form using degrees.

$$-1 + \sqrt{3}i = 2\left(-\frac{1}{2} + \frac{\sqrt{3}}{2}i \right) = 2(\cos 120° + i \sin 120°)$$

So, $r = 2$ and $\theta_0 = 120°$. The three complex cube roots of $-1 + \sqrt{3}i = 2(\cos 120° + i \sin 120°)$ are

$$z_k = \sqrt[3]{2}\left[\cos\left(\frac{120°}{3} + \frac{360°k}{3}\right) + i \sin\left(\frac{120°}{3} + \frac{360°k}{3}\right) \right], \qquad k = 0, 1, 2$$

$$= \sqrt[3]{2}\left[\cos(40° + 120°k) + i \sin(40° + 120°k) \right], \qquad k = 0, 1, 2$$

So,

$$z_0 = \sqrt[3]{2}\left[\cos(40° + 120°\cdot 0) + i \sin(40° + 120°\cdot 0) \right] = \sqrt[3]{2}\,(\cos 40° + i \sin 40°)$$

$$z_1 = \sqrt[3]{2}\left[\cos(40° + 120°\cdot 1) + i \sin(40° + 120°\cdot 1) \right] = \sqrt[3]{2}\,(\cos 160° + i \sin 160°)$$

$$z_2 = \sqrt[3]{2}\left[\cos(40° + 120°\cdot 2) + i \sin(40° + 120°\cdot 2) \right] = \sqrt[3]{2}\,(\cos 280° + i \sin 280°)$$ ∎

> **WARNING:**   Most graphing utilities will only provide the answer z_0 to the calculation $(-1 + \sqrt{3}i)^{\wedge}\left(\frac{1}{3}\right)$. The following paragraph explains how to obtain z_1 and z_2 from z_0. ∎

Notice that each of the three complex roots of $-1 + \sqrt{3}i$ has the same magnitude, $\sqrt[3]{2}$. This means that the points corresponding to each cube root lie the same distance from the origin; that is, the three points lie on a circle with center at the origin and radius $\sqrt[3]{2}$. Furthermore, the arguments of these cube roots are $40°$, $160°$, and $280°$, the difference of consecutive pairs being

$120° = \dfrac{360°}{3}$. This means that the three points are equally spaced on the circle, as shown in Figure 40. These results are not coincidental. In fact, you are asked to show that these results hold for complex nth roots in Problems 53 through 55.

Figure 40

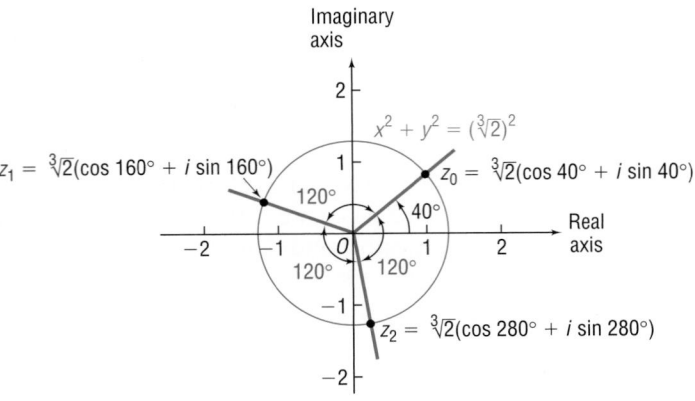

$z_1 = \sqrt[3]{2}(\cos 160° + i \sin 160°)$

$z_0 = \sqrt[3]{2}(\cos 40° + i \sin 40°)$

$x^2 + y^2 = (\sqrt[3]{2})^2$

$z_2 = \sqrt[3]{2}(\cos 280° + i \sin 280°)$

NOW WORK PROBLEM 43.

The Babylonians, Greeks, and Arabs considered square roots of negative quantities to be impossible and equations with complex solutions to be unsolvable. The first hint that there was some connection between real solutions of equations and complex numbers came when Girolamo Cardano (1501–1576) and Tartaglia (1499–1557) found *real* roots of cubic equations by taking cube roots of *complex* quantities. For centuries thereafter, mathematicians worked with complex numbers without much belief in their actual existence. In 1673, John Wallis appears to have been the first to suggest the graphical representation of complex numbers, a truly significant idea that was not pursued further until about 1800. Several people, including Karl Friedrich Gauss (1777–1855), then rediscovered the idea, and the graphical representation helped to establish complex numbers as equal members of the number family. In practical applications, complex numbers have found their greatest uses in the area of alternating current, where they are a commonplace tool, and in the area of subatomic physics.

HISTORICAL PROBLEMS

1. The quadratic formula will work perfectly well if the coefficients are complex numbers. Solve the following using De Moivre's Theorem where necessary.
   [**Hint:** The answers are "nice."]

   (a) $z^2 - (2 + 5i)z - 3 + 5i = 0$
   (b) $z^2 - (1 + i)z - 2 - i = 0$

5.3 Concepts and Vocabulary

In Problems 1–3, fill in the blanks.

1. When a complex number z is written in the polar form $z = r(\cos\theta + i\sin\theta)$, the nonnegative number r is the _____ or _____ of z, and the angle $\theta, 0 \le \theta < 2\pi$, is the _____ of z.

2. _____ Theorem can be used to raise a complex number to a power.

3. A complex number will, in general, have _____ cube roots.

In Problems 4–6, answer True or False to each statement.

4. De Moivre's Theorem is useful for raising a complex number to a positive integer power.

5. Using De Moivre's Theorem, the square of a complex number will have two answers.

6. The polar form of a complex number is unique.

7. Explain how to multiply two complex numbers given in polar form.

8. If $z = r(\cos\theta + i\sin\theta)$, use De Moivre's Theorem to write z^2.

9. If $z = r(\cos\theta + i\sin\theta)$, write down the two complex square roots of z.

5.3 Exercises

In Problems 1–12, plot each complex number in the complex plane and write it in polar form. Express the argument in degrees.

1. $1 + i$ **2.** $-1 + i$ **3.** $\sqrt{3} - i$ **4.** $1 - \sqrt{3}i$

5. $-3i$ **6.** -2 **7.** $4 - 4i$ **8.** $9\sqrt{3} + 9i$

9. $3 - 4i$ **10.** $2 + \sqrt{3}i$ **11.** $-2 + 3i$ **12.** $\sqrt{5} - i$

In Problems 13–22, write each complex number in rectangular form.

13. $2(\cos 120° + i\sin 120°)$ **14.** $3(\cos 210° + i\sin 210°)$ **15.** $4\left(\cos\dfrac{7\pi}{4} + i\sin\dfrac{7\pi}{4}\right)$

16. $2\left(\cos\dfrac{5\pi}{6} + i\sin\dfrac{5\pi}{6}\right)$ **17.** $3\left(\cos\dfrac{3\pi}{2} + i\sin\dfrac{3\pi}{2}\right)$ **18.** $4\left(\cos\dfrac{\pi}{2} + i\sin\dfrac{\pi}{2}\right)$

19. $0.2(\cos 100° + i\sin 100°)$ **20.** $0.4(\cos 200° + i\sin 200°)$ **21.** $2\left(\cos\dfrac{\pi}{18} + i\sin\dfrac{\pi}{18}\right)$

22. $3\left(\cos\dfrac{\pi}{10} + i\sin\dfrac{\pi}{10}\right)$

In Problems 23–30, find zw and $\dfrac{z}{w}$. Leave your answers in polar form.

23. $z = 2(\cos 40° + i\sin 40°)$
$w = 4(\cos 20° + i\sin 20°)$

24. $z = \cos 120° + i\sin 120°$
$w = \cos 100° + i\sin 100°$

25. $z = 3(\cos 130° + i\sin 130°)$
$w = 4(\cos 270° + i\sin 270°)$

26. $z = 2(\cos 80° + i\sin 80°)$
$w = 6(\cos 200° + i\sin 200°)$

27. $z = 2\left(\cos\dfrac{\pi}{8} + i\sin\dfrac{\pi}{8}\right)$
$w = 2\left(\cos\dfrac{\pi}{10} + i\sin\dfrac{\pi}{10}\right)$

28. $z = 4\left(\cos\dfrac{3\pi}{8} + i\sin\dfrac{3\pi}{8}\right)$
$w = 2\left(\cos\dfrac{9\pi}{16} + i\sin\dfrac{9\pi}{16}\right)$

29. $z = 2 + 2i$
$w = \sqrt{3} - i$

30. $z = 1 - i$
$w = 1 - \sqrt{3}i$

In Problems 31–42, write each expression in the standard form $a + bi$. Verify your answer using a graphing utility.

31. $[4(\cos 40° + i\sin 40°)]^3$ **32.** $[3(\cos 80° + i\sin 80°)]^3$ **33.** $\left[2\left(\cos\dfrac{\pi}{10} + i\sin\dfrac{\pi}{10}\right)\right]^5$

34. $\left[\sqrt{2}\left(\cos\dfrac{5\pi}{16} + i\sin\dfrac{5\pi}{16}\right)\right]^4$ **35.** $[\sqrt{3}(\cos 10° + i\sin 10°)]^6$ **36.** $\left[\dfrac{1}{2}(\cos 72° + i\sin 72°)\right]^5$

37. $\left[\sqrt{5}\left(\cos\dfrac{3\pi}{16} + i\sin\dfrac{3\pi}{16}\right)\right]^4$ **38.** $\left[\sqrt{3}\left(\cos\dfrac{5\pi}{18} + i\sin\dfrac{5\pi}{18}\right)\right]^6$ **39.** $(1 - i)^5$

40. $(\sqrt{3} - i)^6$ **41.** $(\sqrt{2} - i)^6$ **42.** $(1 - \sqrt{5}i)^8$

In Problems 43–50, find all the complex roots. Leave your answers in polar form with the argument in degrees.

43. The complex cube roots of $1 + i$

44. The complex fourth roots of $\sqrt{3} - i$

45. The complex fourth roots of $4 - 4\sqrt{3}i$

46. The complex cube roots of $-8 - 8i$

47. The complex fourth roots of $-16i$

48. The complex cube roots of -8

49. The complex fifth roots of i

50. The complex fifth roots of $-i$

---

51. Find the four complex fourth roots of unity (1) and plot each.

52. Find the six complex sixth roots of unity (1) and plot each.

53. Show that each complex nth root of a nonzero complex number w has the same magnitude.

54. Use the result of Problem 53 to draw the conclusion that each complex nth root lies on a circle with center at the origin. What is the radius of this circle?

55. Refer to Problem 54. Show that the complex nth roots of a nonzero complex number w are equally spaced on the circle.

56. Prove formula (6).

PREPARING FOR THIS SECTION

Before getting started, review the following:

✓ Rectangular Coordinates (Section 1.1, pp. 2–3)

✓ Pythagorean Theorem (Section A.2, p. 566)

5.4 VECTORS

OBJECTIVES

1. Graph Vectors
2. Find a Position Vector
3. Add and Subtract Vectors
4. Find a Scalar Product and the Magnitude of a Vector
5. Find a Unit Vector
6. Find a Vector from Its Direction and Magnitude
7. Work with Objects in Static Equilibrium

In simple terms, a **vector** (derived from the Latin *vehere*, meaning "to carry") is a quantity that has both magnitude and direction. It is customary to represent a vector by using an arrow. The length of the arrow represents the **magnitude** of the vector, and the arrowhead indicates the **direction** of the vector.

Many quantities in physics can be represented by vectors. For example, the velocity of an aircraft can be represented by an arrow that points in the direction of movement; the length of the arrow represents speed. If the aircraft speeds up, we lengthen the arrow; if the aircraft changes direction, we introduce an arrow in the new direction. See Figure 41. Based on this representation, it is not surprising that vectors and directed line segments are somehow related.

Figure 41

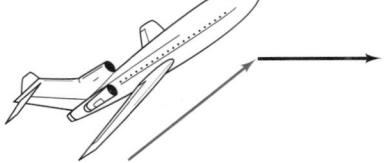

Geometric Vectors

If P and Q are two distinct points in the xy-plane, there is exactly one line containing both P and Q [Figure 42(a)]. The points on that part of the line that joins P to Q, including P and Q, form what is called the **line segment** $\overline{PQ}$ [Figure 42(b)]. If we order the points so that they proceed from P to Q, we have a **directed line segment** from P to Q, or a **geometric vector**, which we

Figure 42

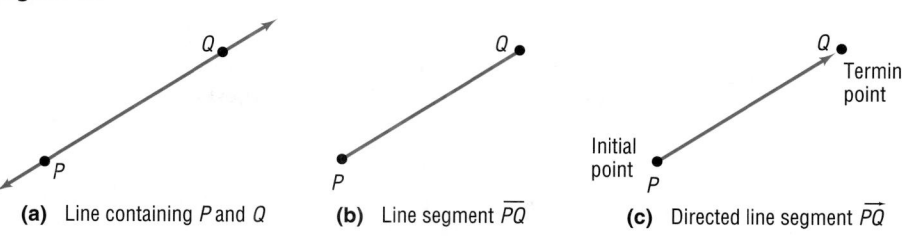

(a) Line containing *P* and *Q*   (b) Line segment $\overline{PQ}$ (c) Directed line segment $\overrightarrow{PQ}$

denote by $\overrightarrow{PQ}$. In a directed line segment $\overrightarrow{PQ}$, we call *P* the **initial point** and *Q* the **terminal point**, as indicated in Figure 42(c).

The magnitude of the directed line segment $\overrightarrow{PQ}$ is the distance from the point *P* to the point *Q*; that is, it is the length of the line segment. The direction of $\overrightarrow{PQ}$ is from *P* to *Q*. If a vector **v*** has the same magnitude and the same direction as the directed line segment $\overrightarrow{PQ}$, we write

$$\mathbf{v} = \overrightarrow{PQ}$$

The vector **v** whose magnitude is 0 is called the **zero vector, 0**. The zero vector is assigned no direction.

Two vectors **v** and **w** are **equal**, written

$$\mathbf{v} = \mathbf{w}$$

if they have the same magnitude and the same direction.

For example, the vectors shown in Figure 43 have the same magnitude and the same direction, so they are equal, even though they have different initial points and different terminal points. As a result, we find it useful to think of a vector simply as an arrow, keeping in mind that two arrows (vectors) are equal if they have the same direction and the same magnitude (length).

Figure 43

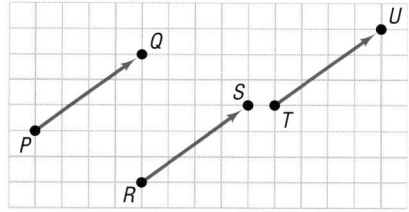

Adding Vectors

Figure 44

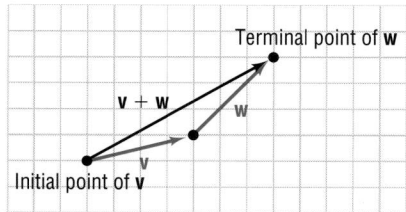

The **sum v + w** of two vectors is defined as follows: We position the vectors **v** and **w** so that the terminal point of **v** coincides with the initial point of **w**, as shown in Figure 44. The vector **v + w** is then the unique vector whose initial point coincides with the initial point of **v** and whose terminal point coincides with the terminal point of **w**.

Vector addition is **commutative**. That is, if **v** and **w** are any two vectors, then

$$\mathbf{v} + \mathbf{w} = \mathbf{w} + \mathbf{v}$$

Figure 45

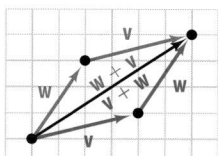

Figure 45 illustrates this fact. (Observe that the commutative property is another way of saying that opposite sides of a parallelogram are equal and parallel.)

Vector addition is also **associative**. That is, if **u**, **v**, and **w** are vectors, then

$$\mathbf{u} + (\mathbf{v} + \mathbf{w}) = (\mathbf{u} + \mathbf{v}) + \mathbf{w}$$

---

*Boldface letters will be used to denote vectors, in order to distinguish them from numbers. For handwritten work, an arrow is placed over the letter to signify a vector.

Figure 46
$(\mathbf{u} + \mathbf{v}) + \mathbf{w} = \mathbf{u} + (\mathbf{v} + \mathbf{w})$

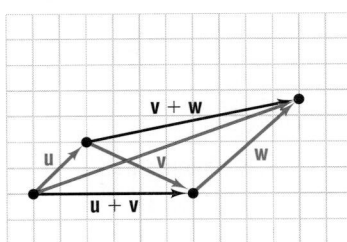

Figure 47

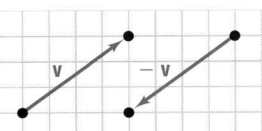

Figure 48

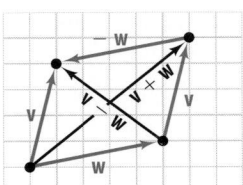

Figure 46 illustrates the associative property for vectors.

The zero vector has the property that

$$\mathbf{v} + \mathbf{0} = \mathbf{0} + \mathbf{v} = \mathbf{v}$$

for any vector $\mathbf{v}$.

If $\mathbf{v}$ is a vector, then $-\mathbf{v}$ is the vector having the same magnitude as $\mathbf{v}$, but whose direction is opposite to $\mathbf{v}$, as shown in Figure 47.

Furthermore,

$$\mathbf{v} + (-\mathbf{v}) = \mathbf{0}$$

If $\mathbf{v}$ and $\mathbf{w}$ are two vectors, we define the **difference $\mathbf{v} - \mathbf{w}$** as

$$\mathbf{v} - \mathbf{w} = \mathbf{v} + (-\mathbf{w})$$

Figure 48 illustrates the relationships among $\mathbf{v}, \mathbf{w}, \mathbf{v} + \mathbf{w}$, and $\mathbf{v} - \mathbf{w}$.

Multiplying Vectors by Numbers

When dealing with vectors, we refer to real numbers as **scalars**. Scalars are quantities that have only magnitude. Examples from physics of scalar quantities are temperature, speed, and time. We now define how to multiply a vector by a scalar.

> If α is a scalar and $\mathbf{v}$ is a vector, the **scalar product** $\alpha\mathbf{v}$ is defined as follows:
>
> 1. If $\alpha > 0$, the product $\alpha\mathbf{v}$ is the vector whose magnitude is α times the magnitude of $\mathbf{v}$ and whose direction is the same as $\mathbf{v}$.
> 2. If $\alpha < 0$, the product $\alpha\mathbf{v}$ is the vector whose magnitude is $|\alpha|$ times the magnitude of $\mathbf{v}$ and whose direction is opposite that of $\mathbf{v}$.
> 3. If $\alpha = 0$ or if $\mathbf{v} = \mathbf{0}$, then $\alpha\mathbf{v} = \mathbf{0}$.

Figure 49

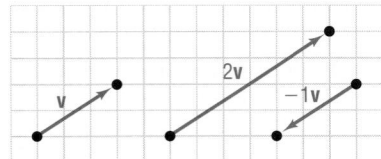

See Figure 49 for some illustrations.

For example, if $\mathbf{a}$ is the acceleration of an object of mass m due to a force $\mathbf{F}$ being exerted on it, then, by Newton's second law of motion, $\mathbf{F} = m\mathbf{a}$. Here, $m\mathbf{a}$ is the product of the scalar m and the vector $\mathbf{a}$.

Scalar products have the following properties:

$$0\mathbf{v} = \mathbf{0} \qquad 1\mathbf{v} = \mathbf{v} \qquad -1\mathbf{v} = -\mathbf{v}$$
$$(\alpha + \beta)\mathbf{v} = \alpha\mathbf{v} + \beta\mathbf{v} \qquad \alpha(\mathbf{v} + \mathbf{w}) = \alpha\mathbf{v} + \alpha\mathbf{w}$$
$$\alpha(\beta\mathbf{v}) = (\alpha\beta)\mathbf{v}$$

① EXAMPLE 1   Graphing Vectors

Figure 50

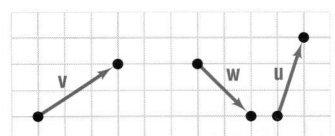

Use the vectors illustrated in Figure 50 to graph each of the following vectors:

(a) $\mathbf{v} - \mathbf{w}$ (b) $2\mathbf{v} + 3\mathbf{w}$ (c) $2\mathbf{v} - \mathbf{w} + \mathbf{u}$

Solution   Figure 51 illustrates each graph.

Figure 51

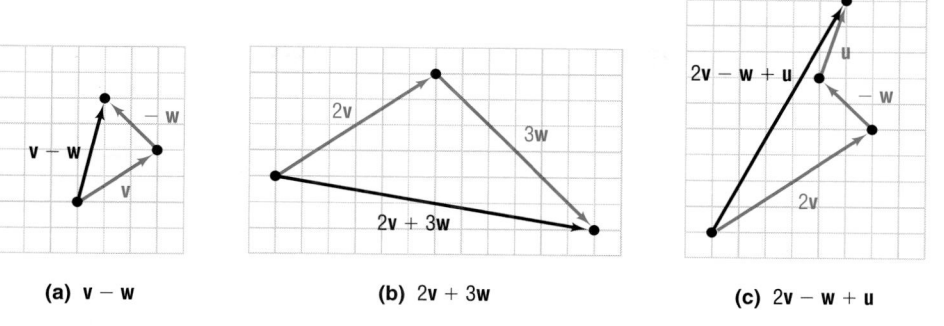

(a) **v** − **w**   (b) **2v** + **3w**   (c) **2v** − **w** + **u**

NOW WORK PROBLEMS 1 AND 3.

Magnitudes of Vectors

If **v** is a vector, we use the symbol $\|\mathbf{v}\|$ to represent the **magnitude** of **v**. Since $\|\mathbf{v}\|$ equals the length of a directed line segment, it follows that $\|\mathbf{v}\|$ has the following properties:

Theorem

Properties of $\|\mathbf{v}\|$

If **v** is a vector and if α is a scalar, then

(a) $\|\mathbf{v}\| \geq 0$ (b) $\|\mathbf{v}\| = 0$ if and only if $\mathbf{v} = \mathbf{0}$

(c) $\|-\mathbf{v}\| = \|\mathbf{v}\|$ (d) $\|\alpha\mathbf{v}\| = |\alpha|\,\|\mathbf{v}\|$

Property (a) is a consequence of the fact that distance is a nonnegative number. Property (b) follows, because the length of the directed line segment $\overrightarrow{PQ}$ is positive unless P and Q are the same point, in which case the length is 0. Property (c) follows because the length of the line segment $\overline{PQ}$ equals the length of the line segment $\overline{QP}$. Property (d) is a direct consequence of the definition of a scalar product.

A vector **u** for which $\|\mathbf{u}\| = 1$ is called a **unit vector**.

To compute the magnitude and direction of a vector, we need an algebraic way of representing vectors.

Algebraic Vectors

2 An **algebraic vector v** is represented as

$$\mathbf{v} = \langle a, b \rangle$$

where a and b are real numbers (scalars) called the **components** of the vector **v**.

We use a rectangular coordinate system to represent algebraic vectors in the plane. If $\mathbf{v} = \langle a, b \rangle$ is an algebraic vector whose initial point is at the origin, then **v** is called a **position vector**. See Figure 52. Notice that the terminal point of the position vector $\mathbf{v} = \langle a, b \rangle$ is $P = (a, b)$.

Figure 52

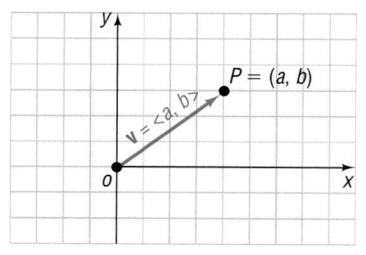

The next result states that any vector whose initial point is not at the origin is equal to a unique position vector.

Theorem

Suppose that $\mathbf{v}$ is a vector with initial point $P_1 = (x_1, y_1)$, not necessarily the origin, and terminal point $P_2 = (x_2, y_2)$. If $\mathbf{v} = \overrightarrow{P_1 P_2}$, then $\mathbf{v}$ is equal to the position vector

$$\mathbf{v} = \langle x_2 - x_1, y_2 - y_1 \rangle \qquad (1)$$

To see why this is true, look at Figure 53. Triangle OPA and triangle $P_1 P_2 Q$ are congruent. (Do you see why? The line segments have the same magnitude, so $d(O, P) = d(P_1, P_2)$; and they have the same direction, so $\angle POA = \angle P_2 P_1 Q$. Since the triangles are right triangles, we have angle-side-angle.) It follows that corresponding sides are equal. As a result, $x_2 - x_1 = a$ and $y_2 - y_1 = b$, so $\mathbf{v}$ may be written as

$$\mathbf{v} = \langle a, b \rangle = \langle x_2 - x_1, y_2 - y_1 \rangle$$

Figure 53
$\langle a, b \rangle = \langle x_2 - x_1, y_2 - y_1 \rangle$

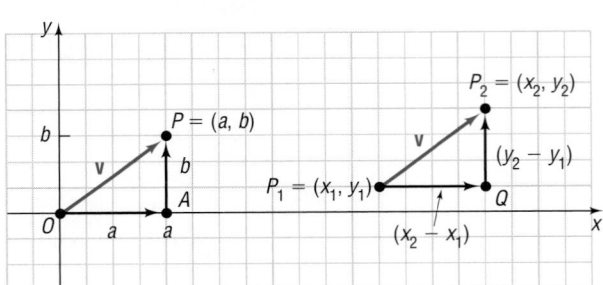

Because of this result, we can replace any algebraic vector by a unique position vector, and vice versa. This flexibility is one of the main reasons for the wide use of vectors. Unless otherwise specified, from now on the term *vector* will mean the unique position vector equal to it.

EXAMPLE 2 **Finding a Position Vector**

Figure 54

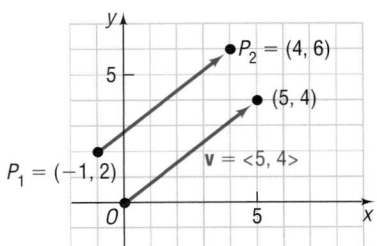

Find the position vector of the vector $\mathbf{v} = \overrightarrow{P_1 P_2}$ if $P_1 = (-1, 2)$ and $P_2 = (4, 6)$.

Solution By equation (1), the position vector equal to $\mathbf{v}$ is

$$\mathbf{v} = \langle 4 - (-1), 6 - 2 \rangle = \langle 5, 4 \rangle$$

See Figure 54.

Two position vectors $\mathbf{v}$ and $\mathbf{w}$ are equal if and only if the terminal point of $\mathbf{v}$ is the same as the terminal point of $\mathbf{w}$. This leads to the following result:

Theorem

Equality of Vectors

Two vectors **v** and **w** are equal if and only if their corresponding components are equal. That is,

$$\text{If } \mathbf{v} = \langle a_1, b_1 \rangle \quad \text{and} \quad \mathbf{w} = \langle a_2, b_2 \rangle$$
$$\text{then} \quad \mathbf{v} = \mathbf{w} \quad \text{if and only if} \quad a_1 = a_2 \quad \text{and} \quad b_1 = b_2.$$

Figure 55

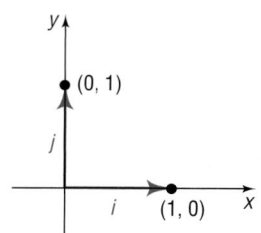

We now present an alternative representation of a vector in the plane that is common in the physical sciences. Let **i** denote the unit vector whose direction is along the positive x-axis; let **j** denote the unit vector whose direction is along the positive y-axis. Then $\mathbf{i} = \langle 1, 0 \rangle$ and $\mathbf{j} = \langle 0, 1 \rangle$, as shown in Figure 55. Any vector $\mathbf{v} = \langle a, b \rangle$ can be written using the unit vectors **i** and **j** as follows:

$$\mathbf{v} = \langle a, b \rangle = a\langle 1, 0 \rangle + b\langle 0, 1 \rangle = a\mathbf{i} + b\mathbf{j}$$

We call a and b the **horizontal** and **vertical components** of **v**, respectively.

─ NOW WORK PROBLEM **21**.

We define addition, subtraction, scalar product, and magnitude in terms of the components of a vector.

Let $\mathbf{v} = a_1\mathbf{i} + b_1\mathbf{j} = \langle a_1, b_1 \rangle$ and $\mathbf{w} = a_2\mathbf{i} + b_2\mathbf{j} = \langle a_2, b_2 \rangle$ be two vectors, and let α be a scalar. Then

$$\mathbf{v} + \mathbf{w} = (a_1 + a_2)\mathbf{i} + (b_1 + b_2)\mathbf{j} = \langle a_1 + a_2, b_1 + b_2 \rangle \quad (2)$$
$$\mathbf{v} - \mathbf{w} = (a_1 - a_2)\mathbf{i} + (b_1 - b_2)\mathbf{j} = \langle a_1 - a_2, b_1 - b_2 \rangle \quad (3)$$
$$\alpha\mathbf{v} = (\alpha a_1)\mathbf{i} + (\alpha b_1)\mathbf{j} = \langle \alpha a_1, \alpha b_1 \rangle \quad (4)$$
$$\|\mathbf{v}\| = \sqrt{a_1^2 + b_1^2} \quad (5)$$

These definitions are compatible with the geometric definitions given earlier in this section. See Figure 56.

Figure 56

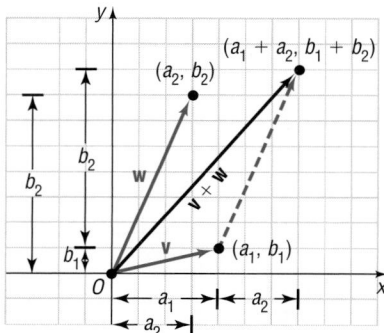

(a) Illustration of property (2)

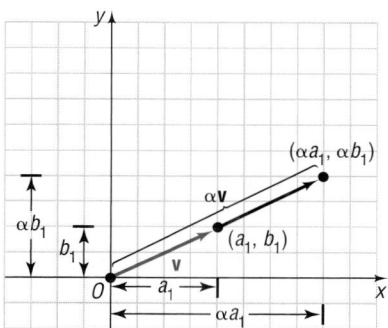

(b) Illustration of property (4), $\alpha > 0$

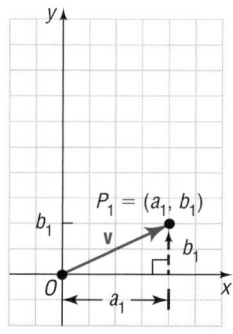

(c) Illustration of property (5):
$\| \mathbf{v} \| = $ Distance from O to P_1
$\| \mathbf{v} \| = \sqrt{a_1^2 + b_1^2}$

To add two vectors, add corresponding components. To subtract two vectors, subtract corresponding components.

---

3   **EXAMPLE 3**   **Adding and Subtracting Vectors**

If $v = 2i + 3j = \langle 2, 3 \rangle$ and $w = 3i - 4j = \langle 3, -4 \rangle$, find:

(a) $v + w$ (b) $v - w$

Solution   (a) $v + w = (2i + 3j) + (3i - 4j) = (2 + 3)i + (3 - 4)j = 5i - j$
or
$v + w = \langle 2, 3 \rangle + \langle 3, -4 \rangle = \langle 2 + 3, 3 + (-4) \rangle = \langle 5, -1 \rangle$

(b) $v - w = (2i + 3j) - (3i - 4j) = (2 - 3)i + [3 - (-4)]j = -i + 7j$
or
$v - w = \langle 2, 3 \rangle - \langle 3, -4 \rangle = \langle 2 - 3, 3 - (-4) \rangle = \langle -1, 7 \rangle$

---

4   **EXAMPLE 4**   **Finding Scalar Products and Magnitudes**

If $v = 2i + 3j = \langle 2, 3 \rangle$ and $w = 3i - 4j = \langle 3, -4 \rangle$, find:

(a) $3v$ (b) $2v - 3w$ (c) $\|v\|$

Solution   (a) $3v = 3(2i + 3j) = 6i + 9j$
or
$3v = 3\langle 2, 3 \rangle = \langle 6, 9 \rangle$

(b) $2v - 3w = 2(2i + 3j) - 3(3i - 4j) = 4i + 6j - 9i + 12j$
$= -5i + 18j$
or
$2v - 3w = 2\langle 2, 3 \rangle - 3\langle 3, -4 \rangle = \langle 4, 6 \rangle - \langle 9, -12 \rangle$
$= \langle 4 - 9, 6 - (-12) \rangle = \langle -5, 18 \rangle$

(c) $\|v\| = \|2i + 3j\| = \sqrt{2^2 + 3^2} = \sqrt{13}$

- **NOW WORK PROBLEMS 27 AND 33.**

For the remainder of the section, we will express a vector v in the form $ai + bj$.

5   Recall that a unit vector u is a vector for which $\|u\| = 1$. In many applications, it is useful to be able to find a unit vector u that has the same direction as a given vector v.

Theorem   **Unit Vector in the Direction of v**

For any nonzero vector v, the vector

$$u = \frac{v}{\|v\|}$$

is a unit vector that has the same direction as v.

Proof Let $\mathbf{v} = a\mathbf{i} + b\mathbf{j}$. Then $\|\mathbf{v}\| = \sqrt{a^2 + b^2}$ and

$$\mathbf{u} = \frac{\mathbf{v}}{\|\mathbf{v}\|} = \frac{a\mathbf{i} + b\mathbf{j}}{\sqrt{a^2 + b^2}} = \frac{a}{\sqrt{a^2 + b^2}}\mathbf{i} + \frac{b}{\sqrt{a^2 + b^2}}\mathbf{j}$$

The vector $\mathbf{u}$ is in the same direction as $\mathbf{v}$, since $\|\mathbf{v}\| > 0$. Furthermore,

$$\|\mathbf{u}\| = \sqrt{\frac{a^2}{a^2 + b^2} + \frac{b^2}{a^2 + b^2}} = \sqrt{\frac{a^2 + b^2}{a^2 + b^2}} = 1$$

Thus, $\mathbf{u}$ is a unit vector in the direction of $\mathbf{v}$. ■

As a consequence of this theorem, if $\mathbf{u}$ is a unit vector in the same direction as a vector $\mathbf{v}$, then $\mathbf{v}$ may be expressed as

$$\mathbf{v} = \|\mathbf{v}\|\mathbf{u} \qquad\qquad (6)$$

This way of expressing a vector is useful in many applications.

EXAMPLE 5 **Finding a Unit Vector**

Find a unit vector in the same direction as $\mathbf{v} = 4\mathbf{i} - 3\mathbf{j}$.

Solution We find $\|\mathbf{v}\|$ first.

$$\|\mathbf{v}\| = \|4\mathbf{i} - 3\mathbf{j}\| = \sqrt{16 + 9} = 5$$

Now we multiply $\mathbf{v}$ by the scalar $\dfrac{1}{\|\mathbf{v}\|} = \dfrac{1}{5}$. A unit vector in the same direction as $\mathbf{v}$ is

$$\frac{\mathbf{v}}{\|\mathbf{v}\|} = \frac{4\mathbf{i} - 3\mathbf{j}}{5} = \frac{4}{5}\mathbf{i} - \frac{3}{5}\mathbf{j}$$

✔ CHECK: This vector is, in fact, a unit vector because

$$\left(\frac{4}{5}\right)^2 + \left(-\frac{3}{5}\right)^2 = \frac{16}{25} + \frac{9}{25} = \frac{25}{25} = 1$$ ■ ■

━ NOW WORK PROBLEM **43**.

Writing a Vector in Terms of Its Magnitude and Direction

⑥ If a vector represents the speed and direction of an object, it is called a **velocity vector**. If a vector represents the direction and amount of a force acting on an object, it is called a **force vector**. In many applications, a vector is described in terms of its magnitude and direction, rather than in terms of its components. For example, a ball thrown with an initial speed of 25 miles per hour at an angle 30° to the horizontal is a velocity vector.

Suppose that we are given the magnitude $\|\mathbf{v}\|$ of a nonzero vector $\mathbf{v}$ and the angle $\alpha, 0° \leq \alpha < 360°$, between $\mathbf{v}$ and $\mathbf{i}$. To express $\mathbf{v}$ in terms of $\|\mathbf{v}\|$ and α, we first find the unit vector $\mathbf{u}$ having the same direction as $\mathbf{v}$.

$$\mathbf{u} = \frac{\mathbf{v}}{\|\mathbf{v}\|} \quad \text{or} \quad \mathbf{v} = \|\mathbf{v}\|\mathbf{u} \tag{7}$$

Figure 57

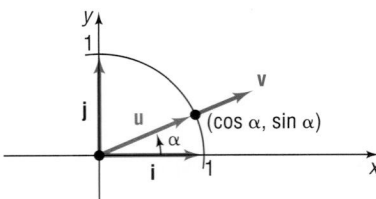

Look at Figure 57. The coordinates of the terminal point of $\mathbf{u}$ are $(\cos \alpha, \sin \alpha)$. Then $\mathbf{u} = \cos \alpha \mathbf{i} + \sin \alpha \mathbf{j}$ and, from (7),

$$\mathbf{v} = \|\mathbf{v}\|(\cos \alpha \mathbf{i} + \sin \alpha \mathbf{j}) \tag{8}$$

where α is the angle between $\mathbf{v}$ and $\mathbf{i}$.

EXAMPLE 6 **Writing a Vector When Its Magnitude and Direction Are Given**

A ball is thrown with an initial speed of 25 miles per hour in a direction that makes an angle of 30° with the positive x-axis. Express the velocity vector $\mathbf{v}$ in terms of $\mathbf{i}$ and $\mathbf{j}$. What is the initial speed in the horizontal direction? What is the initial speed in the vertical direction?

Solution The magnitude of $\mathbf{v}$ is $\|\mathbf{v}\| = 25$ miles per hour, and the angle between the direction of $\mathbf{v}$ and $\mathbf{i}$, the positive x-axis, is $\alpha = 30°$. By equation (8),

$$\mathbf{v} = \|\mathbf{v}\|(\cos \alpha \mathbf{i} + \sin \alpha \mathbf{j}) = 25(\cos 30°\mathbf{i} + \sin 30°\mathbf{j}) = 25\left(\frac{\sqrt{3}}{2}\mathbf{i} + \frac{1}{2}\mathbf{j}\right) = \frac{25\sqrt{3}}{2}\mathbf{i} + \frac{25}{2}\mathbf{j}$$

The initial speed of the ball in the horizontal direction is the horizontal component of $\mathbf{v}$, $\frac{25\sqrt{3}}{2} \approx 21.65$ miles per hour. The initial speed in the vertical direction is the vertical component of $\mathbf{v}$, $\frac{25}{2} = 12.5$ miles per hour. ∎

Application: Static Equilibrium

Figure 58

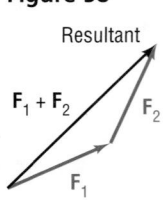

Resultant

$F_1 + F_2$

F_2

F_1

⑦ Because forces can be represented by vectors, two forces "combine" the way that vectors "add." If $\mathbf{F}_1$ and $\mathbf{F}_2$ are two forces simultaneously acting on an object, the vector sum $\mathbf{F}_1 + \mathbf{F}_2$ is the **resultant force**. The resultant force produces the same effect on the object as that obtained when the two forces $\mathbf{F}_1$ and $\mathbf{F}_2$ act on the object. See Figure 58. An application of this concept is *static equilibrium*. An object is said to be in **static equilibrium** if (1) the object is at rest and (2) the sum of all forces acting on the object is zero, that is, if the resultant force is 0.

EXAMPLE 7 **An Object in Static Equilibrium**

A box of supplies that weighs 1200 pounds is suspended by two cables attached to the ceiling, as shown in Figure 59. What is the tension in the two cables?

Figure 59

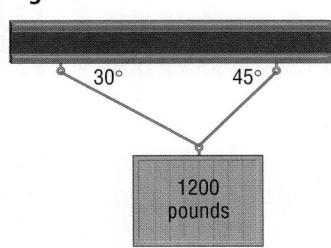

30°   45°

1200
pounds

Solution We draw a force diagram with the vectors drawn as shown in Figure 60. The tensions in the cables are the magnitudes $\|\mathbf{F}_1\|$ and $\|\mathbf{F}_2\|$ of the force vectors $\mathbf{F}_1$ and $\mathbf{F}_2$. The magnitude of the force vector $\mathbf{F}_3$ equals 1200 pounds, the weight of the box. Now write each force vector in terms of the unit vectors $\mathbf{i}$ and $\mathbf{j}$. For $\mathbf{F}_1$ and $\mathbf{F}_2$, we use equation (8). Remember that α is the angle between the vector and the positive x-axis.

Figure 60

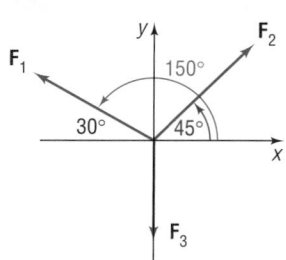

$$\mathbf{F}_1 = \|\mathbf{F}_1\|(\cos 150°\mathbf{i} + \sin 150°\mathbf{j}) = \|\mathbf{F}_1\|\left(-\frac{\sqrt{3}}{2}\mathbf{i} + \frac{1}{2}\mathbf{j}\right) = -\frac{\sqrt{3}}{2}\|\mathbf{F}_1\|\mathbf{i} + \frac{1}{2}\|\mathbf{F}_1\|\mathbf{j}$$

$$\mathbf{F}_2 = \|\mathbf{F}_2\|(\cos 45°\mathbf{i} + \sin 45°\mathbf{j}) = \|\mathbf{F}_2\|\left(\frac{\sqrt{2}}{2}\mathbf{i} + \frac{\sqrt{2}}{2}\mathbf{j}\right) = \frac{\sqrt{2}}{2}\|\mathbf{F}_2\|\mathbf{i} + \frac{\sqrt{2}}{2}\|\mathbf{F}_2\|\mathbf{j}$$

$$\mathbf{F}_3 = -1200\mathbf{j}$$

For static equilibrium, the sum of the force vectors must equal zero.

$$\mathbf{F}_1 + \mathbf{F}_2 + \mathbf{F}_3 = -\frac{\sqrt{3}}{2}\|\mathbf{F}_1\|\mathbf{i} + \frac{1}{2}\|\mathbf{F}_1\|\mathbf{j} + \frac{\sqrt{2}}{2}\|\mathbf{F}_2\|\mathbf{i} + \frac{\sqrt{2}}{2}\|\mathbf{F}_2\|\mathbf{j} - 1200\mathbf{j} = \mathbf{0}$$

The **i** component and **j** component will each equal zero. This results in the two equations

$$-\frac{\sqrt{3}}{2}\|\mathbf{F}_1\| + \frac{\sqrt{2}}{2}\|\mathbf{F}_2\| = 0 \qquad (9)$$

$$\frac{1}{2}\|\mathbf{F}_1\| + \frac{\sqrt{2}}{2}\|\mathbf{F}_2\| - 1200 = 0 \qquad (10)$$

We solve equation (9) for $\|\mathbf{F}_2\|$ and obtain

$$\|\mathbf{F}_2\| = \frac{\sqrt{3}}{\sqrt{2}}\|\mathbf{F}_1\| \qquad (11)$$

Substituting into equation (10) and solving for $\|\mathbf{F}_1\|$, we obtain

$$\frac{1}{2}\|\mathbf{F}_1\| + \frac{\sqrt{2}}{2}\left(\frac{\sqrt{3}}{\sqrt{2}}\|\mathbf{F}_1\|\right) - 1200 = 0$$

$$\frac{1}{2}\|\mathbf{F}_1\| + \frac{\sqrt{3}}{2}\|\mathbf{F}_1\| - 1200 = 0$$

$$\frac{1 + \sqrt{3}}{2}\|\mathbf{F}_1\| = 1200$$

$$\|\mathbf{F}_1\| = \frac{2400}{1 + \sqrt{3}} \approx 878.5 \text{ pounds}$$

Substituting this value into equation (11) yields $\|\mathbf{F}_2\|$.

$$\|\mathbf{F}_2\| = \frac{\sqrt{3}}{\sqrt{2}}\|\mathbf{F}_1\| = \frac{\sqrt{3}}{\sqrt{2}}\frac{2400}{1 + \sqrt{3}} \approx 1075.9 \text{ pounds}$$

The left cable has tension of approximately 878.5 pounds and the right cable has tension of approximately 1075.9 pounds. ∎

HISTORICAL FEATURE

Josiah Gibbs
1839–1903

The history of vectors is surprisingly complicated for such a natural concept. In the *xy*-plane, complex numbers do a good job of imitating vectors. About 1840, mathematicians became interested in finding a system that would do for three dimensions what the complex numbers do for two dimensions. Hermann Grassmann (1809–1877), in Germany, and William Rowan Hamilton (1805–1865), in Ireland, both attempted to find solutions. Hamilton's system was the *quaternions*, which are best thought of as a real number plus a vector, and do for four dimensions what complex numbers do for two dimensions. In this system the order of multiplication matters; that is, $ab \neq ba$. Also, two products of vectors emerged, the scalar (or dot) product and the vector (or cross) product.

Grassmann's abstract style, although easily read today, was almost impenetrable during the previous century, and only a few of his ideas were appreciated. Among those few were the same scalar and vector products that Hamilton had found.

About 1880, the American physicist Josiah Willard Gibbs (1839–1903) worked out an algebra involving only the simplest concepts: the vectors and the two products. He then added some calculus, and the resulting system was simple, flexible, and well adapted to expressing a large number of physical laws. This system remains in use essentially unchanged. Hamilton's and Grassmann's more extensive systems each gave birth to much interesting mathematics, but little of this mathematics is seen at elementary levels.

5.4 Concepts and Vocabulary

In Problems 1–3, fill in the blanks.

1. A vector whose magnitude is 1 is called a(n) _____ vector.

2. The product of a vector by a number is called a _____ product.

3. If $\mathbf{v} = a\mathbf{i} + b\mathbf{j}$, then a is called the _____ component of $\mathbf{v}$ and b is the _____ component of $\mathbf{v}$.

In Problems 4–6, answer True or False to each statement.

4. Vectors are quantities that have magnitude and direction.

5. Force is a physical example of a vector.

6. Mass is a physical example of a vector.

7. Explain in words how to add two vectors drawn on a piece of paper.

5.4 Exercises

In Problems 1–8, use the vectors in the figure at the right to graph each of the following vectors.

1. $\mathbf{v} + \mathbf{w}$	2. $\mathbf{u} + \mathbf{v}$
3. $3\mathbf{v}$	4. $4\mathbf{w}$
5. $\mathbf{v} - \mathbf{w}$	6. $\mathbf{u} - \mathbf{v}$
7. $3\mathbf{v} + \mathbf{u} - 2\mathbf{w}$	8. $2\mathbf{u} - 3\mathbf{v} + \mathbf{w}$

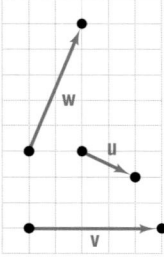

In Problems 9–16, use the figure at the right. Determine whether the given statement is true or false.

9. $\mathbf{A} + \mathbf{B} = \mathbf{F}$	**10.** $\mathbf{K} + \mathbf{G} = \mathbf{F}$
11. $\mathbf{C} = \mathbf{D} - \mathbf{E} + \mathbf{F}$	**12.** $\mathbf{G} + \mathbf{H} + \mathbf{E} = \mathbf{D}$
13. $\mathbf{E} + \mathbf{D} = \mathbf{G} + \mathbf{H}$	**14.** $\mathbf{H} - \mathbf{C} = \mathbf{G} - \mathbf{F}$
15. $\mathbf{A} + \mathbf{B} + \mathbf{K} + \mathbf{G} = \mathbf{0}$	**16.** $\mathbf{A} + \mathbf{B} + \mathbf{C} + \mathbf{H} + \mathbf{G} = \mathbf{0}$

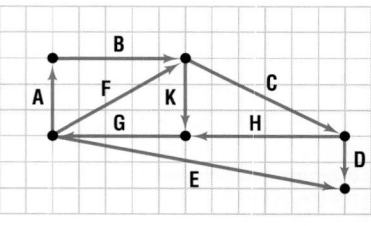

17. If $\|\mathbf{v}\| = 4$, what is $\|3\mathbf{v}\|$?

18. If $\|\mathbf{v}\| = 2$, what is $\|-4\mathbf{v}\|$?

In Problems 19–26, the vector $\mathbf{v}$ has initial point P and terminal point Q. Write $\mathbf{v}$ in the form $a\mathbf{i} + b\mathbf{j}$, that is, find its position vector.

19. $P = (0,0);\quad Q = (3,4)$	**20.** $P = (0,0);\quad Q = (-3,-5)$
21. $P = (3,2);\quad Q = (5,6)$	**22.** $P = (-3,2);\quad Q = (6,5)$
23. $P = (-2,-1);\quad Q = (6,-2)$	**24.** $P = (-1,4);\quad Q = (6,2)$
25. $P = (1,0);\quad Q = (0,1)$	**26.** $P = (1,1);\quad Q = (2,2)$

In Problems 27–32, find $\|\mathbf{v}\|$.

27. $\mathbf{v} = 3\mathbf{i} - 4\mathbf{j}$	**28.** $\mathbf{v} = -5\mathbf{i} + 12\mathbf{j}$	**29.** $\mathbf{v} = \mathbf{i} - \mathbf{j}$
30. $\mathbf{v} = -\mathbf{i} - \mathbf{j}$	**31.** $\mathbf{v} = -2\mathbf{i} + 3\mathbf{j}$	**32.** $\mathbf{v} = 6\mathbf{i} + 2\mathbf{j}$

In Problems 33–38, find each quantity if $\mathbf{v} = 3\mathbf{i} - 5\mathbf{j}$ *and* $\mathbf{w} = -2\mathbf{i} + 3\mathbf{j}$.

33. $2\mathbf{v} + 3\mathbf{w}$ **34.** $3\mathbf{v} - 2\mathbf{w}$ **35.** $\|\mathbf{v} - \mathbf{w}\|$

36. $\|\mathbf{v} + \mathbf{w}\|$ **37.** $\|\mathbf{v}\| - \|\mathbf{w}\|$ **38.** $\|\mathbf{v}\| + \|\mathbf{w}\|$

In Problems 39–44, find the unit vector having the same direction as $\mathbf{v}$.

39. $\mathbf{v} = 5\mathbf{i}$ **40.** $\mathbf{v} = -3\mathbf{j}$ **41.** $\mathbf{v} = 3\mathbf{i} - 4\mathbf{j}$

42. $\mathbf{v} = -5\mathbf{i} + 12\mathbf{j}$ **43.** $\mathbf{v} = \mathbf{i} - \mathbf{j}$ **44.** $\mathbf{v} = 2\mathbf{i} - \mathbf{j}$

45. Find a vector $\mathbf{v}$ whose magnitude is 4 and whose component in the $\mathbf{i}$ direction is twice the component in the $\mathbf{j}$ direction.

46. Find a vector $\mathbf{v}$ whose magnitude is 3 and whose component in the $\mathbf{i}$ direction is equal to the component in the $\mathbf{j}$ direction.

47. If $\mathbf{v} = 2\mathbf{i} - \mathbf{j}$ and $\mathbf{w} = x\mathbf{i} + 3\mathbf{j}$, find all numbers x for which $\|\mathbf{v} + \mathbf{w}\| = 5$.

48. If $P = (-3, 1)$ and $Q = (x, 4)$, find all numbers x such that the vector represented by $\overrightarrow{PQ}$ has length 5.

In Problems 49–54, write the vector $\mathbf{v}$ *in the form* $a\mathbf{i} + b\mathbf{j}$, *given its magnitude* $\|\mathbf{v}\|$ *and the angle* α *it makes with the positive x-axis.*

49. $\|\mathbf{v}\| = 5, \quad \alpha = 60°$ **50.** $\|\mathbf{v}\| = 8, \quad \alpha = 45°$ **51.** $\|\mathbf{v}\| = 14, \quad \alpha = 120°$

52. $\|\mathbf{v}\| = 3, \quad \alpha = 240°$ **53.** $\|\mathbf{v}\| = 25, \quad \alpha = 330°$ **54.** $\|\mathbf{v}\| = 15, \quad \alpha = 315°$

---

55. A child pulls a wagon with a force of 40 pounds. The handle of the wagon makes an angle of 30° with the ground. Express the force vector $\mathbf{F}$ in terms of $\mathbf{i}$ and $\mathbf{j}$.

56. A man pushes a wheelbarrow up an incline of 20° with a force of 100 pounds. Express the force vector $\mathbf{F}$ in terms of $\mathbf{i}$ and $\mathbf{j}$.

57. Resultant Force Two forces of magnitude 40 newtons (N) and 60 newtons act on an object at angles of 30° and $-45°$ with the positive x-axis as shown in the figure. Find the direction and magnitude of the resultant force; that is, find $\mathbf{F}_1 + \mathbf{F}_2$.

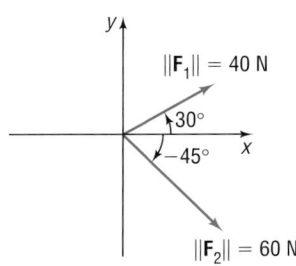

58. Resultant Force Two forces of magnitude 30 newtons (N) and 70 newtons act on an object at angles of 45° and 120° with the positive x-axis as shown in the figure. Find the direction and magnitude of the resultant force; that is, find $\mathbf{F}_1 + \mathbf{F}_2$.

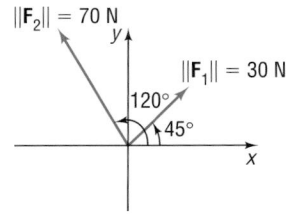

59. Static Equilibrium A weight of 1000 pounds is suspended from two cables as shown in the figure. What is the tension of the two cables?

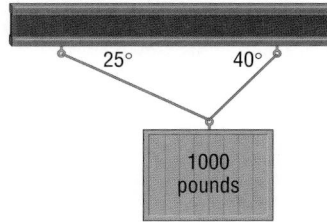

60. Static Equilibrium A weight of 800 pounds is suspended from two cables as shown in the figure. What is the tension of the two cables?

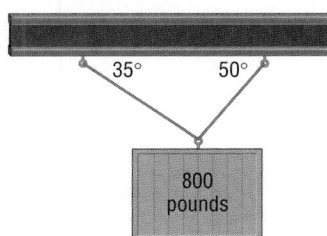

61. Static Equilibrium A tightrope walker located at a certain point deflects the rope as indicated in the figure. If the weight of the tightrope walker is 150 pounds, how much tension is in each part of the rope?

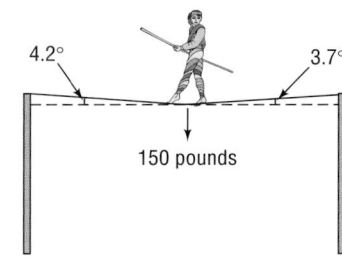

62. **Static Equilibrium**   Repeat Problem 61 if the left angle
    is 3.8°, the right angle is 2.6°, and the weight of the
    tightrope walker is 135 pounds.

63. Show on the following graph the force needed for the ob-
    ject at P to be in static equilibrium.

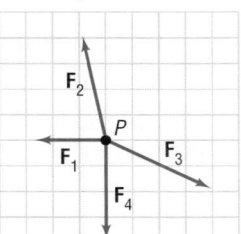

64. Explain in your own words what a vector is. Give an ex-
    ample of a vector.

65. Write a brief paragraph comparing the algebra of com-
    plex numbers and the algebra of vectors.

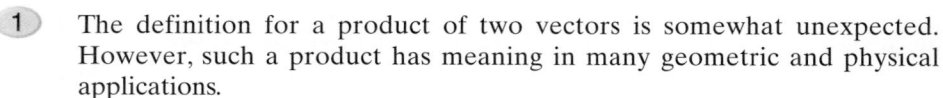

PREPARING FOR THIS SECTION

Before getting started, review the following:

✓ Law of Cosines (Section 4.3, p. 288)

5.5 THE DOT PRODUCT

OBJECTIVES
1. Find the Dot Product of Two Vectors
2. Find the Angle between Two Vectors
3. Determine Whether Two Vectors Are Parallel
4. Determine Whether Two Vectors Are Orthogonal
5. Decompose a Vector into Two Orthogonal Vectors
6. Compute Work

1. The definition for a product of two vectors is somewhat unexpected. However, such a product has meaning in many geometric and physical applications.

> If $\mathbf{v} = a_1\mathbf{i} + b_1\mathbf{j}$ and $\mathbf{w} = a_2\mathbf{i} + b_2\mathbf{j}$ are two vectors, the **dot product** $\mathbf{v} \cdot \mathbf{w}$ is defined as
>
> $$\mathbf{v} \cdot \mathbf{w} = a_1 a_2 + b_1 b_2 \tag{1}$$

EXAMPLE 1   **Finding Dot Products**

If $\mathbf{v} = 2\mathbf{i} - 3\mathbf{j}$ and $\mathbf{w} = 5\mathbf{i} + 3\mathbf{j}$, find:

(a) $\mathbf{v} \cdot \mathbf{w}$ (b) $\mathbf{w} \cdot \mathbf{v}$ (c) $\mathbf{v} \cdot \mathbf{v}$

(d) $\mathbf{w} \cdot \mathbf{w}$ (e) $\|\mathbf{v}\|$ (f) $\|\mathbf{w}\|$

Solution
(a) $\mathbf{v} \cdot \mathbf{w} = 2(5) + (-3)3 = 1$ (b) $\mathbf{w} \cdot \mathbf{v} = 5(2) + 3(-3) = 1$

(c) $\mathbf{v} \cdot \mathbf{v} = 2(2) + (-3)(-3) = 13$ (d) $\mathbf{w} \cdot \mathbf{w} = 5(5) + 3(3) = 34$

(e) $\|\mathbf{v}\| = \sqrt{2^2 + (-3)^2} = \sqrt{13}$ (f) $\|\mathbf{w}\| = \sqrt{5^2 + 3^2} = \sqrt{34}$ ∎

Since the dot product $\mathbf{v} \cdot \mathbf{w}$ of two vectors $\mathbf{v}$ and $\mathbf{w}$ is a real number (scalar), we sometimes refer to it as the **scalar product**.

Properties

The results obtained in Example 1 suggest some general properties.

Theorem

Properties of the Dot Product

If $\mathbf{u}, \mathbf{v}$, and $\mathbf{w}$ are vectors, then

Commutative Property

$$\mathbf{u} \cdot \mathbf{v} = \mathbf{v} \cdot \mathbf{u} \qquad (2)$$

Distributive Property

$$\mathbf{u} \cdot (\mathbf{v} + \mathbf{w}) = \mathbf{u} \cdot \mathbf{v} + \mathbf{u} \cdot \mathbf{w} \qquad (3)$$

$$\mathbf{v} \cdot \mathbf{v} = \|\mathbf{v}\|^2 \qquad (4)$$
$$\mathbf{0} \cdot \mathbf{v} = 0 \qquad (5)$$

Proof   We will prove properties (2) and (4) here and leave properties (3) and (5) as exercises (see Problems 33 and 34).

To prove property (2), we let $\mathbf{u} = a_1\mathbf{i} + b_1\mathbf{j}$ and $\mathbf{v} = a_2\mathbf{i} + b_2\mathbf{j}$. Then

$$\mathbf{u} \cdot \mathbf{v} = a_1a_2 + b_1b_2 = a_2a_1 + b_2b_1 = \mathbf{v} \cdot \mathbf{u}$$

To prove property (4), we let $\mathbf{v} = a\mathbf{i} + b\mathbf{j}$. Then

$$\mathbf{v} \cdot \mathbf{v} = a^2 + b^2 = \|\mathbf{v}\|^2$$

One use of the dot product is to calculate the angle between two vectors.

Angle between Vectors

② Let $\mathbf{u}$ and $\mathbf{v}$ be two vectors with the same initial point A. Then the vectors $\mathbf{u}$, $\mathbf{v}$, and $\mathbf{u} - \mathbf{v}$ form a triangle. The angle θ at vertex A of the triangle is the angle between the vectors $\mathbf{u}$ and $\mathbf{v}$. See Figure 61. We wish to find a formula for calculating the angle θ.

The sides of the triangle have lengths $\|\mathbf{v}\|$, $\|\mathbf{u}\|$, and $\|\mathbf{u} - \mathbf{v}\|$, and θ is the included angle between the sides of length $\|\mathbf{v}\|$ and $\|\mathbf{u}\|$. The Law of Cosines (Section 4.3) can be used to find the cosine of the included angle.

$$\|\mathbf{u} - \mathbf{v}\|^2 = \|\mathbf{u}\|^2 + \|\mathbf{v}\|^2 - 2\|\mathbf{u}\|\|\mathbf{v}\| \cos\theta$$

Now we use property (4) to rewrite this equation in terms of dot products.

$$(\mathbf{u} - \mathbf{v}) \cdot (\mathbf{u} - \mathbf{v}) = \mathbf{u} \cdot \mathbf{u} + \mathbf{v} \cdot \mathbf{v} - 2\|\mathbf{u}\|\|\mathbf{v}\| \cos\theta \qquad (6)$$

Then we apply the distributive property (3) twice on the left side of (6) to obtain

$$\begin{aligned}
(\mathbf{u} - \mathbf{v}) \cdot (\mathbf{u} - \mathbf{v}) &= \mathbf{u} \cdot (\mathbf{u} - \mathbf{v}) - \mathbf{v} \cdot (\mathbf{u} - \mathbf{v}) \\
&= \mathbf{u} \cdot \mathbf{u} - \mathbf{u} \cdot \mathbf{v} - \mathbf{v} \cdot \mathbf{u} + \mathbf{v} \cdot \mathbf{v} \\
&= \mathbf{u} \cdot \mathbf{u} + \mathbf{v} \cdot \mathbf{v} - 2\mathbf{u} \cdot \mathbf{v} \qquad (7)
\end{aligned}$$

↑
Property (2)

Figure 61

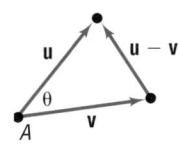

Combining equations (6) and (7), we have

$$\mathbf{u} \cdot \mathbf{u} + \mathbf{v} \cdot \mathbf{v} - 2\mathbf{u} \cdot \mathbf{v} = \mathbf{u} \cdot \mathbf{u} + \mathbf{v} \cdot \mathbf{v} - 2\|\mathbf{u}\|\|\mathbf{v}\| \cos \theta$$
$$\mathbf{u} \cdot \mathbf{v} = \|\mathbf{u}\|\|\mathbf{v}\| \cos \theta$$

We have proved the following result:

Theorem

Angle between Vectors

If $\mathbf{u}$ and $\mathbf{v}$ are two nonzero vectors, the angle $\theta, 0 \le \theta \le \pi$, between $\mathbf{u}$ and $\mathbf{v}$ is determined by the formula

$$\cos \theta = \frac{\mathbf{u} \cdot \mathbf{v}}{\|\mathbf{u}\|\|\mathbf{v}\|} \tag{8}$$

EXAMPLE 2 Finding the Angle θ between Two Vectors

Find the angle θ between $\mathbf{u} = 4\mathbf{i} - 3\mathbf{j}$ and $\mathbf{v} = 2\mathbf{i} + 5\mathbf{j}$.

Solution We compute the quantities $\mathbf{u} \cdot \mathbf{v}$, $\|\mathbf{u}\|$, and $\|\mathbf{v}\|$.

$$\mathbf{u} \cdot \mathbf{v} = 4(2) + (-3)(5) = -7$$
$$\|\mathbf{u}\| = \sqrt{4^2 + (-3)^2} = 5$$
$$\|\mathbf{v}\| = \sqrt{2^2 + 5^2} = \sqrt{29}$$

By formula (8), if θ is the angle between $\mathbf{u}$ and $\mathbf{v}$, then

$$\cos \theta = \frac{\mathbf{u} \cdot \mathbf{v}}{\|\mathbf{u}\|\|\mathbf{v}\|} = \frac{-7}{5\sqrt{29}} \approx -0.26$$

We find that $\theta \approx 105°$. See Figure 62.

Figure 62

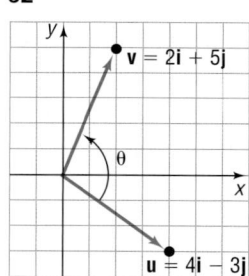

NOW WORK PROBLEMS **1(a)** AND **(b)**.

EXAMPLE 3 Finding the Actual Speed and Direction of an Aircraft

A Boeing 737 aircraft maintains a constant airspeed of 500 miles per hour in the direction due south. The velocity of the jet stream is 80 miles per hour in a northeasterly direction. Find the actual speed and direction of the aircraft relative to the ground.

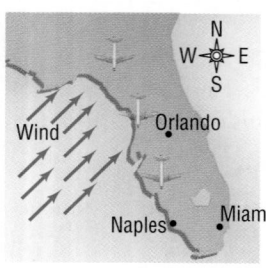

Solution We set up a coordinate system in which north (N) is along the positive y-axis. See Figure 63. Let

$$\mathbf{v}_a = \text{velocity of aircraft relative to the air} = -500\mathbf{j}$$
$$\mathbf{v}_w = \text{velocity of jet stream}$$
$$\mathbf{v}_g = \text{velocity of aircraft relative to ground}$$

The velocity of the jet stream $\mathbf{v}_w$ has magnitude 80 and direction NE (northeast), so $\alpha = 45°$. We express $\mathbf{v}_w$ in terms of $\mathbf{i}$ and $\mathbf{j}$ as

$$\mathbf{v}_w = 80(\cos 45°\mathbf{i} + \sin 45°\mathbf{j}) = 80\left(\frac{\sqrt{2}}{2}\mathbf{i} + \frac{\sqrt{2}}{2}\mathbf{j}\right) = 40\sqrt{2}(\mathbf{i} + \mathbf{j})$$

Figure 63

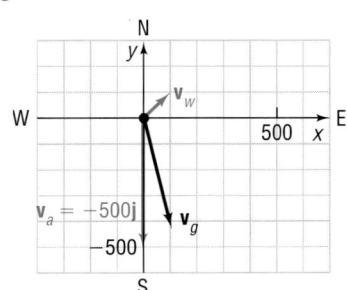

The velocity of the aircraft relative to the ground is

$$\mathbf{v}_g = \mathbf{v}_a + \mathbf{v}_w = -500\mathbf{j} + 40\sqrt{2}(\mathbf{i} + \mathbf{j}) = 40\sqrt{2}\mathbf{i} + (40\sqrt{2} - 500)\mathbf{j}$$

The actual speed of the aircraft is

$$\|\mathbf{v}_g\| = \sqrt{(40\sqrt{2})^2 + (40\sqrt{2} - 500)^2} \approx 447 \text{ miles per hour}$$

The angle θ between $\mathbf{v}_g$ and the vector $\mathbf{v}_a = -500\mathbf{j}$ (the velocity of the aircraft relative to the air) is determined by the equation

$$\cos \theta = \frac{\mathbf{v}_g \cdot \mathbf{v}_a}{\|\mathbf{v}_g\|\|\mathbf{v}_a\|} = \frac{(40\sqrt{2} - 500)(-500)}{(447)(500)} \approx 0.9920$$

$$\theta \approx 7.3°$$

The direction of the aircraft relative to the ground is approximately S7.3°E (about 7.3° east of south).

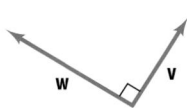 **NOW WORK PROBLEM 19.**

Parallel and Orthogonal Vectors

③ Two vectors $\mathbf{v}$ and $\mathbf{w}$ are said to be **parallel** if there is a nonzero scalar α so that $\mathbf{v} = \alpha\mathbf{w}$. In this case, the angle θ between $\mathbf{v}$ and $\mathbf{w}$ is 0 or π.

EXAMPLE 4 **Determining Whether Vectors Are Parallel**

The vectors $\mathbf{v} = 3\mathbf{i} - \mathbf{j}$ and $\mathbf{w} = 6\mathbf{i} - 2\mathbf{j}$ are parallel, since $\mathbf{v} = \dfrac{1}{2}\mathbf{w}$. Furthermore, since

$$\cos \theta = \frac{\mathbf{v} \cdot \mathbf{w}}{\|\mathbf{v}\|\|\mathbf{w}\|} = \frac{18 + 2}{\sqrt{10}\,\sqrt{40}} = \frac{20}{\sqrt{400}} = 1$$

the angle θ between $\mathbf{v}$ and $\mathbf{w}$ is 0. ∎

④ If the angle θ between two nonzero vectors $\mathbf{v}$ and $\mathbf{w}$ is $\dfrac{\pi}{2}$, the vectors $\mathbf{v}$ and $\mathbf{w}$ are called **orthogonal**.* See Figure 64.

It follows from formula (8) that if $\mathbf{v}$ and $\mathbf{w}$ are orthogonal then $\mathbf{v} \cdot \mathbf{w} = 0$, since $\cos \dfrac{\pi}{2} = 0$.

On the other hand, if $\mathbf{v} \cdot \mathbf{w} = 0$, then either $\mathbf{v} = 0$ or $\mathbf{w} = 0$ or $\cos \theta = 0$. In the latter case, $\theta = \dfrac{\pi}{2}$, and $\mathbf{v}$ and $\mathbf{w}$ are orthogonal. If $\mathbf{v}$ or $\mathbf{w}$ is the zero vector, then, since the zero vector has no specific direction, we adopt the convention that the zero vector is orthogonal to every vector.

Figure 64
v is orthogonal to w

Theorem Two vectors $\mathbf{v}$ and $\mathbf{w}$ are orthogonal if and only if

$$\mathbf{v} \cdot \mathbf{w} = 0$$

Orthogonal, perpendicular, and *normal* are all terms that mean "meet at a right angle." It is customary to refer to two vectors as being *orthogonal*, two lines as being *perpendicular*, and a line and a plane or a vector and a plane as being *normal*.

CHAPTER 5 Polar Coordinates; Vectors

EXAMPLE 5 Determining Whether Two Vectors Are Orthogonal

Figure 65

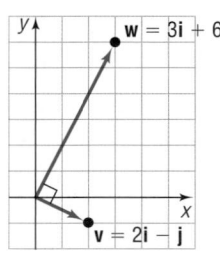

The vectors

$$\mathbf{v} = 2\mathbf{i} - \mathbf{j} \quad \text{and} \quad \mathbf{w} = 3\mathbf{i} + 6\mathbf{j}$$

are orthogonal, since

$$\mathbf{v} \cdot \mathbf{w} = 6 - 6 = 0$$

See Figure 65. ∎

━━ NOW WORK PROBLEM **1(C)**.

Projection of a Vector onto Another Vector

⑤ In many physical applications, it is necessary to find "how much" of a vector is applied in a given direction. Look at Figure 66. The force **F** due to gravity is pulling straight down (toward the center of Earth) on the block. To study the effect of gravity on the block, it is necessary to determine how much of **F** is actually pushing the block down the incline ($\mathbf{F}_1$) and how much is pressing the block against the incline ($\mathbf{F}_2$), at a right angle to the incline. Knowing the **decomposition** of **F** often will allow us to determine when friction is overcome and the block will slide down the incline.

Figure 66

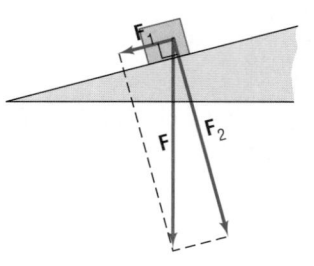

Suppose that **v** and **w** are two nonzero vectors with the same initial point P. We seek to decompose **v** into two vectors: $\mathbf{v}_1$, which is parallel to **w**, and $\mathbf{v}_2$, which is orthogonal to **w**. See Figure 67(a) and (b). The vector $\mathbf{v}_1$ is called the **vector projection of v onto w**.

The vector $\mathbf{v}_1$ is obtained as follows: From the terminal point of **v**, drop a perpendicular to the line containing **w**. The vector $\mathbf{v}_1$ is the vector from P to the foot of this perpendicular. The vector $\mathbf{v}_2$ is given by $\mathbf{v}_2 = \mathbf{v} - \mathbf{v}_1$. Note that $\mathbf{v} = \mathbf{v}_1 + \mathbf{v}_2$, $\mathbf{v}_1$ is parallel to **w**, and $\mathbf{v}_2$ is orthogonal to **w**. This is the decomposition of **v** that we wanted.

Figure 67

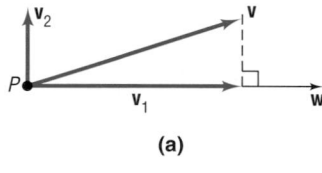

(a)

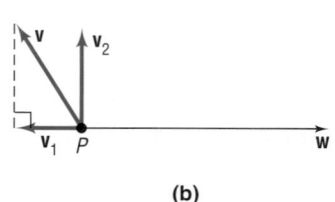

(b)

Now we seek a formula for $\mathbf{v}_1$ that is based on a knowledge of the vectors **v** and **w**. Since $\mathbf{v} = \mathbf{v}_1 + \mathbf{v}_2$, we have

$$\mathbf{v} \cdot \mathbf{w} = (\mathbf{v}_1 + \mathbf{v}_2) \cdot \mathbf{w} = \mathbf{v}_1 \cdot \mathbf{w} + \mathbf{v}_2 \cdot \mathbf{w} \qquad (9)$$

Since $\mathbf{v}_2$ is orthogonal to **w**, we have $\mathbf{v}_2 \cdot \mathbf{w} = 0$. Since $\mathbf{v}_1$ is parallel to **w**, we have $\mathbf{v}_1 = \alpha\mathbf{w}$ for some scalar α. Equation (9) can be written as

$$\mathbf{v} \cdot \mathbf{w} = \alpha\mathbf{w} \cdot \mathbf{w} = \alpha\|\mathbf{w}\|^2$$

$$\alpha = \frac{\mathbf{v} \cdot \mathbf{w}}{\|\mathbf{w}\|^2}$$

Then

$$\mathbf{v}_1 = \alpha\mathbf{w} = \frac{\mathbf{v} \cdot \mathbf{w}}{\|\mathbf{w}\|^2}\mathbf{w}$$

Theorem   If **v** and **w** are two nonzero vectors, the vector projection of **v** onto **w** is

$$\mathbf{v}_1 = \frac{\mathbf{v} \cdot \mathbf{w}}{\|\mathbf{w}\|^2}\mathbf{w} \qquad (10)$$

The decomposition of **v** into $\mathbf{v}_1$ and $\mathbf{v}_2$, where $\mathbf{v}_1$ is parallel to **w** and $\mathbf{v}_2$ is perpendicular to **w**, is

$$\mathbf{v}_1 = \frac{\mathbf{v} \cdot \mathbf{w}}{\|\mathbf{w}\|^2}\mathbf{w} \qquad \mathbf{v}_2 = \mathbf{v} - \mathbf{v}_1 \qquad (11)$$

EXAMPLE 6   Decomposing a Vector into Two Orthogonal Vectors

Figure 68

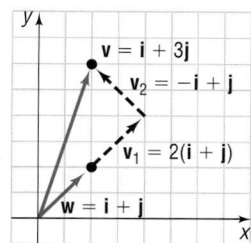

Find the vector projection of $\mathbf{v} = \mathbf{i} + 3\mathbf{j}$ onto $\mathbf{w} = \mathbf{i} + \mathbf{j}$. Decompose $\mathbf{v}$ into two vectors $\mathbf{v}_1$ and $\mathbf{v}_2$, where $\mathbf{v}_1$ is parallel to $\mathbf{w}$ and $\mathbf{v}_2$ is orthogonal to $\mathbf{w}$.

Solution   We use formulas (10) and (11).

$$\mathbf{v}_1 = \frac{\mathbf{v} \cdot \mathbf{w}}{\|\mathbf{w}\|^2}\mathbf{w} = \frac{1 + 3}{(\sqrt{2})^2}\mathbf{w} = 2\mathbf{w} = 2(\mathbf{i} + \mathbf{j})$$

$$\mathbf{v}_2 = \mathbf{v} - \mathbf{v}_1 = (\mathbf{i} + 3\mathbf{j}) - 2(\mathbf{i} + \mathbf{j}) = -\mathbf{i} + \mathbf{j}$$

See Figure 68.

━━━━━━━━━➤ **NOW WORK PROBLEM 13.**

Work Done by a Constant Force

⑥ In elementary physics, the **work** W done by a constant force $\mathbf{F}$ in moving an object from a point A to a point B is defined as

$$W = (\text{magnitude of force})(\text{distance}) = \|\mathbf{F}\|\|\overrightarrow{AB}\|$$

Figure 69

Work is commonly measured in foot-pounds or in Newton-meters (joules).

In this definition, it is assumed that the force $\mathbf{F}$ is applied along the line of motion. If the constant force $\mathbf{F}$ is not along the line of motion, but, instead, is at an angle θ to the direction of motion, as illustrated in Figure 69, then the **work** W **done by** $\mathbf{F}$ in moving an object from A to B is defined as

$$W = \mathbf{F} \cdot \overrightarrow{AB} \qquad (12)$$

This definition is compatible with the force times distance definition given above, since

$$W = (\text{amount of force in the direction of } \overrightarrow{AB})(\text{distance})$$

$$= \|\text{projection of } \mathbf{F} \text{ on } \overrightarrow{AB}\|\|\overrightarrow{AB}\| = \frac{\mathbf{F} \cdot \overrightarrow{AB}}{\|\overrightarrow{AB}\|^2}\|\overrightarrow{AB}\|\|\overrightarrow{AB}\| = \mathbf{F} \cdot \overrightarrow{AB}$$

EXAMPLE 7   Computing Work

Figure 70(a) shows a girl pulling a wagon with a force of 50 pounds. How much work is done in moving the wagon 100 feet if the handle makes an angle of 30° with the ground?

Figure 70

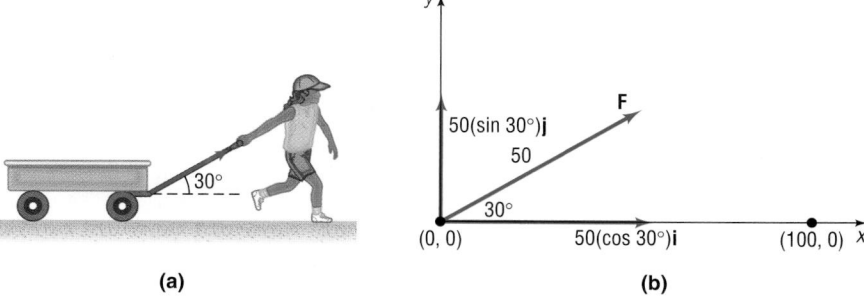

(a)   (b)

Solution   We position the vectors in a coordinate system in such a way that the wagon is moved from $(0, 0)$ to $(100, 0)$. The motion is from $A = (0, 0)$ to $B = (100, 0)$, so $\overrightarrow{AB} = 100\mathbf{i}$. The force vector $\mathbf{F}$, as shown in Figure 70(b), is

$$\mathbf{F} = 50(\cos 30°\mathbf{i} + \sin 30°\mathbf{j}) = 50\left(\frac{\sqrt{3}}{2}\mathbf{i} + \frac{1}{2}\mathbf{j}\right) = 25(\sqrt{3}\mathbf{i} + \mathbf{j})$$

By formula (12), the work done is

$$W = \mathbf{F} \cdot \overrightarrow{AB} = 25(\sqrt{3}\mathbf{i} + \mathbf{j}) \cdot 100\mathbf{i} = 2500\sqrt{3} \text{ foot-pounds} \qquad \blacksquare$$

NOW WORK PROBLEM **29.**

HISTORICAL FEATURE

1. We stated in an earlier Historical Feature that complex numbers were used as vectors in the plane before the general notion of a vector was clarified. Suppose that we make the correspondence

Vector ↔ Complex number
$a\mathbf{i} + b\mathbf{j} \leftrightarrow a + bi$
$c\mathbf{i} + d\mathbf{j} \leftrightarrow c + di$

Show that
$$(a\mathbf{i} + b\mathbf{j}) \cdot (c\mathbf{i} + d\mathbf{j}) = \text{real part}[\overline{(a + bi)}(c + di)]$$

This is how the dot product was found originally. The imaginary part is also interesting. It is a determinant and represents the area of the parallelogram whose edges are the vectors. This is close to some of Hermann Grassmann's ideas and is also connected with the scalar triple product of three-dimensional vectors.

5.5 Concepts and Vocabulary

In Problems 1–3, fill in the blanks.

1. If the angle between two vectors is $\dfrac{\pi}{2}$, the dot product of the two vectors equals _____.

2. If $\mathbf{v} \cdot \mathbf{w} = 0$, then the two vectors $\mathbf{v}$ and $\mathbf{w}$ are _____.

3. If $\mathbf{v} = 3\mathbf{w}$, then the two vectors $\mathbf{v}$ and $\mathbf{w}$ are _____.

In Problems 4–6, answer True or False to each statement.

4. If $\mathbf{v}$ and $\mathbf{w}$ are parallel vectors, then $\mathbf{v} \cdot \mathbf{w} = 0$.

5. Given two nonzero vectors $\mathbf{v}$ and $\mathbf{w}$, it is always possible to decompose $\mathbf{v}$ into two vectors, one parallel to $\mathbf{w}$ and the other perpendicular to $\mathbf{w}$.

6. Work is a physical example of a vector.

7. What does the word orthogonal mean?

5.5 Exercises

In Problems 1–10, (a) find the dot product $\mathbf{v} \cdot \mathbf{w}$; (b) find the angle between $\mathbf{v}$ and $\mathbf{w}$; (c) state whether the vectors are parallel, orthogonal, or neither.

1. $\mathbf{v} = \mathbf{i} - \mathbf{j}, \quad \mathbf{w} = \mathbf{i} + \mathbf{j}$

2. $\mathbf{v} = \mathbf{i} + \mathbf{j}, \quad \mathbf{w} = -\mathbf{i} + \mathbf{j}$

3. $\mathbf{v} = 2\mathbf{i} + \mathbf{j}, \quad \mathbf{w} = \mathbf{i} + 2\mathbf{j}$

4. $\mathbf{v} = 2\mathbf{i} + 2\mathbf{j}, \quad \mathbf{w} = \mathbf{i} + 2\mathbf{j}$

5. $\mathbf{v} = \sqrt{3}\mathbf{i} - \mathbf{j}, \quad \mathbf{w} = \mathbf{i} + \mathbf{j}$

6. $\mathbf{v} = \mathbf{i} + \sqrt{3}\mathbf{j}, \quad \mathbf{w} = \mathbf{i} - \mathbf{j}$

7. $\mathbf{v} = 3\mathbf{i} + 4\mathbf{j}, \quad \mathbf{w} = 4\mathbf{i} + 3\mathbf{j}$

8. $\mathbf{v} = 3\mathbf{i} - 4\mathbf{j}, \quad \mathbf{w} = 4\mathbf{i} - 3\mathbf{j}$

9. $\mathbf{v} = 4\mathbf{i}, \quad \mathbf{w} = \mathbf{j}$

10. $\mathbf{v} = \mathbf{i}, \quad \mathbf{w} = -3\mathbf{j}$

11. Find a so that the vectors $\mathbf{v} = \mathbf{i} - a\mathbf{j}$ and $\mathbf{w} = 2\mathbf{i} + 3\mathbf{j}$ are orthogonal.

12. Find b so that the vectors $\mathbf{v} = \mathbf{i} + \mathbf{j}$ and $\mathbf{w} = \mathbf{i} + b\mathbf{j}$ are orthogonal.

In Problems 13–18, decompose $\mathbf{v}$ into two vectors $\mathbf{v}_1$ and $\mathbf{v}_2$, where $\mathbf{v}_1$ is parallel to $\mathbf{w}$ and $\mathbf{v}_2$ is orthogonal to $\mathbf{w}$.

13. $\mathbf{v} = 2\mathbf{i} - 3\mathbf{j}, \quad \mathbf{w} = \mathbf{i} - \mathbf{j}$

14. $\mathbf{v} = -3\mathbf{i} + 2\mathbf{j}, \quad \mathbf{w} = 2\mathbf{i} + \mathbf{j}$

15. $\mathbf{v} = \mathbf{i} - \mathbf{j}, \quad \mathbf{w} = \mathbf{i} + 2\mathbf{j}$

16. $\mathbf{v} = 2\mathbf{i} - \mathbf{j}, \quad \mathbf{w} = \mathbf{i} - 2\mathbf{j}$

17. $\mathbf{v} = 3\mathbf{i} + \mathbf{j}, \quad \mathbf{w} = -2\mathbf{i} - \mathbf{j}$

18. $\mathbf{v} = \mathbf{i} - 3\mathbf{j}, \quad \mathbf{w} = 4\mathbf{i} - \mathbf{j}$

19. Finding the Actual Speed and Direction of an Aircraft
A DC-10 jumbo jet maintains an airspeed of 550 miles per hour in a southwesterly direction. The velocity of the jet stream is a constant 80 miles per hour from the west. Find the actual speed and direction of the aircraft.

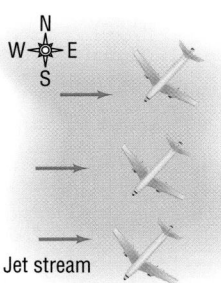

Jet stream

20. Finding the Correct Compass Heading The pilot of an aircraft wishes to head directly east, but is faced with a wind speed of 40 miles per hour from the northwest. If the pilot maintains an airspeed of 250 miles per hour, what compass heading should be maintained? What is the actual speed of the aircraft?

21. Correct Direction for Crossing a River A river has a constant current of 3 kilometers per hour. At what angle to a boat dock should a motorboat, capable of maintaining a constant speed of 20 kilometers per hour, be headed in order to reach a point directly opposite the dock? If the river is $\frac{1}{2}$ kilometer wide, how long will it take to cross?

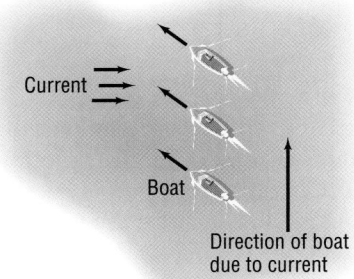

Current

Boat

Direction of boat due to current

22. Correct Direction for Crossing a River Repeat Problem 21 if the current is 5 kilometers per hour.

23. Braking Load A Toyota Sienna with a gross weight of 5300 pounds is parked on a street with a slope of 8°. See the figure. Find the force required to keep the Sienna from rolling down the hill. What is the force perpendicular to the hill?

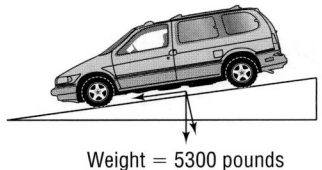

Weight = 5300 pounds

24. Braking Load A Pontiac Bonneville with a gross weight of 4500 pounds is parked on a street with a slope of 10°. Find the force required to keep the Bonneville from rolling down the hill. What is the force perpendicular to the hill?

25. Ground Speed and Direction of an Airplane An airplane has an airspeed of 500 kilometers per hour bearing N45°E. The wind velocity is 60 kilometers per hour in the direction N30°W. Find the resultant vector representing the path of the plane relative to the ground. What is the ground speed of the plane? What is its direction?

26. Ground Speed and Direction of an Airplane An airplane has an airspeed of 600 kilometers per hour bearing S30°E. The wind velocity is 40 kilometers per hour in the direction S45°E. Find the resultant vector representing the path of the plane relative to the ground. What is the ground speed of the plane? What is its direction?

27. Crossing a River A small motorboat in still water maintains a speed of 20 miles per hour. In heading directly across a river (that is, perpendicular to the current) whose current is 3 miles per hour, find a vector representing the speed and direction of the motorboat. What is the true speed of the motorboat? What is its direction?

28. Crossing a River A small motorboat in still water maintains a speed of 10 miles per hour. In heading directly across a river (that is, perpendicular to the current) whose current is 4 miles per hour, find a vector representing the speed and direction of the motorboat. What is the true speed of the motorboat? What is its direction?

29. Computing Work Find the work done by a force of 3 pounds acting in the direction 60° to the horizontal in moving an object 2 feet from $(0, 0)$ to $(2, 0)$.

30. Computing Work Find the work done by a force of 1 pound acting in the direction 45° to the horizontal in moving an object 5 feet from $(0, 0)$ to $(5, 0)$.

31. Computing Work A wagon is pulled horizontally by exerting a force of 20 pounds on the handle at an angle of 30° with the horizontal. How much work is done in moving the wagon 100 feet?

32. Find the acute angle that a constant unit force vector makes with the positive x-axis if the work done by the force in moving a particle from $(0, 0)$ to $(4, 0)$ equals 2.

33. Prove the distributive property:
$$\mathbf{u} \cdot (\mathbf{v} + \mathbf{w}) = \mathbf{u} \cdot \mathbf{v} + \mathbf{u} \cdot \mathbf{w}$$

34. Prove property (5), $\mathbf{0} \cdot \mathbf{v} = 0$.

35. If $\mathbf{v}$ is a unit vector and the angle between $\mathbf{v}$ and $\mathbf{i}$ is α, show that $\mathbf{v} = \cos \alpha \mathbf{i} + \sin \alpha \mathbf{j}$.

36. Suppose that $\mathbf{v}$ and $\mathbf{w}$ are unit vectors. If the angle between $\mathbf{v}$ and $\mathbf{i}$ is α and if the angle between $\mathbf{w}$ and $\mathbf{i}$ is β, use the idea of the dot product $\mathbf{v} \cdot \mathbf{w}$ to prove that
$$\cos(\alpha - \beta) = \cos \alpha \cos \beta + \sin \alpha \sin \beta$$

37. Show that the projection of $\mathbf{v}$ onto $\mathbf{i}$ is $(\mathbf{v} \cdot \mathbf{i})\mathbf{i}$. In fact, show that we can always write a vector $\mathbf{v}$ as
$$\mathbf{v} = (\mathbf{v} \cdot \mathbf{i})\mathbf{i} + (\mathbf{v} \cdot \mathbf{j})\mathbf{j}$$

38. (a) If **u** and **v** have the same magnitude, show that **u** + **v** and **u** − **v** are orthogonal.
   (b) Use this to prove that an angle inscribed in a semi-circle is a right angle (see the figure).

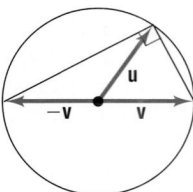

39. Let **v** and **w** denote two nonzero vectors. Show that the vector **v** − α**w** is orthogonal to **w** if $\alpha = \dfrac{\mathbf{v} \cdot \mathbf{w}}{\|\mathbf{w}\|^2}$.

40. Let **v** and **w** denote two nonzero vectors. Show that the vectors $\|\mathbf{w}\|\mathbf{v} + \|\mathbf{v}\|\mathbf{w}$ and $\|\mathbf{w}\|\mathbf{v} - \|\mathbf{v}\|\mathbf{w}$ are orthogonal.

41. In the definition of work given in this section, what is the work done if **F** is orthogonal to $\overrightarrow{AB}$?

42. Prove the **polarization identity**,
$$\|\mathbf{u} + \mathbf{v}\|^2 - \|\mathbf{u} - \mathbf{v}\|^2 = 4(\mathbf{u} \cdot \mathbf{v})$$

43. Make up an application different from any found in the text that requires the dot product.

PREPARING FOR THIS SECTION

Before getting started, review the following:

✓ Distance Formula (Section 1.1, p. 5)

5.6 VECTORS IN SPACE

OBJECTIVES
1. Find the Distance between Two Points
2. Find Position Vectors
3. Perform Operations on Vectors
4. Find the Dot Product
5. Find the Angle between Two Vectors
6. Find the Direction Angles of a Vector

Rectangular Coordinates in Space

In the plane, each point is associated with an ordered pair of real numbers. In space, each point is associated with an ordered triple of real numbers. Through a fixed point, the **origin**, O, draw three mutually perpendicular lines, the x-axis, the y-axis, and the z-axis. On each of these axes, select an appropriate scale and the positive direction. See Figure 71.

The direction chosen for the positive z-axis in Figure 71 makes the system *right-handed*. This conforms to the *right-hand rule*, which states that, if the index finger of the right hand points in the direction of the positive x-axis and the middle finger points in the direction of the positive y-axis, then the thumb will point in the direction of the positive z-axis. See Figure 72.

Figure 71

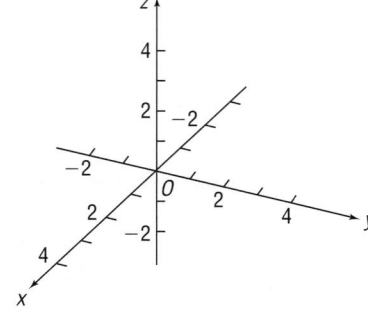

Figure 72

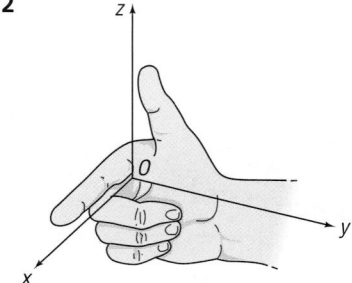

Figure 73

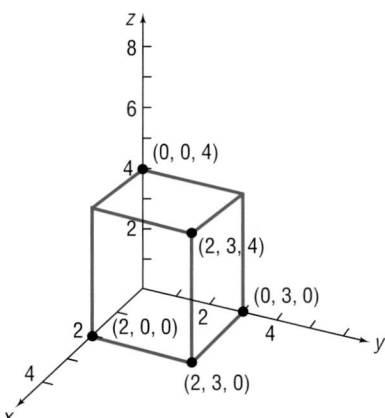

We associate with each point P an ordered triple (x, y, z) of real numbers, the **coordinates of P**. For example, the point $(2, 3, 4)$ is located by starting at the origin and moving 2 units along the positive x-axis, 3 units in the direction of the positive y-axis, and 4 units in the direction of the positive z-axis. See Figure 73.

Figure 73 also shows the location of the points $(2, 0, 0)$, $(0, 3, 0)$, $(0, 0, 4)$, and $(2, 3, 0)$. Points of the form $(x, 0, 0)$ lie on the x-axis, while points of the form $(0, y, 0)$ and $(0, 0, z)$ lie on the y-axis and z-axis, respectively. Points of the form $(x, y, 0)$ lie in a plane, called the **xy-plane**. Its equation is $z = 0$. Similarly, points of the form $(x, 0, z)$ lie in the **xz-plane** (equation $y = 0$) and points of the form $(0, y, z)$ lie in the **yz-plane** (equation $x = 0$). See Figure 74(a). By extension of these ideas, all points obeying the equation $z = 3$ will lie in a plane parallel to and 3 units above the xy-plane. The equation $y = 4$ represents a plane parallel to the xz-plane and 4 units to the right of the plane $y = 0$. See Figure 74(b).

Figure 74

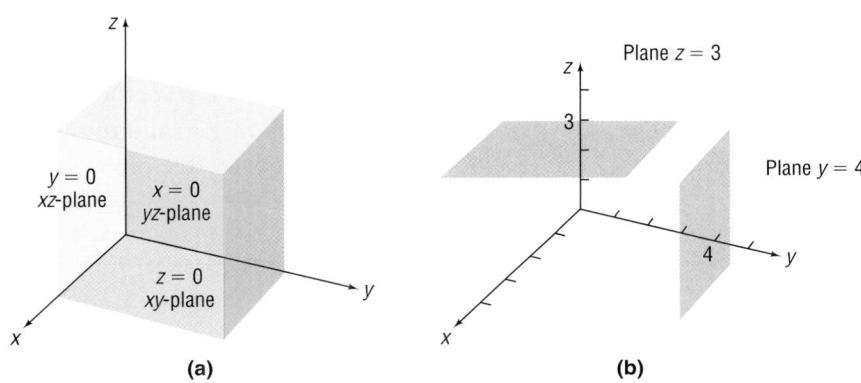

(a)   (b)

NOW WORK PROBLEM 3.

1

The formula for the distance between two points in space is an extension of the Distance Formula for points in the plane given in Chapter 1.

Theorem

Distance Formula in Space

If $P_1 = (x_1, y_1, z_1)$ and $P_2 = (x_2, y_2, z_2)$ are two points in space, the distance d from P_1 to P_2 is

$$d = \sqrt{(x_2 - x_1)^2 + (y_2 - y_1)^2 + (z_2 - z_1)^2} \qquad (1)$$

The proof, which we omit, utilizes a double application of the Pythagorean Theorem.

EXAMPLE 1   **Using the Distance Formula**

Find the distance from $P_1 = (-1, 3, 2)$ to $P_2 = (4, -2, 5)$.

Solution $d = \sqrt{[4 - (-1)]^2 + [-2 - 3]^2 + [5 - 2]^2} = \sqrt{25 + 25 + 9} = \sqrt{59}$

NOW WORK PROBLEM 9.

Representing Vectors in Space

2

To represent vectors in space, we introduce the unit vectors, $\mathbf{i}, \mathbf{j}$, and $\mathbf{k}$ whose directions are along the positive x-axis, positive y-axis, and positive z-axis, respectively. See Figure 75. If $\mathbf{v}$ is a vector with initial point at the origin O and terminal point at $P = (a, b, c)$, then we can represent $\mathbf{v}$ in terms of the vectors $\mathbf{i}, \mathbf{j}$, and $\mathbf{k}$ as

Figure 75

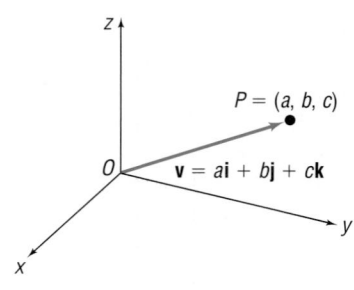

$$\mathbf{v} = a\mathbf{i} + b\mathbf{j} + c\mathbf{k}$$

The scalars a, b, and c are called the **components** of the vector $\mathbf{v} = a\mathbf{i} + b\mathbf{j} + c\mathbf{k}$, with a being the component in the direction $\mathbf{i}$, b the component in the direction $\mathbf{j}$, and c the component in the direction $\mathbf{k}$.

A vector whose initial point is at the origin is called a **position vector**. The next result states that any vector whose initial point is not at the origin is equal to a unique position vector.

Theorem

Suppose that $\mathbf{v}$ is a vector with initial point $P_1 = (x_1, y_1, z_1)$, not necessarily the origin, and terminal point $P_2 = (x_2, y_2, z_2)$. If $\mathbf{v} = \overrightarrow{P_1 P_2}$, then $\mathbf{v}$ is equal to the position vector

$$\mathbf{v} = (x_2 - x_1)\mathbf{i} + (y_2 - y_1)\mathbf{j} + (z_2 - z_1)\mathbf{k} \qquad (2)$$

Figure 76 illustrates this result.

Figure 76

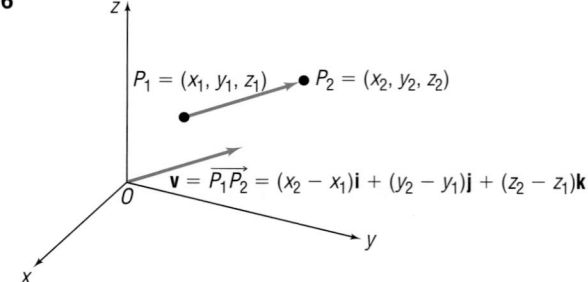

EXAMPLE 2   **Finding a Position Vector**

Find the position vector of the vector $\mathbf{v} = \overrightarrow{P_1 P_2}$ if $P_1 = (-1, 2, 3)$ and $P_2 = (4, 6, 2)$.

Solution   By equation (2), the position vector equal to $\mathbf{v}$ is

$$\mathbf{v} = [4 - (-1)]\mathbf{i} + (6 - 2)\mathbf{j} + (2 - 3)\mathbf{k} = 5\mathbf{i} + 4\mathbf{j} - \mathbf{k}$$

➡ **NOW WORK PROBLEM 23.**

3

Next, we define equality, addition, subtraction, scalar product, and magnitude in terms of the components of a vector.

Let $\mathbf{v} = a_1\mathbf{i} + b_1\mathbf{j} + c_1\mathbf{k}$ and $\mathbf{w} = a_2\mathbf{i} + b_2\mathbf{j} + c_2\mathbf{k}$ be two vectors, and let α be a scalar. Then

$$\mathbf{v} = \mathbf{w} \quad \text{if and only if } a_1 = a_2, b_1 = b_2, \text{ and } c_1 = c_2$$

$$\mathbf{v} + \mathbf{w} = (a_1 + a_2)\mathbf{i} + (b_1 + b_2)\mathbf{j} + (c_1 + c_2)\mathbf{k}$$

$$\mathbf{v} - \mathbf{w} = (a_1 - a_2)\mathbf{i} + (b_1 - b_2)\mathbf{j} + (c_1 - c_2)\mathbf{k}$$

$$\alpha\mathbf{v} = (\alpha a_1)\mathbf{i} + (\alpha b_1)\mathbf{j} + (\alpha c_1)\mathbf{k}$$

$$\|\mathbf{v}\| = \sqrt{a_1^2 + b_1^2 + c_1^2}$$

These definitions are compatible with the geometric ones given earlier in Section 5.4.

EXAMPLE 3   **Adding and Subtracting Vectors**

If $\mathbf{v} = 2\mathbf{i} + 3\mathbf{j} - 2\mathbf{k}$ and $\mathbf{w} = 3\mathbf{i} - 4\mathbf{j} + 5\mathbf{k}$, find:

(a) $\mathbf{v} + \mathbf{w}$ (b) $\mathbf{v} - \mathbf{w}$

Solution   (a) $\mathbf{v} + \mathbf{w} = (2\mathbf{i} + 3\mathbf{j} - 2\mathbf{k}) + (3\mathbf{i} - 4\mathbf{j} + 5\mathbf{k})$
$$= (2 + 3)\mathbf{i} + (3 - 4)\mathbf{j} + (-2 + 5)\mathbf{k}$$
$$= 5\mathbf{i} - \mathbf{j} + 3\mathbf{k}$$

(b) $\mathbf{v} - \mathbf{w} = (2\mathbf{i} + 3\mathbf{j} - 2\mathbf{k}) - (3\mathbf{i} - 4\mathbf{j} + 5\mathbf{k})$
$$= (2 - 3)\mathbf{i} + [3 - (-4)]\mathbf{j} + [-2 - 5]\mathbf{k}$$
$$= -\mathbf{i} + 7\mathbf{j} - 7\mathbf{k}$$

EXAMPLE 4   **Finding Scalar Products and Magnitudes**

If $\mathbf{v} = 2\mathbf{i} + 3\mathbf{j} - 2\mathbf{k}$ and $\mathbf{w} = 3\mathbf{i} - 4\mathbf{j} + 5\mathbf{k}$, find:

(a) $3\mathbf{v}$ (b) $2\mathbf{v} - 3\mathbf{w}$ (c) $\|\mathbf{v}\|$

Solution   (a) $3\mathbf{v} = 3(2\mathbf{i} + 3\mathbf{j} - 2\mathbf{k}) = 6\mathbf{i} + 9\mathbf{j} - 6\mathbf{k}$

(b) $2\mathbf{v} - 3\mathbf{w} = 2(2\mathbf{i} + 3\mathbf{j} - 2\mathbf{k}) - 3(3\mathbf{i} - 4\mathbf{j} + 5\mathbf{k})$
$$= 4\mathbf{i} + 6\mathbf{j} - 4\mathbf{k} - 9\mathbf{i} + 12\mathbf{j} - 15\mathbf{k} = -5\mathbf{i} + 18\mathbf{j} - 19\mathbf{k}$$

(c) $\|\mathbf{v}\| = \|2\mathbf{i} + 3\mathbf{j} - 2\mathbf{k}\| = \sqrt{2^2 + 3^2 + (-2)^2} = \sqrt{17}$

── **NOW WORK PROBLEMS 27 AND 33.**

Recall that a unit vector $\mathbf{u}$ is one for which $\|\mathbf{u}\| = 1$. In many applications, it is useful to be able to find a unit vector $\mathbf{u}$ that has the same direction as a given vector $\mathbf{u}$.

Theorem   **Unit Vector in the Direction of *v***

For any nonzero vector $\mathbf{v}$, the vector

$$\mathbf{u} = \frac{\mathbf{v}}{\|\mathbf{v}\|}$$

is a unit vector that has the same direction as $\mathbf{v}$.

As a consequence of this theorem, if **u** is a unit vector in the same direction as a vector **v**, then **v** may be expressed as

$$\mathbf{v} = \|\mathbf{v}\|\mathbf{u}$$

This way of expressing a vector is useful in many applications.

EXAMPLE 5   Finding a Unit Vector

Find a unit vector in the same direction as $\mathbf{v} = 2\mathbf{i} - 3\mathbf{j} - 6\mathbf{k}$.

Solution   We find $\|\mathbf{v}\|$ first.

$$\|\mathbf{v}\| = \|2\mathbf{i} - 3\mathbf{j} - 6\mathbf{k}\| = \sqrt{4 + 9 + 36} = \sqrt{49} = 7$$

Now we multiply **v** by the scalar $\dfrac{1}{\|\mathbf{v}\|} = \dfrac{1}{7}$. The result is the unit vector

$$\mathbf{u} = \frac{\mathbf{v}}{\|\mathbf{v}\|} = \frac{2\mathbf{i} - 3\mathbf{j} - 6\mathbf{k}}{7} = \frac{2}{7}\mathbf{i} - \frac{3}{7}\mathbf{j} - \frac{6}{7}\mathbf{k}$$

■

NOW WORK PROBLEM 41.

Dot Product

④ The definition of *dot product* is an extension of the definition given for vectors in the plane.

If $\mathbf{v} = a_1\mathbf{i} + b_1\mathbf{j} + c_1\mathbf{k}$ and $\mathbf{w} = a_2\mathbf{i} + b_2\mathbf{j} + c_2\mathbf{k}$ are two vectors, the **dot product v · w** is defined as

$$\mathbf{v} \cdot \mathbf{w} = a_1a_2 + b_1b_2 + c_1c_2 \qquad (3)$$

EXAMPLE 6   Finding Dot Products

If $\mathbf{v} = 2\mathbf{i} - 3\mathbf{j} + 6\mathbf{k}$ and $\mathbf{w} = 5\mathbf{i} + 3\mathbf{j} - \mathbf{k}$, find:

(a) **v · w**   (b) **w · v**   (c) **v · v**
(d) **w · w**   (e) $\|\mathbf{v}\|$ (f) $\|\mathbf{w}\|$

Solution   (a) $\mathbf{v} \cdot \mathbf{w} = 2(5) + (-3)3 + 6(-1) = -5$

(b) $\mathbf{w} \cdot \mathbf{v} = 5(2) + 3(-3) + (-1)(6) = -5$

(c) $\mathbf{v} \cdot \mathbf{v} = 2(2) + (-3)(-3) + 6(6) = 49$

(d) $\mathbf{w} \cdot \mathbf{w} = 5(5) + 3(3) + (-1)(-1) = 35$

(e) $\|\mathbf{v}\| = \sqrt{2^2 + (-3)^2 + 6^2} = \sqrt{49} = 7$

(f) $\|\mathbf{w}\| = \sqrt{5^2 + 3^2 + (-1)^2} = \sqrt{35}$

■

The dot product in space has the same properties as the dot product in the plane.

Theorem   **Properties of the Dot Product**

If **u**, **v**, and **w** are vectors, then

Commutative Property

$$\mathbf{u} \cdot \mathbf{v} = \mathbf{v} \cdot \mathbf{u}$$

Distributive Property

$$\mathbf{u} \cdot (\mathbf{v} + \mathbf{w}) = \mathbf{u} \cdot \mathbf{v} + \mathbf{u} \cdot \mathbf{w}$$

$$\mathbf{v} \cdot \mathbf{v} = \|\mathbf{v}\|^2$$
$$\mathbf{0} \cdot \mathbf{v} = 0$$

⑤   The angle θ between two vectors in space follows the same formula as for two vectors in the plane.

Theorem   **Angle between Vectors**

If **u** and **v** are two nonzero vectors, the angle $\theta, 0 \le \theta \le \pi$, between **u** and **v** is determined by the formula

$$\cos \theta = \frac{\mathbf{u} \cdot \mathbf{v}}{\|\mathbf{u}\|\|\mathbf{v}\|} \qquad (4)$$

EXAMPLE 7   **Finding the Angle θ between Two Vectors**

Find the angle θ between $\mathbf{u} = 2\mathbf{i} - 3\mathbf{j} + 6\mathbf{k}$ and $\mathbf{v} = 2\mathbf{i} + 5\mathbf{j} - \mathbf{k}$.

Solution   We compute the quantities $\mathbf{u} \cdot \mathbf{v}$, $\|\mathbf{u}\|$, and $\|\mathbf{v}\|$.

$$\mathbf{u} \cdot \mathbf{v} = 2(2) + (-3)(5) + 6(-1) = -17$$
$$\|\mathbf{u}\| = \sqrt{2^2 + (-3)^2 + 6^2} = \sqrt{49} = 7$$
$$\|\mathbf{v}\| = \sqrt{2^2 + 5^2 + (-1)^2} = \sqrt{30}$$

By formula (4), if θ is the angle between **u** and **v**, then

$$\cos \theta = \frac{\mathbf{u} \cdot \mathbf{v}}{\|\mathbf{u}\|\|\mathbf{v}\|} = \frac{-17}{7\sqrt{30}} \approx -0.443$$

We find that $\theta \approx 116.3°$.

 NOW WORK PROBLEM 45.

Direction Angles of Vectors in Space

6   A nonzero vector **v** in space can be described by specifying its magnitude and its three **direction angles** α, β, and γ. These direction angles are defined as

α = angle between **v** and **i**, the positive x-axis, $0 \le \alpha \le \pi$
β = angle between **v** and **j**, the positive y-axis, $0 \le \beta \le \pi$
γ = angle between **v** and **k**, the positive z-axis, $0 \le \gamma \le \pi$

See Figure 77.

Figure 77

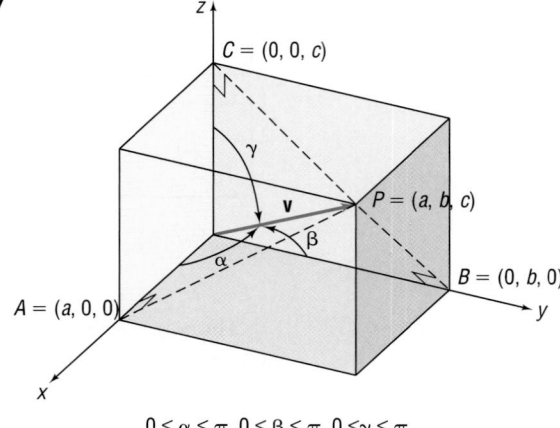

$0 \le \alpha \le \pi, 0 \le \beta \le \pi, 0 \le \gamma \le \pi$

Our first goal is to find an expression for α, β, and γ in terms of the components of a vector. Let $\mathbf{v} = a\mathbf{i} + b\mathbf{j} + c\mathbf{k}$ denote a nonzero vector. The angle α between **v** and **i**, the positive x-axis, obeys

$$\cos \alpha = \frac{\mathbf{v} \cdot \mathbf{i}}{\|\mathbf{v}\|\|\mathbf{i}\|} = \frac{a}{\|\mathbf{v}\|}$$

Similarly,

$$\cos \beta = \frac{b}{\|\mathbf{v}\|} \qquad \cos \gamma = \frac{c}{\|\mathbf{v}\|}$$

Since $\|\mathbf{v}\| = \sqrt{a^2 + b^2 + c^2}$, we have the following result:

Theorem

Direction Angles

If $\mathbf{v} = a\mathbf{i} + b\mathbf{j} + c\mathbf{k}$ is a nonzero vector in space, the direction angles α, β, and γ obey

$$\cos \alpha = \frac{a}{\sqrt{a^2 + b^2 + c^2}} = \frac{a}{\|\mathbf{v}\|} \qquad \cos \beta = \frac{b}{\sqrt{a^2 + b^2 + c^2}} = \frac{b}{\|\mathbf{v}\|}$$
$$\cos \gamma = \frac{c}{\sqrt{a^2 + b^2 + c^2}} = \frac{c}{\|\mathbf{v}\|} \tag{5}$$

The numbers $\cos \alpha$, $\cos \beta$, and $\cos \gamma$ are called the **direction cosines** of the vector **v**. They play the same role in space as slope does in the plane.

EXAMPLE 8 Finding the Direction Angles of a Vector

Find the direction angles of $\mathbf{v} = -3\mathbf{i} + 2\mathbf{j} - 6\mathbf{k}$.

Solution
$$\|\mathbf{v}\| = \sqrt{(-3)^2 + 2^2 + (-6)^2} = \sqrt{49} = 7$$

Using the Theorem on Direction Angles, we get

$$\cos\alpha = \frac{-3}{7} \qquad \cos\beta = \frac{2}{7} \qquad \cos\gamma = \frac{-6}{7}$$

$$\alpha \approx 115.4° \qquad \beta \approx 73.4° \qquad \gamma \approx 149.0°$$

Theorem

Property of Direction Cosines

If α, β, and γ are the direction angles of a nonzero vector $\mathbf{v}$ in space, then

$$\cos^2\alpha + \cos^2\beta + \cos^2\gamma = 1 \qquad\qquad (6)$$

The proof is a direct consequence of equations (5).

Based on equation (6), when two direction cosines are known, the third is determined up to its sign. Knowing two direction cosines is not sufficient to uniquely determine the direction of a vector in space.

EXAMPLE 9 Finding the Direction Angle of a Vector

The vector $\mathbf{v}$ makes an angle of $\alpha = \dfrac{\pi}{3}$ with the positive x-axis, an angle of $\beta = \dfrac{\pi}{3}$ with the positive y-axis, and an acute angle γ with the positive z-axis. Find γ.

Solution By equation (6), we have

$$\cos^2\left(\frac{\pi}{3}\right) + \cos^2\left(\frac{\pi}{3}\right) + \cos^2\gamma = 1 \qquad\qquad 0 \le \gamma \le \pi$$

$$\left(\frac{1}{2}\right)^2 + \left(\frac{1}{2}\right)^2 + \cos^2\gamma = 1$$

$$\cos^2\gamma = \frac{1}{2}$$

$$\cos\gamma = -\frac{\sqrt{2}}{2} \qquad \cos\gamma = \frac{\sqrt{2}}{2} \qquad \text{or}$$

$$\gamma = \frac{3\pi}{4} \qquad\qquad \gamma = \frac{\pi}{4} \qquad \text{or}$$

Since we are requiring that γ be acute, the answer is $\gamma = \dfrac{\pi}{4}$.

The direction cosines of a vector give information about only the direction of the vector; they provide no information about its magnitude. For example, *any* vector parallel to the *xy*-plane and making an angle of $\frac{\pi}{4}$ radian with the positive *x*- and *y*-axes has direction cosines

$$\cos \alpha = \frac{\sqrt{2}}{2} \qquad \cos \beta = \frac{\sqrt{2}}{2} \qquad \cos \gamma = 0$$

However, if the direction angles *and* the magnitude of a vector are known, then the vector is uniquely determined.

EXAMPLE 10 | **Writing a Vector in Terms of Its Magnitude and Direction Cosines**

Show that any nonzero vector **v** in space can be written in terms of its magnitude and direction cosines as

$$\mathbf{v} = \|\mathbf{v}\|[(\cos \alpha)\mathbf{i} + (\cos \beta)\mathbf{j} + (\cos \gamma)\mathbf{k}] \qquad (7)$$

Solution Let $\mathbf{v} = a\mathbf{i} + b\mathbf{j} + c\mathbf{k}$. From equation (5), we see that

$$a = \|\mathbf{v}\| \cos \alpha \qquad b = \|\mathbf{v}\| \cos \beta \qquad c = \|\mathbf{v}\| \cos \gamma$$

Substituting, we find that

$$\mathbf{v} = a\mathbf{i} + b\mathbf{j} + c\mathbf{k} = \|\mathbf{v}\|(\cos \alpha)\mathbf{i} + \|\mathbf{v}\|(\cos \beta)\mathbf{j} + \|\mathbf{v}\|(\cos \gamma)\mathbf{k}$$
$$= \|\mathbf{v}\|[(\cos \alpha)\mathbf{i} + (\cos \beta)\mathbf{j} + (\cos \gamma)\mathbf{k}] \quad \blacksquare$$

━ **NOW WORK PROBLEM 53.**

Example 10 shows that the direction cosines of a vector **v** are also the components of the unit vector in the direction of **v**.

5.6 Concepts and Vocabulary

In Problems 1–3, fill in the blanks.

1. In space, points of the form $(x, y, 0)$ lie in a plane called the _____.

2. If $\mathbf{v} = a\mathbf{i} + b\mathbf{j} + c\mathbf{k}$ is a vector in space, the scalars a, b, c are called the _____ of **v**.

3. The sum of the squares of the direction cosines of a vector in space add up to _____.

In Problems 4–6, answer True or False to each statement.

4. The formula for the distance between two points in space is an extension of the distance formula in the plane.

5. In space, the dot product of two vectors is a positive number.

6. A vector in space may be described by specifying its magnitude and its direction angles.

7. Describe the direction angles of a vector in space.

5.6 Exercises

In Problems 1–8, describe the set of points (x, y, z) *defined by the equation.*

1. $y = 0$

2. $x = 0$

3. $z = 2$

4. $y = 3$

5. $x = -4$

6. $z = -3$

7. $x = 1$ and $y = 2$

8. $x = 3$ and $z = 1$

In Problems 9–14, find the distance from P_1 *to* P_2.

9. $P_1 = (0, 0, 0)$ and $P_2 = (4, 1, 2)$

10. $P_1 = (0, 0, 0)$ and $P_2 = (1, -2, 3)$

11. $P_1 = (-1, 2, -3)$ and $P_2 = (0, -2, 1)$

12. $P_1 = (-2, 2, 3)$ and $P_2 = (4, 0, -3)$

13. $P_1 = (4, -2, -2)$ and $P_2 = (3, 2, 1)$

14. $P_1 = (2, -3, -3)$ and $P_2 = (4, 1, -1)$

In Problems 15–20, opposite vertices of a rectangular box whose edges are parallel to the coordinate axes given. List the coordinates of the other six vertices of the box.

15. $(0, 0, 0)$; $(2, 1, 3)$

16. $(0, 0, 0)$; $(4, 2, 2)$

17. $(1, 2, 3)$; $(3, 4, 5)$

18. $(5, 6, 1)$; $(3, 8, 2)$

19. $(-1, 0, 2)$; $(4, 2, 5)$

20. $(-2, -3, 0)$; $(-6, 7, 1)$

In Problems 21–26, the vector **v** *has initial point P and terminal point Q. Write* **v** *in the form a***i** + b**j** + c**k***; that is, find its position vector.*

21. $P = (0, 0, 0)$; $Q = (3, 4, -1)$

22. $P = (0, 0, 0)$; $Q = (-3, -5, 4)$

23. $P = (3, 2, -1)$; $Q = (5, 6, 0)$

24. $P = (-3, 2, 0)$; $Q = (6, 5, -1)$

25. $P = (-2, -1, 4)$; $Q = (6, -2, 4)$

26. $P = (-1, 4, -2)$; $Q = (6, 2, 2)$

In Problems 27–32, find $\|\mathbf{v}\|$.

27. $\mathbf{v} = 3\mathbf{i} - 6\mathbf{j} - 2\mathbf{k}$

28. $\mathbf{v} = -6\mathbf{i} + 12\mathbf{j} + 4\mathbf{k}$

29. $\mathbf{v} = \mathbf{i} - \mathbf{j} + \mathbf{k}$

30. $\mathbf{v} = -\mathbf{i} - \mathbf{j} + \mathbf{k}$

31. $\mathbf{v} = -2\mathbf{i} + 3\mathbf{j} - 3\mathbf{k}$

32. $\mathbf{v} = 6\mathbf{i} + 2\mathbf{j} - 2\mathbf{k}$

In Problems 33–38, find each quantity if $\mathbf{v} = 3\mathbf{i} - 5\mathbf{j} + 2\mathbf{k}$ *and* $\mathbf{w} = -2\mathbf{i} + 3\mathbf{j} - 2\mathbf{k}$.

33. $2\mathbf{v} + 3\mathbf{w}$

34. $3\mathbf{v} - 2\mathbf{w}$

35. $\|\mathbf{v} - \mathbf{w}\|$

36. $\|\mathbf{v} + \mathbf{w}\|$

37. $\|\mathbf{v}\| - \|\mathbf{w}\|$

38. $\|\mathbf{v}\| + \|\mathbf{w}\|$

In Problems 39–44, find the unit vector having the same direction as **v**.

39. $\mathbf{v} = 5\mathbf{i}$

40. $\mathbf{v} = -3\mathbf{j}$

41. $\mathbf{v} = 3\mathbf{i} - 6\mathbf{j} - 2\mathbf{k}$

42. $\mathbf{v} = -6\mathbf{i} + 12\mathbf{j} + 4\mathbf{k}$

43. $\mathbf{v} = \mathbf{i} + \mathbf{j} + \mathbf{k}$

44. $\mathbf{v} = 2\mathbf{i} - \mathbf{j} + \mathbf{k}$

In Problems 45–52, find the dot product $\mathbf{v} \cdot \mathbf{w}$ *and the angle between* **v** *and* **w**.

45. $\mathbf{v} = \mathbf{i} - \mathbf{j}$, $\mathbf{w} = \mathbf{i} + \mathbf{j} + \mathbf{k}$

46. $\mathbf{v} = \mathbf{i} + \mathbf{j}$, $\mathbf{w} = -\mathbf{i} + \mathbf{j} - \mathbf{k}$

47. $\mathbf{v} = 2\mathbf{i} + \mathbf{j} - 3\mathbf{k}$, $\mathbf{w} = \mathbf{i} + 2\mathbf{j} + 2\mathbf{k}$

48. $\mathbf{v} = 2\mathbf{i} + 2\mathbf{j} - \mathbf{k}$, $\mathbf{w} = \mathbf{i} + 2\mathbf{j} + 3\mathbf{k}$

49. $\mathbf{v} = 3\mathbf{i} - \mathbf{j} + 2\mathbf{k}$, $\mathbf{w} = \mathbf{i} + \mathbf{j} - \mathbf{k}$

50. $\mathbf{v} = \mathbf{i} + 3\mathbf{j} + 2\mathbf{k}$, $\mathbf{w} = \mathbf{i} - \mathbf{j} + \mathbf{k}$

51. $\mathbf{v} = 3\mathbf{i} + 4\mathbf{j} + \mathbf{k}$, $\mathbf{w} = 6\mathbf{i} + 8\mathbf{j} + 2\mathbf{k}$

52. $\mathbf{v} = 3\mathbf{i} - 4\mathbf{j} + \mathbf{k}$, $\mathbf{w} = 6\mathbf{i} - 8\mathbf{j} + 2\mathbf{k}$

In Problems 53–60, find the direction angles of each vector. Write each vector in the form of equation (7).

53. $\mathbf{v} = 3\mathbf{i} - 6\mathbf{j} - 2\mathbf{k}$

54. $\mathbf{v} = -6\mathbf{i} + 12\mathbf{j} + 4\mathbf{k}$

55. $\mathbf{v} = \mathbf{i} + \mathbf{j} + \mathbf{k}$

56. $\mathbf{v} = \mathbf{i} - \mathbf{j} - \mathbf{k}$

57. $\mathbf{v} = \mathbf{i} + \mathbf{j}$

58. $\mathbf{v} = \mathbf{j} + \mathbf{k}$

59. $\mathbf{v} = 3\mathbf{i} - 5\mathbf{j} + 2\mathbf{k}$

60. $\mathbf{v} = 2\mathbf{i} + 3\mathbf{j} - 4\mathbf{k}$

61. The Sphere   In space, the collection of all points that are the same distance from some fixed point is called a **sphere**. See the illustration. The constant distance is called the **radius**, and the fixed point is the **center** of the sphere. Show that the equation of a sphere with center at (x_0, y_0, z_0) and radius r is

$$(x - x_0)^2 + (y - y_0)^2 + (z - z_0)^2 = r^2$$

[**Hint:** Use the Distance Formula (1).]

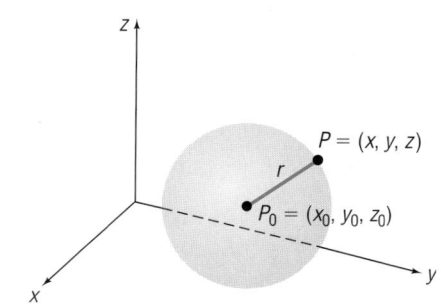

In Problems 62–64, find the equation of a sphere with radius r and center P_0.

62. $r = 1;$ $P_0 = (3, 1, 1)$ **63.** $r = 2;$ $P_0 = (1, 2, 2)$ **64.** $r = 3;$ $P_0 = (-1, 1, 2)$

In Problems 65–70, find the radius and center of each sphere. [**Hint:** *Complete the square in each variable.*]

65. $x^2 + y^2 + z^2 + 2x - 2y = 2$ **66.** $x^2 + y^2 + z^2 + 2x - 2z = -1$

67. $x^2 + y^2 + z^2 - 4x + 4y + 2z = 0$ **68.** $x^2 + y^2 + z^2 - 4x = 0$

69. $2x^2 + 2y^2 + 2z^2 - 8x + 4z = -1$ **70.** $3x^2 + 3y^2 + 3z^2 + 6x - 6y = 3$

The **work** *W done by a constant force* **F** *in moving an object from a point A in space to a point B in space is defined as* $W = \mathbf{F} \cdot \overrightarrow{AB}$. *Use this definition in Problems 71–73.*

71. Work   Find the work done by a force of 3 Newtons acting in the direction $2\mathbf{i} + \mathbf{j} + 2\mathbf{k}$ in moving an object 2 meters from $(0, 0, 0)$ to $(0, 2, 0)$.

72. Work   Find the work done by a force of 1 Newton acting in the direction $2\mathbf{i} + 2\mathbf{j} + \mathbf{k}$ in moving an object 3 meters from $(0, 0, 0)$ to $(1, 2, 2)$.

73. Work   Find the work done in moving an object along a vector $\mathbf{u} = 3\mathbf{i} + 2\mathbf{j} - 5\mathbf{k}$ if the applied force is $\mathbf{F} = 2\mathbf{i} - \mathbf{j} - \mathbf{k}$.

5.7 THE CROSS PRODUCT

OBJECTIVES
1. Find the Cross Product of Two Vectors
2. Know Algebraic Properties of the Cross Product
3. Know Geometric Properties of the Cross Product
4. Find a Vector Orthogonal to Two Given Vectors
5. Find the Area of a Parallelogram

1   For vectors in space, and only for vectors in space, a second product of two vectors is defined, called the *cross product*. The cross product of two vectors in space is, in fact, also a vector that has applications in both geometry and physics.

If $\mathbf{v} = a_1\mathbf{i} + b_1\mathbf{j} + c_1\mathbf{k}$ and $\mathbf{w} = a_2\mathbf{i} + b_2\mathbf{j} + c_2\mathbf{k}$ are two vectors in space, the **cross product** $\mathbf{v} \times \mathbf{w}$ is defined as the vector

$$\mathbf{v} \times \mathbf{w} = (b_1c_2 - b_2c_1)\mathbf{i} - (a_1c_2 - a_2c_1)\mathbf{j} + (a_1b_2 - a_2b_1)\mathbf{k} \quad (1)$$

Notice that the cross product $\mathbf{v} \times \mathbf{w}$ of two vectors is a vector. Because of this, it is sometimes referred to as the **vector product.**

EXAMPLE 1   **Finding Cross Products Using Equation (1)**

If $\mathbf{v} = 2\mathbf{i} + 3\mathbf{j} + 5\mathbf{k}$ and $\mathbf{w} = \mathbf{i} + 2\mathbf{j} + 3\mathbf{k}$, then an application of equation (1) gives

$$\begin{aligned}
\mathbf{v} \times \mathbf{w} &= (3 \cdot 3 - 2 \cdot 5)\mathbf{i} - (2 \cdot 3 - 1 \cdot 5)\mathbf{j} + (2 \cdot 2 - 1 \cdot 3)\mathbf{k} \\
&= (9 - 10)\mathbf{i} - (6 - 5)\mathbf{j} + (4 - 3)\mathbf{k} \\
&= -\mathbf{i} - \mathbf{j} + \mathbf{k}
\end{aligned}$$

Determinants may be used as an aid in computing cross products. A **2 by 2 determinant**, symbolized by

$$\begin{vmatrix} a_1 & b_1 \\ a_2 & b_2 \end{vmatrix}$$

has the value $a_1b_2 - a_2b_1$; that is,

$$\begin{vmatrix} a_1 & b_1 \\ a_2 & b_2 \end{vmatrix} = a_1b_2 - a_2b_1$$

A **3 by 3 determinant** has the value

$$\begin{vmatrix} A & B & C \\ a_1 & b_1 & c_1 \\ a_2 & b_2 & c_2 \end{vmatrix} = \begin{vmatrix} b_1 & c_1 \\ b_2 & c_2 \end{vmatrix} A - \begin{vmatrix} a_1 & c_1 \\ a_2 & c_2 \end{vmatrix} B + \begin{vmatrix} a_1 & b_1 \\ a_2 & b_2 \end{vmatrix} C$$

EXAMPLE 2   Evaluating Determinants

(a) $\begin{vmatrix} 2 & 3 \\ 1 & 2 \end{vmatrix} = 2 \cdot 2 - 1 \cdot 3 = 4 - 3 = 1$

(b) $\begin{vmatrix} A & B & C \\ 2 & 3 & 5 \\ 1 & 2 & 3 \end{vmatrix} = \begin{vmatrix} 3 & 5 \\ 2 & 3 \end{vmatrix} A - \begin{vmatrix} 2 & 5 \\ 1 & 3 \end{vmatrix} B + \begin{vmatrix} 2 & 3 \\ 1 & 2 \end{vmatrix} C$

$$= (9 - 10)A - (6 - 5)B + (4 - 3)C$$
$$= -A - B + C$$

— **NOW WORK PROBLEM 1.**

The cross product of the vectors $\mathbf{v} = a_1\mathbf{i} + b_1\mathbf{j} + c_1\mathbf{k}$, and $\mathbf{w} = a_2\mathbf{i} + b_2\mathbf{j} + c_2\mathbf{k}$, that is,

$$\mathbf{v} \times \mathbf{w} = (b_1c_2 - b_2c_1)\mathbf{i} - (a_1c_2 - a_2c_1)\mathbf{j} + (a_1b_2 - a_2b_1)\mathbf{k},$$

may be written symbolically using determinants as

$$\mathbf{v} \times \mathbf{w} = \begin{vmatrix} \mathbf{i} & \mathbf{j} & \mathbf{k} \\ a_1 & b_1 & c_1 \\ a_2 & b_2 & c_2 \end{vmatrix} = \begin{vmatrix} b_1 & c_1 \\ b_2 & c_2 \end{vmatrix} \mathbf{i} - \begin{vmatrix} a_1 & c_1 \\ a_2 & c_2 \end{vmatrix} \mathbf{j} + \begin{vmatrix} a_1 & b_1 \\ a_2 & b_2 \end{vmatrix} \mathbf{k}$$

EXAMPLE 3   Using Determinants to Find Cross Products

If $\mathbf{v} = 2\mathbf{i} + 3\mathbf{j} + 5\mathbf{k}$ and $\mathbf{w} = \mathbf{i} + 2\mathbf{j} + 3\mathbf{k}$, find:

(a) $\mathbf{v} \times \mathbf{w}$ (b) $\mathbf{w} \times \mathbf{v}$ (c) $\mathbf{v} \times \mathbf{v}$ (d) $\mathbf{w} \times \mathbf{w}$

Solution (a) $\mathbf{v} \times \mathbf{w} = \begin{vmatrix} \mathbf{i} & \mathbf{j} & \mathbf{k} \\ 2 & 3 & 5 \\ 1 & 2 & 3 \end{vmatrix} = \begin{vmatrix} 3 & 5 \\ 2 & 3 \end{vmatrix} \mathbf{i} - \begin{vmatrix} 2 & 5 \\ 1 & 3 \end{vmatrix} \mathbf{j} + \begin{vmatrix} 2 & 3 \\ 1 & 2 \end{vmatrix} \mathbf{k} = -\mathbf{i} - \mathbf{j} + \mathbf{k}$

(b) $\mathbf{w} \times \mathbf{v} = \begin{vmatrix} \mathbf{i} & \mathbf{j} & \mathbf{k} \\ 1 & 2 & 3 \\ 2 & 3 & 5 \end{vmatrix} = \begin{vmatrix} 2 & 3 \\ 3 & 5 \end{vmatrix} \mathbf{i} - \begin{vmatrix} 1 & 3 \\ 2 & 5 \end{vmatrix} \mathbf{j} + \begin{vmatrix} 1 & 2 \\ 2 & 3 \end{vmatrix} \mathbf{k} = \mathbf{i} + \mathbf{j} - \mathbf{k}$

(c) $\quad \mathbf{v} \times \mathbf{v} = \begin{vmatrix} \mathbf{i} & \mathbf{j} & \mathbf{k} \\ 2 & 3 & 5 \\ 2 & 3 & 5 \end{vmatrix}$

$\qquad = \begin{vmatrix} 3 & 5 \\ 3 & 5 \end{vmatrix} \mathbf{i} - \begin{vmatrix} 2 & 5 \\ 2 & 5 \end{vmatrix} \mathbf{j} + \begin{vmatrix} 2 & 3 \\ 2 & 3 \end{vmatrix} \mathbf{k} = 0\mathbf{i} - 0\mathbf{j} + 0\mathbf{k} = \mathbf{0}$

(d) $\quad \mathbf{w} \times \mathbf{w} = \begin{vmatrix} \mathbf{i} & \mathbf{j} & \mathbf{k} \\ 1 & 2 & 3 \\ 1 & 2 & 3 \end{vmatrix}$

$\qquad = \begin{vmatrix} 2 & 3 \\ 2 & 3 \end{vmatrix} \mathbf{i} - \begin{vmatrix} 1 & 3 \\ 1 & 3 \end{vmatrix} \mathbf{j} + \begin{vmatrix} 1 & 2 \\ 1 & 2 \end{vmatrix} \mathbf{k} = 0\mathbf{i} - 0\mathbf{j} + 0\mathbf{k} = \mathbf{0}$ ∎

NOW WORK PROBLEM **9.**

Algebraic Properties of the Cross Product

2   Notice in Example 3(a) and 3(b) that $\mathbf{v} \times \mathbf{w}$ and $\mathbf{w} \times \mathbf{v}$ are negatives of one another. From Examples 3(c) and 3(d), we might conjecture that the cross product of a vector with itself is the zero vector. These and other algebraic properties of cross product are given next.

Theorem

Algebraic Properties of the Cross Product

If $\mathbf{u}$, $\mathbf{v}$, and $\mathbf{w}$ are vectors in space and if α is a scalar, then

$$\mathbf{u} \times \mathbf{u} = \mathbf{0} \qquad (2)$$
$$\mathbf{u} \times \mathbf{v} = -(\mathbf{v} \times \mathbf{u}) \qquad (3)$$
$$\alpha(\mathbf{u} \times \mathbf{v}) = (\alpha\mathbf{u}) \times \mathbf{v} = \mathbf{u} \times (\alpha\mathbf{v}) \qquad (4)$$
$$\mathbf{u} \times (\mathbf{v} + \mathbf{w}) = (\mathbf{u} \times \mathbf{v}) + (\mathbf{u} \times \mathbf{w}) \qquad (5)$$

Proof   We will prove properties (2) and (4) here and leave properties (3) and (5) as exercises (see Problems 49 and 50 at the end of this section).

To prove property (2), we let $\mathbf{u} = a_1\mathbf{i} + b_1\mathbf{j} + c_1\mathbf{k}$. Then

$$\mathbf{u} \times \mathbf{u} = \begin{vmatrix} \mathbf{i} & \mathbf{j} & \mathbf{k} \\ a_1 & b_1 & c_1 \\ a_1 & b_1 & c_1 \end{vmatrix} = \begin{vmatrix} b_1 & c_1 \\ b_1 & c_1 \end{vmatrix} \mathbf{i} - \begin{vmatrix} a_1 & c_1 \\ a_1 & c_1 \end{vmatrix} \mathbf{j} + \begin{vmatrix} a_1 & b_1 \\ a_1 & b_1 \end{vmatrix} \mathbf{k}$$

$$= 0\mathbf{i} - 0\mathbf{j} + 0\mathbf{k} = \mathbf{0}$$

To prove property (4), we let $\mathbf{u} = a_1\mathbf{i} + b_1\mathbf{j} + c_1\mathbf{k}$ and $\mathbf{v} = a_2\mathbf{i} + b_2\mathbf{j} + c_2\mathbf{k}$. Then

$$\alpha(\mathbf{u} \times \mathbf{v}) = \underset{\underset{\text{\scriptsize Apply (1)}}{\uparrow}}{\alpha[(b_1c_2 - b_2c_1)\mathbf{i} - (a_1c_2 - a_2c_1)\mathbf{j} + (a_1b_2 - a_2b_1)\mathbf{k}]}$$

$$= \alpha(b_1c_2 - b_2c_1)\mathbf{i} - \alpha(a_1c_2 - a_2c_1)\mathbf{j} + \alpha(a_1b_2 - a_2b_1)\mathbf{k} \quad (6)$$

Since $\alpha\mathbf{u} = \alpha a_1\mathbf{i} + \alpha b_1\mathbf{j} + \alpha c_1\mathbf{k}$, we have

$$(\alpha\mathbf{u}) \times \mathbf{v} = (\alpha b_1c_2 - b_2\alpha c_1)\mathbf{i} - (\alpha a_1c_2 - a_2\alpha c_1)\mathbf{j} + (\alpha a_1b_2 - a_2\alpha b_1)\mathbf{k}$$

$$= \alpha(b_1c_2 - b_2c_1)\mathbf{i} - \alpha(a_1c_2 - a_2c_1)\mathbf{j} + \alpha(a_1b_2 - a_2b_1)\mathbf{k} \quad (7)$$

Based on equations (6) and (7), the first part of property (4) follows. The second part can be proved in like fashion. ∎

NOW WORK PROBLEM **11.**

③ Geometric Properties of the Cross Product

The cross product has several interesting geometric properties.

Theorem

> **Geometric Properties of the Cross Product**
>
> Let **u** and **v** be vectors in space.
>
> $$\mathbf{u} \times \mathbf{v} \text{ is orthogonal to both } \mathbf{u} \text{ and } \mathbf{v}. \tag{8}$$
>
> $$\|\mathbf{u} \times \mathbf{v}\| = \|\mathbf{u}\|\|\mathbf{v}\| \sin \theta, \text{ where } \theta \text{ is the angle between } \mathbf{u} \text{ and } \mathbf{v}. \tag{9}$$
>
> $$\|\mathbf{u} \times \mathbf{v}\| \text{ is the area of the parallelogram having } \mathbf{u} \neq \mathbf{0} \text{ and } \mathbf{v} \neq \mathbf{0} \text{ as adjacent sides.} \tag{10}$$
>
> $$\mathbf{u} \times \mathbf{v} = \mathbf{0} \text{ if and only if } \mathbf{u} \text{ and } \mathbf{v} \text{ are parallel.} \tag{11}$$

Proof of Property (8)   Let $\mathbf{u} = a_1\mathbf{i} + b_1\mathbf{j} + c_1\mathbf{k}$ and $\mathbf{v} = a_2\mathbf{i} + b_2\mathbf{j} + c_2\mathbf{k}$. Then

$$\mathbf{u} \times \mathbf{v} = (b_1c_2 - b_2c_1)\mathbf{i} - (a_1c_2 - a_2c_1)\mathbf{j} + (a_1b_2 - a_2b_1)\mathbf{k}$$

Now we compute the dot product $\mathbf{u} \cdot (\mathbf{u} \times \mathbf{v})$.

$$\mathbf{u} \cdot (\mathbf{u} \times \mathbf{v}) = (a_1\mathbf{i} + b_1\mathbf{j} + c_1\mathbf{k}) \cdot [(b_1c_2 - b_2c_1)\mathbf{i} - (a_1c_2 - a_2c_1)\mathbf{j} + (a_1b_2 - a_2b_1)\mathbf{k}]$$

$$= a_1(b_1c_2 - b_2c_1) - b_1(a_1c_2 - a_2c_1) + c_1(a_1b_2 - a_2b_1) = 0$$

Since two vectors are orthogonal if their dot product is zero, it follows that **u** and $\mathbf{u} \times \mathbf{v}$ are orthogonal. Similarly, $\mathbf{v} \cdot (\mathbf{u} \times \mathbf{v}) = 0$, so **v** and $\mathbf{u} \times \mathbf{v}$ are orthogonal.

④   As long as the vectors **u** and **v** are not parallel, they will form a plane in space. See Figure 78. Based on property (8), the vector $\mathbf{u} \times \mathbf{v}$ is normal to this plane. As Figure 78 illustrates, there are two vectors normal to the plane containing **u** and **v**. It can be shown that the vector $\mathbf{u} \times \mathbf{v}$ is the one determined by the thumb of the right hand when the other fingers of the right hand are cupped so that they point in a direction from **u** to **v**. See Figure 79.*

Figure 78

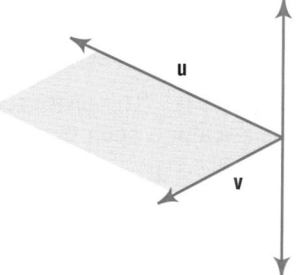

Figure 79

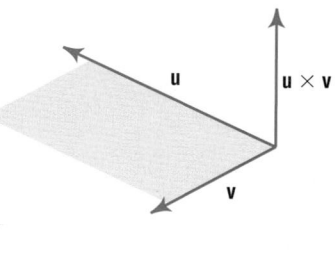

E X A M P L E   4   **Finding a Vector Orthogonal to Two Given Vectors**

Find a vector that is orthogonal to $\mathbf{u} = 3\mathbf{i} - 2\mathbf{j} + \mathbf{k}$ and $\mathbf{v} = -\mathbf{i} + 3\mathbf{j} - \mathbf{k}$.

---

*This is a consequence of using a right-handed coordinate system.

Solution   Based on property (8), such a vector is $\mathbf{u} \times \mathbf{v}$.

$$\mathbf{u} \times \mathbf{v} = \begin{vmatrix} \mathbf{i} & \mathbf{j} & \mathbf{k} \\ 3 & -2 & 1 \\ -1 & 3 & -1 \end{vmatrix} = (2-3)\mathbf{i} - [-3-(-1)]\mathbf{j} + (9-2)\mathbf{k} = -\mathbf{i} + 2\mathbf{j} + 7\mathbf{k}$$

The vector $-\mathbf{i} + 2\mathbf{j} + 7\mathbf{k}$ is orthogonal to both $\mathbf{u}$ and $\mathbf{v}$.

✔ CHECK: Two vectors are orthogonal if their both product is zero.

$$\mathbf{u} \cdot (\mathbf{u} \times \mathbf{v}) = (3\mathbf{i} - 2\mathbf{j} + \mathbf{k}) \cdot (-\mathbf{i} + 2\mathbf{j} + 7\mathbf{k}) = -3 - 4 + 7 = 0$$
$$\mathbf{v} \cdot (\mathbf{u} \times \mathbf{v}) = (-\mathbf{i} + 3\mathbf{j} - \mathbf{k}) \cdot (-\mathbf{i} + 2\mathbf{j} + 7\mathbf{k}) = 1 + 6 - 7 = 0$$

■ ■

✏ **NOW WORK PROBLEM 35.**

Figure 80

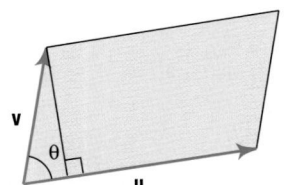

The proof of property (9) is left as an exercise. See Problem 52.

Proof of Property (10)   Suppose that $\mathbf{u}$ and $\mathbf{v}$ are adjacent sides of a parallelogram. See Figure 80. Then the lengths of these sides are $\|\mathbf{u}\|$ and $\|\mathbf{v}\|$. If θ is the angle between $\mathbf{u}$ and $\mathbf{v}$, then the height of the parallelogram is $\|\mathbf{v}\| \sin \theta$ and its area is

$$\text{Area of parallelogram} = \text{Base} \times \text{Height} = \|\mathbf{u}\| [\|\mathbf{v}\| \sin \theta] = \underset{\uparrow}{\|\mathbf{u} \times \mathbf{v}\|}$$

$$\text{Property (9)}$$

■

⑤   **EXAMPLE 5**   **Finding the Area of a Parallelogram**

Find the area of the parallelogram whose vertices are $P_1 = (0,0,0)$, $P_2 = (3, -2, 1)$, $P_3 = (-1, 3, -1)$, and $P_4 = (2, 1, 0)$.

Solution   Two adjacent sides* of this parallelogram are

$$\mathbf{u} = \overrightarrow{P_1 P_2} = 3\mathbf{i} - 2\mathbf{j} + \mathbf{k} \quad \text{and} \quad \mathbf{v} = \overrightarrow{P_1 P_3} = -\mathbf{i} + 3\mathbf{j} - \mathbf{k}$$

Since $\mathbf{u} \times \mathbf{v} = -\mathbf{i} + 2\mathbf{j} + 7\mathbf{k}$ (Example 4), the area of the parallelogram is

$$\text{Area of parallelogram} = \|\mathbf{u} \times \mathbf{v}\| = \sqrt{1 + 4 + 49} = \sqrt{54} = 3\sqrt{6} \quad ■$$

✏ **NOW WORK PROBLEM 43.**

Proof of Property (11)   The proof requires two parts. If $\mathbf{u}$ and $\mathbf{v}$ are parallel, then there is a scalar α such that $\mathbf{u} = \alpha \mathbf{v}$. Then

$$\mathbf{u} \times \mathbf{v} = \underset{\uparrow}{(\alpha \mathbf{v})} \times \mathbf{v} = \alpha \underset{\uparrow}{(\mathbf{v} \times \mathbf{v})} = \mathbf{0}$$

$$\text{Property (4)} \qquad \text{Property (2)}$$

If $\mathbf{u} \times \mathbf{v} = \mathbf{0}$, then, by property (9), we have

$$\|\mathbf{u} \times \mathbf{v}\| = \|\mathbf{u}\| \|\mathbf{v}\| \sin \theta = 0$$

Since $\mathbf{u} \neq \mathbf{0}$ and $\mathbf{v} \neq \mathbf{0}$, then we must have $\sin \theta = 0$, so $\theta = 0$ or $\theta = \pi$. In either case, since θ is the angle between $\mathbf{u}$ and $\mathbf{v}$, then $\mathbf{u}$ and $\mathbf{v}$ are parallel. ■

*__Be careful!__ Not all pairs of vertices give rise to a side. For example, $\overrightarrow{P_1 P_4}$ is a diagonal of the parallelogram since $\overrightarrow{P_1 P_3} + \overrightarrow{P_3 P_4} = \overrightarrow{P_1 P_4}$. Also $\overrightarrow{P_1 P_3}$ and $\overrightarrow{P_2 P_4}$ are not adjacent sides; they are parallel sides.

5.7   Concepts and Vocabulary

In Problems 1–6, answer True or False to each statement.

1. If **u** and **v** are parallel vectors, then **u** × **v** = 0.

2. For any vector **v**, **v** × **v** = 0.

3. If **u** and **v** are vectors, then **u** × **v** + **v** × **u** = 0.

4. **u** × **v** is a vector that is parallel to both **u** and **v**.

5. ‖**u** × **v**‖ = ‖**u**‖‖**v**‖ cos θ, where θ is the angle between **u** and **v**.

6. The area of the parallelogram having **u** and **v** as adjacent sides is the magnitude of the cross product of **u** and **v**.

5.7   Exercises

In Problems 1–8, find the value of each determinant.

1. $\begin{vmatrix} 3 & 4 \\ 1 & 2 \end{vmatrix}$

2. $\begin{vmatrix} -2 & 5 \\ 2 & -3 \end{vmatrix}$

3. $\begin{vmatrix} 6 & 5 \\ -2 & -1 \end{vmatrix}$

4. $\begin{vmatrix} -4 & 0 \\ 5 & 3 \end{vmatrix}$

5. $\begin{vmatrix} A & B & C \\ 2 & 1 & 4 \\ 1 & 3 & 1 \end{vmatrix}$

6. $\begin{vmatrix} A & B & C \\ 0 & 2 & 4 \\ 3 & 1 & 3 \end{vmatrix}$

7. $\begin{vmatrix} A & B & C \\ -1 & 3 & 5 \\ 5 & 0 & -2 \end{vmatrix}$

8. $\begin{vmatrix} A & B & C \\ 1 & -2 & -3 \\ 0 & 2 & -2 \end{vmatrix}$

*In Problems 9–16, find (a) **v** × **w**, (b) **w** × **v**, (c) **w** × **w**, and (d) **v** × **v**.*

9. **v** = 2**i** − 3**j** + **k**
   w = 3**i** − 2**j** − **k**

10. **v** = −**i** + 3**j** + 2**k**
   w = 3**i** − 2**j** − **k**

11. **v** = **i** + **j**
   w = 2**i** + **j** + **k**

12. **v** = **i** − 4**j** + 2**k**
   w = 3**i** + 2**j** + **k**

13. **v** = 2**i** − **j** + 2**k**
   w = **j** − **k**

14. **v** = 3**i** + **j** + 3**k**
   w = **i** − **k**

15. **v** = **i** − **j** − **k**
   w = 4**i** − 3**k**

16. **v** = 2**i** − 3**j**
   w = 3**j** + 2**k**

*In Problems 17–38, use the vectors **u**, **v**, and **w** given next to find each expression.*

$$\mathbf{u} = 2\mathbf{i} - 3\mathbf{j} + \mathbf{k} \qquad \mathbf{v} = -3\mathbf{i} + 3\mathbf{j} + 2\mathbf{k} \qquad \mathbf{w} = \mathbf{i} + \mathbf{j} + 3\mathbf{k}$$

17. **u** × **v**

18. **v** × **w**

19. **v** × **u**

20. **w** × **v**

21. **v** × **v**

22. **w** × **w**

23. (3**u**) × **v**

24. **v** × (4**w**)

25. **u** × (2**v**)

26. (−3**v**) × **w**

27. **u** · (**u** × **v**)

28. **v** · (**v** × **w**)

29. **u** · (**v** × **w**)

30. (**u** × **v**) · **w**

31. **v** · (**u** × **w**)

32. (**v** × **u**) · **w**

33. **u** × (**v** × **v**)

34. (**w** × **w**) × **v**

35. Find a vector orthogonal to both **u** and **v**.

36. Find a vector orthogonal to both **u** and **w**.

37. Find a vector orthogonal to both **u** and **i** + **j**.

38. Find a vector orthogonal to both **u** and **j** + **k**.

In Problems 39–42, find the area of the parallelogram with one corner at P_1 and adjacent sides $\overrightarrow{P_1P_2}$ and $\overrightarrow{P_1P_3}$.

39. $P_1 = (0,0,0)$, $P_2 = (1,2,3)$, $P_3 = (-2,3,0)$

40. $P_1 = (0,0,0)$, $P_2 = (2,3,1)$, $P_3 = (-2,4,1)$

41. $P_1 = (1,2,0)$, $P_2 = (-2,3,4)$, $P_3 = (0,-2,3)$

42. $P_1 = (-2,0,2)$, $P_2 = (2,1,-1)$, $P_3 = (2,-1,2)$

In Problems 43–46, find the area of the parallelogram with vertices P_1, P_2, P_3, and P_4.

43. $P_1 = (1,1,2)$, $P_2 = (1,2,3)$, $P_3 = (-2,3,0)$,
   $P_4 = (-2,4,1)$

44. $P_1 = (2,1,1)$, $P_2 = (2,3,1)$, $P_3 = (-2,4,1)$,
   $P_4 = (-2,6,1)$

45. $P_1 = (1,2,-1)$, $P_2 = (4,2,-3)$, $P_3 = (6,-5,2)$,
   $P_4 = (9,-5,0)$,

46. $P_1 = (-1,1,1)$, $P_2 = (-1,2,2)$, $P_3 = (-3,4,-5)$,
   $P_4 = (-3,5,-4)$,

47. Find a unit vector normal to the plane containing **v** = **i** + 3**j** − 2**k** and **w** = −2**i** + **j** + 3**k**.

48. Find a unit vector normal to the plane containing **v** = 2**i** + 3**j** − **k** and **w** = −2**i** − 4**j** − 3**k**.

49. Prove property (3).

50. Prove property (5).

51. Prove for vectors **u** and **v** that
$$\|\mathbf{u} \times \mathbf{v}\|^2 = \|\mathbf{u}\|^2\|\mathbf{v}\|^2 - (\mathbf{u} \cdot \mathbf{v})^2.$$

   [Hint: Proceed as in the proof of property (4), computing first the left side and then the right side.]

52. Prove property (9).

   [Hint: Use the result of Problem 51 and the fact that if θ is the angle between **u** and **v** then **u** · **v** = ‖**u**‖‖**v**‖ cos θ.]

53. Show that if **u** and **v** are orthogonal then
$$\|\mathbf{u} \times \mathbf{v}\| = \|\mathbf{u}\|\|\mathbf{v}\|.$$

54. Show that if **u** and **v** are orthogonal unit vectors then so is **u** × **v**.

55. If **u** · **v** = 0 and **u** × **v** = 0, what can you conclude about **u** and **v**?

Chapter Review

Things to Know

Relationship between polar coordinates (r, θ) and

rectangular coordinates (x, y) (pp. 321 and 324)

$x = r \cos \theta, y = r \sin \theta$

$r^2 = x^2 + y^2, \tan \theta = \dfrac{y}{x}, \quad x \neq 0$

Polar form of a complex number (p. 346)

If $z = x + yi$, then $z = r(\cos \theta + i \sin \theta)$,

where $r = |z| = \sqrt{x^2 + y^2}$, $\sin \theta = \dfrac{y}{r}$, $\cos \theta = \dfrac{x}{r}$, $0 \leq \theta < 2\pi$

De Moivre's Theorem (p. 349)

If $z = r(\cos \theta + i \sin \theta)$, then
$z^n = r^n[\cos(n\theta) + i \sin(n\theta)]$, where $n \geq 1$ is a positive integer

nth root of a complex number $z = r(\cos \theta_0 + i \sin \theta_0)$ (p. 350)

$\sqrt[n]{z} = \sqrt[n]{r}\left[\cos\left(\dfrac{\theta_0}{n} + \dfrac{2k\pi}{n} \right) + i \sin\left(\dfrac{\theta_0}{n} + \dfrac{2k\pi}{n} \right) \right], \quad k = 0, 1, \ldots, n-1$,

where $n \geq 2$ is an integer.

Vector (p. 354)

Quantity having magnitude and direction; equivalent to a directed line segment $\overrightarrow{PQ}$

Position vector (p. 357)

Vector whose initial point is at the origin

Unit vector (pp. 357 and 360)

Vector whose magnitude is 1

Dot product (pp. 366 and 378)

If $\mathbf{v} = a_1\mathbf{i} + b_1\mathbf{j}$ and $\mathbf{w} = a_2\mathbf{i} + b_2\mathbf{j}$, then $\mathbf{v} \cdot \mathbf{w} = a_1a_2 + b_1b_2$.

If $\mathbf{v} = a_1\mathbf{i} + b_1\mathbf{j} + c_1\mathbf{k}$ and $\mathbf{w} = a_2\mathbf{i} + b_2\mathbf{j} + c_2\mathbf{k}_2$ then $\mathbf{v} \cdot \mathbf{w} = a_1a_2 + b_1b_2 + c_1c_2$.

Angle θ between two nonzero vectors $\mathbf{u}$ and $\mathbf{v}$ (pp. 368 and 379)

$\cos \theta = \dfrac{\mathbf{u} \cdot \mathbf{v}}{\|\mathbf{u}\|\|\mathbf{v}\|}$

Direction angles of vectors in space (p. 380)

If $\mathbf{v} = a\mathbf{i} + b\mathbf{j} + c\mathbf{k}$, then $\mathbf{v} = \|\mathbf{v}\|\left[(\cos \alpha)\mathbf{i} + (\cos \beta)\mathbf{j} + (\cos \gamma)\mathbf{k} \right]$,

where $\cos \alpha = \dfrac{a}{\|\mathbf{v}\|}$, $\cos \beta = \dfrac{b}{\|\mathbf{v}\|}$, $\cos \gamma = \dfrac{c}{\|\mathbf{v}\|}$.

Cross Product (p. 384)

If $\mathbf{v} = a_1\mathbf{i} + b_1\mathbf{j} + c_1\mathbf{k}$ and $\mathbf{w} = a_2\mathbf{i} + b_2\mathbf{j} + c_2\mathbf{k}$,
then $\mathbf{v} \times \mathbf{w} = \left[b_1c_2 - b_2c_1 \right]\mathbf{i} - \left[a_1c_2 - a_2c_1 \right]\mathbf{j} + \left[a_1b_2 - a_2b_1 \right]\mathbf{k}$

Area of parallelogram (p. 387)

$\|\mathbf{u} \times \mathbf{v}\| = \|\mathbf{u}\|\|\mathbf{v}\| \sin \theta$, where θ is the angle between $\mathbf{u}$ and $\mathbf{v}$.

Objectives

Section	You should be able to:	Review Exercises
5.1	① Plot points using polar coordinates (p. 318)	1–6
	② Convert from polar coordinates to rectangular coordinates (p. 321)	1–6
	③ Convert from rectangular coordinates to polar coordinates (p. 322)	7–12
5.2	① Graph and identify polar equations by converting to rectangular equations (p. 327)	13–18
	② Graph polar equations using a graphing utility (p. 329)	19–22
	③ Test polar equations for symmetry (p. 333)	23–28
	④ Graph polar equations by plotting points (p. 334)	23–28
5.3	① Convert a complex number from rectangular form to polar form (p. 346)	29–32
	② Plot points in the complex plane (p. 347)	33–38
	③ Find products and quotients of complex numbers in polar form (p. 347)	39–44

Review Exercises

Blue problem numbers indicate the authors' suggestions for use in a Practice Test.

In Problems 1–6, plot each point given in polar coordinates, and find its rectangular coordinates.

1. $\left(3, \dfrac{\pi}{6}\right)$ 2. $\left(4, \dfrac{2\pi}{3}\right)$ 3. $\left(-2, \dfrac{4\pi}{3}\right)$

4. $\left(-1, \dfrac{5\pi}{4}\right)$ 5. $\left(-3, -\dfrac{\pi}{2}\right)$ 6. $\left(-4, -\dfrac{\pi}{4}\right)$

In Problems 7–12, the rectangular coordinates of a point are given. Find two pairs of polar coordinates (r, θ) for each point, one with $r > 0$ and the other with $r < 0$. Express θ in radians.

7. $(-3, 3)$ 8. $(1, -1)$ 9. $(0, -2)$ 10. $(2, 0)$ 11. $(3, 4)$ 12. $(-5, 12)$

In Problems 13–18, the letters r and θ represent polar coordinates. Write each polar equation as an equation in rectangular coordinates (x, y). Identify the equation and graph it.

13. $r = 2\sin\theta$ 14. $3r = \sin\theta$ 15. $r = 5$

16. $\theta = \dfrac{\pi}{4}$ 17. $r\cos\theta + 3r\sin\theta = 6$ 18. $r^2 + 4r\sin\theta - 8r\cos\theta = 5$

In Problems 19–22, graph each polar equation using a graphing utility.

19. $r = 2 + \sin \theta$ **20.** $r = \sin^2 \theta$ **21.** $r \tan \theta = 2$ **22.** $(r + 5) \sin^2 \theta = 2$

In Problems 23–28, sketch the graph of each polar equation. Be sure to test for symmetry. Verify your graph using a graphing utility.

23. $r = 4 \cos \theta$ **24.** $r = 3 \sin \theta$ **25.** $r = 3 - 3 \sin \theta$

26. $r = 2 + \cos \theta$ **27.** $r = 4 - \cos \theta$ **28.** $r = 1 - 2 \sin \theta$

In Problems 29–32, write each complex number in polar form. Express each argument in degrees.

29. $-1 - i$ **30.** $-\sqrt{3} + i$ **31.** $4 - 3i$ **32.** $3 - 2i$

In Problems 33–38, write each complex number in the standard form $a + bi$ and plot each in the complex plane.

33. $2(\cos 150° + i \sin 150°)$ **34.** $3(\cos 60° + i \sin 60°)$ **35.** $3\left(\cos \dfrac{2\pi}{3} + i \sin \dfrac{2\pi}{3}\right)$

36. $4\left(\cos \dfrac{3\pi}{4} + i \sin \dfrac{3\pi}{4}\right)$ **37.** $0.1(\cos 350° + i \sin 350°)$ **38.** $0.5(\cos 160° + i \sin 160°)$

In Problems 39–44, find zw and $\dfrac{z}{w}$. Leave your answers in polar form.

39. $z = \cos 80° + i \sin 80°$
$w = \cos 50° + i \sin 50°$

40. $z = \cos 205° + i \sin 205°$
$w = \cos 85° + i \sin 85°$

41. $z = 3\left(\cos \dfrac{9\pi}{5} + i \sin \dfrac{9\pi}{5}\right)$
$w = 2\left(\cos \dfrac{\pi}{5} + i \sin \dfrac{\pi}{5}\right)$

42. $z = 2\left(\cos \dfrac{5\pi}{3} + i \sin \dfrac{5\pi}{3}\right)$
$w = 3\left(\cos \dfrac{\pi}{3} + i \sin \dfrac{\pi}{3}\right)$

43. $z = 5(\cos 10° + i \sin 10°)$
$w = \cos 355° + i \sin 355°$

44. $z = 4(\cos 50° + i \sin 50°)$
$w = \cos 340° + i \sin 340°$

In Problems 45–52, write each expression in the standard form $a + bi$.

45. $[3(\cos 20° + i \sin 20°)]^3$ **46.** $[2(\cos 50° + i \sin 50°)]^3$

47. $\left[\sqrt{2}\left(\cos \dfrac{5\pi}{8} + i \sin \dfrac{5\pi}{8}\right)\right]^4$ **48.** $\left[2\left(\cos \dfrac{5\pi}{16} + i \sin \dfrac{5\pi}{16}\right)\right]^4$

49. $(1 - \sqrt{3}i)^6$ **50.** $(2 - 2i)^8$ **51.** $(3 + 4i)^4$ **52.** $(1 - 2i)^4$

53. Find all the complex cube roots of 27.   **54.** Find all the complex fourth roots of -16.

In Problems 55–58, use the figure to graph each of the following:

55. $\mathbf{u} + \mathbf{v}$ **56.** $\mathbf{v} + \mathbf{w}$ **57.** $2\mathbf{u} + 3\mathbf{v}$ **58.** $5\mathbf{v} - 2\mathbf{w}$

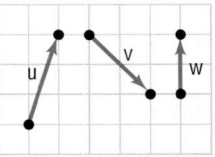

In Problems 59–62, the vector $\mathbf{v}$ is represented by the directed line segment $\overrightarrow{PQ}$. Write $\mathbf{v}$ in the form $a\mathbf{i} + b\mathbf{j}$ and find $\|\mathbf{v}\|$.

59. $P = (1, -2)$; $Q = (3, -6)$ **60.** $P = (-3, 1)$; $Q = (4, -2)$

61. $P = (0, -2)$; $Q = (-1, 1)$ **62.** $P = (3, -4)$; $Q = (-2, 0)$

In Problems 63–72, use the vectors $\mathbf{v} = -2\mathbf{i} + \mathbf{j}$ and $\mathbf{w} = 4\mathbf{i} - 3\mathbf{j}$ to find:

63. $\mathbf{v} + \mathbf{w}$ **64.** $\mathbf{v} - \mathbf{w}$ **65.** $4\mathbf{v} - 3\mathbf{w}$ **66.** $-\mathbf{v} + 2\mathbf{w}$

67. $\|\mathbf{v}\|$ **68.** $\|\mathbf{v} + \mathbf{w}\|$ **69.** $\|\mathbf{v}\| + \|\mathbf{w}\|$ **70.** $\|2\mathbf{v}\| - 3\|\mathbf{w}\|$

71. Find a unit vector in the same direction as $\mathbf{v}$.   **72.** Find a unit vector in the opposite direction of $\mathbf{w}$.

73. Find the vector **v** with magnitude 3 if the angle between **v** and **i** is $60°$.

74. Find the vector **v** with magnitude 5 if the angle between **v** and **i** is $150°$.

75. Find the distance from $P_1 = (1, 3, -2)$ to $P_2 = (4, -2, 1)$.

76. Find the distance from $P_1 = (0, -4, 3)$ to $P_2 = (6, -5, -1)$.

77. A vector **v** has initial point $P = (1, 3, -2)$ and terminal point $Q = (4, -2, 1)$. Write **v** in the form $\mathbf{v} = a\mathbf{i} + b\mathbf{j} + c\mathbf{k}$.

78. A vector **v** has initial point $P = (0, -4, 3)$ and terminal point $Q = (6, -5, -1)$. Write **v** in the form $\mathbf{v} = a\mathbf{i} + b\mathbf{j} + c\mathbf{k}$.

In Problems 79–86, use the vectors $\mathbf{v} = 3\mathbf{i} + \mathbf{j} - 2\mathbf{k}$ and $\mathbf{w} = -3\mathbf{i} + 2\mathbf{j} - \mathbf{k}$ to find each expression.

79. $4\mathbf{v} - 3\mathbf{w}$ **80.** $-\mathbf{v} + 2\mathbf{w}$ **81.** $\|\mathbf{v} - \mathbf{w}\|$ **82.** $\|\mathbf{v} + \mathbf{w}\|$

83. $\|\mathbf{v}\| - \|\mathbf{w}\|$ **84.** $\|\mathbf{v}\| + \|\mathbf{w}\|$ **85.** $\mathbf{v} \times \mathbf{w}$ **86.** $\mathbf{v} \cdot (\mathbf{v} \times \mathbf{w})$

87. Find a unit vector in the same direction as **v** and then in the opposite direction of **v**.

88. Find a unit vector orthogonal to both **v** and **w**.

*In Problems 89–96, find the dot product $\mathbf{v} \cdot \mathbf{w}$ and the angle between **v** and **w**.*

89. $\mathbf{v} = -2\mathbf{i} + \mathbf{j}, \quad \mathbf{w} = 4\mathbf{i} - 3\mathbf{j}$

90. $\mathbf{v} = 3\mathbf{i} - \mathbf{j}, \quad \mathbf{w} = \mathbf{i} + \mathbf{j}$

91. $\mathbf{v} = \mathbf{i} - 3\mathbf{j}, \quad \mathbf{w} = -\mathbf{i} + \mathbf{j}$

92. $\mathbf{v} = \mathbf{i} + 4\mathbf{j}, \quad \mathbf{w} = 3\mathbf{i} - 2\mathbf{j}$

93. $\mathbf{v} = \mathbf{i} + \mathbf{j} + \mathbf{k}, \quad \mathbf{w} = \mathbf{i} - \mathbf{j} + \mathbf{k}$

94. $\mathbf{v} = \mathbf{i} - \mathbf{j} + \mathbf{k}, \quad \mathbf{w} = 2\mathbf{i} + \mathbf{j} + \mathbf{k}$

95. $\mathbf{v} = 4\mathbf{i} - \mathbf{j} + 2\mathbf{k}, \quad \mathbf{w} = \mathbf{i} - 2\mathbf{j} - 3\mathbf{k}$

96. $\mathbf{v} = -\mathbf{i} - 2\mathbf{j} + 3\mathbf{k}, \quad \mathbf{w} = 5\mathbf{i} + \mathbf{j} + \mathbf{k}$

*In Problems 97–102, determine whether **v** and **w** are parallel, orthogonal, or neither.*

97. $\mathbf{v} = 2\mathbf{i} + 3\mathbf{j}; \quad \mathbf{w} = -4\mathbf{i} - 6\mathbf{j}$ **98.** $\mathbf{v} = -2\mathbf{i} - \mathbf{j}; \quad \mathbf{w} = 2\mathbf{i} + \mathbf{j}$ **99.** $\mathbf{v} = 3\mathbf{i} - 4\mathbf{j}; \quad \mathbf{w} = -3\mathbf{i} + 4\mathbf{j}$

100. $\mathbf{v} = -2\mathbf{i} + 2\mathbf{j}; \quad \mathbf{w} = -3\mathbf{i} + 2\mathbf{j}$ **101.** $\mathbf{v} = 3\mathbf{i} - 2\mathbf{j}; \quad \mathbf{w} = 4\mathbf{i} + 6\mathbf{j}$ **102.** $\mathbf{v} = -4\mathbf{i} + 2\mathbf{j}; \quad \mathbf{w} = 2\mathbf{i} + 4\mathbf{j}$

*In Problems 103 and 104, decompose **v** into two vectors, one parallel to **w** and the other orthogonal to **w**.*

103. $\mathbf{v} = 2\mathbf{i} + \mathbf{j}; \quad \mathbf{w} = -4\mathbf{i} + 3\mathbf{j}$

104. $\mathbf{v} = -3\mathbf{i} + 2\mathbf{j}; \quad \mathbf{w} = -2\mathbf{i} + \mathbf{j}$

105. Find the vector projection of $\mathbf{v} = 2\mathbf{i} + 3\mathbf{j}$ onto $\mathbf{w} = 3\mathbf{i} + \mathbf{j}$.

106. Find the vector projection of $\mathbf{v} = -\mathbf{i} + 2\mathbf{j}$ onto $\mathbf{w} = 3\mathbf{i} - \mathbf{j}$.

107. Find the direction angles of the vector $\mathbf{v} = 3\mathbf{i} - 4\mathbf{j} + 2\mathbf{k}$.

108. Find the direction angles of the vector $\mathbf{v} = \mathbf{i} - \mathbf{j} + 2\mathbf{k}$.

109. Find the area of the parallelogram with vertices $P_1 = (1, 1, 1), \quad P_2 = (2, 3, 4), \quad P_3 = (6, 5, 2),$ and $P_4 = (7, 7, 5)$.

110. Find the area of the parallelogram with vertices $P_1 = (2, -1, 1), \quad P_2 = (5, 1, 4), \quad P_3 = (0, 1, 1),$ and $P_4 = (3, 3, 4)$.

111. If $\mathbf{u} \times \mathbf{v} = 2\mathbf{i} - 3\mathbf{j} + \mathbf{k}$, what is $\mathbf{v} \times \mathbf{u}$?

112. Suppose $\mathbf{u} = 3\mathbf{v}$. What is $\mathbf{u} \times \mathbf{v}$?

113. Actual Speed and Direction of a Swimmer A swimmer can maintain a constant speed of 5 miles per hour. If the swimmer heads directly across a river that has a current moving at the rate of 2 miles per hour, what is the actual speed of the swimmer? (See the figure.) If the river is 1 mile wide, how far downstream will the swimmer end up from the point directly across the river from the starting point?

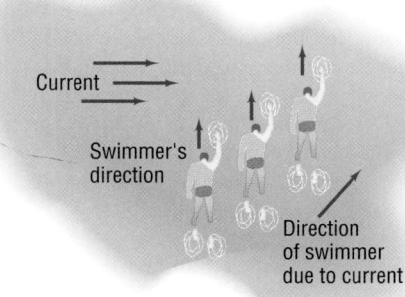

114. Actual Speed and Direction of an Airplane An airplane has an airspeed of 500 kilometers per hour in a northerly direction. The wind velocity is 60 kilometers per hour in a southeasterly direction. Find the actual speed and direction of the plane relative to the ground.

115. Static Equilibrium A weight of 2000 pounds is suspended from two cables as shown in the figure. What are the tensions of each cable?

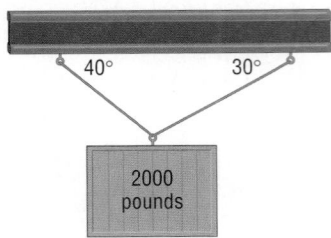

40°   30°

2000 pounds

116. Actual Speed and Distance of a Motorboat A small motorboat is moving at a true speed of 11 miles per hour in a southerly direction. The current is known to be from the northeast at 3 miles per hour. What is the speed of the motorboat relative to the water? In what direction does the compass indicate that the boat is headed?

117. Computing Work Find the work done by a force of 5 pounds acting in the direction 60° to the horizontal in moving an object 20 feet from $(0, 0)$ to $(20, 0)$.

Chapter Projects

1. Mandelbrot Sets

(a) Let $z = x + yi$ be a complex number. We can plot complex numbers using a coordinate system called the complex plane. The x-axis will be referred to as the real axis, because any point that lies on the real axis is of the form $z = x + 0i = x$, a real number. The y-axis is called the imaginary axis, because any point that lies on it is of the form $z = 0 + yi = yi$, a pure imaginary number. To plot the complex number $z = x + yi$, plot the ordered pair (x, y) where x is the signed distance from the imaginary axis and y is the signed distance from the real axis. Draw a complex plane and plot the points $z_1 = 3 + 4i$, $z_2 = -2 + i$, $z_3 = 0 - 2i$, $z_4 = -2$.

(b) Consider the expression $a_n = (a_{n-1})^2 + z$, where z is some complex number (called the **seed**) and $a_0 = z$. Compute $a_1(= a_0^2 + z)$, $a_2(= a_1^2 + z)$, $a_3(= a_2^2 + z)$, a_4, a_5, a_6 for the following seeds: $z_1 = 0.1 - 0.4i$, $z_2 = 0.5 + 0.8i$, $z_3 = -0.9 + 0.7i$, $z_4 = -1.1 + 0.1i$, $z_5 = 0 - 1.3i$, and $z_6 = 1 + 1i$.

(c) The dark portion of the graph represents the set of all values $z = x + yi$ that are in the Mandelbrot Set. Determine which complex numbers in (b) are in this set by plotting them on the graph. Do the complex numbers that are not in the Mandelbrot Set have any common characteristics regarding the values of a_6 found in (b)?

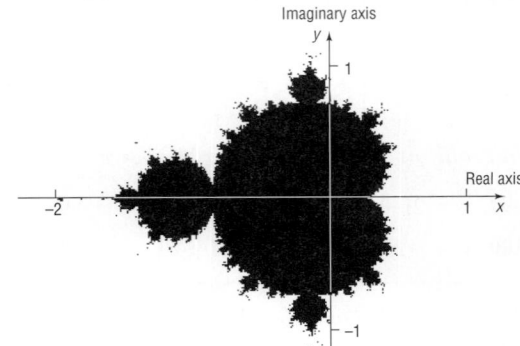

(d) Compute $|z| = \sqrt{x^2 + y^2}$ for each of the complex numbers in (b). Now compute $|a_6|$ for each of the complex numbers in (b). For which complex numbers is $|a_6| \geq |z|$ and $|z| > 2$? Conclude that the criterion for a complex number to be in the Mandelbrot Set is that $|a_n| \geq |z|$ and $|z| > 2$.

2. Compound Interest
In Example 13 on page 340, we graphed a logarithmic spiral. If we write the logarithmic spiral in the form $A = Pe^{r\theta}$ we have the formula for continuously compounded interest where A is the amount resulting from investing a principal P at an annual interest rate r for θ years. Set your graphing calculator to radian and polar coordinate mode.

(a) Graph the amount A that results from a $100 deposit earning 5% compounded continuously using θstep $= 0.1$.

(b) Using TRACE, determine the amount of time it takes (to the nearest tenth of a year) to double your investment.

(c) Using TRACE, determine the amount of time it takes (to the nearest tenth of a year) to triple your investment.

(d) Graph the amount A that results from a \$100 deposit earning 10% compounded continuously using θstep $= 0.1$.

(e) Using TRACE, determine the amount of time it takes (to the nearest tenth of a year) to double your investment.

(f) Using TRACE, determine the amount of time it takes (to the nearest tenth of a year) to triple your investment.

(g) Does doubling the interest rate reduce the time it takes the investment to double by half? Support your answer.

3. **A.** The equation $e^{i\pi} + 1 = 0$ is sometimes called **Euler's equation**, named after Leonhard Euler, a Swiss mathematician from the 18th century. Euler was instrumental in establishing a connection between complex numbers and trigonometric functions. One of his identities is $e^{ix} = \cos x + i \sin x$.

(a) Verify Euler's equation, using the Euler identity stated above.

(b) Verify the other two Euler identities:
$$\sin x = \frac{e^{ix} - e^{-ix}}{2i} \quad \text{and} \quad \cos x = \frac{e^{ix} + e^{-ix}}{2}.$$

(c) Use the Euler identities to write $\sin(1 + i)$ in the standard form $a + bi$. Round to three decimal places.

(d) Prove the theorems for the product and quotient of two complex numbers using Euler's identities.

B. Gauss established a proof of the Fundamental Theorem of Algebra in his doctoral dissertation (at the age of 20). In it, he graphed complex numbers as was shown in Section 5.3 to help establish the existence of complex solutions.

(a) Solve $ix^3 + 8 = 0$ graphically by answering the following:
1. Make the substitution $x = u + iv$ and simplify.
2. Equate the real parts and the imaginary parts on both sides of the equation.
3. Graph each real equation resulting in part 2 on the same set of axes. (Use your graphing calculator.)
4. Find the points of intersection (u, v) by using the INTERSECT feature on your calculator. What are the solutions of the equation? How many should there be? Is that the number you found? Are they complex (but not real) or real? Do they make sense?

(b) Solve the equation by using DeMoivre's Theorem. Do the solutions that you found in part (a) match those that you found here?

Source: Carl B. Boyer, *A History of Mathematics,* 2nd edition. John Wiley & Sons, 1989. Part a–pg. 520; Part b–pgs 560, 561, 596.

Cumulative Review

1. Find the real solutions, if any, of the equation: $3 \sin x + \cos^2 x = 5$.

2. Find an equation for the line containing the origin that makes an angle of $30°$ with the positive x-axis.

3. Find an equation for the circle with center at the point $(0, 1)$ and radius 3. Graph this circle.

4. What is the domain of the function $f(x) = \tan\left(x + \frac{\pi}{4}\right)$?

5. Test the equation $x^2 + y^3 = 2x^4$ for symmetry with respect to the x-axis, the y-axis, and the origin.

6. Solve the triangle $\alpha = 30°$, $\beta = 80°$, and $a = 4$. What is the area of the triangle(s)?

7. Graph the function $y = |\sin x|$.

8. Graph the function $y = \sin|x|$.

9. Find the exact value of $\sin^{-1}\left(-\frac{1}{2}\right)$.

10. Graph the equations $x = 3$ and $y = 4$ using the same set of rectangular coordinates.

11. Graph the equations $r = 2$ and $\theta = \frac{\pi}{3}$ using the same set of polar coordinates.

ANALYTIC GEOMETRY

Pluto resumes 'farthest orbit'

WASHINGTON (AP)—Mere days after surviving attacks on its status as a planet, diminutive Pluto is resuming its traditional spot farthest from the sun.

Pluto was on course to cross the orbit of Neptune at 5:08 a.m. Thursday, NASA reported.

Normally the most distant planet from the sun. Pluto has a highly elliptical orbit that occasionally brings it inside the orbit of Neptune.

That last took place on Feb. 7, 1979, and since then Neptune had been the most distant planet.

Now Pluto once again becomes the farthest planet from the sun, where it will remain for 228 years. It takes pluto 248 years to circle the sun.

SOURCE: *Naples Daily News*, February 11, 1999.

SEE CHAPTER PROJECT 1.

OUTLINE

 For additional study help, go to

www.prenhall.com/sullivanegu3e

Materials include:

- Graphing Calculator Help
- Chapter Quiz
- Chapter Test
- PowerPoint Downloads
- Chapter Projects
- Student Tips

A Look Back, A Look Forward

In Chapter 1, we introduced rectangular coordinates and showed how geometry problems can be solved algebraically. We defined a circle geometrically and then used the distance formula and rectangular coordinates to obtain an equation for a circle. In this chapter we give geometric definitions for the conics and use the distance formula and rectangular coordinates to obtain their equations.

Historically, Apollonius (200 B.C.) was among the first to study *conics* and discover some of their interesting properties. Today, conics are still studied because of their many uses. *Paraboloids of revolution* (parabolas rotated about their axes of symmetry) are used as signal collectors (the satellite dishes used with radar and cable TV, for example), as solar energy col-

lectors, and as reflectors (telescopes, light projection, and so on). The planets circle the Sun in approximately *elliptical* orbits. Elliptical surfaces can be used to reflect signals such as light and sound from one place to another. And *hyperbolas* can be used to determine the positions of ships at sea.

The Greeks used the methods of Euclidean geometry to study conics. However, we shall use the more powerful methods of analytic geometry, bringing to bear both algebra and geometry, for our study of conics.

The chapter concludes with sections on equations of conics in polar coordinates and plane curves and parametric equations.

6.1 CONICS

OBJECTIVES   ① Know the Names of the Conics

Figure 1

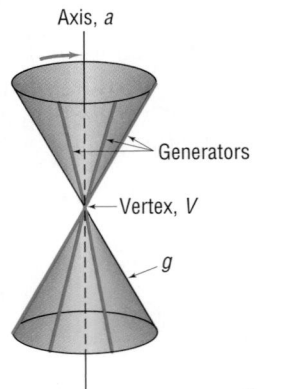

Axis, *a*

Generators

Vertex, *V*

g

① The word *conic* derives from the word *cone*, which is a geometric figure that can be constructed in the following way: Let *a* and *g* be two distinct lines that intersect at a point *V*. Keep the line *a* fixed. Now rotate the line *g* about *a* while maintaining the same angle between *a* and *g*. The collection of points swept out (generated) by the line *g* is called a **(right circular) cone**. See Figure 1. The fixed line *a* is called the **axis** of the cone; the point *V* is called its **vertex**; the lines that pass through *V* and make the same angle with *a* as *g* are called **generators** of the cone. Each generator is a line that lies entirely on the cone. The cone consists of two parts, called **nappes**, that intersect at the vertex.

Conics, an abbreviation for **conic sections**, are curves that result from the intersection of a (right circular) cone and a plane. The conics we shall study arise when the plane does not contain the vertex, as shown in Figure 2. These conics are **circles** when the plane is perpendicular to the axis of the

Figure 2

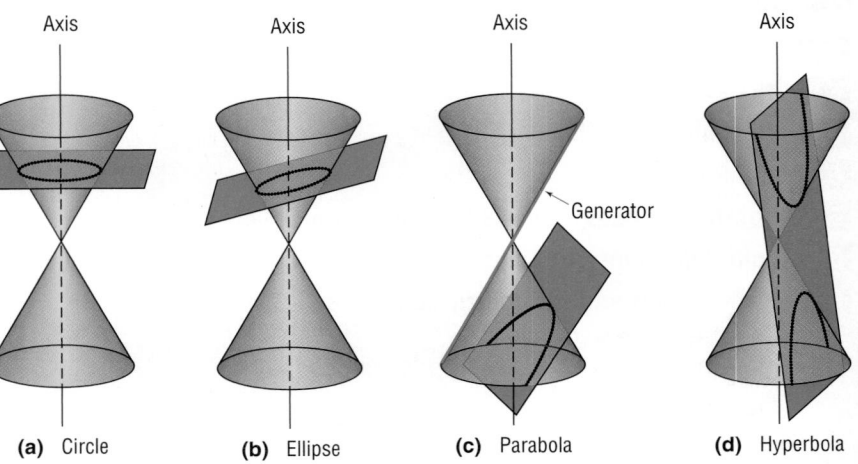

(a) Circle   **(b)** Ellipse   **(c)** Parabola   **(d)** Hyperbola

cone and intersects each generator; **ellipses** when the plane is tilted slightly so that it intersects each generator, but intersects only one nappe of the cone; **parabolas** when the plane is tilted farther so that it is parallel to one (and only one) generator and intersects only one nappe of the cone; and **hyperbolas** when the plane intersects both nappes.

If the plane does contain the vertex, the intersection of the plane and the cone is a point, a line, or a pair of intersecting lines. These are usually called **degenerate conics**.

PREPARING FOR THIS SECTION

Before getting started, review the following:

✓ Distance Formula (Section 1.1, p. 5)

✓ Symmetry (Section 1.3, pp. 20–22)

✓ Square Root Method (Section A.3, p. 578)

✓ Completing the Square (Section A.3, pp. 579–580)

✓ Graphing Techniques: Transformations (Section 1.6, pp. 64–73)

6.2 THE PARABOLA

OBJECTIVES
1. Find the Equation of a Parabola
2. Graph Parabolas
3. Discuss the Equation of a Parabola
4. Work with Parabolas with Vertex at (h, k)
5. Solve Applied Problems Involving Parabolas

The graph of a quadratic function is a parabola. In this section, we begin with a geometric definition of a parabola and use it to obtain an equation.

A **parabola** is the collection of all points P in the plane that are the same distance from a fixed point F as they are from a fixed line D. The point F is called the **focus** of the parabola, and the line D is its **directrix**. As a result, a parabola is the set of points P for which

$$d(F, P) = d(P, D) \tag{1}$$

Figure 3

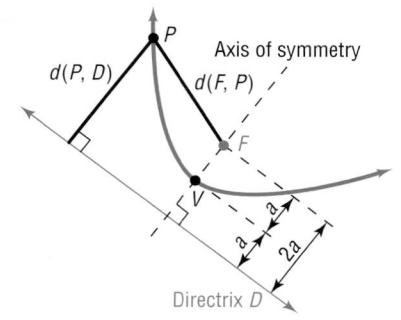

① Figure 3 shows a parabola. The line through the focus F and perpendicular to the directrix D is called the **axis of symmetry** of the parabola. The point of intersection of the parabola with its axis of symmetry is called the **vertex** V.

Because the vertex V lies on the parabola, it must satisfy equation (1): $d(F, V) = d(V, D)$. The vertex is midway between the focus and the directrix. We shall let a equal the distance $d(F, V)$ from F to V. Now we are ready to derive an equation for a parabola. To do this, we use a rectangular system of coordinates positioned so that the vertex V, focus F, and directrix D of the parabola are conveniently located. If we choose to locate the vertex V at the origin $(0, 0)$, then we can conveniently position the focus F on either the x-axis or the y-axis.

Figure 4
$y^2 = 4ax$

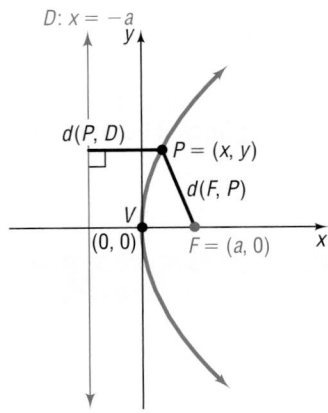

First, we consider the case where the focus F is on the positive x-axis, as shown in Figure 4. Because the distance from F to V is a, the coordinates of F will be $(a, 0)$ with $a > 0$. Similarly, because the distance from V to the directrix D is also a and because D must be perpendicular to the x-axis (since the x-axis is the axis of symmetry), the equation of the directrix D must be $x = -a$. Now, if $P = (x, y)$ is any point on the parabola, then P must obey equation (1):

$$d(F, P) = d(P, D)$$

So we have

$$\sqrt{(x - a)^2 + y^2} = |x + a| \qquad \text{Use the distance formula.}$$
$$(x - a)^2 + y^2 = (x + a)^2 \qquad \text{Square both sides.}$$
$$x^2 - 2ax + a^2 + y^2 = x^2 + 2ax + a^2 \qquad \text{Remove parentheses.}$$
$$y^2 = 4ax \qquad \text{Simplify.}$$

Theorem

Equation of a Parabola; Vertex at $(0, 0)$, Focus at $(a, 0)$, $a > 0$

The equation of a parabola with vertex at $(0, 0)$, focus at $(a, 0)$, and directrix $x = -a, a > 0$, is

$$y^2 = 4ax \qquad\qquad (2)$$

② **EXAMPLE 1**   **Finding the Equation of a Parabola and Graphing It by Hand**

Find an equation of the parabola with vertex at $(0, 0)$ and focus at $(3, 0)$. Graph the equation by hand.

Figure 5
$y^2 = 12x$

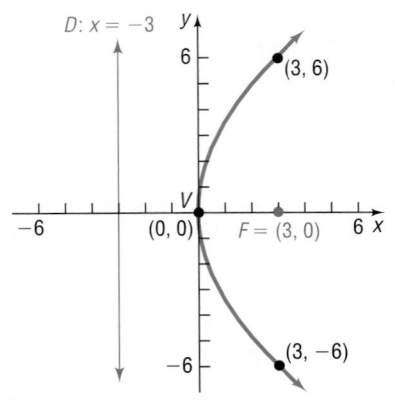

Solution   The distance from the vertex $(0, 0)$ to the focus $(3, 0)$ is $a = 3$. Based on equation (2), the equation of this parabola is

$$y^2 = 4ax$$
$$y^2 = 12x \qquad a = 3$$

To graph this parabola by hand, it is helpful to plot the two points on the graph above and below the focus. To locate them, we let $x = 3$. Then

$$y^2 = 12x = 12(3) = 36$$
$$y = \pm 6 \qquad \text{Solve for } y.$$

The points on the parabola above and below the focus are $(3, 6)$ and $(3, -6)$. These points help in graphing the parabola because they determine the "opening." See Figure 5.

In general, the points on a parabola $y^2 = 4ax$ that lie above and below the focus $(a, 0)$ are each at a distance $2a$ from the focus. This follows from the fact that if $x = a$ then $y^2 = 4ax = 4a^2$ so $y = \pm 2a$. The line segment joining these two points is called the **latus rectum**; its length is $4a$.

✏️ **NOW WORK PROBLEM 15.**

EXAMPLE 2   **Graphing a Parabola Using a Graphing Utility**

Graph the parabola $y^2 = 12x$.

Figure 6
$y^2 = 12x$

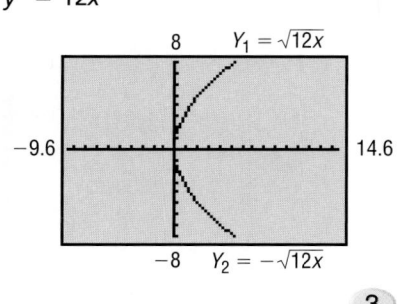

Solution   To graph the parabola $y^2 = 12x$, we need to graph the two functions $Y_1 = \sqrt{12x}$ and $Y_2 = -\sqrt{12x}$ on a square screen. Figure 6 shows the graph of $y^2 = 12x$. Notice that the graph fails the vertical line test, so $y^2 = 12x$ is not a function.

■

By reversing the steps we used to obtain equation (2), it follows that the graph of an equation of the form of equation (2), $y^2 = 4ax$, is a parabola; its vertex is at $(0, 0)$, its focus is at $(a, 0)$, its directrix is the line $x = -a$, and its axis of symmetry is the x-axis.

③   For the remainder of this section, the direction "Discuss the equation" will mean to find the vertex, focus, and directrix of the parabola and graph it.

EXAMPLE 3   **Discussing the Equation of a Parabola**

Discuss the equation: $y^2 = 8x$

Solution   The equation $y^2 = 8x$ is of the form $y^2 = 4ax$, where $4a = 8$ so that $a = 2$. Consequently, the graph of the equation is a parabola with vertex at $(0, 0)$ and focus on the positive x-axis at $(2, 0)$. The directrix is the vertical line $x = -2$. The two points defining the latus rectum are obtained by letting $x = 2$. Then $y^2 = 16$ so $y = \pm 4$. See Figure 7(a) for the graph drawn by hand. Figure 7(b) shows the graph obtained using a graphing utility.

Figure 7
$y^2 = 8x$

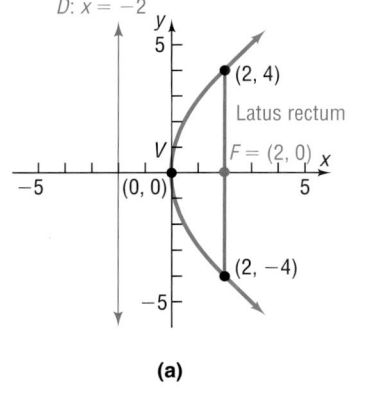

(a)

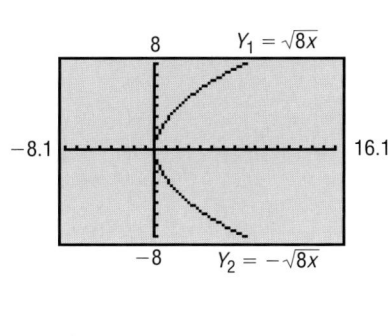

(b)

■

Recall that we arrived at equation (2) after placing the focus on the positive x-axis. If the focus is placed on the negative x-axis, positive y-axis, or negative y-axis, a different form of the equation for the parabola results. The four forms of the equation of a parabola with vertex at $(0, 0)$ and focus on a coordinate axis a distance a from $(0, 0)$ are given in Table 1, and their graphs are given in Figure 8. Notice that each graph is symmetric with respect to its axis of symmetry.

TABLE 1
Equations of a Parabola: Vertex at $(0, 0)$; Focus on an Axis; $a > 0$

Vertex	Focus	Directrix	Equation	Description
$(0, 0)$	$(a, 0)$	$x = -a$	$y^2 = 4ax$	Parabola, axis of symmetry is the x-axis, opens to right
$(0, 0)$	$(-a, 0)$	$x = a$	$y^2 = -4ax$	Parabola, axis of symmetry is the x-axis, opens to left
$(0, 0)$	$(0, a)$	$y = -a$	$x^2 = 4ay$	Parabola, axis of symmetry is the y-axis, opens up
$(0, 0)$	$(0, -a)$	$y = a$	$x^2 = -4ay$	Parabola, axis of symmetry is the y-axis, opens down

Figure 8

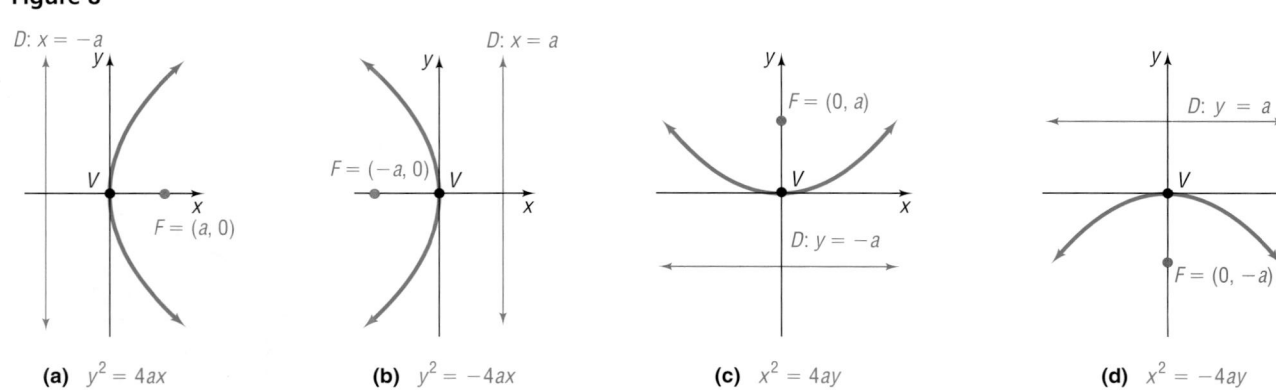

(a) $y^2 = 4ax$ (b) $y^2 = -4ax$ (c) $x^2 = 4ay$ (d) $x^2 = -4ay$

EXAMPLE 4   **Discussing the Equation of a Parabola**

Discuss the equation: $x^2 = -12y$

Solution   The equation $x^2 = -12y$ is of the form $x^2 = -4ay$, with $a = 3$. Consequently, the graph of the equation is a parabola with vertex at $(0, 0)$, focus at $(0, -3)$, and directrix the line $y = 3$. The parabola opens down, and its axis of symmetry is the y-axis. To obtain the points defining the latus rectum, let $y = -3$. Then $x^2 = 36$ so $x = \pm 6$. See Figure 9(a) for the graph drawn by hand. Figure 9(b) shows the graph obtained using a graphing utility.

Figure 9
$x^2 = -12y$

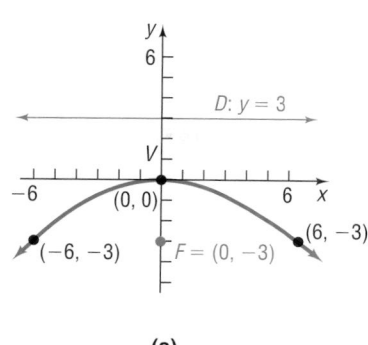

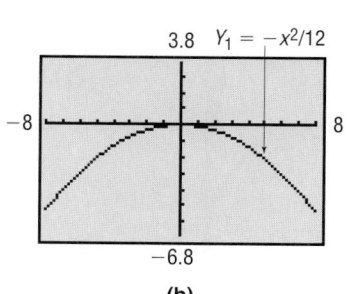

(a)   (b)

NOW WORK PROBLEM **33.**

EXAMPLE 5   Finding the Equation of a Parabola

Find the equation of the parabola with focus at $(0, 4)$ and directrix the line $y = -4$. Graph the equation by hand.

Solution   A parabola whose focus is at $(0, 4)$ and whose directrix is the horizontal line $y = -4$ will have its vertex at $(0, 0)$. (Do you see why? The vertex is midway between the focus and the directrix.) Since the focus is on the positive y-axis at $(0, 4)$, the equation of this parabola is of the form $x^2 = 4ay$, with $a = 4$;

$$x^2 = 4ay = 4(4)y = 16y$$

$$\uparrow$$
$$a = 4$$

Figure 10 shows the graph of $x^2 = 16y$.

Figure 10
$x^2 = 16y$

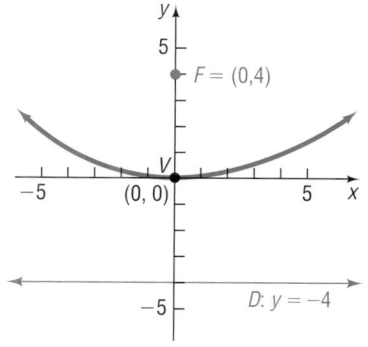

EXAMPLE 6   Finding the Equation of a Parabola

Find the equation of a parabola with vertex at $(0, 0)$ if its axis of symmetry is the x-axis and its graph contains the point $\left(-\dfrac{1}{2}, 2\right)$. Find its focus and directrix, and graph the equation by hand.

Solution   The vertex is at the origin, the axis of symmetry is the x-axis, and the graph contains a point in the second quadrant, so the parabola opens to the left. We see from Table 1 that the form of the equation is

$$y^2 = -4ax$$

Figure 11
$y^2 = -8x$

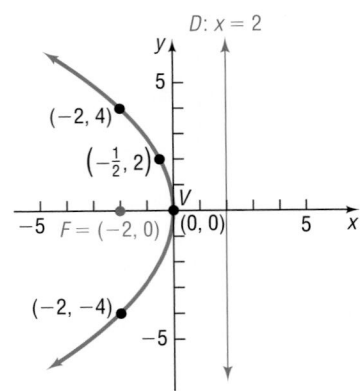

Because the point $\left(-\dfrac{1}{2}, 2\right)$ is on the parabola, the coordinates $x = -\dfrac{1}{2}, y = 2$ must satisfy the equation. Substituting $x = -\dfrac{1}{2}$ and $y = 2$ into the equation, we find that

$$4 = -4a\left(-\dfrac{1}{2}\right) \qquad y^2 = -4ax, x = -\dfrac{1}{2}, y = 2$$

$$a = 2$$

The equation of the parabola is

$$y^2 = -4(2)x = -8x$$

The focus is at $(-2, 0)$ and the directrix is the line $x = 2$. Letting $x = -2$, we find $y^2 = 16$ so $y = \pm 4$. The points $(-2, 4)$ and $(-2, -4)$ define the latus rectum. See Figure 11. ■

NOW WORK PROBLEM **25**.

Vertex at (h, k)

④   If a parabola with vertex at the origin and axis of symmetry along a coordinate axis is shifted horizontally h units and then vertically k units, the result is a parabola with vertex at (h, k) and axis of symmetry parallel to a coordinate axis. The equations of such parabolas have the same forms as those in Table 1, but with x replaced by $x - h$ (the horizontal shift) and y replaced by $y - k$ (the vertical shift). Table 2 gives the forms of the equations of such parabolas. Figure 12(a)–(d) illustrates the graphs for $h > 0, k > 0$.

Figure 12

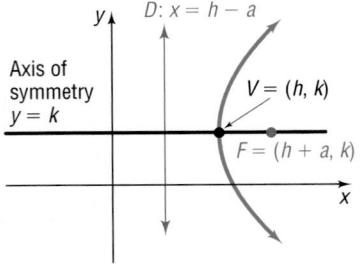

(a) $(y - k)^2 = 4a(x - h)$

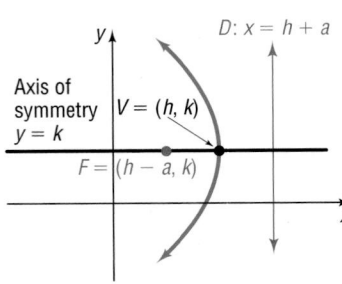

(b) $(y - k)^2 = -4a(x - h)$

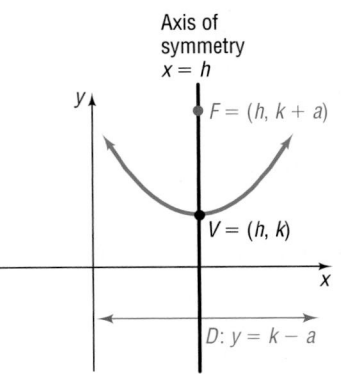

(c) $(x - h)^2 = 4a(y - k)$

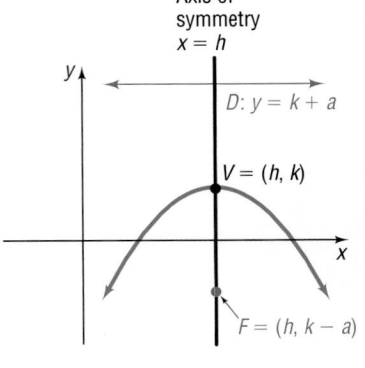

(d) $(x - h)^2 = -4a(y - k)$

TABLE 2
Parabolas with Vertex at (h, k); Axis of Symmetry Parallel to a Coordinate Axis, $a > 0$

Vertex	Focus	Directrix	Equation	Description
(h, k)	$(h + a, k)$	$x = h - a$	$(y - k)^2 = 4a(x - h)$	Parabola, axis of symmetry parallel to x-axis, opens to right
(h, k)	$(h - a, k)$	$x = h + a$	$(y - k)^2 = -4a(x - h)$	Parabola, axis of symmetry parallel to x-axis, opens to left
(h, k)	$(h, k + a)$	$y = k - a$	$(x - h)^2 = 4a(y - k)$	Parabola, axis of symmetry parallel to y-axis, opens up
(h, k)	$(h, k - a)$	$y = k + a$	$(x - h)^2 = -4a(y - k)$	Parabola, axis of symmetry parallel to y-axis, opens down

EXAMPLE 7   **Finding the Equation of a Parabola, Vertex Not at Origin**

Find an equation of the parabola with vertex at $(-2, 3)$ and focus at $(0, 3)$. Graph the equation by hand.

Figure 13
$(y - 3)^2 = 8(x + 2)$

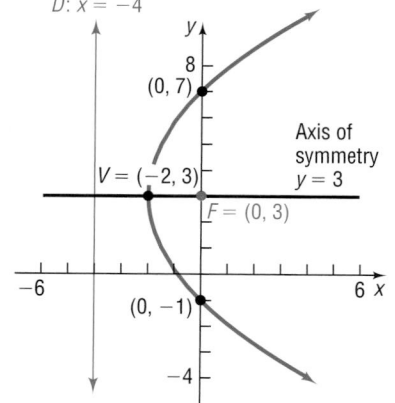

Solution   The vertex $(-2, 3)$ and focus $(0, 3)$ both lie on the horizontal line $y = 3$ (the axis of symmetry). The distance a from the vertex $(-2, 3)$ to the focus $(0, 3)$ is $a = 2$. Also, because the focus lies to the right of the vertex, we know that the parabola opens to the right. Consequently, the form of the equation is

$$(y - k)^2 = 4a(x - h)$$

where $(h, k) = (-2, 3)$ and $a = 2$. Therefore, the equation is

$$(y - 3)^2 = 4 \cdot 2[x - (-2)]$$
$$(y - 3)^2 = 8(x + 2)$$

If $x = 0$, then $(y - 3)^2 = 16$. Then, $y - 3 = \pm 4$ so $y = -1$ or $y = 7$. The points $(0, -1)$ and $(0, 7)$ define the latus rectum; the line $x = -4$ is the directrix. See Figure 13.

━ **NOW WORK PROBLEM 23.**

EXAMPLE 8   **Using a Graphing Utility to Graph a Parabola, Vertex Not at Origin**

Figure 14

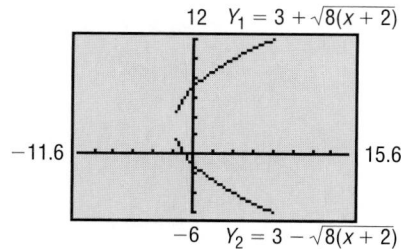

Using a graphing utility, graph the equation $(y - 3)^2 = 8(x + 2)$.

Solution   First, we must solve the equation for y.

$$(y - 3)^2 = 8(x + 2)$$
$$y - 3 = \pm\sqrt{8(x + 2)} \qquad \text{Use the Square Root Method.}$$
$$y = 3 \pm \sqrt{8(x + 2)} \qquad \text{Add 3 to both sides.}$$

Figure 14 shows the graphs of the equations $Y_1 = 3 + \sqrt{8(x + 2)}$ and $Y_2 = 3 - \sqrt{8(x + 2)}$.

━ **NOW WORK PROBLEM 39.**

Polynomial equations define parabolas whenever they involve two variables that are quadratic in one variable and linear in the other. To discuss this type of equation, we first complete the square of the variable that is quadratic.

EXAMPLE 9   Discussing the Equation of a Parabola

Discuss the equation: $x^2 + 4x - 4y = 0$

Solution To discuss the equation $x^2 + 4x - 4y = 0$, we complete the square involving the variable x.

$$x^2 + 4x - 4y = 0$$
$$x^2 + 4x = 4y \qquad \text{Isolate the terms involving x on the left side.}$$
$$x^2 + 4x + 4 = 4y + 4 \qquad \text{Complete the square on the left side.}$$
$$(x + 2)^2 = 4(y + 1) \qquad \text{Factor.}$$

This equation is of the form $(x - h)^2 = 4a(y - k)$, with $h = -2, k = -1$, and $a = 1$. The graph is a parabola with vertex at $(h, k) = (-2, -1)$ that opens up. The focus is at $(-2, 0)$, and the directrix is the line $y = -2$. See Figure 15. ■

Figure 15
$x^2 + 4x - 4y = 0$

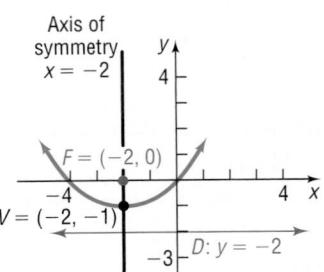

NOW WORK PROBLEM **41.**

⑤   Parabolas find their way into many applications. For example, suspension bridges have cables in the shape of a parabola. Another property of parabolas that is used in applications is their reflecting property.

Reflecting Property

Suppose that a mirror is shaped like a **paraboloid of revolution**, a surface formed by rotating a parabola about its axis of symmetry. If a light (or any other emitting source) is placed at the focus of the parabola, all the rays emanating from the light will reflect off the mirror in lines parallel to the axis of symmetry. This principle is used in the design of searchlights, flashlights, certain automobile headlights, and other such devices. See Figure 16.

Conversely, suppose that rays of light (or other signals) emanate from a distant source so that they are essentially parallel. When these rays strike the surface of a parabolic mirror whose axis of symmetry is parallel to these rays, they are reflected to a single point at the focus. This principle is used in the design of some solar energy devices, satellite dishes, and the mirrors used in some types of telescopes. See Figure 17.

Figure 16
Searchlight

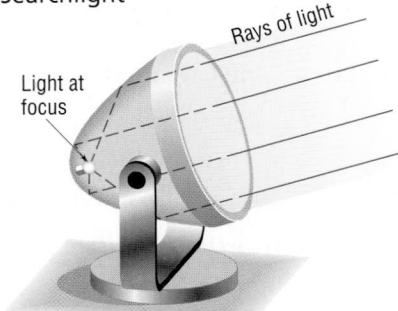

Figure 17
Telescope

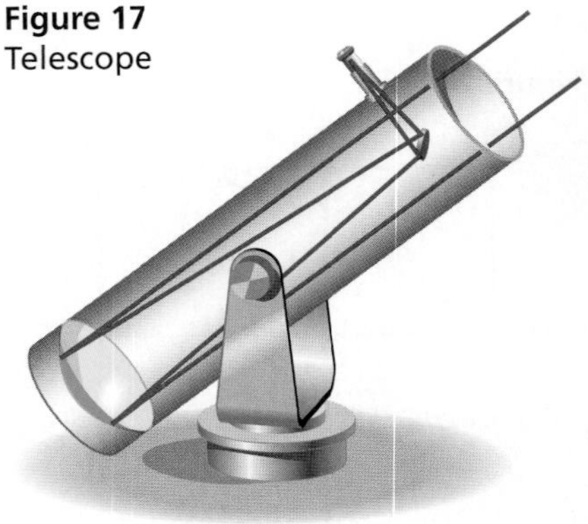

EXAMPLE 10   **Satellite Dish**

A satellite dish is shaped like a paraboloid of revolution. The signals that emanate from a satellite strike the surface of the dish and are reflected to a single point, where the receiver is located. If the dish is 8 feet across at its opening and is 3 feet deep at its center, at what position should the receiver be placed?

Solution   Figure 18(a) shows the satellite dish. We draw the parabola used to form the dish on a rectangular coordinate system so that the vertex of the parabola is at the origin and its focus is on the positive y-axis. See Figure 18(b).

Figure 18

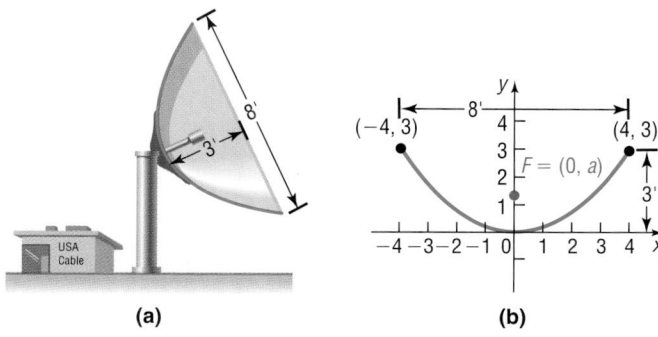

(a)   (b)

The form of the equation of the parabola is

$$x^2 = 4ay$$

and its focus is at $(0, a)$. Since $(4, 3)$ is a point on the graph, we have

$$4^2 = 4a(3)$$

$$a = \frac{4}{3}$$

The receiver should be located $1\frac{1}{3}$ feet from the base of the dish, along its axis of symmetry.   ■

 NOW WORK PROBLEM **57.**

6.2 Concepts and Vocabulary

In Problems 1–3, fill in the blanks.

1. A(n) _____ is the collection of all points in the plane such that the distance from each point to a fixed point equals its distance to a fixed line.

2. The surface formed by rotating a parabola about its axis of symmetry is called a _____ _____ _____.

3. The line segment joining the two points on a parabola above and below its focus is called the _____ _____.

In Problems 4–6, answer True or False to each statement.

4. The vertex of a parabola is a point on the parabola that also is on its axis of symmetry.

5. If a light is placed at the focus of a parabola, all the rays reflected off the parabola will be parallel to the axis of symmetry.

6. The graph of a quadratic function is a parabola.

7. Write down the four equations that are parabolas with vertex at the origin and axis along a coordinate axis.

8. Draw a parabola, labeling its vertex, axis of symmetry, focus, and directrix.

6.2 Exercises

In Problems 1–8, the graph of a parabola is given. Match each graph to its equation.

A. $y^2 = 4x$ C. $y^2 = -4x$ E. $(y - 1)^2 = 4(x - 1)$ G. $(y - 1)^2 = -4(x - 1)$

B. $x^2 = 4y$ D. $x^2 = -4y$ F. $(x + 1)^2 = 4(y + 1)$ H. $(x + 1)^2 = -4(y + 1)$

1.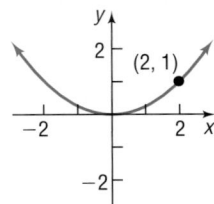

2.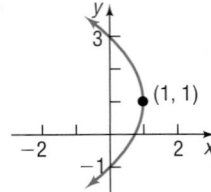

3.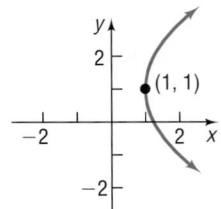

4.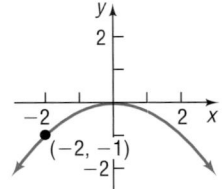

5.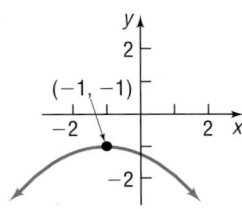

6.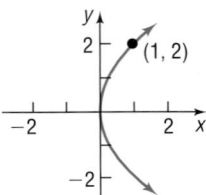

7.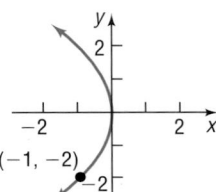

8.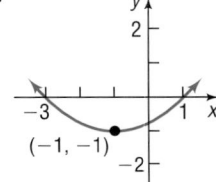

In Problems 9–14, the graph of a parabola is given. Match each graph to its equation.

A. $x^2 = 6y$ B. $y^2 = 6x$ C. $y^2 = -6x$

D. $(x + 2)^2 = -6(y - 2)$ E. $(y - 2)^2 = 6(x + 2)$ F. $(x + 2)^2 = 6(y - 2)$

9.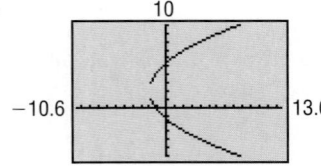

10.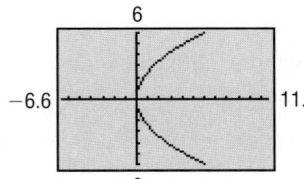

11.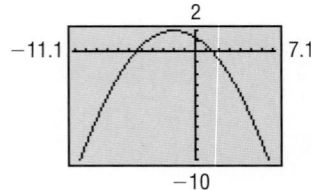

12.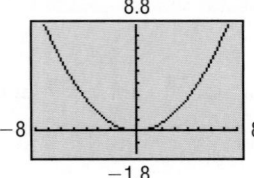

13.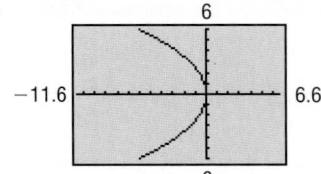

14.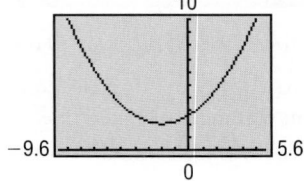

In Problems 15–30, find the equation of the parabola described. Find the two points that define the latus rectum, and graph the equation by hand.

15. Focus at $(4, 0)$;   vertex at $(0, 0)$

16. Focus at $(0, 2)$;   vertex at $(0, 0)$

17. Focus at $(0, -3)$;   vertex at $(0, 0)$

18. Focus at $(-4, 0)$;   vertex at $(0, 0)$

19. Focus at $(-2, 0)$;   directrix the line $x = 2$

20. Focus at $(0, -1)$;   directrix the line $y = 1$

21. Directrix the line $y = -\dfrac{1}{2}$;   vertex at $(0, 0)$

22. Directrix the line $x = -\dfrac{1}{2}$;   vertex at $(0, 0)$

23. Vertex at $(2, -3)$;   focus at $(2, -5)$

24. Vertex at $(4, -2)$;   focus at $(6, -2)$

25. Vertex at $(0, 0)$;   axis of symmetry the y-axis; containing the point $(2, 3)$

26. Vertex at $(0, 0)$;   axis of symmetry the x-axis; containing the point $(2, 3)$

27. Focus at $(-3, 4)$;   directrix the line $y = 2$

28. Focus at $(2, 4)$;   directrix the line $x = -4$

29. Focus at $(-3, -2)$;   directrix the line $x = 1$

30. Focus at $(-4, 4)$;   directrix the line $y = -2$

In Problems 31–48, find the vertex, focus, and directrix of each parabola. Graph the equation (a) by hand and (b) by using a graphing utility.

31. $x^2 = 4y$

32. $y^2 = 8x$

33. $y^2 = -16x$

34. $x^2 = -4y$

35. $(y - 2)^2 = 8(x + 1)$

36. $(x + 4)^2 = 16(y + 2)$

37. $(x - 3)^2 = -(y + 1)$

38. $(y + 1)^2 = -4(x - 2)$

39. $(y + 3)^2 = 8(x - 2)$

40. $(x - 2)^2 = 4(y - 3)$

41. $y^2 - 4y + 4x + 4 = 0$

42. $x^2 + 6x - 4y + 1 = 0$

43. $x^2 + 8x = 4y - 8$

44. $y^2 - 2y = 8x - 1$

45. $y^2 + 2y - x = 0$

46. $x^2 - 4x = 2y$

47. $x^2 - 4x = y + 4$

48. $y^2 + 12y = -x + 1$

In Problems 49–56, write an equation for each parabola.

49.
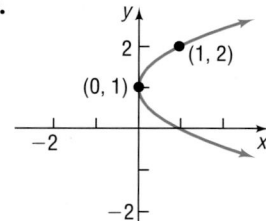

50.
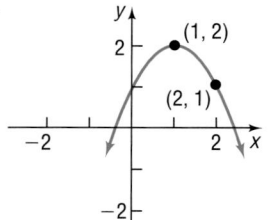

51.
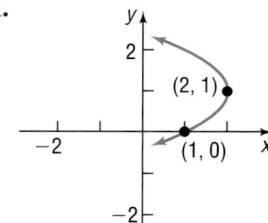

52.
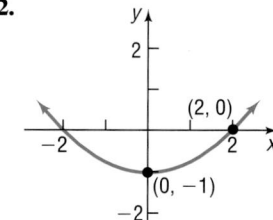

53.
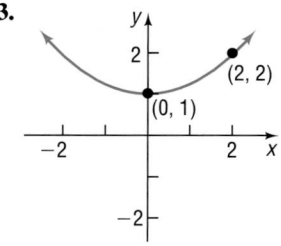

54.
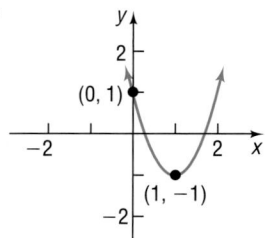

55.

56.
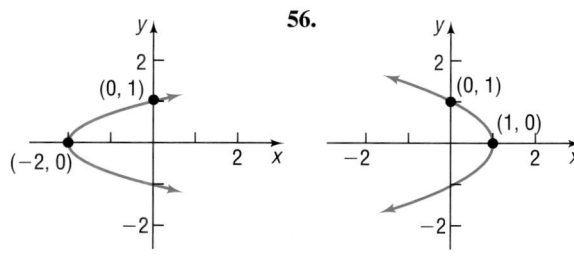

57. Satellite Dish   A satellite dish is shaped like a paraboloid of revolution. The signals that emanate from a satellite strike the surface of the dish and are reflected to a single point, where the receiver is located. If the dish is 10 feet across at its opening and is 4 feet deep at its center, at what position should the receiver be placed?

58. Constructing a TV Dish   A cable TV receiving dish is in the shape of a paraboloid of revolution. Find the location

of the receiver, which is placed at the focus, if the dish is 6 feet across at its opening and 2 feet deep.

59. Constructing a Flashlight   The reflector of a flashlight is in the shape of a paraboloid of revolution. Its diameter is 4 inches and its depth is 1 inch. How far from the vertex should the light bulb be placed so that the rays will be reflected parallel to the axis?

60. Constructing a Headlight A sealed-beam headlight is in the shape of a paraboloid of revolution. The bulb, which is placed at the focus, is 1 inch from the vertex. If the depth is to be 2 inches, what is the diameter of the headlight at its opening?

61. Suspension Bridge The cables of a suspension bridge are in the shape of a parabola, as shown in the figure. The towers supporting the cable are 600 feet apart and 80 feet high. If the cables touch the road surface midway between the towers, what is the height of the cable at a point 150 feet from the center of the bridge?

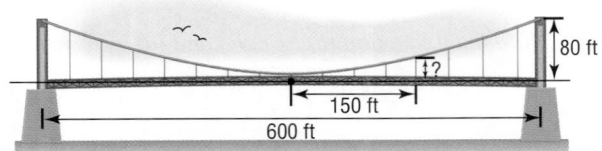

62. Suspension Bridge The cables of a suspension bridge are in the shape of a parabola. The towers supporting the cable are 400 feet apart and 100 feet high. If the cables are at a height of 10 feet midway between the towers, what is the height of the cable at a point 50 feet from the center of the bridge?

63. Searchlight A searchlight is shaped like a paraboloid of revolution. If the light source is located 2 feet from the base along the axis of symmetry and the opening is 5 feet across, how deep should the searchlight be?

64. Searchlight A searchlight is shaped like a paraboloid of revolution. If the light source is located 2 feet from the base along the axis of symmetry and the depth of the searchlight is 4 feet, what should the width of the opening be?

65. Solar Heat A mirror is shaped like a paraboloid of revolution and will be used to concentrate the rays of the sun at its focus, creating a heat source. (See the figure.) If the mirror is 20 feet across at its opening and is 6 feet deep, where will the heat source be concentrated?

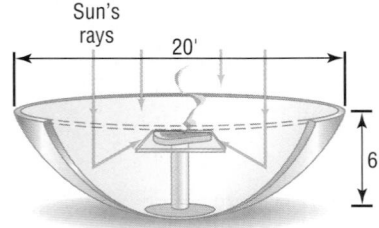

66. Reflecting Telescope A reflecting telescope contains a mirror shaped like a paraboloid of revolution. If the mirror is 4 inches across at its opening and is 3 feet deep, where will the collected light be concentrated?

67. Parabolic Arch Bridge A bridge is built in the shape of a parabolic arch. The bridge has a span of 120 feet and a maximum height of 25 feet. See the illustration. Choose a suitable rectangular coordinate system and find the height of the arch at distances of 10, 30, and 50 feet from the center.

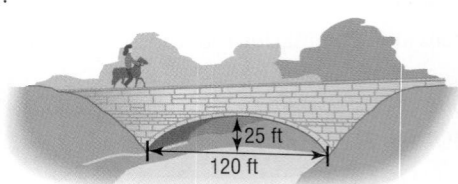

68. Parabolic Arch Bridge A bridge is to be built in the shape of a parabolic arch and is to have a span of 100 feet. The height of the arch a distance of 40 feet from the center is to be 10 feet. Find the height of the arch at its center.

69. Show that an equation of the form
$$Ax^2 + Ey = 0, \qquad A \neq 0, E \neq 0$$
is the equation of a parabola with vertex at $(0,0)$ and axis of symmetry the y-axis. Find its focus and directrix.

70. Show that an equation of the form
$$Cy^2 + Dx = 0, \quad C \neq 0, D \neq 0$$
is the equation of a parabola with vertex at $(0,0)$ and axis of symmetry the x-axis. Find its focus and directrix.

71. Show that the graph of an equation of the form
$$Ax^2 + Dx + Ey + F = 0, \quad A \neq 0$$
(a) Is a parabola if $E \neq 0$.
(b) Is a vertical line if $E = 0$ and $D^2 - 4AF = 0$.
(c) Is two vertical lines if $E = 0$ and $D^2 - 4AF > 0$.
(d) Contains no points if $E = 0$ and $D^2 - 4AF < 0$.

72. Show that the graph of an equation of the form
$$Cy^2 + Dx + Ey + F = 0, \quad C \neq 0$$
(a) Is a parabola if $D \neq 0$.
(b) Is a horizontal line if $D = 0$ and $E^2 - 4CF = 0$.
(c) Is two horizontal lines if $D = 0$ and $E^2 - 4CF > 0$.
(d) Contains no points if $D = 0$ and $E^2 - 4CF < 0$.

PREPARING FOR THIS SECTION

Before getting started, review the following:

✓ Distance Formula (Section 1.1, p. 5)

✓ Completing the Square (Section A.3, pp. 579–580)

✓ Intercepts (Section 1.2, pp. 17–18)

✓ Square Root Method (Section A.3, p. 578)

✓ Symmetry (Section 1.3, pp. 20–22)

✓ Circles (Section 1.3, pp. 26–30)

✓ Graphing Techniques: Transformations (Section 1.6, pp. 64–73)

6.3 THE ELLIPSE

OBJECTIVES
1. Find the Equation of an Ellipse
2. Graph Ellipses
3. Discuss the Equation of an Ellipse
4. Work with Ellipses with Center at (h, k)
5. Solve Applied Problems Involving Ellipses

An **ellipse** is the collection of all points in the plane the sum of whose distances from two fixed points, called the **foci**, is a constant.

The definition actually contains within it a physical means for drawing an ellipse. Find a piece of string (the length of this string is the constant referred to in the definition). Then take two thumbtacks (the foci) and stick them on a piece of cardboard so that the distance between them is less than the length of the string. Now attach the ends of the string to the thumbtacks and, using the point of a pencil, pull the string taut. See Figure 19. Keeping the string taut, rotate the pencil around the two thumbtacks. The pencil traces out an ellipse, as shown in Figure 19.

In Figure 19, the foci are labeled F_1 and F_2. The line containing the foci is called the **major axis**. The midpoint of the line segment joining the foci is called the **center** of the ellipse. The line through the center and perpendicular to the major axis is called the **minor axis**.

The two points of intersection of the ellipse and the major axis are the **vertices**, V_1 and V_2, of the ellipse. The distance from one vertex to the other is called the **length of the major axis**. The ellipse is symmetric with respect to its major axis and with respect to its minor axis.

Figure 19

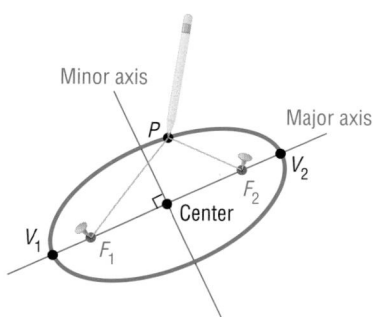

Figure 20
$d(F_1, P) + d(F_2, P) = 2a$

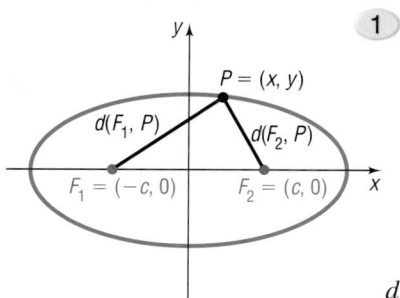

With these ideas in mind, we are now ready to find the equation of an ellipse in a rectangular coordinate system. First, we place the center of the ellipse at the origin. Second, we position the ellipse so that its major axis coincides with a coordinate axis. Suppose that the major axis coincides with the x-axis, as shown in Figure 20. If c is the distance from the center to a focus, then one focus will be at $F_1 = (-c, 0)$ and the other at $F_2 = (c, 0)$. As we shall see, it is convenient to let $2a$ denote the constant distance referred to in the definition. Then, if $P = (x, y)$ is any point on the ellipse, we have

$$d(F_1, P) + d(F_2, P) = 2a$$
Sum of the distances from P to the foci equals a constant, $2a$.

$$\sqrt{(x + c)^2 + y^2} + \sqrt{(x - c)^2 + y^2} = 2a$$
Use the distance formula.

$$\sqrt{(x + c)^2 + y^2} = 2a - \sqrt{(x - c)^2 + y^2}$$
Isolate one radical.

$$(x + c)^2 + y^2 = 4a^2 - 4a\sqrt{(x - c)^2 + y^2} + (x - c)^2 + y^2$$
Square both sides.

$$x^2 + 2cx + c^2 + y^2 = 4a^2 - 4a\sqrt{(x - c)^2 + y^2} + x^2 - 2cx + c^2 + y^2$$
Remove parentheses.

$$4cx - 4a^2 = -4a\sqrt{(x - c)^2 + y^2}$$
Simplify; Isolate the radical.

$$cx - a^2 = -a\sqrt{(x - c)^2 + y^2}$$
Divide each side by 4.

$$(cx - a^2)^2 = a^2[(x - c)^2 + y^2]$$
Square both sides again.

$$c^2x^2 - 2a^2cx + a^4 = a^2(x^2 - 2cx + c^2 + y^2)$$ Remove parentheses.

$$(c^2 - a^2)x^2 - a^2y^2 = a^2c^2 - a^4$$ Rearrange the terms.

$$(a^2 - c^2)x^2 + a^2y^2 = a^2(a^2 - c^2) \qquad \text{(1)}$$ Multiply each side by -1; factor a^2 on the right side.

To obtain points on the ellipse off the x-axis, it must be that $a > c$. To see why, look again at Figure 20.

$$d(F_1, P) + d(F_2, P) > d(F_1, F_2)$$ The sum of the lengths of two sides of a triangle is greater than the length of the third side.

$$2a > 2c$$ $d(F_1, P) + d(F_2, P) = 2a; d(F_1, F_2) = 2c.$

$$a > c$$

Since $a > c$, we also have $a^2 > c^2$, so $a^2 - c^2 > 0$. Let $b^2 = a^2 - c^2, b > 0$. Then $a > b$ and equation (1) can be written as

$$b^2x^2 + a^2y^2 = a^2b^2$$

$$\frac{x^2}{a^2} + \frac{y^2}{b^2} = 1$$ Divide each side by a^2b^2.

Theorem

Equation of an Ellipse; Center at $(0,0)$; Foci at $(\pm c, 0)$; Major Axis along the x-Axis

An equation of the ellipse with center at $(0,0)$, foci at $(-c, 0)$ and $(c, 0)$, and vertices at $(-a, 0)$ and $(a, 0)$ is

$$\frac{x^2}{a^2} + \frac{y^2}{b^2} = 1, \qquad \text{where } a > b > 0 \text{ and } b^2 = a^2 - c^2 \quad \text{(2)}$$

The major axis is the x-axis.

As you can verify, the ellipse defined by equation (2) is symmetric with respect to the x-axis, y-axis, and origin.

To find the vertices of the ellipse defined by equation (2), let $y = 0$. The vertices satisfy the equation $\frac{x^2}{a^2} = 1$, the solutions of which are $x = \pm a$. Consequently, the vertices of the ellipse given by equation (2) are $V_1 = (-a, 0)$ and $V_2 = (a, 0)$. The y-intercepts of the ellipse, found by letting $x = 0$, have coordinates $(0, -b)$ and $(0, b)$. These four intercepts, $(a, 0)$, $(-a, 0)$, $(0, b)$, and $(0, -b)$, are used to graph the ellipse. See Figure 21.

Notice in Figure 21 the right triangle formed with the points $(0, 0)$, $(c, 0)$, and $(0, b)$. Because $b^2 = a^2 - c^2$ (or $b^2 + c^2 = a^2$), the distance from the focus at $(c, 0)$ to the point $(0, b)$ is a.

Figure 21

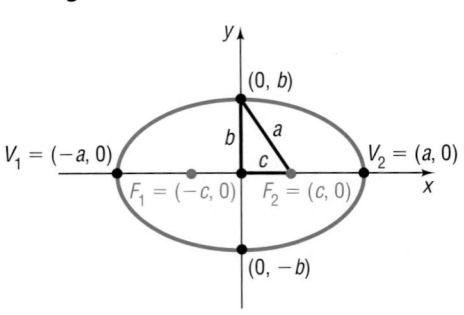

EXAMPLE 1 Finding an Equation of an Ellipse

Find an equation of the ellipse with center at the origin, one focus at $(3, 0)$, and a vertex at $(-4, 0)$. Graph the equation by hand.

Figure 22

$$\frac{x^2}{16} + \frac{y^2}{7} = 1$$

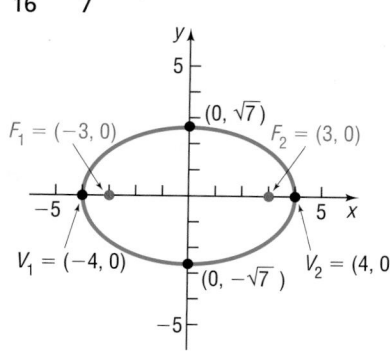

$F_1 = (-3, 0)$ $(0, \sqrt{7})$ $F_2 = (3, 0)$

$V_1 = (-4, 0)$ $(0, -\sqrt{7})$ $V_2 = (4, 0)$

Solution The ellipse has its center at the origin and, since the given focus and vertex lie on the x-axis, the major axis is the x-axis. The distance from the center, $(0, 0)$, to one of the foci, $(3, 0)$, is $c = 3$. The distance from the center, $(0, 0)$, to one of the vertices, $(-4, 0)$, is $a = 4$. From equation (2), it follows that

$$b^2 = a^2 - c^2 = 16 - 9 = 7$$

so an equation of the ellipse is

$$\frac{x^2}{16} + \frac{y^2}{7} = 1$$

Figure 22 shows the graph drawn by hand.

Notice in Figure 22 how we used the intercepts of the equation to graph the ellipse. Following this practice will make it easier for you to obtain an accurate graph of an ellipse when graphing by hand. It also tells you how to set the initial viewing window when using a graphing utility.

EXAMPLE 2 **Graphing an Ellipse Using a Graphing Utility**

Use a graphing utility to graph the ellipse $\dfrac{x^2}{16} + \dfrac{y^2}{7} = 1$.

Figure 23

$3 \quad Y_1 = \sqrt{7\left(1 - \frac{x^2}{16}\right)}$

$-4.5 \qquad\qquad 4.5$

-3

$Y_2 = -\sqrt{7\left(1 - \frac{x^2}{16}\right)}$

Solution First, we must solve $\dfrac{x^2}{16} + \dfrac{y^2}{7} = 1$ for y.

$$\frac{y^2}{7} = 1 - \frac{x^2}{16} \qquad \text{Subtract } \tfrac{x^2}{16} \text{ from each side.}$$

$$y^2 = 7\left(1 - \frac{x^2}{16}\right) \qquad \text{Multiply both sides by 7.}$$

$$y = \pm\sqrt{7\left(1 - \frac{x^2}{16}\right)} \qquad \text{Apply the Square Root Method.}$$

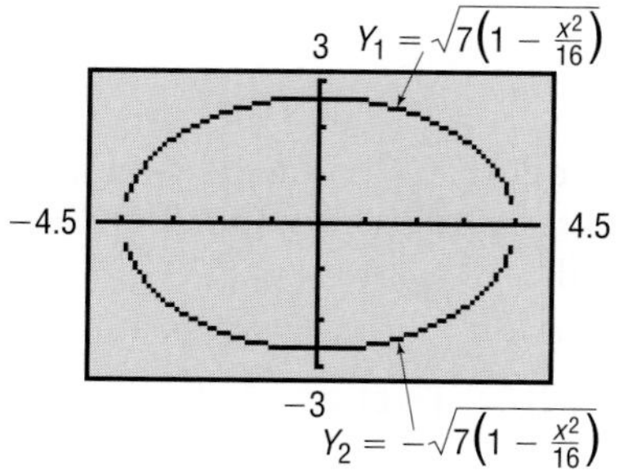

Figure 23* shows the graphs of $Y_1 = \sqrt{7\left(1 - \dfrac{x^2}{16}\right)}$ and $Y_2 = -\sqrt{7\left(1 - \dfrac{x^2}{16}\right)}$.

Notice in Figure 23 that we used a square screen. As with circles and parabolas, this is done to avoid a distorted view of the graph.

─ NOW WORK PROBLEM 19.

An equation of the form of equation (2), with $a > b$, is the equation of an ellipse with center at the origin, foci on the x-axis at $(-c, 0)$ and $(c, 0)$, where $c^2 = a^2 - b^2$, and major axis along the x-axis.

③   For the remainder of this section, the direction "Discuss the equation" will mean to find the center, major axis, foci, and vertices of the ellipse and graph it.

*The initial viewing window selected was Xmin $= -4$, Xmax $= 4$, Ymin $= -3$, Ymax $= 3$. Then we used the ZOOM-SQUARE option to obtain the window shown.

EXAMPLE 3   **Discussing the Equation of an Ellipse**

Discuss the equation: $\dfrac{x^2}{25} + \dfrac{y^2}{9} = 1$

Solution   The given equation is of the form of equation (2), with $a^2 = 25$ and $b^2 = 9$. The equation is that of an ellipse with center $(0,0)$ and major axis along the x-axis. The vertices are at $(\pm a, 0) = (\pm 5, 0)$. Because $b^2 = a^2 - c^2$, we find that

$$c^2 = a^2 - b^2 = 25 - 9 = 16$$

The foci are at $(\pm c, 0) = (\pm 4, 0)$. Figure 24(a) shows the graph drawn by hand. Figure 24(b) shows the graph obtained using a graphing utility.

Figure 24
$$\dfrac{x^2}{25} + \dfrac{y^2}{9} = 1$$

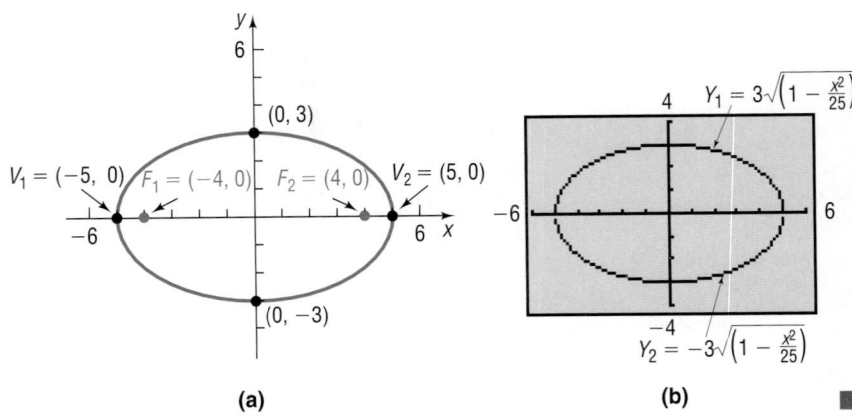

(a)   (b)

NOW WORK PROBLEM **9**.

If the major axis of an ellipse with center at $(0,0)$ lies on the y-axis, then the foci are at $(0, -c)$ and $(0, c)$. Using the same steps as before, the definition of an ellipse leads to the following result:

Theorem   **Equation of an Ellipse; Center at $(0,0)$; Foci at $(0, \pm c)$; Major Axis along the y-Axis**

An equation of the ellipse with center at $(0,0)$, foci at $(0, -c)$ and $(0, c)$, and vertices at $(0, -a)$ and $(0, a)$ is

$$\dfrac{x^2}{b^2} + \dfrac{y^2}{a^2} = 1, \qquad \text{where } a > b > 0 \text{ and } b^2 = a^2 - c^2 \quad (3)$$

The major axis is the y-axis.

Figure 25

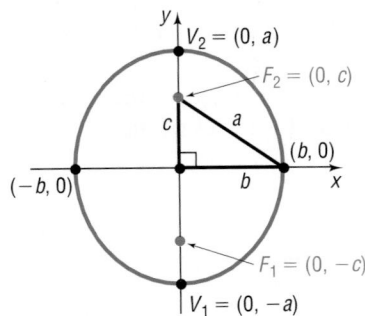

Figure 25 illustrates the graph of such an ellipse. Again, notice the right triangle with the points at $(0,0)$, $(b,0)$, and $(0,c)$.

Look closely at equations (2) and (3). Although they may look alike, there is a difference! In equation (2), the larger number, a^2, is in the denominator of the x^2-term, so the major axis of the ellipse is along the x-axis.

In equation (3), the larger number, a^2, is in the denominator of the y^2-term, so the major axis is along the y-axis.

EXAMPLE 4 Discussing the Equation of an Ellipse

Discuss the equation: $9x^2 + y^2 = 9$

Solution To put the equation in proper form, we divide each side by 9.

$$x^2 + \frac{y^2}{9} = 1$$

The larger number, 9, is in the denominator of the y^2-term so, based on equation (3), this is the equation of an ellipse with center at the origin and major axis along the y-axis. Also, we conclude that $a^2 = 9, b^2 = 1$, and $c^2 = a^2 - b^2 = 9 - 1 = 8$. The vertices are at $(0, \pm a) = (0, \pm 3)$, and the foci are at $(0, \pm c) = (0, \pm 2\sqrt{2})$. Figure 26(a) shows the graph drawn by hand. Figure 26(b) shows the graph obtained using a graphing utility.

Figure 26

$$x^2 + \frac{y^2}{9} = 1$$

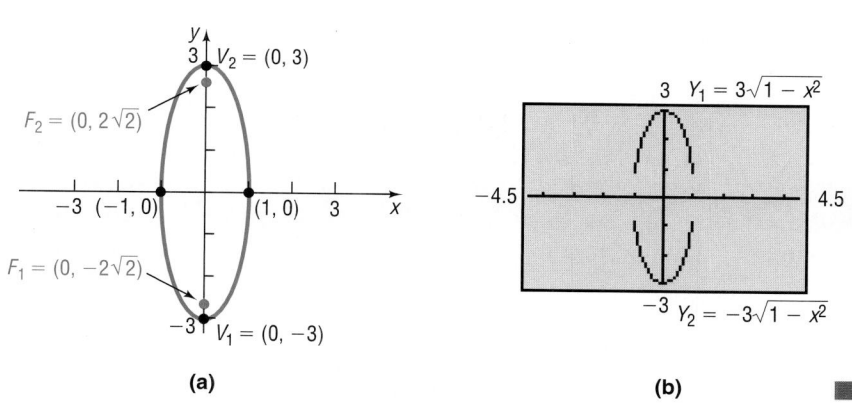

(a)   (b)

NOW WORK PROBLEM 13.

EXAMPLE 5 Finding an Equation of an Ellipse

Find an equation of the ellipse having one focus at $(0, 2)$ and vertices at $(0, -3)$ and $(0, 3)$. Graph the equation by hand.

Figure 27
$$\frac{x^2}{5} + \frac{y^2}{9} = 1$$

Solution Because the vertices are at $(0, -3)$ and $(0, 3)$, the center of this ellipse is at their midpoint, the origin. Also, its major axis lies on the y-axis. The distance from the center, $(0, 0)$, to one of the foci, $(0, 2)$, is $c = 2$. The distance from the center, $(0, 0)$, to one of the vertices, $(0, 3)$, is $a = 3$. So $b^2 = a^2 - c^2 = 9 - 4 = 5$. The form of the equation of this ellipse is given by equation (3).

$$\frac{x^2}{b^2} + \frac{y^2}{a^2} = 1$$

$$\frac{x^2}{5} + \frac{y^2}{9} = 1$$

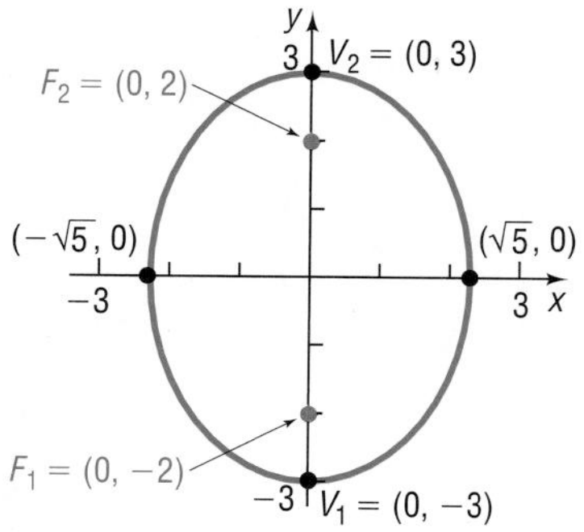

Figure 27 shows the graph.

NOW WORK PROBLEM 21.

The circle may be considered a special kind of ellipse. To see why, let $a = b$ in equation (2) or (3). Then

$$\frac{x^2}{a^2} + \frac{y^2}{a^2} = 1$$
$$x^2 + y^2 = a^2$$

This is the equation of a circle with center at the origin and radius a. The value of c is

$$c^2 = a^2 - b^2 = 0$$

We conclude that the closer the two foci of an ellipse are to the center, the more the ellipse will look like a circle.

Center at (h, k)

④   If an ellipse with center at the origin and major axis coinciding with a coordinate axis is shifted horizontally h units and then vertically k units, the result is an ellipse with center at (h, k) and major axis parallel to a coordinate axis. The equations of such ellipses have the same forms as those given in equations (2) and (3), except that x is replaced by $x - h$ (the horizontal shift) and y is replaced by $y - k$ (the vertical shift). Table 3 gives the forms of the equations of such ellipses, and Figure 28 shows their graphs.

TABLE 3 Ellipses with Center at (h, k) and Major Axis Parallel to a Coordinate Axis

Center	Major Axis	Foci	Vertices	Equation
(h, k)	Parallel to x-axis	$(h + c, k)$	$(h + a, k)$	$\dfrac{(x - h)^2}{a^2} + \dfrac{(y - k)^2}{b^2} = 1,$
		$(h - c, k)$	$(h - a, k)$	$a > b$ and $b^2 = a^2 - c^2$
(h, k)	Parallel to y-axis	$(h, k + c)$	$(h, k + a)$	$\dfrac{(x - h)^2}{b^2} + \dfrac{(y - k)^2}{a^2} = 1,$
		$(h, k - c)$	$(h, k - a)$	$a > b$ and $b^2 = a^2 - c^2$

Figure 28

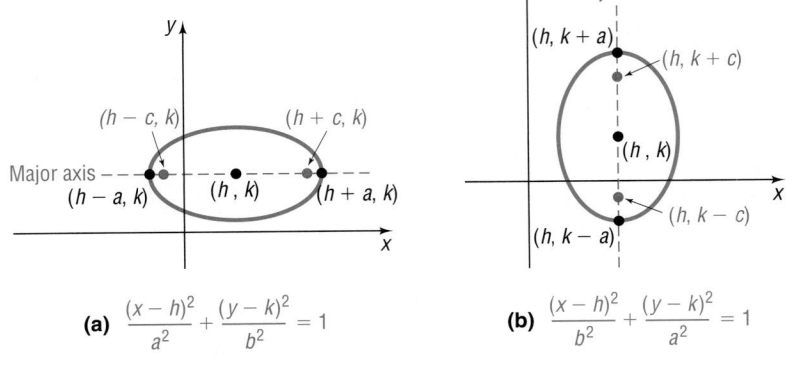

(a) $\dfrac{(x - h)^2}{a^2} + \dfrac{(y - k)^2}{b^2} = 1$ (b) $\dfrac{(x - h)^2}{b^2} + \dfrac{(y - k)^2}{a^2} = 1$

EXAMPLE 6   Finding an Equation of an Ellipse, Center Not at the Origin

Find an equation for the ellipse with center at $(2, -3)$, one focus at $(3, -3)$, and one vertex at $(5, -3)$. Graph the equation by hand.

Solution   The center is at $(h, k) = (2, -3)$, so $h = 2$ and $k = -3$. Since the center, focus, and vertex all lie on the line $y = -3$, the major axis is parallel to the x-axis. The distance from the center $(2, -3)$ to a focus $(3, -3)$ is $c = 1$; the distance from the center $(2, -3)$ to a vertex $(5, -3)$ is $a = 3$. Then, $b^2 = a^2 - c^2 = 9 - 1 = 8$. The form of the equation is

$$\frac{(x - h)^2}{a^2} + \frac{(y - k)^2}{b^2} = 1, \qquad \text{where } h = 2, k = -3, a = 3, b = 2\sqrt{2}$$

$$\frac{(x - 2)^2}{9} + \frac{(y + 3)^2}{8} = 1$$

Figure 29

$$\frac{(x - 2)^2}{9} + \frac{(y + 3)^2}{8} = 1$$

To graph the equation, we use the center $(h, k) = (2, -3)$ to locate the vertices. The major axis is parallel to the x-axis, so the vertices are $a = 3$ units left and right of the center $(2, -3)$. Therefore, the vertices are

$$V_1 = (2 - 3, -3) = (-1, -3) \quad \text{and} \quad V_2 = (2 + 3, -3) = (5, -3)$$

Since $c = 1$ and the major axis is parallel to the x-axis, the foci are 1 unit left and right of the center. Therefore, the foci are

$$F_1 = (2 - 1, -3) = (1, -3) \quad \text{and} \quad F_2 = (2 + 1, -3) = (3, -3)$$

Finally, we use the value of $b = 2\sqrt{2}$ to find the two points above and below the center.

$$(2, -3 - 2\sqrt{2}) \quad \text{and} \quad (2, -3 + 2\sqrt{2})$$

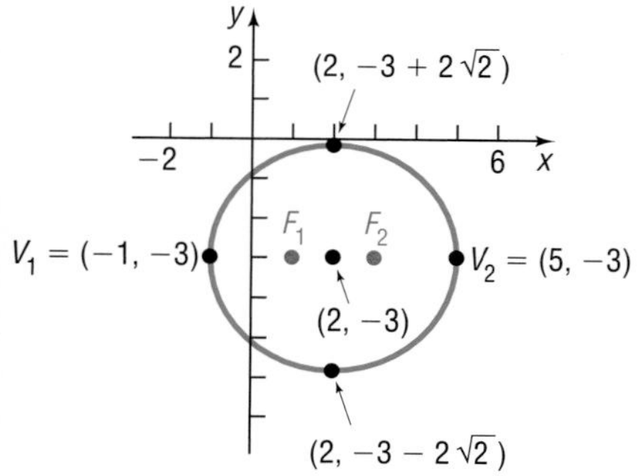

Figure 29 shows the graph.

NOW WORK PROBLEM 45.

EXAMPLE 7   Using a Graphing Utility to Graph an Ellipse, Center Not at the Origin

Using a graphing utility, graph the ellipse: $\dfrac{(x - 2)^2}{9} + \dfrac{(y + 3)^2}{8} = 1$

Solution   First, we must solve the equation $\dfrac{(x - 2)^2}{9} + \dfrac{(y + 3)^2}{8} = 1$ for y.

$$\frac{(y + 3)^2}{8} = 1 - \frac{(x - 2)^2}{9} \qquad \text{Subtract } \frac{(x - 2)^2}{9} \text{ from each side.}$$

$$(y + 3)^2 = 8\left[1 - \frac{(x - 2)^2}{9}\right] \qquad \text{Multiply each side by 8.}$$

$$y + 3 = \pm\sqrt{8\left[1 - \frac{(x - 2)^2}{9}\right]} \qquad \text{Apply the Square Root Method.}$$

$$y = -3 \pm \sqrt{8\left[1 - \frac{(x - 2)^2}{9}\right]} \qquad \text{Subtract 3 from each side.}$$

Figure 30

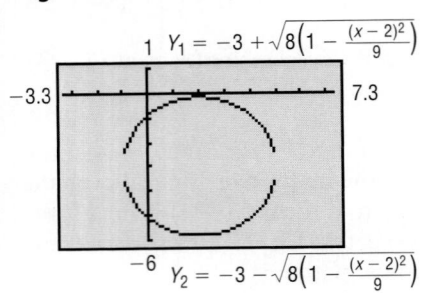

Figure 30 shows the graphs of $Y_1 = -3 + \sqrt{8\left[1 - \dfrac{(x-2)^2}{9}\right]}$ and $Y_2 = -3 - \sqrt{8\left[1 - \dfrac{(x-2)^2}{9}\right]}$.

NOW WORK PROBLEM 33.

EXAMPLE 8   **Discussing the Equation of an Ellipse**

Discuss the equation: $4x^2 + y^2 - 8x + 4y + 4 = 0$

Solution   We proceed to complete the squares in x and in y.

$$4x^2 + y^2 - 8x + 4y + 4 = 0$$
$$4x^2 - 8x + y^2 + 4y = -4 \qquad \text{\small Group like variables; place the constant on the right side.}$$
$$4(x^2 - 2x) + (y^2 + 4y) = -4 \qquad \text{\small Factor out 4 from the first two terms.}$$
$$4(x^2 - 2x + 1) + (y^2 + 4y + 4) = -4 + 4 + 4 \qquad \text{\small Complete each square.}$$
$$4(x - 1)^2 + (y + 2)^2 = 4 \qquad \text{\small Factor.}$$
$$(x - 1)^2 + \frac{(y + 2)^2}{4} = 1 \qquad \text{\small Divide each side by 4.}$$

This is the equation of an ellipse with center at $(1, -2)$ and major axis parallel to the y-axis. Since $a^2 = 4$ and $b^2 = 1$, we have $c^2 = a^2 - b^2 = 4 - 1 = 3$. The vertices are at $(h, k \pm a) = (1, -2 \pm 2)$ or $(1, 0)$ and $(1, -4)$. The foci are at $(h, k \pm c) = (1, -2 \pm \sqrt{3})$ or $(1, -2 - \sqrt{3})$ and $(1, -2 + \sqrt{3})$. Figure 31(a) shows the graph drawn by hand. Figure 31(b) shows the graph obtained using a graphing utility.

Figure 31

$$(x - 1)^2 + \frac{(y + 2)^2}{4} = 1$$

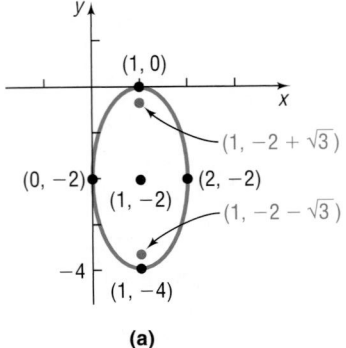

(a)

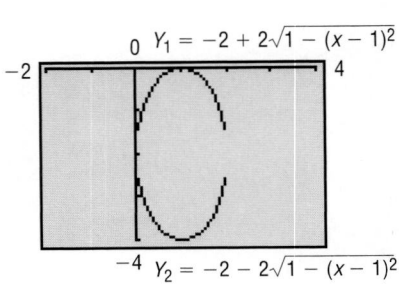

(b)

NOW WORK PROBLEM 37.

Applications

⑤   Ellipses are found in many applications in science and engineering. For example, the orbits of the planets around the Sun are elliptical, with the Sun's position at a focus. See Figure 32.

Figure 32

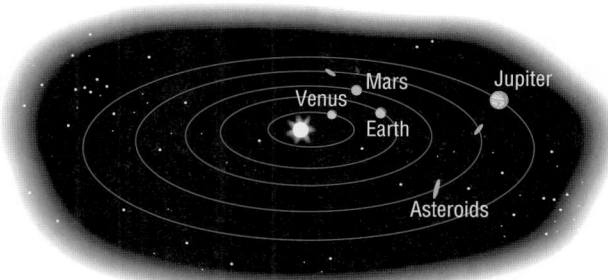

Stone and concrete bridges are often shaped as semielliptical arches. Elliptical gears are used in machinery when a variable rate of motion is required.

Ellipses also have an interesting reflection property. If a source of light (or sound) is placed at one focus, the waves transmitted by the source will reflect off the ellipse and concentrate at the other focus. This is the principle behind *whispering galleries*, which are rooms designed with elliptical ceilings. A person standing at one focus of the ellipse can whisper and be heard by a person standing at the other focus, because all the sound waves that reach the ceiling are reflected to the other person.

EXAMPLE 9 | **Whispering Galleries**

Figure 33 shows the specifications for an elliptical ceiling in a hall designed to be a whispering gallery. In a whispering gallery, a person standing at one focus of the ellipse can whisper and be heard by another person standing at the other focus, because all the sound waves that reach the ceiling from one focus are reflected to the other focus. Where are the foci located in the hall?

Solution   We set up a rectangular coordinate system so that the center of the ellipse is at the origin and the major axis is along the *x*-axis. See Figure 34. The equation of the ellipse is

$$\frac{x^2}{a^2} + \frac{y^2}{b^2} = 1$$

where $a = 25$ and $b = 20$.

Figure 33

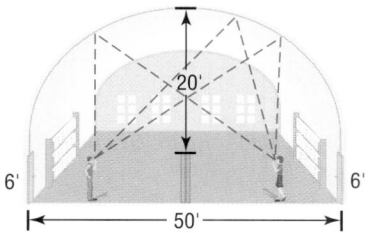

Figure 34

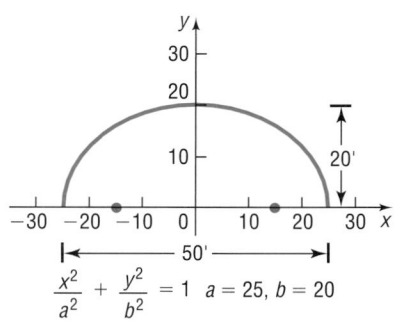

$\dfrac{x^2}{a^2} + \dfrac{y^2}{b^2} = 1$ $a = 25, b = 20$

Then, since

$$c^2 = a^2 - b^2 = 25^2 - 20^2 = 625 - 400 = 225$$

we have $c = 15$. The foci are located 15 feet from the center of the ellipse along the major axis. ■

6.3   Concepts and Vocabulary

In Problems 1–3, fill in the blanks.

1. A(n) _____ is the collection of all points in the plane the sum of whose distances from two fixed points is a constant.

2. For an ellipse, the foci lie on a line called the _____ axis.

3. For the ellipse $\dfrac{x^2}{4} + \dfrac{y^2}{25} = 1$, the vertices are the points _____ and _____.

In Problems 4–6, answer True or False to each statement.

4. The foci, vertices, and center of an ellipse lie on a line called the axis of symmetry.

5. If the center of an ellipse is at the origin and the foci lie on the y-axis, the ellipse is symmetric with respect to the x-axis, the y-axis, and the origin.

6. A circle is a certain type of ellipse.

7. Explain how to draw an ellipse using two thumbtacks and a piece of string.

8. Make up an equation for an ellipse with center at the origin and foci on the x-axis.

9. Make up an equation for an ellipse with center at the origin and foci on the y-axis.

6.3   Exercises

In Problems 1–4, the graph of an ellipse is given. Match each graph to its equation.

A. $\dfrac{x^2}{4} + y^2 = 1$ B. $x^2 + \dfrac{y^2}{4} = 1$ C. $\dfrac{x^2}{16} + \dfrac{y^2}{4} = 1$ D. $\dfrac{x^2}{4} + \dfrac{y^2}{16} = 1$

1.
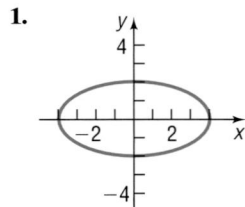

2.
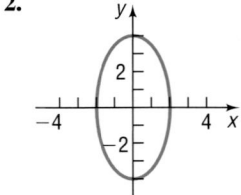

3.
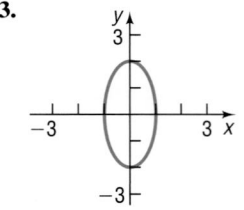

4.
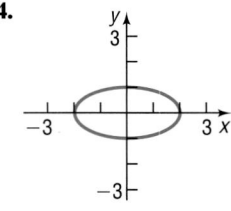

In Problems 5–8, the graph of an ellipse is given. Match each graph to its equation.

A. $\dfrac{(x + 1)^2}{4} + \dfrac{(y - 1)^2}{9} = 1$ B. $\dfrac{(x - 1)^2}{4} + \dfrac{(y + 1)^2}{9} = 1$

C. $\dfrac{(x - 1)^2}{9} + \dfrac{(y + 1)^2}{4} = 1$ D. $\dfrac{(x + 1)^2}{9} + \dfrac{(y - 1)^2}{4} = 1$

5.

6.

7.

8.

In Problems 9–18, find the vertices and foci of each ellipse. Graph each equation (a) by hand and (b) by using a graphing utility.

9. $\dfrac{x^2}{25} + \dfrac{y^2}{4} = 1$

10. $\dfrac{x^2}{9} + \dfrac{y^2}{4} = 1$

11. $\dfrac{x^2}{9} + \dfrac{y^2}{25} = 1$

12. $x^2 + \dfrac{y^2}{16} = 1$

13. $4x^2 + y^2 = 16$

14. $x^2 + 9y^2 = 18$

15. $4y^2 + x^2 = 8$

16. $4y^2 + 9x^2 = 36$

17. $x^2 + y^2 = 16$

18. $x^2 + y^2 = 4$

In Problems 19–28, find an equation for each ellipse. Graph the equation by hand.

19. Center at $(0,0)$; focus at $(3,0)$; vertex at $(5,0)$

20. Center at $(0,0)$; focus at $(-1,0)$; vertex at $(3,0)$

21. Center at $(0,0)$; focus at $(0,-4)$; vertex at $(0,5)$

22. Center at $(0,0)$; focus at $(0,1)$; vertex at $(0,-2)$

23. Foci at $(\pm 2,0)$; length of the major axis is 6

24. Focus at $(0,-4)$; vertices at $(0,\pm 8)$

25. Foci at $(0,\pm 3)$; x-intercepts are ± 2

26. Foci at $(0,\pm 2)$; length of the major axis is 8

27. Center at $(0,0)$; vertex at $(0,4)$; $b = 1$

28. Vertices at $(\pm 5,0)$; $c = 2$

In Problems 29–32, write an equation for each ellipse.

29.

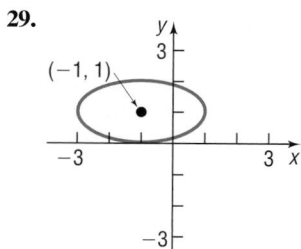

30.

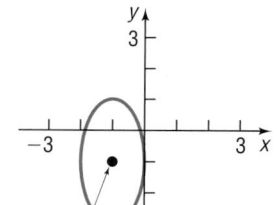

31.

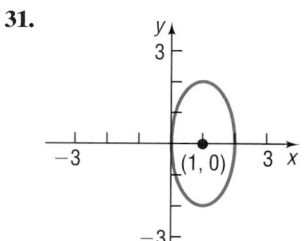

32.

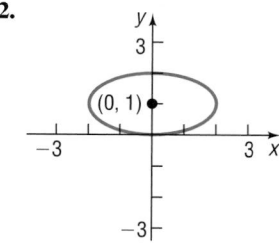

In Problems 33–44, discuss each equation, that is, find the center, foci, and vertices of each ellipse. Graph each equation (a) by hand and (b) by using a graphing utility.

33. $\dfrac{(x-3)^2}{4} + \dfrac{(y+1)^2}{9} = 1$

34. $\dfrac{(x+4)^2}{9} + \dfrac{(y+2)^2}{4} = 1$

35. $(x+5)^2 + 4(y-4)^2 = 16$

36. $9(x-3)^2 + (y+2)^2 = 18$

37. $x^2 + 4x + 4y^2 - 8y + 4 = 0$

38. $x^2 + 3y^2 - 12y + 9 = 0$

39. $2x^2 + 3y^2 - 8x + 6y + 5 = 0$

40. $4x^2 + 3y^2 + 8x - 6y = 5$

41. $9x^2 + 4y^2 - 18x + 16y - 11 = 0$

42. $x^2 + 9y^2 + 6x - 18y + 9 = 0$

43. $4x^2 + y^2 + 4y = 0$

44. $9x^2 + y^2 - 18x = 0$

In Problems 45–54, find an equation for each ellipse. Graph the equation by hand.

45. Center at $(2,-2)$; vertex at $(7,-2)$; focus at $(4,-2)$

46. Center at $(-3,1)$; vertex at $(-3,3)$; focus at $(-3,0)$

47. Vertices at $(4,3)$ and $(4,9)$; focus at $(4,8)$

48. Foci at $(1,2)$ and $(-3,2)$; vertex at $(-4,2)$

49. Foci at $(5,1)$ and $(-1,1)$; length of the major axis is 8

50. Vertices at $(2,5)$ and $(2,-1)$; $c = 2$

51. Center at $(1,2)$; focus at $(4,2)$; contains the point $(1,3)$

52. Center at $(1,2)$; focus at $(1,4)$; contains the point $(2,2)$

53. Center at $(1,2)$; vertex at $(4,2)$; contains the point $(1,3)$

54. Center at $(1,2)$; vertex at $(1,4)$; contains the point $(2,2)$

In Problems 55–58, graph each function by hand. Use a graphing utility to verify your results.

[**Hint:** Notice that each function is half an ellipse.]

55. $f(x) = \sqrt{16 - 4x^2}$ **56.** $f(x) = \sqrt{9 - 9x^2}$ **57.** $f(x) = -\sqrt{64 - 16x^2}$ **58.** $f(x) = -\sqrt{4 - 4x^2}$

---

59. Semielliptical Arch Bridge   An arch in the shape of the upper half of an ellipse is used to support a bridge that is to span a river 20 meters wide. The center of the arch is 6 meters above the center of the river (see the figure). Write an equation for the ellipse in which the *x*-axis coincides with the water level and the *y*-axis passes through the center of the arch.

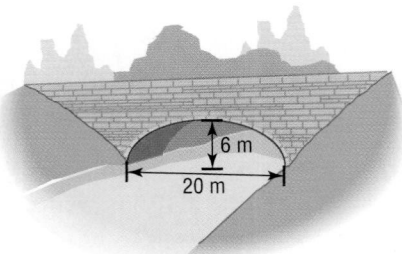

60. Semielliptical Arch Bridge   The arch of a bridge is a semiellipse with a horizontal major axis. The span is 30 feet, and the top of the arch is 10 feet above the major axis. The roadway is horizontal and is 2 feet above the top of the arch. Find the vertical distance from the roadway to the arch at 5-foot intervals along the roadway.

61. Whispering Gallery   A hall 100 feet in length is to be designed as a whispering gallery. If the foci are located 25 feet from the center, how high will the ceiling be at the center?

62. Whispering Gallery   Jim, standing at one focus of a whispering gallery, is 6 feet from the nearest wall. His friend is standing at the other focus, 100 feet away. What is the length of this whispering gallery? How high is its elliptical ceiling at the center?

63. Semielliptical Arch Bridge   A bridge is built in the shape of a semielliptical arch. The bridge has a span of 120 feet and a maximum height of 25 feet. Choose a suitable rectangular coordinate system and find the height of the arch at distances of 10, 30, and 50 feet from the center.

64. Semielliptical Arch Bridge   A bridge is to be built in the shape of a semielliptical arch and is to have a span of 100 feet. The height of the arch, at a distance of 40 feet from the center, is to be 10 feet. Find the height of the arch at its center.

65. Semielliptical Arch   An arch in the form of half an ellipse is 40 feet wide and 15 feet high at the center. Find the height of the arch at intervals of 10 feet along its width.

66. Semielliptical Arch Bridge   An arch for a bridge over a highway is in the form of half an ellipse. The top of the arch is 20 feet above the ground level (the major axis). The highway has four lanes, each 12 feet wide; a center safety strip 8 feet wide; and two side strips, each 4 feet wide. What should the span of the bridge be (the length of its major axis) if the height 28 feet from the center is to be 13 feet?

---

*In Problems 67–70, use the fact that the orbit of a planet about the Sun is an ellipse, with the Sun at one focus. The **aphelion** of a planet is its greatest distance from the Sun, and the **perihelion** is its shortest distance. The **mean distance** of a planet from the Sun is the length of the semimajor axis of the elliptical orbit. See the illustration.*

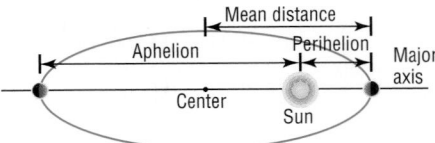

67. Earth   The mean distance of Earth from the Sun is 93 million miles. If the aphelion of Earth is 94.5 million miles, what is the perihelion? Write an equation for the orbit of Earth around the Sun.

68. Mars   The mean distance of Mars from the Sun is 142 million miles. If the perihelion of Mars is 128.5 million miles, what is the aphelion? Write an equation for the orbit of Mars about the Sun.

69. Jupiter   The aphelion of Jupiter is 507 million miles. If the distance from the Sun to the center of its elliptical orbit is 23.2 million miles, what is the perihelion? What is the mean distance? Write an equation for the orbit of Jupiter around the Sun.

70. Pluto   The perihelion of Pluto is 4551 million miles, and the distance of the Sun from the center of its elliptical orbit is 897.5 million miles. Find the aphelion of Pluto. What is the mean distance of Pluto from the Sun? Write an equation for the orbit of Pluto about the Sun.

71. Racetrack Design   Consult the figure. A racetrack is in the shape of an ellipse, 100 feet long and 50 feet wide. What is the width 10 feet from a vertex?

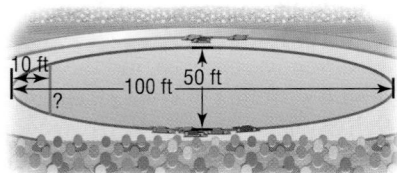

72. Racetrack Design   A racetrack is in the shape of an ellipse 80 feet long and 40 feet wide. What is the width 10 feet from a vertex?

73. Show that an equation of the form

$$Ax^2 + Cy^2 + F = 0, \qquad A \neq 0, C \neq 0, F \neq 0$$

where A and C are of the same sign and F is of opposite sign,
(a) Is the equation of an ellipse with center at $(0,0)$ if $A \neq C$.
(b) Is the equation of a circle with center $(0,0)$ if $A = C$.

74. Show that the graph of an equation of the form

$$Ax^2 + Cy^2 + Dx + Ey + F = 0, \qquad A \neq 0, C \neq 0$$

where A and C are of the same sign,
(a) Is an ellipse if $\dfrac{D^2}{4A} + \dfrac{E^2}{4C} - F$ is the same sign as A.
(b) Is a point if $\dfrac{D^2}{4A} + \dfrac{E^2}{4C} - F = 0$.
(c) Contains no points if $\dfrac{D^2}{4A} + \dfrac{E^2}{4C} - F$ is of opposite sign to A.

75. The **eccentricity** e of an ellipse is defined as the number $\dfrac{c}{a}$, where a and c are the numbers given in equation (2). Because $a > c$, it follows that $e < 1$. Write a brief paragraph about the general shape of each of the following ellipses. Be sure to justify your conclusions.
(a) Eccentricity close to 0
(b) Eccentricity = 0.5
(c) Eccentricity close to 1

PREPARING FOR THIS SECTION

Before getting started, review the following:

✓ Distance Formula (Section 1.1, p. 5)

✓ Completing the Square (Section A.3, pp. 579–580)

✓ Symmetry (Section 1.3, pp. 20–22)

✓ Graphing Techniques: Transformations (Section 1.6, pp. 64–73)

✓ Square Root Method (Section A.3, p. 578)

6.4   THE HYPERBOLA

OBJECTIVES

1   Find the Equation of a Hyperbola
2   Graph Hyperbolas
3   Discuss the Equation of a Hyperbola
4   Find the Asymptotes of a Hyperbola
5   Work with Hyperbolas with Center at (h, k)
6   Solve Applied Problems Involving Hyperbolas

A **hyperbola** is the collection of all points in the plane the difference of whose distances from two fixed points, called the **foci**, is a constant.

1   Figure 35 illustrates a hyperbola with foci F_1 and F_2. The line containing the foci is called the **transverse axis**. The midpoint of the line segment joining the foci is called the **center** of the hyperbola. The line through the center

Figure 35

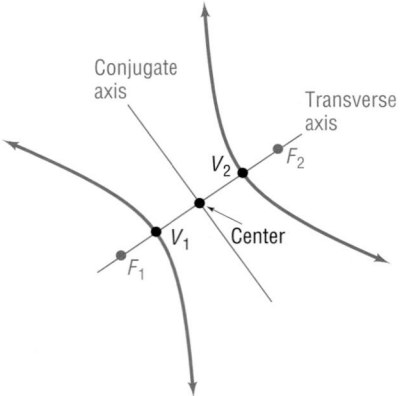

and perpendicular to the transverse axis is called the **conjugate axis**. The hyperbola consists of two separate curves, called **branches**, that are symmetric with respect to the transverse axis, conjugate axis, and center. The two points of intersection of the hyperbola and the transverse axis are the **vertices**, V_1 and V_2, of the hyperbola.

With these ideas in mind, we are now ready to find the equation of a hyperbola in a rectangular coordinate system. First, we place the center at the origin. Next, we position the hyperbola so that its transverse axis coincides with a coordinate axis. Suppose that the transverse axis coincides with the x-axis, as shown in Figure 36.

If c is the distance from the center to a focus, then one focus will be at $F_1 = (-c, 0)$ and the other at $F_2 = (c, 0)$. Now we let the constant difference of the distances from any point $P = (x, y)$ on the hyperbola to the foci F_1 and F_2 be denoted by $\pm 2a$. (If P is on the right branch, the $+$ sign is used; if P is on the left branch, the $-$ sign is used.) The coordinates of P must satisfy the equation

Figure 36
$d(F_1, P) - d(F_2, P) = \pm 2a$

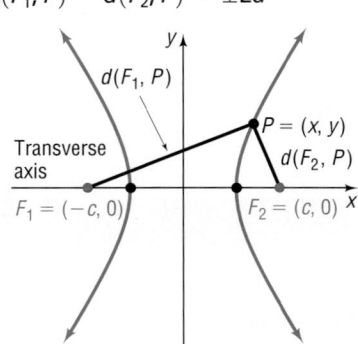

$$d(F_1, P) - d(F_2, P) = \pm 2a \qquad \text{\textit{Difference of the distances from P to the foci equals} } \pm 2a.$$

$$\sqrt{(x + c)^2 + y^2} - \sqrt{(x - c)^2 + y^2} = \pm 2a \qquad \text{\textit{Use the distance formula.}}$$

$$\sqrt{(x + c)^2 + y^2} = \pm 2a + \sqrt{(x - c)^2 + y^2} \qquad \text{\textit{Isolate one radical.}}$$

$$(x + c)^2 + y^2 = 4a^2 \pm 4a\sqrt{(x - c)^2 + y^2} \qquad \text{\textit{Square both sides.}}$$
$$+ (x - c)^2 + y^2$$

Next, we remove the parentheses.

$$x^2 + 2cx + c^2 + y^2 = 4a^2 \pm 4a\sqrt{(x - c)^2 + y^2} + x^2 - 2cx + c^2 + y^2$$

$$4cx - 4a^2 = \pm 4a\sqrt{(x - c)^2 + y^2} \qquad \text{\textit{Simplify; isolate the radical.}}$$

$$cx - a^2 = \pm a\sqrt{(x - c)^2 + y^2} \qquad \text{\textit{Divide each side by 4.}}$$

$$(cx - a^2)^2 = a^2[(x - c)^2 + y^2] \qquad \text{\textit{Square both sides.}}$$

$$c^2x^2 - 2ca^2x + a^4 = a^2(x^2 - 2cx + c^2 + y^2) \qquad \text{\textit{Simplify.}}$$

$$c^2x^2 + a^4 = a^2x^2 + a^2c^2 + a^2y^2 \qquad \text{\textit{Remove parentheses and simplify.}}$$

$$(c^2 - a^2)x^2 - a^2y^2 = a^2c^2 - a^4 \qquad \text{\textit{Rearrange terms.}}$$

$$(c^2 - a^2)x^2 - a^2y^2 = a^2(c^2 - a^2) \qquad \text{\textit{Factor } a^2 \text{ on the right side.}} \qquad (1)$$

To obtain points on the hyperbola off the x-axis, it must be that $a < c$. To see why, look again at Figure 36.

$$d(F_1, P) < d(F_2, P) + d(F_1, F_2) \qquad \text{\textit{Use triangle } F_1PF_2.}$$
$$d(F_1, P) - d(F_2, P) < d(F_1, F_2) \qquad \text{\textit{P is on the right branch, so} } d(F_1, P) - d(F_2, P) = 2a.$$

$$2a < 2c$$
$$a < c$$

Since $a < c$, we also have $a^2 < c^2$, so $c^2 - a^2 > 0$. Let $b^2 = c^2 - a^2$, $b > 0$. Then equation (1) can be written as

$$b^2 x^2 - a^2 y^2 = a^2 b^2$$

$$\frac{x^2}{a^2} - \frac{y^2}{b^2} = 1 \qquad \text{\textit{Divide each side by } } a^2 b^2.$$

To find the vertices of the hyperbola defined by this equation, let $y = 0$. The vertices satisfy the equation $\frac{x^2}{a^2} = 1$, the solutions of which are $x = \pm a$. Consequently, the vertices of the hyperbola are $V_1 = (-a, 0)$ and $V_2 = (a, 0)$. Notice that the distance from the center $(0, 0)$ to either vertex is a.

Theorem

Equation of a Hyperbola; Center at $(0, 0)$; Foci at $(\pm c, 0)$; Vertices at $(\pm a, 0)$; Transverse Axis along the x-Axis

An equation of the hyperbola with center at $(0, 0)$, foci at $(-c, 0)$ and $(c, 0)$, and vertices at $(-a, 0)$ and $(a, 0)$ is

$$\frac{x^2}{a^2} - \frac{y^2}{b^2} = 1, \qquad \text{where } b^2 = c^2 - a^2 \qquad (2)$$

The transverse axis is the x-axis.

Figure 37

$\frac{x^2}{a^2} - \frac{y^2}{b^2} = 1, b^2 = c^2 - a^2$

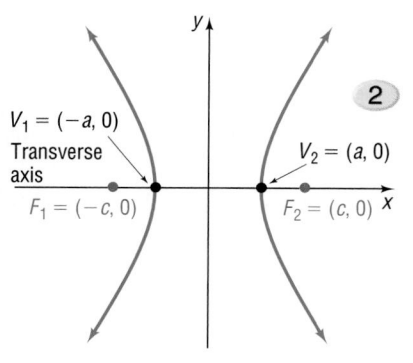

$V_1 = (-a, 0)$
Transverse axis
$F_1 = (-c, 0)$
$V_2 = (a, 0)$
$F_2 = (c, 0)$

As you can verify, the hyperbola defined by equation (2) is symmetric with respect to the x-axis, y-axis, and origin. To find the y-intercepts, if any, let $x = 0$ in equation (2). This results in the equation $\frac{y^2}{b^2} = -1$, which has no real solution. We conclude that the hyperbola defined by equation (2) has no y-intercepts. In fact, since $\frac{x^2}{a^2} - 1 = \frac{y^2}{b^2} \geq 0$, it follows that $\frac{x^2}{a^2} \geq 1$. There are no points on the graph for $-a < x < a$. See Figure 37.

EXAMPLE 1

Finding and Graphing an Equation of a Hyperbola

Find an equation of the hyperbola with center at the origin, one focus at $(3, 0)$, and one vertex at $(-2, 0)$. Graph the equation by hand.

Solution

The hyperbola has its center at the origin, and the transverse axis coincides with the x-axis. One focus is at $(c, 0) = (3, 0)$, so $c = 3$. One vertex is at $(-a, 0) = (-2, 0)$, so $a = 2$. From equation (2), it follows that $b^2 = c^2 - a^2 = 9 - 4 = 5$, so an equation of the hyperbola is

$$\frac{x^2}{4} - \frac{y^2}{5} = 1$$

To graph a hyperbola, it is helpful to locate and plot other points on the graph. For example, to find the points above and below the foci, we let $x = \pm 3$. Then

$$\frac{x^2}{4} - \frac{y^2}{5} = 1$$

$$\frac{(\pm3)^2}{4} - \frac{y^2}{5} = 1 \qquad x = \pm3$$

$$\frac{9}{4} - \frac{y^2}{5} = 1$$

$$\frac{y^2}{5} = \frac{5}{4}$$

$$y^2 = \frac{25}{4}$$

$$y = \pm\frac{5}{2}$$

Figure 38

$$\frac{x^2}{4} - \frac{y^2}{5} = 1$$

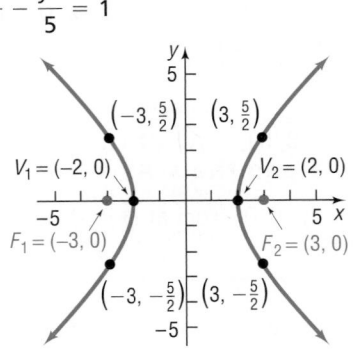

The points above and below the foci are $\left(\pm3, \frac{5}{2}\right)$ and $\left(\pm3, -\frac{5}{2}\right)$. These points help because they determine the "opening" of the hyperbola. See Figure 38. ∎

EXAMPLE 2   Using a Graphing Utility to Graph a Hyperbola

Using a graphing utility, graph the hyperbola: $\dfrac{x^2}{4} - \dfrac{y^2}{5} = 1$

Figure 39

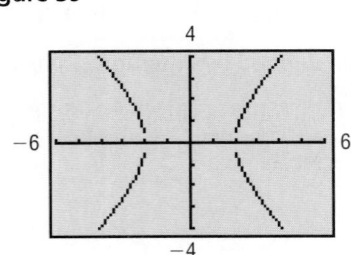

Solution   To graph the hyperbola $\dfrac{x^2}{4} - \dfrac{y^2}{5} = 1$, we need to graph the two functions $Y_1 = \sqrt{5}\sqrt{\dfrac{x^2}{4} - 1}$ and $Y_2 = -\sqrt{5}\sqrt{\dfrac{x^2}{4} - 1}$. As with graphing circles, parabolas, and ellipses on a graphing utility, we use a square screen setting so that the graph is not distorted. Figure 39 shows the graph of the hyperbola. ∎

✏ **NOW WORK PROBLEM 9.**

An equation of the form of equation (2) is the equation of a hyperbola with center at the origin; foci on the x-axis at $(-c, 0)$ and $(c, 0)$, where $c^2 = a^2 + b^2$; and transverse axis along the x-axis.

③   For the remainder of this section, the direction "Discuss the equation" will mean to find the center, transverse axis, vertices, and foci of the hyperbola and graph it.

EXAMPLE 3   Discussing the Equation of a Hyperbola

Discuss the equation: $\dfrac{x^2}{16} - \dfrac{y^2}{4} = 1$

Solution   The given equation is of the form of equation (2), with $a^2 = 16$ and $b^2 = 4$. The graph of the equation is a hyperbola with center at $(0, 0)$ and transverse axis along the x-axis. Also, we know that $c^2 = a^2 + b^2 = 16 + 4 = 20$. The vertices are at $(\pm a, 0) = (\pm4, 0)$, and the foci are at $(\pm c, 0) = (\pm2\sqrt{5}, 0)$.

To locate the points on the graph above and below the foci, we let $x = \pm2\sqrt{5}$. Then

$$\frac{x^2}{16} - \frac{y^2}{4} = 1$$

$$\frac{(\pm 2\sqrt{5})^2}{16} - \frac{y^2}{4} = 1 \qquad x = \pm 2\sqrt{5}$$

$$\frac{20}{16} - \frac{y^2}{4} = 1$$

$$\frac{5}{4} - \frac{y^2}{4} = 1$$

$$\frac{y^2}{4} = \frac{1}{4}$$

$$y = \pm 1$$

The points above and below the foci are $(\pm 2\sqrt{5}, 1)$ and $(\pm 2\sqrt{5}, -1)$. See Figure 40(a) for the graph drawn by hand. Figure 40(b) shows the graph obtained using a graphing utility.

Figure 40

$$\frac{x^2}{16} - \frac{y^2}{4} = 1$$

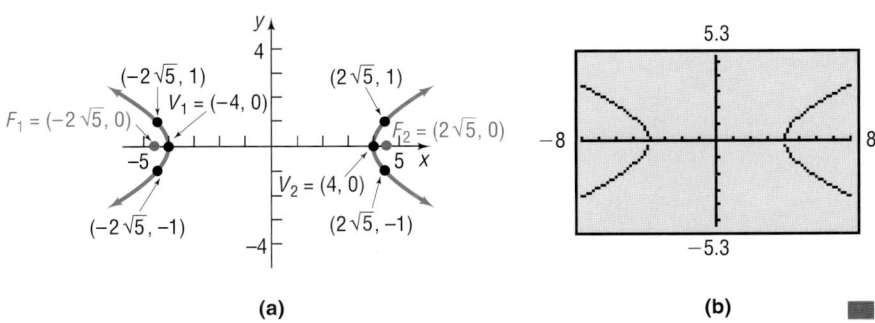

(a)

(b)

NOW WORK PROBLEM **19**.

The next result gives the form of the equation of a hyperbola with center at the origin and transverse axis along the y-axis.

Theorem

Equation of a Hyperbola; Center at $(0, 0)$; Foci at $(0, \pm c)$; Vertices at $(0, \pm a)$; Transverse Axis along the y-Axis

An equation of the hyperbola with center at $(0, 0)$, foci at $(0, -c)$ and $(0, c)$, and vertices at $(0, -a)$ and $(0, a)$ is

$$\frac{y^2}{a^2} - \frac{x^2}{b^2} = 1, \qquad \text{where } b^2 = c^2 - a^2 \tag{3}$$

The transverse axis is the y-axis.

Figure 41

$$\frac{y^2}{a^2} - \frac{x^2}{b^2} = 1, b^2 = c^2 - a^2$$

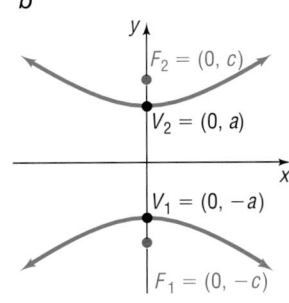

Figure 41 shows the graph of a typical hyperbola defined by equation (3). An equation of the form of equation (2), $\dfrac{x^2}{a^2} - \dfrac{y^2}{b^2} = 1$, is the equation of a hyperbola with center at the origin, foci on the x-axis at $(-c, 0)$ and $(c, 0)$, where $c^2 = a^2 + b^2$, and transverse axis along the x-axis.

An equation of the form of equation (3), $\dfrac{y^2}{a^2} - \dfrac{x^2}{b^2} = 1$, is the equation of a hyperbola with center at the origin, foci on the y-axis at $(0, -c)$ and $(0, c)$, where $c^2 = a^2 + b^2$, and transverse axis along the y-axis.

Notice the difference in the forms of equations (2) and (3). When the y^2-term is subtracted from the x^2-term, the transverse axis is the x-axis. When the x^2-term is subtracted from the y^2-term, the transverse axis is the y-axis.

EXAMPLE 4 **Discussing the Equation of a Hyperbola**

Discuss the equation: $y^2 - 4x^2 = 4$

Solution To put the equation in proper form, we divide each side by 4:

$$\frac{y^2}{4} - x^2 = 1$$

Since the x^2-term is subtracted from the y^2-term, the equation is that of a hyperbola with center at the origin and transverse axis along the y-axis. Also, comparing the above equation to equation (3), we find $a^2 = 4$, $b^2 = 1$, and $c^2 = a^2 + b^2 = 5$. The vertices are at $(0, \pm a) = (0, \pm 2)$, and the foci are at $(0, \pm c) = (0, \pm\sqrt{5})$.

To locate other points on the graph, we let $x = \pm 2$. Then

$$y^2 - 4x^2 = 4$$
$$y^2 - 4(\pm 2)^2 = 4 \qquad x = \pm 2$$
$$y^2 - 16 = 4$$
$$y^2 = 20$$
$$y = \pm 2\sqrt{5}$$

Four other points on the graph are $(\pm 2, 2\sqrt{5})$ and $(\pm 2, -2\sqrt{5})$. See Figure 42(a) for the graph drawn by hand. Figure 42(b) shows the graph obtained using a graphing utility.

Figure 42

$\dfrac{y^2}{4} - x^2 = 1$

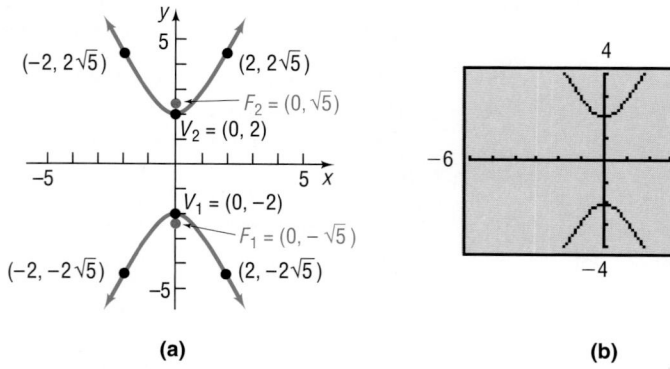

(a)

(b)

EXAMPLE 5 **Finding an Equation of a Hyperbola**

Find an equation of the hyperbola having one vertex at $(0, 2)$ and foci at $(0, -3)$ and $(0, 3)$. Graph the equation by hand.

Figure 43

$$\frac{y^2}{4} - \frac{x^2}{5} = 1$$

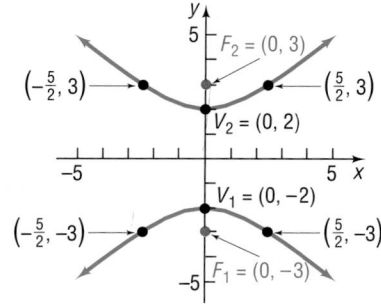

Solution

Since the foci are at $(0, -3)$ and $(0, 3)$, the center of the hyperbola is at their midpoint, the origin. Also, the transverse axis is along the y-axis. The given information also reveals that $c = 3, a = 2$, and $b^2 = c^2 - a^2 = 9 - 4 = 5$. The form of the equation of the hyperbola is given by equation (3):

$$\frac{y^2}{a^2} - \frac{x^2}{b^2} = 1$$

$$\frac{y^2}{4} - \frac{x^2}{5} = 1$$

Let $y = \pm 3$ to obtain points on the graph across from the foci. See Figure 43. ∎

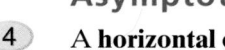 **NOW WORK PROBLEM 13.**

Look at the equations of the hyperbolas in Examples 3 and 5. For the hyperbola in Example 3, $a^2 = 16$ and $b^2 = 4$, so $a > b$; for the hyperbola in Example 5, $a^2 = 4$ and $b^2 = 5$, so $a < b$. We conclude that, for hyperbolas, there are no requirements involving the relative sizes of a and b. Contrast this situation to the case of an ellipse, in which the relative sizes of a and b dictate which axis is the major axis. Hyperbolas have another feature to distinguish them from ellipses and parabolas: Hyperbolas have asymptotes.

Asymptotes

④ A **horizontal or oblique asymptote** of a graph is a line with the property that the distance from the line to points on the graph approaches 0 as $x \to -\infty$ or as $x \to \infty$. The asymptotes provide information about the end behavior of the graph of a hyperbola.

Theorem

Asymptotes of a Hyperbola

The hyperbola $\dfrac{x^2}{a^2} - \dfrac{y^2}{b^2} = 1$ has the two oblique asymptotes

$$y = \frac{b}{a}x \quad \text{and} \quad y = -\frac{b}{a}x \qquad (4)$$

∎

Proof We begin by solving for y in the equation of the hyperbola.

$$\frac{x^2}{a^2} - \frac{y^2}{b^2} = 1$$

$$\frac{y^2}{b^2} = \frac{x^2}{a^2} - 1$$

$$y^2 = b^2\left(\frac{x^2}{a^2} - 1\right)$$

Since $x \neq 0$, we can rearrange the right side in the form

$$y^2 = \frac{b^2 x^2}{a^2}\left(1 - \frac{a^2}{x^2}\right)$$

$$y = \pm\frac{bx}{a}\sqrt{1 - \frac{a^2}{x^2}}$$

Now, as $x \to -\infty$ or as $x \to \infty$, the term $\dfrac{a^2}{x^2}$ approaches 0, so the expression under the radical approaches 1. Thus, as $x \to -\infty$ or as $x \to \infty$, the value of y approaches $\pm\dfrac{bx}{a}$, that is, the graph of the hyperbola approaches the lines

$$y = -\frac{b}{a}x \quad \text{and} \quad y = \frac{b}{a}x$$

These lines are oblique asymptotes of the hyperbola. ∎

The asymptotes of a hyperbola are not part of the hyperbola, but they do serve as a guide for graphing a hyperbola. For example, suppose that we want to graph the equation

$$\frac{x^2}{a^2} - \frac{y^2}{b^2} = 1$$

We begin by plotting the vertices $(-a, 0)$ and $(a, 0)$. Then we plot the points $(0, -b)$ and $(0, b)$ and use these four points to construct a rectangle, as shown in Figure 44. The diagonals of this rectangle have slopes $\dfrac{b}{a}$ and $-\dfrac{b}{a}$, and their extensions are the asymptotes $y = \dfrac{b}{a}x$ and $y = -\dfrac{b}{a}x$ of the hyperbola. If we graph the asymptotes, we can use them to establish the "opening" of the hyperbola and avoid plotting other points.

Figure 44
$$\frac{x^2}{a^2} - \frac{y^2}{b^2} = 1$$

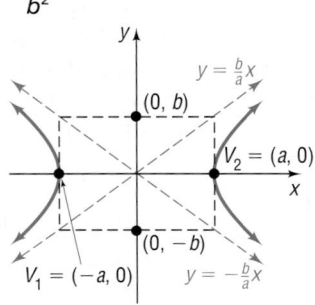

Theorem

Asymptotes of a Hyperbola

The hyperbola $\dfrac{y^2}{a^2} - \dfrac{x^2}{b^2} = 1$ has the two oblique asymptotes

$$y = \frac{a}{b}x \quad \text{and} \quad y = -\frac{a}{b}x \tag{5}$$

∎

You are asked to prove this result in Problem 64.

For the remainder of this section, the direction "Discuss the equation" will mean to find the center, transverse axis, vertices, foci, and asymptotes of the hyperbola and graph it.

EXAMPLE 6   Discussing the Equation of a Hyperbola

Discuss the equation: $9x^2 - 4y^2 = 36$

Solution   Divide each side of the equation by 36 to put the equation in proper form.

$$\frac{x^2}{4} - \frac{y^2}{9} = 1$$

We now proceed to analyze the equation. The center of the hyperbola is the origin. Since the x^2-term is first in the equation, we know that the transverse axis is along the x-axis and the vertices and foci will lie on the x-axis. Using equation (2), we find $a^2 = 4$, $b^2 = 9$, and $c^2 = a^2 + b^2 = 13$. The vertices are $a = 2$ units left and right of the center at $(\pm a, 0) = (\pm 2, 0)$; the foci are $c = \sqrt{13}$ units left and right of the center at $(\pm c, 0) = (\pm\sqrt{13}, 0)$; and the asymptotes have the equations

$$y = \frac{b}{a}x = \frac{3}{2}x \quad \text{and} \quad y = -\frac{b}{a}x = -\frac{3}{2}x$$

To graph the hyperbola by hand, form the rectangle containing the points $(\pm a, 0)$ and $(0, \pm b)$, that is, $(-2, 0)$, $(2, 0)$, $(0, -3)$, and $(0, 3)$. The extensions of the diagonals of this rectangle are the asymptotes. See Figure 45(a) for the graph drawn by hand. Figure 45(b) shows the graph obtained using a graphing utility.

Figure 45

$$\frac{x^2}{4} - \frac{y^2}{9} = 1$$

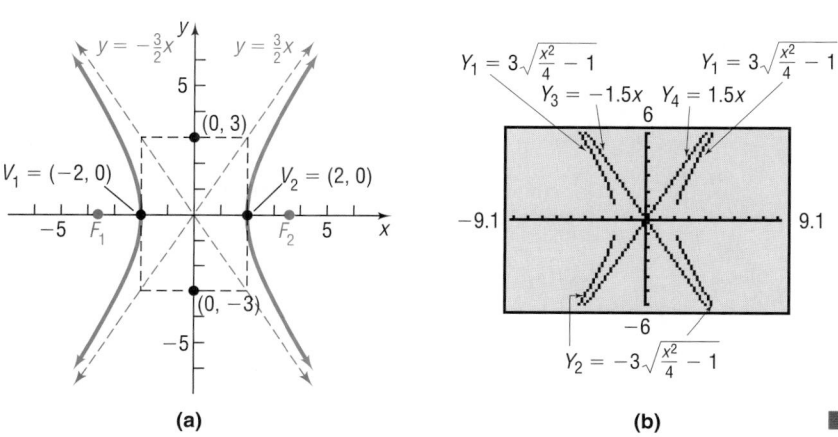

(a)   (b)

NOW WORK PROBLEM **21**.

SEEING THE CONCEPT   Refer to Figure 45(b). Create a TABLE using Y_1 and Y_4 with $x = 10$, 100, 1000, and 10,000. Compare the values of Y_1 and Y_4. Repeat for Y_1 and Y_3, Y_2 and Y_3, and Y_2 and Y_4.

Center at (h, k)

⑤ If a hyperbola with center at the origin and transverse axis coinciding with a coordinate axis is shifted horizontally h units and then vertically k units, the result is a hyperbola with center at (h, k) and transverse axis parallel to a coordinate axis. The equations of such hyperbolas have the same forms as those given in equations (2) and (3), except that x is replaced by $x - h$ (the horizontal shift) and y is replaced by $y - k$ (the vertical shift). Table 4 gives the forms of the equations of such hyperbolas. See Figure 46 for the graphs.

TABLE 4
Hyperbolas With Center at (h, k), and Transverse Axis Parallel to a Coordinate Axis

Center	Transverse Axis	Foci	Vertices	Equation		Asymptotes
(h, k)	Parallel to x-axis	$(h \pm c, k)$	$(h \pm a, k)$	$\dfrac{(x - h)^2}{a^2} - \dfrac{(y - k)^2}{b^2} = 1$,	$b^2 = c^2 - a^2$	$y - k = \pm\dfrac{b}{a}(x - h)$
(h, k)	Parallel to y-axis	$(h, k \pm c)$	$(h, k \pm a)$	$\dfrac{(y - k)^2}{a^2} - \dfrac{(x - h)^2}{b^2} = 1$,	$b^2 = c^2 - a^2$	$y - k = \pm\dfrac{a}{b}(x - h)$

Figure 46

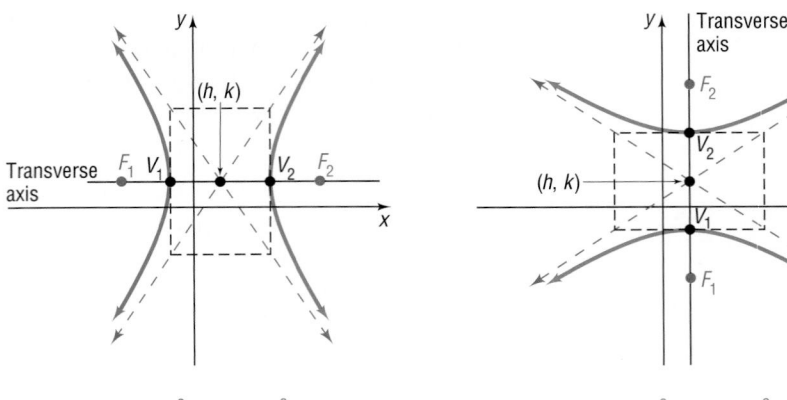

(a) $\dfrac{(x-h)^2}{a^2} - \dfrac{(y-k)^2}{b^2} = 1$

(b) $\dfrac{(y-k)^2}{a^2} - \dfrac{(x-h)^2}{b^2} = 1$

EXAMPLE 7 **Finding an Equation of a Hyperbola, Center Not at the Origin**

Find an equation for the hyperbola with center at $(1, -2)$, one focus at $(4, -2)$, and one vertex at $(3, -2)$. Graph the equation by hand.

Figure 47

$\dfrac{(x-1)^2}{4} - \dfrac{(y+2)^2}{5} = 1$

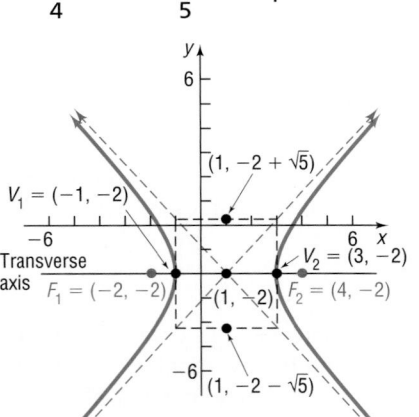

Solution   The center is at $(h, k) = (1, -2)$, so $h = 1$ and $k = -2$. Since the center, focus, and vertex all lie on the line $y = -2$, the transverse axis is parallel to the x-axis. The distance from the center $(1, -2)$ to the focus $(4, -2)$ is $c = 3$; the distance from the center $(1, -2)$ to the vertex $(3, -2)$ is $a = 2$. Thus, $b^2 = c^2 - a^2 = 9 - 4 = 5$. The equation is

$$\frac{(x-h)^2}{a^2} - \frac{(y-k)^2}{b^2} = 1$$

$$\frac{(x-1)^2}{4} - \frac{(y+2)^2}{5} = 1$$

See Figure 47.

▬

NOW WORK PROBLEM **31**.

EXAMPLE 8 **Discussing the Equation of a Hyperbola**

Discuss the equation: $-x^2 + 4y^2 - 2x - 16y + 11 = 0$

Solution   We complete the squares in x and in y.

$$-x^2 + 4y^2 - 2x - 16y + 11 = 0$$

$$-(x^2 + 2x) + 4(y^2 - 4y) = -11 \qquad \text{\textit{Group terms.}}$$

$$-(x^2 + 2x + 1) + 4(y^2 - 4y + 4) = -11 - 1 + 16 \qquad \text{\textit{Complete each square.}}$$

$$-(x+1)^2 + 4(y-2)^2 = 4$$

$$(y-2)^2 - \frac{(x+1)^2}{4} = 1 \qquad \text{\textit{Divide each side by 4.}}$$

This is the equation of a hyperbola with center at $(-1, 2)$ and transverse axis parallel to the y-axis. Also, $a^2 = 1$ and $b^2 = 4$, so $c^2 = a^2 + b^2 = 5$. Since the transverse axis is parallel to the y-axis, the vertices and foci are located a and c units above and below the center, respectively. The vertices are at $(h, k \pm a) = (-1, 2 \pm 1)$, or $(-1, 1)$ and $(-1, 3)$. The foci are at $(h, k \pm c) = (-1, 2 \pm \sqrt{5})$. The asymptotes are $y - 2 = \frac{1}{2}(x + 1)$ and $y - 2 = -\frac{1}{2}(x + 1)$. Figure 48(a) shows the graph drawn by hand. Figure 48(b) shows the graph obtained using a graphing utility.

Figure 48

$$(y - 2)^2 - \frac{(x + 1)^2}{4} = 1$$

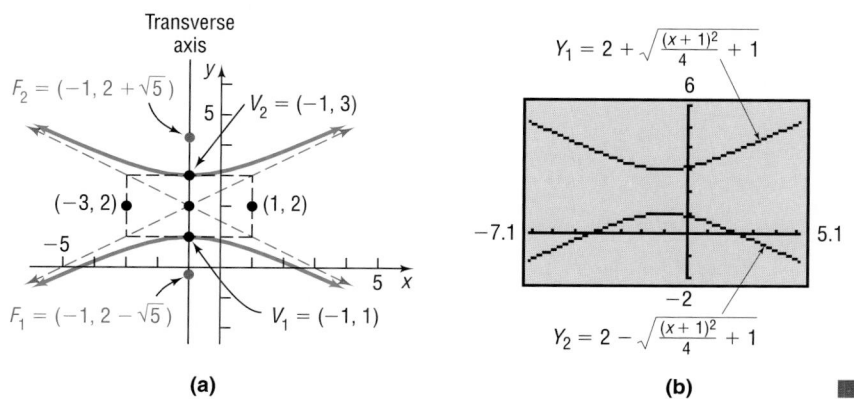

(a)

(b)

────── **NOW WORK PROBLEM 45.**

Applications

6 See Figure 49. Suppose that a gun is fired from an unknown source S. An observer at O_1 hears the report (sound of gun shot) 1 second after another observer at O_2. Because sound travels at about 1100 feet per second, it follows that the point S must be 1100 feet closer to O_2 than to O_1. S lies on one branch of a hyperbola with foci at O_1 and O_2. (Do you see why? The difference of the distances from S to O_1 and from S to O_2 is the constant 1100.) If a third observer at O_3 hears the same report 2 seconds after O_1 hears it, then S will lie on a branch of a second hyperbola with foci at O_1 and O_3. The intersection of the two hyperbolas will pinpoint the location of S.

Figure 49

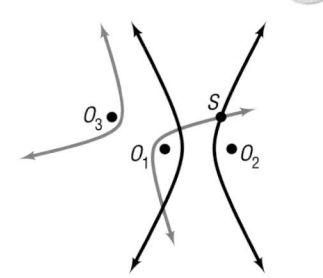

Loran

In the LOng RAnge Navigation system (LORAN), a master radio sending station and a secondary sending station emit signals that can be received by a ship at sea. See Figure 50. Because a ship monitoring the two signals will usually be nearer to one of the two stations, there will be a difference in the distance that the two signals travel, which will register as a slight time difference between the signals. As long as the time difference remains constant, the difference of the two distances will also be constant. If the ship follows a path corresponding to the fixed time difference, it will follow the path of a hyperbola whose foci are located at the positions of the two sending stations. So for each time difference a different hyperbolic path results, each bringing the ship to a different shore location. Navigation charts show the various hyperbolic paths corresponding to different time differences.

Figure 50

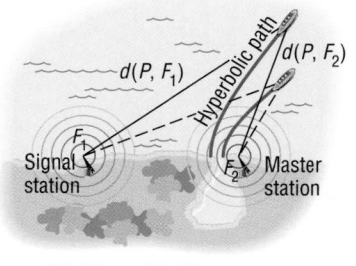

$d(P, F_1) - d(P, F_2) = \text{constant}$

| EXAMPLE 9 | LORAN |

Two LORAN stations are positioned 250 miles apart along a straight shore.

(a) A ship records a time difference of 0.00054 second between the LORAN signals. Set up an appropriate rectangular coordinate system to determine where the ship would reach shore if it were to follow the hyperbola corresponding to this time difference.

(b) If the ship wants to enter a harbor located between the two stations 25 miles from the master station, what time difference should it be looking for?

(c) If the ship is 80 miles offshore when the desired time difference is obtained, what is the approximate location of the ship?

[**Note:** The speed of each radio signal is 186,000 miles per second.]

Solution

(a) We set up a rectangular coordinate system so that the two stations lie on the x-axis and the origin is midway between them. See Figure 51. The ship lies on a hyperbola whose foci are the locations of the two stations. The reason for this is that the constant time difference of the signals from each station results in a constant difference in the distance of the ship from each station. Since the time difference is 0.00054 second and the speed of the signal is 186,000 miles per second, the difference of the distances from the ship to each station (foci) is

Figure 51

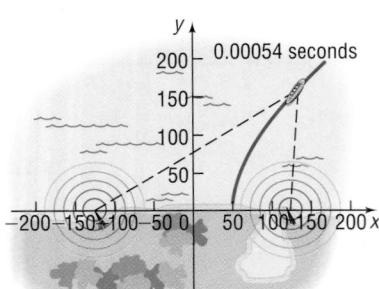

$$\text{Distance} = \text{Speed} \times \text{Time} = 186{,}000 \times 0.00054 \approx 100 \text{ miles}$$

The difference of the distances from the ship to each station, 100, equals $2a$, so $a = 50$ and the vertex of the corresponding hyperbola is at $(50, 0)$. Since the focus is at $(125, 0)$, following this hyperbola the ship would reach shore 75 miles from the master station.

(b) To reach shore 25 miles from the master station, the ship would follow a hyperbola with vertex at $(100, 0)$. For this hyperbola, $a = 100$, so the constant difference of the distances from the ship to each station is $2a = 200$. The time difference that the ship should look for is

$$\text{Time} = \frac{\text{Distance}}{\text{Speed}} = \frac{200}{186{,}000} \approx 0.001075 \text{ second}$$

(c) To find the approximate location of the ship, we need to find the equation of the hyperbola with vertex at $(100, 0)$ and a focus at $(125, 0)$. The form of the equation of this hyperbola is

$$\frac{x^2}{a^2} - \frac{y^2}{b^2} = 1$$

where $a = 100$. Since $c = 125$, we have

$$b^2 = c^2 - a^2 = 125^2 - 100^2 = 5625$$

The equation of the hyperbola is

$$\frac{x^2}{100^2} - \frac{y^2}{5625} = 1$$

Figure 52

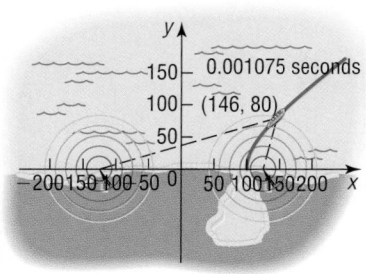

Since the ship is 80 miles from shore, we use $y = 80$ in the equation and solve for x.

$$\frac{x^2}{100^2} - \frac{80^2}{5625} = 1$$

$$\frac{x^2}{100^2} = 1 + \frac{80^2}{5625} \approx 2.14$$

$$x^2 \approx 100^2(2.14)$$

$$x \approx 146$$

The ship is at the position $(146, 80)$. See Figure 52.

NOW WORK PROBLEM 57.

6.4 Concepts and Vocabulary

In Problems 1–3, fill in the blanks.

1. A(n) _____ is the collection of points in the plane the difference of whose distances from two fixed points is a constant.

2. For a hyperbola, the foci lie on a line called the _____ _____.

3. The asymptotes of the hyperbola $\dfrac{x^2}{4} - \dfrac{y^2}{9} = 1$ are _____ and _____.

In Problems 4–6, answer True or False to each statement.

4. The foci of a hyperbola lie on a line called the axis of symmetry.

5. Hyperbolas always have asymptotes.

6. A hyperbola will never intersect its transverse axis.

7. Explain how the asymptotes of a hyperbola are helpful in obtaining its graph.

8. Make up an equation for a hyperbola with center at the origin and transverse axis along the x-axis.

9. Make up an equation for a hyperbola with center at the origin and transverse axis along the y-axis.

6.4 Exercises

In Problems 1–4, the graph of a hyperbola is given. Match each graph to its equation.

A. $\dfrac{x^2}{4} - y^2 = 1$ **B.** $x^2 - \dfrac{y^2}{4} = 1$ **C.** $\dfrac{y^2}{4} - x^2 = 1$ **D.** $y^2 - \dfrac{x^2}{4} = 1$

1.

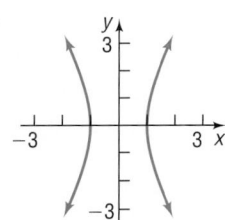

2.

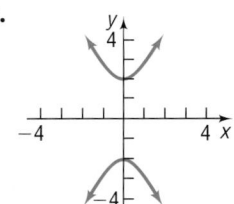

3.

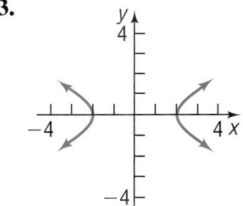

4.

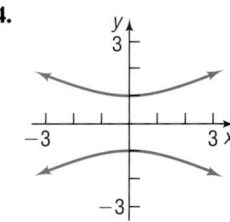

In Problems 5–8, the graph of a hyperbola is given. Match each graph to its equation.

A. $\dfrac{x^2}{16} - \dfrac{y^2}{9} = 1$ B. $\dfrac{x^2}{9} - \dfrac{y^2}{16} = 1$ C. $\dfrac{y^2}{16} - \dfrac{x^2}{9} = 1$ D. $\dfrac{y^2}{9} - \dfrac{x^2}{16} = 1$

5.
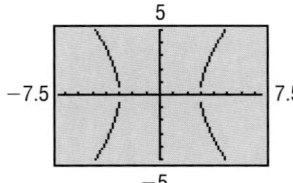

6.
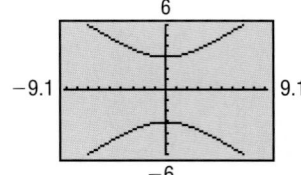

7.
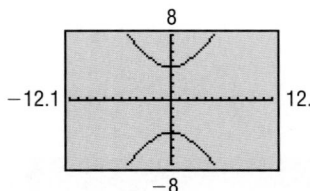

8.
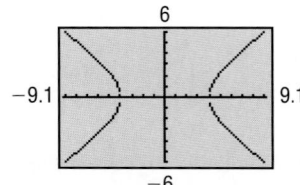

In Problems 9–18, find an equation for the hyperbola described. Graph the equation by hand.

9. Center at $(0,0)$; focus at $(3,0)$; vertex at $(1,0)$
10. Center at $(0,0)$; focus at $(0,5)$; vertex at $(0,3)$
11. Center at $(0,0)$; focus at $(0,-6)$; vertex at $(0,4)$
12. Center at $(0,0)$; focus at $(-3,0)$; vertex at $(2,0)$
13. Foci at $(-5,0)$ and $(5,0)$; vertex at $(3,0)$
14. Focus at $(0,6)$; vertices at $(0,-2)$ and $(0,2)$
15. Vertices at $(0,-6)$ and $(0,6)$; asymptote the line $y = 2x$
16. Vertices at $(-4,0)$ and $(4,0)$; asymptote the line $y = 2x$
17. Foci at $(-4,0)$ and $(4,0)$; asymptote the line $y = -x$
18. Foci at $(0,-2)$ and $(0,2)$; asymptote the line $y = -x$

In Problems 19–26, find the center, transverse axis, vertices, foci, and asymptotes. Graph each equation (a) by hand and (b) by using a graphing utility.

19. $\dfrac{x^2}{25} - \dfrac{y^2}{9} = 1$ 20. $\dfrac{y^2}{16} - \dfrac{x^2}{4} = 1$ 21. $4x^2 - y^2 = 16$ 22. $4y^2 - x^2 = 16$
23. $y^2 - 9x^2 = 9$ 24. $x^2 - y^2 = 4$ 25. $y^2 - x^2 = 25$ 26. $2x^2 - y^2 = 4$

In Problems 27–30, write an equation for each hyperbola.

27.
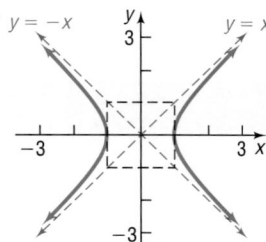

28.
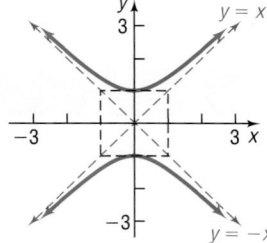

29.
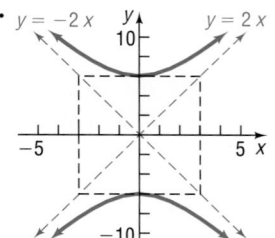

30.
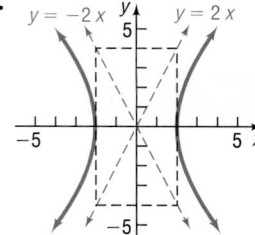

In Problems 31–38, find an equation for the hyperbola described. Graph the equation by hand.

31. Center at $(4,-1)$; focus at $(7,-1)$; vertex at $(6,-1)$
32. Center at $(-3,1)$; focus at $(-3,6)$; vertex at $(-3,4)$
33. Center at $(-3,-4)$; focus at $(-3,-8)$; vertex at $(-3,-2)$
34. Center at $(1,4)$; focus at $(-2,4)$; vertex at $(0,4)$
35. Foci at $(3,7)$ and $(7,7)$; vertex at $(6,7)$
36. Focus at $(-4,0)$; vertices at $(-4,4)$ and $(-4,2)$
37. Vertices at $(-1,-1)$ and $(3,-1)$; asymptote the line $y + 1 = \dfrac{3}{2}(x - 1)$
38. Vertices at $(1,-3)$ and $(1,1)$; asymptote the line $y + 1 = \dfrac{3}{2}(x - 1)$

In Problems 39–52, find the center, transverse axis, vertices, foci, and asymptotes. Graph each equation (a) by hand and (b) by using a graphing utility.

39. $\dfrac{(x-2)^2}{4} - \dfrac{(y+3)^2}{9} = 1$

40. $\dfrac{(y+3)^2}{4} - \dfrac{(x-2)^2}{9} = 1$

41. $(y-2)^2 - 4(x+2)^2 = 4$

42. $(x+4)^2 - 9(y-3)^2 = 9$

43. $(x+1)^2 - (y+2)^2 = 4$

44. $(y-3)^2 - (x+2)^2 = 4$

45. $x^2 - y^2 - 2x - 2y - 1 = 0$

46. $y^2 - x^2 - 4y + 4x - 1 = 0$

47. $y^2 - 4x^2 - 4y - 8x - 4 = 0$

48. $2x^2 - y^2 + 4x + 4y - 4 = 0$

49. $4x^2 - y^2 - 24x - 4y + 16 = 0$

50. $2y^2 - x^2 + 2x + 8y + 3 = 0$

51. $y^2 - 4x^2 - 16x - 2y - 19 = 0$

52. $x^2 - 3y^2 + 8x - 6y + 4 = 0$

In Problems 53–56, graph each function.

[**Hint:** Notice that each function is half a hyperbola.]

53. $f(x) = \sqrt{16 + 4x^2}$

54. $f(x) = -\sqrt{9 + 9x^2}$

55. $f(x) = -\sqrt{-25 + x^2}$

56. $f(x) = \sqrt{-1 + x^2}$

---

57. LORAN Two LORAN stations are positioned 200 miles apart along a straight shore.

(a) A ship records a time difference of 0.00038 second between the LORAN signals. Set up an appropriate rectangular coordinate system to determine where the ship would reach shore if it were to follow the hyperbola corresponding to this time difference.

(b) If the ship wants to enter a harbor located between the two stations 20 miles from the master station, what time difference should it be looking for?

(c) If the ship is 50 miles offshore when the desired time difference is obtained, what is the approximate location of the ship?

[*Note:* The speed of each radio signal is 186,000 miles per second.]

58. LORAN Two LORAN stations are positioned 100 miles apart along a straight shore.

(a) A ship records a time difference of 0.00032 second between the LORAN signals. Set up an appropriate rectangular coordinate system to determine where the ship would reach shore if it were to follow the hyperbola corresponding to this time difference.

(b) If the ship wants to enter a harbor located between the two stations 10 miles from the master station, what time difference should it be looking for?

(c) If the ship is 20 miles offshore when the desired time difference is obtained, what is the approximate location of the ship?

[*Note:* The speed of each radio signal is 186,000 miles per second.]

59. Calibrating Instruments In a test of their recording devices, a team of seismologists positioned two of the devices 2000 feet apart, with the device at point A to the west of the device at point B. At a point between the devices and 200 feet from point B, a small amount of explosive was detonated and a note made of the time at which the sound reached each device. A second explosion is to be carried out at a point directly north of point B.

(a) How far north should the site of the second explosion be chosen so that the measured time difference recorded by the devices for the second detonation is the same as that recorded for the first detonation?

(b) Explain why this experiment can be used to calibrate the instruments.

60. Explain in your own words the LORAN system of navigation.

61. The **eccentricity** e of a hyperbola is defined as the number $\dfrac{c}{a}$, where a and c are the numbers given in equation (2). Because $c > a$, it follows that $e > 1$. Describe the general shape of a hyperbola whose eccentricity is close to 1. What is the shape if e is very large?

62. A hyperbola for which $a = b$ is called an **equilateral hyperbola**. Find the eccentricity e of an equilateral hyperbola.

[*Note:* The eccentricity of a hyperbola is defined in Problem 61.]

63. Two hyperbolas that have the same set of asymptotes are called **conjugate**. Show that the hyperbolas

$$\frac{x^2}{4} - y^2 = 1 \quad \text{and} \quad y^2 - \frac{x^2}{4} = 1$$

are conjugate. Graph each hyperbola on the same set of coordinate axes.

64. Prove that the hyperbola

$$\frac{y^2}{a^2} - \frac{x^2}{b^2} = 1$$

has the two oblique asymptotes

$$y = \frac{a}{b}x \quad \text{and} \quad y = -\frac{a}{b}x$$

65. Show that the graph of an equation of the form

$$Ax^2 + Cy^2 + F = 0, \qquad A \neq 0, C \neq 0, F \neq 0$$

where A and C are of opposite sign, is a hyperbola with center at $(0, 0)$.

66. Show that the graph of an equation of the form

$$Ax^2 + Cy^2 + Dx + Ey + F = 0, \qquad A \neq 0, C \neq 0$$

where A and C are of opposite sign,

(a) Is a hyperbola if $\dfrac{D^2}{4A} + \dfrac{E^2}{4C} - F \neq 0$.

(b) Is two intersecting lines if

$$\frac{D^2}{4A} + \frac{E^2}{4C} - F = 0$$

PREPARING FOR THIS SECTION

Before getting started, review the following:

✓ Sum Formulas for Sine and Cosine (Section 3.4, pp. 222 and 225)

✓ Half-Angle Formulas for Sine and Cosine (Section 3.4, p. 237)

✓ Double-Angle Formulas for Sine and Cosine (Section 3.4, p. 234)

6.5 ROTATION OF AXES; GENERAL FORM OF A CONIC

OBJECTIVES

1. Identify a Conic
2. Use a Rotation of Axes to Transform Equations
3. Discuss an Equation Using a Rotation of Axes
4. Identify Conics without a Rotation of Axes

In this section, we show that the graph of a general second-degree polynomial containing two variables x and y, that is, an equation of the form

$$Ax^2 + Bxy + Cy^2 + Dx + Ey + F = 0 \qquad (1)$$

where A, B, and C are not simultaneously 0, is a conic. We shall not concern ourselves here with the degenerate cases of equation (1), such as $x^2 + y^2 = 0$, whose graph is a single point $(0, 0)$; or $x^2 + 3y^2 + 3 = 0$, whose graph contains no points; or $x^2 - 4y^2 = 0$, whose graph is two lines, $x - 2y = 0$ and $x + 2y = 0$.

We begin with the case where $B = 0$. In this case, the term containing xy is not present, so equation (1) has the form

$$Ax^2 + Cy^2 + Dx + Ey + F = 0$$

where either $A \neq 0$ or $C \neq 0$.

① We have already discussed the procedure for identifying the graph of this kind of equation; we complete the squares of the quadratic expressions in x or y, or both. Once this has been done, the conic can be identified by comparing it to one of the forms studied in Sections 6.2 through 6.4.

In fact, though, we can identify the conic directly from the equation without completing the squares.

Theorem	**Identifying Conics without Completing the Squares**

Excluding degenerate cases, the equation

$$Ax^2 + Cy^2 + Dx + Ey + F = 0 \qquad (2)$$

where A and C cannot both equal zero:

(a) Defines a parabola if $AC = 0$.
(b) Defines an ellipse (or a circle) if $AC > 0$.
(c) Defines a hyperbola if $AC < 0$.

■

Proof

(a) If $AC = 0$, then either $A = 0$ or $C = 0$, but not both, so the form of equation (2) is either

$$Ax^2 + Dx + Ey + F = 0, \qquad A \neq 0$$

or

$$Cy^2 + Dx + Ey + F = 0, \qquad C \neq 0$$

Using the results of Problems 71 and 72 in Exercise 6.2, it follows that, except for the degenerate cases, the equation is a parabola.

(b) If $AC > 0$, then A and C are of the same sign. Using the results of Problems 73 and 74 in Exercise 6.3, except for the degenerate cases, the equation is an ellipse if $A \neq C$ or a circle if $A = C$.

(c) If $AC < 0$, then A and C are of opposite sign. Using the results of Problems 65 and 66 in Exercise 6.4, except for the degenerate cases, the equation is a hyperbola. ■

We will not be concerned with the degenerate cases of equation (2). However, in practice, you should be alert to the possibility of degeneracy.

EXAMPLE 1	**Identifying a Conic without Completing the Squares**

Identify each equation without completing the squares.

(a) $3x^2 + 6y^2 + 6x - 12y = 0$ (b) $2x^2 - 3y^2 + 6y + 4 = 0$
(c) $y^2 - 2x + 4 = 0$

Solution   (a) We compare the given equation to equation (2) and conclude that $A = 3$ and $C = 6$. Since $AC = 18 > 0$, the equation is an ellipse.

(b) Here, $A = 2$ and $C = -3$, so $AC = -6 < 0$. The equation is a hyperbola.

(c) Here, $A = 0$ and $C = 1$, so $AC = 0$. The equation is a parabola. ■

NOW WORK PROBLEM 1.

Although we can now identify the type of conic represented by any equation of the form of equation (2) without completing the squares, we will still need to complete the squares if we desire additional information about a conic.

Now we turn our attention to equations of the form of equation (1), where $B \neq 0$. To discuss this case, we first need to investigate a new procedure: *rotation of axes*.

Rotation of Axes

Figure 53

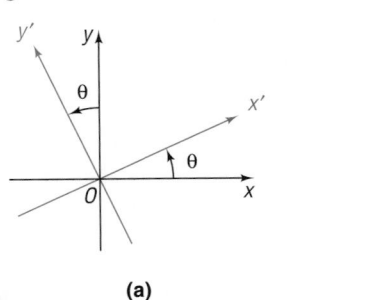

(a)

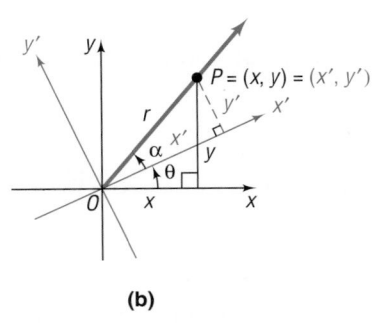

(b)

(2) In a **rotation of axes**, the origin remains fixed while the x-axis and y-axis are rotated through an angle θ to a new position; the new positions of the x- and y-axes are denoted by x' and y', respectively, as shown in Figure 53(a).

Now look at Figure 53(b). There the point P has the coordinates (x, y) relative to the xy-plane, while the same point P has coordinates (x', y') relative to the $x'y'$-plane. We seek relationships that will enable us to express x and y in terms of x', y', and θ.

As Figure 53(b) shows, r denotes the distance from the origin O to the point P, and α denotes the angle between the positive x'-axis and the ray from O through P. Then, using the definitions of sine and cosine, we have

$$x' = r \cos \alpha \qquad y' = r \sin \alpha \qquad (3)$$

$$x = r \cos(\theta + \alpha) \qquad y = r \sin(\theta + \alpha) \qquad (4)$$

Now

$$\begin{aligned}
x &= r \cos(\theta + \alpha) \\
&= r(\cos \theta \cos \alpha - \sin \theta \sin \alpha) && \text{\small Sum formula for cosine} \\
&= (r \cos \alpha)(\cos \theta) - (r \sin \alpha)(\sin \theta) \\
&= x' \cos \theta - y' \sin \theta && \text{\small By equation (3)}
\end{aligned}$$

Similarly,

$$\begin{aligned}
y &= r \sin(\theta + \alpha) \\
&= r(\sin \theta \cos \alpha + \cos \theta \sin \alpha) \\
&= x' \sin \theta + y' \cos \theta
\end{aligned}$$

Theorem

Rotation Formulas

If the x- and y-axes are rotated through an angle θ, the coordinates (x, y) of a point P relative to the xy-plane and the coordinates (x', y') of the same point relative to the new x'- and y'-axes are related by the formulas

$$x = x' \cos \theta - y' \sin \theta \qquad y = x' \sin \theta + y' \cos \theta \qquad (5)$$

EXAMPLE 2   **Rotating Axes**

Express the equation $xy = 1$ in terms of new $x'y'$-coordinates by rotating the axes through a 45° angle. Discuss the new equation.

Solution   Let $\theta = 45°$ in equation (5). Then

$$x = x' \cos 45° - y' \sin 45° = x' \frac{\sqrt{2}}{2} - y' \frac{\sqrt{2}}{2} = \frac{\sqrt{2}}{2}(x' - y')$$

$$y = x' \sin 45° + y' \cos 45° = x' \frac{\sqrt{2}}{2} + y' \frac{\sqrt{2}}{2} = \frac{\sqrt{2}}{2}(x' + y')$$

Substituting these expressions for x and y in $xy = 1$ gives

$$\left[\frac{\sqrt{2}}{2}(x' - y')\right]\left[\frac{\sqrt{2}}{2}(x' + y')\right] = 1$$

$$\frac{1}{2}(x'^2 - y'^2) = 1$$

$$\frac{x'^2}{2} - \frac{y'^2}{2} = 1$$

Figure 54

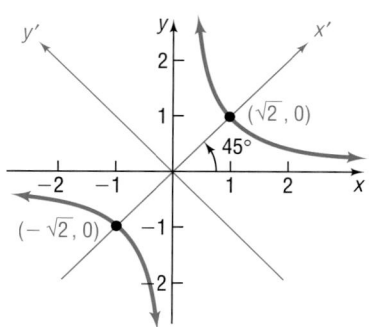

This is the equation of a hyperbola with center at $(0,0)$ and transverse axis along the x'-axis. The vertices are at $(\pm\sqrt{2}, 0)$ on the x'-axis; the asymptotes are $y' = x'$ and $y' = -x'$ (which correspond to the original x- and y-axes). See Figure 54 for the graph drawn by hand. ■

As Example 2 illustrates, a rotation of axes through an appropriate angle can transform a second-degree equation in x and y containing an xy-term into one in x' and y' in which no $x'y'$-term appears. In fact, we will show that a rotation of axes through an appropriate angle will transform any equation of the form of equation (1) into an equation in x' and y' without an $x'y'$-term.

To find the formula for choosing an appropriate angle θ through which to rotate the axes, we begin with equation (1),

$$Ax^2 + Bxy + Cy^2 + Dx + Ey + F = 0, \qquad B \neq 0$$

Next we rotate through an angle θ using rotation formulas (5).

$$A(x' \cos\theta - y' \sin\theta)^2 + B(x' \cos\theta - y' \sin\theta)(x' \sin\theta + y' \cos\theta)$$
$$+ C(x' \sin\theta + y' \cos\theta)^2 + D(x' \cos\theta - y' \sin\theta)$$
$$+ E(x' \sin\theta + y' \cos\theta) + F = 0$$

By expanding and collecting like terms, we obtain

$$(A\cos^2\theta + B\sin\theta\cos\theta + C\sin^2\theta)x'^2 + [B(\cos^2\theta - \sin^2\theta) + 2(C - A)(\sin\theta\cos\theta)]x'y'$$
$$+ (A\sin^2\theta - B\sin\theta\cos\theta + C\cos^2\theta)y'^2$$
$$+ (D\cos\theta + E\sin\theta)x'$$
$$+ (-D\sin\theta + E\cos\theta)y' + F = 0 \qquad (6)$$

In equation (6), the coefficient of $x'y'$ is

$$2(C - A)(\sin\theta\cos\theta) + B(\cos^2\theta - \sin^2\theta)$$

Since we want to eliminate the $x'y'$-term, we select an angle θ so that

$$2(C - A)(\sin\theta\cos\theta) + B(\cos^2\theta - \sin^2\theta) = 0$$
$$(C - A)\sin(2\theta) + B\cos(2\theta) = 0 \quad \text{Double-angle formulas}$$
$$B\cos(2\theta) = (A - C)\sin(2\theta)$$
$$\cot(2\theta) = \frac{A - C}{B}, \qquad B \neq 0$$

Theorem

To transform the equation

$$Ax^2 + Bxy + Cy^2 + Dx + Ey + F = 0, \qquad B \neq 0$$

into an equation in x' and y' without an $x'y'$-term, rotate the axes through an angle θ that satisfies the equation

$$\cot(2\theta) = \frac{A - C}{B} \qquad\qquad (7)$$

Equation (7) has an infinite number of solutions for θ. We shall adopt the convention of choosing the acute angle θ that satisfies (7). Then we have the following two possibilities:

If $\cot(2\theta) \geq 0$, then $0° < 2\theta \leq 90°$ so that $0° < \theta \leq 45°$.

If $\cot(2\theta) < 0$, then $90° < 2\theta < 180°$ so that $45° < \theta < 90°$.

Each of these results in a counterclockwise rotation of the axes through an acute angle θ.*

WARNING Be careful if you use a calculator to solve equation (7).

1. If $\cot(2\theta) = 0$, then $2\theta = 90°$ and $\theta = 45°$.
2. If $\cot(2\theta) \neq 0$, first find $\cos(2\theta)$. Then use the inverse cosine function key(s) to obtain $2\theta, 0° < 2\theta < 180°$. Finally, divide by 2 to obtain the correct acute angle θ. ∎

---

③ **EXAMPLE 3** **Discussing an Equation Using a Rotation of Axes**

Discuss the equation: $x^2 + \sqrt{3}xy + 2y^2 - 10 = 0$

Solution Since an xy-term is present, we must rotate the axes. Using $A = 1, B = \sqrt{3}$, and $C = 2$ in equation (7), the appropriate acute angle θ through which to rotate the axes satisfies the equation

$$\cot(2\theta) = \frac{A - C}{B} = \frac{-1}{\sqrt{3}} = -\frac{\sqrt{3}}{3}, \qquad 0° < 2\theta < 180°$$

Since $\cot(2\theta) = -\dfrac{\sqrt{3}}{3}$, we find $2\theta = 120°$, so $\theta = 60°$. Using $\theta = 60°$ in rotation formulas (5), we find

$$x = x' \cos 60° - y' \sin 60° = \frac{1}{2}x' - \frac{\sqrt{3}}{2}y' = \frac{1}{2}(x' - \sqrt{3}y')$$

$$y = x' \sin 60° + y' \cos 60° = \frac{\sqrt{3}}{2}x' + \frac{1}{2}y' = \frac{1}{2}(\sqrt{3}x' + y')$$

---

*Any rotation (clockwise or counterclockwise) through an angle θ that satisfies $\cot(2\theta) = \dfrac{A - C}{B}$ will eliminate the $x'y'$-term. However, the final form of the transformed equation may be different (but equivalent), depending on the angle chosen.

Substituting these values into the original equation and simplifying, we have

$$x^2 + \sqrt{3}xy + 2y^2 - 10 = 0$$

$$\frac{1}{4}(x' - \sqrt{3}y')^2 + \sqrt{3}\left[\frac{1}{2}(x' - \sqrt{3}y')\right]\left[\frac{1}{2}(\sqrt{3}x' + y')\right] + 2\left[\frac{1}{4}(\sqrt{3}x' + y')^2\right] = 10$$

Multiply both sides by 4 and expand to obtain

$$x'^2 - 2\sqrt{3}x'y' + 3y'^2 + \sqrt{3}(\sqrt{3}x'^2 - 2x'y' - \sqrt{3}y'^2) + 2(3x'^2 + 2\sqrt{3}x'y' + y'^2) = 40$$

$$10x'^2 + 2y'^2 = 40$$

$$\frac{x'^2}{4} + \frac{y'^2}{20} = 1$$

Figure 55

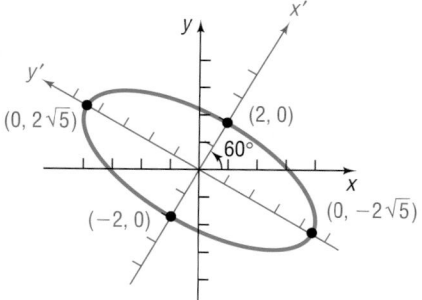

This is the equation of an ellipse with center at $(0,0)$ and major axis along the y'-axis. The vertices are at $(0, \pm 2\sqrt{5})$ on the y'-axis. See Figure 55 for the graph drawn by hand.

To graph the equation $x^2 + \sqrt{3}xy + 2y^2 - 10 = 0$ using a graphing utility, we need to solve the equation for y. Rearranging the terms we observe the equation is quadratic in the variable y: $2y^2 + \sqrt{3}xy + (x^2 - 10) = 0$. We can solve the equation for y using the quadratic formula with $a = 2$, $b = \sqrt{3}x$, and $c = x^2 - 10$.

$$Y_1 = \frac{-\sqrt{3}x + \sqrt{(\sqrt{3}x)^2 - 4(2)(x^2 - 10)}}{2(2)} = \frac{-\sqrt{3}x + \sqrt{-5x^2 + 80}}{4}$$

and

$$Y_2 = \frac{-\sqrt{3}x - \sqrt{(\sqrt{3}x)^2 - 4(2)(x^2 - 10)}}{2(2)} = \frac{-\sqrt{3}x - \sqrt{-5x^2 + 80}}{4}$$

Figure 56 shows the graph of Y_1 and Y_2. ■

Figure 56

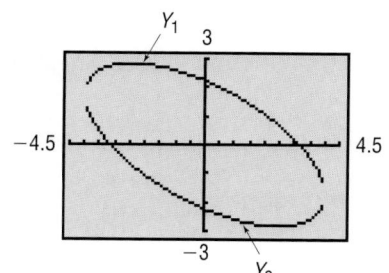

NOW WORK PROBLEM **21.**

In Example 3, the acute angle θ through which to rotate the axes was easy to find because of the numbers that we used in the given equation. In general, the equation $\cot(2\theta) = \dfrac{A - C}{B}$ will not have such a "nice" solution. As the next example shows, we can still find the appropriate rotation formulas without using a calculator approximation by applying half-angle formulas.

EXAMPLE 4   **Discussing an Equation Using a Rotation of Axes**

Discuss the equation: $4x^2 - 4xy + y^2 + 5\sqrt{5}x + 5 = 0$

Solution   Letting $A = 4$, $B = -4$, and $C = 1$ in equation (7), the appropriate angle θ through which to rotate the axes satisfies

$$\cot(2\theta) = \frac{A - C}{B} = \frac{3}{-4} = -\frac{3}{4}$$

In order to use rotation formulas (5), we need to know the values of $\sin\theta$ and $\cos\theta$. Since we seek an acute angle θ, we know that $\sin\theta > 0$ and $\cos\theta > 0$. We use the half-angle formulas in the form

$$\sin\theta = \sqrt{\frac{1 - \cos(2\theta)}{2}} \qquad \cos\theta = \sqrt{\frac{1 + \cos(2\theta)}{2}}$$

Now we need to find the value of $\cos(2\theta)$. Since $\cot(2\theta) = -\dfrac{3}{4}$ and $90° < 2\theta < 180°$ (Do you know why?), it follows that $\cos(2\theta) = -\dfrac{3}{5}$. Then

$$\sin\theta = \sqrt{\frac{1 - \cos(2\theta)}{2}} = \sqrt{\frac{1 - \left(-\dfrac{3}{5}\right)}{2}} = \sqrt{\frac{4}{5}} = \frac{2}{\sqrt{5}} = \frac{2\sqrt{5}}{5}$$

$$\cos\theta = \sqrt{\frac{1 + \cos(2\theta)}{2}} = \sqrt{\frac{1 + \left(-\dfrac{3}{5}\right)}{2}} = \sqrt{\frac{1}{5}} = \frac{1}{\sqrt{5}} = \frac{\sqrt{5}}{5}$$

With these values, the rotation formulas (5) are

$$x = \frac{\sqrt{5}}{5}x' - \frac{2\sqrt{5}}{5}y' = \frac{\sqrt{5}}{5}(x' - 2y')$$

$$y = \frac{2\sqrt{5}}{5}x' + \frac{\sqrt{5}}{5}y' = \frac{\sqrt{5}}{5}(2x' + y')$$

Substituting these values in the original equation and simplifying, we obtain

$$4x^2 - 4xy + y^2 + 5\sqrt{5}x + 5 = 0$$

$$4\left[\frac{\sqrt{5}}{5}(x' - 2y')\right]^2 - 4\left[\frac{\sqrt{5}}{5}(x' - 2y')\right]\left[\frac{\sqrt{5}}{5}(2x' + y')\right]$$
$$+ \left[\frac{\sqrt{5}}{5}(2x' + y')\right]^2 + 5\sqrt{5}\left[\frac{\sqrt{5}}{5}(x' - 2y')\right] = -5$$

Multiply both sides by 5 and expand to obtain

$$4(x'^2 - 4x'y' + 4y'^2) - 4(2x'^2 - 3x'y' - 2y'^2)$$
$$+ 4x'^2 + 4x'y' + y'^2 + 25(x' - 2y') = -25$$

$$25y'^2 - 50y' + 25x' = -25 \quad \text{Combine like terms.}$$
$$y'^2 - 2y' + x' = -1 \quad \text{Divide by 25.}$$
$$y'^2 - 2y' + 1 = -x' \quad \text{Complete the square in } y'.$$
$$(y' - 1)^2 = -x'$$

Figure 57

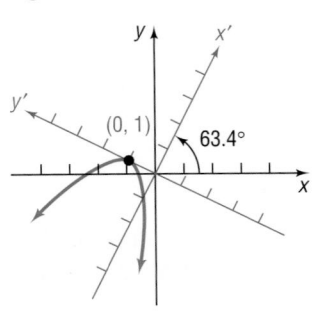

This is the equation of a parabola with vertex at $(0, 1)$ in the $x'y'$-plane. The axis of symmetry is parallel to the x'-axis. Using a calculator to solve $\sin\theta = \dfrac{2\sqrt{5}}{5}$, we find that $\theta \approx 63.4°$. See Figure 57 for the graph drawn by hand.

To graph the equation $4x^2 - 4xy + y^2 + 5\sqrt{5}x + 5 = 0$ using a graphing utility, we need to solve the equation for y. Rearranging the terms we observe the equation is quadratic in the variable y: $y^2 - 4xy + (4x^2 + 5\sqrt{5}x + 5) = 0$. We can solve the equation for y using the quadratic formula with $a = 1, b = -4x$, and $c = 4x^2 + 5\sqrt{5}x + 5$.

Figure 58

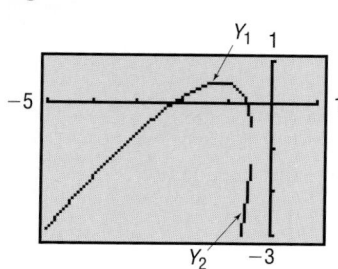

$$Y_1 = \frac{-(-4x) + \sqrt{(-4x)^2 - 4(1)(4x^2 + 5\sqrt{5}x + 5)}}{2(1)} = 2x + \sqrt{-5(\sqrt{5}x + 1)}$$

$$Y_2 = \frac{-(-4x) - \sqrt{(-4x)^2 - 4(1)(4x^2 + 5\sqrt{5}x + 5)}}{2(1)} = 2x - \sqrt{-5(\sqrt{5}x + 1)}$$

Figure 58 shows the graph of Y_1 and Y_2.

NOW WORK PROBLEM 27.

Identifying Conics without a Rotation of Axes

④ Suppose that we are required only to identify (rather than discuss) an equation of the form

$$Ax^2 + Bxy + Cy^2 + Dx + Ey + F = 0, \qquad B \neq 0 \qquad (8)$$

If we apply rotation formulas (5) to this equation, we obtain an equation of the form

$$A'x'^2 + B'x'y' + C'y'^2 + D'x' + E'y' + F' = 0 \qquad (9)$$

where A', B', C', D', E', and F' can be expressed in terms of A, B, C, D, E, F, and the angle θ of rotation (see Problem 43). It can be shown that the value of $B^2 - 4AC$ in equation (8) and the value of $B'^2 - 4A'C'$ in equation (9) are equal no matter what angle θ of rotation is chosen (see Problem 45). In particular, if the angle θ of rotation satisfies equation (7), then $B' = 0$ in equation (9), and $B^2 - 4AC = -4A'C'$. Since equation (9) then has the form of equation (2),

$$A'x'^2 + C'y'^2 + D'x' + E'y' + F' = 0$$

we can identify it without completing the squares, as we did in the beginning of this section. In fact, now we can identify the conic described by any equation of the form of equation (8) without a rotation of axes.

Theorem   **Identifying Conics without a Rotation of Axes**

Except for degenerate cases, the equation

$$Ax^2 + Bxy + Cy^2 + Dx + Ey + F = 0$$

(a) Defines a parabola if $B^2 - 4AC = 0$.
(b) Defines an ellipse (or a circle) if $B^2 - 4AC < 0$.
(c) Defines a hyperbola if $B^2 - 4AC > 0$.

You are asked to prove this theorem in Problem 46.

EXAMPLE 5   **Identifying a Conic without a Rotation of Axes**

Identify the equation: $8x^2 - 12xy + 17y^2 - 4\sqrt{5}x - 2\sqrt{5}y - 15 = 0$

Solution   Here $A = 8$, $B = -12$, and $C = 17$, so $B^2 - 4AC = -400$. Since $B^2 - 4AC < 0$, the equation defines an ellipse.

NOW WORK PROBLEM 33.

6.5 Concepts and Vocabulary

In Problems 1–3, fill in the blanks.

1. To transform the equation

$$Ax^2 + Bxy + Cy^2 + Dx + Ey + F = 0, \qquad B \neq 0$$

into one in x' and y' without an $x'y'$-term, rotate the axes through an acute angle θ that satisfies the equation

_____.

2. Identify the conic: $x^2 - 2y^2 + x - y - 18 = 0$ _____.

3. Identify the conic: $x^2 + 2xy + 3y^2 - 2x + 4y + 10 = 0$ _____.

In Problems 4–6, answer True or False to each statement.

4. The equation $ax^2 + 6y^2 - 12y = 0$ defines an ellipse if $a > 0$.

5. The equation $3x^2 + bxy + 12y^2 = 10$ defines a parabola if $b = -12$.

6. To eliminate the xy-term from the equation $x^2 - 2xy + y^2 - 2x + 3y + 5 = 0$, rotate the axes through an angle θ, where $\cot \theta = B^2 - 4AC$.

7. Write down an equation of the form $Ax^2 + Bxy + Cy^2 = 10$ that is a parabola.

8. Write down an equation of the form $Ax^2 + Bxy + Cy^2 = 10$ that is an ellipse.

9. Write down an equation of the form $Ax^2 + Bxy + Cy^2 = 10$ that is a hyperbola.

6.5 Exercises

In Problems 1–10, identify each equation without completing the squares.

1. $x^2 + 4x + y + 3 = 0$

2. $2y^2 - 3y + 3x = 0$

3. $6x^2 + 3y^2 - 12x + 6y = 0$

4. $2x^2 + y^2 - 8x + 4y + 2 = 0$

5. $3x^2 - 2y^2 + 6x + 4 = 0$

6. $4x^2 - 3y^2 - 8x + 6y + 1 = 0$

7. $2y^2 - x^2 - y + x = 0$

8. $y^2 - 8x^2 - 2x - y = 0$

9. $x^2 + y^2 - 8x + 4y = 0$

10. $2x^2 + 2y^2 - 8x + 8y = 0$

In Problems 11–20, determine the appropriate rotation formulas to use so that the new equation contains no xy-term.

11. $x^2 + 4xy + y^2 - 3 = 0$

12. $x^2 - 4xy + y^2 - 3 = 0$

13. $5x^2 + 6xy + 5y^2 - 8 = 0$

14. $3x^2 - 10xy + 3y^2 - 32 = 0$

15. $13x^2 - 6\sqrt{3}xy + 7y^2 - 16 = 0$

16. $11x^2 + 10\sqrt{3}xy + y^2 - 4 = 0$

17. $4x^2 - 4xy + y^2 - 8\sqrt{5}x - 16\sqrt{5}y = 0$

18. $x^2 + 4xy + 4y^2 + 5\sqrt{5}y + 5 = 0$

19. $25x^2 - 36xy + 40y^2 - 12\sqrt{13}x - 8\sqrt{13}y = 0$

20. $34x^2 - 24xy + 41y^2 - 25 = 0$

In Problems 21–32, rotate the axes so that the new equation contains no xy-term. Discuss and graph the new equation by hand. Refer to Problems 11–20 for Problems 21–30.

21. $x^2 + 4xy + y^2 - 3 = 0$

22. $x^2 - 4xy + y^2 - 3 = 0$

23. $5x^2 + 6xy + 5y^2 - 8 = 0$

24. $3x^2 - 10xy + 3y^2 - 32 = 0$

25. $13x^2 - 6\sqrt{3}xy + 7y^2 - 16 = 0$

26. $11x^2 + 10\sqrt{3}xy + y^2 - 4 = 0$

27. $4x^2 - 4xy + y^2 - 8\sqrt{5}x - 16\sqrt{5}y = 0$

28. $x^2 + 4xy + 4y^2 + 5\sqrt{5}y + 5 = 0$

29. $25x^2 - 36xy + 40y^2 - 12\sqrt{13}x - 8\sqrt{13}y = 0$

30. $34x^2 - 24xy + 41y^2 - 25 = 0$

31. $16x^2 + 24xy + 9y^2 - 130x + 90y = 0$

32. $16x^2 + 24xy + 9y^2 - 60x + 80y = 0$

In Problems 33–42, identify each equation without applying a rotation of axes.

33. $x^2 + 3xy - 2y^2 + 3x + 2y + 5 = 0$

34. $2x^2 - 3xy + 4y^2 + 2x + 3y - 5 = 0$

35. $x^2 - 7xy + 3y^2 - y - 10 = 0$

36. $2x^2 - 3xy + 2y^2 - 4x - 2 = 0$

37. $9x^2 + 12xy + 4y^2 - x - y = 0$

38. $10x^2 + 12xy + 4y^2 - x - y + 10 = 0$

39. $10x^2 - 12xy + 4y^2 - x - y - 10 = 0$

40. $4x^2 + 12xy + 9y^2 - x - y = 0$

41. $3x^2 - 2xy + y^2 + 4x + 2y - 1 = 0$

42. $3x^2 + 2xy + y^2 + 4x - 2y + 10 = 0$

In Problems 43–46, apply rotation formulas (5) to

$$Ax^2 + Bxy + Cy^2 + Dx + Ey + F = 0$$

to obtain the equation

$$A'x'^2 + B'x'y' + C'y'^2 + D'x' + E'y' + F' = 0$$

43. Express A', B', C', D', E', and F' in terms of A, B, C, D, E, F, and the angle θ of rotation.

    [**Hint:** Refer to Equation (6)].

44. Show that $A + C = A' + C'$, and thus show that $A + C$ is **invariant;** that is, its value does not change under a rotation of axes.

45. Refer to Problem 44. Show that $B^2 - 4AC$ is invariant.

46. Prove that, except for degenerate cases, the equation

$$Ax^2 + Bxy + Cy^2 + Dx + Ey + F = 0$$

  (a) Defines a parabola if $B^2 - 4AC = 0$.
  (b) Defines an ellipse (or a circle) if $B^2 - 4AC < 0$.
  (c) Defines a hyperbola if $B^2 - 4AC > 0$.

47. Use rotation formulas (5) to show that distance is invariant under a rotation of axes. That is, show that the distance from $P_1 = (x_1, y_1)$ to $P_2 = (x_2, y_2)$ in the xy-plane equals the distance from $P_1 = (x_1', y_1')$ to $P_2 = (x_2', y_2')$ in the $x'y'$-plane.

48. Show that the graph of the equation $x^{\frac{1}{2}} + y^{\frac{1}{2}} = a^{\frac{1}{2}}$ is part of the graph of a parabola.

49. Formulate a strategy for discussing and graphing an equation of the form

$$Ax^2 + Cy^2 + Dx + Ey + F = 0$$

How does your strategy change if the equation is of the form

$$Ax^2 + Bxy + Cy^2 + Dx + Ey + F = 0$$

PREPARING FOR THIS SECTION

Before getting started, review the following:

✓ Polar Coordinates (Section 5.1, pp. 318–325)

6.6 POLAR EQUATIONS OF CONICS

OBJECTIVES   ① Discuss and Graph Polar Equations of Conics
    ② Convert a Polar Equation of a Conic to a Rectangular Equation

① In Sections 6.2 through 6.4, we gave separate definitions for the parabola, ellipse, and hyperbola based on geometric properties and the distance formula. In this section, we present an alternative definition that simultaneously defines all these conics. As we shall see, this approach is well suited to polar coordinate representation. (Refer to Section 5.1.)

Let D denote a fixed line called the **directrix;** let F denote a fixed point called the **focus**, which is not on D; and let e be a fixed positive number called the **eccentricity**. A **conic** is the set of points P in the plane such that the ratio of the distance from F to P to the distance from D to P equals e. That is, a conic is the collection of points P for which

$$\frac{d(F,P)}{d(D,P)} = e \qquad (1)$$

If $e = 1$, the conic is a **parabola**.

If $e < 1$, the conic is an **ellipse**.

If $e > 1$, the conic is a **hyperbola**.

Observe that if $e = 1$ the definition of a parabola in equation (1) is exactly the same as the definition used earlier in Section 6.2.

In the case of an ellipse, the **major axis** is a line through the focus perpendicular to the directrix. In the case of a hyperbola, the **transverse axis** is a line through the focus perpendicular to the directrix. For both an ellipse and a hyperbola, the eccentricity e satisfies

$$e = \frac{c}{a} \tag{2}$$

where c is the distance from the center to the focus and a is the distance from the center to a vertex.

Just as we did earlier using rectangular coordinates, we derive equations for the conics in polar coordinates by choosing a convenient position for the focus F and the directrix D. The focus F is positioned at the pole, and the directrix D is either parallel to the polar axis or perpendicular to it.

Suppose that we start with the directrix D perpendicular to the polar axis at a distance p units to the left of the pole (the focus F). See Figure 59.

If $P = (r, \theta)$ is any point on the conic, then, by equation (1),

Figure 59

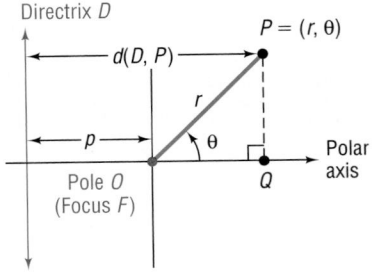

Directrix D

$P = (r, \theta)$

$d(D, P)$

r

p

θ

Pole O (Focus F)

Q

Polar axis

$$\frac{d(F, P)}{d(D, P)} = e \quad \text{or} \quad d(F, P) = e \cdot d(D, P) \tag{3}$$

Now we use the point Q obtained by dropping the perpendicular from P to the polar axis to calculate $d(D, P)$.

$$d(D, P) = p + d(O, Q) = p + r \cos \theta$$

Using this expression and the fact that $d(F, P) = d(O, P) = r$ in equation (3), we get

$$d(F, P) = e \cdot d(D, P)$$
$$r = e(p + r \cos \theta)$$
$$r = ep + er \cos \theta$$
$$r - er \cos \theta = ep$$
$$r(1 - e \cos \theta) = ep$$
$$r = \frac{ep}{1 - e \cos \theta}$$

Theorem

Polar Equation of a Conic; Focus at Pole; Directrix Perpendicular to Polar Axis a Distance p to the Left of the Pole

The polar equation of a conic with focus at the pole and directrix perpendicular to the polar axis at a distance p to the left of the pole is

$$r = \frac{ep}{1 - e \cos \theta} \tag{4}$$

where e is the eccentricity of the conic.

EXAMPLE 1 Discussing and Graphing the Polar Equation of a Conic

Discuss and graph the equation: $r = \dfrac{4}{2 - \cos\theta}$

Solution The given equation is not quite in the form of equation (4), since the first term in the denominator is 2 instead of 1. We divide the numerator and denominator by 2 to obtain

$$r = \dfrac{2}{1 - \dfrac{1}{2}\cos\theta} \qquad r = \dfrac{ep}{1 - e\cos\theta}$$

This equation is in the form of equation (4), with

$$e = \dfrac{1}{2} \quad \text{and} \quad ep = \dfrac{1}{2}p = 2, \quad p = 4$$

We conclude that the conic is an ellipse, since $e = \dfrac{1}{2} < 1$. One focus is at the pole, and the directrix is perpendicular to the polar axis, a distance of $p = 4$ units to the left of the pole. It follows that the major axis is along the polar axis. To find the vertices, we let $\theta = 0$ and $\theta = \pi$. The vertices of the ellipse are $(4, 0)$ and $\left(\dfrac{4}{3}, \pi\right)$. The midpoint of the vertices, $\left(\dfrac{4}{3}, 0\right)$ in polar coordinates, is the center of the ellipse. [Do you see why? The vertices $(4, 0)$ and $\left(\dfrac{4}{3}, \pi\right)$ in polar coordinates are $(4, 0)$ and $\left(-\dfrac{4}{3}, 0\right)$ in rectangular coordinates. The midpoint in rectangular coordinates is $\left(\dfrac{4}{3}, 0\right)$, which is also $\left(\dfrac{4}{3}, 0\right)$ in polar coordinates.] Then $a = $ distance from the center to a vertex $= \dfrac{8}{3}$. Using $a = \dfrac{8}{3}$ and $e = \dfrac{1}{2}$ in equation (2), $e = \dfrac{c}{a}$, we find $c = \dfrac{4}{3}$. Finally, using $a = \dfrac{8}{3}$ and $c = \dfrac{4}{3}$ in $b^2 = a^2 - c^2$, we have

$$b^2 = a^2 - c^2 = \dfrac{64}{9} - \dfrac{16}{9} = \dfrac{48}{9}$$

$$b = \dfrac{4\sqrt{3}}{3}$$

Figure 60(a) shows the graph drawn by hand.

Figures 60(b) shows the graph of the equation obtained using a graphing utility in POLar mode with $\theta\text{min} = 0, \theta\text{max} = 2\pi$ and $\theta\text{step} = \dfrac{\pi}{24}$. ■

Figure 60

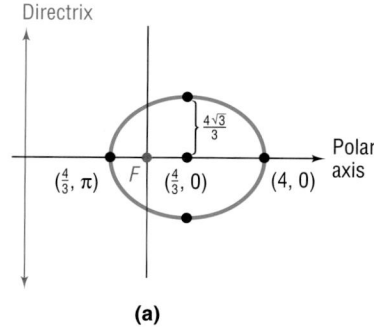

(a)

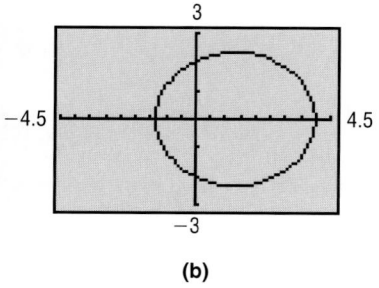

(b)

EXPLORATION Graph $r_1 = \dfrac{4}{2 + \cos\theta}$ and compare the result with Figure 60(b). What do you conclude? Clear the screen and graph $r_1 = \dfrac{4}{2 - \sin\theta}$ and then $r_1 = \dfrac{4}{2 + \sin\theta}$. Compare each of these graphs with Figure 60(b). What do you conclude?

NOW WORK PROBLEM 5.

Equation (4) was obtained under the assumption that the directrix was perpendicular to the polar axis at a distance p units to the left of the pole. A similar derivation (see Problem 37), in which the directrix is perpendicular to the polar axis at a distance p units to the right of the pole, results in the equation

$$r = \frac{ep}{1 + e \cos \theta}$$

In Problems 38 and 39 you are asked to derive the polar equations of conics with focus at the pole and directrix parallel to the polar axis. Table 5 summarizes the polar equations of conics.

TABLE 5 Polar Equations of Conics (Focus at the Pole, Eccentricity e)

Equation	Description
(a) $r = \dfrac{ep}{1 - e \cos \theta}$	Directrix is perpendicular to the polar axis at a distance p units to the left of the pole.
(b) $r = \dfrac{ep}{1 + e \cos \theta}$	Directrix is perpendicular to the polar axis at a distance p units to the right of the pole.
(c) $r = \dfrac{ep}{1 + e \sin \theta}$	Directrix is parallel to the polar axis at a distance p units above the pole.
(d) $r = \dfrac{ep}{1 - e \sin \theta}$	Directrix is parallel to the polar axis at a distance p units below the pole.

Eccentricity

If $e = 1$, the conic is a parabola; the axis of symmetry is perpendicular to the directrix.

If $e < 1$, the conic is an ellipse; the major axis is perpendicular to the directrix.

If $e > 1$, the conic is a hyperbola; the transverse axis is perpendicular to the directrix.

EXAMPLE 2 Discussing and Graphing the Polar Equation of a Conic

Discuss and graph the equation: $r = \dfrac{6}{3 + 3 \sin \theta}$

Solution To place the equation in proper form, we divide the numerator and denominator by 3 to get

$$r = \frac{2}{1 + \sin \theta}$$

Figure 61

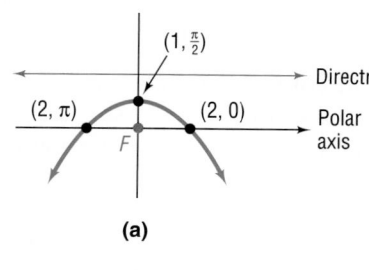

(a)

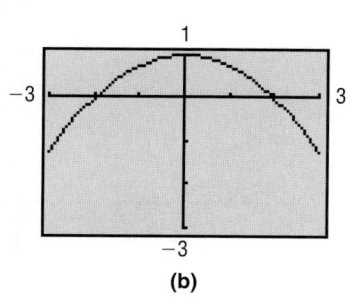

(b)

Referring to Table 5, we conclude that this equation is in the form of equation (c) with

$$e = 1 \quad \text{and} \quad ep = 2$$

$$p = 2$$

The conic is a parabola with focus at the pole. The directrix is parallel to the polar axis at a distance 2 units above the pole; the axis of symmetry is perpendicular to the polar axis. The vertex of the parabola is at $\left(1, \dfrac{\pi}{2}\right)$. (Do you see why?) See Figure 61(a) for the graph drawn by hand. Notice that we plotted two additional points, $(2, 0)$ and $(2, \pi)$, to assist in graphing.

Figure 61(b) shows the graph of the equation using a graphing utility in POLar mode with θmin $= 0$, θmax $= 2\pi$, and θstep $= \dfrac{\pi}{24}$. ∎

NOW WORK PROBLEM 7.

EXAMPLE 3   Discussing and Graphing the Polar Equation of a Conic

Discuss and graph the equation: $r = \dfrac{3}{1 + 3\cos\theta}$

Solution   This equation is in the form of equation (b) in Table 5. We conclude that

$$e = 3 \quad \text{and} \quad ep = 3p = 3$$

$$p = 1$$

This is the equation of a hyperbola with a focus at the pole. The directrix is perpendicular to the polar axis, 1 unit to the right of the pole. The transverse axis is along the polar axis. To find the vertices, we let $\theta = 0$ and $\theta = \pi$. The vertices are $\left(\dfrac{3}{4}, 0\right)$ and $\left(-\dfrac{3}{2}, \pi\right)$. The center, which is at the midpoint of $\left(\dfrac{3}{4}, 0\right)$ and $\left(-\dfrac{3}{2}, \pi\right)$, is $\left(\dfrac{9}{8}, 0\right)$. Then c = distance from the center to a focus $= \dfrac{9}{8}$. Since $e = 3$, it follows from equation (2), $e = \dfrac{c}{a}$, that $a = \dfrac{3}{8}$. Finally, using $a = \dfrac{3}{8}$ and $c = \dfrac{9}{8}$ in $b^2 = c^2 - a^2$, we find

$$b^2 = c^2 - a^2 = \frac{81}{64} - \frac{9}{64} = \frac{72}{64} = \frac{9}{8}$$

$$b = \frac{3}{2\sqrt{2}} = \frac{3\sqrt{2}}{4}$$

Figure 62(a) shows the graph drawn by hand. Notice that we plotted two additional points, $\left(3, \dfrac{\pi}{2}\right)$ and $\left(3, \dfrac{3\pi}{2}\right)$, on the left branch and used symmetry to obtain the right branch. The asymptotes of this hyperbola were found in the usual way by constructing the rectangle shown.

Figures 62(b) and (c) show the graph of the equation using a graphing utility in POLar mode with θmin $= 0$, θmax $= 2\pi$, and θstep $= \dfrac{\pi}{24}$, using both dot mode and connected mode. Notice the extraneous asymptotes in the connected mode.

Figure 62

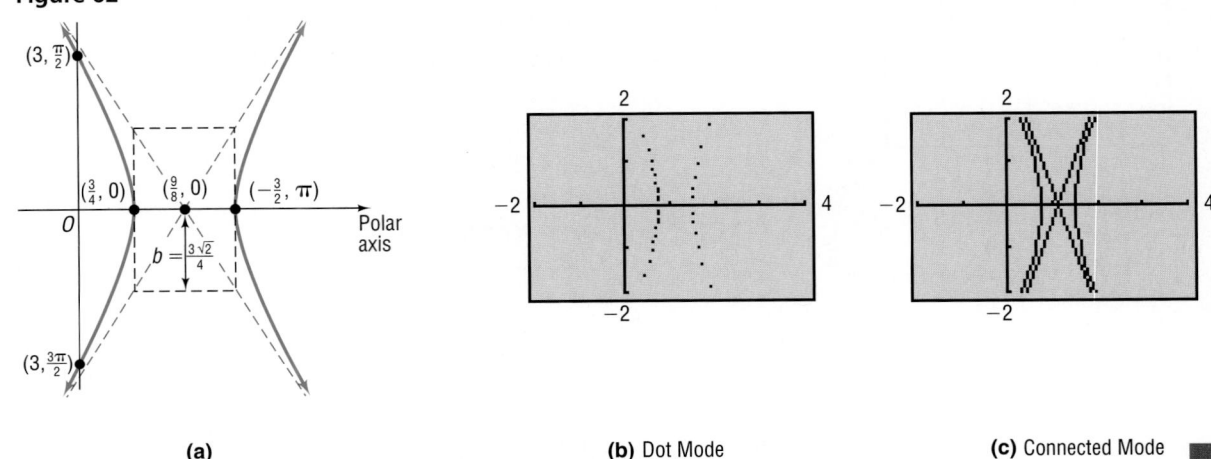

(a)   **(b)** Dot Mode   **(c)** Connected Mode

NOW WORK PROBLEM 11.

② **EXAMPLE 4**   **Converting a Polar Equation to a Rectangular Equation**

Convert the polar equation

$$r = \frac{1}{3 - 3\cos\theta}$$

to a rectangular equation.

Solution   The strategy here is first to rearrange the equation and square each side before using the transformation equations.

$$r = \frac{1}{3 - 3\cos\theta}$$

$$3r - 3r\cos\theta = 1$$

$$3r = 1 + 3r\cos\theta \qquad \text{Rearrange the equation.}$$

$$9r^2 = (1 + 3r\cos\theta)^2 \qquad \text{Square each side.}$$

$$9(x^2 + y^2) = (1 + 3x)^2 \qquad x^2 + y^2 = r^2; x = r\cos\theta$$

$$9x^2 + 9y^2 = 9x^2 + 6x + 1$$

$$9y^2 = 6x + 1$$

This is the equation of a parabola in rectangular coordinates.   ■

NOW WORK PROBLEM 19.

6.6 Concepts and Vocabulary

In Problems 1 and 2, fill in the blanks.

1. The polar equation $r = \dfrac{8}{4 - 2\sin\theta}$ is a conic whose eccentricity is _____. It is a(n) _____ whose directrix is _____ to the polar axis at a distance _____ units _____ the pole.

2. The eccentricity e of a parabola is _____, of an ellipse it is _____, and of a hyperbola it is _____.

In Problems 3 and 4, answer True or False to each statement.

3. If (r, θ) are polar coordinates, the equation $r = \dfrac{2}{2 + 3\sin\theta}$ defines a hyperbola.

4. The eccentricity of any parabola is 1.

5. Write down an equation of the form $r = \dfrac{3e}{1 + e\cos\theta}$ that is a parabola.

6. Write down an equation of the form $r = \dfrac{3e}{1 + e\cos\theta}$ that is an ellipse.

7. Write down an equation of the form $r = \dfrac{3e}{1 + e\cos\theta}$ that is a hyperbola.

6.6 Exercises

In Problems 1–6, identify the conic that each polar equation represents. Also, give the position of the directrix.

1. $r = \dfrac{1}{1 + \cos\theta}$

2. $r = \dfrac{3}{1 - \sin\theta}$

3. $r = \dfrac{4}{2 - 3\sin\theta}$

4. $r = \dfrac{2}{1 + 2\cos\theta}$

5. $r = \dfrac{3}{4 - 2\cos\theta}$

6. $r = \dfrac{6}{8 + 2\sin\theta}$

In Problems 7–18, discuss each equation and graph it by hand. Verify the graph using a graphing utility.

7. $r = \dfrac{1}{1 + \cos\theta}$

8. $r = \dfrac{3}{1 - \sin\theta}$

9. $r = \dfrac{8}{4 + 3\sin\theta}$

10. $r = \dfrac{10}{5 + 4\cos\theta}$

11. $r = \dfrac{9}{3 - 6\cos\theta}$

12. $r = \dfrac{12}{4 + 8\sin\theta}$

13. $r = \dfrac{8}{2 - \sin\theta}$

14. $r = \dfrac{8}{2 + 4\cos\theta}$

15. $r(3 - 2\sin\theta) = 6$

16. $r(2 - \cos\theta) = 2$

17. $r = \dfrac{6\sec\theta}{2\sec\theta - 1}$

18. $r = \dfrac{3\csc\theta}{\csc\theta - 1}$

In Problems 19–30, convert each polar equation to a rectangular equation.

19. $r = \dfrac{1}{1 + \cos\theta}$

20. $r = \dfrac{3}{1 - \sin\theta}$

21. $r = \dfrac{8}{4 + 3\sin\theta}$

22. $r = \dfrac{10}{5 + 4\cos\theta}$

23. $r = \dfrac{9}{3 - 6\cos\theta}$

24. $r = \dfrac{12}{4 + 8\sin\theta}$

25. $r = \dfrac{8}{2 - \sin\theta}$

26. $r = \dfrac{8}{2 + 4\cos\theta}$

27. $r(3 - 2\sin\theta) = 6$

28. $r(2 - \cos\theta) = 2$

29. $r = \dfrac{6\sec\theta}{2\sec\theta - 1}$

30. $r = \dfrac{3\csc\theta}{\csc\theta - 1}$

In Problems 31–36, find a polar equation for each conic. For each, a focus is at the pole.

31. $e = 1$;   directrix is parallel to the polar axis 1 unit above the pole

32. $e = 1$;   directrix is parallel to the polar axis 2 units below the pole

33. $e = \dfrac{4}{5}$;   directrix is perpendicular to the polar axis 3 units to the left of the pole

34. $e = \dfrac{2}{3}$;   directrix is parallel to the polar axis 3 units above the pole

35. $e = 6$;   directrix is parallel to the polar axis 2 units below the pole

36. $e = 5$;   directrix is perpendicular to the polar axis 5 units to the right of the pole

37. Derive equation (b) in Table 5:

$$r = \frac{ep}{1 + e \cos \theta}$$

38. Derive equation (c) in Table 5:

$$r = \frac{ep}{1 + e \sin \theta}$$

39. Derive equation (d) in Table 5:

$$r = \frac{ep}{1 - e \sin \theta}$$

40. Orbit of Mercury   The planet Mercury travels around the Sun in an elliptical orbit given approximately by

$$r = \frac{(3.442)10^7}{1 - 0.206 \cos \theta}$$

where r is measured in miles and the Sun is at the pole. Find the distance from Mercury to the Sun at *aphelion* (greatest distance from the Sun) and at *perihelion* (shortest distance from the Sun). See the figure. Use the aphelion and perihelion to graph the orbit of Mercury using a graphing utility.

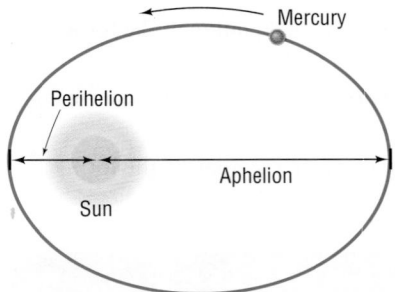

PREPARING FOR THIS SECTION

Before getting started, review the following:

✓ Amplitude and Period of Sinusoidal Graphs (Section 2.4, p. 161)

6.7   PLANE CURVES AND PARAMETRIC EQUATIONS

OBJECTIVES
1. Graph Parametric Equations by Hand
2. Graph Parametric Equations Using a Graphing Utility
3. Find a Rectangular Equation for a Curve Defined Parametrically
4. Use Time as a Parameter in Parametric Equations
5. Find Parametric Equations for Curves Defined by Rectangular Equations

Equations of the form $y = f(x)$, where f is a function, have graphs that are intersected no more than once by any vertical line. The graphs of many of the conics and certain other, more complicated graphs do not have this characteristic. Yet each graph, like the graph of a function, is a collection of points (x, y) in the xy-plane; that is, each is a *plane curve*. In this section, we discuss another way of representing such graphs.

Let $x = f(t)$ and $y = g(t)$, where f and g are two functions whose common domain is some interval I. The collection of points defined by

$$(x, y) = (f(t), g(t))$$

is called a **plane curve**. The equations

$$x = f(t) \qquad y = g(t)$$

where t is in I, are called **parametric equations** of the curve. The variable t is called a **parameter.**

①   Parametric equations are particularly useful in describing movement along a curve. Suppose that a curve is defined by the parametric equations

$$x = f(t), \qquad y = g(t), \qquad a \le t \le b$$

where f and g are each defined over the interval $a \le t \le b$. For a given value of t, we can find the value of $x = f(t)$ and $y = g(t)$, obtaining a point (x, y) on the curve. In fact, as t varies over the interval from $t = a$ to $t = b$, successive values of t give rise to a directed movement along the curve; that is, the curve is traced out in a certain direction by the corresponding succession of points (x, y). See Figure 63. The arrows show the direction, or **orientation**, along the curve as t varies from a to b.

Figure 63

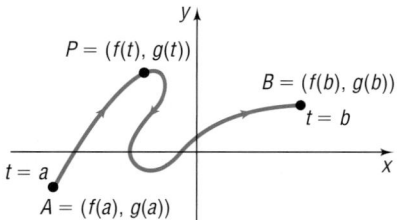

EXAMPLE 1   Discussing a Curve Defined by Parametric Equations

Discuss the curve defined by the parametric equations

$$x = 3t^2, \qquad y = 2t, \qquad -2 \le t \le 2 \qquad (1)$$

Solution   For each number $t, -2 \le t \le 2$, there corresponds a number x and a number y. For example, when $t = -2$, then $x = 12$ and $y = -4$. When $t = 0$, then $x = 0$ and $y = 0$. Indeed, we can set up a table listing various choices of the parameter t and the corresponding values for x and y, as shown in Table 6. Plotting these points and connecting them with a smooth curve leads to Figure 64. The arrows in Figure 64 are used to indicate the orientation.

TABLE 6

t	x	y	(x, y)
-2	12	-4	$(12, -4)$
-1	3	-2	$(3, -2)$
0	0	0	$(0, 0)$
1	3	2	$(3, 2)$
2	12	4	$(12, 4)$

Figure 64

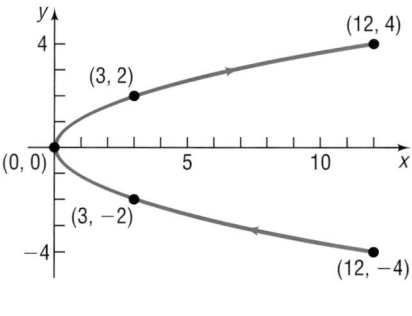

2   Most graphing utilities have the capability of graphing parametric equations. The following steps are usually required in order to obtain the graph of parametric equations. Check your owner's manual to see how yours works.

> ### Graphing Parametric Equations Using a Graphing Utility
>
> **STEP 1:**   Set the mode to PARametric. Enter $x(t)$ and $y(t)$.
>
> **STEP 2:**   Select the viewing window. In addition to setting Xmin, Xmax, Xscl, and so on, the viewing window in parametric mode requires setting minimum and maximum values for the parameter t and an increment setting for t (Tstep).
>
> **STEP 3:**   Graph.

EXAMPLE 2   **Graphing a Curve Defined by Parametric Equations Using a Graphing Utility**

Graph the curve defined by the parametric equations

$$x = 3t^2 \qquad y = 2t \qquad -2 \le t \le 2$$

Solution   **STEP 1:**   Enter the equations $x(t) = 3t^2$, $y(t) = 2t$ with the graphing utility in PARametric mode.

STEP 2:   Select the viewing window. The interval I is $-2 \le t \le 2$, so we select the following square viewing window:

$$T\text{min} = -2 \qquad X\text{min} = 0 \qquad Y\text{min} = -5$$
$$T\text{max} = 2 \qquad X\text{max} = 15 \qquad Y\text{max} = 5$$
$$T\text{step} = 0.1 \qquad X\text{scl} = 1 \qquad Y\text{scl} = 1$$

We choose $T\text{min} = -2$ and $T\text{max} = 2$ because $-2 \le t \le 2$. Finally, the choice for Tstep will determine the number of points the graphing utility will plot. For example, with Tstep at 0.1, the graphing utility will evaluate x and y at $t = -2, -1.9, -1.8$, and so on. The smaller the Tstep, the more points the graphing utility will plot. The reader is encouraged to experiment with different values of Tstep to see how the graph is affected.

STEP 3:   Graph. Notice the direction the graph is drawn. This direction shows the orientation of the curve.

The graph shown in Figure 65 is complete.   ■

Figure 65

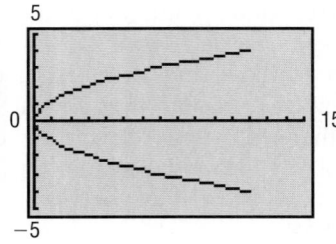

EXPLORATION   Graph the following parametric equations using a graphing utility with $X\text{min} = 0$, $X\text{max} = 15$, $Y\text{min} = -5$, $Y\text{max} = 5$, and $T\text{step} = 0.1$:

1. $x = \dfrac{3t^2}{4}, y = t, -4 \le t \le 4$

2. $x = 3t^2 + 12t + 12, y = 2t + 4, -4 \le t \le 0$

3. $x = 3t^{\frac{2}{3}}, y = 2\sqrt[3]{t}, -8 \le t \le 8$

Compare these graphs to the graph in Figure 65. Conclude that parametric equations defining a curve are not unique, that is, different parametric equations can represent the same graph.

③   The curve given in Examples 1 and 2 should be familiar. To identify it accurately, we find the corresponding rectangular equation by eliminating the parameter t from the parametric equations (1) given in Example 1:

$$x = 3t^2 \qquad y = 2t, \qquad -2 \le t \le 2$$

Noting that we can readily solve for t in $y = 2t$, obtaining $t = \dfrac{y}{2}$, we substitute this expression in the other equation:

$$x = 3t^2 = 3\left(\frac{y}{2}\right)^2 = \frac{3y^2}{4}$$
$$\uparrow$$
$$t = \frac{y}{2}$$

This equation, $x = \dfrac{3y^2}{4}$, is the equation of a parabola with vertex at $(0, 0)$ and axis of symmetry along the x-axis.

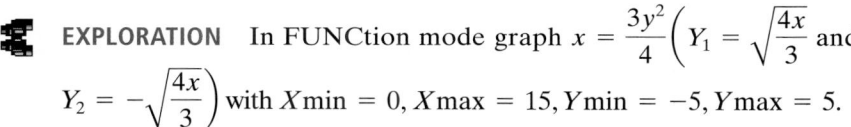 **EXPLORATION**   In FUNCtion mode graph $x = \dfrac{3y^2}{4}\left(Y_1 = \sqrt{\dfrac{4x}{3}}\right.$ and

$Y_2 = -\sqrt{\dfrac{4x}{3}}\Bigg)$ with $X\min = 0, X\max = 15, Y\min = -5, Y\max = 5.$

Compare this graph with Figure 65. Why do the graphs differ?

Note that the parameterized curve defined by equation (1) and shown in Figure 64 (or 65) is only a part of the parabola $x = \dfrac{3y^2}{4}$. The graph of the rectangular equation obtained by eliminating the parameter will, in general, contain more points than the original parameterized curve. Care must therefore be taken when a parameterized curve is sketched by hand after eliminating the parameter. Even so, the process of eliminating the parameter t of a parameterized curve in order to identify it accurately is sometimes a better approach than merely plotting points. However, the elimination process sometimes requires a little ingenuity.

EXAMPLE 3   **Finding the Rectangular Equation of a Curve Defined Parametrically**

Find the rectangular equation of the curve whose parametric equations are

$$x = a \cos t \qquad y = a \sin t$$

where $a > 0$ is a constant. By hand, graph this curve, indicating its orientation.

Solution   The presence of sines and cosines in the parametric equations suggests that we use a Pythagorean identity. In fact, since

$$\cos t = \frac{x}{a} \qquad \sin t = \frac{y}{a}$$

we find that

$$\cos^2 t + \sin^2 t = 1$$

$$\left(\frac{x}{a}\right)^2 + \left(\frac{y}{a}\right)^2 = 1$$

$$x^2 + y^2 = a^2$$

Figure 66

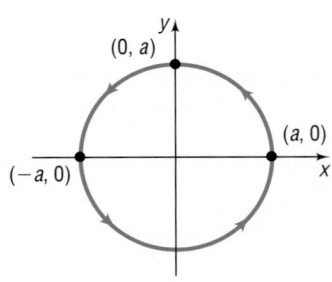

The curve is a circle with center at $(0,0)$ and radius a. As the parameter t increases, say from $t = 0$ [the point $(a, 0)$] to $t = \frac{\pi}{2}$ [the point $(0, a)$] to $t = \pi$ [the point $(-a, 0)$], we see that the corresponding points are traced in a counterclockwise direction around the circle. The orientation is as indicated in Figure 66. ■

━━ N O W   W O R K   P R O B L E M S   **1**   A N D   **1 3 .**

Let's discuss the curve in Example 3 further. The domain of each parametric equation is $-\infty < t < \infty$. Thus, the graph in Figure 66 is actually being repeated each time that t increases by 2π.

If we wanted the curve to consist of exactly 1 revolution in the counterclockwise direction, we could write

$$x = a \cos t, \qquad y = a \sin t, \qquad 0 \le t \le 2\pi$$

This curve starts at $t = 0$ [the point $(a, 0)$] and, proceeding counterclockwise around the circle, ends at $t = 2\pi$ [also the point $(a, 0)$].

If we wanted the curve to consist of exactly three revolutions in the counterclockwise direction, we could write

$$x = a \cos t, \qquad y = a \sin t, \qquad -2\pi \le t \le 4\pi$$

or

$$x = a \cos t, \qquad y = a \sin t, \qquad 0 \le t \le 6\pi$$

or

$$x = a \cos t, \qquad y = a \sin t, \qquad 2\pi \le t \le 8\pi$$

EXAMPLE 4   **Describing Parametric Equations**

Find rectangular equations for and graph the following curves defined by parametric equations.

(a) $x = a \cos t, \qquad y = a \sin t, \qquad 0 \le t \le \pi, \qquad a > 0$

(b) $x = -a \sin t, \qquad y = -a \cos t, \qquad 0 \le t \le \pi, \qquad a > 0$

Solution   (a) We eliminate the parameter t using a Pythagorean identity.

$$\left(\frac{x}{a}\right)^2 + \left(\frac{y}{a}\right)^2 = \cos^2 t + \sin^2 t = 1$$

$$x^2 + y^2 = a^2$$

The curve defined by these parametric equations is a circle, with radius a and center at $(0,0)$. The circle begins at the point $(a, 0), t = 0$; passes through the point $(0, a), t = \frac{\pi}{2}$; and ends at the point $(-a, 0), t = \pi$.

Figure 67

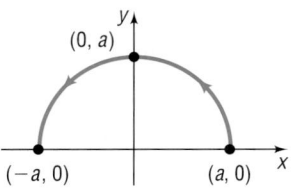

The parametric equations define an upper semicircle of radius a with a counterclockwise orientation. See Figure 67. The rectangular equation is

$$y = a\sqrt{1 - \left(\frac{x}{a}\right)^2}, \qquad -a \le x \le a$$

(b) We eliminate the parameter t using a Pythagorean identity.

$$\left(\frac{x}{-a}\right)^2 + \left(\frac{y}{-a}\right)^2 = \sin^2 t + \cos^2 t = 1$$

$$x^2 + y^2 = a^2$$

Figure 68

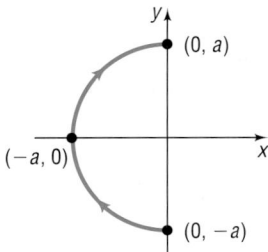

The curve defined by these parametric equations is a circle, with radius a and center at $(0, 0)$. The circle begins at the point $(0, -a), t = 0$; passes through the point $(-a, 0), t = \frac{\pi}{2}$; and ends at the point $(0, a), t = \pi$. The parametric equations define a left semicircle of radius a with a clockwise orientation. See Figure 68. The rectangular equation is

$$x = -a\sqrt{1 - \left(\frac{y}{a}\right)^2}, \qquad -a \le y \le a$$

■

Example 4 illustrates the versatility of parametric equations for replacing complicated rectangular equations, while providing additional information about orientation. These characteristics make parametric equations very useful in applications, such as projectile motion.

SEEING THE CONCEPT   Graph $x = \cos t, y = \sin t$ for $0 \le t \le 2\pi$. Compare to Figure 66. Graph $x = \cos t, y = \sin t$ for $0 \le t \le \pi$. Compare to Figure 67. Graph $x = -\sin t, y = -\cos t$ for $0 \le t \le \pi$. Compare to Figure 68.

Time as a Parameter: Projectile Motion; Simulated Motion

If we think of the parameter t as time, then the parametric equations $x = f(t)$ and $y = g(t)$ of a curve C specify how the x- and y-coordinates of a moving point vary with time.

For example, we can use parametric equations to describe the motion of an object, sometimes referred to as **curvilinear motion**. Using parametric equations, we can specify not only where the object travels, that is, its location (x, y), but also when it gets there, that is, the time t.

When an object is propelled upward at an inclination θ to the horizontal with initial speed v_0, the resulting motion is called **projectile motion.** See Figure 69(a) on page 460.

In calculus it is shown that the parametric equations of the path of a projectile fired at an inclination θ to the horizontal, with an initial speed v_0, from a height h above the horizontal are

$$x = (v_0 \cos \theta)t \qquad y = -\frac{1}{2}gt^2 + (v_0 \sin \theta)t + h \qquad (2)$$

where t is the time and g is the constant acceleration due to gravity (approximately 32 ft/sec/sec or 9.8 m/sec/sec). See Figure 69(b) on page 460.

Figure 69

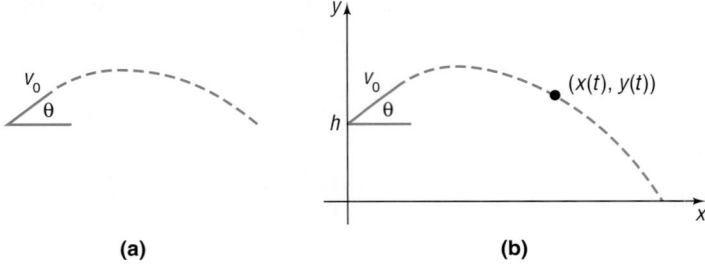

(a)   (b)

EXAMPLE 5   **Projectile Motion**

Figure 70

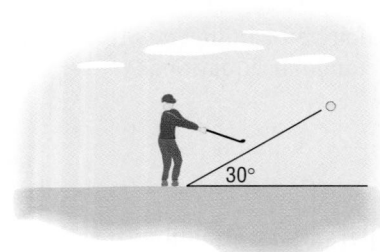

Suppose that Jim hit a golf ball with an initial velocity of 150 feet per second at an angle of 30° to the horizontal. See Figure 70.

(a) Find parametric equations that describe the position of the ball as a function of time.

(b) How long is the golf ball in the air?

(c) When is the ball at its maximum height? Determine the maximum height of the ball.

(d) Determine the distance that the ball traveled.

(e) Using a graphing utility, simulate the motion of the golf ball by simultaneously graphing the equations found in part (a).

Solution   (a) We have $v_0 = 150, \theta = 30°, h = 0$ (the ball is on the ground), and $g = 32$ (since units are in feet and seconds). Substituting these values into equations (2), we find that

$$x = (v_0 \cos \theta)t = (150 \cos 30°)t = 75\sqrt{3}t$$

$$y = -\frac{1}{2}gt^2 + (v_0 \sin \theta)t + h = -\frac{1}{2}(32)t^2 + (150 \sin 30°)t + 0$$

$$= -16t^2 + 75t$$

(b) To determine the length of time that the ball is in the air, we solve the equation $y = 0$.

$$-16t^2 + 75t = 0$$

$$t(-16t + 75) = 0$$

$$t = 0 \sec \quad \text{or} \quad t = \frac{75}{16} = 4.6875 \sec$$

The ball will strike the ground after 4.6875 seconds.

(c) Notice that the height y of the ball is a quadratic function of t, so the maximum height of the ball can be found by determining the vertex of $y = -16t^2 + 75t$. The value of t at the vertex is

$$t = \frac{-b}{2a} = \frac{-75}{-32} = 2.34375 \sec$$

The ball is at its maximum height after 2.34375 seconds. The maximum height of the ball is found by evaluating the function y at $t = 2.34375$ seconds.

Maximum height $= -16(2.34375)^2 + (75)2.34375 \approx 87.89$ feet

Figure 71

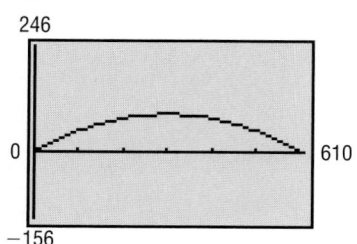

(d) Since the ball is in the air for 4.6875 seconds, the horizontal distance that the ball travels is
$$x = (75\sqrt{3})4.6875 \approx 608.92 \text{ feet}$$

(e) We enter the equations from part (a) into a graphing utility with $T\text{min} = 0$, $T\text{max} = 4.7$, and $T\text{step} = 0.1$. We use ZOOM-SQUARE to avoid any distortion to the angle of elevation. See Figure 71.

 EXPLORATION Simulate the motion of a ball thrown straight up with an initial speed of 100 feet per second from a height of 5 feet above the ground. Use PARametric mode with $T\text{min} = 0$, $T\text{max} = 6.5$, $T\text{step} = 0.1$, $X\text{min} = 0$, $X\text{max} = 5$, $Y\text{min} = 0$, and $Y\text{max} = 180$. What happens to the speed with which the graph is drawn as the ball goes up and then comes back down? How do you interpret this physically? Repeat the experiment using other values for $T\text{step}$. How does this affect the experiment?

[**Hint:** In the projectile motion equations, let $\theta = 90°$, $v_0 = 100$, $h = 5$, and $g = 32$. We use $x = 3$ instead of $x = 0$ to see the vertical motion better.]

RESULT See Figure 72. In Figure 72(a) the ball is going up. In Figure 72(b) the ball is near its highest point. Finally, in Figure 72(c) the ball is coming back down.

Notice that, as the ball goes up, its speed decreases, until at the highest point it is zero. Then the speed increases as the ball comes back down.

Figure 72

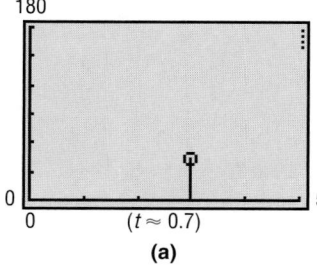

(a)

(b)

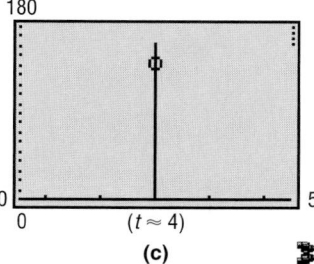

(c)

NOW WORK PROBLEM **27**.

A graphing utility can be used to simulate other kinds of motion as well.

EXAMPLE 6 **Simulating Motion**

Tanya, who is a long distance runner, runs at an average velocity of 8 miles per hour. Two hours after Tanya leaves your house, you leave in your Honda and follow the same route. If your average velocity is 40 miles per hour, how long will it be before you catch up to Tanya? See Figure 73. Use a simulation of the two motions to verify the answer.

Figure 73

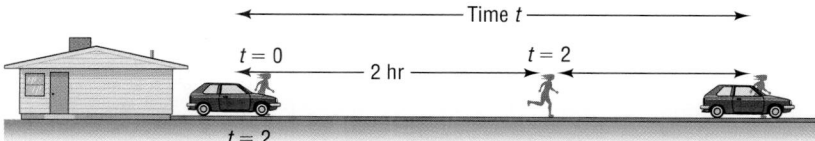

Solution We begin with two sets of parametric equations: one to describe Tanya's motion, the other to describe the motion of the Honda. We choose time $t = 0$ to be when Tanya leaves the house. If we choose $y_1 = 2$ as Tanya's path, then we can use $y_2 = 4$ as the parallel path of the Honda. The horizontal distances traversed in time t (Distance = Velocity × Time) are

$$\text{Tanya:}\quad x_1 = 8t \qquad \text{Honda:}\quad x_2 = 40(t - 2)$$

The Honda catches up to Tanya when $x_1 = x_2$.

$$8t = 40(t - 2)$$
$$8t = 40t - 80$$
$$-32t = -80$$
$$t = \frac{-80}{-32} = 2.5$$

The Honda catches up to Tanya 2.5 hours after Tanya leaves the house. In PARametric mode with Tstep = 0.01, we simultaneously graph

$$\text{Tanya:}\quad x_1 = 8t \qquad\qquad \text{Honda:}\quad x_2 = 40(t - 2)$$
$$y_1 = 2 \qquad\qquad\qquad\qquad y_2 = 4$$

for $0 \le t \le 3$.

Figure 74 shows the relative position of Tanya and the Honda for $t = 0, t = 2, t = 2.25, t = 2.5,$ and $t = 2.75$.

Figure 74

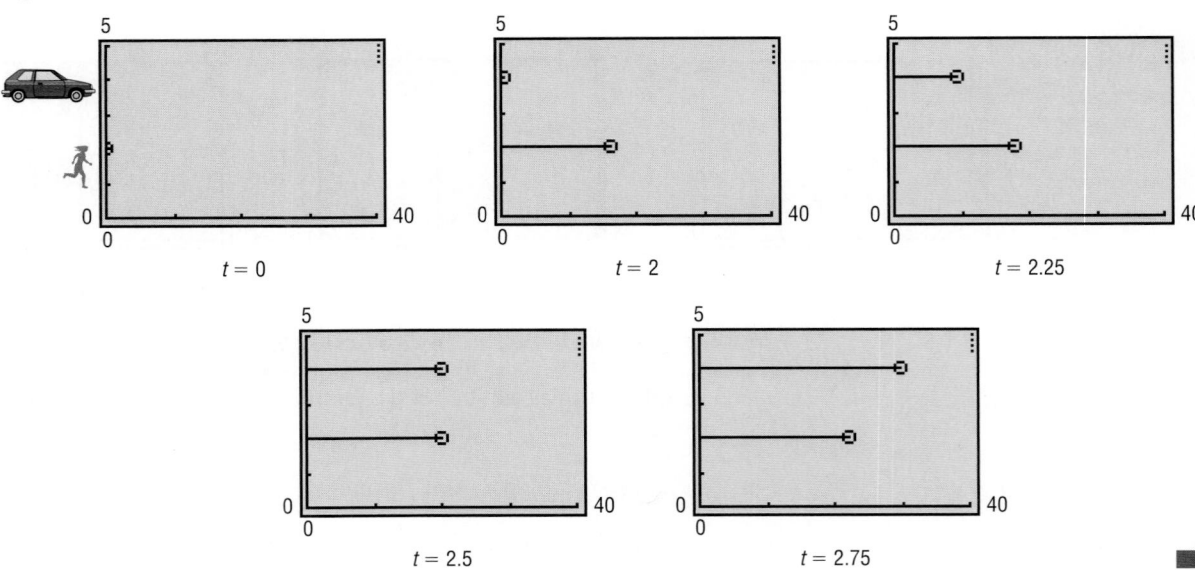

$t = 0$ $t = 2$ $t = 2.25$

$t = 2.5$ $t = 2.75$

Finding Parametric Equations

We now take up the question of how to find parametric equations of a given curve.

⑤ If a curve is defined by the equation $y = f(x)$, where f is a function, one way of finding parametric equations is to let $x = t$. Then $y = f(t)$ and

$$x = t, \quad y = f(t), \qquad t \text{ in the domain of } f$$

are parametric equations of the curve.

EXAMPLE 7 **Finding Parametric Equations for a Curve Defined by a Rectangular Equation**

Find parametric equations for the equation $y = x^2 - 4$.

Solution Let $x = t$. Then the parametric equations are

$$x = t, \quad y = t^2 - 4, \quad -\infty < t < \infty$$

Another less obvious approach to Example 7 is to let $x = t^3$. Then the parametric equations become

$$x = t^3, \quad y = t^6 - 4, \quad -\infty < t < \infty$$

Care must be taken when using this approach, since the substitution for x must be a function that allows x to take on all the values stipulated by the domain of f. For example, letting $x = t^2$ so that $y = t^4 - 4$ does not result in equivalent parametric equations for $y = x^2 - 4$, since only points for which $x \geq 0$ are obtained.

EXAMPLE 8 **Finding Parametric Equations for an Object in Motion**

Find parametric equations for the ellipse

$$x^2 + \frac{y^2}{9} = 1$$

where the parameter t is time (in seconds) and

(a) The motion around the ellipse is clockwise, begins at the point $(0, 3)$, and requires 1 second for a complete revolution.

(b) The motion around the ellipse is counterclockwise, begins at the point $(1, 0)$, and requires 2 seconds for a complete revolution.

Solution (a) See Figure 75. Since the motion begins at the point $(0, 3)$, we want $x = 0$ and $y = 3$ when $t = 0$. Furthermore, since the given equation is an ellipse, we begin by letting

Figure 75

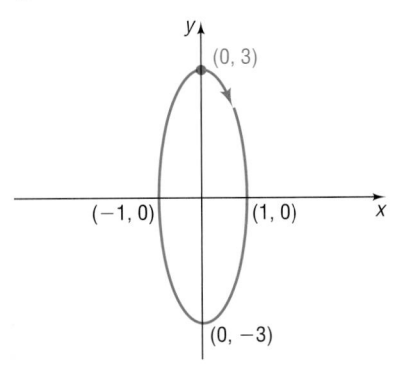

$$x = \sin(\omega t) \qquad \frac{y}{3} = \cos(\omega t)$$

for some constant ω. These parametric equations satisfy the equation of the ellipse. Furthermore, with this choice, when $t = 0$, we have $x = 0$ and $y = 3$.

For the motion to be clockwise, the motion will have to begin with the value of x increasing and y decreasing as t increases. This requires that $\omega > 0$. [Do you know why? If $\omega > 0$, then $x = \sin(\omega t)$ is increasing when $t > 0$ is near zero and $y = 3 \cos(\omega t)$ is decreasing when $t > 0$ is near zero]. See the red part of the graph in Figure 75.

Finally, since 1 revolution requires 1 second, the period $\dfrac{2\pi}{\omega} = 1$, so $\omega = 2\pi$. Parametric equations that satisfy the conditions stipulated are

$$x = \sin(2\pi t), \qquad y = 3\cos(2\pi t), \qquad 0 \leq t \leq 1 \qquad (3)$$

Figure 76

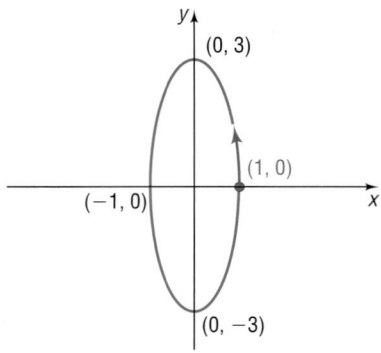

(b) See Figure 76. Since the motion begins at the point $(1, 0)$, we want $x = 1$ and $y = 0$ when $t = 0$. Furthermore, since the given equation is an ellipse, we begin by letting

$$x = \cos(\omega t) \qquad \frac{y}{3} = \sin(\omega t)$$

for some constant ω. These parametric equations satisfy the equation of the ellipse. Furthermore, with this choice, when $t = 0$, we have $x = 1$ and $y = 0$.

For the motion to be counterclockwise, the motion will have to begin with the value of x decreasing and y increasing as t increases. This requires that $\omega > 0$. [Do you know why?] Finally, since 1 revolution requires 2 seconds, the period is $\dfrac{2\pi}{\omega} = 2$, so $\omega = \pi$. The parametric equations that satisfy the conditions stipulated are

$$x = \cos(\pi t), \qquad y = 3\sin(\pi t), \qquad 0 \le t \le 2 \qquad (4) \quad \blacksquare$$

Either of equations (3) or (4) can serve as parametric equations for the ellipse $x^2 + \dfrac{y^2}{9} = 1$ given in Example 8. The direction of the motion, the beginning point, and the time for 1 revolution merely serve to help us arrive at a particular parametric representation.

 **NOW WORK PROBLEM 43.**

The Cycloid

Suppose that a circle of radius a rolls along a horizontal line without slipping. As the circle rolls along the line, a point P on the circle will trace out a curve called a **cycloid** (see Figure 77). We now seek parametric equations* for a cycloid.

Figure 77

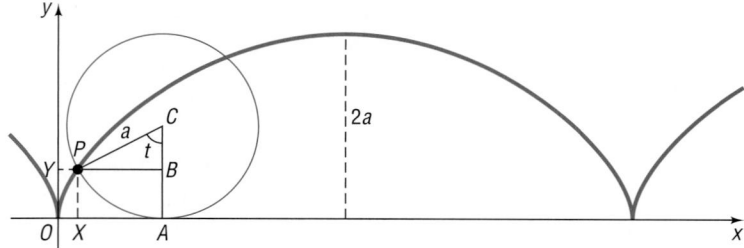

We begin with a circle of radius a and take the fixed line on which the circle rolls as the x-axis. Let the origin be one of the points at which the point P comes in contact with the x-axis. Figure 77 illustrates the position of this point P after the circle has rolled somewhat. The angle t (in radians) measures the angle through which the circle has rolled.

---

*Any attempt to derive the rectangular equation of a cycloid would soon demonstrate how complicated the task is.

Since we require no slippage, it follows that

$$\text{Arc } AP = d(O, A)$$

The length of the arc AP is given by $s = r\theta$, where $r = a$ and $\theta = t$ radians. Then,

$$at = d(O, A) \qquad {\scriptstyle s = r\theta, \text{ where } r = a \text{ and } \theta = t}$$

The x-coordinate of the point P is

$$d(O, X) = d(O, A) - d(X, A) = at - a \sin t = a(t - \sin t)$$

The y-coordinate of the point P is equal to

$$d(O, Y) = d(A, C) - d(B, C) = a - a \cos t = a(1 - \cos t)$$

The parametric equations of the cycloid are

$$x = a(t - \sin t) \qquad y = a(1 - \cos t) \qquad\qquad (5)$$

 EXPLORATION   Graph $x = t - \sin t$, $y = 1 - \cos t$, $0 \le t \le 3\pi$, using your graphing utility with Tstep $= \dfrac{\pi}{36}$ and a square screen. Compare your results with Figure 77.

Applications to Mechanics

If a is negative in equation (5), we obtain an inverted cycloid, as shown in Figure 78(a). The inverted cycloid occurs as a result of some remarkable applications in the field of mechanics. We shall mention two of them: the *brachistochrone* and the *tautochrone*.*

Figure 78

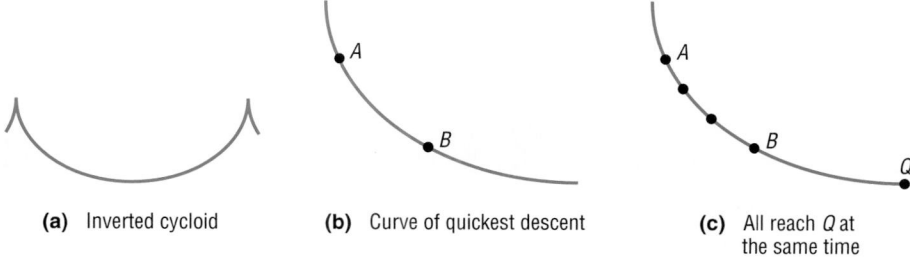

(a) Inverted cycloid   **(b)** Curve of quickest descent   **(c)** All reach Q at the same time

 The **brachistochrone** is the curve of quickest descent. If a particle is constrained to follow some path from one point A to a lower point B (not on the same vertical line) and is acted on only by gravity, the time needed to make the descent is least if the path is an inverted cycloid. See Figure 78(b). This remarkable discovery, which is attributed to many famous mathematicians (including Johann Bernoulli and Blaise Pascal), was a significant step in creating the branch of mathematics known as the *calculus of variations*.

To define the **tautochrone**, let Q be the lowest point on an inverted cycloid. If several particles placed at various positions on an inverted cycloid simultaneously begin to slide down the cycloid, they will reach the point Q at the same time, as indicated in Figure 78(c). The tautochrone property of

---

*In Greek, *brachistochrone* means "the shortest time" and *tautochrone* means "equal time."

Figure 79

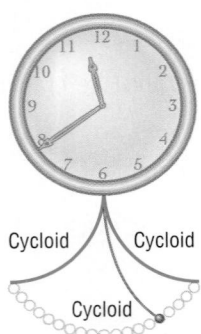

Cycloid   Cycloid

Cycloid

the cycloid was used by Christiaan Huygens (1629–1695), the Dutch mathematician, physicist, and astronomer, to construct a pendulum clock with a bob that swings along a cycloid (see Figure 79). In Huygen's clock, the bob was made to swing along a cycloid by suspending the bob on a thin wire constrained by two plates shaped like cycloids. In a clock of this design, the period of the pendulum is independent of its amplitude.

6.7 Concepts and Vocabulary

In Problems 1–3, fill in the blanks.

1. Let $x = f(t)$ and $y = g(t)$, where f and g are two functions whose common domain is some interval I. The collection of points defined by $(x, y) = (f(t), g(t))$ is called a _____ _____. The variable t is called a _____.

2. The parametric equations $x = 2\sin t$, $y = 3\cos t$ define a(n) _____.

3. If a circle rolls along a horizontal line without slippage, a point P on the circle will trace out a curve called a _____.

In Problems 4 and 5, answer True or False to each statement.

4. Parametric equations defining a curve are unique.

5. Curves defined using parametric equations have an orientation.

6. Make up parametric equations that define an ellipse.

7. Explain what a brachistochrone is.

6.7 Exercises

In Problems 1–20, graph by hand the curve whose parametric equations are given and show its orientation. Verify the graph using a graphing utility. Find the rectangular equation of each curve.

1. $x = 3t + 2$, $y = t + 1$; $0 \le t \le 4$

2. $x = t - 3$, $y = 2t + 4$; $0 \le t \le 2$

3. $x = t + 2$, $y = \sqrt{t}$; $t \ge 0$

4. $x = \sqrt{2t}$, $y = 4t$; $t \ge 0$

5. $x = t^2 + 4$, $y = t^2 - 4$; $-\infty < t < \infty$

6. $x = \sqrt{t} + 4$, $y = \sqrt{t} - 4$; $t \ge 0$

7. $x = 3t^2$, $y = t + 1$; $-\infty < t < \infty$

8. $x = 2t - 4$, $y = 4t^2$; $-\infty < t < \infty$

9. $x = 2e^t$, $y = 1 + e^t$; $t \ge 0$

10. $x = e^t$, $y = e^{-t}$; $t \ge 0$

11. $x = \sqrt{t}$, $y = t^{\frac{3}{2}}$; $t \ge 0$

12. $x = t^{\frac{3}{2}} + 1$, $y = \sqrt{t}$; $t \ge 0$

13. $x = 2\cos t$, $y = 3\sin t$; $0 \le t \le 2\pi$

14. $x = 2\cos t$, $y = 3\sin t$; $0 \le t \le \pi$

15. $x = 2\cos t$, $y = 3\sin t$; $-\pi \le t \le 0$

16. $x = 2\cos t$, $y = \sin t$; $0 \le t \le \dfrac{\pi}{2}$

17. $x = \sec t$, $y = \tan t$; $0 \le t \le \dfrac{\pi}{4}$

18. $x = \csc t$, $y = \cot t$; $\dfrac{\pi}{4} \le t \le \dfrac{\pi}{2}$

19. $x = \sin^2 t$, $y = \cos^2 t$; $0 \le t \le 2\pi$

20. $x = t^2$, $y = \ln t$; $t > 0$

21. Projectile Motion   Bob throws a ball straight up with an initial speed of 50 feet per second from a height of 6 feet.
  (a) Find parametric equations that describe the motion of the ball as a function of time.
  (b) How long is the ball in the air?
  (c) When is the ball at its maximum height? Determine the maximum height of the ball.
  (d) Simulate the motion of the ball by simultaneously graphing the equations found in part (a).

22. Projectile Motion   Alice throws a ball straight up with an initial speed of 40 feet per second from a height of 5 feet.
  (a) Find parametric equations that describe the motion of the ball as a function of time.
  (b) How long is the ball in the air?
  (c) When is the ball at its maximum height? Determine the maximum height of the ball.
  (d) Simulate the motion of the ball by simultaneously graphing the equations found in part (a).

23. Catching a Train   Bill's train leaves at 8:06 AM and accelerates at the rate of 2 meters per second per second. Bill, who can run 5 meters per second, arrives at the train station 5 seconds after the train has left.
  (a) Find parametric equations that describe the motion of the train and Bill as a function of time.

  [**Hint:** The position s at time t of an object having acceleration a is $s = \dfrac{1}{2}at^2$.]

  (b) Determine algebraically whether Bill will catch the train. If so, when?
  (c) Simulate the motion of the train and Bill by simultaneously graphing the equations found in part (a).

24. Catching a Bus   Jodi's bus leaves at 5:30 pm and accelerates at the rate of 3 meters per second per second. Jodi, who can run 5 meters per second, arrives at the bus station 2 seconds after the bus has left.
  (a) Find parametric equations that describe the motion of the bus and Jodi as a function of time.

  [**Hint:** The position s at time t of an object having acceleration a is $s = \dfrac{1}{2}at^2$.]

  (b) Determine algebraically whether Jodi will catch the bus. If so, when?
  (c) Simulate the motion of the bus and Jodi by simultaneously graphing the equations found in part (a).

25. Projectile Motion   Nolan Ryan throws a baseball with an initial speed of 145 feet per second at an angle of 20° to the horizontal. The ball leaves Nolan Ryan's hand at a height of 5 feet.
  (a) Find parametric equations that describe the position of the ball as a function of time.
  (b) How long is the ball in the air?
  (c) When is the ball at its maximum height? Determine the maximum height of the ball.
  (d) Determine the distance that the ball traveled.
  (e) Using a graphing utility, simultaneously graph the equations found in part (a).

26. Projectile Motion   Mark McGwire hit a baseball with an initial speed of 180 feet per second at an angle of 40° to the horizontal. The ball was hit at a height of 3 feet off the ground.
  (a) Find parametric equations that describe the position of the ball as a function of time.
  (b) How long is the ball in the air?
  (c) When is the ball at its maximum height? Determine the maximum height of the ball.
  (d) Determine the distance that the ball traveled.
  (e) Using a graphing utility, simultaneously graph the equations found in part (a).

27. Projectile Motion   Suppose that Adam throws a tennis ball off a cliff 300 meters high with an initial speed of 40 meters per second at an angle of 45° to the horizontal.
  (a) Find parametric equations that describe the position of the ball as a function of time.
  (b) How long is the ball in the air?
  (c) When is the ball at its maximum height? Determine the maximum height of the ball.
  (d) Determine the distance that the ball traveled.
  (e) Using a graphing utility, simultaneously graph the equations found in part (a).

28. Projectile Motion   Suppose that Adam throws a tennis ball off a cliff 300 meters high with an initial speed of 40 meters per second at an angle of 45° to the horizontal on the Moon (gravity on the Moon is one-sixth of that on Earth).
  (a) Find parametric equations that describe the position of the ball as a function of time.
  (b) How long is the ball in the air?
  (c) When is the ball at its maximum height? Determine the maximum height of the ball.
  (d) Determine the distance that the ball traveled.
  (e) Using a graphing utility, simultaneously graph the equations found in part (a).

29. Uniform Motion   A Toyota Paseo (traveling east at 40 mph) and Pontiac Bonneville (traveling north at 30 mph) are heading toward the same intersection. The Paseo is 5 miles from the intersection when the Bonneville is 4 miles from the intersection. See the figure.

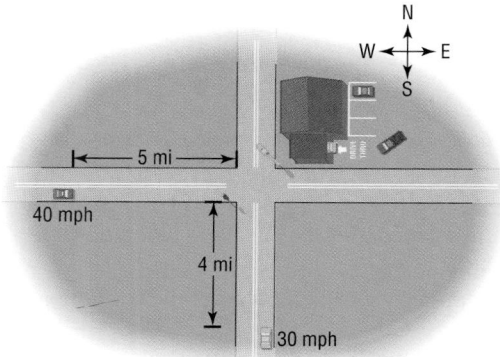

  (a) Find parametric equations that describe the motion of the Paseo and Bonneville.
  (b) Find a formula for the distance between the cars as a function of time.

(c) Graph the function in part (b) using a graphing utility.

(d) What is the minimum distance between the cars? When are the cars closest?

(e) Simulate the motion of the cars by simultaneously graphing the equations found in part (a).

30. Uniform Motion A Cessna (heading south at 120 mph) and a Boeing 747 (heading west at 600 mph) are flying toward the same point at the same altitude. The Cessna is 100 miles from the point where the flight patterns intersect and the 747 is 550 miles from this intersection point. See the figure.

(a) Find parametric equations that describe the motion of the Cessna and 747.

(b) Find a formula for the distance between the planes as a function of time.

(c) Graph the function in part (b) using a graphing utility.

(d) What is the minimum distance between the planes? When are the planes closest?

(e) Simulate the motion of the planes by simultaneously graphing the equations found in part (a).

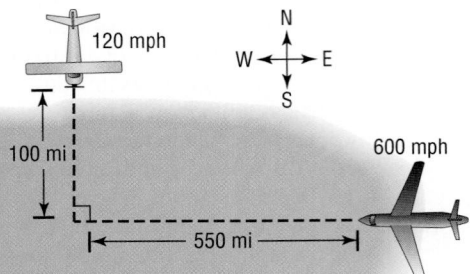

In Problems 31–38, find two different parametric equations for each rectangular equation.

31. $y = 4x - 1$

32. $y = -8x + 3$

33. $y = x^2 + 1$

34. $y = -2x^2 + 1$

35. $y = x^3$

36. $y = x^4 + 1$

37. $x = y^{3/2}$

38. $x = \sqrt{y}$

In Problems 39–42, find parametric equations that define the curve shown.

39.

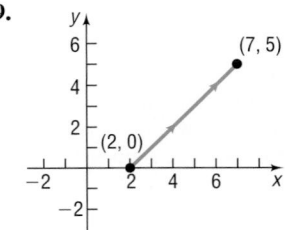

40.

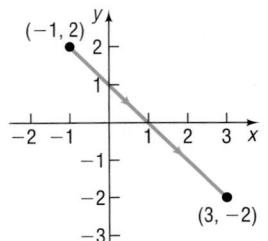

41.

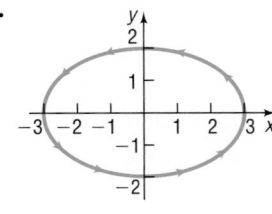

42.

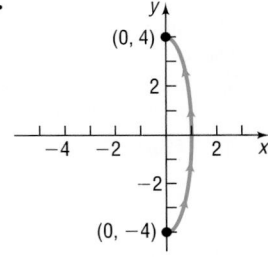

In Problems 43–46, find parametric equations for an object that moves along the ellipse $\dfrac{x^2}{4} + \dfrac{y^2}{9} = 1$ with the motion described.

43. The motion begins at $(2, 0)$, is clockwise, and requires 2 seconds for a complete revolution.

44. The motion begins at $(0, 3)$, is counterclockwise, and requires 1 second for a complete revolution.

45. The motion begins at $(0, 3)$, is clockwise, and requires 1 second for a complete revolution.

46. The motion begins at $(2, 0)$, is counterclockwise, and requires 3 seconds for a complete revolution.

In Problems 47 and 48, the parametric equations of four curves are given. By hand, graph each of them, indicating the orientation.

47. C_1: $x = t$, $y = t^2$; $-4 \le t \le 4$

C_2: $x = \cos t$, $y = 1 - \sin^2 t$; $0 \le t \le \pi$

C_3: $x = e^t$, $y = e^{2t}$; $0 \le t \le \ln 4$

C_4: $x = \sqrt{t}$, $y = t$; $0 \le t \le 16$

48. C_1: $x = t$, $y = \sqrt{1 - t^2}$; $-1 \le t \le 1$

C_2: $x = \sin t$, $y = \cos t$; $0 \le t \le 2\pi$

C_3: $x = \cos t$, $y = \sin t$; $0 \le t \le 2\pi$

C_4: $x = \sqrt{1 - t^2}$, $y = t$; $-1 \le t \le 1$

49. Show that the parametric equations for a line passing through the points (x_1, y_1) and (x_2, y_2) are

$$x = (x_2 - x_1)t + x_1$$
$$y = (y_2 - y_1)t + y_1, \quad -\infty < t < \infty$$

What is the orientation of this line?

50. Projectile Motion   The position of a projectile fired with an initial velocity v_0 feet per second and at an angle θ to the horizontal at the end of t seconds is given by the parametric equations

$$x = (v_0 \cos \theta)t \qquad y = (v_0 \sin \theta)t - 16t^2$$

See the following illustration.

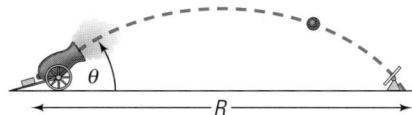

(a) Obtain the rectangular equation of the trajectory and identify the curve.

(b) Show that the projectile hits the ground $(y = 0)$ when $t = \dfrac{1}{16}v_0 \sin \theta$.

(c) How far has the projectile traveled (horizontally) when it strikes the ground? In other words, find the range R.

(d) Find the time t when $x = y$. Then find the horizontal distance x and the vertical distance y traveled by the projectile in this time. Then compute $\sqrt{x^2 + y^2}$. This is the distance R, the range, that the projectile travels up a plane inclined at $45°$ to the horizontal $(x = y)$. See the following illustration. (See also Problem 107 in Exercise 2.4.)

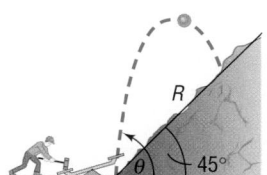

In Problems 51–54, use a graphing utility to graph the curve defined by the given parametric equations.

51. $x = t \sin t, \quad y = t \cos t$

53. $x = 4 \sin t - 2 \sin(2t)$
    $y = 4 \cos t - 2 \cos(2t)$

52. $x = \sin t + \cos t, \quad y = \sin t - \cos t$

54. $x = 4 \sin t + 2 \sin(2t)$
    $y = 4 \cos t + 2 \cos(2t)$

55. Hypocycloid   The hypocycloid is a curve defined by the parametric equations

$$x(t) = \cos^3 t, \qquad y(t) = \sin^3 t, \qquad 0 \le t \le 2\pi$$

(a) Graph the hypocycloid using a graphing utility.

(b) Find rectangular equations of the hypocycloid.

56. In Problem 55, we graphed the hypocycloid. Now graph the rectangular equations of the hypocycloid. Did you obtain a complete graph? If not, experiment until you do.

57. Look up the curves called *hypocycloid* and *epicycloid*. Write a report on what you find. Be sure to draw comparisons with the cycloid.

Chapter Review

Things to Know

Equations

Parabola	See Tables 1 and 2 (pp. 402 and 405).
Ellipse	See Table 3 (p. 416).
Hyperbola	See Table 4 (p. 431).

General equation of a conic (p. 445)	$Ax^2 + Bxy + Cy^2 + Dx + Ey + F = 0$	Parabola if $B^2 - 4AC = 0$
		Ellipse (or circle) if $B^2 - 4AC < 0$
		Hyperbola if $B^2 - 4AC > 0$
Conic in polar coordinates (p. 447)	$\dfrac{d(F, P)}{d(P, D)} = e$	Parabola if $e = 1$
		Ellipse if $e < 1$
		Hyperbola if $e > 1$
Polar equations of a conic with focus at the pole	See Table 5 (p. 450).	
Parametric equations of a curve (p. 455)	$x = f(t), y = g(t), t$ is the parameter	

Definitions

Parabola (p. 399)	Set of points P in the plane for which $d(F, P) = d(P, D)$, where F is the focus and D is the directrix
Ellipse (p. 411)	Set of points P in the plane, the sum of whose distances from two fixed points (the foci) is a constant
Hyperbola (p. 423)	Set of points P in the plane, the difference of whose distances from two fixed points (the foci) is a constant
Conic in polar coordinates (p. 447)	$\dfrac{d(F, P)}{d(P, D)} = e$ Parabola if $e = 1$ Ellipse if $e < 1$ Hyperbola if $e > 1$

Formulas

Rotation formulas (p. 440)	$x = x' \cos\theta - y' \sin\theta$ $y = x' \sin\theta + y' \cos\theta$
Angle θ of rotation that eliminates the $x'y'$-term (p. 442)	$\cot(2\theta) = \dfrac{A - C}{B}, \quad 0° < \theta < 90°$

▨ Objectives

Review Exercises

Blue problem numbers indicate the authors' suggestions for use in a Practice Test.

In Problems 1–20, identify each equation. If it is a parabola, gives its vertex, focus, and directrix; if it is an ellipse, give its center, vertices, and foci; if it is a hyperbola, give its center, vertices, foci, and asymptotes.

1. $y^2 = -16x$

2. $16x^2 = y$

3. $\dfrac{x^2}{25} - y^2 = 1$

4. $\dfrac{y^2}{25} - x^2 = 1$

5. $\dfrac{y^2}{25} + \dfrac{x^2}{16} = 1$

6. $\dfrac{x^2}{9} + \dfrac{y^2}{16} = 1$

7. $x^2 + 4y = 4$

8. $3y^2 - x^2 = 9$

9. $4x^2 - y^2 = 8$

10. $9x^2 + 4y^2 = 36$

11. $x^2 - 4x = 2y$

12. $2y^2 - 4y = x - 2$

13. $y^2 - 4y - 4x^2 + 8x = 4$

14. $4x^2 + y^2 + 8x - 4y + 4 = 0$

15. $4x^2 + 9y^2 - 16x - 18y = 11$

16. $4x^2 + 9y^2 - 16x + 18y = 11$

17. $4x^2 - 16x + 16y + 32 = 0$

18. $4y^2 + 3x - 16y + 19 = 0$

19. $9x^2 + 4y^2 - 18x + 8y = 23$

20. $x^2 - y^2 - 2x - 2y = 1$

In Problems 21–36, obtain an equation of the conic described. Graph the equation by hand.

21. Parabola; focus at $(-2, 0)$; directrix the line $x = 2$

22. Ellipse; center at $(0, 0)$; focus at $(0, 3)$; vertex at $(0, 5)$

23. Hyperbola; center at $(0, 0)$; focus at $(0, 4)$; vertex at $(0, -2)$

24. Parabola; vertex at $(0, 0)$; directrix the line $y = -3$

25. Ellipse; foci at $(-3, 0)$ and $(3, 0)$; vertex at $(4, 0)$

26. Hyperbola; vertices at $(-2, 0)$ and $(2, 0)$; focus at $(4, 0)$

27. Parabola; vertex at $(2, -3)$; focus at $(2, -4)$

28. Ellipse; center at $(-1, 2)$; focus at $(0, 2)$; vertex at $(2, 2)$

29. Hyperbola; center at $(-2, -3)$; focus at $(-4, -3)$; vertex at $(-3, -3)$

30. Parabola; focus at $(3, 6)$; directrix the line $y = 8$

31. Ellipse; foci at $(-4, 2)$ and $(-4, 8)$; vertex at $(-4, 10)$

32. Hyperbola; vertices at $(-3, 3)$ and $(5, 3)$; focus at $(7, 3)$

33. Center at $(-1, 2)$; $a = 3$; $c = 4$; transverse axis parallel to the x-axis

34. Center at $(4, -2)$; $a = 1$; $c = 4$; transverse axis parallel to the y-axis

35. Vertices at $(0, 1)$ and $(6, 1)$; asymptote the line $3y + 2x = 9$

36. Vertices at $(4, 0)$ and $(4, 4)$; asymptote the line $y + 2x = 10$

In Problems 37–46, identify each conic without completing the squares and without applying a rotation of axes.

37. $y^2 + 4x + 3y - 8 = 0$

38. $2x^2 - y + 8x = 0$

39. $x^2 + 2y^2 + 4x - 8y + 2 = 0$

40. $x^2 - 8y^2 - x - 2y = 0$

41. $9x^2 - 12xy + 4y^2 + 8x + 12y = 0$

42. $4x^2 + 4xy + y^2 - 8\sqrt{5}x + 16\sqrt{5}y = 0$

43. $4x^2 + 10xy + 4y^2 - 9 = 0$

44. $4x^2 - 10xy + 4y^2 - 9 = 0$

45. $x^2 - 2xy + 3y^2 + 2x + 4y - 1 = 0$

46. $4x^2 + 12xy - 10y^2 + x + y - 10 = 0$

In Problems 47–52, rotate the axes so that the new equation contains no xy-term. Discuss and graph the new equation.

47. $2x^2 + 5xy + 2y^2 - \dfrac{9}{2} = 0$

48. $2x^2 - 5xy + 2y^2 - \dfrac{9}{2} = 0$

49. $6x^2 + 4xy + 9y^2 - 20 = 0$

50. $x^2 + 4xy + 4y^2 + 16\sqrt{5}x - 8\sqrt{5}y = 0$

51. $4x^2 - 12xy + 9y^2 + 12x + 8y = 0$

52. $9x^2 - 24xy + 16y^2 + 80x + 60y = 0$

In Problems 53–58, identify the conic that each polar equation represents and graph it.

53. $r = \dfrac{4}{1 - \cos\theta}$

54. $r = \dfrac{6}{1 + \sin\theta}$

55. $r = \dfrac{6}{2 - \sin\theta}$

56. $r = \dfrac{2}{3 + 2\cos\theta}$

57. $r = \dfrac{8}{4 + 8\cos\theta}$

58. $r = \dfrac{10}{5 + 20\sin\theta}$

In Problems 59–62, convert each polar equation to a rectangular equation.

59. $r = \dfrac{4}{1 - \cos\theta}$

60. $r = \dfrac{6}{2 - \sin\theta}$

61. $r = \dfrac{8}{4 + 8\cos\theta}$

62. $r = \dfrac{2}{3 + 2\cos\theta}$

In Problems 63–68, by hand graph the curve whose parametric equations are given and show its orientation. Find the rectangular equation of each curve. Verify your results using a graphing utility.

63. $x = 4t - 2$, $y = 1 - t$; $-\infty < t < \infty$

64. $x = 2t^2 + 6$, $y = 5 - t$; $-\infty < t < \infty$

65. $x = 3\sin t$, $y = 4\cos t + 2$; $0 \le t \le 2\pi$

66. $x = \ln t$, $y = t^3$; $t > 0$

67. $x = \sec^2 t$, $y = \tan^2 t$; $0 \le t \le \dfrac{\pi}{4}$

68. $x = t^{\frac{3}{2}}$, $y = 2t + 4$; $t \ge 0$

In Problems 69 and 70, find two different parametric equations for each rectangular equation.

69. $y = -2x + 4$

70. $y = 2x^2 - 8$

In Problems 71 and 72, find parametric equations for an object that moves along the ellipse $\dfrac{x^2}{16} + \dfrac{y^2}{9} = 1$ with the motion described.

71. The motion begins at $(4, 0)$, is counterclockwise, and requires 4 seconds for a complete revolution.

72. The motion begins at $(0, 3)$, is clockwise, and requires 5 seconds for a complete revolution.

---

73. Find an equation of the hyperbola whose foci are the vertices of the ellipse $4x^2 + 9y^2 = 36$ and whose vertices are the foci of this ellipse.

74. Find an equation of the ellipse whose foci are the vertices of the hyperbola $x^2 - 4y^2 = 16$ and whose vertices are the foci of this hyperbola.

75. Describe the collection of points in a plane so that the distance from each point to the point $(3, 0)$ is three-fourths of its distance from the line $x = \dfrac{16}{3}$.

76. Describe the collection of points in a plane so that the distance from each point to the point $(5, 0)$ is five-fourths of its distance from the line $x = \dfrac{16}{5}$.

77. Mirrors A mirror is shaped like a paraboloid of revolution. If a light source is located 1 foot from the base along the axis of symmetry and the opening is 2 feet across, how deep should the mirror be?

78. Parabolic Arch Bridge A bridge is built in the shape of a parabolic arch. The bridge has a span of 60 feet and a maximum height of 20 feet. Find the height of the arch at distances of 5, 10, and 20 feet from the center.

79. Semi-elliptical Arch Bridge A bridge is built in the shape of a semi-elliptical arch. The bridge has a span of 60 feet and a maximum height of 20 feet. Find the height of the arch at distances of 5, 10, and 20 feet from the center.

80. Whispering Galleries The figure shows the specifications for an elliptical ceiling in a hall designed to be a whispering gallery. Where are the foci located in the hall?

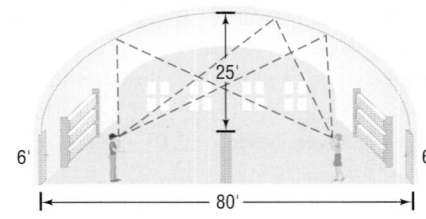

81. LORAN Two LORAN stations are positioned 150 miles apart along a straight shore.
  (a) A ship records a time difference of 0.00032 second between the LORAN signals. Set up an appropriate rectangular coordinate system to determine where the ship would reach shore if it were to follow the hyperbola corresponding to this time difference.
  (b) If the ship wants to enter a harbor located between the two stations 15 miles from the master station, what time difference should it be looking for?
  (c) If the ship is 20 miles offshore when the desired time difference is obtained, what is the approximate location of the ship?
  [*Note:* The speed of each radio signal is 186,000 miles per second.]

82. Uniform Motion Mary's train leaves at 7:15 AM and accelerates at the rate of 3 meters per second per second. Mary, who can run 6 meters per second, arrives at the train station 2 seconds after the train has left.
  (a) Find parametric equations that describe the motion of the train and Mary as a function of time.
  [*Hint:* The position s at time t of an object having acceleration a is $s = \dfrac{1}{2}at^2$.]
  (b) Determine algebraically whether Mary will catch the train. If so, when?
  (c) Simulate the motion of the train and Mary by simultaneously graphing the equations found in part (a).

83. Projectile Motion   Drew Bledsoe throws a football with an initial speed of 100 feet per second at an angle of 35° to the horizontal. The ball leaves Drew Bledsoe's hand at a height of 6 feet.

(a) Find parametric equations that describe the position of the ball as a function of time.

(b) How long is the ball in the air?

(c) When is the ball at its maximum height? Determine the maximum height of the ball.

(d) Determine the distance that the ball travels.

(e) Using a graphing utility, simultaneously graph the equations found in part (a).

84. Formulate a strategy for discussing and graphing an equation of the form

$$Ax^2 + Bxy + Cy^2 + Dx + Ey + F = 0$$

Chapter Projects

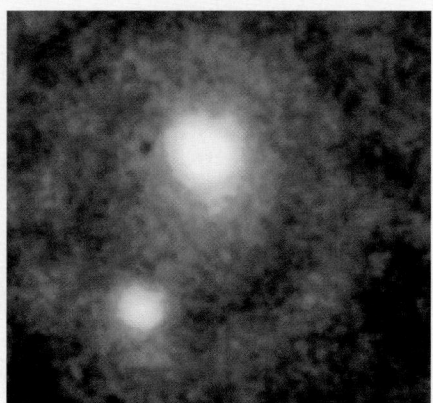

1. **The Orbits of Neptune and Pluto**   The orbit of a planet about the Sun is an ellipse, with the Sun at one focus. The **aphelion** of a planet is its greatest distance from the Sun and the **perihelion** is its shortest distance. The **mean distance** of a planet from the Sun is the length of the semi-major axis of the elliptical orbit. See the illustration.

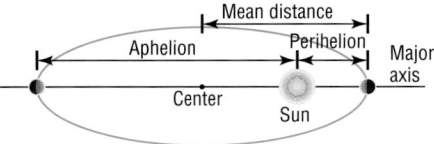

(a) The aphelion of Neptune is 4532.2×10^6 km and its perihelion is 4458.0×10^6 km. Find the equation for the orbit of Neptune around the Sun.

(b) The aphelion of Pluto is 7381.2×10^6 km and its perihelion is 4445.8×10^6 km. Find the equation for the orbit of Pluto around the Sun.

(c) Graph the orbits of Pluto and Neptune on a graphing utility. The orbits of the planets do not intersect! But the orbits do intersect. What is the explanation?

(d) The graphs of the orbits have the same center, so their foci lie in different locations. To see an accurate representation, the location of the Sun (a focus) needs to be the same for both graphs. This can be accomplished by shifting Pluto's orbit to the left. The shift amount is equal to Pluto's distance from the center [in the graph in part (c)] to the Sun minus Neptune's distance from the center to the Sun. Find the new equation representing the orbit of Pluto.

(e) Graph the equation for the orbit of Pluto found in part (d) along with the equation of the orbit of Neptune. Do you see that Pluto's orbit is sometimes inside Neptune's?

(f) Find the point(s) of intersection of the two orbits.

(g) Do you think two planets ever collide?

2. **Constructing a Bridge Over the East River**   A new bridge is to be constructed over the East River in New York City. The space between the supports needs to be 1050 feet; the height at the center of the arch needs to be 350 feet. Two structural possibilities exist: the support could be in the shape of a parabola or the support could be in the shape of a semiellipse.

An empty tanker needs a 280 foot clearance to pass beneath the bridge. The width of the channel for each of the two plans must be determined to verify that the tanker can pass through the bridge.

(a) Determine the equation of a parabola with these characteristics.
[*Hint:* Place the vertex of the parabola at the origin to simplify calculations.]

(b) How wide is the channel that the tanker can pass through?

(c) Determine the equation of a semiellipse with these characteristics.
[*Hint:* Place the center of the semiellipse at the origin to simplify calculations.]

(d) How wide is the channel that the tanker can pass through?

(e) If the river were to flood and rise 10 feet, how would the clearances of the two bridges be affected? Does this affect your decision as to which design to choose? Why?

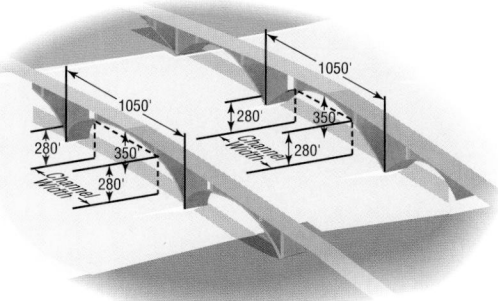

3. **Systems of Parametric Equations**   Consider the following systems of parametric equations:

   I. $x_1 = 4t - 2$, $y_1 = 1 - t$, $-\infty < t < \infty$

   $x_2 = \sec^2 t$, $y_2 = \tan^2 t$, $0 \le t \le \dfrac{\pi}{4}$

   II. $x_1 = \ln t$, $y_2 = t^3$, $t > 0$

   $x_2 = t^{\frac{3}{2}}$, $y_2 = 2t + 4$, $t \ge 0$

   III. $x_1 = 3 \sin t$, $y_1 = 4 \cos t + 2$, $0 \le t \le 2\pi$

   $x_2 = 2 \cos t$, $y_2 = 4 \sin t$, $0 \le t \le 2\pi$

   (a) For system I, set $x_1 = x_2$ and $y_1 = y_2$ and solve each equation for t. If you can solve the resulting equations algebraically, do so. If they cannot be solved algebraically, solve them graphically, using your graphing calculator. Remember that the value of t must be the same for both the x and y equations in order to have a solution for the system.

   (b) Now, graph the system of parametric equations using your graphing calculator and find the point(s) of intersection, if there are any. (You will need to use the TRACE feature to do this. Make sure the same value of t gives any points of intersection of each curve.) What did you notice?

   (c) Did any solutions you found in part (b) match any of those that you found in part (a)? Why or why not? Explain.

   (d) Convert the parametric equations in system I to rectangular coordinates and state the domain and range for each equation. Find the solution to the system either algebraically or graphically. How does this solution compare to what you found in part (c)?

   (e) Repeat parts (a)–(d) for System II.

   (f) Repeat parts (a)–(d) for System III.

   (g) Which method is more efficient—solving in the parametric form or solving in rectangular form? Does this depend on the equations? What must you watch for when solving systems of parametric equations? Explain.

Cumulative Review

1. Find all the solutions of the equation $\sin(2\theta) = 0.5$.
2. Find a polar equation for the line containing the origin that makes an angle of $30°$ with the positive x-axis.
3. Find a polar equation for the circle with center at the point $(0, 4)$ and radius 4. Graph this circle.
4. What is the domain of the function $f(x) = \dfrac{3}{\sin x + \cos x}$?
5. Find the exact value of $\sin\left(\cos^{-1}\dfrac{3}{5} + \tan^{-1} 1\right)$
6. Find the exact value of $\sin \dfrac{\pi}{12}$ using a half-angle-formula.

   Find the exact value using a difference formlua
7. Graph $y = -2\sin(\pi x)$.
8. Graph $y = \cos(2x + 1) - 3$
9. Solve the equation $\cot(2\theta) = 1$, where $0° < \theta < 90°$.
10. Find an equation for each of the following graphs:

    (a) Line:

    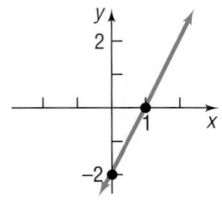

    (b) Circle:

    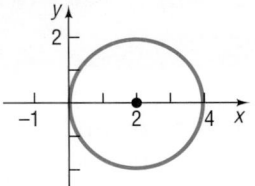

    (c) Ellipse:

    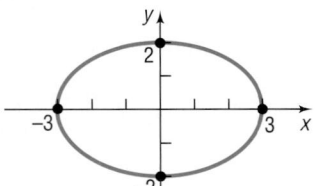

    (d) Parabola:

    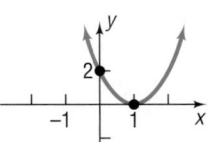

    (e) Hyperbola:

    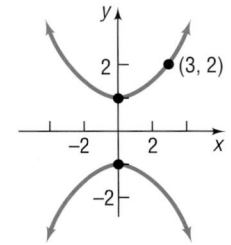

EXPONENTIAL AND LOGARITHMIC FUNCTIONS

CHAPTER

7

The McDonald's Scalding Coffee Case

April 3, 1996

There is a lot of hype about the McDonald's scalding coffee case. No one is in favor of frivolous cases or outlandish results; however, it is important to understand some points that were not reported in most of the stories about the case. McDonald's coffee was not only hot, it was scalding, capable of almost instantaneous destruction of skin, flesh and muscle.

Plaintiff's expert, a scholar in thermodynamics applied to human skin burns, testified that liquids, at 180 degrees, will cause a full thickness burn to human skin in two to seven seconds. Other testimony showed that as the temperature decreases toward 155 degrees, the extent of the burn relative to that temperature decreases exponentially. Thus, if (the) spill had involved coffee at 155 degrees, the liquid would have cooled and given her time to avoid a serious burn.

Miller, Norman, & Associates, Ltd., Attorneys, Moorhead, MN.

SEE CHAPTER PROJECT 1.

OUTLINE

7.1 **Exponential Functions**
7.2 **Logarithmic Functions**
7.3 **Properties of Logarithms**
7.4 **Logarithmic and Exponential Equations**
7.5 **Compound Interest**
7.6 **Growth and Decay**
7.7 **Exponential, Logarithmic, and Logistic Models**
   Chapter Review
   Chapter Projects

 For additional study help, go to

www.prenhall.com/sullivanegu3e

Materials include:

- Graphing Calculator Help
- Chapter Quiz
- Chapter Test
- PowerPoint Downloads
- Chapter Projects
- Student Tips

475

A Look Back, A Look Forward

Algebraic functions are functions that can be expressed in terms of sums, differences, products, quotients, powers, or roots of polynomials. Functions that are not algebraic are termed **transcendental** (they transcend, or go beyond, algebraic functions).

In this chapter, we study two transcendental functions: the *exponential* and *logarithmic* functions. The functions occur frequently in a wide variety of applications, such as biology, chemistry, economics, and psychology.

PREPARING FOR THIS SECTION

Before getting started, review the following:

✓ Exponents (Section A.1, pp. 561–562 and Section A.6, pp. 606–608)

✓ Solving Equations (Section A.3, pp. 571–583)

✓ Graphing Techniques: Transformations (Section 1.6, pp. 64–73)

7.1 EXPONENTIAL FUNCTIONS

OBJECTIVES

1. Evaluate Exponential Functions
2. Graph Exponential Functions
3. Define the Number *e*
4. Solve Exponential Equations

① In Section A.6, we give a definition for raising a real number *a* to a rational power. Based on that discussion, we gave meaning to expressions of the form

$$a^r$$

where the base *a* is a positive real number and the exponent *r* is a rational number.

 But what is the meaning of a^x, where the base *a* is a positive real number and the exponent *x* is an irrational number? Although a rigorous definition requires methods discussed in calculus, the basis for the definition is easy to follow: Select a rational number *r* that is formed by truncating (removing) all but a finite number of digits from the irrational number *x*. Then it is reasonable to expect that

$$a^x \approx a^r$$

For example, take the irrational number $\pi = 3.14159\ldots$. Then, an approximation to a^π is

$$a^\pi \approx a^{3.14}$$

where the digits after the hundredths position have been removed from the value for π. A better approximation would be

$$a^\pi \approx a^{3.14159}$$

where the digits after the hundred-thousandths position have been removed. Continuing in this way, we can obtain approximations to a^π to any desired degree of accuracy.

Graphing calculators can easily evaluate expressions of the form a^x as follows. Enter the base *a*, press the caret key $\boxed{\wedge}$, enter the exponent *x*, and press $\boxed{\text{ENTER}}$.

EXAMPLE 1 **Using a Calculator to Evaluate Powers of 2**

Using a calculator, evaluate:

(a) $2^{1.4}$ (b) $2^{1.41}$ (c) $2^{1.414}$ (d) $2^{1.4142}$ (e) $2^{\sqrt{2}}$

Figure 1

Solution Figure 1 shows the solution to part (a) using a TI-83 graphing calculator.

2^1.4
        2.639015822

(a) $2^{1.4} \approx 2.639015822$

(b) $2^{1.41} \approx 2.657371628$

(c) $2^{1.414} \approx 2.66474965$

(d) $2^{1.4142} \approx 2.665119089$

(e) $2^{\sqrt{2}} \approx 2.665144143$ ∎

  NOW WORK PROBLEM 1 .

It can be shown that the familiar laws for rational exponents hold for real exponents.

Theorem **Laws of Exponents**

If s, t, a, and b are real numbers with $a > 0$ and $b > 0$, then

$$a^s \cdot a^t = a^{s+t} \qquad (a^s)^t = a^{st} \qquad (ab)^s = a^s \cdot b^s$$

$$1^s = 1 \qquad a^{-s} = \frac{1}{a^s} = \left(\frac{1}{a}\right)^s \qquad a^0 = 1 \qquad (1)$$

∎

We are now ready for the following definition:

An **exponential function** is a function of the form

$$f(x) = a^x$$

where a is a positive real number $(a > 0)$ and $a \neq 1$. The domain of f is the set of all real numbers.

We exclude the base $a = 1$, because this function is simply the constant function $f(x) = 1^x = 1$. We also need to exclude bases that are negative, because, otherwise, we would have to exclude many values of x from the domain, such as $x = \dfrac{1}{2}$ and $x = \dfrac{3}{4}$. [Recall that $(-2)^{1/2} = \sqrt{-2}$, $(-3)^{3/4} = \sqrt[4]{(-3)^3} = \sqrt[4]{-27}$, and so on, are not defined in the system of real numbers.]

Some examples of exponential functions are

$$f(x) = 2^x, \qquad F(x) = \left(\frac{1}{3}\right)^x$$

Graphs of Exponential Functions

② First, we graph the exponential function $f(x) = 2^x$.

EXAMPLE 2   **Graphing an Exponential Function**

Graph the exponential function: $f(x) = 2^x$

Solution   The domain of $f(x) = 2^x$ consists of all real numbers. We begin by locating some points on the graph of $f(x) = 2^x$, as listed in Table 1.

Since $2^x > 0$ for all x, the range of f is $(0, \infty)$. From this, we conclude that the graph has no x-intercepts, and, in fact, the graph will lie above the x-axis for all x. As Table 1 indicates, the y-intercept is 1. Table 1 also indicates that as $x \to -\infty$ the value of $f(x) = 2^x$ gets closer and closer to 0. We conclude that the x-axis ($y = 0$) is a horizontal asymptote to the graph as $x \to -\infty$. This gives us the end behavior for x large and negative.

To determine the end behavior for x large and positive, look again at Table 1. As $x \to \infty$, $f(x) = 2^x$ grows very quickly, causing the graph of $f(x) = 2^x$ to rise very rapidly. It is apparent that f is an increasing function and hence is one-to-one.

Figure 2 shows the graph of $f(x) = 2^x$. Notice that all the conclusions given earlier are confirmed by the graph. ◼

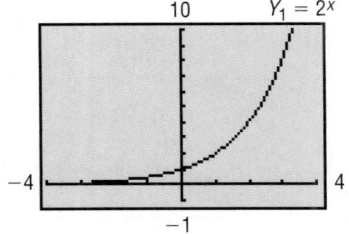

TABLE 1

As we shall see, graphs that look like the one in Figure 2 occur very frequently in a variety of situations. For example, look at the graph in Figure 3, which illustrates the closing price of a share of Harley Davidson stock. Investors might conclude from this graph that the price of Harley Davidson is *behaving exponentially*; that is, the graph exhibits rapid, or exponential, growth. We shall have more to say about situations that lead to exponential growth later in this chapter. For now, we continue to seek properties of the exponential functions.

Figure 2

Figure 3

The graph of $f(x) = 2^x$ in Figure 2 is typical of all exponential functions that have a base larger than 1. Such functions are increasing functions and hence are one-to-one. Their graphs lie above the x-axis, pass through the point $(0, 1)$, and thereafter rise rapidly as $x \to \infty$. As $x \to -\infty$, the x-axis ($y = 0$) is a horizontal asymptote. There are no vertical asymptotes. Finally, the graphs are smooth and continuous, with no corners or gaps.

Figure 4 illustrates the graphs of two more exponential functions whose bases are larger than 1. Notice that for the larger base the graph is steeper when $x > 0$. Figure 5 shows that when $x < 0$ the graph of the equation with the larger base is closer to the x-axis.

Figure 4

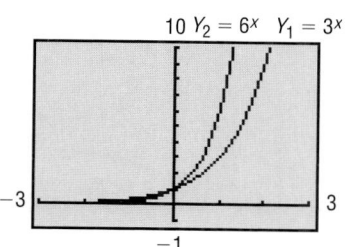

$10\ Y_2 = 6^x\ \ Y_1 = 3^x$

Figure 5

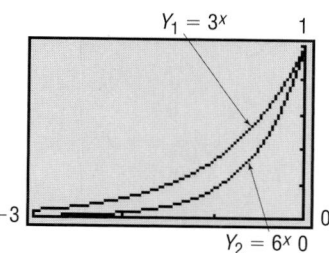

$Y_1 = 3^x$ 1

$Y_2 = 6^x\ 0$

The following display summarizes the information that we have about $f(x) = a^x, a > 1$.

Properties of an Exponential Function $f(x) = a^x, a > 1$

1. The domain is all real numbers; the range is the set of positive real numbers.
2. There are no x-intercepts, the y-intercept is 1.
3. The x-axis ($y = 0$) is a horizontal asymptote as $x \to -\infty$.
4. $f(x) = a^x, a > 1$, is an increasing function and is one-to-one.
5. The graph of f contains the points $(0, 1), (1, a),$ and $\left(-1, \dfrac{1}{a}\right)$.
6. The graph of f is smooth and continuous, with no corners or gaps. See Figure 6.

Figure 6
$f(x) = a^x, a > 1$

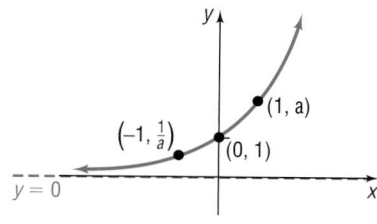

Now we consider $f(x) = a^x$ when $0 < a < 1$.

EXAMPLE 3 **Graphing an Exponential Function**

Graph the exponential function: $f(x) = \left(\dfrac{1}{2}\right)^x$

Solution The domain of $f(x) = \left(\dfrac{1}{2}\right)^x$ consists of all real numbers. As before, we locate some points on the graph, as listed in Table 2. Since $\left(\dfrac{1}{2}\right)^x > 0$ for all x, the range of f is the interval $(0, \infty)$. The graph lies above the x-axis and so has no x-intercepts. The y-intercept is 1. As $x \to -\infty, f(x) = \left(\dfrac{1}{2}\right)^x$ grows very quickly. As $x \to \infty$, the values of $f(x)$ approach 0. The x-axis ($y = 0$) is a horizontal asymptote as $x \to \infty$. It is apparent that f is a decreasing function and hence is one-to-one. Figure 7 illustrates the graph.

TABLE 2

X	Y1
-10	1024
-3	8
-1	2
0	1
1	.5
3	.125
10	9.8E-4

$Y_1 \boxminus (1/2)^{\wedge}X$

Figure 7

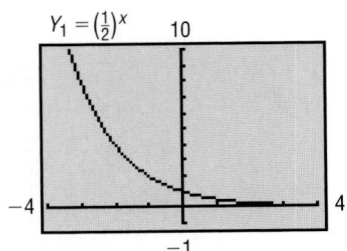

We could have obtained the graph of $y = \left(\dfrac{1}{2}\right)^x$ from the graph of $y = 2^x$ using transformations. The graph of $y = \left(\dfrac{1}{2}\right)^x = \dfrac{1}{2^x} = 2^{-x}$ is a reflection about the y-axis of the graph of $y = 2^x$. Compare Figures 2 and 7.

SEEING THE CONCEPT Using a graphing utility, simultaneously graph:

(a) $Y_1 = 3^x, Y_2 = \left(\dfrac{1}{3}\right)^x$ (b) $Y_1 = 6^x, Y_2 = \left(\dfrac{1}{6}\right)^x$

Conclude that the graph of $Y_2 = \left(\dfrac{1}{a}\right)^x$, for $a > 0$, is the reflection about the y-axis of the graph of $Y_1 = a^x$.

The graph of $f(x) = \left(\dfrac{1}{2}\right)^x$ in Figure 7 is typical of all exponential functions that have a base between 0 and 1. Such functions are decreasing and one-to-one. Their graphs lie above the x-axis and pass through the point $(0, 1)$. The graphs rise rapidly as $x \to -\infty$. As $x \to \infty$, the x-axis $(y = 0)$ is a horizontal asymptote. There are no vertical asymptotes. Finally, the graphs are smooth and continuous, with no corners or gaps.

Figure 8 illustrates the graphs of two more exponential functions whose bases are between 0 and 1. Notice that the smaller base results in a graph that is steeper when $x < 0$. Figure 9 shows that when $x > 0$ the graph of the equation with the smaller base is closer to the x-axis.

Figure 8

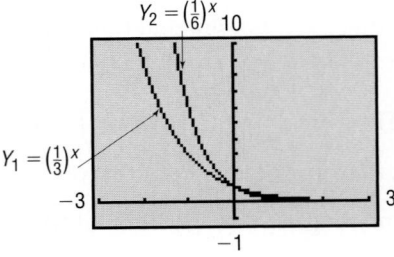

Figure 9

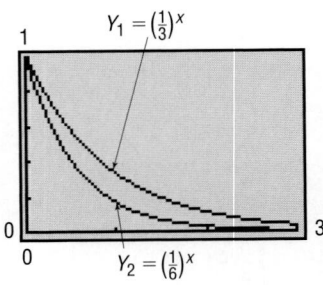

The following display summarizes the information that we have about the function $f(x) = a^x, 0 < a < 1$.

Figure 10
$f(x) = a^x, 0 < a < 1$

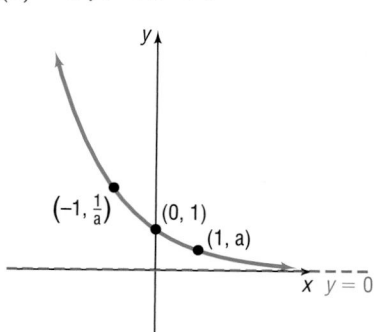

Properties of an Exponential Function $f(x) = a^x, 0 < a < 1$

1. The domain is all real numbers; the range is the set of positive real numbers.
2. There are no x-intercepts; the y-intercept is 1.
3. The x-axis ($y = 0$) is a horizontal asymptote as $x \to \infty$.
4. $f(x) = a^x, 0 < a < 1$, is a decreasing function and is one-to-one.
5. The graph of f contains the points $(0, 1)$, $(1, a)$, and $\left(-1, \dfrac{1}{a}\right)$.
6. The graph of f is smooth and continuous, with no corners or gaps. See Figure 10.

EXAMPLE 4   **Graphing Exponential Functions Using Transformations**

Graph $f(x) = 2^{-x} - 3$ and determine the domain, range, and horizontal asymptote of f.

Solution   We begin with the graph of $y = 2^x$. Figure 11 shows the stages.

Figure 11

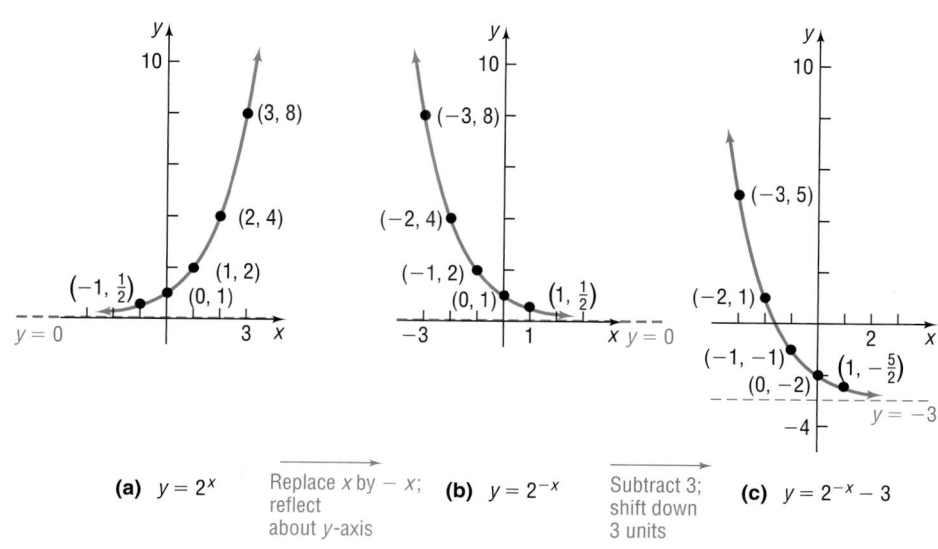

(a) $y = 2^x$ Replace x by $-x$; reflect about y-axis   (b) $y = 2^{-x}$ Subtract 3; shift down 3 units   (c) $y = 2^{-x} - 3$

As Figure 11(c) illustrates, the domain of $f(x) = 2^{-x} - 3$ is the interval $(-\infty, \infty)$ and the range is the interval $(-3, \infty)$. The horizontal asymptote of f is the line $y = -3$.

✔ CHECK: Graph $Y_1 = 2^{-x} - 3$ to verify the graph obtained in Figure 11(c). ■ ■

 **NOW WORK PROBLEM 21.**

The Base e

③ As we shall see shortly, many problems that occur in nature require the use of an exponential function whose base is a certain irrational number, symbolized by the letter e.

Let's look at one way of arriving at this important number e.

The **number e** is defined as the number that the expression

$$\left(1 + \frac{1}{n}\right)^n \tag{2}$$

approaches as $n \to \infty$. In calculus, this is expressed using limit notation as

$$e = \lim_{n \to \infty}\left(1 + \frac{1}{n}\right)^n$$

Table 3 illustrates what happens to the defining expression (2) as n takes on increasingly large values. The last number in the right column in the table is correct to nine decimal places and is the same as the entry given for e on your calculator (if expressed correctly to nine decimal places).

TABLE 3

n	$\dfrac{1}{n}$	$1 + \dfrac{1}{n}$	$\left(1 + \dfrac{1}{n}\right)^n$
1	1	2	2
2	0.5	1.5	2.25
5	0.2	1.2	2.48832
10	0.1	1.1	2.59374246
100	0.01	1.01	2.704813829
1,000	0.001	1.001	2.716923932
10,000	0.0001	1.0001	2.718145927
100,000	0.00001	1.00001	2.718268237
1,000,000	0.000001	1.000001	2.718280469
1,000,000,000	10^{-9}	$1 + 10^{-9}$	2.718281827

TABLE 4

X	Y1
-2	.13534
-1	.36788
0	1
1	2.7183
2	7.3891

Y1 = e^(X)

The exponential function $f(x) = e^x$, whose base is the number e, occurs with such frequency in applications that it is usually referred to as *the* exponential function. Graphing calculators have the key $\boxed{e^x}$ or $\boxed{\exp(x)}$, which may be used to evaluate the exponential function for a given value of x. Use your calculator to find e^x for $x = -2$, $x = -1$, $x = 0$, $x = 1$, and $x = 2$, as we have done to create Table 4. The graph of the exponential function $f(x) = e^x$ is given in Figure 12. Since $2 < e < 3$, the graph of $y = e^x$ is increasing and lies between the graphs of $y = 2^x$ and $y = 3^x$. [See Figure 12(c)].

Figure 12

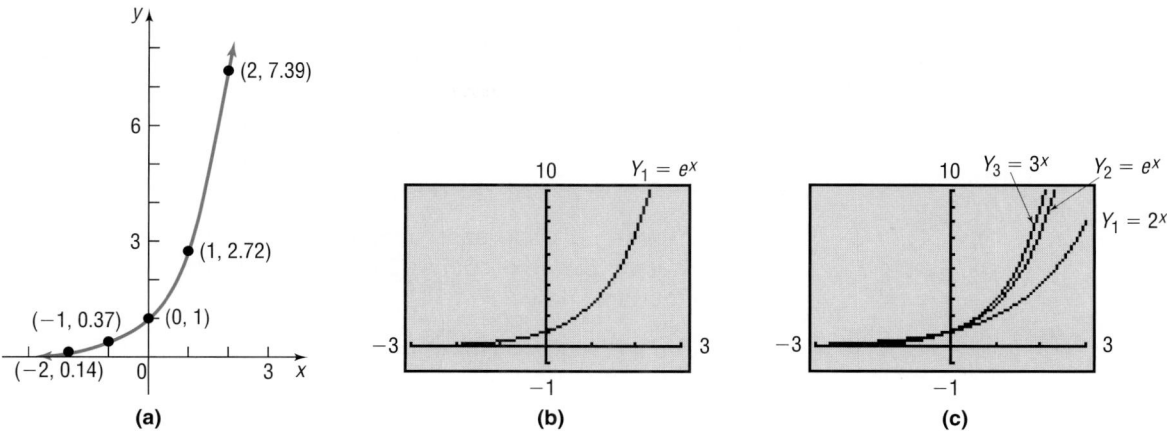

(a)   (b)   (c)

EXAMPLE 5   **Graphing Exponential Functions Using Transformations**

Graph $f(x) = -e^{x-3}$ and determine the domain, range, and horizontal asymptote of f.

Solution   We begin with the graph of $y = e^x$. Figure 13 shows the stages.

Figure 13

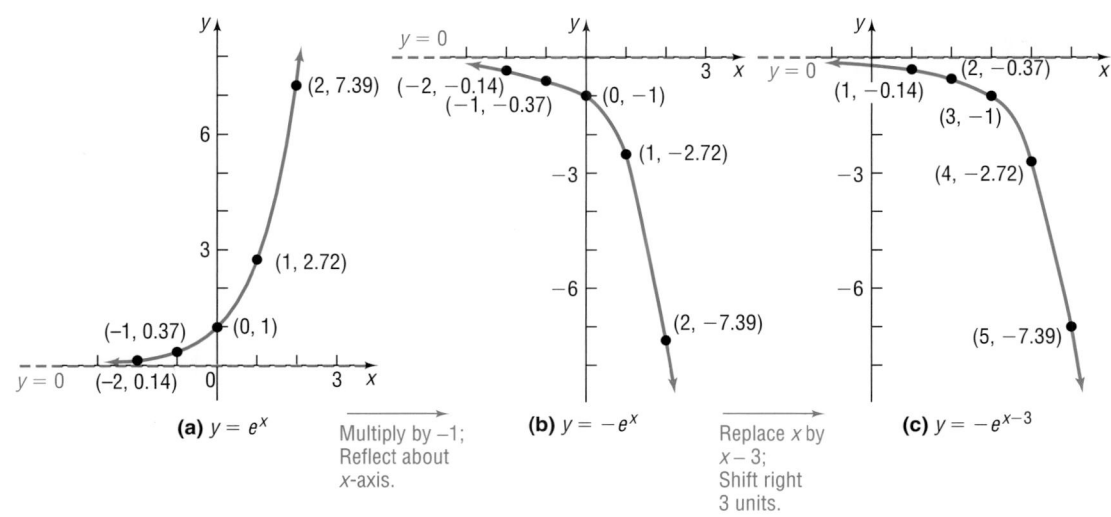

(a) $y = e^x$ Multiply by -1; Reflect about x-axis.   **(b)** $y = -e^x$ Replace x by $x - 3$; Shift right 3 units.   **(c)** $y = -e^{x-3}$

As Figure 13(c) illustrates, the domain of $f(x) = -e^{x-3}$ is the interval $(-\infty, \infty)$ and the range is the interval $(-\infty, 0)$. The horizontal asymptote is the line $y = 0$.

✔ CHECK: Graph $Y_1 = -e^{x-3}$ to verify the graph obtained in Figure 13(c).

NOW WORK PROBLEM **29.**

Exponential Equations

④ Equations that involve terms of the form a^x, $a > 0$, $a \neq 1$, are often referred to as **exponential equations**. Such equations can sometimes be solved by appropriately applying the Laws of Exponents and property (3).

$$\text{If} \quad a^u = a^v, \quad \text{then} \quad u = v \tag{3}$$

Property (3) is a consequence of the fact that exponential functions are one-to-one. To use property (3), each side of the equality must be written with the same base.

EXAMPLE 6   **Solving an Exponential Equation**

Solve: $3^{x+1} = 81$

Solution   Since $81 = 3^4$, we can write the equation as

$$3^{x+1} = 81 = 3^4$$

Figure 14

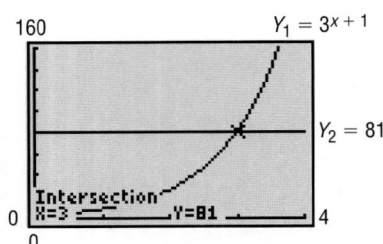

Now we have the same base, 3, on each side, so we can apply property (3) to obtain

$$x + 1 = 4$$
$$x = 3$$

✔ CHECK: We verify the solution by graphing $Y_1 = 3^{x+1}$ and $Y_2 = 81$ to determine where the graphs intersect. See Figure 14. The graphs intersect at $x = 3$.   ■ ■

━━━ NOW WORK PROBLEM **43**.

EXAMPLE 7   **Solving an Exponential Equation**

Solve: $e^{-x^2} = (e^x)^2 \cdot \dfrac{1}{e^3}$

Solution   We use Laws of Exponents first to get the base e on the right side.

$$(e^x)^2 \cdot \frac{1}{e^3} = e^{2x} \cdot e^{-3} = e^{2x-3}$$

As a result,

$$e^{-x^2} = e^{2x-3}$$
$$-x^2 = 2x - 3 \qquad \text{Apply Property (3).}$$
$$x^2 + 2x - 3 = 0 \qquad \text{Place the quadratic equation in standard form.}$$
$$(x + 3)(x - 1) = 0 \qquad \text{Factor.}$$
$$x = -3 \quad \text{or} \quad x = 1 \qquad \text{Use the Zero-Product Property.}$$

The solution set is $\{-3, 1\}$.   ■

You should verify these solutions using a graphing utility.

Many applications involve exponential functions. Let's look at one.

EXAMPLE 8 **Exponential Probability**

Between 9:00 PM and 10:00 PM cars arrive at Burger King's drive-thru at the rate of 12 cars per hour (0.2 car per minute). The following formula from statistics can be used to determine the probability that a car will arrive within t minutes of 9:00 PM.

$$F(t) = 1 - e^{-0.2t}$$

(a) Determine the probability that a car will arrive within 5 minutes of 9 PM (that is, before 9:05 PM).

(b) Determine the probability that a car will arrive within 30 minutes of 9 PM (before 9:30 PM).

(c) Graph F using your graphing utility.

(d) What value does F approach as t becomes unbounded in the positive direction?

Solution (a) The probability that a car will arrive within 5 minutes is found by evaluating $F(t)$ at $t = 5$.

$$F(5) = 1 - e^{-0.2(5)}$$

Figure 15

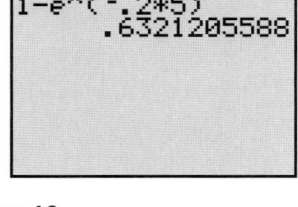

We evaluate this expression in Figure 15. We conclude that there is a 63% probability that a car will arrive within 5 minutes.

(b) The probability that a car will arrive within 30 minutes is found by evaluating $F(t)$ at $t = 30$.

$$F(30) = 1 - e^{-0.2(30)} \approx 0.9975$$

Figure 16

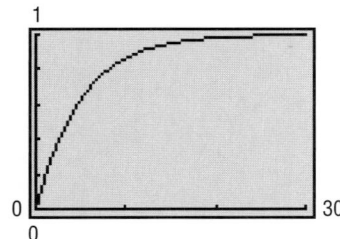

There is a 99.75% probability that a car will arrive within 30 minutes.

(c) See Figure 16 for the graph of F.

(d) As time passes, the probability that a car will arrive increases. The value that F approaches can be found by letting $t \to \infty$. Since $e^{-0.2t} = \dfrac{1}{e^{0.2t}}$, it follows that $e^{-0.2t} \to 0$ as $t \to \infty$. Thus, F approaches 1 as t gets large. ■

✎ **NOW WORK PROBLEM 67.**

SUMMARY **Properties of Exponential Functions**

$f(x) = a^x, \quad a > 1$

Domain: the interval $(-\infty, \infty)$; Range: the interval $(0, \infty)$
x-intercepts: none; y-intercept: 1; horizontal asymptote: x-axis ($y = 0$) as $x \to -\infty$; increasing; one-to-one; smooth; continuous
See Figure 6 for a typical graph.

$f(x) = a^x, \quad 0 < a < 1$

Domain: the interval $(-\infty, \infty)$; Range: the interval $(0, \infty)$
x-intercepts: none; y-intercept: 1; horizontal asymptote: x-axis ($y = 0$) as $x \to \infty$; decreasing; one-to-one; smooth; continuous
See Figure 10 for a typical graph.

If $a^u = a^v$, then $u = v$.

7.1   Concepts and Vocabulary

In Problems 1–3, fill in the blanks.

1. The graph of every exponential function $f(x) = a^x$, $a > 0$, $a \neq 1$, passes through three points: _____ , _____ , and _____ .

2. If the graph of every exponential function $f(x) = a^x$, $a > 0$, $a \neq 1$, is decreasing, then a must be less than _____ .

3. If $3^x = 3^4$, then $x =$ _____ .

In Problems 4–6, answer True or False to each statement.

4. The graph of an exponential function $f(x) = a^x$, $a < 0$, is decreasing.

5. The graphs of $y = 3^x$ and $y = \left(\dfrac{1}{3}\right)^x$ are identical.

6. The range of every exponential function $f(x) = a^x$, $a > 0$, $a \neq 1$, is all real numbers.

7. As the base a of an exponential function $f(x) = a^x$, $a > 1$, increases, what happens to the behavior of its graph for $x > 0$? What happens to the behavior of the graph for $x < 0$?

8. The graphs of $y = a^{-x}$ and $y = \left(\dfrac{1}{a}\right)^x$ are identical. Why?

7.1   Exercises

In Problems 1–10, approximate each number using a calculator. Express your answer rounded to three decimal places.

1. (a) $3^{2.2}$ (b) $3^{2.23}$ (c) $3^{2.236}$ (d) $3^{\sqrt{5}}$

2. (a) $5^{1.7}$ (b) $5^{1.73}$ (c) $5^{1.732}$ (d) $5^{\sqrt{3}}$

3. (a) $2^{3.14}$ (b) $2^{3.141}$ (c) $2^{3.1415}$ (d) 2^{π}

4. (a) $2^{2.7}$ (b) $2^{2.71}$ (c) $2^{2.718}$ (d) 2^{e}

5. (a) $3.1^{2.7}$ (b) $3.14^{2.71}$ (c) $3.141^{2.718}$ (d) π^{e}

6. (a) $2.7^{3.1}$ (b) $2.71^{3.14}$ (c) $2.718^{3.141}$ (d) e^{π}

7. $e^{1.2}$ 8. $e^{-1.3}$ 9. $e^{-0.85}$ 10. $e^{2.1}$

In Problems 11–18, the graph of a transformed exponential function is given. Match each graph to one of the following functions without the aid of a graphing utility.

A. $y = 3^x$ B. $y = 3^{-x}$ C. $y = -3^x$ D. $y = -3^{-x}$

E. $y = 3^x - 1$ F. $y = 3^{x-1}$ G. $y = 3^{1-x}$ H. $y = 1 - 3^x$

11.
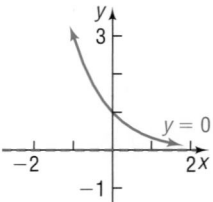

12.
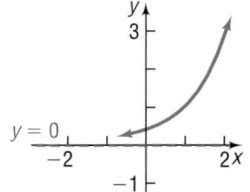

13.
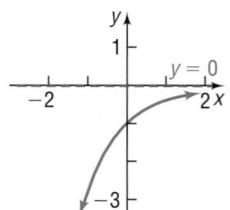

14.
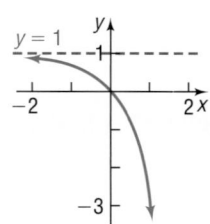

15.
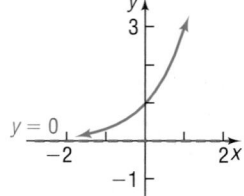

16.
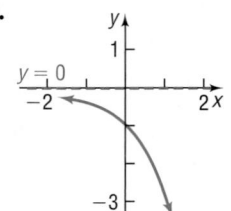

17.
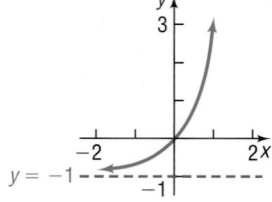

18.
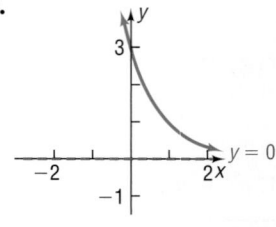

In Problems 19–36, use transformations to graph each function. Determine the domain, range, and horizontal asymptote of each function. Verify your results using a graphing utility.

19. $f(x) = 2^x + 1$

20. $f(x) = 2^{x+2}$

21. $f(x) = 3^{-x} - 2$

22. $f(x) = -3^x + 1$

23. $f(x) = 3(4^x)$

24. $f(x) = \frac{1}{2}(2^x)$

25. $f(x) = 3^{x/2}$

26. $f(x) = -2^{-x/3}$

27. $f(x) = 5 - 2(3^{x+1})$

28. $f(x) = 1 + 3^{x-4}$

29. $f(x) = e^{-x}$

30. $f(x) = -e^x$

31. $f(x) = e^{x+2}$

32. $f(x) = e^x - 1$

33. $f(x) = 5 + e^{-x}$

34. $f(x) = 9 + 3e^x$

35. $f(x) = 2 - e^{x/2}$

36. $f(x) = 7 - 3e^{-2x}$

In Problems 37–42, determine the exponential function whose graph is given.

37.
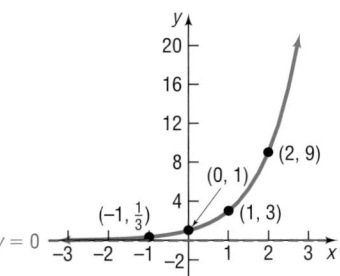

38.
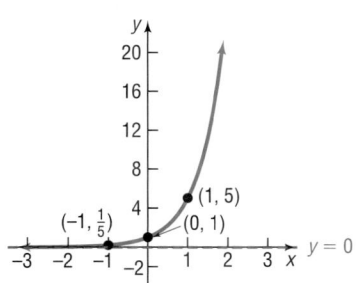

39.
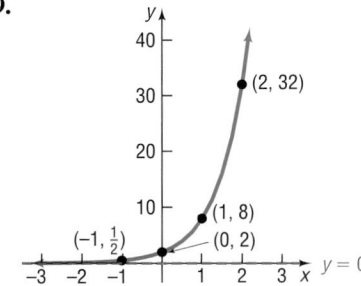

40.
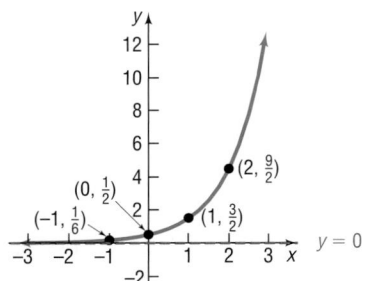

41.
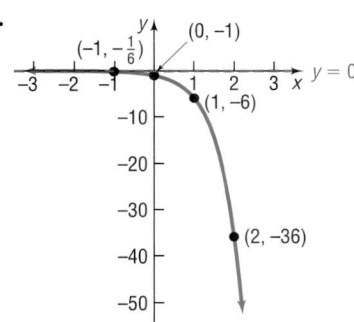

42.
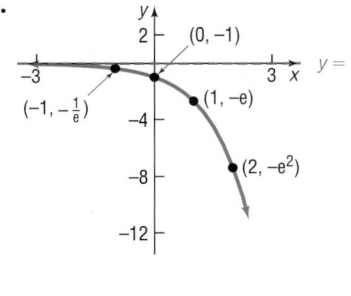

In Problems 43–56, solve each equation. Verify your solution using a graphing utility.

43. $2^{2x+1} = 4$

44. $5^{1-2x} = \frac{1}{5}$

45. $3^{x^3} = 9^x$

46. $4^{x^2} = 2^x$

47. $8^{x^2-2x} = \frac{1}{2}$

48. $9^{-x} = \frac{1}{3}$

49. $2^x \cdot 8^{-x} = 4^x$

50. $\left(\frac{1}{2}\right)^{1-x} = 4$

51. $\left(\frac{1}{5}\right)^{2-x} = 25$

52. $4^x - 2^x = 0$

53. $4^x = 8$

54. $9^{2x} = 27$

55. $e^{x^2} = (e^{3x}) \cdot \frac{1}{e^2}$

56. $(e^4)^x \cdot e^{x^2} = e^{12}$

57. If $4^x = 7$, what does 4^{-2x} equal?

58. If $2^x = 3$, what does 4^{-x} equal?

59. If $3^{-x} = 2$, what does 3^{2x} equal?

60. If $5^{-x} = 3$, what does 5^{3x} equal?

---

61. Optics   If a single pane of glass obliterates 3% of the light passing through it, then the percent p of light that passes through n successive panes is given approximately by the function

$$p(n) = 100e^{-0.03n}$$

   (a) What percent of light will pass through 10 panes?

   (b) What percent of light will pass through 25 panes?

62. Atmospheric Pressure   The atmospheric pressure p on a balloon or plane decreases with increasing height. This

pressure, measured in millimeters of mercury, is related to the height h (in kilometers) above sea level by the function

$$p(h) = 760e^{-0.145h}$$

   (a) Find the atmospheric pressure at a height of 2 kilometers (over a mile).

   (b) What is it at a height of 10 kilometers (over 30,000 feet)?

63. Space Satellites   The number of watts w provided by a space satellite's power supply over a period of d days is given by the function

$$w(d) = 50e^{-0.004d}$$

(a) How much power will be available after 30 days?
(b) How much power will be available after 1 year (365 days)?

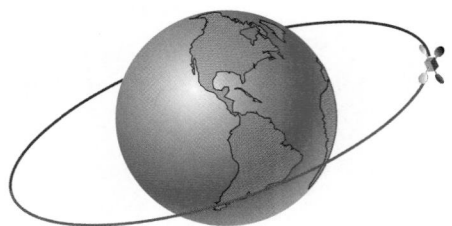

64. Healing of Wounds The normal healing of wounds can be modeled by an exponential function. If A_0 represents the original area of the wound and if A equals the area of the wound after n days, then the function

$$A(n) = A_0 e^{-0.35n}$$

describes the area of a wound on the nth day following an injury when no infection is present to retard the healing. Suppose that a wound initially had an area of 100 square millimeters.
(a) If healing is taking place, how large should the area of the wound be after 3 days?
(b) How large should it be after 10 days?

65. Drug Medication The function

$$D(h) = 5e^{-0.4h}$$

can be used to find the number of milligrams D of a certain drug that is in a patient's bloodstream h hours after the drug has been administered. How many milligrams will be present after 1 hour? After 6 hours?

66. Spreading of Rumors A model for the number of people N in a college community who have heard a certain rumor is

$$N = P(1 - e^{-0.15d})$$

where P is the total population of the community and d is the number of days that have elapsed since the rumor began. In a community of 1000 students, how many students will have heard the rumor after 3 days?

67. Exponential Probability Between 12:00 PM and 1:00 PM, cars arrive at Citibank's drive-thru at the rate of 6 cars per hour (0.1 car per minute). The following formula from statistics can be used to determine the probability that a car will arrive within t minutes of 12:00 PM.

$$F(t) = 1 - e^{-0.1t}$$

(a) Determine the probability that a car will arrive within 10 minutes of 12:00 PM (that is, before 12:10 PM).
(b) Determine the probability that a car will arrive within 40 minutes of 12:00 PM (before 12:40 PM).
(c) Graph F using your graphing utility.
(d) What value does F approach as t becomes unbounded in the positive direction?

68. Exponential Probability Between 5:00 PM and 6:00 PM, cars arrive at Jiffy Lube at the rate of 9 cars per hour (0.15 car per minute). The following formula from statistics can be used to determine the probability that a car will arrive within t minutes of 5:00 PM.

$$F(t) = 1 - e^{-0.15t}$$

(a) Determine the probability that a car will arrive within 15 minutes of 5:00 PM (that is, before 5:15 PM).
(b) Determine the probability that a car will arrive within 30 minutes of 5:00 PM (before 5:30 PM).
(c) Graph F using your graphing utility.
(d) What value does F approach as t becomes unbounded in the positive direction?

69. Poisson Probability Between 5:00 PM and 6:00 PM, cars arrive at McDonald's drive-thru at the rate of 20 cars per hour. The following formula from statistics can be used to determine the probability that x cars will arrive between 5:00 PM and 6:00 PM.

$$P(x) = \frac{20^x e^{-20}}{x!}$$

where

$$x! = x \cdot (x - 1) \cdot (x - 2) \cdots \cdots 3 \cdot 2 \cdot 1$$

(a) Determine the probability that $x = 15$ cars will arrive between 5:00 PM and 6:00 PM.
(b) Determine the probability that $x = 20$ cars will arrive between 5:00 PM and 6:00 PM.

70. Poisson Probability People enter a line for the *Demon Roller Coaster* at the rate of 4 per minute. The following formula from statistics can be used to determine the probability that x people will arrive within the next minute.

$$P(x) = \frac{4^x e^{-4}}{x!}$$

where

$$x! = x \cdot (x - 1) \cdot (x - 2) \cdots \cdots 3 \cdot 2 \cdot 1$$

(a) Determine the probability that $x = 5$ people will arrive within the next minute.
(b) Determine the probability that $x = 8$ people will arrive within the next minute.

71. Relative Humidity The relative humidity is the ratio (expressed as a percent) of the amount of water vapor in the air to the maximum amount that it can hold at a specific temperature. The relative humidity, R, is found using the following formula:

$$R = 10^{\left(\frac{4221}{T+459.4} - \frac{4221}{D+459.4} + 2\right)}$$

where T is the air temperature (in °F) and D is the dew point temperature (in °F).

(a) Determine the relative humidity if the air temperature is 50° Fahrenheit and the dew point temperature is 41° Fahrenheit.
(b) Determine the relative humidity if the air temperature is 68° Fahrenheit and the dew point temperature is 59° Fahrenheit.
(c) What is the relative humidity if the air temperature and the dew point temperature are the same?

72. Learning Curve   Suppose that a student has 500 vocabulary words to learn. If the student learns 15 words after 5 minutes, the function

$$L(t) = 500(1 - e^{-0.0061t})$$

approximates the number of words L that the student will learn after t minutes.
(a) How many words will the student learn after 30 minutes?
(b) How many words will the student learn after 60 minutes?

73. Current in an RL Circuit   The equation governing the amount of current I (in amperes) after time t (in seconds) in a single RL circuit consisting of a resistance R (in ohms), an inductance L (in henrys), and an electromotive force E (in volts) is

$$I = \frac{E}{R}[1 - e^{-(R/L)t}]$$

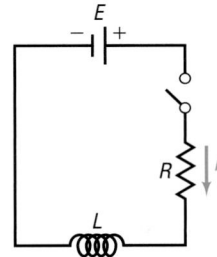

(a) If $E = 120$ volts, $R = 10$ ohms, and $L = 5$ henrys, how much current I_1 is flowing after 0.3 second? After 0.5 second? After 1 second?
(b) What is the maximum current?
(c) Graph this function $I = I_1(t)$, measuring I along the y-axis and t along the x-axis.
(d) If $E = 120$ volts, $R = 5$ ohms, and $L = 10$ henrys, how much current I_2 is flowing after 0.3 second? After 0.5 second? After 1 second?
(e) What is the maximum current?
(f) Graph this function $I = I_2(t)$ on the same viewing window as $I_1(t)$.

74. Current in an RC Circuit   The equation governing the amount of current I (in amperes) after time t (in microseconds) in a single RC circuit consisting of a resistance R (in ohms), a capacitance C (in microfarads), and an electromotive force E (in volts) is

$$I = \frac{E}{R}e^{-t/(RC)}$$

(a) If $E = 120$ volts, $R = 2000$ ohms, and $C = 1.0$ microfarad, how much current I_1 is flowing initially $(t = 0)$? After 1000 microseconds? After 3000 microseconds?
(b) What is the maximum current?
(c) Graph this function $I = I_1(t)$, measuring I along the y-axis and t along the x-axis.
(d) If $E = 120$ volts, $R = 1000$ ohms, and $C = 2.0$ microfarads, how much current I_2 is flowing initially? After 1000 microseconds? After 3000 microseconds?
(e) What is the maximum current?
(f) Graph this function $I = I_2(t)$ on the same viewing window as $I_1(t)$.

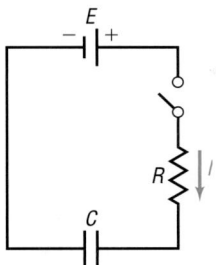

75. Another Formula for e   Use a calculator to compute the values of

$$2 + \frac{1}{2!} + \frac{1}{3!} + \cdots + \frac{1}{n!}$$

for $n = 4, 6, 8,$ and 10. Compare each result with e.
[**Hint:** $1! = 1, 2! = 2 \cdot 1, 3! = 3 \cdot 2 \cdot 1,$
$n! = n(n - 1) \cdots (3)(2)(1)$]

76. Another Formula for e   Use a calculator to compute the various values of the expression. Compare the values to e.

$$2 + \cfrac{1}{1 + \cfrac{1}{2 + \cfrac{2}{3 + \cfrac{3}{4 + 4}}}}$$
etc.

77. Difference Quotient   If $f(x) = a^x$, show that

$$\frac{f(x + h) - f(x)}{h} = a^x\left(\frac{a^h - 1}{h}\right)$$

78. If $f(x) = a^x$, show that $f(A + B) = f(A) \cdot f(B)$.

79. If $f(x) = a^x$, show that $f(-x) = \dfrac{1}{f(x)}$.

80. If $f(x) = a^x$, show that $f(\alpha x) = [f(x)]^{\alpha}$.

Problems 81 and 82 provide definitions for two other transcendental functions.

81. The **hyperbolic sine function**, designated by sinh x, is defined as

$$\sinh x = \frac{1}{2}(e^x - e^{-x})$$

(a) Show that $f(x) = \sinh x$ is an odd function.
(b) Graph $f(x) = \sinh x$ using a graphing utility.

82. The **hyperbolic cosine function**, designated by cosh x, is defined as

$$\cosh x = \frac{1}{2}(e^x + e^{-x})$$

(a) Show that $f(x) = \cosh x$ is an even function.
(b) Graph $f(x) = \cosh x$ using a graphing utility.
(c) Refer to Problem 81. Show that, for every x, $(\cosh x)^2 - (\sinh x)^2 = 1$.

83. Historical Problem Pierre de Fermat (1601–1665) conjectured that the function

$$f(x) = 2^{(2^x)} + 1$$

for $x = 1, 2, 3, \ldots$, would always have a value equal to a prime number. But Leonhard Euler (1707–1783) showed that this formula fails for $x = 5$. Use a calculator to determine the prime numbers produced by f for $x = 1, 2, 3, 4$. Then show that $f(5) = 641 \times 6{,}700{,}417$, which is not prime.

84. The bacteria in a 4-liter container double every minute. After 60 minutes the container is full. How long did it take to fill half the container?

85. Explain in your own words what the number e is. Provide at least two applications that require the use of this number.

86. Do you think that there is a power function that increases more rapidly than an exponential function whose base is greater than 1? Explain.

PREPARING FOR THIS SECTION

Before getting started, review the following:

✓ Solving Inequalities (Section A.5, pp. 598–599)

✓ One-to-one Functions; Inverse Functions (Section 1.7, pp. 77–87)

7.2 LOGARITHMIC FUNCTIONS

OBJECTIVES

1. Change Exponential Expressions to Logarithmic Expressions
2. Change Logarithmic Expressions to Exponential Expressions
3. Evaluate Logarithmic Functions
4. Determine the Domain of a Logarithmic Function
5. Graph Logarithmic Functions
6. Solve Logarithmic Equations

Recall that a one-to-one function $y = f(x)$ has an inverse function that is defined (implicitly) by the equation $x = f(y)$. In particular, the exponential function $y = f(x) = a^x$, $a > 0$, $a \neq 1$, is one-to-one and hence has an inverse function that is defined implicitly by the equation

$$x = a^y, \qquad a > 0, \quad a \neq 1$$

This inverse function is so important that it is given a name, the *logarithmic function*.

The **logarithmic function to the base a**, where $a > 0$ and $a \neq 1$, is denoted by $y = \log_a x$ (read as "y is the logarithm to the base a of x") and is defined by

$$y = \log_a x \quad \text{if and only if} \quad x = a^y$$

The domain of the logarithmic function $y = \log_a x$ is $x > 0$.

EXAMPLE 1 **Relating Logarithms to Exponents**

(a) If $y = \log_3 x$, then $x = 3^y$. For example, $4 = \log_3 81$ is equivalent to $81 = 3^4$.

(b) If $y = \log_5 x$, then $x = 5^y$. For example, $-1 = \log_5\left(\dfrac{1}{5}\right)$ is equivalent to $\dfrac{1}{5} = 5^{-1}$. ∎

1 **EXAMPLE 2** **Changing Exponential Expressions to Logarithmic Expressions**

Change each exponential expression to an equivalent expression involving a logarithm.

(a) $1.2^3 = m$ (b) $e^b = 9$ (c) $a^4 = 24$

Solution We use the fact that $y = \log_a x$ and $x = a^y$, $a > 0, a \neq 1$, are equivalent.

(a) If $1.2^3 = m$, then $3 = \log_{1.2} m$.   (b) If $e^b = 9$, then $b = \log_e 9$.
(c) If $a^4 = 24$, then $4 = \log_a 24$. ∎

━━━ **NOW WORK PROBLEM 1.**

2 **EXAMPLE 3** **Changing Logarithmic Expressions to Exponential Expressions**

Change each logarithmic expression to an equivalent expression involving an exponent.

(a) $\log_a 4 = 5$ (b) $\log_e b = -3$ (c) $\log_3 5 = c$

Solution (a) If $\log_a 4 = 5$, then $a^5 = 4$.   (b) If $\log_e b = -3$, then $e^{-3} = b$.
(c) If $\log_3 5 = c$, then $3^c = 5$. ∎

━━━ **NOW WORK PROBLEM 13.**

3 To find the exact value of a logarithm, we write the logarithm in exponential notation and use the fact that if $a^u = a^v$, then $u = v$.

EXAMPLE 4 **Finding the Exact Value of a Logarithmic Expression**

Find the exact value of:

(a) $\log_2 16$ (b) $\log_3 \dfrac{1}{27}$

Solution (a) $y = \log_2 16$
$2^y = 16$ Change to exponential form.
$2^y = 2^4$ $16 = 2^4$
$y = 4$ Equate exponents.

Therefore, $\log_2 16 = 4$.

(b) $y = \log_3 \dfrac{1}{27}$

$$3^y = \dfrac{1}{27} \qquad \text{Change to exponential form.}$$

$$3^y = 3^{-3} \qquad \dfrac{1}{27} = \dfrac{1}{3^3} = 3^{-3}.$$

$$y = -3 \qquad \text{Equate exponents.}$$

Therefore, $\log_3 \dfrac{1}{27} = -3$.

NOW WORK PROBLEM 25.

Domain of a Logarithmic Function

④ The logarithmic function $y = \log_a x$ has been defined as the inverse of the exponential function $y = a^x$. That is, if $f(x) = a^x$, then $f^{-1}(x) = \log_a x$. Based on the discussion given in Section 1.7 on inverse functions, we know that, for a function f and its inverse f^{-1},

$$\text{Domain of } f^{-1} = \text{Range of } f \quad \text{and} \quad \text{Range of } f^{-1} = \text{Domain of } f$$

Consequently, it follows that

> Domain of logarithmic function = Range of exponential function = $(0, \infty)$
>
> Range of logarithmic function = Domain of exponential function = $(-\infty, \infty)$

In the next box, we summarize some properties of the logarithmic function:

> $y = \log_a x$ (defining equation: $x = a^y$)
>
> Domain: $0 < x < \infty$ Range: $-\infty < y < \infty$

Notice that the domain of a logarithmic function consists of the *positive* real numbers. The argument of a logarithmic function must be greater than zero.

EXAMPLE 5 **Finding the Domain of a Logarithmic Function**

Find the domain of each logarithmic function.

(a) $F(x) = \log_2(x + 3)$ 　　　(b) $g(x) = \log_5\left(\dfrac{1 + x}{1 - x}\right)$

(c) $h(x) = \log_{1/2}|x|$

Solution (a) The domain of F consists of all x for which $x + 3 > 0$, that is, $x > -3$ or, using interval notation, $(-3, \infty)$.

(b) The domain of g is restricted to

$$\dfrac{1 + x}{1 - x} > 0$$

Solving this inequality, we find that the domain of g consists of all x between -1 and 1, that is, $-1 < x < 1$, or using interval notation, $(-1, 1)$.

(c) Since $|x| > 0$, provided that $x \neq 0$, the domain of h consists of all real numbers except zero or, using interval notation, $(-\infty, 0)$ or $(0, \infty)$.   ■

NOW WORK PROBLEMS 39 AND 45.

Graphs of Logarithmic Functions

⑤   Since exponential functions and logarithmic functions are inverses of each other, the graph of a logarithmic function $y = \log_a x$ is the reflection about the line $y = x$ of the graph of the exponential function $y = a^x$, as shown in Figure 17.

Figure 17

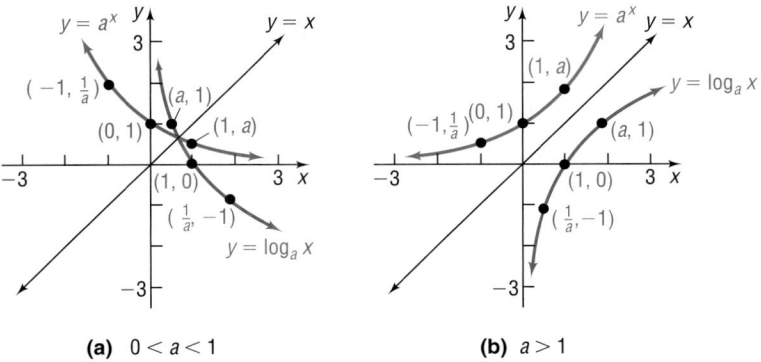

(a) $0 < a < 1$ (b) $a > 1$

Properties of a Logarithmic Function $f(x) = \log_a x$

1. The domain is the set of positive real numbers; the range is all real numbers.
2. The x-intercept of the graph is 1. There is no y-intercept.
3. The y-axis ($x = 0$) is a vertical asymptote of the graph.
4. A logarithmic function is decreasing if $0 < a < 1$ and increasing if $a > 1$.
5. The graph of f contains the points $(1, 0)$, $(a, 1)$, and $\left(\dfrac{1}{a}, -1\right)$.
6. The graph is smooth and continuous, with no corners or gaps.

If the base of a logarithmic function is the number e, then we have the **natural logarithm function**. This function occurs so frequently in applications that it is given a special symbol, **ln** (from the Latin, *logarithmus naturalis*). Thus,

$$y = \ln x \quad \text{if and only if} \quad x = e^y \qquad\qquad (1)$$

Since $y = \ln x$ and the exponential function $y = e^x$ are inverse functions, we can obtain the graph of $y = \ln x$ by reflecting the graph of $y = e^x$ about the line $y = x$. See Figure 18.

Figure 18

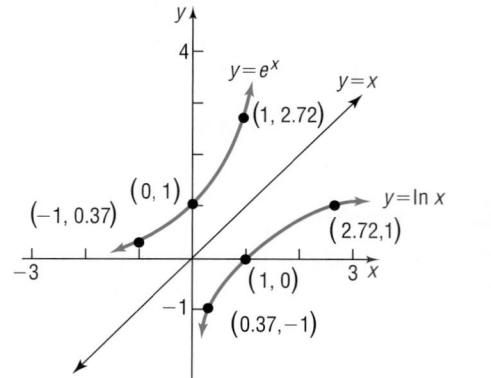

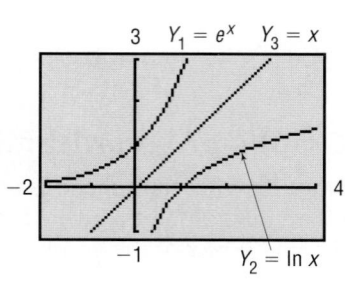

Table 5 displays other points on the graph of $f(x) = \ln x$. Notice for $x \le 0$ that we obtain an error message. Do you recall why?

EXAMPLE 6 — Graphing Logarithmic Functions Using Transformations

Graph $f(x) = -\ln(x + 2)$ by starting with the graph of $y = \ln x$. Determine the domain, range, and vertical asymptote.

Solution The domain consists of all x for which

$$x + 2 > 0 \quad \text{or} \quad x > -2$$

To obtain the graph of $y = -\ln(x + 2)$, we use the steps illustrated in Figure 19.

Figure 19

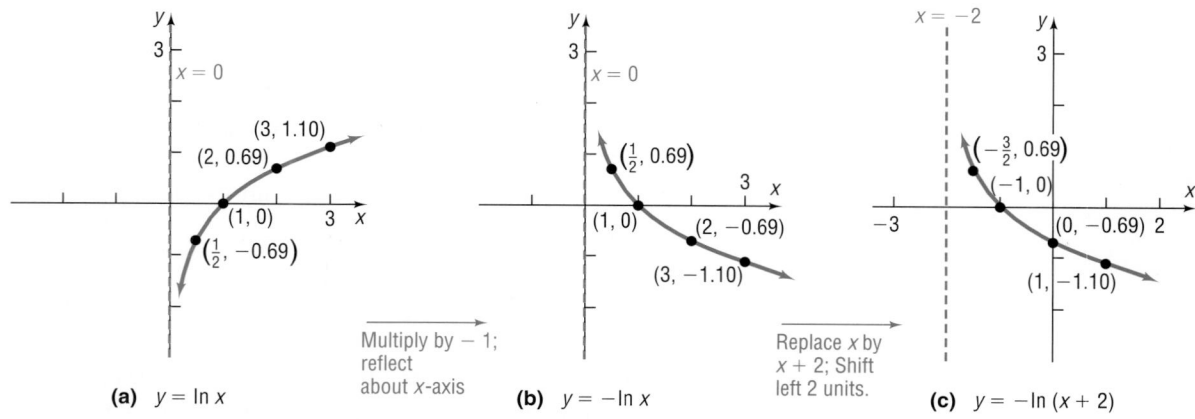

The range of $f(x) = -\ln(x + 2)$ is the interval $(-\infty, \infty)$, and the vertical asymptote is $x = -2$. [Do you see why? The original asymptote $(x = 0)$ is shifted to the left 2 units.]

✔ CHECK: Graph $Y_1 = -\ln(x + 2)$ using a graphing utility to verify Figure 19(c). ■ ▣

──── NOW WORK PROBLEM **61**.

If the base of a logarithmic function is the number 10, then we have the **common logarithm function**. If the base a of the logarithmic function is not indicated, it is understood to be 10. Thus,

$$y = \log x \quad \text{if and only if} \quad x = 10^y$$

Since $y = \log x$ and the exponential function $y = 10^x$ are inverse functions, we can obtain the graph of $y = \log x$ by reflecting the graph of $y = 10^x$ about the line $y = x$. See Figure 20.

Figure 20

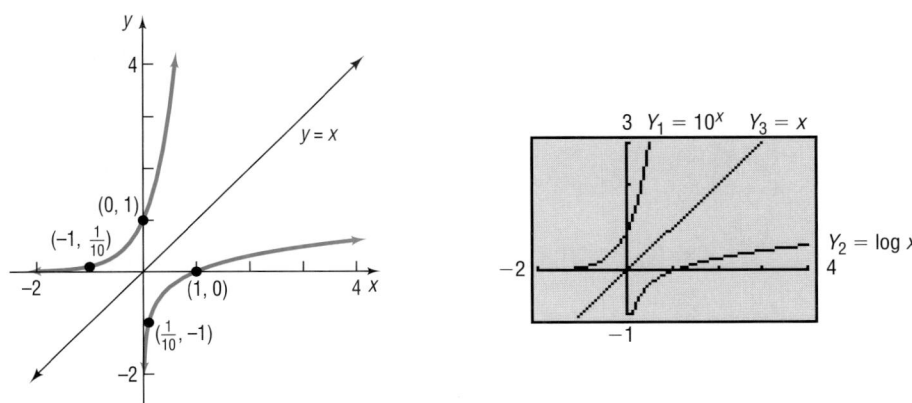

EXAMPLE 7   Graphing Logarithmic Functions Using Transformations

Graph $f(x) = 3 \log(x - 1)$. Determine the domain, range, and vertical asymptote of f.

Solution   The domain consists of all x for which

$$x - 1 > 0 \quad \text{or} \quad x > 1$$

To obtain the graph of $y = 3 \log(x - 1)$, we use the steps illustrated in Figure 21.

Figure 21

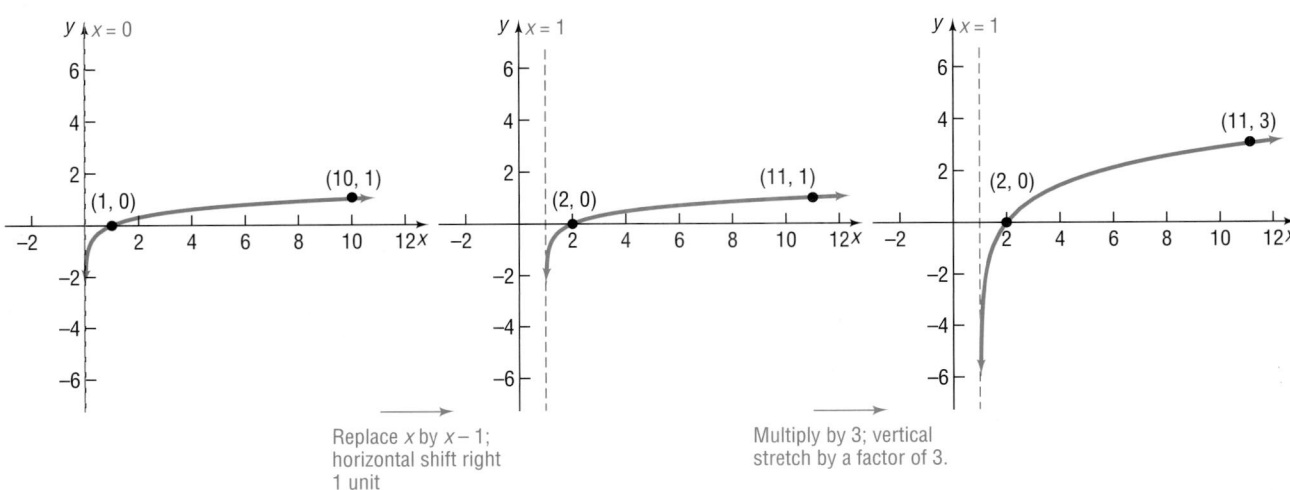

Replace x by $x - 1$; horizontal shift right 1 unit

Multiply by 3; vertical stretch by a factor of 3.

(a) $y = \log x$ **(b)** $y = \log (x - 1)$ **(c)** $y = 3 \log (x - 1)$

The range of $f(x) = 3\log(x - 1)$ is the interval $(-\infty, \infty)$, and the vertical asymptote is $x = 1$.

✔ CHECK: Graph $Y_1 = 3\log(x - 1)$ using a graphing utility to verify Figure 21(c).   ■ ■

NOW WORK PROBLEM 75.

Logarithmic Equations

6 Equations that contain logarithms are called **logarithmic equations**. Care must be taken when solving logarithmic equations algebraically. Be sure to check each apparent solution in the original equation and discard any that are extraneous. In the expression $\log_a M$, remember that a and M are positive and $a \neq 1$.

Some logarithmic equations can be solved by changing from a logarithmic expression to an exponential expression.

EXAMPLE 8   Solving a Logarithmic Equation

Solve: (a) $\log_3(4x - 7) = 2$ (b) $\log_x 64 = 2$

Solution   (a) We can obtain an exact solution by changing the logarithm to exponential form.

$$\log_3(4x - 7) = 2$$
$$4x - 7 = 3^2 \quad \text{Change to exponential form.}$$
$$4x - 7 = 9$$
$$4x = 16$$
$$x = 4$$

(b) We can obtain an exact solution by changing the logarithm to exponential form.

$$\log_x 64 = 2$$
$$x^2 = 64 \quad \text{Change to exponential form.}$$
$$x = \pm\sqrt{64} = \pm 8$$

The base of a logarithm is always positive. As a result, we discard -8; the only solution is 8.   ■

EXAMPLE 9   Using Logarithms to Solve Exponential Equations

Solve: $e^{2x} = 5$

Solution   We can obtain an exact solution by changing the exponential equation to logarithmic form.

$$e^{2x} = 5$$
$$\ln 5 = 2x \quad \text{Change to a logarithmic expression using (1).}$$
$$x = \frac{\ln 5}{2} \approx 0.805$$

■

NOW WORK PROBLEMS 85 AND 97.

EXAMPLE 10   Alcohol and Driving

The concentration of alcohol in a person's blood is measurable. Recent medical research suggests that the risk R (given as a percent) of having an accident while driving a car can be modeled by the equation

$$R = 6e^{kx}$$

where x is the variable concentration of alcohol in the blood and k is a constant.

(a) Suppose that a concentration of alcohol in the blood of 0.04 results in a 10% risk ($R = 10$) of an accident. Find the constant k in the equation. Graph $R = 6e^{kx}$ using this value of k.

(b) Using this value of k, what is the risk if the concentration is 0.17?

(c) Using the same value of k, what concentration of alcohol corresponds to a risk of 100%?

(d) If the law asserts that anyone with a risk of having an accident of 20% or more should not have driving privileges, at what concentration of alcohol in the blood should a driver be arrested and charged with a DUI (Driving Under the Influence)?

Solution   (a) For a concentration of alcohol in the blood of 0.04 and a risk of 10%, we let $x = 0.04$ and $R = 10$ in the equation and solve for k.

$$R = 6e^{kx}$$

$$10 = 6e^{k(0.04)} \qquad\qquad R = 10; x = 0.04.$$

$$\frac{10}{6} = e^{0.04k} \qquad\qquad \text{Divide both sides by 6.}$$

$$0.04k = \ln\frac{10}{6} = 0.5108256 \qquad\qquad \text{Change to a logarithmic expression.}$$

$$k = 12.77 \qquad\qquad \text{Solve for } k.$$

Figure 22

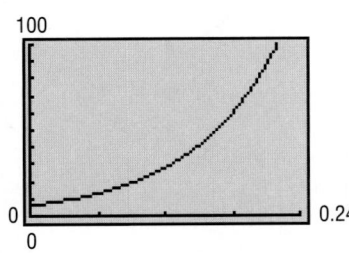

See Figure 22 for the graph of $R = 6e^{12.77x}$.

(b) Using $k = 12.77$ and $x = 0.17$ in the equation, we find the risk R to be

$$R = 6e^{kx} = 6e^{(12.77)(0.17)} = 52.6$$

For a concentration of alcohol in the blood of 0.17, the risk of an accident is about 52.6%. Verify this answer using eVALUEate.

(c) Using $k = 12.77$ and $R = 100$ in the equation, we find the concentration x of alcohol in the blood to be

$$R = 6e^{kx}$$

$$100 = 6e^{12.77x} \qquad\qquad R = 100; k = 12.77.$$

$$\frac{100}{6} = e^{12.77x} \qquad\qquad \text{Divide both sides by 6.}$$

$$12.77x = \ln\frac{100}{6} = 2.8134 \qquad\qquad \text{Change to a logarithmic expression.}$$

$$x = 0.22 \qquad\qquad \text{Solve for } x.$$

For a concentration of alcohol in the blood of 0.22, the risk of an accident is 100%. Verify this answer using a graphing utility.

(d) Using $k = 12.77$ and $R = 20$ in the equation, we find the concentration x of alcohol in the blood to be

$$R = 6e^{kx}$$

$$20 = 6e^{12.77x}$$

$$\frac{20}{6} = e^{12.77x}$$

$$12.77x = \ln\frac{20}{6} = 1.204$$

$$x = 0.094$$

A driver with a concentration of alcohol in the blood of 0.094 or more should be arrested and charged with DUI.   ■

Note: Most states use 0.08 or 0.10 as the blood alcohol content at which a DUI citation is given.

SUMMARY   **Properties of the Logarithmic Function**

$f(x) = \log_a x, \quad a > 1$
$(y = \log_a x \text{ means } x = a^y)$

Domain: the interval $(0, \infty)$; Range: the interval $(-\infty, \infty)$;
x-intercept: 1; y-intercept: none; vertical asymptote: $x = 0$ (y-axis);
increasing; one-to-one
See Figure 23(a) for a typical graph.

$f(x) = \log_a x, \quad 0 < a < 1$
$(y = \log_a x \text{ means } x = a^y)$

Domain: the interval $(0, \infty)$; Range: the interval $(-\infty, \infty)$;
x-intercept: 1; y-intercept: none; vertical asymptote: $x = 0$ (y-axis);
decreasing; one-to-one
See Figure 23(b) for a typical graph.

Figure 23

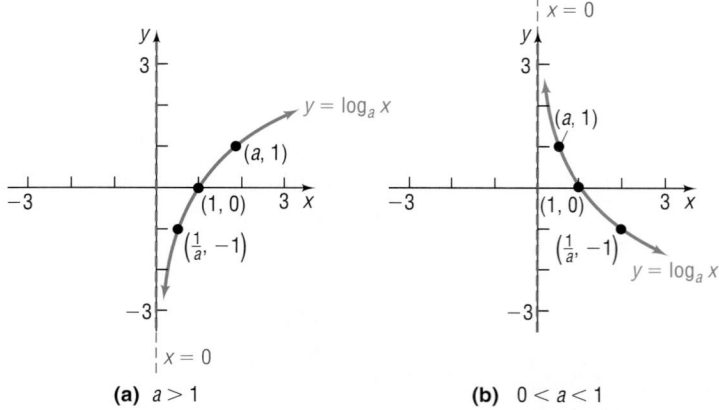

(a) $a > 1$ (b) $0 < a < 1$

7.2   Concepts and Vocabulary

In Problems 1–3, fill in the blanks.

1. The domain of the logarithmic function $f(x) = \log_a x$ is _____.

2. The graph of every logarithmic function $f(x) = \log_a x$, $a > 0$, $a \neq 1$, passes through three points: _____, _____, and _____.

3. If the graph of a logarithmic function $f(x) = \log_a x$, $a > 0$, $a \neq 1$, is increasing, then its base must be larger than _____.

In Problems 4 and 5, answer True or False to each statement.

4. If $y = \log_a x$, then $y = a^x$.

5. The graph of every logarithmic function $f(x) = \log_a x$, $a > 0$, $a \neq 1$, will contain the points $(1, 0)$, $(a, 1)$ and $\left(\dfrac{1}{a}, -1\right)$.

6. In the definition of the logarithmic function, the base a is not allowed to equal 1. Why?

7. If the domain of a logarithmic function is the interval $(1, \infty)$, what is the range of its inverse?

7.2 Exercises

In Problems 1–12, change each exponential expression to an equivalent expression involving a logarithm.

1. $9 = 3^2$
2. $16 = 4^2$
3. $a^2 = 1.6$
4. $a^3 = 2.1$

5. $1.1^2 = M$
6. $2.2^3 = N$
7. $2^x = 7.2$
8. $3^x = 4.6$

9. $x^{\sqrt{2}} = \pi$
10. $x^\pi = e$
11. $e^x = 8$
12. $e^{2.2} = M$

In Problems 13–24, change each logarithmic expression to an equivalent expression involving an exponent.

13. $\log_2 8 = 3$
14. $\log_3\left(\dfrac{1}{9}\right) = -2$
15. $\log_a 3 = 6$
16. $\log_b 4 = 2$

17. $\log_3 2 = x$
18. $\log_2 6 = x$
19. $\log_2 M = 1.3$
20. $\log_3 N = 2.1$

21. $\log_{\sqrt{2}} \pi = x$
22. $\log_\pi x = \dfrac{1}{2}$
23. $\ln 4 = x$
24. $\ln x = 4$

In Problems 25–36, find the exact value of each logarithm without using a calculator.

25. $\log_2 1$
26. $\log_8 8$
27. $\log_5 25$
28. $\log_3\left(\dfrac{1}{9}\right)$

29. $\log_{1/2} 16$
30. $\log_{1/3} 9$
31. $\log_{10} \sqrt{10}$
32. $\log_5 \sqrt[3]{25}$

33. $\log_{\sqrt{2}} 4$
34. $\log_{\sqrt{3}} 9$
35. $\ln \sqrt{e}$
36. $\ln e^3$

In Problems 37–46, find the domain of each function.

37. $f(x) = \ln(x - 3)$
38. $g(x) = \ln(x - 1)$
39. $F(x) = \log_2 x^2$

40. $H(x) = \log_5 x^3$
41. $h(x) = \log_{1/2}(x^2 - 2x + 1)$
42. $G(x) = \log_{1/2}(x^2 - 1)$

43. $f(x) = \ln\left(\dfrac{1}{x + 1}\right)$
44. $g(x) = \ln\left(\dfrac{1}{x - 5}\right)$
45. $g(x) = \log_5\left(\dfrac{x + 1}{x}\right)$

46. $h(x) = \log_3\left(\dfrac{x}{x - 1}\right)$

In Problems 47–50, use a calculator to evaluate each expression. Round your answer to three decimal places.

47. $\ln \dfrac{5}{3}$
48. $\dfrac{\ln 5}{3}$
49. $\dfrac{\ln(10/3)}{0.04}$
50. $\dfrac{\ln(2/3)}{-0.1}$

51. Find a so that the graph of $f(x) = \log_a x$ contains the point $(2, 2)$.

52. Find a so that the graph of $f(x) = \log_a x$ contains the point $\left(\dfrac{1}{2}, -4\right)$.

In Problems 53–60, the graph of a logarithmic function is given. Match each graph to one of the following functions:

A. $y = \log_3 x$
B. $y = \log_3(-x)$
C. $y = -\log_3 x$
D. $y = -\log_3(-x)$

E. $y = \log_3 x - 1$
F. $y = \log_3(x - 1)$
G. $y = \log_3(1 - x)$
H. $y = 1 - \log_3 x$

53.
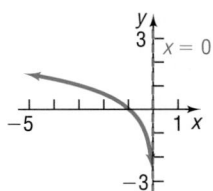

54.
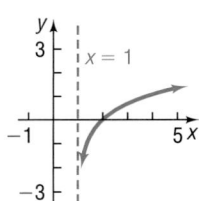

55.
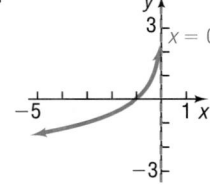

56.

57. 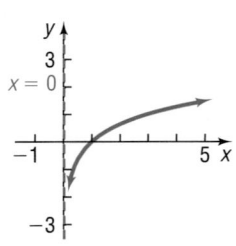   **58.** 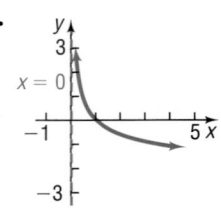   **59.** 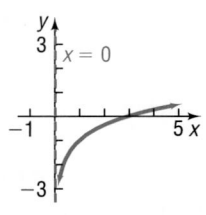   **60.**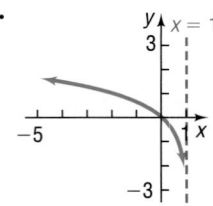

In Problems 61–84, use transformations to graph each function. Determine the domain, range, and vertical asymptote of each function. Verify your results using a graphing utility.

61. $f(x) = \ln(x + 4)$ **62.** $f(x) = \ln(x - 3)$ **63.** $f(x) = 2 + \ln x$ **64.** $f(x) = -\ln(-x)$

65. $g(x) = \ln(2x)$ **66.** $h(x) = \ln\left(\frac{1}{2}x\right)$ **67.** $f(x) = 3 \ln x$ **68.** $f(x) = -2 \ln x$

69. $g(x) = \ln(3 - x)$ **70.** $h(x) = \ln(4 - x)$ **71.** $f(x) = -\ln(x - 1)$ **72.** $f(x) = 2 - \ln x$
73. $f(x) = \log(x - 4)$ **74.** $f(x) = \log(x + 5)$ **75.** $h(x) = 4 \log x$ **76.** $g(x) = -3 \log x$
77. $F(x) = \log(2x)$ **78.** $G(x) = \log(5x)$ **79.** $f(x) = 2 \log(x + 3)$ **80.** $f(x) = -3 \log(x - 2)$
81. $h(x) = 3 + \log(x + 2)$ **82.** $g(x) = 2 - \log(x + 1)$ **83.** $F(x) = \log(2 - x)$ **84.** $G(x) = 3 + \log(4 - x)$

In Problems 85–104, solve each equation.

85. $\log_3 x = 2$ **86.** $\log_5 x = 3$ **87.** $\log_2(2x + 1) = 3$ **88.** $\log_3(3x - 2) = 2$

89. $\log_x 4 = 2$ **90.** $\log_x\left(\frac{1}{8}\right) = 3$ **91.** $\ln e^x = 5$ **92.** $\ln e^{-2x} = 8$

93. $\log_4 64 = x$ **94.** $\log_5 625 = x$ **95.** $\log_3 243 = 2x + 1$ **96.** $\log_6 36 = 5x + 3$

97. $e^{3x} = 10$ **98.** $e^{-2x} = \frac{1}{3}$ **99.** $e^{2x+5} = 8$ **100.** $e^{-2x+1} = 13$

101. $\log_3(x^2 + 1) = 2$ **102.** $\log_5(x^2 + x + 4) = 2$ **103.** $\log_2 8^x = -3$ **104.** $\log_3 3^x = -1$

In Problems 105–108, (a) graph each function and state its domain, range, and asymptote; (b) determine the inverse function; (c) use the graph obtained in (a) to graph the inverse and state its domain, range, and asymptote.

105. $f(x) = 2^x$ **106.** $f(x) = 5^x$ **107.** $f(x) = 2^{x+3}$ **108.** $f(x) = 5^x - 2$

---

109. Optics   If a single pane of glass obliterates 10% of the light passing through it, then the percent P of light that passes through n successive panes is given approximately by the equation

$$P = 100e^{-0.1n}$$

(a) How many panes are necessary to block at least 50% of the light?

(b) How many panes are necessary to block at least 75% of the light?

110. Chemistry   The pH of a chemical solution is given by the formula

$$pH = -\log_{10}[H^+]$$

where $[H^+]$ is the concentration of hydrogen ions in moles per liter. Values of pH range from 0 (acidic) to 14 (alkaline).

(a) What is the pH of a solution for which $[H^+]$ is 0.1?
(b) What is the pH of a solution for which $[H^+]$ is 0.01?
(c) What is the pH of a solution for which $[H^+]$ is 0.001?
(d) What happens to pH as the hydrogen ion concentration decreases?

(e) Determine the hydrogen ion concentration of an orange (pH = 3.5).
(f) Determine the hydrogen ion concentration of human blood (pH = 7.4).

111. Space Satellites   The number of watts w provided by a space satellite's power supply after d days is given by the formula

$$w = 50e^{-0.004d}$$

(a) How long will it take for the available power to drop to 30 watts?
(b) How long will it take for the available power to drop to only 5 watts?

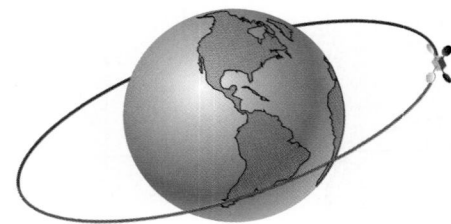

112. **Healing of Wounds**   The normal healing of wounds can be modeled by an exponential function. If A_0 represents the original area of the wound and if A equals the area of the wound after n days, then the formula

$$A = A_0 e^{-0.35n}$$

describes the area of a wound on the nth day following an injury when no infection is present to retard the healing. Suppose that a wound initially had an area of 100 square millimeters.
   (a) If healing is taking place, how many days should pass before the wound is one-half its original size?
   (b) How long before the wound is 10% of its original size?

113. **Exponential Probability**   Between 12:00 PM and 1:00 PM, cars arrive at Citibank's drive-thru at the rate of 6 cars per hour (0.1 car per minute). The following formula from statistics can be used to determine the probability that a car will arrive within t minutes of 12:00 PM.

$$F(t) = 1 - e^{-0.1t}$$

   (a) Determine how many minutes are needed for the probability to reach 50%.
   (b) Determine how many minutes are needed for the probability to reach 80%.
   (c) Is it possible for the probability to equal 100%? Explain.

114. **Exponential Probability**   Between 5:00 PM and 6:00 PM, cars arrive at Jiffy Lube at the rate of 9 cars per hour (0.15 car per minute). The following formula from statistics can be used to determine the probability that a car will arrive within t minutes of 5:00 PM.

$$F(t) = 1 - e^{-0.15t}$$

   (a) Determine how many minutes are needed for the probability to reach 50%.
   (b) Determine how many minutes are needed for the probability to reach 80%.

115. **Drug Medication**   The formula

$$D = 5e^{-0.4h}$$

can be used to find the number of milligrams D of a certain drug that is in a patient's bloodstream h hours after the drug has been administered. When the number of milligrams reaches 2, the drug is to be administered again. What is the time between injections?

116. **Spreading of Rumors**   A model for the number of people N in a college community who have heard a certain rumor is

$$N = P(1 - e^{-0.15d})$$

where P is the total population of the community and d is the number of days that have elapsed since the rumor began. In a community of 1000 students, how many days will elapse before 450 students have heard the rumor?

117. **Current in an *RL* Circuit**   The equation governing the amount of current I (in amperes) after time t (in seconds) in a simple RL circuit consisting of a resistance R (in ohms), an inductance L (in henrys), and an electromotive force E (in volts) is

$$I = \frac{E}{R}[1 - e^{-(R/L)t}]$$

If $E = 12$ volts, $R = 10$ ohms, and $L = 5$ henrys, how long does it take to obtain a current of 0.5 ampere? Of 1.0 ampere? Graph the equation.

118. **Learning Curve**   Psychologists sometimes use the function

$$L(t) = A(1 - e^{-kt})$$

to measure the amount L learned at time t. The number A represents the amount to be learned, and the number k measures the rate of learning. Suppose that a student has an amount A of 200 vocabulary words to learn. A psychologist determines that the student learned 20 vocabulary words after 5 minutes.
   (a) Determine the rate of learning k.
   (b) Approximately how many words will the student have learned after 10 minutes?
   (c) After 15 minutes?
   (d) How long does it take for the student to learn 180 words?

Loudness of Sound   *Problems 119–122 use the following discussion: The* **loudness** $L(x)$, *measured in decibels, of a sound of intensity x, measured in watts per square meter, is defined as* $L(x) = 10 \log \dfrac{x}{I_0}$, *where* $I_0 = 10^{-12}$ *watt per square meter is the least intense sound that a human ear can detect. Determine the loudness, in decibels, of each of the following sounds.*

119. Normal conversation: intensity of $x = 10^{-7}$ watt per square meter.

120. Heavy city traffic: intensity of $x = 10^{-3}$ watt per square meter.

121. Amplified rock music: intensity of 10^{-1} watt per square meter.

122. Diesel truck traveling 40 miles per hour 50 feet away: intensity 10 times that of a passenger car traveling 50 miles per hour 50 feet away whose loudness is 70 decibels.

Problems 123 and 124 use the following discussion: The **Richter scale** *is one way of converting seismographic readings into numbers that provide an easy reference for measuring the magnitude M of an earthquake. All earthquakes are compared to a* **zero-level earthquake** *whose seismographic reading measures 0.001 millimeter at a distance of 100 kilometers from the epicenter. An earthquake whose seismographic reading measures x millimeters has* **magnitude** *M(x) given by*

$$M(x) = \log\left(\frac{x}{x_0}\right)$$

where $x_0 = 10^{-3}$ is the reading of a zero-level earthquake the same distance from its epicenter. Determine the magnitude of the following earthquakes.

123. Magnitude of an Earthquake Mexico City in 1985: seismographic reading of 125,892 millimeters 100 kilometers from the center.

124. Magnitude of an Earthquake San Francisco in 1906: seismographic reading of 7943 millimeters 100 kilometers from the center.

125. Alcohol and Driving The concentration of alcohol in a person's blood is measurable. Suppose that the risk R (given as a percent) of having an accident while driving a car can be modeled by the equation

$$R = 3e^{kx}$$

where x is the variable concentration of alcohol in the blood and k is a constant.

(a) Suppose that a concentration of alcohol in the blood of 0.06 results in a 10% risk ($R = 10$) of an accident. Find the constant k in the equation.

(b) Using this value of k, what is the risk if the concentration is 0.17?

(c) Using the same value of k, what concentration of alcohol corresponds to a risk of 100%?

(d) If the law asserts that anyone with a risk of having an accident of 15% or more should not have driving privileges, at what concentration of alcohol in the blood should a driver be arrested and charged with a DUI?

 (e) Compare this situation with that of Example 10. If you were a lawmaker, which situation would you support? Give your reasons.

126. Is there any function of the form $y = x^{\alpha}, 0 < \alpha < 1$, that increases more slowly than a logarithmic function whose base is greater than 1? Explain.

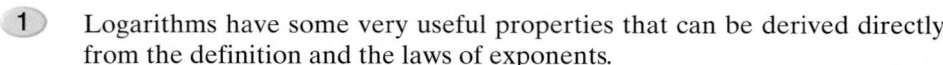

PREPARING FOR THIS SECTION

Before getting started, review the following:

✓ Inverse Functions (Section 1.7, pp. 80–87)

7.3 PROPERTIES OF LOGARITHMS

OBJECTIVES

1. Work with the Properties of Logarithms
2. Write a Logarithmic Expression as a Sum or Difference of Logarithms
3. Write a Logarithmic Expression as a Single Logarithm
4. Evaluate Logarithms Whose Base Is Neither 10 nor e
5. Graph Logarithmic Functions Whose Base Is Neither 10 nor e

1. Logarithms have some very useful properties that can be derived directly from the definition and the laws of exponents.

EXAMPLE 1 **Establishing Properties of Logarithms**

(a) Show that $\log_a 1 = 0$.   (b) Show that $\log_a a = 1$.

Solution (a) This fact was established when we graphed $y = \log_a x$ (see Figure 28). To show the result algebraically, let $y = \log_a 1$. Then

$$y = \log_a 1$$
$$a^y = 1 \qquad \text{\textit{Change to an exponent.}}$$
$$a^y = a^0 \qquad a^0 = 1$$
$$y = 0 \qquad \text{\textit{Solve for y.}}$$
$$\log_a 1 = 0 \qquad y = \log_a 1$$

(b) Let $y = \log_a a$. Then

$$y = \log_a a$$
$$a^y = a \qquad \text{\textit{Change to an exponent.}}$$
$$a^y = a^1 \qquad a^1 = a$$
$$y = 1 \qquad \text{\textit{Solve for y.}}$$
$$\log_a a = 1 \qquad y = \log_a a$$

To summarize:

$$\log_a 1 = 0 \qquad \log_a a = 1$$

Theorem

Properties of Logarithms

In the properties given next, M and a are positive real numbers, with $a \neq 1$, and r is any real number.

The number $\log_a M$ is the exponent to which a must be raised to obtain M. That is,

$$a^{\log_a M} = M \tag{1}$$

The logarithm to the base a of a raised to a power equals that power. That is,

$$\log_a a^r = r \tag{2}$$

The proof uses the fact that $y = a^x$ and $y = \log_a x$ are inverses.

Proof of Property (1)   For inverse functions,
$$f(f^{-1}(x)) = x$$
Using $f(x) = a^x$ and $f^{-1}(x) = \log_a x$, we find
$$f(f^{-1}(x)) = f(\log_a x) = a^{\log_a x} = x$$
Now let $x = M$ to obtain $a^{\log_a M} = M$.

Proof of Property (2)   For inverse functions,
$$f^{-1}(f(x)) = x$$
Using $f(x) = a^x$ and $f^{-1}(x) = \log_a x$, we find
$$f^{-1}(f(x)) = f^{-1}(a^x) = \log_a a^x = x$$
Now let $x = r$ to obtain $\log_a a^r = r$.

| EXAMPLE 2 | **Using Properties (1) and (2)** |

(a) $2^{\log_2 \pi} = \pi$ (b) $\log_{0.2} 0.2^{-\sqrt{2}} = -\sqrt{2}$ (c) $\ln e^{kt} = kt$ ■

NOW WORK PROBLEM 3.

Other useful properties of logarithms are given next.

Theorem

Properties of Logarithms

In the following properties, M, N, and a are positive real numbers, with $a \neq 1$, and r is any real number.

The Log of a Product Equals the Sum of the Logs

$$\log_a(MN) = \log_a M + \log_a N \qquad (3)$$

The Log of a Quotient Equals the Difference of the Logs

$$\log_a\left(\frac{M}{N}\right) = \log_a M - \log_a N \qquad (4)$$

The Log of a Power Equals the Product of the Power and the Log

$$\log_a M^r = r \log_a M \qquad (5)$$

■

We shall derive properties (3) and (5) and leave the derivation of property (4) as an exercise (see Problem 97).

Proof of Property (3) Let $A = \log_a M$ and let $B = \log_a N$. These expressions are equivalent to the exponential expressions

$$a^A = M \quad \text{and} \quad a^B = N$$

Now

$$\log_a(MN) = \log_a(a^A a^B) = \log_a a^{A+B} \qquad \text{Law of Exponents}$$
$$= A + B \qquad \text{Property (2) of logarithms}$$
$$= \log_a M + \log_a N \qquad ■$$

Proof of Property (5) Let $A = \log_a M$. This expression is equivalent to

$$a^A = M$$

Now

$$\log_a M^r = \log_a(a^A)^r = \log_a a^{rA} \qquad \text{Law of Exponents}$$
$$= rA \qquad \text{Property (2) of logarithms}$$
$$= r \log_a M \qquad ■$$

NOW WORK PROBLEM 7.

Logarithms can be used to transform products into sums, quotients into differences, and powers into factors. Such transformations prove useful in certain types of calculus problems.

EXAMPLE 3   **Writing a Logarithmic Expression as a Sum of Logarithms**

Write $\log_a\left(x\sqrt{x^2+1}\right)$ as a sum of logarithms. Express all powers as factors.

Solution $\begin{aligned}\log_a\left(x\sqrt{x^2+1}\right) &= \log_a x + \log_a \sqrt{x^2+1} \qquad \text{Property (3)}\\ &= \log_a x + \log_a(x^2+1)^{1/2}\\ &= \log_a x + \frac{1}{2}\log_a(x^2+1) \qquad \text{Property (5)}\end{aligned}$ ∎

EXAMPLE 4   **Writing a Logarithmic Expression as a Difference of Logarithms**

Write

$$\ln\frac{x^2}{(x-1)^3}$$

as a difference of logarithms. Express all powers as factors.

Solution $\ln\dfrac{x^2}{(x-1)^3} \underset{\substack{\uparrow\\ \text{Property (4)}}}{=} \ln x^2 - \ln(x-1)^3 \underset{\substack{\uparrow\\ \text{Property (5)}}}{=} 2\ln x - 3\ln(x-1)$ ∎

EXAMPLE 5   **Writing a Logarithmic Expression as a Sum and Difference of Logarithms**

Write

$$\log_a\frac{x^3\sqrt{x^2+1}}{(x+1)^4}$$

as a sum and difference of logarithms. Express all powers as factors.

Solution $\begin{aligned}\log_a\frac{x^3\sqrt{x^2+1}}{(x+1)^4} &= \log_a\left(x^3\sqrt{x^2+1}\right) - \log_a(x+1)^4 \qquad \text{Property (4)}\\ &= \log_a x^3 + \log_a\sqrt{x^2+1} - \log_a(x+1)^4 \qquad \text{Property (3)}\\ &= \log_a x^3 + \log_a(x^2+1)^{1/2} - \log_a(x+1)^4\\ &= 3\log_a x + \frac{1}{2}\log_a(x^2+1) - 4\log_a(x+1) \qquad \text{Property (5)}\end{aligned}$ ∎

✏ **NOW WORK PROBLEM 39.**

Another use of properties (3) through (5) is to write sums and/or differences of logarithms with the same base as a single logarithm. This skill will be needed to solve certain logarithmic equations discussed in the next section.

EXAMPLE 6

Writing Expressions as a Single Logarithm

Write each of the following as a single logarithm.

(a) $\log_a 7 + 4 \log_a 3$ (b) $\dfrac{2}{3} \ln 8 - \ln(3^4 - 8)$

(c) $\log_a x + \log_a 9 + \log_a(x^2 + 1) - \log_a 5$

Solution (a) $\log_a 7 + 4 \log_a 3 = \log_a 7 + \log_a 3^4$ Property (5)

$= \log_a 7 + \log_a 81$

$= \log_a(7 \cdot 81)$ Property (3)

$= \log_a 567$

(b) $\dfrac{2}{3} \ln 8 - \ln(3^4 - 8) = \ln 8^{2/3} - \ln(81 - 8)$ Property (5)

$= \ln 4 - \ln 73$ $8^{2/3} = \sqrt[3]{8^2} = \sqrt[3]{64} = 4$

$= \ln\left(\dfrac{4}{73}\right)$ Property (4)

(c) $\log_a x + \log_a 9 + \log_a(x^2 + 1) - \log_a 5 = \log_a(9x) + \log_a(x^2 + 1) - \log_a 5$

$= \log_a[9x(x^2 + 1)] - \log_a 5$

$= \log_a\left[\dfrac{9x(x^2 + 1)}{5}\right]$ ∎

WARNING: A common error made by some students is to express the logarithm of a sum as the sum of logarithms.

$\log_a(M + N)$ is not equal to $\log_a M + \log_a N$

Correct statement $\log_a(MN) = \log_a M + \log_a N$ Property (3)

Another common error is to express the difference of logarithms as the quotient of logarithms.

$\log_a M - \log_a N$ is not equal to $\dfrac{\log_a M}{\log_a N}$

Correct statement $\log_a M - \log_a N = \log_a\left(\dfrac{M}{N}\right)$ Property (4)

A third common error is to express a logarithm raised to a power as the product of the power times the logarithm.

$(\log_a M)^r$ is not equal to $r \log_a M$

Correct statement $\log_a M^r = r \log_a M$ Property (5) ∎

NOW WORK PROBLEM 45.

Two other properties of logarithms that we need to know are consequences of the fact that the logarithmic function $y = \log_a x$ is one-to-one.

Theorem

In the following properties, M, N, and a are positive real numbers, with $a \neq 1$.

If $M = N$, then $\log_a M = \log_a N$.	(6)
If $\log_a M = \log_a N$, then $M = N$.	(7)

When property (6) is used, we start with the equation $M = N$ and say "take the logarithm of both sides" to obtain $\log_a M = \log_a N$.

Properties (6) and (7) are useful for solving exponential and logarithmic equations discussed in the next section.

Using a Calculator to Evaluate Logarithms with Bases Other Than 10 or e

④ Common logarithms, that is, logarithms to the base 10, were used to facilitate arithmetic computations before the widespread use of calculators. (See the Historical Feature at the end of this section.) Natural logarithms, that is, logarithms whose base is the number e, remain very important because they arise frequently in the study of natural phenomena.

Common logarithms are usually abbreviated by writing **log**, with the base understood to be 10 just as natural logarithms are abbreviated by **ln**, with the base understood to be e.

Most calculators have both $\boxed{\log}$ and $\boxed{\ln}$ keys to calculate the common logarithm and natural logarithm of a number. Let's look at an example to see how to approximate logarithms having a base other than 10 or e.

EXAMPLE 7   **Approximating Logarithms Whose Base Is Neither 10 Nor e**

Approximate $\log_2 7$.
Round the answer to four decimal places.

Solution   Let $y = \log_2 7$. Then $2^y = 7$, so

$$2^y = 7$$
$$\ln 2^y = \ln 7 \qquad \text{Property (6)}$$
$$y \ln 2 = \ln 7 \qquad \text{Property (5)}$$
$$y = \frac{\ln 7}{\ln 2} \qquad \text{Solve for } y.$$
$$y \approx 2.8074 \qquad \text{Use your calculator (}\boxed{\ln}\text{ key) and round to four decimal places.} \quad \blacksquare$$

Example 7 shows how to approximate a logarithm whose base is 2 by changing to logarithms involving the base e. In general, we use the **Change-of-Base Formula**.

Theorem   **Change-of-Base Formula**

If $a \ne 1, b \ne 1$, and M are positive real numbers, then

$$\log_a M = \frac{\log_b M}{\log_b a} \qquad (8)$$

Proof   We derive this formula as follows: Let $y = \log_a M$. Then

$$a^y = M$$

$$\log_b a^y = \log_b M \qquad \text{\textit{Property (6)}}$$

$$y \log_b a = \log_b M \qquad \text{\textit{Property (5)}}$$

$$y = \frac{\log_b M}{\log_b a} \qquad \text{\textit{Solve for y.}}$$

$$\log_a M = \frac{\log_b M}{\log_b a} \qquad \text{\textit{y = log}}_a \text{\textit{M}}$$ ■

Since calculators have only keys for $\boxed{\text{log}}$ and $\boxed{\text{ln}}$, in practice, the Change-of-Base Formula uses either $b = 10$ or $b = e$. Thus,

$$\log_a M = \frac{\log M}{\log a} \quad \text{and} \quad \log_a M = \frac{\ln M}{\ln a} \qquad (9)$$

EXAMPLE 8   Using the Change-of-Base Formula

Approximate:

(a) $\log_5 89$

(b) $\log_{\sqrt{2}} \sqrt{5}$

Round answers to four decimal places.

Solution   (a) $\log_5 89 = \dfrac{\log 89}{\log 5} \approx \dfrac{1.94939}{0.69897} = 2.7889$

or

$$\log_5 89 = \frac{\ln 89}{\ln 5} \approx \frac{4.4886}{1.6094} = 2.7889$$

(b) $\log_{\sqrt{2}} \sqrt{5} = \dfrac{\log \sqrt{5}}{\log \sqrt{2}} \approx 2.3219$

or

$$\log_{\sqrt{2}} \sqrt{5} = \frac{\ln \sqrt{5}}{\ln \sqrt{2}} \approx 2.3219$$ ■

NOW WORK PROBLEMS **11** AND **61**.

⑤   We also use the Change-of-Base Formula to graph logarithmic functions whose base is neither 10 nor e.

EXAMPLE 9   Graphing a Logarithmic Function Whose Base Is Neither 10 Nor e

Use a graphing utility to graph $y = \log_2 x$.

Solution   Since graphing utilities only have logarithms with the base 10 or the base e, we need to use the Change-of-Base Formula to express $y = \log_2 x$ in terms

Figure 24

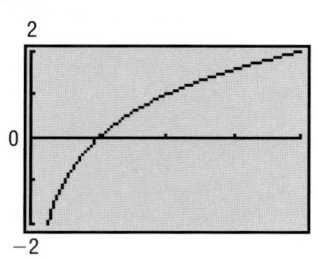

of logarithms with base 10 or base e. We can graph either $y = \dfrac{\ln x}{\ln 2}$ or $y = \dfrac{\log x}{\log 2}$ to obtain the graph of $y = \log_2 x$. See Figure 24.

✔ CHECK: Verify that $y = \dfrac{\ln x}{\ln 2}$ and $y = \dfrac{\log x}{\log 2}$ result in the same graph by graphing each on the same screen. ▪ ▪

NOW WORK PROBLEM **69.**

SUMMARY	Properties of Logarithms

In the list that follows, $a > 0, a \neq 1$, and $b > 0, b \neq 1$; also, $M > 0$ and $N > 0$.

Definition

$y = \log_a x$ means $x = a^y$

Properties of logarithms

$\log_a 1 = 0; \quad \log_a a = 1$

$a^{\log_a M} = M; \quad \log_a a^r = r$

$\log_a(MN) = \log_a M + \log_a N$

$\log_a\left(\dfrac{M}{N}\right) = \log_a M - \log_a N$

$\log_a M^r = r \log_a M$

If $M = N$, then $\log_a M = \log_a N$.

If $\log_a M = \log_a N$, then $M = N$.

Change-of-Base Formula

$\log_a M = \dfrac{\log_b M}{\log_b a}$

7.3 Concepts and Vocabulary

In Problems 1–3, fill in the blanks.

1. The logarithm of a product equals the _____ of the logarithms.

2. If $\log_8 M = \dfrac{\log_5 7}{\log_5 8}$, then $M =$ _____.

3. $\log_a M^r =$ _____.

In Problems 4–6, answer True or False to each statement.

4. $\ln(x + 3) - \ln(2x) = \dfrac{\ln(x + 3)}{\ln(2x)}$

5. $\log_2(3x^4) = 4\log_2(3x)$

6. $\log_2 16 = \dfrac{\ln 16}{\ln 2}$

7. Graph $Y_1 = \log(x^2)$ and $Y_2 = 2\log(x)$ on your graphing utility. Are they equivalent? What might account for any differences in the two functions?

7.3 Exercises

In Problems 1–16, use properties of logarithms to find the exact value of each expression. Do not use a calculator.

1. $\log_3 3^{71}$
2. $\log_2 2^{-13}$
3. $\ln e^{-4}$
4. $\ln e^{\sqrt{2}}$

5. $2^{\log_2 7}$
6. $e^{\ln 8}$
7. $\log_8 2 + \log_8 4$
8. $\log_6 9 + \log_6 4$

9. $\log_6 18 - \log_6 3$
10. $\log_8 16 - \log_8 2$
11. $\log_2 6 \cdot \log_6 4$
12. $\log_3 8 \cdot \log_8 9$

13. $3^{\log_3 5 - \log_3 4}$
14. $5^{\log_5 6 + \log_5 7}$
15. $e^{\log_{e^2} 16}$
16. $e^{\log_{e^2} 9}$

In Problems 17–24, suppose that $\ln 2 = a$ and $\ln 3 = b$. Use properties of logarithms to write each logarithm in terms of a and b.

17. $\ln 6$
18. $\ln \dfrac{2}{3}$
19. $\ln 1.5$
20. $\ln 0.5$

21. $\ln 8$
22. $\ln 27$
23. $\ln \sqrt[5]{6}$
24. $\ln \sqrt[4]{\dfrac{2}{3}}$

In Problems 25–44, write each expression as a sum and/or difference of logarithms. Express powers as factors.

25. $\log_5(25x)$
26. $\log_3 \dfrac{x}{9}$
27. $\log_2 z^3$
28. $\log_7(x^5)$

29. $\ln(ex)$
30. $\ln \dfrac{e}{x}$
31. $\ln(xe^x)$
32. $\ln \dfrac{x}{e^x}$

33. $\log_a(u^2 v^3)$
34. $\log_2\left(\dfrac{a}{b^2}\right)$
35. $\ln(x^2\sqrt{1 - x})$
36. $\ln\left(x\sqrt{1 + x^2}\right)$

37. $\log_2\left(\dfrac{x^3}{x - 3}\right)$
38. $\log_5\left(\dfrac{\sqrt[3]{x^2 + 1}}{x^2 - 1}\right)$
39. $\log\left[\dfrac{x(x + 2)}{(x + 3)^2}\right]$
40. $\log\left[\dfrac{x^3\sqrt{x + 1}}{(x - 2)^2}\right]$

41. $\ln\left[\dfrac{x^2 - x - 2}{(x + 4)^2}\right]^{1/3}$
42. $\ln\left[\dfrac{(x - 4)^2}{x^2 - 1}\right]^{2/3}$
43. $\ln \dfrac{5x\sqrt{1 - 3x}}{(x - 4)^3}$
44. $\ln\left[\dfrac{5x^2\sqrt[3]{1 - x}}{4(x + 1)^2}\right]$

In Problems 45–58, write each expression as a single logarithm.

45. $3 \log_5 u + 4 \log_5 v$

46. $2 \log_3 u - \log_3 v$

47. $\log_3 \sqrt{x} - \log_3 x^3$

48. $\log_2\left(\dfrac{1}{x}\right) + \log_2\left(\dfrac{1}{x^2}\right)$

49. $\log_4(x^2 - 1) - 5 \log_4(x + 1)$

50. $\log(x^2 + 3x + 2) - 2 \log(x + 1)$

51. $\ln\left(\dfrac{x}{x-1}\right) + \ln\left(\dfrac{x+1}{x}\right) - \ln(x^2 - 1)$

52. $\log\left(\dfrac{x^2 + 2x - 3}{x^2 - 4}\right) - \log\left(\dfrac{x^2 + 7x + 6}{x + 2}\right)$

53. $8 \log_2 \sqrt{3x - 2} - \log_2\left(\dfrac{4}{x}\right) + \log_2 4$

54. $21 \log_3 \sqrt[3]{x} + \log_3(9x^2) - \log_3 9$

55. $2 \log_a(5x^3) - \dfrac{1}{2} \log_a(2x + 3)$

56. $\dfrac{1}{3} \log(x^3 + 1) + \dfrac{1}{2} \log(x^2 + 1)$

57. $2 \log_2(x + 1) - \log_2(x + 3) - \log_2(x - 1)$

58. $3 \log_5(3x + 1) - 2 \log_5(2x - 1) - \log_5 x$

59. Write the exponential model $y = ab^x$ as a linear model. [**Hint:** Take the logarithm of both sides.]

60. Write the power model $y = ax^b$ as a linear model.

In Problems 61–68, use the Change-of-Base Formula and a calculator to evaluate each logarithm. Round your answer to three decimal places.

61. $\log_3 21$

62. $\log_5 18$

63. $\log_{1/3} 71$

64. $\log_{1/2} 15$

65. $\log_{\sqrt{2}} 7$

66. $\log_{\sqrt{5}} 8$

67. $\log_\pi e$

68. $\log_\pi \sqrt{2}$

In Problems 69–74, graph each function using a graphing utility and the Change-of-Base Formula.

69. $y = \log_4 x$

70. $y = \log_5 x$

71. $y = \log_2(x + 2)$

72. $y = \log_4(x - 3)$

73. $y = \log_{x-1}(x + 1)$

74. $y = \log_{x+2}(x - 2)$

In Problems 75–84, express y as a function of x. The constant C is a positive number.

75. $\ln y = \ln x + \ln C$

76. $\ln y = \ln(x + C)$

77. $\ln y = \ln x + \ln(x + 1) + \ln C$

78. $\ln y = 2 \ln x - \ln(x + 1) + \ln C$

79. $\ln y = 3x + \ln C$

80. $\ln y = -2x + \ln C$

81. $\ln(y - 3) = -4x + \ln C$

82. $\ln(y + 4) = 5x + \ln C$

83. $3 \ln y = \dfrac{1}{2} \ln(2x + 1) - \dfrac{1}{3} \ln(x + 4) + \ln C$

84. $2 \ln y = -\dfrac{1}{2} \ln x + \dfrac{1}{3} \ln(x^2 + 1) + \ln C$

85. Find the value of $\log_2 3 \cdot \log_3 4 \cdot \log_4 5 \cdot \log_5 6 \cdot \log_6 7 \cdot \log_7 8$.

86. Find the value of $\log_2 4 \cdot \log_4 6 \cdot \log_6 8$.

87. Find the value of $\log_2 3 \cdot \log_3 4 \cdot \ldots \cdot \log_n(n + 1) \cdot \log_{n+1} 2$.

88. Find the value of $\log_2 2 \cdot \log_2 4 \cdot \ldots \cdot \log_2 2^n$.

89. Show that $\log_a\left(x + \sqrt{x^2 - 1}\right) + \log_a\left(x - \sqrt{x^2 - 1}\right) = 0$.

90. Show that $\log_a\left(\sqrt{x} + \sqrt{x - 1}\right) + \log_a\left(\sqrt{x} - \sqrt{x - 1}\right) = 0$.

91. Show that $\ln(1 + e^{2x}) = 2x + \ln(1 + e^{-2x})$.

92. Difference Quotient   If $f(x) = \log_a x$, show that $\dfrac{f(x + h) - f(x)}{h} = \log_a\left(1 + \dfrac{h}{x}\right)^{1/h}, h \neq 0$.

93. If $f(x) = \log_a x$, show that $-f(x) = \log_{1/a} x$.

94. If $f(x) = \log_a x$, show that $f(AB) = f(A) + f(B)$.

95. If $f(x) = \log_a x$, show that $f\left(\dfrac{1}{x}\right) = -f(x)$.

96. If $f(x) = \log_a x$, show that $f(x^\alpha) = \alpha f(x)$.

97. Show that $\log_a\left(\dfrac{M}{N}\right) = \log_a M - \log_a N$, where a, M, and N are positive real numbers, with $a \neq 1$.

98. Show that $\log_a\left(\dfrac{1}{N}\right) = -\log_a N$, where a and N are positive real numbers, with $a \neq 1$.

PREPARING FOR THIS SECTION

Before getting started, review the following:

✓ Solving Equations Using a Graphing Utility (Section A.3, pp. 572–573)

7.4 LOGARITHMIC AND EXPONENTIAL EQUATIONS

OBJECTIVES
1. Solve Logarithmic Equations Using the Properties of Logarithms
2. Solve Exponential Equations
3. Solve Logarithmic and Exponential Equations Using a Graphing Utility

Logarithmic Equations

① In Section 7.2 we solved logarithmic equations by changing a logarithm to exponential form. Often, however, some manipulation of the equation (usually using the properties of logarithms) is required before we can change to exponential form.

Our practice will be to solve equations, whenever possible, by finding exact solutions using algebraic methods. In such cases, we will also verify the solution obtained by using a graphing utility. When algebraic methods cannot be used, approximate solutions will be obtained using a graphing utility. The reader is encouraged to pay particular attention to the form of equations for which exact solutions are possible.

EXAMPLE 1 **Solving a Logarithmic Equation**

Solve: $2 \log_5 x = \log_5 9$

Solution Because each logarithm is to the same base, 5, we can obtain an exact solution as follows:

$$2 \log_5 x = \log_5 9$$
$$\log_5 x^2 = \log_5 9 \qquad \text{\small $\log_a M^r = r \log_a M$}$$
$$x^2 = 9 \qquad \text{\small If $\log_a M = \log_a N$, then $M = N$}$$
$$x = 3 \quad \text{or} \quad \cancel{x = -3} \qquad \text{\small Recall that logarithms of negative numbers are not defined, so, in the expression $2 \log_5 x$, x must be positive. Therefore -3 is extraneous and we discard it.}$$

Figure 25

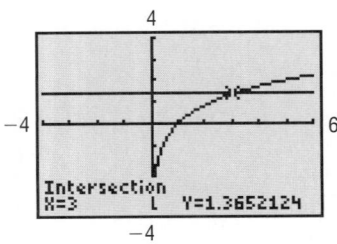

The equation has only one solution, 3.

✔ CHECK: To verify that 3 is the only solution using a graphing utility, graph $Y_1 = 2 \log_5 x = \dfrac{2 \log x}{\log 5}$ and $Y_2 = \log_5 9 = \dfrac{\log 9}{\log 5}$, and determine the point of intersection. See Figure 25. ■ ■

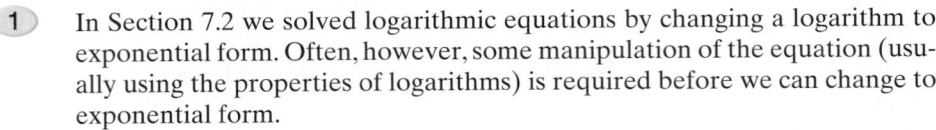

NOW WORK PROBLEM **5**.

EXAMPLE 2   **Solving a Logarithmic Equation**

Solve: $\log_4(x + 3) + \log_4(2 - x) = 1$

Solution   To obtain an exact solution, we need to express the left side as a single logarithm. Then we will change the expression to exponential form.

$$\log_4(x + 3) + \log_4(2 - x) = 1$$
$$\log_4[(x + 3)(2 - x)] = 1 \qquad \log_a M + \log_a N = \log_a (MN)$$
$$(x + 3)(2 - x) = 4^1 = 4 \qquad \text{Change to an exponential expression.}$$
$$-x^2 - x + 6 = 4 \qquad \text{Simplify.}$$
$$x^2 + x - 2 = 0 \qquad \text{Place the quadratic equation in standard form.}$$
$$(x + 2)(x - 1) = 0 \qquad \text{Factor.}$$
$$x = -2 \quad \text{or} \quad x = 1 \qquad \text{Zero-Product Property}$$

Since the arguments of each logarithmic expression in the equation are positive for both $x = -2$ and $x = 1$, neither is extraneous. The solution set is $\{-2, 1\}$.

You should verify that both of these are solutions using a graphing utility.

◼

— **NOW WORK PROBLEM 9.**

Exponential Equations

② In Sections 7.1 and 7.2 we solved certain exponential equations by expressing each side of the equation with the same base. However, many exponential equations cannot be rewritten so each side has the same base. In such cases, sometimes properties of logarithms along with algebraic techniques can be used to obtain a solution.

EXAMPLE 3   **Solving an Exponential Equation**

Solve: $4^x - 2^x - 12 = 0$

Solution   We note that $4^x = (2^2)^x = 2^{2x} = (2^x)^2$, so the equation is actually quadratic in form, and we can rewrite it as

$$(2^x)^2 - 2^x - 12 = 0 \qquad \text{Let } u = 2^x; \text{ then } u^2 - u - 12 = 0.$$

Now we can factor as usual.

$$(2^x - 4)(2^x + 3) = 0 \qquad (u - 4)(u + 3) = 0$$
$$2^x - 4 = 0 \quad \text{or} \quad 2^x + 3 = 0 \qquad u - 4 = 0 \quad \text{or} \quad u + 3 = 0$$
$$2^x = 4 \qquad\qquad 2^x = -3 \qquad u = 2^x = 4 \qquad u = 2^x = -3$$

The equation on the left has the solution $x = 2$, since $2^x = 4 = 2^2$; the equation on the right has no solution, since $2^x > 0$ for all x. The only solution is 2.

◼

In the preceding example, we were able to write the exponential expression using the same base after utilizing some algebra, obtaining an exact solution to the equation. When this is not possible, logarithms can sometimes be used to obtain the solution.

EXAMPLE 4   Solving an Exponential Equation

Solve: $2^x = 5$.

Solution   Since 5 cannot be written as an integral power of 2, we write the exponential equation as the equivalent logarithmic equation.

$$2^x = 5$$

$$x = \log_2 5 \underset{\underset{\text{Change-of-Base Formula (9), Section 7.3}}{\uparrow}}{=} \frac{\ln 5}{\ln 2}$$

Alternatively, we can solve the equation $2^x = 5$ by taking the natural logarithm (or common logarithm) of each side. Taking the natural logarithm,

$$2^x = 5$$
$$\ln 2^x = \ln 5 \qquad \text{If } M = N, \text{ then } \ln M = \ln N$$
$$x \ln 2 = \ln 5 \qquad \ln M^r = r \ln M$$
$$x = \frac{\ln 5}{\ln 2} \qquad \text{Solve for } x.$$

Using a calculator, the solution, rounded to three decimal places, is

$$x = \frac{\ln 5}{\ln 2} \approx 2.322$$

◼

━━━━ **NOW WORK PROBLEM 17.**

EXAMPLE 5   Solving an Exponential Equation

Solve: $8 \cdot 3^x = 5$

Solution
$$8 \cdot 3^x = 5$$
$$3^x = \frac{5}{8} \qquad \text{Solve for } 3^x.$$

$$x = \log_3\left(\frac{5}{8}\right) = \frac{\ln\left(\frac{5}{8}\right)}{\ln 3} \qquad \text{Solve for } x.$$

The solution, rounded to three decimal places, is

$$x = \frac{\ln\left(\frac{5}{8}\right)}{\ln 3} \approx -0.428$$

◼

EXAMPLE 6   Solving an Exponential Equation

Solve: $5^{x-2} = 3^{3x+2}$

Solution   Because the bases are different, we first apply Property (6), Section 7.3 (take the natural logarithm of each side) and then use appropriate properties of logarithms. The result is an equation in x that we can solve.

$$5^{x-2} = 3^{3x+2}$$

$$\ln 5^{x-2} = \ln 3^{3x+2} \qquad \text{If } M = N, \ln M = \ln N$$

$$(x-2)\ln 5 = (3x+2)\ln 3 \qquad \ln M^r = r \ln M$$

$$(\ln 5)x - 2\ln 5 = (3\ln 3)x + 2\ln 3 \qquad \text{Distribute.}$$

$$(\ln 5)x - (3\ln 3)x = 2\ln 3 + 2\ln 5 \qquad \begin{array}{l}\text{Place terms involving } x \text{ on}\\ \text{the left.}\end{array}$$

$$(\ln 5 - 3\ln 3)x = 2(\ln 3 + \ln 5) \qquad \text{Factor.}$$

$$x = \frac{2(\ln 3 + \ln 5)}{\ln 5 - 3\ln 3} \approx -3.212 \qquad \text{Solve for } x.$$

NOW WORK PROBLEM 25.

Graphing Utility Solutions

3   The techniques introduced in this section apply only to certain types of logarithmic and exponential equations. Solutions for other types are usually studied in calculus, using numerical methods. However, we can use a graphing utility to approximate the solution.

EXAMPLE 7   Solving Equations Using a Graphing Utility

Solve: $\log_3 x + \log_4 x = 4$
Express the solution(s) rounded to two decimal places.

Solution   The solution is found by graphing

Figure 26

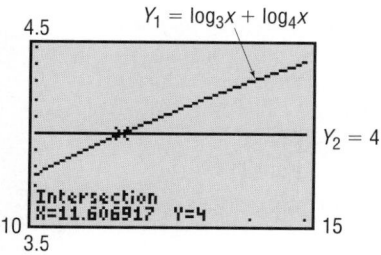

$$Y_1 = \log_3 x + \log_4 x = \frac{\log x}{\log 3} + \frac{\log x}{\log 4} \quad \text{and} \quad Y_2 = 4$$

(Remember that you must use the Change-of-Base Formula to graph Y_1.) Y_1 is an increasing function (do you know why?), and so there is only one point of intersection for Y_1 and Y_2. Figure 26 shows the graphs of Y_1 and Y_2. Using the INTERSECT command, the solution is 11.61, rounded to two decimal places.

EXPLORATION   Can you discover an algebraic solution to Example 7? [**Hint:** Factor $\log x$ from Y_1.]

EXAMPLE 8   Solving Equations Using a Graphing Utility

Solve: $x + e^x = 2$
Express the solution(s) rounded to two decimal places.

Figure 27

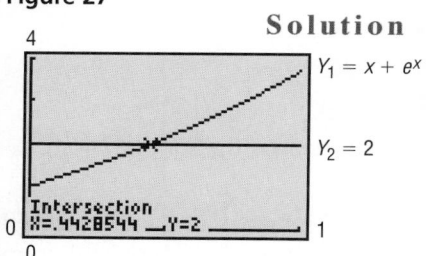

Solution   The solution is found by graphing $Y_1 = x + e^x$ and $Y_2 = 2$. Y_1 is an increasing function (do you know why?), and so there is only one point of intersection for Y_1 and Y_2. Figure 27 shows the graphs of Y_1 and Y_2. Using the INTERSECT command, the solution is 0.44 rounded to two decimal places.

NOW WORK PROBLEM 49.

7.4 Exercises

In Problems 1–44, solve each equation. Verify your solution using a graphing utility.

1. $\log_4(x + 2) = \log_4 8$

2. $\log_5(2x + 3) = \log_5 3$

3. $\frac{1}{2}\log_3 x = 2\log_3 2$

4. $-2\log_4 x = \log_4 9$

5. $2\log_5 x = 3\log_5 4$

6. $3\log_2 x = -\log_2 27$

7. $3\log_2(x - 1) + \log_2 4 = 5$

8. $2\log_3(x + 4) - \log_3 9 = 2$

9. $\log x + \log(x + 15) = 2$

10. $\log_4 x + \log_4(x - 3) = 1$

11. $\ln x + \ln(x + 2) = 4$

12. $\ln(x + 1) - \ln x = 2$

13. $2^{2x} + 2^x - 12 = 0$

14. $3^{2x} + 3^x - 2 = 0$

15. $3^{2x} + 3^{x+1} - 4 = 0$

16. $2^{2x} + 2^{x+2} - 12 = 0$

17. $2^x = 10$

18. $3^x = 14$

19. $8^{-x} = 1.2$

20. $2^{-x} = 1.5$

21. $3^{1-2x} = 4^x$

22. $2^{x+1} = 5^{1-2x}$

23. $\left(\frac{3}{5}\right)^x = 7^{1-x}$

24. $\left(\frac{4}{3}\right)^{1-x} = 5^x$

25. $1.2^x = (0.5)^{-x}$

26. $0.3^{1+x} = 1.7^{2x-1}$

27. $\pi^{1-x} = e^x$

28. $e^{x+3} = \pi^x$

29. $5(2^{3x}) = 8$

30. $0.3(4^{0.2x}) = 0.2$

31. $\log_a(x - 1) - \log_a(x + 6) = \log_a(x - 2) - \log_a(x + 3)$

32. $\log_a x + \log_a(x - 2) = \log_a(x + 4)$

33. $\log_{1/3}(x^2 + x) - \log_{1/3}(x^2 - x) = -1$

34. $\log_4(x^2 - 9) - \log_4(x + 3) = 3$

35. $\log_2(x + 1) - \log_4 x = 1$
   [**Hint:** Change $\log_4 x^2$ to base 2.]

36. $\log_2(3x + 2) - \log_4 x = 3$

37. $\log_{16} x + \log_4 x + \log_2 x = 7$

38. $\log_9 x + 3\log_3 x = 14$

39. $(\sqrt[3]{2})^{2-x} = 2^{x^2}$

40. $\log_2 x^{\log_2 x} = 4$

41. $\dfrac{e^x + e^{-x}}{2} = 1$

42. $\dfrac{e^x + e^{-x}}{2} = 3$

43. $\dfrac{e^x - e^{-x}}{2} = 2$

44. $\dfrac{e^x - e^{-x}}{2} = -2$

   [**Hint:** Multiply each side by e^x.]

In Problems 45–60, use a graphing utility to solve each equation. Express your answer rounded to two decimal places.

45. $\log_5 x + \log_3 x = 1$

46. $\log_2 x + \log_6 x = 3$

47. $\log_5(x + 1) - \log_4(x - 2) = 1$

48. $\log_2(x - 1) - \log_6(x + 2) = 2$

49. $e^x = -x$

50. $e^{2x} = x + 2$

51. $e^x = x^2$

52. $e^x = x^3$

53. $\ln x = -x$

54. $\ln(2x) = -x + 2$

55. $\ln x = x^3 - 1$

56. $\ln x = -x^2$

57. $e^x + \ln x = 4$

58. $e^x - \ln x = 4$

59. $e^{-x} = \ln x$

60. $e^{-x} = -\ln x$

7.5 COMPOUND INTEREST

OBJECTIVES

1   Determine the Future Value of a Lump Sum of Money
2   Calculate Effective Rates of Return
3   Determine the Present Value of a Lump Sum of Money
4   Determine the Time Required to Double a Lump Sum of Money

1   Interest is money paid for the use of money. The total amount borrowed (whether by an individual from a bank in the form of a loan or by a bank from

an individual in the form of a savings account) is called the **principal**. The **rate of interest**, expressed as a percent, is the amount charged for the use of the principal for a given period of time, usually on a yearly (that is, per annum) basis.

Simple Interest Formula

If a principal of P dollars is borrowed for a period of t years at a per annum interest rate r, expressed as a decimal, the interest I charged is

$$I = Prt \qquad\qquad (1)$$

Interest charged according to formula (1) is called **simple interest**.

In working with problems involving interest, we use the term **payment period** as follows:

Annually	Once per year	Monthly	12 times per year
Semiannually	Twice per year	Daily	365 times per year*
Quarterly	Four times per year		

When the interest due at the end of a payment period is added to the principal so that the interest computed at the end of the next payment period is based on this new principal amount (old principal + interest), the interest is said to have been **compounded. Compound interest** is interest paid on previously earned interest.

EXAMPLE 1   **Computing Compound Interest**

A credit union pays interest of 8% per annum compounded quarterly on a certain savings plan. If $1000 is deposited in such a plan and the interest is left to accumulate, how much is in the account after 1 year?

Solution   We use the simple interest formula, $I = Prt$. The principal P is $1000 and the rate of interest is 8% $= 0.08$. After the first quarter of a year, the time t is $\frac{1}{4}$ year, so the interest earned is

$$I = Prt = (\$1000)(0.08)\left(\frac{1}{4}\right) = \$20$$

The new principal is $P + I = \$1000 + \$20 = \$1020$. At the end of the second quarter, the interest on this principal is

$$I = (\$1020)(0.08)\left(\frac{1}{4}\right) = \$20.40$$

---

*Most banks use a 360-day "year." Why do you think they do?

At the end of the third quarter, the interest on the new principal of $1020 + $20.40 = $1040.40 is

$$I = (\$1040.40)(0.08)\left(\frac{1}{4}\right) = \$20.81$$

Finally, after the fourth quarter, the interest is

$$I = (\$1061.21)(0.08)\left(\frac{1}{4}\right) = \$21.22$$

After 1 year the account contains $1082.43. ∎

The pattern of the calculations performed in Example 1 leads to a general formula for compound interest. To fix our ideas, let P represent the principal to be invested at a per annum interest rate r that is compounded n times per year. (For computing purposes, r is expressed as a decimal.) The interest earned after each compounding period is the principal times $\frac{r}{n}$. The amount A after one compounding period is

$$A = P + P\left(\frac{r}{n}\right) = P\left(1 + \frac{r}{n}\right)$$

After two compounding periods, the amount A, based on the new principal $P\left(1 + \frac{r}{n}\right)$, is

$$A = \underbrace{P\left(1 + \frac{r}{n}\right)}_{\substack{\text{New} \\ \text{principal}}} + \underbrace{P\left(1 + \frac{r}{n}\right)\left(\frac{r}{n}\right)}_{\substack{\text{Interest on} \\ \text{new principal}}} = P\left(1 + \frac{r}{n}\right)\left(1 + \frac{r}{n}\right) = P\left(1 + \frac{r}{n}\right)^2$$

After three compounding periods,

$$A = P\left(1 + \frac{r}{n}\right)^2 + P\left(1 + \frac{r}{n}\right)^2\left(\frac{r}{n}\right) = P\left(1 + \frac{r}{n}\right)^2\left(1 + \frac{r}{n}\right) = P\left(1 + \frac{r}{n}\right)^3$$

Continuing this way, after n compounding periods (1 year),

$$A = P\left(1 + \frac{r}{n}\right)^n$$

Because t years will contain $n \cdot t$ compounding periods, after t years we have

$$A = P\left(1 + \frac{r}{n}\right)^{nt}$$

Theorem   **Compound Interest Formula**

The amount A after t years due to a principal P invested at an annual interest rate r compounded n times per year is

$$A = P\left(1 + \frac{r}{n}\right)^{nt} \tag{2}$$

For example, to rework Example 1, we would use $P = \$1000$, $r = 0.08$, $n = 4$ (quarterly compounding), and $t = 1$ year to obtain

$$A = 1000\left(1 + \frac{0.08}{4}\right)^{4(1)} = \$1082.43$$

In equation (2), the amount A is typically referred to as the **accumulated value or future value** of the account, while P is called the **present value**.

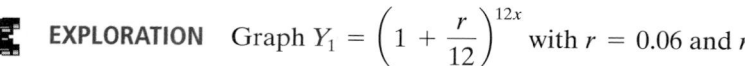 **EXPLORATION**   Graph $Y_1 = \left(1 + \frac{r}{12}\right)^{12x}$ with $r = 0.06$ and $r = 0.12$

for $0 \le x \le 30$. What is the future value of $1 in 30 years when the interest rate per annum is $r = 0.06$ (6%)? What is the future value of $1 in 30 years when the interest rate per annum is $r = 0.12$ (12%)? Does doubling the interest rate double the future value?

Note: In using your calculator, be sure to use stored values, rather than approximations, in order to avoid round-off errors. At the final step, round money to the nearest cent.

NOW WORK PROBLEM 1.

EXAMPLE 2   Comparing Investments Using Different Compounding Periods

Investing $1000 at an annual rate of 10% compounded annually, quarterly, monthly, and daily will yield the following amounts after 1 year:

Annual compounding: $A = P(1 + r)$
$$= (\$1000)(1 + 0.10) = \$1100.00$$

Quarterly compounding: $A = P\left(1 + \frac{r}{4}\right)^4$
$$= (\$1000)(1 + 0.025)^4 = \$1103.81$$

Monthly compounding: $A = P\left(1 + \frac{r}{12}\right)^{12}$
$$= (\$1000)(1 + 0.00833)^{12} = \$1104.71$$

Daily compounding: $A = P\left(1 + \frac{r}{365}\right)^{365}$
$$= (\$1000)(1 + 0.000274)^{365} = \$1105.16$$

From Example 2 we can see that the effect of compounding more frequently is that the amount after 1 year is higher: $1000 compounded 4 times a year at 10% results in $1103.81; $1000 compounded 12 times a year at 10% results in $1104.71; and $1000 compounded 365 times a year at 10% results in $1105.16. This leads to the following question: What would happen to the amount after 1 year if the number of times that the interest is compounded were increased without bound?

Let's find the answer. Suppose that P is the principal, r is the per annum interest rate, and n is the number of times that the interest is compounded each year. The amount after 1 year is

$$A = P\left(1 + \frac{r}{n}\right)^n$$

Now suppose that the number n of times that the interest is compounded per year gets larger and larger; that is, suppose that $n \rightarrow \infty$. Then

$$A = P\left(1 + \frac{r}{n}\right)^n = P\left[1 + \frac{1}{\frac{n}{r}}\right]^n = P\left(\left[1 + \frac{1}{\frac{n}{r}}\right]^{\frac{n}{r}}\right)^r \underset{h = \frac{n}{r}}{=} P\left[\left(1 + \frac{1}{h}\right)^h\right]^r \qquad (3)$$

In (3), as $n \rightarrow \infty$, then $h = \dfrac{n}{r} \rightarrow \infty$, and the expression in brackets equals e. [Refer to (2) on p. 482]. That is, $A \rightarrow Pe^r$.

Table 6 compares $\left(1 + \dfrac{r}{n}\right)^n$, for large values of n, to e^r for $r = 0.05$, $r = 0.10$, $r = 0.15$, and $r = 1$. The larger that n gets, the closer $\left(1 + \dfrac{r}{n}\right)^n$ gets to e^r. No matter how frequent the compounding, the amount after 1 year has the definite ceiling Pe^r.

TABLE 6

$\left(1 + \dfrac{r}{n}\right)^n$			
	$n = 100$	$n = 1000$	$n = 10{,}000$	e^r
$r = 0.05$	1.0512579	1.0512698	1.051271	1.0512711
$r = 0.10$	1.1051157	1.1051654	1.1051703	1.1051709
$r = 0.15$	1.1617037	1.1618212	1.1618329	1.1618342
$r = 1$	2.7048138	2.7169239	2.7181459	2.7182818

When interest is compounded so that the amount after 1 year is Pe^r, we say the interest is **compounded continuously**.

Theorem

Continuous Compounding

The amount A after t years due to a principal P invested at an annual interest rate r compounded continuously is

$$A = Pe^{rt} \qquad (4)$$

EXAMPLE 3 **Using Continuous Compounding**

The amount A that results from investing a principal P of $1000 at an annual rate r of 10% compounded continuously for a time t of 1 year is

$$A = \$1000e^{0.10(1)} = (\$1000)(1.10517) = \$1105.17$$

━ **NOW WORK PROBLEM 9.**

2   The **effective rate of interest** is the equivalent annual simple rate of interest that would yield the same amount as compounding after 1 year. For example, based on Example 3, a principal of $1000 will result in $1105.17 at a rate of 10% compounded continuously. To get this same amount using a simple rate of interest would require that interest of $1105.17 − $1000.00 = $105.17 be earned on the principal. Since $105.17 is 10.517% of $1000, a simple rate of interest of 10.517% is needed to equal 10% compounded continuously. The effective rate of interest of 10% compounded continuously is 10.517%.

Based on the results of Examples 2 and 3, we find the following comparisons:

	Annual Rate	Effective Rate
Annual compounding	10%	10%
Quarterly compounding	10%	10.381%
Monthly compounding	10%	10.471%
Daily compounding	10%	10.516%
Continuous compounding	10%	10.517%

EXAMPLE 4   Computing the Value of an IRA

On January 2, 2002, $2000 is placed in an Individual Retirement Account (IRA) that will pay interest of 10% per annum compounded continuously.
(a) What will the IRA be worth on January 1, 2022?
(b) What is the effective rate of interest?

Solution   (a) The amount A after 20 years is

$$A = Pe^{rt} = \$2000e^{(0.10)(20)} = \$14{,}778.11$$

(b) First, we compute the interest earned on $2000 at $r = 10\%$ compounded continuously for 1 year.

$$A = 2000e^{0.10(1)}$$
$$= \$2210.34$$

So, the interest earned is $2210.34 − $2000.00 = $210.34. Use the simple interest formula, $I = Prt$, with $I = \$210.34$, $P = \$2000$, and $t = 1$, and solve for r, the effective rate of interest.

$$\$210.34 = \$2000 \cdot r \cdot 1$$
$$r = \frac{\$210.34}{\$2000} = 0.10517$$

The effective rate of interest is 10.517%.   ■

NOW WORK PROBLEM **21.**

EXPLORATION How long will it be until A = $4000? $6000? [**Hint:** Graph $Y_1 = 2000e^{0.1x}$ and $Y_2 = 4000$. Use INTERSECT to find x.]

Time is money

3 When people engaged in finance speak of the "time value of money," they are usually referring to the *present value* of money. The **present value** of A dollars to be received at a future date is the principal that you would need to invest now so that it would grow to A dollars in the specified time period. The present value of money to be received at a future date is always less than the amount to be received, since the amount to be received will equal the present value (money invested now) *plus* the interest accrued over the time period.

We use the compound interest formula (2) to get a formula for present value. If P is the present value of A dollars to be received after t years at a per annum interest rate r compounded n times per year, then, by formula (2),

$$A = P\left(1 + \frac{r}{n}\right)^{nt}$$

To solve for P, we divide both sides by $\left(1 + \dfrac{r}{n}\right)^{nt}$, and the result is

$$\frac{A}{\left(1 + \dfrac{r}{n}\right)^{nt}} = P \quad \text{or} \quad P = A\left(1 + \frac{r}{n}\right)^{-nt}$$

Theorem

Present Value Formulas

The present value P of A dollars to be received after t years, assuming a per annum interest rate r compounded n times per year, is

$$P = A\left(1 + \frac{r}{n}\right)^{-nt} \tag{5}$$

If the interest is compounded continuously, then

$$P = Ae^{-rt} \tag{6}$$

To prove (6), solve formula (4) for P.

EXAMPLE 5   **Computing the Value of a Zero-Coupon Bond**

A zero-coupon (noninterest-bearing) bond can be redeemed in 10 years for $1000. How much should you be willing to pay for it now if you want a return of:

(a) 8% compounded monthly?   (b) 7% compounded continuously?

Solution   (a) We are seeking the present value of $1000. We use formula (5) with $A = \$1000$, $n = 12$, $r = 0.08$, and $t = 10$.

$$P = A\left(1 + \frac{r}{n}\right)^{-nt}$$

$$= \$1000\left(1 + \frac{0.08}{12}\right)^{-12(10)}$$

$$= \$450.52$$

For a return of 8% compounded monthly, you should pay $450.52 for the bond.

(b) Here we use formula (6) with $A = \$1000$, $r = 0.07$, and $t = 10$.

$$P = Ae^{-rt}$$

$$= \$1000e^{-(0.07)(10)}$$

$$= \$496.59$$

For a return of 7% compounded continuously, you should pay $496.59 for the bond. ∎

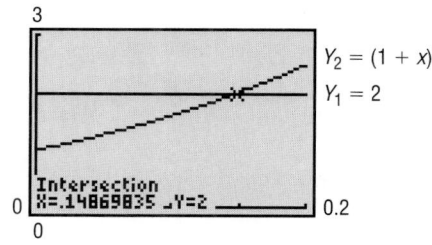 NOW WORK PROBLEM **11**.

---

④ **EXAMPLE 6**   **Rate of Interest Required to Double an Investment**

What annual rate of interest compounded annually should you seek if you want to double your investment in 5 years?

Algebraic Solution   If P is the principal and we want P to double, the amount A will be $2P$. We use the compound interest formula with $n = 1$ and $t = 5$ to find r.

$$2P = P(1 + r)^5$$
$$2 = (1 + r)^5$$
$$1 + r = \sqrt[5]{2}$$
$$r = \sqrt[5]{2} - 1 = 1.148698 - 1 = 0.148698$$

The annual rate of interest needed to double the principal in 5 years is 14.87%.

Graphing Solution   We solve the equation

$$2 = (1 + r)^5$$

for r by graphing the two functions $Y_1 = 2$ and $Y_2 = (1 + x)^5$. The x-coordinate of their point of intersection is the rate r that we seek. See Figure 28.

Figure 28

3

$Y_2 = (1 + x)^5$

$Y_1 = 2$

Intersection
X=.14869835 Y=2

0   0.2

0

Using the INTERSECT command, we find that the point of intersection of Y_1 and Y_2 is $(0.14869835, 2)$, so $r = 0.148698 \approx 14.87\%$ as before. ■

▄▃ ━━━━━ **NOW WORK PROBLEM 23.**

| EXAMPLE 7 | Doubling Time for an Investment |

How long will it take for an investment to double in value if it earns 5% compounded continuously?

Algebraic Solution   If P is the initial investment and we want P to double, the amount A will be $2P$. We use formula (4) for continuously compounded interest with $r = 0.05$. Then

$$A = Pe^{rt}$$
$$2P = Pe^{0.05t} \qquad \text{Let } A = 2P$$
$$2 = e^{0.05t} \qquad \text{Divide both sides by } P$$
$$0.05t = \ln 2 \qquad \text{Rewrite as a logarithm}$$
$$t = \frac{\ln 2}{0.05} = 13.86 \qquad \text{Solve for } t.$$

It will take about 14 years to double the investment.

Graphing Solution   We solve the equation

$$2 = e^{0.05t}$$

for t by graphing the two functions $Y_1 = 2$ and $Y_2 = e^{0.05x}$. Their point of intersection is $(13.86, 2)$. See Figure 29.

Figure 29

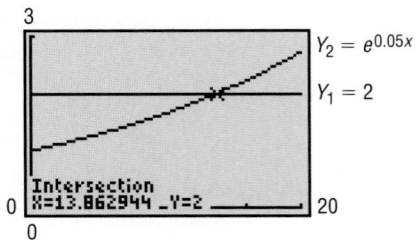

▄▃ ━━━━━ **NOW WORK PROBLEM 29.**

7.5 Exercises

In Problems 1–10, find the amount that results from each investment.

1. $100 invested at 4% compounded quarterly after a period of 2 years

2. $50 invested at 6% compounded monthly after a period of 3 years

3. $500 invested at 8% compounded quarterly after a period of $2\frac{1}{2}$ years

4. $300 invested at 12% compounded monthly after a period of $1\frac{1}{2}$ years

5. $600 invested at 5% compounded daily after a period of 3 years

6. $700 invested at 6% compounded daily after a period of 2 years

7. $10 invested at 11% compounded continuously after a period of 2 years

8. $40 invested at 7% compounded continuously after a period of 3 years

9. $100 invested at 10% compounded continuously after a period of $2\frac{1}{4}$ years

10. $100 invested at 12% compounded continuously after a period of $3\frac{3}{4}$ years

In Problems 11–20, find the principal needed now to get each amount; that is, find the present value.

11. To get $100 after 2 years at 6% compounded monthly

12. To get $75 after 3 years at 8% compounded quarterly

13. To get $1000 after $2\frac{1}{2}$ years at 6% compounded daily

14. To get $800 after $3\frac{1}{2}$ years at 7% compounded monthly

15. To get $600 after 2 years at 4% compounded quarterly

16. To get $300 after 4 years at 3% compounded daily

17. To get $80 after $3\frac{1}{4}$ years at 9% compounded continuously

18. To get $800 after $2\frac{1}{2}$ years at 8% compounded continuously

19. To get $400 after 1 year at 10% compounded continuously

20. To get $1000 after 1 year at 12% compounded continuously

21. Find the effective rate of interest for $5\frac{1}{4}$% compounded quarterly.

22. What interest rate compounded quarterly will give an effective interest rate of 7%?

23. What annual rate of interest is required to double an investment in 3 years? Verify your answer using a graphing utility.

24. What annual rate of interest is required to double an investment in 10 years? Verify your answer using a graphing utility.

In Problems 25–28, which of the two rates would yield the larger amount in 1 year?
[**Hint:** Start with a principal of $10,000 in each instance.]

25. 6% compounded quarterly or $6\frac{1}{4}$% compounded annually

26. 9% compounded quarterly or $9\frac{1}{4}$% compounded annually

27. 9% compounded monthly or 8.8% compounded daily

28. 8% compounded semiannually or 7.9% compounded daily

---

29. How long does it take for an investment to double in value if it is invested at 8% per annum compounded monthly? Compounded continuously? Verify your answer using a graphing utility.

30. How long does it take for an investment to double in value if it is invested at 10% per annum compounded monthly? Compounded continuously? Verify your answer using a graphing utility.

31. If Tanisha has $100 to invest at 8% per annum compounded monthly, how long will it be before she has $150? If the compounding is continuous, how long will it be?

32. If Angela has $100 to invest at 10% per annum compounded monthly, how long will it be before she has $175? If the compounding is continuous, how long will it be?

33. How many years will it take for an initial investment of $10,000 to grow to $25,000? Assume a rate of interest of 6% compounded continuously.

34. How many years will it take for an initial investment of $25,000 to grow to $80,000? Assume a rate of interest of 7% compounded continuously.

35. What will a $90,000 house cost 5 years from now if the inflation rate over that period averages 3% compounded annually?

36. Sears charges 1.25% per month on the unpaid balance for customers with charge accounts (interest is compounded monthly). A customer charges $200 and does not pay her bill for 6 months. What is the bill at that time?

37. Jerome will be buying a used car for $15,000 in 3 years. How much money should he ask his parents for now so that, if he invests it at 5% compounded continuously, he will have enough to buy the car?

38. John will require $3000 in 6 months to pay off a loan that has no prepayment privileges. If he has the $3000 now, how much of it should he save in an account paying 3% compounded monthly so that in 6 months he will have exactly $3000?

39. George is contemplating the purchase of 100 shares of a stock selling for $15 per share. The stock pays no dividends. The history of the stock indicates that it should grow at an annual rate of 15% per year. How much will the 100 shares of stock be worth in 5 years?

40. Tracy is contemplating the purchase of 100 shares of a stock selling for $15 per share. The stock pays no dividends. Her broker says that the stock will be worth $20 per share in 2 years. What is the annual rate of return on this investment?

41. A business purchased for $650,000 in 1997 is sold in 2000 for $850,000. What is the annual rate of return for this investment?

42. Tanya has just inherited a diamond ring appraised at $5000. If diamonds have appreciated in value at an annual rate of 8%, what was the value of the ring 10 years ago when the ring was purchased?

43. Jim places $1000 in a bank account that pays 5.6% compounded continuously. After 1 year, will he have enough money to buy a computer system that costs $1060? If another bank will pay Jim 5.9% compounded monthly, is this a better deal?

44. On January 1, Kim places $1000 in a certificate of deposit that pays 6.8% compounded continuously and matures in 3 months. Then Kim places the $1000 and the interest in a passbook account that pays 5.25% compounded monthly. How much does Kim have in the passbook account on May 1?

45. Will invests $2000 in a bond trust that pays 9% interest compounded semiannually. His friend Henry invests $2000 in a certificate of deposit (CD) that pays $8\frac{1}{2}\%$ compounded continuously. Who has more money after 20 years, Will or Henry?

46. Suppose that April has access to an investment that will pay 10% interest compounded continuously. Which is better: To be given $1000 now so that she can take advantage of this investment opportunity or to be given $1325 after 3 years?

47. Colleen and Bill have just purchased a house for $150,000, with the seller holding a second mortgage of $50,000. They promise to pay the seller $50,000 plus all accrued interest 5 years from now. The seller offers them three interest options on the second mortgage:
(a) Simple interest at 12% per annum
(b) $11\frac{1}{2}\%$ interest compounded monthly
(c) $11\frac{1}{4}\%$ interest compounded continuously
    Which option is best; that is, which results in the least interest on the loan?

48. The First National Bank advertises that it pays interest on savings accounts at the rate of 4.25% compounded daily. Find the effective rate if the bank uses (a) 360 days or (b) 365 days in determining the daily rate.

Problems 49–52 involve zero-coupon bonds. A zero-coupon bond is a bond that is sold now at a discount and will pay its face value at some time when it matures; no interest payments are made.

49. A zero-coupon bond can be redeemed in 20 years for $10,000. How much should you be willing to pay for it now if you want a return of:
(a) 10% compounded monthly?
(b) 10% compounded continuously?

50. A child's grandparents are considering buying a $40,000 face value zero-coupon bond at birth so that she will have enough money for her college education 17 years later. If they want a rate of return of 8% compounded annually, what should they pay for the bond?

51. How much should a $10,000 face value zero-coupon bond, maturing in 10 years, be sold for now if its rate of return is to be 8% compounded annually?

52. If Pat pays $12,485.52 for a $25,000 face value zero-coupon bond that matures in 8 years, what is his annual rate of return?

53. Explain in your own words what the term *compound interest* means. What does *continuous compounding* mean?

54. Explain in your own words the meaning of *present value*.

55. Time to Double or Triple an Investment   The formula

$$y = \frac{\ln m}{n \ln\left(1 + \dfrac{r}{n}\right)}$$

can be used to find the number of years y required to multiply an investment m times when r is the per annum interest rate compounded n times a year.

(a) How many years will it take to double the value of an IRA that compounds annually at the rate of 12%?
(b) How many years will it take to triple the value of a savings account that compounds quarterly at an annual rate of 6%?
(c) Give a derivation of this formula.

56. Time to Reach an Investment Goal   The formula

$$y = \frac{\ln A - \ln P}{r}$$

can be used to find the number of years y required for an investment P to grow to a value A when compounded continuously at an annual rate r.

(a) How long will it take to increase an initial investment of $1000 to $8000 at an annual rate of 10%?
(b) What annual rate is required to increase the value of a $2000 IRA to $30,000 in 35 years?
(c) Give a derivation of this formula.

57. Critical Thinking   You have just contracted to buy a house and will seek financing in the amount of $100,000. You go to several banks. Bank 1 will lend you $100,000 at the rate of 8.75% amortized over 30 years with a loan origination fee of 1.75%. Bank 2 will lend you $100,000 at the rate of 8.375% amortized over 15 years with a loan origination fee of 1.5%. Bank 3 will lend you $100,000 at the rate of 9.125% amortized over 30 years with no loan origination fee. Bank 4 will lend you $100,000 at the rate of 8.625% amortized over 15 years with no loan origination fee. Which loan would you take? Why? Be sure to have sound reasons for your choice. Use the information in the table to assist you. If the amount of the monthly payment does not matter to you, which loan would you take? Again, have sound reasons for your choice. Compare your final decision with others in the class. Discuss.

	Monthly Payment	Loan Origination Fee
Bank 1	$786.70	$1,750.00
Bank 2	$977.42	$1,500.00
Bank 3	$813.63	$0.00
Bank 4	$990.68	$0.00

7.6 GROWTH AND DECAY

OBJECTIVES

1. Find Equations of Populations That Obey the Law of Uninhibited Growth
2. Find Equations of Populations That Obey the Law of Decay
3. Use Newton's Law of Cooling
4. Use Logistic Growth Models

1 Many natural phenomena have been found to follow the law that an amount A varies with time t according to

$$A = A_0 e^{kt} \tag{1}$$

Here A_0 is the original amount $(t = 0)$ and $k \neq 0$ is a constant.

If $k > 0$, then equation (1) states that the amount A is increasing over time; if $k < 0$, the amount A is decreasing over time. In either case, when an amount A varies over time according to equation (1), it is said to follow the **exponential law** or the **law of uninhibited growth** $(k > 0)$ **or decay** $(k < 0)$. See Figure 30.

Figure 30

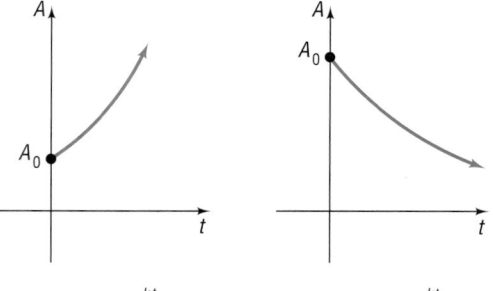

(a) $A(t) = A_0 e^{kt}$, $k > 0$ (b) $A(t) = A_0 e^{kt}$, $k < 0$

For example, we saw in Section 7.5 that continuously compounded interest follows the law of uninhibited growth. In this section we shall look at three additional phenomena that follow the exponential law.

Uninhibited Growth

Cell division is the growth process of many living organisms, such as amoebas, plants, and human skin cells. Based on an ideal situation in which no cells die and no by-products are produced, the number of cells present at a given time follows the law of uninhibited growth. Actually, however, after enough time has passed, growth at an exponential rate will cease due to the influence of factors such as lack of living space and dwindling food supply. The law of uninhibited growth accurately reflects only the early stages of the cell division process.

The cell division process begins with a culture containing N_0 cells. Each cell in the culture grows for a certain period of time and then divides into two identical cells. We assume that the time needed for each cell to divide in two is constant and does not change as the number of cells increases. These new cells then grow, and eventually each divides in two, and so on.

Uninhibited Growth of Cells

A model that gives the number N of cells in the culture after a time t has passed (in the early stages of growth) is

$$N(t) = N_0 e^{kt}, \quad k > 0 \qquad (2)$$

where N_0 is the initial number of cells and k is a positive constant that represents the growth rate of the cells.

In using formula (2) to model the growth of cells, we are using a function that yields positive real numbers, even though we are counting the number of cells, which must be an integer. This is a common practice in many applications.

EXAMPLE 1   **Bacterial Growth**

A colony of bacteria grows according to the law of uninhibited growth $N(t) = 100e^{0.045t}$, where N is measured in grams and t is measured in days.

(a) Determine the initial amount of bacteria.

(b) What is the growth rate of the bacteria?

(c) Graph the function using a graphing utility.

(d) What is the population after 5 days?

(e) How long will it take for the population to reach 140 grams?

(f) What is the doubling time for the population?

Solution (a) The initial amount of bacteria, N_0, is obtained when $t = 0$, so
$N_0 = N(0) = 100e^{0.045(0)} = 100$ grams.

(b) Compare $N(t) = 100e^{0.045t}$ to $N(t) = 100e^{kt}$. The value of k, 0.045, indicates a growth rate of 4.5%.

(c) Figure 31 shows the graph of $N(t) = 100e^{0.045t}$.

(d) The population after 5 days is $N(5) = 100e^{0.045(5)} = 125.2$ grams.

(e) We solve the equation $N(t) = 140$.

Figure 31

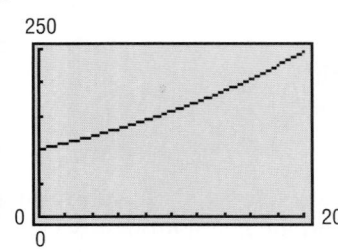

$$
\begin{aligned}
100e^{0.045t} &= 140 \\
e^{0.045t} &= 1.4 && \text{Divide both sides of the equation by 100.} \\
0.045t &= \ln 1.4 && \text{Rewrite as a logarithm.} \\
t &= \frac{\ln 1.4}{0.045} && \text{Divide both sides of the equation by 0.045.} \\
&\approx 7.5 \text{ days}
\end{aligned}
$$

(f) The population doubles when $N(t) = 200$ grams, so we find the doubling time by solving the equation $200 = 100e^{0.045t}$ for t.

$$200 = 100e^{0.045t}$$

$$2 = e^{0.045t} \qquad \text{Divide both sides of the equation by 100.}$$

$$\ln 2 = 0.045t \qquad \text{Rewrite as a logarithm.}$$

$$t = \frac{\ln 2}{0.045} \qquad \text{Divide both sides of the equation by 0.045.}$$

$$\approx 15.4 \text{ days}$$

NOW WORK PROBLEM 1.

EXAMPLE 2   **Bacterial Growth**

A colony of bacteria increases according to the law of uninhibited growth.

(a) If the number of bacteria doubles in 3 hours, find the function that gives the number of cells in the culture.

(b) How long will it take for the size of the colony to triple?

(c) How long will it take for the population to double a second time (that is, increase four times)?

Solution   (a) Using formula (2), the number N of cells at time t is

$$N(t) = N_0 e^{kt}$$

where N_0 is the initial number of bacteria present and k is a positive number. We first seek the number k. The number of cells doubles in 3 hours, so we have

$$N(3) = 2N_0$$

But $N(3) = N_0 e^{k(3)}$, so

$$N_0 e^{k(3)} = 2N_0$$

$$e^{3k} = 2 \qquad \text{Divide both sides by } N_0.$$

$$3k = \ln 2 \qquad \text{Write the exponential equation as a logarithm.}$$

$$k = \frac{1}{3}\ln 2 \approx \frac{1}{3}(0.6931) = 0.2310$$

Formula (2) for this growth process is therefore

$$N(t) = N_0 e^{\left(\frac{1}{3}\ln 2\right)t}$$

(b) The time t needed for the size of the colony to triple requires that $N = 3N_0$. We substitute $3N_0$ for N to get

$$3N_0 = N_0 e^{\left(\frac{1}{3}\ln 2\right)t}$$

$$3 = e^{\left(\frac{1}{3}\ln 2\right)t}$$

$$\left(\frac{1}{3}\ln 2\right)t = \ln 3$$

$$t = \frac{3\ln 3}{\ln 2} \approx 4.755 \text{ hours}$$

It will take about 4.755 hours or 4 hours, 45 minutes for the size of the colony to triple.

 (c) If a population doubles in 3 hours, it will double a second time in 3 more hours, for a total time of 6 hours. ∎

Radioactive Decay

② Radioactive materials follow the law of uninhibited decay.

Uninhibited Radioactive Decay

The amount A of a radioactive material present at time t is given by the following model:

$$A(t) = A_0 e^{kt}, \qquad k < 0 \tag{3}$$

where A_0 is the original amount of radioactive material and k is a negative number that represents the rate of decay.

All radioactive substances have a specific **half-life**, which is the time required for half of the radioactive substance to decay. In **carbon dating**, we use the fact that all living organisms contain two kinds of carbon, carbon 12 (a stable carbon) and carbon 14 (a radioactive carbon, with a half-life of 5600 years). While an organism is living, the ratio of carbon 12 to carbon 14 is constant. But when an organism dies, the original amount of carbon 12 present remains unchanged, whereas the amount of carbon 14 begins to decrease. This change in the amount of carbon 14 present relative to the amount of carbon 12 present makes it possible to calculate when an organism died.

EXAMPLE 3 | **Estimating the Age of Ancient Tools**

Traces of burned wood along with ancient stone tools in an archaeological dig in Chile were found to contain approximately 1.67% of the original amount of carbon 14.

(a) If the half-life of carbon 14 is 5600 years, approximately when was the tree cut and burned?

(b) Using a graphing utility, graph the relation between the percentage of carbon 14 remaining and time.

(c) Determine the time that elapses until half of the carbon 14 remains. This answer should equal the half-life of carbon 14.

(d) Use a graphing utility to verify the answer found in part (a).

Solution (a) Using formula (3), the amount A of carbon 14 present at time t is

$$A(t) = A_0 e^{kt}$$

where A_0 is the original amount of carbon 14 present and k is a negative number. We first seek the number k. To find it, we use the fact that after 5600 years half of the original amount of carbon 14 remains, so $A(5600) = \dfrac{1}{2} A_0$. Thus,

$$\frac{1}{2}A_0 = A_0 e^{k(5600)}$$

$$\frac{1}{2} = e^{5600k} \qquad \text{Divide both sides of the equation by } A_0.$$

$$5600k = \ln\frac{1}{2} \qquad \text{Rewrite as a logarithm.}$$

$$k = \frac{1}{5600}\ln\frac{1}{2} \approx -0.000124$$

Formula (3) therefore becomes

$$A(t) = A_0 e^{\frac{\ln\frac{1}{2}}{5600}t}$$

If the amount A of carbon 14 now present is 1.67% of the original amount, it follows that

$$0.0167 A_0 = A_0 e^{\frac{\ln\frac{1}{2}}{5600}t}$$

$$0.0167 = e^{\frac{\ln\frac{1}{2}}{5600}t} \qquad \text{Divide both sides of the equation by } A_0.$$

$$\frac{\ln\frac{1}{2}}{5600}t = \ln 0.0167 \qquad \text{Rewrite as a logarithm.}$$

$$t = \frac{5600}{\ln\frac{1}{2}}\ln 0.0167 \approx 33{,}062 \text{ years}$$

The tree was cut and burned about 33,062 years ago. Some archaeologists use this conclusion to argue that humans lived in the Americas 33,000 years ago, much earlier than is generally accepted.

Figure 32

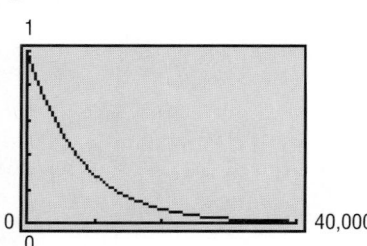

(b) Figure 32 shows the graph of $y = e^{\frac{\ln\frac{1}{2}}{5600}x}$, where y is the fraction of carbon 14 present and x is the time.

(c) By graphing $Y_1 = 0.5$ and $Y_2 = e^{\frac{\ln\frac{1}{2}}{5600}x}$, where x is time, and using INTERSECT, we find that it takes 5600 years until half the carbon 14 remains. The half-life of carbon 14 is 5600 years.

(d) By graphing $Y_1 = 0.0167$ and $Y_2 = e^{\frac{\ln\frac{1}{2}}{5600}x}$, where x is time, and using INTERSECT, we find that it takes 33,062 years until 1.67% of the carbon 14 remains.

━ **NOW WORK PROBLEM 3.**

Newton's Law of Cooling

③ **Newton's Law of Cooling*** states that the temperature of a heated object decreases exponentially over time toward the temperature of the surrounding medium.

> ### Newton's Law of Cooling
>
> The temperature u of a heated object at time t can be modeled by the function
>
> $$u(t) = T + (u_0 - T)e^{kt} \qquad k < 0 \qquad (4)$$
>
> where T is the constant temperature of the surrounding medium, u_0 is the initial temperature of the heated object, and k is a negative constant.

EXAMPLE 4 | **Using Newton's Law of Cooling**

An object is heated to 100°C (degrees Celsius) and is then allowed to cool in a room whose air temperature is 30°C.

(a) If the temperature of the object is 80°C after 5 minutes, when will its temperature be 50°C?
(b) Using a graphing utility, graph the relation found between the temperature and time.
(c) Using a graphing utility, verify that after 18.6 minutes the temperature is 50°C.
(d) Using a graphing utility, determine the elapsed time before the object is 35°C.
(e) What do you notice about the temperature as time passes?

Solution (a) Using formula (4) with $T = 30$ and $u_0 = 100$, the temperature (in degrees Celsius) of the object at time t (in minutes) is

$$u(t) = 30 + (100 - 30)e^{kt} = 30 + 70e^{kt} \qquad (5)$$

where k is a negative constant. To find k, we use the fact that $u = 80$ when $t = 5$. Then

$$80 = 30 + 70e^{k(5)}$$
$$50 = 70e^{5k}$$
$$e^{5k} = \frac{50}{70}$$
$$5k = \ln\frac{5}{7}$$
$$k = \frac{1}{5}\ln\frac{5}{7} \approx -0.0673$$

Formula (4) therefore becomes

$$u(t) = 30 + 70e^{\frac{\ln\frac{5}{7}}{5}t}$$

*Named after Sir Isaac Newton (1642–1727), one of the cofounders of calculus.

We want to find t when $u = 50°C$, so

$$50 = 30 + 70e^{\frac{\ln\frac{5}{7}}{5}t}$$

$$20 = 70e^{\frac{\ln\frac{5}{7}}{5}t}$$

$$e^{\frac{\ln\frac{5}{7}}{5}t} = \frac{20}{70}$$

$$\frac{\ln\frac{5}{7}}{5}t = \ln\frac{2}{7}$$

$$t = \frac{5}{\ln\frac{5}{7}}\ln\frac{2}{7} \approx 18.6\text{ minutes}$$

The temperature of the object will be 50°C after about 18.6 minutes or 18 minutes, 37 seconds.

(b) Figure 33 shows the graph of $y = 30 + 70e^{\frac{\ln\frac{5}{7}}{5}x}$, where y is the temperature and x is the time.

Figure 33

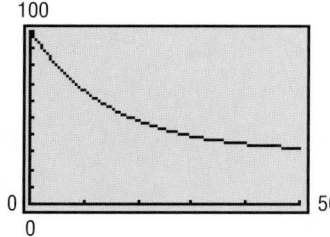

(c) By graphing $Y_1 = 50$ and $Y_2 = 30 + 70e^{\frac{\ln\frac{5}{7}}{5}x}$, where x is time, and using INTERSECT, we find that it takes $x = 18.6$ minutes (18 minutes, 37 seconds) for the temperature to cool to 50°C.

(d) By graphing $Y_1 = 35$ and $Y_2 = 30 + 70e^{\frac{\ln\frac{5}{7}}{5}x}$, where x is time, and using INTERSECT, we find that it takes $x = 39.22$ minutes (39 minutes, 13 seconds) for the temperature to cool to 35°C.

(e) As x increases, the value of $e^{\frac{\ln\frac{5}{7}}{5}x}$ approaches zero, so the value of y approaches 30°C. The temperature of the object approaches 30°C. ∎

Logistic Models

4 The exponential growth model $A(t) = A_0 e^{kt}, k > 0$, assumes uninhibited growth, meaning that the value of the function grows without limit. Recall we stated that cell division could be modeled using this function, assuming that no cells die and no by-products are produced. However, cell division would eventually be limited by factors such as living space and food supply. The **logistic growth model** is an exponential function that can model situations where the growth of the dependent variable is limited.

Other situations that lead to a logistic growth model include population growth and the sales of a product due to advertising. See Problems 21 through 25. The logistic growth model is given next.

Logistic Growth Model

$$P(t) = \frac{c}{1 + ae^{-bt}}$$

where a, b, and c are constants with $c > 0$ and $b > 0$.

The number c is called the **carrying capacity** because the value $P(t)$ approaches c as t approaches infinity; that is, $\lim\limits_{t \to \infty} P(t) = c$. The number b is the growth rate.

EXAMPLE 5   **Fruit Fly Population**

Fruit flies are placed in a half-pint milk bottle with a banana (for food) and yeast plants (for food and to provide a stimulus to lay eggs). Suppose that the fruit fly population after t days is given by

$$P(t) = \frac{230}{1 + 56.5e^{-0.37t}}$$

(a) State the carrying capacity and the growth rate.

(b) Determine the initial population.

(c) Use a graphing utility to graph $P(t)$.

(d) What is the population after 5 days?

(e) How long does it take for the population to reach 180?

(f) How long does it take for the population to reach one-half of the carrying capacity?

Solution (a) As $t \to \infty$, $e^{-0.37t} \to 0$ and $P(t) \to 230/1$. The carrying capacity of the half-pint bottle is 230 fruit flies. The growth rate is $0.37 = 37\%$.

(b) To find the initial number of fruit flies in the half-pint bottle, we evaluate $P(0)$.

$$P(0) = \frac{230}{1 + 56.5e^{-0.37(0)}}$$

$$= \frac{230}{1 + 56.5}$$

$$= 4$$

So initially there were four fruit flies in the half-pint bottle.

(c) See Figure 34 for the graph of $P(t)$.

(d) To find the number of fruit flies in the half-pint bottle after 5 days, we evaluate $P(5)$.

$$P(5) = \frac{230}{1 + 56.5e^{-0.37(5)}} \approx 23 \text{ fruit flies}$$

After 5 days, there are 23 fruit flies in the bottle.

Figure 34

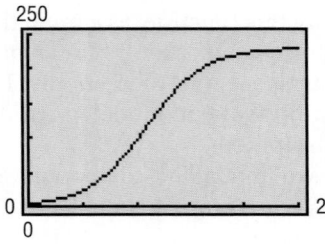

(e) To determine when the population of fruit flies will be 180, we solve the equation

$$\frac{230}{1 + 56.5e^{-0.37t}} = 180$$

$$230 = 180(1 + 56.5e^{-0.37t})$$

$$1.2778 = 1 + 56.5e^{-0.37t} \qquad \text{Divide both sides by 180.}$$

$$0.2778 = 56.5e^{-0.37t} \qquad \text{Subtract 1 from both sides.}$$

$$0.0049 = e^{-0.37t} \qquad \text{Divide both sides by 56.5.}$$

$$\ln(0.0049) = -0.37t \qquad \text{Rewrite as a logarithmic expression.}$$

$$t \approx 14.4 \text{ days} \qquad \text{Divide both sides by } -0.37.$$

It will take approximately 14.4 days (14 days, 9 hours) for the population to reach 180 fruit flies.

We could also solve this problem by graphing $Y_1 = \dfrac{230}{1 + 56.5e^{-0.37t}}$ and $Y_2 = 180$, using INTERSECT to find the solution shown in Figure 35.

Figure 35

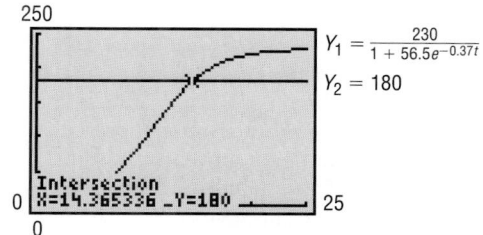

(f) One-half of the carrying capacity is 115 fruit flies. We solve $P(t) = 115$ by graphing $Y_1 = \dfrac{230}{1 + 56.5e^{-0.37t}}$ and $Y_2 = 115$ and using INTERSECT. See Figure 36. The population will reach one-half of the carrying capacity in about 10.9 days (10 days, 22 hours).

Figure 36

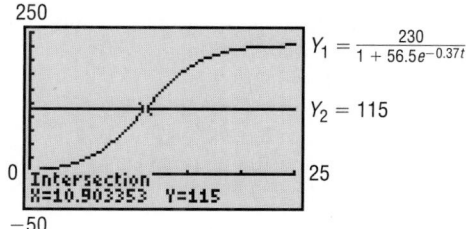

Look back at Figure 36. Notice the point where the graph reaches 115 fruit flies (one-half of the carrying capacity); the graph changes from being curved upward to being curved downward. Using the language of calculus, we say the graph changes from increasing at an increasing rate to increasing at a decreasing rate. For any logistic growth function, when the population reaches one-half the carrying capacity, the population growth starts to slow down.

 EXPLORATION   On the same viewing rectangle, graph $Y_1 = \dfrac{500}{1 + 24e^{-0.03t}}$ and $Y_2 = \dfrac{500}{1 + 24e^{-0.08t}}$. What effect does the growth rate b have on the logistic growth function?

7.6 Exercises

1. **Growth of an Insect Population** The size P of a certain insect population at time t (in days) obeys the function $P(t) = 500e^{0.02t}$.
   (a) Determine the number of insects at $t = 0$ days.
   (b) What is the growth rate of the insect population?
   (c) Graph the function using a graphing utility.
   (d) What is the population after 10 days?
   (e) When will the insect population reach 800?
   (f) When will the insect population double?

2. **Growth of Bacteria** The number N of bacteria present in a culture at time t (in hours) obeys the function $N(t) = 1000e^{0.01t}$.
   (a) Determine the number of bacteria at $t = 0$ hours.
   (b) What is the growth rate of the bacteria?
   (c) Graph the function using a graphing utility.
   (d) What is the population after 4 hours?
   (e) When will the number of bacteria reach 1700?
   (f) When will the number of bacteria double?

3. **Radioactive Decay** Strontium 90 is a radioactive material that decays according to the function $A(t) = A_0 e^{-0.0244t}$, where A_0 is the initial amount present and A is the amount present at time t (in years). Assume that a scientist has a sample of 500 grams of strontium 90.
   (a) What is the decay rate of strontium 90?
   (b) Graph the function using a graphing utility.
   (c) How much strontium 90 is left after 10 years?
   (d) When will 400 grams of strontium 90 be left?
   (e) What is the half-life of strontium 90?

4. **Radioactive Decay** Iodine 131 is a radioactive material that decays according to the function $A(t) = A_0 e^{-0.087t}$ where A_0 is the initial amount present and A is the amount present at time t (in days). Assume that a scientist has a sample of 100 grams of iodine 131.
   (a) What is the decay rate of iodine 131?
   (b) Graph the function using a graphing utility.
   (c) How much iodine 131 is left after 9 days?
   (d) When will 70 grams of iodine 131 be left?
   (e) What is the half-life of iodine 131?

5. **Growth of a Colony of Mosquitoes** The population of a colony of mosquitoes obeys the law of uninhibited growth. If there are 1000 mosquitoes initially and there are 1800 after 1 day, what is the size of the colony after 3 days? How long is it until there are 10,000 mosquitoes?

6. **Bacterial Growth** A culture of bacteria obeys the law of uninhibited growth. If 500 bacteria are present initially and there are 800 after 1 hour, how many will be present in the culture after 5 hours? How long is it until there are 20,000 bacteria?

7. **Population Growth** The population of a southern city follows the exponential law. If the population doubled in size over an 18-month period and the current population is 10,000, what will the population be 2 years from now?

8. **Population Decline** The population of a midwestern city follows the exponential law. If the population decreased from 900,000 to 800,000 from 1993 to 1995, what will the population be in 1997?

9. **Radioactive Decay** The half-life of radium is 1690 years. If 10 grams is present now, how much will be present in 50 years?

10. **Radioactive Decay** The half-life of radioactive potassium is 1.3 billion years. If 10 grams is present now, how much will be present in 100 years? In 1000 years?

11. **Estimating the Age of a Tree** A piece of charcoal is found to contain 30% of the carbon 14 that it originally had.
    (a) When did the tree from which the charcoal came die? Use 5600 years as the half-life of carbon 14.
    (b) Using a graphing utility, graph the relation between the percentage of carbon 14 remaining and time.
    (c) Using INTERSECT, determine the time that elapses until half of the carbon 14 remains.
    (d) Verify the answer found in part (a).

12. **Estimating the Age of a Fossil** A fossilized leaf contains 70% of its normal amount of carbon 14.
    (a) How old is the fossil?
    (b) Using a graphing utility, graph the relation between the percentage of carbon 14 remaining and time.
    (c) Using INTERSECT, determine the time that elapses until half of the carbon 14 remains.
    (d) Verify the answer found in part (a).

13. **Cooling Time of a Pizza** A pizza baked at 450°F is removed from the oven at 5:00 PM into a room that is a constant 70°F. After 5 minutes, the pizza is at 300°F.
    (a) At what time can you begin eating the pizza if you want its temperature to be 135°F?
    (b) Using a graphing utility, graph the relation between temperature and time.
    (c) Using INTERSECT, determine the time that needs to elapse before the pizza is 160°F.
    (d) TRACE the function for large values of time. What do you notice about y, the temperature?

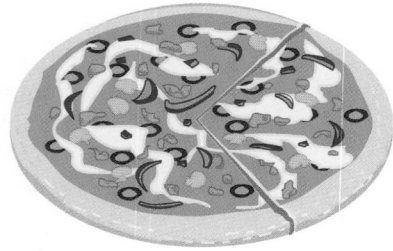

14. **Newton's Law of Cooling** A thermometer reading 72°F is placed in a refrigerator where the temperature is a constant 38°F.

(a) If the thermometer reads 60°F after 2 minutes, what will it read after 7 minutes?

(b) How long will it take before the thermometer reads 39°F?

(c) Using a graphing utility, graph the relation between temperature and time.

(d) Using INTERSECT, determine the time needed to elapse before the thermometer reads 45°F.

(e) TRACE the function for large values of time. What do you notice about y, the temperature?

15. **Newton's Law of Cooling**   A thermometer reading 8°C is brought into a room with a constant temperature of 35°C.

(a) If the thermometer reads 15°C after 3 minutes, what will it read after being in the room for 5 minutes? For 10 minutes?

(b) Graph the relation between temperature and time. TRACE to verify that your answers are correct.

   [**Hint:** You need to construct a formula similar to equation (4).]

16. **Thawing Time of a Steak**   A frozen steak has a temperature of 28°F. It is placed in a room with a constant temperature of 70°F. After 10 minutes, the temperature of the steak has risen to 35°F. What will the temperature of the steak be after 30 minutes? How long will it take the steak to thaw to a temperature of 45°F? [See the hint given for Problem 15.] Graph the relation between temperature and time. TRACE to verify that your answer is correct.

17. **Decomposition of Salt in Water**   Salt (NaCl) decomposes in water into sodium (NA^+) and chloride (Cl^-) ions according to the law of uninhibited decay. If the initial amount of salt is 25 kilograms and, after 10 hours, 15 kilograms of salt is left, how much salt is left after 1 day? How long does it take until $\frac{1}{2}$ kilogram of salt is left?

18. **Voltage of a Conductor**   The voltage of a certain conductor decreases over time according to the law of uninhibited decay. If the initial voltage is 40 volts, and 2 seconds later it is 10 volts, what is the voltage after 5 seconds?

19. **Radioactivity from Chernobyl**   After the release of radioactive material into the atmosphere from a nuclear power plant at Chernobyl (Ukraine) in 1986, the hay in Austria was contaminated by iodine 131 (half-life 8 days). If it is all right to feed the hay to cows when 10% of the iodine 131 remains, how long do the farmers need to wait to use this hay?

20. **Pig Roasts**   The hotel Bora-Bora is having a pig roast. At noon, the chef put the pig in a large earthen oven. The pig's original temperature was 75°F. At 2:00 PM the chef checked the pig's temperature and was upset because it had reached only 100°F. If the oven's temperature remains a constant 325°F, at what time may the hotel serve its guests, assuming that pork is done when it reaches 175°F?

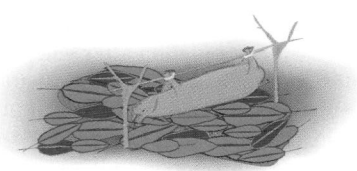

21. **Proportion of the Population That Owns a VCR**   The logistic growth model

$$P(t) = \frac{0.9}{1 + 6e^{-0.32t}}$$

relates the proportion of U.S. households that own a VCR to the year. Let $t = 0$ represent 1984, $t = 1$ represent 1985, and so on.

(a) Determine the maximum percentage of households that will own a VCR.

(b) What percentage of households owned a VCR in 1984 ($t = 0$)?

(c) Use a graphing utility to graph $P(t)$.

(d) What percentage of households owned a VCR in 1999 ($t = 15$)?

(e) When will 0.8 (80%) of U.S. households own a VCR?

(f) How long will it be before 0.45 (45%) of the population owns a VCR?

22. **Market Penetration of Intel's Coprocessor**   The logistic growth model

$$P(t) = \frac{0.90}{1 + 3.5e^{-0.339t}}$$

relates the proportion of new personal computers (PCs) sold at Best Buy that have Intel's latest coprocessor t months after it has been introduced.

(a) Determine the maximum percentage of PCs sold at Best Buy that will have Intel's latest coprocessor.

(b) What percentage of computers sold at Best Buy will have Intel's latest coprocessor when it is first introduced ($t = 0$)?

(c) Use a graphing utility to graph $P(t)$.

(d) What percentage of PCs sold will have Intel's latest coprocessor $t = 4$ months after it is introduced?

(e) When will 0.75 (75%) of PCs sold by Best Buy have Intel's latest coprocessor?

(f) How long will it be before 0.45 (45%) of the PCs sold by Best Buy have Intel's latest coprocessor?

23. **Population of a Bacteria Culture**   The logistic growth model

$$P(t) = \frac{1000}{1 + 32.33e^{-0.439t}}$$

represents the population (in grams) of a bacterium after t hours.

(a) Determine the carrying capacity of the environment. What is the growth rate of the bacteria?

(b) Determine the initial population size.

(c) Use a graphing utility to graph $P(t)$.
(d) What is the population after 9 hours?
(e) When will the population be 700 grams?
(f) How long does it take for the population to reach one-half of the carrying capacity?

24. **Population of a Endangered Species**   Often environmentalists will capture an endangered species and transport the species to a controlled environment where the species can produce offspring and regenerate its population. Suppose that six American bald eagles are captured, transported to Montana, and set free. Based on experience, the environmentalists expect the population to grow according to the model

$$P(t) = \frac{500}{1 + 83.33e^{-0.162t}}$$

where t is measured in years.
(a) Determine the carrying capacity of the environment. What is the growth rate of the bald eagle?
(b) Use a graphing utility to graph $P(t)$.
(c) What is the population after 3 years?
(d) When will the population be 300 eagles?
(e) How long does it take for the population to reach one-half of the carrying capacity?

25. **The *Challenger* Disaster***   After the *Challenger* disaster in 1986, a study of the 23 launches that preceded the fatal flight was made. A mathematical model was developed involving the relationship between the Fahrenheit temperature x around the O-rings and the number y of eroded or leaky primary O-rings. The model stated that

$$y = \frac{6}{1 + e^{-(5.085-0.1156x)}}$$

where the number 6 indicates the 6 primary O-rings on the spacecraft.
(a) What is the predicted number of eroded or leaky primary O-rings at a temperature of 100°F?
(b) What is the predicted number of eroded or leaky primary O-rings at a temperature of 60°F?
(c) What is the predicted number of eroded or leaky primary O-rings at a temperature of 30°F?
(d) Graph the equation. At what temperature is the predicted number of eroded or leaky O-rings 1? 3? 5?

*Linda Tappin, "Analyzing Data Relating to the *Challenger* Disaster," *Mathematics Teacher*, Vol. 87, No. 6, September 1994, pp. 423–426.

PREPARING FOR THIS SECTION

Before getting started, review the following:

✓ Scatter Diagrams; Linear Curve Fitting (Section A.8, pp. 628–631)

7.7 EXPONENTIAL, LOGARITHMIC, AND LOGISTIC MODELS

OBJECTIVES   **1** Use a Graphing Utility to Fit an Exponential Function to Data
2 Use a Graphing Utility to Fit a Logarithmic Function to Data
3 Use a Graphing Utility to Fit a Logistic Function to Data

In Section A.8 we discussed how to find the linear function of best fit $(y = ax + b)$.

In this section we will discuss how to use a graphing utility to find equations of best fit that describe the relation between two variables when the relation is thought to be exponential ($y = ab^x$), logarithmic ($y = a + b \ln x$), or logistic $\left(y = \dfrac{c}{1 + ae^{-bx}} \right)$. As before, we draw a scatter diagram of the data to help determine the appropriate model to use.

Figure 37 shows scatter diagrams that will typically be observed for the three models. Below each scatter diagram are any restrictions on the values of the parameters.

Figure 37

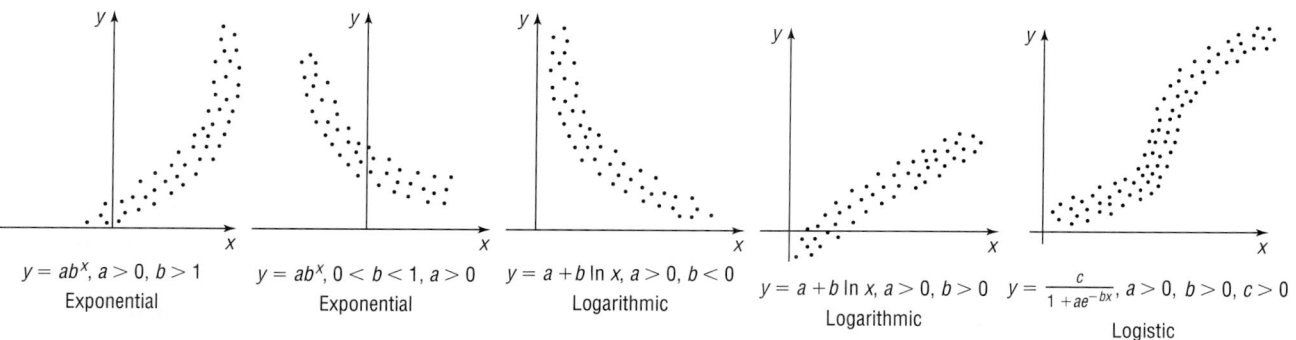

$y = ab^x, a > 0, b > 1$
Exponential

$y = ab^x, 0 < b < 1, a > 0$
Exponential

$y = a + b \ln x, a > 0, b < 0$
Logarithmic

$y = a + b \ln x, a > 0, b > 0$
Logarithmic

$y = \dfrac{c}{1 + ae^{-bx}}, a > 0, b > 0, c > 0$
Logistic

Most graphing utilities have REGression options that fit data to a specific type of curve. Once the data have been entered and a scatter diagram obtained, the type of curve that you want to fit to the data is selected. Then that REGression option is used to obtain the curve of "best fit" of the type selected.

The correlation coefficient r will appear only if the model can be written as a linear expression. Thus, r will appear for the linear, power, exponential, and logarithmic models, since these models can be written as a linear expression (see Problems 59 and 60 in Exercise 7.3). Remember, the closer $|r|$ is to 1, the better the fit.

Let's look at some examples.

Exponential Models

① We saw in Section 7.5 that the future value of money behaves exponentially, and we saw in Section 7.6 that growth and decay models also behave exponentially. The next example shows how data can lead to an exponential model.

EXAMPLE 1 **Fitting an Exponential Function to Data**

Beth is interested in finding a function that explains the closing price of Harley Davidson stock at the end of each month. She obtains the following data:

Year, *x*	Closing Price, *y*
1987 (*x* = 1)	0.392
1988 (*x* = 2)	0.7652
1989 (*x* = 3)	1.1835
1990 (*x* = 4)	1.1609
1991 (*x* = 5)	2.6988
1992 (*x* = 6)	4.5381
1993 (*x* = 7)	5.3379
1994 (*x* = 8)	6.8032
1995 (*x* = 9)	7.0328
1996 (*x* = 10)	11.5585
1997 (*x* = 11)	13.4799
1998 (*x* = 12)	23.5424
1999 (*x* = 13)	31.9342
2000 (*x* = 14)	39.7277

Source: http://finance.yahoo.com

(a) Using a graphing utility, draw a scatter diagram with year as the independent variable.

(b) Using a graphing utility, fit an exponential function to the data.

(c) Express the function found in part (b) in the form $A = A_0 e^{kt}$.

(d) Graph the exponential function found in part (b) or (c) on the scatter diagram.

(e) Using the solution to part (b) or (c), predict the closing price of Harley Davidson stock at the end of 2001.

 (f) Interpret the value of k found in part (c).

Solution

(a) Enter the data into the graphing utility, letting 1 represent 1987, 2 represent 1998, and so on. We obtain the scatter diagram shown in Figure 38.

Figure 38

45

0 ————————— 15
0

(b) A graphing utility fits the data in Figure 38 to an exponential function of the form $y = ab^x$ by using the EXPonential REGression option. See Figure 39. Thus, $y = ab^x = 0.394503(1.40276)^x$.

(c) To express $y = ab^x$ in the form $A = A_0 e^{kt}$, where $x = t$ and $y = A$, we proceed as follows:

$$ab^x = A_0 e^{kt}, \quad x = t$$

Figure 39

```
ExpReg
y=a*b^x
a=.394502732
b=1.402763529
```

When $x = t = 0$, we find $a = A_0$. This leads to

$$a = A_0, \qquad b^x = e^{kt}$$
$$b^x = (e^k)^t$$
$$b = e^k \qquad {\scriptstyle x = t}$$

Since $y = ab^x = 0.394503(1.40276)^x$, we find that $a = 0.394503$ and $b = 1.40276$, so

$$a = A_0 = 0.394503 \quad \text{and} \quad b = 1.40276 = e^k$$

We want to find k, so we rewrite $1.40276 = e^k$ as a logarithm and obtain

$$k = \ln(1.40276) \approx 0.33844$$

As a result, $A = A_0 e^{kt} = 0.394503 e^{0.33844t}$.

Figure 40

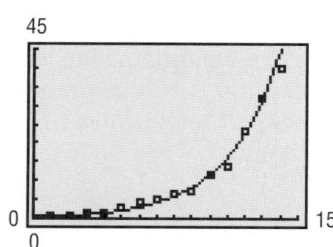

(d) See Figure 40 for the graph of the exponential function of best fit.

(e) Let $t = 15$ (end of 2001) in the function found in part (c). The predicted closing price of Harley Davidson stock at the end of 2001 is

$$A = 0.394503e^{0.33844(15)} = \$63.21$$

(f) The value of k represents the annual interest rate compounded continuously.

$$A = A_0 e^{kt} = 0.394503e^{0.33844t}$$
$$= Pe^{rt} \qquad \text{Equation (4), Section 4.6}$$

The price of Harley Davidson stock has grown at an annual rate of 33.844% (compounded continuously) between 1987 and 2000.   ■

 **N O W   W O R K   P R O B L E M   1 .**

Logarithmic Models

2   Many relations between variables do not follow an exponential model, but, instead, the independent variable is related to the dependent variable using a logarithmic model.

EXAMPLE 2   Fitting a Logarithmic Function to Data

Jodi, a meteorologist, is interested in finding a function that explains the relation between the height of a weather balloon (in kilometers) and the atmospheric pressure (measured in millimeters of mercury) on the balloon. She collects the data shown in the table.

Atmospheric Pressure, p	Height, h
760	0
740	0.184
725	0.328
700	0.565
650	1.079
630	1.291
600	1.634
580	1.862
550	2.235

(a) Using a graphing utility, draw a scatter diagram of the data with atmospheric pressure as the independent variable.

(b) Using a graphing utility, fit a logarithmic function to the data.

(c) Draw the logarithmic function found in part (b) on the scatter diagram.

(d) Use the function found in part (b) to predict the height of the weather balloon if the atmospheric pressure is 560 millimeters of mercury.

Solution   (a) After entering the data into the graphing utility, we obtain the scatter diagram shown in Figure 41.

(b) A graphing utility fits the data in Figure 41 to a logarithmic function of the form $y = a + b \ln x$ by using the Logarithm REGression option. See Figure 42. The logarithmic function of best fit to the data is

$$h(p) = 45.7863 - 6.9025 \ln p$$

where h is the height of the weather balloon and p is the atmospheric pressure. Notice that $|r|$ is close to 1, indicating a good fit.

(c) Figure 43 shows the graph of $h(p) = 45.7863 - 6.9025 \ln p$ on the scatter diagram.

(d) Using the function found in part (b), Jodi predicts the height of the weather balloon when the atmospheric pressure is 560 to be

$$h(560) = 45.7863 - 6.9025 \ln 560$$

$$\approx 2.108 \text{ kilometers}$$

Figure 41

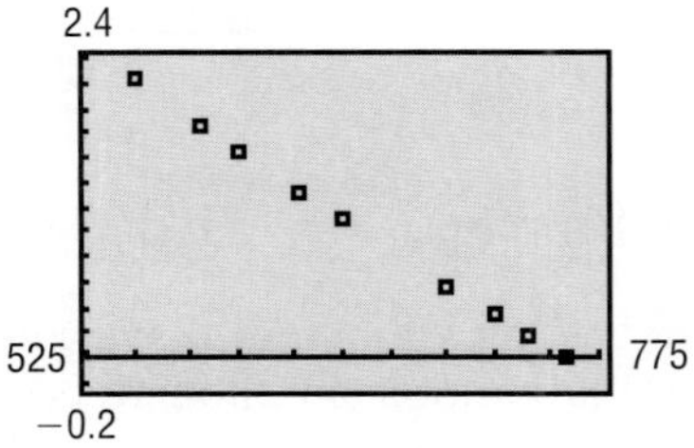

2.4

525   775

−0.2

Figure 42

Figure 43

2.4

525   775

−0.2

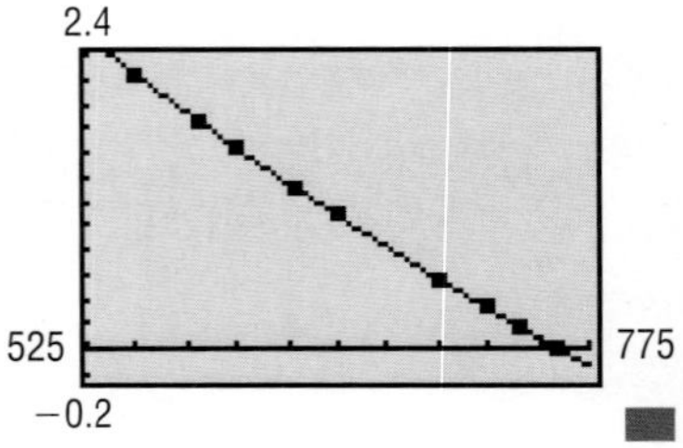

NOW WORK PROBLEM **7.**

Logistic Models

③ Logistic growth models can be used to model situations where the value of the dependent variable is limited. Many real-world situations conform to this scenario. For example, the population of the human race is limited by the availability of natural resources such as food and shelter. When the value of the dependent variable is limited, a logistic growth model is often appropriate.

EXAMPLE 3   **Fitting a Logistic Function to Data**

The data in the table on page 543, obtained from Tor Carlson (Über Geschwindigkeit und Grösse der Hefevermehrung in Würze, *Biochemishe Zeitschrift*, Bd. 57, pp. 313–334, 1913), represent the amount of yeast biomass after *t* hours in a culture.

(a) Using a graphing utility, draw a scatter diagram of the data with time as the independent variable.

(b) Using a graphing utility, fit a logistic function to the data.

(c) Using a graphing utility, graph the function found in part (b) on the scatter diagram.

Time (in hours)	Yeast Biomass
0	9.6
1	18.3
2	29.0
3	47.2
4	71.1
5	119.1
6	174.6
7	257.3
8	350.7
9	441.0
10	513.3
11	559.7
12	594.8
13	629.4
14	640.8
15	651.1
16	655.9
17	659.6
18	661.8

(d) What is the predicted carrying capacity of the culture?
(e) Use the function found in part (b) to predict the population of the culture at $t = 19$ hours.

Solution
(a) See Figure 44 for a scatter diagram of the data.

Figure 44

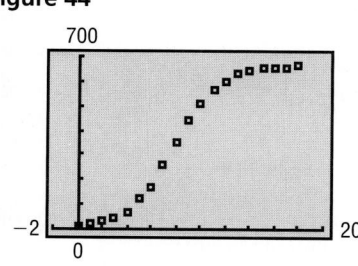

(b) A graphing utility fits a logistic growth model of the form $y = \dfrac{c}{1 + ae^{-bx}}$ by using the LOGISTIC regression option. See Figure 45. The logistic function of best fit to the data is

$$y = \frac{663.0}{1 + 71.6e^{-0.5470x}}$$

where y is the amount of yeast biomass in the culture and x is the time.

(c) See Figure 46 for the graph of the logistic function of best fit.

(d) Based on the logistic growth function found in part (b), the carrying capacity of the culture is 663.

(e) Using the logistic growth function found in part (b), the predicted amount of yeast biomass at $t = 19$ hours is

$$y = \frac{663.0}{1 + 71.6e^{-0.5470(19)}} = 661.5$$

Figure 45

Figure 46

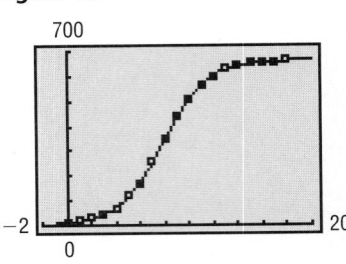

NOW WORK PROBLEM **9.**

7.7 Exercises

1. **Biology**   A strain of E-coli Beu 397-recA441 is placed into a petri dish at 30° Celsius and allowed to grow. The following data are collected. Theory states that the number of bacteria in the petri dish will initially grow according to the law of uninhibited growth. The population is measured using an optical device in which the amount of light that passes through the petri dish is measured.

Time (hours), x	Population, y
0	0.09
2.5	0.18
3.5	0.26
4.5	0.35
6	0.50

Source: Dr. Polly Lavery, Joliet Junior College

(a) Draw a scatter diagram treating time as the predictor variable.
(b) Using a graphing utility, fit an exponential function to the data.
(c) Express the function found in part (b) in the form $N(t) = N_0 e^{kt}$.
(d) Graph the exponential function found in part (b) or (c) on the scatter diagram.
(e) Use the exponential function from part (b) or (c) to predict the population at $x = 7$ hours.
(f) Use the exponential function from part (b) or (c) to predict when the population will reach 0.75.

2. **Biology**   A strain of E-coli SC18del-recA718 is placed into a petri dish at 30° Celsius and allowed to grow. The following data are collected. Theory states that the number of bacteria in the petri dish will initially grow according to the law of uninhibited growth. The population is measured using an optical device in which the amount of light that passes through the petri dish is measured.

Time (hours), x	Population, y
2.5	0.175
3.5	0.38
4.5	0.63
4.75	0.76
5.25	1.20

Source: Dr. Polly Lavery, Joliet Junior College

(a) Draw a scatter diagram treating time as the predictor variable.
(b) Using a graphing utility, fit an exponential function to the data.
(c) Express the function found in part (b) in the form $N(t) = N_0 e^{kt}$.
(d) Graph the exponential function found in part (b) or (c) on the scatter diagram.
(e) Use the exponential function from part (b) or (c) to predict the population at $x = 6$ hours.
(f) Use the exponential function from part (b) or (c) to predict when the population will reach 2.1.

3. **Chemistry**   A chemist has a 100-gram sample of a radioactive material. He records the amount of radioactive material every week for 6 weeks and obtains the following data:

Week	Weight (in Grams)
0	100.0
1	88.3
2	75.9
3	69.4
4	59.1
5	51.8
6	45.5

(a) Using a graphing utility, draw a scatter diagram with week as the independent variable.
(b) Using a graphing utility, fit an exponential function to the data.
(c) Express the function found in part (b) in the form $A(t) = A_0 e^{kt}$.
(d) Graph the exponential function found in part (b) or (c) on the scatter diagram.
(e) From the result found in part (b), determine the half-life of the radioactive material.
(f) How much radioactive material will be left after 50 weeks?
(g) When will there be 20 grams of radioactive material?

4. **Chemistry**   A chemist has a 1000-gram sample of a radioactive material. She records the amount of radioactive material remaining in the sample every day for a week and obtains the following data:

Day	Weight (in Grams)
0	1000.0
1	897.1
2	802.5
3	719.8
4	651.1
5	583.4
6	521.7
7	468.3

(a) Using a graphing utility, draw a scatter diagram with day as the independent variable.
(b) Using a graphing utility, fit an exponential function to the data.
(c) Express the function found in part (b) in the form $A(t) = A_0 e^{kt}$.
(d) Graph the exponential function found in part (b) or (c) on the scatter diagram.
(e) From the result found in part (b), find the half-life of the radioactive material.
(f) How much radioactive material will be left after 20 days?
(g) When will there be 200 grams of radioactive material?

5. **Finance**   The following data represent the amount of money an investor has in an investment account each year for 10 years. She wishes to determine the average annual rate of return on her investment.
(a) Using a graphing utility, draw a scatter diagram with time as the independent variable and the value of the account as the dependent variable.
(b) Using a graphing utility, fit an exponential function to the data.

Year	Value of Account
1991	$10,000
1992	$10,573
1993	$11,260
1994	$11,733
1995	$12,424
1996	$13,269
1997	$13,968
1998	$14,823
1999	$15,297
2000	$16,539

(c) Based on the answer in part (b), what was the average annual rate of return from this account over the past 10 years?
(d) If the investor plans on retiring in 2021, what will the predicted value of this account be?
(e) When will the account be worth $50,000?

6. **Finance**   The following data show the amount of money an investor has in an investment account each year for 7 years. He wishes to determine the average annual rate of return on his investment.

Year	Value of Account
1994	$20,000
1995	$21,516
1996	$23,355
1997	$24,885
1998	$27,434
1999	$30,053
2000	$32,622

(a) Using a graphing utility, draw a scatter diagram with time as the independent variable and the value of the account as the dependent variable.
(b) Using a graphing utility, fit an exponential function to the data.
(c) Based on the answer to part (b), what was the average annual rate of return from this account over the past 7 years?
(d) If the investor plans on retiring in 2020, what will the predicted value of this account be?
(e) When will the account be worth $80,000?

7. **Economics and Marketing**   The following data represent the price and quantity demanded in 2000 for IBM personal computers at Best Buy.

Price ($/Computer)	Quantity Demanded
2300	152
2000	159
1700	164
1500	171
1300	176
1200	180
1000	189

(a) Using a graphing utility, draw a scatter diagram of the data with price as the dependent variable.
(b) Using a graphing utility, fit a logarithmic function to the data.
(c) Using a graphing utility, draw the logarithmic function found in part (b) on the scatter diagram.
(d) Use the function found in part (b) to predict the number of IBM personal computers that would be demanded if the price were $1650.

8. Economics and Marketing The following data represent the price and quantity supplied in 2000 for IBM personal computers.

Price ($/Computer)	Quantity Supplied
2300	180
2000	173
1700	160
1500	150
1300	137
1200	130
1000	113

(a) Using a graphing utility, draw a scatter diagram of the data with price as the dependent variable.
(b) Using a graphing utility, fit a logarithmic function to the data.
(c) Using a graphing utility, draw the logarithmic function found in part (b) on the scatter diagram.
(d) Use the function found in part (b) to predict the number of IBM personal computers that would be supplied if the price were $1650.

9. Population Model The following data obtained from the U.S. Census Bureau represent the population of the United States. An ecologist is interested in finding a function that describes the population of the United States.

Year	Population
1900	76,212,168
1910	92,228,496
1920	106,021,537
1930	123,202,624
1940	132,164,569
1950	151,325,798
1960	179,323,175
1970	203,302,031
1980	226,542,203
1990	248,709,873
2000	281,421,906

(a) Using a graphing utility, draw a scatter diagram of the data using the year as the independent variable and population as the dependent variable.

(b) Using a graphing utility, fit a logistic function to the data.
(c) Using a graphing utility, draw the function found in part (b) on the scatter diagram.
(d) Based on the function found in part (b), what is the carrying capacity of the United States?
(e) Use the function found in part (b) to predict the population of the United States in 2001.
(f) When will the United States population be 300,000,000?

10. Population Model The following data obtained from the U.S. Census Bureau represent the world population. An ecologist is interested in finding a function that describes the world population.

Year	Population (in Billions)
1993	5.531
1994	5.611
1995	5.691
1996	5.769
1997	5.847
1998	5.925
1999	6.003
2000	6.080
2001	6.157

(a) Using a graphing utility, draw a scatter diagram of the data using year as the independent variable and population as the dependent variable.
(b) Using a graphing utility, fit a logistic function to the data.
(c) Using a graphing utility, draw the function found in part (b) on the scatter diagram.
(d) Based on the function found in part (b), what is the carrying capacity of the world?
(e) Use the function found in part (b) to predict the population of the world in 2002.
(f) When will world population be 7 billion?

11. Population Model The data on page 547 obtained from the U.S. Census Bureau represent the population of Illinois. An urban economist is interested in finding a model that describes the population of Illinois.
(a) Using a graphing utility, draw a scatter diagram of the data using year as the independent variable and population as the dependent variable.
(b) Using a graphing utility, fit a logistic function to the data.
(c) Using a graphing utility, draw the function found in part (b) on the scatter diagram.
(d) Based on the function found in part (b), what is the carrying capacity of Illinois?

Year	Population
1900	4,821,550
1910	5,638,591
1920	6,485,280
1930	7,630,654
1940	7,897,241
1950	8,712,176
1960	10,081,158
1970	11,110,285
1980	11,427,409
1990	11,430,602
2000	12,419,293

Year	Population
1900	6,302,115
1910	7,665,111
1920	8,720,017
1930	9,631,350
1940	9,900,180
1950	10,498,012
1960	11,319,366
1970	11,800,766
1980	11,864,720
1990	11,881,643
2000	12,281,054

(e) Use the function found in part (b) to predict the population of Illinois in 2010.

12. Population Model   The following data obtained from the U.S. Census Bureau represent the population of Pennsylvania. An urban economist is interested in finding a model that describes the population of Pennsylvania.

(a) Using a graphing utility, draw a scatter diagram of the data using year as the independent variable and population as the dependent variable.
(b) Using a graphing utility, fit a logistic function to the data.
(c) Using a graphing utility, draw the function found in part (b) on the scatter diagram.
(d) Based on the function found in part (b), what is the carrying capacity of Pennsylvania?
(e) Use the function found in part (b) to predict the population of Pennsylvania in 2010.

Chapter Review

Things to Know

Properties of the exponential function (pp. 479 and 481)	$f(x) = a^x, \quad a > 1$	Domain: the interval $(-\infty, \infty)$; Range: the interval $(0, \infty)$; x-intercepts: none; y-intercept: 1; horizontal asymptote: x-axis $(y = 0)$ as $x \to -\infty$; increasing; one-to-one; smooth; continuous See Figure 6 for a typical graph.
	$f(x) = a^x, \quad 0 < a < 1$	Domain: the interval $(-\infty, \infty)$; Range: the interval $(0, \infty)$; x-intercepts: none; y-intercept: 1; horizontal asymptote: x-axis $(y = 0)$ as $x \to \infty$; decreasing; one-to-one; smooth; continuous See Figure 10 for a typical graph.
Number e (p. 482)	Value approached by the expression $\left(1 + \dfrac{1}{n}\right)^n$ as $n \to \infty$; that is, $\displaystyle\lim_{n \to \infty} \left(1 + \dfrac{1}{n}\right)^n = e.$	
Property of exponents (p. 484)	If $a^u = a^v$, then $u = v$.	

Properties of the logarithmic functions (p. 493)	$f(x) = \log_a x, \quad a > 1$ $(y = \log_a x \text{ means } x = a^y)$	Domain: the interval $(0, \infty)$; Range: the interval $(-\infty, \infty)$; x-intercept: 1; y-intercept: none; vertical asymptote: $x = 0$ (y-axis); increasing; one-to-one; smooth; continuous See Figure 17(b) for a typical graph.
	$f(x) = \log_a x, \quad 0 < a < 1$ $(y = \log_a x \text{ means } x = a^y)$	Domain: the interval $(0, \infty)$; Range: the interval $(-\infty, \infty)$; x-intercept: 1; y-intercept: none; vertical asymptote: $x = 0$ (y-axis); decreasing; one-to-one; smooth; continuous See Figure 17(a) for a typical graph.

Natural logarithm (p. 493) $y = \ln x$ means $x = e^y$.

Properties of logarithms (pp. 503–504)

$$\log_a 1 = 0 \qquad \log_a a = 1 \qquad a^{\log_a M} = M \qquad \log_a a^r = r$$

$$\log_a(MN) = \log_a M + \log_a N \qquad \log_a\left(\frac{M}{N}\right) = \log_a M - \log_a N$$

$$\log_a M^r = r \log_a M$$

(p. 506)   If $M = N$, then $\log_a M = \log_a N$.
If $\log_a M = \log_a N$, then $M = N$.

Formulas

Change-of-Base Formula (p. 507)	$\log_a M = \dfrac{\log_b M}{\log_b a}$
Compound Interest Formula (p. 518)	$A = P\left(1 + \dfrac{r}{n}\right)^{nt}$
Continuous compounding (p. 520)	$A = Pe^{rt}$
Present Value Formula (p. 522)	$P = A\left(1 + \dfrac{r}{n}\right)^{-nt} \quad \text{or} \quad P = Ae^{-rt}$
Growth and Decay (p. 527)	$A(t) = A_0 e^{kt}$
Newton's Law of Cooling (p. 532)	$u(t) = T + (u_0 - T)e^{kt}, \quad k < 0$
Logistic Growth Model (p. 534)	$P(t) = \dfrac{c}{1 + ae^{-bt}}$

Objectives

Review Exercises

Blue problem numbers indicate the authors' suggestions for use in a Practice Test.

In Problems 1 and 2, suppose that $f(x) = 3^x$ and $g(x) = \log_3 x$.

1. Evaluate the following: (a) $f(4)$ (b) $g(9)$ (c) $f(-2)$ (d) $g\left(\dfrac{1}{27}\right)$

2. Evaluate the following: (a) $f(1)$ (b) $g(81)$ (c) $f(-4)$ (d) $g\left(\dfrac{1}{243}\right)$

In Problems 3 and 4, convert each exponential expression to an equivalent expression involving a logarithm. In Problems 5 and 6, convert each logarithmic expression to an equivalent expression involving an exponent.

3. $5^2 = z$ **4.** $a^5 = m$ **5.** $\log_5 u = 13$ **6.** $\log_a 4 = 3$

In Problems 7–10, find the domain of each logarithmic function.

7. $\log(3x - 2)$ **8.** $\log_5(2x + 1)$ **9.** $\log_2(x^2 - 3x + 2)$ **10.** $\ln(x^2 - 9)$

In Problems 11–16, evaluate each expression. Do not use a graphing utility.

11. $\log_2\left(\dfrac{1}{8}\right)$ **12.** $\log_3 81$ **13.** $\ln e^{\sqrt{2}}$

14. $e^{\ln 0.1}$ **15.** $2^{\log_2 0.4}$ **16.** $\log_2 2^{\sqrt{3}}$

In Problems 17–22, write each expression as the sum and/or difference of logarithms. Express powers as factors.

17. $\log_3\left(\dfrac{uv^2}{w}\right)$ **18.** $\log_2(a^2\sqrt{b})^4$ **19.** $\log(x^2\sqrt{x^3 + 1})$

20. $\log_5\left(\dfrac{x^2 + 2x + 1}{x^2}\right)$

21. $\ln\left(\dfrac{x\sqrt[3]{x^2 + 1}}{x - 3}\right)$

22. $\ln\left(\dfrac{2x + 3}{x^2 - 3x + 2}\right)^2$

In Problems 23–28, write each expression as a single logarithm.

23. $3\log_4 x^2 + \dfrac{1}{2}\log_4 \sqrt{x}$

24. $-2\log_3\left(\dfrac{1}{x}\right) + \dfrac{1}{3}\log_3 \sqrt{x}$

25. $\ln\left(\dfrac{x - 1}{x}\right) + \ln\left(\dfrac{x}{x + 1}\right) - \ln(x^2 - 1)$

26. $\log(x^2 - 9) - \log(x^2 + 7x + 12)$

27. $2\log 2 + 3\log x - \dfrac{1}{2}[\log(x + 3) + \log(x - 2)]$

28. $\dfrac{1}{2}\ln(x^2 + 1) - 4\ln\dfrac{1}{2} - \dfrac{1}{2}[\ln(x - 4) + \ln x]$

In Problems 29 and 30, use the Change-of-Base Formula and a calculator to evaluate each logarithm. Round your answer to three decimal places.

29. $\log_4 19$

30. $\log_2 21$

In Problems 31 and 32, graph each function using a graphing utility and the Change-of-Base Formula.

31. $y = \log_3 x$

32. $y = \log_7 x$

In Problems 33–42, use transformations to graph each function. Determine the domain, range, and any asymptotes. Verify your results using a graphing utility.

33. $f(x) = 2^{x-3}$

34. $f(x) = -2^x + 3$

35. $f(x) = \dfrac{1}{2}(3^{-x})$

36. $f(x) = 1 + 3^{2x}$

37. $f(x) = 1 - e^x$

38. $f(x) = 3 + \ln x$

39. $f(x) = 3e^x$

40. $f(x) = \dfrac{1}{2}\ln x$

41. $f(x) = 3 - e^{-x}$

42. $f(x) = 4 - \ln(-x)$

In Problems 43–62, solve each equation. Verify your result using a graphing utility.

43. $4^{1-2x} = 2$

44. $8^{6+3x} = 4$

45. $3^{x^2+x} = \sqrt{3}$

46. $4^{x-x^2} = \dfrac{1}{2}$

47. $\log_x 64 = -3$

48. $\log_{\sqrt{2}} x = -6$

49. $5^x = 3^{x+2}$

50. $5^{x+2} = 7^{x-2}$

51. $9^{2x} = 27^{3x-4}$

52. $25^{2x} = 5^{x^2-12}$

53. $\log_3 \sqrt{x - 2} = 2$

54. $2^{x+1} \cdot 8^{-x} = 4$

55. $8 = 4^{x^2} \cdot 2^{5x}$

56. $2^x \cdot 5 = 10^x$

57. $\log_6(x + 3) + \log_6(x + 4) = 1$

58. $\log_{10}(7x - 12) = 2\log_{10} x$

59. $e^{1-x} = 5$

60. $e^{1-2x} = 4$

61. $2^{3x} = 3^{2x+1}$

62. $2^{x^3} = 3^{x^2}$

In Problems 63 and 64, use the following result: If x is the atmospheric pressure (measured in millimeters of mercury), then the formula for the altitude h(x) (measured in meters above sea level) is

$$h(x) = (30T + 8000)\log\left(\dfrac{P_0}{x}\right)$$

where T is the temperature (in degrees Celsius) and P_0 is the atmospheric pressure at sea level, which is approximately 760 millimeters of mercury.

63. Finding the Altitude of an Airplane At what height is a Piper Cub whose instruments record an outside temperature of 0°C and a barometric pressure of 300 millimeters of mercury?

64. Finding the Height of a Mountain How high is a mountain if instruments placed on its peak record a temperature of 5°C and a barometric pressure of 500 millimeters of mercury?

65. Amplifying Sound   An amplifier's power output P (in watts) is related to its decibel voltage gain d by the formula $P = 25e^{0.1d}$.

(a) Find the power output for a decibel voltage gain of 4 decibels.
(b) For a power output of 50 watts, what is the decibel voltage gain?

66. Limiting Magnitude of a Telescope   A telescope is limited in its usefulness by the brightness of the star it is aimed at and by the diameter of its lens. One measure of a star's brightness is its *magnitude*: the dimmer the star, the larger its magnitude. A formula for the limiting magnitude L of a telescope, that is, the magnitude of the dimmest star that it can be used to view, is given by

$$L = 9 + 5.1 \log d$$

where d is the diameter (in inches) of the lens.
(a) What is the limiting magnitude of a 3.5-inch telescope?
(b) What diameter is required to view a star of magnitude 14?

67. Salvage Value   The number of years n for a piece of machinery to depreciate to a known salvage value can be found using the formula

$$n = \frac{\log s - \log i}{\log (1 - d)}$$

where s is the salvage value of the machinery, i is its initial value, and d is the annual rate of depreciation.
(a) How many years will it take for a piece of machinery to decline in value from $90,000 to $10,000 if the annual rate of depreciation is 0.20 (20%)?
(b) How many years will it take for a piece of machinery to lose half of its value if the annual rate of depreciation is 15%?

68. Funding a College Education   A child's grandparents purchase a $10,000 bond fund that matures in 18 years to be used for her college education. The bond fund pays 4% interest compounded semiannually. How much will the bond fund be worth at maturity? What is the effective rate of interest? How long would it take the bond to double in value under these terms?

69. Funding a College Education   A child's grandparents wish to purchase a bond fund that matures in 18 years to be used for her college education. The bond fund pays 4% interest compounded semiannually. How much should they purchase so that the bond fund will be worth $85,000 at maturity?

70. Funding an IRA   First Colonial Bankshares Corporation advertised the following IRA investment plans.

Target IRA Plans

For each $5000 Maturity Value Desired	
Deposit:	At a Term of:
$620.17	20 Years
$1045.02	15 Years
$1760.92	10 Years
$2967.26	5 Years

(a) Assuming continuous compounding, what was the annual rate of interest that they offered?
(b) First Colonial Bankshares claims that $4000 invested today will have a value of over $32,000 in 20 years. Use the answer found in part (a) to find the actual value of $4000 in 20 years. Assume continuous compounding.

71. Estimating the Date That a Prehistoric Man Died   The bones of a prehistoric man found in the desert of New Mexico contain approximately 5% of the original amount of carbon 14. If the half-life of carbon 14 is 5600 years, approximately how long ago did the man die?

72. Temperature of a Skillet   A skillet is removed from an oven whose temperature is 450°F and placed in a room whose temperature is 70°F. After 5 minutes, the temperature of the skillet is 400°F. How long will it be until its temperature is 150°F?

73. World Population   According to the U.S. Census Bureau, the growth rate of the world's population in 1997 was $k = 1.33\% = 0.0133$. The population of the world in 1997 was 5,840,445,216. Letting $t = 0$ represent 1997, use the uninhibited growth model to predict the world's population in the year 2000.

74. Radioactive Decay   The half-life of radioactive cobalt is 5.27 years. If 100 grams of radioactive cobalt is present now, how much will be present in 20 years? In 40 years?

75. Logistic Growth   The logistic growth model

$$P(t) = \frac{0.8}{1 + 1.67e^{-0.16t}}$$

represents the proportion of new computers sold that utilize the Microsoft Windows 2000 operating system. Let $t = 0$ represent 2000, $t = 1$ represent 2001, and so on.
(a) What proportion of new computers sold in 2000 utilized Windows 2000?
(b) Determine the maximum proportion of new computers sold that will utilize Windows 2000.
(c) Using a graphing utility, graph $P(t)$.
(d) When will 75% of new computers sold utilize Windows 2000?

76. CBL Experiment   The following data were collected by placing a temperature probe in a portable heater, removing the probe, and then recording temperature over time.
According to Newton's Law of Cooling, these data should follow an exponential model.

(a) Using a graphing utility, draw a scatter diagram for the data.
(b) Using a graphing utility, fit an exponential function to the data.
(c) Graph the exponential function found in part (b) on the scatter diagram.
(d) Predict how long it will take for the probe to reach a temperature of 110°F.

Time	Temperature (F°)
0	165.07
1	164.77
2	163.99
3	163.22
4	162.82
5	161.96
6	161.20
7	160.45
8	159.35
9	158.61
10	157.89
11	156.83
12	156.11
13	155.08
14	154.40
15	153.72

87. Wind Chill Factor The following data represent the wind speed (mph) and wind chill factor at an air temperature of 15°F.

Wind Speed (mph)	Wind Chill Factor
5	12
10	−3
15	−11
20	−17
25	−22
30	−25
35	−27

Source: Information Please Almanac

(a) Using a graphing utility, draw a scatter diagram with wind speed as the independent variable.
(b) Using a graphing utility, fit a logarithmic function to the data.
(c) Using a graphing utility, draw the logarithmic function found in part (b) on the scatter diagram.
(d) Use the function found in part (b) to predict the wind chill factor if the air temperature is 15°F and the wind speed is 23 mph.

88. Spreading of a Disease Jack and Diane live in a small town of 50 people. Unfortunately, Jack and Diane both have a cold. Those who come in contact with someone who has this cold will themselves catch the cold. The following data represent the number of people in the small town who have caught the cold after *t* days.

Days, *t*	Number of People with Cold, *C*
0	2
1	4
2	8
3	14
4	22
5	30
6	37
7	42
8	44

(a) Using a graphing utility, draw a scatter diagram of the data. Comment on the type of relation that appears to exist between the days and number of people with a cold.
(b) Using a graphing utility, fit a logistic function to the data.
(c) Graph the function found in part (b) on the scatter diagram.
(d) According to the function found in part (b), what is the maximum number of people who will catch the cold? In reality, what is the maximum number of people who could catch the cold?
(e) Sometime between the second and third day, 10 people in the town had a cold. According to the model found in part (b), when did 10 people have a cold?
(f) How long will it take for 46 people to catch the cold?

Chapter Projects

Year	Value
$t = 0$	
$t = 1$	
$t = 2$	
$t = 3$	
$t = 4$	
$t = 5$	

1. **Hot Coffee**   A fast-food restaurant wants a special container to hold coffee. The restaurant wishes the container to quickly cool the coffee from 200°F to 130°F and keep the liquid between 110° and 130°F as long as possible. The restaurant has three containers to select from.

   (1) The CentiKeeper Company has a container that reduces the temperature of a liquid from 200°F to 100°F in 30 minutes by maintaining a constant temperature of 70°F.

   (2) The TempControl Company has a container that reduces the temperature of a liquid from 200°F to 110°F in 25 minutes by maintaining a constant temperature of 60°F.

   (3) The Hot'n'Cold Company has a container that reduces the temperature of a liquid from 200°F to 120°F in 20 minutes by maintaining a constant temperature of 65°F.

   You need to recommend which container the restaurant should purchase.

   (a) Use Newton's Law of Cooling to find a function relating the temperature of the liquid over time.

   (b) Graph each function using a graphing utility.

   (c) How long does it take each container to lower the coffee temperature from 200°F to 130°F?

   (d) How long will the coffee temperature remain between 110°F and 130°F? This temperature is considered the optimal drinking temperature.

   (e) Which company would you recommend to the restaurant? Why?

   (f) How might the cost of the container affect your decision?

2. **Depreciation of a New Car**   In buying a new car, one consideration might be how well the price of the car holds up over time. Different makes of cars have different depreciation rates. We will discuss one way of computing the depreciation rate for a car. Look up the value of your favorite car for the past 5 years in the NADA ("blue book") available at your local bank, public library, or via the Internet. Let $t = 0$ represent the cost of this car if you purchased it new, $t = 1$ represent the cost of the same car when it is one year old, and so on. Fill in the following table:

   (a) Using a graphing utility, draw a scatter diagram with time as the independent variable and value as the dependent variable.

   (b) Find the exponential function of best fit.

   (c) Express this function in the form $A = A_0 e^{rt}$.

   (d) What is the value of A_0? Compare this value to the purchase price of the vehicle.

   (e) What is the value of r, the depreciation rate?

   (f) Compare your value of r to others in your class. Which car has the lowest rate of depreciation? The highest?

   (g) How might depreciation factor into your decision regarding a new car purchase?

3. **CBL Experiment**   The following data were collected by placing a temperature probe in a cup of hot water, removing the probe, and then recording temperature over time. According to Newton's Law of Cooling, these data should follow an exponential model.

   (a) Using a graphing utility, draw a scatter diagram for the data.

   (b) Using a graphing utility, fit an exponential function to the data.

   (c) Graph the exponential function found in part (b) on the scatter diagram.

   (d) Predict how long it will take for the water to reach a temperature of 110°F.

   (e) Write the model found in part (b) in the form of equation (4) on page 532.

Time	Temperature (F°)	Time	Temperature (F°)
0	175.69	13	148.50
1	173.52	14	146.84
2	171.21	15	145.33
3	169.07	16	143.83
4	166.59	17	142.38
5	164.21	18	141.22
6	161.89	19	140.09
7	159.66	20	138.69
8	157.86	21	137.59
9	155.75	22	136.78
10	153.70	23	135.70
11	151.93	24	134.91
12	150.08	25	133.86

REVIEW

PREPARING FOR THIS BOOK

Before getting started, read the following:

✓ Read the Preface to the Student, page xvi

A.1 ALGEBRA REVIEW

OBJECTIVES

1. Evaluate Algebraic Expressions
2. Determine the Domain of a Variable
3. Graph Inequalities
4. Find Distance on the Real Number Line
5. Use the Laws of Exponents
6. Evaluate Square Roots

Sets

When we want to treat a collection of similar but distinct objects as a whole, we use the idea of a **set.** For example, the set of *digits* consists of the collection of numbers $0, 1, 2, 3, 4, 5, 6, 7, 8,$ and 9. If we use the symbol D to denote the set of digits, then we can write

$$D = \{0, 1, 2, 3, 4, 5, 6, 7, 8, 9\}$$

In this notation, the braces $\{\ \ \}$ are used to enclose the objects, or **elements,** in the set. This method of denoting a set is called the **roster method.** A second way to denote a set is to use **set-builder notation,** where the set D of digits is written as

$$D = \{\quad x \quad | \quad x \text{ is a digit}\}$$

Read as "D is the set of all x such that x is a digit."

555

| EXAMPLE 1 | Using Set-builder Notation and the Roster Method |

(a) $E = \{x|x \text{ is an even digit}\} = \{0, 2, 4, 6, 8\}$

(b) $O = \{x|x \text{ is an odd digit}\} = \{1, 3, 5, 7, 9\}$ ∎

In listing the elements of a set, we do not list an element more than once because the elements of a set are distinct. Also, the order in which the elements are listed is not relevant. For example, $\{2, 3\}$ and $\{3, 2\}$ both represent the same set.

If every element of a set A is also an element of a set B, then we say that A is a **subset** of B. If two sets A and B have the same elements, then we say that A **equals** B. For example, $\{1, 2, 3\}$ is a subset of $\{1, 2, 3, 4, 5\}$; and $\{1, 2, 3\}$ equals $\{2, 3, 1\}$.

Finally, if a set has no elements, it is called the **empty set**, or the **null set**, and it is denoted by the symbol ∅.

Real Numbers

Real numbers are represented by symbols such as

$$25, \quad 0, \quad -3, \quad \frac{1}{2}, \quad -\frac{5}{4}, \quad 0.125, \quad \sqrt{2}, \quad \pi, \quad \sqrt[3]{-2}, \quad 0.666\ldots$$

The set of **counting numbers**, or **natural numbers**, is the set $\{1, 2, 3, 4, \ldots\}$. (The three dots, called an **ellipsis**, indicate that the pattern continues indefinitely.) The set of **integers** is the set $\{\ldots, -3, -2, -1, 0, 1, 2, 3, \ldots\}$. A **rational number** is a number that can be expressed as a *quotient* $\frac{a}{b}$ of two integers, where the integer b cannot be 0. Examples of rational numbers are $\frac{3}{4}, \frac{5}{2}, \frac{0}{4}$, and $-\frac{2}{3}$. Since $\frac{a}{1} = a$ for any integer a, every integer is also a rational number. Real numbers that are not rational are called **irrational**. Examples of irrational numbers are $\sqrt{2}$ and π (the Greek letter pi), which equals the constant ratio of the circumference to the diameter of a circle. See Figure 1.

Real numbers can be represented as **decimals**. Rational real numbers have decimal representations that either **terminate** or are nonterminating with **repeating** blocks of digits. For example, $\frac{3}{4} = 0.75$ which terminates; and $\frac{2}{3} = 0.666\ldots$, in which the digit 6 repeats indefinitely. Irrational real numbers have decimal representations that neither repeat nor terminate. For example, $\sqrt{2} = 1.414213\ldots$ and $\pi = 3.14159\ldots$. In practice, irrational numbers are generally represented by approximations. We use the symbol $\approx$ (read as "approximately equal to") to write $\sqrt{2} \approx 1.4142$ and $\pi \approx 3.1416$.

Figure 1

$$\pi = \frac{c}{d}$$

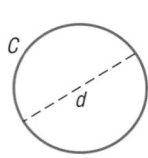

▬▬▬ NOW WORK PROBLEM 3.

Two properties of real numbers that we shall use often are given next. Suppose that a, b, and c are real numbers.

Distributive Property

$$a \cdot (b + c) = ab + ac$$

Zero-Product Property

If $ab = 0$, then either $a = 0$ or $b = 0$ or both equal 0.

The Distributive Property can be used to remove parentheses: $2(x + 3) = 2x + 2\cdot3 = 2x + 6$.

The Zero-Product Property will be used to solve equations (Section A.3).

For example, to solve the equation $2x = 0$, we use the Zero-Product Property as follows: if $2x = 0$, then $2 = 0$ or $x = 0$. Since $2 \neq 0$, it follows that $x = 0$.

Constants and Variables

In algebra we use letters to represent numbers. If the letter used is to represent *any* number from a given set of numbers it is called a **variable.** A **constant** is either a fixed number, such as 5 or $\sqrt{3}$, or a letter that represents a fixed (possibly unspecified) number.

Constants and variables are combined using the operations of addition, subtraction, multiplication, and division to form *algebraic expressions*. Examples of algebraic expressions include

$$x + 3 \qquad \frac{3}{1-t} \qquad 7x - 2y$$

1 To evaluate an algebraic expression, substitute for each variable its numerical value.

EXAMPLE 2   **Evaluating an Algebraic Expression**

Evaluate each expression if $x = 3$ and $y = -1$.

(a) $x + 3y$ (b) $5xy$ (c) $\dfrac{3y}{2 - 2x}$

Solution   (a) Substitute 3 for x and -1 for y in the expression $x + 3y$.

$$x + 3y = 3 + 3(-1) = 3 + (-3) = 0$$
$$\underset{x=3,\,y=-1}{\uparrow}$$

(b) If $x = 3$ and $y = -1$, then

$$5xy = 5(3)(-1) = -15$$

(c) If $x = 3$ and $y = -1$, then

$$\frac{3y}{2 - 2x} = \frac{3(-1)}{2 - 2(3)} = \frac{-3}{2 - 6} = \frac{-3}{-4} = \frac{3}{4}$$

Graphing calculators can be used to evaluate algebraic expressions. Figure 2 shows the results of Example 2 using a TI-83.

Figure 2

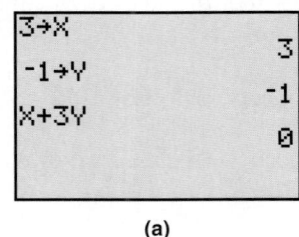
(a)

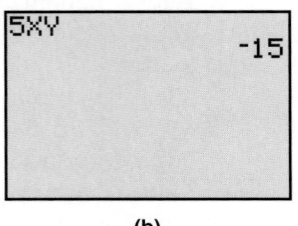
(b)

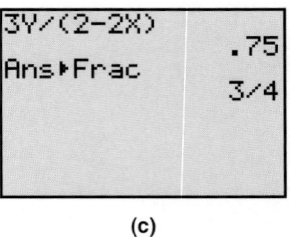
(c)

✏ **NOW WORK PROBLEM 7.**

2   In working with expressions or formulas involving variables, the variables may be allowed to take on values from only a certain set of numbers. For example, in the formula for the area A of a circle of radius r, $A = \pi r^2$, the variable r is necessarily restricted to the positive real numbers. In the expression $\dfrac{1}{x}$, the variable x cannot equal 0, since division by 0 is not defined.

The set of values that a variable in an expression may assume is called the **domain of the variable**.

EXAMPLE 3   Finding the Domain of a Variable

The domain of the variable x in the rational expression

$$\frac{5}{x - 2}$$

is $\{x \mid x \neq 2\}$, since, if $x = 2$, the denominator becomes 0, which is not defined.   ∎

EXAMPLE 4   Circumference of a Circle

In the formula for the circumference C of a circle of radius r,

$$C = 2\pi r$$

the domain of the variable r, representing the radius of the circle, is the set of positive real numbers. The domain of the variable C, representing the circumference of the circle, is also the set of positive real numbers.   ∎

In describing the domain of a variable, we may use either set notation or words, whichever is more convenient.

✏ **NOW WORK PROBLEM 15.**

The Real Number Line

The real numbers can be represented by points on a line called the **real number line**. There is a one-to-one correspondence between real numbers and points on a line. That is, every real number corresponds to a point on the line, and each point on the line has a unique real number associated with it.

Pick a point on the line somewhere in the center, and label it O. This point, called the **origin**, corresponds to the real number 0. See Figure 3. The point 1 unit to the right of O corresponds to the number 1. The distance between 0 and 1 determines the scale of the number line. For example, the point associated with the number 2 is twice as far from O as 1 is. Notice that an arrowhead on the right end of the line indicates the direction in which the numbers increase. Figure 3 also shows the points associated with the irrational numbers $\sqrt{2}$ and π. Points to the left of the origin correspond to the real numbers -1, -2, and so on.

Figure 3
Real number line.

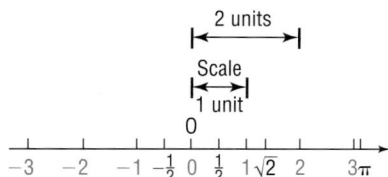

The real number associated with a point P is called the **coordinate** of P, and the line whose points have been assigned coordinates is called the **real number line**.

NOW WORK PROBLEM 27.

The real number line consists of three classes of real numbers, as shown in Figure 4.

Figure 4

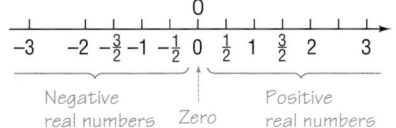

1. The **negative real numbers** are the coordinates of points to the left of the origin O.
2. The real number **zero** is the coordinate of the origin O.
3. The **positive real numbers** are the coordinates of points to the right of the origin O.

Inequalities

③ An important property of the real number line follows from the fact that, given two numbers (points) a and b, either a is to the left of b, a is at the same location as b, or a is to the right of b. See Figure 5.

If a is to the left of b, we say that "a is less than b" and write $a < b$. If a is to the right of b, we say that "a is greater than b" and write $a > b$. If a is at the same location as b, then $a = b$. If a is either less than or equal to b, we write $a \leq b$. Similarly, $a \geq b$ means that a is either greater than or equal to b. Collectively, the symbols $<$, $>$, $\leq$, and $\geq$ are called **inequality symbols**.

Note that $a < b$ and $b > a$ mean the same thing. It does not matter whether we write $2 < 3$ or $3 > 2$.

Furthermore, if $a < b$ or if $b > a$, then the difference $b - a$ is positive. Do you see why?

An **inequality** is a statement in which two expressions are related by an inequality symbol. The expressions are referred to as the **sides** of the inequality. Statements of the form $a < b$ or $b > a$ are called **strict inequalities,** while statements of the form $a \leq b$ or $b \geq a$ are called **nonstrict inequalities**.

Figure 5

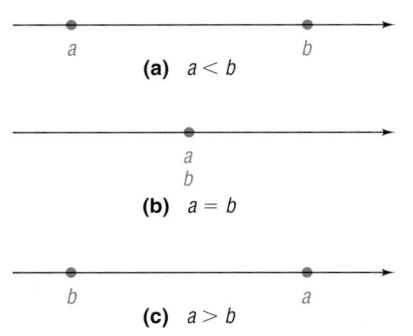

Based on the discussion thus far, we conclude that

$a > 0$ is equivalent to	a is positive
$a < 0$ is equivalent to	a is negative

We sometimes read $a > 0$ by saying that "a is positive." If $a \geq 0$, then either $a > 0$ or $a = 0$, and we may read this as "a is nonnegative."

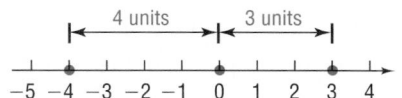 **NOW WORK PROBLEMS 31 AND 41.**

(3) We shall find it useful in later work to graph inequalities on the real number line.

EXAMPLE 5 | **Graphing Inequalities**

(a) On the real number line, graph all numbers x for which $x > 4$.
(b) On the real number line, graph all numbers x for which $x \leq 5$.

Figure 6

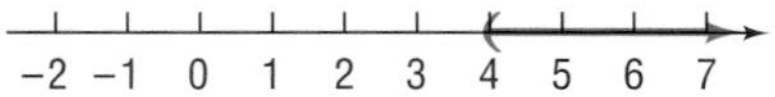

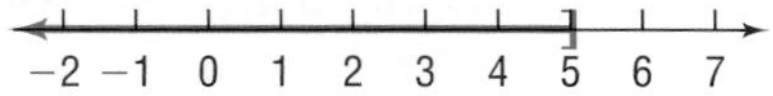

Solution

(a) See Figure 6. Notice that we use a left parenthesis to indicate that the number 4 is not part of the graph.

Figure 7

(b) See Figure 7. Notice that we use a right bracket to indicate that the number 5 is part of the graph. ∎

NOW WORK PROBLEM 47.

Absolute Value

Figure 8

The *absolute value* of a number a is the distance from 0 to a on the number line. For example, -4 is 4 units from 0; and 3 is 3 units from 0. See Figure 8. Thus, the absolute value of -4 is 4, and the absolute value of 3 is 3.

A more formal definition of absolute value is given next.

The **absolute value** of a real number a, denoted by the symbol $|a|$, is defined by the rules

$$|a| = a \quad \text{if } a \geq 0 \qquad \text{and} \qquad |a| = -a \quad \text{if } a < 0$$

For example, since $-4 < 0$, the second rule must be used to get $|-4| = -(-4) = 4$.

EXAMPLE 6 | **Computing Absolute Value**

(a) $|8| = 8$ (b) $|0| = 0$ (c) $|-15| = -(-15) = 15$ ∎

NOW WORK PROBLEM 49.

(4) Look again at Figure 8. The distance from -4 to 3 is 7 units. This distance is the difference $3 - (-4,)$ obtained by subtracting the smaller coordinate from the larger. However, since $|3 - (-4)| = |7| = 7$ and $|-4 - 3| = |-7| = 7$, we can use absolute value to calculate the distance between two points without being concerned about which is smaller.

If P and Q are two points on a real number line with coordinates a and b, respectively, the **distance between P and Q,** denoted by $d(P,Q)$, is

$$d(P,Q) = |b - a|$$

Since $|b - a| = |a - b|$, it follows that $d(P,Q) = d(Q,P)$.

EXAMPLE 7   **Finding Distance on a Number Line**

Let $P, Q,$ and R be points on a real number line with coordinates $-5, 7,$ and -3, respectively. Find the distance

(a) between P and Q (b) between Q and R

Solution   See Figure 9.

Figure 9

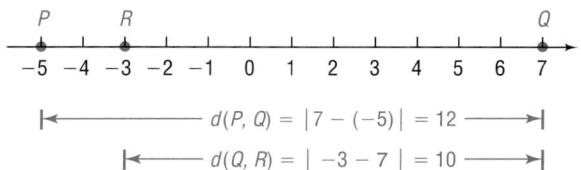

(a) $d(P,Q) = |7 - (-5)| = |12| = 12$
(b) $d(Q,R) = |-3 - 7| = |-10| = 10$

⟍▬▬▬ **NOW WORK PROBLEM 63.**

Exponents

⑤   Integer exponents provide a shorthand device for representing repeated multiplications of a real number.

If a is a real number and n is a positive integer, then the symbol a^n represents the product of n factors of a. That is,

$$a^n = \underbrace{a \cdot a \cdot \ldots \cdot a}_{n \text{ factors}}$$

where it is understood that $a^1 = a$. Then, $a^2 = a \cdot a$, $a^3 = a \cdot a \cdot a$, and so on. In the expression a^n, a is called the **base** and n is called the **exponent,** or **power.** We read a^n as "a raised to the power n" or as "a to the nth power." We usually read a^2 as "a squared" and a^3 as "a cubed."

Care must be taken when parentheses are used in conjunction with exponents. For example, $-2^4 = -(2 \cdot 2 \cdot 2 \cdot 2) = -16$, whereas $(-2)^4 = (-2) \cdot (-2) \cdot (-2) \cdot (-2) = 16$. Notice the difference: The exponent applies only to the number or parenthetical expression immediately preceding it.

If $a \neq 0$, we define

$$a^0 = 1 \qquad \text{if } a \neq 0$$

If $a \neq 0$ and if n is a positive integer, then we define

$$a^{-n} = \frac{1}{a^n} \qquad \text{if } a \neq 0$$

With these definitions, the symbol a^n is defined for any integer n.

The following properties, called the **Laws of Exponents**, can be proved using the preceding definitions. In the list, a and b are real numbers, and m and n are integers.

Laws of Exponents

$$a^m a^n = a^{m+n} \qquad (a^m)^n = a^{mn} \qquad (ab)^n = a^n b^n$$

$$\frac{a^m}{a^n} = a^{m-n} = \frac{1}{a^{n-m}}, \quad \text{if } a \neq 0 \qquad \left(\frac{a}{b}\right)^n = \frac{a^n}{b^n}, \quad \text{if } b \neq 0$$

EXAMPLE 8   **Using the Laws of Exponents**

Write each expression so that all exponents are positive.

(a) $\dfrac{x^5 y^{-2}}{x^3 y}, \quad x \neq 0, y \neq 0$ (b) $\left(\dfrac{x^{-3}}{3y^{-1}}\right)^{-2}, \quad x \neq 0, y \neq 0$

Solution   (a) $\dfrac{x^5 y^{-2}}{x^3 y} = \dfrac{x^5}{x^3} \cdot \dfrac{y^{-2}}{y} = x^{5-3} \cdot y^{-2-1} = x^2 y^{-3} = x^2 \cdot \dfrac{1}{y^3} = \dfrac{x^2}{y^3}$

(b) $\left(\dfrac{x^{-3}}{3y^{-1}}\right)^{-2} = \dfrac{(x^{-3})^{-2}}{(3y^{-1})^{-2}} = \dfrac{x^6}{3^{-2}(y^{-1})^{-2}} = \dfrac{x^6}{\frac{1}{9}y^2} = \dfrac{9x^6}{y^2}$ ∎

NOW WORK PROBLEMS **69** AND **79**.

Square Roots

6   A real number is squared when it is raised to the power 2. The inverse of squaring is finding a **square root**. For example, since $6^2 = 36$ and $(-6)^2 = 36$, the numbers 6 and -6 are square roots of 36.

The symbol $\sqrt{}$, called a **radical sign**, is used to denote the **principal**, or nonnegative, square root. Thus, $\sqrt{36} = 6$.

In general, if a is a nonnegative real number, the nonnegative number b such that $b^2 = a$ is **the principal square root** of a and is denoted by $b = \sqrt{a}$.

The following comments are noteworthy:

1. Negative numbers do not have square roots (in the real number system), because the square of any real number is *nonnegative*. For example, $\sqrt{-4}$ is not a real number, because there is no real number whose square is -4.

2. The principal square root of 0 is 0, since $0^2 = 0$. That is, $\sqrt{0} = 0$.

3. The principal square root of a positive number is positive.

4. If $c \geq 0$, then $(\sqrt{c})^2 = c$. For example, $(\sqrt{2})^2 = 2$ and $(\sqrt{3})^2 = 3$.

EXAMPLE 9   **Evaluating Square Roots**

(a) $\sqrt{64} = 8$ (b) $\sqrt{\dfrac{1}{16}} = \dfrac{1}{4}$

(c) $(\sqrt{1.4})^2 = 1.4$ (d) $\sqrt{(-3)^2} = |-3| = 3$ ■

Examples 9(a) and (b) are examples of square roots of perfect squares, since $64 = 8^2$ and $\dfrac{1}{16} = \left(\dfrac{1}{4}\right)^2$.

Notice the need for the absolute value in Example 9(d). Since $a^2 \geq 0$, the principal square root of a^2 is defined whether $a > 0$ or $a < 0$. However, since the principal square root is nonnegative, we need the absolute value to ensure the nonnegative result.

In general, we have

$$\sqrt{a^2} = |a| \tag{1}$$

EXAMPLE 10   **Using Equation (1)**

(a) $\sqrt{(2.3)^2} = |2.3| = 2.3$ (b) $\sqrt{(-2.3)^2} = |-2.3| = 2.3$

(c) $\sqrt{x^2} = |x|$ ■

✏ **NOW WORK PROBLEM 75.**

A.1 Exercises

In Problems 1–6, list the numbers in each set that are (a) natural numbers, (b) integers, (c) rational numbers, (d) irrational numbers, (e) real numbers.

1. $A = \left\{-6, \dfrac{1}{2}, -1.333\ldots \text{(the 3's repeat)}, \pi, 2, 5\right\}$

2. $B = \left\{-\dfrac{5}{3}, 2.060606\ldots \text{(the block 06 repeats)}, 1.25, 0, 1, \sqrt{5}\right\}$

3. $C = \left\{0, 1, \dfrac{1}{2}, \dfrac{1}{3}, \dfrac{1}{4}\right\}$ **4.** $D = \{-1, -1.1, -1.2, -1.3\}$

5. $E = \left\{\sqrt{2}, \pi, \sqrt{2}+1, \pi + \dfrac{1}{2}\right\}$ **6.** $F = \left\{-\sqrt{2}, \pi + \sqrt{2}, \dfrac{1}{2} + 10.3\right\}$

In Problems 7–14, find the value of each expression if $x = -2$ and $y = 3$. Verify your results using a graphing utility.

7. $x + 2y$ **8.** $3x + y$ **9.** $5xy + 2$ **10.** $-2x + xy$

11. $\dfrac{2x}{x-y}$ **12.** $\dfrac{x+y}{x-y}$ **13.** $\dfrac{3x+2y}{2+y}$ **14.** $\dfrac{2x-3}{y}$

In Problems 15–22, determine which of the value(s) given below, if any, must be excluded from the domain of the variable in each expression.

| (a) $x = 3$ | (b) $x = 1$ | (c) $x = 0$ | (d) $x = -1$ |

15. $\dfrac{x^2 - 1}{x}$ **16.** $\dfrac{x^2 + 1}{x}$ **17.** $\dfrac{x}{x^2 - 9}$ **18.** $\dfrac{x}{x^2 + 9}$

19. $\dfrac{x^2}{x^2 + 1}$ **20.** $\dfrac{x^3}{x^2 - 1}$ **21.** $\dfrac{x^2 + 5x - 10}{x^3 - x}$ **22.** $\dfrac{-9x^2 - x + 1}{x^3 + x}$

In Problems 23–26, determine the domain of the variable x in each expression.

23. $\dfrac{4}{x-5}$ **24.** $\dfrac{-6}{x+4}$ **25.** $\dfrac{x}{x+4}$ **26.** $\dfrac{x-2}{x-6}$

27. On the real number line, label the points with coordinates $0, 1, -1, \dfrac{5}{2}, -2.5, \dfrac{3}{4}$, and 0.25.

28. Repeat Problem 27 for the coordinates $0, -2, 2, -1.5, \dfrac{3}{2}, \dfrac{1}{3}$, and $\dfrac{2}{3}$.

In Problems 29–38, replace the question mark by $<, >,$ or $=$, whichever is correct.

29. $\dfrac{1}{2}\,?\,0$ **30.** $5\,?\,6$ **31.** $-1\,?\,-2$ **32.** $-3\,?\,-\dfrac{5}{2}$ **33.** $\pi\,?\,3.14$

34. $\sqrt{2}\,?\,1.41$ **35.** $\dfrac{1}{2}\,?\,0.5$ **36.** $\dfrac{1}{3}\,?\,0.33$ **37.** $\dfrac{2}{3}\,?\,0.67$ **38.** $\dfrac{1}{4}\,?\,0.25$

In Problems 39–44, write each statement as an inequality.

39. x is positive   **40.** z is negative   **41.** x is less than 2   **42.** y is greater than -5

43. x is less than or equal to 1   **44.** x is greater than or equal to 2

In Problems 45–48, graph the numbers x on the real number line.

45. $x \geq -2$ **46.** $x < 4$ **47.** $x > -1$ **48.** $x \leq 7$

In Problems 49–58, find the value of each expression if $x = 3$ and $y = -2$.

49. $|x+y|$ **50.** $|x-y|$ **51.** $|x|+|y|$ **52.** $|x|-|y|$ **53.** $\dfrac{|x|}{x}$

54. $\dfrac{|y|}{y}$ **55.** $|4x-5y|$ **56.** $|3x+2y|$ **57.** $||4x|-|5y||$ **58.** $3|x|+2|y|$

In Problems 59–64, use the real number line below to compute each distance.

$$
\begin{array}{c}
A \quad\ \ B\ \ C\ \ D\quad\ \ E \\
\overset{\longleftarrow\!\!-\!\!-\!\!-\!\!-\!\!-\!\!-\!\!-\!\!-\!\!-\!\!-\!\!\longrightarrow}{} \\
-4\ \ -3\ \ -2\ \ -1\ \ 0\ \ 1\ \ 2\ \ 3\ \ 4\ \ 5\ \ 6
\end{array}
$$

59. $d(C, D)$ **60.** $d(C, A)$ **61.** $d(D, E)$ **62.** $d(C, E)$ **63.** $d(A, E)$ **64.** $d(D, B)$

In Problems 65–76, simplify each expression.

65. $(-4)^2$ **66.** -4^2 **67.** 4^{-2} **68.** -4^{-2} **69.** $3^{-6} \cdot 3^4$ **70.** $4^{-2} \cdot 4^3$

71. $(3^{-2})^{-1}$ **72.** $(2^{-1})^{-3}$ **73.** $\sqrt{25}$ **74.** $\sqrt{36}$ **75.** $\sqrt{(-4)^2}$ **76.** $\sqrt{(-3)^2}$

In Problems 77–86, simplify each expression. Express the answer so that all exponents are positive. Whenever an exponent is 0 or negative, we assume that the base is not 0.

77. $(8x^3)^2$ **78.** $(-4x^2)^{-1}$ **79.** $(x^2y^{-1})^2$ **80.** $(x^{-1}y)^3$ **81.** $\dfrac{x^2y^3}{xy^4}$

82. $\dfrac{x^{-2}y}{xy^2}$ **83.** $\dfrac{(-2)^3x^4(yz)^2}{3^2xy^3z}$ **84.** $\dfrac{4x^{-2}(yz)^{-1}}{2^3x^4y}$ **85.** $\left(\dfrac{3x^{-1}}{4y^{-1}}\right)^{-2}$ **86.** $\left(\dfrac{5x^{-2}}{6y^{-2}}\right)^{-3}$

In Problems 87–98, express each statement as an equation involving the indicated variables.

87. Area of a Rectangle   The area A of a rectangle is the product of its length l and its width w.

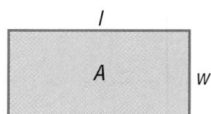

88. Perimeter of a Rectangle   The perimeter P of a rectangle is twice the sum of its length l and its width w.

89. Circumference of a Circle   The circumference C of a circle is the product of π and its diameter d.

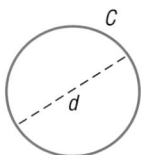

90. Area of a Triangle   The area A of a triangle is one-half the product of its base b and its height h.

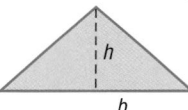

91. Area of an Equilateral Triangle   The area A of an equilateral triangle is $\dfrac{\sqrt{3}}{4}$ times the square of the length x of one side.

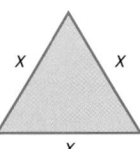

92. Perimeter of an Equilateral Triangle   The perimeter P of an equilateral triangle is 3 times the length x of one side.

93. Volume of a Sphere   The volume V of a sphere is $\dfrac{4}{3}$ times π times the cube of the radius r.

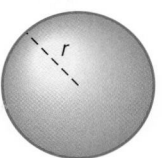

94. Surface Area of a Sphere   The surface area S of a sphere is 4 times π times the square of the radius r.

95. Volume of a Cube   The volume V of a cube is the cube of the length x of a side.

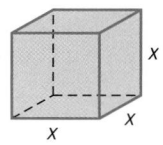

96. Surface Area of a Cube   The surface area S of a cube is 6 times the square of the length x of a side.

97. U.S. Voltage   In the United States, normal household voltage is 115 volts. It is acceptable for the actual voltage x to differ from normal by at most 5 volts. A formula that describes this is

$$|x - 115| \le 5$$

(a) Show that a voltage of 113 volts is acceptable.
(b) Show that a voltage of 109 volts is not acceptable.

98. Foreign Voltage   In other countries, normal household voltage is 220 volts. It is acceptable for the actual voltage x to differ from normal by at most 8 volts. A formula that describes this is

$$|x - 220| \le 8$$

(a) Show that a voltage of 214 volts is acceptable.
(b) Show that a voltage of 209 volts is not acceptable.

A.2   GEOMETRY REVIEW

OBJECTIVES
1 Use the Pythagorean Theorem and Its Converse
2 Know Geometry Formulas

In this section we review some topics studied in geometry that we shall need for our study of algebra.

Figure 10

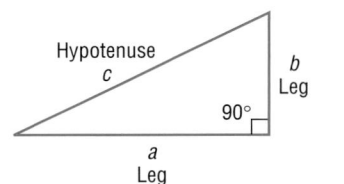

Pythagorean Theorem

1   The *Pythagorean Theorem* is a statement about *right triangles*. A **right triangle** is one that contains a **right angle**, that is, an angle of 90°. The side of the triangle opposite the 90° angle is called the **hypotenuse**; the remaining two sides are called **legs**. In Figure 10 we have used c to represent the length of the hypotenuse and a and b to represent the lengths of the legs. Notice the use of the symbol ⌐ to show the 90° angle. We now state the Pythagorean Theorem.

Pythagorean Theorem

In a right triangle, the square of the length of the hypotenuse is equal to the sum of the squares of the lengths of the legs. That is, in the right triangle shown in Figure 10,

$$c^2 = a^2 + b^2 \tag{1}$$

EXAMPLE 1   **Finding the Hypotenuse of a Right Triangle**

In a right triangle, one leg is of length 4 and the other is of length 3. What is the length of the hypotenuse?

Solution   Since the triangle is a right triangle, we use the Pythagorean Theorem with $a = 4$ and $b = 3$ to find the length c of the hypotenuse. From equation (1), we have

$$c^2 = a^2 + b^2$$
$$c^2 = 4^2 + 3^2 = 16 + 9 = 25$$
$$c = \sqrt{25} = 5$$

 **NOW WORK PROBLEM 3.**

The converse of the Pythagorean Theorem is also true.

Converse of the Pythagorean Theorem

In a triangle, if the square of the length of one side equals the sum of the squares of the lengths of the other two sides, then the triangle is a right triangle. The 90° angle is opposite the longest side.

EXAMPLE 2   **Verifying That a Triangle Is a Right Triangle**

Show that a triangle whose sides are of lengths 5, 12, and 13 is a right triangle. Identify the hypotenuse.

Solution   We square the lengths of the sides.

$$5^2 = 25, \quad 12^2 = 144, \quad 13^2 = 169$$

Figure 11

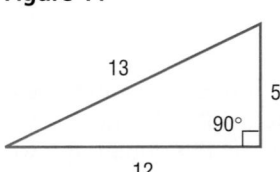

Notice that the sum of the first two squares (25 and 144) equals the third square (169). Hence, the triangle is a right triangle. The longest side, 13, is the hypotenuse. See Figure 11.

 **NOW WORK PROBLEM 11.**

| EXAMPLE 3 | Applying the Pythagorean Theorem |

The tallest inhabited building in the world is the Sears Tower in Chicago.* If the observation tower is 1450 feet above ground level, how far can a person standing in the observation tower see (with the aid of a telescope)? Use 3960 miles for the radius of Earth. See Figure 12.

Figure 12

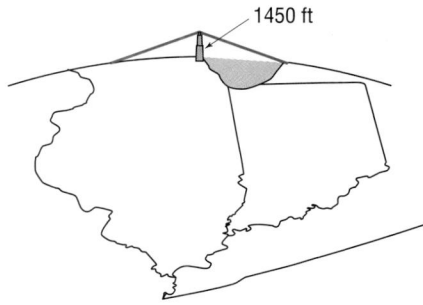
1450 ft

[*Note:* 1 mile = 5280 feet]

Figure 13

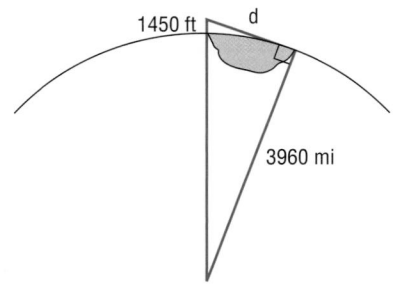
1450 ft d
3960 mi

Solution From the center of Earth, draw two radii: one through the Sears Tower and the other to the farthest point a person can see from the tower. See Figure 13. Apply the Pythagorean Theorem to the right triangle.

Since 1450 feet $= \dfrac{1450}{5280}$ miles, we have

$$d^2 + (3960)^2 = \left(3960 + \frac{1450}{5280}\right)^2$$

$$d^2 = \left(3960 + \frac{1450}{5280}\right)^2 - (3960)^2 \approx 2175.08$$

$$d \approx 46.64$$

A person can see about 47 miles from the observation tower. ■

  NOW WORK PROBLEM 37.

Geometry Formulas

② Certain formulas from geometry are useful in solving algebra problems. We list some of these formulas next.

For a rectangle of length l and width w,

w
l

Area $= lw$	Perimeter $= 2l + 2w$

*Source: Council on Tall Buildings and Urban Habitat (1997): Sears Tower No. 1 for tallest roof (1450 ft) and tallest occupied floor (1431 ft).

For a triangle with base b and altitude h,

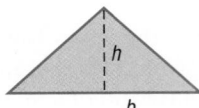

$$\text{Area} = \frac{1}{2}bh$$

For a circle of radius r (diameter $d = 2r$),

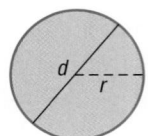

$$\text{Area} = \pi r^2 \qquad \text{Circumference} = 2\pi r = \pi d$$

For a rectangular box of length l, width w, and height h,

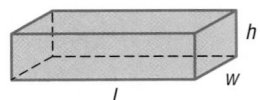

$$\text{Volume} = lwh$$

For a sphere of radius r,

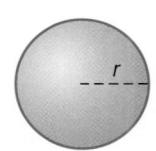

$$\text{Volume} = \frac{4}{3}\pi r^3 \qquad \text{Surface area} = 4\pi r^2$$

For a right circular cylinder of height h and radius r,

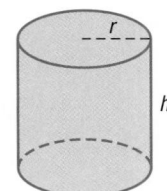

$$\text{Volume} = \pi r^2 h$$

⟍ **NOW WORK PROBLEM 19.**

EXAMPLE 4   **Using Geometry Formulas**

A Christmas tree ornament is in the shape of a semicircle on top of a triangle. How many square centimeters (cm) of copper are required to make the ornament if the height of the triangle is 6 cm and the base is 4 cm?

Solution   See Figure 14. The amount of copper required equals the shaded area. This area is the sum of the area of the triangle and the semicircle. The triangle has height $h = 6$ and base $b = 4$. The semicircle has diameter $d = 4$, so its radius is $r = 2$.

Figure 14

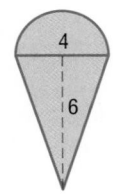

$$\text{Area} = \text{Area of triangle} + \text{Area of semicircle}$$

$$= \frac{1}{2}bh + \frac{1}{2}\pi r^2 = \frac{1}{2}(4)(6) + \frac{1}{2}\pi 2^2 \qquad b = 4; h = 6; r = 2.$$

$$= 12 + 2\pi \approx 18.28 \text{ cm}^2$$

About 18.28 cm^2 of copper are required.   ∎

⟍ **NOW WORK PROBLEM 33.**

A.2 Exercises

In Problems 1–6, the lengths of the legs of a right triangle are given. Find the hypotenuse.

1. $a = 5, b = 12$

2. $a = 6, b = 8$

3. $a = 10, b = 24$

4. $a = 4, b = 3$

5. $a = 7, b = 24$

6. $a = 14, b = 48$

In Problems 7–14, the lengths of the sides of a triangle are given. Determine if the triangle is a right triangle. If it is, identify the hypotenuse.

7. $3, 4, 5$

8. $6, 8, 10$

9. $4, 5, 6$

10. $2, 2, 3$

11. $7, 24, 25$

12. $10, 24, 26$

13. $6, 4, 3$

14. $5, 4, 7$

15. Find the area A of a rectangle with length 4 inches and width 2 inches.

16. Find the area A of a rectangle with length 9 centimeters and width 4 centimeters.

17. Find the area A of a triangle with height 4 inches and base 2 inches.

18. Find the area A of a triangle with height 9 centimeters and base 4 centimeters.

19. Find the area A and circumference C of a circle of radius 5 meters.

20. Find the area A and circumference C of a circle of radius 2 feet.

21. Find the volume V of a rectangular box with length 8 feet, width 4 feet, and height 7 feet.

22. Find the volume V of a rectangular box with length 9 inches, width 4 inches, and height 8 inches.

23. Find the volume V and surface area S of a sphere of radius 4 centimeters.

24. Find the volume V and surface area S of a sphere of radius 3 feet.

25. Find the volume V of a right circular cylinder with radius 9 inches and height 8 inches.

26. Find the volume V of a right circular cylinder with radius 8 inches and height 9 inches.

In Problems 27–30, find the area of the shaded region.

27.
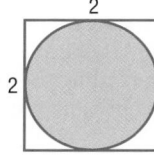

28.
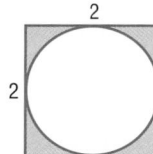

29.
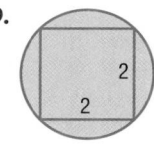

30.
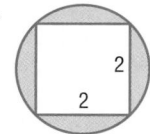

---

31. How many feet does a wheel with a diameter of 16 inches travel after four revolutions?

32. How many revolutions will a circular disk with a diameter of 4 feet have completed after it has rolled 20 feet?

33. In the figure shown, $ABCD$ is a square, with each side of length 6 feet. The width of the border (shaded portion) between the outer square $EFGH$ and $ABCD$ is 2 feet. Find the area of the border.

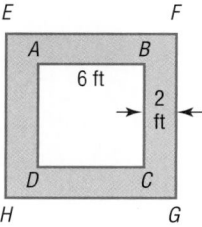

34. Refer to the figure. Square $ABCD$ has an area of 100 square feet; square $BEFG$ has an area of 16 square feet. What is the area of the triangle CGF?

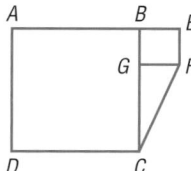

35. Architecture A Norman window consists of a rectangle surmounted by a semicircle. Find the area of the Norman window shown in the illustration. How much wood frame is needed to enclose the window?

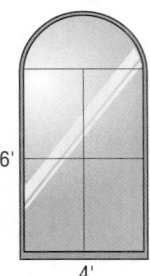

6'

4'

36. Construction A circular swimming pool, 20 feet in diameter, is enclosed by a wooden deck that is 3 feet wide. What is the area of the deck? How much fence is required to enclose the deck?

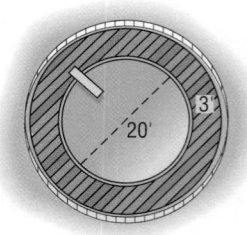

3' 20'

In Problems 37–39, use the facts that the radius of Earth is 3960 miles and 1 mile = 5280 feet.

37. How Far Can You See? The conning tower of the U.S.S. *Silversides*, a World War II submarine now permanently stationed in Muskegon, Michigan, is approximately 20 feet above sea level. How far can you see from the conning tower?

38. How Far Can You See? A person who is 6 feet tall is standing on the beach in Fort Lauderdale, Florida, and looks out onto the Atlantic Ocean. Suddenly, a ship appears on the horizon. How far is the ship from shore?

39. How Far Can You See? The deck of a destroyer is 100 feet above sea level. How far can a person see from the deck? How far can a person see from the bridge, which is 150 feet above sea level?

40. Suppose that m and n are positive integers with $m > n$. If $a = m^2 - n^2$, $b = 2mn$, and $c = m^2 + n^2$, show that $a, b,$ and c are the lengths of the sides of a right triangle. (This formula can be used to find the sides of a right triangle that are integers, such as 3, 4, 5; 5, 12, 13; and so on. Such triplets of integers are called **Pythagorean triples**.)

41. You have 1000 feet of flexible pool siding and wish to construct a swimming pool. Experiment with rectangular-shaped pools with perimeters of 1000 feet. How do their areas vary? What is the shape of the rectangle with the largest area? Now compute the area enclosed by a circular pool with a perimeter (circumference) of 1000 feet. What would be your choice of shape for the pool? If rectangular, what is your preference for dimensions? Justify your choice. If your only consideration is to have a pool that encloses the most area, what shape should you use?

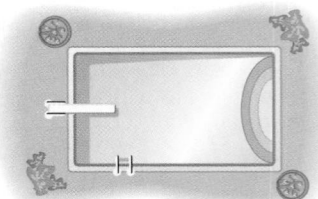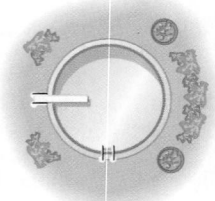

42. The Gibb's Hill Lighthouse, Southampton, Bermuda, in operation since 1846, stands 117 feet high on a hill 245 feet high, so its beam of light is 362 feet above sea level. A brochure states that the light itself can be seen on the horizon about 26 miles distant. Verify the correctness of this information. The brochure further states that ships 40 miles away can see the light and planes flying at 10,000 feet can see it 120 miles away. Verify the accuracy of these statements. What assumption did the brochure make about the height of the ship?

120 miles

40 miles

PREPARING FOR THIS SECTION

Before getting started, review the following:

✓ Domain of a Variable (Section A.1, p. 558)   ✓ Square Roots (Section A.1, pp. 562–563)

✓ Zero-Product Property (Section A.1, p. 557)

A.3 SOLVING EQUATIONS

OBJECTIVES
1. Solve Equations Using a Graphing Utility
2. Solve Equations Algebraically; Equivalent Equations
3. Solve Linear Equations
4. Solve Quadratic Equations by Factoring
5. Solve Equations Using the Square Root Method
6. Solve Quadratic Equations by Completing the Square
7. Obtain the Quadratic Formula
8. Solve Quadratic Equations Using the Quadratic Formula

An **equation in one variable** is a statement in which two expressions, at least one containing the variable, are equal. The expressions are called the **sides** of the equation. Since an equation is a statement, it may be true or false, depending on the value of the variable. Unless otherwise restricted, the admissible values of the variable are those in the domain of the variable. Those admissible values of the variable, if any, that result in a true statement are called **solutions**, or **roots**, of the equation. To **solve an equation** means to find all the solutions of the equation.

For example, the following are all equations in one variable, x:

$$x + 5 = 9 \qquad x^2 + 5x = 2x - 2 \qquad \frac{x^2 - 4}{x + 1} = 0 \qquad x^2 + 9 = 5$$

The first of these statements, $x + 5 = 9$, is true when $x = 4$ and false for any other choice of x. We say that 4 is a solution of the equation $x + 5 = 9$ or that 4 **satisfies** the equation $x + 5 = 9$, because, when we substitute 4 for x, a true statement results.

Sometimes an equation will have more than one solution. For example, the equation

$$\frac{x^2 - 4}{x + 1} = 0$$

has $x = -2$ and $x = 2$ as solutions.

Usually, we will write the solution of an equation in set notation. This set is called the **solution set** of the equation. For example, the solution set of the equation $x^2 - 9 = 0$ is $\{-3, 3\}$.

Unless indicated otherwise, we will limit ourselves to real solutions, that is, solutions that are real numbers. Some equations have no real solution. For example, $x^2 + 9 = 5$ has no real solution, because there is no real number whose square when added to 9 equals 5.

An equation that is satisfied for every choice of the variable for which both sides are defined is called an **identity**. For example, the equation

$$3x + 5 = x + 3 + 2x + 2$$

is an identity, because this statement is true for any real number x.

Solving Equations Using a Graphing Utility

1 In this text, we present two methods for solving equations: algebraic and graphical. We shall see as we proceed through this book that some equations can be solved using algebraic techniques that result in *exact* solutions. For other equations, however, there are no algebraic techniques that lead to an exact solution. For such equations, a graphing utility can often be used to investigate possible solutions. When a graphing utility is used to solve an equation, usually *approximate* solutions are obtained.

One goal of this text is to determine when equations can be solved algebraically. If an algebraic method for solving an equation exists, we shall use it to obtain an exact solution. A graphing utility can then be used to support the algebraic result. However, if an equation must be solved for which no algebraic techniques are available, a graphing utility will be used to obtain approximate solutions. Unless otherwise stated, we shall follow the practice of giving approximate solutions as decimals *rounded to two decimal places*.

The ZERO (or ROOT) feature of a graphing utility can be used to find the solutions of an equation when one side of the equation is 0. In using this feature to solve equations, we make use of the fact that the x-intercepts (or zeros) of the graph of an equation are found by letting $y = 0$ and solving the equation for x. Solving an equation for x when one side of the equation is 0 is equivalent to finding where the graph of the corresponding equation crosses or touches the x-axis.

EXAMPLE 1   **Using ZERO (or ROOT) to Approximate Solutions of an Equation**

Find the solution(s) of the equation $x^3 - x + 1 = 0$. Round answers to two decimal places.

Solution   The solutions of the equation $x^3 - x + 1 = 0$ are the same as the x-intercepts of the graph of $Y_1 = x^3 - x + 1$. We begin by graphing the equation.

Figure 15 shows the graph. From the graph there appears to be one x-intercept (solution to the equation) between -2 and -1.

Using the ZERO (or ROOT) feature of our graphing utility, we determine that the x-intercept, and thus the solution to the equation, is $x = -1.32$ rounded to two decimal places. See Figure 16.

Figure 15   **Figure 16**

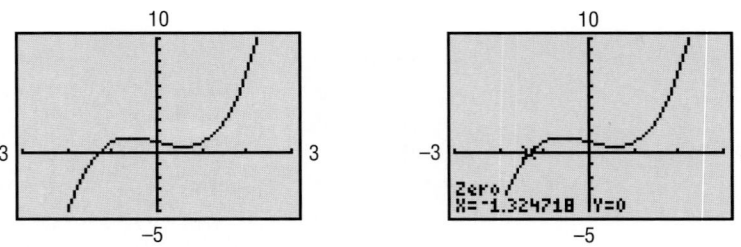

NOW WORK PROBLEM **1**.

A second method for solving equations using a graphing utility involves the INTERSECT feature of the graphing utility. This feature is used most effectively when one side of the equation is not 0.

EXAMPLE 2   Using INTERSECT to Approximate Solutions of an Equation

Find the solution(s) to the equation $4x^4 - 3 = 2x + 1$. Round answers to two decimal places.

Solution   We begin by graphing each side of the equation as follows: graph $Y_1 = 4x^4 - 3$ and $Y_2 = 2x + 1$. See Figure 17. At the point of intersection of the graphs, the value of the y-coordinate is the same. Thus, the x-coordinate of the point of intersection represents the solution to the equation. Do you see why?

The INTERSECT feature on a graphing utility determines the point of intersection of the graphs. Using this feature, we find that the graphs intersect at $(-0.87, -0.73)$ and $(1.12, 3.23)$ rounded to two decimal places. See Figure 18(a) and (b). The solutions of the equation are $x = -0.87$ and $x = 1.12$, rounded to two decimal places.

Figure 17

Figure 18

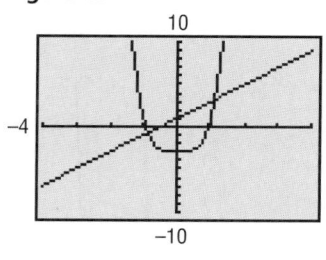

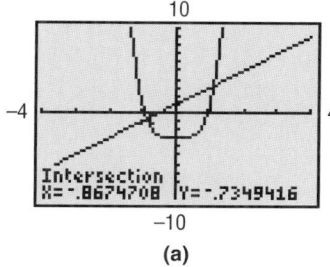

(a)   (b)

NOW WORK PROBLEM **3**.

The steps to follow for approximating solutions of equations are given next.

Steps for Approximating Solutions of Equations Using ZERO (or ROOT)

STEP 1:   Write the equation in the form {expression in x} $= 0$.

STEP 2:   Graph $Y_1 =$ {expression in x}.

STEP 3:   Use ZERO (or ROOT) to determine each x-intercept of the graph.

Steps for Approximating Solutions of Equations Using INTERSECT

STEP 1:   Graph $Y_1 =$ {expression in x on left side of equation};
$Y_2 =$ {expression in x on right side of equation}.

STEP 2:   Use INTERSECT to determine the x-coordinate of each point of intersection.

We now discuss equations that can be solved algebraically to obtain exact solutions. A graphing utility will be used to verify the solution.

Solving Equations Algebraically

2   Two or more equations that have precisely the same solutions are called **equivalent equations**.

Often equations can be solved by finding simpler equivalent equations. For example, all the following equations are equivalent, because each has only the solution $x = 5$:

$$2x + 3 = 13$$
$$2x = 10$$
$$x = 5$$

These three equations illustrate one method for solving many types of equations: Replace the original equation by an equivalent equation, and continue until an equation with an obvious solution, such as $x = 5$, is reached.

The question, though, is "How do I obtain an equivalent equation?" In general, there are five ways to do so.

Procedures That Result in Equivalent Equations

1. Interchange the two sides of the equation:

   Replace $\qquad$ $3 = x$ $\quad$ by $\quad$ $x = 3$

2. Simplify the sides of the equation by combining like terms, eliminating parentheses, and so on:

   Replace $\qquad$ $x + 2 + 6 = 2x + 3(x + 1)$

   by $\qquad$ $x + 8 = 5x + 3$

3. Add or subtract the same expression on both sides of the equation:

   Replace $\qquad$ $3x - 5 = 4$

   by $\qquad$ $(3x - 5) + 5 = 4 + 5$

4. Multiply or divide both sides of the equation by the same nonzero expression:

   Replace $\qquad$ $\dfrac{3x}{x - 1} = \dfrac{6}{x - 1}$ $\qquad$ $x \neq 1$

   by $\qquad$ $\dfrac{3x}{x - 1} \cdot (x - 1) = \dfrac{6}{x - 1} \cdot (x - 1)$

5. If one side of the equation is 0 and the other side can be factored, then we may use the Zero-Product Property and set each factor equal to 0:

   Replace $\qquad$ $x(x - 3) = 0$

   by $\qquad$ $x = 0$ $\quad$ or $\quad$ $x - 3 = 0$

WARNING: Squaring both sides of an equation does not necessarily lead to an equivalent equation. ∎

Whenever it is possible to solve an equation in your head, do so. For example:

The solution of $2x = 8$ is $x = 4$.
The solution of $3x - 15 = 0$ is $x = 5$.

In the next examples, we use the Zero-Product Property (procedure 5, listed above in the box).

EXAMPLE 3   **Solving Equations by Factoring**

Solve the equations: (a) $x^2 = 4x$ (b) $x^3 - x^2 - 4x + 4 = 0$

Solution   (a) We begin by collecting all terms on one side. This results in 0 on one side and an expression to be factored on the other.

$$x^2 = 4x$$
$$x^2 - 4x = 0$$
$$x(x - 4) = 0 \qquad \text{Factor.}$$
$$x = 0 \quad \text{or} \quad x - 4 = 0 \qquad \text{Apply the Zero-Product property.}$$
$$x = 4$$

The solution set is $\{0, 4\}$.

✔ CHECK: $x = 0:$ $0^2 = 4 \cdot 0$ So 0 is a solution.
   $x = 4:$ $4^2 = 4 \cdot 4$ So 4 is a solution.   ■

(b) Do you recall the method of factoring by grouping? We group the terms of $x^3 - x^2 - 4x + 4 = 0$ as follows:

$$(x^3 - x^2) - (4x - 4) = 0$$

Factor out x^2 from the first grouping and 4 from the second.

$$x^2(x - 1) - 4(x - 1) = 0$$

This reveals the common factor $(x - 1)$, so we have

$$(x^2 - 4)(x - 1) = 0$$
$$(x - 2)(x + 2)(x - 1) = 0 \qquad \text{Factor again.}$$
$$x - 2 = 0 \qquad \text{or} \qquad x + 2 = 0 \qquad \text{or} \qquad x - 1 = 0 \qquad \text{Set each factor equal to 0.}$$
$$x = 2 \qquad\qquad x = -2 \qquad\qquad x = 1 \qquad \text{Solve.}$$

The solution set is $\{-2, 1, 2\}$.

✔ CHECK: $x = -2:$ $(-2)^3 - (-2)^2 - 4(-2) + 4 = -8 - 4 + 8 + 4 = 0$ −2 is a solution.
   $x = 1:$ $1^3 - 1^2 - 4(1) + 4 = 1 - 1 - 4 + 4 = 0$ 1 is a solution.
   $x = 2:$ $2^3 - 2^2 - 4(2) + 4 = 8 - 4 - 8 + 4 = 0$ 2 is a solution.   ■ ■

Steps for Solving Equations Algebraically

STEP 1:   List any restrictions on the domain of the variable.

STEP 2:   Simplify the equation by replacing the original equation by a succession of equivalent equations following the procedures listed on page 574.

STEP 3:   If the result of Step 2 is a product of factors equal to 0, use the Zero-Product Property and set each factor equal to 0 (procedure 5).

STEP 4:   Check your solution(s).

NOW WORK PROBLEMS **31** AND **33**.

Linear Equations

3 *Linear equations* are equations such as

$$3x + 12 = 0, \qquad \frac{3}{4}x - \frac{1}{5} = 0, \qquad 0.62x - 0.3 = 0$$

> A **linear equation in one variable** is equivalent to an equation of the form
>
> $$ax + b = 0$$
>
> where a and b are real numbers and $a \neq 0$.

Sometimes a linear equation is called a **first-degree equation**, because the left side is a polynomial in x of degree 1.

EXAMPLE 4   Solving a Linear Equation

Solve the equation: $3(x - 2) = 5(x - 1)$

Algebraic Solution

$$\begin{aligned}
3(x - 2) &= 5(x - 1) \\
3x - 6 &= 5x - 5 && \text{Use the Distributive Property.} \\
3x - 6 - 5x &= 5x - 5 - 5x && \text{Subtract 5x from each side.} \\
-2x - 6 &= -5 && \text{Simplify.} \\
-2x - 6 + 6 &= -5 + 6 && \text{Add 6 to each side.} \\
-2x &= 1 && \text{Simplify.} \\
\frac{-2x}{-2} &= \frac{1}{-2} && \text{Divide each side by } -2. \\
x &= -\frac{1}{2} && \text{Simplify.}
\end{aligned}$$

✔ CHECK: Let $x = -\dfrac{1}{2}$ in the expression in x on the left side of the equation and simplify. Let $x = -\dfrac{1}{2}$ in the expression in x on the right side of the equation and simplify. If the two expressions are equal, the solution checks.

$$3(x - 2) = 3\left(-\frac{1}{2} - 2 \right) = 3\left(-\frac{5}{2} \right) = -\frac{15}{2}$$

$$5(x - 1) = 5\left(-\frac{1}{2} - 1 \right) = 5\left(-\frac{3}{2} \right) = -\frac{15}{2}$$

Since the two expressions are equal, the solution $x = -\dfrac{1}{2}$ checks. ∎

Graphing Solution Graph $Y_1 = 3(x - 2)$ and $Y_2 = 5(x - 1)$. See Figure 19. Using INTERSECT, we find the point of intersection to be $(-0.5, -7.5)$. The solution of the equation is $x = -0.5$. ∎

Figure 19

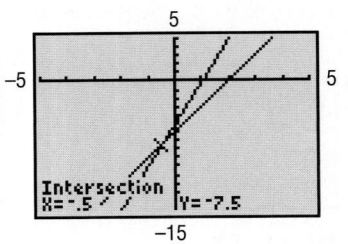

- NOW WORK PROBLEM **15.**

Quadratic Equations

Quadratic equations are equations such as

$$2x^2 + x + 8 = 0, \qquad 3x^2 - 5x = 0, \qquad x^2 - 9 = 0$$

A general definition is given next.

A **quadratic equation** is an equation equivalent to one of the form

$$ax^2 + bx + c = 0 \qquad (1)$$

where a, b, and c are real numbers and $a \neq 0$.

A quadratic equation written in the form $ax^2 + bx + c = 0$ is said to be in **standard form**.

Sometimes, a quadratic equation is called a **second-degree equation**, because the left side is a polynomial of degree 2. We discuss here three algebraic ways of solving quadratic equations: by factoring, by completing the square, and by using the quadratic formula.

④   When a quadratic equation is written in standard form, $ax^2 + bx + c = 0$, it may be possible to factor the expression on the left side as the product of two first-degree polynomials. Then, by setting each factor equal to 0 and solving the resulting linear equations, we obtain the *exact* solutions of the quadratic equation.

Let's look at an example.

EXAMPLE 5   **Solving a Quadratic Equation by Factoring and by Graphing**

Solve the equation: $x^2 = 12 - x$

Algebraic Solution   We put the equation in standard form by adding $x - 12$ to each side:

$$x^2 = 12 - x$$
$$x^2 + x - 12 = 0$$

The left side of the equation may now be factored as

$$(x + 4)(x - 3) = 0$$

so that

$$x + 4 = 0 \qquad \text{or} \qquad x - 3 = 0$$
$$x = -4 \qquad\qquad\qquad x = 3$$

The solution set is $\{-4, 3\}$.

Graphing Solution   Graph $Y_1 = x^2$ and $Y_2 = 12 - x$. See Figure 20(a). From the graph it appears that there are two points of intersection: one near -4; the other near 3. Using INTERSECT, the points of intersection are $(-4, 16)$ and $(3, 9)$, so the solutions of the equation are $x = -4$ and $x = 3$. See Figure 20(b) and (c).

Figure 20

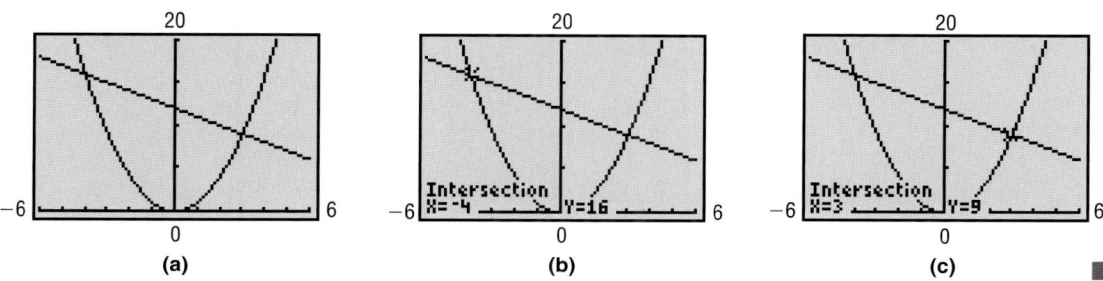

(a)   (b)   (c)

NOW WORK PROBLEMS 19 AND 63.

The Square Root Method

⑤ Suppose that we wish to solve the quadratic equation

$$x^2 = p \qquad (2)$$

where $p \geq 0$ is a nonnegative number. We proceed by factoring the expression $x^2 - p$ and using the Zero-Product Property:

$$x^2 - p = 0 \qquad \text{Put in standard form.}$$
$$(x - \sqrt{p})(x + \sqrt{p}) = 0 \qquad \text{Factor (over the real numbers).}$$
$$x = \sqrt{p} \quad \text{or} \quad x = -\sqrt{p} \qquad \text{Solve.}$$

We have the following result:

$$\text{If } x^2 = p \text{ and } p \geq 0, \text{ then } x = \sqrt{p} \text{ or } x = -\sqrt{p}. \qquad (3)$$

When statement (3) is used, it is called the **Square Root Method**. In statement (3), note that if $p > 0$ the equation $x^2 = p$ has two solutions, $x = \sqrt{p}$ and $x = -\sqrt{p}$. We usually abbreviate these solutions as $x = \pm\sqrt{p}$, read "x equals plus or minus the square root of p." For example, the two solutions of the equation

$$x^2 = 4$$

are

$$x = \pm\sqrt{4}$$

and since $\sqrt{4} = 2$, we have

$$x = \pm 2$$

The solution set is $\{-2, 2\}$.

EXAMPLE 6   **Solving Quadratic Equations by Using the Square Root Method**

Solve each equation.

(a) $x^2 = 5$ (b) $(x - 2)^2 = 16$

Solution (a) $x^2 = 5$

$$x = \pm\sqrt{5} \qquad \text{Use the Square Root Method}$$
$$x = \sqrt{5} \quad \text{or} \quad x = -\sqrt{5}$$

The solution set is $\{-\sqrt{5}, \sqrt{5}\}$.

(b) $(x - 2)^2 = 16$

$$x - 2 = \pm\sqrt{16} \qquad \text{Use the Square Root Method}$$
$$x - 2 = \sqrt{16} \quad \text{or} \quad x - 2 = -\sqrt{16}$$
$$x - 2 = 4 \qquad\qquad\quad x - 2 = -4$$
$$x = 6 \qquad\qquad\qquad x = -2$$

The solution set is $\{-2, 6\}$. ∎

━━━ **NOW WORK PROBLEM 39.**

Completing the Square

We now introduce the method of **completing the square**. The idea behind this method is to "adjust" the left side of a quadratic equation, $ax^2 + bx + c = 0$, so that it becomes a perfect square, that is, the square of a first-degree polynomial. For example, $x^2 + 6x + 9$ and $x^2 - 4x + 4$ are perfect squares because

$$x^2 + 6x + 9 = (x + 3)^2 \quad \text{and} \quad x^2 - 4x + 4 = (x - 2)^2$$

How do we "adjust" the left side? We do it by adding the appropriate number to create a perfect square. For example, to make $x^2 + 6x$ a perfect square, we add 9.

Let's look at several examples of completing the square when the coefficient of x^2 is 1.

Start	Add	Result
$x^2 + 4x$	4	$x^2 + 4x + 4 = (x + 2)^2$
$x^2 + 12x$	36	$x^2 + 12x + 36 = (x + 6)^2$
$x^2 - 6x$	9	$x^2 - 6x + 9 = (x - 3)^2$
$x^2 + x$	$\dfrac{1}{4}$	$x^2 + x + \dfrac{1}{4} = \left(x + \dfrac{1}{2}\right)^2$

Do you see the pattern? Provided that the coefficient of x^2 is 1, we complete the square by adding the square of one-half the coefficient of x.

Start	Add	Result
$x^2 + mx$	$\left(\dfrac{m}{2}\right)^2$	$x^2 + mx + \left(\dfrac{m}{2}\right)^2 = \left(x + \dfrac{m}{2}\right)^2$

 **NOW WORK PROBLEM 43.**

⑥ The next example illustrates how the procedure of completing the square can be used to solve a quadratic equation.

EXAMPLE 7

Solving a Quadratic Equation by Completing the Square

Solve by completing the square: $x^2 + 5x + 4 = 0$

Solution We always begin this procedure by rearranging the equation so that the constant is on the right side.

$$x^2 + 5x + 4 = 0$$
$$x^2 + 5x = -4$$

Since the coefficient of x^2 is 1, we can complete the square on the left side by adding $\left(\dfrac{1}{2}\cdot 5\right)^2 = \dfrac{25}{4}$. Of course, in an equation, whatever we add to the left side must also be added to the right side. We add $\dfrac{25}{4}$ to *both* sides.

$$x^2 + 5x + \frac{25}{4} = -4 + \frac{25}{4} \qquad \text{Add } \frac{25}{4} \text{ to both sides.}$$

$$\left(x + \frac{5}{2}\right)^2 = \frac{9}{4} \qquad \text{Factor; simplify.}$$

$$x + \frac{5}{2} = \pm\sqrt{\frac{9}{4}} \qquad \text{Use the Square Root Method.}$$

$$x + \frac{5}{2} = \pm\frac{3}{2}$$

$$x = -\frac{5}{2} \pm \frac{3}{2}$$

$$x = -\frac{5}{2} + \frac{3}{2} = -1 \quad \text{or} \quad x = -\frac{5}{2} - \frac{3}{2} = -4$$

The solution set is $\{-4, -1\}$. ∎

NOW WORK PROBLEM **47**.

The Quadratic Formula

⑦ We can use the method of completing the square to obtain a general formula for solving the quadratic equation

$$ax^2 + bx + c = 0, \qquad a \neq 0$$

Note: There is no loss in generality to assume that $a > 0$, since if $a < 0$ we can multiply both sides by -1 to obtain an equivalent equation with a positive leading coefficient.

As in Example 7, we begin by rearranging the terms as

$$ax^2 + bx = -c, \qquad a > 0$$

Since $a > 0$, we can divide both sides by a to get

$$x^2 + \frac{b}{a}x = -\frac{c}{a}$$

Now the coefficient of x^2 is 1. To complete the square on the left side, add the square of $\dfrac{1}{2}$ the coefficient of x; that is, add

$$\left(\frac{1}{2}\cdot\frac{b}{a}\right)^2 = \frac{b^2}{4a^2}$$

to each side. Then

$$x^2 + \frac{b}{a}x + \frac{b^2}{4a^2} = \frac{b^2}{4a^2} - \frac{c}{a}$$

$$\left(x + \frac{b}{2a}\right)^2 = \frac{b^2 - 4ac}{4a^2} \qquad \frac{b^2}{4a^2} - \frac{c}{a} = \frac{b^2}{4a^2} - \frac{4ac}{4a^2} = \frac{b^2 - 4ac}{4a^2} \qquad (4)$$

Provided that $b^2 - 4ac \geq 0$, we now can use the Square Root Method to get

$$x + \frac{b}{2a} = \pm\sqrt{\frac{b^2 - 4ac}{4a^2}}$$

$$x + \frac{b}{2a} = \frac{\pm\sqrt{b^2 - 4ac}}{2a} \qquad a > 0 \text{ so } \sqrt{4a^2} = 2a$$

$$x = -\frac{b}{2a} \pm \frac{\sqrt{b^2 - 4ac}}{2a} \qquad \text{Add } -\frac{b}{2a} \text{ to both sides.}$$

$$x = \frac{-b \pm \sqrt{b^2 - 4ac}}{2a} \qquad \text{Combine the quotients on the right.}$$

What if $b^2 - 4ac$ is negative? Then equation (4) states that the left expression (a real number squared) equals the right expression (a negative number). Since this occurrence is impossible for real numbers, we conclude that if $b^2 - 4ac < 0$ the quadratic equation has no *real* solution.*

We now state the *quadratic formula.*

Theorem

Quadratic Formula

Consider the quadratic equation

$$ax^2 + bx + c = 0 \qquad a \neq 0$$

If $b^2 - 4ac < 0$, this equation has no real solution.
If $b^2 - 4ac \geq 0$, the real solution(s) of this equation is (are) given by the **quadratic formula**:

$$x = \frac{-b \pm \sqrt{b^2 - 4ac}}{2a}$$

The quantity $b^2 - 4ac$ is called the **discriminant** of the quadratic equation, because its value tells us whether the equation has real solutions. In fact, it also tells us how many solutions to expect.

Discriminant of a Quadratic Equation

For a quadratic equation $ax^2 + bx + c = 0$:

1. If $b^2 - 4ac > 0$, there are two unequal real solutions.
2. If $b^2 - 4ac = 0$, there is a repeated real solution, a double root.
3. If $b^2 - 4ac < 0$, there is no real solution.

When asked to find the real solutions, if any, of a quadratic equation, always evaluate the discriminant first to see how many real solutions there are.

---

*We consider quadratic equations with $b^2 - 4ac < 0$ in Section A.4.

8   **EXAMPLE 8**   Solving a Quadratic Equation by Using the Quadratic Formula and by Graphing

Find the real solutions, if any, of the equation $3x^2 - 5x + 1 = 0$.

Algebraic Solution   The equation is in standard form, so we compare it to $ax^2 + bx + c = 0$ to find a, b, and c.

$$3x^2 - 5x + 1 = 0$$
$$ax^2 + bx + c = 0, \qquad a = 3, b = -5, c = 1$$

With $a = 3$, $b = -5$, and $c = 1$, we evaluate the discriminant $b^2 - 4ac$.

$$b^2 - 4ac = (-5)^2 - 4(3)(1) = 25 - 12 = 13$$

Since $b^2 - 4ac > 0$, there are two unequal real solutions.
We use the quadratic formula with $a = 3, b = -5, c = 1$, and $b^2 - 4ac = 13$.

$$x = \frac{-b \pm \sqrt{b^2 - 4ac}}{2a} = \frac{-(-5) \pm \sqrt{13}}{2(3)} = \frac{5 \pm \sqrt{13}}{6}$$

The solution set is $\left\{ \dfrac{5 - \sqrt{13}}{6}, \dfrac{5 + \sqrt{13}}{6} \right\}$. These solutions are exact.

Graphing Solution   Figure 21 shows the graph of the equation $Y_1 = 3x^2 - 5x + 1$.

Figure 21

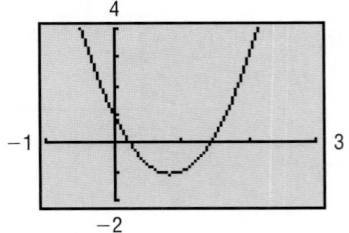

As expected, we see that there are two x-intercepts: one between 0 and 1, the other between 1 and 2. The solutions to the equation are 0.23 and 1.43, rounded to two decimal places. These solutions are approximate. ■

━ **NOW WORK PROBLEM 73.**

EXAMPLE 9   Solving a Quadratic Equation by Using the Quadratic Formula and by Graphing

Find the real solutions, if any, of the equation

$$3x^2 + 2 = 4x$$

Algebraic Solution   The equation, as given, is not in standard form.

$$3x^2 + 2 = 4x$$
$$3x^2 - 4x + 2 = 0 \qquad \text{Put in standard form.}$$
$$ax^2 + bx + c = 0 \qquad \text{Compare to standard form.}$$

With $a = 3, b = -4$, and $c = 2$, we find that

$$b^2 - 4ac = (-4)^2 - 4(3)(2) = 16 - 24 = -8$$

Since $b^2 - 4ac < 0$, the equation has no real solution.

Graphing Solution   We use the standard form of the equation and graph

$$Y_1 = 3x^2 - 4x + 2$$

See Figure 22. We see that there are no x-intercepts, so the equation has no real solution, as expected.

Figure 22

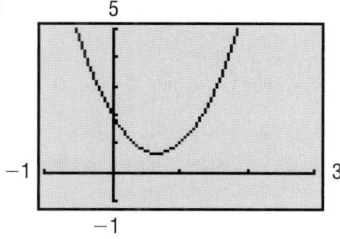

NOW WORK PROBLEM 75.

A.3 Exercises

In Problems 1–12, use a graphing utility to approximate the real solutions, if any, of each equation rounded to two decimal places. All solutions lie between -10 and 10.

1. $x^3 - 4x + 2 = 0$
2. $x^3 - 8x + 1 = 0$
3. $-2x^4 + 5 = 3x - 2$
4. $-x^4 + 1 = 2x^2 - 3$
5. $x^4 - 2x^3 + 3x - 1 = 0$
6. $3x^4 - x^3 + 4x^2 - 5 = 0$
7. $-x^3 - \dfrac{5}{3}x^2 + \dfrac{7}{2}x + 2 = 0$
8. $-x^4 + 3x^3 + \dfrac{7}{3}x^2 - \dfrac{15}{2}x + 2 = 0$
9. $-\dfrac{2}{3}x^4 - 2x^3 + \dfrac{5}{2}x = -\dfrac{2}{3}x^2 + \dfrac{1}{2}$
10. $\dfrac{1}{4}x^3 - 5x = \dfrac{1}{5}x^2 - 4$
11. $x^4 - 5x^2 + 2x + 11 = 0$
12. $-3x^4 + 8x^2 - 2x - 9 = 0$

In Problems 13–18, solve each equation algebraically. Verify your solution using a graphing utility.

13. $3x + 2 = 6$
14. $2x + 7 = 5$
15. $6 - x = 2x + 12$
16. $3 - 2x = 8 - x$
17. $2(3 + 2x) = 3(x - 4)$
18. $3(2 - x) = 2x - 1$

In Problems 19–36, find the real solutions of each equation by factoring. Verify your solution using a graphing utility.

19. $x^2 - 7x + 12 = 0$
20. $x^2 - x - 6 = 0$
21. $2x^2 + 5x - 3 = 0$
22. $3x^2 + 5x - 2 = 0$
23. $x^2 - 9 = 0$
24. $x^2 - 16 = 0$
25. $x^3 + x^2 - 20x = 0$
26. $x^3 + 6x^2 - 7x = 0$
27. $4x^2 = x$
28. $x^2 = 4x$
29. $x^3 = 9x$
30. $4x^3 = x$
31. $x^4 = x^2$
32. $x^4 = 4x^2$
33. $x^3 + x^2 + x + 1 = 0$
34. $x^3 + 4x^2 + x + 4 = 0$
35. $x^3 - 2x^2 - 4x + 8 = 0$
36. $x^3 - 3x^2 - x + 3 = 0$

In Problems 37–42, solve each equation by the Square Root Method. Verify your solution using a graphing utility.

37. $x^2 = 25$
38. $x^2 = 36$
39. $(x - 1)^2 = 4$
40. $(x + 2)^2 = 1$
41. $(2x + 3)^2 = 9$
42. $(3x - 2)^2 = 4$

In Problems 43–46, what number should be added to complete the square of each expression?

43. $x^2 + 8x$

44. $x^2 - 4x$

45. $x^2 - \dfrac{1}{2}x$

46. $x^2 + \dfrac{1}{3}x$

In Problems 47–52, solve each equation by completing the square. Verify your solution using a graphing utility.

47. $x^2 + 4x = 21$

48. $x^2 - 6x = 13$

49. $x^2 - \dfrac{1}{2}x - \dfrac{3}{16} = 0$

50. $x^2 + \dfrac{2}{3}x - \dfrac{1}{3} = 0$

51. $3x^2 + x - \dfrac{1}{2} = 0$

52. $2x^2 - 3x - 1 = 0$

In Problems 53–68, solve each equation by factoring. Verify your solution using a graphing utility.

53. $x^2 - 9x = 0$

54. $x^2 + 4x = 0$

55. $z^2 + z - 6 = 0$

56. $v^2 + 7v + 6 = 0$

57. $2x^2 - 5x - 3 = 0$

58. $3x^2 + 5x + 2 = 0$

59. $3t^2 - 48 = 0$

60. $2y^2 - 50 = 0$

61. $x(x + 8) + 12 = 0$

62. $x(x - 4) = 12$

63. $4x^2 + 9 = 12x$

64. $25x^2 + 16 = 40x$

65. $2x^2 - x = 15$

66. $6x^2 + x = 2$

67. $\dfrac{4(x - 2)}{x - 3} + \dfrac{3}{x} = \dfrac{-3}{x(x - 3)}$

68. $\dfrac{5}{x + 4} = 4 + \dfrac{3}{x - 2}$

In Problems 69–80, find the real solutions, if any, of each equation. Use the quadratic formula. Verify your solution using a graphing utility.

69. $x^2 - 4x + 2 = 0$

70. $x^2 + 4x + 2 = 0$

71. $x^2 - 4x - 1 = 0$

72. $x^2 + 6x + 1 = 0$

73. $2x^2 - 5x + 3 = 0$

74. $2x^2 + 5x + 3 = 0$

75. $4y^2 - y + 2 = 0$

76. $4t^2 + t + 1 = 0$

77. $4x^2 = 1 - 2x$

78. $2x^2 = 1 - 2x$

79. $4x^2 = 9x + 2$

80. $5x = 4x^2 + 5$

In Problems 81–84, solve each equation. The letters a, b, and c are constants.

81. $ax - b = c, \quad a \neq 0$

82. $1 - ax = b, \quad a \neq 0$

83. $\dfrac{x}{a} + \dfrac{x}{b} = c, \quad a \neq 0, b \neq 0, a \neq -b$

84. $\dfrac{a}{x} + \dfrac{b}{x} = c, \quad c \neq 0$

Problems 85–88 list some formulas that occur in applications. Solve each formula for the indicated variable.

85.   Electricity $\dfrac{1}{R} = \dfrac{1}{R_1} + \dfrac{1}{R_2}$ for R

86.   Finance $A = P(1 + rt)$ for r

87.   Mechanics $F = \dfrac{mv^2}{R}$ for R

88.   Chemistry $PV = nRT$ for T

89. One of the steps in the following list contains an error. Identify it and explain what is wrong.

$$x = 2 \tag{1}$$
$$3x - 2x = 2 \tag{2}$$
$$3x = 2x + 2 \tag{3}$$
$$x^2 + 3x = x^2 + 2x + 2 \tag{4}$$
$$x^2 + 3x - 10 = x^2 + 2x - 8 \tag{5}$$
$$(x - 2)(x + 5) = (x - 2)(x + 4) \tag{6}$$
$$x + 5 = x + 4 \tag{7}$$
$$1 = 0 \tag{8}$$

90. Which of the following pairs of equations are equivalent? Explain.
  (a) $x^2 = 9;\quad x = 3$
  (b) $x = \sqrt{9};\quad x = 3$
  (c) $(x - 1)(x - 2) = (x - 1)^2;\quad x - 2 = x - 1$

91. The equation

$$\dfrac{5}{x + 3} + 3 = \dfrac{8 + x}{x + 3}$$

has no solution, yet when we go through the process of solving it we obtain $x = -3$. Write a brief paragraph to explain what causes this to happen.

92. Make up an equation that has no solution and give it to a fellow student to solve. Ask the fellow student to write a critique of your equation.

93. The word *quadratic* seems to imply four (*quad*), yet a quadratic equation is an equation that involves a polynomial of degree 2. Investigate the origin of the term *quadratic* as it is used in the expression *quadratic equation*. Write a brief essay on your findings.

PREPARING FOR THIS SECTION

Before getting started, review the following:

✓ Quadratic Equations (Section A.3, pp. 580–583).

A.4 COMPLEX NUMBERS; QUADRATIC EQUATIONS WITH A NEGATIVE DISCRIMINANT

OBJECTIVES

1 Add, Subtract, Multiply, and Divide Complex Numbers

2 Solve Quadratic Equations with a Negative Discriminant

One property of a real number is that its square is nonnegative. For example, there is no real number x for which

$$x^2 = -1$$

To remedy this situation, we introduce a number called the **imaginary unit**, which we denote by i and whose square is -1; that is,

$$i^2 = -1$$

This should not surprise you. If our universe were to consist only of integers, there would be no number x for which $2x = 1$. This unfortunate circumstance was remedied by introducing numbers such as $\frac{1}{2}$ and $\frac{2}{3}$, the *rational numbers*. If our universe were to consist only of rational numbers, there would be no x whose square equals 2. That is, there would be no number x for which $x^2 = 2$. To remedy this, we introduced numbers such as $\sqrt{2}$ and $\sqrt[3]{5}$, the *irrational numbers*. The *real numbers*, you will recall, consist of the rational numbers and the irrational numbers. Now, if our universe were to consist only of real numbers, then there would be no number x whose square is -1. To remedy this, we introduce a number i, whose square is -1.

In the progression outlined, each time that we encountered a situation that was unsuitable, we introduced a new number system to remedy this situation. And each new number system contained the earlier number system as a subset. The number system that results from introducing the number i is called the **complex number system**.

Complex numbers are numbers of the form $a + bi$, where a and b are real numbers. The real number a is called the **real part** of the number $a + bi$; the real number b is called the **imaginary part** of $a + bi$.

For example, the complex number $-5 + 6i$ has the real part -5 and the imaginary part 6.

When a complex number is written in the form $a + bi$, where a and b are real numbers, we say it is in **standard form**. However, if the imaginary part of a complex number is negative, such as in the complex number $3 + (-2)i$, we agree to write it instead in the form $3 - 2i$.

Also, the complex number $a + 0i$ is usually written merely as a. This serves to remind us that the real numbers are a subset of the complex numbers. The complex number $0 + bi$ is usually written as bi. Sometimes the complex number bi is called a **pure imaginary number**.

Equality, addition, subtraction, and multiplication of complex numbers are defined so as to preserve the familiar rules of algebra for real numbers. Thus, two complex numbers are equal if and only if their real parts are equal and their imaginary parts are equal. That is,

Equality of Complex Numbers

$$a + bi = c + di \quad \text{if and only if } a = c \text{ and } b = d \qquad (1)$$

Two complex numbers are added by forming the complex number whose real part is the sum of the real parts and whose imaginary part is the sum of the imaginary parts. That is,

Sum of Complex Numbers

$$(a + bi) + (c + di) = (a + c) + (b + d)i \qquad (2)$$

To subtract two complex numbers, we use this rule:

Difference of Complex Numbers

$$(a + bi) - (c + di) = (a - c) + (b - d)i \qquad (3)$$

EXAMPLE 1   Adding and Subtracting Complex Numbers

(a) $(3 + 5i) + (-2 + 3i) = [3 + (-2)] + (5 + 3)i = 1 + 8i$
(b) $(6 + 4i) - (3 + 6i) = (6 - 3) + (4 - 6)i = 3 + (-2)i = 3 - 2i$

Figure 23

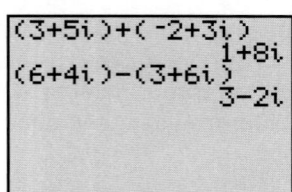

Some graphing calculators have the capability of handling complex numbers.* For example, Figure 23 shows the results of Example 1 using a TI-83 graphing calculator.

NOW WORK PROBLEM 5.

Products of complex numbers are calculated as illustrated in Example 2.

EXAMPLE 2   Multiplying Complex Numbers

$$(5 + 3i) \cdot (2 + 7i) = 5 \cdot (2 + 7i) + 3i(2 + 7i) \quad \text{Distributive property}$$
$$= 10 + 35i + 6i + 21i^2 \quad \text{Distributive property}$$
$$= 10 + 41i + 21(-1) \quad i^2 = -1$$
$$= -11 + 41i$$

*Consult your user's manual for the appropriate keystrokes.

Based on the procedure of Example 2, we define the **product** of two complex numbers by the following formula:

Product of Complex Numbers

$$(a + bi) \cdot (c + di) = (ac - bd) + (ad + bc)i \qquad (4)$$

Do not bother to memorize formula (4). Instead, whenever it is necessary to multiply two complex numbers, follow the usual rules for multiplying two binomials, as in Example 2, remembering that $i^2 = -1$. For example,

$$(2i)(2i) = 4i^2 = -4$$
$$(2 + i)(1 - i) = 2 - 2i + i - i^2 = 3 - i$$

Graphing calculators may also be used to multiply complex numbers. Figure 24 shows the result obtained in Example 2 using a TI-83 graphing calculator.

Figure 24

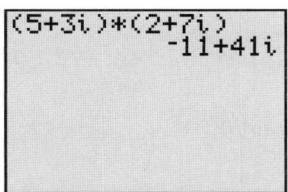

⟢ NOW WORK PROBLEM **11.**

Algebraic properties for addition and multiplication, such as the commutative, associative, and distributive properties, hold for complex numbers. However, the property that every nonzero complex number has a multiplicative inverse, or reciprocal, requires a closer look.

Conjugates

If $z = a + bi$ is a complex number, then its **conjugate**, denoted by $\bar{z}$, is defined as

$$\bar{z} = \overline{a + bi} = a - bi$$

For example, $\overline{2 + 3i} = 2 - 3i$ and $\overline{-6 - 2i} = -6 + 2i$.

EXAMPLE 3   **Multiplying a Complex Number by Its Conjugate**

Find the product of the complex number $z = 3 + 4i$ and its conjugate $\bar{z}$.

Solution   Since $\bar{z} = 3 - 4i$, we have

$$z\bar{z} = (3 + 4i)(3 - 4i) = 9 + 12i - 12i - 16i^2 = 9 + 16 = 25 \quad ■$$

The result obtained in Example 3 has an important generalization.

Theorem   The product of a complex number and its conjugate is a nonnegative real number. That is, if $z = a + bi$, then

$$z\bar{z} = a^2 + b^2 \qquad (5)$$

■

Proof If $z = a + bi$, then

$$z\bar{z} = (a + bi)(a - bi) = a^2 - (bi)^2 = a^2 - b^2i^2 = a^2 + b^2 \qquad \blacksquare$$

To express the reciprocal of a nonzero complex number z in standard form, multiply the numerator and denominator of $\dfrac{1}{z}$ by its conjugate $\bar{z}$. That is, if $z = a + bi$ is a nonzero complex number, then

$$\frac{1}{a + bi} = \frac{1}{z} = \frac{1}{z} \cdot \frac{\bar{z}}{\bar{z}} = \underset{\underset{\text{Use (5).}}{\uparrow}}{\frac{\bar{z}}{z\bar{z}}} = \frac{a - bi}{a^2 + b^2} = \frac{a}{a^2 + b^2} - \frac{b}{a^2 + b^2}i$$

EXAMPLE 4 **Writing the Reciprocal of a Complex Number in Standard Form**

Write $\dfrac{1}{3 + 4i}$ in standard form $a + bi$; that is, find the reciprocal of $3 + 4i$.

Solution The idea is to multiply the numerator and denominator by the conjugate of $3 + 4i$, that is, the complex number $3 - 4i$. The result is

$$\frac{1}{3 + 4i} = \frac{1}{3 + 4i} \cdot \frac{3 - 4i}{3 - 4i} = \frac{3 - 4i}{9 + 16} = \frac{3}{25} - \frac{4}{25}i \qquad \blacksquare$$

Figure 25

```
1/(3+4i)▶Frac
          3/25-4/25i
```

A graphing calculator can be used to verify the result of Example 4. See Figure 25.

To express the quotient of two complex numbers in standard form, we multiply the numerator and denominator of the quotient by the conjugate of the denominator.

EXAMPLE 5 **Writing the Quotient of Complex Numbers in Standard Form**

Write each of the following in standard form.

(a) $\dfrac{1 + 4i}{5 - 12i}$ (b) $\dfrac{2 - 3i}{4 - 3i}$

Solution (a) $\dfrac{1 + 4i}{5 - 12i} = \dfrac{1 + 4i}{5 - 12i} \cdot \dfrac{5 + 12i}{5 + 12i} = \dfrac{5 + 20i + 12i + 48i^2}{25 + 144}$

$$= \frac{-43 + 32i}{169} = \frac{-43}{169} + \frac{32}{169}i$$

(b) $\dfrac{2 - 3i}{4 - 3i} = \dfrac{2 - 3i}{4 - 3i} \cdot \dfrac{4 + 3i}{4 + 3i} = \dfrac{8 - 12i + 6i - 9i^2}{16 + 9}$

$$= \frac{17 - 6i}{25} = \frac{17}{25} - \frac{6}{25}i \qquad \blacksquare$$

NOW WORK PROBLEM 19.

EXAMPLE 6 **Writing Other Expressions in Standard Form**

If $z = 2 - 3i$ and $w = 5 + 2i$, write each of the following expressions in standard form.

(a) $\dfrac{z}{w}$ (b) $\overline{z + w}$ (c) $z + \bar{z}$

Solution (a) $\dfrac{z}{w} = \dfrac{z \cdot \bar{w}}{w \cdot \bar{w}} = \dfrac{(2 - 3i)(5 - 2i)}{(5 + 2i)(5 - 2i)} = \dfrac{10 - 15i - 4i + 6i^2}{25 + 4}$

$= \dfrac{4 - 19i}{29} = \dfrac{4}{29} - \dfrac{19}{29}i$

(b) $\overline{z + w} = \overline{(2 - 3i) + (5 + 2i)} = \overline{7 - i} = 7 + i$

(c) $z + \bar{z} = (2 - 3i) + (2 + 3i) = 4$ ∎

The conjugate of a complex number has certain general properties that we shall find useful later.

For a real number $a = a + 0i$, the conjugate is $\bar{a} = \overline{a + 0i} = a - 0i = a$. That is,

Theorem   The conjugate of a real number is the real number itself.

∎

Other properties that are direct consequences of the definition of the conjugate are given next. In each statement, z and w represent complex numbers.

Theorem   The conjugate of the conjugate of a complex number is the complex number itself.

$$\left(\overline{\bar{z}}\right) = z \tag{6}$$

The conjugate of the sum of two complex numbers equals the sum of their conjugates.

$$\overline{z + w} = \bar{z} + \bar{w} \tag{7}$$

The conjugate of the product of two complex numbers equals the product of their conjugates.

$$\overline{z \cdot w} = \bar{z} \cdot \bar{w} \tag{8}$$

∎

We leave the proofs of equations (6), (7), and (8) as exercises.

Powers of i

The **powers of i** follow a pattern that is useful to know.

$$i^1 = i \qquad\qquad\qquad i^5 = i^4 \cdot i = 1 \cdot i = i$$
$$i^2 = -1 \qquad\qquad\quad i^6 = i^4 \cdot i^2 = -1$$
$$i^3 = i^2 \cdot i = -i \qquad\quad i^7 = i^4 \cdot i^3 = -i$$
$$i^4 = i^2 \cdot i^2 = (-1)(-1) = 1 \qquad i^8 = i^4 \cdot i^4 = 1$$

And so on. The powers of i repeat with every fourth power.

EXAMPLE 7 **Evaluating Powers of i**

(a) $i^{27} = i^{24} \cdot i^3 = (i^4)^6 \cdot i^3 = 1^6 \cdot i^3 = -i$

(b) $i^{101} = i^{100} \cdot i^1 = (i^4)^{25} \cdot i = 1^{25} \cdot i = i$ ∎

EXAMPLE 8 **Writing the Power of a Complex Number in Standard Form**

Write $(2 + i)^3$ in standard form.

Solution We use the special product formula for $(x + a)^3$.

$$(x + a)^3 = x^3 + 3ax^2 + 3a^2x + a^3$$

Using this special product formula,

$$(2 + i)^3 = 2^3 + 3 \cdot i \cdot 2^2 + 3 \cdot i^2 \cdot 2 + i^3$$
$$= 8 + 12i + 6(-1) + (-i)$$
$$= 2 + 11i$$ ∎

✏ **NOW WORK PROBLEMS 25 AND 33.**

Quadratic Equations with a Negative Discriminant

2 Quadratic equations with a negative discriminant have no real number solution. However, if we extend our number system to allow complex numbers, quadratic equations will always have a solution. Since the solution to a quadratic equation involves the square root of the discriminant, we begin with a discussion of square roots of negative numbers.

If N is a positive real number, we define the **principal square root of** $-N$, denoted by $\sqrt{-N}$, as

$$\boxed{\sqrt{-N} = \sqrt{N}i}$$

where i is the imaginary unit and $i^2 = -1$.

EXAMPLE 9 **Evaluating the Square Root of a Negative Number**

(a) $\sqrt{-1} = \sqrt{1}i = i$ (b) $\sqrt{-4} = \sqrt{4}i = 2i$

(c) $\sqrt{-8} = \sqrt{8}i = 2\sqrt{2}i$ ∎

EXAMPLE 10 **Solving Equations**

Solve each equation in the complex number system.

(a) $x^2 = 4$ (b) $x^2 = -9$

Solution (a) $x^2 = 4$

$$x = \pm\sqrt{4} = \pm 2$$

The equation has two solutions, -2 and 2.

(b) $x^2 = -9$

$$x = \pm\sqrt{-9} = \pm\sqrt{9}i = \pm3i$$

The equation has two solutions, $-3i$ and $3i$. ■

──── **NOW WORK PROBLEM 45.**

WARNING: When working with square roots of negative numbers, do not set the square root of a product equal to the product of the square roots (which can be done with positive numbers). To see why, look at this calculation: We know that $\sqrt{100} = 10$. However, it is also true that $100 = (-25)(-4)$, so

$$10 = \sqrt{100} = \sqrt{(-25)(-4)} \neq \sqrt{-25}\,\sqrt{-4} = (\sqrt{25}i)(\sqrt{4}i) = (5i)(2i) = 10i^2 = -10$$

↑

Here is the error.

■

Because we have defined the square root of a negative number, we can now restate the quadratic formula without restriction.

Theorem   In the complex number system, the solutions of the quadratic equation $ax^2 + bx + c = 0$, where a, b, and c are real numbers and $a \neq 0$, are given by the formula

$$x = \frac{-b \pm \sqrt{b^2 - 4ac}}{2a} \qquad (9)$$

■

EXAMPLE 11   **Solving Quadratic Equations in the Complex Number System**

Solve the equation $x^2 - 4x + 8 = 0$ in the complex number system.

Solution   Here $a = 1, b = -4, c = 8$, and $b^2 - 4ac = 16 - 4(1)(8) = -16$. Using equation (9), we find that

$$x = \frac{-(-4) \pm \sqrt{-16}}{2(1)} = \frac{4 \pm \sqrt{16}i}{2} = \frac{4 \pm 4i}{2} = 2 \pm 2i$$

The equation has the solution set $\{2 - 2i, 2 + 2i\}$.

✔ CHECK:

$$2 + 2i: \quad (2 + 2i)^2 - 4(2 + 2i) + 8 = 4 + 8i + 4i^2 - 8 - 8i + 8$$
$$= 4 - 4 = 0$$

$$2 - 2i: \quad (2 - 2i)^2 - 4(2 - 2i) + 8 = 4 - 8i + 4i^2 - 8 + 8i + 8$$
$$= 4 - 4 = 0$$

■ ■

Figure 26

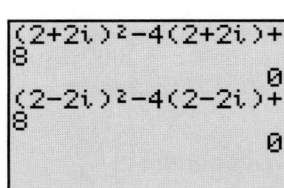

Figure 26 shows the check of the solution using a TI-83 graphing calculator.

──── **NOW WORK PROBLEM 51.**

The discriminant $b^2 - 4ac$ of a quadratic equation still serves as a way to determine the character of the solutions.

Character of the Solutions of a Quadratic Equation

In the complex number system, consider a quadratic equation $ax^2 + bx + c = 0$ with real coefficients.

1. If $b^2 - 4ac > 0$, the equation has two unequal real solutions.
2. If $b^2 - 4ac = 0$, the equation has a repeated real solution, a double root.
3. If $b^2 - 4ac < 0$, the equation has two complex solutions that are not real. The solutions are conjugates of each other.

The third conclusion in the display is a consequence of the fact that if $b^2 - 4ac = -N < 0$ then, by the quadratic formula, the solutions are

$$x = \frac{-b + \sqrt{b^2 - 4ac}}{2a} = \frac{-b + \sqrt{-N}}{2a} = \frac{-b + \sqrt{N}i}{2a} = \frac{-b}{2a} + \frac{\sqrt{N}}{2a}i$$

and

$$x = \frac{-b - \sqrt{b^2 - 4ac}}{2a} = \frac{-b - \sqrt{-N}}{2a} = \frac{-b - \sqrt{N}i}{2a} = \frac{-b}{2a} - \frac{\sqrt{N}}{2a}i$$

which are conjugates of each other.

EXAMPLE 12 **Determining the Character of the Solutions of a Quadratic Equation**

Without solving, determine the character of the solutions of each equation.

(a) $3x^2 + 4x + 5 = 0$
(b) $2x^2 + 4x + 1 = 0$
(c) $9x^2 - 6x + 1 = 0$

Solution (a) Here $a = 3, b = 4$, and $c = 5$, so $b^2 - 4ac = 16 - 4(3)(5) = -44$. The solutions are complex numbers that are not real and are conjugates of each other.

(b) Here $a = 2, b = 4$, and $c = 1$, so $b^2 - 4ac = 16 - 8 = 8$. The solutions are two unequal real numbers.

(c) Here $a = 9, b = -6$, and $c = 1$, so $b^2 - 4ac = 36 - 4(9)(1) = 0$. The solution is a repeated real number, that is, a double root. ■

NOW WORK PROBLEM 65.

A.4 Exercises

In Problems 1–38, write each expression in the standard form $a + bi$. Verify your results using a graphing utility.

1. $(2 - 3i) + (6 + 8i)$

2. $(4 + 5i) + (-8 + 2i)$

3. $(-3 + 2i) - (4 - 4i)$

4. $(3 - 4i) - (-3 - 4i)$

5. $(2 - 5i) - (8 + 6i)$

6. $(-8 + 4i) - (2 - 2i)$

7. $3(2 - 6i)$

8. $-4(2 + 8i)$

9. $2i(2 - 3i)$

10. $3i(-3 + 4i)$

11. $(3 - 4i)(2 + i)$

12. $(5 + 3i)(2 - i)$

13. $(-6 + i)(-6 - i)$

14. $(-3 + i)(3 + i)$

15. $\dfrac{10}{3 - 4i}$

16. $\dfrac{13}{5 - 12i}$

17. $\dfrac{2 + i}{i}$

18. $\dfrac{2 - i}{-2i}$

19. $\dfrac{6 - i}{1 + i}$

20. $\dfrac{2 + 3i}{1 - i}$

21. $\left(\dfrac{1}{2} + \dfrac{\sqrt{3}}{2}i\right)^2$

22. $\left(\dfrac{\sqrt{3}}{2} - \dfrac{1}{2}i\right)^2$

23. $(1 + i)^2$

24. $(1 - i)^2$

25. i^{23}

26. i^{14}

27. i^{-15}

28. i^{-23}

29. $i^6 - 5$

30. $4 + i^3$

31. $6i^3 - 4i^5$

32. $4i^3 - 2i^2 + 1$

33. $(1 + i)^3$

34. $(3i)^4 + 1$

35. $i^7(1 + i^2)$

36. $2i^4(1 + i^2)$

37. $i^6 + i^4 + i^2 + 1$

38. $i^7 + i^5 + i^3 + i$

In Problems 39–44, perform the indicated operations and express your answer in the form $a + bi$.

39. $\sqrt{-4}$

40. $\sqrt{-9}$

41. $\sqrt{-25}$

42. $\sqrt{-64}$

43. $\sqrt{(3 + 4i)(4i - 3)}$

44. $\sqrt{(4 + 3i)(3i - 4)}$

In Problems 45–64, solve each equation in the complex number system. Check your results using a graphing utility.

45. $x^2 + 4 = 0$

46. $x^2 - 4 = 0$

47. $x^2 - 16 = 0$

48. $x^2 + 25 = 0$

49. $x^2 - 6x + 13 = 0$

50. $x^2 + 4x + 8 = 0$

51. $x^2 - 6x + 10 = 0$

52. $x^2 - 2x + 5 = 0$

53. $8x^2 - 4x + 1 = 0$

54. $10x^2 + 6x + 1 = 0$

55. $5x^2 + 2x + 1 = 0$

56. $13x^2 + 6x + 1 = 0$

57. $x^2 + x + 1 = 0$

58. $x^2 - x + 1 = 0$

59. $x^3 - 8 = 0$

60. $x^3 + 27 = 0$

61. $x^4 - 16 = 0$

62. $x^4 - 1 = 0$

63. $x^4 + 13x^2 + 36 = 0$

64. $x^4 + 3x^2 - 4 = 0$

In Problems 65–70, without solving, determine the character of the solutions of each equation. Verify your answer using a graphing utility.

65. $3x^2 - 3x + 4 = 0$

66. $2x^2 - 4x + 1 = 0$

67. $2x^2 + 3x - 4 = 0$

68. $x^2 + 2x + 6 = 0$

69. $9x^2 - 12x + 4 = 0$

70. $4x^2 + 12x + 9 = 0$

71. $2 + 3i$ is a solution of a quadratic equation with real coefficients. Find the other solution.

72. $4 - i$ is a solution of a quadratic equation with real coefficients. Find the other solution.

In Problems 73–76, $z = 3 - 4i$ and $w = 8 + 3i$. Write each expression in the standard form $a + bi$.

73. $z + \bar{z}$

74. $w - \bar{w}$

75. $z\bar{z}$

76. $\overline{z - w}$

---

77. Use $z = a + bi$ to show that $z + \bar{z} = 2a$ and $z - \bar{z} = 2bi$.

78. Use $z = a + bi$ to show that $\left(\bar{\bar{z}}\right) = z$.

79. Use $z = a + bi$ and $w = c + di$ to show that $\overline{z + w} = \bar{z} + \bar{w}$.

80. Use $z = a + bi$ and $w = c + di$ to show that $\overline{z \cdot w} = \bar{z} \cdot \bar{w}$.

81. Explain to a friend how you would add two complex numbers and how you would multiply two complex numbers. Explain any differences in the two explanations.

82. Write a brief paragraph that compares the method used to rationalize denominators and the method used to write the quotient of two complex numbers in standard form.

83. Use an Internet search engine to investigate the origins of complex numbers. Write a paragraph describing what you find and present it to the class.

PREPARING FOR THIS SECTION

Before getting started, review the following:

✓ Inequalities (Section A.1, pp. 559–560)　　　✓ Absolute value (Section A.1, pp. 560–561)

A.5 INEQUALITIES

OBJECTIVES

  ① Use Interval Notation
  ② Use Properties of Inequalities
  ③ Solve Linear Inequalities Algebraically and Graphically
  ④ Solve Combined Inequalities Algebraically and Graphically
  ⑤ Solve Absolute Value Inequalities Algebraically and Graphically

Suppose that a and b are two real numbers and $a < b$. We shall use the notation $a < x < b$ to mean that x is a number *between* a and b. Thus, the expression $a < x < b$ is equivalent to the two inequalities $a < x$ and $x < b$. Similarly, the expression $a \leq x \leq b$ is equivalent to the two inequalities $a \leq x$ and $x \leq b$. The remaining two possibilities, $a \leq x < b$ and $a < x \leq b$, are defined similarly.

Although it is acceptable to write $3 \geq x \geq 2$, it is preferable to reverse the inequality symbols and write instead $2 \leq x \leq 3$ so that, as you read from left to right, the values go from smaller to larger.

A statement such as $2 \leq x \leq 1$ is false because there is no number x for which $2 \leq x$ and $x \leq 1$. Finally, we never mix inequality symbols, as in $2 \leq x \geq 3$.

Intervals

① Let a and b represent two real numbers with $a < b$:

> A **closed interval**, denoted by **[a, b]**, consists of all real numbers x for which $a \leq x \leq b$.
>
> An **open interval**, denoted by **(a, b)**, consists of all real numbers x for which $a < x < b$.
>
> The **half-open**, or **half-closed, intervals** are **(a, b]**, consisting of all real numbers x for which $a < x \leq b$, and **[a, b)**, consisting of all real numbers x for which $a \leq x < b$.

In each of these definitions, a is called the **left endpoint** and b the **right endpoint** of the interval.

The symbol ∞ (read as "infinity") is not a real number but a notational device used to indicate unboundedness in the positive direction. The symbol $-\infty$ (read as "minus infinity" or "negative infinity") also is not a real number, but a notational device used to indicate unboundedness in the negative direction. Using the symbols ∞ and $-\infty$, we can define five other kinds of intervals:

[a, ∞)　　consists of all real numbers x for which $x \geq a$ $(a \leq x < \infty)$

(a, ∞)　　consists of all real numbers x for which $x > a$ $(a < x < \infty)$

(−∞, a]　　consists of all real numbers x for which $x \leq a$ $(-\infty < x \leq a)$

$(-\infty, a)$ consists of all real numbers x for which $x < a \, (-\infty < x < a)$
$(-\infty, \infty)$ consists of all real numbers $x \, (-\infty < x < \infty)$

Note that ∞ and $-\infty$ are never included as endpoints since they are not real numbers.

Table 1 summarizes interval notation, corresponding inequality notation, and their graphs.

TABLE 1

Interval	Inequality	Graph
The open interval (a, b)	$a < x < b$	
The closed interval $[a, b]$	$a \le x \le b$	
The half-open interval $[a, b)$	$a \le x < b$	
The half-open interval $(a, b]$	$a < x \le b$	
The interval $[a, \infty)$	$x \ge a$	
The interval (a, ∞)	$x > a$	
The interval $(-\infty, a]$	$x \le a$	
The interval $(-\infty, a)$	$x < a$	
The interval $(-\infty, \infty)$	All real numbers	

EXAMPLE 1   Writing Inequalities Using Interval Notation

Write each inequality using interval notation.

(a) $1 \le x \le 3$ (b) $-4 < x < 0$ (c) $x > 5$ (d) $x \le 1$

Solution   (a) $1 \le x \le 3$ describes all numbers x between 1 and 3, inclusive. In interval notation, we write $[1, 3]$.

(b) In interval notation, $-4 < x < 0$ is written $(-4, 0)$.

(c) $x > 5$ consists of all numbers x greater than 5. In interval notation, we write $(5, \infty)$.

(d) In interval notation, $x \le 1$ is written $(-\infty, 1]$.   ■

EXAMPLE 2   Writing Intervals Using Inequality Notation

Write each interval as an inequality involving x.

(a) $[1, 4)$ (b) $(2, \infty)$ (c) $[2, 3]$ (d) $(-\infty, -3]$

Solution   (a) $[1, 4)$ consists of all numbers x for which $1 \le x < 4$.

(b) $(2, \infty)$ consists of all numbers x for which $x > 2 \, (2 < x < \infty)$.

(c) $[2, 3]$ consists of all numbers x for which $2 \le x \le 3$.

(d) $(-\infty, -3]$ consists of all numbers x for which $x \le -3 \, (-\infty < x \le -3)$.   ■

━ NOW WORK PROBLEMS 1, 15, AND 23.

Properties of Inequalities

2 The product of two positive real numbers is positive, the product of two negative real numbers is positive, and the product of 0 and 0 is 0. For any real number a, the value of a^2 is 0 or positive; that is, a^2 is nonnegative. This is called the **nonnegative property**.

For any real number a, we have

Nonnegative Property

$$a^2 \geq 0 \tag{1}$$

If we add the same number to both sides of an inequality, we obtain an equivalent inequality. For example, since $3 < 5$, then $3 + 4 < 5 + 4$ or $7 < 9$. This is called the **addition property** of inequalities.

Addition Property of Inequalities

$$\text{If } a < b, \text{ then } a + c < b + c \tag{2a}$$
$$\text{If } a > b, \text{ then } a + c > b + c \tag{2b}$$

The addition property states that the sense, or direction, of an inequality remains unchanged if the same number is added to each side. Figure 27 illustrates the addition property (2a). In Figure 27(a), we see that a lies to the left of b. If c is positive, then $a + c$ and $b + c$ each lie c units to the right of a and b, respectively. Consequently, $a + c$ must lie to the left of $b + c$; that is, $a + c < b + c$. Figure 27(b) illustrates the situation if c is negative.

Figure 27

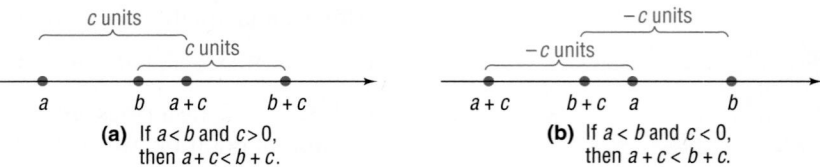

(a) If $a < b$ and $c > 0$, then $a + c < b + c$.

(b) If $a < b$ and $c < 0$, then $a + c < b + c$.

DRAW AN ILLUSTRATION SIMILAR TO FIGURE **27** THAT ILLUSTRATES THAT THE ADDITION PROPERTY **(2b)**.

EXAMPLE 3   **Addition Property of Inequalities**

(a) If $x < -5$, then $x + 5 < -5 + 5$ or $x + 5 < 0$.
(b) If $x > 2$, then $x + (-2) > 2 + (-2)$ or $x - 2 > 0$.   ■

NOW WORK PROBLEM **29**.

We will use two examples to arrive at our next property.

EXAMPLE 4   **Multiplying an Inequality by a Positive Number**

Express as an inequality the result of multiplying each side of the inequality $3 < 7$ by 2.

Solution   We begin with

$$3 < 7$$

Multiplying each side by 2 yields the numbers 6 and 14, so we have

$$6 < 14 \qquad \blacksquare$$

EXAMPLE 5   **Multiplying an Inequality by a Negative Number**

Express as an inequality the result of multiplying each side of the inequality $9 > 2$ by -4.

Solution   We begin with

$$9 > 2$$

Multiplying each side by -4 yields the numbers -36 and -8, so we have

$$-36 < -8 \qquad \blacksquare$$

Note that the effect of multiplying both sides of $9 > 2$ by the negative number -4 is that the direction of the inequality symbol is reversed.

Examples 4 and 5 illustrate the following general **multiplication properties** for inequalities:

Multiplication Properties for Inequalities

> If $a < b$ and if $c > 0$, then $ac < bc$.   (3a)
> If $a < b$ and if $c < 0$, then $ac > bc$.
>
> If $a > b$ and if $c > 0$, then $ac > bc$.   (3b)
> If $a > b$ and if $c < 0$, then $ac < bc$.

The multiplication properties state that the sense, or direction, of an inequality *remains the same* if each side is multiplied by a *positive* real number, while the direction is *reversed* if each side is multiplied by a *negative* real number.

EXAMPLE 6   **Multiplication Property of Inequalities**

(a) If $2x < 6$, then $\dfrac{1}{2}(2x) < \dfrac{1}{2}(6)$ or $x < 3$.

(b) If $\dfrac{x}{-3} > 12$, then $-3\left(\dfrac{x}{-3}\right) < -3(12)$ or $x < -36$.

(c) If $-4x > -8$, then $\dfrac{-4x}{-4} < \dfrac{-8}{-4}$ or $x < 2$.

(d) If $-x < 8$, then $(-1)(-x) > (-1)(8)$ or $x > -8$. $\blacksquare$

✏ **NOW WORK PROBLEM 35.**

Solving Inequalities

An **inequality in one variable** is a statement involving two expressions, at least one containing the variable, separated by one of the inequality symbols, $<$, $\leq$, $>$, or $\geq$. To **solve an inequality** means to find all values of the variable for which the statement is true. These values are called **solutions** of the inequality.

For example, the following are all inequalities involving one variable, x:

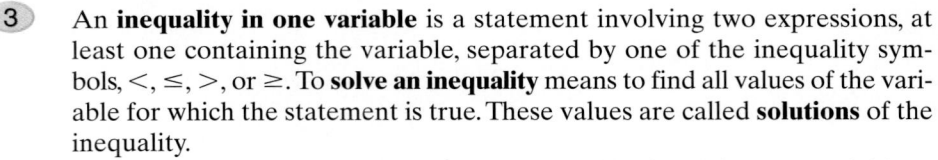

$$x + 5 < 8, \qquad 2x - 3 \geq 4, \qquad x^2 - 1 \leq 3, \qquad \frac{x + 1}{x - 2} > 0$$

Two inequalities having exactly the same solution set are called **equivalent inequalities**. As with equations, one method for solving an inequality is to replace it by a series of equivalent inequalities until an inequality with an obvious solution, such as $x < 3$, is obtained. We obtain equivalent inequalities by applying some of the same operations as those used to find equivalent equations. The addition property and the multiplication properties form the basis for the following procedures.

Procedures That Leave the Inequality Symbol Unchanged

1. Simplify both sides of the inequality by combining like terms and eliminating parentheses:

    Replace $x + 2 + 6 > 2x + 5(x + 1)$
       by $x + 8 > 7x + 5$

2. Add or subtract the same expression on both sides of the inequality:

    Replace $3x - 5 < 4$
       by $(3x - 5) + 5 < 4 + 5$

3. Multiply or divide both sides of the inequality by the same *positive* expression:

    Replace $4x > 16$ by $\dfrac{4x}{4} > \dfrac{16}{4}$

Procedures That Reverse the Sense or Direction of the Inequality Symbol

1. Interchange the two sides of the inequality:

    Replace $3 < x$ by $x > 3$

2. Multiply or divide both sides of the inequality by the same *negative* expression:

    Replace $-2x > 6$ by $\dfrac{-2x}{-2} < \dfrac{6}{-2}$

To solve an inequality using a graphing utility, we follow these steps:

Steps for Solving Inequalities Graphically

STEP 1: Write the inequality in one of the following forms:
$$Y_1 < Y_2, \qquad Y_1 > Y_2, \qquad Y_1 \le Y_2, \qquad Y_1 \ge Y_2$$

STEP 2: Graph Y_1 and Y_2 on the same screen.

STEP 3: If the inequality is of the form $Y_1 < Y_2$, determine on what intervals Y_1 is below Y_2.

If the inequality is of the form $Y_1 > Y_2$, determine on what intervals Y_1 is above Y_2.

If the inequality is not strict ($\le$ or $\ge$), include the x-coordinates of the points of intersection in the solution.

As the examples that follow illustrate, we solve inequalities using many of the same steps that we would use to solve equations. In writing the solution of an inequality, we may use either set notation or interval notation, whichever is more convenient.

EXAMPLE 7 Solving an Inequality

Solve the inequality $4x + 7 \ge 2x - 3$, and graph the solution set.

Algebraic Solution

$$4x + 7 \ge 2x - 3$$
$$4x + 7 - 7 \ge 2x - 3 - 7 \qquad \text{Subtract 7 from both sides.}$$
$$4x \ge 2x - 10 \qquad \text{Simplify.}$$
$$4x - 2x \ge 2x - 10 - 2x \qquad \text{Subtract 2x from both sides.}$$
$$2x \ge -10 \qquad \text{Simplify.}$$
$$\frac{2x}{2} \ge \frac{-10}{2} \qquad \text{Divide both sides by 2. (The direction of the inequality symbol is unchanged.)}$$
$$x \ge -5 \qquad \text{Simplify.}$$

The solution set is $\{x \mid x \ge -5\}$ or, using interval notation, all numbers in the interval $[-5, \infty)$.

Graphing Solution

We graph $Y_1 = 4x + 7$ and $Y_2 = 2x - 3$ on the same screen. See Figure 28. Using the INTERSECT command, we find that Y_1 and Y_2 intersect at $x = -5$. The graph of Y_1 is above that of Y_2, $Y_1 > Y_2$, to the right of the point of intersection. Since the inequality is not strict, the solution set is $\{x \mid x \ge -5\}$ or, using interval notation, $[-5, \infty)$.

Figure 28

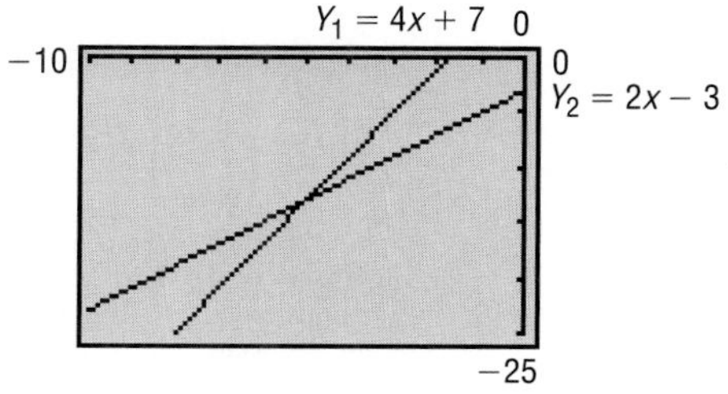

$Y_1 = 4x + 7$

$Y_2 = 2x - 3$

Figure 29

$x \ge -5$ or $[-5, \infty)$

See Figure 29 for the graph of the solution set.

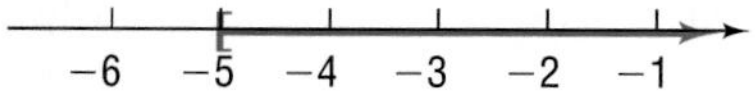

NOW WORK PROBLEM **43.**

④ | **EXAMPLE 8**   **Solving Combined Inequalities**

Solve the inequality $-5 < 3x - 2 < 1$ and draw a graph to illustrate the solution.

Algebraic Solution   Recall that the inequality

$$-5 < 3x - 2 < 1$$

is equivalent to the two inequalities

$$-5 < 3x - 2 \quad \text{and} \quad 3x - 2 < 1$$

We will solve each of these inequalities separately.

$-5 < 3x - 2$		$3x - 2 < 1$
$-5 + 2 < 3x - 2 + 2$	Add 2 to both sides	$3x - 2 + 2 < 1 + 2$
$-3 < 3x$	Simplify.	$3x < 3$
$\dfrac{-3}{3} < \dfrac{3x}{3}$	Divide both sides by 3.	$\dfrac{3x}{3} < \dfrac{3}{3}$
$-1 < x$	Simplify.	$x < 1$

The solution set of the original pair of inequalities consists of all x for which

$$-1 < x \quad \text{and} \quad x < 1$$

This may be written more compactly as $\{x \mid -1 < x < 1\}$. In interval notation, the solution is $(-1, 1)$.

Graphing Solution   To solve a combined inequality, we graph each part: $Y_1 = -5$, $Y_2 = 3x - 2$, and $Y_3 = 1$. We seek the values of x for which the graph of Y_2 is between the graphs of Y_1 and Y_3. See Figure 30. The point of intersection of Y_1 and Y_2 is $(-1, -5)$, and the point of intersection of Y_2 and Y_3 is $(1, 1)$. The inequality is true for all values of x between these two intersection points. Since the inequality is strict, the solution set is $\{x \mid -1 < x < 1\}$ or, using interval notation, $(-1, 1)$.

Figure 30

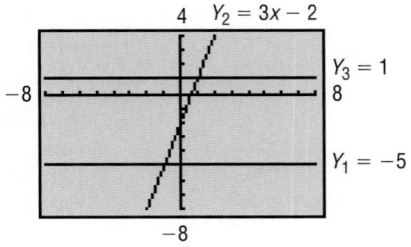

$4 \quad Y_2 = 3x - 2$

$Y_3 = 1$

-8 8

$Y_1 = -5$

-8

Figure 31
$-1 < x < 1$ or $(-1, 1)$

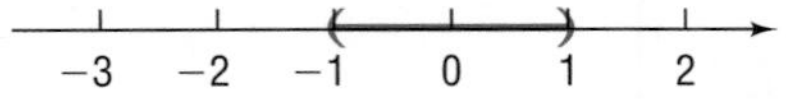

$$\begin{array}{ccccccc} & & (& &) & & \\ \hline -3 & -2 & -1 & 0 & 1 & 2 \end{array}$$

See Figure 31 for the graph of the solution set.   ■

We observe in the algebraic solution of Example 8 that the two inequalities that we solved required exactly the same steps. A shortcut to solving the original inequality algebraically is to deal with the two inequalities at the same time, as follows:

$$-5 < \quad 3x - 2 \quad < 1$$
$$-5 + 2 < 3x - 2 + 2 < 1 + 2 \qquad \text{Add 2 to each part.}$$
$$-3 < \quad 3x \quad < 3 \qquad \text{Simplify.}$$
$$\frac{-3}{3} < \quad \frac{3x}{3} \quad < \frac{3}{3} \qquad \text{Divide each part by 3.}$$
$$-1 < \quad x \quad < 1 \qquad \text{Simplify.}$$

 NOW WORK PROBLEM 63.

5 Let's look at an inequality involving absolute value.

EXAMPLE 9   Solving an Inequality Involving Absolute Value

Solve the inequality: $|x| < 4$

Algebraic Solution   We are looking for all points whose coordinate x is a distance less than 4 units from the origin. See Figure 32 for an illustration. Because any x between -4 and 4 satisfies the condition $|x| < 4$, the solution set consists of all numbers x for which $-4 < x < 4$, that is, all x in the interval $(-4, 4)$.

Figure 32
$-4 < x < 4$ or $(-4, 4)$

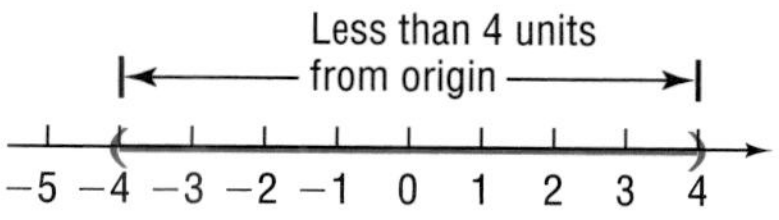

Less than 4 units from origin

$-5 \ -4 \ -3 \ -2 \ -1 \quad 0 \quad 1 \quad 2 \quad 3 \quad 4$

Graphing Solution   We graph $Y_1 = |x|$ and $Y_2 = 4$ on the same screen. See Figure 33. Using the INTERSECT command (twice), we find that Y_1 and Y_2 intersect at $x = -4$ and at $x = 4$. The graph of Y_1 is below that of Y_2, $Y_1 < Y_2$, between the points of intersection. Since the inequality is strict, the solution set is $\{x | -4 < x < 4\}$ or, using interval notation, $(-4, 4)$.

Figure 33

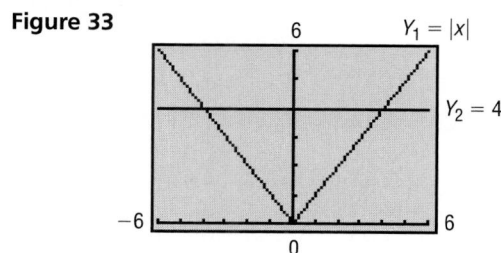

We are led to the following results:

Inequalities Involving Absolute Value

If a is any positive number and if u is any algebraic expression, then

$|u| < a$	is equivalent to	$-a < u < a$	(4)
$|u| \leq a$	is equivalent to	$-a \leq u \leq a$	(5)

In other words, $|u| < a$ is equivalent to $-a < u$ and $u < a$.

| EXAMPLE 10 | Solving an Inequality Involving Absolute Value |

Solve the inequality $|2x + 4| \le 3$, and graph the solution set.

Algebraic Solution

$$|2x + 4| \le 3$$ This follows the form of statement (5); the expression $u = 2x + 4$ is inside the absolute value bars.

$$-3 \le \quad 2x + 4 \quad \le 3$$ Apply statement (5).

$$-3 - 4 \le 2x + 4 - 4 \le 3 - 4$$ Subtract 4 from each part.

$$-7 \le \quad 2x \quad \le -1$$ Simplify.

$$\frac{-7}{2} \le \quad \frac{2x}{2} \quad \le \frac{-1}{2}$$ Divide each part by 2.

$$-\frac{7}{2} \le \quad x \quad \le -\frac{1}{2}$$ Simplify.

The solution set is $\left\{ x \middle| -\dfrac{7}{2} \le x \le -\dfrac{1}{2} \right\}$, that is, all x in the interval $\left[-\dfrac{7}{2}, -\dfrac{1}{2} \right]$.

Graphing Solution We graph $Y_1 = |2x + 4|$ and $Y_2 = 3$ on the same screen. See Figure 34. Using the INTERSECT command (twice), we find that Y_1 and Y_2 intersect at $x = -3.5$ and at $x = -0.5$. The graph of Y_1 is below that of Y_2, $Y_1 < Y_2$, between the points of intersection. Since the inequality is not strict, the solution set is $\{x | -3.5 \le x \le -0.5\}$ or, using interval notation, $[-3.5, -0.5]$.

Figure 34

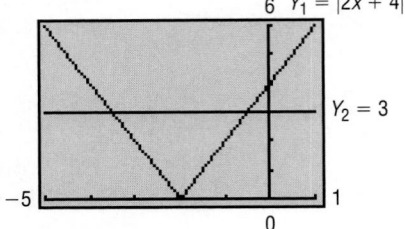

Figure 35

$$-\frac{7}{2} \le x \le -\frac{1}{2} \text{ or } \left[-\frac{7}{2}, -\frac{1}{2} \right]$$

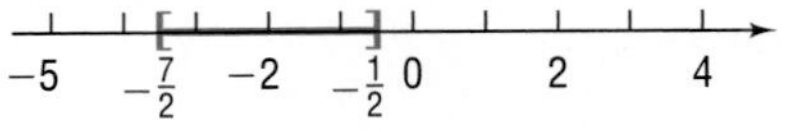

See Figure 35 for a graph of the solution set.

NOW WORK PROBLEM 81.

| EXAMPLE 11 | Solving an Inequality Involving Absolute Value |

Solve the inequality $|x| > 3$, and graph the solution set.

Algebraic Solution

Figure 36

$x < -3$ or $x > 3$

$(-\infty, -3)$ or $(3, \infty)$

We are looking for all points whose coordinate x is a distance greater than 3 units from the origin. Figure 36 illustrates the situation. We conclude that any x less than -3 or greater than 3 satisfies the condition $|x| > 3$. Consequently, the solution set consists of all numbers x for which $x < -3$ or $x > 3$, that is, all x in the intervals $(-\infty, -3)$ or $(3, \infty)$.

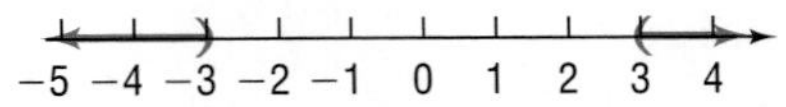

Graphing Solution We graph $Y_1 = |x|$ and $Y_2 = 3$ on the same screen. See Figure 37. Using the INTERSECT command (twice), we find that Y_1 and Y_2 intersect at $x = -3$ and at $x = 3$. The graph of Y_1 is above that of Y_2, $Y_1 > Y_2$, to the left of $x = -3$ and to the right of $x = 3$. Since the inequality is strict, the solution set is $\{x | x < -3$ or $x > 3\}$. Using interval notation, the solution is $(-\infty, -3)$ or $(3, \infty)$.

Figure 37

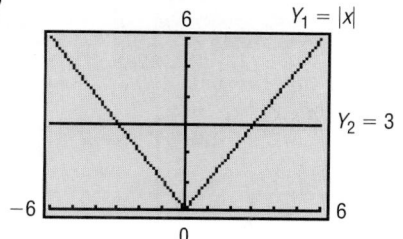

We are led to the following results:

Inequalities Involving Absolute Value

If a is any positive number and u is any algebraic expression, then

$\lvert u \rvert > a$	is equivalent to	$u < -a$ or $u > a$	(6)
$\lvert u \rvert \geq a$	is equivalent to	$u \leq -a$ or $u \geq a$	(7)

EXAMPLE 12   Solving an Inequality Involving Absolute Value

Solve the inequality $\lvert 2x - 5 \rvert > 3$, and graph the solution set.

Algebraic Solution

$\lvert 2x - 5 \rvert > 3$ This follows the form of statement (6); the expression $u = 2x - 5$ is inside the absolute value bars.

$$
\begin{array}{lcl}
2x - 5 < -3 & \text{or} & 2x - 5 > 3 \\
2x - 5 + 5 < -3 + 5 & \text{or} & 2x - 5 + 5 > 3 + 5 \\
2x < 2 & \text{or} & 2x > 8 \\
\dfrac{2x}{2} < \dfrac{2}{2} & \text{or} & \dfrac{2x}{2} > \dfrac{8}{2} \\
x < 1 & \text{or} & x > 4
\end{array}
$$

Apply statement (6).
Add 5 to each part.
Simplify.
Divide each part by 2.
Simplify.

The solution set is $\{x \mid x < 1 \text{ or } x > 4\}$, that is, all x in the intervals $(-\infty, 1)$ or $(4, \infty)$.

Graphing Solution

Figure 38

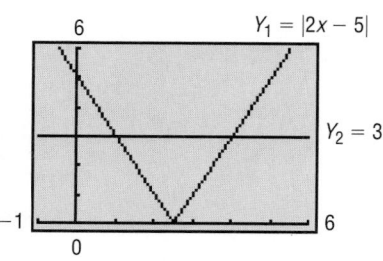

We graph $Y_1 = \lvert 2x - 5 \rvert$ and $Y_2 = 3$ on the same screen. See Figure 38. Using the INTERSECT command (twice), we find that Y_1 and Y_2 intersect at $x = 1$ and at $x = 4$. The graph of Y_1 is above that of Y_2, $Y_1 > Y_2$, to the left of $x = 1$ and to the right of $x = 4$. Since the inequality is strict, the solution set is $\{x \mid x < 1 \text{ or } x > 4\}$. Using interval notation, the solution is $(-\infty, 1)$ or $(4, \infty)$. See Figure 39 for a graph of the solution set.

Figure 39
$x < 1$ or $x > 4$; $(-\infty, 1)$ or $(4, \infty)$

$$\xleftarrow{\hspace{1.5cm}} \quad -2 \;\; -1 \;\; 0 \;\; 1 \;\; 2 \;\; 3 \;\; 4 \;\; 5 \;\; 6 \;\; 7 \quad \xrightarrow{\hspace{1.5cm}}$$

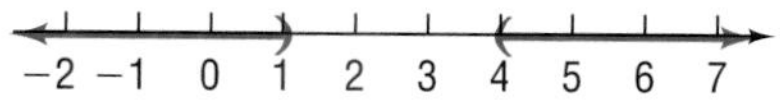

WARNING: A common error to be avoided is to attempt to write the solution $x < 1$ or $x > 4$ as $1 > x > 4$, which is incorrect, since there are no numbers x for which $1 > x$ and $x > 4$. Another common error is to "mix" the symbols and write $1 < x > 4$, which makes no sense.

NOW WORK PROBLEM 85.

A.5 Exercises

In Problems 1–6, express the graph shown in color using interval notation. Also express each as an inequality involving x.

1.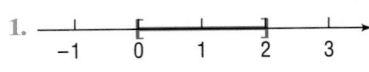
(number line: −1, 0, 1, 2, 3)

2. (number line: −1, 0, 1, 2, 3)

3.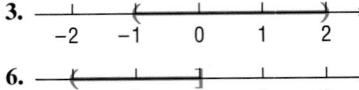
(number line: −2, −1, 0, 1, 2)

4. (number line: −2, −1, 0, 1, 2)

5. (number line: −1, 0, 1, 2, 3)

6. (number line: −1, 0, 1, 2, 3)

In Problems 7–12, an inequality is given. Write the inequality obtained by:

(a) *Adding 3 to each side of the given inequality.*
(b) *Subtracting 5 from each side of the given inequality.*
(c) *Multiplying each side of the given inequality by 3.*
(d) *Multiplying each side of the given inequality by −2.*

7. $3 < 5$

8. $2 > 1$

9. $4 > -3$

10. $-3 > -5$

11. $2x + 1 < 2$

12. $1 - 2x > 5$

In Problems 13–20, write each inequality using interval notation, and illustrate each inequality using the real number line.

13. $0 \le x \le 4$

14. $-1 < x < 5$

15. $4 \le x < 6$

16. $-2 < x < 0$

17. $x \ge 4$

18. $x \le 5$

19. $x < -4$

20. $x > 1$

In Problems 21–28, write each interval as an inequality involving x, and illustrate each inequality using the real number line.

21. $[2, 5]$

22. $(1, 2)$

23. $(-3, -2)$

24. $[0, 1)$

25. $[4, \infty)$

26. $(-\infty, 2]$

27. $(-\infty, -3)$

28. $(-8, \infty)$

In Problems 29–42, fill in the blank with the correct inequality symbol.

29. If $x < 5$, then $x - 5$ _____ 0.

30. If $x < -4$, then $x + 4$ _____ 0.

31. If $x > -4$, then $x + 4$ _____ 0.

32. If $x > 6$, then $x - 6$ _____ 0.

33. If $x \ge -4$, then $3x$ _____ −12.

34. If $x \le 3$, then $2x$ _____ 6.

35. If $x > 6$, then $-2x$ _____ −12.

36. If $x > -2$, then $-4x$ _____ 8.

37. If $x \ge 5$, then $-4x$ _____ −20.

38. If $x \le -4$, then $-3x$ _____ 12.

39. If $2x < 6$, then x _____ 3.

40. If $3x \le 12$, then x _____ 4.

41. If $-\dfrac{1}{2}x \le 3$, then x _____ −6.

42. If $-\dfrac{1}{4}x > 1$, then x _____ −4.

In Problems 43–92, solve each inequality algebraically. Express your answer using set notation or interval notation. Graph the solution set. Verify your solution using a graphing utility.

43. $x + 1 < 5$

44. $x - 6 < 1$

45. $1 - 2x \le 3$

46. $2 - 3x \le 5$

47. $3x - 7 > 2$

48. $2x + 5 > 1$

49. $3x - 1 \ge 3 + x$

50. $2x - 2 \ge 3 + x$

51. $-2(x + 3) < 8$

52. $-3(1 - x) < 12$

53. $4 - 3(1 - x) \le 3$

54. $8 - 4(2 - x) \le -2x$

55. $\dfrac{1}{2}(x - 4) > x + 8$

56. $3x + 4 > \dfrac{1}{3}(x - 2)$

57. $\dfrac{x}{2} \ge 1 - \dfrac{x}{4}$

58. $\dfrac{x}{3} \ge 2 + \dfrac{x}{6}$

59. $0 \le 2x - 6 \le 4$

60. $4 \le 2x + 2 \le 10$

61. $-5 \le 4 - 3x \le 2$

62. $-3 \le 3 - 2x \le 9$

63. $-3 < \dfrac{2x - 1}{4} < 0$

64. $0 < \dfrac{3x + 2}{2} < 4$

65. $1 < 1 - \dfrac{1}{2}x < 4$

66. $0 < 1 - \dfrac{1}{3}x < 1$

67. $(x + 2)(x - 3) > (x - 1)(x + 1)$

68. $(x - 1)(x + 1) > (x - 3)(x + 4)$

69. $x(4x + 3) \le (2x + 1)^2$

70. $x(9x - 5) \le (3x - 1)^2$

71. $\dfrac{1}{2} \le \dfrac{x + 1}{3} < \dfrac{3}{4}$

72. $\dfrac{1}{3} < \dfrac{x + 1}{2} \le \dfrac{2}{3}$

73. $|x| < 6$

74. $|x| \le 9$

75. $|x| \ge 4$

76. $|x| > 1$

77. $|2x| < 8$

78. $|3x| < 15$

79. $|3x| > 12$

80. $|2x| > 6$

81. $|x - 2| + 2 < 3$

82. $|x + 4| + 3 < 5$

83. $|3t - 2| \le 4$

84. $|2u + 5| \le 7$

85. $|x - 3| \geq 2$

86. $|x + 4| \geq 2$

87. $|1 - 4x| - 7 < -2$

88. $|1 - 2x| - 4 < -1$

89. $|1 - 2x| > |-3|$

90. $|2 - 3x| > |-1|$

91. $|2x + 1| < -1$

92. $|3x - 4| \geq 0$

---

93. Express the fact that x differs from 2 by less than $\frac{1}{2}$ as an inequality involving an absolute value. Solve for x.

94. Express the fact that x differs from -1 by less than 1 as an inequality involving an absolute value. Solve for x.

95. Express the fact that x differs from -3 by more than 2 as an inequality involving an absolute value. Solve for x.

96. Express the fact that x differs from 2 by more than 3 as an inequality involving an absolute value. Solve for x.

97. A young adult may be defined as someone older than 21, but less than 30 years of age. Express this statement using inequalities.

98. Middle-aged may be defined as being 40 or more and less than 60. Express the statement using inequalities.

99. **Body Temperature**   Normal human body temperature is 98.6°F. If a temperature x that differs from normal by at least 1.5° is considered unhealthy, write the condition for an unhealthy temperature x as an inequality involving an absolute value, and solve for x.

100. **Household Voltage**   In the United States, normal household voltage is 115 volts. However, it is not uncommon for actual voltage to differ from normal voltage by at most 5 volts. Express this situation as an inequality involving an absolute value. Use x as the actual voltage and solve for x.

101. **Life Expectancy**   Metropolitan Life Insurance Co. reported that an average 25-year-old male in 1996 could expect to live at least 48.4 more years, and an average 25-year-old female in 1996 could expect to live at least 54.7 more years.

(a) To what age can an average 25-year-old male expect to live? Express your answer as an inequality.

(b) To what age can an average 25-year-old female expect to live? Express your answer as an inequality.

(c) Who can expect to live longer, a male or a female? By how many years?

102. **General Chemistry**   For a certain ideal gas, the volume V in cubic centimeters equals 20 times the temperature T in Kelvins (K). If the temperature varies from 353 to 393 K, inclusive, what is the corresponding range of the volume of the gas?

103. **Real Estate**   A real estate agent agrees to sell a large apartment complex according to the following commission schedule: $45,000 plus 25% of the selling price in excess of $900,000. Assuming that the complex will sell at some price between $900,000 and $1,100,000, inclusive, over what range does the agent's commission vary? How does the commission vary as a percent of selling price?

104. **Sales Commission**   A used car salesperson is paid a commission of $25 plus 40% of the selling price in excess of owner's cost. The owner claims that used cars typically sell for at least owner's cost plus $70 and at most owner's cost plus $300. For each sale made, over what range can the salesperson expect the commission to vary?

105. **Federal Tax Withholding**   The percentage method of withholding for federal income tax (1998)* states that a single person whose weekly wages, after subtracting withholding allowances, are over $517, but not over $1105, shall have $69.90 plus 28% of the excess over $517 withheld. Over what range does the amount withheld vary if the weekly wages vary from $525 to $600, inclusive?

106. **Federal Tax Withholding**   Rework Problem 105 if the weekly wages vary from $600 to $700, inclusive.

107. **Electricity Rates**   Commonwealth Edison Company's summer charge for electricity is 10.494¢ per kilowatt-hour.† In addition, each monthly bill contains a customer charge of $9.36. If last summer's bills ranged from a low of $80.24 to a high of $271.80, over what range did usage vary (in kilowatt-hours)?

108. **Water Bills**   The Village of Oak Lawn charges homeowners $21.60 per quarter-year plus $1.70 per 1000 gallons for water usage in excess of 12,000 gallons.‡ In 2000, one homeowner's quarterly bill ranged from a high of $65.75 to a low of $28.40. Over what range did water usage vary?

109. **Markup of a New Car**   The markup over dealer's cost of a new car ranges from 12% to 18%. If the sticker price is $8800, over what range will the dealer's cost vary?

110. **IQ Tests**   A standard intelligence test has an average score of 100. According to statistical theory, of the people who take the test, the 2.5% with the highest scores will have scores of more than 1.96σ above the average, where σ (sigma, a number called the *standard deviation*) depends on the nature of the test. If $\sigma = 12$ for this test and there is (in principle) no upper limit to the score possible on the test, write the interval of possible test scores of the people in the top 2.5%.

Source: *Employer's Tax Guide.* Department of the Treasury, Internal Revenue Service, 1998.
†*Source:* Commonwealth Edison Co., Chicago, Illinois, 2000.
‡*Source:* Village of Oak Lawn, Illinois, 2000.

111. Computing Grades In your Economics 101 class, you have scores of 68, 82, 87, and 89 on the first four of five tests. To get a grade of B, the average of the first five test scores must be greater than or equal to 80 and less than 90. Solve an inequality to find the range of the score that you need on the last test to get a B.

What do I need to get a B?

112. Computing Grades Repeat Problem 111 if the fifth test counts double.

113. A car that averages 25 miles per gallon has a tank that holds 20 gallons of gasoline. After a trip that covered at least 300 miles, the car ran out of gasoline. What is the range of the amount of gasoline (in gallons) that was in the tank at the start of the trip?

114. Repeat Problem 113 if the same car runs out of gasoline after a trip of no more than 250 miles.

115. Arithmetic Mean If $a < b$, show that $a < \dfrac{a+b}{2} < b$.
The number $\dfrac{a+b}{2}$ is called the **arithmetic mean** of a and b.

116. Refer to Problem 115. Show that the arithmetic mean of a and b is equidistant from a and b.

117. Geometric Mean If $0 < a < b$, show that $a < \sqrt{ab} < b$. The number $\sqrt{ab}$ is called the **geometric mean** of a and b.

118. Refer to Problems 115 and 117. Show that the geometric mean of a and b is less than the arithmetic mean of a and b.

119. Harmonic Mean For $0 < a < b$, let h be defined by
$$\frac{1}{h} = \frac{1}{2}\left(\frac{1}{a} + \frac{1}{b}\right)$$
Show that $a < h < b$. The number h is called the **harmonic mean** of a and b.

120. Refer to Problems 115, 117, and 119. Show that the harmonic mean of a and b equals the geometric mean squared divided by the arithmetic mean.

121. Make up an inequality that has no solution. Make up one that has exactly one solution.

122. How would you explain to a fellow student the underlying reason for the multiplication property for inequalities (page 597); that is, the sense or direction of an inequality remains the same if each side is multiplied by a positive real number, while the direction is reversed if each side is multiplied by a negative real number.

PREPARING FOR THIS SECTION

Before getting started, review the following:

✓ Exponents, Square Roots (Section A.1, pp. 561–563)

A.6 NTH ROOTS; RATIONAL EXPONENTS

OBJECTIVES

1. Work with *n*th Roots
2. Simplify Radicals
3. Rationalize Denominators
4. Simplify Expressions with Rational Exponents

*n*th Roots

The **principal *n*th root of a number** a, symbolized by $\sqrt[n]{a}$, where $n \geq 2$ is an integer, is defined as follows:

$$\sqrt[n]{a} = b \quad \text{means} \quad a = b^n$$

where $a \geq 0$ and $b \geq 0$ if $n \geq 2$ is even, and a, b are any real numbers if $n \geq 3$ is odd.

Notice that if a is negative and n is even then $\sqrt[n]{a}$ is not defined. When it is defined, the principal *n*th root of a number is unique.

1   The symbol $\sqrt[n]{a}$ for the principal *n*th root of *a* is sometimes called a **radical**; the integer *n* is called the **index**, and *a* is called the **radicand**. If the index of a radical is 2, we call $\sqrt[n]{a}$ the **square root** of *a* and omit the index 2 by simply writing $\sqrt{a}$. If the index is 3, we call $\sqrt[3]{a}$ the **cube root** of *a*.

EXAMPLE 1   **Evaluating Principal *n*th Roots**

(a) $\sqrt[3]{8} = \sqrt[3]{2^3} = 2$ (b) $\sqrt[3]{-64} = \sqrt[3]{(-4)^3} = -4$

(c) $\sqrt[4]{\dfrac{1}{16}} = \sqrt[4]{\left(\dfrac{1}{2}\right)^4} = \dfrac{1}{2}$ (d) $\sqrt[6]{(-2)^6} = |-2| = 2$

These are examples of **perfect roots**, since each simplifies to a rational number. Notice the absolute value in Example 1(d). If *n* is even, the principal *n*th root must be nonnegative.

In general if $n \geq 2$ is a positive integer and *a* is a real number, we have

$$\sqrt[n]{a^n} = a, \qquad \text{if } n \geq 3 \text{ is odd} \qquad \text{(1a)}$$
$$\sqrt[n]{a^n} = |a|, \qquad \text{if } n \geq 2 \text{ is even} \qquad \text{(1b)}$$

╌ **NOW WORK PROBLEM 1.**

Properties of Radicals

Let $n \geq 2$ and $m \geq 2$ denote positive integers, and let *a* and *b* represent real numbers. Assuming that all radicals are defined, we have the following properties:

$$\sqrt[n]{ab} = \sqrt[n]{a}\,\sqrt[n]{b} \qquad \text{(2a)}$$
$$\sqrt[n]{\dfrac{a}{b}} = \dfrac{\sqrt[n]{a}}{\sqrt[n]{b}} \qquad \text{(2b)}$$
$$\sqrt[n]{a^m} = \left(\sqrt[n]{a}\right)^m \qquad \text{(2c)}$$

2   When used in reference to radicals, the direction to "simplify" will mean to remove from the radicals any perfect roots that occur as factors. Let's look at some examples of how the preceding rules are applied to simplify radicals.

EXAMPLE 2   **Simplifying Radicals**

(a) $\sqrt{32} = \sqrt{16 \cdot 2} = \sqrt{16} \cdot \sqrt{2} = 4\sqrt{2}$
   ↑
   16 is a perfect square.

(b) $\sqrt[3]{16} = \sqrt[3]{8 \cdot 2} = \sqrt[3]{8} \cdot \sqrt[3]{2} = 2\sqrt[3]{2}$
   ↑   ↑
   8 is a perfect cube.   (2a)

(c) $\sqrt[3]{-16x^4} = \sqrt[3]{-8 \cdot 2 \cdot x^3 \cdot x} = \sqrt[3]{(-8x^3)(2x)}$
   ↑   ↑
   Factor perfect   Combine perfect
   cubes inside radical.   cubes.

$\qquad = \sqrt[3]{(-2x)^3 \cdot 2x} = \sqrt[3]{(-2x)^3} \cdot \sqrt[3]{2x}$
   ↑
   $\qquad = -2x\sqrt[3]{2x}$ (2a)

╌ **NOW WORK PROBLEM 7.**

EXAMPLE 3   **Combining Like Radicals**

(a) $-8\sqrt{12} + \sqrt{3} = -8\sqrt{4\cdot 3} + \sqrt{3} = -8\cdot\sqrt{4}\sqrt{3} + \sqrt{3}$
$$= -16\sqrt{3} + \sqrt{3} = -15\sqrt{3}$$

(b) $\sqrt[3]{8x^4} + \sqrt[3]{-x} + 4\sqrt[3]{27x} = \sqrt[3]{2^3x^3x} + \sqrt[3]{-1\cdot x} + 4\sqrt[3]{3^3x}$
$$= \sqrt[3]{(2x)^3}\cdot\sqrt[3]{x} + \sqrt[3]{-1}\cdot\sqrt[3]{x} + 4\sqrt[3]{3^3}\cdot\sqrt[3]{x}$$
$$= 2x\sqrt[3]{x} - 1\cdot\sqrt[3]{x} + 12\sqrt[3]{x}$$
$$= (2x + 11)\sqrt[3]{x}$$

NOW WORK PROBLEM **25.**

Rationalizing

3   When radicals occur in quotients, it is customary to rewrite the quotient so that the denominator contains no square roots. This process is referred to as **rationalizing the denominator**.

The idea is to multiply by an appropriate expression so that the new denominator contains no radicals. For example:

If Denominator Contains the Factor	Multiply By	To Obtain Denominator Free of Radicals
$\sqrt{3}$	$\sqrt{3}$	$(\sqrt{3})^2 = 3$
$\sqrt{3} + 1$	$\sqrt{3} - 1$	$(\sqrt{3})^2 - 1^2 = 3 - 1 = 2$
$\sqrt{2} - 3$	$\sqrt{2} + 3$	$(\sqrt{2})^2 - 3^2 = 2 - 9 = -7$
$\sqrt{5} - \sqrt{3}$	$\sqrt{5} + \sqrt{3}$	$(\sqrt{5})^2 - (\sqrt{3})^2 = 5 - 3 = 2$
$\sqrt[3]{4}$	$\sqrt[3]{2}$	$\sqrt[3]{4}\cdot\sqrt[3]{2} = \sqrt[3]{8} = 2$

In rationalizing the denominator of a quotient, be sure to multiply both the numerator and the denominator by the expression.

EXAMPLE 4   **Rationalizing Denominators**

Rationalize the denominator of each expression.

(a) $\dfrac{4}{\sqrt{2}}$ (b) $\dfrac{\sqrt{3}}{\sqrt[3]{2}}$ (c) $\dfrac{\sqrt{x} - 2}{\sqrt{x} + 2}, x \geq 0$

Solution   (a) $\dfrac{4}{\sqrt{2}} = \dfrac{4}{\sqrt{2}}\cdot\dfrac{\sqrt{2}}{\sqrt{2}} = \dfrac{4\sqrt{2}}{(\sqrt{2})^2} = \dfrac{4\sqrt{2}}{2} = 2\sqrt{2}$

Multiply by $\dfrac{\sqrt{2}}{\sqrt{2}}$.

(b) $\dfrac{\sqrt{3}}{\sqrt[3]{2}} = \dfrac{\sqrt{3}}{\sqrt[3]{2}}\cdot\dfrac{\sqrt[3]{4}}{\sqrt[3]{4}} = \dfrac{\sqrt{3}\sqrt[3]{4}}{\sqrt[3]{8}} = \dfrac{\sqrt{3}\sqrt[3]{4}}{2}$

Multiply by $\dfrac{\sqrt[3]{4}}{\sqrt[3]{4}}$.

(c) $\dfrac{\sqrt{x} - 2}{\sqrt{x} + 2} = \dfrac{\sqrt{x} - 2}{\sqrt{x} + 2}\cdot\dfrac{\sqrt{x} - 2}{\sqrt{x} - 2} = \dfrac{(\sqrt{x} - 2)^2}{(\sqrt{x})^2 - 2^2}$

$$= \dfrac{(\sqrt{x})^2 - 4\sqrt{x} + 4}{x - 4} = \dfrac{x - 4\sqrt{x} + 4}{x - 4}$$

NOW WORK PROBLEM **33.**

Rational Exponents

4 Radicals are used to define rational exponents.

If *a* is a real number and $n \geq 2$ is an integer, then

$$a^{1/n} = \sqrt[n]{a} \qquad (3)$$

provided that $\sqrt[n]{a}$ exists.

EXAMPLE 5 **Using Equation (3)**

(a) $4^{1/2} = \sqrt{4} = 2$ (b) $(-27)^{1/3} = \sqrt[3]{-27} = -3$

(c) $8^{1/2} = \sqrt{8} = 2\sqrt{2}$ (d) $16^{1/3} = \sqrt[3]{16} = 2\sqrt[3]{2}$ ∎

If *a* is a real number and *m* and *n* are integers containing no common factors with $n \geq 2$, then

$$a^{m/n} = \sqrt[n]{a^m} = \left(\sqrt[n]{a}\right)^m \qquad (4)$$

provided that $\sqrt[n]{a}$ exists.

We have two comments about equation (4):

1. The exponent m/n must be in lowest terms and *n* must be positive.
2. In simplifying $a^{m/n}$, either $\sqrt[n]{a^m}$ or $\left(\sqrt[n]{a}\right)^m$ may be used. Generally, taking the root first, as in $\left(\sqrt[n]{a}\right)^m$, is easier.

EXAMPLE 6 **Using Equation (4)**

(a) $4^{3/2} = (\sqrt{4})^3 = 2^3 = 8$ (b) $(-8)^{4/3} = \left(\sqrt[3]{-8}\right)^4 = (-2)^4 = 16$

(c) $(32)^{-2/5} = \left(\sqrt[5]{32}\right)^{-2} = 2^{-2} = \dfrac{1}{4}$ ∎

✏ **NOW WORK PROBLEM 41.**

It can be shown that the laws of exponents hold for rational exponents.

EXAMPLE 7 **Simplifying Expressions with Rational Exponents**

Simplify each expression. Express your answer so that only positive exponents occur. Assume that the variables are positive.

(a) $\left(\dfrac{2x^{1/3}}{y^{2/3}}\right)^{-3}$ (b) $(x^{2/3}y)(x^{-2}y)^{1/2}$

Solution (a) $\left(\dfrac{2x^{1/3}}{y^{2/3}}\right)^{-3} = \left(\dfrac{y^{2/3}}{2x^{1/3}}\right)^3 = \dfrac{(y^{2/3})^3}{(2x^{1/3})^3} = \dfrac{y^2}{2^3(x^{1/3})^3} = \dfrac{y^2}{8x}$

(b) $(x^{2/3}y)(x^{-2}y)^{1/2} = (x^{2/3}y)[(x^{-2})^{1/2}y^{1/2}]$

$= x^{2/3}yx^{-1}y^{1/2} = (x^{2/3}x^{-1})(y \cdot y^{1/2})$

$= x^{-1/3}y^{3/2} = \dfrac{y^{3/2}}{x^{1/3}}$ ∎

✏ **NOW WORK PROBLEM 57.**

 The next two examples illustrate some algebra that you will need to know for certain calculus problems.

EXAMPLE 8 **Writing an Expression as a Single Quotient**

Write the following expression as a single quotient in which only positive exponents appear.

$$(x^2 + 1)^{1/2} + x \cdot \frac{1}{2}(x^2 + 1)^{-1/2} \cdot 2x$$

Solution $(x^2 + 1)^{1/2} + x \cdot \dfrac{1}{2}(x^2 + 1)^{-1/2} \cdot 2x = (x^2 + 1)^{1/2} + \dfrac{x^2}{(x^2 + 1)^{1/2}}$

$= \dfrac{(x^2 + 1)^{1/2}(x^2 + 1)^{1/2} + x^2}{(x^2 + 1)^{1/2}}$

$= \dfrac{(x^2 + 1) + x^2}{(x^2 + 1)^{1/2}}$

$= \dfrac{2x^2 + 1}{(x^2 + 1)^{1/2}}$ ∎

✏ **NOW WORK PROBLEM 61.**

EXAMPLE 9 **Factoring an Expression Containing Rational Exponents**

Factor: $4x^{1/3}(2x + 1) + 2x^{4/3}$

Solution We begin by looking for factors that are common to the two terms. Notice that 2 and $x^{1/3}$ are common factors. Then,

$$4x^{1/3}(2x + 1) + 2x^{4/3} = 2x^{1/3}[2(2x + 1) + x]$$

$$= 2x^{1/3}(5x + 2)$$ ∎

✏ **NOW WORK PROBLEM 67.**

A.6 Exercises

In Problems 1–28, simplify each expression. Assume that all variables are positive when they appear.

1. $\sqrt[3]{27}$ 2. $\sqrt[4]{16}$ 3. $\sqrt[3]{-8}$ 4. $\sqrt[3]{-1}$

5. $\sqrt{8}$ 6. $\sqrt[3]{54}$ 7. $\sqrt[3]{-8x^4}$ 8. $\sqrt[4]{48x^5}$

9. $\sqrt[4]{x^{12}y^8}$ 10. $\sqrt[5]{x^{10}y^5}$ 11. $\sqrt[4]{\dfrac{x^9 y^7}{xy^3}}$ 12. $\sqrt[3]{\dfrac{3xy^2}{81x^4y^2}}$

13. $\sqrt{36x}$ 14. $\sqrt{9x^5}$ 15. $\sqrt{3x^2}\,\sqrt{12x}$ 16. $\sqrt{5x}\,\sqrt{20x^3}$

17. $\left(\sqrt{5}\,\sqrt[3]{9}\right)^2$ 18. $\left(\sqrt[3]{3}\,\sqrt{10}\right)^4$ 19. $(3\sqrt{6})(2\sqrt{2})$ 20. $(5\sqrt{8})(-3\sqrt{3})$

21. $(\sqrt{3}+3)(\sqrt{3}-1)$ 22. $(\sqrt{5}-2)(\sqrt{5}+3)$ 23. $(\sqrt{x}-1)^2$ 24. $(\sqrt{x}+\sqrt{5})^2$

25. $3\sqrt{2}-4\sqrt{8}$ 26. $\sqrt[3]{-x^4}+\sqrt[3]{8x}$ 27. $\sqrt[3]{16x^4}-\sqrt[3]{2x}$ 28. $\sqrt[4]{32x}+\sqrt[4]{2x^5}$

In Problems 29–40, rationalize the denominator of each expression. Assume that all variables are positive when they appear.

29. $\dfrac{1}{\sqrt{2}}$ 30. $\dfrac{6}{\sqrt[3]{4}}$ 31. $\dfrac{-\sqrt{3}}{\sqrt{5}}$ 32. $\dfrac{-\sqrt[3]{3}}{\sqrt{8}}$

33. $\dfrac{\sqrt{3}}{5-\sqrt{2}}$ 34. $\dfrac{\sqrt{2}}{\sqrt{7}+2}$ 35. $\dfrac{2-\sqrt{5}}{2+3\sqrt{5}}$ 36. $\dfrac{\sqrt{3}-1}{2\sqrt{3}+3}$

37. $\dfrac{5}{\sqrt[3]{2}}$ 38. $\dfrac{-2}{\sqrt[3]{9}}$ 39. $\dfrac{\sqrt{x+h}-\sqrt{x}}{\sqrt{x+h}+\sqrt{x}}$ 40. $\dfrac{\sqrt{x+h}+\sqrt{x-h}}{\sqrt{x+h}-\sqrt{x-h}}$

In Problems 41–52, simplify each expression.

41. $8^{2/3}$ 42. $4^{3/2}$ 43. $(-27)^{1/3}$ 44. $16^{3/4}$ 45. $16^{3/2}$ 46. $64^{3/2}$

47. $9^{-3/2}$ 48. $25^{-5/2}$ 49. $\left(\dfrac{9}{8}\right)^{3/2}$ 50. $\left(\dfrac{27}{8}\right)^{2/3}$ 51. $\left(\dfrac{8}{9}\right)^{-3/2}$ 52. $\left(\dfrac{8}{27}\right)^{-2/3}$

In Problems 53–60, simplify each expression. Express your answer so that only positive exponents occur. Assume that the variables are positive.

53. $x^{3/4}x^{1/3}x^{-1/2}$ 54. $x^{2/3}x^{1/2}x^{-1/4}$ 55. $(x^3y^6)^{1/3}$ 56. $(x^4y^8)^{3/4}$

57. $(x^2y)^{1/3}(xy^2)^{2/3}$ 58. $(xy)^{1/4}(x^2y^2)^{1/2}$ 59. $(16x^2y^{-1/3})^{3/4}$ 60. $(4x^{-1}y^{1/3})^{3/2}$

In Problems 61–66, write each expression as a single quotient in which only positive exponents and/or radicals appear.

61. $\dfrac{x}{(1+x)^{1/2}}+2(1+x)^{1/2}$ 62. $\dfrac{1+x}{2x^{1/2}}+x^{1/2}$ 63. $\dfrac{\sqrt{1+x}-x\cdot\dfrac{1}{2\sqrt{1+x}}}{1+x}$

64. $\dfrac{\sqrt{x^2+1}-x\cdot\dfrac{2x}{2\sqrt{x^2+1}}}{x^2+1}$ 65. $\dfrac{(x+4)^{1/2}-2x(x+4)^{-1/2}}{x+4}$ 66. $\dfrac{(9-x^2)^{1/2}+x^2(9-x^2)^{-1/2}}{9-x^2}$

In Problems 67–72, factor each expression.

67. $(x+1)^{3/2}+x\cdot\dfrac{3}{2}(x+1)^{1/2}$ 68. $(x^2+4)^{4/3}+x\cdot\dfrac{4}{3}(x^2+4)^{1/3}\cdot2x$

69. $6x^{1/2}(x^2+x)-8x^{3/2}-8x^{1/2}$ 70. $6x^{1/2}(2x+3)+x^{3/2}\cdot8$

71. $x\left(\dfrac{1}{2}\right)(8-x^2)^{-1/2}(-2x)+(8-x^2)^{1/2}$ 72. $2x(1-x^2)^{3/2}+x^2\left(\dfrac{3}{2}\right)(1-x^2)^{1/2}(-2x)$

A.7 LINES

OBJECTIVES

1. Calculate and Interpret the Slope of a Line
2. Graph Lines Given a Point and the Slope
3. Find the Equation of Vertical Lines
4. Use the Point-Slope Form of a Line; Identify Horizontal Lines
5. Find the Equation of a Line Given Two Points
6. Write the Equation of a Line in Slope–Intercept Form
7. Identify the Slope and y-intercept of a Line from Its Equation
8. Write the Equation of a Line in General Form
9. Define Parallel Lines
10. Find Equations of Parallel Lines
11. Define Perpendicular Lines
12. Find Equations of Perpendicular Lines

In this section we study a certain type of equation that contains two variables, called a *linear equation*, and its graph, a *line*.

Slope of a Line

1. Consider the staircase illustrated in Figure 40. Each step contains exactly the same horizontal **run** and the same vertical **rise**. The ratio of the rise to the run, called the *slope*, is a numerical measure of the steepness of the staircase. For example, if the run is increased and the rise remains the same, the staircase becomes less steep. If the run is kept the same, but the rise is increased, the staircase becomes more steep. This important characteristic of a line is best defined using rectangular coordinates.

Figure 40

Line

} Rise

Run

Let $P = (x_1, y_1)$ and $Q = (x_2, y_2)$ be two distinct points. If $x_1 \neq x_2$, the **slope** m of the nonvertical line L containing P and Q is defined by the formula

$$m = \frac{y_2 - y_1}{x_2 - x_1}, \qquad x_1 \neq x_2 \qquad (1)$$

If $x_1 = x_2$, L is a **vertical line** and the slope m of L is **undefined** (since this results in division by 0).

Figure 41(a) provides an illustration of the slope of a nonvertical line; Figure 41(b) illustrates a vertical line.

Figure 41

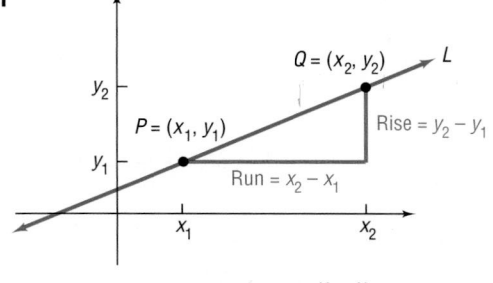

(a) Slope of L is $m = \dfrac{y_2 - y_1}{x_2 - x_1}$

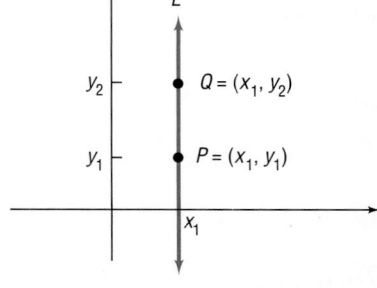

(b) Slope is undefined; L is vertical

As Figure 41(a) illustrates, the slope m of a nonvertical line may be viewed as

$$m = \frac{y_2 - y_1}{x_2 - x_1} = \frac{\text{Rise}}{\text{Run}}$$

We can also express the slope m of a nonvertical line as

$$m = \frac{y_2 - y_1}{x_2 - x_1} = \frac{\text{Change in } y}{\text{Change in } x} = \frac{\Delta y}{\Delta x}$$

That is, the slope m of a nonvertical line L measures the amount that y changes as x changes from x_1 to x_2. This is called the **average rate of change** of y with respect to x.

Two comments about computing the slope of a nonvertical line may prove helpful:

1. Any two distinct points on the line can be used to compute the slope of the line. (See Figure 42 for justification.)

Figure 42
Triangles *ABC* and *PQR* are similar (equal angles). Hence, ratios of corresponding sides are proportional so that

Slope using *P* and *Q* $= \dfrac{y_2 - y_1}{x_2 - x_1}$

$=$ Slope using *A* and *B* $= \dfrac{d(B, C)}{d(A, C)}$

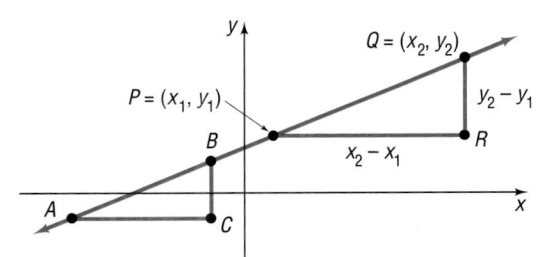

2. The slope of a line may be computed from $P = (x_1, y_1)$ to $Q = (x_2, y_2)$ or from Q to P because

$$\frac{y_2 - y_1}{x_2 - x_1} = \frac{y_1 - y_2}{x_1 - x_2}$$

EXAMPLE 1   Finding and Interpreting the Slope of a Line Containing Two Points

The slope m of the line containing the points $(1, 2)$ and $(5, -3)$ may be computed as

$$m = \frac{-3 - 2}{5 - 1} = \frac{-5}{4} = -\frac{5}{4} \quad \text{or as} \quad m = \frac{2 - (-3)}{1 - 5} = \frac{5}{-4} = -\frac{5}{4}$$

For every 4 unit change in x, y will change by -5 units. That is, if x increases by 4 units, then y will decrease by 5 units. The average rate of change of y with respect to x is $-\dfrac{5}{4}$.

NOW WORK PROBLEMS 1 AND 7.

To get a better idea of the meaning of the slope m of a line, consider the following:

SEEING THE CONCEPT   On the same square screen graph the following equations:

$Y_1 = 0$	*Slope of line is 0.*
$Y_2 = \dfrac{1}{4}x$	*Slope of line is $\dfrac{1}{4}$.*
$Y_3 = \dfrac{1}{2}x$	*Slope of line is $\dfrac{1}{2}$.*
$Y_4 = x$	*Slope of line is 1.*
$Y_5 = 2x$	*Slope of line is 2.*
$Y_6 = 6x$	*Slope of line is 6.*

Figure 43

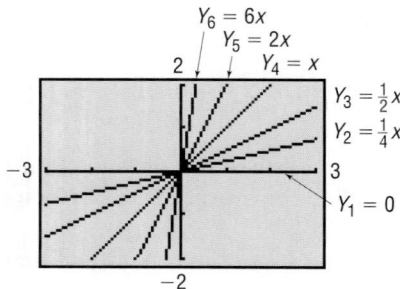

See Figure 43.

SEEING THE CONCEPT   On the same square screen, graph the following equations:

$Y_1 = 0$	*Slope of line is 0.*
$Y_2 = -\dfrac{1}{4}x$	*Slope of line is $-\dfrac{1}{4}$.*
$Y_3 = -\dfrac{1}{2}x$	*Slope of line is $-\dfrac{1}{2}$.*
$Y_4 = -x$	*Slope of line is −1.*
$Y_5 = -2x$	*Slope of line is −2.*
$Y_6 = -6x$	*Slope of line is −6.*

Figure 44

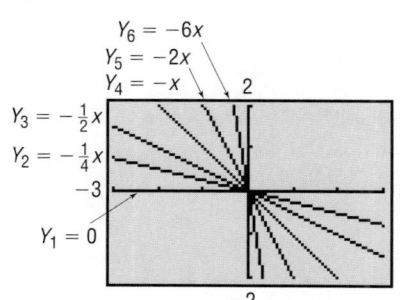

See Figure 44.

Figures 43 and 44 illustrate the following facts:

1. When the slope of a line is positive, the line slants upward from left to right.
2. When the slope of a line is negative, the line slants downward from left to right.
3. When the slope is 0, the line is horizontal.

Figures 43 and 44 also illustrate that the closer the line is to the vertical position, the greater the magnitude of the slope.

2   The next example illustrates how the slope of a line can be used to graph the line.

EXAMPLE 2   **Graphing a Line Given a Point and a Slope**

Draw a graph of the line that contains the point $(3, 2)$ and has a slope of:

(a) $\dfrac{3}{4}$ (b) $-\dfrac{4}{5}$

Solution   (a) Slope $= \dfrac{\text{Rise}}{\text{Run}}$. The fact that the slope is $\dfrac{3}{4}$ means that for every horizontal movement (run) of 4 units to the right there will be a vertical

Figure 45

Slope $= \dfrac{3}{4}$

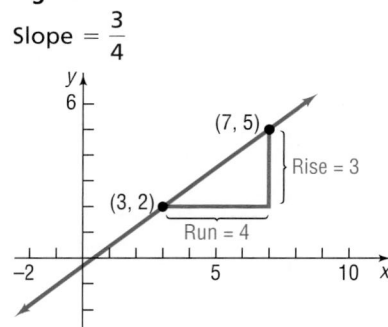

movement (rise) of 3 units. If we start at the given point $(3, 2)$ and move 4 units to the right and 3 units up, we reach the point $(7, 5)$. By drawing the line through this point and the point $(3, 2)$, we have the graph. See Figure 45.

(b) The fact that the slope is

$$-\frac{4}{5} = \frac{-4}{5} = \frac{\text{Rise}}{\text{Run}}$$

means that for every horizontal movement of 5 units to the right there will be a corresponding vertical movement of -4 units (a downward movement). If we start at the given point $(3, 2)$ and move 5 units to the right and then 4 units down, we arrive at the point $(8, -2)$. By drawing the line through these points, we have the graph. See Figure 46.

Figure 46

Slope $= -\dfrac{4}{5}$

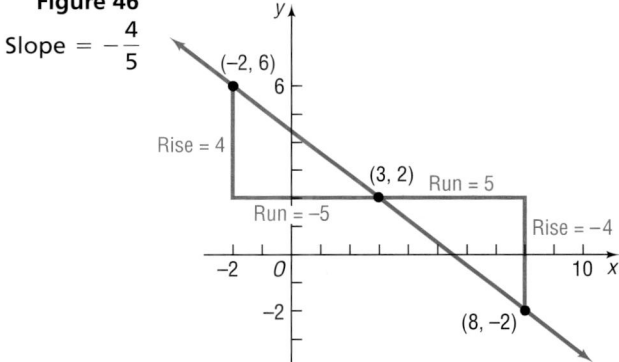

Alternatively, we can set

$$-\frac{4}{5} = \frac{4}{-5} = \frac{\text{Rise}}{\text{Run}}$$

so that for every horizontal movement of -5 units (a movement to the left) there will be a corresponding vertical movement of 4 units (upward). This approach brings us to the point $(-2, 6)$, which is also on the graph shown in Figure 46.

NOW WORK PROBLEM 13.

Equations of Lines

③ Now that we have discussed the slope of a line, we are ready to derive equations of lines. As we shall see, there are several forms of the equation of a line. Let's start with an example.

EXAMPLE 3	Graphing a Line

Graph the equation: $x = 3$

Solution To graph $x = 3$ by hand, recall that we are looking for all points (x, y) in the plane for which $x = 3$. No matter what y-coordinate is used, the corresponding x-coordinate always equals 3. Consequently, the graph of the equation $x = 3$ is a vertical line with x-intercept 3 and undefined slope. See Figure 47(a).

Figure 47
$x = 3$

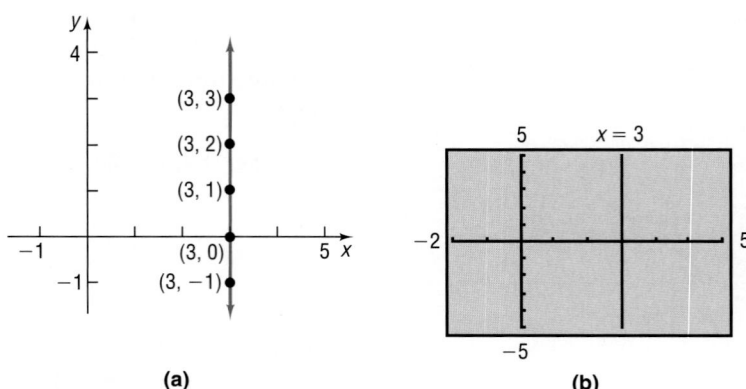

(a)

(b)

To use a graphing utility, we need to express the equation in the form $y = \{\text{expression in } x\}$. But $x = 3$ cannot be put into this form, so an alternative method must be used. Consult your manual to determine the methodology required to draw vertical lines. Figure 47(b) shows the graph that you should obtain. ■

As suggested by Example 3, we have the following result:

Theorem

Equation of a Vertical Line

A vertical line is given by an equation of the form

$$x = a$$

where a is the x-intercept.

■

Figure 48

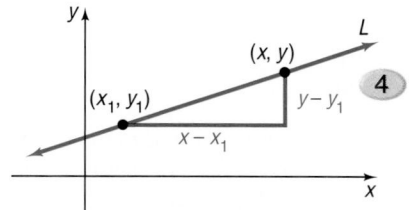

Now let L be a nonvertical line with slope m and containing the point (x_1, y_1). See Figure 48. For any other point (x, y) on L, we have

$$m = \frac{y - y_1}{x - x_1} \quad \text{or} \quad y - y_1 = m(x - x_1)$$

Theorem

Point–Slope Form of an Equation of a Line

An equation of a nonvertical line of slope m that contains the point (x_1, y_1) is

$$y - y_1 = m(x - x_1) \qquad (2)$$

■

EXAMPLE 4

Using the Point–Slope Form of a Line

An equation of the line with slope 4 and containing the point $(1, 2)$ can be found by using the point–slope form with $m = 4$, $x_1 = 1$, and $y_1 = 2$.

$$y - y_1 = m(x - x_1)$$
$$y - 2 = 4(x - 1) \qquad m = 4, x_1 = 1, y_1 = 2$$
$$y = 4x - 2$$

See Figure 49.

Figure 49

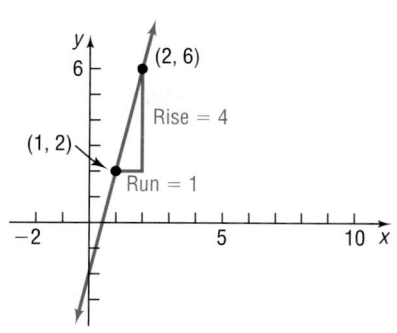

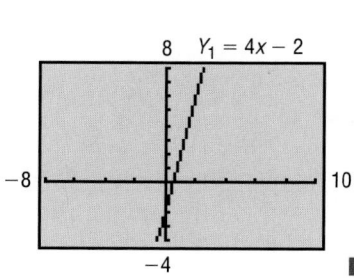

EXAMPLE 5   **Finding the Equation of a Horizontal Line**

Find an equation of the horizontal line containing the point $(3, 2)$.

Solution   Because all the y-coordinates of points on a horizontal line are equal, the slope of a horizontal line is 0. To get an equation, we use the point–slope form with $m = 0$, $x_1 = 3$, and $y_1 = 2$.

$$y - y_1 = m(x - x_1)$$
$$y - 2 = 0 \cdot (x - 3) \qquad \text{\scriptsize{$m = 0, x_1 = 3,$ and $y_1 = 2$}}$$
$$y - 2 = 0$$
$$y = 2$$

See Figure 50 for the graph.

Figure 50
$y = 2$

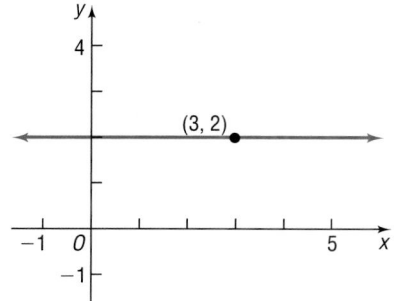

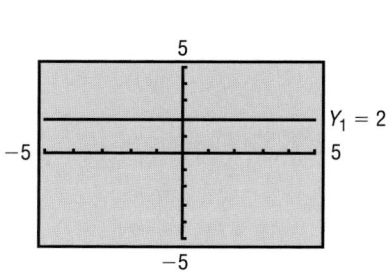

As suggested by Example 5, we have the following result:

Theorem   **Equation of a Horizontal Line**

A horizontal line is given by an equation of the form

$$y = b$$

where b is the y-intercept.

EXAMPLE 6   **Finding an Equation of a Line Given Two Points**

Find an equation of the line L containing the points $(2, 3)$ and $(-4, 5)$. Graph the line L.

Figure 51

$$y - 3 = -\frac{1}{3}(x - 2)$$

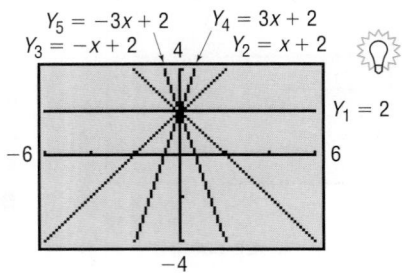

Solution We first compute the slope of the line.

$$m = \frac{5 - 3}{-4 - 2} = \frac{2}{-6} = -\frac{1}{3}$$

We use the point $(2, 3)$ and the slope $m = -\frac{1}{3}$ to get the point–slope form of the equation of the line.

$$y - 3 = -\frac{1}{3}(x - 2)$$

See Figure 51 for the graph. ∎

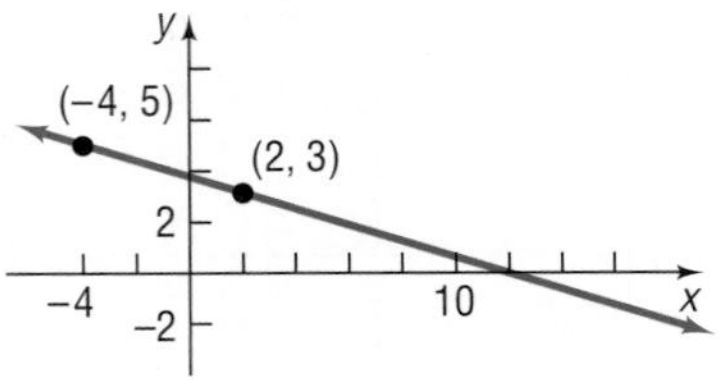

In the solution to Example 6, we could have used the other point, $(-4, 5)$, instead of the point $(2, 3)$. The equation that results, although it looks different, is equivalent to the equation that we obtained in the example. (Try it for yourself.)

➤ **NOW WORK PROBLEM 27.**

6 Another useful equation of a line is obtained when the slope m and y-intercept b are known. In this event, we know both the slope m of the line and a point $(0, b)$ on the line; thus, we may use the point–slope form, equation (2), to obtain the following equation:

$$y - b = m(x - 0) \quad \text{or} \quad y = mx + b$$

Theorem

> **Slope–Intercept Form of an Equation of a Line**
>
> An equation of a line L with slope m and y-intercept b is
>
> $$y = mx + b \qquad (3)$$

∎

Figure 52

$$y = mx + 2$$

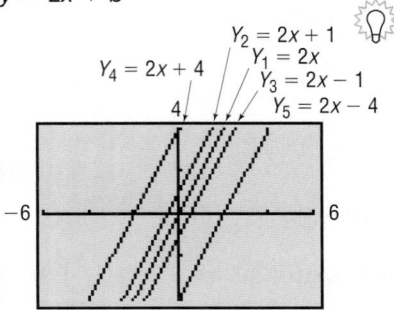

SEEING THE CONCEPT To see the role that the slope m plays, graph the following lines on the same square screen.

$$Y_1 = 2$$
$$Y_2 = x + 2$$
$$Y_3 = -x + 2$$
$$Y_4 = 3x + 2$$
$$Y_5 = -3x + 2$$

See Figure 52. What do you conclude about the lines $y = mx + 2$?

SEEING THE CONCEPT To see the role of the y-intercept b, graph the following lines on the same square screen.

$$Y_1 = 2x$$
$$Y_2 = 2x + 1$$
$$Y_3 = 2x - 1$$
$$Y_4 = 2x + 4$$
$$Y_5 = 2x - 4$$

Figure 53

$$y = 2x + b$$

See Figure 53. What do you conclude about the lines $y = 2x + b$?

7　When the equation of a line is written in slope–intercept form, it is easy to find the slope m and y-intercept b of the line. For example, suppose that the equation of a line is

$$y = -2x + 3$$

Compare it to $y = mx + b$.

$$y = -2x + 3$$
$$\qquad\quad\uparrow\qquad\uparrow$$
$$y = \quad mx \; + \; b$$

The slope of this line is -2 and its y-intercept is 3.

NOW WORK PROBLEM **59**.

EXAMPLE 7　Finding the Slope and *y*-Intercept

Find the slope m and y-intercept b of the equation $2x + 4y = 8$. Graph the equation.

Solution　To obtain the slope and y-intercept, we transform the equation into its slope–intercept form by solving for y.

$$2x + 4y = 8$$
$$4y = -2x + 8$$
$$y = -\frac{1}{2}x + 2 \qquad y = mx + b$$

Figure 54
$2x + 4y = 8$

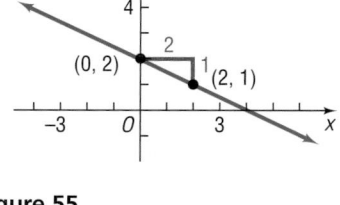

The coefficient of x, $-\dfrac{1}{2}$, is the slope, and the y-intercept is 2. We can graph the line in two ways:

1. Use the fact that the y-intercept is 2 and the slope is $-\dfrac{1}{2}$. Then, starting at the point $(0, 2)$, go to the right 2 units and then down 1 unit to the point $(2, 1)$. See Figure 54.

Or:

2. Locate the intercepts. Because the y-intercept is 2, we know that one intercept is $(0, 2)$. To obtain the x-intercept, let $y = 0$ and solve for x. When $y = 0$, we have

$$2x + 4 \cdot 0 = 8$$
$$2x = 8$$
$$x = 4$$

Figure 55
$2x + 4y = 8$

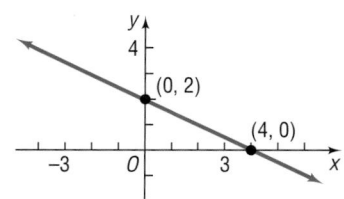

The intercepts are $(4, 0)$ and $(0, 2)$. See Figure 55. ∎

NOW WORK PROBLEM **65**.

8　The form of the equation of the line in Example 7, $2x + 4y = 8$, is called the *general form*.

> The equation of a line L is in **general form** when it is written as
>
> $$Ax + By = C \qquad\qquad (4)$$
>
> where A, B, and C are real numbers and A and B are not both 0.

Every line has an equation that is equivalent to an equation written in general form. For example, a vertical line whose equation is

$$x = a$$

can be written in the general form

$$1 \cdot x + 0 \cdot y = a \qquad A = 1, B = 0, C = a$$

A horizontal line whose equation is

$$y = b$$

can be written in the general form

$$0 \cdot x + 1 \cdot y = b \qquad A = 0, B = 1, C = b$$

Lines that are neither vertical nor horizontal have general equations of the form

$$Ax + By = C \qquad A \neq 0 \text{ and } B \neq 0$$

Because the equation of every line can be written in general form, any equation equivalent to (4) is called a **linear equation**.

Parallel and Perpendicular Lines

9 When two lines (in the plane) do not intersect (that is, they have no points in common), they are said to be **parallel**. Look at Figure 56. There we have drawn two lines and have constructed two right triangles by drawing sides parallel to the coordinate axes. These lines are parallel if and only if the right triangles are similar. (Do you see why? Two angles are equal.) And the triangles are similar if and only if the ratios of corresponding sides are equal. This suggests the following result:

Figure 56

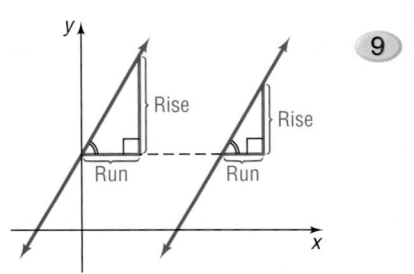

Theorem

Criterion for Parallel Lines

Two nonvertical lines are parallel if and only if their slopes are equal and they have different y-intercepts.

■

The use of the words "if and only if" in the preceding theorem means that actually two statements are being made, one the converse of the other.

If two nonvertical lines are parallel, then their slopes are equal and they have different y-intercepts.

If two nonvertical lines have equal slopes and they have different y-intercepts, then they are parallel.

EXAMPLE 8

Showing That Two Lines Are Parallel

Show that the lines given by the following equations are parallel:

$$L_1: \quad 2x + 3y = 6, \qquad L_2: \quad 4x + 6y = 0$$

Solution   To determine whether these lines have equal slopes and different y-intercepts, we write each equation in slope–intercept form:

$$L_1: \quad 2x + 3y = 6 \qquad\qquad L_2: \quad 4x + 6y = 0$$
$$3y = -2x + 6 \qquad\qquad\qquad 6y = -4x$$
$$y = -\frac{2}{3}x + 2 \qquad\qquad\qquad y = -\frac{2}{3}x$$

$$\text{Slope} = -\frac{2}{3}; y\text{-intercept} = 2 \qquad \text{Slope} = -\frac{2}{3}; y\text{-intercept} = 0$$

Because these lines have the same slope, $-\dfrac{2}{3}$, but different y-intercepts, the lines are parallel. See Figure 57.

Figure 57
Parallel lines

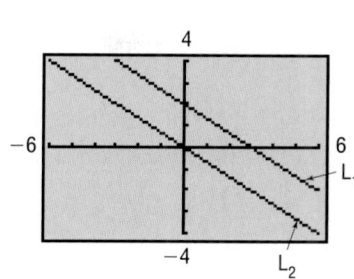

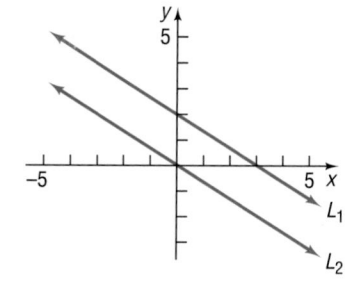

10 **EXAMPLE 9** **Finding a Line That Is Parallel to a Given Line**

Find an equation for the line that contains the point $(2, -3)$ and is parallel to the line $2x + y = 6$.

Solution The slope of the line that we seek equals the slope of the line $2x + y = 6$, since the two lines are to be parallel. We begin by writing the equation of the line $2x + y = 6$ in slope–intercept form.

$$2x + y = 6$$
$$y = -2x + 6$$

The slope is -2. Since the line that we seek contains the point $(2, -3)$, we use the point–slope form to obtain

$$y - y_1 = m(x - x_1)$$ Point-slope form

$$y - (-3) = -2(x - 2)$$ $m = -2, x_1 = 2, y_1 = -3$

$$y + 3 = -2x + 4$$

$$y = -2x + 1$$ Slope–Intercept form

$$2x + y = 1$$ General form

This line is parallel to the line $2x + y = 6$ and contains the point $(2, -3)$. See Figure 58.

Figure 58

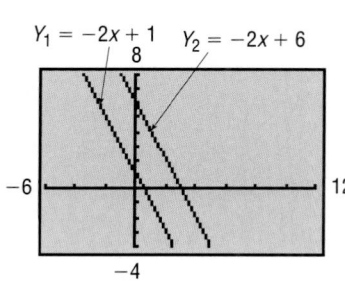

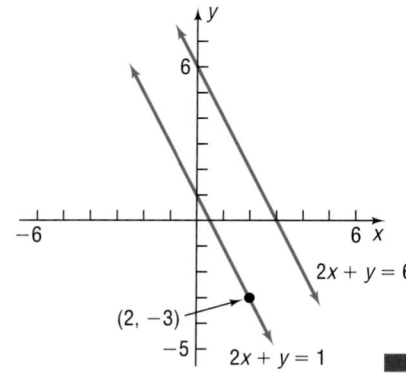

NOW WORK PROBLEM 47.

Figure 59
Perpendicular lines

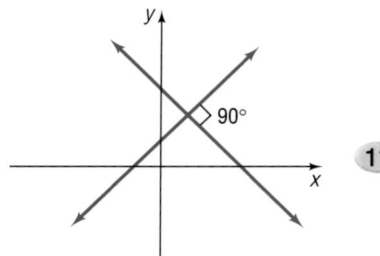

11 When two lines intersect at a right angle ($90°$), they are said to be **perpendicular**. See Figure 59.

The following result gives a condition, in terms of their slopes, for two lines to be perpendicular.

Theorem

Criterion for Perpendicular Lines

Two nonvertical lines are perpendicular if and only if the product of their slopes is -1.

Here we shall prove the "only if" part of the statement:

If two nonvertical lines are perpendicular, then the product of their slopes is -1.

You are asked to prove the "if" part of the theorem; that is:

If two nonvertical lines have slopes whose product is -1, then the lines are perpendicular.

Proof   Let m_1 and m_2 denote the slopes of the two lines. There is no loss in generality (that is, neither the angle nor the slopes are affected) if we situate the lines so that they meet at the origin. See Figure 60. The point $A = (1, m_2)$ is on the line having slope m_2, and the point $B = (1, m_1)$ is on the line having slope m_1. (Do you see why this must be true?)

Figure 60

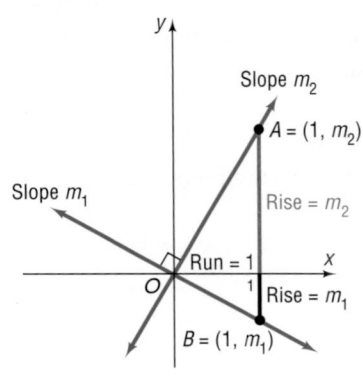

Suppose that the lines are perpendicular. Then triangle OAB is a right triangle. As a result of the Pythagorean Theorem, it follows that

$$[d(O, A)]^2 + [d(O, B)]^2 = [d(A, B)]^2 \qquad (5)$$

By the distance formula, we can write the squares of these distances as

$$[d(O, A)]^2 = (1 - 0)^2 + (m_2 - 0)^2 = 1 + m_2^2$$
$$[d(O, B)]^2 = (1 - 0)^2 + (m_1 - 0)^2 = 1 + m_1^2$$
$$[d(A, B)]^2 = (1 - 1)^2 + (m_2 - m_1)^2 = m_2^2 - 2m_1m_2 + m_1^2$$

Using these facts in equation (5), we get

$$(1 + m_2^2) + (1 + m_1^2) = m_2^2 - 2m_1m_2 + m_1^2$$

which, upon simplification, can be written as

$$m_1m_2 = -1$$

If the lines are perpendicular, the product of their slopes is -1.

You may find it easier to remember the condition for two nonvertical lines to be perpendicular by observing that the equality $m_1m_2 = -1$ means that m_1 and m_2 are negative reciprocals of each other; that is, either

$$m_1 = -\frac{1}{m_2} \text{ or } m_2 = -\frac{1}{m_1}.$$

EXAMPLE 10

Finding the Slope of a Line Perpendicular to Another Line

If a line has slope $\dfrac{3}{2}$, any line having slope $-\dfrac{2}{3}$ is perpendicular to it.

12

EXAMPLE 11 Finding the Equation of a Line Perpendicular to a Given Line

Find an equation of the line that contains the point $(1, -2)$ and is perpendicular to the line $x + 3y = 6$. Graph the two lines.

Solution We first write the equation of the given line in slope–intercept form to find its slope.

$$x + 3y = 6$$
$$3y = -x + 6 \qquad \text{Proceed to solve for } y.$$
$$y = -\frac{1}{3}x + 2 \qquad \text{Place in the form } y = mx + b.$$

The given line has slope $-\dfrac{1}{3}$. Any line perpendicular to this line will have slope 3. Because we require the point $(1, -2)$ to be on this line with slope 3, we use the point–slope form of the equation of a line.

$$y - y_1 = m(x - x_1) \qquad \text{Point–slope form}$$
$$y - (-2) = 3(x - 1) \qquad m = 3, x_1 = 1, y_1 = -2$$

To obtain other forms of the line, we simplify this equation:

$$y - (-2) = 3(x - 1)$$
$$y + 2 = 3x - 3 \qquad \text{Simplify}$$
$$y = 3x - 5 \qquad \text{Slope–intercept form}$$
$$3x - y = 5 \qquad \text{General form}$$

Figure 61 shows the graphs.

Figure 61

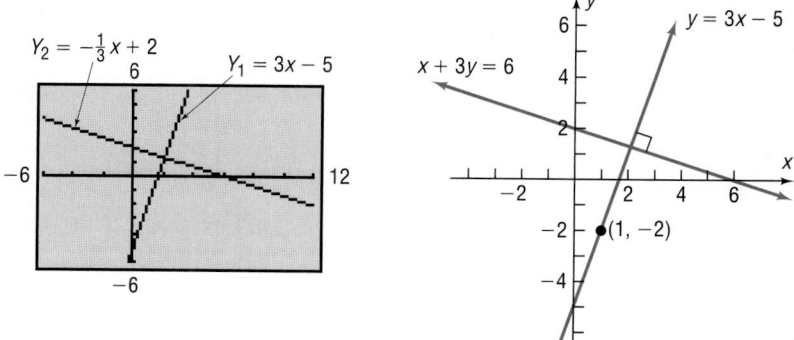

$Y_2 = -\frac{1}{3}x + 2$

$Y_1 = 3x - 5$

$x + 3y = 6$

$y = 3x - 5$

$(1, -2)$

WARNING: Be sure to use a square screen when you graph perpendicular lines. Otherwise, the angle between the two lines will appear distorted. ■

⟶ **NOW WORK PROBLEM 53.**

Overview

The discussion in Section 1.3 about circles and here in Section A.7 about lines dealt with two main types of problems that can be generalized as follows:

1. Given an equation, classify it and graph it.
2. Given a graph, or information about a graph, find its equation.

This text deals with both types of problems. We shall study various equations, classify them, and graph them. Although the second type of problem is usually more difficult to solve than the first, in many instances a graphing utility can be used to solve such problems.

A.7 Exercises

In Problems 1–4, (a) find the slope of the line and (b) interpret the slope.

1.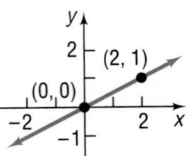

2.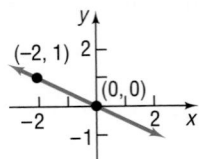

3.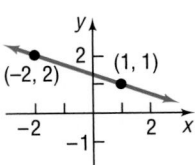

4.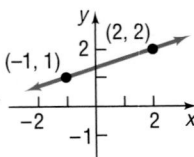

In Problems 5–12, plot each pair of points and determine the slope of the line containing them. Graph the line.

5. $(2, 3); (4, 0)$

6. $(4, 2); (3, 4)$

7. $(-2, 3); (2, 1)$

8. $(-1, 1); (2, 3)$

9. $(-3, -1); (2, -1)$

10. $(4, 2); (-5, 2)$

11. $(-1, 2); (-1, -2)$

12. $(2, 0); (2, 2)$

In Problems 13–20, graph, by hand, the line containing the point P and having slope m.

13. $P = (1, 2);\quad m = 3$

14. $P = (2, 1);\quad m = 4$

15. $P = (2, 4);\quad m = -\dfrac{3}{4}$

16. $P = (1, 3);\quad m = -\dfrac{2}{5}$

17. $P = (-1, 3);\quad m = 0$

18. $P = (2, -4);\quad m = 0$

19. $P = (0, 3)$; slope undefined

20. $P = (-2, 0)$; slope undefined

In Problems 21–26, the slope and a point on a line are given. Use this information to locate three additional points on the line. Answers may vary. [**Hint:** It is not necessary to find the equation of the line. See Example 2.]

21. Slope 4; point $(1, 2)$

22. Slope 2; point $(-2, 3)$

23. Slope $-\dfrac{3}{2}$; point $(2, -4)$

24. Slope $\dfrac{4}{3}$; point $(-3, 2)$

25. Slope -2; point $(-2, -3)$

26. Slope -1; point $(4, 1)$

In Problems 27–34, find an equation of the line L.

27.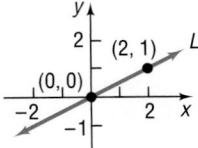

28.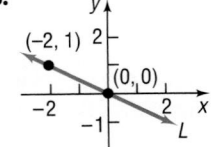

29.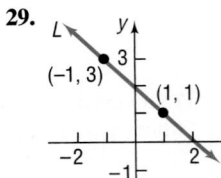

30.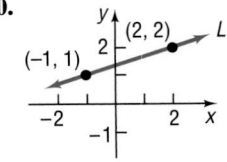

31.

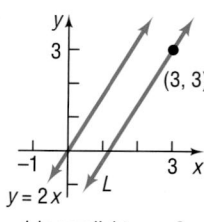

$y = 2x$

L is parallel to $y = 2x$

32.

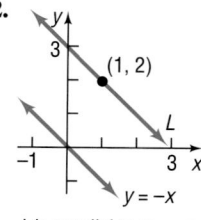

$y = -x$

L is parallel to $y = -x$

33.

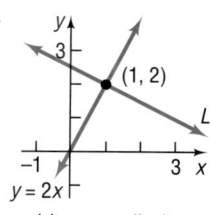

$y = 2x$

L is perpendicular to $y = 2x$

34.

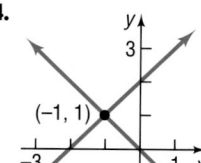

$y = -x$

L is perpendicular to $y = -x$

In Problems 35–58, find an equation for the line with the given properties. Express your answer using either the general form or the slope–intercept form of the equation of a line, whichever you prefer.

35. Slope $= 3$; containing the point $(-2, 3)$

36. Slope $= 2$; containing the point $(4, -3)$

37. Slope $= -\dfrac{2}{3}$; containing the point $(1, -1)$

38. Slope $= \dfrac{1}{2}$; containing the point $(3, 1)$

39. Containing the points $(1, 3)$ and $(-1, 2)$

40. Containing the points $(-3, 4)$ and $(2, 5)$

41. Slope $= -3$; y-intercept $= 3$

42. Slope $= -2$; y-intercept $= -2$

43. x-intercept $= 2$; y-intercept $= -1$

44. x-intercept $= -4$; y-intercept $= 4$

45. Slope undefined; containing the point $(2, 4)$

46. Slope undefined; containing the point $(3, 8)$

47. Parallel to the line $y = 2x$; containing the point $(-1, 2)$

48. Parallel to the line $y = -3x$; containing the point $(-1, 2)$

49. Parallel to the line $2x - y = -2$; containing the point $(0, 0)$

50. Parallel to the line $x - 2y = -5$; containing the point $(0, 0)$

51. Parallel to the line $x = 5$; containing the point $(4, 2)$

52. Parallel to the line $y = 5$; containing the point $(4, 2)$

53. Perpendicular to the line $y = \dfrac{1}{2}x + 4$; containing the point $(1, -2)$

54. Perpendicular to the line $y = 2x - 3$; containing the point $(1, -2)$

55. Perpendicular to the line $2x + y = 2$; containing the point $(-3, 0)$

56. Perpendicular to the line $x - 2y = -5$; containing the point $(0, 4)$

57. Perpendicular to the line $x = 8$; containing the point $(3, 4)$

58. Perpendicular to the line $y = 8$; containing the point $(3, 4)$

In Problems 59–78, find the slope and y-intercept of each line. Graph the line by hand. Check your graph using a graphing utility.

59. $y = 2x + 3$

60. $y = -3x + 4$

61. $\dfrac{1}{2}y = x - 1$

62. $\dfrac{1}{3}x + y = 2$

63. $y = \dfrac{1}{2}x + 2$

64. $y = 2x + \dfrac{1}{2}$

65. $x + 2y = 4$

66. $-x + 3y = 6$

67. $2x - 3y = 6$

68. $3x + 2y = 6$

69. $x + y = 1$

70. $x - y = 2$

71. $x = -4$

72. $y = -1$

73. $y = 5$

74. $x = 2$

75. $y - x = 0$

76. $x + y = 0$

77. $2y - 3x = 0$

78. $3x + 2y = 0$

79. Find an equation of the x-axis

80. Find an equation of the y-axis

In Problems 81–84, match each graph with the correct equation:

(a) $y = x$

(b) $y = 2x$

(c) $y = \dfrac{x}{2}$

(d) $y = 4x$

81.

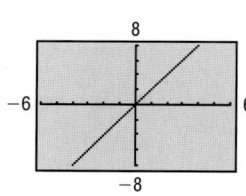

82.

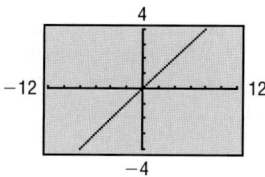

83.

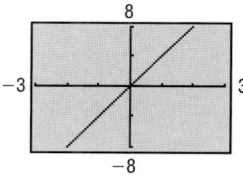

84.

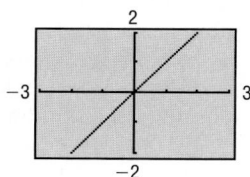

In Problems 85–88, write an equation of each line. Express your answer using either the general form or the slope–intercept form of the equation of a line, whichever you prefer.

85.
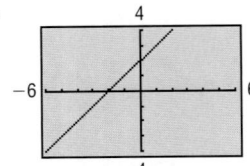

86.
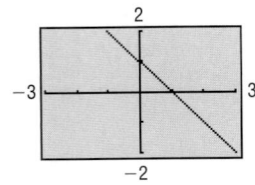

87.
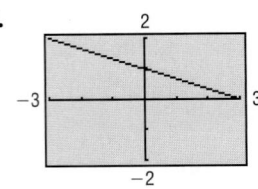

88.
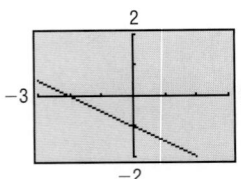

89. Truck Rentals A truck rental company rents a moving truck for one day by charging $29 plus $0.07 per mile. Write a linear equation that relates the cost C, in dollars, of renting the truck to the number x of miles driven. What is the cost of renting the truck if the truck is driven 110 miles? 230 miles?

90. Cost Equation The **fixed costs** of operating a business are the costs incurred regardless of the level of production. Fixed costs include rent, fixed salaries, and costs of buying machinery. The **variable costs** of operating a business are the costs that change with the level of output. Variable costs include raw materials, hourly wages, and electricity. Suppose that a manufacturer of jeans has fixed costs of $500 and variable costs of $8 for each pair of jeans manufactured. Write a linear equation that relates the cost C, in dollars, of manufacturing the jeans to the number x pairs of jeans manufactured. What is the cost of manufacturing 400 pairs of jeans? 740 pairs?

91. Measuring Temperature The relationship between Celsius (°C) and Fahrenheit (°F) degrees of measuring temperature is linear. Find an equation relating °C and °F if 0°C corresponds to 32°F and 100°C corresponds to 212°F. Use the equation to find the Celsius measure of 70°F.

92. Measuring Temperature The Kelvin (K) scale for measuring temperature is obtained by adding 273 to the Celsius temperature.
(a) Write an equation relating K and °C.
(b) Write an equation relating K and °F (see Problem 91).

93. Business: Computing Profit Each Sunday, a newspaper agency sells x copies of a certain newspaper for $1.00 per copy. The cost to the agency of each newspaper is $0.50. The agency pays a fixed cost for storage, delivery, and so on, of $100 per Sunday.
(a) Write an equation that relates the profit P, in dollars, to the number x of copies sold. Graph this equation.
(b) What is the profit to the agency if 1000 copies are sold?
(c) What is the profit to the agency if 5000 copies are sold?

94. Business: Computing Profit Repeat Problem 93 if the cost to the agency is $0.45 per copy and the fixed cost is $125 per Sunday.

95. Cost of Electricity In 2001, Florida Power and Light Company supplied electricity in the summer months to residential customers for a monthly customer charge of $5.65 plus 6.543¢ per kilowatt-hour supplied in the month for the first 750 kilowatt-hours used.* Write an equation

Source: Florida Power and Light Co., Miami, Florida, 2001.

that relates the monthly charge, C, in dollars, to the number x of kilowatt-hours used in the month. Graph this equation. What is the monthly charge for using 300 kilowatt-hours? For using 750 kilowatt-hours?

96. Show that the line containing the points (a, b) and (b, a), $a \neq b$, is perpendicular to the line $y = x$. Also show that the midpoint of (a, b) and (b, a) lies on the line $y = x$.

97. The equation $2x - y = C$ defines a **family of lines**, one line for each value of C. On one set of coordinate axes, graph the members of the family when $C = -4, C = 0$, and $C = 2$. Can you draw a conclusion from the graph about each member of the family?

98. Rework Problem 97 for the family of lines $Cx + y = -4$.

99. The **tangent line** to a circle may be defined as the line that intersects the circle in a single point, called the **point of tangency** (see the figure).

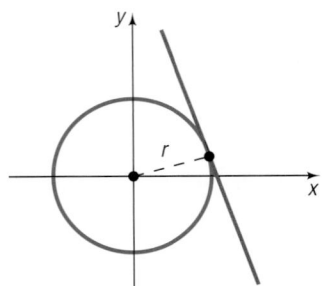

If the equation of the circle is $x^2 + y^2 = r^2$ and the equation of the tangent line is $y = mx + b$, show that:
(a) $r^2(1 + m^2) = b^2$

[**Hint:** The quadratic equation
$x^2 + (mx + b)^2 = r^2$ has exactly one solution.]

(b) The point of tangency is $\left(-r^2 \dfrac{m}{b}, \dfrac{r^2}{b}\right)$.

(c) The tangent line is perpendicular to the line containing the center of the circle and the point of tangency.

100. The Greek method for finding the equation of the tangent line to a circle used the fact that at any point on a circle the lines containing the center and the tangent line are perpendicular (see Problem 99). Use this method to find an equation of the tangent line to the circle $x^2 + y^2 = 9$ at the point $(1, 2\sqrt{2})$.

101. Use the Greek method described in Problem 100 to find an equation of the tangent line to the circle $x^2 + y^2 - 4x + 6y + 4 = 0$ at the point $(3, 2\sqrt{2} - 3)$.

102. Refer to Problem 99. The line $x - 2y + 4 = 0$ is tangent to a circle at $(0, 2)$. The line $y = 2x - 7$ is tangent to the same circle at $(3, -1)$. Find the center of the circle.

103. Find an equation of the line containing the centers of the two circles

$$x^2 + y^2 - 4x + 6y + 4 = 0$$

and

$$x^2 + y^2 + 6x + 4y + 9 = 0$$

104. If a circle of radius 2 is made to roll along the x-axis, what is an equation for the path of the center of the circle?

105. Which of the following equations might have the graph shown? (More than one answer is possible.)
(a) $2x + 3y = 6$ (e) $x - y = -1$
(b) $-2x + 3y = 6$ (f) $y = 3x - 5$
(c) $3x - 4y = -12$ (g) $y = 2x + 3$
(d) $x - y = 1$ (h) $y = -3x + 3$

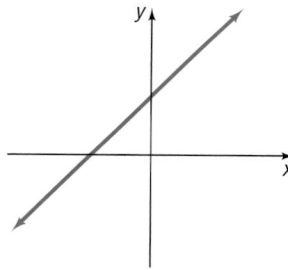

106. Which of the following equations might have the graph shown? (More than one answer is possible.)
(a) $2x + 3y = 6$ (e) $x - y = -1$
(b) $2x - 3y = 6$ (f) $y = -2x - 1$

(c) $3x + 4y = 12$ (g) $y = -\dfrac{1}{2}x + 10$

(d) $x - y = 1$ (h) $y = x + 4$

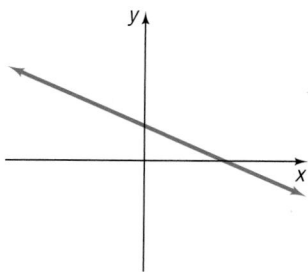

107. The figure that follows shows the graph of two parallel lines. Which of the following pairs of equations might have such a graph?
(a) $x - 2y = 3$ (c) $x - y = -2$
    $x + 2y = 7$ $x - y = 1$
(b) $x + y = 2$ (d) $x - y = -2$
    $x + y = -1$ $2x - 2y = -4$

(e) $x + 2y = 2$
    $x + 2y = -1$

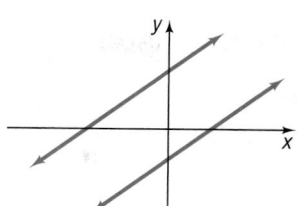

108. The figure below shows the graph of two perpendicular lines. Which of the following pairs of equations might have such a graph?
(a) $y - 2x = 2$ (d) $y - 2x = 2$
    $y + 2x = -1$ $x + 2y = -1$
(b) $y - 2x = 0$ (e) $2x + y = -2$
    $2y + x = 0$ $2y + x = -2$
(c) $2y - x = 2$
    $2y + x = -2$

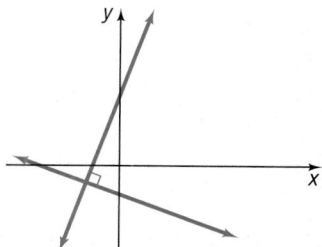

109. The accepted symbol used to denote the slope of a line is the letter m. Investigate the origin of this symbolism. Begin by consulting a French dictionary and looking up the French word *monter*. Write a brief essay on your findings.

110. The term *grade* is used to describe the inclination of a road. How does this term relate to the notion of slope of a line? Is a 4% grade very steep? Investigate the grades of some mountainous roads and determine their slopes. Write a brief essay on your findings.

111. Carpentry Carpenters use the term *pitch* to describe the steepness of staircases and roofs. How does pitch relate to slope? Investigate typical pitches used for stairs and for roofs. Write a brief essay on your findings.

A.8 SCATTER DIAGRAMS; LINEAR CURVE FITTING

OBJECTIVES
1. Draw and Interpret Scatter Diagrams
2. Distinguish between Linear and Nonlinear Relations
3. Use a Graphing Utility to Find the Line of Best Fit

Scatter Diagrams

1. A **relation** is a correspondence between two sets. If x and y are two elements and a relation exists between x and y, then we say that x **corresponds to** y or that y **depends on** x and write $x \rightarrow y$. We may also write $x \rightarrow y$ as the ordered pair (x, y). In this sense, y is referred to as the **dependent** variable and x is called the **independent** variable.

Often we are interested in specifying the type of relation (such as an equation) that might exist between two variables. The first step in finding this relation is to plot the ordered pairs using rectangular coordinates. The resulting graph is called a **scatter diagram.**

EXAMPLE 1   Drawing a Scatter Diagram

The data listed in Table 2 represent the apparent temperature versus the relative humidity in a room whose actual temperature is 72° Fahrenheit.

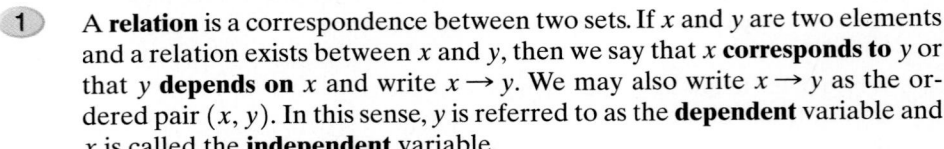

TABLE 2					
Relative Humidity (%), x	Apparent Temperature, y	(x, y)	Relative Humidity (%), x	Apparent Temperature, y	(x, y)
0	64	(0, 64)	60	72	(60, 72)
10	65	(10, 65)	70	73	(70, 73)
20	67	(20, 67)	80	74	(80, 74)
30	68	(30, 68)	90	75	(90, 75)
40	70	(40, 70)	100	76	(100, 76)
50	71	(50, 71)			

(a) Draw a scatter diagram by hand.
(b) Use a graphing utility to draw a scatter diagram.
(c) Describe what happens to the apparent temperature as the relative humidity increases.

Solution   (a) To draw a scatter diagram by hand, we plot the ordered pairs listed in Table 2, with the relative humidity as the x-coordinate and the apparent temperature as the *y*-coordinate. See Figure 62(a). Notice that the points in a scatter diagram are not connected.

(b) Figure 62(b) shows a scatter diagram using a graphing utility.*

(c) We see from the scatter diagrams that, as the relative humidity increases, the apparent temperature increases.

Figure 62

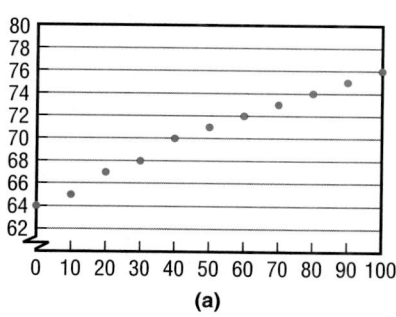

(a)

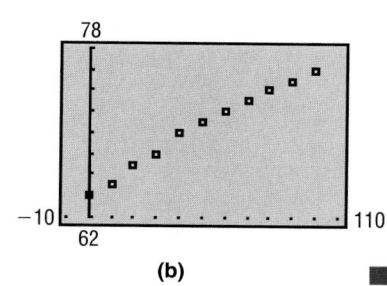

(b)

NOW WORK PROBLEM 7(a).

Curve Fitting

2   Scatter diagrams are used to help us to see the type of relation that exists between two variables. In this text, we will discuss a variety of different relations that may exist between two variables. For now, we concentrate on distinguishing between linear and nonlinear relations. See Figure 63.

Figure 63

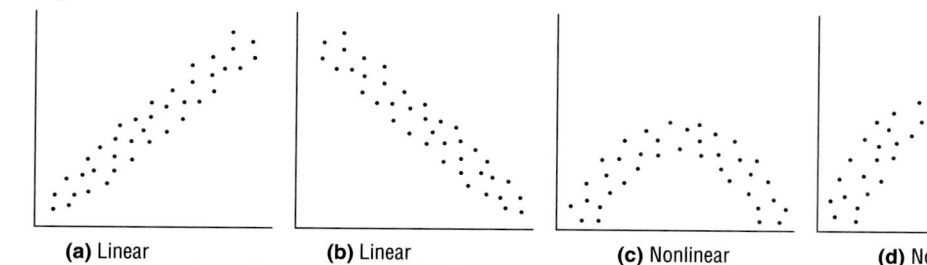

(a) Linear
 $y = mx + b, m > 0$

(b) Linear
 $y = mx + b, m < 0$

(c) Nonlinear

(d) Nonlinear

(e) Nonlinear

EXAMPLE 2   **Distinguishing between Linear and Nonlinear Relations**

Determine whether the relation between the two variables in Figure 64 is linear or nonlinear.

Figure 64

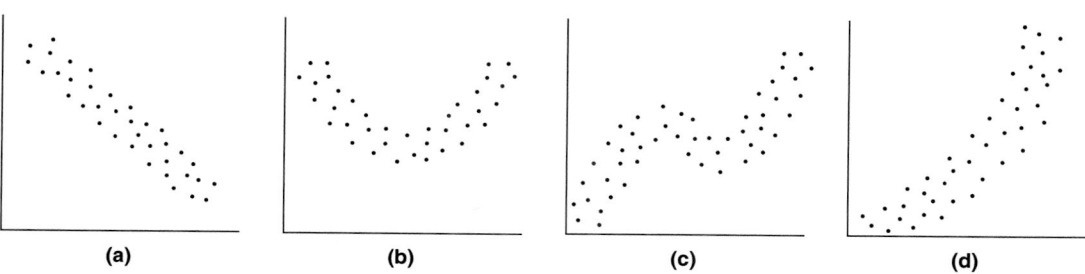

(a)

(b)

(c)

(d)

* Consult your owner's manual for the appropriate keystrokes or visit
www.prenhall.com/sullivanegu3e

Solution (a) Linear   (b) Nonlinear   (c) Nonlinear   (d) Nonlinear ■

━━━━━━ **NOW WORK PROBLEM 1.**

In this section we will study data whose scatter diagrams imply that a linear relation exists between the two variables. Nonlinear data will be discussed later.

Suppose that the scatter diagram of a set of data appears to be linearly related as in Figure 63(a) or (b). We might wish to find an equation of a line that relates the two variables. One way to obtain an equation for such data is to draw a line through two points on the scatter diagram and determine the equation of the line.

EXAMPLE 3 Finding an Equation for Linearly Related Data

Using the data in Table 2 from Example 1:

(a) Select two points and find an equation of the line containing the points.
(b) Graph the line on the scatter diagram obtained in Example 1(b).

Solution (a) Select two points, say $(10, 65)$ and $(70, 73)$. (You should select your own two points and complete the solution.) The slope of the line joining the points $(10, 65)$ and $(70, 73)$ is

$$m = \frac{73 - 65}{70 - 10} = \frac{8}{60} = \frac{2}{15}$$

The equation of the line with slope $\frac{2}{15}$ and passing through $(10, 65)$ is found using the point–slope form with $m = \frac{2}{15}$, $x_1 = 10$, and $y_1 = 65$.

Figure 65

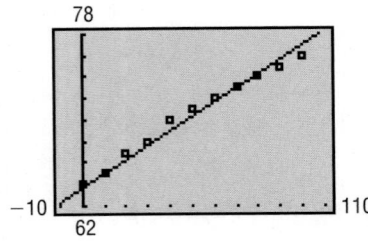

$$y - y_1 = m(x - x_1)$$

$$y - 65 = \frac{2}{15}(x - 10)$$

$$y = \frac{2}{15}x + \frac{191}{3}$$

(b) Figure 65 shows the scatter diagram with the graph of the line found in part (a). ■

━━━━━━ **NOW WORK PROBLEMS 7(b) AND (c).**

Line of Best Fit

④ The line obtained in Example 3 depends on the selection of points, which will vary from person to person. So the line that we found might be different from the line that you found. Although the line we found in Example 3 appears to fit the data well, there may be a line that "fits it better." Do you think your line fits the data better? Is there a line of *best fit*? As it turns out, there is a method for finding the line that best fits linearly related data (called the *line of best fit*).*

───────────

*We shall not discuss the underlying mathematics of lines of best fit. Most books in statistics and many in linear algebra discuss this topic.

EXAMPLE 4   **Finding the Line of Best Fit**

Using the data in Table 2 from Example 1:

(a) Find the line of best fit using a graphing utility.

(b) Graph the line of best fit on the scatter diagram obtained in Example 1(b).

(c) Interpret the slope.

(d) Use the line of best fit to predict the apparent temperature of a room whose temperature is 72°F and relative humidity is 45%.

Solution   (a) Graphing utilities contain built-in programs that find the line of best fit for a collection of points in a scatter diagram. (Look in your owner's manual for details on how to execute the program.) Upon executing the LINear REGression program, we obtain the results shown in Figure 66. The output that the utility provides shows us the equation $y = ax + b$, where a is the slope of the line and b is the y-intercept. The line of best fit that relates relative humidity to apparent temperature may be expressed as the line $y = 0.121x + 64.409$.

(b) Figure 67 shows the graph of the line of best fit, along with the scatter diagram.

Figure 66

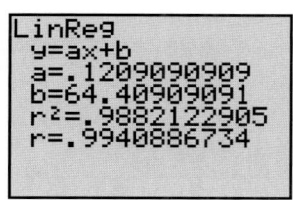

Figure 67

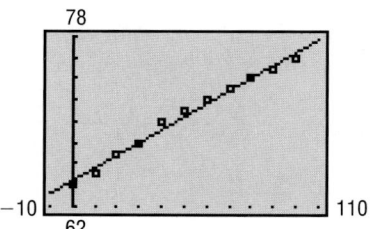

(c) The slope of the line of best fit is 0.121, which means that, for every 1% increase in the relative humidity, apparent room temperature increases 0.121°F.

(d) Letting $x = 45$ in the equation of the line of best fit, we obtain $y = 0.121(45) + 64.409 \approx 70°F$, which is the apparent temperature in the room. ■

━━━━━ NOW WORK PROBLEMS **7(d)** AND **(e)**.

Does the line of best fit appear to be a good fit? In other words, does the line appear to accurately describe the relation between temperature and relative humidity?

And just how "good" is this line of best fit? The answers are given by what is called the *correlation coefficient*. Look again at Figure 66. The last line of output is $r = 0.994$. This number, called the **correlation coefficient, r,** $-1 \leq r \leq 1$, is a measure of the strength of the *linear relation* that exists between two variables. The closer that $|r|$ is to 1, the more perfect the linear relationship is. If r is close to 0, there is little or no *linear* relationship between the variables. A negative value of r, $r < 0$, indicates that as x increases y decreases; a positive value of r, $r > 0$, indicates that as x increases y does also. The data given in Table 2, having a correlation coefficient of 0.994, are indicative of a strong linear relationship with positive slope.

A.8 Exercises

In Problems 1–6, examine the scatter diagram and determine whether the type of relation, if any, that may exist is linear or nonlinear.

1.
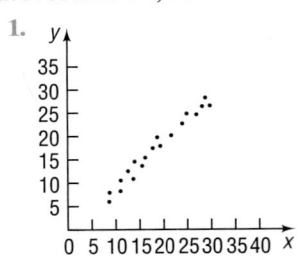

2.

3.
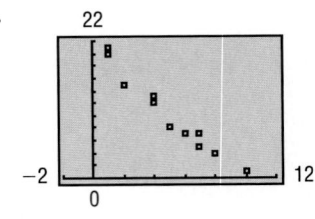

4.
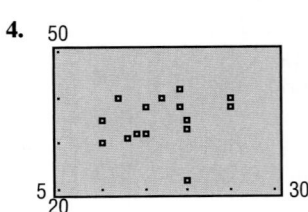

5
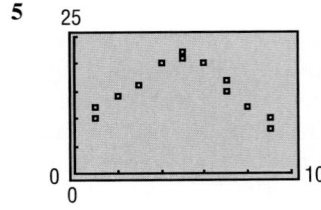

6.
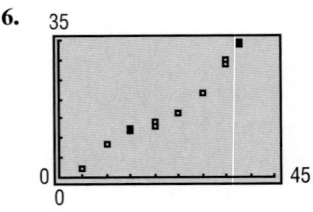

In Problems 7–12:

    (a) *Draw a scatter diagram.*
    (b) *Select two points from the scatter diagram and find the equation of the line containing the points selected.* *
    (c) *Graph the line found in part (b) on the scatter diagram.*
    (d) *Use a graphing utility to find the line of best fit.*
    (e) *Use a graphing utility to graph the line of best fit on the scatter diagram.*

7.
x	3	4	5	6	7	8	9
y	4	6	7	10	12	14	16

8.
x	3	5	7	9	11	13
y	0	2	3	6	9	11

9.
x	-2	-1	0	1	2
y	-4	0	1	4	5

10.
x	-2	-1	0	1	2
y	7	6	3	2	0

11.
x	-20	-17	-15	-14	-10
y	100	120	118	130	140

12.
x	-30	-27	-25	-20	-14
y	10	12	13	13	18

13. **Per Capita Disposable Income versus Consumption** An economist wishes to estimate a linear function that relates per capita consumption expenditures C and disposable income I. Both C and I are measured in dollars. The following data represent the per capita disposable income (income after taxes) and per capita consumption in the United States for 1990 to 1998.

Year	Per Capita Disposable Income (I)	Per Capita Consumption (C)
1990	15,695	14,547
1991	16,700	15,400
1992	17,346	16,035
1993	18,153	16,951
1994	19,711	18,419
1995	20,316	19,061
1996	21,127	19,938
1997	21,871	20,807
1998	22,212	21,385

Source: U.S. Department of Commerce.

Let I represent the independent variable and C the dependent variable.
(a) Use a graphing utility to draw a scatter diagram.
(b) Use a graphing utility to find the line of best fit to the data.
(c) Interpret the slope. The slope of this line is called the **marginal propensity to consume.**
(d) Predict the consumption of a family whose disposable income is $21,500.

14. **Height versus Head Circumference** A pediatrician wanted to estimate a linear function that relates a child's height, H, to their head circumference, C. She randomly selects 9 children from her practice, measures their height and head circumference, and obtains the following data.

* Answers will vary. We will use the first and last data points in the area answer section.

Height, H (inches)	Head Circumference, C (inches)
25.25	16.4
25.75	16.9
25	16.9
27.75	17.6
26.5	17.3
27	17.5
26.75	17.3
26.75	17.5
27.5	17.5

Source: Denise Slucki, Student at Joliet Junior College.

Let H represent the independent variable and C the dependent variable.
(a) Use a graphing utility to draw a scatter diagram.
(b) Use a graphing utility to find the line of best fit to the data.
(c) Interpret the slope.
(d) Predict the head circumference of a child that is 26 inches tall.

15. **Gestation Period versus Life Expectancy** A researcher would like to estimate the linear function relating the gestation period of an animal, G, and its life expectancy, L. She collects the following data.

Animal	Gestation (or incubation) Period, G (days)	Life Expectancy, L (years)
Cat	63	11
Chicken	22	7.5
Dog	63	11
Duck	28	10
Goat	151	12
Lion	108	10
Parakeet	18	8
Pig	115	10
Rabbit	31	7
Squirrel	44	9

Source: Time Almanac 2000.

Let G represent the independent variable and L the dependent variable.
(a) Use a graphing utility to draw a scatter diagram.
(b) Use a graphing utility to find the line of best fit to the data.
(c) Interpret the slope.
(d) Predict the life expectancy of an animal whose gestation period is 89 days.

16. **Mortgage Qualification** The amount of money that a lending institution will allow you to borrow mainly depends on the interest rate and your annual income. The following data represent the annual income, I, required by a bank in order to lend L dollars at an interest rate of 7.5% for 30 years.

Annual Income, I ($)	Loan Amount, L ($)
15,000	44,600
20,000	59,500
25,000	74,500
30,000	89,400
35,000	104,300
40,000	119,200
45,000	134,100
50,000	149,000
55,000	163,900
60,000	178,800
65,000	193,700
70,000	208,600

Source: Information Please Almanac, 1999.

Let I represent the independent variable and L the dependent variable.
(a) Use a graphing utility to draw a scatter diagram of the data.
(b) Use a graphing utility to find the line of best fit to the data.
(c) Graph the line of best fit on the scatter diagram drawn in part (a).
(d) Interpret the slope of the line of best fit.
(e) Determine the loan amount that an individual would qualify for if her income is $42,000.

17. **Demand for Jeans** The marketing manager at Levi–Strauss wishes to find a function that relates the demand D for men's jeans and p, the price of the jeans. The following data were obtained based on a price history of the jeans.

Price ($/Pair), p	Demand (Pairs of Jeans Sold per Day), D
20	60
22	57
23	56
23	53
27	52
29	49
30	44

(a) Does the relation defined by the set of ordered pairs (p, D) represent a function?
(b) Draw a scatter diagram of the data.
(c) Using a graphing utility, find the line of best fit relating price and quantity demanded.
(d) Interpret the slope.
(e) Express the relationship found in part (c) using function notation.
(f) What is the domain of the function?
(g) How many jeans will be demanded if the price is $28 a pair?

18. **Advertising and Sales Revenue** A marketing firm wishes to find a function that relates the sales S of a product and A, the amount spent on advertising the product. The data are obtained from past experience. Advertising and sales are measured in thousands of dollars.

Advertising Expenditures, A	Sales, S
20	335
22	339
22.5	338
24	343
24	341
27	350
28.3	351

(a) Does the relation defined by the set of ordered pairs (A, S) represent a function?
(b) Draw a scatter diagram of the data.
(c) Using a graphing utility, find the line of best fit relating advertising expenditures and sales.
(d) Interpret the slope.
(e) Express the relationship found in part (c) using function notation.
(f) What is the domain of the function?
(g) Predict sales if advertising expenditures are $25,000.

CHAPTER 1   Functions and Their Graphs

1.1 Concepts and Vocabulary *(page 9)*

1. abscissa; ordinate   **2.** quadrants   **3.** midpoint   **4.** F   **5.** F   **6.** T   **7.** Determining Xmin, Xmax, Xscl, Ymin, Ymax, Yscl

1.1 Exercises *(page 9)*

1. (a) Quadrant II   **(b)** Positive x-axis   **(c)** Quadrant III   **3.** The points will be on a vertical line that
(d) Quadrant I   **(e)** Negative y-axis   **(f)** Quadrant IV   is 2 units to the right of the y-axis

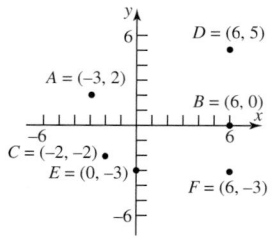
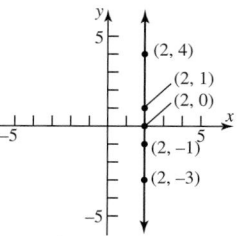

5. $(-1, 4)$; Quadrant II   **7.** $(3, 1)$; Quadrant I   **9.** Xmin $= -11$, Xmax $= 5$, Xscl $= 1$, Ymin $= -3$, Ymax $= 6$, Yscl $= 1$
11. Xmin $= -30$, Xmax $= 50$, Xscl $= 10$, Ymin $= -90$, Ymax $= 50$, Yscl $= 10$ **13.** Xmin $= -10$, Xmax $= 110$, Xscl $= 10$,
Ymin $= -10$, Ymax $= 160$, Yscl $= 10$ **15.** Xmin $= -6$, Xmax $= 6$, Xscl $= 2$, Ymin $= -4$, Ymax $= 4$, Yscl $= 2$
17. Xmin $= -6$, Xmax $= 6$, Xscl $= 2$, Ymin $= -1$, Ymax $= 3$, Yscl $= 1$ **19.** Xmin $= 3$, Xmax $= 9$, Xscl $= 1$, Ymin $= 2$,
Ymax $= 10$, Yscl $= 2$ **21.** $\sqrt{5}$ **23.** $\sqrt{10}$ **25.** $2\sqrt{5}$ **27.** $\sqrt{85}$ **29.** $\sqrt{53}$ **31.** $\sqrt{6.89} \approx 2.62$ **33.** $\sqrt{a^2 + b^2}$ **35.** $4\sqrt{10}$ **37.** $2\sqrt{65}$

39. $d(A, B) = \sqrt{13}$ **41.** $d(A, B) = \sqrt{130}$ **43.** $d(A, B) = 4$
$\ \ d(B, C) = \sqrt{13}$ $\ \ d(B, C) = \sqrt{26}$ $\ \ d(A, C) = 5$
$\ \ d(A, C) = \sqrt{26}$ $\ \ d(A, C) = 2\sqrt{26}$ $\ \ d(B, C) = \sqrt{41}$
$\ \ (\sqrt{13})^2 + (\sqrt{13})^2 = (\sqrt{26})^2$ $(\sqrt{26})^2 + (2\sqrt{26})^2 = (\sqrt{130})^2$ $4^2 + 5^2 = 16 + 25 = (\sqrt{41})^2$
$\ \ \text{Area} = \dfrac{13}{2} \text{ square units}$ $\text{Area} = 26 \text{ square units}$ $\text{Area} = 10 \text{ square units}$

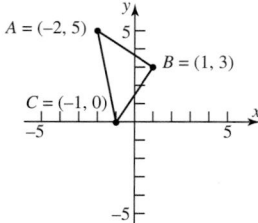

45. $(2, 2); (2, -4)$ **47.** $(0, 0); (8, 0)$ **49.** $(4, 3)$ **51.** $\left(\dfrac{3}{2}, 1\right)$ **53.** $(5, -1)$ **55.** $(1.05, 0.7)$ **57.** $\left(\dfrac{a}{2}, \dfrac{b}{2}\right)$ **59.** $\sqrt{17}; 2\sqrt{5}; \sqrt{29}$

61. $d(P_1, P_2) = 6; d(P_2, P_3) = 4; d(P_1, P_3) = 2\sqrt{13}$; right triangle
63. $d(P_1, P_2) = 2\sqrt{17}; d(P_2, P_3) = \sqrt{34}; d(P_1, P_3) = \sqrt{34}$; isosceles right triangle   **65.** $90\sqrt{2} \approx 127.28$ ft
67. (a) $(90, 0), (90, 90), (0, 90)$ **(b)** $5\sqrt{2161} \approx 232.43$ ft   **(c)** $30\sqrt{149} \approx 366.20$ ft   **69.** $d = 50t$

1.2 Concepts and Vocabulary *(page 19)*

1. intercepts   **2.** zeros; roots   **3.** T   **4.** F   **5.** A complete graph presents enough of the illustration so that a viewer of the graph can
visualize the rest of the graph as an obvious continuation.   **6.** Xmin $= -10$, Xmax $= 10$, Xscl $= 1$, Ymin $= -10$, Ymax $= 10$, Yscl $= 1$
7. Answers will vary. One example is shown in Exercise 14 on page 19.

1.2 Exercises *(page 19)*

1. $(0, 0)$ is on the graph. **3.** $(0, 3)$ is on the graph. **5.** $(0, 2)$ and $(\sqrt{2}, \sqrt{2})$ are on the graph. **7.** $(-1, 0), (1, 0)$

9. $\left(-\dfrac{\pi}{2}, 0\right), (0, 1), \left(\dfrac{\pi}{2}, 0\right)$ **11.** $(0, 0)$ **13.** $(-4, 0), (-1, 0), (0, -3), (4, 0)$ **15.** $-\dfrac{2}{5}$ **17.** $2a + 3b = 6$

19. x-intercept: -2;
y-intercept: 2

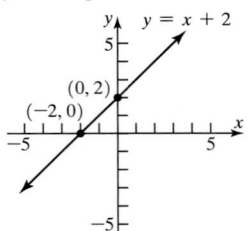

21. x-intercept: -4;
y-intercept: 8

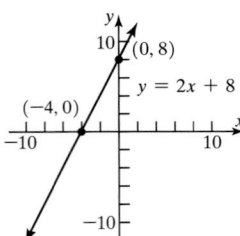

23. x-intercepts: $-1, 1$;
y-intercept: -1

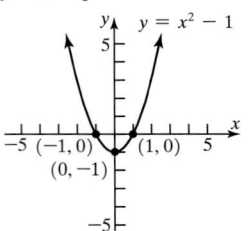

25. x-intercepts: $-2, 2$;
y-intercept: 4

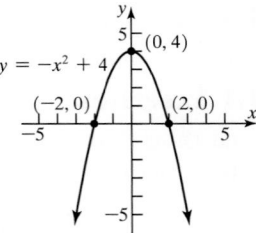

27. x-intercept: 3;
y-intercept: 2

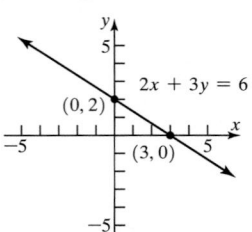

29. x-intercepts: $-2, 2$;
y-intercept: 9

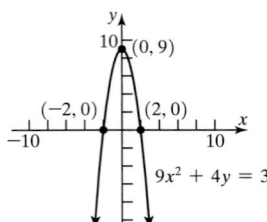

31. x-intercept: 6.5;
y-intercept: -13

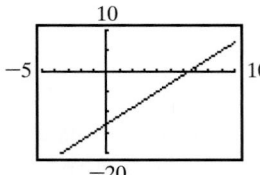

33. x-intercepts: $-2.74, 2.74$;
y-intercept: -15

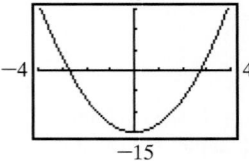

35. x-intercept: 14.33;
y-intercept: -21.5

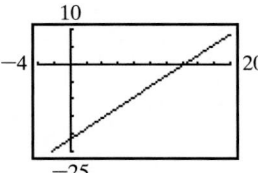

37. x-intercepts: $-2.72, 2.72$;
y-intercept: 12.33

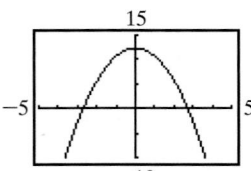

1.3 Concepts and Vocabulary *(page 30)*

1. y-axis **2.** origin **3.** radius **4.** T **5.** T **6.** F **7.** -6

1.3 Exercises *(page 30)*

1. (a) $(3, -4)$ **(b)** $(-3, 4)$ **(c)** $(-3, -4)$ **3. (a)** $(-2, -1)$ **(b)** $(2, 1)$ **(c)** $(2, -1)$ **5. (a)** $(1, -1)$ **(b)** $(-1, 1)$ **(c)** $(-1, -1)$
7. (a) $(-3, 4)$ **(b)** $(3, -4)$ **(c)** $(3, 4)$ **9. (a)** $(0, 3)$ **(b)** $(0, -3)$ **(c)** $(0, 3)$ **11.** x-axis, y-axis, and origin **13.** y-axis **15.** x-axis
17. No symmetry
19.

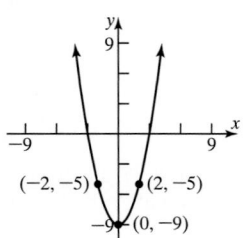

21.

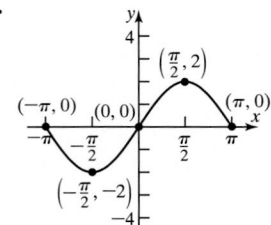

23. Symmetric with respect to the y-axis
25. Symmetric with respect to the origin
27. Symmetric with respect to the y-axis
29. No symmetry
31. No symmetry
33. Symmetric with respect to the origin

35.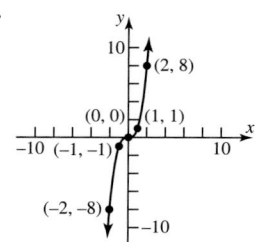

37.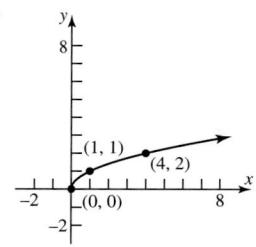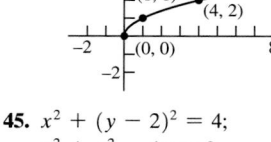

39. Center $(2, 1)$; Radius 2; $(x - 2)^2 + (y - 1)^2 = 4$

41. Center $\left(\dfrac{5}{2}, 2\right)$; Radius $\dfrac{3}{2}$; $\left(x - \dfrac{5}{2}\right)^2 + (y - 2)^2 = \dfrac{9}{4}$
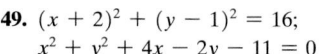

43. $x^2 + y^2 = 4$;
$x^2 + y^2 - 4 = 0$
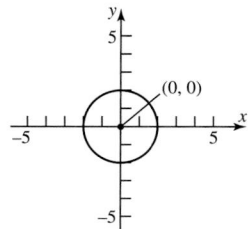

45. $x^2 + (y - 2)^2 = 4$;
$x^2 + y^2 - 4y = 0$
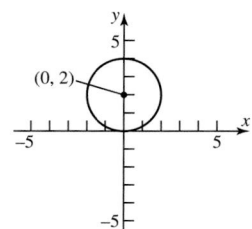

47. $(x - 4)^2 + (y + 3)^2 = 25$;
$x^2 + y^2 - 8x + 6y = 0$
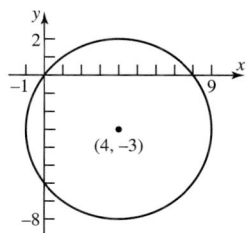

49. $(x + 2)^2 + (y - 1)^2 = 16$;
$x^2 + y^2 + 4x - 2y - 11 = 0$
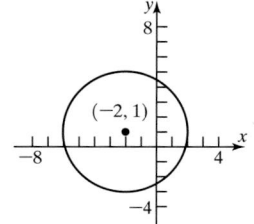

51. $\left(x - \dfrac{1}{2}\right)^2 + y^2 = \dfrac{1}{4}$;
$x^2 + y^2 - x = 0$
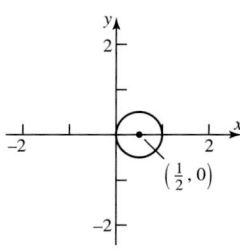

53. $(h, k) = (0, 0); r = 5$
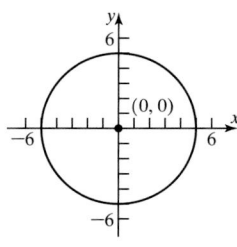

55. $(h, k) = (2, 0); r = 2$
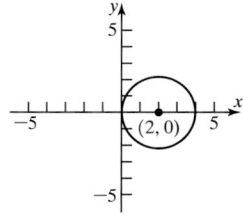

57. $(h, k) = (-2, 2); r = 3$
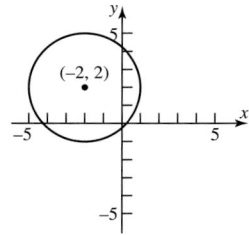

59. $(h, k) = \left(\dfrac{1}{2}, -1\right); r = \dfrac{1}{2}$
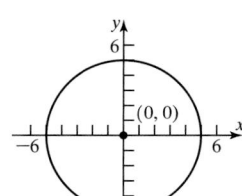

61. $(h, k) = (3, -2); r = 5$
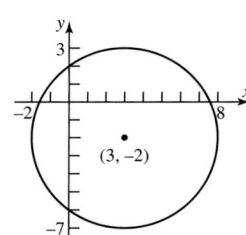

63. $x^2 + y^2 - 13 = 0$

65. $x^2 + y^2 - 4x - 6y + 4 = 0$

67. $x^2 + y^2 + 2x - 6y + 5 = 0$

69. (c)   **71.** (b)

73. $(x + 3)^2 + (y - 1)^2 = 16$

75. $(x - 2)^2 + (y - 2)^2 = 9$

77. (b), (c), (e), (g)

79. $x^2 + y^2 + 2x + 4y - 4168.16 = 0$

1.4 Concepts and Vocabulary *(page 45)*

1. independent; dependent   **2.** vertical   **3.** range   **4.** F   **5.** T   **6.** T   **8.** No limit; at most 1
9. Yes; yes, for point (a, b) the equation $(x - a)^2 + (y - b)^2 = 0$ works, with $y = f(x)$.

1.4 Exercises *(page 46)*

1. Function; Domain: {Dad, Colleen, Kaleigh, Marissa}, Range: {January 8, March 15, September 17}   **3.** Not a function
5. Not a function   **7.** Function; Domain: {1, 2, 3, 4}, Range: {3}   **9.** Not a function   **11.** Function; Domain: {−2, −1, 0, 1}, Range: {0, 1, 4}
13. (a) 5   (b) 7   (c) 3   (d) $-2x + 5$ (e) $-2x - 5$ (f) $2x + 7$ (g) $4x + 5$ (h) $2x + 2h + 5$ **15.** (a) −4   (b) 1   (c) −3
(d) $3x^2 - 2x - 4$ (e) $-3x^2 - 2x + 4$ (f) $3x^2 + 8x + 1$ (g) $12x^2 + 4x - 4$ (h) $3x^2 + 6xh + 3h^2 + 2x + 2h - 4$
17. (a) 0   (b) $\dfrac{1}{2}$ (c) $-\dfrac{1}{2}$ (d) $\dfrac{-x}{x^2 + 1}$ (e) $\dfrac{-x}{x^2 + 1}$ (f) $\dfrac{x + 1}{x^2 + 2x + 2}$ (g) $\dfrac{2x}{4x^2 + 1}$ (h) $\dfrac{x + h}{x^2 + 2xh + h^2 + 1}$
19. (a) 4   (b) 5   (c) 5   (d) $|x| + 4$ (e) $-|x| - 4$ (f) $|x + 1| + 4$ (g) $2|x| + 4$ (h) $|x + h| + 4$ **21.** Function   **23.** Function
25. Not a function   **27.** Not a function   **29.** Function   **31.** Not a function   **33.** All real numbers   **35.** All real numbers

37. $\{x|x \neq -4, x \neq 4\}$ **39.** $\{x|x \neq 0\}$ **41.** $\{x|x \geq 4\}$ **43.** $\{x|x > 9\}$ **45.** $\{x|x > 1\}$ **47. (a)** $f(0) = 3; f(-6) = -3$
(b) $f(6) = 0; f(11) = 1$ **(c)** Positive **(d)** Negative **(e)** $-3, 6$, and 10 **(f)** $-3 < x < 6, 10 < x \leq 11$ **(g)** $\{x|-6 \leq x \leq 11\}$
(h) $\{y|-3 \leq y \leq 4\}$ **(i)** $-3, 6, 10$ **(j)** 3 **(k)** 3 times **(l)** Once **(m)** $0, 4$ **(n)** $-5, 8$ **49.** Not a function

51. Function **(a)** Domain: $\{x|-\pi \leq x \leq \pi\}$; Range: $\{y|-1 \leq y \leq 1\}$ **(b)** $\left(-\dfrac{\pi}{2}, 0\right), \left(\dfrac{\pi}{2}, 0\right), (0, 1)$ **53.** Not a function

55. Function **(a)** Domain: $\{x|x > 0\}$; Range: all real numbers **(b)** $(1, 0)$
57. Function **(a)** Domain: all real numbers; Range: $\{y|y \leq 2\}$ **(b)** $(-3, 0), (3, 0), (0, 2)$
59. Function **(a)** Domain: all real numbers; Range: $\{y|y \geq -3\}$ **(b)** $(1, 0), (3, 0), (0, 9)$

61. (a) Yes **(b)** $f(-2) = 9; (-2, 9)$ **(c)** $0, \dfrac{1}{2}; (0, -1), \left(\dfrac{1}{2}, -1\right)$ **(d)** All real numbers **(e)** $-\dfrac{1}{2}, 1$ **(f)** -1

63. (a) No **(b)** $f(4) = -3; (4, -3)$ **(c)** $14; (14, 2)$ **(d)** $\{x|x \neq 6\}$ **(e)** -2 **(f)** $-\dfrac{1}{3}$ **65. (a)** Yes **(b)** $f(2) = \dfrac{8}{17}; \left(2, \dfrac{8}{17}\right)$
(c) $-1, 1; (-1, 1), (1, 1)$ **(d)** All real numbers **(e)** 0 **(f)** 0 **67.** $C = -3$ **69.** $A = -4$ **71.** $A = 8$; undefined at $x = 3$ **73.** 4
75. $2x + h - 1$ **77.** $3x^2 + 3xh + h^2$ **79. (a)** III **(b)** IV **(c)** I **(d)** V **(e)** II
81.

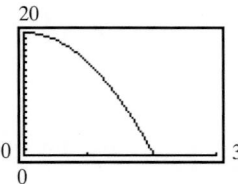

83. (a) 2 hr elapsed during which Kevin was between 0 and 3 mi from home.
(b) 0.5 hr elapsed during which Kevin was 3 mi from home.
(c) 0.3 hr elapsed during which Kevin was between 0 and 3 mi from home.
(d) 0.2 hr elapsed during which Kevin was 0 mi from home.
(e) 0.9 hr elapsed during which Kevin was between 0 and 2.8 mi from home.
(f) 0.3 hr elapsed during which Kevin was 2.8 mi from home.
(g) 1.1 hr elapsed during which Kevin was between 0 and 2.8 mi from home.
(h) 3 mi **(i)** 2 times

85. (a)

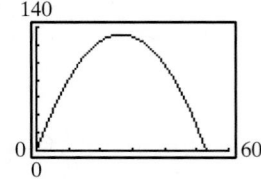

(b) About 15.1 m, 14.07 m, 12.94 m, 11.72 m
(c) After about 1.01 s, 1.43 s, 1.75 s
(d) After about 2.02 s

87. $A(x) = \dfrac{1}{2}x^2$ **89.** $G(x) = 10x$

91. (a) About 81.07 ft **(b)** About 129.59 ft **(c)** About 26.63 ft
(d)

(e) $\dfrac{4225 \pm 65\sqrt{1345}}{16} \approx 115.07$ feet and 413.05 ft
(f) 275 ft; maximum height shown in the table is 131.8 ft
(g) 264 ft **(h)** $\{x|0 \leq x \leq 528.125\}$

93. (a) $222 **(b)** $225 **(c)** $220 **(d)** $230
(e)

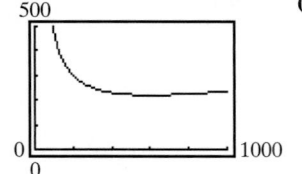

(f) 600 mi/hr

95. Only $h(x) = 2x$
97. No; $f(x)$ has a domain of all real numbers, while $g(x)$ has a domain of $\{x|x \neq -1\}$.

1.5 Concepts and Vocabulary *(page 60)*

1. slope **2.** increasing **3.** even; odd **4.** T **5.** T **6.** F **9.** Yes; the constant function $f(x) = 0$ is both even and odd.

1.5 Exercises *(page 61)*

1. C **3.** E **5.** B **7.** F

9.

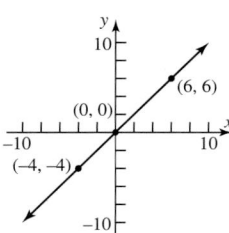

11.

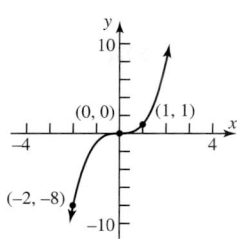

13.

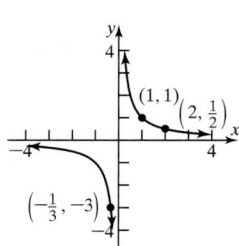

15. Yes **17.** No **19.** $(-8, -2); (0, 2); (5, 10)$ **21.** Yes; $f(2) = 10$ **23.** -2 and 2; $f(-2) = 6$ and $f(2) = 10$

25. (a) $(-2, 0), (0, 3), (2, 0)$ **(b)** Domain: $\{x | -4 \leq x \leq 4\}$ or $[-4, 4]$; Range: $\{y | 0 \leq y \leq 3\}$ or $[0, 3]$
(c) Increasing on $(-2, 0)$ and $(2, 4)$; Decreasing on $(-4, -2)$ and $(0, 2)$ **(d)** Even

27. (a) $(0, 1)$ **(b)** Domain: all real numbers or $(-\infty, \infty)$; Range: $\{y | y > 0\}$ or $(0, \infty)$. **(c)** Increasing on $(-\infty, \infty)$ **(d)** Neither

29. (a) $(-\pi, 0), (0, 0), (\pi, 0)$ **(b)** Domain: $\{x | -\pi \leq x \leq \pi\}$ or $[-\pi, \pi]$; Range: $\{y | -1 \leq y \leq 1\}$ or $[-1, 1]$.
(c) Increasing on $\left(-\dfrac{\pi}{2}, \dfrac{\pi}{2}\right)$; Decreasing on $\left(-\pi, -\dfrac{\pi}{2}\right)$ and on $\left(\dfrac{\pi}{2}, \pi\right)$ **(d)** Odd

31. (a) $\left(0, \dfrac{1}{2}\right), \left(\dfrac{1}{2}, 0\right), \left(\dfrac{5}{2}, 0\right)$ **(b)** Domain: $\{x | -3 \leq x \leq 3\}$ or $[-3, 3]$; Range: $\{y | -1 \leq y \leq 2\}$ or $[-1, 2]$
(c) Increasing on $(2, 3)$; Decreasing on $(-1, 1)$; Constant on $(-3, -1)$ and on $(1, 2)$. **(d)** Neither

33. (a) $0; f(0) = 3$ **(b)** -2 and 2; $f(-2) = 0$ and $f(2) = 0$

35. (a) $\dfrac{\pi}{2}; f\left(\dfrac{\pi}{2}\right) = 1$ **(b)** $-\dfrac{\pi}{2}; f\left(-\dfrac{\pi}{2}\right) = -1$ **37.** Odd **39.** Even **41.** Odd **43.** Neither **45.** Even **47.** Odd **49.** At most one

51.

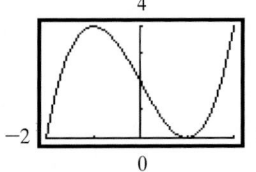

Increasing: $(-2, -1), (1, 2)$
Decreasing: $(-1, 1)$
Local maximum: $(-1, 4)$
Local minimum: $(1, 0)$

53.

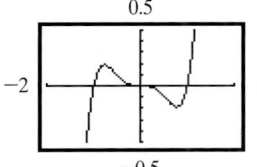

Increasing: $(-2, -0.77), (0.77, 2)$
Decreasing: $(-0.77, 0.77)$
Local maximum: $(-0.77, 0.19)$
Local minimum: $(0.77, -0.19)$

55.

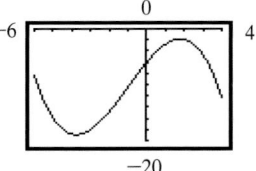
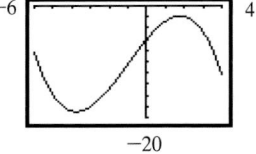

Increasing: $(-3.77, 1.77)$
Decreasing: $(-6, -3.77), (1.77, 4)$
Local maximum: $(1.77, -1.91)$
Local minimum: $(-3.77, -18.89)$

57.

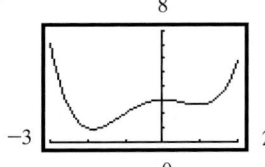
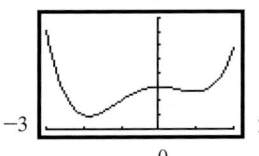
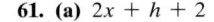

Increasing: $(-1.87, 0), (0.97, 2)$
Decreasing: $(-3, -1.87), (0, 0.97)$
Local maximum: $(0, 3)$
Local minima: $(-1.87, 0.95), (0.97, 2.65)$

59. (a) 2 **(b)** $2; 2; 2; m_{\text{sec}} = 2$ **(c)** $y = 2x + 5$
(d)

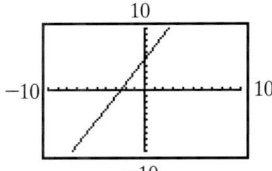

61. (a) $2x + h + 2$ **(b)** $4.5; 4.1; 4.01; m_{\text{sec}} = 4$
(c) $y = 4.01x - 1.01$
(d)

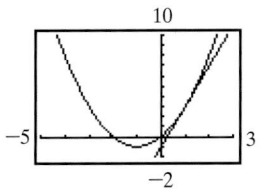

63. (a) $4x + 2h - 3$ **(b)** $2; 1.2; 1.02; m_{\text{sec}} = 1$
(c) $y = 1.02x - 1.02$
(d)

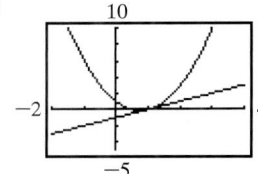

65. (a) $-\dfrac{1}{x(x+h)}$ **(b)** $-\dfrac{2}{3}; -\dfrac{10}{11}; -\dfrac{100}{101}; m_{\sec} = -1$

(c) $y = -\dfrac{100}{101}x + \dfrac{201}{101}$

(d)

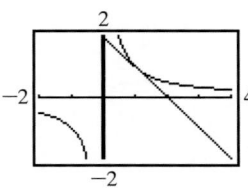

67.

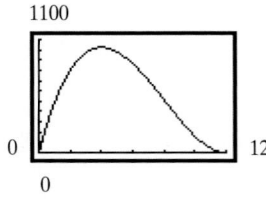

The volume is largest at $x = 4$ inches.

69. (a)

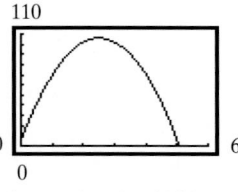

(b) 2.5 seconds   **(c)** 106 feet

71. Each graph is that of $y = x^2$, but shifted vertically. If $y = x^2 + k, k > 0$, the shift is up k units; if $y = x^2 - k, k > 0$, the shift is down k units.

73. Each graph is that of $y = |x|$, but either compressed or stretched. If $y = k|x|$ and $k > 1$, the graph is stretched vertically; if $y = k|x|, 0 < k < 1$, the graph is compressed vertically.

75. The graph of $y = f(-x)$ is the reflection about the y-axis of the graph of $y = f(x)$.

77. They are all ∪-shaped and open upward. All three go through the points $(-1, 1), (0, 0)$, and $(1, 1)$. As the exponent increases, the steepness of the curve increases (except between -1 and 1).

1.6 Concepts and Vocabulary *(page 73)*

1. horizontal; right   **2.** y **3.** compression; $\dfrac{1}{3}$ **4.** T   **5.** F   **6.** T   **7.** $4f(x)$ is a vertical stretch; $f(4x)$ is a horizontal compression.

1.6 Exercises *(page 74)*

1. B   **3.** H   **5.** I   **7.** L   **9.** F   **11.** G   **13.** C   **15.** B   **17.** $y = (x-4)^3$ **19.** $y = x^3 + 4$ **21.** $y = -x^3$ **23.** $y = 4x^3$

25. (1) $y = \sqrt{x} + 2$; (2) $y = -(\sqrt{x} + 2)$; (3) $y = -(\sqrt{-x} + 2)$ **27.** (1) $y = -\sqrt{x}$; (2) $y = -\sqrt{x} + 2$; (3) $y = -\sqrt{x+3} + 2$

29. (c)   **31.** (c)

33.

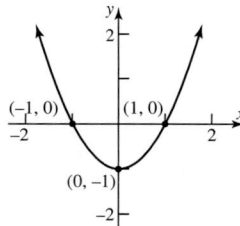

35.

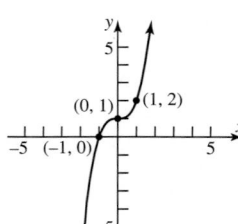

37.

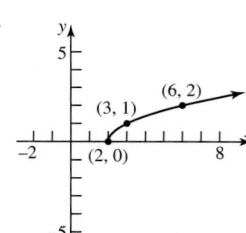

39.

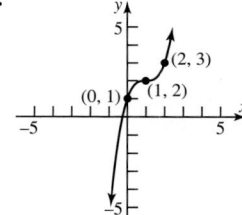

41.

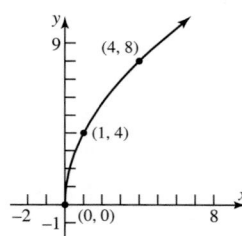

43.

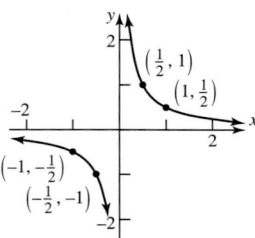

45.

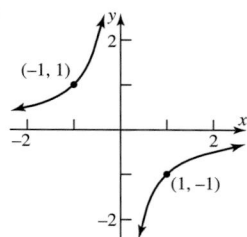

47.

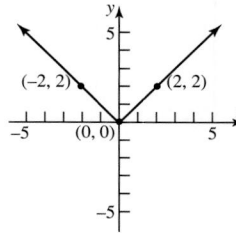

49.

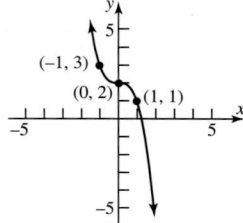

51.

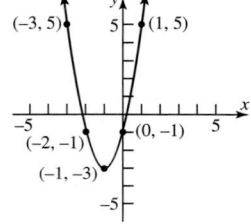

53.

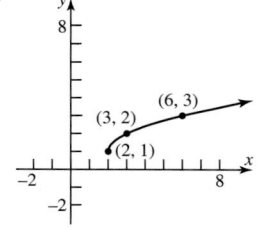

55.

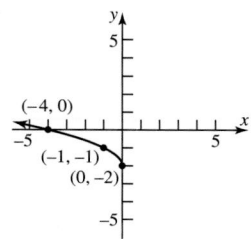

57.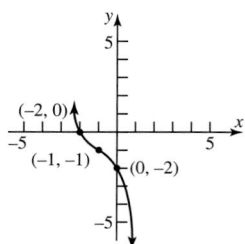

59.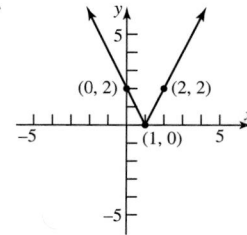

61.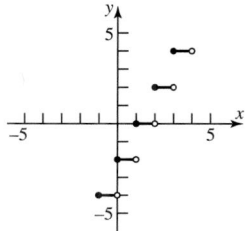

63. (a) $F(x) = f(x) + 3$ **(b)** $G(x) = f(x + 2)$ **(c)** $P(x) = -f(x)$ **(d)** $H(x) = f(x + 1) - 2$

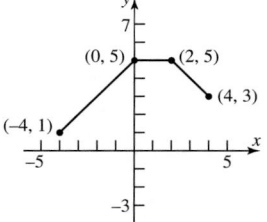   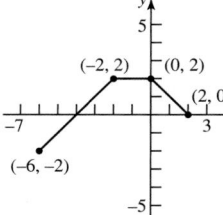   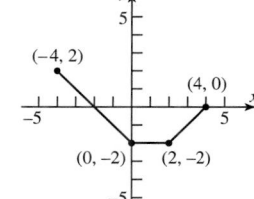   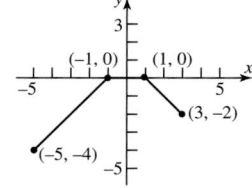

(e) $Q(x) = \frac{1}{2}f(x)$ **(f)** $g(x) = f(-x)$ **(g)** $h(x) = f(2x)$

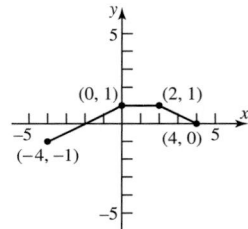   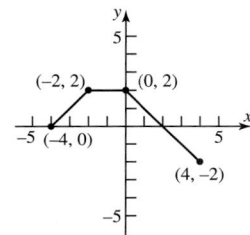   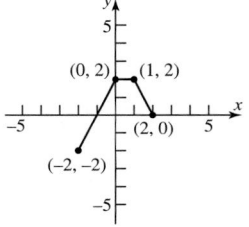

65. (a) $F(x) = f(x) + 3$ **(b)** $G(x) = f(x + 2)$ **(c)** $P(x) = -f(x)$ **(d)** $H(x) = f(x + 1) - 2$

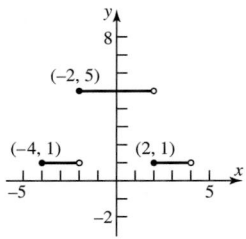   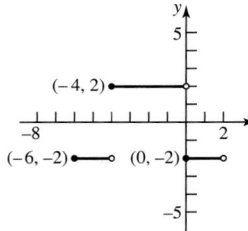   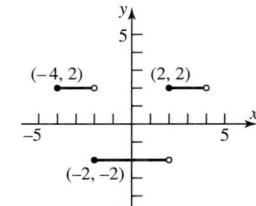   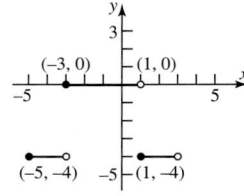

(e) $Q(x) = \frac{1}{2}f(x)$ **(f)** $g(x) = f(-x)$ **(g)** $h(x) = f(2x)$

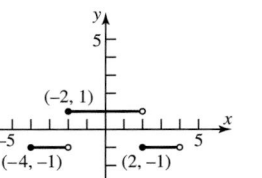   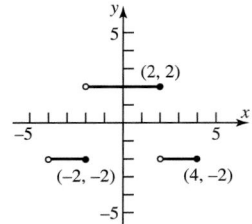   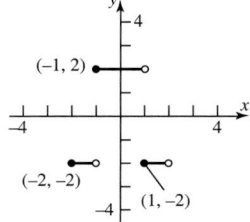

67. (a) $F(x) = f(x) + 3$

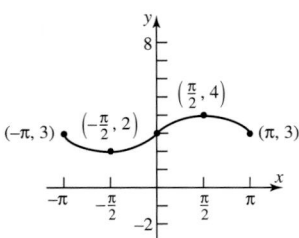

(b) $G(x) = f(x + 2)$

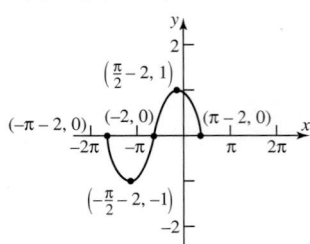

(c) $P(x) = -f(x)$

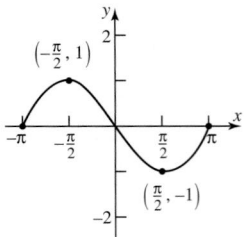

(d) $H(x) = f(x + 1) - 2$

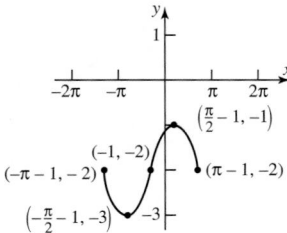

(e) $Q(x) = \frac{1}{2}f(x)$

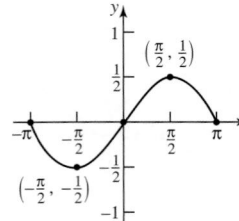

(f) $g(x) = f(-x)$

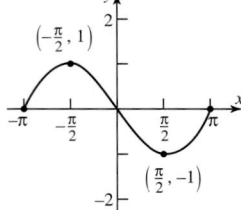

(g) $h(x) = f(2x)$

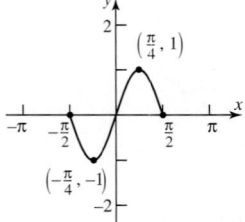

69. (a) $y = |x + 1|$

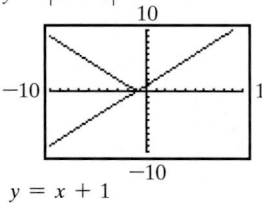

$y = x + 1$

(b) $y = |4 - x^2|$

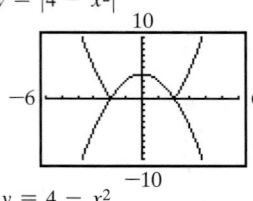

$y = 4 - x^2$

(c) $y = |x^3 + x|$

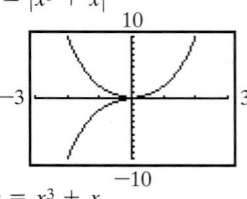

$y = x^3 + x$

(d) Any part of the graph of $y = f(x)$ that lies below the x-axis is reflected about the x-axis to obtain the graph of $y = |f(x)|$.

71. (a)

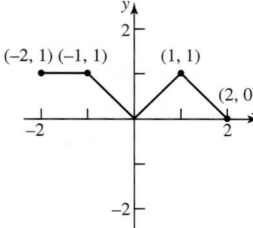

(b)

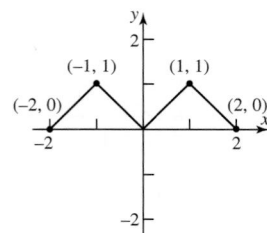

73. $f(x) = (x + 1)^2 - 1$

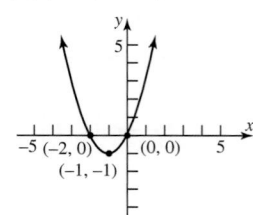

75. $f(x) = (x - 4)^2 - 15$

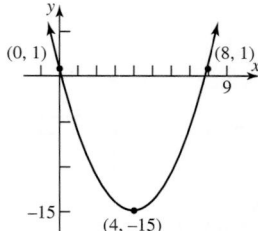

77. $f(x) = \left(x + \frac{1}{2}\right)^2 + \frac{3}{4}$

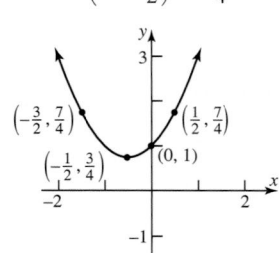

79.

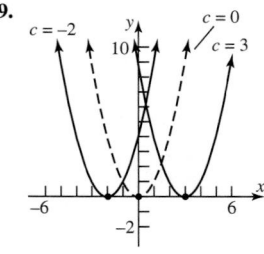

81.
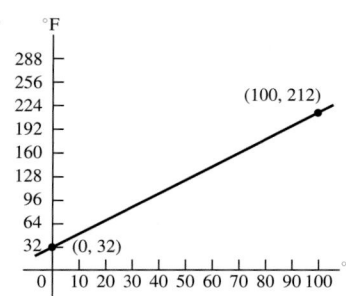

83. (a)
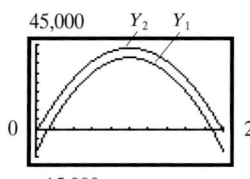

(b) 10% tax
(c) Y_1 is the graph of $p(x)$ shifted down vertically 10,000 units.
Y_2 is the graph of $p(x)$ vertically compressed by a factor of 0.9.
(d) 10% tax

1.7 Concepts and Vocabulary *(page 87)*

1. one-to-one **2.** $y = x$ **3.** $[4, \infty)$ **4.** F **5.** T **6.** Yes, if the domain is $\{x \mid x = 0\}$. **7.** No **8.** On the line $y = x$; No; No

1.7 Exercises *(page 87)*

1. (a)
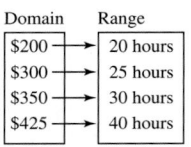
(b) Inverse is a function

3. (a)
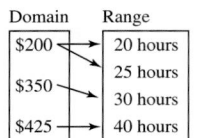
(b) Inverse is not a function

5. (a) $\{(6, 2), (6, -3), (9, 4), (10, 1)\}$
(b) Inverse is not a function
7. (a) $\{(0, 0), (1, 1), (16, 2), (81, 3)\}$
(b) Inverse is a function
9. One-to-one **11.** Not one-to-one
13. One-to-one

15.
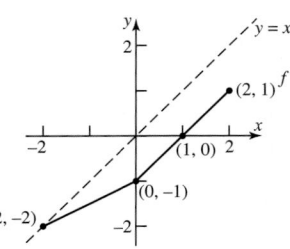

17.
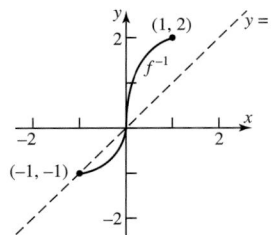

19.
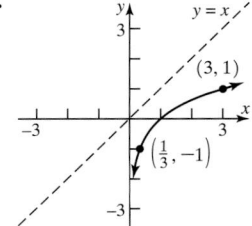

21. $f(g(x)) = f\left(\frac{1}{3}(x - 4)\right) = 3\left[\frac{1}{3}(x - 4)\right] + 4 = x$

$g(f(x)) = g(3x + 4) = \frac{1}{3}[(3x + 4) - 4] = x$

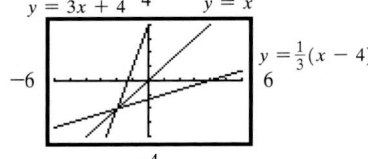

23. $f(g(x)) = 4\left[\frac{x}{4} + 2\right] - 8 = x$

$g(f(x)) = \frac{4x - 8}{4} + 2 = x$

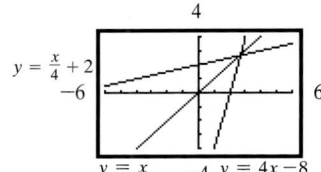

25. $f(g(x)) = (\sqrt[3]{x+8})^3 - 8 = x$
$g(f(x)) = \sqrt[3]{(x^3 - 8) + 8} = x$

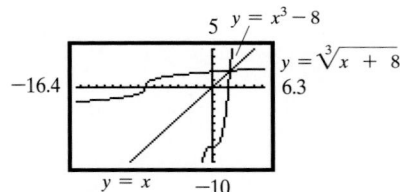

27. $f(g(x)) = \dfrac{1}{\left(\dfrac{1}{x}\right)} = x$

$g(f(x)) = \dfrac{1}{\left(\dfrac{1}{x}\right)} = x$

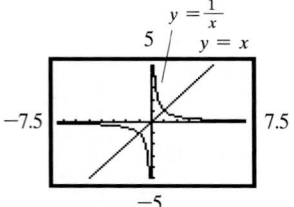

29. $f(g(x)) = \dfrac{2\left(\dfrac{4x-3}{2-x}\right) + 3}{\dfrac{4x-3}{2-x} + 4} = x$

$g(f(x)) = \dfrac{4\left(\dfrac{2x+3}{x+4}\right) - 3}{2 - \dfrac{2x+3}{x+4}} = x$

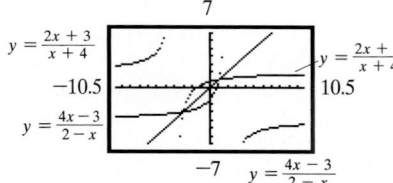

31. $f^{-1}(x) = \dfrac{1}{3}x$

$f(f^{-1}(x)) = 3\left(\dfrac{1}{3}x\right) = x$

$f^{-1}(f(x)) = \dfrac{1}{3}(3x) = x$

Domain f = Range $f^{-1} = (-\infty, \infty)$
Range f = Domain $f^{-1} = (-\infty, \infty)$

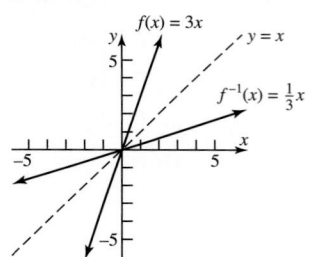

33. $f^{-1}(x) = \dfrac{x}{4} - \dfrac{1}{2}$

$f(f^{-1}(x)) = 4\left(\dfrac{x}{4} - \dfrac{1}{2}\right) + 2 = x$

$f^{-1}(f(x)) = \dfrac{4x+2}{4} - \dfrac{1}{2} = x$

Domain f = Range $f^{-1} = (-\infty, \infty)$
Range f = Domain $f^{-1} = (-\infty, \infty)$

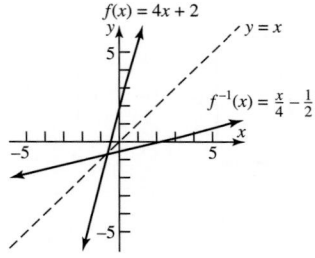

35. $f^{-1}(x) = \sqrt[3]{x+1}$
$f(f^{-1}(x)) = (\sqrt[3]{x+1})^3 - 1 = x$
$f^{-1}(f(x)) = \sqrt[3]{(x^3 - 1) + 1} = x$
Domain f = Range $f^{-1} = (-\infty, \infty)$
Range f = Domain $f^{-1} = (-\infty, \infty)$

37. $f^{-1}(x) = \sqrt{x - 4}$

$f(f^{-1}(x)) = (\sqrt{x - 4})^2 + 4 = x$

$f^{-1}(f(x)) = \sqrt{(x^2 + 4) - 4} = \sqrt{x^2} = |x|$

$\qquad\qquad = x$

Domain f = Range $f^{-1} = [0, \infty)$

Range f = Domain $f^{-1} = [4, \infty)$

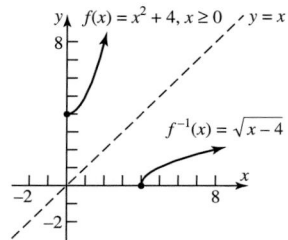

39. $f^{-1}(x) = \dfrac{4}{x}$

$f(f^{-1}(x)) = \dfrac{4}{\dfrac{4}{x}} = x$

$f^{-1}(f(x)) = \dfrac{4}{\dfrac{4}{x}} = x$

Domain f = Range f^{-1} = all real numbers except 0

Range f = Domain f^{-1} = all real numbers except 0

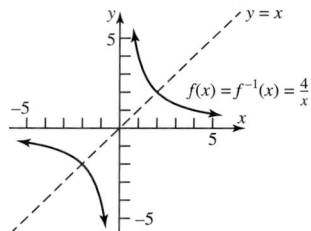

41. $f^{-1}(x) = \dfrac{2x + 1}{x}$

$f(f^{-1}(x)) = \dfrac{1}{\dfrac{2x + 1}{x} - 2} = x$

$f^{-1}(f(x)) = \dfrac{2\left(\dfrac{1}{x - 2}\right) + 1}{\dfrac{1}{x - 2}} = x$

Domain f = Range f^{-1} = all real numbers except 2

Range f = Domain f^{-1} = all real numbers except 0

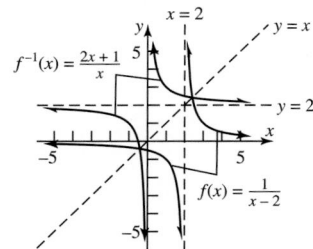

43. $f^{-1}(x) = \dfrac{2 - 3x}{x}$

$f(f^{-1}(x)) = \dfrac{2}{3 + \dfrac{2 - 3x}{x}} = x$

$f^{-1}(f(x)) = \dfrac{2 - 3\left(\dfrac{2}{3 + x}\right)}{\dfrac{2}{3 + x}} = x$

Domain f = all real numbers except -3

Range f = Domain f^{-1} = all real numbers except 0

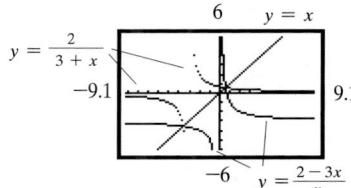

45. $f^{-1}(x) = \sqrt{x} - 2$

$f(f^{-1}(x)) = (\sqrt{x} - 2 + 2)^2 = x$

$f^{-1}(f(x)) = \sqrt{(x + 2)^2} - 2 = |x + 2| - 2 = x, x \geq -2$

Domain $f = [-2, \infty)$

Range f = Domain $f^{-1} = [0, \infty)$

$y = (x + 2)^2, x \geq -2$ 6

$y = \sqrt{x} - 2$

-9.1 9.1

-6

47. $f^{-1}(x) = \dfrac{x}{x - 2}$

$$f(f^{-1}(x)) = \dfrac{2\left(\dfrac{x}{x-2}\right)}{\dfrac{x}{x-2} - 1} = x$$

$$f^{-1}(f(x)) = \dfrac{\dfrac{2x}{x-1}}{\dfrac{2x}{x-1} - 2} = x$$

Domain f = all real numbers except 1
Range f = Domain f^{-1} = all real numbers except 2

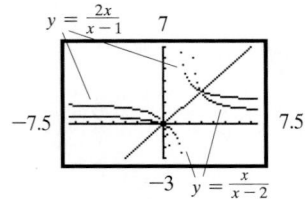

$y = \dfrac{2x}{x-1}$, 7, -7.5, 7.5, -3, $y = \dfrac{x}{x-2}$

49. $f^{-1}(x) = \dfrac{3x + 4}{2x - 3}$

$$f(f^{-1}(x)) = \dfrac{3\left(\dfrac{3x+4}{2x-3}\right) + 4}{2\left(\dfrac{3x+4}{2x-3}\right) - 3} = x$$

$$f^{-1}(f(x)) = \dfrac{3\left(\dfrac{3x+4}{2x-3}\right) + 4}{2\left(\dfrac{3x+4}{2x-3}\right) - 3} = x$$

Domain f = all real numbers except $\dfrac{3}{2}$

Range f = Domain f^{-1} = all real numbers except $\dfrac{3}{2}$

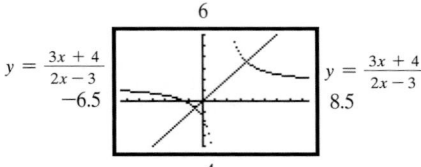

$y = \dfrac{3x+4}{2x-3}$, 6, -6.5, 8.5, -4, $y = \dfrac{3x+4}{2x-3}$

51. $f^{-1}(x) = \dfrac{-2x + 3}{x - 2}$

$$f(f^{-1}(x)) = \dfrac{2\left(\dfrac{-2x+3}{x-2}\right) + 3}{\dfrac{-2x+3}{x-2} + 2} = x$$

$$f^{-1}(f(x)) = \dfrac{-2\left(\dfrac{2x+3}{x+2}\right) + 3}{\dfrac{2x+3}{x+2} - 2} = x$$

Domain f = all real numbers except -2
Range f = Domain f^{-1} = all real numbers except 2

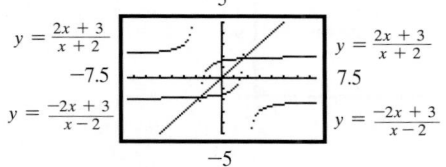

$y = \dfrac{2x+3}{x+2}$, 5, -7.5, 7.5, $y = \dfrac{2x+3}{x+2}$, $y = \dfrac{-2x+3}{x-2}$, $y = \dfrac{-2x+3}{x-2}$, -5

53. $f^{-1}(x) = \dfrac{x^3}{8}$

$$f(f^{-1}(x)) = 2\sqrt[3]{\dfrac{x^3}{8}} = x$$

$$f^{-1}(f(x)) = \dfrac{(2\sqrt[3]{x})^3}{8} = x$$

Domain f = $(-\infty, \infty)$
Range f = Domain f^{-1} = $(-\infty, \infty)$

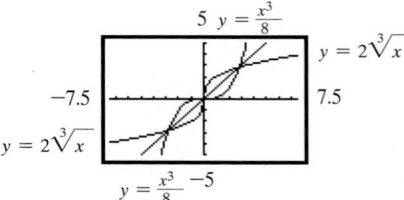

5, $y = \dfrac{x^3}{8}$, $y = 2\sqrt[3]{x}$, -7.5, 7.5, $y = 2\sqrt[3]{x}$, $y = \dfrac{x^3}{8}$, -5

55. $f^{-1}(x) = \dfrac{1}{m}(x - b), m \neq 0$ **57.** Quadrant I   **59.** Possible answer: $f(x) = |x|, x \geq 0$, is one-to-one; $f^{-1}(x) = x, x \geq 0$

61. $f(g(x)) = \dfrac{9}{5}\left[\dfrac{5}{9}(x - 32)\right] + 32 = x; g(f(x)) = \dfrac{5}{9}\left[\left(\dfrac{9}{5}x + 32\right) - 32\right] = x$ **63.** $l(T) = \dfrac{gT^2}{4\pi^2}, T > 0$

65. Yes; The graph must be symmetric about the line $y = x; y = \dfrac{1}{x}$ is an example.

Review Exercises *(page 92)*

1. $2\sqrt{26}; (2, 3)$

3.

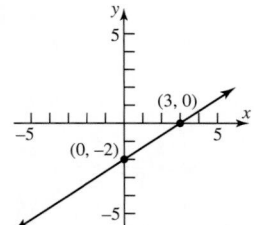

5.

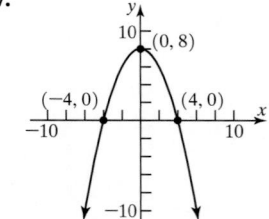

7.

9. x-axis
11. x-axis, y-axis, origin
13. y-axis
15. No symmetry

17.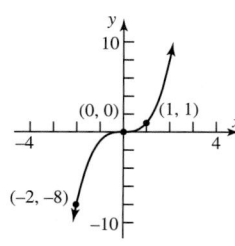

19. $(x - 1)^2 + (y - 3)^2 = 4$

21. $(x + 3)^2 + (y - 2)^2 = 9$

23. Center: $(1, -2)$; radius $= 3$

25. Center: $(1, -2)$; radius $= \sqrt{5}$

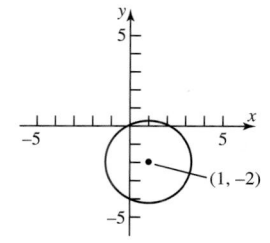

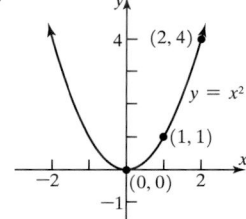

27. Yes

29. $f(x) = -2x + 3$

31. $A = 11$

33. b, c, d

35. (a) $f(-x) = \dfrac{-3x}{x^2 - 4}$ **(b)** $-f(x) = \dfrac{-3x}{x^2 - 4}$ **(c)** $f(x + 2) = \dfrac{3x + 6}{x^2 + 4x}$ **(d)** $f(x - 2) = \dfrac{3x - 6}{x^2 - 4x}$ **37. (a)** $f(-x) = \sqrt{x^2 - 4}$

(b) $-f(x) = -\sqrt{x^2 - 4}$ **(c)** $f(x + 2) = \sqrt{x^2 + 4x}$ **(d)** $f(x - 2) = \sqrt{x^2 - 4x}$ **39. (a)** $f(-x) = \dfrac{x^2 - 4}{x^2}$ **(b)** $-f(x) = -\dfrac{x^2 - 4}{x^2}$

(c) $f(x + 2) = \dfrac{x^2 + 4x}{x^2 + 4x + 4}$ **(d)** $f(x - 2) = \dfrac{x^2 - 4x}{x^2 - 4x + 4}$ **41.** $\{x | x \neq -3, x \neq 3\}$ **43.** $\{x | x \leq 2\}$ **45.** $\{x | x > 0\}$

47. $\{x | x \neq -3, x \neq 1\}$

49. (a) Domain: $\{x | -4 \leq x \leq 4\}$ or $[-4, 4]$
    Range: $\{y | -3 \leq y \leq 1\}$ or $[-3, 1]$
(b) Increasing on $(-4, -1)$ and $(3, 4)$;
    Decreasing on $(-1, 3)$
(c) Local maximum is 1 and occurs at $x = -1$;
    Local minimum is -3 and occurs at $x = 3$
(d) No symmetry **(e)** Neither
(f) $(-2, 0), (0, 0), (4, 0)$

51.

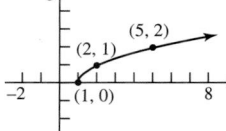

53. Odd **55.** Even **57.** Neither **59.** Odd

61.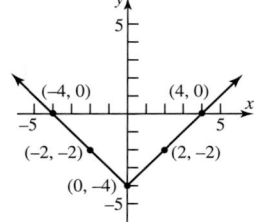

Intercepts: $(-4, 0), (4, 0), (0, -4)$
Domain: all real numbers
Range: $\{y | y \geq -4\}$

63.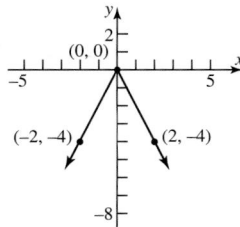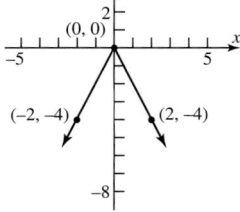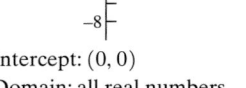

Intercept: $(0, 0)$
Domain: all real numbers
Range: $\{y | y \leq 0\}$

65.

Intercept: $(1, 0)$
Domain: $\{x | x \geq 1\}$
Range: $\{y | y \geq 0\}$

67.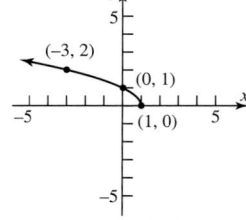

Intercepts: $(0, 1), (1, 0)$
Domain: $\{x | x \leq 1\}$
Range: $\{y | y \geq 0\}$

69.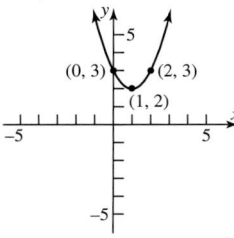

Intercept: $(0, 3)$

Domain: all real numbers
Range: $\{y | y \geq 2\}$

71.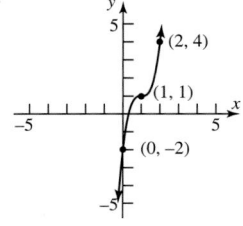

Intercepts: $\left(\sqrt[3]{-\dfrac{1}{3}} + 1, 0 \right), (0, -2)$

Domain: all real numbers
Range: all real numbers

73.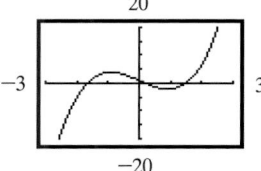

Local maximum: $(-0.91, 4.04)$
Local minimum: $(0.91, -2.04)$
Increasing: $(-3, -0.91); (0.91, 3)$
Decreasing: $(-0.91, 0.91)$

75.

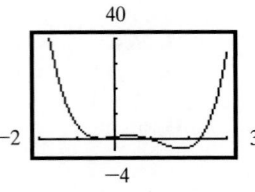

Local maximum: $(0.41, 1.53)$
Local minimum: $(-0.34, 0.54); (1.80, -3.56)$
Increasing: $(-0.34, 0.41); (1.80, 3)$
Decreasing: $(-2, -0.34); (0.41, 1.80)$

77. (a) $\{(2, 1), (5, 3), (8, 5), (10, 6)\}$
  (b) Inverse is a function

79.

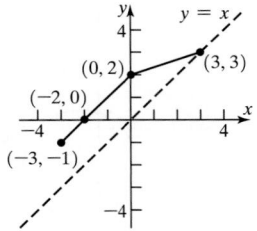

81. $f^{-1}(x) = \dfrac{2x + 3}{5x - 2}$

$$f(f^{-1}(x)) = \dfrac{2\left(\dfrac{2x + 3}{5x - 2}\right) + 3}{5\left(\dfrac{2x + 3}{5x - 2}\right) - 2} = x$$

$$f^{-1}(f(x)) = \dfrac{2\left(\dfrac{2x + 3}{5x - 2}\right) + 3}{5\left(\dfrac{2x + 3}{5x - 2}\right) - 2} = x$$

Domain f = Range f^{-1} = all real numbers except $\dfrac{2}{5}$

Range f = Domain f^{-1} = all real numbers except $\dfrac{2}{5}$

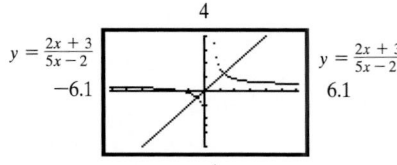

83. $f^{-1}(x) = \dfrac{x + 1}{x}$

$$f(f^{-1}(x)) = \dfrac{1}{\dfrac{x + 1}{x} - 1} = x$$

$$f^{-1}(f(x)) = \dfrac{\dfrac{1}{x - 1} + 1}{\dfrac{1}{x - 1}} = x$$

Domain f = Range f^{-1} = all real numbers except 1
Range f = Domain f^{-1} = all real numbers except 0

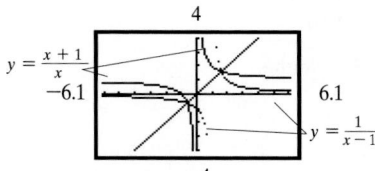

85. $f^{-1}(x) = \dfrac{27}{x^3}$

$$f(f^{-1}(x)) = \dfrac{3}{\left(\dfrac{27}{x^3}\right)^{1/3}} = x$$

$$f^{-1}(f(x)) = \dfrac{27}{\left(\dfrac{3}{x^{1/3}}\right)^3} = x$$

Domain f = Range f^{-1} = all real numbers except 0
Range f = Domain f^{-1} = all real numbers except 0

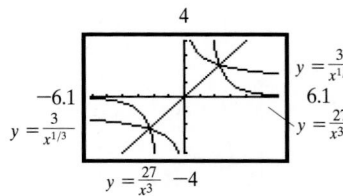

87. (a)

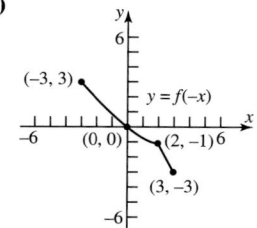

(b)

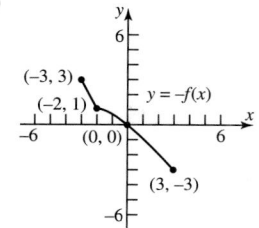

(c)

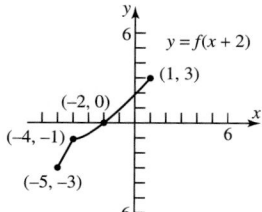

(d)

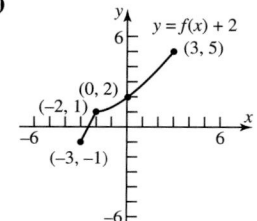

(e)

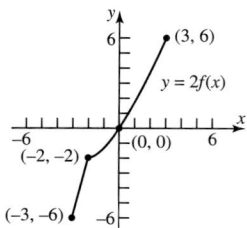

(f)

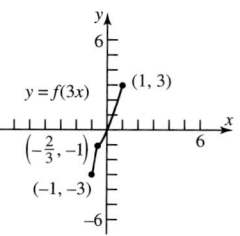

89. $d(A, B) = \sqrt{13}; d(B, C) = \sqrt{13}$ **91.** Center: $(1, -2)$; radius $= 4\sqrt{2}; x^2 + y^2 - 2x + 4y - 27 = 0$

C H A P T E R   2   Trigonometric Functions

2.1 Concepts and Vocabulary *(page 108)*

1. Standard position   **2.** $r\theta; \dfrac{1}{2}r^2\theta$ **3.** $\dfrac{s}{t}; \dfrac{\theta}{t}$ **4.** F   **5.** T   **6.** T   **7.** On a circle of radius r, if the length of arc subtended by a central angle is also r, then the measure of the angle is 1 radian.   **8.** 1 radian is larger.   **9.** When an object travels around a circle, linear speed measures the length of the arc traveled per unit time; angular speed measures the angle swept out per unit time.

2.1 Exercises *(page 109)*

1. 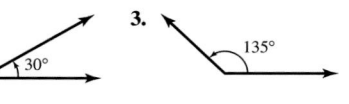   **3.** 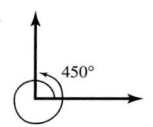   **5.** 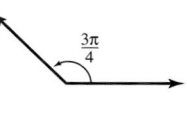   **7.** 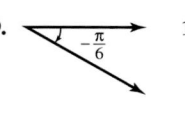   **9.** 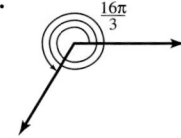   **11.**

13. $\dfrac{\pi}{6}$ **15.** $\dfrac{4\pi}{3}$ **17.** $-\dfrac{\pi}{3}$ **19.** π **21.** $-\dfrac{3\pi}{4}$ **23.** $-\dfrac{\pi}{2}$ **25.** $60°$ **27.** $-225°$ **29.** $90°$ **31.** $15°$ **33.** $-90°$ **35.** $-30°$ **37.** 5 m   **39.** 6 ft

41. 0.6 radian   **43.** $\dfrac{\pi}{3} \approx 1.047$ in.   **45.** 25 m^2 **47.** $2\sqrt{3} \approx 3.464$ ft   **49.** 0.24 radian   **51.** $\dfrac{\pi}{3} \approx 1.047$ in^2 **53.** $s = 2.094$ ft; $A = 2.094$ ft^2

55. $s = 14.661$ yd; $A = 87.965$ yd^2 **57.** 0.30   **59.** -0.70 **61.** 2.18   **63.** $179.91°$ **65.** $114.59°$ **67.** $362.11°$ **69.** $40.17°$ **71.** $1.03°$

73. $9.15°$ **75.** $40°19'12''$ **77.** $18°15'18''$ **79.** $19°59'24''$ **81.** $3\pi \approx 9.42$ in.; $5\pi \approx 15.71$ in.   **83.** $2\pi \approx 6.28$ m^2 **85.** $\dfrac{675\pi}{2} \approx 1060.29$ ft^2

87. $\omega = \dfrac{1}{60}$ radian/sec; $v = \dfrac{1}{12}$ cm/sec   **89.** Approximately 452.49 rpm   **91.** Approximately 359.40 mi   **93.** Approximately 897.84 mi/hr

95. Approximately 2291.94 mi/hr   **97.** $\dfrac{3}{4}$ rpm   **99.** Approximately 2.86 mi/hr   **101.** Approximately 31.47 rpm

103. Approximately 1036.73 mi/hr   **105.** $v_1 = r_1\omega_1, v_2 = r_2\omega_2$, and $v_1 = v_2$ so $r_1\omega_1 = r_2\omega_2 \Rightarrow \dfrac{r_1}{r_2} = \dfrac{\omega_2}{\omega_1}$.

2.2 Concepts and Vocabulary *(page 120)*

1. Complementary   **2.** Cosine   **3.** $62°$ **4.** F   **5.** T   **6.** F   **7.** $\csc\theta = \dfrac{1}{\sin\theta}; \sec\theta = \dfrac{1}{\cos\theta}; \cot\theta = \dfrac{1}{\tan\theta}$ **10.** Tangent, cotangent

2.2 Exercises *(page 121)*

1. $\sin\theta = \dfrac{5}{13}; \cos\theta = \dfrac{12}{13}; \tan\theta = \dfrac{5}{12}; \csc\theta = \dfrac{13}{5}; \sec\theta = \dfrac{13}{12}; \cot\theta = \dfrac{12}{5}$ **3.** $\sin\theta = \dfrac{2\sqrt{13}}{13}; \cos\theta = \dfrac{3\sqrt{13}}{13}; \tan\theta = \dfrac{2}{3}; \csc\theta = \dfrac{\sqrt{13}}{2};$

$\sec\theta = \dfrac{\sqrt{13}}{3}; \cot\theta = \dfrac{3}{2}$ **5.** $\sin\theta = \dfrac{\sqrt{3}}{2}; \cos\theta = \dfrac{1}{2}; \tan\theta = \sqrt{3}; \csc\theta = \dfrac{2\sqrt{3}}{3}; \sec\theta = 2; \cot\theta = \dfrac{\sqrt{3}}{3}$ **7.** $\sin\theta = \dfrac{\sqrt{6}}{3}; \cos\theta = \dfrac{\sqrt{3}}{3};$

$\tan\theta = \sqrt{2}; \csc\theta = \dfrac{\sqrt{6}}{2}; \sec\theta = \sqrt{3}; \cot\theta = \dfrac{\sqrt{2}}{2}$ **9.** $\sin\theta = \dfrac{\sqrt{5}}{5}; \cos\theta = \dfrac{2\sqrt{5}}{5}; \tan\theta = \dfrac{1}{2}; \csc\theta = \sqrt{5}; \sec\theta = \dfrac{\sqrt{5}}{2}; \cot\theta = 2$

11. $\tan\theta = \dfrac{\sqrt{3}}{3}; \csc\theta = 2; \sec\theta = \dfrac{2\sqrt{3}}{3}; \cot\theta = \sqrt{3}$ **13.** $\tan\theta = \dfrac{2\sqrt{5}}{5}; \csc\theta = \dfrac{3}{2}; \sec\theta = \dfrac{3\sqrt{5}}{5}; \cot\theta = \dfrac{\sqrt{5}}{2}$

15. $\cos\theta = \dfrac{\sqrt{2}}{2}; \tan\theta = 1; \csc\theta = \sqrt{2}; \sec\theta = \sqrt{2}; \cot\theta = 1$ **17.** $\sin\theta = \dfrac{2\sqrt{2}}{3}; \tan\theta = 2\sqrt{2}; \csc\theta = \dfrac{3\sqrt{2}}{4}; \sec\theta = 3; \cot\theta = \dfrac{\sqrt{2}}{4}$

19. $\sin\theta = \dfrac{\sqrt{5}}{5}; \cos\theta = \dfrac{2\sqrt{5}}{5}; \csc\theta = \sqrt{5}; \sec\theta = \dfrac{\sqrt{5}}{2}; \cot\theta = 2$ **21.** $\sin\theta = \dfrac{2\sqrt{2}}{3}; \cos\theta = \dfrac{1}{3}; \tan\theta = 2\sqrt{2}; \csc\theta = \dfrac{3\sqrt{2}}{4}; \cot\theta = \dfrac{\sqrt{2}}{4}$

23. $\sin\theta = \dfrac{\sqrt{6}}{3}; \cos\theta = \dfrac{\sqrt{3}}{3}; \csc\theta = \dfrac{\sqrt{6}}{2}; \sec\theta = \sqrt{3}; \cot\theta = \dfrac{\sqrt{2}}{2}$ **25.** $\sin\theta = \dfrac{1}{2}; \cos\theta = \dfrac{\sqrt{3}}{2}; \tan\theta = \dfrac{\sqrt{3}}{3}; \sec\theta = \dfrac{2\sqrt{3}}{3}; \cot\theta = \sqrt{3}$

27. 1   **29.** 1   **31.** 0   **33.** 0   **35.** 1   **37.** 0   **39.** 0   **41.** 1   **43.** 1   **45. (a)** $\dfrac{1}{2}$ **(b)** $\dfrac{3}{4}$ **(c)** 2   **(d)** 2   **47. (a)** 17   **(b)** $\dfrac{1}{4}$ **(c)** 4   **(d)** $\dfrac{17}{16}$

49. (a) $\dfrac{1}{4}$ **(b)** 15   **(c)** 4   **(d)** $\dfrac{16}{15}$ **51. (a)** 0.78   **(b)** 0.79   **(c)** 1.27   **(d)** 1.28   **(e)** 1.61   **(f)** 0.78   **(g)** 0.62   **(h)** 1.27   **53.** 0.6   **55.** 20°

57. (a) 10 min   **(b)** 20 min   **(c)** $T(\theta) = 5\left(1 - \dfrac{1}{3\tan\theta} + \dfrac{1}{\sin\theta}\right)$

(d) Approximately 15.81 min   **(e)** Approximately 10.40 min

(f)

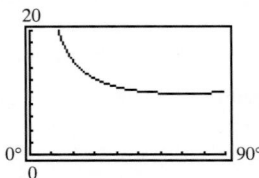

Approximately 70.53°

Approximately 176.78 ft

Approximately 9.71 min

59. (a) $|OA| = |OC| = 1$; angle OAC = angle OCA

angle OAC + angle OCA + $(180° - \theta) = 180°$

$2(\text{angle } OAC) - \theta = 0$; angle $OAC = \dfrac{\theta}{2}$

(b) $\sin\theta = \dfrac{|CD|}{|OC|} = |CD|; \cos\theta = \dfrac{|OD|}{|OC|} = |OD|$

(c) $\tan\dfrac{\theta}{2} = \dfrac{|CD|}{|AD|} = \dfrac{\sin\theta}{1 + |OD|} = \dfrac{\sin\theta}{1 + \cos\theta}$

61. $h = x\tan\theta$ and $h = (1 - x)\tan(n\theta)$; thus, $x\tan\theta = (1 - x)\tan(n\theta)$ so $x = \dfrac{\tan(n\theta)}{\tan\theta + \tan(n\theta)}$.

63. (a) Area $\triangle OAC = \dfrac{1}{2}|AC|\,|OC| = \dfrac{1}{2}\cdot\dfrac{|AC|}{1}\cdot\dfrac{|OC|}{1} = \dfrac{1}{2}\sin\alpha\cos\alpha$

(b) Area $\triangle OCB = \dfrac{1}{2}|BC|\,|OC| = \dfrac{1}{2}|OB|^2\dfrac{|BC|}{|OB|}\cdot\dfrac{|OC|}{|OB|} = \dfrac{1}{2}|OB|^2\sin\beta\cos\beta$

(c) Area $\triangle OAB = \dfrac{1}{2}|BD|\,|OA| = \dfrac{1}{2}|OB|\dfrac{|BD|}{|OB|} = \dfrac{1}{2}|OB|\sin(\alpha + \beta)$ **(d)** $\dfrac{\cos\alpha}{\cos\beta} = \dfrac{1}{\dfrac{|OC|}{|OB|}} = |OB|$

(e) Area $\triangle OAB$ = area $\triangle OAC$ + area $\triangle OCB$

$\dfrac{1}{2}|OB|\sin(\alpha + \beta) = \dfrac{1}{2}\sin\alpha\cos\alpha + \dfrac{1}{2}|OB|^2\sin\beta\cos\beta$

$\sin(\alpha + \beta) = \dfrac{\sin\alpha\cos\alpha + |OB|^2\sin\beta\cos\beta}{|OB|}$

$\sin(\alpha + \beta) = \dfrac{\sin\alpha(|OB|\cos\beta) + |OB|^2\sin\beta\left(\dfrac{\cos\alpha}{|OB|}\right)}{|OB|}$

$\sin(\alpha + \beta) = \sin\alpha\cos\beta + \cos\alpha\sin\beta$

65. $\sin \alpha = \tan \alpha \cos \alpha = \cos \beta \cos \alpha = \cos \beta \tan \beta = \sin \beta$;

$$\sin^2\alpha + \cos^2 \alpha = 1$$
$$\sin^2 \alpha + \tan^2 \beta = 1$$
$$\sin^2 \alpha + \frac{\sin^2 \beta}{\cos^2 \beta} = 1$$
$$\sin^2 \alpha + \frac{\sin^2 \alpha}{1 - \sin^2 \alpha} = 1$$
$$\sin^2 \alpha - \sin^4 \alpha + \sin^2 \alpha = 1 - \sin^2 \alpha$$
$$\sin^4 \alpha - 3 \sin^2 \alpha + 1 = 0$$
$$\sin^2 \alpha = \frac{3 \pm \sqrt{5}}{2}$$
$$\sin^2 \alpha = \frac{3 - \sqrt{5}}{2}$$
$$\sin \alpha = \sqrt{\frac{3 - \sqrt{5}}{2}}$$

Note that $\dfrac{3 + \sqrt{5}}{2} > 1.$

67. Assume (a, b) is a point on the terminal side of θ.
Since θ is acute, $a > 0$ and $b > 0$.
Since $a^2 + b^2 = c^2$, then $0 < b^2 < c^2$,
so $0 < b < c.$
Thus, $0 < \dfrac{b}{c} < 1$ and $0 < \sin \theta < 1.$

2.3 Concepts and Vocabulary *(page 129)*

1. $\dfrac{3}{2}$ **2.** 0.91 **3.** T **4.** F

2.3 Exercises *(page 129)*

1. $\sin 45° = \dfrac{\sqrt{2}}{2}$; $\cos 45° = \dfrac{\sqrt{2}}{2}$; $\tan 45° = 1$; $\csc 45° = \sqrt{2}$; $\sec 45° = \sqrt{2}$; $\cot 45° = 1$ **3.** $\dfrac{\sqrt{3}}{2}$ **5.** $\dfrac{1}{2}$ **7.** $\dfrac{3}{4}$ **9.** $\sqrt{3}$ **11.** $\dfrac{\sqrt{3}}{4}$ **13.** $\sqrt{2}$

15. 2 **17.** $\sqrt{2} + \dfrac{4\sqrt{3}}{3}$ **19.** $-\dfrac{8}{3}$ **21.** $\dfrac{1}{2}$ **23.** 0 **25.** 0.47 **27.** 0.38 **29.** 1.33 **31.** 0.31 **33.** 3.73 **35.** 1.04 **37.** 0.84 **39.** 0.02

41. 0.31 **43.** $R \approx 310.56$ ft; $H \approx 77.64$ ft **45.** $R \approx 19{,}541.95$ m; $H \approx 2278.14$ m **47. (a)** 0.60 sec **(b)** 0.79 sec **(c)** 1.04 sec

49. (a) $T(\theta) = 1 + \dfrac{2}{3 \sin \theta} - \dfrac{1}{4 \tan \theta}$
(b) 1.90 hr; 0.57 hr
(c) 1.69 hr; 0.75 hr
(d) 1.63 hr; 0.86 hr
(e) 1.67 hr **(f)** 2.75 hr

51.

θ	0.5	0.4	0.2	0.1	0.01	0.001	0.0001	0.00001
$\sin \theta$	0.4794	0.3894	0.1987	0.0998	0.0100	0.0010	0.0001	0.00001
$\dfrac{\sin \theta}{\theta}$	0.9589	0.9735	0.9933	0.9983	1.0000	1.0000	1.0000	1.0000

$\dfrac{(\sin \theta)}{\theta}$ approaches 1 as $\theta \to 0.$

(g)

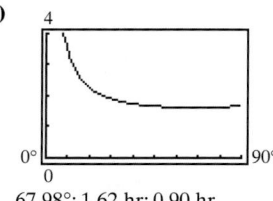

67.98°; 1.62 hr; 0.90 hr

53. 1 **55.** $\dfrac{\sqrt{2}}{2}$

2.4 Concepts and Vocabulary *(page 140)*

1. Tangent; cotangent **2.** Coterminal **3.** 60° **4.** F **5.** T **6.** T **8.** I, IV **9.** $\dfrac{\pi}{2}, \dfrac{3\pi}{2}$ **10.** 50°

2.4 Exercises *(page 141)*

1. $\sin \theta = \dfrac{4}{5}$; $\cos \theta = -\dfrac{3}{5}$; $\tan \theta = -\dfrac{4}{3}$; $\csc \theta = \dfrac{5}{4}$; $\sec \theta = -\dfrac{5}{3}$; $\cot \theta = -\dfrac{3}{4}$ **3.** $\sin \theta = -\dfrac{3\sqrt{13}}{13}$; $\cos \theta = \dfrac{2\sqrt{13}}{13}$; $\tan \theta = -\dfrac{3}{2}$;

$\csc \theta = -\dfrac{\sqrt{13}}{3}$; $\sec \theta = \dfrac{\sqrt{13}}{2}$; $\cot \theta = -\dfrac{2}{3}$ **5.** $\sin \theta = -\dfrac{\sqrt{2}}{2}$; $\cos \theta = -\dfrac{\sqrt{2}}{2}$; $\tan \theta = 1$; $\csc \theta = -\sqrt{2}$; $\sec \theta = -\sqrt{2}$; $\cot \theta = 1$

7. $\sin \theta = \dfrac{1}{2}$; $\cos \theta = \dfrac{\sqrt{3}}{2}$; $\tan \theta = \dfrac{\sqrt{3}}{3}$; $\csc \theta = 2$; $\sec \theta = \dfrac{2\sqrt{3}}{3}$; $\cot \theta = \sqrt{3}$ **9.** $\sin \theta = -\dfrac{\sqrt{2}}{2}$; $\cos \theta = \dfrac{\sqrt{2}}{2}$; $\tan \theta = -1$; $\csc \theta = -\sqrt{2}$;

$\sec \theta = \sqrt{2}$; $\cot \theta = -1$ **11.** II **13.** IV **15.** IV **17.** III **19.** 30° **21.** 60° **23.** 30° **25.** $\dfrac{\pi}{4}$ **27.** $\dfrac{\pi}{3}$ **29.** 45° **31.** $\dfrac{\pi}{3}$ **33.** 80°

35. $\dfrac{\pi}{4}$ **37.** $\dfrac{\sqrt{2}}{2}$ **39.** 1 **41.** 1 **43.** $\sqrt{3}$ **45.** $\dfrac{\sqrt{2}}{2}$ **47.** 0 **49.** $\sqrt{2}$ **51.** $\dfrac{\sqrt{3}}{3}$ **53.** $\dfrac{1}{2}$ **55.** $\dfrac{\sqrt{2}}{2}$ **57.** -2 **59.** $-\sqrt{3}$ **61.** $\dfrac{\sqrt{2}}{2}$ **63.** $\sqrt{3}$

65. $\dfrac{1}{2}$ **67.** $-\dfrac{\sqrt{3}}{2}$ **69.** $-\sqrt{3}$ **71.** $\sqrt{2}$ **73.** 0 **75.** 0 **77.** -1 **79.** $\cos\theta = -\dfrac{5}{13}$; $\tan\theta = -\dfrac{12}{5}$; $\csc\theta = \dfrac{13}{12}$; $\sec\theta = -\dfrac{13}{5}$; $\cot\theta = -\dfrac{5}{12}$

81. $\sin\theta = -\dfrac{3}{5}$; $\tan\theta = \dfrac{3}{4}$; $\csc\theta = -\dfrac{5}{3}$; $\sec\theta = -\dfrac{5}{4}$; $\cot\theta = \dfrac{4}{3}$ **83.** $\cos\theta = -\dfrac{12}{13}$; $\tan\theta = -\dfrac{5}{12}$; $\csc\theta = \dfrac{13}{5}$; $\sec\theta = -\dfrac{13}{12}$; $\cot\theta = -\dfrac{12}{5}$

85. $\sin\theta = -\dfrac{2\sqrt{2}}{3}$; $\tan\theta = 2\sqrt{2}$; $\csc\theta = -\dfrac{3\sqrt{2}}{4}$; $\sec\theta = -3$; $\cot\theta = \dfrac{\sqrt{2}}{4}$ **87.** $\cos\theta = -\dfrac{\sqrt{5}}{3}$; $\tan\theta = -\dfrac{2\sqrt{5}}{5}$; $\csc\theta = \dfrac{3}{2}$; $\sec\theta = -\dfrac{3\sqrt{5}}{5}$;

$\cot\theta = -\dfrac{\sqrt{5}}{2}$ **89.** $\sin\theta = -\dfrac{\sqrt{3}}{2}$; $\cos\theta = \dfrac{1}{2}$; $\tan\theta = -\sqrt{3}$; $\csc\theta = -\dfrac{2\sqrt{3}}{3}$; $\cot\theta = -\dfrac{\sqrt{3}}{3}$ **91.** $\sin\theta = -\dfrac{3}{5}$; $\cos\theta = -\dfrac{4}{5}$; $\csc\theta = -\dfrac{5}{3}$;

$\sec\theta = -\dfrac{5}{4}$; $\cot\theta = \dfrac{4}{3}$ **93.** $\sin\theta = \dfrac{\sqrt{10}}{10}$; $\cos\theta = -\dfrac{3\sqrt{10}}{10}$; $\csc\theta = \sqrt{10}$; $\sec\theta = -\dfrac{\sqrt{10}}{3}$; $\cot\theta = -3$ **95.** $\sin\theta = -\dfrac{1}{2}$; $\cos\theta = -\dfrac{\sqrt{3}}{2}$;

$\tan\theta = \dfrac{\sqrt{3}}{3}$; $\sec\theta = -\dfrac{2\sqrt{3}}{3}$; $\cot\theta = \sqrt{3}$ **97.** 0 **99.** -0.2 **101.** 3 **103.** 5 **105.** 0 **107. (a)** Approximately 16.56 ft

(b) 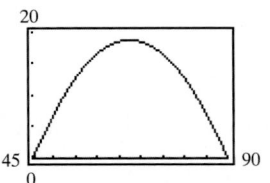 **(c)** $67.50°$

2.5 Concepts and Vocabulary (page 151)

1. 2π; π **2.** All real numbers except odd multiples of $\dfrac{\pi}{2}$ **3.** All real numbers from -1 to 1 inclusive **4.** T **5.** F **6.** F

2.5 Exercises (page 152)

1. $\sin t = -\dfrac{1}{2}$; $\cos t = \dfrac{\sqrt{3}}{2}$; $\tan t = -\dfrac{\sqrt{3}}{3}$; $\csc t = -2$; $\sec t = \dfrac{2\sqrt{3}}{3}$; $\cot t = -\sqrt{3}$ **3.** $\sin t = -\dfrac{\sqrt{2}}{2}$; $\cos t = -\dfrac{\sqrt{2}}{2}$; $\tan t = 1$;

$\csc t = -\sqrt{2}$; $\sec t = -\sqrt{2}$; $\cot t = 1$ **5.** $\sin t = \dfrac{2}{3}$; $\cos t = \dfrac{\sqrt{5}}{3}$; $\tan t = \dfrac{2\sqrt{5}}{5}$; $\csc t = \dfrac{3}{2}$; $\sec t = \dfrac{3\sqrt{5}}{5}$; $\cot t = \dfrac{\sqrt{5}}{2}$

7. $\sin\theta = -\dfrac{4}{5}$; $\cos\theta = \dfrac{3}{5}$; $\tan\theta = -\dfrac{4}{3}$; $\csc\theta = -\dfrac{5}{4}$; $\sec\theta = \dfrac{5}{3}$; $\cot\theta = -\dfrac{3}{4}$ **9.** $\sin\theta = \dfrac{3\sqrt{13}}{13}$; $\cos\theta = -\dfrac{2\sqrt{13}}{13}$; $\tan\theta = -\dfrac{3}{2}$;

$\csc\theta = \dfrac{\sqrt{13}}{3}$; $\sec\theta = -\dfrac{\sqrt{13}}{2}$; $\cot\theta = -\dfrac{2}{3}$ **11.** $\sin\theta = -\dfrac{\sqrt{2}}{2}$; $\cos\theta = -\dfrac{\sqrt{2}}{2}$; $\tan\theta = 1$; $\csc\theta = -\sqrt{2}$; $\sec\theta = -\sqrt{2}$; $\cot\theta = 1$

13. $\dfrac{\sqrt{2}}{2}$ **15.** 1 **17.** 1 **19.** $\sqrt{3}$ **21.** $\dfrac{\sqrt{2}}{2}$ **23.** 0 **25.** $\sqrt{2}$ **27.** $\dfrac{\sqrt{3}}{3}$ **29.** $-\dfrac{\sqrt{3}}{2}$ **31.** $-\dfrac{\sqrt{3}}{3}$ **33.** 2 **35.** -1 **37.** -1 **39.** $\dfrac{\sqrt{2}}{2}$ **41.** 0

43. $-\sqrt{2}$ **45.** $\dfrac{2\sqrt{3}}{3}$ **47.** -1 **49.** -2 **51.** $\dfrac{2-\sqrt{2}}{2}$ **53.** All real numbers **55.** Odd multiples of $\dfrac{\pi}{2}$ **57.** Odd multiples of $\dfrac{\pi}{2}$

59. $[-1,1]$ **61.** $(-\infty,\infty)$ **63.** $(-\infty,-1]$ or $[1,\infty)$ **65.** Odd; yes; origin **67.** Odd; yes; origin **69.** Even; yes; y-axis **71.** 0.9 **73.** 9

75. (a) $-\dfrac{1}{3}$ **(b)** 1 **77. (a)** -2 **(b)** 6 **79. (a)** -4 **(b)** -12

81. Let $P = (x,y)$ be the point on the unit circle that corresponds to t. Consider the equation $\tan t = \dfrac{y}{x} = a$. Then $y = ax$. But

$x^2 + y^2 = 1$ so that $x^2 + a^2x^2 = 1$. Thus, $x = \pm\dfrac{1}{\sqrt{1+a^2}}$ and $y = \pm\dfrac{a}{\sqrt{1+a^2}}$; that is, for any real number a, there is a point

$P = (x,y)$ on the unit circle for which $\tan t = a$. In other words, $-\infty < \tan t < \infty$, and the range of the tangent function is the set
of all real numbers.

83. Suppose there is a number p, $0 < p < 2\pi$, for which $\sin(\theta + p) = \sin\theta$ for all θ. If $\theta = 0$, then $\sin(0 + p) = \sin p = \sin 0 = 0$;

so that $p = \pi$. If $\theta = \dfrac{\pi}{2}$, then $\sin\left(\dfrac{\pi}{2} + p\right) = \sin\left(\dfrac{\pi}{2}\right)$. But $p = \pi$. Thus, $\sin\left(\dfrac{3\pi}{2}\right) = -1 = \sin\left(\dfrac{\pi}{2}\right) = 1$. This is impossible.

Therefore, the smallest positive number p for which $\sin(\theta + p) = \sin\theta$ for all θ is $p = 2\pi$.

85. $\sec\theta = \dfrac{1}{\cos\theta}$; since $\cos\theta$ has period 2π, so does $\sec\theta$.

87. If $P = (a, b)$ is the point on the unit circle corresponding to θ, then $Q = (-a, -b)$ is the point on the unit circle corresponding to $\theta + \pi$. Thus, $\tan(\theta + \pi) = \dfrac{-b}{-a} = \dfrac{b}{a} = \tan \theta$. If there exists a number $p, 0 < p < \pi$, for which $\tan(\theta + p) = \tan \theta$ for all θ, then when $\theta = 0, \tan(p) = \tan 0 = 0$. But this means that p is a multiple of π. Since no multiple of π exists in the interval $(0, \pi)$, this is impossible. Therefore, the fundamental period of $f(\theta) = \tan \theta$ is π.

89. $m = \dfrac{\sin \theta - 0}{\cos \theta - 0} = \dfrac{\sin \theta}{\cos \theta} = \tan \theta$

2.6 Concepts and Vocabulary *(page 166)*

1. $1; \ldots -\dfrac{3\pi}{2}, \dfrac{\pi}{2}, \dfrac{5\pi}{2}, \ldots$ **2.** $\pm 3; \pi$ **3.** $3; \dfrac{\pi}{3}$ **4.** T   **5.** F   **6.** T

2.6 Exercises *(page 166)*

1. 0   **3.** $-\dfrac{\pi}{2} \le x \le \dfrac{\pi}{2}$ **5.** 1   **7.** $0, \pi, 2\pi$ **9.** $\sin x = 1$ for $x = -\dfrac{3\pi}{2}, \dfrac{\pi}{2}; \sin x = -1$ for $x = -\dfrac{\pi}{2}, \dfrac{3\pi}{2}$ **11.** B, C, F

13.
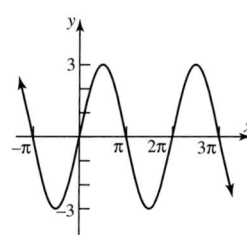

15.
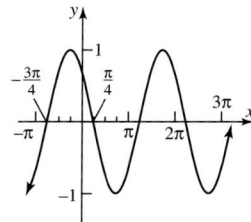

17.
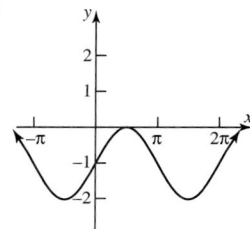

19.
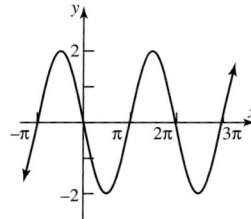

21.
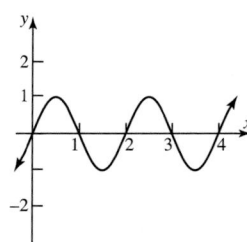

23.
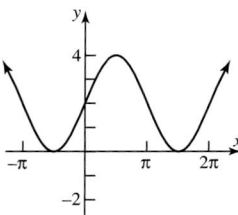

25.
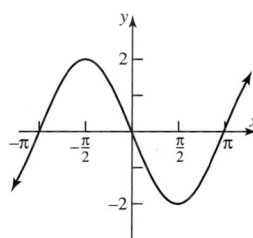

27.
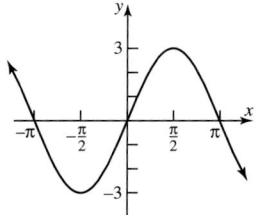

29. Amplitude = 2; period = 2π **31.** Amplitude = 4; period = π **33.** Amplitude = 6; period = 2

35. Amplitude = $\dfrac{1}{2}$; period = $\dfrac{4\pi}{3}$ **37.** Amplitude = $\dfrac{5}{3}$; period = 3   **39.** F   **41.** A   **43.** H   **45.** C   **47.** J   **49.** A   **51.** B

53.
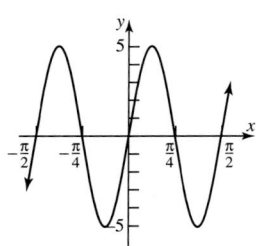

55.
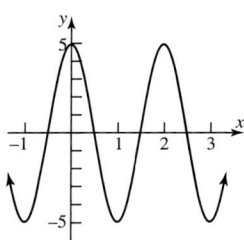

57.
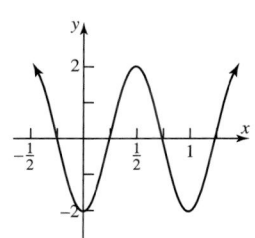

59.
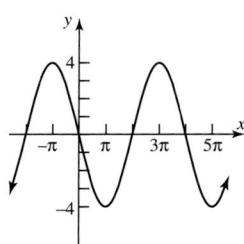

61.
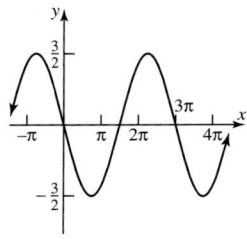

63. $y = \pm 3 \sin(2x)$ **65.** $y = \pm 3 \sin(\pi x)$

67. $y = -3 \cos\left(\dfrac{1}{2}x\right)$ **69.** $y = \dfrac{3}{4} \sin(2\pi x)$ **71.** $y = -\sin\left(\dfrac{3}{2}x\right)$

73. $y = -2 \cos\left(\dfrac{3\pi}{2}x\right)$ **75.** $y = 3 \sin\left(\dfrac{\pi}{2}x\right)$ **77.** $y = -4 \cos(3x)$

79. Period $= \dfrac{1}{30}$, amplitude $= 220$

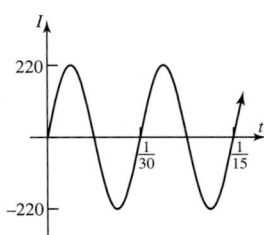

81. (a) Amplitude $= 220$, period $= \dfrac{1}{60}$

(b), (e)

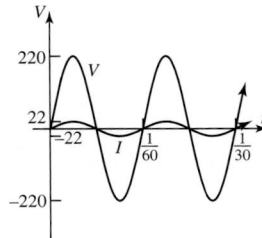

(c) $I = 22 \sin(120\pi t)$

(d) Amplitude $= 22$, period $= \dfrac{1}{60}$

83. (a) $P = \dfrac{[V_0 \sin(2\pi ft)]^2}{R} = \dfrac{V_0^2}{R} \sin^2[2\pi ft]$

(b) Since the graph of P has amplitude $\dfrac{V_0^2}{2R}$,

period $\dfrac{1}{2f}$, and is of the form $y = A \cos(\omega t) + B$,

then $A = -\dfrac{V_0^2}{2R}$ and $B = \dfrac{V_0^2}{2R}$. Since $\dfrac{1}{2f} = \dfrac{2\pi}{\omega}$,

then $\omega = 4\pi f$. Therefore,

$P = -\dfrac{V_0^2}{2R} \cos(4\pi ft) + \dfrac{V_0^2}{2R} = \dfrac{V_0^2}{2R}[1 - \cos(4\pi ft)]$.

85.

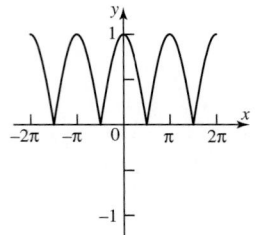

2.7 Concepts and Vocabulary *(page 175)*

1. Origin; odd multiples of $\dfrac{\pi}{2}$ **2.** y-axis; odd multiples of $\dfrac{\pi}{2}$ **3.** $y = \cos x$ **4.** T

2.7 Exercises *(page 175)*

1. 0   **3.** 1   **5.** $\sec x = 1$ for $x = -2\pi, 0, 2\pi$; $\sec x = -1$ for $x = -\pi, \pi$ **7.** $-\dfrac{3\pi}{2}, -\dfrac{\pi}{2}, \dfrac{\pi}{2}, \dfrac{3\pi}{2}$ **9.** $-\dfrac{3\pi}{2}, -\dfrac{\pi}{2}, \dfrac{\pi}{2}, \dfrac{3\pi}{2}$ **11.** D   **13.** B

15.

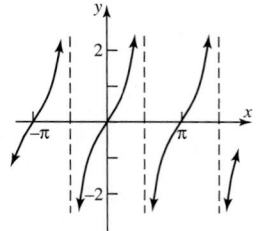

17.

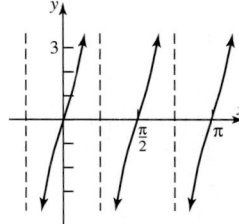

19.

21.

23.

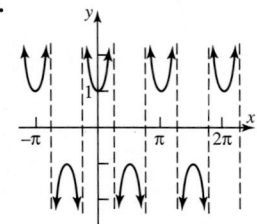

25.

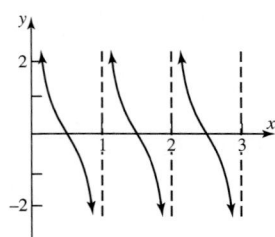

27.

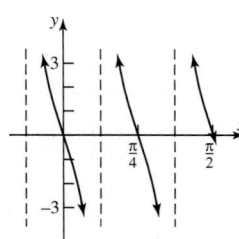

29.

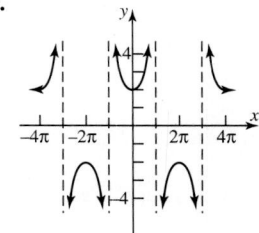

31.

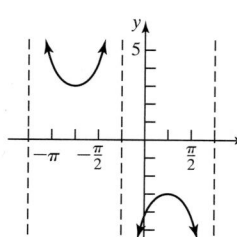

33.

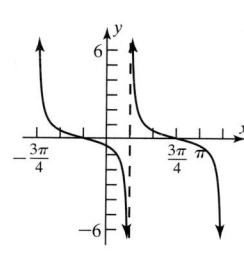

35. (a) $L(\theta) = \dfrac{3}{\cos\theta} + \dfrac{4}{\sin\theta} = 3\sec\theta + 4\csc\theta$

(b)

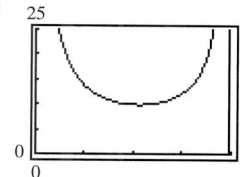

(c) 0.83
(d) 9.87 ft

2.8 Concepts and Vocabulary *(page 185)*

1. Phase shift   **2.** F

2.8 Exercises *(page 186)*

1. Amplitude = 4

Period = π

Phase shift = $\dfrac{\pi}{2}$

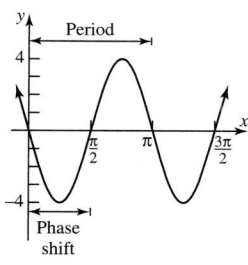

3. Amplitude = 2

Period = $\dfrac{2\pi}{3}$

Phase shift = $-\dfrac{\pi}{6}$

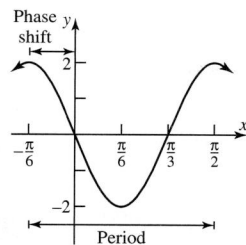

5. Amplitude = 3

Period = π

Phase shift = $-\dfrac{\pi}{4}$

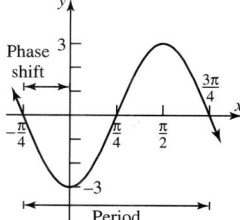

7. Amplitude = 4

Period = 2

Phase shift = $-\dfrac{2}{\pi}$

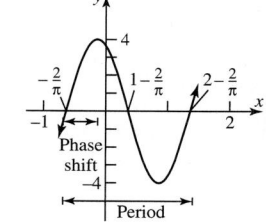

9. Amplitude = 3

Period = 2

Phase shift = $\dfrac{2}{\pi}$

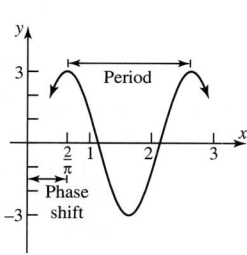

11. Amplitude = 3

Period = π

Phase shift = $\dfrac{\pi}{4}$

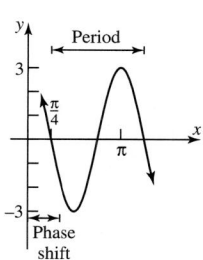

13. $y = \pm 2 \sin\left[2\left(x - \dfrac{1}{2}\right)\right]$ or $y = \pm 2 \sin(2x - 1)$

15. $y = \pm 3 \sin\left[\dfrac{2}{3}\left(x + \dfrac{1}{3}\right)\right]$ or $y = \pm 3 \sin\left(\dfrac{2}{3}x + \dfrac{2}{9}\right)$

17. Period = $\dfrac{1}{15}$; amplitude = 120; phase shift = $\dfrac{1}{90}$

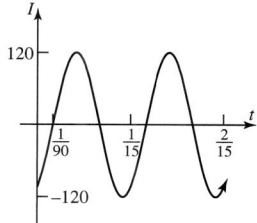

19. (a)

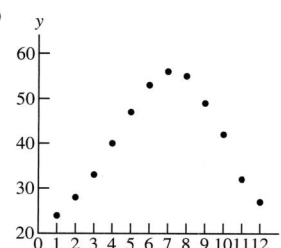

(b) $y = 15.9 \sin\left(\dfrac{\pi}{6}x - \dfrac{2\pi}{3}\right) + 40.1$

(c)

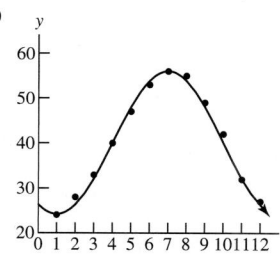

(d) $y = 15.62 \sin(0.517x - 2.096) + 40.377$

(e)

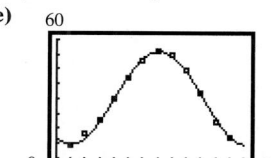

21. (a)

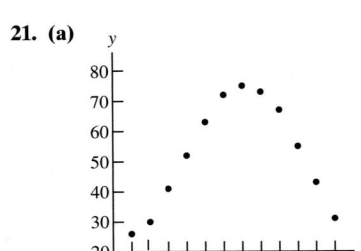

(b) $y = 24.95 \sin\left(\dfrac{\pi}{6}x - \dfrac{2\pi}{3}\right) + 50.45$

(c)

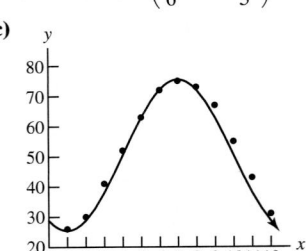

(d) $y = 25.693 \sin(0.476x - 1.814) + 49.854$

(e)

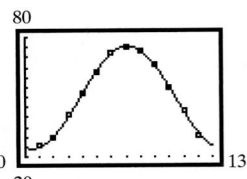

23. (a) 4:08 P.M.

(b) $y = 4.4 \sin\left(\dfrac{4\pi}{25}x - 6.66\right) + 3.8$

(c)

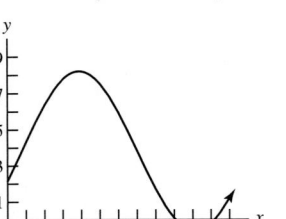

(d) 8.2 ft

25. (a) $y = 1.0835 \sin\left(\dfrac{2\pi}{365}x - 2.45\pi\right) + 11.6665$

(b)

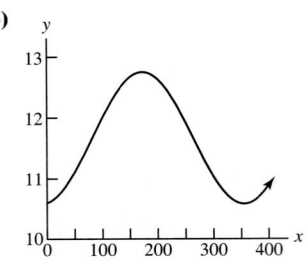

(c) 11.83 hr

27. (a) $y = 5.3915 \sin\left(\dfrac{2\pi}{365}x - 2.45\pi\right) + 10.8415$

(b)

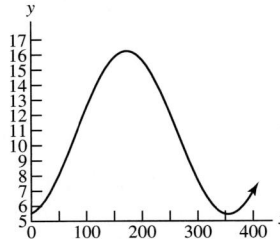

(c) 11.66 hr

Review Exercises *(page 191)*

1. $\dfrac{3\pi}{4}$ **3.** $\dfrac{\pi}{10}$ **5.** $135°$ **7.** $-450°$ **9.** $\dfrac{1}{2}$ **11.** $\dfrac{3\sqrt{2}}{2} - \dfrac{4\sqrt{3}}{3}$ **13.** $-3\sqrt{2} - 2\sqrt{3}$ **15.** 3   **17.** 0   **19.** 0   **21.** 1   **23.** 1   **25.** 1   **27.** -1

29. 1   **31.** $\cos\theta = \dfrac{3}{5}; \tan\theta = \dfrac{4}{3}; \csc\theta = \dfrac{5}{4}; \sec\theta = \dfrac{5}{3}; \cot\theta = \dfrac{3}{4}$ **33.** $\sin\theta = -\dfrac{12}{13}; \cos\theta = -\dfrac{5}{13}; \csc\theta = -\dfrac{13}{12}; \sec\theta = -\dfrac{13}{5}; \cot\theta = \dfrac{5}{12}$

35. $\sin\theta = \dfrac{3}{5}; \cos\theta = -\dfrac{4}{5}; \tan\theta = -\dfrac{3}{4}; \csc\theta = \dfrac{5}{3}; \cot\theta = -\dfrac{4}{3}$ **37.** $\cos\theta = -\dfrac{5}{13}; \tan\theta = -\dfrac{12}{5}; \csc\theta = \dfrac{13}{12}; \sec\theta = -\dfrac{13}{5}; \cot\theta = -\dfrac{5}{12}$

39. $\cos\theta = \dfrac{12}{13}; \tan\theta = -\dfrac{5}{12}; \csc\theta = -\dfrac{13}{5}; \sec\theta = \dfrac{13}{12}; \cot\theta = -\dfrac{12}{5}$ **41.** $\sin\theta = -\dfrac{\sqrt{10}}{10}; \cos\theta = -\dfrac{3\sqrt{10}}{10}; \csc\theta = -\sqrt{10};$

$\sec\theta = -\dfrac{\sqrt{10}}{3}; \cot\theta = 3$ **43.** $\sin\theta = -\dfrac{2\sqrt{2}}{3}; \cos\theta = \dfrac{1}{3}; \tan\theta = -2\sqrt{2}; \csc\theta = -\dfrac{3\sqrt{2}}{4}; \cot\theta = -\dfrac{\sqrt{2}}{4}$

45. $\sin\theta = \dfrac{\sqrt{5}}{5}; \cos\theta = -\dfrac{2\sqrt{5}}{5}; \tan\theta = -\dfrac{1}{2}; \csc\theta = \sqrt{5}; \sec\theta = -\dfrac{\sqrt{5}}{2}$

47.

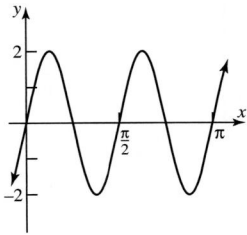

49.

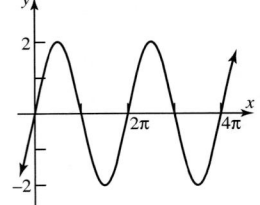

51.

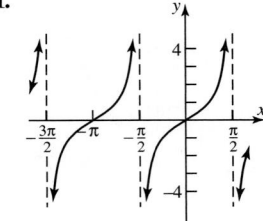

53.

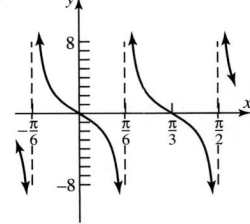

55.
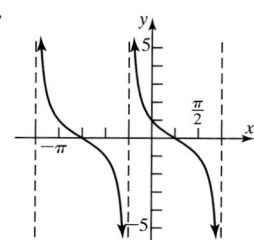

57.
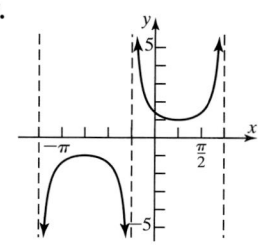

59. Amplitude $= 4$; period $= 2\pi$
61. Amplitude $= 8$; period $= 4$

63. Amplitude $= 4$

Period $= \dfrac{2\pi}{3}$

Phase shift $= 0$

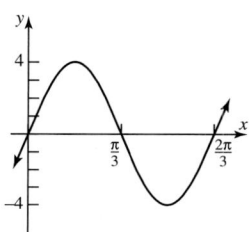

65. Amplitude $= 2$

Period $= \pi$

Phase Shift $= \dfrac{\pi}{2}$

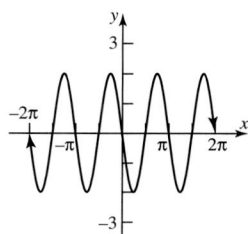

67. Amplitude $= \dfrac{1}{2}$

Period $= \dfrac{4\pi}{3}$

Phase shift $= \dfrac{2\pi}{3}$

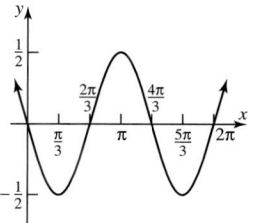

69. Amplitude $= \dfrac{2}{3}$

Period $= 2$

Phase shift $= \dfrac{6}{\pi}$

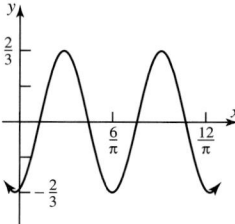

71. $y = 5 \cos \dfrac{x}{4}$ **73.** $y = -6 \cos\left(\dfrac{\pi}{4}x\right)$ **75.** $\sin\theta = \dfrac{5}{13}, \cos\theta = \dfrac{12}{13}, \tan\theta = \dfrac{5}{12}, \csc\theta = \dfrac{13}{5}, \sec\theta = \dfrac{13}{12}, \cot\theta = \dfrac{12}{5}$

77. $\sin\theta = -\dfrac{4}{5}, \cos\theta = \dfrac{3}{5}, \tan\theta = -\dfrac{4}{3}, \csc\theta = -\dfrac{5}{4}, \sec\theta = \dfrac{5}{3}, \cot\theta = -\dfrac{3}{4}$ **79.** $\dfrac{\pi}{5}$

81. Domain: $\left\{x \,\middle|\, x \neq \text{odd multiple of } \dfrac{\pi}{2}\right\}$; range: $\{y \,|\, |y| \geq 1\}$ **83.** $\dfrac{\pi}{3} \approx 1.05$ ft; $\dfrac{\pi}{3} \approx 1.05$ ft^2 **85.** Approximately 114.59 revolutions/hr

87. 0.1 revolution/sec $= \dfrac{\pi}{5}$ radian/sec   **89. (a)** 120   **(b)** $\dfrac{1}{60}$ **(c)**

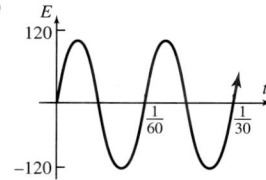

91. (a)

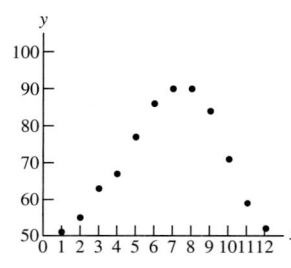

(b) $y = 19.5 \sin\left(\dfrac{\pi}{6}x - \dfrac{2\pi}{3}\right) + 70.5$

(c)

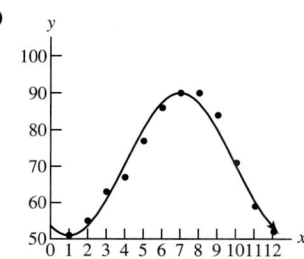

(d) $y = 19.52 \sin(0.54x - 2.28) + 71.01$

(e)

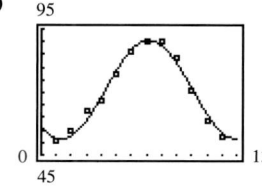

93. (a) $y = 1.85 \sin\left(\dfrac{2\pi}{365}x - 2.45\pi\right) + 11.517$ **(b)**   **(c)** 11.80 hr

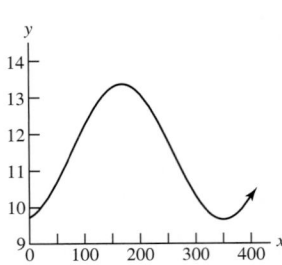

Cumulative Review Exercises *(page 196)*

1. $\left\{-1, \dfrac{1}{2}\right\}$ **2.** $y - 5 = -3(x + 2)$ or $y = -3x - 1$ **3.** $x^2 + (y + 2)^2 = 16$

4.

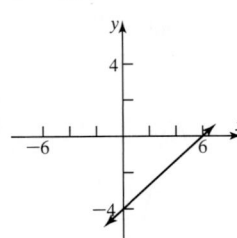

5.

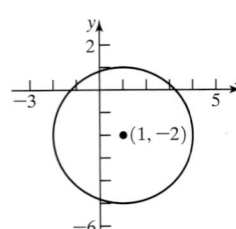

6.

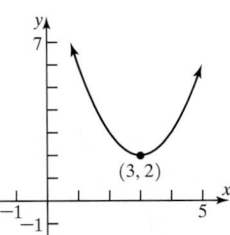

7. (a)

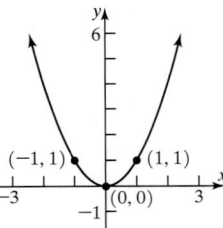

(b)

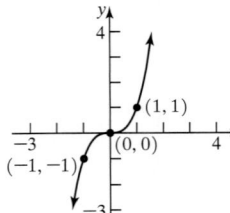

(c)

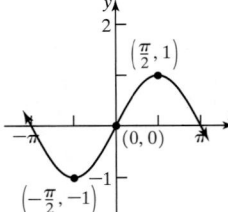

(d)

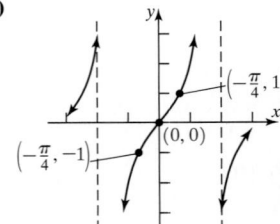

8. $f^{-1}(x) = \dfrac{1}{3}(x + 2)$ **9.** -2 **10.**

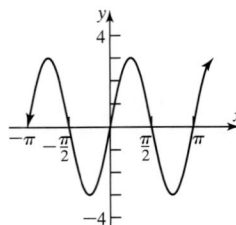

11. $3 - \dfrac{3\sqrt{3}}{2}$ **12.** $y = 3\cos\left(\dfrac{\pi}{6}x\right)$

C H A P T E R   3   Analytic Trigonometry

3.1 Concepts and Vocabulary *(page 208)*

1. $x = \sin y$ **2.** 0   **3.** $\dfrac{\pi}{5}$ **4.** F   **5.** T   **6.** T   **8.** $-\dfrac{\pi}{2} \le x \le \dfrac{\pi}{2}$ **9.** $-1 \le x \le 1$

3.1 Exercises *(page 208)*

1. 0   **3.** $-\dfrac{\pi}{2}$ **5.** 0   **7.** $\dfrac{\pi}{4}$ **9.** $\dfrac{\pi}{3}$ **11.** $\dfrac{5\pi}{6}$ **13.** 0.10   **15.** 1.37   **17.** 0.51   **19.** -0.38 **21.** -0.12 **23.** 1.08   **25.** 0.54   **27.** $\dfrac{4\pi}{5}$

29. -3.5 **31.** $-\dfrac{3\pi}{7}$ **33.** Yes; $-\dfrac{\pi}{6}$ lies in the interval $\left[-\dfrac{\pi}{2}, \dfrac{\pi}{2}\right]$.   **35.** No; 2 is not in the domain of $\sin^{-1} x$.

37. No; $-\dfrac{\pi}{6}$ does not lie in the interval $[0, \pi]$.   **39.** Yes, $-\dfrac{1}{2}$ is in the domain of $\cos^{-1} x$.   **41.** Yes; $-\dfrac{\pi}{3}$ lies in the interval $\left(-\dfrac{\pi}{2}, \dfrac{\pi}{2}\right)$.

43. Yes; 2 is in the domain of $\tan^{-1} x$.   **45. (a)** 13.92 hr or 13 hr, 55 min   **(b)** 12 hr   **(c)** 13.85 hr or 13 hr, 51 min

47. (a) 13.3 hr or 13 hr, 18 min   **(b)** 12 hr   **(c)** 13.26 hr or 13 hr, 15 min   **49. (a)** 12 hr   **(b)** 12 hr   **(c)** 12 hr   **(d)** It's 12 hr.

51. 3.35 min

3.2 Concepts and Vocabulary *(page 214)*

1. $x = \sec y; \geq 1; 0; \pi$ **2.** $\dfrac{\sqrt{2}}{2}$ **3.** F   **4.** T   **5.** T

3.2 Exercises *(page 214)*

1. $\dfrac{\sqrt{2}}{2}$ **3.** $-\dfrac{\sqrt{3}}{3}$ **5.** 2   **7.** $\sqrt{2}$ **9.** $-\dfrac{\sqrt{2}}{2}$ **11.** $\dfrac{2\sqrt{3}}{3}$ **13.** $\dfrac{3\pi}{4}$ **15.** $\dfrac{\pi}{6}$ **17.** $\dfrac{\sqrt{2}}{4}$ **19.** $\dfrac{\sqrt{5}}{2}$ **21.** $-\dfrac{\sqrt{14}}{2}$ **23.** $-\dfrac{3\sqrt{10}}{10}$ **25.** $\sqrt{5}$ **27.** $-\dfrac{\pi}{4}$

29. $\dfrac{\pi}{6}$ **31.** $-\dfrac{\pi}{2}$ **33.** $\dfrac{\pi}{6}$ **35.** $\dfrac{2\pi}{3}$ **37.** 1.32   **39.** 0.46   **41.** -0.34 **43.** 0.42   **45.** -0.73 **47.** 2.55

49.

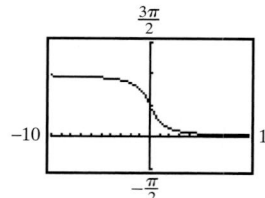

51.

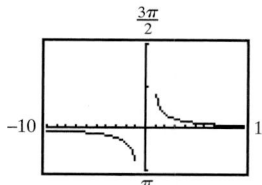

3.3 Concepts and Vocabulary *(page 220)*

1. Identity; conditional   **2.** -1 **3.** 0   **4.** T   **5.** F   **6.** T   **7.** $\sin^2\theta + \cos^2\theta = 1; \tan^2\theta + 1 = \sec^2\theta; 1 + \cot^2\theta = \csc^2\theta$

3.3 Exercises *(page 220)*

1. $\csc\theta \cdot \cos\theta = \dfrac{1}{\sin\theta}\cdot\cos\theta = \dfrac{\cos\theta}{\sin\theta} = \cot\theta$ **3.** $1 + \tan^2(-\theta) = 1 + (-\tan\theta)^2 = 1 + \tan^2\theta = \sec^2\theta$

5. $\cos\theta(\tan\theta + \cot\theta) = \cos\theta\left(\dfrac{\sin\theta}{\cos\theta} + \dfrac{\cos\theta}{\sin\theta}\right) = \cos\theta\left(\dfrac{\sin^2\theta + \cos^2\theta}{\cos\theta\sin\theta}\right) = \cos\theta\left(\dfrac{1}{\cos\theta\sin\theta}\right) = \dfrac{1}{\sin\theta} = \csc\theta$

7. $\tan\theta\cot\theta - \cos^2\theta = \dfrac{\sin\theta}{\cos\theta}\cdot\dfrac{\cos\theta}{\sin\theta} - \cos^2\theta = 1 - \cos^2\theta = \sin^2\theta$ **9.** $(\sec\theta - 1)(\sec\theta + 1) = \sec^2\theta - 1 = \tan^2\theta$

11. $(\sec\theta + \tan\theta)(\sec\theta - \tan\theta) = \sec^2\theta - \tan^2\theta = 1$

13. $\cos^2\theta(1 + \tan^2\theta) = \cos^2\theta + \cos^2\theta\tan^2\theta = \cos^2\theta + \cos^2\theta\cdot\dfrac{\sin^2\theta}{\cos^2\theta} = \cos^2\theta + \sin^2\theta = 1$

15. $(\sin\theta + \cos\theta)^2 + (\sin\theta - \cos\theta)^2 = \sin^2\theta + 2\sin\theta\cos\theta + \cos^2\theta + \sin^2\theta - 2\sin\theta\cos\theta + \cos^2\theta = \sin^2\theta + \cos^2\theta + \sin^2\theta + \cos^2\theta$
$= 1 + 1 = 2$

17. $\sec^4\theta - \sec^2\theta = \sec^2\theta(\sec^2\theta - 1) = (1 + \tan^2\theta)\tan^2\theta = \tan^4\theta + \tan^2\theta$

19. $\sec\theta - \tan\theta = \dfrac{1}{\cos\theta} - \dfrac{\sin\theta}{\cos\theta} = \dfrac{1 - \sin\theta}{\cos\theta}\cdot\dfrac{1 + \sin\theta}{1 + \sin\theta} = \dfrac{1 - \sin^2\theta}{\cos\theta(1 + \sin\theta)} = \dfrac{\cos^2\theta}{\cos\theta(1 + \sin\theta)} = \dfrac{\cos\theta}{1 + \sin\theta}$

21. $3\sin^2\theta + 4\cos^2\theta = 3\sin^2\theta + 3\cos^2\theta + \cos^2\theta = 3(\sin^2\theta + \cos^2\theta) + \cos^2\theta = 3 + \cos^2\theta$

23. $1 - \dfrac{\cos^2\theta}{1 + \sin\theta} = 1 - \dfrac{1 - \sin^2\theta}{1 + \sin\theta} = 1 - \dfrac{(1 + \sin\theta)(1 - \sin\theta)}{1 + \sin\theta} = 1 - (1 - \sin\theta) = \sin\theta$

25. $\dfrac{1 + \tan\theta}{1 - \tan\theta} = \dfrac{1 + \dfrac{1}{\cot\theta}}{1 - \dfrac{1}{\cot\theta}} = \dfrac{\dfrac{\cot\theta + 1}{\cot\theta}}{\dfrac{\cot\theta - 1}{\cot\theta}} = \dfrac{\cot\theta + 1}{\cot\theta - 1}$ **27.** $\dfrac{\sec\theta}{\csc\theta} + \dfrac{\sin\theta}{\cos\theta} = \dfrac{\dfrac{1}{\cos\theta}}{\dfrac{1}{\sin\theta}} + \tan\theta = \dfrac{\sin\theta}{\cos\theta} + \tan\theta = \tan\theta + \tan\theta = 2\tan\theta$

29. $\dfrac{1 + \sin\theta}{1 - \sin\theta} = \dfrac{1 + \dfrac{1}{\csc\theta}}{1 - \dfrac{1}{\csc\theta}} = \dfrac{\dfrac{\csc\theta + 1}{\csc\theta}}{\dfrac{\csc\theta - 1}{\csc\theta}} = \dfrac{\csc\theta + 1}{\csc\theta - 1}$

31. $\dfrac{1 - \sin\theta}{\cos\theta} + \dfrac{\cos\theta}{1 - \sin\theta} = \dfrac{(1 - \sin\theta)^2 + \cos^2\theta}{\cos\theta(1 - \sin\theta)} = \dfrac{1 - 2\sin\theta + \sin^2\theta + \cos^2\theta}{\cos\theta(1 - \sin\theta)} = \dfrac{2 - 2\sin\theta}{\cos\theta(1 - \sin\theta)} = \dfrac{2(1 - \sin\theta)}{\cos\theta(1 - \sin\theta)} = \dfrac{2}{\cos\theta}$
$= 2\sec\theta$

33. $\dfrac{\sin\theta}{\sin\theta - \cos\theta} = \dfrac{1}{\dfrac{\sin\theta - \cos\theta}{\sin\theta}} = \dfrac{1}{1 - \dfrac{\cos\theta}{\sin\theta}} = \dfrac{1}{1 - \cot\theta}$

35. $(\sec\theta - \tan\theta)^2 = \sec^2\theta - 2\sec\theta\tan\theta + \tan^2\theta = \dfrac{1}{\cos^2\theta} - \dfrac{2\sin\theta}{\cos^2\theta} + \dfrac{\sin^2\theta}{\cos^2\theta} = \dfrac{1 - 2\sin\theta + \sin^2\theta}{\cos^2\theta} = \dfrac{(1 - \sin\theta)^2}{1 - \sin^2\theta}$
$= \dfrac{(1 - \sin\theta)^2}{(1 - \sin\theta)(1 + \sin\theta)} = \dfrac{1 - \sin\theta}{1 + \sin\theta}$

37. $\dfrac{\cos\theta}{1-\tan\theta}+\dfrac{\sin\theta}{1-\cot\theta}=\dfrac{\cos\theta}{1-\dfrac{\sin\theta}{\cos\theta}}+\dfrac{\sin\theta}{1-\dfrac{\cos\theta}{\sin\theta}}=\dfrac{\cos\theta}{\dfrac{\cos\theta-\sin\theta}{\cos\theta}}+\dfrac{\sin\theta}{\dfrac{\sin\theta-\cos\theta}{\sin\theta}}=\dfrac{\cos^2\theta}{\cos\theta-\sin\theta}+\dfrac{\sin^2\theta}{\sin\theta-\cos\theta}$

$=\dfrac{\cos^2\theta-\sin^2\theta}{\cos\theta-\sin\theta}=\dfrac{(\cos\theta-\sin\theta)(\cos\theta+\sin\theta)}{\cos\theta-\sin\theta}=\sin\theta+\cos\theta$

39. $\tan\theta+\dfrac{\cos\theta}{1+\sin\theta}=\dfrac{\sin\theta}{\cos\theta}+\dfrac{\cos\theta}{1+\sin\theta}=\dfrac{\sin\theta(1+\sin\theta)+\cos^2\theta}{\cos\theta(1+\sin\theta)}=\dfrac{\sin\theta+\sin^2\theta+\cos^2\theta}{\cos\theta(1+\sin\theta)}=\dfrac{\sin\theta+1}{\cos\theta(1+\sin\theta)}=\dfrac{1}{\cos\theta}=\sec\theta$

41. $\dfrac{\tan\theta+\sec\theta-1}{\tan\theta-\sec\theta+1}=\dfrac{\tan\theta+(\sec\theta-1)}{\tan\theta-(\sec\theta-1)}\cdot\dfrac{\tan\theta+(\sec\theta-1)}{\tan\theta+(\sec\theta-1)}=\dfrac{\tan^2\theta+2\tan\theta(\sec\theta-1)+\sec^2\theta-2\sec\theta+1}{\tan^2\theta-(\sec^2\theta-2\sec\theta+1)}$

$=\dfrac{\sec^2\theta-1+2\tan\theta(\sec\theta-1)+\sec^2\theta-2\sec\theta+1}{\sec^2\theta-1-\sec^2\theta+2\sec\theta-1}=\dfrac{2\sec^2\theta-2\sec\theta+2\tan\theta(\sec\theta-1)}{-2+2\sec\theta}$

$=\dfrac{2\sec\theta(\sec\theta-1)+2\tan\theta(\sec\theta-1)}{2(\sec\theta-1)}=\dfrac{2(\sec\theta-1)(\sec\theta+\tan\theta)}{2(\sec\theta-1)}=\tan\theta+\sec\theta$

43. $\dfrac{\tan\theta-\cot\theta}{\tan\theta+\cot\theta}=\dfrac{\dfrac{\sin\theta}{\cos\theta}-\dfrac{\cos\theta}{\sin\theta}}{\dfrac{\sin\theta}{\cos\theta}+\dfrac{\cos\theta}{\sin\theta}}=\dfrac{\dfrac{\sin^2\theta-\cos^2\theta}{\cos\theta\sin\theta}}{\dfrac{\sin^2\theta+\cos^2\theta}{\cos\theta\sin\theta}}=\dfrac{\sin^2\theta-\cos^2\theta}{1}=\sin^2\theta-\cos^2\theta$

45. $\dfrac{\tan\theta-\cot\theta}{\tan\theta+\cot\theta}+1=\dfrac{\dfrac{\sin\theta}{\cos\theta}-\dfrac{\cos\theta}{\sin\theta}}{\dfrac{\sin\theta}{\cos\theta}+\dfrac{\cos\theta}{\sin\theta}}+1=\dfrac{\dfrac{\sin^2\theta-\cos^2\theta}{\cos\theta\sin\theta}}{\dfrac{\sin^2\theta+\cos^2\theta}{\cos\theta\sin\theta}}+1=\sin^2\theta-\cos^2\theta+1=\sin^2\theta+(1-\cos^2\theta)=2\sin^2\theta$

47. $\dfrac{\sec\theta+\tan\theta}{\cot\theta+\cos\theta}=\dfrac{\dfrac{1}{\cos\theta}+\dfrac{\sin\theta}{\cos\theta}}{\dfrac{\cos\theta}{\sin\theta}+\cos\theta}=\dfrac{\dfrac{1+\sin\theta}{\cos\theta}}{\dfrac{\cos\theta+\cos\theta\sin\theta}{\sin\theta}}=\dfrac{1+\sin\theta}{\cos\theta}\cdot\dfrac{\sin\theta}{\cos\theta(1+\sin\theta)}=\dfrac{\sin\theta}{\cos\theta}\cdot\dfrac{1}{\cos\theta}=\tan\theta\sec\theta$

49. $\dfrac{1-\tan^2\theta}{1+\tan^2\theta}+1=\dfrac{1-\tan^2\theta}{\sec^2\theta}+1=\dfrac{1}{\sec^2\theta}-\dfrac{\tan^2\theta}{\sec^2\theta}+1=\cos^2\theta-\dfrac{\dfrac{\sin^2\theta}{\cos^2\theta}}{\dfrac{1}{\cos^2\theta}}+1=\cos^2\theta-\sin^2\theta+1$

$=\cos^2\theta+(1-\sin^2\theta)=2\cos^2\theta$

51. $\dfrac{\sec\theta-\csc\theta}{\sec\theta\csc\theta}=\dfrac{\sec\theta}{\sec\theta\csc\theta}-\dfrac{\csc\theta}{\sec\theta\csc\theta}=\dfrac{1}{\csc\theta}-\dfrac{1}{\sec\theta}=\sin\theta-\cos\theta$

53. $\sec\theta-\cos\theta-\sin\theta\tan\theta=\left(\dfrac{1}{\cos\theta}-\cos\theta\right)-\sin\theta\cdot\dfrac{\sin\theta}{\cos\theta}=\dfrac{1-\cos^2\theta}{\cos\theta}-\dfrac{\sin^2\theta}{\cos\theta}=\dfrac{\sin^2\theta}{\cos\theta}-\dfrac{\sin^2\theta}{\cos\theta}=0$

55. $\dfrac{1}{1-\sin\theta}+\dfrac{1}{1+\sin\theta}=\dfrac{1+\sin\theta+1-\sin\theta}{(1+\sin\theta)(1-\sin\theta)}=\dfrac{2}{1-\sin^2\theta}=\dfrac{2}{\cos^2\theta}=2\sec^2\theta$

57. $\dfrac{\sec\theta}{1-\sin\theta}=\dfrac{\sec\theta}{1-\sin\theta}\cdot\dfrac{1+\sin\theta}{1+\sin\theta}=\dfrac{\sec\theta(1+\sin\theta)}{1-\sin^2\theta}=\dfrac{\sec\theta(1+\sin\theta)}{\cos^2\theta}=\dfrac{1+\sin\theta}{\cos^3\theta}$

59. $\dfrac{(\sec\theta-\tan\theta)^2+1}{\csc\theta(\sec\theta-\tan\theta)}=\dfrac{\sec^2\theta-2\sec\theta\tan\theta+\tan^2\theta+1}{\dfrac{1}{\sin\theta}\left(\dfrac{1}{\cos\theta}-\dfrac{\sin\theta}{\cos\theta}\right)}=\dfrac{2\sec^2\theta-2\sec\theta\tan\theta}{\dfrac{1}{\sin\theta}\left(\dfrac{1-\sin\theta}{\cos\theta}\right)}=\dfrac{\dfrac{2}{\cos^2\theta}-\dfrac{2\sin\theta}{\cos^2\theta}}{\dfrac{1-\sin\theta}{\sin\theta\cos\theta}}=\dfrac{2-2\sin\theta}{\cos^2\theta}\cdot\dfrac{\sin\theta\cos\theta}{1-\sin\theta}$

$=\dfrac{2(1-\sin\theta)}{\cos\theta}\cdot\dfrac{\sin\theta}{1-\sin\theta}=\dfrac{2\sin\theta}{\cos\theta}=2\tan\theta$

61. $\dfrac{\sin\theta+\cos\theta}{\cos\theta}-\dfrac{\sin\theta-\cos\theta}{\sin\theta}=\dfrac{\sin\theta}{\cos\theta}+1-1+\dfrac{\cos\theta}{\sin\theta}=\dfrac{\sin^2\theta+\cos^2\theta}{\cos\theta\sin\theta}=\dfrac{1}{\cos\theta\sin\theta}=\sec\theta\csc\theta$

63. $\dfrac{\sin^3\theta+\cos^3\theta}{\sin\theta+\cos\theta}=\dfrac{(\sin\theta+\cos\theta)(\sin^2\theta-\sin\theta\cos\theta+\cos^2\theta)}{\sin\theta+\cos\theta}=\sin^2\theta+\cos^2\theta-\sin\theta\cos\theta=1-\sin\theta\cos\theta$

65. $\dfrac{\cos^2\theta-\sin^2\theta}{1-\tan^2\theta}=\dfrac{\cos^2\theta-\sin^2\theta}{1-\dfrac{\sin^2\theta}{\cos^2\theta}}=\dfrac{\cos^2\theta-\sin^2\theta}{\dfrac{\cos^2\theta-\sin^2\theta}{\cos^2\theta}}=\cos^2\theta$

67. $\dfrac{(2\cos^2\theta-1)^2}{\cos^4\theta-\sin^4\theta}=\dfrac{[2\cos^2\theta-(\sin^2\theta+\cos^2\theta)]^2}{(\cos^2\theta-\sin^2\theta)(\cos^2\theta+\sin^2\theta)}=\dfrac{(\cos^2\theta-\sin^2\theta)^2}{\cos^2\theta-\sin^2\theta}=\cos^2\theta-\sin^2\theta=(1-\sin^2\theta)-\sin^2\theta=1-2\sin^2\theta$

69. $\dfrac{1 + \sin\theta + \cos\theta}{1 + \sin\theta - \cos\theta} = \dfrac{(1 + \sin\theta) + \cos\theta}{(1 + \sin\theta) - \cos\theta} \cdot \dfrac{(1 + \sin\theta) + \cos\theta}{(1 + \sin\theta) + \cos\theta} = \dfrac{1 + 2\sin\theta + \sin^2\theta + 2(1 + \sin\theta)(\cos\theta) + \cos^2\theta}{1 + 2\sin\theta + \sin^2\theta - \cos^2\theta}$

$= \dfrac{1 + 2\sin\theta + \sin^2\theta + 2(1 + \sin\theta)(\cos\theta) + (1 - \sin^2\theta)}{1 + 2\sin\theta + \sin^2\theta - (1 - \sin^2\theta)} = \dfrac{2 + 2\sin\theta + 2(1 + \sin\theta)(\cos\theta)}{2\sin\theta + 2\sin^2\theta}$

$= \dfrac{2(1 + \sin\theta) + 2(1 + \sin\theta)(\cos\theta)}{2\sin\theta(1 + \sin\theta)} = \dfrac{2(1 + \sin\theta)(1 + \cos\theta)}{2\sin\theta(1 + \sin\theta)} = \dfrac{1 + \cos\theta}{\sin\theta}$

71. $(a\sin\theta + b\cos\theta)^2 + (a\cos\theta - b\sin\theta)^2 = a^2\sin^2\theta + 2ab\sin\theta\cos\theta + b^2\cos^2\theta + a^2\cos^2\theta - 2ab\sin\theta\cos\theta + b^2\sin^2\theta$
$= a^2(\sin^2\theta + \cos^2\theta) + b^2(\cos^2\theta + \sin^2\theta) = a^2 + b^2$

73. $\dfrac{\tan\alpha + \tan\beta}{\cot\alpha + \cot\beta} = \dfrac{\tan\alpha + \tan\beta}{\dfrac{1}{\tan\alpha} + \dfrac{1}{\tan\beta}} = \dfrac{\tan\alpha + \tan\beta}{\dfrac{\tan\beta + \tan\alpha}{\tan\alpha\tan\beta}} = (\tan\alpha + \tan\beta) \cdot \dfrac{\tan\alpha\tan\beta}{\tan\alpha + \tan\beta} = \tan\alpha\tan\beta$

75. $(\sin\alpha + \cos\beta)^2 + (\cos\beta + \sin\alpha)(\cos\beta - \sin\alpha) = (\sin^2\alpha + 2\sin\alpha\cos\beta + \cos^2\beta) + (\cos^2\beta - \sin^2\alpha)$
$= 2\cos^2\beta + 2\sin\alpha\cos\beta = 2\cos\beta(\cos\beta + \sin\alpha)$

77. $\ln|\sec\theta| = \ln|\cos\theta|^{-1} = -\ln|\cos\theta|$

79. $\ln|1 + \cos\theta| + \ln|1 - \cos\theta| = \ln(|1 + \cos\theta||1 - \cos\theta|) = \ln|1 - \cos^2\theta| = \ln|\sin^2\theta| = 2\ln|\sin\theta|$

81. Let $\theta = \tan^{-1} v$. Then $\tan\theta = v, -\dfrac{\pi}{2} < \theta < \dfrac{\pi}{2}$, so $\sec\theta > 0$ and $\tan^2\theta + 1 = \sec^2\theta$. Thus, $\sec(\tan^{-1} v) = \sec\theta = \sqrt{1 + v^2}$.

83. Let $\theta = \cos^{-1} v$. Then $\cos\theta = v, 0 \le \theta \le \pi$, so $\sin\theta \ge 0$ and $\sin\theta = \sqrt{1 - \cos^2\theta} = \sqrt{1 - v^2}$.

Thus, $\tan(\cos^{-1} v) = \tan\theta = \dfrac{\sin\theta}{\cos\theta} = \dfrac{\sqrt{1 - v^2}}{v}$.

85. Let $\theta = \sin^{-1} v$. Then $\sin\theta = v, -\dfrac{\pi}{2} \le \theta \le \dfrac{\pi}{2}$, so $\cos\theta \ge 0$ and $\cos\theta = \sqrt{1 - \sin^2\theta} = \sqrt{1 - v^2}$.

Thus, $\cos(\sin^{-1} v) = \cos\theta = \sqrt{1 - v^2}$.

3.4 Concepts and Vocabulary *(page 231)*

1. −   **2.** −   **3.** F   **4.** F   **5.** F   **6.** $\frac{1}{4}(\sqrt{6} + \sqrt{2})$ **7.** $\frac{1}{4}(\sqrt{2} + \sqrt{6})$ **8.** 0

3.4 Exercises *(page 231)*

1. $\frac{1}{4}(\sqrt{6} + \sqrt{2})$ **3.** $\frac{1}{4}(\sqrt{2} - \sqrt{6})$ **5.** $-\frac{1}{4}(\sqrt{2} + \sqrt{6})$ **7.** $2 - \sqrt{3}$ **9.** $-\frac{1}{4}(\sqrt{6} + \sqrt{2})$ **11.** $\sqrt{6} - \sqrt{2}$ **13.** $\frac{1}{2}$ **15.** 0   **17.** 1

19. −1   **21.** $\frac{1}{2}$ **23. (a)** $\dfrac{2\sqrt{5}}{25}$ **(b)** $\dfrac{11\sqrt{5}}{25}$ **(c)** $\dfrac{2\sqrt{5}}{5}$ **(d)** 2   **25. (a)** $\dfrac{4 - 3\sqrt{3}}{10}$ **(b)** $\dfrac{-3 - 4\sqrt{3}}{10}$ **(c)** $\dfrac{4 + 3\sqrt{3}}{10}$ **(d)** $\dfrac{25\sqrt{3} + 48}{39}$

27. (a) $-\dfrac{5 + 12\sqrt{3}}{26}$ **(b)** $\dfrac{12 - 5\sqrt{3}}{26}$ **(c)** $-\dfrac{5 - 12\sqrt{3}}{26}$ **(d)** $\dfrac{-240 + 169\sqrt{3}}{69}$ **29. (a)** $-\dfrac{2\sqrt{2}}{3}$ **(b)** $\dfrac{-2\sqrt{2} + \sqrt{3}}{6}$ **(c)** $\dfrac{-2\sqrt{2} + \sqrt{3}}{6}$

(d) $\dfrac{9 - 4\sqrt{2}}{7}$ **31.** $\sin\left(\dfrac{\pi}{2} + \theta\right) = \sin\dfrac{\pi}{2}\cos\theta + \cos\dfrac{\pi}{2}\sin\theta = 1 \cdot \cos\theta + 0 \cdot \sin\theta = \cos\theta$

33. $\sin(\pi - \theta) = \sin\pi\cos\theta - \cos\pi\sin\theta = 0 \cdot \cos\theta - (-1)\sin\theta = \sin\theta$

35. $\sin(\pi + \theta) = \sin\pi\cos\theta + \cos\pi\sin\theta = 0 \cdot \cos\theta + (-1)\sin\theta = -\sin\theta$

37. $\tan(\pi - \theta) = \dfrac{\tan\pi - \tan\theta}{1 + \tan\pi\tan\theta} = \dfrac{0 - \tan\theta}{1 + 0 \cdot \tan\theta} = -\tan\theta$

39. $\sin\left(\dfrac{3\pi}{2} + \theta\right) = \sin\dfrac{3\pi}{2}\cos\theta + \cos\dfrac{3\pi}{2}\sin\theta = (-1)\cos\theta + 0 \cdot \sin\theta = -\cos\theta$

41. $\sin(\alpha + \beta) + \sin(\alpha - \beta) = \sin\alpha\cos\beta + \cos\alpha\sin\beta + \sin\alpha\cos\beta - \cos\alpha\sin\beta = 2\sin\alpha\cos\beta$

43. $\dfrac{\sin(\alpha + \beta)}{\sin\alpha\cos\beta} = \dfrac{\sin\alpha\cos\beta + \cos\alpha\sin\beta}{\sin\alpha\cos\beta} = \dfrac{\sin\alpha\cos\beta}{\sin\alpha\cos\beta} + \dfrac{\cos\alpha\sin\beta}{\sin\alpha\cos\beta} = 1 + \cot\alpha\tan\beta$

45. $\dfrac{\cos(\alpha + \beta)}{\cos\alpha\cos\beta} = \dfrac{\cos\alpha\cos\beta - \sin\alpha\sin\beta}{\cos\alpha\cos\beta} = \dfrac{\cos\alpha\cos\beta}{\cos\alpha\cos\beta} - \dfrac{\sin\alpha\sin\beta}{\cos\alpha\cos\beta} = 1 - \tan\alpha\tan\beta$

47. $\dfrac{\sin(\alpha + \beta)}{\sin(\alpha - \beta)} = \dfrac{\sin\alpha\cos\beta + \cos\alpha\sin\beta}{\sin\alpha\cos\beta - \cos\alpha\sin\beta} = \dfrac{\dfrac{\sin\alpha\cos\beta + \cos\alpha\sin\beta}{\cos\alpha\cos\beta}}{\dfrac{\sin\alpha\cos\beta - \cos\alpha\sin\beta}{\cos\alpha\cos\beta}} = \dfrac{\dfrac{\sin\alpha\cos\beta}{\cos\alpha\cos\beta} + \dfrac{\cos\alpha\sin\beta}{\cos\alpha\cos\beta}}{\dfrac{\sin\alpha\cos\beta}{\cos\alpha\cos\beta} - \dfrac{\cos\alpha\sin\beta}{\cos\alpha\cos\beta}} = \dfrac{\tan\alpha + \tan\beta}{\tan\alpha - \tan\beta}$

49. $\cot(\alpha + \beta) = \dfrac{\cos(\alpha + \beta)}{\sin(\alpha + \beta)} = \dfrac{\cos \alpha \cos \beta - \sin \alpha \sin \beta}{\sin \alpha \cos \beta + \cos \alpha \sin \beta} = \dfrac{\dfrac{\cos \alpha \cos \beta - \sin \alpha \sin \beta}{\sin \alpha \sin \beta}}{\dfrac{\sin \alpha \cos \beta + \cos \alpha \sin \beta}{\sin \alpha \sin \beta}} = \dfrac{\dfrac{\cos \alpha \cos \beta}{\sin \alpha \sin \beta} - \dfrac{\sin \alpha \sin \beta}{\sin \alpha \sin \beta}}{\dfrac{\sin \alpha \cos \beta}{\sin \alpha \sin \beta} + \dfrac{\cos \alpha \sin \beta}{\sin \alpha \sin \beta}} = \dfrac{\cot \alpha \cot \beta - 1}{\cot \beta + \cot \alpha}$

51. $\sec(\alpha + \beta) = \dfrac{1}{\cos(\alpha + \beta)} = \dfrac{1}{\cos \alpha \cos \beta - \sin \alpha \sin \beta} = \dfrac{\dfrac{1}{\sin \alpha \sin \beta}}{\dfrac{\cos \alpha \cos \beta - \sin \alpha \sin \beta}{\sin \alpha \sin \beta}} = \dfrac{\dfrac{1}{\sin \alpha} \cdot \dfrac{1}{\sin \beta}}{\dfrac{\cos \alpha \cos \beta}{\sin \alpha \sin \beta} - \dfrac{\sin \alpha \sin \beta}{\sin \alpha \sin \beta}} = \dfrac{\csc \alpha \csc \beta}{\cot \alpha \cot \beta - 1}$

53. $\sin(\alpha - \beta) \sin(\alpha + \beta) = (\sin \alpha \cos \beta - \cos \alpha \sin \beta)(\sin \alpha \cos \beta + \cos \alpha \sin \beta) = \sin^2 \alpha \cos^2 \beta - \cos^2 \alpha \sin^2 \beta$
$= (\sin^2 \alpha)(1 - \sin^2 \beta) - (1 - \sin^2 \alpha)(\sin^2 \beta) = \sin^2 \alpha - \sin^2 \beta$

55. $\sin(\theta + k\pi) = \sin \theta \cos k\pi + \cos \theta \sin k\pi = (\sin \theta)(-1)^k + (\cos \theta)(0) = (-1)^k \sin \theta, k$ any integer

57. $\dfrac{\sqrt{3}}{2}$ **59.** $-\dfrac{24}{25}$ **61.** $-\dfrac{33}{65}$ **63.** $\dfrac{63}{65}$ **65.** $\dfrac{48 + 25\sqrt{3}}{39}$ **67.** $\dfrac{4}{3}$ **69.** $u\sqrt{1 - v^2} - v\sqrt{1 - u^2}$

71. $\dfrac{u\sqrt{1 - v^2} - v}{\sqrt{1 + u^2}}$ **73.** $\dfrac{uv - \sqrt{1 - u^2}\sqrt{1 - v^2}}{v\sqrt{1 - u^2} + u\sqrt{1 - v^2}}$

75. Let $\alpha = \sin^{-1} v$ and $\beta = \cos^{-1} v$. Then $\sin \alpha = \cos \beta = v$, and since $\sin \alpha = \cos\left(\dfrac{\pi}{2} - \alpha\right)$, $\cos\left(\dfrac{\pi}{2} - \alpha\right) = \cos \beta$.

If $v \geq 0$, then $0 \leq \alpha \leq \dfrac{\pi}{2}$, so that $\left(\dfrac{\pi}{2} - \alpha\right)$ and β both lie in $\left[0, \dfrac{\pi}{2}\right]$. If $v < 0$, then $-\dfrac{\pi}{2} \leq \alpha < 0$, so that $\left(\dfrac{\pi}{2} - \alpha\right)$ and β both lie in

$\left(\dfrac{\pi}{2}, \pi\right]$. Either way, $\cos\left(\dfrac{\pi}{2} - \alpha\right) = \cos \beta$ implies $\dfrac{\pi}{2} - \alpha = \beta$, or $\alpha + \beta = \dfrac{\pi}{2}$.

77. Let $\alpha = \tan^{-1}\dfrac{1}{v}$, and $\beta = \tan^{-1} v$. Because $v \neq 0$, α, $\beta \neq 0$. Then $\tan \alpha = \dfrac{1}{v} = \dfrac{1}{\tan \beta} = \cot \beta$, and since

$\tan \alpha = \cot\left(\dfrac{\pi}{2} - \alpha\right)$, $\cot\left(\dfrac{\pi}{2} - \alpha\right) = \cot \beta$. Because $v > 0, 0 < \alpha < \dfrac{\pi}{2}$ and so $\left(\dfrac{\pi}{2} - \alpha\right)$ and β both lie in $\left(0, \dfrac{\pi}{2}\right)$.

Thus $\cot\left(\dfrac{\pi}{2} - \alpha\right) = \cot \beta$ implies $\dfrac{\pi}{2} - \alpha = \beta$, so $\alpha = \dfrac{\pi}{2} - \beta$, or $\tan^{-1}\dfrac{1}{v} = \dfrac{\pi}{2} - \tan^{-1} v$.

79. $\sin(\sin^{-1} v + \cos^{-1} v) = \sin(\sin^{-1} v) \cos(\cos^{-1} v) + \cos(\sin^{-1} v) \sin(\cos^{-1} v) = (v)(v) + \sqrt{1 - v^2}\sqrt{1 - v^2} = v^2 + 1 - v^2 = 1$

81. $\dfrac{\sin(x + h) - \sin x}{h} = \dfrac{\sin x \cos h + \cos x \sin h - \sin x}{h} = \dfrac{\cos x \sin h - \sin x(1 - \cos h)}{h} = \cos x \cdot \dfrac{\sin h}{h} - \sin x \cdot \dfrac{1 - \cos h}{h}$

83. $\tan\dfrac{\pi}{2}$ is not defined; $\tan\left(\dfrac{\pi}{2} - \theta\right) = \dfrac{\sin\left(\dfrac{\pi}{2} - \theta\right)}{\cos\left(\dfrac{\pi}{2} - \theta\right)} = \dfrac{\cos \theta}{\sin \theta} = \cot \theta$ **85.** $\tan \theta = \tan(\theta_2 - \theta_1) = \dfrac{\tan \theta_2 - \tan \theta_1}{1 + \tan \theta_1 \tan \theta_2} = \dfrac{m_2 - m_1}{1 + m_1 m_2}$

87. No; $\tan\dfrac{\pi}{2}$ is undefined.

3.5 Concepts and Vocabulary *(page 241)*

1. $\sin^2 \theta; 2 \cos^2 \theta; 2 \sin^2 \theta$ **2.** $1 - \cos \theta$ **3.** $\sin \theta$ **4.** T **5.** F **6.** $2 \cos^2 \theta - 1$

7. $\sin\dfrac{30°}{2} = \dfrac{\sqrt{2 - \sqrt{3}}}{2}$ **8.** $\sin(45° - 30°) = \dfrac{1}{4}(\sqrt{6} - \sqrt{2})$

3.5 Exercises *(page 241)*

1. (a) $\dfrac{24}{25}$ **(b)** $\dfrac{7}{25}$ **(c)** $\dfrac{\sqrt{10}}{10}$ **(d)** $\dfrac{3\sqrt{10}}{10}$ **3. (a)** $\dfrac{24}{25}$ **(b)** $-\dfrac{7}{25}$ **(c)** $\dfrac{2\sqrt{5}}{5}$ **(d)** $-\dfrac{\sqrt{5}}{5}$ **5. (a)** $-\dfrac{2\sqrt{2}}{3}$ **(b)** $\dfrac{1}{3}$ **(c)** $\sqrt{\dfrac{3 + \sqrt{6}}{6}}$

(d) $\sqrt{\dfrac{3 - \sqrt{6}}{6}}$ **7. (a)** $\dfrac{4\sqrt{2}}{9}$ **(b)** $-\dfrac{7}{9}$ **(c)** $\dfrac{\sqrt{3}}{3}$ **(d)** $\dfrac{\sqrt{6}}{3}$ **9. (a)** $-\dfrac{4}{5}$ **(b)** $\dfrac{3}{5}$ **(c)** $\sqrt{\dfrac{5 + 2\sqrt{5}}{10}}$ **(d)** $\sqrt{\dfrac{5 - 2\sqrt{5}}{10}}$

11. (a) $-\dfrac{3}{5}$ **(b)** $-\dfrac{4}{5}$ **(c)** $\dfrac{1}{2}\sqrt{\dfrac{10 - \sqrt{10}}{5}}$ **(d)** $-\dfrac{1}{2}\sqrt{\dfrac{10 + \sqrt{10}}{5}}$ **13.** $\dfrac{\sqrt{2 - \sqrt{2}}}{2}$ **15.** $1 - \sqrt{2}$ **17.** $-\dfrac{\sqrt{2 + \sqrt{3}}}{2}$

19. $\dfrac{2}{\sqrt{2 + \sqrt{2}}} = (2 - \sqrt{2})\sqrt{2 + \sqrt{2}}$ **21.** $-\dfrac{\sqrt{2 - \sqrt{2}}}{2}$

23. $\sin^4 \theta = (\sin^2 \theta)^2 = \left(\dfrac{1 - \cos(2\theta)}{2}\right)^2 = \dfrac{1}{4}(1 - 2 \cos(2\theta) + \cos^2(2\theta)) = \dfrac{1}{4} - \dfrac{1}{2} \cos(2\theta) + \dfrac{1}{4} \cos^2(2\theta)$

$= \dfrac{1}{4} - \dfrac{1}{2} \cos(2\theta) + \dfrac{1}{4}\left(\dfrac{1 + \cos(4\theta)}{2}\right) = \dfrac{1}{4} - \dfrac{1}{2} \cos(2\theta) + \dfrac{1}{8} + \dfrac{1}{8} \cos(4\theta) = \dfrac{3}{8} - \dfrac{1}{2} \cos(2\theta) + \dfrac{1}{8} \cos(4\theta)$

25. $\sin(4\theta) = \sin[2(2\theta)] = 2\sin(2\theta)\cos(2\theta) = (4\sin\theta\cos\theta)(1 - 2\sin^2\theta) = 4\sin\theta\cos\theta - 8\sin^3\theta\cos\theta = (\cos\theta)(4\sin\theta - 8\sin^3\theta)$

27. $\sin(5\theta) = 16\sin^5\theta - 20\sin^3\theta + 5\sin\theta$ **29.** $\cos^4\theta - \sin^4\theta = (\cos^2\theta + \sin^2\theta)(\cos^2\theta - \sin^2\theta) = \cos(2\theta)$

31. $\cot(2\theta) = \dfrac{1}{\tan(2\theta)} = \dfrac{1}{\dfrac{2\tan\theta}{1 - \tan^2\theta}} = \dfrac{1 - \tan^2\theta}{2\tan\theta} = \dfrac{1 - \dfrac{1}{\cot^2\theta}}{2\left(\dfrac{1}{\cot\theta}\right)} = \dfrac{\dfrac{\cot^2\theta - 1}{\cot^2\theta}}{\dfrac{2}{\cot\theta}} = \dfrac{\cot^2\theta - 1}{\cot^2\theta} \cdot \dfrac{\cot\theta}{2} = \dfrac{\cot^2\theta - 1}{2\cot\theta}$

33. $\sec(2\theta) = \dfrac{1}{\cos(2\theta)} = \dfrac{1}{2\cos^2\theta - 1} = \dfrac{1}{\dfrac{2}{\sec^2\theta} - 1} = \dfrac{1}{\dfrac{2 - \sec^2\theta}{\sec^2\theta}} = \dfrac{\sec^2\theta}{2 - \sec^2\theta}$

35. $\cos^2(2\theta) - \sin^2(2\theta) = \cos[2(2\theta)] = \cos(4\theta)$

37. $\dfrac{\cos(2\theta)}{1 + \sin(2\theta)} = \dfrac{\cos^2\theta - \sin^2\theta}{1 + 2\sin\theta\cos\theta} = \dfrac{(\cos\theta - \sin\theta)(\cos\theta + \sin\theta)}{\sin^2\theta + \cos^2\theta + 2\sin\theta\cos\theta} = \dfrac{(\cos\theta - \sin\theta)(\cos\theta + \sin\theta)}{(\sin\theta + \cos\theta)(\sin\theta + \cos\theta)} = \dfrac{\cos\theta - \sin\theta}{\cos\theta + \sin\theta}$

$= \dfrac{\dfrac{\cos\theta - \sin\theta}{\sin\theta}}{\dfrac{\cos\theta + \sin\theta}{\sin\theta}} = \dfrac{\dfrac{\cos\theta}{\sin\theta} - \dfrac{\sin\theta}{\sin\theta}}{\dfrac{\cos\theta}{\sin\theta} + \dfrac{\sin\theta}{\sin\theta}} = \dfrac{\cot\theta - 1}{\cot\theta + 1}$

39. $\sec^2\dfrac{\theta}{2} = \dfrac{1}{\cos^2\left(\dfrac{\theta}{2}\right)} = \dfrac{1}{\dfrac{1 + \cos\theta}{2}} = \dfrac{2}{1 + \cos\theta}$

41. $\cot^2\dfrac{\theta}{2} = \dfrac{1}{\tan^2\left(\dfrac{\theta}{2}\right)} = \dfrac{1}{\dfrac{1 - \cos\theta}{1 + \cos\theta}} = \dfrac{1 + \cos\theta}{1 - \cos\theta} = \dfrac{1 + \dfrac{1}{\sec\theta}}{1 - \dfrac{1}{\sec\theta}} = \dfrac{\dfrac{\sec\theta + 1}{\sec\theta}}{\dfrac{\sec\theta - 1}{\sec\theta}} = \dfrac{\sec\theta + 1}{\sec\theta - 1} \cdot \dfrac{\sec\theta}{\sec\theta - 1} = \dfrac{\sec\theta + 1}{\sec\theta - 1}$

43. $\dfrac{1 - \tan^2\left(\dfrac{\theta}{2}\right)}{1 + \tan^2\left(\dfrac{\theta}{2}\right)} = \dfrac{1 - \dfrac{1 - \cos\theta}{1 + \cos\theta}}{1 + \dfrac{1 - \cos\theta}{1 + \cos\theta}} = \dfrac{\dfrac{1 + \cos\theta - (1 - \cos\theta)}{1 + \cos\theta}}{\dfrac{1 + \cos\theta + 1 - \cos\theta}{1 + \cos\theta}} = \dfrac{2\cos\theta}{1 + \cos\theta} \cdot \dfrac{1 + \cos\theta}{2} = \cos\theta$

45. $\dfrac{\sin(3\theta)}{\sin\theta} - \dfrac{\cos(3\theta)}{\cos\theta} = \dfrac{\sin(3\theta)\cos\theta - \cos(3\theta)\sin\theta}{\sin\theta\cos\theta} = \dfrac{\sin(3\theta - \theta)}{\dfrac{1}{2}(2\sin\theta\cos\theta)} = \dfrac{2\sin(2\theta)}{\sin(2\theta)} = 2$

47. $\tan(3\theta) = \tan(\theta + 2\theta) = \dfrac{\tan\theta + \tan(2\theta)}{1 - \tan\theta\tan(2\theta)} = \dfrac{\tan\theta + \dfrac{2\tan\theta}{1 - \tan^2\theta}}{1 - \dfrac{\tan\theta(2\tan\theta)}{1 - \tan^2\theta}} = \dfrac{\tan\theta - \tan^3\theta + 2\tan\theta}{1 - \tan^2\theta - 2\tan^2\theta} = \dfrac{3\tan\theta - \tan^3\theta}{1 - 3\tan^2\theta}$

49. $\dfrac{1}{2}(\ln|1 - \cos(2\theta)| - \ln 2) = \ln\left(\dfrac{|1 - \cos(2\theta)|}{2}\right)^{1/2} = \ln|\sin^2\theta|^{1/2} = \ln|\sin\theta|$

51. $\dfrac{\sqrt{3}}{2}$ **53.** $\dfrac{7}{25}$ **55.** $\dfrac{24}{7}$ **57.** $\dfrac{24}{25}$ **59.** $\dfrac{1}{5}$ **61.** $\dfrac{25}{7}$ **63.** $\sin(2\theta) = \dfrac{4x}{4 + x^2}$ **65.** $-\dfrac{1}{4}$

67. $\dfrac{2z}{1 + z^2} = \dfrac{2\tan\left(\dfrac{\alpha}{2}\right)}{1 + \tan^2\left(\dfrac{\alpha}{2}\right)} = \dfrac{2\tan\left(\dfrac{\alpha}{2}\right)}{\sec^2\left(\dfrac{\alpha}{2}\right)} = \dfrac{\dfrac{2\sin\left(\dfrac{\alpha}{2}\right)}{\cos\left(\dfrac{\alpha}{2}\right)}}{\dfrac{1}{\cos^2\left(\dfrac{\alpha}{2}\right)}} = 2\sin\left(\dfrac{\alpha}{2}\right)\cos\left(\dfrac{\alpha}{2}\right) = \sin\left(2 \cdot \dfrac{\alpha}{2}\right) = \sin\alpha$

69. $A = \dfrac{1}{2}h(\text{base}) = h\left(\dfrac{1}{2}\text{base}\right) = s\cos\dfrac{\theta}{2} \cdot s\sin\dfrac{\theta}{2} = \dfrac{1}{2}s^2\sin\theta$

71.

73. $\sin\dfrac{\pi}{24} = \dfrac{\sqrt{2}}{4}\sqrt{4 - \sqrt{6} - \sqrt{2}}$; $\cos\dfrac{\pi}{24} = \dfrac{\sqrt{2}}{4}\sqrt{4 + \sqrt{6} + \sqrt{2}}$

75. $\sin^3 \theta + \sin^3(\theta + 120°) + \sin^3(\theta + 240°) = \sin^3 \theta + (\sin \theta \cos 120° + \cos \theta \sin 120°)^3 + (\sin \theta \cos 240° + \cos \theta \sin 240°)^3$

$= \sin^3 \theta + \left(-\frac{1}{2}\sin \theta + \frac{\sqrt{3}}{2}\cos \theta\right)^3 + \left(-\frac{1}{2}\sin \theta - \frac{\sqrt{3}}{2}\cos \theta\right)^3$

$= \sin^3 \theta + \frac{1}{8}(3\sqrt{3}\cos^3 \theta - 9\cos^2 \theta \sin \theta + 3\sqrt{3}\cos \theta \sin^2 \theta - \sin^3 \theta) - \frac{1}{8}(\sin^3 \theta + 3\sqrt{3}\sin^2 \theta \cos \theta + 9\sin \theta \cos^2 \theta + 3\sqrt{3}\cos^3 \theta)$

$= \frac{3}{4}\sin^3 \theta - \frac{9}{4}\cos^2 \theta \sin \theta = \frac{3}{4}[\sin^3 \theta - 3\sin \theta(1 - \sin^2 \theta)] = \frac{3}{4}(4\sin^3 \theta - 3\sin \theta) = -\frac{3}{4}\sin(3\theta)$ (from Example 2b)

77. (a) $R = \frac{v_0^2 \sqrt{2}}{16}(\sin \theta \cos \theta - \cos^2 \theta)$ **(b)**   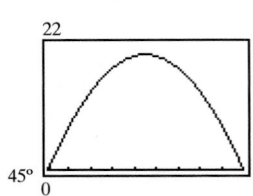   **(c)** $\theta = 67.5°$ makes R largest

$= \frac{v_0^2 \sqrt{2}}{16}\left(\frac{1}{2}\sin(2\theta) - \frac{1 + \cos(2\theta)}{2}\right)$

$= \frac{v_0^2 \sqrt{2}}{32}[\sin(2\theta) - \cos(2\theta) - 1]$

3.6 Exercises *(page 245)*

1. $\frac{1}{2}[\cos(2\theta) - \cos(6\theta)]$ **3.** $\frac{1}{2}[\sin(6\theta) + \sin(2\theta)]$ **5.** $\frac{1}{2}[\cos(2\theta) + \cos(8\theta)]$ **7.** $\frac{1}{2}[\cos \theta - \cos(3\theta)]$ **9.** $\frac{1}{2}[\sin(2\theta) + \sin \theta]$

11. $2\sin \theta \cos(3\theta)$ **13.** $2\cos(3\theta)\cos \theta$ **15.** $2\sin(2\theta)\cos \theta$ **17.** $2\sin \theta \sin\frac{\theta}{2}$ **19.** $\frac{\sin \theta + \sin(3\theta)}{2\sin(2\theta)} = \frac{2\sin(2\theta)\cos \theta}{2\sin(2\theta)} = \cos \theta$

21. $\frac{\sin(4\theta) + \sin(2\theta)}{\cos(4\theta) + \cos(2\theta)} = \frac{2\sin(3\theta)\cos \theta}{2\cos(3\theta)\cos \theta} = \frac{\sin(3\theta)}{\cos(3\theta)} = \tan(3\theta)$ **23.** $\frac{\cos \theta - \cos(3\theta)}{\sin \theta + \sin(3\theta)} = \frac{2\sin(2\theta)\sin \theta}{2\sin(2\theta)\cos \theta} = \frac{\sin \theta}{\cos \theta} = \tan \theta$

25. $\sin \theta[\sin \theta + \sin(3\theta)] = \sin \theta[2\sin(2\theta)\cos \theta] = \cos \theta[2\sin(2\theta)\sin \theta] = \cos \theta\left[2 \cdot \frac{1}{2}[\cos \theta - \cos(3\theta)]\right] = \cos \theta[\cos \theta - \cos(3\theta)]$

27. $\frac{\sin(4\theta) + \sin(8\theta)}{\cos(4\theta) + \cos(8\theta)} = \frac{2\sin(6\theta)\cos(2\theta)}{2\cos(6\theta)\cos(2\theta)} = \frac{\sin(6\theta)}{\cos(6\theta)} = \tan(6\theta)$

29. $\frac{\sin(4\theta) + \sin(8\theta)}{\sin(4\theta) - \sin(8\theta)} = \frac{2\sin(6\theta)\cos(-2\theta)}{2\sin(-2\theta)\cos(6\theta)} = \frac{\sin(6\theta)}{\cos(6\theta)} \cdot \frac{\cos(2\theta)}{-\sin(2\theta)} = \tan(6\theta)[-\cot(2\theta)] = -\frac{\tan(6\theta)}{\tan(2\theta)}$

31. $\frac{\sin \alpha + \sin \beta}{\sin \alpha - \sin \beta} = \frac{2\sin\frac{\alpha + \beta}{2}\cos\frac{\alpha - \beta}{2}}{2\sin\frac{\alpha - \beta}{2}\cos\frac{\alpha + \beta}{2}} = \frac{\sin\frac{\alpha + \beta}{2}}{\cos\frac{\alpha + \beta}{2}} \cdot \frac{\cos\frac{\alpha - \beta}{2}}{\sin\frac{\alpha - \beta}{2}} = \tan\frac{\alpha + \beta}{2}\cot\frac{\alpha - \beta}{2}$

33. $\frac{\sin \alpha + \sin \beta}{\cos \alpha + \cos \beta} = \frac{2\sin\frac{\alpha + \beta}{2}\cos\frac{\alpha - \beta}{2}}{2\cos\frac{\alpha + \beta}{2}\cos\frac{\alpha - \beta}{2}} = \frac{\sin\frac{\alpha + \beta}{2}}{\cos\frac{\alpha + \beta}{2}} = \tan\frac{\alpha + \beta}{2}$

35. $1 + \cos(2\theta) + \cos(4\theta) + \cos(6\theta) = [1 + \cos(6\theta)] + [\cos(2\theta) + \cos(4\theta)] = 2\cos^2(3\theta) + 2\cos(3\theta)\cos(-\theta)$
$= 2\cos(3\theta)[\cos(3\theta) + \cos \theta] = 2\cos(3\theta)[2\cos(2\theta)\cos \theta] = 4\cos \theta \cos(2\theta)\cos(3\theta)$

37. (a) $y = 2\sin(2061\pi t)\cos(357\pi t)$ **(b)** $y_{max} = 2$ **(c)**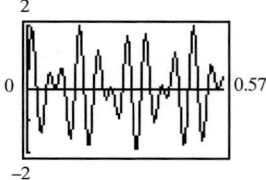

39. $\sin(2\alpha) + \sin(2\beta) + \sin(2\gamma) = 2\sin(\alpha + \beta)\cos(\alpha - \beta) + \sin(2\gamma) = 2\sin(\alpha + \beta)\cos(\alpha - \beta) + 2\sin \gamma \cos \gamma$
$= 2\sin(\pi - \gamma)\cos(\alpha - \beta) + 2\sin \gamma \cos \gamma = 2\sin \gamma \cos(\alpha - \beta) + 2\sin \gamma \cos \gamma = 2\sin \gamma[\cos(\alpha - \beta) + \cos \gamma]$
$= 2\sin \gamma\left(2\cos\frac{\alpha - \beta + \gamma}{2}\cos\frac{\alpha - \beta - \gamma}{2}\right) = 4\sin \gamma \cos\frac{\pi - 2\beta}{2}\cos\frac{2\alpha - \pi}{2} = 4\sin \gamma \cos\left(\frac{\pi}{2} - \beta\right)\cos\left(\alpha - \frac{\pi}{2}\right)$
$= 4\sin \gamma \sin \beta \sin \alpha$

41.
$$\sin(\alpha - \beta) = \sin \alpha \cos \beta - \cos \alpha \sin \beta$$
$$\sin(\alpha + \beta) = \sin \alpha \cos \beta + \cos \alpha \sin \beta$$
$$\sin(\alpha - \beta) + \sin(\alpha + \beta) = 2 \sin \alpha \cos \beta$$
$$\sin \alpha \cos \beta = \frac{1}{2}[\sin(\alpha + \beta) + \sin(\alpha - \beta)]$$

43. $2 \cos \dfrac{\alpha + \beta}{2} \cos \dfrac{\alpha - \beta}{2} = 2 \cdot \dfrac{1}{2}\left[\cos\left(\dfrac{\alpha + \beta}{2} + \dfrac{\alpha - \beta}{2} \right) + \cos\left(\dfrac{\alpha + \beta}{2} - \dfrac{\alpha - \beta}{2} \right) \right] = \cos \dfrac{2\alpha}{2} + \cos \dfrac{2\beta}{2} = \cos \alpha + \cos \beta$

3.7 Concepts and Vocabulary *(page 251)*

1. $\left\{ \dfrac{\pi}{6}, \dfrac{5\pi}{6} \right\}$ **2.** $\left\{ \theta \middle| \theta = \dfrac{\pi}{6} + 2\pi k, \theta = \dfrac{5\pi}{6} + 2\pi k, k \text{ any integer} \right\}$ **3.** F **4.** T **5.** F

3.7 Exercises *(page 251)*

1. $\left\{ \theta \middle| \theta = \dfrac{\pi}{6} + 2k\pi, \theta = \dfrac{5\pi}{6} + 2k\pi \right\}; \dfrac{\pi}{6}, \dfrac{5\pi}{6}, \dfrac{13\pi}{6}, \dfrac{17\pi}{6}, \dfrac{25\pi}{6}, \dfrac{29\pi}{6}$ **3.** $\left\{ \theta \middle| \theta = \dfrac{5\pi}{6} + k\pi \right\}; \dfrac{5\pi}{6}, \dfrac{11\pi}{6}, \dfrac{17\pi}{6}, \dfrac{23\pi}{6}, \dfrac{29\pi}{6}, \dfrac{35\pi}{6}$

5. $\left\{ \theta \middle| \theta = \dfrac{\pi}{2} + 2k\pi, \theta = \dfrac{3\pi}{2} + 2k\pi \right\}; \dfrac{\pi}{2}, \dfrac{3\pi}{2}, \dfrac{5\pi}{2}, \dfrac{7\pi}{2}, \dfrac{9\pi}{2}, \dfrac{11\pi}{2}$ **7.** $\left\{ \theta \middle| \theta = \dfrac{\pi}{3} + k\pi, \theta = \dfrac{2\pi}{3} + k\pi \right\}; \dfrac{\pi}{3}, \dfrac{2\pi}{3}, \dfrac{4\pi}{3}, \dfrac{5\pi}{3}, \dfrac{7\pi}{3}, \dfrac{8\pi}{3}$

9. $\left\{ \theta \middle| \theta = \dfrac{8\pi}{3} + 4k\pi, \theta = \dfrac{10\pi}{3} + 4k\pi \right\}; \dfrac{8\pi}{3}, \dfrac{10\pi}{3}, \dfrac{20\pi}{3}, \dfrac{22\pi}{3}, \dfrac{32\pi}{3}, \dfrac{34\pi}{3}$ **11.** $\left\{ \dfrac{7\pi}{6}, \dfrac{11\pi}{6} \right\}$ **13.** $\left\{ \dfrac{\pi}{3}, \dfrac{2\pi}{3}, \dfrac{4\pi}{3}, \dfrac{5\pi}{3} \right\}$ **15.** $\left\{ \dfrac{\pi}{4}, \dfrac{3\pi}{4}, \dfrac{5\pi}{4}, \dfrac{7\pi}{4} \right\}$

17. $\left\{ \dfrac{\pi}{2}, \dfrac{7\pi}{6}, \dfrac{11\pi}{6} \right\}$ **19.** $\left\{ \dfrac{\pi}{3}, \dfrac{2\pi}{3}, \dfrac{4\pi}{3}, \dfrac{5\pi}{3} \right\}$ **21.** $\left\{ \dfrac{4\pi}{9}, \dfrac{8\pi}{9}, \dfrac{16\pi}{9} \right\}$ **23.** $\left\{ \dfrac{3\pi}{4}, \dfrac{7\pi}{4} \right\}$ **25.** $\left\{ \dfrac{11\pi}{6} \right\}$ **27.** $\left\{ \dfrac{7\pi}{6}, \dfrac{11\pi}{6} \right\}$ **29.** $\left\{ \dfrac{3\pi}{4}, \dfrac{7\pi}{4} \right\}$

31. $\left\{ \dfrac{2\pi}{3}, \dfrac{4\pi}{3} \right\}$ **33.** $\left\{ \dfrac{3\pi}{4}, \dfrac{5\pi}{4} \right\}$ **35.** $\{0.41, 2.73\}$ **37.** $\{1.37, 4.51\}$ **39.** $\{2.69, 3.59\}$ **41.** $\{1.82, 4.46\}$ **43.** $28.90°$

45. Yes; it varies from 1.28 to 1.34 **47.** 1.47

49. If θ is the original angle of incidence and ϕ is the angle of refraction, then $\dfrac{\sin \theta}{\sin \phi} = n_2$. The angle of incidence of the emerging beam is also ϕ, and the index of refraction is $\dfrac{1}{n_2}$. Thus, θ is the angle of refraction of the emerging beam.

3.8 Exercises *(page 258)*

1. $\left\{ \dfrac{\pi}{2}, \dfrac{2\pi}{3}, \dfrac{4\pi}{3}, \dfrac{3\pi}{2} \right\}$ **3.** $\left\{ \dfrac{\pi}{2}, \dfrac{7\pi}{6}, \dfrac{11\pi}{6} \right\}$ **5.** $\left\{ 0, \dfrac{\pi}{4}, \dfrac{5\pi}{4} \right\}$ **7.** $\left\{ \dfrac{\pi}{2}, \dfrac{2\pi}{3}, \dfrac{4\pi}{3}, \dfrac{3\pi}{2} \right\}$ **9.** $\{\pi\}$ **11.** $\left\{ \dfrac{\pi}{3}, \dfrac{2\pi}{3}, \dfrac{4\pi}{3}, \dfrac{5\pi}{3} \right\}$ **13.** $\left\{ \dfrac{\pi}{4}, \dfrac{5\pi}{4} \right\}$

15. $\left\{ 0, \dfrac{\pi}{3}, \pi, \dfrac{5\pi}{3} \right\}$ **17.** $\left\{ \dfrac{\pi}{2}, \dfrac{3\pi}{2} \right\}$ **19.** $\left\{ 0, \dfrac{2\pi}{3}, \dfrac{4\pi}{3} \right\}$ **21.** $\left\{ 0, \dfrac{\pi}{3}, \dfrac{\pi}{2}, \dfrac{2\pi}{3}, \pi, \dfrac{4\pi}{3}, \dfrac{3\pi}{2}, \dfrac{5\pi}{3} \right\}$ **23.** $\left\{ 0, \dfrac{\pi}{5}, \dfrac{2\pi}{5}, \dfrac{3\pi}{5}, \dfrac{4\pi}{5}, \pi, \dfrac{6\pi}{5}, \dfrac{7\pi}{5}, \dfrac{8\pi}{5}, \dfrac{9\pi}{5} \right\}$

25. $\left\{ \dfrac{\pi}{6}, \dfrac{5\pi}{6}, \dfrac{3\pi}{2} \right\}$ **27.** $\left\{ \dfrac{\pi}{3}, \dfrac{5\pi}{3} \right\}$ **29.** No real solutions **31.** No real solutions **33.** $\left\{ \dfrac{\pi}{2}, \dfrac{7\pi}{6} \right\}$ **35.** $\left\{ 0, \dfrac{\pi}{3}, \pi, \dfrac{5\pi}{3} \right\}$ **37.** $\left\{ \dfrac{\pi}{4} \right\}$

39. $\{-1.29, 0\}$ **41.** $\{-2.24, 0, 2.24\}$ **43.** $\{-0.82, 0.82\}$

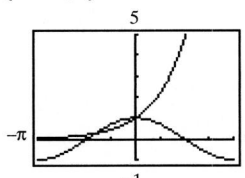

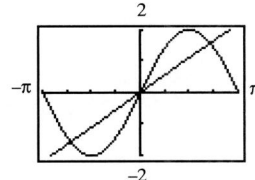

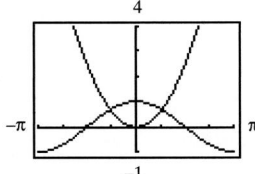

45. $\{-1.31, 1.98, 3.84\}$ **47.** $\{0.52\}$ **49.** $\{1.26\}$ **51.** $\{-1.02, 1.02\}$ **53.** $\{0, 2.15\}$ **55.** $\{0.76, 1.35\}$

57. (a) $60°$ **(b)** $60°$ **(c)** $A(60°) = 12\sqrt{3} \approx 20.78 \text{ in}^2$ **(d)**   **59.** $2.03, 4.91$

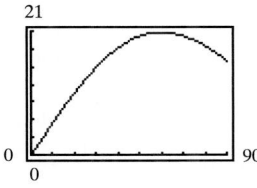

$\theta_{\text{max}} = 60°$

Maximum Area $\approx 20.78 \text{ in}^2$

61. (a) $\approx 29.99°$ or $\approx 60.01°$ **(b)** ≈ 123.58 m   **(c)**

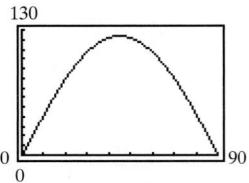

Review Exercises *(page 262)*

1. $\dfrac{\pi}{2}$ **3.** $\dfrac{\pi}{4}$ **5.** $\dfrac{5\pi}{6}$ **7.** $\dfrac{\pi}{4}$ **9.** $-\sqrt{3}$ **11.** $\dfrac{2\sqrt{3}}{3}$ **13.** $\dfrac{3}{5}$ **15.** $-\dfrac{4}{3}$ **17.** $-\dfrac{\pi}{6}$ **19.** $-\dfrac{\pi}{4}$ **21.** $\tan\theta\cot\theta - \sin^2\theta = 1 - \sin^2\theta = \cos^2\theta$

23. $\cos^2\theta(1 + \tan^2\theta) = \cos^2\theta\sec^2\theta = 1$ **25.** $4\cos^2\theta + 3\sin^2\theta = \cos^2\theta + 3(\cos^2\theta + \sin^2\theta) = 3 + \cos^2\theta$

27. $\dfrac{1 - \cos\theta}{\sin\theta} + \dfrac{\sin\theta}{1 - \cos\theta} = \dfrac{(1 - \cos\theta)^2 + \sin^2\theta}{\sin\theta(1 - \cos\theta)} = \dfrac{1 - 2\cos\theta + \cos^2\theta + \sin^2\theta}{\sin\theta(1 - \cos\theta)} = \dfrac{2(1 - \cos\theta)}{\sin\theta(1 - \cos\theta)} = 2\csc\theta$

29. $\dfrac{\cos\theta}{\cos\theta - \sin\theta} = \dfrac{\dfrac{\cos\theta}{\cos\theta}}{\dfrac{\cos\theta - \sin\theta}{\cos\theta}} = \dfrac{1}{1 - \dfrac{\sin\theta}{\cos\theta}} = \dfrac{1}{1 - \tan\theta}$

31. $\dfrac{\csc\theta}{1 + \csc\theta} = \dfrac{\dfrac{1}{\sin\theta}}{1 + \dfrac{1}{\sin\theta}} = \dfrac{1}{1 + \sin\theta} = \dfrac{1}{1 + \sin\theta}\cdot\dfrac{1 - \sin\theta}{1 - \sin\theta} = \dfrac{1 - \sin\theta}{1 - \sin^2\theta} = \dfrac{1 - \sin\theta}{\cos^2\theta}$

33. $\csc\theta - \sin\theta = \dfrac{1}{\sin\theta} - \sin\theta = \dfrac{1 - \sin^2\theta}{\sin\theta} = \dfrac{\cos^2\theta}{\sin\theta} = \cos\theta\cdot\dfrac{\cos\theta}{\sin\theta} = \cos\theta\cot\theta$

35. $\dfrac{1 - \sin\theta}{\sec\theta} = \cos\theta(1 - \sin\theta)\cdot\dfrac{1 + \sin\theta}{1 + \sin\theta} = \dfrac{\cos\theta(1 - \sin^2\theta)}{1 + \sin\theta} = \dfrac{\cos^3\theta}{1 + \sin\theta}$

37. $\cot\theta - \tan\theta = \dfrac{\cos\theta}{\sin\theta} - \dfrac{\sin\theta}{\cos\theta} = \dfrac{\cos^2\theta - \sin^2\theta}{\sin\theta\cos\theta} = \dfrac{1 - 2\sin^2\theta}{\sin\theta\cos\theta}$

39. $\dfrac{\cos(\alpha + \beta)}{\cos\alpha\sin\beta} = \dfrac{\cos\alpha\cos\beta - \sin\alpha\sin\beta}{\cos\alpha\sin\beta} = \dfrac{\cos\alpha\cos\beta}{\cos\alpha\sin\beta} - \dfrac{\sin\alpha\sin\beta}{\cos\alpha\sin\beta} = \cot\beta - \tan\alpha$

41. $\dfrac{\cos(\alpha - \beta)}{\cos\alpha\cos\beta} = \dfrac{\cos\alpha\cos\beta + \sin\alpha\sin\beta}{\cos\alpha\cos\beta} = \dfrac{\cos\alpha\cos\beta}{\cos\alpha\cos\beta} + \dfrac{\sin\alpha\sin\beta}{\cos\alpha\cos\beta} = 1 + \tan\alpha\tan\beta$

43. $(1 + \cos\theta)\left(\tan\dfrac{\theta}{2}\right) = \left(2\cos^2\dfrac{\theta}{2}\right)\dfrac{\sin\left(\dfrac{\theta}{2}\right)}{\cos\left(\dfrac{\theta}{2}\right)} = 2\sin\dfrac{\theta}{2}\cos\dfrac{\theta}{2} = \sin\theta$

45. $2\cot\theta\cot 2\theta = 2\left(\dfrac{\cos\theta}{\sin\theta}\right)\left(\dfrac{\cos 2\theta}{\sin 2\theta}\right) = \dfrac{2\cos\theta(\cos^2\theta - \sin^2\theta)}{2\sin^2\theta\cos\theta} = \dfrac{\cos^2\theta - \sin^2\theta}{\sin^2\theta} = \cot^2\theta - 1$

47. $1 - 8\sin^2\theta\cos^2\theta = 1 - 2(2\sin\theta\cos\theta)^2 = 1 - 2\sin^2(2\theta) = \cos(4\theta)$ **49.** $\dfrac{\sin(2\theta) + \sin(4\theta)}{\cos(2\theta) + \cos(4\theta)} = \dfrac{2\sin(3\theta)\cos(-\theta)}{2\cos(3\theta)\cos(-\theta)} = \tan(3\theta)$

51. $\dfrac{\cos(2\theta) - \cos(4\theta)}{\cos(2\theta) + \cos(4\theta)} - \tan\theta\tan(3\theta) = \dfrac{-2\sin(3\theta)\sin(-\theta)}{2\cos(3\theta)\cos(-\theta)} - \tan\theta\tan(3\theta) = \tan(3\theta)\tan\theta - \tan\theta\tan(3\theta) = 0$

53. $\dfrac{1}{4}(\sqrt{6} - \sqrt{2})$ **55.** $\dfrac{1}{4}(\sqrt{6} - \sqrt{2})$ **57.** $\dfrac{1}{2}$ **59.** $\sqrt{2} - 1$ **61. (a)** $-\dfrac{33}{65}$ **(b)** $-\dfrac{56}{65}$ **(c)** $-\dfrac{63}{65}$ **(d)** $\dfrac{33}{56}$ **(e)** $\dfrac{24}{25}$ **(f)** $\dfrac{119}{169}$ **(g)** $\dfrac{5\sqrt{26}}{26}$

(h) $\dfrac{2\sqrt{5}}{5}$ **63. (a)** $-\dfrac{16}{65}$ **(b)** $-\dfrac{63}{65}$ **(c)** $-\dfrac{56}{65}$ **(d)** $\dfrac{16}{63}$ **(e)** $\dfrac{24}{25}$ **(f)** $\dfrac{119}{169}$ **(g)** $\dfrac{\sqrt{26}}{26}$ **(h)** $-\dfrac{\sqrt{10}}{10}$ **65. (a)** $-\dfrac{63}{65}$ **(b)** $\dfrac{16}{65}$ **(c)** $\dfrac{33}{65}$

(d) $-\dfrac{63}{16}$ **(e)** $\dfrac{24}{25}$ **(f)** $-\dfrac{119}{169}$ **(g)** $\dfrac{2\sqrt{13}}{13}$ **(h)** $-\dfrac{\sqrt{10}}{10}$ **67. (a)** $\dfrac{-\sqrt{3} - 2\sqrt{2}}{6}$ **(b)** $\dfrac{1 - 2\sqrt{6}}{6}$ **(c)** $\dfrac{-\sqrt{3} + 2\sqrt{2}}{6}$ **(d)** $\dfrac{8\sqrt{2} + 9\sqrt{3}}{23}$

(e) $-\dfrac{\sqrt{3}}{2}$ **(f)** $-\dfrac{7}{9}$ **(g)** $\dfrac{\sqrt{3}}{3}$ **(h)** $\dfrac{\sqrt{3}}{2}$ **69. (a)** 1 **(b)** 0 **(c)** $-\dfrac{1}{9}$ **(d)** Not defined **(e)** $\dfrac{4\sqrt{5}}{9}$ **(f)** $-\dfrac{1}{9}$ **(g)** $\dfrac{\sqrt{30}}{6}$

(h) $-\dfrac{\sqrt{6}\sqrt{3} - \sqrt{5}}{6}$ **71.** $\dfrac{4 + 3\sqrt{3}}{10}$ **73.** $-\dfrac{48 + 25\sqrt{3}}{39}$ **75.** $-\dfrac{24}{25}$ **77.** $\left\{\dfrac{\pi}{3}, \dfrac{5\pi}{3}\right\}$ **79.** $\left\{\dfrac{3\pi}{4}, \dfrac{5\pi}{4}\right\}$ **81.** $\left\{\dfrac{3\pi}{4}, \dfrac{7\pi}{4}\right\}$ **83.** $\left\{0, \dfrac{\pi}{2}, \pi, \dfrac{3\pi}{2}\right\}$

85. $\left\{\dfrac{\pi}{3}, \dfrac{2\pi}{3}, \dfrac{4\pi}{3}, \dfrac{5\pi}{3}\right\}$ **87.** $\{0, \pi\}$ **89.** $\left\{0, \dfrac{2\pi}{3}, \pi, \dfrac{4\pi}{3}\right\}$ **91.** $\left\{0, \dfrac{\pi}{6}, \dfrac{5\pi}{6}\right\}$ **93.** $\left\{\dfrac{\pi}{6}, \dfrac{\pi}{2}, \dfrac{5\pi}{6}\right\}$ **95.** $\left\{\dfrac{\pi}{3}, \dfrac{5\pi}{3}\right\}$ **97.** $\left\{\dfrac{\pi}{4}, \dfrac{\pi}{2}, \dfrac{3\pi}{4}, \dfrac{3\pi}{2}\right\}$

99. $\left\{\dfrac{\pi}{2}, \pi\right\}$ **101.** 0.78 **103.** -1.11 **105.** 1.23 **107.** $\{1.11\}$ **109.** $\{0.87\}$ **111.** $\{2.22\}$

Cumulative Review Exercises *(page 265)*

1. $\left\{\dfrac{-1 - \sqrt{13}}{6}, \dfrac{-1 + \sqrt{13}}{6}\right\}$ **2.** $y + 1 = -1(x - 4)$ or $x + y = 3; 6\sqrt{2}; (1, 2)$ **3.** x-axis symmetry; $(0, -3), (0, 3), (3, 0)$

4.

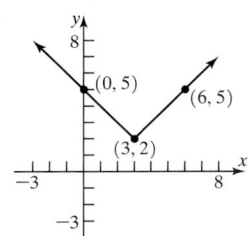

5.

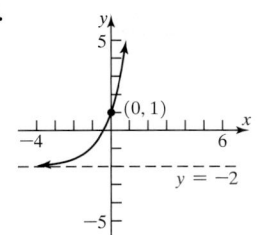

6.

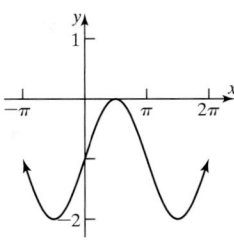

7. (a)

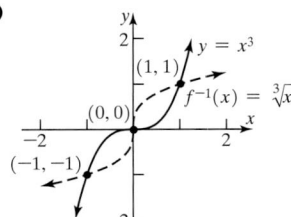

(b)

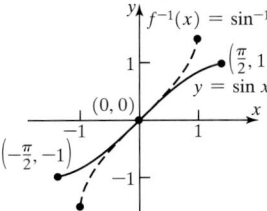

(c)

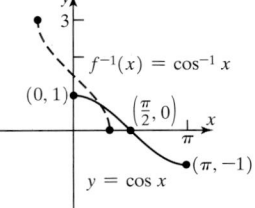

8. (a) $-\dfrac{2\sqrt{2}}{3}$ **(b)** $\dfrac{\sqrt{2}}{4}$ **(c)** $\dfrac{4\sqrt{2}}{9}$ **(d)** $\dfrac{7}{9}$ **(e)** $\sqrt{\dfrac{3 + 2\sqrt{2}}{6}}$ **(f)** $\sqrt{\dfrac{3 - 2\sqrt{2}}{6}}$ **9.** $\dfrac{\sqrt{5}}{5}$

10. (a) $-\dfrac{2\sqrt{2}}{3}$ **(b)** $-\dfrac{2\sqrt{2}}{3}$ **(c)** $\dfrac{7}{9}$ **(d)** $\dfrac{4\sqrt{2}}{9}$ **(e)** $\dfrac{\sqrt{6}}{3}$ **11.** 7.28

C H A P T E R   4   Applications of Trigonometric Functions

4.1 Concepts and Vocabulary *(page 273)*

1. Angle of elevation   **2.** Angle of depression   **3.** T   **4.** F

4.1 Exercises *(page 273)*

1. $a \approx 13.74, c \approx 14.62, \alpha = 70°$ **3.** $b \approx 5.03, c \approx 7.83, \alpha = 50°$ **5.** $a \approx 0.71, c \approx 4.06, \beta = 80°$ **7.** $b \approx 10.72, c \approx 11.83, \beta = 65°$
9. $b \approx 3.08, a \approx 8.46, \alpha = 70°$ **11.** $c \approx 5.83, \alpha = 59.0°, \beta = 31.0°$ **13.** $b \approx 4.58, \alpha \approx 23.6°, \beta = 66.4°$ **15.** 4.59 in., 6.55 in.
17. 5.52 in. or 11.83 in.   **19.** 23.6° and 66.4°   **21.** 70.02 ft   **23.** 985.91 ft   **25.** 137.37 m   **27.** 20.67 ft   **29.** 15.9°   **31.** 60.27 ft
33. 530.18 ft   **35.** 554.52 ft   **37. (a)** 111.96 ft/sec or 76.3 mi/hr   **(b)** 82.42 ft/sec or 56.2 mi/hr   **(c)** 18.8° or less   **39.** S76.6°E
41. 14.9°   **43.** 3.83 mi   **45.** No; Move the tripod back about 1 ft.

47. (a) $A(\theta) = 2 \sin \theta \cos \theta$ **(b)** From double-angle formula, since $2 \sin \theta \cos \theta = \sin(2\theta)$ **(c)** $\theta = 45°$ **(d)** $\dfrac{\sqrt{2}}{2}$ by $\sqrt{2}$

4.2 Concepts and Vocabulary *(page 284)*

1. Oblique   **2.** $\dfrac{\sin \alpha}{a} = \dfrac{\sin \beta}{b} = \dfrac{\sin \gamma}{c}$ **3.** F   **4.** T   **5.** F

4.2 Exercises *(page 284)*

1. $a \approx 3.23, b \approx 3.55, \alpha = 40°$ **3.** $a \approx 3.25, c \approx 4.23, \beta = 45°$ **5.** $\gamma = 95°, c \approx 9.86, a \approx 6.36$ **7.** $\alpha = 40°, a = 2, c \approx 3.06$
9. $\gamma = 120°, b \approx 1.06, c \approx 2.69$ **11.** $\alpha = 100°, a \approx 5.24, c \approx 0.92$ **13.** $\beta = 40°, a \approx 5.64, b \approx 3.86$ **15.** $\gamma = 100°, a \approx 1.31, b \approx 1.31$
17. One triangle; $\beta \approx 30.7°, \gamma \approx 99.3°, c \approx 3.86$ **19.** One triangle; $\gamma \approx 36.2°, \alpha \approx 43.8°, a \approx 3.51$ **21.** No triangle
23. Two triangles; $\gamma_1 \approx 30.9°, \alpha_1 \approx 129.1°, a_1 \approx 9.07$ or $\gamma_2 \approx 149.1°, \alpha_2 \approx 10.9°, a_2 \approx 2.20$ **25.** No triangle   **27.** Two triangles; $a_1 \approx 57.7°$,
$\beta_1 \approx 97.3°, b_1 \approx 2.35$ or $\alpha_2 \approx 122.3°, \beta_2 \approx 32.7°, b_2 \approx 1.28$ **29. (a)** Station Able is about 143.33 mi from the ship; Station Baker is about
135.58 mi from the ship.   **(b)** Approximately 41 min   **31.** 1490.48 ft   **33.** 381.69 ft   **35. (a)** 169.18 mi   **(b)** 161.3°   **37.** 84.7°; 183.72 ft
39. 2.64 mi   **41.** 1.88 mi or 1.53 mi   **43.** 449.36 ft   **45.** 39.39 ft   **47.** 29.97 ft

49. $\dfrac{a - b}{c} = \dfrac{a}{c} - \dfrac{b}{c} = \dfrac{\sin \alpha}{\sin \gamma} - \dfrac{\sin \beta}{\sin \gamma} = \dfrac{\sin \alpha - \sin \beta}{\sin \gamma} = \dfrac{2 \sin\left(\dfrac{\alpha - \beta}{2}\right) \cos\left(\dfrac{\alpha + \beta}{2}\right)}{2 \sin \dfrac{\gamma}{2} \cos \dfrac{\gamma}{2}} = \dfrac{\sin\left(\dfrac{\alpha - \beta}{2}\right) \cos\left(\dfrac{\pi}{2} - \dfrac{\gamma}{2}\right)}{\sin \dfrac{\gamma}{2} \cos \dfrac{\gamma}{2}} = \dfrac{\sin\left(\dfrac{\alpha - \beta}{2}\right)}{\cos \dfrac{\gamma}{2}}$

51. $\dfrac{a-b}{a+b} = \dfrac{\dfrac{a-b}{c}}{\dfrac{a+b}{c}} = \dfrac{\dfrac{\sin\left[\frac{1}{2}(\alpha-\beta)\right]}{\cos\frac{\gamma}{2}}}{\dfrac{\cos\left[\frac{1}{2}(\alpha-\beta)\right]}{\sin\frac{\gamma}{2}}} = \dfrac{\tan\left[\frac{1}{2}(\alpha-\beta)\right]}{\cot\frac{\gamma}{2}} = \dfrac{\tan\left[\frac{1}{2}(\alpha-\beta)\right]}{\tan\left(\frac{\pi}{2}-\frac{\gamma}{2}\right)} = \dfrac{\tan\left[\frac{1}{2}(\alpha-\beta)\right]}{\tan\left[\frac{1}{2}(\alpha+\beta)\right]}$

4.3 Concepts and Vocabulary *(page 291)*

1. Cosines **2.** Sines **3.** Cosines **4.** F **5.** F **6.** T

4.3 Exercises *(page 292)*

1. $b \approx 2.95$, $\alpha \approx 28.7°$, $\gamma \approx 106.3°$ **3.** $c \approx 3.75$, $\alpha \approx 32.1°$, $\beta \approx 52.9°$ **5.** $\alpha \approx 48.5°$, $\beta \approx 38.6°$, $\gamma \approx 92.9°$
7. $\alpha \approx 127.2°$, $\beta \approx 32.1°$, $\gamma \approx 20.7°$ **9.** $c \approx 2.57$, $\alpha \approx 48.6°$, $\beta \approx 91.4°$ **11.** $a \approx 2.99$, $\beta \approx 19.2°$, $\gamma \approx 80.8°$
13. $b \approx 4.14$, $\alpha \approx 43.0°$, $\gamma \approx 27.0°$ **15.** $c \approx 1.69$, $\alpha \approx 65.0°$, $\beta \approx 65.0°$ **17.** $\alpha \approx 67.4°$, $\beta = 90°$, $\gamma \approx 22.6°$
19. $\alpha = 60°$, $\beta = 60°$, $\gamma = 60°$ **21.** $\alpha \approx 33.6°$, $\beta \approx 62.2°$, $\gamma \approx 84.3°$ **23.** $\alpha \approx 97.9°$, $\beta \approx 52.4°$, $\gamma \approx 29.7°$ **25.** 70.75 ft
27. (a) $12.0°$ **(b)** 220.8 mph **29. (a)** 63.7 ft **(b)** 66.8 ft **(c)** 92.8° **31. (a)** 492.6 ft **(b)** 269.3 ft **33.** 342.33 ft

35. Using the Law of Cosines:
$L^2 = x^2 + r^2 - 2rx \cos\theta$
$x^2 - 2rx \cos\theta + r^2 - L^2 = 0$
Then, using the quadratic formula:
$x = r\cos\theta + \sqrt{r^2 \cos^2\theta + L^2 - r^2}$

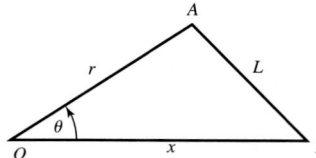

37. $\cos\dfrac{\gamma}{2} = \sqrt{\dfrac{1+\cos\gamma}{2}} = \sqrt{\dfrac{1+\dfrac{a^2+b^2-c^2}{2ab}}{2}} = \sqrt{\dfrac{2ab+a^2+b^2-c^2}{4ab}} = \sqrt{\dfrac{(a+b)^2-c^2}{4ab}} = \sqrt{\dfrac{(a+b+c)(a+b-c)}{4ab}}$
$= \sqrt{\dfrac{2s(2s-2c)}{4ab}} = \sqrt{\dfrac{s(s-c)}{ab}}$

39. $\dfrac{\cos\alpha}{a} + \dfrac{\cos\beta}{b} + \dfrac{\cos\gamma}{c} = \dfrac{b^2+c^2-a^2}{2abc} + \dfrac{a^2+c^2-b^2}{2abc} + \dfrac{a^2+b^2-c^2}{2abc} = \dfrac{b^2+c^2-a^2+a^2+c^2-b^2+a^2+b^2-c^2}{2abc}$
$= \dfrac{a^2+b^2+c^2}{2abc}$

4.4 Concepts and Vocabulary *(page 297)*

1. Heron's **2.** F **3.** T

4.4 Exercises *(page 298)*

1. 2.83 **3.** 2.99 **5.** 14.98 **7.** 9.56 **9.** 3.86 **11.** 1.48 **13.** 2.82 **15.** 1.53 **17.** 30 **19.** 1.73 **21.** 19.90 **23.** 19.81 **25.** 9.03 sq ft
27. $5446.38 **29.** 9.26 sq cm **31.** $A = \dfrac{1}{2}ab\sin\gamma = \dfrac{1}{2}a\sin\gamma\left(\dfrac{a\sin\beta}{\sin\alpha}\right) = \dfrac{a^2\sin\beta\sin\gamma}{2\sin\alpha}$ **33.** 0.92 **35.** 2.27 **37.** 5.44

39. $A = \dfrac{1}{2}r^2(\theta + \sin\theta)$ **41.** 31,145.15 sq ft

43. $h_1 = 2\dfrac{K}{a}$, $h_2 = 2\dfrac{K}{b}$, $h_3 = 2\dfrac{K}{c}$. Then $\dfrac{1}{h_1} + \dfrac{1}{h_2} + \dfrac{1}{h_3} = \dfrac{a}{2K} + \dfrac{b}{2K} + \dfrac{c}{2K} = \dfrac{a+b+c}{2K} = \dfrac{2s}{2K} = \dfrac{s}{K}$.

45. Angle AOB measures $180° - \left(\dfrac{\alpha}{2} + \dfrac{\beta}{2}\right) = 180° - \dfrac{1}{2}(180° - \gamma) = 90° + \dfrac{\gamma}{2}$, and $\sin\left(90° + \dfrac{\gamma}{2}\right) = \cos\left(-\dfrac{\gamma}{2}\right) = \cos\dfrac{\gamma}{2}$ since cosine is
an even function. Therefore, $r = \dfrac{c\sin\dfrac{\alpha}{2}\sin\dfrac{\beta}{2}}{\sin\left(90° + \dfrac{\gamma}{2}\right)} = \dfrac{c\sin\dfrac{\alpha}{2}\sin\dfrac{\beta}{2}}{\cos\dfrac{\gamma}{2}}$.

47. $\cot\dfrac{\alpha}{2} + \cot\dfrac{\beta}{2} + \cot\dfrac{\gamma}{2} = \dfrac{s-a}{r} + \dfrac{s-b}{r} + \dfrac{s-c}{r} = \dfrac{3s-(a+b+c)}{r} = \dfrac{3s-2s}{r} = \dfrac{s}{r}$

4.5 Concepts and Vocabulary *(page 307)*

1. Simple harmonic; amplitude   **2.** Simple harmonic; damped   **3.** T

4.5 Exercises *(page 308)*

1. $d = -5\cos(\pi t)$ **3.** $d = -6\cos(2t)$ **5.** $d = -5\sin(\pi t)$ **7.** $d = -6\sin(2t)$ **9. (a)** Simple harmonic   **(b)** 5 m

(c) $\dfrac{2\pi}{3}$ sec   **(d)** $\dfrac{3}{2\pi}$ oscillation/sec   **11. (a)** Simple harmonic   **(b)** 6 m   **(c)** 2 sec   **(d)** $\dfrac{1}{2}$ oscillation/sec   **13. (a)** Simple harmonic

(b) 3 m   **(c)** 4π sec   **(d)** $\dfrac{1}{4\pi}$ oscillation/sec   **15. (a)** Simple harmonic   **(b)** 2 m   **(c)** 1 sec   **(d)** 1 oscillation/sec

17. (a) $d = -10e^{-0.7t/50}\cos\left(\sqrt{\dfrac{4\pi^2}{25} - \dfrac{0.49}{2500}}\,t\right)$ **19. (a)** $d = -18e^{-0.6t/60}\cos\left(\sqrt{\dfrac{\pi^2}{4} - \dfrac{0.36}{3600}}\,t\right)$ **21. (a)** $d = -5e^{-0.8t/20}\cos\left(\sqrt{\dfrac{4\pi^2}{9} - \dfrac{0.64}{400}}\right)$

(b)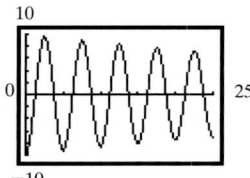

(b)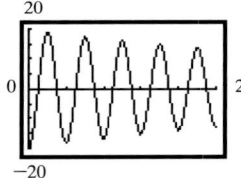

(b)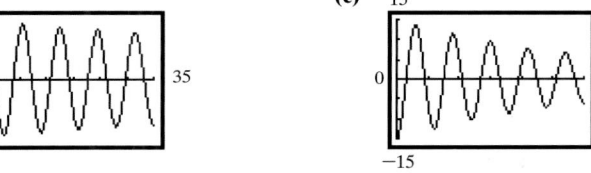

23. (a) The motion is damped.
The bob has mass $m = 20$ kg
with a damping factor of 0.7 kg/sec.
(b) 20 m downward
(c)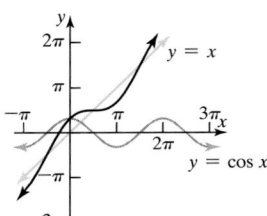
(d) 18.33 m   **(e)** $d \to 0$

25. (a) The motion is damped.
The bob has mass $m = 40$ kg
with a damping factor of 0.6 kg/sec.
(b) 30 m downward
(c)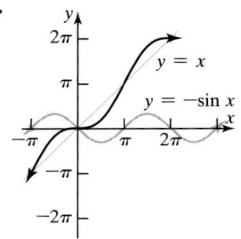
(d) 28.47 m   **(e)** $d \to 0$

27. (a) The motion is damped.
The bob has mass $m = 15$ kg
with a damping factor of 0.9 kg/sec.
(b) 15 m downward
(c)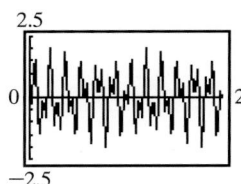
(d) 12.53 m   **(e)** $d \to 0$

29.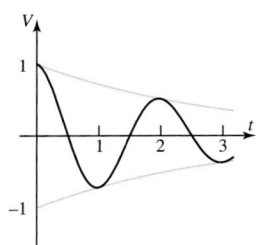

31.

33.

35.

37. (a)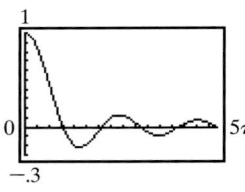
(b) At $t = 0, 2$; at $t = 1, t = 3$
(c) During the approximate intervals
$0.35 < t < 0.67, 1.29 < t < 1.75$,
and $2.19 < t \le 3$

39.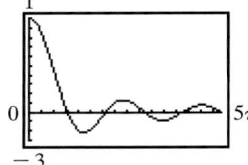

41.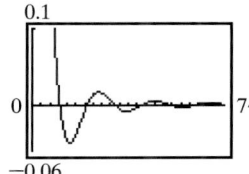

43. $y = \dfrac{1}{x}\sin x$ $y = \dfrac{1}{x^2}\sin x$ $y = \dfrac{1}{x^3}\sin x$

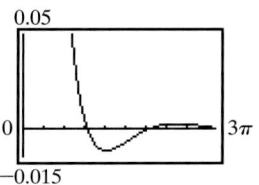

Review Exercises *(page 311)*

1. $\alpha = 70°, b \approx 3.42, a \approx 9.40$ **3.** $a \approx 4.58, \alpha = 66.4°, \beta \approx 23.6°$ **5.** $\gamma = 100°, b \approx 0.65, c \approx 1.29$ **7.** $\beta \approx 56.8°, \gamma \approx 23.2°, b \approx 4.25$
9. No triangle   **11.** $b \approx 3.32, \alpha \approx 62.8°, \gamma \approx 17.2°$ **13.** No triangle   **15.** $c \approx 2.32, \alpha \approx 16.1°, \beta \approx 123.9°$ **17.** $\beta = 36.2°, \gamma = 63.8°,$
$c = 4.55$ **19.** $\alpha = 39.6°, \beta = 18.6°, \gamma = 121.9°$ **21.** Two triangles: $\beta_1 \approx 13.4°, \gamma_1 \approx 156.6°, c_1 \approx 6.86$ or $\beta_2 \approx 166.6°, \gamma_2 \approx 3.4°, c_2 \approx 1.02$
23. $a = 5.23, \beta = 46.0°, \gamma = 64.0°$ **25.** 1.93   **27.** 18.79   **29.** 6   **31.** 3.80   **33.** 0.32   **35.** 839.10 ft   **37.** 23.32 ft   **39.** 2.15 mi
41. 204.07 mi   **43. (a)** 2.59 mi   **(b)** 2.92 mi   **(c)** 2.53 mi   **45. (a)** 131.78 mi   **(b)** 23.1°   **(c)** 0.21 hr   **47.** 8798.67 sq ft

49. 1.92 sq in.   **51.** 76.94 in.   **53. (a)** Simple harmonic   **(b)** 6 ft   **(c)** π sec   **(d)** $\dfrac{1}{\pi}$ oscillation/sec

55. (a) Simple harmonic   **(b)** 2 ft   **(c)** 2 sec   **(d)** $\dfrac{1}{2}$ oscillation/sec   **57.** $d = -4\cos\left(\dfrac{2\pi}{3}t\right)$

59. (a) $d = -15e^{-0.75t/80}\cos\left(\sqrt{\dfrac{4\pi^2}{25} - \dfrac{0.5625}{6400}}\,t\right)$

(b)
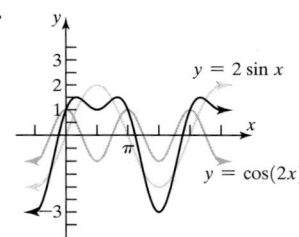

61. (a) The motion is damped. The bob has mass
$m = 20$ kg with a damping factor of 0.6 kg/sec.
(b) 15 m downward
(c)
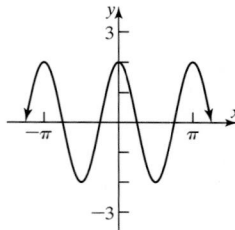
(d) 13.92 m   **(e)** $d \to 0$

63.
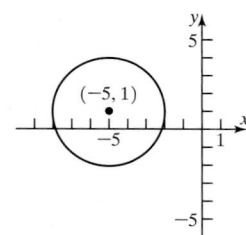

Cumulative Review Exercises *(page 316)*

1. $\left\{\dfrac{1}{3}, 1\right\}$ **2.** $(x+5)^2 + (y-1)^2 = 9$ **3.** $\{x \mid x \le -1 \text{ or } x \ge 4\}$ **4.**

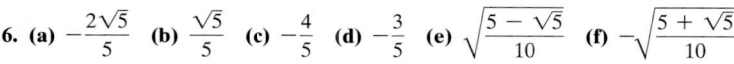

5.

6. (a) $-\dfrac{2\sqrt{5}}{5}$ **(b)** $\dfrac{\sqrt{5}}{5}$ **(c)** $-\dfrac{4}{5}$ **(d)** $-\dfrac{3}{5}$ **(e)** $\sqrt{\dfrac{5-\sqrt{5}}{10}}$ **(f)** $-\sqrt{\dfrac{5+\sqrt{5}}{10}}$

7. (a)
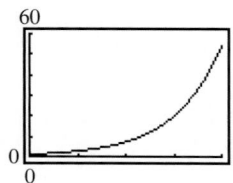
(b)
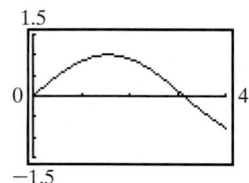
(c)
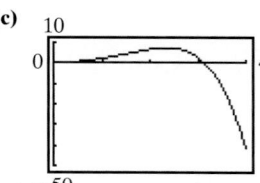
(d)
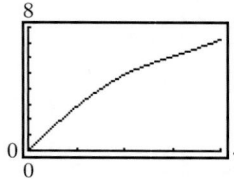

8. (a)

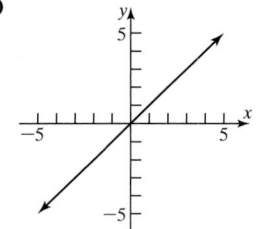

(b)

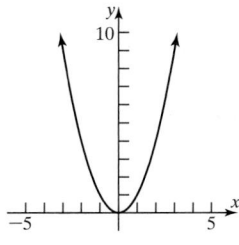

(c)

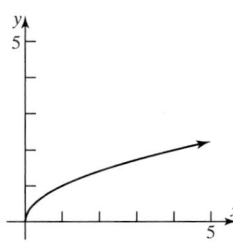

(d)

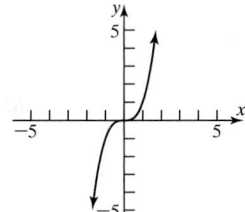

(e)

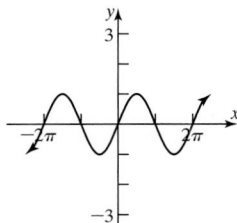

(f)

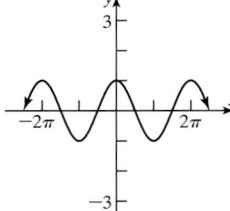

(g)

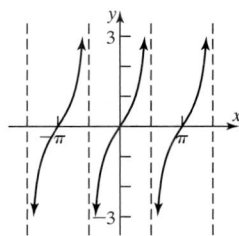

9. Two triangles: $\alpha_1 \approx 59.0°, \beta_1 = 81.0°, b_1 \approx 23.05$ or $\alpha_2 \approx 121.0°, \beta_2 \approx 19.0°, b_2 \approx 7.59$

C H A P T E R   5   Polar Coordinates; Vectors

5.1 Concepts and Vocabulary *(page 325)*

1. pole; polar axis   **2.** -2 **3.** $(-\sqrt{3}, -1)$ **4.** F   **5.** T   **6.** T   **7.** $x = r \cos \theta; y = r \sin \theta$

5.1 Exercises *(page 326)*

1. *A*   **3.** *C*   **5.** *B*   **7.** *A*

9.

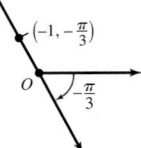

$(3, 90°)$
$90°$
O

11.

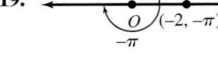

$(-2, 0)$ O

13.

$(6, \frac{\pi}{6})$
$\frac{\pi}{6}$
O

15.

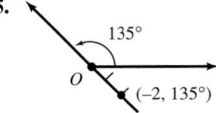

$135°$
O
$(-2, 135°)$

17.

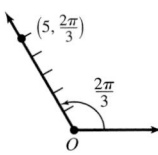

$(-1, -\frac{\pi}{3})$
O
$-\frac{\pi}{3}$

19.

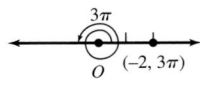

O $(-2, -\pi)$
$-\pi$

21.

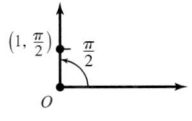

$(5, \frac{2\pi}{3})$
$\frac{2\pi}{3}$
O

(a) $\left(5, -\frac{4\pi}{3}\right)$

(b) $\left(-5, \frac{5\pi}{3}\right)$

(c) $\left(5, \frac{8\pi}{3}\right)$

23.

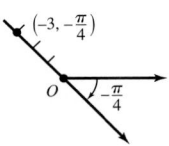

3π
O $(-2, 3\pi)$

(a) $(2, -2\pi)$

(b) $(-2, \pi)$

(c) $(2, 2\pi)$

25.

$(1, \frac{\pi}{2})$
$\frac{\pi}{2}$
O

(a) $\left(1, -\frac{3\pi}{2}\right)$

(b) $\left(-1, \frac{3\pi}{2}\right)$

(c) $\left(1, \frac{5\pi}{2}\right)$

27.

$(-3, -\frac{\pi}{4})$
O
$-\frac{\pi}{4}$

(a) $\left(3, -\frac{5\pi}{4}\right)$

(b) $\left(-3, \frac{7\pi}{4}\right)$

(c) $\left(3, \frac{11\pi}{4}\right)$

29. $(0, 3)$ **31.** $(-2, 0)$ **33.** $(-3\sqrt{3}, 3)$ **35.** $(\sqrt{2}, -\sqrt{2})$ **37.** $\left(-\frac{1}{2}, \frac{\sqrt{3}}{2}\right)$ **39.** $(2, 0)$ **41.** $(-2.57, 7.05)$ **43.** $(-4.98, -3.85)$ **45.** $(3, 0)$

47. $(1, \pi)$ **49.** $\left(\sqrt{2}, -\dfrac{\pi}{4}\right)$ **51.** $\left(2, \dfrac{\pi}{6}\right)$ **53.** $(2.47, -1.02)$ **55.** $(9.3, 0.47)$ **57.** $r^2 = \dfrac{3}{2}$ **59.** $r\cos^2\theta - 4\sin\theta = 0$ **61.** $r^2\sin 2\theta = 1$

63. $r\cos\theta = 4$ **65.** $x^2 + y^2 - x = 0$ or $\left(x - \dfrac{1}{2}\right)^2 + y^2 = \dfrac{1}{4}$ **67.** $(x^2 + y^2)^{3/2} - x = 0$ **69.** $x^2 + y^2 = 4$ **71.** $y^2 = 8(x + 2)$

73. $d = \sqrt{(r_2\cos\theta_2 - r_1\cos\theta_1)^2 + (r_2\sin\theta_2 - r_1\sin\theta_1)^2}$

$\quad = \sqrt{(r_2^2\cos^2\theta_2 - 2r_2\cos\theta_2\, r_1\cos\theta_1 + r_1^2\cos^2\theta_1) + (r_2^2\sin^2\theta_2 - 2r_2\sin\theta_2\, r_1\sin\theta_1 + r_1^2\sin^2\theta_1)}$

$\quad = \sqrt{r_1^2 + r_2^2 - 2r_1r_2(\cos\theta_2\cos\theta_1 + \sin\theta_2\sin\theta_1)}$

$\quad = \sqrt{r_1^2 + r_2^2 - 2r_1r_2\cos(\theta_2 - \theta_1)}$

5.2 Concepts and Vocabulary *(page 343)*

1. polar equation **2.** $r = 2\cos\theta$ **3.** $-r$ **4.** F **5.** F **6.** F **7.** Circle **8.** Line **9.** Heart-shaped **10.** Airplane propeller

5.2 Exercises *(page 343)*

1. $x^2 + y^2 = 16$;
Circle, radius 4, center at pole

3. $y = \sqrt{3}x$; Line through pole,
making an angle of $\dfrac{\pi}{3}$ with polar axis

5. $y = 4$; Horizontal line 4 units
above the pole

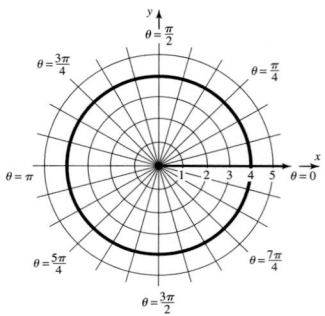
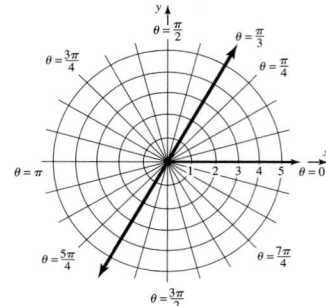
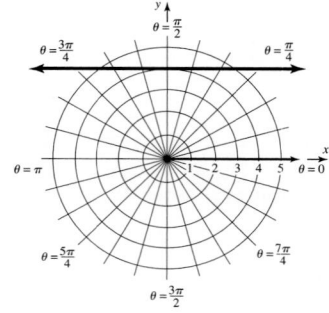

7. $x = -2$; Vertical line 2 units to the
left of the pole

9. $(x - 1)^2 + y^2 = 1$; Circle, radius 1,
center $(1, 0)$ in rectangular coordinates

11. $x^2 + (y + 2)^2 = 4$;
Circle, radius 2, center at $(0, -2)$ in
rectangular coordinates

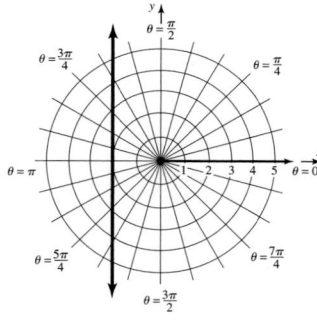
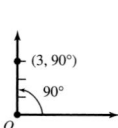

(3, 90°)

90°

O

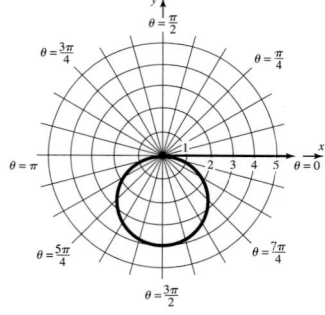

13. $(x - 2)^2 + y^2 = 4$;
Circle, radius 2, center at $(2, 0)$
in rectangular coordinates

15. $x^2 + (y + 1)^2 = 1$;
Circle, radius 1, center at $(0, -1)$ in
rectangular coordinates

17. E **19.** F **21.** H
23. D **25.** D **27.** F **29.** A

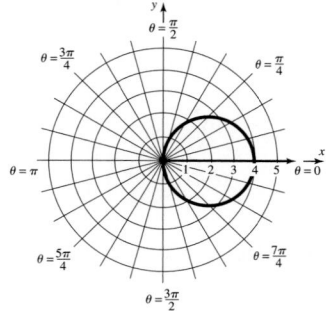
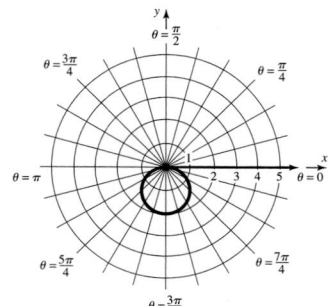

31. Cardioid

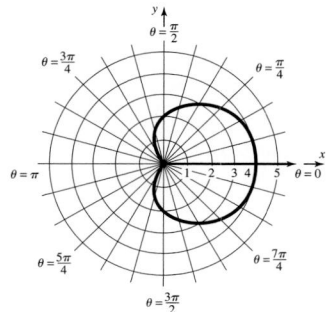

33. Cardioid

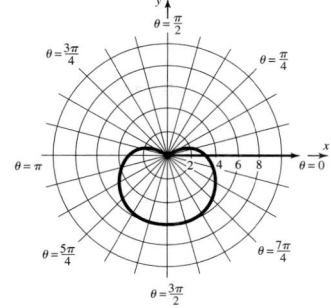

35. Limaçon without inner loop

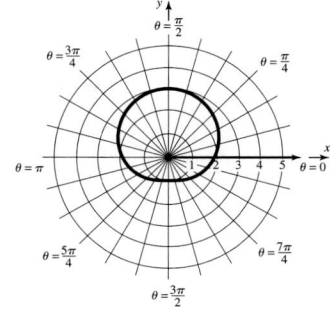

37. Limaçon without inner loop

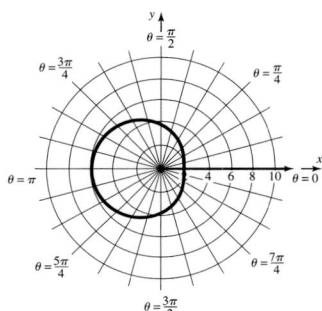

39. Limaçon with inner loop

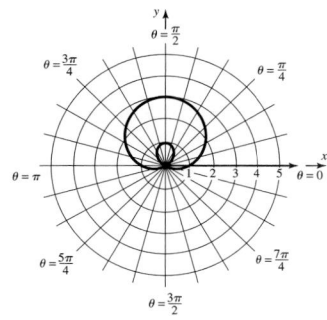

41. Limaçon with inner loop

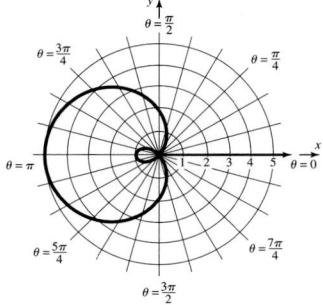

43. Rose

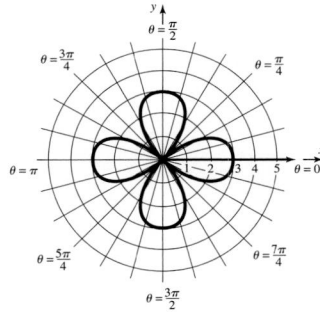

45. Rose

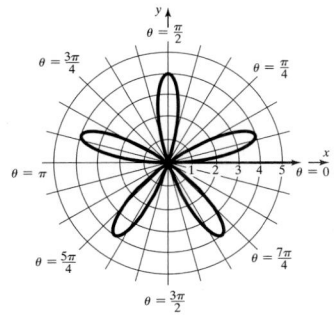

47. Lemniscate

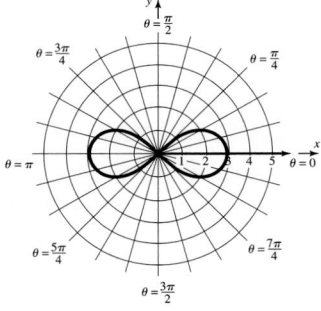

49. Spiral

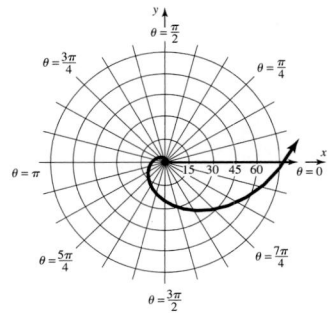

51. Cardioid

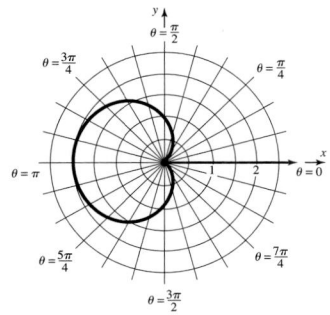

53. Limaçon with inner loop

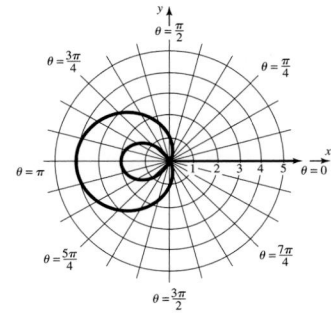

55.

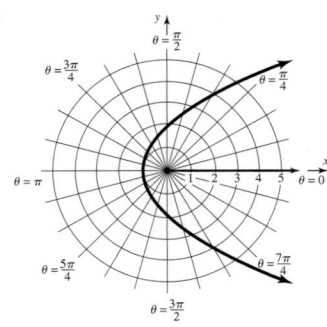

57.

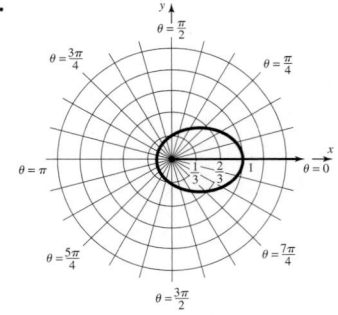

59.

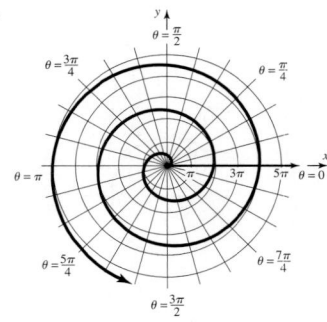

61.

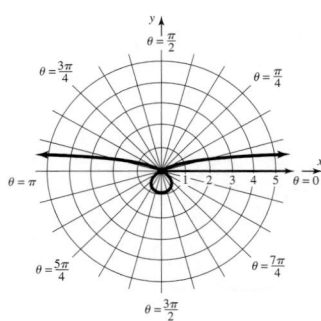

63.

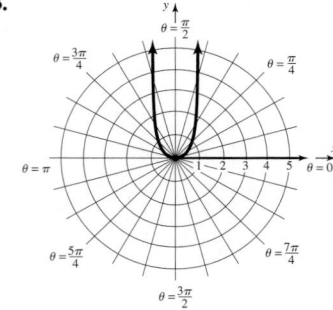

65. $r \sin \theta = a$

$\qquad y = a$

67. $r = 2a \sin \theta$

$\qquad r^2 = 2ar \sin \theta$

$\qquad\qquad x^2 + y^2 = 2ay$

$\qquad x^2 + y^2 - 2ay = 0$

$\qquad x^2 + (y - a)^2 = a^2$

Circle, radius a, center at $(0, a)$
in rectangular coordinates

69. $\qquad r = 2a \cos \theta$

$\qquad\qquad r^2 = 2ar \cos \theta$

$\qquad\qquad x^2 + y^2 = 2ax$

$\qquad x^2 - 2ax + y^2 = 0$

$\qquad (x - a)^2 + y^2 = a^2$

Circle, radius a, center at $(a, 0)$
in rectangular coordinates

71. (a) $r^2 = \cos \theta : r^2 = \cos(\pi - \theta)$

$\qquad r^2 = -\cos \theta$

Not equivalent; test fails.

$\qquad (-r)^2 = \cos(-\theta)$

$\qquad\qquad r^2 = \cos \theta$

New test works.

(b) $r^2 = \sin \theta : r^2 = \sin(\pi - \theta)$

$\qquad r^2 = \sin \theta$

Test works.

$\qquad (-r)^2 = \sin(-\theta)$

$\qquad\qquad r^2 = -\sin \theta$

Not equivalent; new test fails.

Historical Problems (page 352)

1. (a) $1 + 4i, 1 + i$ **(b)** $-1, 2 + i$

5.3 Concepts and Vocabulary (page 352)

1. magnitude; modulus; argument **2.** DeMoivre's **3.** three **4.** T **5.** F **6.** T **8.** $z^2 = r^2[\cos(2\theta) + i \sin(2\theta)]$

9. $\sqrt{r}\left[\cos\left(\dfrac{\theta}{2} + \pi k\right) + i \sin\left(\dfrac{\theta}{2} + \pi k\right)\right]; k = 0, 1$

5.3 Exercises (page 353)

1.

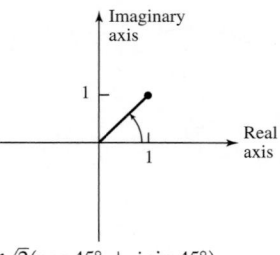

$\sqrt{2}(\cos 45° + i \sin 45°)$

3.

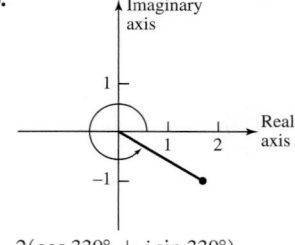

$2(\cos 330° + i \sin 330°)$

5.

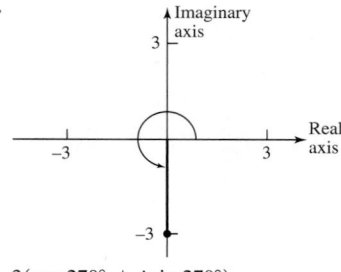

$3(\cos 270° + i \sin 270°)$

7.

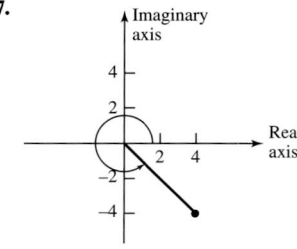

$4\sqrt{2}(\cos 315° + i \sin 315°)$

9.

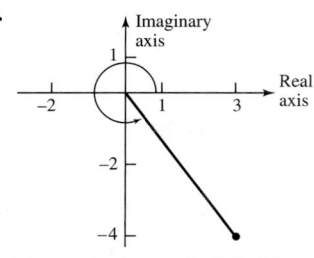

$5(\cos 306.87° + i \sin 306.87°)$

11.

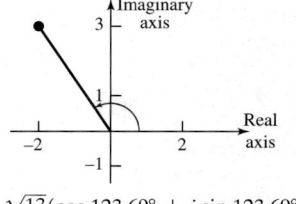

$\sqrt{13}(\cos 123.69° + i \sin 123.69°)$

13. $-1 + \sqrt{3}i$ **15.** $2\sqrt{2} - 2\sqrt{2}i$ **17.** $-3i$ **19.** $-0.03 + 0.2i$ **21.** $1.97 + 0.35i$

23. $zw = 8(\cos 60° + i \sin 60°); \dfrac{z}{w} = \dfrac{1}{2}(\cos 20° + i \sin 20°)$ **25.** $zw = 12(\cos 40° + i \sin 40°); \dfrac{z}{w} = \dfrac{3}{4}(\cos 220° + i \sin 220°)$

27. $zw = 4\left(\cos \dfrac{9\pi}{40} + i \sin \dfrac{9\pi}{40}\right); \dfrac{z}{w} = \cos \dfrac{\pi}{40} + i \sin \dfrac{\pi}{40}$ **29.** $zw = 4\sqrt{2}(\cos 15° + i \sin 15°); \dfrac{z}{w} = \sqrt{2}(\cos 75° + i \sin 75°)$

31. $-32 + 32\sqrt{3}i$ **33.** $32i$ **35.** $\dfrac{27}{2} + \dfrac{27\sqrt{3}}{2}i$ **37.** $-\dfrac{25\sqrt{2}}{2} + \dfrac{25\sqrt{2}}{2}i$ **39.** $-4 + 4i$ **41.** $-23 + 10\sqrt{2}i \approx -23 + 14.14i$

43. $\sqrt[6]{2}(\cos 15° + i \sin 15°), \sqrt[6]{2}(\cos 135° + i \sin 135°), \sqrt[6]{2}(\cos 255° + i \sin 255°)$

45. $\sqrt[4]{8}(\cos 75° + i \sin 75°), \sqrt[4]{8}(\cos 165° + i \sin 165°), \sqrt[4]{8}(\cos 255° + i \sin 255°), \sqrt[4]{8}(\cos 345° + i \sin 345°)$

47. $2(\cos 67.5° + i \sin 67.5°), 2(\cos 157.5° + i \sin 157.5°), 2(\cos 247.5° + i \sin 247.5°), 2(\cos 337.5° + i \sin 337.5°)$

49. $\cos 18° + i \sin 18°, \cos 90° + i \sin 90°, \cos 162° + i \sin 162°, \cos 234° + i \sin 234°, \cos 306° + i \sin 306°$

51. $1, i, -1, -i$

53. Look at formula (8); $|z_k| = \sqrt[n]{r}$ for all k.

55. Look at formula (8). The z_k are spaced apart by an angle of $\dfrac{2\pi}{n}$.

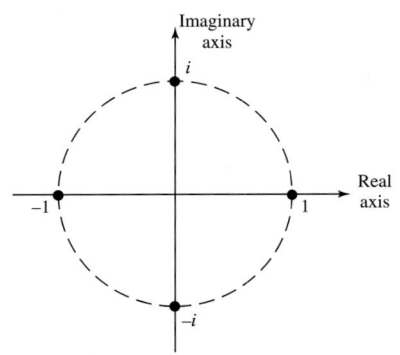

5.4 Concepts and Vocabulary *(page 364)*

1. unit   **2.** scalar   **3.** horizontal; vertical   **4.** T   **5.** T   **6.** F

5.4 Exercises *(page 364)*

1.

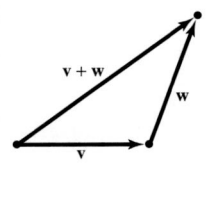

3.

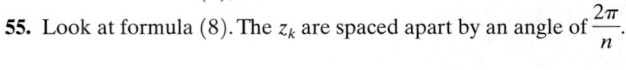

5.

7.

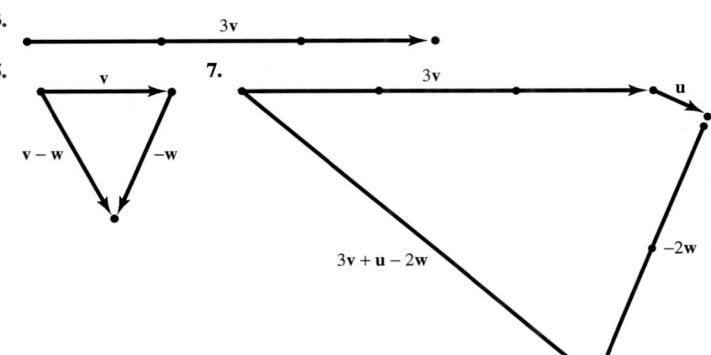

9. T   **11.** F   **13.** F
15. T   **17.** 12
19. $\mathbf{v} = 3\mathbf{i} + 4\mathbf{j}$
21. $\mathbf{v} = 2\mathbf{i} + 4\mathbf{j}$

23. $\mathbf{v} = 8\mathbf{i} - \mathbf{j}$ **25.** $\mathbf{v} = -\mathbf{i} + \mathbf{j}$ **27.** 5   **29.** $\sqrt{2}$ **31.** $\sqrt{13}$ **33.** $-\mathbf{j}$ **35.** $\sqrt{89}$ **37.** $\sqrt{34} - \sqrt{13}$ **39.** $\mathbf{i}$ **41.** $\dfrac{3}{5}\mathbf{i} - \dfrac{4}{5}\mathbf{j}$

43. $\dfrac{\sqrt{2}}{2}\mathbf{i} - \dfrac{\sqrt{2}}{2}\mathbf{j}$ **45.** $\mathbf{v} = \dfrac{8\sqrt{5}}{5}\mathbf{i} + \dfrac{4\sqrt{5}}{5}\mathbf{j}$ or $\mathbf{v} = -\dfrac{8\sqrt{5}}{5}\mathbf{i} - \dfrac{4\sqrt{5}}{5}\mathbf{j}$ **47.** $\{-2 + \sqrt{21}, -2 - \sqrt{21}\}$ **49.** $\mathbf{v} = \dfrac{5}{2}\mathbf{i} + \dfrac{5\sqrt{3}}{2}\mathbf{j}$

51. $\mathbf{v} = -7\mathbf{i} + 7\sqrt{3}\mathbf{j}$ **53.** $\mathbf{v} = \dfrac{25\sqrt{3}}{2}\mathbf{i} - \dfrac{25}{2}\mathbf{j}$ **55.** $\mathbf{F} = 20(\sqrt{3}\mathbf{i} + \mathbf{j})$ **57.** $\mathbf{F} = (20\sqrt{3} + 30\sqrt{2})\mathbf{i} + (20 - 30\sqrt{2})\mathbf{j}$;

magnitude ≈ 80.26 N; direction: $-16.22°$ **59.** Tension in right cable: 1000 lb; Tension in left cable: 845.24 lb
61. Tension in right part: 1088.42 lb; Tension in left part: 1089.07 lb   **63.**

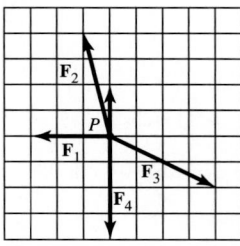

Historical Problems (page 372)

1. $(a\mathbf{i} + b\mathbf{j}) \cdot (c\mathbf{i} + d\mathbf{j}) = ac + bd$

real part $[(\overline{a + bi})(c + di)] = $ real part$[(a - bi)(c + di)] = $ real part$[ac + adi - bci - bdi^2] = ac + bd$

5.5 Concepts and Vocabulary (page 372)

1. zero   **2.** orthogonal   **3.** parallel   **4.** F   **5.** T   **6.** F   **7.** The vectors are perpendicular.

5.5 Exercises (page 372)

1. 0; 90°; orthogonal   **3.** 4; 36.87°; neither   **5.** $\sqrt{3} - 1$; 75°; neither   **7.** 24; 16.26°; neither   **9.** 0; 90°; orthogonal

11. $\dfrac{2}{3}$ **13.** $\mathbf{v}_1 = \dfrac{5}{2}\mathbf{i} - \dfrac{5}{2}\mathbf{j}, \mathbf{v}_2 = -\dfrac{1}{2}\mathbf{i} - \dfrac{1}{2}\mathbf{j}$ **15.** $\mathbf{v}_1 = -\dfrac{1}{5}\mathbf{i} - \dfrac{2}{5}\mathbf{j}, \mathbf{v}_2 = \dfrac{6}{5}\mathbf{i} - \dfrac{3}{5}\mathbf{j}$ **17.** $\mathbf{v}_1 = \dfrac{14}{5}\mathbf{i} + \dfrac{7}{5}\mathbf{j}, \mathbf{v}_2 = \dfrac{1}{5}\mathbf{i} - \dfrac{2}{5}\mathbf{j}$

19. 496.66 mi/hr; 38.46° west of south   **21.** 8.63° off direct heading across the current, upstream; 1.52 min
23. Force required to keep Sienna from rolling down the hill: 737.62 lb; Force perpendicular to the hill: 5248.42 lb.
25. $\mathbf{v} = (250\sqrt{2} - 30)\mathbf{i} + (250\sqrt{2} + 30\sqrt{3})\mathbf{j}$; 518.78 km/hr; N38.59°E   **27.** $\mathbf{v} = 3\mathbf{i} + 20\mathbf{j}$; 20.22 mi/hr;
N8.53°E (Assuming boat traveling north and current traveling east.)   **29.** 3 ft-lb   **31.** $1000\sqrt{3}$ ft-lb ≈ 1732.05 ft-lb
33. Let $\mathbf{u} = a_1\mathbf{i} + b_1\mathbf{j}, \mathbf{v} = a_2\mathbf{i} + b_2\mathbf{j}, \mathbf{w} = a_3\mathbf{i} + b_3\mathbf{j}$. Compute $\mathbf{u} \cdot (\mathbf{v} + \mathbf{w})$ and $\mathbf{u} \cdot \mathbf{v} + \mathbf{u} \cdot \mathbf{w}$.

35. $\cos \alpha = \dfrac{\mathbf{v} \cdot \mathbf{i}}{\|\mathbf{v}\| \, \|\mathbf{i}\|} = \mathbf{v} \cdot \mathbf{i}$; if $\mathbf{v} = x\mathbf{i} + y\mathbf{j}$, then $\mathbf{v} \cdot \mathbf{i} = x = \cos \alpha$ and $\mathbf{v} \cdot \mathbf{j} = y = \cos\left(\dfrac{\pi}{2} - \alpha\right) = \sin \alpha$.

37. $\mathbf{v} = a\mathbf{i} + b\mathbf{j}$; the vector projection of $\mathbf{v}$ onto $\mathbf{i}$ is $\dfrac{\mathbf{v} \cdot \mathbf{i}}{\|\mathbf{i}\|^2}\mathbf{i} = (\mathbf{v} \cdot \mathbf{i})\mathbf{i}; \mathbf{v} \cdot \mathbf{i} = a, \mathbf{v} \cdot \mathbf{j} = b$, so $\mathbf{v} = (\mathbf{v} \cdot \mathbf{i})\mathbf{i} + (\mathbf{v} \cdot \mathbf{j})\mathbf{j}$.

39. $(\mathbf{v} - \alpha\mathbf{w}) \cdot \mathbf{w} = \mathbf{v} \cdot \mathbf{w} - \alpha\mathbf{w} \cdot \mathbf{w} = \alpha\|\mathbf{w}\|^2 - \alpha\|\mathbf{w}\|^2 = 0$ since the dot product of any vector with itself equals the square of its magnitude.
41. $W = \mathbf{F} \cdot \overrightarrow{AB} = 0$ when $\mathbf{F}$ is orthogonal to $\overrightarrow{AB}$.

5.6 Concepts and Vocabulary (page 382)

1. xy-plane   **2.** components   **3.** 1   **4.** T   **5.** F   **6.** T

5.6 Exercises (page 383)

1. All points of the form $(x, 0, z)$ **3.** All points of the form $(x, y, 2)$ **5.** All points of the form $(-4, y, z)$ **7.** All points of the form $(1, 2, z)$
9. $\sqrt{21}$ **11.** $\sqrt{33}$ **13.** $\sqrt{26}$ **15.** $(2, 0, 0); (2, 1, 0); (0, 1, 0); (2, 0, 3); (0, 1, 3); (0, 0, 3)$ **17.** $(1, 4, 3); (3, 2, 3); (3, 4, 3); (3, 2, 5); (1, 4, 5); (1, 2, 5)$
19. $(-1, 2, 2); (4, 0, 2); (4, 2, 2); (-1, 2, 5); (4, 0, 5); (-1, 0, 5)$ **21.** $\mathbf{v} = 3\mathbf{i} + 4\mathbf{j} - \mathbf{k}$ **23.** $\mathbf{v} = 2\mathbf{i} + 4\mathbf{j} + \mathbf{k}$ **25.** $\mathbf{v} = 8\mathbf{i} - \mathbf{j}$ **27.** $2\sqrt{11}$

29. $\sqrt{3}$ **31.** $\sqrt{22}$ **33.** $-\mathbf{j} - 2\mathbf{k}$ **35.** $\sqrt{105}$ **37.** $\sqrt{38} - \sqrt{17}$ **39.** $\mathbf{i}$ **41.** $\dfrac{3}{7}\mathbf{i} - \dfrac{6}{7}\mathbf{j} - \dfrac{2}{7}\mathbf{k}$ **43.** $\dfrac{\sqrt{3}}{3}\mathbf{i} + \dfrac{\sqrt{3}}{3}\mathbf{j} + \dfrac{\sqrt{3}}{3}\mathbf{k}$

45. $\mathbf{v} \cdot \mathbf{w} = 0; \theta = 90°$ **47.** $\mathbf{v} \cdot \mathbf{w} = -2, \theta \approx 100.3°$ **49.** $\mathbf{v} \cdot \mathbf{w} = 0; \theta = 90°$ **51.** $\mathbf{v} \cdot \mathbf{w} = 52; \theta = 0°$ **53.** $\alpha \approx 64.6°; \beta \approx 149°$;
$\gamma \approx 106.6°; \mathbf{v} = 7(\cos 64.6°\mathbf{i} + \cos 149°\mathbf{j} + \cos 106.6°\mathbf{k})$ **55.** $\alpha = \beta = \gamma \approx 54.7°; \mathbf{v} = \sqrt{3}(\cos 54.7°\mathbf{i} + \cos 54.7°\mathbf{j} + \cos 54.7°\mathbf{k})$
57. $\alpha = \beta = 45°; \gamma = 90°; \mathbf{v} = \sqrt{2}(\cos 45°\mathbf{i} + \cos 45°\mathbf{j} + \cos 90°\mathbf{k})$
59. $\alpha \approx 60.9°; \beta \approx 144.2°; \gamma \approx 71.1°; \mathbf{v} = \sqrt{38}(\cos 60.9°\mathbf{i} + \cos 144.2°\mathbf{j} + \cos 71.1°\mathbf{k})$
61. If the point $P = (x, y, z)$ is on the sphere with center $C = (x_0, y_0, z_0)$ and radius r, then the distance between P and C is
$r = \sqrt{(x - x_0)^2 + (y - y_0)^2 + (z - z_0)^2}$. Therefore, the equation for a sphere is $(x - x_0)^2 + (y - y_0)^2 + (z - z_0)^2 = r^2$.
63. $(x - 1)^2 + (y - 2)^2 + (z - 2)^2 = 4$ **65.** radius $= 2$, center $(-1, 1, 0)$ **67.** radius $= 3$, center $(2, -2, -1)$
69. radius $= \dfrac{3\sqrt{2}}{2}$, center $(2, 0, -1)$ **71.** 2 joules   **73.** 9

5.7 Concepts and Vocabulary *(page 389)*

1. T   **2.** T   **3.** T   **4.** F   **5.** F   **6.** T

5.7 Exercises *(page 389)*

1. 2   **3.** 4   **5.** $-11A + 2B + 5C$ **7.** $6A + 27B - 15C$ **9.** (a) $5\mathbf{i} + 5\mathbf{j} + 5\mathbf{k}$ (b) $-5\mathbf{i} - 5\mathbf{j} - 5\mathbf{k}$ (c) 0   (d) 0   **11.** (a) $\mathbf{i} - \mathbf{j} - \mathbf{k}$
(b) $-\mathbf{i} + \mathbf{j} + \mathbf{k}$ (c) 0   (d) 0   **13.** (a) $-\mathbf{i} + 2\mathbf{j} + 2\mathbf{k}$ (b) $\mathbf{i} - 2\mathbf{j} - 2\mathbf{k}$ (c) 0   (d) 0   **15.** (a) $3\mathbf{i} - \mathbf{j} + 4\mathbf{k}$ (b) $-3\mathbf{i} + \mathbf{j} - 4\mathbf{k}$
(c) 0   (d) 0   **17.** $-9\mathbf{i} - 7\mathbf{j} - 3\mathbf{k}$ **19.** $9\mathbf{i} + 7\mathbf{j} + 3\mathbf{k}$ **21.** 0   **23.** $-27\mathbf{i} - 21\mathbf{j} - 9\mathbf{k}$ **25.** $-18\mathbf{i} - 14\mathbf{j} - 6\mathbf{k}$ **27.** 0   **29.** -25 **31.** 25
33. 0   **35.** Any vector of the form $c(-9\mathbf{i} - 7\mathbf{j} - 3\mathbf{k})$, where c is a scalar   **37.** Any vector of the form $c(-\mathbf{i} + \mathbf{j} + 5\mathbf{k})$, where c is a

nonzero scalar   **39.** $\sqrt{166}$ **41.** $\sqrt{555}$ **43.** $\sqrt{34}$ **45.** $\sqrt{998}$ **47.** $\dfrac{11\sqrt{19}}{57}\mathbf{i} + \dfrac{\sqrt{19}}{57}\mathbf{j} + \dfrac{7\sqrt{19}}{57}\mathbf{k}$ or $-\dfrac{11\sqrt{19}}{57}\mathbf{i} - \dfrac{\sqrt{19}}{57}\mathbf{j} - \dfrac{7\sqrt{19}}{57}\mathbf{k}$

49. $\mathbf{u} \times \mathbf{v} = \begin{vmatrix} \mathbf{i} & \mathbf{j} & \mathbf{k} \\ a_1 & b_1 & c_1 \\ a_2 & b_2 & c_2 \end{vmatrix} = (b_1c_2 - b_2c_1)\mathbf{i} - (a_1c_2 - a_2c_1)\mathbf{j} + (a_1b_2 - a_2b_1)\mathbf{k} = -[(b_2c_1 - b_1c_2)\mathbf{i} - (a_2c_1 - a_1c_2)\mathbf{j} + (a_2b_1 - a_1b_2)\mathbf{k}]$

$\qquad = -\begin{vmatrix} \mathbf{i} & \mathbf{j} & \mathbf{k} \\ a_2 & b_2 & c_2 \\ a_1 & b_1 & c_1 \end{vmatrix} = -(\mathbf{v} \times \mathbf{u})$

51. $\mathbf{u} \times \mathbf{v} = \begin{vmatrix} \mathbf{i} & \mathbf{j} & \mathbf{k} \\ a_1 & b_1 & c_1 \\ a_2 & b_2 & c_2 \end{vmatrix} = (b_1c_2 - b_2c_1)\mathbf{i} - (a_1c_2 - a_2c_1)\mathbf{j} + (a_1b_2 - a_2b_1)\mathbf{k}$

$\quad \|\mathbf{u} \times \mathbf{v}\|^2 = (\sqrt{(b_1c_2 - b_2c_1)^2 + (a_1c_2 - a_2c_1)^2 + (a_1b_2 - a_2b_1)^2})^2$
$\qquad\qquad = b_1^2c_2^2 - 2b_1b_2c_1c_2 + b_2^2c_1^2 + a_1^2c_2^2 - 2a_1a_2c_1c_2 + a_2^2c_1^2 + a_1^2b_2^2 - 2a_1a_2b_1b_2 + a_2^2b_1^2$
$\quad \|\mathbf{u}\|^2 = a_1^2 + b_1^2 + c_1^2, \|\mathbf{v}\|^2 = a_2^2 + b_2^2 + c_2^2$
$\quad \|\mathbf{u}\|^2\|\mathbf{v}\|^2 = (a_1^2 + b_1^2 + c_1^2)(a_2^2 + b_2^2 + c_2^2) = a_1^2a_2^2 + a_1^2b_2^2 + a_1^2c_2^2 + b_1^2a_2^2 + b_1^2b_2^2 + b_1^2c_2^2 + c_1^2a_2^2 + c_1^2b_2^2 + c_1^2c_2^2$
$\quad (\mathbf{u} \cdot \mathbf{v})^2 = (a_1a_2 + b_1b_2 + c_1c_2)^2 = (a_1a_2 + b_1b_2 + c_1c_2)(a_1a_2 + b_1b_2 + c_1c_2)$
$\qquad\qquad = a_1^2a_2^2 + a_1a_2b_1b_2 + a_1a_2c_1c_2 + b_1b_2a_1a_2 + b_1^2b_2^2 + b_1b_2c_1c_2 + c_1c_2a_1a_2 + c_1c_2b_1b_2 + c_1^2c_2^2$
$\qquad\qquad = a_1^2a_2^2 + b_1^2b_2^2 + c_1^2c_2^2 + 2a_1a_2b_1b_2 + 2b_1b_2c_1c_2 + 2a_1a_2c_1c_2$
$\quad \|\mathbf{u}\|^2\|\mathbf{v}\|^2 - (\mathbf{u} \cdot \mathbf{v})^2 = a_1^2b_2^2 + a_1^2c_2^2 + b_1^2a_2^2 + b_1^2c_2^2 + c_1^2a_2^2 + c_1^2b_2^2 - 2a_1a_2b_1b_2 - 2b_1b_2c_1c_2 - 2a_1a_2c_1c_2$, which equals $\|\mathbf{u} \times \mathbf{v}\|^2$.

53. We know for any two vectors that $\|\mathbf{u} \times \mathbf{v}\| = \|\mathbf{u}\|\|\mathbf{v}\| \sin\theta$, where θ is the angle between $\mathbf{u}$ and $\mathbf{v}$, so that if $\mathbf{u}$ and $\mathbf{v}$ are orthogonal, then
$\theta = 90°$, and so the result follows.

Review Exercises *(page 391)*

1. $\left(\dfrac{3\sqrt{3}}{2}, \dfrac{3}{2}\right)$ **3.** $(1, \sqrt{3})$ **5.** $(0, 3)$ **7.** $\left(3\sqrt{2}, \dfrac{3\pi}{4}\right), \left(-3\sqrt{2}, -\dfrac{\pi}{4}\right)$ **9.** $\left(2, -\dfrac{\pi}{2}\right), \left(-2, \dfrac{\pi}{2}\right)$

11. $(5, 0.93), (-5, 4.07)$

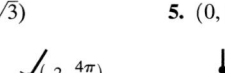

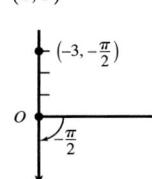

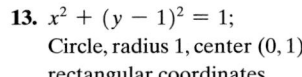

13. $x^2 + (y - 1)^2 = 1$;
Circle, radius 1, center $(0, 1)$ in
rectangular coordinates

15. $x^2 + y^2 = 25$;
Circle, radius 5, center at pole

17. $x + 3y = 6$;
Line through $(6, 0)$ and $(0, 2)$ in
rectangular coordinates

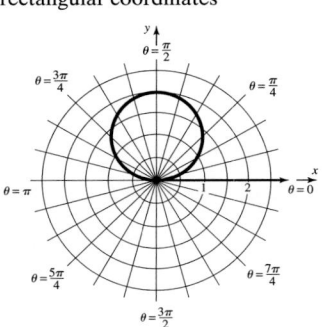

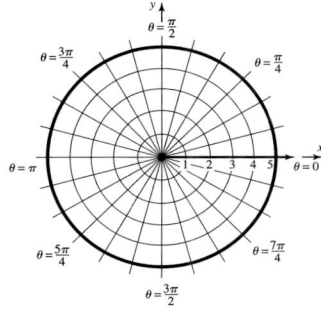

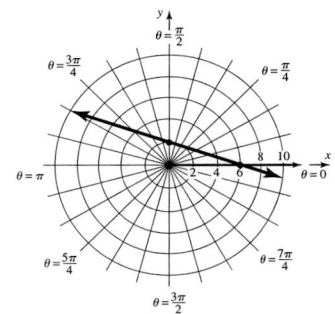

19.

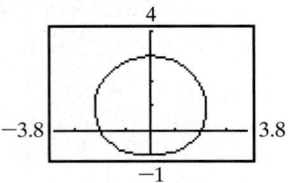

21.

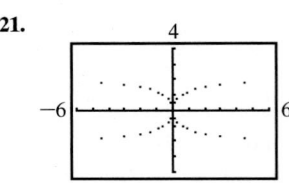

23. Circle; radius 2, center at $(2, 0)$ in rectangular coordinates; symmetric with respect to the polar axis

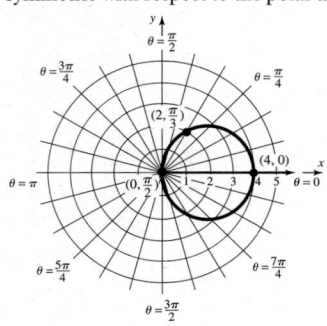

25. Cardioid; symmetric with respect to the line $\theta = \dfrac{\pi}{2}$

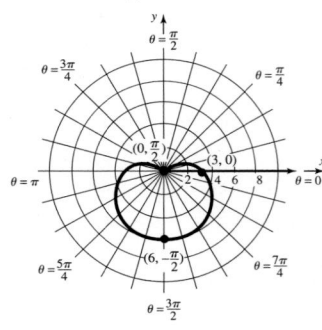

27. Limaçon without inner loop; symmetric with respect to the polar axis

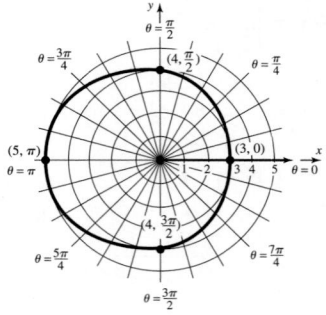

29. $\sqrt{2}(\cos 225° + i \sin 225°)$ **31.** $5(\cos 323.1° + i \sin 323.1°)$

33. $-\sqrt{3} + i$

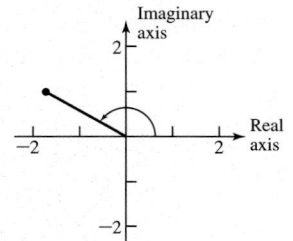

35. $-\dfrac{3}{2} + \dfrac{3\sqrt{3}}{2}i$

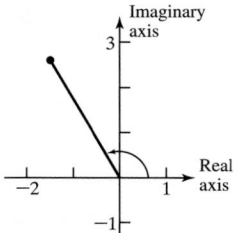

37. $0.10 - 0.02i$

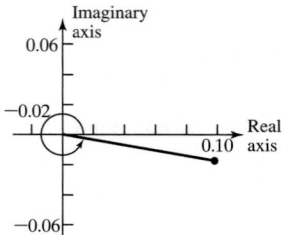

39. $zw = \cos 130° + i \sin 130°$; $\dfrac{z}{w} = \cos 30° + i \sin 30°$ **41.** $zw = 6(\cos 0 + i \sin 0) = 6$; $\dfrac{z}{w} = \dfrac{3}{2}\left(\cos \dfrac{8\pi}{5} + i \sin \dfrac{8\pi}{5}\right)$

43. $zw = 5(\cos 5° + i \sin 5°)$; $\dfrac{z}{w} = 5(\cos 15° + i \sin 15°)$ **45.** $\dfrac{27}{2} + \dfrac{27\sqrt{3}}{2}i$ **47.** $4i$ **49.** 64 **51.** $-527 - 336i$

53. $3, 3(\cos 120° + i \sin 120°), 3(\cos 240° + i \sin 240°)$

55.

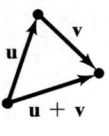

57.

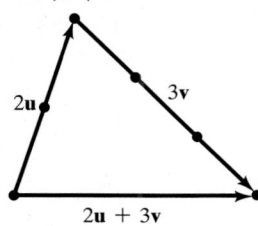

59. $\mathbf{v} = 2\mathbf{i} - 4\mathbf{j}$; $\|\mathbf{v}\| = 2\sqrt{5}$ **61.** $\mathbf{v} = -\mathbf{i} + 3\mathbf{j}$; $\|\mathbf{v}\| = \sqrt{10}$ **63.** $2\mathbf{i} - 2\mathbf{j}$

65. $-20\mathbf{i} + 13\mathbf{j}$ **67.** $\sqrt{5}$ **69.** $\sqrt{5} + 5 \approx 7.24$ **71.** $-\dfrac{2\sqrt{5}}{5}\mathbf{i} + \dfrac{\sqrt{5}}{5}\mathbf{j}$

73. $\dfrac{3}{2} + \dfrac{3\sqrt{3}}{2}\mathbf{i}$ **75.** $\sqrt{43} \approx 6.56$ **77.** $\mathbf{v} = 3\mathbf{i} - 5\mathbf{j} + 3\mathbf{k}$ **79.** $21\mathbf{i} - 2\mathbf{j} - 5\mathbf{k}$

81. $\sqrt{38}$ **83.** 0 **85.** $3\mathbf{i} + 9\mathbf{j} + 9\mathbf{k}$

87. $\dfrac{3\sqrt{14}}{14}\mathbf{i} + \dfrac{\sqrt{14}}{14}\mathbf{j} - \dfrac{\sqrt{14}}{7}\mathbf{k}$; $-\dfrac{3\sqrt{14}}{14}\mathbf{i} - \dfrac{\sqrt{14}}{14}\mathbf{j} + \dfrac{\sqrt{14}}{7}\mathbf{k}$

89. $\mathbf{v} \cdot \mathbf{w} = -11$; $\theta \approx 169.7°$ **91.** $\mathbf{v} \cdot \mathbf{w} = -4$; $\theta \approx 153.4°$ **93.** $\mathbf{v} \cdot \mathbf{w} = 1$; $\theta \approx 70.5°$ **95.** $\mathbf{v} \cdot \mathbf{w} = 0$; $\theta = 90°$ **97.** Parallel **99.** Parallel

101. Orthogonal **103.** $\mathbf{v}_1 = \dfrac{4}{5}\mathbf{i} - \dfrac{3}{5}\mathbf{j}$; $\mathbf{v}_2 = \dfrac{6}{5}\mathbf{i} + \dfrac{8}{5}\mathbf{j}$ **105.** $\mathbf{v}_1 = \dfrac{9}{10}(3\mathbf{i} + \mathbf{j})$ **107.** $\alpha \approx 56.1°$; $\beta \approx 138°$; $\gamma \approx 68.2°$ **109.** $2\sqrt{83}$ sq units

111. $-2\mathbf{i} + 3\mathbf{j} - \mathbf{k}$ **113.** $\sqrt{29} \approx 5.39$ mi/hr; 0.4 mi **115.** Left cable: 1843.21 lb; right cable: 1630.41 lb **117.** 50 foot-pounds

Cumulative Review Exercises *(page 395)*

1. ∅   **2.** $y = \dfrac{\sqrt{3}}{3}x$

3. $x^2 + (y - 1)^2 = 9$ **4.** $\left\{x \,\middle|\, x \neq \dfrac{\pi}{4} + \text{multiples of } \pi\right\}$ **5.** Symmetry with respect to the *y*-axis

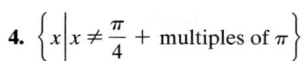

6. $\gamma = 70°, b \approx 7.88, c \approx 7.52$; area ≈ 14.81 **7.**

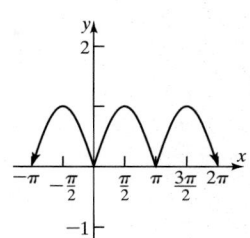

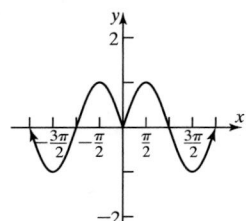

8.   **9.** $-\dfrac{\pi}{6}$ **10.**   **11.**

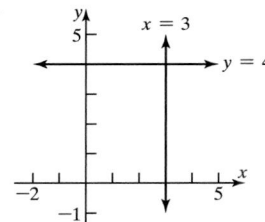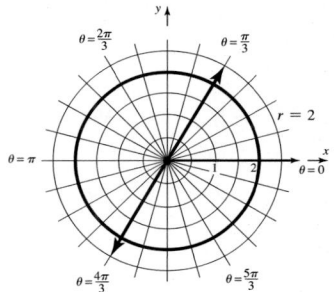

CHAPTER 6   Analytic Geometry

6.2 Concepts and Vocabulary *(page 407)*

1. parabola   **2.** paraboloid   **3.** latus rectum   **4.** T   **5.** F   **6.** T

6.2 Exercises *(page 408)*

1. B   **3.** E   **5.** H   **7.** C   **9.** E   **11.** D   **13.** C

15. $y^2 = 16x$ **17.** $x^2 = -12y$ **19.** $y^2 = -8x$ **21.** $x^2 = 2y$

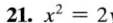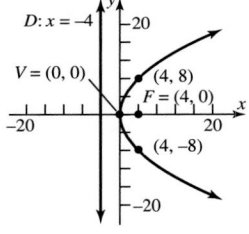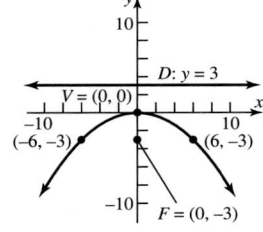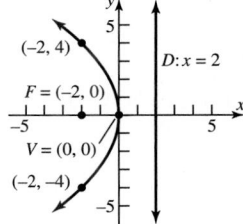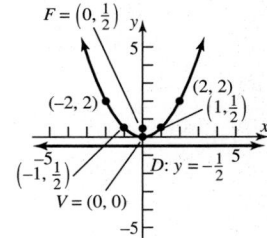

23. $(x - 2)^2 = -8(y + 3)$ **25.** $x^2 = \dfrac{4}{3}y$ **27.** $(x + 3)^2 = 4(y - 3)$ **29.** $(y + 2)^2 = -8(x + 1)$

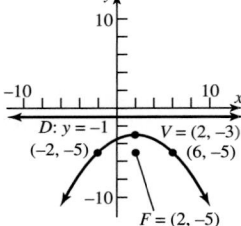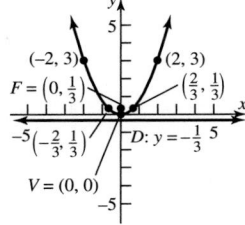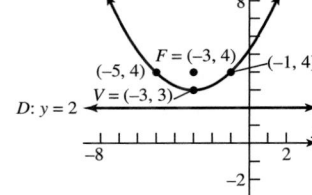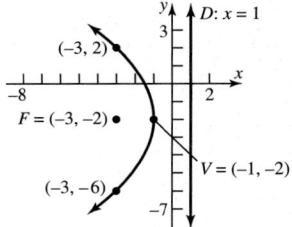

31. Vertex: $(0, 0)$; Focus: $(0, 1)$;
Directrix: $y = -1$

(a)

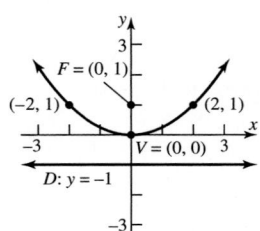

(b)

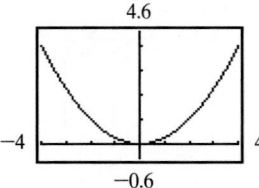

33. Vertex: $(0, 0)$; Focus: $(-4, 0)$;
Directrix: $x = 4$

(a)

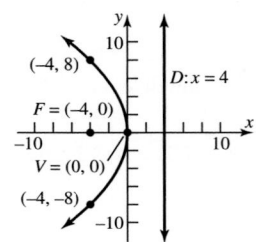

(b)

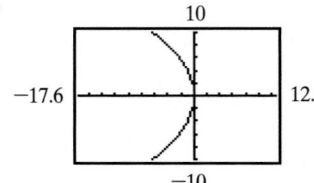

35. Vertex: $(-1, 2)$; Focus: $(1, 2)$;
Directrix: $x = -3$

(a)

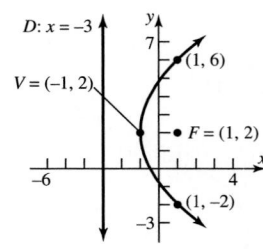

(b)

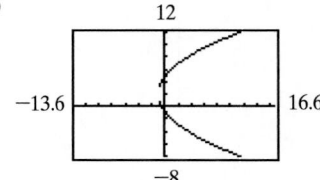

37. Vertex: $(3, -1)$; Focus: $\left(3, -\dfrac{5}{4}\right)$;

Directrix: $y = -\dfrac{3}{4}$

(a)

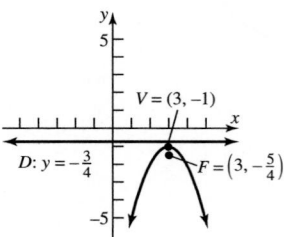
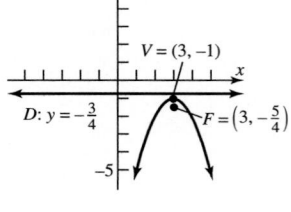

(b)

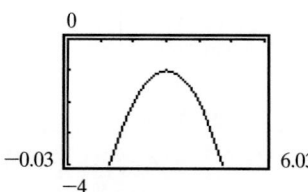

39. Vertex: $(2, -3)$; Focus: $(4, -3)$;

Directrix: $x = 0$

(a)

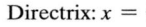
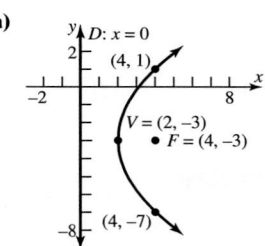

(b)

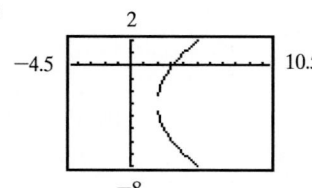

41. Vertex: $(0, 2)$; Focus: $(-1, 2)$;

Directrix: $x = 1$

(a)

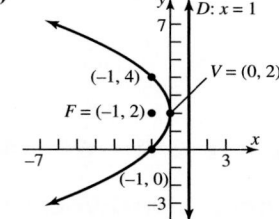

(b)

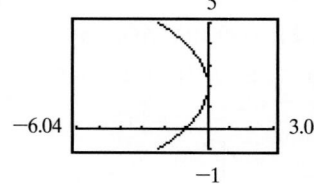

43. Vertex: $(-4, -2)$; Focus: $(-4, -1)$; Directrix: $y = -3$

(a)

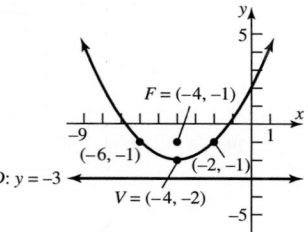

(b)

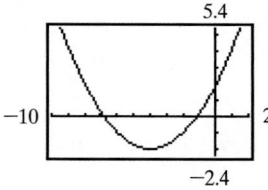

45. Vertex: $(-1, -1)$; Focus: $\left(-\dfrac{3}{4}, -1\right)$; Directrix: $x = -\dfrac{5}{4}$

(a) 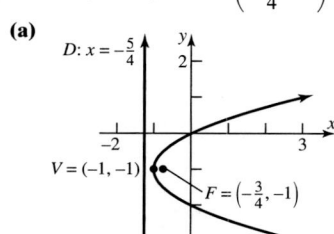   **(b)**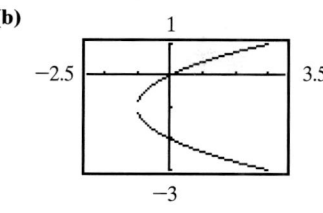

47. Vertex: $(2, -8)$; Focus: $\left(2, -\dfrac{31}{4}\right)$; Directrix: $y = -\dfrac{33}{4}$

(a) 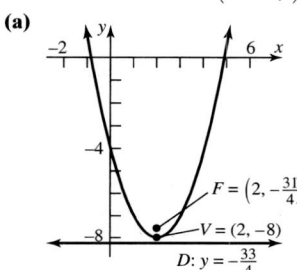   **(b)**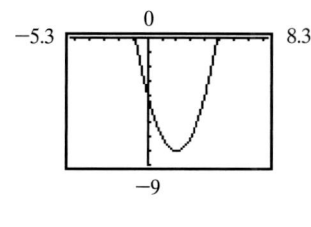

49. $(y - 1)^2 = x$ **51.** $(y - 1)^2 = -(x - 2)$ **53.** $x^2 = 4(y - 1)$ **55.** $y^2 = \dfrac{1}{2}(x + 2)$

57. 1.5625 ft from the base of the dish, along the axis of symmetry   **59.** 1 in. from the vertex   **61.** 20 ft   **63.** 0.78125 ft
65. 4.17 ft from the base along the axis of symmetry   **67.** 24.31 ft, 18.75 ft, 7.64 ft

69. $Ax^2 + Ey = 0, A \neq 0, E \neq 0$ This is the equation of a parabola with vertex at $(0, 0)$ and axis of symmetry the y-axis.

$$Ax^2 = -Ey$$ The focus is $\left(0, -\dfrac{E}{4A}\right)$; the directrix is the line $y = \dfrac{E}{4A}$. The parabola opens up if $-\dfrac{E}{A} > 0$

$$x^2 = -\dfrac{E}{A}y$$ and down if $-\dfrac{E}{A} < 0$.

71. $Ax^2 + Dx + Ey + F = 0, A \neq 0$

$$Ax^2 + Dx = -Ey - F$$

$$x^2 + \dfrac{D}{A}x = -\dfrac{E}{A}y - \dfrac{F}{A}$$

$$\left(x + \dfrac{D}{2A}\right)^2 = -\dfrac{E}{A}y - \dfrac{F}{A} + \dfrac{D^2}{4A^2}$$

$$\left(x + \dfrac{D}{2A}\right)^2 = -\dfrac{E}{A}y + \dfrac{D^2 - 4AF}{4A^2}$$

(a) If $E \neq 0$, then the equation may be written as

$$\left(x + \dfrac{D}{2A}\right)^2 = -\dfrac{E}{A}\left(y - \dfrac{D^2 - 4AF}{4AE}\right)$$

This is the equation of a parabola with vertex at

$\left(-\dfrac{D}{2A}, \dfrac{D^2 - 4AF}{4AE}\right)$ and axis of symmetry parallel to the y-axis.

(b)–(d) If $E = 0$, the graph of the equation contains no points if

$D^2 - 4AF < 0$, is a single vertical line if $D^2 - 4AF = 0$, and is
two vertical lines if $D^2 - 4AF > 0$.

6.3 Concepts and Vocabulary *(page 420)*

1. ellipse   **2.** major   **3.** $(0, -5)$; $(0, 5)$ **4.** F   **5.** T   **6.** T

6.3 Exercises *(page 420)*

1. C   **3.** B   **5.** C   **7.** D

9. Vertices: $(-5, 0)$, $(5, 0)$

Foci: $(-\sqrt{21}, 0)$, $(\sqrt{21}, 0)$

(a)

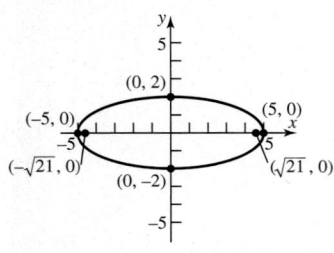

(b)

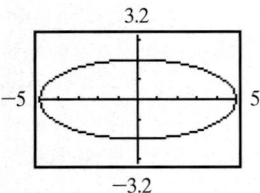

11. Vertices: $(0, -5)$, $(0, 5)$

Foci: $(0, -4)$, $(0, 4)$

(a)

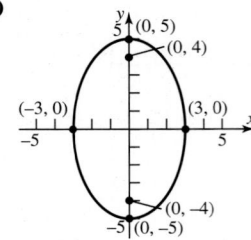

(b)

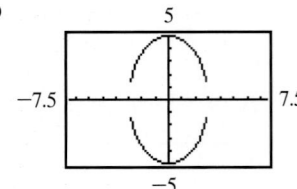

13. $\dfrac{x^2}{4} + \dfrac{y^2}{16} = 1$

Vertices: $(0, -4)$, $(0, 4)$

Foci: $(0, -2\sqrt{3})$, $(0, 2\sqrt{3})$

(a)

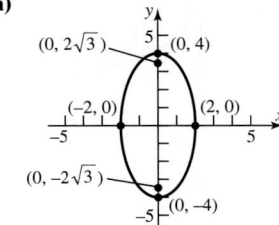

(b)

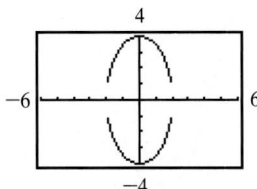

15. $\dfrac{x^2}{8} + \dfrac{y^2}{2} = 1$

Vertices: $(-2\sqrt{2}, 0)$, $(2\sqrt{2}, 0)$; Foci: $(-\sqrt{6}, 0)$, $(\sqrt{6}, 0)$

(a)

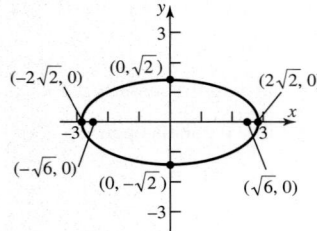

(b)

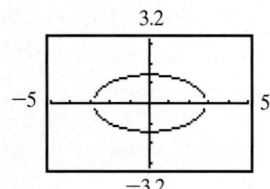

17. $\dfrac{x^2}{16} + \dfrac{y^2}{16} = 1$

Vertices: $(-4, 0)$, $(4, 0)$, $(0, -4)$, $(0, 4)$; Focus: $(0, 0)$

(a)

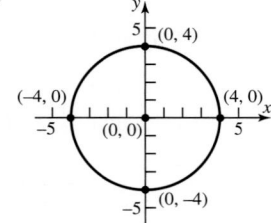

(b)

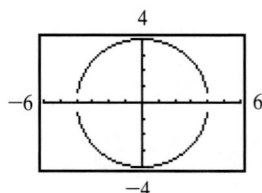

19. $\dfrac{x^2}{25} + \dfrac{y^2}{16} = 1$

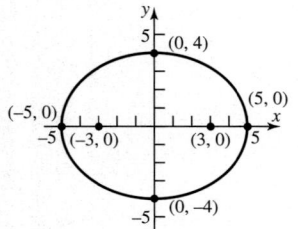

21. $\dfrac{x^2}{9} + \dfrac{y^2}{25} = 1$

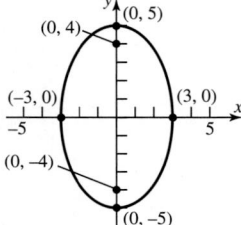

23. $\dfrac{x^2}{9} + \dfrac{y^2}{5} = 1$

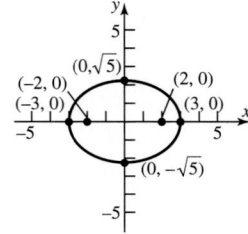

25. $\dfrac{x^2}{4} + \dfrac{y^2}{13} = 1$

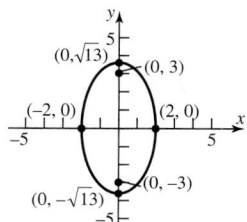

27. $x^2 + \dfrac{y^2}{16} = 1$

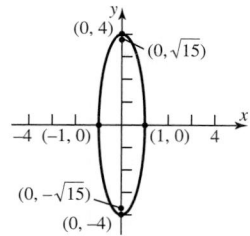

29. $\dfrac{(x+1)^2}{4} + (y-1)^2 = 1$

31. $(x-1)^2 + \dfrac{y^2}{4} = 1$

33. Center: $(3, -1)$; Vertices: $(3, -4)$, $(3, 2)$

Foci: $(3, -1 - \sqrt{5})$, $(3, -1 + \sqrt{5})$

(a)

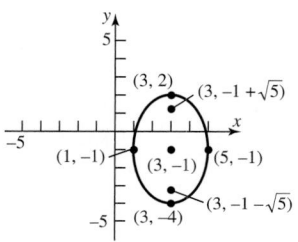

(b)

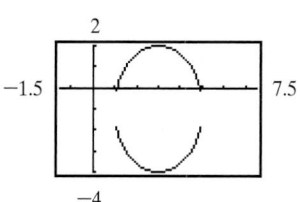

35. $\dfrac{(x+5)^2}{16} + \dfrac{(y-4)^2}{4} = 1$

Center: $(-5, 4)$; Vertices: $(-9, 4)$, $(-1, 4)$

Foci: $(-5 - 2\sqrt{3}, 4)$, $(-5 + 2\sqrt{3}, 4)$

(a)

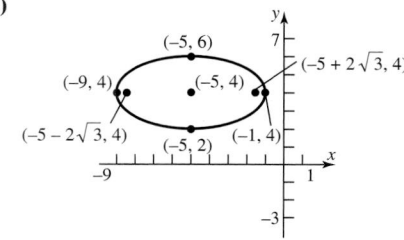

(b)

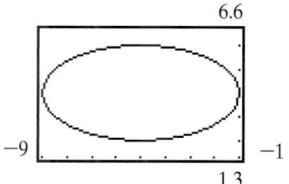

37. $\dfrac{(x+2)^2}{4} + (y-1)^2 = 1$

Center: $(-2, 1)$; Vertices: $(-4, 1)$, $(0, 1)$

Foci: $(-2 - \sqrt{3}, 1)$, $(-2 + \sqrt{3}, 1)$

(a)

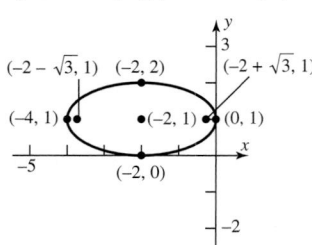

(b)

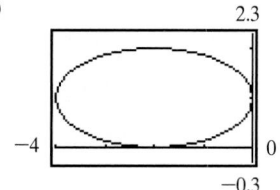

39. $\dfrac{(x-2)^2}{3} + \dfrac{(y+1)^2}{2} = 1$

Center: $(2, -1)$; Vertices: $(2 - \sqrt{3}, -1)$,

$(2 + \sqrt{3}, -1)$; Foci: $(1, -1)$, $(3, -1)$

(a)

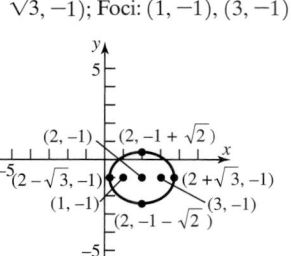

(b)

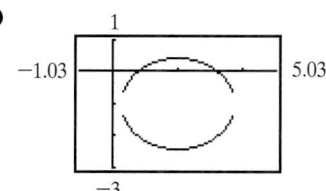

41. $\dfrac{(x-1)^2}{4} + \dfrac{(y+2)^2}{9} = 1$

Center: $(1, -2)$; Vertices: $(1, -5)$, $(1, 1)$

Foci: $(1, -2 - \sqrt{5})$, $(1, -2 + \sqrt{5})$

(a)

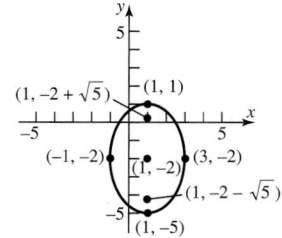

(b)

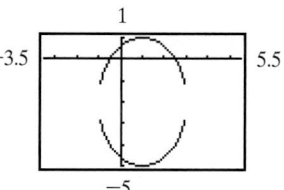

43. $x^2 + \dfrac{(y + 2)^2}{4} = 1$

Center: $(0, -2)$; Vertices: $(0, -4)$, $(0, 0)$

Foci: $(0, -2 - \sqrt{3})$, $(0, -2 + \sqrt{3})$

(a)

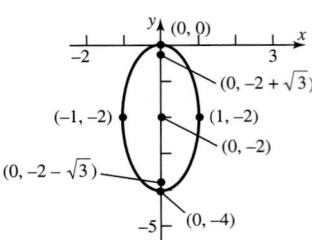

(b)

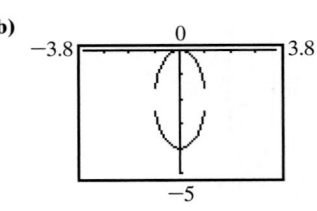

45. $\dfrac{(x - 2)^2}{25} + \dfrac{(y + 2)^2}{21} = 1$

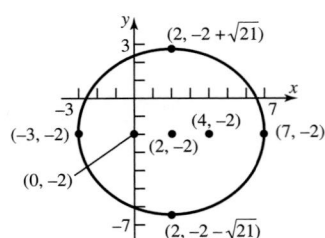

47. $\dfrac{(x - 4)^2}{5} + \dfrac{(y - 6)^2}{9} = 1$

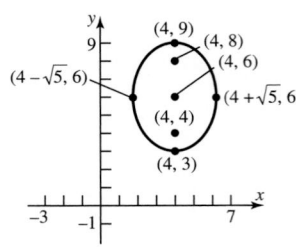

49. $\dfrac{(x - 2)^2}{16} + \dfrac{(y - 1)^2}{7} = 1$

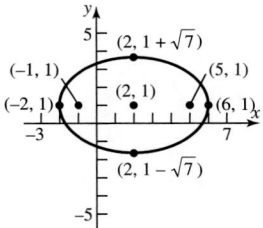

51. $\dfrac{(x - 1)^2}{10} + (y - 2)^2 = 1$

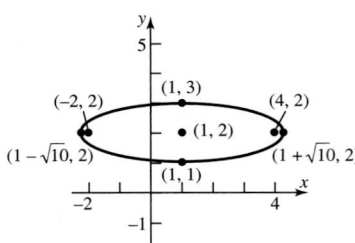

53. $\dfrac{(x - 1)^2}{9} + (y - 2)^2 = 1$

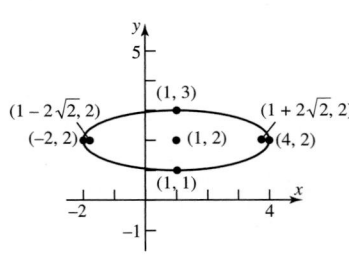

55.

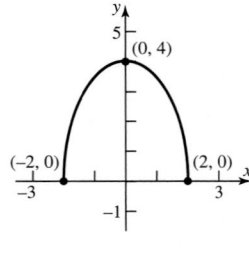

57.

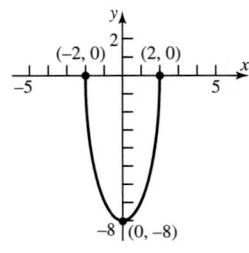

59. $\dfrac{x^2}{100} + \dfrac{y^2}{36} = 1$ **61.** 43.3 ft **63.** 24.65 ft, 21.65 ft, 13.82 ft

65. 0 ft, 12.99 ft, 15 ft, 12.99 ft, 0 ft **67.** 91.5 million mi; $\dfrac{x^2}{(93)^2} + \dfrac{y^2}{8646.75} = 1$

69. perihelion: 460.6 million mi; mean distance: 483.8 million mi; $\dfrac{x^2}{483.8^2} + \dfrac{y^2}{233{,}524.2} = 1$

71. 30 ft

73. (a) $Ax^2 + Cy^2 + F = 0$ If A and C are of the same sign and F is of opposite sign, then the equation takes the form

$$Ax^2 + Cy^2 = -F \qquad \dfrac{x^2}{\left(-\dfrac{F}{A}\right)} + \dfrac{y^2}{\left(-\dfrac{F}{C}\right)} = 1, \text{where } -\dfrac{F}{A} \text{ and } -\dfrac{F}{C} \text{ are positive. This is the equation of an ellipse}$$

with center at $(0, 0)$.

(b) If $A = C$, the equation may be written as $x^2 + y^2 = -\dfrac{F}{A}$.

This is the equation of a circle with center at $(0, 0)$ and radius equal to $\sqrt{-\dfrac{F}{A}}$.

6.4 Concepts and Vocabulary *(page 435)*

1. hyperbola   **2.** transverse axis   **3.** $y = \frac{3}{2}x; y = -\frac{3}{2}x$ **4.** F   **5.** T   **6.** F

6.4 Exercises *(page 435)*

1. B   **3.** A   **5.** B   **7.** C

9. $x^2 - \frac{y^2}{8} = 1$

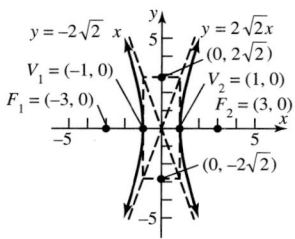

11. $\frac{y^2}{16} - \frac{x^2}{20} = 1$

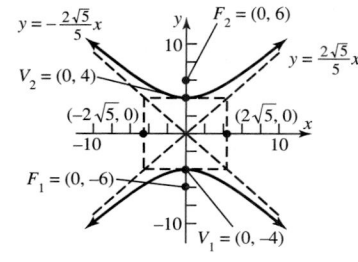

13. $\frac{x^2}{9} - \frac{y^2}{16} = 1$

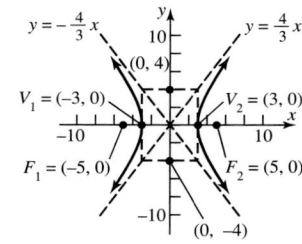

15. $\frac{y^2}{36} - \frac{x^2}{9} = 1$

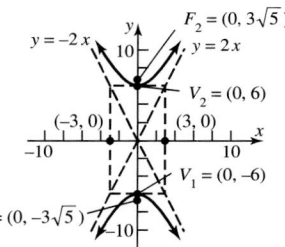

17. $\frac{x^2}{8} - \frac{y^2}{8} = 1$

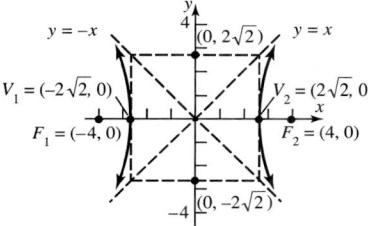

19. $\frac{x^2}{25} - \frac{y^2}{9} = 1$

Center: $(0, 0)$
Transverse axis: x-axis
Vertices: $(-5, 0)$, $(5, 0)$
Foci: $(-\sqrt{34}, 0)$, $(\sqrt{34}, 0)$
Asymptotes: $y = \pm\frac{3}{5}x$

(a)

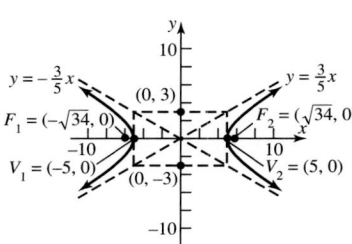

(b)

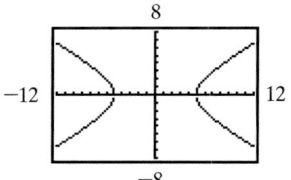

21. $\frac{x^2}{4} - \frac{y^2}{16} = 1$

Center: $(0, 0)$
Transverse axis: x-axis
Vertices: $(-2, 0)$, $(2, 0)$
Foci: $(-2\sqrt{5}, 0)$, $(2\sqrt{5}, 0)$
Asymptotes: $y = \pm 2x$

(a)

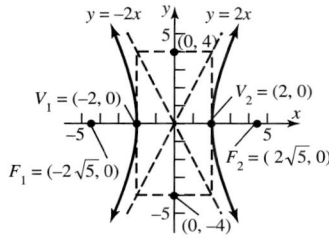

(b)

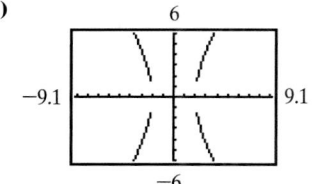

23. $\frac{y^2}{9} - x^2 = 1$

Center: $(0, 0)$
Transverse axis: y-axis
Vertices: $(0, -3)$, $(0, 3)$
Foci: $(0, -\sqrt{10})$, $(0, \sqrt{10})$
Asymptotes: $y = \pm 3x$

(a)

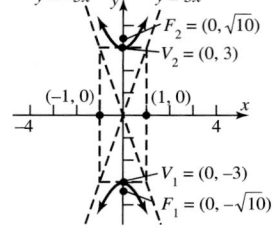

(b)

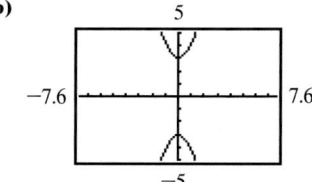

25. $\dfrac{y^2}{25} - \dfrac{x^2}{25} = 1$; Center: $(0, 0)$; Transverse axis: y-axis; Vertices: $(0, -5)$, $(0, 5)$;

Foci: $(0, -5\sqrt{2})$, $(0, 5\sqrt{2})$; Asymptotes: $y = \pm x$

(a) 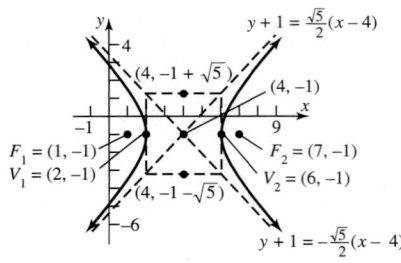   **(b)**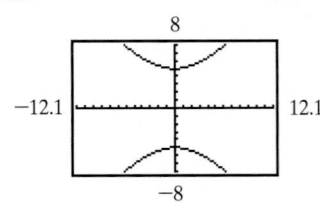

27. $x^2 - y^2 = 1$

29. $\dfrac{y^2}{36} - \dfrac{x^2}{9} = 1$

31. $\dfrac{(x-4)^2}{4} - \dfrac{(y+1)^2}{5} = 1$

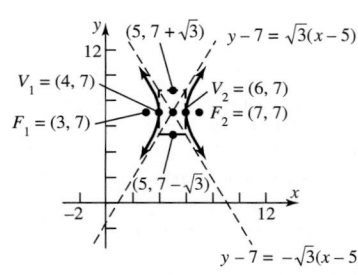

33. $\dfrac{(y+4)^2}{4} - \dfrac{(x+3)^2}{12} = 1$

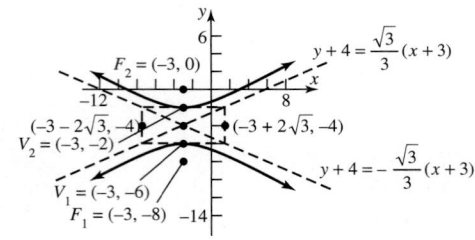

35. $(x-5)^2 - \dfrac{(y-7)^2}{3} = 1$

37. $\dfrac{(x-1)^2}{4} - \dfrac{(y+1)^2}{9} = 1$

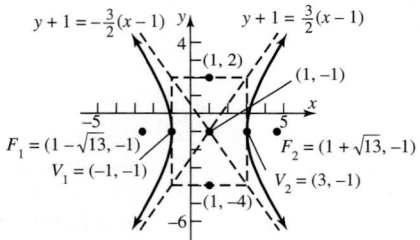

39. $\dfrac{(x-2)^2}{4} - \dfrac{(y+3)^2}{9} = 1$

Center: $(2, -3)$

Transverse axis: Parallel to x-axis

Vertices: $(0, -3)$, $(4, -3)$

Foci: $(2 - \sqrt{13}, -3)$, $(2 + \sqrt{13}, -3)$

Asymptotes: $y + 3 = \pm\dfrac{3}{2}(x - 2)$

(a) 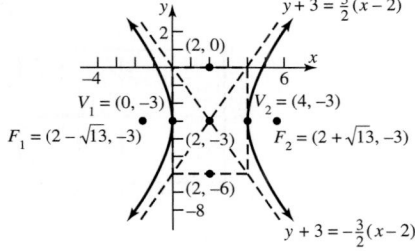   **(b)**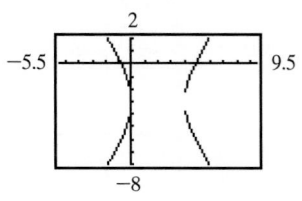

41. $\dfrac{(y-2)^2}{4} - (x+2)^2 = 1$

Center: $(-2, 2)$

Transverse axis: Parallel to y-axis

Vertices: $(-2, 0), (-2, 4)$

Foci: $(-2, 2 - \sqrt{5}), (-2, 2 + \sqrt{5})$

Asymptotes: $y - 2 = \pm 2(x + 2)$

(a)

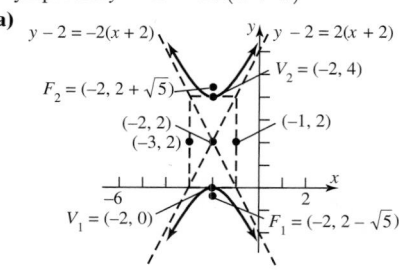

(b)

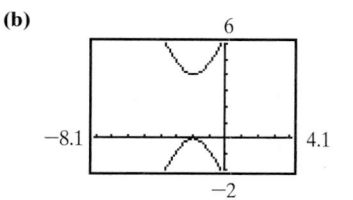

43. $\dfrac{(x+1)^2}{4} - \dfrac{(y+2)^2}{4} = 1$

Center: $(-1, -2)$

Transverse axis: Parallel to x-axis

Vertices: $(-3, -2), (1, -2)$

Foci: $(-1 - 2\sqrt{2}, -2), (-1 + 2\sqrt{2}, -2)$

Asymptotes: $y + 2 = \pm(x + 1)$

(a)

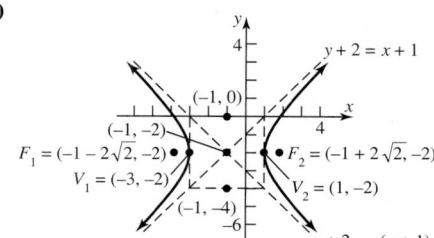

(b)

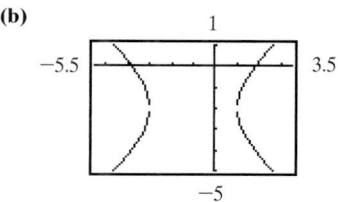

45. $(x-1)^2 - (y+1)^2 = 1$

Center: $(1, -1)$

Transverse axis: Parallel to x-axis

Vertices: $(0, -1), (2, -1)$

Foci: $(1 - \sqrt{2}, -1), (1 + \sqrt{2}, -1)$

Asymptotes: $y + 1 = \pm(x - 1)$

(a)

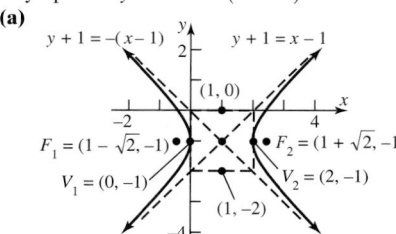

(b)

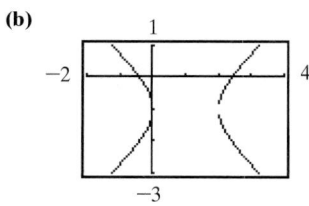

47. $\dfrac{(y-2)^2}{4} - (x+1)^2 = 1$

Center: $(-1, 2)$

Transverse axis: Parallel to y-axis

Vertices: $(-1, 0), (-1, 4)$

Foci: $(-1, 2 - \sqrt{5}), (-1, 2 + \sqrt{5})$

Asymptotes: $y - 2 = \pm 2(x + 1)$

(a)

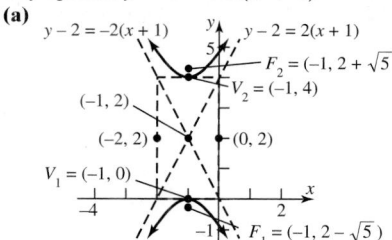

(b)

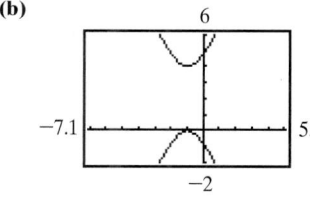

49. $\dfrac{(x-3)^2}{4} - \dfrac{(y+2)^2}{16} = 1$

Center: $(3, -2)$

Transverse axis: Parallel to x-axis

Vertices: $(1, -2)$, $(5, -2)$

Foci: $(3 - 2\sqrt{5}, -2)$, $(3 + 2\sqrt{5}, -2)$

Asymptotes: $y + 2 = \pm 2(x - 3)$

(a)

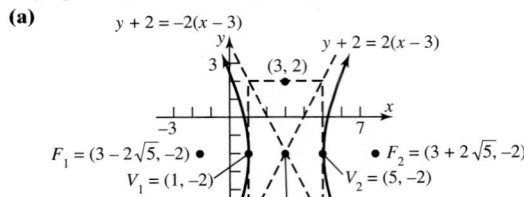

(b)

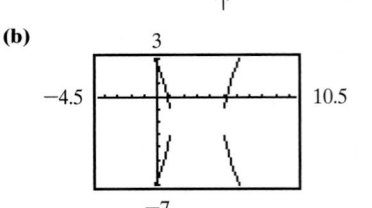

51. $\dfrac{(y-1)^2}{4} - (x+2)^2 = 1$

Center: $(-2, 1)$

Transverse axis: Parallel to y-axis

Vertices: $(-2, -1)$, $(-2, 3)$

Foci: $(-2, 1 - \sqrt{5})$, $(-2, 1 + \sqrt{5})$

Asymptotes: $y - 1 = \pm 2(x + 2)$

(a)

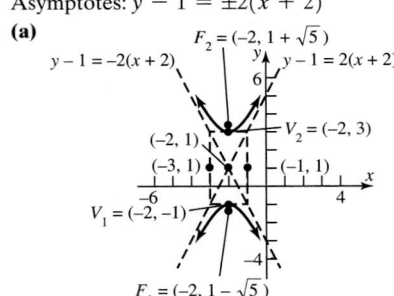

(b)

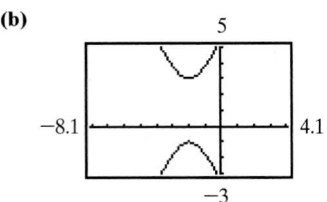

53.

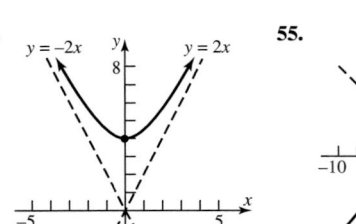

55.

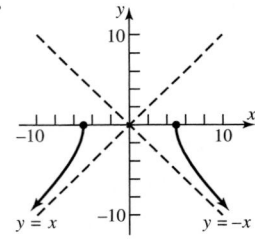

57. **(a)** The ship will reach shore at a point 64.66 mi from the master station.

(b) 0.00086 s **(c)** $(104, 50)$

59. **(a)** 450 ft

61. If e is close to 1, narrow hyperbola; if e is very large, wide hyperbola

63. $\dfrac{x^2}{4} - y^2 = 1$; asymptotes $y = \pm\dfrac{1}{2}x$, $y^2 - \dfrac{x^2}{4} = 1$; asymptotes $y = \pm\dfrac{1}{2}x$

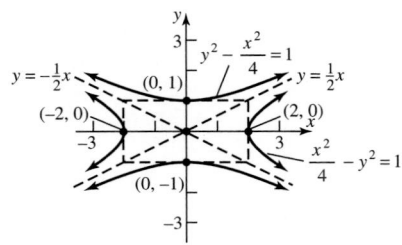

65. $Ax^2 + Cy^2 + F = 0$ If A and C are of opposite sign and $F \neq 0$, this equation may be written as $\dfrac{x^2}{\left(-\dfrac{F}{A}\right)} + \dfrac{y^2}{\left(-\dfrac{F}{C}\right)} = 1$,

$Ax^2 + Cy^2 = -F$ where $-\dfrac{F}{A}$ and $-\dfrac{F}{C}$ are opposite in sign. This is the equation of a hyperbola with center $(0, 0)$.

The transverse axis is the x-axis if $-\dfrac{F}{A} > 0$; the transverse axis is the y-axis if $-\dfrac{F}{A} < 0$.

6.5 Concepts and Vocabulary *(page 446)*

1. $\cot(2\theta) = \dfrac{A - C}{B}$ **2.** Hyperbola   **3.** Ellipse   **4.** T   **5.** T   **6.** F

6.5 Exercises *(page 446)*

1. Parabola   **3.** Ellipse   **5.** Hyperbola   **7.** Hyperbola   **9.** Circle

11. $x = \dfrac{\sqrt{2}}{2}(x' - y'),\, y = \dfrac{\sqrt{2}}{2}(x' + y')$ **13.** $x = \dfrac{\sqrt{2}}{2}(x' - y'),\, y = \dfrac{\sqrt{2}}{2}(x' + y')$ **15.** $x = \dfrac{1}{2}(x' - \sqrt{3}y'),\, y = \dfrac{1}{2}(\sqrt{3}x' + y')$

17. $x = \dfrac{\sqrt{5}}{5}(x' - 2y'),\, y = \dfrac{\sqrt{5}}{5}(2x' + y')$ **19.** $x = \dfrac{\sqrt{13}}{13}(3x' - 2y'),\, y = \dfrac{\sqrt{13}}{13}(2x' + 3y')$

21. $\theta = 45°$ (see Problem 11)

$x'^2 - \dfrac{y'^2}{3} = 1$

Hyperbola
Center at origin
Transverse axis is the x'-axis.
Vertices at $(\pm 1, 0)$

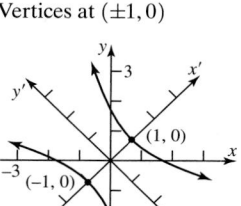

23. $\theta = 45°$ (see Problem 13)

$x'^2 + \dfrac{y'^2}{4} = 1$

Ellipse
Center at $(0, 0)$
Major axis is the y'-axis.
Vertices at $(0, \pm 2)$

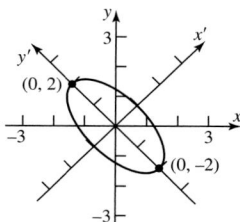

25. $\theta = 60°$ (see Problem 15)

$\dfrac{x'^2}{4} + y'^2 = 1$

Ellipse
Center at $(0, 0)$
Major axis is the x'-axis.
Vertices at $(\pm 2, 0)$

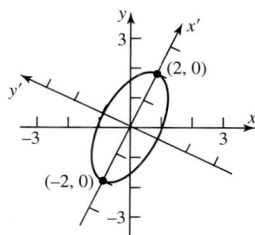

27. $\theta \approx 63°$ (see Problem 17)

$y'^2 = 8x'$

Parabola
Vertex at $(0, 0)$
Focus at $(2, 0)$

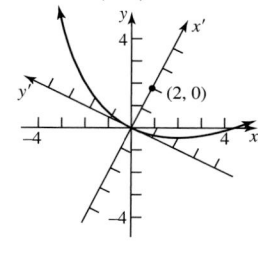

29. $\theta \approx 34°$ (see Problem 19)

$\dfrac{(x' - 2)^2}{4} + y'^2 = 1$

Ellipse

Center at $(2, 0)$

Major axis is the x'-axis.

Vertices at $(4, 0)$ and $(0, 0)$

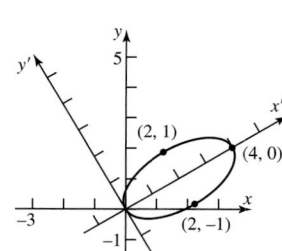

31. $\cot(2\theta) = \dfrac{7}{24}$;

$\theta = \sin^{-1}\left(\dfrac{3}{5}\right) \approx 37°$

$(x' - 1)^2 = -6\left(y' - \dfrac{1}{6}\right)$

Parabola

Vertex at $\left(1, \dfrac{1}{6}\right)$

Focus at $\left(1, -\dfrac{4}{3}\right)$

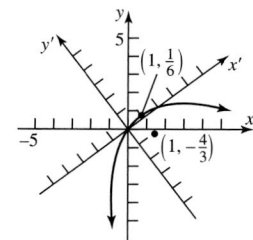

33. Hyperbola
35. Hyperbola
37. Parabola
39. Ellipse
41. Ellipse

43. Refer to equation (6):
$A' = A\cos^2\theta + B\sin\theta\cos\theta + C\sin^2\theta$
$B' = B(\cos^2\theta - \sin^2\theta) + 2(C - A)(\sin\theta\cos\theta)$
$C' = A\sin^2\theta - B\sin\theta\cos\theta + C\cos^2\theta$
$D' = D\cos\theta + E\sin\theta$
$E' = -D\sin\theta + E\cos\theta$
$F' = F$

45. Use Problem 43 to find $B'^2 - 4A'C'$.
After much cancellation, $B'^2 - 4A'C' = B^2 - 4AC$.

47. The distance between P_1 and P_2 in the $x'y'$-plane equals $\sqrt{(x_2' - x_1')^2 + (y_2' - y_1')^2}$.
Assuming $x' = x\cos\theta - y\sin\theta$ and $y' = x\sin\theta + y\cos\theta$, then $(x_2' - x_1')^2 = (x_2\cos\theta - y_2\sin\theta - x_1\cos\theta + y_1\sin\theta)^2$
$= \cos^2\theta(x_2 - x_1)^2 - 2\sin\theta\cos\theta(x_2 - x_1)(y_2 - y_1) + \sin^2\theta(y_2 - y_1)^2$, and
$(y_2' - y_1')^2 = (x_2\sin\theta + y_2\cos\theta - x_1\sin\theta - y_1\cos\theta)^2 = \sin^2\theta(x_2 - x_1)^2 + 2\sin\theta\cos\theta(x_2 - x_1)(y_2 - y_1) + \cos^2\theta(y_2 - y_1)^2$.
Therefore, $(x_2' - x_1')^2 + (y_2' - y_1')^2 = \cos^2\theta(x_2 - x_1)^2 + \sin^2\theta(x_2 - x_1)^2 + \sin^2\theta(y_2 - y_1)^2 + \cos^2\theta(y_2 - y_1)^2$
$= (x_2 - x_1)^2(\cos^2\theta + \sin^2\theta) + (y_2 - y_1)^2(\sin^2\theta + \cos^2\theta) = (x_2 - x_1)^2 + (y_2 - y_1)^2$.

6.6 Concepts and Vocabulary *(page 453)*

1. $\frac{1}{2}$; ellipse; parallel; 4; below   **2.** 1; < 1; > 1 **3.** T   **4.** T

6.6 Exercises *(page 453)*

1. Parabola; directrix is perpendicular to the polar axis 1 unit to the right of the pole.

3. Hyperbola; directrix is parallel to the polar axis $\frac{4}{3}$ units below the pole.

5. Ellipse; directrix is perpendicular to the polar axis $\frac{3}{2}$ units to the left of the pole.

7. Parabola; directrix is perpendicular to the polar axis 1 unit to the right of the pole; vertex is at $\left(\frac{1}{2}, 0\right)$.

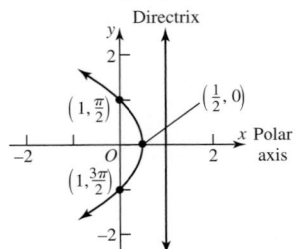

9. Ellipse; directrix is parallel to the polar axis $\frac{8}{3}$ units above the pole; vertices are at $\left(\frac{8}{7}, \frac{\pi}{2}\right)$ and $\left(8, \frac{3\pi}{2}\right)$.

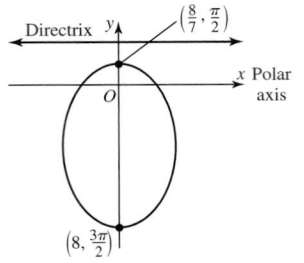

11. Hyperbola; directrix is perpendicular to the polar axis $\frac{3}{2}$ units to the left of the pole; vertices are at $(-3, 0)$ and $(1, \pi)$.

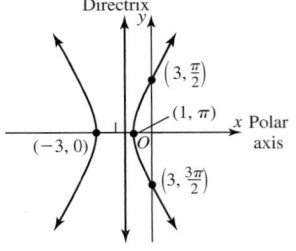

13. Ellipse; directrix is parallel to the polar axis 8 units below the pole; vertices are at $\left(8, \frac{\pi}{2}\right)$ and $\left(\frac{8}{3}, \frac{3\pi}{2}\right)$.

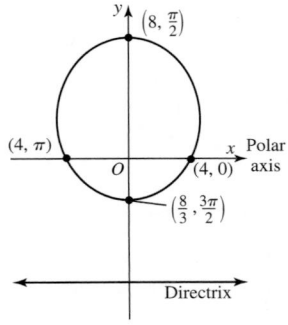

15. Ellipse; directrix is parallel to the polar axis 3 units below the pole; vertices are at $\left(6, \frac{\pi}{2}\right)$ and $\left(\frac{6}{5}, \frac{3\pi}{2}\right)$.

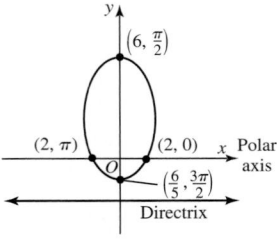

17. Ellipse; directrix is perpendicular to the polar axis 6 units to the left of the pole; vertices are at $(6, 0)$ and $(2, \pi)$.

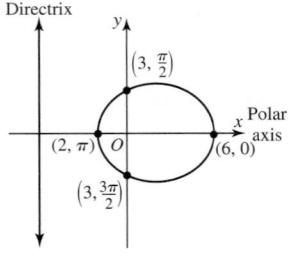

19. $y^2 + 2x - 1 = 0$ **21.** $16x^2 + 7y^2 + 48y - 64 = 0$ **23.** $3x^2 - y^2 + 12x + 9 = 0$ **25.** $4x^2 + 3y^2 - 16y - 64 = 0$

27. $9x^2 + 5y^2 - 24y - 36 = 0$ **29.** $3x^2 + 4y^2 - 12x - 36 = 0$ **31.** $r = \dfrac{1}{1 + \sin\theta}$ **33.** $r = \dfrac{12}{5 - 4\cos\theta}$ **35.** $r = \dfrac{12}{1 - 6\sin\theta}$

37. Use $d(D, P) = p - r\cos\theta$ in the derivation of equation (a) in Table 5.

39. Use $d(D, P) = p + r\sin\theta$ in the derivation of equation (a) in Table 5.

6.7 Concepts and Vocabulary *(page 466)*

1. plane curve; parameter   **2.** ellipse   **3.** cycloid   **4.** F   **5.** T

6.7 Exercises *(page 466)*

1.

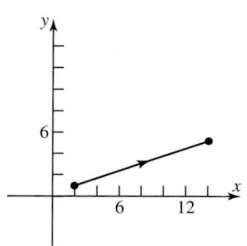

$x - 3y + 1 = 0$

3.

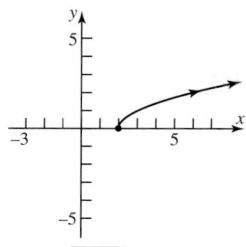

$y = \sqrt{x - 2}$

5.

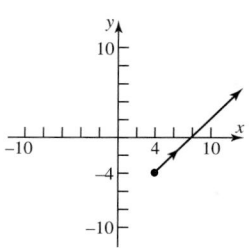

$x = y + 8$

7.

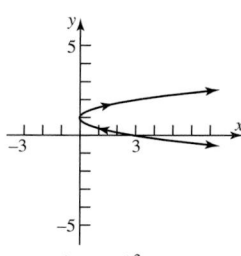

$x = 3(y - 1)^2$

9.

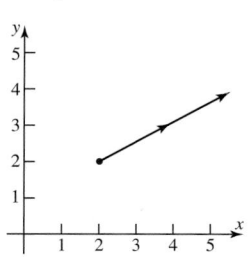

$2y = 2 + x$

11.

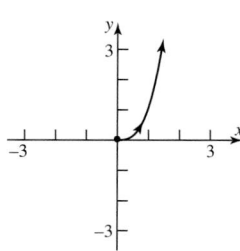

$y = x^3$

13.

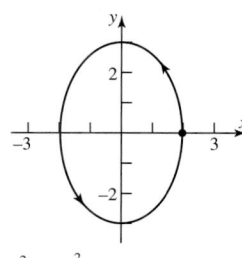

$\dfrac{x^2}{4} + \dfrac{y^2}{9} = 1$

15.

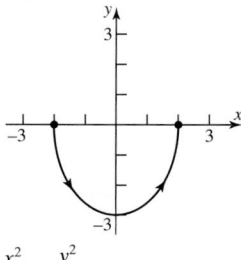

$\dfrac{x^2}{4} + \dfrac{y^2}{9} = 1$

17.

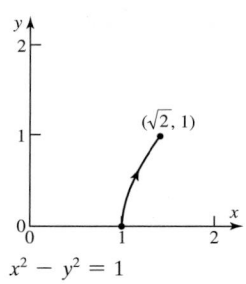

$x^2 - y^2 = 1$

19.

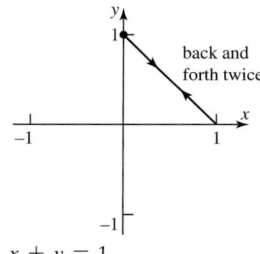

back and forth twice

$x + y = 1$

21. **(a)** $x = 3; y = -16t^2 + 50t + 6$
(b) 3.24 sec   **(c)** 1.5625 sec; 45.0625 ft
(d)

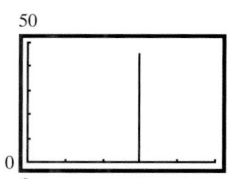

23. **(a)** Train: $x_1 = t^2, y_1 = 1$;
Bill: $x_2 = 5(t - 5), y_2 = 3$
(b) Bill won't catch the train.
(c)

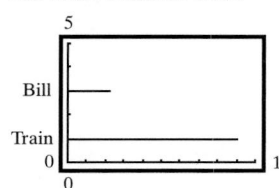

25. **(a)** $x = (145 \cos 20°)t$
$y = -16t^2 + (145 \sin 20°)t + 5$
(b) 3.20 sec
(c) 1.55 sec; 43.43 ft
(d) 435.65 ft
(e)

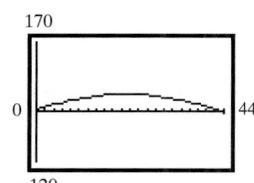

27. **(a)** $x = (40 \cos 45°)t$
$y = -4.9t^2 + (40 \sin 45°)t + 300$
(b) 11.23 sec
(c) 2.89 sec; 340.82 m
(d) 317.52 m
(e)

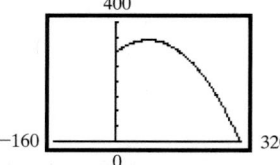

29. **(a)** Paseo: $x = 40t - 5, y = 0$; Bonneville: $x = 0, y = 30t - 4$ **(b)** $d = \sqrt{(40t - 5)^2 + (30t - 4)^2}$
(c)

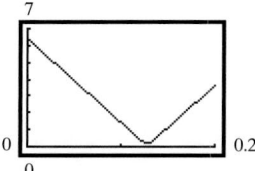

(d) 0.2 mi; 7.68 min   **(e)** Turn axes off to see the graph:

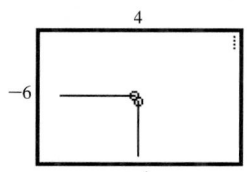

31. $x = t$ or $x = \dfrac{t+1}{4}$
$y = 4t - 1$ or $y = t$

33. $x = t$ or $x = t^3$
$y = t^2 + 1$ or $y = t^6 + 1$

35. $x = t$ or $x = \sqrt[3]{t}$
$y = t^3$ or $y = t$

37. $x = t^4$ or $x = t^6$
$y = t^6$ or $y = t^9$

39. $x = t + 2, y = t, 0 \le t \le 5$

41. $x = 3 \cos t, y = 2 \sin t, 0 \le t \le 2\pi$

43. $x = 2 \cos(\pi t), y = -3 \sin(\pi t), 0 \le t \le 2$ **45.** $x = 2 \sin(2\pi t), y = 3 \cos(2\pi t), 0 \le t \le 1$

47.
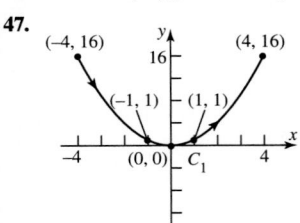 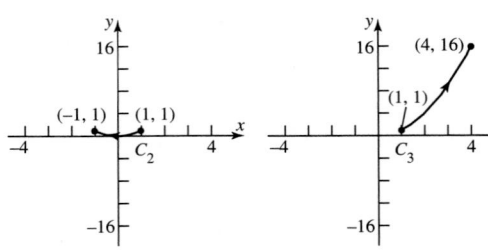 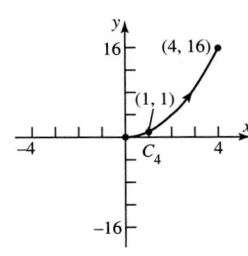

49. The orientation is from (x_1, y_1) to (x_2, y_2).

51. 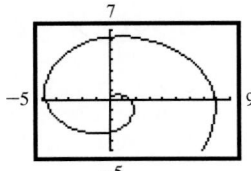   **53.** 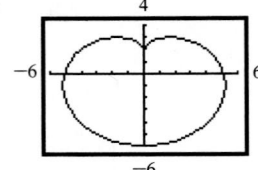   **55. (a)**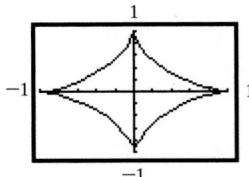

(b) $x^{2/3} + y^{2/3} = 1$

Review Exercises *(page 471)*

1. Parabola; vertex $(0, 0)$, focus $(-4, 0)$, directrix $x = 4$

3. Hyperbola; center $(0, 0)$, vertices $(5, 0)$ and $(-5, 0)$, foci $(\sqrt{26}, 0)$ and $(-\sqrt{26}, 0)$, asymptotes $y = \dfrac{1}{5}x$ and $y = -\dfrac{1}{5}x$

5. Ellipse; center $(0, 0)$, vertices $(0, 5)$ and $(0, -5)$, foci $(0, 3)$ and $(0, -3)$

7. $x^2 = -4(y - 1)$: Parabola; vertex $(0, 1)$, focus $(0, 0)$, directrix $y = 2$

9. $\dfrac{x^2}{2} - \dfrac{y^2}{8} = 1$: Hyperbola; center $(0, 0)$, vertices $(\sqrt{2}, 0)$ and $(-\sqrt{2}, 0)$, foci $(\sqrt{10}, 0)$ and $(-\sqrt{10}, 0)$, asymptotes $y = 2x$ and $y = -2x$

11. $(x - 2)^2 = 2(y + 2)$: Parabola; vertex $(2, -2)$, focus $\left(2, -\dfrac{3}{2}\right)$, directrix $y = -\dfrac{5}{2}$

13. $\dfrac{(y - 2)^2}{4} - (x - 1)^2 = 1$: Hyperbola; center $(1, 2)$, vertices $(1, 4)$ and $(1, 0)$, foci $(1, 2 + \sqrt{5})$ and $(1, 2 - \sqrt{5})$, asymptotes $y - 2 = \pm 2(x - 1)$

15. $\dfrac{(x - 2)^2}{9} + \dfrac{(y - 1)^2}{4} = 1$: Ellipse; center $(2, 1)$, vertices $(5, 1)$ and $(-1, 1)$, foci $(2 + \sqrt{5}, 1)$ and $(2 - \sqrt{5}, 1)$

17. $(x - 2)^2 = -4(y + 1)$: Parabola; vertex $(2, -1)$, focus $(2, -2)$, directrix $y = 0$

19. $\dfrac{(x - 1)^2}{4} + \dfrac{(y + 1)^2}{9} = 1$: Ellipse; center $(1, -1)$, vertices $(1, 2)$ and $(1, -4)$, foci $(1, -1 + \sqrt{5})$ and $(1, -1 - \sqrt{5})$

21. $y^2 = -8x$ **23.** $\dfrac{y^2}{4} - \dfrac{x^2}{12} = 1$ **25.** $\dfrac{x^2}{16} + \dfrac{y^2}{7} = 1$

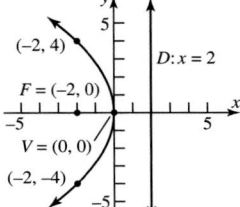   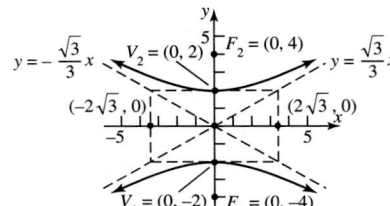   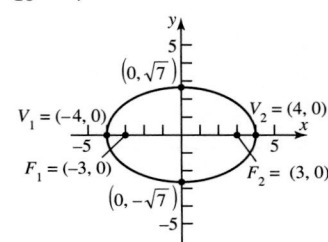

27. $(x - 2)^2 = -4(y + 3)$

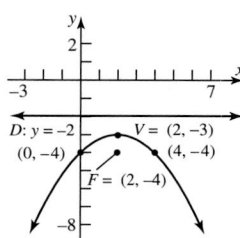

29. $(x + 2)^2 - \dfrac{(y + 3)^2}{3} = 1$

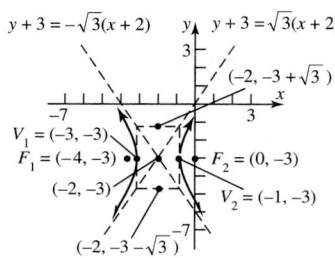

31. $\dfrac{(x + 4)^2}{16} + \dfrac{(y - 5)^2}{25} = 1$

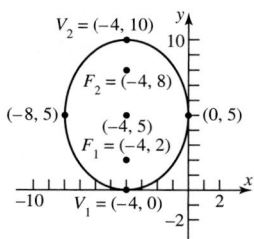

33. $\dfrac{(x + 1)^2}{9} - \dfrac{(y - 2)^2}{7} = 1$

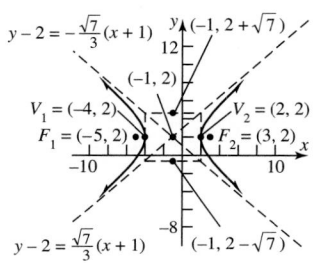

35. $\dfrac{(x - 3)^2}{9} - \dfrac{(y - 1)^2}{4} = 1$

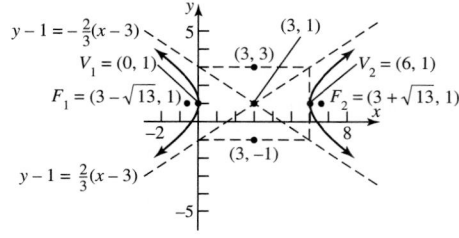

37. Parabola

39. Ellipse

41. Parabola

43. Hyperbola

45. Ellipse

47. $x'^2 - \dfrac{y'^2}{9} = 1$

Hyperbola
Center at the origin
Transverse axis the x'-axis
Vertices at $(\pm 1, 0)$

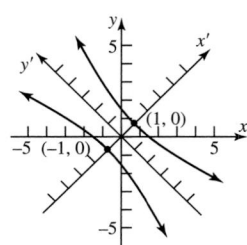

49. $\dfrac{x'^2}{2} + \dfrac{y'^2}{4} = 1$

Ellipse
Center at origin
Major axis the y'-axis
Vertices at $(0, \pm 2)$

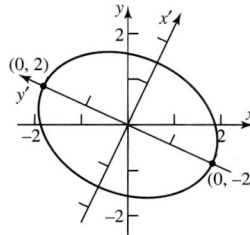

51. $y'^2 = -\dfrac{4\sqrt{13}}{13}x'$

Parabola
Vertex at the origin
Focus on the x'-axis at $\left(-\dfrac{\sqrt{13}}{13}, 0\right)$

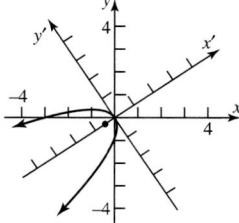

53. Parabola; directrix is perpendicular to the polar axis 4 units to the left of the pole; vertex is $(2, \pi)$.

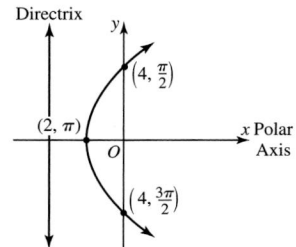

55. Ellipse; directrix is parallel to the polar axis 6 units below the pole; vertices are $\left(6, \dfrac{\pi}{2}\right)$ and $\left(2, \dfrac{3\pi}{2}\right)$.

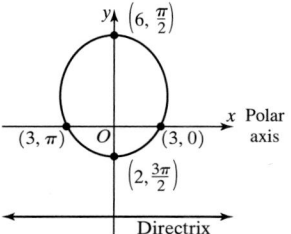

57. Hyperbola; directrix is perpendicular to the polar axis 1 unit to the right of the pole; vertices are $\left(\dfrac{2}{3}, 0\right)$ and $(-2, \pi)$.

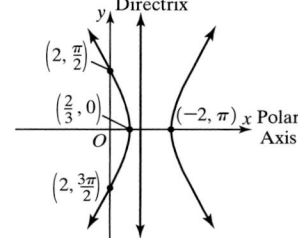

59. $y^2 - 8x - 16 = 0$ **61.** $3x^2 - y^2 - 8x + 4 = 0$

63.

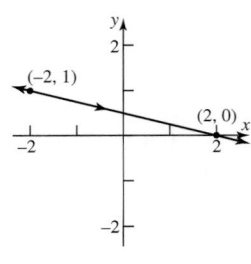

$x + 4y = 2$

65.

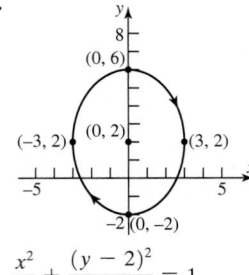

$\dfrac{x^2}{9} + \dfrac{(y-2)^2}{16} = 1$

67.

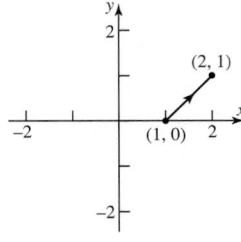

$1 + y = x$

69. $x = t,\ y = -2t + 4,\ -\infty < t < \infty;\ x = \dfrac{t-4}{-2},\ y = t,\ -\infty < t < \infty$ **71.** $x = 4\cos\left(\dfrac{\pi}{2}t\right),\ y = 3\sin\left(\dfrac{\pi}{2}t\right),\ 0 \le t \le 4$

73. $\dfrac{x^2}{5} - \dfrac{y^2}{4} = 1$ **75.** The ellipse $\dfrac{x^2}{16} + \dfrac{y^2}{7} = 1$ **77.** $\dfrac{1}{4}$ ft or 3 in. **79.** 19.72 ft, 18.86 ft, 14.91 ft

81. (a) 45.24 mi from the Master Station **(b)** 0.000645 sec **(c)** $(66, 20)$

83. (a) $x = (100\cos 35°)t;\ y = -16t^2 + (100\sin 35°)t + 6$

(b) 3.6866 s **(c)** 1.7924 s; 57.4 ft **(d)** 302 ft **(e)**

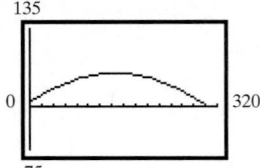

Cumulative Review Exercises *(page 474)*

1. $\theta = \dfrac{\pi}{12} \pm \pi k,\ k$ is any integer; $\theta = \dfrac{5\pi}{12} \pm \pi k,\ k$ is any integer **2.** $\theta = \dfrac{\pi}{6}$

3. $r = 8\sin\theta$

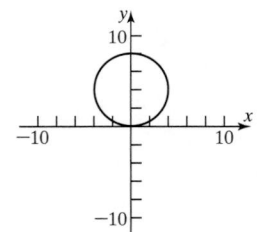

4. $\left\{x \,\middle|\, x \ne \dfrac{3\pi}{4} \pm \pi k,\ k \text{ is an integer}\right\}$ **5.** $\dfrac{7\sqrt{2}}{10}$

6. $\sin\dfrac{\pi}{12} = \sqrt{\dfrac{1 - \cos\dfrac{\pi}{6}}{2}} = \dfrac{\sqrt{2 - \sqrt{3}}}{2}$;

$\sin\dfrac{\pi}{12} = \sin\left(\dfrac{\pi}{3} - \dfrac{\pi}{4}\right) = \sin\dfrac{\pi}{3}\cos\dfrac{\pi}{4} - \cos\dfrac{\pi}{3}\sin\dfrac{\pi}{4} = \dfrac{\sqrt{6} - \sqrt{2}}{4}$

7.

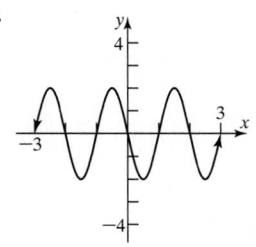

8.

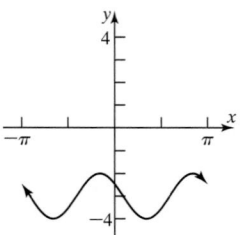

9. $\theta = 22.5°$ **10. (a)** $y = 2x - 2$ **(b)** $(x - 2)^2 + y^2 = 4$ **(c)** $\dfrac{x^2}{9} + \dfrac{y^2}{4} = 1$ **(d)** $y = 2(x - 1)^2$ **(e)** $\dfrac{y^2}{1} - \dfrac{x^2}{3} = 1$

C H A P T E R   7   Exponential and Logarithmic Functions

7.1 Concepts and Vocabulary *(page 486)*

1. $\left(-1, \dfrac{1}{a}\right), (0,1), (1,a)$ **2.** 1   **3.** 4   **4.** F   **5.** F   **6.** F   **7.** Becomes steeper; Lies closer to the x-axis

8. Because $y = a^{-x} = (a^{-1})^x = \left(\dfrac{1}{a}\right)^x$.

7.1 Exercises *(page 486)*

1. (a) 11.212   **(b)** 11.587   **(c)** 11.664   **(d)** 11.665   **3. (a)** 8.815   **(b)** 8.821   **(c)** 8.824   **(d)** 8.825
5. (a) 21.217   **(b)** 22.217   **(c)** 22.440   **(d)** 22.459   **7.** 3.320   **9.** 0.427   **11.** *B*   **13.** *D*   **15.** *A*   **17.** *E*

19.

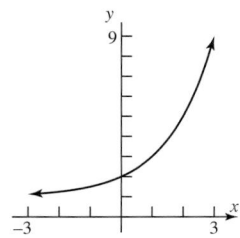

Domain: $(-\infty, \infty)$ or all real numbers
Range: $(1, \infty)$ or $\{y|y > 1\}$
Horizontal asymptote: $y = 1$

21.

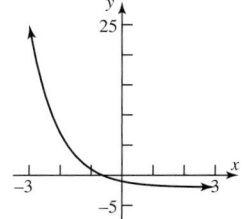

Domain: $(-\infty, \infty)$ or all real numbers
Range: $(-2, \infty)$ or $\{y|y > -2\}$
Horizontal asymptote: $y = -2$

23.

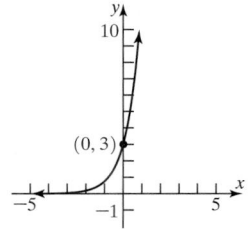

Domain: $(-\infty, \infty)$ or all real numbers
Range: $(0, \infty)$ or $\{y|y > 0\}$
Horizontal asymptote: $y = 0$

25.

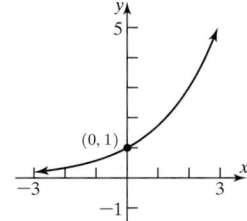

Domain: $(-\infty, \infty)$ or all real numbers
Range: $(0, \infty)$ or $\{y|y > 0\}$
Horizontal asymptote: $y = 0$

27.

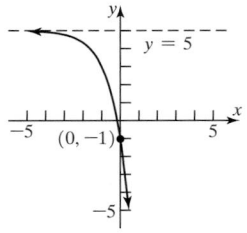

Domain: $(-\infty, \infty)$ or all real numbers
Range: $(-\infty, 5)$ or $\{y|y < 5\}$
Horizontal asymptote: $y = 5$

29.

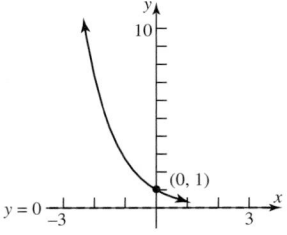

Domain: $(-\infty, \infty)$ or all real numbers
Range: $(0, \infty)$ or $\{y|y > 0\}$
Horizontal asymptote: $y = 0$

31.

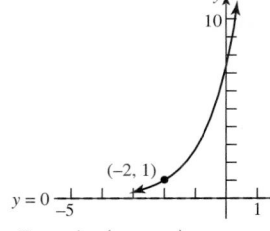

Domain: $(-\infty, \infty)$ or all real numbers
Range: $(0, \infty)$ or $\{y|y > 0\}$
Horizontal asymptote: $y = 0$

33.

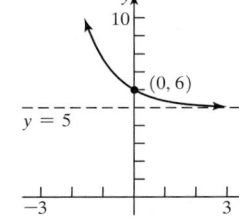

Domain: $(-\infty, \infty)$ or all real numbers
Range: $(5, \infty)$ or $\{y|y > 5\}$
Horizontal asymptote: $y = 5$

35.

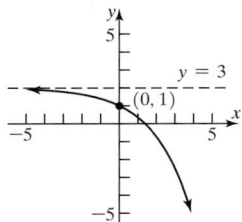

Domain: $(-\infty, \infty)$ or all real numbers
Range: $(-\infty, 2)$ or $\{y|y < 2\}$
Horizontal asymptote: $y = 2$

37. $f(x) = 3^x$ **39.** $f(x) = 2(4^x)$ **41.** $f(x) = -6^x$

43. $\left\{\dfrac{1}{2}\right\}$ **45.** $\{-\sqrt{2}, 0, \sqrt{2}\}$ **47.** $\left\{1 - \dfrac{\sqrt{6}}{3}, 1 + \dfrac{\sqrt{6}}{3}\right\}$

49. $\{0\}$ **51.** $\{4\}$ **53.** $\left\{\dfrac{3}{2}\right\}$ **55.** $\{1, 2\}$ **57.** $\dfrac{1}{49}$ **59.** $\dfrac{1}{4}$

61. (a) 74%   **(b)** 47%   **63. (a)** 44 watts   **(b)** 11.6 watts
65. 3.35 milligrams; 0.45 milligrams
67. (a) 0.632   **(b)** 0.982   **(c)**

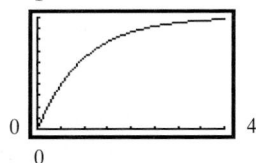

(d) 1

69. (a) 0.052   **(b)** 0.089   **71. (a)** 70.95%   **(b)** 72.62%   **(c)** 100%

73. (a) 5.414 amperes, 7.585 amperes, 10.376 amperes   **(b)** 12 amperes
(d) 3.343 amperes, 5.309 amperes, 9.443 amperes   **(e)** 24 amperes
(c), (f)

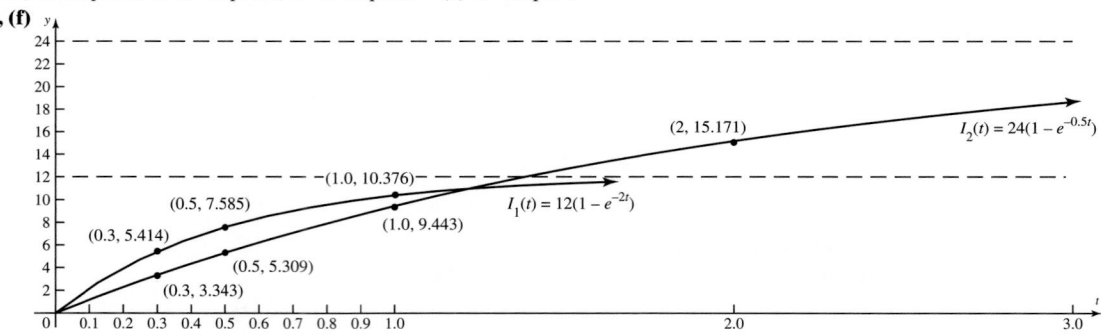

75. $n = 4: 2.7083; n = 6: 2.7181; n = 8: 2.7182788; n = 10: 2.7182818$

77. $\dfrac{f(x + h) - f(x)}{h} = \dfrac{a^{x+h} - a^x}{h} = \dfrac{a^x a^h - a^x}{h} = \dfrac{a^x(a^h - 1)}{h}$ **79.** $f(-x) = a^{-x} = \dfrac{1}{a^x} = \dfrac{1}{f(x)}$

81. (a) $f(-x) = \dfrac{1}{2}(e^{-x} - e^{-(-x)}) = \dfrac{1}{2}(e^{-x} - e^{x})$ **(b)**
$$= -\dfrac{1}{2}(e^{x} - e^{-x}) = -f(x)$$

$y = \frac{1}{2}(e^x - e^{-x})$

83. $f(1) = 5, f(2) = 17, f(3) = 257, f(4) = 65{,}537, f(5) = 4{,}294{,}967{,}297 = 641 \times 6{,}700{,}417$

7.2 Concepts and Vocabulary *(page 498)*

1. $\{x | x > 0\}$ or $(0, \infty)$ **2.** $(1, 0), (a, 1), \left(\dfrac{1}{a}, -1\right)$ **3.** 1   **4.** F   **5.** T
6. Because $y = \log_1 x$ means $1^y = 1 = x$, which cannot be true for $x \neq 1$.   **7.** $(1, \infty)$

7.2 Exercises *(page 499)*

1. $2 = \log_3 9$ **3.** $2 = \log_a 1.6$ **5.** $2 = \log_{1.1} M$ **7.** $x = \log_2 7.2$ **9.** $\sqrt{2} = \log_x \pi$ **11.** $x = \ln 8$ **13.** $2^3 = 8$ **15.** $a^6 = 3$ **17.** $3^x = 2$
19. $2^{1.3} = M$ **21.** $(\sqrt{2})^x = \pi$ **23.** $e^x = 4$ **25.** 0   **27.** 2   **29.** -4 **31.** $\dfrac{1}{2}$ **33.** 4   **35.** $\dfrac{1}{2}$ **37.** $\{x | x > 3\}; (3, \infty)$
39. $\{x | x \neq 0\}; (-\infty, 0)$ or $(0, \infty)$ **41.** $\{x | x \neq 1\}; (-\infty, 1)$ or $(1, \infty)$ **43.** $\{x | x > -1\}; (-1, \infty)$
45. $\{x | x < -1$ or $x > 0\}; (-\infty, -1)$ or $(0, \infty)$ **47.** 0.511   **49.** 30.099   **51.** $\sqrt{2}$ **53.** B   **55.** D   **57.** A   **59.** E

61.

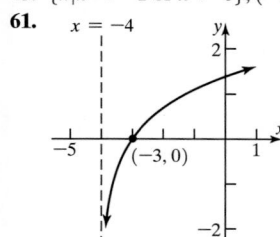

Domain: $(-4, \infty)$
Range: $(-\infty, \infty)$
Vertical asymptote: $x = -4$

63.

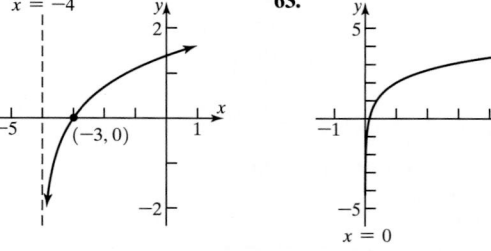

Domain: $(0, \infty)$
Range: $(-\infty, \infty)$
Vertical asymptote: $x = 0$

65.

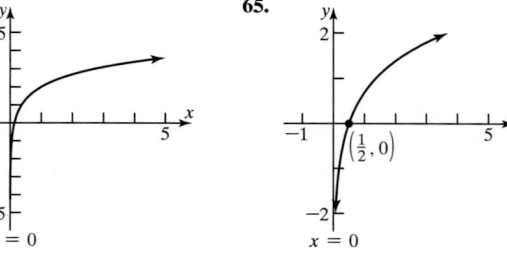

Domain: $(0, \infty)$
Range: $(-\infty, \infty)$
Vertical asymptote: $x = 0$

67.

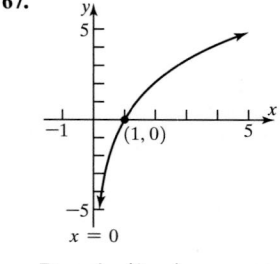

Domain: $(0, \infty)$
Range: $(-\infty, \infty)$
Vertical asymptote: $x = 0$

69.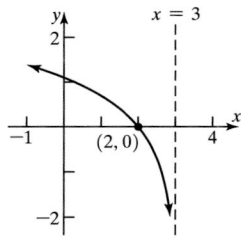
Domain: $(-\infty, 3)$
Range: $(-\infty, \infty)$
Vertical asymptote: $x = 3$

71.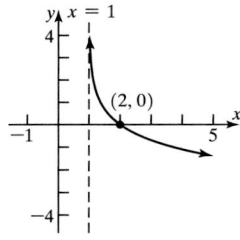
Domain: $(1, \infty)$
Range: $(-\infty, \infty)$
Vertical asymptote: $x = 1$

73.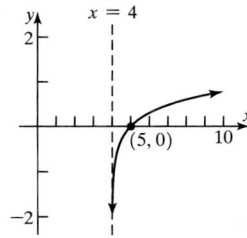
Domain: $(4, \infty)$
Range: $(-\infty, \infty)$
Vertical asymptote: $x = 4$

75.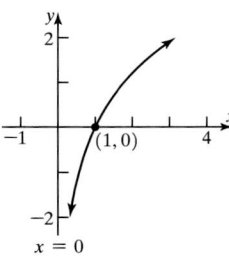
Domain: $(0, \infty)$
Range: $(-\infty, \infty)$
Vertical asymptote: $x = 0$

77.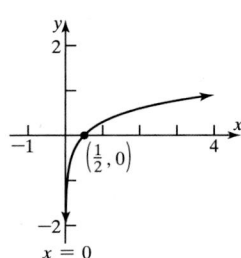
Domain: $(0, \infty)$
Range: $(-\infty, \infty)$
Vertical asymptote: $x = 0$

79.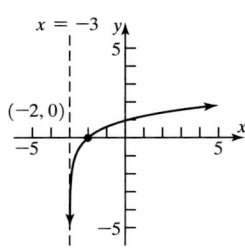
Domain: $(-3, \infty)$
Range: $(-\infty, \infty)$
Vertical asymptote: $x = -3$

81.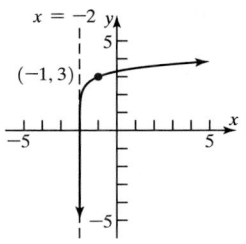
Domain: $(-2, \infty)$
Range: $(-\infty, \infty)$
Vertical asymptote: $x = -2$

83.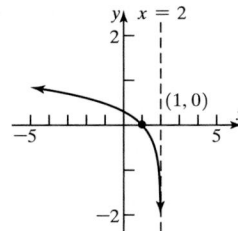
Domain: $(-\infty, 2)$
Range: $(-\infty, \infty)$
Vertical asymptote: $x = 2$

85. $\{9\}$ **87.** $\left\{\dfrac{7}{2}\right\}$ **89.** $\{2\}$ **91.** $\{5\}$ **93.** $\{3\}$ **95.** $\{2\}$ **97.** $\left\{\dfrac{\ln 10}{3}\right\}$ **99.** $\left\{\dfrac{\ln 8 - 5}{2}\right\}$ **101.** $\{-2\sqrt{2}, 2\sqrt{2}\}$ **103.** $\{-1\}$

105. (a)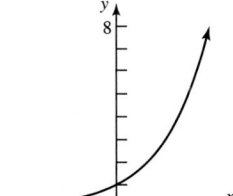
Domain: $(-\infty, \infty)$
Range: $(0, \infty)$
Horizontal asymptote: $y = 0$

(b) $f^{-1}(x) = \log_2 x$

(c)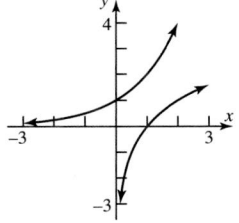
Domain of f^{-1} = Range of $f = (0, \infty)$
Range of f^{-1} = Domain of $f = (-\infty, \infty)$
Vertical asymptote of f^{-1}: $x = 0$

107. (a)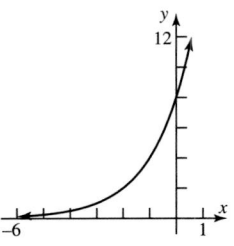
Domain: $(-\infty, \infty)$
Range: $(0, \infty)$
Horizontal asymptote: $y = 0$

(b) $f^{-1}(x) = \log_2 x - 3$

(c)
Domain of f^{-1} = Range of $f = (0, \infty)$
Range of f^{-1} = Domain of $f = (-\infty, \infty)$
Vertical asymptote of f^{-1}: $x = 0$

109. (a) $n \approx 6.93$ so 7 panes are necessary
(b) $n \approx 13.86$ so 14 panes are necessary

111. (a) $d \approx 127.7$ so it takes about 128 days
(b) $d \approx 575.6$ so it takes about 576 days

113. (a) 6.93 min   **(b)** 16.09 min
(c) No, since $F(t)$ can never equal one.

115. $h \approx 2.29$ so the time between injections is about
2 hours, 17 minutes

117. 0.2695 seconds
0.8959 seconds

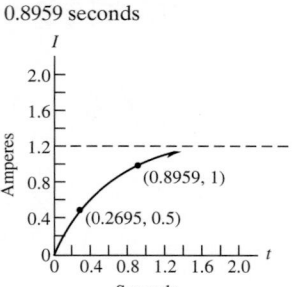

119. 50 decibels **121.** 110 decibels **123.** 8.1
125. (a) $k = 20.07$ **(b)** 91% **(c)** 0.175 **(d)** 0.08

7.3 Concepts and Vocabulary *(page 510)*

1. Sum **2.** 7 **3.** $r \log_a M$ **4.** F **5.** F **6.** T

7.3 Exercises *(page 510)*

1. 71 **3.** -4 **5.** 7 **7.** 1 **9.** 1 **11.** 2 **13.** $\dfrac{5}{4}$ **15.** 4 **17.** $a + b$ **19.** $b - a$ **21.** $3a$ **23.** $\dfrac{1}{5}(a + b)$ **25.** $2 + \log_5 x$ **27.** $3 \log_2 z$

29. $1 + \ln x$ **31.** $\ln x + x$ **33.** $2 \log_a u + 3 \log_a v$ **35.** $2 \ln x + \dfrac{1}{2} \ln(1 - x)$ **37.** $3 \log_2 x - \log_2(x - 3)$

39. $\log x + \log(x + 2) - 2 \log(x + 3)$ **41.** $\dfrac{1}{3} \ln(x - 2) + \dfrac{1}{3} \ln(x + 1) - \dfrac{2}{3} \ln(x + 4)$ **43.** $\ln 5 + \ln x + \dfrac{1}{2} \ln(1 - 3x) - 3 \ln(x - 4)$

45. $\log_5 u^3 v^4$ **47.** $-\dfrac{5}{2} \log_3 x$ **49.** $\log_4 \dfrac{x - 1}{(x + 1)^4}$ **51.** $-2 \ln(x - 1)$ **53.** $\log_2[x(3x - 2)^4]$ **55.** $\log_a\left(\dfrac{25x^6}{\sqrt{2x + 3}}\right)$

57. $\log_2 \dfrac{(x + 1)^2}{(x + 3)(x - 1)}$ **59.** $\log y = \log a + x \log b$ **61.** 2.771 **63.** -3.880 **65.** 5.615 **67.** 0.874

69. $y = \dfrac{\log x}{\log 4}$

71. $y = \dfrac{\log(x + 2)}{\log 2}$

73. $y = \dfrac{\log(x + 1)}{\log(x - 1)}$

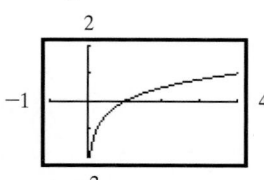

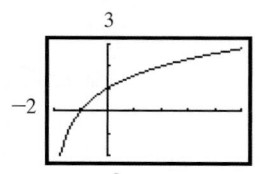

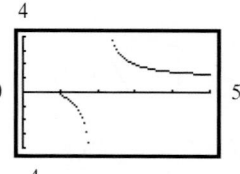

75. $y = Cx$ **77.** $y = Cx(x + 1)$ **79.** $y = Ce^{3x}$ **81.** $y = Ce^{-4x} + 3$ **83.** $y = \dfrac{\sqrt[3]{C}(2x + 1)^{1/6}}{(x + 4)^{1/9}}$ **85.** 3 **87.** 1

89. $\log_a(x + \sqrt{x^2 - 1}) + \log_a(x - \sqrt{x^2 - 1}) = \log_a[(x + \sqrt{x^2 - 1})(x - \sqrt{x^2 - 1})]$
$= \log_a[x^2 - (x^2 - 1)] = \log_a 1 = 0$

91. $\ln(1 + e^{2x}) = \ln(e^{2x}(e^{-2x} + 1)) = \ln e^{2x} + \ln(e^{-2x} + 1) = 2x + \ln(1 + e^{-2x})$

93. $y = f(x) = \log_a x; a^y = x$ implies $a^y = \left(\dfrac{1}{a}\right)^{-y} = x$, so $-y = \log_{1/a} x = -f(x)$.

95. $f(x) = \log_a x; f\left(\dfrac{1}{x}\right) = \log_a \dfrac{1}{x} = \log_a 1 - \log_a x = -f(x)$

97. $\log_a \dfrac{M}{N} = \log_a (M \cdot N^{-1}) = \log_a M + \log_a N^{-1} = \log_a M - \log_a N$,
since $a^{\log_a N^{-1}} = N^{-1}$ implies $a^{-\log_a N^{-1}} = N$, i.e., $\log_a N = -\log_a N^{-1}$

7.4 Exercises *(page 516)*

1. 6 **3.** 16 **5.** 8 **7.** 3 **9.** 5 **11.** $-1 + \sqrt{1 + e^4} \approx 6.456$ **13.** $\dfrac{\ln 3}{\ln 2} \approx 1.585$ **15.** 0 **17.** $\dfrac{\ln 10}{\ln 2} \approx 3.322$ **19.** $-\dfrac{\ln 1.2}{\ln 8} \approx -0.088$

21. $\dfrac{\ln 3}{2 \ln 3 + \ln 4} \approx 0.307$ **23.** $\dfrac{\ln 7}{\ln 0.6 + \ln 7} \approx 1.356$ **25.** 0 **27.** $\dfrac{\ln \pi}{1 + \ln \pi} \approx 0.534$ **29.** $\dfrac{\ln 1.6}{3 \ln 2} \approx 0.226$ **31.** $\dfrac{9}{2}$ **33.** 2 **35.** 1 **37.** 16

39. $-1, \dfrac{2}{3}$ **41.** 0 **43.** $\ln(2 + \sqrt{5}) \approx 1.444$ **45.** 1.92 **47.** 2.79 **49.** -0.57 **51.** -0.70 **53.** 0.57 **55.** 0.39, 1.00 **57.** 1.32 **59.** 1.31

7.5 Exercises *(page 524)*

1. $108.29 **3.** $609.50 **5.** $697.09 **7.** $12.46 **9.** $125.23 **11.** $88.72 **13.** $860.72 **15.** $554.09 **17.** $59.71 **19.** $361.93 **21.** 5.35%

23. 26% **25.** $6\frac{1}{4}$% compounded annually **27.** 9% compounded monthly **29.** 104.32 months; 103.97 months **31.** 61.02 months; 60.82 months

33. 15.27 years **35.** $104,335 **37.** $12,910.62 **39.** About $30.17 per share or $3017 **41.** 9.35%

43. Not quite. Jim will have $1057.60. The second bank gives a better deal, since Jim will have $1060.62 after 1 year.

45. Will has $11,632.73; Henry has $10,947.89. **47. (a)** Interest is $30,000 **(b)** Interest is $38,613.59

(c) Interest is $37,752.73. Simple interest at 12% is best. **49. (a)** $1364.62 **(b)** $1353.35 **51.** $4631.93

55. (a) 6.12 years **(b)** 18.45 years **(c)** $mP = P\left(1 + \dfrac{r}{n}\right)^{nt}$

$$m = \left(1 + \frac{r}{n}\right)^{nt}$$

$$\ln m = \ln\left(1 + \frac{r}{n}\right)^{nt} = nt \ln\left(1 + \frac{r}{n}\right)$$

$$t = \frac{\ln m}{n \ln\left(1 + \dfrac{r}{n}\right)}$$

7.6 Exercises *(page 536)*

1. (a) 500 insects
 (b) $0.02 = 2\%$
 (c)

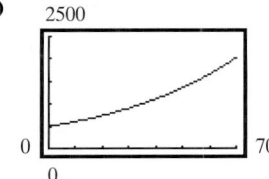

 (d) ≈ 611 insects
 (e) After about 23.5 days
 (f) After about 34.7 days

13. (a) 5:18 PM
 (b)

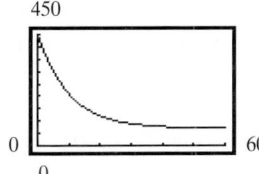

 (c) About 14.3 minutes
 (d) The temperature of the pizza approaches 70°F.

23. (a) 1000 g, 43.9% **(b)** 30 g
 (c)

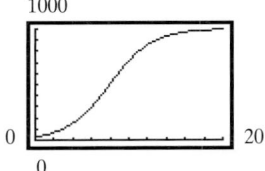

 (d) 616.6 g **(e)** After 9.85 hours
 (f) About 7.9 hours

3. (a) $-0.0244 = -2.44\%$
 (b)

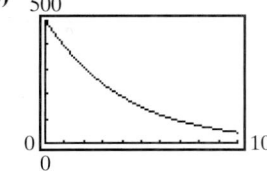

 (c) About 391.7 grams
 (d) After about 9.1 years
 (e) 28.4 years

15. (a) 18.63°C; 25.1°C
 (b)

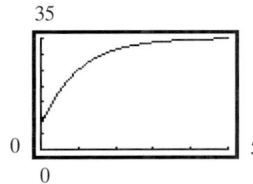

25. (a) 9.23×10^{-3}, or about 0 **(b)** 0.81, or about 1
 (c) 5.01, or about 5 **(d)** 57.91°, 43.99°, 30.07°

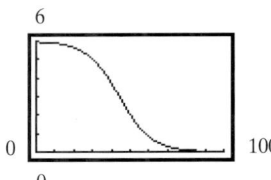

5. 5832; 3.9 days
7. 25,198 **9.** 9.797 grams
11. (a) 9727 years ago
 (b)

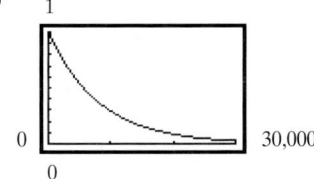

 (c) 5600 years

17. 7.34 kg; 76.6 hours
19. 26.6 days
21. (a) 90% **(b)** 12.86%
 (c)

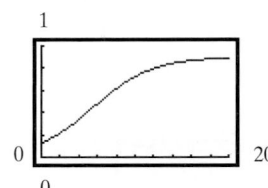

 (d) 85.77% **(e)** 1996
 (f) About 5.6 years

7.7 Exercises *(page 544)*

1. (a)

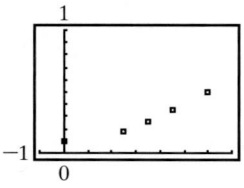

(b) $y = 0.0903(1.3384)^x$
(c) $N(t) = 0.0903e^{0.2915t}$
(d)

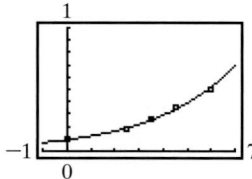

(e) 0.69
(f) After about 7.26 hours

3. (a)

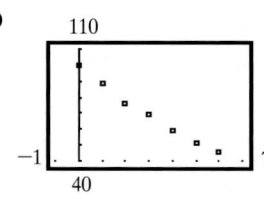

(b) $y = 100.326(0.8769)^x$
(c) $A = 100.326e^{-0.1314t}$
(d)

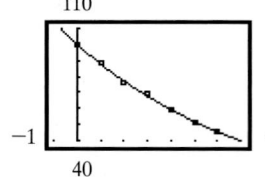

(e) 5.3 weeks   **(f)** 0.14 grams
(g) After about 12.3 weeks

5. (a)

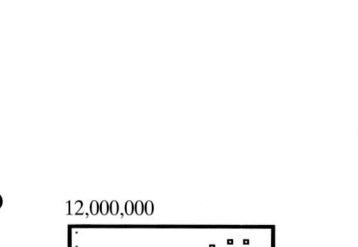

(b) value $= 2.7018 \times 10^{-44}(1.056554737)^{\text{year}}$
(c) 5.66%
(d) $52,166
(e) In 2029

7. (a)

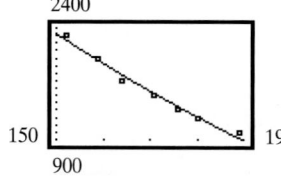

(b) $y = 32,741.02 - 6070.96 \ln x$
(c)

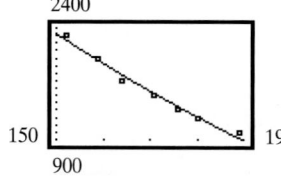

(d) Approximately 168 computers

9. (a)

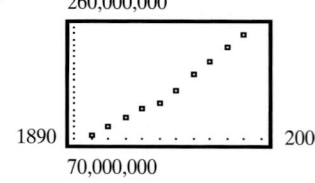

(b) $y = \dfrac{799,475,916.5}{1 + 1.56344 \times 10^{14}e^{-0.0160x}}$
(c)

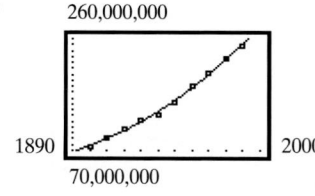

(d) 799,475,917
(e) 283,391,335   **(f)** 2007

11. (a)

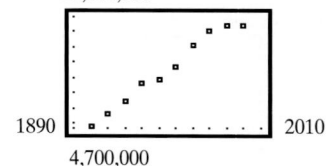

(b) $y = \dfrac{14,471,245.24}{1 + 3.860 \times 10^{20}\,e^{-0.0246x}}$
(c)

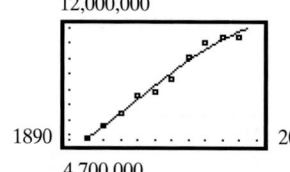

(d) 14,471,245
(e) Approximately 12,750,816

Review Exercises *(page 549)*

1. (a) 81   **(b)** 2   **(c)** $\dfrac{1}{9}$ **(d)** -3 **3.** $\log_5 z = 2$ **5.** $5^{13} = u$ **7.** $\left\{x \middle| x > \dfrac{2}{3}\right\}; \left(\dfrac{2}{3}, \infty\right)$ **9.** $\left\{x \middle| x < 1 \text{ or } x > 2\right\}; (-\infty, 1) \text{ or } (2, \infty)$

11. -3 **13.** $\sqrt{2}$ **15.** 0.4   **17.** $\log_3 u + 2\log_3 v - \log_3 w$ **19.** $2\log x + \dfrac{1}{2}\log(x^3 + 1)$ **21.** $\ln x + \dfrac{1}{3}\ln(x^2 + 1) - \ln(x - 3)$

23. $\dfrac{25}{4}\log_4 x$ **25.** $-2\ln(x + 1)$ **27.** $\log\left(\dfrac{4x^3}{[(x + 3)(x - 2)]^{1/2}}\right)$ **29.** 2.124   **31.**

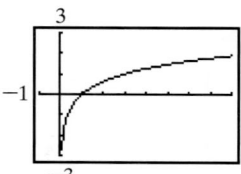

33.

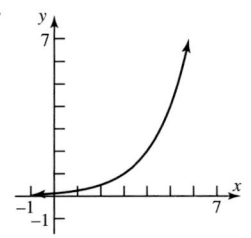

Domain: $(-\infty, \infty)$
Range: $(0, \infty)$
Horizontal asymptote: $y = 0$

35.

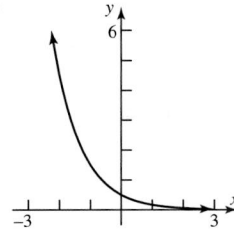

Domain: $(-\infty, \infty)$
Range: $(0, \infty)$
Horizontal asymptote: $y = 0$

37.

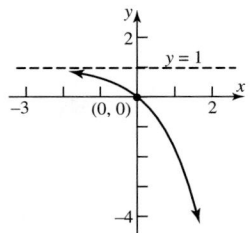

Domain: $(-\infty, \infty)$
Range: $(-\infty, 1)$
Horizontal asymptote: $y = 1$

39.

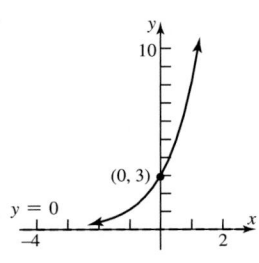

Domain: $(-\infty, \infty)$
Range: $(0, \infty)$
Horizontal asymptote: $y = 0$

41.

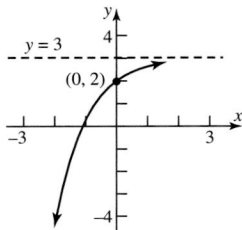

Domain: $(-\infty, \infty)$
Range: $(-\infty, 3)$
Horizontal asymptote: $y = 3$

43. $\left\{\dfrac{1}{4}\right\}$ **45.** $\left\{\dfrac{-1 - \sqrt{3}}{2}, \dfrac{-1 + \sqrt{3}}{2}\right\}$ **47.** $\left\{\dfrac{1}{4}\right\}$

49. $\left\{\dfrac{2 \ln 3}{\ln 5 - \ln 3} \approx 4.301\right\}$ **51.** $\left\{\dfrac{12}{5}\right\}$ **53.** $\{83\}$

55. $\left\{\dfrac{1}{2}, -3\right\}$ **57.** $\{-1\}$ **59.** $\{1 - \ln 5 \approx -0.609\}$

61. $\left\{\dfrac{\ln 3}{3 \ln 2 - 2 \ln 3} \approx -9.327\right\}$ **63.** 3229.5 meters

65. (a) 37.3 watts (b) 6.9 decibels **67.** (a) 9.85 years (b) 4.27 years **69.** \$41,668.97 **71.** 24,203 years ago **73.** 6,078,190,457

75. (a) 0.3 (b) 0.8 (c)   (d) About 20.1 years

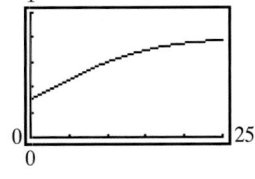

77. (a)

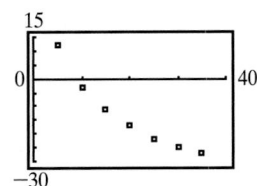

(b) Wind chill $= 44.198 - 20.331 \ln$ (wind speed)
(c)

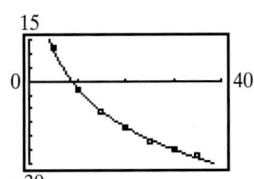

(d) Approximately -20 degrees Fahrenheit

A P P E N D I X   A   Review

A.1 Exercises *(page 563)*

1. (a) $\{2, 5\}$ (b) $\{-6, 2, 5\}$ (c) $\left\{-6, \dfrac{1}{2}, -1.333..., 2, 5\right\}$ (d) $\{\pi\}$ (e) $\left\{-6, \dfrac{1}{2}, -1.333..., \pi, 2, 5\right\}$ **3.** (a) $\{1\}$ (b) $\{0, 1\}$

(c) $\left\{0, 1, \dfrac{1}{2}, \dfrac{1}{3}, \dfrac{1}{4}\right\}$ (d) None (e) $\left\{0, 1, \dfrac{1}{2}, \dfrac{1}{3}, \dfrac{1}{4}\right\}$ **5.** (a) None (b) None (c) None (d) $\left\{\sqrt{2}, \pi, \sqrt{2} + 1, \pi + \dfrac{1}{2}\right\}$

(e) $\left\{\sqrt{2}, \pi, \sqrt{2} + 1, \pi + \dfrac{1}{2}\right\}$ **7.** 4 **9.** -28 **11.** $\dfrac{4}{5}$ **13.** 0 **15.** $x = 0$ **17.** $x = 3$ **19.** None **21.** $x = 1, x = 0, x = -1$

23. $\{x \mid x \neq 5\}$ **25.** $\{x \mid x \neq -4\}$

27.

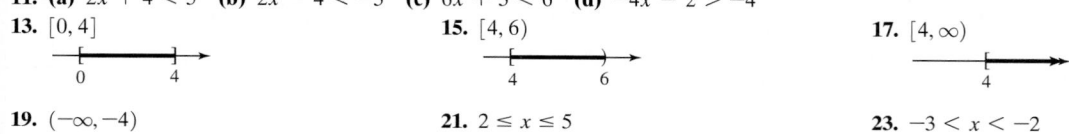

29. $>$ **31.** $>$ **33.** $>$ **35.** $=$ **37.** $<$ **39.** $x > 0$ **41.** $x < 2$ **43.** $x \le 1$

45. (number line with -2)   **47.** (number line with -1)   **49.** 1   **51.** 5   **53.** 1   **55.** 22   **57.** 2   **59.** 1   **61.** 2   **63.** 6   **65.** 16

67. $\dfrac{1}{16}$ **69.** $\dfrac{1}{9}$ **71.** 9   **73.** 5   **75.** 4   **77.** $64x^6$ **79.** $\dfrac{x^4}{y^2}$ **81.** $\dfrac{x}{y}$ **83.** $-\dfrac{8x^3z}{9y}$ **85.** $\dfrac{16x^2}{9y^2}$

87. $A = lw$; all the variables are positive real numbers.   **89.** $C = \pi d$; d and C are positive real numbers.

91. $A = \dfrac{\sqrt{3}}{4}x^2$; x and A are positive real numbers.   **93.** $V = \dfrac{4}{3}\pi r^3$; r and V are positive real numbers.

95. $V = x^3$; x and V are positive real numbers.   **97. (a)** $2 \le 5$ **(b)** $6 > 5$

A.2 Exercises *(page 569)*

1. 13   **3.** 26   **5.** 25   **7.** Right triangle; 5   **9.** Not a right triangle   **11.** Right triangle; 25   **13.** Not a right triangle   **15.** 8 in²   **17.** 4 in²

19. $A = 25\pi$ m²; $C = 10\pi$ m   **21.** 224 ft³   **23.** $V = \dfrac{256}{3}\pi$ cm³; $S = 64\pi$ cm²   **25.** 648π in³   **27.** π square units   **29.** 2π square units

31. $\dfrac{16}{3}\pi \approx 16.76$ ft   **33.** 64 ft²   **35.** $24 + 2\pi \approx 30.28$ ft²; $16 + 2\pi \approx 22.28$ ft   **37.** About 5.48 mi   **39.** About 12.25 mi; about 15.00 mi

A.3 Exercises *(page 583)*

1. $\{-2.21, 0.54, 1.68\}$ **3.** $\{-1.55, 1.15\}$ **5.** $\{-1.12, 0.36\}$ **7.** $\{-2.69, -0.49, 1.51\}$ **9.** $\{-2.86, -1.34, 0.20, 1.00\}$ **11.** No real solutions

13. $\left\{\dfrac{4}{3}\right\}$ **15.** $\{-2\}$ **17.** $\{-18\}$ **19.** $\{3, 4\}$ **21.** $\left\{-3, \dfrac{1}{2}\right\}$ **23.** $\{-3, 3\}$ **25.** $\{-5, 0, 4\}$ **27.** $\left\{0, \dfrac{1}{4}\right\}$ **29.** $\{-3, 0, 3\}$ **31.** $\{-1, 0, 1\}$

33. $\{-1\}$ **35.** $\{-2, 2\}$ **37.** $\{-5, 5\}$ **39.** $\{-1, 3\}$ **41.** $\{-3, 0\}$ **43.** 16   **45.** $\dfrac{1}{16}$ **47.** $\{-7, 3\}$ **49.** $\left\{-\dfrac{1}{4}, \dfrac{3}{4}\right\}$

51. $\left\{\dfrac{-1 - \sqrt{7}}{6}, \dfrac{-1 + \sqrt{7}}{6}\right\}$ **53.** $\{0, 9\}$ **55.** $\{-3, 2\}$ **57.** $\left\{-\dfrac{1}{2}, 3\right\}$ **59.** $\{-4, 4\}$ **61.** $\{-6, -2\}$ **63.** $\left\{\dfrac{3}{2}\right\}$ **65.** $\left\{-\dfrac{5}{2}, 3\right\}$

67. $\left\{-\dfrac{3}{4}, 2\right\}$ **69.** $\{2 - \sqrt{2}, 2 + \sqrt{2}\}$ **71.** $\{2 - \sqrt{5}, 2 + \sqrt{5}\}$ **73.** $\left\{1, \dfrac{3}{2}\right\}$ **75.** No real solutions   **77.** $\left\{\dfrac{-1 - \sqrt{5}}{4}, \dfrac{-1 + \sqrt{5}}{4}\right\}$

79. $\left\{\dfrac{9 - \sqrt{113}}{8}, \dfrac{9 + \sqrt{113}}{8}\right\}$ **81.** $x = \dfrac{b + c}{a}$ **83.** $x = \dfrac{abc}{a + b}$ **85.** $R = \dfrac{R_1 R_2}{R_1 + R_2}$ **87.** $R = \dfrac{mv^2}{F}$

A.4 Exercises *(page 593)*

1. $8 + 5i$ **3.** $-7 + 6i$ **5.** $-6 - 11i$ **7.** $6 - 18i$ **9.** $6 + 4i$ **11.** $10 - 5i$ **13.** 37

15. $\dfrac{6}{5} + \dfrac{8}{5}i$ **17.** $1 - 2i$ **19.** $\dfrac{5}{2} - \dfrac{7}{2}i$ **21.** $-\dfrac{1}{2} + \dfrac{\sqrt{3}}{2}i$ **23.** $2i$ **25.** $-i$ **27.** i **29.** -6 **31.** $-10i$ **33.** $-2 + 2i$ **35.** 0   **37.** 0   **39.** $2i$

41. $5i$ **43.** $5i$ **45.** $\{-2i, 2i\}$ **47.** $\{-4, 4\}$ **49.** $\{3 - 2i, 3 + 2i\}$ **51.** $\{3 - i, 3 + i\}$ **53.** $\left\{\dfrac{1}{4} - \dfrac{1}{4}i, \dfrac{1}{4} + \dfrac{1}{4}i\right\}$

55. $\left\{-\dfrac{1}{5} - \dfrac{2}{5}i, -\dfrac{1}{5} + \dfrac{2}{5}i\right\}$ **57.** $\left\{-\dfrac{1}{2} - \dfrac{\sqrt{3}}{2}i, -\dfrac{1}{2} + \dfrac{\sqrt{3}}{2}i\right\}$ **59.** $\{2, -1 - \sqrt{3}i, -1 + \sqrt{3}i\}$ **61.** $\{-2, 2, -2i, 2i\}$ **63.** $\{-3i, -2i, 2i, 3i\}$

65. Two complex solutions   **67.** Two unequal real solutions   **69.** A repeated real solution   **71.** $2 - 3i$ **73.** 6   **75.** 25

77. $z + \bar{z} = (a + bi) + (a - bi) = 2a$; $z - \bar{z} = (a + bi) - (a - bi) = 2bi$

79. $\overline{z + w} = \overline{(a + bi) + (c + di)} = \overline{(a + c) + (b + d)i} = (a + c) - (b + d)i = (a - bi) + (c - di) = \bar{z} + \bar{w}$

A.5 Exercises *(page 604)*

1. $[0, 2]; 0 \le x \le 2$ **3.** $(-1, 2); -1 < x < 2$ **5.** $[0, 3); 0 \le x < 3$ **7. (a)** $6 < 8$ **(b)** $-2 < 0$ **(c)** $9 < 15$ **(d)** $-6 > -10$

9. (a) $7 > 0$ **(b)** $-1 > -8$ **(c)** $12 > -9$ **(d)** $-8 < 6$

11. (a) $2x + 4 < 5$ **(b)** $2x - 4 < -3$ **(c)** $6x + 3 < 6$ **(d)** $-4x - 2 > -4$

13. $[0, 4]$ (number line 0 to 4)   **15.** $[4, 6)$ (number line 4 to 6)   **17.** $[4, \infty)$ (number line from 4)

19. $(-\infty, -4)$ (number line to -4)   **21.** $2 \le x \le 5$ (number line 2 to 5)   **23.** $-3 < x < -2$ (number line -3 to -2)

25. $x \geq 4$

27. $x < -3$

29. $<$ **31.** $>$ **33.** $\geq$ **35.** $<$
37. $\leq$ **39.** $<$ **41.** $\geq$

43. $\{x \mid x < 4\}; (-\infty, 4)$

45. $\{x \mid x \geq -1\}; [-1, \infty)$

47. $\{x \mid x > 3\}; (3, \infty)$

49. $\{x \mid x \geq 2\}; [2, \infty)$

51. $\{x \mid x > -7\}; (-7, \infty)$

53. $\left\{x \mid x \leq \dfrac{2}{3}\right\}; \left(-\infty, \dfrac{2}{3}\right]$

55. $\{x \mid x < -20\}; (-\infty, -20)$

57. $\left\{x \mid x \geq \dfrac{4}{3}\right\}; \left[\dfrac{4}{3}, \infty\right)$

59. $\{x \mid 3 \leq x \leq 5\}; [3, 5]$

61. $\left\{x \mid \dfrac{2}{3} \leq x \leq 3\right\}; \left[\dfrac{2}{3}, 3\right]$

63. $\left\{x \mid -\dfrac{11}{2} < x < \dfrac{1}{2}\right\}; \left(-\dfrac{11}{2}, \dfrac{1}{2}\right)$

65. $\{x \mid -6 < x < 0\}; (-6, 0)$

67. $\{x \mid x < -5\}; (-\infty, -5)$

69. $\{x \mid x \geq -1\}; [-1, \infty)$

71. $\left\{x \mid \dfrac{1}{2} \leq x < \dfrac{5}{4}\right\}; \left[\dfrac{1}{2}, \dfrac{5}{4}\right)$

73. $\{x \mid -6 < x < 6\}; (-6, 6)$

75. $\{x \mid x \leq -4 \text{ or } x \geq 4\};$
$(-\infty, -4] \text{ or } [4, \infty)$

77. $\{x \mid -4 < x < 4\}; (-4, 4)$

79. $\{x \mid x < -4 \text{ or } x > 4\};$
$(-\infty, -4) \text{ or } (4, \infty)$

81. $\{x \mid 1 < x < 3\}; (1, 3)$

83. $\left\{t \mid -\dfrac{2}{3} \leq t \leq 2\right\}; \left[-\dfrac{2}{3}, 2\right]$

85. $\{x \mid x \leq 1 \text{ or } x \geq 5\}; (-\infty, 1] \text{ or } [5, \infty)$

87. $\left\{x \mid -1 < x < \dfrac{3}{2}\right\}; \left(-1, \dfrac{3}{2}\right)$

89. $\{x \mid x < -1 \text{ or } x > 2\}; (-\infty, -1) \text{ or } (2, \infty)$

91. No real solutions; $\varnothing$

93. $\left| x - 2 \right| < \dfrac{1}{2}; \left\{x \mid \dfrac{3}{2} < x < \dfrac{5}{2}\right\}$

95. $|x + 3| > 2; \{x \mid x < -5 \text{ or } x > -1\}$

97. $21 < \text{age} < 30$ **99.** $|x - 98.6| \geq 1.5; \{x \mid x \leq 97.1 \text{ or } x \geq 100.1\}$
101. (a) Male ≥ 73.4 **(b)** Female ≥ 79.7 **(c)** A female can expect to live at least 6.3 years longer.
103. The agent's commission ranges from $45,000 to $95,000, inclusive. As a percent of selling price, the commission ranges from 5% to approximately 8.6%, inclusive.   **105.** The amount withheld varies from $72.14 to $93.14, inclusive.   **107.** The usage varied from approximately 675.43 to 2500.86 kilowatt-hours, inclusive.   **109.** The dealer's cost varies from $7457.63 to $7857.14, inclusive.
111. You need at least a 74 on the last test.   **113.** The amount of gasoline ranged from 12 to 20 gal, inclusive.

115. $\dfrac{a + b}{2} - a = \dfrac{a + b - 2a}{2} = \dfrac{b - a}{2} > 0; \text{therefore}, a < \dfrac{a + b}{2}.$

$b - \dfrac{a + b}{2} = \dfrac{2b - a - b}{2} = \dfrac{b - a}{2} > 0; \text{therefore}, b > \dfrac{a + b}{2}.$

117. $(\sqrt{ab})^2 - a^2 = ab - a^2 = a(b - a) > 0$; thus, $(\sqrt{ab})^2 > a^2$ and $\sqrt{ab} > a$

$b^2 - (\sqrt{ab})^2 = b^2 - ab = b(b - a) > 0$; thus $b^2 > (\sqrt{ab})^2$ and $b > \sqrt{ab}$

119. $h = \dfrac{1}{\dfrac{1}{2}\left(\dfrac{1}{a} + \dfrac{1}{b}\right)} = \dfrac{2}{\dfrac{b}{ab} + \dfrac{a}{ab}} = \dfrac{2ab}{a + b}$

$h - a = \dfrac{2ab}{a + b} - a = \dfrac{ab - a^2}{a + b} = \dfrac{a(b - a)}{a + b} > 0$; thus, $h > a$.

$b - h = b - \dfrac{2ab}{a + b} = \dfrac{b^2 - ab}{a + b} = \dfrac{b(b - a)}{a + b} > 0$; thus $h < b$.

A.6 Exercises *(page 606)*

1. 3 **3.** -2 **5.** $2\sqrt{2}$ **7.** $-2x\sqrt[3]{x}$ **9.** x^3y^2 **11.** x^2y **13.** $6\sqrt{x}$ **15.** $6x\sqrt{x}$ **17.** $15\sqrt[3]{3}$ **19.** $12\sqrt{3}$ **21.** $2\sqrt{3}$ **23.** $x - 2\sqrt{x} + 1$

25. $-5\sqrt{2}$ **27.** $(2x - 1)\sqrt[3]{2x}$ **29.** $\dfrac{\sqrt{2}}{2}$ **31.** $-\dfrac{\sqrt{15}}{5}$ **33.** $\dfrac{\sqrt{3}(5 + \sqrt{2})}{23}$ **35.** $\dfrac{-19 + 8\sqrt{5}}{41}$ **37.** $\dfrac{5\sqrt[3]{4}}{2}$ **39.** $\dfrac{2x + h - 2\sqrt{x(x + h)}}{h}$

41. 4 **43.** -3 **45.** 64 **47.** $\dfrac{1}{27}$ **49.** $\dfrac{27\sqrt{2}}{32}$ **51.** $\dfrac{27\sqrt{2}}{32}$ **53.** $x^{7/12}$ **55.** xy^2 **57.** $x^{4/3}y^{5/3}$ **59.** $\dfrac{8x^{3/2}}{y^{1/4}}$ **61.** $\dfrac{3x + 2}{(1 + x)^{1/2}}$ **63.** $\dfrac{2 + x}{2(1 + x)^{3/2}}$

65. $\dfrac{4 - x}{(x + 4)^{3/2}}$ **67.** $\dfrac{1}{2}(5x + 2)(x + 1)^{1/2}$ **69.** $2x^{1/2}(3x - 4)(x + 1)$ **71.** $\dfrac{2(2 - x)(2 + x)}{(8 - x^2)^{1/2}}$

A.7 Exercises *(page 624)*

1. **(a)** $\dfrac{1}{2}$ **(b)** If x increases by 2 units, y will increase by 1 unit. **3.** **(a)** $-\dfrac{1}{3}$ **(b)** If x increases by 3 units, y will decrease by 1 unit.

5. Slope $= -\dfrac{3}{2}$

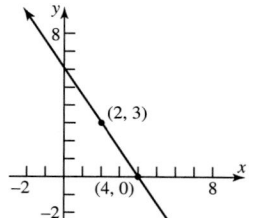

7. Slope $= -\dfrac{1}{2}$

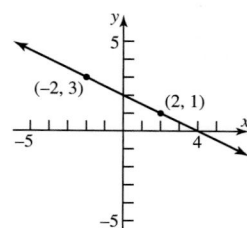

9. Slope $= 0$

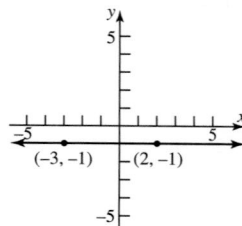

11. Slope undefined

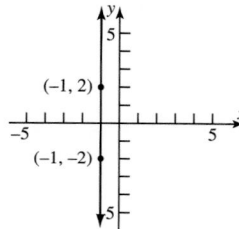

13.

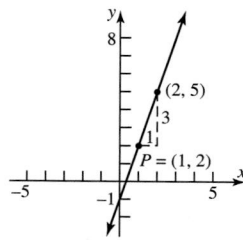

15.

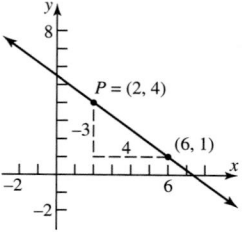

17.

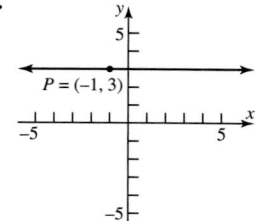

19.

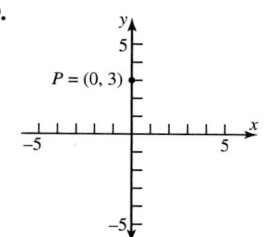

21. $(2, 6); (3, 10); (4, 14)$
23. $(4, -7); (6, -10); (8, -13)$
25. $(-1, -5); (0, -7); (1, -9)$
27. $x - 2y = 0$ or $y = \dfrac{1}{2}x$
29. $x + y = 2$ or $y = -x + 2$
31. $2x - y = 3$ or $y = 2x - 3$
33. $x + 2y = 5$ or $y = -\dfrac{1}{2}x + \dfrac{5}{2}$

35. $3x - y = -9$ or $y = 3x + 9$ **37.** $2x + 3y = -1$ or $y = -\frac{2}{3}x - \frac{1}{3}$ **39.** $x - 2y = -5$ or $y = \frac{1}{2}x + \frac{5}{2}$

41. $3x + y = 3$ or $y = -3x + 3$ **43.** $x - 2y = 2$ or $y = \frac{1}{2}x - 1$ **45.** $x = 2$; no slope-intercept form

47. $2x - y = -4$ or $y = 2x + 4$ **49.** $2x - y = 0$ or $y = 2x$ **51.** $x = 4$; no slope–intercept form

53. $2x + y = 0$ or $y = -2x$ **55.** $x - 2y = -3$ or $y = \frac{1}{2}x + \frac{3}{2}$ **57.** $y = 4$

59. Slope $= 2$; y-intercept $= 3$

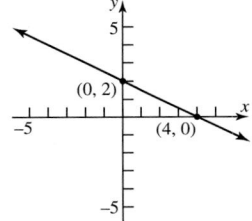

61. $y = 2x - 2$; Slope $= 2$; y-intercept $= -2$

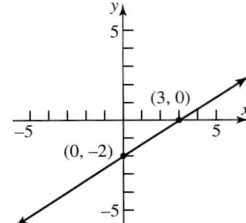

63. Slope $= \frac{1}{2}$; y-intercept $= 2$

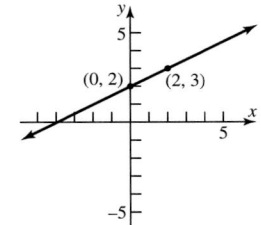

65. $y = -\frac{1}{2}x + 2$; Slope $= -\frac{1}{2}$; y-intercept $= 2$

67. $y = \frac{2}{3}x - 2$; Slope $= \frac{2}{3}$; y-intercept $= -2$

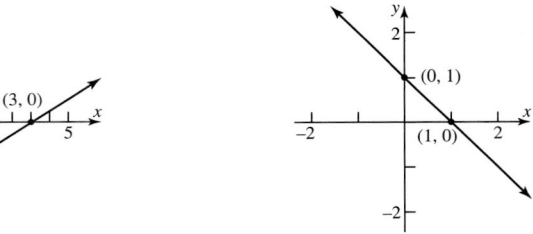

69. $y = -x + 1$; Slope $= -1$; y-intercept $= 1$

71. Slope undefined; no y-intercept

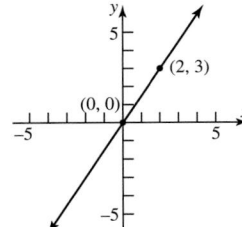

73. Slope $= 0$; y-intercept $= 5$

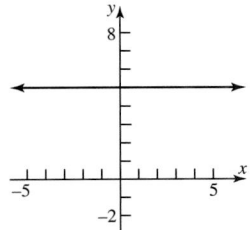

75. $y = x$; Slope $= 1$; y-intercept $= 0$

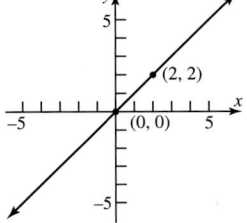

77. $y = \frac{3}{2}x$; Slope $= \frac{3}{2}$; y-intercept $= 0$

79. $y = 0$ **81.** (b)   **83.** (d)   **85.** $x - y = -2$ or $y = x + 2$

87. $x + 3y = 3$ or $y = -\frac{1}{3}x + 1$

89. $C = 0.07x + 29$; \$36.70; \$45.10

91. $^\circ C = \frac{5}{9}(^\circ F - 32)$; approximately $21.11^\circ C$

93. (a) $P = 0.5x - 100$
(b) \$400   **(c)** \$2400

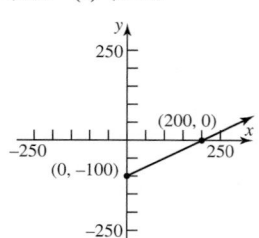

95. $C = 0.06543x + 5.65$; \$25.28; \$54.72

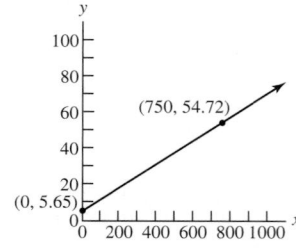

97. All have the same slope, 2; the lines are parallel.

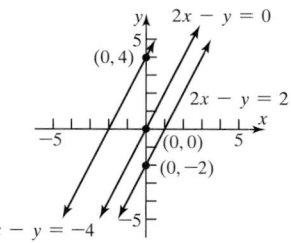

99. (a) $x^2 + (mx + b)^2 = r^2$

$(1 + m^2)x^2 + 2mbx + b^2 - r^2 = 0$

One solution if and only if discriminant $= 0$

$(2mb)^2 - 4(1 + m^2)(b^2 - r^2) = 0$

$-4b^2 + 4r^2 + 4m^2r^2 = 0$

$r^2(1 + m^2) = b^2$

(b) $x = \dfrac{-2mb}{2(1 + m^2)} = \dfrac{-2mb}{2b^2/r^2} = -\dfrac{r^2m}{b}$

$y = m\left(-\dfrac{r^2m}{b}\right) + b = -\dfrac{r^2m^2}{b} + b = \dfrac{-r^2m^2 + b^2}{b} = \dfrac{r^2}{b}$

(c) Slope of tangent line $= m$

Slope of line joining center to point of tangency $= \dfrac{r^2/b}{-r^2m/b} = -\dfrac{1}{m}$

101. $\sqrt{2}x + 4y - 11\sqrt{2} + 12 = 0$

103. $x + 5y + 13 = 0$

105. (b), (c), (e), (g)

107. (c)

A.8 Exercises *(page 632)*

1. Linear relation   **3.** Linear relation   **5.** Nonlinear relation

7. (a)

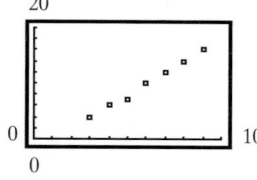

(b) Answers will vary. Using $(3, 4)$ and $(9, 16)$: $y = 2x - 2$

(c)

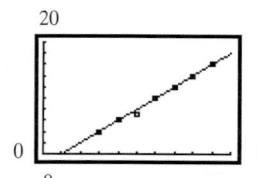

(d) $y = 2.0357x - 2.3571$

(e)

9. (a)

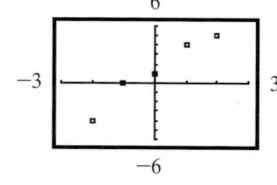

(b) Answers will vary. Using $(-2, -4)$ and $(2, 5)$: $y = \dfrac{9}{4}x + \dfrac{1}{2}$

(c)

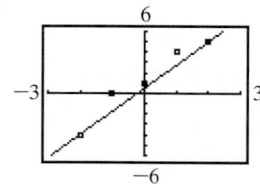

(d) $y = 2.2x + 1.2$

(e)

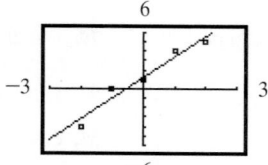

11. (a)

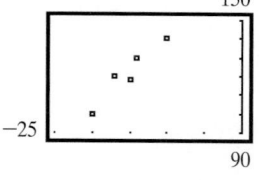

(b) Answers will vary. Using $(-20, 100)$ and $(-10, 140)$: $y = 4x + 180$

(c)

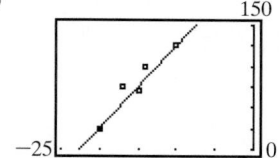

(d) $y = 3.8613x + 180.292$

(e)

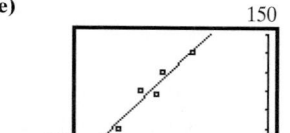
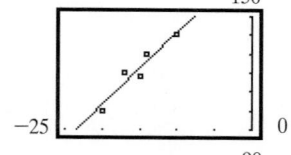

13. (a)

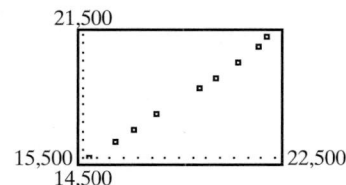

(b) $C(I) = 1.03748I - 1897.5071$

(c) If income increases by \$1, then consumption increases by about \$1.04.

(d) About \$20,408

15. (a)

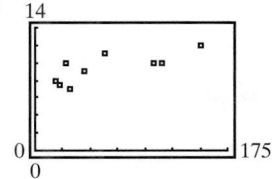

(b) $L(G) = 0.0261G + 7.8738$

(c) If gestation period increases by 1 day, then life expectancy increases by about 0.0261 years.

(d) About 10.2 years

17. (a) No   **(b)**

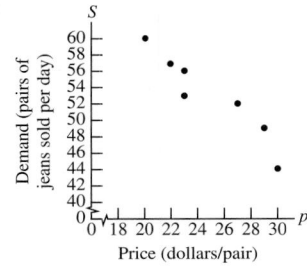

(c) $D = -1.3355p + 86.1974$

(d) If the price increases by \$1, the quantity sold per day decreases by about 1.34 pairs of jeans.

(e) $D(p) = -1.3355p + 86.1974$

(f) $\{p \mid p > 0\}$

(g) About 49 pairs

CONICS

Parabola

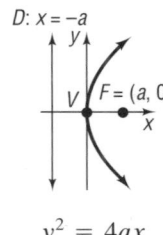

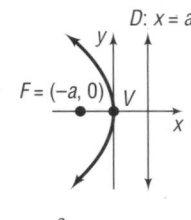

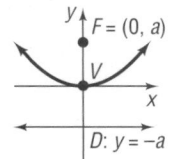

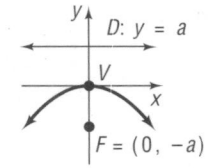

$$y^2 = 4ax \qquad y^2 = -4ax \qquad x^2 = 4ay \qquad x^2 = -4ay$$

Ellipse

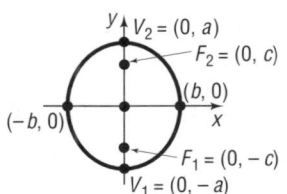

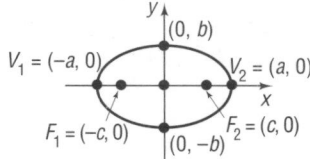

$$\frac{x^2}{a^2} + \frac{y^2}{b^2} = 1, \quad a > b, \quad c^2 = a^2 - b^2 \qquad\qquad \frac{x^2}{b^2} + \frac{y^2}{a^2} = 1, \quad a > b, \quad c^2 = a^2 - b^2$$

Hyperbola

$$\frac{x^2}{a^2} - \frac{y^2}{b^2} = 1, \quad c^2 = a^2 + b^2 \qquad\qquad \frac{y^2}{a^2} - \frac{x^2}{b^2} = 1, \quad c^2 = a^2 + b^2$$

$$\text{Asymptotes:} \quad y = \frac{b}{a}x, \quad y = -\frac{b}{a}x \qquad\qquad \text{Asymptotes:} \quad y = \frac{a}{b}x, \quad y = -\frac{a}{b}x$$

PROPERTIES OF LOGARITHMS

$$\log_a (MN) = \log_a M + \log_a N$$

$$\log_a\!\left(\frac{M}{N}\right) = \log_a M - \log_a N$$

$$\log_a M^r = r \log_a M$$

$$\log_a M = \frac{\log M}{\log a} = \frac{\ln M}{\ln a}$$

THE
Norton
Anthology
OF
American
Literature

RONALD GOTTESMAN
University of Southern California

FRANCIS MURPHY
Smith College

LAURENCE B. HOLLAND

HERSHEL PARKER
University of Delaware

DAVID KALSTONE
Rutgers, The State University of New Jersey

WILLIAM H. PRITCHARD
Amherst College

Shorter Edition

W • W • NORTON & COMPANY

NEW YORK • LONDON

W. W. Norton & Company, Inc. 500 Fifth Avenue, New York, N.Y. 10110

*Since this page cannot legibly accommodate all the copyright notices,
the three pages following constitute an extension of the copyright page.*

Library of Congress Cataloging in Publication Data
Main entry under title:
The Norton anthology of American literature. Shorter Edition.
    Includes bibliographies and index.
    1. American literature.   I. Gottesman, Ronald.
PS507.N65   1980   810'.8   79–23969

                    4 5 6 7 8 9 0

        ISBN 0-393-95119-7 CL

        ISBN 0-393-95112-X PBK

Contents

American Literature 1820–1865

American Literature 1865–1914

American Literature
between the Wars 1914–1945   1229

Contemporary American Prose 1945–

Contemporary American Poetry 1945–

and contemporary poetry, and to stimulate them to carry on their reading after they have finished the course.

In making their selections for the Shorter Edition, the editors were guided by general criteria of historical importance and thematic relevance. More important, however, was the literary worth and the accessibility of each author or work, especially for students of the present day. The editors have represented the work of major authors as diversely and copiously as possible: there are, for example, more essays by Emerson, more stories by Hawthorne, more poems by Whitman as well as by many of the important poets between the wars than may be found in any other short anthology of American Literature. In addition to these major authors, the editors have included a number of authors whose importance, even in a short survey of the literature, ought not to go unrecognized—such figures, for example, as Margaret Fuller, Gertrude Stein, Frank O'Hara, and John Ashbery. And many selections have been made that, although omitted in other anthologies, are central to an understanding of their authors: John Winthrop's sermon, for example; Thoreau's *Life without Principle*; Kate Chopin's hitherto unanthologized story *The Storm*; Pound's *Mauberley 1920*; and very recent work by almost all of the contemporary poets. The editors have also undertaken, as in the two-volume edition, to redress the long neglect of women writers in America.

This latest of the "Norton anthologies" incorporates the features that have established a new standard in literary texts for the classroom. The format is that of a book to be read for pleasure: there are no forbidding double columns of prose and verse, and the text is inviting to the eye. The editorial materials—introductions, headnotes, footnotes—are terse but full, and designed to give the student the information needed, without pre-empting the interpretative function either of the student or the instructor. The "Selected Bibliographies" at the end of the volume provide guides to further reading and research. All these editorial aids make the anthology self-sufficient, so that a student is able to understand each selection and to place it in its historical and biographical context without the need for recourse to a collection of reference books.

Care has been taken to provide the most accurate available versions of each work that is represented. Indeed, several of the major texts—the excerpts from Franklin's *Autobiography*, some of the material by Mark Twain, Howells's *Novel-Writing and Novel-Reading*—have been newly edited from the original manuscripts. And each text is printed in the form that accords, as closely as it is possible to determine, to the intentions of its author. On the principle that nonfunctional features such as archaic spellings and typography pose unnecessary problems for beginning students, we have modernized the spelling and (very sparingly) the punctuation in early American literature texts. The one exception is Franklin's *Autobiography*, which was left unchanged because it is a new edition from manuscript. We have used square brackets to indicate titles supplied by the editors, and have, whenever a portion of a text has been omitted, indicated that omission by three asterisks.

The editors of this anthology were selected on the basis of their expertness in their individual areas, and also because they combine respect for the best that has been thought and said about literature in the past with an alertness (as participants, as well as observers) to the altering interests, procedures, and evaluations in contemporary scholarship and criticism. Each editor was given ultimate responsibility for his own period, but all collaborated in the total enterprise.

In preparing *The Norton Anthology of American Literature* we have incurred obligations to hundreds of teachers throughout the country who generously answered our questions, made suggestions, or prepared critiques. Many of them are listed in the two-volume edition; here we take the opportunity to thank them warmly again for their invaluable aid. Thanks are also due to the editors' assistants, and to many members of the publisher's staff. All have helped us in the difficult undertaking to represent, in a single volume, the extraordinary variety and quality of the American literary heritage.

Early American Literature 1620–1820

L ONG BEFORE Captain John Smith established Jamestown in 1607, the European imagination had been entranced by rumors of the New World's plenty. But it was probably Captain Smith, rather than any other, who convinced English readers that there was an earthly paradise not far from their shores. In his *A Description of New England* (1616) he wrote that "Here nature and liberty afford us that freely which in England we want [lack], or it costs us dearly." What greater satisfaction is there, he asked, than hauling in one's supper by dropping a hook and line into any plentiful river or stream; is it not "pretty sport" to "pull up two pence, six pence, and twelve pence" as fast as you can let out a line? One hundred twenty-five years later another Virginia planter, William Byrd, would add to the fabled accounts of the place in his *History of the Dividing Line*, and it is significant that Thomas Jefferson's one book, *Notes on the State of Virginia* (1785, 1787), was written in response to inquiries made by a French naturalist concerning the geography and resources of his state. European readers were anxious to sort American fable from fact, but as Smith's *Description* convinced them, the facts themselves were fabulous.

THE PURITAN EXPERIMENT: PLYMOUTH PLANTATION

Although those separatists from the Church of England whom we call "Pilgrims" were familiar with Captain Smith's *Description* and followed his map of the Atlantic coast, they were not sympathetic to his proposal that he join their emigration to the New World; for Smith was primarily an adventurer, explorer, and trader, and while this group was not composed entirely of "reborn" Christians (only about twenty-seven of the one hundred persons aboard the *Mayflower* were Puritans), and even those were not indifferent to the material well-being of their venture, their leaders had more in mind than mercantile success. These pilgrims thought of themselves as soldiers in a war against Satan—the Arch-Enemy—who planned to ruin the Kingdom of God on earth by sowing discord among those who professed to be Christians. This small band of believers saw no hope of reforming a national church and its Anglican hierarchy from within. In 1608, five years after the death of Queen Elizabeth and with an enemy of Puritanism, James Stuart, on the throne, they left England and settled in Holland, where, William Bradford tells us, they saw "fair and beautiful cities" and the "grisly face of poverty" confronting them. Isolated by their language, and unable to farm, they turned to mastering trades (Bradford himself became a weaver). Later, fearing that they would eventually lose their identity as a religious community living as strangers in a foreign land, they applied for a charter to settle in the Virginia Plantation—a vast tract of land which included what is now New England. Sponsored by merchants who were anxious to receive repayment in goods from the New World, they sailed from Southampton, England, in September, 1620. Sixty-six days later, taken by strong winds much farther north than they had anticipated, they dropped anchor at Cape Cod and established their colony at Plymouth.

In spite of the fact that their separatism does not make them representative of the large number of emigrants who came to these shores in the seventeenth century (Plymouth was eventually absorbed into the Massachusetts Bay Colony in 1691 when a new charter was negotiated), their story has become an integral part of our literature. Bradford's account of a chosen people, exiles in a "howling wilderness," who struggled against all adversity to bring into being the City of God on earth, is ingrained in our national consciousness. Both in the nineteenth century and in the twentieth, Americans have seen themselves as a "redeemer nation," without, of course, possessing Bradford's Christian ideals. What gives Bradford's book its great strength, in spite of his obvious prejudices, is his ability to keep the ideals of the Pilgrims before us as he describes the harsh reality of their struggle against not only the external forces of nature but the even more damaging corruption of worldliness within the community.

THE PURITAN EXPERIMENT: THE MASSACHUSETTS BAY COLONY

Far more representative in attitude toward the Church of England were the Puritans who joined the Massachusetts Bay Colony under the leadership of John Winthrop. They were dissenting but nonseparating—although it might be argued that geographical distance from London, and a charter which located the seat of their colony in Boston, left them nonseparating in theory rather than practice. Whatever their difference with respect to the Church of England, however, the basic beliefs of both groups were identical: both held with Martin Luther that no Pope or Bishop had a right to impose any law upon a Christian soul without consent and, following John Calvin, that God chose freely those He would save and those He would damn eternally.

Too much can be made of this doctrine of election; those who have not read the actual Puritan sermons often come away from secondary sources with the mistaken notion that Puritans talked about nothing but damnation. Puritans did indeed hold that God had chosen, before their birth, those whom He wished to save; but it does not follow that the Puritans considered most of us to be born damned. While Puritans argued that Adam broke the "Covenant of Works" (the promise God made to Adam that he was immortal and could live in Paradise forever as long as he obeyed God's commandments) when he disobeyed and ate of the tree of knowledge of good and evil, thereby bringing sin and death into the world, their central doctrine was the new "Covenant of Grace," a binding agreement Christ made with all men who believed in Him, and which He sealed with His crucifixion, promising them eternal life. Puritans thus addressed themselves not to the hopelessly unregenerate but to the indifferent, and they addressed the heart more often than the mind, always distinguishing between "historical" or rational understanding and heartfelt "saving faith." There is more joy in Puritan life and thought than we often credit, and this joy is the direct result of meditation on the doctrine of Christ's redeeming power. Edward Taylor is not alone in making his rapturous litany of Christ's attributes: "He is altogether lovely in everything, lovely in His person, lovely in His natures, lovely in His properties, lovely in His offices, lovely in His titles, lovely in His practice, lovely in His purchases and lovely in His relations." All of Taylor's art is a meditation on the miraculous gift of the Incarnation, and, in this respect, his sensibility is typically Puritan. Anne Bradstreet, who is remarkably frank about confessing her religious doubts, told her children that it was "upon this rock Christ Jesus" that she built her faith.

Their lives, however, were hard. Anne Bradstreet's father told people in England to come over and join them if their lives were "endued with grace," but that others were "not fitted for this business"; that there was not a house where one had not died, and that if they survived the terrible winter they had to face the devastating infections that were the result of summer heat. Bradford's account of what he called "the starving time" are among the most moving in his history, and nothing in Captain Smith's *Discovery* had hinted at how oppressive daily life might be. Sarah Kemble Knight of Boston provides a healthy antidote to any sentimental notion we might have that life on the frontier was invigorating. Puritan letters, diaries, histories, and poetry all attest to their faith in a larger plan, a "noble design" as Cotton Mather put it, which made daily life bearable.

In this Christocentric world it is not surprising that Puritans held to the strictest requirements regarding communion or, as they preferred to call it, the Lord's Supper. It was the most important of the two sacraments they recognized (baptism being the other), and they guarded it with a zeal which set them apart from all other dissenting churches. In the beginning, communion was taken only by church members—those who had stood before their minister and elders and given an account (or "relation") of their conversion—and was regarded as a sign of election. This insistence on challenging their members made these New England churches more rigorous than any others, and confirmed the feeling that they were a special few. Thus, when John Winthrop addressed the immigrants to the Bay Colony aboard the flagship *Arbella* in 1630, he told them that the eyes of the world were upon them, and that they would be an example for all, a "city upon a hill."

PURITAN HISTORIOGRAPHY

Puritans held the writing of history in high regard; for, as heirs of Renaissance thought, they believed that lasting truths were to be gained by studying the lives of noble men. Cotton Mather urged students of the ministry to read not only early church historians but the classical historians Xenophon, Livy, Tacitus, and Plutarch as well. Puritans saw all of human time as a progression toward the fulfillment of God's design on earth. Therefore, pre-Christian history could be read as a preparation for Christ's entry into the world. They learned this lesson from medieval Biblical scholars, who interpreted figures in the Old Testament as foreshadowings of Christ. This method of comparison, called "typology," was an ingrained habit of Puritan thinking, and it made them compare themselves, as a chosen people, to the Israelites of old, who had been given the promise of a new land. Cotton Mather said that John Winthrop was the Puritan Moses whose education had prepared him to fulfill the "noble design of carrying a colony of chosen people into an American wilderness."

Puritans believed that God's hand was present in every human event and that He rewarded good and punished bad. History, therefore, revealed what God approved of or condemned, and if God looked favorably upon a nation, His approval could be evidenced in its success. Puritans had enough confidence in God's design to believe that no facts were too small or insignificant to be included in that design; everything could emblemize something. In writing about Anne Bradstreet, Adrienne Rich observes that seventeenth-century Puritan life was perhaps "the most self-conscious ever lived"; that "faith underwent its hourly testing, the domestic mundanities were episodes in the drama; the piecemeal thoughts of a woman stirring a pot, clues to her 'justification' in Christ." John Winthrop in his diary records a struggle between a snake and a mouse and is surprised to see the seemingly weaker emerge the victor. His Boston friend, Mr. Wilson, however, saw the event as a battle between Satan and a "poor contemptible people, which God hath brought hither, which should overcome Satan here and dispossess him of his kingdom." When a young sailor on board the *Mayflower* mocked those Puritans who were sick, Bradford found it fitting that the sailor should himself succumb to a "grievous disease." This sense of the universal significance of all things meant that drama was present in every believer's life and that individual lives could be as symbolic as the life of a nation.

The greatest of all the Puritan historians was Cotton Mather, and in his *Magnalia Christi Americana* (1702) the myth of a chosen people took on its fullest resonance of meaning. By the time Mather undertook his history, the original Puritan community had vanished, leaving behind heirs to its lands and fortunes but not to its spirituality. Mather saw himself as one of the last defenders of the "old New England way," and all the churches as under attack from new forces of secularism. As a historian, Mather solved his problem by not focusing on the dissolution of the Puritan community but writing "saints' lives" instead, each of which would serve as an example of the progress of the individual Christian soul and an allegory of the potential American hero. Under Mather's artistry, Winthrop's vision of a community of saints living in mutual concern and sympathy became an ideal rather than a historical reality. The words "New England" would symbolize the effort to realize the City of God on earth, and "whether New England may live anywhere else or no," he said, "it must live in our history."

AN EXPANDING UNIVERSE

It should come as no surprise to learn that Cotton Mather was defensively retrospective in his ecclesiastical history of New England; for the enormous changes—economic, social, philosophical, and scientific—which occurred between Mather's birth in 1663 and the publication of the *Magnalia* in 1702 inevitably affected the influence and authority of Congregational churches. In 1686 Mather himself joined with Boston merchants in jailing their colonial Governor, Sir Edmund Andros, and was successful in getting him sent back to England. It was a rare occasion when church and trade saw eye to eye; the Puritan clergy disliked Andros's Anglicanism as much as the merchants hated his taxes. It was an act celebrated annually in Boston until it was replaced by celebrations honoring American independence.

The increase in population alone would account for a greater diversity of opinion in the matter of churches. In 1670, for example, the population of the colonies numbered approximately 111,000. Thirty years later the colonies contained more than a quarter of a million persons; by 1760, if one included Georgia, they numbered 1,600,000, and the settled area had tripled. The demand for and price of colonial goods increased in England, and vast fortunes were to be made in New England with any business connected with shipbuilding: especially timber, tar, and pitch. Virginia planters became rich in tobacco, and rice and indigo from the Carolinas were in constant demand.

New England towns were full of acrimonious debate between first settlers and newcomers. Town histories are full of accounts of splinter groups and the establishment of the "Second" church. In the beginning land was apportioned to settlers and allotted free, but by 1713 speculators in land were hard at work, buying as much as possible for as little as possible and selling high. The idea of a "community" of mutually helpful souls was fast disappearing. Life in the colonies was not easy, but the hardships and dangers the first settlers faced were mostly overcome, and compared to crowded cities like London, it was healthier, cheaper, and more hopeful. Those who could arrange their passage came in great numbers. Boston almost doubled in size from 1700 to 1720. It is also important to note that the great emigration to America which occurred in the first half of the eighteenth century was not primarily English. Dutch and Germans came in large numbers and so did French Protestants. Jewish merchants and craftsmen were well known in New York and Philadelphia.

By 1750 Philadelphia had become the unofficial capital of the colonies and was second only to London as a city of commerce. In 1681 the Quaker William Penn exchanged a large claim against the Crown for land in the New World. He was named Proprietor (rather than Governor, since he actually owned the territory) of Pennsylvania and immediately opened the land to settlement by people of all faiths. Penn had the genius to bestow the privilege of self-government on the people of Pennsylvania and in his "Frame of Government" told them that "Liberty without obedience is confusion, and obedience without liberty is slavery." These thousands of emigrants did not think of themselves as displaced Englishmen; they thought of themselves as Americans. In 1702 no one would have dreamed of an independent union of colonies, but by 1762, fifty years later, it was a distinct possibility.

THE ENLIGHTENMENT

Great challenges to seventeenth-century beliefs were posed by scientists and philosophers, and it has sometimes been suggested that the "modern" period dates from 1662 and the founding of the British scientific academy known, because of the patronage of King Charles II, as the Royal Society. The greatest scientists of the age like Sir Isaac Newton (1642–1727) and philosophers like John Locke (1632–1704) saw no conflict between their discoveries and traditionally held Christian truths. They saw nothing heretical in arguing that the universe was an orderly system and that by the application of reason mankind would comprehend its laws. But the inevitable result of their inquiries was to make the universe seem more rational and benevolent than it had been represented by Puritan doctrine. Because the world seemed more comprehensible, people paid less attention to re-

vealed religion, and a number of seventeenth-century modes of thought—
Bradford and Winthrop's penchant for the allegorical and emblematic, seeing every natural and human event as a message from God, for instance—
seemed almost medieval and decidedly quaint. These new scientists and
philosophers were called "Deists"; they deduced the existence of a Supreme
Being from the construction of the universe itself rather than from the
Bible. "A creation," as one distinguished historian has put it, "presupposes
a creator." People were less interested in the metaphysical wit of introspective divines than in the progress of ordinary men as they made their
way in the world. They assumed that men were naturally good, and dwelt
on neither the Fall nor the Incarnation. A harmonious universe proclaimed
the beneficence of God, and Deists argued that man himself should be as
generous. They were not interested in theology but in man's own nature.
Americans as well as Englishmen knew Alexander Pope's famous couplet:

> Know then thyself, presume not God to scan,
> The proper study of mankind is man.

Locke said that "our business" here on earth "is not to know all things, but
those which concern our conduct." In suggesting that we are not born with
a set of innate ideas of good or evil and that the mind is rather like a blank
wax tablet upon which experiences are inscribed (a *tabula rasa*), Locke
qualified traditional belief.

THE GREAT AWAKENING

A conservative reaction against the world view of the new science was
bound to follow, and the first half of the eighteenth century witnessed a
number of religious revivals in both England and America. They were
sometimes desperate efforts to reassert the old values in the face of the
new and, oddly enough, were themselves the direct product of the new cult
of feeling, a philosophy which argued that man's greatest pleasure was
derived from the good he did for others and that his sympathetic emotions
(his joy as well as his tears) should not be contained. Phillis Wheatley,
whose poem on the death of the Methodist George Whitefield (1714–70)
made her famous, said that Whitefield prayed that "grace in every heart
might dwell," and longed to see "America excell." Whitefield's revival
meetings along the Atlantic seaboard were a great personal triumph; but
they were no more famous than the "extraordinary circumstances" which
occurred in Northampton, Massachusetts, under the leadership of Jonathan
Edwards in the 1730s and which have come to be synonymous with "the
Great Awakening."

Edwards also read his Locke, but he wished to liberate human beings
from their senses, not define them by those senses. Edwards was fond of
pointing out that the five senses are what we share with beasts, and that
if our ultimate goal were merely a heightened sensibility, feverish sickness
is the condition where the senses are most acute. Edwards was interested
in *supernatural* concerns, but he was himself influenced by Locke in arguing that true belief is something which we feel and do not merely comprehend intellectually. Edwards took the one doctrine most difficult for
eighteenth-century minds to accept—election—and persuaded his congregation that God's sovereignty was not only the most reasonable doctrine,
but that it was the most "delightful," and appeared to him (using adjectives
which suggest that the best analogy is to what can be apprehended sensu-

ally) "exceeding pleasant, bright, and sweet." In carefully reasoned, calmly argued prose, as harmonious and as ordered as anything the age produced, Edwards brought his great intellect to bear on doctrines that had been current the century before. Most people, when they think about the Puritans, remember Edwards's sermon *Sinners in the Hands of an Angry God*, forgetting that one hundred years had lapsed between that sermon and Winthrop's *Model of Christian Charity*. When Edwards tried to re-assert "the old New England way" and demanded accounts of conversion before admission to church membership, he was accused of being a re-actionary who thrived on hysteria, was removed from his pulpit, and was effectively silenced. He spent his last years as a missionary to the Indians in Stockbridge, Massachusetts, a town forty miles to the west of North-ampton. There he remained until invited to become president of the Col-lege of New Jersey. His death in Princeton was the direct result of his willingness to be vaccinated against smallpox and so to set an example for his frightened and superstitious students; it serves as a vivid reminder of how complicated in any one individual the reponse to the "new science" could become.

THE AMERICAN CRISIS

On June 7, 1776, at the second Continental Congress, Richard Henry Lee of Virginia moved that "these united colonies are, and of a right ought to be, free and independent states." A committee was duly appointed to pre-pare a declaration of independence, and it was approved on July 4. Although these motions and their swiftness took some delegates by surprise—the purpose of the Congress had, after all, not been to declare independence but to protest the usurpation of rights by King and Parliament and to effect a compromise with the mother country—others saw them as the inevitable consequence of the events of the decade preceding. The Stamp Act of 1764, taxing all newspapers, legal documents, and licenses, had in-furiated Bostonians and resulted in the burning of the Governor's palace; in Virginia, Patrick Henry had taken the occasion to speak impassionedly against taxation without representation. In 1770 a Boston mob had been fired upon by British soldiers, and three years later the famous "Tea Party" occurred, an act which drew hard lines in the matter of acceptable limits of British rule. The news of the April confrontation with the British in Concord and Lexington, Massachusetts, was still on everyone's tongue in Philadelphia when the second Continental Congress convened in May of 1775.

Although the drama of these events cannot be underestimated, most historians agree that it was Thomas Paine's *Common Sense*, published in January, 1776, that gave the needed push for revolution. In the course of two months it was read by almost every American. In arguing that separa-tion from England was the only reasonable course and that "the Almighty" had planted these feelings in us "for good and wise purposes," Paine was appealing to basic tenets of the Enlightenment. His clarion call to those that "love mankind," those "that dare oppose not only the tyranny but the tyrant, stand forth!" did not go unheeded. Americans needed an apol-ogist for the Revolution, and in December of 1776, when Washington's troops were at their most demoralized, it was, again, Paine's first *Crisis* paper—popularly called *The American Crisis*—which was read to all the regiments and was said to have inspired their future success.

Paine first came to America in 1774 with a note from Benjamin Franklin recommending him to publishers and editors. He was only one of a number of young writers who were able to take advantage of the times. This was, in fact, the great age of the newspaper and the moral essay; Franklin tells us that he modeled his own style on the clarity, good sense, and simplicity of the English essayists Joseph Addison and Richard Steele. The first newspaper in the colonies appeared in 1704, but by the time of the Revolution there were almost fifty papers and forty magazines. The great cry was for a "national literature" (meaning anti-British), and the political events of the 1770s were advantageous for a career. Philip Freneau made his first success as a writer as a satirist of the British, and after the publication of his *Poems Written Chiefly during the Late War* (1786) he turned to newspaper work, editing the New York *Daily Advertiser* and writing anti-Federalist party essays, making himself an enemy of Alexander Hamilton in the process. Freneau's career was marked by restlessness and indecision, although in his case financial necessity came between his life and his art. The first American writer able to live exclusively by his craft was Washington Irving.

The crisis in American life caused by the Revolution made artists self-conscious about American subjects. It would be another fifty years before writers discovered ways of being American without compromising their integrity. One of the ironies of our history is that the Revolution itself has rarely proved to be a usable subject for American literature and art.

THE PURSUIT OF HAPPINESS

When John Winthrop described his "model" for a Christian community, he envisioned a group of men and women working together for the common good, each one of whom knew his place in the social structure and accepted God's disposition of goods. At all times, he said, "some must be rich, some poor, some high and eminent in power and dignity," others low and "in subjection." Ideally, it was to be a community of love, each made equal by their fallen nature and their concern for the salvation of their souls; but it was to be a stable community, and Winthrop would not have imagined very much social change. One hundred forty years later John Quincy Adams, our second President, envisioned a model community, decreed by higher laws, when he said that the American colonies were a part of a "grand scheme and design in Providence for the illumination of the ignorant and the emancipation of the slavish part of mankind all over the earth." Adams witnessed social mobility of a kind and number, however, that no European before him would have dreamed possible. As historians have observed, European critics of America in the eighteenth and nineteenth centuries never understood that great social change was possible without social upheaval primarily because there was no feudal hierarchy to overthrow. When Crèvecoeur wanted to distinguish America from Europe, it was the medievalism of the latter that he wished to stress. The visitor to America, he said, "views not the hostile castle, and the haughty mansion, contrasted with the clay-built hut and miserable cabin, where cattle and men help to keep each other warm, and dwell in meanness, smoke and indigence." Of course, not everyone was free. Some of our founding fathers, like Thomas Jefferson, were large slaveowners, and it was still not possible to vote without owning property. Women had hardly any rights at all: they could not vote, and young women were educated at home, excluded by their studies from anything other than domestic employment. Nevertheless, the same forces that were undermining church authority in New England (in New York and Philadelphia no such hierarchy existed) were effecting social change. The two assumptions held to be true by most eighteenth-century Americans were, as Russel Nye once put it, "the perfectability of man, and the prospect of his future progress." Much of the imaginative energy of the second half of the eighteenth century was expanded in correcting institutional injustices: the tyranny of monarchy, the tolerance of slavery, the misuse of prisons. Few doubted that with the application of intelligence the human lot could be improved; and writers like Freneau, Franklin, and Crèvecoeur argued that, if it were not too late, the white man might learn something about brotherhood and manners from noble savages rather than from rude white settlers, slaveowners, and backwoodsmen.

In many ways it is Franklin who best represents the spirit of the Enlightenment in America: self-educated, social, assured, a man of the world, ambitious and public-spirited, speculative about the nature of the universe, but in matters of religion content to observe the actual conduct of men rather than to debate supernatural matters which are unprovable. When Ezra Stiles asked him about his religion, he said he believed in the "creator of the universe" but he doubted the "divinity of Jesus." He would never dogmatize about it, however, because he expected soon "an opportunity of knowing the truth with less trouble." Franklin always presents himself as a man depending on firsthand experience, too worldly-wise to be caught off guard. His posture, however, belies one side of the eighteenth century which can be accounted for neither by the inheritance of Calvin nor by the empiricism of Locke: those idealistic assumptions which underlie the great public documents of the American Revolution, especially the Declaration of Independence. There are truths which, Joel Barlow once said, were "as perceptible when first presented to the mind as age or world of experience could make them." Given the representative nature of Franklin's character, it seems right that of the documents most closely associated with the formation of the American Republic—the Declaration of Independence, the treaty of alliance with France, the Treaty of Paris, and the Constitution—only he should have signed all four.

The fact that Americans in the last quarter of the eighteenth century would hold that "certain truths are self-evident, that all men are created equal, that they are endowed by their Creator with certain unalienable Rights, that among these are Life, Liberty and the pursuit of Happiness," is the result, as both Leon Howard and Gary Wills have argued, of their reading the Scottish philosophers, particularly Francis Hutcheson and Lord Kames (Henry Home), who argued that all men in all places possess a sense common to all—a moral sense—which contradicted the notion of the mind as an empty vessel awaiting experience. This idealism paved the way for writers like Bryant, Emerson, Thoreau, and Whitman, but in the 1770s its presence is found chiefly in politics and ethics. The assurance of a universal sense of right and wrong made possible both the overthrow of tyrants and the restoration of order; and it allowed men to make new earthly covenants, not, as was the case with Bradford and Winthrop, for the glory of God, but, as Thomas Jefferson argued, for man's right to happiness on earth.

JOHN WINTHROP
1588–1649

John Winthrop was born in Groton, England, on an estate which his father purchased from Henry VIII. It was a prosperous farm, and Winthrop had all the advantages which his father's social and economic position would allow. He went to Cambridge University for two years and married at the age of seventeen. It was probably at Cambridge University that Winthrop was exposed to Puritan ideas. Unlike Bradford and the Pilgrims, however, Winthrop was not a separatist; that is, he wished to reform the national church from within, purging it of everything that harked back to Rome, especially the hierarchy of the clergy and all the traditional Catholic rituals. For a time Winthrop thought of becoming a clergyman himself, but instead he turned to the practice of law.

In the 1620s severe economic depression in England made Winthrop realize that he could not depend upon the support of his father's estate. The ascension of Charles I to the throne—who was known to be sympathetic to Roman Catholicism and impatient with Puritan reformers—was also taken as an ominous sign for Puritans, and Winthrop was not alone in predicting that "God will bring some heavy affliction upon the land, and that speedily." Winthrop came to realize that he could not antagonize the King by expressing openly the Puritan cause without losing all that he possessed. The only recourse seemed to be to obtain the King's permission to emigrate. In March of 1629 a group of enterprising merchants, all sympathetic believers, were able to get a charter from the Council for New England for land in the New World. They called themselves "The Company of Massachusetts Bay in New England."

From four candidates, Winthrop was chosen Governor in October, 1629; for the next twenty years most of the responsibility for the Colony rested in his hands. On April 8, 1630, an initial group of some seven hundred emigrants sailed from England. The ship carrying Winthrop was called the *Arbella*. Somewhere in the middle of the Atlantic ocean Winthrop delivered his sermon *A Model of Christian Charity*. It set out clearly and eloquently the ideals of a harmonious Christian community, and reminded all those on board that they would stand as an example to the world either of the triumph or else the failure of this Christian enterprise. When Cotton Mather wrote his history of New England some fifty years after Winthrop's death, he chose Winthrop as his model of the perfect earthly ruler. Although the actual history of the Colony showed that Winthrop's ideal of a perfectly selfless community was impossible to realize in fact, Winthrop emerges from the story as a man of unquestioned integrity and deep humanity.

From A Model of Christian Charity[1]
I
A MODEL HEREOF

God Almighty in His most holy and wise providence, hath so dis-

---

1. The text is from Old South Leaflets, Old South Association, Old South Meet- inghouse, Boston, Massachusetts, No. 207, edited by Samuel Eliot Morison. The

posed of the condition of mankind, as in all times some must be rich, some poor, some high and eminent in power and dignity; others mean and in subjection.

THE REASON HEREOF

First, to hold conformity with the rest of His works, being delighted to show forth the glory of His wisdom in the variety and difference of the creatures; and the glory of His power, in ordering all these differences for the preservation and good of the whole; and the glory of His greatness, that as it is the glory of princes to have many officers, so this great King will have many stewards, counting Himself more honored in dispensing His gifts to man by man than if He did it by His own immediate hands.

Secondly, that He might have the more occasion to manifest the work of His Spirit: first upon the wicked in moderating and restraining them, so that the rich and mighty should not eat up the poor, nor the poor and despised rise up against their superiors and shake off their yoke; secondly in the regenerate, in exercising His graces, in them, as in the great ones, their love, mercy, gentleness, temperance, etc.; in the poor and inferior sort, their faith patience, obedience etc.

Thirdly, that every man might have need of other, and from hence they might be all knit more nearly together in the bonds of brotherly affection. From hence it appears plainly that no man is made more honorable than another or more wealthy, etc., out of any particular and singular respect to himself, but for the glory of his creator and the common good of the creature, man. Therefore God still reserves the property of these gifts to Himself as [in] Ezekiel 16.17. He there calls wealth His gold and His silver.[2] [In] Proverbs 3.9, he claims their service as His due, honor the Lord with thy riches etc.[3] All men being thus (by divine providence) ranked into two sorts, rich and poor; under the first are comprehended all such as are able to live comfortably by their own meanes duly improved; and all others are poor according to the former distribution.

There are two rules whereby we are to walk one towards another: justice and mercy. These are always distinguished in their act and in their object, yet may they both concur in the same subject in each respect; as sometimes there may be an occasion of showing mercy to a rich man in some sudden danger of distress, and also doing of mere justice to a poor man in regard of some particular contract, etc.

---

original manuscript for Winthrop's sermon is lost; but a copy made during Winthrop's lifetime was published by the Massachusetts Historical Society in 1838.
2. "Thou hast also taken thy fair jewels of my gold and of my silver, which I had given thee, and madest to thyself images of men, and didst commit whoredom with them."
3. "Honor the Lord with thy substance, and with the firstfruits of all thine increase: So shall thy barns be filled with plenty, and thy presses burst out with new wine."

There is likewise a double law by which we are regulated in our conversation one towards another in both the former respects: the law of nature and the law of grace, or the moral law or the law of the Gospel, to omit the rule of justice as not properly belonging to this purpose otherwise than it may fall into consideration in some particular cases. By the first of these laws man as he was enabled so withal [is] commanded to love his neighbor as himself.[4] Upon this ground stands all the precepts of the moral law, which concerns our dealings with men. To apply this to the works of mercy, this law requires two things: first, that every man afford his help to another in every want or distress; secondly, that he performed this out of the same affection which makes him careful of his own goods, according to that of our Savior. Matthew: "Whatsoever ye would that men should do to you."[5] This was practiced by Abraham and Lot in entertaining the Angels and the old man of Gibeah.[6]

The law of grace or the Gospel hath some difference from the former, as in these respects: First, the law of nature was given to man in the estate of innocency; this of the Gospel in the estate of regeneracy.[7] Secondly, the former propounds one man to another, as the same flesh and image of God; this as a brother in Christ also, and in the communion of the same spirit and so teacheth us to put a difference between Christians and others. *Do good to all, especially to the household of faith*: Upon this ground the Israelites were to put a difference between the brethren of such as were strangers though not of Canaanites.[8] Third, the law of nature could give no rules for dealing with enemies, for all are to be considered as friends in the state of innocency, but the Gospel commands love to an enemy. Proof. If thine Enemy hunger, feed him; Love your Enemies, do good to them that hate you. Matthew: 5.44.

* * *

II

* * *

Thus stands the cause between God and us. We are entered into covenant[9] with Him for this work. We have taken out a commission, the Lord hath given us leave to draw our own articles. We

4. Matthew 5.43; 19.19.
5. "All things therefore whatsoever ye would that men should do unto you even so do ye also unto them: for this is the law of the prophets" (Matthew 7.12).
6. Abraham entertains the angels in Genesis 18: "And the Lord appeared unto him in the plains of Mamre: and he sat in the tent door in the heat of the day; And he lifted up his eyes and looked, and, lo, three men stood by him: and when he saw them, he ran to meet them * * *" (Genesis 18.1–2). Lot was Abraham's nephew, and he escaped the destruction of the city of Sodom because he defended two angels who were his guests from a

mob (Genesis 19.1–14). In Judges, 19.16–21, an old citizen of Gibeah offered shelter to a traveling priest or Levite and defended him from enemies in a neighboring city.
7. Men lost their natural innocence when Adam fell; that state is called unregenerate. When Christ came to ransom man for Adam's sin, he offered salvation for those who believed in Him and became regenerate, or saved.
8. One who lived in Canaan, the Land of Promise for the Israelites.
9. A legal contract; the Israelites entered into a covenant with God in which He promised to protect them if they kept His word and were faithful to Him.

have professed to enterprise these actions, upon these and those ends, we have hereupon besought Him of favour and blessing. Now if the Lord shall please to hear us, and bring us in peace to the place we desire, then hath He ratified this covenant and sealed our commission, [and] will expect a strict performance of the articles contained in it; but if we shall neglect the observation of these articles which are the ends we have propounded, and, dissembling with our God, shall fall to embrace this present world and prosecute our carnal intentions, seeking great things for ourselves and our posterity, the Lord will surely break out in wrath against us; be revenged of such a perjured people and make us know the price of the breach of such a covenant.

Now the only way to avoid this shipwreck, and to provide for our posterity, is to follow the counsel of Micah,[1] to do justly, to love mercy, to walk humbly with our God. For this end, we must be knit together in this work as one man. We must entertain each other in brotherly affection, we must be willing to abridge ourselves of our superfluities, for the supply of other's necessities. We must uphold a familiar commerce together in all meekness, gentleness, patience and liberality. We must delight in each other, make other's conditions our own, rejoice together, mourn together, labour and suffer together, always having before our eyes our commission and community in the work, our community as members of the same body. So shall we keep the unity of the spirit in the bond of peace. The Lord will be our God, and delight to dwell among us as His own people, and will command a blessing upon us in all our ways, so that we shall see much more of His wisdom, power, goodness and truth, than formerly we have been acquainted with. We shall find that the God of Israel is among us, when ten of us shall be able to resist a thousand of our enemies; when He shall make us a praise and glory that men shall say of succeeding plantations, "the lord make it like that of NEW ENGLAND." For we must consider that we shall be as a city upon a hill.[2] The eyes of all people are upon us, so that if we shall deal falsely with our God in this work we have undertaken, and so cause him to withdraw His present help from us, we shall be made a story and a by-word through the world. We shall open the mouths of enemies to speak evil of the ways of God, and all professors for God's sake. We shall shame the faces of many of God's worthy servants, and cause their prayers to be turned into curses upon us 'til we be consumed out of the good land whither we are agoing.

1. The Book of Micah preserves the words of this eighth-century-B.C. prophet. Micah speaks continually of the judgment of God on His people and the necessity to hope for salvation: "I will bear the indignation of the Lord, because I have sinned against him, until he plead my cause, and execute judgment for me: he will bring me forth to the light, and I shall behold his righteousness" (Micah 7.9).

2. "Ye are the light of the world. A city that is set on a hill cannot be hid. Neither do men light a candle, and put it under a bushel, but on a candlestick; and it giveth light unto all that are in the house" (Matthew 5.14–15).

And to shut up this discourse with that exhortation of Moses, that faithful servant of the Lord, in his last farewell to Israel, Deuteronomy 30.[3] Beloved, there is now set before us life and good, death and evil, in that we are commanded this day to love the Lord our God, and to love one another, to walk in His ways and to keep His commandments and His ordinance and His laws, and the articles of our covenant with Him, that we may live and be multiplied, and that the Lord our God may bless us in the land whither we go to possess it. But if our hearts shall turn away, so that we will not obey, but shall be seduced, and worship other Gods, our pleasures and profits, and serve them; it is propounded unto us this day, we shall surely perish out of the good land whither we pass over this vast sea to possess it.

Therefore let us choose life,
that we and our seed
may live by obeying His
voice and cleaving to Him,
for He is our life and
our prosperity.

1630   1838

3. "And it shall come to pass, when all these things are come upon thee, the blessing and the curse, which I have set before thee, and thou shalt call them to mind among all the nations, whither the Lord thy God hath driven thee, And shalt return unto the Lord thy God, and shalt obey his voice according to all that I command thee this day, thou and thy children, with all thine heart, and with all thy soul; That then the Lord thy God will turn thy captivity, and have compassion upon thee, and will return and gather thee from all the nations, whither the Lord thy God hath scattered thee" (Deuteronomy 30.1–3).

WILLIAM BRADFORD
1590–1657

William Bradford epitomizes the spirit of determination and self-sacrifice that seems to us characteristic of our first "Pilgrims," a word Bradford himself used to describe the community of believers who sailed from Southampton, England, on the *Mayflower* and settled in Plymouth, Massachusetts, in 1620. For Bradford, as well as for the other members of this community, the decision to settle at Plymouth was the last step in a long march of exile from England, and the hardships they suffered in the new land were tempered with the knowledge that they were in a place they had chosen for themselves, where they were safe from persecution. Shortly after their arrival Bradford was elected Governor. His duties involved more than that title might imply today: he was chief judge and jury, superintended agriculture and trade, and made allotments of land. It would be hard to imagine a historian better prepared to write the history of this colony.

Bradford's own life provides a model of the life of the community as a whole. He was born in Yorkshire, in the town of Austerfield, of parents who were modestly well off. Bradford's father died when he was a child,

and he was brought up by his grandparents and uncles. He did not receive
a university education; instead, he was taught the arts of farming. When he
was only twelve or thirteen, he heard the sermons of the nonconformist
minister Richard Clyfton, who preached in a neighboring parish; these sermons changed Bradford's life. For Clyfton was the religious guide of a
small community of believers who met at the house of William Brewster in
Scrooby, Nottinghamshire, and it was with this group, in 1606, that Bradford wished to be identified. Much against the opposition of uncles and
grandparents, he left home and joined them. They were known as "separatists," because unlike the majority of Puritans, they saw no hope of reforming the Church of England from within. They wished to follow Calvin's
model and to set up "particular" churches, each one founded on a formal
covenant, entered into by those who professed their faith and swore to the
covenant. Their model was the Old Testament covenant God made with
Adam and which Christ renewed. In their covenanted churches God offered
Himself as a contractual partner to each believer; it was a contract freely
initiated but perpetually binding. They were not sympathetic to the idea of
a national church. Separating was, however, by English law an act of treason, and many believers paid a high price for their dreams of purity. Sick of
the hidden life that the Church of England forced upon them, the Scrooby
community took up residence in Holland. Bradford joined them in 1609
and there learned to be a weaver. When he came into his inheritance he
went into business for himself.

Living in a foreign land was not easy, and eventually the Scrooby community petitioned for a grant of land in the New World. Their original grant
was for land in the Virginia territory, but high seas prevented them from
reaching those shores and they settled at Plymouth, Massachusetts, instead.
In the second book of Bradford's history he describes the signing of the
"Mayflower Compact," a civil covenant designed to allow the temporal
state to serve the godly citizen. It was the first of a number of "plantation
covenants" designed to protect the rights of citizens beyond the reach of
established governments.

Bradford was a self-educated man, deeply committed to the Puritan
cause. In his ecclesiastical history of New England, Cotton Mather
describes him as "a person for study as well as action; and hence notwithstanding the difficulties which he passed in his youth, he attained unto a
notable skill in languages * * * but the Hebrew he most of all studied,
because, he said, he would see with his own eyes the ancient oracles of God
in their native beauty * * * the crown of all his life was his holy, prayerful,
watchful and fruitful walk with God, wherein he was exemplary." Bradford
served as Governor for all but five of the remaining years of his life.

The manuscript of Bradford's *History*, although known to early historians,
disappeared from Boston after the Revolution. The first book (through Chapter IX) had been copied into the Plymouth church records and was thus
preserved, but the second book was assumed lost. The manuscript was found
in the residence of the Bishop of London and published for the first time in
1856. In 1897 it was returned to this country by ecclesiastical decree and was
deposited in the State House in Boston.

From Of Plymouth Plantation[1]

From *Book I, Chapter I.*
[*The Separatist Interpretation of the Reformation in England, 1550–1607*]

* * * When as by the travail and diligence of some godly and zealous preachers, and God's blessing on their labors, as in other places of the land, so in the North parts,[2] many became enlightened by the Word of God and had their ignorance and sins discovered[3] unto them, and began by His grace to reform their lives and make conscience of their ways; the work of God was no sooner manifest in them but presently they were both scoffed and scorned by the profane[4] multitude; and the ministers urged with the yoke of subscription,[5] or else must be silenced. And the poor people were so vexed with apparitors and pursuivants and the commissary courts,[6] as truly their affliction was not small. Which, notwithstanding, they bore sundry years with much patience, till they were occasioned by the continuance and increase of these troubles, and other means which the Lord raised up in those days, to see further into things by the light of the Word of God. How not only these base and beggarly ceremonies were unlawful, but also that the lordly and tyrannous power of the prelates ought not to be submitted unto; which thus, contrary to the freedom of the gospel, would load and burden men's consciences and by their compulsive power make a profane mixture of persons and things in the worship of God. And that their offices and callings, courts and canons, etc. were unlawful and antichristian; being such as have no warrant in the Word of God, but the same that were used in popery and still retained. Of which a famous author thus writeth in his Dutch commentaries, at the coming of King James into England:

> The new king (saith he) found there established the reformed religion according to the reformed religion of King Edward VI, retaining or keeping still the spiritual state of the bishops, etc. after the old manner, much varying and differing from the reformed churches in Scotland, France and the Netherlands, Emden, Geneva, etc., whose reformation is cut, or shapen much nearer the first Christian churches, as it was used in the Apostles' times.[7]

So many, therefore, of these professors as saw the evil of these things in these parts, and whose hearts the Lord had touched with

---

1. The text is from *Of Plymouth Plantation*, ed. Samuel Eliot Morison (New York: Knopf, 1953), pp. 8–10, 58–63, 73–79, 204–10, 252–54.
2. I.e., of England and Scotland.
3. Revealed.
4. Unholy.
5. I.e., to subscribe to the tenets of the Church of England; "urged": threatened.
6. I.e., vexed with officers and summoners of the Church of England, and the

court of a Bishop's jurisdiction.
7. From Emanuel van Meteren's *General History of the Netherlands* (1608). King James (1566–1625) ascended the throne in 1603. Most Puritans preferred the model of the Calvinist system in Geneva or the Church of Scotland, which replaced a hierarchy of Archbishops, Bishops, and priests with a national assembly and a parish presbytery consisting of ministers and elders.

heavenly zeal for His truth, they shook off this yoke of antichristian bondage, and as the Lord's free people joined themselves (by a covenant[8] of the Lord) into a church estate, in the fellowship of the gospel, to walk in all His ways made known, or to be made known unto them, according to their best endeavours, whatsoever it should cost them, the Lord assisting them. And that it cost them something this ensuing history will declare.

These people became two distinct bodies or churches, and in regard of distance of place did congregate severally; for they were of sundry towns and villages, some in Nottinghamshire, some of Lincolnshire, and some of Yorkshire where they border nearest together. In one of these churches (besides others of note) was Mr. John Smith,[9] a man of able gifts and a good preacher, who afterwards was chosen their pastor. But these afterwards falling into some errors in the Low Countries,[1] there (for the most part) buried themselves and their names.

But in this other church (which must be the subject of our discourse) besides other worthy men, was Mr. Richard Clyfton, a grave and reverend preacher, who by his pains and diligence had done much good, and under God had been a means of the conversion of many. And also that famous and worthy man Mr. John Robinson, who afterwards was their pastor for many years, till the Lord took him away by death. Also Mr. William Brewster a reverend man, who afterwards was chosen an elder of the church and lived with them till old age.

But after these things they could not long continue in any peaceable condition, but were hunted and persecuted on every side, so as their former afflictions were but as flea-bitings in comparison of these which now came upon them. For some were taken and clapped up in prison, others had their houses beset and watched night and day, and hardly escaped their hands; and the most were fain to flee and leave their houses and habitations, and the means of their livelihood.

Yet these and many other sharper things which afterward befell them, were no other than they looked for, and therefore were the better prepared to bear them by the assistance of God's grace and Spirit.

Yet seeing themselves thus molested, and that there was no hope of their continuance there, by a joint consent they resolved to go into the Low Countries, where they heard was freedom of religion for all men; as also how sundry from London and other parts of the land had been exiled and persecuted for the same cause, and were

8. A solemn agreement between the members of a church to act together in harmony with the precepts of the Gospel.
9. A Cambridge University graduate who seceded from the Church of England in 1605. Richard Clyfton and John Robinson were also Cambridge University graduates who were separatists. William Brewster (1576–1644) was a church leader of the Pilgrims in both Leyden and Plymouth.
1. Holland.

gone thither, and lived at Amsterdam and in other places of the land. So after they had continued together about a year, and kept their meetings every Sabbath in one place or other, exercising the worship of God amongst themselves, notwithstanding all the diligence and malice of their adversaries, they seeing they could no longer continue in that condition, they resolved to get over into Holland as they could. Which was in the year 1607 and 1608; of which more at large in the next chapter.

Book I, Chapter IX. Of Their Voyage and How They Passed the Sea; and of Their Safe Arrival at Cape Cod

September 6. These troubles[2] being blown over, and now all being compact together in one ship, they put to sea again with a prosperous wind, which continued divers days together, which was some encouragement unto them; yet, according to the usual manner, many were afflicted with seasickness. And I may not omit here a special work of God's providence. There was a proud and very profane young man, one of the seamen, of a lusty,[3] able body, which made him the more haughty; he would alway be contemning the poor people in their sickness and cursing them daily with grievous execrations; and did not let[4] to tell them that he hoped to help to cast half of them overboard before they came to their journey's end, and to make merry with what they had; and if he were by any gently reproved, he would curse and swear most bitterly. But it pleased God before they came half seas over, to smite this young man with a grievous disease, of which he died in a desperate manner, and so was himself the first that was thrown overboard. Thus his curses light on his own head, and it was an astonishment to all his fellows for they noted it to be the just hand of God upon him.

After they had enjoyed fair winds and weather for a season, they were encountered many times with cross winds and met with many fierce storms with which the ship was shroudly[5] shaken, and her upper works made very leaky; and one of the main beams in the midships was bowed and cracked, which put them in some fear that the ship could not be able to perform the voyage. So some of the chief of the company, perceiving the mariners to fear the sufficiency of the ship as appeared by their mutterings, they entered into serious consultation with the master and other officers of the ship, to consider in time of the danger, and rather to return than to cast themselves into a desperate and inevitable peril. And truly there was great distraction and difference of opinion amongst the mariners themselves; fain would they do what could be done for their wages' sake (being now near half the seas over) and on the other hand

2. Some of the Scrooby community originally sailed from Delftshaven about August 1, 1620, on board the *Speedwell*, but it proved unseaworthy and it was necessary to transfer everything to the *Mayflower*.
3. Strong, energetic.
4. Hesitate.
5. Shrewdly, in its original sense of wickedly.

they were loath to hazard their lives too desperately. But in examining of all opinions, the master and others affirmed they knew the ship to be strong and firm under water; and for the buckling of the main beam, there was a great iron screw the passengers brought out of Holland, which would raise the beam into his place; the which being done, the carpenter and master affirmed that with a post put under it, set firm in the lower deck and otherways bound, he would make it sufficient. And as for the decks and upper works, they would caulk them as well as they could, and though with the working of the ship they would not long keep staunch,[6] yet there would otherwise be no great danger, if they did not overpress her with sails. So they committed themselves to the will of God and resolved to proceed.

In sundry of these storms the winds were so fierce and the seas so high, as they could not bear a knot of sail, but were forced to hull[7] for divers days together. And in one of them, as they thus lay at hull in a mighty storm, a lusty young man called John Howland, coming upon some occasion above the gratings was, with a seele[8] of the ship, thrown into sea; but it pleased God that he caught hold of the topsail halyards which hung overboard and ran out at length. Yet he held his hold (though he was sundry fathoms under water) till he was hauled up by the same rope to the brim of the water, and then with a boat hook and other means got into the ship again and his life saved. And though he was something ill with it, yet he lived many years after and became a profitable member both in church and commonwealth. In all this voyage there died but one of the passengers, which was William Butten, a youth, servant to Samuel Fuller, when they drew near the coast.

But to omit other things (that I may be brief) after long beating at sea they fell with that land which is called Cape Cod; the which being made and certainly known to be it, they were not a little joyful. After some deliberation had amongst themselves and with the master of the ship, they tacked about and resolved to stand for the southward (the wind and weather being fair) to find some place about Hudson's River for their habitation. But after they had sailed that course about half the day, they fell amongst dangerous shoals and roaring breakers, and they were so far entangled therewith as they conceived themselves in great danger; and the wind shrinking upon them withal, they resolved to bear up again for the Cape and thought themselves happy to get out of those dangers before night overtook them, as by God's good providence they did. And the next day they got into the Cape Harbor[9] where they rid in safety.

A word or two by the way of this cape. It was thus first named by

6. **Watertight.**
7. **Drift with the wind under short sail.**
8. **Roll.**
9. **Cape Harbor is now Provincetown**

Harbor; they arrived on November 11, 1620, the journey from England having taken 65 days.

Captain Gosnold and his company, Anno 1602, and after by Captain Smith was called Cape James; but it retains the former name amongst seamen. Also, that point which first showed those dangerous shoals unto them they called Point Care, and Tucker's Terror; but the French and Dutch to this day call it Malabar[1] by reason of those perilous shoals and the losses they have suffered there.

Being thus arrived in a good harbor, and brought safe to land, they fell upon their knees and blessed the God of Heaven who had brought them over the vast and furious ocean, and delivered them from all the perils and miseries thereof, again to set their feet on the firm and stable earth, their proper element. And no marvel if they were thus joyful, seeing wise Seneca was so affected with sailing a few miles on the coast of his own Italy, as he affirmed, that he had rather remain twenty years on his way by land than pass by sea to any place in a short time, so tedious and dreadful was the same unto him.[2]

But here I cannot but stay and make a pause, and stand half amazed at this poor people's present condition; and so I think will the reader, too, when he well considers the same. Being thus passed the vast ocean, and a sea of troubles before in their preparation (as may be remembered by that which went before), they had now no friends to welcome them nor inns to entertain or refresh their weatherbeaten bodies; no houses or much less towns to repair to, to seek for succor. It is recorded in Scripture as a mercy to the Apostle and his shipwrecked company, that the barbarians showed them no small kindness in refreshing them,[3] but these savage barbarians, when they met with them (as after will appear) were readier to fill their sides full of arrows than otherwise. And for the season it was winter, and they that know the winters of that country know them to be sharp and violent, and subject to cruel and fierce storms, dangerous to travel to known places, much more to search an unknown coast. Besides, what could they see but a hideous and desolate wilderness, full of wild beasts and wild men—and what multitudes there might be of them they knew not. Neither could they, as it were, go up to the top of Pisgah[4] to view from this wilderness a more goodly country to feed their hopes; for which way soever they turned their eyes (saved upward to the heavens) they could have little solace or content in respect of any outward objects. For summer being done, all things stand upon them with a weatherbeaten face, and the whole country, full of woods and thickets, represented a wild and savage hue. If they looked behind them,

1. The prefix "mal" means bad; the reference here is to the dangerous sandbars.
2. Bradford notes that this *remark may be found in the Moral Epistles to Lucilius*, line 5, of the Roman Stoic philosopher (4? B.C.–A.D. 65).
3. "And when they were escaped, then they knew that the island was called Melita. And the barbarous people showed us no little kindness: for they kindled a fire, and received us every one, because of the present rain, and because of the cold" (Acts 28.1–2).
4. Mountain from which Moses saw the Promised Land (Deuteronomy 34.1–4).

there was the mighty ocean which they had passed and was now as a main bar and gulf to separate them from all the civil parts of the world. If it be said they had a ship to succor them, it is true; but what heard they daily from the master and company? But that with speed they should look out a place (with their shallop)[5] where they would be, at some near distance; for the season was such as he would not stir from thence till a safe harbor was discovered by them, where they would be, and he might go without danger; and that victuals consumed apace but he must and would keep sufficient for themselves and their return. Yea, it was muttered by some that if they got not a place in time, they would turn them and their goods ashore and leave them. Let it also be considered what weak hopes of supply and succor they left behind them, that might bear up their minds in this sad condition and trials they were under; and they could not but be very small. It is true, indeed, the affections and love of their brethren at Leyden[6] was cordial and entire towards them, but they had little power to help them or themselves; and how the case stood between them and the merchants at their coming away hath already been declared.

What could now sustain them but the Spirit of God and His grace? May not and ought not the children of these fathers rightly say: "Our fathers were Englishmen which came over this great ocean, and were ready to perish in this wilderness; but they cried unto the Lord, and He heard their voice and looked on their adversity,"[7] etc. "Let them therefore praise the Lord, because He is good: and His mercies endure forever." "Yea, let them which have been redeemed of the Lord, show how He hath delivered them from the hand of the oppressor. When they wandered in the desert wilderness out of the way, and found no city to dwell in, both hungry and thirsty, their soul was overwhelmed in them. Let them confess before the Lord His loving kindness and His wonderful works before the sons of men."[8]

From *Book II, Chapter XI*.[9] *The Remainder of Anno 1620*

[THE MAYFLOWER COMPACT]

I shall a little return back, and begin with a combination[1] made by them before they came ashore; being the first foundation of their government in this place. Occasioned partly by the discon-

---

5. Small boat.
6. In Holland. A substantial number of separatists remained in the Netherlands.
7. "And the Egyptians evil entreated us, and afflicted us, and laid upon us hard bondage: And we cried unto the Lord God of our fathers, the Lord heard our voice, and looked on our affliction, and our labor and our oppression: And the Lord brought us forth out of Egypt with a mighty hand * * *" (Deuteronomy 26.6–8).

8. "O give thanks unto the Lord, for he is good: for his mercy endureth for ever. Let the redeemed of the Lord say so, whom he hath redeemed from the hand of the enemy; And gathered them out of the lands, from the east, and from the west, from the north, and from the south" (Psalm 107.1–5).
9. Bradford numbered only the first 10 chapters of his manuscript.
1. A form of union, a joining together.

tented and mutinous speeches that some of the strangers[2] amongst them had let fall from them in the ship: That when they came ashore they would use their own liberty, for none had power to command them, the patent they had being for Virginia and not for New England, which belonged to another government, with which the Virginia Company had nothing to do. And partly that such an act by them done, this their condition considered, might be as firm as any patent,[3] and in some respects more sure.

The form was as followeth:
IN THE NAME OF GOD, AMEN.

We whose names are underwritten, the loyal subjects of our dread Sovereign Lord King James, by the Grace of God of Great Britain, France, and Ireland King, Defender of the Faith, etc.

Having undertaken, for the glory of God and advancement of the Christian Faith and Honour of our King and Country, a Voyage to plant the First Colony in the Northern Parts of Virginia, do by these presents solemnly and mutually in the presence of God and one of another, Covenant and Combine ourselves together into a Civil Body Politic, our better ordering and preservation and furtherance of the ends aforesaid; and by virtue hereof to enact, constitute and frame such just and equal Laws, Ordinances, Acts, Constitutions and Offices, from time to time, as shall be thought most meet and covenient for the general good of the Colony, unto which we promise all due submission and obedience. In witness whereof we have hereunder subscribed our names at Cape Cod, the 11th of November, in the year of the reign of our Sovereign Lord King James, of England, France and Ireland the eighteenth, and of Scotland the fifty-fourth. Anno Domini 1620.

After this they chose, or rather confirmed, Mr. John Carver[4] (a man godly and well approved amongst them) their Governor for that year. And after they had provided a place for their goods, or common store (which were long in unlading[5] for want of boats, foulness of the winter weather and sickness of divers) and begun some small cottages for their habitation; as time would admit, they met and consulted of laws and orders, both for their civil and military government as the necessity of their condition did require, still adding thereunto as urgent occasion in several times, and as cases did require.

In these hard and difficult beginnings they found some discontents and murmurings arise amongst some, and mutinous speeches

---

2. Puritans called themselves "saints" and those outside their churches "strangers." Many of those who came to Plymouth with them were not church members but adventurers looking forward to business success and making new lives in the New World.

3. A document signed by a sovereign granting privileges to those named in it.

4. Carver (c. 1576–1621), like Bradford, was an original member of the group who went to Holland and, like Bradford, a tradesman. Bradford was elected Governor after Carver's death.

5. Unloading.

the church must also be divided, and those that had lived so long together in Christian and comfortable fellowship must now part and suffer many divisions. First, those that lived on their lots on the other side of the Bay, called Duxbury, they could not long bring their wives and children to the public worship and church meetings here, but with such burthen as, growing to some competent number, they sued to be dismissed and become a body of themselves. And so they were dismissed about this time, though very unwillingly. But to touch this sad matter, and handle things together that fell out afterward; to prevent any further scattering from this place and weakening of the same, it was thought best to give out some good farms to special persons that would promise to live at Plymouth, and likely to be helped to the church or commonwealth, and so tie the lands to Plymouth as farms for the same; and there they might keep their cattle and tillage by some servants and retain their dwellings here. And so some special lands were granted at a place general called Green's Harbor, where no allotments had been in the former division, a place very well meadowed and fit to keep and rear cattle good store. But alas, this remedy proved worse than the disease; for within a few years those that had thus got footing there rent themselves away, partly by force and partly wearing the rest with importunity and pleas of necessity, so as they must either suffer them to go or live in continual opposition and contention. And other still, as they conceived themselves straitened[3] or to want accommodation, broke away under one pretence or other, thinking their own conceived necessity and the example of others a warrant sufficient for them. And this I fear will be the ruin of New England, at least of the churches of God there, and will provoke the Lord's displeasure against them.* * *

1630–50   1856

3. Financially restricted.

ANNE BRADSTREET
c. 1612–1672

Anne Bradstreet's father, Thomas Dudley, was the manager of the country estate of the Puritan Earl of Lincoln, and his daughter was very much the apple of his eye. When she was about seven she had eight tutors in languages, music, and dancing and her father took great care to see that she received an education superior to that of most young women of the time. When she was only sixteen she married a young man, Simon Bradstreet, a recent graduate of *Cambridge University, who was* associated with her father in conducting the affairs of the Earl of Lincoln's estate. He also shared her father's Puritan beliefs. A year after the marriage her husband

was appointed to assist in the preparations of the Company of the Massachusetts Bay, and the following year the Bradstreets and the Dudleys sailed with Winthrop's fleet. Anne Bradstreet tells us that when she first "came into this country" she "found a new world and new manners," at which her "heart rose" in resistance, "But after I was convinced it was the way of God, I submitted to it and joined the church at Boston."

We know very little of Anne Bradstreet's daily life, except that it was a hard existence. The wilderness, Samuel Eliot Morison once observed, "made men stern and silent, children unruly, servants insolent." William Bradford's wife, Dorothy, staring at the barren dunes of Cape Cod, said that she preferred the surety of drowning to the unknown life ashore. Added to the hardship of daily living was the fact that Anne Bradstreet was never very strong. She had rheumatic fever as a child and as a result suffered recurrent periods of severe fatigue; nevertheless, she risked death by childbirth eight times. Her husband was secretary to the Company and later Governor of the Bay Colony; he was always involved in the colony's diplomatic missions; and in 1661 he went to England to renegotiate the Bay Company charter with Charles II. All of Simon's tasks must have added to her responsibilities at home. And like any good Puritan she added to the care of daily life the examination of her conscience. She tells us in one of the prose "Meditations" written for her children that she was troubled many times about the truth of the Scriptures; that she never saw any convincing miracles, and that she always wondered if those of which she read "were feigned." What proved to her finally that God exists was not her reading but the evidence of her own eyes. She is the first in a long line of American poets who took their consolation not from theology but from the "wonderous works," as she wrote, "that I see, the vast frame of the heaven and the earth, the order of all things, night and day, summer and winter, spring and autumn, the daily providing for this great household upon the earth, the preserving and directing of all to its proper end."

When Anne Bradstreet was a young girl she had written poems to please her father, and he made much of their reading them together. After her marriage she continued writing. Quite unknown to her, her brother-in-law, John Woodbridge, pastor of the Andover church, brought with him to London a manuscript collection of her poetry and had it printed there in 1650. It was the first published volume of poems written by a resident in the New World and was widely read. Reverend Edward Taylor, also a poet, and living in the frontier community of Westfield, Massachusetts, had a copy of the second edition of Bradstreet's poems (1678) in his library. Although she herself probably took greatest pride in her long meditative poems on the ages of man and on the seasons, the poems which have attracted present-day readers are the more intimate ones which reflect her concern for her family and home and the pleasures she took in everyday life rather than in the life to come.

The text used is the *Works of Anne Bradstreet*, ed. Jeannine Hensley (Cambridge: Harvard University Press, 1967).

The Prologue

1

To sing of wars, of captains, and of kings,
Of cities founded, commonwealths begun,
For my mean[1] pen are too superior things:
Or how they all, or each their dates have run
Let poets and historians set these forth,   5
My obscure lines shall not so dim their worth.

2

But when my wond'ring eyes and envious heart
Great Bartas'[2] sugared lines do but read o'er,
Fool[3] I do grudge the Muses[4] did not part
'Twixt him and me that overfluent store;   10
A Bartas can do what a Bartas will
But simple I according to my skill.

3

From schoolboy's tongue no rhet'ric we expect,
Nor yet a sweet consort[5] from broken strings,
Nor perfect beauty where's a main defect: 15
My foolish, broken, blemished Muse so sings,
And this to mend, alas, no art is able,
'Cause nature made it so irreparable.

4

Nor can I, like that fluent sweet tongued Greek,
Who lisped at first, in future times speak plain.[6]   20
By art he gladly found what he did seek,
A full requital of his striving pain.
Art can do much, but this maxim's most sure:
A weak or wounded brain admits no cure.

5

I am obnoxious to each carping tongue   25
Who says my hand a needle better fits,
A poet's pen all scorn I should thus wrong,
For such despite they cast on female wits:
If what I do prove well, it won't advance,
They'll say it's stol'n, or else it was by chance.   30

6

But sure the antique Greeks were far more mild
Else of our sex, why feigned they those nine
And poesy made Calliope's[7] own child;
So 'mongst the rest they placed the arts divine:
But this weak knot they will full soon untie.   35
The Greeks did nought, but play the fools and lie.

---

1. Humble.
2. Guillaume du Bartas (1544–90) was a French writer much admired by the Puritans. He was most famous as the author of *The Divine Weeks,* an epic poem translated by Joshua Sylvester and intended to recount the great moments in Christian history.
3. I.e., like a fool.
4. In Greek mythology, the nine goddesses of the arts and sciences.
5. Accord, harmony of sound.
6. The Greek orator Demosthenes (c. 383–322 B.C.) conquered a speech defect.
7. The muse of epic poetry.

7

Let Greeks be Greeks, and women what they are;
Men have precedency and still excel,
It is but vain unjustly to wage war;
Men can do best, and women know it well.   40
Preeminence in all and each is yours;
Yet grant some small acknowledgement of ours.

8

And oh ye high flown quills[8] that soar the skies,
And ever with your prey still catch your praise,
If e'er you deign these lowly lines your eyes,   45
Give thyme or parsley wreath, I ask no bays;[9]
This mean and unrefined ore of mine
Will make your glist'ring gold but more to shine.

1650

The Author to Her Book[1]

Thou ill-formed offspring of my feeble brain,
Who after birth didst by my side remain,
Till snatched from thence by friends, less wise than true,
Who thee abroad, exposed to public view,
Made thee in rags, halting to th' press to trudge,   5
Where errors were not lessened (all may judge).
At thy return my blushing was not small,
My rambling brat (in print) should mother call,
I cast thee by as one unfit for light,
Thy visage was so irksome in my sight;   10
Yet being mine own, at length affection would
Thy blemishes amend, if so I could:
I washed thy face, but more defects I saw,
And rubbing off a spot still made a flaw.
I stretched thy joints to make thee even feet,[2]   15
Yet still thou run'st more hobbling than is meet;
In better dress to trim thee was my mind,
But nought save homespun cloth i' th' house I find.
In this array 'mongst vulgars[3] may'st thou roam.
In critic's hands beware thou dost not come,   20
And take thy way where yet thou art not known;
If for thy father asked, say thou hadst none;
And for thy mother, she alas is poor,
Which caused her thus to send thee out of door.

1678

Before the Birth of One of Her Children

All things within this fading world hath end,
Adversity doth still our joys attend;

---

8. Pens.
9. Garlands of laurel, used to crown the head of a poet.
1. *The Tenth Muse* was published in 1650 without Anne Bradstreet's knowledge. She is thought to have written this poem in 1666 when a second edition was contemplated.
2. I.e., metrical feet; to smooth out the lines.
3. The common people.

No ties so strong, no friends so dear and sweet,
But with death's parting blow is sure to meet.
The sentence past is most irrevocable,   5
A common thing, yet oh, inevitable.
How soon, my Dear, death may my steps attend,
How soon't may be thy lot to lose thy friend,
We both are ignorant, yet love bids me
These farewell lines to recommend to thee,   10
That when that knot's untied that made us one,
I may seem thine, who in effect am none.
And if I see not half my days that's due,
What nature would, God grant to yours and you;
The many faults that well you know I have   15
Let be interred in my oblivious grave;
If any worth or virtue were in me,
Let that live freshly in thy memory
And when thou feel'st no grief, as I no harms,
Yet love thy dead, who long lay in thine arms.   20
And when thy loss shall be repaid with gains
Look to my little babes, my dear remains.
And if thou love thyself, or loved'st me,
These O protect from stepdame's[1] injury.
And if chance to thine eyes shall bring this verse,   25
With some sad sighs honor my absent hearse;
And kiss this paper for thy love's dear sake,
Who with salt tears this last farewell did take.

                                                                    1678

To My Dear and Loving Husband

If ever two were one, then surely we.
If ever man were loved by wife, then thee;
If ever wife was happy in a man,
Compare with me, ye women, if you can.
I prize thy love more than whole mines of gold   5
Or all the riches that the East doth hold.
My love is such that rivers cannot quench,
Nor ought but love from thee, give recompense.
Thy love is such I can no way repay,
The heavens reward thee manifold, I pray.   10
Then while we live, in love let's so persevere
That when we live no more, we may live ever.

                                                                    1678

A Letter to Her Husband,
Absent upon Public Employment

My head, my heart, mine eyes, my life, nay, more,
My joy, my magazine[1] of earthly store,
If two be one, as surely thou and I,
How stayest thou there, whilst I at Ipswich[2] lie?

---

1. I.e., stepmother's.
1. Warehouse, storehouse.

2. Ipswich, Massachusetts, is north of Boston.

So many steps, head from the heart to sever, 5
If but a neck, soon should we be together.
I, like the Earth this season, mourn in black,
My Sun is gone so far in's zodiac,
Whom whilst I 'joyed, nor storms, nor frost I felt,
His warmth such frigid colds did cause to melt. 10
My chilled limbs now numbed lie forlorn;
Return; return, sweet Sol, from Capricorn;[3]
In this dead time, alas, what can I more
Than view those fruits which through thy heat I bore?
Which sweet contentment yield me for a space, 15
True living pictures of their father's face.
O strange effect! now thou art southward gone,
I weary grow the tedious day so long;
But when thou northward to me shalt return,
I wish my Sun may never set, but burn 20
Within the Cancer[4] of my glowing breast,
The welcome house of him my dearest guest.
Where ever, ever stay, and go not thence,
Till nature's sad decree shall call thee hence;
Flesh of thy flesh, bone of thy bone,
I here, thou there, yet both but one. 25

1678

In Memory of My Dear Grandchild Elizabeth Bradstreet, Who Deceased August, 1665, Being a Year and Half Old

Farewell dear babe, my heart's too much content,
Farewell sweet babe, the pleasure of mine eye,
Farewell fair flower that for a space was lent,
Then ta'en away unto eternity.
Blest babe, why should I once bewail thy fate, 5
Or sigh thy days so soon were terminate,
Sith[5] thou art settled in an everlasting state.

2

By nature trees do rot when they are grown,
And plums and apples thoroughly ripe do fall,
And corn and grass are in their season mown, 10
And time brings down what is both strong and tall.
But plants new set to be eradicate,
And buds new blown to have so short a date,
Is by His hand alone that guides nature and fate.

1678

3. Capricorn, the 10th sign of the zodiac, represents winter; "Sol": sun.
4. Cancer, the fourth sign of the zodiac, represents summer.
5. Since.

Here Follows Some Verses upon the Burning of Our House July 10th, 1666

Copied Out of a Loose Paper

In silent night when rest I took
For sorrow near I did not look
I wakened was with thund'ring noise
And piteous shrieks of dreadful voice.
That fearful sound of "Fire!" and "Fire!"   5
Let no man know is my desire.
I, starting up, the light did spy,
And to my God my heart did cry
To strengthen me in my distress
And not to leave me succorless.   10
Then, coming out, beheld a space
The flame consume my dwelling place.
And when I could no longer look,
I blest His name that gave and took,[1]
That laid my goods now in the dust.   15
Yea, so it was, and so 'twas just.
It was His own, it was not mine,
Far be it that I should repine;
He might of all justly bereft
But yet sufficient for us left.   20
When by the ruins oft I past
My sorrowing eyes aside did cast,
And here and there the places spy
Where oft I sat and long did lie:
Here stood that trunk, and there that chest,   25
There lay that store I counted best.
My pleasant things in ashes lie,
And them behold no more shall I.
Under thy roof no guest shall sit,
Nor at thy table eat a bit.   30
No pleasant tale shall e'er be told,
Nor things recounted done of old.
No candle e'er shall shine in thee,
Nor bridegroom's voice e'er heard shall be.
In silence ever shall thou lie,   35
Adieu, Adieu, all's vanity.
Then straight I 'gin my heart to chide,
And did thy wealth on earth abide?
Didst fix thy hope on mold'ring dust?
The arm of flesh didst make thy trust?   40
Raise up thy thoughts above the sky
That dunghill mists away may fly.
Thou hast an house on high erect,
Framed by that mighty Architect,
With glory richly furnished,   45

1. "The Lord gave, and the Lord hath taken away; blessed be the name of the Lord" (Job 1.21).

Stands permanent though this be fled.
It's purchaséd and paid for too
By Him who hath enough to do.
A price so vast as is unknown
Yet by His gift is made thine own;   50
There's wealth enough, I need no more,
Farewell, my pelf,[2] farewell my store.
The world no longer let me love,
My hope and treasure lies above.

1867

2. Possessions, usually in the sense of being falsely gained.

EDWARD TAYLOR

c. 1642–1729

Given the importance of Edward Taylor's role in the town in which he lived for fifty-eight years, it is curious that we should know so little about his life. Taylor was probably born in Sketchly, Leicestershire County, England; his father was a "yeoman farmer"—that is, he was not a "gentleman" with large estates, but an independent landholder with title to his farm. Although his poetry contains no images which reflect his boyhood in Leicestershire, the dialect of that farming country is ever-present and gives his verse an air of provincial charm, but also, it must be admitted, makes it difficult and complex for the modern reader. Taylor did not enter Harvard until he was twenty-nine years old, and he stayed only three years. It is assumed, therefore, that he had some university education in England, but it is not known where. We do know that he taught school and that he left his family and sailed to New England in 1668 because he would not sign an oath of loyalty to the Church of England. Rather than compromise his religious principles as a Puritan, he preferred exile in what he once called a "howling wilderness." It was at Harvard that he must have decided to leave teaching and prepare himself for the ministry.

In 1671 a delegation from the frontier town of Westfield, Massachusetts, asked Taylor to join them as their minister, and after a good deal of soul-searching he journeyed with them the hundred miles west to Westfield, where he remained the rest of his life. As by far the most educated member of that community, he served as minister, physician, and public servant. Taylor married twice and had fourteen children, many of whom died in infancy. A rigorous observer of all churchly functions, Taylor did not shy away from the religious controversies of the period. He was a strict observer of the "old" New England way, demanding a public account of conversion before admission to church membership and the right to partake of the sacrament of communion.

Taylor was a learned man as well as a pious one. Like most Harvard ministers, he knew Latin, Hebrew, and Greek. He had a passion for books and copied out in his own hand volumes that he borrowed from his college roommate Samuel Sewall. He was known to Sewall and others as a good preacher, and on occasion he sent poems and letters to Boston friends, some parts of which were published during his lifetime. But Taylor's work as a poet was generally unknown until, in the 1930s, Thomas H. Johnson discovered that most of Taylor's poems had been deposited in the Yale University Library by Taylor's grandson, Ezra Stiles, a former president of Yale. It was one of the major literary discoveries of the twentieth century, and revealed a body of work by a Puritan divine that was remarkable both in its quantity and quality.

Taylor's interest in poetry was lifelong, and he tried his hand at a variety of poetic genres: elegies on the death of public figures; lyrics in the manner of Elizabethan songs; a long poem, *God's Determinations*, in the tradition of the medieval "debate"; and an almost unreadable five-hundred-page *Metrical History of Christianity*, primarily a book of martyrs. But Taylor's best verse is to be found in a series called *Preparatory Meditations*. These poems, written for his own pleasure and never a part of any religious service, followed upon his preparation for a sermon to be delivered at monthly communion. They gave the poet an occasion to summarize the emotional and intellectual content of his sermon and to speak directly and fervently to God. Sometimes these poems are gnarled and difficult to follow, but they also reveal a unique voice, unmistakably Taylor's. They are written in an idiom which harkens back to the verse Taylor must have known as a child in England—the metaphysical lyrics of John Donne and George Herbert—and so delight in puns and paradoxes and a rich profusion of metaphors and images. Nothing previously discovered about Puritan literature had suggested that there was a writer in New England who had sustained such a long-term love affair with poetry.

From PREPARATORY MEDITATIONS[1]

Prologue

Lord, can a crumb of dust the earth outweigh,
  Outmatch all mountains, nay, the crystal sky?
embosom in't designs that shall display
  And trace into the boundless deity?
  Yea, hand a pen whose moisture doth guide o'er   5
  Eternal glory with a glorious glore.[2]

If it its pen had of an angel's quill,
  And sharpened on a precious stone ground tight,
And dipped in liquid gold, and moved by skill
  In Crystal leaves should golden letters write,   10
  It would but blot and blur, yea, jag, and jar
  Unless thou mak'st the pen, and scrivener.

1. The full title is *Preparatory Meditations before my Approach to the Lord's Supper. Chiefly upon the Doctrine preached upon the Day of Administration [of Communion]*. Taylor administered communion once a month to those members of his congregation who had made a declaration of their faith. He wrote these meditations in private; they are the result of his contemplation of the Biblical texts which served as the basis for the communion sermon. One hundred ninety-five of these poems survive, dating from 1682 to 1725. The text used here is from *Poems of Edward Taylor*, ed. Donald E. Stanford (New Haven, Conn.: Yale University Press, 1960).
2. Scottish form of "glory."

> I am this crumb of dust which is designed
>   To make my pen unto Thy praise alone,
> And my dull fancy[3] I would gladly grind   15
>   Unto an edge on Zion's[4] precious stone.
>   And write in liquid gold upon Thy name
>   My letters till Thy glory forth doth flame.
>
> Let not th'attempts break down my dust, I pray,
>   Nor laugh Thou them to scorn but pardon give.   20
> Inspire this crumb of dust 'til it display
>   Thy glory through't: and then thy dust shall live.
>   Its failings then thou'lt overlook, I trust,
>   They being slips slipped from Thy crumb of dust.
>
> Thy crumb of dust breathes two words from its breast,   25
>   That Thou wilt guide its pen to write aright
> To prove Thou art, and that Thou art the best
>   And show Thy properties to shine most bright.
>   And then Thy works will shine as flowers on stems
>   Or as in jewelry shops, do gems.   30

c. 1682   1939

Meditation 8 (First Series)

John 6.51. I am the living bread.[1]

> I kenning through astronomy divine
>   The world's bright battlement,[2] wherein I spy
> A golden path my pencil cannot line,
>   From that bright throne unto my threshold lie.
>   And while my puzzled thoughts about it pour   5
>   I find the bread of life in't at my door.
>
> When that this bird of paradise[3] put in
>   This wicker cage (my corpse) to tweedle praise
> Had pecked the fruit forbade: and so did fling
>   Away its food; and lost its golden days;   10
>   It fell into celestial famine sore:
>   And never could attain a morsel more.
>
> Alas! alas! poore bird, what wilt thou do?
>   The creatures field no food for souls e'er gave.
> And if thou knock at angels' doors they show   15

3. I.e., imagination.
4. A hill in Jerusalem where the Jews built their temple; the city of God on earth.
1. "The Jews then murmured at him, because he said, I am the bread which came down from heaven. And they said, Is not this Jesus, the son of Joseph, whose father and mother we know? how *is it* then that he saith, I came down from heaven? *Jesus therefore* answered * * * Verily, verily, I say unto *you*, He that believeth on me hath everlasting life. I am that bread of life" (John 6.41–51). Christ offers a "New Covenant of Faith" in place of the "Old Covenant of Works" which Adam broke when he disobeyed God's commandment.
2. I.e., discerning, by means of "divine astronomy," the towers of heaven. Taylor goes on to suggest that there is an invisible golden path from this world to the gates of Heaven.
3. I.e., the soul, which is like a bird kept in the body's cage.

An empty barrel: they no soul bread have.
Alas! poor bird, the world's white loaf is done.
And cannot yield thee here the smallest crumb.

In this sad state, God's tender bowels[4] run
　　Out streams of grace: and He to end all strife　　　　　20
The purest wheat in heaven His dear-dear son
　　Grinds, and kneads up into this bread of life.
　　Which bread of life from Heaven down came and stands
　　Dished on thy table up by Angels hands.

Did God mold up this bread in heaven, and bake,　　　　　25
　　Which from His table came, and to thine goeth?
Doth He bespeak thee thus, this soul bread take.
　　Come eat thy fill of this thy God's white loaf?
　　It's food too fine for angels, yet come, take
　　And eat thy fill. It's heaven's sugar cake.　　　　　30

What grace is this knead in this loaf? This thing
　　Souls are but petty things it to admire.
Ye angels, help: This fill would to the brim
　　Heav'ns whelmed-down[5] crystal meal bowl, yea and higher
　　This bread of life dropped in thy mouth, doth cry.　　35
　　Eat, eat me, Soul, and thou shalt never die.
June 8, 1684　　　　　　　　　　　　　　　　　　　1939

Meditation 22 (First Series)

Philippians 2.9. God hath highly exalted him.[1]

When thy bright beams, my Lord, do strike mine eye,
　　Methinks I then could truly chide outright
My hide-bound soul that stands so niggardly
　　That scarce a thought gets glorified by't.
　　My quaintest[2] metaphors are ragged stuff,　　　　　5
　　Making the sun seem like a mullipuff.[3]

It's my desire, Thou shouldst be glorified:
　　But when Thy glory shines before mine eye,
I pardon crave, lest my desire be pride.
　　Or bed Thy glory in a cloudy sky.　　　　　　　　10

4. Here used in the sense of the interior of the body, the "seat of the tender and sympathetic emotions," the heart.
5. Turned over. The *Oxford English Dictionary* quotes a passage from Dryden which is relevant: "That the earth is like a trencher and the Heavens a dish whelmed over it."
1. "Let this mind be in you, which was also in Christ Jesus: Who, being in the form of God, thought it not robbery to be equal with God: But made himself no reputation, and took upon him the form of a servant, and was made in the likeness of men: And being found in fashion as a man, he humbled himself, and became obedient unto death, even the death of the cross. Wherefore God also hath highly exalted him, and given him a name which is above every name: That at the name of Jesus every knee should bow, of things in heaven, and things in earth, and things under the earth; And that every tongue should confess that Jesus Christ is Lord, to the glory of God the Father" (Philippians 2.5–11).
2. Most skilled, wise.
3. Fuzz ball.

The sun grows wan; and angels palefaced shrink,
Before Thy shine, which I besmear with ink.

But shall the bird sing forth Thy praise, and shall
  The little bee present her thankful hum?
But I who see Thy shining glory fall   15
  Before mine eyes, stand blockish, dull, and dumb?
  Whether I speak, or speechless stand, I spy,
  I fail Thy glory: therefore pardon cry.

But this I find; my rhymes do better suit
  Mine own dispraise than tune forth praise to Thee.   20
Yet being chid, whether consonant,[4] or mute,
  I force my tongue to tattle, as You see.
  That I Thy glorious praise may trumpet right,
  Be thou my song, and make, Lord, me thy pipe.

This shining sky will fly away apace,   25
  When Thy bright glory splits the same to make
Thy majesty a pass, whose fairest face
  Too foul a path is for Thy feet to take.
  What glory then, shall tend Thee through the sky
  Draining the heaven much of angels dry?   30

What light then flame will in Thy judgment seat,
  'Fore which all men and angels shall appear?
How shall Thy glorious righteousness them treat,
  Rend'ring to each after his works done here?
  Then saints with angels thou wilt glorify: 35
  And burn lewd[5] men, and devils gloriously.

One glimpse, my Lord, of Thy bright judgment day,
  And glory piercing through, like fiery darts,
All devils, doth me make for grace to pray,
  For filling grace had I ten thousand hearts.   40
  I'd through ten hells to see Thy judgment day
  Wouldst Thou but gild my soul with Thy bright ray.

June 12, 1687   1960

Meditation 38 (First Series)

1 John 2.1. An Advocate with the Father.[1]

Oh! what a thing is man? Lord, who am I?
  That Thou shouldst give him law[2] (Oh! golden line)
To regulate his thoughts, words, life thereby.
  And judge him wilt thereby too in Thy time.

---

4. Talkative, making sounds.
5. Worthless, fallen.
1. "And if any man sin, we have an advocate with the Father, Jesus Christ the righteous: And he is the propitiation for our sins; and not for ours only, but also for the sins of the whole world" (1 John 2.1–2).
2. I.e., the Ten Commandments set forth in the Old Testament prescribing our behavior. This inheritance is our "golden line," or lineage.

A court of justice Thou in heaven hold'st   5
To try his case while he's here housed on mold.[3]

How do Thy angels lay before Thine eye
    My deeds both white and black I daily do?
How doth Thy court Thou Panelist[4] there them try?
    But flesh complains. What right for this? let's know.   10
    For right, or wrong I can't appear unto't.
    And shall a sentence pass on such a suite?

Soft; blemish not this golden bench, or place.
    Here is no bribe, nor colorings[5] to hide
Nor pettifogger[6] to befog the case   15
    But justice hath her glory here well tried.
    Her spotless law all spotted cases tends.
    Without respect or disrespect them ends.

God's judge Himself: and Christ attorney is,
    The Holy Ghost regesterer[7] is found.   20
Angels the sergeants[8] are, all creatures kiss
    The book, and do as Evidences[9] abound.
    All cases pass according to pure law
    And in the sentence is no fret,[1] nor flaw.

What sayst, my soul? Here all thy deeds are tried.   25
    Is Christ thy advocate to plead thy cause?
Art thou His client? Such shall never slide.
    He never lost His case: He pleads such laws
    As carry do the same, nor doth refuse
    The vilest sinner's case that doth Him choose.   30

This is His honour, not dishonor: nay,
    No habeas-corpus[2] against His clients came
For all their fines His purse doth make down pay.
    He non-suits Satan's suit or casts[3] the same.
    He'll plead thy case, and not accept a fee.   35
    He'll plead sub forma pauperis[4] for thee.

My case is bad. Lord, be my advocate.
    My sin is red: I'm under God's arrest.
    Thou hast the hint of pleading; plead my state.

---

3. I.e., the body will decay; only the soul is immortal.
4. I.e., impanel, as a jury.
5. Deceitful appearances.
6. Lawyer who handles trivial cases and is given to professional tricks and quibblings.
7. I.e., registrar, court recorder and keeper of records.
8. Attendants at court who maintain order.
9. Witnesses.
1. Malice, ill will.
2. "Thou shalt have the body"; i.e., no person may be kept in jail without the charge against him being quickly brought before a judge. Christ's clients are brought to trial immediately; He has Himself paid for man's sins with His crucifixion and covered their fines with His own blood.
3. I.e., stops a suit because of insufficient evidence; "casts": dismisses.
4. "According to form of poverty," a procedure in which an impoverished person is able to sue another without threat of costs in the event that he should lose.

Although it's bad Thy plea will make it best. 40
If Thou wilt plead my case before the King:
I'll wagonloads of love and glory bring.

July 6, 1690   1939

From GOD'S DETERMINATIONS[1]

The Preface

Infinity, when all things it beheld
In nothing, and of nothing all did build,
Upon what base was fixed the lath wherein
He turned this globe, and riggalled[2] so trim?
Who blew the bellows of His furnace vast?   5
Or held the mold wherein the world was cast?
Who laid its corner stone?[3] Or whose command?
Where stand the pillars upon which it stands?
Who laced and filleted[4] the earth so fine,
With rivers like green ribbons smaragdine?[5]   10
Who made the seas its selvage,[6] and it locks
Like a quilt ball[7] within a silver box?
Who spread its canopy? Or curtains spun?
Who in this bowling alley bowled the sun?
Who made it always when it rises set   15
To go at once both down, and up to get?
Who th' curtain rods made for this tapestry?
Who hung the twinkling lanterns in the sky?
Who? who did this? or who is He? Why, know
It's only might almighty this did do.   20
His hand hath made this noble work which stands
His glorious handiwork not made by hands.
Who spake all things from nothing; and with ease
Can speak all things to nothing, if He please.
Whose little finger at His pleasure can   25
Out mete[8] ten thousand worlds with half a span:
Whose might almighty can by half a looks

1. The subject of this "debate" poem is made clear in the full title: *God's determinations touching His elect: and the Elect's combat in their conversion, and coming up to God in Christ, together with the comfortable effects thereof.* In this group of poems Taylor explores the progress of the human soul from the creation of the world and the fall of man to the redemption of the Christian soul through Christ's crucifixion: Christ's mercy triumphs over justice—the punishment that man deserves for his disobedience—and the soul is finally carried to Heaven to share *in the joys of the* resurrection. The text used here is from *Poems of Edward Taylor*, ed. Donald E. Stanford (New Haven, Conn.: Yale University Press, 1960).

2. Grooved.

3. "Where wast thou when I laid the foundations of the earth? declare if thou hast understanding. Who hath laid the measures thereof, if thou knowest? or who hath stretched the line upon it? Whereupon are the foundations thereof fastened? or who laid the corner stone thereof; When the morning stars sang together, and all the songs of God shouted for joy? Or who shut up the sea with doors, when it brake forth, as if it had issued out of the womb?" (Job 38.4–8).

4. Encircled, bound around.

5. Emerald green.

6. The border of woven material which prevents unraveling.

7. A ball of wool which would unravel if it were not kept in a box.

8. Outmeasure.

Root up the rocks and rock the hills by th'roots.
Can take this mighty world up in His hand,
And shake it like a squitchen[9] or a wand.   30
Whose single frown will make heavens shake
Like as an wind makes quake.
Oh! what a might is this whose single frown
Doth shake the world as it would shake it down?
Which all from nothing fet,[1] from nothing, all: 35
Hath all on nothing set, lets nothing fall.
Gave all to nothing man indeed, whereby
Through nothing man all might Him glorify.
In nothing then embossed the brightest gem
More precious than all preciousness in them.   40
But nothing man did throw down all by sin:
And darkened that lightsome gem in him.
    That now his brightest diamond is grown
    Darker by far than any coalpit stone.

c. 1685   1939

The Soul's Groan to Christ for Succor

Good Lord, behold this dreadful enemy
    Who makes me tremble with his fierce assaults;
I dare not trust, yet fear to give the lie,
    For in my soul, my soul finds many faults.
    And though I justify myself to's face: 5
    I do condemn myself before Thy grace.

He strives to mount my sins, and them advance
    Above Thy merits, pardons, or good will
Thy grace to lessen, and Thy wrath t'enhance
    As if Thou couldst not pay the sinner's bill.   10
    He chiefly injures Thy rich grace, I find,
    Though I confess my heart to sin inclined.

Those graces which Thy grace enwrought in me,
    He makes as nothing but a pack of sins.
He maketh grace no grace, but cruelty;   15
    Is grace's honeycomb a comb of stings?
    This makes me ready leave Thy grace and run.
    Which if I do, I find I am undone.

I know he is Thy cur, therefore I be
    Perplexed lest I from Thy pasture stray.   20
He bays, and barks so veh'mently at me.
    Come rate[2] this cur, Lord, break his teeth, I pray.
    Remember me, I humbly pray Thee first.
    Then halter up this cur that is so cursed.

c. 1685   1939

9. Switch.
1. Made.
                    2. Give reproof, specifically, to a dog.

Christ's Reply

Peace, Peace, my honey, do not cry,
My little darling, wipe thine eye,
    Oh cheer, cheer up, come see.
Is anything too dear,[1] my dove,
Is anything too good, my love,   5
    To get or give for thee?

If in the several[2] thou art,
This yelper fierce will at thee bark:
    That thou art Mine this shows.
As Spot barks back the sheep again   10
Before they to the pound are ta'en,
    So he and hence 'way goes.

But yet this cur that bay so sore
Is broken-toothed, and muzzled sure,
    Fear not, my pretty heart.   15
His barking is to make thee cling
Close underneath thy Savior's wing.
    Why did my sweeten[3] start?

And if he run an inch too far,
I'll check his chain, and rate[4] the cur.
    My chick, keep close to me.   20
The poles shall sooner kiss, and greet
And parallels shall sooner meet
    Than thou shalt harmed be.

He seeks to aggravate thy sin   25
And screw them to the highest pin,
    To make thy faith to quail.
Yet mountain sins like mites should show
And then these mites for naught should go
    Could he but once prevail.   30

I smote thy sins upon the head.
They deadened are, though not quite dead:
    And shall not rise again.
I'll put away the guilt thereof,
And purge its filthiness clear off: 35
    My blood doth out the stain.

And though thy judgment was remiss
Thy headstrong will too wilful is.
    I will renew the same.
And though thou do too frequently   40
Offend as heretofore hereby,
    I'll not severely blame.

1. Difficult to get; expensive.
2. Divided, in the sense that the soul is torn between Christ and the devil.
3. Beloved.
4. Give reproof; specifically, to a dog.

And though thy senses do inveigle
Thy noble soul to tend the beagle,
    That t'hunt her games forth go,   45
I'll lure her back to me, and change
Those fond affections that do range
    As yelping beagles do.

Although thy sins increase their race,
And though when thou hast sought for grace,   50
    Thou fallst more than before,
If thou by true repentence rise,
And faith makes me thy sacrifice,
    I'll pardon all, though more.

Though Satan strive to block thy way   55
By all his stratagems he may:
    Come, come though through the fire.
For hell that gulf of fire for sins,
Is not so hot as t'burn thy shins.
    Then credit not the liar.   60

Those cursed vermin sins that crawl
All o'er thy soul, both great and small,
    Are only satan's own:
Which he in his malignity
Unto thy soul's true sanctity   65
    In at the doors hath thrown.

And though they be rebellion high,
Atheism or apostasy;
    Though blasphemy it be:
Unto what quality, or size   70
Excepting one, so e'er it rise.
    Repent, I'll pardon thee.

Although thy soul was once a stall[5]
Rich hung with Satan's nicknacks all;
    If thou repent thy sin,   75
A tabernacle in't I'll place
Filled with God's spirit, and His grace.
    Oh, comfortable thing!

I dare the world therefore to show
A God like Me, to anger slow: 80
    Whose wrath is full of grace.
Doth hate all sins both great and small:
Yet when repented, pardons all.
    Frowns with a smiling face.

As for thy outward postures each,   85
Thy gestures, actions, and thy speech,

5. A small booth in which things are sold.

I eye and eying spare,
If thou repent. My grace is more
Ten thousand times still trebled o'er
    Than thou canst want, or wear.   90

As for the wicked charge he makes,
That he of every dish first takes
    Of all thy holy things,
It's false, deny the same, and say,
That which he had he stole away   95
    Out of thy offerings.[6]

Though to thy grief, poor heart, thou find
In prayer too oft a wandering mind,
    In sermons spirits dull,
Though faith in fiery furnace flags,   100
And zeal in chilly seasons lags,
    Temptation's powerful.

These faults are his, and none of thine
So far as thou dost them decline.
    Come, then, receive my grace.   105
And when he buffets thee therefore,
If thou My aid and grace implore,
    I'll show a pleasant face.

But still look for temptations deep,
Whilst that thy noble spark doth keep   110
    Within a mudwalled cote.[7]
These white frosts and the showers that fall
Are but to whiten thee withall,
    Not rot the web they smote.

If in the fire where gold is tried   115
Thy soul is put, and purified,
    Wilt thou lament thy loss?
If silver-like this fire refine
Thy soul and make it brighter shine:
    Wilt thou bewail the dross?   120

Oh! fight My field: no colors fear:
I'll be thy front, I'll be thy rear.
    Fail not: My battles fight.
Defy the tempter, and his Mock.
Anchor thy heart on Me thy rock.   125
    I do in thee delight.

c. 1685   1939

6. Satan argues that man does nothing in a disinterested way for the love of God, but only from the fear of hell. Christ argues that the soul should affirm its generous impulses and put Satan in his place.
7. Cottage.

The Joy of Church Fellowship
Rightly Attended[1]

In heaven soaring up, I dropped an ear
  On Earth: and oh! sweet melody!
And listening, found it was the saints[2] who were
  Encoached for heaven that sang for joy.
    For in Christ's coach they sweetly sing;   5
    As they to glory ride therein.

Oh! joyous hearts! enfired with holy flame!
  Is speech thus tasseléd[3] with praise?
Will not your inward fire of joy contain:
  That it in open flames doth blaze?   10
    For in christ's coach saints sweetly sing,
    As they to glory ride therein.

And if a string do slip, by chance, they soon
  Do screw it up again; whereby
They set it in a more melodious tune   15
  And a diviner harmony.
    For in Christ's coach they sweetly sing
    As they to Glory ride therein.

In all their acts, public, and private, nay
  And secret, too, they praise impart.   20
But in their acts divine and worship, they
  With hymns do offer up their heart.
    Thus in Christ's coach they sweetly sing
    As they to glory ride therein.

Some few not in;[4] and some whose time and place   25
  Block up this coach's way do go
As travelers afoot, and so do trace
  The road that gives them right thereto;
    While in this coach these sweetly sing
    As they to glory ride therein.   30

c. 1685

1939

Huswifery[1]

Make me, O Lord, Thy spining wheel complete.
Thy holy word my distaff make for me.

1. The final poem in *God's Determinations*.
2. I.e., "visible saints," those who were church members when alive.
3. Ornamented, in the sense in which tassels might be applied to a piece of fabric.
4. I.e., those who are saved but are outside the church.
1. Housekeeping; used here to mean weaving. In Taylor's *Treatise Concerning the Lord's Supper* he considers the sig-nificance of the sacrament of communion and takes as his text a passage from the New Testament: "And he saith unto him, Friend, how camest thou in hither not having a wedding garment? And he was speechless" (Matthew 22.12). Taylor argues that the wedding garment is the proper sign of the regenerate Christian. The text used here is from *Poems of Edward Taylor*, ed. Donald E. Stanford (New Haven, Conn.: Yale University Press, 1960).

Make mine affections Thy swift flyers neat
    And make my soul Thy holy spool to be.
    My conversation make to be Thy reel   5
    And reel the yarn thereon spun of Thy wheel.[2]

Make me Thy loom then, knit therein this twine:
    And make Thy holy spirit, Lord, wind quills:[3]
Then weave the web Thyself. The yarn is fine.
    Thine ordinances make my fulling mills.[4]   10
    Then dye the same in heavenly colors choice,
    All pinked with varnished[5] flowers of paradise.

Then clothe therewith mine understanding, will,
    Affections, judgment, conscience, memory,
My words, and actions, that their shine may fill   15
    My ways with glory and Thee glorify.
    Then mine apparel shall display before Ye
    That I am clothed in holy robes for glory.

1939

2. In the lines above Taylor refers to the working parts of a spinning wheel: the "distaff" holds the raw wool or flax; the "flyers" regulate the spinning; the "spool" twists the yarn; and the "reel" takes up the finished thread.

3. I.e., be like a spool or bobbin.
4. Where cloth is beaten and cleansed with fuller's earth, or soap.
5. "Pinked": adorned; "varnished": glossy, sparkling.

COTTON MATHER
1663–1728

Cotton Mather, as the eldest son of Increase Mather and the grandson of Richard Mather and John Cotton, was the heir apparent to the Congregational hierarchy which had dominated the churches of New England for almost fifty years. Like his father before him, Cotton Mather attended Harvard College. He was admitted at the age of twelve, and when he graduated in 1678, President Urian Oakes told the commencement audience that his hope was great that "in this youth, Cotton and Mather shall, in fact as well as name, joint together and once more appear in life." He was expected by his family to excel and did not disappoint them, but there is no doubt he had to pay a price for his ambition: he stammered badly when young, so much so that it was assumed he could never be a preacher, and he was subject all his life to nervous disorders which drove him alternatively to ecstasy and despair. His enemies often complained that he was vain and aggressive. But he was also a genius of sorts, competent in the natural sciences and gifted in the study of ancient languages. He possessed a strong mind, and by the time he had stopped writing he could boast that he had published more than four hundred separate works. A worthy successor to his father's position as pastor of the Second Church of Boston, he remained connected with that church from 1685, when he was ordained, until his death.

Like Benjamin Franklin, Mather found great satisfaction in doing good works, and organized societies for building churches, supported schools for the children of slaves, and worked to establish funds for indigent clergy. But for all his worldly success, Mather's life was darkened by disappointment and tragedy. He lost two wives and saw his third wife go insane, and of his fifteen children, only two lived until his death. More than one of his contemporaries observed that he never overcame his bitterness at being rejected for the presidency of Harvard. It was the one thing his father had achieved that he could not succeed in doing.

Although he was a skillful preacher and an eminent theologian, it is his work as a historian which has earned Mather a significant place in American literature. No one has described more movingly the hopes of the first generation of Puritans, and what gives Mather's best writing its urgency is the sense that the Puritan community as he knew it was fading away. By the time that Mather was writing his history of New England, the issues which seemed most pressing to his parishioners were political and social rather than theological. In his diary of 1700 he noted that "there was hardly any but my father and myself to appear in defense of our invaded churches." Everything that Mather wrote can be seen as a call to defend the old order of church authority against the encroachment of an increasingly secular world. As an apologist for the "old New England way" there is no doubt that Mather left himself open to attack, and by the end of the seventeenth century he had become a scapegoat for the worst in Puritan culture. He is often blamed for the Salem witch trials, for example, but he never actually attended one of them; his greatest crime was in not speaking out against those who he knew had exceeded the limits of authority. Mather saw the devil's presence in Salem as a final effort to undermine and destroy religious community.

In spite of its rambling and sometimes self-indulgent nature, the *Magnalia Christi Americana* (the title may be translated as "A History of the Wonderful Works of Christ in America") remains Mather's most impressive work. It is described on the title page as an "ecclesiastical history of New England," and in the course of its seven books, Mather attempts to record for future readers not only a history of the New England churches and the college (Harvard) where its ministers were trained, but representative biographies of "saint's" lives. Although it is true that Mather was so caught up in his vision of a glorious past that he was sometimes quite blind to the suffocating realities of the world in which he lived, no one has set forth more clearly the history of a people who transformed a wilderness into a garden and the ideal of a harmonious community which has been characterized time and again as the "American dream." It is, however, in Mather's biographical sketches—his lives of Bradford, Winthrop, Eliot, and Phips—that the *Magnalia* is most arresting; for it is in his account of a particular saint's reconciliation with God on earth that New England's story is most eloquently realized.

Much has been said about Mather's style, mostly by those who hate his pedantry. But Mather did not favor any one manner of writing. He is fond of paradoxes and repetition, and he sometimes displays his learning shamelessly, but his prose can be quite straightforward when the occasion demands it. Mather also tolerated a variety of prose styles, but clearly liked the allusive style best. He tells us in his guide to young ministers (*Manu-*

ductio ad Ministerium, 1726) that the prose which he likes best is that in which "there is not only a vigor sensible in every sentence, but the paragraph is embellished with profitable references, even to something beyond what is directly spoken. Formal and painful quotations are not studied; yet all that could be learnt from them is insinuated. The writer pretends not unto reading, yet he could not have written as he does if he had not read very much in his time; and his composures are not only a cloth of gold, but also stuck with as many jewels as the gown of a Russian ambassador."

From The Wonders of the Invisible World[1]

[*A People of God in the Devil's Territories*]

The New Englanders are a people of God settled in those, which were once the devil's territories; and it may easily be supposed that the devil was exceedingly disturbed, when he perceived such a people here accomplishing the promise of old made unto our blessed Jesus, that He should have the utmost parts of the earth for His possession.[2] There was not a greater uproar among the Ephesians,[3] when the Gospel was first brought among them, than there was among the powers of the air (after whom those Ephesians walked) when first the silver trumpets of the Gospel here made the joyful sound. The devil thus irritated, immediately tried all sorts of methods to overturn this poor plantation: and so much of the church, as was fled into this wilderness, immediately found the serpent cast out of his mouth a flood for the carrying of it away. I believe that never were more satanical devices used for the unsettling of any people under the sun, than what have been employed for the extirpation of the vine which God has here planted, casting out the heathen, and preparing a room before it, and causing it to take deep root, and fill the land, so that it sent its boughs unto the Atlantic Sea eastward, and its branches unto the Connecticut River westward, and the hills were covered with the shadow thereof. But all those attempts of hell have hitherto been abortive, many an Ebenezer[4] has been erected unto the praise of God, by his poor people here; and having obtained help from God, we continue to

1. In May, 1692, Governor William Phips of Massachusetts appointed a court to "hear and determine" the cases against some 19 persons in Salem, Massachusetts, accused of witchcraft. Mather had long been interested in the subject of witchcraft, and in this work, written at the request of the judges, he describes the case against the accused. Mather, like many others, saw the evidence of witchcraft as the devil's work, a last-ditch effort to undermine the Puritan ideal. Mather was himself skeptical of much of the *evidence* used against the *accused,* especially as the trials proceeded in the summer of 1692, but like a number of prominent persons in the community, he made no public protest.

First published in 1693, the text used here is taken from the reprint by John Russell Smith (London, 1862).
2. After Jesus was baptized he went into the desert to fast for 40 days; it was there that the devil tempted him and offered him the world. See Luke 4.
3. Ephesus was an ancient city of Ionia in west Asia Minor and famous for its temples to the goddess Diana. When St. Paul preached there he received hostile treatment, and riots followed the sermons of missionaries who attempted to convert the Ephesians.
4. Literally a stone of help; a commemorative monument like the one Samuel erected to commemorate victory over the Philistines (1 Samuel 7.12).

this day Wherefore the devil is now making one attempt more upon us; an attempt more difficult, more surprising, more snarled with unintelligible circumstances than any that we have hitherto encountered; an attempt so critical, that if we get well through, we shall soon enjoy halcyon days with all the vultures of hell trodden under our feet. He has wanted his incarnate legions to persecute us, as the people of God have in the other hemisphere been persecuted: he has therefore drawn forth his more spiritual ones to make an attack upon us. We have been advised by some credible Christians yet alive, that a malefactor, accused of witchcraft as well as murder, and executed in this place more than forty years ago, did then give notice of an horrible plot against the country by witchcraft, and a foundation of witchcraft then laid, which if it were not seasonably discovered, would probably blow up, and pull down all the churches in the country. And we have now with horror seen the discovery of such a witchcraft! An army of devils is horribly broke in upon the place which is the center, and after a sort, the first-born of our English settlements: and the houses of the good people there are filled with the doleful shrieks of their children and servants, tormented by invisible hands, with tortures altogether preternatural. After the mischiefs there endeavored, and since in part conquered, the terrible plague of evil angels hath made its progress into some other places, where other persons have been in like manner diabolically handled. These our poor afflicted neighbors, quickly after they become infected and infested with these demons, arrive to a capacity of discerning those which they conceive the shapes of their troublers; and notwithstanding the great and just suspicion that the demons might impose the shapes of innocent persons in their spectral exhibitions upon the sufferers (which may perhaps prove no small part of the witch-plot in the issue), yet many of the persons thus represented, being examined, several of them have been convicted of a very damnable witchcraft: yea, more than one twenty have confessed, that they have signed unto a book, which the devil showed them, and engaged in his hellish design of bewitching and ruining our land. We know not, at least I know not, how far the delusions of Satan may be interwoven into some circumstances of the confessions; but one would think all the rules of understanding human affairs are at an end, if after so many most voluntary harmonious confessions, made by intelligent persons of all ages, in sundry towns, at several times, we must not believe the main strokes wherein those confessions all agree: especially when we have a thousand preternatural things every day before our eyes, wherein the confessors do acknowledge their concernment, and give demonstration of their being so concerned. If the devils now can strike the minds of men with any poisons of so fine a composition and operation, that scores of innocent people shall unite, in confessions of a crime, which we see actually committed, it is a thing prodigious, beyond the wonders of the former ages, and it threatens no less than a sort of a dissolution upon the world. Now, by these confessions 'tis agreed that the Devil has made a dreadful knot of witches in the country, and by the help of witches has dreadfully increased that knot: that these witches have driven a trade of commissioning their confederate spirits to do all sorts of mischiefs to the neighbors, whereupon there have ensued such mischievous consequences upon the bodies and estates of the neighborhood, as could not otherwise be accounted for: yea, that at prodigious witch-meetings, the wretches have proceeded so far as to concert and consult the methods of rooting out the Christian religion from this country, and setting up instead of it perhaps a more gross diabolism than ever the world saw before. And yet it will be a thing little short of miracle, if in so spread a business as this, the Devil should not get in some of his juggles,[5] to confound the discovery of all the rest.

* * *

But I shall no longer detain my reader from his expected entertainment, in a brief account of the trials which have passed upon some of the malefactors lately executed at Salem, for the witchcrafts whereof they stood convicted. For my own part, I was not present at any of them; nor ever had I any personal prejudice at the persons thus brought upon the stage; much less at the surviving relations of those persons, with and for whom I would be as hearty a mourner as any man living in the world: The Lord comfort them! But having received a command[6] so to do, I can do no other than shortly relate the chief matters of fact, which occurred in the trials of some that were executed, in an abridgment collected out of the court papers on this occasion put into my hands. You are to take the truth, just as it was; and the truth will hurt no good man. There might have been more of these, if my book would not thereby have swollen too big; and if some other worthy hands did not perhaps intend something further in these collections; for which cause I have only singled out four or five, which may serve to illustrate the way of dealing, wherein witchcrafts use to be concerned; and I report matters not as an advocate, but as an historian.

* * *

The Trial of Martha Carrier

AT THE COURT OF OYER AND TERMINER,[7]
HELD BY ADJOURNMENT AT SALEM, AUGUST 2. 1692

I. Martha Carrier was indicted for the bewitching certain persons, according to the form usual in such cases, pleading not guilty to her indictment; there were first brought in a considerable number of the bewitched persons *who* not only made the court sensible[8] of an

5. Tricks.
6. I.e., the request by the judges of the Salem trials to explain the sentencing of
people accused of witchcraft.
7. To hear and determine.
8. Aware.

horrid witchcraft committed upon them, but also deposed that it was Martha Carrier, or her shape, that grievously tormented them, by biting, pricking, pinching and choking of them. It was further deposed that while this Carrier was on her examination before the magistrates, the poor people were so tortured that every one expected their death upon the very spot, but that upon the binding of Carrier they were eased. Moreover the look of Carrier then laid the afflicted people for dead; and her touch, if her eye at the same time were off them, raised them again: which things were also now seen upon her trial. And it was testified that upon the mention of some having their necks twisted almost round, by the shape of this Carrier, she replied, "It's no matter though their necks had been twisted quite off."

II. Before the trial of this prisoner, several of her own children had frankly and fully confessed not only that they were witches themselves, but that this their mother had made them so. This confession they made with great shows of repentance, and with much demonstration of truth. They related place, time, occasion; they gave an account of journeys, meetings and mischiefs by them performed, and were very credible in what they said. Nevertheless, this evidence was not produced against the prisoner at the bar,[9] inasmuch as there was other evidence enough to proceed upon.

III. Benjamin Abbot gave his testimony that last March was a twelvemonth, this Carrier was very angry with him, upon laying out some land near her husband's: her expressions in this anger were that she would stick as close to Abbot as the bark stuck to the tree; and that he should repent of it afore seven years came to an end, so as Doctor Prescot should never cure him. These words were heard by others besides Abbot himself; who also heard her say, she would hold his nose as close to the grindstone as ever it was held since his name was Abbot. Presently after this, he was taken with a swelling in his foot, and then with a pain in his side, and exceedingly tormented. It bred into a sore, which was lanced[1] by Doctor Prescot, and several gallons of corruption[2] ran out of it. For six weeks it continued very bad, and then another sore bred in the groin, which was also lanced by Doctor Prescot. Another sore than bred in his groin, which was likewise cut, and put him to very great misery: he was brought unto death's door, and so remained until Carrier was taken, and carried away by the constable, from which very day he began to mend, and so grew better every day, and is well ever since.

Sarah Abbot also, his wife, testified that her husband was not only all this while afflicted in his body, but also that strange, extraordinary and unaccountable calamities befell his cattle; their death being such as they could guess at no natural reason for.

IV. Allin Toothaker testified that Richard, the son of Martha

9. Court.
1. Cut open.
2. Pus; infected matter.

Carrier, having some difference with him, pulled him down by the hair of the head. When he rose again he was going to strike at Richard Carrier but fell down flat on his back to the ground, and had not power to stir hand or foot, until he told Carrier he yielded; and then he saw the shape of Martha Carrier go off his breast.

This Toothaker had received a wound in the wars; and he now testified that Martha Carrier told him he should never be cured. Just afore the apprehending of Carrier, he could thrust a knitting needle into his wound four inches deep; but presently after her being seized, he was thoroughly healed.

He further testified that when Carrier and he some times were at variance, she would clap her hands at him, and say he should get nothing by it; whereupon he several times lost his cattle, by strange deaths, whereof no natural causes could be given.

V. John Rogger also testified that upon the threatening words of this malicious Carrier, his cattle would be strangely bewitched; as was more particularly then described.

VI. Samuel Preston testified that about two years ago, having some difference with Martha Carrier, he lost a cow in a strange, preternatural, unusual manner; and about a month after this, the said Carrier, having again some difference with him, she told him he had lately lost a cow, and it should not be long before he lost another; which accordingly came to pass; for he had a thriving and well-kept cow, which without any known cause quickly fell down and died.

VII. Phebe Chandler testified that about a fortnight before the apprehension of Martha Carrier, on a Lordsday, while the Psalm was singing in the Church, this Carrier then took her by the shoulder and shaking her, asked her, where she lived: she made her no answer, although as Carrier, who lived next door to her father's house, could not in reason but know who she was. Quickly after this, as she was at several times crossing the fields, she heard a voice, that she took to be Martha Carrier's, and it seemed as if it was over her head. The voice told her she should within two or three days be poisoned. Accordingly, within such a little time, one half of her right hand became greatly swollen and very painful; as also part of her face: whereof she can give no account how it came. It continued very bad for some days; and several times since she has had a great pain in her breast; and been so seized on her legs that she has hardly been able to go. She added that lately, going well to the house of God, Richard, the son of Martha Carrier, looked very earnestly upon her, and immediately her hand, which had formerly been poisoned, as is abovesaid, began to pain her greatly, and she had a strange burning at her stomach; but was then struck deaf, so that she could not hear any of the prayer, or singing, till the two or three last words of the Psalm.

VIII. One Foster, who confessed her own share in the witchcraft for which the prisoner stood indicted, affirmed that she had seen the prisoner at some of their witch-meetings, and that it was this Carrier, who persuaded her to be a witch. She confessed that the Devil carried them on a pole to a witch-meeting; but the pole broke, and she hanging about Carrier's neck, they both fell down, and she then received an hurt by the fall, whereof she was not at this very time recovered.

IX. One Lacy, who likewise confessed her share in this witch-craft, now testified, that she and the prisoner were once bodily present at a witch-meeting in Salem Village; and that she knew the prisoner to be a witch, and to have been at a diabolical sacrament, and that the prisoner was the undoing of her and her children by enticing them into the snare of the devil.

X. Another Lacy, who also confessed her share in this witchcraft, now testified, that the prisoner was at the witch-meeting, in Salem Village, where they had bread and wine administered unto them.

XI. In the time of this prisoner's trial, one Susanna Sheldon in open court had her hands unaccountably tied together with a wheel-band[3] so fast that without cutting it, it could not be loosed: it was done by a specter; and the sufferer affirmed it was the prisoner's.

Memorandum. This rampant hag, Martha Carrier, was the person of whom the confessions of the witches, and of her own children among the rest, agreed that the devil had promised her she should be Queen of Hebrews.

1692, 1693

From MAGNALIA CHRISTI AMERICANA[1]

Galeacius Secundus:[2]

The Life of William Bradford, Esq., Governor of Plymouth Colony

Omnium somnos illius vigilantia defendit; omnium otium, illius labor; omnium delitias, illius industria; omnium vacationem, illius occupatio.[3]

It has been a matter of some observation, that although York-shire be one of the largest shires in England; yet for all the fires of

3. A band or strap that goes around a wheel.

1. "A History of the Wonderful Works of Christ in America." Mather's book is subtitled *The ecclesiastical History of New England from its first planting, in the year 1620, unto the year of our Lord, 1698.* The *Magnalia* contains seven books. The first book is concerned with the discovery of America and the founding and history of the New England settlements. The second book contains lives of Governors of New England, and the life of William Bradford may be found there; the third book contains lives of 60 famous "Divines, by whose ministry the churches of New England have been planted and continued." Other books contain a history of Harvard University, a record of church ordinances passed in synods, and a record of "illustrious" and "wonderous" events which have been witnessed by people in New England. First published in London in 1702, the text used here is taken from the reprint, London, 1852.

2. "The second shield-bearer." Bradford was the second Governor of Plymouth and elected after the death of John Carver.

3. "His vigilance defends the sleep of all; his labor, their rest; his industry, their pleasures; and his diligence, their leisure."

martyrdom which were kindled in the days of Queen Mary,[4] it afforded no more fuel than one poor leaf; namely, John Leaf, an apprentice, who suffered for the doctrine of the Reformation at the same time and stake with the famous John Bradford.[5] But when the reign of Queen Elizabeth[6] would not admit the reformation of worship to proceed unto those degrees, which were proposed and pursued by no small number of the faithful in those days, Yorkshire was not the least of the shires in England that afforded suffering witnesses thereunto. The churches there gathered were quickly molested with such a raging persecution, that if the spirit of separa-tion in them did carry them unto a further extreme than it should have done, one blamable cause thereof will be found in the extrem-ity of that persecution. Their troubles made that cold country too hot for them, so that they were under a necessity to seek a retreat in the Low Countries;[7] and yet the watchful malice and fury of their adversaries rendered it almost impossible for them to find what they sought. For them to leave their native soil, their lands and their friends, and go into a strange place, where they must hear foreign language, and live meanly[8] and hardly, and in other employ-ments than that of husbandry, wherein they had been educated, these must needs have been such discouragements as could have been conquered by none, save those who sought first the kingdom of God, and the righteousness thereof. But that which would have made these discouragements the more unconquerable unto an ordi-nary faith, was the terrible zeal of their enemies to guard all ports, and search all ships, that none of them should be carried off. I will not relate all the sad things of this kind then seen and felt by this people of God; but only exemplify those trials with one short story. Diverse of this people having hired a Dutchman, then lying at Hull, to carry them over to Holland, he promised faithfully to take them in, between Grimsby and Hull; but they coming to the place a day or two too soon, the appearance of such a multitude alarmed the officers of the town adjoining, who came with a great body of sol-diers to seize upon them. Now it happened that one boat full of men had been carried aboard, while the women were yet in a bark that lay aground in a creek at low water. The Dutchman perceiving the storm that was thus beginning ashore, swore by the sacrament that he would stay no longer for any of them; and so taking the advantage of a fair wind then blowing, he put out to sea for Zeeland.[9] The women thus left near Grimsby-common, bereaved of their husbands, who had been hurried from them, and forsaken of

4. During the reign of Mary Tudor (1553–58) an effort was made to restore Roman Catholicism to the position of a national church, and a number of Prot-estants were executed.
5. John Bradford (1510?–55) was burned at the stake with Leaf on July 1, 1555. Their story was well known from John Foxe's *Book of Martyrs*.

6. Elizabeth followed Mary Tudor to the throne (1558–1603) and by virtue of the Act of Uniformity proscribed traditional church ritual.
7. A number of English Puritans went to Holland to form their own churches without threat of civil disobedience.
8. In poverty.
9. I.e., the Netherlands.

their neighbors, of whom none durst in this fright stay with them, were a very rueful spectacle; some crying for fear, some shaking for cold, all dragged by troops of armed and angry men from one Justice to another, till not knowing what to do with them, they even dismissed them to shift as well as they could for themselves. But by their singular afflictions, and by their Christian behaviors, the cause for which they exposed themselves did gain considerably. In the meantime, the men at sea found reason to be glad that their families were not with them, for they were surprised with an horrible tempest, which held them for fourteen days together, in seven whereof they saw not sun, moon or star, but were driven upon the coast of Norway. The mariners often despaired of life, and once with doleful shrieks gave over all, as thinking the vessel was founded: but the vessel rose again, and when the mariners with sunk hearts often cried out, "We sink! we sink!" the passengers, without such distraction of mind, even while the water was running into their mouths and ears, would cheerfully shout, "Yet, Lord, thou canst save! Yet, Lord, thou canst save!" And the Lord accordingly brought them at last safe unto their desired haven: and not long after helped their distressed relations thither after them, where indeed they found upon almost all accounts a new world, but a world in which they found that they must live like strangers and pilgrims.

Among those devout people was our William Bradford, who was born Anno 1588, in an obscure village called Austerfield, where the people were as unacquainted with the Bible, as the Jews do seem to have been with part of it in the days of Josiah;[1] a most ignorant and licentious people, and like unto their priest. Here, and in some other places, he had a comfortable inheritance left him of his honest parents, who died while he was yet a child, and cast him on the education,[2] first of his grandparents, and then of his uncles, who devoted him, like his ancestors, unto the affairs of husbandry. Soon a long sickness kept him, as he would afterwards thankfully say, from the vanities of youth, and made him the fitter for what he was afterwards to undergo. When he was about a dozen years old, the reading of the Scriptures began to cause great impressions upon him; and those impressions were much assisted and improved, when he came to enjoy Mr. Richard Clifton's[3] illuminating ministry, not far from his abode; he was then also further befriended, by being brought into the company and fellowship of such as were then called professors;[4] though the young man that brought him into it did after become a profane and wicked apostate.[5] Nor could the wrath of his uncles, nor the scoff of his neighbors, now turned upon

---

1. **King of Judah** (638?–608? B.C.). Josiah was ignorant of the book of the law of the God of Israel and worshiped false gods. See 2 Kings 22 ff.
2. I.e., made his education dependent upon.
3. A Puritan minister in the town of Scrooby, who also settled in Amsterdam with the Scrooby separatists. He died in 1616.
4. I.e., those who declared their faith.
5. One who denies what he formerly professed.

him, as one of the Puritans, divert him from his pious inclinations.

At last, beholding how fearfully the evangelical and apostolical church-form, whereinto the churches of the primitive times were cast by the good spirit of God, had been deformed by the apostacy of the succeeding times; and what little progress the Reformation had yet made in many parts of Christendom towards its recovery, he set himself by reading, by discourse, by prayer, to learn whether it was not his duty to withdraw from the communion of the parish-assemblies, and engage with some society of the faithful, that should keep close unto the written word of God, as the rule of their worship. And after many distresses of mind concerning it, he took up a very deliberate and understanding resolution, of doing so; which resolution he cheerfully prosecuted, although the provoked rage of his friends tried all the ways imaginable to reclaim him from it, unto all whom his answer was:

> Were I like to endanger my life, or consume my estate by any ungodly courses, your counsels to me were very seasonable; but you know that I have been diligent and provident in my calling, and not only desirous to augment what I have, but also to enjoy it in your company; to part from which will be as great a cross as can befall me. Nevertheless, to keep a good conscience, and walk in such a way as God has prescribed in His Word; is a thing which I must prefer before you all, and above life itself. Wherefore, since 'tis for a good cause that I am like to suffer the disasters which you lay before me, you have no cause to be either angry with me, or sorry for me; yea, I am not only willing to part with every thing that is dear to me in this world for this cause, but I am also thankful that God has given me an heart so to do, and will accept me so to suffer for Him.

Some lamented him, some derided him, all dissuaded him: nevertheless, the more they did it, the more fixed he was in his purpose to seek the ordinances of the gospel, where they should be dispensed with most of the commanded purity; and the sudden deaths of the chief relations which thus lay at him,[6] quickly after convinced him what a folly it had been to have quitted his profession, in expectation of any satisfaction from them. So to Holland he attempted a removal.

Having with a great company of Christians hired a ship to transport them for Holland, the master perfidiously betrayed them into the hands of those persecutors, who rifled and ransacked their goods, and clapped their persons into prison at Boston,[7] where they lay for a month together. But Mr. Bradford being a young man of about eighteen, was dismissed sooner than the rest, so that within a while he had opportunity with some others to get over to Zeeland,

6. I.e., struck out at him and tried to make him change his beliefs.
7. Boston, England.

through perils, both by land and sea not inconsiderable; where he was not long ashore ere a viper seized on his hand—that is, an officer—who carried him unto the magistrates, unto whom an envious passenger had accused him as having fled out of England. When the magistrates understood the true cause of his coming thither, they were well satisfied with him; and so he repaired joyfully unto his brethren at Amsterdam, where the difficulties to which he afterwards stooped in learning and serving of a Frenchman at the working of silks, were abundantly compensated by the delight wherewith he sat under the shadow of our Lord, in His purely dispensed ordinances. At the end of two years, he did, being of age to do it, convert his estate in England into money; but setting up for himself, he found some of his designs by the providence of God frowned upon, which he judged a correction bestowed by God upon him for certain decays of internal piety, whereinto he had fallen; the consumption of his estate he thought came to prevent a consumption in his virtue. But after he had resided in Holland about half a score years, he was one of those who bore a part in that hazardous and generous enterprise of removing into New England, with part of the English church at Leyden, where, at their first landing, his dearest consort[8] accidentally falling overboard, was drowned in the harbor; and the rest of his days were spent in the services, and the temptations, of that American wilderness.

Here was Mr. Bradford, in the year 1621, unanimously chosen the governor of the plantation; the difficulties whereof were such, that if he had not been a person of more than ordinary piety, wisdom and courage, he must have sunk under them. He had, with a laudable industry, been laying up a treasure of experiences, and he had now occasion to use it; indeed, nothing but an experienced man could have been suitable to the necessities of the people. The potent nations of the Indians, into whose country they were come, would have cut them off, if the blessing of God upon his conduct had not quelled them; and if his prudence, justice and moderation had not overruled them, they had been ruined by their own distempers. One specimen of his demeanor is to this day particularly spoken of. A company of young fellows that were newly arrived were very unwilling to comply with the governor's order for working abroad on the public account; and therefore on Christmas Day, when he had called upon them, they excused themselves, with a pretense that it was against their conscience to work such a day.[9] The governor gave them no answer, only that he would spare them till they were better informed; but by and by he found them all at play in the street, sporting themselves with various diversions; whereupon commanding the instruments of their games to be taken from them, he effectually gave them to understand that it was

8 Wife.
9. Puritans did not observe Christmas as a holiday.

against his conscience that they should play whilst others were at work, and that if they had any devotion to the day, they should show it at home in the exercises of religion, and not in the streets with pastime and frolics; and this gentle reproof put a final stop to all such disorders for the future.

For two years together after the beginning of the colony, whereof he was now governor, the poor people had a great experiment of "man's not living by bread alone";[1] for when they were left all together without one morsel of bread for many months, one after another, still the good providence of God relieved them, and supplied them, and this for the most part out of the sea. In this low condition of affairs, there was no little exercise for the prudence and patience of the governor, who cheerfully bore his part in all; and, that industry might not flag, he quickly set himself to settle propriety[2] among the new planters, foreseeing that while the whole country labored upon a common stock,[3] the husbandry and business of the plantation could not flourish, as Plato[4] and others long since dreamed that it would if a community were established. Certainly, if the spirit which dwelt in the old Puritans, had not inspired these new planters, they had sunk under the burden of these difficulties; but our Bradford had a double portion of that spirit.

The plantation was quickly thrown into a storm that almost overwhelmed it, by the unhappy actions of a minister sent over from England by the adventurers[5] concerned for the plantation; but by the blessing of Heaven on the conduct of the governor, they weathered out that storm. Only the adventurers, hereupon breaking to pieces, threw up all their concernments with the infant colony; whereof they gave this as one reason, that the planters dissembled with his Majesty and their friends in their petition, wherein they declared for a church discipline, agreeing with the French and others of the reforming churches in Europe.[6] Whereas 'twas now urged, that they had admitted into their communion a person who at his admission utterly renounced the Churches of England, (which person, by the way, was that very man who had made the complaints against them) and therefore, though they denied the name of Brownists,[7] yet they were the thing. In answer hereunto, the very words written by the governor were these:

> Whereas you tax us with dissembling about the French discipline, you do us wrong, for we both hold and practice the discipline of the French and other Reformed Churches (as

1. Luke 4.4.
2. I.e., property.
3. Property held in common.
4. *Greek philosopher* (427?–347 B.C.).
5. English investors.
6. In Europe states were declared as either Protestant or Catholic; in France the Edict of Nantes (1598) provided liberty of conscience for all without denying the authority of the Crown.
7. Robert Browne (c. 1550–1633) was a separatist clergyman and identified with Congregationalism, a system whereby each church is independent of any national church.

they have published the same in the Harmony of Confessions) according to our means, in effect and substance. But whereas you would tie us up to the French discipline in every circumstance, you derogate from the liberty we have in Christ Jesus. The Apostle Paul would have none to follow him in any thing, but wherein he follows Christ; much less ought any Christian or church in the world to do it. The French may err, we may err, and other churches may err, and doubtless do in many circumstances. That honor therefore belongs only to the infallible Word of God, and pure Testament of Christ, to be propounded and followed as the only rule and pattern for direction herein to all churches and Christians. And it is too great arrogancy for any man or church to think that he or they have so sounded the Word of God unto the bottom, as precisely to set down the church's discipline without error in substance or circumstance, that no other without blame may digress or differ in any thing from the same. And it is not difficult to show that the reformed churches differ in many circumstances among themselves.

By which words it appears how far he was free from that rigid spirit of separation, which broke to pieces the separatists themselves in the Low Countries, unto the great scandal of the reforming churches.[8] He was indeed a person of a well-tempered spirit, or else it had been scarce possible for him to have kept the affairs of Plymouth in so good a temper for thirty-seven years together; in every one of which he was chosen their governor, except the three years wherein Mr. Winslow,[9] and the two years wherein Mr. Prince,[1] at the choice of the people, took a turn with him.

The leader of a people in a wilderness had need be a Moses;[2] and if a Moses had not led the people of Plymouth Colony, where this worthy person was the governor, the people had never with so much unanimity and importunity still called him to lead them. Among many instances thereof, let this one piece of self-denial be told for a memorial of him, wheresoever this history shall be considered: the patent of the colony was taken in his name, running in these terms: "To William Bradford, his heirs, associates, and assigns," but when the number of the freemen[3] was much increased, and many new townships erected, the General Court there desired of Mr. Bradford that he would make a surrender of the same into their hands, which he willingly and presently assented unto, and confirmed it according to their desire by his hand and seal, reserving no more for himself than was his proportion, with

8. By the time the movement for reform had come to an end, the movement for separating ended in dissension and mutual recrimination, with particular churches arguing they were more pure than others. Two English Puritans baptized themselves on the grounds that there were no pure churches to baptize them.

9. Edward Winslow (1595–1655).
1. Thomas Prince (1600–73).
2. The Hebrew lawgiver and prophet who led the Israelites out of Egypt.
3. I.e., those who are not indentured servants and able to work for themselves.

others, by agreement. But as he found the providence of heaven many ways recompensing his many acts of self-denial, so he gave this testimony to the faithfulness of the divine promises: that he had forsaken friends, houses and lands for the sake of the gospel, and the Lord gave them him again. Here he prospered in his estate; and besides a worthy son which he had by a former wife, he had also two sons and a daughter by another, whom he married in this land.

He was a person for study as well as action; and hence, notwithstanding the difficulties through which he passed in his youth, he attained unto a notable skill in languages: the Dutch tongue was become almost as vernacular to him as the English; the French tongue he could also manage; the Latin and the Greek he had mastered; but the Hebrew he most of all studied, because he said he would see with his own eyes the ancient oracles of God in their native beauty. He was also well skilled in history, in antiquity, and in philosophy; and for theology he became so versed in it, that he was an irrefragable disputant against the errors, especially those of Anabaptism,[4] which with trouble he saw rising in his colony; wherefore he wrote some significant things for the confutation of those errors. But the crown of all was his holy, prayerful, watchful, and fruitful walk with God, wherein he was very exemplary.

At length he fell into an indisposition of body, which rendered him unhealthy for a whole winter; and as the spring advanced, his health yet more declined; yet he felt himself not what he counted sick, till one day, in the night after which, the God of heaven so filled his mind with ineffable consolations, that he seemed little short of Paul, rapt up unto the unutterable entertainments of Paradise.[5] The next morning he told his friends that the good spirit of God had given him a pledge of his happiness in another world, and the first fruits of his eternal glory; and on the day following he died, May 9, 1657, in the 69th year of his age—lamented by all the colonies of New England as a common blessing and father to them all.

O mihi si Similis Contingat Clausula Vitae![6]

Plato's brief description of a governor, is all that I will now leave as his character, in an

EPITAPH.

Νομευς Τροψος ἀγελης ανθρωπινης.[7]

Men are but flocks: Bradford beheld their need,
And long did them at once both rule and feed.

1702

4. *Anabaptists* opposed the baptism of children and advocated separation of church and state.
5. Paul was converted on the road to Damascus when a light from heaven transfixed him. See Acts 9.3–5.
6. "Oh, that such an end of life might come to me."
7. "Shepherd and provider of the human flock."

SARAH KEMBLE KNIGHT
1666–1727

Like a number of the classics of early American literature, *The Private Journal* of Mrs. Sarah Kemble Knight was not published until the nineteenth century; it at once found an enthusiastic audience eager to read documents of social history from the American past. Mrs. Knight's shorthand journal, as transcribed and edited by Theodore Dwight, provided a healthy antidote to the soul-searching journals of Mrs. Knight's contemporaries, and revealed an earthiness and ready wit, an appetite for living, and a frankness not often found in colonial literature. Mrs. Knight was a keen observer of provincial America and a woman who did not suffer fools gladly. Something of Mrs. Knight's tough-mindedness was undoubtedly the result of the fact that she had to make her way in the world with considerable ingenuity and that early on in her life she displayed a gift for managing other people's affairs.

Sarah Kemble Knight was the daughter of a Boston merchant. She married a man much older who was a sometime sea captain and London agent for an American company. After her father died in 1689, Mrs. Knight assumed full responsibility for being head of the household. While her husband was abroad, she kept a boarding house and taught school (hence the title "Madam Knight") and supposedly could number Benjamin Franklin and the Mather children among her pupils. Mrs. Knight taught penmanship, made copies of court records, and wrote letters for people having business with the courts. She trained herself in the ways of the law and had a reputation for settling estates with skill. In 1704, while her husband was abroad, Mrs. Knight took upon herself the task of settling her cousin Caleb Trowbridge's estate on behalf of his young widow. Mrs. Knight set out for New Haven, Connecticut, on Monday, October 2, 1704. From there she went to New York and returned home to Boston in March, 1705. It was a hazardous journey, one not undertaken lightly in those years, and almost unprecedented for a woman traveling alone.

In 1706 Mrs. Knight apparently became a widow; at least her husband is not mentioned again after that date. In 1714 she followed her married daughter to Connecticut. Her last years were spent in New London, where she ran an inn and, true to her genius, made shrewd investments in property.

The text used here is from *The Journal of Madame Knight*, ed. George P. Winship (Boston, 1920; reprinted by Peter Smith, New York, 1935).

From The Private Journal of a Journey
from Boston to New York

✳ ✳ ✳

Tuesday, October the Third

About 8 in the morning, I with the post proceeded forward without observing anything remarkable; and about two, afternoon, arrived at the post's second stage,[1] where the western post met him and exchanged letters. Here, having called for something to eat, the woman brought in a twisted thing like a cable, but something[2] whiter; and laying it on the board, tugged for life to bring it into a capacity to spread; which having with great pains accomplished, she served in a dish of pork and cabbage, I suppose the remains of Dinner. The sauce was of a deep purple, which I thought was boiled in her dye kettle; the bread was Indian and everything on the table service agreeable to these. I, being hungry, got a little down; but my stomach was soon cloyed, and what cabbage I swallowed served me for a cud the whole day after.

Having here discharged the ordinary[3] for self and guide (as I understood was the custom), about three, afternoon, went on with my third guide, who rode very hard; and having crossed Providence ferry, we come to a river which they generally ride through. But I dare not venture; so the post got a lad and canoe to carry me to t'other side, and he rid through and led my horse. The canoe was very small and shallow, so that when we were in, she seemed ready to take in water, which greatly terrified me, and caused me to be very circumspect, sitting with my hands fast on each side, my eyes steady, not daring so much as to lodge my tongue a hair's breadth more on one side of my mouth than t'other nor so much as think on Lot's wife,[4] for a wry thought would have overset our wherry:[5] but was soon put out of this pain, by feeling the canoe on shore, which I as soon almost saluted with my feet; and rewarding my sculler, again mounted and made the best of our way forwards. The road here was very even and the day pleasant, it being now near sunset. But the post told me we had near 14 miles to ride to the next stage (where we were to lodge). I asked him of the rest of the road, foreseeing we must travail in the night. He told me there was a bad river we were to ride through, which was so very fierce a horse could sometimes hardly stem it: but it was but narrow, and we should soon be over. I cannot express the concern of mind this relation set me in: no thoughts but those of the dangerous river could entertain my imagination, and they were as formidable as various, still tormenting me with blackest ideas of my approaching fate—sometimes seeing myself drowning, otherwhiles drowned, and at the best, like a holy sister just come out of a spiritual bath in dripping garments.

Now was the glorious luminary, with his swift coursers arrived at his stage,[6] leaving poor me with the rest of this part of the lower world in darkness, with which we were soon surrounded. The only

1. A stopping place where mail was left from the previous stage of a journey and then forwarded to the next. People who traveled "post" accompanied the driver to the next stop.
2. Somewhat.
3. Paid for the meal, an "ordinary" is a meal served at an inn or public house.

4. I.e., not daring to look back. When Sodom was being destroyed, Lot's wife looked back and was turned into a pillar of salt (Genesis 19.26).
5. Small boat.
6. I.e., it was night; in Greek mythology the sun god Apollo rides across the sky in a chariot. "Coursers": horses.

glimmering we now had was from the spangled skies, whose imperfect reflections rendered every object formidable. Each lifeless trunk, with its shattered limbs, appeared an armed enemy; and every little stump like a ravenous devourer. Nor could I so much as discern my guide, when at any distance, which added to the terror.

Thus, absolutely lost in thought, and dying with the very thoughts of drowning, I come up with the post, who I did not see 'til even with his horse: he told me he stopped for me; and we rode on very deliberately a few paces, when we entered a thicket of trees and shrubs, and I perceived by the horse's going, we were on the descent of a hill, which, as we come nearer the bottom, 'twas totally dark with the trees that surrounded it. But I knew by the going of the horse we had entered the water, which my guide told me was the hazardous river he had told me of; and he, riding up close to my side, bid me not fear—we should be over immediately. I now rallied all the courage I was mistress of, knowing that I must either venture my fate of drowning, or be left like the children in the wood.[7] So, as the post bid me, I gave reins to my nag; and sitting as steady as just before in the canoe, in a few minutes got safe to the other side, which he told me was the Narragansett country.

Here we found great difficulty in travailing, the way being very narrow, and on each side the trees and bushes gave us very unpleasant welcomes with their branches and boughs, which we could not avoid, it being so exceeding dark. My guide, as before so now, put on harder than I, with my weary bones, could follow; so left me and the way behind him. Now returned my distressed apprehensions of the place where I was: the dolesome woods, my company next to none, going I knew not whither, and encompassed with terrifying darkness; The least of which was enough to startle a more masculine courage. Added to which the reflections, as in the afternoon of the day that my call[8] was very questionable, which, til then I had not so prudently as I ought considered. Now, coming to the foot of a hill, I found great difficulty in ascending; but being got to the top, was there amply recompensed with the friendly appearance of the kind conductress of the night, just then advancing above the horizontal line. The raptures which the sight of that fair planet produced in me, caused me, for the moment, to forget my present weariness and past toils; and inspired me for most of the remaining way with very diverting thoughts, some of which, with other occurrences of the day, I reserved to note down when I should come to my stage. My thoughts on the sight of the moon were to this purpose:

> Fair Cynthia,[9] all the homage that I may
> Unto a creature, unto thee I pay;

7. The phrase "children in the wood" or "babes in the woods" refers to a ballad in which two children are taken out to be murdered and instead are left in the woods where they die during the night.
8. Mission.
9. The moon personified.

In lonesome woods to meet so kind a guide,
To me's more worth than all the world beside.
Some joy I felt just now, when safe got o'er
Yon surly river to this rugged shore,
Deeming rough welcomes from these clownish trees
Better than lodgings with Nereides.[1]
Yet swelling fears surprise; all dark appears—
Nothing but light can dissipate those fears.
My fainting vitals can't lend strength to say,
But softly whisper, O I wish 'twere day.
The murmur hardly warmed the ambient air,
E're thy bright aspect rescues from despair:
Makes the old hag[2] her sable mantle loose,
And a bright joy does through my soul diffuse.
The boisterous trees now lend a passage free,
And pleasant prospects thou giv'st light to see.

From hence we kept on, with more ease than before: the way
being smooth and even, the night warm and serene, and the tall
and thick trees at a distance, especially when the moon glared light
through the branches, filled my imagination with the pleasant delusion of a sumptuous city, filled with famous buildings and churches,
with their spiring steeples, balconies, galleries and I know not what:
grandeurs which I had heard of, and which the stories of foreign
countries had given me the idea of.

Here stood a lofty church—there is a steeple,
And there the grand parade—O see the people!
That famous castle there, were I but nigh,
To see the mote and bridge and walls so high—
They're very fine! says my deluded eye.

Being thus agreeably entertained without a thought of anything but
thoughts themselves, I on a sudden was roused from these pleasing
imaginations, by the post's sounding his horn, which assured me he
was arrived at the stage, where we were to lodge: and that music was
then most musical and agreeable to me.

Being come to Mr. Havens', I was very civilly received, and courteously entertained, in a clean comfortable house; and the good
woman was very active in helping off my riding clothes, and
then asked what I would eat. I told her I had some chocolate,
if she would prepare it; which with the help of some milk, and
a little clean brass kettle, she soon effected to my satisfaction.
I then betook me to my apartment, which was a little room
parted from the kitchen by a single board partition; where, after
I had noted the occurrences of the past day, I went to bed,

---

1. Sea nymphs, daughters of Nereus.   2. I.e., night.

which, though pretty hard, yet neat and handsome. But I could get no sleep, because of the clamor of some of the town topers in next room, who were entered into a strong debate concerning the signification of the name of their country (*viz.*), *Narragansett*. One said it was named so by the Indians, because there grew a brier there, of a prodigious height and bigness, the like hardly ever known, called by the Indians narragansett; and quotes an Indian of so barbarous a name for his author, that I could not write it. His antagonist replied no—it was from a spring it had its name, which he well knew where it was, which was extreme cold in summer, and as hot as could be imagined in the winter, which was much resorted to by the natives, and by them called Narragansett (hot and cold), and that was the original of their place's name—with a thousand impertinances not worth notice, which he uttered with such a roaring voice and thundering blows with the fist of wickedness on the table, that it pierced my very head. I heartily fretted, and wished 'um tonguetied; but with as little success as a friend of mine once, who was (as she said) kept a whole night awake, on a journey, by a country left. and a sergeant, insigne[3] and a deacon, contriving how to bring a triangle into a square. They kept calling for t'other gill,[4] which while they were swallowing, was some intermission; but presently, like oil to fire, increased the flame. I set my candle on a chest by the bedside, and setting up, fell to my old way of composing my resentments, in the following manner:

> I ask thy aid, O potent rum!
> To charm these wrangling topers dumb.
> Thou hast their giddy brains possessed—
> The man confounded with the beast—
> And I, poor I, can get no rest.
> Intoxicate them with thy fumes:
> O still their tongues' til morning comes!

And I know not but my wishes took effect; for the dispute soon ended with t'other dram; and so good night!

* * *

January the Sixth

Being now well recruited and fit for business I discoursed the person I was concerned with, that we might finish in order to my return to Boston. They delayed as they had hitherto done hoping to tire my patience. But I was resolute to stay and see an end of the matter let it be never so much to my disadvantage—So January 9th they come again and promise the Wednesday following to go through with the distribution of the estate which they delayed 'til

---

3. "Ensigne": lowest grade of commissioned officer. "Left.": lieutenant.
4. A measure of wine.

Thursday and then come with new amusements.[1] But at length by the mediation of that holy good gentleman, the Rev. Mr. James Pierpont, the minister of New Haven, and with the advice and assistance of other our good friends we come to an accommodation and distribution, which having finished though not 'til February, the man that waited on me to York taking the charge of me I set out for Boston. We went from New Haven upon the ice (the ferry being not passable thereby) and the Rev. Mr. Pierpont with Madam Prout, cousin Trowbridge and divers others were taking leave, we went onward without anything remarkable 'til we come to New London and lodged again at Mr. Saltonstalls—and here I dismissed my guide, and my generous entertainer provided me Mr. Samuel Rogers of that place to go home with me—I stayed a day here longer than intended by the commands of the honorable Governor Winthrop to stay and take a supper with him whose wonderful civility I may not omit. The next morning I crossed the ferry to Groton, having had the honor of the company, of Madam Livingston (who is the governor's daughter) and Mary Christophers and divers others to the boat—and that night lodged at Stonington and had roast beef and pumpkin sauce for supper. The next night at Haven's and had roast fowl, and the next day we come to a river which by reason of the freshets coming down was swelled so high we feared it impassable and the rapid stream was very terrifying—However we must over and that in a small canoe. Mr. Rogers assuring me of his good conduct,[2] I after a stay of near an hour on the shore for consultation, went into the canoe, and Mr. Rogers paddled about 100 yards up the creek by the shore side, turned into the swift stream and dexterously steering her in a moment we come to the other side as swiftly passing as an arrow shot out of the bow by a strong arm. I stayed on the shore 'til he returned to fetch our horses, which he caused to swim over, himself bringing the furniture in the canoe. But it is past my skill to express the exceeding fright all their transactions formed in me. We were now in the colony of the Massachusetts and taking lodgings at the first inn we come to, had a pretty difficult passage the next day which was the second of March by reason of the sloughy[3] ways then thawed by the sun. Here I met Capt. John Richards of Boston who was going home, so being very glad of his company we rode something harder than hitherto, and missing my way in going up a very steep hill, my horse dropped down under me as dead; this new surprise no little hurt me, meeting it just at the entrance into Dedham from whence we intended to reach home that night. But was now obliged to get another horse there and leave my own, resolving for Boston that night if possible. But in going over the causeway at Dedham the Bridge being overflowed by the high waters coming down, I very narrowly escaped

1. I.e., trifles which caused further delays.   3. Muddy.
2. Company for safekeeping.

falling over into the river horse and all which 'twas almost a miracle I did not—now it grew late in the afternoon and the people having very much discouraged us about the sloughy way which they said we should find very difficult and hazardous, it so wrought on me being tired and dispirited and disappointed of my desires of going home, that I agreed to lodge there that night which we did at the house of one Draper, and the next day being March 3d we got safe home to Boston, where I found my aged and tender mother and my dear and only child in good health with open arms ready to receive me, and my kind relations and friends flocking in to welcome me and hear the story of my transactions and travails, I having this day been five months from home and now I cannot fully express my joy and satisfaction. But desire sincerely to adore my great Benefactor for thus graciously carrying forth and returning in safety his unworthy handmaid.

1704–5   1825

WILLIAM BYRD
1674–1744

William Byrd was the son of a wealthy Virginia planter, merchant, and Indian trader. He was educated in England and spent the formative years of his life in London, where he was trained to assume the responsibility of managing his father's estate. Byrd was urbane and witty and traveled in sophisticated London circles. He loved the theater and numbered Wycherley and Congreve among his many friends. The diaries which he kept during his London years tell us little about his private thoughts but a great deal about his social life as a London man about town. Byrd recorded his habits of eating and prayer with the same detachment which compelled him to keep track of his whoring and his reading.

More than half of Byrd's life was spent in England, but in 1726, after many years of alternation, he returned to Virginia to stay. In a letter written in that year to his friend Charles Boyle, Earl of Orrery, Byrd described the satisfactions to be derived from the role of the Virginia planter:

> Besides the advantage of pure air, we abound in all kinds of provisions without expense (I mean we who have plantations). I have a large family of my own, and my doors are open to everybody, yet I have no bills to pay, and half-a-crown will rest undisturbed in my pockets for many moons altogether. Like one of the patriarchs, I have my flock and herds, my bondmen and bondwomen, and every sort of trade amongst my own servants, so that I live in a kind of independence on everyone but Providence. * * * Thus, my Lord, we are very happy in our Canaans if we could but forget the onions and fleshpots of Egypt.

Byrd's "New Canaan," unlike its New England counterpart, and in spite of his nod to Divine Providence, is a self-sufficient, earthly garden, combining the best of civilized and rural life. Byrd's house at Westover epitomized his ideal of the perfect plantation, and he spared no expense to fashion it. The house he rebuilt is still standing, one of the most beautiful examples of eighteenth-century architecture in America. It housed a library of some 3,600 volumes and rivaled Cotton Mather's library in Boston; its gardens were the envy of all who saw them. Byrd had a passion for land, and before he died he escalated the 26,000 acres left to him by his father into a vast holding of 179,000 acres. The cities of Richmond and Petersburg, Virginia, were created from Byrd's landholdings.

Given Byrd's position in the affairs of Virginia, it is not surprising that he was asked to accept a number of public commissions. In 1728 he accepted a commission to survey the much-disputed boundary line between Virginia and North Carolina. The diary which Byrd kept on this trip served as the sourcebook for his *History of the Dividing Line*. Although not published until 1841, Byrd's manuscript was circulated among his London friends, and its existence was known to later naturalists such as Thomas Jefferson. In its wealth of natural detail (some of it—like his description of the possum catching and eating little birds with its claws—quite fantastic) and Indian lore, it helped to satisfy London curiosity for all things American.

From History of the Dividing Line[1]

From *October*[2]

1. There was a white frost this morning on the ground, occasioned by a northwest wind, which stood our friend in dispersing all aguish[3] damps and making the air wholesome at the same time that it made it cold. Encouraged, therefore, by the weather, our surveyors got to work early and, by the benefit of clear woods and level ground, drove the line twelve miles and twelve poles.[4]

At a small distance from our camp we crossed Great Creek and about seven miles farther Nutbush Creek, so called from the many hazel trees growing upon it. By good luck, many branches of these creeks were full of reeds, to the great comfort of our horses. Near five miles from thence we encamped on a branch that runs into Nutbush Creek, where those reeds flourished more than ordinary. The land we marched over was for the most part broken and stony and in some places covered over with thickets almost impenetrable.

At night the surveyors, taking advantage of a very clear sky, made a third trial of the variation and found it still something less than three degrees; so that it did not diminish by advancing toward the west or by approaching the mountains, nor yet by increasing our distance from the sea, but remained much the same we had found it at Currituck Inlet.

One of our Indians killed a large fawn, which was very welcome,

1. *The* text used here is taken from *The Prose Works of William Byrd of Westover*, ed. Louis B. Wright (Cambridge, Mass.: Harvard University Press, 1966).
2. The boundary commission began its expedition on March 5, 1728. By October 1, they were close to present-day Bristol, Tennessee.
3. Feverish.
4. A linear measure equal to 5.5 yards.

though, like Hudibras'[5] horse, it had hardly flesh enough to cover its bones.

In the low grounds the Carolina gentlemen showed us another plant, which they said was used in their country to cure the bite of the rattlesnake. It put froth several leaves in figure like a heart and was clouded so like the common Asarabacca that I conceived it to be of that family.

2. So soon as the horses could be found, we hurried away the surveyors, who advanced the line 9 miles and 254 poles. About three miles from the camp they crossed a large creek, which the Indians called Massamony, signifying in their language "Paint Creek," because of the great quantity of red ocher found in its banks. This in every fresh tinges the water, just as the same mineral did formerly, and to this day continues to tinge, the famous river Adonis in Phoenicia,[6] by which there hangs a celebrated fable.

Three miles beyond that we passed another water with difficulty called Yapatsco or Beaver Creek. Those industrious animals had dammed up the water so high that we had much ado to get over. 'Tis hardly credible how much work of this kind they will do in the space of one night. They bite young saplings into proper lengths with their foreteeth, which are exceeding strong and sharp, and afterwards drag them to the place where they intend to stop the water. Then they know how to join timber and earth together with so much skill that their work is able to resist the most violent flood that can happen. In this they are qualified to instruct their betters, it being certain their dams will stand firm when the strongest that are made by men will be carried down the stream. We observed very broad, low grounds upon this creek, with a growth of large trees and all the other signs of fertility, but seemed subject to be everywhere overflowed in a fresh. The certain way to catch these sagacious animals is this: squeeze all the juice out of the large pride[1] of the beaver and six drops out of the small pride. Powder the inward bark of sassafras and mix it with this juice; then bait therewith a steel trap and they will eagerly come to it and be taken.

About three miles and an half farther we came to the banks of another creek, called the the Saponi language Ohimpamony, signifying "Jumping Creek," from the frequent jumping of fish during the spring season.

Here we encamped, and by the time the horses were hobbled our hunters brought us no less than a brace and an half of deer, which made great plenty and consequently great content in our quarters. Some of our people had shot a great wildcat, which was that fatal moment making a comfortable meal upon a fox squirrel, and an amibitious sportsman of our company claimed the merit of killing

5. In Samuel Butler's satire *Hudibras* (1663–78), a Presbyterian minister sets out on a journey upon a starving horse.
6. A river which turns red each spring, supposedly with the blood of Adonis, who, a mortal, was beloved of Aphrodite and wounded while hunting.
1. Testis.

this monster after it was dead. The wildcat is as big again as any household cat and much the fiercest inhabitant of the woods. Whenever it is disabled, it will tear its own flesh for madness. Although a panther will run away from a man, wildcat will only make a surly retreat, now and then facing about if he be too closely pursued, and will even pursue in his turn if he observe the least sign of fear or even of caution in those that pretend to follow him. The flesh of this beast, as well as of the panther, is as white as veal and altogether as sweet and delicious.

3. We got to work early this morning and carried the line 8 miles and 160 poles. We forded several runs of excellent water and afterwards traversed a large level of high land, full of lofty walnut, poplar, and white oak trees, which are certain proofs of a fruitful soil. This level was near two miles in length and of an unknown breadth, quite out of danger of being overflowed, which is a misfortune most of the low grounds are liable to in those parts. As we marched along, we saw many buffalo tracks and abundance of their dung very fresh but could not have the pleasure of seeing them. They either smelt us out, having that sense very quick,[2] or else were alarmed at the noise that so many people must necessarily make in marching along. At the sight of a man they will snort and grunt, cock up their ridiculous short tails, and tear up the ground with a sort of timorous fury. These wild cattle hardly ever range alone but herd together like those that are tame. They are seldom seen so far north as forty degrees of latitude, delighting much in canes and reeds which grow generally more southerly.

We quartered on the banks of a creek that the inhabitants call Tewahominy or Tuskarooda[3] Creek, because one of that nation had been killed thereabouts and his body thrown into the creek.

Our people had the fortune to kill a brace of does, one of which we presented to the Carolina gentlemen, who were glad to partake of the bounty of Providence at the same time that they sneered at us for depending upon it.

4. We hurried away the surveyors about nine this morning, who extended the line 7 miles and 160 poles, notwithstanding the ground was exceedingly uneven. At the distance of five miles we forded a stream to which we gave the name of Bluewing Creek because of the great number of those fowls[4] that then frequented it. About two and a half miles beyond that, we came upon Sugartree Creek, so called from the many trees of that kind that grow upon it.[5] By tapping this tree in the first warm weather in February, one may get from twenty to forty gallons of liquor, very sweet to the taste and agreeable to the stomach. This may be boiled into molasses first and afterwards into very good sugar, allowing about ten gallons of liquor to make a pound. There is no doubt,

2. Acutely.
3. Now Tuscarora.
4. I.e., blue-winged teal.
5. I.e., sugar maples.

too, that a very fine spirit may be distilled from the molasses, at least as good as rum. The sugar tree delights only in rich ground, where it grows very tall, and by the softness and sponginess of the wood should be a quick grower. Near this creek we discovered likewise several spice trees, the leaves of which are fragrant and the berries they bear are black when dry and of a hot taste, not much unlike pepper. The low grounds upon the creek are very wide, sometimes on one side, sometimes on the other, though most commonly upon the opposite shore the high land advances close to the bank, only on the north side of the line it spread itself into a great breadth of rich low ground on both sides the creek for four miles together, as far as this stream runs into Hyco River, whereof I shall presently make mention. One of our men spied three buffaloes, but his piece being loaded only with goose shot, he was able to make no effectual impression on their thick hides; however, this disappointment was made up by a brace of bucks and as many wild turkeys killed by the rest of the company. Thus Providence was very bountiful to our endeavors, never disappointing those that faithfully rely upon it and pray heartily for their daily bread.

5. This day we met with such uneven grounds and thick underwoods that with all our industry we were able to advance the line but 4 miles and 312 poles. In this small distance it intersected a large stream four times, which our Indian at first mistook for the south branch of Roanoke River; but, discovering his error soon after, he assured us 'twas a river called Hycootonmony, or Turkey Buzzard River, from the great number of those unsavory birds that roost on the tall trees growing near its banks.

Early in the afternoon, to our very great surprise, the commissioners of Carolina acquainted us with their resolution to return home. This declaration of theirs seemed the more abrupt because they had not been so kind as to prepare us by the least hint of their intention to desert us. We therefore let them understand they appeared to us to abandon the business they came about with too much precipitation, this being but the fifteenth day since we came out the last time. But although we were to be so unhappy as to lose the assistance of their great abilities, yet we, who were concerned for Virginia, determined, by the grace of God, not to do our work by halves but, all deserted as we were like to be, should think it our duty to push the line quite to the mountains; and if their government should refuse to be bound by so much of the line as we run without their commissioners, yet at least it would bind Virginia and stand as a direction how far His Majesty's lands extend to the southward. In short, these gentlemen were positive, and the most we could agree upon was to subscribe plats[6] of our work as far as we had acted together; though at the same time we insisted these plats

6. Maps.

should be got ready by Monday noon at farthest, when we on the part of Virginia intended, if we were alive, to move forward without farther loss of time, the season being then too far advanced to admit of any unnecessary or complaisant delays.

6. We lay still this day, being Sunday, on the bank of Hyco River and had only prayers, our chaplain not having spirits enough to preach. The gentlemen of Carolina assisted not at our public devotions, because they were taken up all the morning in making a formidable protest against our proceeding on the line without them. When the divine service was over, the surveyors set about making the plats of so much of the line as we had run this last campaign. Our pious friends of Carolina assisted in this work with some seeming scruple, pretending it was a violation of the Sabbath, which we were the more surprised at because it happened to be the first qualm of conscience they had ever been troubled with during the whole journey. They had made no bones of staying from prayers to hammer out an unnecessary protest, though divine service was no sooner over but an unusual fit of godliness made them fancy that finishing the plats, which was now matter of necessity, was a profanation of the day. However, the expediency of losing no time, for us who thought it our duty to finish what we had undertaken, make such a labor pardonable.

In the afternoon, Mr. Fitzwilliam, one of the commissioners for Virginia, acquainted his colleagues it was his opinion that by His Majesty's order they could not proceed farther on the line but in conjunction with the commissioners of Carolina; for which reason he intended to retire the next morning with those gentlemen. This looked a little odd in our brother commissioner; though, in justice to him as well as to our Carolina friends, they stuck by us as long as our good liquor lasted and were so kind to us as to drink our good journey to the mountains in the last bottle we had left.

7. The duplicates of the plats could not be drawn fair this day before noon, where they were countersigned by the commissioners of each government. Then those of Carolina delivered their protest, which was by this time licked into form and signed by them all. And we have been so just to them as to set it down at full length in the Appendix, that their reasons for leaving us may appear in their full strength. After having thus adjusted all our affairs with the Carolina commissioners and kindly supplied them with bread to carry them back, which they hardly deserved at our hands, we took leave both of them and our colleague, Mr. Fitzwilliam. This gentlemen had still a stronger reason for hurrying him back to Williamsburg, which was that neither the General Court might lose an able judge nor himself a double salary, not despairing in the least but he should have the whole pay of commissioner into the bargain, though he did not half the work. This, to be sure, was relying more on interest of his friends than on the justice of his cause; in which, however, he

had the misfortune to miscarry when it came to be fairly considered.

It was two o'clock in the afternoon before these arduous affairs could be dispatched, and then, all forsaken as we were, we held on our course toward the west. But it was our misfortune to meet with so many thickets in this afternoon's work that we could advance no further than 2 miles and 260 poles. In this small distance we crossed the Hyco the fifth time and quartered near Buffalo Creek, so named from the frequent tokens we discovered of that American behemoth. Here the bushes were so intolerably thick that we were obliged to cover the bread bags with our deerskins, otherwise the joke of one of the Indians must have happened to us in good earnest: that in a few days we must cut up our house to make bags for the bread and so be forced to expose our backs in compliment to our bellies. We computed we had then biscuit enough left to last us, with good management, seven weeks longer; and this being our chief dependence, it imported us to be very careful both in the carriage and the distribution of it.

We had now no other drink but what Adam drank in Paradise, though to our comfort we found the water excellent, by the help of which we perceived our appetites to mend, our slumbers to sweeten, the stream of life to run cool and peaceably in our veins, and if ever we dreamt of women, they were kind.

Our men killed a very fat buck and several turkeys. These two kinds of meat boiled together, with the addition of a little rice or French barley, made excellent soup, and, what happens rarely in other good things, it never cloyed, no more than an engaging wife would do, by being a constant dish. Our Indian was very superstitious in this matter and told us, with a face full of concern, that if we continued to boil venison and turkey together we should for the future kill nothing, because the spirit that presided over the woods would drive all the game out of our sight. But we had the happiness to find this an idle superstition, and though his argument could not convince us, yet our repeated experience at last, with much ado, convinced him.

We observed abundance of coltsfoot and maidenhair in many places and nowhere larger quantity than here. They are both excellent pectoral[7] plants and seem to have greater virtues much in this part of the world than in more northern climates; and I believe it may pass for a rule in botanics that where any vegetable is planted by the hand of Nature it has more virtue than in places whereto it is transplanted by the curiosity of man.

* * *

12. We were so cruelly entangled with bushes and grapevines all day that we could advance the line no farther than five miles and twenty-eight poles. The vines grew very thick in these woods, twin-

7. Good for diseases of the chest.

ing lovingly round the trees almost everywhere, especially to the saplings. This makes it evident how natural both the soil and climate of this country are to vines, though I believe most to our own vines. The grapes we commonly met with were black, though there be two or three kinds of white grapes that grow wild. The black are very sweet but small, because the strength of the vine spends itself in wood, though without question a proper culture would make the same grapes both larger and sweeter. But, with all these disadvantages, I have drunk tolerable good wine pressed from them, though made without skill. There is then good reason to believe it might admit of great improvement if rightly managed.

Our Indian killed a bear, two years old, that was feasting on these grapes. He was very fat, as they generally are in that season of the year. In the fall the flesh of this animal has a high relish different from that of other creatures, though inclining nearest to that of pork, or rather of wild boar. A true woodsman prefers this sort of meat to that of the fattest vension, not only for the *haut goût*,[8] but also because the fat of it is well tasted and never rises in the stomach. Another proof of the goodness of this meat is that it is less apt to corrupt than any other we are acquainted with.

As agreeable as such rich diet was to the men, yet we who were not accustomed to it tasted it at first with some sort of squeamishness, that animal being of the dog kind, though a little use soon reconciled us to this American venison. And that its being of the dog kind might give us the less disgust, we had the example of that ancient and polite people, the Chinese, who reckon dog's flesh too good for any under the quality of a mandarin. This beast is in truth a very clean feeder, living, while the season lasts, upon acorns, chestnuts, and chinquapins, wild honey and wild grapes. They are naturally not carnivorous, unless hunger constrain them to it after the mast[9] is all gone and the product of the woods quite exhausted. They are not provident enough to lay up any hoard like the squirrels, nor can they, after all, live very long upon licking their paws, as Sir John Mandeville[1] and some travelers tell us, but are forced in the winter months to quit the mountains and visit the inhabitants. Their errand is then to surprise a poor hog at a pinch to keep them from starving. And to show that they are not flesh eaters by trade, they devour their prey very awkwardly. They don't kill it right out and feast upon its blood and entrails, like other ravenous beasts, but, having, after a fair pursuit, seized it with their paws, they begin first upon the rump and so devour one collop after another till they come to the vitals, the poor animal crying all the while for several minutes together. However, in so doing, Bruin acts a little imprudently, because the dismal outcry of the hog alarms the neighbor-

8. Gamy flavor.
9. A collective name for the fruit of the beech, chestnut, oak, and other forest trees.
1. Ostensible author of a mid-14th-century book of travels.

hood, and 'tis odds but he pays the forfeit with his life before he can secure his retreat.

But bears soon grow weary of this unnatural diet, and about January, when there is nothing to be gotten in the woods, they retire into some cave or hollow tree, where they sleep away two or three months very comfortably. But then they quit their holes in March, when the fish begin to run up the rivers, on which they are forced to keep Lent till some fruit or berry comes in season. But bears are fondest of chestnuts, which grow plentifully toward the mountains, upon very large trees, where the soil happens to be rich. We were curious to know how it happened that many of the outward branches of those trees came to be broke off in that solitary place and were informed that the bears are so discreet as not to trust their unwieldly bodies on the smaller limbs of the tree that would not bear their weight, but after venturing as far as is safe, which they can judge to an inch, they bite off the end of the branch, which falling down, they are content to finish their repast upon the ground. In the same cautious manner they secure the acorns that grow on the weaker limbs of the oak. And it must be allowed that in these instances a bear carries instinct a great way and acts more reasonably than many of his betters, who indiscreetly venture upon frail projects that won't bear them.

13. This being Sunday, we rested from our fatigue and had leisure to reflect on the signal mercies of Providence.

The great plenty of meat wherewith Bearskin[2] furnished us in these lonely woods made us once more shorten the men's allowance of bread from five to four pounds of biscuit a week. This was the more necessary because we knew not yet how long our business might require us to be out.

In the afternoon our hunters went forth and returned triumphantly with three brace of wild turkeys. They told us they could see mountains distinctly from every eminence, though the atmosphere was so thick with smoke that they appeared at a greater distance than they really were.

In the evening we examined our friend Bearskin concerning the religion of his country, and he explained it to us without any of that reserve to which his nation is subject. He told us he believed there was one supreme god, who had several subaltern deities under him. And that this master god made the world a long time ago. That he told the sun, the moon, and stars their business in the beginning, which they, with good looking-after, have faithfully performed ever since. That the same power that made all things at first has taken care to keep them in the same method and motion ever since. He believed that God had formed many worlds before he formed this, but that those worlds either grew old and ruinous or

2. Their Indian guide.

were destroyed for the dishonesty of the inhabitants. That God is very just and very good, ever well pleased with those men who possess those godlike qualities. That he takes good people into his safe protection, makes them very rich, fills their bellies plentifully, preserves them from sickness and from being surprised or overcome by their enemies. But all such as tell lies and cheat those they have dealings with he never fails to punish with sickness, poverty, and hunger and, after all that, suffers them to be knocked on the head and scalped by those that fight against them.

He believed that after death both good and bad people are conducted by a strong guard into a great road, in which departed souls travel together for some time till at a certain distance this road forks into two paths, the one extremely level and the other stony and mountainous. Here the good are parted from the bad by a flash of lightning, the first being hurried away to the right, the other to the left. The right-hand road leads to a charming, warm country, where the spring is everlasting and every month is May; and as the year is always in its youth, so are the people, and particularly the women are bright as stars and never scold. That in this happy climate there are deer, turkeys, elks, and buffaloes innumerable, perpetually fat and gentle, while the trees are loaded with delicious fruit quite throughout the four seasons. That the soil brings forth corn spontaneously, without the curse of labor, and so very wholesome that none who have the happiness to eat of it are ever sick, grow old, or die. Near the entrance into this blessed land sits a venerable old man on a mat richly woven, who examines strictly all that are brought before him, and if they have behaved well, the guards are ordered to open the crystal gate and let them enter into the land of delight. The left-hand path is very rugged and uneven leading to a dark and barren country where it is always winter. The ground is the whole year round covered with snow, and nothing is to be seen upon the trees but icicles. All the people are hungry yet have not a morsel of anything to eat except a bitter kind of potato, that gives them the dry gripes[3] and fills their whole body with loathsome ulcers that stink and are insupportably painful. Here all the women are old and ugly, having claws like a panther with which they fly upon the men that slight their passion. For it seems these haggard old furies are intolerably fond and expect a vast deal of cherishing. They talk much and exceedingly shrill, giving exquisite pain to the drum of the ear, which in that place of the torment is so tender that every sharp note wounds it to the quick. At the end of this path sits a dreadful old woman on a monstrous toadstool, whose head is covered with rattlesnakes instead of tresses, with glaring white eyes that strike a terror unspeakable into all that behold her. This hag pronounces sentence of woe upon all the miserable wretches that hold up their hands at her tribunal. After this they

3. Dry heaves; vomiting.

are delivered over to huge turkey buzzards, like harpies, that fly away with them to the place above-mentioned. Here, after they have been tormented a certain number of years according to their several degrees of guilt, they are driven back into this world to try if they will mend their manners and merit a place the next time in the regions of bliss.

This was the substance of Bearskin's religion and was as much to the purpose as could be expected from a mere state of nature, without one glimpse of revelation or philosophy. It contained, however, the three great articles of natural religion: the belief of a god, the moral distinction betwixt good and evil, and the expectation of rewards and punishments in another world. Indeed, the Indian notion of a future happiness is a little gross and sensual, like Mahomet's Paradise. But how can it be otherwise in a people that are contented with Nature as they find her and have no other lights but what they receive from purblind tradition?

1728   *   *   *   1841

JONATHAN EDWARDS
1703–1758

Although it is certainly true that, as Perry Miller once put it, the true life of Jonathan Edwards is the life of a mind, the circumstances surrounding Edwards's career are not without their drama, and his rise to eminence and fall from power remain one of the most moving stories in American literature.

Edwards was born in East Windsor, Connecticut, a town not far from Hartford, the son of the Reverend Timothy Edwards and Esther Stoddard Edwards. There was little doubt from the beginning as to his career. Edwards's mother was the daughter of the Reverend Solomon Stoddard of Northampton, Massachusetts, one of the most influential and independent figures in the religious life of New England. Western Massachusetts clergymen were so anxious for his approval, that he was sometimes called the "Pope of the Connecticut Valley," and his gifted grandson, the only male child in a family of eleven children, was groomed to be his heir.

Edwards was a studious and dutiful child, and from an early age showed remarkable gifts of observation and exposition. When he was eleven he wrote an essay on the flying spider which is still very readable. Most of Edwards's early education he received at home. In 1716, when he was thirteen, Edwards was admitted to Yale College; he stayed on to read theology in New Haven for two years after his graduation in 1720. Like Benjamin Franklin, Edwards determined to perfect himself, and in one of his early notebooks he resolved "never to lose one moment of time, but to improve it in the most profitable way" he could. As a student he always rose at four in the morning, studied thirteen hours a day, and reserved part of each day for walking. It was a routine that Edwards varied little, even when, after spending two years in New York, he came to Northampton to assist his grandfather in his church. He married in 1727. In 1729 Solomon Stoddard died, and Edwards was named to succeed him. In the twenty-four years that Edwards lived in Northampton he managed to tend his duties as pastor of a growing congregation and deliver brilliant sermons, to write some of his most important books—concerned primarily with defining the nature of true religious experience—and watch his five children grow up. Until the mid-1740s his relations with the town seemed enviable.

In spite of the awesome—even imposing—quality of Jonathan Edwards's mind, all of his work is of a piece and, in essence, readily graspable. What Edwards was trying to do was to restore to his congregation and to his read ers that original sense of religious commitment which he felt had been lost since the first days of the Puritan exodus, and he wanted to do this by transforming his congregation from mere believers who understood the logic of Christian doctrine to converted Christians who were genuinely moved by the principles of their belief. Edwards says that he read the work of the English philosopher John Locke (1632–1704) with more pleasure "than the greedy miser finds when gathering up handfuls of silver and gold, from some newly discovered treasure." For Locke confirmed Edwards's conviction that we must do more than comprehend religious ideas; we must be *moved* by them, we must know them experientially: the difference, as he says, is like that between reading the word "fire" and actually being burned. Basic to this newly felt belief is the recognition that nothing that man can do warrants his salvation—that man is totally dependent on God, and that he is saved solely by God's grace. In his progress as a Christian, Edwards says that he experienced several steps toward conversion, but that his true conversion came only when he had achieved a "full and constant sense of the absolute sovereignty of God, and a delight in that sovereignty." The word "delight" reminds us that Edwards is trying to inculcate and describe a religious feeling that approximates a physical sensation, recognizing always that supernatural feelings and natural ones are actually very different. In his patient and lucid prose Edwards became a master at the art of persuading his congregation that they could—and *must*—possess this intense awareness of their precarious condition. The exaltation which his parishioners felt when they experienced delight in God's sovereignty was the characteristic fervid emotion of religious revivalism.

For fifteen years, beginning in 1734, this spirit of revivalism transformed complacent believers all along the eastern seaboard. This period of new religious fervor has been called "The Great Awakening," and in the early years Edwards could do no wrong. His meetinghouse was filled with newly converted believers, and the details of the spiritual life of Edwards and his congregation were the subject of inquiry by Christian believers everywhere. But in his attempt to restore the church to the position of authority it held in the years of his grandfather's reign, Edwards went too far. When he named backsliders from his pulpit—including the children and parents of the best families in town—and tried to return to the old order of communion, permitting the sacrament to be taken only by those who had publicly declared themselves to be saved, the people of the town turned against him. Residents of the Connecticut Valley everywhere were tired of religious controversy, and the hysterical behavior of a few fanatics turned many against the spirit of revivalism. On June 22, 1750, by a vote of two hundred to twenty, Edwards was dismissed from his church and effectively silenced. Although the congregation had difficulty naming a successor to Edwards, they preferred to have no sermons rather than let Edwards preach. For the next seven years he served as missionary to the Indians in Stockbridge, Massachusetts, a town thirty-five miles to the west of Northampton. There he wrote his monumental treatises debating the doctrine of the freedom of the will and defining the nature of true virtue: "that consent, propensity and union of heart to Being in general, that is immediately exercised in a general good will." It was in Stockbridge that Edwards received, very reluctantly, a call to become president of the College of New Jersey (later called Princeton). Three months after his arrival in Princeton, Edwards died of smallpox, the result of the inoculation taken to prevent infection.

Personal Narrative[1]

I had a variety of concerns and exercises[2] about my soul from my childhood, but had two more remarkable seasons of awakening[3] before I met with that change by which I was brought to those new dispositions and that new sense of things that I have since had. The first time was when I was a boy, some years before I went to college, at a time of remarkable awakening in my father's congregation. I was then very much affected[4] for many months and concerned about the things of religion and my soul's salvation and was abundant in duties. I used to pray five times a day in secret, and to spend much time in religious talk with other boys and used to meet with them to pray together. I experienced I know not what kind of delight in religion. My mind was much engaged in it, and had much self-righteous pleasure; and it was my delight to abound in religious duties. I, with some of my schoolmates, joined together and built a booth in a swamp, in a very secret and retired place, for a place of prayer. And besides, I had particular secret places of my own in the woods, where I used to retire by myself, and used to be from time to time much affected. My affections seemed to be lively and easily moved, and I seemed to be in my element, when engaged in religious duties. And I am ready to think, many are deceived with such affections and such a kind of delight, as I then had in religion, and mistake it for grace.

But in process of time, my convictions and affections wore off; and I entirely lost all those affections and delights, and left off

---

1. Because of Edwards's reference to an evening in January, 1739, this essay must have been written after that date. Edwards's reasons for writing it are not known, and it was not published in his lifetime. After his death his friend Samuel Hopkins had access to his manuscripts and prepared *The Life and Character of the Late Rev. Mr. Jonathan Edwards*, which was published in 1765. In that volume the *Personal Narrative* ap- peared in Section IV as a chapter entitled "An account of his conversion, experiences, and religious exercises, given by himself."

2. Religious thoughts, specifically for improving one's relations to God.

3. I.e., spiritual awakenings, renewals.

4. Emotionally aroused, as opposed to merely understanding rationally the arguments for Christian faith.

secret prayer, at least as to any constant performance of it, and returned like a dog to his vomit, and went on in ways of sin.[5]

Indeed, I was at some times very uneasy, especially towards the latter part of the time of my being at college.[6] 'Til it pleased God, in my last year at college, at a time when I was in the midst of many uneasy thoughts about the state of my soul, to seize me with a pleurisy;[7] in which he brought me nigh to the grave, and shook me over the pit of hell.

But yet, it was not long after my recovery before I fell again into my old ways of sin. But God would not suffer me to go on with any quietness; but I had great and violent inward struggles: 'til after many conflicts with wicked inclinations and repeated resolutions and bonds that I laid myself under by a kind of vows to God, I was brought wholly to break off all former wicked ways and all ways of known outward sin, and to apply myself to seek my salvation and practice the duties of religion, but without that kind of affection and delight that I had formerly experienced. My concern now wrought more by inward struggles and conflicts and self-reflections. I made seeking my salvation the main business of my life. But yet it seems to me I sought after a miserable manner, which has made me sometimes since to question whether ever it issued in that which was saving,[8] being ready to doubt, whether such miserable seeking was ever succeeded. But yet I was brought to seek salvation in a manner that I never was before. I felt a spirit to part with all things in the world for an interest in Christ. My concern continued and prevailed, with many exercising thoughts and inward struggles; but yet it never seemed to be proper to express my concern that I had, by the name of terror.

From my childhood up, my mind had been wont to be full of objections against the doctrine of God's sovereignty, in choosing whom He would to eternal life and rejecting whom He pleased, leaving them eternally to perish and be everlastingly tormented in hell. It used to appear like a horrible doctrine to me. But I remember the time very well when I seemed to be convinced, and fully satisfied, as to this sovereignty of God and His justice in thus eternally disposing of men according to His sovereign pleasure. But never could give an account how or by what means I was thus convinced; not in the least imagining, in the time of it nor a long time after, that there was any extraordinary influence of God's spirit in it; but only that now I saw further, and my reason apprehended the justice and reasonableness of it. However, my mind rested in it; and it put an end to all those cavils and objections, that had 'til then abode with me, all the preceding part of my life. And there has been a

---

5. "As a dog returneth to his vomit, so a fool returneth to his folly" (Proverbs 26.11).
6. Edwards was an undergraduate at Yale from 1716 to 1720 and a divinity student from 1720 to 1722.
7. A respiratory disorder.
8. I.e., truly redeeming, capable of making the penitent a "saint."

wonderful alteration in my mind, with respect to the doctrine of God's sovereignty, from that day to this; so that I scarce ever have found so much as the rising of an objection against God's sovereignty, in the most absolute sense, in showing mercy to whom He will show mercy and hardening and eternally damning whom He will.[9] God's absolute sovereignty and justice, with respect to salvation and damnation, is what my mind seems to rest assured of, as much as of anything that I see with my eyes; at least it is so at times. But I have oftentimes since that first conviction had quite another kind of sense of God's sovereignty than I had then. I have often since not only had a conviction, but a delightful conviction. The doctrine of God's sovereignty has very often appeared an exceeding pleasant, bright and sweet doctrine to me; and absolute sovereignty is what I love to ascribe to God. But my first conviction was not with this.

The first that I remember that ever I found anything of that sort of inward, sweet delight in God and divine things, that I have lived much in since, was on reading those words, 1 Timothy 1.17, "Now unto the king eternal, immortal, invisible, the only wise God, be honor and glory for ever and ever, Amen." As I read the words, there came into my soul, and was as it were diffused through it, a sense of the glory of the Divine Being, a new sense, quite different from anything I ever experienced before. Never any words of scripture seemed to me as these words did. I thought with myself, how excellent a being that was, and how happy I should be if I might enjoy that God and be rapt[1] up to God in Heaven, and be as it were swallowed up in Him. I kept saying, and as it were singing over these words of scripture to myself; and went to prayer to pray to God that I might enjoy Him; and prayed in a manner quite different from what I used to do, with a new sort of affection. But it never came into my thought that there was anything spiritual or of a saving nature in this.

From about that time I began to have a new kind of apprehensions and ideas of Christ, and the work of redemption, and the glorious way of salvation by Him. I had an inward, sweet sense of these things, that at times came into my heart; and my soul was led away in pleasant views and contemplations of them. And my mind was greatly engaged to spend my time in reading and meditating on Christ, and the beauty and excellency of His person, and the lovely way of salvation, by free grace in Him. I found no books so delightful to me as those that treated of these subjects. Those words Canticles 2.1, used to be abundantly with me: "I am the Rose of Sharon, the lily of the valleys." The words seemed to me, sweetly to represent the loveliness and beauty of Jesus Christ. And the whole book of Canticles[2] used to be pleasant to me; and I used to be

9. "Therefore hath he mercy on whom he will have mercy, and whom he will be hardeneth" (Romans 9.18).

1. Lifted.
2. I.e., Song of Solomon.

much in reading it, about that time. And found, from time to time, an inward sweetness that used, as it were, to carry me away in my contemplations, in what I know not how to express otherwise, than by a calm, sweet abstraction of soul from all the concerns of this world, and a kind of vision, or fixed ideas and imaginations, of being alone in the mountains or some solitary wilderness, far from all mankind, sweetly conversing with Christ, and rapt and swallowed up in God. The sense I had of divine things would often of a sudden as it were, kindle up a sweet burning in my heart, an ardor of my soul, that I know not how to express.

Not long after I first began to experience these things, I gave an account to my father of some things that had passed in my mind. I was pretty much affected by the discourse we had together. And when the discourse was ended, I walked abroad alone, in a solitary place in my father's pasture, for contemplation. And as I was walking there, and looked up on the sky and clouds; there came into my mind a sweet sense of the glorious majesty and grace of God that I know not how to express. I seemed to see them both in a sweet conjunction, majesty and meekness joined together. It was a sweet and gentle, and holy majesty; and also a majestic meekness; an awful sweetness; a high, and great, and holy gentleness.

After this my sense of divine things gradually increased, and became more and more lively, and had more of that inward sweetness. The appearance of everything was altered: there seemed to be, as it were, a calm, sweet cast, or appearance of divine glory, in almost everything. God's excellency, His wisdom, His purity and love, seemed to appear in everything: in the sun, moon and stars; in the clouds, and blue sky; in the grass, flowers, trees; in the water, and all nature; which used greatly to fix my mind. I often used to sit and view the moon for a long time, and so in the daytime spent much time in viewing the clouds and sky to behold the sweet glory of God in these things, in the meantime, singing forth with a low voice my contemplations of the Creator and Redeemer. And scarce anything, among all the works of nature, was so sweet to me as thunder and lightning. Formerly, nothing had been so terrible to me. I used to be a person uncommonly terrified with thunder, and it used to strike me with terror when I saw a thunderstorm rising. But now, on the contrary, it rejoiced me. I felt God at the first appearance of a thunderstorm. And used to take the opportunity at such times to fix myself to view the clouds, and see the lightnings play, and hear the majestic and awful voice of God's thunder, which often times was exceeding entertaining, leading me to sweet contemplations of my great and glorious God. And while I viewed, used to spend my time, as it always seemed natural to me, to sing or chant forth my meditations, to speak my thoughts in soliloquies, and speak with a singing voice.

I felt then a great satisfaction as to my good estate.[3] But that did not content me. I had vehement longings of soul after God and Christ, and after more holiness, wherewith my heart seemed to be full and ready to break: which often brought to my mind the words of the psalmist, Psalm 119.28: "My soul breaketh for the longing it hath." I often felt a mourning and lamenting in my heart that I had not turned to God sooner, that I might have had more time to grow in grace. My mind was greatly fixed on divine things; I was almost perpetually in the contemplation of them. Spent most of my time in thinking of divine things, year after year. And used to spend abundance of my time in walking alone in the woods and solitary places for meditation, soliloquy and prayer, and converse with God. And it was always my manner, at such times, to sing forth my contemplations. And was almost constantly in ejaculatory prayer, wherever I was. Prayer seemed to be natural to me, as the breath by which the inward burnings of my heart had vent.

The delights which I now felt in things of religion were of an exceeding different kind from those forementioned, that I had when I was a boy. They were totally of another kind; and what I then had no more notion or idea of, than one born blind has of pleasant and beautiful colors. They were of a more inward, pure, soul-animating and refreshing nature. Those former delights never reached the heart, and did not arise from any sight of the divine excellency of the things of God or any taste of the soul-satisfying and life-giving good there is in them.

My sense of divine things seemed gradually to increase, 'til I went to preach at New York, which was about a year and a half after they began. While I was there, I felt them, very sensibly,[4] in a much higher degree, than I had done before. My longings after God and holiness, were much increased. Pure and humble, holy and heavenly Christianity appeared exceeding amiable to me. I felt in me a burning desire to be in everything a complete Christian, and conformed to the blessed image of Christ, and that I might live in all things, according to the pure, sweet and blessed rules of the gospel. I had an eager thirsting after progress in these things. My longings after it put me upon pursuing and pressing after them. It was my continual strife day and night, and constant inquiry, how I should be more holy, and live more holily, and more becoming a child of God, and disciple of Christ. I sought an increase of grace and holiness, and that I might live an holy life with vastly more earnestness than ever I sought grace, before I had it. I used to be continually examining myself, and studying and contriving for likely ways and means how I should live holily with far greater diligence

3. Conditions of being.
4. Edwards was in New York from August, 1722, to April, 1723, assisting at a Presbyterian church; "sensibly": feelingly.

and earnestness than ever I pursued anything in my life; but with too great a dependence on my own strength, which afterwards proved a great damage to me. My experience had not then taught me, as it has done since, my extreme feebleness and impotence, every manner of way, and the innumerable and bottomless depths of secret corruption and deceit that there was in my heart. However, I went on with my eager pursuit after more holiness, and sweet conformity to Christ.

The Heaven I desired was a heaven of holiness, to be with God, and to spend my eternity in divine love, and holy communion with Christ. My mind was very much taken up with contemplations on heaven, and the enjoyments of those there, and living there in perfect holiness, humility and love. And it used at that time to appear a great part of the happiness of heaven that there the saints could express their love to Christ. It appeared to me a great clog and hindrance and burden to me that what I felt within I could not express to God and give vent to as I desired. The inward ardor of my soul seemed to be hindered and pent up, and could not freely flame out as it would. I used often to think how in heaven this sweet principle should freely and fully vent and express itself. Heaven appeared to me exceeding delightful as a world of love. It appeared to me that all happiness consisted in living in pure, humble, heavenly, divine love.

I remember the thoughts I used then to have of holiness. I remember I then said sometimes to myself, "I do certainly know that I love holiness such as the gospel prescribes." It appeared to me there was nothing in it but what was ravishingly lovely. It appeared to me to be the highest beauty and amiableness, above all other beauties, that it was a divine beauty, far purer than anything here upon earth; and that everything else, was like mire, filth and defilement in comparison of it.

Holiness, as I then wrote down some of my contemplations on it, appeared to me to be of a sweet, pleasant, charming, serene, calm nature. It seemed to me it brought an inexpressible purity, brightness, peacefulness and ravishment to the soul, and that it made the soul like a field or garden of God, with all manner of pleasant flowers; that is, all pleasant, delightful and undisturbed, enjoying a sweet calm, and the gently vivifying beams of the sun. The soul of a true Christian, as I then wrote my meditations, appeared like such a little white flower as we see in the spring of the year, low and humble on the ground, opening its bosom, to receive the pleasant beams of the sun's glory, rejoicing, as it were, in a calm rapture, diffusing around a sweet fragrancy, standing peacefully and lovingly in the midst of other flowers round about, all in like manner opening their bosoms, to drink in the light of the sun.

There was no part of creature holiness that I then, and at other times, had so great a sense of the loveliness of, as humility, brokenness of heart and poverty of spirit, and there was nothing that I had such a spirit to long for. My heart, as it were, panted after this to lie low before God, and in the dust; that I might be nothing, and that God might be all; that I might become as a little child.[5]

While I was there at New York, I sometimes was much affected with reflections on my past life, considering how late it was, before I began to be truly religious and how wickedly I had lived 'til then; and once so as to weep abundantly, and for a considerable time together.

On January 12, 1722–3 I made a solemn dedication of myself to God, and wrote it down; giving up myself, and all that I had to God; to be for the future in no respect my own; to act as one that had no right to himself, in any respect. And solemnly vowed to take God for my whole portion and felicity, looking on nothing else as any part of my happiness, nor acting as if it were: and His law for the constant rule of my obedience, engaging to fight with all my might against the world, the flesh and the devil, to the end of my life. But have reason to be infinitely humbled, when I consider, how much I have failed of answering my obligation.

I had then abundance of sweet religious conversation in the family where I lived, with Mr. John Smith, and his pious mother. My heart was knit in affection to those in whom were appearances of true piety, and I could bear the thoughts of no other companions but such as were holy, and the disciples of the blessed Jesus.

I had great longings for the advancement of Christ's kingdom in the world. My secret prayer used to be in great part taken up in praying for it. If I heard the least hint of anything that happened in any part of the world that appeared to me in some respect or other, to have a favorable aspect on the interest of Christ's kingdom, my soul eagerly catched at it; and it would much animate and refresh me. I used to be earnest to read public newsletters, mainly for that end, to see if I could not find some news favorable to the interest of religion in the world.

I very frequently used to retire into a solitary place, on the banks of Hudson's river, at some distance from the city, for contemplation on divine things and secret converse with God, and had many sweet hours there. Sometimes Mr. Smith and I walked there together to converse of the things of God, and our conversation used much to turn on the advancement of Christ's kingdom in the world, and the glorious things that God would accomplish for His church in the latter days.

I had then, and at other times, the greatest delight in the holy Scriptures, of any book whatsoever. Oftentimes in reading it, every word seemed to touch my heart. I felt an harmony between something in my heart, and those sweet and powerful words. I seemed

5. "Verily I say unto you, Whosoever shall not receive the kingdom of God as a little child, he shall not enter therein" (Mark 10.15).

often to see so much light exhibited by every sentence, and such a refreshing ravishing food communicated, that I could not get along in reading. Used oftentimes to dwell long on one sentence, to see the wonders contained in it; and yet almost every sentence seemed to be full of wonders.

I came away from New York in the month of April, 1723, and had a most bitter parting with Madam Smith and her son. My heart seemed to sink within me, at leaving the family and city, where I had enjoyed so many sweet and pleasant days. I went from New York to Weathersfield[6] by water. As I sailed away, I kept sight of the city as long as I could; and when I was out of sight of it, it would affect me much to look that way, with a kind of melancholy mixed with sweetness. However, that night after this sorrowful parting, I was greatly comforted in God at Westchester, where we went ashore to lodge, and had a pleasant time of it all the voyage to Saybrook.[7] It was sweet to me to think of meeting dear Christians in heaven, where we should never part more. At Saybrook we went ashore to lodge on Saturday, and there kept sabbath where I had a sweet and refreshing season, walking alone in the fields.

After I came home to Windsor, remained much in a like frame of my mind as I had been in at New York, but only sometimes felt my heart ready to sink with the thoughts of my friends at New York. And my refuge and support was in contemplations on the heavenly state, as I find in my diary of May 1, 1723. It was my comfort to think of that state where there is fulness of joy; where reigns heavenly, sweet, calm and delightful love, without alloy; where there are continually the dearest expressions of this love; where is the enjoyment of the persons loved without ever parting; where these persons that appear so lovely in this world will really be inexpressibly more lovely, and full of love to us. And how sweetly will the mutual lovers join together to sing the praises of God and the Lamb![8] How full will it fill us with joy to think that this enjoyment, these sweet exercises will never cease or come to an end, but will last to all eternity!

Continued much in the same frame in the general that I had been in at New York, 'til I went to New Haven to live there as tutor of the college, having some special seasons of uncommon sweetness; particularly once at Boston in a journey from Boston, walking out alone in the fields. After I went to New Haven, I sunk in religion, my mind being diverted from my eager and violent pursuits after holiness by some affairs that greatly perplexed and distracted my mind.

In September, 1725, was taken ill at New Haven, and, endeavoring to go home to Windsor, was so ill at the North Village that I

6. Wethersfield, Connecticut, is very near his father's home in Windsor.
7. Westchester and Saybrook are in New York and Connecticut, respectively.
8. In Revelation the symbol of Christ.

could go no further, where I lay sick for about a quarter of a year. And in this sickness, God was pleased to visit me again with the sweet influences of His spirit. My mind was greatly engaged there on divine, pleasant contemplations and longings of soul. I observed that those who watched with me would often be looking out for the morning, and seemed to wish for it. Which brought to my mind those words of the psalmist, which my soul with sweetness made its own language: "My soul waitest for the Lord, more than they that watch for the morning, I say, more than they that watch for the morning."[9] And when the light of the morning came, and the beams of the sun came in at the windows, it refreshed my soul from one morning to another. It seemed to me to be some image of the sweet light of God's glory.

I remember, about that time, I used greatly to long for the conversion of some that I was concerned with. It seemed to me I could gladly honor them, and with delight be a servant to them, and lie at their feet, if they were but truly holy.

But sometime after this, I was again greatly diverted in my mind with some temporal concerns that exceedingly took up my thoughts, greatly to the wounding of my soul, and went on through various exercises, that it would be tedious to relate, that gave me much more experience of my own heart than ever I had before.

Since I came to this town,[1] I have often had sweet complacency in God, in views of His glorious perfections and the excellency of Jesus Christ. God has appeared to me a glorious and lovely Being, chiefly on the account of His holiness. The holiness of God has always appeared to me the most lovely of all His attributes. The doctrines of God's absolute sovereignty and free grace in showing mercy to whom He would show mercy, and man's absolute dependence on the operations of God's Holy Spirit, have very often appeared to me as sweet and glorious doctrines. These doctrines have been much my delight. God's sovereignty has ever appeared to me as great part of His glory. It has often been sweet to me to go to God and adore Him as a sovereign God, and ask sovereign mercy of Him.

I have loved the doctrines of the gospel; they have been to my soul like green pastures. The gospel has seemed to me to be the richest treasure, the treasure that I have most desired and longed that it might dwell richly in me. The way of salvation by Christ has appeared in a general way glorious and excellent, and most pleasant and beautiful. It has often seemed to me that it would in a great measure spoil heaven to receive it in any other way. That text has often been affecting and delightful to me, Isaiah 32.2: "A man shall be an hiding place from the wind, and a covert from the tempest, etc."

9. Psalm 130.6.
1. Northampton, Massachusetts, where, in 1726, Edwards came to help his grand-father in conducting the affairs of his parish; "complacency": contentment.

It has often appeared sweet to me to be united to Christ; to have Him for my head, and to be a member of His body; and also to have Christ for my teacher and prophet. I very often think with sweetness and longings and pantings of soul, of being a little child, taking hold of Christ, to be led by Him through the wilderness of this world. That text, Matthew 18.3 at the beginning, has often been sweet to me, "Except ye be converted, and become as little children, etc." I love to think of coming to Christ, to receive salvation of Him, poor in spirit, and quite empty of self; humbly exalting Him alone; cut entirely off from my own root, and to grow into and out of Christ; to have God in Christ to be all in all; and to live by faith on the Son of God, a life of humble, unfeigned confidence in Him. That Scripture has often been sweet to me, Psalm 115.1: "Not unto us, O Lord, not unto us, but unto Thy name give glory, for Thy mercy, and for Thy truth's sake." And those words of Christ, Luke 10.21: "In that hour Jesus rejoiced in spirit, and said, I thank thee, O Father, Lord of heaven and earth, that Thou hast hid these things from the wise and prudent, and hast revealed them unto babes: Even so Father, for so it seemed good in Thy sight." That sovereignty of God that Christ rejoiced in seemed to me to be worthy to be rejoiced in, and that rejoicing of Christ seemed to me to show the excellency of Christ, and the spirit that He was of.

Sometimes only mentioning a single word causes my heart to burn within me, or only seeing the name of Christ or the name of some attribute of God. And God has appeared glorious to me on account of the Trinity. It has made me have exalting thoughts of God, that He subsists in three persons: Father, Son, and Holy Ghost.

The sweetest joys and delights I have experienced have not been those that have arisen from a hope of my own good estate,[2] but in a direct view of the glorious things of the gospel. When I enjoy this sweetness it seems to carry me above the thoughts of my own safe estate. It seems at such times a loss that I cannot bear, to take off my eye from the glorious, pleasant object I behold without me, to turn my eye in upon myself, and my own good estate.

My heart has been much on the advancement of Christ's kingdom in the world. The histories of the past advancement of Christ's kingdom have been sweet to me. When I have read histories of past ages, the pleasantest thing in all my reading has been to read of the kingdom of Christ being promoted. And when I have expected in my reading to come to any such thing, I have lotted[3] upon it all the way as I read. And my mind has been much entertained and delighted with the Scripture promises and prophecies of the future glorious advancement of Christ's kingdom on earth.

---

2. Condition of being.   3. Rejoiced.

I have sometimes had a sense of the excellent fullness of Christ, and His meetness and suitableness as a Savior; whereby He has appeared to me, far above all, the chief of ten thousands.[4] And His blood and atonement has appeared sweet, and His righteousness sweet; which is always accompanied with an ardency of spirit, and inward strugglings and breathings and groanings, that cannot be uttered, to be emptied of myself, and swallowed up in Christ.

Once, as I rid out into the woods for my health, Anno[5] 1737, and having lit from my horse in a retired place, as my manner commonly has been, to walk for divine contemplation and prayer, I had a view, that for me was extraordinary, of the glory of the Son of God, as mediator between God and man, and His wonderful, great, full, pure and sweet grace and love, and meek and gentle condescension. This grace, that appeared to me so calm and sweet, appeared great above the heavens. The person of Christ appeared ineffably excellent, with an excellency great enough to swallow up all thought and conception, which continued, as near as I can judge, about an hour, which kept me, the bigger part of the time, in a flood of tears, and weeping aloud. I felt withal an ardency of soul to be, what I know not otherwise how to express, than to be emptied and annihilated; to lie in the dust, and to be full of Christ alone; to love Him with a holy and pure love; to trust in Him; to live upon Him; to serve and follow Him, and to be totally wrapt up in the fullness of Christ; and to be perfectly sanctified and made pure with a divine and heavenly purity. I have several other times had views very much of the same nature and that have had the same effects.

I have many times had a sense of the glory of the third person in the Trinity in His office of sanctifier; in His holy operations communicating divine light and life to the soul. God in the communications of His Holy Spirit has appeared as an infinite fountain of divine glory and sweetness, being full and sufficient to fill and satisfy the soul, pouring forth itself in sweet communications, like the sun in its glory, sweetly and pleasantly diffusing light and life.

I have sometimes had an affecting sense of the excellency of the word of God, as a word of life; as the light of life; a sweet, excellent, life-giving word, accompanied with a thirsting after that word, that it might dwell richly in my heart.

I have often, since I lived in this town, had very affecting views of my own sinfulness and vileness; very frequently so as to hold me in a kind of loud weeping, sometimes for a considerable time together, so that I have often been forced to shut myself up.[6] I have had a vastly greater sense of my wickedness, and the badness of my heart, since my conversion, than ever I had before. It has often appeared to me, that if God should mark iniquity against me,

4. "My beloved is white and ruddy, the chiefest among ten thousand" (Song of Solomon 5.10).
5. In the year.
6. I.e., retire to his study.

I should appear the very worst of all mankind, of all that have been since the beginning of the world of this time, and that I should have by far the lowest place in hell. When others that have come to talk with me about their soul concerns have expressed the sense they have had of their own wickedness by saying that it seemed to them that they were as bad as the devil himself, I thought their expressions seemed exceeding faint and feeble to represent my wickedness. I thought I should wonder that they should content themselves with such expressions as these, if I had any reason to imagine that their sin bore any proportion to mine. It seemed to me I should wonder at myself if I should express my wickedness in such feeble terms as they did.

My wickedness, as I am in myself, has long appeared to me perfectly ineffable and infinitely swallowing up all thought and imagination, like an infinite deluge or infinite mountains over my head. I know not how to express better what my sins appear to me to be than by heaping infinite upon infinite, and multiplying infinite by infinite. I go about very often, for this many years, with these expressions in my mind and in my mouth, "Infinite upon infinite. Infinite upon infinite!" When I look into my heart and take a view of my wickedness, it looks like an abyss infinitely deeper than hell. And it appears to me that were it not for free grace, exalted and raised up to the infinite height of all the fullness and glory of the great Jehovah,[7] and the arm of His power and grace stretched forth, in all the majesty of His power and in all the glory of His sovereignty, I should appear sunk down in my sins infinitely below hell itself, far beyond sight of everything but the piercing eye of God's grace, that can pierce even down to such a depth and to the bottom of such an abyss.

And yet I be not in the least inclined to think that I have a greater conviction of sin than ordinary. It seems to me my conviction of sin is exceeding small and faint. It appears to me enough to amaze me that I have no more sense of my sin. I know certainly that I have very little sense of my sinfulness. That my sins appear to me so great don't seem to me to be because I have so much more conviction of sin than other Christians, but because I am so much worse and have so much more wickedness to be convinced of. When I have had these turns of weeping and crying for my sins, I thought I knew in the time of it that my repentance was nothing to my sin.

I have greatly longed of late for a broken heart and to lie low before God. And when I ask for humility of God, I can't bear the thoughts of being no more humble than other Christians. It seems to me that though their degrees of humility may be suitable for them, yet it would be a vile self-exaltation in me not to be the

7. The God of the Old Testament.

lowest in humility of all mankind. Others speak of their longing to be humbled to the dust. Though that may be a proper expression for them I always think for myself that I ought to be humbled down below hell. 'Tis an expression that it has long been natural for me to use in prayer to God. I ought to lie infinitely low before God.

It is affecting to me to think how ignorant I was, when I was a young Christian, of the bottomless, infinite depths of wickedness, pride, hypocrisy and deceit left in my heart.

I have vastly a greater sense of my universal, exceeding dependence on God's grace and strength and mere good pleasure, of late, than I used formerly to have, and have experienced more of an abhorrence of my own righteousness. The thought of any comfort or joy, arising in me, on any consideration or reflection on my own amiableness, or any of my performances or experiences, or any goodness of heart or life is nauseous and detestable to me. And yet I am greatly afflicted with a proud and self-righteous spirit, much more sensibly than I used to be formerly. I see that serpent rising and putting forth its head, continually, everywhere, all around me.

Though it seems to me that in some respects I was a far better Christian for two or three years after my first conversion than I am now, and lived in a more constant delight and pleasure, yet of late years I have had a more full and constant sense of the absolute sovereignty of God and a delight in that sovereignty, and have had more of a sense of the glory of Christ as a mediator as revealed in the gospel. On one Saturday night in particular, had a particular discovery of the excellency of the gospel of Christ, above all other doctrines, so that I could not but say to myself, "This is my chosen light, my chosen doctrine," and of Christ, "This is my chosen prophet." It appeared to me to be sweet beyond all expression to follow Christ and to be taught and enlightened and instructed by Him, to learn of Him, and live to Him.

Another Saturday night, January, 1738–9, had such a sense how sweet and blessed a thing it was to walk in the way of duty, to do that which was right and meet to be done and agreeable to the holy mind of God, that it caused me to break forth into a kind of a loud weeping, which held me some time, so that I was forced to shut myself up, and fasten the doors. I could not but as it were cry out, "How happy are they which do that which is right in the sight of God! They are blessed indeed, they are the happy ones!" I had at the same time, a very affecting sense how meet and suitable it was that God should govern the world, and order all things according to His own pleasure, and I rejoiced in it, and God reigned, and that His will was done.

c. 1740   1765

[Sarah Pierrepont] [8]

They say there is a young lady in [New Haven] who is beloved of that Great Being, who made and rules the world, and that there are certain seasons in which this Great Being, in some way or other invisible, comes to her and fills her mind with exceeding sweet delight, and that she hardly cares for anything, except to meditate on Him—that she expects after a while to be received up where He is, to be raised up out of the world and caught up into heaven; being assured that He loves her too well to let her remain at a distance from Him always. There she is to dwell with Him, and to be ravished with His love and delight forever. Therefore, if you present all the world before her, with the richest of its treasures, she disregards it and cares not for it, and is unmindful of any pain or affliction. She has a strange sweetness in her mind, and singular purity in her affections; is most just and conscientious in all her conduct; and you could not persuade her to do anything wrong or sinful, if you would give her all the world, lest she should offend this Great Being. She is of a wonderful sweetness, calmness and universal benevolence of mind; especially after this Great God has manifested Himself to her mind. She will sometimes go about from place to place, singing sweetly; and seems to be always full of joy and pleasure; and no one knows for what. She loves to be alone, walking in the fields and groves, and seems to have some one invisible always conversing with her.

1723   1829

Sinners in the Hands of an Angry God [1]

Deuteronomy 32.35

Their foot shall slide in due time. [2]

In this verse is threatened the vengeance of God on the wicked unbelieving Israelites, who were God's visible people, and who lived under the means of grace,[3] but who, notwithstanding all God's

8. This tribute to Sarah Pierrepont was written, according to S. E. Dwight, in 1723, when she was 13 years old and Edwards 20. He married her in 1727. The manuscript appeared in a blank leaf in a book, and no holograph exists. It was first published in Sereno E. Dwight's *The Life of President Edwards* (New York, 1829).

1. Edwards delivered this sermon in Enfield, Connecticut, a town about 30 miles south of Northampton, on Sunday, July 8, 1741. In Benjamin Trumbull's *A Complete History of Connecticut* (1797, 1818) we are told that Edwards read his sermon in a level voice with his sermon book in his left hand, and in spite of his calm "there was such a

breathing of distress, and weeping, that the preacher was obliged to speak to the people and desire silence, that he might be heard." The text is from Sereno E. Dwight, ed., *The Works of Jonathan Edwards*, Vol. VII (New York, 1829–30).

2. "To me belongeth vengeance, and recompense; their foot shall slide in due time: for the day of their calamity is at hand, and the things that shall come upon them make haste."

3. I.e., the Ten Commandments. For Protestants following the Westminster Confession (1646), the "means of grace" consist of "preaching of the word and the administration of the sacraments of baptism and the Lord's Supper."

wonderful works towards them, remained (as in verse 28.)[4] void of counsel, having no understanding in them. Under all the cultivations of heaven, they brought forth bitter and poisonous fruit, as in the two verses next preceding the text.[5] The expression I have chosen for my text, "Their foot shall slide in due time," seems to imply the following things, relating to the punishment and destruction to which these wicked Israelites were exposed.

1. That they were always exposed to destruction; as one that stands or walks in slippery places is always exposed to fall. This is implied in the manner of their destruction coming upon them, being represented by their foot sliding. The same is expressed, Psalm 73.18: "Surely thou didst set them in slippery places; thou castedst them down into destruction."

2. It implies that they were always exposed to sudden unexpected destruction. As he that walks in slippery places is every moment liable to fall, he cannot foresee one moment whether he shall stand or fall the next; and when he does fall, he falls at once without warning. Which is also expressed in Psalm 73. 18–19: "Surely thou didst set them in slippery places; thou castedst them down into destruction: How are they brought into desolation as in a moment!"

3. Another thing implied is, that they are liable to fall of themselves, without being thrown down by the hand of another; as he that stands or walks on slippery ground needs nothing but his own weight to throw him down.

4. That the reason why they are not fallen already, and do not fall now, is only that God's appointed time is not come. For it is said, that when that due time or appointed times comes, their foot shall slide. Then they shall be left to fall, as they are inclined by their own weight. God will not hold them up in these slippery places any longer, but will let them go; and then, at that very instant, they shall fall into destruction; as he that stands on such slippery declining ground, on the edge of a pit, he cannot stand alone, when he is let go he immediately falls and is lost.

The observation from the words that I would now insist upon is this. "There is nothing that keeps wicked men at any one moment out of hell, but the mere pleasure of God." By the mere pleasure of God, I mean His sovereign pleasure, His arbitrary will, restrained by no obligation, hindered by no manner of difficulty, any more than if nothing else but God's mere will had in the least degree, or in any respect whatsoever, any hand in the preservation of wicked men one moment. The truth of this observation may appear by the following considerations.

4. "They are a nation void of counsel, neither is there any understanding in them" (Deuteronomy 32.28).

5. "For their vine is of the vine of Sodom, and the fields of Gomorrah: their grapes are grapes of gall, their clusters are bitter: their wine is the poison of dragons, and the cruel venom of asps" (Deuteronomy 32.32–33). Sodom and Gomorrah were wicked cities destroyed by a rain of fire and sulphur from heaven (Genesis 19.28).

1. There is no want of power in God to cast wicked men into hell at any moment. Men's hands cannot be strong when God rises up. The strongest have no power to resist Him, nor can any deliver[6] out of His hands. He is not only able to cast wicked men into hell, but He can most easily do it. Sometimes an earthly prince meets with a great deal of difficulty to subdue a rebel, who has found means to fortify himself, and has made himself strong by the numbers of his followers. But it is not so with God. There is no fortress that is any defense from the power of God. Though hand join in hand, and vast multitudes of God's enemies combine and associate themselves, they are easily broken in pieces. They are as great heaps of light chaff before the whirlwind; or large quantities of dry stubble before devouring flames. We find it easy to tread on and crush a worm that we see crawling on the earth; so it is easy for us to cut or singe a slender thread that any thing hangs by: thus easy is it for God, when he pleases, to cast His enemies down to hell. What are we, that we should think to stand before him, at whose rebuke the earth trembles, and before whom the rocks are thrown down?

2. They deserve to be cast into hell; so that divine justice never stands in the way, it makes no objection against God's using His power at any moment to destroy them. Yea, on the contrary, justice calls aloud for an infinite punishment of their sins. Divine justice says of the tree that brings forth such grapes of Sodom, "Cut it down, why cumbereth it the ground? Luke 13.7. The sword of divine justice is every moment brandished over their heads, and it is nothing but the hand of arbitrary mercy, and God's will, that holds it back.

3. They are already under a sentence of condemnation to hell. They do not only justly deserve to be cast down thither, but the sentence of the law of God, that eternal and immutable rule of righteousness that God has fixed between Him and mankind, is gone out against them, and stands against them; so that they are bound over already to hell. John 3.18: "He that believeth not is condemned already." So that every unconverted man properly belongs to hell; that is his place; from thence he is, John 8.23: "Ye are from beneath." And thither he is bound; it is the place that justice, and God's word, and the sentence of his unchangeable law assign to him.

4. They are now the objects of that very same anger and wrath of God that is expressed in the torments of hell. And the reason why they do not go down to hell at each moment is not because God, in whose power they are, is not then very angry with them as He is with many miserable creatures now tormented in hell, who there feel and bear the fierceness of His wrath. Yea, God is a great deal more angry with great numbers that are now on earth: yea, doubtless,

---

6. I.e., rescue others.

with many that are now in this congregation, who it may be are at ease, than He is with many of those who are now in the flames of hell.

So that it is not because God is unmindful of their wickedness, and does not resent it, that He does not let loose His hand and cut them off. God is not altogether such an one as themselves, though they may imagine Him to be so. The wrath of God burns against them, their damnation does not slumber; the pit is prepared, the fire is made ready, the furnace is now hot, ready to receive them; the flames do now rage and glow. The glittering sword is whet,[7] and held over them, and the pit hath opened its mouth under them.

5. The devil stands ready to fall upon them, and seize them as his own, at what moment God shall permit him. They belong to him; he has their souls in his possession, and under his dominion. The scripture represents them as his goods, Luke 11.12.[8] The devils watch them; they are ever by them at their right hand; they stand waiting for them, like greedy hungry lions that see their prey, and expect to have it, but are for the present kept back. If God should withdraw His hand, by which they are restrained, they would in one moment fly upon their poor souls. The old serpent is gaping for them; hell opens it mouth wide to receive them; and if God should permit it, they would be hastily swallowed up and lost.

6. There are in the souls of wicked men those hellish principles reigning that would presently kindle and flame out into hell fire, if it were not for God's restraints. There is laid in the very nature of carnal men a foundation for the torments of hell. There are those corrupt principles, in reigning power in them, and in full possession of them, that are seeds of hell fire. These principles are active and powerful, exceeding violent in their nature, and if it were not for the restraining hand of God upon them, they would soon break out, they would flame out after the same manner as the same corruptions, the same enmity does in the hearts of damned souls, and would beget the same torments as they do in them. The souls of the wicked are in scripture compared to the troubled sea, Isaiah 57.20[9] For the present, God restrains their wickedness by His mighty power, as He does the raging waves of the troubled sea, saying, "Hitherto shalt thou come, but no further;"[1] but if God should withdraw that restraining power, it would soon carry all before it. Sin is the ruin and misery of the soul; it is destructive in its nature; and if God should leave it without restraint, there would need nothing else to make the soul perfectly miserable. The corruption of the heart of man is immoderate and boundless in its fury; and while wicked men live here, it is like fire pent up by God's restraints,

7. Sharpened.
8. "Or if he shall ask an egg, will he offer him a scorpion?"
9. "But the wicked are like the troubled sea, when it cannot rest, whose waters cast up mire and dirt."
1. Job 38.11.

whereas if it were let loose, it would set on fire the course of nature; and as the heart is now a sink of sin, so if sin was not restrained, it would immediately turn the soul into a fiery oven, or a furnace of fire and brimstone.

7. It is no security to wicked men for one moment that there are no visible means of death at hand. It is no security to a natural man that he is now in health and that he does not see which way he should now immediately go out of the world by any accident, and that there is no visible danger in any respect in his circumstances. The manifold and continual experience of the world in all ages, shows this is no evidence that a man is not on the very brink of eternity, and that the next step will not be into another world. The unseen, unthought-of ways and means of persons going suddenly out of the world are innumerable and inconceivable. Unconverted men walk over the pit of hell on a rotten covering, and there are innumerable places in this covering so weak that they will not bear their weight, and these places are not seen. The arrows of death fly unseen at noonday;[2] the sharpest sight cannot discern them. God has so many different unsearchable ways of taking wicked men out of the world and sending them to hell, that there is nothing to make it appear that God had need to be at the expense of a miracle, or go out of the ordinary course of His providence, to destroy any wicked man at any moment. All the means that there are of sinners going out of the world are so in God's hands, and so universally and absolutely subject to His power and determination, that it does not depend at all the less on the mere will of God whether sinners shall at any moment go to hell than if means were never made use of or at all concerned in the case.

8. Natural men's prudence and care to preserve their own lives, or the care of others to preserve them, do not secure them a moment. To this, divine providence and universal experience do also bear testimony. There is this clear evidence that men's own wisdom is no security to them from death; that if it were otherwise we should see some difference between the wise and politic men of the world, and others, with regard to their liableness to early and unexpected death: but how is it in fact? Ecclesiastes 2.16: "How dieth the wise man? even as the fool."

9. All wicked men's pains and contrivance which they use to escape hell, while they continue to reject Christ, and so remain wicked men, do not secure them from hell one moment. Almost every natural[3] man that hears of hell, flatters himself that he shall escape it; he depends upon himself for his own security; he flatters himself in what he has done, in what he is now doing, or what he intends to do. Every one lays out matters in his own mind how he shall avoid damnation, and flatters himself that he contrives well for

2. "Thou shalt not be afraid for the terror by night; nor for the arrow that flieth by day" (Psalm 91.5).
3. I.e., unregenerate, unsaved.

himself, and that his schemes will not fail. They hear indeed that there are but few saved, and that the greater part of men that have died heretofore are gone to hell; but each one imagines that he lays out matters better for his own escape than others have done. He does not intend to come to that place of torment; he says within himself that he intends to take effectual care, and to order matters so for himself as not to fail.

But the foolish children of men miserably delude themselves in their own schemes, and in confidence in their own strength and wisdom; they trust to nothing but a shadow. The greater part of those who heretofore have lived under the same means of grace, and are now dead, are undoubtedly gone to hell; and it was not because they were not as wise as those who are now alive: it was not because they did not lay out matters as well for themselves to secure their own escape. If we could speak with them, and inquire of them, one by one, whether they expected, when alive, and when they used to hear about hell, ever to be the subjects of that misery, we doubtless, should hear one and another reply, "No, I never intended to come here: I had laid out matters otherwise in my mind; I thought I should contrive well for myself: I thought my scheme good. I intended to take effectual care; but it came upon me unexpected; I did not look for it at that time, and in that manner; it came as a thief: Death outwitted me: God's wrath was too quick for me. Oh, my cursed foolishness! I was flattering myself, and pleasing myself with vain dreams of what I would do hereafter; and when I was saying, peace and safety, then suddenly destruction came upon me."

10. God has laid Himself under no obligation by any promise to keep any natural man out of hell one moment. God certainly has made no promises either of eternal life or of any deliverance or preservation from eternal death but what are contained in the covenant of grace,[4] the promises that are given in Christ, in whom all the promises are yea and amen. But surely they have no interest in the promises of the covenant of grace who are not the children of the covenant, who do not believe in any of the promises, and have no interest in the Mediator of the covenant.[5]

So that, whatever some have imagined and pretended[6] about promises made to natural men's earnest seeking and knocking, it is plain and manifest that whatever pains a natural man takes in religion, whatever prayers he makes, till he believes in Christ, God is under no manner of obligation to keep him a moment from eternal destruction.

So that, thus it is that natural men are held in the hand of God, over the pit of hell; they have deserved the fiery pit, and are already

4. The original covenant God made with Adam is called the Covenant of Works; the second covenant Christ made with fallen man—declaring that if he believed in Him he would be saved—is called the Covenant of Grace.
5. I.e., Christ, who took upon Himself man's sins and suffered for them.
6. Claimed.

sentenced to it; and God is dreadfully provoked, His anger is as great towards them as to those that are actually suffering the executions of the fierceness of His wrath in hell, and they have done nothing in the least to appease or abate that anger, neither is God in the least bound by any promise to hold them up one moment; the devil is waiting for them, hell is gaping for them, the flames gather and flash about them, and would fain lay hold on them, and swallow them up; the fire pent up in their own hearts is struggling to break out: and they have no interest in any Mediator, there are no means within reach that can be any security to them. In short, they have no refuge, nothing to take hold of; all that preserves them every moment is the mere arbitrary will, and uncovenanted, unobliged forbearance of an incensed God.

Application

The use of this awful[7] subject may be for awakening unconverted persons in this congregation. This that you have heard is the case of every one of you that are out of Christ. That world of misery, that lake of burning brimstone is extended abroad under you. There is the dreadful pit of the glowing flames of the wrath of God; there is hell's wide gaping mouth open; and you have nothing to stand upon, nor any thing to take hold of; there is nothing between you and hell but the air; it is only the power and mere pleasure of God that holds you up.

You probably are not sensible[8] of this; you find you are kept out of hell, but do not see the hand of God in it; but look at other things, as the good state of your bodily constitution, your care of your own life, and the means you use for your own preservation. But indeed these things are nothing; if God should withdraw His hand, they would avail no more to keep you from falling, than the thin air to hold up a person that is suspended in it.

Your wickedness makes you as it were heavy as lead, and to tend downwards with great weight and pressure towards hell; and if God should let you go, you would immediately sink and swiftly descend and plunge into the bottomless gulf, and your healthy constitution, and your own care and prudence, and best contrivance, and all your righteousness, would have no more influence to uphold you and keep you out of hell, than a spider's web would have to stop a fallen rock. Were it not for the sovereign pleasure of God, the earth would not bear you one moment; for you are a burden to it; the creation groans with you; the creature is made subject to the bondage of your corruption, not willingly; the sun does not willingly shine upon you to give you light to serve sin and Satan; the earth does not willingly yield her increase to satisfy your lusts; nor is it willingly a stage for your wickedness to be acted upon; the air does not will-

7. Awesome.   8. Aware.

ingly serve you for breath to maintain the flame of life in your vitals, while you spend your life in the service of God's enemies. God's creatures are good, and were made for men to serve God with, and do not willingly subserve to any other purpose, and groan when they are abused to purposes so directly contrary to their nature and end. And the world would spew you out, were it not for the sovereign hand of Him who hath subjected it in hope. There are black clouds of God's wrath now hanging directly over your heads, full of the dreadful storm, and big with thunder; and were it not for the restraining hand of God, it would immediately burst forth upon you. The sovereign pleasure of God, for the present, stays His rough wind; otherwise it would come with fury, and your destruction would come like a whirlwind, and you would be like the chaff of the summer threshing floor.

The wrath of God is like great waters that are dammed for the present; they increase more and more, and rise higher and higher, till an outlet is given; and the longer the stream is stopped, the more rapid and mighty is its course when once it is let loose. It is true that judgment against your evil works has not been executed hitherto; the floods of God's vengeance have been withheld; but your guilt in the meantime is constantly increasing, and you are every day treasuring up more wrath; the waters are constantly rising, and waxing more and more mighty; and there is nothing but the mere pleasure of God that holds the waters back, that are unwilling to be stopped, and press hard to go forward. If God should only withdraw His hand from the floodgate, it would immediately fly open, and the fiery floods of the fierceness and wrath of God, would rush forth with inconceivable fury, and would come upon you with omnipotent power; and if your strength were ten thousand times greater than it is, yea, ten thousand times greater than the strength of the stoutest, sturdiest devil in hell, it would be nothing to withstand or endure it.

The bow of God's wrath is bent, and the arrow made ready on the string, and justice bends the arrow at your heart, and strains the bow; and it is nothing but the mere pleasure of God, and that of an angry God, without any promise or obligation at all, that keeps the arrow one moment from being made drunk with your blood. Thus all you that never passed under a great change of heart, by the mighty power of the Spirit of God upon your souls, all you that were never born again, and made new creatures, and raised from being dead in sin, to a state of new, and before altogether unexperienced light and life, are in the hands of an angry God. However you may have reformed your life in many things, and may have had religious affections,[9] and may keep up a form of religion in your families and closets,[1] and in the house of God, it is nothing but

9. Feelings.　　　　　　　　　1. Studies; rooms for meditation.

His mere pleasure that keeps you from being this moment swallowed up in everlasting destruction. However unconvinced you may now be of the truth of what you hear, by and by you will be fully convinced of it. Those that are gone from being in the like circumstances with you see that it was so with them; for destruction came suddenly upon most of them; when they expected nothing of it and while they were saying, peace and safety: now they see that those things on which they depended for peace and safety, were nothing but thin air and empty shadows.

The God that holds you over the pit of hell, much as one holds a spider or some loathsome insect over the fire, abhors you, and is dreadfully provoked: His wrath towards you burns like fire; He looks upon you as worthy of nothing else but to be cast into the fire; He is of purer eyes than to bear to have you in His sight; you are ten thousand times more abominable in His eyes than the most hateful venomous serpent is in ours. You have offended Him infinitely more than ever a stubborn rebel did his prince; and yet it is nothing but His hand that holds you from falling into the fire every moment. It is to be ascribed to nothing else, that you did not go to hell the last night; that you was suffered to awake again in this world, after you closed your eyes to sleep. And there is no other reason to be given, why you have not dropped into hell since you arose in the morning, but that God's hand has held you up. There is no other reason to be given why you have not gone to hell, since you have sat here in the house of God, provoking His pure eyes by your sinful wicked manner of attending His solemn worship. Yea, there is nothing else that is to be given as a reason why you do not this very moment drop down into hell.

O sinner! Consider the fearful danger you are in: it is a great furnace of wrath, a wide and bottomless pit, full of the fire of wrath, that you are held over in the hand of that God, whose wrath is provoked and incensed as much against you, as against many of the damned in hell. You hang by a slender thread, with the flames of divine wrath flashing about it, and ready every moment to singe it, and burn it asunder; and you have no interest in any Mediator, and nothing to lay hold of to save yourself, nothing to keep off the flames of wrath, nothing of your own, nothing that you ever have done, nothing that you can do, to induce God to spare you one moment. And consider here more particularly,

1. Whose wrath it is? It is the wrath of the infinite God. If it were only the wrath of man, though it were of the most potent prince, it would be comparatively little to be regarded. The wrath of kings is very much dreaded, especially of absolute monarchs, who have the possessions and lives of their subjects wholly in their power, to be disposed of at their mere will. Proverbs 20.2: "The fear of a king is as the roaring of a lion: Whoso provoketh him to anger, sinneth against his own soul." The subject that very much

enrages an arbitrary prince, is liable to suffer the most extreme torments that human art can invent, or human power can inflict. But the greatest earthly potentates in their greatest majesty and strength, and when clothed in their greatest terrors, are but feeble, despicable worms of the dust, in comparison of the great and almighty Creator and King of heaven and earth. It is but little that they can do, when most enraged, and when they have exerted the utmost of their fury. All the kings of the earth, before God, are as grasshoppers; they are nothing, and less than nothing: both their love and their hatred is to be despised. The wrath of the great King of kings, is as much more terrible than theirs, as His majesty is greater. Luke 12.4–5: "And I say unto you, my friends, Be not afraid of them that kill the body, and after that, have no more that they can do. But I will forewarn you whom you shall fear: fear him, which after he hath killed, hath power to cast into hell: yea, I say unto you, Fear him."

2. It is the fierceness of His wrath that you are exposed to. We often read of the fury of God; as in Isaiah 59.18: "According to their deeds, accordingly he will repay fury to his adversaries." So Isaiah 66.15: "For behold, the Lord will come with fire, and with his chariots like a whirlwind, to render his anger with fury, and his rebuke with flames of fire." And in many other places. So, Revelation 19.15: we read of "the wine press of the fierceness and wrath of Almighty God."[2] The words are exceeding terrible. If it had only been said, "the wrath of God," the words would have implied that which is infinitely dreadful: but it is "the fierceness and wrath of God." The fury of God! the fierceness of Jehovah![3] Oh, how dreadful must that be! Who can utter or conceive what such expressions carry in them! But it is also "the fierceness and wrath of Almighty God." As though there would be a very great manifestation of His almighty power in what the fierceness of His wrath should inflict, as though omnipotence should be as it were enraged, and exerted, as men are wont to exert their strength in the fierceness of their wrath. Oh! then, what will be the consequence! What will become of the poor worms that shall suffer it! Whose hands can be strong? And whose heart can endure? To what a dreadful, inexpressible, inconceivable depth of misery must the poor creature be sunk who shall be the subject of this!

Consider this, you that are here present that yet remain in an unregenerate state. That God will execute the fierceness of His anger implies that He will inflict wrath without any pity. When God beholds the ineffable extremity of your case, and sees your torment to be so vastly disproportioned to your strength, and sees how your poor soul is crushed, and sinks down, as it were, into an infinite gloom; He will have no compassion upon you, He will not forbear

2. "He treadeth the winepress of the fierceness and wrath of Almighty God."

3. The God of the Old Testament, the Hebrew God.

the executions of His wrath, or in the least lighten His hand; there shall be no moderation or mercy, nor will God then at all stay His rough wind; He will have no regard to your welfare, nor be at all careful lest you should suffer too much in any other sense, than only that you shall not suffer beyond what strict justice requires. Nothing shall be withheld because it is so hard for you to bear. Ezekiel 8.18: "Therefore will I also deal in fury: mine eye shall not spare, neither will I have pity; and though they cry in mine ears with a loud voice, yet I will not hear them." Now God stands ready to pity you; this is a day of mercy; you may cry now with some encouragement of obtaining mercy. But when once the day of mercy is past, your most lamentable and dolorous cries and shrieks will be in vain; you will be wholly lost and thrown away of God as to any regard to your welfare. God will have no other use to put you to, but to suffer misery; you shall be continued in being to no other end; for you will be a vessel of wrath fitted to destruction; and there will be no other use of this vessel, but to be filled full of wrath. God will be so far from pitying you when you cry to Him, that it is said He will only "laugh and mock." Proverbs 1.25–26, etc.[4]

How awful are those words, Isaiah 63.3, which are the words of the great God: "I will tread them in mine anger, and will trample them in my fury, and their blood shall be sprinkled upon my garments, and I will stain all my raiment." It is perhaps impossible to conceive of words that carry in them greater manifestations of these three things, viz., contempt, and hatred, and fierceness of indignation. If you cry to God to pity you, He will be so far from pitying you in your doleful case, or showing you the least regard or favor, that instead of that, He will only tread you under foot. And though He will know that you cannot bear the weight of omnipotence treading upon you, yet He will not regard that, but He will crush you under His feet without mercy; He will crush out your blood, and make it fly and it shall be sprinkled on His garments, so as to stain all His raiment. He will not only hate you, but He will have you in the utmost contempt: no place shall be thought fit for you, but under His feet to be trodden down as the mire of the streets.

3. The misery you are exposed to is that which God will inflict to that end, that He might show what that wrath of Jehovah is. God hath had it on His heart to show to angels and men both how excellent His love is, and also how terrible His wrath is. Sometimes earthly kings have a mind to show how terrible their wrath is, by the extreme punishments they would execute on those that would provoke them. Nebuchadnezzar, that mighty and haughty monarch of the Chaldean empire, was willing to show his wrath when enraged with Shadrach, Meshech, and Abednego; and accordingly gave orders that the burning fiery furnace should be heated seven

---

4. "But ye have set at nought all my counsel, and would none of my reproof: I also will laugh at your calamity; I will mock you when your fear cometh."

times hotter than it was before; doubtless, it was raised to the utmost degree of fierceness that human art could raise it.[5] But the great God is also willing to show His wrath, and magnify His awful majesty and mighty power in the extreme sufferings of His enemies. Romans 9.22: "What if God, willing to show his wrath, and to make his power known, endure with much long-suffering the vessels of wrath fitted to destruction?" And seeing this in His design, and what He has determined, even to show how terrible the restrained wrath, the fury and fierceness of Jehovah is, He will do it to effect. There will be something accomplished and brought to pass that will be dreadful with a witness. When the great and angry God hath risen up and executed His awful vengeance on the poor sinner, and the wretch is actually suffering the infinite weight and power of His indignation, then will God call upon the whole universe to behold that awful majesty and mighty power that is to be seen in it. Isaiah 33.12–14: "And the people shall be as the burnings of lime, as thorns cut up shall they be burnt in the fire. Hear ye that are far off, what I have done; ye that are near, acknowledge my might. The sinners in Zion are afraid; fearfulness hath surprised the hypocrites," etc.

Thus it will be with you that are in an unconverted state, if you continue in it; the infinite might, and majesty, and terribleness of the omnipotent God shall be magnified upon you, in the ineffable strength of your torments. You shall be tormented in the presence of the holy angels, and in the presence of the Lamb; and when you shall be in this state of suffering, the glorious inhabitants of heaven shall go forth and look on the awful spectacle, that they may see what the wrath and fierceness of the Almighty is; and when they have seen it, they will fall down and adore that great power and majesty. Isaiah 66.23–24: "And it shall come to pass, that from one new moon to another, and from one sabbath to another, shall flesh come to worship before me, saith the Lord. And they shall go forth and look upon the carcasses of the men that have transgressed against me; for their worm shall not die, neither shall their fire be quenched, and they shall be an abhorring unto all flesh."

4. It is everlasting wrath. It would be dreadful to suffer this fierceness and wrath of Almighty God one moment; but you must suffer it to all eternity. There will be no end to this exquisite horrible misery. When you look forward, you shall see a long forever, a boundless duration before you, which will swallow up your thoughts, and amaze your soul; and you will absolutely despair of ever having any deliverance, any end, any mitigation, any rest at all. You will know certainly that you must wear out long ages, millions of millions of ages, in wrestling and conflicting with this almighty merciless vengeance; and then when you have so done, when so many ages have actually been spent by you in this manner, you will

5. See Daniel 3.1–30.

know that all is but a point to what remains. So that your punishment will indeed be infinite. Oh, who can express what the state of a soul in such circumstances is! All that we can possibly say about it gives but a very feeble, faint representation of it; it is inexpressible and inconceivable: For "who knows the power of God's anger?"[6]

How dreadful is the state of those that are daily and hourly in the danger of this great wrath and infinite misery! But this is the dismal case of every soul in this congregation that has not been born again, however moral and strict, sober and religious, they may otherwise be. Oh that you would consider it, whether you be young or old! There is reason to think that there are many in this congregaton now hearing this discourse that will actually be the subjects of this very misery to all eternity. We know not who they are, or in what seats they sit, or what thoughts they now have. It may be they are now at ease, and hear all these things without much disturbance, and are now flattering themselves that they are not the persons, promising themselves that they shall escape. If they knew that there was one person, and but one, in the whole congregation, that was to be the subject of this misery, what an awful thing would it be to think of! If we knew who it was, what an awful sight would it be to see such a person! How might all the rest of the congregation lift up a lamentable and bitter cry over him! But, alas! instead of one, how many is it likely will remember this discourse in hell? And it would be a wonder, if some that are now present should not be in hell in a very short time, even before this year is out. And it would be no wonder if some persons, that now sit here, in some seats of this meetinghouse, in health, quiet and secure, should be there before tomorrow morning. Those of you that finally continue in a natural condition, that shall keep out of hell longest will be there in a little time! your damnation does not slumber; it will come swiftly, and, in all probability, very suddenly upon many of you. You have reason to wonder that you are not already in hell. It is doubtless the case of some whom you have seen and known, that never deserved hell more than you, and that heretofore appeared as likely to have been now alive as you. Their case is past all hope; they are crying in extreme misery and perfect despair; but here you are in the land of the living and in the house of God, and have an opportunity to obtain salvation. What would not those poor damned hopeless souls give for one day's opportunity such as you now enjoy!

And now you have an extraordinary opportunity, a day wherein Christ has thrown the door of mercy wide open, and stands in calling and crying with a loud voice to poor sinners; a day wherein many are flocking to Him, and pressing into the kingdom of God. Many are daily coming from the east, west, north and south; many

6. "Who knoweth the power of thine anger? even according to thy fear, so is thy wrath" (Psalm 90.11).

that were very lately in the same miserable condition that you are in are now in a happy state, with their hearts filled with love to Him who has loved them, and washed them from their sins in His own blood, and rejoicing in hope of the glory of God. How awful is it to be left behind at such a day! To see so many others feasting, while you are pining and perishing! To see so many rejoicing and singing for joy of heart, while you have cause to mourn for sorrow of heart, and howl for vexation of spirit! How can you rest one moment in such a condition? Are not your souls as precious as the souls of the people at Suffield,[7] where they are flocking from day to day to Christ?

Are there not many here who have lived long in the world, and are not to this day born again? and so are aliens from the commonwealth of Israel,[8] and have done nothing ever since they have lived, but treasure up wrath against the day of wrath? Oh, sirs, your case, in an especial manner, is extremely dangerous. Your guilt and hardness of heart is extremely great. Do you not see how generally persons of your years are passed over and left, in the present remarkable and wonderful dispensation of God's mercy? You had need to consider yourselves, and awake thoroughly out of sleep. You cannot bear the fierceness and wrath of the infinite God. And you, young men, and young women, will you neglect this precious season which you now enjoy, when so many others of your age are renouncing all youthful vanities, and flocking to Christ? You especially have now an extraordinary opportunity; but if you neglect it, it will soon be with you as with those persons who spent all the precious days in youth in sin, and are now come to such a dreadful pass in blindness and hardness. And you, children, who are unconverted, do not you know that you are going down to hell, to bear the dreadful wrath of that God, who is now angry with you every day and every night? Will you be content to be the children of the devil, when so many other children in the land are converted, and are become the holy and happy children of the King of kings?

And let every one that is yet of Christ, and hanging over the pit of hell, whether they be old men and women, or middle-aged, or young people, or little children, now hearken to the loud calls of God's word and providence. This acceptable year of the Lord, a day of such great favors to some, will doubtless be a day of as remarkable vengeance to others. Men's hearts harden, and their guilt increases apace at such a day as this, if they neglect their souls; and never was there so great danger of such person being given up to hardness of heart and blindness of mind. God seems now to be hastily gathering in His elect in all parts of the land; and probably the greater part of adult persons that ever shall be saved, will be brought in now in a little time, and that it will be as it was on the

---

7. "A town in the neighborhood" [Edwards's note].

8. I.e., not among the chosen people and, therefore, saved.

great outpouring of the Spirit upon the Jews in the apostles' days;[9] the election will obtain, and the rest will be blinded. If this should be the case with you, you will eternally curse this day, and will curse the day that ever you was born, to see such a season of the pouring out of God's Spirit, and will wish that you had died and gone to hell before you had seen it. Now undoubtedly it is, as it was in the days of John the Baptist, the axe is in an extraordinary manner laid at the root of the trees,[1] that every tree which brings not forth good fruit, may be hewn down and cast into the fire.

Therefore, let everyone that is out of Christ, now awake and fly from the wrath to come. The wrath of Almighty God is now undoubtedly hanging over a great part of this congregation: Let everyone fly out of Sodom: "Haste and escape for your lives, look not behind you, escape to the mountain, lest you be consumed."[2]

1741

[The Beauty of the World][1]

The beauty of the world consists wholly of sweet mutual consents,[2] either within itself or with the Supreme Being. As to the corporeal world, though there are many other sorts of consents, yet the sweetest and most charming beauty of it is its resemblance of spiritual beauties. The reason is that spiritual beauties are infinitely the greatest, and bodies being but the shadows of beings, they must be so much the more charming as they shadow forth spiritual beauties. This beauty is peculiar to natural things, it surpassing the art of man.

Thus there is the resemblance of a decent trust, dependence and acknowledgment in the planets continually moving around the sun, receiving his influences by which they are made happy, bright and beautiful: a decent attendance in the secondary planets, an image of majesty, power, glory, and beneficence in the sun in the midst of all, and so in terrestrial things, as I have shown in another place.

It is very probable that that wonderful suitableness of green for the grass and plants, the blues of the skie, the white of the clouds, the colors of flowers, consists in a complicated proportion that these colors make one with another, either in their magnitude of the rays, the number of vibrations that are caused in the atmosphere, or some other way. So there is a great suitableness between the objects

9. In Acts 2 the apostle Peter admonishes a crowd to repent and be converted, saying, "Save yourselves from this untoward generation. Then they that gladly received his word were baptized: and the same day there were added unto them about three thousand souls" (Acts 2.40–41).
1. "And now also the axe is laid unto the root of the trees: therefore every tree which brings not forth good fruit is hewn down, and cast into the fire" (Matthew 3.10).
2. Genesis 19.17.
1. This fragment was found among Edwards's papers and first published, with this title, by Perry Miller in *Images or Shadows of Divine Things* (New Haven, Conn.: Yale University Press, 1948). The editorial emendations are by Perry Miller.
2. Agreements.

of different sense, as between sounds, colors, and smells; as between colors of the woods and flowers and the smells and the singing of birds, which it is probable consist in a certain proportion of the vibrations that are made in the different organs. So there are innumerable other agreeablenesses of motions, figures, etc. The gentle motions of waves, of [the] lily, etc., as it is agreeable to other things that represent calmness, gentleness, and benevolence, etc. the fields and woods seem to rejoice, and how joyful do the birds seem to be in it. How much a resemblance is there of every grace in the field covered with plants and flowers when the sun shines serenely and undisturbedly upon them, how a resemblance, I say, of every grace and beautiful disposition of mind, of an inferior towards a superior cause, preserver, benevolent benefactor, and a fountain of happiness.

How great a resemblance of a holy and virtuous soul is a calm, serene day. What an infinite number of such like beauties is there in that one thing, the light, and how complicated an harmony and proportion is it probable belongs to it.

There are beauties that are more palpable and explicable, and there are hidden and secret beauties. The former pleases, and we can tell why; we can explain the particular point for the agreement that renders the thing pleasing. Such are all artificial regularities; we can tell wherein the regularity lies that affects us. [The] latter sort are those beauties that delight us and we cannot tell why. Thus, we find ourselves pleased in beholding the color of the violets, but we know not what secret regularity or harmony it is that creates that pleasure in our minds. These hidden beauties are commonly by far the greatest, because the more complex a beauty is, the more hidden is it. In this latter fact consists principally the beauty of the world, and very much in light and colors. Thus mere light is pleasing to the mind. If it be to the degree of effulgence, it is very sensible, and mankind have agreed in it: they all represent glory and extraordinary beauty by brightness. The reason of it is either that light or our organ of seeing is so contrived that an harmonious motion is excited in the animal spirits and propogated to the brain. That mixture we call white is a proportionate mixture that is harmonious, as Sir Isaac Newton[3] has shown, to each particular simple color, and contains in it some harmony or other that is delightful. And each sort of rays play a distinct tune to the soul, besides those lovely mixtures that are found in nature. Those beauties, how lovely is the green of the face of the earth in all manner of colors, in flowers the color of the skies, and lovely tinctures of the morning and evening.

Corollary:[4] Hence the reason why almost all men, and those that

3. Sir Isaac Newton (1642–1727), in his *Opticks* (1704), explained the phenomena of color, proving that differences in color are caused by differing degrees of refrangibility.

4. Something which naturally follows from a proved proposition.

seem to be very miserable, love life, because they cannot bear to lose sight of such a beautiful and lovely world. The ideas, that every moment whilst we live have a beauty that we take not distinct notice of, brings a pleasure that, when we come to the trial, we had rather live in much pain and misery than lose.

1948

BENJAMIN FRANKLIN
1706–1790

Benjamin Franklin was born on Milk Street in Boston, the tenth son in a family of fifteen children. His father, Josiah, was a tallow chandler and soap boiler who came to Boston in 1682 from Ecton in Northamptonshire, England, and was proud of his Protestant ancestors. He married Abiah Folger, whose father was a teacher to the Indians. Josiah talked of offering his son Benjamin as his "tithe" to the church and enrolled him in Boston Grammar School as a preparation for the study of the ministry, but his plans were too ambitious and Benjamin was forced to leave school and work for his father. He hated his father's occupation and threatened to run away to sea. A compromise was made, and when Benjamin was twelve he was apprenticed to his brother, a printer. He must have been a natural student of the printing trade; he loved books and reading, he learned quickly, and he liked to write. His brother unwittingly published Benjamin's first essay when he printed an editorial left on his desk signed "Silence Dogood." When his brother was imprisoned in 1722 for offending Massachusetts officials, Franklin carried on publication of the paper by himself.

In 1723 Franklin broke with his brother and ran away to Philadelphia. It was a serious act for an apprentice, and his brother was justly indignant and angry. But the break was inevitable; for Franklin was proud and independent by nature, and too clever for his brother by far. At seventeen, with little money in his pocket but already an expert printer, he proceeded to make his way in the world, subject to the usual "errata," as he liked to call his mistakes, but confident that he could profit from lessons learned and not repeat them. His most serious error was in trusting a foolish man who wanted to be important to everyone. As a result of Governor Keith's "favors," Benjamin found himself alone and without employment in London in 1724. He returned to the colonies two years later.

Franklin had an uncanny instinct for success. He taught himself French, Spanish, Italian, and Latin, yet was shrewd enough to know that people did not like to do business with merchants who were smarter than they. He dressed plainly and sometimes carried his own paper in a wheelbarrow through Philadelphia streets to assure future customers that he was hardworking and not above doing things for himself. By the time he was twenty-four he was the sole owner of a successful printing shop and editor and publisher of the *Pennsylvania Gazette*. He offered his *Poor Richard's Almanac* for sale in 1733 and made it an American institution, filling it with maxims for achieving wealth and preaching hard work and thrift. In 1730 he married Deborah Read, the daughter of his first landlady, and they had two children. Franklin had two illegitimate children, and Deborah took Franklin's son William into the household. He was later to become Governor of New Jersey and a Loyalist during the Revolution; Franklin addressed the first part of his *Autobiography* to him. Before he retired from business at the age of forty-two, Franklin had founded a library, invented a stove, established a fire company, subscribed to an academy which was to become the University of Pennsylvania, and served as secretary to the American Philosophical Society. It was his intention when he retired to devote himself to public affairs and his lifelong passion for the natural sciences, especially the phenomena of sound, vapors, earthquakes, and electricity.

Franklin's observations on electricity were published in London in 1751 and, despite his disclaimers in the *Autobiography*, brought him the applause of British scientists. Science was Franklin's great passion, the only thing, the American historian Charles Beard once said, about which Franklin was not ironic. His inquiring mind was challenged most by the mechanics of the ordinary phenomena of the world, and he was convinced that man's rational powers would enable him to solve riddles that had puzzled mankind for centuries. Franklin believed that man was naturally innocent, that all the mysteries which charmed the religious mind could be explained to our advantage, and that education, properly undertaken, would transform our lives and set us free from the tyrannies of church and monarchy. Franklin had no illusions about the "errata" of mankind, but his metaphor suggests that we can change and alter our past in a way that the word *sins* does not.

Franklin's remaining years, however, were not spent in a laboratory, but at the diplomatic table in London, Paris, and Philadelphia, where his gift for irony served him well. For he was a born diplomat, detached, adaptable, witty, urbane, charming, and clever, and of the slightly more than forty years left to him after his retirement, more than half were spent abroad. In 1757 he went to England to represent the colonies and stayed for five years, returning in 1763. It was in England in 1768 that Franklin first noted his growing sense of alienation from England and the impossibility of compromise with the mother country. Parliament can make *all* laws for the colonies or *none*, he said, and "I think the arguments for the latter more numerous and weighty, than those for the former." When he returned to Philadelphia in May, 1775, he was chosen as a representative to the second Continental Congress, and he served on the committee to draft the Declaration of Independence. In October, 1776, he was appointed Minister to France, where he successfully negotiated a treaty of allegiance and became something of a cult hero. In 1781 he was a member of the American delegation to the Paris peace conference, and he signed the Treaty of Paris which brought the Revolutionary War to an end. Franklin protested his too long stay in Europe and returned to Philadelphia in 1785, serving as a delegate to the Constitutional Convention. When he died in 1790, he was one of the most beloved Americans. Twenty thousand people attended his funeral.

This hero of the eighteenth century, however, has not universally charmed our own. For a number of readers Franklin has been identified as a garrulous but insensitive man of the world, too adaptable for a man of integrity and too willing to please. D. H. Lawrence is only one of a number of Franklin's critics who have charged him with insensitivity and indifference to the darker recesses of the soul. There is no question but that Franklin, like Emerson, has been reduced by his admirers—the hero of those who seek only the way to wealth. But such critics often ignore Franklin's own comments on the follies of humankind.

The Way to Wealth[1]

Preface to Poor Richard Improved

Courteous Reader,

I have heard that nothing gives an author so great pleasure, as to find his works respectfully quoted by other learned authors. This pleasure I have seldom enjoyed; for though I have been, if I may say it without vanity, an eminent author of almanacs annually now a full quarter of a century, my brother authors in the same way, for what reason I know not, have ever been very sparing in their applauses, and no other author has taken the least notice of me, so that did not my writings produce me some solid pudding, the great deficiency of praise would have quite discouraged me.

I concluded at length, that the people were the best judges of my merit; for they buy my works; and besides, in my rambles, where I am not personally known, I have frequently heard one or other of my adages repeated, with "as Poor Richard says" at the end on 't; this gave me some satisfaction, as it showed not only that my instructions were regarded, but discovered likewise some respect for my authority; and I own, that to encourage the practice of remembering and repeating those wise sentences, I have sometimes quoted myself with great gravity.

Judge, then, how much I must have been gratified by an incident I am going to relate to you. I stopped my horse lately where a great number of people were collected at a vendue[2] of merchant goods. The hour of sale not being come, they were conversing on the badness of the times and one of the company called to a plain clean old man, with white locks, "Pray, Father Abraham, what think you of the times? Won't these heavy taxes quite ruin the country? How shall we be ever able to pay them? What would you advise us to?" Father Abraham stood up, and replied, "If you'd have my advice, I'll give it you in short, for a *word to the wise is enough, and many words won't fill a bushel,* as Poor Richard says." They joined in desiring him to speak his mind, and gathering round him, he proceeded as follows:

"Friends," says he, "and neighbors, the taxes are indeed very

---

1. Franklin composed this essay for the 25th anniversary issue of his *Almanac,* the first issue of which, under the fictitious editorship of "Richard Saunders," appeared in 1733. For this essay Franklin brought together the best of his maxims in the guise of a speech by Father Abraham. It is frequently reprinted as *The Way to Wealth,* but is also known by earlier titles: *Poor Richard Improved* and *Father Abraham's Speech.* The text used here is from *The Writings of Benjamin Franklin,* Vol. III, ed. Albert Henry Smyth (New York: Macmillan, 1907).

2. Auction or sale.

heavy, and if those laid on by the government were the only ones we had to pay, we might more easily discharge them; but we have many others, and much more grievous to some of us. We are taxed twice as much by our idleness, three times as much by our pride, and four times as much by our folly; and from these taxes the commissioners cannot ease or deliver us by allowing an abatement. However, let us hearken to good advice, and something may be done for us; *God helps them that help themselves,* as Poor Richard says, in his Almanack of 1733.

"It would be thought a hard government that should tax its people one-tenth part of their time, to be employed in its service. But idleness taxes many of us much more, if we reckon all that is spent in absolute sloth, or doing of nothing, with that which is spent in idle employments or amusements, that amount to nothing. Sloth, by bringing on diseases, absolutely shortens life. *Sloth, like rust, consumes faster than labor wears; while the used key is always bright,* as Poor Richard says. *But dost thou love life, then do not squander time, for that's the stuff life is made of,* as Poor Richard says. How much more than is necessary do we spend in sleep, forgetting that *the sleeping fox catches no poultry* and that *there will be sleeping enough in the grave,* as Poor Richard says.

"*If time be of all things the most precious, wasting time must be,* as Poor Richard says, *the greatest prodigality;* since, as he elsewhere tells us, *lost time is never found again; and what we call time enough, always proves little enough:* let us then up and be doing, and doing to the purpose; so by diligence shall we do more with less perplexity. *Sloth makes all things difficult, but industry all easy,* as Poor Richard says; *and he that riseth late must trot all day, and shall scarce overtake his business at night;* while *laziness travels so slowly, that poverty soon overtakes him,* as we read in Poor Richard, who adds, *drive thy business, let not that drive thee,* and *early to bed, and early to rise, makes a man healthy, wealthy, and wise.*

"So what signifies wishing and hoping for better times. We may make these times better, if we bestir ourselves. *Industry need not wish,* as Poor Richard says, *and he that lives upon hope will die fasting. There are no gains without pains; then help hands, for I have no lands,* or if I have, they are smartly taxed. And, as Poor Richard likewise observes, *he that hath a trade hath an estate; and he that hath a calling, hath an office of profit and honor;* but then the trade must be worked at, and the calling well followed, or neither the estate nor the office will enable us to pay our taxes. If we are industrious, we shall never starve; for, as Poor Richard says, *at the workingman's house hunger looks in, but dares not enter.* Nor will the bailiff or the constable enter, for *industry pays debts, while despair increaseth them,* says Poor Richard. What though you have found no treasure, nor has any rich relation left you a legacy, *diligence is the mother of goodluck,* as Poor Richard says, and

God gives all things to industry. Then plow deep, while sluggards sleep, and you shall have corn to sell and to keep, says Poor Dick. Work while it is called today, for you know not how much you may be hindered tomorrow, which makes Poor Richard say, *one today is worth two tomorrows,* and farther, *have you somewhat to do tomorrow, do it today.* If you were a servant, would you not be ashamed that a good master should catch you idle? Are you then your own master, *be ashamed to catch yourself idle,* as Poor Dick says. When there is so much to be done for yourself, your family, your country, and your gracious king, be up by peep of day; *let not the sun look down and say, inglorious here he lies.* Handle your tools without mittens; remember that *the cat in gloves catches no mice,* as Poor Richard says. 'Tis true there is much to be done, and perhaps you are weak-handed, but stick to it steadily; and you will see great effects, for *constant dropping wears away stones,* and *by diligence and patience the mouse ate in two the cable;* and *little strokes fell great oaks,* as Poor Richard says in his Almanack, the year I cannot just now remember.

"Methinks I hear some of you say, "must a man afford himself no leisure?" I will tell thee, my friend, what Poor Richard says, *employ thy time well, if thou meanest to gain leisure; and, since thou art not sure of a minute, throw not away an hour.* Leisure is time for doing something useful; this leisure the diligent man will obtain, but the lazy man never; so that, as Poor Richard says *a life of leisure and a life of laziness are two things.* Do you imagine that sloth will afford you more comfort than labour? No, for as Poor Richard says, *trouble springs from idleness, and grievous toil from needless ease. Many without labor, would live by their wits only, but they break for want of stock.* Whereas industry gives comfort, and plenty, and respect: *fly pleasures, and they'll follow you. The diligent spinner has a large shift;*[3] *and now I have a sheep and a cow, everybody bids me good morrow;* all which is well said by Poor Richard.

"But with our industry, we must likewise be steady, settled, and careful, and oversee our own affairs with our own eyes, and not trust too much to others; for, as Poor Richard says

> *I never saw an oft-removed tree,*
> *Nor yet an oft-removed family,*
> *That throve so well as those that settled be.*

And again, *three removes*[4] *is as bad as a fire;* and again, *keep thy shop, and thy shop will keep thee;* and again, *if you would have your business done, go; if not, send.* And again,

> *He that by the plough would thrive,*
> *Himself must either hold or drive.*

---

3. Wardrobe.   4. Moves.

And again, *the eye of a master will do more work than both his hands*; and again, *want of care does us more damage than want of knowledge*; and again, *not to oversee workmen is to leave them your purse open.* Trusting too much to others' care is the ruin of many; for, as the Almanack says, *in the affairs of this world, men are saved, not by faith, but by the want of it*; but a man's own care is profitable; for, saith Poor Dick, *learning is to the studious*, and *riches to the careful*, as well as *power to the bold*, and *heaven to the virtuous*, and farther, *if you would have a faithful servant, and one that you like, serve yourself.* And again, he adviseth to circumspection and care, even in the smallest matters, because sometimes *a little neglect may breed great mischief*; adding, *for want of a nail the shoe was lost; for want of a shoe the horse was lost; and for want of a horse the rider was lost*, being overtaken and slain by the enemy; *all for want of care about a horseshoe nail.*

"So much for industry, my friends, and attention to one's own business; but to these we must add frugality, if we would make our industry more certainly successful. A man may, if he knows not how to save as he gets, keep his nose all his life to the grindstone, and die not worth a groat[5] at last. *A fat kitchen makes a lean will*, as Poor Richard says; and

> *Many estates are spent in the getting,*
> *Since women for tea forsook spinning and knitting,*
> *And men for punch forsook hewing and splitting.*

If you would be wealthy, says he, in another Almanack, *think of saving as well as of getting: the Indies have not made Spain rich, because her outgoes are greater than her incomes.*

"Away then with your expensive follies, and you will not then have so much cause to complain of hard times, heavy taxes, and chargeable families; for, as Poor Dick says,

> *Women and wine, game and deceit,*
> *Make the wealth small and the wants great.*

And farther, *what maintains one vice would bring up two children.* You may think perhaps, that a little tea, or a little punch now and then, diet a little more costly, clothes a little finer, and a little entertainment now and then, can be no great matter; but remember what Poor Richard says, *many a little makes a mickle*;[6] and farther, *Beware of little expenses; a small leak will sink a great ship*; and again, *who dainties love shall beggars prove*; and moreover, *fools make feasts, and wise men eat them.*

"Here you are all got together at this vendue of fineries and knick-nacks. You call them goods; but if you do not take care, they will prove evils to some of you. You expect they will be sold cheap, and perhaps they may for less than they cost; but if you have no occasion

---

5. A silver coin worth about four pence.   6. Lot.

for them, they must be dear to you. Remember what Poor Richard says; *buy what thou hast no need of, and ere long thou shalt sell thy necessaries.* And again, *at a great pennyworth pause a while*: he means, that perhaps the cheapness is apparent only, and not real; or the bargain, by straightening thee in thy business, may do thee more harm than good. For in another place he says, *many have been ruined by buying good pennyworths*. Again, Poor Richard says, *'tis foolish to lay out money in a purchase of repentance*; and yet this folly is practiced every day at vendues, for want of minding the Almanack. *Wise men*, as Poor Dick says, *learn by others' harms, fools scarcely by their own*; but *felix quem faciunt aliena pericula cautum*.[7] Many a one, for the sake of finery on the back, have gone with a hungry belly, and half-starved their families. *Silks and satins, scarlet and velvets*, as Poor Richard says, *put out the kitchen fire*.

"These are not the necessaries of life; they can scarcely be called the conveniences; and yet only because they look pretty, how many want to have them! The artificial wants of mankind thus become more numerous than the natural; and, as Poor Dick says, *for one poor person, there are an hundred indigent*. By these, and other extravagancies, the genteel are reduced to poverty, and forced to borrow of those whom they formerly despised, but who through industry and frugality have maintained their standing; in which case it appears plainly, that *a plowman on his legs is higher than a gentleman on his knees*, as Poor Richard says. Perhaps they have had a small estate left them, which they knew not the getting of; they think, " 'Tis day, and will never be night"; that a little to be spent out of so much is not worth minding; *a child and a fool*, as Poor Richard says, *imagine twenty shillings and twenty years can never be spent* but, *always taking out of the meal-tub, and never putting in, soon comes to the bottom*; as Poor Dick says, *when the well's dry, they know the worth of water*. But this they might have known before, if they had taken his advice; *if you would know the value of money, go and try to borrow some*; for, *he that goes a-borrowing goes a-sorrowing*; and indeed so does he that lends to such people, when he goes to get it in again. Poor Dick farther advises, and says,

> *Fond pride of dress is sure a very curse;*
> *E'er fancy you consult, consult your purse.*

And again, *pride is as loud a beggar as want, and a great deal more saucy*. When you have bought one fine thing, you must buy ten more, that your appearance may be all of a piece; but Poor Dick says, *'tis easier to suppress the first desire, than to satisfy all that follow it*. And 'tis as truly folly for the poor to ape the rich, as for the frog to swell, in order to equal the ox.

---

7. A Latin version of the proverb just quoted.

> *Great estates may venture more,*
> *But little boats should keep near shore.*

'Tis, however, a folly soon punished; for *pride that dines on vanity supps on contempt*, as Poor Richard says. And in another place, *pride breakfasted with plenty, dined with poverty, and supped with infamy*. And after all, of what use is this pride of appearance, for which so much is risked so much is suffered? It cannot promote health, or ease pain; it makes no increase of merit in the person, it creates envy, it hastens misfortune.

> *What is a butterfly? At best*
> *He's but a caterpillar dressed*
> *The gaudy fop's his picture just,*

as Poor Richard says.

"But what madness must it be to run in debt for these super-fluities! We are offered, by the terms of this vendue, *six months' credit*; and that perhaps has induced some of us to attend it, because we cannot spare the ready money, and hope now to be fine without it. But, ah, think what you do when you run in debt; you give to another power over your liberty. If you cannot pay at the time, you will be ashamed to see your creditor; you will be in fear when you speak to him; you will make poor pitiful sneaking excuses, and by degrees come to lose your veracity, and sink into base downright lying; for, as Poor Richard says, *the second vice is lying, the first is running in debt*. And again, to the same purpose, *lying rides upon debt's back*. Whereas a free-born Englishman ought not to be ashamed or afraid to see or speak to any man living. But poverty often deprives a man of all spirit and virtue: *'tis hard for an empty bag to stand upright*, as Poor Richard truly says.

"What would you think of that prince, or that government, who should issue an edict forbidding you to dress like a gentleman or a gentlewoman, on pain of imprisonment or servitude? Would you not say, that you were free, have a right to dress as you please, and that such an edict would be a breach of your privileges, and such a government tyrannical? And yet you are about to put yourself under that tyranny, when you run in debt for such dress! Your creditor has authority, at his pleasure to deprive you of your liberty, by con-fining you in gaol[8] for life, or to sell you for a servant, if you should not be able to pay him! When you have got your bargain, you may, perhaps, think little of payment; but *creditors*, Poor Richard tells us, *have better memories than debtors*; and in another place says, *creditors are a superstitious sect, great observers of set days and times*. The day comes round before you are aware, and the demand is made before you are prepared to satisfy it, or if you bear your debt in mind, the term which at first seemed so long will, as it lessens, appear extremely short. Time will seem to have added wings to his

8. Jail.

heels as well as shoulders. *Those have a short Lent*, saith Poor Richard, *who owe money to be paid at Easter*. Then since, as he says, *The borrower is a slave to the lender, and the debtor to the creditor*, disdain the chain, preserve your freedom; and maintain your independency: be industrious and free; be frugal and free. At present, perhaps, you may think yourself in thriving circumstances, and that you can bear a little extravagance without injury; but,

> *For age and want, save while you may;*
> *No morning sun lasts a whole day,*

as Poor Richard says. Gain may be temporary and uncertain, but ever while you live, expense is constant and certain; and *'tis easier to build two chimneys than to keep one in fuel*, as Poor Richard says. So, *rather go to bed supperless than rise in debt*.

> *Get what you can, and what you get hold;*
> *'Tis the stone that will turn all your lead into gold,*

as Poor Richard says. And when you have got the philosopher's stone,[9] sure you will no longer complain of bad times, or the difficulty of paying taxes.

"This doctrine, my friends, is reason and wisdom; but after all, do not depend too much upon your own industry, and frugality, and prudence, though excellent things, for they may all be blasted without the blessing of heaven; and therefore, ask that blessing humbly, and be not uncharitable to those that at present seem to want it, but comfort and help them. Remember, Job[1] suffered, and was afterwards prosperous.

"And now to conclude, *experience keeps a dear[2] school, but fools will learn in no other, and scarce in that*; for it is true, *we may give advice, but we cannot give conduct*, as Poor Richard says: however, remember this, *they that won't be counseled, can't be helped*, as Poor Richard says: and farther, that, *if you will not hear reason, she'll surely rap your knuckles*."

Thus the old gentleman ended his harangue. The people heard it, and approved the doctrine, and immediately practiced the contrary, just as if it had been a common sermon; for the vendue opened, and they began to buy extravangantly, notwithstanding, his cautions and their own fear of taxes. I found the good man had thoroughly studied my almanacs, and digested all I had dropped on these topics during the course of five and twenty years. The frequent mention he made of me must have tired any one else, but my vanity was wonderfully delighted with it, though I was conscious that not a tenth part of the wisdom was my own, which he ascribed to me, but rather the gleanings I had made of the sense of all ages and nations. However, I resolved to be the better for the echo of it; and

9. A substance thought to transform base metals into gold, much sought after by alchemists.

1. The Old Testament patriarch who suffered with faith.

2. Expensive.

though I had at first determined to buy stuff for a new coat, I went away resolved to wear my old one a little longer. Reader, if thou wilt do the same, thy profit will be as great as mine. I am, as ever, thine to serve thee,

Richard Saunders
July 7, 1757

1757   1758

The Speech of Polly Baker[1]

The speech of Miss Polly Baker before a court of judicature, at Connecticut near Boston in New England; where she was prosecuted the fifth time, for having a bastard child: Which influenced the court to dispense with her punishment, and which induced one of her judges to marry her the next day—by whom she had fifteen children.

"May it please the honorable bench to indulge me in a few words: I am a poor, unhappy woman, who have no money to fee[2] lawyers to plead for me, being hard put to it to get a living. I shall not trouble your honors with long speeches; for I have not the presumption to expect that you may, by any means, be prevailed on to deviate in your sentence from the law, in my favor. All I humbly hope is, that your honors would charitably move the governor's goodness on my behalf, that my fine may be remitted. This is the fifth time, gentlemen, that I have been dragged before your court on the same account; twice I have paid heavy fines, and twice have been brought to public punishment, for want[3] of money to pay those fines. This may have been agreeable to the laws, and I don't dispute it; but since laws are sometimes unreasonable in themselves, and therefore repealed; and others bear too hard on the subject in particular circumstances, and therefore there is left a power somewhere to dispense with the execution of them; I take the liberty to say, that I think this law, by which I am punished, both unreasonable in itself, and particularly severe with regard to me, who have always lived an inoffensive life in the neighborhood where I was born, and defy my enemies (if I have any) to say I ever wronged any man, woman, or child. Abstracted from the law, I cannot conceive (may it please your honors) what the nature of my offense is. I have brought five fine children into the world, at the risk of my life; I have maintained them well by my own industry, without burdening the township, and would have done it better, if it had not

1. In his later years, Franklin admitted to the authorship of *The Speech of Polly Baker* in a conversation with Thomas Jefferson. It was probably first printed in Philadelphia, but it does not appear in *The Pennsylvania Gazette*. The first known printing is *The Gentleman's Maga-* *zine*, London, 1747. The text used here is from *The Writings of Benjamin Franklin*, Vol. II, ed. Albert Henry Smyth (New York: Macmillan, 1907).
2. Pay.
3. Lack.

been for the heavy charges and fines I have paid. Can it be a crime (in the nature of things, I mean) to add to the king's subjects, in a new country, that really wants people? I own[4] it, I should think it rather a praiseworthy than a punishable action. I have debauched no other woman's husband, nor enticed any other youth; these things I never was charged with; nor has anyone the least cause of complaint against me, unless, perhaps, the ministers of justice, because I have had children without being married, by which they have missed a wedding fee. But can this be a fault of mine? I appeal to your honors. You are pleased to allow I don't want sense; but I must be stupefied to the last degree, not to prefer the honorable state of wedlock to the condition I have lived in. I always was, and still am willing to enter into it; and doubt not my behaving well in it, having all the industry, frugality, fertility, and skill in economy[5] appertaining to a good wife's character. I defy any one to say I ever refused an offer of that sort: on the contrary, I readily consented to the only proposal of marriage that ever was made me, which was when I was a virgin, but too easily confiding in the person's sincerity that made it, I unhappily lost my honor by trusting to his; for he got me with child, and then forsook me.

"That very person, you all know, he is now become a magistrate of this country; and I had hopes he would have appeared this day on the bench, and have endeavored to moderate the court in my favor; then I should have scorned to have mentioned it; but I must now complain of it, as unjust and unequal, that my betrayer and undoer, the first cause of all my faults and miscarriages (if they must be deemed such), should be advanced to honor and power in this government that punishes my misfortunes with stripes and infamy. I should be told, 'tis like, that were there no act of assembly[6] in the case, the precepts of religion are violated by my transgressions. If mine is a religious offense, leave it to religious punishments. You have already excluded me from the comforts of your church communion. Is not that sufficient? You believe I have offended heaven, and must suffer eternal fire. Will not that be sufficient? What need is there then of your additional fines and whipping? I own I do not think as you do, for, if I thought what you call a sin was really such, I could not presumptuously commit it. But, how can it be believed that heaven is angry at my having children, when to the little done by me towards it, God has been pleased to add His divine skill and admirable workmanship in the formation of their bodies, and crowned the whole by furnishing them with rational and immortal souls?

"Forgive me, gentlemen, if I talk a little extravagantly on these matters; I am no divine,[7] but if you, gentlemen, must be making laws, do not turn natural and useful actions into crimes by your pro-

---

4. Acknowledge.
5. Household management.

6. I.e., legislative assembly.
7. Clergyman.

hibitions. But take into your wise consideration the great and growing number of bachelors in the country, many of whom, from the
mean fear of the expenses of a family, have never sincerely and hon-
orably courted a woman in their lives; and by their manner of living
leave unproduced (which is little better than murder) hundreds of
their posterity to the thousandth generation. Is not this a greater
offense against the public good than mine? Compel them, then, by
law, either to marriage, or to pay double the fine of fornication
every year. What must poor young women do, whom customs and
nature forbid to solicit the men, and who cannot force themselves
upon husbands, when the laws take no care to provide them any,
and yet severely punish them if they do their duty without them;
the duty of the first and great command of nature and nature's
God, *increase and multiply*; a duty, from the steady performance of
which nothing has been able to deter me, but for its sake I have
hazarded the loss of the public esteem, and have frequently endured
public disgrace and punishment; and therefore ought, in my humble
opinion, instead of a whipping, to have a statue erected to my
memory."

1747

Remarks Concerning the Savages of North America[1]

Savages we call them, because their manners differ from ours,
which we think the perfection of civility; they think the same of
theirs.

Perhaps, if we could examine the manners of different nations
with impartiality, we should find no people so rude, as to be without any rules of politeness; nor any so polite, as not to have some
remains of rudeness.

The Indian men, when young, are hunters and warriors; when
old, counselors; for all their government is by counsel of the sages;
there is no force, there are no prisons, no officers to compel obedience, or inflict punishment. Hence they generally study oratory,
the best speaker having the most influence. The Indian women till
the ground, dress the food, nurse and bring up the children, and
preserve and hand down to posterity the memory of public transactions. These employments of men and women are accounted natural
and honorable. Having few artificial wants, they have abundance of
leisure for improvement by conversation. Our laborious manner of
life, compared with theirs, they esteem slavish and base; and the
learning, on which we value ourselves, they regard as frivolous and
useless. An instance of this occurred at the Treaty of Lancaster, in
Pennsylvania, *anno* 1744, between the government of Virginia and

1. The text used here is from *The Writ-
ings of Benjamin Franklin*, Vol. X, ed.   Albert Henry Smyth (New York: Macmillan, 1907).

the Six Nations.[2] After the principal business was settled, the commissioners from Virginia acquainted the Indians by a speech, that there was at Williamsburg a college, with a fund for educating Indian youth; and that, if the Six Nations would send down half a dozen of their young lads to that college, the government would take care that they should be well provided for, and instructed in all the learning of the white people. It is one of the Indian rules of politeness not to answer a public proposition the same day that it is made; they think it would be treating it as a light matter, and that they show it respect by taking time to consider it, as of a matter important. They therefore deferred their answer till the day following; when their speaker began, by expressing their deep sense of the kindness of the Virginia government, in making them that offer; "for we know," says he, "that you highly esteem the kind of learning taught in those Colleges, and that the maintenance of our young men, while with you, would be very expensive to you. We are convinced, therefore, that you mean to do us good by your proposal; and we thank you heartily. But you, who are wise, must know that different nations have different conceptions of things; and you will therefore not take it amiss, if our ideas of this kind of education happen not to be the same with yours. We have had some experience of it; several of our young people were formerly brought up at the colleges of the northern provinces; they were instructed in all your sciences; but, when they came back to us, they were bad runners, ignorant of every means of living in the woods, unable to bear either cold or hunger, knew neither how to build a cabin, take a deer, or kill an enemy, spoke our language imperfectly, were therefore neither fit for hunters, warriors, nor counselors; they were totally good for nothing. We are however not the less obliged by your kind offer, though we decline accepting it; and, to show our grateful sense of it, if the gentlemen of Virginia will send us a dozen of their sons, we will take great care of their education, instruct them in all we know, and make *men* of them."

Having frequent occasions to hold public councils, they have acquired great order and decency in conducting them. The old men sit in the foremost ranks, the warriors in the next, and the women and children in the hindmost. The business of the women is to take exact notice of what passes, imprint it in their memories (for they have no writing), and communicate it to their children. They are the records of the council, and they preserve traditions of the stipulations in treaties 100 years back; which, when we compare with our writings, we always find exact. He that would speak, rises. The rest observe a profound silence. When he has finished and sits down, they leave him 5 or 6 minutes to recollect, that, if he has omitted anything he intended to say, or has anything to add, he may rise

2. A confederation of Iroquois tribes: Seneca, Cayuga, Oneida, Onondaga, Mohawk, and Tuscarora.

again and deliver it. To interrupt another, even in common conversation, is reckoned highly indecent. How different this from the conduct of a polite British House of Commons, where scarce a day passes without some confusion, that makes the speaker hoarse in calling *to order*; and how different from the mode of conversation in many polite companies of Europe, where, if you do not deliver your sentence with great rapidity, you are cut off in the middle of it by the impatient loquacity of those you converse with, and never suffered to finish it!

The politeness of these savages in conversation is indeed carried to excess, since it does not permit them to contradict or deny the truth of what is asserted in their presence. By this means they indeed avoid disputes; but then it becomes difficult to know their minds, or what impression you make upon them. The missionaries who have attempted to convert them to Christianity, all complain of this as one of the great difficulties of their mission. The Indians hear with patience the truths of the Gospel explained to them, and give their usual tokens of assent and approbation; you would think they were convinced. No such matter. It is mere civility.

A Swedish minister, having assembled the chiefs of the Susquehanah Indians, made a sermon to them, acquainting them with the principal historical facts on which our religion is founded; such as the fall of our first parents by eating an apple, the coming of Christ to repair the mischief, His miracles and suffering, &c. When he had finished, an Indian orator stood up to thank him. "What you have told us," says he, "is all very good. It is indeed bad to eat apples. It is better to make them all into cider. We are much obliged by your kindness in coming so far, to tell us these things which you have heard from your mothers. In return, I will tell you some of those we have heard from ours. In the beginning, our fathers had only the flesh of animals to subsist on; and if their hunting was unsuccessful, they were starving. Two of our young hunters, having killed a deer, made a fire in the woods to broil some part of it. When they were about to satisfy their hunger, they beheld a beautiful young woman descend from the clouds, and seat herself on that hill, which you see yonder among the blue mountains. They said to each other, it is a spirit that has smelled our broiling vension, and wishes to eat of it; let us offer some to her. They presented her with the tongue; she was pleased with the taste of it, and said, 'Your kindness shall be rewarded; come to this place after thirteen moons, and you shall find something that will be of great benefit in nourishing you and your children to the latest generations.' They did so, and, to their surprise, found plants they had never seen before; but which, from that ancient time, have been constantly cultivated among us, to our great advantage. Where her right hand had touched the ground, they found maize; where her left hand had touch it, they found kidney-beans; and where her backside had sat on it, they found tobacco." The good missionary, disgusted with this idle tale, said, "What I delivered to you were sacred truths; but what you tell me is mere fable, fiction, and falsehood." The Indian, offended, replied, "My brother, it seems your friends have not done you justice in your education; they have not well instructed you in the rules of common civility. You saw that we, who understand and practice those rules, believed all your stories; why do you refuse to believe ours?"

When any of them come into our towns, our people are apt to crowd round them, gaze upon them, and incommode them, where they desire to be private; this they esteem great rudeness, and the effect of the want of instruction in the rules of civility and good manners. "We have," say they, "as much curiosity as you, and when you come into our towns, we wish for opportunities of looking at you; but for this purpose we hide ourselves behind bushes, where you are to pass, and never intrude ourselves into your company."

Their manner of entering one another's village has likewise its rules. It is reckoned uncivil in traveling strangers to enter a village abruptly, without giving notice of their approach. Therefore, as soon as they arrive within hearing, they stop and hollow,[3] remaining there till invited to enter. Two old men usually come out to them, and lead them in. There is in every village a vacant dwelling, called *the strangers' house.* Here they are placed, while the old men go round from hut to hut, acquainting the inhabitants, that strangers are arrived, who are probably hungry and weary; and every one sends them what he can spare of victuals, and skins to repose on. When the strangers are refreshed, pipes and tobacco are brought; and then, but not before, conversation begins, with inquiries who they are, whither bound, what news, &c.; and it usually ends with offers of service, if the strangers have occasion of guides, or any neccessaries for continuing their journey; and nothing is exacted for the entertainment.

The same hospitality, esteemed among them as a principal virtue, is practiced by private persons; of which Conrad Weiser, our interpreter, gave me the following instances. He had been naturalized among the Six Nations, and spoke well the Mohawk language. In going through the Indian country, to carry a message from our Governor to the Council at Onondaga, he called at the habitation of Canassatego, an old acquaintance, who embraced him, spread furs for him to sit on, placed before him some boiled beans and venison, and mixed some rum and water for his drink. When he was well refreshed, and had lit his pipe, Canassatego began to converse with him; asked how he had fared the many years since they had seen each other; whence he then came; what occasioned the journey, &c.

3. Cry out; announce themselves.

Conrad answered all his questions; and when the discourse began to flag, the Indian, to continue it, said, "Conrad, you have lived long among the white people, and know something of their customs; I have been sometimes at Albany, and have observed, that once in seven days they shut up their shops, and assemble all in the great house; tell me what it is for? What do they do there?" "They meet there," says Conrad, "to hear and learn *good things.*" "I do not doubt," says the Indian, "that they tell you so; they have told me the same; but I doubt the truth of what they say, and I will tell you my reasons. I went lately to Albany to sell my skins and buy blankets, knives, powder, rum, &c. You know I used generally to deal with Hans Hanson; but I was a little inclined this time to try some other merchant. However, I called first upon Hans, and asked him what he would give for beaver. He said he could not give any more than four shillings a pound; 'but,' says he, 'I cannot talk on business now; this is the day when we meet together to learn *good things,* and I am going to the meeting.' So I thought to myself, 'Since we cannot do any business today, I may as well go to the meeting too,' and I went with him. There stood up a man in black, and began to talk to the people very angrily. I did not understand what he said; but, perceiving that he looked much at me and at Hanson, I imagined he was angry at seeing me there; so I went out, sat down near the house, struck fire, and lit my pipe, waiting till the meeting should break up. I thought too, that the man had mentioned something of beaver, and I suspected it might be the subject of their meeting. So, when they came out, I accosted my merchant. 'Well, Hans,' says I, 'I hope you have agreed to give more than four shillings a pound.' 'No,' says he, 'I cannot give so much; I cannot give more than three shillings and sixpence.' I then spoke to several other dealers, but they all sung the same song,—three and sixpence, —three and sixpence. This made it clear to me, that my suspicion was right; and, that whatever they pretended of meeting to learn *good things,* the real purpose was to consult how to cheat Indians in the price of beaver. Consider but a little, Conrad, and you must be of my opinion. If they met so often to learn *good things,* they would certainly have learned some before this time. But they are still ignorant. You know our practice. If a white man, in traveling through our country, enters one of our cabins, we all treat him as I treat you; we dry him if he is wet, we warm him if he is cold, we give him meat and drink, that he may allay his thirst and hunger; and we spread soft furs for him to rest and sleep on; we demand nothing in return. But, if I go into a white man's house at Albany, and ask for victuals and drink, they say, 'Where is your money?' and if I have none, they say, 'Get out, you Indian dog.' You see they have not yet learned those little *good things,* that we need no meetings to be instructed in, because our mothers taught them to us when we were children; and therefore it is impossible their meetings should be, as they say, for any such purpose, or have any such effect; they are only to contrive *the cheating of Indians in the price of beaver.*"[4]

1784

Letter to Ezra Stiles[1]

Reverend and Dear Sir,   Philadelphia, March 9, 1790

I received your kind letter of January 28, and am glad you have at length received the portrait of Governor Yale[2] from his family, and deposited it in the college library. He was a great and good man, and had the merit of doing infinite service to your country by his munificence to that institution. The honor you propose doing me by placing mine in the same room with his is much too great for my deserts; but you always had a partiality for me, and to that it must be ascribed. I am however too much obliged to Yale College, the first learned society that took notice of me[3] and adorned me with its honors, to refuse a request that comes from it through so esteemed a friend. But I do not think any one of the portraits you mention, as in my possession, worthy of the place and company you propose to place it in. You have an excellent artist lately arrived. If he will undertake to make one for you, I shall cheerfully pay the expense; but he must not delay setting about it, or I may slip through his fingers, for I am now in my eighty-fifth year, and very infirm.[4]

I send with this a very learned work, as it seems to me, on the ancient Samaritan coins, lately printed in Spain, and at least curious for the beauty of the impression. Please to accept it for your college library. I have subscribed for the Encyclopædia[5] now printing here, with the intention of presenting it to the college. I shall probably

4. "It is remarkable that in all ages and countries hospitality has been allowed as the virtue of those whom the civilized were pleased to call barbarians. The Greeks celebrated the Scythians for it. The Saracens possessed it eminently, and it is to this day the reigning virtue of the wild Arabs. St. Paul, too, in the relation of his voyage and shipwreck on the island of Melité says the barbarous people showed us no little kindness; for they kindled a fire, and received us every one, because of the present rain, and because of the cold" [Franklin's note]. Saint Paul's account of his visit to Melita may be found in Acts 28. The Scythians were nomadic tribes of southeastern Europe known for their plundering.

1. Ezra Stiles (1727–95) was the grandson of the poet Edward Taylor and president of Yale College. He wrote to Franklin on January 28, 1780, asking Franklin to provide him with some information about his "religious sentiments" and his opinion "concerning Jesus of Nazareth." Stiles hoped Franklin would not think his inquiry an "impertinence" because he revered Franklin with an affection "bordering on adoration." The text used here is from *The Writings of Benjamin Franklin*, Vol. X, ed. Albert Henry Smyth (New York: Macmillan, 1907).

2. Elihu Yale (1649–1721) was an official in the East India Company, and Governor of Fort Saint George. Yale College was named for him after the receipt from London of books and goods in 1714 and 1718.

3. Franklin was awarded an honorary master's degree from Yale in 1753.

4. Franklin died on April 17, 1790.

5. The third edition of the *Encyclopaedia Britannica*, printed in the United States for the first time. It was customary in undertaking expensive publishing ventures to get customers to subscribe for future volumes and assure the success of the undertaking.

depart before the work is finished, but shall leave directions for its continuance to the end. With this you will receive some of the first numbers.

You desire to know something of my religion. It is the first time I have been questioned upon it. But I cannot take your curiosity amiss, and shall endeavor in a few words to gratify it. Here is my creed. I believe in one God, Creator of the Universe. That He governs it by His providence. That He ought to be worshiped. That the most acceptable service we render to Him is doing good to His other children. That the soul of man is immortal, and will be treated with justice in another life respecting its conduct in this. These I take to be the fundamental principles of all sound religion, and I regard them as you do in whatever sect I meet with them.

As to Jesus of Nazareth, my opinion of whom you particularly desire, I think the system of morals and His religion, as He left them to us, the best the world ever saw or is likely to see; but I apprehend it has received various corrupting changes, and I have with most of the present dissenters in England, some doubts as to His divinity; though it is question I do not dogmatize upon, having never studied it, and think it needless to busy myself with it now, when I expect soon an opportunity of knowing the truth with less trouble. I see no harm, however, in its being believed, if that belief has the good consequence, as probably it has, of making His doctrines more respected and better observed; especially as I do not perceive, that the Supreme takes it amiss, by distinguishing the unbelievers in His government of the world with any peculiar marks of His displeasure.

I shall only add, respecting myself, that, having experienced the goodness of that Being in conducting me prosperously through a long life, I have no doubt of its continuance in the next, though without the smallest conceit of meriting such goodness. My sentiments on this head you will see in the copy of an old letter enclosed, which I wrote in answer to one from a zealous religionist, whom I had relieved in a paralytic case by electricity, and who, being afraid I should grow proud upon it, sent me his serious though rather impertinent caution. I send you also the copy of another letter,[6] which will show something of my disposition relating to religion. With great and sincere esteem and affection, I am, your obliged old friend and most obedient humble servant

B. Franklin

P.S. Had not your college some present of books from the King of France? Please to let me know, if you had an expectation given you of more, and the nature of that expectation? I have a reason for the inquiry.

I confide, that you will not expose me to criticism and censure by publishing any part of his communication to you. I have ever let

6. It has been suggested that this letter was addressed to Thomas Paine.

others enjoy their religious sentiments, without reflecting on them
for those that appeared to me unsupportable and even absurd. All
sects here, and we have a great variety, have experienced my good
will in assisting them with subscriptions for building their new
places of worship; and, as I have never opposed any of their doctrines, I hope to go out of the world in peace with them all.

1790   1840

The Autobiography   Franklin turned to the manuscript of *The Autobiography* on four different occasions over a period of nineteen years.
The first part, addressed to his son William Franklin (c. 1731–1813),
who was Governor of New Jersey when Franklin was writing this
section, was composed while Franklin was visiting the country home of
Bishop Jonathan Shipley at Twyford, a village about fifty miles from London. It was begun on July 30 and concluded on or about August 13,
1771. Franklin did not work on the manuscript again until he was living in
France and was Minister of the newly formed United States, about thirteen
years later. The last two sections were written in August, 1788, and the
winter of 1789–90, when Franklin stopped because of illness. Before he
died he carried his life up to the year 1758. The account ends, therefore,
before Franklin's great triumphs as a diplomat and public servant.

The text for *The Autobiography* here reprinted is entirely new; it is the
first to have been taken directly from the manuscript itself (all other editors
have merely corrected earlier printed texts). The text was established by J.
A. Leo Lemay and Paul Zall for their Norton Critical Edition of *The Autobiography* (in preparation), and is here reprinted with their kind permission. Because this text differs significantly from all published versions, we
have departed from our usual policy of modernizing early texts in this one
instance in order to reprint exactly what Franklin wrote. The editors note
that they have omitted short dashes which Franklin often wrote after sentences, and punctuation marks which "have been clearly superseded by revisions or additions." Careless slips have been corrected silently but may be
found in their section on emendations in their complete text. Professors
Lemay and Zall have been very generous in letting us see their footnotes
and biographical sketches. Every student of Franklin's *Autobiography* must
also acknowledge the extremely helpful edition of Leonard W. Labaree *et
al.* (New Haven, Conn.: Yale University Press, 1964).

From The Autobiography

Dear Son,   Twyford, at the Bishop of St. Asaph's 1771

I have ever had a Pleasure in obtaining any little Anecdotes of
my Ancestors. You may remember the Enquiries I made among the
Remains[1] of my Relations when you were with me in England;

THE AUTOBIOGRAPHY OF BEN-
JAMIN FRANKLIN: A Norton Critical
Edition, edited by J. A. Leo Lemay and
Paul M. Zall. Copyright © 1979 by W. W.
Norton & Company, Inc. Reprinted by
permission of the editors and W. W.
Norton & Company, Inc.

1. I.e., the remaining representatives of a
family. Franklin and his son toured England in 1758 and visited ancestral homes
at Ecton and Banbury, Northamptonshire, England.

and the Journey I took for that purpose. Now imagining it may be equally agreable to you to know the Circumstances of *my* Life, many of which you are yet unacquainted with; and expecting a Weeks uninterrupted Leisure in my present Country Retirement, I sit down to write them for you. To which I have besides some other Inducements. Having emerg'd from the Poverty and Obscurity in which I was born and bred, to a State of Affluence and some Degree of Reputation in the World, and having gone so far thro' Life with a considerable Share of Felicity, the conducing Means I made use of, which, with the Blessing of God, so well succeeded, my Posterity may like to know, as they may find some of them suitable to their own Situations, and therefore fit to be imitated. That Felicity, when I reflected on it, has induc'd me sometimes to say, that were it offer'd to my Choice, I should have no Objection to a Repetition of the same Life from its Beginning, only asking the Advantage Authors have in a second Edition to correct some Faults of the first. So would I if I might, besides correcting the Faults, change some sinister Accidents and Events of it for others more favourable, but tho' this were deny'd, I should still accept the Offer. However, since such a Repetition is not to be expected, the Thing most like living one's Life over again, seems to be a *Recollection* of that Life; and to make that Recollection as durable as possible, the putting it down in Writing. Hereby, too, I shall indulge the Inclination so natural in old Men, to be talking of themselves and their own past Actions, and I shall indulge it, without being troublesome to others who thro' respect to Age might think themselves oblig'd to give me a Hearing, since this may be read or not as any one pleases. And lastly, (I may as well confess it, since my Denial of it will be believ'd by no body) perhaps I shall a good deal gratify my own *Vanity*. Indeed I scarce ever heard or saw the introductory Words, *Without Vanity I may say*, etc. but some vain thing immediately follow'd. Most People dislike Vanity in others whatever Share they have of it themselves, but I give it fair Quarter wherever I meet with it, being persuaded that it is often productive of Good to the Possessor and to others that are within his Sphere of Action: And therefore in many Cases it would not be quite absurd if a Man were to thank God for his Vanity among the other Comforts of Life.

And now I speak of thanking God, I desire with all Humility to acknowledge, that I owe the mention'd Happiness of my past life to his kind Providence, which led me to the Means I us'd and gave them Success. My Belief of This, induces me to *hope*, tho' I must not *presume*, that the same Goodness will still be exercis'd towards me in continuing that Happiness, or in enabling me to bear a fatal Reverso,[2] which I may experience as others have done, the Com-

---

2. I.e., a backhanded stroke, a word used in dueling with rapiers.

plexion of my future Fortune being known to him only: and in whose Power it is to bless us even our Afflictions.

* * *

At Ten Years old, I was taken home to assist my Father in his Business, which was that of a Tallow Chandler and Sope-Boiler.[3] A Business he was not bred to, but had assumed on his Arrival in New England and on finding his Dying Trade would not maintain his Family, being in little Request. Accordingly I was employed in cutting Wick for the Candles, filling the Dipping Mold, and the Molds for cast Candles, attending the Shop, going of Errands, etc. I dislik'd the Trade and had a strong Inclination for the Sea; but my Father declar'd against it; however, living near the Water, I was much in and about it, learnt early to swim well, and to manage Boats, and when in a Boat or Canoe with other Boys I was commonly allow'd to govern,[4] especially in any case of Difficulty; and upon other Occasions I was generally a Leader among the Boys, and sometimes led them into Scrapes, of which I will mention one Instance, as it shows an early projecting public Spirit tho' not then justly conducted. There was a Salt Marsh that bounded part of the Mill Pond, on the Edge of which at Highwater, we us'd to stand to fish for Minews. By much Trampling, we had made it a mere Quagmire. My Proposal was to build a Wharf there fit for us to stand upon, and I show'd my Comrades a large Heap of Stones which were intended for a new House near the Marsh, and which would very well suit our Purpose. Accordingly in the Evening when the Workmen were gone, I assembled a Number of my Playfellows, and working with them diligently like so many Emmets,[5] sometimes two or three to a Stone, we brought them all away and built our little Wharff. The next Morning the Workmen were surpriz'd at Missing the Stones; which were found in our Wharff; Enquiry was made after the Removers; we were discovered and complain'd of; several of us were corrected by our Fathers; and tho' I pleaded the Usefulness of the Work, mine convinc'd me that nothing was useful which was not honest.

* * *

* * * I continu'd thus employ'd in my Father's Business for two Years, that is till I was 12 Years old; and my Brother John,[6] who was bred to that Business having left my Father, married and set up for himself at Rhodeisland, there was all Appearance that I was destin'd to supply his Place and be a Tallow Chandler. But my Dislike to the Trade continuing, my Father was under Apprehensions that if he did not find one for me more agreable, I should

---

3. Maker of candles and soap.
4. Steer.
5. Ants.

6. John Franklin (1690–1756), Franklin's favorite brother; he was to become Postmaster of Boston.

break away and get to Sea, as his Son Josiah had done to his great Vexation. He therefore sometimes took me to walk with him, and see Joiners, Bricklayers, Turners, Braziers,[7] etc. at their Work, that he might observe my Inclination, and endeavour to fix it on some Trade or other on Land. It has ever since been a Pleasure to me to see good Workmen handle their Tools; and it has been useful to me, having learnt so much by it, as to be able to do little Jobs my self in my House, when a Workman could not readily be got; and to construct little Machines for my Experiments while the Intention of making the Experiment was fresh and warm in my Mind. My Father at last fix'd upon the Cutler's Trade, and my Uncle Benjamin's Son Samuel who was bred to that Business in London being about that time established in Boston, I was sent to be with him some time on liking. But his Expectations of a Fee with me displeasing my Father, I was taken home again.

From a Child I was fond of Reading, and all the little Money that came into my Hands was ever laid out in Books. Pleas'd with the Pilgrim's Progress, my first Collection was of John Bunyan's[8] Works, in separate little Volumes. I afterwards sold them to enable me to buy R. Burton's[9] Historical Collections; they were small Chapmen's Books[1] and cheap, 40 or 50 in all. My Father's little Library consisted chiefly of Books in polemic Divinity, most of which I read, and have since often regretted, that at a time when I had such a Thirst for Knowledge, more proper Books had not fallen in my Way, since it was now resolv'd I should not be a Clergyman. Plutarch's Lives[2] there was, in which I read abundantly, and I still think that time spent to great Advantage. There was also a Book of Defoe's[3] called an Essay on Projects and another of Dr. Mather's[4] call'd Essays to do Good, which perhaps gave me a Turn of Thinking that had an Influence on some of the principal future Events of my Life.

This Bookish Inclination at length determin'd my Father to make me a Printer, tho' he had already one Son, (James) of that Profession. In 1717 my Brother James return'd from England with a Press and Letters[1] to set up his Business in Boston. I lik'd it much better than that of my Father, but still had a Hankering for the Sea. To prevent the apprehended Effect of such an Inclination, my Father was impatient to have me bound[2] to my Brother. I stood out some time, but at last was persuaded and signed the

---

7. Woodworkers, bricklayers, latheworkers, brassworkers.
8. John Bunyan (1628–88) published *Pilgrim's Progress* in 1678; his works were enormously popular and available in cheap one-shilling editions.
9. Burton was a pseudonym for Nathaniel Crouch (1632?–1725), a popularizer of British history.
1. Peddlers' books, hence inexpensive.

2. Plutarch (A.D. 46?–120?), Greek biographer who wrote *Parallel Lives* of noted Greek and Roman figures.
3. Daniel Defoe's *Essay on Projects* (1697) proposed remedies for economic improvement.
4. Cotton Mather published *Bonifacius: An Essay upon the Good* in 1710.
1. Type.
2. Apprenticed.

Indentures,[3] when I was yet but 12 Years old. I was to serve as an Apprentice till I was 21 Years of Age, only I was to be allow'd Journeyman's Wages[4] during the last Year. In a little time I made great Proficiency in the Business, and became a useful Hand to my Brother. I now had Access to better Books. An Acquaintance with the Apprentices of Booksellers, enabled me sometimes to borrow a small one, which I was careful to return soon and clean. Often I sat up in my Room reading the greatest Part of the Night, when the Book was borrow'd in the Evening and to be return'd early in the Morning lest it should be miss'd or wanted. And after some time an ingenious Tradesman[5] who had a pretty Collection of Books, and who frequented our Printing House, took Notice of me, invited me to his Library, and very kindly lent me such Books as I chose to read. I now took a Fancy to Poetry, and made some little Pieces. My Brother, thinking it might turn to account encourag'd me, and put me on composing two occasional Ballads. One was called the *Light House Tragedy*, and contain'd an Account of the drowning of Capt. Worthilake with his Two Daughters; the other was a Sailor Song on the Taking of *Teach* or Blackbeard the Pirate.[6] They were wretched Stuff, in the Grubstreet Ballad Stile,[7] and when they were printed he sent me about the Town to sell them. The first sold wonderfully, the Event being recent, having made a great Noise. This flatter'd my Vanity. But my Father discourag'd me, by ridiculing my Performances, and telling me Verse-makers were generally Beggars; so I escap'd being a Poet, most probably a very bad one. But as Prose Writing has been of great Use to me in the Course of my Life, and was a principal Means of my Advancement, I shall tell you how in such a Situation I acquir'd what little Ability I have in that Way.

There was another Bookish Lad in the Town, John Collins by Name, with whom I was intimately acquainted. We sometimes disputed, and very fond we were of Argument, and very desirous of confuting one another. Which disputacious Turn, by the way, is apt to become a very bad Habit, making People often extreamly disagreable in Company, by the Contradiction that is necessary to bring it into Practice, and thence, besides souring and spoiling the Conversation, is productive of Disgusts and perhaps Enmities where you may have occasion for Friendship. I had caught it by reading my Father's Books of Dispute about Religion. Persons of good Sense, I have since observ'd, seldom fall into it, except Lawyers, University

3. A contract binding him to work for his brother for nine years. James Franklin (1697–1735) had learned the printer's trade in England.
4. I.e., be paid for each day's work, having served his apprenticeship.
5. "Mr. Matthew Adams" [Franklin's note]. "Pretty": exceptionally fine.
6. The full texts of these ballads cannot be found; George Worthylake, lighthouse keeper on Beacon Island, Boston Harbor, and his wife and daughter were drowned on November 3, 1718; the pirate Blackbeard, Edward Teach, was killed off the Carolina coast on November 22, 1718.
7. Grub Street in London was inhabited by poor literary hacks who capitalized on poems of topical interest.

Men, and Men of all Sorts that have been bred at Edinborough.[8] A Question was once some how or other started between Collins and me, of the Propriety of educating the Female Sex in Learning, and their Abilities for Study. He was of Opinion that it was improper; and that they were naturally unequal to it. I took the contrary Side, perhaps a little for Dispute sake. He was naturally more eloquent, had a ready Plenty of Words, and sometimes as I thought bore me down more by his Fluency than by the Strength of his Reasons. As we parted without settling the Point, and were not to see one another again for some time, I sat down to put my Arguments in Writing, which I copied fair and sent to him. He answer'd and I reply'd. Three or four Letters of a Side had pass'd, when my Father happen'd to find my Papers, and read them. Without entring into the Discussion, he took occasion to talk to me about the Manner of my Writing, observ'd that tho' I had the Advantage of my Antagonist in correct Spelling and pointing[9] (which I ow'd to the Printing House) I fell far short in elegance of Expression, in Method and in Perspicuity, of which he convinc'd me by several Instances. I saw the Justice of his Remarks, and thence grew more attentive to the *Manner* in Writing, and determin'd to endeavour at Improvement.

About this time I met with an odd Volume of the Spectator.[1] I had never before seen any of them. I bought it, read it over and over, and was much delighted with it. I thought the Writing excellent, and wish'd if possible to imitate it. With that View, I took some of the Papers, and making short Hints of the Sentiment in each Sentence, laid them by a few Days, and then without looking at the Book, try'd to compleat the Papers again, by expressing each hinted Sentiment at length and as fully as it had been express'd before, in any suitable Words that should come to hand.

Then I compar'd my Spectator with the Original, discover'd some of my Faults and corrected them. But I found I wanted a Stock of Words or a Readiness in recollecting and using them, which I thought I should have acquir'd before that time, if I had gone on making Verses, since the continual Occasion for Words of the same Import but of different Length, to suit the Measure,[2] or of different Sound for the Rhyme, would have laid me under a constant Necessity of searching for Variety, and also have tended to fix that Variety in my Mind, and make me Master of it. Therefore I took some of the Tales and turn'd them into Verse: And after a time, when I had pretty well forgotten the Prose, turn'd them back again. I also sometimes jumbled my Collections of Hints into Confusion, and

---

8. Scottish Presbyterians were noted for their argumentative nature.
9. Punctuation. Spelling and punctuation were not standardized in this period.
1. An English periodical published daily from March 1, 1711, to December 6, 1712, and revived in 1714. It contained essays by Joseph Addison (1672–1719) and Richard Steele (1672–1729). It addressed itself primarily to matters of literature and morality.
2. Meter.

after some Weeks, endeavour'd to reduce them into the best Order, before I began to form the full Sentences, and compleat the Paper. This was to teach me Method in the Arrangement of Thoughts. By comparing my Work afterwards with the original, I discover'd many faults and amended them; but I sometimes had the Pleasure of Fancying that in certain Particulars of small Import, I had been lucky enough to improve the Method or the Language and this encourag'd me to think I might possibly in time come to be a tolerable English Writer, of which I was extreamly ambitious.

* * *

My Brother had in 1720 or 21, begun to print a Newspaper. It was the second[3] that appear'd in America, and was called *The New-England Courant*. The only one before it, was *the Boston News Letter*. I remember his being dissuaded by some of his Friends from the Undertaking, as not likely to succeed, one Newspaper being in their Judgment enough for America. At this time 1771 there are not less than five and twenty. He went on however with the Undertaking, and after having work'd in composing the Types and printing off the Sheets I was employ'd to carry the Papers thro' the Streets to the Customers. He had some ingenious Men among his Friends who amus'd themselves by writing little Pieces for this Paper, which gain'd it Credit, and made it more in Demand; and these Gentlemen often visited us. Hearing their Conversations, and their Accounts of the Approbation their Papers were receiv'd with, I was excited to try my Hand among them. But being still a Boy, and suspecting that my Brother would object to printing any Thing of mine in his Paper if he knew it to be mine, I contriv'd to disguise my Hand, and writing an anonymous Paper I put it in at Night under the Door of the Printing House.

It was found in the Morning and communicated to his Writing Friends when they call'd in as Usual. They read it, commented on it in my Hearing, and I had the exquisite Pleasure, of finding it met with their Approbation, and that in their different Guesses at the Author none were named but Men of some Character among us for Learning and Ingenuity. I suppose now that I was rather lucky in my Judges: And that perhaps they were not really so very good ones as I then esteem'd them. Encourag'd however by this, I wrote and convey'd in the same Way to the Press several more Papers,[4] which were equally approv'd, and I kept my Secret till my small Fund of Sense for such Performances was pretty well exhausted, and then I discovered[5] it; when I began to be considered a little more by my Brother's Acquaintance, and in a manner that did not quite please him, as he thought, probably with reason, that it tended to make me too vain. And perhaps this might be one Occasion of the Differ-

3. Actually the fifth; James Franklin's paper appeared on August 7, 1721.
4. *The Silence Dogood Letters* (April 12–October 8, 1722) were the earliest essay series in America.
5. Revealed.

ences that we began to have about this Time. Tho' a Brother, he considered himself as Master, and me as his Apprentice; and accordingly expected the same Services from me as he would from another; while I thought he demean'd me too much in some he requir'd of me, who from a Brother expected more Indulgence. Our Disputes were often brought before our Father, and I fancy I was either generally in the right, or else a better Pleader, because the Judgment was generally in my favour: But my Brother was passionate and had often beaten me, which I took extreamly amiss; and thinking my Apprenticeship very tedious, I was continually wishing for some Opportunity of shortening it, which at length offered in a manner unexpected.[6]

One of the Pieces in our News-Paper, on some political Point which I have now forgotten, gave Offence to the Assembly.[7] He was taken up, censur'd and imprison'd for a Month by the Speaker's Warrant, I suppose because he would not discover his Author. I too was taken up and examin'd before the Council; but tho' I did not give them any Satisfaction, they contented themselves with admonishing me, and dismiss'd me; considering me perhaps as an Apprentice who was bound to keep his Master's Secrets. During my Brother's Confinement, which I resented a good deal, notwithstanding our private Differences, I had the Management of the Paper, and I made bold to give our Rulers some Rubs[1] in it, which my Brother took very kindly, while others began to consider me in an unfavourable Light, as a young Genius that had a Turn for Libelling and Satyr.[2] My Brother's Discharge was accompany'd with an Order of the House, (a very odd one) *that James Franklin should no longer print the Paper called the New England Courant*. There was a Consultation held in our Printing House among his Friends what he should do in this Case. Some propos'd to evade the Order by changing the Name of the Paper; but my Brother seeing Inconveniences in that, it was finally concluded on as a better Way, to let it be printed for the future under the Name of *Benjamin Franklin*. And to avoid the Censure of the Assembly that might fall on him, as still printing it by his Apprentice, the Contrivance was, that my old Indenture should be return'd to me with a full Discharge on the Back of it, to be shown on Occasion; but to secure to him the Benefit of my Service I was to sign new Indentures for the Remainder of the Term, which were to be kept private. A very flimsy Scheme it was, but however it was *immediately* executed, and the Paper went

---

6. "I fancy his harsh and tyrannical Treatment of me, might be a means of impressing me with that Aversion to arbitrary Power that has stuck to me thro' my whole Life" [Franklin's note].
7. On June 11, 1722, the *Courant* hinted that there was collusion between local authorities and pirates raiding off Boston

Harbor. James Franklin was jailed from June 12 to July 7. "Assembly": Massachusetts legislative body; the lower house, elected by towns of the Massachusetts General Court.
1. Insults; annoyances.
2. Satire.

on accordingly under my Name for several Months.[3] At length a fresh Difference arising between my Brother and me, I took upon me to assert my Freedom, presuming that he would not venture to produce the new Indentures. It was not fair in me to take this Advantage, and this I therefore reckon one of the first Errata[4] of my Life: But the Unfairness of it weigh'd little with me, when under the *Impressions of* Resentment, for the Blows his Passion too often urg'd him to bestow upon me. Tho' He was otherwise not an ill-natur'd Man: Perhaps I was too saucy and provoking.

When he found I would leave him, he took care to prevent my getting Employment in any other Printing-House of the Town, by going round and speaking to every Master, who accordingly refus'd to give me Work. I then thought of going to New York as the nearest Place where there was a Printer: and I was the rather inclin'd to leave Boston, when I reflected that I had already made my self a little obnoxious to the governing Party; and from the arbitrary Proceedings of the Assembly in my Brother's Case it was likely I might if I stay'd soon bring my self into Scrapes; and farther that my indiscrete Disputations about Religion began to make me pointed at with Horror by good People, as an Infidel or Atheist; I determin'd on the Point; but my Father now siding with my brother, I was sensible that if I attempted to go openly, Means would be used to prevent me. My Friend Collins therefore undertook to manage a little for me. He agreed with the Captain of a New York Sloop for my Passage, under the Notion of my being a young Acquaintance of his that had got a naughty Girl with Child, whose Friends would compel me to marry her, and therefore I could not appear or come away publickly. So I sold some of my Books to raise a little Money, was taken on board privately, and as we had a fair Wind, in three Days I found my self in New York near 300 Miles from home, a Boy of but 17, without the least Recommendation to or Knowledge of any Person in the Place, and with very little Money in my Pocket.

My Inclinations for the Sea, were by this time worne out, or I might now have gratify'd them. But having a Trade, and supposing my self a pretty good Workman, I offer'd my Service to the Printer of the Place, old Mr. William Bradford.[5] He could give me no Employment, having little to do, and Help enough already: But, says he, my Son at Philadelphia has lately lost his principal Hand, Aquila Rose, by Death. If you go thither I believe he may employ you. Philadelphia was 100 Miles farther. I set out, however, in a Boat for Amboy;[6] leaving my Chest and Things to follow me round by Sea. ⁕ ⁕ ⁕

3. The paper continued under Franklin's name until 1726, nearly three years after he left Boston.
4. Printer's term for errors.
5. William Bradford (1663–1752), one of the first American printers and father of Andrew Bradford (1686–1742), Franklin's future competitor in Philadelphia.
6. Perth Amboy, New Jersey.

I have been the more particular in this Description of my Journey, and shall be so of my first Entry into that City, that you may in your Mind compare such unlikely Beginning with the Figure I have since made there. I was in my working Dress, my best Cloaths being to come round by Sea. I was dirty from my Journey; my Pockets were stuff'd out with Shirts and Stockings; I knew no Soul, nor where to look for Lodging. I was fatigu'd with Travelling, Rowing and Want of Rest. I was very hungry, and my whole Stock of Cash consisted of a Dutch Dollar and about a Shilling in Copper. The latter I gave the People of the Boat for my Passage, who at first refus'd it on Account of my Rowing; but I insisted on their taking it, a Man being sometimes more generous when he has but a little Money than when he has plenty, perhaps thro' Fear of being thought to have but little. Then I walk'd up the Street, gazing about, till near the Market House I met a Boy with Bread. I had made many a Meal on Bread, and inquiring where he got it, I went immediately to the Baker's he directed me to in second Street; and ask'd for Bisket, intending such as we had in Boston, but they it seems were not made in Philadelphia, then I ask'd for a three-penny Loaf, and was told they had none such: so not considering or knowing the Difference of Money and the greater Cheapness nor the Names of his Bread, I bad him give me three pennyworth of any sort. He gave me accordingly three great Puffy Rolls; I was surpriz'd at the Quantity, but took it, and having no Room in my Pockets, walk'd off, with a Roll under each Arm, and eating the other. Thus I went up Market Street as far as fourth Street, passing by the Door of Mr. Read, my future Wife's Father, when she standing at the Door saw me, and thought I made as I certainly did a most awkward ridiculous Appearance. Then I turn'd and went down Chestnut Street and part of Walnut Street, eating my Roll all the Way, and coming round found my self again at Market street Wharff, near the Boat I came in, to which I went for a Draught of the River Water, and being fill'd with one of my Rolls, gave the other two to a Woman and her Child that came down the River in the Boat with us and were waiting to go farther. Thus refresh'd I walk'd again up the Street, which by this time had many clean dress'd People in it who were all walking the same Way; I join'd them, and thereby was led into the great Meeting House of the Quakers near the Market. I sat down among them, and after *looking round a while and* hearing nothing said, being very drowzy thro' Labour and want of Rest the preceding Night, I fell fast asleep, and continu'd so till the Meeting broke up, when one was kind enough to rouse me. This was therefore the first House I was in or slept in, in Philadelphia.

Walking again down towards the River, and looking in the Faces of People, I met a young Quaker Man whose Countenance I lik'd, and accosting him requested he would tell me where a Stranger could get Lodging. We were then near the Sign of the Three Mariners. Here, says he, is one Place that entertains Strangers, but it is not a reputable House; if thee wilt walk with me, I'll show thee a better. He brought me to the Crooked Billet in Water-Street. Here I got a Dinner. And while I was eating it, several sly Questions were ask'd me, as it seem'd to be suspected from my youth and Appearance, that I might be some Runaway. After Dinner my Sleepiness return'd: and being shown to a Bed, I lay down without undressing, and slept till Six in the Evening; was call'd to Supper; went to Bed again very early and slept soundly till the next Morning. Then I made my self as tidy as I could, and went to Andrew Bradford the Printer's. I found in the Shop the old Man his Father, whom I had seen at New York, and who travelling on horse back had got to Philadelphia before me. He introduc'd me to his Son, who receiv'd me civilly, gave me a Breakfast, but told me he did not at present want a Hand, being lately supply'd with one. But there was another Printer in town lately set up, one Keimer,[1] who perhaps might employ me; if not, I should be welcome to lodge at his House, and he would give me a little Work to do now and then till fuller Business should offer.

The old Gentleman said, he would go with me to the new Printer: And when we found him, Neighbour, says Bradford, I have brought to see you a young Man of your Business, perhaps you may want such a One. He ask'd me a few Questions, put a Composing Stick[2] in my Hand to see how I work'd, and then said he would employ me soon, tho' he had just then nothing for me to do. And taking old Bradford whom he had never seen before, to be one of the Towns People that had a Good Will for him, enter'd into a Conversation on his present Undertaking and Prospects; while Bradford not discovering that he was the other Printer's Father; on Keimer's Saying he expected soon to get the greatest Part of the Business into his own Hands, drew him on by artful Questions and starting little Doubts, to explain all his Views, what Interest he rely'd on, and in what manner he intended to proceed. I who stood by and heard all, saw immediately that one of them was a crafty old Sophister,[3] and the other a mere Novice. Bradford left me with Keimer, who was greatly surpriz'd when I told him who the old Man was.

Keimer's Printing House I found, consisted of an old shatter'd Press, and one small worn-out Fount of English,[4] which he was then using himself, composing in it an Elegy on Aquila Rose[5] before-mentioned, an ingenious young Man of excellent Character

1. Samuel Keimer (c. 1688–1742) was a *printer in London before coming to* Philadelphia.
2. An instrument of adjustable width in which type is set before being put on a galley.
3. Trickster, rationalizer.
4. An oversized type, not practicable for books and newspapers.
5. Aquila Rose (c. 1695–1723), journeyman printer for Andrew Bradford; his son Joseph apprenticed with Franklin.

much respected in the Town, Clerk of the Assembly,[6] and a pretty Poet. Keimer made Verses, too, but very indifferently. He could not be said to write them, for his Manner was to compose them in the Types directly out of his Head; so there being no Copy, but one Pair of Cases,[7] and the Elegy likely to require all the Letter, no one could help him. I endeavour'd to put his Press (which he had not yet us'd, and of which he understood nothing) into Order fit to be work'd with; and promising to come and print off his Elegy as soon as he should have got it ready, I return'd to Bradford's who gave me a little Job to do for the present, and there I lodg'd and dieted.[8] A few Days after Keimer sent for me to print off the Elegy. And now he had got another Pair of Cases, and a Pamphlet to reprint, on which he set me to work.

These two Printers I found poorly qualified for their Business. Bradford had not been bred to it, and was very illiterate; and Keimer tho' something of a Scholar, was a mere Compositor, knowing nothing of Presswork. He had been one of the French Prophets[9] and could act their enthusiastic Agitations. At this time he did not profess any particular Religion, but something of all on occasion; was very ignorant of the World, and had, as I afterwards found, a good deal of the Knave in his Composition. He did not like my Lodging at Bradford's while I work'd with him. He had a House indeed, but without Furniture, so he could not lodge me: But he got me a Lodging at Mr. Read's before-mentioned, who was the Owner of his House. And my Chest and Clothes being come by this time, I made rather a more respectable Appearance in the Eyes of Miss Read, than I had done when she first happen'd to see me eating my Roll in the Street.

I began now to have some Acquaintance among the young People of the Town, that were Lovers of Reading with whom I spent my Evenings very pleasantly and gaining Money by my Industry and Frugality, I lived very agreably, forgetting Boston as much as I could, and not desiring that any there should know where I resided except my Friend Collins who was in my Secret, and kept it when I wrote to him. At length an Incident happened that sent me back again much sooner than I had intended.

I had a Brother-in-law, Robert Holmes,[1] Master of a Sloop, that traded between Boston and Delaware. He being at NewCastle 40 Miles below Philadelphia, heard there of me, and wrote me a Letter, mentioning the Concern of my Friends in Boston at my abrupt Departure, assuring me of their Goodwill to me, and that every thing would be accommodated to my Mind if I would return,

---

6. One who has charge of the records, documents, and correspondence of any organized body; here, the Pennsylvania legislative council.
7. Two shallow trays which contain upper-case and lower-case type.

8. Boarded.
9. An English sect which preached doomsday and cultivated emotional fits.
1. Robert Homes (d. before 1743), husband of Franklin's sister Mary, and a ship's captain.

to which he exhorted me very earnestly. I wrote an Answer to his Letter, thank'd him for his Advice, but stated my Reason for quitting Boston fully, and in such a Light as to convince him I was not so wrong as he had apprehended. Sir William Keith[2] Governor of the Province, was then at New Castle, and Captain Holmes happening to be in Company with him when my Letter came to hand, spoke to him of me, and show'd him the Letter. The Governor read it, and seem'd surpriz'd when he was told my Age. He said I appear'd a young Man of promising Parts, and therefore should be encouraged: The Printers at Philadelphia were wretched ones, and if I would set up there, he made no doubt I should succeed; for his Part, he would procure me the publick Business, and do me every other Service in his Power. This my Brother-in-Law afterwards told me in Boston. But I knew as yet nothing of it; when one Day Keimer and I being at Work together near the Window, we saw the Governor and other Gentleman (which prov'd to be Colonel French, of New Castle) finely dress'd, come directly across the Street to our House, and heard them at the Door.

Keimer ran down immediately, thinking it a Visit to him. But the Governor enquir'd for me, came up, and with a Condescension and Politeness I had been quite unus'd to, made me many Compliments, desired to be acquainted with me, blam'd me kindly for not having made my self known to him when I first came to the Place, and would have me away with him to the Tavern where he was going with Colonel French to taste as he said some excellent Madeira. I was not a little surpriz'd, and Keimer star'd like a Pig poison'd. I went however with the Governor and Colonel French, to a Tavern the Corner of Third Street, and over the Madeira he propos'd my Setting up my Business, laid before me the Probabilities of Success, and both he and Colonel French assur'd me I should have their Interest and Influence in procuring the Publick-Business of both Governments. On my doubting whether my Father would assist me in it, Sir William said he would give me a Letter to him, in which he would state the Advantages, and he did not doubt of prevailing with him. So it was concluded I should return to Boston in the first Vessel with the Governor's Letter recommending me to my Father.

In the mean time the Intention was to be kept secret, and I went on working with Keimer as usual, the Governor sending for me now and then to dine with him, a very great Honour I thought it, and conversing with me in the most affable, familiar, and friendly manner imaginable. About the End of April 1724, a little Vessel offer'd for Boston. I took Leave of Keimer as going to see my Friends. The Governor gave me an ample Letter, saying many flattering things of me to my Father, and strongly recommending the Project of my setting up at Philadelphia, as a Thing that must

2. Sir William Keith (1680–1749), Governor of Pennsylvania 1717–26; he fled to England in 1728 to escape debtor's prison.

make my Fortune. We struck on a Shoal in going down the Bay
and sprung a Leak, we had a blustring time at Sea, and were oblig'd
to pump almost continually, at which I took my Turn. We arriv'd
safe however at Boston in about a Fortnight. I had been absent
Seven Months and my Friends had heard nothing of me, for my
Brother Holmes was not yet return'd; and had not written about
me. My unexpected Appearance surpriz'd the Family; all were however very glad to see me and made me Welcome, except my
Brother.

I went to see him at his Printing-House: I was better dress'd than
ever while in his Service, having a genteel new Suit from Head to
foot, a Watch, and my Pockets lin'd with near Five Pounds Sterling
in Silver. He receiv'd me not very frankly, look'd me all over, and
turn'd to his Work again. The Journey-Men were inquisitive where
I had been, what sort of a Country it was, and how I lik'd it? I
prais'd it much, and the happy life I led in it; expressing strongly
my Intention of returning to it; and one of them asking what kind
of Money we had there, I produc'd a handful of Silver and spread it
before them, which was a kind of Raree-Show[3] they had not been
us'd to, Paper being the Money of Boston. Then I took an Opportunity of letting them see my Watch: and lastly, (my Brother still
grum and sullen) I gave them a Piece of Eight to drink[4] and took
my Leave. This Visit of mine offended him extreamly. For when
my Mother some time after spoke to him of a Reconciliation, and
of her Wishes to see us on good Terms together, and that we might
live for the future as Brothers, he said, I had insulted him in such a
Manner before his People that he could never forget or forgive it.
In this however he was mistaken.

My Father receiv'd the Governor's Letter with some apparent
Surprize; but said little of it to me for some Days; when Captain
Holmes returning, he show'd it to him, ask'd if he knew Keith, and
what kind of a Man he was: Adding his Opinion that he must be
of small Discretion, to think of setting a Boy up in Business who
wanted yet 3 Years of being at Man's Estate. Holmes said what he
could in favor of the Project; but my Father was clear in the Impropriety of it; and at last gave a flat Denial to it. Then he wrote a
civil Letter to Sir William thanking him for the Patronage he had
so kindly offered me, but declining to assist me as yet in Setting up,
I being in his Opinion too young to be trusted with the Management of a Business so important; and for which the Preparation
must be so expensive.

* * *

Keimer and I liv'd on a pretty good familiar Footing and agreed
tolerably well: for he suspected nothing of my Setting up. He

---

3. A sidewalk peep show; silver coins
were rare in the colonies.

4. A Spanish dollar for them to buy
drinks with.

retain'd a great deal of his old Enthusiasms, and lov'd an Argumentation. We therefore had many Disputations. I us'd to work him so with my Socratic Method, and had trapann'd[5] him so often by Questions apparently so distant from any Point we had in hand, and yet by degrees led to the Point, and brought him into Difficulties and Contradictions, that at last he grew ridiculously cautious, and would hardly answer me the most common Question, without asking first, *What do you intend to infer from that?* However it gave him so high an Opinion of my Abilities in the Confuting Way, that he seriously propos'd my being his Colleague in a Project he had of setting up a new Sect. He was to preach the Doctrines, and I was to confound all Opponents. When he came to explain with me upon the Doctrines, I found several Conundrums[6] which I objected to, unless I might have my Way a little too, and introduce some of mine. Keimer wore his Beard at full length, because somewhere in the Mosaic Law it is said, *thou shalt not mar the Corners of thy Beard.*[7] He likewise kept the seventhday Sabbath; and these two Points were Essentials with him. I dislik'd both, but agreed to admit them upon Condition of his adopting the Doctrine of using no animal Food. I doubt, says he, my Constitution will not bear that. I assur'd him it would, and that he would be the better for it. He was usually a great Glutton, and I promis'd my self some Diversion in half-starving him. He agreed to try the Practice if I would keep him Company. I did so and we held it for three Months. We had our Victuals dress'd and brought to us regularly by a Woman in the Neighborhood, who had from me a List of 40 Dishes to be prepar'd for us at different times, in all which there was neither Fish Flesh nor Fowl, and the Whim suited me the better at this time from the Cheapness of it, not costing above 18 Pence Sterling each, per Week. I have since kept several Lents most strictly, Leaving the common Diet for that, and that for the common, abruptly, without the least Inconvenience: So that I think there is little in the Advice of making those Changes by easy Gradations. I went on pleasantly, but Poor Keimer suffer'd grievously, tir'd of the Project, long'd for the Flesh Pots of Egypt,[1] and order'd a roast Pig; He invited me and two Women Friends to dine with him, but it being brought too soon upon Table, he could not resist the Temptation, and ate it all up before we came.

I had made some Courtship during this time to Miss Read. I had a great Respect and Affection for her, and had some Reason to believe she had the same for me: but as I was about to take a long

5. Trapped.
6. Puzzles; difficult questions.
7. "Ye shall not round the corners of your heads, neither shalt thou mar the corners of thy beard" (Leviticus 19.27). Keimer probably also wore his hair long.
1. "And the whole congregation of the children of Israel murmured against Moses and Aaron in the wilderness: And the children of Israel said unto them, Would to God we had died by the hand of the Lord in the land of Egypt, when we sat by the flesh pots, and when we did eat bread to the full" (Exodus 16.2–3).

Voyage, and we were both very young, only a little above 18, it was thought most prudent by her Mother to prevent our going too far at present, as a Marriage if it was to take place would be more convenient after my Return, when I should be as I expected set up in my Business. Perhaps too she thought my Expectations not so well founded as I imagined them to be.

My chief Acquaintances at this time were, Charles Osborne, Joseph Watson, and James Ralph;[2] All Lovers of Reading. The two first were Clerks to an eminent Scrivener or Conveyancer[3] in the Town, Charles Brogden; the other was Clerk to a Merchant. Watson was a pious sensible young Man, of great Integrity. The others rather more lax in their Principles of Religion, particularly Ralph, who as well as Collins had been unsettled by me, for which they both made me suffer. Osborne was sensible, candid, frank, sincere, and affectionate to his Friends; but in litterary matters too fond of Criticising. Ralph, was ingenious, genteel in his Manners, and extremely eloquent; I think I never knew a prettier Talker. Both of them great Admirers of Poetry, and began to try their Hands in little Pieces. Many pleasant Walks we four had together, on Sundays into the Woods near Skuylkill,[4] where we read to one another and conferr'd on what we read. Ralph was inclin'd to pursue the Study of Poetry, not doubting but he might become eminent in it and make his Fortune by it, alledging that the best Poets must when they first began to write, make as many Faults as he did. Osborne dissuaded him, assur'd him he had no Genius for Poetry, and advis'd him to think of nothing beyond the Business he was bred to; that in the mercantile way tho' he had no Stock, he might by his Diligence and Punctuality recommend himself to Employment as a Factor,[5] and in time acquire wherewith to trade on his own Account. I approv'd the amusing one's Self with Poetry now and then, so far as to improve one's Language, but no farther. On this it was propos'd that we should each of us at our next Meeting produce a Piece of our own Composing, in order to improve by our mutual Observations, Criticisms and Corrections. As Language and Expression was what we had in View, we excluded all Considerations of Invention,[6] by agreeing that the Task should be a Version of the 18th Palsm, which describes the Descent of a Diety.[7] When the Time of our Meeting drew nigh, Ralph call'd on me first, and let me know his Piece was ready. I told him I had been busy, and having little Inclination had done nothing. He then show'd me his Piece for my Opinion; and I much aprov'd it, as it appear'd to me to have great Merit. Now, says he, Osborne will never allow the

2. Charles Osborne's dates are unknown; Joseph Watson died about 1728; James Ralph (c. 1695–1762) became well known as a political journalist.
3. One who draws up leases and deeds; Charles Brockden (1683–1769) came to Philadelphia in 1706.
4. Schuylkill River, at Philadelphia.
5. Business agent.
6. I.e., originality.
7. "He bowed the heavens also, and came down: and darkness was under his feet" (Psalm 18.9).

least Merit in any thing of mine, but makes 1000 Criticisms out of mere Envy. He is not so jealous of you. I wish therefore you would take this Piece, and produce it as yours. I will pretend not to have had time, and so produce nothing: We shall then see what he will say to it. It was agreed, and I immediately transcrib'd it that it might appear in my own hand. We met.

Watson's Performance was read: there were some Beauties in it: but many Defects. Osborne's was read: It was better. Ralph did it Justice, remark'd some Faults, but applauded the Beauties. He himself had nothing to produce. I was backward, seem'd desirous of being excus'd, had not had sufficient Time to correct; etc. but no Excuse could be admitted, produce I must. It was read and repeated; Watson and Osborne gave up the Contest; and join'd in applauding it immoderately. Ralph only made some Criticisms and propos'd some Amendments, but I defended my Text. Osborne was against Ralph, and told him he was no better a Critic than Poet; so he dropt the Argument. As they two went home together, Osborne express'd himself still more strongly in favour of what he thought my Production, having restrain'd himself before as he said, lest I should think it Flattery. But who would have imagin'd, says he, that Franklin had been capable of such a Performance; such Painting, such Force! such Fire! He has even improv'd the Original! In his common Conversation, he seems to have no Choice of Words; he hesitates and blunders; and yet, good God, how he writes!

When we next met, Ralph discover'd the Trick, we had plaid him, and Osborne was a little laught at. This Transaction fix'd Ralph in his Resolution of becoming a Poet. I did all I could to dissuade him from it, but He continu'd scribbling Verses, till *Pope*[8] cur'd him. He became however a pretty good Prose Writer. More of him hereafter. But as I may not have occasion again to mention the other two, I shall just remark here, that Watson died in my Arms a few Years after, much lamented, being the best of our Set. Osborne went to the West Indies, where he became an eminent Lawyer and made Money, but died young. He and I had made a serious Agreement, that the one who happen'd first to die, should if possible make a friendly Visit to the other, and acquaint him how he found things in that separate State. But he never fulfill'd his Pomise.

The Governor, seeming to like my Company, had me frequently to his House; and his Setting me up was always mention'd as a fix'd thing. I was to take with me Letters recommendatory to a Number of his Friends, besides the Letter of Credit to furnish me with the necessary Money for purchasing the Press and Types, Paper, etc.

8. In the second edition of the *Dunciad* (1728), a poem which attacks ignorance of all kinds, Alexander Pope responded to Ralph's slur against him in *Sawney*: "Silence, ye Wolves: while Ralph to Cynthia howls, / And makes Night hideous—Answer him ye Owls" (Book 3, lines 159–60). In the 1742 edition Pope included another dig at Ralph: "And see: The very Gazeteers give o'er, / Ev'n Ralph repents" (Book 1, lines 215–16).

For these Letters I was appointed to call at different times, when they were to be ready, but a future time was still[9] named. Thus we went on till the Ship whose Departure too had been several times postponed was on the Point of sailing. Then when I call'd to take my Leave and receive the Letters, his Secretary, Dr. Bard,[1] came out to me and said the Governor was extreamly busy, in writing, but would be down at Newcastle[2] before the Ship, and there the Letters would be delivered to me.

Ralph, tho' married and having one Child, had determined to accompany me in this Voyage. It was thought he intended to establish a Correspondence, and obtain Goods to sell on Commission. But I found afterwards, that thro' some Discontent with his Wifes Relations, he purposed to leave her on their Hands, and never return again. Having taken leave of my Friends, and interchang'd some Promises with Miss Read, I left Philadelphia in the Ship, which anchor'd at Newcastle. The Governor was there. But when I went to his Lodging, the Secretary came to me from him with the civillest Message in the World, that he could not then see me being engag'd in Business of the utmost Importance; but should send the Letters to me on board, wish'd me heartily a good Voyage and a speedy Return, etc. I return'd on board, a little puzzled, but still not doubting.

Mr. Andrew Hamilton,[3] a famous Lawyer of Philadelphia, had taken Passage in the same Ship for himself and Son: and with Mr. Denham[4] a Quaker Merchant, and Messrs. Onion and Russel Masters of an Iron Work in Maryland, had engag'd the Great Cabin; so that Ralph and I were forc'd to take up with a Birth in the Steerage: And none on board knowing us, were considered as ordinary Persons. But Mr. Hamilton and his Son (it was James, since Governor) return'd from New Castle to Philadelphia, the Father being recall'd by a great Fee to plead for a seized Ship. And just before we sail'd Colonel French coming on board, and showing me great Respect, I was more taken Notice of, and with my Friend Ralph invited by the other Gentlemen to come into the Cabin, there being now Room. Accordingly we remov'd thither.

Understanding that Colonel French had brought on board the Governor's Dispatches, I ask'd the Captain for those Letters that were to be under my Care. He said all were put into the Bag together; and he could not then come at them; but before we landed in England, I should have an Opportunity of picking them out. So I was satisfy'd for the present, and we proceeded on our Voyage. We had a sociable Company in the Cabin, and lived

9. Always.
1. Patrick Bard or Baird was resident in Philadelphia as Port Physician after 1720.
2. Delaware.
3. Andrew Hamilton (c. 1678–1741); his son James Hamilton (c. 1710–83) was Governor of Pennsylvania four times between 1748 and 1773.
4. Thomas Denham (d. 1728), merchant and benefactor, left Bristol, England, in 1715.

uncommonly well, having the Addition of all Mr. Hamilton's Stores, who had laid in plentifully. In this Passage Mr. Denham contracted a Friendship for me that continued during his Life. The Voyage was otherwise not a pleasant one, as we had a great deal of bad Weather.

When we came into the Channel, the Captain kept his Word with me, and gave me an Opportunity of examining the Bag for the Governor's Letters. I found none upon which my Name was put, as under my Care; I pick'd out 6 or 7 that by the Handwriting I thought might be the promis'd Letters, especially as one of them was directed to Basket[5] the King's Printer, and another to some Stationer. We arrived in London the 24th of December, 1724. I waited upon the Stationer who came first in my Way, delivering the Letter as from Governor Keith. I don't know such a Person, says he: but opening the Letter, O, this is from Riddlesden;[6] I have lately found him to be a compleat Rascal, and I will have nothing to do with him, nor receive any Letters from him. So putting the Letter into my Hand, he turn'd on his Heel and left me to serve some Customer. I was surprized to find these were not the Governor's Letters. And after recollecting and comparing Circumstances, I began to doubt his Sincerity. I found my Friend Denham, and opened the whole Affair to him. He let me into Keith's Character, told me there was not the least Probability that he had written any Letters for me, that no one who knew him had the smallest Dependence on him, and he laught at the Notion of the Governor's giving me a Letter of Credit, having as he said no Credit to give. On my expressing some Concern about what I should do: He advis'd me to endeavour getting some Employment in the Way of my Business. Among the Printers here, says he, you will improve yourself; and when you return to America, you will set up to greater Advantage.

* * *

Thus I spent about 18 Months in London. Most Part of the Time, I work'd hard at my Business, and spent but little upon my self except in seeing Plays, and in Books. My Friend Ralph had kept me poor. He owed me about 27 Pounds; which I was now never likely to receive; a great Sum out of my small Earnings. I lov'd him notwithstanding, for he had many amiable Qualities Tho' I had by no means improv'd my Fortune. But I had pick'd up some very ingenious Acquaintance whose Conversation was of great Advantage to me, and I had read considerably.

We sail'd from Gravesend on the 23d of July 1726. For The Incidents of the Voyage, I refer you to my Journal, where you will find them all minutely related. Perhaps the most important Part of

5. John Baskett (d. 1742).
6. William Riddlesden (d. before 1733), well known in Maryland as a man of "infamy."

that Journal is the *Plan*[7] to be found in it which I formed at Sea, for regulating my future Conduct in Life. It is the more remarkable, as being form'd when I was so young, and yet being pretty faithfully adhered to quite thro' to old Age. We landed in Philadelphia the 11th of October, where I found sundry Alterations. Keith was no longer Governor, being superceded by Major Gordon:[8] I met him walking the Streets as a common Citizen. He seem'd a little asham'd at seeing me, but pass'd without saying any thing. I should have been as much asham'd at seeing Miss Read, had not her Friends despairing with Reason of my Return, after the Receipt of my Letter, persuaded her to marry another, one Rogers, a Potter, which was done in my Absence. With him however she was never happy, and soon parted from him, refusing to cohabit with him, or bear his Name, It being now said that he had another Wife. He was a worthless Fellow tho' an excellent Workman which was the Temptation to her Friends. He got into Debt, and ran away in 1727 or 28, went to the West Indies, and died there. Keimer had got a better House, a Shop well supply'd with Stationary, plenty of new Types, a number of Hands tho' none good, and seem'd to have a great deal of Business.

Mr. Denham took a Store in Water Street, where we open'd our Goods. I attended the Business diligently, studied Accounts, and grew in a little Time expert at selling. We lodg'd and boarded together, he counsell'd me as a Father, having a sincere Regard for me: I respected and lov'd him: and we might have gone on together very happily: But in the Beginning of February 1726/7 when I had just pass'd my 21st Year, we both were taken ill. My Distemper was a Pleurisy,[1] which very nearly carried me off: I suffered a good deal, gave up the Point[2] in my own mind, and was rather disappointed when I found my self recovering; regretting in some degree that I must now sometime or other have all that disagreable Work to do over again. I forget what his Distemper was. It held him a long time, and at length carried him off. He left me a small Legacy in a nuncupative Will,[3] as a Token of his Kindness for me, and he left me once more to the wide World. For the Store was taken into the Care of his Executors, and my Employment under him ended: My Brother-in-law Homes, being now at Philadelphia, advis'd my Return to my Business. And Keimer tempted me with an Offer of large Wages by the Year to come and take the Management of his Printing-House that he might better attend his Stationer's Shop. I had heard a bad Character of him in London, from his Wife and her Friends, and was not fond of having any more to do with him. I try'd for farther Employment as a Mer-

7. Only the "Outline" and "Preamble" of Franklin's *"Plan"* survive.
8. Patrick Gordon (1644–1736), Governor of Pennsylvania from 1726 to 1736.

1. A disease of the lungs.
2. End; i.e., resigned himself to death.
3. An oral will.

chant's Clerk; but not readily meeting with any, I clos'd again with Keimer.

*   *   *

Before I enter upon my public Appearance in Business, it may be well to let you know the then State of my Mind, with regard to my Principles and Morals, that you may see how far those influenc'd the Future Events of my Life. My Parents had early given me religious Impressions, and brought me through my Childhood piously in the Dissenting Way.[4] But I was scarce 15 when, after doubting by turns of several Points as I found them disputed in the different Books I read, I began to doubt of Revelation it self. Some Books against Deism fell into my Hands; they were said to be the Substance of Sermons preached at Boyle's Lectures.[5] It happened that they wrought an Effect on me quite contrary to what was intended by them: For the Arguments of the Deists which were quoted to be refuted, appeared to me much Stronger than the Refutations. In short I soon became a thorough Deist. My Arguments perverted some others, particularly Collins and Ralph: but each of them having afterwards wrong'd me greatly without the least Compunction, and recollecting Keith's Conduct towards me, (who was another Freethinker) and my own towards Vernon and Miss Read which at Times gave me great Trouble, I began to suspect that this Doctrine tho' it might be true, was not very useful. My London Pamphlet, which had for its Motto those Lines of Dryden

> *Whatever is, is right—*
> *Tho' purblind Man/ Sees but a Part of*
> *the Chain, the nearest Link,*
> *His Eyes not carrying to the equal Beam,*
> *That poizes all, above.*[6]

And from the Attributes of God, his infinite Wisdom, Goodness and Power concluded that nothing could possibly be wrong in the World, and that Vice and Virtue were empty Distinctions, no such Things existing: appear'd now not so clever a Performance as I once thought it; and I doubted whether some Error had not insinuated itself unperceiv'd into my Argument, so as to infect all that follow'd, as is common in metaphysical Reasonings. I grew convinc'd that *Truth, Sincerity* and *Integrity* in Dealings between Man and Man, were of the utmost Importance to the Felicity of Life,

4. I.e., the Protestant way.
5. Robert Boyle (1627–91), English physicist and chemist, endowed annual lectures for preaching eight sermons a year against "infidels." Deism accepts a Supreme Being as the author of finite existence, but denies Christian doctrines of revelation and supernaturalism.

6. The first line is not from John Dryden (1631–1700) but from Alexander Pope's *Essay on Man* (1733), Epistle I, line 294, but Dryden's line is close: "Whatever is, is in its Causes just." The rest of the poem is recalled accurately from Dryden's *Oedipus*, III.i.244–48.

and I form'd written Resolutions, (which still remain in my Journal Book) to practise them ever while I lived. Revelation had indeed no weight with me as such; but I entertain'd an Opinion, that tho' certain Actions might not be bad *because* they were forbidden by it, or good *because* it commanded them; yet probably those Actions might be forbidden *because* they were bad for us, or commanded *because* they were beneficial to us, in their own Natures, all the Circumstances of things considered. And this Persuasion, with the kind hand of Providence, or some guardian angel, or accidental favourable Circumstances and Situations, or all together, perserved me (thro' this dangerous Time of Youth and the hazardous Situations I was sometimes in among Strangers, remote from the Eye and Advice of my Father), without any *wilful* gross Immorality or Injustice that might have been expected from my Want of Religion. I say *wilful*, because the Instances I have mentioned, had something of *Necessity* in them, from my Youth, Inexperience, and the Knavery of others. I had therefore a tolerable Character to begin the World with, I valued it properly, and determin'd to preserve it.

* * *

CONTINUATION OF THE ACCOUNT OF MY LIFE BEGUN AT PASSY 1784

It is some time since I receiv'd the above Letters,[7] but I have been too busy till now to think of complying with the Request they contain. It might too be much better done if I were at home among my Papers, which would aid my Memory, and help to ascertain Dates. But my Return being uncertain, and having just now a little Leisure, I will endeavour to recollect and write what I can; if I live to get home, it may there be corrected and improv'd.

Not having any Copy here of what is already written, I know not whether an Account is given of the means I used to establish the Philadelphia publick Library, which from a small Beginning is now become so considerable, though I remember to have come down to near the Time of that Transaction, 1730. I will therefore begin here, with an Account of it, which may be struck out if found to have been already given.

At the time I establish'd my self in Pensylvania, there was not a good Bookseller's Shop in any of the Colonies to the Southward of Boston. In New-York and Philadelphia the Printers were indeed Stationers, they sold only Paper, etc., Almanacks, Ballads, and a few common School Books. Those who lov'd Reading were oblig'd to send for their Books from England. The Members of the Junto had each a few. We had left the Alehouse where we first met, and hired a Room to hold our Club in. I propos'd that we should all of us bring our Books to that Room, where they would not only be ready to consult in our Conferences, but become a common Benefit, each

---

7. The second part of *The Autobiography* begins with letters from two friends, James Abel and Benjamin Vaughan, urging Franklin to continue his memoirs.

of us being at Liberty to borrow such as he wish'd to read at home. This was accordingly done, and for some time contented us. Finding the Advantage of this little Collection, I propos'd to render the Benefit from Books more common by commencing a Public Subscription Library. I drew a Sketch of the Plan and Rules that would be necessary, and got a skilful Conveyancer Mr. Charles Brockden[8] to put the whole in Form of Articles of Agreement to be subscribed, by which each Subscriber engag'd to pay a certain Sum down for the first Purchase of Books and an annual Contribution for encreasing them. So few were the Readers at that time in Philadelphia, and the Majority of us so poor, that I was not able with great Industry to find more than Fifty Persons, mostly young Tradesmen, willing to pay down for this purpose Forty shillings each, and Ten Shillings per Annum. On this little Fund we began. The Books were imported. The Library was open one Day in the Week for lending them to the Subscribers, on their Promisory Notes to pay Double the Value if not duly returned. The Institution soon manifested its Utility, was imitated by other Towns and in other Provinces, the Librarys were augmented by Donations, Reading became fashionable, and our People having no publick Amusements to divert their Attention from Study became better acquainted with Books, and in a few Years were observ'd by Strangers to be better instructed and more intelligent than People of the same Rank generally are in other Countries.

When we were about to sign the above-mentioned Articles, which were to be binding on us, our Heirs, etc. for fifty Years, Mr. Brockden, the Scrivener, said to us, "You are young Men, but it is scarce probable that any of you will live to see the Expiration of the Term fix'd in this Instrument." A Number of us, however, are yet living: But the Instrument was after a few Years rendred null by a Charter that incorporated and gave Perpetuity to the Company.

The Objections, and Reluctances I met with in Soliciting the Subscriptions, made me soon feel the Impropriety of presenting one's self as the Proposer of any useful Project that might be suppos'd to raise one's Reputation in the smallest degree above that of one's Neighbours, when one has need of their Assistance to accomplish that Project. I therefore put my self as much as I could out of sight, and stated it as a Scheme of a *Number of Friends*, who had requested me to go about and propose it to such as they thought Lovers of Reading. In this way my Affair went on more smoothly, and I ever after practis'd it on such Occasions; and from my frequent Successes, can heartily recommend it. The present little Sacrifice of your Vanity will afterwards be amply repaid. If it remains a while uncertain to whom the Merit belongs, some one more vain

8. Charles Brockden (1683–1769) was Philadelphia's leading drafter of legal documents; "Conveyancer": an attorney who specializes in the transfer of real estate and property.

than yourself will be encourag'd to claim it, and then even Envy will be dispos'd to do you Justice, by plucking those assum'd Feathers, and restoring them to their right Owner.

This Library afforded me the Means of Improvement by constant Study, for which I set apart an Hour or two each Day; and thus repair'd in some Degree the Loss of the Learned Education my Father once intended for me. Reading was the only Amusement I allow'd my self. I spent no time in Taverns, Games, or Frolicks of any kind. And my Industry in my Business continu'd as indefatigable as it was necessary. I was in debt for my Printing-house, I had a young Family coming on to be educated,[9] and I had to contend with for Business two Printers who were establish'd in the Place before me. My Circumstances however grew daily easier: my original Habits of Frugality continuing. And My Father having among his Instructions to me when a Boy, frequently repeated a Proverb of Solomon, "*Seest thou a Man diligent in his Calling, he shall stand before Kings, he shall not stand before mean Men.*"[1] I from thence consider'd Industry as a Means of obtaining Wealth and Distinction, which encourag'd me; tho' I did not think that I should ever literally stand before Kings, which however has since happened; for I have stood before five,[2] and even had the honour of sitting down with one, the King of Denmark, to Dinner.

We have an English Proverb that says,

> He that would thrive
> Must ask his Wife;[3]

it was lucky for me that I had one as much dispos'd to Industry and Frugality as my self. She assisted me chearfully in my Business, folding and stitching Pamphlets, tending Shop, purchasing old Linen Rags for the Paper-makers, etc. etc. We kept no idle Servants, our Table was plain and simple, our Furniture of the cheapest. For instance my Breakfast was a long time Bread and Milk, (no Tea,) and I ate it out of a twopenny earthen Porringer[4] with a Pewter Spoon. But mark how Luxury will enter Families, and make a Progress, in Spite of Principle. Being Call'd one Morning to Breakfast, I found it in a China[5] Bowl with a Spoon of Silver. They had been bought for me without my Knowledge by my Wife, and had cost her the enormous Sum of three and twenty Shillings, for which she had no other Excuse or Apology to make, but that she thought *her* Husband deserv'd a Silver Spoon and China Bowl as well as any of his Neighbours. This was the first Appearance of Plate[6] and China in our House, which afterwards in a Course of

---

9. Franklin had three children: William, born c. 1731; Francis, born in 1732; Sarah, born in 1743.
1. Proverbs 22.29.
2. Louis XV and Louis XVI of France, George II and George III of England,

and Christian VI of Denmark.
3. More commonly: "He that will thrive must ask leave of his wife."
4. Bowl.
5. I.e., porcelain.
6. Silver.

Years as our Wealth encreas'd, augmented gradually to several Hundred Pounds in Value.

I had been religiously educated as a Presbyterian; and tho' some of the Dogmas of that Persuasion, such as the Eternal Decrees of God, Election, Reprobation,[7] etc. appear'd to me unintelligible, others doubtful, and I early absented myself from the Public Assemblies of the Sect, Sunday being my Studying-Day, I never was without some religious Principles; I never doubted, for instance, the Existance of the Deity, that he made the World, and govern'd it by his Providence; that the most acceptable Service of God was the doing Good to Man; that our Souls are immortal; and that all Crime will be punished and Virtue rewarded either here or hereafter; these I esteem'd the Essentials of every Religion, and being to be found in all the Religions we had in our Country I respected them all, tho' with different degrees of Respect as I found them more or less mix'd with other Articles which without any Tendency to inspire, promote or confirm Morality, serv'd principally to divide us and make us unfriendly to one another. This Respect to all, with an Opinion that the worst had some good Effects, induc'd me to avoid all Discourse that might tend to lessen the good Opinion another might have of his own Religion; and as our Province increas'd in People and new Places of worship were continually wanted, and generally erected by voluntary Contribution, my Mite[8] for such purpose, whatever might be the Sect, was never refused.

Tho' I seldom attended any Public Worship, I had still an Opinion of its Propriety, and of its Utility when rightly conducted, and I regularly paid my annual subscription for the Support of the only Presbyterian Minister or Meeting we had in Philadelphia. He us'd to visit me sometimes as a Friend, and admonish me to attend his Administrations, and I was now and then prevail'd on to do so, once for five Sundays successively. Had he been, *in my Opinion*, a good Preacher perhaps I might have continued, notwithstanding the occasion I had for the Sunday's Leisure in my Course of Study: But his Doctrines were chiefly either polemic Arguments, or Explications of the peculiar Doctrines of our Sect, and were all to me very dry, uninteresting and unedifying, since not a single moral Principle was inculcated or enforc'd, their Aim seeming to be rather to make us Presbyterians than good Citizens. At length he took for his Text that Verse of the 4th Chapter of Philippians, *Finally, Brethren, Whatsoever Things are true, honest, just, pure, lovely, or of good report, if there be any virtue, or any praise, think on these Things;*[9] and I imagin'd in a Sermon on such a Text, we could not miss of having some Morality: But he confin'd himself to five Points only as meant by the Apostle, viz: 1. Keeping holy the Sab-

---

7. Punishment; "Election": God's choosing who is to be saved and who is to be damned.

8. Small contribution.

9. A paraphrase of Philippians 4.8.

bath Day. 2. Being diligent in Reading the Holy Scriptures. 3. Attending duly the Publick Worship. 4. Partaking of the Sacrament. 5. Paying a due Respect to God's Ministers.—These might be all good Things, but as they were not the kind of good Things that I expected from that Text, I despaired of ever meeting with them from any other, was disgusted, and attended his Preaching no more. I had some Years before compos'd a little Liturgy or Form of Prayer for my own private Use, viz, in 1728, entitled, *Articles of Belief and Acts of Religion*.[1] I return'd to the Use of this, and went no more to the public Assemblies. My Conduct might be blameable, but I leave it without attempting farther to excuse it, my present purpose being to relate Facts, and not to make Apologies for them.

It was about this time that I conceiv'd the bold and arduous Project of arriving at moral Perfection. I wish'd to live without committing any Fault at any time; I would conquer all that either Natural Inclination, Custom, or Company might lead me into. As I knew, or thought I knew, what was right and wrong, I did not see why I might not *always* do the one and avoid the other. But I soon found I had undertaken a Task of more Difficulty than I had imagined: While my Care was employ'd in guarding against one Fault, I was often surpriz'd by another. Habit took the Advantage of Inattention. Inclination was sometimes too strong for Reason. I concluded at length, that the mere speculative Conviction that it was our Interest to be compleatly virtuous, was not sufficient to prevent our Slipping, and that the contrary Habits must be broken and good Ones acquired and established, before we can have any Dependance on a steady uniform Rectitude of Conduct. For this purpose I therefore contriv'd the following Method.

In the various Enumerations of the moral Virtues I had met with in my Reading, I found the Catalogue more or less numerous, as different Writers included more or fewer Ideas under the same Name. Temperance, for Example, was by some confin'd to Eating and Drinking, while by others it was extended to mean the moderating every other Pleasure, Appetite, Inclination or Passion, bodily or mental, even to our Avarice and Ambition. I propos'd to myself, for the sake of Clearness, to use rather more Names with fewer Ideas annex'd to each, than a few Names with more Ideas; and I included after Thirteen Names of Virtues all that at that time occurr'd to me as necessary or desirable, and annex'd to each a short Precept, which fully express'd the Extent I gave to its Meaning.

These Names of Virtues with their Precepts were

1. *Temperance.* Eat not to Dulness. Drink not to Elevation.

1. Only the first part of Franklin's *Articles of Belief and Acts of Religion* survives. It can be found in *The Papers of* *Benjamin Franklin*, Vol. I, ed. Leonard W. Labaree *et al.* (New Haven, Conn.: Yale University Press, 1964).

2. *Silence.* Speak not but what may benefit others or your self. Avoid trifling Conversation.

3. *Order.* Let all your Things have their Places. Let each Part of your Business have its Time.

4. *Resolution.* Resolve to perform what you ought. Perform without fail what you resolve.

5. *Frugality.* Make no Expence but to do good to others or yourself: i.e. Waste nothing.

6. *Industry.* Lose no Time. Be always employ'd in something useful. Cut off all unnecessary Actions.

7. *Sincerity.* Use no hurtful Deceit. Think innocently and justly; and, if you speak; speak accordingly.

8. *Justice.* Wrong none, by doing Injuries or omitting the Benefits that are your Duty.

9. *Moderation.* Avoid Extreams. Forbear resenting Injuries so much as you think they deserve.

10. *Cleanliness.* Tolerate no Uncleanness in Body, Cloaths or Habitation.

11. *Tranquility.* Be not disturbed at Trifles, or at Accidents common or unavoidable.

12. *Chastity.* Rarely use Venery but for Health or Offspring; Never to Dulness, Weakness, or the Injury of your own or another's Peace or Reputation.

13. *Humility.* Imitate Jesus and Socrates.

My intention being to acquire the *Habitude*[2] of all these Virtues, I judg'd it would be well not to distract my Attention by attempting the whole at once, but to fix it on one of them at a time, and when I should be Master of that, then to proceed to another, and so on till I should have gone thro' the thirteen. And as the previous Acquisition of some might facilitate the Acquisition of certain others, I arrang'd them with that View as they stand above. *Temperance* first, as it tends to procure that Coolness and Clearness of Head, which is so necessary where constant Vigilance was to be kept up, and Guard maintained, against the unremitting Attraction of ancient Habits, and the Force of perpetual Temptations. This being acquir'd and establish'd, *Silence* would be more easy, and my Desire being to gain Knowledge at the same time that I improv'd in Virtue, and considering that in Conversation it was obtain'd rather by the Use of the Ears than of the Tongue, and therefore wishing to break a Habit I was getting into of Prattling, Punning and Joking, which only made me acceptable to trifling Company, I gave *Silence* the second Place. This, and the next, *Order*, I expected would allow me more Time for attending to my Project and my Studies; RESOLUTION once become habitual, would keep me firm in

2. I.e., making these virtues an integral part of his nature.

my Endeavours to obtain all the subsequent Virtues; *Frugality* and *Industry*, by freeing me from my remaining Debt, and producing Affluence and Independance would make more easy the Practice of *Sincerity* and *Justice*, etc. etc. Conceiving then that agreable to the Advice of Pythagoras[3] in his Golden Verses, daily examination would be necessary, I contriv'd the following Method for conducting that Examination.

I made a little Book in which I allotted a Page for each of the Virtues. I rul'd each Page with red Ink so as to have seven Columns, one for each Day of the Week, marking each Column with a Letter for the Day. I cross'd these Columns with thirteen red Lines, marking the Beginning of each Line with the first Letter of one of the Virtues, on which Line and in its proper Column I might mark by a little black Spot every Fault I found upon Examination, to have been committed respecting that Virtue upon that Day.

Form of the Pages

TEMPERANCE.						
Eat not to Dulness. *Drink not to Elevation.*						

	S	M	T	W	T	F	S
T							
S	••	•		•		•	
O	•	•	•		•	•	•
R			•			•	
F		•			•		
I		•					
S							
J							
M							
Cl.							
T							
Ch.							
H							

I determined to give a Week's strict Attention to each of the Virtues successively. Thus in the first Week my great Guard was to avoid every the least Offence against Temperance, leaving the other Virtues to their ordinary Chance, only marking every Evening the Faults of the Day. Thus if in the first Week I could keep my first Line marked T clear of Spots, I suppos'd the Habit of that Virtue so much strengthen'd and its opposite weaken'd, that I might venture extending my Attention to include the next, and for the following Week keep both Lines clear of Spots. Proceeding thus to the

3. Pythagoras (sixth century B.C.) was a Greek philosopher and mathematician. Franklin added a note here: "Insert those Lines that direct it in a Note," and wished to include verses translated: "Let sleep not close your eyes till you have thrice examined the transactions of the day: where have I strayed, what have I done, what good have I omitted?"

last, I could go thro' a Course compleat in Thirteen Weeks, and
four Courses in a Year. And like him who having a Garden to weed,
does not attempt to eradicate all the bad Herbs at once, which
would exceed his Reach and his Strength, but works on one of the
Beds at a time, and having accomplish'd the first proceeds to a
second; so I should have, (I hoped) the encouraging Pleasure of
seeing on my Pages the Progress I made in Virtue, by clearing successively my Lines of their Spots, till in the End by a Number of
Courses, I should be happy in viewing a clean Book after a thirteen
Weeks daily Examination.

This my little Book had for its Motto these Lines from *Addison's
Cato;*[4]

> *Here will I hold: If there is a* Pow'r *above us,*
> *(And that there is, all Nature cries aloud*
> *Thro' all her Works) he must delight in Virtue,*
> *And that which he delights in must be happy.*

Another from *Cicero.*[5]

> *O Vitae Philosophia Dux! O Virtutum indagatrix, expultrixque
> vitiorum! Unus dies bene, et ex preceptis tuis actus, peccanti
> immortalitati est anteponendus.*

Another from the Proverbs of Solomon speaking of Wisdom or
Virtue;

> Length of Days is in her right hand, and in her Left Hand
> Riches and Honours; Her Ways are Ways of Pleasantness, and
> all her Paths are Peace.
>
> —III, 16, 17

And conceiving God to be the Fountain of Wisdom, I thought it
right and necessary to solicit his Assistance for obtaining it; to this
End I form'd the following little Prayer, which was prefix'd to my
Tables of Examination; for daily Use.

> *O Powerful Goodness! bountiful Father! merciful Guide! In-
> crease in me that Wisdom which discovers my truest Interests;
> Strengthen my Resolutions to perform what that Wisdom dictates.
> Accept my kind Offices to thy other Children, as the only Return
> in my Power for thy continual Favours to me.*

4. Joseph Addison, *Cato, a Tragedy*
(1713), V.i.15–18. Franklin also used
these lines as an epigraph for his *Articles of Belief and Acts of Religion.*
5. Marcus Tullius Cicero (106–43 B.C.),
Roman philosopher and orator. The quotation is from *Tusculan Disputations,*
V.ii.5, but several lines are omitted after
vitiorum: "Oh, philosophy, guide of life:
Oh, searcher out of virtues and expeller
of vices! * * * One day lived well and
according to thy precepts is to be preferred to an eternity of sin."

I us'd also sometimes a little Prayer which I took from *Thomson's*[6] Poems, viz

> *Father of Light and Life, thou Good supreme,*
> *O teach me what is good, teach me thy self!*
> *Save me from Folly, Vanity and Vice,*
> *From every low Pursuit, and fill my Soul*
> *With Knowledge, conscious Peace, and Virtue pure,*
> *Sacred, substantial, neverfading Bliss!*

The Precept of *Order* requiring that *every Part of my Business should have its allotted Time,* one Page in my little Book contain'd the following Scheme of Employment for the Twenty-four Hours of a natural Day,

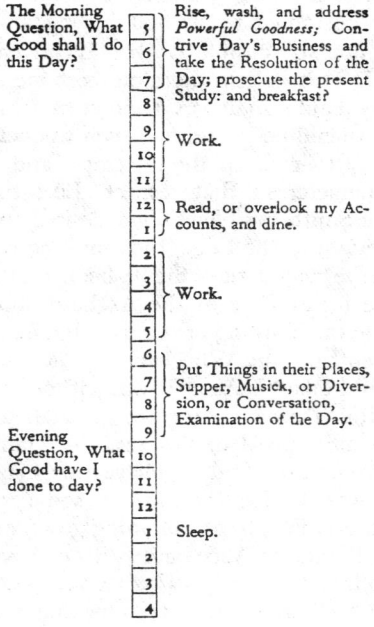

The Morning Question, What Good shall I do this Day?	5 6 7 8	Rise, wash, and address *Powerful Goodness;* Contrive Day's Business and take the Resolution of the Day; prosecute the present Study: and breakfast?
	9 10 11	Work.
	12 1	Read, or overlook my Accounts, and dine.
	2 3 4 5	Work.
	6 7 8 9	Put Things in their Places, Supper, Musick, or Diversion, or Conversation, Examination of the Day.
Evening Question, What Good have I done to day?	10 11 12	
	1 2 3 4	Sleep.

I enter'd upon the Execution of this Plan for Self Examination, and continu'd it with occasional Intermissions for some time. I was surpriz'd to find myself so much fuller of Faults than I had imagined, but I had the Satisfaction of seeing them diminish. To avoid the Trouble of renewing now and then my little Book, which by scraping out the Marks on the Paper of old Faults to make room for new Ones in a new Course, became full of Holes: I transferr'd my Tables and Precepts to the Ivory Leaves of a Memorandum

6. James Thomson (1700–48), *The Seasons,* "Winter" (1726), lines 218–23.

Book, on which the Lines were drawn with red Ink that made a durable Stain, and on those Lines I mark'd my Faults with a black Lead Pencil, which Marks I could easily wipe out with a wet Sponge. After a while I went thro' one Course only in a Year, and afterwards only one in several Years; till at length I omitted them entirely, being employ'd in Voyages and Business abroad with a Multiplicity of Affairs, that interfered. But I always carried my little Book with me.

My Scheme of ORDER, gave me the most Trouble, and I found, that tho' it might be practicable where a Man's Business was such as to leave him the Disposition of his Time, that of a Journey-man Printer for instance, it was not possible to be exactly observ'd by a Master, who must mix with the World, and often receive People of Business at their own Hours. *Order* too, with regard to Places for Things, Papers, etc. I found extremely difficult to acquire. I had not been early accustomed to it, and having an exceeding good Memory, I was not so sensible of the Inconvenience attending Want of Method. This Article therefore cost me so much painful Attention and my Faults in it vex'd me so much, and I made so little Progress in Amendment, and had such frequent Relapses, that I was almost ready to give up the Attempt, and content my self with a faulty Character in that respect. Like the Man who in buying an Ax of a Smith my Neighbour, desired to have the whole of its Surface as bright as the Edge; the Smith consented to grind it bright for him if he would turn the Wheel. He turn'd while the Smith press'd the broad Face of the Ax hard and heavily on the Stone, which made the Turning of it very fatiguing. The Man came every now and then from the Wheel to see how the Work went on; and at length would take his Ax as it was without farther Grinding. No, says the Smith, Turn on, turn on; we shall have it bright by and by; as yet 'tis only speckled. Yes, says the Man; but—*I think I like a speckled Ax best.*— And I believe this may have been the Case with many who having for want of some such Means as I employ'd found the Difficulty of obtaining good, and breaking bad Habits, in other Points of Vice and Virtue, have given up the Struggle, and concluded that *a speckled Ax was best*. For something that pretended to be Reason was every now and then suggesting to me, that such extream Nicety as I exacted of my self might be a kind of Foppery in Morals, which if it were known would make me ridiculous; that a perfect Character might be attended with the Inconvenience of being envied and hated; and that a benevolent Man should allow a few Faults in himself, to keep his Friends in Countenance.

In Truth I found myself incorrigible with respect to *Order*; and now I am grown old, and my Memory bad, I feel very sensibly the want of it. But on the whole, tho' I never arrived at the Perfection I had been so ambitious of obtaining, but fell far short of it, yet I was by the Endeavour made a better and a happier Man than I otherwise should have been, if I had not attempted it; As those who aim at perfect Writing by imitating the engraved Copies,[7] tho' they never reach the wish'd for Excellence of those Copies, their Hand is mended by the Endeavour, and is tolerable while it continues fair and legible.

And it may be well my Posterity should be informed, that to this little Artifice, with the Blessing of God, their Ancestor ow'd the constant Felicity of his Life down to his 79th Year in which this is written. What Reverses may attend the Remainder is in the Hand of Providence: But if they arrive the Reflection on past Happiness enjoy'd ought to help his Bearing them with more Resignation. To *Temperance* he ascribes his long-continu'd Health, and what is still left to him of a good Constitution. To *Industry* and *Frugality* the early Easiness of his Circumstances, and Acquisition of his Fortune, with all that knowledge which enabled him to be an useful Citizen, and obtain'd for him some Degree of Reputation among the Learned. To *Sincerity* and *Justice* the Confidence of his Country, and the honourable Employs it conferr'd upon him. And to the joint Influence of the whole Mass of the Virtues, even in their imperfect State he was able to acquire them, all that Evenness of Temper, and that Chearfulness in Conversation which makes his Company still sought for, and agreable even to his younger Acquaintance. I hope therefore that some of my Descendants may follow the Example and reap the Benefit.

It will be remark'd[8] that, tho my Scheme was not wholly without Religion there was in it no Mark of any of the distinguishing Tenets of any particular Sect. I had purposely avoided them; for being fully persuaded of the Utility and Excellency of my Method, and that it might be serviceable to People in all Religions, and intending some time or other to publish it, I would not have any thing in it that should prejudice any one of any Sect against it. I purposed writing a little Comment on each Virtue, in which I would have shown the Advantages of possessing it, and the Mischiefs attending its opposite Vice; and I should have called my Book the ART *of Virtue*, because it would have shown the *Means and Manner* of obtaining Virtue; which would have distinguish'd it from the mere Exhortation to be good, that does not instruct and indicate the Means; but is like the Apostle's Man of verbal Charity, who only, without showing to the Naked and the Hungry *how* or where they might get Cloaths or Victuals, exhorted them to be fed and clothed. *James* II, 15, 16.[9]

But it so happened that my Intention of writing and publishing

---

7. I.e., the models in the printed book.
8. Observed.
9. "If a brother or sister be naked, and destitute of daily food, And one of you say unto them, Depart in peace, be ye warmed and filled; notwithstanding ye give them not those things which are needful to the body; what doth it profit?"

this Comment was never fulfilled. I did indeed, from time to time put down short Hints of the Sentiments, Reasonings, etc. to be made use of in it; some of which I have still by me: But the necessary close Attention to private Business in the earlier part of Life, and public Business since, have occasioned my postponing it. For it being connected in my Mind with *a great and extensive Project* that required the whole Man to execute, and which an unforeseen Succession of Employs prevented my attending to, it has hitherto remain'd unfinish'd.

In this Piece it was my Design to explain and enforce this Doctrine, that vicious Actions are not hurtful because they are forbidden, but forbidden because they are hurtful, the Nature of Man alone consider'd: That it was therefore every ones Interest to be virtuous, who wish'd to be happy even in this World. And I should from this Circumstance, there being always in the World a Number of rich Merchants, Nobility, States and Princes, who have need of honest Instruments for the Management of their Affairs, and such being so rare have endeavoured to convince young Persons, that no Qualities were so likely to make a poor Man's Fortune as those of Probity and Integrity.

My List of Virtues contain'd at first but twelve: But a Quaker Friend having kindly inform'd me that I was generally thought proud; that my Pride show'd itself frequently in Conversation; that I was not content with being in the right when discussing any Point, but was overbearing and rather insolent; of which he convinc'd me by mentioning several Instances; I determined endeavouring to cure myself if I could of this Vice or Folly among the rest, and I added *Humility* to my List, giving an extensive Meaning to the Word. I cannot boast of much Success in acquiring the *Reality* of this Virtue; but I had a good deal with regard to the *Appearance* of it. I made it a Rule to forbear all direct Contradiction to Sentiments of others, and all positive Assertion of my own. I even forbid myself, agreable to the old Laws of our Junto, the Use of every Word or Expression in the Language that imported[1] a fix'd Opinion; such as *certainly, undoubtedly*, etc. and I adopted instead of them, *I conceive, I apprehend*, or *I imagine* a thing to be so or so, or it so appears to me at present. When another asserted something that I thought an Error, I deny'd my self the Pleasure of contradicting him abruptly, and of showing immediately some Absurdity in his Proposition; and in answering I began by observing that in certain Cases or Circumstances his Opinion would be right, but that in the present case there *appear'd* or *seem'd* to me some Difference, etc. I soon found the Advantage of this Change in my Manners. The Conversations I engag'd in went on more pleasantly. The modest way in which I propos'd my Opinions, procur'd them a readier Reception and less

1. Suggested.

Contradiction; I had less Mortification when I was found to be in
the wrong, and I more easily prevail'd with others to give up their
Mistakes and join with me when I happen'd to be in the right. And
this Mode, which I at first put on, with some violence to natural
Inclination, became at length so easy and so habitual to me, that
perhaps for these Fifty Years past no one has ever heard a dogmati-
cal Expression escape me. And to this Habit (after my Character of
Integrity) I think it principally owing, that I had early so much
Weight with my Fellow Citizens, when I proposed new Institutions, or Alterations in the old; and so much Influence in public
Councils when I became a Member. For I was but a bad Speaker,
never eloquent, subject to much Hesitation in my choice of Words,
hardly correct in Language, and yet I generally carried my Points.

In reality there is perhaps no one of our natural Passions so hard
to subdue as *Pride*. Disguise it, struggle with it, beat it down, stifle
it, mortify it as much as one pleases, it is still alive, and will every
now and then peep out and show itself. You will see it perhaps
often in this History. For even if I could conceive that I had compleatly overcome it, I should probably be proud of my Humility.

*   *   *

1771–90   1868

ST. JEAN DE CRÈVECOEUR
1735–1813

Crèvecoeur was a man with a mysterious past, and a number of details of
his life have puzzled his biographers. He was born Michel-Guillaume-Jean
de Crèvecoeur in Caen, Normandy, in 1735. When he was nineteen, he
left home and sailed to England, where he took up residence with distant
relatives. He planned to marry, but his fiancée died before the ceremony
took place, and in 1755 he went to Canada; he enlisted in the Canadian
militia, served the government as a surveyor and cartographer, and was
wounded in the defense of Quebec. His military career came to an end in
1759, and for the next ten years Crèvecoeur traveled extensively in the colonies as a surveyor and Indian trader. In 1769 he bought land in Orange
County, New York, and, newly married, settled into the life of an American farmer.

Given the history of Crèvecoeur's restlessness, it is hard to know whether
or not he would have been happy forever at Pine Hill, but the advent of
the American Revolution and his Tory sympathies were enough to determine his return to France. He claimed that he wished to re-establish ownership of family lands, and it is ironic, given his political sympathies, that he
was arrested and imprisoned as a rebel spy when he tried to sail from the
port of New York. Not until 1780, did Crèvecoeur succeed in reaching
London. He remained in France until 1783, when he returned as French
consul to New York, Connecticut, and New Jersey, only to learn that his

farm had been burned in an Indian attack, his wife was dead, and his children were housed with strangers.

Crèvecoeur was a great success as a diplomat—he was made an honorary citizen of a number of American cities, and the town of St. Johnsbury, Vermont, was named in his honor—but he did not remain long in America. He returned to France in 1785 and after 1790 remained there permanently, first living in Paris and retiring, after 1793, to Normandy.

The first year that Crèvecoeur spent at Pine Hill he began to write a series of essays about America based on his travels and experience as a farmer. He brought them to London in 1780 and, suppressing those essays most unsympathetic to the American cause, sold them to the bookseller Thomas Davies. *Letters from an American Farmer* appeared in 1782 and was an immediate success. Crèvecoeur found himself a popular hero when the expanded French edition appeared in 1783. Its publication followed close enough upon the American Revolution to satisfy an almost insatiable demand for things American and confirmed, for most readers, a vision of a new land, rich and promising, where industry prevailed over class and fashion. George Washington said the book was "too flattering" to be true, but more careful readers of these twelve letters will take note of a more ambiguous attitude throughout: Crèvecoeur's hymn to the land does not make him blind to the ignorant frontiersman or the calculating slaveholder. His final letter, "Distresses of a Frontiersman," affirms the possibility of a harmonious relationship with Nature, but he writes from an Indian village and with no successful historical models in mind.

From Letters from an American Farmer[1]
From *Letter III. What Is an American*

I wish I could be acquainted with the feelings and thoughts which must agitate the heart and present themselves to the mind of an enlightened Englishman, when he first lands on this continent. He must greatly rejoice that he lived at a time to see this fair country discovered and settled; he must necessarily feel a share of national pride, when he views the chain of settlements which embellishes these extended shores. When he says to himself, this is the work of my countrymen, who, when convulsed by factions,[2] afflicted by a variety of miseries and wants, restless and impatient, took refuge here. They brought along with them their national genius,[3] to which they principally owe what liberty they enjoy, and what substance they possess. Here he sees the industry of his native country displayed in a new manner, and traces in their works the embryos of all the arts, sciences, and ingenuity which flourish in Europe. Here he beholds fair cities, substantial villages, extensive fields, an immense country filled with decent houses, good roads, orchards, meadows, and bridges, where an hundred years ago all was wild, woody, and uncultivated! What a train of pleasing ideas this fair spectacle must suggest; it is a prospect which must inspire a good citizen with the most heartfelt

1. From *Letters from an American Farmer*, ed. Albert Boni and Charles Boni (New York, 1925).

2. Disputes.

3. Spirit; distinctive national character.

pleasure. The difficulty consists in the manner of viewing so extensive a scene. He is arrived on a new continent; a modern society
offers itself to his contemplation, different from what he had hitherto seen. It is not composed, as in Europe, of great lords who possess everything, and of a herd of people who have nothing. Here
are no aristocratical families, no courts, no kings, no bishops, no
ecclesiastical dominion, no invisible power giving to a few a very visible one; no great manufacturers employing thousands, no great
refinements of luxury. The rich and the poor are not so far removed
from each other as they are in Europe. Some few towns excepted,
we are all tillers of the earth, from Nova Scotia to West Florida.
We are a people of cultivators, scattered over an immense territory,
communicating with each other by means of good roads and navigable rivers, united by the silken bands of mild government, all
respecting the laws, without dreading their power, because they are
equitable. We are all animated with the spirit of an industry which
is unfettered and unrestrained, because each person works for himself. If he travels through our rural districts he views not the hostile
castle, and the haughty mansion, contrasted with the clay-built hut
and miserable cabin, where cattle and men help to keep each other
warm, and dwell in meanness, smoke, and indigence. A pleasing
uniformity of decent competence appears throughout our habitations. The meanest of our log-houses is a dry and comfortable habitation. Lawyer or merchant are the fairest titles our towns afford;
that of a farmer is the only appellation of the rural inhabitants of
our country. It must take some time ere he can reconcile himself
to our dictionary, which is but short in words of dignity, and names
of honor. There, on a Sunday, he sees a congregation of respectable
farmers and their wives, all clad in neat homespun, well mounted,
or riding in their own humble wagons. There is not among them an
esquire, saving the unlettered magistrate. There he sees a parson as
simple as his flock, a farmer who does not riot[4] on the labor of
others. We have no princes, for whom we toil, starve, and bleed: we
are the most perfect society now existing in the world. Here man is
free as he ought to be; nor is this pleasing equality so transitory as
many others are. Many ages will not see the shores of our great
lakes replenished with inland nations, nor the unknown bounds of
North America entirely peopled. Who can tell how far it extends?
Who can tell the millions of men whom it will feed and contain?
for no European foot has as yet traveled half the extent of this
mighty continent!

The next wish of this traveler will be to know whence came all
these people? They are a mixture of English, Scotch, Irish, French,
Dutch, Germans, and Swedes. From this promiscuous breed, that
race now called Americans have arisen. The eastern provinces[5] must

---

4. I.e., indulge himself.   5. New England.

indeed be excepted, as being the unmixed descendants of Englishmen. I have heard many wish that they had been more intermixed
also: for my part, I am no wisher, and think it much better as it has
happened. They exhibit a most conspicuous figure in this great and
variegated picture; they too enter for a great share in the pleasing
perspective displayed in these thirteen provinces. I know it is fashionable to reflect on them, but respect them for what they have
done; for the accuracy and wisdom with which they have settled
their territory; for the decency of their manners; for their early love
of letters; their ancient college,[6] the first in this hemisphere; for
their industry, which to me who am but a farmer is the criterion of
everything. There never was a people, situated as they are, who with
so ungrateful a soil have done more in so short a time. Do you
think that the monarchical ingredients which are more prevalent in
other governments have purged them from all foul stains? Their histories assert the contrary.

In this great American asylum, the poor of Europe have by some
means met together, and in consequence of various causes; to what
purpose should they ask one another what countrymen they are?
Alas, two thirds of them had no country. Can a wretch who wanders about, who works and starves, whose life is a continual scene of
sore affliction or pinching penury, can that man call England or any
other kingdom his country? A country that had no bread for him,
whose fields procured him no harvest, who met with nothing but
the frowns of the rich, the severity of the laws, with jails and punishments; who owned not a single foot of the extensive surface of
this planet? No! Urged by a variety of motives, here they came.
Everything has tended to regenerate them; new laws, a new mode
of living, a new social system; here they are become men: in Europe
they were as so many useless plants, wanting vegetative mold and
refreshing showers; they withered, and were mowed down by want,
hunger, and war; but now by the power of transplantation, like all
other plants they have taken root and flourished! Formerly they
were not numbered in any civil lists[7] of their country, except in
those of the poor; here they rank as citizens. By what invisible
power has this surprising metamorphosis been performed? By that
of the laws and that of their industry. The laws, the indulgent laws,
protect them as they arrive, stamping on them the symbol of adoption; they receive ample rewards for their labors; these accumulated rewards procure them lands; those lands confer on them the
title of freemen, and to that title every benefit is affixed which men
can possibly require. This is the great operation daily performed by
our laws. From whence proceed these laws? From our government.
Whence the government? It *is* derived from the original genius and

6. Harvard College was founded in 1636.
7. Recognized employees of the civil government: ambassadors, judges, secre-
taries, etc.

strong desire of the people ratified and confirmed by the crown. This is the great chain which links us all, this is the picture which every province exhibits, Nova Scotia excepted. There the crown has done all;[8] either there were no people who had genius, or it was not much attended to: the consequence is that the province is very thinly inhabited indeed; the power of the crown in conjunction with the mosquitoes has prevented men from settling there. Yet some parts of it flourished once, and it contained a mild, harmless set of people. But for the fault of a few leaders, the whole were banished. The greatest political error the crown ever committed in America was to cut off men from a country which wanted nothing but men!

What attachment can a poor European emigrant have for a country where he had nothing? The knowledge of the language, the love of a few kindred as poor as himself, were the only cords that tied him: his country is now that which gives him land, bread, protection, and consequence: *Ubi panis ibi patria*[9] is the motto of all emigrants. What then is the American, this new man? He is either a European, or the descendant of a European, hence that strange mixture of blood, which you will find in no other country. I could point out to you a family whose grandfather was an Englishman, whose wife was Dutch, whose son married a French woman, and whose present four sons have now four wives of different nations. *He* is an American, who, leaving behind him all his ancient prejudices and manners, receives new ones from the new mode of life he has embraced, the new government he obeys, and the new rank he holds. He becomes an American by being received in the broad lap of our great *Alma Mater*.[1] Here individuals of all nations are melted into a new race of men, whose labors and posterity will one day cause great changes in the world. Americans are the western pilgrims, who are carrying along with them that great mass of arts, sciences, vigor, and industry which began long since in the east; they will finish the great circle. The Americans were once scattered all over Europe; here they are incorporated into one of the finest systems of population which has ever appeared, and which will hereafter become distinct by the power of the different climates they inhabit. The American ought therefore to love this country much better than that wherein either he or his forefathers were born. Here the rewards of his industry follow with equal steps the progress of his labor; his labor is founded on the basis of nature, *self-interest*; can it want a stronger allurement? Wives and children, who before in vain demanded of him a morsel of bread, now, fat and frolicsome, gladly help their father to clear those fields whence exuberant crops are to arise to feed and to clothe them all; without any

---

8. In 1755 the French Acadians were banished from Nova Scotia by the British, who took it in 1710.

9. "Where there is bread, there is one's fatherland."

1. Literally, dear mother.

part being claimed, either by a despotic prince, a rich abbot, or a mighty lord. Here religion demands but little of him; a small voluntary salary to the minister, and gratitude to God; can he refuse these? The American is a new man, who acts upon new principles; he must therefore entertain new ideas, and form new opinions. From involuntary idleness, servile dependence, penury, and useless labor, he has passed to toils of a very different nature, rewarded by ample subsistence.—This is an American.

British America is divided into many provinces, forming a large association, scattered along a coast 1,500 miles extent and about 200 wide. This society I would fain examine, at least such as it appears in the middle provinces; if it does not afford that variety of tinges and gradations which may be observed in Europe, we have colors peculiar to ourselves. For instance, it is natural to conceive that those who live near the sea must be very different from those who live in the woods; the intermediate space will afford a separate and distinct class.

Men are like plants; the goodness and flavor of the fruit proceeds from the peculiar soil and exposition in which they grow. We are nothing but what we derive from the air we breathe, the climate we inhabit, the government we obey, the system of religion we profess, and the nature of our employment. Here you will find but few crimes; these have acquired as yet no root among us. I wish I was able to trace all my ideas; if my ignorance prevents me from describing them properly, I hope I shall be able to delineate a few of the outlines, which are all I propose.

Those who live near the sea feed more on fish than on flesh, and often encounter that boisterous element. This renders them more bold and enterprising; this leads them to neglect the confined occupations of the land. They see and converse with a variety of people, their intercourse with mankind becomes extensive. The sea inspires them with a love of traffic, a desire of transporting produce from one place to another; and leads them to a variety of resources which supply the place of labor. Those who inhabit the middle settlements, by far the most numerous, must be very different; the simple cultivation of the earth purifies them, but the indulgences of the government, the soft remonstrances of religion, the rank of independent freeholders, must necessarily inspire them with sentiments, very little known in Europe among people of the same class. What do I say? Europe has no such class of men; the early knowledge they acquire, the early bargains they make, give them a great degree of sagacity. As freemen they will be litigious; pride and obstinacy are often the cause of lawsuits; the nature of our laws and governments may be another. As citizens it is easy to imagine that they will carefully read the newspapers, enter into every political disquisition, freely blame or censure governors and others. As farmers they will be careful and anxious to get as much as they can, because what they get is their own. As northern men they will love the cheerful cup. As Christians, religion curbs them not in their opinions; the general indulgence leaves everyone to think for themselves in spiritual matters; the laws inspect our actions, our thoughts are left to God. Industry, good living, selfishness, litigiousness, country politics, the pride of freemen, religious indifference are their characteristics. If you recede still farther from the sea, you will come into more modern settlements; they exhibit the same strong lineaments, in a ruder appearance. Religion seems to have still less influence, and their manners are less improved.

Now we arrive near the great woods, near the last inhabited districts;[2] there men seem to be placed still farther beyond the reach of government, which in some measure leaves them to themselves. How can it pervade every corner; as they were driven there by misfortunes, necessity of beginnings, desire of acquiring large tracts of land, idleness, frequent want of economy,[3] ancient debts; the reunion of such people does not afford a very pleasing spectacle. When discord, want of unity and friendship; when either drunkenness or idleness prevail in such remote districts; contention, inactivity, and wretchedness must ensue. There are not the same remedies to these evils as in a long-established community. The few magistrates they have are in general little better than the rest; they are often in a perfect state of war; that of man against man, sometimes decided by blows, sometimes by means of the law; that of man against every wild inhabitant of these venerable woods, of which they are come to dispossess them. There men appear to be no better than carnivorous animals of a superior rank, living on the flesh of wild animals when they can catch them, and when they are not able, they subsist on grain. He who would wish to see America in its proper light, and have a true idea of its feeble beginnings and barbarous rudiments, must visit our extended line of frontiers where the last settlers dwell, and where he may see the first labors of settlement, the mode of clearing the earth, in all their different appearances; where men are wholly left dependent on their native tempers and on the spur of uncertain industry, which often fails when not sanctified by the efficacy of a few moral rules. There, remote from the power of example and check of shame, many families exhibit the most hideous parts of our society. They are a kind of forlorn hope, preceding by ten or twelve years the most respectable army of veterans which come after them. In that space, prosperity will polish some, vice and the law will drive off the rest, who uniting again with others like themselves will recede still farther; making room for more industrious people, who will finish their improvements, convert the loghouse into a convenient habitation,

---

2. I.e., the frontier; the land west of the original colonies and east of the Mississippi.

3. I.e., they were improvident and spent beyond their means.

and rejoicing that the first heavy labors are finished, will change in a few years that hitherto barbarous country into a fine fertile, well-regulated district. Such is our progress, such is the march of the Europeans toward the interior parts of this continent. In all societies there are offcasts; this impure part serves as our precursors or pioneers; my father himself was one of that class,[4] but he came upon honest principles, and was therefore one of the few who held fast; by good conduct and temperance, he transmitted to me his fair inheritance, when not above one in fourteen of his contemporaries had the same good fortune.

Forty years ago this smiling country was thus inhabited; it is now purged, a general decency of manners prevails throughout, and such has been the fate of our best countries.

Exclusive of those general characteristics, each province has its own, founded on the government, climate, mode of husbandry, customs, and peculiarity of circumstances. Europeans submit insensibly to these great powers, and become, in the course of a few generations, not only Americans in general, but either Pennsylvanians, Virginians, or provincials under some other name. Whoever traverses the continent must easily observe those strong differences, which will grow more evident in time. The inhabitants of Canada, Massachusetts, the middle provinces, the southern ones will be as different as their climates; their only points of unity will be those of religion and language.

As I have endeavored to show you how Europeans become Americans, it may not be disagreeable to show you likewise how the various Christian sects introduced wear out, and how religious indifference becomes prevalent. When any considerable number of a particular sect happen to dwell contiguous to each other, they immediately erect a temple, and there worship the Divinity agreeably to their own peculiar ideas. Nobody disturbs them. If any new sect springs up in Europe it may happen that many of its professors[5] will come and settle in America. As they bring their zeal with them, they are at liberty to make proselytes if they can, and to build a meeting and to follow the dictates of their consciences; for neither the government nor any other power interferes. If they are peaceable subjects, and are industrious, what is it to their neighbors how and in what manner they think fit to address their prayers to the Supreme Being? But if the sectaries are not settled close together, if they are mixed with other denominations, their zeal will cool for want of fuel, and will be extinguished in a little time. Then the Americans become as to religion what they are as to country, allied to all. In them the name of Englishman, Frenchman, and European is lost, and in like manner, the strict modes of Christianity as practiced in Europe are lost also. This effect will extend itself still far-

---

4. His father never came to America.   5. Believers.

ther hereafter, and though this may appear to you as a strange idea, yet it is a very true one. I shall be able perhaps hereafter to explain myself better; in the meanwhile, let the following example serve as my first justification.

Let us suppose you and I to be traveling; we observe that in this house, to the right, lives a Catholic, who prays to God as he has been taught, and believes in transubstantiation;[6] he works and raises wheat, he has a large family of children, all hale and robust; his belief, his prayers offend nobody. About one mile farther on the same road, his next neighbor may be a good honest plodding German Lutheran, who addresses himself to the same God, the God of all, agreeably to the modes he has been educated in, and believes in consubstantiation;[7] by so doing he scandalizes nobody; he also works in his fields, embellishes the earth, clears swamps, etc. What has the world to do with his Lutheran principles? He perse-cutes nobody, and nobody persecutes him, he visits his neighbors, and his neighbors visit him. Next to him lives a seceder,[8] the most enthusiastic of all sectaries;[9] his zeal is hot and fiery, but separated as he is from others of the same complexion, he has no congrega-tion of his own to resort to, where he might cabal and mingle reli-gious pride with worldly obstinacy. He likewise raises good crops, his house is handsomely painted, his orchard is one of the fairest in the neighborhood. How does it concern the welfare of the country, or of the province at large, what this man's religious sentiments are, or really whether he has any at all? He is a good farmer, he is a sober, peaceable, good citizen: William Penn[1] himself would not wish for more. This is the visible character, the invisible one is only guessed at, and is nobody's business. Next again lives a Low Dutch-man, who implicitly believes the rules laid down by the synod of Dort.[2] He conceives no other idea of a clergyman than that of a hired man; if he does his work well he will pay him the stipulated sum; if not he will dismiss him, and do without his sermons, and let his church be shut up for years. But notwithstanding this coarse idea, you will find his house and farm to be the neatest in all the country; and you will judge by his wagon and fat horses that he thinks more of the affairs of this world than of those of the next. He is sober and laborious, therefore he is all he ought to be as to the affairs of this life; as for those of the next, he must trust to the great Creator. Each of these people instruct their children as well as

6. The doctrine followed by Roman Catholics that the bread and wine used in the sacrament of communion have been changed into the real presence of Christ's body and blood.
7. As distinguished from transubstan-tiation; the doctrine which affirms that Christ's body is not present in or under the elements of bread and wine, but that the bread and wine are signs of Christ's presence through faith.

8. One who has withdrawn from any re-ligious body.
9. One who dissents from the established church.
1. William Penn (1644–1718), English Quaker and founder of Philadelphia.
2. The Synod of Dort met in Holland in 1618 and attempted to settle disputes be-tween Protestant Reformed Churches; "Low Dutchman": someone from Hol-land, not Belgium.

they can, but these instructions are feeble compared to those which are given to the youth of the poorest class in Europe. Their children will therefore grow up less zealous and more indifferent in matters of religion than their parents. The foolish vanity, or rather the fury of making proselytes, is unknown here; they have no time, the seasons call for all their attention, and thus in a few years, this mixed neighborhood will exhibit a strange religious medley, that will be neither pure Catholicism nor pure Calvinism. A very perceptible indifference, even in the first generation, will become apparent; and it may happen that the daughter of the Catholic will marry the son of the seceder, and settle by themselves at a distance from their parents. What religious education will they give their children? A very imperfect one. If there happens to be in the neighborhood any place of worship, we will suppose a Quaker's meeting; rather than not show their fine clothes, they will go to it, and some of them may perhaps attach themselves to that society. Others will remain in a perfect state of indifference; the children of these zealous parents will not be able to tell what their religious principles are, and their grandchildren still less. The neighborhood of a place of worship generally leads them to it, and the action of going thither is the strongest evidence they can give of their attachment to any sect. The Quakers are the only people who retain a fondness for their own mode of worship; for be they ever so far separated from each other, they hold a sort of communion with the society, and seldom depart from its rules, at least in this country. Thus all sects are mixed as well as all nations; thus religious indifference is imperceptibly disseminated from one end of the continent to the other; which is at present one of the strongest characteristics of the Americans. Where this will reach no one can tell, perhaps it may leave a vacuum fit to receive other systems. Persecution, religious pride, the love of contradiction are the food of what the world commonly calls religion. These motives have ceased here; zeal in Europe is confined; here it evaporates in the great distance it has to travel; there it is a grain of powder inclosed, here it burns away in the open air, and consumes without effect.

But to return to our back settlers. I must tell you that there is something in the proximity of the woods which is very singular. It is with men as it is with the plants and animals that grow and live in the forests; they are entirely different from those that live in the plains. I will candidly tell you all my thoughts but you are not to expect that I shall advance any reasons. By living in or near the woods, their actions are regulated by the wildness of the neighborhood. The deer often come to eat their grain, the wolves to destroy their sheep, the bears to kill their hogs, the foxes to catch their poultry. This surrounding hostility immediately puts the gun into their hands; they watch these animals, they kill some; and thus by defending their property, they soon become professed hunters; this is the progress; once hunters, farewell to the plow. The chase renders them ferocious, gloomy, and unsociable; a hunter wants no neighbor, he rather hates them, because he dreads the competition. In a little time their success in the woods makes them neglect their tillage. They trust to the natural fecundity of the earth, and therefore do little; carelessness in fencing often exposes what little they sow to destruction; they are not at home to watch; in order therefore to make up the deficiency, they go oftener to the woods. That new mode of life brings along with it a new set of manners, which I cannot easily describe. These new manners, being grafted on the old stock, produce a strange sort of lawless profligacy, the impressions of which are indelible. The manners of the Indian natives are respectable, compared with this European medley. Their wives and children live in sloth and inactivity; and having no proper pursuits, you may judge what education the latter receive. Their tender minds have nothing else to contemplate but the example of their parents; like them they grow up a mongrel breed, half civilized, half savage, except nature stamps on them some constitutional propensities. That rich, that voluptuous sentiment is gone that struck them so forcibly; the possession of their freeholds[3] no longer conveys to their minds the same pleasure and pride. To all these reasons you must add their lonely situation, and you cannot imagine what an effect on manners the great distances they live from each other has! Consider one of the last settlements in its first view: of what is it composed? Europeans who have not that sufficient share of knowledge they ought to have, in order to prosper; people who have suddenly passed from oppression, dread of government, and fear of laws into the unlimited freedom of the woods. This sudden change must have a very great effect on most men, and on that class particularly. Eating of wild meat, whatever you may think, tends to alter their temper: though all the proof I can adduce is that I have seen it: and having no place of worship to resort to, what little society this might afford is denied them. The Sunday meetings, exclusive of religious benefits, were the only social bonds that might have inspired them with some degree of emulation in neatness. Is it then surprising to see men thus situated, immersed in great and heavy labors, degenerate a little? It is rather a wonder the effect is not more diffusive. The Moravians[4] and the Quakers are the only instances in exception to what I have advanced. The first never settle singly, it is a colony of the society which emigrates; they carry with them their forms, worship, rules, and decency: the others never begin so hard, they are always able to buy improvements, in which there is a great advantage, for by that time the country is recovered

3. Land held outright for a specified period of time.
4. The Moravians were followers of Jacob Hutter, who was executed in 1536; they were Christian family communities who gave up private property and were noted for their industry and thrift. They suffered a number of persecutions in the 17th century and emigrated to other lands.

from its first barbarity. Thus our bad people are those who are half cultivators and half hunters; and the worst of them are those who have degenerated altogether into the hunting state. As old plowmen and new men of the woods, as Europeans and new-made Indians, they contract the vices of both; they adopt the moroseness and ferocity of a native, without his mildness, or even his industry at home. If manners are not refined, at least they are rendered simple and inoffensive by tilling the earth; all our wants are supplied by it, our time is divided between labor and rest, and leaves none for the commission of great misdeeds. As hunters it is divided between the toil of the chase, the idleness of repose, or the indulgence of inebriation. Hunting is but a licentious idle life, and if it does not always pervert good dispositions; yet, when it is united with bad luck, it leads to want: want stimulates that propensity to rapacity and injustice, too natural to needy men, which is the fatal gradation. After this explanation of the effects which follow by living in the woods, shall we yet vainly flatter ourselves with the hope of converting the Indians? We should rather begin with converting our back-settlers; and now if I dare mention the name of religion, its sweet accents would be lost in the immensity of these woods. Men thus placed are not fit either to receive or remember its mild instructions; they want[5] temples and ministers, but as soon as men cease to remain at home, and begin to lead an erratic life, let them be either tawny or white, they cease to be its disciples.

* * *

Europe contains hardly any other distinctions but lords and tenants; this fair country alone is settled by freeholders, the possessors of the soil they cultivate, members of the government they obey, and the framers of their own laws, by means of their representatives. This is a thought which you have taught me to cherish; our difference from Europe, far from diminishing, rather adds to our usefulness and consequence as men and subjects. Had our forefathers remained there, they would only have crowded it, and perhaps prolonged those convulsions which had shook it so long. Every industrious European who transports himself here may be compared to a sprout growing at the foot of a great tree; it enjoys and draws but a little portion of sap; wrench it from the parent roots, transplant it, and it will become a tree bearing fruit also. Colonists are therefore entitled to the consideration due to the most useful subjects; a hundred families barely existing in some parts of Scotland will here in six years cause an annual exportation of 10,000 bushels of wheat; 100 bushels being but a common quantity for an industrious family to sell, if they cultivate good land. It is here then that the idle may be employed, the useless become useful, and the poor become rich; but by riches I do not mean gold and silver, we

5. Lack.

have but little of those metals; I mean a better sort of wealth, cleared lands, cattle, good houses, good clothes, and an increase of people to enjoy them.

There is no wonder that this country has so many charms, and presents to Europeans so many temptations to remain in it. A traveler in Europe becomes a stranger as soon as he quits his own kingdom; but it is otherwise here. We know, properly speaking, no strangers; this is every person's country; the variety of our soils, situations, climates, governments, and produce hath something which must please everybody. No sooner does a European arrive, no matter of what condition, than his eyes are opened upon the fair prospect; he hears his language spoken, he retraces many of his own country manners, he perpetually hears the names of families and towns with which he is acquainted; he sees happiness and prosperity in all places disseminated; he meets with hospitality, kindness, and plenty everywhere; he beholds hardly any poor; he seldom hears of punishments and executions; and he wonders at the elegance of our towns, those miracles of industry and freedom. He cannot admire enough our rural districts, our convenient roads, good taverns, and our many accommodations; he involuntarily loves a country where everything is so lovely.

* * *

After a foreigner from any part of Europe is arrived, and become a citizen, let him devoutly listen to the voice of our great parent, which says to him, "Welcome to my shores, distressed European; bless the hour in which thou didst see my verdant fields, my fair navigable rivers, and my green mountains!—If thou wilt work, I have bread for thee; if thou wilt be honest, sober, and industrious, I have greater rewards to confer on thee—ease and independence. I will give thee fields to feed and clothe thee; a comfortable fireside to sit by, and tell thy children by what means thou hast prospered; and a decent bed to repose on. I shall endow thee beside with the immunities of a freeman. If thou wilt carefully educate thy children, teach them gratitude to God, and reverence to that government, that philanthropic government, which has collected here so many men and made them happy. I will also provide for thy progeny; and to every good man this ought to be the most holy, the most powerful, the most earnest wish he can possibly form, as well as the most consolatory prospect when he dies. Go thou and work and till; thou shalt prosper, provided thou be just, grateful, and industrious."

c. 1769–80   1782

THOMAS PAINE
1737–1809

The author of two of the most popular books in eighteenth-century America, and the most persuasive rhetorician of the cause for independence that our country has ever known, Thomas Paine was born in England in 1737, the son of a Quaker father and an Anglican mother, and did not come to America until he was thirty-seven years old. Paine's early years prepared him to be a supporter of the Revolution. The discrepancy between his high intelligence and the limitations imposed upon him by poverty and caste made him long for a new social order. He once said that a sermon he heard at the age of eight impressed him with the cruelty inherent in Christianity and made him a rebel forever. When he arrived in Philadelphia with letters of introduction from Benjamin Franklin, recommending him as an "ingenious, worthy young man," he had already had a remarkably full life. Until he was thirteen he went to grammar school, and then was apprenticed in his father's corset shop; at nineteen he ran away from home to go to sea. From 1757 to 1774 he was a corsetmaker, a tobacconist and grocer, a schoolteacher, and an exciseman (a government employee who taxed goods). His efforts to organize the excisemen and make Parliament raise their salary was unprecedented. He lost his job when he admitted he had stamped as examined goods which had not been opened. His first wife died less than a year after his marriage, and he was separated from his second wife after three years. Scandals about his private life and questions about his integrity while employed as an exciseman provided his critics with ammunition for the rest of his life. Franklin was right, however, in recognizing Paine's genius; for, like Franklin himself, he was a remarkable man, self-taught and curious about everything, from the philosophy of law to natural science.

In Philadelphia he seemed to find himself as a journalist, and he made his way quickly in that city, first as a spokesman against slavery and then as the anonymous author of *Common Sense*, the first pamphlet published in this country to urge immediate independence from Britain. Paine was obviously the right man in the right place at the right time. Relations with England were at their lowest ebb: Boston was under siege, and the Second Continental Congress had convened in Philadelphia. *Common Sense* sold almost half a million copies, and its authorship (followed by the charge of traitor) could not be kept a secret for long. Paine enlisted in the Revolutionary Army and served as an aide-de-camp in battles in New York, New Jersey, and Pennsylvania. He followed his triumph of *Common Sense* with the first of sixteen pamphlets entitled *Crisis*. The first *Crisis* paper ("These are the times that try men's souls") was read to Washington's troops at Trenton and did much to shore up the spirits of the Revolutionary soldiers.

Paine received a number of political appointments as rewards for his services as a writer for the American cause, but he misused his privileges and lost the most lucrative offices. He was too indiscreet and hot-tempered for public employment. In 1787 he returned to England, determined to get financial assistance to construct an iron bridge for which he had devised plans. It came to nothing. But in England he wrote his second most successful work, his *Rights of Man* (1791–92), an impassioned plea against hereditary monarchy, the traditional institution Paine never tired of arguing against. Paine was charged with treason and fled to France, where he was made a citizen and lionized as a spokesman for revolution. The horrors of the French Revolution, however, brought home to Paine the fact that the mere overthrow of monarchy did not usher in light and order. When he protested the execution of Louis XVI, he was accused of sympathy with the crown and imprisoned. He was saved from trial by the American ambassador, James Monroe, who offered him an American citizenship and safe passage back to New York.

Paine spent the last years of his life in New York City and in New Rochelle, New York. They were unhappy, impoverished years, and his reputation suffered enormously as a result of *The Age of Reason* (1794). Paine's attempt to define his beliefs was viewed as an attack on Christianity and, by extension, on conventional society. He was ridiculed and despised. Even George Washington, who had supported Paine's early writing, thought English criticism of him was "not a bad thing." Paine had clearly outlived his time. He was buried on his farm at New Rochelle after his request for a Quaker grave site was refused. Ten years later an enthusiastic admirer exhumed his bones with the intention of having him reburied in England. The admirer's plans came to nothing, and the whereabouts of Paine's grave is, at present, unknown.

Paine's great gift as a stylist was "plainness." He said he needed no "ceremonious expressions." "It is my design," he wrote, "to make those who can scarcely read understand," to put his arguments in a language "as plain as the alphabet," and to shape everything "to fit the powers of thinking and the turn of language to the subject, so as to bring out a clear conclusion that shall hit the point in question and nothing else."

From Common Sense[1]

Introduction

Perhaps the sentiments contained in the following pages are not yet sufficiently fashionable to procure them general favor; a long habit of not thinking a thing wrong, gives it a superficial appearance of being right, and raises at first a formidable outcry in defence of custom. But the tumult soon subsides. Time makes more converts than reason.

As a long and violent abuse of power is generally the means of calling the right of it in question (and in matters too which might never have been thought of, had not the sufferers been aggravated into the inquiry), and as the King of England hath undertaken in his own right, to support the Parliament in what he calls theirs, and as the good people of this country are grievously oppressed by the combination, they have an undoubted privilege to inquire into the

---

1. The full title is *Common Sense: Addressed to the Inhabitants of America, on the following Interesting Subjects: viz.: I. Of the Origin and Design of Government in General; with Concise Remarks on the English Constitution. II. Of Monarchy and Hereditary Succession. III.* *Thoughts on the Present State of American Affairs. IV. Of the Present Ability of America; with some Miscellaneous Reflections.* The text used here is from *The Writings of Thomas Paine,* Vol. 1, ed. M. D. Conway (New York, 1894–96).

pretensions of both, and equally to reject the usurpation of either.

In the following sheets, the author hath studiously avoided everything which is personal among ourselves. Compliments as well as censure to individuals make no part thereof. The wise and the worthy need not the triumph of a pamphlet; and those whose sentiments are injudicious or unfriendly will cease of themselves, unless too much pains is bestowed upon their conversions.

The cause of America is in a great measure the cause of all mankind. Many circumstances have, and will, arise which are not local, but universal, and through which the principles of all lovers of mankind are affected, and in the event of which their affections are interested. The laying a country desolate with fire and sword, declaring war against the natural rights of all mankind, and extirpating the defenders thereof from the face of the earth, is the concern of every man to whom nature hath given the power of feeling; of which class, regardless of party censure, is

The Author

From *III. Thoughts on the Present State of American Affairs*

In the following pages I offer nothing more than simple facts, plain arguments, and common sense: and have no other preliminaries to settle with the reader, than that he will divest himself of prejudice and prepossession, and suffer his reason and his feelings to determine for themselves: that he will put on, or rather that he will not put off, the true character of a man, and generously enlarge his views beyond the present day.

Volumes have been written on the subject of the struggle between England and America. Men of all ranks have embarked in the controversy, from different motives, and with various designs; but all have been ineffectual, and the period of debate is closed. Arms as the last resource decide the contest; the appeal was the choice of the King, and the continent has accepted the challenge.

It hath been reported of the late Mr. Pelham[2] (who though an able minister was not without his faults) that on his being attacked in the House of Commons on the score that his measures were only of a temporary kind, replied, "they will last my time." Should a thought so fatal and unmanly possess the colonies in the present contest, the name of ancestors will be remembered by future generations with detestation.

The sun never shined on a cause of greater worth. 'Tis not the affair of a city, a county, a province, or a kingdom; but of a continent—of at least one eighth part of the habitable globe. 'Tis not the concern of a day, a year, or an age; posterity are virtually involved in the contest, and will be more or less affected even to the end of time, by the proceedings now. Now is the seed time of conti-

---

2. Prime Minister of Britain (1743–54).

nental union, faith and honor. The least fracture now will be like a name engraved with the point of a pin on the tender rind of a young oak; the wound would enlarge with the tree, and posterity read it in full grown characters.

By referring the matter from argument to arms, a new era for politics is struck—a new method of thinking hath arisen. All plans, proposals, etc. prior to the nineteenth of April, i.e., to the commencement of hostilities,[3] are like the almanacs of the last year; which though proper then, are superceded and useless now. Whatever was advanced by the advocates on either side of the question then, terminated in one and the same point, viz., a union with Great Britain; the only difference between the parties was the method of effecting it; the one proposing force, the other friendship; but it hath so far happened that the first hath failed, and the second hath withdrawn her influence.

As much hath been said of the advantages of reconciliation, which, like an agreeable dream, hath passed away and left us as we were, it is but right that we should examine the contrary side of the argument, and inquire into some of the many material injuries which these colonies sustain, and always will sustain, by being connected with and dependent on Great Britain. To examine that connection and dependence, on the principles of nature and common sense, to see what we have to trust to, if separated, and what we are to expect, if dependent.

I have heard it asserted by some, that as America has flourished under her former connection with Great Britain, the same connection is necessary towards her future happiness, and will always have the same effect. Nothing can be more fallacious than this kind of argument. We may as well assert that because a child has thrived upon milk, that it is never to have meat, or that the first twenty years of our lives is to become a precedent for the next twenty. But even this is admitting more than is true; for I answer roundly, that America would have flourished as much, and probably much more, had no European power taken any notice of her. The commerce by which she hath enriched herself are the necessaries of life, and will always have a market while eating is the custom of Europe.

But she has protected us, say some. That she hath engrossed[4] us is true, and defended the continent at our expense as well as her own, is admitted; and she would have defended Turkey from the same motive, viz., for the sake of trade and dominion.

Alas! we have been long led away by ancient *prejudices* and made large sacrifices to superstition. We have boasted the protection of Great Britain without considering that her motive was interest not attachment; and that she did not protect us from our enemies on

---

3. The "Minutemen" of Lexington, Massachusetts, defended their ammunition stores against the British on April 19, 1775, and engaged in the first armed conflict of the American Revolution.
4. Dominated.

our account; but from her enemies on her own account, from those who had no quarrel with us on any other account, and who will always be our enemies on the same account. Let Britain waive her pretensions to the continent, or the continent throw off the dependence, and we should be at peace with France and Spain, were they at war with Britain. The miseries of Hanover's last war[5] ought to warn us against connections.

It hath lately been asserted in Parliament, that the colonies have no relation to each other but through the parent country, i.e., that Pennsylvania and the Jerseys,[6] and so on for the rest, are sister colonies by the way of England; this is certainly a very roundabout way of proving relationship, but it is the nearest and only true way of proving enmity (or enemyship, if I may so call it). France and Spain never were, nor perhaps ever will be, our enemies as Americans, but as our being the subjects of Great Britain.

But Britain is the parent country, say some. Then the more shame upon her conduct. Even brutes do not devour their young, nor savages make war upon their families; Wherefore, the assertion, if true, turns to her reproach; but it happens not to be true, or only partly so, and the phrase parent or mother country hath been jesuitically[7] adopted by the King and his parasites, with a low papistical design of gaining an unfair bias on the credulous weakness of our minds. Europe, and not England, is the parent country of America. This new world hath been the asylum for the persecuted lovers of civil and religious liberty from every part of Europe. Hither have they fled, not from the tender embraces of the mother, but from the cruelty of the monster; and it is so far true of England, that the same tyranny which drove the first emigrants from home, pursues their descendants still.

In this extensive quarter of the globe, we forget the narrow limits of three hundred and sixty miles (the extent of England) and carry our friendship on a larger scale; we claim brotherhood with every European Christian, and triumph in the generosity of the sentiment.

It is pleasant to observe by what regular gradations we surmount the force of local prejudices, as we enlarge our acquaintance with the world. A man born in any town in England divided into parishes, will naturally associate most with his fellow parishioners (because their interests in many cases will be common) and distinguish him by the name of neighbor; if he meet him but a few miles from home, he drops the narrow idea of a street, and salutes

5. King George III of Great Britain was a descendant of the Prussian House of Hanover; Paine is referring to the Seven Years' War (1756–63), which originally involved Prussia and Austria and grew to involve all the major European powers. American losses in the French and Indian campaigns were heavy, even though the war was settled in Britain's favor.
6. The colony was divided into East and West Jersey.
7. I.e., cunningly.

him by the name of townsman; if he travel out of the county and meet him in any other, he forgets the minor divisions of street and town, and calls him countryman, i.e., countyman: but if in their foreign excursions they should associate in France, or any other part of Europe, their local remembrance would be enlarged into that of Englishmen. And by a just parity of reasoning, all Europeans meeting in America, or any other quarter of the globe, are countrymen; for England, Holland, Germany, or Sweden, when compared with the whole, stand in the same places on the larger scale, which the divisions of street, town, and county do on the smaller ones; distinctions too limited for continental minds. Not one third of the inhabitants, even of this province,[8] are of English descent. Wherefore, I reprobate the phrase of parent or mother country applied to England only, as being false, selfish, narrow and ungenerous.

But, admitting that we were all of English descent, what does it amount to? Nothing. Britain, being now an open enemy, extinguishes every other name and title: and to say that reconciliation is our duty is truly farcical. The first King of England of the present line (William the Conqueror) was a Frenchman, and half the peers of England are descendants from the same country; wherefore, by the same method of reasoning, England ought to be governed by France.

Much hath been said of the united strength of Britain and the colonies, that in conjunction they might bid defiance to the world: but this is mere presumption; the fate of war is uncertain, neither do the expressions mean anything; for this continent would never suffer itself to be drained of inhabitants to support the British arms in either Asia, Africa, or Europe.

Besides, what have we to do with setting the world at defiance? Our plan is commerce, and that, well attended to, will secure us the peace and friendship of all Europe; because it is the interest of all Europe to have America a free port. Her trade will always be a protection, and her barrenness of gold and silver secure her from invaders.

I challenge the warmest advocate for reconciliation to show a single advantage that this continent can reap by being connected with Great Britain. I repeat the challenge; not a single advantage is derived. Our corn will fetch its price in any market in Europe, and our imported goods must be paid for buy them where we will.

But the injuries and disadvantages which we sustain by that connection, are without number; and our duty to mankind at large, as well as to ourselves, instruct us to renounce the alliance: because, any submission to, or dependence on, Great Britain tends directly to involve this continent in European wars and quarrels, and set us at variance with nations who would otherwise seek our friendship,

8. I.e., Pennsylvania.

176 * Thomas Paine

and against whom we have neither anger nor complaint. As Europe is our market for trade, we ought to form no partial connection with any part of it. It is the true interest of America to steer clear of European contentions, which she never can do, while, by her dependence on Britain, she is made the makeweight in the scale of British politics.

Europe is too thickly planted with kingdoms to be long at peace, and whenever a war breaks out between England and any foreign power, the trade of America goes to ruin, because of her connection with Britain. The next war may not turn out like the last,[9] and should it not, the advocates for reconciliation now will be wishing for separation then, because neutrality in that case would be a safer convoy than a man of war. Everything that is right or reasonable pleads for separation. The blood of the slain, the weeping voice of nature cries, " 'Tis time to part." Even the distance at which the Almighty hath placed England and America is a strong and natural proof that the authority of the one over the other was never the design of Heaven. The time likewise at which the continent was discovered adds weight to the argument, and the manner in which it was peopled increases the force of it. The Reformation was preceded by the discovery of America: as if the Almighty graciously meant to open a sanctuary to the persecuted in future years, when home should afford neither friendship nor safety.

The authority of Great Britain over this continent is a form of government which sooner or later must have an end: and a serious mind can draw no true pleasure by looking forward, under the painful and positive conviction that what he calls "the present constitution" is merely temporary. As parents, we can have no joy, knowing that this government is not sufficiently lasting to insure anything which we may bequeath to posterity: and by a plain method of argument, as we are running the next generation into debt, we ought to do the work of it, otherwise we use them meanly and pitifully. In order to discover the line of our duty rightly, we should take our children in our hand, and fix our station a few years farther into life; that eminence will present a prospect which a few present fears and prejudices conceal from our sight.

Though I would carefully avoid giving unnecessary offense, yet I am inclined to believe, that all those who espouse the doctrine of reconciliation, may be included within the following descriptions.

Interested men who are not to be trusted, weak men who cannot see, prejudiced men who will not see, and a certain set of moderate men who think better of the European world than it deserves; and this last class, by an ill-judged deliberation, will be the cause of more calamities to this continent than all the other three.

---

9. The Seven Years' War concluded with the Treaty of Paris (1763), and Britain gained all the French territory in North America.

It is the good fortune of many to live distant from the scene of present sorrow; the evil is not sufficiently brought to their doors to make them feel the precariousness with which all American property is possessed. But let our imaginations transport us a few moments to Boston; that seat of wretchedness will teach us wisdom, and instruct us forever to renounce a power in whom we can have no trust.[1] The inhabitants of that unfortunate city, who but a few months ago were in ease and affluence, have now no other alternative than to stay and starve, or turn out to beg. Endangered by the fire of their friends if they continue within the city, and plundered by the soldiery if they leave it, in their present situation they are prisoners without the hope of redemption, and in a general attack for their relief they would be exposed to the fury of both armies.

Men of passive tempers look somewhat lightly over the offenses of Great Britain, and, still hoping for the best, are apt to call out, "Come, come, we shall be friends again for all this." But examine the passions and feelings of mankind: bring the doctrine of reconciliation to the touchstone of nature, and then tell me whether you can hereafter love, honor, and faithfully serve the power that hath carried fire and sword into your land? If you cannot do all these, then are you only deceiving yourselves, and by your delay bringing ruin upon posterity. Your future connection with Britain, whom you can neither love nor honor, will be forced and unnatural, and, being formed only on the plan of present convenience, will in a little time fall into a relapse more wretched than the first. But if you say, you can still pass the violations over, then I ask, hath your house been burnt? Hath your property been destroyed before your face? Are your wife and children destitute of a bed to lie on, or bread to live on? Have you lost a parent or a child by their hands, and yourself the ruined and wretched survivor? If you have not, then are you not a judge of those who have. But if you have, and can still shake hands with the murderers, then are you unworthy the name of husband, father, friend, or lover, and whatever may be your rank or title in life, you have the heart of a coward, and the spirit of a sycophant.

This is not inflaming or exaggerating matters, but trying them by those feelings and affections which nature justifies, and without which we should be incapable of discharging the social duties of life, or enjoying the felicities of it. I mean not to exhibit horror for the purpose of provoking revenge, but to awaken us from fatal and unmanly slumbers, that we may pursue determinately some fixed object. 'Tis not in the power of Britain or of Europe to conquer America, if she doth not conquer herself by delay and timidity. The present winter is worth an age if rightly employed, but if lost or neglected the whole continent will partake of the misfortune; and

1. Boston was under British military occupation and blockaded for six months.

there is no punishment which that man doth not deserve, be he who, or what, or where he will, that may be the means of sacrificing a season so precious and useful.

'Tis repugnant to reason, to the universal order of things, to all examples from former ages, to suppose that this continent can long remain subject to any external power. The most sanguine in Britain doth not think so. The utmost stretch of human wisdom cannot, at this time, compass a plan, short of separation, which can promise the continent even a year's security. Reconciliation is now a fallacious dream. Nature hath deserted the connection, and art cannot supply her place. For, as Milton wisely expresses, "never can true reconcilement grow where wounds of deadly hate have pierced so deep."[2]

* * *

A government of our own is our natural right: and when a man seriously reflects on the precariousness of human affairs, he will become convinced that it is infinitely wiser and safer to form a constitution of our own in a cool deliberate manner, while we have it in our power, than to trust such an interesting event to time and chance. If we omit it now, some Massanello[3] may hereafter arise, who, laying hold of popular disquietudes, may collect together the desperate and the discontented, and by assuming to themselves the powers of government, finally sweep away the liberties of the continent like a deluge. Should the government of America return again into the hands of Britain, the tottering situation of things will be a temptation for some desperate adventurer to try his fortune; and in such a case, what relief can Britain give? Ere she could hear the news, the fatal business might be done; and ourselves suffering like the wretched Britons under the oppression of the conqueror. Ye that oppose independence now, ye know not what ye do: ye are opening a door to eternal tyranny by keeping vacant the seat of government. There are thousands and tens of thousands, who would think it glorious to expel from the continent that barbarous and hellish power, which hath stirred up the Indians and the Negroes to destroy us; the cruelty hath a double guilt: it is dealing brutally by us, and treacherously by them.

To talk of friendship with those in whom our reason forbids us to have faith, and our affections wounded through a thousand pores instruct us to detest, is madness and folly. Every day wears out the little remains of kindred between us and them; and can there be any reason to hope, that as the relationship expires, the affection will increase, or that we shall agree better when we have ten times more and greater concerns to quarrel over than ever?

2. *Paradise Lost*, IV.98–99.
3. "Thomas Anello, otherwise Massanello, a fisherman of Naples, who after spiriting up his countrymen in the public market place, against the oppression of the Spaniards, to whom the place was then subject, prompted them to revolt, and in the space of a day became King" [Paine's note].

Ye that tell us of harmony and reconciliation, can ye restore to us the time that is past? Can ye give to prostitution its former innocence? Neither can ye reconcile Britian and America. The last cord now is broken, the people of England are presenting addresses against us. There are injuries which nature cannot forgive; she would cease to be nature if she did. As well can the lover forgive the ravisher of his mistress, as the continent forgive the murders of Britain. The Almighty hath implanted in us these unextinguishable feelings for good and wise purposes. They are the guardians of His image in our hearts. They distinguish us from the herd of common animals. The social compact would dissolve, and justice be extirpated from the earth, or have only a casual existence were we callous to the touches of affection. The robber and the murderer would often escape unpunished, did not the injuries which our tempers sustain provoke us into justice.

O! ye that love mankind! Ye that dare oppose not only the tyranny but the tyrant, stand forth! Every spot of the old world is overrun with oppression. Freedom hath been hunted round the globe. Asia and Africa have long expelled her. Europe regards her like a stranger, and England hath given her warning to depart. O! receive the fugitive, and prepare in time an asylum for mankind.

1776

The Crisis,[1] No. 1

These are the times that try men's souls. The summer soldier and the sunshine patriot will, in this crisis, shrink from the service of their country; but he that stands it now, deserves the love and thanks of man and woman. Tyranny, like hell, is not easily conquered; yet we have this consolation with us, that the harder the conflict, the more glorious the triumph. What we obtain too cheap, we esteem too lightly: it is dearness only that gives everything its value. Heaven knows how to put a proper price upon its goods; and it would be strange indeed if so celestial an article as freedom should not be highly rated. Britain, with an army to enforce her tyranny, has declared that she has a right (not only to tax) but "to bind us in all cases whatsoever," and if being bound in that manner is not slavery, then is there not such a thing as slavery upon earth. Even the expression is impious; for so unlimited a power can belong only to God.

Whether the independence of the continent was declared too soon, or delayed too long, I will not now enter into as an argument; my own simple opinion is, that had it been eight months earlier, it would have been much better. We did not make a proper use of

---

1. The first of 16 pamphlets which appeared under this title. Paine sometimes referred to this particular essay as *The American Crisis*. There were three pamphlet editions in one week: one undated, one dated December 19, and the one reprinted here, dated December 23. The text used here is from *The Writings of Thomas Paine*, Vol. 1, ed. M. D. Conway (New York, 1894–96).

last winter, neither could we, while we were in a dependent state. However, the fault, if it were one, was all our own;[2] we have none to blame but ourselves. But no great deal is lost yet. All that Howe[3] has been doing for this month past is rather a ravage than a conquest, which the spirit of the Jerseys,[4] a year ago, would have quickly repulsed, and which time and a little resolution will soon recover.

I have as little superstition in me as any man living, but my secret opinion has ever been, and still is, that God Almighty will not give up a people to military destruction, or leave them unsupportedly to perish, who have so earnestly and so repeatedly sought to avoid the calamities of war, by every decent method which wisdom could invent. Neither have I so much of the infidel in me as to suppose that He has relinquished the government of the world, and given us up to the care of devils; and as I do not, I cannot see on what grounds the King of Britain can look up to heaven for help against us: a common murderer, a highwayman, or a housebreaker has as good a pretense as he.

'Tis surprising to see how rapidly a panic will sometimes run through a country. All nations and ages have been subject to them: Britain has trembled like an ague[5] at the report of a French fleet of flat-bottomed boats; and in the fourteenth[6] century the whole English army, after ravaging the kingdom of France, was driven back like men petrified with fear; and this brave exploit was performed by a few broken forces collected and headed by a woman, Joan of Arc. Would that heaven might inspire some Jersey maid to spirit up her countrymen, and save her fair fellow sufferers from ravage and ravishment! Yet panics, in some cases, have their uses; they produce as much good as hurt. Their duration is always short; the mind soon grows through them, and acquires a firmer habit than before. But their peculiar advantage is that they are the touchstones of sincerity and hypocrisy, and bring things and men to light, which might otherwise have lain forever undiscovered. In fact, they have the same effect on secret traitors, which an imaginary apparition would have upon a private murderer. They sift out the hidden thoughts of man, and hold them up in public to the world. Many a disguised tory[7] has lately shown his head, that shall penitentially solemnize with curses the day on which Howe arrived upon the Delaware.

As I was with the troops at Fort Lee, and marched with them to the edge of Pennsylvania, I am well acquainted with many circum-

2. "The present winter is worth an age, if rightly employed; but, if lost or neglected, the whole continent will partake of the evil; and there is no punishment that man does not deserve, be he who, or what, or where he will, that may be the means of sacrificing a season so precious and useful" [Paine's note, taken from *Common Sense*]. Paine wanted an immediate declaration of independence, uniting the colonies and enlisting the aid

of France and Spain.
3. Lord William Howe (1729–1814) was Commander of the British army in America from 1775 to 1778.
4. The colony was divided into East and West Jersey.
5. I.e., like one who is chilled.
6. Properly, the 15th century. Joan of Arc led the French to victory over the English in 1429.
7. I.e., supporter of the King.

stances, which those who live at a distance know but little or nothing of. Our situation there was exceedingly cramped, the place
being a narrow neck of land between the North River[8] and the
Hackensack. Our force was inconsiderable, being not one fourth so
great as Howe could bring against us. We had no army at hand to
have relieved the garrison, had we shut ourselves up and stood on
our defense. Our ammunition, light artillery, and the best part of
our stores had been removed on the apprehension that Howe would
endeavor to penetrate the Jerseys, in which case Fort Lee could be of
no use to us; for it must occur to every thinking man, whether in the
army or not, that these kind of field forts are only for temporary
purposes, and last in use no longer than the enemy directs his force
against the particular object, which such forts are raised to defend.
Such was our situation and condition at Fort Lee on the morning
of the 20th of November, when an officer arrived with information
that the enemy with 200 boats had landed about seven miles
above: Major General Green,[9] who commanded the garrison,
immediately ordered them under arms, and sent express to General
Washington at the town of Hackensack, distant, by the way of the
ferry, six miles. Our first object was to secure the bridge over the
Hackensack, which laid up the river between the enemy and us,
about six miles from us, and three from them. General Washington
arrived in about three quarters of an hour, and marched at the head
of the troops towards the bridge, which place I expected we should
have a brush for; however, they did not choose to dispute it
with us, and the greatest part of our troops went over the bridge,
the rest over the ferry, except some which passed at a mill on a
small creek, between the bridge and the ferry, and made their way
through some marshy grounds up to the town of Hackensack, and
there passed the river. We brought off as much baggage as the
wagons could contain, the rest was lost. The simple object was to
bring off the garrison, and march them on till they could be
strengthened by the Jersey or Pennsylvania militia, so as to be
enabled to make a stand. We staid four days at Newark, collected
our outposts with some of the Jersey militia, and marched out twice
to meet the enemy, on being informed that they were advancing,
though our numbers were greatly inferior to theirs. Howe, in my
little opinion, committed a great error in generalship in not throwing a body of forces off from Staten Island through Amboy, by
which means he might have seized all our stores at Brunswick, and
intercepted our march into Pennsylvania; but if we believe the
power of hell to be limited, we must likewise believe that their
agents are under some providential control.[1]

8. I.e., the Hudson River.
9. Paine was aide-de-camp to Major General Nathanael Greene (1742–86).
1. The American losses were larger than
Paine implies. General Howe took 3,000
prisoners and a large store of military
supplies when he captured Fort Lee.
Paine wrote *The Crisis, No. 1* while
serving with Washington's army as it
retreated through New Jersey.

I shall not now attempt to give all the particulars of our retreat to the Delaware; suffice it for the present to say, that both officers and men, though greatly harassed and fatigued, frequently without rest, covering, or provision, the inevitable consequences of a long retreat, bore it with a manly and martial spirit. All their wishes centered in one, which was that the country would turn out and help them to drive the enemy back. Voltaire has remarked that King William never appeared to full advantage but in difficulties and in action;[2] the same remark may be made on General Washington, for the character fits him. There is a natural firmness in some minds which cannot be unlocked by trifles, but which, when unlocked, discovers a cabinet[3] of fortitude; and I reckon it among those kind of public blessings, which we do not immediately see, that God hath blessed him with uninterrupted health, and given him a mind that can even flourish upon care.

I shall conclude this paper with some miscellaneous remarks on the state of our affairs; and shall begin with asking the following question: Why is it that the enemy have left the New England provinces, and made these middle ones the seat of war? The answer is easy: New England is not infested with tories, and we are. I have been tender in raising the cry against these men, and used numberless arguments to show them their danger, but it will not do to sacrifice a world either to their folly or their baseness. The period is now arrived, in which either they or we must change our sentiments, or one or both must fall. And what is a tory? Good God! what is he? I should not be afraid to go with a hundred whigs[4] against a thousand tories, were they to attempt to get into arms. Every tory is a coward; for servile, slavish, self-interested fear is the foundation of toryism; and a man under such influence, though he may be cruel, never can be brave.

But, before the line of irrecoverable separation be drawn between us, let us reason the matter together: Your conduct is an invitation to the enemy, yet not one in a thousand of you has heart enough to join him. Howe is as much deceived by you as the American cause is injured by you. He expects you will all take up arms, and flock to his standard, with muskets on your shoulders. Your opinions are of no use to him, unless you support him personally, for 'tis soldiers, and not tories, that he wants.

I once felt all that kind of anger, which a man ought to feel, against the mean principles that are held by the tories: a noted one, who kept a tavern at Amboy,[5] was standing at his door, with as pretty a child in his hand, about eight or nine years old, as I ever saw, and after speaking his mind as freely as he thought was pru-

2. Voltaire (1694–1778) made this remark about King William III of England (1650–1702) in his *History of Louis the Fourteenth* (1751).
3. Storehouse.
4. Supporters of the Revolution.
5. Paine was stationed at Perth Amboy, New Jersey, while in the Continental Army.

dent, finished with this unfatherly expression, "Well! give me peace in my day." Not a man lives on the continent but fully believes that a separation must some time or other finally take place, and a generous parent should have said, "If there must be trouble, let it be in my day, that my child may have peace"; and this single reflection, well applied, is sufficient to awaken every man to duty. Not a place upon earth might be so happy as America. Her situation is remote from all the wrangling world, and she has nothing to do but to trade with them. A man can distinguish himself between temper and principle, and I am as confident, as I am that God governs the world, that America will never be happy till she gets clear of foreign dominion. Wars, without ceasing, will break out till that period arrives, and the continent must in the end be conqueror; for though the flame of liberty may sometimes cease to shine, the coal can never expire.

America did not, nor does not, want force; but she wanted a proper application of that force. Wisdom is not the purchase of a day, and it is no wonder that we should err at the first setting off. From an excess of tenderness, we were unwilling to raise an army, and trusted our cause to the temporary defense of a well-meaning militia. A summer's experience has now taught us better; yet with those troops, while they were collected, we were able to set bounds to the progress of the enemy, and thank God! they are again assembling. I always considered militia as the best troops in the world for a sudden exertion, but they will not do for a long campaign. Howe, it is probable, will make an attempt on this city;[6] should he fail on this side the Delaware, he is ruined: if he succeeds, our cause is not ruined. He stakes all on his side against a part on ours; admitting he succeeds, the consequences will be that armies from both ends of the continent will march to assist their suffering friends in the middle states; for he cannot go everywhere, it is impossible. I consider Howe as the greatest enemy the tories have; he is bringing a war into their country, which, had it not been for him and partly for themselves, they had been clear of. Should he now be expelled, I wish with all the devotion of a Christian, that the names of whig and tory may never more be mentioned; but should the tories give him encouragement to come, or assistance if he come, I as sincerely wish that our next year's arms may expel them from the continent, and the congress appropriate their possessions to the relief of those who have suffered in well-doing. A single successful battle next year will settle the whole. America could carry on a two years' war by the confiscation of the property of disaffected persons, and be made happy by their expulsion. Say not that this is revenge; call it rather the soft resentment of a suffering people, who, having no object in view but the good of all, have staked their own all upon a seemingly doubtful

6. Philadelphia.

event. Yet it is folly to argue against determined hardness; eloquence may strike the ear, and the language of sorrow draw forth the tear of compassion, but nothing can reach the heart that is steeled with prejudice.

Quitting this class of men, I turn with the warm ardor of a friend to those who have nobly stood, and are yet determined to stand the matter out: I call not upon a few, but upon all: not on this state or that state, but on every state: up and help us; lay your shoulders to the wheel; better have too much force than too little, when so great an object is at stake. Let it be told to the future world that in the depth of winter, when nothing but hope and virtue could survive, that the city and the country, alarmed at one common danger, came forth to meet and to repulse it. Say not that thousands are gone, turn out your tens of thousands;[7] throw not the burden of the day upon Providence, but "show your faith by your works"[8] that God may bless you. It matters not where you live, or what rank of life you hold, the evil or the blessing will reach you all. The far and the near, the home counties and the back,[9] the rich and poor will suffer or rejoice alike. The heart that feels not now is dead: the blood of his children will curse his cowardice who shrinks back at a time when a little might have saved the whole, and made them happy. I love the man that can smile in trouble, that can gather strength from distress, and grow brave by reflection. 'Tis the business of little minds to shrink; but he whose heart is firm, and whose conscience approves his conduct, will pursue his principles unto death. My own line of reasoning is to myself as straight and clear as a ray of light. Not all the treasures of the world, so far as I believe, could have induced me to support an offensive war, for I think it murder; but if a thief breaks into my house, burns and destroys my property, and kills or threatens to kill me, or those that are in it, and to "bind me in all cases whatsoever"[1] to his absolute will, am I to suffer it? What signifies it to me, whether he who does it is a king or a common man; my countryman or not my countryman; whether it be done by an individual villain, or an army of them? If we reason to the root of things we shall find no difference; neither can any just cause be assigned why we should punish in the one case and pardon in the other. Let them call me rebel, and welcome, I feel no concern from it; but I should suffer the misery of devils were I to make a whore of my soul by swearing allegiance to one whose character is that of a sottish, stupid, stubborn, worthless, brutish man. I conceive likewise a horrid idea in receiving mercy from a being, who at the last day shall be shrieking to the rocks and mountains to cover him, and fleeing with terror from the orphan, the widow, and the slain of America.

There are cases which cannot be overdone by language, and this is one. There are persons, too, who see not the full extent of the evil which threatens them; they solace themselves with hopes that the enemy, if he succeed, will be merciful. It is the madness of folly to expect mercy from those who have refused to do justice; and even mercy, where conquest is the object, is only a trick of war; the cunning of the fox is as murderous as the violence of the wolf, and we ought to guard equally against both. Howe's first object is, partly by threats and partly by promises, to terrify or seduce the people to deliver up their arms and receive mercy. The ministry recommended the same plan to Gage,[2] and this is what the tories call making their peace, "a peace which passeth all understanding" indeed![3] A peace which would be the immediate forerunner of a worse ruin than any we have yet thought of. Ye men of Pennsylvania, do reason upon these things! Were the back counties to give up their arms, they would fall an easy prey to the Indians, who are all armed: this perhaps is what some tories would not be sorry for. Were the home counties to deliver up their arms, they would be exposed to the resentment of the back counties, who would then have it in their power to chastise their defection at pleasure. And were any one state to give up its arms, that state must be garrisoned by all Howe's army of Britons and Hessians[4] to preserve it from the anger of the rest. Mutual fear is the principal link in the chain of mutual love, and woe be to that state that breaks the compact. Howe is mercifully inviting you to barbarous destruction, and men must be either rogues or fools that will not see it. I dwell not upon the vapors of imagination: I bring reason to your ears, and, in language as plain as A, B, C, hold up truth to your eyes.

I thank God that I fear not. I see no real cause for fear. I know our situation well, and can see the way out of it. While our army was collected, Howe dared not risk a battle; and it is no credit to him that he decamped from the White Plains,[5] and waited a mean opportunity to ravage the defenceless Jerseys; but it is great credit to us, that, with a handful of men, we sustained an orderly retreat for near an hundred miles, brought off our ammunition, all our field pieces, the greatest part of our stores, and had four rivers to pass. None can say that our retreat was precipitate, for we were near three weeks in performing it, that the country[6] might have time to come in. Twice we marched back to meet the enemy, and remained out till dark. The sign of fear was not seen in our camp, and had

2. General Thomas Gage, who commanded the British armies in America from 1763 to 1775, prior to Howe.
3. "And the peace of God, which passeth all understanding, shall keep your hearts and minds through Christ Jesus" (Philippians 4.7).
4. German mercenaries.
5. At White Plains, New York, on October 28, 1776, General Howe successfully overcame Washington's troops, but failed to take full advantage of his victory.
6. I.e., the local volunteers.

not some of the cowardly and disaffected inhabitants spread false alarms through the country, the Jerseys had never been ravaged. Once more we are again collected and collecting; our new army at both ends of the continent is recruiting fast, and we shall be able to open the next campaign with sixty thousand men, well armed and clothed. This is our situation, and who will may know it. By perseverance and fortitude we have the prospect of a glorious issue; by cowardice and submission, the sad choice of a variety of evils—a ravaged country—a depopulated city—habitations without safety, and slavery without hope—our homes turned into barracks and bawdyhouses for Hessians, and a future race to provide for, whose fathers we shall doubt of. Look on this picture and weep over it! and if there yet remains one thoughtless wretch who believes it not, let him suffer it unlamented.

Common Sense
1776

THOMAS JEFFERSON
1743–1826

President of the United States, first Secretary of State, and Minister to France, Governor of Virginia, and Congressman, Thomas Jefferson once said that he wished to be remembered for only three things: drafting the Declaration of Independence, writing and supporting the Virginia Statute for Religious Freedom (1786), and founding the University of Virginia. Jefferson might well have included a number of other accomplishments in this list: he was a remarkable architect and designed the Virginia state capital, his residence Monticello, and the original buildings for the University of Virginia; he farmed thousands of acres and built one of the most beautiful plantations in America; he had a library of some 10,000 volumes, which served as the basis for the Library of Congress, and a collection of paintings and sculpture that made him America's greatest patron of the arts; and was known the world over for his spirit of scientific inquiry and as the creator of a number of remarkable inventions. The three acts for which he wished to be remembered, however, have this in common: they all testify to Jefferson's lifelong passion to liberate the human mind from tyranny, whether imposed by the state, the church, or our own ignorance.

Jefferson was born at Shadwell, in what is now Albemarle County, Virginia. Peter Jefferson, his father, was a county official and surveyor. He made the first accurate map of Virginia, something of which Jefferson was always proud. When his father died Thomas was only fourteen. Peter Jefferson left his son 2,750 acres of land, and Jefferson added to this acreage until he died; at one time he owned almost 10,000 acres. Jefferson tells us in his *Autobiography* that his father's education had been "quite neglected" but that he was always "eager after information" and determined to improve himself. In 1760, when Jefferson entered William and Mary College in Williamsburg, Virginia, he had mastered Latin and Greek, played the vi-

olin respectably, and was a skilled horseman. He was tall and a bit awkward
looking, but a good companion. Williamsburg was the capital of Virginia
as well as a college town, and Jefferson was fortunate enough to make the
acquaintance of three men who strongly influenced his life: Governor
Francis Fauquier, a Fellow of the Royal Society; George Wythe, one of
the best teachers of law in the country; and Dr. William Small, an emigrant from Scotland who taught mathematics and philosophy and who introduced Jefferson, as Garry Wills has put it, "to the invigorating realm of
the Scottish Enlightenment," especially the work of Francis Hutcheson,
author of *An Inquiry into the Original of Our Ideas of Beauty and Virtue*
(1725), and Lord Kames (Henry Home), author of *Essays on the Principles of Morality and Natural Religion* (1751). Jefferson flourished in Williamsburg, and it is hard to imagine a city in America where his natural interests and talents could have been more sympathetically encouraged.

Jefferson stayed on in Williamsburg to read law after graduation, and was
admitted to the bar. In 1769 he was elected to the Virginia House of Bur-
gesses and began a distinguished career in the legislature. In 1774 he wrote
an influential and daring pamphlet on *A Summary View of the Rights of
British America*, denying all parliamentary authority over America and arguing that ties to the British monarchy were voluntary and not irrevocable.
Jefferson's reputation as a writer preceded him to Philadelphia, where he
was a delegate to the second Continental Congress, and on June 11, 1776,
he was elected to join Benjamin Franklin, John Adams, Roger Sherman,
and Robert Livingston in drafting a declaration of independence. Although
committee members made suggestions, the draft was very much Jefferson's
own. As Garry Wills has recently shown, Jefferson was unhappy with the
changes made by Congress to his draft, and rightly so; for congressional
changes went contrary to some of his basic arguments. Jefferson wished to
place the British *people* on record as the ultimate cause of the Revolution,
because they tolerated a corrupt Parliament and King; and he wished to include a strong statement against slavery. Congress tolerated neither passage.
Jefferson was justified, however, in asking that he be remembered as the author of the Declaration. It was, as Dumas Malone, Jefferson's biographer,
once put it, a "dangerous but glorious opportunity." Whether as the result
of these frustrations, or merely Jefferson's wish to be nearer his family, he
left the Congress in September, 1776, and entered the Virginia House of
Delegates. In 1779 he was elected Governor, and although re-elected the
following year, Jefferson's term of office came to an ignominious end when
he resigned. After the British captured Richmond in 1781, Jefferson and
the legislature moved to Charlottesville, and he and the legislators barely escaped imprisonment when the pursuing British army descended on them at
Monticello. Jefferson's resignation and the lack of preparations for the defense of the city were held against him, and it was some time before he regained the confidence of Virginians.

From 1781 to 1784 Jefferson withdrew from public life and remained at
Monticello, completing his only book, *Notes on the State of Virginia*. In
1784 he was appointed Minister to France and served with Benjamin
Franklin on the commission which signed the Treaty of Paris, ending the
Revolutionary War. He returned to Monticello in 1789, and in 1790
Washington appointed him the first Secretary of State under the newly
adopted Constitution. After three years he announced his plans for retire-

ment once again and withdrew to Monticello, where he rotated his crops and built a grist mill. But Jefferson's political blood was too thick for retirement, and in 1796 he ran for the office of President, losing to John Adams and taking the office of Vice-President instead. In 1800 he was elected President, the first to be inaugurated in Washington. He named Benjamin Latrobe surveyor of public buildings, and he worked with Latrobe in planning a great city.

When Jefferson returned to Monticello in 1809, he knew that this time his public life was over. For the final seventeen years of his life he kept a watchful eye on everything that grew in Monticello. But Jefferson was never far from the world. He rose every morning to attack his voluminous correspondence. The Library of Congress holds more than 55,000 Jefferson manuscripts and letters, and the most recent edition of his writings will run to sixty volumes. Jefferson left no treatise on political philosophy and, in a sense, was no political thinker. He was always more interested in the practical consequences of ideas. He remained an agrarian aristocrat all his life, and it is to the liberty of mind and the values of the land that he always returned. As Dumas Malone puts it: he was a "homely aristocrat in manner of life and personal tastes; he distrusted all rulers and feared the rise of an industrial proletariat, but more than any of his eminent contemporaries, he trusted the common man, if measurably enlightened and kept in rural virtue." Jefferson died a few hours before John Adams on the Fourth of July, 1826.

From The Autobiography of Thomas Jefferson[1]
From *The Declaration of Independence*

* * *

It appearing in the course of these debates, that the colonies of New York, New Jersey, Pennsylvania, Delaware, Maryland, and South Carolina were not yet matured for falling from the parent stem, but that they were fast advancing to that state, it was thought most prudent to wait a while for them, and to postpone the final decision to July 1st; but, that this might occasion as little delay as possible, a committee was appointed to prepare a Declaration of Independence. The committee were John Adams, Dr. Franklin, Roger Sherman, Robert R. Livingston, and myself. Committees were also appointed, at the same time, to prepare a plan of confederation for the colonies, and to state the terms proper to be proposed for foreign alliance. The committee for drawing the Decla-

1. On June 7, 1776, Richard Henry Lee of Virginia proposed to the second Continental Congress, meeting in Philadelphia, that "these united Colonies are, and of a right ought to be, free and independent states." On June 11, a committee of five—John Adams of Massachusetts, Benjamin Franklin of Pennsylvania, Roger Sherman of Connecticut, Robert Livingston of New York, and Thomas Jefferson of Virginia—was instructed to draft a declaration of independence. The draft presented to Congress on June 28 was primarily the work of Jefferson. Lee's resolution was passed on July 2, and the Declaration was adopted on July 4 with the changes noted by Jefferson in this text, taken from his *Autobiography*. On August 2 a copy in parchment was signed by all the delegates but three, and they signed later. The text used here is from *The Writings of Thomas Jefferson*, ed. A. A. Lipscomb and A. E. Bergh (1903).

ration of Independence, desired me to do it. It was accordingly done, and being approved by them, I reported it to the House on Friday, the 28th of June, when it was read, and ordered to lie on the table. On Monday, the 1st of July, the House resolved itself into a committee of the whole, and resumed the consideration of the original motion made by the delegates of Virginia, which, being again debated through the day, was carried in the affirmative by the votes of New Hampshire, Connecticut, Massachusetts, Rhode Island, New Jersey, Maryland, Virginia, North Carolina and Georgia. South Carolina and Pennsylvania voted against it. Delaware had but two members present, and they were divided. The delegates from New York declared they were for it themselves, and were assured their constituents were for it; but that their instructions having been drawn near a twelve-month before, when reconciliation was still the general object, they were enjoined by them to do nothing which should impede that object. They, therefore, thought themselves not justifiable in voting on either side, and asked leave to withdraw from the question; which was given them. The committee rose and reported their resolution to the House. Mr. Edward Rutledge, of South Carolina, then requested the determination might be put off to the next day, as he believed his colleagues, though they disapproved of the resolution, would then join in it for the sake of unanimity. The ultimate question, whether the House would agree to the resolution of the committee, was accordingly postponed to the next day, when it was again moved, and South Carolina concurred in voting for it. In the meantime, a third member had come post[2] from the Delaware counties, and turned the vote of that colony in favor of the resolution. Members of a different sentiment attending that morning from Pennsylvania also, her vote was changed, so that the whole twelve colonies who were authorized to vote at all, gave their voices for it; and, within a few days, the convention of New York approved of it, and thus supplied the void occasioned by the withdrawing of her delegates from the vote.

Congress proceeded the same day to consider the Declaration of Independence, which had been reported and lain on the table the Friday preceding, and on Monday referred to a committee of the whole. The pusillanimous idea that we had friends in England worth keeping terms with, still haunted the minds of many. For this reason, those passages which conveyed censures on the people of England were struck out, lest they should give them offense. The clause too, reprobating the enslaving the inhabitants of Africa, was struck out in complaisance to South Carolina and Georgia, who had never attempted to restrain the importation of slaves, and who, on the contrary, still wished to continue it. Our northern brethren also, I believe, felt a little tender under those censures; for though their

2. I.e., by stagecoach.

people had very few slaves themselves, yet they had been pretty considerable carriers of them to others. The debates, having taken up the greater parts of the 2d, 3d, and 4th days of July, were, on the evening of the last, closed; the Declaration was reported by the committee, agreed to by the House, and signed by every member present, except Mr. Dickinson.[3] As the sentiments of men are known not only by what they receive, but what they reject also, I will state the form of the Declaration as originally reported. The parts struck out by Congress shall be distinguished by a black line drawn under them, and those inserted by them shall be placed in the margin, or in a concurrent column.

A DECLARATION BY THE REPRESENTATIVES OF THE UNITED
STATES OF AMERICA, IN GENERAL CONGRESS ASSEMBLED.

When, in the course of human events, it becomes necessary for one people to dissolve the political bands which have connected them with another, and to assume among the powers of the earth the separate and equal station to which the laws of nature and of nature's God entitle them, a decent respect to the opinions of mankind requires that they should declare the causes which impel them to the separation.

We hold these truths to be self evident: that all men are created equal;[4] that they are endowed by their Creator with inherent and inalienable rights; that certain among these are life, liberty, and the pursuit of happiness;[5] that to secure these rights, governments are instituted among men, deriving their just powers from the consent of the governed; that whenever any form of government becomes destructive of these ends, it is the right of the people to alter or to abolish it, and to institute new government, laying its foundation on such principles, and organizing its powers in such form, as to them shall seem most likely to effect their safety and happiness. Prudence, indeed, will dictate that governments long established should not be changed for light

3. John Dickinson of Pennsylvania, who opposed it.

4. Garry Wills, in his study of the Declaration (*Inventing America*, 1978), tells us that Jefferson means "equal" in possessing a moral sense: "The moral sense is not only man's *highest* faculty, but the one that is *equal* to all men."

5. In his *Second Treatise on Government* (1689) John Locke defined man's natural rights to "life, liberty, and property." Jefferson's substitution of "pursuit of happiness" has puzzled a number of critics. Wills suggests that Jefferson was less influenced by Locke than the Scottish philosophers, particularly Francis

Hutcheson and his *Inquiry into the Original of Our Ideas of Beauty and Virtue* (1725). Wills tells us that "the pursuit of happiness is a phenomenon both obvious and paradoxical. It supplies us with the ground of human right and the goal of human virtue. It is the basic drive of the self, and the only means given for transcending the self. * * * Men in the eighteenth century felt they could become conscious of their freedom only by discovering how they were bound: When they found what they *must* pursue, they knew they had a *right* to pursue it."

and transient causes; and accordingly all experience hath shown that mankind are more disposed to suffer while evils are sufferable, than to right themselves by abolishing the forms to which they are accustomed. But when a long train of abuses and usurpations, begun at a distinguished[6] period and pursuing invariably the same object, evinces a design to reduce them under absolute despotism, it is their right, it is their duty to throw off such government, and to provide new guards for their future security. Such has been the patient sufferance of these colonies; and such is now the necessity which constrains them to expunge their former systems *alter* of government. The history of the present king of Great Britain[7] is a history of unremitting injuries and usurpa- *repeated* tions, among which appears no solitary fact to contra- *all having* dict the uniform tenor of the rest, but all have in direct object the establishment of an absolute tyranny over these states. To prove this, let facts be submitted to a candid world for the truth of which we pledge a faith yet unsullied by falsehood.

He has refused his assent to laws the most wholesome and necessary for the public good.

He has forbidden his governors to pass laws of immediate and pressing importance, unless suspended in their operation till his assent should be obtained; and, when so suspended, he has utterly neglected to attend to them.

He has refused to pass other laws for the accommodation of large districts of people, unless those people would relinquish the right of representation in the legislature, a right inestimable to them, and formidable to tyrants only.

He has called together legislative bodies at places unusual, uncomfortable, and distant from the depository of their public records, for the sole purpose of fatiguing them into compliance with his measures.

He has dissolved representative houses repeatedly and continually for opposing with manly firmness his invasions on the rights of the people.

He has refused for a long time after such dissolutions to cause others to be elected, whereby the legislative powers, incapable of annihilation, have returned to the people at large for their exercise, the state remaining, in the meantime, exposed to all the dangers of invasion from without and convulsions within.

6. I.e., discernible.　　　7. King George III (1738–1820).

He has endeavored to prevent the population of these states; for that purpose obstructing the laws for naturalization of foreigners, refusing to pass others to encourage their migrations hither, and raising the conditions of new appropriations of lands.

He has suffered the administration of justice totally to cease in some of these states refusing his assent to laws for establishing judiciary powers. *[margin: obstructed by]*

He has made our judges dependent on his will alone for the tenure of their offices, and the amount and payment of their salaries.

He has erected a multitude of new offices, by a self-assumed power and sent hither swarms of new officers to harass our people and eat out their substance.

He has kept among us in times of peace standing armies and ships of war without the consent of our legislatures.

He has affected to render the military independent of, and superior to, the civil power.

He has combined with others[8] to subject us to a jurisdiction foreign to our constitutions and unacknowledged by our laws, giving his assent to their acts of pretended legislation for quartering large bodies of armed troops among us; for protecting them by a mock trial from punishment for any murders which they should commit on the inhabitants of these states; for cutting off our trade with all parts of the world; for imposing taxes on us without our consent; for depriving us [ ] of the benefits of trial by jury; for transporting us beyond seas to be tried for pretended offenses; for abolishing the free system of English laws in a neighboring province,[9] establishing therein an arbitrary government, and enlarging its boundaries, so as to render it at once an example and fit instrument for introducing the same absolute rule into these states; for taking away our charters, abolishing our most valuable laws, and altering fundamentally the forms of our governments; for suspending our own legislatures, and declaring themselves invested with power to legislate for us in all cases whatsoever. *[margin: in many cases] [margin: colonies]*

He has abdicated government here withdrawing his governors, and declaring us out of his allegiance and protection. *[margin: by declaring us out of his protection, and waging war against us.]*

8. I.e., the British Parliament.
9. The Quebec Act of 1774 recognized the Roman Catholic religion in Quebec and extended the borders of the province to the Ohio River: it restored French civil law and thus angered the New England colonies. It was often referred to as one of the "intolerable acts."

He has plundered our seas, ravaged our coasts, burnt our towns, and destroyed the lives of our people.

He is at this time transporting large armies of foreign mercenaries to complete the works of death, desolation and tyranny already begun with circumstances of cruelty and perfidy [ ] unworthy the head of a civilized nation.[1] *scarcely paralleled in the most barbarous ages, and totally*

He has constrained our fellow citizens taken captive on the high seas, to bear arms against their country, to become the executioners of their friends and brethren, or to fall themselves by their hands.

He has [ ] endeavored to bring on the inhabitants of our frontiers, the merciless Indian savages, whose known rule of warfare is an undistinguished destruction of all ages, sexes and conditions of existence. *excited domestic insurrection among us, and has*

He has incited treasonable insurrections of our fellow citizens, with the allurements of forfeiture and confiscation of our property.

He has waged cruel war against human nature itself, violating its most sacred rights of life and liberty in the persons of a distant people who never offended him, captivating and carrying them into slavery in another hemisphere, or to incur miserable death in their transportation thither. This piratical warfare, the opprobrium of INFIDEL powers, is the warfare of the CHRISTIAN king of Great Britain. Determined to keep open a market where MEN should be bought and sold, he has prostituted his negative for suppressing every legislative attempt to prohibit or to restrain this execrable commerce. And that this assemblage of horrors might want no fact of distinguished die, he is now exciting those very people to rise in arms among us, and to purchase that liberty of which he has deprived them, by murdering the people on whom he also obtruded them: thus paying off former crimes committed against the LIBERTIES of one people, with crimes which he urges them to commit against the LIVES of another.

In every stage of these oppressions we have petitioned for redress in the most humble terms: our repeated petitions have been answered only by repeated injuries.

A prince whose character is thus marked by every act which may define a tyrant is unfit to be the ruler of a [ ] people who mean to be free. Future ages will *free* scarcely believe that the hardiness of one man adventured, within the short compass of twelve years only,

1. German soldiers hired by the King for colonial service.

to lay a foundation so broad and so undisguised for
tyranny over a people fostered and fixed in principles of
freedom.

Nor have we been wanting in attentions to our British brethren. We have warned them from time to time
of attempts by their legislature to extend a jurisdiction
over these our states. We have reminded them of the
circumstances of our emigration and settlement here,
no one of which could warrant so strange a pretension:
that these were effected at the expense of our own
blood and treasure, unassisted by the wealth or the
strength of Great Britain: that in constituting indeed
our several forms of government, we had adopted one
common king, thereby laying a foundation for perpetual
league and amity with them: but that submission to
their parliament was no part of our constitution, nor
ever in idea, if history may be credited: and, we [   ]
appealed to their native justice and magnanimity as
well as to the ties of our common kindred to disavow
these usurpations which were likely to interrupt our
connection and correspondence. They too have been
deaf to the voice of justice and of consanguinity, and
when occasions have been given them, by the regular
course of their laws, of removing from their councils
the disturbers of our harmony, they have, by their free
election, reestablished them in power. At this very time
too, they are permitting their chief magistrate to send
over not only soldiers of our common blood, but Scotch
and foreign mercenaries to invade and destroy us. These
facts have given the last stab to agonizing affection, and
manly spirit bids us to renounce forever these unfeeling
brethren. We must endeavor to forget our former love
for them, and hold them as we hold the rest of mankind, enemies in war, in peace friends. We might have
been a free and a great people together; but a communication of grandeur and of freedom, it seems, is below
their dignity. Be it so, since they will have it. The road
to happiness and to glory is open to us, too. We will
tread it apart from them, and acquiesce in the necessity
which denounces[2] our eternal separation [   ]!

Margin notes:
an
unwarrantable
us

have

and we have
conjured
them by
would
inevitably

We must
therefore

and hold them
as we hold the
rest of man-
kind, enemies
in war, in
peace friends.

2. Proclaims.

We therefore the representatives of the United States of America in General Congress assembled, do in the name, and by the authority of the good people of these states reject and renounce all allegiance and subjection to the kings of Great Britain and all others who may hereafter claim by, through or under them; we utterly dissolve all political connection which may heretofore have subsisted between us and the people or parliament of Great Britain: and finally we do assert and declare these colonies to be free and independent states, and that as free and independent states, they have full power to levy war, conclude peace, contract alliances, establish commerce, and to do all other acts and things which independent states may of right do.

And for the support of this declaration, we mutually pledge to each other our lives, our fortunes, and our sacred honor.

We, therefore, the representatives of the United States of America in General Congress assembled, appealing to the supreme judge of the world for the rectitude of our intentions, do in the name, and by the authority of the good people of these colonies, solemnly publish and declare, that these united colonies are, and of right ought to be free and independent states; that they are absolved from all allegiance to the British crown, and that all political connection between them and the state of Great Britain is, and ought to be, totally dissolved; and that as free and independent states, they have full power to levy war, conclude peace, contract alliances, establish commerce, and to do all other acts and things which independent states may of right do.

And for the support of this declaration, with a firm reliance on the protection of divine providence, we mutually pledge to each other our lives, our fortunes, and our sacred honor.

The Declaration thus signed on the 4th, on paper, was engrossed on parchment, and signed again on the 2d of August.

1821   1829

From Notes on the State of Virginia[1]
From *Query VI. Productions Mineral, Vegetable, and Animal*

* * *

The opinion advanced by the Count de Buffon,[2] is 1. That the animals common both to the old and new world, are smaller in the latter. 2. That those peculiar to the new, are on a smaller scale. 3. That those which have been domesticated in both, have degenerated in America. and 4. That on the whole it exhibits fewer species. And the reason he thinks is, that the heats of America are less; that more waters are spread over its surface by nature, and fewer of these drained off by the hand of man. In other words, that *heat* is friendly, and *moisture* adverse to the production and development of large quadrupeds. I will not meet this hypothesis on its first doubtful ground, whether the climate of America be comparatively more humid? Because we are not furnished with observations sufficient to decide this question. And though, till it be decided, we are as free to deny, as others are to affirm, the fact, yet for a moment let it be supposed. The hypothesis, after this supposition, proceeds to another; that *moisture* is unfriendly to animal growth. The truth of this is inscrutable to us by reasonings a priori.[3] Nature has hidden from us her modus agendi.[4] Our only appeal on such questions is to experience; and I think that experience is against the supposition. It is by the assistance of *heat* and *moisture* that vegetables are elaborated from the elements of earth, air, water, and fire. We accordingly see the more humid climates produce the greater quantity of vegetables. Vegetables are mediately or immediately the food of every animal: and in proportion to the quantity of food, we see animals not only multiplied in their numbers, but improved in their bulk, as far as the laws of their nature will admit. Of this opinion is the Count de Buffon himself in another part of his work: "in general it seems that somewhat cold countries are better suited to our oxen than hot countries, and they are the heavier and bigger in proportion as the climate is damper and more abounding in pasture lands. The oxen of Denmark, of Podolie,[5] of the Ukraine, and of Tartary which is inhabited by the Calmouques,[6] are the largest

---

1. In 1781, the year Jefferson retired as Governor of Virginia, he received a request from the Marquis de Barbé-Marbois, secretary of the French legation at Philadelphia, to answer 23 questions concerning the geographical boundaries, the ecology, and the social history of Virginia. Jefferson took the occasion to make some observations on slavery, manufacturing, and government. He wanted especially to counter the notion, prevalent among European naturalists, that species in North America had degenerated and were inferior to Old World types. Jefferson's replies were published privately in 1784-85. The threat of an unauthorized French translation prompted Jefferson to publish an authorized edition in London in 1787. The text used here is from the Norton edition, edited by William Peden (1954).
2. Georges Louis Leclerc de Buffon (1707-88), French naturalist and keeper of the Royal Gardens. Buffon suggested that North American species were degenerate in his *Natural History* (1749-88).
3. Assumptions; previously held ideas.
4. The mode by which a thing acts or operates.
5. A village in northeast India.
6. More commonly spelled "Kalmucks," a nomadic Mongol tribe.

of all." Here then a race of animals, and one of the largest too, has been increased in its dimensions by *cold* and *moisture,* in direct opposition to the hypothesis, which supposes that these two circumstances diminish animal bulk, and that it is their contraries *heat* and dryness which enlarge it. But when we appeal to experience, we are not to rest satisfied with a single fact. Let us therefore try our question on more general ground. Let us take two portions of the earth, Europe and America for instance, sufficiently extensive to give operation to general causes; let us consider the circumstances peculiar to each, and observe their effect on animal nature. America, running through the torrid as well as temperate zone, has more *heat,* collectively taken, than Europe. But Europe, according to our hypothesis, is the dryest. They are equally adapted then to animal productions; each being endowed with one of those causes which befriend animal growth, and with one which opposes it. If it be thought unequal to compare Europe with America, which is so much larger, I answer, not more so than to compare America with the whole world. Besides, the purpose of the comparison is to try an hypothesis, which makes the size of animals depend on the *heat* and *moisture* of climate. If therefore we take a region, so extensive as to comprehend a sensible distinction of climate, and so extensive too as that local accidents, or the intercourse of animals on its borders, may not materially affect the size of those in its interior parts, we shall comply with those conditions which the hypothesis may reasonably demand. The objection would be the weaker in the present case, because any intercourse of animals which may take place on the confines of Europe and Asia, is to the advantage of the former, Asia producing certainly larger animals than Europe. * * *

Hitherto I have considered this hypothesis as applied to brute animals only, and not in its extension to the man of America, whether aboriginal or transplanted. It is the opinion of Mons. de Buffon that the former furnishes no exception to it: "Although the savage of the new world is about the same height as man in our world, this does not suffice for him to constitute an exception to the general fact that all living nature has become smaller on that continent. The savage is feeble, and has small organs of generation; he has neither hair nor beard, and no ardor whatever for his female; although swifter than the European because he is better accustomed to running, he is, on the other hand, less strong in body; he is also less sensitive, and yet more timid and cowardly; he has no vivacity, no activity of mind; the activity of his body is less an exercise, a voluntary motion, than a necessary action caused by want; relieve him of hunger and thirst, and you deprive him of the active principle of all his movements; he will rest stupidly upon his legs or lying down entire days. There is no need for seeking further the cause of the isolated mode of life of these savages and their repugnance for society: the most precious spark of the fire of nature has been

refused to them; they lack ardor for their females, and consequently
have no love for their fellow men: not knowing this strongest and
most tender of all affections, their other feelings are also cold and
languid; they love their parents and children but little; the most intimate of all ties, the family connection, binds them therefore but
loosely together; between family and family there is no tie at all;
hence they have no communion, no commonwealth, no state of
society. Physical love constitutes their only morality; their heart is
icy, their society cold, and their rule harsh. They look upon their
wives only as servants for all work, or as beasts of burden, which
they load without consideration with the burden of their hunting,
and which they compel without mercy, without gratitude, to perform tasks which are often beyond their strength. They have only
few children, and they take little care of them. Everywhere the original defect appears: they are indifferent because they have little
sexual capacity, and this indifference to the other sex is the fundamental defect which weakens their nature, prevents its development,
and—destroying the very germs of life—uproots society at the same
time. Man is here no exception to the general rule. Nature, by refusing him the power of love, has treated him worse and lowered him
deeper than any animal." An afflicting picture indeed, which, for
the honor of human nature, I am glad to believe has no original. Of
the Indian of South America I know nothing; for I would not
honor with the appellation of knowledge, what I derive from the
fables published of them. These I believe to be just as true as the
fables of Aesop.[1] This belief is founded on what I have seen of
man, white, red, and black, and what has been written of him by
authors, enlightened themselves, and writing amidst an enlightened
people. The Indian of North America being more within our reach,
I can speak of him somewhat from my own knowledge, but more
from the information of others better acquainted with him, and on
whose truth and judgment I can rely. From these sources I am able
to say, in contradiction to this representation, that he is neither
more defective in ardor, nor more impotent with his female, than
the white reduced to the same diet and exercise: that he is brave,
when an enterprise depends on bravery; education with him making
the point of honor consist in the destruction of an enemy by stratagem, and in the preservation of his own person free from injury; or
perhaps this is nature; while it is education which teaches us to
honor force more than finesse; that he will defend himself against
an host of enemies, always choosing to be killed, rather than to surrender, though it be to the whites, who he knows will treat him
well: that in other situations also he meets death with more deliberation, and endures tortures with a firmness unknown almost to religious enthusiasm with us: that he is affectionate to his children,

1. A Greek slave (c. 620–560 B.C.), reported to be the author of fables.

careful of them, and indulgent in the extreme: that his affections comprehend his other connections, weakening, as with us, from circle to circle, as they recede from the center: that his friendships are strong and faithful to the uttermost extremity: that his sensibility is keen, even the warriors weeping most bitterly on the loss of their children, though in general they endeavor to appear superior to human events: that his vivacity and activity of mind is equal to ours in the same situation; hence his eagerness for hunting, and for games of chance. The women are submitted to unjust drudgery. This I believe is the case with every barbarous people. With such, force is law. The stronger sex therefore imposes on the weaker. It is civilization alone which replaces women in the enjoyment of their natural equality. That first teaches us to subdue the selfish passions, and to respect those rights in others which we value in ourselves. Were we in equal barbarism, our females would be equal drudges. The man with them is less strong than with us, but their woman stronger than ours; and both for the same obvious reason; because our man and their woman is habituated to labor, and formed by it. With both races the sex which is indulged with ease is least athletic. An Indian man is small in the hand and wrist for the same reason for which a sailor is large and strong in the arms and shoulders, and a porter in the legs and thighs.—They raise fewer children than we do. The causes of this are to be found, not in a difference of nature, but of circumstance. The women very frequently attending the men in their parties of war and of hunting, childbearing becomes extremely inconvenient to them. It is said, therefore, that they have learnt the practice of procuring abortion by the use of some vegetable; and that it even extends to prevent conception for a considerable time after. During these parties they are exposed to numerous hazards, to excessive exertions, to the greatest extremities of hunger. Even at their homes the nation depends for food, through a certain part of every year, on the gleanings of the forest: that is, they experience a famine once in every year. With all animals, if the female be badly fed, or not fed at all, her young perish: and if both male and female be reduced to like want, generation becomes less active, less productive. To the obstacles then of want and hazard, which nature has opposed to the multiplication of wild animals, for the purpose of restraining their numbers within certain bounds, those of labor and of voluntary abortion are added with the Indian. No wonder then if they multiply less than we do. Where food is regularly supplied, a single farm will show more of cattle, than a whole country of forests can of buffaloes. The same Indian women, when married to white traders, who feed them and their children plentifully and regularly, who exempt them from excessive drudgery, who keep them stationary and unexposed to accident, produce and raise as many children as the white women. Instances are known, under these circumstances, of their rearing a dozen children.

An inhuman practice once prevailed in this country of making slaves of the Indians. (This practice commenced with the Spaniards with the first discovery of America). It is a fact well known with us, that the Indian women so enslaved produced and raised as numerous families as either the whites or blacks among whom they lived.—It has been said, that Indians have less hair than the whites, except on the head. But this is a fact of which fair proof can scarcely be had. With them it is disgraceful to be hairy on the body. They say it likens them to hogs. They therefore pluck the hair as fast as it appears. But the traders who marry their women, and prevail on them to discontinue this practice, say, that nature is the same with them as with the whites. Nor, if the fact be true, is the consequence necessary which has been drawn from it. Negroes have notoriously less hair than the whites; yet they are more ardent. But if cold and moisture be the agents of nature for diminishing the races of animals, how comes she all at once to suspend their operation as to the physical man of the new world, whom the Count acknowledges to be "about the same size as the man of our hemisphere," and to let loose their influence on his moral faculties? How has this "combination of the elements and other physical causes, so contrary to the enlargement of animal nature in this new world, these obstacles to the development and formation of great germs," been arrested and suspended, so as to permit the human body to acquire its just dimensions, and by what inconceivable process has their action been directed on his mind alone? To judge of the truth of this, to form a just estimate of their genius and mental powers, more facts are wanting, and great allowance to be made for those circumstances of their situation which call for a display of particular talents only. This done, we shall probably find that they are formed in mind as well as in body, on the same module with the "Homo sapiens Europaeus."[2] The principles of their society forbidding all compulsion, they are to be led to duty and to enterprise by personal influence and persuasion. Hence eloquence in council, bravery and address in war, become the foundations of all consequences with them. To these acquirements all their faculties are directed. Of their bravery and address in war we have multiplied proofs, because we have been the subjects on which they were exercised. Of their eminence in oratory we have fewer examples, because it is displayed chiefly in their own councils. Some, however, we have of very superior luster. I may challenge the whole orations of Demosthenes and Cicero,[3] and of any more eminent orator, if Europe has furnished more eminent, to produce a single passage, superior to the speech of Logan, a Mingo chief, to Lord Dunmore,[4] when governor of this state. And, as a testimony of their talents in

2. European man.
3. Demosthenes (385?–322 B.C.) was an Athenian orator; Cicero (106–43 B.C.) was a Roman orator and statesman.
4. John Murray, Earl of Dunmore (1732–1809), was the colonial Governor of Virginia from 1771 to 1775; "Mingo": an Iroquois tribe.

this line, I beg leave to introduce it, first stating the incidents necessary for understanding it. In the spring of the year 1774, a robbery was committed by some Indians on certain land-adventurers on the river Ohio. The whites in that quarter, according to their custom, undertook to punish this outrage in a summary way. Captain Michael Cresap,[5] and a certain Daniel Great-house, leading on these parties, surprised, at different times, traveling and hunting parties of the Indians, having their women and children with them, and murdered many. Among these were unfortunately the family of Logan, a chief celebrated in peace and war, and long distinguished as the friend of the whites. This unworthy return provoked his vengeance. He accordingly signalized himself in the war which ensued. In the autumn of the same year a decisive battle was fought at the mouth of the Great Kanhaway, between the collected forces of the Shawanese, Mingoes, and Delawares, and a detachment of the Virginia militia. The Indians were defeated, and sued for peace. Logan however disdained to be seen among the suppliants. But, lest the sincerity of a treaty should be distrusted, from which so distinguished a chief absented himself, he sent by a messenger the following speech to be delivered to Lord Dunmore.

"I appeal to any white man to say, if ever he entered Logan's cabin hungry, and he gave him not meat; if ever he came cold and naked, and he clothed him not. During the course of the last long and bloody war, Logan remained idle in his cabin, an advocate for peace. Such was my love for the whites, that my countrymen pointed as they passed, and said, 'Logan is the friend of white men.' I had even thought to have lived with you, but for the injuries of one man. Col. Cresap, the last spring, in cold blood, and unprovoked, murdered all the relations of Logan, not sparing even my women and children. There runs not a drop of my blood in the veins of any living creature. This called on me for revenge. I have sought it: I have killed many: I have fully glutted my vengeance. For my country, I rejoice at the beams of peace. But do not harbor a thought that mine is the joy of fear. Logan never felt fear. He will not turn on his heel to save his life. Who is there to mourn for Logan?—Not one."

Before we condemn the Indians of this continent as wanting genius,[6] we must consider that letters have not yet been introduced among them. Were we to compare them in their present state with the Europeans North of the Alps, when the Roman arms and arts first crossed those mountains, the comparison would be unequal, because, at that time, those parts of Europe were swarming with numbers; because numbers produce emulation, and multiply the chances of improvement, and one improvement begets another. Yet I may safely ask, How many good poets, how many able mathemati-

5. Michael Cresap (1742–75) was a Maryland soldier and frontiersman.

6. Intelligence.

cians, how many great inventors in arts or sciences had Europe North of the Alps then produced? And it was sixteen centuries after this before a Newton[7] could be formed. I do not mean to deny, that there are varieties in the race of man, distinguished by their powers both of body and mind. I believe there are, as I see to be the case in the races of other animals. I only mean to suggest a doubt, whether the bulk and faculties of animals depend on the side of the Atlantic on which their food happens to grow, or which furnishes the elements of which they are compounded? Whether nature has enlisted herself as a Cis[8] or transatlantic partisan? I am induced to suspect, there has been more eloquence than sound reasoning displayed in support of this theory; that it is one of those cases where the judgment has been seduced by a glowing pen: and whilst I render every tribute of honor and esteem to the celebrated zoologist, who has added, and is still adding, so many precious things to the treasures of science, I must doubt whether in this instance he has not cherished error also, by lending her for a moment his vivid imagination and bewitching language. * * *

1780–81   1787

Letter to John Adams[1]
[*The Natural Aristocrat*]

Monticello, October 28, 1813

Dear Sir,—According to the reservation between us, of taking up one of the subjects of our correspondence at a time, I turn to your letters of August the 16th and September the 2d. * * * I agree with you that there is a natural aristocracy among men. The grounds of this are virtue and talents. Formerly bodily powers gave place among the aristoi.[2] But since the invention of gunpowder has armed the weak as well as the strong with missile death, bodily strength, like beauty, good humor, politeness and other accomplishments, has become but an auxiliary ground of distinction. There is also an artificial aristocracy, founded on wealth and birth, without either virtue or talents; for with these it would belong to the first class. The natural aristocracy I consider as the most precious gift of nature for the instruction, the trusts, and government of society. And indeed, it would have been inconsistent in creation to have

7. Sir Isaac Newton (1642–1727), English philosopher and mathematician, most frequently identified with his theory of gravitation.
8. On this side.
1. Thomas Jefferson and John Adams (1735–1826) became estranged when Adams was elected second President in 1796. Adams's Federalist positions were *opposed by Jefferson*, who succeeded him as President in 1801. In 1812 they began to correspond and were able to debate their differences. The text used here is from *The Writings of Thomas*

Jefferson, Vol. 13, ed. A. A. Lipscomb and A. E. Bergh (1903).
2. The best. On July 9, 1813, Adams wrote to Jefferson that he recalled a maxim from the work of Theognis which said that " 'nobility in men is worth as much as it is in horses, asses, or rams; but the meanest [i.e., poorest] blooded puppy *in the world*, if he gets a little money is as good a man as the best of them.' Yet birth and wealth together have prevailed over virtue and talents in all ages. The many will acknowledge no other *aristoi*."

formed man for the social state, and not to have provided virtue and wisdom enough to manage the concerns of the society. May we not even say that that form of government is the best which provides the most effectually for a pure selection of these natural aristoi into the offices of government? The artificial aristocracy is a mischievous ingredient in government, and provision should be made to prevent its ascendency. On the question, what is the best provision, you and I differ; but we differ as rational friends, using the free exercise of our own reason, and mutually indulging its errors. You think it best to put the pseudo-aristoi into a separate chamber of legislation, where they may be hindered from doing mischief by their co-ordinate branches, and where, also, they may be a protection to wealth against the agrarian and plundering enterprises of the majority of the people. I think that to give them power in order to prevent them from doing mischief is arming them for it, and increasing instead of remedying the evil. For if the co-ordinate branches can arrest their action, so may they that of the co-ordinates. Mischief may be done negatively as well as positively. Of this, a cabal in the Senate of the United States has furnished many proofs. Nor do I believe them necessary to protect the wealthy; because enough of these will find their way into every branch of the legislation, to protect themselves. From fifteen to twenty legislatures of our own, in action for thirty years past, have proved that no fears of an equalization of property are to be apprehended from them. I think the best remedy is exactly that provided by all our constitutions, to leave to the citizens the free election and separation of the aristoi from the pseudo-aristoi, of the wheat from the chaff. In general they will elect the really good and wise. In some instances, wealth may corrupt, and birth blind them; but not in sufficient degree to endanger the society.

It is probable that our difference of opinion may, in some measure, be produced by a difference of character in those among whom we live. From what I have seen of Massachusetts and Connecticut myself, and still more from what I have heard, and the character given of the former by yourself,[3] who know them so much better, there seems to be in those two states a traditionary reverence for certain families, which has rendered the offices of the government nearly hereditary in those families. I presume that from an early period of your history, members of those families happening to possess virtue and talents, have honestly exercised them for the good of the people, and by their services have endeared their names to them. In coupling Connecticut with you, I mean it politically only, not morally. For having made the Bible the common law of their land, they seem to have modeled their morality on the story of

---

3. "Vol. 1, page 111" [Jefferson's note]. A reference to Adams's *Defense of the Constitutions of Government of the* *United States of America,* 3 vols. (Philadelphia, 1797). This work was first published in 1787.

Jacob and Laban.[4] But although this hereditary succession to office with you, may, in some degree, be founded in real family merit, yet in a much higher degree, it has proceeded from your strict alliance of Church and State. These families are canonized in the eyes of the people on common principles, "you tickle me, and I will tickle you." In Virginia we have nothing of this. Our clergy, before the Revolution, having been secured against rivalship by fixed salaries, did not give themselves the trouble of acquiring influence over the people. Of wealth, there were great accumulations in particular families, handed down from generation to generation, under the English law of entails.[5] But the only object of ambition for the wealthy was a seat in the King's Council.[6] All their court then was paid to the crown and its creatures; and they philippized[7] in all collisions between the King and the people. Hence they were unpopular; and that unpopularity continues attached to their names. A Randolph, a Carter, or a Burwell[8] must have great personal superiority over a common competitor to be elected by the people even at this day. At the first session of our legislature after the Declaration of Independence, we passed a law abolishing entails. And this was followed by one abolishing the privilege of primogeniture, and dividing the lands of intestates[9] equally among all their children, or other representatives. These laws, drawn by myself, laid the axe to the foot of pseudo-aristocracy. And had another which I prepared been adopted by the legislature, our work would have been complete. It was a bill for the more general diffusion of learning. This proposed to divide every county into wards of five or six miles square, like your townships; to establish in each ward a free school for reading, writing and common arithmetic; to provide for the annual selection of the best subjects from these schools, who might receive, at the public expense, a higher degree of education at a district school; and from these district schools to select a certain number of the most promising subjects, to be completed at an university, where all the useful sciences should be taught. Worth and genius would thus have been sought out from every condition of life, and completely prepared by education for defeating the competition of wealth and birth for public trusts. My proposition had, for a further object, to impart to these wards those portions of self-government for which they are best qualified, by confiding to them the care of their poor, their roads, police, elections, the nomination of jurors, administration of justice in small cases, elementary exercises of militia; in short, to have made them little republics, with a warden at the head of each,

4. I.e., a dynastic family, founded on the marital relations between the daughters of Jacob and Laban (Genesis 24–31).
5. An estate which cannot be willed but *must pass from* a proscribed list of successors.
6. The Privy Council, a select group of advisors, appointed by the King.
7. Argued against liberty for the people; spoke corrupted by their desire to please the King.
8. John Randolph, Landon Carter, and Lewis Burwell were all Virginia aristocrats.
9. Those who died without wills; "primogeniture": a law which gave estates to the eldest son.

for all those concerns which, being under their eye, they would better manage than the larger republics of the county or state. A general call of ward meetings by their wardens on the same day through the state, would at any time produce the genuine sense of the people on any required point, and would enable the state to act in mass, as your people have so often done, and with so much effect by their town meetings. The law for religious freedom,[1] which made a part of this system, having put down the aristocracy of the clergy, and restored to the citizen the freedom of the mind, and those of entails and descents nurturing an equality of condition among them, this on education would have raised the mass of the people to the high ground of moral respectability necessary to their own safety, and to orderly government; and would have completed the great object of qualifying them to select the veritable aristoi, for the trusts of government, to the exclusion of the pseudalists; and the same Theognis who has furnished the epigraphs of your two letters, assures us that "Ουδεμιαν πω, Κυρν', αγαθοι πολιν ωλεσαν ανδρες."[2] Although this law has not yet been acted on but in a small and inefficient degree, it is still considered as before the legislature, with other bills of the revised code, not yet taken up, and I have great hope that some patriotic spirit will, at a favorable moment, call it up, and make it the keystone of the arch of our government.

With respect to aristocracy, we should further consider, that before the establishment of the American states, nothing was known to history but the man of the old world, crowded within limits either small or overcharged, and steeped in the vices which that situation generates. A government adapted to such men would be one thing; but a very different one, that for the man of these states. Here every one may have land to labor for himself, if he chooses; or, preferring the exercise of any other industry, may exact for it such compensation as not only to afford a comfortable subsistence, but wherewith to provide for a cessation from labor in old age. Every one, by his property, or by his satisfactory situation, is interested in the support of law and order. And such men may safely and advantageously reserve to themselves a wholesome control over their public affairs, and a degree of freedom, which, in the hands of the canaille[3] of the cities of Europe, would be instantly perverted to the demolition and destruction of everything public and private. The history of the last twenty-five years of France,[4] and of the last forty years in America, nay of its last two hundred years, proves the truth of both parts of this observation.

But even in Europe a change has sensibly taken place in the mind of man. Science had liberated the ideas of those who read and

1. Passed in 1786.
2. "Curnis, good men have never harmed any city." Theognis was a Greek elegiac poet of the sixth century B.C.
3. Mob.
4. I.e., since the French Revolution (1789).

reflect, and the American example had kindled feelings of right in the people. An insurrection has consequently begun, of science, talents, and courage, against rank and birth, which have fallen into contempt. It has failed in its first effort, because the mobs of the cities, the instrument used for its accomplishment, debased by ignorance, poverty, and vice, could not be restrained to rational action. But the world will recover from the panic of this first catastrophe. Science is progressive, and talents and enterprise on the alert. Resort may be had to the people of the country, a more governable power from their principles and subordination; and rank, and birth, and tinsel-aristocracy will finally shrink into insignificance, even there. This, however, we have no right to meddle with. It suffices for us, if the moral and physical condition of our own citizens qualifies them to select the able and good for the direction of their government, with a recurrence of elections at such short periods as will enable them to displace an unfaithful servant, before the mischief he meditates may be irremediable.

I have thus stated my opinion on a point on which we differ, not with a view to controversy, for we are both too old to change opinions which are the result of a long life of inquiry and reflection; but on the suggestions of a former letter of yours, that we ought not to die before we have explained ourselves to each other. We acted in perfect harmony, through a long and perilous contest for our liberty and independence. A constitution has been acquired, which, though neither of us thinks perfect, yet both consider as competent to render our fellow citizens the happiest and the securest on whom the sun has ever shone. If we do not think exactly alike as to its imperfections, it matters little to our country, which, after devoting to it long lives of disinterested labor, we have delivered over to our successors in life, who will be able to take care of it and of themselves.

Of the pamphlet on aristocracy which has been sent to you, or who may be its author, I have heard nothing but through your letter. If the person you suspect, it may be known from the quaint, mystical, and hyperbolical ideas, involved in affected, newfangled and pedantic terms which stamp his writings. Whatever it be, I hope your quiet is not to be affected at this day by the rudeness or intemperance of scribblers; but that you may continue in tranquility to live and to rejoice in the prosperity of our country, until it shall be your own wish to take your seat among the aristoi who have gone before you. Ever and affectionately yours.

PHILIP FRENEAU
1752–1832

Philip Freneau had all the advantages that wealth and social position could bestow, and the Freneau household in Manhattan was frequently visited by well-known writers and painters. Philip received a good education at the hands of tutors and at fifteen entered the sophomore class at the College of New Jersey (now Princeton University). There he became fast friends with his roommate, James Madison, a future President, and a classmate, Hugh Henry Brackenridge, who was to become a successful novelist. In their senior year Freneau and Brackenridge composed an ode on *The Rising Glory of America*, and Brackenridge read the poem at commencement. It establishes early in Freneau's career his recurrent vision of a glorious future in which America would fulfill the collective hope of humankind:

> Paradise anew
> Shall flourish, by no second Adam lost,
> No dangerous tree with deadly fruit shall grow,
> No tempting serpent to allure the soul
> From native innocence. . . . The lion and the lamb
> In mutual friendship linked, shall browse the shrub,
> And timorous deer with softened tigers stray
> O'er mead, or lofty hill, or grassy plain * * *

For a short time Freneau taught school and hoped to make a career as a writer, but it was an impractical wish. When he was offered a position as secretary on a plantation in the West Indies in 1776, he sailed to Santa Cruz and remained there almost three years. It was in that country, where "Sweet orange groves in lonely valleys rise," that Freneau wrote some of his most sensuous lyrics; but as he tells us in *To Sir Toby*, he could not talk of "blossoms" and an "endless spring" forever in a land which abounded in poverty and misery and where the owners grew wealthy on a slave economy. In 1778 he returned home and enlisted as a seaman on a blockade runner; two years later he was captured at sea and imprisoned on the British ship *Scorpion*, anchored in New York Harbor. He was treated brutally, and when he was exchanged from the hospital ship *Hunter* his family feared for his life.

Freneau was to spend ten more years of his life at sea, first as a master of a merchant ship in 1784, and again in 1803, but immediately after he regained his health he moved to Philadelphia to work in the post office, and it was in that city that he gained his reputation as a satirist, journalist, and poet. As editor of the *Freeman's Journal*, Freneau wrote impassioned verse in support of the American Revolution and turned all his rhetorical gifts against anyone who was thought to be in sympathy with the British monarchy. It was during this period in his life that he became identified as the "Poet of the American Revolution." In 1791, after he returned from duties at sea, Jefferson, as Secretary of State, offered him a position as translator in his department, understanding that Freneau would have plenty of free time to devote to his newspaper, *The National Gazette*. Like Thomas Paine, Freneau was a strong supporter of the French Revolution, and he had a sharp eye for anyone not sympathetic to the democratic cause. He had a special grudge against Alexander Hamilton, Secretary of the Treasury, as chief spokesman for the Federalists. President Washington thought it was ironic that "that rascal Freneau" should be employed by his administration when he attacked it so outspokenly.

The *National Gazette* ceased publication in 1793, and after Jefferson resigned his office, Freneau left Philadelphia for good, alternating between ship's captain and newspaper editor in New York and New Jersey. He spent his last years on his New Jersey farm, unable to make it supporting and with no hope of further employment. Year after year he sold all the land he inherited from his father and was finally reduced to applying for a pension as a veteran of the American Revolution. He died impoverished and unknown, lost in a blizzard.

Freneau's biographer, Lewis Leary, subtitled his book *A Study in Literary Failure* and began that work by observing that "Philip Freneau failed in almost everything he attempted." Freneau's most sympathetic readers have always felt that he was born in a time not ripe for poetry, and that his genuine lyric gifts were always in conflict with his political pamphleteering. Had he been born fifty years later, perhaps he could have joined Cooper and Irving in a life devoted exclusively to letters. There is no doubt that he did much to pave the way for these later writers. Freneau is not "the Father of American Poetry" (as his readers, anxious for a spokesman for a national literary consciousness, liked to call him), but his obsession with the beautiful, transient things of nature, and the conflict in his art between the sensuous and the didactic, are central to the concerns of American poetry.

Texts used are *The Poems of Philip Freneau*, ed. F. L. Pattee (Princeton: The University Library, 1902), and *The Poems of Freneau*, ed. H. H. Clark (New York: Harcourt, Brace, 1929).

On the Emigration to America and Peopling the Western Country

To western woods, and lonely plains,
Palemon[1] from the crowd departs,
Where Nature's wildest genius reigns,
To tame the soil, and plant the arts—
What wonders there shall freedom show,   5
What mighty states successive grow!

From Europe's proud, despotic shores
Hither the stranger takes his way,
And in our new found world explores
A happier soil, a milder sway,   10
Where no proud despot holds him down,
No slaves insult him with a crown.

What charming scenes attract the eye,
On wild Ohio's savage stream!

---

1. Conventionally, any young man setting out on a journey. Palemon appears in Chaucer's *Knight's Tale,* an adaptation of Boccaccio's *Teseide.*

There Nature reigns, whose works outvie   15
The boldest pattern art can frame;
There ages past have rolled away,
And forests bloomed but to decay.

From these fair plains, these rural seats,
So long concealed, so lately known,   20
The unsocial Indian far retreats,
To make some other clime his own,
When other streams, less pleasing flow,
And darker forests round him grow.

Great sire[2] of floods! whose varied wave   25
Through climes and countries takes its way,
To whom creating Nature gave
Ten thousand streams to swell thy sway!
No longer shall they useless prove,
Nor idly through the forests rove;   30

Nor longer shall your princely flood
From distant lakes be swelled in vain,
Nor longer through a darksome wood
Advance, unnoticed, to the main,[3]
Far other ends, the heavens decree—   35
And commerce plans new freights for thee.

While virtue warms the generous breast,
There heaven-born freedom shall reside,
Nor shall the voice of war molest,
Nor Europe's all-aspiring pride—   40
There Reason shall new laws devise,
And order from confusion rise.

Forsaking kings and regal state,
With all their pomp and fancied bliss,
The traveler owns,[4] convinced though late,   45
No realm so free, so blessed as this—
The east is half to slaves consigned,
Where kings and priests enchain the mind.

O come the time, and haste the day,
When man shall man no longer crush,   50
When Reason shall enforce her sway,
Nor these fair regions raise our blush,
Where still the African complains,
And mourns his yet unbroken chains.

Far brighter scenes a future age,   55
The muse predicts, these states will hail,

2. "Mississippi" [Freneau's note].   4. Admits.
3. Ocean.

Whose genius may the world engage,
Whose deeds may over death prevail,
And happier systems bring to view,
Than all the eastern sages knew.   60

                                                        1785

The Wild Honey Suckle

Fair flower, that dost so comely grow,
Hid in this silent, dull retreat,
Untouched thy honeyed blossoms blow,[1]
Unseen thy little branches greet:
   No roving foot shall crush thee here,   5
   No busy hand provoke a tear.

By Nature's self in white arrayed,
She bade thee shun the vulgar[2] eye,
And planted here the guardian shade,
And sent soft waters murmuring by;   10
   Thus quietly thy summer goes,
   Thy days declining to repose.

Smit with those charms, that must decay,
I grieve to see your future doom;
They died—nor were those flowers more gay,   15
The flowers that did in Eden bloom;
   Unpitying frosts, and Autumn's power
   Shall leave no vestige of this flower.

From morning suns and evening dews
At first thy little being came: 20
If nothing once, you nothing lose,
For when you die you are the same;
   The space between, is but an hour,
   The frail duration of a flower.

                                                        1786

The Indian Burying Ground

In spite of all the learned have said,
I still my old opinion keep;
The posture, that we give the dead,
Points out the soul's eternal sleep.

Not so the ancients of these lands—   5
The Indian, when from life released,
Again is seated with his friends,
And shares again the joyous feast.[1]

1. Bloom.
2. Common; unfeeling.
1. "The North American Indians bury their dead in a sitting posture; decorating the corpse with wampum, the images of birds, quadrupeds, &: And (if that of a warrior) with bows, arrows, tomhawks and other military weapons" [Freneau's note].

His imaged birds, and painted bowl,
And vension, for a journey dressed,   10
Bespeak the nature of the soul,
Activity, that knows no rest.

His bow, for action ready bent,
And arrows, with a head of stone,
Can only mean that life is spent,   15
And not the old ideas gone.

Thou, stranger, that shalt come this way,
No fraud upon the dead commit—
Observe the swelling turf, and say
They do not lie, but here they sit.   20

Here still a lofty rock remains,
On which the curious eye may trace
(Now wasted, half, by wearing rains)
The fancies of a ruder race.

Here still an agéd elm aspires,   25
Beneath whose far-projecting shade
(And which the shepherd still admires)
The children of the forest played!

There oft a restless Indian queen
(Pale Sheba,[2] with her braided hair)   30
And many a barbarous form is seen
To chide the man that lingers there.

By midnight moons, o'er moistening dews,
In habit for the chase arrayed,
The hunter still the deer pursues,   35
The hunter and the deer, a shade!

And long shall timorous fancy see
The painted chief, and pointed spear,
And Reason's self shall bow the knee
To shadows and delusions here.   40

                                     1788

On the Religion of Nature

The power, that gives with liberal hand
    The blessings man enjoys, while here,
And scatters through a smiling land
    Abundant products of the year;
        That power of nature, ever blessed,   5
        Bestowed religion with the rest.

2. **Sheba** is the Queen who visited Solomon to test his wisdom. 1 Kings 10.1–13

Born with ourselves, her early sway
  Inclines the tender mind to take
The path of right, fair virtue's way
  Its own felicity to make.   10
    This universally extends
    And leads to no mysterious ends.

Religion, such as nature taught,
  With all divine perfection suits;
Had all mankind this sytem sought   15
  Sophists[1] would cease their vain disputes,
    And from this source would nations know
    All that can make their heaven below.

This deals not curses on mankind,
  Or dooms them to perpetual grief,   20
If from its aid no joys they find,
  It damns them not for unbelief;
    Upon a more exalted plan
    Creatress nature dealt with man—

Joy to the day, when all agree   25
  On such grand systems to proceed,
From fraud, design, and error free,
  And which to truth and goodness lead:
    Then persecution will retreat
    And man's religion be complete.   30

                                                         1815

1. Teachers of philosophy.

PHILLIS WHEATLEY
c. 1753–1784

Phillis Wheatley was only nineteen years old when her *Poems on Various Subjects, Religious and Moral* was published in London in 1773. At the time of their publication she was the object of considerable public attention because, in addition to being a child prodigy, Phillis Wheatley was a black slave. She had been born in Africa, probably somewhere in present-day Senegal or Gambia, and was brought to Boston in 1761 when she was about seven years of age. She was purchased by a wealthy tailor, John Wheatley, for his wife, Susannah, probably as a companion. Phillis Wheatley was fortunate in her surroundings for Mrs. Wheatley had a sympathetic heart. Phillis was both very frail and remarkably intelligent, and the Wheatley household seems to have been alert to her gifts. In an age when few white women were given an education, she was taught to read and write, and in a very short time began also to read Latin writers. She came to know the Bible well; and three English poets—Milton, Pope, and Gray—touched her deeply and exerted a strong influence on her verse. It is clear that the

Wheatleys were deeply committed Christians and thought of her as, first of all, a soul in need of salvation. Her poem in 1770 on the death of the Reverend George Whitefield, the great English evangelical preacher who frequently toured New England, made her famous. She had been taken to London partly for reasons of health, but also to meet a number of distinguished persons (the Countess of Huntington and the Lord Mayor of London) who knew her poetry and wanted to assist her in its publication. Her literary gifts, intelligence, and piety were a striking example to her English and American audience of the triumph of native human capacities over the circumstances of birth. Phillis returned to Boston sooner than had been planned, when she received news of Mrs. Wheatley's fatal illness.

After the death of the Wheatleys, Phillis was freed and in 1778 married John Peters, a freedman. Almost nothing is known about Peters other than the fact that the Wheatley family disliked him, that he may have been a spokesman for Negro rights, that he moved from job to job and was imprisoned for debt in 1784, and was probably in prison when Phillis died. She spent her last years alone and in great poverty in a wreck of a house in Boston. She had lost two children; and when she was on her deathbed her third child lay ill beside her and died soon after she did. Both mother and child were buried together in an unmarked grave.

Until the 1830s, when Massachusetts abolitionists reprinted her poetry, Phillis Wheatley's verse remained forgotten. Hers was a thoroughly conventional poetic talent, tied too strongly to Miltonic cadences and the balanced couplets of Alexander Pope; but given the stringencies of the time and of her situation, it would be unrealistic to expect anything more. The only hint of injustice found in her poetry is in the line "Some view our sable race with scornful eye." It would be almost a hundred years before a black American writer could drop the mask of convention and write about the formation of his own unique sensibility. Nevertheless, Phillis Wheatley expresses the popular sentiments of the age in matters of poetic taste, religious feeling, and national identity. She is the first black writer of consequence in America; and her life constitutes a deeply moving account of unfulfilled promise.

The text used here is from *The Poems of Phillis Wheatley*, ed. Julian D. Mason, Jr. (Chapel Hill: University of North Carolina Press, 1966).

On Being Brought from Africa to America

'Twas mercy brought me from my pagan land,
Taught my benighted soul to understand
That there's a God, that there's a Savior too:
Once I redemption neither sought nor knew.
Some view our sable[1] race with scornful eye,   5
"Their color is a diabolic dye."
Remember, Christians, Negroes, black as Cain,[2]
May be refined, and join the angelic train.

1773

1. Black.
2. Cain slew his brother Abel and was "marked" by God for doing so. This mark has sometimes been taken to be the origin of the Negro (Genesis 4.1–15).

To the University of Cambridge, in New England[1]

While an intrinsic ardor prompts to write,
The muses promise to assist my pen;
'Twas not long since I left my native shore
The land of errors,[2] and Egyptian gloom.[3]
Father of mercy, 'twas Thy gracious hand   5
Brought me in safety from those dark abodes.

    Students, to you 'tis given to scan the heights
Above, to traverse the ethereal space,
And mark the systems of revolving worlds.
Still more, ye sons of science[4] ye receive   10
The blissful news by messengers from heav'n,
How Jesus' blood for your redemption flows.
See Him with hands outstretched upon the cross;
Immense compassion in His bosom glows;
He hears revilers, nor resents their scorn: 15
What matchless mercy in the Son of God!
When the whole human race by sin had fall'n,
He deigned to die that they might rise again,
And share with Him in the sublimest skies,
Life without death, and glory without end.   20

    Improve your privileges while they stay,
Ye pupils, and each hour redeem, that bears
Or good or bad report of you to heav'n.
Let sin, that baneful evil to the soul,
By you be shunned, nor once remit your guard;   25
Suppress the deadly serpent in its egg.
Ye blooming plants of human race divine,
An Ethiop[5] tells you 'tis your greatest foe;
Its transient sweetness turns to endless pain,
And in immense perdition sinks the soul.   30

1767   1773

On the Death of the
Rev. Mr. George Whitefield, 1770[1]

Hail, happy saint, on thine immortal throne,
Possessed of glory, life, and bliss unknown;
We hear no more the music of thy tongue,
Thy wonted[2] auditories cease to throng.

1. Harvard.
2. I.e., theological errors, since Africa was unconverted.
3. "And Moses stretched forth his hand toward heaven; and there was a thick darkness in all the land of Egypt three *days*" (*Exodus 10.22*).
4. Knowledge.
5. Ethiopian. In Wheatley's time "Ethiopian" was a conventional name for the

Negro peoples of Africa.
1. George Whitefield (1714–70), an English follower of John Wesley, was the best-known revivalist in the 18th century. He made several visits to America and died in Newberyport, Massachusetts. This was Wheatley's first published poem, and it made her famous.
2. Accustomed.

Thy sermons in unequaled accents flowed,   5
And every bosom with devotion glowed;
Thou didst in strains of eloquence refined
Inflame the heart, and captivate the mind.
Unhappy we the setting sun deplore,
So glorious once, but ah! it shines no more.   10

Behold the prophet in his towering flight!
He leaves the earth for heav'n's unmeasured height,
And worlds unknown receive him from our sight.
There Whitefield wings with rapid course his way,
And sails to Zion³ through vast seas of day.   15
Thy prayers, great saint, and thine incessant cries
Have pierced the bosom of thy native skies.
Thou moon hast seen, and all the stars of light,
How he has wrestled with his God by night.
He prayed that grace in every heart might dwell,   20
He longed to see America excel;
He charged⁴ its youth that every grace divine
Should with full luster in their conduct shine;
That Savior, which his soul did first receive,
The greatest gift that ev'n a God can give,   25
He freely offered to the numerous throng,
That on his lips with listening pleasure hung.

"Take Him, ye wretched, for your only good,
Take Him my dear Americans," he said,
Ye thirsty, come to this life-giving stream,   30
Ye preachers, take Him for your joyful theme;
Take Him my dear Americans," he said,
"Be your complaints on His kind bosom laid:
Take Him, ye Africans, He longs for you,
Impartial Savior is His title due: 35
Washed in the fountain of redeeming blood,
You shall be sons, and kings, and priests to God."

Great *Countess*,⁵ we Americans revere
Thy name, and mingle in thy grief sincere;
New England deeply feels, the orphans mourn,   40
Their more than father will no more return.

But, though arrested by the hand of death,
Whitefield no more exerts his laboring breath,
Yet let us view him in the eternal skies,
Let every heart to this bright vision rise;   45
While the tomb safe retains its sacred trust,
Till life divine re-animates his dust.

1770   1770, 1773

3. Here, the heavenly city of God.
4. Exhorted.
5. Selina Shirley Hastings (1707–91), Countess of Huntington, was a strong supporter of George Whitefield and active in Methodist church affairs. Wheatley visited her in England in 1773.

To S. M.,[1] a Young African Painter, on Seeing His Works

To show the laboring bosom's deep intent,
And thought in living characters to paint,
When first thy pencil did those beauties give,
And breathing figures learnt from thee to live,
How did those prospects give my soul delight,   5
A new creation rushing on my sight?
Still, wondrous youth! each noble path pursue,
On deathless glories fix thine ardent view:
Still may the painter's and the poet's fire
To aid thy pencil, and thy verse conspire!   10
And may the charms of each seraphic[2] theme
Conduct thy footsteps to immortal fame!
High to the blissful wonders of the skies
Elate thy soul, and raise thy wishful eyes.
Thrice happy, when exalted to survey   15
That splendid city, crowned with endless day,
Whose twice six gates[3] on radiant hinges ring:
Celestial Salem[4] blooms in endless spring.

Calm and serene thy moments glide along,
And may the muse inspire each future song!   20
Still, with the sweets of contemplation blest,
May peace with balmy wings your soul invest!
But when these shades of time are chased away,
And darkness ends in everlasting day,
On what seraphic pinions shall we move,   25
And view the landscapes in the realms above?
There shall thy tongue in heavenly murmurs flow,
And there my muse with heavenly transport glow:
No more to tell of Damon's[5] tender sighs,
Or rising radiance of Aurora's[6] eyes,   30
For nobler themes demand a nobler strain,
And purer language on the ethereal plain.
Cease, gentle muse! the solemn gloom of night
Now seals the fair creation from my sight.

1773

---

1. Scipio Moorhead, a servant to the *Reverend John Moorhead* of Boston.
2. Angelic.
3. Heaven, like the city of Jerusalem, is thought to have had 12 gates (as many gates as tribes of Israel).
4. Heavenly Jerusalem.
5. In classical mythology Damon pledged his life for his friend Pythias.
6. The Roman goddess of the dawn.

American Literature
1820–1865

THE INFLUENCE OF WASHINGTON IRVING

IN A PAINTING popular during the late nineteenth century, Christian Schussele reverentially depicted *Washington Irving and His Literary Friends at Sunnyside*. Working in 1863, four years after Irving's death, Schussele portrayed an astonishing number of elegantly clad notables in Irving's snug study in his Gothic cottage-castle on the Hudson River, north of New York City. Among them were several writers in this anthology: Irving himself, Oliver Wendell Holmes, Nathaniel Hawthorne, Henry Wadsworth Longfellow, Ralph Waldo Emerson, William Cullen Bryant, and James Fenimore Cooper. Intermingled with these men were poets and novelists now seldom read: William Gilmore Simms, Fitz-Greene Halleck, Nathaniel Parker Willis, James Kirke Paulding, John Pendleton Kennedy, and Henry T. Tuckerman, along with the historians William H. Prescott and George Bancroft. The Schussele painting was a pious hoax, for these guests never assembled together at one time, at Sunnyside or anywhere else, and while a few of those depicted were indeed among Irving's friends, he barely knew some of them and never met others at all. But in several ways the scene is profoundly true to American literary history.

As Schussele's painting suggests, Irving, beloved by ordinary readers and by most of his fellow writers, was the central figure in the American literary world between 1809 (the year of his parody *History of New York*) and the Civil War, especially after he demonstrated in *The Sketch Book* (1819–20) that memorable fiction—*Rip Van Winkle* and *The Legend of Sleepy Hollow*—could be set in the United States; he also proved, by the book's international success, that an American writer could win a British and Continental audience. Irving's legion of imitators included several of the persons in the painting, and among his fellow writers Irving's reputation was enhanced by his generosity, as in his gallantly relinquishing the subject of the conquest of Mexico to Prescott or in urging the publisher George P. Putnam to bring out an American edition of the first book by the unknown Herman Melville. Although James Fenimore Cooper's fame as a fiction writer rivaled Irving's in the 1820s and 1830s, his influence never approached the breadth of Irving's. Nor did the influence of Ralph Waldo Emerson, despite his profoundly provocative effects on such writers as Margaret Fuller, Henry David Thoreau, Walt Whitman, Herman Melville, and

Emily Dickinson—effects that make modern literary historians see him as the seminal writer of the century.

Mentioning the names of Fuller, Thoreau, Melville, and Dickinson suggests still another way the Schussele painting is exemplary. Since the painter set out to depict representative literary men as much as to depict genuine intimates of Irving, it is striking that he omitted writers who now seem among the most important of the century: Edgar Allan Poe, Thoreau, Whitman, Melville, John Greenleaf Whittier (who was frowned upon as a militant abolitionist until 1866, when *Snow-Bound* made him seem a safe poet to admire), and Dickinson (in the 1860s an unpublished recluse). The painter would probably have considered Thomas Bangs Thorpe and other southwestern humorists to be subliterary, despite the fact that Irving had influenced such writing and had delighted in reading it.

THE SMALL WORLD OF AMERICAN WRITERS

Perhaps most important, paintings like the one by Schussele (and the similar wishful fad of depicting famous literary people in cozy association through the then-new technique of composite photography) capture the fact that in the nineteenth century the American literary world was very small indeed, so small that most of the writers in this period knew each other, often intimately, or else knew much about each other. They lived, if not in each other's pockets, at least in each other's houses, or boardinghouses: Lemuel Shaw, from 1830 to 1860 Chief Justice of the Massachusetts Supreme Court, and Herman Melville's father-in-law after 1847, for a time stayed in a Boston boardinghouse run by Ralph Waldo Emerson's widowed mother; the Longfellows summered in the 1840s at the Pittsfield boardinghouse run by Melville's cousin, a house where Melville had stayed in his early teens; in Pittsfield and Lenox, Hawthorne and Melville paid each other overnight visits; in Concord the Hawthornes rented the Old Manse, the Emerson ancestral home, and later bought a house there from the educator Bronson Alcott; in Concord the Emersons welcomed many guests, including Margaret Fuller, and when the master was away Thoreau sometimes stayed in the house to help Mrs. Emerson with the children and the property.

Many of the writers of this period came together casually for dining and drinking, the hospitality at the editor Evert A. Duyckinck's house in New York being famous, open to southerners like Simms as well as New Yorkers like Melville and Bostonians like the elder Richard Henry Dana. In the late 1850s a Bohemian group of newspaper and theater people and writers drank together at Pfaff's saloon on Broadway above Bleecker Street; for a time Whitman was a fixture there. Of the clubs formed by writers, artists, and other notables (usually male), the three most memorable are the Bread and Cheese Club, which Cooper organized in 1824 in the back room of his publisher's Manhattan bookstore; the Transcendental Club, started in Boston in 1836 and lasting four years; and the Saturday Club, a more convivial Boston group formed in 1856. Members of the Bread and Cheese Club included the poet William Cullen Bryant, Samuel F. B. Morse (the painter who later invented the telegraph), the poet Fitz-Greene Halleck, and Thomas Cole, the English-born painter of the American landscape. Emerson was the leading spirit of the Transcendental Club, but other members included Bronson Alcott, later Margaret Fuller, and George Ripley, the organizer of the Transcendental commune at Brook Farm, near Roxbury.

Among the members of the Saturday Club were Emerson, James Russell Lowell, Henry Wadsworth Longfellow, Oliver Wendell Holmes, and the historians John Lothrop Motley and William H. Prescott; Nathaniel Hawthorne attended some meetings.

THE SMALL COUNTRY

Such intimacy was inevitable in a country which had only a few literary and publishing centers, all of them along the Atlantic seaboard. Despite the acquisition of the Louisiana Territory from France in 1803 and the vast Southwest from Mexico in 1848, most of the writers we still read lived all their lives in the original thirteen states, except for trips abroad, and their practical experience was of a compact country: in 1840 the "northwestern" states were those covered by the Northwest Ordinance of 1787 (Ohio, Indiana, Illinois, and Michigan; Wisconsin was still a territory), while the "southwestern" humor writers such as George Washington Harris, Thomas Bangs Thorpe, and Johnson Jones Hooper wrote in the region bounded by Georgia, Louisiana, and Tennessee.

Improvements in transportation were shrinking the country even while territorial gains were enlarging it. When Irving went from Manhattan to Albany in 1800, steamboats had not yet been invented; the Hudson voyage was slow and dangerous; and in 1803 the wagons of his Canada-bound party barely made it through the bogs beyond Utica. The Erie Canal, completed in 1825, changed things: in the 1830s and 1840s Hawthorne, Melville, and Fuller took the canal boats in safety, suffering only from crowded and stuffy sleeping conditions. When Irving went buffalo hunting in Indian territory (now Oklahoma) in 1832 he left the steamboat at St. Louis and went on horseback, camping out at night except when his party reached one of the line of missions built to accommodate whites who were Christianizing the Indians. By the 1840s railroads had replaced stagecoaches between many eastern towns, although to get to New Orleans in 1848 Whitman had to change from railroad to stagecoach to steamboat. Despite frequent train wrecks, steamboat explosions, and Atlantic shipwrecks, by the 1850s travel had ceased to be the hazardous adventure it had been. But the few American writers who saw much of the country were still provincials in their practical attitude toward their literary careers, for their publishers and purchasers were concentrated mainly in or near New York, Philadelphia, and Boston.

And the New York, Philadelphia, and Boston of this period were themselves tiny in comparison to their modern size. The site of Brook Farm, now long since a victim of urban sprawl, was chosen because it was nine miles remote from Boston and two miles away from the nearest farm. The population of New York City at the start of the 1840s was only a third of a million and was concentrated in Lower Manhattan: Union Square was the edge of town. Horace Greeley, the editor of the New York *Tribune*, escaped the bustle of the city by living on a ten-acre farm up the East River on Turtle Bay, where the East Fifties are now; there he and his wife provided a bucolic retreat for Margaret Fuller when she was his literary critic and metropolitan reporter. In 1853 the Crystal Palace, an exposition of arts, crafts, and sciences, created in imitation of the great Crystal Palace at the London World's Fair of 1851, failed—largely because it was too far out of town, up west of the new Croton Water Reservoir that had recently brought running water to the city. The reservoir was on the spot where the New York Public Library now stands, at Forty-second Street and Fifth Avenue, and the Crystal Palace was on the site of the modern Bryant Park, named for the nature poet but now long inured to sadder urban visitors than those who made their way to the Fair in the 1850s.

THE ECONOMICS OF AMERICAN LETTERS

Geography and modes of transportation bore directly upon publishing procedures in the United States of this period. For a long time writers who wanted to publish a book carried the manuscript to a local printer and paid job rates to have it printed and bound. Longfellow worked in this fashion with a firm in Brunswick, Maine, when he printed his translation of *Elements of French Grammar* and other textbooks during his first years as a teacher. Fiction was also sometimes sent to a local printer, as when Johnson Jones Hooper paid a firm in Tuscaloosa, Alabama, to print *A Ride with Old Kit Kuncker* before having it brought out the next year by a regular Philadelphia publisher. However, the true publishing centers were major seaports which could receive the latest British books by the fastest ships and, hastily reprinting them, distribute them inland by river traffic as well as in coastal cities. After 1820 the leading publishing towns were New York and Philadelphia, with the Erie Canal soon giving New York an advantage in the Ohio trade. Boston remained only a provincial publishing center until after 1850, when publishers realized the value of the new railroad connections to the West. Despite the aggressive merchandizing techniques of a few firms, the creation of a national book-buying market for literature, especially American literature, was long delayed.

The problem was that the economic interests of American publisher-booksellers were antithetical to the interests of American writers. A national copyright law became effective in the United States in 1790, but it was 1891 before American writers had international protection and foreign writers received protection in the United States. Until the end of the century, American printers routinely pirated English writers, paying nothing to Sir Walter Scott or Charles Dickens for their novels, which were rushed into print and sold very cheaply in New York, Philadelphia, and other cities. American readers benefited from the situation, for they could buy the best British—and Continental—writings cheaply, but American writers suffered, since if they were to receive royalties, their books had to be priced above the prices charged for works of the most famous British writers. American publishers were willing to carry a few native novelists and poets as prestige items for a while, but they were businessmen, not philanthropists.

To compound the problem, Irving's apparent conquest of the British publishing system, by which he received large sums for *The Sketch Book* and succeeding volumes, proved delusory. Cooper and others followed in Irving's track and were paid by magnanimous British publishers under a system whereby works first printed in Great Britain were presumed to hold a British copyright. But this practice was ruled illegal by a British judge in 1849, and the British market dried up for American writers.

Throughout this period, making a serious American contribution to the literature of the world was no guarantee at all of monetary rewards. Except possibly for a few authors of sentimental best sellers, including what Hawthorne jealously called "that damned mob of scribbling women," the United States was not a country in which one could make a living by writing. It was not even a place where the best authors could always publish what they wrote. The only writers who could consistently find a publisher were Irving and Cooper, who kept their appeal on the basis of early success (though more copies had to be sold in order to make the same profit) and the magazine or newspaper editors who could fill some of their own columns when they wanted. These editors included (for various periods of time) Poe, Harris, Thorpe, Hooper, Lowell, and four other notable examples: Fuller, who for several years reported for the New York *Tribune* at home and from Europe; Whitman, who for much of the 1840s and 1850s was free to editorialize in one Brooklyn or Manhattan newspaper or another; Whittier, who for more than two decades before the Civil War was corresponding editor of the Washington *National Era*; and, most conspicuously, Bryant, long-time owner of the New York *Evening Post*. Whitman was his own publisher for most editions of *Leaves of Grass*, and filled mail orders himself, as Thoreau also did when an occasional request came for one of the 700 copies of his first book which the publisher had turned back to him. At crucial moments in his career Melville was balked from writing what he wanted to write, as when he sacrificed his literary aspirations after the failure of *Mardi* and wrote *Redburn* and *White-Jacket*, which he regarded as mere drudgery; and at other times he was "prevented from publishing" works he had written, including at least one which was subsequently destroyed. Ironically, the writer freest to pursue literary greatness in this period was probably Emily Dickinson, whose "letter to the world" remained unmailed during her lifetime.

THE QUEST FOR AN AMERICAN LITERARY DESTINY

In the first half of the nineteenth century, lobbying for the existence of an American literature in magazines seemed to take up more space than the literature itself. Especially after the War of 1812 confirmed American independence, theorists called for a great literature which would match the emerging political greatness of the nation. Huckstering critics soon developed specific notions as to the subjects which would-be writers should choose: preferably the distant colonial past (the nearest we could hope to come to the medieval settings which were serving Sir Walter Scott so well), or possibly Indian legends, or, still less desirable (because too near the mundane present), subjects from the recent Revolutionary past. Such exhortations were the stock-in-trade of commencement speakers and literary critics in the 1820s and 1830s. But in *The Poet* (1842) Emerson boldly called for a poet who would write of the United States as it was, not as it might have been:

We have yet had no genius in America, with tyrannous eye, which knew the value of our incomparable materials, and saw, in the barbarism and materialism of the times, another carnival of the same gods whose picture he so much admires in Homer; then in the middle age; then in Calvinism. Banks and tariffs, the newspaper and caucus, methodism and unitarianism, are flat and dull to dull people, but rest on the same foundations of wonder as the town of Troy, and the temple of Delphos, and are as swiftly passing away. Our logrolling, our stumps and their politics, our fisheries, our Negroes, and Indians, our boasts, and our repudiations, the wrath of rogues, and the pusillanimity of honest men, the northern trade, the southern planting, the western clearing, Oregon, and Texas, are yet unsung. Yet America is a poem in our eyes; its ample geography dazzles the imagination, and it will not wait long for metres.

Later Whitman was to say that he had remained simmering, simmering, until Emerson brought him to a boil.

During the 1840s Evert A. Duyckinck and other New York literary men (primarily through the columns of the *Democratic Review* and the *Literary World*) mustered a squad of promoters of the great literature that was to come. The propagandists perfected the rhetorical strategy of linking literary destiny to geography and political destiny: the "great nation of futurity" must have a literature to match Niagara Falls and the Rocky Mountains. Herman Melville for several years was associated with Duyckinck's magazines, and he half-champions and half-spoofs the chauvinistic rhetoric in the essay on Hawthorne which he wrote for the *Literary World* in 1850. An American, he proclaimed, was "bound to carry republican progressiveness into Literature, as well as into Life," even to the point of believing that sooner or later American writers would rival Shakespeare, whom a generation of Bardolators regarded as unapproachable. This was literary manifest destiny with a vengeance, warranted only because as he wrote the essay Melville had already written his way well into what he later titled *Moby-Dick*.

None of the American writers of the period was chauvinistic enough to think that a great American literature could be written without reference to past English and European literature. As Cooper protested in *Notions of the Americans* (1828), writers in the United States possessed the same literary heritage that writers in Great Britain did. Shakespeare, Spenser, Milton, Bunyan, Addison, Pope, Fielding, Johnson, and Burns, along with many others (especially some now neglected writers of the eighteenth century) were the possession of all educated Americans born in the late eighteenth century or early in the nineteenth. Americans were not long behind the British in responding to the Romantics Wordsworth and Coleridge, then to Byron, Moore, and Scott. By the 1830s Carlyle was a force in the lives of several American writers through his translations of recent German philosophical works and his own jeremiads against contemporary British values. Americans had access to the latest British and Continental discussions of art, religion, politics, and science, for British magazines, especially the quarterly reviews, were imported promptly and widely reprinted. Nineteenth-century American writing reveals its full meanings only in the light of European influences and parallel developments.

THE NEW AMERICANNESS OF AMERICAN LITERATURE

Despite the cultural cross-connections with Europe, the best literature that emerged in the United States was distinctively new, and a few perceptive critics very early began trying to define its special quality. This analysis from the review of *The Whale* (the English title of *Moby-Dick*) in the London *Leader* had currency in America as well, for the popular *Harper's New Monthly Magazine* quoted it approvingly:

Want [lack] of originality has long been the just and standing reproach to American literature; the best of its writers were but second-hand Englishmen. Of late some have given evidence of originality; not *absolute* originality, but such genuine outcomings of the American intellect as can be safely called national. Edgar Poe, Nathaniel Hawthorne, Herman Melville are assuredly no British offshoots; nor is Emerson—the *German* American that he is! The observer of this commencement of an American literature, properly so called, will notice as significant that these writers have a wild and mystic love of the supersensual, peculiarly their own. To move a horror skilfully, with something of the earnest faith in the Unseen, and with weird imagery to shape these Phantasms so vividly that the most incredulous mind is hushed, absorbed—to do this no European pen has apparently any longer the power—to do this American literature is without a rival. What *romance* writer can be named with Hawthorne? Who knows the terrors of the seas like Herman Melville?"

Plainly, this was meant as praise, but to employ "weird imagery" in order to "move a horror skilfully" was hardly the ambition of any American writer of the period besides Poe; for their part, Hawthorne and Melville were not concerned with the supernatural except as stage devices for heightning their psychological analyses.

But literary historians have not improved much on the reviewer in the *Leader* in deciding what was American about American literature. American writers were not achieving originality in form: Irving's sentences were accepted as models of English prose style precisely because they were themselves modeled upon the sentences of Addison and Goldsmith, long the prime exemplars of decorous English prose. Melville's sentences often looked like those of whatever powerful master of the English language he had most recently been reading—Shakespeare, Burton, Taylor, Sterne, De Quincey, Carlyle. Nor was the content of the best American writing of this period original in anything like an "absolute" sense. Modern scholars have shown that in his most "American" stories, *Rip Van Winkle* and *The Legend of Sleepy Hollow*, Irving drew upon, and even closely translated, parts of German tales. In *Moby-Dick* Melville's metaphysics are recognizably of the generation of Goethe, Byron, and Carlyle. Thoreau's recurrent ideas came mainly from Emerson (at least Emerson himself insisted they did), but Emerson had picked them up from dozens of ancient and modern philosophers.

Yet, as everyone in the country sensed by the 1850s, there was some elusive quality about its new literature that was *American.* Irving's German-influenced stories were profoundly moving to Americans, who knew more than most Britons what it was to feel the trauma of rapid change, especially to experience repeated physical uprootings, and Americans found in the ne'er-do-well Rip a model for making a success of failure. In Cooper's novels was a sense of the immensity of physical nature and the power of human beings to destroy nature that most European writers could experience only vicariously. In Melville's *Moby-Dick* was a sense of the grandeur of the physical universe and man's role in it long suppressed in European consciousness. In *Leaves of Grass* Whitman undertook another elemental task—to become the national poet of a new people on a new continent. What proved most enduringly "American" about Emerson was his wide streak of Yankee individualism best displayed in *Self-Reliance*, which became an inspiration to thousands of Americans who were determined to hitch their wagons, as Emerson said, to a star. Even Thoreau's *Walden,* which many contemporaries took merely as an American counterpart of the English naturalist Gilbert White's *Natural History and Antiquities of Selbourne*, was in fact consciously an American counterscripture, a Franklinesque retort to Poor Richard, a how-to book on getting a living by working at what you love. At a time when grandiloquence in political rhetoric was often taken for eloquence, Abraham Lincoln mastered both the majestic cadences of the King James Version of the Bible and the extravagant toughness of backwoods tall talk. Dickinson's poems in their minute intensity were as ambitious as Whitman's, magnificent attempts to define her experience at whatever cost in wrenched syntax and rhyme. At best, beyond question, American writers were accomplishing things yet unattempted in the English language.

THE AESTHETICS OF A NATIONAL LITERATURE

The great writers of the period for the most part defined their aesthetic problems by themselves, though Emerson's *The Poet* aided some of the others. The primary difficulty of how to keep from being secondhand English writers had not been squarely faced by the theorists of nationality in literature, who most often seemed to think that adoption of an American setting or, more vaguely, the infusion of an American "spirit" guaranteed Americanness. Insofar as the issues had been addressed by Americans before the 1840s, it was primarily by painters and sculptors, the most prominent of whom had received their training abroad, then had found it impossible to reconcile their European notions of noble subject and style with Americanness. Only the Hudson River school of painters, led by Thomas Cole, found a pantheistic majesty in American landscapes not anticipated by the history-filled landscapes of European painting. But despite all the theorizing and actual creating, the major writers were pretty much on their own in confronting their aesthetic problems—such as Hawthorne's attempts to strike a balance between the allegorical and the realistic, Emerson's difficulty in achieving unity from the mutually repellent particles of his thought, Thoreau's attempts to unite the Transcendentalist and the naturalist in himself, Whitman's struggle to domesticate the epic catalogue without falling into self-parody, Melville's attempt to create a tragedy in a democracy, and Dickinson's attempts to walk the hairline between mere coyness and psychological precision.

THE WRITERS AND THEIR AMERICA

When the great American writers of the mid-nineteenth century took stock of their country, they sometimes caught the contagion of an ebullient, expansionist mood that struck many observers as the dominant one of the time, and even Thoreau, the most relentless critic of the values of his society, insisted that to some extent he counted himself among "those who find their encouragement and inspiration in precisely the present condition of things, and cherish it with the fondness and enthusiasm of lovers." But often they felt a profound alienation. Emerson was a preacher who had renounced his pulpit, and the other great writers—also preachers without pulpits—devoted much of their artistic effort to analyzing conditions of life in America and to exhorting their fellow citizens to live more wisely.

Conformity and Materialism

It seemed to some of the writers that Americans, even while deluding themselves that they were the most self-reliant populace in the world, were systematically selling out their individuality. Emerson sounded the alarm: "Society everywhere is in conspiracy against the manhood of every one of its members. Society is a joint-stock company in which the members agree for the better securing of his bread to each shareholder, to surrender the liberty and culture of the eater. The virtue in most request is conformity." In *The Celestial Railroad* Hawthorne satirically described the condition at the Vanity Fair of modern America, where there was a "species of machine for the wholesale manufacture of individual morality." He went on: "This excellent result is effected by societies for all manner of virtuous purposes; with which a man has merely to connect himself, throwing, as it were, his quota of virtue into the common stock; and the president and directors will take care that the aggregate amount be well applied." Thoreau repeatedly satirized America as a nation of joiners which tried to force every newcomer "to belong to their desperate odd-fellow society"; to Thoreau, members of the Odd Fellows and other social organizations were simply not odd *enough*, not individual enough.

But none of the writers found anything comical in the wholesale loss of Yankee individualism as both men and women deserted worn-out farms for factories, where many began to feel what Emerson called "the disproportion between their faculties and the work offered them." Far too often, the search for a better life had degenerated into a desire to possess factory-made objects. "Things are in the saddle," Emerson said sweepingly, "and ride mankind." In elaboration of that accusation, Thoreau wrote *Walden* as a treatise on expanding the spiritual life by simplifying material wants. Informing Thoreau's outrage at the materialism of his time was the bitter knowledge that even the most impoverished were being led to waste their money (and, therefore, their lives) on trumpery. In a vocabulary echoing Benjamin Franklin, he condemned the emerging consumer economy which was devoted, even in the infancy of advertising, to the creation of "artificial wants" for things which were unneeded or outright pernicious. And to counter the loss of an archetypal Yankee virtue, he made himself into a Jack-of-all-trades and strong master of one, the art of writing.

Sex and Sexual Roles

At a time when sex was banished from the magazines and from almost all books except medical treatises, Whitman alone called for a healthy sense of the relation between body and soul and created a forum for discussing sexual joy and anguish. The other male writers made no challenge to conventional sexual roles; when Emerson, for instance, said that society "is in conspiracy against the manhood of every one of its members," he meant "*man*hood," not "manhood and womanhood." Only Whitman among the male authors regularly employed what we would call nonsexist language, and only Whitman rejected the opinion that woman's proper "sphere" was a limited, subservient, supportive one. While the attitudes of most male—and female—writers of the time reflected and embodied the prevailing sexism, Whitman rejected the "empty dish, gallantry" as a degraded attitude: "This tepid wash, this diluted deferential love, as in songs, fictions, and so forth, is enough to make a man vomit." Instead, he insisted on equality: "Women in These States approach the day of that organic equality with men, without which, I see, men cannot have organic equality among themselves." Of the other writers only Margaret Fuller thought so deeply about sexual roles. Ironically, as the mother of a tardily acknowledged child (and perhaps not the wife of its Italian father), Fuller was an incalculable threat to the little Boston literary society in the months before her death by shipwreck prevented her arrival home. Of the women writers of the time, Dickinson, who never married, was the most bitterly ironic observer of the sacrifices marriage often required of a woman, as in her depiction of the bride who "rose to His Requirement—dropt / The Playthings of Her Life / To take the honorable Work / Of Woman, and of Wife." But women had no monopoly on sexual anguish. Melville, who as a young man had known the pagan Eden of the South Seas, found that the claims of his intellect and imagination, his pursuit of a literary career, could be acknowledged only by denying the claims of his wife and children. And Whitman, the only writer of the period to advance a "Programme" for honest depiction of sex in literature, himself endured torments over his homoerotic longings.

Nature

In "a new country," Thoreau said, "fuel is an encumbrance," and his generation acted as if trees existed to be burned (and mountains to be graded and wild animals to be slaughtered). But while Thoreau faced the possibility that like villains we might grub our forests all up, "poaching on our own national domains," he had no deep anxiety that primeval nature like the Maine woods would be destroyed. Melville was likewise sure that the whale would not perish: "hunted from the savannas and glades of the middle seas, the whale-bone whales can at last resort to their Polar citadels, and diving under the ultimate glassy barriers and walls there, come up among icy fields and floes; and in a charmed circle of everlasting December, bid defiance to all pursuit from man." Of the major writers of the period, Emerson, Thoreau, and Whitman felt an intensity of communion with nature that warrants their being called nature-mystics, and Dickinson, bounded by town lots and fields near her house in Amherst, found a profoundly un-Christian, "Druidic difference" which enhanced nature for her with a sense of harmony between it and human beings. The writers diverged in their wider views of the universe, Melville describing in *Moby-Dick* the maddening of a cabin-boy abandoned in the immensity of ocean, and Thoreau, by contrast, insisting that he was not lonely at Walden: "Why should I feel lonely? is not our planet in the Milky Way?" But whatever their sense of man's place in the cosmos, they all found nature a force in their lives in ways out of keeping with the times, when the Romantic sense of nature as restorer and healer of mankind seemed to persist, as Thoreau pointed out, in the absurd form of uneasy rest-day strollers anxious to pass their allotted time in the woods and return to town.

Orthodox Religion and Transcendentalism

All the major writers found themselves at odds with the dominant religion of their time, a nominal Protestant Christianity which exerted practical control over what could be printed in books and magazines. This church, Emerson said, acted "as if God were dead." Whitman was more bitter still: "The churches are one vast lie; the people do not believe them, and they do not believe themselves." The writers all came of Protestant backgrounds where Calvinism was more or less watered down (less in the cases of Melville and Dickinson), but they tended to apply absolute standards toward what passed for Christianity. In *The Celestial Railroad* Hawthorne memorably satirized the American urge to be progressive and liberal in theology as well as in politics, and Melville extended the satire throughout an entire book, *The Confidence-Man*.

Awareness of the fact of religious ecstasy was not at issue. Emerson, for instance, showed in *The Over-Soul* a clinical sense of the varieties of religious experience, the "varying forms of that shudder of awe and delight with which the individual soul always mingles with the universal soul." Similarly, Thoreau acknowledged the validity of the "second birth and peculiar religious experience" available to the "solitary hired man on a farm in the outskirts of Concord," but felt that any religious denomination in America would pervert that mystical experience into something available only under its auspices and something to be brought into line with its particular doctrines. Like Thoreau, Whitman saw all religious ecstasy as equally valid, and came forth in *Song of Myself* outbidding "the old cautious hucksters" like Jehovah, Kronos, Zeus, and Hercules, gods who held too low an estimate of the value of men and women. Among these writers Melville was alone in his anguish at the realization that Christianity was impracticable. Melville also felt the brutal power of the Calvinistic Jehovah with special keenness: mankind was "god-bullied" even as the hull of the *Pequod* was in *Moby-Dick*, and the best way man had of demonstrating his own divinity lay in defying the omnipotent tyrant. To Dickinson also God was a bully—a "Mastiff," whom subservience might, or might not, appease.

Transcendentalism in the late 1830s and early 1840s was treated in newspapers and magazines as something between a national laughingstock and a clear menace to organized religion. The running journalistic joke, which Hawthorne echoed in *The Celestial Railroad*, was that no one could define the term, other than that it was highfalutin, foreign, and obscurely dangerous. The conservative Christian view is well represented by a passage which appeared in Harriet Beecher Stowe's newspaper serialization of *Uncle Tom's Cabin* (1851) but was omitted from the book version, a sarcastic indictment of the reader who might find it hard to believe that Tom could be stirred by a passage in the Bible: "I mention this, of course, philosophic friend, as a psychological phenomenon. Very likely it would do no such a thing for you, because you are an enlightened man, and have out-grown the old myths of past centuries. But then you have Emerson's Essays and Carlyle's Miscellanies, and other productions of the latter day, suited to your advanced development." Such early observers understood well enough that Transcendentalism was more pantheistic than Christian. The "defiant Pantheism" infusing Thoreau's shorter pieces helped keep them out of the magazines, and James Russell Lowell for the *Atlantic Monthly* publication of a section of *The Maine Woods* censored a sentence in which Thoreau declared that a pine tree was as immortal as he was, and perchance would "go to as high a heaven."

Melville was also at least once kept from publication by the religious scruples of the magazines, and often he was harshly condemned for what he had managed to publish. For years he bore the wrath of reviewers such as the one who denounced him for writing *Moby-Dick* and the Harpers for publishing it: "The Judgment day will hold him liable for not turning his talents to better account, when, too, both authors and publishers of injurious books will be conjointly answerable for the influence of those books upon the wide circle of immortal minds on which they have written their mark. The book-maker and the book-publisher had better do their work with a view to the trial it must undergo at the bar of God." This was extreme, but Emerson, Thoreau, and Whitman all suffered in comparable ways for transgressing the code of the Doctors of Divinity (Thoreau said he wished it were not the D.D.'s but the chickadee-dees who acted as censors).

Immigration and Xenophobia

However threatened conservative Protestants felt by Transcendentalism and by religious speculations like Melville's, they felt far more threatened by Catholicism when refugees from the Napoleonic wars were followed by refugees from oppressed and famine-struck Ireland. In Boston, Lyman Beecher, father of Harriet Beecher Stowe, thundered out antipapist sermons, then professed dismay when in 1834 a mob in Charlestown, across the Charles River from Boston, burned the Ursuline Convent School where daughters of many wealthy families were educated. Through the 1830s and 1840s and long afterward, the country was saturated with lurid books and pamphlets purporting to reveal the truth about sexual practices in nunneries and monasteries (accounts of how priests and nuns disposed of their babies were specially prized) and about the Pope's schemes to take over the Mississippi Valley (Samuel F. B. Morse and others warned that Jesuits were prowling the Ohio Valley, in disguise). An extreme of xenophobia was reached in the summer of 1844, when rioters in Philadelphia (the city, everyone pointed out, of brotherly love) burned Catholic churches and a seminary. Melville was replying to the current hostility when he followed a description of the pestilent conditions of steerage-passengers in emigrant ships with this plea: "Let us waive that agitated national topic, as to whether such multitudes of foreign poor should be landed on our American shores; let us waive it, with the one only thought, that if they can get here, they have God's right to come; though they bring all Ireland and her miseries with them. For the whole world is the patrimony of the whole world; there is no telling who does not own a stone in the Great Wall of China."

For all his humanitarian eloquence, Melville, like the other writers, realized that the new immigrants were changing the country from the cozy, homogeneous land it had been, or had seemed to be. By the end of the Civil War many native Amercians shared Stowe's profound nostalgia for the days before the railroads, before the influx of Catholics, before the even more alien influx of immigrants from southern and eastern Europe, few of whom spoke English and many of whom were not Christian at all.

Politics and Wars

The major writers of the period lived with the anguishing paradox that the most idealistic nation in the world was implicated in continuing national sins: the near-genocide of the American Indians (whole tribes in colonial times had already become, in Melville's phrase for the Massachusetts Pequods, as extinct as the ancient Medes), the enslavement of blacks, and (partly a by-product of slavery) the staged "Executive's War" against Mexico, started by President Polk before being declared by Congress. Emerson was an exception, but most writers were silent about the successive removal of Indian tribes to less desirable lands: American destiny plainly required a little practical callousness, most whites felt, in a secular version of the colonial notion that God had willed the extirpation of the red men. The imperialistic Mexican War was so gaudily exotic—and so distant— that only a small minority of American writers voiced more than perfunctory opposition; an exception was Thoreau, who spent a night in the Concord jail in symbolic protest against being taxed to support the war.

It was Negro slavery, what Melville called "man's foulest crime," which most stirred the consciences of the white writers; and in describing his own enslavement the fugitive Frederick Douglass developed a notable capacity to stir readers as well as audiences in the lecture halls. When the Fugitive Slave Law was enforced in Boston in 1851 (by Melville's father-in-law, Chief Justice Shaw), Thoreau worked his outrage into his journals; then after another famous case in 1854 he combined the experiences into his most scathing speech, *Slavery in Massachusetts*, for delivery at a Fourth of July countercelebration at which a copy of the Constitution was burned because slavery was written into it. In that speech Thoreau summed up the disillusionment that many of his generation shared. He had felt a vast but indefinite loss after the 1854 case, he said: "I did not know at first what ailed me. At last it occurred to me that what I had lost was a country." (Successive generations of American writers would experience the same trauma: Howells, Twain, and others when the United States turned from savior to conqueror in the Philippines after the Spanish-American War, Robert Lowell and many others after it became clear that the involvement of the United States in Vietnam was not purely a gesture of compassion toward a grateful, beleaguered nation.) More obliquely than Thoreau, Melville explored black slavery in *Benito Cereno* as an index to the emerging national character. At his bitterest, he felt in the mid-1850s that "free Ameriky" was "intrepid, unprincipled, reckless, predatory, with boundless ambition, civilized in externals but a savage at heart."

John Brown's raid on Harpers Ferry in 1859, immediately repudiated by the new Republican party, drew from the now tubercular Thoreau a passionate defense. During the Civil War itself (1861–65), Lincoln found the genius to suit diverse occasions with right language and length of utterance, but the major writers fell silent—Thoreau dying, Hawthorne dying, Melville living as a man without a profession and free to witness the war briefly from camp and follow it by telegraph bulletins and newspapers, Whitman nursing the wounded in Washington hospitals. After the end of the war two volumes of poetry, Whitman's *Drum-Taps* (to which he appended the great elegy on Lincoln, *When Lilacs Last in the Dooryard Bloom'd*) and Melville's *Battle-Pieces*, summed up the national experience. Both writers looked ahead as well as backward, Whitman calling "reconciliation" the "word over all," and Melville urging in a prose "Supplement" that the victorious North "be Christians toward our fellow-whites, as well as philanthropists toward the blacks, our fellow-men." Later in *Specimen Days* Whitman made a memorable attempt to do the impossible—to put the real war realistically into a book.

Both Whitman and Melville, especially in their later years, saw American politics cease to be concerned with great national struggles over momentous issues; rather, politics meant corruption, on a petty or a grand scale. Melville lived out the Gilded Age as an employee at the notoriously corrupt Custom House in New York City. In *Clarel*, foreseeing a descent from the present "civic barbarism" to "the Dark Ages of Democracy," he portrayed his American pilgrims to the Holy Land as recognizing sadly that the time might come to honor the God of Limitations in what had been

the Land of Opportunity, a time when Americans might cry: "To Terminus build fanes! / Columbus ended earth's romance: / No New World
to mankind remains!"

THE HEROISM OF AMERICAN WRITERS

Against a society which often lost sight of principles, whether aesthetic,
social, or political, Emerson offered the challenge which the other great
writers took up: "Let us affront and reprimand the smooth mediocrity and
squalid contentment of the times, and hurl in the face of custom, and
trade, and office, the fact which is the upshot of all history, that there is a
great responsible Thinker and Actor moving wherever moves a man; that
a true man belongs to no other time or place, but is the centre of things."
In the same spirit Melville looked bravely at the risks which lay beyond the
imitation of Irving:

> But the graceful writer, who perhaps of all Americans has received the
> most plaudits from his own country for his productions,—that very popu-
> lar and amiable writer, however good, and self-reliant in many things,
> perhaps owes his chief reputation to the self-acknowledged imitation of
> a foreign model, and to the studied avoidance of all topics but smooth
> ones. But it is better to fail in originality, than to succeed in imitation.
> He who has never failed somewhere, that man can not be great. Failure
> is the true test of greatness.

In the same spirit Whitman commanded his readers: "re-examine all you
have been told at school or church or in any book, dismiss whatever insults your own soul, and your very flesh shall be a great poem and have
the richest fluency not only in its words but in the silent lines of its lips
and face and between the lashes of your eyes and in every motion and
joint of your body." As Emerson had warned them they must, the great
writers of this time relinquished "display and immediate fame" in order
to wrestle, in Melville's phrase, "with the angel—Art," making their writings into classics from which later generations, and sometimes even their
own, would date eras in their lives. As the selections in this volume demonstrate, all of Emerson's great fellow writers fervently shared his conviction
that "nothing is of any value in books, excepting the transcendental and
extraordinary."

WASHINGTON IRVING
1783–1859

Washington Irving, the first American to achieve an international literary reputation, was born in New York City on April 3, 1783, the last of eleven children of a Scottish-born father and English-born mother. Well into his thirties his brothers routinely tried to make plans for him, and his own devotion to his family was a dominant emotion throughout his life. He read widely in English literature at home, modeling his early prose upon the graceful *Spectator* papers by Joseph Addison, but delighted by many other writers, including Shakespeare, Oliver Goldsmith, and Laurence Sterne. His brothers enjoyed writing poems and essays as pleasant, companionable recreation, and at nineteen Irving wrote a series of satirical essays on the theater and New York society for his brother Peter's newspaper, the *Morning Courier*.

When Irving showed signs of tuberculosis in 1804, his brothers sent him abroad for a two-year tour of Europe, where in his notebooks he steadily became an acute observer and felicitous recorder of what he witnessed. On his return, he began studying law with Judge Josiah Hoffman, but, more important for his career, he and his brother William (along with William's brother-in-law, James Kirke Paulding) started an anonymous satirical magazine, *Salmagundi* (the name of a spicy hash), which ran through 1807 with sketches and poems on politics and drama as well as familiar essays on a great range of topics. Then in 1808 Irving began work on *A History of New York*, at first conceiving it as a parody of Samuel Latham Mitchill's pompously titled *The Picture of New-York; or The Traveller's Guide through the Commercial Metropolis of the United States*, then taking on a variety of satiric targets, including President Jefferson, whom he portrayed as an early Dutch Governor of New Amsterdam, William the Testy. Exuberant, broadly comic, the *History* spoofed historians' pedantries but was itself the result of many months of antiquarian reading in local libraries, where his researches gave Irving refuge from grief over the sudden death of Judge Hoffman's daughter Mitilda, to whom he had become engaged. Then the *History* was launched by a charming publicity campaign. First a newspaper noted the disappearance of a "small elderly gentleman, dressed in an old black coat and cocked hat, by the name of KNICKERBOCKER," adding that there were "some reasons for believing he is not entirely in his right mind." After further "news" items the old man's fictitious landlord announced that he had found in Knickerbocker's room a *"very curious kind of a written book"* which he intended to dispose of to pay the bill that was owed him, and the book at last appeared, ascribed to Diedrich Knickerbocker. With its publication Irving became an American celebrity. Reprinted in England, the *History* reached Sir Walter Scott, who declared that it made his sides hurt from laughter. Like all but the rarest of topical satires, however, it has become increasingly inaccessible to later generations of readers, who can hardly comprehend Irving's strategies and targets without precisely the sort of antiquarian footnotes he found delight in mocking.

During the War of 1812 Irving was editor of the *Analectic Magazine*, which he filled mainly with essays from British periodicals but where he printed his own timely series of patriotic biographical sketches of American naval heroes. Toward the end of the war he was made a colonel in the New York State Militia. Then in May, 1815, a major break occurred in his life: he left for Europe and stayed away for seventeen years. At first he worked in Liverpool with his brother Peter, an importer of English hardware. In 1818 Peter went bankrupt, shortly after their mother died in New York; profoundly grieved and shamed, Irving once again took refuge in writing. During his work on *The Sketch Book* he met Scott, who buoyed him by admiration for the *History* and helpfully directed Irving's attention to the wealth of unused literary material in German folk tales; there, as scholars have shown, Irving found the source for *Rip Van Winkle*, some passages of which are close paraphrases of the original. In 1819 Irving began sending *The Sketch Book* to the United States for publication in installments. When the full version was printed in England the next year, it made Irving famous and brought him the friendship of many of the leading British writers of the time. His new pseudonym, Geoffrey Crayon, became universally recognized, and over the next years selections from *The Sketch Book* entered the classroom as models of English prose just as selections from Addison had long been used. As Irving knew, part of his British success derived from general astonishment that a man born in the United States could write in such an English way about English scenes: Addison lay behind the sketches of English country life, just as Oliver Goldsmith's essays on the Boar's-head Tavern in Eastcheap and on Westminster Abbey lay behind Irving's on the same topics. But in among the graceful, tame tributes to English scenes and characters were two vigorous tales set in rural New York, *Rip Van Winkle* and *The Legend of Sleepy Hollow*. Everyone who read them knew instantly that they were among the literary treasures of the language, and it very soon became hard to remember that they had not always been among the English classics.

Irving's next book, *Bracebridge Hall* (1822), a worshipful tribute to old-fashioned English country life, was, as the author realized, a feeble follow-up, and *Tales of a Traveller* (1824) was widely taken as a sign that he had written himself out. At a loss to sustain his career, Irving gambled on accepting an invitation from an acquaintance, the American Minister to Spain: he was to come to Spain as an attaché of the legation (a device for giving him entree into manuscript collections) and translate Martín Fernández de Navarrete's new compilation of accounts of the voyages of Columbus, including Columbus's own lost journals as copied by an earlier historian. Helped by the American consul in Madrid, Obadiah Rich, who owned a magnificent collection of books and manuscripts on Spanish and Latin American history, Irving worked intensely and in 1828 published *The Life and Voyages of Christopher Columbus*, not a translation of Navarrete (though the Spaniard's volume supplied most of the facts) but a biography of Irving's own, shaped by his skill at evocative re-creation of history. Out of these Spanish years came also *The Conquest of Granada* (1829), *Voyages and Discoveries of the Companions of Columbus* (1831), and *The Alhambra* (1832), which became known as "the Spanish *Sketch Book*."

In 1829 Irving was appointed Secretary to the American legation in London, where he became a competent, hard-working diplomat, aided by his access to the highest levels of British society. No longer the latest rage,

Irving by now was a solidly established author. On his return to the United States in 1832 his reputation was in need of redemption from a different charge—that of becoming too Europeanized. As if in an effort to make amends, Irving turned to three studies of the American West: *A Tour on the Prairies* (1835), based on his horseback journey into what is now Oklahoma; *Astoria* (1836), an account of John Jacob Astor's fur-trading colony in Oregon, written in Astor's own library and based on published accounts as well as research in Astor's archives (in which task Irving was assisted by his nephew Peter); and *The Adventures of Captain Bonneville, U.S.A.* (1837), an account of a Frenchman's explorations in the Rockies and the Far West.

In the late 1830s Irving bought and began refurbishing a house near Tarrytown, along the Hudson north of New York City, just where he had dreamed of settling down in *The Legend of Sleepy Hollow*. At Sunnyside he made a home for several members of his family, including as many as five nieces at a time, but he wrote little. From this somewhat purposeless stage of his life he was rescued by appointment as Minister to Spain in 1842; he served four years in Madrid with great success. After his return he arranged with G. P. Putnam to publish a collected edition of his writings and took the occasion to revise some of them. Using essays he had written years before, he also prepared for the edition a derivative biography of Oliver Goldsmith (1849), after which critics more than ever compared him to the Irish prince of hack writers. Irving's main work after 1851 was his long-contemplated life of George Washington. He worked in libraries, read old newspapers, studied government records, and visited battlefields, but once again he drew very heavily on published biographies, especially the recent one by Jared Sparks. He forced himself, in the most heroic effort of his career, to complete the successive five volumes, the first of which was published in 1855. Just after finishing the last he collapsed, and died a few months later, on November 28, 1859.

Decades before his death, Irving had achieved the status of a classic writer; in his own country he had no rival as a stylist. As schoolboys, Hawthorne and Longfellow were inspired by the success of *The Sketch Book*, and their prose, as well as that of a horde of now-unread writers, owed much to Irving. Although Melville, in his essay on Hawthorne's *Mosses from an Old Manse*, declared his preference for creative geniuses over adept imitators like Irving, he could not escape Irving's influence, which emerges both in his short stories and in a late poem, *Rip Van Winkle's Lilacs*, which showed he saw Rip as an archetypal artist figure. (Melville's debt was even more tangible, for early in 1846 Irving had passed the word to Putnam that *Typee* was worth reprinting in New York; but then Irving had been generous to younger writers all his life, as in his supervision of the London publication of Bryant's poems in 1832.) The southwestern humorists of the 1840s, whom Irving read and enjoyed, were much more robust than Irving in his mature years, yet they learned from him that realistic details of rural life in America could be worked memorably into fiction. From the beginning, Americans identified with Rip as a counter-hero, an anti-Franklinian who made a success of failure, and successive generations have responded profoundly to Irving's pervasive theme of mutability, especially as localized in his portrayal of the bewildering and destructive rapidity of change in American life.

Rip Van Winkle[1]

[The following Tale was found among the papers of the late
Diedrich Knickerbocker, an old gentleman of New-York, who was
very curious in the Dutch history of the province, and the manners
of the descendants from its primitive settlers. His historical researches, however, did not lay so much among books, as among
men; for the former are lamentably scanty on his favourite topics;
whereas he found the old burghers, and still more, their wives, rich
in that legendary lore, so invaluable to true history. Whenever,
therefore, he happened upon a genuine Dutch family, snugly shut
up in its low-roofed farm house, under a spreading sycamore, he
looked upon it as a little clasped volume of black-letter,[2] and studied it with the zeal of a bookworm.

The result of all these researches was a history of the province,
during the reign of the Dutch governors, which he published some
years since. There have been various opinions as to the literary
character of his work, and, to tell the truth, it is not a whit better
than it should be. Its chief merit is its scrupulous accuracy, which,
indeed, was a little questioned, on its first appearance, but has since
been completely established;[3] and it is now admitted into all historical collections, as a book of unquestionable authority.

The old gentleman died shortly after the publication of his work,
and now, that he is dead and gone, it cannot do much harm to his
memory, to say, that his time might have been much better employed in weightier labours. He, however, was apt to ride his hobby
his own way; and though it did now and then kick up the dust a
little in the eyes of his neighbours, and grieve the spirit of some
friends, for whom he felt the truest deference and affection; yet his
errors and follies are remembered "more in sorrow than in anger,"[4]
and it begins to be suspected, that he never intended to injure or
offend. But however his memory may be appreciated by critics, it is
still held dear among many folk, whose good opinion is well worth
having; particularly certain biscuit bakers, who have gone so far as
to imprint his likeness on their new year cakes, and have thus given

1. Irving sent a tentative batch of
sketches from England to his brothers in
New York in March, 1819. This group,
which included *Rip Van Winkle*, was
published in New York (by C. S. Van
Winkle) in May, 1819, as *The Sketch
Book of Geoffrey Crayon, Gent.*, No. I,
the source of the text printed here. In
1820, after a total of seven installments
had been printed separately, the publisher bound them together.
2. Type face in early printed books, resembling *medieval* script; such books, because of their value, were often equipped
with clasps so they could be shut tightly

and even locked.
3. Irving knew that most of his first
readers would remember with delight the
wildly inaccurate Knickerbocker *History*.
He is also echoing Cervantes's humorous
assurance of accuracy at the outset of
Don Quixote.
4. Shakespeare's *Hamlet*, 1.1.231–32. To
this quotation Irving appended the following footnote: "Vide [see] the excellent discourse of G. C. Verplanck, Esq.
before the New-York Historical Society."
If Irving's friend Gulian C. Verplanck
ever made such an address, it was in
fun.

him a chance for immortality, almost equal to being stamped on a Waterloo medal, or a Queen Anne's farthing.[5]]

Rip Van Winkle
A Posthumous Writing of Diedrich Knickerbocker

> By Woden, God of Saxons,
> From whence comes Wensday, that is Wodensday,
> Truth is a thing that ever I will keep
> Unto thylke day in which I creep into
> My sepulchre——
>
> —CARTWRIGHT[6]

Whoever has made a voyage up the Hudson, must remember the Kaatskill mountains. They are a dismembered branch of the great Appalachian family, and are seen away to the west of the river, swelling up to a noble height, and lording it over the surrounding country. Every change of season, every change of weather, indeed, every hour of the day, produces some change in the magical hues and shapes of these mountains, and they are regarded by all the good wives, far and near, as perfect barometers. When the weather is fair and settled, they are clothed in blue and purple, and print their bold outlines on the clear evening sky; but some times, when the rest of the landscape is cloudless, they will gather a hood of gray vapours about their summits, which, in the last rays of the setting sun, will glow and light up like a crown of glory.

At the foot of these fairy mountains, the voyager may have descried the light smoke curling up from a village, whose shingle roofs gleam among the trees, just where the blue tints of the upland melt away into the fresh green of the nearer landscape. It is a little village of great antiquity, having been founded by some of the Dutch colonists, in the early times of the province, just about the beginning of the government of the good Peter Stuyvesant,[7] (may he rest in peace!) and there were some of the houses of the original settlers standing within a few years, with lattice windows, gable fronts surmounted with weathercocks, and built of small yellow bricks brought from Holland.

In that same village, and in one of these very houses, (which, to tell the precise truth, was sadly time worn and weather beaten,) there lived many years since, while the country was yet a province of Great Britain, a simple good natured fellow, of the name of Rip Van Winkle. He was a descendant of the Van Winkles who figured

5. Irving's irony cuts in different directions: Waterloo medals were minted liberally after the defeat of Napoleon in 1815, while farthings (tiny coins) from the reign of Queen Anne of England (1702–14) were commonly, though wrongly, considered rare, one story saying only three were minted.
6. In this quotation from *The Ordinary*, 3.1.1050–54, a play by the English writer William Cartwright (1611–43), the speaker is a pedant named Moth. "Woden": Norse god of war.
7. Peter Stuyvesant (1592–1672), last Governor of the Dutch province of New Netherlands, in 1655 (as mentioned below) defeated Swedish colonists at Fort Christina, near what is now Wilmington, Delaware.

so gallantly in the chivalrous days of Peter Stuyvesant, and accompanied him to the siege of Fort Christina. He inherited, however, but little of the martial character of his ancestors. I have observed that he was a simple good natured man; he was moreover a kind neighbour, and an obedient, henpecked husband. Indeed, to the latter circumstance might be owing that meekness of spirit which gained him such universal popularity; for those men are most apt to be obsequious and conciliating abroad, who are under the discipline of shrews at home. Their tempers, doubtless, are rendered pliant and malleable in the fiery furnace of domestic tribulation, and a curtain lecture[8] is worth all the sermons in the world for teaching the virtues of patience and long suffering. A termagant wife may, therefore, in some respects, be considered a tolerable blessing; and if so, Rip Van Winkle was thrice blessed.

Certain it is, that he was a great favourite among all the good wives of the village, who, as usual with the amiable sex, took his part in all family squabbles, and never failed, whenever they talked those matters over in their evening gossippings, to lay all the blame on Dame Van Winkle. The children of the village, too, would shout with joy whenever he approached. He assisted at their sports, made their playthings, taught them to fly kites and shoot marbles, and told them long stories of ghosts, witches, and Indians. Whenever he went dodging about the village, he was surrounded by a troop of them, hanging on his skirts, clambering on his back, and playing a thousand tricks on him with impunity; and not a dog would bark at him throughout the neighbourhood.

The great error in Rip's composition was an insuperable aversion to all kinds of profitable labour. It could not be for the want of assiduity or perseverance; for he would sit on a wet rock, with a rod as long and heavy as a Tartar's lance, and fish all day without a murmur, even though he should not be encouraged by a single nibble. He would carry a fowling piece on his shoulder, for hours together, trudging through woods and swamps, and up hill and down dale, to shoot a few squirrels or wild pigeons. He would never even refuse to assist a neighbour in the roughest toil, and was a foremost man at all country frolicks for husking Indian corn, or building stone fences; the women of the village, too, used to employ him to run their errands, and to do such little odd jobs as their less obliging husbands would not do for them;—in a word, Rip was ready to attend to any body's business but his own; but as to doing family duty, and keeping his farm in order, it was impossible.

In fact, he declared it was no use to work on his farm; it was the most pestilent little piece of ground in the whole country; every thing about it went wrong, and would go wrong, in spite of him. His fences were continually falling to pieces; his cow would either go

8. Tirade delivered by a wife after the curtains around the four-poster bed have been drawn for the night.

astray, or get among the cabbages; weeds were sure to grow quicker in his fields than any where else; the rain always made a point of setting in just as he had some out-door work to do. So that though his patrimonial estate had dwindled away under his management, acre by acre, until there was little more left than a mere patch of Indian corn and potatoes, yet it was the worst conditioned farm in the neighbourhood.

His children, too, were as ragged and wild as if they belonged to nobody. His son Rip, an urchin begotten in his own likeness, promised to inherit the habits, with the old clothes of his father. He was generally seen trooping like a colt at his mother's heels, equipped in a pair of his father's cast-off galligaskins,[9] which he had much ado to hold up with one hand, as a fine lady does her train in bad weather.

Rip Van Winkle, however, was one of those happy mortals, of foolish, well-oiled dispositions, who take the world easy, eat white bread or brown, which ever can be got with least thought or trouble, and would rather starve on a penny than work for a pound. If left to himself, he would have whistled life away, in perfect content-ment; but his wife kept continually dinning in his ears about his idleness, his carelessness, and the ruin he was bringing on his fam-ily. Morning, noon, and night, her tongue was incessantly going, and every thing he said or did was sure to produce a torrent of household eloquence. Rip had but one way of replying to all lec-tures of the kind, and that, by frequent use, had grown into a habit. He shrugged his shoulders, shook his head, cast up his eyes, but said nothing. This, however, always provoked a fresh volley from his wife, so that he was fain to draw off his forces, and take to the outside of the house—the only side which, in truth, belongs to a henpecked husband.

Rip's sole domestic adherent was his dog Wolf, who was as much henpecked as his master; for Dame Van Winkle regarded them as companions in idleness, and even looked upon Wolf with an evil eye, as the cause of his master's so often going astray. True it is, in all points of spirit befitting an honourable dog, he was as coura-geous an animal as ever scoured the woods—but what courage can withstand the ever-during and all-besetting terrors of a woman's tongue? The moment Wolf entered the house, his crest fell, his tail drooped to the ground, or curled between his legs, he sneaked about with a gallows air, casting many a sidelong glance at Dame Van Winkle, and at the least flourish of a broomstick or ladle, would fly to the door with yelping precipitation.

Times grew worse and worse with Rip Van Winkle as years of matrimony rolled on; a tart temper never mellows with age, and a sharp tongue is the only edge tool that grows keener by constant use. For a long while he used to console himself, when driven from

9. Loose, wide breeches.

home, by frequenting a kind of perpetual club of the sages, phi-losophers, and other idle personages of the village, that held its sessions on a bench before a small inn, designated by a rubicund portrait of his majesty George the Third. Here they used to sit in the shade, of a long lazy summer's day, talk listlessly over village gossip, or tell endless sleepy stories about nothing. But it would have been worth any statesman's money to have heard the profound discussions that sometimes took place, when by chance an old newspaper fell into their hands, from some passing traveller. How solemnly they would listen to the contents, as drawled out by Der-rick Van Bummel, the schoolmaster, a dapper learned little man, who was not to be daunted by the most gigantic word in the dic-tionary; and how sagely they would deliberate upon public events some months after they had taken place.

The opinions of this junto[1] were completely controlled by Nicho-las Vedder, a patriarch of the village, and landlord of the inn, at the door of which he took his seat from morning till night, just moving sufficiently to avoid the sun, and keep in the shade of a large tree; so that the neighbours could tell the hour by his movements as accurately as by a sun dial. It is true, he was rarely heard to speak, but smoked his pipe incessantly. His adherents, however, (for every great man has his adherents,) perfectly understood him, and knew how to gather his opinions. When any thing that was read or related displeased him, he was observed to smoke his pipe vehemently, and send forth short, frequent, and angry puffs; but when pleased, he would inhale the smoke slowly and tranquilly, and emit it in light and placid clouds, and sometimes taking the pipe from his mouth, and letting the fragrant vapour curl about his nose, would gravely nod his head in token of perfect approbation.

From even this strong hold the unlucky Rip was at length routed by his termagant wife, who would suddenly break in upon the tranquillity of the assemblage, call the members all to nought, nor was that august personage, Nicholas Vedder himself, sacred from the daring tongue of this terrible virago, who charged him outright with encouraging her husband in habits of idleness.

Poor Rip was at last reduced almost to despair; and his only alternative to escape from the labour of the farm and the clamour of his wife, was to take gun in hand, and stroll away into the woods. Here he would sometimes seat himself at the foot of a tree, and share the contents of his wallet[2] with Wolf, with whom he sym-pathised as a fellow sufferer in persecution. "Poor Wolf," he would say, "thy mistress leads thee a dogs' life of it; but never mind, my lad, while I live thou shalt never want a friend to stand by thee!" Wolf would wag his tail, look wistfully in his master's face, and if dogs can feel pity, I verily believe he reciprocated the sentiment with all his heart.

1. Ruling committee.   2. Knapsack.

In a long ramble of the kind on a fine autumnal day, Rip had unconsciously scrambled to one of the highest parts of the Kaatskill mountains. He was after his favourite sport of squirrel shooting, and the still solitudes had echoed and re-echoed with the reports of his gun. Panting and fatigued, he threw himself, late in the afternoon, on a green knoll, covered with mountain herbage, that crowned the brow of a precipice. From an opening between the trees, he could overlook all the lower country for many a mile of rich woodland. He saw at a distance the lordly Hudson, far, far below him, moving on its silent but majestic course, the reflection of a purple cloud, or the sail of a lagging bark, here and there sleeping on its glassy bosom, and at last losing itself in the blue highlands.

On the other side he looked down into a deep mountain glen, wild, lonely, and shagged, the bottom filled with fragments from the impending cliffs, and scarcely lighted by the reflected rays of the setting sun. For some time Rip lay musing on this scene, evening was gradually advancing, the mountains began to throw their long blue shadows over the valleys, he saw that it would be dark long before he could reach the village, and he heaved a heavy sigh when he thought of encountering the terrors of Dame Van Winkle.

As he was about to descend, he heard a voice from a distance, hallooing, "Rip Van Winkle! Rip Van Winkle!" He looked around, but could see nothing but a crow winging its solitary flight across the mountain. He thought his fancy must have deceived him, and turned again to descend, when he heard the same cry ring through the still evening air; "Rip Van Winkle! Rip Van Winkle!"—at the same time Wolf bristled up his back, and giving a low growl, skulked to his master's side, looking fearfully down into the glen. Rip now felt a vague apprehension stealing over him; he looked anxiously in the same direction, and perceived a strange figure slowly toiling up the rocks, and bending under the weight of something he carried on his back. He was surprised to see any human being in this lonely and unfrequented place, but supposing it to be some one of the neighbourhood in need of his assistance, he hastened down to yield it.

On nearer approach, he was still more surprised at the singularity of the stranger's appearance. He was a short square built old fellow, with thick bushy hair, and a grizzled beard. His dress was of the antique Dutch fashion—a cloth jerkin[3] strapped round the waist— several pair of breeches, the outer one of ample volume, decorated with rows of buttons down the sides, and bunches at the knees. He bore on his shoulder a stout keg, that seemed full of liquor, and made signs for Rip to approach and assist him with the load. Though rather shy and distrustful of this new acquaintance, Rip complied with his usual alacrity, and mutually relieving each other, they clambered up a narrow gully, apparently the dry bed of a

3. Jacket fitted tightly at the waist.

mountain torrent. As they ascended, Rip every now and then heard long rolling peals, like distant thunder, that seemed to issue out of a deep ravine, or rather cleft between lofty rocks, toward which their rugged path conducted. He paused for an instant, but supposing it to be the muttering of one of those transient thunder showers which often take place in mountain heights, he proceeded. Passing through the ravine, they came to a hollow, like a small amphitheatre, surrounded by perpendicular precipices, over the brinks of which impending trees shot their branches, so that you only caught glimpses of the azure sky, and the bright evening cloud. During the whole time, Rip and his companion had laboured on in silence; for though the former marvelled greatly what could be the object of carrying a keg of liquor up this wild mountain, yet there was something strange and incomprehensible about the unknown, that inspired awe, and checked familiarity.

On entering the amphitheatre, new objects of wonder presented themselves. On a level spot in the centre was a company of odd-looking personages playing at nine-pins. They were dressed in a quaint, outlandish fashion: some wore short doublets,[4] others jerkins, with long knives in their belts, and most had enormous breeches, of similar style with that of the guide's. Their visages, too, were peculiar: one had a large head, broad face, and small piggish eyes; the face of another seemed to consist entirely of nose, and was surmounted by a white sugarloaf hat, set off with a little red cockstail. They all had beards, of various shapes and colours. There was one who seemed to be the commander. He was a stout old gentleman, with a weather-beaten countenance; he wore a laced doublet, broad belt and hanger,[5] high crowned hat and feather, red stockings, and high heeled shoes, with roses in them. The whole group reminded Rip of the figures in an old Flemish painting, in the parlour of Dominie[6] Van Schaick, the village parson, and which had been brought over from Holland at the time of the settlement.

What seemed particularly odd to Rip, was, that though these folks were evidently amusing themselves, yet they maintained the gravest faces, the most mysterious silence, and were, withal, the most melancholy party of pleasure he had ever witnessed. Nothing interrupted the stillness of the scene, but the noise of the balls, which, whenever they were rolled, echoed along the mountains like rumbling peals of thunder.

As Rip and his companion approached them, they suddenly desisted from their play, and stared at him with such fixed statue-like gaze, and such strange, uncouth, lack lustre countenances, that his heart turned within him, and his knees smote together. His companion now emptied the contents of the keg into large flagons, and made signs to him to wait upon the company. He obeyed with fear

---

4. Male garment covering from neck to upper thighs, where it hooked to hose.
5. Short, curved sword.
6. Minister.

and trembling; they quaffed the liquor in profound silence, and then returned to their game.

By degrees, Rip's awe and apprehension subsided. He even ventured, when no eye was fixed upon him, to taste the beverage, which he found had much of the flavour of excellent Hollands.[7] He was naturally a thirsty soul, and was soon tempted to repeat the draught. One taste provoked another, and he reiterated his visits to the flagon so often, that at length his senses were overpowered, his eyes swam in his head, his head gradually declined, and he fell into a deep sleep.

On awaking, he found himself on the green knoll from whence he had first seen the old man of the glen. He rubbed his eyes—it was a bright sunny morning. The birds were hopping and twittering among the bushes, and the eagle was wheeling aloft, and breasting the pure mountain breeze. "Surely," thought Rip, "I have not slept here all night." He recalled the occurrences before he fell asleep. The strange man with the keg of liquor—the mountain ravine—the wild retreat among the rocks—the wo-begone party at nine-pins—the flagon—"Oh! that flagon! that wicked flagon!" thought Rip—"what excuse shall I make to Dame Van Winkle?"

He looked round for his gun, but in place of the clean well-oiled fowling-piece, he found an old firelock lying by him, the barrel encrusted with rust, the lock falling off, and the stock worm-eaten. He now suspected that the grave roysters of the mountain had put a trick upon him, and having dosed him with liquor, had robbed him of his gun. Wolf, too, had disappeared, but he might have strayed away after a squirrel or partridge. He whistled after him, shouted his name, but all in vain; the echoes repeated his whistle and shout, but no dog was to be seen.

He determined to revisit the scene of the last evening's gambol, and if he met with any of the party, to demand his dog and gun. As he arose to walk he found himself stiff in the joints, and wanting in his usual activity. "These mountain beds do not agree with me," thought Rip, "and if this frolick should lay me up with a fit of the rheumatism, I shall have a blessed time with Dame Van Winkle." With some difficulty he got down into the glen: he found the gully up which he and his companion had ascended the preceding evening, but to his astonishment a mountain stream was now foaming down it, leaping from rock to rock, and filling the glen with babbling murmurs. He, however, made shift to scramble up its sides, working his toilsome way through thickets of birch, sassafras, and witch hazel, and sometimes tripped up or entangled by the wild grape vines that twisted their coils and tendrils from tree to tree, and spread a kind of network in his path.

At length he reached to where the ravine had opened through the cliffs, to the amphitheatre; but no traces of such opening remained.

7. Kind of gin.

The rocks presented a high impenetrable wall, over which the torrent came tumbling in a sheet of feathery foam, and fell into a broad deep basin, black from the shadows of the surrounding forest. Here, then, poor Rip was brought to a stand. He again called and whistled after his dog; he was only answered by the cawing of a flock of idle crows, sporting high in air about a dry tree that overhung a sunny precipice; and who, secure in their elevation, seemed to look down and scoff at the poor man's perplexities. What was to be done? the morning was passing away, and Rip felt famished for his breakfast. He grieved to give up his dog and gun; he dreaded to meet his wife; but it would not do to starve among the mountains. He shook his head, shouldered the rusty firelock, and, with a heart full of trouble and anxiety, turned his steps homeward.

As he approached the village, he met a number of people, but none that he knew, which somewhat surprised him, for he had thought himself acquainted with every one in the country round. Their dress, too, was of a different fashion from that to which he was accustomed. They all stared at him with equal marks of surprise, and whenever they cast eyes upon him, invariably stroked their chins. The constant recurrence of this gesture, induced Rip, involuntarily, to do the same, when, to his astonishment, he found his beard had grown a foot long!

He had now entered the skirts of the village. A troop of strange children ran at his heels, hooting after him, and pointing at his gray beard. The dogs, too, not one of which he recognized for his old acquaintances, barked at him as he passed. The very village seemed altered: it was larger and more populous. There were rows of houses which he had never seen before, and those which had been his familiar haunts had disappeared. Strange names were over the doors—strange faces at the windows—every thing was strange. His mind now began to misgive him, that both he and the world around him were bewitched. Surely this was his native village, which he had left but the day before. There stood the Kaatskill mountains—there ran the silver Hudson at a distance—there was every hill and dale precisely as it had always been—Rip was sorely perplexed—"That flagon last night," thought he, "has addled my poor head sadly!"

It was with some difficulty he found the way to his own house, which he approached with silent awe, expecting every moment to hear the shrill voice of Dame Van Winkle. He found the house gone to decay—the roof fallen in, the windows shattered, and the doors off the hinges. A half starved dog, that looked like Wolf, was skulking about it. Rip called him by name, but the cur snarled, showed his teeth, and passed on. This was an unkind cut indeed—"My very dog," sighed poor Rip, "has forgotten me!"

He entered the house, which, to tell the truth, Dame Van Winkle had always kept in neat order. It was empty, forlorn, and apparently abandoned. This desolateness overcame all his connubial fears—he

called loudly for his wife and children—the lonely chambers rung
for a moment with his voice, and then all again was silence.

He now hurried forth, and hastened to his old resort, the little
village inn—but it too was gone. A large ricketty wooden building
stood in its place, with great gaping windows, some of them broken,
and mended with old hats and petticoats, and over the door was
painted, "The Union Hotel, by Jonathan Doolittle." Instead of the
great tree that used to shelter the quiet little Dutch inn of yore,
there now was reared a tall naked pole, with something on top that
looked like a red night cap,[8] and from it was fluttering a flag, on
which was a singular assemblage of stars and stripes—all this was
strange and incomprehensible. He recognised on the sign, however,
the ruby face of King George, under which he had smoked so many
a peaceful pipe, but even this was singularly metamorphosed. The
red coat was changed for one of blue and buff,[9] a sword was stuck
in the hand instead of a sceptre, the head was decorated with a
cocked hat, and underneath was painted in large characters,
GENERAL WASHINGTON.

There was, as usual, a crowd of folk about the door, but none
that Rip recollected. The very character of the people seemed
changed. There was a busy, bustling, disputatious tone about it,
instead of the accustomed phlegm and drowsy tranquillity. He
looked in vain for the sage Nicholas Vedder, with his broad face,
double chin, and fair long pipe, uttering clouds of tobacco smoke
instead of idle speeches; or Van Bummel, the schoolmaster, doling
forth the contents of an ancient newspaper. In place of these, a lean
bilious looking fellow, with his pockets full of handbills, was
haranguing vehemently about rights of citizens—election—members
of congress—liberty—Bunker's hill—heroes of seventy-six—and
other words, that were a perfect Babylonish jargon[1] to the bewildered Van Winkle.

The appearance of Rip, with his long grizzled beard, his rusty
fowling piece, his uncouth dress, and the army of women and
children that had gathered at his heels, soon attracted the attention
of the tavern politicians. They crowded around him, eyeing him
from head to foot, with great curiosity. The orator bustled up to
him, and drawing him partly aside, inquired "which side he voted?"
Rip stared in vacant stupidity. Another short but busy little fellow
pulled him by the arm, and raising on tiptoe, inquired in his ear,
"whether he was Federal or Democrat."[2] Rip was equally at a loss

---

8. Limp, close-fitting cap adopted during
the French Revolution as a symbol of
liberty; the pole is a "liberty pole"—
i.e., a tall flagstaff topped by a liberty
cap.
9. Colors of the Revolutionary uniform.
Irving's joke is that the new proprietor,
being a Yankee, is so parsimonious that
he will only touch up the sign, not replace it with a true portrait of Washington.
1. Cf. Genesis 11.1–9, Babel being confused with Babylon.
2. Political parties which developed in
George Washington's administrations,
Alexander Hamilton leading the Federalists and Thomas Jefferson the Democrats.

to comprehend the question; when a knowing, self-important old gentleman, in a sharp cocked hat, made his way through the crowd, putting them to the right and left with his elbows as he passed, and planting himself before Van Winkle, with one arm akimbo, the other resting on his cane, his keen eyes and sharp hat penetrating, as it were, into his very soul, demanded, in an austere tone, "what brought him to the election with a gun on his shoulder, and a mob at his heels, and whether he meant to breed a riot in the village?" "Alas! gentlemen," cried Rip, somewhat dismayed, "I am a poor quiet man, a native of the place, and a loyal subject of the King, God bless him!"

Here a general shout burst from the bystanders—"A tory! a tory! a spy! a refugee! hustle him! away with him!" It was with great difficulty that the self-important man in the cocked hat restored order; and having assumed a tenfold austerity of brow, demanded again of the unknown culprit, what he came there for, and whom he was seeking. The poor man humbly assured them that he meant no harm; but merely came there in search of some of his neighbours, who used to keep about the tavern.

"Well—who are they?—name them."

Rip bethought himself a moment, and inquired, "where's Nicholas Vedder?"

There was a silence for a little while, when an old man replied, in a thin piping voice, "Nicholas Vedder? why he is dead and gone these eighteen years! There was a wooden tombstone in the church yard that used to tell all about him, but that's rotted and gone too."

"Where's Brom Dutcher?"

"Oh he went off to the army in the beginning of the war; some say he was killed at the battle of Stoney-Point—others say he was drowned in a squall, at the foot of Antony's Nose.[3] I don't know—he never came back again."

"Where's Van Bummel, the schoolmaster?"

"He went off to the wars too, was a great militia general, and is now in Congress."

Rip's heart died away, at hearing of these sad changes in his home and friends, and finding himself thus alone in the world. Every answer puzzled him, too, by treating of such enormous lapses of time, and of matters which he could not understand: war—congress—Stoney-Point;—he had no courage to ask after any more friends, but cried out in despair, "does nobody here know Rip Van Winkle?"

"Oh, Rip Van Winkle!" exclaimed two or three, "Oh, to be sure! that's Rip Van Winkle yonder, leaning against the tree."

3. "Stoney Point": on the west bank of the Hudson south of West Point, captured by General Anthony Wayne (1745– 96) during the Revolution; "Antony's Nose": mountain near West Point.

Rip looked, and beheld a precise counterpart of himself, as he went up the mountain: apparently as lazy, and certainly as ragged. The poor fellow was now completely confounded. He doubted his own identity, and whether he was himself or another man. In the midst of his bewilderment, the man in the cocked hat demanded who he was, and what was his name?

"God knows," exclaimed he, at his wit's end; "I'm not myself— I'm somebody else—that's me yonder—no—that's somebody else, got into my shoes—I was myself last night, but I fell asleep on the mountain, and they've changed my gun, and every thing's changed, and I'm changed, and I can't tell what's my name, or who I am!"

The bystanders began now to look at each other, nod, wink significantly, and tap their fingers against their foreheads. There was a whisper, also, about securing the gun, and keeping the old fellow from doing mischief. At the very suggestion of which, the self-important man in the cocked hat retired with some precipitation. At this critical moment a fresh likely woman pressed through the throng to get a peep at the graybearded man. She had a chubby child in her arms, which, frightened at his looks, began to cry. "Hush, Rip," cried she, "hush, you little fool, the old man wont hurt you." The name of the child, the air of the mother, the tone of her voice, all awakened a train of recollections in his mind.

"What is your name, my good woman?" asked he.

"Judith Gardenier."

"And your father's name?"

"Ah, poor man, his name was Rip Van Winkle; it's twenty years since he went away from home with his gun, and never has been heard of since—his dog came home without him; but whether he shot himself, or was carried away by the Indians, nobody can tell. I was then but a little girl."

Rip had but one question more to ask; but he put it with a faltering voice:

"Where's your mother?"

Oh, she too had died but a short time since; she broke a blood vessel in a fit of passion at a New-England pedlar.

There was a drop of comfort, at least, in this intelligence. The honest man could contain himself no longer.—He caught his daughter and her child in his arms.—"I am your father!" cried he—"Young Rip Van Winkle once—old Rip Van Winkle now!— Does nobody know poor Rip Van Winkle!"

All stood amazed, until an old woman, tottering out from among the crowd, put her hand to her brow, and peering under it in his face for a moment, exclaimed, "Sure enough! it is Rip Van Winkle —it is himself. Welcome home again, old neighbour—Why, where have you been these twenty long years?"

Rip's story was soon told, for the whole twenty years had been to him but as one night. The neighbours stared when they heard it;

some were seen to wink at each other, and put their tongues in their cheeks; and the self-important man in the cocked hat, who, when the alarm was over, had returned to the field, screwed down the corners of his mouth, and shook his head—upon which there was a general shaking of the head throughout the assemblage.

It was determined, however, to take the opinion of old Peter Vanderdonk, who was seen slowly advancing up the road. He was a descendant of the historian of that name,[4] who wrote one of the earliest accounts of the province. Peter was the most ancient inhabitant of the village, and well versed in all the wonderful events and traditions of the neighbourhood. He recollected Rip at once, and corroborated his story in the most satisfactory manner. He assured the company that it was a fact, handed down from his ancestor the historian, that the Kaatskill mountains had always been haunted by strange beings. That it was affirmed that the great Hendrick Hudson,[5] the first discoverer of the river and country, kept a kind of vigil there every twenty years, with his crew of the Half-moon, being permitted in this way to revisit the scenes of his enterprize, and keep a guardian eye upon the river, and the great city called by his name. That his father had once seen them in their old Dutch dresses playing at nine pins in a hollow of the mountain; and that he himself had heard, one summer afternoon, the sound of their balls, like long peals of thunder.

To make a long story short, the company broke up, and returned to the more important concerns of the election. Rip's daughter took him home to live with her; she had a snug, well-furnished house, and a stout cheery farmer for a husband, whom Rip recollected for one of the urchins that used to climb upon his back. As to Rip's son and heir, who was the ditto of himself, seen leaning against the tree, he was employed to work on the farm; but evinced an hereditary disposition to attend to any thing else but his business.

Rip now resumed his old walks and habits; he soon found many of his former cronies, though all rather the worse for the wear and tear of time; and preferred making friends among the rising generation, with whom he soon grew into great favour.

Having nothing to do at home, and being arrived at that happy age when a man can do nothing with impunity, he took his place once more on the bench, at the inn door, and was reverenced as one of the patriarchs of the village, and a chronicle of the old times "before the war." It was some time before he could get into the regular track of gossip, or could be made to comprehend the strange events that had taken place during his torpor. How that there had been a revolutionary war—that the country had thrown off the yoke

---

4. Adriaen Van der Donck (1620–55?) wrote a history of New Netherlands (Amsterdam, 1655).
5. Henry Hudson (d. 1611), English navigator in the service of the Dutch;

"great city" is ironic, for the town named for him on the east bank of the Hudson River was flourishing but not a metropolis.

of old England—and that, instead of being a subject of his Majesty George the Third, he was now a free citizen of the United States. Rip, in fact, was no politician; the changes of states and empires made but little impression on him. But there was one species of despotism under which he had long groaned, and that was—petticoat government. Happily, that was at an end; he had got his neck out of the yoke of matrimony, and could go in and out whenever he pleased, without dreading the tyranny of Dame Van Winkle. Whenever her name was mentioned, however, he shook his head, shrugged his shoulders, and cast up his eyes; which might pass either for an expression of resignation to his fate, or joy at his deliverance.

He used to tell his story to every stranger that arrived at Mr. Doolittle's hotel. He was observed, at first, to vary on some points every time he told it, which was, doubtless, owing to his having so recently awaked. It at last settled down precisely to the tale I have related, and not a man, woman, or child in the neighbourhood, but knew it by heart. Some always pretended to doubt the reality of it, and insisted that Rip had been out of his head, and that this was one point on which he always remained flighty. The old Dutch inhabitants, however, almost universally gave it full credit. Even to this day they never hear a thunder storm of a summer afternoon, about the Kaatskill, but they say Hendrick Hudson and his crew are at their game of nine pins; and it is a common wish of all henpecked husbands in the neighbourhood, when life hangs heavy on their hands, that they might have a quieting draught out of Rip Van Winkle's flagon.

NOTE

The foregoing tale, one would suspect, had been suggested to Mr. Knickerbocker by a little German superstition about Charles V. and the Kypphauser mountain;[6] the subjoined note, however, which he had appended to the tale, shows that it is an absolute fact, narrated with his usual fidelity:

"The story of Rip Van Winkle may seem incredible to many, but nevertheless I give it my full belief, for I know the vicinity of our old Dutch settlements to have been very subject to marvellous events and appearances. Indeed, I have heard many stranger stories than this, in the villages along the Hudson; all of which were too well authenticated to admit of a doubt. I have even talked with Rip Van Winkle myself, who, when last I saw him, was a very venerable old man, and so perfectly rational and consistent on every other point, that I think no conscientious person could refuse to take this

---

6. Later Irving changed "Charles V." (Holy Roman Emperor 1519–56) to "The Emperor Frederick *der Rothbart*" (i.e., Frederick Barbarossa, Holy Roman Emperor 1152–90). ("Rothbart" and "Barbarossa" both mean "redbeard.") In either form, the allusion is a red herring, a disarming way of suggesting indebtedness to a German source while concealing the most specific source, the story of "Peter Klaus" in the folk tales of J. C. C. N. Otmar.

into the bargain; nay, I have seen a certificate on the subject taken before a country justice, and signed with a cross, in the justice's own hand writing. The story, therefore, is beyond the possibility of doubt.   D.K."

1819

JAMES FENIMORE COOPER
1789–1851

James Fenimore Cooper, the first successful American novelist, was born on September 15, 1789, in Burlington, New Jersey, but taken in infancy to Cooperstown, on Otsego Lake in central New York, where his wealthy father owned great tracts of land. A few years before, the region had been wilderness, but during Cooper's boyhood there were few backwoodsmen left, and fewer Indians; in his novels the information about Indians came from older people and from books. In 1801 his father sent him to study in Albany in preparation for Yale, where he spent two years in his mid-teens before being expelled for pranks, thereby acquiring a lifelong distaste for New Englanders. He became a sailor in 1806 then two years later a midshipman in the navy. At twenty he inherited a fortune from his father and married Susan De Lancey, whose family had lost possessions by siding with the British in the Revolution but still owned lands in Westchester County. For several years Cooper and his wife wavered between Scarsdale and Otsego as a permanent home. Wherever they settled, Cooper seemed certain to live as a landed gentleman. His first book, *Precaution* (1820), a novel dealing with English high society, was the result of his casual bet with his wife that he could write a better book than the one he had been reading to her. Following that insignificant start, he wrote *The Spy* (1821), the first important historical romance of the Revolution, and on its success he moved to New York City to take up his new career. From the first his faults (such as syntactical awkwardness, arbitrary plotting, and heavy-handed attempts at humor) were obvious enough, but so were his genuine achievements in opening up new American scenes and themes for fiction. Founding the Bread and Cheese Club, he became the center of a circle which included notable painters of the Hudson River school as well as writers (William Cullen Bryant among them) and professional men. In 1823 he published *The Pioneers*, the first of what eventually consisted of five books about Natty Bumppo, known collectively as the *Leather-Stocking Tales*; the second, *The Last of the Mohicans*, followed in 1826. Cooper has other claims to fame—the virtual creation of the sea novel (starting with *The Pilot*, 1824), authorship of the first serious American novels of manners and the first American sociopolitical novels—but with Natty Bumppo, the aged hunter, he had created one of the most popular characters in world literature.

In 1826, at the height of his fame, Cooper sailed for Europe. In Paris, where he became intimate with the aged Lafayette, he wrote *The Prairie*

(1827) and *Notions of the Americans* (1828), a defense of the United States against the attacks of European travelers. Smarting under the half-complimentary, half-patronizing epithet of "The American Scott," he wrote three historical novels set in medieval Europe as a realistic corrective to Sir Walter Scott's glorifications of the past. On his return to the United States in 1833 Cooper was so stung by a review of one of these novels that he renounced novel writing in an angry *Letter to His Countrymen* (1834). Then at Cooperstown he gave notice that a point of land on Otsego Lake where the townspeople had been picnicking was private property and not to be used without permission. Newspapers began attacking him as a would-be aristocrat poisoned by his residence abroad, and for years Cooper embroiled himself in lawsuits designed not to gain damages for the journalistic libels but to tame the irresponsible press. Legally in the right, Cooper sacrificed his peace of mind to establish the principle that reviewers must work within the bounds of truth when they deal with the author rather than the book. Even as he was becoming the great national scold of his time, Cooper managed to write book after book—social and political satires growing out of his experiences with the press, a reactionary primer on *The American Democrat* (1838), and, despite his avowal in 1834, a series of sociopolitical novels and two more *Leather-Stocking Tales, The Pathfinder* (1840) and *The Deerslayer* (1841). His monumental *History of the Navy of the United States of America* (1839) became the focus of new quarrels and a new lawsuit.

When Cooper died on September 14, 1851, a day before his sixty-second birthday, he was a byword for litigiousness and social pretentiousness. A lifelong defender of American democracy as he knew it in his youth against European aristocracy and then against what American democracy had become, he was out of step with his countrypeople. Yet throughout the century and into the next his *Leather-Stocking Tales* had an incalculable vogue in the United States and abroad. In his own time and shortly afterward, major European writers as diverse as Honoré de Balzac and Leo Tolstoy were profoundly moved by *The Pioneers* and the subsequent Natty Bumppo novels, but gradually the *Leather-Stocking Tales* became something only schoolchildren read. Not until the 1920s did scholars begin to see Cooper's value as the country's first great social critic. It now seems clear that no revolution in taste will lead to widespread admiration of Cooper as a literary artist, but he will always be a major source for the student of ideas in America. Some of his opinions now seem hopelessly reactionary, as when he defends American slavery as legal and, after all, mild ("physical suffering cannot properly be enumerated among its evils") or when he deplores the dangers of universal manhood suffrage and argues for restricting voting on certain issues to property-owners, who have the greater stake in society. What most appeals to modern readers are his profoundly ambivalent dramatizations of such enduring American conflicts as natural right versus legal right, order versus change, primeval wilderness versus civilization. And new readers will always encounter the *Leather-Stocking Tales* with a sense of something long known and loved, for if Cooper is no longer read even by children, everyone has read books—and seen films—which are directly and indirectly influenced by his grand conception of Natty Bumppo.

From The Pioneers
[*The Slaughter of the Pigeons*][1]

"Men, boys, and girls,
Desert th' unpeopled village; and wild crowds
Spread o'er the plain, by the sweet frenzy driven."
—SOMERVILLE

From this time to the close of April, the weather continued to be
a succession of great and rapid changes. One day, the soft airs of
spring would seem to be stealing along the valley, and, in unison
with an invigorating sun, attempting, covertly, to rouse the dormant
powers of the vegetable world; while on the next, the surly blasts
from the north would sweep across the lake, and erase every impression left by their gentle adversaries. The snow, however, finally
disappeared, and the green wheat fields were seen in every direction,
spotted with the dark and charred stumps that had, the preceding
season, supported some of the proudest trees of the forest.[3] Ploughs
were in motion, wherever those useful implements could be used,
and the smokes of the sugar-camps[4] were no longer seen issuing
from the summits of the woods of maple. The lake had lost all the
characteristic beauty of a field of ice, but still a dark and gloomy
covering concealed its waters, for the absence of currents left them
yet hid under a porous crust, which, saturated with the fluid, barely
retained enough of its strength to preserve the contiguity of its
parts. Large flocks of wild geese were seen passing over the country,
which would hover, for a time, around the hidden sheet of water,
apparently searching for an opening, where they might obtain a
resting-place; and then, on finding themselves excluded by the chill
covering, would soar away to the north, filling the air with their

1. *The Pioneers, or The Sources of the*
Susquehanna; A Descriptive Tale (New
York: Charles Wiley, 1823) is the first
of five Cooper novels in which Natty
Bumppo is the major character. The text
is that of the first edition, Vol. II, Chapter 3 (Chapter 22 in later one-volume
editions). *The Pioneers* begins in December, 1793, at the settlement of Templeton (modeled on Cooperstown) at
Otsego Lake in central New York, some
50 miles west of Albany. The episode reprinted here occurs in the spring of 1794.
Natty Bumppo is in his early 70s, six
feet tall (then a great height), gray-eyed,
with lank, sandy hair, sunburned, robust,
but thin almost to emaciation. One yellow tooth survives in his enormous
mouth, and he gives forth a remarkable
kind of inward laugh. He wears a fox-
skin hat and is clad in deerskin—coat,
moccasins, and even the leggings which
fasten over the keees of his buckskin
breeches and give him the nickname of
"Leather-Stocking." For his old and unusually long rifle he carries gunpowder in
an enormous ox horn slung over his
shoulder by a strap of deerskin. This
was the unprepossessing figure who captured the imagination of the United
States and Europe.
2. From *The Chace*, II, 197–99, by the
English poet William Somerville (1675–
1742). The last word should be "seized,"
not "driven."
3. The practice was to chop timber down
in the spring, let it dry through the summer, then burn the cleared area so that
only blackened logs and stumps remained. Nothing was salvaged except
some ashes used as the basis for potash.
4. Where sugar was made from maple
sap.

discordant screams, as if venting their complaints at the tardy operations of nature.

For a week, the dark covering of the Otsego was left to the undisturbed possession of two eagles, who alighted on the centre of its field, and sat proudly eyeing the extent of their undisputed territory. During the presence of these monarchs of the air, the flocks of migrating birds avoided crossing the plain of ice, by turning into the hills, and apparently seeking the protection of the forests, while the white and bald heads of the tenants of the lake were turned upward, with a look of majestic contempt, as if penetrating to the very heavens, with the acuteness of their vision. But the time had come, when even these kings of birds were to be dispossessed. An opening had been gradually increasing, at the lower extremity of the lake, and around the dark spot where the current of the river had prevented the formation of ice, during even the coldest weather; and the fresh southerly winds, that now breathed freely up the valley, obtained an impression on the waters. Mimic waves begun to curl over the margin of the frozen field, which exhibited an outline of crystallizations, that slowly receded towards the north. At each step the power of the winds and the waves increased, until, after a struggle of a few hours, the turbulent little billows succeeded in setting the whole field in an undulating motion, when it was driven beyond the reach of the eye, with a rapidity, that was as magical as the change produced in the scene by this expulsion of the lingering remnant of winter. Just as the last sheet of agitated ice was disappearing in the distance, the eagles rose over the border of crystals, and soared with a wide sweep far above the clouds, while the waves tossed their little caps of snow into the air, as if rioting in their release from a thraldom of five months duration.

The following morning Elizabeth[5] was awakened by the exhilarating sounds of the martins, who were quarrelling and chattering around the little boxes which were suspended above her windows, and the cries of Richard,[6] who was calling, in tones as animating as the signs of the season itself—

"Awake! awake! my lady fair! the gulls are hovering over the lake already, and the heavens are alive with the pigeons. You may look an hour before you can find a hole, through which, to get a peep at the sun. Awake! awake! lazy ones! Benjamin[7] is overhauling the ammunition, and we only wait for our breakfasts, and away for the mountains and pigeon-shooting."

5. Elizabeth Temple, daughter of Judge Marmaduke Temple, the founder of Templeton and its chief landowner; at the outset of the story she returns from four years at school.

6. Richard (Dickon) Jones, the sheriff, a cousin of Judge Temple; he superintends "all the minor concerns of Temple's business."

7. Benjamin Penguillan (called Ben Pump), a Cornishman and former sailor, "major-domo" or steward under Jones. In the next paragraph Pump is called "the ex-steward" because he had been a steward to the captain in his seagoing years. One of his charges at the Templeton house is to keep the stove in the parlor hot in winter.

There was no resisting this animated appeal, and in a few minutes Miss Temple and her friend[8] descended to the parlour. The doors of the hall were thrown open, and the mild, balmy air of a clear spring morning was ventilating the apartment, where the vigilance of the ex-steward had been so long maintaining an artificial heat, with such unremitted diligence. All of the gentlemen, we do not include Monsieur Le Quoi,[9] were impatiently waiting their morning's repast, each being equipt in the garb of a sportsman. Mr. Jones made many visits to the southern door, and would cry—

"See, cousin Bess! see, 'duke![1] the pigeon-roosts of the south have broken up! They are growing more thick every instant. Here is a flock that the eye cannot see the end of. There is food enough in it to keep the army of Xerxes[2] for a month, and feathers enough to make beds for the whole county. Xerxes, Mr. Edwards,[3] was a Grecian king, who—no, he was a Turk, or a Persian, who wanted to conquer Greece, just the same as these rascals will overrun our wheat-fields, when they come back in the fall.—Away! away! Bess; I long to pepper them from the mountain."

In this wish both Marmaduke and young Edwards seemed equally to participate, for really the sight was most exhilarating to a sportsman; and the ladies soon dismissed the party, after a hasty breakfast.

If the heavens were alive with pigeons, the whole village seemed equally in motion, with men, women, and children. Every species of fire-arms, from the French ducking-gun, with its barrel of near six feet in length, to the common horseman's pistol, was to be seen in the hands of the men and boys; while bows and arrows, some made of the simple stick of a walnut sapling, and others in a rude imitation of the ancient cross-bows, were carried by many of the latter.

The houses, and the signs of life apparent in the village, drove the alarmed birds from the direct line of their flight, towards the mountains, along the sides and near the bases of which they were glancing in dense masses, that were equally wonderful by the rapidity of their motion, as by their incredible numbers.

We have already said, that across the inclined plane which fell from the steep ascent of the mountain to the banks of the Susquehanna, ran the highway, on either side of which a clearing of many acres had been made, at a very early day. Over those clearings, and up the eastern mountain, and along the dangerous path that was cut into its side, the different individuals posted themselves, as suited their inclinations; and in a few moments the attack commenced.

Amongst the sportsmen was to be seen the tall, gaunt form of

8. Louisa Grant, daughter of the Episcopal minister.
9. Once a West Indian planter, now a refugee because of the French Revolution.

1. Short for "Marmaduke," the judge.
2. Xerxes the Great (519?–465 B.C.) was King of Persia (486–465 B.C.).
3. Oliver Edwards, a mysterious young stranger.

Leather-stocking,[4] who was walking over the field, with his rifle hanging on his arm, his dogs following close at his heels, now scenting the dead or wounded birds, that were beginning to tumble from the flocks, and then crouching under the legs of their master, as if they participated in his feelings, at this wasteful and unsportsmanlike execution.

The reports of the fire-arms became rapid, whole volleys rising from the plain, as flocks of more than ordinary numbers darted over the opening, covering the field with darkness, like an interposing cloud; and then the light smoke of a single piece would issue from among the leafless bushes on the mountain, as death was hurled on the retreat of the affrighted birds, who would rise from a volley, for many feet into the air, in a vain effort to escape the attacks of man. Arrows, and missiles of every kind, were seen in the midst of the flocks; and so numerous were the birds, and so low did they take their flight, that even long poles, in the hands of those on the sides of the mountain, were used to strike them to the earth.

During all this time, Mr. Jones, who disdained the humble and ordinary means of destruction used by his companions, was busily occupied, aided by Benjamin, in making arrangements for an assault of a more than ordinarily fatal character. Among the relics of the old military excursions, that occasionally are discovered throughout the different districts of the western part of New-York, there had been found in Templeton, at its settlement, a small swivel,[5] which would carry a ball of a pound weight. It was thought to have been deserted by a war-party of the whites, in one of their inroads into the Indian settlements, when, perhaps, their convenience or their necessities induced them to leave such an encumbrance to the rapidity of their march, behind them in the woods. This miniature cannon had been released from the rust, and mounted on little wheels, in a state for actual service. For several years, it was the sole organ for extraordinary rejoicings that was used in those mountains. On the mornings of the Fourth of July, it would be heard, with its echoes ringing among the hills, and telling forth its sounds, for thirteen times, with all the dignity of a two-and-thirty pounder; and even Captain Hollister,[6] who was the highest authority in that part of the country on all such occasions, affirmed that, considering its dimensions, it was no despicable gun for a salute. It was somewhat the worse for the service it had performed, it is true, there being but a trifling difference in size between the touch-hole and the muzzle.[7] Still, the grand conceptions of Richard had suggested the importance of such an instrument, in hurling

---

4. I.e., Natty Bumppo.
5. Small cannon capable of being swung higher or lower.
6. The landlord of the major village inn, The Bold Dragoon; his rank comes from his having been an early commander of local militia.
7. Ordinarily the muzzle (or mouth) would be considerably larger than the touch-hole, the vent by which fire is communicated to the powder.

death at his nimble enemies. The swivel was dragged by a horse into a part of the open space, that the sheriff thought most eligible for planting a battery of the kind, and Mr. Pump proceeded to load it. Several handfuls of duck-shot were placed on top of the powder, and the Major-domo soon announced that his piece was ready for service.

The sight of such an implement collected all the idle spectators to the spot, who, being mostly boys, filled the air with their cries of exultation and delight. The gun was pointed on high, and Richard, holding a coal of fire in a pair of tongs, patiently took his seat on a stump, awaiting the appearance of a flock that was worthy of his notice.

So prodigious was the number of the birds, that the scattering fire of the guns, with the hurling of missiles, and the cries of the boys, had no other effect than to break off small flocks from the immense masses that continued to dart along the valley, as if the whole creation of the feathered tribe were pouring through that one pass. None pretended to collect the game, which lay scattered over the fields in such profusion, as to cover the very ground with the fluttering victims.

Leather-stocking was a silent, but uneasy spectator of all these proceedings, but was able to keep his sentiments to himself until he saw the introduction of the swivel into the sports.

"This comes of settling a country!" he said—"here have I known the pigeons to fly for forty long years, and, till you made your clearings, there was nobody to scare or to hurt them. I loved to see them come into the woods, for they were company to a body; hurting nothing; being, as it was, as harmless as a garter-snake. But now it gives me sore thoughts when I hear the frighty things whizzing through the air, for I know it's only a motion to bring out all the brats in the village at them. Well! the Lord won't see the waste of his creaters for nothing, and right will be done to the pigeons, as well as others, by-and-by.—There's Mr. Oliver, as bad as the rest of them, firing into the flocks as if he was shooting down nothing but the Mingo[8] warriors."

Among the sportsmen was Billy Kirby,[9] who, armed with an old musket, was loading, and, without even looking into the air, was firing, and shouting as his victims fell even on his own person. He heard the speech of Natty, and took upon himself to reply—

"What's that, old Leather-stocking!" he cried; "grumbling at the loss of a few pigeons! If you had to sow your wheat twice, and three times, as I have done, you wouldn't be so massyfully[1] feeling'd to'ards the divils.—Hurrah, boys! scatter the feathers. This is better

8. In the *Leather-Stocking* novels set in New York, the Mingos (Iroquois) are made out to be the "bad Indians" while the Delawares are the "good Indians."
9. A woodchopper.
1. Mercifully.

than shooting at a turkey's head and neck, old fellow."[2]

"It's better for you, maybe, Billy Kirby," returned the indignant old hunter, "and all them as don't know how to put a ball down a rifle-barrel, or how to bring it up ag'in with a true aim; but it's wicked to be shooting into flocks in this wastey manner; and none do it, who know how to knock over a single bird. If a body has a craving for pigeon's flesh, why! it's made the same as all other creaters, for man's eating, but not to kill twenty and eat one. When I want such a thing, I go into the woods till I find one to my liking, and then I shoot him off the branches without touching a feather of another, though there might be a hundred on the same tree. But you couldn't do such a thing, Billy Kirby—you couldn't do it if you tried."

"What's that you say, you old, dried cornstalk! you sapless stub!" cried the wood-chopper. "You've grown mighty boasting, sin[3] you killed the turkey; but if you're for a single shot, here goes at that bird which comes on by himself."

The fire from the distant part of the field had driven a single pigeon below the flock to which it had belonged, and, frightened with the constant reports of the muskets, it was approaching the spot where the disputants stood, darting first from one side, and then to the other, cutting the air with the swiftness of lightning, and making a noise with its wings, not unlike the rushing of a bullet. Unfortunately for the wood-chopper, notwithstanding his vaunt, he did not see his bird until it was too late for him to fire as it approached, and he pulled his trigger at the unlucky moment when it was darting immediately over his head. The bird continued its course with incredible velocity.

Natty had dropped his piece from his arm, when the challenge was made, and, waiting a moment, until the terrified victim had got in a line with his eyes, and had dropped near the bank of the lake, he raised his rifle with uncommon rapidity, and fired. It might have been chance, or it might have been skill, that produced the result; it was probably a union of both; but the pigeon whirled over in the air, and fell into the lake, with a broken wing. At the sound of his rifle, both his dogs started from his feet, and in a few minutes the "slut"[4] brought out the bird, still alive.

The wonderful exploit of Leather-stocking was noised through the field with great rapidity, and the sportsmen gathered in to learn the truth of the report.

"What," said young Edwards, "have you really killed a pigeon on the wing, Natty, with a single ball?"

"Haven't I killed loons before now, lad, that dive at the flash?"

2. In an earlier chapter Natty Bumppo   3. Since.
had beaten Kirby in a turkey-shooting   4. Bitch, female dog.
contest.

returned the hunter. "It's much better to kill only such as you want, without wasting your powder and lead, than to be firing into God's creaters in such a wicked manner. But I come out for a bird, and you know the reason why I like small game, Mr. Oliver, and now I have got one I will go home, for I don't like to see these wasty ways that you are all practysing, as if the least thing was not made for use, and not to destroy."

"Thou sayest well, Leather-stocking," cried Marmaduke, "and I begin to think it time to put an end to this work of destruction."

"Put an ind, Judge, to your clearings. An't the woods his work as well as the pigeons? Use, but don't waste. Wasn't the woods made for the beasts and birds to harbour in? and when man wanted their flesh, their skins, or their feathers, there's the place to seek them. But I'll go to the hut with my own game, for I wouldn't touch one of the harmless things that kiver the ground here, looking up with their eyes at me, as if they only wanted tongues to say their thoughts."

With this sentiment in his mouth, Leather-stocking threw his rifle over his arm, and, followed by his dogs, stepped across the clearing with great caution, taking care not to tread on one, of the hundreds of the wounded birds that lay in his path. He soon entered the bushes on the margin of the lake, and was hid from view.

Whatever might be the impression the morality of Natty made on the Judge, it was utterly lost on Richard. He availed himself of the gathering of the sportsmen, to lay a plan for one "fell swoop"[5] of destruction. The musket-men were drawn up in battle array, in a line extending on each side of his artillery, with orders to await the signal of firing from himself.

"Stand by, my lads," said Benjamin, who acted as an aid-de-camp on this momentous occasion, "stand by, my hearties, and when Squire Dickens heaves out the signal for to begin the firing, d'ye see, you may open upon them in a broadside. Take care and fire low, boys, and you'll be sure to hull the flock."

"Fire low!" shouted Kirby—"hear the old fool! If we fire low, we may hit the stumps, but not ruffle a pigeon."

"How should you know, you lubber?"[6] cried Benjamin, with a very unbecoming heat, for an officer on the eve of battle—"how should you know, you grampus? Havn't I sailed aboard of the Boadishy[7] for five years? and wasn't it a standing order to fire low, and to hull your enemy? Keep silence at your guns, boys, and mind the order that is passed."

The loud laughs of the musketmen were silenced by the authori-

---

5. Shakespeare's *Macbeth*, 4.3.219, in Macduff's lament for his dead wife and children.
6. Landlubber, clumsy fellow.
7. "Grampus": variety of small whale, used here as a term of contempt; "Boadishy": the *Boadicea*, a ship named for the British Queen who led a rebellion against the Roman rulers in A.D. 62.

tative voice of Richard, who called to them for attention and obedience to his signals.

Some millions of pigeons were supposed to have already passed, that morning, over the valley of Templeton; but nothing like the flock that was now approaching had been seen before. It extended from mountain to mountain in one solid blue mass, and the eye looked in vain over the southern hills to find its termination. The front of this living column was distinctly marked by a line, but very slightly indented, so regular and even was the flight. Even Marmaduke forgot the morality of Leather-stocking as it approached, and, in common with the rest, brought his musket to his shoulder.

"Fire!" cried the Sheriff, clapping his coal to the priming of the cannon. As half of Benjamin's charge escaped through the touch-hole, the whole volley of the musketry preceded the report of the swivel. On receiving this united discharge of small-arms, the front of the flock darted upward, while, at the same instant, myriads of those in their rear rushed with amazing rapidity into their places, so that when the column of white smoke gushed from the mouth of the little cannon, an accumulated mass of objects was gliding over its point of direction. The roar of the gun echoed along the mountains, and died away to the north, like distant thunder, while the whole flock of alarmed birds seemed, for a moment, thrown into one disorderly and agitated mass. The air was filled with their irregular flights, layer rising over layer, far above the tops of the highest pines, none daring to advance beyond the dangerous pass; when, suddenly, some of the leaders of the feathered tribe shot across the valley, taking their flight directly over the village, and the hundreds of thousands in their rear followed their example, deserting the eastern side of the plain to their persecutors and the fallen.

"Victory!" shouted Richard, "victory! we have driven the enemy from the field."

"Not so, Dickon," said Marmaduke; "the field is covered with them; and, like the Leather-stocking, I see nothing but eyes, in every direction, as the innocent sufferers turn their heads in terror, to examine my movements. Full one half of those that have fallen are yet alive: and I think it is time to end the sport; if sport it be."

"Sport!" cried the Sheriff; "it is princely sport. There are some thousands of the blue-coated boys on the ground, so that every old woman in the village may have a pot-pie for the asking."

"Well, we have happily frightened the birds from this pass," said Marmaduke, "and our carnage must of necessity end, for the present.—Boys, I will give thee sixpence a hundred for the pigeons' heads only; so go to work, and bring them into the village, when I will pay thee."

This expedient produced the desired effect, for every urchin on the ground went industriously to work to wring the necks of the wounded birds. Judge Temple retired towards his dwelling with that kind of feeling, that many a man has experienced before him, who discovers, after the excitement of the moment has passed, that he has purchased pleasure at the price of misery to others. Horses were loaded with the dead; and, after this first burst of sporting, the shooting of pigeons became a business, for the remainder of the season, more in proportion to the wants of the people.[8] Richard, however, boasted for many a year, of his shot with the "cricket;"[9] and Benjamin gravely asserted, that he thought that they killed nearly as many pigeons on that day, as there were Frenchmen destroyed on the memorable occasion of Rodney's victory.[1]

1823

8. The pigeons described in this chapter —the passenger pigeons—are extinct, the last known specimen dying in 1914 at the Cincinnati Zoological Garden.
9. I.e., the little cannon.
1. The British admiral George Brydges,

Baron Rodney (1719–92), defeated the French off Dominica, in the West Indies, in April, 1782. Penguillan's nickname comes from his tall tale about manning the pumps to keep the ship from sinking after Rodney's victory.

WILLIAM CULLEN BRYANT
1794–1878

William Cullen Bryant was born in the backwoods of Massachusetts, at Cummington, but his father was a physician who loved the classics, and Cullen, as the boy was called, was trained early in Greek and Latin. For religion he was taught a harsh Calvinism which held that the Fall of Man had brought about the Fall of Nature as well. But Bryant's first published poem was political, not about nature and religion: when he wrote an anti-Jefferson lampoon, *The Embargo*, his Federalist father printed it as a pamphlet (1808). Bryant entered Williams College in 1810 but dropped out after a few months with the expectation of entering Yale. Dr. Bryant could not afford that expense, and instead Cullen read for the law, being admitted to practice in 1815. Meanwhile, in 1813 or 1814, Bryant wrote the first, shorter version of *Thanatopsis*, the poem by which he is best remembered. Since his early teens Bryant had been reading the melancholy and sometimes scarifying meditations of the British "graveyard poets" of the previous decades, especially Robert Blair (*The Grave*), Thomas Gray (*Elegy Written in a Country Churchyard*), Bishop Beilby Porteus (*Death*), and various poems by Henry Kirke White. Such poems by their luxurious sonorousness tempered the Calvinism instilled in the boy, but they often poeticized religious doctrine, as in Blair's account of the resurrection at the Judgment Day: "the time draws on / When not a single spot of burial-earth, / Whether on land, or on the spacious sea, / But must give back its long-committed dust / Inviolate" (*The Grave*). In 1810 or soon afterward Bryant read *Lyrical Ballads* and responded strongly to Wordsworth's near-pantheistic view of Nature. *Thanatopsis* as published in the *North American Review* in 1817 is nondoctrinally meditative. The fuller version of 1821 concludes with a fervent injunction to trust in something or someone who remains unspecified: Bryant's Calvinistic earnestness was outliving his commitment to particular doctrines. (Symptomatically, a reference to the Fall of Nature in the first version of *The Prairies*, 1834, was later removed.) *Thanatopsis* won Bryant immediate acknowledgment in 1817, but a full-time career as a poet was economically impossible. In 1820, the year his father died, Bryant was appointed as justice of the peace in Berkshire County. Early in 1821 he married Frances Fairchild in Great Barrington, Massachusetts, and later that year published a very slim volume of *Poems*.

Stirred by the conflict between his literary ambition and his need to support his family, Bryant in 1825 chanced a move to New York City as an editor of the *New-York Review and Atheneum Magazine*. He was welcomed as a literary celebrity and quickly fitted into metropolitan life, becoming an early member of James Fenimore Cooper's Bread and Cheese Club. His magazine failed, as almost all periodicals did at that time, but Bryant stayed on in New York as editorial assistant on the *Evening Post* (1826), then soon became part owner and editor-in-chief. Bryant was not immune to the pettier temptations of the then brawling occupation of journalism, but over the decades he made the *Evening Post* one of the most respected newspapers in the country, mainly through editorials in which he argued out his position on many momentous issues. A deeply committed Jacksonian Democrat, despite his youthful Federalism, Bryant rarely let party loyalty interfere with principle. He led the antislavery Free-Soil movement within the Democratic party as long as this seemed a feasible way of achieving his ends, then helped to form the Republican party. In 1860 he was an influential advocate of Abraham Lincoln.

As he prospered with his newspaper, Bryant became a great traveler, at home and abroad, and through his letters to the *Evening Post* he helped to shape a sense of the world for his countrypeople. *Letters of a Traveller* appeared in 1850; *Letters of a Traveller, Second Series*, in 1859; and *Letters from the East* (that is, the Mid-East) in 1869. His community service took many forms, most tangibly in his campaign for the creation of Central Park. His private life was happy. In 1844 he moved his family to a fine old farmhouse on the Sound in then rural Long Island, and for many years relieved his strenuous urban activity with peaceful respites at his estate, Cedarmere. Left a widower in 1866, Bryant continued to work at the *Evening Post*. Blessed with patriarchal fame and great wealth, as well as astonishing health, which owed much to a daily set of vigorous exercises, Bryant in his seventies undertook the remarkably ambitious task of translating Homer. His version of the *Iliad* was published in 1870, and that of the *Odyssey* two years later. Together with the 1876 printing of his *Poems* (a new accumulation of many old and a few new verses), these translations crowned his career.

Bryant died of the consequences of a fall suffered after he gave a speech at the unveiling of a statue of the Italian patriot Joseph Mazzini in Central Park. In New York City flags were lowered to half mast, and he was mourned throughout the country as a great poet and editor.

Bryant had, in fact, written very little poetry; his translations from Homer are many times as long as his own verses. And his poems, early and late, are for the most part limited to a few subjects treated in ways that soon become predictable. His collected poetry consists of accurately rhymed or sonorously unrhymed blank verse on landscapes, flora, meteorological phenomena, historical personages and events, friends, Indian legends, and a few other topics. Yet the country's response showed plainly that he was providing what it needed at a time of national self-consciousness about the scarcity of talented poets—a loftiness of diction that at best seemed securely Miltonic, a way of making American landscapes and subjects as worthy of celebration as Old World scenes and topics, and a moral stance which blended ecumenical vagueness with didactic earnestness. Bryant's fame as a poet was accurately analyzed by his early biographer, W. A. Bradley: "He appeared much more remarkable to his early contemporaries than he ever can to us, because of the contrast which he presented with what had gone before. And later, after a period in which he suffered somewhat of an eclipse through the rise of new schools and new poets to contest with him the palm of supremacy, his great age, the traditions of an earlier day which he represented, his personality which so perfectly embodied the prophetic and seer-like aspect of the poetic ideal, and finally local pride in the possession of a poet whom New York could produce to oppose the claims of its rival, Boston, to literary supremacy,—all these tended to create a regard for Bryant that was rather personal than literary." On the strength of a few—mainly very early—poems and a notable public life, Bryant passed into what seemed, to his own time, literary immortality. Only as historians describe his part in the great political issues of his time is he passing into a perhaps truer immortality as a man who may not have been a great American poet but who led a great American life.

Thanatopsis[1]

To him who in the love of Nature holds
Communion with her visible forms, she speaks
A various language; for his gayer hours
She has a voice of gladness, and a smile
And eloquence of beauty, and she glides   5
Into his darker musings, with a mild
And gentle sympathy, that steals away
Their sharpness, ere he is aware. When thoughts
Of the last bitter hour come like a blight
Over thy spirit, and sad images   10
Of the stern agony, and shroud, and pall,
And breathless darkness, and the narrow house,
Make thee to shudder, and grow sick at heart;—
Go forth under the open sky, and list
To Nature's teachings, while from all around—   15
Earth and her waters, and the depths of air,—
Comes a still voice—Yet a few days, and thee

1. The text is that of the first full printing, in Bryant's *Poems* (Cambridge: Hilliard and Metcalf, 1821). The title ("meditation on death") was supplied by an editor for the central section of the poem (lines 17–73) when that section was printed in the *North American Review* (September, 1817).

The all-beholding sun shall see no more
In all his course; nor yet in the cold ground,
Where thy pale form was laid, with many tears,   20
Nor in the embrace of ocean shall exist
Thy image. Earth, that nourished thee, shall claim
Thy growth, to be resolv'd to earth again;
And, lost each human trace, surrend'ring up
Thine individual being, shalt thou go   25
To mix forever with the elements,
To be a brother to th' insensible rock
And to the sluggish clod, which the rude swain[2]
Turns with his share, and treads upon. The oak
Shall send his roots abroad, and pierce thy mould.   30
Yet not to thy eternal resting place
Shalt thou retire alone—nor couldst thou wish
Couch more magnificent. Thou shalt lie down
With patriarchs of the infant world—with kings
The powerful of the earth—the wise, the good,   35
Fair forms, and hoary seers of ages past,
All in one mighty sepulchre.—The hills
Rock-ribb'd and ancient as the sun,—the vales
Stretching in pensive quietness between;
The venerable woods—rivers that move   40
In majesty, and the complaining brooks
That make the meadows green; and pour'd round all,
Old ocean's grey and melancholy waste,—
Are but the solemn decorations all
Of the great tomb of man. The golden sun,   45
The planets, all the infinite host of heaven,
Are shining on the sad abodes of death,
Through the still lapse of ages. All that tread
The globe are but a handful to the tribes
That slumber in its bosom.—Take the wings   50
Of morning—and the Barcan desert[3] pierce,
Or lose thyself in the continuous woods
Where rolls the Oregan, and hears no sound,
Save his own dashings—yet—the dead are there,
And millions in those solitudes, since first   55
The flight of years began, have laid them down
In their last sleep—the dead reign there alone.—
So shalt thou rest—and what if thou shalt fall
Unnoticed by the living—and no friend
Take note of thy departure? All that breathe   60
Will share thy destiny. The gay will laugh
When thou art gone, the solemn brood of care
Plod on, and each one as before will chase
His favourite phantom; yet all these shall leave

2. Farmer; "share": plowshare.
3. In Barca (northeast Libya); "Oregan" (early variant spelling of Oregon): now the Columbia River. (For his distant examples Bryant ranges across the Atlantic and then westward across the North American continent.)

Their mirth and their employments, and shall come, 65
And make their bed with thee. As the long train
Of ages glide away, the sons of men,
The youth in life's green spring, and he who goes
In the full strength of years, matron, and maid,
The bow'd with age, the infant in the smiles 70
And beauty of its innocent age cut off,—
Shall one by one be gathered to thy side,
By those, who in their turn shall follow them.
So live, that when thy summons comes to join
The innumerable caravan, that moves 75
To the pale realms of shade, where each shall take
His chamber in the silent halls of death,
Thou go not, like the quarry-slave at night,
Scourged to his dungeon, but sustain'd and sooth'd
By an unfaltering trust, approach thy grave, 80
Like one who wraps the drapery of his couch
About him, and lies down to pleasant dreams.

c. 1814   1821

RALPH WALDO EMERSON
1803–1882

Emerson was a man who had no personal excesses such as doomed Poe, no
mysterious decade such as lent glamor to Hawthorne, no exotic adventures
such as Melville founded his career upon, no dramatic struggles for artistic
recognition such as Whitman waged, no local notoriety as a crank and
extremist such as Thoreau acquired. He led a respectable, conventional life
as a family man and decent, solid citizen. Yet in both literature and philosophy this man of conventional life became the American writer with
whom every other significant writer of his time had to come to terms. At
one extreme, Melville reacted so hostilely to the optimistic side of Emerson's
thought that he satirized him in *The Confidence-Man* as a great American
philosophical con man. At the other extreme, without Emerson's inspiration the writings of Thoreau are all but unthinkable and Whitman's great
poetry might never have been written. Emerson's persisting influence upon
twentieth-century American writers is evident in astonishing permutations,
on writers as diverse as Theodore Dreiser, Robert Frost, Wallace Stevens,
his namesake Ralph Waldo Ellison, and A. R. Ammons.

Emerson was born in Boston, Massachusetts, on May 25, 1803, son of
a Unitarian minister and the second of five surviving boys. He was eight
years old when the death of his father left the family to the meager charity
of the church. Determined to send four sons to Harvard (another son,
mentally retarded, was cared for by rural relatives), Emerson's mother kept

a succession of boardinghouses. Emerson grew up in the city, protected from the lower-class "rough boys" in his early years and sent at nine to the Boston Public Latin School. So poorly clothed that two brothers had to make do at times with one coat, the boys were encouraged by a brilliant eccentric aunt, Mary Moody Emerson, to regard deprivation as ecstatic self-denial. Emerson showed no remarkable literary promise either in his early prose exercises or in his adolescent satires in imitation of Alexander Pope. His Harvard years, 1817–21, were frugal, industrious, and undistinguished. After graduation he served, he said, as "a hopeless Schoolmaster," unable to impose his authority upon his pupils. Escaping into the study of theology in 1825, he began preaching in October, 1826, and early in 1829 was ordained as junior pastor of Boston's Second Church where Increase Mather and Cotton Mather had preached a century and more before.

Biographers have pointed out that Emerson's dedication to the ministry at the age of twenty-one was to a life of public service through eloquence, not to a life of preserving and disseminating religious dogma. In any case, Boston was no longer a Puritan stronghold. Boston Unitarianism, led in the 1820s by William Ellery Channing, still accepted the Bible as the revelation of God's intentions for humankind, but no longer held that human beings were innately depraved or that Jesus was more than the highest type of mankind. Emerson's skepticism toward Christianity was strengthened by his exposure to the German "higher criticism," which heretically interpreted Biblical miracles in the light of comparable stories in other cultures. Emerson was gradually developing a faith greater in individual moral sentiment than in revealed religion. Around 1830–31 his reading of Samuel Taylor Coleridge's *Aids to Reflection* provided him with a basic terminology in his postulation of an intuitive "Reason," which is superior to the mere "Understanding," or ordinary rationality operating on the materials of sense experience. Undogmatic about Christianity as he became, Emerson nevertheless seems to have undergone an intense religious experience around these same years, 1830 or 1831, something comparable to the sweet inward burning which the Calvinist Jonathan Edwards had delighted in describing. Emerson's knowledge of this emotion is clear from his later essay *The Over-Soul*, but he felt no impulse to account for it according to the tenets of a particular church.

In the year of his ordination, Emerson married a young woman from New Hampshire, Ellen Tucker. She died sixteen months later of tuberculosis, the disease which had already infected Emerson and others in his family. Early in 1832 Emerson notified his church that he had become so skeptical of the validity of the Lord's Supper that he could no longer administer it. A few months later he resigned, keeping the sober good will of many in his flock, and embarked on a leisurely European tour which constituted a postgraduate education in art and natural science. In the custom of that time, he called upon literary men, meeting Walter Savage Landor in Italy, listening to Coleridge converse with such cogent volubility that he seemed to be reading aloud, and hearing William Wordsworth recite his poetry. Most important for his intellectual growth and for his reputation was his visit to Thomas Carlyle at Craigenputtock in Scotland, beginning a lifelong alliance in which each helped to publish and create an audience for the other.

In 1834 Emerson drifted into a quiet retreat at Concord, Massachusetts, where generations of his ancestors had been ministers. That year he received the first installment of his wife's legacy. Soon he was assured of more than a thousand dollars annually, enough so that he did not need to hold a steady job again. He continued to preach occasionally, and began lecturing at New England lyceums, the public halls which brought a variety of speakers and performers both to the cities and to smaller towns. In 1835, after a prudent courtship, he married Lydia Jackson of Plymouth, having explained to her his work and the conditions under which he must pursue it. One condition was that he must live in rural Concord rather than move into the bustle of Plymouth. His assessment of his accomplishments was focused on his being a "poet," even when writing prose: "I am a poet, of a low class without doubt yet a poet. That is my nature & vocation. My singing be sure is very 'husky,' & is for the most part in prose. Still am I a poet in the sense of a perceiver & dear lover of the harmonies that are in the soul & in matter, & specially of the correspondences between these & those. A sunset, a forest, a snow storm, a certain river-view, are more to me than many friends & do ordinarily divide my day with books. Wherever I go therefore I guard & study my rambling propensities with a care that is ridiculous to people, but to me is the care of my high calling."

At this time, before *Nature* was published and before his essays were written, Emerson may well have hoped to gain his literary fame by his verse, and in fact the poems he had written thus far were not so husky or unmelodious as he implied. His main problem as a poet was not the huskiness he complained of but the more serious failure first to arrive at, then to apply, his great insight: that in true poetry the thought creates its own meter, the content creates its own form. Emerson's first little book, *Nature* (1836), did not establish him as an important American writer (for one thing, it was anonymous, and not every reviewer was in on what became an open secret around Boston), but it did confirm his future as a prose writer, however poetic that prose might be. One reviewer noticed the influence of Wordsworth and Coleridge in the tendency to "look on Nature with the spiritual eye," so that one creates Nature in perceiving it. Another found that the author had adopted "the Berkeleyan system" of philosophy which denies "the outward and real existence" of Nature, and noticed also that the new school of philosophy called Transcendentalism was "a revival of the Old Platonic school" in rejecting a scientific attitude toward Nature. Yet another reviewer stressed the influence of Wordsworth's *Immortality* ode and of Coleridge's *Dejection: An Ode* in the author's concept of Nature. Another hailed the book as revealing a mind cognate with Thomas Carlyle's, "however inferior in energies and influences," and defined the philosophy of the book as "an Idealistic Pantheism" like that of Carlyle's *Sartor Resartus* (1834). A Swedenborgian writer in London took *Nature* as self-evidently the work of an American Swedenborgian, especially in "the beautiful and heart-cheering doctrine of correspondences" between Nature's moods and man's. As all these reviewers understood, *Nature* was not a Christian book but one influenced by a range of idealistic philosophies, ancient and very modern, Transcendentalism being merely the latest name for an old way of thinking. Although the favorable reception of the book in England encouraged some American journalists, hitherto skeptical, to take Emerson more seriously as a force in modern thought, Emerson's immediate reward was having the book become the unofficial manifesto for "the Symposium" or the Transcendental Club, which held its first meeting only a few days after *Nature* was published.

The Transcendental Club had influences, on Emerson and on the intellectual life of the country, out of proportion to its small membership and its short life of four years. It was composed mainly of ministers who were repelled by John Locke's views that the mind is a passive receiver of sense impressions and enthusiastic about Coleridge's alternative view of the mind as creative in perception. Among the members were the educator Bronson Alcott, the abolitionist and Unitarian minister Theodore Parker, and the Unitarian minister Orestes A. Brownson (later a major force in American Catholicism). Such friends were welcome, for during the early 1830s deaths broke up the close-knit band of Emerson brothers. Emerson himself had gone South to recover from tuberculosis in 1826–27; weakened by the same disease, Edward Emerson became mentally deranged in 1827 and died in 1834; and the youngest brother, Charles, died in 1836. There was compensation for Emerson in the circle of admirers who began forming about him at the time of his second marriage and the publication of *Nature*. Alcott, Margaret Fuller, and others sought him out, some paying him frequent and prolonged visits or even settling in Concord to be near him.

Nature reached a smaller audience than did many of his lectures, which were often reported by newspapers in substantial part; his formal Harvard addresses to the Phi Beta Kappa Society in 1837 on the American scholar and to the Divinity School graduates in 1838 on the state of Christianity were both printed as pamphlets, according to the custom of the time. The second of these speeches occasioned a brief, virulent series of attacks in the press for its heresies, giving Emerson a notoriety that barred him from speaking at Harvard for three decades. His unsigned contributions to the Transcendentalists' magazine *The Dial* did not enhance his reputation; indeed, he sometimes was attacked in newspapers as the author of Alcott's *Orphic Sayings*, which jocular contemporaries took as the ultimate of transcendental gibberish. Only with the publication of *Essays* (1841) did Emerson's lasting reputation begin. Far more than *Nature*, this book was directed to a popular audience. The essays had been tried out, in whole or large part, in his lectures, so that their final form was shaped by the responses of many audiences.

By the early 1840s, Emerson's life had settled into its enduring routine. He gave intermittent lectures in Boston and made lecture tours in the Northeast and, later, in the Middle Atlantic states to supplement the income from his legacy. Early in 1842 his first son, Waldo, died at the age of five, the last of the untimely deaths in Emerson's immediate family; after that Emerson devoted himself more and more to the personal problems of his circle of family and friends. His editing of *The Dial* from 1842 till 1844, for instance, was undertaken mainly to support his friends, especially Margaret Fuller. He worked steadily at a succession of essays, usually derived from his extensive journals by way of one or more intervening lectures. *Essays* (1841) was followed by *Essays: Second Series* (1844). The second collection demonstrated even more thoroughly than the first that Emerson's intellect had sharpened in the years since *Nature*. In *The Poet* especially his grappling with aesthetic problems was more incisive; he spoke from practical experience as well as theoretical speculation in defining the present state of literature in America, and he brilliantly foretold the nature of the great national poets to come. In *Experience* and other essays he resolutely and realistically faced the conflict between idealism and ordinary life. The deferential minister and the once-tentative lecturer had become a confident American prophet. Emerson slowly gained recognition for his poems, which he collected at the end of 1846. A second trip to Europe (1847–48) capped his secure middle age. He became something of a country squire, buying up many pieces of property in and around Concord. Always aware that there was a certain coldness in his disposition, he deliberately set out to make himself into a more sociable man, taking part in Boston club life (and smoking cigars to mask his diffidence). As his reputation expanded, he widened his lecture tours into the Midwest. His newer books, among them *Representative Men* (1849) and *Conduct of Life* (1860), were less forceful than his earlier ones, though they sold better because of the enlarged market and his established fame. After long resisting attempts by reformers to gain his support for various social issues, Emerson became a fervent advocate in the 1850s for abolitionism, though his efforts were too late and too local to make him a national leader. The rest of Emerson's writing, like the rest of his life, was a slow anticlimax to the intellectual ferment of the years between the mid-1830s and the mid-1840s, though it was only during these later decades that his earlier work first won general recognition. Emerson's memory began to fail more than ten years before his death, and he declined into a benign senility during which the English-speaking world, and even many who read him in translation, continued to honor the intellectual liberator that he had been in his middle life.

Although Emerson's contemporary reputation rested on his essays, he had all along been writing another masterpiece, his journals, which were not published in full until the 1960s and 1970s under the title *Journals and Miscellaneous Notebooks*. It will take time before readers fully grasp the importance of these writings as the historical record of a response to people and events, the most thorough documentation we possess of the growth of a nineteenth-century American writer, and the remarkable account of a spiritual life. A large number of important journal entries is reprinted here, interspersed with some of Emerson's more memorable letters so as to present a picture of his day-to-day life.

A critic said that Emerson wanted to get his whole philosophy into each essay; more than that, he got as much of it as he could into everything he wrote. Emerson's point of view may shift from one pole of a subject to the other even within a single work, for his mind moved like that of a dramatist who embodies felt or imagined moods in various characters, but the subject remains Emersonian. In this there is challenge for the new reader to find pattern in diversity. And for those who have already cherished Emerson through the various stages of life there is the warmth of familiarity, however unsettling this mild-mannered man always remains to any receptive reader.

From Nature

*"Nature is but an image or imitation of wisdom, the last thing of the
soul; nature being a thing which doth only do, but not know."*

—PLOTINUS

Introduction

Our age is retrospective. It builds the sepulchres of the fathers. It
writes biographies, histories, and criticism. The foregoing generations beheld God and nature face to face; we, through their eyes.
Why should not we also enjoy an original relation to the universe?
Why should not we have a poetry and philosophy of insight and not
of tradition, and a religion by revelation to us, and not the history
of theirs? Embosomed for a season in nature, whose floods of life
stream around and through us, and invite us by the powers they
supply, to action proportioned to nature, why should we grope
among the dry bones of the past, or put the living generation into
masquerade out of its faded wardrobe? The sun shines to-day also.
There is more wool and flax in the fields. There are new lands, new
men, new thoughts. Let us demand our own works and laws and
worship.

Undoubtedly we have no questions to ask which are unanswerable. We must trust the perfection of the creation so far, as to
believe that whatever curiosity the order of things has awakened in
our minds, the order of things can satisfy. Every man's condition is
a solution in hieroglyphic to those inquiries he would put. He acts it
as life, before he apprehends it as truth. In like manner, nature is
already, in its forms and tendencies, describing its own design. Let
us interrogate the great apparition, that shines so peacefully around
us. Let us inquire, to what end is nature?

All science has one aim, namely, to find a theory of nature. We
have theories of races and of functions, but scarcely yet a remote
approximation to an idea of creation. We are now so far from the
road to truth, that religious teachers dispute and hate each other,
and speculative men are esteemed unsound and frivolous. But to a
sound judgment, the most abstract truth is the most practical.
Whenever a true theory appears, it will be its own evidence. Its test

1. Emerson found the motto from the
Roman philosopher Plotinus (205?–270?)
in his copy of Ralph Cudworth's *The
True Intellectual System of the Universe.*
Since its anonymous publication as a
little book (Boston, 1836), the source of
the Introduction and first chapter printed
here, *Nature* has been recognized as a
major document in American Romanticism and Transcendentalism. In *Emerson's "Nature"—Origin, Growth, Mean-*
ing, Merton M. Sealts, Jr., and Alfred
R. Ferguson point out "how wide the
divergence has been over how to read
such a work—whether as doctrine or
mysticism, philosophy or poetry." The
excerpts printed here afford perhaps the
best epitome of Emerson's earnest faith
in the opportunities of the present time
and in the kinship between man's spirit
and external nature.

is, that it will explain all phenomena. Now many are thought not only unexplained but inexplicable; as language, sleep, dreams, beasts, sex.

Philosophically considered, the universe is composed of Nature and the Soul. Strictly speaking, therefore, all that is separate from us, all which Philosophy distinguishes as the NOT ME, that is, both nature and art, all other men and my own body, must be ranked under this name, NATURE. In enumerating the values of nature and casting up their sum, I shall use the word in both senses;—in its common and in its philosophical import. In inquiries so general as our present one, the inaccuracy is not material; no confusion of thought will occur. *Nature*, in the common sense, refers to essences unchanged by man; space, the air, the river, the leaf. *Art* is applied to the mixture of his will with the same things, as in a house, a canal, a statue, a picture. But his operations taken together are so insignificant, a little chipping, baking, patching, and washing, that in an impression so grand as that of the world on the human mind, they do not vary the result.

Chapter I

To go into solitude, a man needs to retire as much from his chamber as from society. I am not solitary whilst I read and write, though nobody is with me. But if a man would be alone, let him look at the stars. The rays that come from those heavenly worlds, will separate between him and vulgar things. One might think the atmosphere was made transparent with this design, to give man, in the heavenly bodies, the perpetual presence of the sublime. Seen in the streets of cities, how great they are! If the stars should appear one night in a thousand years, how would men believe and adore; and preserve for many generations the remembrance of the city of God which had been shown! But every night come out these preachers of beauty, and light the universe with their admonishing smile.

The stars awaken a certain reverence, because though always present, they are always inaccessible; but all natural objects make a kindred impression, when the mind is open to their influence. Nature never wears a mean appearance. Neither does the wisest man extort all her secret, and lose his curiosity by finding out all her perfection. Nature never became a toy to a wise spirit. The flowers, the animals, the mountains, reflected all the wisdom of his best hour, as much as they had delighted the simplicity of his childhood.

When we speak of nature in this manner, we have a distinct but most poetical sense in the mind. We mean the integrity of impression made by manifold natural objects. It is this which distinguishes the stick of timber of the wood-cutter, from the tree of the poet. The charming landscape which I saw this morning, is indubitably made up of some twenty or thirty farms. Miller owns this field, Locke that, and Manning the woodland beyond. But none of them owns the landscape. There is a property in the horizon which no man has but he whose eye can integrate all the parts, that is, the poet. This is the best part of these men's farms, yet to this their land-deeds give them no title.

To speak truly, few adult persons can see nature. Most persons do not see the sun. At least they have a very superficial seeing. The sun illuminates only the eye of the man, but shines into the eye and the heart of the child. The lover of nature is he whose inward and outward senses are still truly adjusted to each other; who has re-tained the spirit of infancy even into the era of manhood. His intercourse with heaven and earth, becomes part of his daily food. In the presence of nature, a wild delight runs through the man, in spite of real sorrows. Nature says,—he is my creature, and maugre[2] all his impertinent griefs, he shall be glad with me. Not the sun or the summer alone, but every hour and season yields its tribute of delight; for every hour and change corresponds to and authorizes a different state of the mind, from breathless noon to grimmest mid-night. Nature is a setting that fits equally well a comic or a mourn-ing piece. In good health, the air is a cordial of incredible virtue. Crossing a bare common, in snow puddles, at twilight, under a clouded sky, without having in my thoughts any occurrence of special good fortune, I have enjoyed a perfect exhilaration. Almost I fear to think how glad I am. In the woods too, a man casts off his years, as the snake his slough, and at what period soever of life, is always a child. In the woods, is perpetual youth. Within these plantations of God, a decorum and sanctity reign, a perennial festi-val is dressed, and the guest sees not how he should tire of them in a thousand years. In the woods, we return to reason and faith. There I feel that nothing can befal me in life,—no disgrace, no calamity, (leaving me my eyes,) which nature cannot repair. Standing on the bare ground,—my head bathed by the blithe air, and uplifted into infinite space,—all mean egotism vanishes. I become a transparent eye-ball.[3] I am nothing. I see all. The currents of the Universal Being circulate through me; I am part or particle of God. The name of the nearest friend sounds then foreign and accidental. To be brothers, to be acquaintances,—master or servant, is then a trifle and a disturbance. I am the lover of uncontained and immortal beauty. In the wilderness, I find something more dear and connate[4] than in streets or villages. In the tranquil landscape, and especially

2. In spite of.
3. This remarkable image, satirized in Emerson's time and learnedly explicated in ours, characterizes the paradoxical state of being in which one is merged into nature (or "Universal Being") even while retaining a unique perception of the experience.
4. Related.

in the distant line of the horizon, man beholds somewhat as beautiful as his own nature.

The greatest delight which the fields and woods minister, is the suggestion of an occult relation between man and the vegetable. I am not alone and unacknowledged. They nod to me and I to them. The waving of the boughs in the storm, is new to me and old. It takes me by surprise, and yet is not unknown. Its effect is like that of a higher thought or a better emotion coming over me, when I deemed I was thinking justly or doing right.

Yet it is certain that the power to produce this delight, does not reside in nature, but in man, or in a harmony of both. It is necessary to use these pleasures with great temperance. For, nature is not always tricked in holiday attire, but the same scene which yesterday breathed perfume and glittered as for the frolic of the nymphs, is overspread with melancholy today. Nature always wears the colors of the spirit. To a man laboring under calamity, the heat of his own fire hath sadness in it. Then, there is a kind of contempt of the landscape felt by him who has just lost by death a dear friend. The sky is less grand as it shuts down over less worth in the population.

1836

The American Scholar[1]

Mr. President, and Gentlemen,

I greet you on the re-commencement of our literary year. Our anniversary is one of hope, and, perhaps, not enough of labor. We do not meet for games of strength or skill, for the recitation of histories, tragedies and odes, like the ancient Greeks; for parliaments of love and poesy, like the Troubadours;[2] nor for the advancement of science, like our cotemporaries in the British and European capitals. Thus far, our holiday has been simply a friendly sign of the survival of the love of letters amongst a people too busy to give to letters any more. As such, it is precious as the sign of an indestructible instinct. Perhaps the time is already come, when it ought to be, and will be something else; when the sluggard intellect of this continent will look from under its iron lids and fill the postponed expectation of the world with something better than the exertions of mechanical skill. Our day of dependence, our long apprenticeship to the learning of other lands, draws to a close. The millions that around us are rushing into life, cannot always be fed on the sere remains of foreign harvests. Events, actions arise, that must be sung, that will sing themselves. Who can doubt that poetry will revive and lead in a new age, as the star in the constellation

1. The text printed here is that of the first publication (Boston, 1837) as a pamphlet entitled *An Oration, Delivered before the Phi Beta Kappa Society, at Cambridge, August 31, 1837.* By altering the title to *The American Scholar* when he republished it in *Essays* (1841), Emerson expanded the application to all American college students and all others who dedicate themselves to thought.
2. Courtly poets of southern France, especially Provence, in the 12th and 13th centuries.

Harp which now flames in our zenith, astronomers announce, shall one day be the pole-star for a thousand years.

In the light of this hope, I accept the topic which not only usage, but the nature of our association, seem to prescribe to this day,— the AMERICAN SCHOLAR. Year by year, we come up hither to read one more chapter of his biography. Let us inquire what new lights, new events and more days have thrown on his character, his duties and his hopes.

It is one of those fables, which out of an unknown antiquity, convey an unlooked for wisdom, that the gods, in the beginning, divided Man into men, that he might be more helpful to himself;[3] just as the hand was divided into fingers, the better to answer its end.

The old fable covers a doctrine ever new and sublime; that there is One Man,—present to all particular men only partially, or through one faculty; and that you must take the whole society to find the whole man. Man is not a farmer, or a professor, or an engineer, but he is all. Man is priest, and scholar, and statesman, and producer, and soldier. In the *divided* or social state, these functions are parcelled out to individuals, each of whom aims to do his stint of the joint work, whilst each other performs his. The fable implies that the individual to possess himself, must sometimes return from his own labor to embrace all the other laborers. But unfortunately, this original unit, this fountain of power, has been so distributed to multitudes, has been so minutely subdivided and peddled out, that it is spilled into drops, and cannot be gathered. The state of society is one in which the members have suffered amputation from the trunk, and strut about so many walking monsters,—a good finger, a neck, a stomach, an elbow, but never a man.

Man is thus metamorphosed into a thing, into many things. The planter, who is Man sent out into the field to gather food, is seldom cheered by any idea of the true dignity of his ministry. He sees his bushel and his cart, and nothing beyond, and sinks into the farmer, instead of Man on the farm. The tradesman scarcely ever gives an ideal worth to his work, but is ridden by the routine of his craft, and the soul is subject to dollars. The priest becomes a form; the attorney, a statute-book; the mechanic, a machine; the sailor, a rope of a ship.

In this distribution of functions, the scholar is the delegated intellect. In the right state, he is, *Man Thinking*. In the degenerate state, when the victim of society, he tends to become a mere thinker, or, still worse, the parrot of other men's thinking.

In this view of him, as Man Thinking, the whole theory of his office[4] is contained. Him nature solicits, with all her placid, all her

3. One such fable Emerson knew from Plato's *Symposium*.
4. Function.

monitory pictures. Him the past instructs. Him the future invites. Is not, indeed, every man a student, and do not all things exist for the student's behoof? And, finally, is not the true scholar the only true master? But, as the old oracle said, "All things have two handles. Beware of the wrong one." In life, too often, the scholar errs with mankind and forfeits his privilege. Let us see him in his school, and consider him in reference to the main influences he receives.

I. The first time and the first in importance of the influences upon the mind is that of nature. Every day, the sun; and, after sunset, night and her stars. Ever the winds blow; ever the grass grows. Every day, men and women, conversing, beholding and beholden. The scholar must needs stand wistful and admiring before this great spectacle. He must settle its value in his mind. What is nature to him? There is never a beginning, there is never an end to the inexplicable continuity of this web of God, but always circular power returning into itself. Therein it resembles his own spirit, whose beginning, whose ending he never can find—so entire, so boundless. Far, too, as her splendors shine, system on system shooting like rays, upward, downward, without centre, without circumference,—in the mass and in the particle nature hastens to render account of herself to the mind. Classification begins. To the young mind, every thing is individual, stands by itself. By and by, it finds how to join two things, and see in them one nature; then three, then three thousand; and so, tyrannized over by its own unifying instinct, it goes on tying things together, diminishing anomalies, discovering roots running under ground, whereby contrary and remote things cohere, and flower out from one stem. It presently learns, that, since the dawn of history, there has been a constant accumulation and classifying of facts. But what is classification but the perceiving that these objects are not chaotic, and are not foreign, but have a law which is also a law of the human mind? The astronomer discovers that geometry, a pure abstraction of the human mind, is the measure of planetary motion. The chemist finds proportions and intelligible method throughout matter: and science is nothing but the finding of analogy, identity in the most remote parts. The ambitious soul sits down before each refractory fact; one after another, reduces all strange constitutions, all new powers, to their class and their law, and goes on forever to animate the last fibre of organization, the outskirts of nature, by insight.

Thus to him, to this school-boy under the bending dome of day, is suggested, that he and it proceed from one root; one is leaf and one is flower; relation, sympathy, stirring in every vein. And what is that Root? Is not that the soul of his soul?—A thought too bold—a dream *too wild*. Yet when this spiritual light shall have revealed the law of more earthly natures,—when he has learned to worship the soul, and to see that the natural philosophy that now is, is only the first gropings of its gigantic hand, he shall look forward to an ever expanding knowledge as to a becoming creator. He shall see that nature is the opposite of the soul, answering to it part for part. One is seal, and one is print. Its beauty is the beauty of his own mind. Its laws are the laws of his own mind. Nature then becomes to him the measure of his attainments. So much of nature as he is ignorant of, so much of his own mind does he not yet possess. And, in fine, the ancient precept, "Know thyself," and the modern precept, "Study nature," become at last one maxim.

II. The next great influence[5] into the spirit of the scholar, is, the mind of the Past,—in whatever form, whether of literature, of art, of institutions, that mind is inscribed. Books are the best type of the influence of the past, and perhaps we shall get at the truth—learn the amount of this influence more conveniently—by considering their value alone.

The theory of books is noble. The scholar of the first age received into him the world around; brooded thereon; gave it the new arrangement of his own mind, and uttered it again. It came into him—life; it went out from him—truth. It came to him—short-lived actions; it went out from him—immortal thoughts. It came to him —business; it went from him—poetry. It was—dead fact; now, it is quick[6] thought. It can stand, and it can go. It now endures, it now flies, it now inspires.[7] Precisely in proportion to the depth of mind from which it issued, so high does it soar, so long does it sing.

Or, I might say, it depends on how far the process had gone, of transmuting life into truth. In proportion to the completeness of the distillation, so will the purity and imperishableness of the product be. But none is quite perfect. As no air-pump can by any means make a perfect vacuum, so neither can any artist entirely exclude the conventional, the local, the perishable from his book, or write a book of pure thought that shall be as efficient, in all respects, to a remote posterity, as to cotemporaries, or rather to the second age. Each age, it is found, must write its own books; or rather, each generation for the next succeeding. The books of an older period will not fit this.

Yet hence arises a grave mischief. The sacredness which attaches to the act of creation,—the act of thought,—is instantly transferred to the record. The poet chanting, was felt to be a divine man. Henceforth the chant is divine also. The writer was a just and wise spirit. Henceforward it is settled, the book is perfect; as love of the hero corrupts into worship of his statue. Instantly, the book becomes noxious. The guide is a tyrant. We sought a brother, and lo, a governor. The sluggish and perverted mind of the multitude, always slow to open to the incursions of Reason, having once so opened, having once received this book, stands upon it, and makes an outcry, if it is disparaged. Colleges are built on it. Books are

5. Inflowing.
6. "Business": busyness, activity; "quick":
living.
7. Breathes in; "go": walk.

written on it by thinkers, not by Man Thinking; by men of talent, that is, who start wrong, who set out from accepted dogmas, not from their own sight of principles. Meek young men grow up in libraries, believing it their duty to accept the views which Cicero, which Locke, which Bacon[8] have given, forgetful that Cicero, Locke and Bacon were only young men in libraries when they wrote these books.

Hence, instead of Man Thinking, we have the bookworm. Hence, the book-learned class, who value books, as such; not as related to nature and the human constitution, but as making a sort of Third Estate[9] with the world and the soul. Hence, the restorers of readings, the emendators, the bibliomaniacs of all degrees.

This is bad; this is worse than it seems. Books are the best of things, well used; abused, among the worst. What is the right use? What is the one end which all means go to effect? They are for nothing but to inspire. I had better never see a book than to be warped by its attraction clean out of my own orbit, and made a satellite instead of a system. The one thing in the world of value, is, the active soul,—the soul, free, sovereign, active. This every man is entitled to; this every man contains within him, although in almost all men, obstructed, and as yet unborn. The soul active sees absolute truth; and utters truth, or creates. In this action, it is genius; not the privilege of here and there a favorite, but the sound estate of every man. In its essence, it is progressive. The book, the college, the school of art, the institution of any kind, stop with some past utterance of genius. This is good, say they,—let us hold by this. They pin me down. They look backward and not forward. But genius always looks forward. The eyes of man are set in his forehead, not in his hindhead. Man hopes. Genius creates. To create,—to create,—is the proof of a divine presence. Whatever talents may be, if the man create not, the pure efflux[1] of the Deity is not his:— cinders and smoke, there may be, but not yet flame. There are creative manners, there are creative actions, and creative words; manners, actions, words, that is, indicative of no custom or authority, but springing spontaneous from the mind's own sense of good and fair.

On the other part, instead of being its own seer, let it receive always from another mind its truth, though it were in torrents of light, without periods of solitude, inquest and self-recovery, and a fatal disservice is done. Genius is always sufficiently the enemy of

8. These examples are not especially apt, since none of the three wrote books at an unusually precocious age. As a young man Marcus Tullius Cicero (106–43 B.C.), Roman statesman, was best known for his oratory; John Locke (1632–1704), English philosopher and political thinker, wrote *Essay Concerning Human Understanding* (1690); before he was

40. Sir Francis Bacon (1561–1626), English statesman and philosopher, wrote his *Essays*.
9. On the analogy of the three-part division of estates of the realm, in which the third estate is the common people while the first estate is the nobility and the second the clergy.
1. Flowing forth.

genius by over-influence. The literature of every nation bear me witness. The English dramatic poets have Shakspearized now for two hundred years.

.. Undoubtedly there is a right way of reading,—so it be sternly subordinated. Man Thinking must not be subdued by his instruments. Books are for the scholar's idle times. When he can read God directly, the hour is too precious to be wasted in other mens' transcripts of their readings. But when the intervals of darkness come, as come they must,—when the soul seeth not, when the sun is hid, and the stars withdraw their shining,—we repair to the lamps which were kindled by their ray to guide our steps to the East again, where the dawn is. We hear that we may speak. The Arabian proverb says, "A fig tree looking on a fig tree, becometh fruitful."

It is remarkable, the character of the pleasure we derive from the best books. They impress us ever with the conviction that one nature wrote and the same reads. We read the verses of one of the great English poets, of Chaucer, of Marvell, of Dryden, with the most modern joy,—with a pleasure, I mean, which is in great part caused by the abstraction of all *time* from their verses. There is some awe mixed with the joy of our surprise, when this poet, who lived in some past world, two or three hundred years ago, says that which lies close to my own soul, that which I also had well nigh thought and said. But for the evidence thence afforded to the philosophical doctrine of the identity of all minds, we should suppose some pre-established harmony, some foresight of souls that were to be, and some preparation of stores for their future wants, like the fact observed in insects, who lay up food before death for the young grub they shall never see.

I would not be hurried by any love of system, by any exaggeration of instincts, to underrate the Book. We all know, that as the human body can be nourished on any food, though it were boiled grass and the broth of shoes, so the human mind can be fed by any knowledge. And great and heroic men have existed, who had almost no other information than by the printed page. I only would say, that it needs a strong head to bear that diet. One must be an inventor to read well. As the proverb says, "He that would bring home the wealth of the Indies, must carry out the wealth of the Indies." There is then creative reading, as well as creative writing. When the mind is braced by labor and invention, the page of whatever book we read becomes luminous with manifold allusion. Every sentence is doubly significant, and the sense of our author is as broad as the world. We then see, what is always true, that as the seer's hour of vision is short and rare among heavy days and months, so is its record, perchance, the least part of his volume. The discerning will read in his Plato or Shakspeare, only that least part,—only the authentic utterances of the oracle,—and all the rest he rejects, were it never so many times Plato's and Shakspeare's.

Of course, there is a portion of reading quite indispensable to a
wise man. History and exact science he must learn by laborious
reading. Colleges, in like manner, have their indispensable office,—
to teach elements. But they can only highly serve us, when they aim
not to drill, but to create; when they gather from far every ray of
various genius to their hospitable halls, and, by the concentrated
fires, set the hearts of their youth on flame. Thought and knowledge
are natures in which apparatus and pretension avail nothing.
Gowns, and pecuniary foundations, though of towns of gold, can
never countervail the least sentence or syllable of wit. Forget this,
and our American colleges will recede in their public importance
whilst they grow richer every year.

III. There goes in the world a notion that the scholar should be a
recluse, a valetudinarian,—as unfit for any handiwork or public
labor, as a penknife for an axe. The so called "practical men" sneer
at speculative men, as if, because they speculate or *see*, they could
do nothing. I have heard it said that the clergy,—who are always
more universally than any other class, the scholars of their day,—
are addressed as women: that the rough, spontaneous conversation
of men they do not hear, but only a mincing and diluted speech.
They are often virtually disfranchised; and, indeed, there are advocates for their celibacy. As far as this is true of the studious classes,
it is not just and wise. Action is with the scholar subordinate, but it
is essential. Without it, he is not yet man. Without it, thought can
never ripen into truth. Whilst the world hangs before the eye as a
cloud of beauty, we can not even see its beauty. Inaction is cowardice, but there can be no scholar without the heroic mind. The
preamble of thought, the transition through which it passes from
the unconscious to the conscious, is action. Only so much do I
know, as I have lived. Instantly we know whose words are loaded
with life, and whose not.

The world,—this shadow of the soul, or *other me*, lies wide
around. Its attractions are the keys which unlock my thoughts and
make me acquainted with myself. I launch eagerly into this resounding tumult. I grasp the hands of those next me, and take my place in
the ring to suffer and to work, taught by an instinct that so shall the
dumb abyss be vocal with speech. I pierce its order; I dissipate its
fear; I dispose of it within the circuit of my expanding life. So much
only of life as I know by experience, so much of the wilderness have
I vanquished and planted, or so far have I extended my being, my
dominion. I do not see how any man can afford, for the sake of his
nerves and his nap, to spare any action in which he can partake. It
is pearls and rubies to his discourse. Drudgery, calamity, exasperation, want, are instructers in eloquence and wisdom. The true
scholar grudges every opportunity of action past by, as a loss of
power.

It is the raw material out of which the intellect moulds her splendid products. A strange process too, this, by which experience is converted into thought, as a mulberry leaf is converted into satin. The manufacture goes forward at all hours.

The actions and events of our childhood and youth are now matters of calmest observation. They lie like fair pictures in the air. Not so with our recent actions,—with the business which we now have in hand. On this we are quite unable to speculate. Our affections as yet circulate through it. We no more feel or know it, than we feel the feet, or the hand, or the brain of our body. The new deed is yet a part of life,—remains for a time immersed in our unconscious life. In some contemplative hour, it detaches itself from the life like a ripe fruit, to become a thought of the mind. Instantly, it is raised, transfigured; the corruptible has put on incorruption.[2] Always now it is an object of beauty, however base its origin and neighborhood. Observe, too, the impossibility of antedating this act. In its grub state, it cannot fly, it cannot shine,—it is a dull grub. But suddenly, without observation, the selfsame thing unfurls beautiful wings, and is an angel of wisdom. So is there no fact, no event, in our private history, which shall not, sooner or later, lose its adhesive inert form, and astonish us by soaring from our body into the empyrean.[3] Cradle and infancy, school and playground, the fear of boys, and dogs, and ferules,[4] the love of little maids and berries, and many another fact that once filled the whole sky, are gone already; friend and relative, profession and party, town and country, nation and world, must also soar and sing.

Of course, he who has put forth his total strength in fit actions, has the richest return of wisdom. I will not shut myself out of this globe of action and transplant an oak into a flower pot, there to hunger and pine; nor trust the revenue of some single faculty, and exhaust one vein of thought, much like those Savoyards,[5] who, getting their livelihood by carving shepherds, shepherdesses, and smoking Dutchmen, for all Europe, went out one day to the mountain to find stock, and discovered that they had whittled up the last of their pine trees. Authors we have in numbers,[6] who have written out their vein, and who, moved by a commendable prudence, sail for Greece or Palestine, follow the trapper into the prairie, or ramble round Algiers to replenish their merchantable stock.

If it were only for a vocabulary the scholar would be covetous of action. Life is our dictionary. Years are well spent in country labors; in town—in the insight into trades and manufactures; in frank intercourse with many men and women; in science; in art; to the one end of mastering in all their facts a language, by which to

2. "For this corruptible must put on incorruption, and this mortal must put on immortality" (1 Corinthians 15.53).
3. The highest reaches of heaven.
4. Rods used for punishing children.
5. Savoy is in the western Alps, where France, Italy, and Switzerland converge.

6. Emerson's contemporaries would have understood a reference to writers now unread, such as Nathaniel Parker Willis, as well as to two still-famous writers, James Fenimore Cooper, author of *The Prairie* (1827), and Washington Irving, author of *A Tour on the Prairies* (1835).

278 ★ Ralph Waldo Emerson

illustrate and embody our perceptions. I learn immediately from any speaker how much he has already lived, through the poverty or the splendor of his speech. Life lies behind us as the quarry from whence we get tiles and copestones for the masonry of to-day. This is the way to learn grammar. Colleges and books only copy the language which the field and the workyard made.

But the final value of action, like that of books, and better than books, is, that it is a resource. That great principle of Undulation in nature, that shows itself in the inspiring and expiring of the breath; in desire and satiety; in the ebb and flow of the sea, in day and night, in heat and cold, and as yet more deeply ingrained in every atom and every fluid, is known to us under the name of Polarity,— these "fits of easy transmission and reflection," as Newton called them,[7] are the law of nature because they are the law of spirit.

The mind now thinks; now acts; and each fit reproduces the other. When the artist has exhausted his materials, when the fancy no longer paints, when thoughts are no longer apprehended, and books are a weariness,—he has always the resource *to live*. Character is higher than intellect. Thinking is the function. Living is the functionary. The stream retreats to its source. A great soul will be strong to live, as well as strong to think. Does he lack organ or medium to impart his truths? He can still fall back on this elemental force of living them. This is a total act. Thinking is a partial act. Let the grandeur of justice shine in his affairs. Let the beauty of affection cheer his lowly roof. Those "far from fame" who dwell and act with him, will feel the force of his constitution in the doings and passages of the day better than it can be measured by any public and designed display. Time shall teach him that the scholar loses no hour which the man lives. Herein he unfolds the sacred germ of his instinct screened from influence. What is lost in seemliness is gained in strength. Not out of those on whom systems of education have exhausted their culture, comes the helpful giant to destroy the old or to build the new, but out of unhandselled[8] savage nature, out of terrible Druids and Berserkirs, come at last Alfred and Shakspear.[9]

I hear therefore with joy whatever is beginning to be said of the dignity and necessity of labor to every citizen. There is virtue yet in the hoe and the spade, for learned as well as for unlearned hands. And labor is every where welcome; always we are invited to work; only be this limitation observed, that a man shall not for the sake of wider activity sacrifice any opinion to the popular judgments and modes of action.

I have now spoken of the education of the scholar by nature, by books, and by action. It remains to say somewhat of his duties.

7. From the *Optics* (1704) of Sir Isaac Newton (1642–1727), English scientist and mathematician.
8. A handsel is a gift to express good wishes at the outset of some enterprise; apparently Emerson uses the word to mean something like unauspicious.
9. Uncivilized Celts and Anglo-Saxons; Alfred was the enlightened ninth-century King of the West Saxons.

They are such as become Man Thinking. They may all be comprised in self-trust. The office of the scholar is to cheer, to raise, and to guide men by showing them facts amidst appearances. He plies the slow, unhonored, and unpaid task of observation. Flamsteed and Herschel,[1] in their glazed observatory, may catalogue the stars with the praise of all men, and, the results being splendid and useful, honor is sure. But he, in his private observatory, cataloguing obscure and nebulous stars of the human mind, which as yet no man has thought of as such,—watching days and months, sometimes, for a few facts; correcting still his old records;—must relinquish display and immediate fame. In the long period of his preparation, he must betray often an ignorance and shiftlessness in popular arts, incurring the disdain of the able who shoulder him aside. Long he must stammer in his speech; often forego the living for the dead. Worse yet, he must accept—how often! poverty and solitude. For the ease and pleasure of treading the old road, accepting the fashions, the education, the religion of society, he takes the cross of making his own, and, of course, the self accusation, the faint heart, the frequent uncertainty and loss of time which are the nettles and tangling vines in the way of the self-relying and self-directed; and the state of virtual hostility in which he seems to stand to society, and especially to educated society. For all this loss and scorn, what offset? He is to find consolation in exercising the highest functions of human nature. He is one who raises himself from private considerations, and breathes and lives on public and illustrious thoughts. He is the world's eye. He is the world's heart. He is to resist the vulgar prosperity that retrogrades ever to barbarism, by preserving and communicating heroic sentiments, noble biographies, melodious verse, and the conclusions of history. Whatsoever oracles the human heart in all emergencies, in all solemn hours has uttered as its commentary on the world of actions,—these he shall receive and impart. And whatsoever new verdict Reason from her inviolable seat pronounces on the passing men and events of to-day,—this he shall hear and promulgate.

These being his functions, it becomes him to feel all confidence in himself, and to defer never to the popular cry. He and he only knows the world. The world of any moment is the merest appearance. Some great decorum, some fetish of a government, some ephemeral trade, or war, or man, is cried up by half mankind and cried down by the other half, as if all depended on this particular up or down. The odds are that the whole question is not worth the poorest thought which the scholar has lost in listening to the controversy. Let him not quit his belief that a popgun is a popgun, though the ancient and honorable of the earth affirm it to be the crack of doom. In silence, in steadiness, in severe abstraction, let

1. John Flamsteed (1646–1719), English astronomer, first royal astronomer at Greenwich observatory; Sir William Her- schel (1738–1822), German-born English astronomer, founder of sidereal astronomy. "Glazed": glass-roofed.

him hold by himself; add observation to observation; patient of neglect, patient of reproach, and bide his own time,—happy enough if he can satisfy himself alone that this day he has seen something truly. Success treads on every right step. For the instinct is sure that prompts him to tell his brother what he thinks. He then learns that in going down into the secrets of his own mind, he has descended into the secrets of all minds. He learns that he who has mastered any law in his private thoughts, is master to that extent of all men whose language he speaks, and of all into whose language his own can be translated. The poet in utter solitude remembering his spontaneous thoughts and recording them, is found to have recorded that which men in "cities vast" find true for them also. The orator distrusts at first the fitness of his frank confessions,—his want of knowledge of the persons he addresses,—until he finds that he is the complement of his hearers;—that they drink his words because he fulfils for them their own nature; the deeper he dives into his privatest secretest presentiment,—to his wonder he finds, this is the most acceptable, most public, and universally true. The people delight in it; the better part of every man feels, This is my music: this is myself.

In self-trust, all the virtues are comprehended. Free should the scholar be,—free and brave. Free even to the definition of freedom, "without any hindrance that does not arise out of his own constitution." Brave; for fear is a thing which a scholar by his very function puts behind him. Fear always springs from ignorance. It is a shame to him if his tranquillity, amid dangerous times, arise from the presumption that like children and women, his is a protected class; or if he seek a temporary peace by the diversion of his thoughts from politics or vexed questions, hiding his head like an ostrich in the flowering bushes, peeping into microscopes, and turning rhymes, as a boy whistles to keep his courage up. So is the danger a danger still: so is the fear worse. Manlike let him turn and face it. Let him look into its eye and search its nature, inspect its origin—see the whelping of this lion,—which lies no great way back; he will then find in himself a perfect comprehension of its nature and extent; he will have made his hands meet on the other side, and can henceforth defy it, and pass on superior. The world is his who can see through its pretension. What deafness, what stone-blind custom, what overgrown error you behold, is there only by sufferance,—by your sufferance. See it to be a lie, and you have already dealt it its mortal blow.

Yes, we are the cowed,—we the trustless. It is a mischievous notion that we are come late into nature; that the world was finished a long time ago. As the world was plastic and fluid in the hands of God, so it is ever to so much of his attributes as we bring to it. To ignorance and sin, it is flint. They adapt themselves to it as they may; but in proportion as a man has anything in him divine, the firmament flows before him, and takes his signet[2] and form. Not he is great who can alter matter, but he who can alter my state of mind. They are the kings of the world who give the color of their present thought to all nature and all art, and persuade men by the cheerful serenity of their carrying the matter, that this thing which they do, is the apple which the ages have desired to pluck, now at last ripe, and inviting nations to the harvest. The great man makes the great thing. Wherever Macdonald sits, there is the head of the table.[3] Linnæus makes botany the most alluring of studies and wins it from the farmer and the herb-woman. Davy, chemistry: and Cuvier, fossils.[4] The day is always his, who works in it with serenity and great aims. The unstable estimates of men crowd to him whose mind is filled with a truth, as the heaped waves of the Atlantic follow the moon.

For this self-trust, the reason is deeper than can be fathomed,— darker than can be enlightened. I might not carry with me the feeling of my audience in stating my own belief. But I have already shown the ground of my hope, in adverting to the doctrine that man is one. I believe man has been wronged: he has wronged himself. He has almost lost the light that can lead him back to his prerogatives. Men are become of no account. Men in history, men in the world of to-day are bugs, are spawn, and are called "the mass" and "the herd." In a century, in a millenium, one or two men; that is to say —one or two approximations to the right state of every man. All the rest behold in the hero or the poet their own green and crude being —ripened; yes, and are content to be less, so *that* may attain to its full stature. What a testimony—full of grandeur, full of pity, is borne to the demands of his own nature, by the poor clansman, the poor partisan, who rejoices in the glory of his chief. The poor and the low find some amends to their immense moral capacity, for their acquiescence in a political and social inferiority. They are content to be brushed like flies from the path of a great person, so that justice shall be done by him to that common nature which it is the dearest desire of all to see enlarged and glorified. They sun themselves in the great man's light, and feel it to be their own element. They cast the dignity of man from their downtrod selves upon the shoulders of a hero, and will perish to add one drop of blood to make that great heart beat, those giant sinews combat and conquer. He lives for us, and we live in him.

Men such as they are, very naturally seek money or power; and power because it is as good as money,—the "spoils," so called, "of office." And why not? for they aspire to the highest, and this, in their sleep-walking, they dream is highest. Wake them, and they

2. Seal.
3. An old proverb says, "Where Macgregor sits, there is the head of the table"; Emerson substitutes another type name for a Scottish chief.

4. Carolus Linnaeus (1707–78), Swedish botanist; Sir Humphry Davy (1778–1829), English chemist; Georges Cuvier (1769–1832), French pioneer in comparative anatomy and paleontology.

shall quit the false good and leap to the true, and leave government to clerks and desks. This revolution is to be wrought by the gradual domestication of the idea of Culture. The main enterprise of the world for splendor, for extent, is the upbuilding of a man. Here are the materials strown along the ground. The private life of one man shall be a more illustrious monarchy,—more formidable to its enemy, more sweet and serene in its influence to its friend, than any kingdom in history. For a man, rightly viewed, comprehendeth the particular natures of all men. Each philosopher, each bard, each actor, has only done for me, as by a delegate, what one day I can do for myself. The books which once we valued more than the apple of the eye, we have quite exhausted. What is that but saying that we have come up with the point of view which the universal mind took through the eyes of that one scribe; we have been that man, and have passed on. First, one; then, another; we drain all cisterns, and waxing greater by all these supplies, we crave a better and more abundant food. The man has never lived that can feed us ever. The human mind cannot be enshrined in a person who shall set a barrier on any one side to this unbounded, unboundable empire. It is one central fire which flaming now out of the lips of Etna, lightens the capes of Sicily; and now out of the throat of Vesuvius, illuminates the towers and vineyards of Naples.[5] It is one light which beams out of a thousand stars. It is one soul which animates all men.

But I have dwelt perhaps tediously upon this abstraction of the Scholar. I ought not to delay longer to add what I have to say, of nearer reference to the time and to this country.

Historically, there is thought to be a difference in the ideas which predominate over successive epochs, and there are data for marking the genius of the Classic, of the Romantic, and now of the Reflective or Philosophical age.[6] With the views I have intimated of the oneness or the identity of the mind through all individuals, I do not much dwell on these differences. In fact, I believe each individual passes through all three. The boy is a Greek; the youth, romantic; the adult, reflective. I deny not, however, that a revolution in the leading idea may be distinctly enough traced.

Our age is bewailed as the age of Introversion. Must that needs be evil? We, it seems, are critical. We are embarrassed with second thoughts. We cannot enjoy any thing for hankering to know whereof the pleasure consists. We are lined with eyes. We see with our feet. The time is infected with Hamlet's unhappiness,—

"Sicklied o'er with the pale cast of thought."[7]

Is it so bad then? Sight is the last thing to be pitied. Would we be

---

5. Active volcanoes in eastern Sicily and western Italy.
6. Emerson proceeds to refute the self-excusing notion that his age was merely a time for criticism, not for genuinely creative achievements.
7. *Hamlet*, 3.1.85–87.

blind? Do we fear lest we should outsee nature and God, and drink truth dry? I look upon the discontent of the literary class as a mere announcement of the fact that they find themselves not in the state of mind of their fathers, and regret the coming state as untried; as a boy dreads the water before he has learned that he can swim. If there is any period one would desire to be born in,—is it not the age of Revolution; when the old and the new stand side by side, and admit of being compared; when the energies of all men are searched by fear and by hope; when the historic glories of the old, can be compensated by the rich possibilities of the new era? This time, like all times, is a very good one, if we but know what to do with it.

I read with joy some of the auspicious signs of the coming days as they glimmer already through poetry and art, through philosophy and science, through church and state.

One of these signs is the fact that the same movement which effected the elevation of what was called the lowest class in the state, assumed in literature a very marked and as benign an aspect. Instead of the sublime and beautiful, the near, the low, the common, was explored and poetised. That which had been negligently trodden under foot by those who were harnessing and provisioning themselves for long journies into far countries, is suddenly found to be richer than all foreign parts. The literature of the poor, the feelings of the child, the philosophy of the street, the meaning of household life, are the topics of the time. It is a great stride. It is a sign—is it not? of new vigor, when the extremities are made active, when currents of warm life run into the hands and the feet. I ask not for the great, the remote, the romantic; what is doing in Italy or Arabia; what is Greek art, or Provencal Minstrelsy; I embrace the common, I explore and sit at the feet of the familiar, the low. Give me insight into to-day, and you may have the antique and future worlds. What would we really know the meaning of? The meal in the firkin; the milk in the pan; the ballad in the street; the news of the boat; the glance of the eye; the form and the gait of the body;— show me the ultimate reason of these matters;—show me the sublime presence of the highest spiritual cause lurking, as always it does lurk, in these suburbs and extremities of nature; let me see every trifle bristling with the polarity that ranges it instantly on an eternal law; and the shop, the plough, and the ledger, referred to the like cause by which light undulates and poets sing;—and the world lies no longer a dull miscellany and lumber room,[8] but has form and order; there is no trifle; there is no puzzle; but one design unites and animates the farthest pinnacle and the lowest trench.

This idea has inspired the genius of Goldsmith, Burns, Cowper, and, in a newer time, of Goethe, Wordsworth, and Carlyle. This idea they have differently followed and with various success. In

8. Junk room.

contrast with their writing, the style of Pope, of Johnson, of Gibbon, looks cold and pedantic.[9] This writing is blood-warm. Man is surprised to find that things near are not less beautiful and wondrous than things remote. The near explains the far. The drop is a small ocean. A man is related to all nature. This perception of the worth of the vulgar, is fruitful in discoveries. Goethe, in this very thing the most modern of the moderns, has shown us, as none ever did, the genius of the ancients.

There is one man of genius who has done much for this philosophy of life, whose literary value has never yet been rightly estimated; —I mean Emanuel Swedenborg.[1] The most imaginative of men, yet writing with the precision of a mathematician, he endeavored to engraft a purely philosophical Ethics on the popular Christianity of his time. Such an attempt, of course, must have difficulty which no genius could surmount. But he saw and showed the connexion between nature and the affections of the soul. He pierced the emblematic or spiritual character of the visible, audible, tangible world. Especially did his shade-loving muse hover over and interpret the lower parts of nature; he showed the mysterious bond that allies moral evil to the foul material forms, and has given in epical parables a theory of insanity, of beasts, of unclean and fearful things.

Another sign of our times, also marked by an analogous political movement is, the new importance given to the single person. Every thing that tends to insulate the individual,—to surround him with barriers of natural respect, so that each man shall feel the world is his, and man shall treat with man as a sovereign state with a sovereign state;—tends to true union as well as greatness. "I learned," said the melancholy Pestalozzi,[2] "that no man in God's wide earth is either willing or able to help any other man." Help must come from the bosom alone. The scholar is that man who must take up into himself all the ability of the time, all the contributions of the past, all the hopes of the future. He must be an university of knowledges. If there be one lesson more than another which should pierce his ear, it is, The world is nothing, the man is all; in yourself is the law of all nature, and you know not yet how a globule of sap ascends; in yourself slumbers the whole of Reason; it is for you to know all, it is for you to dare all. Mr. President and Gentlemen, this confidence in the unsearched might of man, belongs by all motives, by all prophecy, by all preparation, to the American

9. Himself nurtured on such "cold and pedantic" writers as Alexander Pope, Samuel Johnson, and Edward Gibbon, in this passage Emerson conventionally contrasts them with the so-called pre-Romantics like Oliver Goldsmith, Robert Burns, and William Cowper and Romantics like Goethe, Wordsworth, and Carlyle, supposedly marked by greater attention to aspects of ordinary life.
1. No important critic after Emerson has taken up this advocacy of literary greatness in Emanuel Swedenborg (1688–1772), Swedish scientist and theologian. He was a passion of Emerson's because of intellectual and spiritual affinities, not intrinsic literary merit.
2. Johann Heinrich Pestalozzi (1746–1827), Swiss educator and benefactor of poor children, whose ideas on education influenced Emerson's friends Bronson Alcott and Elizabeth Peabody.

Scholar. We have listened too long to the courtly muses of Europe. The spirit of the American freeman is already suspected to be timid, imitative, tame. Public and private avarice make the air we breathe thick and fat. The scholar is decent, indolent, complaisant.[3] See already the tragic consequence. The mind of this country taught to aim at low objects, eats upon itself. There is no work for any but the decorous and the complaisant. Young men of the fairest promise, who begin life upon our shores, inflated by the mountain winds, shined upon by all the stars of God, find the earth below not in unison with these,—but are hindered from action by the disgust which the principles on which business is managed inspire, and turn drudges, or die of disgust,—some of them suicides. What is the remedy? They did not yet see, and thousands of young men as hopeful now crowding to the barriers for the career, do not yet see, that if the single man plant himself indomitably on his instincts, and there abide, the huge world will come round to him. Patience— patience;—with the shades of all the good and great for company; and for solace, the perspective of your own infinite life; and for work, the study and the communication of principles, the making those instincts prevalent, the conversion of the world. Is it not the chief disgrace in the world, not to be an unit;—not to be reckoned one character;—not to yield that peculiar fruit which each man was created to bear, but to be reckoned in the gross, in the hundred, or the thousand, of the party, the section, to which we belong; and our opinion predicted geographically, as the north, or the south. Not so, brothers and friends,—please God, ours shall not be so. We will walk on our own feet; we will work with our own hands; we will speak our own minds. Then shall man be no longer a name for pity, for doubt, and for sensual indulgence. The dread of man and the love of man shall be a wall of defence and a wreath of love around all. A nation of men will for the first time exist, because each believes himself inspired by the Divine Soul which also inspires all men.

1837

The Divinity School Address[1]

In this refulgent summer it has been a luxury to draw the breath of life. The grass grows, the buds burst, the meadow is spotted with fire and gold in the tint of flowers. The air is full of birds, and sweet

3. Too ready to please others.
1. *An Address Delivered before the Senior Class in Divinity College, Cambridge, Sunday Evening 15 July, 1838* was published as a pamphlet in Boston soon after it was given. That original text is followed here, though with the title used in *Essays* (1841). Outraged attacks appeared in newspapers and pamphlets, and Emerson cautioned himself in his journal to remain "steady."

(Perry Miller, in *The Transcendentalists* [1950], reprints some of the documents in this brief furor, including the most notorious attack on Emerson, Andrews Norton's *The Latest Form of Infidelity*.) Emerson retracted nothing privately or publicly and was not invited back to Harvard for three decades, after the university had become more secular and Emerson's own international reputation had muted the charges against him.

with the breath of the pine, the balm-of-Gilead,[2] and the new hay. Night brings no gloom to the heart with its welcome shade. Through the transparent darkness pour the stars their almost spiritual rays. Man under them seems a young child, and his huge globe a toy. The cool night bathes the world as with a river, and prepares his eyes again for the crimson dawn. The mystery of nature was never displayed more happily. The corn and the wine have been freely dealt to all creatures, and the never-broken silence with which the old bounty goes forward, has not yielded yet one word of explanation. One is constrained to respect the perfection of this world, in which our senses converse. How wide; how rich; what invitation from every property it gives to every faculty of man! In its fruitful soils; in its navigable sea; in its mountains of metal and stone; in its forests of all woods; in its animals; in its chemical ingredients; in the powers and path of light, heat, attraction, and life, is it well worth the pith and heart of great men to subdue and enjoy it. The planters, the mechanics, the inventors, the astronomers, the builders of cities, and the captains, history delights to honor.

But the moment the mind opens, and reveals the laws which traverse the universe, and make things what they are, then shrinks the great world at once into a mere illustration and fable of this mind. What am I? and What is? asks the human spirit with a curiosity new-kindled, but never to be quenched. Behold these outrunning laws, which our imperfect apprehension can see tend this way and that, but not come full circle. Behold these infinite relations, so like, so unlike; many, yet one. I would study, I would know, I would admire forever. These works of thought have been the entertainments of the human spirit in all ages.

A more secret, sweet, and overpowering beauty appears to man when his heart and mind open to the sentiment of virtue. Then instantly he is instructed in what is above him. He learns that his being is without bound; that, to the good, to the perfect, he is born, low as he now lies in evil and weakness. That which he venerates is still his own, though he has not realized it yet. *He ought.* He knows the sense of that grand word, though his analysis fails entirely to render account of it. When in innocency, or when by intellectual perception, he attains to say,—'I love the Right; Truth is beautiful within and without, forevermore. Virtue, I am thine: save me: use me: thee will I serve, day and night, in great, in small, that I may be not virtuous, but virtue;'—then is the end of the creation answered, and God is well pleased.

The sentiment of virtue is a reverence and delight in the presence of certain divine laws. It perceives that this homely game of life we

2. An aromatic evergreen tree, named for the curative resin associated with Gilead in Jeremiah 8.22: "Is there no balm in Gilead; is there no physician there?"

play, covers, under what seem foolish details, principles that astonish. The child amidst his baubles, is learning the action of light,
motion, gravity, muscular force; and in the game of human life,
love, fear, justice, appetite, man, and God, interact. These laws
refuse to be adequately stated. They will not by us or for us be
written out on paper, or spoken by the tongue. They elude, evade
our persevering thought, and yet we read them hourly in each
other's faces, in each other's actions, in our own remorse. The
moral traits which are all globed into every virtuous act and
thought,—in speech, we must sever, and describe or suggest by
painful enumeration of many particulars. Yet, as this sentiment is
the essence of all religion, let me guide your eyes to the precise
objects of the sentiment, by an enumeration of some of those classes
of facts in which this element is conspicuous.

The intuition of the moral sentiment is an insight of the perfection of the laws of the soul. These laws execute themselves. They
are out of time, out of space, and not subject to circumstance.
Thus; in the soul of man there is a justice whose retributions are
instant and entire. He who does a good deed, is instantly ennobled
himself. He who does a mean deed, is by the action itself contracted. He who puts off impurity, thereby puts on purity. If a man
is at heart just, then in so far is he God; the safety of God, the
immortality of God, the majesty of God do enter into that man with
justice. If a man dissemble, deceive, he deceives himself, and goes
out of acquaintance with his own being. A man in the view of
absolute goodness, adores, with total humility. Every step so downward, is a step upward. The man who renounces himself, comes to
himself by so doing.

See how this rapid intrinsic energy worketh everywhere, righting
wrongs, correcting appearances, and bringing up facts to a harmony
with thoughts. Its operation in life, though slow to the senses, is, at
last, as sure as in the soul. By it, a man is made the Providence to
himself, dispensing good to his goodness, and evil to his sin. Character is always known. Thefts never enrich; alms never impoverish;
murder will speak out of stone walls. The least admixture of a
lie,—for example, the smallest mixture of vanity, the least attempt
to make a good impression, a favorable appearance,—will instantly
vitiate the effect. But speak the truth, and all nature and all spirits
help you with unexpected furtherance. Speak the truth, and all
things alive or brute are vouchers, and the very roots of the grass
underground there, do seem to stir and move to bear you witness.
See again the perfection of the Law as it applies itself to the affections, and becomes the law of society. As we are, so we associate.
The good, by affinity, seek the good; the vile, by affinity, the vile.
Thus of their own volition, souls proceed into heaven, into hell.

These facts have always suggested to man the sublime creed, that
the world is not the product of manifold power, but of one will, of

one mind; and that one mind is everywhere, in each ray of the star, in each wavelet of the pool, active; and whatever opposes that will, is everywhere baulked and baffled, because things are made so, and not otherwise. Good is positive. Evil is merely privative,[3] not absolute. It is like cold, which is the privation of heat. All evil is so much death or nonentity. Benevolence is absolute and real. So much benevolence as a man hath, so much life hath he. For all things proceed out of this same spirit, which is differently named love, justice, temperance, in its different applications, just as the ocean receives different names on the several shores which it washes. All things proceed out of the same spirit, and all things conspire with it. Whilst a man seeks good ends, he is strong by the whole strength of nature. In so far as he roves from these ends, he bereaves himself of power, of auxiliaries; his being shrinks out of all remote channels, he becomes less and less, a mote, a point, until absolute badness is absolute death.

The perception of this law of laws always awakens in the mind a sentiment which we call the religious sentiment, and which makes our highest happiness. Wonderful is its power to charm and to command. It is a mountain air. It is the embalmer of the world. It is myrrh and storax, and chlorine and rosemary.[4] It makes the sky and the hills sublime, and the silent song of the stars is it. By it, is the universe made safe and habitable, not by science or power. Thought may work cold and intransitive in things, and find no end or unity. But the dawn of the sentiment of virtue on the heart, gives and is the assurance that Law is sovereign over all natures; and the worlds, time, space, eternity, do seem to break out into joy.

This sentiment is divine and deifying. It is the beatitude of man. It makes him illimitable. Through it, the soul first knows itself. It corrects the capital mistake of the infant man, who seeks to be great by following the great, and hopes to derive advantages *from another*,—by showing the fountain of all good to be in himself, and that he, equally with every man, is a door into the deeps of Reason. When he says, "I ought;" when love warms him; when he chooses, warned from on high, the good and great deed; then, deep melodies wander through his soul from Supreme Wisdom. Then he can worship, and be enlarged by his worship; for he can never go behind this sentiment. In the sublimest flights of the soul, rectitude is never surmounted, love is never outgrown.

This sentiment lies at the foundation of society, and successively creates all forms of worship. The principle of veneration never dies

3. I.e., not an active power, but the absence of a power.
4. One of the gifts the wise men brought to Jesus was myrrh, a perfume made from aromatic resins. Storax is also an aromatic resin. Chlorine, in this sense, is a greenish-yellow gas used for purification. Rosemary is an aromatic evergreen shrub of southern Europe and Asia Minor, used in cookery and perfumery.

out. Man fallen into superstition, into sensuality, is never wholly without the visions of the moral sentiment. In like manner, all the expressions of this sentiment are sacred and permanent in proportion to their purity. The expressions of this sentiment affect us deeper, greatlier, than all other compositions. The sentences of the oldest time, which ejaculate this piety, are still fresh and fragrant. This thought dwelled always deepest in the minds of men in the devout and contemplative East; not alone in Palestine, where it reached its purest expression, but in Egypt, in Persia, in India, in China. Europe has always owed to oriental genius, its divine impulses. What these holy bards said, all sane men found agreeable and true. And the unique impression of Jesus upon mankind, whose name is not so much written as ploughed into the history of this world, is proof of the subtle virtue of this infusion.

Meantime, whilst the doors of the temple stand open, night and day, before every man, and the oracles of this truth cease never, it is guarded by one stern condition; this, namely; It is an intuition. It cannot be received at second hand. Truly speaking, it is not instruction, but provocation, that I can receive from another soul. What he announces, I must find true in me, or wholly reject; and on his word, or as his second, be he who he may, I can accept nothing. On the contrary, the absence of this primary faith is the presence of degradation. As is the flood so is the ebb. Let this faith depart, and the very words it spake, and the things it made, become false and hurtful. Then falls the church, the state, art, letters, life. The doctrine of the divine nature being forgotten, a sickness infects and dwarfs the constitution. Once man was all; now he is an appendage, a nuisance. And because the indwelling Supreme Spirit cannot wholly be got rid of, the doctrine of it suffers this perversion, that the divine nature is attributed to one or two persons, and denied to all the rest, and denied with fury. The doctrine of inspiration is lost; the base doctrine of the majority of voices, usurps the place of the doctrine of the soul. Miracles, prophecy, poetry, the ideal life, the holy life, exist as ancient history merely; they are not in the belief, nor in the aspiration of society; but, when suggested, seem ridiculous. Life is comic or pitiful, as soon as the high ends of being fade out of sight, and man becomes near-sighted, and can only attend to what addresses the senses.

These general views, which, whilst they are general, none will contest, find abundant illustration in the history of religion, and especially in the history of the Christian church. In that, all of us have had our birth and nurture. The truth contained in that, you, my young friends, are now setting forth to teach. As the Cultus, or established worship of the civilized world, it has great historical interest for us. Of its blessed words, which have been the consolation of humanity, you need not that I should speak. I shall endeavor to discharge my duty to you, on this occasion, by pointing out two errors in its administration, which daily appear more gross from the point of view we have just now taken.

Jesus Christ belonged to the true race of prophets. He saw with open eye the mystery of the soul. Drawn by its severe harmony, ravished with its beauty, he lived in it, and had his being there. Alone in all history, he estimated the greatness of man. One man was true to what is in you and me. He saw that God incarnates himself in man, and evermore goes forth anew to take possession of his world. He said, in this jubilee of sublime emotion, 'I am divine. Through me, God acts; through me, speaks. Would you see God, see me; or, see thee, when thou also thinkest as I now think.' But what a distortion did his doctrine and memory suffer in the same, in the next, and the following ages! There is no doctrine of the Reason which will bear to be taught by the Understanding.[5] The understanding caught this high chant from the poet's lips, and said, in the next age, 'This was Jehovah come down out of heaven. I will kill you, if you say he was a man.' The idioms of his language, and the figures of his rhetoric, have usurped the place of his truth; and churches are not built on his principles, but on his tropes. Christianity became a Mythus,[6] as the poetic teaching of Greece and of Egypt, before. He spoke of miracles; for he felt that man's life was a miracle, and all that man doth, and he knew that this daily miracle shines, as the man is diviner. But the very word Miracle, as pronounced by Christian churches, gives a false impression; it is Monster. It is not one with the blowing clover and the falling rain.

He felt respect for Moses and the prophets; but no unfit tenderness at postponing their initial revelations, to the hour and the man that now is; to the eternal revelation in the heart. Thus was he a true man. Having seen that the law in us is commanding, he would not suffer it to be commanded. Boldly, with hand, and heart, and life, he declared it was God. Thus was he a true man. Thus is he, as I think, the only soul in history who has appreciated the worth of a man.

1. In thus contemplating Jesus, we become very sensible of the first defect of historical Christianity. Historical Christianity has fallen into the error that corrupts all attempts to communicate religion. As it appears to us, and as it has appeared for ages, it is not the doctrine of the soul, but an exaggeration of the personal, the positive, the ritual. It has dwelt, it dwells, with noxious exaggeration about the *person* of Jesus. The soul knows no persons. It invites every man to expand to the full circle of the universe, and will have no preferences but those of spontaneous love. But by this eastern

---

5. Emerson reverses the common meaning of Reason, using it in the sense of intuitive, suprarational knowledge, while by Understanding he means knowledge arrived at through a logical reasoning process.
6. A cult deliberately fostered.

monarchy of a Christianity, which indolence and fear have built, the friend of man is made the injurer of man. The manner in which his name is surrounded with expressions, which were once sallies of admiration and love, but are now petrified into official titles, kills all generous sympathy and liking. All who hear me, feel, that the language that describes Christ to Europe and America, is not the style of friendship and enthusiasm to a good and noble heart, but is appropriated and formal,—paints a demigod, as the Orientals or the Greeks would describe Osiris or Apollo.[7] Accept the injurious impositions of our early catachetical instruction, and even honesty and self-denial were but splendid sins, if they did not wear the Christian name. One would rather be

'A pagan suckled in a creed outworn,'[8]

than to be defrauded of his manly right in coming into nature, and finding not names and places, not land and professions, but even virtue and truth foreclosed and monopolized. You shall not be a man even. You shall not own the world; you shall not dare, and live after the infinite Law that is in you, and in company with the infinite Beauty which heaven and earth reflect to you in all lovely forms; but you must subordinate your nature to Christ's nature; you must accept our interpretations; and take his portrait as the vulgar draw it.

That is always best which gives me to myself. The sublime is excited in me by the great stoical doctrine, Obey thyself. That which shows God in me, fortifies me. That which shows God out of me, makes me a wart and a wen. There is no longer a necessary reason for my being. Already the long shadows of untimely oblivion creep over me, and I shall decease forever.

The divine bards are the friends of my virtue, of my intellect, of my strength. They admonish me, that the gleams which flash across my mind, are not mine, but God's; that they had the like, and were not disobedient to the heavenly vision.[9] So I love them. Noble provocations go out from them, inviting me also to emancipate myself; to resist evil; to subdue the world; and to Be. And thus by his holy thoughts, Jesus serves us, and thus only. To aim to convert a man by miracles, is a profanation of the soul. A true conversion, a true Christ, is now, as always, to be made, by the reception of beautiful sentiments. It is true that a great and rich soul, like his, falling among the simple, does so preponderate, that, as his did, it names the world. The world seems to them to exist for him, and they have not yet drunk so deeply of his sense, as to see that only by coming again to themselves, or to God in themselves, can they grow

---

7. Emerson associates Egypt (where Osiris was a fertility god) with the Orient and associates Greece (where Apollo was the god of the sun) with European culture.

8. From Wordsworth's sonnet *The World Is Too Much with Us.*
9. "I was not disobedient unto the heavenly vision" (Acts 26.19).

forevermore. It is a low benefit to give me something; it is a high benefit to enable me to do somewhat of myself. The time is coming when all men will see, that the gift of God to the soul is not a vaunting, overpowering, excluding sanctity, but a sweet, natural goodness, a goodness like thine and mine, and that so invites thine and mine to be and to grow.

The injustice of the vulgar tone of preaching is not less flagrant to Jesus, than it is to the souls which it profanes. The preachers do not see that they make his gospel not glad, and shear him of the locks of beauty and the attributes of heaven. When I see a majestic Epaminondas,[1] or Washington; when I see among my contemporaries, a true orator, an upright judge, a dear friend; when I vibrate to the melody and fancy of a poem; I see beauty that is to be desired. And so lovely, and with yet more entire consent of my human being, sounds in my ear the severe music of the bards that have sung of the true God in all ages. Now do not degrade the life and dialogues of Christ out of the circle of this charm, by insulation and peculiarity. Let them lie as they befel, alive and warm, part of human life, and of the landscape, and of the cheerful day.

2. The second defect of the traditionary and limited way of using the mind of Christ is a consequence of the first; this, namely; that the Moral Nature, that Law of laws, whose revelations introduce greatness,—yea, God himself, into the open soul, is not explored as the fountain of the established teaching in society. Men have come to speak of the revelation as somewhat long ago given and done, as if God were dead. The injury to faith throttles the preacher; and the goodliest of institutions becomes an uncertain and inarticulate voice.

It is very certain that it is the effect of conversation with the beauty of the soul, to beget a desire and need to impart to others the same knowledge and love. If utterance is denied, the thought lies like a burden on the man. Always the seer is a sayer. Somehow his dream is told. Somehow he publishes it with solemn joy. Sometimes with pencil on canvas; sometimes with chisel on stone; sometimes in towers and aisles of granite, his soul's worship is builded; sometimes in anthems of indefinite music; but clearest and most permanent, in words.

The man enamored of this excellency, becomes its priest or poet. The office is coeval with the world. But observe the condition, the spiritual limitation of the office. The spirit only can teach. Not any profane man, not any sensual, not any liar, not any slave can teach, but only he can give, who has; he only can create, who is. The man on whom the soul descends, through whom the soul speaks, alone can teach. Courage, piety, love, wisdom, can teach; and every man can open his door to these angels, and they shall bring him the gift

1. Theban general (418?–362 B.C.) whose military innovations helped end Sparta's dominance in Greece.

of tongues. But the man who aims to speak as books enable, as synods use, as the fashion guides, and as interest commands, babbles. Let him hush.

To this holy office, you propose to devote yourselves. I wish you may feel your call in throbs of desire and hope. The office is the first in the world. It is of that reality, that it cannot suffer the deduction of any falsehood. And it is my duty to say to you, that the need was never greater of new revelation than now. From the views I have already expressed, you will infer the sad conviction, which I share, I believe, with numbers, of the universal decay and now almost death of faith in society. The soul is not preached. The Church seems to totter to its fall, almost all life extinct. On this occasion, any complaisance, would be criminal, which told you, whose hope and commission it is to preach the faith of Christ, that the faith of Christ is preached.

It is time that this ill-suppressed murmur of all thoughtful men against the famine of our churches; this moaning of the heart because it is bereaved of the consolation, the hope, the grandeur, that come alone out of the culture of the moral nature; should be heard through the sleep of indolence, and over the din of routine. This great and perpetual office of the preacher is not discharged. Preaching is the expression of the moral sentiment in application to the duties of life. In how many churches, by how many prophets, tell me, is man made sensible that he is an infinite Soul; that the earth and heavens are passing into his mind; that he is drinking forever the soul of God? Where now sounds the persuasion, that by its very melody imparadises my heart, and so affirms its own origin in heaven? Where shall I hear words such as in elder ages drew men to leave all and follow,—father and mother, house and land, wife and child?[2] Where shall I hear these august laws of moral being so pronounced, as to fill my ear, and I feel ennobled by the offer of my uttermost action and passion? The test of the true faith, certainly, should be its power to charm and command the soul, as the laws of nature control the activity of the hands,—so commanding that we find pleasure and honor in obeying. The faith should blend with the light of rising and of setting suns, with the flying cloud, the singing bird, and the breath of flowers. But now the priest's Sabbath has lost the splendor of nature; it is unlovely; we are glad when it is done; we can make, we do make, even sitting in our pews, a far better, holier, sweeter, for ourselves.

Whenever the pulpit is usurped by a formalist, then is the worshipper defrauded and disconsolate. We shrink as soon as the

2. See Matthew 19.28–29: "And Jesus said unto them, Verily I say unto you, That ye which have followed me, in the regeneration when the Son of man shall sit in the throne of his glory, ye also shall sit upon twelve thrones, judging the twelve tribes of Israel. And every one that hath forsaken houses, or brethren, or sisters, or father, or mother, or wife, or children, or lands, for my name's sake, shall receive an hundredfold, and shall inherit everlasting life."

prayers begin, which do not uplift, but smite and offend us. We are
fain to wrap our cloaks about us, and secure, as best we can, a
solitude that hears not. I once heard a preacher who sorely tempted
me to say, I would go to church no more. Men go, thought I, where
they are wont to go, else had no soul entered the temple in the
afternoon. A snowstorm was falling around us. The snowstorm was
real; the preacher merely spectral; and the eye felt the sad contrast
in looking at him, and then out of the window behind him, into the
beautiful meteor of the snow. He had lived in vain. He had no one
word intimating that he had laughed or wept, was married or in
love, had been commended, or cheated, or chagrined. If he had ever
lived and acted, we were none the wiser for it. The capital secret of
his profession, namely, to convert life into truth, he had not
learned. Not one fact in all his experience, had he yet imported into
his doctrine. This man had ploughed, and planted, and talked, and
bought, and sold; he had read books; he had eaten and drunken; his
head aches; his heart throbs; he smiles and suffers; yet was there not
a surmise, a hint, in all the discourse, that he had ever lived at all.
Not a line did he draw out of real history. The true preacher can
always be known by this, that he deals out to the people his life,—
life passed through the fire of thought. But of the bad preacher, it
could not be told from his sermon, what age of the world he fell in;
whether he had a father or a child; whether he was a freeholder or a
pauper; whether he was a citizen or a countryman; or any other fact
of his biography.

It seemed strange that the people should come to church. It
seemed as if their houses were very unentertaining, that they should
prefer this thoughtless clamor. It shows that there is a commanding
attraction in the moral sentiment, that can lend a faint tint of light
to dulness and ignorance, coming in its name and place. The good
hearer is sure he has been touched sometimes; is sure there is
somewhat to be reached, and some word that can reach it. When he
listens to these vain words, he comforts himself by their relation to
his remembrance of better hours, and so they clatter and echo
unchallenged.

I am not ignorant that when we preach unworthily, it is not
always quite in vain. There is a good ear, in some men, that draws
supplies to virtue out of very indifferent nutriment. There is poetic
truth concealed in all the common-places of prayer and of sermons,
and though foolishly spoken, they may be wisely heard; for, each is
some select expression that broke out in a moment of piety from
some stricken or jubilant soul, and its excellency made it remembered. The prayers and even the dogmas of our church, are like the
zodiac of Denderah,[3] and the astronomical monuments of the Hindoos, wholly insulated from anything now extant in the life and

3. At Dandarah, a village in Upper
Egypt, the ceiling of a ruined ancient
temple is sculpted with astronomical
scenes.

business of the people. They mark the height to which the waters once rose. But this docility is a check upon the mischief from the good and devout. In a large portion of the community, the religious service gives rise to quite other thoughts and emotions. We need not chide the negligent servant. We are struck with pity, rather, at the swift retribution of his sloth. Alas for the unhappy man that is called to stand in the pulpit, and *not* give bread of life. Everything that befals, accuses him. Would he ask contributions for the missions, foreign or domestic? Instantly his face is suffused with shame, to propose to his parish, that they should send money a hundred or a thousand miles, to furnish such poor fare as they have at home, and would do well to go the hundred or the thousand miles, to escape. Would he urge people to a godly way of living;—and can he ask a fellow creature to come to Sabbath meetings, when he and they all know what is the poor uttermost they can hope for therein? Will he invite them privately to the Lord's Supper? He dares not. If no heart warm this rite, the hollow, dry, creaking formality is too plain, than that he can face a man of wit and energy, and put the invitation without terror. In the street, what has he to say to the bold village blasphemer? The village blasphemer sees fear in the face, form, and gait of the minister.

Let me not taint the sincerity of this plea by any oversight of the claims of good men. I know and honor the purity and strict conscience of numbers of the clergy. What life the public worship retains, it owes to the scattered company of pious men, who minister here and there in the churches, and who, sometimes accepting with too great tenderness the tenet of the elders, have not accepted from others, but from their own heart, the genuine impulses of virtue, and so still command our love and awe, to the sanctity of character. Moreover, the exceptions are not so much to be found in a few eminent preachers, as in the better hours, the truer inspirations of all,—nay, in the sincere moments of every man. But with whatever exception, it is still true, that tradition characterizes the preaching of this country; that it comes out of the memory, and not out of the soul; that it aims at what is usual, and not at what is necessary and eternal; that thus, historical Christianity destroys the power of preaching, by withdrawing it from the exploration of the moral nature of man, where the sublime is, where are the resources of astonishment and power. What a cruel injustice it is to that Law, the joy of the whole earth, which alone can make thought dear and rich; that Law whose fatal sureness the astronomical orbits poorly emulate, that it is travestied and depreciated, that it is behooted and behowled, and not a trait, not a word of it articulated. The pulpit in losing sight of this Law, loses all its inspiration, and gropes after it knows not what. And for want of this culture, the soul of the community is sick and faithless. It wants nothing so much as a stern, high, stoical, Christian discipline, to make it know itself and the divinity that speaks through it. Now man is ashamed of himself; he skulks and sneaks through the world, to be tolerated, to be pitied, and scarcely in a thousand years does any man dare to be wise and good, and so draw after him the tears and blessings of his kind.

Certainly there have been periods when, from the inactivity of the intellect on certain truths, a greater faith was possible in names and persons. The Puritans in England and America, found in the Christ of the Catholic Church, and in the dogmas inherited from Rome, scope for their austere piety, and their longings for civil freedom. But their creed is passing away, and none arises in its room. I think no man can go with his thoughts about him, into one of our churches, without feeling that what hold the public worship had on men, is gone or going. It has lost its grasp on the affection of the good, and the fear of the bad. In the country,—neighborhoods, half parishes are *signing off*,—to use the local term. It is already beginning to indicate character and religion to withdraw from the religious meetings. I have heard a devout person, who prized the Sabbath, say in bitterness of heart, "On Sundays, it seems wicked to go to church." And the motive, that holds the best there, is now only a hope and a waiting. What was once a mere circumstance, that the best and the worst men in the parish, the poor and the rich, the learned and the ignorant, young and old, should meet one day as fellows in one house, in sign of an equal right in the soul,—has come to be a paramount motive for going thither.

My friends, in these two errors, I think, I find the causes of that calamity of a decaying church and a wasting unbelief, which are casting malignant influences around us, and making the hearts of good men sad. And what greater calamity can fall upon a nation, than the loss of worship? Then all things go to decay. Genius leaves the temple, to haunt the senate, or the market. Literature becomes frivolous. Science is cold. The eye of youth is not lighted by the hope of other worlds, and age is without honor. Society lives to trifles, and when men die, we do not mention them.

And now, my brothers, you will ask, What in these desponding days can be done by us? The remedy is already declared in the ground of our complaint of the Church. We have contrasted the Church with the Soul. In the soul, then, let the redemption be sought. In one soul, in your soul, there are resources for the world. Wherever a man comes, there comes revolution. The old is for slaves. When a man comes, all books are legible, all things transparent, all religions are forms. He is religious. Man is the wonder-worker. He is seen amid miracles. All men bless and curse. He saith yea and nay, only. The stationariness of religion; the assumption that the age of inspiration is past, that the Bible is closed; the fear of degrading the character of Jesus by representing him as a man; indicate with sufficient clearness the falsehood of our theology. It is the office of a true teacher to show us that God is, not was; that He speaketh, not spake. The true Christianity,—a faith like Christ's in the infinitude of man,—is lost. None believeth in the soul of man, but only in some man or person old and departed. Ah me! no man goeth alone. All men go in flocks to this saint or that poet, avoiding the God who seeth in secret. They cannot see in secret; they love to be blind in public. They think society wiser than their soul, and know not that one soul, and their soul, is wiser than the whole world. See how nations and races flit by on the sea of time, and leave no ripple to tell where they floated or sunk, and one good soul shall make the name of Moses, or of Zeno, or of Zoroaster,[4] reverend forever. None assayeth the stern ambition to be the Self of the nation, and of nature, but each would be an easy secondary to some Christian scheme, or sectarian connexion, or some eminent man. Once leave´ your own knowledge of God, your own sentiment, and take secondary knowledge, as St. Paul's, or George Fox's, or Swedenborg's,[5] and you get wide from God with every year this secondary form lasts, and if, as now, for centuries,—the chasm yawns to that breadth, that men can scarcely be convinced there is in them anything divine.

Let me admonish you, first of all, to go alone; to refuse the good models, even those most sacred in the imagination of men, and dare to love God without mediator or veil. Friends enough you shall find who will hold up to your emulation Wesleys and Oberlins,[6] Saints and Prophets. Thank God for these good men, but say, 'I also am a man.' Imitation cannot go above its model. The imitator dooms himself to hopeless mediocrity. The inventor did it, because it was natural to him, and so in him it has a charm. In the imitator, something else is natural, and he bereaves himself of his own beauty, to come short of another man's.

Yourself a newborn bard of the Holy Ghost,—cast behind you all conformity, and acquaint men at first hand with Deity. Be to them a man. Look to it first and only, that you are such; that fashion, custom, authority, pleasure, and money are nothing to you,—are not bandages over your eyes, that you cannot see,—but live with the privilege of the immeasurable mind. Not too anxious to visit periodically all families and each family in your parish connexion, —when you meet one of these men or women, be to them a divine man; be to them thought and virtue; let their timid aspirations find in you a friend; let their trampled instincts be genially tempted out

4. "Moses": Hebrew lawgiver who led the exodus from Egypt; "Zeno": Greek philosopher (342?–270? B.C.), founder of Stoicism; "Zoroaster": Iranian religious reformer (sixth century B.C.), founder of religion still practiced by the Parsees.
5. St. Paul, the apostle to the Gentiles, hero of the book of Acts, and author of other books of the New Testament; George Fox (1624–91), English founder of the Society of Friends (Quakers);

Emanuel Swedenborg (1688–1772), Swedish scientist and theologian.
6. John Wesley (1703–91) and his brother Charles (1707–88) founded the Methodist movement in the Church of England; Jean Frédéric Oberlin (1740–1826), Alsatian Lutheran clergyman and philanthropist, innnovator in children's education, honored by the naming of the town and college in Ohio.

in your atmosphere; let their doubts know that you have doubted, and their wonder feel that you have wondered. By trusting your own soul, you shall gain a greater confidence in other men. For all our penny-wisdom, for all our soul-destroying slavery to habit, it is not to be doubted, that all men have sublime thoughts; that all men do value the few real hours of life; they love to be heard; they love to be caught up into the vision of principles. We mark with light in the memory the few interviews, we have had in the dreary years of routine and of sin, with souls that made our souls wiser; that spoke what we thought; that told us what we knew; that gave us leave to be what we inly were. Discharge to men the priestly office, and, present or absent, you shall be followed with their love as by an angel.

And, to this end, let us not aim at common degrees of merit. Can we not leave, to such as love it, the virtue that glitters for the commendation of society, and ourselves pierce the deep solitudes of absolute ability and worth? We easily come up to the standard of goodness in society. Society's praise can be cheaply secured, and almost all men are content with those easy merits; but the instant effect of conversing with God, will be, to put them away. There are sublime merits; persons who are not actors, not speakers, but influences; persons too great for fame, for display; who disdain eloquence; to whom all we call art and artist, seems too nearly allied to show and by-ends, to the exaggeration of the finite and selfish, and loss of the universal. The orators, the poets, the commanders encroach on us only as fair women do, by our allowance and homage. Slight them by preoccupation of mind, slight them, as you can well afford to do, by high and universal aims, and they instantly feel that you have right, and that it is in lower places that they must shine. They also feel your right; for they with you are open to the influx of the all-knowing Spirit, which annihilates before its broad noon the little shades and gradations of intelligence in the compositions we call wiser and wisest.

In such high communion, let us study the grand strokes of rectitude: a bold benevolence, an independence of friends, so that not the unjust wishes of those who love us, shall impair our freedom, but we shall resist for truth's sake the freest flow of kindness, and appeal to sympathies far in advance; and,—what is the highest form in which we know this beautiful element,—a certain solidity of merit, that has nothing to do with opinion, and which is so essentially and manifestly virtue, that it is taken for granted, that the right, the brave, the generous step will be taken by it, and nobody thinks of commending it. You would compliment a coxcomb doing a good act, but you would not praise an angel. The silence that accepts merit as the most natural thing in the world, is the highest applause. Such souls, when they appear, are the Imperial Guard of Virtue, the perpetual reserve, the dictators of fortune. One needs not praise their courage,—they are the heart and soul of nature. O my friends, there are resources in us on which we have not drawn. There are men who rise refreshed on hearing a threat; men to whom a crisis which intimidates and paralyzes the majority—demanding not the faculties of prudence and thrift, but comprehension, immovableness, the readiness of sacrifice,—comes graceful and beloved as a bride. Napoleon said of Massena,[7] that he was not himself until the battle began to go against him; then, when the dead began to fall in ranks around him, awoke his powers of combination, and he put on terror and victory as a robe. So it is in rugged crises, in unweariable endurance, and in aims which put sympathy out of question, that the angel is shown. But these are heights that we can scarce remember and look up to, without contrition and shame. Let us thank God that such things exist.

And now let us do what we can to rekindle the smouldering, nigh quenched fire on the altar. The evils of that church that now is, are manifest. The question returns, What shall we do? I confess, all attempts to project and establish a Cultus with new rites and forms, seem to me vain. Faith makes us, and not we it, and faith makes its own forms. All attempts to contrive a system, are as cold as the new worship introduced by the French to the goddess of Reason,[8]—to-day, pasteboard and fillagree, and ending to-morrow in madness and murder. Rather let the breath of new life be breathed by you through the forms already existing. For, if once you are alive, you shall find they shall become plastic[9] and new. The remedy to their deformity is, first, soul, and second, soul, and evermore, soul. A whole popedom[1] of forms, one pulsation of virtue can uplift and vivify. Two inestimable advantages Christianity has given us; first; the Sabbath, the jubilee of the whole world; whose light dawns welcome alike into the closet of the philosopher, into the garret of toil, and into prison cells, and everywhere suggests, even to the vile, a thought of the dignity of spiritual being. Let it stand forevermore, a temple, which new love, new faith, new sight shall restore to more than its first splendor to mankind. And secondly, the institution of preaching;—the speech of man to men,—essentially the most flexible of all organs, of all forms. What hinders that now, everywhere, in pulpits, in lecture-rooms, in houses, in fields, wherever the invitation of men or your own occasions lead you, you speak the very truth, as your life and conscience teach it, and cheer the waiting, fainting hearts of men with new hope and new revelation.

I look for the hour when that supreme Beauty, which ravished the souls of those Eastern men, and chiefly of those Hebrews, and

---

7. André Masséna (1758–1817), marshal of the Empire under Napoleon; the anecdote is taken from Barry Edward O'Meara's *Napoleon in Exile* (Boston, 1823).

8. A reference to the French "worship of Reason" promulgated in 1793 during the Reign of Terror.

9. Receptive to influences, capable of receiving new shapes.

1. I.e., rigid hierarchy.

through their lips spoke oracles to all time, shall speak in the West
also. The Hebrew and Greek Scriptures contain immortal sentences,
that have been bread of life to millions. But they have no epical
integrity; are fragmentary; are not shown in their order to the intellect. I look for the new Teacher, that shall follow so far those
shining laws, that he shall see them come full circle; shall see their
rounding complete grace; shall see the world to be the mirror of the
soul; shall see the identity of the law of gravitation with purity of
heart; and shall show that the Ought, that Duty, is one thing with
Science, with Beauty, and with Joy.

1838, 1841

Self-Reliance[1]

Ne te quæsiveiris extra.[2]

> *"Man is his own star, and the soul that can*
> *Render an honest and a perfect man,*
> *Command all light, all influence, all fate,*
> *Nothing to him falls early or too late.*
> *Our acts our angels are, or good or ill,*
> *Our fatal shadows that walk by us still."*
> —EPILOGUE TO BEAUMONT AND FLETCHER'S *Honest Man's Fortune*[3]

> *Cast the bantling*[4] *on the rocks,*
> *Suckle him with the she-wolf's teat:*
> *Wintered with the hawk and fox,*
> *Power and speed be hands and feet.*

I read the other day some verses written by an eminent painter
which were original and not conventional. Always the soul hears an
admonition in such lines, let the subject be what it may. The sentiment they instil is of more value than any thought they may contain. To believe your own thought, to believe that what is true for
you in your private heart, is true for all men,—that is genius. Speak
your latent conviction and it shall be the universal sense; for always
the inmost becomes the outmost,—and our first thought is rendered
back to us by the trumpets of the Last Judgment. Familiar as the
voice of the mind is to each, the highest merit we ascribe to Moses,
Plato, and Milton, is that they set at naught books and traditions,
and spoke not what men but what they thought. A man should
learn to detect and watch that gleam of light which flashes across
his mind from within, more than the lustre of the firmament of
bards and sages. Yet he dismisses without notice his thought, because it is his. In every work of genius we recognise our own
rejected thoughts: they come back to us with a certain alienated
majesty. Great works of art have no more affecting lesson for us

1. *Self-Reliance*, first published in *Essays* (Boston, 1841), the source of the pres-
ent text, is even more than most of Emerson's essays a collection of thoughts from his journals, often by way of various lectures over a period of years. The earliest of the journal entries reused in this essay is from 1832, the year Emerson re-
nounced his pulpit, and numerous other reused journal entries and lecture rework-
ings are from the years 1838–40, when *The Divinity School Address* provided a major test of his own self-trust.
2. Persius, *Satire,* I.7, "Do not search outside yourself," meaning "Do not imitate."
3. 1613.
4. Baby. The stanza is Emerson's.

than this. They teach us to abide by our spontaneous impression with good humored inflexibility then most when the whole cry of voices is on the other side. Else, to-morrow a stranger will say with masterly good sense precisely what we have thought and felt all the time, and we shall be forced to take with shame our own opinion from another.

There is a time in every man's education when he arrives at the conviction that envy is ignorance; that imitation is suicide; that he must take himself for better, for worse, as his portion; that though the wide universe is full of good, no kernel of nourishing corn can come to him but through his toil bestowed on that plot of ground which is given to him to till. The power which resides in him is new in nature, and none but he knows what that is which he can do, nor does he know until he has tried. Not for nothing one face, one character, one fact makes much impression on him, and another none. It is not without preëstablished harmony, this sculpture in the memory. The eye was placed where one ray should fall, that it might testify of that particular ray. Bravely let him speak the utmost syllable of his confession. We but half express ourselves, and are ashamed of that divine idea which each of us represents. It may be safely trusted as proportionate and of good issues, so it be faithfully imparted, but God will not have his work made manifest by cowards. It needs a divine man to exhibit any thing divine. A man is relieved and gay when he has put his heart into his work and done his best; but what he has said or done otherwise, shall give him no peace. It is a deliverance which does not deliver. In the attempt his genius deserts him; no muse befriends; no invention, no hope.

Trust thyself: every heart vibrates to that iron string. Accept the place the divine Providence has found for you; the society of your contemporaries, the connexion of events. Great men have always done so and confided themselves childlike to the genius of their age, betraying their perception that the Eternal was stirring at their heart, working through their hands, predominating in all their being. And we are now men, and must accept in the highest mind the same transcendent destiny; and not pinched in a corner, not cowards fleeing before a revolution, but redeemers and benefactors, pious aspirants to be noble clay plastic under the Almighty effort, let us advance and advance on Chaos and the Dark.

What pretty oracles nature yields us on this text in the face and behavior of children, babes and even brutes. That divided and rebel mind, that distrust of a sentiment because our arithmetic has computed the strength and means opposed to our purpose, these have not. Their mind being whole, their eye is as yet unconquered, and when we look in their faces, we are disconcerted. Infancy conforms to nobody: all conform to it, so that one babe commonly makes four or five out of the adults who prattle and play to it. So God has armed youth and puberty and manhood no less with its own piquancy and charm, and made it enviable and gracious and its claims not to be put by, if it will stand by itself. Do not think the youth has no force because he cannot speak to you and me. Hark! in the next room, who spoke so clear and emphatic? Good Heaven! it is he! it is that very lump of bashfulness and phlegm which for weeks has done nothing but eat when you were by, that now rolls out these words like bell-strokes. It seems he knows how to speak to his contemporaries. Bashful or bold, then, he will know how to make us seniors very unnecessary.

The nonchalance of boys who are sure of a dinner, and would disdain as much as a lord to do or say aught to conciliate one, is the healthy attitude of human nature. How is a boy the master of society; independent, irresponsible, looking out from his corner on such people and facts as pass by, he tries and sentences them on their merits, in the swift summary way of boys, as good, bad, interesting, silly, eloquent, troublesome. He cumbers himself never about consequences, about interests: he gives an independent, genuine verdict. You must court him: he does not court you. But the man is, as it were, clapped into jail by his consciousness. As soon as he has once acted or spoken with eclat, he is a committed person, watched by the sympathy or the hatred of hundreds whose affections must now enter into his account. There is no Lethe[5] for this. Ah, that he could pass again into his neutral, godlike independence! Who can thus lose all pledge, and having observed, observe again from the same unaffected, unbiased, unbribable, unaffrighted innocence, must always be formidable, must always engage the poet's and the man's regards. Of such an immortal youth the force would be felt. He would utter opinions on all passing affairs, which being seen to be not private but necessary, would sink like darts into the ear of men, and put them in fear.

These are the voices which we hear in solitude, but they grow faint and inaudible as we enter into the world. Society everywhere is in conspiracy against the manhood of every one of its members. Society is a joint-stock company[6] in which the members agree for the better securing of his bread to each shareholder, to surrender the liberty and culture of the eater. The virtue in most request is conformity. Self-reliance is its aversion. It loves not realities and creators, but names and customs.

Whoso would be a man must be a nonconformist. He who would gather immortal palms must not be hindered by the name of goodness, but must explore if it be goodness. Nothing is at last sacred but the integrity of our own mind. Absolve you to yourself, and you shall have the suffrage of the world. I remember an answer which when quite young I was prompted to make to a valued adviser who

5. Oblivion-producing water from the river of the underworld in Greek mythology.

6. Business where the capital is held by its joint owners in transferable shares.

was wont to importune me with the dear old doctrines of the church. On my saying, What have I to do with the sacredness of traditions, if I live wholly from within? my friend suggested—"But these impulses may be from below, not from above." I replied, 'They do not seem to me to be such; but if I am the devil's child, I will live then from the devil.' No law can be sacred to me but that of my nature. Good and bad are but names very readily transferable to that or this; the only right is what is after my constitution, the only wrong what is against it. A man is to carry himself in the presence of all opposition as if every thing were titular and ephemeral but he. I am ashamed to think how easily we capitulate to badges and names, to large societies and dead institutions. Every decent and well-spoken individual affects and sways me more than is right. I ought to go upright and vital, and speak the rude truth in all ways. If malice and vanity wear the coat of philanthropy, shall that pass? If an angry bigot assumes this bountiful cause of Abolition, and comes to me with his last news from Barbadoes,[7] why should I not say to him, 'Go love thy infant; love thy wood-chopper: be good-natured and modest: have that grace; and never varnish your hard, uncharitable ambition with this incredible tenderness for black folk a thousand miles off. Thy love afar is spite at home.' Rough and graceless would be such greeting, but truth is handsomer than the affectation of love. Your goodness must have some edge to it—else it is none. The doctrine of hatred must be preached as the counteraction of the doctrine of love when that pules and whines. I shun father and mother and wife and brother, when my genius calls me.[8] I would write on the lintels of the doorpost, *Whim.*[9] I hope it is somewhat better than whim at last, but we cannot spend the day in explanation. Expect me not to show cause why I seek or why I exclude company. Then, again, do not tell me, as a good man did to-day, of my obligation to put all poor men in good situations. Are they *my* poor? I tell thee, thou foolish philanthropist, that I grudge the dollar, the dime, the cent I give to such men as do not belong to me and to whom I do not belong. There is a class of persons to whom by all spiritual affinity I am bought and sold; for them I will go to prison, if need be; but your miscellaneous popular charities; the education at college of fools; the building of meeting-houses to the vain end to which many now stand; alms to sots; and the thousandfold Relief Societies;—though I confess with shame I sometimes succumb and give the dollar, it is a wicked dollar which by-and-by I shall have the manhood to withhold.

7. Island in the eastern Caribbean where slavery was officially abolished in 1834 and all slaves freed by 1838.
8. For shunning family in order to obey a divine command see Matthew 10.34–37.
9. See Exodus 12 for God's instructions to Moses on marking with blood the "two side posts" and the "upper door post" (or lintel) of houses so that God would spare those within when he passed through to "smite all the firstborn in the land of Egypt, both man and beast." Emerson equates importunate distractions from other people as a death threat to his intellectual and spiritual life.

Virtues are in the popular estimate rather the exception than the rule. There is the man *and* his virtues. Men do what is called a good action, as some piece of courage or charity, much as they would pay a fine in expiation of daily non-appearance on parade. Their works are done as an apology or extenuation of their living in the world,— as invalids and the insane pay a high board. Their virtues are penances. I do not wish to expiate, but to live. My life is not an apology, but a life. It is for itself and not for a spectacle. I much prefer that it should be of a lower strain, so it be genuine and equal, than that it should be glittering and unsteady. I wish it to be sound and sweet, and not to need diet and bleeding.[1] My life should be unique; it should be an alms, a battle, a conquest, a medicine. I ask primary evidence that you are a man, and refuse this appeal from the man to his actions. I know that for myself it makes no difference whether I do or forbear those actions which are reckoned excellent. I cannot consent to pay for a privilege where I have intrinsic right. Few and mean as my gifts may be, I actually am, and do not need for my own assurance or the assurance of my fellows any secondary testimony.

What I must do, is all that concerns me, not what the people think. This rule, equally arduous in actual and in intellectual life, may serve for the whole distinction between greatness and meanness. It is the harder, because you will always find those who think they know what is your duty better than you know it. It is easy in the world to live after the world's opinion; it is easy in solitude to live after our own; but the great man is he who in the midst of the crowd keeps with perfect sweetness the independence of solitude.

The objection to conforming to usages that have become dead to you, is, that it scatters your force. It loses your time and blurs the impression of your character. If you maintain a dead church, contribute to a dead Bible-Society, vote with a great party either for the Government or against it, spread your table like base housekeepers, —under all these screens, I have difficulty to detect the precise man you are. And, of course, so much force is withdrawn from your proper life. But do your thing, and I shall know you. Do your work, and you shall reinforce yourself. A man must consider what a blindman's-buff is this game of conformity. If I know your sect, I anticipate your argument. I hear a preacher announce for his text and topic the expediency of one of the institutions of his church. Do I not know beforehand that not possibly can he say a new and spontaneous word? Do I not know that with all this ostentation of examining the grounds of the institution, he will do no such thing? Do I not know that he is pledged to himself not to look but at one side; the permitted side, not as a man, but as a parish minister? He is a retained attorney, and these airs of the bench are the emptiest affectation. Well, most men have bound their eyes with one or

---

1. The old medical treatment of bloodletting.

another handkerchief, and attached themselves to some one of these communities of opinion. This conformity makes them not false in a few particulars, authors of a few lies, but false in all particulars. Their every truth is not quite true. Their two is not the real two, their four not the real four: so that every word they say chagrins us, and we know not where to begin to set them right. Meantime nature is not slow to equip us in the prison-uniform of the party to which we adhere. We come to wear one cut of face and figure, and acquire by degrees the gentlest asinine expression. There is a mortifying experience in particular which does not fail to wreak itself also in the general history; I mean, "the foolish face of praise,"[2] the forced smile which we put on in company where we do not feel at ease in answer to conversation which does not interest us. The muscles, not spontaneously moved, but moved by a low usurping wilfulness, grow tight about the outline of the face and make the most disagreeable sensation, a sensation of rebuke and warning which no brave young man will suffer twice.

For non-conformity the world whips you with its displeasure. And therefore a man must know how to estimate a sour face. The bystanders look askance on him in the public street or in the friend's parlor. If this aversation had its origin in contempt and resistance like his own, he might well go home with a sad countenance; but the sour faces of the multitude, like their sweet faces, have no deep cause,—disguise no god, but are put on and off as the wind blows, and a newspaper directs. Yet is the discontent of the multitude more formidable than that of the senate and the college. It is easy enough for a firm man who knows the world to brook the rage of the cultivated classes. Their rage is decorous and prudent, for they are timid as being very vulnerable themselves. But when to their feminine rage the indignation of the people is added, when the ignorant and the poor are aroused, when the unintelligent brute force that lies at the bottom of society is made to growl and mow, it needs the habit of magnanimity and religion to treat it godlike as a trifle of no concernment.

The other terror that scares us from self-trust is our consistency; a reverence for our past act or word, because the eyes of others have no other data for computing our orbit than our past acts, and we are loath to disappoint them.

But why should you keep your head over your shoulder? Why drag about this monstrous corpse of your memory, lest you contradict somewhat you have stated in this or that public place? Suppose you should contradict yourself; what then? It seems to be a rule of wisdom never to rely on your memory alone, scarcely even in acts of pure memory, but bring the past for judgment into the thousand-eyed present, and live ever in a new day. Trust your emotion. In your metaphysics you have denied personality to the Diety: yet

2. Alexander Pope, *Epistle to Dr. Arbuthnot*, line 212.

when the devout motions of the soul come, yield to them heart and life, though they should clothe God with shape and color. Leave your theory as Joseph his coat in the hand of the harlot, and flee.[3]

A foolish consistency is the hobgoblin of little minds, adored by little statesmen and philosophers and divines. With consistency a great soul has simply nothing to do. He may as well concern himself with his shadow on the wall. Out upon your guarded lips! Sew them up with packthread, do. Else, if you would be a man, speak what you think to-day in words as hard as cannon balls, and to-morrow speak what to-morrow thinks in hard words again, though it contradict every thing you said to-day. Ah, then, exclaim the aged ladies, you shall be sure to be misunderstood. Misunderstood! It is a right fool's word. Is it so bad then to be misunderstood? Pythagoras was misunderstood, and Socrates, and Jesus, and Luther, and Copernicus, and Galileo, and Newton, and every pure and wise spirit that ever took flesh. To be great is to be misunderstood.

I suppose no man can violate his nature. All the sallies of his will are rounded in by the law of his being as the inequalities of Andes and Himmaleh[4] are insignificant in the curve of the sphere. Nor does it matter how you guage and try him. A character is like an acrostic or Alexandrian stanza;[5] —read it forward, backward, or across, it still spells the same thing. In this pleasing contrite wood-life which God allows me, let me record day by day my honest thought without prospect or retrospect, and, I cannot doubt, it will be found symmetrical, though I mean it not, and see it not. My book should smell of pines and resound with the hum of insects. The swallow over my window should interweave that thread or straw he carries in his bill into my web also. We pass for what we are. Character teaches above our wills. Men imagine that they communicate their virtue or vice only by overt actions and do not see that virtue or vice emit a breath every moment.

Fear never but you shall be consistent in whatever variety of actions, so they be each honest and natural in their hour. For of one will, the actions will be harmonious, however unlike they seem. These varieties are lost sight of when seen at a little distance, at a little height of thought. One tendency unites them all. The voyage of the best ship is a zigzag line of a hundred tacks. This is only microscopic criticism. See the line from a sufficient distance, and it straightens itself to the average tendency. Your genuine action will explain itself and will explain your other genuine actions. Your conformity explains nothing. Act singly, and what you have already done singly, will justify you now. Greatness always appeals to the

3. The story of Joseph and Potiphar's wife is in Genesis 39.
4. Mountain ranges in South America and Asia, the latter (now spelled Him- alayas) separating India from China.
5. A palindrome, reading the same backward as forward.

future. If I can be great enough now to do right and scorn eyes, I
must have done so much right before, as to defend me now. Be it
how it will, do right now. Always scorn appearances, and you
always may. The force of character is cumulative. All the foregone
days of virtue work their health into this. What makes the majesty
of the heroes of the senate and the field, which so fills the imagination? The consciousness of a train of great days and victories behind. There they all stand and shed an united light on the advancing
actor. He is attended as by a visible escort of angels to every man's
eye. That is it which throws thunder into Chatham's voice, and
dignity into Washington's port, and America into Adams's eye.[6]
Honor is venerable to us because it is no ephemeris. It is always
ancient virtue. We worship it to-day, because it is not of to-day. We
love it and pay it homage, because it is not a trap for our love and
homage, but is self-dependent, self-derived, and therefore of an old
immaculate pedigree, even if shown in a young person.

I hope in these days we have heard the last of conformity and
consistency. Let the words be gazetted[7] and ridiculous henceforward. Instead of the gong for dinner, let us hear a whistle from the
Spartan fife.[8] Let us bow and apologize never more. A great man is
coming to eat at my house. I do not wish to please him: I wish that
he should wish to please me. I will stand here for humanity, and
though I would make it kind, I would made it true. Let us affront
and reprimand the smooth mediocrity and squalid contentment of
the times, and hurl in the face of custom, and trade, and office, the
fact which is the upshot of all history, that there is a great responsible Thinker and Actor moving wherever moves a man; that a true
man belongs to no other time or place, but is the centre of things.
Where he is, there is nature. He measures you, and all men, and all
events. You are constrained to accept his standard. Ordinarily every
body in society reminds us of somewhat else or of some other
person. Character, reality, reminds you of nothing else. It takes
place of the whole creation. The man must be so much that he must
make all circumstances indifferent,—put all means into the shade.
This all great men are and do. Every true man is a cause, a country,
and an age; requires infinite spaces and numbers and time fully to
accomplish his thought;—and posterity seem to follow his steps as a
procession. A man Cæsar is born, and for ages after, we have a
Roman Empire. Christ is born, and millions of minds so grow and
cleave to his genius, that he is confounded with virtue and the

6. William Pitt, first Earl of Chatham
(1708–78), English statesman and great
orator; George Washington (1732–99),
first President of the United States (port
means carriage or physical bearing);
Adams may be Samuel Adams (1722–
1803), leader of Revolutionary movement in Massachusetts, or, more likely,
his younger relative John Quincy Adams

(1767–1848), sixth President of the
United States and, afterward, long-time
member of the House of Representatives,
known as "Old Man Eloquence."
7. Labeled in public as not to be used
henceforth.
8. Emerson is associating the gong with
lax ease and the Spartan fife with disciplined alertness.

possible of man. An institution is the lengthened shadow of one
man; as, the Reformation, of Luther; Quakerism, of Fox; Methodism, of Wesley; Abolition, of Clarkson.[9] Scipio, Milton called "the
height of Rome;"[1] and all history resolves itself very easily into the
biography of a few stout and earnest persons.

Let a man then know his worth, and keep things under his feet.
Let him not peep or steal, or skulk up and down with the air of a
charity-boy, a bastard, or an interloper, in the world which exists
for him. But the man in the street finding no worth in himself which
corresponds to the force which built a tower or sculptured a marble
god, feels poor when he looks on these. To him a palace, a statue,
or a costly book have an alien and forbidding air, much like a gay
equipage, and seem to say like that, 'Who are you, sir?' Yet they all
are his, suitors for his notice, petitioners to his faculties that they
will come out and take possession. The picture waits for my verdict:
it is not to command me, but I am to settle its claims to praise. That
popular fable of the sot who was picked up dead drunk in the street,
carried to the duke's house, washed and dressed and laid in the
duke's bed, and, on his waking, treated with all obsequious ceremony like the duke, and assured that he had been insane,[2]—owes
its popularity to the fact, that it symbolizes so well the state of man,
who is in the world a sort of sot, but now and then wakes up,
exercises his reason, and finds himself a true prince.

Our reading is mendicant and sycophantic. In history, our imagination makes fools of us, plays us false. Kingdom and lordship,
power and estate are a gaudier vocabulary than private John and
Edward in a small house and common day's work: but the things of
life are the same to both: the sum total of both is the same. Why all
this deference to Alfred, and Scanderbeg, and Gustavus?[3] Suppose
they were virtuous: did they wear out virtue? As great a stake
depends on your private act to-day, as followed their public and
renowned steps. When private men shall act with vast views, the
lustre will be transferred from the actions of kings to those of gentlemen.

The world has indeed been instructed by its kings, who have so
magnetized the eyes of nations. It has been taught by this colossal
symbol the mutual reverence that is due from man to man. The
joyful loyalty with which men have every where suffered the king,
the noble, or the great proprietor to walk among them by a law of
his own, make his own scale of men and things, and reverse theirs,
pay for benefits not with money but with honor, and represent the

9. These founders are Martin Luther (1483–1546); George Fox (1624–91); John Wesley (1703–91); Thomas Clarkeson (1760–1846).
1. Scipio Africanus (237–183 B.C.), the conqueror of Carthage.
2. The fable is best known in the version in the "Induction" to Shakespeare's *The Taming of the Shrew.*
3. National heroes of England (849–899), Albania (1404?–68), and Sweden (1594–1632).

Law in his person, was the hieroglyphic by which they obscurely signified their consciousness of their own right and comeliness, the right of every man.

The magnetism which all original action exerts is explained when we inquire the reason of self-trust. Who is the Trustee? What is the aboriginal Self on which a universal reliance may be grounded? What is the nature and power of that science-baffling star, without parallax,[4] without calculable elements, which shoots a ray of beauty even into trivial and impure actions, if the least mark of independence appear? The inquiry leads us to that source, at once the essence of genius, the essence of virtue, and the essence of life, which we call Spontaneity or Instinct. We denote this primary wisdom as Intuition, whilst all later teachings are tuitions. In that deep force, the last fact behind which analysis cannot go, all things find their common origin. For the sense of being which in calm hours rises, we know not how, in the soul, is not diverse from things, from space, from light, from time, from man, but one with them, and proceedeth obviously from the same source whence their life and being also proceedeth. We first share the life by which things exist, and afterwards see them as appearances in nature, and forget that we have shared their cause. Here is the fountain of action and the fountain of thought. Here are the lungs of that inspiration which giveth man wisdom, of that inspiration of man which cannot be denied without impiety and atheism. We lie in the lap of immense intelligence, which makes us organs of its activity and receivers of its truth. When we discern justice, when we discern truth, we do nothing of ourselves, but allow a passage to its beams. If we ask whence this comes, if we seek to pry into the soul that causes,—all metaphysics, all philosophy is at fault. Its presence or its absence is all we can affirm. Every man discerns between the voluntary acts of his mind, and his involuntary perceptions. And to his involuntary perceptions, he knows a perfect respect is due. He may err in the expression of them, but he knows that these things are so, like day and night, not to be disputed. All my wilful actions and acquisitions are but roving;—the most trivial reverie, the faintest native emotion are domestic and divine. Thoughtless people contradict as readily the statement of perceptions as of opinions, or rather much more readily; for, they do not distinguish between perception and notion. They fancy that I choose to see this or that thing. But perception is not whimsical, but fatal. If I see a trait, my children will see it after me, and in course of time, all mankind,—although it may chance that no one has seen it before me. For my perception of it is as much a fact as the sun.

The relations of the soul to the divine spirit are so pure that it is profane to seek to interpose helps. It must be that when God speaketh, he should communicate not one thing, but all things; should fill the world with his voice; should scatter forth light, nature, time, souls from the centre of the present thought; and new date and new create the whole. Whenever a mind is simple, and receives a divine wisdom, then old things pass away,—means, teachers, texts, temples fall; it lives now and absorbs past and future into the present hour. All things are made sacred by relation to it,—one thing as much as another. All things are dissolved to their centre by their cause, and in the universal miracle petty and particular miracles disappear. This is and must be. If, therefore, a man claims to know and speak of God, and carries you backward to the phraseology of some old mouldered nation in another country, in another world, believe him not. Is the acorn better than the oak which is its fulness and completion? Is the parent better than the child into whom he has cast his ripened being? Whence then this worship of the past? The centuries are conspirators against the sanity and majesty of the soul. Time and space are but physiological colors which the eye maketh, but the soul is light; where it is, is day; where it was, is night; and history is an impertinence and an injury, if it be anything more than a cheerful apologue or parable of my being and becoming.

Man is timid and apologetic. He is no longer upright. He dares not say 'I think,' 'I am,' but quotes some saint or sage. He is ashamed before the blade of grass or the blowing rose. These roses under my window make no reference to former roses or to better ones; they are for what they are; they exist with God to-day. There is no time to them. There is simply the rose; it is perfect in every moment of its existence. Before a leaf-bud has burst, its whole life acts; in the full-blown flower, there is no more; in the leafless root, there is no less. Its nature is satisfied, and it satisfies nature, in all moments alike. There is no time to it. But man postpones or remembers; he does not live in the present, but with reverted eye laments the past, or, heedless of the riches that surround him, stands on tiptoe to foresee the future. He cannot be happy and strong until he too lives with nature in the present, above time.

This should be plain enough. Yet see what strong intellect dare not yet hear God himself, unless he speak the phraseology of I know not what David, or Jeremiah, Paul.[5] We shall not always set so great a price on a few texts, on a few lines. We are like children who repeat by rote the sentences of grandames and tutors, and, as they grow older, of the men of talents and character they chance to see,—painfully recollecting the exact words they spoke; afterwards, when they come into the point of view which those had who uttered these sayings, they understand them, and are willing to

5. Biblical authors of the Book of Psalms, the Book of Jeremiah, and various New Testament Epistles.

let the words go; for, at any time, they can use words as good, when occasion comes. So was it with us, so will it be, if we proceed. If we live truly, we shall see truly. It is as easy for the strong man to be strong, as it is for the weak to be weak. When we have new perception, we shall gladly disburthen the memory of its hoarded treasures as old rubbish. When a man lives with God, his voice shall be as sweet as the murmur of the brook and the rustle of the corn.

And now at last the highest truth on this subject remains unsaid; probably, cannot be said; for all that we say is the far off remembering of the intuition. That thought, by what I can now nearest approach to say it, is this. When good is near you, when you have life in yourself,—it is not by any known or appointed way; you shall not discern the foot-prints of any other; you shall not see the face of man; you shall not hear any name;—the way, the thought, the good shall be wholly strange and new. It shall exclude all other being. You take the way from man not to man. All persons that ever existed are its fugitive ministers. There shall be no fear in it. Fear and hope are alike beneath it. It asks nothing. There is somewhat low even in hope. We are then in vision. There is nothing that can be called gratitude nor properly joy. The soul is raised over passion. It seeth identity and eternal causation. It is a perceiving that Truth and Right are. Hence it becomes a Tranquillity out of the knowing that all things go well. Vast spaces of nature; the Atlantic Ocean, the South Sea; vast intervals of time, years, centuries, are of no account. This which I think and feel, underlay that former state of life and circumstances, as it does underlie my present, and will always all circumstance, and what is called life, and what is called death.

Life only avails, not the having lived. Power ceases in the instant of repose; it resides in the moment of transition from a past to a new state; in the shooting of the gulf; in the darting to an aim. This one fact the world hates, that the soul *becomes*; for, that forever degrades the past; turns all riches to poverty; all reputation to a shame; confounds the saint with the rogue; shoves Jesus and Judas equally aside. Why then do we prate of self-reliance? Inasmuch as the soul is present, there will be power not confident but agent. To talk of reliance, is a poor external way of speaking. Speak rather of that which relies, because it works and is. Who has more soul than I, masters me, though he should not raise his finger. Round him I must revolve by the gravitation of spirits; who has less, I rule with like facility. We fancy it rhetoric when we speak of eminent virtue. We do not yet see that virtue is Height, and that a man or a company of men plastic and permeable to principles, by the law of nature must overpower and ride all cities, nations, kings, rich men, poets, who are not.

This is the ultimate fact which we so quickly reach on this as on every topic, the resolution of all into the ever blessed ONE. Virtue is the governor, the creator, the reality. All things real are so by so much of virtue as they contain. Hardship, husbandry, hunting, whaling, war, eloquence, personal weight, are somewhat, and engage my respect as examples of the soul's presence and impure action. I see the same law working in nature for conservation and growth. The poise of a planet, the bended tree recovering itself from the strong wind, the vital resources of every vegetable and animal, are also demonstrations of the self-sufficing, and therefore self-relying soul. All history from its highest to its trivial passages is the various record of this power.

Thus all concentrates; let us not rove; let us sit at home with the cause. Let us stun and astonish the intruding rabble of men and books and institutions by a simple declaration of the divine fact. Bid them take the shoes from off their feet,[6] for God is here within. Let our simplicity judge them, and our docility to our own law demonstrate the poverty of nature and fortune beside our native riches.

But now we are a mob. Man does not stand in awe of man, nor is the soul admonished to stay at home, to put itself in communication with the internal ocean, but it goes abroad to beg a cup of water of the urns of men. We must go alone. Isolation must precede true society. I like the silent church before the service begins, better than any preaching. How far off, how cool, how chaste the persons look, begirt each one with a precinct or sanctuary. So let us always sit. Why should we assume the faults of our friend, or wife, or father, or child, because they sit around our hearth, or are said to have the same blood? All men have my blood, and I have all men's. Not for that will I adopt their petulance or folly, even to the extent of being ashamed of it. But your isolation must not be mechanical, but spiritual, that is, must be elevation. At times the whole world seems to be in conspiracy to importune you with emphatic trifles. Friend, client, child, sickness, fear, want, charity, all knock at once at thy closet door and say, 'Come out unto us.'—Do not spill thy soul; do not all descend; keep thy state; stay at home in thine own heaven; come not for a moment into their facts, into their hubbub of conflicting appearances, but let in the light of thy law on their confusion. The power men possess to annoy me, I give them by a weak curiosity. No man can come near me but through my act. "What we love that we have, but by desire we bereave ourselves of the love."

If we cannot at once rise to the sanctities of obedience and faith, let us at least resist our temptations, let us enter into the state of war, and wake Thor and Woden,[7] courage and constancy in our Saxon breasts. This is to be done in our smooth times by speaking the truth. Check this lying hospitality and lying affection. Live no

6. Exodus 3.5.
7. Norse gods, here taken as ancestral gods of the Anglo-Saxon as well, asso-ciated respectively with courage and en-durance. Emerson took seriously the idea of racial traits.

longer to the expectation of these deceived and deceiving people with whom we converse. Say to them, O father, O mother, O wife, O brother, O friend, I have lived with you after appearances hitherto. Henceforward I am the truth's. Be it known unto you that henceforward I obey no law less than the eternal law. I will have no covenants but proximities. I shall endeavor to nourish my parents, to support my family, to be the chaste husband of one wife,—but these relations I must fill after a new and unprecedented way. I appeal from your customs. I must be myself. I cannot break myself any longer for you, or you. If you can love me for what I am, we shall be the happier. If you cannot, I will still seek to deserve that you should. I must be myself. I will not hide my tastes or aversions. I will so trust that what is deep is holy, that I will do strongly before the sun and moon whatever inly rejoices me, and the heart appoints. If you are noble, I will love you; if you are not, I will not hurt you and myself by hypocritical attentions. If you are true, but not in the same truth with me, cleave to your companions; I will seek my own. I do this not selfishly, but humbly and truly. It is alike your interest and mine and all men's, however long we have dwelt in lies, to live in truth. Does this sound harsh to-day? You will soon love what is dictated by your nature as well as mine, and if we follow the truth, it will bring us out safe at last.—but so you may give these friends pain. Yes, but I cannot sell my liberty and my power, to save their sensibility. Besides, all persons have their moments of reason when they look out into the region of absolute truth; then will they justify me and do the same thing.

The populace think that your rejection of popular standards is a rejection of all standard, and mere antinomianism;[8] and the bold sensualist will use the name of philosophy to gild his crimes. But the law of consciousness abides. There are two confessionals, in one or the other of which we must be shriven. You may fulfil your round of duties by clearing yourself in the *direct*, or, in the *reflex* way. Consider whether you have satisfied your relations to father, mother, cousin, neighbor, town, cat, and dog; whether any of these can upbraid you. But I may also neglect this reflex standard, and absolve me to myself. I have my own stern claims and perfect circle. It denies the name of duty to many offices that are called duties. But if I can discharage its debts, it enables me to dispense with the popular code. If any one imagines that this law is lax, let him keep its commandment one day.

And truly it demands something godlike in him who has cast off the common motives of humanity, and has ventured to trust himself for a task-master. High be his heart, faithful his will, clear his sight, that he may in good earnest be doctrine, society, law to himself, that a simple purpose may be to him as strong as iron necessity is to others.

8. Rejection of moral and religious laws.

If any man consider the present aspects of what is called by distinction *society*, he will see the need of these ethics. The sinew and heart of man seem to be drawn out, and we are become timorous desponding whimperers. We are afraid of truth, afraid of fortune, afraid of death, and afraid of each other. Our age yields no great and perfect persons. We want men and women who shall renovate life and our social state, but we see that most natures are insolvent; cannot satisfy their own wants, have an ambition out of all proportion to their practical force, and so do lean and beg day and night continually. Our housekeeping is mendicant, our arts, our occupations, our marriages, our religion we have not chosen, but society has chosen for us. We are parlor soldiers. The rugged battle of fate, where strength is born, we shun.

If our young men miscarry in their first enterprizes, they lose all heart. If the young merchant fails, men say he is *ruined*. If the finest genius studies at one of our colleges, and is not installed in an office within one year afterwards in the cities or suburbs of Boston or New York, it seems to his friends and to himself that he is right in being disheartened and in complaining the rest of his life. A sturdy lad from New Hampshire or Vermont, who in turn tries all the professions, who *teams it, farms it, peddles,* keeps a school, preaches, edits a newspaper, goes to Congress, buys a township, and so forth, in successive years, and always, like a cat, falls on his feet, is worth a hundred of these city dolls.[9] He walks abreast with his days, and feels no shame in not 'studying a profession,' for he does not postpone his life, but lives already. He has not one chance, but a hundred chances. Let a stoic arise who shall reveal the resources of man, and tell men they are not leaning willows, but can and must detach themselves; that with the exercise of self-trust, new powers shall appear; that a man is the word made flesh, born to shed healing to the nations, that he should be ashamed of our compassion, and that the moment he acts from himself, tossing the laws, the books, idolatries, and customs out of the window,—we pity him no more but thank and revere him,—and that teacher shall restore the life of man to splendor, and make his name dear to all History.

It is easy to see that a greater self-reliance,—a new respect for the divinity in man,—must work a revolution in all the offices and relations of men; in their religion; in their education; in their pursuits; their modes of living; their association; in their property; in their speculative views.

1. In what prayers do men allow themselves! That which they call a holy office, is not so much as brave and manly. Prayer looks abroad and asks for some foreign addition to come through some foreign virtue, and loses itself in endless mazes of natural and supernatural, and mediatorial and miraculous. Prayer that craves a

9. Emerson's journal entry for May 27, 1839, shows that this passage was originally modeled on the young Thoreau.

particular commodity—any thing less than all good, is vicious. Prayer is the contemplation of the facts of life from the highest point of view. It is the soliloquy of a beholding and jubilant soul. It is the spirit of God pronouncing his works good. But prayer as a means to effect a private end, is theft and meanness. It supposes dualism and not unity in nature and consciousness. As soon as the man is at one with God, he will not beg. He will then see prayer in all action. The prayer of the farmer kneeling in his field to weed it, the prayer of the rower kneeling with the stroke of his oar, are true prayers heard throughout nature, though for cheap ends. Caratach, in Fletcher's Bonduca, when admonished to inquire the mind of the god Audate, replies,

> "His hidden meaning lies in our endeavors,
> Our valors are our best gods."[1]

Another sort of false prayers are our regrets. Discontent is the want of self-reliance: it is infirmity of will. Regret calamities, if you can thereby help the sufferer; if not, attend your own work, and already the evil begins to be repaired. Our sympathy is just as base. We come to them who weep foolishly, and sit down and cry for company, instead of imparting to them truth and health in rough electric shocks, putting them once more in communication with the soul. The secret of fortune is joy in our hands. Welcome evermore to gods and men is the self-helping man. For him all doors are flung wide. Him all tongues greet, all honors crown, all eyes follow with desire. Our love goes out to him and embraces him, because he did not need it. We solicitously and apologetically caress and celebrate him, because he held on his way and scorned our disapprobation. The gods love him because men hated him. "To the persevering mortal," said Zoroaster,[2] "the blessed Immortals are swift."

As men's prayers are a disease of the will, so are their creeds a disease of the intellect. They say with those foolish Israelites, 'Let not God speak to us, lest we die. Speak thou, speak any man with us, and we will obey.'[3] Everywhere I am bereaved of meeting God in my brother, because he has shut his own temple doors, and recites fables merely of his brother's, or his brother's brother's God. Every new mind is a new classification. If it prove a mind of uncommon activity and power, a Locke, a Lavoisier, a Hutton, a Bentham, a Spurzheim,[4] it imposes its classification on other men,

1. Lines 1294–95, slightly misquoted.
2. Religious prophet of ancient Persia.
3. See the fearful words of the Hebrews after God has given Moses the Ten Commandments, Exodus 20.19: "And they said unto Moses, Speak thou with us, and we will hear: but let not God speak with us, lest we die."
4. These innovators are John Locke (1632–1704), English philosopher; Antoine Lavoisier (1726–97), French chemist; James Hutton (1726–97), Scottish geologist; Jeremy Bentham (1748–1832), English philosopher; Johann Kaspar Spurzheim (1776–1832), German physician whose work led to the pseudoscience of phrenology, reading character by the bumps on the skull.

and lo! a new system. In proportion always to the depth of the thought, and so to the number of the objects it touches and brings within reach of the pupil, is his complacency. But chiefly is this apparent in creeds and churches, which are also classifications of some powerful mind acting on the great elemental thought of Duty, and man's relation to the Highest. Such is Calvinism, Quakerism, Swedenborgianism.[5] The pupil takes the same delight in subordinating every thing to the new terminology that a girl does who has just learned botany, in seeing a new earth and new seasons thereby. It will happen for a time, that the pupil will feel a real debt to the teacher,—will find his intellectual power has grown by the study of his writings. This will continue until he has exhausted his master's mind. But in all unbalanced minds, the classification is idolized, passes for the end, and not for a speedily exhaustible means, so that the walls of the system blend to their eye in the remote horizon with the walls of the universe; the luminaries of heaven seem to them hung on the arch their master built. They cannot imagine how you aliens have any right to see,—how you can see; 'It must be somehow that you stole the light from us.' They do not yet perceive, that, light unsystematic, indomitable, will break into any cabin, even into theirs. Let them chirp awhile and call it their own. If they are honest and do well, presently their neat new pinfold will be too strait and low, will crack, will lean, will rot and vanish, and the immortal light, all young and joyful, million-orbed, million-colored, will beam over the universe as on the first morning.

2. It is for want of self-culture that the idol of Travelling, the idol of Italy, of England, of Egypt, remains for all educated Americans. They who made England, Italy, or Greece venerable in the imagination, did so not by rambling round creation as a moth round a lamp, but by sticking fast where they were, like an axis of the earth. In manly hours, we feel that duty is our place, and that the merrymen of circumstance should follow as they may. The soul is no traveller: the wise man stays at home with the soul, and when his necessities, his duties, on any occasion call him from his house, or into foreign lands, he is at home still, and is not gadding abroad from himself, and shall make men sensible by the expression of his countenance, that he goes the missionary of wisdom and virtue, and visits cities and men like a sovereign, and not like an interloper or a valet.

I have no churlish objection to the circumnavigation of the globe, for the purposes of art, of study, and benevolence, so that the man is first domesticated, or does not go abroad with the hope of finding somewhat greater than he knows. He who travels to be amused, or

---

5. Three widely varying religious movements founded by or based on the teachings of John Calvin (1509–64), French theologian; George Fox (1624–91), English clergyman; and Emanuel Swedenborg (1688–1772), Swedish scientist and theologian.

to get somewhat which he does not carry, travels away from himself, and grows old even in youth among old things. In Thebes, in Palmyra, his will and mind have become old and dilapidated as they. He carries ruins to ruins.

Travelling is a fool's paradise. We owe to our first journeys the discovery that place is nothing. At home I dream that at Naples, at Rome, I can be intoxicated with beauty, and lose my sadness. I pack my trunk, embrace my friends, embark on the sea, and at last wake up in Naples, and there beside me is the stern Fact, the sad self, unrelenting, identical, that I fled from. I seek the Vatican, and the palaces. I affect to be intoxicated with sights and suggestions, but I am not intoxicated. My giant goes with me wherever I go.

3. But the rage of travelling is itself only a symptom of a deeper unsoundness affecting the whole intellectual action. The intellect is vagabond, and the universal system of education fosters restlessness. Our minds travel when our bodies are forced to stay at home. We imitate; and what is imitation but the travelling of the mind? Our houses are built with foreign taste; our shelves are garnished with foreign ornaments; our opinions, our tastes, our whole minds lean, and follow the Past and the Distant, as the eyes of a maid follow her mistress. The soul created the arts wherever they have flourished. It was in his own mind that the artist sought his model. It was an application of his own thought to the thing to be done and the conditions to be observed. And why need we copy the Doric or the Gothic mode?[6] Beauty, convenience, grandeur of thought, and quaint expression are as near to us as to any, and if the American artist will study with hope and love the precise thing to be done by him, considering the climate, the soil, the length of the day, the wants of the people, the habit and form of the government, he will create a house in which all these will find themselves fitted, and taste and sentiment will be satisfied also.

Insist on yourself; never imitate. Your own gifts you can present every moment with the cumulative force of a whole life's cultivation; but of the adopted talent of another, you have only an extemporaneous, half possession. That which each can do best, none but his Maker can teach him. No man yet knows what it is, nor can, till that person has exhibited it. Where is the master who could have taught Shakspeare? Where is the master who could have instructed Franklin, or Washington, or Bacon, or Newton? Every great man is an unique. The Scipionism[7] of Scipio is precisely that part he could not borrow. If any body will tell me whom the great man imitates in the original crisis when he performs a great act, I will tell him who else than himself can teach him. Shakespeare will never be made by the study of Shakespeare. Do that which is assigned thee, and thou canst not hope too much or dare too much.

6. I.e., Greek or medieval architecture.   7. I.e., the essence of the man.

There is at this moment, there is for me an utterance bare and grand as that of the colossal chisel of Phidias,[8] or trowel of the Egyptians, or the pen of Moses, or Dante, but different from all these. Now possibly will the soul all rich, all eloquent, with thousand-cloven tongue, deign to repeat itself; but if I can hear what these patriarchs say, surely I can reply to them in the same pitch of voice: for the ear and the tongue are two organs of one nature. Dwell up there in the simple and noble regions of thy life, obey thy heart, and thou shalt reproduce the Foreworld again.

4. As our Religion, our Education, our Art look abroad, so does our spirit of society. All men plume themselves on the improvement of society, and no man improves.

Society never advances. It recedes as fast on one side as it gains on the other. Its progress is only apparent, like the workers of a treadmill. It undergoes continual changes: it is barbarous, it is civilized, it is christianized, it is rich, it is scientific; but this change is not amelioration. For every thing that is given, something is taken. Society acquires new arts and loses old instincts. What a contrast between the well-clad, reading, writing, thinking American, with a watch, a pencil, and a bill of exchange in his pocket, and the naked New Zealander, whose property is a club, a spear, a mat, and an undivided twentieth of a shed to sleep under. But compare the health of the two men, and you shall see that his aboriginal strength the white man has lost. If the traveller tell us truly, strike the savage with a broad axe, and in a day or two the flesh shall unite and heal as if you struck the blow into soft pitch, and the same blow shall send the white to his grave.

The civilized man has built a coach, but has lost the use of his feet. He is supported on crutches, but looses so much support of muscle. He has got a fine Geneva watch, but he has lost the skill to tell the hour by the sun. A Greenwich nautical almanac he has, and so being sure of the information when he wants it, the man in the street does not know a star in the sky. The solstice he does not observe; the equinox he knows as little; and the whole bright calendar of the year is without a dial in his mind. His notebooks impair his memory; his libraries overload his wit; the insurance office increases the number of accidents; and it may be a question whether machinery does not encumber; whether we have not lost by refinement some energy, by a chistianity entrenched in establishments and forms, some vigor of wild virtue. For every stoic was a stoic;[9] but in Christendom where is the Christian?

There is no more deviation in the moral standard than in the

8. Greek sculptor of fifth century B.C.
9. Emerson refers particularly to the Stoics, members of the Greek school of philosophy founded by Zeno about 308 B.C. It taught the ideal of a calm, passionless existence in which any occurrence is accepted as inevitable.

standard of height or bulk. No greater men are now than ever were. A singular equality may be observed between the great men of the first and of the last ages; nor can all the science, art, religion and philosophy of the nineteenth century avail to educate greater men than Plutarch's heroes,[1] three or four and twenty centuries ago. Not in time is the race progressive. Phocion, Socrates, Anaxagoras, Diogenes,[2] are great men, but they leave no class. He who is really of their class will not be called by their name, but be wholly his own man, and, in his turn the founder of a sect. The arts and inventions of each period are only its costume, and do not invigorate men. The harm of the improved machinery may compensate its good. Hudson and Behring[3] accomplished so much in their fishing-boats, as to astonish Parry and Franklin,[4] whose equipment exhausted the resources of science and art. Galileo, with an opera-glass, discovered a more splendid series of facts than any one since. Columbus found the New World in an undecked boat. It is curious to see the periodical disuse and perishing of means and machinery which were introduced with loud laudation, a few years or centuries before. The great genius returns to essential man. We reckoned the improvements of the art of war among the triumphs of science, and yet Napoleon conquered Europe by the Bivouac, which consisted of falling back on naked valor, and disencumbering it of all aids. The Emperor held it impossible to make a perfect army, says Las Cases, "without abolishing our arms, magazines, commissaries, and carriages, until in imitation of the Roman custom, the soldier should receive his supply of corn, grind it in his hand-mill, and bake his bread himself."[5]

Society is a wave. The wave moves onward, but the water of which it is composed, does not. The same particle does not rise from the valley to the ridge. Its unity is only phenomenal. The persons who make up a nation to-day, next year die, and their experience with them.

And so the reliance on Property, including the reliance on governments which protect it, is the want of self-reliance. Men have looked away from themselves and at things so long, that they have come to esteem what they call the soul's progress, namely, the religious, learned, and civil institutions, as guards of property, and they deprecate assaults on these, because they feel them to be assaults on property. They measure their esteem of each other, by

1. The lives of famous Greeks and Romans written by Plutarch (46?–120?), Greek biographer.
2. Four Greek philosophers, 402?–317 B.C., 470?–399 B.C., 500?–428 B.C., and 412?–323 B.C.
3. Henry Hudson (d. 1611) English navigator (sometimes in service of the Dutch); Vitus Jonassen Bering (1680–

1741), Danish navigator who explored the northern Pacific Ocean.
4. Sir William Edward Parry (1790–1855) and Sir John Franklin (1786–1847), English explorers of the arctic.
5. Comte Emmanuel de Las Cases (1766–1842), author of a book recording his conversations with the exiled Napoleon at Saint Helena.

what each has, and not by what each is. But a cultivated man becomes ashamed of his property, ashamed of what he has, out of new respect for his being. Especially he hates what he has, if he see that it is accidental,—came to him by inheritance, or gift, or crime; then he feels that it is not having; it does not belong to him, has no root in him, and merely lies there, because no revolution or no robber takes it away. But that which a man is, does always by necessity acquire, and what the man acquires is permanent and living property, which does not wait the beck of rulers, or mobs, or revolutions, or fire, or storm, or bankruptcies, but perpetually renews itself wherever the man is put. "Thy lot or portion of life," said the Caliph Ali, "is seeking after thee; therefore be at rest from seeking after it."[6] Our dependence on these foreign goods leads us to our slavish respect for numbers. The political parties meet in numerous conventions; the greater the concourse, and with each new uproar of announcement, The delegation from Essex![7] The Democrats from New Hampshire! The Whigs of Maine! the young patriot feels himself stronger than before by a new thousand of eyes and arms. In like manner the reformers summon conventions, and vote and resolve in multitude. But not so, O friends! will the God deign to enter and inhabit you, but by a method precisely the reverse. It is only as a man puts off from himself all external support, and stands alone, that I see him to be strong and to prevail. He is weaker by every recruit to his banner. Is not a man better than a town? Ask nothing of men, and in the endless mutation, thou only firm column must presently appear the upholder of all that surrounds thee. He who knows that power is in the soul, that he is weak only because he has looked for good out of him and elsewhere, and so perceiving, throws himself unhesitatingly on his thought, instantly rights himself, stands in the erect position, commands his limbs, works miracles; just as a man who stands on his feet is stronger than a man who stands on his head.

So use all that is called Fortune. Most men gamble with her, and gain all, and lose all, as her wheel rolls. But do thou leave as unlawful these winnings, and deal with Cause and Effect, the chancellors of God. In the Will work and acquire, and thou hast chained the wheel of Chance, and shalt always drag her after thee. A political victory, a rise of rents, the recovery of your sick, or the return of your absent friend, or some other quite external event, raises your spirits, and you think good days are preparing for you. Do not believe it. It can never be so. Nothing can bring you peace but yourself. Nothing can bring you peace but the triumph of principles.

1841

6. Fourth Moslem caliph of Mecca (602?–661).
7. County in Massachusetts.

The Over-Soul[1]

There is a difference between one and another hour of life, in their authority and subsequent effect. Our faith comes in moments; our vice is habitual. Yet is there a depth in those brief moments, which constrains us to ascribe more reality to them than to all other experiences. For this reason, the argument, which is always forthcoming to silence those who conceive extraordinary hopes of man, namely, the appeal to experience, is forever invalid and vain. A mightier hope abolishes despair. We give up the past to the objector, and yet we hope. He must explain this hope. We grant that human life is mean; but how did we find out that it was mean? What is the ground of this uneasiness of ours; of this old discontent? What is the universal sense of want and ignorance, but the fine inuendo[2] by which the great soul makes its enormous claim? Why do men feel that the natural history of man has never been written, but always he is leaving behind what you have said of him, and it becomes old, and books of metaphysics worthless? The philosophy of six thousand years has not searched the chambers and magazines of the soul. In its experiments there has always remained, in the last analysis, a residuum it could not resolve. Man is a stream whose source is hidden. Always our being is descending into us from we know not whence. The most exact calculator has no prescience that somewhat incalculable may not baulk the very next moment. I am constrained every moment to acknowledge a higher origin for events than the will I call mine.

As with events, so it is with thoughts. When I watch that flowing river, which, out of regions I see not, pours for a season its streams into me,—I see that I am a pensioner,—not a cause, but a surprised spectator of this ethereal water; that I desire and look up, and put myself in the attitude of reception, but from some alien energy the visions come.

The Supreme Critic on all the errors of the past and the present, and the only prophet of that which must be, is that great nature in which we rest, as the earth lies in the soft arms of the atmosphere; that Unity, that Over-Soul, within which every man's particular being is contained and made one with all other; that common heart, of which all sincere conversation is the worship, to which all right action is submission; that overpowering reality which confutes our tricks and talents, and constrains every one to pass for what he is,

---

1. *The Over-Soul*, reprinted here from its first appearance in *Essays* (Boston, 1841), is Emerson's most comprehensive and sensitive analysis of the varieties of religious experience. In his *Journals*, Vol. 7, 412, Emerson gives a partial definition of the term which serves as title of the essay: *Oversoul or The Highest Thou Always Unknown.* Another definition is in the third paragraph of the essay. Had Emerson used the word *God*, his readers would have interpreted him as meaning merely the Jehovah of Judaism and Christianity, not the power which lies behind all religions.
2. An acceptable spelling in Emerson's day.

322 ★ Ralph Waldo Emerson

and to speak from his character and not from his tongue; and which evermore tends and aims to pass into our thought and hand, and become wisdom, and virtue, and power, and beauty. We live in succession, in division, in parts, in particles. Meantime within man is the soul of the whole; the wise silence; the universal beauty, to which every part and particle is equally related; the eternal ONE. And this deep power in which we exist, and whose beatitude is all accessible to us, is not only self-sufficing and perfect in every hour, but the act of seeing, and the thing seen, the seer and the spectacle, the subject and the object, are one. We see the world piece by piece, as the sun, the moon, the animal, the tree; but the whole, of which these are the shining parts, is the soul. It is only by the vision of that Wisdom, that the horoscope of the ages can be read, and it is only by falling back on our better thoughts, by yielding to the spirit of prophecy which is innate in every man, that we can know what it saith. Every man's words, who speaks from that life, must sound vain to those who do not dwell in the same thought on their own part. I dare not speak for it. My words do not carry its august sense; they fall short and cold. Only itself can inspire whom it will, and behold! their speech shall be lyrical, and sweet, and universal as the rising of the wind. Yet I desire, even by profane words, if sacred I may not use, to indicate the heaven of this deity, and to report what hints I have collected of the transcendent simplicity and energy of the Highest Law.

If we consider what happens in conversation, in reveries, in remorse, in times of passion, in surprises, in the instructions of dreams wherein often we see ourselves in masquerade,—the droll disguises only magnifying and enhancing a real element, and forcing it on our distinct notice,—we shall catch many hints that will broaden and lighten into knowledge of the secret of nature. All goes to show that the soul in man is not an organ, but animates and exercises all the organs; is not a function, like the power of memory, of calculation, of comparison,—but uses these as hands and feet; is not a faculty, but a light; is not the intellect or the will, but the master of the intellect and the will;—is the vast back-ground of our being, in which they lie,—an immensity not possessed and that cannot be possessed. From within or from behind, a light shines through us upon things, and makes us aware that we are nothing, but the light is all. A man is the façade of a temple wherein all wisdom and all good abide. What we commonly call man, the eating, drinking, planting, counting man, does not, as we know him, represent himself, but misrepresents himself. Him we do not respect, but the soul, whose organ he is, would he let it appear through his action, would make our knees bend. When it breathes through his intellect, it is genius; when it breathes through his will, it is virtue; when it flows through his affection, it is love. And the blindness of the intellect begins, when it would be something of itself. The weakness of the will begins when the individual would be something of himself. All reform aims, in some one particular, to let the great soul have its way through us; in other words, to engage us to obey.

Of this pure nature every man is at some time sensible. Language cannot paint it with his colors. It is too subtle. It is undefinable, unmeasureable, but we know that it pervades and contains us. We know that all spiritual being is in man. A wise old proverb says, "God comes to see us without bell:" that is, as there is no screen or ceiling between our heads and the infinite heavens, so is there no bar or wall in the soul where man, the effect, ceases, and God, the cause, begins. The walls are taken away. We lie open on one side to the deeps of spiritual nature, to all the attributes of God. Justice we see and know, Love, Freedom, Power. These natures no man ever got above, but always they tower over us, and most in the moment when our interests tempt us to wound them.

The sovereignty of this nature whereof we speak, is made known by its independency of those limitations which circumscribe us on every hand. The soul circumscribeth all things. As I have said, it contradicts all experience. In like manner it abolishes time and space. The influence of the senses has, in most men, overpowered the mind to that degree, that the walls of time and space have come to look solid, real and insurmountable; and to speak with levity of these limits, is, in the world, the sign of insanity. Yet time and space are but inverse measures of the force of the soul. A man is capable of abolishing them both. The spirit sports with time—

> "Can crowd eternity into an hour,
> Or stretch an hour to eternity."[3]

We are often made to feel that there is another youth and age than that which is measured from the year of our natural birth. Some thoughts always find us young and keep us so. Such a thought is the love of the universal and eternal beauty. Every man parts from that contemplation with the feeling that it rather belongs to ages than to mortal life. The least activity of the intellectual powers redeems us in a degree from the influences of time. In sickness, in languor, give us a strain of poetry or a profound sentence, and we are refreshed; or produce a volume of Plato, or Shakspeare, or remind us of their names, and instantly we come into a feeling of longevity. See how the deep, divine thought demolishes centuries, and millenniums, and makes itself present through all ages. Is the teaching of Christ less effective now than it was when first his mouth was opened? The emphasis of facts and persons to my soul

3. Adapted from William Blake's *Auguries of Innocence*: "To see a World in a Grain of Sand/And a Heaven in a Wild Flower,/Hold Infinity in the palm of your hand/And Eternity in an hour."

has nothing to do with time. And so, always, the soul's scale is one; the scale of the senses and the understanding is another. Before the great revelations of the soul, Time, Space and Nature shrink away. In common speech, we refer all things to time, as we habitually refer the immensely sundered stars to one concave sphere. And so we say that the Judgment is distant or near, that the Millennium[4] approaches, that a day of certain political, moral, social reforms is at hand, and the like, when we mean, that in the nature of things, one of the facts we contemplate is external and fugitive, and the other is permanent and connate with the soul. The things we now esteem fixed, shall, one by one, detach themselves, like ripe fruit, from our experience, and fall. The wind shall blow them none knows whither.[5] The landscape, the figures, Boston, London, are facts as fugitive as any institution past, or any whiff of mist or smoke, and so is society, and so is the world. The soul looketh steadily forwards, creating a world alway before her, and leaving worlds alway behind her. She has no dates, nor rites, nor persons, nor specialties, nor men. The soul knows only the soul. All else is idle weeds[6] for her wearing.

After its own law and not by arithmetic is the rate of its progress to be computed. The soul's advances are not made by gradation, such as can be represented by motion in a straight line; but rather by ascension of state, such as can be represented by metamorphosis, —from the egg to the worm, from the worm to the fly. The growths of genius are of a certain *total* character, that does not advance the elect individual first over John, then Adam, then Richard, and give to each the pain of discovered inferiority, but by every throe of growth, the man expands there where he works, passing, at each pulsation, classes, populations of men. With each divine impulse the mind rends the thin rinds of the visible and finite, and comes out into eternity, and inspires and expires its air. It converses with truths that have always been spoken in the world, and becomes conscious of a closer sympathy with Zeno and Arrian,[7] than with persons in the house.

This is the law of moral and of mental gain. The simple rise as by specific levity, not into a particular virtue, but into the region of all the virtues. They are in the spirit which contains them all. The soul

4. From the reference in Revelation 20.2 to the "thousand years" during which Satan will be bound and Jesus will reign on earth; Emerson has in mind the extraordinary publicity—and hysteria—created by the prediction of the American preacher William Miller (1782–1849) that the Second Coming of Jesus would occur in 1843.
5. "The wind bloweth where it listeth, and thou hearest the sound thereof, but canst not tell whence it cometh, and whither it goeth: so is every one that is born of the Spirit" (John 3.8).
6. Black mourning garments.
7. Emerson's point is not only that these are men of the distant past and, therefore, difficult to feel affinity with, but also that as the founder of Stoicism and the biographer of another Stoic, Epictetus, these two men would take a philosophical stand against the cultivation of a close, sympathetic relationship.

is superior to all the particulars of merit. The soul requires purity, but purity is not it; requires justice, but justice is not that; requires beneficence, but is somewhat better: so that there is a kind of descent and accommodation felt when we leave speaking of moral nature, to urge a virtue which it enjoins. For, to the soul in her pure action, all the virtues are natural, and not painfully acquired. Speak to his heart, and the man becomes suddenly virtuous.

Within the same sentiment is the germ of intellectual growth, which obeys the same law. Those who are capable of humility, of justice, of love, of aspiration, are already on a platform that commands the sciences and arts, speech and poetry, action and grace. For whoso dwells in this moral beatitude, does already anticipate those special powers which men prize so highly; just as love does justice to all the gifts of the object beloved. The lover has no talent, no skill, which passes for quite nothing with his enamored maiden, however little she may possess of related faculty. And the heart, which abandons itself to the Supreme Mind, finds itself related to all its works and will travel a royal road to particular knowledges and powers. For, in ascending to this primary and aboriginal sentiment, we have come from our remote station on the circumference instantaneously to the centre of the world, where, as in the closet of God, we see causes, and anticipate the universe, which is but a slow effect.

One mode of the divine teaching is the incarnation of the spirit in a form,—in forms, like my own. I live in society; with persons who answer to thoughts in my own mind, or outwardly express to me a certain obedience to the great instincts to which I live. I see its presence to them. I am certified of a common nature; and so these other souls, these separated selves, draw me as nothing else can. They stir in me the new emotions we call passion; of love, hatred, fear, admiration, pity; thence comes conversation, competition, persuasion, cities, and war. Persons are supplementary to the primary teaching of the soul. In youth we are mad for persons. Childhood and youth see all the world in them. But the larger experience of man discovers the identical nature appearing through them all. Persons themselves acquaint us with the impersonal. In all conversation between two persons, tacit reference is made as to a third party, to a common nature. That third party or common nature is not social; it is impersonal; is God. And so in groups where debate is earnest, and especially on great questions of thought, the company become aware of their unity; aware that the thought rises to an equal height in all bosoms, that all have a spiritual property in what was said, as well as the sayer. They all wax wiser than they were. It arches over them like a temple, this unity of thought, in which every heart beats with nobler sense of power and duty, and thinks and acts with unusual solemnity. All are conscious of attaining to a higher self-possession. It shines for all. There is a certain wisdom of humanity which is common to the greatest men with the lowest, and which our ordinary education often labors to silence and obstruct. The mind is one, and the best minds who love truth for its own sake, think much less of property in truth. Thankfully they accept it everywhere, and do not label or stamp it with any man's name, for it is theirs long beforehand. It is theirs from eternity. The learned and the studious of thought have no monopoly of wisdom. Their violence of direction in some degree disqualifies them to think truly. We owe many valuable observations to people who are not very acute or profound, and who say the thing without effort, which we want and have long been hunting in vain. The action of the soul is oftener in that which is felt and left unsaid, than in that which is said in any conversation. It broods over every society, and they unconsciously seek for it in each other. We know better than we do. We do not yet possess ourselves, and we know at the same time that we are much more. I feel the same truth how often in my trivial conversation with my neighbors, that somewhat higher in each of us overlooks this by-play, and Jove nods to Jove from behind each of us.

Men descend to meet. In their habitual and mean service to the world, for which they forsake their native nobleness, they resemble those Arabian Sheikhs, who dwell in mean houses and affect an external poverty, to escape the rapacity of the Pacha, and reserve all their display of wealth for their interior and guarded retirements.

As it is present in all persons, so it is in every period of life. It is adult already in the infant man. In my dealing with my child, my Latin and Greek, my accomplishments and my money, stead me nothing. They are all lost on him: but as much soul as I have, avails. If I am merely wilful, he gives me a Rowland for an Oliver,[8] sets his will against mine, one for one, and leaves me, if I please, the degradation of beating him by my superiority of strength. But if I renounce my will, and act for the soul, setting that up as umpire between us two, out of his young eyes looks the same soul; he reveres and loves with me.

The soul is the perceiver and revealer of truth. We know truth when we see it, let skeptic and scoffer say what they choose. Foolish people ask you, when you have spoken what they do not wish to hear, 'How do you know it is truth, and not an error of your own?' We know truth when we see it, from opinion, as we know when we are awake that we are awake. It was a grand sentence of Emanuel Swedenborg,[9] which would alone indicate the greatness of that

8. Emerson is playing on the proverbial expression to give "a Rowland for an Oliver," to match one incredible lie with another. (Roland and Oliver were among Charlemagne's Twelve Peers and became heroes of medieval romances.)

9. Swedish theologian and statesman (1688–1772), Emerson's favorite example of a mystic.

man's perception,—"It is no proof of a man's understanding to be able to affirm whatever he pleases, but to be able to discern that what is true is true, and that what is false is false, this is the mark and character of intelligence." In the book I read, the good thought returns to me, as every truth will, the image of the whole soul. To the bad thought which I find in it, the same soul becomes a discerning, separating sword and lops it away. We are wiser than we know. If we will not interfere with our thought, but will act entirely, or see how the thing stands in God, we know the particular thing, and every thing, and every man. For, the Maker of all things and all persons, stands behind us, and casts his dread omniscience through us over things.

But beyond this recognition of its own in particular passages of the individual's experience, it also reveals truth. And here we should seek to reinforce ourselves by its very presence, and to speak with a worthier, loftier strain of that advent. For the soul's communication of truth is the highest event in nature, for it then does not give somewhat from itself, but it gives itself, or passes into and becomes that man whom it enlightens; or in proportion to that truth he receives, it takes him to itself.

We distinguish the announcements of the soul, its manifestations of its own nature, by the term *Revelation*. These are always attended by the emotion of the sublime. For this communication is an influx of the Divine mind into our mind. It is an ebb of the individual rivulet before the flowing surges of the sea of life. Every distinct apprehension of this central commandment agitates men with awe and delight. A thrill passes through all men at the reception of new truth, or at the performance of a great action, which comes out of the heart of nature. In these communications, the power to see, is not separated from the will to do, but the insight proceeds from obedience, and the obedience proceeds from a joyful perception. Every moment when the individual feels himself invaded by it, is memorable. Always, I believe, by the necessity of our constitution, a certain enthusiasm attends the individual's consciousness of that divine presence. The character and duration of this enthusiasm varies with the state of the individual, from an extasy and trance and prophetic inspiration,—which is its rarer appearance, to the faintest glow of virtuous emotion, in which form it warms, like our household fires, all the families and associations of men, and makes society possible. A certain tendency to insanity has always attended the opening of the religious sense of men, as if "blasted with excess of light."[1] The trances of Socrates; the "union" of Plotinus; the vision of Porphyry; the conversion of Paul; the aurora of Behmen; the convulsions of George Fox and his Quakers;

1. From *The Progress of Poesy*, III.ii.7, by Thomas Gray (1716–71), English poet.

the illumination of Swedenborg; are of this kind.[2] What was in the case of these remarkable persons a ravishment, has, in innumerable instances in common life, been exhibited in less striking manner. Everywhere the history of religion betrays a tendency to enthusiasm. The rapture of the Moravian and Quietist; the opening of the internal sense of the Word, in the language of the New Jerusalem Church; the revival of the Calvinistic Churches; the experiences of the Methodists,[3] are varying forms of that shudder of awe and delight with which the individual soul always mingles with the universal soul.

The nature of these revelations is always the same: they are perceptions of the absolute law. They are solutions of the soul's own questions. They do not answer the questions which the understanding asks. The soul answers never by words, but by the thing itself that is inquired after.

Revelation is the disclosure of the soul. The popular notion of a revelation, is, that it is a telling of fortunes. In past oracles of the soul, the understanding seeks to find answers to sensual questions, and undertakes to tell from God how long men shall exist, what their hands shall do, and who shall be their company, adding even mames, and dates and places. But we must pick no locks. We must check this low curiosity. An answer in words is delusive; it is really no answer to the questions you ask. Do not ask a description of the countries towards which you sail. The description does not describe them to you, and to-morrow you arrive there, and know them by inhabiting them. Men ask of the immortality of the soul, and the employments of heaven, and the state of the sinner, and so forth. They even dream that Jesus has left replies to precisely these interrogatories. Never a moment did that sublime spirit speak in their *patois*.[4] To truth, justice, love, the attributes of the soul, the idea of immutableness is essentially associated. Jesus, living in these moral sentiments, heedless of sensual fortunes, heeding only the manifestations of these, never made the separation of the idea of duration from the essence of these attributes; never uttered a syllable concerning the duration of the soul. It was left to his disciples to sever duration from the moral elements and to teach the immortal-

2. I.e., whatever the particular vocabulary, the same psychological phenomenon unites the experiences of the pagan Greek philosopher, two Neoplatonic Greek philosophers of the early Christian era, the apostle Paul, the German theosophist Behmen or Böhme (1575–1624), the English 17th-century founder of Quakerism, and 18th-century Swedish mystic.
3. The Moravians were exiles from Czechoslovakia, members of a Protestant sect in Saxony in the 18th century; Quietists were 17th-century believers in per-

fection through passivity of the soul, condemned as heretics by the Roman Catholic church; members of the New Jerusalem Church in Emerson's time were followers of Swedenborg; Calvinist churches, including the Presbyterian and Congregational, would emphasize innate depravity; Methodists began as an 18th-century offshoot of the Church of England. Once again Emerson's point is that whatever the terminology, the experience is the same.
4. Dialect; here, within their circumscribed limits.

ity of the soul as a doctrine, and maintain it by evidences. The moment the doctrine of the immortality is separately taught, man is already fallen. In the flowing of love, in the adoration of humility, there is no question of continuance. No inspired man ever asks this question, or condescends to these evidences. For the soul is true to itself, and the man in whom it is shed abroad, cannot wander from the present, which is infinite, to a future, which would be finite.

These questions which we lust to ask about the future, are a confession of sin. God has no answer for them. No answer in words can reply to a question of things. It is not in an arbitrary "decree of God," but in the nature of man that a veil shuts down on the facts of to-morrow: for the soul will not have us read any other cipher but that of cause and effect. By this veil, which curtains events, it instructs the children of men to live in to-day. The only mode of obtaining an answer to these questions of the senses, is, to forego all low curiosity, and, accepting the tide of being which floats us into the secret of nature, work and live, work and live, and all unawares, the advancing soul has built and forged for itself a new condition, and the question and the answer are one.

Thus is the soul the perceiver and revealer of truth. By the same fire, serene, impersonal, perfect, which burns until it shall dissolve all things into the waves and surges of an ocean of light,—we see and know each other, and what spirit each is of. Who can tell the grounds of his knowledge of the character of the several individuals in his circle of friends? No man. Yet their acts and words do not disappoint him. In that man, though he knew no ill of him, he put no trust. In that other, though they had seldom met, authentic signs had yet passed, to signify that he might be trusted as one who had an interest in his own character. We know each other very well,— which of us has been just to himself, and whether that which we teach or behold, is only an aspiration, or is our honest effort also.

We are all discerners of spirits. That diagnosis lies aloft in our life or unconscious power, not in the understanding. The whole intercourse of society, its trade, its religion, its friendships, its quarrels, —is one wide, judicial investigation of character. In full court, or in small committee, or confronted face to face, accuser and accused, men offer themselves to be judged. Against their will they exhibit those decisive trifles by which character is read. But who judges? and what? Not our understanding. We do not read them by learning or craft. No; the wisdom of the wise man consists herein, that he does not judge them; he lets them judge themselves, and merely reads and records their own verdict.

By virtue of this inevitable nature, private will is overpowered, and, maugre[5] our efforts, or our imperfections, your genius will speak from you, and mine from me. That which we are, we shall

5. Malgré, despite.

teach, not voluntarily, but involuntarily. Thoughts come into our minds by avenues which we never left open, and thoughts go out of our minds through avenues which we never voluntarily opened. Character teaches over our head. The infallible index of true progress is found in the tone the man takes. Neither his age, nor his breeding, nor company, nor books, nor actions, nor talents, nor all together, can hinder him from being deferential to a higher spirit than his own. If he have not found his home in God, his manners, his forms of speech, the turn of his sentences, the build, shall I say, of all his opinions will involuntarily confess it, let him brave it out how he will. If he have found his centre, the Deity will shine through him, through all the disguises of ignorance, of ungenial temperament, of unfavorable circumstance. The tone of seeking, is one, and the tone of having is another.

The great distinction between teachers sacred or literary; between poets like Herbert, and poets like Pope; between philosophers like Spinoza, Kant, and Coleridge,—and philosophers like Locke, Paley, Mackintosh, and Stewart;[6] between men of the world who are reckoned accomplished talkers, and here and there a fervent mystic, prophesying half-insane under the infinitude of his thought, is, that one class speak *from within*, or from experience, as parties and possessors of the fact; and the other class, *from without*, as spectators merely, or perhaps as acquainted with the fact, on the evidence of third persons. It is of no use to preach to me from without. I can do that too easily myself. Jesus speaks always from within, and in a degree that transcends all others. In that, is the miracle. That includes the miracle. My soul believes beforehand that it ought so to be. All men stand continually in the expectation of the appearance of such a teacher. But if a man do not speak from within the veil, where the word is one with that it tells of, let him lowly confess it.

The same Omniscience flows into the intellect, and makes what we call genius. Much of the wisdom of the world is not wisdom, and the most illuminated class of men are no doubt superior to literary fame, and are not writers. Among the multitude of scholars and authors, we feel no harrowing presence; we are sensible of a knack and skill rather than of inspiration; they have a light, and know not whence it comes, and call it their own: their talent is

6. Emerson's oppositions pit the 17th-century religious poet against the more worldly 18th-century poet, then the metaphysically minded 17th-century Dutch philosopher Baruch Spinoza, the 18th-century German philosopher Immanuel Kant, and the English poet-philosopher S. T. Coleridge (1772–1834) against the rational, pragmatic English writers John Locke (1632–1704), author of *Essay Concerning Human Understanding* (1690), which argued that knowledge comes only from sense experience and subsequent reflection upon such experience, and William Paley (1743–1805), who in *Principles of Moral and Political Philosophy* (1785) took a utilitarian view of Christianity, as well as two Scottish philosophers, Sir James Mackintosh (1765–1832), advocate of a version of utilitarianism, and Dugald Stewart (1753–1828), a member of the Scottish Common-Sense school of philosophy.

some exaggerated faculty, some overgrown member, so that their strength is a disease. In these instances, the intellectual gifts do not make the impression of virtue, but almost of vice; and we feel that a man's talents stand in the way of his advancement in truth. But genius is religious. It is a larger imbibing of the common heart. It is not anomalous, but more like, and not less like other men. There is in all great poets, a wisdom of humanity, which is superior to any talents they exercise. The author, the wit, the partisan, the fine gentleman, does not take place of the man. Humanity shines in Homer, in Chaucer, in Spenser, in Shakspeare, in Milton. They are content with truth. They use the positive degree. They seem frigid and phlegmatic to those who have been spiced with the frantic passion and violent coloring of inferior, but popular writers. For, they are poets by the free course which they allow to the informing soul, which, though their eyes beholdeth again, and blesseth the things which it hath made. The soul is superior to its knowledge; wiser than any of its works. The great poet makes us feel our own wealth, and then we think less of his compositions. His greatest communication to our mind, is, to teach us to despise all he has done. Shakspeare carries us to such a lofty strain of intelligent activity, as to suggest a wealth which beggars his own; and we then feel that the splendid works which he has created, and which in other hours, we extol as a sort of self-existent poetry, take no stronger hold of real nature than the shadow of a passing traveller on the rock. The inspiration which uttered itself in Hamlet and Lear, could utter things as good from day to day, forever. Why then should I make account of Hamlet and Lear, as if we had not the soul from which they fell as syllables from the tongue?

This energy does not descend into individual life, on any other condition than entire possession. It comes to the lowly and simple; it comes to whomsoever will put off what is foreign and proud; it comes as insight; it comes as serenity and grandeur. When we see those whom it inhabits, we are apprised of new degrees of greatness. From that inspiration the man comes back with a changed tone. He does not talk with men, with an eye to their opinion. He tries them. It requires of us to be plain and true. The vain traveller attempts to embellish his life by quoting my Lord, and the Prince, and the Countess, who thus said or did to *him*. The ambitious vulgar,[7] show you their spoons, and brooches, and rings, and preserve their cards and compliments. The more cultivated, in their account of their own experience, cull out the pleasing poetic circumstance; the visit to Rome; the man of genius they saw; the brilliant friend they know; still further on, perhaps, the gorgeous landscape, the mountain lights, the mountain thoughts, they enjoyed yesterday,—and so seek to throw a romantic color over their life. But the soul that

7. Common people.

ascendeth to worship the great God, is plain and true; has no rose color; no fine friends; no chivalry; no adventures; does not want admiration; dwells in the hour that now is, in the earnest experience of the common day,—by reason of the present moment, and the mere trifle having become porous to thought, and bibulous of the sea of light.

Converse with a mind that is grandly simple, and literature looks like word-catching. The simplest utterances are worthiest to be written, yet are they so cheap, and so things of course, that in the infinite riches of the soul, it is like gathering a few pebbles off the ground, or bottling a little air in a phial, when the whole earth, and the whole atmosphere are ours. The mere author, in such society, is like a pickpocket among gentlemen, who has come in to steal a gold button or a pin. Nothing can pass there, or make you one of the circle, but the casting aside your trappings, and dealing man to man in naked truth, plain confession and omniscient affirmation.

Souls, such as these, treat you as gods would; walk as gods in the earth, accepting without any admiration, your wit, your bounty, your virtue, even, say rather your act of duty, for your virtue they own as their proper blood, royal as themselves, and over-royal, and the father of the gods. But what rebuke their plain fraternal bearing casts on the mutual flattery with which authors solace each other, and wound themselves! These flatter not. I do not wonder that these men go to see Cromwell, and Christina, and Charles II., and James I., and the Grand Turk.[8] For they are in their own elevation, the fellows of kings, and must feel the servile tone of conversation in the world. They must always be a godsend to princes, for they confront them, a king to a king, without ducking or concession, and give a high nature the refreshment and satisfaction of resistance, of plain humanity, of even companionship, and of new ideas. They leave them wiser and superior men. Souls like these make us feel that sincerity is more excellent than flattery. Deal so plainly with man and woman, as to constrain the utmost sincerity, and destroy all hope of trifling with you. It is the highest compliment you can pay. Their "highest praising," said Milton, "is not flattery, and their plainest advice is a kind of praising."[9]

Ineffable is the union of man and God in every act of the soul. The simplest person, who in his integrity worships God, becomes God; yet forever and ever the influx of this better and universal self is new and unsearchable. Ever it inspires awe and astonishment. How dear, how soothing to man, arises the idea of God, peopling the lonely place, effacing the scars of our mistakes and disappointments! When we have broken our god of tradition, and ceased from

8. As the peculiar unchronological order shows, Emerson primarily means any rulers, not only rulers especially receptive to unusual common people, although he may have been thinking of Descartes's welcome at the court of Queen Christina of Sweden or Milton's services to Cromwell.

9. The *Areopagitica* (1644), third paragraph, slightly adapted.

our god of rhetoric, then may God fire the heart with his presence. It is the doubling of the heart itself, nay, the infinite enlargement of the heart with a power of growth to a new infinity on every side. It inspires in man an infallible trust. He has not the conviction, but the sight that the best is the true, and may in that thought easily dismiss all particular uncertainties and fears, and adjourn to the sure revelation of time, the solution of his private riddles. He is sure that his welfare is dear to the heart of being. In the presence of law to his mind, he is overflowed with a reliance so universal, that it sweeps away all cherished hopes and the most stable projects of mortal condition in its flood. He believes that he cannot escape from his good. The things that are really for thee, gravitate to thee. You are running to seek your friend. Let your feet run, but your mind need not. If you do not find him, will you not acquiesce that it is best you should not find him? for there is a power, which, as it is in you, is in him also, and could therefore very well bring you together, if it were for the best. You are preparing with eagerness to go and render a service to which your talent and your taste invite you, the love of men, and the hope of fame. Has it not occurred to you, that you have no right to go, unless you are equally willing to be prevented from going? O believe, as thou livest, that every sound that is spoken over the round world, which thou oughtest to hear, will vibrate on thine ear. Every proverb, every book, every by-word that belongs to thee for aid or comfort, shall surely come home through open or winding passages. Every friend whom not thy fantastic will, but the great and tender heart in thee craveth, shall lock thee in his embrace. And this, because the heart in thee is the heart of all; not a valve, not a wall, not an intersection is there any where in nature, but one blood rolls uninterruptedly, an endless circulation through all men, as the water of the globe is all one sea, and, truly seen, its tide is one.

Let man then learn the revelation of all nature, and all thought to his heart; this, namely; that the Highest dwells with him; that the sources of nature are in his own mind, if the sentiment of duty is there. But if he would know what the great God speaketh, he must 'go into his closet and shut the door,' as Jesus said.[1] God will not make himself manifest to cowards. He must greatly listen to himself, withdrawing himself from all the accents of other men's devotion. Their prayers even are hurtful to him, until he have made his own. The soul makes no appeal from itself. Our religion vulgarly stands on numbers of believers. Whenever the appeal is made,—no matter how indirectly,—to numbers, proclamation is then and there made, that religion is not. He that finds God a sweet, enveloping thought to him, never counts his company. When I sit in that

1. Matthew 6.6.

presence, who shall dare to come in? When I rest in perfect humility, when I burn with pure love,—what can Calvin or Swedenborg say?

It makes no difference whether the appeal is to numbers or to one. The faith that stands on authority is not faith. The reliance on authority, measures the decline of religion, the withdrawal of the soul. The position men have given to Jesus, now for many centuries of history, is a position of authority. It characterizes themselves. It cannot alter the eternal facts. Great is the soul, and plain. It is no flatterer, it is no follower; it never appeals from itself. It always believes in itself. Before the immense possibilities of man, all mere experience, all past biography, however spotless and sainted, shrinks away. Before that holy heaven which our presentiments foreshow us, we cannot easily praise any form of life we have seen or read of. We not only affirm that we have few great men, but absolutely speaking, that we have none; that we have no history, no record of any character or mode of living, that entirely contents us. The saints and demigods whom history worships, we are constrained to accept with a grain of allowance. Though in our lonely hours, we draw a new strength out of their memory, yet pressed on our attention, as they are by the thoughtless and customary, they fatigue and invade. The soul gives itself alone, original, and pure, to the Lonely, Original and Pure, who, on that condition, gladly inhabits, leads, and speaks through it. Then it is glad, young, and nimble. It is not wise, but it sees through all things. It is not called religious, but it is innocent. It calls the light its own, and feels that the grass grows, and the stone falls by a law inferior to, and dependent on its nature. Behold, it saith, I am born into the great, the universal mind. I the imperfect, adore my own Perfect. I am somehow receptive of the great soul, and thereby I do overlook the sun and the stars, and feel them to be but the fair accidents and effects which change and pass. More and more the surges of everlasting nature enter into me, and I become public and human in my regards and actions. So come I to live in thoughts, and act with energies which are immortal. Thus revering the soul, and learning, as the ancient said, that "its beauty is immense," man will come to see that the world is the perennial miracle which the soul worketh, and be less astonished at particular wonders; he will learn that there is no profane history; that all history is sacred; that the universe is represented in an atom, in a moment of time. He will weave no longer a spotted life of shreds and patches, but he will live with a divine unity. He will cease from what is base and frivolous in his own life, and be content with all places and any service he can render. He will calmly front the morrow in the negligency of that trust which carries God with it, and so hath already the whole future in the bottom of the heart.

1841

The Poet[1]

A moody child and wildly wise
Pursued the game with joyful eyes,
Which chose, like meteors, their way,
And rived the dark with private ray:
They overleapt the horizon's edge,
Searched with Apollo's privilege;
Through man, and woman, and sea, and star,
Saw the dance of nature forward far;
Through worlds, and races, and terms, and times,
Saw musical order, and pairing rhymes.

Olympian bards who sung
Divine ideas below,
Which always find us young,
And always keep us so.

Those who are esteemed umpires of taste, are often persons who have acquired some knowledge of admired pictures or sculptures, and have an inclination for whatever is elegant; but if you inquire whether they are beautiful souls, and whether their own acts are like fair pictures, you learn that they are selfish and sensual. Their cultivation is local, as if you should rub a log of dry wood in one spot to produce fire, all the rest remaining cold. Their knowledge of the fine arts is some study of rules and particulars, or some limited judgment of color or form, which is exercised for amusement or for show. It is a proof of the shallowness of the doctrine of beauty, as it lies in the minds of our amateurs, that men seem to have lost the perception of the instant dependence of form upon soul. There is no doctrine of forms in our philosophy. We were put into our bodies, as fire is put into a pan, to be carried about; but there is no accurate adjustment between the spirit and the organ, much less is the latter the germination of the former. So in regard to other forms, the intellectual men do not believe in any essential dependence of the material world on thought and volition. Theologians think it a pretty air-castle to talk of the spiritual meaning of a ship or a cloud, of a city or a contract, but they prefer to come again to the solid ground of historical evidence; and even the poets are contented with a civil and conformed manner of living, and to write poems from the fancy, at a safe distance from their own experience. But the highest minds of the world have never ceased to explore the double meaning, or, shall I say, the quadruple, or the centuple, or much more manifold meaning, of every sensuous fact: Orpheus, Emped-ocles, Heraclitus, Plato, Plutarch, Dante, Swedenborg,[2] and the masters of sculpture, picture, and poetry. For we are not pans and

1. First published in *Essays, Second Series* (1844), the source of the present text, *The Poet* contains the fullest elaboration of Emerson's aesthetic ideas and his most incisive comments on contemporary poetry and criticism. The first prefatory poem is from one of Emerson's own uncompleted poems, and the second is from his *Ode to Beauty*.

2. Emerson mentions a legendary Greek poet, three Greek philosophers (of the fifth, sixth, and fourth centuries B.C.), a first-century-A.D. Greek biographer, a medieval Italian poet, and an 18th century Swedish scientist and mystic.

barrows, nor even porters of the fire and torchbearers, but children of the fire, made of it, and only the same divinity transmuted, and at two or three removes, when we know least about it. And this hidden truth, that the fountains when all this river of Time, and its creatures, floweth, are intrinsically ideal and beautiful, draws us to the consideration of the nature and functions of the Poet, or the man of Beauty, to the means and materials he uses, and to the general aspect of the art in the present time.

The breadth of the problem is great, for the poet is representative. He stands among partial men for the complete man, and apprises us not of his wealth, but of the commonwealth. The young man reveres men of genius, because, to speak truly, they are more himself than he is. They receive of the soul as he also receives, but they more. Nature enhances her beauty, to the eye of loving men, from their belief that the poet is beholding her shows at the same time. He is isolated among his contemporaries, by truth and by his art, but with this consolation in his pursuits, that they will draw all men sooner or later. For all men live by truth, and stand in need of expression. In love, in art, in avarice, in politics, in labor, in games, we study to utter our painful secret. The man is only half himself, the other half is his expression.

Notwithstanding this necessity to be published, adequate expression is rare. I know not how it is that we need an interpreter; but the great majority of men seem to be minors, who have not yet come into possession of their own, or mutes, who cannot report the conversation they have had with nature. There is no man who does not anticipate a supersensual utility in the sun, and stars, earth, and water. These stand and wait to render him a peculiar service. But there is some obstruction, or some excess of phlegm in our constitution, which does not suffer them to yield the due effect. Too feeble fall the impressions of nature on us to make us artists. Every touch should thrill. Every man should be so much an artist, that he could report in conversation what had befallen him. Yet, in our experience, the rays or appulses have sufficient force to arrive at the senses, but not enough to reach the quick, and compel the reproduction of themselves in speech. The poet is the person in whom these powers are in balance, the man without impediment, who sees and handles that which others dream of, traverses the whole scale of experience, and its representative of man, in virtue of being the largest power to receive and to impart.

For the Universe has three children, born at one time, which reappear, under different names, in every system of thought, whether they be called cause, operation, and effect; or, more poetically, Jove, Pluto, Neptune; or, theologically, the Father, the Spirit, and the Son; but which we will call here, the Knower, the Doer, and the Sayer. These stand respectively for the love of truth, for the love of good, and for the love of beauty. These three are equal. Each is that which he is essentially, so that he cannot be surmounted or analyzed, and each of these three has the power of the others latent in him, and his own patent.

The poet is the sayer, the namer, and represents beauty. He is a sovereign, and stands on the centre. For the world is not painted, or adorned, but is from the beginning beautiful; and God has not made some beautiful things, but Beauty is the creator of the universe. Therefore the poet is not any permissive potentate, but is emperor in his own right. Criticism is infested with a cant of materialism, which assumes that manual skill and activity is the first merit of all men, and disparages such as say and do not, overlooking the fact, that some men, namely, poets, are natural sayers, sent into the world to the end of expression, and confounds them with those whose province is action, but who quit it to imitate the sayers. But Homer's words are as costly and admirable to Homer, as Agamemnon's victories are to Agamemnon.[3] The poet does not wait for the hero or the sage, but, as they act and think primarily, so he writes primarily what will and must be spoken, reckoning the others, though primaries also, yet, in respect to him, secondaries and servants; as sitters or models in the studio of a painter, or as assistants who bring building materials to an architect.

For poetry was all written before time was, and whenever we are so finely organized that we can penetrate into that region where the air is music, we hear those primal warblings, and attempt to write them down, but we lose ever and anon a word, or a verse, and substitute something of our own, and thus miswrite the poem. The men of more delicate ear write down these cadences more faithfully, and these transcripts, though imperfect, become the songs of the nations. For nature is as truly beautiful as it is good, or as it is reasonable, and must as much appear, as it must be done, or be known. Words and deeds are quite indifferent modes of the divine energy. Words are also actions, and actions are a kind of words.

The sign and credentials of the poet are, that he announces that which no man foretold. He is the true and only doctor;[4] he knows and tells; he is the only teller of news, for he was present and privy to the appearance which he describes. He is a beholder of ideas, and an utterer of the necessary and causal. For we do not speak now of men of poetical talents, or of industry and skill in metre, but of the true poet. I took part in a conversation the other day, concerning a recent writer of lyrics, a man of subtle mind, whose head appeared to be a music-box of delicate tunes and rhythms, and whose skill, and command of language, we could not sufficiently praise. But when the question arose, whether he was not only a lyrist, but a poet, we were obliged to confess that he is plainly a contemporary,

3. Emerson is comparing the author (Homer) with his character (Agamem- non, in the *Iliad*).
4. Teacher.

not an eternal man. He ·does not stand out of our low limitations, like a Chimborazo under the line,[5] running up from the torrid base through all the climates of the globe, with belts of the herbage of every latitude on its high and mottled sides; but this genius is the landscape-garden of a modern house, adorned with fountains and statues, with well-bred men and women standing and sitting in the walks and terraces. We hear, through all the varied music, the ground-tone of conventional life. Our poets are men of talents who sing, and not the children of music. The argument is secondary, the finish of the verses is primary.

For it is not metres, but a metre-making argument, that makes a poem,—a thought so passionate and alive, that, like the spirit of a plant or an animal, it has an architecture of its own, and adorns nature with a new thing. The thought and the form are equal in the order of time, but in the order of genesis the thought is prior to the form. The poet has a new thought: he has a whole new experience to unfold; he will tell us how it was with him, and all men will be the richer in his fortune. For, the experience of each new age requires a new confession, and the world seems always waiting for its poet. I remember, when I was young, how much I was moved one morning by tidings that genius had appeared in a youth who sat near me at table. He had left his work, and gone rambling none knew whither, and had written hundreds of lines, but could not tell whether that which was in him was therein told: he could tell nothing but that all was changed,—man, beast, heaven, earth, and sea. How gladly we listened! how credulous! Society seemed to be compromised. We sat in the aurora of a sunrise which was to put out all the stars. Boston seemed to be at twice the distance it had the night before, or was much farther than that. Rome,—what was Rome? Plutarch and Shakspeare were in the yellow leaf, and Homer no more should be heard of. It is much to know that poetry has been written this very day, under this very roof, by your side. What! that wonderful spirit has not expired! these stony moments are still sparkling and animated! I had fancied that the oracles were all silent, and nature had spent her fires, and behold! all night, from every pore, these fine auroras have been streaming. Every one has some interest in the advent of the poet, and no one knows how much it may concern him. We know that the secret of the world is profound, but who or what shall be our interpreter, we know not. A mountain ramble, a new style of face, a new person, may put the key into our hands. Of course, the value of genius to us is in the veracity of its report. Talent may frolic and juggle; genius realizes and adds. Mankind, in good earnest, have availed so far in under-standing themselves and their work, that the foremost watchman on the peak announces his news. It is the truest word ever spoken, and

5. "Chimborazo": a mountain in Ecuador; "line": equator.

the phrase will be the fittest, most musical, and the unerring voice of the world for that time.

All that we call sacred history attests that the birth of a poet is the principal event in chronology. Man, never so often deceived, still watches for the arrival of a brother who can hold him steady to a truth, until he has made it his own. With what joy I begin to read a poem, which I confide in as an inspiration! And now my chains are to be broken; I shall mount above these clouds and opaque airs in which I live,—opaque, though they seem transparent,—and from the heaven of truth I shall see and comprehend my relations. That will reconcile me to life, and renovate nature, to see trifles animated by a tendency, and to know what I am doing. Life will no more be a noise; now I shall see men and women, and know the signs by which they may be discerned from fools and satans. This day shall be better than my birth-day: then I became an animal: now I am invited into the science of the real. Such is the hope, but the fruition is postponed. Oftener it falls, that this winged man, who will carry me into the heaven, whirls me into the clouds, then leaps and frisks about with me from cloud to cloud, still affirming that he is bound heavenward; and I, being myself a novice, am slow in perceiving that he does not know the way into the heavens, and is merely bent that I should admire his skill to rise, like a fowl or a flying fish, a little way from the ground or the water; but the all-piercing, all-feeding, and ocular[6] air of heaven, that man shall never inhabit. I tumble down again soon into my old nooks, and lead the life of exaggerations as before, and have lost my faith in the possibility of any guide who can lead me thither where I would be.

But leaving these victims of vanity, let us, with new hope, observe how nature, by worthier impulses, has ensured the poet's fidelity to his office of announcement and affirming, namely, by the beauty of things, which becomes a new, and higher beauty, when expressed. Nature offers all her creatures to him as a picture-language. Being used as a type, a second wonderful value appears in the object, far better than its old value, as the carpenter's stretched cord, if you hold your ear close enough, is musical in the breeze. "Things more excellent than every image," says Jamblichus,[7] "are expressed through images." Things admit of being used as symbols, because nature is a symbol, in the whole, and in every part. Every line we can draw in the sand, has expression; and there is no body without its spirit or genius. All form is an effect of character; all condition, of the quality of the life; all harmony, of health; (and, for this reason, a perception of beauty should be sympathetic, or proper only to the good.) The beautiful rests on the foundations of the necessary. The soul makes the body, as the wise Spenser teaches:—

6. Visible.
7. Neoplatonic philosopher of the fourth century A.D. (Neoplatonism: a mystical religious system combining features of Platonic and other Greek philosophies with features of Eastern religions, primarily Judaism and Christianity.)

"So every spirit, as it is most pure,
And hath in it the more of heavenly light,
So it the fairer body doth procure
To habit in, and it more fairly dight,
With cheerful grace and amiable sight.
For, of the soul, the body form doth take,
For soul is form, and doth the body make."[8]

Here we find ourselves, suddenly, not in a critical speculation, but in a holy place, and should go very warily and reverently. We stand before the secret of the world, there where Being passes into Appearance, and Unity into Variety.

The Universe is the externisation of the soul. Wherever the life is, that bursts into appearance around it. Our science is sensual, and therefore superficial. The earth, and the heavenly bodies, physics, and chemistry, we sensually treat, as if they were self-existent; but these are the retinue of that Being we have. "The mighty heaven," said Proclus,[9] "exhibits, in its transfigurations, clear images of the splendor of intellectual perceptions; being moved in conjunction with the unapparent periods of intellectual natures." Therefore, science always goes abreast with the just elevation of the man, keeping step with religion and metaphysics; or, the state of science is an index of our self-knowledge. Since everything in nature answers to a moral power, if any phenomenon remains brute and dark, it is that the corresponding faculty in the observer is not yet active.

No wonder, then, if these waters be so deep, that we hover over them with a religious regard. The beauty of the fable proves the importance of the sense; to the poet, and to all others; or, if you please, every man is so far a poet as to be susceptible of these enchantments of nature: for all men have the thoughts whereof the universe is the celebration. I find that the fascination resides in the symbol. Who loves nature? Who does not? Is it only poets, and men of leisure and cultivation, who live with her? No; but also hunters, farmers, grooms, and butchers, though they express their affection in their choice of life, and not in their choice of words. The writer wonders what the coachman or the hunter values in riding, in horses, and dogs. It is not superficial qualities. When you talk with him, he holds these at as slight a rate as you. His worship is sympathetic; he has no definitions, but he is commanded in nature, by the living power which he feels to be there present. No imitation, or playing of these things, would content him; he loves the earnest of the northwind, of rain, of stone, and wood, and iron. A beauty not explicable, is dearer than a beauty which we can see to the end

---

8. *An Hymn in Honour of Beauty* (1596), by the English poet Edmund Spenser (1552–99).

9. Greek Neoplatonic philosopher (411–485).

of. It is nature the symbol, nature certifying the supernatural, body overflowed by life, which he worships, with coarse, but sincere rites.

The inwardness, and mystery, of this attachment, drives men of every class to the use of emblems. The schools of poets, and philosophers, are not more intoxicated with their symbols, than the populace with theirs. In our political parties, compute the power of badges and emblems. See the great ball which they roll from Baltimore to Bunker hill![1] In the political processions, Lowell goes in a loom, and Lynn in a shoe, and Salem in a ship. Witness the cider-barrel, the log-cabin, the hickory-stick, the palmetto, and all the cognizances of party. See the power of national emblems. Some stars, lilies, leopards, a crescent, a lion, an eagle, or other figure, which came into credit God knows how, on an old rag of bunting, blowing in the wind, on a fort, at the ends of the earth, shall make the blood tingle under the rudest, or the most conventional exterior. The people fancy they hate poetry, and they are all poets and mystics!

Beyond this universality of the symbolic language, we are apprised of the divineness of this superior use of things, whereby the world is a temple, whose walls are covered with emblems, pictures, and commandments of the Deity, in this, that there is no fact in nature which does not carry the whole sense of nature; and the distinctions which we make in events, and in affairs, of low and high, honest and base, disappear when nature is used as a symbol. Thought makes every thing fit for use. The vocabulary of an omniscient man would embrace words and images excluded from polite conversation. What would be base, or even obscene, to the obscene, becomes illustrious, spoken in a new connexion of thought. The piety of the Hebrew prophets purges their grossness. The circumcision is an example of the power of poetry to raise the low and offensive. Small and mean things serve as well as great symbols. The meaner the type by which a law is expressed, the more pungent it is, and the more lasting in the memories of men: just as we choose the smallest box, or case, in which any needful utensil can be carried. Bare lists of words are found suggestive, to an imaginative and excited mind; as it is related of Lord Chatham, that he was accustomed to read in Bailey's Dictionary,[2] when he was preparing to speak in Parliament. The poorest experience is rich enough for all the purposes of expressing thought. Why covet a knowledge of new facts? Day and night, house and garden, a few books, a few actions, serve us as well as would all trades and all spectacles. We are far from having exhausted the significance of the few symbols we use.

1. This passage mentions a recent political stunt; in the next sentence towns are symbolized by major products.
2. William Pitt (1708–78), English statesman, famous for his oratory; Nathan (or Nathaniel) Bailey (d. 1742) published *An University Etymological English Dictionary* (1721), which ran through many editions.

342 ★ Ralph Waldo Emerson

We can come to use them yet with a terrible simplicity. It does not need that a poem should be long. Every word was once a poem. Every new relation is a new word. Also, we use defects and deformities to a sacred purpose, so expressing our sense that the evils of the world are such only to the evil eye. In the old mythology, mythologists observe, defects are ascribed to divine natures, as lameness to Vulcan, blindness to Cupid, and the like, to signify exuberances.

For, as it is dislocation and detachment from the life of God, that makes things ugly, the poet, who re-attaches things to nature and the Whole,—re-attaching even artificial things, and violations of nature, to nature, by a deeper insight,—disposes very easily of the most disagreeable facts. Readers of poetry see the factory-village, and the railway, and fancy that the poetry of the landscape is broken up by these; for these works of art are not yet consecrated in their reading; but the poet sees them fall within the great Order not less than the bee-hive, or the spider's geometrical web. Nature adopts them very fast into her vital circles, and the gliding train of cars she loves like her own. Besides, in a centred mind, it signifies nothing how many mechanical inventions you exhibit. Though you add millions, and never so surprising, the fact of mechanics has not gained a grain's weight. The spiritual fact remains unalterable, by many or by few particulars; as no mountain is of any appreciable height to break the curve of the sphere. A shrewd country-boy goes to the city for the first time, and the complacent citizen is not satisfied with his little wonder. It is not that he does not see all the fine houses, and know that he never saw such before, but he disposes of them as easily as the poet finds place for the railway. The chief value of the new fact, is to enhance the great and constant fact of Life, which can dwarf any and every circumstance, and to which the belt of wampum, and the commerce of America, are alike:

The world being thus put under the mind for verb and noun, the poet is he who can articulate it. For, though life is great, and fascinates, and absorbs,—and though all men are intelligent of the symbols through which it is named,—yet they cannot originally use them. We are symbols, and inhabit symbols; workmen, work, and tools, words and things, birth and death, all are emblems; but we sympathize with the symbols, and, being infatuated with the economical uses of things, we do not know that they are thoughts. The poet, by an ulterior intellectual perception, gives them a power which makes their old use forgotten, and puts eyes, and a tongue, into every dumb and inanimate object. He perceives the independence of the thought on the symbol, the stability of the thought, the accidency and fugacity of the symbol. As the eyes of Lyncæus[3] were said to see through the earth, so the poet turns the world to

3. In Greek mythology, the keenest-sighted crewman on the *Argo*, in which Jason sailed in search of the Golden Fleece.

glass, and shows us all things in their right series and procession. For, through that better perception, he stands one step nearer to things, and sees the flowing or metamorphosis; perceives that thought is multiform; that within the form of every creature is a force impelling it to ascend into a higher form; and, following with his eyes the life, uses the forms which express that life, and so his speech flows with the flowing of nature. All the facts of the animal economy, sex, nutriment, gestation, birth, growth, are symbols of the passage of the world into the soul of man, to suffer there a change, and reappear a new and higher fact. He uses forms according to the life, and not according to the form. This is true science. The poet alone knows astronomy, chemistry, vegetation, and animation, for he does not stop at these facts, but employs them as signs. He knows why the plain, or meadow of space, was strown with these flowers we call suns, and moons, and stars; why the great deep is adorned with animals, with men, and gods; for, in every word he speaks he rides on them as the horses of thought.

By virtue of this science the poet is the Namer, or Language-maker, naming things sometimes after their appearance, sometimes after their essence, and giving to every one its own name and not another's, thereby rejoicing the intellect, which delights in detachment or boundary. The poets made all the words, and therefore language is the archives of history, and, if we must say it, a sort of tomb of the muses. For, though the origin of most of our words is forgotten, each word was at first a stroke of genius, and obtained currency, because for the moment it symbolized the world to the first speaker and to the hearer. The etymologist finds the deadest word to have been once a brilliant picture. Language is fossil poetry. As the limestone of the continent consists of infinite masses of the shells of animalcules, so language is made up of images, or tropes, which now, in their secondary use, have long ceased to remind us of their poetic origin. But the poet names the thing because he sees it, or comes one step nearer to it than any other. This expression, or naming, is not art, but a second nature, grown out of the first, as a leaf out of a tree. What we call nature, is a certain self-regulated motion, or change; and nature does all things by her own hands, and does not leave another to baptise her, but baptises herself; and this through the metamorphosis again. I remember that a certain poet[4] described it to me thus:

> Genius is the activity which repairs the decays of things, whether wholly or partly of a material and finite kind. Nature, through all her kingdoms, insures herself. Nobody cares for planting the poor fungus: so she shakes down from the gills of one agaric countless spores, any one of which, being preserved, transmits new billions of spores to-morrow or next day. The

4. A private joke: the poet is Emerson himself.

new agaric of this hour has a chance which the old one had not. This atom of seed is thrown into a new place, not subject to the accidents which destroyed its parent two rods off. She makes a man; and having brought him to ripe age, she will no longer run the risk of losing this wonder at a blow, but she detaches from him a new self, that the kind may be safe from accidents to which the individual is exposed. So when the soul of the poet has come to ripeness of thought, she detaches and sends away from it its poems or songs,—a fearless, sleepless, deathless progeny, which is not exposed to the accidents of the weary kingdom of time: a fearless, vivacious offspring, clad with wings (such was the virtue of the soul out of which they came), which carry them fast and far, and infix them irrecoverably into the hearts of men. These wings are the beauty of the poet's soul. The songs, thus flying immortal from their mortal parent, are pursued by clamorous flights of censures, which swarm in far greater numbers, and threaten to devour them; but these last are not winged. At the end of a very short leap they fall plump down, and rot, having received from the souls out of which they came no beautiful wings. But the melodies of the poet ascend, and leap, and pierce into the deeps of infinite time.

So far the bard taught me, using his freer speech. But nature has a higher end, in the production of new individuals, than security, namely, *ascension*, or, the passage of the soul into higher forms. I knew, in my younger days, the sculptor who made the statue of the youth which stands in the public garden. He was, as I remember, unable to tell directly, what made him happy, or unhappy, but by wonderful indirections he could tell. He rose one day, according to his habit, before the dawn, and saw the morning break, grand as the eternity out of which it came, and, for many days after, he strove to express this tranquillity, and, lo! his chisel had fashioned out of marble the form of a beautiful youth, Phosphorus,[5] whose aspect is such, that, it is said, all persons who look on it become silent. The poet also resigns himself to his mood, and that thought which agitated him is expressed, but *alter idem*,[6] in a manner totally new. The expression is organic, or, the new type which things themselves take when liberated. As, in the sun, objects paint their images on the retina of the eye, so they, sharing the aspiration of the whole universe, tend to paint a far more delicate copy of their essence in his mind. Like the metamorphosis of things into higher organic forms, is their change into melodies. Over everything stands its dæmon, or soul, and, as the form of the thing is reflected by the eye, so the soul of the thing is reflected by a melody. The sea, the mountain-ridge, Niagara, and every flower-bed, pre-exist, or super-exist, in pre-cantations, which sail like odors in the air, and when

5. The Greek god associated with the morning star.
6. The same, yet different.

any man goes by with an ear sufficiently fine, he overhears them, and endeavors to write down the notes, without diluting or depraving them. And herein is the legitimation of criticism, in the mind's faith, that the poems are a corrupt version of some text in nature, with which they ought to be made to tally. A rhyme in one of our sonnets should not be less pleasing than the iterated nodes of a seashell, or the resembling difference of a group of flowers. The pairing of the birds is an idyl, not tedious as our idyls are; a tempest is a rough ode, without falsehood or rant: a summer, with its harvest sown, reaped, and stored, is an epic song, subordinating how many admirably executed parts. Why should not the symmetry and truth that modulate these, glide into our spirits, and we participate the invention of nature?

This insight, which expresses itself by what is called Imagination, is a very high sort of seeing, which does not come by study, but by the intellect being where and what it sees, by sharing the path, or circuit of things through forms, and so making them translucid to others. The path of things is silent. Will they suffer a speaker to go with them? A spy they will not suffer; a lover, a poet, is the transcendency of their own nature,—him they will suffer. The condition of true naming, on the poet's part, is his resigning himself to the divine *aura*[7] which breathes through forms, and accompanying that.

It is a secret which every intellectual man quickly learns, that, beyond the energy of his possessed and conscious intellect, he is capable of a new energy (as of an intellect doubled on itself), by abandonment to the nature of things; that, beside his privacy of power as an individual man, there is a great public power, on which he can draw, by unlocking, at all risks, his human doors, and suffering the ethereal tides to roll and circulate through him: then he is caught up into the life of the Universe, his speech is thunder, his thought is law, and his words are universally intelligible as the plants and animals. The poet knows that he speaks adequately, then, only when he speaks somewhat wildly, or, "with the flower of the mind;" not with the intellect, used as an organ, but with the intellect released from all service, and suffered to take its direction from its celestial life; or, as the ancients were wont to express themselves, not with intellect alone, but with the intellect inebriated by nectar. As the traveller who has lost his way, throws his reins on his horse's neck, and trusts to the instinct of the animal to find his road, so must we do with the divine animal who carries us through this world. For if in any manner we can stimulate this instinct, new passages are opened for us into nature, the mind flows into and through things hardest and highest, and the metamorphosis is possible.

7. Distinctive quality.

This is the reason why bards love wine, mead, narcotics, coffee, tea, opium, the fumes of sandal-wood and tobacco, or whatever other species of animal exhilaration. All men avail themselves of such means as they can, to add this extraordinary power to their normal powers; and to this end they prize conversation, music, pictures, sculpture, dancing, theatres, travelling, war, mobs, fires, gaming, politics, or love, or science, or animal intoxication, which are several coarser or finer *quasi*-mechanical substitutes for the true nectar, which is the ravishment of the intellect by coming nearer to the fact. These are auxiliaries to the centrifugal tendency of a man, to his passage out into free space, and they help him to escape the custody of that body in which he is pent up, and of that jail-yard of individual relations in which he is enclosed. Hence a great number of such as were professionally expressors of Beauty, as painters, poets, musicians, and actors, have been more than others wont to lead a life of pleasure and indulgence; all but the few who received the true nectar; and, as it was a spurious mode of attaining freedom, as it was an emancipation not into the heavens, but into the freedom of baser places, they were punished for that advantage they won, by a dissipation and deterioration. But never can any advantage be taken of nature by a trick. The spirit of the world, the great calm presence of the creator, comes not forth to the sorceries of opium or of wine. The sublime vision comes to the pure and simple soul in a clean and chaste body. That is not an inspiration which we owe to narcotics, but some counterfeit excitement and fury. Milton says, that the lyric poet may drink wine and live generously, but the epic poet, he who shall sing of the gods, and their descent unto men, must drink water out of a wooden bowl.[8] For poetry is not 'Devil's wine,' but God's wine. It is with this as it is with toys. We fill the hands and nurseries of our children with all manner of dolls, drums, and horses, withdrawing their eyes from the plain face and sufficing objects of nature, the sun, and moon, the animals, the water, and stones, which should be their toys. So the poet's habit of living should be set on a key so low and plain, that the common influences should delight him. His cheerfulness should be the gift of the sunlight; the air should suffice for his inspiration, and he should be tipsy with water. That spirit which suffices quiet hearts, which seems to come forth to such from every dry knoll of sere grass, from every pine-stump, and half-imbedded stone, on which the dull March sun shines, comes forth to the poor and hungry, and such as are of simple taste. If thou fill thy brain with Boston and New York, with fashion and covetousness, and wilt stimulate thy jaded senses with wine and French coffee, thou shalt find no radiance of wisdom in the lonely waste of the pinewoods.

If the imagination intoxicates the poet, it is not inactive in other

8. In Milton's *Sixth Latin Elegy*.

men. The metamorphosis excites in the beholder an emotion of joy. The use of symbols has a certain power of emancipation and exhilaration for all men. We seem to be touched by a wand, which makes us dance and run about happily, like children. We are like persons who come out of a cave or cellar into the open air. This is the effect on us of tropes,[9] fables, oracles, and all poetic forms. Poets are thus liberating gods. Men have really got a new sense, and found within their world, another world, or nest of worlds; for, the metamorphosis once seen, we divine that it does not stop. I will not now consider how much this makes the charm of algebra and the mathematics, which also have their tropes, but it is felt in every definition; as, when Aristotle defines *space* to be an immovable vessel, in which things are contained;—or, when Plato defines a *line* to be a flowing point; or, *figure* to be a bound of solid; and many the like. What a joyful sense of freedom we have, when Vitruvius announces the old opinion of artists, that no architect can build any house well, who does not know something of anatomy. When Socrates, in Charmides, tells us that the soul is cured of its maladies by certain incantations, and that these incantations are beautiful reasons, from which temperance is generated in souls; when Plato calls the world an animal; and Timæus affirms that the plants also are animals; or affirms a man to be a heavenly tree, growing with his root, which is his head, upward; and, as George Chapman, following him, writes,—

> "So in our tree of man, whose nervie root
> Springs in his top;"

when Orpheus speaks of hoariness as "that white flower which marks extreme old age;" when Proclus calls the universe the statue of the intellect; when Chaucer, in his praise of 'Gentilesse,' compares good blood in mean condition to fire, which, though carried to the darkest house betwixt this and the mount of Caucasus, will yet hold its natural office, and burn as bright as if twenty thousand men did it behold; when John saw, in the apocalypse, the ruin of the world through evil, and the stars fall from heaven, as the figtree casteth her untimely fruit; when Æsop reports the whole catalogue of common daily relations through the masquerade of birds and beasts;—we take the cheerful hint of the immortality of our essence, and its versatile habit and escapes, as when the gypsies say, "it is in vain to hang them, they cannot die."[1]

9. Figures of speech.
1. Emerson's freewheeling allusiveness embodies the liberation he is celebrating. *Charmides* is one of Plato's dialogues, *Timaeus* another; the Chapman quotation is from his dedication to his translation of Homer; Chaucer's praise of gentilesse is in the "Wife of Bath's Tale"; John's vision is in Revelation 6.13; the Greek Aesop in the sixth century B.C. wrote beast fables which commented on human foibles. The saying attributed to gypsies is unlocated.

The poets are thus liberating gods. The ancient British bards had for the title of their order, "Those who are free throughout the world." They are free, and they make free. An imaginative book renders us much more service at first, by stimulating us through its tropes, than afterward, when we arrive at the precise sense of the author. I think nothing is of any value in books, excepting the transcendental and extraordinary. If a man is inflamed and carried away by his thought, to that degree that he forgets the authors and the public, and heeds only this one dream, which holds him like an insanity, let me read his paper, and you may have all the arguments and histories and criticism. All the value which attaches to Pythagoras, Paracelsus, Cornelius Agrippa, Cardan, Kepler, Swedenborg, Schelling, Oken,[2] or any other who introduces questionable facts into his cosmogony, as angels, devils, magic, astrology, palmistry, mesmerism,[3] and so on, is the certificate we have of departure from routine, and that here is a new witness. That also is the best success in conversation, the magic of liberty, which puts the world, like a ball, in our hands. How cheap even the liberty then seems; how mean to study, when an emotion communicates to the intellect the power to sap and upheave nature: how great the perspective! nations, times, systems, enter and disappear, like threads in tapestry of large figure and many colors; dream delivers us to dream, and, while the drunkenness lasts, we will sell our bed, our philosophy, our religion, in our opulence.

There is good reason why we should prize this liberation. The fate of the poor shepherd, who, blinded and lost in the snowstorm, perishes in a drift within a few feet of his cottage door, is an emblem of the state of man. On the brink of the waters of life and truth, we are miserably dying. The inaccessibleness of every thought but that we are in, is wonderful. What if you come near to it,—you are as remote, when you are nearest, as when you are farthest. Every thought is also a prison; every heaven is also a prison. Therefore we love the poet, the inventor, who in any form, whether in an ode, or in an action, or in looks and behavior, has yielded us a new thought. He unlocks our chains, and admits us to a new scene.

This emancipation is dear to all men, and the power to impart it, as it must come from greater depth and scope of thought, is a measure of intellect. Therefore all books of the imagination endure, all which ascend to that truth, that the writer sees nature beneath him, and uses it as his exponent.[4] Every verse or sentence, possessing this virtue, will take care of its own immortality. The religions

2. Emerson combines a Greek mathematician and mystic philosopher of the sixth century B.C., a 16th-century Swiss alchemist, a 16th-century German physician, a 16th-century Italian mathematician, a 17th-century German astronomer, an 18th-century Swedish statesman and mystic, a 19th-century German philosopher, and a 19th-century German naturalist.

3. Hypnotism.

4. Means of expounding his beliefs.

of the world are the ejaculations[5] of a few imaginative men.

But the quality of the imagination is to flow, and not to freeze. The poet did not stop at the color, or the form, but read their meaning; neither may he rest in this meaning, but he makes the same objects exponents of his new thought. Here is the difference betwixt the poet and the mystic, that the last nails a symbol to one sense, which was a true sense for a moment, but soon becomes old and false. For all symbols are fluxional; all language is vehicular and transitive, and is good, as ferries and horses are, for conveyance, not as farms and houses are, for homestead. Mysticism consists in the mistake of an accidental and individual symbol for an universal one. The morning-redness happens to be the favorite meteor to the eyes of Jacob Behmen,[6] and comes to stand to him for truth and faith; and he believes should stand for the same realities to every reader. But the first reader prefers as naturally the symbol of a mother and child, or a gardener and his bulb, or a jeweller polishing a gem. Either of these, or of a myriad more, are equally good to the person to whom they are significant. Only they must be held lightly, and be very willingly translated into the equivalent terms which others use. And the mystic must be steadily told,—All that you say is just as true without the tedious use of that symbol as with it. Let us have a little algebra, instead of this trite rhetoric,—universal signs, instead of these village symbols,—and we shall both be gainers. The history of hierarchies seems to show, that all religious error consisted in making the symbol too stark and solid, and, at last, nothing but an excess of the organ of language.

Swedenborg, of all men in the recent ages, stands eminently for the translator of nature into thought. I do not know the man in history to whom things stood so uniformly for words. Before him the metamorphosis continually plays. Everything on which his eye rests, obeys the impulses of moral nature. The figs become grapes whilst he eats them. When some of his angels affirmed a truth, the laurel twig which they held blossomed in their hands. The noise which, at a distance, appeared like gnashing and thumping, on coming nearer was found to be the voice of disputants. The men, in one of his visions, seen in heavenly light, appeared like dragons, and seemed in darkness: but, to each other, they appeared as men, and, when the light from heaven shone into their cabin, they complained of the darkness, and were compelled to shut the window that they might see.

There was this perception in him, which makes the poet or seer, an object of awe and terror, namely, that the same man, or society of men, may wear one aspect to themselves and their companions, and a different aspect to higher intelligences. Certain priests, whom

---

5. Throwings forth. 6. German mystic (1575–1624).

he describes as conversing very learnedly together, appeared to the children, who were at some distance, like dead horses: and many the like misappearances. And instantly the mind inquires, whether these fishes under the bridge, yonder oxen in the pasture, those dogs in the yard, are immutably fishes, oxen, and dogs, or only so appear to me, and perchance to themselves appear upright men; and whether I appear as a man to all eyes. The Bramins and Pythagoras propounded the same question, and if any poet has witnessed the transformation, he doubtless found it in harmony with various experiences. We have all seen changes as considerable in wheat and caterpillars. He is the poet, and shall draw us with love and terror, who sees, through the flowing vest, the firm nature, and can declare it.

I look in vain for the poet whom I describe. We do not, with sufficient plainness, or sufficient profoundness, address ourselves to life, nor dare we chaunt our own times and social circumstance. If we filled the day with bravery, we should not shrink from celebrating it. Time and nature yield us many gifts, but not yet the timely man, the new religion, the reconciler, whom all things await. Dante's praise is, that he dared to write his autobiography in colossal cipher, or into universality. We have yet had no genius in America, with tyrannous eye, which knew the value of our incomparable materials, and saw, in the barbarism and materialism of the times, another carnival of the same gods whose picture he so much admires in Homer; then in the middle age; then in Calvinism. Banks and tariffs, the newspaper and caucus, methodism and unitarianism, are flat and dull to dull people, but rest on the same foundations of wonder as the town of Troy, and the temple of Delphos,[7] and are as swiftly passing away. Our logrolling, our stumps and their politics,[8] our fisheries, our Negroes, and Indians, our boasts, and our repudiations, the wrath of rogues, and the pusillanimity of honest men, the northern trade, the southern planting, the western clearing, Oregon, and Texas, are yet unsung. Yet America is a poem in our eyes; its ample geography dazzles the imagination, and it will not wait long for metres. If I have not found that excellent combination of gifts in my countrymen which I seek, neither could I aid myself to fix the idea of the poet by reading now and then in Chalmers's collection of five centuries of English poets.[9] These are wits, more than poets,

7. "Troy": the site of the Trojan War in Asia Minor; "Delphos": the home of the Delphic oracle, or prophetess, in Greece.
8. "Logrolling" seems to be used in the metaphorical sense of exchanging political favors; "stumps" refers to the practice political orators had of addressing audiences from any makeshift platform, even a tree stump. Just below,

"boasts" is the common correction for the first edition's "boats"; Emerson is contrasting the optimism of the states as they sold bonds here and abroad with their seemingly blithe repudiation of states' debts when grandiose projects fell through.
9. A commonly used set compiled by Alexander Chalmers (1759–1834), Scottish journalist and biographer.

though there have been poets among them. But when we adhere to the ideal of the poet, we have our difficulties even with Milton and Homer. Milton is too literary, and Homer too literal and historical.

But I am not wise enough for a national criticism, and must use the old largeness a little longer, to discharge my errand from the muse to the poet concerning his art.

Art is the path of the creator to his work. The paths, or methods, are ideal and eternal, though few men ever see them, not the artist himself for years, or for a lifetime, unless he come into the conditions. The painter, the sculptor, the composer, the epic rhapsodist, the orator, all partake one desire, namely, to express themselves symmetrically and abundantly, not dwarfishly and fragmentarily. They found or put themselves in certain conditions, as, the painter and sculptor before some impressive human figures; the orator, into the assembly of the people; and the others, in such scenes as each has found exciting to his intellect; and each presently feels the new desire. He hears a voice, he sees a beckoning. Then he is apprised, with wonder, what herds of dæmons hem him in. He can no more rest; he says, with the old painter, "By God, it is in me, and must go forth of me." He pursues a beauty, half seen, which flies before him. The poet pours out verses in every solitude. Most of the things he says are conventional, no doubt; but by and by he says something which is original and beautiful. That charms him. He would say nothing else but such things. In our way of talking, we say, 'That is yours, this is mine;' but the poet knows well that it is not his; that it is as strange and beautiful to him as to you; he would fain hear the like eloquence at length. Once having tasted this immortal ichor,[1] he cannot have enough of it, and, as an admirable creative power exists in these intellections, it is of the last importance that these things get spoken. What a little of all we know is said! What drops of all the sea of our science are baled up! and by what accident it is that these are exposed, when so many secrets sleep in nature! Hence the necessity of speech and song; hence these throbs and heart-beatings in the orator, at the door of the assembly, to the end, namely, that thought may be ejaculated as Logos, or Word.

Doubt not, O poet, but persist. Say, 'It is in me, and shall out.' Stand there, baulked and dumb, stuttering and stammering, hissed and hooted, stand and strive, until, at last, rage draw out of thee that *dream*-power which every night shows thee is thine own; a power transcending all limit and privacy, and by virtue of which a man is the conductor of the whole river of electricity. Nothing walks, or creeps, or grows, or exists, which must not in turn arise and walk before him as exponent of his meaning. Comes he to that

1. In Greek myth, blood of the gods; but Emerson may mean nectar, the drink of the gods.

power, his genius is no longer exhaustible. All the creatures, by pairs and by tribes, pour into his mind as into a Noah's ark, to come forth again to people a new world. This is like the stock of air for our respiration, or for the combustion of our fireplace, not a measure of gallons, but the entire atmosphere if wanted. And therefore the rich poets, as Homer, Chaucer, Shakspeare, and Raphael, have obviously no limits to their works, except the limits of their lifetime, and resemble a mirror carried through the street, ready to render an image of every created thing.

O poet! a new nobility is conferred in groves and pastures, and not in castles, or by the sword-blade, any longer. The conditions are hard, but equal. Thou shalt leave the world, and know the muse only. Thou shalt not know any longer the times, customs, graces, politics, or opinions of men, but shalt take all from the muse. For the time of towns is tolled from the world by funereal chimes, but in nature the universal hours are counted by succeeding tribes of animals and plants, and by growth of joy on joy. God wills also that thou abdicate a manifold and duplex life, and that thou be content that others speak for thee. Others shall be thy gentlemen, and shall represent all courtesy and worldly life for thee; others shall do the great and resounding actions also. Thou shalt lie close hid with nature, and canst not be afforded to the Capitol or the Exchange.[2] The world is full of renunciations and apprenticeships, and this is thine: thou must pass for a fool and a churl for a long season. This is the screen and sheath in which Pan[3] has protected his well-beloved flower, and thou shalt be known only to thine own, and they shall console thee with tenderest love. And thou shalt not be able to rehearse the names of thy friends in thy verse, for an old shame before the holy ideal. And this is the reward: that the ideal shall be real to thee, and the impressions of the actual world shall fall like summer rain, copious, but not troublesome, to thy invulnerable essence. Thou shalt have the whole land for thy park and manor, the sea for thy bath and navigation, without tax and without envy; the woods and the rivers thou shalt own; and thou shalt possess that wherein others are only tenants and boarders. Thou true land-lord! sea-lord! air-lord! Wherever snow falls, or water flows, or birds fly, wherever day and night meet in twilight, wherever the blue heaven is hung by clouds, or sown with stars, wherever are forms with transparent boundaries, wherever are outlets into celestial space, wherever is danger, and awe, and love, there is Beauty, plenteous as rain, shed for thee, and though thou shouldest walk the world over, thou shalt not be able to find a condition inopportune or ignoble.

1844

2. Stock exchange.
3. In Greek myth, the god of woods and

fields, represented with goat's legs, horns, and ears.

Experience

The lords of life, the lords of life,—
I saw them pass,
In their own guise,
Like and unlike,
Portly and grim,
Use and Surprise,
Surface and Dream,
Succession swift, and spectral Wrong,
Temperament without a tongue,
And the inventor of the game
Omnipresent without name;—
Some to see, some to be guessed,
They marched from east to west:
Little man, least of all,
Among the legs of his guardians tall,
Walked about with puzzled look:—
Him by the hand dear nature took;
Dearest nature, strong and kind,
Whispered, 'Darling, never mind!
Tomorrow they will wear another face,
The founder thou! these are thy race!"[1]

Where do we find ourselves? In a series of which we do not know the extremes, and believe that it has none. We wake and find ourselves on a stair; there are stairs below us, which we seem to have ascended; there are stairs above us, many a one, which go upward and out of sight. But the Genius[2] which, according to the old belief, stands at the door by which we enter, and gives us the lethe[3] to drink, that we may tell no tales, mixed the cup too strongly, and we cannot shake off the lethargy now at noonday. Sleep lingers all our lifetime about our eyes, as night hovers all day in the boughs of the fir-tree. All things swim and glitter. Our life is not so much threatened as our perception. Ghostlike we glide through nature, and should not know our place again. Did our birth fall in some fit of indigence and frugality in nature, that she was so sparing of her fire and so liberal of her earth, that it appears to us that we lack the affirmative principle, and though we have health and reason, yet we have no superfluity of spirit for new creation? We have enough to live and bring the year about, but not an ounce to impart or to invest. Ah that our Genius were a little more of a genius! We are like millers on the lower levels of a stream, when the factories above them have exhausted the water. We too fancy that the upper people must have raised their dams.

If any of us knew what we were doing, or where we are going, then when we think we best know! We do not know today whether we are busy or idle. In times when we thought ourselves indolent, we have afterwards discovered, that much was accomplished, and

1. First published in *Essays, Second Series* (Boston, 1844), *Experience* emerged rapidly from Emerson's broodings following the death of his young son Waldo in January, 1842, rather than being built up from scattered reflections over a period of years, as most of his essays were. The consequent intensity of concentration and ruthless determination to tell the truth as he saw it give the essay a strong claim to being Emerson's masterpiece. The epigraph is by Emerson.
2. Governing or guardian spirit.
3. Water from the river of forgetfulness in the underworld of Greek myth.

much was begun in us. All our days are so unprofitable while they pass, that 'tis wonderful where or when we ever got anything of this which we call wisdom, poetry, virtue. We never got it on any date calendar day. Some heavenly days must have been intercalated somewhere, like those that Hermes won with dice of the Moon, that Osiris[4] might be born. It is said, all martyrdoms looked mean when they were suffered. Every ship is a romantic object, except that we sail in. Embark, and the romance quits our vessel, and hangs on every other sail in the horizon. Our life looks trivial, and we shun to record it. Men seem to have learned of the horizon the art of perpetual retreating and reference. 'Yonder uplands are rich pasturage, and my neighbor has fertile meadow, but my field,' says the querulous farmer, 'only holds the world together.' I quote another man's saying; unluckily, that other withdraws himself in the same way, and quotes me. 'Tis the trick of nature thus to degrade today; a good deal of buzz, and somewhere a result slipped magically in. Every roof is agreeable to the eye, until it is lifted; then we find tragedy and moaning women, and hard-eyed husbands, and deluges of lethe, and the men ask, 'What's the news?' as if the old were so bad. How many individuals can we count in society? how many actions? how many opinions? So much of our time is preparation, so much is routine, and so much retrospect, that the pith of each man's genius contracts itself to a very few hours. The history of literature—take the net result of Tiraboschi, Warton, or Schlegel,[5] —is a sum of very few ideas, and of very few original tales,—all the rest being variation of these. So in this great society wide lying around us, a critical analysis would find very few spontaneous actions. It is almost all custom and gross sense. There are even few opinions, and these seem organic in the speakers, and do not disturb the universal necessity.

What opium is instilled into all disaster! It shows formidable as we approach it, but there is at last no rough rasping friction, but the most slippery sliding surfaces. We fall soft on a thought. Ate Dea[6] is gentle,

> "Over men's heads walking aloft,
> With tender feet treading so soft."[7]

People grieve and bemoan themselves, but it is not half so bad with them as they say. There are moods in which we court suffering, in the hope that here, at least, we shall find reality, sharp peaks and

4. Chief Egyptian god. The following story is told in Plutarch's *Morals*: the sun god forbade his wife Rhea to give birth on any day of the year, but Hermes won five new days from the moon, during which Osiris could be born.
5. Girolamo Tiraboschi (1731–94); Thomas Warton (1728–90), and either Friedrich von Schlegel (1772–1829) or his brother August Wilhelm von Schlegel (1767–1845), historians, respectively, of Italian, British, and European literature.
6. The goddess of mischief or fatal recklessness.
7. The *Iliad*, Book 19.

edges of truth. But it turns out to be scene-painting and counterfeit. The only thing grief has taught me, is to know how shallow it is. That, like all the rest, plays about the surface, and never introduces me into the reality, for contact with which, we would even pay the costly price of sons and lovers. Was it Boscovich[8] who found out that bodies never come in contact? Well, souls never touch their objects. An innavigable sea washes with silent waves between us and the things we aim at and converse with. Grief too will make us idealists. In the death of my son, now more than two years ago, I seem to have lost a beautiful estate,—no more. I cannot get it nearer to me. If tomorrow I should be informed of the bankruptcy of my principal debtors, the loss of my property would be a great inconvenience to me, perhaps, for many years; but it would leave me as it found me,—neither better nor worse. So is it with this calamity: it does not touch me: some thing which I fancied was a part of me, which could not be torn away without tearing me, nor enlarged without enriching me, falls off from me, and leaves no scar. It was caducous.[9] I grieve that grief can teach me nothing, nor carry me one step into real nature. The Indian who was laid under a curse, that the wind should not blow on him, nor water flow to him, nor fire burn him, is a type of us all. The dearest events are summer-rain, and we the Para coats[1] that shed every drop. Nothing is left us now but death. We look to that with a grim satisfaction, saying, there at least is reality that will not dodge us.

I take this evanescence and lubricity of all objects, which lets them slip through our fingers then when we clutch hardest, to be the most unhandsome part of our condition. Nature does not like to be observed, and likes that we should be her fools and playmates. We may have the sphere for our cricket-ball, but not a berry for our philosophy. Direct strokes she never gave us power to make; all our blows glance, all our hits are accidents. Our relations to each other are oblique and casual.

Dream delivers us to dream, and there is no end to illusion. Life is a train of moods like a string of beads, and, as we pass through them, they prove to be many-colored lenses which paint the world their own hue, and each shows only what lies in its focus. From the mountain you see the mountain. We animate what we can, and we see only what we animate. Nature and books belong to the eyes that see them. It depends on the mood of the man, whether he shall see the sunset or the fine poem. There are always sunsets, and there is always genius; but only a few hours so serene that we can relish nature or criticism. The more or less depends on structure or

8. Ruggiero Giuseppe Boscovich (1711–87), Italian physicist who advanced a molecular theory of matter.
9. Not long-lasting.
1. Rubber overcoats.

temperament. Temperament is the iron wire on which the beads are strung. Of what use is fortune or talent to a cold and defective nature? Who cares what sensibility or discrimination a man has at some time shown, if he falls asleep in his chair? or if he laugh and giggle? or if he apologize? or is affected with egotism? or thinks of his dollar? or cannot go by food? or has gotten a child in his boyhood? Of what use is genius, if the organ is too convex or too concave, and cannot find a focal distance within the actual horizon of human life? Of what use, if the brain is too cold or too hot, and the man does not care enough for results, to stimulate him to experiment, and hold him up in it? or if the web is too finely woven, too irritable by pleasure and pain, so that life stagnates from too much reception, without due outlet? Of what use to make heroic vows of amendment, if the same old law-breaker is to keep them? What cheer can the religious sentiment yield, when that is suspected to be secretly dependent on the seasons of the year, and the state of the blood? I knew a witty physician who found theology in the biliary duct, and used to affirm that if there was disease in the liver, the man became a Calvinist, and if that organ was sound, he became a Unitarian.[2] Very mortifying is the reluctant experience that some unfriendly excess or imbecility neutralizes the promise of genius. We see young men who owe us a new world, so readily and lavishly they promise, but they never acquit the debt; they die young and dodge the account: or if they live, they lose themselves in the crowd.

Temperament also enters fully into the system of illusions, and shuts us in a prison of glass which we cannot see. There is an optical illusion about every person we meet. In truth, they are all creatures of given temperament, which will appear in a given character, whose boundaries they will never pass: but we look at them, they seem alive, and we presume there is impulse in them. In the moment it seems impulse; in the year, in the lifetime, it turns out to be a certain uniform tune which the revolving barrel of the music-box must play. Men resist the conclusion in the morning, but adopt it as the evening wears on, that temper prevails over everything of time, place, and condition, and is inconsumable in the flames of religion. Some modifications the moral sentiment avails to impose, but the individual texture holds its dominion, if not to bias the moral judgments, yet to fix the measure of activity and of enjoyment.

I thus express the law as it is read from the platform of ordinary life, but must not leave it without noticing the capital exception. For temperament is a power which no man willingly hears any one praise but himself. On the platform of physics, we cannot resist the

2. I.e., the Calvinistic sense of Original Sin is seen as an intellectual manifestation of a bodily disease; the Unitarian view of man has none of the Calvinistic preoccupation with eternal damnation for all but the select few, the elect.

contracting influences of so-called science. Temperament puts all divinity to rout. I know the mental proclivity of physicians. I hear the chuckle of the phrenologists.[3] Theoretic kidnappers and slave-drivers, they esteem each man the victim of another, who winds him round his finger by knowing the law of his being, and by such cheap signboards as the color of his beard, or the slope of his occiput, reads the inventory of his fortunes and character. The grossest ignorance does not disgust like this impudent knowingness. The physicians say, they are not materialists; but they are:—Spirit is matter reduced to an extreme thinness: O *so* thin!—But the definition of *spiritual* should be, *that which is its own evidence.* What notions do they attach to love! what to religion! One would not willingly pronounce these words in their hearing, and give them the occasion to profane them. I saw a gracious gentleman who adapts his conversation to the form of the head of the man he talks with! I had fancied that the value of life lay in its inscrutable possibilities; in the fact that I never know, in addressing myself to a new individual, what may befall me. I carry the keys of my castle in my hand, ready to throw them at the feet of my lord, whenever and in what disguise so ever he shall appear. I know he is in the neighborhood hidden among vagabonds. Shall I preclude my future, by taking a high seat, and kindly adapting my conversation to the shape of heads? When I come to that, the doctors shall buy me for a cent.——'But, sir, medical history; the report to the Institute; the proven facts!'—I distrust the facts and the inferences. Temperament is the veto or limitation-power in the constitution, very justly applied to restrain an opposite excess in the constitution, but absurdly offered as a bar to original equity. When virtue is in presence, all subordinate powers sleep. On its own level, or in view of nature, temperament is final. I see not, if one be once caught in this trap of so-called sciences, any escape for the man from the links of the chain of physical necessity. Given such an embryo, such a history must follow. On this platform, one lives in a sty of sensualism, and would soon come to suicide. But it is impossible that the creative power should exclude itself. Into every intelligence there is a door which is never closed, through which the creator passes. The intellect, seeker of absolute truth, or the heart, lover of absolute good, intervenes for our succor, and at one whisper of these high powers, we awake from ineffectual struggles with this nightmare. We hurl it into its own hell, and cannot again contract ourselves to so base a state.

The secret of the illusoriness is in the necessity of a succession of moods or objects. Gladly we would anchor, but the anchorage is quicksand. This onward trick of nature is too strong for us: *Pero si*

3. Pseudoscientists who claimed to read character by the formation of the skull.

muove.[4] When, at night, I look at the moon and stars, I seem stationary, and they to hurry. Our love of the real draws us to permanence, but health of body consists in circulation, and sanity of mind in variety or facility of association. We need change of objects. Dedication to one thought is quickly odious. We house with the insane, and must humor them; then conversation dies out. Once I took such delight in Montaigne, that I thought I should not need any other book; before that, in Shakspeare; then in Plutarch; then in Plotinus; at one time in Bacon; afterwards in Goethe; even in Bettine;[5] but now I turn the pages of either of them languidly, whilst I still cherish their genius. So with pictures; each will bear an emphasis of attention once, which it cannot retain, though we fain would continue to be pleased in that manner. How strongly I have felt of pictures, that when you have seen one well, you must take your leave of it; you shall never see it again. I have had good lessons from pictures, which I have since seen without emotion or remark. A deduction must be made from the opinion, which even the wise express of a new book or occurrence. Their opinion gives me tidings of their mood, and some vague guess at the new fact, but is nowise to be trusted as the lasting relation between that intellect and that thing. The child asks, 'Mamma, why don't I like the story as well as when you told it me yesterday?' Alas, child, it is even so with the oldest cherubim of knowledge. But will it answer thy question to say, Because thou wert born to a whole, and this story is a particular? The reason of the pain this discovery causes us (and we make it late in respect to works of art and intellect), is the plaint of tragedy which murmurs from it in regard to persons, to friendship and love.

That immobility and absence of elasticity which we find in the arts, we find with more pain in the artist. There is no power of expansion in men. Our friends early appear to us as representatives of certain ideas, which they never pass or exceed. They stand on the brink of the ocean of thought and power, but they never take the single step that would bring them there. A man is like a bit of Labrador spar,[6] which has no lustre as you turn it in your hand, until you come to a particular angle; then it shows deep and beautiful colors. There is no adaptation or universal applicability in men, but each has his special talent, and the mastery of successful men consists in adroitly keeping themselves where and when that turn shall be oftenest to be practised. We do what we must, and call it by

4. "It moves, all the same"—Galileo's muttered protest after the Inquisition (tribunal of the Roman Catholic church charged with suppressing heresy) had forced him to retract the idea that the earth revolves around the sun.
5. Michel de Montaigne (1533–92), French essayist; Plutarch (46?–120?), Greek biographer of famous Greeks and Romans; Plotinus (205?–270?), Egyp-

tian-born Roman Neoplatonist philosopher; Sir Francis Bacon (1561–1626), English essayist, philosopher, and statesman; Johann Wolfgang von Goethe (1749–1832), German poet and dramatist; Elizabeth ("Bettine") von Arnim (1785–1859), whose purported correspondence with Goethe was published in 1835.
6. Labradorite, crystalline rock.

the best names we can, and would fain have the praise of having intended the result which ensues. I cannot recall any form of man who is not superfluous sometimes. But is not this pitiful? Life is not worth the taking, to do tricks in.

Of course, it needs the whole society, to give the symmetry we seek. The parti-colored wheel must revolve very fast to appear white. Something is learned too by conversing with so much folly and defect. In fine, whoever loses, we are always of the gaining party. Divinity is behind our failures and follies also. The plays of children are nonsense, but very educative nonsense. So it is with the largest and solemnest things, with commerce, government, church, marriage, and so with the history of every man's bread, and the ways by which he is to come by it. Like a bird which alights nowhere, but hops perpetually from bough to bough, is the Power which abides in no man and in no woman, but for a moment speaks from this one, and for another moment from that one.

But what help from these fineries or pedantries? What help from thought? Life is not dialectics. We, I think, in these times, have had lessons enough of the futility of criticism. Our young people have thought and written much on labor and reform, and for all that they have written, neither the world nor themselves have got on a step. Intellectual tasting of life will not supersede muscular activity. If a man should consider the nicety of the passage of a piece of bread down his throat, he would starve. At Education-Farm,[7] the noblest theory of life sat on the noblest figures of young men and maidens, quite powerless and melancholy. It would not rake or pitch a ton of hay; it would not rub down a horse; and the men and maidens it left pale and hungry. A political orator wittily compared our party promises to western roads, which opened stately enough, with planted trees on either side, to tempt the traveller, but soon became narrow and narrower, and ended in a squirrel-track, and ran up a tree. So does culture with us; it ends in head-ache. Unspeakably sad and barren does life look to those, who a few months ago were dazzled with the splendor of the promise of the times. "There is now no longer any right course of action, nor any self-devotion left among the Iranis."[8] Objections and criticism we have had our fill of. There are objections to every course of life and action, and the practical wisdom infers an indifferency, from the omnipresence of objection. The whole frame of things preaches indifferency. Do not craze yourself with thinking, but go about your business anywhere. Life is not intellectual or critical, but sturdy. Its chief good is for well-mixed people who can enjoy what they find, without question. Nature hates peeping and our mothers speak her very sense

7. Brook Farm, the Transcendentalist commune at West Roxbury, Mass.
8. In the Persian *Desatir*, ancient scriptures credited to Zoroaster, sixth-century-B.C. founder of the Parsee religion.

when they say, "Children, eat your victuals, and say no more of it."
To fill the hour,—that is happiness; to fill the hour, and leave no
crevice for a repentance or an approval. We live amid surfaces, and
the true art of life is to skate well on them. Under the oldest mouldi-
est conventions, a man of native force prospers just as well as in the
newest world, and that by skill of handling and treatment. He can
take hold anywhere. Life itself is a mixture of power and form, and
will not bear the least excess of either. To finish the moment, to find
the journey's end in every stop of the road, to live the greatest number of good hours, is wisdom. It is not the part of men, but of fanatics, or of mathematicians, if you will, to say, that, the shortness of
life considered, it is not worth caring whether for so short a duration
we were sprawling in want, or sitting high. Since our office is with
moments, let us husband them. Five minutes of today are worth as
much to me, as five minutes in the next millennium. Let us be
poised, and wise, and our own, today. Let us treat the men and
women well: treat them as if they were real: perhaps they are. Men
live in their fancy, like drunkards whose hands are too soft and
tremulous for successful labor. It is a tempest of fancies, and the
only ballast I know, is a respect to the present hour. Without any
shadow of doubt, amidst this vertigo of shows and politics, I settle
myself ever the firmer in the creed, that we should not postpone and
refer and wish, but do broad justice where we are, by whomsoever
we deal with, accepting our actual companions and circumstances,
however humble or odious, as the mystic officials to whom the universe has delegated its whole pleasure for us. If these are mean and
malignant, their contentment, which is the last victory of justice, is
a more satisfying echo to the heart, than the voice of poets and the
casual sympathy of admirable persons. I think that however a
thoughtful man may suffer from the defects and absurdities of his
company, he cannot with affection deny to any set of men and
women, a sensibility to extraordinary merit. The coarse and frivolous
have an instinct of superiority, if they have not a sympathy, and
honor it in their blind capricious way with sincere homage.

The fine young people despise life, but in me, and in such as with
me are free from dyspepsia, and to whom a day is a sound and solid
good, it is a great excess of politeness to look scornful and to cry
for company. I am grown by sympathy a little eager and sentimental, but leave me alone, and I should relish every hour and what
it brought me, the potluck of the day, as heartily as the oldest gossip
in the bar-room. I am thankful for small mercies. I compared notes
with one of my friends who expects everything of the universe, and
is disappointed when anything is less than the best, and I found that
I begin at the other extreme, expecting nothing, and am always full
of thanks for moderate goods. I accept the clangor and jangle of
contrary tendencies. I find my account in sots and bores also. They
give a reality to the circumjacent picture, which such a vanishing

meteorous appearance can ill spare. In the morning I awake, and find the old world, wife, babes, and mother, Concord and Boston, the dear old spiritual world, and even the dear old devil not far off. If we will take the good we find, asking no questions, we shall have heaping measures. The great gifts are not got by analysis. Everything good is on the highway. The middle region of our being is the temperate zone. We may climb into the thin and cold realm of pure geometry and lifeless science, or sink into that of sensation. Between these extremes is the equator of life, of thought, of spirit, of poetry,—a narrow belt. Moreover, in popular experience, everything good is on the highway. A collector peeps into all the picture-shops of Europe, for a landscape of Poussin, a crayon-sketch of Salvator; but the Transfiguration, the Last Judgment, the Communion of St. Jerome, and what are as transcendent as these, are on the walls of the Vatican, the Uffizii, or the Louvre, where every footman may see them;[9] to say nothing of nature's pictures in every street, of sunsets and sunrises every day, and the sculpture of the human body never absent. A collector recently bought at public auction, in London, for one hundred and fifty-seven guineas,[1] an autograph of Shakspeare: but for nothing a school-boy can read Hamlet, and can detect secrets of highest concernment yet unpublished therein. I think I will never read any but the commonest books,—the Bible, Homer, Dante, Shakspeare, and Milton. Then we are impatient of so public a life and planet, and run hither and thither for nooks and secrets. The imagination delights in the woodcraft of Indians, trappers, and bee-hunters. We fancy that we are strangers, and not so intimately domesticated in the planet as the wild man, and the wild beast and bird. But the exclusion reaches them also; reaches the climbing, flying, gliding, feathered and four-footed man. Fox and woodchuck, hawk and snipe, and bittern, when nearly seen, have no more root in the deep world than many, and are just such superficial tenants of the globe. Then the new molecular philosophy shows astronomical interspaces betwixt atom and atom, shows that the world is all outside: it has no inside.

The mid-world is best. Nature, as we know her, is no saint. The lights of the church, the ascetics, Gentoos and Grahamites,[2] she does not distinguish by any favor. She comes eating and drinking and sinning. Her darlings, the great, the strong, the beautiful, are not children of our law, do not come out of the Sunday School, nor weigh their food, nor punctually keep the commandments. If we

9. Nicolas Poussin (1594–1665), French painter; Salvator Rosa (1615–73), Italian painter of wild landscapes. The *Transfiguration* is that by Raphael, in Rome; the *Last Judgment* is Michelangelo's, in Florence; and the *Communion of St. Jerome* is that by Il Domenichino, in Paris. The collector hunts for minor paintings in out-of-the-way shops while the great paintings are in museums where anyone may see them.
1. British gold coin worth one shilling more than a pound.
2. Hindu sectarians or contemporary food-faddists (from Sylvester Graham [1794–1851], vegetarian whose efforts at food reform are memorialized in the graham cracker).

will be strong with her strength, we must not harbor such disconsolate consciences, borrowed too from the consciences of other nations. We must set up the strong present tense against all the rumors of wrath, past or to come. So many things are unsettled which it is of the first importance to settle,—and, pending their settlement, we will do as we do. Whilst the debate goes forward on the equity of commerce, and will not be closed for a century or two, New and Old England may keep shop. Law of copyright and international copyright[3] is to be discussed, and, in the interim, we will sell our books for the most we can. Expediency of literature, reason of literature, lawfulness of writing down a thought, is questioned; much is to say on both sides, and, while the fight waxes hot, thou, dearest scholar, stick to thy foolish task, add a line every hour, and between whiles add a line. Right to hold land, right of property, is disputed, and the conventions convene, and before the vote is taken, dig away in your garden, and spend your earnings as a waif or godsend to all serene and beautiful purposes. Life itself is a bubble and a skepticism, and a sleep within a sleep. Grant it, and as much more as they will,—but thou, God's darling! heed thy private dream: thou wilt not be missed in the scorning and skepticism: there are enough of them: stay there in thy closet, and toil, until the rest are agreed what to do about it. Thy sickness, they say, and thy puny habit, require that thou do this or avoid that, but know that thy life is a flitting state, a tent for a night, and do thou, sick or well, finish that stint. Thou art sick, but shalt not be worse, and the universe, which holds thee dear, shall be the better.

Human life is made up of the two elements, power and form, and the proportion must be invariably kept, if we would have it sweet and sound. Each of these elements in excess makes a mischief as hurtful as its defect. Everything runs to excess: every good quality is noxious, if unmixed, and, to carry the danger to the edge of ruin, nature causes each man's peculiarity to superabound. Here, among the farms, we adduce the scholars as examples of this treachery. They are nature's victims of expression. You who see the artist, the orator, the poet, too near, and find their life no more excellent than that of mechanics or farmers, and themselves victims of partiality, very hollow and haggard, and pronounce them failures,—not heroes, but quacks,—conclude very reasonably, that these arts are not for man, but are disease. Yet nature will not bear you out. Irresistible nature made men such, and makes legions more of such, every day. You love the boy reading in a book, gazing at a drawing, or a cast: yet what are these millions who read and behold, but incipient writers and sculptors? Add a little more of that quality which now reads and sees, and they will seize the pen and chisel. And if one remembers how innocently he began to be an artist, he perceives that nature joined with his enemy. A man is a golden impossibility.

3. Not passed by the American Congress until 1891.

The line he must walk is a hair's breadth. The wise through excess of wisdom is made a fool.

How easily, if fate would suffer it, we might keep forever these beautiful limits, and adjust ourselves, once for all, to the perfect calculation of the kingdom of known cause and effect. In the street and in the newspapers, life appears so plain a business, that manly resolution and adherence to the multiplication-table through all weathers, will insure success. But ah! presently comes a day, or is it only a half-hour, with its angel-whispering,—which discomfits the conclusions of nations and of years! Tomorrow again, everything looks real and angular, the habitual standards are reinstated, common sense is as rare as genius,—is the basis of genius, and experience is hands and feet to every enterprise;—and yet, he who should do his business on this understanding, would be quickly bankrupt. Power keeps quite another road than the turnpikes of choice and will, namely, the subterranean and invisible tunnels and channels of life. It is ridiculous that we are diplomatists, and doctors, and considerate people: there are no dupes like these. Life is a series of surprises, and would not be worth taking or keeping, if it were not. God delights to isolate us every day, and hide from us the past and the future. We would look about us, but with grand politeness he draws down before us an impenetrable screen of purest sky, and another behind us of purest sky. 'You will not remember,' he seems to say, 'and you will not expect.' All good conversation, manners, and action, come from a spontaneity which forgets usages, and makes the moment great. Nature hates calculators; her methods are saltatory and impulsive. Man lives by pulses; our organic movements are such; and the chemical and ethereal agents are undulatory and alternate; and the mind goes antagonizing on, and never prospers but by fits. We thrive by casualties. Our chief experiences have been casual. The most attractive class of people are those who are powerful obliquely, and not by the direct stroke: men of genius, but not yet accredited: one gets the cheer of their light, without paying too great a tax. Theirs is the beauty of the bird, or the morning light, and not of art. In the thought of genius there is always a surprise; and the moral sentiment is well called "the newness," for it is never other; as new to the oldest intelligence as to the young child,—"the kingdom that cometh without observation."[4] In like manner, for practical success, there must not be too much design. A man will not be observed in doing that which he can do best. There is a certain magic about his properest action, which stupefies your powers of observation, so that though it is done before you, you wist not of it. The art of life has a pudency, and will not be exposed. Every man is an impossibility, until he is born; every thing impossible, until we see a success. The ardors of piety agree at last with the coldest skepticism,—that nothing is of us or

4. Luke 17.20.

our works,—that all is of God. Nature will not spare us the smallest leaf of laurel. All writing comes by the grace of God, and all doing and having. I would gladly be moral, and keep due metes and bounds, which I dearly love, and allow the most to the will of man, but I have set my heart on honesty in this chapter, and I can see nothing at last, in success or failure, than more or less of vital force supplied from the Eternal. The results of life are uncalculated and uncalculable. The years teach much which the days never know. The persons who compose our company, converse, and come and go, and design and execute many things, and somewhat comes of it all, but an unlooked for result. The individual is always mistaken. He designed many things, and drew in other persons as coadjutors, quarrelled with some or all, blundered much, and something is done; all are a little advanced, but the individual is always mistaken. It turns out somewhat new, and very unlike what he promised himself.

The ancients, struck with this irreducibleness of the elements of human life to calculation, exalted Chance into a divinity, but that is to stay too long at the spark,—which glitters truly at one point,— but the universe is warm with the latency of the same fire. The miracle of life which will not be expounded, but will remain a miracle, introduces a new element. In the growth of the embryo, Sir Everard Home,[5] I think, noticed that the evolution was not from one central point, but co-active from three or more points. Life has no memory. That which proceeds in succession might be remembered, but that which is co-existent, or ejaculated from a deeper cause, as yet far from being conscious, knows not its own tendency. So is it with us, now skeptical, or without unity, because immersed in forms and effects all seeming to be of equal yet hostile value, and now religious, whilst in the reception of spiritual law. Bear with these distractions, with this coetaneous growth of the parts: they will one day be *members*, and obey one will. On that one will, on that secret cause, they nail our attention and hope. Life is hereby melted into an expectation or a religion. Underneath the inharmonious and trivial particulars, is a musical perfection, the Ideal journeying always with us, the heaven without rent or seam. Do but observe the mode of our illumination. When I converse with a profound mind, or if at any time being alone I have good thoughts, I do not at once arrive at satisfactions, as when, being thirsty, I drink water, or go to the fire, being cold: no! but I am at first apprised of my vicinity to a new and excellent region of life. By persisting to read or to think, this region gives further sign of itself, as it were in flashes of light, in sudden discoveries of its profound beauty and repose, as if the clouds that covered it parted at intervals, and showed the approaching traveller the inland mountains,

5. Scottish surgeon (1756–1832).

with the tranquil eternal meadows spread at their base, whereon flocks graze, and shepherds pipe and dance. But every insight from this realm of thought is felt as initial, and promises a sequel. I do not make it; I arrive there, and behold what was there already. I make! O no! I clap my hands in infantine joy and amazement, before the first opening to me of this august magnificence, old with the love and homage of innumerable ages, young with the life of life, the sunbright Mecca of the desert. And what a future it opens! I feel a new heart beating with the love of the new beauty. I am ready to die out of nature, and be born again into this new yet unapproachable America I have found in the West.

> "Since neither now nor yesterday began
> These thoughts, which have been ever, nor yet can
> A man be found who their first entrance knew."[6]

If I have described life as a flux of moods, I must now add, that there is that in us which changes not, and which ranks all sensations and states of mind. The consciousness in each man is a sliding scale, which identifies him now with the First Cause, and now with the flesh of his body; life above life, in infinite degrees. The sentiment from which it sprung determines the dignity of any deed, and the question ever is, not, what you have done or forborne, but, at whose command you have done or forborne it.

Fortune, Minerva,[7] Muse, Holy Ghost,—these are quaint names, too narrow to cover this unbounded substance. The baffled intellect must still kneel before this cause, which refuses to be named,—ineffable cause, which every fine genius has essayed to represent by some emphatic symbol, as, Thales by water, Anaximenes by air, Anaxagoras by ($Nov\varsigma$) thought, Zoroaster by fire, Jesus and the moderns by love:[8] and the metaphor of each has become a national religion. The Chinese Mencius[9] has not been the least successful in his generalization. "I fully understand language," he said, "and nourish well my vast-flowing vigor."—"I beg to ask what you call vast-flowing vigor?"—said his companion. "The explanation," replied Mencius, "is difficult. This vigor is supremely great, and in the highest degree unbending. Nourish it correctly, and do it no injury, and it will fill up the vacancy between heaven and earth. This vigor accords with and assists justice and reason, and leaves no hunger." —In our more correct writing, we give to this generalization the name of Being, and thereby confess that we have arrived as far as we can go. Suffice it for the joy of the universe, that we have not arrived at a wall, but at interminable oceans. Our life seems not

---

6. A free translation of the conclusion of one of the heroine's speeches in Sophocles' *Antigone* (lines 456–57).
7. Roman goddess of wisdom.
8. The first three are Greek philosophers of, respectively, the seventh, sixth, and fifth centuries B.C.; Zoroaster is the sixth-century-B.C. Persian founder of the fire worship of the Parsees.
9. Meng-tsu, third-century-B.C. compiler of doctrines of Confucianism.

present, so much as prospective; not for the affairs on which it is
wasted, but as a hint of this vast-flowing vigor. Most of life seems to
be mere advertisement of faculty: information is given us not to sell
ourselves cheap; that we are very great. So, in particulars, our
greatness is always in a tendency or direction, not in an action. It is
for us to believe in the rule, not in the exception. The noble are thus
known from the ignoble. So in accepting the leading of the sentiments, it is not what we believe concerning the immortality of the
soul, or the like, but *the universal impulse to believe,* that is the
material circumstance, and is the principal fact in the history of the
globe. Shall we describe this cause as that which works directly?
The spirit is not helpless or needful of mediate organs. It has plentiful powers and direct effects. I am explained without explaining, I
am felt without acting, and where I am not. Therefore all just
persons are satisfied with their own praise. They refuse to explain
themselves, and are content that new actions should do them that
office. They believe that we communicate without speech, and
above speech, and that no right action of ours is quite unaffecting to
our friends, at whatever distance; for the influence of action is not
to be measured by miles. Why should I fret myself, because a
circumstance has occurred, which hinders my presence where I was
expected? If I am not at the meeting, my presence where I am,
should be as useful to the commonwealth of friendship and wisdom,
as would be my presence in that place. I exert the same quality of
power in all places. Thus journeys the mighty Ideal before us; it
never was known to fall into the rear. No man ever came to an
experience which was satiating, but his good is tidings of a better.
Onward and onward! In liberated moments, we know that a new
picture of life and duty is already possible; the elements already
exist in many minds around you, of a doctrine of life which shall
transcend any written record we have. The new statement will comprise the skepticisms, as well as the faiths of society, and out of
unbeliefs a creed shall be formed. For, skepticisms are not gratuitous or lawless, but are limitations of the affirmative statement, and
the new philosophy must take them in, and make affirmations outside of them, just as much as it must include the oldest beliefs.

It is very unhappy, but too late to be helped, the discovery we
have made, that we exist. That discovery is called the Fall of Man.
Ever afterwards, we suspect our instruments. We have learned that
we do not see directly, but mediately, and that we have no means of
correcting these colored and distorting lenses which we are, or of
computing the amount of their errors. Perhaps these subject-lenses
have a creative power; perhaps there are no objects. Once we lived
in what we saw; now, the rapaciousness of this new power, which
threatens to absorb all things, engages us. Nature, art, persons,
letters, religions,—objects, successively tumble in, and God is but

one of its ideas. Nature and literature are subjective phenomena; every evil and every good thing is a shadow which we cast. The street is full of humiliations to the proud. As the fop contrived to dress his bailiffs in his livery, and make them wait on his guests at table, so the chagrins[1] which the bad heart gives off as bubbles, at once take form as ladies and gentlemen in the street, shopmen or barkeepers in hotels, and threaten or insult whatever is threatenable and insultable in us. 'Tis the same with our idolatries. People forget that it is the eye which makes the horizon, and the rounding mind's eye which makes this or that man a type of representative of humanity with the name of hero or saint. Jesus the "providential man," is a good man on whom many people are agreed that these optical laws shall take effect. By love on one part, and by forbearance to press objection on the other part, it is for a time settled, that we will look at him in the centre of the horizon, and ascribe to him the properties that will attach to any man so seen. But the longest love or aversion has a speedy term. The great and crescive[2] self, rooted in absolute nature, supplants all relative existence, and ruins the kingdom of mortal friendship and love. Marriage (in what is called the spiritual world) is impossible, because of the inequality between every subject and every object. The subject is the receiver of Godhead, and at every comparison must feel his being enhanced by that cryptic might. Though not in energy, yet by presence, this magazine[3] of substance cannot be otherwise than felt: nor can any force of intellect attribute to the object the proper deity which sleeps or wakes forever in every subject. Never can love make consciousness and ascription equal in force. There will be the same gulf between every me and thee, as between the original and the picture. The universe is the bride of the soul. All private sympathy is partial. Two human beings are like globes, which can touch only in a point, and, whilst they remain in contact, all other points of each of the spheres are inert; their turn must also come, and the longer a particular union lasts, the more energy of appetency[4] the parts not in union acquire.

Life will be imaged, but cannot be divided nor doubled. Any invasion of its unity would be chaos. The soul is not twin-born, but the only begotten, and though revealing itself as child in time, child in appearance, is of a fatal and universal power, admitting no co-life. Every day, every act betrays the ill-concealed deity. We believe in ourselves, as we do not believe in others. We permit all things to ourselves, and that which we call sin in others, is experiment for us. It is an instance of our faith in ourselves, that men never speak of crime as lightly as they think: or, every man thinks a latitude safe for himself, which is nowise to be indulged to another. The act

1. Ill-humored feelings.
2. Increasing.
3. Stored supply.
4. Strong impulse toward union.

looks very differently on the inside, and on the outside; in its quality, and in its consequences. Murder in the murderer is no such ruinous thought as poets and romancers will have it; it does not unsettle him, or fright him from his ordinary notice of trifles: it is an act quite easy to be contemplated, but in its sequel, it turns out to be a horrible jangle and confounding of all relations. Especially the crimes that spring from love, seem right and fair from the actor's point of view, but, when acted, are found destructive of society. No man at last believes that he can be lost, nor that the crime in him is as black as in the felon. Because the intellect qualifies in our own case the moral judgments. For there is no crime to the intellect. That is antinomian or hypernomian,[5] and judges law as well as fact. "It is worse than a crime, it is a blunder," said Napoleon, speaking the language of the intellect. To it, the world is a problem in mathematics or the science of quantity, and it leaves out praise and blame, and all weak emotions. All stealing is comparative. If you come to absolutes, pray who does not steal? Saints are sad, because they behold sin, (even when they speculate,) from the point of view of the conscience, and not of the intellect; a confusion of thought. Sin seen from the thought, is a diminution or *less*: seen from the conscience or will, it is pravity or *bad*. The intellect names it shade, absence of light, and no essence. The conscience must feel it as essence, essential evil. This it is not: it has an objective existence, but no subjective.

Thus inevitably does the universe wear our color, and every object fall successively into the subject itself. The subject exists, the subject enlarges; all things sooner or later fall into place. As I am, so I see; use what language we will, we can never say anything but what we are; Hermes, Cadmus, Columbus, Newton, Buonaparte, are the mind's ministers.[6] Instead of feeling a poverty when we encounter a great man, let us treat the new comer like a travelling geologist, who passes through our estate, and shows us good slate, or limestone, or anthracite, in our brush pasture. The partial action of each strong mind in one direction, is a telescope for the objects on which it is pointed. But every other part of knowledge is to be pushed to the same extravagance, ere the soul attains her due sphericity. Do you see that kitten chasing so prettily her own tail? If you could look with her eyes, you might see her surrounded with hundreds of figures performing complex dramas, with tragic and comic issues, long conversations, many characters, many ups and downs of fate,—and meantime it is only puss and her tail. How

5. Against or beyond the control of law.
6. I.e., great gods or men of legend and history are servants of the human mind because our subjectivity uses them to light up areas of our own being. (Hermes is the Greek god of invention; Cadmus the mythical inventor of the alphabet and creator of the Thebans by sowing dragon's teeth; Columbus discovered America; Newton discovered the law of gravity; and Napoleon Bonaparte in Emerson's childhood was the conqueror of much of Europe.)

long before our masquerade will end its noise of tamborines, laughter, and shouting, and we shall find it was a solitary performance?—A subject and an object,—it takes so much to make the galvanic circuit complete, but magnitude adds nothing. What imports it whether it is Kepler[7] and the sphere; Columbus and America; a reader and his book; or puss with her tail?

It is true that all the muses and love and religion hate these developments, and will find a way to punish the chemist, who publishes in the parlor the secrets of the laboratory. And we cannot say too little of our constitutional necessity of seeing things under private aspects, or saturated with our humors. And yet is the God the native of these bleak rocks. That need makes in morals the capital virtue of self-trust. We must hold hard to this poverty, however scandalous, and by more vigorous self-recoveries, after the sallies of action, possess our axis more firmly. The life of truth is cold, and so far mournful; but it is not the slave of tears, contritions, and perturbations. It does not attempt another's work, nor adopt another's facts. It is a main lesson of wisdom to know your own from another's. I have learned that I cannot dispose of other people's facts; but I possess such a key to my own, as persuades me against all their denials, that they also have a key to theirs. A sympathetic person is placed in the dilemma of a swimmer among drowning men, who all catch at him, and if he gives so much as a leg or a finger, they will drown him. They wish to be saved from the mischiefs of their vices, but not from their vices. Charity would be wasted on this poor waiting on the symptoms. A wise and hardy physician will say, *Come out of that*, as the first condition of advice.

In this our talking America, we are ruined by our good nature and listening on all sides. This compliance takes away the power of being greatly useful. A man should not be able to look other than directly and forthright. A preoccupied attention is the only answer to the importunate frivolity of other people: an attention, and to an aim which makes their wants frivolous. This is a divine answer, and leaves no appeal, and no hard thoughts. In Flaxman's drawing of the Eumenides of Æschylus, Orestes supplicates Apollo, whilst the Furies sleep on the threshold.[8] The face of the god expresses a shade of regret and compassion, but calm with the conviction of the irreconcilableness of the two spheres. He is born into other politics, into the eternal and beautiful. The man at his feet asks for his interest in turmoils of the earth, into which his nature cannot enter.

7. Johannes Kepler, 16th-century German physicist and pioneer in laws of planetary motion.
8. John Flaxman (1755–1826), English illustrator. In the clearer modern usage, the title of Aeschylus's play *The Eume-* *nides* would be italicized; in the scene depicted by Flaxman, the Furies, or Eumenides, who have pursued Orestes since his murder of his adulterous mother, are temporarily lulled by the power of Apollo, who sanctioned the murder.

And the Eumenides there lying express pictorially this disparity.
The god is surcharged with his divine destiny.

Illusion, Temperament, Succession, Surface, Surprise, Reality,
Subjectiveness,—these are threads on the loom of time, these are
the lords of life. I dare not assume to give their order, but I name
them as I find them in my way. I know better than to claim any
completeness for my picture. I am a fragment, and this is a fragment
of me. I can very confidently announce one or another law, which
throws itself into relief and form, but I am too young yet by some
ages to compile a code. I gossip for my hour concerning the eternal
politics. I have seen many fair pictures not in vain. A wonderful time
I have lived in. I am not the novice I was fourteen, nor yet seven
years ago. Let who will ask, where is the fruit? I find a private fruit
sufficient. This is a fruit,—that I should not ask for a rash effect
from meditations, counsels, and the hiving of truths. I should feel it
pitiful to demand a result on this town and county, an overt effect
on the instant month and year. The effect is deep and secular[9] as the
cause. It works on periods in which mortal lifetime is lost. All I
know is reception; I am and I have: but I do not get, and when I
have fancied I had gotten anything, I found I did not. I worship
with wonder the great Fortune. My reception has been so large, that
I am not annoyed by receiving this or that superabundantly. I say to
the Genius, if he will pardon the proverb, *In for a mill, in for a
million.* When I receive a new gift, I do not macerate my body to
make the account square, for, if I should die, I could not make the
account square. The benefit overran the merit the first day, and has
overran the merit ever since. The merit itself, so-called, I reckon part
of the receiving.

Also, that hankering after an overt or practical effect seems to me
an apostasy. In good earnest, I am willing to spare this most unnecessary deal of doing. Life wears to me a visionary face. Hardest,
roughest action is visionary also. It is but a choice between soft and
turbulent dreams. People disparage knowing and the intellectual
life, and urge doing. I am very content with knowing, if only I
could know. That is an august entertainment, and would suffice me
a great while. To know a little, would be worth the expense of this
world. I hear always the law of Adrastia,[1] "that every soul which
had acquired any truth, should be safe from harm until another
period."

I know that the world I converse with in the city and in the
farms, is not the world I *think*. I observe that difference, and shall
observe it. One day, I shall know the value and law of this discrep-
ance. But I have not found that much was gained by manipular

9. Lasting from century to century.
1. Another name for Nemesis or Destiny. The quotation is from the *Phaedrus*
by Plato.

attempts to realize the world of thought. Many eager persons successively make an experiment in this way, and make themselves ridiculous. They acquire democratic manners, they foam at the mouth, they hate and deny. Worse, I observe, that, in the history of mankind, there is never a solitary example of success,—taking their own tests of success. I say this polemically, or in reply to the inquiry, why not realize your world? But far be from me the despair which prejudges the law by a paltry empiricism,—since there never was a right endeavor, but it succeeded. Patience and patience, we shall win at the last. We must be very suspicious of the deceptions of the element of time. It takes a good deal of time to eat or to sleep, or to earn a hundred dollars, and a very little time to entertain a hope and an insight which becomes the light of our life. We dress our garden, eat our dinners, discuss the household with our wives, and these things make no impression, are forgotten next week; but in the solitude to which every man is always returning, he has a sanity and revelations, which in his passage into new worlds he will carry with him. Never mind the ridicule, never mind the defeat: up again, old heart!—it seems to say,—there is victory yet for all justice; and the true romance which the world exists to realize, will be the transformation of genius into practical power.

1844

Hymn Sung at the Completion of the Concord Monument, April 19, 1836

By the rude bridge that arched the flood,
　Their flag to April's breeze unfurled,
Here once the embattled farmers stood,
　And fired the shot heard round the world.

The foe long since in silence slept;　　　　　　5
　Alike the conqueror silent sleeps;
And Time the ruined bridge has swept
　Down the dark stream which seaward creeps.

On this green bank, by this soft stream,
　We set to-day a votive stone;　　　　　　　10
That memory may their deed redeem,
　When, like our sires, our sons are gone.

Spirit, that made those heroes dare
　To die, or leave their children free,
Bid Time and Nature gently spare　　　　　　15
　The shaft we raise to them and thee.

1836

Each and All

Little thinks, in the field, yon red-cloaked clown,[1]
Of thee from the hill-top looking down;
The heifer that lows in the upland farm,
Far-heard, lows not thine ear to charm;
The sexton, tolling his bell at noon,   5
Deems not that great Napoleon
Stops his horse, and lists with delight,
Whilst his files sweep round yon Alpine height;
Nor knowest thou what argument
Thy life to thy neighbor's creed has lent.   10
All are needed by each one;
Nothing is fair or good alone.
I thought the sparrow's note from heaven,
Singing at dawn on the alder bough;
I brought him home, in his nest, at even;   15
He sings the song, but it cheers not now,
For I did not bring home the river and sky;—
He sang to my ear,—they sang to my eye.
The delicate shells lay on the shore;
The bubbles of the latest wave   20
Fresh pearls to their enamel gave;
And the bellowing of the savage sea
Greeted their safe escape to me.
I wiped away the weeds and foam,
I fetched my sea-born treasures home;   25
But the poor, unsightly, noisome things
Had left their beauty on the shore,
With the sun, and the sand, and the wild uproar.
The lover watched his graceful maid,
As 'mid the virgin train she strayed,   30
Nor knew her beauty's best attire
Was woven still by the snow-white choir.
At last she came to his hermitage,
Like the bird from the woodlands to the cage;—
The gay enchantment was undone,   35
A gentle wife, but fairy none.
Then I said, 'I covet truth;
Beauty is unripe childhood's cheat;
I leave it behind with the games of youth.'—
As I spoke, beneath my feet   40
The ground-pine curled its pretty wreath,
Running over the club-moss burrs;
I inhaled the violet's breath;
Around me stood the oaks and firs;
Pine-cones and acorns lay on the ground,   45
Over me soared the eternal sky,
Full of light and of deity;
Again I saw, again I heard,

---

1. Peasant.

The rolling river, the morning bird;—
Beauty through my senses stole;   50
I yielded myself to the perfect whole.

1846

The Problem

I like a church; I like a cowl;
I love a prophet of the soul;
And on my heart monastic aisles
Fall like sweet strains, or pensive smiles;
Yet not for all his faith can see   5
Would I that cowled churchman be.

Why should the vest[1] on him allure,
Which I could not on me endure?

Not from a vain or shallow thought
His awful Jove young Phidias[2] brought;   10
Never from lips of cunning fell
The thrilling Delphic oracle;[3]
Out from the heart of nature rolled
The burdens of the Bible old;
The litanies of nations came,   15
Like the volcano's tongue of flame,
Up from the burning core below,—
The canticles of love and woe;
The hand that rounded Peter's dome,
And groined the aisles of Christian Rome,[4]   20
Wrought in a sad sincerity;
Himself from God he could not free;
He builded better than he knew;—
The conscious stone to beauty grew.

Know'st thou what wove yon woodbird's nest   25
Of leaves, and feathers from her breast?
Or how the fish outbuilt her shell,
Painting with morn each annual cell?
Or how the sacred pine-tree adds
To her old leaves new myriads?   30
Such and so grew these holy piles,
Whilst love and terror laid the tiles.
Earth proudly wears the Parthenon,
As the best gem upon her zone;
And Morning opes with haste her lids,   35
To gaze upon the Pyramids;
O'er England's abbeys bends the sky,

1. Vestment.
2. Greek sculptor of the fifth century
B.C.
3. The prophetess at the temple of
Apollo at Delphos, in Greece.

4. Late in life the great Renaissance
artist Michelangelo (1475–1564) became
the chief architect of St. Peter's Cathedral
in Rome.

374 ★ Ralph Waldo Emerson

As on its friends, with kindred eye;
For, out of Thought's interior sphere,
These wonders rose to upper air;   40
And Nature gladly gave them place,
Adopted them into her race,
And granted them an equal date
With Andes and with Ararat.[5]

These temples grew as grows the grass;   45
Art might obey, but not surpass.
The passive Master lent his hand
To the vast soul that o'er him planned;
And the same power that reared the shrine,
Bestrode the tribes that knelt within.   50
Ever the fiery Pentecost[6]
Girds with one flame the countless host,
Trances the heart through chanting choirs,
And through the priest the mind inspires.
The word unto the prophet spoken   55
Was writ on tables yet unbroken;
The word by seers or sibyls told,
In groves of oak, or fanes[7] of gold,
Still floats upon the morning wind,
Still whispers to the willing mind.   60
One accent of the Holy Ghost
The heedless world hath never lost.
I know what say the fathers wise,—
The Book itself before me lies,
Old *Chrysostom*, best Augustine,[8]   65
And he who blent both in his line,
The younger *Golden Lips* or mines,
Taylor, the Shakspeare of divines.[9]
His words are music in my ear,
I see his cowled portrait dear;   70
And yet, for all his faith could see,
I would not the good bishop be.

1846

Uriel[1]

It fell in the ancient periods,
Which the brooding soul surveys,
Or ever the wild Time coined itself
Into calendar months and days.

5. "Andes": mountain range in South America; "Ararat": mountain in Asia Minor where Noah's ark landed after the flood.
6. The Holy Spirit, whose descent is described in Acts 2.
7. Temples.
8. Saint John Chrysostom (345?–407), Patriarch of Constantinople, renowned for eloquence (his surname meaning "golden-mouthed"), a major force in early Christianity; Saint Augustine (354–430), Bishop of Hippo in Roman Africa, greatest religious thinker of early Christianity, in early life a teacher of rhetoric.
9. Jeremy Taylor (1613–67), Anglican clergyman and author, famous from early manhood for his eloquence; "divines": preachers.
1. Borrowed from the name of Milton's angel of the sun (*Paradise Lost*, III.648–54).

This was the lapse of Uriel,   5
Which in Paradise befell.
Once, among the Pleiads walking,
SAID[2] overheard the young gods talking;
And the treason, too long pent,
To his ears was evident.   10
The young deities discussed
Laws of form, and metre just,
Orb, quintessence, and sunbeams,
What subsisteth, and what seems.
One, with low tones that decide,   15
And doubt and reverend use defied,
With a look that solved the sphere,
And stirred the devils everywhere,
Gave his sentiment divine
Against the being of a line.   20
'Line in nature is not found;
Unit and universe are round;
In vain produced, all rays return;
Evil will bless, and ice will burn.'
As Uriel spoke with piercing eye,   25
A shudder ran around the sky;
The stern old war-gods shook their heads;
The seraphs frowned from myrtle-beds;
Seemed to the holy festival
The rash word boded ill to all;   30
The balance-beam of Fate was bent;
The bounds of good and ill were rent;
Strong Hades could not keep his own,
But all slid to confusion.
A sad self-knowledge, withering, fell   35
On the beauty of Uriel;
In heaven once eminent, the god
Withdrew, that hour, into his cloud;
Whether doomed to long gyration
In the sea of generation,   40
Or by knowledge grown too bright
To hit the nerve of feebler sight.
Straightway, a forgetting wind
Stole over the celestial kind,
And their lips the secret kept,   45
If in ashes the fire-seed slept.
But now and then, truth-speaking things
Shamed the angels' veiling wings;
And, shrilling from the solar course,
Or from fruit of chemic force,   50
Procession of a soul in matter,
Or the speeding change of water,

2. "Pleiades": A cluster of seven stars named in Greek mythology for the daughters of Atlas and Pleione. "Said," in later editions changed to "Seyd," is adapted from the Persian poet Saadi, whom Emerson had read.

Or out of the good of evil born,
Came Uriel's voice of cherub scorn,
And a blush tinged the upper sky,   55
And the gods shook, they knew not why.

1846

The Snow-Storm

Announced by all the trumpets of the sky,
Arrives the snow, and, driving o'er the fields,
Seems nowhere to alight: the whited air
Hides hills and woods, the river, and the heaven,
And veils the farm-house at the garden's end.   5
The sled and traveller stopped, the courier's feet
Delayed, all friends shut out, the housemates sit
Around the radiant fireplace, enclosed
In a tumultuous privacy of storm.

Come see the north wind's masonry.   10
Out of an unseen quarry evermore
Furnished with tile, the fierce artificer
Curves his white bastions with projected roof
Round every windward stake, or tree, or door.
Speeding, the myriad-handed, his wild work   15
So fanciful, so savage, nought cares he
For number or proportion. Mockingly,
On coop or kennel he hangs Parian wreaths;[1]
A swan-like form invests the hidden thorn;
Fills up the farmer's lane from wall to wall,   20
Maugre[2] the farmer's sighs; and, at the gate,
A tapering turret overtops the work.
And when his hours are numbered, and the world
Is all his own, retiring, as he were not,
Leaves, when the sun appears, astonished Art   25
To mimic in slow structures, stone by stone,
Built in an age, the mad wind's night-work,
The frolic architecture of the snow.

1846

Ode, Inscribed to W. H. Channing[1]

Though loath to grieve
The evil time's sole patriot,
I cannot leave
My honied thought
For the priest's cant,   5
Or statesman's rant.

---

1. I.e., sculpted in white, from the marble quarried in the island of Paros, in the Aegean sea, for sculptors of classical Greece.
2. Despite.
1. A young clergyman, nephew of the famous Unitarian minister William Ellery Channing; as the poem makes clear, he had urged Emerson to take an overt political role in resisting the war waged by the United States against Mexico.

If I refuse
My study for their politique,
Which at the best is trick,
The angry Muse   10
Puts confusion in my brain.

But who is he that prates
Of the culture of mankind,
Of better arts and life?
Go, blindworm, go,   15
Behold the famous States
Harrying Mexico
With rifle and with knife!

Or who, with accent bolder,
Dare praise the freedom-loving mountaineer?   20
I found by thee, O rushing Contoocook!
And in thy valleys, Agiochook![2]
The jackals of the negro-holder.

The God who made New Hampshire
Taunted the lofty land   25
With little men;—
Small bat and wren

House in the oak:—
If earth-fire cleave
The upheaved land, and bury the folk,   30
The southern crocodile would grieve.

Virtue palters; Right is hence;
Freedom praised, but hid;
Funeral eloquence
Rattles the coffin-lid.   35

What boots thy zeal,
O glowing friend,
That would indignant rend
The northland from the south?
Wherefore? to what good end?   40
Boston Bay and Bunker Hill[3]
Would serve things still;—
Things are of the snake.

The horseman serves the horse,
The neatherd serves the neat,[4]   45

2. New Hampshire had gone Democratic, which meant in effect proslavery, reason enough for Emerson to taunt the debased inhabitants of that state with the majesty of their rivers like Contoocook and mountains like Agiochook.
3. Emerson does not spare his own state from accusation of materialism, whatever its heroic past memorialized in the Bunker Hill Monument. One manifestation of that materialism was a commercial alliance between the South which grew cotton and the North which shipped and manufactured it.
4. The cowherd serves the cow.

The merchant serves the purse,
The eater serves his meat;
'Tis the day of the chattel,
Web to weave, and corn to grind;
Things are in the saddle,   50
And ride mankind.

There are two laws discrete,
Not reconciled,—
Law for man, and law for thing;
The last builds town and fleet,   55
But it runs wild,
And doth the man unking.

'Tis fit the forest fall,
The steep be graded,
The mountain tunnelled,   60
The sand shaded,
The orchard planted,
The glebe[5] tilled,
The prairie granted,
The steamer built.   65

Let man serve law for man;
Live for friendship, live for love,
For truth's and harmony's behoof;
The state may follow how it can,
As Olympus follows Jove.[6]   70

    Yet do not I invite[7]
The wrinkled shopman to my sounding woods,
Nor bid the unwilling senator
Ask votes of thrushes in the solitudes.
Every one to his chosen work;—   75
Foolish hands may mix and mar;
Wise and sure the issues are.
Round they roll till dark is light,
Sex to sex, and even to odd;—
The over-god   80
Who marries Right to Might,
Who peoples, unpeoples,—
He who exterminates
Races by stronger races,
Black by white faces,—   85
Knows to bring honey
Out of the lion;[8]
Grafts gentlest scion
On pirate and Turk.

5. Soil.
6. As minor gods follow the supreme
god.
7. Later texts read "implore."
8. A reference to the riddle (Judges 14)
which Samson propounded after finding
the carcass of a lion in which bees had
made honey: "Out of the eater came
forth meat, and out of the strong came
forth sweetness."

The Cossack eats Poland,   90
Like stolen fruit;[9]
Her last noble is ruined,
Her last poet mute:
Straight, into double band
The victors divide;   95
Half for freedom strike and stand;—
The astonished Muse finds thousands at her side.

          1846

Merlin[1]

I

Thy trivial harp will never please
Or fill my craving ear;
Its chords should ring as blows the breeze,
Free, peremptory, clear.
No jingling serenader's art,   5
Nor tinkle of piano strings,
Can make the wild blood start
In its mystic springs.
The kingly bard
Must smite the chords rudely and hard,   10
As with hammer or with mace;
That they may render back
Artful thunder, which conveys
Secrets of the solar track,
Sparks of the supersolar blaze.   15
Merlin's blows are strokes of fate,
Chiming with the forest tone,
When boughs buffet boughs in the wood;
Chiming with the gasp and moan
Of the ice-imprisoned flood;   20
With the pulse of manly hearts;
With the voice of orators;
With the din of city arts;
With the cannonade of wars;
With the marches of the brave;   25
And prayers of might from martyrs' cave.

Great is the art,
Great be the manners, of the bard.
He shall not his brain encumber
With the coil of rhythm and number;   30
But, leaving rule and pale forethought,
He shall aye[2] climb

9. Poland had been partitioned three times in the late 18th century, with Russia (the Cossack) getting the lion's share.

1. Here, the type of a great poet (not the magician-prophet of Arthurian legend).
2. Always.

For his rhyme.
'Pass in, pass in,' the angels say,
'In to the upper doors,   35
Nor count compartments of the floors,
But mount to paradise
By the stairway of surprise.'

Blameless master of the games,
King of sport that never shames,   40
He shall daily joy dispense
Hid in song's sweet influence.
Things[3] more cheerly live and go,
What time the subtle mind
Sings aloud the tune whereto   45
Their pulses beat,
And march their feet,
And their members are combined.

By Sybarites[4] beguiled,
He shall no task decline;   50
Merlin's mighty line
Extremes of nature reconciled,—
Bereaved a tyrant of his will,
And made the lion mild.
Songs can the tempest still,   55
Scattered on the stormy air,
Mould the year to fair increase,
And bring in poetic peace.

He shall not seek to weave,
In weak, unhappy times,   60
Efficacious rhymes;
Wait his returning strength.
Bird, that from the nadir's floor
To the zenith's top can soar,
The soaring orbit of the muse exceeds that
        journey's length.   65
Nor profane affect to hit
Or compass that, by meddling wit,
Which only the propitious mind
Publishes when 'tis inclined.
There are open hours   70
When the God's will sallies free,
And the dull idiot might see
The flowing fortunes of a thousand years;—
Sudden, at unawares,
Self-moved, fly-to the doors   75
Nor sword of angels could reveal
What they conceal.

3. Later texts read "Forms."
4. Like people from Sybaris, the Greek
city in Italy notorious for wealth and
hedonistic indulgence.

II

The rhyme of the poet
Modulates the king's affairs;
Balance-loving Nature   80
Made all things in pairs.
To every foot its antipode;
Each color with its counter glowed;
To every tone beat answering tones,
Higher or graver;   85
Flavor gladly blends with flavor;
Leaf answers leaf upon the bough;
And match the paired cotyledons.
Hands to hands, and feet to feet,
Coeval grooms and brides;[5]   90
Eldest rite, two married sides
In every mortal meet.
Light's far furnace shines,
Smelting balls and bars,
Forging double stars,   95
Glittering twins and trines.
The animals are sick with love,
Lovesick with rhyme;
Each with all propitious time
Into chorus wove.   100

Like the dancers' ordered band,
Thoughts come also hand in hand;
In equal couples mated,
Or else alternated;
Adding by their mutual gage,   105
One to other, health and age.
Solitary fancies go
Short-lived wandering to and fro,
Most like to bachelors,
Or an ungiven maid,   110
Not ancestors,
With no posterity to make the lie afraid,
Or keep truth undecayed.
Perfect-paired as eagle's wings,
Justice is the rhyme of things;   115
Trade and counting use
The self-same tuneful muse;
And Nemesis,[6]
Who with even matches odd,
Who athwart space redresses   120
The partial wrong,
Fills the just period,
And finishes the song.

5. Later texts read "In one body grooms and brides."
6. Here, the imposer of order, not the personification of divine disapproval and punishment.

Subtle rhymes, with ruin rife,
Murmur in the house of life,   125
Sung by the Sisters⁷ as they spin;
In perfect time and measure they
Build and unbuild our echoing clay,
As the two twilights of the day
Fold us music-drunken in.   130

                                                1846

Days

Daughters of Time, the hypocritic Days,
Muffled and dumb, like barefoot dervishes,¹
And marching single in an endless file,
Bring diadems and fagots² in their hands.
To each they offer gifts, after his will,—   5
Bread, kingdoms, stars, or sky that holds them all.
I, in my pleached³ garden, watched the pomp,
Forgot my morning wishes, hastily
Took a few herbs and apples, and the Day
Turned and departed silent. I, too late,   10
Under her solemn fillet⁴ saw the scorn.

                                                1857

7. The Fates, in Greek mythology.   3. Shaded with interlaced branches or
1. Moslem ascetics.   vines.
2. Crowns and sticks, unequal portions.   4. Headband.

NATHANIEL HAWTHORNE
1804–1864

Hawthorne was born on Independence Day, 1804, in Salem, Massachusetts,
a descendant of Puritan immigrants; one ancestor had been a judge in the
Salem witchcraft trials. The family, like the seaport town, was on the decline. When his sea-captain father died in Dutch Guiana in 1808, his
mother's brothers took responsibility for his education. In his early teens
he lived three years as free as "a bird of the air" at Sebago Lake, in Maine
(then still a part of Massachusetts), acquiring a love of tramping which he
always kept. By his midteens he was reading eighteenth-century novelists
like Henry Fielding, Tobias Smollett, and Horace Walpole as well as contemporary writers like William Godwin and Sir Walter Scott, and forming
an ambition to be a writer himself. At Bowdoin College shyness caused him
to try to evade the obligatory public declamations, but in social clubs he

formed smoking, card-playing, and drinking friendships; two fellow members of the Democratic literary society, Horatio Bridge and Franklin Pierce, later President, became lifelong friends; Longfellow, another classmate, belonged to the rival Federalist society. Hawthorne kept outdoors a good deal at the bucolic college but managed, as he later said, to read "desultorily right and left." At the graduation ceremonies in 1825, Longfellow spoke optimistically on the possibility that "Our Native Writers" could achieve lasting fame. Hawthorne went home to Salem and became a writer, but he was agonizingly slow in winning acclaim.

Hawthorne's years between 1825 and 1837 have fascinated his biographers and critics. Hawthorne himself took pains to propagate the notion that he had lived as a hermit who left his upstairs room only for nighttime walks and hardly communicated even with his mother and sisters. Twentieth-century scholars have shown that although in fact Hawthorne was intensely committed to his writing and was steeping himself in colonial history more than the political issues of his time, he socialized in Salem, had several more or less serious flirtations, kept in touch with Pierce and Bridge, among others, and spent most of the summers knocking about all over New England (an uncle owned stage-lines). He even got as far as Detroit one year. Often called his apprenticeship, these dozen years in fact encompassed as well his period of finest creativity. The first surviving piece of his true apprenticework is the historical novel *Fanshawe*, which Hawthorne paid to have published in 1828 and then quickly suppressed.

Over the next several years Hawthorne tried unsuccessfully to find a publisher for collections of the tales he was writing. In chagrin he burned *Seven Tales of My Native Land* (including one or two stories of witchcraft) though at least one of the seven, *Alice Doane's Appeal*, survives in an altered form. By 1829 he was negotiating—again fruitlessly—for the publication of a volume called *Provincial Tales*, which included *The Gentle Boy*, as well, apparently, as *Roger Malvin's Burial* and *My Kinsman, Major Molineux*. In tales like these he had found his special—though highly unsatisfactory—outlets for publication: magazines and the literary annuals that were issued each fall as genteel Christmas gifts. For his tales Hawthorne got a few dollars each and no fame at all, since publication in the annuals was anonymous. He continued to strive to interest a bookseller in his tales, offering what could have been a remarkable volume called *"The Story Teller,"* in which the title character wandered about New England telling his stories in dramatic settings and circumstances. One story, *Mr. Higginbotham's Catastrophe*, reached print in its narrative frame, but the editor of the *New-England Magazine* scrapped the frame for *Young Goodman Brown* and others which are now known as isolated items instead of interrelated elements in a larger whole. The biographer Randall Stewart plausibly suggests that *"The Story Teller"* "would have united in one work Hawthorne's imaginative and reportorial faculties as none of his published writings quite do." In 1836 Hawthorne turned to literary hackwork, making an encyclopedia for the Boston publisher Samuel G. Goodrich, whose annual, *The Token*, had become the regular market for his tales. In the same year Bridge secretly persuaded Goodrich to publish a collection of Hawthorne's tales by promising to repay any losses. *Twice-Told Tales* appeared in March, 1837, with Hawthorne's name on the title page; the title was a self-deprecating allusion to Shakespeare's *King John* 3.4: "Life is as tedious as a twice-told tale / Vexing the dull eare of a drowsie man." The book was reviewed in England as well as the United States, and opened up what Hawthorne called "an intercourse with the world." A notebook entry written sometime in 1836 was only a little premature: "In this dismal and sordid chamber FAME was won."

Throughout the early stories, both those collected in *Twice-Told Tales* and those he left for later gleaning, Hawthorne mused obsessively over a small range of psychological themes: the consequences of pride, selfishness, and secret guilt; the conflict between lighthearted and somber attitudes toward life; the difficulty of preventing isolation from leading to coldness of heart; the impingement of the past (especially the Puritan past) upon the present; the futility of comprehensive social reforms; the impossibility of eradicating sin from the human heart. Above all, his theme was curiosity about the recesses of other men's and women's beings. About this theme he was always ambivalent, for he knew that his success as a writer depended upon his keen psychological analysis of people he met, while he could never forget that invasion of the sanctity of another's personality may harden the heart even as it enriches the mind. He knew that there was "something of the hawk-eye" about him, and that the line was vague between prurient curiosity and legitimate artistic study of character. At his best, he was a master of psychological insight, and some of his power of psychological burrowing remained with him throughout his career, even in the romances which were left unfinished at his death.

The year 1837 was the start of Howthorne's public literary career; it also marked the end of his single-minded dedication to his work. In the fall of 1838 Elizabeth Peabody, a Salemite who was to become a major force in American educational reform, sought out the new local celebrity. When Hawthorne met her sister Sophia, twenty-nine and an invalid, his life abruptly changed course. Within a few months he and Sophia were engaged. To save money for marriage, Hawthorne worked as salt and coal measurer in the Boston Custom House during 1839 and 1840, then the next year invested in the utopian community Brook Farm, more as a business venture than as a philosophical gesture; the only return, however, was the locale he later used for *The Blithedale Romance* (1852). During his engagement, Hawthorne's main literary productions were letters to Sophia —full of ironical self-deprecation, satirical reportage, and romantic effusions. In December, 1841, he wrote Evert A. Duyckinck and Cornelius Mathews, New York magazine editors, that his early stories had grown out of quietude and seclusion, the lack of which would probably prevent him from writing any more. Marriage, not literature, became Hawthorne's new career long before the actual ceremony in July, 1842. As he rather severely put it, "when a man has taken upon himself to beget children, he has no longer any right to a life of his own."

The first three years of marriage, spent at the Old Manse in Concord, the home of Emerson's ancestors, seemed idyllic to the Hawthornes, but a hoped-for novel never materialized. By now comfortably familiar with accounts of the Puritan and Revolutionary past, he wrote a child's history of colonial and revolutionary New England, *Grandfather's Chair* (1841) and four years later a rewriting of Bridge's *Journal of an African Cruiser*. *Mosses from an Old Manse* (1846) consisted mainly of new tales, but among the early ones first collected in it were *Roger Malvin's Burial* and *Young Goodman Brown*. His literary earnings were not rising, but his reputation was, partly through his own shrewd creation of a public persona. Knowing that certain readers who delight in realism would be disturbed by the shadowiness of some of his stories, he anticipated the worst that could be said, declaring in the whimsical survey of his career in the headnote to *Rappaccini's Daughter* that "M. de l'Aubépine" (French for "Hawthorne") had "an inveterate love of allegory, which is apt to invest his plots and characters with the aspect of scenery and people in the clouds, and to steal away the human warmth out of his conceptions." Any hostility, of course, was disarmed by such self-criticism, and in the introductory essay for *Mosses from an Old Manse* Hawthorne pursued his strategy of evoking for himself an equivalent of the Miltonic fit audience though few, yet enlarging that audience without letting his readers feel they were part of a mob. Hawthorne insisted winningly both on his ultimate reserve—his refusal to "serve up" his own heart delicately fried—and on his eagerness to communicate with his chosen audience. Even after he attained a large readership, he knew the value of trading on his own early obscurity so as to make a reader feel like a special discoverer of a rarity yet unshared by the many. In the 1851 edition of *Twice-Told Tales*, Hawthorne observed that the author, "on the internal evidence of his sketches, came to be regarded as a mild, shy, gentle, melancholic, exceedingly sensitive, and not very forcible man, hiding his blushes under an assumed name, the quaintness of which was supposed, somehow or other, to symbolize his personal and literary traits." While summarizing the image critics had conceived of him, he helped fix that image for a century and more as *the* Hawthorne.

Through long service to the local Democrats, Hawthorne was named Surveyor of the Port of Salem in 1846. The office was something of a sinecure, but his forenoons—always his most productive hours—had to be spent at the Custom House, and he wrote little. Hawthorne was thrown out of office by the new Whig administration in June, 1849, amid a furious controversy in the newspapers. He then spent a summer of "great diversity and severity" of emotion climaxed by his mother's death. In September he was at work upon *The Scarlet Letter*, which he planned as a long tale to make up half a volume called "*Old Time Legends; together with Sketches, Experimental and Ideal.*" Besides the long introduction, *The Custom House*, which was Howthorne's means of revenging himself on the Salem Whigs who had ousted him, he planned to include some still uncollected tales. James Fields, the young associate of the publisher William D. Ticknor, persuaded him that a longer fiction would sell better than another collection of stories, and Hawthorne obligingly expanded his manuscript. Although it was frequently denounced as licentious or morbid, *The Scarlet Letter* (1850) was nevertheless a literary sensation in the United States and Great Britain, and Hawthorne was proclaimed as the finest American romancer. There had already been many novels set in Puritan New England, and many more followed, but *The Scarlet Letter* remains the single classic of the group, appealing to tastes of changing generations in different ways; perhaps the most powerful appeal has not changed at all: the remarkable way Hawthorne manages to evoke emotional sympathy for the heroine even when he is condemning her actions.

During a year and a half in the Berkshires of western Massachusetts, where Melville became his "not-too-distant neighbor," Hawthorne wrote *The House of the Seven Gables* (1851), assembled *The Snow Image,* mainly from very early pieces, and wrote for children A *Wonder-Book* (1852). Escaping from the rigors of the Berkshire winters, he wrote *The Blithedale Romance* (1852) in West Newton; then in the first home he had owned, the Wayside at Concord, he put together a political biography of his friend Franklin Pierce for the campaign of 1852 and worked up *The Tanglewood Tales* (1853), prettified stories from mythology. This productivity was broken when President Pierce appointed him American consul at Liverpool. The consulship came as a blessing despite the disruption of his new life at Concord, for his literary income was not enough to support his family, which now included a son and two daughters.

At Liverpool (1853–57) Hawthorne was an uncommonly industrious consul; he had always been more comfortable among businessmen and politicians than among literary people. Tireless in sightseeing among ancient inns, castles, and other public buildings, he also set himself a rigorous course of gallery-going and elaborately recorded his observations in his notebooks. Exposed to great museums for the first time, Hawthorne surprised himself with his affinity for the seventeenth-century Dutch masters of genre painting, deciding that those painters "accomplish all they aim at, —a praise, methinks, which can be given to no other men since the world began." He forced himself—fortified with liquor—to make the required public speeches, and late in his consulship let himself be lionized during an extended trip to London. A stay in Italy—starting in the miserably cold first months of 1858—ate deeply into the more than $30,000 he had earned at Liverpool, and malaria nearly killed his daughter Una. Except during her illness, he kept up his minutely detailed tourist's account, as well as a record of the family's contacts with the English and American colony of painters, sculptors, and writers. Many pages of the notebooks went nearly verbatim into a book which he began in Florence in 1858 and finished late in 1859, after his return to England. This romance, suggested by the statue of a faun attributed to the classical Greek sculptor Praxiteles, was published in London (1860) as *Transformation* and in the United States under Hawthorne's preferred title, *The Marble Faun.*

The Hawthornes came home in June, 1860, during the general acclaim of the new romance, and set about fitting up the Wayside; this project was a considerable drain on Hawthorne's savings, which were already depleted by prolonged residence abroad after resigning his consulship and by generous, though unwise, loans to friends. His literary stature made even his abolitionist neighbors respectful toward him, but Hawthorne was keenly aware that his sympathy for the South ran counter to the mood of neighbors such as Emerson and Thoreau. For the *Atlantic Monthly* Fields solicited a series of sketches which Hawthorne adapted from his English notebooks. Fields paid well, but he was pressing Hawthorne into overwork. Despite short excursions designed to restore his vigor, Hawthorne's physical and psychic energies waned steadily. Humiliated by his weakness, he intermittently forced himself to work on his literary projects, especially the English sketches which he published as *Our Old Home* (1863), loyally dedicating it to Pierce, who because of his southern sympathies was now anathema to many northerners. Hawthorne began four romances, overlapping attempts to grapple with two major themes, an American claimant to an ancestral English estate and the search for an elixir of life. He finished none of them before his death in May, 1864, while traveling in New Hampshire with Pierce. He was buried in the Sleepy Hollow Cemetery at Concord. Alcott, Emerson, Fields, Holmes, Longfellow, and Lowell were among his pallbearers.

My Kinsman, Major Molineux[1]

After the kings of Great Britain had assumed the right of appointing the colonial governors,[2] the measures of the latter seldom met with the ready and general approbation, which had been paid to those of their predecessors, under the original charters. The people looked with most jealous scrutiny to the exercise of power, which did not emanate from themselves, and they usually rewarded the rulers with slender gratitude, for the compliances, by which, in softening their instructions from beyond the sea, they had incurred the reprehension of those who gave them. The annals of Massachusetts Bay will inform us, that of six governors, in the space of about forty years from the surrender of the old charter, under James II., two were imprisoned by a popular insurrection; a third, as Hutchinson[3] inclines to believe, was driven from the province by the whizzing of a musket ball; a fourth, in the opinion of the same historian, was hastened to his grave by continual bickerings with the house of representatives; and the remaining two, as well as their successors, till the Revolution, were favored with few and brief intervals of peaceful sway. The inferior members of the court party,[4] in times of high political excitement, led scarcely a more desirable life. These remarks may serve as preface to the following adventures, which chanced upon a summer night, not far from a hundred years ago. The reader, in order to avoid a long and dry detail of colonial affairs, is requested to dispense with an account of the train of circumstances, that had caused much temporary inflammation of the popular mind.

It was near nine o'clock of a moonlight evening, when a boat crossed the ferry with a single passenger, who had obtained his conveyance, at that unusual hour, by the promise of an extra fare.

---

1. The text here is that of the first printing in *The Token* for 1832, where the story is identified as being "By the Author of 'Sights from a Steeple.'"
2. I.e., after 1684, when the British government annulled the Massachusetts charter.
3. The particular annals, or year-by-year histories, that Hawthorne has in mind are

The History of the Colony and Province of Massachusetts-Bay (1764, 1767) by the last Royal governor, Thomas Hutchinson (1711–80). James II (1633–1701) reigned briefly (1685–88) before being exiled to France in the Glorious Revolution.
4. The pro-Crown party.

While he stood on the landing-place, searching in either pocket for the means of fulfilling his agreement, the ferryman lifted a lantern, by the aid of which, and the newly risen moon, he took a very accurate survey of the stranger's figure. He was a youth of barely eighteen years, evidently country-bred, and now, as it should seem, upon his first visit to town. He was clad in a coarse grey coat, well worn, but in excellent repair; his under garments were durably constructed of leather, and sat tight to a pair of serviceable and well-shaped limbs; his stockings of blue yarn, were the incontrovertible handiwork of a mother or a sister; and on his head was a three-cornered hat, which in its better days had perhaps sheltered the graver brow of the lad's father. Under his left arm was a heavy cudgel, formed of an oak sapling, and retaining a part of the hardened root; and his equipment was completed by a wallet,[5] not so abundantly stocked as to incommode the vigorous shoulders on which it hung. Brown, curly hair, well-shaped features, and bright, cheerful eyes, were nature's gifts, and worth all that art could have done for his adornment.

The youth, one of whose names was Robin, finally drew from his pocket the half of a little province-bill[6] of five shillings, which, in the depreciation of that sort of currency, did but satisfy the ferryman's demand, with the surplus of a sexangular piece of parchment valued at three pence. He then walked forward into the town, with as light a step, as if his day's journey had not already exceeded thirty miles, and with as eager an eye, as if he were entering London city, instead of the little metropolis of a New England colony. Before Robin had proceeded far, however, it occurred to him, that he knew not whither to direct his steps; so he paused, and looked up and down the narrow street, scrutinizing the small and mean wooden buildings, that were scattered on either side.

'This low hovel cannot be my kinsman's dwelling,' thought he, 'nor yonder old house, where the moonlight enters at the broken casement; and truly I see none hereabouts that might be worthy of him. It would have been wise to inquire my way of the ferryman, and doubtless he would have gone with me, and earned a shilling from the Major for his pains. But the next man I meet will do as well.'

He resumed his walk, and was glad to perceive that the street now became wider, and the houses more respectable in their appearance. He soon discerned a figure moving on moderately in advance, and hastened his steps to overtake it. As Robin drew nigh, he saw that the passenger was a man in years, with a full periwig of grey hair, a wide-skirted coat of dark cloth, and silk stockings rolled about his knees. He carried a long and polished cane, which he struck down perpendicularly before him, at every step; and at regular intervals he uttered two successive hems, of a peculiarly

5. Knapsack.   6. Local paper money.

solemn and sepulchral intonation. Having made these observations, Robin laid hold of the skirt of the old man's coat, just when the light from the open door and windows of a barber's shop, fell upon both their figures.

'Good evening to you, honored Sir,' said he, making a low bow, and still retaining his hold of the skirt. 'I pray you to tell me whereabouts is the dwelling of my kinsman, Major Molineux?'

The youth's question was uttered very loudly; and one of the barbers, whose razor was descending on a well-soaped chin, and another who was dressing a Ramillies wig,[7] left their occupations, and came to the door. The citizen, in the meantime, turned a long favored countenance upon Robin, and answered him in a tone of excessive anger and annoyance. His two sepulchral hems, however, broke into the very centre of his rebuke, with most singular effect, like a thought of the cold grave obtruding among wrathful passions.

'Let go my garment, fellow! I tell you. I know not the man you speak of. What! I have authority, I have—hem, hem—authority; and if this be the respect you show your betters, your feet shall be brought acquainted with the stocks,[8] by daylight, tomorrow morning!'

Robin released the old man's skirt, and hastened away, pursued by an ill-mannered roar of laughter from the barber's shop. He was at first considerably surprised by the result of his question, but, being a shrewd youth, soon thought himself able to account for the mystery.

'This is some country representative,' was his conclusion, 'who has never seen the inside of my kinsman's door, and lacks the breeding to answer a stranger civilly. The man is old, or verily— I might be tempted to turn back and smite him on the nose. Ah, Robin, Robin! even the barber's boys laugh at you, for choosing such a guide! You will be wiser in time, friend Robin.'

He now became entangled in a succession of crooked and narrow streets, which crossed each other, and meandered at no great distance from the water-side. The smell of tar was obvious to his nostrils, the masts of vessels pierced the moonlight above the tops of the buildings, and the numerous signs, which Robin paused to read, informed him that he was near the centre of business. But the streets were empty, the shops were closed, and lights were visible only in the second stories of a few dwelling-houses. At length, on the corner of a narrow lane, through which he was passing, he beheld the broad countenance of a British hero swinging before the door of an inn, whence proceeded the voices of many guests. The casement of one of the lower windows was thrown back, and a very thin curtain permitted Robin to distinguish a party at supper, round a well-

7. Elaborately plaited wig named for Ramillies, Belgium.
8. Instrument of punishment having a

heavy wooden frame with holes for confining the ankles and sometimes the wrists as well.

furnished table. The fragrance of the good cheer steamed forth into the outer air, and the youth could not fail to recollect, that the last remnant of his travelling stock of provision had yielded to his morning appetite, and that noon had found, and left him, dinnerless.

'Oh, that a parchment three-penny might give me a right to sit down at yonder table,' said Robin, with a sigh. 'But the Major will make me welcome to the best of his victuals; so I will even step boldly in, and inquire my way to his dwelling.'

He entered the tavern, and was guided by the murmur of voices, and fumes of tobacco, to the public room. It was a long and low apartment, with oaken walls, grown dark in the continual smoke, and a floor, which was thickly sanded, but of no immaculate purity. A number of persons, the larger part of whom appeared to be mariners, or in some way connected with the sea, occupied the wooden benches, or leather-bottomed chairs, conversing on various matters, and occasionally lending their attention to some topic of general interest. Three or four little groups were draining as many bowls of punch, which the great West India trade had long since made a familiar drink in the colony. Others, who had the aspect of men who lived by regular and laborious handicraft, preferred the insulated bliss of an unshared potation, and became more taciturn under its influence. Nearly all, in short, evinced a predilection for the Good Creature[9] in some of its various shapes, for this is a vice, to which, as the Fast-day[1] sermons of a hundred years ago will testify, we have a long hereditary claim. The only guests to whom Robin's sympathies inclined him, were two or three sheepish countrymen, who were using the inn somewhat after the fashion of a Turkish Caravansary;[2] they had gotten themselves into the darkest corner of the room, and, heedless of the Nicotian[3] atmosphere, were supping on the bread of their own ovens, and the bacon cured in their own chimney-smoke. But though Robin felt a sort of brotherhood with these strangers, his eyes were attracted from them, to a person who stood near the door, holding whispered conversation with a group of ill-dressed associates. His features were separately striking almost to grotesqueness, and the whole face left a deep impression in the memory. The forehead bulged out into a double prominence, with a vale between; the nose came boldly forth in an irregular curve, and its bridge was of more than a finger's breadth; the eyebrows were deep and shaggy, and the eyes glowed beneath them like fire in a cave.

While Robin deliberated of whom to inquire respecting his kins-

9. Hawthorne is playing upon the warning against food fanatics in 1 Timothy 4.4: "For every creature of God is good, and nothing to be refused, if it be received with thanksgiving."
1. Days set apart for public penitence.

2. An inn built around a court for accommodating caravans.
3. Heavy with tobacco fumes (from Jean Nicot, who introduced tobacco into France when he was French ambassador at Lisbon).

man's dwelling, he was accosted by the innkeeper, a little man in a stained white apron, who had come to pay his professional welcome to the stranger. Being in the second generation from a French protestant, he seemed to have inherited the courtesy of his parent nation; but no variety of circumstance was ever known to change his voice from the one shrill note in which he now addressed Robin.

'From the country, I presume, Sir?' said he, with a profound bow. 'Beg to congratulate you on your arrival, and trust you intend a long stay with us. Fine town here, Sir, beautiful buildings, and much that may interest a stranger. May I hope for the honor of your commands in respect to supper?'

'The man sees a family likeness! the rogue has guessed that I am related to the Major!' thought Robin, who had hitherto experienced little superfluous civility.

All eyes were now turned on the country lad, standing at the door, in his worn three-cornered hat, grey coat, leather breeches, and blue yarn stockings, leaning on an oaken cudgel, and bearing a wallet on his back. Robin replied to the courteous innkeeper, with such an assumption of consequence, as befitted the Major's relative.

'My honest friend,' he said, 'I shall make it a point to patronise your house on some occasion, when—' here he could not help lowering his voice—'I may have more than a parchment three-pence in my pocket. My present business,' continued he, speaking with lofty confidence, 'is merely to inquire the way to the dwelling of my kinsman, Major Molineux.'

There was a sudden and general movement in the room, which Robin interpreted as expressing the eagerness of each individual to become his guide. But the innkeeper turned his eyes to a written paper on the wall, which he read, or seemed to read, with occasional recurrences to the young man's figure.

'What have we here?' said he, breaking his speech into little dry fragments. "Left the house of the subscriber, bounden servant,[4] Hezekiah Mudge—had on when he went away, grey coat, leather breeches, master's third best hat. One pound currency reward to whoever shall lodge him in any jail in the province." 'Better trudge, boy, better trudge.'

Robin had began to draw his hand towards the lighter end of the oak cudgel, but a strange hostility in every countenance, induced him to relinquish his purpose of breaking the courteous innkeeper's head. As he turned to leave the room, he encountered a sneering glance from the bold-featured personage whom he had before noticed; and no sooner was he beyond the door, than he heard a general laugh, in which the innkeeper's voice might be distinguished, like the dropping of small stones into a kettle.

'Now is it not strange,' thought Robin, with his usual shrewdness,

4. A person bound by contract to servitude for seven years (or another set period), usually in repayment for transportation to the colonies.

'is it not strange, that the confession of an empty pocket, should outweigh the name of my kinsman, Major Molineux? Oh, if I had one of these grinning rascals in the woods, where I and my oak sapling grew up together, I would teach him that my arm is heavy, though my purse be light!'

On turning the corner of the narrow lane, Robin found himself in a spacious street, with an unbroken line of lofty houses on each side, and a steepled building at the upper end, whence the ringing of a bell announced the hour of nine. The light of the moon, and the lamps from numerous shop windows, discovered people promenading on the pavement, and amongst them, Robin hoped to recognise his hitherto inscrutable relative. The result of his former inquiries made him unwilling to hazard another, in a scene of such publicity, and he determined to walk slowly and silently up the street, thrusting his face close to that of every elderly gentleman, in search of the Major's lineaments. In his progress, Robin encountered many gay and gallant figures. Embroidered garments, of showy colors, enormous periwigs, gold-laced hats, and silver hilted swords, glided past him and dazzled his optics. Travelled youths, imitators of the European fine gentlemen of the period, trod jauntily along, half-dancing to the fashionable tunes which they hummed, and making poor Robin ashamed of his quiet and natural gait. At length, after many pauses to examine the gorgeous display of goods in the shop windows, and after suffering some rebukes for the impertinence of his scrutiny into people's faces, the Major's kinsman found himself near the steepled building, still unsuccessful in his search. As yet, however, he had seen only one side of the thronged street; so Robin crossed, and continued the same sort of inquisition down the opposite pavement, with stronger hopes than the philosopher seeking an honest man,[5] but with no better fortune. He had arrived about midway towards the lower end, from which his course began, when he overheard the approach of some one, who struck down a cane on the flag-stones at every step, uttering, at regular intervals, two sepulchral hems.

'Mercy on us!' quoth Robin, recognising the sound.

Turning a corner, which chanced to be close at his right hand, he hastened to pursue his researches, in some other part of the town. His patience was now wearing low, and he seemed to feel more fatigue from his rambles since he crossed the ferry, than from his journey of several days on the other side. Hunger also pleaded loudly within him, and Robin began to balance the propriety of demanding, violently and with lifted cudgel, the necessary guidance from the first solitary passenger, whom he should meet. While a resolution to this effect was gaining stength, he entered a street of mean appearance, on either side of which, a row of ill-built houses

---

5. Diogenes, the Greek philosopher (412?–323 B.C.), carried a lantern about in daytime in his search for an honest man.

was straggling towards the harbor. The moonlight fell upon no passenger along the whole extent, but in the third domicile which Robin passed, there was a half-opened door, and his keen glance detected a woman's garment within.

'My luck may be better here,' said he to himself.

Accordingly, he approached the door, and beheld it shut closer as he did so; yet an open space remained, sufficing for the fair occupant to observe the stranger, without a corresponding display on her part. All that Robin could discern was a strip of scarlet petticoat, and the occasional sparkle of an eye, as if the moonbeams were trembling on some bright thing.

'Pretty mistress,'—for I may call her so with a good conscience, thought the shrewd youth, since I know nothing to the contrary—'my sweet pretty mistress, will you be kind enough to tell me whereabouts I must seek the dwelling of my kinsman, Major Molineux?'

Robin's voice was plaintive and winning, and the female, seeing nothing to be shunned in the handsome country youth, thrust open the door, and came forth into the moonlight. She was a dainty little figure, with a white neck, round arms, and a slender waist, at the extremity of which her scarlet petticoat jutted out over a hoop, as if she were standing in a balloon. Moreover, her face was oval and pretty, her hair dark beneath the little cap, and her bright eyes possessed a sly freedom, which triumphed over those of Robin.

'Major Molineux dwells here,' said this fair woman.

Now her voice was the sweetest Robin had heard that night, the airy counterpart of a stream of melted silver; yet he could not help doubting whether that sweet voice spoke gospel truth. He looked up and down the mean street, and then surveyed the house before which they stood. It was a small, dark edifice of two stories, the second of which projected over the lower floor; and the front apartment had the aspect of a shop for petty commodities.

'Now truly I am in luck,' replied Robin, cunningly, 'and so indeed is my kinsman, the Major, in having so pretty a housekeeper. But I prithee trouble him to step to the door; I will deliver him a message from his friends in the country, and then go back to my lodgings at the inn.'

'Nay, the Major has been a-bed this hour or more, said the lady of the scarlet petticoat; 'and it would be to little purpose to disturb him to night, seeing his evening draught was of the strongest. But he is a kind-hearted man, and it would be as much as my life's worth, to let a kinsman of his turn away from the door. You are the good old gentleman's very picture, and I could swear that was his rainy-weather hat. Also, he has garments very much resembling those leather—But come in, I pray, for I bid you hearty welcome in his name.'

So saying, the fair and hospitable dame took our hero by the hand; and though the touch was light, and the force was gentleness,

394   ★   *Nathaniel Hawthorne*

and though Robin read in her eyes what he did not hear in her
words, yet the slender waisted woman, in the scarlet petticoat,
proved stronger than the athletic country youth. She had drawn his
half-willing footsteps nearly to the threshold, when the opening of a
door in the neighborhood, startled the Major's housekeeper, and,
leaving the Major's kinsman, she vanished speedily into her own
domicile. A heavy yawn preceded the appearance of a man, who,
like the Moonshine of Pyramus and Thisbe, carried a lantern,[6]
needlessly aiding his sister luminary in the heavens. As he walked
sleepily up the street, he turned his broad, dull face on Robin, and
displayed a long staff, spiked at the end.

'Home, vagabond, home!' said the watchman, in accents that
seemed to fall asleep as soon as they were uttered. 'Home, or we'll
set you in the stocks by peep of day!'

'This is the second hint of the kind,' thought Robin. 'I wish they
would end my difficulties, by setting me there to-night.'

Nevertheless, the youth felt an instinctive antipathy towards the
guardian of midnight order, which at first prevented him from asking his usual question. But just when the man was about to vanish
behind the corner, Robin resolved not to lose the opportunity, and
shouted lustily after him—

'I say, friend! will you guide me to the house of my kinsman,
Major Molineux?'

The watchman made no reply, but turned the corner and was
gone; yet Robin seemed to hear the sound of drowsy laughter stealing along the solitary street. At that moment, also, a pleasant titter
saluted him from the open window above his head; he looked up,
and caught the sparkle of a saucy eye; a round arm beckoned to
him, and next he heard light footsteps descending the staircase
within. But Robin, being of the household of a New England
clergyman, was a good youth, as well as a shrewd one; so he
resisted temptation, and fled away.

He now roamed desperately, and at random, through the town,
almost ready to believe that a spell was on him, like that, by which
a wizard of his country, had once kept three pursuers wandering, a
whole winter night, within twenty paces of the cottage which they
sought. The streets lay before him, strange and desolate, and the
lights were extinguished in almost every house. Twice, however,
little parties of men, among whom Robin distinguished individuals
in outlandish attire, came hurrying along, but though on both occasions they paused to address him, such intercourse did not at all
enlighten his perplexity. They did but utter a few words in some
language of which Robin knew nothing, and perceiving his inability
to answer, bestowed a curse upon him in plain English, and hastened away. Finally, the lad determined to knock at the door of every

6. In Shakespeare's *Midsummer Night's Dream*, 5.1, the craftsmen's play within a
play.

mansion that might appear worthy to be occupied by his kinsman, trusting that perseverance would overcome the fatality which had hitherto thwarted him. Firm in this resolve, he was passing beneath the walls of a church, which formed the corner of two streets, when, as he turned into the shade of its steeple, he encountered a bulky stranger, muffled in a cloak. The man was proceeding with the speed of earnest business, but Robin planted himself full before him, holding the oak cudgel with both hands across his body, as a bar to further passage.

'Halt, honest man, and answer me a question,' said he, very resolutely. 'Tell me, this instant, whereabouts is the dwelling of my kinsman, Major Molineux?'

'Keep your tongue between your teeth, fool, and let me pass,' said a deep, gruff voice, which Robin partly remembered. 'Let me pass, I say, or I'll strike you to the earth!'

'No, no, neighbor!' cried Robin, flourishing his cudgel, and then thrusting its larger end close to the man's muffled face. 'No, no, I'm not the fool you take me for, nor do you pass, till I have an answer to my question. Whereabouts is the dwelling of my kinsman, Major Molineux?'

The stranger, instead of attempting to force his passage, stept back into the moonlight, unmuffled his own face and stared full into that of Robin.

'Watch here an hour, and Major Molineux will pass by,' said he.

Robin gazed with dismay and astonishment, on the unprecedented physiognomy of the speaker. The forehead with its double prominence, the broad-hooked nose, the shaggy eyebrows, and fiery eyes, were those which he had noticed at the inn, but the man's complexion had undergone a singular, or more properly, a two-fold change. One side of the face blazed of an intense red, while the other was black as midnight, the division line being in the broad bridge of the nose; and a mouth, which seemed to extend from ear to ear, was black or red, in contrast to the color of the cheek. The effect was as if two individual devils, a fiend of fire and a fiend of darkness, had united themselves to form this infernal visage. The stranger grinned in Robin's face, muffled his party-colored features, and was out of sight in a moment.

'Strange things we travellers see!' ejaculated Robin.

He seated himself, however, upon the steps of the church-door, resolving to wait the appointed time for his kinsman's appearance. A few moments were consumed in philosophical speculations, upon the species of the *genus homo*, who had just left him, but having settled this point shrewdly, rationally, and satisfactorily, he was compelled to look elsewhere for amusement. And first he threw his eyes along the street; it was of more respectable appearance than most of those into which he had wandered, and the moon, 'creating, like the imaginative power, a beautiful strangeness in familiar objects,' gave something of romance to a scene, that might not have possessed it in the light of day. The irregular, and often quaint architecture of the houses, some of whose roofs were broken into numerous little peaks; while others ascended, steep and narrow, into a single point; and others again were square; the pure milk-white of some of their complexions, the aged darkness of others, and the thousand sparklings, reflected from bright substances in the plastered walls of many; these matters engaged Robin's attention for awhile, and then began to grow wearisome. Next he endeavored to define the forms of distant objects, starting away with almost ghostly indistinctness, just as his eye appeared to grasp them; and finally he took a minute survey of an edifice, which stood on the opposite side of the street, directly in front of the church-door, where he was stationed. It was a large square mansion, distinguished from its neighbors by a balcony, which rested on tall pillars, and by an elaborate gothic window, communicating therewith.

'Perhaps this is the very house I have been seeking,' thought Robin.

Then he strove to speed away the time, by listening to a murmur, which swept continually along the street, yet was scarcely audible, except to an unaccustomed ear like his; it was a low, dull, dreamy sound, compounded of many noises, each of which was at too great a distance to be separately heard. Robin marvelled at this snore of a sleeping town, and marvelled more, whenever its continuity was broken, by now and then a distant shout, apparently loud where it originated. But altogether it was a sleep-inspiring sound, and to shake off its drowsy influence, Robin arose, and climbed a window-frame, that he might view the interior of the church. There the moonbeams came trembling in, and fell down upon the deserted pews, and extended along the quiet aisles. A fainter, yet more awful radiance, was hovering round the pulpit, and one solitary ray had dared to rest upon the opened page of the great bible. Had Nature, in that deep hour, become a worshipper in the house, which man had builded? Or was that heavenly light the visible sanctity of the place, visible because no earthly and impure feet were within the walls? The scene made Robin's heart shiver with a sensation of loneliness, stronger than he had ever felt in the remotest depths of his native woods; so he turned away, and sat down again before the door. There were graves around the church, and now an uneasy thought obtruded into Robin's breast. What if the object of his search, which had been so often and so strangely thwarted, were all the time mouldering in his shroud? What if his kinsman should glide through yonder gate, and nod and smile to him in passing dimly by?

'Oh, that any breathing thing were here with me!' said Robin.

Recalling his thoughts from this uncomfortable track, he sent them over forest, hill, and stream, and attempted to imagine how that evening of ambiguity and weariness, had been spent by his father's household. He pictured them assembled at the door, beneath the tree, the great old tree, which had been spared for its huge twisted trunk, and venerable shade, when a thousand leafy brethren fell. There, at the going down of the summer sun, it was his father's custom to perform domestic worship, that the neighbors might come and join with him like brothers of the family, and that the wayfaring man might pause to drink at that fountain, and keep his heart pure by freshening the memory of home. Robin distinguished the seat of every individual of the little audience; he saw the good man in the midst, holding the scriptures in the golden light that shone from the western clouds; he beheld him close the book, and all rise up to pray. He heard the old thanksgivings for daily mercies, the old supplications for their continuance, to which he had so often listened in weariness, but which were now among his dear remembrances. He perceived the slight inequality of his father's voice when he came to speak of the Absent One; he noted how his mother turned her face to the broad and knotted trunk, how his elder brother scorned, because the beard was rough upon his upper lip, to permit his features to be moved; how his younger sister drew down a low hanging branch before her eyes; and how the little one of all, whose sports had hitherto broken the decorum of the scene, understood the prayer for her playmate, and burst into clamorous grief. Then he saw them go in at the door; and when Robin would have entered also, the latch tinkled into its place, and he was excluded from his home.

'Am I here, or there?' cried Robin, starting; for all at once, when his thoughts had become visible and audible in a dream, the long, wide, solitary street shone out before him.

He aroused himself, and endeavored to fix his attention steadily upon the large edifice which he had surveyed before. But still his mind kept vibrating between fancy and reality; by turns, the pillars of the balcony lengthened into the tall, bare stems of pines, dwindled down to human figures, settled again in their true shape and size, and then commenced a new succession of changes. For a single moment, when he deemed himself awake, he could have sworn that a visage, one which he seemed to remember, yet could not absolutely name as his kinsman's, was looking towards him from the Gothic window. A deeper sleep wrestled with, and nearly overcame him, but fled at the sound of footsteps along the opposite pavement. Robin rubbed his eyes, discerned a man passing at the foot of the balcony, and addressed him in a loud, peevish, and lamentable cry.

'Halloo, friend! must I wait here all night for my kinsman, Major Molineux?'

The sleeping echoes awoke, and answered the voice; and the passenger, barely able to discern a figure sitting in the oblique shade of the steeple, traversed the street to obtain a nearer view. He was himself a gentleman in his prime, of open, intelligent, cheerful, and altogether prepossessing countenance. Perceiving a country youth, apparently homeless and without friends, he accosted him in a tone of real kindness, which had become strange to Robin's ears.

'Well, my good lad, why are you sitting here?' inquired he. 'Can I be of service to you in any way?'

'I am afraid not, Sir,' replied Robin, despondingly; 'yet I shall take it kindly, if you'll answer me a single question. I've been searching half the night for one Major Molineux; now, Sir, is there really such a person in these parts, or am I dreaming?'

'Major Molineux! The name is not altogether strange to me,' said the gentleman, smiling. 'Have you any objection to telling me the nature of your business with him?'

Then Robin briefly related that his father was a clergyman, settled on a small salary, at a long distance back in the country, and that he and Major Molineux were brothers' children. The Major, having inherited riches, and acquired civil and military rank, had visited his cousin in great pomp a year or two before; had manifested much interest in Robin and an elder brother, and, being childless himself, had thrown out hints respecting the future establishment of one of them in life. The elder brother was destined to succeed to the farm, which his father cultivated, in the interval of sacred duties; it was therefore determined that Robin should profit by his kinsman's generous intentions, especially as he had seemed to be rather the favorite, and was thought to possess other necessary endowments.

'For I have the name of being a shrewd youth,' observed Robin, in this part of his story.

'I doubt not you deserve it,' replied his new friend, good naturedly; 'but pray proceed.'

'Well, Sir, being nearly eighteen years old, and well grown, as you see,' continued Robin, raising himself to his full height, 'I thought it high time to begin the world. So my mother and sister put me in handsome trim, and my father gave me half the remnant of his last year's salary, and five days ago I started for this place, to pay the Major a visit. But would you believe it, Sir? I crossed the ferry a little after dusk, and have yet found nobody that would show me the way to his dwelling; only an hour or two since, I was told to wait here, and Major Molineux would pass by.'

'Can you describe the man who told you this?' inquired the gentleman.

'Oh, he was a very ill-favored fellow, Sir,' replied Robin, 'with two great bumps on his forehead, a hook nose, fiery eyes, and, what struck me as the strangest, his face was of two different colors. Do you happen to know such a man, Sir?'

'Not intimately,' answered the stranger, 'but I chanced to meet him a little time previous to your stopping me. I believe you may trust his word, and that the Major will very shortly pass through this street. In the mean time, as I have a singular curiosity to witness your meeting, I will sit down here upon the steps, and bear you company.'

He seated himself accordingly, and soon engaged his companion in animated discourse. It was but of brief continuance, however, for a noise of shouting, which had long been remotely audible, drew so much nearer, that Robin inquired its cause.

'What may be the meaning of this uproar?' asked he. 'Truly, if your town be always as noisy, I shall find little sleep, while I am an inhabitant.'

'Why, indeed, friend Robin, there do appear to be three or four riotous fellows abroad to-night,' replied the gentleman. 'You must not expect all the stillness of your native woods, here in our streets. But the watch will shortly be at the heels of these lads, and—'

'Aye, and set them in the stocks by peep of day,' interrupted Robin, recollecting his own encounter with the drowsy lantern-bearer. 'But, dear Sir, if I may trust my ears, an army of watchmen would never make head against such a multitude of rioters. There were at least a thousand voices went to make up that one shout.'

'May not one man have several voices, Robin, as well as two complexions?' said his friend.

'Perhaps a man may; but heaven forbid that a woman should!' responded the shrewd youth, thinking of the seductive tones of the Major's housekeeper.

The sounds of a trumpet in some neighboring street, now became so evident and continual, that Robin's curiosity was strongly excited. In addition to the shouts, he heard frequent bursts from many instruments of discord, and a wild and confused laughter filled up the intervals. Robin rose from the steps, and looked wistfully towards a point, whither several people seemed to be hastening.

'Surely some prodigious merrymaking is going on,' exclaimed he. 'I have laughed very little since I left home, Sir, and should be sorry to lose an opportunity. Shall we just step round the corner by that darkish house, and take our share of the fun?'

'Sit down again, sit down, good Robin,' replied the gentleman, laying his hand on the skirt of the grey coat. 'You forget that we must wait here for your kinsman; and there is reason to believe that he will pass by, in the course of a very few moments.'

The near approach of the uproar had now disturbed the neighborhood; windows flew open on all sides; and many heads, in the attire of the pillow, and confused by sleep suddenly broken, were protruded to the gaze of whoever had leisure to observe them. Eager voices hailed each other from house to house, all demanding the explanation, which not a soul could give. Half-dressed men hurried towards the unknown commotion, stumbling as they went over the stone steps, that thrust themselves into the narrow foot-walk. The shouts, the laughter, and the tuneless bray, the antipodes of music, came onward with increasing din, till scattered individuals, and then denser bodies, began to appear round a corner, at the distance of a hundred yards.

'Will you recognise your kinsman, Robin, if he passes in this crowd?' inquired the gentleman.

'Indeed, I can't warrant it, Sir; but I'll take my stand here, and keep a bright look out,' answered Robin, descending to the outer edge of the pavement.

A mighty stream of people now emptied into the street, and came rolling slowly towards the church. A single horseman wheeled the corner in the midst of them, and close behind him came a band of fearful wind-instruments, sending forth a fresher discord, now that no intervening buildings kept it from the ear. Then a redder light disturbed the moonbeams, and a dense multitude of torches shone along the street, concealing by their glare whatever object they illuminated. The single horseman, clad in a military dress, and bearing a drawn sword, rode onward as the leader, and, by his fierce and variegated countenance, appeared like war personified; the red of one cheek was an emblem of fire and sword; the blackness of the other betokened the mourning which attends them. In his train, were wild figures in the Indian dress, and many fantastic shapes without a model, giving the whole march a visionary air, as if a dream had broken forth from some feverish brain, and were sweeping visibly through the midnight streets. A mass of people, inactive, except as applauding spectators, hemmed the procession in, and several women ran along the sidewalks, piercing the confusion of heavier sounds, with their shrill voices of mirth of terror.

'The double-faced fellow has his eye upon me,' muttered Robin, with an indefinite but uncomfortable idea, that he was himself to bear a part in the pageantry.

The leader turned himself in the saddle, and fixed his glance full upon the country youth, as the steed went slowly by. When Robin had freed his eyes from those fiery ones, the musicians were passing before him, and the torches were close at hand; but the unsteady brightness of the latter formed a veil which he could not penetrate. The rattling of wheels over the stones sometimes found its way to his ear, and confused traces of a human form appeared at intervals, and then melted into the vivid light. A moment more, and the leader thundered a command to halt; the trumpets vomited a horrid breath, and held their peace; the shouts and laughter of the people died away, and there remained only an universal hum, nearly allied to silence. Right before Robin's eyes was an uncovered cart. There the torches blazed the brightest, there the moon shone out like day, and there, in tar-and-feathery dignity, sate his kinsman, Major Molineux!

He was an elderly man, of large and majestic person, and strong, square features, betokening a steady soul; but steady as it was, his enemies had found the means to shake it. His face was pale as death, and far more ghastly; the broad forehead was contracted in his agony, so that the eyebrows formed one dark grey line; his eyes were red and wild, and the foam hung white upon his quivering lip. His whole frame was agitated by a quick, and continual tremor, which his pride strove to quell, even in those circumstances of overwhelming humiliation. But perhaps the bitterest pang of all was when his eyes met those of Robin; for he evidently knew him on the instant, as the youth stood witnessing the foul disgrace of a head that had grown grey in honor. They stared at each other in silence, and Robin's knees shook, and his hair bristled, with a mixture of pity and terror. Soon, however, a bewildering excitment began to seize upon his mind; the preceding adventures of the night, the un-expected appearance of the crowd, the torches, the confused din, and the hush that followed, the spectre of his kinsman reviled by that great multitude, all this, and more than all, a perception of tremen-dous ridicule in the whole scene, affected him with a sort of mental inebriety. At that moment a voice of sluggish merriment saluted Robin's ears; he turned instinctively, and just behind the corner of the church stood the lantern-bearer, rubbing his eyes, and drowsily enjoying the lad's amazement. Then he heard a peal of laughter like the ringing of silvery bells; a woman twitched his arm, a saucy eye met his, and he saw the lady of the scarlet petticoat. A sharp, dry cachinnation appealed to his memory, and, standing on tiptoe in the crowd, with his white apron over his head, he beheld the courteous little innkeeper. And lastly, there sailed over the heads of the multi-tude a great, broad laugh, broken in the midst by two deep sepul-chral hems; thus—

'Haw, haw, haw—hem, hem—haw, haw, haw, haw!'

The sound proceeded from the balcony of the opposite edifice, and thither Robin turned his eyes. In front of the Gothic window stood the old citizen, wrapped in a wide gown, his grey periwig exchanged for a nightcap, which was thrust back from his forehead, and his silk stockings hanging down about his legs. He supported himself on his polished cane in a fit of convulsive merriment, which manifested itself on his solemn old features, like a funny inscription on a tomb-stone. Then Robin seemed to hear the voices of the barbers; of the guests of the inn; and of all who had made sport of him that night. The contagion was spreading among the multitude, when, all at once, it seized upon Robin, and he sent forth a shout of laughter that echoed through the street; every man shook his sides, every man emptied his lungs, but Robin's shout was the loudest there. The cloud-spirits peeped from their silvery islands, as the congregated mirth went roaring up the sky! The Man in the Moon heard the far bellow; 'Oho,' quoth he, 'the old Earth is frolicsome to-night!'

When there was a momentary calm in that tempestuous sea of sound, the leader gave the sign, and the procession resumed its march. On they went, like fiends that throng in mockery round some dead potentate, mighty no more, but majestic still in his agony. On they went, in counterfeited pomp, in senseless uproar, in frenzied merriment, trampling all on an old man's heart. On swept the tumult, and left a silent street behind.

. . . . .

'Well, Robin, are you dreaming?' inquired the gentleman, laying his hand on the youth's shoulder.

Robin started, and withdrew his arm from the stone post, to which he had instinctively clung, while the living stream rolled by him. His cheek was somewhat pale, and his eye not quite so lively as in the earlier part of the evening.

'Will you be kind enough to show me the way to the Ferry?' said he, after a moment's pause.

'You have then adopted a new subject of inquiry?' observed his companion, with a smile.

'Why, yes, Sir,' replied Robin, rather dryly. 'Thanks to you, and to my other friends, I have at last met my kinsman, and he will scarce desire to see my face again. I begin to grow weary of a town life, Sir. Will you show me the way to the Ferry?'

'No, my good friend Robin, not to-night, at least,' said the gentleman. 'Some few days hence, if you continue to wish it, I will speed you on your journey. Or, if you prefer to remain with us, perhaps, as you are a shrewd youth, you may rise in the world, without the help of your kinsman, Major Molineux.'

1832, 1837

Roger Malvin's Burial[1]

One of the few incidents of Indian warfare, naturally susceptible of the moonlight of romance, was that expedition, undertaken, for the defence of the frontiers, in the year 1725, which resulted in the well-remembered 'Lovell's Fight.'[2] Imagination, by casting certain circumstances judiciously into the shade, may see much to admire in the heroism of a little band, who gave battle to twice their number in the heart of the enemy's country. The open bravery displayed by both parties was in accordance with civilized ideas of valor, and chivalry itself might not blush to record the deeds of one or two individuals. The battle, though so fatal to those who fought,

---

1. The text is that of the first printing, in *The Token* for 1832.
2. An incident in the Penobscot War in

Maine (then part of Massachusetts) during 1725.

was not unfortunate in its consequences to the country; for it broke the strength of a tribe, and conduced to the peace which subsisted during several ensuing years. History and tradition are unusually minute in their memorials of this affair; and the captain of a scouting party of frontier-men has acquired as actual a military renown, as many a victorious leader of thousands. Some of the incidents contained in the following pages will be recognised, notwithstanding the substitution of fictitious names, by such as have heard, from old men's lips, the fate of the few combatants who were in a condition to retreat, after 'Lovell's Fight.'

.   .   .   .   .

The early sunbeams hovered cheerfully upon the tree-tops, beneath which two weary and wounded men had stretched their limbs the night before. Their bed of withered oak leaves was strewn upon the small level space, at the foot of a rock, situated near the summit of one of the gentle swells, by which the face of the country is there diversified. The mass of granite, rearing its smooth, flat surface, fifteen or twenty feet above their heads, was not unlike a gigantic grave-stone, upon which the veins seemed to form an inscription in forgotten characters. On a tract of several acres around this rock, oaks and other hard-wood trees had supplied the place of the pines, which were the usual growth of the land; and a young and vigorous sapling stood close beside the travellers.

The severe wound of the elder man had probably deprived him of sleep; for, so soon as the first ray of sunshine rested on the top of the highest tree, he reared himself painfully from his recumbent posture, and sat erect. The deep lines of his countenance, and the scattered grey of his hair, marked him as past the middle age; but his muscular frame would, but for the effects of his wound, have been as capable of sustaining fatigue, as in the early vigor of life. Languor and exhaustion now sat upon his haggard features, and the despairing glance which he sent forward through the depths of the forest, proved his own conviction that his pilgrimage was at an end. He next turned his eyes to the companion, who reclined by his side. The youth, for he had scarcely attained the years of manhood, lay, with his head upon his arm, in the embrace of a unquiet sleep, which a thrill of pain from his wounds seemed each moment on the point of breaking. His right hand grasped a musket, and, to judge from the violent action of his features, his slumbers were bringing back a vision of the conflict, of which he was one of the few survivors. A shout,—deep and loud to his dreaming fancy,—found its way in an imperfect murmur to his lips, and, starting even at the slight sound of his own voice, he suddenly awoke. The first act of reviving recollection, was to make anxious inquiries respecting the condition of his wounded fellow traveller. The latter shook his head.

'Reuben, my boy,' said he, 'this rock, beneath which we sit, will serve for an old hunter's grave-stone. There is many and many a long mile of howling wilderness before us yet; nor would it avail me anything, if the smoke of my own chimney were but on the other side of that swell of land. The Indian bullet was deadlier than I thought.'

'You are weary with our three days' travel,' replied the youth, 'and a little longer rest will recruit you. Sit you here, while I search the woods for the herbs and roots, that must be our sustenance; and having eaten, you shall lean on me, and we will turn our faces homeward. I doubt not, that, with my help, you can attain to some one of the frontier garrisons.'

'There is not two days' life in me, Reuben,' said the other, calmly, 'and I will no longer burthen you with my useless body, when you can scarcely support your own. Your wounds are deep, and your strength is failing fast; yet, if you hasten onward alone, you may be preserved. For me there is no hope; and I will await death here.'

'If it must be so, I will remain and watch by you,' said Reuben, resolutely.

'No, my son, no,' rejoined his companion. 'Let the wish of a dying man have weight with you; give me one grasp of your hand, and get you hence. Think you that my last moments will be eased by the thought, that I leave you to die a more lingering death? I have loved you like a father, Reuben, and, at a time like this, I should have something of a father's authority. I charge you to be gone, that I may die in peace.'

'And because you have been a father to me, should I therefore leave you to perish, and to lie unburied in the wilderness?' exclaimed the youth. 'No; if your end be in truth approaching, I will watch by you, and receive your parting words. I will dig a grave here by the rock, in which, if my weakness overcome me, we will rest together; or, if Heaven gives me strength, I will seek my way home.'

'In the cities, and wherever men dwell,' replied the other, 'they bury their dead in the earth; they hide them from the sight of the living; but here, where no step may pass, perhaps for a hundred years, wherefore should I not rest beneath the open sky, covered only by the oak-leaves, when the autumn winds shall strew them? And for a monument, here is this grey rock, on which my dying hand shall carve the name of Roger Malvin: and the traveller in days to come will know, that here sleeps a hunter and a warrior. Tarry not, then, for a folly like this, but hasten away, if not for your own sake, for hers who will else be desolate.'

Malvin spoke the last few words in a faultering voice, and their effect upon his companion was strongly visible. They reminded him that there were other, and less questionable duties, than that of sharing the fate of a man whom his death could not benefit. Nor can it be affirmed that no selfish feeling strove to enter Reuben's heart, though the consciousness made him more earnestly resist his companion's entreaties.

'How terrible, to wait the slow approach of death, in this solitude!' exclaimed he. 'A brave man does not shrink in the battle, and, when friends stand round the bed, even women may die composedly; but here'—

'I shall not shrink, even here, Reuben Bourne;' interrupted Malvin, 'I am a man of no weak heart; and, if I were, there is a surer support than that of earthly friends. You are young, and life is dear to you. Your last moments will need comfort far more than mine; and when you have laid me in the earth, and are alone, and night is settling on the forest, you will feel all the bitterness of the death that may now be escaped. But I will urge no selfish motive to your generous nature. Leave me for my sake; that, having said a prayer for your safety, I may have space to settle my account, undisturbed by worldly sorrows.'

'And your daughter! How shall I dare to meet her eye?' exclaimed Reuben. 'She will ask the fate of her father, whose life I vowed to defend with my own. Must I tell her, that he travelled three days' march with me from the field of battle, and that then I left him to perish in the wilderness? Were it not better to lie down and die by your side, than to return safe, and say this to Dorcas?'

'Tell my daughter,' said Roger Malvin, 'that, though yourself sore wounded, and weak, and weary, you led my tottering footsteps many a mile, and left me only at my earnest entreaty, because I would not have your blood upon my soul. Tell her, that through pain and danger you were faithful, and that, if your life-blood could have saved me, it would have flowed to its last drop. And tell her, that you will be something dearer than a father, and that my blessing is with you both, and that my dying eyes can see a long and pleasant path, in which you will journey together.'

As Malvin spoke, he almost raised himself from the ground, and the energy of his concluding words seemed to fill the wild and lonely forest with a vision of happiness. But when he sank exhausted upon his bed of oak-leaves, the light, which had kindled in Reuben's eye, was quenched. He felt as if it were both sin and folly to think of happiness at such a moment. His companion watched his changing countenance, and sought, with generous art, to wile him to his own good.

'Perhaps I deceive myself in regard to the time I have to live,' he resumed. 'It may be, that, with speedy assistance, I might recover my wound. The foremost fugitives must, ere this, have carried tidings of our fatal battle to the frontiers, and parties will be out to succour those in like condition with ourselves. Should you meet one of these, and guide them hither, who can tell but that I may sit by my own fireside again?'

A mournful smile strayed across the features of the dying man, as

he insinuated that unfounded hope; which, however, was not without its effect on Reuben. No merely selfish motive, nor even the
desolate condition of Dorcas, could have induced him to desert his
companion, at such a moment. But his wishes seized upon the
thought, that Malvin's life might be preserved, and his sanguine
nature heightened, almost to certainty, the remote possibility of
procuring human aid.

'Surely there is reason, weighty reason, to hope that friends are
not far distant;' he said, half aloud. 'There fled one coward, un-
wounded, in the beginning of the fight, and most probably he made
good speed. Every true man on the frontier would shoulder his
musket, at the news; and though no party may range so far into the
woods as this, I shall perhaps encounter them in one day's march.
Counsel me faithfully,' he added, turning to Malvin, in distrust of
his own motives. 'Were your situation mine, would you desert me
while life remained?'

'It is now twenty years,' replied Roger Malvin, sighing, however,
as he secretly acknowledged the wide dissimilarity between the two
cases,—'it is now twenty years, since I escaped, with one dear
friend, from Indian captivity, near Montreal. We journeyed many
days through the woods, till at length, overcome with hunger and
weariness, my friend lay down, and besought me to leave him; for
he knew, that, if I remained, we both must perish. And, with but
little hope of obtaining succour, I heaped a pillow of dry leaves
beneath his head, and hastened on.'

'And did you return in time to save him?' asked Reuben, hanging
on Malvin's words, as if they were to be prophetic of his own
success.

'I did,' answered the other, 'I came upon the camp of a hunting
party, before sunset of the same day. I guided them to the spot
where my comrade was expecting death; and he is now a hale and
hearty man, upon his own farm, far within the frontiers, while I lie
wounded here, in the depths of the wilderness.'

This example, powerful in effecting Reuben's decision, was aided,
unconsciously to himself, by the hidden strength of many another
motive. Roger Malvin perceived that the victory was nearly won.

'Now go, my son, and Heaven prosper you!' he said. 'Turn not
back with our friends, when you meet them, lest your wounds and
weariness overcome you; but send hitherward two or three, that
may be spared, to search for me. And believe me, Reuben, my heart
will be lighter with every step you take towards home.' Yet there
was perhaps a change, both in his countenance and voice, as he
spoke thus; for, after all, it was a ghastly fate, to be left expiring in
the wilderness.

Reuben Bourne, but half convinced that he was acting rightly, at
length raised himself from the ground, and prepared for his depar-

ture. And first, though contrary to Malvin's wishes, he collected a stock of roots and herbs, which had been their only food during the last two days. This useless supply he placed within reach of the dying man, for whom, also, he swept together a fresh bed of dry oak-leaves. Then, climbing to the summit of the rock, which on one side was rough and broken, he bent the oak-sapling downwards, and bound his handkerchief to the topmost branch. This precaution was not unnecessary, to direct any who might come in search of Malvin; for every part of the rock, except its broad, smooth front, was concealed, at a little distance, by the dense undergrowth of the forest. The handkerchief had been the bandage of a wound upon Reuben's arm; and, as he bound it to the tree, he vowed, by the blood that stained it, that he would return, either to save his companion's life, or to lay his body in the grave. He then descended, and stood, with downcast eyes, to receive Roger Malvin's parting words.

The experience of the latter suggested much and minute advice, respecting the youth's journey through the trackless forest. Upon this subject he spoke with calm earnestness, as if he were sending Reuben to the battle or the chase, while he himself remained secure at home; and not as if the human countenance, that was about to leave him, were the last he would ever behold. But his firmness was shaken, before he concluded.

'Carry my blessing to Dorcas, and say that my last prayer shall be for her and you. Bid her have no hard thoughts because you left me here'—Reuben's heart smote him—'for that your life would not have weighed with you, if its sacrifice could have done me good. She will marry you, after she has mourned a little while for her father; and Heaven grant you long and happy days! and may your children's children stand round your death-bed! And, Reuben,' added he, as the weakness of mortality made its way at last, 'return, when your wounds are healed and your weariness refreshed, return to this wild rock, and lay my bones in the grave, and say a prayer over them.'

An almost superstitious regard, arising perhaps from the customs of the Indians, whose war was with the dead, as well as the living, was paid by the frontier inhabitants to the rites of sepulture; and there are many instances of the sacrifice of life, in the attempt to bury those who had fallen by the 'sword of the wilderness.' Reuben, therefore, felt the full importance of the promise, which he most solemnly made, to return, and perform Roger Malvin's obsequies. It was remarkable, that the latter, speaking his whole heart in his parting words, no longer endeavored to persuade the youth, that even the speediest succour might avail to the preservation of his life. Reuben was internally convinced, that he should see Malvin's living face no more. His generous nature would fain have delayed him, at whatever risk, till the dying scene were past; but the desire of existence, and the hope of happiness had strengthened in his heart, and he was unable to resist them.

'It is enough,' said Roger Malvin, having listened to Reuben's promise. 'Go, and God speed you!'

The youth pressed his hand in silence, turned, and was departing. His slow and faultering steps, however, had borne him but a little way, before Malvin's voice recalled him.

'Reuben, Reuben,' said he, faintly; and Reuben returned and knelt down by the dying man.

'Raise me, and let me lean against the rock,' was his last request. 'My face will be turned towards home, and I shall see you a moment longer, as you pass among the trees.'

Reuben, having made the desired alteration in his companion's posture, again began his solitary pilgrimage. He walked more hastily at first, than was consistent with his strength; for a sort of guilty feeling, which sometimes torments men in their most justifiable acts, caused him to seek concealment from Malvin's eyes. But, after he had trodden far upon the rustling forest-leaves, he crept back, impelled by a wild and painful curiosity, and, sheltered by the earthy roots of an uptorn tree, gazed earnestly at the desolate man. The morning sun was unclouded, and the trees and shrubs imbibed the sweet air of the month of May; yet there seemed a gloom on Nature's face, as if she sympathized with mortal pain and sorrow. Roger Malvin's hands were uplifted in a fervent prayer, some of the words of which stole through the stillness of the woods, and entered Reuben's heart, torturing it with an unutterable pang. They were the broken accents of a petition for his own happiness and that of Dorcas; and, as the youth listened, conscience, or something in its similitude, pleaded strongly with him to return, and lie down again by the rock. He felt how hard was the doom of the kind and generous being whom he had deserted in his extremity. Death would come, like the slow approach of a corpse, stealing gradually towards him through the forest, and showing its ghastly and motionless features from behind a nearer, and yet a nearer tree. But such must have been Reuben's own fate, had he tarried another sunset; and who shall impute blame to him, if he shrank from so useless a sacrifice? As he gave a parting look, a breeze waved the little banner upon the sapling-oak, and reminded Reuben of his vow.

．　．　．　．　．

Many circumstances contributed to retard the wounded traveller, in his way to the frontiers. On the second day, the clouds, gathering densely over the sky, precluded the possibility of regulating his course by the position of the sun; and he knew not but that every effort of his almost exhausted strength, was removing him farther from the home he sought. His scanty sustenance was supplied by the berries, and other spontaneous products of the forest. Herds of deer, it is true, sometimes bounded past him, and partridges frequently whirred up before his footsteps; but his ammunition had been expended in the fight, and he had no means of slaying them. His wounds, irritated by the constant exertion in which lay the only hope of life, wore away his strength, and at intervals confused his reason. But, even in the wanderings of intellect, Reuben's young heart clung strongly to existence, and it was only through absolute incapacity of motion, that he at last sank down beneath a tree, compelled there to await death. In this situation he was discovered by a party, who, upon the first intelligence of the fight, had been despatched to the relief of the survivors. They conveyed him to the nearest settlement, which chanced to be that of his own residence.

Dorcas, in the simplicity of the olden time, watched by the bedside of her wounded lover, and administered all those comforts, that are in the sole gift of woman's heart and hand. During several days, Reuben's recollection strayed drowsily among the perils and hardships through which he had passed, and he was incapable of returning definite answers to the inquiries, with which many were eager to harass him. No authentic particulars of the battle had yet been circulated; nor could mothers, wives, and children tell, whether their loved ones were detained by captivity, or by the stronger chain of death. Dorcas nourished her apprehensions in silence, till one afternoon, when Reuben awoke from an unquiet sleep, and seemed to recognise her, more perfectly than at any previous time. She saw that his intellect had become composed, and she could no longer restrain her filial anxiety.

'My father, Reuben?' she began; but the change in her lover's countenance made her pause.

The youth shrank, as if with a bitter pain, and the blood gushed vividly into his wan and hollow cheeks. His first impulse was to cover his face; but, apparently with a desperate effort, he half raised himself, and spoke vehemently, defending himself against an imaginary accusation.

'Your father was sore wounded in the battle, Dorcas, and he bade me not burthen myself with him, but only to lead him to the lakeside, that he might quench his thirst and die. But I would not desert the old man in his extremity, and, though bleeding myself, I supported him; I gave him half my strength, and led him away with me. For three days we journeyed on together, and your father was sustained beyond my hopes; but, awaking at sunrise on the fourth day, I found him faint and exhausted,—he was unable to proceed, —his life had ebbed away fast,—and'—

'He died!' exclaimed Dorcas, faintly.

Reuben felt it impossible to acknowledge, that his selfish love of life had hurried him away, before her father's fate was decided. He spoke not; he only bowed his head; and, between shame and exhaustion, sank back and hid his face in the pillow. Dorcas wept, when her fears were thus confirmed; but the shock, as it had been long anticipated, was on that account the less violent.

'You dug a grave for my poor father, in the wilderness, Reuben?' was the question by which her filial piety manifested itself.

'My hands were weak, but I did what I could,' replied the youth in a smothered tone. 'There stands a noble tomb-stone above his head, and I would to Heaven I slept as soundly as he!'

Dorcas, perceiving the wildness of his latter words, inquired no farther at that time; but her heart found ease in the thought, that Roger Malvin had not lacked such funeral rites as it was possible to bestow. The tale of Reuben's courage and fidelity lost nothing, when she communicated it to her friends; and the poor youth, tottering from his sick chamber to breathe the sunny air, experienced from every tongue the miserable and humiliating torture of unmerited praise. All acknowledged that he might worthily demand the hand of the fair maiden, to whose father he had been 'faithful unto death;' and, as my tale is not of love, it shall suffice to say, that, in the space of two years, Reuben became the husband of Dorcas Malvin. During the marriage ceremony, the bride was covered with blushes, but the bridegroom's face was pale.

There was now in the breast of Reuben Bourne an incommunicable thought; something which he was to conceal most heedfully from her whom he most loved and trusted. He regretted, deeply and bitterly, the moral cowardice that had restrained his words, when he was about to disclose the truth to Dorcas; but pride, the fear of losing her affection, the dread of universal scorn, forbade him to rectify this falsehood. He felt, that, for leaving Roger Malvin, he deserved no censure. His presence, the gratuitous sacrifice of his own life, would have added only another, and a needless agony to the last moments of the dying man. But concealment had imparted to a justifiable act, much of the secret effect of guilt; and Reuben, while reason told him that he had done right, experienced in no small degree, the mental horrors, which punish the perpetrator of undiscovered crime. By a certain association of ideas, he at times almost imagined himself a murderer. For years, also, a thought would occasionally recur, which, though he perceived all its folly and extravagance, he had not power to banish from his mind; it was a haunting and torturing fancy, that his father-in-law was yet sitting at the foot of the rock, on the withered forest-leaves, alive, and awaiting his pledged assistance. These mental deceptions, however, came and went, nor did he ever mistake them for realities; but in the calmest and clearest moods of his mind, he was conscious that he had a deep vow unredeemed, and that an unburied corpse was calling to him, out of the wilderness. Yet, such was the consequence of his prevarication, that he could not obey the call. It was now too late to require the assistance of Roger Malvin's friends, in performing his long-deferred sepulture; and superstitious fears, of which none were more susceptible than the people of the outward settlements, forbade Reuben to go alone. Neither did he know where, in the pathless and illimitable forest, to seek that smooth and lettered rock, at the base of which the body lay; his remembrance of every portion of his travel thence was indistinct, and the latter part had left no impression upon his mind. There was, however, a continual impulse, a voice audible only to himself, commanding him to go forth and redeem his vow; and he had a strange impression, that, were he to make the trial, he would be led straight to Malvin's bones. But, year after year, that summons, unheard but felt, was disobeyed. His one secret thought, became like a chain, binding down his spirit, and, like a serpent, gnawing into his heart; and he was transformed into a sad and downcast, yet irritable man.

In the course of a few years after their marriage, changes began to be visible in the external prosperity of Reuben and Dorcas. The only riches of the former had been his stout heart and strong arm; but the latter, her father's sole heiress, had made her husband master of a farm, under older cultivation, larger, and better stocked than most of the frontier establishments. Reuben Bourne, however, was a neglectful husbandman; and while the lands of the other settlers became annually more fruitful, his deteriorated in the same proportion. The discouragements to agriculture were greatly lessened by the cessation of Indian war, during which men held the plough in one hand, and the musket in the other; and were fortunate if the products of their dangerous labor were not destroyed, either in the field or in the barn, by the savage enemy. But Reuben did not profit by the altered condition of the country; nor can it be denied, that his intervals of industrious attention to his affairs were but scantily rewarded with success. The irritability, by which he had recently become distinguished, was another cause of his declining prosperity, as it occasioned frequent quarrels, in his unavoidable intercourse with the neighboring settlers. The results of these were innumerable law-suits; for the people of New England, in the earliest stages and wildest circumstances of the country, adopted, whenever attainable, the legal mode of deciding their differences. To be brief, the world did not go well with Reuben Bourne, and, though not till many years after his marriage, he was finally a ruined man, with but one remaining expedient against the evil fate that had pursued him. He was to throw sunlight into some deep recess of the forest, and seek subsistence from the virgin bosom of the wilderness.

The only child of Reuben and Dorcas was a son, now arrived at the age of fifteen years, beautiful in youth, and giving promise of a glorious manhood. He was peculiarly qualified for, and already began to excel in, the wild accomplishments of frontier life. His foot was fleet, his aim true, his apprehension quick, his heart glad and high; and all, who anticipated the return of Indian war, spoke of

Cyrus Bourne as a future leader in the land. The boy was loved by his father, with a deep and silent strength, as if whatever was good and happy in his own nature had been transferred to his child, carrying his affections with it. Even Dorcas, though loving and beloved, was far less dear to him; for Reuben's secret thoughts and insulated emotions had gradually made him a selfish man; and he could no longer love deeply, except where he saw, or imagined, some reflection or likeness of his own mind. In Cyrus he recognised what he had himself been in other days; and at intervals he seemed to partake of the boy's spirit, and to be revived with a fresh and happy life. Reuben was accompanied by his son in the expedition, for the purpose of selecting a tract of land, and felling and burning the timber, which necessarily preceded the removal of the household gods.[3] Two months of autumn were thus occupied; after which Reuben Bourne and his young hunter returned, to spend their last winter in the settlements.

. . . . .

It was early in the month of May, that the little family snapped asunder whatever tendrils of affection had clung to inanimate objects, and bade farewell to the few, who, in the blight of fortune, called themselves their friends. The sadness of the parting moment had, to each of the pilgrims, its peculiar alleviations. Reuben, a moody man, and misanthropic because unhappy, strode onward, with his usual stern brow and downcast eye, feeling few regrets, and disdaining to acknowledge any. Dorcas, while she wept abundantly over the broken ties by which her simple and affectionate nature had bound itself to everything, felt that the inhabitants of her inmost heart moved on with her, and that all else would be supplied wherever she might go. And the boy dashed one tear-drop from his eye, and thought of the adventurous pleasures of the untrodden forest. Oh! who, in the enthusiasm of a day-dream, has not wished that he were a wanderer in a world of summer wilderness, with one fair and gentle being hanging lightly on his arm? In youth, his free and exulting step would know no barrier but the rolling ocean or the snow-topt mountains; calmer manhood would choose a home, where Nature had strewn a double wealth, in the vale of some transparent stream; and when hoary age, after long, long years of that pure life, stole on and found him there, it would find him the father of a race, the patriarch of a people, the founder of a mighty nation yet to be. When death, like the sweet sleep which we welcome after a day of happiness, came over him, his far descendants would mourn over the venerated dust. Enveloped by tradition in mysterious attributes, the men of future generations would call him

---

3. I.e., prized possessions, because of the value placed on personal idols in many cultures. In Genesis 31.19 Rachel, without telling her husband Jacob, steals her father's household gods.

godlike; and remote posterity would see him standing, dimly glorious, far up the valley of a hundred centuries!

The tangled and gloomy forest, through which the personages of my tale were wandering, differed widely from the dreamer's Land of Fantasië; yet there was something in their way of life that Nature asserted as her own; and the gnawing cares, which went with them from the world, were all that now obstructed their happiness. One stout and shaggy steed, the bearer of all their wealth, did not shrink from the added weight of Dorcas; although her hardy breeding sustained her, during the larger part of each day's journey, by her husband's side. Reuben and his son, their muskets on their shoulders, and their axes slung behind them, kept an unwearied pace, each watching with a hunter's eye for the game that supplied their food. When hunger bade, they halted and prepared their meal on the bank of some unpolluted forest-brook, which, as they knelt down with thirsty lips to drink, murmured a sweet unwillingness, like a maiden, at love's first kiss. They slept beneath a hut of branches, and awoke at peep of light, refreshed for the toils of another day. Dorcas and the boy went on joyously, and even Reuben's spirit shone at intervals with an outward gladness; but inwardly there was a cold, cold sorrow, which he compared to the snow-drifts, lying deep in the glens and hollows of the rivulets, while the leaves were brightly green above.

Cyrus Bourne was sufficiently skilled in the travel of the woods, to observe, that his father did not adhere to the course they had pursued, in their expedition of the preceding autumn. They were now keeping farther to the north, striking out more directly from the settlements, and into a region, of which savage beasts and savage men were as yet the sole possessors. The boy sometimes hinted his opinions upon the subject, and Reuben listened attentively, and once or twice altered the direction of their march in accordance with his son's counsel. But having so done, he seemed ill at ease. His quick and wandering glances were sent forward, apparently in search of enemies lurking behind the tree-trunks; and seeing nothing there, he would cast his eyes backward, as if in fear of some pursuer. Cyrus, perceiving that his father gradually resumed the old direction, forbore to interfere; nor, though something began to weigh upon his heart, did his adventurous nature permit him to regret the increased length and the mystery of their way.

On the afternoon of the fifth day, they halted and made their simple encampment, nearly an hour before sunset. The face of the country, for the last few miles, had been diversified by swells of land, resembling huge waves of a petrified sea; and in one of the corresponding hollows, a wild and romantic spot, had the family reared their hut, and kindled their fire. There is something chilling, and yet heart-warming, in the thought of three, united by strong bands of love, and insulated from all that breathe beside. The dark and gloomy pines looked down upon them, and, as the wind swept through their tops, a pitying sound was heard in the forest; or did those old trees groan, in fear that men were come to lay the axe to their roots at last? Reuben and his son, while Dorcas made ready their meal, proposed to wander out in search of game, of which that day's march had afforded no supply. The boy, promising not to quit the vicinity of the encampment, bounded off with a step as light and elastic as that of the deer he hoped to slay; while his father, feeling a transient happiness as he gazed after him, was about to pursue an opposite direction. Dorcas, in the meanwhile, had seated herself near their fire of fallen branches, upon the moss-grown and mouldering trunk of a tree, uprooted years before. Her employment, diversified by an occasional glance at the pot, now beginning to simmer over the blaze, was the perusal of the current year's Massachusetts Almanac, which, with the exception of an old black-letter[4] Bible, comprised all the literary wealth of the family. None pay a greater regard to arbitrary divisions of time, than those who are excluded from society; and Dorcas mentioned, as if the information were of importance, that it was now the twelfth of May. Her husband started.

'The twelfth of May! I should remember it well,' muttered he, while many thoughts occasioned a momentary confusion in his mind. 'Where am I? Whither am I wandering? Where did I leave him?'

Dorcas, too well accustomed to her husband's wayward moods to note any peculiarity of demeanor, now laid aside the Almanac, and addressed him in that mournful tone, which the tender-hearted appropriate to griefs long cold and dead.

'It was near this time of the month, eighteen years ago, that my poor father left this world for a better. He had a kind arm to hold his head, and a kind voice to cheer him, Reuben, in his last moments; and the thought of the faithful care you took of him, has comforted me, many a time since. Oh! death would have been awful to a solitary man, in a wild place like this!'

'Pray Heaven, Dorcas,' said Reuben, in a broken voice, 'pray Heaven, that neither of us three die solitary, and lie unburied, in this howling wilderness!' And he hastened away, leaving her to watch the fire, beneath the gloomy pines.

Reuben Bourne's rapid pace gradually slackened, as the pang, unintentionally inflicted by the words of Dorcas, became less acute. Many strange reflections, however, thronged upon him; and, straying onward, rather like a sleep-walker than a hunter, it was attributable to no care of his own, that his devious course kept him in the vicinity of the encampment. His steps were imperceptibly led almost

4. Printed in early type resembling the shapes of letters used by medieval and early Renaissance scribes.

in a circle, nor did he observe that he was on the verge of a tract of land heavily timbered, but not with pine-trees. The place of the latter was here supplied by oaks, and other of the harder woods; and around their roots clustered a dense and bushy undergrowth, leaving, however, barren spaces between the trees, thick-strewn with withered leaves. Whenever the rustling of the branches, or the creaking of the trunks made a sound, as if the forest were waking from slumber, Reuben instinctively raised the musket that rested on his arm, and cast a quick, sharp glance on every side; but, convinced by a partial observation that no animal was near, he would again give himself up to his thoughts. He was musing on the strange influence, that had led him away from his premeditated course, and so far into the depths of the wilderness. Unable to penetrate to the secret place of his soul, where his motives lay hidden, he believed that a supernatural voice had called him onward, and that a supernatural power had obstructed his retreat. He trusted that it was Heaven's intent to afford him an opportunity of expiating his sin; he hoped that he might find the bones, so long unburied; and that, having laid the earth over them, peace would throw its sunlight into the sepulchre of his heart. From these thoughts he was aroused by a rustling in the forest, at some distance from the spot to which he had wandered. Perceiving the motion of some object behind a thick veil of undergrowth, he fired, with the instinct of a hunter, and the aim of a practised marksman. A low moan, which told his success, and by which even animals can express their dying agony, was unheeded by Reuben Bourne. What were the recollections now breaking upon him?

The thicket, into which Reuben had fired, was near the summit of a swell of land, and was clustered around the base of a rock, which, in the shape and smoothness of one of its surfaces, was not unlike a gigantic gravestone. As if reflected in a mirror, its likeness was in Reuben's memory. He even recognised the veins which seemed to form an inscription in forgotten characters; everything remained the same, except that a thick covert of bushes shrouded the lower part of the rock, and would have hidden Roger Malvin, had he still been sitting there. Yet, in the next moment, Reuben's eye was caught by another change, that time had effected, since he last stood, where he was now standing again, behind the earthy roots of the uptorn tree. The sapling, to which he had bound the blood-stained symbol of his vow, had increased and strengthened into an oak, far indeed from its maturity, but with no mean spread of shadowy branches. There was one singularity, observable in this tree, which made Reuben tremble. The middle and lower branches were in luxuriant life, and an excess of vegetation had fringed the trunk, almost to the ground; but a blight had apparently stricken the upper part of the oak, and the very topmost bough was withered, sapless, and utterly dead. Reuben remembered how the little banner had fluttered on that topmost bough, when it was green and lovely, eighteen years before. Whose guilt had blasted it?

. . . . .

Dorcas, after the departure of the two hunters, continued her preparations for their evening repast. Her sylvan table was the moss-covered trunk of a large fallen tree, on the broadest part of which she had spread a snow-white cloth, and arranged what were left of the bright pewter vessels, that had been her pride in the settlements. It had a strange aspect—that one little spot of homely comfort, in the desolate heart of Nature. The sunshine yet lingered upon the higher branches of the trees that grew on rising ground; but the shades of evening had deepened into the hollow, where the encampment was made; and the fire-light began to redden as it gleamed up the tall trunks of the pines, or hovered on the dense and obscure mass of foliage, that circled round the spot. The heart of Dorcas was not sad; for she felt that it was better to journey in the wilderness, with two whom she loved, than to be a lonely woman in a crowd that cared not for her. As she busied herself in arranging seats of mouldering wood, covered with leaves, for Reuben and her son, her voice danced through the gloomy forest, in the measure of a song that she had learned in youth. The rude melody, the production of a bard who won no name, was descriptive of a winter evening in a frontier-cottage, when, secured from savage inroad by the high-piled snow-drifts, the family rejoiced by their own fire-side. The whole song possessed that nameless charm, peculiar to unborrowed thought; but four continually-recurring lines shone out from the rest, like the blaze of the hearth whose joys they celebrated. Into them, working magic with a few simple words, the poet had instilled the very essence of domestic love and household happiness, and they were poetry and picture joined in one. As Dorcas sang, the walls of her forsaken home seemed to encircle her; she no longer saw the gloomy pines, nor heard the wind, which still, as she began each verse, sent a heavy breath through the branches, and died away in a hollow moan, from the burthen of the song. She was aroused by the report of a gun, in the vicinity of the encampment; and either the sudden sound, or her loneliness by the glowing fire, caused her to tremble violently. The next moment, she laughed in the pride of a mother's heart.

'My beautiful young hunter! my boy has slain a deer!' she exclaimed, recollecting that, in the direction whence the shot proceeded, Cyrus had gone to the chase.

She waited a reasonable time, to hear her son's light step bounding over the rustling leaves, to tell of his success. But he did not immediately appear, and she sent her cheerful voice among the trees, in search of him.

'Cyrus! Cyrus!'

His coming was still delayed, and she determined, as the report of the gun had apparently been very near, to seek for him in person. Her assistance, also, might be necessary in bringing home the venison, which she flattered herself he had obtained. She therefore set forward, directing her steps by the long-past sound, and singing as she went, in order that the boy might be aware of her approach, and run to meet her. From behind the trunk of every tree, and from every hiding place in the thick foliage of the undergrowth, she hoped to discover the countenance of her son, laughing with the sportive mischief that is born of affection. The sun was now beneath the horizon, and the light that came down among the trees was sufficiently dim to create many illusions in her expecting fancy. Several times she seemed indistinctly to see his face gazing out from among the leaves; and once she imagined that he stood beckoning to her, at the base of a craggy rock. Keeping her eyes on this object, however, it proved to be no more than the trunk of an oak, fringed to the very ground with little branches, one of which, thrust out farther than the rest, was shaken by the breeze. Making her way round the foot of the rock, she suddenly found herself close to her husband, who had approached in another direction. Leaning upon the butt of his gun, the muzzle of which rested upon the withered leaves, he was apparently absorbed in the contemplation of some object at his feet.

'How is this, Reuben? Have you slain the deer, and fallen asleep over him?' exclaimed Dorcas, laughing cheerfully, on her first slight observation of his posture and appearance.

He stirred not, neither did he turn his eyes towards her; and a cold, shuddering fear, indefinite in its source and object, began to creep into her blood. She now perceived that her husband's face was ghastly pale, and his features were rigid, as if incapable of assuming any other expression than the strong despair which had hardened upon them. He gave not the slightest evidence that he was aware of her approach.

'For the love of Heaven, Reuben, speak to me!' cried Dorcas, and the strange sound of her own voice affrighted her even more than the dead silence.

Her husband started, stared into her face; drew her to the front of the rock, and pointed with his finger.

Oh! there lay the boy, asleep, but dreamless, upon the fallen forest-leaves! His cheek rested upon his arm, his curled locks were thrown back from his brow, his limbs were slightly relaxed. Had a sudden weariness overcome the youthful hunter? Would his mother's voice arouse him? She knew that it was death.

'This broad rock is the grave-stone of your near kindred, Dorcas,' said her husband. 'Your tears will fall at once over your father and your son.'

She heard him not. With one wild shriek, that seemed to force its way from the sufferer's inmost soul, she sank insensible by the side

of her dead boy. At that moment, the withered topmost bough of
the oak loosened itself, in the stilly air, and fell in soft, light fragments upon the rock, upon the leaves, upon Reuben, upon his wife
and child, and upon Roger Malvin's bones. Then Reuben's heart
was stricken, and the tears gushed out like water from a rock. The
vow that the wounded youth had made, the blighted man had come
to redeem. His sin was expiated, the curse was gone from him; and,
in the hour, when he had shed blood dearer to him than his own, a
prayer, the first for years, went up to Heaven from the lips of
Reuben Bourne.

1832   1846

Young Goodman Brown[1]

Young goodman[2] Brown came forth, at sunset, into the street of
Salem village, but put his head back, after crossing the threshold, to
exchange a parting kiss with his young wife. And Faith, as the wife
was aptly named, thrust her own pretty head into the street, letting
the wind play with the pink ribbons of her cap, while she called to
goodman Brown.

'Dearest heart,' whispered she, softly and rather sadly, when her
lips were close to his ear, 'pr'y thee, put off your journey until
sunrise, and sleep in your own bed to-night. A lone woman is
troubled with such dreams and such thoughts, that she's afeard of
herself, sometimes. Pray, tarry with me this night, dear husband, of
all nights in the year!'

'My love and my Faith,' replied young goodman Brown, 'of all
nights in the year, this one night must I tarry away from thee. My
journey, as thou callest it, forth and back again, must needs be done
'twixt now and sunrise. What, my sweet, pretty wife, dost thou
doubt me already, and we but three months married!'

'Then, God bless you!' said Faith, with the pink ribbons, 'and
may you find all well, when you come back.'

'Amen!' cried goodman Brown. 'Say thy prayers, dear Faith, and
go to bed at dusk, and no harm will come to thee.'

So they parted; and the young man pursued his way, until, being
about to turn the corner by the meeting-house, he looked back, and
saw the head of Faith still peeping after him, with a melancholy air,
in spite of her pink ribbons.

'Poor little Faith!' thought he, for his heart smote him. 'What a
wretch am I, to leave her on such an errand! She talks of dreams,
too. Methought, as she spoke, there was trouble in her face, as if a
dream had warned her what work is to be done to-night. But, no,

1. The text followed here is that of the
first publication, in the *New-England
Magazine* (April, 1835); the story was
ascribed to "the author of 'The Gray
Champion,'" which had appeared in the
same magazine three months earlier.

2. Hawthorne puns on the title used to
address a man of humble birth and the
moral implications of "good man"; what
with "Brown" as a surname, the hero is
equivalent to Young Mister Anybody.

no! 't would kill her to think it. Well; she's a blessed angel on earth; and after this one night, I'll cling to her skirts and follow her to Heaven.'

With this excellent resolve for the future, goodman Brown felt himself justified in making more haste on his present evil purpose. He had taken a dreary road, darkened by all the gloomiest trees of the forest, which barely stood aside to let the narrow path creep through, and closed immediately behind. It was all as lonely as could be; and there is this peculiarity in such a solitude, that the traveler knows not who may be concealed by the innumerable trunks and the thick boughs overhead; so that, with lonely foot-steps, he may yet be passing through an unseen multitude.

'There may be a devilish Indian behind every tree,' said goodman Brown, to himself; and he glanced fearfully behind him, as he added, 'What if the devil himself should be at my very elbow!'

His head being turned back, he passed a crook of the road, and looking forward again, beheld the figure of a man, in grave and decent attire, seated at the foot of an old tree. He arose, at goodman Brown's approach, and walked onward, side by side with him.

'You are late, goodman Brown,' said he. 'The clock of the Old South was striking as I came through Boston; and that is full fifteen minutes agone.'[3]

'Faith kept me back awhile,' replied the young man, with a tremor in his voice, caused by the sudden appearance of his companion, though not wholly unexpected.

It was now deep dusk in the forest, and deepest in that part of it where these two were journeying. As nearly as could be discerned, the second traveler was about fifty years old, apparently in the same rank of life as goodman Brown, and bearing a considerable resemblance to him, though perhaps more in expression than features. Still, they might have been taken for father and son. And yet, though the elder person was as simply clad as the younger, and as simple in manner too, he had an indescribable air of one who knew the world, and would not have felt abashed at the governor's dinner-table, or in king William's[4] court, were it possible that his affairs should call him thither. But the only thing about him, that could be fixed upon as remarkable, was his staff, which bore the likeness of a great black snake, so curiously wrought, that it might almost be seen to twist and wriggle itself, like a living serpent. This, of course, must have been an ocular deception, assisted by the uncertain light.

'Come, goodman Brown!' cried his fellow-traveler, 'this is a dull pace for the beginning of a journey. Take my staff, if you are so soon weary.'

'Friend,' said the other, exchanging his slow pace for a full stop, 'having kept covenant by meeting thee here, it is my purpose now to

3. This speed could only be supernatural.
4. William of Orange, first cousin and husband of Queen Mary II, with whom he jointly ruled England, 1689–1702.

return whence I came. I have scruples, touching the matter thou wot'st of.'

'Sayest thou so?' replied he of the serpent, smiling apart. 'Let us walk on, nevertheless, reasoning as we go, and if I convince thee not, thou shalt turn back. We are but a little way in the forest, yet.'

'Too far, too far!' exclaimed the goodman, unconsciously resuming his walk. 'My father never went into the woods on such an errand, nor his father before him. We have been a race of honest men and good Christians, since the days of the martyrs.[5] And shall I be the first of the name of Brown, that ever took this path, and kept'—

'Such company, thou wouldst say,' observed the elder person, interpreting his pause. 'Good, goodman Brown! I have been as well acquainted with your family as with ever a one among the Puritans; and that's no trifle to say. I helped your grandfather, the constable, when he lashed the Quaker woman so smartly through the streets of Salem. And it was I that brought your father a pitch-pine knot, kindled at my own hearth, to set fire to an Indian village, in king Philip's[6] war. They were my good friends, both; and many a pleasant walk have we had along this path, and returned merrily after midnight. I would fain be friends with you, for their sake.'

'If it be as thou sayest,' replied goodman Brown, 'I marvel they never spoke of these matters. Or, verily, I marvel not, seeing that the least rumor of the sort would have driven them from New-England. We are a people of prayer, and good works, to boot, and abide no such wickedness.'

'Wickedness or not,' said the traveler with the twisted staff, 'I have a very general acquaintance here in New-England. The deacons of many a church have drunk the communion wine with me; the selectmen, of divers towns, make me their chairman; and a majority of the Great and General Court[7] are firm supporters of my interest. The governor and I, too—but these are state-secrets.'

'Can this be so!' cried goodman Brown, with a stare of amazement at his undisturbed companion. 'Howbeit, I have nothing to do with the governor and council; they have their own ways, and are no rule for a simple husbandman,[8] like me. But, were I to go on with thee, how should I meet the eye of that good old man, our minister, at Salem village? Oh, his voice would make me tremble, both Sabbath-day and lecture-day!'[9]

---

5. I.e., during the reign of the Catholic Mary Tudor of England (1553–58), called "Bloody Mary" for her persecution of Protestants. Common reading in New England was John Foxe's *Acts and Monuments* (1563), soon known as the *Book of Martyrs*; it concluded with horrifically detailed accounts of martyrdoms under Mary.

6. Indian leader of the Wampanoags who waged war (1675–76) against the New England colonists.
7. The legislature.
8. Usually, farmer; here, man of ordinary status.
9. Midweek sermon day, Wednesday or Thursday.

Thus far, the elder traveler had listened with due gravity, but now burst into a fit of irrepressible mirth, shaking himself so violently, that his snake-like staff actually seemed to wriggle in sympathy.

'Ha! ha! ha!' shouted he, again and again; then composing himself, 'Well, go on, goodman Brown, go on; but, pr'y thee, don't kill me with laughing!'

'Well, then, to end the matter at once,' said goodman Brown, considerably nettled, 'there is my wife, Faith. It would break her dear little heart; and I'd rather break my own!'

'Nay, if that be the case,' answered the other, 'e'en go thy ways, goodman Brown. I would not, for twenty old women like the one hobbling before us, that Faith should come to any harm.'

As he spoke, he pointed his staff at a female figure on the path, in whom goodman Brown recognized a very pious and exemplary dame, who had taught him his catechism, in youth, and was still his moral and spiritual adviser, jointly with the minister and deacon Gookin.

'A marvel, truly, that goody Cloyse[1] should be so far in the wilderness, at night-fall!' said he. 'But, with your leave, friend, I shall take a cut through the woods, until we have left this Christian woman behind. Being a stranger to you, she might ask whom I was consorting with, and whither I was going.'

'Be it so,' said his fellow-traveler. 'Betake you to the woods, and let me keep the path.'

Accordingly, the young man turned aside, but took care to watch his companion, who advanced softly along the road, until he had come within a staff's length of the old dame. She, meanwhile, was making the best of her way, with singular speed for so aged a woman, and mumbling some indistinct words, a prayer, doubtless, as she went. The traveler put forth his staff, and touched her withered neck with what seemed the serpent's tail.

'The devil!' screamed the pious old lady.

'Then goody Cloyse knows her old friend?' observed the traveler, confronting her, and leaning on his writhing stick.

'Ah, forsooth, and is it your worship, indeed?' cried the good dame. 'Yea, truly is it, and in the very image of my old gossip, goodman Brown, the grandfather of the silly fellow that now is. But, would your worship believe it? my broomstick hath strangely disappeared, stolen, as I suspect, by that unhanged witch, goody Cory, and that, too, when I was all anointed with the juice of smallage and cinque-foil and wolf's-bane'—[2]

'Mingled with fine wheat and the fat of a new-born babe,' said the shape of old goodman Brown.

---

1. Hawthorne uses historical names of people involved in the Salem witchcraft trials. "Goody" means "goodwife" and was a polite title for a married woman of humble rank.

2. Plants associated with witchcraft: wild celery or parsley; a five-lobed plant of the rose family (from the Latin for "five fingers"); hooded, poisonous plant known as monkshood ("bane" means "poison").

'Ah, your worship knows the receipt,' cried the old lady, cackling aloud. 'So, as I was saying, being all ready for the meeting, and no horse to ride on, I made up my mind to foot it; for they tell me, there is a nice young man to be taken into communion to-night. But now your good worship will lend me your arm, and we shall be there in a twinkling.'

'That can hardly be,' answered her friend. 'I may not spare you my arm, goody Cloyse, but here is my staff, if you will.'

So saying, he threw it down at her feet, where, perhaps, it assumed life, being one of the rods which its owner had formerly lent to the Egyptian Magi.[3] Of this fact, however, goodman Brown could not take cognizance. He had cast up his eyes in astonishment, and looking down again, beheld neither goody Cloyse nor the serpentine staff, but his fellow-traveler alone, who waited for him as calmly as if nothing had happened.

'That old woman taught me my catechism!' said the young man; and there was a world of meaning in this simple comment.

They continued to walk onward, while the elder traveler exhorted his companion to make good speed and persevere in the path, discoursing so aptly, that his arguments seemed rather to spring up in the bosom of his auditor, than to be suggested by himself. As they went, he plucked a branch of maple, to serve for a walking-stick, and began to strip it of the twigs and little boughs, which were wet with evening dew. The moment his fingers touched them, they became strangely withered and dried up, as with a week's sunshine. Thus the pair proceeded, at a good free pace, until suddenly, in a gloomy hollow of the road, goodman Brown sat himself down on the stump of a tree, and refused to go any farther.

'Friend,' said he, stubbornly, 'my mind is made up. Not another step will I budge on this errand. What if a wretched old woman do choose to go to the devil, when I thought she was going to Heaven! Is that any reason why I should quit my dear Faith, and go after her?'

'You will think better of this, by-and-by,' said his acquaintance, composedly. 'Sit here and rest yourself awhile; and when you feel like moving again, there is my staff to help you along.'

Without more words, he threw his companion the maple stick, and was as speedily out of sight, as if he had vanished into the deepening gloom. The young man sat a few moments, by the roadside, applauding himself greatly, and thinking with how clear a conscience he should meet the minister, in his morning-walk, nor shrink from the eye of good old deacon Gookin. And what calm sleep would be his, that very night, which was to have been spent so wickedly, but purely and sweetly now, in the arms of Faith! Amidst

3. See Exodus 7.11 for the magicians of Egypt who duplicated Aaron's feat of casting down his rod before Pharaoh and making it turn into a serpent.

these pleasant and praiseworthy meditations, goodman Brown heard the tramp of horses along the road, and deemed it advisable to conceal himself within the verge of the forest, conscious of the guilty purpose that had brought him thither, though now so happily turned from it.

On came the hoof-tramps and the voices of the riders, two grave old voices, conversing soberly as they drew near. These mingled sounds appeared to pass along the road, within a few yards of the young man's hiding-place; but owing, doubtless, to the depth of the gloom, at that particular spot, neither the travelers nor their steeds were visible. Though their figures brushed the small boughs by the way-side, it could not be seen that they intercepted, even for a moment, the faint gleam from the strip of bright sky, athwart which they must have passed. Goodman Brown alternately crouched and stood on tip-toe, pulling aside the branches, and thrusting forth his head as far as he durst, without discerning so much as a shadow. It vexed him the more, because he could have sworn, were such a thing possible, that he recognized the voices of the minister and deacon Gookin, jogging along quietly, as they were wont to do, when bound to some ordination or ecclesiastical council. While yet within hearing, one of the riders stopped to pluck a switch.

'Of the two, reverend Sir,' said the voice like the deacon's, 'I had rather miss an ordination-dinner than to-night's meeting. They tell me that some of our community are to be here from Falmouth[4] and beyond, and others from Connecticut and Rhode-Island; besides several of the Indian powows,[5] who, after their fashion, know almost as much deviltry as the best of us. Moreover, there is a goodly young woman to be taken into communion.'

'Mighty well, deacon Gookin!' replied the solemn old tones of the minister. 'Spur up, or we shall be late. Nothing can be done, you know, until I get on the ground.'

The hoofs clattered again, and the voices, talking so strangely in the empty air, passed on through the forest, where no church had ever been gathered, nor solitary Christian prayed. Whither, then, could these holy men be journeying, so deep into the heathen wilderness? Young goodman Brown caught hold of a tree, for support, being ready to sink down on the ground, faint and over-burthened with the heavy sickness of his heart. He looked up to the sky, doubting whether there really was a Heaven above him. Yet, there was the blue arch, and the stars brightening in it.

'With Heaven above, and Faith below, I will yet stand firm against the devil!' cried goodman Brown.

While he still gazed upward, into the deep arch of the firmament,

4. Town on Cape Cod, about 70 miles from Salem.
5. Medicine men. Usually spelled "pow- wow" and later used to refer to any conference or gathering.

and had lifted his hands to pray, a cloud, though no wind was stirring, hurried across the zenith, and hid the brightening stars. The blue sky was still visible, except directly overhead, where this black mass of cloud was sweeping swiftly northward. Aloft in the air, as if from the depths of the cloud, came a confused and doubtful sound of voices. Once, the listener fancied that he could distinguish the accents of town's-people of his own, men and women, both pious and ungodly, many of whom he had met at the communion-table, and had seen others rioting at the tavern. The next moment, so indistinct were the sounds, he doubted whether he had heard aught but the murmur of the old forest, whispering without a wind. Then came a stronger swell of those familiar tones, heard daily in the sunshine, at Salem village, but never, until now, from a cloud of night. There was one voice, of a young woman, uttering lamentations, yet with an uncertain sorrow, and entreating for some favor, which, perhaps, it would grieve her to obtain. And all the unseen multitude, both saints and sinners, seemed to encourage her onward.

'Faith!' shouted goodman Brown, in a voice of agony and desperation; and the echoes of the forest mocked him, crying—'Faith! Faith!' as if bewildered wretches were seeking her, all through the wilderness.

The cry of grief, rage, and terror, was yet piercing the night, when the unhappy husband held his breath for a response. There was a scream, drowned immediately in a louder murmur of voices, fading into far-off laughter, as the dark cloud swept away, leaving the clear and silent sky above goodman Brown. But something fluttered lightly down through the air, and caught on the branch of a tree. The young man seized it, and beheld a pink ribbon.

'My Faith is gone!' cried he, after one stupefied moment. 'There is no good on earth; and sin is but a name. Come, devil! for to thee is this world given.'

And maddened with despair, so that he laughed loud and long, did goodman Brown grasp his staff and set forth again, at such a rate, that he seemed to fly along the forest-path, rather than to walk or run. The road grew wilder and drearier, and more faintly traced, and vanished at length, leaving him in the heart of the dark wilderness, still rushing onward, with the instinct that guides mortal man to evil. The whole forest was peopled with frightful sounds; the creaking of the trees, the howling of wild beasts, and the yell of Indians; while, sometimes, the wind tolled like a distant church-bell, and sometimes gave a broad roar around the traveler, as if all Nature were laughing him to scorn. But he was himself the chief horror of the scene, and shrank not from its other horrors.

'Ha! ha! ha!' roared goodman Brown, when the wind laughed at him. 'Let us hear which will laugh loudest! Think not to frighten me with your deviltry! Come witch, come wizard, come Indian powow,

come devil himself! and here comes goodman Brown. You may as
well fear him as he fear you!'

In truth, all through the haunted forest, there could be nothing
more frightful than the figure of goodman Brown. On he flew,
among the black pines, brandishing his staff with frenzied gestures,
now giving vent to an inspiration of horrid blasphemy, and now
shouting forth such laughter, as set all the echoes of the forest
laughing like demons around him. The fiend in his own shape is less
hideous, than when he rages in the breast of man. Thus sped the
demoniac on his course, until, quivering among the trees, he saw a
red light before him, as when the felled trunks and branches of a
clearing have been set on fire, and throw up their lurid blaze against
the sky, at the hour of midnight. He paused, in a lull of the tempest
that had driven him onward, and heard the swell of what seemed a
hymn, rolling solemnly from a distance, with the weight of many
voices. He knew the tune; it was a familiar one in the choir of the
village meeting-house. The verse died heavily away, and was lengthened by a chorus, not of human voices, but of all the sounds of the
benighted wilderness, pealing in awful harmony together. Goodman
Brown cried out; and his cry was lost to his own ear, by its unison
with the cry of the desert.

In the interval of silence, he stole forward, until the light glared
full upon his eyes. At one extremity of an open space, hemmed in
by the dark wall of the forest, arose a rock, bearing some rude,
natural resemblance either to an altar or a pulpit, and surrounded
by four blazing pines, their tops a flame, their stems untouched, like
candles at an evening meeting. The mass of foliage, that had overgrown the summit of the rock, was all on fire, blazing high into the
night, and fitfully illuminating the whole field. Each pendent twig
and leafy festoon was in a blaze. As the red light arose and fell, a
numerous congregation alternately shone forth, then disappeared in
shadow, and again grew, as it were, out of the darkness, peopling
the heart of the solitary woods at once.

'A grave and dark-clad company!' quoth goodman Brown.

In truth, they were such. Among them, quivering to-and-fro,
between gloom and splendor, appeared faces that would be seen,
next day, at the council-board of the province, and others which,
Sabbath after Sabbath, looked devoutly heavenward, and benig-
nantly over the crowded pews, from the holiest pulpits in the land.
Some affirm, that the lady of the governor was there. At least, there
were high dames well known to her, and wives of honored husbands, and widows, a great multitude, and ancient maidens, all of
excellent repute, and fair young girls, who trembled, lest their
mothers should espy them. Either the sudden gleams of light, flashing over the obscure field, bedazzled goodman Brown, or he recognized a score of the church-members of Salem village, famous for
their especial sanctity. Good old deacon Gookin had arrived, and

waited at the skirts of that venerable saint, his revered pastor. But, irreverently consorting with these grave, reputable, and pious people, these elders of the church, these chaste dames and dewy virgins, there were men of dissolute lives and women of spotted fame, wretches given over to all mean and filthy vice, and suspected even of horrid crimes. It was strange to see, that the good shrank not from the wicked, nor were the sinners abashed by the saints. Scattered, also, among their pale-faced enemies, were the Indian priests, or powows, who had often scared their native forest with more hideous incantations than any known to English witchcraft.

'But, where is Faith?' thought goodman Brown; and, as hope came into his heart, he trembled.

Another verse of the hymn arose, a slow and solemn strain, such as the pious love, but joined to words which expressed all that our nature can conceive of sin, and darkly hinted at far more. Unfathomable to mere mortals is the lore of fiends. Verse after verse was sung, and still the chorus of the desert swelled between, like the deepest tone of a mighty organ. And, with the final peal of that dreadful anthem, there came a sound, as if the roaring wind, the rushing streams, the howling beasts, and every other voice of the unconverted wilderness, were mingling and according with the voice of guilty man, in homage to the prince of all. The four blazing pines threw up a loftier flame, and obscurely discovered shapes and visages of horror on the smoke-wreaths, above the impious assembly. At the same moment, the fire on the rock shot redly forth, and formed a glowing arch above its base, where now appeared a figure. With reverence be it spoken, the apparition bore no slight similitude, both in garb and manner, to some grave divine of the New-England churches.

'Bring forth the converts!' cried a voice, that echoed through the field and rolled into the forest.

At the word, goodman Brown stept forth from the shadow of the trees, and approached the congregation, with whom he felt a loathful brotherhood, by the sympathy of all that was wicked in his heart. He could have well nigh sworn, that the shape of his own dead father beckoned him to advance, looking downward from a smoke-wreath, while a woman, with dim features of despair, threw out her hand to warn him back. Was it his mother? But he had no power to retreat one step, nor to resist, even in thought, when the minister and good old deacon Gookin, seized his arms, and led him to the blazing rock. Thither came also the slender form of a veiled female, led between Goody Cloyse, that pious teacher of the catechism, and Martha Carrier, who had received the devil's promise to be queen of hell. A rampant hag was she! And there stood the proselytes, beneath the canopy of fire.

'Welcome, my children,' said the dark figure, 'to the communion of your race!⁶ Ye have found, thus young, your nature and your destiny. My children, look behind you!"

They turned; and flashing forth, as it were, in a sheet of flame, the fiend-worshippers were seen; the smile of welcome gleamed darkly on every visage.

'There,' resumed the sable form, 'are all whom ye have reverenced from youth. Ye deemed them holier than yourselves, and shrank from your own sin, contrasting it with their lives of righteousness, and prayerful aspirations heavenward. Yet, here are they all, in my worshipping assembly! This night it shall be granted you to know their secret deeds; how hoary-bearded elders of the church have whispered wanton words to the young maids of their households; how many a woman, eager for widow's weeds, has given her husband a drink at bed-time, and let him sleep his last sleep in her bosom; how beardless youths have made haste to inherit their fathers' wealth; and how fair damsels—blush not, sweet ones!—have dug little graves in the garden, and bidden me, the sole guest, to an infant's funeral. By the sympathy of your human hearts for sin, ye shall scent out all the places—whether in church, bedchamber, street, field, or forest—where crime has been committed, and shall exult to behold the whole earth one stain of guilt, one mighty blood-spot. Far more than this! It shall be your's to penetrate, in every bosom, the deep mystery of sin, the fountain of all wicked arts, and which, inexhaustibly supplies more evil impulses than human power—than my power, at its utmost!—can make manifest in deeds. And now, my children, look upon each other.'

They did so; and, by the blaze of the hell-kindled torches, the wretched man beheld his Faith, and the wife her husband, trembling before that unhallowed altar.

'Lo! there ye stand, my children,' said the figure, in a deep and solemn tone, almost sad, with its despairing awfulness, as if his once angelic nature could yet mourn for our miserable race. 'Depending upon one another's hearts, ye had still hoped, that virtue were not all a dream. Now are ye undeceived! Evil is the nature of mankind. Evil must be your only happiness. Welcome, again, my children, to the communion of your race!'

'Welcome!' repeated the fiend-worshippers, in one cry of despair and triumph.

And there they stood, the only pair, as it seemed, who were yet hesitating on the verge of wickedness, in this dark world. A basin was hollowed, naturally, in the rock. Did it contain water, reddened by the lurid light? or was it blood? or, perchance, a liquid flame? Herein did the Shape of Evil dip his hand, and prepare to lay the mark of baptism upon their foreheads, that they might be partakers

---

6. The *New-England Magazine* erroneously printed "grave," corrected to "race" in *Mosses from an Old Manse* (1846).

of the mystery of sin, more conscious of the secret guilt of others, both in deed and thought, than they could now be of their own. The husband cast one look at his pale wife, and Faith at him. What polluted wretches would the next glance shew them to each other, shuddering alike at what they disclosed and what they saw!

'Faith! Faith!' cried the husband. 'Look up to Heaven, and resist the Wicked One!'

Whether Faith obeyed, he knew not. Hardly had he spoken, when he found himself amid calm night and solitude, listening to a roar of the wind, which died heavily away through the forest. He staggered against the rock and felt it chill and damp, while a hanging twig, that had been all on fire, besprinkled his cheek with the coldest dew.

The next morning, young goodman Brown came slowly into the street of Salem village, staring around him like a bewildered man. The good old minister was taking a walk along the graveyard, to get an appetite for breakfast and meditate his sermon, and bestowed a blessing, as he passed, on goodman Brown. He shrank from the venerable saint, as if to avoid an anathema. Old deacon Gookin was at domestic worship, and the holy words of his prayer were heard through the open window. 'What God doth the wizard pray to?' quoth goodman Brown. Goody Cloyse, that excellent old Christian, stood in the early sunshine, at her own lattice, catechising a little girl, who had brought her a pint of morning's milk. Goodman Brown snatched away the child, as from the grasp of the fiend himself. Turning the corner by the meeting-house, he spied the head of Faith, with the pink ribbons, gazing anxiously forth, and bursting into such joy at sight of him, that she skipt along the street, and almost kissed her husband before the whole village. But, goodman Brown looked sternly and sadly into her face, and passed on without a greeting.

Had goodman Brown fallen asleep in the forest, and only dreamed a wild dream of a witch-meeting?

Be it so, if you will. But, alas! it was a dream of evil omen for young goodman Brown. A stern, a sad, a darkly meditative, a distrustful, if not a desperate man, did he become, from the night of that fearful dream. On the Sabbath-day, when the congregation were singing a holy psalm, he could not listen, because an anthem of sin rushed loudly upon his ear, and drowned all the blessed strain. When the minister spoke from the pulpit, with power and fervid eloquence, and, with his hand on the open bible, of the sacred truths of our religion, and of saint-like lives and triumphant deaths, and of future bliss or misery unutterable, then did goodman Brown turn pale, dreading, lest the roof should thunder down upon the gray blasphemer and his hearers. Often, awakening suddenly at midnight, he shrank from the bosom of Faith, and at morning or eventide, when the family knelt down at prayer, he scowled, and muttered to himself, and gazed sternly at his wife, and turned away. And when he had lived long, and was borne to his grave, a hoary corpse, followed by Faith, an aged woman, and children and grand-children, a goodly procession, besides neighbors, not a few, they carved no hopeful verse upon his tomb-stone; for his dying hour was gloom.

1835

The May-Pole of Merry Mount[1]

There is an admirable foundation for a philosophic romance, in the curious history of the early settlement of Mount Wallaston, or Merry Mount. In the slight sketch here attempted, the facts, recorded on the grave pages of our New England annalists, have wrought themselves, almost spontaneously, into a sort of allegory. The masques, mummeries, and festive customs, described in the text, are in accordance with the manners of the age. Authority, on these points may be found in Strutt's Book of English Sports and Pastimes.[2]

Bright were the days at Merry Mount, when the May-Pole[3] was the banner-staff of that gay colony! They who reared it, should their banner be triumphant, were to pour sun-shine over New England's rugged hills, and scatter flower-seeds throughout the soil. Jollity and gloom were contending for an empire. Midsummer eve[4] had come, bringing deep verdure to the forest, and roses in her lap, of a more vivid hue than the tender buds of Spring. But May, or her mirthful spirit, dwelt all the year round at Merry Mount, sporting with the Summer months, and revelling with Autumn, and basking in the glow of Winter's fireside. Through a world of toil and care, she flitted with a dreamlike smile, and came hither to find a home among the lightsome hearts of Merry Mount.

Never had the May-Pole been so gaily decked as at sunset on mid-summer eve. This venerated emblem was a pine tree, which had preserved the slender grace of youth, while it equalled the loftiest height of the old wood monarchs. From its top streamed a silken banner, colored like the rainbow. Down nearly to the ground, the pole was dressed with birchen boughs, and others of the liveliest green, and some with silvery leaves, fastened by ribbons that flut-tered in fantastic knots of twenty different colors, but no sad ones. Garden flowers, and blossoms of the wilderness, laughed gladly forth amid the verdure, so fresh and dewy, that they must have grown by magic on that happy pine tree. Where this green and flowery splendor terminated, the shaft of the May-Pole was stained with the seven brilliant hues of the banner at its top. On the lowest

1. The text is that of the first printing in *The Token* (1836), where the story is ascribed to "the Author of 'The Gentle Boy.' "
2. Joseph Strutt, *The Sports and Pastimes of the People of England* (London, 1801). Hawthorne also knew Nathaniel Morton's *New England Memorial* (1669), which drew on William Bradford's manuscript history *Of Plymouth Plantation.*

3. In English tradition the tall pole placed in a prominent place in a village where on the first of May flower-bedecked young people could dance around it after a night of gathering new vegetation and blossoms in the woods. Puritans con-demned the custom as a sexual orgy.
4. The 20th of June, the day before the longest day of the year.

green bough hung an abundant wreath of roses, some that had been gathered in the sunniest spots of the forest, and others, of still richer blush, which the colonists had reared from English seed. Oh, people of the Golden Age, the chief of your husbandry, was to raise flowers!

But what was the wild throng that stood hand in hand about the May-Pole? It could not be, that the Fauns and Nymphs, when driven from their classic groves and homes of ancient fable, had sought refuge, as all the persecuted did, in the fresh woods of the West. These were Gothic monsters, though perhaps of Grecian ancestry. On the shoulders of a comely youth, uprose the head and branching antlers of a stag; a second, human in all other points, had the grim visage of a wolf; a third, still with the trunk and limbs of a mortal man, showed the beard and horns of a venerable he-goat. There was the likeness of a bear erect, brute in all but his hind legs, which were adorned with pink silk stockings. And here again, almost as wondrous, stood a real bear of the dark forest, lending each of his fore paws to the grasp of a human hand, and as ready for the dance as any in that circle. This inferior nature rose half-way, to meet his companions as they stooped. Other faces wore the similitude of man or woman, but distorted or extravagant, with red noses pendulous before their mouths, which seemed of awful depth, and stretched from ear to ear in an eternal fit of laughter. Here might be seen the Salvage Man,[5] well known in heraldry, hairy as a baboon, and girdled with green leaves. By his side, a nobler figure, but still a counterfeit, appeared an Indian hunter, with feathery crest and wampum belt. Many of this strange company wore fools-caps, and had little bells appended to their garments, tinkling with a silvery sound, responsive to the inaudible music of their gleesome spirits. Some youths and maidens were of soberer garb, yet well maintained their places in the irregular throng, by the expression of wild revelry upon their features. Such were the colonists of Merry Mount, as they stood in the broad smile of sunset, round their venerated May-Pole.

Had a wanderer, bewildered in the melancholy forest, heard their mirth, and stolen a half-affrighted glance, he might have fancied them the crew of Comus,[6] some already transformed to brutes, some midway between man and beast, and the others rioting in the flow of tipsey jollity that foreran the change. But a band of Puritans, who watched the scene, invisible themselves, compared the masques to those devils and ruined souls, with whom their superstition peopled the black wilderness.

Within the ring of monsters, appeared the two airiest forms, that had ever trodden on any more solid footing than a purple and

---

5. Person clad in foliage to represent a savage, as in medieval and Renaissance pageantry.

6. The god of revelry, here associated with Milton's *Comus* (1634).

golden cloud. One was a youth, in glistening apparel, with a scarf of the rainbow pattern crosswise on his breast. His right hand held a gilded staff, the ensign[7] of high dignity among the revellous, and his left grasped the slender fingers of a fair maiden, not less gaily decorated than himself. Bright roses glowed in contrast with the dark and glossy curls of each, and were scattered round their feet, or had sprung up spontaneously there. Behind this lightsome couple, so close to the May-Pole that its boughs shaded his jovial face, stood the figure of an English priest, canonically dressed, yet decked with flowers, in Heathen fashion, and wearing a chaplet of the native vine leaves. By the riot of his rolling eye, and the pagan decorations of his holy garb, he seemed the wildest monster there, and the very Comus of the crew.

'Votaries of the May-Pole,' cried the flower-decked priest, 'merrily, all day long, have the woods echoed to your mirth. But be this your merriest hour, my hearts! Lo, here stand the Lord and Lady of the May, whom I, a clerk[8] of Oxford, and high priest of Merry Mount, am presently to join in holy matrimony. Up with your nimble spirits, ye morrice-dancers, green-men, and glee-maidens,[9] bears and wolves, and horned gentlemen! Come; a chorus now, rich with the old mirth of Merry England, and the wilder glee of this fresh forest; and then a dance, to show the youthful pair what life is made of, and how airily they should go through it! All ye that love the May-Pole, lend your voices to the nuptial song of the Lord and Lady of the May!'

This wedlock was more serious than most affairs of Merry Mount, where jest and delusion, trick and fantasy, kept up a continual carnival. The Lord and Lady of the May, though their titles must be laid down at sunset, were really and truly to be partners for the dance of life, beginning the measure that same bright eve. The wreath of roses, that hung from the lowest green bough of the May-Pole, had been twined for them, and would be thrown over both their heads, in symbol of their flowery union. When the priest had spoken, therefore, a riotous uproar burst from the rout of monstrous figures.

'Begin you the stave,[1] reverend Sir,' cried they all; 'and never did the woods ring to such a merry peal, as we of the May-Pole shall send up!'

Immediately a prelude of pipe, cittern,[2] and viol, touched with practised minstrelsy, began to play from a neighboring thicket, in such a mirthful cadence, that the boughs of the May-Pole quivered to the sound. But the May Lord, he of the gilded staff, chancing to look into his Lady's eyes, was wonderstruck at the almost pensive glance that met his own.

---

7. Sign, token.
8. In Anglican usage, lay minister who assists the parish clergyman.
9. Participants in an English folk dance

(originally "Moorish dance"); men bedecked in greenery; girl singers.
1. Stanza.
2. Guitar with pear-shaped body.

'Edith, sweet Lady of the May,' whispered he, reproachfully, 'is your wreath of roses a garland to hang above our graves, that you look so sad? Oh, Edith, this is our golden time! Tarnish it not by any pensive shadow of the mind; for it may be, that nothing of futurity will be brighter than the mere remembrance of what is now passing.'

'That was the very thought that saddened me! How came it in your mind too?' said Edith, in a still lower tone than he; for it was high treason to be sad at Merry Mount. 'Therefore do I sigh amid this festive music. And besides, dear Edgar, I struggle as with a dream, and fancy that these shapes of our jovial friends are visionary, and their mirth unreal, and that we are no true Lord and Lady of the May. What is the mystery in my heart?'

Just then, as if a spell had loosened them, down came a little shower of withering rose leaves from the May-Pole. Alas, for the young lovers! No sooner had their hearts glowed with real passion, than they were sensible of something vague and unsubstantial in their former pleasures, and felt a dreary presentiment of inevitable change. From the moment that they truly loved, they had subjected themselves to earth's doom of care, and sorrow, and troubled joy, and had no more a home at Merry Mount. That was Edith's mystery. Now leave we the priest to marry them, and the masquers to sport round the May-Pole, till the last sunbeam be withdrawn from its summit, and the shadows of the forest mingle gloomily in the dance. Meanwhile, we may discover who these gay people were.

Two hundred years ago, and more, the old world and its inhabitants became mutually weary of each other. Men voyaged by thousands to the West; some to barter glass beads, and such like jewels, for the furs of the Indian hunter; some to conquer virgin empires; and one stern band to pray. But none of these motives had much weight with the colonists of Merry Mount. Their leaders were men who had sported so long with life, that when Thought and Wisdom came, even these unwelcome guests were led astray, by the crowd of vanities which they should have put to flight. Erring Thought and perverted Wisdom were made to put on masques, and play the fool. The men of whom we speak, after losing the heart's fresh gaiety, imagined a wild philosophy of pleasure, and came hither to act out their latest day-dream. They gathered followers from all that giddy tribe, whose whole life is like the festal days of soberer men. In their train were minstrels, not unknown in London streets; wandering players, whose theatres had been the halls of noblemen; mummeries, rope-dancers, and mountebanks,[3] who would long be missed at wakes, church-ales, and fairs; in a word, mirth-makers of every sort, such as abounded in that age, but now began to be discountenanced by the rapid growth of Puritanism. Light had their

---

3. Masked actors; tightrope walkers; showmen who "climb on a bench" to hawk medicines or (as here) to tell stories or do tricks.

footsteps been on land, and as lightly they came across the sea. Many had been maddened by their previous troubles into a gay despair; others were as madly gay in the flush of youth, like the May Lord and his Lady; but whatever might be the quality of their mirth, old and young were gay at Merry Mount. The young deemed themselves happy. The elder spirits, if they knew that mirth was but the counterfeit of happiness, yet followed the false shadow wilfully, because at least her garments glittered brightest. Sworn triflers of a life-time, they would not venture among the sober truths of life, not even to be truly blest.

All the hereditary pastimes of Old England were transplanted hither. The King of Christmas was duly crowned, and the Lord of Misrule[4] bore potent sway. On the eve of Saint John,[5] they felled whole acres of the forest to make bonfires, and danced by the blaze all night, crowned with garlands, and throwing flowers into the flame. At harvest time, though their crop was of the smallest, they made an image with the sheaves of Indian corn, and wreathed it with autumnal garlands, and bore it home triumphantly. But what chiefly characterized the colonists of Merry Mount, was their veneration for the May-Pole. It has made their true history a poet's tale. Spring decked the hallowed emblem with young blossoms and fresh green boughs; Summer brought roses of the deepest blush, and the perfected foliage of the forest; Autumn enriched it with that red and yellow gorgeousness, which converts each wildwood leaf into a painted flower; and Winter silvered it with sleet, and hung it round with icicles, till it flashed in the cold sunshine, itself a frozen sunbeam. Thus each alternate season did homage to the May-Pole, and paid it a tribute of its own richest splendor. Its votaries danced round it, once, at least, in every month; sometimes they called it their religion, or their altar; but always, it was the banner-staff of Merry Mount.

Unfortunately, there were men in the new world, of a sterner faith than these May-Pole worshippers. Not far from Merry Mount was a settlement of Puritans, most dismal wretches, who said their prayers before daylight, and then wrought in the forest or the cornfield, till evening made it prayer time again. Their weapons were always at hand, to shoot down the straggling savage. When they met in conclave, it was never to keep up the old English mirth, but to hear sermons three hours long, or to proclaim bounties on the heads of wolves and the scalps of Indians. Their festivals were fast-days, and their chief pastime the singing of psalms. Woe to the youth or maiden, who did but dream of a dance! The selectman nodded to the constable; and there sat the light-heeled reprobate in the stocks; or if he danced, it was round the whipping-post, which might be termed the Puritan May-Pole.

4. Master of the traditional Christmas revelry.
5. Midsummer eve.

A party of these grim Puritans, toiling through the difficult woods, each with a horse-load of iron armor to burthen his footsteps, would sometimes draw near the sunny precincts of Merry Mount. There were the silken colonists, sporting round their May-Pole; perhaps teaching a bear to dance, or striving to communicate their mirth to the grave Indian; or masquerading in the skins of deer and wolves, which they had hunted for that especial purpose. Often, the whole colony were playing at blindman's bluff, magistrates and all with their eyes bandaged, except a single scape-goat, whom the blinded sinners pursued by the tinkling of the bells at his garments. Once, it is said, they were seen following a flower-decked corpse, with merriment and festive music, to his grave. But did the dead man laugh? In their quietest times, they sang ballads and told tales, for the edification of their pious visiters; or perplexed them with juggling tricks; or grinned at them through horse-collars; and when sport itself grew wearisome, they made game of their own stupidity, and began a yawning match. At the very least of these enormities, the men of iron shook their heads and frowned so darkly, that the revellers looked up, imagining that a momentary cloud had overcast the sunshine, which was to be perpetual there. On the other hand, the Puritans affirmed, that, when a psalm was pealing from their place of worship, the echo, which the forest sent them back, seemed often like the chorus of a jolly catch, closing with a roar of laughter. Who but the fiend, and his fond slaves, the crew of Merry Mount, had thus disturbed them! In due time, a feud arose, stern and bitter on one side, and as serious on the other as any thing could be, among such light spirits as had sworn allegiance to the May-Pole. The future complexion of New England was involved in this important quarrel. Should the grisly saints establish their jurisdiction over the gay sinners, then would their spirits darken all the clime, and make it a land of clouded visages, of hard toil, of sermon and psalm, forever. But should the banner-staff of Merry Mount be fortunate, sunshine would break upon the hills, and flowers would beautify the forest, and late posterity do homage to the May-Pole!

After these authentic passages from history, we return to the nuptials of the Lord and Lady of the May. Alas! we have delayed too long, and must darken our tale too suddenly. As we glanced again at the May-Pole, a solitary sun-beam is fading from the summit, and leaves only a faint golden tinge, blended with the hues of the rain bow banner. Even that dim light is now withdrawn, relinquishing the whole domain of Merry Mount to the evening gloom, which has rushed so instantaneously from the black surrounding woods. But some of these black shadows have rushed forth in human shape.

Yes: with the setting sun, the last day of mirth had passed from Merry Mount. The ring of gay masquers was disordered and broken; the stag lowered his antlers in dismay; the wolf grew weaker than a lamb; the bells of the morrice dancers tinkled with tremulous affright. The Puritans had played a characteristic part in the May-Pole mummeries. Their darksome figures were intermixed with the wild shapes of their foes, and made the scene a picture of the moment, when waking thoughts start up amid the scattered fantasies of a dream. The leader of the hostile party stood in the centre of the circle, while the rout of monsters cowered around him, like evil spirits in the presence of a dread magician. No fantastic foolery could look him in the face. So stern was the energy of his aspect, that the whole man, visage, frame, and soul, seemed wrought of iron, gifted with life and thought, yet all of one substance with his head-piece and breast-plate. It was the Puritan of Puritans; it was Endicott[6] himself!

'Stand off, priest of Baal!'[7] said he, with a grim frown, and laying no reverent hand upon the surplice. 'I know thee, Claxton![8] Thou art the man, who couldst not abide the rule even of thine own corrupted church,[9] and hast come hither to preach iniquity, and to give example of it in thy life. But now shall it be seen that the Lord hath sanctified this wilderness for his peculiar people. Woe unto them that would defile it! And first for this flower-decked abomination, the altar of thy worship!'

And with his keen sword, Endicott assaulted the hallowed May-Pole. Nor long did it resist his arm. It groaned with a dismal sound; it showered leaves and rose-buds upon the remorseless enthusiast; and finally, with all its green boughs, and ribbons, and flowers, symbolic of departed pleasures, down fell the banner-staff of Merry Mount. As it sank, tradition says, the evening sky grew darker, and the woods threw forth a more sombre shadow.

'There,' cried Endicott, looking triumphantly on his work, 'there lies the only May-Pole in New-England! The thought is strong within me, that, by its fall, is shadowed forth the fate of light and idle mirth-makers, amongst us and our posterity. Amen, saith John Endicott!'

'Amen!' echoed his followers.

But the votaries of the May-Pole gave one groan for their idol. At the sound, the Puritan leader glanced at the crew of Comus, each a figure of broad mirth, yet, at this moment, strangely expressive of sorrow and dismay.

'Valiant captain,' quoth Peter Palfrey, the Ancient[1] of the band, 'what order shall be taken with the prisoners?'

'I thought not to repent me of cutting down a May-Pole,' replied Endicott, 'yet now I could find in my heart to plant it again, and

6. John Endicott (1589?–1665), several times Governor of the Massachusetts colony.
7. For the slaying of the prophets of the fertility god Baal, see 1 Kings 18.
8. "Did Governor Endicott speak less positively, we should suspect a mistake here. The Reverend Mr. Claxton, though an eccentric, is not known to have been an immoral man. We rather doubt his identity with the priest of Merry Mount" [Hawthorne's note].
9. That is, the Anglican Church.
1. Lieutenant.

give each of these bestial pagans one other dance round their idol. It would have served rarely for a whipping-post!'

'But there are pine trees enow,' suggested the lieutenant.

'True, good Ancient,' said the leader. 'Wherefore, bind the heathen crew, and bestow on them a small matter of stripes apiece, as earnest of our future justice. Set some of the rogues in the stocks to rest themselves, so soon as Providence shall bring us to one of our own well-ordered settlements, where such accommodations may be found. Further penalties, such as branding and cropping of ears, shall be thought of hereafter.'

'How many stripes for the priest?' inquired Ancient Palfrey.

'None as yet,' answered Endicott, bending his iron frown upon the culprit. 'It must be for the Great and General Court[2] to determine, whether stripes and long imprisonment, and other grievous penalty, may atone for his transgressions. Let him look to himself! For such as violate our civil order, it may be permitted us to show mercy. But woe to the wretch that troubleth our religion!'

'And this dancing bear,' resumed the officer. 'Must he share the stripes of his fellows?'

'Shoot him through the head!' said the energetic Puritan. 'I suspect witchcraft in the beast.'

'Here be a couple of shining ones,' continued Peter Palfrey, pointing his weapon at the Lord and Lady of the May. 'They seem to be of high station among these mis-doers. Methinks their dignity will not be fitted with less than a double share of stripes.'

Endicott rested on his sword, and closely surveyed the dress and aspect of the hapless pair. There they stood, pale, downcast, and apprehensive. Yet there was an air of mutual support, and of pure affection, seeking aid and giving it, that showed them to be man and wife, with the sanction of a priest upon their love. The youth, in the peril of the moment, had dropped his gilded staff, and thrown his arm about the Lady of the May, who leaned against his breast, too lightly to burthen him, but with weight enough to express that their destinies were linked together, for good or evil. They looked first at each other, and then into the grim captain's face. There they stood, in the first hour of wedlock, while the idle pleasures, of which their companions were the emblems, had given place to the sternest cares of life, personified by the dark Puritans. But never had their youthful beauty seemed so pure and high, as when its glow was chastened by adversity.

'Youth,' said Endicott, 'ye stand in an evil case, thou and thy maiden wife. Make ready presently; for I am minded that ye shall both have a token to remember your wedding-day!'

'Stern man,' exclaimed the May Lord, 'How can I move thee? Were the means at hand, I would resist to the death. Being power-

2. Massachusetts legislature.

less, I entreat! Do with me as thou wilt; but let Edith go untouched!'

'Not so,' replied the immitigable zealot. 'We are not wont to show an idle courtesy to that sex, which requireth the stricter discipline. What sayest thou, maid? Shall thy silken bridegroom suffer thy share of the penalty, besides his own?'

'Be it death,' said Edith, 'and lay it all on me!'

Truly, as Endicott had said, the poor lovers stood in a woeful case. Their foes were triumphant, their friends captive and abased, their home desolate, the benighted wilderness around them, and a rigorous destiny, in the shape of the Puritan leader, their only guide. Yet the deepening twilight could not altogether conceal, that the iron man was softened; he smiled, at the fair spectacle of early love; he almost sighed, for the inevitable blight of early hopes.

'The troubles of life have come hastily on this young couple,' observed Endicott. 'We will see how they comport themselves under their present trials, ere we burthen them with greater. If, among the spoil, there be any garments of a more decent fashion, let them be put upon this May Lord and his Lady, instead of their glistening vanities. Look to it, some of you.'

'And shall not the youth's hair be cut?' asked Peter Palfrey, looking with abhorrence at the love-lock and long glossy curls of the young man.

'Crop it forthwith, and that in the true pumpkin shell fashion,'[3] answered the captain. 'Then bring them along with us, but more gently than their fellows. There be qualities in the youth, which may make him valiant to fight, and sober to toil, and pious to pray; and in the maiden, that may fit her to become a mother in our Israel,[4] bringing up babes in better nurture than her own hath been. Nor think ye, young ones, that they are the happiest, even in our lifetime of a moment, who misspend it in dancing round a May-Pole!'

And Endicott, the severest Puritan of all who laid the rock-foundation of New England, lifted the wreath of roses from the ruin of the May-Pole, and threw it, with his own gauntleted hand, over the heads of the Lord and Lady of the May. It was a deed of prophecy. As the moral gloom of the world overpowers all systematic gaiety, even so was their home of wild mirth made desolate amid the sad forest. They returned to it no more. But, as their flowery garland was wreathed of the brightest roses that had grown there, so, in the tie that united them, were intertwined all the purest and best of their early joys. They went heavenward, supporting each other along the difficult path which it was their lot to tread, and never wasted one regretful thought on the vanities of Merry Mount.

1835

---

3. Roundhead style, close-cropped in Puritan fashion.
4. Endicott makes the standard 17th-century Puritan identification of the New England settlers with the Jews, another persecuted, God-chosen minority.

Wakefield[1]

In some old magazine or newspaper, I recollect a story, told as truth, of a man—let us call him Wakefield—who absented himself for a long time, from his wife. The fact, thus abstractedly stated, is not very uncommon, nor—without a proper distinction of circumstances—to be condemned either as naughty or nonsensical. Howbeit, this, though far from the most aggravated, is perhaps the strangest instance, on record, of marital delinquency; and, moreover, as remarkable a freak as may be found in the whole list of human oddities. The wedded couple lived in London. The man, under pretence of going a journey, took lodgings in the next street to his own house, and there, unheard of by his wife or friends, and without the shadow of a reason for such self-banishment, dwelt upwards of twenty years. During that period, he beheld his home every day, and frequently the forlorn Mrs. Wakefield. And after so great a gap in his matrimonial felicity—when his death was reckoned certain, his estate settled, his name dismissed from memory, and his wife, long, long ago, resigned to her autumnal widowhood—he entered the door one evening, quietly, as from a day's absence, and became a loving spouse until death.

This outline is all that I remember. But the incident, though of the purest originality, unexampled, and probably never to be repeated, is one, I think, which appeals to the general sympathies of mankind. We know, each for himself, that none of us would perpetrate such a folly, yet feel as if some other might. To my own contemplations, at least, it has often recurred, always exciting wonder, but with a sense that the story must be true, and a conception of its hero's character. Whenever any subject so forcibly affects the mind, time is well spent in thinking of it. If the reader choose, let him do his own meditation; or if he prefer to ramble with me through the twenty years of Wakefield's vagary, I bid him welcome; trusting that there will be a pervading spirit and a moral, even should we fail to find them, done up neatly, and condensed into the final sentence. Thought has always its efficacy, and every striking incident its moral.

What sort of a man was Wakefield? We are free to shape out our own idea, and call it by his name. He was now in the meridian of life; his matrimonial affections, never violent, were sobered into a calm, habitual sentiment; of all husbands, he was likely to be the most constant, because a certain sluggishness would keep his heart at rest, wherever it might be placed. He was intellectual, but not actively so; his mind occupied itself in long and lazy musings, that tended to no purpose, or had not vigor to attain it; his thoughts were seldom so energetic as to seize hold of words. Imagination, in

1. The text is from the first publication, in the *New-England Magazine* (May, 1835).

the proper meaning of the term, made no part of Wakefield's gifts. With a cold, but not depraved nor wandering heart, and a mind never feverish with riotous thoughts, nor perplexed with originality, who could have anticipated, that our friend would entitle himself to a foremost place among the doers of eccentric deeds? Had his acquaintances been asked, who was the man in London, the surest to perform nothing to-day which should be remembered on the morrow, they would have thought of Wakefield. Only the wife of his bosom might have hesitated. She, without having analyzed his character, was partly aware of a quiet selfishness, that had rusted into his inactive mind—of a peculiar sort of vanity, the most uneasy attribute about him—of a disposition to craft, which had seldom produced more positive effects than the keeping of petty secrets, hardly worth revealing—and, lastly, of what she called a little strangeness, sometimes, in the good man. This latter quality is indefinable, and perhaps non-existent.

Let us now imagine Wakefield bidding adieu to his wife. It is the dusk of an October evening. His equipment is a drab great-coat, a hat covered with an oil-cloth, top-boots, an umbrella in one hand and a small portmanteau in the other. He has informed Mrs. Wakefield that he is to take the night-coach into the country. She would fain inquire the length of his journey, its object, and the probable time of his return; but, indulgent to his harmless love of mystery, interrogates him only by a look. He tells her not to expect him positively by the return coach, nor to look alarmed should he tarry three or four days; but, at all events, to look for him at supper on Friday evening. Wakefield himself, be it considered, has no suspicion of what is before him. He holds out his hand; she gives her own, and meets his parting kiss, in the matter-of-course way of a ten years' matrimony; and forth goes the middle-aged Mr. Wakefield, almost resolved to perplex his good lady by a whole week's absence. After the door has closed behind him, she perceives it thrust partly open, and a vision of her husband's face, through the aperture, smiling on her, and gone in a moment. For the time, this little incident is dismissed without a thought. But, long afterwards, when she has been more years a widow than a wife, that smile recurs, and flickers across all her reminiscences of Wakefield's visage. In her many musings, she surrounds the original smile with a multitude of fantasies, which make it strange and awful; as, for instance, if she imagines him in a coffin, that parting look is frozen on his pale features; or, if she dreams of him in Heaven, still his blessed spirit wears a quiet and crafty smile. Yet, for its sake, when all others have given him up for dead, she sometimes doubts whether she is a widow.

But, our business is with the husband. We must hurry after him, along the street, ere he lose his individuality, and melt into the great mass of London life. It would be vain searching for him there. Let us follow close at his heels, therefore, until, after several superfluous turns and doublings, we find him comfortably established by the fireside of a small apartment, previously bespoken. He is in the next street to his own, and at his journey's end. He can scarcely trust his good fortune, in having got thither unperceived—recollecting that, at one time, he was delayed by the throng, in the very focus of a lighted lantern; and, again, there were foot-steps, that seemed to tread behind his own, distinct from the multitudinous tramp around him; and, anon, he heard a voice shouting afar, and fancied that it called his name. Doubtless, a dozen busy-bodies had been watching him, and told his wife the whole affair. Poor Wakefield! Little knowest thou thine own insignificance in this great world! No mortal eye but mine has traced thee. Go quietly to thy bed, foolish man; and, on the morrow, if thou wilt be wise, get thee home to good Mrs. Wakefield, and tell her the truth. Remove not thyself, even for a little week, from thy place in her chaste bosom. Were she, for a single moment, to deem thee dead, or lost, or lastingly divided from her, thou wouldst be woefully conscious of a change in thy true wife, forever after. It is perilous to make a chasm in human affections; not that they gape so long and wide—but so quickly close again!

Almost repenting of his frolic, or whatever it may be termed, Wakefield lies down betimes, and starting from his first nap, spreads forth his arms into the wide and solitary waste of the unaccustomed bed. 'No'—thinks he, gathering the bed-clothes about him—'I will not sleep alone another night.'

In the morning, he rises earlier than usual, and sets himself to consider what he really means to do. Such are his loose and rambling modes of thought, that he has taken this very singular step, with the consciousness of a purpose, indeed, but without being able to define it sufficiently for his own contemplation. The vagueness of the project, and the convulsive effort with which he plunges into the execution of it, are equally characteristic of a feeble-minded man. Wakefield sifts his ideas, however, as minutely as he may, and finds himself curious to know the progress of matters at home—how his exemplary wife will endure her widowhood, of a week; and, briefly, how the little sphere of creatures and circumstances, in which he was a central object, will be affected by his removal. A morbid vanity, therefore, lies nearest the bottom of the affair. But, how is he to attain his ends? Not, certainly, by keeping close in this comfortable lodging, where, though he slept and awoke in the next street to his home, he is as effectually abroad, as if the stage-coach had been whirling him away all night. Yet, should he reappear, the whole project is knocked in the head. His poor brains being hopelessly puzzled with this dilemma, he at length ventures out, partly resolving to cross the head of the street, and send one hasty glance towards his forsaken domicile. Habit—for he is a man of habits—

takes him by the hand, and guides him, wholly unaware, to his own door, where, just at the critical moment, he is aroused by the scraping of his foot upon the step. Wakefield! whither are you going?

At that instant, his fate was turning on the pivot. Little dreaming of the doom to which his first backward step devotes him, he hurries away, breathless with agitation hitherto unfelt, and hardly dares turn his head, at the distant corner. Can it be, that nobody caught sight of him? Will not the whole household—the decent Mrs. Wakefield, the smart maid-servant, and the dirty little foot-boy—raise a hue-and-cry, through London streets, in pursuit of their fugitive lord and master? Wonderful escape! He gathers courage to pause and look homeward, but is perplexed with a sense of change about the familiar edifice, such as affects us all, when, after a separation of months or years, we again see some hill or lake, or work of art, with which we were friends, of old. In ordinary cases, this indescribable impression is caused by the comparison and contrast between our imperfect reminiscences and the reality. In Wakefield, the magic of a single night has wrought a similar transformation, because, in that brief period, a great moral change has been effected. But this is a secret from himself. Before leaving the spot, he catches a far and momentary glimpse of his wife, passing athwart the front window, with her face turned towards the head of the street. The crafty nincompoop takes to his heels, scared with the idea, that, among a thousand such atoms of mortality, her eye must have detected him. Right glad is his heart, though his brain be somewhat dizzy, when he finds himself by the coal-fire of his lodgings.

So much for the commencement of this long whim-wham. After the critical conception, and the stirring up of the man's sluggish temperament to put it in practice, the whole matter evolves itself in a natural train. We may suppose him, as the result of deep deliberation, buying a new wig, of reddish hair, and selecting sundry garments, in a fashion unlike his customary suit of brown, from a Jew's old-clothes bag. It is accomplished. Wakefield is another man. The new system being now established, a retrograde movement to the old would be almost as difficult as the step that placed him in his unparalleled position. Furthermore, he is rendered obstinate by a sulkiness, occasionally incident to his temper, and brought on, at present, by the inadequate sensation which he conceived to have been produced in the bosom of Mrs. Wakefield. He will not go back until she be frightened half to death. Well; twice or thrice has she passed before his sight, each time with a heavier step, a paler cheek, and more anxious brow; and, in the third week of his non-appearance, he detects a portent of evil entering the house, in the guise of an apothecary. Next day, the knocker is muffled. Towards night-fall, comes the chariot of a physician, and deposits its big-wigged and solemn burthen at Wakefield's door, whence, after a quarter of an hour's visit, he emerges, perchance the herald of a funeral. Dear woman! Will she die? By this time, Wakefield is excited to something like energy of feeling, but still lingers away from his wife's bedside, pleading with his conscience, that she must not be disturbed at such a juncture. If aught else restrains him, he does not know it. In the course of a few weeks, she gradually recovers; the crisis is over; her heart is sad, perhaps, but quiet; and, let him return soon or late, it will never be feverish for him again. Such ideas glimmer through the mist of Wakefield's mind, and render him indistinctly conscious that an almost impassible gulf divides his hired apartment from his former home. 'It is but in the next street!' he sometimes says. Fool! it is in another world. Hitherto, he has put off his return from one particular day to another; henceforward, he leaves the precise time undetermined. Not to-morrow—probably next week—pretty soon. Poor man! The dead have nearly as much chance of re-visiting their earthly homes, as the self-banished Wakefield.

Would that I had a folio to write, instead of a brief article in the New-England! Then might I exemplify how an influence, beyond our control, lays its strong hand on every deed which we do, and weaves its consequences into an iron tissue of necessity. Wakefield is spell-bound. We must leave him, for ten years or so, to haunt around his house, without once crossing the threshold, and to be faithful to his wife, with all the affection of which his heart is capable, while he is slowly fading out of hers. Long since, it must be remarked, he has lost the perception of singularity in his conduct.

Now for a scene! Amid the throng of a London street, we distinguish a man, now waxing elderly, with few characteristics to attract careless observers, yet bearing, in his whole aspect, the hand-writing of no common fate, for such as have the skill to read it. He is meagre; his low and narrow forehead is deeply wrinkled; his eyes, small and lustreless, sometimes wander apprehensively about him, but often seem to look inward. He bends his head, and moves with an indescribable obliquity of gait, as if unwilling to display his full front to the world. Watch him, long enough to see what we have described, and you will allow, that circumstances—which often produce remarkable men from nature's ordinary handiwork—have produced one such here. Next, leaving him to sidle along the foot-walk, cast your eyes in the opposite direction, where a portly female, considerably in the wane of life, with a prayer-book in her hand, is proceeding to yonder church. She has the placid mien of settled widowhood. Her regrets have either died away, or have become so essential to her heart, that they would be poorly exchanged for joy. Just as the lean man and well conditioned woman are passing, a slight obstruction occurs, and brings these two figures directly in contact. Their hands touch; the pressure of the crowd forces her bosom against his shoulder; they stand, face to face, staring into each other's eyes. After a ten years' separation, thus Wakefield meets his wife!

The throng eddies away, and carries them asunder. The sober widow, resuming her former pace, proceeds to church, but pauses in the portal, and throws a perplexed glance along the street. She passes in, however, opening her prayer-book as she goes. And the man? With so wild a face, that busy and selfish London stands to gaze after him, he hurries to his lodgings, bolts the door, and throws himself upon the bed. The latent feelings of years break out; his feeble mind acquires a brief energy from their strength; all the miserable strangeness of his life is revealed to him at a glance; and he cries out, passionately—'Wakefield! Wakefield! You are mad!'

Perhaps he was so. The singularity of his situation must have so moulded him to itself, that, considered in regard to his fellow-creatures and the business of life, he could not be said to possess his right mind. He had contrived, or rather he had happened, to dissever himself from the world—to vanish—to give up his place and privileges with living men, without being admitted among the dead. The life of a hermit is nowise parallel to his. He was in the bustle of the city, as of old; but the crowd swept by, and saw him not; he was, we may figuratively say, always beside his wife, and at his hearth, yet must never feel the warmth of the one, nor the affection of the other. It was Wakefield's unprecedented fate, to retain his original share of human sympathies, and to be still involved in human interests, while he had lost his reciprocal influence on them. It would be a most curious speculation, to trace out the effect of such circumstances on his heart and intellect, separately, and in unison. Yet, changed as he was, he would seldom be conscious of it, but deem himself the same man as ever; glimpses of the truth, indeed, would come, but only for the moment; and still he would keep saying—'I shall soon go back!'—nor reflect, that he had been saying so for twenty years.

I conceive, also, that these twenty years would appear, in the retrospect, scarcely longer than the week to which Wakefield had at first limited his absence. He would look on the affair as no more than an interlude in the main business of his life. When, after a little while more, he should deem it time to re-enter his parlor, his wife would clap her hands for joy, on beholding the middle-aged Mr. Wakefield. Alas, what a mistake! Would Time but await the close of our favorite follies, we should be young men, all of us, and till Doom's Day.

One evening, in the twentieth year since he vanished, Wakefield is taking his customary walk towards the dwelling which he still calls his own. It is a gusty night of autumn, with frequent showers, that patter down upon the pavement, and are gone, before a man can put up his umbrella. Pausing near the house, Wakefield discerns, through the parlor-windows of the second floor, the red glow, and the glimmer and fitful flash, of a comfortable fire. On the ceiling, appears a grotesque shadow of good Mrs. Wakefield. The cap, the nose and chin, and the broad waist, form an admirable caricature, which dances, moreover, with the up-flickering and down-sinking blaze, almost too merrily for the shade of an elderly widow. At this instant, a shower chances to fall, and is driven, by the unmannerly gust, full into Wakefield's face and bosom. He is quite penetrated with its autumnal chill. Shall he stand, wet and shivering here, when his own hearth has a good fire to warm him, and his own wife will run to fetch the gray coat and small-clothes, which, doubtless, she has kept carefully in the closet of their bed-chamber? No! Wakefield is no such fool. He ascends the steps—heavily!—for twenty years have stiffened his legs, since he came down—but he knows it not. Stay, Wakefield! Would you go to the sole home that is left you? Then step into your grave! The door opens. As he passes in, we have a parting glimpse of his visage, and recognize the crafty smile, which was the precursor of the little joke, that he has ever since been playing off at his wife's expense. How unmercifully has he quizzed the poor woman! Well; a good night's rest to Wakefield!

This happy event—supposing it to be such—could only have occurred at an unpremeditated moment. We will not follow our friend across the threshold. He has left us much food for thought, a portion of which shall lend its wisdom to a moral, and be shaped into a figure. Amid the seeming confusion of our mysterious world, individuals are so nicely adjusted to a system, and systems to one another, and to a whole, that, by stepping aside for a moment, a man exposes himself to a fearful risk of losing his place forever. Like Wakefield, he may become, as it were, the Outcast of the Universe.

1835

The Minister's Black Veil

A Parable[1]

BY THE AUTHOR OF 'SIGHTS FROM A STEEPLE'

The sexton stood in the porch of Milford meeting-house, pulling lustily at the bell-rope. The old people of the village came stooping along the street. Children, with bright faces, tript merrily beside their parents, or mimicked a graver gait, in the conscious dignity of their sunday clothes. Spruce bachelors looked sidelong at the pretty maidens, and fancied that the sabbath sunshine made them prettier than on week-days. When the throng had mostly streamed into the

---

1. "Another clergyman in New-England, Mr. Joseph Moody, of York, Maine, who died about eighty years since, made himself remarkable by the same eccentricity that is here related of the Reverend Mr. Hooper. In his case, however, the symbol had a different import. In early life he had accidentally killed a beloved friend; and from that day till the hour of his own death, he hid his face from men" [Hawthorne's note]. The text is that of the first printing in *The Token* (1836).

porch, the sexton began to toll the bell, keeping his eye on the Reverend Mr. Hooper's door. The first glimpse of the clergyman's figure was the signal for the bell to cease its summons.

'But what has good Parson Hooper got upon his face?' cried the sexton in astonishment.

All within hearing immediately turned about, and beheld the semblance of Mr. Hooper, pacing slowly his meditative way towards the meeting-house. With one accord they started, expressing more wonder than if some strange minister were coming to dust the cushions of Mr. Hooper's pulpit.

'Are you sure it is our parson?' inquired Goodman Gray of the sexton.

'Of a certainty it is good Mr. Hooper,' replied the sexton. 'He was to have exchanged pulpits with Parson Shute of Westbury; but Parson Shute sent to excuse himself yesterday, being to preach a funeral sermon.'

The cause of so much amazement may appear sufficiently slight. Mr. Hooper, a gentlemanly person of about thirty, though still a bachelor, was dressed with due clerical neatness, as if a careful wife had starched his band, and brushed the weekly dust from his Sunday's garb. There was but one thing remarkable in his appearance. Swathed about his forehead, and hanging down over his face, so low as to be shaken by his breath, Mr. Hooper had on a black veil. On a nearer view, it seemed to consist of two folds of crape, which entirely concealed his features, except the mouth and chin, but probably did not intercept his sight, farther than to give a darkened aspect to all living and inanimate things. With this gloomy shade before him, good Mr. Hooper walked onward, at a slow and quiet pace, stooping somewhat and looking on the ground, as is customary with abstracted men, yet nodding kindly to those of his parishioners who still waited on the meeting-house steps. But so wonderstruck were they, that his greeting hardly met with a return.

'I can't really feel as if good Mr. Hooper's face was behind that piece of crape,' said the sexton.

'I don't like it,' muttered an old woman, as she hobbled into the meeting-house. 'He has changed himself into something awful, only by hiding his face.'

'Our parson has gone mad!' cried Goodman Gray, following him across the threshhold.

A rumor of some unaccountable phenomenon had preceded Mr. Hooper into the meeting-house, and set all the congregation astir. Few could refrain from twisting their heads towards the door; many stood upright, and turned directly about; while several little boys clambered upon the seats, and came down again with a terrible racket. There was a general bustle, a rustling of the women's gowns and shuffling of the men's feet, greatly at variance with that hushed repose which should attend the entrance of the minister. But Mr.

Hooper appeared not to notice the perturbation of his people. He entered with an almost noiseless step, bent his head mildly to the pews on each side, and bowed as he passed his oldest parishioner, a white-haired great-grandsire, who occupied an arm-chair in the centre of the aisle. It was strange to observe, how slowly this venerable man became conscious of something singular in the appearance of his pastor. He seemed not fully to partake of the prevailing wonder, till Mr. Hooper had ascended the stairs, and showed himself in the pulpit, face to face with his congregation, except for the black veil. That mysterious emblem was never once withdrawn. It shook with his measured breath as he gave out the psalm; it threw its obscurity between him and the holy page, as he read the Scriptures; and while he prayed, the veil lay heavily on his uplifted countenance. Did he seek to hide it from the dread Being whom he was addressing?

Such was the effect of this simple piece of crape, that more than one woman of delicate nerves was forced to leave the meetinghouse. Yet perhaps the pale-faced congregation was almost as fearful a sight to the minister, as his black veil to them.

Mr. Hooper had the reputation of a good preacher, but not an energetic one: he strove to win his people heavenward, by mild persuasive influences, rather than to drive them thither, by the thunders of the Word. The sermon which he now delivered, was marked by the same characteristics of style and manner, as the general series of his pulpit oratory. But there was something, either in the sentiment of the discourse itself, or in the imagination of the auditors, which made it greatly the most powerful effort that they had ever heard from their pastor's lips. It was tinged, rather more darkly than usual, with the gentle gloom of Mr. Hooper's temperament. The subject had reference to secret sin, and those sad mysteries which we hide from our nearest and dearest, and would fain conceal from our own consciousness, even forgetting that the Omniscient can detect them. A subtle power was breathed into his words. Each member of the congregation, the most innocent girl, and the man of hardened breast, felt as if the preacher had crept upon them, behind his awful veil, and discovered their hoarded iniquity of deed or thought. Many spread their clasped hands on their bosoms. There was nothing terrible in what Mr. Hooper said; at least, no violence; and yet, with every tremor of his melancholy voice, the hearers quaked. An unsought pathos came hand in hand with awe. So sensible were the audience of some unwonted attribute in their minister, that they longed for a breath of wind to blow aside the veil, almost believing that a stranger's visage would be discovered, though the form, gesture, and voice were those of Mr. Hooper.

At the close of the services, the people hurried out with indecorous confusion, eager to communicate their pent-up amazement, and conscious of lighter spirits, the moment they lost sight of the black veil. Some gathered in little circles, huddled closely together, with their mouths all whispering in the centre; some went homeward alone, wrapt in silent meditation; some talked loudly, and profaned the Sabbath-day with ostentatious laughter. A few shook their sagacious heads, intimating that they could penetrate the mystery; while one or two affirmed that there was no mystery at all, but only that Mr. Hooper's eyes were so weakened by the midnight lamp, as to require a shade. After a brief interval, forth came good Mr. Hooper also, in the rear of his flock. Turning his veiled face from one group to another, he paid due reverence to the hoary heads, saluted the middle-aged with kind dignity, as their friend and spiritual guide, greeted the young with mingled authority and love, and laid his hands on the little children's heads to bless them. Such was always his custom on the Sabbath-day. Strange and bewildered looks repaid him for his courtesy. None, as on former occasions, aspired to the honor of walking by their pastor's side. Old Squire Saunders, doubtless by an accidental lapse of memory, neglected to invite Mr. Hooper to his table, where the good clergyman had been wont to bless the food, almost every Sunday since his settlement. He returned, therefore, to the parsonage, and, at the moment of closing the door, was observed to look back upon the people, all of whom had their eyes fixed upon the minister. A sad smile gleamed faintly from beneath the black veil, and flickered about his mouth, glimmering as he disappeared.

'How strange,' said a lady, 'that a simple black veil, such as any woman might wear on her bonnet, should become such a terrible thing on Mr. Hooper's face!'

'Something must surely be amiss with Mr. Hooper's intellects,' observed her husband, the physician of the village. 'But the strangest part of the affair is the effect of this vagary, even on a sober-minded man like myself. The black veil, though it covers only our pastor's face, throws its influence over his whole person, and makes him ghost-like from head to foot. Do you not feel it so?'

'Truly do I,' replied the lady; 'and I would not be alone with him for the world. I wonder he is not afraid to be alone with himself!'

'Men sometimes are so,' said her husband.

The afternoon service was attended with similar circumstances. At its conclusion, the bell tolled for the funeral of a young lady. The relatives and friends were assembled in the house, and the more distant acquaintances stood about the door, speaking of the good qualities of the deceased, when their talk was interrupted by the appearance of Mr. Hooper, still covered with his black veil. It was now an appropriate emblem. The clergyman stepped into the room where the corpse was laid, and bent over the coffin, to take a last farewell of his deceased parishioner. As he stooped, the veil hung straight down from his forehead, so that, if her eye-lids had not been closed for ever, the dead maiden might have seen his face. Could Mr. Hooper be fearful of her glance, that he so hastily caught back the black veil? A person, who watched the interview between the dead and living, scrupled not to affirm, that, at the instant when the clergyman's features were disclosed, the corpse had slightly shuddered, rustling the shroud and muslin cap, though the countenance retained the composure of death. A superstitious old woman was the only witness of this prodigy. From the coffin, Mr. Hooper passed into the chambers of the mourners, and thence to the head of the staircase, to make the funeral prayer. It was a tender and heart-dissolving prayer, full of sorrow, yet so imbued with celestial hopes, that the music of a heavenly harp, swept by the fingers of the dead, seemed faintly to be heard among the saddest accents of the minister. The people trembled, though they but darkly understood him, when he prayed that they, and himself, and all of mortal race, might be ready, as he trusted this young maiden had been, for the dreadful hour that should snatch the veil from their faces. The bearers went heavily forth, and the mourners followed, saddening all the street, with the dead before them, and Mr. Hooper in his black veil behind.

'Why do you look back?' said one in the procession to his partner.

'I had a fancy,' replied she, 'that the minister and the maiden's spirit were walking hand in hand.'

'And so had I, at the same moment,' said the other.

That night, the handsomest couple in Milford village were to be joined in wedlock. Though reckoned a melancholy man, Mr. Hooper had a placid cheerfulness for such occasions, which often excited a sympathetic smile, where livelier merriment would have been thrown away. There was no quality of his disposition which made him more beloved than this. The company at the wedding awaited his arrival with impatience, trusting that the strange awe, which had gathered over him throughout the day, would now be dispelled. But such was not the result. When Mr. Hooper came, the first thing that their eyes rested on was the same horrible black veil, which had added deeper gloom to the funeral, and could portend nothing but evil to the wedding. Such was its immediate effect on the guests, that a cloud seemed to have rolled duskily from beneath the black crape, and dimmed the light of the candles. The bridal pair stood up before the minister. But the bride's cold fingers quivered in the tremulous hand of the bridegroom, and her death-like paleness caused a whisper, that the maiden who had been buried a few hours before, was come from her grave to be married. If ever another wedding were so dismal, it was that famous one, where they tolled the wedding-knell.[2] After performing the cere-

2. A reference to Hawthorne's own *The Wedding Knell*, which appeared in *The Token* for 1836 along with this story.

mony, Mr. Hooper raised a glass of wine to his lips, wishing happiness to the new-married couple, in a strain of mild pleasantry that ought to have brightened the features of the guests, like a cheerful gleam from the hearth. At that instant, catching a glimpse of his figure in the looking-glass, the black veil involved his own spirit in the horror with which it overwhelmed all others. His frame shuddered—his lips grew white—he spilt the untasted wine upon the carpet—and rushed forth into the darkness. For the Earth, too, had on her Black Veil.

The next day, the whole village of Milford talked of little else than Parson Hooper's black veil. That, and the mystery concealed behind it, supplied a topic for discussion between acquaintances meeting in the street, and good women gossiping at their open windows. It was the first item of news that the tavern-keeper told to his guests. The children babbled of it on their way to school. One imitative little imp covered his face with an old black handkerchief, thereby so affrighting his playmates, that the panic seized himself, and he well nigh lost his wits by his own waggery.

It was remarkable, that, of all the busy-bodies and impertinent people in the parish, not one ventured to put the plain question to Mr. Hooper, wherefore he did this thing. Hitherto, whenever there appeared the slightest call for such interference, he had never lacked advisers, nor shown himself averse to be guided by their judgment. If he erred at all, it was by so painful a degree of self-distrust, that even the mildest censure would lead him to consider an indifferent action as a crime. Yet, though so well acquainted with this amiable weakness, no individual among his parishioners chose to make the black veil a subject of friendly remonstrance. There was a feeling of dread, neither plainly confessed nor carefully concealed, which caused each to shift the responsibility upon another, till at length it was found expedient to send a deputation of the church, in order to deal with Mr. Hooper about the mystery, before it should grow into a scandal. Never did an embassy so ill discharge its duties. The minister received them with friendly courtesy, but became silent, after they were seated, leaving to his visitors the whole burthen[3] of introducing their important business. The topic, it might be supposed, was obvious enough. There was the black veil, swathed round Mr. Hooper's forehead, and concealing every feature above his placid mouth, on which, at times, they could perceive the glimmering of a melancholy smile. But that piece of crape, to their imagination, seemed to hang down before his heart, the symbol of a fearful secret between him and them. Were the veil but cast aside, they might speak freely of it, but not till then. Thus they sat a considerable time, speechless, confused, and shrinking uneasily from Mr. Hooper's eye, which they felt to be fixed upon them with an invisible glance. Finally, the deputies returned abashed to their

3. Burden.

constituents, pronouncing the matter too weighty to be handled, except by a council of the churches, if, indeed, it might not require a general synod.

But there was one person in the village, unappalled by the awe with which the black veil had impressed all beside herself. When the deputies returned without an explanation, or even venturing to demand one, she, with the calm energy of her character, determined to chase away the strange cloud that appeared to be settling round Mr. Hooper, every moment more darkly than before. As his plighted wife, it should be her privilege to know what the black veil concealed. At the minister's first visit, therefore, she entered upon the subject, with a direct simplicity, which made the task easier both for him and her. After he had seated himself, she fixed her eyes steadfastly upon the veil, but could discern nothing of the dreadful gloom that had so overawed the multitude: it was but a double fold of crape, hanging down from his forehead to his mouth, and slightly stirring with his breath.

'No,' said she aloud, and smiling, 'there is nothing terrible in this piece of crape, except that it hides a face which I am always glad to look upon. Come, good sir, let the sun shine from behind the cloud. First lay aside your black veil: then tell me why you put it on.'

Mr. Hooper's smile glimmered faintly.

'There is an hour to come,' said he, 'when all of us shall cast aside our veils. Take it not amiss, beloved friend, if I wear this piece of crape till then.'

'Your words are a mystery too,' returned the young lady. 'Take away the veil from them, at least.'

'Elizabeth, I will,' said he, 'so far as my vow may suffer me. Know, then, this veil is a type[4] and a symbol, and I am bound to wear it ever, both in light and darkness, in solitude and before the gaze of multitudes, and as with strangers, so with my familiar friends. No mortal eye will see it withdrawn. This dismal shade must separate me from the world: even you, Elizabeth, can never come behind it!'

'What grievous affliction hath befallen you,' she earnestly inquired, 'that you should thus darken your eyes for ever?'

'If it be a sign of mourning,' replied Mr. Hooper, 'I, perhaps, like most other mortals, have sorrows dark enough to be typified by a black veil.'

'But what if the world will not believe that it is the type of an innocent sorrow?' urged Elizabeth. 'Beloved and respected as you are, there may be whispers, that you hide your face under the consciousness of secret sin. For the sake of your holy office, do away this scandal!'

The color rose into her cheeks, as she intimated the nature of the rumors that were already abroad in the village. But Mr. Hooper's

4. Symbol (the phrase "a type and a symbol" is redundant).

mildness did not forsake him. He even smiled again—that same sad smile, which always appeared like a faint glimmering of light, proceeding from the obscurity beneath the veil.

'If I hide my face for sorrow, there is cause enough,' he merely replied; 'and if I cover it for secret sin, what mortal might not do the same?'

And with this gentle, but unconquerable obstinacy, did he resist all her entreaties. At length Elizabeth sat silent. For a few moments she appeared lost in thought, considering, probably, what new methods might be tried, to withdraw her lover from so dark a fantasy, which, if it had no other meaning, was perhaps a symptom of mental disease. Though of a firmer character than his own, the tears rolled down her cheeks. But, in an instant, as it were, a new feeling took the place of sorrow: her eyes were fixed insensibly on the black veil, when, like a sudden twilight in the air, its terrors fell around her. She arose, and stood trembling before him.

'And do you feel it then at last?' said he mournfully.

She made no reply, but covered her eyes with her hand, and turned to leave the room. He rushed forward and caught her arm.

'Have patience with me, Elizabeth!' cried he passionately. 'Do not desert me, though this veil must be between us here on earth. Be mine, and hereafter there shall be no veil over my face, no darkness between our souls! It is but a mortal veil—it is not for eternity! Oh, you know not how lonely I am and how frightened to be alone behind my black veil. Do not leave me in this miserable obscurity for ever!'

'Lift the veil but once, and look me in the face,' said she.

'Never! It cannot be!' replied Mr. Hooper.

'Then, farewell!' said Elizabeth.

She withdrew her arm from his grasp, and slowly departed, pausing at the door, to give one long, shuddering gaze, that seemed almost to penetrate the mystery of the black veil. But, even amid his grief, Mr. Hooper smiled to think that only a material emblem had separated him from happiness, though the horrors which it shadowed forth, must be drawn darkly between the fondest of lovers.

From that time no attempts were made to remove Mr. Hooper's black veil, or, by a direct appeal, to discover the secret which it was supposed to hide. By persons who claimed a superiority to popular prejudice, it was reckoned merely an eccentric whim, such as often mingles with the sober actions of men otherwise rational, and tinges them all with its own semblance of insanity. But with the multitude, good Mr. Hooper was irreparably a bugbear.[5] He could not walk the street with any peace of mind, so conscious was he that the gentle and timid would turn aside to avoid him, and that others would make it a point of hardihood to throw themselves in his way.

5. Object of dread.

The impertinence of the latter class compelled him to give up his customary walk, at sunset, to the burial ground; for when he leaned pensively over the gate, there would always be faces behind the grave-stones, peeping at his black veil. A fable went the rounds, that the stare of the dead people drove him thence. It grieved him, to the very depth of his kind heart, to observe how the children fled from his approach, breaking up their merriest sports, while his melancholy figure was yet afar off. Their instinctive dread caused him to feel, more strongly than aught else, that a preternatural horror was interwoven with the threads of the black crape. In truth, his own antipathy to the veil was known to be so great, that he never willingly passed before a mirror, nor stooped to drink at a still fountain, lest, in its peaceful bosom, he should be affrighted by himself. This was what gave plausibility to the whispers, that Mr. Hooper's conscience tortured him for some great crime, too horrible to be entirely concealed, or otherwise than so obscurely intimated. Thus, from beneath the black veil, there rolled a cloud into the sunshine, an ambiguity of sin or sorrow, which enveloped the poor minister, so that love or sympathy could never reach him. It was said, that ghost and fiend consorted with him there. With self-shudderings and outward terrors, he walked continually in its shadow, groping darkly within his own soul, or gazing through a medium that saddened the whole world. Even the lawless wind, it was believed, respected his dreadful secret, and never blew aside the veil. But still good Mr. Hooper sadly smiled, at the pale visages of the worldly throng as he passed by.

Among all its bad influences, the black veil had the one desirable effect, of making its wearer a very efficient clergyman. By the aid of his mysterious emblem—for there was no other apparent cause—he became a man of awful power, over souls that were in agony for sin. His converts always regarded him with a dread peculiar to themselves, affirming, though but figuratively, that, before he brought them to celestial light, they had been with him behind the black veil. Its gloom, indeed, enabled him to sympathize with all dark affections. Dying sinners cried aloud for Mr. Hooper, and would not yield their breath till he appeared; though ever, as he stooped to whisper consolation, they shuddered at the veiled face so near their own. Such were the terrors of the black veil, even when death had bared his visage! Strangers came long distances to attend service at his church, with the mere idle purpose of gazing at his figure, because it was forbidden them to behold his face. But many were made to quake ere they departed! Once, during Governor Belcher's[6] administration, Mr. Hooper was appointed to preach the election sermon. Covered with his black veil, he stood before the

6. Jonathan Belcher (1682–1757) was Governor of Massachusetts and New Hampshire (1730–41); an Election Sermon was preached at the installing of each new Governor (in this case, at one of Belcher's installations for a new term).

chief magistrate, the council, and the representatives, and wrought so deep an impression, that the legislative measures of that year, were characterized by all the gloom and piety of our earliest ancestral sway.

In this manner Mr. Hooper spent a long life, irreproachable in outward act, yet shrouded in dismal suspicions; kind and loving, though unloved, and dimly feared; a man apart from men, shunned in their health and joy, but ever summoned to their aid in mortal anguish. As years wore on, shedding their snows above his sable veil, he acquired a name throughout the New-England churches, and they called him Father Hooper. Nearly all his parishioners, who were of mature age when he was settled, had been borne away by many a funeral: he had one congregation in the church, and a more crowded one in the church-yard; and having wrought so late into the evening, and done his work so well, it was now good Father Hooper's turn to rest.

Several persons were visible by the shaded candlelight, in the death-chamber of the old clergyman. Natural connections he had none. But there was the decorously grave, though unmoved physician, seeking only to mitigate the last pangs of the patient whom he could not save. There were the deacons, and other eminently pious members of his church. There, also, was the Reverend Mr. Clark, of Westbury, a young and zealous divine, who had ridden in haste to pray by the bed-side of the expiring minister. There was the nurse, no hired handmaiden of death, but one whose calm affection had endured thus long, in secresy, in solitude, amid the chill of age, and would not perish, even at the dying hour. Who, but Elizabeth! And there lay the hoary head of good Father Hooper upon the death-pillow, with the black veil still swathed about his brow and reaching down over his face, so that each more difficult gasp of his faint breath caused it to stir. All through life that piece of crape had hung between him and the world: it had separated him from cheerful brotherhood and woman's love, and kept him in that saddest of all prisons, his own heart; and still it lay upon his face, as if to deepen the gloom of his darksome chamber, and shade him from the sunshine of eternity.

For some time previous, his mind had been confused, wavering doubtfully between the past and the present, and hovering forward, as it were, at intervals, into the indistinctness of the world to come. There had been feverish turns, which tossed him from side to side, and wore away what little strength he had. But in his most convulsive struggles, and in the wildest vagaries of his intellect, when no other thought retained its sober influence, he still showed an awful solicitude lest the black veil should slip aside. Even if his bewildered soul could have forgotten, there was a faithful woman at his pillow, who, with averted eyes, would have covered that aged face, which she had last beheld in the comeliness of manhood. At length the death-stricken old man lay quietly in the torpor of mental and bodily exhaustion, with an imperceptible pulse, and breath that grew fainter and fainter, except when a long, deep, and irregular inspiration seemed to prelude the flight of his spirit.

The minister of Westbury approached the bedside.

'Venerable Father Hooper,' said he, 'the moment of your release is at hand. Are you ready for the lifting of the veil, that shuts in time from eternity?'

Father Hooper at first replied merely by a feeble motion of his head; then, apprehensive, perhaps, that his meaning might be doubtful, he exerted himself to speak.

'Yea,' said he, in faint accents, 'my soul hath a patient weariness until that veil be lifted.'

'And is it fitting,' resumed the Reverend Mr. Clark, 'that a man so given to prayer, of such a blameless example, holy in deed and thought, so far as mortal judgment may pronounce; is it fitting that a father in the church should leave a shadow on his memory, that may seem to blacken a life so pure? I pray you, my venerable brother, let not this thing be! Suffer us to be gladdened by your triumphant aspect, as you go to your reward. Before the veil of eternity be lifted, let me cast aside this black veil from your face!'

And thus speaking, the reverend Mr. Clark bent forward to reveal the mystery of so many years. But, exerting a sudden energy, that made all the beholders stand aghast, Father Hooper snatched both his hands from beneath the bed-clothes, and pressed them strongly on the black veil, resolute to struggle, if the minister of Westbury would contend with a dying man.

'Never!' cried the veiled clergyman. 'On earth, never!'

'Dark old man!' exclaimed the affrighted minister, 'with what horrible crime upon your soul are you now passing to the judgment?'

Father Hooper's breath heaved; it rattled in his throat; but, with a mighty effort, grasping forward with his hands, he caught hold of life, and held it back till he should speak. He even raised himself in bed; and there he sat, shivering with the arms of death around him, while the black veil hung down, awful, at that last moment, in the gathered terrors of a life-time. And yet the faint, sad smile, so often there, now seemed to glimmer from its obscurity, and linger on Father Hooper's lips.

'Why do you tremble at me alone?' cried he, turning his veiled face round the circle of pale spectators. 'Tremble also at each other! Have men avoided me, and women shown no pity, and children screamed and fled, only for my black veil? What, but the mystery which it obscurely typifies, has made this piece of crape so awful? When the friend shows his inmost heart to his friend; the lover to his best-beloved; when man does not vainly shrink from the eye of his Creator, loathsomely treasuring up the secret of his sin; then deem me a monster, for the symbol beneath which I have lived, and die! I look around me, and lo! on every visage a black veil!'

While his auditors shrank from one another, in mutual affright, Father Hooper fell back upon his pillow, a veiled corpse, with a faint smile lingering on the lips. Still veiled, they laid him in his coffin, and a veiled corpse they bore him to the grave. The grass of many years has sprung up and withered on that grave, the burial-stone is moss-grown, and good Mr. Hooper's face is dust; but awful is still the thought, that it mouldered beneath the black veil!

1836

HENRY WADSWORTH LONGFELLOW
1807–1882

Longfellow was born in Portland, Maine (then still a part of Massachusetts), on February 27, 1807, and died on March 24, 1882, in Cambridge, Massachusetts, the most beloved American poet of his time. His father sent him to Bowdoin, thinking that he would become a lawyer. Instead, Longfellow became so proficient a student of languages that Bowdoin created for him a professorship of modern languages, then one of only a handful in the country. With support from his father, Longfellow studied languages in Europe for three years before taking up his work at Bowdoin in 1829. In Spain, Washington Irving was hospitable, and Longfellow's prose romance *Outre-Mer* (1833–35) was in loving imitation of *The Sketch-Book*. Having concentrated on the Romance languages during his first European stay, Longfellow returned to perfect himself in Germanic languages, a condition for his becoming a professor at Harvard late in 1836. He took teaching seriously, although in the early years he spent most of his time instilling the rudiments of foreign languages (for which he wrote and published his own textbooks) without being able to teach the literatures. Later, at Harvard, he taught an extraordinary range of European literatures of many periods and thereby became an incalculable force in American cultural life. It would be hard, also, to overestimate the importance of his anthology *The Poets and Poetry of Europe* (1845) in bringing home to the ordinary reader the rich variety of European literatures. His own poetry became a means of teaching readers of his day something of the possible range of poetic subject matter and techniques, ancient, medieval, and modern. Irving had been notably successful in domesticating European subject matter while employing a British prose style; now Longfellow domesticated European meters, as in his adaptation of classical Greek meters to tell the story of Evangeline Belle-fontaine, set in the recent North American past, or in using Finnish folk meter in his celebration of American Indian legends in *Hiawatha*. Longfellow became a great teacher of the masses. If his worst fault is that he made poetry seem so easy to write that anyone could do it, his greatest virtue is that he made poetry seem worth reading and worth writing.

Longfellow married in 1831, during his professorship at Bowdoin, but in 1835, during his second European trip, his wife died after miscarrying.

Longfellow stayed on, fulfilling his commitment to Harvard, and before he returned home he had met Fanny Appleton, the Boston heiress who was to become his second wife. She was slow to return his affection, and he embarrassed her by the transparent account of their meeting in the prose romance *Hyperion* (1839), but after their marriage in 1843 their life was idyllic. Longfellow's father-in-law bought the couple, as a wedding gift, Craigie House in Cambridge, a mansion George Washington had used as headquarters and where Longfellow himself had been renting rooms. Their life was elegant. Emerson, who lived amply enough in Concord, was intimidated: "If Socrates were here, we could go and talk with him; but Longfellow we cannot go and talk with; there is a palace, and servants, and a row of bottles of different colored wines, and wine glasses, and fine coats." But the sumptuousness proved supportive and encouraging to Longfellow's poetry and to his work at Harvard until he resigned in 1854. Popular as he was, Longfellow could not make a living from his poetry. In 1855 and 1856, for instance, the phenomenal sales of *Hiawatha* brought his total earnings from poetry to around $3,700 and $7,400, but in the 1840s and 1850s his average annual income from poetry hardly exceeded his Harvard salary of $1,500 ($1,800 after 1845). In 1861 Mrs. Longfellow was fatally burned as she was sealing up locks of her daughters' hair. In his grief Longfellow turned to translating the entire *Divine Comedy* of Dante, making the labor the occasion for regular meetings with friends such as James Russell Lowell and the young William Dean Howells, who had lived in Venice. Longfellow's last decades were uneventful, except for one final visit to Europe in 1868–69, during which Queen Victoria gave him a private audience. His seventy-fifth birthday was celebrated nationally. Of his death his brother and official biographer wrote: "The long, busy, blameless life was ended. The loneliness of separation was over. He was dead. But the world was better and happier for his having lived."

My Lost Youth[1]

Often I think of the beautiful town
    That is seated by the sea;
Often in thought go up and down
The pleasant streets of that dear old town,
    And my youth comes back to me.   5
      And a verse of a Lapland song
      Is haunting my memory still:
    "A boy's will is the wind's will,
And the thoughts of youth are long, long thoughts."

I can see the shadowy lines of its trees,   10
    And catch, in sudden gleams,
The sheen of the far-surrounding seas,

1. Longfellow wrote this poem about his hometown of Portland, Maine, in March, 1855, at Cambridge, deriving the refrain from lines in John Scheffer's *The History of Lapland* (1674): "A Youth's desire is the desire of the wind, / All his essaies / Are long delaies, / No issue can they find." ("Essaies" or "essays": attempts; "issue": outlet.) The text is that of the first printing in *Putnam's Monthly Magazine*, Vol. 6 (August, 1855). It was reprinted in *The Courtship of Miles Standish and Other Poems* (Boston: Ticknor and Fields, 1858).

And islands that were the Hesperides[2]
   Of all my boyish dreams.
      And the burden of that old song,   15
      It murmurs and whispers still:
"A boy's will is the wind's will,
And the thoughts of youth are long, long thoughts."

I remember the black wharves and the slips,
   And the sea-tides tossing free;   20
And Spanish sailors with bearded lips,
And the beauty and mystery of the ships,
   And the magic of the sea.
      And the voice of that wayward song
      Is singing and saying still: 25
"A boy's will is the wind's will,
And the thoughts of youth are long, long thoughts."

I remember the bulwarks by the shore,
And the fort upon the hill;
The sun-rise gun, with its hollow roar,   30
The drum-beat repeated o'er and o'er,
   And the bugle wild and shrill.
      And the music of that old song
      Throbs in my memory still:
"A boy's will is the wind's will,   35
And the thoughts of youth are long, long thoughts."

I remember the sea-fight far away,
   How it thundered o'er the tide![3]
And the dead captains, as they lay
In their graves, o'erlooking the tranquil bay,   40
   Where they in battle died.
      And the sound of that mournful song
      Goes through me with a thrill:
"A boy's will is the wind's will,
And the thoughts of youth are long, long thoughts."   45

I can see the breezy dome of groves,
   The shadows of Deering's Woods;
And the friendships old and the early loves
Come back with a Sabbath sound, as of doves
   In quiet neighborhoods.   50
      And the verse of that sweet old song,
      It flutters and murmurs still:
"A boy's will is the wind's will,
And the thoughts of youth are long, long thoughts."

I remember the gleams and glooms that dart   55
   Across the schoolboy's brain;

2. In Greek mythology, fabled islands where the golden apples grew.
3. The American *Enterprise* and the British *Boxer* fought near Portland in 1813. Both captains were killed and carried ashore for burial.

The song and the silence in the heart,
That in part are prophecies, and in part
   Are longings wild and vain.
     And the voice of that fitful song   60
     Sings on, and is never still:
"A boy's will is the wind's will,
And the thoughts of youth are long, long thoughts."

There are things of which I may not speak;
   There are dreams that cannot die;   65
There are thoughts that make the strong heart weak,
And bring a pallor into the cheek,
   And a mist before the eye.
     And the words of that fatal song
     Come over me like a chill: 70
"A boy's will is the wind's will,
And the thoughts of youth are long, long thoughts."

Strange to me now are the forms I meet
   When I visit the dear old town;
But the native air is pure and sweet,   75
And the trees that o'ershadow each well-known street,
   As they balance up and down,
     Are singing the beautiful song,
     Are sighing and whispering still:
"A boy's will is the wind's will,   80
And the thoughts of youth are long, long thoughts."

And Deering's Woods are fresh and fair,
   And with joy that is almost pain
My heart goes back to wander there,
And among the dreams of the days that were,   85
   I find my lost youth again.
     And the strange and beautiful song,
     The groves are repeating it still:
"A boy's will is the wind's will,
And the thoughts of youth are long, long thoughts." 90
1855   1855

JOHN GREENLEAF WHITTIER
1807–1892

John Greenleaf Whittier was born on December 17, 1807, on a farm near Haverhill, Massachusetts, of a Quaker family. No longer persecuted in New England, Quakers were still a people apart, and Whittier grew up with a sense of being different from most of his neighbors. Labor on the debt-ridden farm overstrained his health in adolescence, and thereafter through-

out his long life he suffered from intermittent physical collapses. At fourteen, having had only meager education in a household suspicious of non-
Quaker literature, he found in the Scottish poet Robert Burns a model for
imitation, one using a regional dialect, dealing with homely subjects, and
displaying a strong social conscience. His first poem was published in 1826
in a local newspaper run by another young man, William Lloyd Garrison,
whose dedication to the antislavery movement was to affect Whittier's life
profoundly. In 1827 Garrison helped persuade Whittier's father that the
young poet deserved more education, and Whittier supported himself
through two terms at Haverhill Academy. During this time and later
Whittier was near serious courtships, but like many of his relatives, he
never married; among the obstacles were his Quakerism, his poverty, and
his commitment to abolitionism. In 1836, six years after his father's death,
Whittier and his mother and sisters moved from the farm to the house in
nearby Amesbury, Massachusetts, which he owned until his death.

In his twenties Whittier became editor of various newspapers, some of
regional importance. He was elected for a term to the Massachusetts legislature (1835) and became a behind-the-scenes force in the Whig party, and
later in the antislavery Liberty party, which he helped to found in 1839.
The turning point in his career came in 1833 with the publication of his
abolitionist manifesto *Justice and Expediency*, in which Whittier concluded that there was only one practicable and just scheme of emancipation: "Immediate abolition of slavery; an immediate acknowledgment of
the great truth, that man cannot hold property in man; an immediate surrender of baneful prejudice to Christian love; an immediate practical obedience to the command of Jesus Christ: 'Whatsoever ye would that men
should do unto you, do ye even so to them.'" Over the next three decades
Whittier paid for his principles in many ways, some subtle, some as overt
as being mobbed and stoned in 1835. The climactic danger came in 1838
when Whittier, in disguise, joined a mob to save some of his papers as his
office was ransacked and burned.

From the 1830s through the 1850s Whittier was a working editor associated with abolitionist papers, becoming the sort of man he was to
describe in *The Tent on the Beach* (1867): "a dreamer born, / Who, with
a mission to fulfil, / Had left the Muses' haunts to turn / The crank of
an opinion-mill, / Making his rustic reed of song / A weapon in the war
with wrong." Yet he continued to write about his own region, one legacy
from his family being a rich oral history. His first book, *Legends of New
England* (1831), had included stories in both prose and poetry. His first
book of poetry was *Lays of My Home* (1843), and the prose *Supernaturalism of New England* followed in 1847. *Leaves from Margaret Smith's Journal* (1849) is a fictional re-creation of colonial life in the form of the diary
of a young woman. Through his fictional and historical prose and through
his poetry Whittier was setting a very early example of faithful treatment
of American village and rural life which later local colorists and regionalists
were to follow: the elderly Whittier's paternal interest in the career of
Sarah Orne Jewett epitomizes this influence. But from the beginning a crucial problem for Whittier had been how to be true to the occasional beauty
of rural life without portraying it in the sentimental manner which prevailed at the time. Whittier succeeded best in some late poems, especially
Snow-Bound (1866) and the "Prelude" to *Among the Hills* (1868).

460 ★ John Greenleaf Whittier

Whittier's reputation began undergoing a change in the late 1850s, when abolitionism had ceased to be almost as much abhorred in the North as in the South; partly the new favor he received was a result of the founding in 1857 of the *Atlantic Monthly*, which was always hospitable to his poems, humorous folk legends as well as militant odes. *Snow-Bound* brought Whittier extraordinary acclaim and immediate financial security; although one of the themes of the poem was his sense of his own approaching death, Whittier ironically lived another quarter century during which he was revered as a great American poet. He died on September 7, 1892.

Prelude to *Among the Hills*[1]

Along the roadside, like the flowers of gold
That tawny Incas[2] for their gardens wrought,
Heavy with sunshine droops the golden-rod,
And the red pennons of the cardinal-flowers
Hang motionless upon their upright staves.   5
The sky is hot and hazy, and the wind,
Wing-weary with its long flight from the south,
Unfelt; yet, closely scanned, yon maple leaf
With faintest motion, as one stirs in dreams,
Confesses it. The locust by the wall   10
Stabs the noon-silence with his sharp alarm.
A single hay-cart down the dusty road
Creaks slowly, with its driver fast asleep
On the load's top. Against the neighboring hill,
Huddled along the stone wall's shady side,   15
The sheep show white, as if a snow-drift still
Defied the dog-star.[3] Through the open door
A drowsy smell of flowers—gray heliotrope,
And white sweet-clover, and shy mignonette—
Comes faintly in, and silent chorus lends   20
To the pervading symphony of peace.

No time is this for hands long overworn
To task their strength; and (unto Him be praise
Who giveth quietness!) the stress and strain
Of years that did the work of centuries   25
Have ceased, and we can draw our breath once more
Freely and full. So, as yon harvesters
Make glad their nooning underneath the elms
With tale and riddle and old snatch of song,
I lay aside grave themes, and idly turn   30
The leaves of Memory's sketch-book, dreaming o'er
Old summer pictures of the quiet hills,
And human life, as quiet, at their feet.

1. From the first printing, in *Among the Hills, and Other Poems* (Boston: Fields, Osgood, & Co., 1869).
2. An allusion to the belief that gold was so plentiful among the Inca Indians of Peru that they fashioned golden ornamental flowers for their gardens.
3. Sirius, star visible near the sun at dawn during the torrid "dog days" of August.

And yet not idly all. A farmer's son,
Proud of field-lore and harvest craft, and feeling   35
All their fine possibilities, how rich
And restful even poverty and toil
Become when beauty, harmony, and love
Sit at their humble hearth as angels sat
At evening in the patriarch's tent, when man   40
Makes labor noble, and his farmer's frock
The symbol of a Christian chivalry
Tender and just and generous to her
Who clothes with grace all duty; still, I know
Too well the picture has another side,—   45
How wearily the grind of toil goes on
Where love is wanting, how the eye and ear
And heart are starved amidst the plenitude
Of nature, and how hard and colorless
Is life without an atmosphere. I look   50
Across the lapse of half a century,
And call to mind old homesteads, where no flower
Told that the spring had come, but evil weeds,
Nightshade and rough-leaved burdock in the place
Of the sweet doorway greeting of the rose   55
And honeysuckle, where the house walls seemed
Blistering in sun, without a tree or vine
To cast the tremulous shadow of its leaves
Across the curtainless windows from whose panes
Fluttered the signal rags of shiftlessness;   60
Within, the cluttered kitchen-floor, unwashed
(Broom-clean I think they called it); the best room
Stifling with cellar damp, shut from the air
In hot midsummer, bookless, pictureless
Save the inevitable sampler hung   65
Over the fireplace, or a mourning-piece,[4]
A green-haired woman, peony-cheeked, beneath
Impossible willows; the wide-throated hearth
Bristling with faded pine-boughs half concealing
The piled-up rubbish at the chimney's back;   70
And, in sad keeping with all things about them,
Shrill, querulous women, sour and sullen men,
Untidy, loveless, old before their time,
With scarce a human interest save their own
Monotonous round of small economies,[5]   75
Or the poor scandal of the neighborhood;
Blind to the beauty everywhere revealed,
Treading the May-flowers with regardless feet;
For them the song-sparrow and the bobolink
Sang not, nor winds made music in the leaves;   80
For them in vain October's holocaust
Burned, gold and crimson, over the hills,

4. A piece of art in memory of a de-   5. Management of domestic affairs, par-
parted relative.   ticularly those involving the budget.

The sacramental mystery of the woods.
Church-goers, fearful of the unseen Powers,
But grumbling over pulpit-tax and pew-rent,[6]   85
Saving, as shrewd economists, their souls
And winter pork with the least possible outlay
Of salt and sanctity; in daily life
Showing as little actual comprehension
Of Christian charity and love and duty,   90
As if the Sermon on the Mount[7] had been
Outdated like a last year's almanac:
Rich in broad woodlands and in half-tilled fields,
And yet so pinched and bare and comfortless,
The veriest straggler limping on his rounds,   95
The sun and air his sole inheritance,
Laughed at a poverty that paid its taxes,
And hugged his rags in self-complacency!

Not such should be the homesteads of a land
Where whoso wisely wills and acts may dwell   100
As king and lawgiver, in broad-acred state,
With beauty, art, taste, culture, books, to make
His hour of leisure richer than a life
Of fourscore to the barons of old time,
Our yeoman[8] should be equal to his home   105
Set in the fair, green valleys, purple walled,
A man to match his mountains, not to creep
Dwarfed and abased below them. I would fain
In this light way (of which I needs must own
With the knife-grinder of whom Canning[9] sings,   110
"Story, God bless you! I have none to tell you!")
Invite the eye to see and heart to feel
The beauty and the joy within their reach,—
Home, and home loves, and the beatitudes
Of nature free to all. Haply in years   115
That wait to take the places of our own,
Heard where some breezy balcony looks down
On happy homes, or where the lake in the moon
Sleeps dreaming of the mountains, fair as Ruth,
In the old Hebrew pastoral, at the feet   120
Of Boaz,[1] even this simple lay of mine
May seem the·burden of a prophecy,

6. Fees to support the minister and pay for the use of a pew.
7. Matthew 5–7, Jesus' fullest statement of the absolute behavior he expects of his followers in contrast to the conventional ways of this world.
8. Farmer.
9. During the 1790s, as a way of turning English public opinion against the French Revolution, the statesman George Canning (1770–1827) wrote for *The Anti-Jacobin*, a paper, as its title says, opposed to the most radical French fac- tion. Canning's *The Friend of Humanity and the Knife-Grinder*, extremely popular in Whittier's time, is a satire of misplaced humanitarianism and bleeding-heart liberalism. The line Whittier quotes is the drink-loving knife-grinder's brusque retort to the torrential address of the would-be philanthropist.
1. See Ruth 3 for the story of how the young widow, an ancestress of David, reminded Boaz of his family obligation to marry her.

Finding its late fulfilment in a change
Slow as the oak's growth, lifting manhood up
Through broader culture, finer manners, love,   125
And reverence, to the level of the hills.

O Golden Age, whose light is of the dawn,
And not of sunset, forward, not behind,
Flood the new heavens and earth, and with thee bring
All the old virtues, whatsoever things   130
Are pure and honest and of good repute,
But add thereto whatever bard has sung
Or seer has told of when in trance and dream
They saw the Happy Isles of prophecy!
Let Justice hold her scale, and Truth divide   135
Between the right and wrong; but give the heart
The freedom of its fair inheritance;
Let the poor prisoner, cramped and starved so long,
At Nature's table feast his ear and eye
With joy and wonder; let all harmonies   140
Of sound, form, color, motion, wait upon
The princely guest, whether in soft attire
Of leisure clad, or the coarse frock of toil.
And, lending life to the dead form of faith,
Give human nature reverence for the sake   145
Of One who bore it, making it divine
With the ineffable tenderness of God;
Let common need, the brotherhood of prayer,
The heirship of an unknown destiny,
The unsolved mystery round about us, make   150
A man more precious than the gold of Ophir.[2]
Sacred, inviolate, unto whom all things
Should minister, as outward types and signs
Of the eternal beauty which fulfils
The one great purpose of creation, Love,   155
The sole necessity of Earth and Heaven!

                                                                1869

2. Source of treasures of gold brought to King Solomon (1 Kings 10.11).

EDGAR ALLAN POE
1809–1849

The life of Poe is the most melodramatic of any of the major American writers of his generation. Determining the facts has proved difficult, since lurid legend became entwined with fact even before he died. Some legends were spread by Poe himself. Given to claiming that he was born in 1811 or 1813 and had written certain poems far earlier than he had, Poe also exaggerated the length of his attendance at the University of Virginia and, in imitation of Lord Byron, fabricated a "quixotic expedition to join the

Greeks, then struggling for liberty." Two days after Poe's death his supposed friend Rufus Griswold, a prominent anthologizer of American literature, began a campaign of character assassination in which he ultimately rewrote Poe's correspondence so as to alienate many of his friends who could only assume that Poe had treacherously maligned them behind their backs. Griswold's forgeries went unexposed for many years, poisoning every biographer's image of Poe, and legend still feeds upon half truth in much writing on him.

Yet biographers now possess a great deal of factual evidence about most periods of Poe's life. His mother, Elizabeth Arnold, had been an actress, prominent among the wandering seaport players in a profession which was then considered disreputable. She was a teenage widow when she married David Poe, Jr., in 1806. Poe, also an actor, worked up to choice supporting roles before liquor destroyed his career. Edgar, the Poes' second child, was born in Boston on January 19, 1809; a year later David Poe deserted the family. In December, 1811, Elizabeth Poe died at twenty-four while acting in Richmond, Virginia, and her husband disappeared, probably dying soon afterward at the age of twenty-seven.

The disruptions of Poe's first two years were followed by apparent security, for John Allan, a young Richmond merchant, took him in as the children were parceled out. As "Master Allan," Poe accompanied the family to England in 1815, where he attended good schools. On their return in 1820 the boy continued in school, but under his own last name. During Poe's adolescence uncertainty about his future and shameful certainty about his past affected his feelings—and those of his prosperous playmates. Around 1824, Allan's attitude toward the boy changed; one rumor suggests that Edgar took the side of his foster mother in a quarrel. Poe spent most of 1826 at the new University of Virginia, doing well in his studies, although he was already drinking. Under the pretext that Allan had not provided him an adequate allowance, he gambled, and lost some $2,000— "debts of honor" which a gentleman must repay. Allan had just inherited a fortune of several hundred thousand dollars (with purchasing power of several million today), but he refused to pay Poe's debts. After a quarrel with Allan in March, 1827, Poe looked up his father's relatives in Baltimore and then went on to his birthplace, where he paid for the printing of *Tamerlane and Other Poems*, "By a Bostonian." Before its publication, "Edgar A. Perry" had joined the army. Poe was partially reconciled with Allan in March, 1829, just after Mrs. Allan died. Released from the army with the rank of sergeant major, Poe sought Allan's influence to gain him an appointment to West Point, although he was past the age limit for admission.

While he was waiting for the appointment, Poe shortened *Tamerlane*, revised other poems, and added new ones to make up a second volume, *Al Aaraaf, Tamerlane, and Minor Poems*, published at Baltimore in December, 1829. He entered West Point in June, 1830, but felt he could not fit into life at the academy without supplemental income, and Allan was interested in his own life, not Poe's. Just after Poe went to West Point a woman in Richmond bore Allan twin sons. In October, 1830, Allan married again and within a month his new wife was pregnant. Losing any remaining hope that if he dutifully pursued a military career he might become Allan's heir, Poe got himself expelled by missing classes and roll calls. Supportive friends among the cadets made up a subscription for his *Poems*, published in May,

1831. In this third volume Poe revised some earlier poems and for the first time included versions of both *To Helen* (the famous "Helen, thy beauty is to me," not a later, inferior poem of the same title) and *Israfel*.

Poe's mature career—from his twenty-first year to his death in his fortieth year—was spent in four literary centers: Baltimore, Richmond, Philadelphia, and New York. The Baltimore years—mid-1831 to late 1835—were marked by great industry and comparative sobriety. Poe lived in sordid poverty among his once-prosperous relatives, including his poetaster brother who died in 1831; his Grandmother Poe, whose death in 1835 cut off a Revolutionary widow's pension of $240 per annum on which the household relied; his aunt Maria Poe Clemm; and her daughter Virginia, whom Poe secretly married in 1835, when she was thirteen. Poe's first story, *Metzengerstein* (later subtitled *In Imitation of the German*) was published in the Philadelphia *Saturday Courier*, anonymously, in January, 1832, and other stories appeared in the same paper through the year. By early 1833, Poe was projecting a volume of eleven stories, *Tales of the Folio Club*, never published under that title. In May, 1833, he sent the *New-England Magazine* one of a set of *Eleven Tales of the Arabesque*—apparently the same eleven; a postscript added to the manuscript said simply, "I am poor." With his *Tales of the Folio Club*, Poe impressed all three judges of a contest in the Baltimore *Saturday Visiter*. One judge, the novelist John P. Kennedy, became a loyal mentor, offering timely money and advice.

Poe returned to Richmond in 1835, twenty-six years old, as assistant editor of T. L. White's new *Southern Literary Messenger*, at a salary of $540 a year, subsistence wages even in the 1830s. Allan was dead, survived by three small legitimate sons, and Poe had no contact with the widow. From the start, White deplored what he called Poe's tendency to "sip the juice," and gave him editorial duties without commensurate recognition or authority, even though the circulation of the magazine rose swiftly under Poe's guidance. The *Messenger* published stories by Poe, but it was through his critical pieces that he gained a national reputation as a reviewer in the virulently sarcastic British manner—a literary hatchetman.

Fired from the *Messenger* early in 1837, Poe took his aunt and his wife (whom he had publicly remarried in May, 1836) to New York City, where for two years he lived hand to mouth on the fringes of the publishing world, selling a few stories and reviews. He had written a short novel, *The Narrative of Arthur Gordon Pym*, in Richmond, where White ran two installments in the *Messenger* early in 1837. *Harper's* finally brought it out in July, 1838, but it earned him neither money nor reputation, since it purported only to be edited by Poe. In 1838 Poe moved to Philadelphia, where for weeks the family survived on bread and molasses. But he continued writing, and *Ligeia* appeared in the Baltimore *American Museum* in September, 1838, where other stories and poems followed. Resorting to literary hackwork just as Hawthorne was doing, Poe put his name on *The Conchologist's First Book* (1839). In May, 1839, he got his first steady job in over two years, as co-editor of *Burton's Gentleman's Magazine*. There he published book reviews and stories, among them *The Fall of the House of Usher* and *William Wilson*. Late in 1839, a Philadelphia firm published *Tales of the Grotesque and Arabesque*, but it sold badly. Poe was now at the height of his powers as a writer of tales, though his personal life continued unstable, as did his career as an editor. William Burton fired him for drinking in May, 1840, but recommended him to George Graham, who carried on Burton's magazine as *Graham's*. Throughout 1841, Poe was with *Graham's* as co-editor, courting subscribers by articles on cryptography and on character as revealed in handwriting. In January, 1842, Virginia Poe, not yet twenty, burst a blood vessel in her throat (she lived only five more years). Leaving *Graham's* in some unhappiness, Poe revived a project for his own magazine, now to be called *The Stylus*. In 1843 he worked at times for the Philadelphia weekly *Saturday Museum*. On a trip to Washington seeking a patronage job (and subscriptions to *The Stylus*) he reportedly was so drunk when he called on President Tyler that he wore his cloak inside out.

In April, 1844, Poe moved his family to New York City, where he wrote for newspapers and worked as subeditor on the *Sunday Times*. Poe's most successful year was 1845. The February issue of *Graham's* contained James Russell Lowell's complimentary article on Poe, and *The Raven* appeared in the February *American Review* after advance publication in the New York *Evening Mirror*. Capitalizing on the sensation the poem created, Poe lectured on "Poets of America" and became a principal reviewer for the new weekly, the *Broadway Journal*. *The Raven* won him entrée into the literary life of New York. One new literary acquaintance, Evert A. Duyckinck, soon to be Melville's friend also, selected a dozen of Poe's stories for a collection brought out by Wiley & Putnam in June and arranged for the same firm to publish *The Raven and Other Poems* in November. Having acquired critical clout despite a growing number of enemies, Poe had great hopes for the *Broadway Journal*, of which he became sole owner; but it failed early in 1846. Meanwhile Poe was marring his new opportunities by drinking.

With fame the tempo of Poe's life spun into a blur of literary feuds, flirtations with literary ladies, and drinking bouts which ended in quarrels. Virginia's death in January, 1847, slowed the tempo: during much of that year Poe was seriously ill himself—perhaps with a brain lesion—and drinking steadily. He worked away at *Eureka*, a prose statement of a theory of the universe, and soon after Virginia's death he wrote *Ulalume*. The year 1848 was frenetic, culminating in a brief engagement to Helen Power Whitman of Providence; his letters to her are effusively hysterical. He flirted with Mrs. Nancy Richmond of Lowell, Massachusetts, in equally desperate letters, and may—as he wrote her—have tried to commit suicide by taking laudanum. He managed to write a little still, the story *Hop-Frog* and the poem *Annabel Lee*. While headed South in June, 1849, he drank on the train and got off in Philadelphia to seek asylum, he said, from two men who were trying to kill him. In Richmond he spent two improbably happy months, being received into society by his childhood friends and becoming engaged to the sweetheart of his teens, the now-widowed Elmira Royster Shelton. He gave lectures and readings, and joined the Sons of Temperance. On the way to accept a hundred dollars for editing the poems of a Philadelphia woman, he stopped off in Baltimore, broke his temperance pledge, and was found senseless near a polling place on an Election Day (October 3). Taken to a hospital, he died on October 7, 1849, "of congestion of the brain."

If Poe had disappeared from the American literary scene after publishing his third volume of poems in 1831, a literary historian grubbing among

privately printed nineteenth-century collections of poetry would have classified him (once his authorship of the anonymous *Tamerlane* had been
established) as an odd American imitator of major British Romantics like
Lord Byron and Percy Bysshe Shelley as well as then-popular ones like
Thomas Moore. In both form and content Poe's early poetry is typically
Romantic, although of an unusually limited range. Well before his twenty-
first birthday he had earned the right to call himself a poet, but by British
standards he was not an important one.

It was the handful of poems which Poe wrote a decade and a half later
that made him famous as a poet. *The Raven* brought him international
celebrity, and poems like *Ulalume* and *The Bells* soon enhanced that fame
among Poe's constantly enlarging posthumous audience. These poems became standard declamation pieces in schools and remained so well into the
present century. In subject matter they progress little beyond the Romantic
gothicism of Poe's early years, but in technique they are remarkable. Innumerable young people have learned to love poetry from them and have
continued to love poetry even after they stopped loving only Poe. There
could be worse fates for a man who started out as a belated, second-rate
imitator of first- and second-rate British Romantics.

But the bulk of Poe's collected writings consists of his criticism, and his
most abiding ambition was to become a powerful critic. Just as he had
modeled his poems and first tales upon British examples (or British imitations of the German), he took his critical concepts from treatises on aesthetics by late eighteenth-century Scottish Common-Sense philosophers
(later modified by his borrowings from A. W. Schlegel and Coleridge) and
took his stance as a reviewer from the slashing critics of the British quarterlies. Poe's employers were often uneasy about their reviewer, both because
his virulence brought reproaches (though it was good for business) and because they suspected that for all his stress on aesthetic principles, Poe's reviews were apt to be unjust to writers he was jealous of and laudatory
toward others he wished to curry favor with. But Poe's basic critical principles were consistent enough, however he deviated from them in his reviewing. He thought poetry should appeal only to the sense of beauty, not truth;
informational poetry, poetry of ideas, or any sort of didactic poetry was
illegitimate. Holding that the true poetic emotion was a vague sensory state,
he set himself against realistic details in poetry, although the prose tale,
with truth as one object, could profit from the discreet use of specifics.
Both poems and tales should be short enough to be read in one sitting;
otherwise the unity of effect would be dissipated. In Poe's view, good
writers calculate their effects precisely. At a time when even famous poets
such as Longfellow rarely wrote a poem of sustained coherence, Poe's reaction, with the stress on forethought, seems understandable. But his criticism
is often dogmatic and self-serving, weakened partly because it was applied
to some of the most wretched writing a reviewer ever had to discuss, for
Poe never had the luxury of reviewing only worthwhile volumes.

Poe's first tales have proved hard to classify—are they burlesques of
popular kinds of fiction or serious attempts at contributing to or somehow
altering those genres? Poe's own comments tend to becloud his intentions
rather than to clarify them. In 1836 his benefactor John P. Kennedy wrote
him: "Some of your *bizarreries* have been mistaken for satire—and admired
too in that character. *They* deserved it, but *you* did not, for you did not

intend them so. I like your grotesque—it is of the very best stamp; and I am sure you will do wonders for yourself in the comic—I mean the serio-tragicomic." Poe's reply is tantalizing: "You are nearly, but not altogether right in relation to the satire of some of my Tales. Most of them were *intended* for half banter, half satire—although I might not have fully acknowledged this to be their aim even to myself." The problem of determining the nature of a given work—imitation? satire? spoof? hoax?—is crucial in Poe criticism.

At the core of Poe's defenses of his stories is the hardheadedness of a professional writer who wanted to crack the popular market. Such stories, he claimed, were the products of superior minds disciplining themselves to the task at hand, not the indulgences of Romantic genius. Poe worked hard at structuring his tales of aristocratic madmen, self-tormented murderers, neurasthenic necrophiliacs, and other deviant types so as to produce the greatest possible horrific effects on the reader. In the detective story, which Poe created when he was thirty-two, with all its major conventions complete, the structuring was equally contrived, although the effect desired was one of awe at the brilliance of his preternatural logician-hero. Seriously as he took the writing of his tales, Poe never claimed that prose writing was for him, as he said poetry was, a "passion," not merely a "purpose."

Other American writers, from Poe's time to ours, have often been uneasy about him. The "jingle man," Ralph Waldo Emerson is supposed to have called him, and Henry James thought that enthusiasm for Poe was "the mark of a decidedly primitive stage of reflection," while T. S. Eliot said Poe's intellect was that of "a highly gifted young person before puberty." Yet no other American writer, except possibly Mark Twain, has been so thoroughly absorbed by later writers—writers as diverse as E. A. Robinson, Frank Norris, Theodore Dreiser, William Faulkner, as well as the great Russian-American player of complex Poesque games, Vladimir Nabokov. Some American literary critics and historians have always been hard pressed to understand why foreign writers like Charles Baudelaire and Stéphane Mallarmé could idolize Poe and translate his works lovingly, why the French Symbolist poets could draw on him for their aesthetic ideas, how August Strindberg could fantasize that because he was born in 1849 Poe's spirit had passed to him, how the influence of someone so childish could seem profound when it came back to English indirectly, through foreigners Poe had influenced. Some American critics have often felt reproached when British writers such as Dante Gabriel Rossetti, Algernon Swinburne, Robert Louis Stevenson, Arthur Conan Doyle, and George Bernard Shaw expressed delight in Poe or indebtedness to him. More than a century and a quarter after his death, American critics are still taking sides about Poe, hailing him as a pioneering aesthetician, psychological investigator, and literary technician, or else reviling him as an absurd fraud, a subliterary vulgarian. But whatever his influence on artists of the past and present, and whatever his status with literary critics and historians, Poe's reputation with the reading public—through the whole range of literacy—is more assured than that of any other major American writer of his century, again with the possible exception of Mark Twain. For the professional writer that Poe struggled to be, that is probably a fate even better than being precisely understood and logically classified.

To Helen[1]

Helen, thy beauty is to me
  Like those Nicéan barks[2] of yore,
That gently, o'er a perfumed sea,
  The weary, way-worn wanderer bore
  To his own native shore.   5

On desperate seas long wont to roam,
  Thy hyacinth hair, thy classic face,
Thy Naiad[3] airs have brought me home
  To the glory that was Greece,
And the grandeur that was Rome.   10

Lo! in yon brilliant window-niche
  How statue-like I see thee stand,
  The agate lamp within thy hand!
Ah, Psyche,[4] from the regions which
  Are Holy-Land!   15

                        1831, 1845

Israfel[1]

In Heaven a spirit doth dwell
  "Whose heart-strings are a lute;"
None sing so wildly well
As the angel Israfel,
And the giddy stars (so legends tell)   5
Ceasing their hymns, attend the spell
  Of his voice, all mute.

Tottering above
  In her highest noon,
  The enamoured moon   10
Blushes with love,
  While, to listen, the red levin
  (With the rapid Pleiads,[2] even,
  Which were seven,)
  Pauses in Heaven.   15

1. The text is that of 1845, with two errors of indentation corrected. The poem was first published in 1831 where, among other differences, lines 9 and 10 read: "To the beauty of fair Greece,/And the grandeur of old Rome."
2. Variously annotated by Poe scholars, the Nicéan boats are more important for their musicality and vaguely classical suggestiveness than for their vaguely Mediterranean reference.
3. Nymphlike, fairylike.
4. Goddess of the soul.
1. "And the angel Israfel, whose heart-strings are a lute, and who has the sweet-est voice of all God's creatures.—KORAN" [Poe's note]. A version of this poem appeared in the 1831 volume; the present text is from 1845. In 1831 the footnote to the title read (correctly): "And the angel Israfel, who has the sweetest voice of all God's creatures.—KORAN." Poe later expanded the quotation. Parallels among English Romantic poems include Coleridge's *Kubla Khan*, where glimpses of heavenly song also inspire but ultimately frustrate the speaker.
2. In Greek mythology the seven daughters of Atlas became stars, making up a constellation.

And they say (the starry choir
   And the other listening things)
That Israfeli's fire
Is owing to that lyre
   By which he sits and sings—   20
The trembling living wire
   Of those unusual strings.

But the skies that angel trod,
   Where deep thoughts are a duty—
Where Love's a grown-up God—   25
   Where the Houri[3] glances are
Imbued with all the beauty
   Which we worship in a star.

Therefore, thou art not wrong,
   Israfeli, who despisest   30
An unimpassioned song;
To thee the laurels belong,
   Best bard, because the wisest!
Merrily live, and long!

The ecstasies above   35
   With thy burning measures suit—
Thy grief, thy joy, thy hate, thy love,
   With the fervour of thy lute—
   Well may the stars be mute!

Yes, Heaven is thine; but this   40
   Is a world of sweets and sours;
   Our flowers are merely—flowers,
And the shadow of thy perfect bliss
   Is the sunshine of ours.

If I could dwell   45
Where Israfel
   Hath dwelt, and he where I,
He might not sing so wildly well
   A mortal melody,
While a bolder note than this might swell   50
   From my lyre within the sky.

1831, 1845

The City in the Sea[1]

Lo! Death has reared himself a throne
In a strange city lying alone
Far down within the dim West,

---

3. **Beautiful virgin waiting in paradise for the devout Mohammedan.**

1. The text is that of 1845; it was first published in 1831 as *The Doomed City*.

Where the good and the bad and the worst
   and the best
Have gone to their eternal rest.   5
There shrines and palaces and towers
(Time-eaten towers that tremble not!)
Resemble nothing that is ours.
Around, by lifting winds forgot,
Resignedly beneath the sky   10
The melancholy waters lie.

No rays from the holy heaven come down
On the long night-time of that town;
But light from out the lurid sea
Streams up the turrets silently—   15
Gleams up the pinnacles far and free—
Up domes—up spires—up kingly halls—
Up fanes[2]—up Babylon-like walls—
Up shadowy long-forgotten bowers
Of sculptured ivy and stone flowers—   20
Up many and many a marvellous shrine
Whose wreathéd friezes intertwine
The viol, the violet, and the vine.
Resignedly beneath the sky
The melancholy waters lie.   25
So blend the turrets and shadows there
That all seem pendulous in air,
While from a proud tower in the town
Death looks gigantically down.

There open fanes and gaping graves   30
Yawn level with the luminous waves;
But not the riches there that lie
In each idol's diamond eye—
Not the gaily-jewelled dead
Tempt the waters from their bed;   35
For no ripples curl, alas!
Along that wilderness of glass—
No swellings tell that winds may be
Upon some far-off happier sea—
No heavings hint that winds have been   40
On seas less hideously serene.

But lo, a stir is in the air!
The wave—there is a movement there!
As if the towers had thrust aside,
In slightly sinking, the dull tide—   45
As if their tops had feebly given
A void within the filmy Heaven.
The waves have now a redder glow—

2. Temples.

472 Edgar Allan Poe

The hours are breathing faint and low—
And when, amid no earthly moans,   50
Down, down that town shall settle hence,
Hell, rising from a thousand thrones,
Shall do it reverence.

                              1831, 1845

Dream-land[1]

By a route obscure and lonely,
Haunted by ill angels only,
Where an Eidolon,[2] named NIGHT,
On a black throne reigns upright,
I have reached these lands but newly   5
From an ultimate dim Thule—[3]
From a wild weird clime, that lieth, sublime,
        Out of SPACE—out of TIME.

Bottomless vales and boundless floods,
And chasms, and caves, and Titan[4] woods,   10
With forms that no man can discover
For the dews that drip all over;
Mountains toppling evermore
Into seas without a shore;
Seas that restlessly aspire,   15
Surging, unto skies of fire;
Lakes that endlessly outspread
Their lone waters, lone and dead,—
Their still waters, still and chilly
With the snows of the lolling lily.   20
By a route obscure and lonely,
Haunted by ill angels only,
Where an Eidolon, named NIGHT,
On a black throne reigns upright,
I have reached my home but newly   25
From this ultimate dim Thule.

By the lakes that thus outspread
Their lone waters, lone and dead,—
Their sad waters, sad and chilly
With the snows of the lolling lily,—   30
By the mountain—near the river
Murmuring lowly, murmuring ever,—
By the gray woods,—by the swamp
Where the toad and the newt encamp,—

1. This version, from *Graham's Maga-*
zine, 25 (June, 1844), 256, contains three
repetitions of the refrain; the 1845 volume text prints those lines only once, as
the last stanza.
2. Phantom.

3. A fabled island located north of Britain by ancient geographers and thought
of as the northernmost habitable region.
4. I.e., enormous, as were the Titans,
the children of Heaven and Earth, deposed by the gods of Olympus.

By the dismal tarns and pools   35
    Where dwell the Ghouls,—
By each spot the most unholy—
In each nook most melancholy,—
There the traveler meets aghast
Sheeted Memories of the Past—   40
Shrouded forms that start and sigh
As they pass the wanderer by—
White-robed forms of friends long given,
In agony, to the worms, and Heaven.

By a route obscure and lonely,   45
Haunted by ill angels only,
Where an Eidolon, named NIGHT,
On a black throne reigns upright,
I have journeyed home but newly
From this ultimate dim Thule.   50

For the heart whose woes are legion
'T is a peaceful, soothing region—
For the spirit that walks in shadow
'T is—oh 't is an Eldorado!⁵
But the traveler, traveling through it,   55
May not—dare not openly view it;
Never its mysteries are exposed
To the weak human eye unclosed;
So wills the King, who hath forbid
The uplifting of the fringéd lid;   60
And thus the sad Soul that here passes
Beholds it but through darkened glasses.

By a route obscure and lonely,
Haunted by ill angels only,
Where an Eidolon, named NIGHT,   65
On a black throne reigns upright,
I have wandered home but newly
From this ultimate dim Thule.

                                              1844, 1845

The Raven¹
By ⸺ Quarles

[*The following lines from a correspondent—besides the deep quaint
strain of the sentiment, and the curious introduction of some ludicrous
touches amidst the serious and impressive, as was doubtless intended by
the author—appear to us one of the most felicitous specimens of unique*

5. Legendary golden country sought by
Spanish conquerors of South America.
1. This printing of Poe's most famous
poem is taken from the *American Review: A Whig Journal of Politics, Literature, Art and Science*, 1 (February,
1845), where it was first set in type; before the issue was distributed, the New
York *Evening Mirror* printed the poem,

on January 29, 1845, from the pages of
the *American Review*. The prefatory
paragraph, signed as if it were by the
editor of the *American Review*, is retained here since Poe most likely had a
hand in it, if he did not write it all.
Many minor variations appear in later
texts.

rhyming which has for some time met our eye. The resources of English rhythm for varieties of melody, measure, and sound, producing corresponding diversities of effect, have been thoroughly studied, much more perceived, by very few poets in the language. While the classic tongues, especially the Greek, possess, by power of accent, several advantages for versification over our own, chiefly through greater abundance of spondaic feet,[2] we have other and very great advantages of sound by the modern usage of rhyme. Alliteration is nearly the only effect of that kind which the ancients had in common with us. It will be seen that much of the melody of "The Raven" arises from alliteration, and the studious use of similar sounds in unusual places. In regard to its measure, it may be noted that if all the verses were like the second, they might properly be placed merely in short lines, producing a not uncommon form; but the presence in all the others of one line—mostly the second in the verse—which flows continuously, with only an aspirate pause in the middle, like that before the short line in the Sapphic Adonic,[3] while the fifth has at the middle pause no similarity of sound with any part beside, gives the versification an entirely different effect. We could wish the capacities of our noble language, in prosody, were better understood.—Ed. Am. Rev.]

Once upon a midnight dreary, while I pondered, weak and weary,
Over many a quaint and curious volume of forgotten lore,
While I nodded, nearly napping, suddenly there came a tapping,
As of some one gently rapping, rapping at my chamber door.
" 'Tis some visiter," I muttered, "tapping at my chamber door—   5
        Only this, and nothing more."

Ah, distinctly I remember it was in the bleak December,
And each separate dying ember wrought its ghost upon the floor.
Eagerly I wished the morrow;—vainly I had tried to borrow
From my books surcease of sorrow—sorrow for the lost Lenore—   10
For the rare and radiant maiden whom the angels name Lenore—
        Nameless here for evermore.

And the silken sad uncertain rustling of each purple curtain
Thrilled me—filled me with fantastic terrors never felt before;
So that now, to still the beating of my heart, I stood repeating   15
" 'Tis some visiter entreating entrance at my chamber door—
Some late visiter entreating entrance at my chamber door;—
        This it is and nothing more."

Presently my soul grew stronger; hesitating then no longer,
"Sir," said I, "or Madam, truly your forgiveness I implore;   20
But the fact is I was napping, and so gently you came rapping,
And so faintly you came tapping, tapping at my chamber door,
That I scarce was sure I heard you"—here I opened wide the door;—
        Darkness there, and nothing more.

Deep into that darkness peering, long I stood there wondering,
        fearing,   25
Doubting, dreaming dreams no mortal ever dared to dream before;
But the silence was unbroken, and the darkness gave no token,
And the only word there spoken was the whispered word, "Lenore!"
This I whispered, and an echo murmured back the word, "Lenore!"
        Merely this, and nothing more.   30

2. A spondee is a metrical foot consisting of two stressed syllables.
3. A Greek lyric form. In prosody an adonic is a dactyl (a foot with one long syllable and two short ones) followed by a spondee.

Then into the chamber turning, all my soul within me burning,
Soon I heard again a tapping somewhat louder than before.
"Surely," said I, "surely that is something at my window lattice;
Let me see, then, what thereat is, and this mystery explore—
Let my heart be still a moment and this mystery explore;—   35
          'Tis the wind, and nothing more!"

Open here I flung the shutter, when, with many a flirt and flutter,
In there stepped a stately raven of the saintly days of yore;
Not the least obeisance made he; not an instant stopped or stayed he;
But, with mien of lord or lady, perched above my chamber
    door—   40
Perched upon a bust of Pallas[4] just above my chamber door—
          Perched, and sat, and nothing more.

Then this ebony bird beguiling my sad fancy into smiling,
By the grave and stern decorum of the countenance it wore,
"Though thy crest be shorn and shaven, thou," I said, "art sure no
    craven,   45
Ghastly grim and ancient raven wandering from the Nightly shore—
Tell me what thy lordly name is on the Night's Plutonian[5] shore!"
          Quoth the raven, "Nevermore."

Much I marvelled this ungainly fowl to hear discourse so plainly,
Though its answer little meaning—little relevancy bore;   50
For we cannot help agreeing that no sublunary[6] being
Ever yet was blessed with seeing bird above his chamber door—
Bird or beast upon the sculptured bust above his chamber door,
          With such name as "Nevermore."

But the raven, sitting lonely on the placid bust, spoke only   55
That one word, as if his soul in that one word he did outpour.
Nothing farther then he uttered—not a feather then he fluttered—
Till I scarcely more than muttered, "Other friends have flown
    before—
On the morrow *he* will leave me, as my hopes have flown before."
          Quoth the raven, "Nevermore."   60

Wondering at the stillness broken by reply so aptly spoken,
"Doubtless," said I, "what it utters is its only stock and store,
Caught from some unhappy master whom unmerciful Disaster
Followed fast and followed faster—so, when Hope he would adjure,
Stern Despair returned, instead of the sweet Hope he dared
    adjure—   65
          That sad answer, "Nevermore!"[7]

4. Athena, the Greek goddess of wisdom and the arts.
5. Black, as in the underworld of Greek mythology.
6. Earthly, beneath the moon.
7. This stanza concluded in the 1845 volume with these lines: "Followed fast and followed faster till his songs one burden bore—/Till the dirges of his Hope that melancholy burden bore of 'Never—nevermore.'"

But the raven still beguiling all my sad soul into smiling,
Straight I wheeled a cushioned seat in front of bird, and bust, and
    door;
Then upon the velvet sinking, I betook myself to linking
Fancy unto fancy, thinking what this ominous bird of yore—   70
What this grim, ungainly, ghastly, gaunt, and ominous bird of yore
                  Meant in croaking "Nevermore."

This I sat engaged in guessing, but no syllable expressing
To the fowl whose fiery eyes now burned into my bosom's core;
This and more I sat divining, with my head at ease reclining   75
On the cushion's velvet lining that the lamplight gloated o'er,
But whose velvet violet lining with the lamplight gloating o'er,
                She shall press, ah, nevermore!

Then, methought, the air grew denser, perfumed from an unseen
    censer
Swung by angels whose faint foot-falls tinkled on the tufted floor. 80
"Wretch," I cried, "thy God hath lent thee—by these angels he
    hath sent thee
Respite—respite and Nepenthe[8] from thy memories of Lenore!
Let me quaff this kind Nepenthe and forget this lost Lenore!"
                Quoth the raven, "Nevermore."

"Prophet!" said I, "thing of evil!—prophet still, if bird **or**
    devil!—   85
Whether Tempter sent, or whether tempest tossed thee here
    ashore,
Desolate, yet all undaunted, on this desert land enchanted—
On this home by Horror haunted—tell me truly, I implore—
Is there—*is* there balm in Gilead?[9]—tell me—tell me, I implore!"
                Quoth the raven, "Nevermore." 90

"Prophet!" said I, "thing of evil!—prophet still, if bird or devil!
By that Heaven that bends above us—by that God we both adore—
Tell this soul with sorrow laden if, within the distant Aidenn,[1]
It shall clasp a sainted maiden whom the angels name Lenore—
Clasp a rare and radiant maiden whom the angels name Lenore." 95
                Quoth the raven, "Nevermore."

"Be that word our sign of parting, bird or fiend!" I shrieked,
    upstarting—
"Get thee back into the tempest and the Night's Plutonian shore!
Leave no black plume as a token of that lie thy soul hath spoken!
Leave my loneliness unbroken!—quit the bust above my door! 100
Take thy beak from out my heart, and take thy form from off
    my door!"
                Quoth the raven, "Nevermore."

8. Drug that induces oblivion.
9. An echo of the ironic words in Jeremiah 8.22: "Is there no balm in Gilead; is there no physician there?" (Gilead is a mountainous area east of the Jordan River between the Sea of Galilee and the Dead Sea; evergreens growing there were an ample source of medicinal resins.)
1. One of Poe's vaguely evocative place names, designed to suggest Eden.

And the raven, never flitting, still is sitting, still is sitting
On the pallid bust of Pallas just above my chamber door;
And his eyes have all the seeming of a demon that is dreaming,   105
And the lamp-light o'er him streaming throws his shadow on the
    floor;
And my soul from out that shadow that lies floating on the floor
                              Shall be lifted—nevermore!
                                                      1845

To _____ _____ _____.[1] Ulalume: A Ballad

The skies they were ashen and sober;
    The leaves they were crispéd and sere—
    The leaves they were withering and sere;
It was night in the lonesome October
    Of my most immemorial year;   5
It was hard by the dim lake of Auber,
    In the misty mid region of Weir[2]—
It was down by the dank tarn[3] of Auber.
    In the ghoul-haunted woodland of Weir.

Here once, through an alley Titanic,[4]   10
    Of cypress, I roamed with my Soul—
    Of cypress, with Psyche, my Soul.
These were days when my heart was volcanic

    As the scoriac rivers[5] that roll—
    As the lavas that restlessly roll   15
Their sulphurous currents down Yaanek
    In the ultimate climes of the pole—
That groan as they roll down Mount Yaanek
    In the realms of the boreal pole.[6]

Our talk had been serious and sober,   20
    But our thoughts they were palsied and sere—
    Our memories were treacherous and sere—
For we knew not the month was October,
    And we marked not the night of the year—
    (Ah, night of all nights in the year!)   25
We noted not the dim lake of Auber—
    (Though once we had journeyed down here)—
We remembered not the dank tarn of Auber,
    Nor the ghoul-haunted woodland of Weir.

1. This is the longer version of the poem; Poe sometimes dropped the tenth stanza. The source is the *American Review*, 6 (December, 1847), 599–600, the first printing.
2. "Auber" and "Weir" are surnames Poe probably knew; as place names they are chosen for their rhyme value and connotative suggestions ("Weir," for instance, suggesting "weird").
3. A small mountain lake.
4. The alley—the pathway—is titanic because the cypress trees on either side are enormous, on a scale to match that of the pre-Olympian Greek gods.
5. Rivers of lava.
6. North pole.

And now, as the night was senescent   30
    And star-dials pointed to morn—
    As the star-dials hinted of morn—
At the end of our path a liquescent
    And nebulous lustre was born,
Out of which a miraculous crescent   35
    Arose with a duplicate horn—
Astarte's[7] bediamonded crescent
    Distinct with its duplicate horn.

And I said—"She is warmer than Dian:[8]
    She rolls through an ether of sighs—   40
    She revels in a region of sighs:
She has seen that the tears are not dry on
    These cheeks, where the worm never dies,
And has come past the stars of the Lion[9]
    To point us the path to the skies—   45
    To the Lethean[1] peace of the skies—
Come up, in despite of the Lion,
    To shine on us with her bright eyes—
Come up through the lair of the Lion
    With Love in her luminous eyes."   50

But Psyche,[2] uplifting her finger,
    Said—"Sadly this star I mistrust—
    Her pallor I strangely mistrust:—
Oh, hasten!—oh, let us not linger!
    Oh, fly!—let us fly!—for we must."   55
In terror she spoke, letting sink her
    Wings till they trailed in the dust—
In agony sobbed, letting sink her
    Plumes till they trailed in the dust—
    Till they sorrowfully trailed in the dust.   60

I replied—"This is nothing but dreaming:
    Let us on by this tremulous light!
    Let us bathe in this crystalline light!
Its Sybillic[3] splendor is beaming
    With Hope and in Beauty to-night:—   65
    See!—it flickers up the sky through the night!
Ah, we safely may trust to its gleaming,
    And be sure it will lead us aright—
We safely may trust to a gleaming
    That cannot but guide us aright,   70
    Since it flickers up to Heaven through the night."

---

7. Phoenician fertility goddess, here described as a moon goddess; the horns are the ends of a new moon.
8. The chaste Roman goddess of the moon.
9. The constellation Leo.

1. Absolute peace, as if bathed in the oblivion-giving waters of Lethe.
2. The soul, imaged as a butterfly.
3. Now spelled "sibyllic"—mysterious prophetic.

Thus I pacified Psyche and kissed her,
    And tempted her out of her gloom—
    And conquered her scruples and gloom:
And we passed to the end of the vista,   75
    And were stopped by the door of a tomb—
    By the door of a legended tomb;
And I said—"What is written, sweet sister,
    On the door of this legended tomb?"
She replied—"Ulalume—Ulalume—   80
'Tis the vault of thy lost Ulalume!"

Then my heart it grew ashen and sober
    As the leaves that were crispéd and sere—
    As the leaves that were withering and sere,
And I cried—"It was surely October   85
    On *this* very night of last year
    That I journeyed—I journeyed down here—
    That I brought a dread burden down here—
    On this night of all nights in the year,
    Oh, what demon has tempted me here?   90
Well I know, now, this dim lake of Auber—
    This misty mid region of Weir—
Well I know, now, this dank tarn of Auber,
    In the ghoul-haunted woodland of Weir."

Said *we*, then—the two, then—"Ah, can it   95
    Have been that the woodlandish ghouls—
    The pitiful, the merciful ghouls—
To bar up our way and to ban it
    From the secret that lies in these wolds—
    From the thing that lies hidden in these wolds—   100
Had drawn up the spectre of a planet
    From the limbo of lunary souls—
This sinfully scintillant[4] planet
    From the Hell of the planetary souls?"

                          1847

Annabel Lee[1]

It was many and many a year ago,
    In a kingdom by the sea
That a maiden there lived whom you may know
    By the name of ANNABEL LEE;
And this maiden she lived with no other thought   5
    Than to love and be loved by me.

I was a child and *she* was a child,
    In this kingdom by the sea;

---

4. Sparkling, shining.
1. The text is that of the first printing, in Rufus Griswold's article in the New York *Tribune* (October 9, 1849), signed "Ludwig."

But we loved with a love that was more than love—
    I and my ANNABEL LEE—
With a love that the wingèd seraphs of heaven
      Coveted her and me.   10

And this was the reason that, long ago,
    In this kingdom by the sea,
A wind blew out of a cloud, chilling   15
    My beautiful ANNABEL LEE;
So that her highborn kinsmen came
    And bore her away from me,
To shut her up in a sepulchre
    In this kingdom by the sea.   20

The angels, not half so happy in heaven,
    Went envying her and me—
Yes!—that was the reason (as all men know,
    In this kingdom by the sea)
That the wind came out of the cloud by night,   25
    Chilling and killing my ANNABEL LEE.

But our love it was stronger by far than the love
    Of those who were older than we—
    Of many far wiser than we—
And neither the angels in heaven above,   30
    Nor the demons down under the sea,
Can ever dissever my soul from the soul
    Of the beautiful ANNABEL LEE:

For the moon never beams, without bringing me dreams
    Of the beautiful ANNABEL LEE;   35
And the stars never rise, but I feel the bright eyes
    Of the beautiful ANNABEL LEE:
And so, all the night tide, I lie down by the side
Of my darling—my darling—my life and my bride,
    In her sepulchre there by the sea—   40
    In her tomb by the sounding sea.

1849

The Fall of the House of Usher[1]

During the whole of a dull, dark, and soundless day in the
autumn of the year, when the clouds hung oppressively low in the
heavens, I had been passing alone, on horseback, through a singularly dreary tract of country; and at length found myself, as the
shades of the evening drew on, within view of the melancholy
House of Usher. I know not how it was—but, with the first glimpse
of the building, a sense of insufferable gloom pervaded my spirit. I
say insufferable; for the feeling was unrelieved by any of that half-

1. The text is that of the first publication   *Amerian Monthly Review*, 5 (September,
in *Burton's Gentleman's Magazine, and*   1839).

pleasurable, because poetic, sentiment, with which the mind usually receives even the sternest natural images of the desolate or terrible. I looked upon the scene before me—upon the mere house, and the simple landscape features of the domain—upon the bleak walls— upon the vacant eye-like windows—upon a few rank sedges—and upon a few white trunks of decayed trees—with an utter depression of soul which I can compare to no earthly sensation more properly than to the after-dream of the reveller upon opium—the bitter lapse into common life—the hideous dropping off of the veil. There was an iciness, a sinking, a sickening of the heart—an unredeemed dreariness of thought which no goading of the imagination could torture into aught of the sublime. What was it—I paused to think— what was it that so unnerved me in the contemplation of the House of Usher? It was a mystery all insoluble; nor could I grapple with the shadowy fancies that crowded upon me as I pondered. I was forced to fall back upon the unsatisfactory conclusion, that while, beyond doubt, there *are* combinations of very simple natural objects which have the power of thus affecting us, still the reason, and the analysis, of this power, lie among considerations beyond our depth. It was possible, I reflected, that a mere different arrangement of the particulars of the scene, of the details of the picture, would be sufficient to modify, or perhaps to annihilate its capacity for sorrowful impression; and, acting upon this idea, I reined my horse to the precipitous brink of a black and lurid tarn[2] that lay in unruffled lustre by the dwelling, and gazed down—but with a shudder even more thrilling than before—upon the re-modelled and inverted images of the gray sedge, and the ghastly tree-stems, and the vacant and eye-like windows.

Nevertheless, in this mansion of gloom I now proposed to myself a sojourn of some weeks. Its proprietor, Roderick Usher, had been one of my boon companions in boyhood; but many years had elapsed since our last meeting. A letter, however, had lately reached me in a distant part of the country—a letter from him—which, in its wildly importunate nature, had admitted of no other than a personal reply. The MS. gave evidence of nervous agitation. The writer spoke of acute bodily illness—of a pitiable mental idiosyncrasy which oppressed him—and of an earnest desire to see me, as his best, and indeed, his only personal friend, with a view of attempting, by the cheerfulness of my society, some alleviation of his malady. It was the manner in which all this, and much more, was said—it was the apparent *heart* that went with his request—which allowed me no room for hesitation—and I accordingly obeyed, what I still considered a very singular summons, forthwith.

Although, as boys, we had been even intimate associates, yet I really knew little of my friend. His reserve had been always excessive and habitual. I was aware, however, that his very ancient

2. A small lake, normally in the mountains.

family had been noted, time out of mind, for a peculiar sensibility of temperament, displaying itself, through long ages, in many works of exalted art, and manifested, of late, in repeated deeds of munificent yet unobtrusive charity, as well as in a passionate devotion to the intricacies, perhaps even more than to the orthodox and easily recognizable beauties, of musical science. I had learned, too, the very remarkable fact, that the stem of the Usher race, all time-honored as it was, had put forth, at no period, any enduring branch; in other words, that the entire family lay in the direct line of descent, and had always, with very trifling and very temporary variation, so lain. It was this deficiency, I considered, while running over in thought the perfect keeping of the character of the premises with the accredited character of the people, and while speculating upon the possible influence which the one, in the long lapse of centuries, might have exercised upon the other—it was this deficiency, perhaps, of collateral issue, and the consequent undeviating transmission, from sire to son, of the patrimony with the name, which had, at length, so identified the two as to merge the original title of the estate in the quaint and equivocal appellation of the "House of Usher"—an appellation which seemed to include, in the minds of the peasantry who used it, both the family and the family mansion.

I have said that the sole effect of my somewhat childish experiment, of looking down within the tarn, had been to deepen the first singular impression. There can be no doubt that the consciousness of the rapid increase of my superstition—for why should I not so term it?—served mainly to accelerate the increase itself. Such, I have long known, is the paradoxical law of all sentiments having terror as a basis. And it might have been for this reason only, that, when I again uplifted my eyes to the house itself, from its image in the pool, there grew in my mind a strange fancy—a fancy so ridiculous, indeed, that I but mention it to show the vivid force of the sensations which oppressed me. I had so worked upon my imagination as really to believe that around about the whole mansion and domain there hung an atmosphere peculiar to themselves and their immediate vicinity—an atmosphere which had no affinity with the air of heaven, but which had reeked up from the decayed trees, and the gray walls, and the silent tarn, in the form of an inelastic vapor or gas—dull, sluggish, faintly discernible, and leaden-hued. Shaking off from my spirit what *must* have been a dream, I scanned more narrowly the real aspect of the building. Its principal feature seemed to be that of an excessive antiquity. The discoloration of ages had been great. Minute fungi overspread the whole exterior, hanging in a fine tangled web-work from the eaves. Yet all this was apart from any extraordinary dilapidation. No portion of the masonry had fallen; and there appeared to be a wild inconsistency between its still perfect adaptation of parts, and the utterly porous, and evidently decayed condition of the individual stones. In this there was much that reminded me of the specious totality of old wood-work which has rotted for long years in some neglected vault, with no disturbance from the breath of the external air. Beyond this indication of extensive decay, however, the fabric gave little token of instability. Perhaps the eye of a scrutinizing observer might have discovered a barely perceptible fissure, which, extending from the roof of the building in front, made its way down the wall in a zigzag direction, until it became lost in the sullen waters of the tarn.

Noticing these things, I rode over a short causeway to the house. A servant in waiting took my horse, and I entered the Gothic archway of the hall. A valet, of stealthy step, thence conducted me, in silence, through many dark and intricate passages in my progress to the studio of his master. Much that I encountered on the way contributed, I know not how, to heighten the vague sentiments of which I have already spoken. While the objects around me— while the carvings of the ceilings, the sombre tapestries of the walls, the ebon blackness of the floors, and the phantasmagoric armorial trophies which rattled as I strode, were but matters to which, or to such as which, I had been accustomed from my infancy—while I hesitated not to acknowledge how familiar was all this—I still wondered to find how unfamiliar were the fancies which ordinary images were stirring up. On one of the staircases, I met the physician of the family. His countenance, I thought, wore a mingled expression of low cunning and perplexity. He accosted me with trepidation and passed on. The valet now threw open a door and ushered me into the presence of his master.

The room in which I found myself was very large and excessively lofty. The windows were long, narrow, and pointed, and at so vast a distance from the black oaken floor as to be altogether inaccessible from within. Feeble gleams of encrimsoned light made their way through the trelliced panes, and served to render sufficiently distinct the more prominent objects around; the eye, however, struggled in vain to reach the remoter angles of the chamber, or the recesses of the vaulted and fretted ceiling. Dark draperies hung upon the walls. The general furniture was profuse, comfortless, antique, and tattered. Many books and musical instruments lay scattered about, but failed to give any vitality to the scene. I felt that I breathed an atmosphere of sorrow. An air of stern, deep, and irredeemable gloom hung over and pervaded all.

Upon my entrance, Usher arose from a sofa upon which he had been lying at full length, and greeted me with a vivacious warmth which had much in it, I at first thought of an overdone cordiality— of the constrained effort of the ennuyé[3] man of the world. A glance, however, at his countenance convinced me of his perfect sincerity. We sat down; and for some moments, while he spoke not,

3. Bored.

I gazed upon him with a feeling half of pity, half of awe. Surely, man had never before so terribly altered, in so brief a period, as had Roderick Usher! It was with difficulty that I could bring myself to admit the identity of the wan being before me with the companion of my early boyhood. Yet the character of his face had been at all times remarkable. A cadaverousness of complexion; an eye large, liquid, and luminous beyond comparison; lips somewhat thin and very pallid, but of a surpassingly beautiful curve; a nose of a delicate Hebrew model, but with a breadth of nostril unusual in similar formations; a finely moulded chin, speaking, in its want of prominence, of a want of moral energy; hair of a more than web-like softness and tenuity; these features, with an inordinate expansion above the regions of the temple, made up altogether a countenance not easily to be forgotten. And now in the mere exaggeration of the prevailing character of these features, and of the expression they were wont to convey, lay so much of change that I doubted to whom I spoke. The now ghastly pallor of the skin, and the now miraculous lustre of the eye, above all things startled and even awed me. The silken hair, too, had been suffered to grow all unheeded, and as, in its wild gossamer texture, it floated rather than fell about the face, I could not, even with effort, connect its arabesque expression with any idea of simple humanity.

In the manner of my friend I was at once struck with an incoherence—an inconsistency; and I soon found this to arise from a series of feeble and futile struggles to overcome an habitual trepidancy, an excessive nervous agitation. For something of this nature I had indeed been prepared, no less by his letter, than by reminiscences of certain boyish traits, and by conclusions deduced from his peculiar physical conformation and temperament. His action was alternately vivacious and sullen. His voice varied rapidly from a tremulous indecision (when the animal spirits seemed utterly in abeyance) to that species of energetic concision—that abrupt, weighty, unhurried, and hollow-sounding enunciation—that leaden, self-balanced and perfectly modulated guttural utterance, which may be observed in the moments of the intensest excitement of the lost drunkard, or the irreclaimable eater of opium.

It was thus that he spoke of the object of my visit, of his earnest desire to see me, and of the solace he expected me to afford him. He entered, at some length, into what he conceived to be the nature of his malady. It was, he said, a constitutional and a family evil, and one for which he despaired to find a remedy—a mere nervous affection, he immediately added, which would undoubtedly soon pass off. It displayed itself in a host of unnatural sensations. Some of these, as he detailed them, interested and bewildered me— although, perhaps, the terms, and the general manner of the narration had their weight. He suffered much from a morbid acuteness of the senses; the most insipid food was alone endurable; he could wear only garments of certain texture; the odors of all flowers were oppressive; his eyes were tortured by even a faint light; and there were but peculiar sounds, and these from stringed instruments, which did not inspire him with horror.

To an anomalous species of terror I found him a bounden slave. "I shall perish," said he, "I *must* perish in this deplorable folly. Thus, thus, and not otherwise, shall I be lost. I dread the events of the future, not in themselves, but in their results. I shudder at the thought of any, even the most trivial, incident, which may operate upon this intolerable agitation of soul. I have, indeed, no abhorrence of danger, except in its absolute effect—in terror. In this unnerved—in this pitiable condition—I feel that I must inevitably abandon life and reason together in my struggles with some fatal demon of fear."

I learned, moreover, at intervals, and through broken and equivocal hints, another singular feature of his mental condition. He was enchained by certain superstitious impressions in regard to the dwelling which he tenanted, and from which, for many years, he had never ventured forth—in regard to an influence whose supposititious force was conveyed in terms too shadowy here to be restated —an influence which some peculiarities in the mere form and substance of his family mansion, had, by dint of long sufferance, he said, obtained over his spirit—an effect which the *physique* of the gray walls and turrets, and of the dim tarn into which they all looked down, had, at length, brought about upon the *morale* of his existence.

He admitted, however, although with hesitation, that much of the peculiar gloom which thus afflicted him could be traced to a more natural and far more palpable origin—to the severe and long-continued illness—indeed to the evidently approaching dissolution —of a tenderly beloved sister; his sole companion for long years— his last and only relative on earth. "Her decease," he said, with a bitterness which I can never forget, "would leave him (him the hopeless and the frail) the last of the ancient race of the Ushers." As he spoke, the lady Madeline (for so was she called) passed slowly through a remote portion of the apartment, and, without having noticed my presence, disappeared. I regarded her with an utter astonishment not unmingled with dread. Her figure, her air, her features—all, in their very minutest development were those— were identically (I can use no other sufficient term) were identically those of the Roderick Usher who sat beside me. A feeling of stupor oppressed me, as my eyes followed her retreating steps. As a door, at length, closed upon her exit, my glance sought instinctively and eagerly the countenance of the brother—but he had buried his face in his hands, and I could only perceive that a far more than ordinary wanness had overspread the emaciated fingers through which trickled many passionate tears.

The disease of the lady Madeline had long baffled the skill of her physicians. A settled apathy, a gradual wasting away of the person, and frequent although transient affections of a partially cataleptical character, were the unusual diagnosis. Hitherto she had steadily borne up against the pressure of her malady, and had not betaken herself finally to bed; but, on the closing in of the evening of my arrival at the house, she succumbed, as her brother told me at night with inexpressible agitation, to the prostrating power of the destroyer—and I learned that the glimpse I had obtained of her person would thus probably be the last I should obtain—that the lady, at least while living, would be seen by me no more.

For several days ensuing, her name was unmentioned by either Usher or myself; and, during this period, I was busied in earnest endeavors to alleviate the melancholy of my friend. We painted and read together—or I listened, as if in a dream, to the wild improvisations of his speaking guitar. And thus, as a closer and still closer intimacy admitted me more unreservedly into the recesses of his spirit, the more bitterly did I perceive the futility of all attempt at cheering a mind from which darkness, as if an inherent positive quality, poured forth upon all objects of the moral and physical universe, in one unceasing radiation of gloom.

I shall ever bear about me, as Moslem in their shrouds at Mecca; a memory of the many solemn hours I thus spent alone with the master of the House of Usher. Yet I should fail in any attempt to convey an idea of the exact character of the studies, or of the occupations, in which he involved me, or led me the way. An excited and highly distempered ideality threw a sulphurous lustre over all. His long improvised dirges will ring for ever in my ears. Among other things, I bear painfully in mind a certain singular perversion and amplification of the wild air of the last waltz of Von Weber.[4] From the paintings over which his elaborate fancy brooded, and which grew, touch by touch, into vaguenesses at which I shuddered the more thrillingly, because I shuddered knowing not, why from these paintings (vivid as their images now are before me) I would in vain endeavor to educe more than a small portion which should lie within the compass of merely written words. By the utter simplicity, by the nakedness, of his designs, he arrested and over-awed attention. If ever mortal painted an idea, that mortal was Roderick Usher. For me at least—in the circumstancs then surrounding me—there arose out of the pure abstractions which the hypochondriac contrived to throw upon his canvas, an intensity of intolerable awe, no shadow of which felt I ever yet in the contemplation of the certainly glowing yet too concrete reveries of Fuseli.[5]

One of the phantasmagoric conceptions of my friend, partaking not so rigidly of the spirit of abstraction, may be shadowed forth,

4. Karl Maria von Weber (1786–1826) established Romanticism in German opera; *The Last Waltz of Von Weber* was composed by Karl Gottlieb Reissiger (1798–1859).

5. Henry Fuseli (1741–1825), Swiss painter who made his reputation in London; noted for his interest in the supernatural.

although feebly, in words. A small picture presented the interior of an immensely long and rectangular vault or tunnel, with low walls, smooth, white, and without interruption or device. Certain accessory points of the design served well to convey the idea that this excavation lay at an exceeding depth below the surface of the earth. No outlet was observed in any portion of its vast extent, and no torch, or other artificial source of light was discernible—yet a flood of intense rays rolled throughout, and bathed the whole in a ghastly and inappropriate splendor.

I have just spoken of that morbid condition of the auditory nerve which rendered all music intolerable to the sufferer, with the exception of certain effects of stringed instruments. It was, perhaps, the narrow limits to which he thus confined himself upon the guitar, which gave birth, in great measure, to the fantastic character of his performances. But the fervid *facility* of his impromptus could not be so accounted for. They must have been, and were, in the notes, as well as in the words of his wild fantasies, (for he not unfrequently accompanied himself with rhymed verbal improvisations,) the result of that intense mental collectedness and concentration to which I have previously alluded as observable only in particular moments of the highest artificial excitement. The words of one of these rhapsodies I have easily borne away in memory. I was, perhaps, the more forcibly impressed with it, as he gave it, because, in the under or mystic current of its meaning, I fancied that I perceived, and for the first time, a full consciousness on the part of Usher, of the tottering of his lofty reason upon her throne. The verses, which were entitled "The Haunted Palace," ran very nearly, if not accurately, thus:[6]

I

In the greenest of our valleys,
　By good angels tenanted,
Once a fair and stately palace—
　Snow-white palace—reared its head.
In the monarch Thought's dominion—
　It stood there!
Never seraph spread a pinion
　Over fabric half so fair.

II

Banners yellow, glorious, golden,
　On its roof did float and flow;
(This—all this—was in the olden
　Time long ago)
And every gentle air that dallied,
　In that sweet day,
Along the ramparts plumed and pallid,
　A winged odor went away.

6. In the original printing this note appeared at the end of the story: "The ballad of 'The Haunted Palace,' introduced in this tale, was published separately, some months ago, in the Baltimore 'Museum.' "

III

Wanderers in that happy valley
    Through two luminous windows saw
Spirits moving musically
    To a lute's well-tunéd law,
Round about a throne, where sitting
    (Porphyrogene[7]!)
In state his glory well befitting,
    The sovereign of the realm was seen.

IV

And all with pearl and ruby glowing
    Was the fair palace door,
Through which came flowing, flowing, flowing,
    And sparkling evermore,
A troop of Echoes whose sole duty
    Was but to sing,
In voices of surpassing beauty,
    The wit and wisdom of their king.

V

But evil things, in robes of sorrow,
    Assailed the monarch's high estate;
(Ah, let us mourn, for never morrow
    Shall dawn upon him, desolate!)
And, round about his home, the glory
    That blushed and bloomed
Is but a dim-remembered story
    Of the old time entombed.

VI

And travellers now within that valley,
    Through the red-litten windows, see
Vast forms that move fantastically
    To a discordant melody;
While, like a rapid ghastly river,
    Through the pale door,
A hideous throng rush out forever,
    And laugh—but smile no more.

I well remember that suggestions arising from this ballad led us into a train of thought wherein there became manifest an opinion of Usher's which I mention not so much on account of its novelty, (for other men have thought thus,) as on account of the pertinacity with which he maintained it. This opinion, in its general form, was that of the sentience of all vegetable things. But, in his disordered fancy, the idea had assumed a more daring character, and trespassed, under certain conditions, upon the kingdom of inorganiza-

---

7. Born to the purple, of royal birth.

tion. I lack words to express the full extent, or the earnest *abandon* of his persuasion. The belief, however, was connected (as I have previously hinted) with the gray stones of the home of his forefathers. The condition of the sentience had been here, he imagined, fulfilled in the method of collocation of these stones—in the order of their arrangement, as well as in that of the many fungi which overspread them, and of the decayed trees which stood around—above all, in the long undisturbed endurance of this arrangement, and in its reduplication in the still waters of the tarn. Its evidence—the evidence of the sentience—was to be seen, he said, (and I here started as he spoke,) in *the gradual yet certain condensation of an atmosphere of their own about the waters and the walls*. The result was discoverable, he added, in that silent, yet importunate and terrible influence which for centuries had moulded the destinies of his family, and which made *him* what I now saw him—what he was. Such opinions need no comment, and I will make none.

Our books—the books which, for years, had formed no small portion of the mental existence of the invalid—were, as might be supposed, in strict keeping with this character of phantasm. We pored together over such works as the Ververt et Chartreuse of Gresset; the Belphegor of Machiavelli; the Selenography of Brewster; the Heaven and Hell of Swedenborg; the Subterranean Voyage of Nicholas Klimm de Holberg; the Chiromancy of Robert Flud, of Jean d'Indaginé, and of De la Chambre; the Journey into the Blue Distance of Tieck; and the City of the Sun of Campanella. One favorite volume was a small octavo edition of the Directorium Inquisitorium, by the Dominican Eymeric de Gironne; and there were passages in Pomponius Mela, about the old African Satyrs and Ægipans, over which Usher would sit dreaming for hours. His chief delight, however, was found in the earnest and repeated perusal of an exceedingly rare and curious book in quarto Gothic—the manual of a forgotten church—the *Vigilae Mortuorum secundum Chorum Ecclesiae Maguntinae*.[8]

---

8. The titles are real, although the way they sound in the narrator's inventory is at least as important as their precise contents. Jean Baptiste Gresset (1709–77) wrote the anticlerical *Vairvert* and *Ma Chartreuse*. In *Belphegor*, by Niccolò Machiavelli (1469–1527), a demon comes to earth to prove that women damn men to Hell. Emanuel Swedenborg (1688–1772), Swedish scientist and mystic, presents a fantastically precise anatomy of living conditions in Heaven and Hell, seeing the two places as mutually attractive opposites. Ludwig Holberg (1684–1754), Danish dramatist and historian, deals with a voyage to the land of death and back. Robert Flud (1574–1637), English physician and noted Rosicrucian (the Rosicrucians then being a new organization of esoteric philosophy and theology which purported to be based on ancient lore from the Middle East), and two Frenchmen, Jean D'Indaginé (fl. early 16th century) and Maria Cireau de la Chambre (1594–1669), all wrote on chiromancy (palm reading). The German Ludwig Tieck (1773–1853) wrote *Das Alte Buch; oder Reise ins Blaue hinein*, which deals with a journey to another world. *The City of the Sun* by the Italian Tommaso Campanella (1568–1639) is a famous utopian work. Nicholas Eymeric de Gerône, who was Inquisitor-General for Castile in 1356, recorded procedures for torturing heretics. Pomponius Mela was a first-century Roman whose widely used book on geography (printed in Italy in 1471) described strange beasts ("oegipans" are African goat-men). A book called *The Vigils of the Dead, According to the Church-Choir of Mayence* was printed at Basel around 1500.

I could not help thinking of the wild ritual of this work, and of its probable influence upon the hypochondriac, when, one evening, having informed me abruptly that the lady Madeline was no more, he stated his intention of preserving her corpse for a fortnight, previously to its final interment, in one of the numerous vaults within the main walls of the building. The wordly reason, however, assigned for this singular proceeding, was one which I did not feel at liberty to dispute. The brother had been led to his resolution (so he told me) by considerations of the unusual character of the malady of the deceased, of certain obtrusive and eager inquiries on the part of her medical men, and of the remote and exposed situation of the burial ground of the family. I will not deny that when I called to mind the sinister countenance of the person whom I met upon the staircase, on the day of my arrival at the house, I had no desire to oppose what I regarded as at best but a harmless, and not by any means an unnatural precaution.[9]

At the request of Usher, I personally aided him in the arrangements for the temporary entombment. The body having been encoffined, we two alone bore it to its rest. The vault in which we placed it (and which had been so long unopened that our torches, half smothered in its oppressive atmosphere, gave us little opportunity for investigation) was small, damp, and utterly without means of admission for light; lying, at great depth, immediately beneath that portion of the building in which was my own sleeping apartment. It had been used, apparently, in remote feudal times, for the worst purposes of a donjon-keep, and, in later days, as a place of deposit for powder, or other highly combustible substance, as a portion of its floor, and the whole interior of a long archway through which we reached it, were carefully sheathed with copper. The door, of massive iron, had been, also, similarly protected. Its immense weight caused an unusually sharp grating sound, as it moved upon its hinges.

Having deposited our mournful burden upon tressels within this region of horror, we partially turned aside the yet unscrewed lid of the coffin, and looked upon the face of the tenant. The exact similitude between the brother and sister even here again startled and confounded me. Usher, divining, perhaps, my thoughts, murmured out some few words from which I learned that the deceased and himself had been twins, and that sympathies of a scarcely intelligible nature had always existed between them. Our glances, however, rested not long upon the dead—for we could not regard her unawed. The disease which had thus entombed the lady in the maturity of youth, had left, as usual in all maladies of a strictly cataleptical character, the mockery of a faint blush upon the bosom

---

9. The shortage of corpses for dissection had led to the new profession of resurrection men, who dug up fresh corpses and sold them to medical students and surgeons.

and the face, and that suspiciously lingering smile upon the lip which is so terrible in death. We replaced and screwed down the lid, and, having secured the door of iron, made our way, with toil, into the scarcely less gloomy apartments of the upper portion of the house.

And now, some days of bitter grief having elapsed, an observable change came over the features of the mental disorder of my friend. His ordinary manner had vanished. His ordinary occupations were neglected or forgotten. He roamed from chamber to chamber with hurried, unequal, and objectless step. The pallor of his countenance had assumed, if possible, a more ghastly hue—but the luminousness of his eye had utterly gone out. The once occasional huskiness of his tone was heard no more; and a tremulous quaver, as if of extreme terror, habitually characterized his utterance.—There were times, indeed, when I thought his unceasingly agitated mind was laboring with an oppressive secret, to divulge which he struggled for the necessary courage. At times, again, I was obliged to resolve all into the mere inexplicable vagaries of madness, as I beheld him gazing upon vacancy for long hours, in an attitude of the profoundest attention, as if listening to some imaginary sound. It was no wonder that his condition terrified—that it infected me. I felt creeping upon me, by slow yet certain degrees, the wild influences of his own fantastic yet impressive superstitions.

It was, most especially, upon retiring to bed late in the night of the seventh or eighth day after the entombment of the lady Madeline, that I experienced the full power of such feelings. Sleep came not near my couch—while the hours waned and waned away. I struggled to reason off the nervousness which had dominion over me. I endeavored to believe that much, if not all of what I felt, was due to phantasmagoric influence of the gloomy furniture of the room—of the dark and tattered draperies, which, tortured into motion by the breath of a rising tempest, swayed fitfully to and fro upon the walls, and rustled uneasily about the decorations of the bed. But my efforts were fruitless. An irrepressible tremor gradually pervaded my frame; and, at length, there sat upon my very heart an incubus[1] of utterly causeless alarm. Shaking this off with a gasp and a struggle, I uplifted myself upon the pillows, and, peering earnestly within the intense darkness of the chamber, harkened—I know not why, except that an instinctive spirit prompted me—to certain low and indefinite sounds which came, through the pauses of the storm, at long intervals, I knew not whence. Overpowered by an intense sentiment of horror, unaccountable yet unendurable, I threw on my clothes with haste, for I felt that I should sleep no more during the night, and endeavored to arouse myself from the pitiable condition into which I had fallen, by pacing rapidly to and fro through the apartment.

1. An evil spirit supposed to lie upon people in their sleep.

I had taken but few turns in this manner, when a light step on an adjoining staircase arrested my attention. I presently recognized it as that of Usher. In an instant afterwards he rapped, with a gentle touch, at my door, and entered, bearing a lamp. His countenance was, as usual, cadaverously wan—but there was a species of mad hilarity in his eyes—an evidently restrained hysteria in his whole demeanor. His air appalled me—but any thing was preferable to the solitude which I had so long endured, and I even welcomed his presence as a relief.

"And you have not seen it?" he said abruptly, after having stared about him for some moments in silence—"you have not then seen it?—but, stay! you shall." Thus speaking, and having carefully shaded his lamp, he hurried to one of the gigantic casements, and threw it freely open to the storm.

The impetuous fury of the entering gust nearly lifted us from our feet. It was, indeed, a tempestuous yet sternly beautiful night, and one wildly singular in its terror and its beauty. A whirlwind had apparently collected its force in our vicinity; for there were frequent and violent alterations in the direction of the wind; and the exceeding density of the clouds (which hung so low as to press upon the turrents of the house) did not prevent our perceiving the life-like velocity with which they flew careering from all points against each other, without passing away into the distance. I say that even their exceeding density did not prevent our perceiving this—yet we had no glimpse of the moon or stars—nor was there any flashing forth of the lightning. But the under surfaces of the huge masses of agitated vapor, as well as all terrestrial objects immediately around us, were glowing in the unnatural light of a faintly luminous and distinctly visible gaseous exhalation which hung about and enshrouded the mansion.

"You must not—you shall not behold this!" said I, shudderingly, to Usher, as I led him, with a gentle violence, from the window to a seat. "These appearances, which bewilder you, are merely electrical phenomena not uncommon—or it may be that they have their ghastly origin in the rank miasma of the tarn. Let us close this casement—the air is chilling and dangerous to your frame. Here is one of your favorite romances. I will read, and you shall listen— and so we will pass away this terrible night together."

The antique volume which I had taken up was the "Mad Trist" of Sir Launcelot Canning[2]—but I had called it a favorite of Usher's more in sad jest than in earnest; for, in truth, there is little in its uncouth and unimaginative prolixity which could have had interest for the lofty and spiritual ideality of my friend. It was, however, the only book immediately at hand; and I indulged a vague hope that the excitement which now agitated the hypochondriac might find

2. Not a real book. "Trist" here means simply meeting, or prearranged or fated encounter, not the lovers' meeting implied in the modern use of "tryst."

relief (for the history of mental disorder is full of similar anomalies) even in the extremeness of the folly which I should read. Could I have judged, indeed, by the wild, overstrained air of vivacity with which he harkened, or apparently harkened, to the words of the tale, I might have well congratulated myself upon the success of my design.

I had arrived at that well-known portion of the story where Ethelred, the hero of the Trist, having sought in vain for peaceable admission into the dwelling of the hermit, proceeds to make good an entrance by force. Here, it will be remembered, the words of the narrative run thus—

"And Ethelred, who was by nature of a doughty heart, and who was now mighty withal, on account of the powerfulness of the wine which he had drunken, waited no longer to hold parley with the hermit, who, in sooth, was of an obstinate and maliceful turn, but, feeling the rain upon his shoulders, and fearing the rising of the tempest, uplifted his mace outright, and, with blows, made quickly room in the plankings of the door for his gauntleted hand, and now pulling therewith sturdily, he so cracked, and ripped, and tore all asunder, that the noise of the dry and hollow-sounding wood alarummed and reverberated throughout the forest."

At the termination of this sentence I started, and, for a moment, paused; for it appeared to me (although I at once concluded that my excited fancy had deceived me)—it appeared to me that, from some very remote portion of the mansion or of its vicinity, there came, indistinctly, to my ears, what might have been, in its exact similarity of character, the echo (but a stifled and dull one certainly) of the very cracking and ripping sound which Sir Launcelot had so particularly described. It was, beyond doubt, the coincidence alone which had arrested my attention; for, amid the rattling of the sashes of the casements, and the ordinary commingled noises of the still increasing storm, the sound, in itself, had nothing, surely, which should have interested or disturbed me. I continued the story.

"But the good champion Ethelred, now entering within the door, was sore enraged and amazed to perceive no signal of the maliceful hermit; but, in the stead thereof, a dragon of scaly and prodigious demeanor, and of a fiery tongue, which sate in guard before a palace of gold, with a floor of silver; and upon the wall there hung a shield of shining brass with this legend enwritten—

> Who entereth herein, a conqueror hath bin,
> Who slayeth the dragon, the shield he shall win.

And Ethelred uplifted his mace, and struck upon the head of the dragon, which fell before him, and gave up his pesty breath, with a shriek so horrid and harsh, and withal so piercing, that Ethelred had fain to close his ears with his hands against the dreadful noise of it, the like whereof was never before heard."

Here again I paused abruptly, and now with a feeling of wild amazement—for there could be no doubt whatever that, in this instance, I did actually hear (although from what direction it proceeded I found it impossible to say) a low and apparently distant, but harsh, protracted, and most unusual screaming or grating sound —the exact counterpart of what my fancy had already conjured up as the sound of the dragon's unnatural shriek as described by the romancer.

Oppressed, as I certainly was, upon the occurrence of this second and most extraordinary coincidence, by a thousand conflicting sensations, in which wonder and extreme terror were predominant, I still retained sufficient presence of mind to avoid exciting, by any observation, the sensitive nervousness of my companion. I was by no means certain that he had noticed the sounds in question; although, assuredly, a strange alteration had, during the last few minutes, taken place in his demeanor. From a position fronting my own, he had gradually brought round his chair, so as to sit with his face to the door of the chamber, and thus I could but partially perceive his features, although I saw that his lips trembled as if he were murmuring inaudibly. His head had dropped upon his breast—yet I knew that he was not asleep, from the wide and rigid opening of the eye, as I caught a glance of it in profile. The motion of his body, too, was at variance with his idea—for he rocked from side to side with a gentle yet constant and uniform sway. Having rapidly taken notice of all this, I resumed the narrative of Sir Launcelot, which thus proceeded:—

"And now, the champion, having escaped from the terrible fury of the dragon, bethinking himself of the brazen shield, and of the breaking up of the enchantment which was upon it, removed the carcass from out of the way before him, and approached valorously over the silver pavement of the castle to where the shield was upon the wall; which in sooth tarried not for his full coming, but fell down at his feet upon the silver floor, with a mighty great and terrible ringing sound."

No sooner had these syllables passed my lips, than—as if a shield of brass had indeed, at the moment, fallen heavily upon a floor of silver—I became aware of a distinct, hollow, metallic, and clangorous, yet apparently muffled reverberation. Completely unnerved, I started convulsively to my feet, but the measured rocking movement of Usher was undisturbed. I rushed to the chair in which he sat. His eyes were bent fixedly before him, and throughout his whole countenance there reigned a more than stony rigidity. But, as I laid my hand upon his shoulder, there came a strong shudder over his frame; a sickly smile quivered about his lips; and I saw that he spoke in a low, hurried, and gibbering murmur, as if unconscious of my presence. Bending closely over his person, I at length drank in the hideous import of his words.

"Not hear it?—yes, I hear it, and *have* heard it. Long—long—long—many minutes, many hours, many days, have I heard it—yet I dared not—oh, pity me, miserable wretch that I am!—I dared not—*I dared* not speak! *We have put her living in the tomb!* Said I not that my senses were acute?—I *now* tell you that I heard her first feeble movements in the hollow coffin. I heard them—many, many days ago—yet I dared not—*I dared not speak!* And now—to-night—Ethelred—ha! ha!—the breaking of the hermit's door, and the death-cry of the dragon, and the clangor of the shield—say, rather, the rending of the coffin, and the grating of the iron hinges, and her struggles within the coppered archway of the vault! Oh whither shall I fly? Will she not be here anon? Is she not hurrying to upbraid me for my haste? Have I not heard her footsteps on the stair? Do I not distinguish that heavy and horrible beating of her heart? Madman!"—here he sprung violently to his feet, and shrieked out his syllables, as if in the effort he were giving up his soul—"Madman! *I tell you that she now stands without the door!*"

As if in the superhuman energy of his utterance there had been found the potency of a spell—the huge antique pannels to which the speaker pointed, threw slowly back, upon the instant, their ponderous and ebony jaws. It was the work of the rushing gust—but then without those doors there *did* stand the lofty and enshrouded figure of the lady Madeline of Usher. There was blood upon her white robes, and the evidence of some bitter struggle upon every portion of her emaciated frame. For a moment she remained trembling and reeling to and fro upon the threshold—then, with a low moaning cry, fell heavily inward upon the person of her brother, and in her horrible and now final death-agonies, bore him to the floor a corpse, and a victim to the terrors he had dreaded.

From that chamber, and from that mansion, I fled aghast. The storm was still abroad in all its wrath as I found myself crossing the old causeway. Suddenly there shot along the path a wild light, and I turned to see whence a gleam so unusual could have issued—for the vast house and its shadows were alone behind me. The radiance was that of the full, setting, and blood-red moon, which now shone vividly through that once barely-discernible fissure, of which I have before spoken, as extending from the roof of the building, in a zig-zag direction, to the base. While I gazed, this fissure rapidly widened—there came a fierce breath of the whirlwind—the entire orb of the satellite burst at once upon my sight—my brain reeled as I saw the mighty walls rushing asunder—there was a long tumultuous shouting sound like the voice of a thousand waters—and the deep and dank tarn at my feet closed sullenly and silently over the fragments of the "*House of Usher.*"

1839

William Wilson. A Tale[1]

What say of it? what say of conscience *grim,*
That spectre in my path?
—CHAMBERLAINE'S PHARRONIDA[2]

Let me call myself, for the present, William Wilson. The fair page now lying before me need not be sullied with my real appellation. This has been already too much an object for the scorn, for the horror, for the detestation of my race. To the uttermost regions of the globe have not the indignant winds bruited its unparalleled infamy? oh, outcast of all outcasts most abandoned! To the earth art thou not for ever dead? to its honours, to its flowers, to its golden aspirations? and a cloud, dense, dismal, and limitless, does it not hang eternally between thy hopes and heaven?

I would not, if I could, here or to-day, embody a record of my later years of unspeakable misery, and unpardonable crime. This epoch—these later years—took unto themselves a sudden elevation in turpitude, whose origin alone it is my present purpose to assign. Men usually grow base by degrees. From me, in an instant, all virtue dropped bodily as a mantle. I shrouded my nakedness in triple guilt. From comparatively trivial wickedness I passed, with the stride of a giant, into more than the enormities of an Elah-Gabalus.[3] What chance, what one event brought this evil thing to pass, bear with me while I relate. Death approaches; and the shadow which fore-runs him has thrown a softening influence over my spirit. I long, in passing through the dim valley, for the sympathy—I had nearly said for the pity—of my fellow-men. I would fain have them believe that I have been, in some measure, the slave of circumstances beyond human control. I would wish them to seek out for me, in the details I am about to give, some little oasis of *fatality* amid a wilderness of error. I would have them allow—what they cannot refrain from allowing—that, although temptation may have erewhile existed as great, man was never *thus*, at least, tempted before—certainly, never *thus* fell. And therefore has he never thus suffered. Have I not indeed been living in a dream? And am I not now dying a victim to the horror and the mystery of the wildest of all sublunary visions?

I am come of a race whose imaginative and easily excitable temperament has at all times rendered them remarkable; and, in my earliest infancy, I gave evidence of having fully inherited the family

---

1. This tale is reprinted from its first appearance in the Philadelphia annual, *The Gift*, dated 1840 but published in September, 1839. Into it Poe worked some memories of the school he attended at Stoke-Newington (Bransby was the name of his own principal there). January 19 is Poe's own birthday as well as William Wilson's, and in different printings of the story Wilson's birthyear appears as 1809, 1811, and 1813, the latter two dates being those Poe also used in autobiographical accounts to make him even more precocious than he was.
2. The epigraph is not in this 1659 poem by William Chamberlayne.
3. Elagabalus (b. 204), boy Emperor of Rome (218–222), murdered by his imperial guards. Among the "enormities" were the imposition of the worship of Baal, the Semitic fertility god, and the favor displayed toward handsome homosexual boys.

character. As I advanced in years it was more strongly developed; becoming, for many reasons, a cause of serious disquietude to my friends, and of positive injury to myself. I grew self-willed, addicted to the wildest caprices, and a prey to the most ungovernable passions. Weak-minded, and beset with constitutional infirmities akin to my own, my parents could do but little to check the evil propensities which distinguished me. Some feeble and ill-directed efforts resulted in complete failure on their part, and of course, in total triumph on mine. Thenceforward my voice was a household law; and at an age when few children have abandoned their leading-strings, I was left to the guidance of my own will, and became, in all but name, the master of my own actions.

My earliest recollections of a school-life are connected with a large, rambling, cottage-built, and somewhat decayed building in a misty-looking village of England, where were a vast number of gigantic and gnarled trees, and where all the houses were excessively ancient and inordinately tall. In truth, it was a dream-like and spirit-soothing place, that venerable old town. At this moment, in fancy, I feel the refreshing chilliness of its deeply-shadowed avenues, inhale the fragrance of its thousand shrubberies, and thrill anew with undefinable delight, at the deep, hollow note of the church-bell, breaking each hour, with sullen and sudden roar, upon the stillness of the dusky atmosphere in which the old, fretted, Gothic steeple lay imbedded and asleep.

It gives me, perhaps, as much of pleasure as I can now in any manner experience, to dwell upon minute recollections of the school and its concerns. Steeped in misery as I am—misery, alas! only too real—I shall be pardoned for seeking relief, however slight and temporary, in the weakness of a few rambling details. These, moreover, utterly trivial, and even ridiculous in themselves, assume, to my fancy, adventitious importance as connected with a period and a locality, when and where I recognise the first ambiguous monitions of the destiny which afterwards so fully overshadowed me. Let me then remember.

The house, I have said, was old, irregular, and cottage-built. The grounds were extensive, and an enormously high and solid brick wall, topped with a bed of mortar and broken glass, encompassed the whole. This prison-like rampart formed the limit of our domain; beyond it we saw but thrice a week—once every Saturday afternoon, when, attended by two ushers,[4] we were permitted to take brief walks in a body through some of the neighbouring fields—and twice during Sunday, when we were paraded in the same formal manner to the morning and evening service in the one church of the village. Of this church the principal of our school was pastor. With how deep a spirit of wonder and perplexity was I wont to regard him from our remote pew in the gallery, as, with step solemn and

4. Assistant schoolmasters.

ᵃe

I

slow he ascended the pulpit! This reverend man, with countenance so demurely benign, with robes so glossy and so clerically flowing, with wig so minutely powdered, so rigid and so vast—could this be he who of late, with sour visage, and in snuffy habiliments, administered, ferule in hand, the Draconian[5] laws of the academy? Oh, gigantic paradox too utterly monstrous for solution!

At an angle of the ponderous wall frowned a more ponderous gate. It was riveted and studded with iron bolts, and surmounted with jagged iron spikes. What impressions of deep awe it inspired! It was never opened save for the three periodical egressions and ingressions already mentioned; then, in every creak of its mighty hinges we found a plenitude of mystery, a world of matter for solemn remark, or for far more solemn meditation.

The extensive enclosure was irregular in form, having many capacious recesses. Of these, three or four of the largest constituted the play-ground. It was level, and covered with fine hard gravel. I well remember it had no trees, nor benches, nor any thing similar within it. Of course it was in the rear of the house. In front lay a small parterre, planted with box and other shrubs; but through this sacred division we passed only upon rare occasions indeed, such as a first advent or final departure from school, or perhaps, when a parent or friend having called for us, we joyfully took our way home for the Christmas or Mid-summer holydays.

But the house—how quaint an old building was this!—to me how veritably a palace of enchantment! There was really no end to its windings, to its incomprehensible sub-divisions. It was impossible, at any given time, to say with certainty upon which of its two stories one happened to be. From each room to every other there were sure to be found three or four steps either in ascent or descent. Then the lateral branches were innumerable—inconceivable, and so returning in upon themselves, that our most exact ideas in regard to the whole mansion were not very far different from those with which we pondered upon infinity. During the five years of my residence here I was never able to ascertain with precision, in what remote locality lay the little sleeping apartment assigned to myself and some eighteen or twenty other scholars.

The school-room was the largest in the house—I could not help thinking, in the world. It was very long, narrow, and dismally low, with pointed Gothic windows and a ceiling of oak. In a remote and terror-inspiring angle was a square enclosure of eight or ten feet, comprising the sanctum, "during hours," of our principal, the Reverend Dr. Bransby. It was a solid structure, with massy door, sooner than open which in the absence of "the Dominie,"[6] we would all have willingly perished by the *peine forte et dure*.[7] In other angles

5. Merciless, from Draco, Athenian lawgiver, whose code (621? B.C.) set death as the penalty for numerous crimes.
6. "Minister" or "schoolteacher" (Brans-
by was both).
7. Pressing to death, as with large flat rocks.

were two other similar boxes, far less reverenced, indeed, but still greatly matters of awe. One of these was the pulpit of "the classical" usher, one of the "English and mathematical." Intesrpersed about the room, crossing and recrossing in endless irregularity, were in-numerable benches and desks, black, ancient, and time-worn, piled desperately with much-bethumbed books, and so beseamed with initial letters, names at full length, meaningless gashes, grotesque figures, and other multiplied efforts of the knife, as to have utterly lost what little of original form might have been their portion in days long departed. A huge bucket with water stood at one extrem-ity of the room, and a clock of stupendous dimensions at the other.

Encompassed by the massy walls of this venerable academy I passed, yet not in tedium or disgust, the years of the third lustrum[8] of my life. The teeming brain of childhood requires no external world of incident to occupy or amuse it, and the apparently dismal monotony of a school, was replete with more intense excitement than my riper youth has derived from luxury, or my full manhood from crime. Yet I must believe that my first mental developement had in it much of the uncommon, even much of the *outré*.[9] Upon mankind at large the events of very early existence rarely leave in mature age any definite impression. All is gray shadow—a weak and irregular remembrance—an indistinct regathering of feeble plea-sures and phantasmagoric pains. With me this is not so. In child-hood I must have felt with the energy of a man what I now find stamped upon memory in lines as vivid, as deep, and as durable as the exergues of the Carthaginian medals.[1]

Yet in fact—in the fact of the world's view—how little was there to remember! The morning's awakening, the nightly summons to bed; the connings,[2] the recitations; the periodical half-holidays and perambulations; the playground, with its broils, its pastimes, its intrigues—these, by a mental sorcery long forgotten, were made to involve a wilderness of sensation, a world of rich incident, an uni-verse of varied emotion, of excitement the most passionate and spirit-stirring. *"Oh, le bon temps, que ce siecle de fer!"*[3]

In truth, the ardency, the enthusiasm, and the imperiousness of my disposition soon rendered me a marked character among my schoolmates, and by slow but natural gradations, gave me an ascen-dency over all not greatly older than myself—over all with one single exception. This exception was found in the person of a scholar, who although no relation, bore the same Christian and surname as myself—a circumstance, in truth, little remarkable, for, notwithstanding a noble descent, mine was one of those every-day

8. Five-year period.
9. Extreme, exaggerated.
1. Perhaps Poe has in mind no particular medal of Carthage (the ancient sea power on the Mediterranean near modern Tunis, defeated by Rome in the second century B.C.); exergues are the spaces beneath

the central design on the reverse of coins.
2. Memorizings.
3. From Voltaire's *Le Mondain* (1736): "Oh, this age of iron is a good time." "Iron" implies dull utilitarianism in con-trast to the fabled heroic age of gold.

appellations which seem, by prescriptive right, to have been, time out of mind, the common property of the mob. In this narrative I have therefore designated myself as William Wilson—a fictitious title not very dissimilar to the real. My namesake alone, of those who in school phraseology constituted "our set," presumed to compete with me in the studies of the class, in the sports and broils of the play-ground—to refuse implicit belief in my assertions, and submission to my will—indeed to interfere with my arbitrary dictation in any respect whatsoever. If there be on earth a supreme and unqualified despotism, it is the despotism of a master mind in boyhood over the less energetic spirits of his companions.

Wilson's rebellion was to me a source of the greatest embarrassment—the more so as, in spite of the bravado with which in public I made a point of treating him and his pretensions, I secretly felt that I feared him, and could not help thinking the equality which he maintained so easily with myself a proof of his true superiority, since not to be overcome cost me a perpetual struggle. Yet this superiority—even this equality—was in truth acknowledged by no one but myself; our companions, by some unaccountable blindness, seemed not even to suspect it. Indeed, his competition, his resistance, and especially his impertinent and dogged interference with my purposes, were not more pointed than private. He appeared to be utterly destitute alike of the ambition which urged, and of the passionate energy of mind which enabled me to excel. In his rivalry he might have been supposed actuated solely by a whimsical desire to thwart, astonish, or mortify myself; although there were times when I could not help observing, with a feeling made up of wonder, abasement, and pique, that he mingled with his injuries, his insults, or his contradictions, a certain most inappropriate, and assuredly most unwelcome *affectionateness* of manner. I could only conceive this singular behaviour to arise from a consummate self-conceit assuming the vulgar airs of patronage and protection.

Perhaps it was this latter trait in Wilson's conduct, conjoined with our identity of name, and the mere accident of our having entered the school upon the same day, which set afloat the notion that we were brothers, among the senior classes in the academy. These do not usually inquire with much strictness into the affairs of their juniors. I have before said, or should have said, that Wilson was not, in the most remote degree, connected with my family. But assuredly if we *had* been brothers we must have been twins, for, since leaving Dr. Bransby's, I casually learned that my namesake— a somewhat remarkable coincidence—was born on the nineteenth of January, 1811—and this is precisely the day of my own nativity.

It may seem strange that in spite of the continual anxiety occasioned me by the rivalry of Wilson, and his intolerable spirit of contradiction, I could not bring myself to hate him altogether. We had, to be sure, nearly every day a quarrel, in which, yielding me publicly the palm of victory, he, in some manner, contrived to make me feel that it was he who had deserved it; yet a sense of pride upon my part, and a veritable dignity upon his own, kept us always upon what are called "speaking terms," while there were many points of strong congeniality in our tempers, operating to awake in me a sentiment which our position alone, perhaps, prevented from ripening into friendship. It is difficult, indeed, to define, or even to describe, my real feelings towards him. They were formed of a heterogeneous mixture—some petulant animosity, which was not yet hatred, some esteem, more respect, much fear, with a world of uneasy curiosity. To the moralist fully acquainted with the minute springs of human action, it will be unnecessary to say, in addition, that Wilson and myself were the most inseparable of companions.

It was no doubt the anomalous state of affairs existing between us which turned all my attacks upon him, and they were many, either open or covert, into the channel of banter or practical joke (giving pain while assuming the aspect of mere fun) rather than into that of a more serious and determined hostility. But my endeavours on this head were by no means uniformly successful, even when my plans were the most wittily concocted; for my namesake had much about him, in character, of that unassuming and quiet austerity which, while enjoying the poignancy of its own jokes, has no heel of Achilles[4] in itself, and absolutely refuses to be laughed at. I could find, indeed, but one vulnerable point, and that, lying in a personal peculiarity arising, perhaps, from constitutional disease, would have been spared by any antagonist less at his wit's end than myself—my rival had a weakness in the faucial or guttural organs which precluded him from raising his voice at any time *above a very low whisper*. Of this defect I did not fail to take what poor advantage lay in my power.

Wilson's retaliations in kind were many, and there was one form of his practical wit that disturbed me beyond measure. How his sagacity first discovered at all that so petty a thing would vex me is a question I never could solve—but, having discovered, he habitually practised the annoyance. I had always felt aversion to my uncourtly patronymic, and its very common, if not plebeian, praenomen. The words were venom in my ears; and when, upon the day of my arrival, a second William Wilson came also to the academy, I felt angry with him for bearing the name, and doubly disgusted with the name because a stranger bore it who would be the cause of its twofold repetition, who would be constantly in my presence, and

4. I.e., no vulnerable spot. The mother of Achilles, the hero of Homer's *Iliad*, tried to made her son immortal by dipping him into the river Styx. But no water touched the heel she held him by, and in that heel he received his death wound.

whose concerns, in the ordinary routine of the school business, must, inevitably, on account of the detestable coincidence, be often confounded with my own.

The feeling of vexation thus engendered, grew stronger with every circumstance tending to show resemblance, moral or phyical, between my rival and myself. I had not then discovered the remarkable fact that we were of the same age; but I saw that we were of the same height, and I perceived that we were not altogether unlike in general contour of person and outline of feature. I was galled, too, by the rumour touching a relationship which had grown current in the upper forms. In a word, nothing could more seriously disturb me, (although I scrupulously concealed such disturbance,) than any allusion to a similarity of mind, person, or condition existing between us. But, in truth, I had no reason to believe that (with the exception of the matter of relationship, and in the case of Wilson himself,) this similarity had ever been made a subject of comment, or even observed at all by our schoolfellows. That *he* observed it in all its bearings, and as fixedly as I, was apparent, but that he could discover in such circumstances so fruitful a field of annoyance for myself can only be attributed, as I said before, to his more than ordinary penetration.

His cue, which was to perfect an imitation of myself, lay both in words and in actions; and most admirably did he play his part. My dress it was an easy matter to copy; my gait and general manner were, without difficulty, appropriated; in spite of his constitutional defect, even my voice did not escape him. My louder tones were, of course, unattempted, but then the key, it was identical; *and his singular whisper, it grew the very echo of my own.*

How greatly this most exquisite portraiture harassed me, (for it could not justly be termed a caricature,) I will not now venture to describe. I had but one consolation—in the fact that the imitation, apparently, was noticed by myself alone, and that I had to endure only the knowing and strangely sarcastic smiles of my namesake himself. Satisfied with having produced in my bosom the intended effect, he seemed to chuckle in secret over the sting he had inflicted, and was characteristically disregardful of the public applause which the success of his witty endeavours might have so easily elicited. That the school, indeed, did not feel his design, perceive its accomplishment, and participate in his sneer, was, for many anxious months, a riddle I could not resolve. Perhaps the *gradation* of his copy rendered it not so readily perceptible, or, more possibly, I owed my security to the masterly air of the copyist, who, disdaining the letter, which in a painting is all the obtuse can see, gave but the full spirit of his original for my individual contemplation and chagrin.

I have already more than once spoken of the disgusting air of patronage which he assumed towards me, and of his frequent offi-

cious interference with my will. This interference often took the
ungracious character of advice; advice not openly given, but hinted
or insinuated. I received it with a repugnance which gained strength
as I grew in years. Yet, at this distant day, let me do him the simple
justice to acknowledge that I can recall no occasion when the suggestions of my rival were on the side of those errors or follies so
usual to his immature age, and seeming inexperience; that his moral
sense, at least, if not his general talents and worldly wisdom, was
far keener than my own; and that I might, to-day, have been a
better, and thus a happier man, had I more seldom rejected the
counsels embodied in those meaning whispers which I then but too
cordially hated, and too bitterly derided.

As it was, I at length grew restive in the extreme, under his
distasteful supervision, and daily resented more and more openly
what I considered his intolerable arrogance. I have said that, in the
first years of our connexion as schoolmates, my feelings in regard to
him might have been easily ripened into friendship; but, in the latter
months of my residence at the academy, although the intrusion of
his ordinary manner had, beyond doubt, in some measure, abated,
my sentiments, in nearly similar proportion, partook very much of
positive hatred. Upon one occasion he saw this, I think, and afterwards avoided, or made a show of avoiding me.

It was about the same period, if I remember aright, that, in an
altercation of violence with him, in which he was more than usually
thrown off his guard, and spoke and acted with an openness of
demeanour rather foreign to his nature, I discovered, or fancied I
discovered, in his accent, his air, and general appearance, a something which first startled, and then deeply interested me, by bringing
to mind dim visions of my earliest infancy; wild, confused, and
thronging memories of a time when memory herself was yet unborn. I cannot better describe the sensation which oppressed me
than by saying that I could with difficulty shake off the belief that
myself and the being who stood before me had been acquainted at
some epoch very long ago; some point of the past even infinitely
remote. The delusion, however, faded rapidly as it came; and I
mention it at all but to define the day of the last conversation I
there held with my singular namesake.

The huge old house, with its countless subdivisions, had several
enormously large chambers communicating with each other, where
slept the greater number of the students. There were, however, as
must necessarily happen in a building so awkwardly planned, many
little nooks or recesses, the odds and ends of the structure; and
these the economic ingenuity of Dr. Bransby had also fitted up as
dormitories—although, being the merest closets, they were capable
of accommodating only a single individual. One of these small
apartments were occupied by Wilson.

It was upon a gloomy and tempestuous night of an early autumn,

about the close of my fifth year at the school, and immediately after
the altercation just mentioned, that, finding every one wrapped in
sleep, I arose from bed, and, lamp in hand, stole through a wilderness of narrow passages from my own bed-room to that of my rival.
I had been long plotting one of those ill-natured pieces of practical
wit at his expense in which I had hitherto been so uniformly unsuccessful. It was my intention, now, to put my scheme in operation,
and I resolved to make him feel the whole extent of the malice with
which I was imbued. Having reached his closet, I noiselessly entered, leaving the lamp with a shade over it, on the outside. I
advanced a step, and listened to the sound of his tranquil breathing.
Assured of his being asleep, I returned, took the light, and with it
again approached the bed. Close curtains were around it, which, in
the prosecution of my plan, I slowly and quietly withdrew, when the
bright rays fell vividly upon the sleeper, and my eyes, at the same
moment upon his countenance. I looked, and a numbness, an iciness of feeling instantly pervaded my frame. My breast heaved, my
knees tottered, my whole spirit became possessed with an objectless
yet intolerable horror. Gasping for breath, I lowered the lamp in
still nearer proximity to the face. Were these—*these* the lineaments
of William Wilson? I saw, indeed, that they were his, but I shook as
with a fit of the ague in fancying they were not. What *was* there
about them to confound me in this manner? I gazed—while my
brain reeled with a multitude of incoherent thoughts. Not thus he
appeared—assuredly not *thus*—in the vivacity of his waking hours.
The same name; the same contour of person; the same day of
arrival at the academy! And then his dogged and meaningless imitation of my gait, my voice, my habits, and my manner! Was it, in
truth, within the bounds of human possibility that *what I now
witnessed* was the result of the habitual practice of this sarcastic
imitation? Awe-stricken, and with a creeping shudder, I extinguished the lamp, passed silently from the chamber, and left, at
once, the halls of that old academy, never to enter them again.

After a lapse of some months, spent at home in mere idleness, I
found myself a student at Eton. The brief interval had been sufficient to enfeeble my remembrance of the events at Dr. Bransby's, or
at least, to effect a material change in the nature of the feelings with
which I remembered them. The truth—the tragedy—of the drama
was no more. I could now find room to doubt the evidence of my
senses; and seldom called up the subject at all but with wonder at
the extent of human credulity, and a smile at the vivid force of the
imagination which I hereditarily possessed. Neither was this species
of scepticism likely to be diminished by the character of the life I
led at Eton. The vortex of thoughtless folly into which I there so
immediately and so recklessly plunged, washed away all but the
froth of my past hours—engulfed, at once, every solid or serious

impression, and left to memory only the veriest levities of a former existence.

I do not wish, however, to trace the course of my miserable profligacy here—a profligacy which set at defiance the laws, while it eluded the vigilance of the institution. Three years of folly, passed without profit, had but given me rooted habits of vice, and added, in a somewhat unusual degree, to my bodily stature, when, after a week of soulless dissipation, I invited a small party of the most dissolute students to a secret carousal in my chamber. We met at a late hour of the night, for our debaucheries were to be faithfully protracted until morning. The wine flowed freely, and there were not wanting other, perhaps more dangerous, seductions; so that the gray dawn had already faintly appeared in the east, while our delirious extravagance was at its height. Madly flushed with cards and intoxication, I was in the act of insisting upon a toast of more than intolerable profanity, when my attention was suddenly diverted by the violent, although partial unclosing of the door of the apartment, and by the eager voice from without of a servant. He said that some person, apparently in great haste, demanded to speak with me in the hall.

Wildly excited with the potent *Vin de Barac*, the unexpected interruption rather delighted than surprised me. I staggered forward at once, and a few steps brought me to the vestibule of the building. In this low and small room there hung no lamp; and now no light at all was admitted, save that of the exceedingly feeble dawn which made its way through a semicircular window. As I put my foot over the threshold I became aware of the figure of a youth about my own height, and (what then peculiarly struck my mad fancy) habited in a white cassimere morning frock, cut in the novel fashion of the one I myself wore at the moment. This the faint light enabled me to perceive—but the features of his face I could not distinguish. Immediately upon my entering he strode hurriedly up to me, and, seizing me by the arm with a gesture of petulant impatience, whispered the words "William Wilson!" in my ear. I grew perfectly sober in an instant.

There was that in the manner of the stranger, and in the tremulous shake of his uplifted finger, as he held it between my eyes and the light, which filled me with unqualified amazement—but it was not this which had so violently moved me. It was the pregnancy of solemn admonition in the singular, low, hissing utterance; and, above all, it was the character, the tone, *the key*, of those few, simple, and familiar, yet whispered, syllables, which came with a thousand thronging memories of by-gone days, and struck upon my soul with the shock of a galvanic battery. Ere I could recover the use of my senses he was gone.

Although this event failed not of a vivid effect upon my disordered imagination, yet was it evanescent as vivid. For some weeks, indeed, I busied myself in earnest inquiry, or was wrapped in a cloud of morbid speculation. I did not pretend to disguise from my perception the identity of the singular individual who thus perseveringly interfered with my affairs, and harassed me with his insinuated counsel. But who and what was this Wilson?—and whence came he?—and what were his purposes? Upon neither of these points could I be satisfied—merely ascertaining, in regard to him, that a sudden accident in his family had caused his removal from Dr. Bransby's Academy on the afternoon of the day in which I myself had eloped. But in a brief period I ceased to think upon the subject; my attention being all absorbed in a contemplated departure for Oxford. Thither I soon went; the uncalculating vanity of my parents furnished me with an outfit, and annual establishment, which would enable me to indulge at will in the luxury already so dear to my heart—to vie in profuseness of expenditure with the haughtiest heirs of the wealthiest earldoms in Great Britain.

Excited by such appliances to vice, my constitutional temperament broke forth with redoubled ardour, and I spurned even the common restraints of decency in the mad infatuation of my revels. But it were absurd to pause in the detail of my extravagance. Let it suffice, that among spendthrifts I out-heroded Herod,[5] and that, giving name to a multitude of novel follies, I added no brief appendix to the long catalogue of vices then usual in the most dissolute university of Europe.

It could hardly be credited, however, that I had, even here, so utterly fallen from the gentlemanly estate as to seek acquaintance with the vilest arts of the gambler by profession, and, having become an adept in his despicable science, to practise it habitually as a means of increasing my already enormous income at the expense of the weak-minded among my fellow-collegians. Such, nevertheless, was the fact. And the very enormity of this offence against all manly and honourable sentiment proved, beyond doubt, the main, if not the sole reason of the impunity with which it was committed. Who, indeed, among my most abandoned associates, would not rather have disputed the clearest evidence of his senses, than have suspected of such courses the gay, the frank, the generous William Wilson—the noblest and most liberal commoner at Oxford—him whose follies (said his parasites) were but the follies of youth and unbridled fancy—whose errors but inimitable whim—whose darkest vice but a careless and dashing extravagance.

I had been now two years successfully busied in this way, when there came to the university a young *parvenu*[6] nobleman, Glendin-

---

5. I.e., exceeded excesses, from Hamlet's advice to the players, *Hamlet*, 3.2. In medieval mystery plays a favorite luridly acted villain was Herod (73–4 B.C.), the cruel King of Judea (see Matthew 2).
6. Upstart, newly rich.

ning—rich, said report, as Herodes Atticus[7]—his riches, too, as easily acquired. I soon found him of weak intellect, and, of course, marked him as a fitting subject for my skill. I frequently engaged him in play, and contrived, with a gambler's usual art, to let him win considerable sums, the more effectually to entangle him in my snares. At length, my schemes being ripe, I met him (with the full intention that this meeting should be final and decisive) at the chambers of a fellow-commoner, (Mr. Preston,) equally intimate with both, but who, to do him justice, entertained not even a remote suspicion of my design. To give to this a better colouring, I had contrived to have assembled a party of some eight or ten, and was solicitously careful that the introduction of cards should appear accidental, and originate in the proposal of my contemplated dupe himself. To be brief upon a vile topic, none of the low finesse was omitted, so customary upon similar occasions that it is a just matter for wonder how any are still found so besotted as to fall its victim.

We had protracted our sitting far into the night, and I had at length effected the manœuvre of getting Glendinning as my sole antagonist. The game, too, was my favourite écarté. The rest of the company, interested in the extent of our play, had abandoned their own cards, and were standing around us as spectators. The *parvenu*, who had been induced by my artifices in the early part of the evening to drink deeply, now shuffled, dealt, or played with a wild nervousness of manner for which his intoxication, I thought, might partially, but could not altogether, account. In a very short period he had become my debtor to a large amount of money, when, having taken a long draught of port, he did precisely what I had been coolly anticipating, proposed to double our already extravagant stakes. With a well feigned show of reluctance, and not until after my repeated refusal had seduced him into some angry words which gave a colour of *pique* to my compliance, did I finally comply. The result, of course, did but prove how entirely the prey was in my toils—in less than a single hour he had quadrupled his debt. For some time his countenance had been losing the florid tinge lent it by the wine—but now, to my astonishment, I perceived that it had grown to a pallor truly fearful. I say to my astonishment. Glendinning had been represented to my eager inquiries as immeasurably wealthy; and the sums which he had as yet lost, although in themselves vast, could not, I supposed, very seriously annoy, much less so violently affect him. That he was overcome by the wine just swallowed, was the idea which most readily presented itself; and, rather with a view to the preservation of my own character in the eyes of my associates, than from any less interested motive, I was about to insist, peremptorily, upon a discontinuance of the play, when some expressions at my elbow from among the company, and

---

7. Athenian rhetorician of the second century, proverbial for his extreme wealth.

an ejaculation evincing utter despair on the part of Glendinning, gave me to understand that I had effected his total ruin under circumstances which, rendering him an object for the pity of all, should have protected him from the ill offices of a fiend.

What now might have been my conduct it is difficult to say. The pitiable condition of my dupe had thrown an air of embarrassed gloom over all, and, for some moments, a profound and unbroken silence was maintained, during which I could not help feeling my cheeks tingle with the many burning glances of scorn or reproach cast upon me by the less abandoned of the party. I will even own that an intolerable weight of anxiety was for a brief instant lifted from my bosom by the sudden and extraordinary interruption which ensued. The wide, heavy folding doors of the apartment were all at once thrown open, to their full extent, with a vigorous and rushing impetuosity that extinguished, as if by magic, every candle in the room. Their light, in dying, enabled us just to perceive that a stranger had entered of about my own height, and closely muffled in a cloak. The darkness, however, was now total; and we could only feel that he was standing in our midst. Before any one of us could recover from the extreme astonishment into which this rudeness had thrown all, we heard the voice of the intruder.

"Gentlemen"—he said, in a low, distinct, and never-to-be-forgotten *whisper* which thrilled to the very marrow of my bones— "Gentlemen, I make no apology for this behaviour, because in thus behaving I am but fulfilling a duty. You are, beyond doubt, uninformed of the true character of the person who has to-night won at écarté a large sum of money from Lord Glendinning. I will therefore put you upon an expeditious and decisive plan of obtaining this very necessary information. Please to examine, at your leisure, the inner linings of the cuff of his left sleeve, and the several little packages which may be found in the somewhat capacious pockets of his embroidered morning wrapper."

While he spoke, so profound was the stillness that one might have heard a pin dropping upon the floor. In ceasing, he at once departed, and as abruptly as he had entered. Can I—shall I describe my sensations?—must I say that I felt all the horrors of the damned? Most assuredly I had but little time given for reflection. Many hands roughly seized me upon the spot, and lights were immediately reprocured. A search ensued. In the lining of my sleeve were found all of the court-cards essential in écarté, and, in the pockets of my wrapper, a number of packs, fac-similes of those used at our sittings, with the single exception that mine were of the species called, technically, *arrondé*; the honours[8] being slightly convex at the ends, the lower cards slightly convex at the sides. In this disposition, the dupe who cuts, as customary, at the breadth of the pack, will invariably find that he cuts his antagonist an honour;

8. Face cards.

while the gambler, cutting at the length, will, as certainly, cut nothing for his victim which may count in the records of the game.

Any outrageous burst of indignation upon this shameful discovery would have affected me less than the silent contempt, or the sarcastic composure with which it was received.

"Mr. Wilson," said our host, stooping to remove from beneath his feet an exceedingly luxurious cloak of rare furs, "Mr. Wilson, this is your property." (The weather was cold; and, upon quitting my own room, I had thrown a cloak over my dressing wrapper, putting it off upon reaching the scene of play.) "I presume it is supererogatory to seek here (eyeing the folds of the garment with a bitter smile,) for any farther evidence of your skill. Indeed we have had enough. You will see the necessity, I hope, of quitting Oxford—at all events, of quitting, instantly, my chambers."

Abased, humbled to the dust as I then was, it is probable that I should have resented this galling language by immediate personal violence, had not my whole attention been immediately arrested, by a fact of the most startling character. The cloak which I had worn was of a rare description of fur; how rare, how extravagantly costly, I shall not venture to say. Its fashion, too, was of my own fantastic invention; for I was fastidious, to a degree of absurd coxcombry, in matters of this frivolous nature. When, therefore, Mr. Preston reached me that which he had picked up upon the floor, and near the folding doors of the apartment, it was with an astonishment nearly bordering upon terror, that I perceived my own already hanging on my arm, (where I had no doubt unwittingly placed it,) and that the one presented me was but its exact counterpart in every, in even the minutest possible particular. The singular being who had so disastrously exposed me, had been muffled, I remembered, in a cloak; and none had been worn at all by any of the members of our party with the exception of myself. Retaining some presence of mind, I took the one offered me by Preston, placed it, unnoticed, over my own, left the apartment with a resolute scowl of defiance, and, next morning ere dawn of day, commenced a hurried journey from Oxford to the continent, in a perfect agony of horror and of shame.

I fled in vain. My evil destiny pursued me as if in exultation, and proved, indeed, that the exercise of its mysterious dominion had as yet only begun. Scarcely had I set foot in Paris ere I had fresh evidence of the detestable interest taken by this Wilson in my concerns. Years flew, while I experienced no relief. Villain!—at Rome, with how untimely, yet with how spectral an officiousness, stepped he in between me and my ambition! At Vienna, too, at Berlin, and at Moscow! Where, in truth, had I *not* bitter cause to curse him within my heart? From his inscrutable tyranny did I at length flee, panic-stricken, as from a pestilence; and to the very ends of the earth *I fled in vain.*

And again, and again, in secret communion with my own spirit, would I demand the questions "Who is he?—whence came he?—and what are his objects?" But no answer was there found. And now I scrutinized, with a minute scrutiny, the forms, and the methods, and the leading traits of his impertinent supervision. But even here there was very little upon which to base a conjecture. It was noticeable, indeed, that, in no one of the multiplied instances in which he had of late crossed my path, had he so crossed it except to frustrate those schemes, or to disturb those actions, which, fully carried out, might have resulted in bitter mischief. Poor justification this, in truth, for an authority so imperiously assumed! Poor indemnity for natural rights of self-agency so pertinaciously, so insultingly denied!

I had also been forced to notice that my tormentor, for a very long period of time, (while scrupulously and with miraculous dexterity maintaining his whim of an identity of apparel with myself,) had so contrived it, in the execution of his varied interference with my will, that I saw not, at any moment, the features of his face. Be Wilson what he might, *this*, at least, was but the veriest of affection, or of folly. Could he, for an instant, have supposed that, in my admonisher at Eton, in the destroyer of my honour at Oxford, in him who thwarted my ambition at Rome, my revenge in Paris, my passionate love at Naples, or what he falsely termed my avarice in Egypt, that in this, my arch-enemy and evil genius, I could fail to recognize the William Wilson of my schoolboy days, the namesake, the companion, the rival, the hated and dreaded rival at Dr. Bransby's? Impossible!—But let me hasten to the last eventful scene of the drama.

Thus far I had succumbed supinely to this imperious domination. The sentiments of deep awe with which I habitually regarded the elevated character, the majestic wisdom, the apparent omnipresence and omnipotence of Wilson, added to a feeling of even terror, with which certain other traits in his nature and assumptions inspired me, had operated, hitherto, to impress me with an idea of my own utter weakness and helplessness, and to suggest an implicit, although bitterly reluctant submission to his arbitrary will. But, of late days, I had given myself up entirely to wine; and its maddening influence upon my hereditary temper rendered me more and more impatient of control. I began to murmur, to hesitate, to resist. And was it only fancy which induced me to believe that, with the increase of my own firmness, that of my tormentor underwent a proportional diminution? Be this as it may, I now began to feel the inspirations of a burning hope, and at length nurtured in my secret thoughts a stern and desperate resolution that I would submit no longer to be enslaved.

It was at Rome, during the carnival of 18—, that I attended a masquerade in the palazzo of the Neapolitan Duke Di Broglio. I

had indulged more freely than usual in the excesses of the wine-
table; and now the suffocating atmosphere of the crowded rooms
irritated me beyond endurance. The difficulty, too, of forcing my
way through the mazes of the company contributed not a little to
the ruffling of my temper; for I was anxiously seeking, let me not
say with what unworthy motive, the young, the gay, the beautiful
wife of the aged and doting Di Broglio. With a too unscrupulous
confidence she had previously communicated to me the secret of the
costume in which she would be habited, and now, having caught a
glimpse of her person, I was hurrying to make my way into her
presence. At this moment I felt a light hand laid upon my shoulder,
and that ever-remembered, low, damnable whisper within my ear.

In a perfect whirlwind of wrath, I turned at once upon him who
had thus interrupted me, and seized him violently by the collar. He
was attired, as I expected, like myself; wearing a large Spanish
cloak, and a mask of black silk which entirely covered his features.

"Scoundrel!" I said, in a voice husky with rage, while every
syllable I uttered seemed as new fuel to my fury, "scoundrel! impostor! accursed villain! you shall not—you *shall not* dog me unto
death! Follow me, or I stab you where I stand," and I broke my
way from the room into a small antechamber adjoining, dragging
him unresistingly with me as I went.

Upon entering, I thrust him furiously from me. He staggered
against the wall, while I closed the door with an oath, and commanded him to draw. He hesitated but for an instant, then, with a
slight sigh, drew in silence, and put himself upon his defence.

The contest was brief indeed. I was frantic with every species of
wild excitement, and felt within my single arm the energy and the
power of a multitude. In a few seconds I forced him by sheer
strength against the wainscoting, and thus, getting him at mercy,
plunged my sword, with brute ferocity, repeatedly through and
through his bosom.

At this instant some person tried the latch of the door. I hastened
to prevent an intrusion, and then immediately returned to my dying
antagonist. But what human language can adequately portray *that*
astonishment, *that* horror which possessed me at the spectacle then
presented to view. The brief moment in which I averted my eyes
had been sufficient to produce, apparently, a material change in the
arrangements at the upper or farther end of the room. A large
mirror, it appeared to me, now stood where none had been perceptible before; and, as I stepped up to it in extremity of terror, mine
own image, but with features all pale and dabbled in blood, advanced, with a feeble and tottering gait, to meet me.

Thus it appeared, I say, but was not. It was my antagonist—it
was Wilson, who then stood before me in the agonies of his dissolution. Not a line in all the marked and singular lineaments of that
face which was not, even identically, mine own! His mask and cloak

lay, where he had thrown them, upon the floor.

It was Wilson, but he spoke no longer in a whisper, and I could have fancied that I myself was speaking while he said—

"*You have conquered, and I yield. Yet, henceforward art thou also dead—dead to the world and its hopes. In me didst thou exist—and, in my death, see by this image, which is thine, how utterly thou hast murdered thyself.*"

1839

The Black Cat[1]

For the most wild, yet most homely narrative which I am about to pen, I neither expect nor solicit belief. Mad indeed would I be to expect it, in a case where my very senses reject their own evidence. Yet, mad am I not—and very surely do I not dream. But to-morrow I die, and to-day I would unburthen my soul. My immediate purpose is to place before the world, plainly, succinctly, and without comment, a series of mere household events. In their consequences, these events have terrified—have tortured—have destroyed me. Yet I will not attempt to expound them. To me, they have presented little but Horror—to many they will seem less terrible than *barroques*.[2] Hereafter, perhaps, some intellect may be found which will reduce my phantasm to the common-place—some intellect more calm, more logical, and far less excitable than my own, which will perceive, in the circumstances I detail with awe, nothing more than an ordinary succession of very natural causes and effects.

From my infancy I was noted for the docility and humanity of my disposition. My tenderness of heart was even so conspicuous as to make me the jest of my companions. I was especially fond of animals, and was indulged by my parents with a great variety of pets. With these I spent most of my time, and never was so happy as when feeding and caressing them. This peculiarity of character grew with my growth, and, in my manhood, I derived from it one of my principal sources of pleasure. To those who have cherished an affection for a faithful and sagacious dog, I need hardly be at the trouble of explaining the nature or the intensity of the gratification thus derivable. There is something in the unselfish and self-sacrificing love of a brute, which goes directly to the heart of him who has had frequent occasion to test the paltry friendship and gossamer fidelity of mere *Man*.

I married early, and was happy to find in my wife a disposition not uncongenial with my own. Observing my partiality for domestic pets, she lost no opportunity of procuring those of the most agree-

1. *The Black Cat* was first published August 19, 1843, in the *United States Saturday Post*, a Philadelphia paper. In a letter to James Russell Lowell (July 2, 1844) Poe listed it as one of his eight best tales. The present text is that of the Wiley & Putnam *Tales* (New York, 1845), pp. 37–46.
2. Baroque, bizarre.

able kind. We had birds, gold-fish, a fine dog, rabbits, a small monkey, and *a cat*.

This latter was a remarkably large and beautiful animal, entirely black, and sagacious to an astonishing degree. In speaking of his intelligence, my wife, who at heart was not a little tinctured with superstition, made frequent allusion to the ancient popular notion, which regarded all black cats as witches in disguise. Not that she was ever *serious* upon this point—and I mention the matter at all for no better reason than that it happens, just now, to be remembered.

Pluto[3]—this was the cat's name—was my favorite pet and playmate. I alone fed him, and he attended me wherever I went about the house. It was even with difficulty that I could prevent him from following me through the streets.

Our friendship lasted, in this manner, for several years, during which my general temperament and character—through the instrumentality of the Fiend Intemperance—had (I blush to confess it) experienced a radical alteration for the worse. I grew, day by day, more moody, more irritable, more regardless of the feelings of others. I suffered myself to use intemperate language to my wife. At length, I even offered her personal violence. My pets, of course, were made to feel the change in my disposition. I not only neglected, but ill-used them. For Pluto, however, I still retained sufficient regard to restrain me from maltreating him, as I made no scruple of maltreating the rabbits, the monkey, or even the dog, when by accident, or through affection, they came in my way. But my disease grew upon me—for what disease is like Alcohol!—and at length even Pluto, who was now becoming old, and consequently somewhat peevish—even Pluto began to experience the effects of my ill temper.

One night, returning home, much intoxicated, from one of my haunts about town, I fancied that the cat avoided my presence. I seized him; when, in his fright at my violence, he inflicted a slight wound upon my hand with his teeth. The fury of a demon instantly possessed me. I knew myself no longer. My original soul seemed, at once, to take its flight from my body; and a more than fiendish malevolence, gin-nurtured, thrilled every fibre of my frame. I took from my waistcoat-pocket a pen-knife, opened it, grasped the poor beast by the throat, and deliberately cut one of its eyes from the socket! I blush, I burn, I shudder, while I pen the damnable atrocity.

When reason returned with the morning—when I had slept off the fumes of the night's debauch—I experienced a sentiment half of horror, half of remorse, for the crime of which I had been guilty;

---

3. Appropriate because in Roman mythology Pluto is the god of the dead and the ruler of the underworld.

but it was, at best, a feeble and equivocal feeling, and the soul remained untouched. I again plunged into excess, and soon drowned in wine all memory of the deed.

In the meantime the cat slowly recovered. The socket of the lost eye presented, it is true, a frightful appearance, but he no longer appeared to suffer any pain. He went about the house as usual, but, as might be expected, fled in extreme terror at my approach. I had so much of my old heart left, as to be at first grieved by this evident dislike on the part of a creature which had once so loved me. But this feeling soon gave place to irritation. And then came, as if to my final and irrevocable overthrow, the spirit of PERVERSENESS.[4] Of this spirit philosophy takes no account. Yet I am not more sure that my soul lives, than I am that perverseness is one of the primitive impulses of the human heart—one of the indivisible primary faculties, or sentiments, which give direction to the character of Man. Who has not, a hundred times, found himself committing a vile or a silly action, for no other reason than because he knows he should *not*? Have we not a perpetual inclination, in the teeth of our best judgment, to violate that which is *Law*, merely because we understand it to be such? This spirit of perverseness, I say, came to my final overthrow. It was this unfathomable longing of the soul *to vex itself*—to offer violence to its own nature—to do wrong for the wrong's sake only—that urged me to continue and finally to consummate the injury I had inflicted upon the unoffending brute. One morning, in cool blood, I slipped a noose about its neck and hung it to the limb of a tree;—hung it with the tears streaming from my eyes, and with the bitterest remorse at my heart;—hung it *because* I knew that it had loved me, and *because* I felt it had given me no reason of offence;—hung it *because* I knew that in so doing I was committing a sin—a deadly sin that would so jeopardize my immortal soul as to place it—if such a thing were possible—even beyond the reach of the infinite mercy of the Most Merciful and Most Terrible God.

On the night of the day on which this cruel deed was done, I was aroused from sleep by the cry of fire. The curtains of my bed were in flames. The whole house was blazing. It was with great difficulty that my wife, a servant, and myself, made our escape from the conflagration. The destruction was complete. My entire worldly wealth was swallowed up, and I resigned myself thenceforward to despair.

I am above the weakness of seeking to establish a sequence of cause and effect, between the disaster and the atrocity. But I am detailing a chain of facts—and wish not to leave even a possible link imperfect. On the day succeeding the fire, I visited the ruins.

4. Poe's story *The Imp of the Perverse* is built around the notion that there is "an innate and primitive principle of human action, a paradoxical something, which we may call *perverseness*, for want of a more characteristic term."

The walls, with one exception, had fallen in. This exception was found in a compartment wall, not very thick, which stood about the middle of the house, and against which had rested the head of my bed. The plastering had here, in great measure, resisted the action of the fire—a fact which I attributed to its having been recently spread. About this wall a dense crowd were collected, and many persons seemed to be examining a particular portion of it with very minute and eager attention. The words "strange!" "singular!" and other similar expressions, excited my curiosity. I approached and saw, as if graven in *bas relief* upon the white surface, the figure of a gigantic *cat*. The impression was given with an accuracy truly marvellous. There was a rope about the animal's neck.

When I first beheld this apparition—for I could scarcely regard it as less—my wonder and my terror were extreme. But at length reflection came to my aid. The cat, I remembered, had been hung in a garden adjacent to the house. Upon the alarm of fire, this garden had been immediately filled by the crowd—by some one of whom the animal must have been cut from the tree and thrown, through an open window, into my chamber. This had probably been done with the view of arousing me from sleep. The falling of other walls had compressed the victim of my cruelty into the substance of the freshly-spread plaster; the lime of which, with the flames, and the *ammonia* from the carcass, had then accomplished the portraiture as I saw it.

Although I thus readily accounted to my reason, if not altogether to my conscience, for the startling fact just detailed, it did not the less fail to make a deep impression upon my fancy. For months I could not rid myself of the phantasm of the cat; and, during this period, there came back into my spirit a half-sentiment that seemed, but was not, remorse. I went so far as to regret the loss of the animal, and to look about me, among the vile haunts which I now habitually frequented, for another pet of the same species, and of somewhat similar appearance, with which to supply its place.

One night as I sat, half stupified, in a den of more than infamy, my attention was suddenly drawn to some black object, reposing upon the head of one of the immense hogsheads[5] of Gin, or of Rum, which constituted the chief furniture of the apartment. I had been looking steadily at the top of this hogshead for some minutes, and what now caused me surprise was the fact that I had not sooner perceived the object thereupon. I approached it, and touched it with my hand. It was a black cat—a very large one—fully as large as Pluto, and closely resembling him in every respect but one. Pluto had not a white hair upon any portion of his body; but this cat had a large, although indefinite splotch of white, covering nearly the whole region of the breast.

5. Large barrels or casks, usually holding 63 gallons.

Upon my touching him, he immediately arose, purred loudly, rubbed against my hand, and appeared delighted with my notice. This, then, was the very creature of which I was in search. I at once offered to purchase it of the landlord; but this person made no claim to it—knew nothing of it—had never seen it before.

I continued my caresses, and, when I prepared to go home, the animal evinced a disposition to accompany me. I permitted it to do so; occasionally stooping and patting it as I proceeded. When it reached the house it domesticated itself at once, and became immediately a great favorite with my wife.

For my own part, I soon found a dislike to it arising within me. This was just the reverse of what I had anticipated; but—I know not how or why it was—its evident fondness for myself rather disgusted and annoyed. By slow degrees, these feelings of disgust and annoyance rose into the bitterness of hatred. I avoided the creature; a certain sense of shame, and the remembrance of my former deed of cruelty, preventing me from physically abusing it. I did not, for some weeks, strike, or otherwise violently ill use it; but gradually—very gradually—I came to look upon it with unutterable loathing, and to flee silently from its odious presence, as from the breath of a pestilence.

What added, no doubt, to my hatred of the beast, was the discovery, on the morning after I brought it home, that, like Pluto, it also had been deprived of one of its eyes. This circumstance, however, only endeared it to my wife, who, as I have already said, possessed, in a high degree, that humanity of feeling which had once been my distinguishing trait, and the source of many of my simplest and purest pleasures.

With my aversion to this cat, however, its partiality for myself seemed to increase. It followed my footsteps with a pertinacity which it would be difficult to make the reader comprehend. Whenever I sat, it would crouch beneath my chair, or spring upon my knees, covering me with its loathsome caresses. If I arose to walk it would get between my feet and thus nearly throw me down, or, fastening its long and sharp claws in my dress, clamber, in this manner, to my breast. At such times, although I longed to destroy it with a blow, I was yet withheld from so doing, partly by a memory of my former crime, but chiefly—let me confess it at once—by absolute *dread* of the beast.

This dread was not exactly a dread of physical evil—and yet I should be at a loss how otherwise to define it. I am almost ashamed to own—yes, even in this felon's cell, I am almost ashamed to own—that the terror and horror with which the animal inspired me, had been heightened by one of the merest chimæras[6] it would be possible to conceive. My wife had called my attention, more than

6. Foolish, impossible fancyings.

once, to the character of the mark of white hair, of which I have spoken, and which constituted the sole visible difference between the strange beast and the one I had destroyed. The reader will remember that this mark, although large, had been originally very indefinite; but, by slow degrees—degrees nearly imperceptible, and which for a long time my Reason struggled to reject as fanciful—it had, at length, assumed a rigorous distinctness of outline. It was now the representation of an object that I shudder to name—and for this, above all, I loathed, and dreaded, and would have rid myself of the monster *had I dared*—it was now, I say, the image of a hideous—of a ghastly thing—of the GALLOWS!—oh, mournful and terrible engine of Horror and of Crime—of Agony and of Death!

And now was I indeed wretched beyond the wretchedness of mere Humanity. And *a brute beast*—whose fellow I had contemptuously destroyed—*a brute beast* to work out for *me*—for me a man, fashioned in the image of the High God—so much of insufferable wo! Alas! neither by day nor by night knew I the blessing of Rest any more! During the former the creature left me no moment alone; and, in the latter, I started, hourly, from dreams of unutterable fear, to find the hot breath of *the thing* upon my face, and its vast weight—an incarnate Night-Mare that I had no power to shake off—incumbent eternally upon my *heart*!

Beneath the pressure of torments such as these, the feeble remnant of the good within me succumbed. Evil thoughts became my sole intimates—the darkest and most evil of thoughts. The moodiness of my usual temper increased to hatred of all things and of all mankind; while, from the sudden, frequent, and ungovernable outbursts of a fury to which I now blindly abandoned myself, my uncomplaining wife, alas! was the most usual and the most patient of sufferers.

One day she accompanied me, upon some household errand, into the cellar of the old building which our poverty compelled us to inhabit. The cat followed me down the steep stairs, and, nearly throwing me headlong, exasperated me to madness. Uplifting an axe, and forgetting, in my wrath, the childish dread which had hitherto stayed my hand, I aimed a blow at the animal which, of course, would have proved instantly fatal had it descended as I wished. But this blow was arrested by the hand of my wife. Goaded, by the interference, into a rage more than demoniacal, I withdrew my arm from her grasp and buried the axe in her brain. She fell dead upon the spot, without a groan.

This hideous murder accomplished, I set myself forthwith, and with entire deliberation, to the task of concealing the body. I knew that I could not remove it from the house, either by day or by night, without the risk of being observed by the neighbors. Many projects entered my mind. At one period I thought of cutting the corpse into minute fragments, and destroying them by fire. At another, I resolved to dig a grave for it in the floor of the cellar. Again, I deliberated about casting it in the well in the yard—about packing it in a box, as if merchandize, with the usual arrangements, and so getting a porter to take it from the house. Finally I hit upon what I considered a far better expedient than either of these. I determined to wall it up in the cellar—as the monks of the middle ages are recorded to have walled up their victims.

For a purpose such as this the cellar was well adapted. Its walls were loosely constructed, and had lately been plastered throughout with a rough plaster, which the dampness of the atmosphere had prevented from hardening. Moreover, in one of the walls was a projection, caused by a false chimney, or fireplace, that had been filled up, and made to resemble the rest of the cellar. I made no doubt that I could readily displace the bricks at this point, insert the corpse, and wall the whole up as before, so that no eye could detect any thing suspicious.

And in this calculation I was not deceived. By means of a crowbar I easily dislodged the bricks, and, having carefully deposited the body against the inner wall, I propped it in that position, while, with little trouble, I re-laid the whole structure as it originally stood. Having procured mortar, sand, and hair, with every possible precaution, I prepared a plaster which could not be distinguished from the old, and with this I very carefully went over the new brick-work. When I had finished, I felt satisfied that all was right. The wall did not present the slightest appearance of having been disturbed. The rubbish on the floor was picked up with the minutest care. I looked around triumphantly, and said to myself—"Here at least, then, my labor has not been in vain."

My next step was to look for the beast which had been the cause of so much wretchedness; for I had, at length, firmly resolved to put it to death. Had I been able to meet with it, at the moment, there could have been no doubt of its fate; but it appeared that the crafty animal had been alarmed at the violence of my previous anger, and forebore to present itself in my present mood. It is impossible to describe, or to imagine, the deep, the blissful sense of relief which the absence of the detested creature occasioned in my bosom. It did not make its appearance during the night—and thus for one night at least, since its introduction into the house, I soundly and tranquilly slept; aye, *slept* even with the burden of murder upon my soul!

The second and the third day passed, and still my tormentor came not. Once again I breathed as a freeman. The monster, in terror, had fled the premises forever! I should behold it no more! My happiness was supreme! The guilt of my dark deed disturbed me but little. Some few inquiries had been made, but these had been readily answered. Even a search had been instituted—but of course nothing was to be discovered. I looked upon my future felicity as secured.

Upon the fourth day of the assassination, a party of the police came, very unexpectedly, into the house, and proceeded again to make rigorous investigation of the premises. Secure, however, in the inscrutability of my place of concealment, I felt no embarrassment whatever. The officers bade me accompany them in their search. They left no nook or corner unexplored. At length, for the third or fourth time, they descended into the cellar. I quivered not in a muscle. My heart beat calmly as that of one who slumbers in innocence. I walked the cellar from end to end. I folded my arms upon my bosom, and roamed easily to and fro. The police were thoroughly satisfied and prepared to depart. The glee at my heart was too strong to be restrained. I burned to say if but one word, by way of triumph, and to render doubly sure their assurance of my guiltlessness.

"Gentlemen," I said at last, as the party ascended the steps, "I delight to have allayed your suspicions. I wish you all health, and a little more courtesy. By the bye, gentlemen, this—this is a very well constructed house." [In the rabid desire to say something easily, I scarcely knew what I uttered at all.]—"I may say an *excellently* well constructed house. These walls—are you going, gentlemen?— these walls are solidly put together;" and here, through the mere phrenzy of bravado, I rapped heavily, with a cane which I held in my hand, upon that very portion of the brick-work behind which stood the corpse of the wife of my bosom.

But may God shield and deliver me from the fangs of the Arch-Fiend! No sooner had the reverberation of my blows sunk into silence, than I was answered by a voice from within the tomb!—by a cry, at first muffled and broken, like the sobbing of a child, and then quickly swelling into one long, loud, and continuous scream, utterly anomalous and inhuman—a howl—a wailing shriek, half of horror and half of triumph, such as might have arisen only out of hell, conjointly from the throats of the damned in their agony and of the demons that exult in the damnation.

Of my own thoughts it is folly to speak. Swooning, I staggered to the opposite wall. For one instant the party upon the stairs remained motionless, through extremity of terror and of awe. In the next, a dozen stout arms were toiling at the wall. It fell bodily. The corpse, already greatly decayed and clotted with gore, stood erect before the eyes of the spectators. Upon its head, with red extended mouth and solitary eye of fire, sat the hideous beast whose craft had seduced me into murder, and whose informing voice had consigned me to the hangman. I had walled the monster up within the tomb!

1843

The Purloined Letter[1]

At Paris, just after dark one gusty evening in the autumn of 18——, I was enjoying the twofold luxury of meditation and a meerschaum, in company with my friend C. Auguste Dupin, in his little back library, or book-closet, *au troisiême*,[2] No. 33, *Rue Dunôt, Faubourg St. Germain*. For one hour at least we had maintained a profound silence; while each, to any casual observer, might have seemed intently and exclusively occupied with the curling eddies of smoke that oppressed the atmosphere of the chamber. For myself, however, I was mentally discussing certain topics which had formed matter for conversation between us at an earlier period of the evening; I mean the affair of the Rue Morgue, and the mystery attending the murder of Marie Roget. I looked upon it, therefore, as something of a coincidence, when the door of our apartment was thrown open and admitted our old acquaintance, Monsieur G——, the Prefect of the Parisian police.

We gave him a hearty welcome; for there was nearly half as much of the entertaining as of the contemptible about the man, and we had not seen him for several years. We had been sitting in the dark, and Dupin now arose for the purpose of lighting a lamp, but sat down again, without doing so, upon G.'s saying that he had called to consult us, or rather to ask the opinion of my friend, about some official business which had occasioned a great deal of trouble.

"If it is any point requiring reflection," observed Dupin, as he forbore to enkindle the wick, "we shall examine it to better purpose in the dark."

"That is another of your odd notions," said the Prefect, who had a fashion of calling every thing "odd" that was beyond his comprehension, and thus lived amid an absolute legion of "oddities."

"Very true," said Dupin, as he supplied his visiter with a pipe, and rolled towards him a very comfortable chair.

"And what is the difficulty now?" I asked. "Nothing more in the assassination way, I hope?"

"Oh no; nothing of that nature. The fact is, the business is *very*

---

1. The text is that of the first publication in *The Gift*, a Philadelphia annual dated 1845 but for sale late in 1844. Historians of detective fiction usually cite Poe's three stories about C. Auguste Dupin as the first of the genre. This is the third Dupin story, the others being *The Murders in the Rue Morgue* (1841) and *The Mystery of Marie Rôget* (1842). Here the criminal is known from the beginning and the solution comes from Dupin's analytical powers; but in *The Murders in the Rue Morgue* Poe is at some pains to stress that Dupin's powers are not of mere "calculation": rather, "the analyst throws himself into the spirit of his opponent, identifies himself there-with, and not unfrequently sees thus, at a glance, the sole methods (sometimes indeed absurdly simple ones) by which he may seduce into error or hurry into miscalculation."

2. Fourth floor (since the French do not count the first, the *rez-de-chaussée*). In *The Murders in the Rue Morgue* the narrator describes his and Dupin's quarters, "a time-eaten and grotesque mansion, long deserted through superstitions," "tottering to its fall in a retired and desolate portion of the Faubourg St. Germain," but meanwhile furnished "in a style which suited the rather fantastic gloom" of their common temperament.

simple indeed, and I make no doubt that we can manage it sufficiently well ourselves; but then I thought Dupin would like to hear the details of it, because it is so excessively *odd*."

"Simple and odd," said Dupin.

"Why, yes; and not exactly that, either. The fact is, we have all been a good deal puzzled because the affair *is* so simple, and yet baffles us altogether."

"Perhaps it is the very simplicity of the thing which puts you at fault," said my friend.

"What nonsense you *do* talk!" replied the Prefect, laughing heartily.

"Perhaps the mystery is a little *too* plain," said Dupin.

"Oh, good heavens! who ever heard of such an idea?"

"A little *too* self-evident."

"Ha! ha! ha!—ha! ha! ha!—ho! ho! ho!" roared out our visiter, profoundly amused, "oh, Dupin, you will be the death of me yet!"

"And what, after all, *is* the matter on hand?" I asked.

"Why, I will tell you," replied the Prefect, as he gave a long, steady, and contemplative puff, and settled himself in his chair. "I will tell you in a few words; but, before I begin, let me caution you that this is an affair demanding the greatest secrecy, and that I should most probably lose the position I now hold, were it known that I confided it to any one."

"Proceed," said I.

"Or not," said Dupin.

"Well, then; I have received personal information, from a very high quarter, that a certain document of the last importance, has been purloined from the royal apartments. The individual who purloined it is known; this beyond a doubt; he was seen to take it. It is known, also, that it still remains in his possession."

"How is this known?" asked Dupin.

"It is clearly inferred," replied the Prefect, "from the nature of the document, and from the non-appearance of certain results which would at once arise from its passing *out* of the robber's possession;—that is to say, from his employing it as he must design in the end to employ it."

"Be a little more explicit," I said.

"Well, I may venture so far as to say that the paper gives its holder a certain power in a certain quarter where such power is immensely valuable." The Prefect was fond of the cant of diplomacy.

"Still I do not quite understand," said Dupin.

"No? Well; the disclosure of the document to a third person, who shall be nameless, would bring in question the honour of a personage of most exalted station; and this fact gives the holder of the document an ascendancy over the illustrious personage whose honour and peace are so jeopardized."

"But this ascendancy," I interposed, "would depend upon the robber's knowledge of the loser's knowledge of the robber. Who would dare—"

"The thief," said G, "is the—Minister D——, who dares all things, those unbecoming as well as those becoming a man. The method of the theft was not less ingenious than bold. The document in question—a letter, to be frank—had been received by the personage robbed while alone in the royal *boudoir*. During its perusal she was suddenly interrupted by the entrance of the other exalted personage from whom especially it was her wish to conceal it. After a hurried and vain endeavour to thrust it in a drawer, she was forced to place it, open as it was, upon a table. The address, however, was uppermost, and the contents thus unexposed, the letter escaped notice. At this juncture enters the Minister D——. His lynx eye immediately perceives the paper, recognises the handwriting of the address, observes the confusion of the personage addressed, and fathoms her secret. After some business transactions, hurried through in his ordinary manner, he produces a letter somewhat similar to the one in question, opens it, pretends to read it, and then places it in close juxtaposition to the other. Again he converses, for some fifteen minutes, upon the public affairs. At length, in taking leave, he takes also from the table the letter to which he had no claim. Its rightful owner saw, but, of course, dared not call attention to the act, in the presence of the third personage who stood at her elbow. The minister decamped; leaving his own letter—one of no importance—upon the table."

"Here, then," said Dupin to me, "you have precisely what you demand to make the ascendancy complete—the robber's knowledge of the loser's knowledge of the robber."

"Yes," replied the Prefect; "and the power thus attained has, for some months past, been wielded, for political purposes, to a very dangerous extent. The personage robbed is more thoroughly convinced, every day, of the necessity of reclaiming her letter. But this, of course, cannot be done openly. In fine, driven to despair, she has committed the matter to me."

"Than whom," said Dupin, amid a perfect whirlwind of smoke, "no more sagacious agent could, I suppose, be desired, or even imagined."

"You flatter me," replied the Prefect; "but it is possible that some such opinion may have been entertained."

"It is clear," said I, "as you observe, that the letter is still in possession of the minister; since it is this possession, and not any employment, of the letter, which bestows the power. With the employment the power departs."

"True," said G——; "and upon this conviction I proceeded. My first care was to make thorough search of the minister's hotel; and here my chief embarrassment lay in the necessity of searching without his knowledge. Beyond all things, I have been warned of the danger which would result from giving him reason to suspect our design."

"But," said I, "you are quite *au fait*[3] in these investigations. The Parisian police have done this thing often before."

"O yes; and for this reason I did not despair. The habits of the minister gave me, too, a great advantage. He is frequently absent from home all night. His servants are by no means numerous. They sleep at a distance from their master's apartments, and, being chiefly Neapolitans, are readily made drunk. I have keys, as you know, with which I can open any chamber or cabinet in Paris. For three months a night has not passed, during the greater part of which I have not been engaged, personally, in ransacking the D—— Hotel. My honour is interested, and, to mention a great secret, the reward is enormous. So I did not abandon the search until I had become fully satisfied that the thief is a more astute man than myself. I fancy that I have investigated every nook and corner of the premises in which it is possible that the paper can be concealed."

"But is it not possible," I suggested, "that although the letter may be in possession of the minister, as it unquestionably is, he may have concealed it elsewhere than upon his own premises?"

"This is barely possible," said Dupin. "The present peculiar condition of affairs at court, and especially of those intrigues in which D—— is known to be involved, would render the instant availability of the document—its susceptibility of being produced at a moment's notice—a point of nearly equal importance with its possession."

"Its susceptibility of being produced?" said I.

"That is to say, of being *destroyed*," said Dupin.

"True," I observed; "the paper is clearly then upon the premises. As for its being upon the person of the minister, we may consider that as out of the question."

"Entirely," said the Prefect. "He has been twice waylaid, as if by footpads, and his person rigorously searched under my own inspection."

"You might have spared yourself this trouble," said Dupin. "D——, I presume, is not altogether a fool, and, if not, must have anticipated these waylayings, as a matter of course."

"Not *altogether* a fool," said G——, "but then he's a poet, which I take to be only one remove from a fool."

"True;" said Dupin, after a long and thoughtful whiff from his meerschaum, "although I have been guilty of certain doggrel myself."

"Suppose you detail," said I, "the particulars of your search."

"Why the fact is, we took our time, and we searched *every where*. I have had long experience in these affairs. I took the entire build-

3. At home, expert.

ing, room by room; devoting the nights of a whole week to each. We examined, first, the furniture of each apartment. We opened every possible drawer; and I presume you know that, to a properly trained police agent, such a thing as a *secret* drawer is impossible. Any man is a dolt who permits a 'secret' drawer to escape him in a search of this kind. The thing is *so* plain. There is a certain amount of bulk—of space—to be accounted for in every cabinet. Then we have accurate rules. The fiftieth part of a line could not escape us. After the cabinets we took the chairs. The cushions we probed with the fine long needles you have seen me employ. From the tables we removed the tops."

"Why so?"

"Sometimes the top of a table, or other similarly arranged piece of furniture, is removed by the person wishing to conceal an article; then the leg is excavated, the article deposited within the cavity, and the top replaced. The bottoms and tops of bedposts are employed in the same way."

"But could not the cavity be detected by sounding?" I asked.

"By no means, if, when the article is deposited, a sufficient wadding of cotton be placed around it. Besides, in our case, we were obliged to proceed without noise."

"But you could not have removed—you could not have taken to pieces *all* articles of furniture in which it would have been possible to make a deposit in the manner you mention. A letter may be compressed into a thin spiral roll, not differing much in shape or bulk from a large knitting-needle, and in this form it might be inserted into the rung of a chair, for example. You did not take to pieces all the chairs?"

"Certainly not; but we did better—we examined the rungs of every chair in the hotel, and, indeed, the jointings of every description of furniture, by the aid of a most powerful microscope.[4] Had there been any traces of recent disturbance we should not have failed to detect it *instanter*.[5] A single grain of gimlet-dust, or saw-dust, for example, would have been as obvious as an apple. Any disorder in the glueing—any unusual gaping in the joints—would have sufficed to insure detection."

"Of course you looked to the mirrors, between the boards and the plates, and you probed the beds and the bed-clothes, as well as the curtains and carpets."

"That of course; and when we had absolutely completed every particle of the furniture in this way, then we examined the house itself. We divided its entire surface into compartments, which we numbered, so that none might be missed; then we scrutinized each individual square inch throughout the premises, including the two houses immediately adjoining, with the microscope, as before."

4. I.e., a powerful magnifying glass.   5. Instantly.

"The two houses adjoining!" I exclaimed; "you must have had a great deal of trouble."

"We had; but the reward offered is prodigious."

"You include the *grounds* about the houses?"

"All the grounds are paved with brick. They gave us comparatively little trouble. We examined the moss between the bricks, and found it undisturbed."

"And the roofs?"

"We surveyed every inch of the external surface, and probed carefully beneath every tile."

"You looked among D——'s papers, of course, and into the books of the library?"

"Certainly; we opened every package and parcel; we not only opened every book, but we turned over every leaf in each volume, not contenting ourselves with a mere shake, according to the fashion of some of our police officers. We also measured the thickness of every book-*cover*, with the most accurate admeasurement, and applied to them the most jealous scrutiny of the microscope. Had any of the bindings been recently meddled with, it would have been utterly impossible that the fact should have escaped observation. Some five or six volumes, just from the hands of the binder, we carefully probed, longitudinally, with the needles."

"You explored the floors beneath the carpets?"

"Beyond doubt. We removed every carpet, and examined the boards with the microscope."

"And the paper on the walls?"

"Yes."

"You looked into the cellars?"

"We did; and, as time and labour were no objects, we dug up every one of them to the depth of four feet."

"Then," I said, "you have been making a miscalculation, and the letter is *not* upon the premises, as you suppose."

"I fear you are right there," said the Prefect. "And now, Dupin, what would you advise me to do?"

"To make a thorough re-search of the premises."

"That is absolutely needless," replied G——. "I am not more sure that I breathe than I am that the letter is not at the Hotel."

"I have no better advice to give you," said Dupin. "You have, of course, an accurate description of the letter?"

"Oh yes!"—And here the Prefect, producing a memorandum-book, proceeded to read aloud a minute account of the internal, and especially of the external, appearance of the missing document. Soon after finishing the perusal of this description, he took his departure, more entirely depressed in spirits than I had ever known the good gentleman before.

In about a month afterwards he paid us another visit, and found us occupied very nearly as before. He took a pipe and a chair, and entered into some ordinary conversation. At length I said,—

"Well, but G——, what of the purloined letter? I presume you have at last made up your mind that there is no such thing as overreaching the Minister?"

"Confound him, say I—yes; I made the re-examination, however, as Dupin suggested—but it was all labour lost, as I knew it would be."

"How much was the reward offered, did you say?" asked Dupin.

"Why, a very great deal—a *very* liberal reward—I don't like to say how much, precisely; but one thing I *will* say, that I wouldn't mind giving my individual check for fifty thousand francs to any one who could obtain me that letter. The fact is, it is becoming of more and more importance every day; and the reward has been lately doubled. If it were trebled, however, I could do no more than I have done."

"Why, yes," said Dupin, drawlingly, between the whiffs of his meerschaum, "I really—think, G——, you have not exerted yourself—to the utmost in this matter. You might—do a little more, I think, eh?"

"How?—in what way?"

"Why—puff, puff—you might—puff, puff—employ counsel in the matter, eh?—puff, puff, puff. Do you remember the story they tell of Abernethy?"

"No; hang Abernethy!"

"To be sure! hang him and welcome. But, once upon a time, a certain rich miser conceived the design of spunging upon this Abernethy for a medical opinion. Getting up, for this purpose, an ordinary conversation in a private company, he insinuated his case to the physician, as that of an imaginary individual.

" 'We will suppose,' said the miser, 'that his symptoms are such and such; now, doctor, what would *you* have directed him to take?'

" 'Take!' said Abernethy, 'why, take *advice*, to be sure.' "

"But," said the Prefect, a little discomposed, "I am *perfectly* willing to take advice, and to pay for it. I would *really* give fifty thousand francs, every *centime* of it, to any one who would aid me in the matter!"

"In that case," replied Dupin, opening a drawer, and producing a check-book, "you may as well fill me up a check for the amount mentioned. When you have signed it, I will hand you the letter."

I was astounded. The Prefect appeared absolutely thunder-stricken. For some minutes he remained speechless and motionless, looking incredulously at my friend with open mouth, and eyes that seemed starting from their sockets; then, apparently recovering himself in some measure, he seized a pen, and after several pauses and vacant stares, finally filled up and signed a check for fifty thousand francs, and handed it across the table to Dupin. The latter examined it carefully and deposited it in his pocket-book; then, unlocking an *escritoire,*[6] took thence a letter and gave it to the Prefect. This functionary grasped it in a perfect agony of joy, opened it with a trembling hand, cast a rapid glance at its contents, and then, scrambling and struggling to the door, rushed at length unceremoniously from the room and from the house, without having uttered a solitary syllable since Dupin had requested him to fill up the check.

When he had gone, my friend entered into some explanations.

"The Parisian police," he said, "are exceedingly able in their way. They are persevering, ingenious, cunning, and thoroughly versed in the knowledge which their duties seem chiefly to demand. Thus, when G—— detailed to us his mode of searching the premises at the Hotel D——, I felt entire confidence in his having made a satisfactory investigation—so far as his labours extended."

"So far as his labours extended?" said I.

"Yes," said Dupin. "The measures adopted were not only the best of their kind, but carried out to absolute perfection. Had the letter been deposited within the range of their search, these fellows would, beyond a question, have found it."

I merely laughed—but he seemed quite serious in all that he said.

"The measures, then," he continued, "were good in their kind, and well executed; their defect lay in their being inapplicable to the case, and to the man. A certain set of highly ingenious resources are, with the Prefect, a sort of Procrustean bed,[7] to which he forcibly adapts his designs. But he perpetually errs by being too deep or too shallow, for the matter in hand; and many a schoolboy is a better reasoner than he. I knew one about eight years of age, whose success at guessing in the game of 'even and odd' attracted universal admiration. This game is simple, and is played with marbles. One player holds in his hand a number of these toys, and demands of another whether that number is even or odd. If the guess is right, the guesser wins one; if wrong, he loses one. The boy to whom I allude won all the marbles of the school. Of course he had some principle of guessing; and this lay in mere observation and admeasurement of the astuteness of his opponents. For example, an arrant simpleton is his opponent, and, holding up his closed hand, asks, 'are they even or odd?' Our schoolboy replies, 'odd,' and loses; but upon the second trial he wins, for he then says to himself, 'the simpleton had them even upon the first trial, and his amount of cunning is just sufficient to make him have them odd upon the second; I will therefore guess odd;'—he guesses odd, and wins. Now, with a simpleton a degree above the first, he would have reasoned thus: 'this fellow finds that in the first instance I guessed

6. **Writing desk (now spelled *écritoire*).**
7. **Procrustes, legendary Greek bandit, made his victims fit the bed he bound** them to, either by streching them to the required length or by hacking off any surplus length in the feet and legs.

odd, and, in the second, he will propose to himself, upon the first impulse, a simple variation from even to odd, as did the first simpleton; but then a second thought will suggest that this is too simple a variation, and finally he will decide upon putting it even as before. I will therefore guess even;'—he guesses even, and wins. Now this mode of reasoning in the schoolboy, whom his fellows termed 'lucky,'—what, in its last analysis, is it?"

"It is merely," I said, "an identification of the reasoner's intellect with that of his opponent."

"It is," said Dupin; "and, upon inquiring of the boy by what means he effected the *thorough* identification in which his success consisted, I received answer as follows: 'When I wish to find out how wise, or how stupid, or how good, or how wicked is any one, or what are his thoughts at the moment, I fashion the expression of my face, as accurately as possible, in accordance with the expression of his, and then wait to see what thoughts or sentiments arise in my mind or heart, as if to match or correspond with the expression.' This response of the schoolboy lies at the bottom of all the spurious profundity which has been attributed to Rochefoucault, to La Bruyère, to Machiavelli, and to Campanella."[8]

"And the identification," I said, "of the reasoner's intellect with that of his opponent, depends, if I understand you aright, upon the accuracy with which the opponent's intellect is admeasured."

"For its practical value it depends upon this," replied Dupin; "and the Prefect and his cohort fail so frequently, first, by default of this identification, and, secondly, by ill-admeasurement, or rather through non-admeasurement, of the intellect with which they are engaged. They consider only their *own* ideas of ingenuity; and, in searching for any thing hidden, advert only to the modes in which *they* would have hidden it. They are right in this much—that their own ingenuity is a faithful representative of that of *the mass*; but when the cunning of the individual felon is diverse in character from their own, the felon foils them, of course. This always happens when it is above their own, and very usually when it is below. They have no variation of principle in their investigations; at best, when urged by some unusual emergency—by some extraordinary reward —they extend or exaggerate their old modes of *practice*, without touching their principles. What, for example, in this case of D——, has been done to vary the principle of action? What is all this boring, and probing, and sounding, and scrutinizing with the microscope, and dividing the surface of the building into registered square inches—what is it all but an exaggeration *of the application* of the one principle or set of principles of search, which are based upon the one set of notions regarding human ingenuity, to which the

8. An oddly assorted group of moralists and political and religious philosophers, all denigrated by Dupin. The original reads "La Bougive," probably a printer's error.

Prefect, in the long routine of his duty, has been accustomed? Do you not see he has taken it for granted that *all* men proceed to conceal a letter,—not exactly in a gimlet-hole bored in a chair-leg—but, at least, in *some* out-of-the-way hole or corner suggested by the same tenor of thought which would urge a man to secrete a letter in a gimlet-hole bored in a chair-leg? And do you not see also, that such *recherches*[9] nooks for concealment are adapted only for ordinary occasions, and would be adopted only by ordinary intellects; for, in all cases of concealment, a disposal of the article concealed—a disposal of it in this *recherché* manner,—is, in the very first instance, presumed and presumable; and thus its discovery depends, not at all upon the acumen, but altogether upon the mere care, patience, and determination of the seekers; and where the case is of importance—or, what amounts to the same thing in the policial eyes, when the reward is of magnitude, the qualities in question have *never* been known to fail. You will now understand what I meant in suggesting that, had the purloined letter been hidden any where within the limits of the Prefect's examination—in other words, had the principle of its concealment been comprehended within the principles of the Prefect—its discovery would have been a matter altogether beyond question. This functionary, however, has been thoroughly mystified; and the remote source of his defeat lies in the supposition that the Minister is a fool, because he has acquired renown as a poet. All fools are poets; this the Prefect *feels*; and he is merely guilty of a *non distributio medii*[1] in thence inferring that all poets are fools."

"But is this really the poet?" I asked. "There are two brothers, I know; and both have attained reputation in letters. The Minister I believe has written learnedly on the Differential Calculus. He is a mathematician, and no poet."

"You are mistaken; I know him well; he is both. As poet *and* mathematician, he would reason well; as poet, profoundly; as mere mathematician, he could not have reasoned at all, and thus would have been at the mercy of the Prefect."

"You surprise me," I said, "by these opinions, which have been contradicted by the voice of the world. You do not mean to set at naught the well-digested idea of centuries. The mathematical reason has been long regarded as *the* reason *par excellence*."

"'Il y a à parièr,' replied Dupin, quoting from Chamfort, 'que toute idée publique, toute convention reçue, est une sottise, car elle a convenue au plus grand nombre.'[2] The mathematicians, I grant

---

9. Out of the ordinary, esoteric (then permissible without the acute accent or with it, as just below).
1. A fallacy in logic in which neither premise of a syllogism "distributes" (that is, conveys information about every member of the class) the middle term. Acording to Dupin, the Prefect does not allow for the possibility that some poets are not fools.
2. "The odds are that every common notion, every accepted convention, is nonsense, precisely because it has suited itself to the majority." Sébastian Roch Nicolas Chamfort (1741–94), author of *Maximes et Pensées*.

you, have done their best to promulgate the popular error to which you allude, and which is none the less an error for its promulgation as truth. With an art worthy a better cause, for example, they have insinuated the term 'analysis' into application to algebra. The French are the originators of this particular deception; but if a term is of any importance—if words derive any value from applicability —then 'analysis' conveys 'algebra' about as much as, in Latin, '*ambitus*' implies 'ambition,' '*religio*' 'religion,' or '*homines honesti*,' a set of *honourable* men."

"You have a quarrel on hand, I see," said I, "with some of the algebraists of Paris; but proceed."

"I dispute the availability, and thus the value, of that reason which is cultivated in any especial form other than the abstractly logical. I dispute, in particular, the reason educed by mathematical study. The mathematics are the science of form and quantity; mathematical reasoning is merely logic applied to observation upon form and quantity. The great error lies in supposing that even the truths of what is called *pure* algebra, are abstract or general truths. And this error is so egregious that I am confounded at the universality with which it has been received. Mathematical axioms are *not* axioms of general truth. What is true of *relation*—of form and quantity—is often grossly false in regard to morals, for example. In this latter science it is very usually *un*true that the aggregated parts are equal to the whole. In chemistry also the axiom fails. In the consideration of motive it fails; for two motives, each of a given value, have not, necessarily, a value when united, equal to the sum of their values apart. There are numerous other mathematical truths which are only truths within the limits of *relation*. But the mathematician argues, from his *finite truths*, through habit, as if they were of an absolutely general applicability—as the world indeed imagines them to be. Bryant, in his very learned 'Mythology,' mentions an analogous source of error, when he says that 'although the Pagan fables are not believed, yet we forget ourselves continually, and make inferences from them as existing realities.'[3] With the algebraist, however, who are Pagans themselves, the 'Pagan fables' *are* believed, and the inferences are made, not so much through lapse of memory, as through an unaccountable addling of the brains. In short, I never yet encountered the mere mathematician who could be trusted out of equal roots, or one who did not clandestinely hold it as a point of his faith x^2+px was absolutely and unconditionally equal to q. Say to one of these gentlemen, by way of experiment, if you please, that you believe occasions may occur where x^2+px is *not* altogether equal to q, and, having made him understand what you mean, get out of his reach as speedily as

---

3. Jacob Bryant (1715–1804), English scholar who wrote *A New System, or an*   *Analysis of Antient Mythology* (1774–76).

convenient, for, beyond doubt, he will endeavour to knock you down.

"I mean to say," continued Dupin, while I merely laughed at his last observations, "that if the Minister had been no more than a mathematician, the Prefect would have been under no necessity of giving me this check. Had he been no more than a poet, I think it probable that he would have foiled us all. I knew him, however, as both mathematician and poet, and my measures were adapted to his capacity, with reference to the circumstances by which he was surrounded. I knew him as a courtier, too, and as a bold *intriguant*. Such a man, I considered, could not fail to be aware of the ordinary policial modes of action. He could not have failed to anticipate— and events have proved that he did not fail to anticipate—the way-layings to which he was subjected. He must have foreseen, I reflected, the secret investigations of his premises. His frequent absences from home at night, which were hailed by the Prefect as certain aids to his success, I regarded only as *ruses*, to afford opportunity for thorough search to the police, and thus the sooner to impress them with the conviction to which G——, in fact, did finally arrive—the conviction that the letter was not upon the premises. I felt, also, that the whole train of thought, which I was at some pains in detailing to you just now, concerning the invariable principle of policial action in searches for articles concealed—I felt that this whole train of thought would necessarily pass through the mind of the Minister. It would imperatively lead him to despise all the ordinary *nooks* of concealment. *He* could not, I reflected, be so weak as not to see that the most intricate and remote recess of his hotel would be as open as his commonest closets to the eyes, to the probes, to the gimlets, and to the microscopes of the Prefect. I saw, in fine, that he would be driven, as a matter of course, to *simplicity*, if not deliberately induced to it as a matter of choice. You will remember, perhaps, how desperately the Prefect laughed when I suggested, upon our first interview, that it was just possible this mystery troubled him so much on account of its being so *very* self-evident."

"Yes," said I, "I remember his merriment well. I really thought he would have fallen into convulsions."

"The material world," continued Dupin, "abounds with very strict analogies to the immaterial; and thus some colour of truth has been given to the rhetorical dogma, that metaphor, or simile, may be made to strengthen an argument, as well as to embellish a description. The principle of the *vis inertiæ*,[4] for example, with the amount of *momentum* proportionate with it and consequent upon it, seems to be identical in physics and metaphysics. It is not more true in the former, that a large body is with more difficulty set in

4. The power of inertia.

motion than a smaller one, and that its subsequent *impetus* is commensurate with this difficulty, than it is, in the latter, that intellects of the vaster capacity, while more forcible, more constant, and more eventful in their movements than those of inferior grade, are yet the less readily moved, and more embarrassed and full of hesitation in the first few steps of their progress. Again: have you ever noticed which of the street signs, over the shop-doors, are the most attractive of attention?"

"I have never given the matter a thought," I said.

"There is a game of puzzles," he resumed, "which is played upon a map. One party playing requires another to find a given word—the name of town, river, state, or empire—any word, in short, upon the motley and perplexed surface of the chart. A novice in the game generally seeks to embarrass his opponents by giving them the most minutely lettered names; but the adept selects such words as stretch, in large characters, from one end of the chart to the other. These, like the over-largely lettered signs and placards of the street, escape obesrvation by dint of being excessively obvious; and here the physical oversight is precisely analogous with the moral inapprehension by which the intellect suffers to pass unnoticed those considerations which are too obtrusively and too palpably self-evident. But this is a point, it appears, somewhat above or beneath the understanding of the Prefect. He never once thought it probable, or possible, that the Minister had deposited the letter immediately beneath the nose of the whole world, by way of best preventing any portion of that world from perceiving it.

"But the more I reflected upon the daring, dashing, and discriminating ingenuity of D——; upon the fact that the document must always have been *at hand*, if he intended to use it to good purpose; and upon the decisive evidence, obtained by the Prefect, that it was not hidden within the limits of that dignitary's ordinary search—the more satisfied I became that, to conceal this letter, the Minister had resorted to the comprehensive and sagacious expedient of not attempting to conceal it at all.

"Full of these ideas, I prepared myself with a pair of green spectacles, and called one fine morning, quite by accident, at the ministerial hotel. I found D—— at home, yawning, lounging, and dawdling as usual, and pretending to be in the last extremity of *ennui*.[5] He is, perhaps, the most really energetic human being now alive—but that is only when nobody sees him.

"To be even with him, I complained of my weak eyes, and lamented the necessity of the spectacles, under cover of which I cautiously and thoroughly surveyed the whole apartment, while seemingly intent only upon the conversation of my host.

"I paid especial attention to a large writing-table near which he sat, and upon which lay confusedly, some miscellanous letters and

5. Boredom.

other papers, with one or two musical instruments and a few books. Here, however, after a long and very deliberate scrutiny, I saw nothing to excite particular suspicion.

"At length my eyes, in going the circuit of the room, fell upon a trumpery fillagree card-rack of pasteboard, that hung dangling by a dirty blue riband, from a little brass knob just beneath the middle of the mantel-piece. In this rack, which had three or four compartments, were five or six visiting-cards, and a solitary letter. This last was much soiled and crumpled. It was torn nearly in two, across the middle—as if a design, in the first instance, to tear it entirely up as worthless, had been altered, or stayed, in the second. It had a large black seal, bearing the D—— cipher *very* conspicuously, and was addressed, in a diminutive female hand, to D——, the minister, himself. It was thrust carelessly, and even, as it seemed, contemptuously, into one of the uppermost divisions of the rack.

"No sooner had I glanced at this letter, than I concluded it to be that of which I was in search. To be sure, it was, to all appearance, radically different from the one of which the Prefect had read us so minute a description. Here the seal was large and black, with the D—— cipher; there, it was small and red, with the ducal arms of the S—— family. Here, the address, to the minister, was diminutive and feminine; there, the superscription, to a certain royal personage, was markedly bold and decided; the size alone formed a point of correspondence. But, then, the *radicalness* of these differences, which was excessive; the dirt, the soiled and torn condition of the paper, so inconsistent with the *true* methodical habits of D——, and so suggestive of a design to delude the beholder into an idea of the worthlessness of the document; these things, together with the hyper-obtrusive situation of this document, full in the view of every visiter, and thus exactly in accordance with the conclusions to which I had previously arrived; these things, I say, were strongly corroborative of suspicion, in one who came with the intention to suspect.

"I protracted my visit as long as possible, and, while I maintained a most animated discussion with the minister, upon a topic which I knew well had never failed to interest and excite him, I kept my attention really riveted upon the letter. In this examination, I committed to memory its external appearance and arrangement in the rack; and also fell, at length, upon a discovery which set at rest whatever trivial doubt I might have entertained. In scrutinizing the edges of the paper, I observed them to be more *chafed* than seemed necessary. They presented the *broken* appearance which is manifested when a stiff paper, having been once folded and pressed with a folder, is refolded in a reversed direction, in the same creases or edges which had formed the original fold. This discovery was sufficient. It was clear to me that the letter had been turned, as a glove, inside out, re-directed, and re-sealed. I bade the minister good morning, and took my departure at once, leaving a gold snuff-box upon the table.

"The next morning I called for the snuff-box, when we resumed, quite eagerly, the conversation of the preceding day. While thus engaged, however, a loud report, as if of a pistol, was heard immediately beneath the windows of the hotel, and was succeeded by a series of fearful screams, and the shoutings of a terrified mob. D—— rushed to a casement, threw it open, and looked out. In the meantime, I stepped to the card-rack, took the letter, put it in my pocket, and replaced it by a *fac-simile*, which I had carefully prepared at my lodgings—imitating the D—— cipher, very readily, by means of a seal formed of bread.

"The disturbance in the street had been occasioned by the frantic behaviour of a man with a musket. He had fired it among a crowd of women and children. It proved, however, to have been without ball, and the fellow was suffered to go his way as a lunatic or a drunkard. When he had gone, D—— came from the window, whither I had followed him immediately upon securing the object in view. Soon afterwards I bade him farewell. The pretended lunatic was a man in my own pay."

"But what purpose had you," I asked, "in replacing the letter by a *fac-simile*? Would it not have been better, at the first visit, to have seized it openly, and departed?"

"D——," replied Dupin, "is a desperate man, and a man of nerve. His hotel, too, is not without attendants devoted to his interests. Had I made the wild attempt you suggest, I should never have left the ministerial presence alive. The good people of Paris would have heard of me no more. But I had an object apart from these considerations. You know my political prepossessions. In this matter, I act as a partisan of the lady concerned. For eighteen months the minister has had her in his power. She has now him in hers—since, being unaware that the letter is not in his possession, he will proceed with his exactions as if it was. Thus will he inevitably commit himself, at once, to his political destruction. His downfall, too, will not be more precipitate than awkward. It is all very well to talk about the *facilis descensus Averni*;[6] but in all kinds of climbing, as Catalini said of singing, it is far more easy to get up than to come down.[7] In the present instance I have no sympathy— at least no pity—for him who descends. He is that *monstrum horrendum*,[8] an unprincipled man of genius. I confess, however, that I should like very well to know the precise character of his thoughts, when, being defied by her whom the Prefect terms 'a certain per-

---

6. Slightly misquoted from Virgil's *Aeneid*, Book VI: "the descent to Avernus [Hell] is easy."
7. Angelica Catalani (1780–1849), Ital-
ian singer.
8. Dreadful monstrosity (Virgil's epithet for Polyphemus, the one-eyed man-eating giant).

sonage,' he is reduced to opening the letter which I left for him in the card-rack."

"How? did you put any thing particular in it?"

"Why—it did not seem altogether right to leave the interior blank —that would have been insulting. To be sure, D——, at Vienna once, did me an evil turn, which I told him, quite good-humouredly, that I should remember. So, as I knew he would feel some curiosity in regard to the identity of the person who had outwitted him, I thought it a pity not to give him a clue. He is well acquainted with my MS., and I just copied into the middle of the blank sheet the words—

"'—— Un dessein si funeste,
S'il n'est digne d'Atrée, est digne de Thyeste.'

They are to be found in Crébillon's 'Atrée.' "⁹

1844

The Philosophy of Composition[1]

Charles Dickens, in a note now lying before me, alluding to an examination I once made of the mechanism of "Barnaby Rudge," says—"By the way, are you aware that Godwin wrote his 'Caleb Williams' backwards? He first involved his hero in a web of difficulties, forming the second volume, and then, for the first, cast about him for some mode of accounting for what had been done."[2]

I cannot think this the *precise* mode of procedure on the part of Godwin—and indeed what he himself acknowledges, is not altogether in accordance with Mr. Dickens' idea—but the author of "Caleb Williams" was too good an artist not to perceive the advantage derivable from at least a somewhat similar process. Nothing is more clear than that every plot, worth the name, must be elaborated to its *dénouement* before any thing be attempted with the pen. It is only with the *dénouement* constantly in view that we can give a plot its indispensable air of consequence, or causation, by making the incidents, and especially the tone at all points, tend to the development of the intention.

9. Prosper Jolyot de Crébillon wrote *Atrée et Thyeste* (1707), in which Thyestes seduces the wife of his brother Atreus, the King of Mycenae; in revenge Atreus murders the sons of Thyestes and serves them to their father at a feast. The quotation reads, "So baneful a scheme, if not worthy of Atreus, is worthy of Thyestes."

1. The title means something like "The Theory of Writing." Poe wrote the work as a lecture in hopes of capitalizing on the success of *The Raven*. For years in his reviews Poe had campaigned for deliberate artistry rather than uncontrolled effusions, and *The Philosophy of Composition* must be regarded as part of that campaign rather than as a factual account of how Poe actually wrote *The Raven*. In a letter of Agust 9, 1846, Poe called the essay his "best specimen of analysis." The text here is that of the first printing, in *Graham's Magazine*, 28 (April, 1846), 163–67.

2. William Godwin makes this claim in his 1832 preface to *Caleb Williams* (first published in 1794).

There is a radical error, I think, in the usual mode of constructing a story. Either history affords a thesis—or one is suggested by an incident of the day—or, at best, the author sets himself to work in the combination of striking events to form merely the basis of his narrative—designing, generally, to fill in with description, dialogue, or authorial comment, whatever crevices of fact, or action, may, from page to page, render themselves apparent.

I prefer commencing with the consideration of an *effect*. Keeping originality *always* in view—for he is false to himself who ventures to dispense with so obvious and so easily attainable a source of interest—I say to myself, in the first place, "Of the innumerable effects, or impressions, of which the heart; the intellect, or (more generally) the soul is susceptible, what one shall I, on the present occasion, select?" Having chosen a novel, first, and secondly a vivid effect, I consider whether it can best be wrought by incident or tone—whether by ordinary incidents and peculiar tone, or the converse, or by peculiarity both of incident and tone—afterward looking about me (or rather within) for such combinations of event, or tone, as shall best aid me in the construction of the effect.

I have often thought how interesting a magazine paper might be written by any author who would—that is to say, who could—detail, step by step, the processes by which any one of his compositions attained its ultimate point of completion. Why such a paper has never been given to the world, I am much at a loss to say—but, perhaps, the autorial vanity has had more to do with the omission than any one other cause. Most writers—poets in especial—prefer having it understood that they compose by a species of fine frenzy[3]—an ecstatic intuition—and would positively shudder at letting the public take a peep behind the scenes, at the elaborate and vacillating crudities of thought—at the true purposes seized only at the last moment—at the innumerable glimpses of idea that arrived not at the maturity of full view—at the fully matured fancies discarded in despair as unmanageable—at the cautious selections and rejections—at the painful erasures and interpolations—in a word, at the wheels and pinions—the tackle for scene-shifting—the step-ladders and demon-traps—the cock's feathers, the red paint and the black patches, which, in ninety-nine cases out of the hundred, constitute the properties of the literary *histrio*.[4]

I am aware, on the other hand, that the case is by no means common, in which an author is at all in condition to retrace the steps by which his conclusions have been attained. In general, suggestions, having arisen pell-mell, are pursued and forgotten in a similar manner.

---

3. Shakespeare's *Midsummer Night's Dream* 5.1.12, in Theseus's description of the poet: "The poet's eye, in a fine frenzy rolling,/Doth glance from heaven to earth, from earth to heaven/And as imagination bodies forth/The forms of things unknown, the poet's pen/Turns them to shapes, and gives to airy nothing/A local habitation and a name."
4. Artist.

For my own part, I have neither sympathy with the repugnance alluded to, nor, at any time, the least difficulty in recalling to mind the progressive steps of any of my compositions; and, since the interest of an analysis, or reconstruction, such as I have considered a *desideratum*,[5] is quite independent of any real or fancied interest in the thing analyzed, it will not be regarded as a breach of decorum on my part to show the *modus operandi*[6] by which some one of my own works was put together. I select "The Raven," as the most generally known. It is my design to render it manifest that no one point in its composition is referrible either to accident or intuition— that the work proceeded, step by step, to its completion with the precision and rigid consequence of a mathematical problem.

Let us dismiss, as irrelevant to the poem *per se*, the circumstance —or say the necessity—which, in the first place, gave rise to the intention of composing *a* poem that should suit at once the popular and the critical taste.

We commence, then, with this intention.

The initial consideration was that of extent. If any literary work is too long to be read at one sitting, we must be content to dispense with the immensely important effect derivable from unity of impression—for, if two sittings be required, the affairs of the world interfere, and every thing like totality is at once destroyed. But since, *ceteris paribus*,[7] no poet can afford to dispense with *any thing* that may advance his design, it but remains to be seen whether there is, in extent, any advantage to counterbalance the loss of unity which attends it. Here I say no, at once. What we term a long poem is, in fact, merely a succession of brief ones—that is to say, of brief poetical effects. It is needless to demonstrate that a poem is such, only inasmuch as it intensely excites, by elevating, the soul; and all intense excitements are, through a psychal necessity, brief. For this reason, at least one half of the "Paradise Lost"[8] is essentially prose —a succession of poetical excitements interspersed, *inevitably*, with corresponding depressions—the whole being deprived, through the extremeness of its length, of the vastly important artistic element, totality, or unity, of effect.

It appears evident, then, that there is a distinct limit, as regards length, to all works of literary art—the limit of a single sitting—and that, although in certain classes of prose composition, such as "Robinson Crusoe,"[9] (demanding no unity,) this limit may be advantageously overpassed, it can never properly be overpassed in a poem. Within this limit, the extent of a poem may be made to bear mathematical relation to its merit—in other words, to the excitement or elevation—again in other words, to the degree of the true

5. Something to be desired.
6. Method of procedure.
7. Other things being equal.
8. The 12-book blank-verse epic by John Milton which contains some 10,500 lines,

more than a hundred times as many lines as Poe considered desirable in a poem.
9. Daniel Defoe's novel of shipwreck in the Caribbean (1719), based on the experiences of Alexander Selkirk.

poetical effect which it is capable of inducing; for it is clear that the brevity must be in direct ratio of the intensity of the intended effect:—this, with one proviso—that a certain degree of duration is absolutely requisite for the production of any effect at all.

Holding in view these considerations, as well as that degree of excitement which I deemed not above the popular, while not below the critical, taste, I reached at once what I conceived the proper *length* for my intended poem—a length of about one hundred lines. It is, in fact, a hundred and eight.

My next thought concerned the choice of an impression, or effect, to be conveyed: and here I may as well observe that, throughout the construction, I kept steadily in view the design of rendering the work *universally* appreciable. I should be carried too far out of my immediate topic were I to demonstrate a point upon which I have repeatedly insisted, and which, with the poetical, stands not in the slightest need of demonstration—the point, I mean, that Beauty is the sole legitimate province of the poem. A few words, however, in elucidation of my real meaning, which some of my friends have evinced a disposition to misrepresent. That pleasure which is at once the most intense, the most elevating, and the most pure, is, I believe, found in the contemplation of the beautiful. When, indeed, men speak of Beauty, they mean, precisely, not a quality, as is supposed, but an effect—they refer, in short, just to that intense and pure elevation of *soul*—*not* of intellect, or of heart—upon which I have commented, and which is experienced in consequence of contemplating "the beautiful." Now I designate Beauty as the province of the poem, merely because it is an obvious rule of Art that effects should be made to spring from direct causes —that objects should be attained through means best adapted for their attainment—no one as yet having been weak enough to deny that the peculiar elevation alluded to, is *most readily* attained in the poem. Now the object, Truth, or the satisfaction of the intellect, and the object Passion, or the excitement of the heart, are, although attainable, to a certain extent, in poetry, far more readily attainable in prose. Truth, in fact, demands a precision, and Passion, a *homeliness* (the truly passionate will comprehend me) which are absolutely antagonistic to that Beauty which, I maintain, is the excitement, or pleasurable elevation, of the soul. It by no means follows from any thing here said, that passion, or even truth, may not be introduced, and even profitably introduced, into a poem—for they may serve in elucidation, or aid the general effect, as do discords in music, by contrast—but the true artist will always contrive, first, to tone them into proper subservience to the predominant aim, and, secondly, to enveil them, as far as possible, in that Beauty which is the atmosphere and the essence of the poem.

Regarding, then, Beauty as my province, my next question referred to the *tone* of its highest manifestation—and all experience has shown that this tone is one of *sadness*. Beauty of whatever kind, in its supreme development, invariably excites the sensitive soul to tears. Melancholy is thus the most legitimate of all the poetical tones.

The length, the province, and the tone, being thus determined, I betook myself to ordinary induction, with the view of obtaining some artistic piquancy which might serve me as a key-note in the construction of the poem—some pivot upon which the whole structure might turn. In carefully thinking over all the usual artistic effects—or more properly *points*, in the theatrical sense—I did not fail to perceive immediately that no one had been so universally employed as that of the *refrain*. The universality of its employment sufficed to assure me of its intrinsic value, and spared me the necessity of submitting it to analysis. I considered it, however, with regard to its susceptibility of improvement, and soon saw it to be in a primitive condition. As commonly used, the *refrain*, or burden, not only is limited to lyric verse, but depends for its impression upon the force of monotone—both in sound and thought. The pleasure is deduced solely from the sense of identity—of repetition. I resolved to diversify, and so vastly heighten, the effect, by adhering, in general, to the monotone of sound, while I continually varied that of thought: that is to say, I determined to produce continuously novel effects, by the variation *of the application* of the *refrain* —the *refrain* itself remaining, for the most part, unvaried.

These points being settled, I next bethought me of the *nature* of my *refrain*. Since its application was to be repeatedly varied, it was clear that the *refrain* itself must be brief, for there would have been an insurmountable difficulty in frequent variations of application in any sentence of length. In proportion to the brevity of the sentence, would, of course, be the facility of the variation. This led me at once to a single word as the best *refrain*.

The question now arose as to the *character* of the word. Having made up my mind to a *refrain*, the division of the poem into stanzas was, of course, a corollary: the *refrain* forming the close to each stanza. That such a close, to have force, must be sonorous and susceptible of protracted emphasis, admitted no doubt: and these considerations inevitably led me to the long *o* as the most sonorous vowel, in connection with *r* as the most producible consonant.

The sound of the *refrain* being thus determined, it became necessary to select a word embodying this sound, and at the same time in the fullest possible keeping with that melancholy which I had predetermined as the tone of the poem. In such a search it would have been absolutely impossible to overlook the word "Nevermore." In fact, it was the very first which presented itself.

The next *desideratum* was a pretext for the continuous use of the one word "nevermore." In observing the difficulty which I at once found in inventing a sufficiently plausible reason for its continuous repetition, I did not fail to perceive that this difficulty arose solely from the pre-assumption that the word was to be so continuously or monotonously spoken by *a human* being—I did not fail to perceive, in short, that the difficulty lay in the reconciliation of this monot-ony with the exercise of reason on the part of the creature repeating the word. Here, then, immediately arose the idea of a *non*-reasoning creature capable of speech; and, very naturally, a parrot, in the first instance, suggested itself, but was superseded forthwith by a Raven, as equally capable of speech, and infinitely more in keeping with the intended *tone*.

I had now gone so far as the conception of a Raven—the bird of ill omen—monotonously repeating the one-word, "Nevermore," at the conclusion of each stanza, in a poem of melancholy tone, and in length about one hundred lines. Now, never losing sight of the object *supremeness*, or perfection, at all points, I asked myself—"Of all melancholy topics, what, according to the *universal* understand-ing of mankind, is the *most* melancholy?" Death—was the obvious reply. "And when," I said, "is this most melancholy of topics most poetical?" From what I have already explained at some length, the answer, here also, is obvious—"When it most closely allies itself to *Beauty*: the death, then, of a beautiful woman is, unquestionably, the most poetical topic in the world—and equally is it beyond doubt that the lips best suited for such topic are those of a bereaved lover."

I had now to combine the two ideas, of a lover lamenting his deceased mistress and a Raven continuously repeating the word "Nevermore"—I had to combine these, bearing in mind my design of varying, at every turn, the *application* of the word repeated; but the only intelligible mode of such combination is that of imagining the Raven employing the word in answer to the queries of the lover. And here it was that I saw at once the opportunity afforded for the effect on which I had been depending—that is to say, the effect of the *variation of application*. I saw that I could make the first query propounded by the lover—the first query to which the Raven should reply "Nevermore"—that I could make this first query a common-place one—the second less so—the third still less, and so on—until at length the lover, startled from his original *nonchalance* by the melancholy character of the word itself—by its frequent repetition —and by a consideration of the ominous reputation of the fowl that uttered it—is at length excited to superstition, and wildly propounds queries of a far different character—queries whose solution he has passionately at heart—propounds them half in superstition and half in that species of despair which delights in self-torture—propounds them not altogether because he believes in the prophetic or demoniac character of the bird (which, reason assures him, is merely repeating a lesson learned by rote) but because he experi-ences a phrenzied pleasure in so modeling his questions as to receive from the *expected* "Nevermore" the most delicious because the most intolerable of sorrow. Perceiving the opportunity thus afforded me —or, more strictly, thus forced upon me in the progress of the construction—I first established in mind the climax, or concluding query—that to which "Nevermore" should be in the last place an answer—that in reply to which this word "Nevermore" should involve the utmost conceivable amount of sorrow and despair.

Here then the poem may be said to have its beginning—at the end, where all works of art should begin—for it was here, at this point of my preconsiderations, that I first put pen to paper in the composition of the stanza:

"Prophet," said I, "thing of evil! prophet still if bird or devil!
By that heaven that bends above us—by that God we both adore,
Tell this soul with sorrow laden, if within the distant Aidenn,
It shall clasp a sainted maiden whom the angels name Lenore—
Clasp a rare and radiant maiden whom the angels name Lenore."
    Quoth the raven "Nevermore."

I composed this stanza, at this point, first that, by establishing the climax, I might the better vary and graduate, as regards seriousness and importance, the preceding queries of the lover—and, secondly, that I might definitely settle the rhythm, the metre, and the length and general arrangement of the stanza—as well as graduate the stanzas which were to precede, so that none of them might surpass this in rhythmical effect. Had I been able, in the subsequent composition, to construct more vigorous stanzas, I should, without scruple, have purposely enfeebled them, so as not to interfere with the climacteric effect.

And here I may as well say a few words of the versification. My first object (as usual) was originality. The extent to which this has been neglected, in versification, is one of the most unaccountable things in the world. Admitting that there is little possibility of variety in mere *rhythm*, it is still clear that the possible varieties of metre and stanza are absolutely infinite—and yet, *for centuries, no man, in verse, has ever done, or ever seemed to think of doing, an original thing.* The fact is, originality (unless in minds of very unusual force) is by no means a matter, as some suppose, of impulse or intuition. In general, to be found, it must be elaborately sought, and although a positive merit of the highest class, demands in its attainment less of invention than negation.

Of course, I pretend to no originality in either the rhythm or metre of the "Raven." The former is trochaic—the latter is octameter acatalectic, alternating with heptameter catalectic repeated in the *refrain* of the fifth verse, and terminating with tetrameter catalectic. Less pedantically—the feet employed throughout (trochees) consist of a long syllable followed by a short: the first line of the stanza consists of eight of these feet—the second of seven and a half (in effect two-thirds)—the third of eight—the fourth of seven and a half—the fifth the same—the sixth three and a half. Now, each of these lines, taken individually, has been employed before, and what originality the "Raven" has, is in their *combination into stanza*; nothing even remotely approaching this combination has ever been attempted. The effect of this originality of combination is aided by other unusual, and some altogether novel effects, arising from an extension of the application of the principles of rhyme and alliteration.

The next point to be considered was the mode of bringing together the lover and the Raven—and the first branch of this consideration was the *locale*. For this the most natural suggestion might seem to be a forest, or the fields—but it has always appeared to me that a close *circumscription of space* is absolutely necessary to the effect of insulated incident:—it has the force of a frame to a picture. It has an indisputable moral power in keeping concentrated the attention, and, of course, must not be confounded with mere unity of place.

I determined, then, to place the lover in his chamber—in a chamber rendered sacred to him by memories of her who had frequented it. The room is represented as richly furnished—this in mere pursuance of the ideas I have already explained on the subject of Beauty, as the sole true poetical thesis.

The *locale* being thus determined, I had now to introduce the bird—and the thought of introducing him through the window, was inevitable. The idea of making the lover suppose, in the first instance, that the flapping of the wings of the bird against the shutter, is a "tapping" at the door, originated in a wish to increase, by prolonging, the reader's curiosity, and in a desire to admit the incidental effect arising from the lover's throwing open the door, finding all dark, and thence adopting the half-fancy that it was the spirit of his mistress that knocked.

I made the night tempestuous, first, to account for the Raven's seeking admission, and secondly, for the effect of contrast with the (physical) serenity within the chamber.

I made the bird alight on the bust of Pallas,[1] also for the effect of contrast between the marble and the plumage—it being understood that the bust was absolutely *suggested* by the bird—the bust of *Pallas* being chosen, first, as most in keeping with the scholarship of the lover, and, secondly, for the sonorousness of the word, Pallas, itself.

About the middle of the poem, also, I have availed myself of the force of contrast, with a view of deepening the ultimate impression. For example, an air of the fantastic—approaching as nearly to the

---

1. Pallas Athena, the Greek goddess of wisdom and the arts.

ludicrous as was admissible—is given to the Raven's entrance. He comes in "with many a flirt and flutter."

Not the *least obeisance made he*—not a moment stopped or stayed he,
But with *mien of lord or lady*, perched above my chamber door.

In the two stanzas which follow, the design is more obviously carried out:—

Then this ebony bird beguiling my sad fancy into smiling
By the *grave and stern decorum of the countenance it wore,*
"Though thy *crest be shorn and shaven* thou," I said, "art sure no craven,
Ghastly grim and ancient Raven wandering from the nightly shore—
Tell me what thy lordly name is on the Night's Plutonian shore!"
Quoth the Raven "Nevermore."
—
Much I marvelled *this ungainly fowl* to hear discourse so plainly,
Though its answer little meaning—little relevancy bore;
For we cannot help agreeing that no living human being
Ever yet was blessed with seeing bird above his chamber door—
Bird or beast upon the sculptured bust above his chamber door,
With such name as "Nevermore."

The effect of the *dénouement* being thus provided for, I immediately drop the fantastic for a tone of the most profound seriousness: —this tone commencing in the stanza directly following the one last quoted, with the line,

But the Raven, sitting lonely on that placid bust, spoke only, etc.

From this epoch the lover no longer jests—no longer sees any thing even of the fantastic in the Raven's demeanor. He speaks of him as a "grim, ungainly, ghastly, gaunt, and ominous bird of yore," and feels the "fiery eyes" burning into his "bosom's core." This revolution of thought, or fancy, on the lover's part, is intended to induce a similar one on the part of the reader—to bring the mind into a proper frame for the *dénouement*—which is now brought about as rapidly and as *directly* as possible.

With the *dénouement* proper—with the Raven's reply, "Nevermore," to the lover's final demand if he shall meet his mistress in another world—the poem, in its obvious phase, that of a simple narrative, may be said to have its completion. So far, every thing is within the limits of the accountable—of the real. A raven, having learned by rote the single word "Nevermore," and having escaped from the custody of its owner, is driven, at midnight, through the violence of a storm, to seek admission at a window from which a light still gleams—the chamber-window of a student, occupied half in poring over a volume, half in dreaming of a beloved mistress deceased. The casement being thrown open at the fluttering of the bird's wings, the bird itself perches on the most convenient seat out of the immediate reach of the student, who, amused by the incident and the oddity of the visiter's demeanor, demands of it, in jest and without looking for a reply, its name. The raven addressed, answers with its customary word, "Nevermore"—a word which finds immediate echo in the melancholy heart of the student, who, giving utterance aloud to certain thoughts suggested by the occasion, is again startled by the fowl's repetition of "Nevermore." The student now guesses the state of the case, but is impelled, as I have before explained, by the human thirst for self-torture, and in part by superstition, to propound such queries to the bird as will bring him, the lover, the most of the luxury of sorrow, through the anticipated answer "Nevermore." With the indulgence, to the utmost extreme, of this self-torture, the narration, in what I have termed its first or obvious phase, has a natural termination, and so far there has been no overstepping of the limits of the real.

But in subjects so handled, however skilfully, or with however vivid an array of incident, there is always a certain hardness or nakedness, which repels the artistical eye. Two things are invariably required—first, some amount of complexity, or more properly, adaptation; and, secondly, some amount of suggestiveness—some under current, however indefinite of meaning. It is this latter, in especial, which imparts to a work of art so much of that *richness* (to borrow from colloquy a forcible term) which we are too fond of confounding with *the ideal*. It is the *excess* of the suggested meaning—it is the rendering this the upper instead of the under current of the theme—which turns into prose (and that of the very flattest kind) the so called poetry of the so called transcendentalists.

Holding these opinions, I added the two concluding stanzas of the poem—their suggestiveness being thus made to pervade all the narrative which has preceded them. The under-current of meaning is rendered first apparent in the lines—

"Take thy beak from out *my heart*, and take thy form from off my
    door!"
                Quoth the Raven "Nevermore!"

It will be observed that the words, "from out my heart," involve the first metaphorical expression in the poem. They, with the answer, "Nevermore," dispose the mind to seek a moral in all that has been previously narrated. The reader begins now to regard the Raven as emblematical—but it is not until the very last line of the very last stanza, that the intention of making him emblematical of

Mournful and Never-ending Remembrance is permitted distinctly to be seen:

And the Raven, never flitting, still is sitting, still is sitting,
On the pallid bust of Pallas just above my chamber door;
And his eyes have all the seeming of a demon's that is dreaming,
And the lamplight o'er him streaming throws his shadow on the
    floor;
And my soul *from out that shadow* that lies floating on the floor
                    Shall be lifted—nevermore.

1846

ABRAHAM LINCOLN
1809–1865

Abraham Lincoln's life and presidency can be seen as affirmative answers to the central question raised by the intellectual and political ferment of the late eighteenth century: Can individuals and nations rule themselves? Lincoln's career as self-made man is a paradigm of the possibilities of individual self-regulation and development within a context of freedom; his unshakable commitment to the preservation of the Union made possible the survival of a self-governing nation devoted to the principles of equality. Only by making himself independent and responsible could Lincoln be the Great Emancipator of others; only by surviving the test of civil war could the United States be the model and hope for democratic nations.

Lincoln was born on February 12, 1809, in a backwoods cabin in Hardin County, Kentucky, to nearly illiterate parents. He attended school only sporadically—probably for no more than a year all told—and was essentially self-taught. Though his access to books was limited, he absorbed and retained what he read of the King James Bible, *Aesop's Fables*, John Bunyan's *Pilgrim's Progress*, Daniel Defoe's *Robinson Crusoe*, and Mason Locke Weems's *A History of the Life and Death, Virtues, and Exploits of General George Washington*. Lincoln never lost his love of reading—adding Shakespeare, John Stuart Mill, Lord Byron, and Robert Burns to his list of favorite authors, and was always, in sensibility and by achievement, a great master of words.

Lincoln spent his impoverished youth in Kentucky and southern Indiana, where his father farmed for a living. His mother died when he was nine, but his stepmother, who soon joined the family with children of her own, seems to have singled out Abraham for special affection; he later spoke of her as his "angel mother." In 1830 the family moved to Illinois; after helping the family settle by splitting rails to fence in a new farm, young Lincoln set out on his own, making a trip to New Orleans as a flatboatman. He soon returned to settle in the tiny village of New Salem, Illinois, where he worked as storekeeper, postmaster, and surveyor. In 1832 he volunteered for service in the Black Hawk War; he was elected captain of his company but, as he later observed, saw more action against mosquitoes than he did against Indians.

Lincoln had considered blacksmithing as a trade, but decided instead in the early 1830s to prepare himself for a career in law. This he did by studying independently the basic law books of the time: Blackstone's *Commentaries*, Chitty's *Pleadings*, Greenleaf's *Evidence*, Story's *Equity* and his *Equity Pleadings*. In 1834 he was elected to the first of four terms in the state legislature, at that time a position of small influence and smaller salary. He passed the state bar examination in 1836 and moved the next year to the new state capital in Springfield. Here he entered a succession of law partnerships, the most enduring with William H. Herndon, later his biographer. By dint of hard work—which included twice-yearly sessions following the court on horseback or buggy as it moved from town to town to reach the people across the Illinois countryside—Lincoln prospered as a lawyer and earned a reputation as a shrewd, sensible, fair, and honest practitioner.

Much has been made of Lincoln's romance with Ann Rutledge, whom he had known in New Salem, but she died at nineteen years of age; so far as the records show, Lincoln's only love was Mary Todd. She came from a well-to-do Kentucky family, and the social aristocracy of Springfield to which she belonged advised her against marrying Lincoln despite his success as a lawyer and his obvious good qualities. He, too, apparently had misgivings about the prospects for the marriage, but they were married in the fall of 1842. The relationship between Mary Todd and Abraham Lincoln has been subject to endless speculation. She was witty and intense, and no doubt her temper and extravagance were often a trial to her husband, especially later in their marriage; but he was often absent from home, absorbed in his flourishing law practice, and was himself moody and sharp-tongued. On balance, the Lincolns seem to have shared as much affection and pleasure in their union as one might reasonably expect. They certainly seem to have joined in affectionate concern for their four boys, only one of whom survived to adulthood.

The network of political and other historical events of the 1840s and '50s that would result in Lincoln's election to the presidency in 1860 is complicated, but the central issue involved in these events is not. Very simply, the question was whether or not slavery would be permitted in the new territories, which eventually would become states. When he was elected to Congress in 1846, Lincoln voted against abolitionist measures but he insisted that the new territories must be kept free as "places for poor people to go and better their condition." He also joined in a vote of censure against President Polk for engaging in the war against Mexico (1848), a war he believed to be both unnecessary and unconstitutional. He did not run for re-election and it seemed that his political career had come to an end.

By 1854 the two major political parties of the time—the Whigs (to which Lincoln belonged) and the Democrats—had reached compromise on the extension of slavery into new territories and states. Strong antislavery elements in both parties established independent organizations, and when, in 1854, the Republican party was organized, Lincoln soon joined it. His new party lost the presidential election of 1856 to the Democrats, but in 1858 Lincoln re-entered political life as the Republican candidate in the senatorial election. He opposed the Democrat Stephen A. Douglas, who had earlier sponsored the Kansas-Nebraska Bill, a bill which would have left it to new territories to establish their status as slave or free when they achieved statehood. Lincoln may have won the famous series of debates with Douglas, but he lost the election. More important for the future, though, he had gained national recognition and he found a theme commensurate with his rapidly intensifying powers of thought and expression. As the "House Divided" speech suggests, Lincoln now added to the often biting satirical humor, and to the logic and natural grace of his earlier utterances, a resonance and wisdom that mark his emergence as a national political leader and as a master of language.

This reputation was enhanced by the "Cooper Union Address" in 1860, and at the Republican convention he won nomination on the third ballot. Lincoln was elected sixteenth President of the United States in November, 1860, but before he took office on March 4, 1861, seven states had seceded from the Union to form the Confederacy. Little more than a month after his inauguration, the Civil War had begun. He devoted himself to the preservation of the Union, without which, he believed, neither individuals nor the nation could live freely and decently. To preserve the Union he had to develop an overall war strategy, devise a workable command system, and find the right personnel to execute his plans. All of this he was to accomplish by trial and error in the early years of the war. At the same time he had to develop popular support for his purposes by using his extraordinary political skills in times of high passion and internal division. And when the war ended, leaving him and the country exhausted, he had immediately to face the monumental problems of healing a traumatized nation.

Only by degrees had Lincoln come to commit himself to the elimination of slavery throughout the country. Initially he wished only to stop the spread of slavery; then he saw that "a house divided against itself cannot stand," and, finally, he took the leading role in the passage of the Thirteenth Amendment, which outlawed slavery everywhere and forever in the United States. Elected to a second term in 1864, he had served scarcely a month of his new term when he was assassinated by the demented Shakespearean actor John Wilkes Booth as he attended a play. He died on April 15, 1865.

The texts of Lincoln's addresses are taken from Roy P. Basler's *Abraham Lincoln: His Speeches and Writings,* pp. 734 and 792–93. Lincoln's spellings have been retained throughout. The "Gettysburg Address" of November 19, 1863, is taken from facsimiles reproduced in W. F. Barton's *Lincoln at Gettysburg,* 1930; the "Second Inaugural" was delivered on March 4, 1865, and is based on photostats of the original manuscript owned by the Abraham Lincoln Association.

[The Presidential Question:]
Speech in the United States House of Representatives
July 27, 1848[1]

* * *

The other day, one of the gentlemen from Georgia, an eloquent man, and a man of learning, so far as I can judge, not being learned myself, came down upon us astonishingly. He spoke in what the

1. The speech from which this excerpt comes argues that General Zachary Taylor, the Whig candidate, is more suitable than the Democrats' candidate, General Lewis Cass, to the majority of Whigs and Democrats alike.

Baltimore American calls the "scathing and withering style." At the end of his second severe flash I was struck blind, and found myself feeling with my fingers for an assurance of my continued physical existence. A little of the bone was left, and I gradually revived. He eulogized Mr. Clay[2] in high and beautiful terms, and then declared that we had deserted all our principles, and had turned Henry Clay out, like an old horse, to root. This is terribly severe. It cannot be answered by argument; at least, I cannot so answer it. I merely wish to ask the gentleman if the Whigs are the only party he can think of, who sometimes turn old horses out to root. Is not a certain Martin Van Buren[3] an old horse, which your own party have turned out to root? and is he not rooting a 'little to your discomfort about now? But in not nominating Mr. Clay, we deserted our principles, you say. Ah! in what? Tell us, ye men of principles, what principle we violated? We say you did violate principle in discarding Van Buren, and we can tell you how. You violated the primary, the cardinal, the one great living principle of all Democratic representative government—the principle that the representative is bound to carry out the known will of his constituents. A large majority of the Baltimore Convention of 1844 were, by their constituents, instructed to procure Van Buren's nomination if they could. In violation, in utter, glaring contempt of this, you rejected him—rejected him, as the gentleman from New York, the other day expressly admitted, for *availability*—that same "general availability" which you charge upon us, and daily chew over here, as something exceedingly odious and unprincipled. But the gentleman from Georgia gave us a second speech yesterday, all well considered and put down in writing, in which Van Buren was scathed and withered a "few" for his present position and movements. I cannot remember the gentleman's precise language, but I do remember he put Van Buren down, down, till he got him where he was finally to "stink" and "rot."

Mr. Speaker, it is no business or inclination of mine to defend Martin Van Buren. In the war of extermination now waging between him and his old admirers, I say, devil take the hindmost—and the foremost. But there is no mistaking the origin of the breach; and if the curse of "stinking" and "rotting" is to fall on the first and greatest violators of principle in the matter, I disinterestedly suggest, that the gentleman from Georgia and his present co-workers are bound to take it upon themselves.

But the gentleman from Georgia further says, we have deserted all our principles, and taken shelter under General Taylor's military coat tail; and he seems to think this is exceedingly degrading. Well, as his faith is, so be it unto him. But can he remember no other military coat tail under which a certain other party have been sheltering for near a quarter of a century? Has he no acquaintance

---

2. American Congressman, Senator, Secretary of State (1777–1852).

3. Eighth President of the United States (1782–1862).

with the ample military coat tail of General Jackson?[4] Does he not know that his own party have run the last five Presidential races under that coat tail, and that they are now running the sixth under that same cover? Yes, sir, that coat tail was used, not only for General Jackson himself, but has been clung to with the grip of death by every Democratic candidate since. You have never ventured, and dare not now venture, from under it. Your campaign papers have constantly been "Old Hickories," with rude likenesses of the old General upon them; hickory poles and hickory brooms your never-ending emblems; Mr. Polk,[5] himself, was "Young Hickory," "Little Hickory," or something so; and even now your campaign paper here is proclaiming that Cass and Butler[6] are of the true "Hickory stripe." No, sir; you dare not give it up. Like a horde of hungry ticks, you have stuck to the tail of the Hermitage lion[7] to the end of his life, and you are still sticking to it, and drawing a loathsome sustenance from it after he is dead. A fellow once advertised that he had made a discovery, by which he could make a new man out of an old one, and have enough of the stuff left to make a little yellow dog. Just such a discovery has General Jackson's popularity been to you. You not only twice made President of him out of it, but you have had enough of the stuff left to make Presidents of several comparatively small men since; and it is your chief reliance now to make still another.

Mr. Speaker, old horses and military coat tails, or tails of any sort, are not figures of speech such as I would be the first to introduce into discussions here; but as the gentleman from Georgia has thought fit to introduce them, he and you are welcome to all you have made, or can make, by them. If you have any more old horses, trot them out; any more tails, just cock them, and come at us.

I repeat, I would not introduce this mode of discussion here; but I wish gentlemen on the other side to understand, that the use of degrading figures is a game at which they may not find themselves able to take all the winnings. [We give it up.] Aye, you give it up, and well you may, but from a very different reason from that which you would have us understand. The point—the power to hurt—of all figures, consists in the *truthfulness* of their application; and understanding this, you may well give it up. They are weapons which hit you, but miss us.

But, in my hurry, I was very near closing on the subject of military tails, before I was done with it. There is one entire article of the sort I have not discussed yet; I mean the military tail you Democrats are now engaged in dovetailing on to the great Michigander. Yes, sir, all his biographers (and they are legion) have him

---

4. Andrew Jackson (1767–1845), seventh President of the United States.
5. James K. Polk (1795–1849), eleventh President of the United States.
6. Benjamin Franklin Butler (1795–1858),
American lawyer and politician.
7. Andrew Jackson's home, near Nashville, Tennessee, was known as the Hermitage.

in hand, tying him to a military tail, like so many mischievous boys tying a dog to a bladder of beans. True, the material they have is very limited; but they drive at it, might and main. He *in*vaded Canada without resistance, and he *out*vaded it without pursuit. As he did both under orders, I suppose there was, to him, neither credit nor discredit in them; but they are made to constitute a large part of the tail. He was not at Hull's surrender, but he was close by. He was volunteer aid to General Harrison on the day of the battle of the Thames; and, as you said in 1840, Harrison was picking whortleberries two miles off, while the battle was fought, I suppose it is a just conclusion, with you, to say Cass was aiding Harrison to pick whortleberries. This is about all, except the mooted question of the broken sword.[8] Some authors say he broke it; some say he threw it away; and some others, who ought to know, say nothing about it. Perhaps it would be a fair historical compromise to say, if he did not break it, he did not do anything else with it.

By the way, Mr. Speaker, did you know I am a military hero? Yes, sir, in the days of the Black Hawk war, I fought, bled, and came away. Speaking of General Cass's career, reminds me of my own. I was not at Stillman's defeat, but I was about as near it as Cass was to Hull's surrender; and, like him, I saw the place very soon afterwards. It is quite certain I did not break my sword, for I had none to break; but I bent a musket pretty badly on one occasion. If Cass broke his sword, the idea is, he broke it in desperation; I bent the musket by accident. If General Cass went in advance of me in picking whortleberries, I guess I surpassed him in charges upon the wild onions. If he saw any live fighting Indians, it was more than I did, but I had a good many bloody struggles with the mosquitoes; and although I never fainted from loss of blood, I can truly say I was often very hungry.

Mr. Speaker, if I should ever conclude to doff whatever our Democratic friends may suppose there is of black-cockade Federalism about me, and, thereupon, they shall take me up as their candidate for the Presidency, I protest they shall not make fun of me, as they have of General Cass, by attempting to write me into a military hero.

* * *

1848

Address Delivered at the Dedication of the Cemetery at Gettysburg
November 19, 1863

Four score and seven years ago our fathers brought forth on this continent, a new nation, conceived in Liberty, and dedicated to the proposition that all men are created equal.

Now we are engaged in a great civil war, testing whether that

8. Cass is said to have angrily broken his sword when he learned of the surrender of Detroit in the War of 1812.

nation, or any nation so conceived and so dedicated, can long endure. We are met on a great battle-field of that war. We have come to dedicate a portion of that field, as a final resting place for those who here gave their lives that that nation might live. It is altogether fitting and proper that we should do this.

But, in a larger sense, we can not dedicate—we can not con-secrate—we can not hallow—this ground. The brave men, living and dead, who struggled here, have consecrated it, far above our poor power to add or detract. The world will little note, nor long remember what we say here, but it can never forget what they did here. It is for us the living, rather, to be dedicated here to the unfinished work which they who fought here have thus far so nobly advanced. It is rather for us to be here dedicated to the great task remaining before us—that from these honored dead we take in-creased devotion to that cause for which they gave the last full measure of devotion—that we here highly resolve that these dead shall not have died in vain—that this nation, under God, shall have a new birth of freedom—and that government of the people, by the people, for the people, shall not perish from the earth.

November 19, 1863   *modest*   Abraham Lincoln
                                                           1863

Second Inaugural Address
March 4, 1865

At this second appearing to take the oath of the presidential office, there is less occasion for an extended address than there was at the first. Then a statement, somewhat in detail, of a course to be pursued, seemed <u>fitting and proper.</u> Now, at the expiration of four years, during which public declarations have been constantly called forth on every point and phase of the great contest which still absorbs the attention, and engrosses the energies of the nation, little that is new could be presented. The progress of our arms, upon which all else chiefly depends, is as well known to the public as to myself; and it is, I trust, reasonably satisfactory and encouraging to all. With high hope for the future, no prediction in regard to it is ventured.

On the occasion corresponding to this four years ago, all thoughts were anxiously directed to an impending civil war. All dreaded it—all sought to avert it. While the inaugural address was being delivered from this place, devoted altogether to *saving* the Union without war, insurgent agents were in the city seeking to *destroy* it without war—seeking to dissol[v]e the Union, and divide effects, by negotiation. Both parties deprecated war; but one of them would *make* war rather than let the nation survive; and the other would *accept* war rather than let it perish. And the war came.

One eighth of the whole population were colored slaves, not distributed generally over the Union, but localized in the Southern part of it. These slaves constituted a peculiar and powerful interest. All knew that this interest was, somehow, the cause of the war. To strengthen, perpetuate, and extend this interest was the object for which the insurgents would rend the Union, even by war; while the government claimed no right to do more than to restrict the territorial enlargement of it. Neither party expected for the war, the magnitude, or the duration, which it has already attained. Neither anticipated that the *cause* of the conflict might cease with, or even before, the conflict itself should cease. Each looked for an easier triumph, and a result less fundamental and astounding. Both read the same Bible, and pray to the same God; and each invokes His aid against the other. It may seem strange that any men should dare to ask a just God's assistance in wringing their bread from the sweat of other men's faces; but let us judge not that we be not judged. The prayers of both could not be answered; that of neither has been answered fully. The Almighty has his own purposes. "Woe unto the world because of offences! for it must needs be that offences come; but woe to that man by whom the offence cometh!" If we shall suppose that American Slavery is one of those offences which, in the providence of God, must needs come, but which, having continued through His appointed time, He now wills to remove, and that He gives to both North and South, this terrible war, as the woe due to those by whom the offence came, shall we discern therein any departure from those divine attributes which the believers in a Living God always ascribe to Him? Fondly do we hope —fervently do we pray—that this mighty scourge of war may speedily pass away. Yet, if God wills that it continue, until all the wealth piled by the bond-man's two hundred and fifty years of unrequited toil shall be sunk, and until every drop of blood drawn with the lash, shall be paid by another drawn with the sword, as was said three thousand years ago, so still it must be said "the judgments of the Lord, are true and righteous altogether."

With malice toward none; with charity for all; with firmness in the right, as God gives us to see the right, let us strive on to finish the work we are in; to bind up the nation's wounds; to care for him who shall have borne the battle, and for his widow, and his orphan —to do all which may achieve and cherish a just and lasting peace, among ourselves, and with all nations.

1865   1865

OLIVER WENDELL HOLMES
1809–1894

Oliver Wendell Holmes was born on August 29, 1809, in Cambridge, Massachusetts, of old, honorable stock—what Holmes himself called the Brahmin caste. At Harvard he was class poet in 1829. He studied law at

Harvard, then went to Paris to study medicine (1833–35). Harvard awarded him an M.D. in 1836, the year his *Poems* was published. Holmes was professor of anatomy at Dartmouth from 1839 to 1840, then moved to Boston. A medical treatise, *Homœopathy and Its Kindred Delusions*, was published in 1843, and the next year the controversial *Contagiousness of Puerperal Fever*, which tellingly surveyed the evidence that childbed fever, then a great danger to new mothers, might be contagious (as in fact it was, being spread by unsanitary doctors and midwives). Holmes's own three children were born in the early 1840s; the first, his namesake, became the distinguished jurist. From 1847 until 1882 Holmes was professor of anatomy at Harvard, phenomenally popular with students. Until well after his retirement, literature was only a secondary interest. Aggressive in arguing his positions in medical journals, Holmes studiously avoided involvement in social causes which agitated so many of his contemporaries, remaining always a respectable, conservative, law-abiding citizen.

Holmes became nationally famous as an essayist and a poet in the late 1850s, when the *Atlantic Monthly* serialized his humorous essays, collected in *The Autocrat of the Breakfast Table* (1858), and printed poems which at once became classroom favorites, especially *The Chambered Nautilus* and *The Deacon's Masterpiece*. In three "medicated novels," as he called them, *Elsie Venner* (1861), *The Guardian Angel* (1867), and *A Mortal Antipathy* (1885), Holmes explored genetic and psychological determinism in a series of disturbed characters and took an analytic attitude toward theological convictions; cautious reviewers warned that the fiction was not entirely safe for the impressionable to read. Modern readers are attracted by Holmes's clinical portrayals, and all three of the novels are valuable for their local color and witty discursive passages. From the 1860s through the 1880s Holmes published several volumes of poems and essays and an unsatisfying biography of Emerson (1885) which ignored his friend's radicalism; his professional writing for medical journals continued to win respect. In his last decades Holmes was an American institution, the most famous after-dinner speaker of his time and the most reliably witty writer of poems for special occasions. He died on October 7, 1894, the last of his literary generation.

The Chambered Nautilus[1]

This is the ship of pearl, which, poets feign,
    Sails the unshadowed main,—
    The venturous bark that flings
On the sweet summer wind its purpled wings
In gulfs enchanted, where the siren sings,   5
    And coral reefs lie bare,
Where the cold sea-maids rise to sun their streaming hair.

Its webs of living gauze no more unfurl;
    Wrecked is the ship of pearl!
    And every chambered cell,   10
Where its dim dreaming life was wont to dwell,

1. From the first printing in the *Atlantic Monthly* (February, 1858). The "nautilus" (from the Greek, "sailor") mollusk is so named because it was thought to have a membrane which served as a sail. The pearly nautilus of this poem is found in the Indian Ocean and the South Pacific Ocean.

As the frail tenant shaped his growing shell,
  Before thee lies revealed,—
Its irised ceiling rent, its sunless crypt unsealed!

Year after year beheld the silent toil   15
    That spread his lustrous coil;
    Still, as the spiral grew,
He left the past year's dwelling for the new,
Stole with soft step its shining archway through,
    Built up its idle door,   20
Stretched in his last-found home, and knew the old no more.

Thanks for the heavenly message brought by thee,
    Child of the wandering sea,
    Cast from her lap, forlorn!
From thy dead lips a clearer note is born   25
Than ever Triton[2] blew from wreathéd horn!
    While on mine ear it rings,
Through the deep caves of thought I hear a voice that sings:—

Build thee more stately mansions, O my soul,
    As the swift seasons roll!   30
    Leave thy low-vaulted past!
Let each new temple, nobler than the last,
Shut thee from heaven with a dome more vast,
    Till thou at length art free,
Leaving thine outgrown shell by life's unresting sea!   35

                                                                      1858

2. In Greek mythology, a sea god who
controlled the waves with a trumpet made
of a conch shell. The line echoes the sonnet *The World Is Too Much with Us*

by the English poet William Wordsworth
(1770–1850): "Or hear old Triton blow
his wreathéd horn."

MARGARET FULLER
1810–1850

Sarah Margaret Fuller was born at Cambridgeport (now part of Cambridge), Massachusetts, on May 23, 1810. Her father supervised her education, making her a prodigy but depriving her of a childhood. After a brief,
traumatic stay at a girls' school in her early teens, she returned to pursue
her rigorous education at home, steeping herself in the classics and in modern languages and literatures, especially German. Accustomed to intense,
lonely study, Fuller nevertheless formed lasting intellectual and emotional
friendships with a few young Harvard scholars, among them her co-biogra-
phers James Freeman Clarke and W. H. Channing. A Cambridge lady,
Eliza Farrar, undertook to instill some of the social graces into the father-
taught Margaret. The death of her father in 1835 burdened Fuller with
the education of younger brothers and sisters. Setting aside her own ambi-

tions (including a planned trip to Europe), she taught for several years, in Boston and Providence. During this time the German novelist and dramatist Goethe became the chief influence on her religion and philosophy, and she tormented herself with the hope that she might have money, time, and ability to write his biography. In 1839 she began leading "Conversation" classes among an elite group of Boston women. Later, men participated also, and during the next years her topics included Greek mythology, the Fine Arts, Ethics, Education, Demonology, Creeds, and the Ideal.

A close friend of Emerson's since she first sought him out in 1836, Fuller edited the Transcendentalists' magazine *The Dial* from 1840 to 1842, meanwhile continuing to translate works by and about Goethe. In 1844 her little *Summer on the Lakes*, an account of a trip to the Midwest, led Horace Greeley to hire her as literary critic for his New York *Tribune*, making her probably the first self-supporting American woman journalist. More than merely a literary reviewer, Fuller wrote a series of reports on public questions, among them the conditions of the blind, of the insane, and of female prisoners. In 1845 Greeley published her *Woman in the Nineteenth Century*, the title article of which was an expansion of a controversial *Dial* essay, *The Great Lawsuit*. This is one of the great neglected documents of American sexual liberation—not merely of feminism, for Fuller recognized that both men and women were imprisoned by social roles, although men at least had the power to make and enforce the definitions of those roles. In 1846 some of her *Tribune* pieces were collected in *Papers on Literature and Art*. In New York she fell in love with James Nathan, a German Jew who, baffled by her mixture of sexual honesty and prudery, fled home in June, 1845, letting the growing spaces between his letters persuade her gradually that he had rejected her.

While still hoping for a reunion with Nathan, Fuller sailed for Europe in August, 1846, intending to support herself as foreign correspondent for the *Tribune*. In England one of her idols, Thomas Carlyle (then in his fifties), disappointed her by his reactionary political views and his insensitivity to the worth of others, especially the Italian revolutionary Joseph Mazzini, who had sought refuge in England. In Paris she met another idol, George Sand, who proved more satisfactory than Carlyle, and another political revolutionary, the exiled Polish poet Adam Mickiewicz. Sand's example of sexually liberated womanhood stirred Fuller profoundly, as did Mickiewicz's blunt speculation that she could not deeply respond to Europe while remaining a virgin—not the sort of comment men like Emerson and Greeley had accustomed her to. Fuller went on to Italy, then not a unified country but a collection of states—some controlled by the Pope, others independent, and, to the north, a third group controlled by Austria. Soon after her arrival in Rome she became the object of courtship by a Roman of the lower nobility, Giovanni Angelo Ossoli, almost eleven years younger than she. When she returned from summering in northern Italy, Rome was undergoing anti-Papal ferment, and her dispatches to the *Tribune* became more and more political. Making use of her connections with varying factions, she began an earnest accumulation of documents concerning the forthcoming revolution—newspapers, pamphlets, leaflets.

And she began a love affair with Ossoli. In December she was pregnant, with no man or woman she could confide in, either in the United States or Europe. At the start of 1848 she wrote guardedly to a friend at home:

"with this year I enter upon a sphere of my destiny so difficult that at present I see no way out except through the gate of death." Marriage seemed out of the question because of the certain opposition of Ossoli's family. Through a dismal rainy season, in which she lived on pennies a day, Fuller covered for the *Tribune* such events as the popular agitation against the Jesuits. She became intimate with the Princess Belgioioso, a leader of the anti-Austrian faction who drew her still more deeply into Italian politics. When cities of northern Italy revolted against the Austrians in March, Fuller described to her New York readers the joyous response of the Roman citizens. The revolutionaries Mickiewicz and Mazzini entered Italy; both kept in touch with Fuller out of their respect for her personal commitment to their goals and their sense of her value in shaping American opinion.

That spring, 1848, Emerson wrote from England urging her to return home with him before war broke out. Still keeping her secret, she withdrew instead to the Abruzzi region to wait out her pregnancy. Ossoli had become a member of the civic guard, but he managed to be with her for the birth of Angelo on September 5. Leaving the baby in Rieti with a wet nurse, Fuller returned to Rome late in November, in time to report the flight of the Pope and, early in 1849, the arrival of the Italian nationalist Giuseppe Garibaldi and the proclamation of the Roman Republic. She shared the triumph of Mazzini's arrival in Rome, but the Republic was short-lived. Anticipating the intervention of the French on behalf of the Pope, Princess Belgioioso urgently wrote Fuller on April 30, 1849: "You are named Regolatrice of the Hospital of the Fate Bene Fratelli"—on an island in the Tiber. Fuller ran the hospital heroically when the French laid siege, despite her concern for Ossoli, who was fighting with the Republican forces, and her uncertainty about the baby, whom she had hardly seen since he was two months old. After Rome fell to the French on the fourth of July she made her way to Rieti, only to find that the nurse, assuming the baby had been abandoned, was allowing him to starve. Retreating to Florence with Ossoli and the baby, Fuller faced down her shocked acquaintances, including Robert and Elizabeth Barrett Browning, and began work on her history of the Roman Republic. While at Florence she may have married Ossoli, as his sister later claimed. In May, 1850, she sailed for the United States with Ossoli and the baby, full of forebodings about the ship and the way they would be received at home. All three died in a shipwreck off Fire Island, New York, on July 19. The body of the baby was washed ashore, as well as a trunk which contained some of Fuller's papers but not the history. Thoreau sought in vain for her body.

Among Fuller's family and friends mourning was mixed with relief, for gossip condemned her conduct as scandalous, and her own frank letters about the disparities between her and her spouse in age and intellectual cultivation had confirmed their worst fears. Emerson, Clarke, and Channing edited her *Memoirs* (1852) in a way that sanitized and trivialized her last years; in the process they made her a prime exhibit of the self-defeating American drive toward self-culture, an emphasis which slighted her life-long activism. In 1903 her friend Julia Ward Howe published Fuller's love letters to James Nathan, thereby sealing the image for much of the twentieth century—Fuller as a would-be intellectual old maid, sex-starved for many years, then sex-crazed, twice in two years flinging herself at handsome young foreigners. Hawthorne's old verdict seemed confirmed: "There never was such a tragedy as her whole story; the sadder and sterner, because so much of the ridiculous was mixed up with it, and because she could bear anything better than to be ridiculous."

Sexist ridicule dies hard, and in Fuller's case its death has been retarded by the inaccessibility of most of her writings. The few twentieth-century reprints of her works have been selective and have quickly passed out of print. Few libraries possess the collections of parts of her reportage for the *Tribune*. Yet perhaps no complete edition of her published writings and surviving papers could establish her as a major figure: she thought much better than she wrote. Such an edition, however, might establish her candidacy for being considered seriously as what Hawthorne said mockingly: "the greatest, wisest, best woman of the age."

From The Great Lawsuit
MAN versus *MEN*. *WOMAN* versus *WOMEN*[1]
[TWO KINDS OF SLAVERY]

It is worthy of remark, that, as the principle of liberty is better understood and more nobly interpreted, a broader protest is made in behalf of woman. As men become aware that all men have not had their fair chance, they are inclined to say that no women have had a fair chance. The French revolution, that strangely disguised angel, bore witness in favor of woman, but interpreted her claims no less ignorantly than those of man. Its idea of happiness did not rise beyond outward enjoyment, unobstructed by the tyranny of others. The title it gave was Citoyen, Citoyenne,[2] and it is not unimportant to woman that even this species of equality was awarded her. Before, she could be condemned to perish on the scaffold for treason, but not as a citizen, but a subject. The right, with which this title then invested a human being, was that of bloodshed and license. The Goddess of Liberty was impure. Yet truth was prophesied in the ravings of that hideous fever induced by long ignorance and abuse. Europe is conning a valued lesson from the blood-stained page. The same tendencies, farther unfolded, will bear good fruit in this country.

Yet, in this country, as by the Jews, when Moses was leading them to the promised land,[3] everything has been done that inherited depravity could, to hinder the promise of heaven from its fulfilment. The cross, here as elsewhere, has been planted only to be blasphemed by cruelty and fraud. The name of the Prince of Peace has been profaned by all kinds of injustice towards the Gentile whom he said he came to save. But I need not speak of what has

1. Reprinted here from the Boston *Dial*, Vol. 4 (July, 1843).
2. Male citizen, female citizen, equal under the law—one of the most promising early achievements of the French Revolution.
3. Fuller emphasizes the irony of Moses'

promulgation of a degraded role for women under Mosaic law even while leading the Israelites, male and female, out of Egypt toward their promised homeland. See Leviticus 12 and Numbers 30.

been done towards the red man, the black man. These deeds are the
scoff of the world; and they have been accompanied by such pious
words, that the gentlest would not dare to intercede with, "Father
forgive them, for they know not what they do."[4]

Here, as elsewhere, the gain of creation consists always in the
growth of individual minds, which live and aspire, as flowers bloom
and birds sing, in the midst of morasses; and in the continual
development of that thought, the thought of human destiny, which
is given to eternity to fulfil, and which ages of failure only seemingly impede. Only seemingly, and whatever seems to the contrary,
this country is as surely destined to elucidate a great moral law, as
Europe was to promote the mental culture of man.

Though the national independence be blurred by the servility of
individuals; though freedom and equality have been proclaimed
only to leave room for a monstrous display of slave dealing and
slave keeping; though the free American so often feels himself free,
like the Roman, only to pamper his appetites and his indolence
through the misery of his fellow beings, still it is not in vain, that
the verbal statement has been made, "All men are born free and
equal."[5] There it stands, a golden certainty, wherewith to encourage
the good, to shame the bad. The new world may be called clearly to
perceive that it incurs the utmost penalty, if it reject the sorrowful
brother. And if men are deaf, the angels hear. But men cannot be
deaf. It is inevitable that an external freedom, such as has been
achieved for the nation, should be so also for every member of it.
That, which has once been clearly conceived in the intelligence,
must be acted out. It has become a law, irrevocable as that of the
Medes in their ancient dominion.[6] Men will privately sin against it,
but the law so clearly expressed by a leading mind of the age,

> "Tutti fatti a sembianza d' un Solo;
> Figli tutti d' un solo riscatto,
> In qual ora, in qual parte del suolo
> Trascorriamo quest' aura vital,
> Siam fratelli, siam stretti ad un patto:
> Maladetto colui che lo infrange,
> Che s' innalza sul fiacco che piange,
> Che contrista uno spirto immortal."[7]

> "All made in the likeness of the One,
> All children of one ransom,
> In whatever hour, in whatever part of the soil
> We draw this vital air,

4. Jesus' words in Luke 23.34, in reference to the Roman soldiers who had
just nailed him to the cross.
5. As in the Declaration of Independence, second paragraph, "all men are
created equal."

6. Media is now part of Iran. In Fuller's
notion, as in Thoreau's, real insight always leads to action.
7. "Manzoni" [Fuller's note]; Alessandro
Manzoni (1785–1873), Italian poet.

We are brothers, we must be bound by one compact,
Accursed he who infringes it,
Who raises himself upon the weak who weep,
Who saddens an immortal spirit."

cannot fail of universal recognition.

We sicken no less at the pomp than at the strife of words. We feel that never were lungs so puffed with the wind of declamation, on moral and religious subjects, as now. We are tempted to implore these "word-heroes," these word-Catos, word-Christs,[8] to beware of cant above all things; to remember that hypocrisy is the most hopeless as well as the meanest of crimes, and that those must surely be polluted by it, who do not keep a little of all this morality and religion for private use.[9] We feel that the mind may "grow black and rancid in the smoke" even of altars. We start up from the harangue to go into our closet and shut the door. But, when it has been shut long enough, we remember that where there is so much smoke, there must be some fire; with so much talk about virtue and freedom must be mingled some desire for them; that it cannot be in vain that such have become the common topics of conversation among men; that the very newspapers should proclaim themselves Pilgrims, Puritans, Heralds of Holiness.[1] The king that maintains so costly a retinue cannot be a mere Count of Carabbas[2] fiction. We have waited here long in the dust; we are tired and hungry, but the triumphal procession must appear at last.

Of all its banners, none has been more steadily upheld, and under none has more valor and willingness for real sacrifices been shown, than that of the champions of the enslaved African. And this band it is, which, partly in consequence of a natural following out of principles, partly because many women have been prominent in that cause, makes, just now, the warmest appeal in behalf of woman.

Though there has been a growing liberality on this point, yet society at large is not so prepared for the demands of this party, but that they are, and will be for some time, coldly regarded as the Jacobins[3] of their day.

"Is it not enough," cries the sorrowful trader, "that you have done all you could to break up the national Union, and thus destroy the prosperity of our country, but now you must be trying to break up family union, to take my wife away from the cradle, and the kitchen hearth, to vote at polls, and preach from a pulpit? Of

---

8. Windy would-be reformers, whether of political and social morality like Marcus Porcius Cato (234–149 B.C.), Roman statesman, or of religion, in fancied imitation of Jesus.
9. "Dr. Johnson's one piece of advice should be written on every door; 'Clear your mind of cant.' But Byron, to whom it was so acceptable, in clearing away the noxious vine, shook down the building too. Stirling's emendation is note-

worthy, 'Realize your cant, not cast it off' " [Fuller's note].
1. Then common names for newspapers in Massachusetts and elsewhere.
2. Type name for a purse-proud nobleman (or fake nobleman, as in the nursery tale *Puss in Boots*).
3. Political radicals of the extreme left, from the political group founded in Paris in 1789 near the church of Saint-Jacques.

course, if she does such things, she cannot attend to those of her own sphere. She is happy enough as she is. She has more leisure than I have, every means of improvement, every indulgence."

"Have you asked her whether she was satisfied with these indulgences?"

"No, but I know she is. She is too amiable to wish what would make me unhappy, and too judicious to wish to step beyond the sphere of her sex. I will never consent to have our peace disturbed by any such discussions."

" 'Consent'—you? it is not consent from you that is in question, it is assent from your wife."

"Am I not the head of my house?"

"You are not the head of your wife. God has given her a mind of her own."

"I am the head and she the heart."

"God grant you play true to one another then. If the head represses no natural pulse of the heart, there can be no question as to your giving your consent. Both will be of one accord, and there needs but to present any question to get a full and true answer. There is no need of precaution, of indulgence, or consent. But our doubt is whether the heart consents with the head, or only acquiesces in its decree; and it is to ascertain the truth on this point, that we propose some liberating measures."

Thus vaguely are these questions proposed and discussed at present. But their being proposed at all implies much thought, and suggests more. Many women are considering within themselves what they need that they have not, and what they can have, if they find they need it. Many men are considering whether women are capable of being and having more than they are and have, and whether, if they are, it will be best to consent to improvement in their condition.

The numerous party, whose opinions are already labelled and adjusted too much to their mind to admit of any new light, strive, by lectures on some model-women of bridal-like beauty and gentleness, by writing or lending little treatises, to mark out with due precision the limits of woman's sphere, and woman's mission, and to prevent other than the rightful shepherd from climbing the wall, or the flock from using any chance gap to run astray.

Without enrolling ourselves at once on either side, let us look upon the subject from that point of view which to-day offers. No better, it is to be feared, than a high house-top. A high hill-top, or at least a cathedral spire, would be desirable.

It is not surprising that it should be the Anti-Slavery party that pleads for woman, when we consider merely that she does not hold property on equal terms with men; so that, if a husband dies without a will, the wife, instead of stepping at once into his place as head of the family, inherits only a part of his fortune, as if she were a child, or ward only, not an equal partner.

We will not speak of the innumerable instances, in which profligate or idle men live upon the earnings of industrious wives; or if the wives leave them and take with them the children, to perform the double duty of mother and father, follow from place to place, and threaten to rob them of the children, if deprived of the rights of a husband, as they call them, planting themselves in their poor lodgings, frightening them into paying tribute by taking from them the children, running into debt at the expense of these otherwise so overtasked helots.[4] Though such instances abound, the public opinion of his own sex is against the man, and when cases of extreme tyranny are made known, there is private action in the wife's favor. But if woman be, indeed, the weaker party, she ought to have legal protection, which would make such oppression impossible.

And knowing that there exists, in the world of men, a tone of feeling towards women as towards slaves, such as is expressed in the common phrase, "Tell that to women and children;" that the infinite soul can only work through them in already ascertained limits; that the prerogative of reason, man's highest portion, is allotted to them in a much lower degree; that it is better for them to be engaged in active labor, which is to be furnished and directed by those better able to think, &c. &c.; we need not go further, for who can review the experience of last week, without recalling words which imply, whether in jest or earnest, these views, and views like these? Knowing this, can we wonder that many reformers think that measures are not likely to be taken in behalf of women, unless their wishes could be publicly represented by women?

That can never be necessary, cry the other side. All men are privately influenced by women; each has his wife, sister, or female friends, and is too much biassed by these relations to fail of representing their interests. And if this is not enough, let them propose and enforce their wishes with the pen. The beauty of home would be destroyed, the delicacy of the sex be violated, the dignity of halls of legislation destroyed, by an attempt to introduce them there. Such duties are inconsistent with those of a mother; and then we have ludicrous pictures of ladies in hysterics at the polls, and senate chambers filled with cradles.

But if, in reply, we admit as truth that woman seems destined by nature rather to the inner circle, we must add that the arrangements of civilized life have not been as yet such as to secure it to her. Her circle, if the duller, is not the quieter. If kept from excitement, she is not from drudgery. Not only the Indian carries the burdens of the camp, but the favorites of Louis the Fourteenth accompany him in his journeys, and the washerwoman stands at her tub and carries home her work at all seasons, and in all states of health.[5]

As to the use of the pen, there was quite as much opposition to

4. Slaves. Fuller fairly describes the legal situation in her time.
5. In Fuller's view courtesans and washerwomen are equally enslaved, however disparate their conditions are.

woman's possessing herself of that help to free-agency as there is
now to her seizing on the rostrum or the desk; and she is likely to
draw, from a permission to plead her cause that way, opposite
inferences to what might be wished by those who now grant it.

As to the possibility of her filling, with grace and dignity, any
such position, we should think those who had seen the great ac-
tresses, and heard the Quaker preachers of modern times, would not
doubt, that woman can express publicly the fulness of thought and
emotion, without losing any of the peculiar beauty of her sex.

As to her home, she is not likely to leave it more than she now
does for balls, theatres, meetings for promoting missions, revival
meetings, and others to which she flies, in hope of an animation for
her existence, commensurate with what she sees enjoyed by men.
Governors of Ladies' Fairs are no less engrossed by such a charge,
than the Governor of the State by his; presidents of Washingtonian
societies,[6] no less away from home than presidents of conventions.
If men look straitly to it, they will find that, unless their own lives
are domestic, those of the women will not be. The female Greek, of
our day, is as much in the street as the male, to cry, What news?[7]
We doubt not it was the same in Athens of old. The women, shut
out from the market-place, made up for it at the religious festivals.
For human beings are not so constituted, that they can live without
expansion; and if they do not get it one way, must another, or
perish.

And, as to men's representing women fairly, at present, while we
hear from men who owe to their wives not only all that is comfortable and graceful, but all that is wise in the arrangement of their
lives, the frequent remark, "You cannot reason with a woman,"
when from those of delicacy, nobleness, and poetic culture, the
contemptuous phrase, "Women and children," and that in no light
sally of the hour, but in works intended to give a permanent statement of the best experiences, when not one man in the million, shall
I say, no, not in the hundred million, can rise above the view that
woman was made *for man,* when such traits as these are daily
forced upon the attention, can we feel that man will always do
justice to the interests of woman? Can we think that he takes a
sufficiently discerning and religious view of her office and destiny,
ever to do her justice, except when prompted by sentiment; accidentally or transiently, that is, for his sentiment will vary according
to the relations in which he is placed. The lover, the poet, the artist,
are likely to view her nobly. The father and the philosopher have
some chance of liberality; the man of the world, the legislator for
expediency, none.

6. *Roughly equivalent in purpose and
reputation to the modern Daughters of
the American Revolution.*
7. Fuller mentions only the ancient
Greek males as targets of satire for their

newsmongering, but it was the Latin
"quid nunc?" ("What now?") which
entered English as a label for male
busybodies, "quidnuncs."

Under these circumstances, without attaching importance in themselves to the changes demanded by the champions of woman, we hail them as signs of the times. We would have every arbitrary barrier thrown down. We would have every path laid open to woman as freely as to man. Were this done, and a slight temporary fermentation allowed to subside, we believe that the Divine would ascend into nature to a height unknown in the history of past ages, and nature, thus instructed, would regulate the spheres not only so as to avoid collision, but to bring forth ravishing harmony.

Yet then, and only then, will human beings be ripe for this, when inward and outward freedom for woman, as much as for man, shall be acknowledged as a right, not yielded as a concession. As the friend of the negro assumes that one man cannot, by right, hold another in bondage, should the friend of woman assume that man cannot, by right, lay even well-meant restrictions on woman. If the negro be a soul, if the woman be a soul, apparelled in flesh, to one master only are they accountable. There is but one law for all souls, and, if there is to be an interpreter of it, he comes not as man, or son of man, but as Son of God.

Were thought and feeling once so far elevated than man should esteem himself the brother and friend, but nowise the lord and tutor of woman, were he really bound with her in equal worship, arrangements as to function and employment would be of no consequence. What woman needs is not as a woman to act or rule, but as a nature to grow, as an intellect to discern, as a soul to live freely, and unimpeded to unfold such powers as were given her when we left our common home. If fewer talents were given her, yet, if allowed the free and full employment of these, so that she may render back to the giver his own with usury, she will not complain, nay, I dare to say she will bless and rejoice in her earthly birth-place, her earthly lot.

Let us consider what obstructions impede this good era, and what signs give reason to hope that it draws near.

1843

T. B. THORPE
1815–1878

Thomas Bangs Thorpe was born on March 1, 1815, at Westfield, Massachusetts, son of a Methodist minister who died when the boy was four. Thorpe spent his childhood in Dutch Albany, with his mother's family, and summers with his paternal grandparents in Connecticut. In 1827 he went with the family to New York City, where in 1830 he began studying with the eccentric painter John Quidor, an early illustrator of Irving. (When Thorpe first exhibited at the American Academy of Fine Arts in 1833, the subject of his painting was Ichabod Crane.) From 1834 to 1836 Thorpe attended Wesleyan University in Middletown, Connecticut, then withdrew in ill health. He went south early in 1837 and recovered his health while painting plantations families in Louisiana and Mississippi. He married in 1838, and the next year became a minor celebrity with a casual sketch of a Louisiana bee hunter published in William T. Porter's *Spirit of the Times* and widely reprinted in the United States and abroad.

During a trip to New York in 1840 Thorpe solidified his friendship with Porter and arranged to write for the *Knickerbocker Magazine*. The next year the *Spirit* published *The Big Bear of Arkansas*, and in the aftermath of this second triumph Thorpe wrote many sketches, mainly about hunting. He continued to paint, frontier scenes and animal paintings as well as portraits, while editing newspapers in Louisiana. In 1845 Porter used Thorpe's most famous story as the title of his anthology of the best southwestern humor writing he had published (*The Big Bear of Arkansas, and Other Sketches*), and later that year Thorpe published a book of his own, *Mysteries of the Backwoods; or, Sketches of the Southwest: Including Character, Scenery, and Rural Sports*. At the outbreak of the Mexican War Thorpe wrote and illustrated *Our Army on the Rio Grande*. Much involved with Whig politics, he failed to establish any long or profitable journalistic connection, and in 1854 returned to New York, where he wrote for *Harper's*. In 1854 he published an enlargement of his *Mysteries* volume as *The Hive of "The Bee-Hunter": A Repository of Sketches, Including Peculiar American Character, Scenery, and Rural Sports*. His wife died in 1855 and he remarried two years later, about the time he went on the staff of *Frank Leslie's Illustrated Newspaper*. From 1859 to 1861, after Porter's death, Thorpe was part owner of the *Spirit*, but he was still painting, notably a mammoth view of Niagara Falls (1860) and a view of Irving's grave (1862). In 1862 he went to occupied New Orleans, where he was put in charge of distributing food to the poor on a massive scale as well as enforcing sanitation. He served as a member of the Union-backed Louisiana constitutional convention in 1864 before returning North, where from 1869 on he worked at the New York City Custom House; in his last years he also wrote for a new magazine, *Appleton's*. Thorpe died of Bright's disease on September 20, 1878.

Restless, nervous (as Porter described him), self-effacing, Thorpe never quite brought his varied powers to fruition. Yet he has a permanent niche in American literary history for a story which gave its name to "the Big Bear School of literature."

The Big Bear of Arkansas[1]

A steamboat on the Mississippi frequently, in making her regular trips, carries, between places varying from one to two thousand miles apart; and as these boats advertise to land passengers and freight at "all intermediate landings," the heterogeneous character of the passengers of one of these up-country boats can scarcely be

1. The text is that of the first printing, in *The Spirit of the Times: A Chronicle of the Turf, Agriculture, Field Sports, Literature and the Stage* (March 27, 1841). The story was reprinted in *The Big Bear of Arkansas, and Other Sketches*, ed. William T. Porter (Philadelphia: Carey and Hart, 1845), and in Thorpe's *The Hive of "The Bee-Hunter"* (New York: D. Appleton, 1854).

imagined by one who has never seen it with his own eyes. Starting from New Orleans in one of these boats, you will find yourself associated with men from every State in the Union, and from every portion of the globe; and a man of observation need not lack for amusement or instruction in such a crowd, if he will take the trouble to read the great book of character so favorably opened before him. Here may be seen jostling together the wealthy Southern planter, and the pedlar of tin-ware from New England—the Northern merchant, and the Southern jockey—a venerable bishop, and a desperate gambler—the land speculator, and the honest farmer—professional men of all creeds and characters—Wolvereens, Suckers, Hoosiers, Buckeyes, and Corncrackers,[2] beside a "plentiful sprinkling" of the half-horse and half-alligator species of men, who are peculiar to "old Mississippi," and who appear to gain a livelihood simply by going up and down the river. In the pursuit of pleasure or business, I have frequently found myself in such a crowd.

On one occasion, when in New Orleans, I had occasion to take a trip of a few miles up the Mississippi, and I hurried on board the well-known, "high-pressure-and-beat-every-thing" steamboat "Invincible," just as the last note of the last bell was sounding, and when the confusion and bustle that is natural to a boat's getting under way had subsided, I discovered that I was associated in as heterogeneous a crowd as was ever got together. As my trip was to be of a few hours duration only, I made no endeavors to become acquainted with my fellow passengers, most of whom would be together many days. Instead of this, I took out of my pocket the "latest paper," and more critically than usual examined its contents; my fellow passengers at the same time disposed of themselves in little groups. While I was thus busily employed in reading, and my companions were more busily still employed in discussing such subjects as suited their humors best, we were startled most unexpectedly by a loud Indian whoop, uttered in the "social hall," that part of the cabin fitted off for a bar; then was to be heard a loud crowing, which would not have continued to have interested us—such sounds being quite common in that *place of spirits*—had not the hero of these windy accomplishments stuck his head into the cabin and hallooed out, "Hurra for the Big Bar of Arkansaw!" and then might be heard a confused hum of voices, unintelligible, save in such broken sentences as "horse," "screamer,"[3] "lightning is slow," &c. As might have been expected, this continued interruption attracted the attention of every one in the cabin; all conversation

2. I.e., people from Michigan, Illinois, Indiana, Ohio, and Kentucky, respectively. The "half-horse and half-alligator species of men" are the heroic, tall-talking raftsmen, the most celebrated of whom was Mike Fink (1770?–1823), known as "the last of the boatmen." See Chapter 16 of *Huckleberry Finn* in this anthology (which restores the usually omitted raftsman's section).
3. Something of unusual size, strength, or speed.

dropped, and in the midst of this surprise the "Big Bar" walked into
the cabin, took a chair, put his feet on the stove, and looking back
over his shoulder, passed the general and familiar salute of "Strangers, how are you?" He then expressed himself as much at home as
if he had been at "the Forks of Cypress," and "prehaps a little more
so." Some of the company at this familiarity looked a little angry,
and some astonished, but in a moment every face was wreathed in a
smile. There was something about the intruder that won the heart
on sight. He appeared to be a man enjoying perfect health and
contentment—his eyes were as sparkling as diamonds, and good
natured to simplicity. Then his perfect confidence in himself was
irresistibly droll. "Prehaps," said he, "gentlemen," running on without a person speaking, "prehaps you have been to New Orleans
often; I never made *the first visit before*, and I don't intend to make
another in a crow's life. I am thrown away in that ar place, and
useless, that ar a fact. Some of the gentlemen thar called me *green*
—well, prehaps I am, said I, *but I arn't so at home*; and if I aint off
my trail much, the heads of them perlite chaps themselves wern't
much the hardest, for according to my notion, they were *real know-
nothings*, green as a pumpkin-vine—couldn't, in farming, I'll bet,
raise a crop of turnips—and as for shooting, they'd miss a barn if
the door was swinging, and that, too, with the best rifle in the
country. And then they talked to me 'bout hunting, and laughed at
my calling the principal game in Arkansaw poker, and high-low-
jack. 'Prehaps,' said I, 'you prefer checkers and rolette;'[4] at this
they laughed harder than ever, and asked me if I lived in the woods,
and didn't know what *game* was? At this I rather think I laughed.
'Yes,' I roared, and says, 'Strangers, if you'd asked me *how we got
our meat* in Arkansaw, I'd a told you at once, and given you a list
of varmints that would make a caravan, beginning with the bar, and
ending off with the cat; that's *meat* though, not game.' Game,
indeed, that's what city folks call it, and with them it means chippen-
birds and shite-pokes;[5] maybe such trash live in my diggings, but I
arn't noticed them yet—a bird any way is too trifling. I never did
shoot at but one, and I'd never forgiven myself for that had it
weighed less than forty pounds; I wouldn't draw a rifle on anything
less than that; and when I meet with another wild turkey of the
same weight I will drap him."

"A wild turkey weighing forty pounds?" exclaimed twenty voices
in the cabin at once.

"Yes, strangers, and wasn't it a whopper? You see, the thing was
so fat that he couldn't fly far, and when he fell out of the tree, after
I shot him, on striking the ground he bust open behind, and the way
the pound gobs of tallow rolled out of the opening was perfectly
beautiful."

4. The reading "checkers" is from the
1845 edition; the *Spirit* has "chickens,"
probably a misreading of the manuscript; "rolette": roulette.
5. Chipping sparrows and herons.

"Where did all that happen?" asked a cynical looking hoosier.

"Happen! happened in Arkansaw; where else could it have happened, but in the creation State, the finishing-up country; a State where the *sile* runs down to the centre of the 'arth, and government gives you a title to every inch of it. Then its airs, just breathe them, and they will make you snort like a horse. It's a State without a fault, it is."

"Excepting mosquitoes," cried the hoosier.

"Well, stranger, except them; for it ar a fact that they are rather *enormous*, and do push themselves in somewhat troublesome. But, stranger, they never stick twice in the same place, and give them a fair chance for a few months, and you will get as much above noticing them as an alligator. They can't hurt my feelings, for they lay under the skin; and I never knew but one case of injury resulting from them, and that was to a Yankee: and they take worse to foreigners anyhow than they do to natives. But the way they used that fellow up! first they punched him until he swelled up and busted, then he sup-per-a-ted, as the doctor called it, until he was raw as beef; then he took the ager,[6] owing to the warm weather, and finally he took a steamboat and left the country. He was the only man that ever took mosquitoes at heart that I know of. But mosquitoes is natur, and I never find fault with her; if they ar large, Arkansaw is large, her varmints ar large, her trees ar large, her rivers ar large, and a small mosquitoe would be of no more use in Arkansaw than preaching in a cane-brake."

This knock-down argument in favor of big mosquitoes used the hoosier up, and the logician started on a new track, to explain how numerous bear were in his "diggings," where he represented them to be "about as plenty as blackberries, and a little plentifuler."

Upon the utterance of this assertion, a timid little man near me enquired if the bear in Arkansaw ever attacked the settlers in numbers.

"No," said our hero, warming with the subject, "no, stranger, for you see it ain't the natur of bar to go in droves, but the way they squander about in pairs and single ones is edifying. And then the way I hunt them—the old black rascals know the crack of my gun as well as they know a pig's squealing. They grow thin in our parts, it frightens them so, and they do take the noise dreadfully, poor things. That gun of mine is perfect *epidemic among bar*—if not watched closely, it will go off as quick on a warm scent as my dog Bowie-knife[7] will; and then that dog, whew! why the fellow thinks that the world is full of bar, he finds them so easy. It's lucky he don't talk as well as think, for with his natural modesty, if he should suddenly learn how much he is acknowledged to be ahead of all other dogs in the universe, he would be astonished to death in two

6. Ague.
7. Named for James Bowie, who was killed at the Alamo in 1836; Bowie rhymes with "who he."

minutes. Strangers, that dog knows a bar's way as well as a horse-jockey knows a woman's; he always barks at the right time—bites at the exact place—and whips without getting a scratch. I never could tell whether he was made expressly to hunt bar, or whether bar was made expressly for him to hunt; any way, I believe they were ordained to go together as naturally as Squire Jones says a man and woman is, when he moralizes in marrying a couple. In fact, Jones once said, said he, 'Marriage according to law is a civil contract of divine origin, it's common to all countries as well as Arkansaw, and people take to it as naturally as Jim Doggett's Bowie-knife takes to bar.' "

"What season of the year do your hunts take place?" enquired a gentlemanly foreigner, who, from some peculiarities of his baggage, I suspected to be an Englishman, on some hunting expedition, probably, at the foot of the Rocky Mountains.

"The season for bar hunting, stranger," said the man of Arkansaw, "is generally all the year round, and the hunts take place about as regular. I read in history that varmints have their fat season, and their lean season. That is not the case in Arkansaw, feeding as they do upon the *spontenacious* productions of the sile, they have one continued fat season the year round—though in winter things in this way is rather more greasy than in summer, I must admit. For that reason bar with us run in warm weather, but in winter they only waddle. Fat, fat! it's an enemy to speed—it tames everything that has plenty of it. I have seen wild turkies, from its influence, as gentle as chickens. Run a bar in this fat condition, and the way it improves the critter for eating is amazing; it sort of mixes the ile up with the meat until you can't tell t'other from which. I've done this often. I recollect one perty morning in particular, of putting an old he fellow on the stretch, and considering the weight he carried, he run well. But the dogs soon tired him down, and when I came up with him wasn't he in a beautiful sweat—I might say fever; and then to see his tongue sticking out of his mouth a feet,[8] and his sides sinking and opening like a bellows, and his cheeks so fat he couldn't look cross. In this fix I blazed at him, and pitch me naked into a briar patch if the steam didn't come out of the bullet hole ten foot in a straight line. The fellow, I reckon, was made on the high-pressure system, and the lead sort of bust his biler."

"That column of steam was rather curious, or else the bear must have been *warm*,"[9] observed the foreigner with a laugh.

"Stranger, as you observe, that bar was WARM, and the blowing off of the steam show'd it, and also how hard the varmint had been run. I have no doubt if he had kept on two miles farther his insides would have been stewed; and I expect to meet with a varmint yet of extra bottom, who will run himself into a skin full of bar's-grease: it

8. Perhaps a misprint for "foot."
9. Some slang usage may be involved, although the foreigner is probably employing humorous understatement.

is possible, much onlikelier things have happened."

"Where abouts are these bear so abundant?" enquired the foreigner, with increasing interest.

"Why, stranger, they inhabit the neighborhood of my settlement, one of the prettiest places on Old Mississippi—a perfect location, and no mistake; a place that had some defects until the river made the 'cut-off' at 'Shirt-tail bend,' and that remedied the evil, as it brought my cabin on the edge of the river—a great advantage in wet weather, I assure you, as you can now roll a barrel of whiskey into my yard in high water, from a boat, as easy as falling off a log; it's a great improvement, as toting it by land in a jug, as I used to do, *evaporated* it too fast, and it became expensive. Just stop with me, stranger, a month or two, or a year if you like and you will appreciate my place. I can give you plenty to eat, for beside hog and hominy, you can have bar ham, and bar sausages, and a mattrass of bar-skins to sleep on, and a wildcat-skin, pulled off hull, stuffed with corn-shucks for a pillow. That bed would put you to sleep if you had the rheumatics in every joint in your body. I call that ar bed a *quietus*.[1] Then look at my land, the government ain't got another such a piece to dispose of. Such timber, and such bottom land, why you can't preserve anything natural you plant in it, unless you pick it young, things thar will grow out of shape so quick. I once planted in those diggings a few potatoes and beets, they took a fine start, and after that an ox team couldn't have kept them from growing. About that time I went off to old Kentuck on bisiness, and did not hear from them things in three months, when I accidentally stumbled on a fellow who had stopped at my place, with an idea of buying me out. 'How did you like things?' said I. 'Pretty well,' said he; 'the cabin is convenient, and the timber land is good; but that bottom land ain't worth the first red cent.' 'Why?' said I. ' 'Cause,' said he. ' 'Cause what?' said I. ' 'Cause it's full of cedar stumps and Indian mounds,' said he, 'and *it can't be cleared.*' 'Lord,' said I, 'them ar "cedar stumps" is beets, and them ar "Indian mounds" ar tater hills,'—as I expected the crop was overgrown and useless; the sile is too rich, *and planting in Arkansaw is dangerous.* I had a good sized sow killed in that same bottom land; the old thief stole an ear of corn, and took it down where she slept at night to eat; well, she left a grain or two on the ground, and lay down on them; before morning the corn shot up, and the percussion killed her dead. I don't plant any more; natur intended Arkansaw for a hunting ground, and I go according to natur."

The questioner, who thus elicited the description of our hero's settlement, seemed to be perfectly satisfied, and said no more; but the "Big Bar of Arkansaw" rambled on from one thing to another with a volubility perfectly astonishing, occasionally disputing with those around him, particularly with a "live sucker" from Illinois,

1. Final discharge from all care.

who had the daring to say that our Arkansaw friend's stories "smelt rather tall."

In this manner the evening was spent, but conscious that my own association with so singular a personage would probably end before morning, I asked him if he would not give me a description of some particular bear hunt—adding that I took great interest in such things, though I was no sportsman. The desire seemed to please him, and he squared himself round towards me, saying, that he could give me an idea of a bar hunt that was never beat in this world, or in any other. His manner was so singular, that half of his story consisted in his excellent way of telling it, the great peculiarity of which was, the happy manner he had of emphasizing the prominent parts of his conversation. As near as I can recollect, I have italicized them, and given the story in his own words.

"Stranger," said he, "in bar hunts *I am numerous*, and which particular one as you say I shall tell puzzles me. There was the old she devil I shot at the hurricane last fall—then there was the old hog thief I popped over at the Bloody Crossing, and then——Yes, I have it, I will give you an idea of a hunt, in which the greatest bar was killed that ever lived, *none excepted*; about an old fellow that I hunted, more or less, for two or three years, and if that ain't a *particular bar hunt*, I ain't got one to tell. But in the first place, stranger, let me say, I am pleased with you, because you ain't ashamed to gain information by asking, and listening, and that's what I say to Countess's pups every day when I'm home—and I have got great hopes of them ar pups, because they are continually *nosing* about, and though they stick it sometimes in the wrong place, they gain experience anyhow, and may learn something useful to boot. Well, as I was saying about this big bar, you see when I and some more first settled in our region, we were drivin to hunting naturally; we soon liked it, and after that we found it an easy matter to make the thing our business. One old chap who had pioneered 'afore us, gave us to understand that we had settled in the right place. He dwelt upon its merits until it was affecting, and showed us, to prove his assertions, more marks on the sassafras trees than I ever saw on a tavern door 'lection time.[2] 'Who keeps that ar reckoning?' said I. 'The bar,' said he. 'What for?' said I. 'Can't tell,' said he, 'but so it is, the bar bite the bark and wood too, at the highest point from the ground they can reach, and you can tell by the marks,' said he, 'the length of the bar to an inch.' 'Enough,' said I, 'I've learned something here a'ready, and I'll put it in practice.' Well, stranger, just one month from that time I killed a bar, and told its exact length before I measured it by those very marks—and when I did that I swelled up considerable—I've been a prouder man ever since. So I went on, larning something every day, until I was

---

2. I.e., reckonings, drinking bills, posted on the doors of inns or taverns; drinking would be extremely heavy during an election.

reckoned a buster,[3] and allowed to be decidedly the best bar hunter in my district; and that is a reputation as much harder to earn than to be reckoned first man in Congress, as an iron ram-rod is harder than a toad-stool. Did the varmints grow over cunning, by being fooled with by green-horn hunters, and by this means get trouble-some, they send for me as a matter of course, and thus I do my own hunting, and most of my neighbors'. I walk into the varmints though, and it has become about as much the same to me as drinking. It is told in two sentences—a bar is started, and he is killed. The thing is somewhat monotonous now—I know just how much they will run, where they will tire, how much they will growl, and what a thundering time I will have in getting them home. I could give you this history of the chase with all particulars at the commencement, I know the signs so well. *Stranger, I'm certain.* Once I met with a match, though, and I will tell you about it, for a common hunt would not be worth relating.

"On a fine fall day, long time ago, I was trailing about for bar, and what should I see but fresh marks on the sassafras trees, about eight inches above any in the forests that I knew of. Says I, them marks is a hoax, or it indicates the d——t bar that was ever grown. In fact, stranger, I couldn't believe it was real, and I went on. Again I saw the same marks, at the same height, and *I knew the thing lived.* That conviction came home to my soul like an earthquake. Says I, here is something a-purpose for me—that bar is mine, or I give up the hunting business. The very next morning what should I see but a number of buzzards hovering over my cornfield. The rascal has been there, said I, for that sign is certain; and, sure enough, on examining, I found the bones of what had been as beautiful a hog the day before, as was ever raised by a Buck-eye. Then I tracked the critter out of the field to the woods, and all the marks he left behind, showed me that he was *the Bar.*

"Well, stranger, the first fair chase I ever had with that big critter, I saw him no less than three distinct times at a distance, the dogs run him over eighteen miles, and broke down, my horse gave out, and I was as nearly used up as a man can be, made on *my* principle, *which is patent.* Before this adventure, such things were unknown to me as possible; but, strange as it was, that bar got me used to it, before I was done with him,—for he got so at last, that he would leave me on a long chase *quite easy.* How he did it, I never could understand. That a bar runs at all, is puzzling; but how this one could tire down, and bust up a pack of hounds and a horse, that were used to overhauling every thing they started after in no time, was past my understanding. Well, stranger, that bar finally got so sassy, that he used to help himself to a hog off my premises when-ever he wanted one;—the buzzards followed after what he left, and so between *bar and buzzard,* I rather think I was *out of pork.* Well,

3. Something stupendous, "busting" all records.

missing that bar so often, took hold of my vitals, and I wasted away. The thing had been carried too far, and it reduced me in flesh faster than an ager. I would see that bar in every thing I did,—*he hunted me*, and that, too, like a devil, which I began to think he was. While in this fix, I made preparations to give him a last brush, and be done with it. Having completed every thing to my satisfaction, I started at sun-rise, and to my great joy, I discovered from the way the dogs run, that they were near him—finding his trail was nothing, for that had become as plain to the pack as a turnpike-road.[4] On we went, and coming to an open country, what should I see but the bar very leisurely ascending a hill, and the dogs close at his heels, either a match for him this time in speed, or else he did not care to get out of their way—I don't know which. But, wasn't he a beauty though? I loved him like a brother. On he went, until coming to a tree, the limbs of which formed a crotch about six feet from the ground,—into this crotch he got and seated himself,—the dogs yelling all around it—and there he sat eyeing them, as quiet as a pond in low water. A green-horn friend of mine, in company, reached shooting distance before me, and blazed away, hitting the critter in the centre of his forehead. The bar shook his head as the ball struck it, and then he walked down from that tree as gently as a lady would from a carriage. 'Twas a beautiful sight to see him do that,—he was in such a rage, that he seemed to be as little afraid of the dogs, as if they had been sucking pigs; and the dogs warn't slow in making a ring around him at a respectful distance, I tell you; even Bowie-knife himself stood off. Then the way his eyes flashed— why the fire of them would have singed a cat's hair; in fact, that bar was in a *wrath all over*. Only one pup came near him, and he was brushed out so totally with the bar's left paw, that he entirely disappeared; and that made the old dogs more cautious still. In the mean time, I came up, and taking deliberate aim as a man should do, at his side, just back of his foreleg, *if my gun did not snap*,[5] call me a coward, and I won't take it personal. Yes, stranger, *it snapped*, and I could not find a cap about my person. While in this predicament, I turned round to my fool friend—says I, 'Bill,' says I, 'you're an ass—you're a fool—you might as well have tried to kill that bar by barking the tree under his belly, as to have done it by hitting him in the head. Your shot has made a tiger of him, and blast me, if a dog gets killed or wounded when they come to blows, I will stick my knife into your liver, I will——' my wrath was up. I had lost my caps, my gun had snapped, the fellow with me had fired at the bar's head, and I expected every moment to see him close in with the

4. A toll road (from the pike or spear-shaped implement which turns to permit entrance after payment).
5. I.e., the hammer fell but the gun mis-fired; "cap": metallic percussion cap filled with fulminating (explosive) powder.

dogs, and kill a dozen of them at least. In this thing I was mistaken, for the bar leaped over the ring formed by the dogs, and giving a fierce growl, was off—the pack of course in full cry after him. The run this time was short, for coming to the edge of a lake the varmint jumped in, and swam to a little island in the lake, which it reached just a moment before the dogs. I'll have him now, said I, for I had found my caps in the *lining of my coat*—so, rolling a log into the lake, I paddled myself across to the island, just as the dogs had cornered the bar in a thicket. I rushed up and fired—at the same time the critter leaped over the dogs and came within three feet of me, running like mad; he jumped into the lake, and tried to mount the log I had just deserted, but every time he got half his body on it, it would roll over and send him under; the dogs, too, got around him, and pulled him about, and finally Bowie-knife clenched with him, and they sunk into the lake together. Stranger, about this time I was excited, and I stripped off my coat, drew my knife, and intended to have taken a part with Bowie-knife myself when the bar rose to the surface. But the varmint staid under—Bowie-knife came up alone, more dead than alive, and with the pack came ashore. Thank God, said I, the old villain has got his deserts at last. Determined to have the body, I cut a grape-vine for a rope, and dove down where I could see the bar in the water, fastened my queer rope to his leg, and fished him, with great difficulty, ashore. Stranger, may I be chawed to death by young alligators, if the thing I looked at wasn't a *she bar, and not the old critter after all.* The way matters got mixed on that island was onaccountably curious, and thinking of it made me more than ever convinced that I was hunting the devil himself. I went home that night and took to my bed—the thing was killing me. The entire team of Arkansaw in bar-hunting, acknowledged himself used up, and the fact sunk into my feelings like a snagged boat will in the Mississippi. I grew as cross as a bar with two cubs and a sore tail. The thing got out 'mong my neighbors, and I was asked how come on that individ-u-al that never lost a bar when once started? and if that same individ-u-al didn't wear telescopes when he turned a she bar, of ordinary size, into an old he one, a little larger than a horse? Prehaps, said I, friends—getting wrathy—prehaps you want to call somebody a liar. Oh, no, said they, we only heard such things as being *rather common* of late, but we don't believe one word of it; oh, no,—and then they would ride off and laugh like so many hyenas over a dead nigger. It was too much, and I determined to catch that bar, go to Texas, or die,—and I made my preparations accordin'. I had the pack shut up and rested. I took my rifle to pieces, and iled it. I put caps in every pocket about my person, *for fear of the lining.* I then told my neighbors that on Monday morning—naming the day—I would start THAT BAR, and bring him home with me, or they might divide my settlement among them, the owner having disappeared. Well, stranger, on the morning previous to the great day of my hunting expedition, I went into the woods near my house, taking my gun and Bowie-knife along, just *from habit,* and there sitting down also from habit, what should I see, getting over my fence, but *the bar!* Yes, the old varmint was within a hundred yards of me, and the way he walked *over that fence,*—stranger, he loomed up like a *black mist,* he seemed so large, and he walked right towards me. I raised myself, took deliberate aim, and fired. Instantly the varmint wheeled, gave a yell, and *walked through the fence* like a falling tree would through a cobweb. I started after, but was tripped up by my inexpressibles, which either from habit, or the excitement of the moment, were about my heels, and before I had really gathered myself up, I heard the old varmint groaning in a thicket near by, like a thousand sinners, and by the time I reached him he was a corpse. Stranger, it took five niggers and myself to put that carcase on a mule's back, and old long ears waddled under his load, as if he was foundered in every leg of his body, and with a common whopper of a bar, he would have trotted off, and enjoyed himself. 'Twould astonish you to know how big he was,—I made a *bed spread of his skin,* and the way it used to cover my bar mattrass, and leave several feet on each side to tuck up, would have delighted you. It was in fact a creation bar, and if it had lived in Sampson's time,[6] and had met him, in a fair fight, it would have licked him in the twinkling of a dice-box. But, stranger, I never liked the way I hunted him, *and missed him.* There is something curious about it, I could never understand,—and I never was satisfied at his giving in so *easy at last.* Prehaps, he had heard of my preparations to hunt him the next day, so he jist come in, like Capt. Scott's coon,[7] to save his wind to grunt with in dying; but that ain't likely. My private opinion is, that that bar was an *unhuntable bar, and died when his time come.*"

When the story was ended, our hero sat some minutes with his auditors in a grave silence; I saw there was a mystery to him connected with the bear whose death he had just related, that had evidently made a strong impression on his mind. It was also evident that there was some superstitious awe connected with the affair,—a feeling common with all "children of the wood," when they meet with any thing out of their every day experience. He was the first one, however, to break the silence, and jumping up he asked all present to "liquor" before going to bed,—a thing which he did, with a number of companions, evidently to his heart's content.

6. Variant spelling of Samson, strongman and judge of Israel (Judges 13–16).
7. In a favorite anecdote of the time a fine marksman is taking aim at a coon when the animal recognizes that his situation is hopeless and cries out "I'm a gone coon!" or "Don't shoot, captain, I'm coming down!" The story may have become attached to the name of General Winfield Scott (1786–1866), hero of the War of 1812.

Long before day, I was put ashore at my place of destination, and I can only follow with the reader, in imagination, our Arkansas friend, in his adventures at the "Forks of Cypress" on the Mississippi.

1841

FREDERICK DOUGLASS

1817–1895

Born a slave in Maryland, Douglass taught himself to read and write, escaped to Massachuetts by disguising himself as a sailor, became one of the most effective orators of his day, an influential newspaper editor, a confidant of the radical abolitionist John Brown, a militant reformer, and a respected diplomat. The first two accounts of his experiences belong to the tradition of fugitive-slave narratives popular in the North before the Civil War; the final volume, published when Douglass was in his mid-sixties, reveals one of the most remarkable and successful lives of the nineteenth century.

Narrative of the Life of Frederick Douglass, an American Slave (1845) told in 125 pages the story of his life from early childhood until he escaped from bondage (and changed his last name from Bailey to Douglass) in 1838. The vivid detail, the dignity of tone, and the sincerity of the writing left no doubt that Douglass had in fact suffered the horrors he had been describing in powerful lectures for several years. In 1855 he published a revised and enlarged version of the *Narrative* under the title *My Bondage and My Freedom*. This work balanced a more detailed account of his life as a slave with the impressive record of his intellectual growth and personal achievement since he had joined forces with the abolitionists in 1841. It told of his intimacy with the Garrisonian wing of the abolitionist movement (which demanded immediate freeing of all slaves on moral grounds), of his successful speaking tour of the British Isles, the purchase of his freedom for $700 by a group of his admirers, and his move to Rochester, New York, where he brought out in December, 1847, the first issue of the increasingly outspoken weekly newspaper he published for the next thirteen years (first as *The North Star*, later as *Frederick Douglass's Weekly* and *Monthly*). The third of Douglass's autobiographies, *The Life and Times of Frederick Douglass* (1881) subsumes the first two and adds to them the events of his career just before, during, and after the Civil War and traces the rising arc of his fame, influence, and ultimately honored recognition of his countrymen, black and white alike.

Wrongly accused of complicity in John Brown's raid on the arsenal at Harpers Ferry in 1859, Douglass was obliged to flee to Canada and thence to England. Once the Civil War began, he took an active role in the campaign to make black men eligible for Union service; he became a successful recruiter of black soldiers, whose ranks soon included two of his own sons. Having helped to enlist these men, Douglass was only acting in character when he took his protests over their unequal pay and treatment directly to President Lincoln.

It was also in character for Douglass to criticize Lincoln's successors over what Douglass felt was an insufficiently prompt and just Reconstruction policy once the war had been won. Douglass was particularly insistent on the necessity for swift passage of the Fifteenth Amendment guaranteeing suffrage to the newly emancipated slaves. Never satisfied with the grudging legal concessions the Civil War yielded, Douglass continued to object to every sign of discrimination—economic, sexual, legal, and social. Even after he had been appointed United States Marshal and then Recorder of Deeds for the District of Columbia, he continued to speak out on such matters as the exploitation of black sharecroppers in the South, to demand antilynching legislation, to protest the exclusion of black people from public accommodations. He was also active in suffrage movements for women, believing firmly in the power of the ballot as one of the necessities of freedom. It would be hard to exaggerate the importance for later black leaders such as Booker T. Washington and W. E. B. DuBois of Douglass's exemplary career as a champion of human rights. His life, in fact, has become the heroic paradigm for all oppressed people.

From My Bondage and My Freedom
Chapter XVII. The Last Flogging[1]

A SLEEPLESS NIGHT—RETURN TO COVEY'S—PURSUED BY COVEY—THE CHASE DEFEATED—VENGEANCE POSTPONED—MUSINGS IN THE WOODS—THE ALTERNATIVE—DEPLORABLE SPECTACLE—NIGHT IN THE WOODS—EXPECTED ATTACK—ACCOSTED BY SANDY, A FRIEND, NOT A HUNTER—SANDY'S HOSPITALITY—THE "ASH CAKE" SUPPER—THE INTERVIEW WITH SANDY—HIS ADVICE—SANDY A CONJURER AS WELL AS A CHRISTIAN—THE MAGIC ROOT—STRANGE MEETING WITH COVEY—HIS MANNER—COVEY'S SUNDAY FACE—AUTHOR'S DEFENSIVE RESOLVE—THE FIGHT—THE VICTORY, AND ITS RESULTS.

Sleep itself does not always come to the relief of the weary in body, and the broken in spirit; especially when past troubles only foreshadow coming disasters. The last hope had been extinguished. My master, who I did not venture to hope would protect me as *a man*, had even now refused to protect me as *his property*; and had cast me back, covered with reproaches and bruises, into the hands

1. From *My Bondage and My Freedom* (New York and Auburn: Miller, Orton, and Mulligan, 1855), Douglass's second of three autobiographies, the first of which was published as the *Narrative of the Life of Frederick Douglass* in 1845. The first 16 chapters of the second autobiography recount Frederick's childhood, his reading lessons from his mistress Miss Sophia (Mrs. Hugh Auld), and his recognition of the implications of enslavement. At the close of Chapter 14, Douglass is bound out for a year by his master Thomas Auld to work for Edward Covey, a noted "slave breaker" who was to improve Frederick's character and conduct as a slave. Chapters 15 and 16 describe Covey's wrathful character and the brutal weekly floggings which Douglass endures at his hand. After six months at Covey's, Douglass becomes ill and is incapable of physical labor. Covey beats Douglass to force him to continue his work, but to no avail; Douglass's wounds are simply multiplied. At the first opportunity, Douglass escapes from Covey into the fields and returns to the home of his master in St. Michael's, seven miles distant. Captain Auld refuses to credit Douglass's fears that Covey might kill him and orders his return to Covey's farm so that he will not forfeit a year's wages. Douglass manages to beg one night's rest at Auld's before his return.

of a stranger to that mercy which was the soul of the religion he professed. May the reader never spend such a night as that allotted to me, previous to the morning which was to herald my return to the den of horrors from which I had made a temporary escape.

I remained all night—sleep I did not—at St. Michael's; and in the morning (Saturday) I started off, according to the order of Master Thomas, feeling that I had no friend on earth, and doubting if I had one in heaven. I reached Covey's about nine o'clock; and just as I stepped into the field, before I had reached the house, Covey, true to his snakish habits, darted out at me from a fence corner, in which he had secreted himself, for the purpose of securing me. He was amply provided with a cowskin and a rope; and he evidently intended to *tie me up*, and to wreak his vengeance on me to the fullest extent. I should have been an easy prey, had he succeeded in getting his hands upon me, for I had taken no refreshment since noon on Friday; and this, together with the pelting, excitement, and the loss of blood, had reduced my strength. I, however, darted back into the woods, before the ferocious hound could get hold of me, and buried myself in a thicket, where he lost sight of me. The corn-field afforded me cover, in getting to the woods. But for the tall corn, Covey would have overtaken me, and made me his captive. He seemed very much chagrined that he did not catch me, and gave up the chase, very reluctantly; for I could see his angry movements, toward the house from which he had sallied, on his foray.

Well, now I am clear of Covey, and of his wrathful lash, for the present. I am in the wood, buried in its somber gloom, and hushed in its solemn silence; hid from all human eyes; shut in with nature and nature's God, and absent from all human contrivances. Here was a good place to pray; to pray for help for deliverance—a prayer I had often made before. But how could I pray? Covey could pray—Capt. Auld could pray—I would fain pray; but doubts (arising partly from my own neglect of the means of grace, and partly from the sham religion which everywhere prevailed, cast in my mind a doubt upon all religion, and led me to the conviction that prayers were unavailing and delusive) prevented my embracing the opportunity, as a religious one. Life, in itself, had almost become burdensome to me. All my outward relations were against me; I must stay here and starve, (I was already hungry,) or go home to Covey's, and have my flesh torn to pieces, and my spirit humbled under the cruel lash of Covey. This was the painful alternative presented to me. The day was long and irksome. My physical condition was deplorable. I was weak, from the toils of the previous day, and from the want of food and rest; and had been so little concerned about my appearance, that I had not yet washed the blood from my garments. I was an object of horror, even to myself. Life, in Baltimore, when most oppressive, was a paradise to this. What had I done, what had my parents done, that such a life as this should be mine? That day, in the woods, I would have exchanged my manhood for the brutehood of an ox.

Night came. I was still in the woods, unresolved what to do. Hunger had not yet pinched me to the point of going home, and I laid myself down in the leaves to rest; for I had been watching for hunters all day, but not being molested during the day, I expected no disturbance during the night. I had come to the conclusion that Covey relied upon hunger to drive me home; and in this I was quite correct—the facts showed that he had made no effort to catch me, since morning.

During the night, I heard the step of a man in the woods. He was coming toward the place where I lay. A person lying still has the advantage over one walking in the woods, in the day time, and this advantage is much greater at night. I was not able to engage in a physical struggle, and I had recourse to the common resort of the weak. I hid myself in the leaves to prevent discovery. But, as the night rambler in the woods drew nearer, I found him to be a *friend*, not an enemy; it was a slave of Mr. William Groomes, of Easton, a kind hearted fellow, named "Sandy." Sandy lived with Mr. Kemp that year, about four miles from St. Michael's. He, like myself, had been hired out by the year; but, unlike myself, had not been hired out to be broken. Sandy was the husband of a free woman, who lived in the lower part of *"Potpie Neck,"* and he was now on his way through the woods, to see her, and to spend the Sabbath with her.

As soon as I had ascertained that the disturber of my solitude was not an enemy, but the good-hearted Sandy—a man as famous among the slaves of the neighborhood for his good nature, as for his good sense—I came out from my hiding place, and made myself known to him. I explained the circumstances of the past two days, which had driven me to the woods, and he deeply compassionated my distress. It was a bold thing for him to shelter me, and I could not ask him to do so; for, had I been found in his hut, he would have suffered the penalty of thirty-nine lashes on his bare back, if not something worse. But, Sandy was too generous to permit the fear of punishment to prevent his relieving a brother bondman from hunger and exposure; and, therefore, on his own motion, I accompanied him to his home, or rather to the home of his wife—for the house and lot were hers. His wife was called up—for it was now about midnight—a fire was made, some Indian meal was soon mixed with salt and water, and an ash cake was baked in a hurry to relieve my hunger. Sandy's wife was not behind him in kindness—both seemed to esteem it a privilege to succor me; for, although I was hated by Covey and by my master, I was loved by the colored people, because *they* thought I was hated for my knowledge, and persecuted because I was feared. I was the *only* slave *now* in that region who could read and write. There had been one other man, belonging to Mr. Hugh Hamilton, who could read, (his name was "Jim,") but he, poor fellow, had, shortly after my coming into the neighborhood, been sold off to the far south. I saw Jim ironed, in the cart, to be carried to Easton for sale,—pinioned like a yearling for the slaughter. My knowledge was now the pride of my brother slaves; and, no doubt, Sandy felt something of the general interest in me on that account. The supper was soon ready, and though I have feasted since, with honorables, lord mayors and aldermen, over the sea, my supper on ash cake and cold water, with Sandy, was the meal, of all my life, most sweet to my taste, and now most vivid in my memory.

Supper over, Sandy and I went into a discussion of what was *possible* for me, under the perils and hardships which now overshadowed my path. The question was, must I go back to Covey, or must I now attempt to run away? Upon a careful survey, the latter was found to be impossible; for I was on a narrow neck of land, every avenue from which would bring me in sight of pursuers. There was the Chesapeake bay to the right, and "Potpie" river to the left, and St. Michael's and its neighborhood occupying the only space through which there was any retreat.

I found Sandy an old adviser. He was not only a religious man, but he professed to believe in a system for which I have no name. He was a genuine African, and had inherited some of the so called magical powers, said to be possessed by African and eastern nations. He told me that he could help me; that, in those very woods, there was an herb, which in the morning might be found, possessing all the powers required for my protection, (I put his thoughts in my own language;) and that, if I would take his advice, he would procure me the root of the herb of which he spoke. He told me further, that if I would take that root and wear it on my right side, it would be impossible for Covey to strike me a blow; that with this root about my person, no white man could whip me. He said he had carried it for years, and that he had fully tested its virtues. He had never received a blow from a slaveholder since he carried it; and he never expected to receive one, for he always meant to carry that root as a protection. He knew Covey well, for Mrs. Covey was the daughter of Mr. Kemp; and he (Sandy) had heard of the barbarous treatment to which I was subjected, and he wanted to do something for me.

Now all this talk about the root, was, to me, very absurd and ridiculous, if not positively sinful. I at first rejected the idea that the simple carrying a root on my right side, (a root, by the way, over which I walked every time I went into the woods,) could possess any such magic power as he ascribed to it, and I was, therefore, not disposed to cumber my pocket with it. I had a positive aversion to all pretenders to "*divination*." It was beneath one of my intelligence to countenance such dealings with the devil, as this power implied. But, with all my learning—it was really precious little—Sandy was more than a match for me. "My book learning," he said, "had not kept Covey off me," (a powerful argument just then,) and he entreated me, with flashing eyes, to try this. If it did me no good, it could do me no harm, and it would cost me nothing, any way. Sandy was so earnest, and so confident of the good qualities of this weed, that, to please him, rather than from any conviction of its excellence, I was induced to take it. He had been to me the good Samaritan,[2] and had, almost providentially, found me, and helped me when I could not help myself; how did I know but that the hand of the Lord was in it? With thoughts of this sort, I took the roots from Sandy, and put them in my right hand pocket.

This was, of course, Sunday morning. Sandy now urged me to go home, with all speed, and to walk up bravely to the house, as though nothing had happened. I saw in Sandy too deep an insight into human nature, with all his superstition, not to have some respect for his advice; and perhaps, too, a slight gleam or shadow of his superstition had fallen upon me. At any rate, I started off toward Covey's, as directed by Sandy. Having, the previous night, poured my griefs into Sandy's ears, and got him enlisted in my behalf, having made his wife a sharer in my sorrows, and having, also, become well refreshed by sleep and food, I moved off, quite courageously, toward the much dreaded Covey's. Singularly enough, just as I entered his yard gate, I met him and his wife, dressed in their Sunday best—looking as smiling as angels—on their way to church. The manner of Covey astonished me. There was something really benignant in his countenance. He spoke to me as never before; told me that the pigs had got into the lot, and he wished me to drive them out; inquired how I was, and seemed an altered man. This extraordinary conduct of Covey, really made me begin to think that Sandy's herb had more virtue in it than I, in my pride, had been willing to allow; and, had the day been other than Sunday, I should have attributed Covey's altered manner solely to the magic power of the root. I suspected, however, that the *Sabbath*, and not the *root*, was the real explanation of Covey's manner. His religion hindered him from breaking the Sabbath, but not from breaking my skin. He had more respect for the *day* than for the *man*, for whom the day was mercifully given; for while he would cut and slash my body during the week, he would not hesitate, on Sunday, to teach me the value of my soul, or the way of life and salvation by Jesus Christ.

All went well with me till Monday morning; and then, whether the root had lost its virtue, or whether my tormentor had gone deeper into the black art than myself, (as was sometimes said of

2. See Luke 10.29–37.

him,) or whether he had obtained a special indulgence, for his faithful Sabbath day's worship, it is not necessary for me to know, or to inform the reader; but, this much I *may* say,—the pious and benignant smile which graced Covey's face on *Sunday*, wholly disappeared on *Monday*. Long before daylight, I was called up to go and feed, rub, and curry the horses. I obeyed the call, and I would have so obeyed it, had it been made at an earlier hour, for I had brought my mind to a firm resolve, during that Sunday's reflection, viz: to obey every order, however unreasonable, if it were possible, and, if Mr. Covey should then undertake to beat me, to defend and protect myself to the best of my ability. My religious views on the subject of resisting my master, had suffered a serious shock, by the savage persecution to which I had been subjected, and my hands were no longer tied by my religion. Master Thomas's indifference had severed the last link. I had now to this extent "backslidden" from this point in the slave's religious creed; and I soon had occasion to make my fallen state known to my Sunday-pious brother, Covey.

Whilst I was obeying his order to feed and get the horses ready for the field, and when in the act of going up the stable loft for the purpose of throwing down some blades, Covey sneaked into the stable, in his peculiar snake-like way, and seizing me suddenly by the leg, he brought me to the stable floor, giving my newly mended body a fearful jar. I now forgot my *roots*, and remembered my pledge to *stand up in my own defense*. The brute was endeavoring skillfully to get a slip-knot on my legs, before I could draw up my feet. As soon as I found what he was up to, I gave a sudden spring, (my two day's rest had been of much service to me,) and by that means, no doubt, he was able to bring me to the floor so heavily. He was defeated in his plan of tying me. While down, he seemed to think he had me very securely in his power. He little thought he was—as the rowdies say—"in" for a "rough and tumble" fight; but such was the fact. Whence came the daring spirit necessary to grapple with a man who, eight-and-forty hours before, could, with his slightest word have made me tremble like a leaf in a storm, I do not know; at any rate, *I was resolved to fight*, and, what was better still, I was actually hard at it. The fighting madness had come upon me, and I found my strong fingers firmly attached to the throat of my cowardly tormentor; as heedless of consequences, at the moment, as though we stood as equals before the law. The very color of the man was forgotten. I felt as supple as a cat, and was ready for the snakish creature at every turn. Every blow of his was parried, though I dealt no blows in turn. I was strictly on the *defensive*, preventing him from injuring me, rather than trying to injure him. I flung him on the ground several times, when he meant to have hurled me there. I held him so firmly by the throat, that his blood followed my nails. He held me, and I held him.

All was fair, thus far, and the contest was about equal. My resistance was entirely unexpected, and Covey was taken all aback by it, for he trembled in every limb. "*Are you going to resist,* you scoundrel?" he said. To which, I returned a polite "*yes sir*"; steadily gazing my interrogator in the eye, to meet the first approach or dawning of the blow, which I expected my answer would call forth. But, the conflict did not long remain thus equal. Covey soon cried out lustily for help; not that I was obtaining any marked advantage over him, or was injuring him, but because he was gaining none over me, and was not able, single handed, to conquer me. He called for his cousin Hughes, to come to his assistance, and now the scene was changed. I was compelled to give blows, as well as to parry them; and, since I was, in any case, to suffer for resistance, I felt (as the musty proverb goes) that "I might as well be hanged for an old sheep as a lamb." I was still *defensive* toward Covey, but *aggressive* toward Hughes; and, at the first approach of the latter, I dealt a blow, in my desperation, which fairly sickened my youthful assailant. He went off, bending over with pain, and manifesting no disposition to come within my reach again. The poor fellow was in the act of trying to catch and tie my right hand, and while flattering himself with success, I gave him the kick which sent him staggering away in pain, at the same time that I held Covey with a firm hand.

Taken completely by surprise, Covey seemed to have lost his usual strength and coolness. He was frightened, and stood puffing and blowing, seemingly unable to command words or blows. When he saw that poor Hughes was standing half bent with pain—his courage quite gone—the cowardly tyrant asked if I "meant to persist in my resistance." I told him "I *did mean to resist, come what might;*" that I had been by him treated like a *brute*, during the last six months; and that I should stand it *no longer*. With that, he gave me a shake, and attempted to drag me toward a stick of wood, that was lying just outside the stable door. He meant to knock me down with it; but, just as he leaned over to get the stick, I seized him with both hands by the collar, and, with a vigorous and sudden snatch, I brought my assailant harmlessly, his full length, on the *not over* clean ground—for we were now in the cow yard. He had selected the place for the fight, and it was but right that he should have all the advantages of his own selection.

By this time, Bill, the hired man, came home. He had been to Mr. Hemsley's, to spend the Sunday with his nominal wife, and was coming home on Monday morning, to go to work. Covey and I had been skirmishing from before daybreak, till now, that the sun was almost shooting his beams over the eastern woods, and we were still at it. I could not see where the matter was to terminate. He evidently was afraid to let me go, lest I should again make off to the woods; otherwise, he would probably have obtained arms from the house, to frighten me. Holding me, Covey called upon Bill for assistance. The scene here, had something comic about it. "Bill," who knew *precisely* what Covey wished him to do, affected ignorance, and pretended he did not know what to do. "What shall I do, Mr. Covey," said Bill. "Take hold of him—take hold of him!" said Covey. With a toss of his head, peculiar to Bill, he said, "indeed, Mr. Covey, I want to go to work." "*This is* your work," said Covey; "take hold of him." Bill replied, with spirit, "My master hired me here, to work, and *not* to help you whip Frederick." It was now my turn to speak. "Bill," said I, "don't put your hands on me." To which he replied, "My God! Frederick, I aint goin' to tech ye," and Bill walked off, leaving Covey and myself to settle our matters as best we might.

But, my present advantage was threatened when I saw Caroline (the slave-woman of Covey) coming to the cow yard to milk, for she was a powerful woman, and could have mastered me very easily, exhausted as I now was. As soon as she came into the yard, Covey attempted to rally her to his aid. Strangely—and, I may add, fortunately—Caroline was in no humor to take a hand in any such sport. We were all in open rebellion, that morning. Caroline answered the command of her master to "*take hold of me*," precisely as Bill had answered, but in *her*, it was at greater peril so to answer; she was the slave of Covey, and he could do what he pleased with her. It was *not* so with Bill, and Bill knew it. Samuel Harris, to whom Bill belonged, did not allow his slaves to be beaten, unless they were guilty of some crime which the law would punish. But, poor Caroline, like myself, was at the mercy of the merciless Covey; nor did she escape the dire effects of her refusal. He gave her several sharp blows.

Covey at length (two hours had elapsed) gave up the contest. Letting me go, he said,—puffing and blowing at a great rate—"now, you scoundrel, go to your work; I would not have whipped you half so much as I have had you not resisted." The fact was, *he had not whipped me at all*. He had not, in all the scuffle, drawn a single drop of blood from me. I had drawn blood from him; and, even without this satisfaction, I should have been victorious, because my aim had not been to injure him, but to prevent him injuring me.

During the whole six months that I lived with Covey, after this transaction, he never laid on me the weight of his finger in anger. He would, occasionally, say he did not want to have to get hold of me again—a declaration which I had no difficulty in believing; and I had a secret feeling, which answered, "you need not wish to get hold of me again, for you will be likely to come off worse in a second fight than you did in the first."

Well, my dear reader, this battle with Mr. Covey,—undignified as it was, and as I fear my narration of it is—was the turning point in my *"life as a slave."* It rekindled in my breast the smouldering embers of liberty; it brought up my Baltimore dreams, and revived a sense of my own manhood. I was a changed being after that fight. I was *nothing* before; I WAS A MAN NOW. It recalled to life my crushed self-respect and my self-confidence, and inspired me with a renewed determination to be A FREEMAN. A man, without force, is without the essential dignity of humanity. Human nature is so constituted, that it cannot *honor* a helpless man, although it can *pity* him; and even this it cannot do long, if the signs of power do not arise.

He only can understand the effect of this combat on my spirit, who has himself incurred something, hazarded something, in re-pelling the unjust and cruel aggressions of a tyrant. Covey was a tyrant, and a cowardly one, withal. After resisting him, I felt as I had never felt before. It was a resurrection from the dark and pestiferous tomb of slavery, to the heaven of comparative freedom. I was no longer a servile coward, trembling under the frown of a brother worm of the dust, but, my long-cowed spirit was roused to an attitude of manly independence. I had reached the point, at which I was *not afraid to die*. This spirit made me a freeman in *fact*, while I remained a slave in *form*. When a slave cannot be flogged he is more than half free. He has a domain as broad as his own manly heart to defend, and he is really *"a power on earth."* While slaves prefer their lives with flogging, to instant death, they will always find christians enough, like unto Covey, to accommodate that preference. From this time, until that of my escape from slavery, I was never fairly whipped. Several attempts were made to whip me, but they were always unsuccessful. Bruises I did get, as I shall hereafter inform the reader; but the case I have been describ-ing, was the end of the brutification to which slavery had subjected me.

The reader will be glad to know why, after I had so grievously offended Mr. Covey, he did not have me taken in hand by the authorities; indeed, why the law of Maryland, which assigns hang-ing to the slave who resists his master, was not put in force against me; at any rate, why I was not taken up, as is usual in such cases, and publicly whipped, for an example to other slaves, and as a means of deterring me from committing the same offense again. I confess, that the easy manner in which I got off, was, for a long time, a surprise to me, and I cannot, even now, fully explain the cause.

The only explanation I can venture to suggest, is the fact, that Covey was, probably, ashamed to have it known and confessed that he had been mastered by a boy of sixteen. Mr. Covey enjoyed the

unbounded and very valuable reputation, of being a first rate overseer and *negro breaker*. By means of this reputation, he was able to
procure his hands for *very trifling* compensation, and with very
great ease. His interest and his pride mutually suggested the wisdom
of passing the matter by, in silence. The story that he had undertaken to whip a lad, and had been resisted, was, of itself, sufficient
to damage him; for his bearing should, in the estimation of slave-
holders, be of that imperial order that should make such an occurrence *impossible*. I judge from these circumstances, that Covey
deemed it best to give me the go-by. It is, perhaps, not altogether
creditable to my natural temper, that, after this conflict with Mr.
Covey, I did, at times, purposely aim to provoke him to an attack,
by refusing to keep with the other hands in the field, but I could
never bully him to another battle. I had made up my mind to do
him serious damage, if he ever again attempted to lay violent hands
on me.

> "Hereditary bondmen, know yet not
> Who would be free, themselves must strike the blow?"[3]

1845, 1855

3. Byron, *Childe Harold*. Canto ii, st. 76.

HENRY DAVID THOREAU
1817–1862

Thoreau won his place in American literature by adventuring at home—
traveling, as he put it, a good deal in Concord. With that kind of paradox
he infuriated and inspired his Massachusetts neighbors and audiences while
he lived; his writings have infuriated and inspired successive generations of
readers since his death.

Of the men and women who made Concord the center of Transcendentalism, only Thoreau was born there. He lived in Concord all his life, except
for a few years in early childhood, his college years at nearby Cambridge,
and several months on Staten Island in 1843. He made numerous short
excursions, including three to northern Maine, four to Cape Cod, others to
New Hampshire, one to Quebec, and a last trip to Minnesota (1861) in a
futile attempt to strengthen his tubercular lungs. Never marrying, and horrified by the one proposal that he received, his most complex personal relationship outside his family was with his older neighbor, Ralph Waldo
Emerson, though the discrepancies between their rarefied ideals of friendship and the realities of social commerce finally left them frustrated with
each other. Aside from Emerson, contemporary writers meant little to him
except for Thomas Carlyle, whom he regarded as one of the great exhorting

prophets of the generation, and Walt Whitman, though he was always a reader of any history of travel and exploration that could suggest possible ways of experimenting with life. He steeped himself in the classics—Greek, Roman, and English—and he knew in translation the sacred writings of the Hindus. He wrote constantly in his journals, which he began at Emerson's suggestion. Ultimately he made them a finished literary form, but in his early career he used them primarily as sources for his lectures, for his essays, and for both of the books that he published, *A Week on the Concord and Merrimack Rivers* (1849) and *Walden* (1854). Through his writings and lectures he attracted admirers, a few of whom must be called disciples. Much effort—and much unwonted tact—went into satisfying their demands upon him while keeping them at an appropriate distance. In the 1850s, as his journals became more and more the record of his observations of nature, his scientific discoveries made him well known to important naturalists such as Louis Agassiz. During the same years, he became one of the most outspoken abolitionists. Although he was never one to affiliate himself with groups, he became known as a reliable abolitionist speaker— not as important as Wendell Phillips, William Lloyd Garrison, or Theodore Parker, but effective enough to be summoned to fill in for Frederick Douglass at a convention in Boston. Thoreau moved into the political forefront only with his defense of John Brown, immediately after the arrests at Harpers Ferry. He was forty-four when he died at Concord on May 6, 1862, in his mother's house. The little national fame he had achieved was as an eccentric Emersonian social experimenter and a firebrand champion of Brown. Emerson, himself famous as the sage of Concord, called Thoreau pre-eminently "*the* man of Concord," a sincere compliment that precisely delimited his sense of his younger friend as ultimately far more provincial than himself.

Thoreau's nonliterary neighbors, whom he taunted in *Walden* in order to compel their attention, knew him as an educated man without an occupation—an affront to a society in which few sons (and no daughters) had the privilege of going to Harvard College. Even Emerson thought that he had drifted into his odd way of life rather than choosing it deliberately. Thoreau might in fact have made a career of his first job as a Concord schoolteacher had he not quickly resigned rather than inflict corporal punishment on his students. He would have taken another teaching job, but in that depression year of 1837 could find none. He and his older brother John started their own progressive school in Concord, but it disbanded when John became ill. John died early in 1842, and Thoreau never went back to teaching. That year he became a handyman at Emerson's house in exchange for room and board, and stayed there intermittently during the 1840s, especially when Emerson was away on long trips. He tried tutoring at the Staten Island home of Emerson's brother William in 1843, but he grew miserably homesick. One long-term advantage was that the job had permitted him some contact with the New York publishing circle. He spent two years on Emerson's property at Walden Pond (1845–47) in a cabin he built himself. He first lectured at the Concord Lyceum in 1838; from the late 1840s onward he occasionally earned twenty-five dollars or so for lecturing in small towns such as New Bedford and Worcester and, less often, in Boston.

Sometimes he charmed his audiences with woodlore and what reviewers called his "comical" and "highfalutin" variety of laconic Yankee wit; sometimes he infuriated them with righteous challenges to the way they lived. No critic, however friendly, claimed that Thoreau had much presence as a public speaker, except during the fury of some of his abolitionist addresses. After 1848 he earned some money now and then by surveying property. He sold a few magazine articles, but earned nothing from his two books. He worked at times in his father's pencil factory, and carried on the business when his father died in 1859, thereby aggravating his tuberculosis with the dust from graphite. His whole life, after the period of uncertainty about an occupation in his early manhood, became a calculated refusal to live by the materialistic values of the neighbors who provided him with a microcosm of the world. By simplifying his needs—an affront to what was already a consumer society devoted to arousing "artificial wants"—he succeeded, with minimal compromises, in living his life rather than wasting it in earning a living.

Among Thoreau's literary acquaintances such as Bronson Alcott, Ellery Channing, and Margaret Fuller, Emerson was his first and most powerful champion. Emerson published many of Thoreau's early poems and essays in *The Dial* between 1842 and 1844, and tried to persuade publishers in Boston and New York to print Thoreau's first book. Emerson saw to it that *Week* and, later, *Walden* were known to his British friends, including Thomas Carlyle. Hawthorne, a sometime Concord resident, liked Thoreau although he thought him "the most unmalleable fellow alive." In 1845 Hawthorne discouraged Evert A. Duyckinck from looking for any popular book from Thoreau except perhaps "a book of simple observation of nature." Later, Hawthorne mentioned Thoreau in the prefaces to *Mosses from an Old Manse* (1846) and *The Scarlet Letter* (1850), and Elizabeth Peabody, Hawthorne's sister-in-law, printed *Resistance to Civil Government* in her *Aesthetic Papers* (1849). Horace Greeley, the vigorous editor of the New York *Tribune*, did more than anyone besides Emerson to make Thoreau a national figure, from 1843 onward mentioning his contributions to *The Dial*, reviewing his books, advertising his lectures, reprinting some of his writings, and aggressively forcing some of Thoreau's essays upon magazine editors in New York and Philadelphia, then dunning them for payment. George William Curtis, who was one of the "raisers" of the Walden cabin, printed three parts of *A Yankee in Canada* (1866) in *Putnam's Monthly* (1853), but Thoreau withdrew the rest when Curtis wanted to modify what Greeley guessed were "very flagrant heresies (like your defiant Pantheism)." Curtis also accepted *Cape Cod* for *Putnam's Monthly*, but held it for three years before starting to print it in 1855, and even then Thoreau had to contend with Curtis's religious scruples. The *Atlantic Monthly*, founded in 1857 by men of abolitionist sympathies, ought to have become a regular outlet, but the editor, James Russell Lowell, had acquired a dislike for Thoreau, either at Harvard or during Lowell's enforced rustication at Concord. Lowell accepted part of what became *The Maine Woods* but deleted a climactic sentence about a pine tree: "It is as immortal as I am, and perchance will go to as high a heaven, there to tower above me still." Thoreau scathingly declared that the expurgation had been made in "a very mean and cowardly manner," and the *Atlantic Monthly* was closed to him until just before his death, when the new editors solicited manuscripts.

When he died, Thoreau was putting many of his works in shape for publication. His last audible words had to do with *The Maine Woods*: "moose" and "Indian." Although Thoreau had published only *Week* and *Walden* in book form, substantial sections of two posthumous books, *The Maine Woods* and *Cape Cod*, had appeared in magazines. Had they appeared as books, these two might have won Thoreau a wider reputation as a conservationist and an acute observer of people and places, but they would hardly have won him a loftier literary fame.

A Week on the Concord and Merrimack Rivers (1849) purports to be the record of a canoe excursion Thoreau and his brother took upriver. They leave Concord on a Saturday; by Thursday they are as far into New Hampshire as their canoe will go; and then they go back downstream to Concord on a Thursday and Friday (really of the next week).The book consists partly of descriptions of the fauna and flora that the brothers see, along with brief mention of people they encounter. Many pages are devoted to local history, plundered from gazetteers; Thoreau even includes a narrative of Indian captivity which he ends with some Hawthornesque sensationalism. Poems by Thoreau and others and fragments of his translations from Greek epics and drama also take up space. The bulk of the small book, however, consists of numerous essays, spliced in hit or miss, on a variety of topics such as rivers, fish and fishing, fables, Christianity, poetry, reading, writing, reformers, Oriental scriptures, canal boats, Anacreon, quackery, pedestrian travel, Persius, the distinction between art and nature, the Concord Cattle Show, Ossian, and Chaucer. The longest essay is on friendship. Much of this, verse as well as prose, was fugitive material salvaged from issues of *The Dial*. First completed in the spring of 1846, *Week* was revised and expanded over the next years (the essay on friendship being added in 1848) before Thoreau published an edition of one thousand copies at his own expense in 1849; the true story of his having to accommodate some seven hundred unsold copies in his attic is one of the more grimly ironic episodes in the history of earning a living in America by writing. Emerson had generously assured one editor in 1846 that *Week* contained the results of years of study. That is even truer of the book in its final form, but it was never worked into a unified whole. Its great merit is that by its disastrous reception Thoreau was forced not to publish *Walden* right away (there were ads for it in *Week*). Instead, he kept the manuscript of *Walden* for several more years, reworking it many times. If *Week* had been even a modest success, *Walden* probably would not have been a literary classic.

As early as 1857, Thoreau made clear his intention to publish *The Maine Woods* as a book, though in his lifetime only the first two parts appeared. In this book there is very little satire and very little of the reflective writing shunned by magazines of the time. The most heightened passages of "Ktaadn" deal with Thoreau's realization that the Maine woods were "primeval, untamed, and forever untameable *Nature*," and his peroration would not have offended even the spoilers of nature, citing as it did the inviolable areas in America still left for exploration. Even passages on conservation are not in the voice of a nature-loving Jeremiah, though "Chesun-cook" ends with the hope that we shall not, like villains, grub the forests all up, "poaching on our own national domains." These two sections, like al-most all of the third, are largely straightforward descriptions of people, places, plants, and animals, with special attention paid to what woodlore could be picked up from lumbermen and Indian guides. The book's modern editor aptly says that as Maine became a favorite hunting and resort area in the 1870s and 1880s *The Maine Woods* served as a backwoods Baedeker. The backwoods have retreated, but the book is a durable record of what a trained and resourceful observer could discover of primeval nature only a short way from Concord, a reminder that Thoreau was a frontiersman, an explorer of the primeval wilderness as well as of the higher latitudes to be found within oneself.

Thoreau also wrote *Cape Cod* as a book, but during his lifetime he was able to publish only the first four chapters. If he had managed to publish it in the early 1850s it might have gone some way toward making him a popular author. At this time Cape Cod was not fashionable, so Thoreau had a subject almost as exotic as Melville's "Encantadas" (which appeared in *Putnam's Monthly* a little earlier than the chapters from *Cape Cod*) with the added piquancy that Thoreau's unknown land was in the backyards of New York and Boston. Much of the book is vivid eyewitness reportage in the punning style of Thoreau's maturity and with a cheeriness none of his other works sustains. Many pages are openly cribbed from local histories; at best, such information is supplemented by fresh stories from the local in-habitants (who, to Thoreau's delight, often turned out to be even more cantankerous than himself) and by Thoreau's own observations. There is no rage in the book, even in the satire; much of the book is joyous tall talk. In the late twentieth century there is poignancy in reading Thoreau's concluding glance at the future of Cape Cod. The last sentence is a power-ful image of Thoreau's repudiation of the worst aspects of his time: "A man may stand there [at Cape Cod] and put all America behind him."

None of the other books which Thoreau published or projected conveys anything like the image of the whole Thoreau that *Walden* does, and even his most representative short work, *Life without Principle*, contains little to suggest his cheerier humor or his love of nature. *Walden* has the medita-tiveness of *Week* without its diffuseness, the natural observation of *The Maine Woods* without its constriction to particular excursions, the atten-tion to quaintnesses of person and place of *Cape Cod* without its some-times smothering admixture of borrowed facts. The meandering of *Week* and the travelogue quality of both *The Maine Woods* and *Cape Cod* are replaced in *Walden* by an account which is both a factual record of a particular experience and a parable of all experience. The parables of *Week* are elaborated more richly and focused more memorably in *Walden*; in it the satiric verve of *Cape Cod* is focused on issues far more momentous; in it the nature study of *The Maine Woods* is infused with Thoreau's Trans-cendentalism. *Week* was the product of diverse impulses; *Cape Cod* and *The Maine Woods* were products of single but limited impulses—perfect of their kind but not belonging to the first order of aspiration or achieve-ment. *Walden* was the product of a single impulse, but one of the strongest

literary impulses ever felt: the determination to write a basic book on how
to live wisely, a book so profoundly liberating that from the reading of it
men and women would date new eras in their lives. In *Walden* Thoreau's
whole character emerges. In it he becomes, in the highest sense, a public
servant, offering the English-speaking public the fruits of his experience,
thought, and artistic dedication.

Thoreau's early writing, even well after his college days, was undistinguished—mere educated prose, less individual, for the most part, than the
thirdhand prose that Melville uneasily employed in *Typee*. Thoreau's prose
ran to clichés even when the topics, such as love and friendship, were those
that were to recur in memorable forms in his later writings. As late as
Week, the writing was often pedestrianly learned, not up to the alertness
of a profoundly educated walker like Thoreau. In *Walden* and a few other
works of his mature years, however, Thoreau's style totally subserves his
main purpose. Throughout *Walden* that purpose is to force the reader to
evaluate the way he has been living and thinking. Whether with his famous
aphoristic sentences, his brief fables or allegories, his thick-strewn puns, or
many other rhetorical devices, Thoreau's intention always is to make the
reader look beyond the obvious, routine sense of an expression to see what
idea once vitalized it. He ultimately wants the reader to re-evaluate any
institution, from the Christian religion to the Constitution of the United
States, but first he makes the reader work up his courage by re-evaluating on
a smaller scale. Thoreau's rhetorical devices afford the hard exercise by
which a reader may learn to think freshly. The prose of *Walden*, in short,
is designed as a practical course in the liberation of the reader.

Recognition of Thoreau as an important writer was slow in coming. Literary people of his own time knew well enough who he was, but the reading
public did not until the publication of *Walden* occasioned comment in
some widely read newspapers and magazines. What became Thoreau's
most famous essay, *Resistance to Civil Government* (the posthumous title
On the Duty of Civil Disobedience, now usually cut to the last two words,
is apparently not authorial), was published anonymously and never attached
to his name in print during his life, though such people as Emerson and
Hawthorne knew Thoreau was the author; many decades passed before
anyone explicitly acted upon the essay's radical advice. Thoreau's early
essays in magazines like the *Democratic Review* and *Graham's* were anonymous, and Greeley did not mention him by name when he printed in the
Tribune for May 25, 1848, a remarkable quotation from a Thoreau letter
that was to become part of the first chapter of *Walden*. Thoreau's *Putnam's
Monthly* and *Atlantic Monthly* pieces were also anonymous, according to
the custom, so that most of his readers probably never knew they were
reading Thoreau. Ironically, his widest-read works published under his name
during his lifetime were not *Week* or even *Walden*, but *Slavery in Massachusetts* (printed in William Lloyd Garrison's *Liberator* and copied in the
Tribune) and *A Plea for Captain John Brown* (printed in the fast-selling
Echoes of Harper's Ferry [1860]).

Between June, 1862 (the month after Thoreau's death), and November,
1863, the *Atlantic Monthly* published *Walking, Autumn Tints, Wild
Apples, Life without Principle*, and *Night and Moonlight* anonymously,
but publicized them as Thoreau's. Ticknor and Fields reissued *Week* and

Walden and quickly got out five new books: *Excursions* (1863), *The Maine Woods* (1864), *Cape Cod* (1864), *Letters to Various Persons* (1865), and *A Yankee in Canada, with Anti-Slavery and Reform Papers* (1866). The expanded form of Emerson's funeral speech, published in the *Atlantic Monthly* for August, 1863, confirmed Thoreau's growing reputation even while unnecessarily stressing some of his less attractive traits, especially his "habit of antagonism." In the *North American Review* for October, 1865, James Russell Lowell—by then the foremost American critic—had his revenge for Thoreau's scorn for his censorship. Reviewing *Letters to Various Persons*, Lowell depicted Thoreau as a mere echoer of Emerson, "surly and stoic," with "a morbid self-consciousness that pronounces the world of men empty and worthless before trying it." Perhaps most damning, Thoreau was a man who "had no humor." Even Robert Louis Stevenson's description (1880) of Thoreau as a "skulker" had a less baneful effect. In American literary histories and classroom anthologies of the next sixty years, Lowell's words were endlessly quoted or paraphrased.

With Thoreau's credit as social philosopher so thoroughly squelched, his friends began emphasizing his role as a student of nature. Channing published *Thoreau: The Poet-Naturalist* (1873), and John Burroughs's essays followed in the 1880s. Capitalizing on this new attention, Thoreau's disciple H. G. O. Blake, who had inherited the journals from Thoreau's sister Sophia, published *Early Spring in Massachusetts* (1881), *Summer* (1884), *Winter* (1887), and *Autumn* (1892). British critics became interested in Thoreau, and in 1890 an important biography was published by the socialist H. S. Salt—just in time to introduce Thoreau to many Fabians and Labour party members. Thoreau was at last becoming widely recognized as a social philosopher as well as a naturalist. In 1906 Mahatma Gandhi, in his African exile, read *Civil Disobedience* and made it—and later *Life without Principle*—major documents in his struggle for Indian independence. The publication of the journals in 1906 in chronological order (Blake had plundered the journals for seasonal passages regardless of the years in which they occurred) gave readers for the first time a nearly full body of evidence for understanding and judging Thoreau. By the 1930s, when for many "Simplify!" had become not a whim but a necessity, Thoreau had attained the status of a major American voice. Scholarly attention in the next decades began to exalt him to a literary rank higher than Emerson's, even while civil-rights leaders such as Martin Luther King, Jr., tested his tactics of civil disobedience throughout the South and sometimes into the North. Today the counterculture's concern with experiments in living and the general American concern for ecological sanity are establishing Thoreau more firmly than ever as a great American prophet, while his potential value to the radical left remains largely untested. Oddly enough, editors and publishers have kept much of his best work from being known by frequently reprinting *Walden* and *Civil Disobedience* together while ignoring his other works. Thoreau has yet to achieve his full recognition as a great prose stylist as well as a lover of nature, a New England mystic, and a powerful social philosopher. He remains the most challenging major writer America has produced. No good reader will ever be entirely pleased with him- or herself or with the current state of culture and civilization while reading any of Thoreau's best works.

Resistance to Civil Government[1]

I heartily accept the motto,—"That government is best which governs least;"[2] and I should like to see it acted up to more rapidly and systematically. Carried out, it finally amounts to this, which also I believe,—"That government is best which governs not at all;" and when men are prepared for it, that will be the kind of government which they will have. Government is at best but an expedient; but most governments are usually, and all governments are sometimes, inexpedient. The objections which have been brought against a standing army, and they are many and weighty, and deserve to prevail, may also at last be brought against a standing government. The standing army is only an arm of the standing government. The government itself, which is only the mode which the people have chosen to execute their will, is equally liable to be abused and perverted before the people can act through it. Witness the present Mexican war, the work of comparatively a few individuals using the standing government as their tool; for, in the outset, the people would not have consented to this measure.[3]

This American government,—what is it but a tradition, though a recent one, endeavoring to transmit itself unimpaired to posterity, but each instant losing some of its integrity? It has not the vitality and force of a single living man; for a single man can bend it to his will. It is a sort of wooden gun to the people themselves; and, if ever they should use it in earnest as a real one against each other, it will surely split. But it is not the less necessary for this; for the people must have some complicated machinery or other, and hear its din, to satisfy that idea of government which they have. Governments show thus how successfully men can be imposed on, even impose on themselves, for their own advantage. It is excellent, we must all

1. *Resistance to Civil Government* is reprinted here from its first appearance, in *Aesthetic Papers* (Boston: Elizabeth P. Peabody, 1849); the editor and publisher was Hawthorne's sister-in-law. Thoreau had delivered the paper (or parts of it) as a lecture in January and again in February, 1848, before the Concord Lyceum, under the title *The Rights and Duties of the Individual in Relation to Government*. After his death it was reprinted in *A Yankee in Canada, with Anti-Slavery and Reform Papers* (1866) as *Civil Disobedience*, the title by which it much later became world-famous. That title, although very commonly used, may well not be authorial, and Thoreauvians are accustoming themselves to the title of the first printing which, as Thoreau indicates, was a play on *Duty of Submission to Civil Government*, the title of one of the chapters in William Paley's *Principles of Moral and Political Philosophy* (1785). Ignored in its own time, in the 20th century the influence of the essay has been profound, most notably in Mahatma Gandhi's struggle for Indian independence and in the American civil-rights movement under the leadership of Martin Luther King, Jr.

2. Associated with Jeffersonianism, these words appeared on the masthead of the *Democratic Review*, the New York magazine that had published two early Thoreau pieces in 1843.

3. The Mexican War, widely criticized by Whigs and many Democrats as an "executive's war" because President Polk commenced hostilities without a congressional declaration of war, ended on February 2, 1848, just after Thoreau first delivered this essay as a lecture. He repeated the lecture after the official ending of the war (or perhaps gave another installment of it), and the next year let it go to press with the out-of-date reference.

allow; yet this government never of itself furthered any enterprise, but by the alacrity with which it got out of its way. *It* does not keep the country free. *It* does not settle the West. *It* does not educate. The character inherent in the American people has done all that has been accomplished; and it would have done somewhat more, if the government had not sometimes got in its way. For government is an expedient by which men would fain succeed in letting one another alone; and, as has been said, when it is most expedient, the governed are most let alone by it. Trade and commerce, if they were not made of India rubber, would never manage to bounce over the obstacles which legislators are continually putting in their way; and, if one were to judge these men wholly by the effects of their actions, and not partly by their intentions, they would deserve to be classed and punished with those mischievous persons who put obstructions on the railroads.

But, to speak practically and as a citizen, unlike those who call themselves no-government men, I ask for, not at once no government, but *at once* a better government. Let every man make known what kind of government would command his respect, and that will be one step toward obtaining it.

After all, the practical reason why, when the power is once in the hands of the people, a majority are permitted, and for a long period continue, to rule, is not because they are most likely to be in the right, nor because this seems fairest to the minority, but because they are physically the strongest. But a government in which the majority rule in all cases cannot be based on justice, even as far as men understand it. Can there not be a government in which majorities do not virtually decide right and wrong, but conscience?—in which majorities decide only those questions to which the rule of expediency is applicable? Must the citizen ever for a moment, or in the least degree, resign his conscience to the legislator? Why has every man a conscience, then? I think that we should be men first, and subjects afterward. It is not desirable to cultivate a respect for the law, so much as for the right. The only obligation which I have a right to assume, is to do at any time what I think right. It is truly enough said,[4] that a corporation has no conscience; but a corporation of conscientious men is a corporation *with* a conscience. Law never made men a whit more just; and, by means of their respect for it, even the well-disposed are daily made the agents of injustice. A common and natural result of an undue respect for law is, that you may see a file of soldiers, colonel, captain, corporal, privates, powder-monkeys and all, marching in admirable order over hill and dale to the wars, against their wills, aye, against their common sense and consciences, which makes it very steep marching indeed, and produces a palpitation of the heart. They have no doubt that it is a

---

4. By Sir Edward Coke, 1612, in a famous legal decision.

damnable business in which they are concerned; they are all peace-ably inclined. Now, what are they? Men at all? or small moveable forts and magazines, at the service of some unscrupulous man in power? Visit the Navy Yard, and behold a marine, such a man as an American government can make, or such as it can make a man with its black arts, a mere shadow and reminiscence of humanity, a man laid out alive and standing, and already, as one may say, buried under arms with funeral accompaniments, though it may be

> "Not a drum was heard, nor a funeral note,
> As his corse to the ramparts we hurried;
> Not a soldier discharged his farewell shot
> O'er the grave where our hero we buried."[5]

The mass of men serve the State thus, not as men mainly, but as machines, with their bodies. They are the standing army, and the militia, jailers, constables, *posse comitatus,*[6] &c. In most cases there is no free exercise whatever of the judgment or of the moral sense; but they put themselves on a level with wood and earth and stones; and wooden men can perhaps be manufactured that will serve the purpose as well. Such command no more respect than men of straw, or a lump of dirt. They have the same sort of worth only as horses and dogs. Yet such as these even are commonly esteemed good citizens. Others, as most legislators, politicians, lawyers, ministers, and office-holders, serve the State chiefly with their heads; and, as they rarely make any moral distinctions, they are as likely to serve the devil, without intending it, as God. A very few, as heroes, patriots, martyrs, reformers in the great sense, and *men,* serve the State with their consciences also, and so necessarily resist it for the most part; and they are commonly treated by it as enemies. A wise man will only be useful as a man, and will not submit to be "clay," and "stop a hole to keep the wind away,"[7] but leave that office to his dust at least:—

> "I am too high-born to be propertied,
> To be a secondary at control,
> Or useful serving-man and instrument
> To any sovereign state throughout the world."[8]

He who gives himself entirely to his fellow-men appears to them useless and selfish; but he who gives himself partially to them is pronounced a benefactor and philanthropist.

How does it become a man to behave toward this American government to-day? I answer that he cannot without disgrace be

5. From Charles Wolfe's *Burial of Sir John Moore at Corunna* (1817), a song Thoreau liked to sing.
6. Sheriff's posse.
7. Shakespeare's *Hamlet,* 5.1.236–37.
8. Shakespeare's *King John,* 5.1.79–82.

associated with it. I cannot for an instant recognize that political organization as *my* government which is the *slave's* government also.

All men recognize the right of revolution; that is, the right to refuse allegiance to and to resist the government, when its tyranny or its inefficiency are great and unendurable. But almost all say that such is not the case now. But such was the case, they think, in the Revolution of '75. If one were to tell me that this was a bad government because it taxed certain foreign commodities brought to its ports, it is most probable that I should not make an ado about it, for I can do without them: all machines have their friction; and possibly this does enough good to counterbalance the evil. At any rate, it is a great evil to make a stir about it. But when the friction comes to have its machine, and oppression and robbery are organized, I say, let us not have such a machine any longer. In other words, when a sixth of the population of a nation which has undertaken to be the refuge of liberty are slaves, and a whole country is unjustly overrun and conquered by a foreign army, and subjected to military law, I think that it is not too soon for honest men to rebel and revolutionize. What makes this duty the more urgent is the fact, that the country so overrun is not our own, but ours is the invading army.

Paley, a common authority with many on moral questions, in his chapter on the "Duty of Submission to Civil Government,"[9] resolves all civil obligation into expediency; and he proceeds to say, "that so long as the interest of the whole society requires it, that is, so long as the established government cannot be resisted or changed without public inconveniency, it is the will of God that the established government be obeyed, and no longer."—"This principle being admitted, the justice of every particular case of resistance is reduced to a computation of the quantity of the danger and grievance on the one side, and of the probability and expense of redressing it on the other." Of this, he says, every man shall judge for himself. But Paley appears never to have contemplated those cases to which the rule of expediency does not apply, in which a people, as well as an individual, must do justice, cost what it may. If I have unjustly wrested a plank from a drowning man, I must restore it to him though I drown myself.[1] This, according to Paley, would be inconvenient. But he that would save his life, in such a case, shall lose it.[2] This people must cease to hold slaves, and to make war on Mexico, though it cost them their existence as a people.

In their practice, nations agree with Paley; but does any one

---

9. The precise title of the chapter in William Paley's *Principles of Moral and Political Philosophy* (1785) is "The Duty of Submission to Civil Government Explained." This book by Paley, English theologian and moralist (1743–1805), was one of Thoreau's Harvard textbooks.

1. A problem in situational ethics cited by Cicero in *De Officiis*, III, which Thoreau had studied.

2. Matthew 10.39; Luke 9.24.

think that Massachusetts does exactly what is right at the present crisis?

> "A drab of state, a cloth-o'-silver slut,
> To have her train borne up, and her soul trail in the dirt."[3]

Practically speaking, the opponents to a reform in Massachusetts are not a hundred thousand politicians at the South, but a hundred thousand merchants and farmers here,[4] who are more interested in commerce and agriculture than they are in humanity, and are not prepared to do justice to the slave and to Mexico, *cost what it may.* I quarrel not with far-off foes, but with those who, near at home, co-operate with, and do the bidding of those far away, and without whom the latter would be harmless. We are accustomed to say, that the mass of men are unprepared; but improvement is slow, because the few are not materially wiser or better than the many. It is not so important that many should be as good as you, as that there be some absolute goodness somewhere; for that will leaven the whole lump.[5] There are thousands who are *in opinion* opposed to slavery and to the war, who yet in effect do nothing to put an end to them; who, esteeming themselves children of Washington and Franklin,[6] sit down with their hands in their pockets, and say that they know not what to do, and do nothing; who even postpone the question of freedom to the question of free-trade, and quietly read the prices-current along with the latest advices from Mexico, after dinner, and, it may be, fall asleep over them both. What is the price-current of an honest man and patriot to-day? They hesitate, and they regret, and sometimes they petition; but they do nothing in earnest and with effect. They will wait, well disposed, for others to remedy the evil, that they may no longer have it to regret. At most, they give only a cheap vote, and a feeble countenance and God-speed, to the right, as it goes by them. There are nine hundred and ninety-nine patrons of virtue to one virtuous man; but it is easier to deal with the real possessor of a thing than with the temporary guardian of it.

All voting is a sort of gaming, like chequers or backgammon, with a slight moral tinge to it, a playing with right and wrong, with moral questions; and betting naturally accompanies it. The character of the voters is not staked. I cast my vote, perchance, as I think right; but I am not vitally concerned that that right should prevail. I am willing to leave it to the majority. Its obligation, therefore, never exceeds that of expediency. Even voting *for the right* is *doing* nothing for it. It is only expressing to men feebly your desire that it should prevail. A wise man will not leave the right to the mercy of chance, nor wish it to prevail through the power of the

---

3. Cyril Tourneur (1575?–1626), *The Revenger's Tragedy*, IV.iv.
4. Thoreau refers to the economic alliance of southern cotton growers with northern shippers and manufacturers.
5. 1 Corinthians 5.6.
6. I.e., children of rebels and revolutionaries.

majority. There is but little virtue in the action of masses of men. When the majority shall at length vote for the abolition of slavery, it will be because they are indifferent to slavery, or because there is but little slavery left to be abolished by their vote. *They* will then be the only slaves. Only *his* vote can hasten the abolition of slavery who asserts his own freedom by his vote.

I hear of a convention to be held at Baltimore, or elsewhere, for the selection of a candidate for the Presidency, made up chiefly of editors, and men who are politicians by profession; but I think, what is it to any independent, intelligent, and respectable man what decision they may come to, shall we not have the advantage of his wisdom and honesty, nevertheless? Can we not count upon some independent votes? Are there not many individuals in the country who do not attend conventions? But no: I find that the respectable man, so called, has immediately drifted from his position, and de-spairs of his country, when his country has more reason to despair of him. He forthwith adopts one of the candidates thus selected as the only *available* one, thus proving that he is himself *available* for any purposes of the demagogue. His vote is of no more worth than that of any unprincipled foreigner or hireling native, who may have been bought. Oh for a man who is a *man*, and, as my neighbor says, has a bone in his back which you cannot pass your hand through! Our statistics are at fault: the population has been returned too large. How many *men* are there to a square thousand miles in this country? Hardly one. Does not America offer any inducement for men to settle here? The American had dwindled into an Odd Fellow,—one who may be known by the development of his organ of gregariousness, and a manifest lack of intellect and cheerful self-reliance;[7] whose first and chief concern, on coming into the world, is to see that the alms-houses are in good repair; and, before yet he has lawfully donned the virile garb,[8] to collect a fund for the support of the widows and orphans that may be; who, in short, ventures to live only by the aid of the mutual insurance company, which has promised to bury him decently.

It is not a man's duty, as a matter of course, to devote himself to the eradication of any, even the most enormous wrong; he may still properly have other concerns to engage him; but it is his duty, at least, to wash his hands of it, and, if he gives it no thought longer, not to give it practically his support. If I devote myself to other pursuits and contemplations, I must first see, at least, that I do not pursue them sitting upon another man's shoulders. I must get off him first, that he may pursue his contemplations too. See what gross inconsistency is tolerated. I have heard some of my townsmen say, "I should like to have them order me out to help put down an

7. The Odd Fellows are a secret fraternal organization, chosen by Thoreau for the satirical value of its name: in his view the archetypal American is not the in-dividualist, the genuine odd fellow, but the conformist.

8. Adult garb allowed a Roman boy on reaching 14.

insurrection of the slaves, or to march to Mexico,—see if I would go;" and yet these very men have each, directly by their allegiance, and so indirectly, at least, by their money, furnished a substitute. The soldier is applauded who refuses to serve in an unjust war by those who do not refuse to sustain the unjust government which makes the war; is applauded by those whose own act and authority he disregards and sets at nought; as if the State were penitent to that degree that it hired one to scourge it while it sinned, but not to that degree that it left off sinning for a moment. Thus, under the name of order and civil government, we are all made at last to pay homage to and support our own meanness. After the first blush of sin, comes its indifference; and from immoral it becomes, as it were, *un*moral, and not quite unnecessary to that life which we have made.

The broadest and most prevalent error requires the most disinterested virtue to sustain it. The slight reproach to which the virtue of patriotism is commonly liable, the noble are most likely to incur. Those who, while they disapprove of the character and measures of a government, yield to it their allegiance and support, are undoubtedly its most conscientious supporters, and so frequently the most serious obstacles to reform. Some are petitioning the State to dissolve the Union, to disregard the requisitions of the President. Why do they not dissolve it themselves,—the union between themselves and the State,—and refuse to pay their quota into its treasury? Do not they stand in the same relation to the State, that the State does to the Union? And have not the same reasons prevented the State from resisting the Union, which have prevented them from resisting the State?

How can a man be satisfied to entertain an opinion merely, and enjoy *it*? Is there any enjoyment in it, if his opinion is that he is aggrieved? If you are cheated out of a single dollar by your neighbor, you do not rest satisfied with knowing that you are cheated, or with saying that you are cheated, or even with petitioning him to pay you your due; but you take effectual steps at once to obtain the full amount, and see that you are never cheated again. Action from principle,—the perception and the performance of right,—changes things and relations; it is essentially revolutionary, and does not consist wholly with any thing which was. It not only divides states and churches, it divides families; aye, it divides the *individual*, separating the diabolical in him from the divine.

Unjust laws exist: shall we be content to obey them, or shall we endeavor to amend them, and obey them until we have succeeded, or shall we transgress them at once? Men generally, under such a government as this, think that they ought to wait until they have persuaded the majority to alter them. They think that, if they should resist, the remedy would be worse than the evil. But it is the fault of the government itself that the remedy *is* worse than the evil. *It* makes it worse. Why is it not more apt to anticipate and provide for reform? Why does it not cherish its wise minority? Why does it cry and resist before it is hurt? Why does it not encourage its citizens to be on the alert to point out its faults, and *do* better than it would have them? Why does it always crucify Christ, and excommunicate Copernicus and Luther,[9] and pronounce Washington and Franklin rebels?

One would think, that a deliberate and practical denial of its authority was the only offence never contemplated by government; else, why has it not assigned its definite, its suitable and proportionate penalty? If a man who has no property refuses but once to earn nine shillings[1] for the State, he is put in prison for a period unlimited by any law that I know, and determined only by the discretion of those who placed him there; but if he should steal ninety times nine shillings from the State, he is soon permitted to go at large again.

If the injustice is part of the necessary friction of the machine of government, let it go, let it go: perchance it will wear smooth,— certainly the machine will wear out. If the injustice has a spring, or a pulley, or a rope, or a crank, exclusively for itself, then perhaps you may consider whether the remedy will not be worse than the evil; but if it is of such a nature that it requires you to be the agent of injustice to another, then, I say, break the law. Let your life be a counter friction to stop the machine. What I have to do is to see, at any rate, that I do not lend myself to the wrong which I condemn.

As for adopting the ways which the State has provided for remedying the evil, I know not of such ways. They take too much time, and a man's life will be gone. I have other affairs to attend to. I came into this world, not chiefly to make this a good place to live in, but to live in it, be it good or bad. A man has not every thing to do, but something; and because he cannot do *every thing*, it is not necessary that he should do *something* wrong. It is not my business to be petitioning the governor or the legislature any more than it is theirs to petition me; and, if they should not hear my petition, what should I do then? But in this case the State has provided no way: its very Constitution is the evil. This may seem to be harsh and stubborn and unconciliatory; but it is to treat with the utmost kindness and consideration the only spirit that can appreciate or deserves it. So is all change for the better, like birth and death which convulse the body.

I do not hesitate to say, that those who call themselves abolitionists should at once effectually withdraw their support, both in per-

---

9. Thoreau uses Copernicus (1473–1543), the Polish astronomer who died too soon after the publication of his new system of astronomy to be excommunicated from the Catholic Church for writing it, and Martin Luther (1483–1546), the German leader of the Protestant Reformation who was excommunicated, as announcers of new truths.
1. The amount of the poll tax Thoreau had refused to pay.

son and property, from the government of Massachusetts, and not wait till they constitute a majority of one, before they suffer the right to prevail through them. I think that it is enough if they have God on their side, without waiting for that other one. Moreover, any man more right than his neighbors, constitutes a majority of one already.[2]

I meet this American government, or its representative the State government, directly, and face to face, once a year, no more, in the person of its tax-gatherer; this is the only mode in which a man situated as I am necessarily meets it; and it then says distinctly, Recognize me; and the simplest, the most effectual, and, in the present posture of affairs, the indispensablest mode of treating with it on this head, of expressing your little satisfaction with and love for it, is to deny it then. My civil neighbor, the tax-gatherer,[3] is the very man I have to deal with,—for it is, after all, with men and not with parchment that I quarrel,—and he has voluntarily chosen to be an agent of the government. How shall he ever know well what he is and does as an officer of the government, or as a man, until he is obliged to consider whether he shall treat me, his neighbor, for whom he has respect, as a neighbor and well-disposed man, or as a maniac and disturber of the peace, and see if he can get over this obstruction to his neighborliness without a ruder and more impetuous thought or speech corresponding with his action? I know this well, that if one thousand, if one hundred, if ten men whom I could name,—if ten *honest* men only,—aye, if *one* HONEST man, in this State of Massachusetts, *ceasing to hold slaves*, were actually to withdraw from this copartnership, and be locked up in the county jail therefor, it would be the abolition of slavery in America. For it matters not how small the beginning may seem to be: what is once well done is done for ever. But we love better to talk about it: that we say is our mission. Reform keeps many scores of newspapers in its service, but not one man. If my esteemed neighbor, the State's ambassador,[4] who will devote his days to the settlement of the question of human rights in the Council Chamber, instead of being threatened with the prisons of Carolina, were to sit down the prisoner of Massachusetts, that State which is so anxious to foist the sin of slavery upon her sister,—though at present she can discover only an act of inhospitality to be the ground of a quarrel with her,—the Legislature would not wholly waive the subject the following winter.

Under a government which imprisons any unjustly, the true place for a just man is also a prison. The proper place to-day, the only place which Massachusetts has provided for her freer and less de-

2. John Knox (1505?–72), the Scottish religious reformer, said that "a man with God is always in the majority."
3. Sam Staples, who sometimes assisted Thoreau in his surveying.
4. Samuel Hoar (1778–1856), local political figure who as agent of the state of Massachusetts had been expelled from Charleston, South Carolina, in 1844 while interceding on behalf of imprisoned Negro seamen from Massachusetts. The South Carolina legislature had voted to ask the Governor to expel Hoar.

sponding spirits, is in her prisons, to be put out and locked out of the State by her own act, as they have already put themselves out by their principles. It is there that the fugitive slave, and the Mexican prisoner on parole, and the Indian come to plead the wrongs of his race, should find them; on that separate, but more free and honorable ground, where the State places those who are not *with* her but *against* her,—the only house in a slave-state in which a free man can abide with honor. If any think that their influence would be lost there, and their voices no longer afflict the ear of the State, that they would not be as an enemy within its walls, they do not know by how much truth is stronger than error, nor how much more eloquently and effectively he can combat injustice who has experienced a little in his own person. Cast your whole vote, not a strip of paper merely, but your whole influence. A minority is powerless while it conforms to the majority; it is not even a minority then; but it is irresistible when it clogs by its whole weight. If the alternative is to keep all just men in prison, or give up war and slavery, the State will not hesitate which to choose. If a thousand men were not to pay their tax-bills this year, that would not be a violent and bloody measure, as it would be to pay them, and enable the State to commit violence and shed innocent blood. This is, in fact, the definition of a peaceable revolution, if any such is possible. If the tax-gatherer, or any other public officer, asks me, as one has done, "But what shall I do?" my answer is, "If you really wish to do any thing, resign your office." When the subject has refused allegiance, and the officer has resigned his office, then the revolution is accomplished. But even suppose blood should flow. Is there not a sort of blood shed when the conscience is wounded? Through this wound a man's real manhood and immortality flow out, and he bleeds to an everlasting death. I see this blood flowing now.

I have contemplated the imprisonment of the offender, rather than the seizure of his goods,—though both will serve the same purpose,—because they who assert the purest right, and consequently are most dangerous to a corrupt State, commonly have not spent much time in accumulating property. To such the State renders comparatively small service, and a slight tax is wont to appear exorbitant, particularly if they are obliged to earn it by special labor with their hands. If there were one who lived wholly without the use of money, the State itself would hesitate to demand it of him. But the rich man—not to make any invidious comparison—is always sold to the institution which makes him rich. Absolutely speaking, the more money, the less virtue; for money comes between a man and his objects, and obtains them for him; and it was certainly no great virtue to obtain it. It puts to rest many questions which he would otherwise be taxed to answer; while the only new question which it puts is the hard but superfluous one, how to spend it. Thus his moral ground is taken from under his feet. The opportunities of living are diminished in proportion as what are called the "means" are increased. The best thing a man can do for his culture when he is rich is to endeavour to carry out those schemes which he entertained when he was poor. Christ answered the Herodians according to their condition. "Show me the tribute-money," said he;—and one took a penny out of his pocket;—If you use money which has the image of Cæsar on it, and which he has made current and valuable, that is, *if you are men of the State*, and gladly enjoy the advantages of Cæsar's government, then pay him back some of his own when he demands it: "Render therefore to Cæsar that which is Cæsar's, and to God those things which are God's,"[5]—leaving them no wiser than before as to which was which; for they did not wish to know.

When I converse with the freest of my neighbors, I perceive that, whatever they may say about the magnitude and seriousness of the question, and their regard for the public tranquillity, the long and the short of the matter is, that they cannot spare the protection of the existing government, and they dread the consequences of disobedience to it to their property and families. For my own part, I should not like to think that I ever rely on the protection of the State. But, if I deny the authority of the State when it presents its tax-bill, it will soon take and waste all my property, and so harass me and my children without end. This is hard. This makes it impossible for a man to live honestly and at the same time comfortably in outward respects. It will not be worth the while to accumulate property; that would be sure to go again. You must hire or squat somewhere, and raise but a small crop, and eat that soon. You must live within yourself, and depend upon yourself, always tucked up and ready for a start, and not have many affairs. A man may grow rich in Turkey even, if he will be in all respects a good subject of the Turkish government. Confucious said,—"If a State is governed by the principles of reason, poverty and misery are subjects of shame; if a State is not governed by the principles of reason, riches and honors are the subjects of shame."[6] No: until I want the protection of Massachusetts to be extended to me in some distant southern port, where my liberty is endangered, or until I am bent solely on building up an estate at home by peaceful enterprise, I can afford to refuse allegiance to Massachusetts, and her right to my property and life. It costs me less in every sense to incur the penalty of disobedience to the State, than it would to obey. I should feel as if I were worth less in that case.

Some years ago, the State met me in behalf of the church, and commanded me to pay a certain sum toward the support of a clergyman whose preaching my father attended, but never I myself. "Pay it," it said, "or be locked up in the jail." I declined to pay. But,

---

5. Matthew 22.16–21. In their attempt to entrap Jesus, the Pharisees (a Jewish sect that held to Mosaic law) were util- izing secular government functionaries of Herod, the Tetrarch or King of Judea. 6. *Analects*, VIII.13.

unfortunately, another man saw fit to pay it. I did not see why the schoolmaster should be taxed to support the priest, and not the priest the schoolmaster; for I was not the State's schoolmaster, but I supported myself by voluntary subscription. I did not see why the lyceum should not present its tax-bill, and have the State to back its demand, as well as the church. However, at the request of the selectmen, I condescended to make some such statement as this in writing:—"Know all men by these presents, that I, Henry Thoreau, do not wish to be regarded as a member of any incorporated society which I have not joined." This I gave to the town-clerk; and he has it. The State, having thus learned that I did not wish to be regarded as a member of that church, has never made a like demand on me since; though it said that it must adhere to its original presumption that time. If I had known how to name them, I should then have signed off in detail from all the societies which I never signed on to; but I did not know where to find a complete list.

I have paid no poll-tax for six years.[7] I was put into a jail once on this account, for one night; and, as I stood considering the walls of solid stone, two or three feet thick, the door of wood and iron, a foot thick, and the iron grating which strained the light, I could not help being struck with the foolishness of that institution which treated me as if I were mere flesh and blood and bones, to be locked up. I wondered that it should have concluded at length that this was the best use it could put me to, and had never thought to avail itself of my services in some way. I saw that, if there was a wall of stone between me and my townsmen, there was a still more difficult one to climb or break through, before they could get to be as free as I was. I did not for a moment feel confined, and the walls seemed a great waste of stone and mortar. I felt as if I alone of all my townsmen had paid my tax. They plainly did not know how to treat me, but behaved like persons who are underbred. In every threat and in every compliment there was a blunder; for they thought that my chief desire was to stand the other side of that stone wall. I could not but smile to see how industriously they locked the door on my meditations, which followed them out again without let or hinderance, and *they* were really all that was dangerous. As they could not reach me, they had resolved to punish my body; just as boys, if they cannot come at some person against whom they have a spite, will abuse his dog. I saw that the State was half-witted, that it was timid as a lone woman with her silver spoons, and that it did not know its friends from its foes, and I lost all my remaining respect for it, and pitied it.

Thus the State never intentionally confronts a man's sense, intellectual or moral, but only his body, his senses. It is not armed with superior wit or honesty, but with superior physical strength. I was

7. Since 1840. The jail mentioned in the next sentence was the Middlesex County jail in Concord, a large three-story building.

not born to be forced. I will breathe after my own fashion. Let us see who is the strongest. What force has a multitude? They only can force me who obey a higher law than I. They force me to become like themselves. I do not hear of *men* being *forced* to live this way or that by masses of men. What sort of life were that to live? When I meet a government which says to me, "Your money or your life,"[8] why should I be in haste to give it my money? It may be in a great strait, and not know what to do: I cannot help that. It must help itself; do as I do. It is not worth the while to snivel about it. I am not responsible for the successful working of the machinery of society. I am not the son of the engineer. I perceive that, when an acorn and a chestnut fall side by side, the one does not remain inert to make way for the other, but both obey their own laws, and spring and grow and flourish as best they can, till one, perchance, over-shadows and destroys the other. If a plant cannot live according to its nature, it dies; and so a man.

The night in prison was novel and interesting enough. The prisoners in their shirt-sleeves were enjoying a chat and the eve-ning air in the door-way, when I entered. But the jailer said, "Come, boys, it is time to lock up;" and so they dispersed, and I heard the sound of their steps returning into the hollow apart-ments. My room-mate was introduced to me by the jailer, as "a first-rate fellow and a clever man." When the door was locked, he showed me where to hang my hat, and how he managed matters there. The rooms were whitewashed once a month; and this one, at least, was the whitest, most simply furnished, and probably the neatest apartment in the town. He naturally wanted to know where I came from, and what brought me there; and, when I had told him, I asked him in my turn how he came there, presuming him to be an honest man, of course; and, as the world goes, I believe he was. "Why," said he, "they accuse me of burning a barn; but I never did it." As near as I could discover, he had probably gone to bed in a barn when drunk, and smoked his pipe there; and so a barn was burnt. He had the reputation of being a clever man, had been there some three months waiting for his trial to come on, and would have to wait as much longer; but he was quite domesticated and con-tented, since he got his board for nothing, and thought that he was well treated.

He occupied one window, and I the other; and I saw, that, if one stayed there long, his principal business would be to look out the window. I had soon read all the tracts that were left there, and examined where former prisoners had broken out, and where a grate had been sawed off, and heard the history of the various occupants of that room; for I found that even here there was a history and a gossip which never circulated beyond the walls of the jail. Probably this is the only house in the town where verses are composed, which are afterward printed in a

8. The cry of the highway robber.

circular form, but not published. I was shown quite a long list of verses which were composed by some young men who had been detected in an attempt to escape, who avenged themselves by singing them.

I pumped my fellow-prisoner as dry as I could, for fear I should never see him again; but at length he showed me which was my bed, and left me to blow out the lamp.

It was like travelling into a far country, such as I had never expected to behold, to lie there for one night. It seemed to me that I never had heard the town-clock strike before, nor the evening sounds of the village; for we slept with the windows open, which were inside the grating. It was to see my native village in the light of the middle ages, and our Concord was turned into a Rhine stream, and visions of knights and castles passed before me. They were the voices of old burghers that I heard in the streets. I was an involuntary spectator and auditor of whatever was done and said in the kitchen of the adjacent village-inn,—a wholly new and rare experience to me. It was a closer view of my native town. I was fairly inside of it. I never had seen its institutions before. This is one of its peculiar institutions; for it is a shire town.[9] I began to comprehend what its inhabitants were about.

In the morning, our breakfasts were put through the hole in the door, in small oblong-square tin pans, made to fit, and holding a pint of chocolate, with brown bread, and an iron spoon. When they called for the vessels again, I was green enough to return what bread I had left; but my comrade seized it, and said that I should lay that up for lunch or dinner. Soon after, he was let out to work at haying in a neighboring field, whither he went every day, and would not be back till noon; so he bade me good-day, saying that he doubted if he should see me again.

When I came out of prison,—for some one interfered, and paid the tax,—I did not perceive that great changes had taken place on the common, such as he observed who went in a youth, and emerged a tottering and gray-headed man; and yet a change had to my eyes come over the scene,—the town, and State, and country,—greater than any that mere time could effect. I saw yet more distinctly the State in which I lived. I saw to what extent the people among whom I lived could be trusted as good neighbors and friends; that their friendship was for summer weather only; that they did not greatly purpose to do right; that they were a distinct race from me by their prejudices and superstitions, as the Chinamen and Malays are; that, in their sacrifices to humanity, they ran no risks, not even to their property; that, after all, they were not so noble but they treated the thief as he had treated them, and hoped, by a certain outward observance and a few prayers, and by walking in a particular straight though useless path from time to time, to save their souls. This may be to judge my neighbors harshly;

9. Comparable to "county seat."

for I believe that most of them are not aware that they have such an institution as the jail in their village.

It was formerly the custom in our village, when a poor debtor came out of jail, for his acquaintances to salute him, looking through their fingers, which were crossed to represent the grating of a jail window, "How do ye do?" My neighbors did not thus salute me, but first looked at me, and then at one another, as if I had returned from a long journey. I was put into jail as I was going to the shoemaker's to get a shoe which was mended. When I was let out the next morning, I proceeded to finish my errand, and, having put on my mended shoe, joined a huckleberry party, who were impatient to put themselves under my conduct; and in half an hour,—for the horse was soon tackled,[1] —was in the midst of a huckleberry field, on one of our highest hills, two miles off; and then the State was nowhere to be seen.

This is the whole history of "My Prisons."[2]

I have never declined paying the highway tax, because I am as desirous of being a good neighbor as I am of being a bad subject; and, as for supporting schools, I am doing my part to educate my fellow-countrymen now. It is for no particular item in the tax-bill that I refuse to pay it. I simply wish to refuse allegiance to the State, to withdraw and stand aloof from it effectually. I do not care to trace the course of my dollar, if I could, till it buys a man, or a musket to shoot one with,—the dollar is innocent,—but I am concerned to trace the effects of my allegiance. In fact, I quietly declare war with the State, after my fashion, though I will still make what use and get what advantage of her I can, as is usual in such cases.

If others pay the tax which is demanded of me, from a sympathy with the State, they do but what they have already done in their own case, or rather they abet injustice to a greater extent than the State requires. If they pay the tax from a mistaken interest in the individual taxed, to save his property or prevent his going to jail, it is because they have not considered wisely how far they let their private feelings interfere with the public good.

This, then, is my position at present. But one cannot be too much on his guard in such a case, lest his action be biassed by obstinacy, or an undue regard for the opinions of men. Let him see that he does only what belongs to himself and to the hour.

I think sometimes, Why, this people mean well; they are only ignorant; they would do better if they knew how: why give your neighbors this pain to treat you as they are not inclined to? But I think, again, this is no reason why I should do as they do, or permit others to suffer much greater pain of a different kind. Again, I sometimes say to myself, When many millions of men, without

1. Harnessed.
2. A wry comparison to the title of a book (1832) by the Italian poet Silvio Pellico (1789–1854) on his years of hard labor in Austrian prisons.

heat, without ill-will, without personal feeling of any kind, demand of you a few shillings only, without the possibility, such is their constitution, of retracting or altering their present demand, and without the possibility, on your side, of appeal to any other millions, why expose yourself to this overwhelming brute force? You do not resist cold and hunger, the winds and the waves, thus obstinately; you quietly submit to a thousand similar necessities. You do not put your head into the fire. But just in proportion as I regard this as not wholly a brute force, but partly a human force, and consider that I have relations to those millions as to so many millions of men, and not of mere brute or inanimate things, I see that appeal is possible, first and instantaneously, from them to the Maker of them, and, secondly, from them to themselves. But, if I put my head deliberately into the fire, there is no appeal to fire or to the Maker of fire, and I have only myself to blame. If I could convince myself that I have any right to be satisfied with men as they are, and to treat them accordingly, and not according, in some respects, to my requisitions and expectations of what they and I ought to be, then, like a good Mussulman[3] and fatalist, I should endeavor to be satisfied with things as they are, and say it is the will of God. And, above all, there is this difference between resisting this and a purely brute or natural force, that I can resist this with some effect; but I cannot expect, like Orpheus,[4] to change the nature of the rocks and trees and beasts.

I do not wish to quarrel with any man or nation. I do not wish to split hairs, to make fine distinctions, or set myself up as better than my neighbors. I seek rather, I may say, even an excuse for conforming to the laws of the land. I am but too ready to conform to them. Indeed I have reason to suspect myself on this head; and each year, as the tax-gatherer comes round, I find myself disposed to review the acts and position of the general and state governments, and the spirit of the people, to discover a pretext for conformity. I believe that the State will soon be able to take all my work of this sort out of my hands, and then I shall be no better a patriot than my fellow-countrymen. Seen from a lower point of view, the Constitution, with all its faults, is very good; the law and the courts are very respectable; even this State and this American government are, in many respects, very admirable and rare things, to be thankful for, such as a great many have described them; but seen from a point of view a little higher, they are what I have described them; seen from a higher still, and the highest, who shall say what they are, or that they are worth looking at or thinking of at all?

However, the government does not concern me much, and I shall bestow the fewest possible thoughts on it. It is not many moments

3. Mohammedan.
4. The son of Calliope, one of the Muses, who gave him the gift of music. Trees and rocks moved to the playing of his lyre. He charmed the three-headed dog Cerberus in an unsuccessful attempt to bring his dead wife Eurydice up from the underworld.

608 ★ *Henry David Thoreau*

that I live under a government, even in this world. If a man is thought-free, fancy-free, imagination-free, that which *is not* never for a long time appearing *to be* to him, unwise rulers or reformers cannot fatally interrupt him.

I know that most men think differently from myself; but those whose lives are by profession devoted to the study of these or kindred subjects, content me as little as any. Statesmen and legislators, standing so completely within the institution, never distinctly and nakedly behold it. They speak of moving society, but have no resting-place without it. They may be men of a certain experience and discrimination, and have no doubt invented ingenious and even useful systems, for which we sincerely thank them; but all their wit and usefulness lie within certain not very wide limits. They are wont to forget that the world is not governed by policy and expediency. Webster[5] never goes behind government, and so cannot speak with authority about it. His words are wisdom to those legislators who contemplate no essential reform in the existing government; but for thinkers, and those who legislate for all time, he never once glances at the subject. I know of those whose serene and wise speculations on this theme would soon reveal the limits of his mind's range and hospitality. Yet, compared with the cheap professions of most reformers, and the still cheaper wisdom and eloquence of politicians in general, his are almost the only sensible and valuable words, and we thank Heaven for him. Comparatively, he is always strong, original, and, above all, practical. Still his quality is not wisdom, but prudence. The lawyer's truth is not Truth, but consistency, or a consistent expediency. Truth is always in harmony with herself, and is not concerned chiefly to reveal the justice that may consist with wrong-doing. He well deserves to be called, as he has been called, the Defender of the Constitution. There are really no blows to be given by him but defensive ones. He is not a leader, but a follower. His leaders are the men of '87.[6] "I have never made an effort," he says, "and never propose to make an effort; I have never countenanced an effort, and never mean to countenance an effort, to disturb the arrangement as originally made, by which the various States came into the Union."[7] Still thinking of the sanction which the Constitution gives to slavery, he says, "Because it was a part of the original compact,—let it stand." Notwithstanding his special acuteness and ability, he is unable to take a fact out of its merely political relations, and behold it as it lies absolutely to be disposed of by the intellect,—what, for instance, it behoves a man to do here in America to-day with regard to slavery, but ventures, or is driven, to make some such desperate answer as the following, while professing to speak absolutely, and as a private man,—from which

5. Daniel Webster (1782–1852), prominent Whig politician of the second quarter of the 19th century.
6. The writers of the Constitution, who convened at Philadelphia in 1787.
7. From Webster's speech on *The Admission of Texas* (December 22, 1845).

what new and singular code of social duties might be inferred?—
"The manner," says he, "in which the government of those States
where slavery exists are to regulate it, is for their own consideration,
under their responsibility to their constituents, to the general laws of
propriety, humanity, and justice, and to God. Associations formed
elsewhere, springing from a feeling of humanity, or any other cause,
have nothing whatever to do with it. They have never received any
encouragement from me, and they never will."[8]

They who know of no purer sources of truth, who have traced up
its stream no higher, stand, and wisely stand, by the Bible and the
Constitution, and drink at it there with reverence and humility; but
they who behold where it comes trickling into this lake or that pool,
gird up their loins once more, and continue their pilgrimage toward
its fountain-head.

No man with a genius for legislation has appeared in America.
They are rare in the history of the world. There are orators, politicians, and eloquent men, by the thousand; but the speaker has not
yet opened his mouth to speak, who is capable of settling the much-
vexed questions of the day. We love eloquence for its own sake, and
not for any truth which it may utter, or any heroism it may inspire.
Our legislators have not yet learned the comparative value of free-
trade and of freedom, of union, and of rectitude, to a nation. They
have no genius or talent for comparatively humble questions of
taxation and finance, commerce and manufactures and agriculture.
If we were left solely to the wordy wit of legislators in Congress for
our guidance, uncorrected by the seasonable experience and the
effectual complaints of the people, America would not long retain
her rank among the nations. For eighteen hundred years, though
perchance I have no right to say it, the New Testament has been
written; yet where is the legislator who has wisdom and practical
talent enough to avail himself of the light which it sheds on the
science of legislation?

The authority of government, even such as I am willing to submit
to,—for I will cheerfully obey those who know and can do better
than I, and in many things even those who neither know nor can do
so well,—is still an impure one: to be strictly just, it must have the
sanction and consent of the governed. It can have no pure right
over my person and property but what I concede to it. The progress
from an absolute to a limited monarchy, from a limited monarchy
to a democracy, is a progress toward a true respect for the individual. Is a democracy, such as we know it, the last improvement
possible in government? Is it not possible to take a step further
towards recognizing and organizing the rights of man? There will
never be a really free and enlightened State, until the State comes to
recognize the individual as a higher and independent power, from

8. "These extracts have been inserted since the Lecture was read" [Thoreau's note; he means the quotation beginning "The manner"].

which all its own power and authority are derived, and treats him accordingly. I please myself with imagining a State at last which can afford to be just to all men, and to treat the individual with respect as a neighbor; which even would not think it inconsistent with its own repose, if a few were to live aloof from it, not meddling with it, nor embraced by it, who fulfilled all the duties of neighbors and fellow-men. A State which bore this kind of fruit, and suffered it to drop off as fast as it ripened, would prepare the way for a still more perfect and glorious State, which also I have imagined, but not yet anywhere seen.

1849, 1866

From Walden, or Life in the Woods[1]

I do not propose to write an ode to dejection, but to brag as lustily as chanticleer in the morning, standing on his roost, if only to wake my neighbors up.

Economy[2]

When I wrote the following pages, or rather the bulk of them, I lived alone, in the woods, a mile from any neighbor, in a house which I had built myself, on the shore of Walden Pond, in Concord, Massachusetts, and earned my living by the labor of my hands only. I lived there two years and two months. At present I am a sojourner in civilized life again.

I should not obtrude my affairs so much on the notice of my readers if very particular inquiries had not been made by my townsmen concerning my mode of life, which some would call impertinent, though they do not appear to me at all impertinent, but, considering the circumstances, very natural and pertinent. Some have asked what I got to eat; if I did not feel lonesome; if I was not afraid; and the like. Others have been curious to learn what portion of my income I devoted to charitable purposes; and some, who have large families, how many poor children I maintained. I will therefore ask those of my readers who feel no particular interest in me to pardon me if I undertake to answer some of these questions in this book. In most books, the *I*, or first person, is omitted; in this it will be retained; that, in respect to egotism, is the main difference. We commonly do not remember that it is, after all, always the first

1. Thoreau began writing *Walden* early in 1846, some months after he began living at Walden Pond, and by late 1847, when he moved back into the village of Concord, he had drafted roughly half the book. Between 1852 and 1854 he rewrote the manuscript several times and substantially enlarged it. The text printed here is that of the first edition (Boston: Ticknor and Fields, 1854), with a few printer's errors corrected on the basis of Thoreau's set of marked proofs, his corrections in his copy of *Walden*, and scholars' comparisons of the printed book and the manuscript drafts, especially the edition by J. Lyndon Shanley (Princeton: Princeton University Press, 1971).

Any annotator of *Walden* is deeply indebted to Walter Harding, editor of *The Variorum Walden* (Boston: Twayne, 1962), and Philip Van Doren Stern, editor of *The Annotated Walden* (New York: Clarkson N. Potter, 1970).

2. As Thoreau explains later in the chapter, the title means something like "philosophy of living."

person that is speaking. I should not talk so much about myself if there were any body else whom I knew as well. Unfortunately, I am confined to this theme by the narrowness of my experience. Moreover, I, on my side, require of every writer, first or last, a simple and sincere account of his own life, and not merely what he has heard of other men's lives; some such account as he would send to his kindred from a distant land; for if he has lived sincerely, it must have been in a distant land to me. Perhaps these pages are more particularly addressed to poor students. As for the rest of my readers, they will accept such portions as apply to them. I trust that none will stretch the seams in putting on the coat, for it may do good service to him whom it fits.

I would fain say something, not so much concerning the Chinese and Sandwich Islanders[3] as you who read these pages, who are said to live in New England; something about your condition, especially your outward condition or circumstances in this world, in this town, what it is, whether it is necessary that it be as bad as it is, whether it cannot be improved as well as not. I have travelled a good deal in Concord; and every where, in shops, and offices, and fields, the inhabitants have appeared to me to be doing penance in a thousand remarkable ways. What I have heard of Brahmins sitting exposed to four fires and looking in the face of the sun; or hanging suspended, with their heads downward, over flames; or looking at the heavens over their shoulders "until it becomes impossible for them to resume their natural position, while from the twist of the neck nothing but liquids can pass into the stomach;" or dwelling, chained for life, at the foot of a tree; or measuring with their bodies, like caterpillars, the breadth of vast empires; or standing on one leg on the tops of pillars,—even these forms of conscious penance are hardly more incredible and astonishing than the scenes which I daily witness.[4] The twelve labors of Hercules[5] were trifling in comparison with those which my neighbors have undertaken; for they were only twelve, and had an end; but I could never see that these men slew or captured any monster or finished any labor. They have no friend Iolas to burn with a hot iron the root of the hydra's head, but as soon as one head is crushed, two spring up.

I see young men, my townsmen, whose misfortune it is to have inherited farms, houses, barns, cattle, and farming tools; for these are more easily acquired than got rid of. Better if they had been born in the open pasture and suckled by a wolf, that they might

3. Hawaiians.
4. Thoreau's source has not been found for this depiction of the religious self-torture of high-caste Hindus in India.
5. Son of Zeus and Alcmene, this half mortal could become a god only by performing 12 labors, each apparently impossible. The second labor, the slaying of the Lernaean hydra, a many-headed sea monster, is referred to just below. (Hercules' friend Iolas helped by searing the stump each time Hercules cut off one of the heads, which otherwise would have regenerated.) The seventh labor, mentioned in the following paragraph, was the cleansing of Augeas's pestilent stables in one day, a feat Hercules accomplished by diverting two nearby rivers through the stables.

612 ★ Henry David Thoreau

have seen with clearer eyes what field they were called to labor in. Who made them serfs of the soil? Why should they eat their sixty acres, when man is condemned to eat only his peck of dirt? Why should they begin digging their graves as soon as they are born? They have got to live a man's life, pushing all these things before them, and get on as well as they can. How many a poor immortal soul have I met well nigh crushed and smothered under its load, creeping down the road of life, pushing before it a barn seventy-five feet by forty, its Augean stables never cleansed, and one hundred acres of land, tillage, mowing, pasture, and wood-lot! The portionless, who struggle with no such unnecessary inherited encumbrances, find it labor enough to subdue and cultivate a few cubic feet of flesh.

But men labor under a mistake. The better part of the man is soon ploughed into the soil for compost. By a seeming fate, commonly called necessity, they are employed, as it says in an old book, laying up treasures which moth and rust will corrupt and thieves break through and steal.[6] It is a fool's life, as they will find when they get to the end of it, if not before. It is said that Deucalion and Pyrrha created men by throwing stones over their heads behind them:—[7]

> Inde genus durum sumus, experiensque laborum,
> Et documenta damus quâ simus origine nati.

Or, as Raleigh rhymes it in his sonorous way,—

> "From thence our kind hard-hearted is, enduring pain and care,
> Approving that our bodies of a stony nature are."

So much for a blind obedience to a blundering oracle, throwing the stones over their heads behind them, and not seeing where they fell.

Most men, even in this comparatively free country, through mere ignorance and mistake, are so occupied with the factitious cares and superfluously coarse labors of life that its finer fruits cannot be plucked by them. Their fingers, from excessive toil, are too clumsy and tremble too much for that. Actually, the laboring man has not leisure for a true integrity day by day; he cannot afford to sustain the manliest relations to men; his labor would be depreciated in the market. He has no time to be any thing but a machine. How can he remember well his ignorance—which his growth requires—who has so often to use his knowledge? We should feed and clothe him gratuitously sometimes, and recruit him with our cordials, before

6. Matthew 6.19.
7. Deucalion and Pyrrha, husband and wife in the Greek analogue to the Biblical legend of Noah and the Flood, repopulated the earth by throwing stones behind them over their shoulders. The stones thrown by Deucalion turned into men, and the stones thrown by Pyrrha turned into women. The quotation is from Ovid's *Metamorphoses*, I.414–15, as translated in Sir Walter Raleigh's *History of the World*.

we judge of him. The finest qualities of our nature, like the bloom on fruits, can be preserved only by the most delicate handling. Yet we do not treat ourselves nor one another thus tenderly.

Some of you, we all know, are poor, find it hard to live, are sometimes, as it were, gasping for breath. I have no doubt that some of you who read this book are unable to pay for all the dinners which you have actually eaten, or for the coats and shoes which are fast wearing or are already worn out, and have come to this page to spend borrowed or stolen time, robbing your creditors of an hour. It is very evident what mean and sneaking lives many of you live, for my sight has been whetted by experience; always on the limits, trying to get into business and trying to get out of debt, a very ancient slough, called by the Latins *æs alienum*, another's brass, for some of their coins were made of brass; still living, and dying, and buried by this other's brass; always promising to pay, promising to pay, to-morrow, and dying to-day, insolvent; seeking to curry favor, to get custom, by how many modes, only not state-prison offences; lying, flattering, voting, contracting yourselves into a nut-shell of civility, or dilating into an atmosphere of thin and vaporous generosity, that you may persuade your neighbor to let you make his shoes, or his hat, or his coat, or his carriage, or import his groceries for him; making yourselves sick, that you may lay up something against a sick day, something to be tucked away in an old chest, or in a stocking behind the plastering, or, more safely, in the brick bank; no matter where, no matter how much or how little.

I sometimes wonder that we can be so frivolous, I may almost say, as to attend to the gross but somewhat foreign form of servitude called Negro Slavery, there are so many keen and subtle masters that enslave both north and south. It is hard to have a southern overseer; it is worse to have a northern one; but worst of all when you are the slave-driver of yourself. Talk of a divinity in man! Look at the teamster on the highway, wending to market by day or night; does any divinity stir within him? His highest duty to. fodder and water his horses! What is his destiny to him compared with the shipping interests? Does not he drive for Squire Make-a-stir?[8] How godlike, how immortal, is he? See how he cowers and sneaks, how vaguely all the day he fears, not being immortal nor divine, but the slave and prisoner of his own opinion of himself, a fame won by his own deeds. Public opinion is a weak tyrant compared with our own private opinion. What a man thinks of himself, that it is which determines, or rather indicates, his fate. Self-emancipation even in the West Indian provinces of the fancy and imagination,—what Wilberforce[9] is there to bring that about? Think, also, of the ladies

8. An allegorical name modeled upon those in John Bunyan's *Pilgrim's Progress*, familiar to almost any reader in Thoreau's time.

9. William Wilberforce (1759–1833), English philanthropist, leading opponent of the slave trade until its abolition in 1807.

of the land weaving toilet cushions against the last day, not to betray too green an interest in their fates! As if you could kill time without injuring eternity.

The mass of men lead lives of quiet desperation. What is called resignation is confirmed desperation. From the desperate city you go into the desperate country, and have to console yourself with the bravery of minks and muskrats. A stereotyped but unconscious despair is concealed even under what are called the games and amusements of mankind. There is no play in them, for this comes after work. But it is a characteristic of wisdom not to do desperate things.

When we consider what, to use the words of the catechism, is the chief end of man,[1] and what are the true necessaries and means of life, it appears as if men had deliberately chosen the common mode of living because they preferred it to any other. Yet they honestly think there is no choice left. But alert and healthy natures remember that the sun rose clear. It is never too late to give up our prejudices. No way of thinking or doing, however ancient, can be trusted without proof. What every body echoes or in silence passes by as true to-day may turn out to be falsehood to-morrow, mere smoke of opinion, which some had trusted for a cloud that would sprinkle fertilizing rain on their fields. What old people say you cannot do you try and find that you can. Old deeds for old people, and new deeds for new. Old people did not know enough once, perchance, to fetch fresh fuel to keep the fire a-going; new people put a little dry wood under a pot, and are whirled round the globe with the speed of birds, in a way to kill old people, as the phrase is. Age is no better, hardly so well, qualified for an instructor as youth, for it has not profited so much as it has lost. One may almost doubt if the wisest man has learned any thing of absolute value by living. Practically, the old have no very important advice to give the young, their own experience has been so partial, and their lives have been such miserable failures, for private reasons, as they must believe; and it may be that they have some faith left which belies that experience, and they are only less young than they were. I have lived some thirty years on this planet, and I have yet to hear the first syllable of valuable or even earnest advice from my seniors. They have told me nothing, and probably cannot tell me any thing, to the purpose. Here is life, an experiment to a great extent untried by me; but it does not avail me that they have tried it. If I have any experience which I think valuable, I am sure to reflect that this my Mentors[2] said nothing about.

One farmer says to me, "You cannot live on vegetable food solely, for it furnishes nothing to make bones with;" and so he

1. From the Shorter Catechism in the *New England Primer*: "What is the chief end of man? Man's chief end is to glorify God and to enjoy him forever."

2. From Mentor, in Homer's *Odyssey*, the friend whom Odysseus entrusted with the education of his son Telemachus.

religiously devotes a part of his day to supplying his system with the raw material of bones; walking all the while he talks behind his oxen, which, with vegetable-made bones, jerk him and his lumbering plough along in spite of every obstacle. Some things are really necessaries of life in some circles, the most helpless and diseased, which in others are luxuries merely, and in others still are entirely unknown.

The whole ground of human life seems to some to have been gone over by their predecessors, both the heights and the valleys, and all things to have been cared for. According to Evelyn, "the wise Solomon prescribed ordinances for the very distances of trees; and the Roman prætors have decided how often you may go into your neighbor's land to gather the acorns which fall on it without trespass, and what share belongs to that neighbor."[3] Hippocrates[4] has even left directions how we should cut our nails; that is, even with the ends of the fingers, neither shorter nor longer. Undoubtedly the very tedium and ennui which presume to have exhausted the variety and the joys of life are as old as Adam. But man's capacities have never been measured; nor are we to judge of what he can do by any precedents, so little has been tried. Whatever have been thy failures hitherto, "be not afflicted, my child, for who shall assign to thee what thou hast left undone?"[5]

We might try our lives by a thousand simple tests; as, for instance, that the same sun which ripens my beans illumines at once a system of earths like ours. If I had remembered this it would have prevented some mistakes. This was not the light in which I hoed them. The stars are the apexes of what wonderful triangles! What distant and different beings in the various mansions of the universe are contemplating the same one at the same moment! Nature and human life are as various as our several constitutions. Who shall say what prospect life offers to another? Could a greater miracle take place than for us to look through each other's eyes for an instant? We should live in all the ages of the world in an hour; ay, in all the worlds of the ages. History, Poetry, Mythology!—I know of no reading of another's experience so startling and informing as this would be.

The greater part of what my neighbors call good I believe in my soul to be bad, and if I repent of any thing, it is very likely to be my good behavior. What demon possessed me that I behaved so well? You may say the wisest thing you can old man,—you who have lived seventy years, not without honor of a kind,—I hear an irresistible voice which invites me away from all that. One generation

3. *Silva: or, a Discourse of Forest-Trees*, by John Evelyn (1620–1706); "praetors": in the Roman Republic, high elected magistrates.
4. Greek physician (460?–377? B.C.), known as the father of medicine.

5. "Be not afflicted, my child, for who shall efface what thou hast formerly done, or shall assign to thee what thou hast left undone?" H. H. Wilson's translation of the *Vishnu Purana* (London, 1840), p. 87.

abandons the enterprises of another like stranded vessels.

I think that we may safely trust a good deal more than we do. We may waive just so much care of ourselves as we honestly bestow elsewhere. Nature is as well adapted to our weakness as to our strength. The incessant anxiety and strain of some is a well nigh incurable form of disease. We are made to exaggerate the importance of what work we do; and yet how much is not done by us! or, what if we had been taken sick? How vigilant we are! determined not to live by faith if we can avoid it; all the day long on the alert, at night we unwillingly say our prayers and commit ourselves to uncertainties. So thoroughly and sincerely are we compelled to live, reverencing our life, and denying the possibility of change. This is the only way, we say; but there are as many ways as there can be drawn radii from one centre. All change is a miracle to contemplate; but it is a miracle which is taking place every instant. Confucius said, "To know that we know what we know, and that we do not know what we do not know, that is true knowledge."[6] When one man has reduced a fact of the imagination to be a fact to his understanding, I foresee that all men will at length establish their lives on that basis.

Let us consider for a moment what most of the trouble and anxiety which I have referred to is about, and how much it is necessary that we be troubled, or, at least, careful. It would be some advantage to live a primitive and frontier life, though in the midst of an outward civilization, if only to learn what are the gross necessaries of life and what methods have been taken to obtain them; or even to look over the old day-books of the merchants, to see what it was that men most commonly bought at the stores, what they stored, that is, what are the grossest groceries. For the improvements of ages have had but little influence on the essential laws of man's existence; as our skeletons, probably, are not to be distinguished from those of our ancestors.

By the words *necessary of life*, I mean whatever, of all that man obtains by his own exertions, has been from the first, or from long use has become, so important to human life that few, if any, whether from savageness, or poverty, or philosophy, ever attempt to do without it. To many creatures there is in this sense but one necessary of life, Food. To the bison of the prairie it is a few inches of palatable grass, with water to drink; unless he seeks the Shelter of the forest or the mountain's shadow. None of the brute creation requires more than Food and Shelter. The necessaries of life for man in this climate may, accurately enough, be distributed under the several heads of Food, Shelter, Clothing, and Fuel; for not till we have secured these are we prepared to entertain the true problems of life with freedom and a prospect of success. Man has

6. *Analects*, II.17.

invented, not only houses, but clothes and cooked food; and possibly from the accidental discovery of the warmth of fire, and the consequent use of it, at first a luxury, arose the present necessity to sit by it. We observe cats and dogs acquiring the same second nature. By proper Shelter and Clothing we legitimately retain our own internal heat; but with an excess of these, or of Fuel, that is, with an external heat greater than our own internal, may not cookery properly be said to begin? Darwin, the naturalist, says of the inhabitants of Tierra del Fuego, that while his own party, who were well clothed and sitting close to a fire, were far from too warm, these naked savages, who were farther off, were observed, to his great surprise, "to be streaming with perspiration at undergoing such a roasting."[7] So, we are told, the New Hollander[8] goes naked with impunity, while the European shivers in his clothes. Is it impossible to combine the hardiness of these savages with the intellectualness of the civilized man? According to Liebig,[9] man's body is a stove, and food the fuel which keeps up the internal combustion in the lungs. In cold weather we eat more, in warm less. The animal heat is the result of a slow combustion, and disease and death take place when this is too rapid; or for want of fuel, or from some defect in the draught, the fire goes out. Of course the vital heat is not to be confounded with fire; but so much for analogy. It appears, therefore, from the above list, that the expression, *animal life*, is nearly synonymous with the expression, *animal heat*; for while Food may be regarded as the Fuel which keeps up the fire within us,—and Fuel serves only to prepare that Food or to increase the warmth of our bodies by addition from without,—Shelter and Clothing also serve only to retain the *heat* thus generated and absorbed.

The grand necessity, then, for our bodies, is to keep warm, to keep the vital heat in us. What pains we accordingly take, not only with our Food, and Clothing, and Shelter, but with our beds, which are our night-clothes, robbing the nests and breasts of birds to prepare this shelter within a shelter, as the mole has its bed of grass and leaves at the end of its burrow! The poor man is wont to complain that this is a cold world; and to cold, no less physical than social, we refer directly a great part of our ails. The summer, in some climates, makes possible to man a sort of Elysian[1] life. Fuel, except to cook his Food, is then unnecessary; the sun is his fire, and many of the fruits are sufficiently cooked by its rays; while Food generally is more various, and more easily obtained, and Clothing and Shelter are wholly or half unnecessary. At the present day, and in this country, as I find by my own experience, a few implements,

7. Charles Darwin, *Journal of * * * the Various Countries Visited by H.M.S. Beagle* (1839).
8. I.e., Australian aborigine.
9. Justus, Baron von Liebig (1803–73),

German chemist, author of *Organic Chemistry.*
1. In Greek mythology, Elysium is the home of the blessed after death.

a knife, an axe, a spade, a wheelbarrow, &c., and for the studious, lamplight, stationery, and access to a few books, rank next to necessaries, and can all be obtained at a trifling cost. Yet some, not wise, go to the other side of the globe, to barbarous and unhealthy regions, and devote themselves to trade for ten or twenty years, in order that they may live,—that is, keep comfortably warm,—and die in New England at last. The luxuriously rich are not simply kept comfortably warm, but unnaturally hot; as I implied before, they are cooked, of course à la mode.

Most of the luxuries, and many of the so called comforts of life, are not only not indispensable, but positive hinderances to the elevation of mankind. With respect to luxuries and comforts, the wisest have ever lived a more simple and meager life than the poor. The ancient philosophers, Chinese, Hindoo, Persian, and Greek, were a class than which none has been poorer in outward riches, none so rich in inward. We know not much about them. It is remarkable that *we* know so much of them as we do. The same is true of the more modern reformers and benefactors of their race. None can be an impartial or wise observer of human life but from the vantage ground of what *we* should call voluntary poverty. Of a life of luxury the fruit is luxury, whether in agriculture, or commerce, or literature, or art. There are nowadays professors of philosophy, but not philosophers. Yet it is admirable to profess because it was once admirable to live. To be a philosopher is not merely to have subtle thoughts, nor even to found a school, but so to love wisdom as to live according to its dictates, a life of simplicity, independence, magnanimity, and trust. It is to solve some of the problems of life, not only theoretically, but practically. The success of great scholars and thinkers is commonly a courtier-like success, not kingly, not manly. They make shift to live merely by conformity, practically as their fathers did, and are in no sense the progenitors of a nobler race of men. But why do men degenerate ever? What makes families run out? What is the nature of the luxury which enervates and destroys nations? Are we sure that there is none of it in our own lives? The philosopher is in advance of his age even in the outward form of his life. He is not fed, sheltered, clothed, warmed, like his contemporaries. How can a man be a philosopher and not maintain his vital heat by better methods than other men?

When a man is warmed by the several modes which I have described, what does he want next? Surely not more warmth of the same kind, as more and richer food, larger and more splendid houses, finer and more abundant clothing, more numerous incessant and hotter fires, and the like. When he has obtained those things which are necessary to life, there is another alternative than to obtain the superfluities; and that is, to adventure on life now, his vacation from humbler toil having commenced. The soil, it appears, is suited to the seed, for it has sent its radicle downward, and it may now send its shoot upward also with confidence. Why has man rooted himself thus firmly in the earth, but that he may rise in the same proportion into the heavens above?—for the nobler plants are valued for the fruit they bear at last in the air and light, far from the ground, and are not treated like the humbler esculents, which, though they may be biennials, are cultivated only till they have perfected their root, and often cut down at top for this purpose, so that most would not know them in their flowering season.

I do not mean to prescribe rules to strong and valiant natures, who will mind their own affairs whether in heaven or hell, and perchance build more magnificently and spend more lavishly than the richest, without ever impoverishing themselves, not knowing how they live,—if, indeed, there are any such, as has been dreamed; nor to those who find their encouragement and inspiration in precisely the present condition of things, and cherish it with the fondness and enthusiasm of lovers,—and, to some extent, I reckon myself in this number; I do not speak to those who are well employed, in whatever circumstances, and they know whether they are well employed or not;—but mainly to the mass of men who are discontented, and idly complaining of the hardness of their lot or of the times, when they might improve them. There are some who complain most energetically and inconsolably of any, because they are, as they say, doing their duty. I also have in my mind that seemingly wealthy, but most terribly impoverished class of all, who have accumulated dross, but know not how to use it, or get rid of it, and thus have forged their own golden or silver fetters.

If I should attempt to tell how I have desired to spend my life in years past, it would probably surprise those of my readers who are somewhat acquainted with its actual history; it would certainly astonish those who know nothing about it. I will only hint at some of the enterprises which I have cherished.

In any weather, at any hour of the day or night, I have been anxious to improve the nick of time, and notch it on my stick too; to stand on the meeting of two eternities, the past and future, which is precisely the present moment; to toe that line. You will pardon some obscurities, for there are more secrets in my trade than in most men's, and yet not voluntarily kept, but inseparable from its very nature. I would gladly tell all that I know about it, and never paint "No Admittance" on my gate.

I long ago lost a hound, a bay horse, and a turtle-dove, and am still on their trail.[2] Many are the travellers I have spoken concern-

2. Thoreau's reply to B. B. Wiley, April 26, 1857, suggests something of the evocative way he wanted this passage interpreted: "If others have their losses, which they are busy repairing, so have I *mine*, & their hound & horse may *perhaps* be the symbols of some of them. But also I have lost, or am in danger of losing, a far finer & more etherial treasure, which commonly no loss of which they are conscious will symbolize—this I answer hastily & with some hesitation, according as I now understand my own words."

ing them, describing their tracks and what calls they answered to. I have met one or two who had heard the hound, and the tramp of the horse, and even seen the dove disappear behind a cloud, and they seemed as anxious to recover them as if they had lost them themselves.

To anticipate, not the sunrise and the dawn merely, but, if possible, Nature herself! How many mornings, summer and winter, before yet any neighbor was stirring about his business, have I been about mine! No doubt, many of my townsmen have met me returning from this enterprise, farmers starting for Boston in the twilight, or woodchoppers going to their work. It is true, I never assisted the sun materially in his rising, but, doubt not, it was of the last importance only to be present at it.

So many autumn, ay, and winter days, spent outside the town, trying to hear what was in the wind, to hear and carry it express! I well-nigh sunk all my capital in it, and lost my own breath into the bargain, running in the face of it. If it had concerned either of the political parties, depend upon it, it would have appeared in the Gazette with the earliest intelligence.[3] At other times watching from the observatory of some cliff or tree, to telegraph any new arrival; or waiting at evening on the hill-tops for the sky to fall, that I might catch something, though I never caught much, and that, manna-wise, would dissolve again in the sun.[4]

For a long time I was reporter to a journal, of no very wide circulation, whose editor has never yet seen fit to print the bulk of my contributions, and, as is too common with writers, I got only my labor for my pains.[5] However, in this case my pains were their own reward.

For many years I was self-appointed inspector of snow storms and rain storms, and did my duty faithfully; surveyor, if not of highways, then of forest paths and all across-lot routes, keeping them open, and ravines bridged and passable at all seasons, where the public heel had testified to their utility.

I have looked after the wild stock of the town, which give a faithful herdsman a good deal of trouble by leaping fences; and I have had an eye to the unfrequented nooks and corners of the farm; though I did not always know whether Jonas or Solomon worked in a particular field to-day; that was none of my business. I have watered the red huckleberry, the sand cherry and the nettle tree, the red pine and the black ash, the white grape and the yellow violet, which might have withered else in dry seasons.

In short, I went on thus for a long time, I may say it without

3. "Gazette": newspaper; "intelligence": news.
4. In Exodus 16 manna is the bread that God rained from heaven so the Israelites could survive in the desert on their way from Egypt to the Promised Land.
5. Thoreau puns on the common usage of "journal" to mean a daily newspaper as well as a diary; Thoreau is the negligent or too demanding editor.

boasting, faithfully minding my business, till it became more and more evident that my townsmen would not after all admit me into the list of town officers, nor make my place a sinecure with a moderate allowance. My accounts, which I can swear to have kept faithfully, I have, indeed, never got audited, still less accepted, still less paid and settled. However, I have not set my heart on that.

Not long since, a strolling Indian went to sell baskets at the house of a well-known lawyer in my neighborhood. "Do you wish to buy any baskets?" he asked. "No, we do not want any," was the reply. "What!" exclaimed the Indian as he went out the gate, "do you mean to starve us?" Having seen his industrious white neighbors so well off,—that the lawyer had only to weave arguments, and by some magic wealth and standing followed, he had said to himself; I will go into business; I will weave baskets; it is a thing which I can do. Thinking that when he had made the baskets he would have done his part, and then it would be the white man's to buy them. He had not discovered that it was necessary for him to make it worth the other's while to buy them, or at least make him think that it was so, or to make something else which it would be worth his while to buy. I too had woven a kind of basket of a delicate texture, but I had not made it worth any one's while to buy them.[6] Yet not the less, in my case, did I think it worth my while to weave them, and instead of studying how to make it worth men's while to buy my baskets, I studied rather how to avoid the necessity of selling them. The life which men praise and regard as successful is but one kind. Why should we exaggerate any one kind at the expense of the others?

Finding that my fellow-citizens were not likely to offer me any room in the court house, or any curacy or living[7] any where else, but I must shift for myself, I turned my face more exclusively than ever to the woods, where I was better known. I determined to go into business at once, and not wait to acquire the usual capital, using such slender means as I had already got. My purpose in going to Walden Pond was not to live cheaply nor to live dearly there, but to transact some private business with the fewest obstacles; to be hindered from accomplishing which for want of a little common sense, a little enterprise and business talent, appeared not so sad as foolish.

I have always endeavored to acquire strict business habits; they are indispensable to every man. If your trade is with the Celestial Empire,[8] then some small counting house on the coast, in some Salem harbor, will be fixture enough. You will export such articles as the country affords, purely native products, much ice and pine timber and a little granite, always in native bottoms. These will be

6. A reference to Thoreau's poorly selling first book, *A Week on the Concord and Merrimack Rivers* (1849).
7. A church office with a fixed, steady income.
8. China, from the belief that the Chinese Emperors were sons of Heaven.

good ventures. To oversee all the details yourself in person; to be at once pilot and captain, and owner and underwriter; to buy and sell and keep the accounts; to read every letter received, and write or read every letter sent; to superintend the discharge of imports night and day; to be upon many parts of the coast almost at the same time;—often the richest freight will be discharged upon a Jersey shore;[9]—to be your own telegraph, unweariedly sweeping the horizon, speaking all passing vessels bound coastwise; to keep up a steady despatch of commodities, for the supply of such a distant and exorbitant market; to keep yourself informed of the state of the markets, prospects of war and peace every where, and anticipate the tendencies of trade and civilization,—taking advantage of the results of all exploring expeditions, using new passages and all improvements in navigation;—charts to be studied, the position of reefs and new lights and buoys to be ascertained, and ever, and ever, the logarithmic tables to be corrected, for by the error of some calculator the vessel often splits upon a rock that should have reached a friendly pier,—there is the untold fate of La Perouse;[1]— universal science to be kept pace with, studying the lives of all great discoverers and navigators, great adventurers and merchants from Hanno[2] and the Phœnicians down to our day; in fine, account of stock to be taken from time to time, to know how you stand. It is a labor to task the faculties of a man,—such problems of profit and loss, of interest, of tare and tret, and gauging of all kinds in it, as demand a universal knowledge.

I have thought that Walden Pond would be a good place for business, not solely on account of the railroad and the ice trade; it offers advantages which it may not be good policy to divulge; it is a good port and a good foundation. No Neva marshes to be filled; though you must every where build on piles of your own driving. It is said that a flood-tide, with a westerly wind, and ice in the Neva, would sweep St. Petersburg[3] from the face of the earth.

As this business was to be entered into without the usual capital, it may not be easy to conjecture where those means, that will still be indispensable to every such undertaking, were to be obtained. As for Clothing, to come at once to the practical part of the question, perhaps we are led oftener by the love of novelty, and a regard for the opinions of men, in procuring it, than by a true utility. Let him who has work to do recollect that the object of clothing is, first, to retain the vital heat, and secondly, in this state of society, to cover nakedness, and he may judge how much of any necessary or impor-

9. I.e., by shipwreck on the way to New York.
1. Jean François de Galaup (1741–88), French explorer of the western Pacific.
2. Carthaginian navigator (sixth and fifth centuries B.C.) credited with opening the coast of west Africa to trade.

"Tret": an allowance to the purchaser for waste or refuse in certain materials, four pounds being thrown in for every 104 pounds of suttle weight, or weight after the "tare" (the weight of the vehicle or smaller container) is deducted.
3. Now Leningrad.

tant work may be accomplished without adding to his wardrobe. Kings and queens who wear a suit but once, though made by some tailor or dress-maker to their majesties, cannot know the comfort of wearing a suit that fits. They are no better than wooden horses to hang the clean clothes on. Every day our garments become more assimilated to ourselves, receiving the impress of the wearer's character, until we hesitate to lay them aside, without such delay and medical appliances and some such solemnity even as our bodies. No man ever stood the lower in my estimation for having a patch in his clothes; yet I am sure that there is greater anxiety, commonly, to have fashionable, or at least clean and unpatched clothes, than to have a sound conscience. But even if the rent is not mended, perhaps the worst vice betrayed is improvidence. I sometimes try my acquaintances by such tests as this;—who could wear a patch, or two extra seams only, over the knee? Most behave as if they believed that their prospects for life would be ruined if they should do it. It would be easier for them to hobble to town with a broken leg than with a broken pantaloon. Often if an accident happens to a gentleman's legs, they can be mended; but if a similar accident happens to the legs of his pantaloons, there is no help for it; for he considers, not what is truly respectable, but what is respected. We know but few men, a great many coats and breeches. Dress a scarecrow in your last shift, you standing shiftless by, who would not soonest salute the scarecrow? Passing a cornfield the other day, close by a hat and coat on a stake, I recognized the owner of the farm. He was only a little more weather-beaten than when I saw him last. I have heard of a dog that barked at every stranger who approached his master's premises with clothes on, but was easily quieted by a naked thief. It is an interesting question how far men would retain their relative rank if they were divested of their clothes. Could you, in such a case, tell surely of any company of civilized men, which belonged to the most respected class? When Madam Pfeiffer, in her adventurous travels round the world, from east to west, had got so near home as Asiatic Russia, she says that she felt the necessity of wearing other than a travelling dress, when she went to meet the authorities, for she "was now in a civilized country, where —— – people are judged of by their clothes."[4] Even in our democratic New England towns the accidental possession of wealth, and its manifestation in dress and equipage alone, obtain for the possessor almost universal respect. But they who yield such respect, numerous as they are, are so far heathen, and need to have a missionary sent to them. Beside, clothes introduced sewing, a kind of work which you may call endless; a woman's dress, at least, is never done.[5]

---

4. Ida Pfeiffer (1797–1858), *A Lady's Voyage round the World* (1852).
5. A play on the saying "Man may work from sun to sun,/But woman's work is never done."

A man who has at length found something to do will not need to get a new suit to do it in; for him the old will do, that has lain dusty in the garret for an indeterminate period. Old shoes will serve a hero longer than they have served his valet,—if a hero ever has a valet,— bare feet are older than shoes, and he can make them do. Only they who go to soirées and legislative halls must have new coats, coats to change as often as the man changes in them. But if my jacket and trousers, my hat and shoes, are fit to worship God in, they will do; will they not? Who ever saw his old clothes,—his old coat, actually worn out, resolved into its primitive elements, so that it was not a deed of charity to bestow it on some poor boy, by him perchance to be bestowed on some poorer still, or shall we say richer, who could do with less? I say, beware of all enterprises that require new clothes, and not rather a new wearer of clothes. If there is not a new man, how can the new clothes be made to fit? If you have any enterprise before you, try it in your old clothes. All men want, not something to *do with*, but something to *do*, or rather something to *be*. Perhaps we should never procure a new suit, however ragged or dirty the old, until we have so conducted, so enterprised or sailed in some way, that we feel like new men in the old, and that to retain it would be like keeping new wine in old bottles.[6] Our moulting season, like that of the fowls, must be a crisis in our lives. The loon retires to solitary ponds to spend it. Thus also the snake casts its slough, and the caterpillar its wormy coat, by an internal industry and expansion; for clothes are but our outmost cuticle and mortal coil. Otherwise we shall be found sailing under false colors, and be inevitably cashiered[7] at last by our own opinion, as well as that of mankind.

We don garment after garment, as if we grew like exogenous plants by addition without. Our outside and often thin and fanciful clothes are our epidermis or false skin, which partakes not of our life, and may be stripped off here and there without fatal injury; our thicker garments, constantly worn, are our cellular integument, or cortex; but our shirts are our liber[8] or true bark, which cannot be removed without girdling and so destroying the man. I believe that all races at some seasons wear something equivalent to the shirt. It is desirable that a man be clad so simply that he can lay his hands on himself in the dark, and that he live in all respects so compactly and preparedly, that, if an enemy take the town, he can, like the old philosopher, walk out the gate empty-handed without anxiety. While one thick garment is, for most purposes, as good as three thin ones, and cheap clothing can be obtained at prices really to suit customers; while a thick coat can be bought for five dollars, which will last as many years, thick pantaloons for two dollars, cowhide

6. "Neither do men put new wine into old bottles: else the bottles break, and the wine runneth out, and the bottles perish: but they put new wine into new bottles, and both are preserved" (Matthew 9.17).
7. Fired.
8. Inner bark.

boots for a dollar and a half a pair, a summer hat for a quarter of a dollar, and a winter cap for sixty-two and a half cents, or a better be made at home at a nominal cost, where is he so poor that, clad in such a suit, *of his own earning*, there will not be found wise men to do him reverence?

When I ask for a garment of a particular form, my tailoress tells me gravely, "They do not make them so now," not emphasizing the "They" at all, as if she quoted an authority as impersonal as the Fates, and I find it difficult to get made what I want, simply because she cannot believe that I mean what I say, that I am so rash. When I hear this oracular sentence, I am for a moment absorbed in thought, emphasizing to myself each word separately that I may come at the meaning of it, that I may find out by what degree of consanguinity *They* are related to *me*, and what authority they may have in an affair which affects me so nearly; and, finally, I am inclined to answer her with equal mystery, and without any more emphasis of the "they,"—"It is true, they did not make them so recently, but they do now." Of what use this measuring of me if she does not measure my character, but only the breadth of my shoulders, as it were a peg to hang the coat on? We worship not the Graces, nor the Parcæ,[9] but Fashion. She spins and weaves and cuts with full authority. The head monkey at Paris puts on a traveller's cap, and all the monkeys in America do the same. I sometimes despair of getting any thing quite simple and honest done in this world by the help of men. They would have to be passed through a powerful press first, to squeeze their old notions out of them, so that they would not soon get upon their legs again, and then there would be some one in the company with a maggot in his head, hatched from an egg deposited there nobody knows when, for not even fire kills these things, and you would have lost your labor. Nevertheless, we will not forget that some Egyptian wheat is said to have been handed down to us by a mummy.

On the whole, I think that it cannot be maintained that dressing has in this or any country risen to the dignity of an art. At present men make shift to wear what they can get. Like shipwrecked sailors, they put on what they can find on the beach, and at a little distance, whether of space or time, laugh at each other's masquerade. Every generation laughs at the old fashions, but follows religiously the new. We are amused at beholding the costume of Henry VIII., or Queen Elizabeth, as much as if it was that of the King and Queen of the Cannibal Islands. All costume off a man is pitiful or grotesque. It is only the serious eye peering from and the sincere life passed within it, which restrain laughter and consecrate the costume of any people. Let Harlequin[1] be taken with a fit of the colic and his trappings will have to serve that mood too. When the soldier is

9. In Roman mythology, the three Fates.
1. A type of comic servant in *commedia* *dell'arte*, dressed in mask and many-colored tights.

hit by a cannon ball rags are as becoming as purple.

The childish and savage taste of men and women for new patterns keeps how many shaking and squinting through kaleidoscopes
that they may discover the particular figure which this generation
requires to-day. The manufacturers have learned that this taste is
merely whimsical. Of two patterns which differ only by a few
threads more or less of a particular color, the one will be sold
readily, the other lie on the shelf, though it frequently happens that
after the lapse of a season the latter becomes the most fashionable.
Comparatively, tattooing is not the hideous custom which it is
called. It is not barbarous merely because the printing is skin-deep
and unalterable.

I cannot believe that our factory system is the best mode by
which men may get clothing. The condition of the operatives is
becoming every day more like that of the English; and it cannot be
wondered at, since, as far as I have heard or observed, the principal
object is, not that mankind may be well and honestly clad, but,
unquestionably, that the corporations may be enriched. In the long
run men hit only what they aim at. Therefore, though they should
fail immediately, they had better aim at something high.

As for a Shelter, I will not deny that this is now a necessary of
life, though there are instances of men having done without it for
long periods in colder countries than this. Samuel Laing says that
"The Laplander in his skin dress, and in a skin bag which he puts
over his head and shoulders, will sleep night after night on the
snow——in a degree of cold which would extinguish the life of one
exposed to it in any woollen clothing." He had seen them asleep
thus. Yet he adds, "They are not hardier than other people."[2] But,
probably, man did not live long on the earth without discovering the
convenience which there is in a house, the domestic comforts,
which phrase may have originally signified the satisfactions of the
house more than of the family; though these must be extremely
partial and occasional in those climates where the house is associated in our thoughts with winter or the rainy season chiefly, and
two thirds of the year, except for a parasol, is unnecessary. In our
climate, in the summer, it was formerly almost solely a covering at
night. In the Indian gazettes[3] a wigwam was the symbol of a day's
march, and a row of them cut or painted on the bark of a tree
signified that so many times they had camped. Man was not made
so large limbed and robust but that he must seek to narrow his
world, and wall in a space such as fitted him. He was at first bare
and out of doors; but though this was pleasant enough in serene and
warm weather, by daylight, the rainy season and the winter, to say
nothing of the torrid sun, would perhaps have nipped his race in the

2. *Journal of a Residence in Norway* (1837).

3. In Indian sign language (in messages equivalent to gazettes or newspapers).

bud if he had not made haste to clothe himself with the shelter of a house. Adam and Eve, according to the fable, wore the bower before other clothes. Man wanted a home, a place of warmth, or comfort, first of physical warmth, then the warmth of the affections.

We may imagine a time when, in the infancy of the human race, some enterprising mortal crept into a hollow in a rock for shelter. Every child begins the world again, to some extent, and loves to stay out doors, even in wet and cold. It plays house, as well as horse, having an instinct for it. Who does not remember the interest with which when young he looked at shelving rocks, or any approach to a cave? It was the natural yearning of that portion of our most primitive ancestor which still survived in us. From the cave we have advanced to roofs of palm leaves, of bark and boughs, of linen woven and stretched, of grass and straw, of boards and shingles, of stones and tiles. At last, we know not what it is to live in the open air, and our lives are domestic in more senses than we think. From the hearth to the field is a great distance. It would be well perhaps if we were to spend more of our days and nights without any obstruction between us and the celestial bodies, if the poet did not speak so much from under a roof, or the saint dwell there so long. Birds do not sing in caves, nor do doves cherish their innocence in dovecots.

However, if one designs to construct a dwelling house, it behooves him to exercise a little Yankee shrewdness, lest after all he find himself in a workhouse, a labyrinth without a clew, a museum, an almshouse, a prison, or a splendid mausoleum instead. Consider first how slight a shelter is absolutely necessary. I have seen Penobscot Indians, in this town, living in tents of thin cotton cloth, while the snow was nearly a foot deep around them, and I thought that they would be glad to have it deeper to keep out the wind. Formerly, when how to get my living honestly, with freedom left for my proper pursuits, was a question which vexed me even more than it does now, for unfortunately I am become somewhat callous, I used to see a large box by the railroad, six feet long by three wide, in which the laborers locked up their tools at night, and it suggested to me that every man who was hard pushed might get such a one for a dollar, and, having bored a few auger holes in it, to admit the air at least, get into it when it rained and at night, and hook down the lid, and so have freedom in his love, and in his soul be free. This did not appear the worst, nor by any means a despicable alternative. You could sit up as late as you pleased, and, whenever you got up, go abroad without any landlord or house-lord dogging you for rent. Many a man is harassed to death to pay the rent of a larger and more luxurious box who would not have frozen to death in such a box as this. I am far from jesting. Economy is a subject which admits of being treated with levity, but it cannot so be disposed of. A comfortable house for a rude and hardy race, that lived mostly out of doors, was once made here almost entirely of such materials as Nature furnished ready to their hands. Gookin, who was super-intendent of the Indians subject to the Massachusetts Colony, writ-ing in 1674, says, "The best of their houses are covered very neatly, tight and warm, with barks of trees, slipped from their bodies at those seasons when the sap is up, and made into great flakes, with pressure of weighty timber, when they are green. . . . The meaner sort are covered with mats which they make of a kind of bulrush, and are also indifferently tight and warm, but not so good as the former. . . . Some I have seen, sixty or a hundred feet long and thirty feet broad. . . . I have often lodged in their wigwams, and found them as warm as the best English houses."[4] He adds, that they were commonly carpeted and lined within with well-wrought embroidered mats, and were furnished with various utensils. The Indians had advanced so far as to regulate the effect of the wind by a mat suspended over the hole in the roof and moved by a string. Such a lodge was in the first instance constructed in a day or two at most, and taken down and put up in a few hours; and every family owned one, or its apartment in one.

In the savage state every family owns a shelter as good as the best, and sufficient for its coarser and simpler wants; but I think that I speak within bounds when I say that, though the birds of the air have their nests, and the foxes their holes,[5] and the savages their wigwams, in modern civilized society not more than one half the families own a shelter. In the large towns and cities, where civiliza-tion especially prevails, the number of those who own a shelter is a very small fraction of the whole. The rest pay an annual tax for this outside garment of all, become indispensable summer and winter, which would buy a village of Indian wigwams, but now helps to keep them poor as long as they live. I do not mean to insist here on the disadvantage of hiring compared with owning, but it is evident that the savage owns his shelter because it costs so little, while the civilized man hires his commonly because he cannot afford to own it; nor can he, in the long run, any better afford to hire. But, answers one, by merely paying this tax the poor civilized man secures an abode which is a palace compared with the savage's. An annual rent of from twenty-five to a hundred dollars, these are the country rates, entitles him to the benefit of the improvements of centuries, spacious apartments, clean paint and paper, Rumford fireplace,[6] back plastering, Venetian blinds, copper pump, spring lock, a commodious cellar, and many other things. But how happens it that he who is said to enjoy these things is so commonly a *poor* civilized man, while the savage, who has them not, is rich as a savage? If it is asserted that civilization is a real advance in the condition of man,

4. Daniel Gookin, *Historical Collections of the Indians in New England* (1792).
5. "The foxes have holes, and the birds of the air have nests; but the Son of man hath not where to lay his head" (Matthew 8.20).
6. Benjamin Thompson, Count Rumford (1753–1814) devised a shelf inside the chimney to prevent smoke from being carried back into a room by downdrafts.

—and I think that it is, though only the wise improve their advantages,—it must be shown that it has produced better dwellings without making them more costly; and the cost of a thing is the amount of what I will call life which is required to be exchanged for it, immediately or in the long run. An average house in this neighborhood costs perhaps eight hundred dollars, and to lay up this sum will take from ten to fifteen years of the laborer's life, even if he is not encumbered with a family;—estimating the pecuniary value of every man's labor at one dollar a day, for if some receive more, others receive less;—so that he must have spent more than half his life commonly before *his* wigwam will be earned. If we suppose him to pay a rent instead, this is but a doubtful choice of evils. Would the savage have been wise to exchange his wigwam for a palace on these terms?

It may be guessed that I reduce almost the whole advantage of holding this superfluous property as a fund in store against the future, so far as the individual is concerned, mainly to the defraying of funeral expenses. But perhaps a man is not required to bury himself. Nevertheless this points to an important distinction between the civilized man and the savage; and, no doubt, they have designs on us for our benefit, in making the life of a civilized people an *institution*, in which the life of the individual is to a great extent absorbed, in order to preserve and perfect that of the race. But I wish to show at what a sacrifice this advantage is at present obtained, and to suggest that we may possibly so live as to secure all the advantage without suffering any of the disadvantage. What mean ye by saying that the poor ye have always with you, or that the fathers have eaten sour grapes, and the children's teeth are set on edge?[7]

"As I live, saith the Lord God, ye shall not have occasion any more to use this proverb in Israel."

"Behold all souls are mine; as the soul of the father, so also the soul of the son is mine: the soul that sinneth it shall die."

When I consider my neighbors, the farmers of Concord, who are at least as well off as the other classes, I find that for the most part they have been toiling twenty, thirty, or forty years, that they may become the real owners of their farms, which commonly they have inherited with encumbrances, or else bought with hired money,—and we may regard one third of that toil as the cost of their houses,—but commonly they have not paid for them yet. It is true,

7. Thoreau is repudiating Jesus' words to his disciples "For ye have the poor always with you; but me ye have not always" (Matthew 26.11) by combining it with God's reproof to Ezekiel for employing a negatively deterministic proverb: "What mean ye, that ye use this proverb concerning the land of Israel, saying, The fathers have eaten sour grapes, and the children's teeth are set on edge?" (Ezekiel 18.2). The two verses within Thoreau's quotation marks are Ezekiel 18.3–4, but Thoreau so truncates the passage that the reader may find it hard to understand that the Biblical intent (as well as Thoreau's own) is optimistic, to reject the notion that the sins of the fathers are visited unto their children.

the encumbrances sometimes outweigh the value of the farm, so that the farm itself becomes one great encumbrance, and still a man is found to inherit it, being well acquainted with it, as he says. On applying to the assessors, I am surprised to learn that they cannot at once name a dozen in the town who own their farms free and clear. If you would know the history of these homesteads, inquire at the bank where they are mortgaged. The man who has actually paid for his farm with labor on it is so rare that every neighbor can point to him. I doubt if there are three such men in Concord. What has been said of the merchants, that a very large majority, even ninety-seven in a hundred, are sure to fail, is equally true of the farmers. With regard to the merchants, however, one of them says pertinently that a great part of their failures are not genuine pecuniary failures, but merely failures to fulfil their engagements, because it is inconvenient; that is, it is the moral character that breaks down. But this puts an infinitely worse face on the matter, and suggests, beside, that probably not even the other three succeed in saving their souls, but are perchance bankrupt in a worse sense than they who fail honestly. Bankruptcy and repudiation are the spring-boards from which much of our civilization vaults and turns its somersets, but the savage stands on the unelastic plank of famine. Yet the Middlesex Cattle Show goes off here with *éclat* annually, as if all the joints of the agricultural machine were suent.[8]

The farmer is endeavoring to solve the problem of a livelihood by a formula more complicated than the problem itself. To get his shoestrings he speculates in herds of cattle. With consummate skill he has set his trap with a hair spring to catch comfort and independence, and then, as he turned away, got his own leg into it. This is the reason he is poor; and for a similar reason we are all poor in respect to a thousand savage comforts, though surrounded by luxuries. As Chapman sings,—[9]

> "The false society of men—
> —for earthly greatness
> All heavenly comforts rarefies to air."

And when the farmer has got his house, he may not be the richer but the poorer for it, and it be the house that has got him. As I understand it, that was a valid objection urged by Momus[1] against the house which Minerva made, that she "had not made it movable, by which means a bad neighborhood might be avoided;" and it may still be urged, for our houses are such unwieldy property that we are often imprisoned rather than housed in them; and the bad neighborhood to be avoided is our own scurvy selves. I know one or two

8. In good working order, broken in.
9. George Chapman (1559?–1634), *Caesar and Pompey*, V.ii.

1. In Greek mythology, the god of pleasantry but also of carping criticism.

families, at least, in this town, who, for nearly a generation, have been wishing to sell their houses in the outskirts and move into the village, but have not been able to accomplish it, and only death will set them free.

Granted that the *majority* are able at last either to own or hire the modern house with all its improvements. While civilization has been improving our houses, it has not equally improved the men who are to inhabit them. It has created palaces, but it was not so easy to create noblemen and kings. And *if the civilized man's pursuits are no worthier than the savage's, if he is employed the greater part of his life in obtaining gross necessaries and comforts merely, why should he have a better dwelling than the former?*

But how do the poor *minority* fare? Perhaps it will be found, that just in proportion as some have been placed in outward circumstances above the savage, others have been degraded below him. The luxury of one class is counterbalanced by the indigence of another. On the one side is the palace, on the other are the almshouse and "silent poor".[2] The myriads who built the pyramids to be the tombs of the Pharaohs were fed on garlic, and it may be were not decently buried themselves. The mason who finishes the cornice of the palace returns at night perchance to a hut not so good as a wigwam. It is a mistake to suppose that, in a country where the usual evidences of civilization exist, the condition of a very large body of the inhabitants may not be as degraded as that of savages. I refer to the degraded poor, not now to the degraded rich. To know this I should not need to look farther than to the shanties which every where border our railroads, that last improvement in civilization; where I see in my daily walks human beings living in sties, and all winter with an open door, for the sake of light, without any visible, often imaginable, wood pile, and the forms of both old and young are permanently contracted by the long habit of shrinking from cold and misery, and the development of all their limbs and faculties is checked. It certainly is fair to look at that class by whose labor the works which distinguish this generation are accomplished. Such too, to a greater or less extent, is the condition of the operatives of every denomination in England, which is the great workhouse of the world. Or I could refer you to Ireland, which is marked as one of the white or enlightened spots on the map.[3] Contrast the physical condition of the Irish with that of the North American Indian, or the South Sea Islander, or any other savage race before it was degraded by contact with the civilized man. Yet I have no doubt that that people's rulers are as wise as the average of civilized rulers. Their condition only proves what squalidness may consist with civilization. I hardly need refer now to the laborers in

2. Harding identifies these as the poor of Concord who received public charity secretly in order to retain their dwellings and not go to the poorhouse.

3. Thoreau refers to the habit some cartographers had of leaving unexplored terrain in a dark color; other cartographers left unexplored areas white.

our Southern States who produce the staple exports of this country, and are themselves a staple production of the South.[4] But to confine myself to those who are said to be in *moderate* circumstances.

Most men appear never to have considered what a house is, and are actually though needlessly poor all their lives because they think that they must have such a one as their neighbors have. As if one were to wear any sort of coat which the tailor might cut out for him, or, gradually leaving off palmleaf hat or cap of woodchuck skin, complain of hard times because he could not afford to buy him a crown! It is possible to invent a house still more convenient and luxurious than we have, which yet all would admit that man could not afford to pay for. Shall we always study to obtain more of these things, and not sometimes to be content with less? Shall the respectable citizen thus gravely teach, by precept and example, the necessity of the young man's providing a certain number of superfluous glow-shoes,[5] and umbrellas, and empty guest chambers for empty guests, before he dies? Why should not our furniture be as simple as the Arab's or the Indian's? When I think of the benefactors of the race, whom we have apotheosized as messengers from heaven, bearers of divine gifts to man, I do not see in my mind any retinue at their heels, any car-load of fashionable furniture. Or what if I were to allow—would it not be a singular allowance?—that our furniture should be more complex than the Arab's, in proportion as we are morally and intellectually his superiors! At present our houses are cluttered and defiled with it, and a good housewife would sweep out the greater part into the dust hole, and not leave her morning's work undone. Morning work! By the blushes of Aurora and the music of Memnon,[6] what should be man's *morning work* in this world? I had three pieces of limestone on my desk, but I was terrified to find that they required to be dusted daily, when the furniture of my mind was all undusted still, and I threw them out the window in disgust. How, then, could I have a furnished house? I would rather sit in the open air, for no dust gathers on the grass, unless where man has broken ground.

It is the luxurious and dissipated who set the fashions which the herd so diligently follow. The traveller who stops at the best houses, so called, soon discovers this, for the publicans presume him to be a Sardanapalus,[7] and if he resigned himself to their tender mercies he would soon be completely emasculated. I think that in the railroad car we are inclined to spend more on luxury than on safety and convenience, and it threatens without attaining these to become no better than a modern drawing room, with its divans, and ottomans,

4. The accusation, denied by many historians, that some plantations, especially in Virginia, were run for the sole purpose of breeding slave children for sale.
5. Galoshes.
6. The Roman goddess of the dawn and her son, an Ethiopian prince who fought for Priam at Troy. Memnon is associated here with the Egyptian colossus near Thebes which in ancient times emitted a sound at dawn, presumably because of the warming of air currents.
7. Effeminate ruler of Assyria in the ninth century B.C.

and sunshades, and a hundred other oriental things, which we are taking west with us, invented for the ladies of the harem and the effeminate natives of the Celestial Empire, which Jonathan[8] should be ashamed to know the names of. I would rather sit on a pumpkin and have it all to myself, than be crowded on a velvet cushion. I would rather ride on earth in an ox cart with a free circulation, than go to heaven in the fancy car of an excursion train and breathe a *malaria* all the way.

The very simplicity and nakedness of man's life in the primitive ages imply this advantage at least, that they left him still but a sojourner in nature. When he was refreshed with food and sleep he contemplated his journey again. He dwelt, as it were, in a tent in this world, and was either threading the valleys, or crossing the plains, or climbing the mountain tops. But lo! men have become the tools of their tools. The man who independently plucked the fruits when he was hungry is become a farmer; and he who stood under a tree for shelter, a housekeeper. We now no longer camp as for a night, but have settled down on earth and forgotten heaven. We have adopted Christianity merely as an improved method of *agri*-culture. We have built for this world a family mansion, and for the next a family tomb. The best works of art are the expression of man's struggle to free himself from this condition, but the effect of our art is merely to make this low state comfortable and that higher state to be forgotten. There is actually no place in this village for a work of *fine* art, if any had come down to us, to stand, for our lives, our houses and streets, furnish no proper pedestal for it. There is not a nail to hang a picture on, nor a shelf to receive the bust of a hero or a saint. When I consider how our houses are built and paid for, or not paid for, and their internal economy managed and sus-tained, I wonder that the floor does not give way under the visitor while he is admiring the gewgaws upon the mantel-piece, and let him through into the cellar, to some solid and honest though earthy foundation. I cannot but perceive that this so called rich and refined life is a thing jumped at, and I do not get on in the enjoyment of the *fine* arts which adorn it, my attention being wholly occupied with the jump; for I remember that the greatest genuine leap, due to human muscles alone, on record, is that of certain wandering Arabs, who are said to have cleared twenty-five feet on level ground. Without factitious support, man is sure to come to earth again beyond that distance. The first question which I am tempted to put to the proprietor of such great impropriety is, Who bolsters you? Are you one of the ninety-seven who fail? or of the three who succeed? Answer me these questions, and then perhaps I may look at your bawbles and find them ornamental. The cart before the horse is neither beautiful nor useful. Before we can adorn our

8. A type name at first applied to New Englanders, then later (as here) to the in-habitants of the entire United States.

houses with beautiful objects the walls must be stripped, and our
lives must be stripped, and beautiful housekeeping and beautiful
living be laid for a foundation: now, a taste for the beautiful is most
cultivated out of doors, where there is no house and no housekeeper.

Old Johnson, in his "Wonder-Working Providence," speaking of
the first settlers of this town, with whom he was contemporary, tells
us that "they burrow themselves in the earth for their first shelter
under some hillside, and, casting the soil aloft upon timber, they
make a smoky fire against the earth, at the highest side." They did
not "provide them houses," says he, "till the earth, by the Lord's
blessing, brought forth bread to feed them," and the first year's crop
was so light that "they were forced to cut their bread very thin for a
long season."[9] The secretary of the Province of New Netherland,
writing in Dutch, in 1650, for the information of those who wished
to take up land there, states more particularly, that "those in New
Netherland, and especially in New England, who have no means to
build farm houses at first according to their wishes, dig a square pit
in the ground, cellar fashion, six or seven feet deep, as long and as
broad as they think proper, case the earth inside with wood all
round the wall, and line the wood with the bark of trees or something else to prevent the caving in of the earth; floor this cellar with
plank, and wainscot it overhead for a ceiling, raise a roof of spars
clear up, and cover the spars with bark or green sods, so that they
can live dry and warm in these houses with their entire families for
two, three, and four years, it being understood that partitions are
run through those cellars which are adapted to the size of the
family. The wealthy and principal men in New England, in the
beginning of the colonies, commenced their first dwelling houses in
this fashion for two reasons; firstly, in order not to waste time in
building, and not to want food the next season; secondly, in order
not to discourage poor laboring people whom they brought over in
numbers from Fatherland. In the course of three or four years,
when the country became adapted to agriculture, they built themselves handsome houses, spending on them several thousands."[1]

In this course which our ancestors took there was a show of
prudence at least, as if their principle were to satisfy the more
pressing wants first. But are the more pressing wants satisfied now?
When I think of acquiring for myself one of our luxurious dwellings, I am deterred, for, so to speak, the country is not yet adapted
to *human* culture, and we are still forced to cut our *spiritual* bread
far thinner than our forefathers did their wheaten. Not that all
architectural ornament is to be neglected even in the rudest periods;
but let our houses first be lined with beauty, where they come in

9. Edward Johnson, *Wonder-working
Providence of Sion's Saviour in New
England* (1654).

1. Edmund Bailey O'Callaghan, *Documentary History of the State of New-
York* (1851).

contact with our lives, like the tenement of the shellfish, and not overlaid with it. But, alas! I have been inside one or two of them, and know what they are lined with.

Though we are not so degenerate but that we might possibly live in a cave or a wigwam or wear skins to-day, it certainly is better to accept the advantages, though so dearly bought, which the invention and industry of mankind offer. In such a neighborhood as this, boards and shingles, lime and bricks, are cheaper and more easily obtained than suitable caves, or whole logs, or bark in sufficient quantities, or even well-tempered clay or flat stones. I speak understandingly on this subject, for I have made myself acquainted with it both theoretically and practically. With a little more wit we might use these materials so as to become richer than the richest now are, and make our civilization a blessing. The civilized man is a more experienced and wiser savage. But to make haste to my own experiment.

Near the end of March, 1845, I borrowed an axe and went down to the woods by Walden Pond, nearest to where I intended to build my house, and began to cut down some tall arrowy white pines, still in their youth, for timber. It is difficult to begin without borrowing, but perhaps it is the most generous course thus to permit your fellow-men to have an interest in your enterprise. The owner of the axe, as he released his hold on it, said that it was the apple of his eye; but I returned it sharper than I received it. It was a pleasant hillside where I worked, covered with pine woods, through which I looked out on the pond, and a small open field in the woods where pines and hickories were springing up. The ice in the pond was not yet dissolved, though there were some open spaces, and it was all dark colored and saturated with water. There were some slight flurries of snow during the days that I worked there; but for the most part when I came out on to the railroad, on my way home, its yellow sand heap stretched away gleaming in the hazy atmosphere, and the rails shone in the spring sun, and I heard the lark and pewee and other birds already come to commence another year with us. They were pleasant spring days, in which the winter of man's discontent was thawing as well as the earth, and the life that had lain torpid began to stretch itself. One day, when my axe had come off and I had cut a green hickory for a wedge, driving it with a stone, and had placed the whole to soak in a pond hole in order to swell the wood, I saw a striped snake run into the water, and he lay on the bottom, apparently without inconvenience, as long as I staid there, or more than a quarter of an hour; perhaps because he had not yet fairly come out of the torpid state. It appeared to me that for a like reason men remain in their present low and primitive condition; but if they should feel the influence of the spring of springs arousing them, they would of necessity rise to a higher and more ethereal life. I had previously seen the snakes in frosty mornings in my path with portions of their bodies still numb and inflexible, waiting for the sun to thaw them. On the 1st of April it rained and melted the ice, and in the early part of the day, which was very foggy, I heard a stray goose groping about over the pond and cackling as if lost, or like the spirit of the fog.

So I went on for some days cutting and hewing timber, and also studs and rafters, all with my narrow axe, not having many communicable or scholar-like thoughts, singing to myself,—[2]

> Men say they know many things;
> But lo! they have taken wings,—
> The arts and sciences,
> And a thousand appliances;
> The wind that blows
> Is all that any body knows.

I hewed the main timbers six inches square, most of the studs on two sides only, and the rafters and floor timbers on one side, leaving the rest of the bark on, so that they were just as straight and much stronger than sawed ones. Each stick was carefully mortised or tenoned by its stump, for I had borrowed other tools by this time. My days in the woods were not very long ones; yet I usually carried my dinner of bread and butter, and read the newspaper in which it was wrapped, at noon, sitting amid the green pine boughs which I had cut off, and to my bread was imparted some of their fragrance, for my hands were covered with a thick coat of pitch. Before I had done I was more the friend than the foe of the pine tree, though I had cut down some of them, having become better acquainted with it. Sometimes a rambler in the wood was attracted by the sound of my axe, and we chatted pleasantly over the chips which I had made.

By the middle of April, for I made no haste in my work, but rather made the most of it, my house was framed and ready for the raising. I had already bought the shanty of James Collins, an Irishman who worked on the Fitchburg Railroad, for boards. James Collins' shanty was considered an uncommonly fine one. When I called to see it he was not at home. I walked about the outside, at first unobserved from within, the window was so deep and high. It was of small dimensions, with a peaked cottage roof, and not much else to be seen, the dirt being raised five feet all around as if it were a compost heap. The roof was the soundest part, though a good deal warped and made brittle by the sun. Door-sill there was none, but a perennial passage for the hens under the door board. Mrs. C. came to the door and asked me to view it from the inside. The hens were

2. Like other poems in *Walden* not enclosed in quotation marks, this poem is Thoreau's.

driven in by my approach. It was dark, and had a dirt floor for the most part, dank, clammy, and aguish, only here a board and there a board which would not bear removal. She lighted a lamp to show me the inside of the roof and the walls, and also that the board floor extended under the bed, warning me not to step into the cellar, a sort of dust hole two feet deep. In her own words, they were "good boards overhead, good boards all around, and a good window,"— of two whole squares originally, only the cat had passed out that way lately. There was a stove, a bed, and a place to sit, an infant in the house where it was born, a silk parasol, gilt-framed looking-glass, and a patent new coffee mill nailed to an oak sapling, all told. The bargain was soon concluded, for James had in the mean while returned. I to pay four dollars and twenty-five cents to-night, he to vacate at five to-morrow morning, selling to nobody else mean-while: I to take possession at six. It were well, he said, to be there early, and anticipate certain indistinct but wholly unjust claims on the score of ground rent and fuel. This he assured me was the only encumbrance. At six I passed him and his family on the road. One large bundle held their all,—bed, coffee-mill, looking-glass, hens,—all but the cat, she took to the woods and became a wild cat, and, as I learned afterward, trod in a trap set for woodchucks, and so became a dead cat at last.

I took down this dwelling the same morning, drawing the nails, and removed it to the pond side by small cartloads, spreading the boards on the grass there to bleach and warp back again in the sun. One early thrush gave me a note or two as I drove along the woodland path. I was informed treacherously by a young Patrick that neighbor Seeley, an Irishman, in the intervals of the carting, transferred the still tolerable, straight, and drivable nails, staples, and spikes to his pocket, and then stood when I came back to pass the time of day, and look freshly up, unconcerned, with spring thoughts, at the devastation; there being a dearth of work, as he said. He was there to represent spectatordom, and help make this seemingly insignificant event one with the removal of the gods of Troy.[3]

I dug my cellar in the side of a hill sloping to the south, where a woodchuck had formerly dug his burrow, down through sumach and blackberry roots, and the lowest stain of vegetation, six feet square by seven deep, to a fine sand where potatoes would not freeze in any winter. The sides were left shelving, and not stoned; but the sun having never shone on them, the sand still keeps its place. It was but two hours' work. I took particular pleasure in this breaking of ground, for in almost all latitudes men dig into the earth for an equable temperature. Under the most splendid house in the city is still to be found the cellar where they store their roots as

3. In Virgil's *Aeneid*, Book II, after the fall of Troy, Aeneas escapes with his father and son and his household gods.

of old, and long after the superstructure has disappeared posterity remark its dent in the earth. The house is still but a sort of porch at the entrance of a burrow.

At length, in the beginning of May, with the help of some of my acquaintances, rather to improve so good an occasion for neighborliness than from any necessity, I set up the frame of my house. No man was ever more honored in the character of his raisers[4] than I. They are destined, I trust, to assist at the raising of loftier structures one day. I began to occupy my house on the 4th of July, as soon as it was boarded and roofed, for the boards were carefully feather-edged[5] and lapped, so that it was perfectly impervious to rain; but before boarding I laid the foundation of a chimney at one end, bringing two cartloads of stones up the hill from the pond in my arms. I built the chimney after my hoeing in the fall, before a fire became necessary for warmth, doing my cooking in the mean while out of doors on the ground, early in the morning: which mode I still think is in some respects more convenient and agreeable than the usual one. When it stormed before my bread was baked, I fixed a few boards over the fire, and sat under them to watch my loaf, and passed some pleasant hours in that way. In those days, when my hands were much employed, I read but little, but the least scraps of paper which lay on the ground, my holder, or table-cloth, afforded me as much entertainment, in fact answered the same purpose as the Iliad.[6]

It would be worth the while to build still more deliberately than I did, considering, for instance, what foundation a door, a window, a cellar, a garret, have in the nature of man, and perchance never raising any superstructure until we found a better reason for it than our temporal necessities even. There is some of the same fitness in a man's building his own house that there is in a bird's building its own nest. Who knows but if men constructed their dwellings with their own hands, and provided food for themselves and families simply and honestly enough, the poetic faculty would be universally developed, as birds universally sing when they are so engaged? But alas! we do like cowbirds and cuckoos, which lay their eggs in nests which other birds have built, and cheer no traveller with their chattering and unmusical notes. Shall we forever resign the pleasure of construction to the carpenter? What does architecture amount to in the experience of the mass of men? I never in all my walks came across a man engaged in so simple and natural an occupation as building his house. We belong to the community. It is not the tailor

---

4. These "raisers" (a pun) included Emerson; Alcott; Ellery Channing; two young brothers who had studied at Brook Farm, Burrill and George William Curtis; and the Concord farmer Edmund Hosmer and his three sons.

5. I.e., on the boards to be nailed horizontally the top and bottom edges were cut at 45-degree angles and overlapped so as to shed rain.
6. Greek epic of the siege of Troy traditionally attributed to Homer.

alone who is the ninth part of a man; it is as much the preacher, and the merchant, and the farmer. Where is this division of labor to end? and what object does it finally serve? No doubt another *may* also think for me; but it is not therefore desirable that he should do so to the exclusion of my thinking for myself.

True, there are architects so called in this country, and I have heard of one at least possessed with the idea of making architectural ornaments have a core of truth, a necessity, and hence a beauty, as if it were a revelation to him.[7] All very well perhaps from his point of view, but only a little better than the common dilettantism. A sentimental reformer in architecture, he began at the cornice, not at the foundation. It was only how to put a core of truth within the ornaments, that every sugar plum in fact might have an almond or caraway seed in it,—though I hold that almonds are most whole- some without the sugar,—and not how the inhabitant, the indweller, might build truly within and without, and let the ornaments take care of themselves. What reasonable man ever supposed that orna- ments were something outward and in the skin merely,—that the tortoise got his spotted shell, or the shellfish its mother-o'-pearl tints, by such a contract as the inhabitants of Broadway their Trinity Church? But a man has no more to do with the style of architecture of his house than a tortoise with that of its shell: nor need the soldier be so idle as to try to paint the precise *color* of his virtue on his standard. The enemy will find it out. He may turn pale when the trial comes. This man seemed to me to lean over the cornice and timidly whisper his half truth to the rude occupants who really knew it better than he. What of architectural beauty I now see, I know has gradually grown from within outward, out of the necessi- ties and character of the indweller, who is the only builder,—out of some unconscious truthfulness, and nobleness, without ever a thought for the appearance; and whatever additional beauty of this kind is destined to be produced will be preceded by a like uncon- scious beauty of life. The most interesting dwellings in this country, as the painter knows, are the most unpretending, humble log huts and cottages of the poor commonly; it is the life of the inhabitants whose shells they are, and not any peculiarity in their surfaces merely, which makes them *picturesque*; and equally interesting will be the citizen's suburban box, when his life shall be as simple and as agreeable to the imagination, and there is as little straining after effect in the style of his dwelling. A great proportion of architec- tural ornaments are literally hollow, and a September gale would strip them off, like borrowed plumes, without injury to the substan- tials. They can do without *architecture* who have no olives nor wines in the cellar. What if an equal ado were made about the

---

7. The sculptor Horatio Greenough (1805–52), whose theories Thoreau knew only imperfectly from a private letter of Greenough's to Emerson. The ideas attributed here are at variance with Greenough's published comments on architecture.

ornaments of style in literature, and the architects of our bibles spent as much time about their cornices as the architects of our churches do? So are made the *belles-lettres* and the *beaux-arts* and their professors. Much it concerns a man, forsooth, how a few sticks are slanted over him or under him, and what colors are daubed upon his box. It would signify somewhat, if, in any earnest sense, *he* slanted them and daubed it; but the spirit having departed out of the tenant, it is of a piece with constructing his own coffin,— the architecture of the grave, and "carpenter" is but another name for "coffin-maker." One man says, in his despair or indifference to life, take up a handful of the earth at your feet, and paint your house that color. Is he thinking of his last and narrow house? Toss up a copper for it as well. What an abundance of leisure he must have! Why do you take up a handful of dirt? Better paint your house your own complexion; let it turn pale or blush for you. An enterprise to improve the style of cottage architecture! When you have got my ornaments ready I will wear them.

Before winter I built a chimney, and shingled the sides of my house, which were already impervious to rain, with imperfect and sappy shingles made of the first slice of the log, whose edges I was obliged to straighten with a plane.

I have thus a tight shingled and plastered house, ten feet wide by fifteen long, and eight-feet posts, with a garret and a closet, a large window on each side, two trap doors, one door at the end, and a brick fireplace opposite. The exact cost of my house, paying the usual price for such materials as I used, but not counting the work, all of which was done by myself, was as follows; and I give the details because very few are able to tell exactly what their houses cost, and fewer still, if any, the separate cost of the various materials which compose them:—

Boards	$8 03½	Mostly shanty boards
Refuse shingles for roof and sides,	4 00	
Laths,	1 25	
Two second-hand windows with glass,	2 43	
One thousand old brick,	4 00	
Two casks of lime,	2 40	That was high
Hair,	0 31	More than I needed
Mantle-tree iron,	0 15	
Nails,	3 90	
Hinges and screws,	0 14	
Latch,	0 10	
Chalk,	0 01	
Transportation,	1 40	I carried a good part on my back
In all,	$28 12½	

These are all the materials excepting the timber stones and sand, which I claimed by squatter's right. I have also a small wood-shed adjoining, made chiefly of the stuff which was left after building the house.

I intend to build me a house which will surpass any on the main street in Concord in grandeur and luxury, as soon as it pleases me as much and will cost me no more than my present one.

I thus found that the student who wishes for a shelter can obtain one for a lifetime at an expense not greater than the rent which he now pays annually. If I seem to boast more than is becoming, my excuse is that I brag for humanity rather than for myself; and my shortcomings and inconsistencies do not affect the truth of my statement. Notwithstanding much cant and hypocrisy,—chaff which I find it difficult to separate from my wheat, but for which I am as sorry as any man,—I will breathe freely and stretch myself in this respect, it is such a relief to both the moral and physical system; and I am resolved that I will not through humility become the devil's attorney. I will endeavor to speak a good word for the truth. At Cambridge College[3] the mere rent of a student's room, which is only a little larger than my own, is thirty dollars each year, though the corporation had the advantage of building thirty-two side by side and under one roof, and the occupant suffers the inconvenience of many and noisy neighbors, and perhaps a residence in the fourth story. I cannot but think that if we had more true wisdom in these respects, not only less education would be needed, because, forsooth, more would already have been acquired, but the pecuniary expense of getting an education would in a great measure vanish. Those conveniences which the student requires at Cambridge or elsewhere cost him or somebody else ten times as great a sacrifice of life as they would with proper management on both sides. Those things for which the most money is demanded are never the things which the student most wants. Tuition, for instance, is an important item in the term bill, while for the far more valuable education which he gets by associating with the most cultivated of his contemporaries no charge is made. The mode of founding a college is, commonly, to get up a subscription of dollars and cents, and then following blindly the principles of a divison of labor to its extreme, a principle which should never be followed but with circumspection, —to call in a contractor who makes this a subject of speculation, and he employs Irishmen or other operatives actually to lay the foundations, while the students that are to be are said to be fitting themselves for it; and for these oversights successive generations have to pay. I think that it would be *better than this*, for the students, or those who desire to be benefited by it, even to lay the foundation themselves. The student who secures his coveted leisure and retirement by systematically shirking any labor necessary to

8. Harvard University.

man obtains but an ignoble and unprofitable leisure, defrauding himself of the experience which alone can make leisure fruitful. "But," says one, "you do not mean that the students should go to work with their hands instead of their heads?" I do not mean that exactly, but I mean something which he might think a good deal like that; I mean that they should not *play* life, or *study* it merely, while the community supports them at this expensive game, but earnestly *live* it from beginning to end. How could youths better learn to live than by at once trying the experiment of living? Methinks this would exercise their minds as much as mathematics. If I wished a boy to know something about the arts and sciences, for instance, I would not pursue the common course, which is merely to send him into the neighborhood of some professor, where any thing is professed and practised but the art of life;—to survey the world through a telescope or a microscope, and never with his natural eye; to study chemistry, and not learn how his bread is made, or mechanics, and not learn how it is earned; to discover new satellites to Neptune, and not detect the motes in his eyes, or to what vagabond he is a satellite himself; or to be devoured by the monsters that swarm all around him, while contemplating the monsters in a drop of vinegar. Which would have advanced the most at the end of the month,—the boy who had made his own jack-knife from the ore which he had dug and smelted, reading as much as would be necessary for this,—or the boy who had attended the lectures on metallurgy at the Institute in the mean while, and had received a Rodgers' penknife from his father? Which would be most likely to cut his fingers?—To my astonishment I was informed on leaving college that I had studied navigation!—why, if I had taken one turn down the harbor I should have known more about it. Even the *poor* student studies and is taught only *political* economy, while that economy of living which is synonymous with philosophy is not even sincerely professed in our colleges. The consequence is, that while he is reading Adam Smith, Ricardo, and Say,[9] he runs his father in debt irretrievably.

As with our colleges, so with a hundred "modern improvements"; there is an illusion about them; there is not always a positive advance. The devil goes on exacting compound interest to the last for his early share and numerous succeeding investments in them. Our inventions are wont to be pretty toys, which distract our attention from serious things. They are but improved means to an unimproved end, an end which it was already but too easy to arrive at; as railroads lead to Boston or New York. We are in great haste to construct a magnetic telegraph from Maine to Texas; but Maine and Texas, it may be, have nothing important to communicate. Either is in such a predicament as the man who was earnest to be

---

9. Three economists, the Scottish Adam Smith (1723–90), the English David Ricardo (1772–1823), and the French Jean Baptiste Say (1767–1832).

introduced to a distinguished deaf woman, but when he was presented, and one end of her ear trumpet was put into his hand, had nothing to say. As if the main object were to talk fast and not to talk sensibly. We are eager to tunnel under the Atlantic and bring the old world some weeks nearer to the new; but perchance the first news that will leak through into the broad, flapping American ear will be that the Princess Adelaide has the whooping cough. After all, the man whose horse trots a mile in a minute does not carry the most important messages; he is not an evangelist, nor does he come round eating locusts and wild honey. I doubt if Flying Childers[1] ever carried a peck of corn to mill.

One says to me, "I wonder that you do not lay up money; you love to travel; you might take the cars and go to Fitchburg to-day and see the country." But I am wiser than that. I have learned that the swiftest traveller is he that goes afoot. I say to my friend, Suppose we try who will get there first. The distance is thirty miles; the fare ninety cents. That is almost a day's wages. I remember when wages were sixty cents a day for laborers on this very road. Well, I start now on foot, and get there before night; I have travelled at that rate by the week together. You will in the mean while have earned your fare, and arrive there some time to-morrow, or possibly this evening, if you are lucky enough to get a job in season. Instead of going to Fitchburg, you will be working here the greater part of the day. And so, if the railroad reached round the world, I think that I should keep ahead of you; and as for seeing the country and getting experience of that kind, I should have to cut your acquaintance altogether.

Such is the universal law, which no man can ever outwit, and with regard to the railroad even we may say it is as broad as it is long. To make a railroad round the world available to all mankind is equivalent to grading the whole surface of the planet. Men have an indistinct notion that if they keep up this activity of joint stocks and spades long enough all will at length ride somewhere, in next to no time, and for nothing; but though a crowd rushes to the depot, and the conductor shouts "All aboard!" when the smoke is blown away and the vapor condensed, it will be perceived that a few are riding, but the rest are run over,—and it will be called, and will be, "A melancholy accident." No doubt they can ride at last who shall have earned their fare, that is, if they survive so long, but they will probably have lost their elasticity and desire to travel by that time. This spending of the best part of one's life earning money in order to enjoy a questionable liberty during the least valuable part of it, reminds me of the Englishman who went to India to make a fortune first, in order that he might return to England and live the life of a poet. He should have gone up garret at once. "What!" exclaim a million Irishmen starting up from all the shanties in the land, "is

1. English race horse.

not this railroad which we have built a good thing?" Yes, I answer, *comparatively* good, that is, you might have done worse; but I wish, as you are brothers of mine, that you could have spent your time better than digging in this dirt.

Before I finished my house, wishing to earn ten or twelve dollars by some honest and agreeable method, in order to meet my unusual expenses, I planted about two acres and a half of light and sandy soil near it chiefly with beans, but also a small part with potatoes, corn, peas, and turnips. The whole lot contains eleven acres, mostly growing up to pines and hickories, and was sold the preceding season for eight dollars and eight cents an acre. One farmer said that it was "good for nothing but to raise cheeping squirrels on." I put no manure on this land, not being the owner, but merely a squatter, and not expecting to cultivate so much again, and I did not quite hoe it all once. I got out several cords of stumps in ploughing, which supplied me with fuel for a long time, and left small circles of virgin mould, easily distinguishable through the summer by the greater luxuriance of the beans there. The dead and for the most part unmerchantable wood behind my house, and the driftwood from the pond, have supplied the remainder of my fuel. I was obliged to hire a team and a man for the ploughing, though I held the plough myself. My farm outgoes for the first season were, for implements, seed, work, &c., $14 72½. The seed corn was given me. This never costs any thing to speak of, unless you plant more than enough. I got twelve bushels of beans, and eighteen bushels of potatoes, beside some peas and sweet corn. The yellow corn and turnips were too late to come to any thing. My whole income from the farm was

$$\begin{array}{lr} & \$23\ 44. \\ \text{Deducting the outgoes,} & 14\ 72\tfrac{1}{2} \\ \hline \text{there are left,} & \$\ 8\ 71\tfrac{1}{2}, \end{array}$$

beside produce consumed and on hand at the time this estimate was made of the value of $4 50,—the amount on hand much more than balancing a little grass which I did not raise. All things considered, that is, considering the importance of a man's soul and of to-day, notwithstanding the short time occupied by my experiment, nay, partly even because of its transient character, I believe that that was doing better than any farmer in Concord did that year.

The next year I did better still, for I spaded up all the land which I required, about a third of an acre, and I learned from the experience of both years, not being in the least awed by many celebrated works on husbandry, Arthur Young[2] among the rest, that if one would live simply and eat only the crop which he raised, and raise

2. Author of *Rural Oeconomy, or Essays on the Practical Parts of Husbandry* (1773).

no more than he ate, and not exchange it for an insufficient quantity of more luxurious and expensive things, he would need to cultivate only a few rods of ground, and that it would be cheaper to spade up that than to use oxen to plough it, and to select a fresh spot from time to time than to manure the old, and he could do all his necessary farm work as it were with his left hand at odd hours in the summer; and thus he would not be tied to an ox, or horse, or cow, or pig, as at present. I desire to speak impartially on this point, and as one not interested in the success or failure of the present economical and social arrangements. I was more independent than any farmer in Concord, for I was not anchored to a house or farm, but could follow the bent of my genius, which is a very crooked one, every moment. Beside being better off than they already, if my house had been burned or my crops had failed, I should have been nearly as well off as before.

I am wont to think that men are not so much the keepers of herds as herds are the keepers of men, the former are so much the freer. Men and oxen exchange work; but if we consider necessary work only, the oxen will be seen to have greatly the advantage, their farm is so much the larger. Man does some of his part of the exchange work in his six weeks of haying, and it is no boy's play. Certainly no nation that lived simply in all respects, that is, no nation of philosophers, would commit so great a blunder as to use the labor of animals. True, there never was and is not likely soon to be a nation of philosophers, nor am I certain it is desirable that there should be. However, *I* should never have broken a horse or bull and taken him to board for any work he might do for me, for fear I should become a horse-man or a herds-man merely; and if society seems to be the gainer by so doing, are we certain that what is one man's gain is not another's loss, and that the stable-boy has equal cause with his master to be satisfied? Granted that some public works would not have been constructed without this aid, and let man share the glory of such with the ox and horse; does it follow that he could not have accomplished works yet more worthy of himself in that case? When men begin to do, not merely unnecessary or artistic, but luxurious and idle work, with their assistance, it is inevitable that a few do all the exchange work with the oxen, or, in other words, become the slaves of the strongest. Man thus not only works for the animal within him, but, for a symbol of this, he works for the animal without him. Though we have many substantial houses of brick or stone, the prosperity of the farmer is still measured by the degree to which the barn overshadows the house. This town is said to have the largest houses for oxen cows and horses hereabouts, and it is not behindhand in its public buildings; but there are very few halls for free worship or free speech in this county. It should not be by their architecture, but why not even by their power of abstract thought, that nations should seek to commemorate themselves? How much more admirable the Bhagvat-Geeta[3] than all the ruins of the East! Towers and temples are the luxury of princes. A simple and independent mind does not toil at the bidding of any prince. Genius is not a retainer to any emperor, nor is its material silver, or gold, or marble, except to a trifling extent. To what end, pray, is so much stone hammered? In Arcadia,[4] when I was there, I did not see any hammering stone. Nations are possessed with an insane ambition to perpetuate the memory of themselves by the amount of hammered stone they leave. What if equal pains were taken to smooth and polish their manners? One piece of good sense would be more memorable than a monument as high as the moon. I love better to see stones in place. The grandeur of Thebes[5] was a vulgar grandeur. More sensible is a rod of stone wall that bounds an honest man's field than a hundred-gated Thebes that has wandered farther from the true end of life. The religion and civilization which are barbaric and heathenish build splendid temples; but what you might call Christianity does not. Most of the stone a nation hammers goes toward its tomb only. It buries itself alive. As for the Pyramids, there is nothing to wonder at in them so much as the fact that so many men could be found degraded enough to spend their lives constructing a tomb for some ambitious booby, whom it would have been wiser and manlier to have drowned in the Nile, and then given his body to the dogs. I might possibly invent some excuse for them and him, but I have no time for it. As for the religion and love of art of the builders, it is much the same all the world over, whether the building be an Egyptian temple or the United States Bank. It costs more than it comes to. The mainspring is vanity, assisted by the love of garlic and bread and butter. Mr. Balcom, a promising young architect, designs it on the back of his Vitruvius,[6] with hard pencil and ruler, and the job is let out to Dobson & Sons, stonecutters. When the thirty centuries begin to look down on it, mankind begin to look up at it. As for your high towers and monuments, there was a crazy fellow once in this town who undertook to dig through to China, and he got so far that, as he said, he heard the Chinese pots and kettles rattle; but I think that I shall not go out of my way to admire the hole which he made. Many are concerned about the monuments of the West and the East,—to know who built them. For my part, I should like to know who in those days did not build them,—who were above such trifling. But to proceed with my statistics.

By surveying, carpentry, and day-labor of various other kinds in the village in the mean while, for I have as many trades as fingers, I had earned $13 34. The expense of food for eight months, namely,

3. A sacred Hindu text.
4. Place epitomizing rustic simplicity and contentment, from the region in Greece celebrated by the bucolic poets.

5. Ancient city in Upper Egypt.
6. Vitruvius Pollio, Roman architect during the reigns of Julius Caesar and Augustus, author of *De Architectura*.

from July 4th to March 1st, the time when these estimates were made, though I lived there more than two years,—not counting potatoes, a little green corn, and some peas, which I had raised, nor considering the value of what was on hand at the last date, was

Rice,	01 73½	
Molasses,	1 73	Cheapest form of the saccharine.
Rye meal,	1 04¾	
Indian meal,	0 99¾	Cheaper than rye.
Pork,	0 22	
Flour,	0 88	Costs more than Indian meal, both money and trouble.
Sugar,	0 80	
Lard,	0 65	
Apples,	0 25	
Dried apple,	0 22	All experiments which failed.
Sweet potatoes,	0 10	
One pumpkin,	0 6	
One watermelon,	0 2	
Salt,	0 3	

Yes, I did eat $8 74, all told; but I should not thus unblushingly publish my guilt, if I did not know that most of my readers were equally guilty with myself, and that their deeds would look no better in print. The next year I sometimes caught a mess of fish for my dinner, and once I went so far as to slaughter a woodchuck which ravaged my bean-field,—effect his transmigration, as a Tartar[7] would say,—and devour him, partly for experiment's sake; but though it afforded me a momentary enjoyment, notwithstanding a musky flavor, I saw that the longest use would not make that a good practice, however it might seem to have your woodchucks ready dressed by the village butcher.

Clothing and some incidental expenses within the same dates, though little can be inferred from this item, amounted to

$$\$8 \ 40¾$$

Oil and some household utensils, 2 00

So that all the pecuniary outgoes, excepting for washing and mending, which for the most part were done out of the house, and their bills have not yet been received,—and these are all and more than all the ways by which money necessarily goes out in this part of the world,—were

7. An inhabitant of Tartary, a broad area of Central Asia overrun by the Tatars (Tartars) in the 12th century.

House,	$28	12½
Farm one year,	14	72½
Food eight months,	8	74
Clothing, &c., eight months,	8	40¾
Oil, &c., eight months,	2	00
In all,	$61	99¾

I address myself now to those of my readers who have a living to get. And to meet this I have for farm produce sold

	$23	44
Earned by day-labor,	13	34
In all,	$36	78,

which subtracted from the sum of the outgoes leaves a balance of $25 21¾ on the one side,—this being very nearly the means with which I started, and the measure of expenses to be incurred,—and on the other, beside the leisure and independence and health thus secured, a comfortable house for me as long as I choose to occupy it.

These statistics, however accidental and therefore uninstructive they may appear, as they have a certain completeness, have a certain value also. Nothing was given me of which I have not rendered some account. It appears from the above estimate, that my food alone cost me in money about twenty-seven cents a week. It was, for nearly two years after this, rye and Indian meal without yeast, potatoes, rice, a very little salt pork, molasses, and salt, and my drink water. It was fit that I should live on rice, mainly, who loved so well the philosophy of India. To meet the objections of some inveterate cavillers, I may as well state, that if I dined out occasionally, as I always had done, and I trust shall have opportunities to do again, it was frequently to the detriment of my domestic arrangements. But the dining out, being, as I have stated, a constant element, does not in the least affect a comparative statement like this.

I learned from my two years' experience that it would cost incredibly little trouble to obtain one's necessary food, even in this latitude; that a man may use as simple a diet as the animals, and yet retain health and strength. I have made a satisfactory dinner, satisfactory on several accounts, simply off a dish of purslane (*Portulaca oleracea*) which I gathered in my cornfield, boiled and salted. I give the Latin on account of the savoriness of the trivial name. And pray what more can a reasonable man desire, in peaceful times, in ordinary noons, than a sufficient number of ears of green sweet-corn boiled, with the addition of salt? Even the little variety which I used was a yielding to the demands of appetite, and not of health. Yet men have come to such a pass that they frequently starve, not for want of necessaries, but for want of luxuries; and I know a good woman who thinks that her son lost his life because he took to drinking water only.

The reader will perceive that I am treating the subject rather from an economic than a dietetic point of view, and he will not venture to put my abstemiousness to the test unless he has a well-stocked larder.

Bread I at first made of pure Indian meal and salt, genuine hoe-cakes, which I baked before my fire out of doors on a shingle or the end of a stick of timber sawed off in building my house; but it was wont to get smoked and to have a piny flavor. I tried flour also; but have at last found a mixture of rye and Indian meal most convenient and agreeable. In cold weather it was no little amusement to bake several small loaves of this in succession, tending and turning them as carefully as an Egyptian his hatching eggs.[8] They were a real cereal fruit which I ripened, and they had to my senses a fragrance like that of other noble fruits, which I kept in as long as possible by wrapping them in cloths. I made a study of the ancient and indispensable art of bread-making, consulting such authorities as offered, going back to the primitive days and first invention of the unleavened kind, when from the wildness of nuts and meats men first reached the mildness and refinement of this diet, and travelling gradually down in my studies through that accidental souring of the dough which, it is supposed, taught the leavening process, and through the various fermentations thereafter, till I came to "good, sweet, wholesome bread," the staff of life. Leaven, which some deem the soul of bread, the *spiritus* which fills its cellular tissue, which is religiously preserved like the vestal fire,— some precious bottle-full, I suppose, first brought over in the May-flower, did the business for America, and its influence is still rising, swelling, spreading, in cerealian billows over the land,—this seed I regularly and faithfully procured from the village, till at length one morning I forgot the rules, and scalded my yeast; by which accident I discovered that even this was not indispensable,—for my discoveries were not by the synthetic but analytic process,—and I have gladly omitted it since, though most housewives earnestly assured me that safe and wholesome bread without yeast might not be, and elderly people prophesied a speedy decay of the vital forces. Yet I find it not to be an essential ingredient, and after going without it for a year am still in the land of the living; and I am glad to escape the trivialness of carrying a bottle-full in my pocket, which would sometimes pop and discharge its contents to my discomfiture. It is simpler and more respectable to omit it. Man is an animal who more than any other can adapt himself to all climates and circumstances. Neither did I put any sal soda, or other acid or alkali, into my bread. It would seem that I made it according to the recipe

8. Egyptians had devised incubators.

which Marcus Porcius Cato gave about two centuries before Christ. "Panem depsticium sic facito. Manus mortariumque bene lavato. Farinam in mortarium indito, aquæ paulatim addito, subigitoque pulchre. Ubi bene subegeris, defingito, coquitoque sub testu."[9] Which I take to mean—"Make kneaded bread thus. Wash your hands and trough well. Put the meal into the trough, add water gradually, and knead it thoroughly. When you have kneaded it well, mould it, and bake it under a cover," that is, in a baking-kettle. Not a word about leaven. But I did not always use this staff of life. At one time, owing to the emptiness of my purse, I saw none of it for more than a month.

Every New Englander might easily raise all his own breadstuffs in this land of rye and Indian corn, and not depend on distant and fluctuating markets for them. Yet so far are we from simplicity and independence that, in Concord, fresh and sweet meal is rarely sold in the shops, and hominy and corn in a still coarser form are hardly used by any. For the most part the farmer gives to his cattle and hogs the grain of his own producing, and buys flour, which is at least no more wholesome, at a greater cost, at the store. I saw that I could easily raise my bushel or two of rye and Indian corn, for the former will grow on the poorest land, and the latter does not require the best, and grind them in a hand-mill, and so do without rice and pork; and if I must have some concentrated sweet, I found by experiment that I could make a very good molasses either of pumpkins or beets, and I knew that I needed only to set out a few maples to obtain it more easily still, and while these were growing I could use various substitutes beside those which I have named, "For," as the Forefathers sang,—

> "we can make liquor to sweeten our lips
> Of pumpkins and parsnips and walnut-tree chips."[1]

Finally, as for salt, that grossest of groceries, to obtain this might be a fit occasion for a visit to the seashore, or, if I did without it altogether, I should probably drink the less water. I do not learn that the Indians ever troubled themselves to go after it.

Thus I could avoid all trade and barter, so far as my food was concerned, and having a shelter already, it would only remain to get clothing and fuel. The pantaloons which I now wear were woven in a farmer's family,—thank Heaven there is so much virtue still in man; for I think the fall from the farmer to the operative as great and memorable as that from the man to the farmer;—and in a new country fuel is an encumbrance. As for a habitat, if I were not permitted still to squat, I might purchase one acre at the same price for which the land I cultivated was sold—namely, eight dollars and eight cents. But as it was, I considered that I enhanced the value of

9. *De agri cultura*, 74.
1. From John Warner Barber's *Historical Collections* (1839).

the land by squatting on it.

There is a certain class of unbelievers who sometimes ask me such questions as, if I think that I can live on vegetable food alone; and to strike at the root of the matter at once,—for the root is faith,—I am accustomed to answer such, that I can live on board nails. If they cannot understand that, they cannot understand much that I have to say. For my part, I am glad to hear of experiments of this kind being tried; as that a young man tried for a fortnight to live on hard raw corn on the ear, using his teeth for all mortar. The squirrel tribe tried the same and succeeded. The human race is interested in these experiments, though a few old women who are incapacitated for them, or who own their thirds in mills, may be alarmed.

My furniture, part of which I made myself, and the rest cost me nothing of which I have not rendered an account, consisted of a bed, a table, a desk, three chairs, a looking-glass three inches in diameter, a pair of tongs and andirons, a kettle, a skillet, and a frying-pan, a dipper, a wash-bowl, two knives and forks, three plates, one cup, one spoon, a jug for oil, a jug for molasses, and a japanned[2] lamp. None is so poor that he need sit on a pumpkin. That is shiftlessness. There is a plenty of such chairs as I like best in the village garrets to be had for taking them away. Furniture! Thank God, I can sit and I can stand without the aid of a furniture warehouse. What man but a philosopher would not be ashamed to see his furniture packed in a cart and going up country exposed to the light of heaven and the eyes of men, a beggarly account of empty boxes? That is Spaulding's furniture.[3] I could never tell from inspecting such a load whether it belonged to a so called rich man or a poor one; the owner always seemed poverty-stricken. Indeed, the more you have of such things the poorer you are. Each load looks as if it contained the contents of a dozen shanties; and if one shanty is poor, this is a dozen times as poor. Pray, for what do we *move* ever but to get rid of our furniture, our *exuviæ*;[4] at last to go from this world to another newly furnished, and leave this to be burned? It is the same as if all these traps were buckled to a man's belt, and he could not move over the rough country where our lines are cast without dragging them,—dragging his trap. He was a lucky fox that left his tail in the trap. The muskrat will gnaw his third leg off to be free. No wonder man has lost his elasticity. How often he is at a dead set! "Sir, if I may be so bold, what do you mean by a dead set?" If you are a seer, whenever you meet a man you will see all that he owns, ay, and much that he pretends to disown, behind him, even to his kitchen furniture and all the trumpery which he saves and will not burn, and he will appear to be harnessed to it and

2. Lacquered with decorative scenes in the Japanese manner.

3. Unidentified.

4. Discarded objects.

making what headway he can. I think that the man is at a dead set who has got through a knot hole or gateway where his sledge load of furniture cannot follow him. I cannot but feel compassion when I hear some trig, compact-looking man, seemingly free, all girded and ready, speak of his "furniture," as whether it is insured or not. "But what shall I do with my furniture?" My gay butterfly is entangled in a spider's web then. Even those who seem for a long while not to have any, if you inquire more narrowly you will find have some stored in somebody's barn. I look upon England to-day as an old gentleman who is travelling with a great deal of baggage, trumpery which has accumulated from long housekeeping, which he has not the courage to burn; great trunk, little trunk, bandbox and bundle. Throw away the first three at least. It would surpass the powers of a well man nowadays to take up his bed and walk, and I should certainly advise a sick one to lay down his bed and run. When I have met an immigrant tottering under a bundle which contained his all,—looking like an enormous wen which had grown out of the nape of his neck,—I have pitied him, not because that was his all, but because he had all *that* to carry. If I have got to drag my trap, I will take care that it be a light one and do not nip me in a vital part. But perchance it would be wisest never to put one's paw into it.

I would observe, by the way, that it costs me nothing for curtains, for I have no gazers to shut out but the sun and moon, and I am willing that they should look in. The moon will not sour milk nor taint meat of mine, nor will the sun injure my furniture or fade my carpet, and if he is sometimes too warm a friend, I find it still better economy to retreat behind some curtain which nature has provided, than to add a single item to the details of housekeeping. A lady once offered me a mat, but as I had no room to spare within the house, nor time to spare within or without to shake it, I declined it, preferring to wipe my feet on the sod before my door. It is best to avoid the beginnings of evil.

Not long since I was present at the auction of a deacon's effects, for his life had not been ineffectual:—

"The evil that men do lives after them."[5]

As usual, a great proportion was trumpery which had begun to accumulate in his father's day. Among the rest was a dried tapeworm. And now, after lying half a century in his garret and other dust holes, these things were not burned; instead of a *bonfire*, or purifying destruction of them, there was an *auction*, or increasing of them.[6] The neighbors eagerly collected to view them, bought them all, and carefully transported them to their garrets and dust holes, to lie there till their estates are settled, when they will start

5. Tag from Antony's speech to the citizens, in Shakespeare's *Julius Caesar*, 3.3.

6. Thoreau puns on the Latin root of *auction*, which means "to increase."

again. When a man dies he kicks the dust.

The customs of some savage nations might, perchance, be profitably imitated by us, for they at least go through the semblance of casting their slough annually; they have the idea of the thing, whether they have the reality or not. Would it not be well if we were to celebrate such a "busk," or "feast of first fruits," as Bartram describes to have been the custom of the Mucclasse Indians?[7] "When a town celebrates the busk," says he, "having previously provided themselves with new clothes, new pots, pans, and other household utensils and furniture, they collect all their worn out clothes and other despicable things, sweep and cleanse their houses, squares, and the whole town, of their filth, which with all the remaining grain and other old provisions they cast together into one common heap, and consume it with fire. After having taken medicine, and fasted for three days, all the fire in the town is extinguished. During this fast they abstain from the gratification of every appetite and passion whatever. A general amnesty is proclaimed; all malefactors may return to their town.——"

"On the fourth morning, the high priest, by rubbing dry wood together, produces new fire in the public square, from whence every habitation in the town is supplied with the new and pure flame."

They then feast on the new corn and fruits and dance and sing for three days, "and the four following days they receive visits and rejoice with their friends from neighboring towns who have in like manner purified and prepared themselves."

The Mexicans also practised a similar purification at the end of every fifty-two years, in the belief that it was time for the world to come to an end.

I have scarcely heard of a truer sacrament, that is, as the dictionary defines it, "outward and visible sign of an inward and spiritual grace," than this, and I have no doubt that they were originally inspired directly from Heaven to do thus, though they have no biblical record of the revelation.

For more than five years I maintained myself thus solely by the labor of my hands, and I found, that by working about six weeks in a year, I could meet all the expenses of living. The whole of my winters, as well as most of my summers, I had free and clear for study. I have thoroughly tried school-keeping, and found that my expenses were in proportion, or rather out of proportion, to my income, for I was obliged to dress and train, not to say think and believe, accordingly, and I lost my time into the bargain. As I did not teach for the good of my fellow-men, but simply for a livelihood, this was a failure. I have tried trade; but I found that it would take ten years to get under way in that, and that then I should probably be on my way to the devil. I was actually afraid that I

7. William Bartram, *Travels through North and South Carolina* (1791).

might by that time be doing what is called a good business. When formerly I was looking about to see what I could do for a living, some sad experience in conforming to the wishes of friends being fresh in my mind to tax my ingenuity, I thought often and seriously of picking huckleberries; that surely I could do, and its small profits might suffice,—for my greatest skill has been to want but little,—so little capital it required, so little distraction from my wonted moods, I foolishly thought. While my acquaintances went unhesitatingly into trade or the professions, I contemplated this occupation as most like theirs; ranging the hills all summer to pick the berries which came in my way, and thereafter carelessly dispose of them; so, to keep the flocks of Admetus.[8] I also dreamed that I might gather the wild herbs, or carry evergreens to such villagers as loved to be reminded of the woods, even to the city, by hay-cart loads. But I have since learned that trade curses every thing it handles; and though you trade in messages from heaven, the whole curse of trade attaches to the business.

As I preferred some things to others, and especially valued my freedom, as I could fare hard and yet succeed well, I did not wish to spend my time in earning rich carpets or other fine furniture, or delicate cookery, or a house in the Grecian or the Gothic style just yet. If there are any to whom it is no interruption to acquire these things, and who know how to use them when acquired, I relinquish to them the pursuit. Some are "industrious," and appear to love labor for its own sake, or perhaps because it keeps them out of worse mischief; to such I have at present nothing to say. Those who would not know what to do with more leisure than they now enjoy, I might advise to work twice as hard as they do,—work till they pay for themselves, and get their free papers. For myself I found that the occupation of a day-laborer was the most independent of any, especially as it required only thirty or forty days in a year to support one. The laborer's day ends with the going down of the sun, and he is then free to devote himself to his chosen pursuit, independent of his labor; but his employer, who speculates from month to month, has no respite from one end of the year to the other.

In short, I am convinced, both by faith and experience, that to maintain one's self on this earth is not a hardship but a pastime, if we will live simply and wisely; as the pursuits of the simpler nations are still the sports of the more artificial. It is not necessary that a man should earn his living by the sweat of his brow, unless he sweats easier than I do.

One young man of my acquaintance, who has inherited some acres, told me that he thought he should live as I did, *if he had the means.* I would not have any one adopt *my* mode of living on any account; for, beside that before he has fairly learned it I may have

---

8. Apollo, Greek god of poetry, tended the flocks of Admetus while banished from Olympus.

found out another for myself, I desire that there may be as many different persons in the world as possible; but I would have each one be very careful to find out and pursue *his own* way, and not his father's or his mother's or his neighbor's instead. The youth may build or plant or sail, only let him not be hindered from doing that which he tells me he would like to do. It is by a mathematical point only that we are wise, as the sailor or the fugitive slave keeps the polestar in his eye; but that is sufficient guidance for all our life. We may not arrive at our port within a calculable period, but we would preserve the true course.

Undoubtedly, in this case, what is true for one is truer still for a thousand, as a large house is not more expensive than a small one in proportion to its size, since one roof may cover, one cellar underlie, and one wall separate several apartments. But for my part, I preferred the solitary dwelling. Moreover, it will commonly be cheaper to build the whole yourself than to convince another of the advantage of the common wall; and when you have done this, the common partition, to be much cheaper, must be a thin one, and that other may prove a bad neighbor, and also not keep his side in repair. The only coöperation which is commonly possible is exceedingly partial and superficial; and what little true coöperation there is, is as if it were not, being a harmony inaudible to men. If a man has faith he will coöperate with equal faith every where; if he has not faith, he will continue to live like the rest of the world, whatever company he is joined to. To coöperate, in the highest as well as the lowest sense, means *to get our living together*. I heard it proposed lately that two young men should travel together over the world, the one without money, earning his means as he went, before the mast and behind the plough, the other carrying a bill of exchange in his pocket. It was easy to see that they could not long be companions or coöperate, since one would not *operate* at all. They would part at the first interesting crisis in their adventures. Above all, as I have implied, the man who goes alone can start today; but he who travels with another must wait till that other is ready, and it may be a long time before they get off.

But all this is very selfish, I have heard some of my townsmen say. I confess that I have hitherto indulged very little in philanthropic enterprises. I have made some sacrifices to a sense of duty, and among others have sacrificed this pleasure also. There are those who have used all their arts to persuade me to undertake the support of some poor family in the town; and if I had nothing to do,—for the devil finds employment for the idle,—I might try my hand at some such pastime as that. However, when I have thought to indulge myself in this respect, and lay their Heaven under an obligation by maintaining certain poor persons in all respects as comfortably as I maintain myself, and have even ventured so far as

to make them the offer, they have one and all unhesitatingly preferred to remain poor. While my townsmen and women are devoted
in so many ways to the good of their fellows, I trust that one at
least may be spared to other and less humane pursuits. You must
have a genius for charity as well as for any thing else. As for Doing-
good, that is one of the professions which are full. Moreover, I have
tried it fairly, and, strange as it may seem, am satisfied that it does
not agree with my constitution. Probably I should not consciously
and deliberately forsake my particular calling to do the good which
society demands of me, to save the universe from annihilation; and
I believe that a like but infinitely greater steadfastness elsewhere is
all that now preserves it. But I would not stand between any man
and his genius; and to him who does this work, which I decline,
with his whole heart and soul and life, I would say, Persevere, even
if the world call it doing evil, as it is most likely they will.

I am far from supposing that my case is a peculiar one; no doubt
many of my readers would make a similar defence. At doing
something,—I will not engage that my neighbors shall pronounce it
good,—I do not hesitate to say that I should be a capital fellow to
hire; but what that is, it is for my employer to find out. What *good* I
do, in the common sense of that word, must be aside from my main
path, and for the most part wholly unintended. Men say, practically, Begin where you are and such as you are, without aiming
mainly to become of more worth, and with kindness aforethought
go about doing good. If I were to preach at all in this strain, I
should say rather, Set about being good. As if the sun should stop
when he had kindled his fires up to the splendor of a moon or a star
of the sixth magnitude, and go about like a Robin Goodfellow,[9]
peeping in at every cottage window, inspiring lunatics, and tainting
meats, and making darkness visible, instead of steadily increasing
his genial heat and beneficence till he is of such brightness that no
mortal can look him in the face, and then, and in the mean while
too, going about the world in his own orbit, doing it good, or rather,
as a truer philosophy has discovered, the world going about him
getting good. When Phaeton,[1] wishing to prove his heavenly birth
by his beneficence, had the sun's chariot but one day, and drove out
of the beaten track, he burned several blocks of houses in the lower
streets of heaven, and scorched the surface of the earth, and dried
up every spring, and made the great desert of Sahara, till at length
Jupiter hurled him headlong to the earth with a thunderbolt, and
the sun, through grief at his death, did not shine for a year.

There is no odor so bad as that which arises from goodness
tainted. It is human, it is divine, carrion. If I knew for a certainty
that a man was coming to my house with the conscious design of

9. Mischievous fairy, known as Puck in
Shakespeare's *A Midsummer Night's
Dream.*
1. In Greek mythology, the son of

Helios. He attempted to drive his father's
chariot, the sun, with disastrous consequences.

doing me good, I should run for my life, as from that dry and
parching wind of the African deserts called the simoom, which fills
the mouth and nose and ears and eyes with dust till you are suffo-
cated, for fear that I should get some of his good done to me,—
some of its virus mingled with my blood. No,—in this case I would
rather suffer evil the natural way. A man is not a good *man* to me
because he will feed me if I should be starving, or warm me if I
should be freezing, or pull me out of a ditch if I should ever fall
into one. I can find you a Newfoundland dog that will do as much.
Philanthropy is not love for one's fellow-man in the broadest sense.
Howard[2] was no doubt an exceedingly kind and worthy man in his
way, and has his reward; but, comparatively speaking, what are a
hundred Howards to *us*, if their philanthropy do not help *us* in our
best estate, when we are most worthy to be helped? I never heard of
a philanthropic meeting in which it was sincerely proposed to do
any good to me, or the like of me.

The Jesuits were quite balked by those Indians who, being burned
at the stake, suggested new modes of torture to their tormentors.[3]
Being superior to physical suffering, it sometimes chanced that they
were superior to any consolation which the missionaries could offer;
and the law to do as you would be done by fell with less persuasiveness on the ears of those, who, for their part, did not care how they
were done by, who loved their enemies after a new fashion, and
came very near freely forgiving them all they did.

Be sure that you give the poor the aid they most need, though it
be your example which leaves them far behind. If you give money,
spend yourself with it, and do not merely abandon it to them. We
make curious mistakes sometimes. Often the poor man is not so
cold and hungry as he is dirty and ragged and gross. It is partly his
taste, and not merely his misfortune. If you give him money, he will
perhaps buy more rags with it. I was wont to pity the clumsy Irish
laborers who cut ice on the pond, in such mean and ragged clothes,
while I shivered in my more tidy and somewhat more fashionable
garments, till, one bitter cold day, one who had slipped into the
water came to my house to warm him, and I saw him strip off three
pairs of pants and two pairs of stockings ere he got down to the
skin, though they were dirty and ragged enough, it is true, and that
he could afford to refuse the *extra* garments which I offered him, he
had so many *intra* ones. This ducking was the very thing he needed.
Then I began to pity myself, and I saw that it would be a greater
charity to bestow on me a flannel shirt than a whole slop-shop on
him. There are a thousand hacking at the branches of evil to one
who is striking at the root, and it may be that he who bestows the
largest amount of time and money on the needy is doing the most

2. John Howard (1726?–90), English
prison reformer.
3. Thoreau's source is unknown, but
Harding cites comparable accounts in
*The Jesuit Relations and Allied Docu-
ments* (1898), Vol. 17, 109.

by his mode of life to produce that misery which he strives in vain to relieve. It is the pious slave-breeder devoting the proceeds of every tenth slave to buy a Sunday's liberty for the rest. Some show their kindness to the poor by employing them in their kitchens. Would they not be kinder if they employed themselves there? You boast of spending a tenth part of your income in charity; may be you should spend the nine tenths so, and done with it. Society recovers only a tenth part of the property then. Is this owing to the generosity of him in whose possession it is found, or to the remissness of the officers of justice?

Philanthropy is almost the only virtue which is sufficiently appreciated by mankind. Nay, it is greatly overrated; and it is our selfishness which overrates it. A robust poor man, one sunny day here in Concord, praised a fellow-townsman to me, because, as he said, he was kind to the poor; meaning himself. The kind uncles and aunts of the race are more esteemed than its true spiritual fathers and mothers. I once heard a reverend lecturer on England, a man of learning and intelligence, after enumerating her scientific, literary, and political worthies, Shakspeare, Bacon, Cromwell, Milton, Newton, and others, speak next of her Christian heroes, whom, as if his profession required it of him, he elevated to a place far above all the rest, as the greatest of the great. They were Penn, Howard, and Mrs. Fry.[4] Every one must feel the falsehood and cant of this. The last were not England's best men and women; only, perhaps, her best philanthropists.

I would not subtract any thing from the praise that is due to philanthropy, but merely demand 'ustice for all who by their lives and works are a blessing to mankind. I do not value chiefly a man's uprightness and benevolence, which are, as it were, his stem and leaves. Those plants of whose greenness withered we make herb tea for the sick, serve but a humble use, and are most employed by quacks. I want the flower and fruit of a man; that some fragrance be wafted over from him to me, and some ripeness flavor our intercourse. His goodness must not be a partial and transitory act, but a constant superfluity, which costs him nothing and of which he is unconscious. This is a charity that hides a multitude of sins. The philanthropist too often surrounds mankind with the remembrance of his own cast-off griefs as an atmosphere, and calls it sympathy. We should impart our courage, and not our despair, our health and ease, and not our disease, and take care that this does not spread by contagion. From what southern plains comes up the voice of wailing? Under what latitudes reside the heathen to whom we would send light? Who is that intemperate and brutal man whom we would redeem? If any thing ail a man, so that he does not perform

---

4. William Penn (1644–1718), Quaker leader and proprietor of Pennsylvania; John Howard (see footnote 2, above); and Elizabeth Fry (1780–1845), English Quaker and prison reformer.

his functions, if he have a pain in his bowels even,—for that is the seat of sympathy,—he forthwith sets about reforming—the world. Being a microcosm himself, he discovers, and it is a true discovery, and he is the man to make it,—that the world has been eating green apples; to his eyes, in fact, the globe itself is a great green apple, which there is danger awful to think of that the children of men will nibble before it is ripe; and straightway his drastic philanthropy seeks out the Esquimaux and the Patagonian, and embraces the populous Indian and Chinese villages; and thus, by a few years of philanthropic activity, the powers in the mean while using him for their own ends, no doubt, he cures himself of his dyspepsia, the globe acquires a faint blush on one or both of its cheeks, as if it were beginning to be ripe, and life loses its crudity and is once more sweet and wholesome to live. I never dreamed of any enormity greater than I have committed. I never knew, and never shall know, a worse man than myself.

I believe that what so saddens the reformer is not his sympathy with his fellows in distress, but, though he be the holiest son of God, is his private ail. Let this be righted, let the spring come to him, the morning rise over his couch, and he will forsake his generous companions without apology. My excuse for not lecturing against the use of tobacco is, that I never chewed it; that is a penalty which reformed tobacco-chewers have to pay; though there are things enough I have chewed, which I could lecture against. If you should ever be betrayed into any of these philanthropies, do not let your left hand know what your right hand does, for it is not worth knowing. Rescue the drowning and tie your shoe-strings. Take your time, and set about some free labor.

Our manners have been corrupted by communication with the saints. Our hymn-books resound with a melodious cursing of God and enduring him forever. One would say that even the prophets and redeemers had rather consoled the fears than confirmed the hopes of man. There is nowhere recorded a simple and irrepressible satisfaction with the gift of life, any memorable praise of God. All health and success does me good, however far off and withdrawn it may appear; all disease and failure helps to make me sad and does me evil, however much sympathy it may have with me or I with it. If, then, we would indeed restore mankind by truly Indian, botanic, magnetic, or natural means, let us first be as simple and well as Nature ourselves, dispel the clouds which hang over our own brows, and take up a little life into our pores. Do not stay to be an overseer of the poor, but endeavor to become one of the worthies of the world.

I read in the Gulistan, or Flower Garden, of Sheik Sadi of Shiraz, that "They asked a wise man, saying; Of the many celebrated trees which the Most High God has created lofty and umbrageous, they call none azad, or free, excepting the cypress, which bears no fruit;

what mystery is there in this? He replied; Each has its appropriate produce, and appointed season, during the continuance of which it is fresh and blooming, and during their absence dry and withered; to neither of which states is the cypress exposed, being always flourishing; and of this nature are the azads, or religious independents.— Fix not thy heart on that which is transitory; for the Dijlah, or Tigris, will continue to flow through Bagdad after the race of caliphs is extinct: if thy hand has plenty, be liberal as the date tree; but if it affords nothing to give away, be an azad, or free man, like the cypress."[5]

Complemental Verses[1]

THE PRETENSIONS OF POVERTY

"Thou dost presume too much, poor needy wretch,
To claim a station in the firmament,
Because thy humble cottage, or thy tub,
Nurses some lazy or pedantic virtue
In the cheap sunshine or by shady springs,
With roots and pot-herbs; where thy right hand,
Tearing those humane passions from the mind,
Upon whose stocks fair blooming virtues flourish,
Degradeth nature, and benumbeth sense,
And, Gorgon-like,[2] turns active men to stone.
We not require the dull society
Of your necessitated temperance,
Or that unnatural stupidity
That knows nor joy nor sorrow; nor your forc'd
Falsely exalted passive fortitude
Above the active. This low abject brood,
That fix their seats in mediocrity,
Become your servile minds; but we advance
Such virtues only as admit excess,
Brave, bounteous acts, regal magnificence,
All-seeing prudence, magnanimity
That knows no bound, and that heroic virtue
For which antiquity hath left no name,
But patterns only, such as Hercules,
Achilles, Theseus. Back to thy loath'd cell;
And when thou seest the new enlightened sphere,
Study to know but what those worthies were."

—T. CAREW

Where I Lived, and What I Lived For

At a certain season of our life we are accustomed to consider every spot as the possible site of a house. I have thus surveyed the country on every side within a dozen miles of where I live. In

---

5. **Muslih-ud-Din** (Saadi) (1184?–1291), *The Gulistan or Rose Garden.*
1. From *Coelum Britanicum* by the English Cavalier poet Thomas Carew (1595?–1645?), offered ironically as a

retort to "Economy."
2. In Greek mythology the Gorgons were three sisters who, with snakes for hair and eyes, turned any beholder into stone.

imagination I have bought all the farms in succession, for all were to be bought and I knew their price. I walked over each farmer's premises, tasted his wild apples, discoursed on husbandry with him, took his farm at his price, at any price, mortgaging it to him in my mind; even put a higher price on it,—took every thing but a deed of it,—took his word for his deed, for I dearly love to talk,—cultivated it, and him too to some extent, I trust, and withdrew when I had enjoyed it long enough, leaving him to carry it on. This experience entitled me to be regarded as a sort of real-estate broker by my friends. Wherever I sat, there I might live, and the landscape radiated from me accordingly. What is a house but a *sedes*, a seat?—better if a country seat. I discovered many a site for a house not likely to be soon improved, which some might have thought too far from the village, but to my eyes the village was too far from it. Well, there I might live, I said; and there I did live, for an hour, a summer and a winter life; saw how I could let the years run off, buffet the winter through, and see the spring come in. The future inhabitants of this region, wherever they may place their houses, may be sure that they have been anticipated. An afternoon sufficed to lay out the land into orchard woodlot and pasture, and to decide what fine oaks or pines should be left to stand before the door, and whence each blasted tree could be seen to the best advantage; and then I let it lie, fallow perchance, for a man is rich in proportion to the number of things which he can afford to let alone.

My imagination carried me so far that I even had the refusal of several farms,—the refusal was all I wanted,—but I never got my fingers burned by actual possession. The nearest that I came to actual possession was when I bought the Hollowell Place, and had begun to sort my seeds, and collected materials with which to make a wheelbarrow to carry it on or off with; but before the owner gave me a deed of it, his wife—every man has such a wife—changed her mind and wished to keep it, and he offered me ten dollars to release him. Now, to speak the truth, I had but ten cents in the world, and it surpassed my arithmetic to tell, if I was that man who had ten cents, or who had a farm, or ten dollars, or all together. However, I let him keep the ten dollars and the farm too, for I had carried it far enough; or rather, to be generous, I sold him the farm for just what I gave for it, and, as he was not a rich man, made him a present of ten dollars, and still had my ten cents, and seeds, and materials for a wheelbarrow left. I found thus that I had been a rich man without any damage to my poverty. But I retained the landscape, and I have since annually carried off what it yielded without a wheelbarrow. With respect to landscapes,—

> "I am monarch of all I *survey*,
> My right there is none to dispute."[1]

1. William Cowper's *Verses Supposed to Be Written by Alexander Selkirk*, with the pun italicized. Selkirk was Daniel Defoe's model for Robinson Crusoe.

I have frequently seen a poet withdraw, having enjoyed the most valuable part of a farm, while the crusty farmer supposed that he had got a few wild apples only. Why, the owner does not know it for many years when a poet has put his farm in rhyme, the most admirable kind of invisible fence, has fairly impounded it, milked it, skimmed it, and got all the cream, and left the farmer only the skimmed milk.

The real attractions of the Hollowell farm, to me, were; its complete retirement, being about two miles from the village, half a mile from the nearest neighbor, and separated from the highway by a broad field; its bounding on the river, which the owner said protected it by its fogs from frosts in the spring, though that was nothing to me; the gray color and ruinous state of the house and barn, and the dilapidated fences, which put such an interval between me and the last occupant; the hollow and lichen-covered apple trees, gnawed by rabbits, showing what kind of neighbors I should have; but above all, the recollection I had of it from my earliest voyages up the river, when the house was concealed behind a dense grove of red maples, through which I heard the house-dog bark. I was in haste to buy it, before the proprietor finished getting out some rocks, cutting down the hollow apple trees, and grubbing up some young birches which had sprung up in the pasture, or, in short, had made any more of his improvements. To enjoy these advantages I was ready to carry it on; like Atlas,[2] to take the world on my shoulders,—I never heard what compensation he received for that,—and do all those things which had no other motive or excuse but that I might pay for it and be unmolested in my possession of it; for I knew all the while that it would yield the most abundant crop of the kind I wanted if I could only afford to let it alone. But it turned out as I have said.

All that I could say, then, with respect to farming on a large scale, (I have always cultivated a garden,) was, that I had had my seeds ready. Many think that seeds improve with age. I have no doubt that time discriminates between the good and the bad; and when at last I shall plant, I shall be less likely to be disappointed. But I would say to my fellows, once for all, As long as possible live free and uncommitted. It makes but little difference whether you are committed to a farm or the county jail.

Old Cato, whose "De Re Rusticâ" is my "Cultivator," says, and the only translation I have seen makes sheer nonsense of the passage, "When you think of getting a farm, turn it thus in your mind, not to buy greedily; nor spare your pains to look at it, and do not think it enough to go round it once. The oftener you go there the *more it will please you, if it is good.*"[3] I think I shall not buy

---

2. A Titan whom Zeus forced to stand upon the earth supporting the heavens on his head and in his hands as punishment for warring against the Olympian gods.
3. *De agri cultura*, 1.1.

greedily, but go round and round it as long as I live, and be buried in it first, that it may please me the more at last.

The present was my next experiment of this kind, which I purpose to describe more at length; for convenience, putting the experience of two years into one. As I have said, I do not propose to write an ode to dejection, but to brag as lustily as chanticleer in the morning, standing on his roost, if only to wake my neighbors up.

When first I took up my abode in the woods, that is, began to spend my nights as well as days there, which, by accident, was on Independence Day, or the fourth of July, 1845, my house was not finished for winter, but was merely a defence against the rain, without plastering or chimney, the walls being of rough weather-stained boards, with wide chinks, which made it cool at night. The upright white hewn studs and freshly planed door and window casings gave it a clean and airy look, especially in the morning, when its timbers were saturated with dew, so that I fancied that by noon some sweet gum would exude from them. To my imagination it retained throughout the day more or less of this auroral character, reminding me of a certain house on a mountain which I had visited the year before. This was an airy and unplastered cabin, fit to entertain a traveling god, and where a goddess might trail her garments. The winds which passed over my dwelling were such as sweep over the ridges of mountains, bearing the broken strains, or celestial parts only, of terrestrial music. The morning wind forever blows, the poem of creation is uninterrupted; but few are the ears that hear it. Olympus is but the outside of the earth every where.

The only house I had been the owner of before, if I except a boat, was a tent, which I used occasionally when making excursions in the summer, and this is still rolled up in my garret; but the boat, after passing from hand to hand, has gone down the stream of time. With this more substantial shelter about me, I had made some progress toward settling in the world. This frame, so slightly clad, was a sort of crystallization around me, and reacted on the builder. It was suggestive somewhat as a picture in outlines. I did not need to go out doors to take the air, for the atmosphere within had lost none of its freshness. It was not so much within doors as behind a door where I sat, even in the rainiest weather. The Harivansa[4] says, "An abode without birds is like a meat without seasoning." Such was not my abode, for I found myself suddenly neighbor to the birds; not by having imprisoned one, but having caged myself near them. I was not only nearer to some of those which commonly frequent the garden and the orchard, but to those wilder and more thrilling songsters of the forest which never, or rarely, serenade a villager,—the wood-thrush, the veery, the scarlet tanager, the field-sparrow, the whippoorwill, and many others.

4. A Hindu epic poem.

664 ★ *Henry David Thoreau*

I was seated by the shore of a small pond, about a mile and a half south of the village of Concord and somewhat higher than it, in the midst of an extensive wood between that town and Lincoln, and about two miles south of that our only field known to fame, Concord Battle Ground;[5] but I was so low in the woods that the opposite shore, half a mile off, like the rest, covered with wood, was my most distant horizon. For the first week, whenever I looked out on the pond it impressed me like a tarn[6] high up on the side of a mountain, its bottom far above the surface of other lakes, and, as the sun arose, I saw it throwing off its nightly clothing of mist, and here and there, by degrees, its soft ripples or its smooth reflecting surface was revealed, while the mists, like ghosts, were stealthily withdrawing in every direction into the woods, as at the breaking up of some nocturnal conventicle. The very dew seemed to hang upon the trees later into the day than usual, as on the sides of mountains.

This small lake was of most value as a neighbor in the intervals of a gentle rain storm in August, when, both air and water being perfectly still, but the sky overcast, mid-afternoon had all the serenity of evening, and the wood-thrush sang around, and was heard from shore to shore. A lake like this is never smoother than at such a time; and the clear portion of the air above it being shallow and darkened by clouds, the water, full of light and reflections, becomes a lower heaven itself so much the more important. From a hill top near by, where the wood had been recently cut off, there was a pleasing vista southward across the pond, through a wide indentation in the hills which form the shore there, where their opposite sides sloping toward each other suggested a stream flowing out in that direction through a wooded valley, but stream there was none. That way I looked between and over the near green hills to some distant and higher ones in the horizon, tinged with blue. Indeed, by standing on tiptoe I could catch a glimpse of some of the peaks of the still bluer and more distant mountain ranges in the north-west, those true-blue coins from heaven's own mint, and also of some portion of the village. But in other directions, even from this point, I could not see over or beyond the woods which surrounded me. It is well to have some water in your neighborhood, to give buoyancy to and float the earth. One value even of the smallest well is, that when you look into it you see that earth is not continent but insular. This is as important as that it keeps butter cool. When I looked across the pond from this peak toward the Sudbury meadows, which in time of flood I distinguished elevated perhaps by a mirage in their seething valley, like a coin in a basin, all the earth beyond the pond appeared like a thin crust insulated and floated even by this small sheet of intervening water, and I was reminded that this on which I dwelt was but *dry land*.

5. The site of battle on the first day of the American Revolution, April 19, 1775.
6. Lake.

Though the view from my door was still more contracted, I did not feel crowded or confined in the least. There was pasture enough for my imagination. The low shrub-oak plateau to which the opposite shore arose, stretched away toward the prairies of the West and the steppes of Tartary, affording ample room for all the roving families of men. "There are none happy in the world but beings who enjoy freely a vast horizon,"—said Damodara,[7] when his herds required new and larger pastures.

Both place and time were changed, and I dwelt nearer to those parts of the universe and to those eras in history which had most attracted me. Where I lived was as far off as many a region viewed nightly by astronomers. We are wont to imagine rare and delectable places in some remote and more celestial corner of the system, behind the constellation of Cassiopeia's Chair, far from noise and disturbance. I discovered that my house actually had its site in such a withdrawn, but forever new and unprofaned, part of the universe. If it were worth the while to settle in those parts near to the Pleiades or the Hyades, to Aldebaran or Altair,[8] then I was really there, or at an equal remoteness from the life which I had left behind, dwindled and twinkling with as fine a ray to my nearest neighbor, and to be seen only in moonless nights by him. Such was that part of creation where I had squatted;—

> "There was a shepherd that did live,
>   And held his thoughts as high
> As were the mounts whereon his flocks
>   Did hourly feed him by."[9]

What should we think of the shepherd's life if his flocks always wandered to higher pastures than his thoughts?

Every morning was a cheerful invitation to make my life of equal simplicity, and I may say innocence, with Nature herself. I have been as sincere a worshipper of Aurora as the Greeks. I got up early and bathed in the pond; that was a religious exercise, and one of the best things which I did. They say that characters were engraven on the bathing tub of king Tching-thang to this effect: "Renew thyself completely each day; do it again, and again, and forever again."[1] I can understand that. Morning brings back the heroic ages. I was as much affected by the faint hum of a mosquito making its invisible and unimaginable tour through my apartment at earliest dawn, when I was sitting with door and windows open, as I could be by

7. Another name for Krishna, the eighth avatar of Vishnu in Hindu mythology; Thoreau translates from a French edition of *Harivansa*.
8. The Pleiades and the Hyades are constellations; Aldebaran, in the constellation Taurus, is one of the brightest stars; Altair is in the constellation Aquila.
9. Anonymous Jacobean verse set to music in *The Muses Garden* (1611) and probably found by Thoreau in Thomas Evans's *Old Ballads* (1810).
1. Confucius, *The Great Learning*, Chapter 1.

any trumpet that ever sang of fame. It was Homer's requiem; itself an Iliad and Odyssey in the air, singing its own wrath and wanderings. There was something cosmical about it; a standing advertisement, till forbidden,[2] of the everlasting vigor and fertility of the world. The morning, which is the most memorable season of the day, is the awakening hour. Then there is least somnolence in us; and for an hour, at least, some part of us awakes which slumbers all the rest of the day and night. Little is to be expected of that day, if it can be called a day, to which we are not awakened by our Genius, but by the mechanical nudgings of some servitor, are not awakened by our own newly-acquired force and aspirations from within, accompanied by the undulations of celestial music, instead of factory bells, and a fragrance filling the air—to a higher life than we fell asleep from; and thus the darkness bear its fruit, and prove itself to be good, no less than the light. That man who does not believe that each day contains an earlier, more sacred, and auroral hour than he has yet profaned, has despaired of life, and is pursuing a descending and darkening way. After a partial cessation of his sensuous life, the soul of man, or its organs rather, are reinvigorated each day, and his Genius tries again what noble life it can make. All memorable events, I should say, transpire in morning time and in a morning atmosphere. The Vedas[3] say, "All intelligences awake with the morning." Poetry and art, and the fairest and most memorable of the actions of men, date from such an hour. All poets and heroes, like Memnon,[4] are the children of Aurora, and emit their music at sunrise. To him whose elastic and vigorous thought keeps pace with the sun, the day is a perpetual morning. It matters not what the clocks say or the attitudes and labors of men. Morning is when I am awake and there is a dawn in me. Moral reform is the effort to throw off sleep. Why is it that men give so poor an account of their day if they have not been slumbering? They are not such poor calculators. If they had not been overcome with drowsiness they would have performed something. The millions are awake enough for physical labor; but only one in a million is awake enough for effective intellectual exertion, only one in a hundred millions to a poetic or divine life. To be awake is to be alive. I have never yet met a man who was quite awake. How could I have looked him in the face?

We must learn to reawaken and keep ourselves awake, not by mechanical aids, but by an infinite expectation of the dawn, which does not forsake us in our soundest sleep. I know of no more encouraging fact than the unquestionable ability of man to elevate

---

2. In newspaper advertisements "TF" signaled to the compositor that an item was to be repeated daily "till forbidden."
3. The Vedas are Hindu scriptures; the quotation has not been located.
4. In Greek mythology, Aurora was the goddess of the dawn, mother of the heroic prince Memnon, slain by Achilles in the Trojan War. As in "Economy," the reference is to the sounds emitted by the ancient temple of Memnon at Thebes.

his life by a conscious endeavor. It is something to be able to paint a particular picture, or to carve a statue, and so to make a few objects beautiful; but it is far more glorious to carve and paint the very atmosphere and medium through which we look, which morally we can do. To affect the quality of the day, that is the highest of arts. Every man is tasked to make his life, even in its details, worthy of the contemplation of his most elevated and critical hour. If we refused, or rather used up, such paltry information as we get, the oracles would distinctly inform us how this might be done.

I went to the woods because I wished to live deliberately, to front only the essential facts of life, and see if I could not learn what it had to teach, and not, when I came to die, discover that I had not lived. I did not wish to live what was not life, living is so dear; nor did I wish to practise resignation, unless it was quite necessary. I wanted to live deep and suck out all the marrow of life, to live so sturdily and Spartan-like as to put to rout all that was not life, to cut a broad swath and shave close, to drive life into a corner, and reduce it to its lowest terms, and, if it proved to be mean, why then to get the whole and genuine meanness of it, and publish its meanness to the world; or if it were sublime, to know it by experience, and be able to give a true account of it in my next excursion. For most men, it appears to me, are in a strange uncertainty about it, whether it is of the devil or of God, and have *somewhat hastily* concluded that it is the chief end of man here to "glorify God and enjoy him forever."[5]

Still we live meanly, like ants; though the fable tells us that we were long ago changed into men;[6] like pygmies we fight with cranes; it is error upon error, and clout upon clout, and our best virtue has for its occasion a superfluous and evitable wretchedness. Our life is frittered away by detail. An honest man has hardly need to count more than his ten fingers, or in extreme cases he may add his ten toes, and lump the rest. Simplicity, simplicity, simplicity! I say, let your affairs be as two or three, and not a hundred or a thousand; instead of a million count half a dozen, and keep your accounts on your thumb nail. In the midst of this chopping sea of civilized life, such are the clouds and storms and quicksands and thousand-and-one items to be allowed for, that a man has to live, if he would not founder and go to the bottom and not make his port at all, by dead reckoning, and he must be a great calculator indeed who succeeds. Simplify, simplify. Instead of three meals a day, if it be necessary eat but one; instead of a hundred dishes, five; and reduce other things in proportion. Our life is like a German Con-

5. From the Shorter Catechism in the *New England Primer*.
6. In a Greek fable Aeacus persuaded Zeus to turn ants into men. The Trojans are compared to cranes fighting with pygmies (*Iliad*, Book 3).

federacy,[7] made up of petty states, with its boundary forever fluctuating, so that even a German cannot tell you how it is bounded at any moment. The nation itself, with all its so called internal improvements, which, by the way, are all external and superficial, is just such an unwieldy and overgrown establishment, cluttered with furniture and tripped up by its own traps, ruined by luxury and heedless expense, by want of calculation and a worthy aim, as the million households in the land; and the only cure for it as for them is in a rigid economy, a stern and more than Spartan simplicity of life and elevation of purpose. It lives too fast. Men think that it is essential that the *Nation* have commerce, and export ice, and talk through a telegraph, and ride thirty miles an hour, without a doubt, whether *they* do or not; but whether we should live like baboons or like men, is a little uncertain. If we do not get out sleepers,[8] and forge rails, and devote days and nights to the work, but go to tinkering upon our *lives* to improve *them*, who will build railroads? And if railroads are not built, how shall we get to heaven in season? But if we stay at home and mind our business, who will want railroads? We do not ride on the railroad; it rides upon us. Did you ever think what those sleepers are that underlie the railroad? Each one is a man, an Irish-man, or a Yankee man. The rails are laid on them, and they are covered with sand, and the cars run smoothly over them. They are sound sleepers, I assure you. And every few years a new lot is laid down and run over; so that, if some have the pleasure of riding on a rail, others have the misfortune to be ridden upon. And when they run over a man that is walking in his sleep, a supernumerary sleeper in the wrong position, and wake him up, they suddenly stop the cars, and make a hue and cry about it, as if this were an exception. I am glad to know that it takes a gang of men for every five miles to keep the sleepers down and level in their beds as it is, for this is a sign that they may sometime get up again.

Why should we live with such hurry and waste of life? We are determined to be starved before we are hungry. Men say that a stitch in time saves nine, and so they take a thousand stitches to-day to save nine to-morrow. As for *work*, we haven't any of any consequence. We have the Saint Vitus' dance,[9] and cannot possibly keep our heads still. If I should only give a few pulls at the parish bell-rope, as for a fire, that is, without setting the bell, there is hardly a man on his farm in the outskirts of Concord, notwithstanding that press of engagements which was his excuse so many times this morning, nor a boy, nor a woman, I might almost say, but would forsake all and follow that sound, not mainly to save property from

---

7. Later in the century Germany was unified under Prince Otto von Bismarck (1815–98), first chancellor of the German Empire.

8. Wooden railroad ties (another pun).
9. Chorea, a severe nervous disorder characterized by jerky motions.

the flames, but, if we will confess the truth, much more to see it burn, since burn it must, and we, be it known, did not set it on fire,—or to see it put out, and have a hand in it, if that is done as handsomely; yes, even if it were the parish church itself. Hardly a man takes a half hour's nap after dinner, but when he wakes he holds up his head and asks, "What's the news?" as if the rest of mankind had stood his sentinels. Some give directions to be waked every half hour, doubtless for no other purpose; and then, to pay for it, they tell what they have dreamed. After a night's sleep the news is as indispensable as the breakfast. "Pray tell me any thing new that has happened to a man any where on this globe",—and he reads it over his coffee and rolls, that a man has had his eyes gouged out this morning on the Wachito River;[1] never dreaming the while that he lives in the dark unfathomed mammoth cave of this world, and has but the rudiment of an eye himself.[2]

For my part, I could easily do without the post-office. I think that there are very few important communications made through it. To speak critically, I never received more than one or two letters in my life—I wrote this some years ago—that were worth the postage. The penny-post is, commonly, an institution through which you seriously offer a man that penny for his thoughts which is so often safely offered in jest. And I am sure that I never read any memorable news in a newspaper. If we read of one man robbed, or murdered, or killed by accident, or one house burned, or one vessel wrecked, or one steamboat blown up, or one cow run over on the Western Railroad, or one mad dog killed, or one lot of grasshoppers in the winter,—we never need read of another. One is enough. If you are acquainted with the principle, what do you care for a myriad instances and applications? To a philosopher all *news*, as it is called, is gossip, and they who edit and read it are old women over their tea. Yet not a few are greedy after this gossip. There was such a rush, as I hear, the other day at one of the offices to learn the foreign news by the last arrival, that several large squares of plate glass belonging to the establishment were broken by the pressure,—news which I seriously think a ready wit might write a twelvemonth or twelve years beforehand with sufficient accuracy. As for Spain, for instance, if you know how to throw in Don Carlos and the Infanta, and Don Pedro and Seville and Granada, from time to time in the right proportions,—they may have changed the names a little since I saw the papers,—and serve up a bull-fight when other entertainments fail, it will be true to the letter, and give us as good an idea of the exact state or ruin of things in Spain as the most succinct and lucid reports under this head in the newspapers: and as for England, almost the last significant scrap of news

---

1. Also spelled Ouachita, a tributary of the Red River; Thoreau refers to a common-enough incident in backwoods brawling.

2. Sightless fish had been found in Kentucky's Mammoth Cave.

from that quarter was the revolution of 1649; and if you have learned the history of her crops for an average year, you never need attend to that thing again, unless your speculations are of a merely pecuniary character. If one may judge who rarely looks into the newspapers, nothing new does ever happen in foreign parts, a French revolution not excepted.

What news! how much more important to know what that is which was never old! "Kieou-pe-yu (great dignitary of the state of Wei) sent a man to Khoung-tseu to know his news. Khoung-tseu caused the messenger to be seated near him, and questioned him in these terms: What is your master doing? The messenger answered with respect: My master desires to diminish the number of his faults, but he cannot accomplish it. The messenger being gone, the philosopher remarked: What a worthy messenger! What a worthy messenger!"[3] The preacher, instead of vexing the ears of drowsy farmers on their day of rest at the end of the week,—for Sunday is the fit conclusion of an ill-spent week, and not the fresh and brave beginning of a new one,—with this one other draggle-tail of a sermon, should shout with thundering voice,—"Pause! Avast! Why so seeming fast, but deadly slow?"[4]

Shams and delusions are esteemed for soundest truths, while reality is fabulous. If men would steadily observe realities only, and not allow themselves to be deluded, life, to compare it with such things as we know, would be like a fairy tale and the Arabian Nights' Entertainments. If we respected only what is inevitable and has a right to be, music and poetry would resound along the streets. When we are unhurried and wise, we perceive that only great and worthy things have any permanent and absolute existence,—that petty fears and petty pleasures are but the shadow of the reality. This is always exhilarating and sublime. By closing the eyes and slumbering, and consenting to be deceived by shows, men establish and confirm their daily life of routine and habit every where, which still is built on purely illusory foundations. Children, who play life, discern its true law and relations more clearly than men, who fail to live it worthily, but who think that they are wiser by experience, that is, by failure. I have read in a Hindoo book, that "there was a king's son, who, being expelled in infancy from his native city, was brought up by a forester, and, growing up to maturity in that state, imagined himself to belong to the barbarous race with which he lived. One of his father's ministers having discovered him, revealed to him what he was, and the misconception of his character was removed, and he knew himself to be a prince. So soul," continues the Hindoo philosopher, "from the circumstances in which it is placed, mistakes its own character, until the truth is revealed to it by some holy teacher,

---

3. Confucius, *Analects*, XIV.
4. Father Taylor of the Seaman's Bethel
in Boston was one such preacher famous for the nautical cast of his sermons.

and then it knows itself to be *Brahme*."[5] I perceive that we inhabitants of New England live this mean life that we do because our vision does not penetrate the surface of things. We think that that *is* which *appears* to be. If a man should walk through this town and see only the reality, where, think you, would the "Mill-dam"[6] go to? If he should give us an account of the realities he beheld there, we should not recognize the place in his description. Look at a meeting-house, or a court-house, or a jail, or a shop, or a dwelling-house, and say what that thing really is before a true gaze, and they would all go to pieces in your account of them. Men esteem truth remote, in the outskirts of the system, behind the farthest star, before Adam and after the last man. In eternity there is indeed something true and sublime. But all these times and places and occasions are now and here. God himself culminates in the present moment, and will never be more divine in the lapse of all the ages. And we are enabled to apprehend at all what is sublime and noble only by the perpetual instilling and drenching of the reality which surrounds us. The universe constantly and obediently answers to our conceptions; whether we travel fast or slow, the track is laid for us. Let us spend our lives in conceiving then. The poet or the artist never yet had so fair and noble a design but some of his posterity at least could accomplish it.

Let us spend one day as deliberately as Nature, and not be thrown off the track by every nutshell and mosquito's wing that falls on the rails. Let us rise early and fast, or break fast, gently and without perturbation; let company come and let company go, let the bells ring and the children cry,—determined to make a day of it. Why should we knock under and go with the stream? Let us not be upset and overwhelmed in that terrible rapid and whirlpool called a dinner, situated in the meridian shallows. Weather this danger and you are safe, for the rest of the way is down hill. With unrelaxed nerves, with morning vigor, sail by it, looking another way, tied to the mast like Ulysses.[7] If the engine whistles, let it whistle till it is hoarse for its pains. If the bell rings, why should we run? We will consider what kind of music they are like. Let us settle ourselves, and work and wedge our feet downward through the mud and slush of opinion, and prejudice, and tradition, and delusion, and appearance, that alluvion[8] which covers the globe, through Paris and London, through New York and Boston and Concord, through church and state, through poetry and philosophy and religion, till we come to a hard bottom and rocks in place, which we can call *reality*, and say, This is, and no mistake; and then begin, having a

---

5. In the Hindu triad Brahma is the divine reality in the aspect of creator, while Vishnu is the preserver and Siva the destroyer.
6. The business center of Concord.
7. A precaution Ulysses (Odysseus) took

to prevent his yielding to the call of the Sirens, sea nymphs whose singing lured ships to destruction.
8. Sediment deposited by flowing water along a shore or bank.

point d'appui,[9] below freshet and frost and fire, a place where you might found a wall or a state, or set a lamp-post safely, or perhaps a gauge, not a Nilometer,[1] but a Realometer, that future ages might know how deep a freshet of shams and appearances had gathered from time to time. If you stand right fronting and face to face to a fact, you will see the sun glimmer on both its surfaces, as if it were a cimeter, and feel its sweet edge dividing you through the heart and marrow, and so you will happily conclude your mortal career. Be it life or death, we crave only reality. If we are really dying, let us hear the rattle in our throats and feel cold in the extremities; if we are alive, let us go about our business.

Time is but the stream I go a-fishing in. I drink at it; but while I drink I see the sandy bottom and detect how shallow it is. Its thin current slides away, but eternity remains. I would drink deeper; fish in the sky, whose bottom is pebbly with stars. I cannot count one. I know not the first letter of the alphabet. I have always been regretting that I was not as wise as the day I was born. The intellect is a cleaver; it discerns and rifts its way into the secret of things. I do not wish to be any more busy with my hands than is necessary. My head is hands and feet. I feel all my best faculties concentrated in it. My instinct tells me that my head is an organ for burrowing, as some creatures use their snout and fore-paws, and with it I would mine and burrow my way through these hills. I think that the richest vein is somewhere hereabouts; so by the divining rod and thin rising vapors I judge; and here I will begin to mine.

Reading

With a little more deliberation in the choice of their pursuits, all men would perhaps become essentially students and observers, for certainly their nature and destiny are interesting to all alike. In accumulating property for ourselves or our posterity, in founding a family or a state, or acquiring fame even, we are mortal; but in dealing with truth we are immortal, and need fear no change nor accident. The oldest Egyptian or Hindoo philosopher raised a corner of the veil from the statue of the divinity; and still the trembling robe remains raised, and I gaze upon as fresh a glory as he did, since it was I in him that was then so bold, and it is he in me that now reviews the vision. No dust has settled on that robe; no time has elapsed since that divinity was revealed. That time which we really improve, or which is improvable, is neither past, present, nor future.

My residence was more favorable, not only to thought, but to serious reading, than a university; and though I was beyond the range of the ordinary circulating library, I had more than ever come within the influence of those books which circulate round the world,

9. Basis, leverage point.
1. Gauge used at Memphis in ancient

times for measuring the height of the Nile.

whose sentences were first written on bark, and are now merely copied from time to time on to linen paper. Says the poet Mîr Camar Uddîn Mast, "Being seated to run through the region of the spiritual world; I have had this advantage in books. To be intoxicated by a single glass of wine; I have experienced this pleasure when I have drunk the liquor of the esoteric doctrines."[1] I kept Homer's Iliad on my table through the summer, though I looked at his page only now and then. Incessant labor with my hands, at first, for I had my house to finish and my beans to hoe at the same time, made more study impossible. Yet I sustained myself by the prospect of such reading in future. I read one or two shallow books of travel in the intervals of my work, till that employment made me ashamed of myself, and I asked where it was then that *I* lived.

The student may read Homer or Æschylus in the Greek without danger of dissipation or luxuriousness, for it implies that he in some measure emulate their heroes, and consecrate morning hours to their pages. The heroic books, even if printed in the character of our mother tongue, will always be in a language dead to degenerate times; and we must laboriously seek the meaning of each word and line, conjecturing a larger sense than common use permits out of what wisdom and valor and generosity we have. The modern cheap and fertile press, with all its translations, has done little to bring us nearer to the heroic writers of antiquity. They seem as solitary, and the letter in which they are printed as rare and curious, as ever. It is worth the expense of youthful days and costly hours, if you learn only some words of an ancient language, which are raised out of the trivialness of the street, to be perpetual suggestions and provocations. It is not in vain that the farmer remembers and repeats the few Latin words which he has heard. Men sometimes speak as if the study of the classics would at length make way for more modern and practical studies; but the adventurous student will always study classics, in whatever language they may be written and however ancient they may be. For what are the classics but the noblest recorded thoughts of man? They are the only oracles which are not decayed, and there are such answers to the most modern inquiry in them as Delphi and Dodona[2] never gave. We might as well omit to study Nature because she is old. To read well, that is, to read true books in a true spirit, is a noble exercise, and one that will task the reader more than any exercise which the customs of the day esteem. It requires a training such as the athletes underwent, the steady intention almost of the whole life to this object. Books must be read as deliberately and reservedly as they were written. It is not enough even to be able to speak the language of that nation by which they are written, for there is a memorable interval between the spoken and the written language, the language heard and the language read.

1. Thoreau knew this 18th-century Hindu poet from a French translation in a his-tory of Hindu literature.
2. Oracles of ancient Greece.

The one is commonly transitory, a sound, a tongue, a dialect merely, almost brutish, and we learn it unconsciously, like the brutes, of our mothers. The other is the maturity and experience of that; if that is our mother tongue, this is our father tongue, a reserved and select expression, too significant to be heard by the ear, which we must be born again in order to speak. The crowds of men who merely *spoke* the Greek and Latin tongues in the middle ages were not entitled by the accident of birth to *read* the works of genius written in those languages; for these were not written in that Greek or Latin which they knew, but in the select language of literature. They had not learned the nobler dialects of Greece and Rome, but the very materials on which they were written were waste paper to them, and they prized instead a cheap contemporary literature. But when the several nations of Europe had acquired distinct though rude written languages of their own, sufficient for the purposes of their rising literatures, then first learning revived, and scholars were enabled to discern from that remoteness the treasures of antiquity. What the Roman and Grecian multitude could not *hear*, after the lapse of ages a few scholars *read*, and a few scholars only are still reading it.

However much we may admire the orator's occasional bursts of eloquence, the noblest written words are commonly as far behind or above the fleeting spoken language as the firmament with its stars is behind the clouds. *There* are the stars, and they who can may read them. The astronomers forever comment on and observe them. They are not exhalations like our daily colloquies and vaporous breath. What is called eloquence in the forum is commonly found to be rhetoric in the study. The orator yields to the inspiration of a transient occasion, and speaks to the mob before him, to those who can *hear* him; but the writer, whose more equable life is his occasion, and who would be distracted by the event and the crowd which inspire the orator, speaks to the intellect and heart of mankind, to all in any age who can *understand* him.

No wonder that Alexander carried the Iliad with him on his expeditions in a precious casket.[3] A written word is the choicest of relics. It is something at once more intimate with us and more universal than any other work of art. It is the work of art nearest to life itself. It may be translated into every language, and not only be read but actually breathed from all human lips;—not be represented on canvas or in marble only, but be carved out of the breath of life itself. The symbol of an ancient man's thought becomes a modern man's speech. Two thousand summers have imparted to the monuments of Grecian literature, as to her marbles, only a maturer golden and autumnal tint, for they have carried their own serene and celestial atmosphere into all lands to protect them against the corrosion of time. Books are the treasured wealth of the world and

3. Plutarch attests to this in his biography of Alexander.

the fit inheritance of generations and nations. Books, the oldest and the best, stand naturally and rightfully on the shelves of every cottage. They have no cause of their own to plead, but while they enlighten and sustain the reader his common sense will not refuse them. Their authors are a natural and irresistible aristocracy in every society, and, more than kings or emperors, exert an influence on mankind. When the illiterate and perhaps scornful trader has earned by enterprise and industry his coveted leisure and independence, and is admitted to the circles of wealth and fashion, he turns inevitably at last to those still higher but yet inaccessible circles of intellect and genius, and is sensible only of the imperfection of his culture and the vanity and insufficiency of all his riches, and further proves his good sense by the pains which he takes to secure for his children that intellectual culture whose want he so keenly feels; and thus it is that he becomes the founder of a family.

Those who have not learned to read the ancient classics in the language in which they were written must have a very imperfect knowledge of the history of the human race; for it is remarkable that no transcript of them has ever been made into any modern tongue, unless our civilization itself may be regarded as such a transcript. Homer has never yet been printed in English, nor Æschylus, nor Virgil even,—works as refined, as solidly done, and as beautiful almost as the morning itself; for later writers, say what we will of their genius, have rarely, if ever, equalled the elaborate beauty and finish and the lifelong and heroic literary labors of the ancients. They only talk of forgetting them who never knew them. It will be soon enough to forget them when we have the learning and the genius which will enable us to attend to and appreciate them. That age will be rich indeed when those relics which we call Classics, and the still older and more than classic but even less known Scriptures of the nations, shall have still further accumulated, when the Vaticans[4] shall be filled with Vedas and Zendavestas and Bibles, with Homers and Dantes and Shakspeares, and all the centuries to come shall have successively deposited their trophies in the forum of the world. By such a pile we may hope to scale heaven at last.

The works of the great poets have never yet been read by mankind, for only great poets can read them. They have only been read as the multitude read the stars, at most astrologically, not astronomically. Most men have learned to read to serve a paltry convenience, as they have learned to cipher in order to keep accounts and not be cheated in trade; but of reading as a noble intellectual exercise they know little or nothing; yet this only is reading, in a high sense, not that which lulls us as a luxury and suffers the nobler faculties to sleep the while, but what we have to stand on tiptoe to read and devote our most alert and wakeful hours to.

4. I.e., libraries.

I think that having learned our letters we should read the best
that is in literature, and not be forever repeating our a b abs, and
words of one syllable, in the fourth or fifth classes, sitting on the
lowest and foremost form all our lives.[5] Most men are satisfied if
they read or hear read, and perchance have been convicted by the
wisdom of one good book, the Bible, and for the rest of their lives
vegetate and dissipate their faculties in what is called easy reading.
There is a work in several volumes in our Circulating Library entitled Little Reading,[6] which I thought referred to a town of that
name which I had not been to. There are those who, like cor-
morants and ostriches, can digest all sorts of this, even after the
fullest dinner of meats and vegetables, for they suffer nothing to be
wasted. If others are the machines to provide this provender, they
are the machines to read it. They read the nine thousandth tale
about Zebulon and Sephronia, and how they loved as none had ever
loved before, and neither did the course of their true love run
smooth,—at any rate, how it did run and stumble, and get up again
and go on! how some poor unfortunate got up onto a steeple, who
had better never have gone up as far as the belfry; and then, having
needlessly got him up there, the happy novelist rings the bell for all
the world to come together and hear, O dear! how he did get down
again! For my part, I think that they had better metamorphose all
such aspiring heroes of universal noveldom into man weathercocks,
as they used to put heroes among the constellations, and let them
swing round there till they are rusty, and not come down at all to
bother honest men with their pranks. The next time the novelist
rings the bell I will not stir though the meeting-house burn down.
"The Skip of the Tip-Toe-Hop, a Romance of the Middle Ages, by
the celebrated author of 'Tittle-Tol-Tan,'[7] to appear in monthly
parts; a great rush; don't all come together." All this they read with
saucer eyes, and erect and primitive curiosity, and with unwearied
gizzard, whose corrugations even yet need no sharpening, just as
some little four-year-old bencher[8] his two-cent gilt-covered edition
of Cinderella,—without any improvement, that I can see, in the
pronunciation, or accent, or emphasis, or any more skill in extract-
ing or inserting the moral. The result is dulness of sight, a stagna-
tion of the vital circulations, and a general deliquium and sloughing
off of all the intellectual faculties. This sort of gingerbread is baked
daily and more sedulously than pure wheat or rye-and-Indian in
almost every oven, and finds a surer market.

The best books are not read even by those who are called good

5. I.e., with the youngest children at the front of a one-room schoolhouse.
6. Harding points out a basis for Thoreau's irony: a book called *Much Instruction from Little Reading* is included in the 1836 *Catalogue of Concord Social Library*.
7. Probably a play on James Fenimore Cooper's novel *The Wept of the Wish-ton-Wish*, which Thoreau would not have wasted his time reading.
8. A child too young to have graduated to a desk.

readers. What does our Concord culture amount to? There is in this town, with a very few exceptions, no taste for the best or for very good books even in English literature, whose words all can read and spell. Even the college-bred and so called liberally educated men here and elsewhere have really little or no acquaintance with the English classics; and as for the recorded wisdom of mankind, the ancient classics and Bibles, which are accessible to all who will know of them, there are the feeblest efforts any where made to become acquainted with them. I know a woodchopper, of middle age, who takes a French paper, not for news as he says, for he is above that, but to "keep himself in practice," he being a Canadian by birth; and when I ask him what he considers the best thing he can do in this world, he says, beside this, to keep up and add to his English. This is about as much as the college bred generally do or aspire to do, and they take an English paper for the purpose. One who has just come from reading perhaps one of the best English books will find how many with whom he can converse about it? Or suppose he comes from reading a Greek or Latin classic in the original, whose praises are familiar even to the so called illiterate; he will find nobody at all to speak to, but must keep silence about it. Indeed, there is hardly the professor in our colleges, who, if he has mastered the difficulties of the language, has proportionally mastered the difficulties of the wit and poetry of a Greek poet, and has any sympathy to impart to the alert and heroic reader; and as for the sacred Scriptures, or Bibles of mankind, who in this town can tell me even their titles? Most men do not know that any nation but the Hebrews have had a scripture. A man, any man, will go considerably out of his way to pick up a silver dollar; but here are golden words, which the wisest men of antiquity have uttered, and whose worth the wise of every succeeding age have assured us of;—and yet we learn to read only as far as Easy Reading, the primers and class-books, and when we leave school, the "Little Reading," and story books, which are for boys and beginners; and our reading, our conversation and thinking, are all on a very low level, worthy only of pygmies and manikins.

I aspire to be acquainted with wiser men than this our Concord soil has produced, whose names are hardly known here. Or shall I hear the name of Plato and never read his book? As if Plato were my townsman and I never saw him,—my next neighbor and I never heard him speak or attended to the wisdom of his words. But how actually is it? His Dialogues, which contain what was immortal in him, lie on the next shelf, and yet I never read them. We are under-bred and low-lived and illiterate; and in this respect I confess I do not make any very broad distinction between the illiterateness of my townsman who cannot read at all, and the illiterateness of him who has learned to read only what is for children and feeble intellects.

We should be as good as the worthies of antiquity, but partly by first knowing how good they were. We are a race of tit-men,[9] and soar but little higher in our intellectual flights than the columns of the daily paper.

It is not all books that are as dull as their readers. There are probably words addressed to our condition exactly, which, if we could really hear and understand, would be more salutary than the morning or the spring to our lives, and possibly put a new aspect on the face of things for us. How many a man has dated a new era in his life from the reading of a book. The book exists for us perchance which will explain our miracles and reveal new ones. The at present unutterable things we may find somewhere uttered. These same questions that disturb and puzzle and confound us have in their turn occurred to all the wise men; not one has been omitted; and each has answered them, according to his ability, by his words and his life. Moreover, with wisdom we shall learn liberality. The solitary hired man on a farm in the outskirts of Concord, who has had his second birth and peculiar religious experience, and is driven as he believes into silent gravity and exclusiveness by his faith, may think it is not true; but Zoroaster, thousands of years ago, travelled the same road and had the same experience; but he, being wise, knew it to be universal, and treated his neighbors accordingly, and is even said to have invented and established worship among men. Let him humbly commune with Zoroaster then, and, through the liberalizing influence of all the worthies, with Jesus Christ himself, and let "our church" go by the board.

We boast that we belong to the nineteenth century and are making the most rapid strides of any nation. But consider how little this village does for its own culture. I do not wish to flatter my townsmen, nor to be flattered by them, for that will not advance either of us. We need to be provoked,—goaded like oxen, as we are, into a trot. We have a comparatively decent system of common schools, schools for infants only; but excepting the half-starved Lyceum[1] in the winter, and latterly the puny beginning of a library suggested by the state, no school for ourselves. We spend more on almost any article of bodily aliment or ailment than on our mental aliment. It is time that we had uncommon schools, that we did not leave off our education when we begin to be men and women. It is time that villages were universities, and their elder inhabitants the fellows of universities, with leisure—if they are indeed so well off—to pursue liberal studies the rest of their lives. Shall the world be confined to one Paris or one Oxford forever? Cannot students be boarded here

9. Runts.
1. Public hall where local citizens and others, often with national reputations, gave lectures on a great variety of topics. Thoreau was one of those in charge of lecture series at Concord for several years, and in 1844–45 divided the town by bringing Wendell Phillips, the abolitionist, for a second controversial lecture. Concord Lyceum was in Thoreau's time one of the more liberal in the nation.

and get a liberal education under the skies of Concord? Can we not hire some Abelard[2] to lecture to us? Alas! what with foddering the cattle and tending the store, we are kept from school too long, and our education is sadly neglected. In this country, the village should in some respects take the place of the nobleman of Europe. It should be the patron of the fine arts. It is rich enough. It wants only the magnanimity and refinement. It can spend money enough on such things as farmers and traders value, but it is thought Utopian to propose spending money for things which more intelligent men know to be of far more worth. This town has spent seventeen thousand dollars on a town-house, thank fortune or politics, but probably it will not spend so much on living wit, the true meat to put into that shell, in a hundred years. The one hundred and twenty-five dollars annually subscribed for a Lyceum in the winter is better spent than any other equal sum raised in the town. If we live in the nineteenth century, why should we not enjoy the advantages which the nineteenth century offers? Why should our life be in any respect provincial? If we will read newspapers, why not skip the gossip of Boston and take the best newspaper in the world at once?—not be sucking the pap of "neutral family" papers, or browsing "Olive-Branches" here in New England. Let the reports of all the learned societies come to us, and we will see if they know any thing. Why should we leave it to Harper & Brothers and Redding & Co.[3] to select our reading? As the nobleman of cultivated taste surrounds himself with whatever conduces to his culture,—genius—learning—wit—books—paintings—statuary—music—philosophical instruments, and the like; so let the village do,—not stop short at a pedagogue, a parson, a sexton, a parish library, and three selectmen, because our pilgrim forefathers got through a cold winter once on a bleak rock with these. To act collectively is according to the spirit of our institutions; and I am confident that, as our circumstances are more flourishing, our means are greater than the nobleman's. New England can hire all the wise men in the world to come and teach her, and board them round the while, and not be provincial at all. That is the *uncommon* school we want. Instead of noblemen, let us have noble villages of men. If it is necessary, omit one bridge over the river, go round a little there, and throw one arch at least over the darker gulf of ignorance which surrounds us.

*   *   *

Conclusion

To the sick the doctors wisely recommend a change of air and scenery. Thank Heaven, here is not all the world. The buck-eye does not grow in New England, and the mocking-bird is rarely heard here. The wild-goose is more of a cosmopolite than we; he

---

2. Peter Abelard (1079–1142) was a great teacher of philosophy and theology in medieval France.

3. Major publishers and booksellers of New York City and Boston, respectively.

breaks his fast in Canada, takes a luncheon in the Ohio, and plumes himself for the night in a southern bayou. Even the bison, to some extent, keeps pace with the seasons, cropping the pastures of the Colorado only till a greener and sweeter grass awaits him by the Yellowstone. Yet we think that if rail-fences are pulled down, and stone-walls piled up on our farms, bounds are henceforth set to our lives and our fates decided. If you are chosen town-clerk, forsooth, you cannot go to Tierra del Fuego[1] this summer: but you may go to the land of infernal fire nevertheless. The universe is wider than our views of it.

Yet we should oftener look over the tafferel of our craft, like curious passengers, and not make the voyage like stupid sailors picking oakum.[2] The other side of the globe is but the home of our correspondent. Our voyaging is only great-circle sailing, and the doctors prescribe for diseases of the skin merely. One hastens to Southern Africa to chase the giraffe; but surely that is not the game he would be after. How long, pray, would a man hunt giraffes if he could? Snipes and woodcocks also may afford rare sport; but I trust it would be nobler game to shoot one's self.—

"Direct your eye sight inward, and you'll find
A thousand regions in your mind
Yet undiscovered. Travel them, and be
Expert in home-cosmography."[3]

What does Africa,—what does the West stand for? Is not our own interior white on the chart? black though it may prove, like the coast, when discovered. Is it the source of the Nile, or the Niger, or the Mississippi, or a North West Passage around this continent, that we would find? Are these the problems which most concern mankind? Is Franklin[4] the only man who is lost, that his wife should be so earnest to find him? Does Mr. Grinnell[5] know where he himself is? Be rather the Mungo Park, the Lewis and Clarke and Frobisher,[6] of your own streams and oceans; explore your own higher latitudes,—with shiploads of preserved meats to support you, if they be necessary; and pile the empty cans sky-high for a sign. Were preserved meats invented to preserve meat merely? Nay, be a Columbus to whole new continents and worlds within you, opening new channels, not of trade, but of thought. Every man is the lord of

1. Thoreau puns on the meaning of the name of the archipelago at the southern tip of South America, "land of fire."
2. Common nautical busywork: picking old rope apart so the pieces of hemp could be tarred and used for calking.
3. William Habington (1605–54), *To My Honoured Friend Sir Ed. P. Knight.*
4. Sir John Franklin (1786–1847), lost on a British expedition to the arctic.
5. Henry Grinnell (1799–1874), a rich New York whale-oil merchant (from a New Bedford family who sponsored two attempts to rescue Sir John Franklin, one in 1850 and another in 1853).
6. I.e., an explorer like Mungo Park (1771–1806), Scottish explorer of Africa; Meriwether Lewis (1774–1809) and William Clark (1770–1838), leaders of the American expedition into the Louisiana Territory (1804–6); Martin Frobisher (1535?–94), English mariner.

a realm beside which the earthly empire of the Czar is but a petty state, a hummock left by the ice. Yet some can be patriotic who have no *self*-respect, and sacrifice the greater to the less. They love the soil which makes their graves, but have no sympathy with the spirit which may still animate their clay. Patriotism is a maggot in their heads. What was the meaning of that South-Sea Exploring Expedition,[7] with all its parade and expense, but an indirect recognition of the fact, that there are continents and seas in the moral world, to which every man is an isthmus or an inlet, yet unexplored by him, but that it is easier to sail many thousand miles through cold and storm and cannibals, in a government ship, with five hundred men and boys to assist one, than it is to explore the private sea, the Atlantic and Pacific Ocean of one's being alone.—

> "Erret, et extremos alter scrutetur Iberos.
> Plus habet hic vitæ, plus habet ille viæ."[8]

Let them wander and scrutinize the outlandish Australians. I have more of God, they more of the road.

It is not worth the while to go round the world to count the cats in Zanzibar.[9] Yet do this even till you can do better, and you may perhaps find some "Symmes' Hole"[1] by which to get at the inside at last. England and France, Spain and Portugal, Gold Coast and Slave Coast, all front on this private sea; but no bark from them has ventured out of sight of land, though it is without doubt the direct way to India. If you would learn to speak all tongues and conform to the customs of all nations, if you would travel farther than all travellers, be naturalized in all climes, and cause the Sphinx to dash her head against a stone,[2] even obey the precept of the old philosopher, and Explore thyself. Herein are demanded the eye and the nerve. Only the defeated and deserters go to the wars, cowards that run away and enlist. Start now on that farthest western way, which does not pause at the Mississippi or the Pacific, nor conduct toward a worn-out China or Japan, but leads on direct a tangent to this sphere, summer and winter, day and night, sun down, moon down, and at last earth down too.

It is said that Mirabeau took to highway robbery "to ascertain

7. The famous expedition to the Pacific antarctic led by Charles Wilkes during 1838–42.
8. Thoreau's journal for May 10, 1841, begins: "A good warning to the restless tourists of these days is contained in the last verses of Claudian's 'Old Man of Verona'"; Thoreau substitutes "Australians" for "Spaniards" in his translation. Claudian: Claudius Claudianus, last of the Latin classic poets (fl. A.D. 395), author of *Epigrammata*, where Thoreau found the passage he translates.
9. Thoreau had read Charles Pickering's *The Races of Man* (1851), which reports on the domestic cats in Zanzibar (Harding).
1. In 1818 Captain John Symmes theorized that the earth was hollow with openings at both North and South Poles.
2. As the Theban Sphinx did when Oedipus guessed her riddle. (Thebes: an ancient Greek city, not the Egyptian city Thoreau has previously referred to.)

what degree of resolution was necessary in order to place one's self in formal opposition to the most sacred laws of society." He declared that "a soldier who fights in the ranks does not require half so much courage as a foot-pad,"—"that honor and religion have never stood in the way of a well-considered and a firm resolve."[3] This was manly, as the world goes; and yet it was idle, if not desperate. A saner man would have found himself often enough "in formal opposition" to what are deemed "the most sacred laws of society," through obedience to yet more sacred laws, and so have tested his resolution without going out of his way. It is not for a man to put himself in such an attitude to society, but to maintain himself in whatever attitude he find himself through obedience to the laws of his being, which will never be one of opposition to a just government, if he should chance to meet with such.

I left the woods for as good a reason as I went there. Perhaps it seemed to me that I had several more lives to live, and could not spare any more time for that one. It is remarkable how easily and insensibly we fall into a particular route, and make a beaten track for ourselves. I had not lived there a week before my feet wore a path from my door to the pond-side; and though it is five or six years since I trod it, it is still quite distinct. It is true, I fear that others may have fallen into it, and so helped to keep it open. The surface of the earth is soft and impressible by the feet of men; and so with the paths which the mind travels. How worn and dusty, then, must be the highways of the world, how deep the ruts of tradition and conformity! I did not wish to take a cabin passage, but rather to go before the mast and on the deck of the world, for there I could best see the moonlight amid the mountains. I do not wish to go below now.

I learned this, at least, by my experiment; that if one advances confidently in the direction of his dreams, and endeavors to live the life which he has imagined, he will meet with a success unexpected in common hours. He will put some things behind, will pass an invisible boundary; new, universal, and more liberal laws will begin to establish themselves around and within him; or the old laws be expanded, and interpreted in his favor in a more liberal sense, and he will live with the license of a higher order of beings. In proportion as he simplifies his life, the laws of the universe will appear less complex, and solitude will not be solitude, nor poverty poverty, nor weakness weakness. If you have built castles in the air, your work need not be lost; that is where they should be. Now put the foundations under them.

It is a ridiculous demand which England and America make, that you shall speak so that they can understand you. Neither men nor toad-stools grow so. As if that were important, and there were not

3. Thoreau encountered this passage by the Comte de Mirabeau (1749–91) in *Harper's*, I (1850), 651.

enough to understand you without them. As if Nature could support but one order of understandings, could not sustain birds as well as quadrupeds, flying as well as creeping things, and *hush* and *who*, which Bright[4] can understand, were the best English. As if there were safety in stupidity alone. I fear chiefly lest my expression may not be *extra-vagant* enough, may not wander far enough beyond the narrow limits of my daily experience, so as to be adequate to the truth of which I have been convinced. *Extra vagance!* it depends on how you are yarded. The migrating buffalo, which seeks new pastures in another latitude, is not extravagant like the cow which kicks over the pail, leaps the cow-yard fence, and runs after her calf, in milking time. I desire to speak somewhere *without* bounds; like a man in a waking moment, to men in their waking moments; for I am convinced that I cannot exaggerate enough even to lay the foundation of a true expression. Who that has heard a strain of music feared then lest he should speak extravagantly any more forever? In view of the future or possible, we should live quite laxly and undefined in front, our outlines dim and misty on that side; as our shadows reveal an insensible perspiration toward the sun. The volatile truth of our words should continually betray the inadequacy of the residual statement. Their truth is instantly *translated*; its literal monument alone remains. The words which express our faith and piety are not definite; yet they are significant and fragrant like frankincense to superior natures.

Why level downward to our dullest perception always, and praise that as common sense? The commonest sense is the sense of men asleep, which they express by snoring. Sometimes we are inclined to class those who are once-and-a-half witted with the half-witted, because we appreciate only a third part of their wit. Some would find fault with the morning-red, if they ever got up early enough. "They pretend," as I hear, "that the verses of Kabir have four different senses; illusion, spirit, intellect, and the exoteric doctrine of the Vedas;"[5] but in this part of the world it is considered a ground for complaint if a man's writings admit of more than one interpretation. While England endeavors to cure the potato-rot, will not any endeavor to cure the brain-rot, which prevails so much more widely and fatally?

I do not suppose that I have attained to obscurity, but I should be proud if no more fatal fault were found with my pages on this score than was found with the Walden ice. Southern customers objected to its blue color, which is the evidence of its purity, as if it were muddy, and preferred the Cambridge ice, which is white, but tastes of weeds. The purity men love is like the mists which envelop the earth, and not like the azure ether beyond.

Some are dinning in our ears that we Americans, and moderns generally, are intellectual dwarfs compared with the ancients, or

4. Name for an ox.
5. M. Garcin de Tassy, *Histoire de la littérature hindoui* (1839), p. 279.

even the Elizabethan men.[6] But what is that to the purpose? A living dog is better than a dead lion.[7] Shall a man go and hang himself because he belongs to the race of pygmies, and not be the biggest pygmy that he can? Let every one mind his own business, and endeavor to be what he was made.

Why should we be in such desperate haste to succeed, and in such desperate enterprises? If a man does not keep pace with his companions, perhaps it is because he hears a different drummer. Let him step to the music which he hears, however measured or far away. It is not important that he should mature as soon as an apple-tree or an oak. Shall he turn his spring into summer? If the condition of things which we were made for is not yet, what were any reality which we can substitute? We will not be shipwrecked on a vain reality. Shall we with pains erect a heaven of blue glass over ourselves, though when it is done we shall be sure to gaze still at the true ethereal heaven far above, as if the former were not?

There was an artist in the city of Kouroo who was disposed to strive after perfection. One day it came into his mind to make a staff. Having considered that in an imperfect work time is an ingredient, but into a perfect work time does not enter, he said to himself, It shall be perfect in all respects, though I should do nothing else in my life. He proceeded instantly to the forest for wood, being resolved that it should not be made of unsuitable material; and as he searched for and rejected stick after stick, his friends gradually deserted him, for they grew old in their works and died, but he grew not older by a moment. His singleness of purpose and resolution, and his elevated piety, endowed him, without his knowledge, with perennial youth. As he made no compromise with Time, Time kept out of his way, and only sighed at a distance because he could not overcome him. Before he had found a stock in all respects suitable the city of Kouroo was a hoary ruin, and he sat on one of its mounds to peel the stick. Before he had given it the proper shape the dynasty of the Candahars was at an end, and with the point of the stick he wrote the name of the last of that race in the sand, and then resumed his work. By the time he had smoothed and polished the staff Kalpa was no longer the pole-star; and ere he had put on the ferule and the head adorned with precious stones, Brahma had awoke and slumbered many times. But why do I stay to mention these things? When the finishing stroke was put to his work, it suddenly expanded before the eyes of the astonished artist into the fairest of all the creations of Brahma. He had made a new system in making a staff, a world with full and fair proportions; in which, though the old cities and dynasties had passed away, fairer and

6. There had been serious as well as satirical speculation as to the debilitating effects of the American climate, a continuation of the older question as to whether or not modern civilization could ever achieve the heights of ancient Greek and Roman civilization.
7. Ecclesiastes 9.4.

more glorious ones had taken their places. And now he saw by the heap of shavings still fresh at his feet, that, for him and his work, the former lapse of time had been an illusion, and that no more time had elapsed than is required for a single scintillation from the brain of Brahma to fall on and inflame the tinder of a mortal brain. The material was pure, and his art was pure; how could the result be other than wonderful?

No face which we can give to a matter will stead us so well at last as the truth. This alone wears well. For the most part, we are not where we are, but in a false position. Through an infirmity of our natures, we suppose a case, and put ourselves into it, and hence are in two cases at the same time, and it is doubly difficult to get out. In sane moments we regard only the facts, the case that is. Say what you have to say, not what you ought. Any truth is better than make-believe. Tom Hyde, the tinker, standing on the gallows, was asked if he had any thing to say. "Tell the tailors," said he, "to remember to make a knot in their thread before they take the first stitch."[8] His companion's prayer is forgotten.

However mean your life is, meet it and live it; do not shun it and call it hard names. It is not so bad as you are. It looks poorest when you are richest. The fault-finder will find faults even in paradise. Love your life, poor as it is. You may perhaps have some pleasant, thrilling, glorious hours, even in a poor-house. The setting sun is reflected from the windows of the alms-house as brightly as from the rich man's abode; the snow melts before its door as early in the spring. I do not see but a quiet mind may live as contentedly there, and have as cheering thoughts, as in a palace. The town's poor seem to me often to live the most independent lives of any. May be they are simply great enough to receive without misgiving. Most think that they are above being supported by the town; but it oftener happens that they are not above supporting themselves by dishonest means, which should be more disreputable. Cultivate poverty like a garden herb, like sage. Do not trouble yourself much to get new things, whether clothes or friends. Turn the old; return to them. Things do not change; we change. Sell your clothes and keep your thoughts. God will see that you do not want society. If I were confined to a corner of a garret all my days, like a spider, the world would be just as large to me while I had my thoughts about me. The philosopher said: "From an army of three divisions one can take away its general, and put it in disorder; from the man the most abject and vulgar one cannot take away his thought."[9] Do not seek so anxiously to be developed, to subject yourself to many influences to be played on; it is all dissipation. Humility like darkness reveals

8. Presumably a reference to the tailors who will sew Hyde's shroud, although some custom may be involved such as that of making the last stitch through the nose in preparing a sailor for burial at sea.
9. Confucius, *Analects*, IX.25.

the heavenly lights. The shadows of poverty and meanness gather around us, "and lo! creation widens to our view."[1] We are often reminded that if there were bestowed on us the wealth of Crœsus,[2] our aims must still be the same, and our means essentially the same. Moreover, if you are restricted in your range by poverty, if you cannot buy books and newspapers, for instance, you are but confined to the most significant and vital experiences; you are compelled to deal with the material which yields the most sugar and the most starch. It is life near the bone where it is sweetest. You are defended from being a trifler. No man loses ever on a lower level by magnanimity on a higher. Superfluous wealth can buy superfluities only. Money is not required to buy one necessary of the soul.

I live in the angle of a leaden wall, into whose composition was poured a little alloy of bell metal. Often, in the repose of my midday, there reaches my ears a confused *tintinnabulum*[3] from without. It is the noise of my contemporaries. My neighbors tell me of their adventures with famous gentlemen and ladies, what notabilities they met at the dinner-table; but I am no more interested in such things than in the contents of the Daily Times. The interest and the conversation are about costume and manners chiefly; but a goose is a goose still, dress it as you will. They tell me of California and Texas, of England and the Indies, of the Hon. Mr.—— of Georgia or of Massachusetts, all transient and fleeting phenomena, till I am ready to leap from their court-yard like the Mameluke bey.[4] I delight to come to my bearings,—not walk in procession with pomp and parade, in a conspicuous place, but to walk even with the Builder of the universe, if I may,—not to live in this restless, nervous, bustling, trivial Nineteenth Century, but stand or sit thoughtfully while it goes by. What are men celebrating? They are all on a committee of arrangements, and hourly expect a speech from somebody. God is only the president of the day, and Webster is his orator.[5] I love to weigh, to settle, to gravitate toward that which most strongly and rightfully attracts me;—not hang by the beam of the scale and try to weigh less,—not suppose a case, but take the case that is; to travel the only path I can, and that on which no power can resist me. It affords me no satisfaction to commence to spring an arch before I have got a solid foundation. Let us not play at kittlybenders.[6] There is a solid bottom every where. We read that the traveller asked the boy if the swamp before him had a hard

1. From the sonnet *To Night* by the British writer Joseph Blanco White (1775–1841).
2. King of Lydia (d. 546 B.C.), fabled as the richest man on earth.
3. Tinkling.
4. A famous romantic exploit: in 1811 the Egyptian Mehemet Ali Pasha attempted to massacre the Mameluke caste; but one bey, or officer, escaped by leaping from a wall onto his horse.
5. Political meetings then had "presidents" (since they presided) rather than "chairmen" or "chairpersons." Thoreau plays on the catch phrase from Mohammedanism "There is no other God than Allah, and Mohammed is his prophet." Thoreau regarded Daniel Webster with contempt.
6. Harding defines this as a "child's game of running out onto thin ice without breaking through."

bottom. The boy replied that it had. But presently the traveller's horse sank in up to the girths, and he observed to the boy, "I thought you said that this bog had a hard bottom." "So it has," answered the latter, "but you have not got half way to it yet." So it is with the bogs and quicksands of society; but he is an old boy that knows it. Only what is thought said or done at a certain rare coincidence is good. I would not be one of those who will foolishly drive a nail into mere lath and plastering; such a deed would keep me awake nights. Give me a hammer, and let me feel for the furring.[7] Do not depend on the putty. Drive a nail home and clinch it so faithfully that you can wake up in the night and think of your work with satisfaction,—a work at which you would not be ashamed to invoke the Muse. So will help you God, and so only. Every nail driven should be as another rivet in the machine of the universe, you carrying on the work.

Rather than love, than money, than fame, give me truth. I sat at a table where were rich food and wine in abundance, and obsequious attendance, but sincerity and truth were not; and I went away hungry from the inhospitable board. The hospitality was as cold as the ices. I thought that there was no need of ice to freeze them. They talked to me of the age of the wine and the fame of the vintage; but I thought of an older, a newer, and purer wine, of a more glorious vintage, which they had not got, and could not buy. The style, the house and grounds and "entertainment" pass for nothing with me. I called on the king, but he made me wait in his hall, and conducted like a man incapacitated for hospitality. There was a man in my neighborhood who lived in a hollow tree. His manners were truly regal. I should have done better had I called on him.

How long shall we sit in our porticoes practising idle and musty virtues, which any work would make impertinent? As if one were to begin the day with long-suffering, and hire a man to hoe his potatoes; and in the afternoon go forth to practise Christian meekness and charity with goodness aforethought! Consider the China[8] pride and stagnant self-complacency of mankind. This generation reclines a little to congratulate itself on being the last of an illustrious line; and in Boston and London and Paris and Rome, thinking of its long descent, it speaks of its progress in art and science and literature with satisfaction. There are the Records of the Philosophical Societies, and the public Eulogies of *Great Men!* It is the good Adam contemplating his own virtue. "Yes, we have done great deeds, and sung divine songs, which shall never die,"—that is, as long as *we* can remember them. The learned societies and great men of Assyria,—where are they? What youthful philosophers and ex-

---

7. Narrow lumber nailed as backing for lath. The first edition reads "furrowing," and one manuscript draft reads "stud."

8. From China's lingering isolationism, despite the China trade so important to the New England economy.

perimentalists we are! There is not one of my readers who has yet lived a whole human life. These may be but the spring months in the life of the race. If we have had the seven-years' itch, we have not seen the seventeen-year locust yet in Concord. We are acquainted with a mere pellicle of the globe on which we live. Most have not delved six feet beneath the surface, nor leaped as many above it. We know not where we are. Beside, we are sound asleep nearly half our time. Yet we esteem ourselves wise, and have an established order on the surface. Truly, we are deep thinkers, we are ambitious spirits! As I stand over the insect crawling amid the pine needles on the forest floor, and endeavoring to conceal itself from my sight, and ask myself why it will cherish those humble thoughts, and hide its head from me who might perhaps be its benefactor, and impart to its race some cheering information, I am reminded of the greater Benefactor and Intelligence that stands over me the human insect.

There is an incessant influx of novelty into the world, and yet we tolerate incredible dulness. I need only suggest what kind of sermons are still listened to in the most enlightened countries. There are such words as joy and sorrow, but they are only the burden of a psalm, sung with a nasal twang, while we believe in the ordinary and mean. We think that we can change our clothes only. It is said that the British Empire is very large and respectable, and that the United States are a first-rate power. We do not believe that a tide rises and falls behind every man which can float the British Empire like a chip, if he should ever harbor it in his mind. Who knows what sort of seventeen-year locust will next come out of the ground? The government of the world I live in was not framed, like that of Britain, in after-dinner conversations over the wine.

The life in us is like the water in the river. It may rise this year higher than man has ever known it, and flood the parched uplands; even this may be the eventful year, which will drown out all our muskrats. It was not always dry land where we dwell. I see far inland the banks which the stream anciently washed, before science began to record its freshets. Every one has heard the story which has gone the rounds of New England, of a strong and beautiful bug which came out of the dry leaf of an old table of apple-tree wood, which had stood in a farmer's kitchen for sixty years, first in Connecticut, and afterward in Massachusetts,—from an egg deposited in the living tree many years earlier still, as appeared by counting the annual layers beyond it; which was heard gnawing out for several weeks, hatched perchance by the heat of an urn.[9] Who does not feel his faith in a resurrection and immortality strengthened by hearing of this? Who knows what beautiful and winged life, whose egg has been buried for ages under many concentric layers of

9. A major account of the incident is in   *land and New York* (1821), II, 398.
Timothy Dwight, *Travels in New Eng-*

woodenness in the dead dry life of society, deposited at first in the alburnum of the green and living tree, which has been gradually converted into the semblance of its well-seasoned tomb,—heard perchance gnawing out now for years by the astonished family of man, as they sat round the festive board,—may unexpectedly come forth from amidst society's most trivial and handselled furniture, to enjoy its perfect summer life at last!

I do not say that John or Jonathan[10] will realize all this; but such is the character of that morrow which mere lapse of time can never make to dawn. The light which puts out our eyes is darkness to us. Only that day dawns to which we are awake. There is more day to dawn. The sun is but a morning star.

THE END

1846, 1850

Life without Principle[1]

At a lyceum, not long since, I felt that the lecturer had chosen a theme too foreign to himself, and so failed to interest me as much as he might have done. He described things not in or near to his heart, but toward his extremities and superficies. There was, in this sense, no truly central or centralizing thought in the lecture. I would have had him deal with his privatest experience, as the poet does. The greatest compliment that was ever paid me was when one asked me what I *thought*, and attended to my answer. I am surprised, as well as delighted, when this happens, it is such a rare use he would make of me, as if he were acquainted with the tool. Commonly, if men want anything of me, it is only to know how many acres I make of their land,—since I am a surveyor,—or, at most, what trivial news I have burdened myself with. They never will go to law for my meat; they prefer the shell. A man once came a considerable distance to ask me to lecture on Slavery; but on conversing with him, I found that he and his clique expected seven-eighths of the lecture to be theirs, and only one-eighth mine; so I declined. I take it for granted, when I am invited to lecture anywhere,—for I have had a little experience in that business,—that there is a desire to

10. John Bull or Brother Jonathan, i.e., England or America. Thoreau is now addressing not the restricted audience of the opening of "Economy" but all readers of the English language.
1. Although *Life without Principle* is little known in comparison to *Resistance to Civil Government*, it is, as Walter Harding says, "unquestionably the favorite of true Thoreau *aficionados*." The piece grew out of journal entries in the early 1850s and in some form was delivered as a lecture in 1854. Some of the titles it had in the next years were *Getting a Living, Misspent Lives, What Shall It Profit* (in allusion to Mark 8.36: "For what shall it profit a man, if he shall gain the whole world, and lose his own soul?"),

and *The Higher Law*. Thoreau was near death when Ticknor and Fields accepted the work for the *Atlantic Monthly* in March, 1862. In response to some criticism from the publishers, Thoreau—too weak to write—dictated a reply: "As for another title for the Higher Law article, I can think of nothing better than, Life without Principle." Under that title it was published in October, 1863, a year and a half after Thoreau's death. It was unsigned, in accordance with contemporary magazine practice. Thought of now as an essay, *Life without Principle* carries still the stringent immediacy of its origins as a lecture which sometimes outraged its auditors.

hear what I *think* on some subject, though I may be the greatest
fool in the country,—and not that I should say pleasant things
merely, or such as the audience will assent to; and I resolve, accordingly, that I will give them a strong dose of myself. They have
sent for me, and engaged to pay for me, and I am determined that
they shall have me, though I bore them beyond all precedent.

So now I would say something similar to you, my readers. Since
you are my readers, and I have not been much of a traveller, I will
not talk about people a thousand miles off, but come as near home
as I can. As the time is short, I will leave out all the flattery, and
retain all the criticism.

Let us consider the way in which we spend our lives.

This world is a place of business.[2] What an infinite bustle! I am
awaked almost every night by the panting of the locomotive. It
interrupts my dreams. There is no sabbath. It would be glorious to
see mankind at leisure for once. It is nothing but work, work, work.
I cannot easily buy a blank-book to write thoughts in; they are
commonly ruled for dollars and cents. An Irishman, seeing me
making a minute[3] in the fields, took it for granted that I was
calculating my wages. If a man was tossed out of a window when
an infant, and so made a cripple for life, or scared out of his wits by
the Indians, it is regretted chiefly because he was thus incapacitated
for—business! I think that there is nothing, not even crime, more
opposed to poetry, to philosophy, ay, to life itself, than this incessant business.

There is a coarse and boisterous money-making fellow in the
outskirts of our town, who is going to build a bank-wall under the
hill along the edge of his meadow. The powers have put this into his
head to keep him out of mischief, and he wishes me to spend three
weeks digging there with him. The result will be that he will perhaps
get some more money to hoard, and leave for his heirs to spend
foolishly. If I do this, most will commend me as an industrious and
hard-working man; but if I choose to devote myself to certain labors
which yield more real profit, though but little money, they may be
inclined to look on me as an idler. Nevertheless, as I do not need
the police of meaningless labor to regulate me, and do not see
anything absolutely praiseworthy in this fellow's undertaking, any
more than in many an enterprise of our own or foreign governments, however amusing it may be to him or them, I prefer to finish
my education at a different school.

If a man walk in the woods for love of them half of each day, he
is in danger of being regarded as a loafer; but if he spends his whole
day as a speculator, shearing off those woods and making earth bald
before her time, he is esteemed an industrious and enterprising
citizen. As if a town had no interest in its forests but to cut them
down!

2. A pun on "busyness." 3. Note or memo.

Most men would feel insulted, if it were proposed to employ them in throwing stones over a wall, and then in throwing them back, merely that they might earn their wages. But many are no more worthily employed now. For instance: just after sunrise, one summer morning, I noticed one of my neighbors walking beside his team, which was slowly drawing a heavy hewn stone swung under the axle, surrounded by an atmosphere of industry,—his day's work begun,—his brow commenced to sweat,—a reproach to all sluggards and idlers,—pausing abreast the shoulders of his oxen, and half turning round with a flourish of his merciful whip, while they gained their length on him. And I thought, Such is the labor which the American Congress exists to protect,—honest, manly toil,— honest as the day is long,—that makes his bread taste sweet, and keeps society sweet,—which all men respect and have consecrated: one of the sacred band, doing the needful, but irksome drudgery. Indeed, I felt a slight reproach, because I observed this from the window, and was not abroad and stirring about a similar business. The day went by, and at evening I passed the yard of another neighbor, who keeps many servants, and spends much money foolishly, while he adds nothing to the common stock, and there I saw the stone of the morning lying beside a whimsical structure intended to adorn this Lord Timothy Dexter's[4] premises, and the dignity forthwith departed from the teamster's labor, in my eyes. In my opinion, the sun was made to light worthier toil than this. I may add, that his employer has since run off, in debt to a good part of the town, and, after passing through Chancery,[5] has settled somewhere else, there to become once more a patron of the arts.

The ways by which you may get money almost without exception lead downward. To have done anything by which you earned money *merely* is to have been truly idle or worse. If the laborer gets no more than the wages which his employer pays him, he is cheated, he cheats himself. If you would get money as a writer or lecturer, you must be popular, which is to go down perpendicularly. Those services which the community will most readily pay for it is most disagreeable to render. You are paid for being something less than a man. The State does not commonly reward a genius any more wisely. Even the poet-laureate would rather not have to celebrate the accidents of royalty. He must be bribed with a pipe[6] of wine; and perhaps another poet is called away from his muse to gauge that very pipe. As for my own business, even that kind of surveying which I could do with most satisfaction my employers do not want. They would prefer that I should do my work coarsely and not too well, ay, not well enough. When I observe that there are different

4. "Lord" Timothy Dexter (1747–1806) was an eccentric wealthy merchant of Newburyport, Massachusetts, who had the fence, lawns, and exotically planted gardens of his "Palace" decorated with dozens of life-size painted wood statues, mostly of famous men, including the first three presidents.
5. Bankruptcy court.
6. Cask containing two hogshead (variously construed as 126 or 280 gallons).

ways of surveying, my employer commonly asks which will give
him the most land, not which is most correct. I once invented a rule
for measuring cord-wood, and tried to introduce it in Boston; but
the measurer there told me that the sellers did not wish to have their
wood measured correctly,—that he was already too accurate for
them, and therefore they commonly got their wood measured in
Charlestown before crossing the bridge.

The aim of the laborer should be, not to get his living, to get "a
good job," but to perform well a certain work; and, even in a
pecuniary sense, it would be economy for a town to pay its laborers
so well that they would not feel that they were working for low
ends, as for a livelihood merely, but for scientific, or even moral
ends. Do not hire a man who does your work for money, but him
who does it for love of it.

It is remarkable that there are few men so well employed, so
much to their minds, but that a little money or fame would commonly buy them off from their present pursuit. I see advertisements
for *active* young men, as if activity were the whole of a young
man's capital. Yet I have been surprised when one has with confidence proposed to me, a grown man, to embark in some enterprise
of his, as if I had absolutely nothing to do, my life having been a
complete failure hitherto. What a doubtful compliment this is to
pay me! As if he had met me half-way across the ocean beating up
against the wind, but bound nowhere, and proposed to me to go
along with him! If I did, what do you think the underwriters would
say? No, no! I am not without employment at this stage of the
voyage. To tell the truth, I saw an advertisement for able-bodied
seamen, when I was a boy, sauntering in my native port, and as
soon as I came of age I embarked.[7]

The community has no bribe that will tempt a wise man. You
may raise money enough to tunnel a mountain, but you cannot
raise money enough to hire a man who is minding *his own* business.
An efficient and valuable man does what he can, whether the
community pay him for it or not. The inefficient offer their inefficiency to the highest bidder, and are forever expecting to be put into
office. One would suppose that they were rarely disappointed.

Perhaps I am more than usually jealous with respect to my freedom. I feel that my connection with and obligation to society are
still very slight and transient. Those slight labors which afford me a
livelihood, and by which it is allowed that I am to some extent
serviceable to my contemporaries, are as yet commonly a pleasure
to me, and I am not often reminded that they are a necessity. So far
I am successful. But I foresee, that, if my wants should be much
increased, the labor required to supply them would become a
drudgery. If I should sell both my forenoons and afternoons to
society, as most appear to do, I am sure, that, for me, there would

7. A brief allegory: what Thoreau embarked on was the voyage of life.

be nothing left worth living for. I trust that I shall never thus sell my birthright for a mess of pottage.[8] I wish to suggest that a man may be very industrious, and yet not spend his time well. There is no more fatal blunderer than he who consumes the greater part of his life getting his living. All great enterprises are self-supporting. The poet, for instance, must sustain his body by his poetry, as a steam planing-mill feeds its boilers with the shavings it makes. You must get your living by loving. But as it is said of the merchants that ninety-seven in a hundred fail, so the life of men generally, tried by this standard, is a failure, and bankruptcy may be surely prophesied.

Merely to come into the world the heir of a fortune is not to be born, but to be still-born, rather. To be supported by the charity of friends, or a government-pension,—provided you continue to breathe,—by whatever fine synonymes you describe these relations, is to go into the almshouse. On Sundays the poor debtor goes to church to take an account of stock, and finds, of course, that his outgoes have been greater than his income. In the Catholic Church, especially, they go into Chancery, make a clean confession, give up all, and think to start again. Thus men will lie on their backs, talking about the fall of man, and never make an effort to get up.

As for the comparative demand which men make on life, it is an important difference between two, that the one is satisfied with a level success, that his marks can all be hit by point-blank shots, but the other, however low and unsuccessful his life may be, constantly elevates his aim, though at a very slight angle to the horizon. I should much rather be the last man,—though, as the Orientals say, "Greatness doth not approach him who is forever looking down; and all those who are looking high are growing poor."

It is remarkable that there is little or nothing to be remembered written on the subject of getting a living: how to make getting a living not merely honest and honorable, but altogether inviting and glorious; for if *getting* a living is not so, then living is not. One would think, from looking at literature, that this question had never disturbed a solitary individual's musings. Is it that men are too much disgusted with their experience to speak of it? The lesson of value which money teaches, which the Author of the Universe has taken so much pains to teach us, we are inclined to skip altogether. As for the means of living, it is wonderful how indifferent men of all classes are about it, even reformers, so called,—whether they inherit, or earn, or steal it. I think that society has done nothing for us in this respect, or at least has undone what she has done. Cold and hunger seem more friendly to my nature than those methods which men have adopted and advise to ward them off.

The title *wise* is, for the most part, falsely applied. How can one be a wise man, if he does not know any better how to live than

8. Genesis 25.34.

other men?—if he is only more cunning and intellectually subtle? Does Wisdom work in a tread-mill? or does she teach how to succeed *by her example?* Is there any such thing as wisdom not applied to life? Is she merely the miller who grinds the finest logic? It is pertinent to ask if Plato got his *living* in a better way or more successfully than his contemporaries,—or did he succumb to the difficulties of life like other men? Did he seem to prevail over some of them merely by indifference, or by assuming grand airs? or find it easier to live, because his aunt remembered him in her will? The ways in which most men get their living, that is, live, are mere makeshifts, and a shirking of the real business of life,—chiefly because they do not know, but partly because they do not mean, any better.

The rush to California,[9] for instance, and the attitude, not merely of merchants, but of philosophers and prophets, so called, in relation to it, reflect the greatest disgrace on mankind. That so many are ready to live by luck, and so get the means of commanding the labor of others less lucky, without contributing any value to society! And that is called enterprise! I know of no more startling development of the immorality of trade, and all the common modes of getting a living. The philosophy and poetry and religion of such a mankind are not worth the dust of a puff-ball. The hog that gets his living by rooting, stirring up the soil so, would be ashamed of such company. If I could command the wealth of all the worlds by lifting my finger, I would not pay *such* a price for it. Even Mahomet knew that God did not make this world in jest.[1] It makes God to be a moneyed gentleman who scatters a handful of pennies in order to see mankind scramble for them. The world's raffle! A subsistence in the domains of Nature a thing to be raffled for! What a comment, what a satire on our institutions! The conclusion will be, that mankind will hang itself upon a tree. And have all the precepts in all the Bibles taught men only this? and is the last and most admirable invention of the human race only an improved muck-rake? Is this the ground on which Orientals and Occidentals meet? Did God direct us so to get our living, digging where we never planted,—and He would, perchance, reward us with lumps of gold?

God gave the righteous man a certificate entitling him to food and raiment, but the unrighteous man found a *facsimile* of the same in God's coffers, and appropriated it, and obtained food and raiment like the former. It is one of the most extensive systems of counterfeiting that the world has seen. I did not know that mankind were suffering for want of gold. I have seen a little of it. I know that it is very malleable, but not so malleable as wit. A grain of gold will gild a great surface, but not so much as a grain of wisdom.

The gold-digger in the ravines of the mountains is as much a gambler as his fellow in the saloons of San Francisco. What differ-

---

9. The gold rush, which began in 1849.   1. The allusion is unlocated.

ence does it make, whether you shake dirt or shake dice? If you win, society is the loser. The gold-digger is the enemy of the honest laborer, whatever checks and compensations there may be. It is not enough to tell me that you worked hard to get your gold. So does the Devil work hard. The way of transgressors may be hard in many respects. The humblest observer who goes to the mines sees and says that gold-digging is of the character of a lottery; the gold thus obtained is not the same thing with the wages of honest toil. But, practically, he forgets what he has seen, for he has seen only the fact, not the principle, and goes into trade there, that is, buys a ticket in what commonly proves another lottery, where the fact is not so obvious.

After reading Howitt's account of the Australian gold-diggings[2] one evening, I had in my mind's eye, all night, the numerous valleys, with their streams, all cut up with foul pits, from ten to one hundred feet deep, and half a dozen feet across, as close as they can be dug, and partly filled with water,—the locality to which men furiously rush to probe for their fortunes,—uncertain where they shall break ground,—not knowing but the gold is under their camp itself,—sometimes digging one hundred and sixty feet before they strike the vein, or then missing it by a foot,—turned into demons, and regardless of each other's rights, in their thirst for riches,— whole valleys, for thirty miles, suddenly honey-combed by the pits of the miners, so that even hundreds are drowned in them,—standing in water, and covered with mud and clay, they work night and day, dying of exposure and disease. Having read this, and partly forgotten it, I was thinking, accidentally, of my own unsatisfactory life, doing as others do; and with that vision of the diggings still before me, I asked myself, why I might not be washing some gold daily, though it were only the finest particles,—why I might not sink a shaft down to the gold within me, and work that mine. *There* is a Ballarat, a Bendigo for you,—what though it were a sulky- gully?[3] At any rate, I might pursue some path, however solitary and narrow and crooked, in which I could walk with love and reverence. Wherever a man separates from the multitude, and goes his own way in this mood, there indeed is a fork in the road, though ordinary travellers may see only a gap in the paling. His solitary path across-lots will turn out the *higher way* of the two.

Men rush to California and Australia as if the true gold were to be found in that direction; but that is to go to the very opposite extreme to where it lies. They go prospecting farther and farther away from the true lead, and are most unfortunate when they think themselves most successful. Is not our *native* soil auriferous?[4] Does

2. William Howitt, *Land, Labour, and Gold; or, Two Years in Victoria*, 2 vols. (London: Longman, 1855).
3. Ballarat and Bendigo were two Australian diggings; Bendigo was the later and more productive one. Many of the diggings had "gully" in their name (e.g., Iron-Bark Gully, Long Gully, and Devil's Gully); a sulky-gully would be hard to extract gold from—here, a recalcitrant mind.
4. Gold-bearing.

not a stream from the golden mountains flow through our native valley? and has not this for more than geologic ages been bringing down the shining particles and forming the nuggets for us? Yet, strange to tell, if a digger steal away, prospecting for this true gold, into the unexplored solitudes around us, there is no danger that any will dog his steps, and endeavor to supplant him. He may claim and undermine the whole valley even, both the cultivated and the un-cultivated portions, his whole life long in peace, for no one will ever dispute his claim. They will not mind his cradles or his toms. He is not confined to a claim twelve feet square, as at Ballarat, but may mine anywhere, and wash the whole wide world in his tom.

Howitt says of the man who found the great nugget which weighed twenty-eight pounds, at the Bendigo diggings in Australia: —"He soon began to drink; got a horse, and rode all about, gener-ally at full gallop, and, when he met people, called out to inquire if they knew who he was, and then kindly informed them that he was 'the bloody wretch that had found the nugget.' At last he rode full speed against a tree, and nearly knocked his brains out." I think, however, there was no danger of that, for he had already knocked his brains out against the nugget. Howitt adds, "He is a hopelessly ruined man."[5] But he is a type of the class. They are all fast men. Hear some of the names of the places where they dig:—"Jackass Flat,"—"Sheep's-Head Gully,"—"Murderer's Bar," etc. Is there no satire in these names? Let them carry their ill-gotten wealth where they will, I am thinking it will still be "Jackass Flat," if not "Mur-derer's Bar," where they live.

The last resource of our energy has been the robbing of grave-yards on the Isthmus of Darien,[6] an enterprise which appears to be but in its infancy; for, according to late accounts, an act has passed its second reading in the legislature of New Granada,[7] regulating this kind of mining; and a correspondent of the "Tribune"[8] writes: —"In the dry season, when the weather will permit of the country being properly prospected, no doubt other rich '*guacas*' [that is, graveyards] will be found." To emigrants he says:—"Do not come before December; take the Isthmus route in preference to the Boca del Toro one; bring no useless baggage, and do not cumber yourself with a tent; but a good pair of blankets will be necessary; a pick, shovel, and axe of good material will be almost all that is required": advice which might have been taken from the "Burker's Guide."[9] And he concludes with this line in Italics and small capitals: "*If you are doing well at home,* STAY THERE," which may fairly be inter-preted to mean, "If you are getting a good living by robbing grave-yards at home, stay there."

5. *Land, Labour, and Gold*, I, 19.
6. Panama.
7. Colombia.
8. The New York *Tribune*, edited by Thoreau's long-time promoter Horace Greeley.

9. I.e., "Murderer's Handbook"—from William Burke (1792–1829), body snatch-er who resorted to murder in order to meet the medical demand for bodies to dissect.

But why go to California for a text? She is the child of New England, bred at her own school and church.

It is remarkable that among all the preachers there are so few moral teachers. The prophets are employed in excusing the ways of men. Most reverend seniors,[1] the *illuminati* of the age, tell me, with a gracious, reminiscent smile, betwixt an aspiration and a shudder, not to be too tender about these things,—to lump all that, that is, make a lump of gold of it. The highest advice I have heard on these subjects was grovelling. The burden of it was,—It is not worth your while to undertake to reform the world in this particular. Do not ask how your bread is buttered; it will make you sick, if you do,—and the like. A man had better starve at once than lose his innocence in the process of getting his bread. If within the sophisticated man there is not an unsophisticated one, then he is but one of the Devil's angels. As we grow old, we live more coarsely, we relax a little in our disciplines, and, to some extent, cease to obey our finest instincts. But we should be fastidious to the extreme of sanity, disregarding the gibes of those who are more unfortunate than ourselves.

In our science and philosophy, even, there is commonly no true and absolute account of things. The spirit of sect and bigotry has planted its hoof amid the stars. You have only to discuss the problem, whether the stars are inhabited or not, in order to discover it. Why must we daub the heavens as well as the earth? It was an unfortunate discovery that Dr. Kane was a Mason, and that Sir John Franklin was another.[2] But it was a more cruel suggestion that possibly that was the reason why the former went in search of the latter. There is not a popular magazine in this country that would dare to print a child's thought on important subjects without comment. It must be submitted to the D. D.s. I would it were the chickadee-dees.[3]

You come from attending the funeral of mankind to attend to a natural phenomenon. A little thought is sexton[4] to all the world.

I hardly know an *intellectual* man, even, who is so broad and truly liberal that you can think aloud in his society. Most with whom you endeavor to talk soon come to a stand against some institution in which they appear to hold stock,—that is, some particular, not universal, way of viewing things. They will continually thrust their own low roof, with its narrow skylight, between you and the sky, when it is the unobstructed heavens you would view. Get out of the way with your cobwebs, wash your windows, I say! In some lyceums they tell me that they have voted to exclude the

1. A sarcastic echo of Shakespeare's *Othello*, 1.3.78.
2. Elisha Kent Kane (1820–57), a U.S. medical officer, died during his second expedition to the arctic to find Sir John Franklin, not knowing that Franklin had died in 1847, before any rescue missions were sent out. Kane had lectured in Boston and other New England towns (1852–53) to raise funds for his second expedition.
3. A contemptuous pun on the timid "D.D.'s" (Doctors of Divinity) who exercised near-total, though unofficial, literary censorship in the mid-19th century.
4. Gravedigger.

subject of religion. But how do I know what their religion is, and when I am near to or far from it? I have walked into such an arena and done my best to make a clean breast of what religion I have experienced, and the audience never suspected what I was about. The lecture was as harmless as moonshine to them. Whereas, if I had read to them the biography of the greatest scamps in history, they might have thought that I had written the lives of the deacons of their church. Ordinarily, the inquiry is, Where did you come from? or, Where are you going? That was a more pertinent question which I overheard one of my auditors put to another once,—"What does he lecture for?" It made me quake in my shoes.

To speak impartially, the best men that I know are not serene, a world in themselves. For the most part, they dwell in forms, and flatter and study effect only more finely than the rest. We select granite for the underpinning of our houses and barns; we build fences of stone; but we do not ourselves rest on an underpinning of granitic truth, the lowest primitive rock. Our sills are rotten. What stuff is the man made of who is not coexistent in our thought with the purest and subtilest truth? I often accuse my finest acquaintances of an immense frivolity; for, while there are manners and compliments we do not meet, we do not teach one another the lessons of honesty and sincerity that the brutes do, or of steadiness and solidity that the rocks do. The fault is commonly mutual, however; for we do not habitually demand any more of each other.

That excitement about Kossuth,[5] consider how characteristic, but superficial, it was!—only another kind of politics or dancing. Men were making speeches to him all over the country, but each expressed only the thought, or the want of thought, of the multitude. No man stood on truth. They were merely banded together, as usual, one leaning on another, and all together on nothing; as the Hindoos made the world rest on an elephant, the elephant on a tortoise, and the tortoise on a serpent, and had nothing to put under the serpent. For all fruit of that stir we have the Kossuth hat.

Just so hollow and ineffectual, for the most part, is our ordinary conversation. Surface meets surface. When our life ceases to be inward and private, conversation degenerates into mere gossip. We rarely meet a man who can tell us any news which he has not read in a newspaper, or been told by his neighbor; and, for the most part, the only difference between us and our fellow is, that he has seen the newspaper, or been out to tea, and we have not. In proportion as our inward life fails, we go more constantly and desperately to the post-office. You may depend on it, that the poor fellow who walks away with the greatest number of letters, proud of his extensive correspondence, has not heard from himself this long while.

5. Lajos Kossuth (1802–94), Hungarian revolutionist. For weeks after his first triumphal arrival in the United States in December, 1851, American newspapers printed little news that did not concern him, and his slouch hat set an immediate fad.

I do not know but it is too much to read one newspaper a week. I have tried it recently, and for so long it seems to me that I have not dwelt in my native region. The sun, the clouds, the snow, the trees say not so much to me. You cannot serve two masters. It requires more than a day's devotion to know and to possess the wealth of a day.

We may well be ashamed to tell what things we have read or heard in our day. I do not know why my news should be so trivial,—considering what one's dreams and expectations are, why the developments should be so paltry. The news we hear, for the most part, is not news to our genius. It is the stalest repetition. You are often tempted to ask, why such stress is laid on a particular experience which you have had,—that, after twenty-five years, you should meet Hobbins, Registrar of Deeds,[6] again on the sidewalk. Have you not budged an inch, then? Such is the daily news. Its facts appear to float in the atmosphere, insignificant as the sporules of fungi, and impinge on some neglected *thallus*, or surface of our minds, which affords a basis for them, and hence a parasitic growth. We should wash ourselves clean of such news. Of what consequence, though our planet explode, if there is no character involved in the explosion? In health we have not the least curiosity about such events. We do not live for idle amusement. I would not run round a corner to see the world blow up.

All summer, and far into the autumn, perchance, you unconsciously went by[7] the newspapers and the news, and now you find it was because the morning and the evening were full of news to you. Your walks were full of incidents. You attended, not to the affairs of Europe, but to your own affairs in Massachusetts fields. If you chance to live and move and have your being[8] in that thin stratum in which the events that make the news transpire,—thinner than the paper on which it is printed,—then these things will fill the world for you; but if you soar above or dive below that plane, you cannot remember nor be reminded of them. Really to see the sun rise or go down every day, so to relate ourselves to a universal fact, would preserve us sane forever. Nations! What are nations? Tartars, and Huns, and Chinamen! Like insects, they swarm. The historian strives in vain to make them memorable. It is for want of a man that there are so many men. It is individuals that populate the world. Any man thinking may say with the Spirit of Loda,—[9]

> "I look down from my height on nations,
> And they become ashes before me;—

6. Thoreau is punning: the imaginary Hobbins is a recollector of *actions*; he judges the person today by something that person did long ago.
7. I.e., passed up, failed to read.
8. Acts 17.28: "For in him [the Lord] we live, and move, and have our being;

as certain also of your own poets have said, For we are also his offspring."
9. The *Atlantic Monthly* reads "Lodin," but Thoreau means the Spirit of Loda, whose speech from *Carric-Thura* in James Macpherson's *Poems of Ossian* follows in slightly compressed form.

> Calm is my dwelling in the clouds;
> Pleasant are the great fields of my rest."

Pray, let us live without being drawn by dogs, Esquimaux-fashion, tearing over hill and dale, and biting each other's ears.

Not without a slight shudder at the danger, I often perceive how near I had come to admitting into my mind the details of some trivial affair,—the news of the street; and I am astonished to observe how willing men are to lumber their minds with such rubbish, —to permit idle rumors and incidents of the most insignificant kind to intrude on ground which should be sacred to thought. Shall the mind be a public arena, where the affairs of the street and the gossip of the tea-table chiefly are discussed? Or shall it be a quarter of heaven itself,—an hypæthral[1] temple, consecrated to the service of the gods? I find it so difficult to dispose of the few facts which to me are significant, that I hesitate to burden my attention with those which are insignificant, which only a divine mind could illustrate. Such is, for the most part, the news in newspapers and conversation. It is important to preserve the mind's chastity in this respect. Think of admitting the details of a single case of the criminal court into our thoughts, to stalk profanely through their very *sanctum sanctorum*[2] for an hour, ay, for many hours! to make a very bar-room of the mind's inmost apartment, as if for so long the dust of the street had occupied us,—the very street itself, with all its travel, its bustle, and filth had passed through our thoughts' shrine! Would it not be an intellectual and moral suicide? When I have been compelled to sit spectator and auditor in a court-room for some hours, and have seen my neighbors, who were not compelled, stealing in from time to time, and tiptoeing about with washed hands and faces, it has appeared to my mind's eye, that, when they took off their hats, their ears suddenly expanded into vast hoppers for sound, between which even their narrow heads were crowded. Like the vanes of windmills, they caught the broad, but shallow stream of sound, which, after a few titillating gyrations in their coggy brains, passed out the other side. I wondered if, when they got home, they were as careful to wash their ears as before their hands and faces. It has seemed to me, at such a time, that the auditors and the witnesses, the jury and the counsel, the judge and the criminal at the bar,—if I may presume him guilty before he is convicted,— were all equally criminal, and a thunderbolt might be expected to descend and consume them all together.

By all kinds of traps and sign-boards, threatening the extreme penalty of the divine law, exclude such trespassers from the only ground which can be sacred to you. It is so hard to forget what it is worse than useless to remember! If I am to be a thoroughfare, I

---

1. Roofless, open-air.   2. Holy of holies, the most sacred spot.

prefer that it be of the mountain-brooks, the Parnassian streams,[3] and not the town-sewers. There is inspiration, that gossip which comes to the ear of the attentive mind from the courts of heaven. There is the profane and stale revelation of the bar-room and the police court. The same ear is fitted to receive both communications. Only the character of the hearer determines to which it shall be open, and to which closed. I believe that the mind can be permanently profaned by the habit of attending to trivial things, so that all our thoughts shall be tinged with triviality. Our very intellect shall be macadamized,[4] as it were,—its foundation broken into fragments for the wheels of travel to roll over; and if you would know what will make the most durable pavement, surpassing rolled stones, spruce blocks, and asphaltum, you have only to look into some of our minds which have been subjected to this treatment so long.

If we have thus desecrated ourselves,—as who has not?—the remedy will be by wariness and devotion to reconsecrate ourselves, and make once more a fane[5] of the mind. We should treat our minds, that is, ourselves, as innocent and ingenuous children, whose guardians we are, and be careful what objects and what subjects we thrust on their attention. Read not the Times.[6] Read the Eternities. Conventionalities are at length as bad as impurities. Even the facts of science may dust the mind by their dryness, unless they are in a sense effaced each morning, or rather rendered fertile by the dews of fresh and living truth. Knowledge does not come to us by details, but in flashes of light from heaven. Yes, every thought that passes through the mind helps to wear and tear it, and to deepen the ruts, which, as in the streets of Pompeii, evince how much it has been used. How many things there are concerning which we might well deliberate, whether we had better know them,—had better let their peddling-carts be driven, even at the slowest trot or walk, over that bridge of glorious span[7] by which we trust to pass at last from the farthest brink of time to the nearest shore of eternity! Have we no culture, no refinement,—but skill only to live coarsely and serve the Devil?—to acquire a little worldly wealth, or fame, or liberty, and make a false show with it, as if we were all husk and shell, with no tender and living kernel to us? Shall our institutions be like those chestnut-burs which contain abortive nuts, perfect only to prick the fingers?

America is said to be the arena on which the battle of freedom is to be fought; but surely it cannot be freedom in a merely political sense that is meant. Even if we grant that the American has freed himself from a political tyrant, he is still the slave of an economical

3. The springs of inspiration, like the springs on Mount Parnassus in Greek mythology.
4. John McAdam (1756–1836), Scottish engineer, pioneered paving roads with broken stones in a tar or asphalt binder.
5. Thoreau puns on the Latin sense of *profane*, some outrage done *in front of* the fane, the temple.
6. A pun on the common use of *Times* as a newspaper name.
7. I.e., the mind.

and moral tyrant. Now that the republic—the *res-publica*—has been settled, it is time to look after the *res-privata*,—the private state,—to see, as the Roman senate charged its consuls, *"ne quid res-*PRIVATA *detrimenti caperet,"* that the *private* state receive no detriment.

Do we call this the land of the free? What is it to be free from King George and continue the slaves of King Prejudice? What is it to be born free and not to live free? What is the value of any political freedom, but as a means to moral freedom? Is it a freedom to be slaves, or a freedom to be free, of which we boast? We are a nation of politicians, concerned about the outmost defences only of freedom. It is our children's children who may perchance be really free. We tax ourselves unjustly. There is a part of us which is not represented. It is taxation without representation. We quarter troops, we quarter fools and cattle of all sorts upon ourselves. We quarter our gross bodies on our poor souls, till the former eat up all the latter's substance.

With respect to a true culture and manhood, we are essentially provincial still, not metropolitan,—mere Jonathans.[8] We are provincial, because we do not find at home our standards,—because we do not worship truth, but the reflection of truth,—because we are warped and narrowed by an exclusive devotion to trade and commerce and manufactures and agriculture and the like, which are but means, and not the end.

So is the English Parliament provincial. Mere country-bumpkins, they betray themselves, when any more important question arises for them to settle, the Irish question, for instance,—the English question why did I not say? Their natures are subdued to what they work in.[9] Their "good breeding" respects only secondary objects. The finest manners in the world are awkwardness and fatuity, when contrasted with a finer intelligence. They appear but as the fashions of past days,—mere courtliness, knee-buckles and small-clothes, out of date. It is the vice, but not the excellence of manners, that they are continually being deserted by the character; they are cast-off clothes or shells, claiming the respect which belonged to the living creature. You are presented with the shells instead of the meat, and it is no excuse generally, that, in the case of some fishes, the shells are of more worth than the meat. The man who thrusts his manners upon me does as if he were to insist on introducing me to his cabinet of curiosities,[1] when I wished to see himself. It was not in this sense that the poet Decker called Christ "the first true gentleman that ever breathed."[2] I repeat that in this sense the most splendid court in Christendom is provincial, having authority to

8. Americans (more specifically, New Englanders).
9. An echo of Shakespeare's Sonnet 111 ("And almost thence my nature is sub-du'd/To what it works in, like the dyer's hand").

1. In Thoreau's time there was a special word, *virtuosi*, for avid collectors and admirers of odd and rare artifacts.
2. Thomas Dekker, *The Honest Whore*, I.xii.

consult about Trans-alpine interests only, and not the affairs of Rome.[3] A prætor or proconsul would suffice to settle the questions which absorb the attention of the English Parliament and the American Congress.

Government and legislation! these I thought were respectable professions. We have heard of heaven-born Numas, Lycurguses, and Solons,[4] in the history of the world, whose *names* at least may stand for ideal legislators; but think of legislating to *regulate* the breeding of slaves, or the exportation of tobacco! What have divine legislators to do with the exportation or the importation of tobacco? what humane ones with the breeding of slaves? Suppose you were to submit the question to any son of God,—and has He no children in the nineteenth century? is it a family which is extinct?—in what condition would you get it again? What shall a State like Virginia say for itself at the last day, in which these have been the principal, the staple productions? What ground is there for patriotism in such a State? I derive my facts from statistical tables which the States themselves have published.

A commerce that whitens[5] every sea in quest of nuts and raisins, and makes slaves of its sailors for this purpose! I saw, the other day, a vessel which had been wrecked, and many lives lost, and her cargo of rags, juniper-berries, and bitter almonds were strewn along the shore. It seemed hardly worth the while to tempt the dangers of the sea between Leghorn and New York for the sake of a cargo of juniper-berries and bitter almonds. America sending to the Old World for her bitters! Is not the sea-brine, is not shipwreck, bitter enough to make the cup of life go down here? Yet such, to a great extent, is our boasted commerce; and there are those who style themselves statesmen and philosophers who are so blind as to think that progress and civilization depend on precisely this kind of interchange and activity,—the activity of flies about a molasses-hogshead. Very well, observes one, if men were oysters. And very well, answer I, if men were mosquitoes.

Lieutenant Herndon,[6] whom our Government sent to explore the Amazon, and, it is said, to extend the area of Slavery, observed that there was wanting there "an industrious and active population, who know what the comforts of life are, and who have artificial wants to draw out the great resources of the country." But what are the "artificial wants" to be encouraged? Not the love of luxuries, like the tobacco and slaves of, I believe, his native Virginia, nor the ice and granite and other material wealth of our native New England; nor are "the great resources of a country" that fertility or barren-

3. That is, the papal court in Rome, which in Thoreau's time was losing some of its secular power at Rome while retaining influence beyond the Alps.
4. Numa, Lycurgus, and Solon were "ideal legislators" of, respectively, Rome, Sparta, and Athens.

5. I.e., with sails.
6. Lieutenant William Lewis Herndon (1813–57) wrote *Exploration of the Valley of the Amazon*, 2 vols. (Washington: Robert Armstrong, Public Printer, 1854).

ness of soil which produces these. The chief want, in every State that I have been into, was a high and earnest purpose in its inhabitants. This alone draws out "the great resources" of Nature, and at last taxes her beyond her resources; for man naturally dies out of her. When we want culture more than potatoes, and illumination more than sugar-plums, then the great resources of a world are taxed and drawn out, and the result, or staple production, is, not slaves, nor operatives,[7] but men,—those rare fruits called heroes, saints, poets, philosophers, and redeemers.

In short, as a snow-drift is formed where there is a lull in the wind, so, one would say, where there is a lull of truth, an institution springs up. But the truth blows right on over it, nevertheless, and at length blows it down.

What is called politics is comparatively something so superficial and inhuman, that, practically, I have never fairly recognized that it concerns me at all. The newspapers, I perceive, devote some of their columns specially to politics or government without charge; and this, one would say, is all that saves it; but, as I love literature, and, to some extent, the truth also, I never read those columns at any rate. I do not wish to blunt my sense of right so much. I have not got to answer for having read a single President's Message. A strange age of the world this, when empires, kingdoms, and republics come a-begging to a private man's door, and utter their complaints at his elbow! I cannot take up a newspaper but I find that some wretched government or other, hard pushed, and on its last legs, is interceding with me, the reader, to vote for it,—more importunate than an Italian beggar; and if I have a mind to look at its certificate, made, perchance, by some benevolent merchant's clerk, or the skipper that brought it over, for it cannot speak a word of English itself, I shall probably read of the eruption of some Vesuvius, or the overflowing of some Po, true or forged, which brought it into this condition. I do not hesitate, in such a case, to suggest work, or the almshouse; or why not keep its castle in silence, as I do commonly? The poor President,[8] what with preserving his popularity and doing his duty, is completely bewildered. The newspapers are the ruling power. Any other government is reduced to a few marines at Fort Independence.[9] If a man neglects to read the Daily Times, Government will go down on its knees to him, for this is the only treason in these days.

Those things which now most engage the attention of men, as politics and the daily routine, are, it is true, vital functions of human society, but should be unconsciously performed, like the corresponding functions of the physical body. They are *infra-*

7. Factory workers.
8. The journal passage from which this part of the essay is derived specifically deals with Franklin Pierce.
9. In Boston Harbor.

human, a kind of vegetation. I sometimes awake to a half-conscious-
ness of them going on about me, as a man may become conscious
of some of the processes of digestion in a morbid state, and so have
the dyspepsia, as it is called. It is as if a thinker submitted himself to
be rasped by the great gizzard of creation. Politics is, as it were, the
gizzard of society, full of grit and gravel, and the two political
parties are its two opposite halves,—sometimes split into quarters, it
may be, which grind on each other. Not only individuals, but States,
have thus a confirmed dyspepsia, which expresses itself, you can
imagine by what sort of eloquence.[1] Thus our life is not altogether
a forgetting,[2] but also, alas! to a great extent, a remembering of
that which we should never have been conscious of, certainly not in
our waking hours. Why should we not meet, not always as dyspep-
tics, to tell our bad dreams, but sometimes as *eupeptics*,[3] to
congratulate each other on the ever glorious morning? I do not
make an exorbitant demand, surely.

1863

1. A scatological joke.
2. An echo of Wordsworth's *Immortal-*
ity ode, line 58.
3. Exuberantly healthy people.

JAMES RUSSELL LOWELL
1819–1891

Lowell was born on February 22, 1819, at Elmwood, the family estate in
Cambridge, Massachusetts, then hardly more than a village. After an indulgent childhood he attended Harvard, where he drew reprimands for
skipping classes and other minor infractions of the rules. He studied law
but never made a living from his practice. Instead, he published ardent,
derivative verse (Wordsworth and Tennyson were prime influences) in
newspapers and magazines, collecting some of them in *A Year's Life*
(1841). Exuberant and idealistic, he founded a short-lived Boston literary
magazine, *The Pioneer*, in 1843, then briefly worked his way into the New
York City literary scene and abolitionist journalism, winning friends and
often alienating some of those friends by his inability to blunt a cruel
satiric barb and by his impatience with anyone not devoted to his current
cause. At the end of 1844 he married Maria White, herself deeply interested in abolitionism and other reform movements. Lowell's breakthrough
to national reputation came in 1848 with the publication of *A Fable for
Critics*, the book form of *The Biglow Papers*, and a Christmas-market edition of *The Vision of Sir Launfal*. The first is a verse satire on contemporary American writers. *The Biglow Papers*, a part-prose, part-verse satire on
American imperialism against Mexico and southern lust for new slave territory, was a major contribution to altering northern consciousness about
slavery, but the topical nature of Lowell's humor and the painstaking representation of New England dialect pronunciation have rendered it nearly

inaccessible to all but specialists. The *Vision,* a didactic narrative of a dream vision experienced by one of King Arthur's knights, was for generations a classroom staple.

Unable to make a living from his poetry and saddened by the successive deaths of two young daughters, Lowell took his family abroad, where a baby son died, leaving only one daughter to outlive Lowell. Mrs. Lowell died in 1853, after their return to Cambridge. Lowell wrote desultorily for the magazines until 1855, when Harvard offered him Longfellow's professorship. Hoping to perfect his German before assuming the post, Lowell passed several lonely months in Dresden, then joined family and friends in Italy. Leon Howard has summed up Lowell's situation in the mid-1850s: "Of his poetic career, little remained except for a few periods of occasional inspiration, a certain amount of philosophy, and an accomplished craftsmanship which enabled him to exploit fully those relics of youthful enthusiasm still left in his notebooks. Instead of writing strong poems in his maturity that could carry his early writings down to posterity on their shoulders, as he had promised, he found that his early verses often had to be used to sustain his later productivity and reputation."

In 1857 Lowell married his daughter's governess, to general amazement, and took up teaching and the editorship of the new *Atlantic Monthly.* Editing proved a good excuse for writing only at intervals and not finishing more ambitious projects such as *The Nooning,* conceived in 1849 as a sort of local *Canterbury Tales.* The *Atlantic* drew him once again into politics, and after the change of ownership removed him as editor he became co-editor of the *North American Review,* which he made into a weapon for political commentary. The Civil War brought him personal tragedy in the loss of nephews and other young friends, and after it ended he remained engaged with issues of Reconstruction, dreading that the new rights of the Negroes would be ignored by the South. But for all his national exposure in the 1850s and 1860s he was aware that his life had not fulfilled its promise.

Always a sociable man, Lowell was one of the founders of the Saturday Club, which also counted as members Emerson, Longfellow, Hawthorne, Holmes, Dana, Agassiz, and other literary men and historians. In the 1860s there were supplemental pleasures in the Dante Club, of which the nucleus was Longfellow, Lowell, and his younger friend Charles Eliot Norton. Lowell led the life of a literary man and professor, his chief new publication being *The Cathedral.* Hard pressed to live on his income, in 1871 Lowell sold almost twenty acres from around his house in Cambridge, and left the next year for Europe, returning in 1874. He became briefly absorbed in American politics, and attacked corruption in stinging poems. President Hayes appointed him Minister to Spain in 1877. The second Mrs. Lowell nearly died in Madrid and was intermittently insane thereafter. Lowell became Minister to England in 1880 and was soon notorious for the frequency of his dining out and for mishandling matters of state. Just after Lowell's wife died in early 1885, President Cleveland removed him from his post, and thereafter for several years he alternated between quiet winters at his daughter's house in Southborough, Massachusetts, and summers in London, where he reveled in invitations to dine out every day. He died at Elmwood of cancer on August 12, 1891.

From A Fable for Critics[1]
[*Emerson, Channing, Thoreau, Hawthorne, Poe, Longfellow, Irving, and Lowell*]

"There comes Emerson first, whose rich words, every one,
Are like gold nails in temples to hang trophies on,[2]
Whose prose is grand verse, while his verse, the Lord knows,   525
Is some of it pr———No, 'tis not even prose;
I'm speaking of metres; some poems have welled
From those rare depths of soul that have ne'er been excelled;
They're not epics, but that doesn't matter a pin,
In creating, the only hard thing's to begin;   530
A grass-blade's no easier to make than an oak,
If you've once found the way, you've achieved the grand stroke;
In the worst of his poems are mines of rich matter,
But thrown in a heap with a crush and a clatter;
Now it is not one thing nor another alone   535
Makes a poem, but rather the general tone,
The something pervading, uniting the whole,
The before unconceived, unconceivable soul,
So that just in removing this trifle or that, you
Take away, as it were, a chief limb of the statue;   540
Roots, wood, bark, and leaves, singly perfect may be,
But, clapt hodge-podge together, they don't make a tree.

     "But, to come back to Emerson, (whom, by the way,
I believe we left waiting,)—his is, we may say,
A Greek head on right Yankee shoulders, whose range   545
Has Olympus for one pole, for t'other the Exchange;[3]
He seems, to my thinking, (although I'm afraid

1. The text is that of the first edition (1848). As Lowell's page-long rhyming subtitles said, this work is "A GLANCE/ AT A FEW OF OUR LITERARY PROGENIES/(Mrs. Malaprop's word)/ FROM/THE TUB OF DIOGENES." (Mrs. Malaprop is the character in Sheridan's *The Rivals*, 1775, who gets her words wrong; the skepticism of the Greek philosopher Diogenes (412?–323 B.C.) toward his contemporaries defines Lowell's stance in the poem toward his own contemporaries.
     The following selections are in quotation marks because they are spoken by Phoebus (or Apollo), Greek god of poetry, whose occasional irritation springs partly from the fact that the magazine editors Evert A. Duyckinck and Cornelius Mathews have been pestering him with autographed copies of books by American writers. Lowell's joke is that Duyckinck, Mathews, and other people working for the establishment of an international copyright law and recognition of American authors often claimed greatness for nonentities. The ultimate joke, which Duyckinck probably never appreciated any more than Lowell did, is that in fact, even in 1848, Duyckinck had at least one authentic literary genius among the American writers he was promoting—Herman Melville, whom Lowell does not mention in *A Fable for Critics*.
     The line numbers given here begin with the first lines of the *Fable* to be set off as verse and do not count the rhymed title page and other prefatory matter (rhymed footnotes are not counted either).
2. Ecclesiastes 12.11: "The words of the wise are as goads, and as nails fastened by the masters of assemblies, which are given from one shepherd." Many of Lowell's readers would have known that the next verse admonishes that "of making many books there is no end"—a deft undercutting of the compliment.
3. Stock market. This is an early noticing of the union of opposites in Emerson—the idealist like the Greek philosopher Plotinus (c. 205–270) and the shrewd realist like Michel de Montaigne (1533–92), French ("Gascon") essayist.

708 ★ James Russell Lowell

The comparison must, long ere this, have been made,)
A Plotinus-Montaigne, where the Egyptian's gold mist
And the Gascon's shrewd wit cheek-by-jowl co-exist;   550
All admire, and yet scarcely six converts he's got
To I don't (nor they either) exactly know what;
For though he builds glorious temples, 'tis odd
He leaves never a doorway to get in a god.
'Tis refreshing to old-fashioned people like me,   555
To meet such a primitive Pagan as he,
In whose mind all creation is duly respected
As parts of himself—just a little projected;
And who's willing to worship the stars and the sun,
A convert to—nothing but Emerson.   560
So perfect a balance there is in his head,
That he talks of things sometimes as if they were dead;
Life, nature, love, God, and affairs of that sort,
He looks at as merely ideas; in short,
As if they were fossils stuck round in a cabinet,[4]   565
Of such vast extent that our Earth's a mere dab in it;
Composed just as he is inclined to conjecture her,
Namely, one part pure earth, ninety-nine parts pure lecturer;
You are filled with delight at his clear demonstration,
Each figure, word, gesture, just fits the occasion,   570
With the quiet precision of science he'll sort 'em,
But you can't help suspecting the whole a *post mortem*.

                                    * * *

    "He has imitators in scores, who omit
No part of the man but his wisdom and wit,—
Who go carefully o'er the sky-blue of his brain,
And when he has skimmed it once, skim it again;   610
If at all they resemble him, you may be sure it is
Because their shoals mirror his mists and obscurities,
As a mud-puddle seems deep as heaven for a minute,
While a cloud that floats o'er is reflected within it.

    "There comes———,[5] for instance; to see him's rare sport,   615
Tread in Emerson's tracks with legs painfully short;
How he jumps, how he strains, and gets red in the face,
To keep step with the mystagogue's natural pace!
He follows as close as a stick to a rocket,[1]
His fingers exploring the prophet's each pocket.   620
Fie, for shame, brother bard; with good fruit of your own,
Can't you let neighbor Emerson's orchards alone?
Besides, 'tis no use, you'll not find e'en a core,—
———[2] has picked up all the windfalls before.
They might strip every tree, and E. never would catch 'em,   625

---

4. Exhibition case.
5. I.e., "Channing," according to Leon
Howard's book on Lowell, *Victorian
Knight-Errant*: William Ellery Channing
(1818–1901) Thoreau's ne'er-do-well poetaster friend.

1. A fireworks device.
2. "Thoreau," according to Howard;
"windfalls": casual, unexpected legacies
(Thoreau was then not as well known
as Channing but he was more often on
the spot at Concord).

His Hesperides[3] have no rude dragon to watch 'em;
When they send him a dish-full, and ask him to try 'em,
He never suspects how the sly rogues came by 'em;
He wonders why 'tis there are none such his trees on,
And thinks 'em the best he has tasted this season.   630

* * *

"There is Hawthorne, with genius so shrinking and rare
That you hardly at first see the strength that is there;
A frame so robust, with a nature so sweet,   995
So earnest, so graceful, so solid, so fleet,
Is worth a descent from Olympus to meet;
'Tis as if a rough oak that for ages had stood,
With his gnarled bony branches like ribs of the wood,
Should bloom, after cycles of struggle and scathe,   1000
With a single anemone trembly and rathe;
His strength is so tender, his wildness so meek,
That a suitable parallel sets one to seek,—
He's a John Bunyan Fouqué, a Puritan Tieck;[4]
When Nature was shaping him, clay was not granted   1005
For making so full-sized a man as she wanted,
So, to fill out her model, a little she spared
From some finer-grained stuff for a woman prepared,
And she could not have hit a more excellent plan
For making him fully and perfectly man.   1010
The success of her scheme gave her so much delight,
That she tried it again, shortly after, in Dwight;[5]
Only, while she was kneading and shaping the clay,
She sang to her work in her sweet childish way,
And found, when she'd put the last touch to his soul,   1015
That the music had somehow got mixed with the whole.

* * *

"There comes Poe with his raven, like Barnaby Rudge,[6]   1295
Three-fifths of him genius and two-fifths sheer fudge,[1]
Who talks like a book of iambs and pentameters,
In a way to make people of common-sense damn metres,
Who has written some things quite the best of their kind,
But the heart somehow seems all squeezed out by the mind,   1300
Who—but hey-day! What's this? Messieurs Mathews[2] and Poe,

3. The garden of the golden apples in the Hesperides, the Isles of the Blest.
4. I.e., Hawthorne is didactic as John Bunyan (1628–88), English Baptist and author of *Pilgrim's Progress*, yet as fanciful as Friedrich Fouqué (1777–1843), German writer of fiction, including the fairy tale *Undine*. The same union of opposites is found in "Puritan Tieck," since Ludwig Tieck (1773–1853), is another famous German Romantic author of fanciful, unpuritanic tales.
5. John Sullivan Dwight (1813–93), Bostonian music critic and minor poet.
6. Halfwit, in Charles Dickens's book of that title (1841), whose companion is a knowing, evil-looking raven.
1. Humbug.

2. Cornelius Mathews (1817–89), New York editor and fiction writer, much ridiculed by some of his contemporaries, including Lowell, who quite unfairly lumps him together here with Poe; Mathews merely thought Longfellow was imitative, but Poe launched a virulent campaign against him as a plagiarist. The literary debates of the 1840s pitted champions of a great, peculiarly American literature against those who thought that the important American literature would come through following English models. At times (as in Melville's *Hawthorne and His Mosses*) the issues were simplified to New York City versus Boston, creativity versus imitation.

You mustn't fling mud-balls at Longfellow so,
Does it make a man worse that his character's such
As to make his friends love him (as you think) too much?
Why, there is not a bard at this moment alive   1305
More willing than he that his fellow should thrive;
While you are abusing him thus, even now
He would help either one of you out of a slough;
You may say that he's smooth and all that till you're hoarse,
But remember that elegance also is force;   1310
After polishing granite as much as you will,
The heart keeps its tough old persistency still;
Deduct all you can that still keeps you at bay,—
Why, he'll live till men weary of Collins and Gray;[3]
I'm not over-fond of Greek metres in English,[4]   1315
To me rhyme's a gain, so it be not too jinglish,
And your modern hexameter verses are no more
Like Greek ones than sleek Mr Pope is like Homer;[5]
As the roar of the sea to the coo of a pigeon is
So, compared to your moderns sounds old Melesigenes;[6]   1320
I may be too partial, the reason, perhaps, o't is
That I've heard the old blind man recite his own rhapsodies,
And my ear with that music impregnate may be,
Like the poor exiled shell with the soul of the sea,
Or as one can't bear Strauss[7] when his nature is cloven   1325
To its deeps within deeps by the stroke of Beethoven;
But, set that aside, and 'tis truth that I speak,
Had Theocritus[8] written in English, not Greek,
I believe that his exquisite sense would scarce change a line
In that rare, tender, virgin-like pastoral Evangeline.   1330
That's not ancient nor modern, its place is apart
Where Time has no sway, in the realm of pure Art,
'Tis a shrine of retreat from Earth's hubbub and strife
As quiet and chaste as the author's own life.

* * *

"What! Irving? thrice welcome, warm heart and fine brain,
You bring back the happiest spirit from Spain,[9]
And the gravest sweet humor, that ever were there
Since Cervantes met death in his gentle despair;   1440
Nay, don't be embarrassed, nor look so beseeching,—
I shan't run directly against my own preaching,
And, having just laughed at their Raphaels and Dantes,[1]

---

3. William Collins (1721–59) and Thomas Gray (1716–71), English poets.
4. Longfellow had attempted to adapt Greek hexameters to English poetry in *Evangeline* (1847).
5. The English poet Alexander Pope (1688–1744) *translated Homer's Iliad and Odyssey* into heroic couplets.
6. Homer (Melos-born).
7. Johann Strauss (1804–49), Austrian composer, known for his waltzes.
8. Greek pastoral poet of the third cen-

tury B.C.
9. Washington Irving had been Minister to Spain in the early 1840s. Miguel de Cervantes Saavedra (1547–1616), Spanish novelist, author of *Don Quixote*, a major influence on Irving's work.
1. "Their Raphaels and Dantes" refers to the string of American writers whom Duyckinck and Mathews have pressed upon Phoebus as creative geniuses, all as great as the greatest Italian painters and writers.

Go to setting you up beside matchless Cervantes;
But allow me to speak what I honestly feel,—   1445
To a true poet-heart add the fun of Dick Steele,
Throw in all of Addison,[2] *minus* the chill,
With the whole of that partnership's stock and good will,
Mix well, and, while stirring, hum o'er, as a spell,
The fine *old* English Gentleman,[3] simmer it well,   1450
Sweeten just to your own private liking, then strain,
That only the finest and clearest remain,
Let it stand out of doors till a soul it receives
From the warm lazy sun loitering down through green leaves,
And you'll find a choice nature, not wholly deserving   1455
A name either English or Yankee,—just Irving.

*   *   *

"There is Lowell, who's striving Parnassus to climb
With a whole bale of *isms* tied together with rhyme,
He might get on alone, spite of brambles and boulders,   1580
But he can't with that bundle he has on his shoulders,
The top of the hill he will ne'er come nigh reaching
Till he learns the distinction 'twixt singing and preaching;
His lyre has some chords that would ring pretty well,
But he'd rather by half make a drum of the shell,   1585
And rattle away till he's old as Methusalem,[4]
At the head of a march to the last New Jerusalem.

*   *   *

1848

2. Irving had plainly formed his prose style in part from reading the essays in *The Spectator* written by Sir Richard Steele (1672–1729), British essayist and dramatist, and Joseph Addison (1672–1719), English poet and essayist.

3. An allusion to an essay in Irving's *Bracebridge Hall.*
4. Methuselah, who lived 969 years (Genesis 5.27). The "last New Jerusalem": the latest utopian scheme for social reform.

WALT WHITMAN
1819–1892

Whitman was born on May 31, 1819, son of a Long Island farmer turned carpenter who moved the family into Brooklyn in 1823 during a building boom. The ancestors were undistinguished, but stories survived of some forceful characters among them, and Whitman's father was acquainted with powerful personalities like the aged Thomas Paine. Whitman left school at eleven to become an office boy in a law firm, then worked for a doctor; already he was enthralled with the novels of Sir Walter Scott. By twelve he was working in the printing office of a newspaper and contributing sentimental items. By fifteen, when his family moved back into the interior of Long Island, Whitman was on his own. Very early he reached full physical maturity, and in his mid-teens was contributing "pieces"—probably correct, conventional poems—to one of the best Manhattan papers, the *Mirror,* and often crossing the ferry from Brooklyn to attend debating societies and to use his journalist's passes at theaters in Manhattan. His rich fantasy life was fueled by numberless romantic novels. By sixteen he was a compositor in Manhattan, a journeyman printer. But two great fires in 1835 disrupted the printing industry, and as he turned seventeen he rejoined his family. For five years he taught intermittently at country and small-town schools, interrupting teaching to start a newspaper of his own in 1838 and to work briefly on another Long Island paper. Although forced into the exile of Long Island, he refused to compromise further with the sort of life he wanted. During his visits home he outraged his father by refusing to do farmwork. Although he was innovative in the classroom, he struck some of the farm families he boarded with as unwilling to fulfill his role of teacher outside school hours; the main charge against him was laziness. He was active in debating societies, however, and already thought of himself as a writer. By early 1840 he had started a series of "Sun-Down Papers from the Desk of a School-Master" for the Long Island *Democrat* and was writing poems. One of his stories prophetically culminated with the dream of writing "a wonderful and ponderous book."

Just before he turned twenty-one he went back to Manhattan, his teaching days over, and began work on Park Benjamin's *New World,* a literary weekly which pirated British novels; he also began a political career by speaking at Democratic rallies. Simultaneously he was publishing stories in the *Democratic Review,* the foremost magazine of the Democratic party. Before he was twenty-three, he became editor of a Manhattan daily, the *Aurora,* and briefly transformed himself into a sartorial dandy while he spiked his editorial columns with his high democratic hopes. He exulted in the extremes of the city, where the violence of street gangs was countered by the lectures of Emerson, and where even a young editor could get to know the poet Bryant (by livelihood, editor of the *Evening Post*). Fired from the *Aurora,* which publicly charged him with laziness, he wrote a temperance novel, *Franklin Evans, or the Inebriate,* for a one-issue extra of the *New World* late in 1842. For the next years he was journalist, hack writer, and a doughty minor politician. In 1845 he returned to Brooklyn, where he became a special contributor to the Long Island *Star,* assigned to Manhattan events including musical and theatrical engagements. Just before he was twenty-seven he took over the editorship of the Brooklyn *Eagle;* for years he had kept to his eccentric daily routine of apparently purposeless walks in which he absorbed metropolitan sights and sounds, and now he formed the habit of a daily swim and shower at a bathhouse. On the *Eagle* he did most of the literary reviews, handling books by Carlyle, Emerson, Melville, Margaret Fuller, George Sand, Goethe, and others. Like most Democrats, he was able to justify the Mexican War, and he hero-worshiped Zachary Taylor (on whom Melville was writing a series of satirical sketches). Linking territorial acquisition to personal and civic betterment, he was, in his nationalistic moods, capable of hailing the great American mission of "peopling the New World with a noble race." Yet by the beginning of 1848 he was fired from the *Eagle* because like Bryant he had become a Free-soiler, opposed to the acquisition of more slave territory. Taking a chance offer of newspaper work, he made a brief but vivid trip to New Orleans, his only extensive journey until late in life, when he made a trip into the West.

By the summer of 1848, Whitman was back in New York, starting experiments with poetry and, in August, serving as delegate to the Buffalo Free-Soil convention. In the next years he was profoundly influenced by his association with a group of Brooklyn artists. All through the 1840s he had attended operas on his newspaperman's passes, hearing the greatest singers of the time. He was most profoundly stirred by the coloratura soprano Marietta Alboni during the season of 1852–53; years later he said that a scene from *Norma* he cherished as one of his life's "rare and blessed bits of hours." He went so far as to say that but for the "emotions, raptures, uplifts" of opera he could never have written *Leaves of Grass*. In an effort to control the disposition of his time, he became a "house builder" around 1851 or 1852, perhaps acting as contractor sometimes, but also simply hiring out as a carpenter. By the early 1850s he had set a durable pattern of having discrete sets of friends simultaneously, the roughs and the artists, moving casually from one set to the other but seldom, if ever, mingling them. Living with his family, now back in Brooklyn, Whitman baffled and outraged them by ignoring regular mealtimes and appearing to loaf away his days in strolls, in reading at libraries, and in writing in the room he shared with a brother. Always self-taught, he undertook a more systematic plan of study. He became something of an expert on Egyptology through his trips to the Egyptian Museum on Broadway and his conversations with its proprietor; Egyptology taught Whitman cultural relativism just as Melville's experiences in the South Seas had taught him a decade before. He became a student of astronomy, attending lectures and reading recent books; much of the information went into the cosmic concepts in *Song of Myself* and other poems. Cutting articles out of the great British quarterlies and monthlies, Whitman annotated them and argued with them in the margins, developing in the process clear ideas about aesthetics for the first time and formulating his notions about pantheism. From Frances Wright's *A Few Days in Athens* (1822), he absorbed a neo-Epicurean sense of religion as the enemy of natural human pleasure and virtue. By about 1853 he had arrived at something like his special poetic form in the little poem *Pictures*. He had given up newspaper work for carpentry; around the end of 1854 he gave up that also, and simply wrote. By the spring of 1855 he was seeing his "wonderful and ponderous book" through the press, probably setting some of the type himself. *Leaves of Grass* was on sale within a day or two of the Fourth of July, 1855.

Facing the title page of this remarkable book was an engraving of a lounging working man, broad-hatted, bearded, shirt open at the neck to reveal a colored undershirt, the right arm akimbo, left hand in pants pocket, weight on the right leg. Such a man would hardly be expected to read verse, much less write it. The title page said simply *"Leaves of Grass"* and gave the place and date of publication as "Brooklyn, New York: 1855." The back of the title page named "Walter Whitman" as the man who had entered the work for copyright; that this was the author was confirmed by a line far down in the first poem (the one later retitled *Song of Myself*; here they were all entitled *Leaves of Grass*).

Following the copyright page was an untitled essay running to ten pages, double column—an intimidating mass of type, most of it in long paragraphs punctuated by sets of what looked like ellipses. The essay was oracular in tone, sweeping in message, the outgrowth of Whitman's long musing on the place of art in American life. One of its starting points was pretty clearly Emerson's *The Poet*, especially the provocative list of some of the incomparable materials awaiting the tyrannous eye of the great American poet. It circled back to its major points much like Emerson's own essays, focusing primarily on the sort of poet America required and the sort of poetry he would write. The reader soon had every reason to suspect that the "Walter Whitman" of the copyright page might consider himself that poet and *Leaves of Grass* an example of that poetry.

In this manifesto Whitman declared that the American poet would not repudiate past beliefs but would incorporate them into newer ones, just as Americans are composed of all peoples. To be commensurate with this new American stock, the poet would incarnate the American geography, occupations, and the people themselves in a new and transcendent poetic form. The great poet would find encouragement and support in the sciences and in branches of history, since his poetry would not be escapist and otherworldly but solidly tied to verifiable knowledge. He would never populate a tree with a hamadryad, much less rue the vanishing of that mythological creature, and never write a poem (like John Milton's *Paradise Lost*) which would show God as "contending against some being or influence." This poet would have a sense of ultimate causality—that cruelty and goodness perpetuate themselves, that "no result exists now without being from its long antecedent result." By his capacity for encouraging and exalting others, the poet would soon replace the priest as servant to the people; then at last every man would be his own priest.

The great American poet would create both new forms and new subject matter for poetry. Rhyme would not be primary, if used at all; uniformity of stanzaic pattern would be abandoned. Whitman was even clearer about the new content. The American poetry would not echo the melancholy complaints of the Graveyard school nor proliferate the moral precepts of didactic writers like Longfellow. Exaggeration of both style and subject would be replaced by "genuineness," by respect for the way things really are.

Whitman was confident in his linguistic resources. He paid tribute to the strength of the English language, calling it "the medium that shall well nigh express the inexpressible." He was also confident that he could broaden the literary class—that small, elite body—to include all the people. He ended with a daring prediction: "The proof of a poet is that his country absorbs him as affectionately as he has absorbed it." Whitman would refine his "programme" over the next three and a half decades, but not alter its basic tenets.

Yet the prefatory essay did not quite prepare the reader for the poems that followed. The recitation of some of the incomparable materials awaiting the American poet suggested the comprehensiveness and brilliance of detail in some of the poetic catalogues (in fact, portions of the preface were later worked into poems); the declarations about science prepared for the sophisticated understanding the poems revealed of geology and astronomy, if not for the poetic uses of that knowledge; and the tributes to the American people partially suggested the profound tenderness which would be manifested toward all people in the poems. Yet not even the occasional ironic passages of the essay suggested that the poems which followed would include master strokes of comedy; nothing adequately prepared for the intense and explicit sexuality and the psychological complexity of the poems; and not even his comments on form suggested that Whitman had already achieved a new form splendidly appropriate to his democratic subject matter and had already become a supreme master of the English language.

The publication of *Leaves of Grass* did not immediately change Whitman's life. His father died just after it appeared, and support of his mother and a feeble-minded brother devolved more and more on Whitman. He sent copies of his book out broadcast, and got an immediate response from Emerson greeting him "at the beginning of a great career, which yet must have had a long foreground somewhere, for such a start." As weeks passed with few reviews, he wrote a few himself to be published anonymously, and in October he let Horace Greeley's *Tribune* print Emerson's letter. Unfazed by his own effrontery, he put clippings of the letter in presentation copies, to Longfellow among others. While Whitman was angling for reviews in England and working on an expansion of the book, Emerson visited him (in December of 1855), and in the fall of 1856 Bronson Alcott, Thoreau, and others came out to his house in Brooklyn. Whitman continued to do miscellaneous journalism, mainly for a weekly magazine called *Life Illustrated*, and wrote a long political tract on *The Eighteenth Presidency!* which he never distributed. The excitement of the election year was reflected in the excessive nationalism of some of the new poems as well as the pomposity of titles, but among the new poems in the second edition (1856) was the great *Sun-Down Poem*, later called *Crossing Brooklyn Ferry*. Whitman flaunted Emerson's praise on the spine of the new edition and in the back printed the whole letter, following it with an open letter to the Master at Concord in which Whitman announced his determination "to meet people and The States face to face, to confront them with an American rude tongue."

In the late 1850s, Whitman's main statements on the national crisis were contained in editorials he wrote during his editorship of the Brooklyn *Times*, from 1857 to 1859, but he was also working on a political, sexual, and poetic programme which he could bring to the people of the states as a wandering lecturer; he envisioned for himself a way of life that would fuse the lecture tours of men like Emerson with his own love of the open road and good comrades. He lectured locally, but nothing came of the grander scheme, and his new poems reflected very little of the national situation. One group of twelve poems, called "Calamus" (those published under that name in 1860 numbered forty-five poems), told a fairly coherent story of homosexual longings and torment from the recognition of such feelings and the difficulty of their being satisfied. During these years Whitman was admitting to himself, perhaps for the first time, the homoerotic nature of much of his sexual urge. The sexual torment of these years was compounded by his increasingly sordid family life. His brother Andrew's widow became a prostitute and neglected her children; an older sister was embittered by marriage to an unsuccessful artist; a brother crazed by syphilis had violent spells in which he threatened their mother; a feeble-minded brother shared a room with Whitman; and his mother, always idealized in Whitman's poems and his recollections, seems to have been a whining, self-pitying nag. This family situation, like the national affairs, went into the new poems only obliquely. By late 1859 Whitman had in hand the "Calamus" and "Children of Adam" poems (then called "Enfans d'Adam") and others such as *A Child's Reminiscence* (*Out of the Cradle Endlessly Rocking*) in nearly final form. Early the next year he received an opportune letter from the Boston firm of Thayer & Eldridge: "We are young men. We 'celebrate' ourselves by acts. Try us. You can do us good. We can do you good—pecuniarily." For the first time, Whitman had a publisher.

Emerson welcomed him in Boston, arguing through a two-hour walk in the cold Common that Whitman should not publish the "Children of Adam" poems. Characteristically, Whitman refused to compromise. When Emerson tried to introduce him into the Boston literary coterie, the Saturday Club, Longfellow, Holmes, and Lowell all objected, and apparently the families of Emerson, Thoreau, and Alcott refused to have him invited to Concord. Yet Whitman met a few kindred spirits during the weeks he labored over the proofs of his third edition, which was printed in May, 1860. Besides the accustomed attacks, there were defenses, including one by the actress Adah Isaacks Menken, and, for the first time, there were parodies, a sure sign of public interest. But Thayer & Eldridge went bankrupt, and Whitman was left again with the book on his own hands.

For society at the turn of the decade Whitman had, as usual, his separate groups of friends, in one group stage drivers (i.e., drivers of the horse-pulled buses on which he was an inveterate passenger) and in another literary, publishing, and theatrical people, especially the bohemian habitués of the famous Pfaff's saloon on lower Broadway (where the rather stuffy young W. D. Howells once went slumming and was warmed by the grasp of Whitman's "mighty fist" at their meeting). For years Whitman had paid cheering visits to prisoners and had made regular visits to sick stage drivers. Almost imperceptibly, his role of visitor of the sick merged into his Civil War services as hospital attendant. Just before Christmas, 1862, Whitman went to Washington to find his brother George, who had been reported wounded. The injury was slight, but Whitman stayed on with his brother in camp (such was the casual fashion of the war), the two sharing a tent with other soldiers. Returning to Washington, he rented a small room, found a job as copyist in the Paymaster's office, and began visiting Brooklyn soldiers in the hospitals. Without premeditation, Whitman yielded to the irresistible appeal of the occasion: his informality, gentleness, resourcefulness, and lack of preachiness were uniquely required. Looking older than his age (only the early forties), he could serve as benevolent father to the men, and they returned his extraordinary sympathy and love; his friends became accustomed to seeing recovered soldiers stop him in the street to hug and kiss him. It has been thought that the hospital experience was a sublimation of Whitman's homosexual feelings, but Gay Wilson Allen has shown that those emotions permeate some of his correspondence with soldiers. As Allen says of the letters to Sergeant Tom Sawyer, they are "not easy for a modern critic to interpret, for they were evidently motivated by a mixture of vicarious paternalism, longing for companionship, and some rather confused erotic impulses that perhaps Whitman himself did not clearly understand." Whatever Whitman's mixture of emotions, his role of wound-dresser was heroic, and it eventually undercut his buoyant physical health.

During the war, Whitman's deepest emotions and energies were reserved for his hospital work, though toward the end he wrote a series of war poems designed to trace his own varying attitudes toward the conflict, from his early near-mindless jingoism to something quite rare in American poetry up to that time, a dedication to simple realism. Whitman later wrote a section in *Specimen Days* showing that "The Real War Will Never Get in the Books," but to incorporate the real war into a book of poetry became one of the dominant impulses of the *Drum-Taps* collection (1865). After Lincoln's assassination Whitman delayed the new volume until it could include in a "sequel" *When Lilacs Last in the Dooryard Bloom'd*, his masterpiece of the 1860s. Even as *Drum-Taps* was being published, Whitman was revising a copy of the Boston edition of *Leaves of Grass* at his desk in the Department of the Interior. The new Secretary read the annotated copy and abruptly fired him. The consequences might have been minor: Whitman's friend William O'Connor quickly got him a new post in the Attorney General's office, and Whitman had been dismissed before without catastrophic reactions, though this was the first time he had been fired because of the sexual passages in *Leaves of Grass*. But the incident turned O'Connor from a devoted friend into a disciple who quickly began writing a book (with Whitman's help in supplying information and documents) called *The Good Gray Poet* (January, 1866). It was a piece of pure hagiography, in which Whitman was identified with Jesus. O'Connor's book, coming out simultaneously with *Drum-Taps*, polarized opinion, with negative immediate effects, but in the long run it strengthened Whitman's determination not to yield to censorship or apologize for his earlier poems, and it set a pattern by which other remarkable men and women would be drawn to Whitman as disciples, seeking to care for his few physical needs—minimal food and shelter—while working for his reputation as a great poet.

For several years Whitman continued as a clerk in the Attorney General's office, living most of the time in a bare, unheated room but gaining access to good lights for nighttime reading in his government office. He continued to rework *Leaves of Grass*, incorporating *Drum-Taps* into it in 1867, and with his friends' help continued to propagandize for its acceptance. After much correspondence involving Whitman, O'Connor, and a newer admirer (John Burroughs, the naturalist), William Michael Rossetti published a volume of Whitman's poems in London early in 1868. Though it was a selected and even an expurgated edition, it created many English admirers. As an ex-newspaperman, Whitman was accustomed to having immediate outlets for his political thoughts and general observations on the American scene. Such ruminations which in earlier years might have gone into editorials went into essays in the *Galaxy* during 1867 and 1868, then (in expanded form) into *Democratic Vistas* (1870), a passionate look to the future of democracy and democratic literature in America, based upon his realistic appraisal of postwar culture. Another prose work, *Specimen Days*, published in book form in 1882, has affinities with Whitman's early editorial records of strolls through the city, but it is even more intensely personal, the record of representative days in the life of an American who had lived in the midst of great national events and who had kept alert to nature and his own mind and body. Through the late 1860s and early 1870s Whitman's compartmentalized life went on. He formed an emotional relationship with Peter Doyle, a young streetcar driver, about the time he was beginning in a small way to be treated as a literary lion. He was still enduring devastating torments over his sexual drives.

The Washington years, a time of slow, faltering growth in reputation, marred by severe setbacks and complications, ended early in 1873, when Whitman suffered a paralytic stroke. His mother died a few months later, and Whitman joined his brother George's household in Camden, New Jersey, intending only a temporary move during his recuperation. His convalescence was bitterly lonely, for his well-meaning brother provided none of the intellectual companionship of a Burroughs nor the emotional response of a Peter Doyle. During the second year of his illness, the government decided not to hold his job for him any longer, and he became dependent on occasional publication in newspapers and magazines—not an easy market, since his genteel enemies either ignored him in print or joined in a cabal to exclude him from some major publishing organs such as *Scribner's*. The 1867 edition of *Leaves of Grass* had involved much reworking and rearrangement, and the fifth edition (1871) continued that process, with many of the original *Drum-Taps* poems being distributed throughout the book and with an assemblage of old and new poems in a large new section, "Passage to India." The "Centennial Edition" of 1876 was a reissue of the 1871 edition and was most notable for the way his English admirers got funds to him by having important literary people subscribe to it. American public opinion was gradually swayed by new evidences that the invalid in Camden could command the respect of Alfred Lord Tennyson, the poet laureate, and many other famous British writers. Not even the most puritanic American critics could hold their ranks firm in the face of such extravagant admiration, however rough and outrageous his poetry. Yet in 1881, when the reputable Boston firm of James R. Osgood & Co. printed the sixth edition of *Leaves of Grass*, the Boston District Attorney threatened to prosecute on the grounds of obscenity, and Whitman found himself with the plates on his hands and no publisher until he had impressions from the Osgood plates made by Philadelphia printers Rees Welsh & Co. in 1882 and David McKay thereafter. The "deathbed" edition of 1891–92 was in fact a reissue of the 1881 edition with the addition of two later groups of poems, "Sands at Seventy" (from *November Boughs*, which Whitman published in 1888) and "Good-bye My Fancy" (from the 1891 collection of that name).

Of all the American writers of the nineteenth century, Whitman offers the most inspiring example of fidelity to his art. While Hawthorne let marriage become his true career, and while Melville ceased writing for a public that would not accept him, Whitman persisted. (James persisted also, but he was equipped with material, educational, and social advantages Whitman lacked.) Outraging his employers and his family by his odd hours and the semblance of mere loafing, outraging his well-wishers by refusing to compromise on minor points that might have gained him fuller acceptance, finagling reviews, reviewing himself, writing admiring accounts of his work for others to sign, shocking some of his followers by refusing to give autographs gratis, Whitman kept on, like what he called some high-and-dry "hard-cased dilapidated grim ancient shell-fish or time-bang'd conch," uncompromising to the end, never bowing to the materialism and puerilities of nineteenth-century America. Appropriately, when he finally accumulated a few worldly belongings about him at Camden, he managed to give a nautical cast to his room, for eccentric as he seemed, crotchety, stubborn, Whitman was a literary equivalent of Melville's Bulkington in *Moby-Dick*, willing to renounce the comforts of the shore, all normal earthly felicity, for a life of the intellect and the imagination. He died at Camden on March 26, 1892, secure in the knowledge that he had held unwaveringly true to his art and to his role as an artist who had made that art prevail.

From INSCRIPTIONS[1]
When I Read the Book

When I read the book, the biography famous,
And is this then (said I) what the author calls a man's life?
And so will some one when I am dead and gone write my life?
(As if any man really knew aught of my life,
Why even I myself I often think know little or nothing of
 my real life,   5
Only a few hints, a few diffused faint clews and indirections
I seek for my own use to trace out here.)

1867, 1871

Beginning My Studies

Beginning my studies the first step pleas'd me so much,
The mere fact consciousness, these forms, the power of motion,
The least insect or animal, the senses, eyesight, love,
The first step I say awed me and pleas'd me so much,
I have hardly gone and hardly wish'd to go any farther,   5
But stop and loiter all the time to sing it in ecstatic songs.

1865, 1871

Song of Myself[1]

1

I celebrate myself, and sing myself,
And what I assume you shall assume,
For every atom belonging to me as good belongs to you.

I loafe and invite my soul,
I lean and loafe at my ease observing a spear of summer grass.   5

My tongue, every atom of my blood, form'd from this soil, this air,
Born here of parents born here from parents the same, and their
 parents the same,

---

1. This title was first used for the opening poems of *Leaves of Grass* in 1871; in 1881 the number of poems was increased from 9 to 24, a result not so much of new composition as of a shifting of the contents of the various sections of the book.

1. Undeniably uncouth yet carefully fashioned, the poem was the outgrowth of years of labor; anticipatory notebook entries survive from the late 1840s and related poetic passages from the early 1850s. It first appeared in the 1855 *Leaves of Grass* without a title and with no internal divisions. In the second edition (1856), the title was *Poem of Walt Whitman, an American*; then it became *Walt Whitman* in 1860 and remained under that title until 1881, when it finally became *Song of Myself*. During all of this the poem itself was changed only

slightly. Perhaps the most important changes made the purpose of the speaker's journeying more explicit: he is absorbing all that he sees in order to write the poem.

For textual information and guidance through Whitman's vocabulary, this edition has often drawn upon the work of Harold W. Blodgett and Sculley Bradley in *"Leaves of Grass": Comprehensive Reader's Edition* (New York: New York University Press, 1965; Norton, 1968), reprinted with corrections and additions in the Bradley and Blodgett Norton Critical Edition of *Leaves of Grass* (New York: Norton, 1973).

Unless otherwise indicated, all texts of Whitman's poems are those of the "Deathbed edition," the green hardbound issue of 1891–92.

I, now thirty-seven years old in perfect health begin,
Hoping to cease not till death.

Creeds and schools in abeyance,   10
Retiring back a while sufficed at what they are, but never forgotten,
I harbor for good or bad, I permit to speak at every hazard,
Nature without check with original energy.

2

Houses and rooms are full of perfumes, the shelves are crowded with
    perfumes,
I breathe the fragrance myself and know it and like it,   15
The distillation would intoxicate me also, but I shall not let it.

The atmosphere is not a perfume, it has no taste of the distillation,
    it is odorless,
It is for my mouth forever, I am in love with it,
I will go to the bank by the wood and become undisguised and
    naked,
I am mad for it to be in contact with me.   20

The smoke of my own breath,
Echoes, ripples, buzz'd whispers, love-root, silk-thread, crotch and
    vine,
My respiration and inspiration, the beating of my heart, the passing
    of blood and air through my lungs,
The sniff of green leaves and dry leaves, and of the shore and dark-
    color'd sea-rocks, and of hay in the barn,
The sound of the belch'd words of my voice loos'd to the eddies of
    the wind,   25
A few light kisses, a few embraces, a reaching around of arms,
The play of shine and shade on the trees as the supple boughs wag,
The delight alone or in the rush of the streets, or along the fields
    and hill-sides,
The feeling of health, the full-noon trill, the song of me rising from
    bed and meeting the sun.

Have you reckon'd a thousand acres much? have you reckon'd the
    earth much?   30
Have you practis'd so long to learn to read?
Have you felt so proud to get at the meaning of poems?

Stop this day and night with me and you shall possess the origin of
    all poems,
You shall possess the good of the earth and sun, (there are millions
    of suns left,)
You shall no longer take things at second or third hand, nor look
    through the eyes of the dead, nor feed on the spectres in
    books,   35
You shall not look through my eyes either, nor take things from me,
You shall listen to all sides and filter them from your self.

3

I have heard what the talkers were talking, the talk of the beginning and the end,
But I do not talk of the beginning or the end.

There was never any more inception than there is now,   40
Nor any more youth or age than there is now,
And will never be any more perfection than there is now,
Nor any more heaven or hell than there is now.

Urge and urge and urge,
Always the procreant urge of the world.

Out of the dimness opposite equals advance, always substance and
 increase, always sex,   45
Always a knit of identity, always distinction, always a breed of life.

To elaborate is no avail, learn'd and unlearn'd feel that it is so.

Sure as the most certain sure, plumb in the uprights, well entretied,[2]
 braced in the beams,
Stout as a horse, affectionate, haughty, electrical,   50
I and this mystery here we stand.

Clear and sweet is my soul, and clear and sweet is all that is not my
 soul.

Lack one lacks both, and the unseen is proved by the seen,
Till that becomes unseen and receives proof in its turn.

Showing the best and dividing it from the worst age vexes age,   55
Knowing the perfect fitness and equanimity of things, while they
 discuss I am silent, and go bathe and admire myself.

Welcome is every organ and attribute of me, and of any man hearty
 and clean,
Not an inch nor a particle of an inch is vile, and none shall be less
 familiar than the rest.

I am satisfied—I see, dance, laugh, sing;
As the hugging and loving bed-fellow sleeps at my side through the
 night, and withdraws at the peep of the day with stealthy
 tread,   60
Leaving me baskets cover'd with white towels swelling the house
 with their plenty,
Shall I postpone my acceptation and realization and scream at my
 eyes,
That they turn from gazing after and down the road,
And forthwith cipher[3] and show me to a cent,

2. Cross-braced.   3. Calculate.

Exactly the value of one and exactly the value of two, and which is
    ahead?   65

4

Trippers and askers surround me,
People I meet, the effect upon me of my early life or the ward and
    city I live in, or the nation,
The latest dates, discoveries, inventions, societies, authors old and
    new,
My dinner, dress, associates, looks, compliments, dues,
The real or fancied indifference of some man or woman I love,   70
The sickness of one of my folks or of myself, or ill-doing or loss or
    lack of money, or depressions or exaltations,
Battles, the horrors of fratricidal war, the fever of doubtful news,
    the fitful events;
These come to me days and nights and go from me again,
But they are not the Me myself.

Apart from the pulling and hauling stands what I am,   75
Stands amused, complacent, compassionating, idle, unitary,
Looks down, is erect, or bends an arm on an impalpable certain rest,
Looking with side-curved head curious what will come next,
Both in and out of the game and watching and wondering at it.

Backward I see in my own days where I sweated through fog with
    linguists and contenders,   80
I have no mockings or arguments, I witness and wait.

5

I believe in you my soul, the other I am must not abase itself to you,
And you must not be abased to the other.

Loafe with me on the grass, loose the stop from your throat,
Not words, not music or rhyme I want, not custom or lecture, not
    even the best,   85
Only the lull I like, the hum of your valvèd voice.

I mind how once we lay such a transparent summer morning,
How you settled your head athwart my hips and gently turn'd over
    upon me,
And parted the shirt from my bosom-bone, and plunged your tongue
    to my bare-stript heart,
And reach'd till you felt my beard, and reach'd till you held my
    feet.   90

Swiftly arose and spread around me the peace and knowledge that
    pass all the argument of the earth,
And I know that the hand of God is the promise of my own,
And I know that the spirit of God is the brother of my own,
And that all the men ever born are also my brothers, and the women
    my sisters and lovers,

And that a kelson[4] of the creation is love,   95
And limitless are leaves stiff or drooping in the fields,
And brown ants in the little wells beneath them,
And mossy scabs of the worm fence, heap'd stones, elder, mullein
and poke-weed.

6

A child said *What is the grass?* fetching it to me with full hands;
How could I answer the child? I do not know what it is any more
than he.   100

I guess it must be the flag of my disposition, out of hopeful green
stuff woven.

Or I guess it is the handkerchief of the Lord,
A scented gift and remembrancer designedly dropt,
Bearing the owner's name someway in the corners, that we may see
and remark, and say *Whose?*

Or I guess the grass is itself a child, the produced babe of the veg-
etation.   105

Or I guess it is a uniform hieroglyphic,
And it means, Sprouting alike in broad zones and narrow zones,
Growing among black folks as among white,
Kanuck, Tuckahoe, Congressman, Cuff,[5] I give them the same, I
receive them the same.

And now it seems to me the beautiful uncut hair of graves.   110

Tenderly will I use you curling grass,
It may be you transpire from the breasts of young men,
It may be if I had known them I would have loved them,
It may be you are from old people, or from offspring taken soon out
of their mothers' laps,
And here you are the mothers' laps.   115

This grass is very dark to be from the white heads of old mothers,
Darker than the colorless beards of old men,
Dark to come from under the faint red roofs of mouths.

O I perceive after all so many uttering tongues,
And I perceive they do not come from the roofs of mouths for
nothing.   120

I wish I could translate the hints about the dead young men and
women,

---

4. A basic structural unit; a keelson or
kelson is a reinforcing timber bolted to
the keel (backbone) of a ship.

5. French Canadian, Virginian (from eaters of an Indian foodplant, tuckahoe),
Negro (from an African word, *cuffee*).

And the hints about old men and mothers, and the offspring taken
    soon out of their laps.

What do you think has become of the young and old men?
And what do you think has become of the women and children?

They are alive and well somewhere,   125
The smallest sprout shows there is really no death,
And if ever there was it led forward life, and does not wait at the
    end to arrest it,
And ceas'd the moment life appear'd.

All goes onward and outward, nothing collapses,
And to die is different from what any one supposed, and luckier. 130

7

Has any one supposed it lucky to be born?
I hasten to inform him or her it is just as lucky to die, and I know it.

I pass death with the dying and birth with the new-wash'd babe, and
    am not contain'd between my hat and boots,
And peruse manifold objects, no two alike and every one good,
The earth good and the stars good, and their adjuncts all good. 135

I am not an earth nor an adjunct of an earth,
I am the mate and companion of people, all just as immortal and
    fathomless as myself,
(They do not know how immortal, but I know.)

Every kind for itself and its own, for me mine male and female,
For me those that have been boys and that love women,   140
For me the man that is proud and feels how it stings to be slighted,
For me the sweet-heart and the old maid, for me mothers and the
    mothers of mothers,
For me lips that have smiled, eyes that have shed tears,
For me children and the begetters of children.

Undrape! you are not guilty to me, nor stale nor discarded,   145
I see through the broadcloth and gingham whether or no,
And am around, tenacious, acquisitive, tireless, and cannot be shaken
    away.

8

The little one sleeps in its cradle,
I lift the gauze and look a long time, and silently brush away flies
    with my hand.

The youngster and the red-faced girl turn aside up the bushy hill,   150
I peeringly view them from the top.

The suicide sprawls on the bloody floor of the bedroom,
I witness the corpse with its dabbled hair, I note where the pistol
    has fallen.

The blab of the pave, tires of carts, sluff of boot-soles, talk of the
    promenaders,
The heavy omnibus, the driver with his interrogating thumb, the
    clank of the shod horses on the granite floor,   155
The snow-sleighs, clinking, shouted jokes, pelts of snow-balls,
The hurrahs for popular favorites, the fury of rous'd mobs,
The flap of the curtain'd litter, a sick man inside borne to the
    hospital,
The meeting of enemies, the sudden oath, the blows and fall,
The excited crowd, the policeman with his star quickly working his
    passage to the centre of the crowd,   160
The impassive stones that receive and return so many echoes,
What groans of over-fed or half-starv'd who fall sunstruck or in fits,
What exclamations of women taken suddenly who hurry home and
    give birth to babes,
What living and buried speech is always vibrating here, what howls
    restrain'd by decorum,
Arrests of criminals, slights, adulterous offers made, acceptances,
    rejections with convex lips,   165
I mind them or the show or resonance of them—I come and I
    depart.

9

The big doors of the country barn stand open and ready,
The dried grass of the harvest-time loads the slow-drawn wagon,
The clear light plays on the brown gray and green intertinged,
The armfuls are pack'd to the sagging mow.   170

I am there, I help, I came stretch'd atop of the load,
I felt its soft jolts, one leg reclined on the other,
I jump from the cross-beams and sieze the clover and timothy,
And roll head over heels and tangle my hair full of wisps.

10

Alone far in the wilds and mountains I hunt,   175
Wandering amazed at my own lightness and glee,
In the late afternoon choosing a safe spot to pass the night,
Kindling a fire and broiling the fresh-kill'd game,
Falling asleep on the gather'd leaves with my dog and gun by my
    side.

The Yankee clipper is under her sky-sails, she cuts the sparkle and
    scud,   180
My eyes settle the land, I bend at her prow or shout joyously from
    the deck.

The boatmen and clam-diggers arose early and stopt for me,
I tuck'd my trowser-ends in my boots and went and had a good
    time;
You should have been with us that day round the chowder-kettle.

I saw the marriage of the trapper in the open air in the far west, the
    bride was a red girl,   185
Her father and his friends sat near cross-legged and dumbly smok-
    ing, they had moccasins to their feet and large thick blankets
    hanging from their shoulders,
On a bank lounged the trapper, he was drest mostly in skins, his
    luxuriant beard and curls protected his neck, he held his bride
    by the hand,
She had long eyelashes, her head was bare, her coarse straight locks
    descended upon her voluptuous limbs and reach'd to her feet.

The runaway slave came to my house and stopt outside,
I heard his motions crackling the twigs of the woodpile,   190
Through the swung half-door of the kitchen I saw him limpsy[6] and
    weak,
And went where he sat on a log and led him in and assured him,
And brought water and fill'd a tub for his sweated body and bruis'd
    feet,
And gave him a room that enter'd from my own, and gave him some
    coarse clean clothes,
And remember perfectly well his revolving eyes and his awkward-
    ness,   195
And remember putting plasters on the galls of his neck and ankles;
He staid with me a week before he was recuperated and pass'd
    north,
I had him sit next me at table, my fire-lock lean'd in the corner.

11

Twenty-eight young men bathe by the shore,
Twenty-eight young men and all so friendly;   200
Twenty-eight years of womanly life and all so lonesome.

She owns the fine house by the rise of the bank,
She hides handsome and richly drest aft the blinds of the window.

Which of the young men does she like the best?
Ah the homeliest of them is beautiful to her.   205

Where are you off to, lady? for I see you,
You splash in the water there, yet stay stock still in your room.

Dancing and laughing along the beach came the twenty-ninth
    bather,
The rest did not see her, but she saw them and loved them.

6. Limping or swaying.

The beards of the young men glisten'd with wet, it ran from their
    long hair,   210
Little streams pass'd all over their bodies.

An unseen hand also pass'd over their bodies,
It descended tremblingly from their temples and ribs.

The young men float on their backs, their white bellies bulge to the
    sun, they do not ask who seizes fast to them,
They do not know who puffs and declines with pendant and bending
    arch,   215
They do not think whom they souse with spray.

12

The butcher-boy puts off his killing-clothes, or sharpens his knife at
    the stall in the market,
I loiter enjoying his repartee and his shuffle and break-down.[7]

Blacksmiths with grimed and hairy chests environ the anvil,
Each has his main-sledge, they are all out, there is a great heat in
    the fire.   220

From the cinder-strew'd threshold I follow their movements,
The lithe sheer of their waists plays even with their massive arms,
Overhand the hammers swing, overhand so slow, overhand so sure,
They do not hasten, each man hits in his place.

13

The negro holds firmly the reins of his four horses, the block swags
    underneath on its tied-over chain,   225
The negro that drives the long dray of the stone-yard, steady and
    tall he stands pois'd on one leg on the string-piece,[8]
His blue shirt exposes his ample neck and breast and loosens over
    his hip-band,
His glance is calm and commanding, he tosses the slouch of his hat
    away from his forehead,
The sun falls on his crispy hair and mustache, falls on the black of
    his polish'd and perfect limbs.

I behold the picturesque giant and love him, and I do not stop
    there,   230
I go with the team also.

In me the caresser of life wherever moving, backward as well as for-
    ward sluing,
To niches aside and junior[9] bending, not a person or object missing,
Absorbing all to myself and for this song.

7. Two favorite minstrel-show dances, the first involving the sliding of feet across the floor, the second faster and noiser.

8. Long, heavy timber used to keep a load in place.
9. Smaller.

Oxen that rattle the yoke and chain or halt in the leafy shade, what
    is that you express in your eyes?   235
It seems to me more than all the print I have read in my life.

My tread scares the wood-drake and wood-duck on my distant and
    day-long ramble,
They rise together, they slowly circle around.

I believe in those wing'd purposes,
And acknowledge red, yellow, white, playing within me,   240
And consider green and violet and the tufted crown[1] intentional,
And do not call the tortoise unworthy because she is not something
    else,
And the jay in the woods never studied the gamut,[2] yet trills pretty
    well to me,
And the look of the bay mare shames silliness out of me.

14

The wild gander leads his flock through the cool night,   245
Ya-honk he says, and sounds it down to me like an invitation,
The pert may suppose it meaningless, but I listening close,
Find its purpose and place up there toward the wintry sky.

The sharp-hoof'd moose of the north, the cat on the house-sill, the
    chickadee, the prairie-dog,
The litter of the grunting sow as they tug at her teats,   250
The brood of the turkey-hen and she with her half-spread wings,
I see in them and myself the same old law.

The press of my foot to the earth springs a hundred affections,
They scorn the best can do to relate them.

I am enamour'd of growing out-doors,   255
Of men that live among cattle or taste of the ocean or woods,
Of the builders and steerers of ships and the wielders of axes and
    mauls, and the drivers of horses,
I can eat and sleep with them week in and week out.

What is commonest, cheapest, nearest, easiest, is Me,
Me going in for my chances, spending for vast returns,   260
Adorning myself to bestow myself on the first that will take me,
Not asking the sky to come down to my good will,
Scattering it freely forever.

15

The pure contralto sings in the organ loft,
The carpenter dresses his plank, the tongue of his foreplane whistles
    its wild ascending lisp,   265

1. Of the wood drake.   2. The series of recognized musical notes.

The married and unmarried children ride home to their Thanks-
giving dinner,
The pilot seizes the king-pin, he heaves down with a strong arm,
The mate stands braced in the whale-boat, lance and harpoon are
ready,
The duck-shooter walks by silent and cautious stretches,
The deacons are ordain'd with cross'd hands at the altar,   270
The spinning-girl retreats and advances to the hum of the big wheel,
The farmer stops by the bars[3] as he walks on a First-day[4] loafe and
looks at the oats and rye,
The lunatic is carried at last to the asylum a confirm'd case,
(He will never sleep any more as he did in the cot in his mother's
bed-room;)
The jour printer[5] with gray head and gaunt jaws works at his
case,   275
He turns his quid of tobacco while his eyes blurr with the manu-
script;
The malform'd limbs are tied to the surgeon's table,
What is removed drops horribly in a pail;
The quadroon girl is sold at the auction-stand, the drunkard nods
by the bar-room stove,
The machinist rolls up his sleeves, the policeman travels his beat,
the gate-keeper marks who pass,   280
The young fellow drives the express-wagon, (I love him, though I
do not know him;)
The half-breed straps on his light boots to compete in the race,
The western turkey-shooting draws old and young, some lean on
their rifles, some sit on logs,
Out from the crowd steps the marksman, takes his position, levels
his piece;
The groups of newly-come immigrants cover the wharf or levee, 285
As the woolly-pates hoe in the sugar-field, the overseer views them
from his saddle,
The bugle calls in the ball-room, the gentlemen run for their part-
ners, the dancers bow to each other,
The youth lies awake in the cedar-roof'd garret and harks to the
musical rain,
The Wolverine[6] sets traps on the creek that helps fill the Huron,
The squaw wrapt in her yellow-hemm'd cloth is offering moccasins
and bead-bags for sale,   290
The connoisseur peers along the exhibition-gallery with half-shut
eyes bent sideways,
As the deck-hands make fast the steamboat the plank is thrown for
the shore-going passengers,
The young sister holds out the skein while the elder sister winds it
off in a ball, and stops now and then for the knots,

3. Of a rail fence.
4. Sunday. Whitman frequently uses the numerical Quaker substitutes for the customary pagan names of days and months.
5. A journeyman printer is one who has passed his apprenticeship and is fully qualified for all professional work; in this usage, Whitman may imply that the man works by the day, without a steady job.
6. Inhabitant of Michigan.

The one-year wife is recovering and happy having a week ago borne
    her first child,
The clean-hair'd Yankee girl works with her sewing-machine or in
    the factory or mill,   295
The paving-man[7] leans on his two-handed rammer, the reporter's
    lead flies swiftly over the note-book, the sign-painter is lettering
    with blue and gold,
The canal boy trots on the tow-path, the book-keeper counts at his
    desk, the shoemaker waxes his thread,
The conductor beats time for the band and all the performers fol-
    low him,
The child is baptized, the convert is making his first professions,
The regatta is spread on the bay, the race is begun, (how the white
    sails sparkle!)   300
The drover watching his drove sings out to them that would stray,
The pedler sweats with his pack on his back, (the purchaser hig-
    gling about the odd cent;)
The bride unrumples her white dress, the minute-hand of the clock
    moves slowly,
The opium-eater reclines with rigid head and just-open'd lips,
The prostitute draggles her shawl, her bonnet bobs on her tipsy and
    pimpled neck,   305
The crowd laugh at her blackguard oaths, the men jeer and wink to
    each other,
(Miserable! I do not laugh at your oaths nor jeer you;)
The President holding a cabinet council is surrounded by the great
    Secretaries,
On the piazza walk three matrons stately and friendly with twined
    arms,
The crew of the fish-smack pack repeated layers of halibut in the
    hold,   310
The Missourian crosses the plains toting his wares and his cattle,
As the fare-collector goes through the train he gives notice by the
    jingling of loose change,
The floor-men are laying the floor, the tinners are tinning the roof,
    the masons are calling for mortar,
In single file each shouldering his hod pass onward the laborers;
Seasons pursuing each other the indescribable crowd is gather'd, it
    is the fourth of Seventh-month, (what salutes of cannon and
    small arms!)   315
Seasons pursuing each other the plougher ploughs, the mower
    mows, and the winter-grain falls in the ground;
Off on the lakes the pike-fisher watches and waits by the hole in the
    frozen surface,
The stumps stand thick round the clearing, the squatter strikes deep
    with his axe,
Flatboatmen make fast towards dusk near the cotton-wood or
    pecan-trees,
Coon-seekers go through the regions of the Red river or through

7. Man building or repairing streets.

those drain'd by the Tennessee, or through those of the Arkansas,   320
Torches shine in the dark that hangs on the Chattahooche or Altamahaw,[8]
Patriarchs sit at supper with sons and grandsons and great-grandsons around them,
In walls of adobie, in canvas tents, rest hunters and trappers after their day's sport,
The city sleeps and the country sleeps,
The living sleep for their time, the dead sleep for their time,   325
The old husband sleeps by his wife and the young husband sleeps by his wife;
And these tend inward to me, and I tend outward to them,
And such as it is to be of these more or less I am,
And of these one and all I weave the song of myself.

16

I am of old and young, of the foolish as much as the wise,   330
Regardless of others, ever regardful of others,
Maternal as well as paternal, a child as well as a man,
Stuff'd with the stuff that is coarse and stuff'd with the stuff that is fine,
One of the Nation of many nations, the smallest the same and the largest the same,
A Southerner soon as a Northerner, a planter nonchalant and hospitable down by the Oconee[9] I live,   335
A Yankee bound my own way ready for trade, my joints the limberest joints on earth and the sternest joints on earth,
A Kentuckian walking the vale of the Elkhorn in my deer-skin leggings, a Louisianian or Georgian,
A boatman over lakes or bays or along coasts, a Hoosier, Badger, Buckeye;[1]
At home on Kanadian[2] snow-shoes or up in the bush, or with fishermen off Newfoundland,
At home in the fleet of ice-boats, sailing with the rest and tacking,   340
At home on the hills of Vermont or in the woods of Maine, or the Texan ranch,
Comrade of Californians, comrade of free North-Westerners, (loving their big proportions,)
Comrade of raftsmen and coalmen, comrade of all who shake hands and welcome to drink and meat,
A learner with the simplest, a teacher of the thoughtfullest,
A novice beginning yet experient of myriads of seasons,   345
Of every hue and caste am I, of every rank and religion,
A farmer, mechanic, artist, gentleman, sailor, quaker,
Prisoner, fancy-man, rowdy, lawyer, physician, priest.

8. Georgia rivers.
9. River in central Georgia.
1. Inhabitants of Indiana, Wisconsin, and Ohio.
2. Apparently Whitman found something muscular in this spelling.

I resist any thing better than my own diversity,
Breathe the air but leave plenty after me,   350
And am not stuck up, and am in my place.

(The moth and the fish-eggs are in their place,
The bright suns I see and the dark suns I cannot see are in their
    place,
The palpable is in its place and the impalpable is in its place.)

17

These are really the thoughts of all men in all ages and lands, they
    are not original with me,   355
If they are not yours as much as mine they are nothing, or next to
    nothing,
If they are not the riddle and the untying of the riddle they are
    nothing,
If they are not just as close as they are distant they are nothing.

This is the grass that grows wherever the land is and the water is,
This the common air that bathes the globe.   360

18

With music strong I come, with my cornets and my drums,
I play not marches for accepted victors only, I play marches for
    conquer'd and slain persons.

Have you heard that it was good to gain the day?
I also say it is good to fall, battles are lost in the same spirit in
    which they are won.

I beat and pound for the dead,   365
I blow through my embouchures[3] my loudest and gayest for them.

Vivas to those who have fail'd!
And to those whose war-vessels sank in the sea!
And to those themselves who sank in the sea!
And to all generals that lost engagements, and all overcome
    heroes!   370
And the numberless unknown heroes equal to the greatest heroes
    known!

19

This is the meal equally set, this the meat for natural hunger,
It is for the wicked just the same as the righteous, I make appoint-
    ments with all,
I will not have a single person slighted or left away,
The kept-woman, sponger, thief, are hereby invited,   375
The heavy-lipp'd slave is invited, the venerealee is invited;
There shall be no difference between them and the rest.

3. Mouthpiece of musical instruments such as the cornet, mentioned above.

This is the press of a bashful hand, this the float and odor of hair,
This the touch of my lips to yours, this the murmur of yearning,
This the far-off depth and height reflecting my own face,   380
This the thoughtful merge of myself, and the outlet again.

Do you guess I have some intricate purpose?
Well I have, for the Fourth-month showers have, and the mica on
    the side of a rock has.

Do you take it I would astonish?
Does the daylight astonish? does the early redstart twittering through
    the woods?   385
Do I astonish more than they?

This hour I tell things in confidence,
I might not tell everybody, but I will tell you.

20

Who goes there? hankering, gross, mystical, nude;
How is it I extract strength from the beef I eat?   390

What is a man anyhow? what am I? what are you?

All I mark as my own you shall offset it with your own,
Else it were time lost listening to me.

I do not snivel that snivel the world over,
That months are vacuums and the ground but wallow and filth.   395

Whimpering and truckling fold with powders⁴ for invalids, con-
    formity goes to the fourth-remov'd,⁵
I wear my hat as I please indoors or out.

Why should I pray? why should I venerate and be ceremonious?

Having pried through the strata, analyzed to a hair, counsel'd with
    doctors and calculated close,
I find no sweeter fat than sticks to my own bones.   400

In all people I see myself, none more and not one a barley-corn⁶ less,
And the good or bad I say of myself I say of them.

I know I am solid and sound,
To me the converging objects of the universe perpetually flow,
All are written to me, and I must get what the writing means.   405

4. A reference to the custom of a physician's wrapping up a dose of medicine
in a piece of paper.
5. Those very remote in relationship,
from the genealogical usage such as third

cousin, fourth removed.
6. The seed or grain of barley, but also
a unit of measure equal to about one-
third inch.

I know I am deathless,
I know this orbit of mine cannot be swept by a carpenter's compass,
I know I shall not pass like a child's carlacue[7] cut with a burnt stick
    at night.

I know I am august,
I do not trouble my spirit to vindicate itself or be understood,   410
I see that the elementary laws never apologize,
(I reckon I behave no prouder than the level I plant my house by,
    after all.)

I exist as I am, that is enough,
If no other in the world be aware I sit content,
And if each and all be aware I sit content.   415

One world is aware and by far the largest to me, and that is myself,
And whether I come to my own to-day or in ten thousand or ten
    million years,
I can cheerfully take it now, or with equal cheerfulness I can wait.

My foothold is tenon'd and mortis'd[8] in granite,
I laugh at what you call dissolution,   420
And I know the amplitude of time.

21

I am the poet of the Body and I am the poet of the Soul,
The pleasures of heaven are with me and the pains of hell are with
    me,
The first I graft and increase upon myself, the latter I translate into
    a new tongue.

I am the poet of the woman the same as the man,   425
And I say it is as great to be a woman as to be a man,
And I say there is nothing greater than the mother of men.

I chant the chant of dilation or pride,
We have had ducking and deprecating about enough,
I show that size is only development.   430

Have you outstript the rest? are you the President?
It is a trifle, they will more than arrive there every one, and still pass
    on.

I am he that walks with the tender and growing night,
I call to the earth and sea half-held by the night.

Press close bare-bosom'd night—press close magnetic nourishing
    night!   435

---

7. Also spelled curlicue, a fancy flourish made with a writing implement, here made in the dark with a lighted stick, and so lasting only a moment.
8. Carpenter's terms for a way of holding boards together: a mortise is a cavity in a piece of wood into which is placed the projection (tenon) from another piece of wood.

Night of south winds—night of the large few stars!
Still nodding night—mad naked summer night.

Smile O voluptuous cool-breath'd earth!
Earth of the slumbering and liquid trees!
Earth of departed sunset—earth of the mountains misty-topt!   440
Earth of the vitreous pour of the full moon just tinged with blue!
Earth of shine and dark mottling the tide of the river!
Earth of the limpid gray of clouds brighter and clearer for my sake!
Far-swooping elbow'd earth—rich apple-blossom'd earth!
Smile, for your lover comes.   445

Prodigal, you have given me love—therefore I to you give love!
O unspeakable passionate love.

22

You sea! I resign myself to you also—I guess what you mean,
I behold from the beach your crooked inviting fingers,
I believe you refuse to go back without feeling of me,   450
We must have a turn together, I undress, hurry me out of sight of
    the land,
Cushion me soft, rock me in billowy drowse,
Dash me with amorous wet, I can repay you.

Sea of stretch'd ground-swells,
Sea breathing broad and convulsive breaths,   455
Sea of the brine of life and of unshovell'd yet always-ready graves,
Howler and scooper of storms, capricious and dainty sea,
I am integral with you, I too am of one phase and of all phases.

Partaker of influx and efflux I, extoller of hate and conciliation,
Extoller of amies[9] and those that sleep in each others' arms.   460

I am he attesting sympathy,
(Shall I make my list of things in the house and skip the house that
    supports them?)

I am not the poet of goodness only, I do not decline to be the poet
    of wickedness also.

What blurt is this about virtue and about vice?
Evil propels me and reform of evil propels me, I stand indif-
    ferent,   465
My gait is no fault-finder's or rejecter's gait,
I moisten the roots of all that has grown.

Did you fear some scrofula out of the unflagging pregnancy?
Did you guess the celestial laws are yet to be work'd over and recti-
    fied?

9. Friends, in Whitman's specialized sense of comrades.

I find one side a balance and the antipodal side a balance,   470
Soft doctrine as steady help as stable doctrine,
Thoughts and deeds of the present our rouse and early start.

This minute that comes to me over the past decillions,
There is no better than it and now.

What behaved well in the past or behaves well to-day is not such a
     wonder,   475
The wonder is always and always how there can be a mean man or
     an infidel.

23

Endless unfolding of words of ages!
And mine a word of the modern, the word En-Masse.

A word of the faith that never balks,
Here or henceforward it is all the same to me, I accept Time abso-
     lutely.   480

It alone is without flaw, it alone rounds and completes all,
That mystic baffling wonder alone completes all.

I accept Reality and dare not question it,
Materialism first and last imbuing.

Hurrah for positive science! long live exact demonstration!   485
Fetch stonecrop[1] mixt with cedar and branches of lilac,
This is the lexicographer, this the chemist, this made a grammar of
     the old cartouches,[2]
These mariners put the ship through dangerous unknown seas,
This is the geologist, this works with the scalpel, and this is a
     mathematician.

Gentlemen, to you the first honors always!   490
Your facts are useful, and yet they are not my dwelling,
I but enter by them to an area of my dwelling.

Less the reminders of properties told my words,
And more the reminders they of life untold, and of freedom and
     extrication,
And make short account of neuters and geldings, and favor men and
     women fully equipt,   495
And beat the gong of revolt, and stop with fugitives and them that
     plot and conspire.

1. A fleshy-leafed plant of the genus *Sedum.*
2. Scroll-like tablet with space for an inscription. Whitman knew the Egyptian cartouches with hieroglyphics, the deciphering of which contributed to knowledge of ancient life.

25

Dazzling and tremendous how quick the sun-rise would kill
me,   560
If I could not now and always send sun-rise out of me.

We also ascend dazzling and tremendous as the sun,
We found our own O my soul in the calm and cool of the day-break.

My voice goes after what my eyes cannot reach,
With the twirl of my tongue I encompass worlds and volumes of
worlds.   565

Speech is the twin of my vision, it is unequal to measure itself,
It provokes me forever, it says sarcastically,
Walt you contain enough, why don't you let it out then?

Come now I will not be tantalized, you conceive too much of artic-
ulation,
Do you not know O speech how the buds beneath you are
folded?   570
Waiting in gloom, protected by frost,
The dirt receding before my prophetical screams,
I underlying causes to balance them at last,
My knowledge my live parts, it keeping tally with the meaning of
all things,
Happiness, (which whoever hears me let him or her set out in search
of this day.)   575

My final merit I refuse you, I refuse putting from me what I really
am,
Encompass worlds, but never try to encompass me,
I crowd your sleekest and best by simply looking toward you.

Writing and talk do not prove me,
I carry the plenum[7] of proof and every thing else in my face,   580
With the hush of my lips I wholly confound the skeptic.

26

Now I will do nothing but listen,
To accrue what I hear into this song, to let sounds contribute
toward it.

I hear bravuras of birds, bustle of growing wheat, gossip of flames,
clack of sticks cooking my meals,
I hear the sound I love, the sound of the human voice,   585
I hear all sounds running together, combined, fused or following,
Sounds of the city and sounds out of the city, sounds of the day
and night,

---

7. Fullness.

Talkative young ones to those that like them, the loud laugh of
    work-people at their meals,
The angry base of disjointed friendship, the faint tones of the sick,
The judge with hands tight to the desk, his pallid lips pronouncing
    a death-sentence,   590
The heave'e'yo of stevedores unlading ships by the wharves, the
    refrain of the anchor-lifters,
The ring of alarm-bells, the cry of fire, the whirr of swift-streaking
    engines and hose-carts with premonitory tinkles and color'd
    lights,
The steam-whistle, the solid roll of the train of approaching cars,
The slow march play'd at the head of the association marching two
    and two,
(They go to guard some corpse, the flag-tops are draped with black
    muslin.)   595

I hear the violoncello, ('tis the young man's heart's complaint,)
I hear the key'd cornet, it glides quickly in through my ears,
It shakes mad-sweet pangs through my belly and breast.

I hear the chorus, it is a grand opera,
Ah this indeed is music—this suits me.   600

A tenor large and fresh as the creation fills me,
The orbic flex of his mouth is pouring and filling me full.

I hear the train'd soprano (what work with hers is this?)
The orchestra whirls me wider than Uranus[8] flies,
It wrenches such ardors from me I did not know I possess'd
    them,   605
It sails me, I dab with bare feet, they are lick'd by the indolent
    waves,
I am cut by bitter and angry hail, I lose my breath,
Steep'd amid honey'd morphine, my windpipe throttled in fakes[9]
    of death,
At length let up again to feel the puzzle of puzzles,
And that we call Being.   610

27

To be in any form, what is that?
(Round and round we go, all of us, and ever come back thither,)
If nothing lay more develop'd the quahaug[1] in its callous shell
    were enough.

Mine is no callous shell,
I have instant conductors all over me whether I pass or stop, 615
They seize every object and lead it harmlessly through me.

8. In our solar system, the seventh   9. Coils of rope.
planet from the sun, then thought to be   1. Edible clam of the Atlantic coast.
the most remote.

I merely stir, press, feel with my fingers, and am happy,
To touch my person to some one else's is about as much as I can
     stand.

28

Is this then a touch? quivering me to a new identity,
Flames and ether making a rush for my veins,   620
Treacherous tip of me reaching and crowding to help them,
My flesh and blood playing out lightning to strike what is hardly
     different from myself,
On all sides prurient provokers stiffening my limbs,
Straining the udder of my heart for its withheld drip,
Behaving licentious toward me, taking no denial,   625
Depriving me of my best as for a purpose,
Unbuttoning my clothes, holding me by the bare waist,
Deluding my confusion with the calm of the sunlight and pasture-
     fields,
Immodestly sliding the fellow-senses away,
They bribed to swap off with touch and go and graze at the edges
     of me,   630
No consideration, no regard for my draining strength or my anger,
Fetching the rest of the herd around to enjoy them a while,
Then all uniting to stand on a headland and worry me.

The sentries desert every other part of me,
They have left me helpless to a red marauder,
They all come to the headland to witness and assist against me. 635

I am given up by traitors,
I talk wildly, I have lost my wits, I and nobody else am the greatest
     traitor,
I went myself first to the headland, my own hands carried me there.

You villain touch! what are you doing? my breath is tight in its
     throat,   640
Unclench your floodgates, you are too much for me.

29

Blind loving wrestling touch, sheath'd hooded sharp-tooth'd touch!
Did it make you ache so, leaving me?

Parting track'd by arriving, perpetual payment of perpetual loan,
Rich showering rain, and recompense richer afterward.   645

Sprouts take and accumulate, stand by the curb prolific and vital,
Landscapes projected masculine, full-sized and golden.

30

All truths wait in all things,
They neither hasten their own delivery nor resist it,
They do not need the obstetric forceps of the surgeon,   650

The insignificant is as big to me as any,
(What is less or more than a touch?)

Logic and sermons never convince,
The damp of the night drives deeper into my soul.

(Only what proves itself to every man and woman is so,   655
Only what nobody denies is so.)

A minute and a drop of me settle my brain,
I believe the soggy clods shall become lovers and lamps,
And a compend[2] of compends is the meat of a man or woman,
And a summit and flower there is the feeling they have for each
    other,   660
And they are to branch boundlessly out of that lesson until it be-
    comes omnific,[3]
And until one and all shall delight us, and we them.

31

I believe a leaf of grass is no less than the journey-work of the stars,
And the pismire[4] is equally perfect, and a grain of sand, and the
    egg of the wren,
And the tree-toad is a chef-d'œuvre for the highest,   665
And the running blackberry would adorn the parlors of heaven,
And the narrowest hinge in my hand puts to scorn all machinery,
And the cow crunching with depress'd head surpasses any statue,
And a mouse is miracle enough to stagger sextillions of infidels.

I find I incorporate gneiss,[5] coal, long-threaded moss, fruits, grains,
    esculent roots,   670
And am stucco'd with quadrupeds and birds all over,
And have distanced what is behind me for good reasons,
But call any thing back again when I desire it.

In vain the speeding or shyness,
In vain the plutonic rocks[6] send their old heat against my
    approach,   675
In vain the mastodon retreats beneath its own powder'd bones,
In vain objects stand leagues off and assume manifold shapes,
In vain the ocean settling in hollows and the great monsters lying
    low,
In vain the buzzard houses herself with the sky,
In vain the snake slides through the creepers and logs,   680
In vain the elk takes to the inner passes of the woods,
In vain the razor-bill'd auk sails far north to Labrador,
I follow quickly, I ascend to the nest in the fissure of the cliff.

---

2. A compendium, where something is
reduced to short, essential summary.
3. All-encompassing.
4. Ant.
5. Metamorphic rock in which minerals
are arranged in layers.
6. Rock of igneus (fire-created) or mag-
matic (molten) origin (from Pluto, ruler
of infernal regions).

32

I think I could turn and live with animals, they are so placid and
    self-contain'd,
I stand and look at them long and long.   685

They do not sweat and whine about their condition,
They do not lie awake in the dark and weep for their sins,
They do not make me sick discussing their duty to God,
Not one is dissatisfied, not one is demented with the mania of own-
    ing things,
Not one kneels to another, nor to his kind that lived thousands of
    years ago,   690
Not one is respectable or unhappy over the whole earth.

So they show their relations to me and I accept them,
They bring me tokens of myself, they evince them plainly in their
    possession.

I wonder where they get those tokens,
Did I pass that way huge times ago and negligently drop them?   695

Myself moving forward then and now and forever,
Gathering and showing more always and with velocity,
Infinite and omnigenous,[7] and the like of these among them,
Not too exclusive toward the reachers of my remembrancers,
Picking out here one that I love, and now go with him on brotherly
    terms.   700

A gigantic beauty of a stallion, fresh and responsive to my caresses,
Head high in the forehead, wide between the ears,
Limbs glossy and supple, tail dusting the ground,
Eyes full of sparkling wickedness, ears finely cut, flexibly moving.

His nostrils dilate as my heels embrace him,   705
His well-built limbs tremble with pleasure as we race around and
    return.

I but use you a minute, then I resign you, stallion,
Why do I need your paces when I myself out-gallop them?
Even as I stand or sit passing faster than you.

33

Space and Time! now I see it is true, what I guessed at,   710
What I guess'd when I loaf'd on the grass,
What I guess'd while I lay alone in my bed,
And again as I walk'd the beach under the paling stars of the
    morning.

7. Belonging to every form of life.

My ties and ballasts leave me, my elbows rest in sea-gaps,[8]
I skirt sierras, my palms cover continents,   715
I am afoot with my vision.

By the city's quadrangular houses—in log huts, camping with lum-
  bermen,
Along the ruts of the turnpike, along the dry gulch and rivulet bed,
Weeding my onion-patch or hoeing rows of carrots and parsnips,
  crossing savannas,[9] trailing in forests,
Prospecting, gold-digging, girdling the trees of a new purchase,   720
Scorch'd ankle-deep by the hot sand, hauling my boat down the
  shallow river,
Where the panther walks to and fro on a limb overhead, where the
  buck turns furiously at the hunter,
Where the rattlesnake suns his flabby length on a rock, where the
  otter is feeding on fish,
Where the alligator in his tough pimples sleeps by the bayou,
Where the black bear is searching for roots or honey, where the
  beaver pats the mud with his paddle-shaped tail;   725
Over the growing sugar, over the yellow-flower'd cotton plant, over
  the rice in its low moist field,
Over the sharp-peak'd farm house, with its scallop'd scum and
  slender shoots from the gutters,[1]
Over the western persimmon, over the long-leav'd corn, over the
  delicate blue-flower flax,
Over the white and brown buckwheat, a hummer and buzzer there
  with the rest,
Over the dusky green of the rye as it ripples and shades in the
  breeze;   730
Scaling mountains, pulling myself cautiously up, holding on by low
  scragged limbs,
Walking the path worn in the grass and beat through the leaves of
  the brush,
Where the quail is whistling betwixt the woods and the wheat-lot,
Where the bat flies in the Seventh-month eve, where the great gold-
  bug drops through the dark,
Where the brook puts out of the roots of the old tree and flows to
  the meadow,   735
Where cattle stand and shake away flies with the tremulous shud-
  dering of their hides,
Where the cheese-cloth hangs in the kitchen, where andirons
  straddle the hearth-slab, where cobwebs fall in festoons from
  the rafters;
Where trip-hammers crash, where the press is whirling its cylinders,
Wherever the human heart beats with terrible throes under its ribs,
Where the pear-shaped balloon is floating aloft, (floating in it my-
  self and looking composedly down,)   740

8. Estuaries or bays.
9. Flat, treeless, tropical grassland.
1. Presumably debris washed or blown down roofs, settling into scalloplike shapes in the gutter and providing nu-triment for grasses or weeds.

Where the life-car[2] is drawn on the slip-noose, where the heat
    hatches pale-green eggs in the dented sand,
Where the she-whale swims with her calf and never forsakes it,
Where the steam-ship trails hind-ways its long pennant of smoke,
Where the fin of the shark cuts like a black chip out of the water,
Where the half-burn'd brig is riding on unknown currents,   745
Where shells grow to her slimy deck, where the dead are corrupt-
    ing below;
Where the dense-starr'd flag is borne at the head of the regiments,
Approaching Manhattan up by the long-stretching island,
Under Niagara, the cataract falling like a veil over my countenance,
Upon a door-step, upon the horse-block of hard wood outside,   750
Upon the race-course, or enjoying picnics or jigs or a good game of
    base-ball,
At he-festivals, with blackguard gibes, ironical license, bull-dances,[3]
    drinking, laughter,
At the cider-mill tasting the sweets of the brown mash, sucking
    the juice through a straw,
At apple-peelings wanting kisses for all the red fruit I find,
At musters,[4] beach-parties, friendly bees,[5] huskings, house-
    raisings;   755
Where the mocking-bird sounds his delicious gurgles, cackles,
    screams, weeps,
Where the hay-rick stands in the barn-yard, where the dry-stalks are
    scatter'd, where the brood-cow waits in the hovel,
Where the bull advances to do his masculine work, where the stud
    to the mare, where the cock is treading the hen,
Where the heifers browse, where geese nip their food with short
    jerks,
Where sun-down shadows lengthen over the limitless and lonesome
    prairie,   760
Where herds of buffalo make a crawling spread of the square miles
    far and near,
Where the humming-bird shimmers, where the neck of the long-
    lived swan is curving and winding,
Where the laughing-gull scoots by the shore, where she laughs her
    near-human laugh,
Where bee-hives range on a gray bench in the garden half hid by the
    high weeds,
Where band-neck'd partridges roost in a ring on the ground with
    their heads out,   765
Where burial coaches enter the arch'd gates of a cemetery,
Where winter wolves bark amid wastes of snow and icicled trees,
Where the yellow-crown'd heron comes to the edge of the marsh at
    night and feeds upon small crabs,

---

2. Water-tight compartment for lowering passengers from a ship when emergency evacuation is required.
3. Rowdy backwoods dances where in
the absence of women men took male
partners.

4. Any assemblage of people, but particularly a gathering of military troops
for drill.
5. Gathering where people work while
socializing with their neighbors.

Where the splash of swimmers and divers cools the warm noon,
Where the katy-did works her chromatic[6] reed on the walnut-tree
    over the well,   770
Through patches of citrons and cucumbers with silver-wired leaves,
Through the salt-lick or orange glade, or under conical firs,
Through the gymnasium, through the curtain'd saloon, through the
    office or public hall;
Pleas'd with the native and pleas'd with the foreign, pleas'd with the
    new and old,
Pleas'd with the homely woman as well as the handsome,   775
Pleas'd with the quakeress as she puts off her bonnet and talks
    melodiously,
Pleas'd with the tune of the choir of the whitewash'd church,
Pleas'd with the earnest words of the sweating Methodist preacher,
    impress'd seriously at the camp-meeting;
Looking in at the shop-windows of Broadway the whole forenoon,
    flatting the flesh of my nose on the thick plate-glass,
Wandering the same afternoon with my face turn'd up to the
    clouds, or down a lane or along the beach,   780
My right and left arms round the sides of two friends, and I in the
    middle;
Coming home with the silent and dark-cheek'd bush-boy, (behind
    me he rides at the drape of the day,)
Far from the settlements studying the print of animals' feet, or the
    moccasin print,
By the cot in the hospital reaching lemonade to a feverish patient,
Nigh the coffin'd corpse when all is still, examining with a candle; 785
Voyaging to every port to dicker and adventure,
Hurrying with the modern crowd as eager and fickle as any,
Hot toward one I hate, ready in my madness to knife him,
Solitary at midnight in my back yard, my thoughts gone from me a
    long while,
Walking the old hills of Judæa with the beautiful gentle God by
    my side,   790
Speeding through space, speeding through heaven and the stars,
Speeding amid the seven satellites[7] and the broad ring, and the
    diameter of eighty thousand miles,
Speeding with tail'd meteors, throwing fire-balls like the rest,
Carrying the crescent child that carries its own full mother in its
    belly,
Storming, enjoying, planning, loving, cautioning,   795
Backing and filling, appearing and disappearing,
I tread day and night such roads.

I visit the orchards of spheres and look at the product,
And look at quintillions ripen'd and look at quintillions green.

I fly those flights of a fluid and swallowing soul,   800
My course runs below the soundings of plummets.

6. Consisting of chords or harmonies based upon nonharmonic tones.
7. The seven then-known planets.

I help myself to material and immaterial,
No guard can shut me off, no law prevent me.

I anchor my ship for a little while only,
My messengers continually cruise away or bring their returns to
    me.   805

I go hunting polar furs and the seal, leaping chasms with a pike-
    pointed staff, clinging to topples[8] of brittle and blue.

I ascend to the foretruck,
I take my place late at night in the crow's-nest,
We sail the arctic sea, it is plenty light enough,
Through the clear atmosphere I stretch around on the wonderful
    beauty,   810
The enormous masses of ice pass me and I pass them, the scenery is
    plain in all directions,
The white-topt mountains show in the distance, I fling out my
    fancies toward them,
We are approaching some great battle-field in which we are soon to
    be engaged,
We pass the colossal outposts of the encampment, we pass with
    still feet and caution,
Or we are entering by the suburbs some vast and ruin'd city,   815
The blocks and fallen architecture more than all the living cities of
    the globe.

I am a free companion, I bivouac by invading watchfires,
I turn the bridegroom out of bed and stay with the bride myself,
I tighten her all night to my thighs and lips.

My voice is the wife's voice, the screech by the rail of the stairs,   820
They fetch my man's body up dripping and drown'd.

I understand the large hearts of heroes,
The courage of present times and all times,
How the skipper saw the crowded and rudderless wreck of the steam-
    ship, and Death chasing it up and down the storm,
How he knuckled tight and gave not back an inch, and was faith-
    ful of days and faithful of nights,   825
And chalk'd in large letters on a board, *Be of good cheer, we will
    not desert you;*
How he follow'd with them and tack'd with them three days and
    would not give it up,
How he saved the drifting company at last,
How the lank loose-gown'd women look'd when boated from the
    side of their prepared graves,
How the silent old-faced infants and the lifted sick, and the sharp-
    lipp'd unshaved men;   830

---

8. Toppled pieces of ice.

All this I swallow, it tastes good, I like it well, it becomes mine,
I am the man, I suffer'd, I was there.

The disdain and calmness of martyrs,
The mother of old, condemn'd for a witch, burnt with dry wood,
    her children gazing on,
The hounded slave that flags in the race, leans by the fence, blowing,
    cover'd with sweat,   835
The twinges that sting like needles his legs and neck, the murderous
    buckshot and the bullets,
All these I feel or am.

I am the hounded slave, I wince at the bite of the dogs,
Hell and despair are upon me, crack and again crack the marksmen,
I clutch the rails of the fence, my gore dribs,⁹ thinn'd with the
    ooze of my skin,   840
I fall on the weeds and stones,
The riders spur their unwilling horses, haul close,
Taunt my dizzy ears and beat me violently over the head with whip-
    stocks.

Agonies are one of my changes of garments,
I do not ask the wounded person how he feels, I myself become the
    wounded person,   845
My hurts turn livid upon me as I lean on a cane and observe.

I am the mash'd fireman with breast-bone broken,
Tumbling walls buried me in their debris,
Heat and smoke I inspired, I heard the yelling shouts of my com-
    rades,
I heard the distant click of their picks and shovels,   850
They have clear'd the beams away, they tenderly lift me forth.

I lie in the night air in my red shirt, the pervading hush is for my
    sake,
Painless after all I lie exhausted but not so unhappy,
White and beautiful are the faces around me, the heads are bared of
    their fire-caps,
The kneeling crowd fades with the light of the torches.   855

Distant and dead resuscitate,
They show as the dial or move as the hands of me, I am the clock
    myself.

I am an old artillerist, I tell of my fort's bombardment,
I am there again.

Again the long roll of the drummers,   860
Again the attacking cannon, mortars,
Again to my listening ears the cannon responsive.

9. Dribbles down, diluted with sweat.

I take part, I see and hear the whole,
The cries, curses, roar, the plaudits for well-aim'd shots,
The ambulanza slowly passing trailing its red drip,   865
Workmen searching after damages, making indispensable repairs,
The fall of grenades through the rent roof, the fan-shaped explosion,
The whizz of limbs, heads, stone, wood, iron, high in the air.

Again gurgles the mouth of my dying general, he furiously waves
    with his hand,
He gasps through the clot *Mind not me—mind—the entrench-
    ments.*   870

34

Now I tell what I knew in Texas in my early youth,
(I tell not the fall of Alamo,
Not one escaped to tell the fall of Alamo,[1]
The hundred and fifty are dumb yet at Alamo,)
'Tis the tale of the murder in cold blood of four hundred and twelve
    young men.   875

Retreating they had form'd in a hollow square with their baggage
    for breastworks,
Nine hundred lives out of the surrounding enemy's, nine times their
    number, was the price they took in advance,
Their colonel was wounded and their ammunition gone,
They treated for an honorable capitulation, receiv'd writing and
    seal, gave up their arms and march'd back prisoners of war.

They were the glory of the race of rangers,   880
Matchless with horse, rifle, song, supper, courtship,
Large, turbulent, generous, handsome, proud, and affectionate,
Bearded, sunburnt, drest in the free costume of hunters,
Not a single one over thirty years of age.

The second First-day morning they were brought out in squads and
    massacred, it was beautiful early summer,   885
The work commenced about five o'clock and was over by eight.

None obey'd the command to kneel,
Some made a mad and helpless rush, some stood stark and straight,
A few fell at once, shot in the temple or heart, the living and dead
    lay together,
The maim'd and mangled dug in the dirt, the new-comers saw them
    there,   890
Some half-kill'd attempted to crawl away,

---

1. The fall of the Alamo during the Mexican War was already well established in the American consciousness. Whitman here celebrates a lesser-known but bloodier massacre, the murder of some 400 "Texans" (most of them new emigrants from southern states) after they surrendered to the Mexicans near Goliad (now in Texas) in late March, 1836, three weeks after the fall of the Alamo (now in San Antonio, Texas).

These were despatch'd with bayonets or batter'd with the blunts of
  muskets,
A youth not seventeen years old seiz'd his assassin till two more
  came to release him,
The three were all torn and cover'd with the boy's blood.

At eleven o'clock began the burning of the bodies;   895
That is the tale of the murder of the four hundred and twelve young
  men.

35

Would you hear of an old-time sea-fight?[2]
Would you learn who won by the light of the moon and stars?
List to the yarn, as my grandmother's father the sailor told it to me.

Our foe was no skulk in his ship I tell you, (said he,)   900
His was the surly English pluck, and there is no tougher or truer, and
  never was, and never will be;
Along the lower'd eve he came horribly raking us.

We closed with him, the yards entangled, the cannon touch'd,
My captain lash'd fast with his own hands.

We had receiv'd some eighteen pound shots under the water,   905
On our lower-gun-deck two large pieces had burst at the first fire,
  killing all around and blowing up overhead.

Fighting at sun-down, fighting at dark,
Ten o'clock at night, the full moon well up, our leaks on the gain,
  and five feet of water reported,
The master-at-arms loosing the prisoners confined in the after-hold
  to give them a chance for themselves.

The transit to and from the magazine[3] is now stopt by the
  sentinels,   910
They see so many strange faces they do not know whom to trust.

Our frigate takes fire,
The other asks if we demand quarter?
If our colors are struck and the fighting done?

Now I laugh content, for I hear the voice of my little captain,   915
We have not struck, he composedly cries, *we have just begun our
  part of the fighting*.

2. The famous Revolutionary sea battle on September 23, 1779, between the American *BonHomme Richard*, commanded by John Paul Jones, and the British *Serapis* off the coast of York-shire. A brilliant account of the battle, based on naval histories, is in Melville's *Israel Potter*, Chapters 19 and 20.
3. Storeroom for ammunition.

Only three guns are in use,
One is directed by the captain himself against the enemy's main-
　　mast,
Two well serv'd with grape and canister[4] silence his musketry and
　　clear his decks.

The tops[5] alone second the fire of this little battery, especially the
　　main-top,　　　　　　　　　　　　　　　　　　　　　　　　　920
They hold out bravely during the whole of the action.

Not a moment's cease,
The leaks gain fast on the pumps, the fire eats toward the powder-
　　magazine.

One of the pumps has been shot away, it is generally thought we are
　　sinking.

Serene stands the little captain,　　　　　　　　　　　　　　925
He is not hurried, his voice is neither high nor low,
His eyes give more light to us than our battle-lanterns.

Toward twelve there in the beams of the moon they surrender to us.

36

Stretch'd and still lies the midnight,
Two great hulls motionless on the breast of the darkness,　　930
Our vessel riddled and slowly sinking, preparations to pass to the
　　one we have conquer'd,
The captain on the quarter-deck coldly giving his orders through a
　　countenance white as a sheet,
Near by the corpse of the child that serv'd in the cabin,
The dead face of an old salt with long white hair and carefully
　　curl'd whiskers,
The flames spite of all that can be done flickering aloft and
　　below,　　　　　　　　　　　　　　　　　　　　　　　　　935
The husky voices of the two or three officers yet fit for duty,
Formless stacks of bodies and bodies by themselves, dabs of flesh
　　upon the masts and spars,
Cut of cordage, dangle of rigging, slight shock of the soothe of
　　waves,
Black and impassive guns, litter of powder-parcels, strong scent,
A few large stars overhead, silent and mournful shining,　　940
Delicate sniffs of sea-breeze, smells of sedgy grass and fields by the
　　shore, death-messages given in charge to survivors,
The hiss of the surgeon's knife, the gnawing teeth of his saw.

---

4. Grapeshot, clusters of small iron
balls, was packed inside a metal cylinder
(canister) and used to charge a cannon.

5. Platforms enclosing the heads of each
mast (here, the sailors manning the
tops).

Wheeze, cluck, swash of falling blood, short wild scream, and long,
    dull, tapering groan,
These so, these irretrievablè.

37

You laggards there on guard! look to your arms!   945
In at the conquer'd doors they crowd! I am possess'd!
Embody all presences outlaw'd or suffering,
See myself in prison shaped like another man,
And feel the dull unintermitted pain.

For me the keepers of convicts shoulder their carbines and keep
    watch,   950
It is I let out in the morning and barr'd at night.

Not a mutineer walks handcuff'd to jail but I am handcuff'd to him
    and walk by his side,
(I am less the jolly one there, and more the silent one with sweat on
    my twitching lips.)

Not a youngster is taken for larceny but I go up too, and am tried
    and sentenced.

Not a cholera patient lies at the last gasp but I also lie at the last
    gasp,   955
My face is ash-color'd, my sinews gnarl, away from me people retreat.

Askers embody themselves in me and I am embodied in them,
I project my hat, sit shame-faced, and beg.

38

Enough! enough! enough!
Somehow I have been stunn'd. Stand back!   960
Give me a little time beyond my cuff'd head, slumbers, dreams,
    gaping,
I discover myself on the verge of a usual mistake.

That I could forget the mockers and insults!
That I could forget the trickling tears and the blows of the blud-
    geons and hammers!
That I could look with a separate look on my own crucifixion and
    bloody crowning.   965

I remember now,
I resume the overstaid fraction,
The grave of rock multiplies what has been confided to it, or to any
    graves,
Corpses rise, gashes heal, fastenings roll from me.

I troop forth replenish'd with supreme power, one of an average un-
    ending procession,   970

Inland and sea-coast we go, and pass all boundary lines,
Our swift ordinances on their way over the whole earth,
The blossoms we wear in our hats the growth of thousands of years.

Eleves,[6] I salute you! come forward!
Continue your annotations, continue your questionings.　975

39

The friendly and flowing savage, who is he?
Is he waiting for civilization, or past it and mastering it?

Is he some Southwesterner rais'd out-doors? is he Kanadian?
Is he from the Mississippi country? Iowa, Oregon, California?
The mountains? prairie-life, bush-life? or sailor from the sea?　980

Wherever he goes men and women accept and desire him,
They desire he should like them, touch them, speak to them, stay
　　with them.

Behavior lawless as snow-flakes, words simple as grass, uncomb'd
　　head, laughter, and naivetè,
Slow-stepping feet, common features, common modes and ema-
　　nations,
They descend in new forms from the tips of his fingers,　985
They are wafted with the odor of his body or breath, they fly out of
　　the glance of his eyes.

40

Flaunt of the sunshine I need not your bask—lie over!
You light surfaces only, I force surfaces and depths also.

Earth! you seem to look for something at my hands,
Say, old top-knot,[7] what do you want?　990

Man or woman, I might tell how I like you, but cannot,
And might tell what it is in me and what it is in you, but cannot,
And might tell that pining I have, that pulse of my nights and days.

Behold, I do not give lectures or a little charity,
When I give I give myself.　995

You there, impotent, loose in the knees,
Open your scarf'd chops[8] till I blow grit within you,
Spread your palms and lift the flaps of your pockets,

6. From the French for "students," but Whitman's usage carries some of the sense of "disciples" or "acolytes" as well.

7. As Blodgett and Bradley say in their note, this "epithet was familiar in frontier humor as a comic, half-affectionate term for an Indian, whose tuft of hair or ornament on top of the head was characteristic of certain tribes."

8. With chops (jaws) tied up in a scarf, as one might do for a toothache, earache, or other ailment. See line 1069.

I am not to be denied, I compel, I have stores plenty and to spare,
And any thing I have I bestow.   1000

I do not ask who you are, that is not important to me,
You can do nothing and be nothing but what I will infold you.

To cotton-field drudge or cleaner of privies I lean,
On his right cheek I put the family kiss,
And in my soul I swear I never will deny him.   1005

On women fit for conception I start bigger and nimbler babes,
(This day I am jetting the stuff of far more arrogant republics.)

To any one dying, thither I speed and twist the knob of the door,
Turn the bed-clothes toward the foot of the bed,
Let the physician and the priest go home.   1010

I seize the descending man and raise him with resistless will,
O despairer, here is my neck,
By God, you shall not go down! hang your whole weight upon me.

I dilate you with tremendous breath, I buoy you up,
Every room of the house do I fill with an arm'd force,   1015
Lovers of me, bafflers of graves.

Sleep—I and they keep guard all night,
Not doubt, not decease shall dare to lay finger upon you,
I have embraced you, and henceforth possess you to myself,
And when you rise in the morning you will find what I tell you is
    so.   1020

41

I am he bringing help for the sick as they pant on their backs,
And for strong upright men I bring yet more needed help.

I heard what was said of the universe,
Heard it and heard it of several thousand years;
It is middling well as far as it goes—but is that all?   1025

Magnifying and applying come I,
Outbidding at the start the old cautious hucksters,[9]

9. The "old cautious hucksters" are gods or priests who made too little of the divinity in man. Among these rough sketches for gods are portraits of Jehovah, the God of the Jews and Christians; Kronos or Cronus, in Greek mythology the Titan who ruled the universe until dethroned by Zeus, his son, the chief of the Olympian gods; Hercules, son of Zeus and the mortal Alcmene, who won immortality by performing 12 supposedly impossible feats; Osiris, the Egyptian god who annually died and was reborn, symbolizing the fertility of nature; Isis, Egyptian goddess of fertility, sister and wife of Osiris; Belus, a legendary god-king of Assyria; Brahma, in Hinduism the divine reality in the role of creator; Buddha, the Indian philosopher Gautama Siddhartha, founder of Buddhism; Manito, nature god of the Algonquian Indians; Allah, the supreme being in the Moslem religion; Odin, the chief Norse god; and Mexitli, an Aztec war god.

Taking myself the exact dimensions of Jehovah,
Lithographing Kronos, Zeus his son, and Hercules his grandson,
Buying drafts of Osiris, Isis, Belus, Brahma, Buddha,   1030
In my portfolio placing Manito loose, Allah on a leaf, the crucifix
    engraved,
With Odin and the hideous-faced Mexitli and every idol and image,
Taking them all for what they are worth and not a cent more,
Admitting they were alive and did the work of their days,
(They bore mites as for unfledg'd birds who have now to rise and
    fly and sing for themselves,)   1035
Accepting the rough deific sketches to fill out better in myself, be-
    stowing them freely on each man and woman I see,
Discovering as much or more in a framer framing a house,
Putting higher claims for him there with his roll'd-up sleeves driving
    the mallet and chisel,
Not objecting to special revelations, considering a curl of smoke or
    a hair on the back of my hand just as curious as any revelation,
Lads ahold of fire-engines and hook-and-ladder ropes no less to me
    than the gods of the antique wars,   1040
Minding their voices peal through the crash of destruction,
Their brawny limbs passing safe over charr'd laths, their white fore-
    heads whole and unhurt out of the flames;
By the mechanic's wife with her babe at her nipple interceding for
    every person born,
Three scythes at harvest whizzing in a row from three lusty angels
    with shirts bagg'd out at their waists,
The snag-tooth'd hostler with red hair redeeming sins past and to
    come,   1045
Selling all he possesses, traveling on foot to fee lawyers for his
    brother and sit by him while he is tried for forgery;
What was strewn in the amplest strewing the square rod about me,
    and not filling the square rod then,
The bull and the bug never worshipp'd half enough,[1]
Dung and dirt more admirable than was dream'd,
The supernatural of no account, myself waiting my time to be one
    of the supremes,   1050
The day getting ready for me when I shall do as much good as the
    best, and be as prodigious;
By my life-lumps![2] becoming already a creator,
Putting myself here and now to the ambush'd womb of the shadows.

42

A call in the midst of the crowd,
My own voice, orotund sweeping and final.   1055

Come my children,
Come my boys and girls, my women, household and intimates,

---

1. As Whitman implies, the bull and the bug had in fact been worshiped in earlier religions, the bull in several, the scarab beetle as an Egyptian symbol of the soul; but they had been worshiped wrongly, as supernatural objects.
2. A felicitously comic way of referring to spurts of semen, here used figuratively.

Now the performer launches his nerve, he has pass'd his prelude on
    the reeds within.

Easily written loose-finger'd chords—I feel the thrum of your climax
    and close.

My head slues round on my neck,   1060
Music rolls, but not from the organ,
Folks are around me, but they are no household of mine.

Ever the hard unsunk ground,
Ever the eaters and drinkers, ever the upward and downward sun,
    ever the air and the ceaseless tides,
Ever myself and my neighbors, refreshing, wicked, real,   1065
Ever the old inexplicable query, ever that thorn'd thumb, that breath
    of itches and thirsts,
Ever the vexer's *hoot! hoot!* till we find where the sly one hides and
    bring him forth,
Ever love, ever the sobbing liquid of life,
Ever the bandage under the chin, ever the trestles[3] of death.

Here and there with dimes on the eyes[4] walking,   1070
To feed the greed of the belly the brains liberally spooning,
Tickets buying, taking, selling, but in to the feast never once going,
Many sweating, ploughing, thrashing, and then the chaff for pay-
    ment receiving,
A few idly owning, and they the wheat continually claiming.

This is the city and I am one of the citizens,   1075
Whatever interests the rest interests me, politics, wars, markets,
    newspapers, schools,
The mayor and councils, banks, tariffs, steamships, factories, stocks,
    stores, real estate and personal estate.

The little plentiful manikins skipping around in collars and tail'd
    coats,
I am aware who they are, (they are positively not worms or fleas,)
I acknowledge the duplicates of myself, the weakest and shallowest
    is deathless with me,   1080
What I do and say the same waits for them,
Every thought that flounders in me the same flounders in them.

I know perfectly well my own egotism,
Know my omnivorous lines and must not write any less,
And would fetch you whoever you are flush with myself.   1085

Not words of routine this song of mine,
But abruptly to question, to leap beyond yet nearer bring;

---

3. Sawhorses or similar supports holding   4. Coins were placed on eyelids to hold
up a coffin.   them closed until burial.

This printed and bound book—but the printer and the printing-
office boy?
The well-taken photographs—but your wife or friend close and solid
in your arms?
The black ship mail'd with iron, her mighty guns in her turrets—
but the pluck of the captain and engineers?   1090
In the houses the dishes and fare and furniture—but the host and
hostess, and the look out of their eyes?
The sky up there—yet here or next door, or across the way?
The saints and sages in history—but you yourself?
Sermons, creeds, theology—but the fathomless human brain,
And what is reason? and what is love? and what is life?   1095

43

I do not despise you priests, all time, the world over,
My faith is the greatest of faiths and the least of faiths,
Enclosing worship ancient and modern and all between ancient
and modern,
Believing I shall come again upon the earth after five thousand
years,
Waiting responses from oracles, honoring the gods, saluting the
sun,   1100
Making a fetich of the first rock or stump, powowing with sticks in
the circle of obis,[5]
Helping the llama or brahmin as he trims the lamps of the idols,
Dancing yet through the streets in a phallic procession, rapt and
austere in the woods a gymnosophist,
Drinking mead from the skull-cup, to Shastas and Vedas admirant,
minding the Koran,
Walking the teokallis, spotted with gore from the stone and knife,
beating the serpent-skin drum,   1105
Accepting the Gospels,[6] accepting him that was crucified, knowing
assuredly that he is divine,
To the mass kneeling or the puritan's prayer rising, or sitting pa-
tiently in a pew,
Ranting and frothing in my insane crisis, or waiting dead-like till
my spirit arouses me,
Looking forth on pavement and land, or outside of pavement and
land,
Belonging to the winders of the circuit of circuits.   1110

One of that centripetal and centrifugal gang I turn and talk like a
man leaving charges before a journey.

5. Witch doctors, either in Africa or among blacks in the New World.
6. "Llama" is Whitman's spelling for lama, a Buddhist monk of Tibet or Mongolia. A brahmin here is also a Buddhist priest. Gymnosophists were members of an ancient Hindu ascetic sect, thought to have forgone clothing, as the name ("naked philosophers") implies. The other worshipers include old Teutonic drinkers of mead, an alcoholic beverage made of fermented honey; admiring or wondering readers of the sastras (or shastras or shasters, books of Hindu law) or of the Vedas, the oldest sacred writings of Hinduism; those attentive to the Koran, the sacred book of Islam, containing Allah's revelations to Mohammed; worshipers walking the teokallis, an ancient Central American temple built upon a pyramidal mound; believers in the New Testament Gospels.

Down-hearted doubters dull and excluded,
Frivolous, sullen, moping, angry, affected, dishearten'd, atheistical,
I know every one of you, I know the sea of torment, doubt, despair
and unbelief.

How the flukes[7] splash!   1115
How they contort rapid as lightning, with spasms and spouts of
blood!

Be at peace bloody flukes of doubters and sullen mopers,
I take my place among you as much as among any,
The past is the push of you, me, all, precisely the same,
And what is yet untried and afterward is for you, me, all, precisely
the same.   1120

I do not know what is untried and afterward,
But I know it will in its turn prove sufficient, and cannot fail.

Each who passes is consider'd, each who stops is consider'd, not a
single one can it fail.

It cannot fail the young man who died and was buried,
Nor the young woman who died and was put by his side,   1125
Nor the little child that peep'd in at the door, and then drew back
and was never seen again,
Nor the old man who has lived without purpose, and feels it with
bitterness worse than gall,
Nor him in the poor house tubercled by rum and the bad disorder,
Nor the numberless slaughter'd and wreck'd, nor the brutish koboo[8]
call'd the ordure of humanity,
Nor the sacs merely floating with open mouths for food to slip
in,   1130
Nor any thing in the earth, or down in the oldest graves of the
earth,
Nor any thing in the myriads of spheres, nor the myriads of myriads
that inhabit them,
Nor the present, nor the least wisp that is known.

44

It is time to explain myself—let us stand up.

What is known I strip away,   1135
I launch all men and women forward with me into the Unknown.

The clock indicates the moment—but what does eternity indicate?

We have thus far exhausted trillions of winters and summers,
There are trillions ahead, and trillions ahead of them.

7. The flat parts on either side of a whale's tail; here used figuratively.
8. Native of Sumatra.

Births have brought us richness and variety,   1140
And other births will bring us richness and variety.

I do not call one greater and one smaller,
That which fills its period and place is equal to any.

Were mankind murderous or jealous upon you, my brother, my
    sister?
I am sorry for you, they are not murderous or jealous upon me, 1145
All has been gentle with me, I keep no account with lamentation,
(What have I to do with lamentation?)

I am an acme of things accomplish'd, and I an encloser of things to
    be.

My feet strike an apex of the apices[9] of the stairs,
On every step bunches of ages, and larger bunches between the
    steps,   1150
All below duly travel'd, and still I mount and mount.

Rise after rise bow the phantoms behind me,
Afar down I see the huge first Nothing, I know I was even there,
I waited unseen and always, and slept through the lethargic mist,
And took my time, and took no hurt from the fetid carbon.[1]   1155

Long I was hugg'd close—long and long.

Immense have been the preparations for me,
Faithful and friendly the arms that have help'd me.

Cycles[2] ferried my cradle, rowing and rowing like cheerful boat-
    men,
For room to me stars kept aside in their own rings,   1160
They sent influences to look after what was to hold me.

Before I was born out of my mother generations guided me,
My embryo has never been torpid, nothing could overlay it.

For it the nebula cohered to an orb,
The long slow strata piled to rest it on,   1165
Vast vegetables gave it sustenance,
Monstrous sauroids transported it in their mouths and deposited it
    with care.

All forces have been steadily employ'd to complete and delight me,
Now on this spot I stand with my robust soul.

9. The highest points (variant plural of
apex).
1. Whitman knew a good deal about geology and pre-Darwinian theories of evolution. Here the periods of lethargic mist
and fetid carbon are prehuman ages,
probably ages far earlier than the period
of the "monstrous sauroids" (line 1166).
2. Centuries.

45

O span of youth! ever-push'd elasticity!   1170
O manhood, balanced, florid and full.

My lovers suffocate me,
Crowding my lips, thick in the pores of my skin,
Jostling me through streets and public halls, coming naked to me at
        night,
Crying by day *Ahoy!* from the rocks of the river, swinging and
        chirping over my head,   1175
Calling my name from flower-beds, vines, tangled underbrush,
Lighting on every moment of my life,
Bussing[3] my body with soft balsamic busses,
Noiselessly passing handfuls out of their hearts and giving them to
        be mine.

Old age superbly rising! O welcome, ineffable grace of dying
        days!   1180

Every condition promulges[4] not only itself, it promulges what grows
        after and out of itself,
And the dark hush promulges as much as any.

I open my scuttle at night and see the far-sprinkled systems,
And all I see multiplied as high as I can cipher edge but the rim of
        the farther systems.

Wider and wider they spread, expanding, always expanding,   1185
Outward and outward and forever outward.

My sun has his sun and round him obediently wheels,
He joins with his partners a group of superior circuit,
And greater sets follow, making specks of the greatest inside them.

There is no stoppage and never can be stoppage,   1190
If I, you, and the worlds, and all beneath or upon their surfaces,
        were this moment reduced back to a pallid float,[5] it would not
        avail in the long run,
We should surely bring up again where we now stand,
And surely go as much farther, and then farther and farther.

A few quadrillions of eras, a few octillions of cubic leagues, do not
        hazard[6] the span or make it impatient,
They are but parts, any thing is but a part.   1195

See ever so far, there is limitless space outside of that,
Count ever so much, there is limitless time around that.

---

3. Kissing.
4. Promulgates, officially announces.
5. That period before the solar system

6. Imperil, make hazardous.
had defined itself.

My rendezvous is appointed, it is certain,
The Lord will be there and wait till I come on perfect terms,
The great Camerado, the lover true for whom I pine will be
    there.   1200

46

I know I have the best of time and space, and was never measured
    and never will be measured.

I tramp a perpetual journey, (come listen all!)
My signs are a rain-proof coat, good shoes, and a staff cut from the
    woods,
No friend of mine takes his ease in my chair,
I have no chair, no church, no philosophy,   1205
I lead no man to a dinner-table, library, exchange,[7]
But each man and each woman of you I lead upon a knoll,
My left hand hooking you round the waist,
My right hand pointing to landscapes of continents and the public
    road.

Not I, not any one else can travel that road for you,   1210
You must travel it for yourself.

It is not far, it is within reach,
Perhaps you have been on it since you were born and did not know,
Perhaps it is everywhere on water and on land.

Shoulder your duds dear son, and I will mine, and let us hasten
    forth,   1215
Wonderful cities and free nations we shall fetch as we go.

If you tire, give me both burdens, and rest the chuff[8] of your hand
    on my hip,
And in due time you shall repay the same service to me,
For after we start we never lie by again.

This day before dawn I ascended a hill and look'd at the crowded
    heaven,   1220
And I said to my spirit *When we become the enfolders of those orbs,
and the pleasure and knowledge of every thing in them, shall
we be fill'd and satisfied then?*
And my spirit said *No, we but level that lift to pass and continue
beyond.*

You are also asking me questions and I hear you,
I answer that I cannot answer, you must find out for yourself.

Sit a while dear son,   1225
Here are biscuits to eat and here is milk to drink,

---

7. Stock exchange.   8. The meaty part of the palm.

But as soon as you sleep and renew yourself in sweet clothes, I kiss
    you with a good-by kiss and open the gate for your egress hence.

Long enough have you dream'd contemptible dreams,
Now I wash the gum from your eyes,
You must habit yourself to the dazzle of the light and of every
    moment of your life.   1230

Long have you timidly waded holding a plank by the shore,
Now I will you to be a bold swimmer,
To jump off in the midst of the sea, rise again, nod to me, shout,
    and laughingly dash with your hair.

47

I am the teacher of athletes,
He that by me spreads a wider breast than my own proves the width
    of my own,   1235
He most honors my style who learns under it to destroy the teacher.

The boy I love, the same becomes a man not through derived power,
    but in his own right,
Wicked rather than virtuous out of conformity or fear,
Fond of his sweetheart, relishing well his steak,
Unrequited love or a slight cutting him worse than sharp steel
    cuts,   1240
First-rate to ride, to fight, to hit the bull's eye, to sail a skiff, to
    sing a song or play on the banjo,
Preferring scars and the beard and faces pitted with small-pox over
    all latherers,
And those well-tann'd to those that keep out of the sun.

I teach straying from me, yet who can stray from me?
I follow you whoever you are from the present hour,   1245
My words itch at your ears till you understand them.

I do not say these things for a dollar or to fill up the time while I
    wait for a boat,
(It is you talking just as much as myself, I act as the tongue of you,
Tied in your mouth, in mine it begins to be loosen'd.)

I swear I will never again mention love or death inside a house, 1250
And I swear I will never translate myself at all, only to him or her
    who privately stays with me in the open air.

If you would understand me go to the heights or water-shore,
The nearest gnat is an explanation, and a drop or motion of waves
    a key,
The maul, the oar, the hand-saw, second my words.

No shutter'd room or school can commune with me,   1255
But roughs and little children better than they.

The young mechanic is closest to me, he knows me well,
The woodman that takes his axe and jug with him shall take me
with him all day,
The farm-boy ploughing in the field feels good at the sound of my
voice,
In vessels that sail my words sail, I go with fishermen and seamen
and love them.   1260

The soldier camp'd or upon the march is mine,
On the night ere the pending battle many seek me, and I do not
fail them,
On that solemn night (it may be their last) those that know me
seek me.

My face rubs to the hunter's face when he lies down alone in his
blanket,
The driver thinking of me does not mind the jolt of his wagon,   1265
The young mother and old mother comprehend me,
The girl and the wife rest the needle a moment and forget where
they are,
They and all would resume what I have told them.

48

I have said that the soul is not more than the body,
And I have said that the body is not more than the soul,   1270
And nothing, not God, is greater to one than one's self is,
And whoever walks a furlong without sympathy walks to his own
funeral drest in his shroud,
And I or you pocketless of a dime may purchase the pick of the
earth,
And to glance with an eye or show a bean in its pod confounds the
learning of all times,
And there is no trade or employment but the young man following
it may become a hero,   1275
And there is no object so soft but it makes a hub for the wheel'd
universe,
And I say to any man or woman, Let your soul stand cool and composed before a million universes.

And I say to mankind, Be not curious about God,
For I who am curious about each am not curious about God,
(No array of terms can say how much I am at peace about God and
about death.)   1280

I hear and behold God in every object, yet understand God not in
the least,
Nor do I understand who there can be more wonderful than myself.

Why should I wish to see God better than this day?
I see something of God each hour of the twenty-four, and each
moment then,

In the faces of men and women I see God, and in my own face in
the glass,   1285
I find letters from God dropt in the street, and every one is sign'd by
God's name,
And I leave them where they are, for I know that wheresoe'er I go,
Others will punctually come for ever and ever.

49

And as to you Death, and you bitter hug of mortality, it is idle to
try to alarm me.

To his work without flinching the accoucheur[9] comes,   1290
I see the elder-hand pressing receiving supporting,
I recline by the sills of the exquisite flexible doors,
And mark the outlet, and mark the relief and escape.

And as to you Corpse I think you are good manure, but that does
not offend me,
I smell the white roses sweet-scented and growing,   1295
I reach to the leafy lips, I reach to the polish'd breasts of melons.

And as to you Life I reckon you are the leavings of many deaths,
(No doubt I have died myself ten thousand times before.)

I hear you whispering there O stars of heaven,
O suns—O grass of graves—O perpetual transfers and pro-
motions,   1300
If you do not say any thing how can I say any thing?

Of the turbid pool that lies in the autumn forest,
Of the moon that descends the steeps of the soughing twilight,
Toss, sparkles of day and dusk—toss on the black stems that decay
in the muck,
Toss to the moaning gibberish of the dry limbs.   1305

I ascend from the moon, I ascend from the night,
I perceive that the ghastly glimmer is noonday sunbeams reflected,
And debouch[1] to the steady and central from the offspring great or
small.

50

There is that in me—I do not know what it is—but I know it is in
me.

Wrench'd and sweaty—calm and cool then my body becomes,
I sleep—I sleep long.   1310

I do not know it—it is without name—it is a word unsaid,
It is not in any dictionary, utterance, symbol.

9. Midwife.   1. Pour forth.

Something it swings on more than the earth I swing on,
To it the creation is the friend whose embracing awakes me.   1315

Perhaps I might tell more. Outlines! I plead for my brothers and
    sisters.

Do you see O my brothers and sisters?
It is not chaos or death—it is form, union, plan—it is eternal life
    —it is Happiness.

51

The past and present wilt—I have fill'd them, emptied them,
And proceed to fill my next fold of the future.   1320

Listener up there! what have you to confide to me?
Look in my face while I snuff the sidle[2] of evening,
(Talk honestly, no one else hears you, and I stay only a minute
    longer.)

Do I contradict myself?
Very well then I contradict myself,   1325
(I am large, I contain multitudes.)

I concentrate toward them that are nigh, I wait on the door-slab.

Who has done his day's work? who will soonest be through with
    his supper?
Who wishes to walk with me?

Will you speak before I am gone? will you prove already too
    late?   1330

52

The spotted hawk swoops by and accuses me, he complains of my
    gab and my loitering.

I too am not a bit tamed, I too am untranslatable,
I sound my barbaric yawp over the roofs of the world.

The last scud of day[3] holds back for me,
It flings my likeness after the rest and true as any on the shadow'd
    wilds,   1335
It coaxes me to the vapor and the dusk.

I depart as air, I shake my white locks at the runaway sun,
I effuse my flesh in eddies, and drift it in lacy jags.

---

2. To snuff is to put out, as in extinguishing a candle; here the light is the hesitant last light of day, sidling or moving along edgeways.
3. Wind-driven clouds, or merely the last rays of the sun.

I bequeath myself to the dirt to grow from the grass I love,
If you want me again look for me under your boot-soles.　　　1340

You will hardly know who I am or what I mean,
But I shall be good health to you nevertheless,
And filter and fibre your blood.

Failing to fetch me at first keep encouraged,
Missing me one place search another,　　　1345
I stop somewhere waiting for you.

　　　　　　　　　　　　　　　　　　1855, 1881

From CHILDREN OF ADAM[1]
From Pent-up Aching Rivers

From pent-up aching rivers,
From that of myself without which I were nothing,
From what I am determin'd to make illustrious, even if I stand sole
　　　among men,
From my own voice resonant, singing the phallus,
Singing the song of procreation,　　　5
Singing the need of superb children and therein superb grown
　　　people,
Singing the muscular urge and the blending,
Singing the bedfellow's song, (O resistless yearning!
O for any and each the body correlative attracting!
O for you whoever you are your correlative body! O it, more than
　　　all else, you delighting!)　　　10
From the hungry gnaw that eats me night and day,
From native moments, from bashful pains, singing them,
Seeking something yet unfound though I have diligently sought it
　　　many a long year,
Singing the true song of the soul fitful at random,
Renascent with grosset Nature or among animals,　　　15
Of that, of them and what goes with them my poems informing,

1. This group of poems celebrating sex first appeared in the 1860 edition of *Leaves of Grass* as "Enfans d'Adam"; later the contents and order were slightly altered until they reached final form in 1871. In their edition Blodgett and Bradley quote a note in which Whitman identifies the relationship of this group to the "Calamus" poems: "Theory of a Cluster of Poems the same *to the passion of Woman-Love* as the 'Calamus-Leaves' are to adhesiveness, manly love. Full of animal-fire, tender, burning,— the tremulous ache, delicious, yet such a torment. The swelling elate and vehement, that will not be denied. Adam, as a central figure and type. One piece presenting a vivid picture (in connection with the spirit) of a fully complete, well-developed man, eld, bearded, swart, fiery,—as a more than rival of the youthful type-hero of novels and love poems" (*Notes and Fragments*, p. 124, No. 142).

From the first the "Children of Adam" poems were controversial. Having seen them in manuscript or proofs early in 1860, Emerson tried to persuade Whitman not to print them in the third edition then being printed in Boston. In "A Memorandum at a Venture" (*North American Review*, June, 1882), Whitman said of the poems that "the sexual passion in itself, while normal and unperverted, is inherently legitimate, creditable, not necessarily an improper theme for poet, as confessedly not for scientist." Honest treatment of the sexual passion was basic "to the whole construction, organicism, and intentions of 'Leaves of Grass.' "

Of the smell of apples and lemons, of the pairing of birds,
Of the wet of woods, of the lapping of waves,
Of the mad pushes of waves upon the land, I them chanting,
The overture lightly sounding, the strain anticipating,   20
The welcome nearness, the sight of the perfect body,
The swimmer swimming naked in the bath, or motionless on his
    back lying and floating,
The female form approaching, I pensive, love-flesh tremulous
    aching,
The divine list for myself or you or for any one making,
The face, the limbs, the index from head to foot, and what it
    arouses,   25
The mystic deliria, the madness amorous, the utter abandonment,
(Hark close and still what I now whisper to you,
I love you, O you entirely possess me,
O that you and I escape from the rest and go utterly off, free and
    lawless,
Two hawks in the air, two fishes swimming in the sea not more law-
    less than we;)
The furious storm through me careering, I passionately
    trembling,   30
The oath of the inseparableness of two together, of the woman
    that loves me and whom I love more than my life, that oath
    swearing,
(O I willingly stake all for you,
O let me be lost if it must be so!
O you and I! what is it to us what the rest do or think?
What is all else to us? only that we enjoy each other and exhaust
    each other if it must be so;)   35
From the master, the pilot I yield the vessel to,
The general commanding me, commanding all, from him permis-
    sion taking,
From time the programme hastening, (I have loiter'd too long as
    it is,)
From sex, from the warp and from the woof,[2]   40
From privacy, from frequent repinings alone,
From plenty of persons near and yet the right person not near,
From the soft sliding of hands over me and thrusting of fingers
    through my hair and beard,
From the long sustain'd kiss upon the mouth or bosom,
From the close pressure that makes me or any man drunk, fainting
    with excess,   45
From what the divine husband knows, from the work of fatherhood,
From exultation, victory and relief, from the bedfellow's embrace
    in the night,
From the act-poems of eyes, hands, hips and bosoms,
From the cling of the trembling arm,
From the bending curve and the clinch,   50
From side by side the pliant coverlet off-throwing,

2. Lengthwise and crosswise threads in a fabric.

From the one so unwilling to have me leave, and me just as unwill-
    ing to leave,
(Yet a moment O tender waiter, and I return,)
From the hour of shining stars and dropping dews,
From the night a moment I emerging flitting out,   55
Celebrate you act divine and you children prepared for,
And you stalwart loins.

                              1860, 1881

Spontaneous Me

Spontaneous me, Nature,
The loving day, the mounting sun, the friend I am happy with,
The arm of my friend hanging idly over my shoulder,
The hillside whiten'd with blossoms of the mountain ash,
The same late in autumn, the hues of red, yellow, drab, purple, and
    light and dark green,   5
The rich coverlet of the grass, animals and birds, the private un-
    trimm'd bank, the primitive apples, the pebble-stones,
Beautiful dripping fragments, the negligent list of one after an-
    other as I happen to call them to me or think of them,
The real poems, (what we call poems being merely pictures,)
The poems of the privacy of the night, and of men like me,
This poem drooping shy and unseen that I always carry, and that
    all men carry,   10
(Know once for all, avow'd on purpose, wherever are men like me,
    are our lusty lurking masculine poems,)
Love-thoughts, love-juice, love-odor, love-yielding, love-climbers,
    and the climbing sap,
Arms and hands of love, lips of love, phallic thumb of love, breasts
    of love, bellies press'd and glued together with love,
Earth of chaste love, life that is only life after love,
The body of my love, the body of the woman I love, the body of the
    man, the body of the earth,   15
Soft forenoon airs that blow from the south-west,
The hairy wild-bee that murmurs and hankers up and down, that
    gripes the full-grown lady-flower, curves upon her with amorous
    firm legs, takes his will of her, and holds himself tremulous and
    tight till he is satisfied;
The wet of woods through the early hours,
Two sleepers at night lying close together as they sleep, one with an
    arm slanting down across and below the waist of the other,
The smell of apples, aromas from crush'd sage-plant, mint, birch-
    bark,   20
The boy's longings, the glow and pressure as he confides to me what
    he was dreaming,
The dead leaf whirling its spiral whirl and falling still and content
    to the ground,
The no-form'd stings that sights, people, objects, sting me with,
The hubb'd sting of myself, stinging me as much as it ever can any
    one,

The sensitive, orbic, underlapp'd brothers, that only privileged feelers
    may be intimate where they are,   25

The curious roamer the hand roaming all over the body, the bashful
    withdrawing of flesh where the fingers soothingly pause and
    edge themselves,

The limpid liquid within the young man,

The vex'd corrosion so pensive and so painful,

The torment, the irritable tide that will not be at rest,

The like of the same I feel, the like of the same in others,   30

The young man that flushes and flushes, and the young woman that
    flushes and flushes,

The young man that wakes deep at night, the hot hand seeking to
    repress what would master him,

The mystic amorous night, the strange half-welcome pangs, visions,
    sweats,

The pulse pounding through palms and trembling encircling fingers,
    the young man all color'd, red, ashamed, angry;

The souse upon me of my lover the sea, as I lie willing and
    naked,   35

The merriment of the twin babes that crawl over the grass in the
    sun, the mother never turning her vigilant eyes from them,

The walnut-trunk, the walnut-husks, and the ripening or ripen'd
    long-round walnuts,

The continence of vegetables, birds, animals,

The consequent meanness of me should I skulk or find myself in-
    decent, while birds and animals never once skulk or find them-
    selves indecent,

The great chastity of paternity, to match the great chastity of
    maternity,   40

The oath of procreation I have sworn, my Adamic and fresh
    daughters,

The greed that eats me day and night with hungry gnaw, till I
    saturate what shall produce boys to fill my place when I am
    through,

The wholesome relief, repose, content,

And this bunch pluck'd at random from myself,

It has done its work—I toss it carelessly to fall where it may.   45

                                          1856, 1867

Once I Pass'd through a Populous City

Once I pass'd through a populous city imprinting my brain for
    future use with its shows, architecture, customs, traditions,

Yet now of all that city I remember only a woman I casually met
    there who detain'd me for love of me,

Day by day and night by night we were together—all else has long
    been forgotten by me,

I remember I say only that woman who passionately clung to me,

Again we wander, we love, we separate again,   5

Again she holds me by the hand, I must not go,

I see her close beside me with silent lips sad and tremulous.

                                          1860, 1861

Facing West from California's Shores

Facing west from California's shores,
Inquiring, tireless, seeking what is yet unfound,
I, a child, very old, over waves, towards the house of maternity, the
    land of migrations, look afar,
Look off the shores of my Western sea, the circle almost circled;
For starting westward from Hindustan, from the vales of Kash-
    mere,   5
From Asia, from the north, from the God, the sage, and the hero,
From the south, from the flowery peninsulas and the spice islands,
Long having wander'd since, round the earth having wander'd,
Now I face home again, very pleas'd and joyous,
(But where is what I started for so long ago?   10
And why is it yet unfound?)

                                                              1860, 1867

From CALAMUS[1]
Scented Herbage of My Breast

Scented herbage of my breast,
Leaves from you I glean, I write, to be perused best afterwards,
Tomb-leaves, body-leaves growing up above me above death,
Perennial roots, tall leaves, O the winter shall not freeze you deli-
    cate leaves,
Every year shall you bloom again, out from where you retired you
    shall emerge again;   5
O I do not know whether many passing by will discover you or
    inhale your faint odor, but I believe a few will;
O slender leaves! O blossoms of my blood! I permit you to tell in
    your own way of the heart that is under you,
O I do not know what you mean there underneath yourselves, you
    are not happiness,
You are often more bitter than I can bear, you burn and sting me,
Yet you are beautiful to me you faint tinged roots, you make me
    think of death,   10
Death is beautiful from you, (what indeed is finally beautiful except
    death and love?)

1. The "Calamus" group first appeared in the third edition of *Leaves of Grass* (1860) with pretty much the contents and order the group finally reached in 1881. Comparisons with the "Children of Adam" sequence are inevitable; Whitman himself saw the first as celebrating "ama-tive" love of men and women and the "Calamus" poems as celebrating "ad-hesive love" of men for men. Blodgett and Bradley quote Whitman's insistence in *Democratic Vistas* that the adhesive love he celebrates was political in na-ture: "It is to the development, identifi-cation, and general prevalence of that fervid comradeship, (the adhesive love, at least rivaling the amative love hitherto possessing imaginative literature, if not going beyond it,) that I look for the counterbalance and offset of our material-istic and vulgar American democracy, and for the spiritualization thereof." For the 1876 preface to *Leaves of Grass* Whitman rewrote this passage as a direct comment on the "Calamus" poems. Whatever the programmatic content thus vouched for, some of the poems are in-fused with a personal torment which strikes most readers as autoerotic or homoerotic. Edwin H. Miller says pro-vocatively that the "Children of Adam" poems express "the public image of the singer of sexuality" while the "Calamus" poems express "the private man and his personal hunger."

O I think it is not for life I am chanting here my chant of lovers, I
    think it must be for death,
For how calm, how solemn it grows to ascend to the atmosphere of
    lovers,
Death or life I am then indifferent, my soul declines to prefer,
(I am not sure but the high soul of lovers welcomes death most,) 15
Indeed O death, I think now these leaves mean precisely the same
    as you mean,
Grow up taller sweet leaves that I may see! grow up out of my
    breast!
Spring away from the conceal'd heart there!
Do not fold yourself so in your pink-tinged roots timid leaves!
Do not remain down there so ashamed, herbage of my breast! 20
Come I am determin'd to unbare this broad breast of mine, I have
    long enough stifled and choked;
Emblematic and capricious blades I leave you, now you serve me
    not,
I will say what I have to say by itself,
I will sound myself and comrades only, I will never again utter a
    call only their call,
I will raise with it immortal reverberations through the States, 25
I will give an example to lovers to take permanent shape and will
    through the States,
Through me shall the words be said to make death exhilarating,
Give me your tone therefore O death, that I may accord with it,
Give me yourself, for I see that you belong to me now above all,
    and are folded inseparably together, you love and death are,
Nor will I allow you to balk me any more with what I was calling
    life,
                                      30
For now it is convey'd to me that you are the purports essential,
That you hide in these shifting forms of life, for reasons, and that
    they are mainly for you,
That you beyond them come forth to remain, the real reality,
That behind the mask of materials you patiently wait, no matter
    how long,
That you will one day perhaps take control of all, 35
That you will perhaps dissipate this entire show of appearance,
That may-be you are what it is all for, but it does not last so very
    long,
But you will last very long.

                                        1860, 1881

Whoever You Are Holding Me Now in Hand

Whoever you are holding me now in hand,
Without one thing all will be useless,
I give you fair warning before you attempt me further,
I am not what you supposed, but far different.

Who is he that would become my follower? 5
Who would sign himself a candidate for my affections?

The way is suspicious, the result uncertain, perhaps destructive,
You would have to give up all else, I alone would expect to be your
    sole and exclusive standard,
Your novitiate would even then be long and exhausting,
The whole past theory of your life and all conformity to the lives
    around you would have to be abandon'd,   10
Therefore release me now before troubling yourself any further,
    let go your hand from my shoulders,
Put me down and depart on your way.

Or else by stealth in some wood for trial,
Or back of a rock in the open air,
(For in any roof'd room of a house I emerge not, nor in com-
    pany,   15
And in libraries I lie as one dumb, a gawk, or unborn, or dead,)
But just possibly with you on a high hill, first watching lest any
    person for miles around approach unawares,
Or possibly with you sailing at sea, or on the beach of the sea or
    some quiet island,
Here to put your lips upon mine I permit you,
With the comrade's long-dwelling kiss or the new husband's
    kiss,   20
For I am the new husband and I am the comrade.

Or if you will, thrusting me beneath your clothing,
Where I may feel the throbs of your heart or rest upon your hip,
Carry me when you go forth over land or sea;
For thus merely touching you is enough, is best,   25
And thus touching you would I silently sleep and be carried
    eternally.

But these leaves conning you con at peril,
For these leaves and me you will not understand,
They will elude you at first and still more afterward, I will certainly
    elude you,
Even while you should think you had unquestionably caught me,
    behold!   30
Already you see I have escaped from you.

For it is not for what I have put into it that I have written this book,
Nor is it by reading it you will acquire it,
Nor do those know me best who admire me and vauntingly praise
    me,
Nor will the candidates for my love (unless at most a very few)
    prove victorious,   35
Nor will my poems do good only, they will do just as much evil,
    perhaps more,
For all is useless without that which you may guess at many times
    and not hit, that which I hinted at;
Therefore release me and depart on your way.

                                     1860, 1881

When I Heard at the Close of the Day

When I heard at the close of the day how my name had been
    receiv'd with plaudits in the capitol, still it was not a happy
    night for me that follow'd,
And else when I carous'd, or when my plans were accomplish'd,
    still I was not happy,
But the day when I rose at dawn from the bed of perfect health,
    refresh'd, singing, inhaling the ripe breath of autumn,
When I saw the full moon in the west grow pale and disappear in
    the morning light,
When I wander'd alone over the beach, and undressing bathed,
    laughing with the cool waters, and saw the sun rise,   5
And when I thought how my dear friend my lover was on his way
    coming, O then I was happy,
O then each breath tasted sweeter, and all that day my food
    nourish'd me more, and the beautiful day pass'd well,
And the next came with equal joy, and with the next at evening
    came my friend,
And that night while all was still I heard the waters roll slowly con-
    tinually up the shores,
I heard the hissing rustle of the liquid and sands as directed to me
    whispering to congratulate me,   10
For the one I love most lay sleeping by me under the same cover in
    the cool night,
In the stillness in the autumn moonbeams his face was inclined
    toward me,
And his arm lay lightly around my breast—and that night I was
    happy.

                  1860, 1867

Trickle Drops

Trickle drops! my blue veins leaving!
O drops of me! trickle, slow drops,
Candid from me falling, drip, bleeding drops,
From wounds made to free you whence you were prison'd,
From my face, from my forehead and lips,   5
From my breast, from within where I was conceal'd, press forth red
    drops, confession drops,
Stain every page, stain every song I sing, every word I say, bloody
    drops,
Let them know your scarlet heat, let them glisten,
Saturate them with yourself all ashamed and wet,
Glow upon all I have written or shall write, bleeding drops,   10
Let it all be seen in your light, blushing drops.

                  1860, 1867

I Saw in Louisiana a Live-Oak Growing

I saw in Louisiana a live-oak growing,
All alone stood it and the moss hung down from the branches,

Without any companion it grew there uttering joyous leaves of dark
    green,
And its look, rude, unbending, lusty, made me think of myself,
But I wonder'd how it could utter joyous leaves standing alone there
    without its friend near, for I knew I could not,   5
And I broke off a twig with a certain number of leaves upon it, and
    twined around it a little moss,
And brought it away, and I have placed it in sight in my room,
It is not needed to remind me as of my own dear friends,
(For I believe lately I think of little else than of them,)
Yet it remains to me a curious token, it makes me think of manly
    love;   10
For all that, and though the live-oak glistens there in Louisiana
    solitary in a wide flat space,
Uttering joyous leaves all its life without a friend a lover near,
I know very well I could not.

                                                    1860, 1867

Here the Frailest Leaves of Me

Here the frailest leaves of me and yet my strongest lasting,
Here I shade and hide my thoughts, I myself do not expose them,
And yet they expose me more than all my other poems.

                                                    1860, 1871

Crossing Brooklyn Ferry[1]

1

Flood-tide below me! I see you face to face!
Clouds of the west—sun there half an hour high—I see you also
    face to face.

Crowds of men and women attired in the usual costumes, how
    curious you are to me!
On the ferry-boats the hundreds and hundreds that cross, returning
    home, are more curious to me than you suppose,
And you that shall cross from shore to shore years hence are more
    to me, and more in my meditations, than you might suppose. 5

2

The impalpable sustenance of me from all things at all hours of
    the day,
The simple, compact, well-join'd scheme, myself disintegrated,
    every one disintegrated yet part of the scheme,
The similitudes of the past and those of the future,
The glories strung like beads on my smallest sights and hearings, on

1. *Crossing Brooklyn Ferry* is one of a dozen poems which follow the "Calamus" section and precede the "Birds of Passage" section; mostly longish poems, like this one, they have no section title. Perhaps the clearest example of Whitman's desire to work by indirection, *Crossing Brooklyn Ferry* succeeds by alluring the reader without his quite knowing why.
    First published as *Sun-Down Poem* in the second edition (1856), *Crossing Brooklyn Ferry* was given its final title in 1860.

the walk in the street and the passage over the river,
The current rushing so swiftly and swimming with me far away,   10
The others that are to follow me, the ties between me and them,
The certainty of others, the life, love, sight, hearing of others.

Others will enter the gates of the ferry and cross from shore to
   shore,
Others will watch the run of the flood-tide,
Others will see the shipping of Manhattan north and west, and the
   heights of Brooklyn to the south and east,   15
Others will see the islands large and small;
Fifty years hence, others will see them as they cross, the sun half an
   hour high,
A hundred years hence, or ever so many hundred years hence, others
   will see them,
Will enjoy the sunset, the pouring-in of the flood-tide, the falling-
   back to the sea of the ebb-tide.

                              3
It avails not, time nor place—distance avails not,   20
I am with you, you men and women of a generation, or ever so
   many generations hence,
Just as you feel when you look on the river and sky, so I felt,
Just as any of you is one of a living crowd, I was one of a crowd,
Just as you are refresh'd by the gladness of the river and the bright
   flow, I was refresh'd,
Just as you stand and lean on the rail, yet hurry with the swift
   current, I stood yet was hurried,   25
Just as you look on the numberless masts of ships and the thick-
   stemm'd pipes of steamboats, I look'd.

I too many and many a time cross'd the river of old,
Watched the Twelfth-month[2] sea-gulls, saw them high in the air
   floating with motionless wings, oscillating their bodies,
Saw how the glistening yellow lit up parts of their bodies and left
   the rest in strong shadow,
Saw the slow-wheeling circles and the gradual edging toward the
   south,   30
Saw the reflection of the summer sky in the water,
Had my eyes dazzled by the shimmering track of beams,
Look'd at the fine centrifugal spokes of light round the shape of my
   head in the sunlit water,
Look'd on the haze on the hills southward and south-westward,
Look'd on the vapor as it flew in fleeces tinged with violet,   35
Look'd toward the lower bay to notice the vessels arriving,
Saw their approach, saw aboard those that were near me,
Saw the white sails of schooners and sloops, saw the ships at anchor,
The sailors at work in the rigging or out astride the spars,
The round masts, the swinging motion of the hulls, the slender ser-
   pentine pennants,   40

2. December.

The large and small steamers in motion, the pilots in their pilot-
houses,
The white wake left by the passage, the quick tremulous whirl of
the wheels,
The flags of all nations, the falling of them at sunset,
The scallop-edged waves in the twilight, the ladled cups, the frolicsome crests and glistening,
The stretch afar growing dimmer and dimmer, the gray walls of the
granite storehouses by the docks,   45
On the river the shadowy group, the big steam-tug closely flank'd on
each side by the barges, the hay-boat, the belated lighter,[3]
On the neighboring shore the fires from the foundry chimneys burn-
ing high and glaringly into the night,
Casting their flicker of black contrasted with wild red and yellow
light over the tops of houses, and down into the clefts of streets.

4

These and all else were to me the same as they are to you,
I loved well those cities, loved well the stately and rapid river,   50
The men and women I saw were all near to me,
Others the same—others who look back on me because I look'd
forward to them,
(The time will come, though I stop here to-day and to-night.)

5

What is it then between us?
What is the count of the scores or hundreds of years between
us?   55

Whatever it is, it avails not—distance avails not, and place avails
not,
I too lived, Brooklyn of ample hills was mine,
I too walk'd the streets of Manhattan island, and bathed in the
waters around it,
I too felt the curious abrupt questionings stir within me,
In the day among crowds of people sometimes they came upon
me,   60
In my walks home late at night or as I lay in my bed they came
upon me,
I too had been struck from the float forever held in solution,
I too had receiv'd identity by my body,
That I was I knew was of my body, and what I should be I knew I
should be of my body.

6

It is not upon you alone the dark patches fall,   65
The dark threw its patches down upon me also,
The best I had done seem'd to me blank and suspicious,
My great thoughts as I supposed them, were they not in reality
meagre?
Nor is it you alone who know what it is to be evil,
I am he who knew what it was to be evil,   70
I too knitted the old knot of contrariety,

3. Barge used to load or unload a cargo ship.

Blabb'd, blush'd, resented, lied, *stole, grudg'd,*
Had guile, anger, lust, hot wishes I dared not speak,
Was wayward, vain, greedy, shallow, sly, cowardly, malignant,
The wolf, the snake, the hog, not wanting in me,   75
The cheating look, the frivolous word, the adulterous wish, not
  wanting,
Refusals, hates, postponements, meanness, laziness, none of these
  wanting,
Was one with the rest, the days and haps of the rest,
Was call'd by my nighest name by clear loud voices of young men
  as they saw me approaching or passing,
Felt their arms on my neck as I stood, or the negligent leaning of
  their flesh against me as I sat,   80
Saw many I loved in the street or ferry-boat or public assembly, yet
  never told them a word,
Lived the same life with the rest, the same old laughing, gnawing,
  sleeping,
Play'd the part that still looks back on the actor or actress,
The same old role, the role that is what we make it, as great as we
  like,
Or as small as we like, or both great and small.   85

7

Closer yet I approach you,
What thought you have of me now, I had as much of you—I laid
  in my stores in advance,
I consider'd long and seriously of you before you were born.

Who was to know what should come home to me?
Who knows but I am enjoying this?   90
Who knows, for all the distance, but I am as good as looking at you
  now, for all you cannot see me?

8

Ah, what can ever be more stately and admirable to me than mast-
  hemm'd Manhattan?
River and sunset and scallop-edg'd waves of flood-tide?
The sea-gulls oscillating their bodies, the hay-boat in the twilight,
  and the belated lighter?
What gods can exceed these that clasp me by the hand, and with
  voices I love call me promptly and loudly by my nighest name
  as I approach?   95
What is more subtle than this which ties me to the woman or man
  that looks in my face?
Which fuses me into you now, and pours my meaning into you?

We understand then do we not?
What I promis'd without mentioning it, have you not accepted?
What the study could not teach—what the preaching could not
  accomplish is accomplish'd, is it not?   100

9

Flow on, river! flow with the flood-tide, and ebb with the ebb-tide!
Frolic on, crested and scallop-edg'd waves!

Gorgeous clouds of the sunset! drench with your splendor me, or
    the men and women generations after me!
Cross from shore to shore, countless crowds of passengers!
Stand up, tall masts of Mannahatta![4] stand up, beautiful hills of
    Brooklyn!   105
Throb, baffled and curious brain! throw out questions and answers!
Suspend here and everywhere, eternal float of solution!
Gaze, loving and thirsting eyes, in the house or street or public
    assembly!
Sound out, voices of young men! loudly and musically call me by
    my nighest name!
Live, old life! play the part that looks back on the actor or
    actress!   110
Play the old role, the role that is great or small according as one
    makes it!
Consider, you who peruse me, whether I may not in unknown ways
    be looking upon you;
Be firm, rail over the river, to support those who lean idly, yet haste
    with the hasting current;
Fly on, sea-birds! fly sideways, or wheel in large circles high in the
    air;
Receive the summer sky, you water, and faithfully hold it till all
    downcast eyes have time to take it from you!   115
Diverge, fine spokes of light, from the shape of my head, or any
    one's head, in the sunlit water!
Come on, ships from the lower bay! pass up or down, white-sail'd
    schooners, sloops, lighters!
Flaunt away, flags of all nations! be duly lower'd at sunset!
Burn high your fires, foundry chimneys! cast black shadows at night-
    fall! cast red and yellow light over the tops of the houses!
Appearances, now or henceforth, indicate what you are,   120
You necessary film, continue to envelop the soul,
About my body for me, and your body for you, be hung our divinest
    aromas,
Thrive, cities—bring your freight, bring your shows, ample and suf-
    ficient rivers,
Expand, being than which none else is perhaps more spiritual,
Keep your places, objects than which none else is more lasting.   125

You have waited, you always wait, you dumb, beautiful ministers,
We receive you with free sense at last, and are insatiate hence-
    forward,
Not you any more shall be able to foil us, or withhold yourselves
    from us,
We use you, and do not cast you aside—we plant you permanently
    within us,
We fathom you not—we love *you*—*there is* perfection in you
    also.   130

---

4. Variant for the Indian word normally spelled Manhattan.

You furnish your parts toward eternity,
Great or small, you furnish your parts toward the soul.

1856, 1881

From SEA-DRIFT[1]
Out of the Cradle Endlessly Rocking

Out of the cradle endlessly rocking,
Out of the mocking-bird's throat, the musical shuttle,
Out of the Ninth-month midnight,
Over the sterile sands and the fields beyond, where the child leaving
    his bed wander'd alone, bareheaded, barefoot,
Down from the shower'd halo,   5
Up from the mystic play of shadows twining and twisting as if they
    were alive,
Out from the patches of briers and blackberries,
From the memories of the bird that chanted to me,
From your memories sad brother, from the fitful risings and fallings
    I heard,
From under that yellow half-moon late-risen and swollen as if with
    tears,   10
From those beginning notes of yearning and love there in the mist,
From the thousand responses of my heart never to cease,
From the myriad thence-arous'd words,
From the word stronger and more delicious than any,
From such as now they start the scene revisiting,   15
As a flock, twittering, rising, or overhead passing,
Borne hither, ere all eludes me, hurriedly,
A man, yet by these tears a little boy again,
Throwing myself on the sand, confronting the waves,
I, chanter of pains and joys, uniter of here and hereafter,   20
Taking all hints to use them, but swiftly leaping beyond them,
A reminiscence sing.

Once Paumanok,[2]
When the lilac-scent was in the air and Fifth-month grass was
    growing,
Up this seashore in some briers,   25
Two feather'd guests from Alabama, two together,

---

1. First published as *A Child's Reminiscence* in the New York *Saturday Press* for December 24, 1859, this poem was incorporated into the 1860 *Leaves of Grass* as *A Word Out of the Sea*. Whitman continued to revise it until it reached the present form in the "Sea-Drift" section of the 1881 edition, a section made up of two new poems, seven poems from *Sea-Shore Memories* in the 1871 "Passage to India" section, and two poems from the 1876 "Two Rivulets" section. *Out of the Cradle Endlessly Rocking* had been the first of the "Sea-Shore Memories" group.

The poem is about the way at a crisis in his adult life the poet remembers (and now fully comprehends) the boyhood experience of the annunciation of Whitman's role as a poet. On the most obvious level, the poem belongs to the Romantic tradition of poems about the revisiting of a spot important to the poet's earlier life: examples are Wordsworth's *Tintern Abbey* and *Wye Revisited* and Longfellow's *My Lost Youth*.
2. Long Island.

And their nest, and four light-green eggs spotted with brown,
And every day the he-bird to and fro near at hand,
And every day the she-bird crouch'd on her nest, silent, with bright eyes,
And every day I, a curious boy, never too close, never disturbing them,   30
Cautiously peering, absorbing, translating.

Shine! shine! shine!
Pour down your warmth, great sun!
While we bask, we two together.

Two together!   35
Winds blow south, or winds blow north,
Day come white, or night come black,
Home, or rivers and mountains from home,
Singing all time, minding no time,
While we two keep together.   40

Till of a sudden,
May-be kill'd, unknown to her mate,
One forenoon the she-bird crouch'd not on the nest,
Nor return'd that afternoon, nor the next
Nor ever appear'd again.   45

And thenceforward all summer in the sound of the sea,
And at night under the full of the moon in calmer weather,
Over the hoarse surging of the sea,
Or flitting from brier to brier by day,
I saw, I heard at intervals the remaining one, the he-bird,   50
The solitary guest from Alabama.

Blow! blow! blow!
Blow up sea-winds along Paumanok's shore;
I wait and I wait till you blow my mate to me.

Yes, when the stars glisten'd,   55
All night long on the prong of a moss-scallop'd stake,
Down almost amid the slapping waves,
Sat the lone singer wonderful causing tears.

He call'd on his mate,
He pour'd forth the meanings which I of all men know.   60

Yes my brother I know,
The rest might not, but I have treasur'd every note,
For more than once dimly down to the beach gliding,
Silent, avoiding the moonbeams, blending myself with the shadows,
Recalling now the obscure shapes, the echoes, the sounds and sights after their sorts,   65
The white arms out in the breakers tirelessly tossing,

I, with bare feet, a child, the wind wafting my hair,
Listen'd long and long.

Listen'd to keep, to sing, now translating the notes,
Following you my brother.   70

Soothe! soothe! soothe!
Close on its wave soothes the wave behind,
And again another behind embracing and lapping, every one close,
But my love soothes not me, not me.

Low hangs the moon, it rose late,   75
It is lagging—O I think it is heavy with love, with love.

O madly the sea pushes upon the land,
With love, with love.

O night! do I not see my love fluttering out among the breakers?
What is that little black thing I see there in the white?   80

Loud! loud! loud!
Loud I call to you, my love!

High and clear I shoot my voice over the waves,
Surely you must know who is here, is here,
You must know who I am, my love.   85

Low-hanging moon!
What is that dusky spot in your brown yellow?
O it is the shape, the shape of my mate!
O moon do not keep her from me any longer.

Land! land! O land!   90
Whichever way I turn, O I think you could give me my mate back
*   again if you only would,*
For I am almost sure I see her dimly whichever way I look.

O rising stars!
Perhaps the one I want so much will rise, will rise with some of you.

O throat! O trembling throat!   95
Sound clearer through the atmosphere!
Pierce the woods, the earth,
Somewhere listening to catch you must be the one I want.

Shake out carols!
Solitary here, the night's carols!   100
Carols of lonesome love! death's carols!
Carols under that lagging, yellow, waning moon!
O under that moon where she droops almost down into the sea!
O reckless despairing carols.

But soft! sink low!   105
Soft! let me just murmur,
And do you wait a moment you husky-nois'd sea,
For somewhere I believe I heard my mate responding to me,
So faint, I must be still, be still to listen,
But not altogether still, for then she might not come immediately
*   to me.*   110

Hither my love!
Here I am! here!
With this just-sustain'd note I announce myself to you,
This gentle call is for you my love, for you.

Do not be decoy'd elsewhere,   115
That is the whistle of the wind, it is not my voice,
That is the fluttering, the fluttering of the spray,
Those are the shadows of leaves.

O darkness! O in vain!
O I am very sick and sorrowful.   120

O brown halo in the sky near the moon, drooping upon the sea!
O troubled reflection in the sea!
O throat! O throbbing heart!
And I singing uselessly, uselessly all the night.

O past! O happy life! O songs of joy!   125
In the air, in the woods, over fields,
Loved! loved! loved! loved! loved!
But my mate no more, no more with me!
We two together no more.

The aria sinking,   130
All else continuing, the stars shining,
The winds blowing, the notes of the bird continuous echoing,
With angry moans the fierce old mother incessantly moaning,
On the sands of Paumanok's shore gray and rustling,
The yellow half-moon enlarged, sagging down, drooping, the face
    of the sea almost touching,   135
The boy ecstatic, with his bare feet the waves, with his hair the
    atmosphere dallying,
The love in the heart long pent, now loose, now at last tumultu-
    ously bursting,
The aria's meaning, the ears, the soul, swiftly depositing,
The strange tears down the cheeks coursing,
The colloquy there, the trio, each uttering,   140
The undertone, the savage old mother incessantly crying,
To the boy's soul's questions sullenly timing, some drown'd secret
    hissing,
To the outsetting bard.

Demon or bird! (said the boy's soul,)
Is it indeed toward your mate you sing? or is it really to me?   145
For I, that was a child, my tongue's use sleeping, now I have heard
      you,
Now in a moment I know what I am for, I awake,
And already a thousand singers, a thousand songs, clearer, louder
      and more sorrowful than yours,
A thousand warbling echoes have started to life within me, never
      to die.

O you singer solitary, singing by yourself, projecting me,   150
O solitary me listening, never more shall I cease perpetuating
      you,
Never more shall I escape, never more the reverberations,
Never more the cries of unsatisfied love be absent from me,
Never again leave me to be the peaceful child I was before what
      there in the night,
By the sea under the yellow and sagging moon,   155
The messenger there arous'd, the fire, the sweet hell within,
The unknown want, the destiny of me.

O give me the clew! (it lurks in the night here somewhere,)
O if I am to have so much, let me have more!

A word then, (for I will conquer it,)   160
The word final, superior to all,
Subtle, sent up—what is it?—I listen;
Are you whispering it, and have been all the time, you sea-waves?
Is that it from your liquid rims and wet sands?

Whereto answering, the sea,   165
Delaying not, hurrying not,
Whisper'd me through the night, and very plainly before daybreak,
Lisp'd to me the low and delicious word death,
And again death, death, death, death,
Hissing melodious, neither like the bird nor like my arous'd child's
      heart,   170
But edging near as privately for me rustling at my feet,
Creeping thence steadily up to my ears and laving me softly all
      over,
Death, death, death, death, death.

Which I do not forget,
But fuse the song of my dusky demon and brother,   175
That he sang to me in the moonlight on Paumanok's gray beach,
With the thousand responsive songs at random,
My own songs awaked from that hour,
And with them the key, the word up from the waves,
The word of the sweetest song and all songs,   180
That strong and delicious word which, creeping to my feet,

(Or like some old crone rocking the cradle, swathed in sweet gar-
     ments, bending aside,)
The sea whisper'd me.

1859, 1881

From BY THE ROADSIDE[1]
A Hand-Mirror

Hold it up sternly—see this it sends back, (who is it? is it you?)
Outside fair costume, within ashes and filth,
No more a flashing eye, no more a sonorous voice or springy step,
Now some slave's eye, voice, hands, step,
A drunkard's breath, unwholesome eater's face, venerealee's flesh,   5
Lungs rotting away piecemeal, stomach sour and cankerous,
Joints rheumatic, bowels clogged with abomination,
Blood circulating dark and poisonous streams,
Words babble, hearing and touch callous,
No brain, no heart left, no magnetism of sex;   10
Such from one look in this looking-glass ere you go hence,
Such a result so soon—and from such a beginning!

1860, 1860

When I Heard the Learn'd Astronomer

When I heard the learn'd astronomer,
When the proofs, the figures, were ranged in columns before me,
When I was shown the charts and diagrams, to add, divide, and
     measure them,
When I sitting heard the astronomer where he lectured with much
     applause in the lecture-room,
How soon unaccountable I became tired and sick,   5
Till rising and gliding out I wander'd off by myself,
In the mystical moist night-air, and from time to time,
Look'd up in perfect silence at the stars.

1865, 1865

To a President

All you are doing and saying is to America dangled mirages,
You have not learn'd of Nature—of the politics of Nature you have
     not learn'd the great amplitude, rectitude, impartiality,
You have not seen that only such as they are for these States,
And that what is less than they must sooner or later lift off from
     these States.

1860, 1860

---

1. "By the Roadside" is the 1881 section title for around two dozen poems, most of which first appeared in the 1860 edition of *Leaves of Grass*. As Blodgett and Bradley say, "The group is truly a me- lange held together by the common bond of the poet's experience as roadside ob- server—passive, but alert and continually recording." Several of the poems are mere jottings of two, three, or four lines.

I Sit and Look Out

I sit and look out upon all the sorrows of the world, and upon all
    oppression and shame,
I hear secret convulsive sobs from young men at anguish with them-
    selves, remorseful after deeds done,
I see in low life the mother misused by her children, dying, ne-
    glected, gaunt, desperate,
I see the wife misused by her husband, I see the treacherous seducer
    of young women,
I mark the ranklings of jealousy and unrequited love attempted to
    be hid, I see these sights on the earth,   5
I see the workings of battle, pestilence, tyranny, I see martyrs and
    prisoners,
I observe a famine at sea, I observe the sailors casting lots who shall
    be kill'd to preserve the lives of the rest,
I observe the slights and degradations cast by arrogant persons upon
    laborers, the poor, and upon negroes, and the like;
All these—all the meanness and agony without end I sitting look out
    upon,
See, hear, and am silent.   10

                                                        1860, 1871

The Dalliance of the Eagles

Skirting the river road, (my forenoon walk, my rest,)
Skyward in air a sudden muffled sound, the dalliance of the eagles,
The rushing amorous contact high in space together,
The clinching interlocking claws, a living, fierce, gyrating wheel,
Four beating wings, two beaks, a swirling mass tight grappling,   5
In tumbling turning clustering loops, straight downward falling,
Till o'er the river pois'd, the twain yet one, a moment's lull,
A motionless still balance in the air, then parting, talons loosing,
Upward again on slow-firm pinions slanting, their separate diverse
    flight,
She hers, he his, pursuing.   10

                                                        1880, 1881

To the States

To Identify the 16th, 17th, or 18th Presentiad[1]

Why reclining, interrogating? why myself and all drowsing?
What deepening twilight—scum floating atop of the waters,
Who are they as bats and night-dogs askant in the capitol?
What a filthy Presidentiad! (O South, your torrid suns! O North,
    your arctic freezings!)

---

1. "Presidentiad" means "presidency."
Whitman is counting two-term Presidents
(Washington, Jefferson, and Jackson)
twice, and counting both Presidents who
died in office and their successors (Harrison and Tyler, Taylor and Fillmore) to
arrive at the 16th Presidentiad for Mil-
ard Fillmore's term (1850–53), the 17th
for Franklin Pierce's term (1853–57),
and the 18th for James Buchanan's term
(1857–61). Whitman's pamphlet *The
Eighteenth Presidency!* expresses the
same scorn for the presidencies of the
1850s.

Are those really Congressmen? are those the great Judges? is that
    the President?   5
Then I will sleep awhile yet, for I see that these States sleep, for
    reasons;
(With gathering murk, with muttering thunder and lambent shoots
    we all duly awake,
South, North, East, West, inland and seaboard, we will surely
    awake.)

                                           1860, 1860

From DRUM-TAPS[1]
Beat! Beat! Drums!

Beat! beat! drums!—blow! bugles! blow!
Through the windows—through doors—burst like a ruthless force,
Into the solemn church, and scatter the congregation,
Into the school where the scholar is studying;
Leave not the bridegroom quiet—no happiness must he have now
    with his bride,   5
Nor the peaceful farmer any peace, ploughing his field or gathering
    his grain,
So fierce you whirr and pound you drums—so shrill you bugles
    blow.

Beat! beat! drums!—blow! bugles! blow!
Over the traffic of cities—over the rumble of wheels in the streets;
Are beds prepared for sleepers at night in the houses? no sleepers
    must sleep in those beds,   10
No bargainers' bargains by day—no brokers or speculators—would
    they continue?
Would the talkers be talking? would the singer attempt to sing?
Would the lawyer rise in the court to state his case before the judge?
Then rattle quicker, heavier drums—you bugles wilder blow.

Beat! beat! drums!—blow! bugles! blow!   15
Make no parley—stop for no expostulation,
Mind not the timid—mind not the weeper or prayer,
Mind not the old man beseeching the young man,

1. The contents of the original *Drum-Taps* (first printed in 1865 as a little book) differed considerably from the contents of the "Drum-Taps" section finally arrived at in the 1881 *Leaves of Grass*. In the final arrangement the poetic purpose shifts throughout, roughly reflecting the chronology of the Civil War and the chronology of the composition of the poems. The first purpose is propagandistic. Indeed, *Beat! Beat! Drums!* served as a kind of recruiting poem when it was first printed (and reprinted) in the fall of 1861, having been composed after the southern victory at the first battle of Bull Run. Later Whitman seems to have understood that the early jingoistic poems had a certain his- torical value that made them worth preserving. The dominant impulse of most of the later poems is realistic—a determination to record the war the way it was, and in the best of the poems the realistic record is achieved through elaborate technical subtleties. The stages of Whitman's own attitudes toward the war are well stated in the epigraph he gave the whole "Drum-Taps" group in 1871 then inserted parenthetically into *The Wound Dresser* in the 1881 edition: "Arous'd and angry, I'd thought to beat the alarum, and urge relentless war,/But soon my fingers fail'd me, my face droop'd and I resign'd myself/To sit by the wounded and soothe them, or silently watch the dead."

Let not the child's voice be heard, nor the mother's entreaties,
Make even the trestles to shake the dead where they lie awaiting
    the hearses,   20
So strong you thump O terrible drums—so loud you bugles blow.

                                  1861, 1867

Cavalry Crossing a Ford

A line in long array where they wind betwixt green islands,
They take a serpentine course, their arms flash in the sun—hark to
    the musical clank,
Behold the silvery river, in it the splashing horses loitering stop to
    drink,
Behold the brown-faced men, each group, each person, a picture, the
    negligent rest on the saddles,
Some emerge on the opposite bank, others are just entering the ford
    —while,   5
Scarlet and blue and snowy white,
The guidon flags flutter gayly in the wind.

                                  1865, 1871

Vigil Strange I Kept on the Field One Night

Vigil strange I kept on the field one night;
When you my son and my comrade dropt at my side that day,
One look I but gave which your dear eyes return'd with a look I
    shall never forget,
One touch of your hand to mine O boy, reach'd up as you lay on
    the ground,
Then onward I sped in the battle, the even-contested battle,   5
Till late in the night reliev'd to the place at last again I made my
    way,
Found you in death so cold dear comrade, found your body son of
    responding kisses, (never again on earth responding,)
Bared your face in the starlight, curious the scene, cool blew the
    moderate night-wind,
Long there and then in vigil I stood, dimly around me the battle-
    field spreading,
Vigil wondrous and vigil sweet there in the fragrant silent night,   10
But not a tear fell, not even a long-drawn sigh, long, long I gazed,
Then on the earth partially reclining sat by your side leaning my
    chin in my hands,
Passing sweet hours, immortal and mystic hours with you dearest
    comrade—not a tear, not a word,
Vigil of silence, love and death, vigil for you my son and my soldier,
As onward silently stars aloft, eastward new ones upward stole,   15
Vigil final for you brave boy, (I could not save you, swift was your
    death,
I faithfully loved you and cared for you living, I think we shall
    surely meet again,)
Till at latest lingering of the night, indeed just as the dawn appear'd,

788 ★ Walt Whitman

My comrade I wrapt in his blanket, envelop'd well his form,
Folded the blanket well, tucking it carefully over head and carefully
    under feet,   20
And there and then and bathed by the rising sun, my son in his
    grave, in his rude-dug grave I deposited,
Ending my vigil strange with that, vigil of night and battle-field
    dim,
Vigil for boy of responding kisses, (never again on earth re-
    sponding,)
Vigil for comrade swiftly slain, vigil I never forget, how as day
    brighten'd,
I rose from the chill ground and folded my soldier well in his
    blanket,   25
And buried him where he fell.

1865, 1867

A March in the Ranks Hard-Prest, and the Road Unknown

A march in the ranks hard-prest, and the road unknown,
A route through a heavy wood with muffled steps in the darkness,
Our army foil'd with loss severe, and the sullen remnant retreating,
Till after midnight glimmer upon us the lights of a dim-lighted
    building,
We come to an open space in the woods, and halt by the dim-
    lighted building,   5
'Tis a large old church at the crossing roads, now an impromptu
    hospital,
Entering but for a minute I see a sight beyond all the pictures and
    poems ever made,
Shadows of deepest, deepest black, just lit by moving candles and
    lamps,
And by one great pitchy torch stationary with wild red flame and
    clouds of smoke,
By these, crowds, groups of forms vaguely I see on the floor, some
    in the pews laid down,   10
At my feet more distinctly a soldier, a mere lad, in danger of bleed-
    ing to death, (he is shot in the abdomen,)
I stanch the blood temporarily, (the youngster's face is white as a
    lily,)
Then before I depart I sweep my eyes o'er the scene fain to absorb
    it all,
Faces, varieties, postures beyond description, most in obscurity,
    some of them dead,
Surgeons operating, attendants holding lights, the smell of ether,
    the odor of blood,   15
The crowd, O the crowd of the bloody forms, the yard outside also
    fill'd,
Some on the bare ground, some on planks or stretchers, some in
    the death-spasm sweating,
An occasional scream or cry, the doctor's shouted orders or calls,

The glisten of the little steel instruments catching the glint of the
    torches,
These I resume as I chant, I see again the forms, I smell the
    odor,   20
Then hear outside the orders given, *Fall in, my men, fall in;*
But first I bend to the dying lad, his eyes open, a half-smile gives he
    me,
Then the eyes close, calmly close, and I speed forth to the darkness,
Resuming, marching, ever in darkness marching, on in the ranks,
The unknown road still marching.   25
1865, 1867

A Sight in Camp in the Daybreak
Gray and Dim

A sight in camp in the daybreak gray and dim,
As from my tent I emerge so early sleepless,
As slow I walk in the cool fresh air the path near by the hospital
    tent,
Three forms I see on stretchers lying, brought out there untended
    lying,
Over each the blanket spread, ample brownish woolen blanket, 5
Gray and heavy blanket, folding, covering all.

Curious I halt and silent stand,
Then with light fingers I from the face of the nearest the first just
    lift the blanket;
Who are you elderly man so gaunt and grim, with well-gray'd hair,
    and flesh all sunken about the eyes?
Who are you my dear comrade?   10

Then to the second I step—and who are you my child and darling?
Who are you sweet boy with cheeks yet blooming?

Then to the third—a face nor child nor old, very calm, as of beauti-
    ful yellow-white ivory;
Young man I think I know you—I think this face is the face of the
    Christ himself,
Dead and divine and brother of all, and here again he lies. 15
1865, 1867

As Toilsome I Wander'd Virginia's Woods

As toilsome I wander'd Virginia's woods,
To the music of rustling leaves kick'd by my feet, (for 'twas
    autumn,)
I mark'd at the foot of a tree the grave of a soldier;
Mortally wounded he and buried on the retreat, (easily all could I
    understand,)
The halt of a mid-day hour, when up! no time to lose—yet this sign
    left,   5

On a tablet scrawl'd and nail'd on the tree by the grave,
Bold, cautious, true, and my loving comrade.

Long, long I muse, then on my way go wandering,
Many a changeful season to follow, and many a scene of life,
Yet at times through changeful season and scene, abrupt, alone, or
    in the crowded street,   10
Comes before me the unknown soldier's grave, comes the inscrip-
    tion rude in Virginia's woods,
Bold, cautious, true, and my loving comrade.

1865, 1867

The Wound-Dresser

1

An old man bending I come among new faces,
Years looking backward resuming in answer to children,
Come tell us old man, as from young men and maidens that love me,
(Arous'd and angry, I'd thought to beat the alarum, and urge relent-
    less war,
But soon my fingers fail'd me, my face droop'd and I resign'd
    myself,   5
To sit by the wounded and soothe them, or silently watch the
    dead;)
Years hence of these scenes, of these furious passions, these
    chances,
Of unsurpass'd heroes, (was one side so brave? the other was
    equally brave;)
Now be witness again, paint the mightiest armies of earth,
Of those armies so rapid so wondrous what saw you to tell us?   10
What stays with you latest and deepest? of curious panics,
Of hard-fought engagements or sieges tremendous what deepest
    remains?

2

O maidens and young men I love and that love me,
What you ask of my days those the strangest and sudden your talk-
    ing recalls,
Soldier alert I arrive after a long march cover'd with sweat and
    dust,   15
In the nick of time I come, plunge in the fight, loudly shout in
    the rush of successful charge,
Enter the captur'd works[1]—yet lo, like a swift-running river they
    fade,
Pass and are gone they fade—I dwell not on soldiers' perils or
    soldiers' joys,
(Both I remember well—many the hardships, few the joys, yet I
    was content.)

But in silence, in dreams' projections,   20
While the world of gain and appearance and mirth goes on,

1. Fortifications.

So soon what is over forgotten, and waves wash the imprints off
the sand,
With hinged knees returning I enter the doors, (while for you up
there,
Whoever you are, follow without noise and be of strong heart.)

Bearing the bandages, water and sponge,   25
Straight and swift to my wounded I go,
Where they lie on the ground after the battle brought in,
Where their priceless blood reddens the grass the ground,
Or to the rows of the hospital tent, or under the roof'd hospital,
To the long rows of cots up and down each side I return,   30
To each and all one after another I draw near, not one do I miss,

An attendant follows holding a tray, he carries a refuse pail,
Soon to be fill'd with clotted rags and blood, emptied, and fill'd
again.

I onward go, I stop,
With hinged knees and steady hand to dress wounds,   35
I am firm with each, the pangs are sharp yet unavoidable,
One turns to me his appealing eyes—poor boy! I never knew you,
Yet I think I could not refuse this moment to die for you, if that
would save you.

3
On, on I go, (open doors of time! open hospital doors!)
The crush'd head I dress, (poor crazed hand tear not the bandage
away,)   40
The neck of the cavalry-man with the bullet through and through
I examine,
Hard the breathing rattles, quite glazed already the eye, yet life
struggles hard,
(Come sweet death! be persuaded O beautiful death!
In mercy come quickly.)

From the stump of the arm, the amputated hand,   45
I undo the clotted lint, remove the slough, wash off the matter and
blood,
Back on his pillow the soldier bends with curv'd neck and side-
falling head,
His eyes are closed, his face is pale, he dares not look on the bloody
stump,
And has not yet look'd on it.

I dress a wound in the side, deep, deep,   50
But a day or two more, for see the frame all wasted and sinking,
And the yellow-blue countenance see.

I dress the perforated shoulder, the foot with the bullet-wound,
Cleanse the one with a gnawing and putrid gangrene, so sickening,
so offensive,

While the attendant stands behind aside me holding the tray and
    pail.   55

I am faithful, I do not give out,
The fractur'd thigh, the knee, the wound in the abdomen,
These and more I dress with impassive hand, (yet deep in my
    breast a fire, a burning flame.)

4

Thus in silence in dreams' projections,
Returning, resuming, I thread my way through the hospitals,   60
The hurt and wounded I pacify with soothing hand,
I sit by the restless all the dark night, some are so young,
Some suffer so much, I recall the experience sweet and sad,
(Many a soldier's loving arms about this neck have cross'd and
    rested,
Many a soldier's kiss dwells on these bearded lips.)   65

        1865, 1881

Ethiopia Saluting the Colors

Who are you dusky woman, so ancient hardly human,
With your woolly-white and turban'd head, and bare bony feet?
Why rising by the roadside here, do you the colors greet?

('Tis while our army lines Carolina's sands and pines,
Forth from thy hovel door thou Ethiopia com'st to me,   5
As under doughty Sherman[1] I march toward the sea.)

Me master years a hundred since from my parents sunder'd,
A little child, they caught me as the savage beast is caught,
Then hither me across the sea the cruel slaver brought.

No further does she say, but lingering all the day,   10
Her high-borne turban'd head she wags, and rolls her darkling eye,
And courtesies to the regiments, the guidons[2] moving by.

What is it fateful woman, so blear, hardly human?
Why wag your head with turban bound, yellow, red and green?
Are the things so strange and marvelous you see or have seen?   15

        1871, 1881

Reconciliation

Word over all, beautiful as the sky,
Beautiful that war and all its deeds of carnage must in time be
    utterly lost,
That the hands of the sisters Death and Night incessantly softly
    wash again, and ever again, this soil'd world;
For my enemy is dead, a man divine as myself is dead,

---

1. William Tecumseh Sherman (1820–
91), Union general who ravaged a huge
swathe of the South in his march from
Atlanta to the sea.
2. Regimental pennants, usually with
forked ends.

I look where he lies white-faced and still in the coffin—I draw near, 5
Bend down and touch lightly with my lips the white face in the
coffin.

1865–66, 1881

Spirit Whose Work Is Done
(Washington City, 1865.)

Spirit whose work is done—spirit of dreadful hours!
Ere departing fade from my eyes your forests of bayonets;
Spirit of gloomiest fears and doubts, (yet onward ever unfaltering
pressing,)
Spirit of many a solemn day and many a savage scene—electric
spirit,
That with muttering voice through the war now closed, like a tire-
less phantom flitted,   5
Rousing the land with breath of flame, while you beat and beat the
drum,
Now as the sound of the drum, hollow and harsh to the last, rever-
berates round me,
As your ranks, your immortal ranks, return, return from the battles,
As the muskets of the young men yet lean over their shoulders,
As I look on the bayonets bristling over their shoulders,   10
As those slanted bayonets, whole forests of them appearing in the
distance, approach and pass on, returning homeward,
Moving with steady motion, swaying to and fro to the right and
left,
Evenly lightly rising and falling while the steps keep time;
Spirit of hours I knew, all hectic red one day, but pale as death
next day,
Touch my mouth ere you depart, press my lips close,   15
Leave me your pulses of rage—bequeath them to me—fill me with
currents convulsive,
Let them scorch and blister out of my chants when you are gone,
Let them identify you to the future in these songs.

1865–66, 1881

From MEMORIES OF PRESIDENT LINCOLN[1]
When Lilacs Last in the Dooryard Bloom'd

1

When lilacs last in the dooryard bloom'd,
And the great star[2] early droop'd in the western sky in the night,
I mourn'd, and yet shall mourn with ever-returning spring.

1. Composed in the months following
Lincoln's assassination on April 14, 1865,
this elegy was printed in the fall of that
year as an appendix to the recently published *Drum-Taps* volume. In the 1881
edition of *Leaves of Grass* it and three
lesser poems were joined to make up the
section "Memories of President Lincoln."

Not simply a poem about the death of
Lincoln, *When Lilacs Last in the Dooryard Bloom'd* is about the stages by
which a poet transmutes his grief into
poetry.
2. Literally, Venus, although it becomes
associated with Lincoln himself.

Ever-returning spring, trinity sure to me you bring,
Lilac blooming perennial and drooping star in the west,   5
And thought of him I love.

2

O powerful western fallen star!
O shades of night—O moody, tearful night!
O great star disappear'd—O the black murk that hides the star!
O cruel hands that hold me powerless—O helpless soul of me!   10
O harsh surrounding cloud that will not free my soul.

3

In the dooryard fronting an old farm-house near the white-wash'd
    palings,
Stands the lilac-bush tall-growing with heart-shaped leaves of rich
    green,
With many a pointed blossom rising delicate, with the perfume
    strong I love,
With every leaf a miracle—and from this bush in the dooryard,   15
With delicate-color'd blossoms and heart-shaped leaves of rich
    green,
A sprig with its flower I break.

4

In the swamp in secluded recesses,
A shy and hidden bird is warbling a song.

Solitary the thrush,
The hermit withdrawn to himself, avoiding the settlements,   20
Sings by himself a song.

Song of the bleeding throat,
Death's outlet song of life, (for well dear brother I know,
If thou wast not granted to sing thou would'st surely die.)   25

5

Over the breast of the spring, the land, amid cities,
Amid lanes and through old woods, where lately the violets peep'd
    from the ground, spotting the gray debris,
Amid the grass in the fields each side of the lanes, passing the end-
    less grass,
Passing the yellow-spear'd wheat, every grain from its shroud in
    the dark-brown fields uprisen,
Passing the apple-tree blows[3] of white and pink in the orchards,   30
Carrying a corpse to where it shall rest in the grave,
Night and day journeys a coffin.

6

Coffin that passes through lanes and streets,
Through day and night with the great cloud darkening the land,
With the pomp of the inloop'd flags with the cities draped in
    black,   35
With the show of the States themselves as of crape-veil'd women
    standing,
With processions long and winding and the flambeaus[4] of the night,

3. Blossoms.   4. Torches.

With the countless torches lit, with the silent sea of faces and the
  unbared heads,
With the waiting depot, the arriving coffin, and the sombre faces,
With dirges through the night, with the thousand voices rising strong
  and solemn,   40
With all the mournful voices of the dirges pour'd around the coffin,
The dim-lit churches and the shuddering organs—where amid these
  you journey,
With the tolling tolling bells' perpetual clang,
Here, coffin that slowly passes,
I give you my sprig of lilac.   45

7

(Nor for you, for one alone,
Blossoms and branches green to coffins all I bring,
For fresh as the morning, thus would I chant a song for you O
  sane and sacred death.

All over bouquets of roses,
O death, I cover you over with roses and early lilies,   50
But mostly and now the lilac that blooms the first,
Copious I break, I break the sprigs from the bushes,
With loaded arms I come, pouring for you,
For you and the coffins all of you O death.)

8

O western orb sailing the heaven,   55
Now I know what you must have meant as a month since I walk'd,
As I walk'd in silence the transparent shadowy night,
As I saw you had something to tell as you bent to me night after
  night,
As you droop'd from the sky low down as if to my side, (while the
  other stars all look'd on,)
As we wander'd together the solemn night, (for something I know
  not what kept me from sleep,)   60
As the night advanced, and I saw on the rim of the west how full
  you were of woe,
As I stood on the rising ground in the breeze in the cool transparent
  night,
As I watch'd where you pass'd and was lost in the netherward black
  of the night,
As my soul in its trouble dissatisfied sank, as where you sad orb,
Concluded, dropt in the night, and was gone.   65

9

Sing on there in the swamp,
O singer bashful and tender, I hear your notes, I hear your call,
I hear, I come presently, I understand you,
But a moment I linger, for the lustrous star has detain'd me,
The star my departing comrade holds and detains me.   70

10

O how shall I warble myself for the dead one there I loved?
And how shall I deck my song for the large sweet soul that has gone?
And what shall my perfume be for the grave of him I love?

Sea-winds blown from east and west,
Blown from the Eastern sea and blown from the Western sea, till
    there on the prairies meeting,   75
These and with these and the breath of my chant,
I'll perfume the grave of him I love.

11

O what shall I hang on the chamber walls?
And what shall the pictures be that I hang on the walls,
To adorn the burial-house of him I love?   80

Pictures of growing spring and farms and homes,
With the Fourth-month[5] eve at sundown, and the gray smoke lucid
    and bright,
With floods of the yellow gold of the gorgeous, indolent, sinking
    sun, burning, expanding the air,
With the fresh sweet herbage under foot, and the pale green leaves
    of the trees prolific,
In the distance the flowing glaze, the breast of the river, with a
    wind-dapple here and there,   85
With ranging hills on the banks, with many a line against the sky,
    and shadows,
And the city at hand with dwellings so dense, and stacks of chim-
    neys,
And all the scenes of life and the workshops, and the workmen
    homeward returning.

12

Lo, body and soul—this land,
My own Manhattan with spires, and the sparkling and hurrying
    tides, and the ships,   90
The varied and ample land, the South and the North in the light,
    Ohio's shores and flashing Missouri,
And ever the far-spreading prairies cover'd with grass and corn.

Lo, the most excellent sun so calm and haughty,
The violet and purple morn with just-felt breezes,
The gentle soft-born measureless light,   95
The miracle spreading bathing all, the fulfill'd noon,
The coming eve delicious, the welcome night and the stars,
Over my cities shining all, enveloping man and land.

13

Sing on, sing on you gray-brown bird,
Sing from the swamps, the recesses, pour your chant from the
    bushes,   100
Limitless out of the dusk, out of the cedars and pines.

Sing on dearest brother, warble your reedy song,
Loud human song, with voice of uttermost woe.

O liquid and free and tender!
O wild and loose to my soul!—O wondrous singer!   105

5. April.

You only I hear—yet the star holds me, (but will soon depart,)
Yet the lilac with mastering odor holds me.

14

Now while I sat in the day and look'd forth,
In the close of the day with its light and the fields of spring, and
    the farmers preparing their crops,
In the large unconscious scenery of my land with its lakes and for-
    ests,   110
In the heavenly aerial beauty, (after the perturb'd winds and the
    storms,)
Under the arching heavens of the afternoon swift passing, and the
    voices of children and women,
The many-moving sea-tides, and I saw the ships how they sail'd,
And the summer approaching with richness, and the fields all busy
    with labor,
And the infinite separate houses, how they all went on, each with
    its meals and minutia of daily usages,   115
And the streets how their throbbings throbb'd, and the cities pent
    —lo, then and there,
Falling upon them all and among them all, enveloping me with the
    rest,
Appear'd the cloud, appear'd the long black trail,
And I knew death, its thought, and the sacred knowledge of death.

Then with the knowledge of death as walking one side of me,   120
And the thought of death close-walking the other side of me,
And I in the middle as with companions, and as holding the hands
    of companions,
I fled forth to the hiding receiving night that talks not,
Down to the shores of the water, the path by the swamp in the
    dimness,
To the solemn shadowy cedars and ghostly pines so still.   125

And the singer so shy to the rest receiv'd me,
The gray-brown bird I know receiv'd us comrades three,
And he sang the carol of death, and a verse for him I love.

From deep secluded recesses,
From the fragrant cedars and the ghostly pines so still,   130
Came the carol of the bird.

And the charm of the carol rapt me,
As I held as if by their hands my comrades in the night,
And the voice of my spirit tallied the song of the bird.

Come lovely and soothing death,   135
Undulate round the world, serenely arriving, arriving,
In the day, in the night, to all, to each,
Sooner or later delicate death.

Prais'd be the fathomless universe,
For life and joy, and for objects and knowledge curious,   140

And for love, sweet love—but praise! praise! praise!
For the sure-enwinding arms of cool-enfolding death.

Dark mother always gliding near with soft feet,
Have none chanted for thee a chant of fullest welcome?
Then I chant it for thee, I glorify thee above all,   145
I bring thee a song that when thou must indeed come, come unfal-
      teringly.

Approach strong deliveress,
When it is so, when thou hast taken them I joyously sing the dead,
Lost in the loving floating ocean of thee,
Laved in the flood of thy bliss O death.   150

From me to thee glad serenades,
Dances for thee I propose saluting thee, adornments and feastings
      for thee,
And the sights of the open landscape and the high-spread sky are
      fitting,
And life and the fields, and the huge and thoughtful night.

The night in silence under many a star,   155
The ocean shore and the husky whispering wave whose voice I know,
And the soul turning to thee O vast and well-veil'd death,
And the body gratefully nestling close to thee.

Over the tree-tops I float thee a song,
Over the rising and sinking waves, over the myriad fields and the
      prairies wide,   160
Over the dense-pack'd cities all and the teeming wharves and ways,
I float this carol with joy, with joy to thee O death.

15

To the tally of my soul,
Loud and strong kept up the gray-brown bird,
With pure deliberate notes spreading filling the night.   165

Loud in the pines and cedars dim,
Clear in the freshness moist and the swamp-perfume,
And I with my comrades there in the night.

While my sight that was bound in my eyes unclosed,
As to long panoramas of visions.   170

And I saw askant[6] the armies,
I saw as in noiseless dreams hundreds of battle-flags,
Borne through the smoke of the battles and pierc'd with missiles
      I saw them,
And carried hither and yon through the smoke, and torn and bloody,
And at last but a few shreds left on the staffs, (and all in
      silence,)   175
And the staffs all splinter'd and broken.

---

6. Sideways, aslant, an appropriate word for introducing a surrealistic vision.

I saw battle-corpses, myriads of them,
And the white skeletons of young men, I saw them,
I saw the debris and debris of all the slain soldiers of the war,
But I saw they were not as was thought,   180
They themselves were fully at rest, they suffer'd not,
The living remain'd and suffer'd, the mother suffer'd,
And the wife and the child and the musing comrade suffer'd,
And the armies that remain'd suffer'd.

16

Passing the visions, passing the night,   185
Passing, unloosing the hold of my comrades' hands,
Passing the song of the hermit bird and the tallying song of my soul,
Victorious song, death's outlet song, yet varying ever-altering song,
As low and wailing, yet clear the notes, rising and falling, flooding
       the night,
Sadly sinking and fainting, as warning and warning, and yet again
       bursting with joy,   190
Covering the earth and filling the spread of the heaven,
As that powerful psalm in the night I heard from recesses,
Passing, I leave thee lilac with heart-shaped leaves,
I leave thee there in the door-yard, blooming, returning with spring.

I cease from my song for thee,   195
From my gaze on thee in the west, fronting the west, communing
       with thee,
O comrade lustrous with silver face in the night.

Yet each to keep and all, retrievements out of the night,
The song, the wondrous chant of the gray-brown bird,
And the tallying chant, the echo arous'd in my soul,   200
With the lustrous and drooping star with the countenance full of
       woe,
With the holders holding my hand nearing the call of the bird,
Comrades mine and I in the midst, and their memory ever to keep,
       for the dead I loved so well,
For the sweetest, wisest soul of all my days and lands—and this for
       his dear sake,
Lilac and star and bird twined with the chant of my soul,   205
There in the fragrant pines and the cedars dusk and dim.

                                                        1865–66, 1881

From AUTUMN RIVULETS
There Was a Child Went Forth[1]

There was a child went forth every day,
And the first object he look'd upon, that object he became,
And that object became part of him for the day or a certain part of
       the day,
Or for many years or stretching cycles of years.

1. This poem was first published in the 1856 edition of *Leaves of Grass* as *Poem of the Child That Went Forth, and Always Goes Forth, Forever and Forever,* then subsequently published under other titles until the present one was reached in the 1871 edition.

The early lilacs became part of this child,   5
And grass and white and red morning-glories, and white and red
    clover, and the song of the phœbe-bird,
And the Third-month lambs and the sow's pink-faint litter, and the
    mare's foal and the cow's calf,
And the noisy brood of the barnyard or by the mire of the pond-side,
And the fish suspending themselves so curiously below there, and
    the beautiful curious liquid,
And the water-plants with their graceful flat heads, all became part
    of him.   10

The field-sprouts of Fourth-month and Fifth-month became part of
    him,
Winter-grain sprouts and those of the light-yellow corn, and the
    esculent roots of the garden,
And the apple-trees cover'd with blossoms and the fruit afterward,
    and wood-berries, and the commonest weeds by the road,
And the old drunkard staggering home from the outhouse of the
    tavern whence he had lately risen,
And the schoolmistress that pass'd on her way to the school,   15
And the friendly boys that pass'd, and the quarrelsome boys,
And the tidy and fresh-cheek'd girls, and the barefoot negro boy and
    girl,
And all the changes of city and country wherever he went.

His own parents, he that had father'd him and she that had conceiv'd
    him in her womb and birth'd him,
They gave this child more of themselves than that,   20
They gave him afterward every day, they became part of him.

The mother at home quietly placing the dishes on the supper-table,
The mother with mild words, clean her cap and gown, a wholesome
    odor falling off her person and clothes as she walks by,
The father, strong, self-sufficient, manly, mean, anger'd, unjust,
The blow, the quick loud word, the tight bargain, the crafty lure,   25
The family usages, the language, the company, the furniture, the
    yearning and swelling heart,
Affection that will not be gainsay'd, the sense of what is real, the
    thought if after all it should prove unreal,
The doubts of day-time and the doubts of night-time, the curious
    whether and how,
Whether that which appears so is so, or is it all flashes and specks?
Men and women crowding fast in the streets, if they are not flashes
    and specks what are they?   30
The streets themselves and the façades of houses, and goods in the
    windows,
Vehicles, teams, the heavy-plank'd wharves, the huge crossing at
    the ferries,
The village on the highland seen from afar at sunset, the river
    between,
Shadows, aureola and mist, the light falling on roofs and gables of

white or brown two miles off,
The schooner near by sleepily dropping down the tide, the little
    boat slack-tow'd astern,   35
The hurrying tumbling waves, quick-broken crests, slapping,
The strata of color'd clouds, the long bar of maroon-tint away soli-
    tary by itself, the spread of purity it lies motionless in,
The horizon's edge, the flying sea-crow, the fragrance of salt marsh
    and shore mud,
These became part of that child who went forth every day, and who
    now goes, and will always go forth every day.

1855, 1871

My Picture-Gallery[1]

In a little house keep I pictures suspended, it is not a fix'd house,
It is round, it is only a few inches from one side to the other;
Yet behold, it has room for all the shows of the world, all memories!
Here the tableaus of life, and here the groupings of death;
Here, do you know this? this is cicerone[2] himself,   5
With finger rais'd he points to the prodigal pictures.

1880, 1881

The Sleepers[1]

1

I wander all night in my vision,
Stepping with light feet, swiftly and noiselessly stepping and
    stopping,
Bending with open eyes over the shut eyes of sleepers,
Wandering and confused, lost to myself, ill-assorted, contradictory,
Pausing, gazing, bending, and stopping.   5

How solemn they look there, stretch'd and still,
How quiet they breathe, the little children in their cradles.

The wretched features of ennuyés,[2] the white features of corpses,
    the livid faces of drunkards, the sick-gray faces of onanists,[3]
The gash'd bodies on battle-fields, the insane in their strong-door'd
    rooms, the sacred idiots, the new-born emerging from gates,
    and the dying emerging from gates,
The night pervades them and infolds them.   10

1. First published in a periodical (1880), then in the 1881 edition of *Leaves of Grass*, but drafted well before the publication of the 1855 edition.
2. Tour guide.
1. This was the fourth poem, untitled, in the 1855 *Leaves of Grass*; it was entitled *Night Poem* in the 1856 edition, *Sleep-Chasings* in 1860 and 1867, and finally *The Sleepers* in 1871 and thereafter. This printing follows the final edition, where it appears with five other poems, mostly long ones, between "Autumn Rivulets" and "Whispers of Heavenly Death." Minor revisions are not indicated here, but footnotes record lines present in 1855 and subsequently deleted. As Blodgett and Bradley say, at least some of the deletions were made "not so much for aesthetic as for discretionary reasons," the lines being too explicitly sexual.
2. The debauched.
3. Masturbators.

The married couple sleep calmly in their bed, he with his palm on
    the hip of the wife, and she with her palm on the hip of the
    husband,
The sisters sleep lovingly side by side in their bed,
The men sleep lovingly side by side in theirs,
And the mother sleeps with her little child carefully wrapt.

The blind sleep, and the deaf and dumb sleep,   15
The prisoner sleeps well in the prison, the runaway son sleeps,
The murderer that is to be hung next day, how does he sleep?
And the murder'd person, how does he sleep?

The female that loves unrequited sleeps,
And the male that loves unrequited sleeps,   20
The head of the money-maker that plotted all day sleeps,
And the enraged and treacherous dispositions, all, all sleep.

I stand in the dark with drooping eyes by the worst-suffering and the
    most restless,
I pass my hands soothingly to and fro a few inches from them,
The restless sink in their beds, they fitfully sleep.   25

Now I pierce the darkness, new beings appear,
The earth recedes from me into the night,
I saw that it was beautiful, and I see that what is not the earth is
    beautiful.

I go from bedside to bedside, I sleep close with the other sleepers
    each in turn,
I dream in my dream all the dreams of the other dreamers,   30
And I become the other dreamers.

I am a dance—play up there! the fit is whirling me fast!

I am the ever-laughing—it is new moon and twilight,
I see the hiding of douceurs,[4] I see nimble ghosts whichever way I
    look,
Cache[5] and cache again deep in the ground and sea, and where it
    is neither ground nor sea.   35

Well do they do their jobs those journeymen divine,
Only from me can they hide nothing, and would not if they could,
I reckon I am their boss and they make me a pet besides,
And surround me and lead me and run ahead when I walk,
To lift their cunning covers to signify me with stretch'd arms, and
    resume the way;   40
Onward we move, a gay gang of blackguards! with mirth-shouting
    music and wild-flapping pennants of joy!

4. Sweetnesses; here, sexual delights.   if it is used as a verb, "hiding place" if
5. A French borrowing meaning "hide"   used as a noun.

I am the actor, the actress, the voter, the politician,
The emigrant and the exile, the criminal that stood in the box,
He who has been famous and he who shall be famous after to-day,
The stammerer, the well-form'd person, the wasted or feeble per-
    son.   45

I am she who adorn'd herself and folded her hair expectantly,
My truant lover has come, and it is dark.

Double yourself and receive me darkness,
Receive me and my lover too, he will not let me go without him.

I roll myself upon you as upon a bed, I resign myself to the dusk.   50

He whom I call answers me and takes the place of my lover,
He rises with me silently from the bed.

Darkness, you are gentler than my lover, his flesh was sweaty and
    panting,
I feel the hot moisture yet that he left me.

My hands are spread forth, I pass them in all directions,   55
I would sound up the shadowy shore to which you are journeying.

Be careful darkness! already what was it touch'd me?
I thought my lover had gone, else darkness and he are one,
I hear the heart-beat, I follow, I fade away.[6]

2

I descend my western course, my sinews are flaccid,   60
Perfume and youth course through me and I am their wake.

It is my face yellow and wrinkled instead of the old woman's,
I sit low in a straw-bottom chair and carefully darn my grandson's
    stockings.

It is I too, the sleepless widow looking out on the winter midnight,
I see the sparkles of starshine on the icy and pallid earth.   65

6. In 1855 these three verse paragraphs followed:
O hotcheeked and blushing! O foolish hectic!
O for pity's sake, no one must see me now! . . . . my clothes were stolen while I was abed,
Now I am thrust forth, where shall I run?

Pier that I saw dimly last night when I looked from the windows,
Pier out from the main, let me catch myself with you and stay . . . . I will not chafe you;
I feel ashamed to go naked about the world,

And am curious to know where my feet stand . . . . and what is this flooding me, childhood or manhood . . . . and the hunger that crosses the bridge between.

The cloth laps a first sweet eating and drinking,
Laps life-swelling yolks . . . . laps ear of rose-corn, milky and just ripened:
The white teeth stay, and the boss-tooth advances in darkness,
And liquor is spilled on lips and bosoms by touching glasses, and the best liquor afterward.

A shroud I see and I am the shroud, I wrap a body and lie in the
   coffin,
It is dark here under ground, it is not evil or pain here, it is blank
   here, for reasons.
(It seems to me that every thing in the light and air ought to be
   happy,
Whoever is not in his coffin and the dark grave let him know he
   has enough.)

                                   3
I see a beautiful gigantic swimmer swimming naked through the
   eddies of the sea,   70
His brown hair lies close and even to his head, he strikes out with
   courageous arms, he urges himself with his legs,
I see his white body, I see his undaunted eyes,
I hate the swift-running eddies that would dash him head-foremost
   on the rocks.

What are you doing you ruffianly red-trickled waves?
Will you kill the courageous giant? will you kill him in the prime of
   his middle age?   75

Steady and long he struggles,
He is baffled, bang'd, bruis'd, he holds out while his strength holds
   out,
The slapping eddies are spotted with his blood, they bear him away,
   they roll him, swing him, turn him,
His beautiful body is borne in the circling eddies, it is continually
   bruis'd on rocks,
Swiftly and out of sight is borne the brave corpse.   80

                                   4
I turn but do not extricate myself,
Confused, a past-reading, another, but with darkness yet.

The beach is cut by the razory ice-wind, the wreck-guns sound,[7]
The tempest lulls, the moon comes floundering through the drifts.

I look where the ship helplessly heads end on, I hear the burst as she
   strikes, I hear the howls of dismay, they grow fainter and
   fainter.   85

I cannot aid with my wringing fingers,
I can but rush to the surf and let it drench me and freeze upon me.

I search with the crowd, not one of the company is wash'd to us
   alive,
In the morning I help pick up the dead and lay them in rows in a
   barn.

                                   5
Now of the older war-days, the defeat at Brooklyn,[8]   90
Washington stands inside the lines, he stands on the intrench'd hills

7. In a signal for help.
8. Washington was defeated at Brooklyn Heights in August, 1776.

amid a crowd of officers,
His face is cold and damp, he cannot repress the weeping drops,
He lifts the glass perpetually to his eyes, the color is blanch'd from
his cheeks,
He sees the slaughter of the southern braves[9] confided to him by
their parents.

The same at last and at last when peace is declared,   95
He stands in the room of the old tavern, the well-belov'd soldiers
all pass through,
The officers speechless and slow draw near in their turns,
The chief encircles their necks with his arm and kisses them on the
cheek,
He kisses lightly the wet cheeks one after another, he shakes hands
and bids good-by to the army.

6

Now what my mother told me one day as we sat at dinner together,   100
Of when she was a nearly grown girl living home with her parents
on the old homestead.

A red squaw came one breakfast-time to the old homestead,
On her back she carried a bundle of rushes for rush-bottoming chairs,
Her hair, straight, shiny, coarse, black, profuse, half-envelop'd her
face,
Her step was free and elastic, and her voice sounded exquisitely as
she spoke.   105

My mother look'd in delight and amazement at the stranger,
She look'd at the freshness of her tall-borne face and full and pliant
limbs,
The more she look'd upon her she loved her,
Never before had she seen such wonderful beauty and purity,
She made her sit on a bench by the jamb of the fireplace, she
cook'd food for her,   110
She had no work to give her, but she gave her remembrance and
fondness.

The red squaw staid all the forenoon, and toward the middle of the
afternoon she went away,
O my mother was loth to have her go away,
All the week she thought of her, she watch'd for her many a month,
She remember'd her many a winter and many a summer,   115
But the red squaw never came nor was heard of there again.[1]

9. Washington's soldiers, largely recruited
in his own Virginia and other southern
colonies.
1. In 1855 these three verse paragraphs
followed:
Now Lucifer was not dead . . . . or if he
     was I am his sorrowful terrible heir;
I have been wronged . . . . I am op-
     pressed . . . . I hate him that
     oppresses me,
I will either destroy him, or he shall
     release me.

Damn him! how he does defile me,
How he informs against my brother and
     sister and takes pay for their blood,
How he laughs when I look down the
     bend after the steamboat that carries
     away my woman.

Now the vast dusk bulk that is the
     whale's bulk . . . . it seems mine,
Warily, sportsman! though I lie so sleepy
     and sluggish, my tap is death.

7

A show of the summer softness—a contact of something unseen—
　　an amour of the light and air,
I am jealous and overwhelm'd with friendliness,
And will go gallivant with the light and air myself.[2]

O love and summer, you are in the dreams and in me,　　　　120
Autumn and winter are in the dreams, the farmer goes with his thrift,
The droves[3] and crops increase, the barns are well-fill'd.

Elements merge in the night, ships make tacks in the dreams,
The sailor sails, the exile returns home,
The fugitive returns unharm'd, the immigrant is back beyond
　　months and years,　　　　125
The poor Irishman lives in the simple house of his childhood with
　　the well-known neighbors and faces,
They warmly welcome him, he is barefoot again, he forgets he is
　　well off,
The Dutchman voyages home, and the Scotchman and Welshman
　　voyage home, and the native of the Mediterranean voyages
　　home,
To every port of England, France, Spain, enter well-fill'd ships,
The Swiss foots it toward his hills, the Prussian goes his way, the
　　Hungarian his way, and the Pole his way,　　　　130
The Swede returns, and the Dane and Norwegian return.

The homeward bound and the outward bound,
The beautiful lost swimmer, the ennuyé, the onanist, the female
　　that loves unrequited, the money-maker,
The actor and actress, those through with their parts and those wait-
　　ing to commence,
The affectionate boy, the husband and wife, the voter, the nominee
　　that is chosen and the nominee that has fail'd.　　　　135
The great already known and the great any time after to-day,
The stammerer, the sick, the perfect-form'd, the homely,
The criminal that stood in the box, the judge that sat and sentenced
　　him, the fluent lawyers, the jury, the audience,
The laugher and weeper, the dancer, the midnight widow, the red
　　squaw,
The consumptive, the erysipalite,[4] the idiot, he that is wrong'd,　140
The antipodes,[5] and every one between this and them in the dark,
I swear they are averaged now—one is no better than the other,
The night and sleep have liken'd them and restored them.

I swear they are all beautiful,
Every one that sleeps is beautiful, every thing in the dim light is
　　beautiful,　　　　145

2. In 1855 "myself" was followed by a
comma and this line concluded the verse
paragraph: "And have an unseen something to be in contact with them also."
3. Herds of cattle, from their being
"driven" to market in a group.
4. Someone with "red skin," a severe inflammation caused by a streptococcus.
5. Any two places on opposite sides of
the earth.

The wildest and bloodiest is over, and all is peace.

Peace is always beautiful,
The myth of heaven indicates peace and night.

The myth of heaven indicates the soul,
The soul is always beautiful, it appears more or it appears less, it
    comes or it lags behind,   150
It comes from its embower'd garden and looks pleasantly on itself
    and encloses the world,
Perfect and clean the genitals previously jetting, and perfect and
    clean the womb cohering,
The head well-grown proportion'd and plumb, and the bowels and
    joints proportion'd and plumb.

The soul is always beautiful,
The universe is duly in order, every thing is in its place,   155
What has arrived is in its place and what waits shall be in its place,
The twisted skull waits, the watery or rotten blood waits,
The child of the glutton or venerealee waits long, and the child of
    the drunkard waits long, and the drunkard himself waits long,
The sleepers that lived and died wait, the far advanced are to go on
    in their turns, and the far behind are to come on in their turns,
The diverse shall be no less diverse, but they shall flow and unite—
    they unite now.   160

8

The sleepers are very beautiful as they lie unclothed,
They flow hand in hand over the whole earth from east to west as
    they lie unclothed,
The Asiatic and African are hand in hand, the European and Ameri-
    can are hand in hand,
Learn'd and unlearn'd are hand in hand, and male and female are
    hand in hand,
The bare arm of the girl crosses the bare breast of her lover, they
    press close without lust, his lips press her neck,   165
The father holds his grown or ungrown son in his arms with mea-
    sureless love, and the son holds the father in his arms with mea-
    sureless love,
The white hair of the mother shines on the white wrist of the
    daughter,
The breath of the boy goes with the breath of the man, friend is
    inarm'd by friend,
The scholar kisses the teacher and the teacher kisses the scholar,
    the wrong'd is made right,
The call of the slave is one with the master's call, and the master
    salutes the slave,   170
The felon steps forth from the prison, the insane becomes sane, the
    suffering of sick persons is reliev'd,
The sweatings and fevers stop, the throat that was unsound is sound,
    the lungs of the consumptive are resumed, the poor distress'd
    head is free,

The joints of the rheumatic move as smoothly as ever, and smoother
    than ever,
Stiflings and passages open, the paralyzed become supple,
The swell'd and convuls'd and congested awake to themselves in
    condition,   175
They pass the invigoration of the night and the chemistry of the
    night, and awake.

I too pass from the night,
I stay a while away O night, but I return to you again and love you.

Why should I be afraid to trust myself to you?
I am not afraid, I have been well brought forward by you,   180
I love the rich running day, but I do not desert her in whom I lay
    so long,
I know not how I came of you and I know not where I go with you,
    but I know I came well and shall go well.

I will stop only a time with the night, and rise betimes,
I will duly pass the day O my mother, and duly return to you.[6]
            1855, 1881

From WHISPERS OF HEAVENLY DEATH
Quicksand Years[7]

Quicksand years that whirl me I know not wither,
Your schemes, politics, fail, lines give way, substances mock and
    elude me,
Only the theme I sing, the great and strong-possess'd soul, eludes
    not,
One's-self must never give way—that is the final substance—that
    out of all is sure,
Out of politics, triumphs, battles, life, what at last finally re-
    mains?   5
When shows break up what but One's-Self is sure?
            1865, 1871

A Noiseless Patient Spider

A noiseless patient spider,
I mark'd where on a little promontory it stood isolated,
Mark'd how to explore the vacant vast surrounding,
It launch'd forth filament, filament, filament, out of itself,
Ever unreeling them, ever tirelessly speeding them.   5

6. In 1855 this line began a verse para-graph (the previous line standing alone) and ended with a semicolon, being fol-lowed by these two lines: "Not you will yield forth the dawn again more surely than you will yield forth me again,/Not the worm yields the babe in its time more surely than I shall be yielded from you in my time."
7. This poem first appeared in *Drum-Taps* (1865), then in the "Whispers of Heavenly Death" section of *Passage to India* (1871), and reached its final posi-tion in the "Whispers of Heavenly Death" section of the 1881 *Leaves of Grass*.

And you O my soul where you stand,
Surrounded, detached, in measureless oceans of space,
Ceaselessly musing, venturing, throwing, seeking the spheres to
    connect them,
Till the bridge you will need be form'd, till the ductile anchor hold,
Till the gossamer thread you fling catch somewhere, O my soul.   10

1868, 1881

From FROM NOON TO STARRY NIGHT
Faces[1]

1

Sauntering the pavement or riding the country by-road, lo, such
    faces!
Faces of friendship, precision, caution, suavity, ideality,
The spiritual-prescient face, the always welcome common benevolent
    face,
The face of the singing of music, the grand faces of natural lawyers
    and judges broad at the back-top,[2]
The faces of hunters and fishers bulged at the brows, the shaved
    blanch'd faces of orthodox citizens,   5
The pure, extravagant, yearning, questioning artist's face,
The ugly face of some beautiful soul, the handsome detested or
    despised face,
The sacred faces of infants, the illuminated face of the mother of
    many children,
The face of an amour, the face of veneration,
The face as of a dream, the face of an immobile rock,   10
The face withdrawn of its good and bad, a castrated face,
A wild hawk, his wings clipp'd by the clipper,
A stallion that yielded at last to the thongs and knife of the gelder.

Sauntering the pavement thus, or crossing the ceaseless ferry, faces
    and faces and faces,
I see them and complain not, and am content with all.   15

2

Do you suppose I could be content with all if I thought them their
    own finalè?

This now is too lamentable a face for a man,
Some abject louse asking leave to be, cringing for it,
Some milk-nosed maggot blessing what lets it wrig[3] to its hole.

This face is a dog's snout sniffing for garbage,   20
Snakes nest in that mouth, I hear the sibilant threat.

This face is a haze more chill than the arctic sea,
Its sleepy and wabbling icebergs crunch as they go.

1. This poem was sixth of the 12 poems
in the 1855 *Leaves of Grass.* In the 1881
edition it became the second poem in
the section "From Noon to Starry

Night."
2. Phrenological, meaning the top of the
back of the head.
3. Wriggle.

This is a face of bitter herbs, this an emetic, they need no label,
And more of the drug-shelf, laudanum, caoutchouc, or hog's-lard.[4] 25

This face is an epilepsy, its wordless tongue gives out the unearthly
  cry,
Its veins down the neck distend, its eyes roll till they show nothing
  but their whites,
Its teeth grit, the palms of the hands are cut by the turn'd-in nails,
The man falls struggling and foaming to the ground, while he speculates well.

This face is bitten by vermin and worms,   30
And this is some murderer's knife with a half-pull'd scabbard.
This face owes to the sexton[5] his dismalest fee,
An unceasing death-bell tolls there.

3
Features of my equals would you trick me with your creas'd and
  cadaverous march?
Well, you cannot trick me.   35

I see your rounded never-erased flow,
I see 'neath the rims of your haggard and mean disguises.

Splay[6] and twist as you like, poke with the tangling fores[7] of fishes
  or rats,
You'll be unmuzzled, you certainly will.

I saw the face of the most smear'd and slobbering idiot they had
  at the asylum,   40
And I knew for my consolation what they knew not,
I knew of the agents that emptied and broke my brother,
The same wait to clear the rubbish from the fallen tenement,
And I shall look again in a score or two of ages,
And I shall meet the real landlord perfect and unharm'd, every inch
  as good as myself.   45

4
The Lord advances, and yet advances,
Always the shadow in front, always the reach'd hand bringing up the
  laggards.

Out of this face emerge banners and horses—O superb! I see what
  is coming,
I see the high pioneer-caps, see staves of runners clearing the way,
I hear victorious drums,   50

This face is a life-boat,
This is the face commanding and bearded, it asks no odds of the
  rest,

4. "Laudanum": a tincture of opium; death bell.
"caoutchouc": natural rubber. 6. Here, to force open.
5. Cemetery caretaker, toller of the 7. Snouts.

This face is flavor'd fruit ready for eating,
This face of a healthy honest boy is the programme of all good.

These faces bear testimony slumbering or awake, 55
They show their descent from the Master himself.

Off the word I have spoken I except not one—red, white, black, are
all deific,
In each house is the ovum, it comes forth after a thousand years.

Spots or cracks at the windows do not disturb me,
Tall and sufficient stand behind and make signs to me, 60
I read the promise and patiently wait.
This is a full-grown lily's face,
She speaks to the limber-hipp'd man near the garden pickets,
Come here she blushingly cries, *Come nigh to me limber-hipp'd
man,*
Stand at my side till I lean as high as I can upon you, 65
Fill me with albescent[8] honey, bend down to me,
Rub to me with your chafing beard, rub to my breast and shoulders.

5

The old face of the mother of many children,
Whist! I am fully content.

Lull'd and late is the smoke of the First-day[9] morning, 70
It hangs low over the rows of trees by the fences,
It hangs thin by the sassafras and wild-cherry and cat-brier under
them.

I saw the rich ladies in full dress at the soiree,
I heard what the singers were singing so long,
Heard who sprang in crimson youth from the white froth and the
water-blue. 75

Behold a woman!
She looks out from her quaker cap, her face is clearer and more
beautiful than the sky.

She sits in an armchair under the shaded porch of the farmhouse,
The sun just shines on her old white head.

Her ample gown is of cream-hued linen, 80
Her grandsons raised the flax, and her grand-daughters spun it with
the distaff and the wheel.

The melodious character of the earth,
The finish beyond which philosophy cannot go and does not wish
to go,
The justified mother of men.

1855, 1881

8. Whitish.   9. Sunday.

To a Locomotive in Winter[1]

Thee for my recitative,
Thee in the driving storm even as now, the snow, the winter-day
    declining,
Thee in thy panoply,[2] thy measur'd dual throbbing and thy beat
    convulsive,
Thy black cylindric body, golden brass and silvery steel,
Thy ponderous side-bars, parallel and connecting rods, gyrating,
    shuttling at thy sides,   5
Thy metrical, now swelling pant and roar, now tapering in the
    distance,
Thy great protruding head-light fix'd in front,
Thy long, pale, floating vapor-pennants, tinged with delicate purple,
The dense and murky clouds out-belching from thy smoke-stack,
Thy knitted frame, thy springs and valves, the tremulous twinkle of
    thy wheels,   10
Thy train of cars behind, obedient, merrily following,
Through gale or calm, now swift, now slack, yet steadily careering;
Type of the modern—emblem of motion and power—pulse of the
    continent,
For once come serve the Muse and merge in verse, even as here I
    see thee,
With storm and buffeting gusts of wind and falling snow,   15
By day thy warning ringing bell to sound its notes,
By night thy silent signal lamps to swing.

Fierce-throated beauty!
Roll through my chant with all thy lawless music, thy swinging
    lamps at night,
Thy madly-whistled laughter, echoing, rumbling like an earthquake,
    rousing all,   20
Law of thyself complete, thine own track firmly holding,
(No sweetness debonair of tearful harp or glib piano thine,)
Thy trills of shrieks by rocks and hills return'd,
Launch'd o'er the prairies wide, across the lakes,
To the free skies unpent and glad and strong.   25

1876, 1881

From SECOND ANNEX: GOOD-BYE MY FANCY[1]

Preface Note to 2d Annex, Concluding L. of G.—1891

Had I not better withhold (in this old age and paralysis of me)
such little tags and fringe-dots (maybe specks, stains,) as follow a
long dusty journey, and witness it afterward? I have probably not

---

1. First printed in the New York *Tribune* on February 19, 1876, as a sample from the forthcoming *Two Rivulets*, this poem was moved to the "From Noon to Starry Night" group in the 1881 *Leaves of Grass*.
2. Suit of armor.

1. This, the second and final "annex" to *Leaves of Grass*, consisting of prose headnote and 31 poems, mostly new, was printed separately in 1891 as *Good-bye My Fancy*, then added to *Leaves of Grass* in the 1891–92 edition. "Fancy": the poetic imagination.

been enough afraid of careless touches, from the first—and am not now—nor of parrot-like repetitions—nor platitudes and the commonplace. Perhaps I am too democratic for such avoidances. Besides, is not the verse-field, as originally plann'd by my theory, now sufficiently illustrated—and full time for me to silently retire?—(indeed amid no loud call or market for my sort of poetic utterance.)

In answer, or rather defiance, to that kind of well-put interrogation, here comes this little cluster, and conclusion of my preceding clusters. Though not at all clear that, as here collated, it is worth printing (certainly I have nothing fresh to write)—I while away the hours of my 72d year—hours of forced confinement in my den—by putting in shape this small old age collation:

> Last droplets of and after spontaneous rain,
> From many limpid distillations and past showers;
> (Will they germinate anything? mere exhalations as they all are
> —the land's and sea's—America's;
> Will they filter to any deep emotion? any heart and brain?)

However that may be, I feel like improving to-day's opportunity and wind up. During the last two years I have sent out, in the lulls of illness and exhaustion, certain chirps—lingering-dying ones probably (undoubtedly)—which now I may as well gather and put in fair type while able to see correctly—(for my eyes plainly warn me they are dimming, and my brain more and more palpably neglects or refuses, month after month, even slight tasks or revisions.)

In fact, here I am these current years 1890 and '91, (each successive fortnight getting stiffer and stuck deeper) much like some hard-cased dilapidated grim ancient shell-fish or time-bang'd conch (no legs, utterly non-locomotive) cast up high and dry on the shore-sands, helpless to move anywhere—nothing left but behave myself quiet, and while away the days yet assign'd, and discover if there is anything for the said grim and time-bang'd conch to be got at last out of inherited good spirits and primal buoyant centre-pulses down there deep somewhere within his gray-blurr'd old shell. . . . .
(Reader, you must allow a little fun here—for one reason there are too many of the following poemets about death, &c., and for another the passing hours (July 5, 1890) are so sunny-fine. And old as I am I feel to-day almost a part of some frolicsome wave, or for sporting yet like a kid or kitten—probably a streak of physical adjustment and perfection here and now. I believe I have it in me perennially anyhow.)

Then behind all, the deep-down consolation (it is a glum one, but I dare not be sorry for the fact of it in the past, nor refrain from dwelling, even vaunting here at the end) that this late-years palsied old shorn and shell-fish condition of me is the indubitable outcome

and growth, now near for 20 years along, of too over-zealous, over-
continued bodily and emotional excitement and action through the
times of 1862, '3, '4 and '5, visiting and waiting on wounded and
sick army volunteers, both sides, in campaigns or contests, or after
them, or in hospitals or fields south of Washington City, or in that
place and elsewhere—those hot, sad, wrenching times—the army
volunteers, all States,—or North or South—the wounded, suffering,
dying—the exhausting, sweating summers, marches, battles, carnage
—those trenches hurriedly heap'd by the corpse-thousands, mainly
unknown—Will the America of the future—will this vast rich
Union ever realize what itself cost, back there after all?—those
hecatombs[2] of battle-deaths—Those times of which, O far-off reader,
this whole book is indeed finally but a reminiscent memorial from
thence by me to you?

Osceola[1]

[*When I was nearly grown to manhood in Brooklyn, New York, (middle of 1838,) I met one of the return'd U. S. Marines from Fort Moultrie, S. C., and had long talks with him—learn'd the occurrence below
described—death of Osceola. The latter was a young, brave, leading
Seminole in the Florida war of that time—was surrender'd to our troops,
imprison'd and literally died of "a broken heart," at Fort Moultrie. He
sicken'd of his confinement—the doctor and officers made every allowance and kindness possible for him; then the close:*]

When his hour for death had come,
He slowly rais'd himself from the bed on the floor,
Drew on his war-dress, shirt, leggings, and girdled the belt around
    his waist,
Call'd for vermilion paint (his looking-glass was held before him,)
Painted half his face and neck, his wrists, and back-hands.   5
Put the scalp-knife carefully in his belt—then lying down, resting a
    moment,
Rose again, half sitting, smiled, gave in silence his extended hand to
    each and all,
Sank faintly low to the floor (tightly grasping the tomahawk handle,)
Fix'd his look on wife and little children—the last:

(And here a line in memory of his name and death.)   10
                                        1890, 1891–92

"The Rounded Catalogue Divine Complete"[2]

[*Sunday, —— — ——.—Went this forenoon to church. A college professor, Rev. Dr. ——, gave us a fine sermon, during which I caught
the above words; but the minister included in his "rounded catalogue"
letter and spirit, only the esthetic things, and entirely ignored what I
name in the following:*]

The devilish and the dark, the dying and diseas'd,
The countless (nineteen-twentieths) low and evil, crude and savage,
The crazed, prisoners in jail, the horrible, rank, malignant,

2. Many hundreds, from the Greek for a
hundred oxen (in the ritual sacrifice of a
hundred oxen at the same time).

1. The bracketed headnote is Whitman's.
2. The bracketed headnote is Whitman's.

Venom and filth, serpents, the ravenous sharks, liars, the dissolute;
(What is the part the wicked and the loathesome bear within earth's
    orbic scheme?)   5
Newts, crawling things in slime and mud, poisons,
The barren soil, the evil men, the slag and hideous rot.

1891, 1891–92

Good-bye My Fancy!

Good-bye my Fancy!
Farewell dear mate, dear love!
I'm going away, I know not where,
Or to what fortune, or whether I may ever see you again,
So Good-bye my Fancy.   5

Now for my last—let me look back a moment;
The slower fainter ticking of the clock is in me,
Exit, nightfall, and soon the heart-thud stopping.

Long have we lived, joy'd, caress'd together;
Delightful!—now separation—Good-bye my Fancy.   10

Yet let me not be too hasty,
Long indeed have we lived, slept, filter'd, become really blended
    into one;
Then if we die we die together, (yes, we'll remain one,)
If we go anywhere we'll go together to meet what happens,
May-be we'll be better off and blither, and learn something,   15
May-be it is yourself now really ushering me to the true songs, (who
    knows?)
May-be it is you the mortal knob really undoing, turning—so now
    finally,
Good-bye—and hail! my Fancy.

1891, 1891–92

From Democratic Vistas [1]
[*American Literature*]

* * *

    What, however, do we more definitely mean by New World liter-
ature? Are we not doing well enough here already? Are not the
United States this day busily using, working, more printer's type,
more presses, than any other country? uttering and absorbing more
publications than any other? Do not our publishers fatten quicker
and deeper? (helping themselves, under shelter of a delusive and
sneaking law, or rather absence of law, to most of their forage,
poetical, pictorial, historical, romantic, even comic, without money
and without price—and fiercely resisting the timidest proposal to

1. The little book *Democratic Vistas*, published late in 1870 but dated 1871, was made up of three essays written for the *Galaxy*: "Democracy" appeared in December, 1867, and "Personalism" in May, 1868, but the magazine rejected the third, "Literature," which was first published in the book form, the source of the present text. *Democratic Vistas* is one of the most neglected of major American literary, political, and philosophical documents.

pay for it.)[2]

Many will come under this delusion—but my purpose is to dispel it. I say that a nation may hold and circulate rivers and oceans of very readable print, journals, magazines, novels, library-books, "poetry," &c.—such as the States to-day possess and circulate—of unquestionable aid and value—hundreds of new volumes annually composed and brought out here, respectable enough, indeed unsurpass'd in smartness and erudition—with further hundreds, or rather millions, (as by free forage or theft aforemention'd,) also thrown into the market,—And yet, all the while, the said nation, land, strictly speaking, may possess no literature at all.

Repeating our inquiry, what, then, do we mean by real literature? especially the American literature of the future? Hard questions to meet. The clues are inferential, and turn us to the past. At best, we can only offer suggestions, comparisons, circuits.

It must still be reiterated, as, for the purpose of these memoranda, the deep lesson of history and time, that all else in the contributions of a nation or age, through its politics, materials, heroic personalities, military eclat, &c., remains crude, and defers, in any close and thorough-going estimate, until vitalized by national, original archetypes in literature. They only put the nation in form, finally tell anything—prove, complete anything—perpetuate anything. Without doubt, some of the richest and most powerful and populous communities of the antique world, and some of the grandest personalities and events, have, to after and present times, left themselves entirely unbequeath'd. Doubtless, greater than any that have come down to us, were among those lands, heroisms, persons, that have not come down to us at all, even by name, date, or location. Others have arrived safely, as from voyages over wide, century-stretching seas. The little ships, the miracles that have buoy'd them, and by incredible chances safely convey'd them, (or the best of them, their meaning and essence,) over long wastes, darkness, lethargy, ignorance, &c., have been a few inscriptions—a few immortal compositions, small in size, yet compassing what measureless values of reminiscence, contemporary portraitures, manners, idioms and beliefs, with deepest inference, hint and thought, to tie and touch forever the old, new body, and the old, new soul! These! and still these! bearing the freight so dear—dearer than pride—dearer than love. All the best experience of humanity, folded, saved, freighted to us here. Some of these tiny ships we call Old and New Testament, Homer, Eschylus, Plato, Juvenal, &c. Precious minims![3] I think, if we were forced to choose, rather than have you, and the likes of you, and what belongs to, and has grown

---

2. From the founding of the United States until 1891, when Congress finally passed an international copyright law, American writers had been victimized by the fact that American publishers could reprint foreign books without payment to the authors: it cost a publisher only printing expenses to publish Dickens in this country, but Cooper or Melville or Clemens had to be paid royalties.

3. In this sense, small containers, treasures being gathered into the small compass of a book.

of you, blotted out and gone, we could better afford, appalling as that would be, to lose all actual ships, this day fasten'd by wharf, or floating on wave, and see them, with all their cargoes, scuttled and sent to the bottom.

Gather'd by geniuses of city, race or age, and put by them in highest of art's forms, namely, the literary form, the peculiar combinations and the outshows of that city, age, or race, its particular modes of the universal attributes and passions, its faiths, heroes, lovers and gods, wars, traditions, struggles, crimes, emotions, joys, (or the subtle spirit of these,) having been pass'd on to us to illumine our own selfhood, and its experiences—what they supply, indispensable and highest, if taken away, nothing else in all the world's boundless storehouses could make up to us, or ever again return.

For us, along the great highways of time, those monuments stand —those forms of majesty and beauty. For us those beacons burn through all the nights. Unknown Egyptians, graving hieroglyphs; Hindus, with hymn and apothegm and endless epic; Hebrew prophet, with spirituality, as in flashes of lightning, conscience like red-hot iron, plaintive songs and screams of vengeance for tyrannies and enslavement; Christ, with bent head, brooding love and peace, like a dove; Greek, creating eternal shapes of physical and esthetic proportion; Roman, lord of satire, the sword, and the codex; —of the figures, some far off and veil'd, others nearer and visible; Dante, stalking with lean form, nothing but fibre, not a grain of superfluous flesh; Angelo,[4] and the great painters, architects, musicians; rich Shakespeare, luxuriant as the sun, artist and singer of feudalism in its sunset, with all the gorgeous colors, owner thereof, and using them at will; and so to such as German Kant and Hegel,[5] where they, though near us, leaping over the ages, sit again, impassive, imperturbable, like the Egyptian gods. Of these, and the like of these, is it too much, indeed, to return to our favorite figure, and view them as orbs and systems of orbs, moving in free paths in the spaces of that other heaven, the kosmic intellect, the soul?

Ye powerful and resplendent ones! ye were, in your atmospheres, grown not for America, but rather for her foes, the feudal and the old—while our genius is democratic and modern. Yet could ye, indeed, but breathe your breath of life into our New World's nostrils—not to enslave us, as now, but, for our needs, to breed a spirit like your own—perhaps, (dare we to say it?) to dominate, even destroy, what you yourselves have left! On your plane, and no less, but even higher and wider, will I mete and measure for our wants to-day and here. I demand races of orbic bards, with uncon-

---

4. Dante Alighieri (1265–1321), Italian poet, author of the *Divine Comedy*; Angelo: then an acceptable form of the name of Michelangelo Buonarroti (1475–1564), Italian artist.

5. Immanuel Kant (1724–1804) and Georg Wilhelm Friedrich Hegel (1770–1831), philosophers whose writings were familiar to Whitman.

ditional uncompromising sway. Come forth, sweet democratic despots of the west!

By points and specimens like these we, in reflection, token what we mean by any land's or people's genuine literature. And thus compared and tested, judging amid the influence of loftiest products only, what do our current copious fields of print, covering in manifold forms, the United States, better, for an analogy, present, then, as in certain regions of the sea, those spreading, undulating masses of squid, through which the whale swimming, with head half out, feeds?

Not but that doubtless our current so-called literature, (like an endless supply of small coin,) performs a certain service, and maybe, too, the service needed for the time, (the preparation-service, as children learn to spell.) Everybody reads, and truly nearly everybody writes, either books, or for the magazines or journals. The matter has magnitude, too, after a sort. There is something impressive about the huge editions of the dailies and weeklies, the mountain-stacks of white paper piled in the press-vaults, and the proud, crashing, ten-cylinder presses, which I can stand and watch any time by the half hour. Then, (though the States in the field of imagination present not a single first-class work, not a single great literatus,) the main objects, to amuse, to titillate, to pass away time, to circulate the news, and rumors of news, to rhyme and read rhyme, are yet attain'd, and on a scale of infinity. To-day, in books, in the rivalry of writers, especially novelists, success, (so-call'd,) is for him or her who strikes the mean flat average, the sensational appetite for stimulus, incident,[6] &c., and depicts, to the common calibre, sensual, exterior life. To such, or the luckiest of them, as we see, the audiences are limitless and profitable; but they cease presently. While this day, or any day, to workmen portraying interior or spiritual life, the audiences were limited, and often laggard—but they last forever.

Compared with the past, our modern science soars, and our journals serve; but ideal and even ordinary romantic literature, does not, I think, substantially advance. Behold the prolific brood of the contemporary novel, magazine-tale, theatre-play, &c. The same endless thread of tangled and superlative love-story, inherited, apparently from the Amadises and Palmerins[7] of the 13th, 14th, and 15th centuries over there in Europe. The costumes and associations are brought down to date, the seasoning is hotter and more varied, the dragons and ogres are left out—but the *thing*, I should say, has not advanced—is just as sensational, just as strain'd—remains about the same, nor more, nor less.

---

6. In the reprinting of *Democratic Vistas* as part of *Specimen Days & Collect* (1882), Whitman expanded the series to "the sensational appetite for stimulus, incident, persiflage, &c."

7. Amadis de Gaul (with Gaul first meaning Wales, then being understood as meaning France) was the hero of various chivalric romances, as was Palmerin, the hero of *Palmerin of England*.

What is the reason our time, our lands, that we see no fresh local courage, sanity, of our own—the Mississippi, stalwart Western men, real mental and physical facts, Southerners, &c., in the body of our literature? especially the poetic part of it. But always, instead, a parcel of dandies and ennuyees, dapper little gentlemen from abroad, who flood us with their thin sentiment of parlors, parasols, piano-songs, tinkling rhymes, the five-hundredth importation—or whimpering and crying about something, chasing one aborted conceit after another, and forever occupied in dyspeptic amours with dyspeptic women.

While, current and novel, the grandest events and revolutions, and stormiest passions of history, are crossing to-day with unparallel'd rapidity and magnificence over the stages of our own and all the continents, offering new materials, opening new vistas, with largest needs, inviting the daring launching forth of conceptions in literature, inspired by them, soaring in highest regions, serving art in its highest, (which is only the other name for serving God, and serving humanity,) where is the man of letters, where is the book, with any nobler aim than to follow in the old track, repeat what has been said before—and, as its utmost triumph, sell well, and be erudite or elegant?

* * *

1870

HERMAN MELVILLE
1819–1891

Melville's life, works, and reputation are the stuff of legend. With almost no formal education, he turned his early South Sea adventuring to literary use, charming readers in Britain and the United States with his first book, *Typee*, the story of his captivity by a Polynesian tribe. Once established as a popular young author, he simultaneously began exploring philosophy and experimenting with literary style and form. His readers were outraged, and for the rest of Melville's brief career he was torn between his own urge toward aesthetic and philosophical adventuring and the public's demand for racy sea stories which did not disturb its opinions on politics, religion, and metaphysics. By his mid-thirties, broken in reputation and health, he ceased writing fiction, gradually passing into a stern and neglected middle age as a deputy customs inspector in Manhattan. During the forty years he lived after publishing *Moby-Dick*, Melville withdrew into the privacy of his family while men like G. W. Curtis, R. H. Stoddard, E. C. Stedman, T. B. Aldrich, E. P. Whipple, and R. W. Gilder reigned over a magazine-dominated literary domain whose intellectual and artistic values formed a counterpart to the prevailing shoddiness of political values in post-Civil War America. Rediscovered by a few English readers just before his death, Melville was all but forgotten for another thirty years. Finally the centennial of his birth brought about a revival of interest; by the 1920s literary and cultural historians began to see Melville as the archetypal artist in a money-grubbing century hostile to all grandeur of intellect and spirit. That was a new distortion, but the "Melville Revival" of the 1920s succeeded in establishing him as one of the greatest American writers, although it took another decade or two for him to gain much space in college textbooks. The facts of his life are as poignant—and as archetypal—as the legends.

Melville began life with everything in his favor: heredity first of all, with two genuine Revolutionary heroes for grandfathers. The Melvill family (the "e" was added in the 1830s) was solidly established in Boston and the Gansevoorts were linked to the greatest Dutch patroon families of New York. Melville's much-traveled father, Allan Melvill, a dry-goods merchant in New York City, took inordinate pride in the genealogy of the Melvills, tracing the line past Scottish Renaissance courtiers to a Queen of Hungary and tracing his mother's family, the Scollays, to the Kings of Norway: "& so it appears we are of a royal line in both sides of the House—after all, it is not only an amusing but a just cause of pride, to resort back through the ages to such ancestry, & should produce a correspondent spirit of emulation in their descendants to the remotest posterity." As the third oldest of eight children born between 1815 and 1830, Herman Melville spent his early childhood in luxury. But Allan Melvill began borrowing from relatives in the 1820s, alternating between overenthusiasm about the future of business in America and dread of an inevitable recession. In 1832 he suddenly fell ill and died in a delirium which some in the family thought of as madness. He was many thousands of dollars in debt, and his family, then living in Albany, became dependent on the conscientious but finely calculated care of the Gansevoorts, especially Melville's Uncle Peter.

Biographers justifiably hold that Melville's mature psychology is best understood as that of the decayed patrician. During his teens, he was distinctly a poor relation. The Princeton-educated Peter Gansevoort hobnobbed with the leading politicians of the day, entertaining President Van Buren at dinner during the years in which his widowed sister, Maria Melville, saw her brilliant oldest son Gansevoort and her more plodding second son Herman make do with what self-improvement they could derive from the Albany debating societies. Dropping out of school when he was twelve, a few months after his father's death, Melville clerked for two years at a bank. He spent several months during 1834 at nearby Pittsfield, Massachusetts, on the farm of his Uncle Thomas Melvill, who had lived for decades in Paris and now cultivated the pose of nobility in exile. Beginning early in 1835, Melville worked two and a half years at his brother Gansevoort's fur-cap store in Albany. Just after he turned eighteen he taught in a country school near Pittsfield, where he boarded with Yankee backwoods families. The next spring he took a course in surveying and engineering at the Lansingburgh Academy, near Albany, but in the aftermath of the Panic of 1837 found no work. He signed on a voyage to and from Liverpool in 1839, the summer he turned twenty, then the next year job-hunted fruitlessly around the Midwest. At twenty-one, in January, 1841, he took the desperate measure of sailing on a whaler for the South Seas. His education, his crucial experience, had begun.

From Peru he wrote, in Gansevoort's paraphrase, that he was "not dissatisfied with his lot"—"The fact of his being one of a crew so much superior in morale and early advantages to the ordinary run of whaling crews affords him constant gratification." Nevertheless, in the summer of 1842 Melville and a shipmate, Toby Greene, jumped ship at Nukahiva, in the Marquesas, and for a few weeks Melville lived with a tribe quite untainted by Western civilization. Picked up by an Australian whaler a month after he deserted, he took part in a comic opera revolt and was imprisoned by the British Consul at Tahiti, along with a learned friend (the "Dr. Long Ghost" of *Omoo*) who became his companion in exploring the flora and, especially, the fauna of Tahiti and Eimeo. Shipping on a Nantucket whaler at Eimeo, Melville was discharged at Honolulu, where he knocked about a few months before signing on the frigate *United States* as an ordinary seaman. After a leisurely cruise in the Pacific, including a revisit to the Marquesas, the *United States* sailed for home, arriving at Boston in October, 1844. On this ship Melville again encountered some remarkably literate, and even literary, sailors. No newspapers welcomed the young sailor home, but that month Democratic papers in New York were hailing the triumphant return of his brother Gansevoort from a splendidly histrionic stump-speaking tour in the West on behalf of Polk's campaign for the presidency. Herman Melville was twenty-five; he later said that from that year, beginning August 1, 1844, he dated his life. He apparently did not look for a job after his return to the States; within weeks or months of his discharge he began writing.

Circumstances were propitious. In the spring of 1845 Gansevoort was rewarded for his services to the Democrats with the secretaryship to the American Legation in London. When he sailed in the summer, he had with him Herman's chaotic manuscript. It purported to be a straight autobiographical account of his detainment "in an indulgent captivity for about the space of four months," but in fact Melville had quadrupled the time he had spent in the valley of the Typees in order to account for the detailed anthropological observations, many of which came from earlier books by sea captains and missionaries. Gansevoort interested John Murray (the son of Lord Byron's friend and publisher) in the book for his Home and Colonial Library, and after it was eked out by additions from Melville and tidied up by a professional "reader," *Typee* was published early in 1846. As the earliest personal account of the South Seas to have the readability and suspense of adventure fiction, it made a great sensation, capturing the imagination of both the literary reviewers and the reading public with the surefire combination of anthropological novelty and what reviewers regularly tagged (remembering *Othello*) as "hair-breadth 'scapes." It was attended by vigorous, sales-stimulating controversy over its authenticity, capped by the emergence of Toby, the long-lost fellow runaway, in the person of Richard Tobias Greene, a housepainter near Buffalo. G. P. Putnam of Wiley & Putnam (he was a cousin of Sophia Hawthorne's) had bought *Typee* in England at the urging of Washington Irving, but his partner, John Wiley, was appalled once he read closely the attacks on missionary operations in the South Seas. Although the American edition was already printed, Wiley demanded expurgations of sexual and political passages as well as of the attacks on the missionaries, and Melville agreed to excise a total of some thirty pages, contenting himself with exclaiming to the New York editor Evert Duyckinck that expurgation was an "odious" word. Melville followed the fortunes of *Typee* with zest and even wanted to manipulate the controversy through a planted newspaper review of his own. In the middle of publicity over *Typee*, Gansevoort died suddenly at the age of thirty. In less than a year the unknown sailor, the unappreciated second son, had become a sensationally newsworthy writer and the head of his family.

Melville immediately turned to the composition of a sequel, *Omoo*, the account—more strictly autobiographical than *Typee*—of his beachcombing in Tahiti and Eimeo. When he offered the book to Murray, Melville wrote exuberantly that a "little experience in this art of book-craft has done wonders." He had in mind the condition of his manuscript, but he might well have said the same of his ability to manage a narrative. *Omoo* lacked the suspense of *Typee*, but it was a more polished performance of a writer far surer of himself. It is a fine, humorous production, full of vivid character sketches and memorable documentation of the evils wrought by the Christianizers. It delighted readers in 1847 and gave great pleasure to later South Sea wanderers like Robert Louis Stevenson and Henry Adams.

In the flush of his success with *Omoo*, Melville married Elizabeth Knapp Shaw on August 4, 1847, three days after his twenty-eighth birthday. Her father, Lemuel Shaw, the Chief Justice of Massachusetts, had been a school friend of Allan Melvill at the turn of the century and had been engaged to one of Allan's sisters who died early of tuberculosis. Allan had taken advantage of Shaw's friendship to borrow from him in the 1820s; Herman's uncle, Thomas Melvill, and Thomas's son Robert had further abused that friendship in the 1830s; then in the early 1840s Gansevoort Melville had sought Shaw out as patron. Melville dedicated *Typee* to Shaw, though it is not clear what their personal acquaintance had been; after the marriage Shaw provided several advances against his daughter's inheritance, the first being $2,000 toward the purchase of a house in New York, where Melville established himself with his bride, his younger brother Allan, and Allan's own bride, his mother, four sisters, and his new manuscript. Melville was well on his way to becoming a minor literary fixture of New York City, a participant in projects of the Duyckinck literary clique such as the short-lived satirical *Yankee Doodle*, a resident authority and reviewer of books on nautical matters and inland exploration, and a reliable dispenser of vigorous, humorous, authentic tales of exotic adventure.

Instead, the Polynesian adventurer discovered the world of the mind and the potentialities of the English language as he worked his way into his third book, *Mardi*, which was finally published in January, 1849, almost two years after he began it. His friends had some baffled inklings at the changes in Melville which could make him call the seventeenth-century writer Sir Thomas Browne a "cracked archangel" because of the speculations in the *Religio Medici* (Melville's new friend Evert Duyckinck wrote his brother, "Was ever any thing of this sort said before by a sailor?"), but for the most part the evidence of the transformation went into the manuscript of *Mardi*. It had begun as a South Sea adventure story like *Typee* and *Omoo*, or as they would have been if they had been written by a man intoxicated wth his discovery of his powers. Gone were the journalistic clichés of *Typee*, with its "umbrageous shades," its scenes "vividly impressed" upon the mind, the "performing" of "ablutions in the stream," and the stock language for moments of horror ("A cold sweat stood upon my brow, and spell-bound with terror I awaited my fate!"). Some clichés like these had undoubtedly been used for deliberate comic effects in *Typee*, but Melville's control of tone was so intermittent that the reader could seldom be sure when he was satirizing his own language. In *Mardi* Melville began using language with ecstatic verve and firm-handed control; from the start the joy of his new powers informs the most trivial passages, such as the narrator's description of Jarl's pouring "needle and thread into the frightful gashes that agonized my hapless nether integuments, which thou callest 'ducks.'" The great Melvillean style—or the first great style, since he forged others out of the tensions between his artistic drives and his financial necessities—was one which compounded the vernacular and technical concreteness of sailor talk with Renaissance rhetoric. Melville read on as he wrote—Robert Burton, François Rabelais, Sir Thomas Browne, and motley dozens more, and *Mardi* altered, becoming, as Elizabeth Foster says, a "continuum of adventure in an open boat and a derelict ship, wild allegorical romance, and fantastic travelogue-satire." In the spring of 1848, after Melville thought he was through with *Mardi*, news of the new European revolutions led him to interpolate a long section of allegorical satire on European and American politics. Sometime in the last year of composition, he bade farewell to the New York literary cliques with another allegorical section on the great poet Lombardo's creation of a masterpiece which puzzled his small-minded contemporaries.

In his solitary expansion of mind Melville had become reckless, admitting in his book that he had "voyaged chartless," and he ultimately foundered in an attempt to persuade Murray that the work, though professedly fiction, would not retroactively impugn the much-challenged authenticity of the first two books. Another London publisher, Richard Bentley, promptly enough took the book, but Murray had been prescient. Many of the reviewers were appalled at the betrayal of their expectations of another *Typee* or *Omoo*, though a discriminating minority recognized what a valuable book they had in hand. It sold poorly, especially in the overpriced three-volume English edition, and deeply damaged Melville's growing reputation except with a few readers. *Mardi* is, in fact, almost unreadable, except for a rarely dedicated lover of antiquarian literary, philosophical, metaphysical, and political hodgepodge—the sort of eccentric scholar who loves Burton's *Anatomy of Melancholy*, Browne's *Vulgar Errors*, and Laurence Sterne's *Tristram Shandy*. Melvilleans find it inexhaustibly fascinating, recognizing in it Melville's exuberant response to his realization that he was—or could become—a great literary genius. *Mardi* was his declaration of literary independence, though he did not fully achieve that independence until *Moby-Dick*, two books and two years later.

Early in 1849, during the interval between completing *Mardi* and its publication, Melville's first son, Malcolm, was born at the Shaw house in Boston, and Melville rested, went to the theater, heard Emerson lecture, and read Shakespeare with full attentiveness for the first time. He spoke hopefully of undertaking a work that would carry him beyond *Mardi*, but the first reviews of that book showed that he could not afford another such luxury. Accepting the responsibilities of a new father, he wrote *Redburn* (1849) and *White-Jacket* (1850) as acts of contrition, both ground out during one four- or five-month period in the 1849 summer swelter of a cholera-ridden New York City. As he promised Richard Bentley, *Redburn* would contain no metaphysics, only cakes and ale. Written

in the first person by the middle-aged, sentimental Wellingborough Redburn, it is the story of the narrator's first voyage, which like Melville's own
was a summer voyage to and from Liverpool, though Redburn is hardly
more than a boy while Melville was twenty. Often as good as *Huckleberry
Finn*, better than such a twentieth-century rival as *Catcher in the Rye*,
Redburn could have been a minor classic if Melville had sustained the
point of view he had established—lovingly satiric toward the boy Redburn,
more pointedly satiric toward the convention-bound narrator. But interest
in his experiment with a limited character's first-person narrative flagged,
and the second half of the book is only intermittently as compelling as the
first. The reviewers and the readers liked it, especially the air of documentary convincingness which reminded them of *Robinson Crusoe* and
other works by Daniel Defoe.

Long before *Redburn* was published, Melville had completed *White-
Jacket*, which was based on his experiences on the man-of-war *United States*
in 1843 and 1844, supplemented by lavish borrowings from earlier nautical
literature. In *White-Jacket* Melville came into something like creative
equilibrium, for his first-person narrator was once again, as in *Typee* and
Omoo, at the same stage of development as the writer, capable of saying
precisely what Melville might at that given moment be capable of saying.
Overshadowed by *Moby-Dick*, slighted by most modern readers because of
its unpromising—"unliterary"—subject matter, *White-Jacket* has been
adequately praised only by its first readers. Melville himself never could
quite regard it as much more than a product of forced labor, like *Redburn*
(that "little nursery tale") the literary equivalent of "sawing wood."

Rather than bargaining with Bentley by mail (as he had just done
for *Redburn*) and having the publisher again cite the new British ruling
on copyright (which now was denied to books by American authors even
if first printed in Great Britain), Melville sailed for London in October,
1849, carrying with him proofs of the Harper edition of *White-Jacket*. The
trip was ample reward for his summer of drudgery, for he went as a triumphant young author (*Redburn* was just being favorably reviewed) to a
country he had last seen as a penniless and anonymous American sailor. In
high spirits he entered into weeks of antiquarian bookbuying, sightseeing,
library- and museum-going, and literary socializing. By his responses to
these experiences he confirmed his new sense of himself as "a pondering
man." An observer described him as warily hawking his book "from Picca-
dilly to Whitechapel, calling upon every publisher in his way," and in fact
Melville had repeatedly met refusal because of the copyright problem, but
he ultimately settled with Bentley upon good terms. Homesick and guilty
about his extended holiday, he cut short his trip after a brief excursion into
France and Germany. Leaving early meant refusing the Duke of Rutland's
"cordial invitation to visit him at his Castle," Melville's one chance to
learn "what the highest English aristocracy really & practically is." Soon
after his return to New York on the first of February, 1850, enthusiastic
reviews of *White-Jacket* began arriving from England, and in March the
American edition was published to similar acclaim. In a buoyant mood,
sure of his powers and sure of his ability to keep an audience, Melville
began his whaling book. (By mid-1851 its working title was *The Whale*,
which remained the title for the English edition; *Moby-Dick* was a last-
minute substitute for the American edition.)

Like *Mardi*, *Moby-Dick* was luxury for Melville, an enormous, slowly written book. "Slowly" deserves qualification: Melville lived with the book some seventeen months, often writing very steadily for many weeks on end, but allowing several lengthy interruptions. By May 1, 1850, Melville was telling Richard Henry Dana, his well-known fellow sea-writer, that he was "half way in the work." Critics have speculated that the book began as a matter-of-fact sea narrative, but Melville's letter makes it clear that from the start the challenge to his art lay in getting poetry from blubber and managing to "throw in a little fancy" without, as he said, resulting in gambols as ungainly as those of the whales. Furthermore, he meant "to give the truth of the thing." None of these intentions clashes with the book he finally completed, though whatever plans he had were later altered to accommodate new literary sources as well as his maturing philosophical and theological preoccupations.

One crucial event during the composition of *Moby-Dick* was Melville's meeting with Nathaniel Hawthorne while he was vacationing at his Uncle Thomas's old place in the Berkshires (now occupied by his cousin Robert as a select boardinghouse where the likes of former President Tyler and the poet Longfellow had stayed). Reading Hawthorne's *Mosses from an Old Manse* just after their meeting may have had some minor stylistic influence on a few passages in *Moby-Dick*; more importantly, Melville undertook for the Duyckinck brothers' *Literary World* a review of *Mosses* in which he articulated many of his deepest attitudes toward the problems and opportunities of American writers. Infusing the whole review is Melville's exultant sense that the day had come when American writers could rival Shakespeare; in praising Hawthorne's achievements, he was honoring what he knew lay in his own manuscript. Furthermore, Melville gave clearer hints at what sort of "truth" he might be trying to give in *Moby-Dick*—dark, "Shakespearean" truths about human nature and the universe which "in this world of lies" can be told only "covertly, and by snatches." Out of his failures with *Mardi* and the slave labor of the next two books, Melville had built a literary theory in which a writer writes simultaneously for two audiences, one composed of the mob, the other of "eagle-eyed" readers who perceive the true meaning of those passages which the author has "directly calculated to deceive—egregiously deceive—the superficial skimmer of pages." *Moby-Dick*, now reported by Evert Duyckinck to be "a new book mostly done—a romantic, fanciful & literal & most enjoyable presentment of the Whale Fishery—something quite new"—was to be such a book. It was the culmination of Melville's reading in great literature from the Bible through Rabelais, Burton, John Milton, Sterne, Lord Byron, Thomas De Quincey, and Thomas Carlyle, yet anchored also in the nautical world of Baron Cuvier, Frederick Debell Bennett, William Scoresby, and Obed Macy, a fusion of aspects of Sir Thomas Browne and the American travel-writer J. Ross Browne, with incidental hints from a multitude of quaint old encyclopedic volumes.

Still exultantly feeling his new powers, Melville moved his family to a farm near Pittsfield late in 1850. By December he had settled again into intense work on his book until the spring chores took him away from it. During 1851 the most stimulating fact of Melville's existence, other than the book he brought to completion and saw through the press, was Hawthorne's presence at Lenox, near enough for several visits to be exchanged

except during the worst of the Berkshire winter. Melville's intense friendship provided him with a desperately needed sense of literary community
as well as a confidant for his metaphysical and philosophical speculations.
His letters to Hawthorne, preserved now mostly in nineteenth-century
transcripts and printings by Hawthorne's descendants, are among the glories of American literature and a priceless record of Melville's state of mind
during his last months with *Moby-Dick*. Uppermost in them is his sense of
kinship with the great writers and thinkers of the world—a sense that would
seem megalomanic if his manuscript had not vindicated him. The recurrent themes of the letters—democracy and aristocracy, the ironic failure of
Christians to be Christian, fame and immortality, the brotherhood of great-
souled mortals—were all recurrent themes of *Moby-Dick*. From his own
situation as a descendant of Kings abandoned to the universe, yet struggling
back to reclaim his rightful majesty, Melville created a hero who dared to
turn God's lightning back against him and whose nature could only be
explained by venturing deep below the antiquities of the earth to question
a titanic captive god. For all Ahab's insanity, which was recognized by the
narrator, Ishmael, Melville's emotional sympathies were with the defiant
Ahab who rejected the slavish values of the shore in order to defy the
malignancy in the universe. That was the world of the mind. But as he
finished *Moby-Dick* Melville was a family man whose household included
his mother and sisters as well as a small child and a pregnant wife. He owed
the Harpers $700 because they had advanced him more than his earlier
books had earned, and in April, 1851, they refused him an advance on his
whaling book. On May 1, Melville borrowed $2,050 from T. D. Stewart,
an old Lansingburgh acquaintance, and a few weeks later he painfully defined his literary-economic dilemma to Hawthorne: "What I feel most
moved to write, that is banned,—it will not pay. Yet, altogether, write the
other way I cannot. So the product is a final hash, and all my books are
botches."

Late in 1851, about the time *Moby-Dick* was published, Melville tried
once again to find a form in which he could write as profoundly as he could
while retaining the popularity he had so easily won with his first two books.
Settling on the gothic novel in its midcentury transmogrification as the
sentimental psychological novel favored by women bookbuyers, he began
Pierre, thinking he could express the agonies of the growth of a human
psyche even while enthralling readers with the romantic and ethical perplexities attending on young Pierre Glendinning's discovery of a dark
maiden who might be his unacknowledged half sister. Melville was relentlessly analyzing both the tragic and the satiric implications of the impracti-
cability of Christianity, for Pierre's calamitous decision was to obey his
heart's idealism and attempt a life in imitation of the "divine unidenti-
fiableness" of Jesus, who required of his followers the rejection of all worldly
kith and kin. After weeks of intense labor Melville took his manuscript to
New York City around New Year's Day, 1852, hoping to publish it as a
taut 360-page book, little more than half the size of *Moby-Dick*. But despite the early sales of the whaling book, Melville was still in debt to the
Harpers, who offered him a punitive contract for *Pierre*—twenty cents on
the dollar after expenses rather than the old rate of fifty cents. Stung, Melville accepted, but his rage and shame over the contract mingled with pain
from the reviews of *Moby-Dick* in the January periodicals: the *Southern*

Quarterly Review, for instance, said a "writ *de lunatico*" was justified against Melville and his characters. Within days of coming to terms with the Harpers, Melville began working into *Pierre* a sometimes wry, sometimes recklessly bitter account of his own literary career, ultimately enlarging the work by 150 printed pages and wrecking whatever chance he had of making the work what he had hoped—as much more profound than *Moby-Dick* as the legendary Krakens are larger than whales.

Pierre would probably have failed with its first readers even if it had been completed and published in its projected shorter form, for the subject matter even in the first half included atheism and incest and the language Melville created as a tool for psychological probing seemed hysterical and artificial to the reviewers. In any case, the *Pierre* that Harpers finally published late in July, 1852 (giving Melville time for a fruitless negotiation with Bentley, who refused to publish it without expurgation), all but ended Melville's career. It was widely denounced as immoral, and one *Pierre*-inspired news account was captioned "HERMAN MELVILLE CRAZY." In panic the family made efforts to gain Melville some government post, preferably foreign, but nothing came of their attempts to call in old favors. Melville stayed on the farm with his expanding household: his second son, Stanwix, had been born in October, 1851, and two daughters, Elizabeth and Frances, were born in 1853 and 1855. After *Pierre* Melville's career seemed stuttering away: one story about a patient Nantucket wife was apparently written and destroyed; a book about tortoise-hunting in the Galapagos Islands was partially written, and some of that part was destroyed. Melville was undergoing a profound psychological crisis which left him more resigned to fate than defiant, and in addition to his older ailment of weak eyes he developed a new set of crippling afflictions diagnosed as sciatica and rheumatism.

Late in 1853 Melville began a new, low-keyed career as writer of short stories for the two major American monthlies, *Harper's* and *Putnam's*. All stories were anonymous, by magazine policy (though authorship was often leaked to editors of newspapers and other magazines), so what Melville published in the next years did not add greatly to his fame. These tales were for the most part apparently innocuous, deliberately tailored to the bland tastes of the upper-middle-class readers, yet many of them concealed covert allegories with religious, sexual, and autobiographical implications. Literature became for Melville a means of expressing his true attitudes by elaborately convoluted aesthetic dodges. Some of his stories manifest a love of allegory as inveterate as Hawthorne's and are written in a shifty, deceptive style appropriate to a once popular writer who had perforce become a literary sleight-of-hand man. One serial, the story of a Revolutionary exile named Israel Potter, stretched out to book length. Offering it to the publisher, Melville promised that it would contain nothing "to shock the fastidious," and in fact he restrained his imagination and his metaphysical and theological compulsions. Straightforward novel that it is, *Israel Potter* contains passages of great historical interest, especially the complex portraits of Benjamin Franklin, John Paul Jones, and Ethan Allen. In 1856, Melville collected the *Putnam's* stories as *The Piazza Tales*, supplying a new prefatory sketch, "The Piazza," which marked his development past his earlier simple admiration for Hawthorne's subjects and techniques. Clear in all of Melville's writings in the mid-fifties is his growing tendency to brood less over his own career and his own relationship with cosmic forces and more over the American national character and the conditions in American life that would allow honest craftsmen like himself to be rejected. Self-pity tinges some of these writings, but more often a wry jocularity, an almost comfortable self-mockery. In them Melville achieved a new sureness of artistic control, even though the power of *Moby-Dick* and parts of *Pierre* was never regained. From this period of physical and psychic suffering and of financial distress emerged a new masterpiece, *The Confidence-Man.*

During the summer Melville was writing *Redburn* and *White-Jacket*, a swindler arrested in New York City caught the imagination of editors and readers alike with the archetypal perfection of his ploy, a request that his victims prove their confidence in him by entrusting him with cash, a watch, or some other valuable. Journalists dubbed the petty crook the "Confidence Man," a term which quickly entered the national vocabulary, used literally and metaphorically in countless newspaper stories of the next years. The Confidence Man surfaced again in 1855, at Albany, under yet another alias, just when Melville was most apt to see allegorical implications in commonplaces and be most deeply concerned about the pervasive shallow optimism in all aspects of American life. In *The Confidence-Man* he projected the confidence game into cosmic terms by having the Original Confidence Man himself, the guller of Eve, come on board the American ship of faith, the *Fidèle*, and in a series of disguises probe the consciences of representative Americans, hoping less to enter souls in the hellish account book of the "Black Rapids Coal Company" than to demonstrate in fair play that no real Christians are aboard. In order to blaspheme with impunity in the America of the mid-fifties, where at least one of his stories had been rejected as too obviously satiric of upper-class New York churchgoers, Melville perfected yet another style, one in which the object is to set up elaborate qualifications where ideas are not so much asserted as hedgingly offered and ambiguously retracted, where examples of litotes such as "not unlikely" and "less unrefined" almost but not quite turn a double negative into a positive, where only the rarest of eagle-eyed readers could perceive that behind the noncommital prose lurked satanic ironies. *The Confidence-Man* was a devastating indictment of national confidence in the form of mingled metaphysical satire and low comedy. It went almost unread in the United States; in England the reviews were more intelligent but the sales were also disappointing, and Melville did not earn a cent from either edition.

By the spring of 1856 Melville may have recovered from most of his mental, spiritual, and physical agonies, but his economic distress was greater than ever. Besides a mortgage on the farm held by the previous owner, Melville still owed the principal and the accumulating interest on the $2,050 he had borrowed in 1851, and the lender was pressing for full repayment. Melville was forced to sell part of the farm, but Judge Shaw met the family's anxieties about Herman's state of mind by providing funds for an extended trip to Europe and the Levant, from October, 1856, to May, 1857. In England Melville told Hawthorne that he did not anticipate much pleasure in his rambles, since "the spirit of adventure" had gone out of him. For upwards of a decade, Melville's adventuring had been inward—philosophical, metaphysical, psychological, and artistic. As Hawthorne hoped, Melville brightened as he went onward, and after a few days he began to keep a journal. Melville's sightseeing and gallery-going were as compulsively American as Hawthorne's own. Many of Melville's observations were predictable responses to the places, palaces, and paintings given largest space in the guidebooks, but what he saw gradually led him to energetically original responses, as in his then unfashionable response to classical statuary, earlier Italian painters like Giotto, and the realistic Dutch and Flemish genre painters. As Howard Horsford, the editor of this journal, says, the entries show Melville's taste in the process of being formed. Horsford points out "the peculiar urgency, the sharpness, vividness, and freshness" of those entries where Melville was most deeply moved: "Many passages, such as those on the Pyramids, or the descriptions of the Jerusalem scene and the Palestinian landscape, are in his finest rhetorical style; many of his comments on people, places, and things display the most cutting edge of his irony and satire, as in his accounts of the Church of the Holy Sepulchre and of the missionaries in the Near East." Horsford also draws a precise contrast between the Melville of the late 1840s and the one of the mid-fifties. When Melville "had embarked on metaphysical speculation at the time of *Mardi*," it had been "a welcome release, an escape into a new freedom from orthodoxy and dogmatism, a mental emancipation." Now Melville's metaphysical speculations had ended in "joyless skepticism." He told Hawthorne that he had "pretty much made up his mind to be annihilated"—to give up any religious hope of immortality; but, Hawthorne decided, "still he does not seem to rest in that anticipation." When he returned home, Melville was more than ever "a pondering man," but he told a young Gansevoort cousin that he was "not going to write any more at present."

That moment stretched on. Melville lectured in the East and Midwest for three seasons (1857–1860) without much profit, speaking in successive years on "Statues in Rome," "The South Seas," and "Traveling." He prepared a volume of poems in 1860 but instead of trying to place it, he sailed on a voyage to San Francisco as passenger on a ship captained by his youngest brother, Thomas, leaving his wife and his brother Allan to seek fruitlessly for a publisher. In 1861 Judge Shaw died, and with some of his wife's inheritance they moved back to New York City two years later. In 1866 Melville published *Battle-Pieces*, a volume of Civil War poems that was casually or disdainfully reviewed and quickly forgotten; now it ranks with Whitman's *Drum-Taps* as the best of hundreds of volumes of poetry to come out of the war. For all his front of nonchalance, Melville was devastated by the loss of his career and further rebuffs when he sought a government job in Washington. As unemployed men do, Melville took out his frustrations on his family, so that for years his wife's half brothers considered him insane as well as financially incompetent; and by early 1867 Melville's wife was also convinced of his insanity. Her sense of loyalty to him and her horror of gossip, however, were strong enough to make her reject her minister's suggestion that she pretend to make a routine visit to Boston and then barricade herself in the house of one of her brothers; but as her family realized, the law was on Melville's side, whatever unrecorded abuses he was guilty of. In 1866 Melville had at last obtained a political job—not as consul in some exotic capital but as a deputy inspector of customs in New York City; ironically, his beat during some years took

830 ★ Herman Melville

him frequently to the pier on Gansevoort Street, named for his mother's heroic father. Gradually the routine of his work reduced some of the tensions in the family, and after Malcolm killed himself late in 1867 at the age of eighteen the Melvilles closed ranks.

As Melville had predicted to Hawthorne, he became known as the "man who lived among the cannibals," holding his place in encyclopedias and literary histories primarily as the author of *Typee* and *Omoo*, all but forgotten by the post-bellum literary world. But for years through the late 1860s and early 1870s Melville worked on a poem about a motley group of American European pilgrims—and tourists—who talked their way through some of the same Palestinian scenes he had visited a decade and more earlier; apparently he carried about pocket-sized slips of paper for writing in odd moments at work as well as during his evenings. This poem, *Clarel*, grew to 18,000 lines and lay unpublished for many months or perhaps even two or three years before it appeared in 1876, paid for by a specific bequest from the dying Peter Gansevoort. It is America's most thoughtful contribution to the conflict of religious faith and Darwinian skepticism which obsessed English contemporaries such as Matthew Arnold and Thomas Hardy. Like *Mardi* it is inexhaustible for what it reveals of Melville's mind and art, but unlike *Mardi* it is plotted with the surety of artistic control which he had learned in the 1850s; but *Mardi* had been read and argued about, and *Clarel* was ignored.

Stanwix, the second Melville son, drifted away without a career, beachcombing for a time in Central America, finally dying in San Francisco in 1886. The first daughter, called Bessie, developed severe arthritis and never married, and died in 1908. Only Frances married, and she lived until 1934, so bitter against her father that she could not recognize him in the words of twentieth-century admirers and flatly refused to talk about him. But through the 1880s Melville and his wife drew closer together. An extraordinary series of legacies came to them in Melville's last years; ironically the wealth was too late to make much change in their lives, but it allowed him to retire from the Custom House at the beginning of 1886 and devote himself to his writing. From time to time after *Clarel* he had written poems which ultimately went into two volumes which he printed privately shortly before his death, except for some that remained unpublished until the 1920s and later. Melville developed the habit of writing prose headnotes to poems, notably some dealing with an imaginary Burgandy Club in which he found consolation for his loneliness. He could relax with the intelligent good-fellows of his imagination as he could never relax among the popular literary men of the 1870s and 1880s who now and then tried to patronize him. In the mid-1880s one poem about a British sailor evoked a headnote which, expanded and re-expanded, was left nearly finished at Melville's death as *Billy Budd, Sailor*, his final study of the ambiguous claims of authority and individuality.

Before Melville's death in 1891, something like a revival of his fame was in progress, especially in England. American newspapers became accustomed to reprinting and briefly commenting on extraordinary items in British periodicals, such as Robert Buchanan's footnote to Melville's name in a poetic tribute to Whitman (1885): "I sought everywhere for this Triton, who is still living somewhere in New York. No one seemed to know anything of the one great imaginative writer fit to stand shoulder to shoulder with Whitman on that continent." The recurrent imagery—used by Melville as well as journalists—was of burial and possible resurrection. To a young British admirer Melville wrote in 1884 that he really wished "that the books you have so patiently disinterred better merited what you say of them," and in a subsequent letter the next year wrote: "—In a former note you mentioned that altho' you had unearthed several of my buried books, yet there was one—'Clarel'—that your spade had not succeeded in getting at. Fearing that you never will get at it by yourself, I have disinterred a copy for you of which I ask your acceptance and mail it with this note. It is the sole presentation-copy of the issue." In yet another letter to the same admirer he expressed one preoccupation of his last decades: "And it must have occurred to you as it has to me, that the further our civilization advances upon its present lines, so much the cheaper sort of thing does 'fame' become, especially of the literary sort."

When he died, Melville had reason to think his reputation would ultimately be established. Just after his death, new editions of *Typee, Omoo, White-Jacket,* and *Moby-Dick* were published both in the United States and England, and Mrs. Melville remained a loyal and alert custodian of his memory until her death in 1906, but interest sputtered away except for small cults of Melville lovers who, as an anonymous British writer said in 1922, came to use *Moby-Dick* (or *The Whale*) as the test of a worthy reader and friend, proffering it without special comment and staking all on the response of the reader. The true Melville revival began with articles on Melville's centennial in 1919. That revival, one of the most curious phenomena of American literary history, swept Melville from the ranks of the lesser American writers—lesser than James Fenimore Cooper and William Gilmore Simms—into the rarefied company of Shakespeare and a few fellow immortals of world literature so that only Whitman, James, and Faulkner are seen as his American equals. Many of the materials for a biography had by then been lost (Melville burned his letters from Hawthorne, family members censored their files from dangerous years like the 1860s), but scholars have found the study of Melville's life and works inexhaustible. Even during the mass consumption of Melville in the classrooms and the spawning of the White Whale in comic books, cartoons, and seafood restaurants, a few lonely cultists are still to be found, tracing his journeys in the South Seas and Manhattan Island, and visiting his grave in the Bronx, faithful to the Melville who speaks to them without the aid of interpreter. That may be the true sign of the rarest literary immortality.

Bartleby, the Scrivener[1]

A Story of Wall-Street

I am a rather elderly man. The nature of my avocations for the last thirty years has brought me into more than ordinary contact with what would seem an interesting and somewhat singular set of men, of whom as yet nothing that I know of has ever been written:

1. The text is from the first printing in the November and December, 1853, issues of *Putnam's Monthly Magazine,* the first work by Melville's to be printed after the disastrous reception of *Pierre* during the summer and fall of 1852. (One work, probably the story of Agatha Robinson, a Nantucket woman who displayed patience, endurance, and resignedness, was apparently destroyed after being rejected by the Harpers.)

—I mean the law-copyists or scriveners. I have known very many of them, professionally and privately, and if I pleased, could relate divers histories, at which good-natured gentlemen might smile, and sentimental souls might weep. But I waive the biographies of all other scriveners for a few passages in the life of Bartleby, who was a scrivener the strangest I ever saw or heard of. While of other law-copyists I might write the complete life, of Bartleby nothing of that sort can be done. I believe that no materials exist for a full and satisfactory biography of this man. It is an irreparable loss to litera-ture. Bartleby was one of those beings of whom nothing is ascer-tainable, except from the original sources, and in his case those are very small. What my own astonished eyes saw of Bartleby, *that* is all I know of him, except, indeed, one vague report which will appear in the sequel.

Ere introducing the scrivener, as he first appeared to me, it is fit I make some mention of myself, my *employées*, my business, my chambers, and general surroundings; because some such description is indispensable to an adequate understanding of the chief character about to be presented.

Imprimis: I am a man who, from his youth upwards, has been filled with a profound conviction that the easiest way of life is the best. Hence, though I belong to a profession proverbially energetic and nervous, even to turbulence, at times, yet nothing of that sort have I ever suffered to invade my peace. I am one of those unam-bitious lawyers who never addresses a jury, or in any way draws down public applause; but in the cool tranquillity of a snug retreat, do a snug business among rich men's bonds and mortgages and title-deeds. All who know me, consider me an eminently *safe* man. The late John Jacob Astor, a personage little given to poetic enthusiasm, had no hesitation in pronouncing my first grand point to be pru-dence; my next, method. I do not speak it in vanity, but simply record the fact, that I was not unemployed in my profession by the late John Jacob Astor; a name which, I admit, I love to repeat, for it hath a rounded and orbicular sound to it, and rings like unto bullion. I will freely add, that I was not insensible to the late John Jacob Astor's good opinion.

Some time prior to the period at which this little history begins, my avocations had been largely increased. The good old office, now extinct in the State of New-York, of a Master in Chancery,[2] had been conferred upon me. It was not a very arduous office, but very pleasantly remunerative. I seldom lose my temper; much more sel-dom indulge in dangerous indignation at wrongs and outrages; but I must be permitted to be rash here and declare, that I consider the

2. The narrator is understandably con-cerned about the abolition of a sinecure, but heirs had cause to rejoice, for chan-cery had kept estates tied up in pro-longed litigation. In a poem written around the 1870s, *At the Hostelry*, Mel-ville says that divided Italy, "Nigh para-lysed, by cowls misguided," was "Locked as in Chancery's numbing hand." New York had adopted a "new Constitution" (see just below) in 1846.

sudden and violent abrogation of the office of Master in Chancery, by the new Constitution, as a——premature act; inasmuch as I had counted upon a life-lease of the profits, whereas I only received those of a few short years. But this is by the way.

My chambers were up stairs at No.—Wall-street. At one end they looked upon the white wall of the interior of a spacious sky-light shaft, penetrating the building from top to bottom. This view might have been considered rather tame than otherwise, deficient in what landscape painters call "life." But if so, the view from the other end of my chambers offered, at least, a contrast, if nothing more. In that direction my windows commanded an unobstructed view of a lofty brick wall, black by age and everlasting shade; which wall required no spy-glass to bring out its lurking beauties, but for the benefit of all near-sighted spectators, was pushed up to within ten feet of my window panes. Owing to the great height of the surrounding buildings, and my chambers being on the second floor, the interval between this wall and mine not a little resembled a huge square cistern.

At the period just preceding the advent of Bartleby, I had two persons as copyists in my employment, and a promising lad as an office-boy. First, Turkey; second, Nippers; third, Ginger Nut. These may seem names, the like of which are not usually found in the Directory. In truth they were nicknames, mutually conferred upon each other by my three clerks, and were deemed expressive of their respective persons or characters. Turkey was a short, pursy[3] Englishman of about my own age, that is, somewhere not far from sixty. In the morning, one might say, his face was of a fine florid hue, but after twelve o'clock, meridian—his dinner hour—it blazed like a grate full of Christmas coals; and continued blazing—but, as it were, with a gradual wane—till 6 o'clock, P.M. or thereabouts, after which I saw no more of the proprietor of the face, which gaining its meridian with the sun, seemed to set with it, to rise, culminate, and decline the following day, with the like regularity and undiminished glory. There are many singular coincidences I have known in the course of my life, not the least among which was the fact, that exactly when Turkey displayed his fullest beams from his red and radiant countenance, just then, too, at that critical moment, began the daily period when I considered his business capacities as seriously disturbed for the remainder of the twenty-four hours. Not that he was absolutely idle, or averse to business then; far from it. The difficulty was, he was apt to be altogether too energetic. There was a strange, inflamed, flurried, flighty recklessness of activity about him. He would be incautious in dipping his pen into his inkstand. All his blots upon my documents, were dropped there after twelve o'clock, meridian. Indeed, not only would he be reckless and sadly given to making blots in the afternoon, but some days

---

3. Shortwinded from obesity.

he went further, and was rather noisy. At such times, too, his face flamed with augmented blazonry, as if cannel coal had been heaped on anthracite. He made an unpleasant racket with his chair; spilled his sand-box; in mending his pens, impatiently split them all to pieces, and threw them on the floor in a sudden passion; stood up and leaned over his table, boxing his papers about in a most indecorous manner, very sad to behold in an elderly man like him. Nevertheless, as he was in many ways a most valuable person to me, and all the time before twelve o'clock, meridian, was the quickest, steadiest creature too, accomplishing a great deal of work in a style not easy to be matched—for these reasons, I was willing to overlook his eccentricities, though indeed, occasionally, I remonstrated with him. I did this very gently, however, because, though the civilest, nay, the blandest and most reverential of men in the morning, yet in the afternoon he was disposed, upon provocation, to be slightly rash with his tongue, in fact, insolent. Now, valuing his morning services as I did, and resolved not to lose them; yet, at the same time made uncomfortable by his inflamed ways after twelve o'clock; and being a man of peace, unwilling by my admonitions to call forth unseemly retorts from him; I took upon me, one Saturday noon (he was always worse on Saturdays), to hint to him, very kindly, that perhaps now that he was growing old, it might be well to abridge his labors; in short, he need not come to my chambers after twelve o'clock, but, dinner over, had best go home to his lodgings and rest himself till tea-time. But no; he insisted upon his afternoon devotions. His countenance became intolerably fervid, as he oratorically assured me—gesticulating with a long ruler at the other end of the room—that if his services in the morning were useful, how indispensable, then, in the afternoon?

"With submission, sir," said Turkey on this occasion, "I consider myself your right-hand man. In the morning I but marshal and deploy my columns; but in the afternoon I put myself at their head, and gallantly charge the foe, thus!"—and he made a violent thrust with the ruler.

"But the blots, Turkey," intimated I.

"True,—but, with submission, sir, behold these hairs! I am getting old. Surely, sir, a blot or two of a warm afternoon is not to be severely urged against gray hairs. Old age—even if it blot the page —is honorable. With submission, sir, we *both* are getting old."

This appeal to my fellow-feeling was hardly to be resisted. At all events, I saw that go he would not. So I made up my mind to let him stay, resolving, nevertheless, to see to it, that during the afternoon he had to do with my less important papers.

Nippers, the second on my list, was a whiskered, sallow, and, upon the whole, rather piratical-looking young man of about five and twenty. I always deemed him the victim of two evil powers— ambition and indigestion. The ambition was evinced by a certain impatience of the duties of a mere copyist, an unwarrantable usurpation of strictly professional affairs, such as the original drawing up of legal documents. The indigestion seemed betokened in an occasional nervous testiness and grinning irritability, causing the teeth to audibly grind together over mistakes committed in copying; unnecessary maledictions, hissed, rather than spoken, in the heat of business; and especially by a continual discontent with the height of the table where he worked. Though of a very ingenious mechanical turn, Nippers could never get this table to suit him. He put chips under it, blocks of various sorts, bits of pasteboard, and at last went so far as to attempt an exquisite adjustment by final pieces of folded blotting paper. But no invention would answer. If, for the sake of easing his back, he brought the table lid at a sharp angle well up towards his chin, and wrote there like a man using the steep roof of a Dutch house for his desk:—then he declared that it stopped the circulation in his arms. If now he lowered the table to his waistbands, and stooped over it in writing, then there was a sore aching in his back. In short, the truth of the matter was, Nippers knew not what he wanted. Or, if he wanted any thing, it was to be rid of a scrivener's table altogether. Among the manifestations of his diseased ambition was a fondness he had for receiving visits from certain ambiguous-looking fellows in seedy coats, whom he called his clients. Indeed I was aware that not only was he, at times, considerable of a ward-politician, but he occasionally did a little business at the Justices' courts, and was not unknown on the steps of the Tombs.[4] I have good reason to believe, however, that one individual who called upon him at my chambers, and who, with a grand air, he insisted was his client, was no other than a dun,[5] and the alleged title-deed, a bill. But with all his failings, and the annoyances he caused me, Nippers, like his compatriot Turkey, was a very useful man to me; wrote a neat, swift hand; and, when he chose, was not deficient in a gentlemanly sort of deportment. Added to this, he always dressed in a gentlemanly sort of way; and so, incidentally, reflected credit upon my chambers. Whereas with respect to Turkey, I had much ado to keep him from being a reproach to me. His clothes were apt to look oily and smell of eating-houses. He wore his pantaloons very loose and baggy in summer. His coats were execrable; his hat not to be handled. But while the hat was a thing of indifference to me, inasmuch as his natural civility and deference, as a dependent Englishman, always led him to doff it the moment he entered the room, yet his coat was another matter. Concerning his coats, I reasoned with him; but with no effect. The truth was, I suppose, that a man with so small an income, could not afford to sport such a lustrous face and a lustrous coat at one and

---

4. I.e., Nippers is suspected of arranging bail for prisoners or other such activities which strike the narrator as unseemly if not nefarious.
5. Bill collector.

the same time. As Nippers once observed, Turkey's money went chiefly for red ink. One winter day I presented Turkey with a highly-respectable looking coat of my own, a padded gray coat, of a most comfortable warmth, and which buttoned straight up from the knee to the neck. I thought Turkey would appreciate the favor, and abate his rashness and obstreperousness of afternoons. But no. I verily believe that buttoning himself up in so downy and blanket-like a coat had a pernicious effect upon him; upon the same principle that too much oats are bad for horses. In fact, precisely as a rash, restive horse is said to feel his oats, so Turkey felt his coat. It made him insolent. He was a man whom prosperity harmed.

Though concerning the self-indulgent habits of Turkey I had my own private surmises, yet touching Nippers I was well persuaded that whatever might be his faults in other respects, he was, at least, a temperate young man. But indeed, nature herself seemed to have been his vintner, and at his birth charged him so thoroughly with an irritable, brandy-like disposition, that all subsequent potations were needless. When I consider how, amid the stillness of my chambers, Nippers would sometimes impatiently rise from his seat, and stooping over his table, spread his arms wide apart, seize the whole desk, and move it, and jerk it, with a grim, grinding motion on the floor, as if the table were a perverse voluntary agent, intent on thwarting and vexing him; I plainly perceive that for Nippers, brandy and water were altogether superfluous.

It was fortunate for me that, owing to its peculiar cause—indigestion—the irritability and consequent nervousness of Nippers, were mainly observable in the morning, while in the afternoon he was comparatively mild. So that Turkey's paroxysms only coming on about twelve o'clock, I never had to do with their eccentricities at one time. Their fits relieved each other like guards. When Nippers' was on, Turkey's was off; and *vice versa*. This was a good natural arrangement under the circumstances.

Ginger Nut, the third on my list, was a lad some twelve years old. His father was a carman,[6] ambitious of seeing his son on the bench instead of a cart, before he died. So he sent him to my office as student at law, errand boy, and cleaner and sweeper, at the rate of one dollar a week. He had a little desk to himself, but he did not use it much. Upon inspection, the drawer exhibited a great array of the shells of various sorts of nuts. Indeed, to this quick-witted youth the whole noble science of the law was contained in a nut-shell. Not the least among the employments of Ginger Nut, as well as one which he discharged with the most alacrity, was his duty as cake and apple purveyor for Turkey and Nippers. Copying law papers being proverbially a dry, husky sort of business, my two scriveners were fain to moisten their mouths very often with Spitzenbergs[7] to be had at the numerous stalls nigh the Custom House and Post Office.

6. Driver, teamster.　　　　7. Red-and-yellow New York apples.

Also, they sent Ginger Nut very frequently for that peculiar cake—small, flat, round, and very spicy—after which he had been named by them. Of a cold morning when business was but dull, Turkey would gobble up scores of these cakes, as if they were mere wafers —indeed they sell them at the rate of six or eight for a penny—the scrape of his pen blending with the crunching of the crisp particles in his mouth. Of all the fiery afternoon blunders and flurried rashnesses of Turkey, was his once moistening a ginger-cake between his lips, and clapping it on to a mortgage for a seal.[8] I came within an ace of dismissing him then. But he mollified me by making an oriental bow, and saying—"With submission, sir, it was generous of me to find you in stationery on my own account."

Now my original business—that of a conveyancer[9] and title hunter, and drawer-up of recondite documents of all sorts—was considerably increased by receiving the master's office. There was now great work for scriveners. Not only must I push the clerks already with me, but I must have additional help. In answer to my advertisement, a motionless young man one morning, stood upon my office threshold, the door being open, for it was summer. I can see that figure now—pallidly neat, pitiably respectable, incurably forlorn! It was Bartleby.

After a few words touching his qualifications, I engaged him, glad to have among my corps of copyists a man of so singularly sedate an aspect, which I thought might operate beneficially upon the flighty temper of Turkey, and the fiery one of Nippers.

I should have stated before that ground glass folding-doors divided my premises into two parts, one of which was occupied by my scriveners, the other by myself. According to my humor I threw open these doors, or closed them. I resolved to assign Bartleby a corner by the folding-doors, but on my side of them, so as to have this quiet man within easy call, in case any trifling thing was to be done. I placed his desk close up to a small side-window in that part of the room, a window which originally had afforded a lateral view of certain grimy back-yards and bricks, but which, owing to subsequent erections, commanded at present no view at all, though it gave some light. Within three feet of the panes was a wall, and the light came down from far above, between two lofty buildings, as from a very small opening in a dome. Still further to a satisfactory arrangement, I procured a high green folding screen, which might entirely isolate Bartleby from my sight, though not remove him from my voice. And thus, in a manner, privacy and society were conjoined.

At first Bartleby did an extraordinary quantity of writing. As if long famishing for something to copy, he seemed to gorge himself

8. The narrator is playing on the resemblance between thin cookies and wax wafers used for sealing documents.
9. Someone who draws up deeds for transferring title to property. "Title hunter": someone who checks records to be sure there are no encumbrances on the title of property to be transferred.

on my documents. There was no pause for digestion. He ran a day and night line, copying by sun-light and by candle-light. I should have been quite delighted with his application, had he been cheerfully industrious. But he wrote on silently, palely, mechanically.

It is, of course, an indispensable part of a scrivener's business to verify the accuracy of his copy, word by word. Where there are two or more scriveners in an office, they assist each other in this examination, one reading from the copy, the other holding the original. It is a very dull, wearisome, and lethargic affair. I can readily imagine that to some sanguine temperaments it would be altogether intolerable. For example, I cannot credit that the mettlesome poet Byron would have contentedly sat down with Bartleby to examine a law document of, say five hundred pages, closely written in a crimpy hand.

Now and then, in the haste of business, it had been my habit to assist in comparing some brief document myself, calling Turkey or Nippers for this purpose. One object I had in placing Bartleby so handy to me behind the screen, was to avail myself of his services on such trivial occasions. It was on the third day, I think, of his being with me, and before any necessity had arisen for having his own writing examined, that, being much hurried to complete a small affair I had in hand, I abruptly called to Bartleby. In my haste and natural expectancy of instant compliance, I sat with my head bent over the original on my desk, and my right hand sideways, and somewhat nervously extended with the copy, so that immediately upon emerging from his retreat, Bartleby might snatch it and proceed to business without the least delay.

In this very attitude did I sit when I called to him, rapidly stating what it was I wanted him to do—namely, to examine a small paper with me. Imagine my surprise, nay, my consternation, when without moving from his privacy, Bartleby in a singularly mild, firm voice, replied, "I would prefer not to."

I sat awhile in perfect silence, rallying my stunned faculties. Immediately it occurred to me that my ears had deceived me, or Bartleby had entirely misunderstood my meaning. I repeated my request in the clearest tone I could assume. But in quite as clear a one came the previous reply, "I would prefer not to."

"Prefer not to," echoed I, rising in high excitement, and crossing the room with a stride. "What do you mean? Are you moon-struck? I want you to help me compare this sheet here—take it," and I thrust it towards him.

"I would prefer not to," said he.

I looked at him steadfastly. His face was leanly composed; his gray eye dimly calm. Not a wrinkle of agitation rippled him. Had there been the least uneasiness, anger, impatience or impertinence in his manner; in other words, had there been any thing ordinarily human about him, doubtless I should have violently dismissed him from the premises. But as it was, I should have as soon thought of turning my pale plaster-of-paris bust of Cicero[1] out of doors. I stood gazing at him awhile, as he went on with his own writing, and then reseated myself at my desk. This is very strange, thought I. What had one best do? But my business hurried me. I concluded to forget the matter for the present, reserving it for my future leisure. So calling Nippers from the other room, the paper was speedily examined.

A few days after this, Bartleby concluded four lengthy documents, being quadruplicates of a week's testimony taken before me in my High Court of Chancery. It became necessary to examine them. It was an important suit, and great accuracy was imperative. Having all things arranged I called Turkey, Nippers and Ginger Nut from the next room, meaning to place the four copies in the hands of my four clerks, while I should read from the original. Accordingly Turkey, Nippers and Ginger Nut had taken their seats in a row, each with his document in hand, when I called to Bartleby to join this interesting group.

"Bartleby! quick, I am waiting."

I heard a slow scrape of his chair legs on the uncarpeted floor, and soon he appeared standing at the entrance of his hermitage.

"What is wanted?" said he mildly.

"The copies, the copies," said I hurriedly. "We are going to examine them. There"—and I held towards him the fourth quadruplicate.

"I would prefer not to," he said, and gently disappeared behind the screen.

For a few moments I was turned into a pillar of salt,[2] standing at the head of my seated column of clerks. Recovering myself, I advanced towards the screen, and demanded the reason for such extraordinary conduct.

"*Why* do you refuse?"

"I would prefer not to."

With any other man I should have flown outright into a dreadful passion, scorned all further words, and thrust him ignominiously from my presence. But there was something about Bartleby that not only strangely disarmed me, but in a wonderful manner touched and disconcerted me. I began to reason with him.

"These are your own copies we are about to examine. It is labor saving to you, because one examination will answer for your four papers. It is common usage. Every copyist is bound to help examine his copy. Is it not so? Will you not speak? Answer!"

"I prefer not to," he replied in a flute-like tone. It seemed to me that while I had been addressing him, he carefully revolved every statement that I made; fully comprehended the meaning; could not

1. Roman orator and statesman (106–42 B.C.).   2. The punishment of Lot's disobedient wife (Genesis 19.26).

gainsay the irresistible conclusion; but, at the same time, some paramount consideration prevailed with him to reply as he did.

"You are decided, then, not to comply with my request—a request made according to common usage and common sense?"

He briefly gave me to understand that on that point my judgment was sound. Yes: his decision was irreversible.

It is not seldom the case that when a man is browbeaten in some unprecedented and violently unreasonable way, he begins to stagger in his own plainest faith. He begins, as it were, vaguely to surmise that, wonderful as it may be, all the justice and all the reason is on the other side. Accordingly, if any disinterested persons are present, he turns to them for some reinforcement for his own faltering mind.

"Turkey," said I, "what do you think of this? Am I not right?"

"With submission, sir," said Turkey, with his blandest tone, "I think that you are."

"Nippers," said I, "what do *you* think of it?"

"I think I should kick him out of the office."

(The reader of nice perceptions will here perceive that, it being morning, Turkey's answer is couched in polite and tranquil terms, but Nippers replies in ill-tempered ones. Or, to repeat a previous sentence, Nippers's ugly mood was on duty, and Turkey's off.)

"Ginger Nut," said I, willing to enlist the smallest suffrage in my behalf, "what do *you* think of it?"

"I think, sir, he's a little *luny*," replied Ginger Nut, with a grin.

"You hear what they say," said I, turning towards the screen, "come forth and do your duty."

But he vouchsafed no reply. I pondered a moment in sore perplexity. But once more business hurried me. I determined again to postpone the consideration of this dilemma to my future leisure. With a little trouble we made out to examine the papers without Bartleby, though at every page or two, Turkey deferentially dropped his opinion that this proceeding was quite out of the common; while Nippers, twitching in his chair with a dyspeptic nervousness, ground out between his set teeth occasional hissing maledictions against the stubborn oaf behind the screen. And for his (Nippers's) part, this was the first and the last time he would do another man's business without pay.

Meanwhile Bartleby sat in his hermitage, oblivious to every thing but his own peculiar business there.

Some days passed, the scrivener being employed upon another lengthy work. His late remarkable conduct led me to regard his ways narrowly. I observed that he never went to dinner; indeed that he never went any where. As yet I had never of my personal knowledge known him to be outside of my office. He was a perpetual sentry in the corner. At about eleven o'clock though, in the morning, I noticed that Ginger Nut would advance toward the opening in Bartleby's screen, as if silently beckoned thither by a gesture invisible to me where I sat. The boy would then leave the office jingling a few pence, and reappear with a handful of ginger-nuts which he delivered in the hermitage, receiving two of the cakes for his trouble.

He lives, then, on ginger-nuts, thought I; never eats a dinner, properly speaking; he must be a vegetarian then; but no; he never eats even vegetables, he eats nothing but ginger-nuts. My mind then ran on in reveries concerning the probable effects upon the human constitution of living entirely on ginger-nuts. Ginger-nuts are so called because they contain ginger as one of their peculiar constituents, and the final flavoring one. Now what was ginger? A hot, spicy thing. Was Bartleby hot and spicy? Not at all. Ginger, then, had no effect upon Bartleby. Probably he preferred it should have none.

Nothing so aggravates an earnest person as a passive resistance. If the individual so resisted be of a not inhumane temper, and the resisting one perfectly harmless in his passivity; then, in the better moods of the former, he will endeavor charitably to construe to his imagination what proves impossible to be solved by his judgment. Even so, for the most part, I regarded Bartleby and his ways. Poor fellow! thought I, he means no mischief; it is plain he intends no insolence; his aspect sufficiently evinces that his eccentricities are involuntary. He is useful to me. I can get along with him. If I turn him away, the chances are he will fall in with some less indulgent employer, and then he will be rudely treated, and perhaps driven forth miserably to starve. Yes. Here I can cheaply purchase a delicious self-approval. To befriend Bartleby; to humor him in his strange wilfulness, will cost me little or nothing, while I lay up in my soul what will eventually prove a sweet morsel for my conscience. But this mood was not invariable with me. The passiveness of Bartleby sometimes irritated me. I felt strangely goaded on to encounter him in new opposition, to elicit some angry spark from him answerable to my own. But indeed I might as well have essayed to strike fire with my knuckles against a bit of Windsor soap.[3] But one afternoon the evil impulse in me mastered me, and the following little scene ensued:

"Bartleby," said I, "when those papers are all copied, I will compare them with you."

"I would prefer not to."

"How? Surely you do not mean to persist in that mulish vagary?" No answer.

I threw open the folding-doors near by, and turning upon Turkey and Nippers, exclaimed in an excited manner—

"He says, a second time, he won't examine his papers. What do you think of it, Turkey?"

3. Brown hand soap.

It was afternoon, be it remembered. Turkey sat glowing like a brass boiler, his bald head steaming, his hands reeling among his blotted papers.

"Think of it?" roared Turkey; "I think I'll just step behind his screen, and black his eyes for him!"

So saying, Turkey rose to his feet and threw his arms into a pugilistic position. He was hurrying away to make good his promise, when I detained him, alarmed at the effect of incautiously rousing Turkey's combativeness after dinner.

"Sit down, Turkey," said I, "and hear what Nippers has to say. What do you think of it, Nippers? Would I not be justified in immediately dismissing Bartleby?"

"Excuse me, that is for you to decide, sir. I think his conduct quite unusual, and indeed unjust, as regards Turkey and myself. But it may only be a passing whim."

"Ah," exclaimed I, "you have strangely changed your mind then—you speak very gently of him now."

"All beer," cried Turkey; "gentleness is effects of beer—Nippers and I dined together to-day. You see how gentle I am, sir. Shall I go and black his eyes?"

"You refer to Bartleby, I suppose. No, not to-day, Turkey," I replied; "pray, put up your fists."

I closed the doors, and again advanced towards Bartleby. I felt additional incentives tempting me to my fate. I burned to be rebelled against again. I remembered that Bartleby never left the office.

"Bartleby," said I, "Ginger Nut is away; just step round to the Post Office, won't you? (it was but a three minutes' walk,) and see if there is any thing for me."

"I would prefer not to."

"You *will* not?"

"I *prefer* not."

I staggered to my desk, and sat there in a deep study. My blind inveteracy returned. Was there any other thing in which I could procure myself to be ignominiously repulsed by this lean, penniless wight?—my hired clerk? What added thing is there, perfectly reasonable, that he will be sure to refuse to do?

"Bartleby!"

No answer.

"Bartleby," in a louder tone.

No answer.

"Bartleby," I roared.

Like a very ghost, agreeably to the laws of magical invocation, at the third summons, he appeared at the entrance of his hermitage.

"Go to the next room, and tell Nippers to come to me."

"I prefer not to," he respectfully and slowly said, and mildly disappeared.

"Very good, Bartleby," said I, in a quiet sort of serenely severe self-possessed tone, intimating the unalterable purpose of some terrible retribution very close at hand. At the moment I half intended something of the kind. But upon the whole, as it was drawing towards my dinner-hour, I thought it best to put on my hat and walk home for the day, suffering much from perplexity and distress of mind.

Shall I acknowledge it? The conclusion of this whole business was, that it soon became a fixed fact of my chambers, that a pale young scrivener, by the name of Bartleby, had a desk there; that he copied for me at the usual rate of four cents a folio (one hundred words); but he was permanently exempt from examining the work done by him, that duty being transferred to Turkey and Nippers, out of compliment doubtless to their superior acuteness; moreover, said Bartleby was never on any account to be dispatched on the most trivial errand of any sort; and that even if entreated to take upon him such a matter, it was generally understood that he would prefer not to—in other words, that he would refuse point-blank.

As days passed on, I became considerably reconciled to Bartleby. His steadiness, his freedom from all dissipation, his incessant industry (except when he chose to throw himself into a standing revery behind his screen), his great stillness, his unalterableness of demeanor under all circumstances, made him a valuable acquisition. One prime thing was this,—*he was always there*;—first in the morning, continually through the day, and the last at night. I had a singular confidence in his honesty. I felt my most precious papers perfectly safe in his hands. Sometimes to be sure I could not, for the very soul of me, avoid falling into sudden spasmodic passions with him. For it was exceeding difficult to bear in mind all the time those strange peculiarities, privileges, and unheard of exemptions, forming the tacit stipulations on Bartleby's part under which he remained in my office. Now and then, in the eagerness of dispatching pressing business, I would inadvertently summon Bartleby, in a short, rapid tone, to put his finger, say, on the incipient tie of a bit of red tape with which I was about compressing some papers. Of course, from behind the screen the usual answer, "I prefer not to," was sure to come; and then, how could a human creature with the common infirmities of our nature, refrain from bitterly exclaiming upon such perverseness—such unreasonableness. However, every added repulse of this sort which I received only tended to lessen the probability of my repeating the inadvertence.

Here it must be said, that according to the custom of most legal gentlemen occupying chambers in densely-populated law buildings, there were several keys to my door. One was kept by a woman residing in the attic, which person weekly scrubbed and daily swept and dusted my apartments. Another was kept by Turkey for convenience sake. The third I sometimes carried in my own pocket. The fourth I knew not who had.

Now, one Sunday morning I happened to go to Trinity Church, to hear a celebrated preacher, and finding myself rather early on the ground, I thought I would walk round to my chambers for a while. Luckily I had my key with me; but upon applying it to the lock, I found it resisted by something inserted from the inside. Quite surprised, I called out; when to my consternation a key was turned from within; and thrusting his lean visage at me, and holding the door ajar, the apparition of Bartleby appeared, in his shirt sleeves, and otherwise in a strangely tattered dishabille, saying quietly that he was sorry, but he was deeply engaged just then, and—preferred not admitting me at present. In a brief word or two, he moreover added, that perhaps I had better walk round the block two or three times, and by that time he would probably have concluded his affairs.

Now, the utterly unsurmised appearance of Bartleby, tenanting my law-chambers of a Sunday morning, with his cadaverously gentlemanly *nonchalance*, yet withal firm and self-possessed, had such a strange effect upon me, that incontinently I slunk away from my own door, and did as desired. But not without sundry twinges of impotent rebellion against the mild effrontery of this unaccountable scrivener. Indeed, it was his wonderful mildness chiefly, which not only disarmed me, but unmanned me, as it were. For I consider that one, for the time, is a sort of unmanned when he tranquilly permits his hired clerk to dictate to him, and order him away from his own premises. Furthermore, I was full of uneasiness as to what Bartleby could possibly be doing in my office in his shirt sleeves, and in an otherwise dismantled condition of a Sunday morning. Was any thing amiss going on? Nay, that was out of the question. It was not to be thought of for a moment that Bartleby was an immoral person. But what could he be doing there?—copying? Nay again, whatever might be his eccentricities, Bartleby was an eminently decorous person. He would be the last man to sit down to his desk in any state approaching to nudity. Besides, it was Sunday; and there was something about Bartleby that forbade the supposition that he would by any secular occupation violate the proprieties of the day.

Nevertheless, my mind was not pacified; and full of a restless curiosity, at last I returned to the door. Without hindrance I inserted my key, opened it, and entered. Bartleby was not to be seen. I looked round anxiously, peeped behind his screen; but it was very plain that he was gone. Upon more closely examining the place, I surmised that for an indefinite period Bartleby must have ate, dressed, and slept in my office, and that too without plate, mirror, or bed. The cushioned seat of a ricketty old sofa in one corner bore the faint impress of a lean, reclining form. Rolled away under his desk, I found a blanket; under the empty grate, a blacking box and brush; on a chair, a tin basin, with soap and a ragged towel; in a newspaper a few crumbs of ginger-nuts and a morsel of cheese. Yes, thought I, it is evident enough that Bartleby has been making his home here, keeping bachelor's hall all by himself. Immediately then the thought came sweeping across me, What miserable friendlessness and loneliness are here revealed! His poverty is great; but his solitude, how horrible! Think of it. Of a Sunday, Wall-street is deserted as Petra;[4] and every night of every day it is an emptiness. This building too, which of week-days hums with industry and life, at nightfall echoes with sheer vacancy, and all through Sunday is forlorn. And here Bartleby makes his home; sole spectator of a solitude which he has seen all populous—a sort of innocent and transformed Marius brooding among the ruins of Carthage![5]

For the first time in my life a feeling of overpowering stinging melancholy seized me. Before, I had never experienced aught but a not-unpleasing sadness. The bond of a common humanity now drew me irresistibly to gloom. A fraternal melancholy! For both I and Bartleby were sons of Adam. I remembered the bright silks and sparkling faces I had seen that day, in gala trim, swan-like sailing down the Mississippi of Broadway; and I contrasted them with the pallid copyist, and thought to myself, Ah, happiness courts the light, so we deem the world is gay; but misery hides aloof, so we deem that misery there is none. These sad fancyings—chimeras, doubtless, of a sick and silly brain—led on to other and more special thoughts, concerning the eccentricities of Bartleby. Presentiments of strange discoveries hovered round me. The scrivener's pale form appeared to me laid out, among uncaring strangers, in its shivering winding sheet.

Suddenly I was attracted by Bartleby's closed desk, the key in open sight left in the lock.

I mean no mischief, seek the gratification of no heartless curiosity, thought I; besides, the desk is mine, and its contents too, so I will make bold to look within. Every thing was methodically arranged, the papers smoothly placed. The pigeon holes were deep, and removing the files of documents, I groped into their recesses. Presently I felt something there, and dragged it out. It was an old bandanna handkerchief, heavy and knotted. I opened it, and saw it was a savings' bank.

I now recalled all the quiet mysteries which I had noted in the man. I remembered that he never spoke but to answer; that though at intervals he had considerable time to himself, yet I had never seen him reading—no, not even a newspaper; that for long periods he would stand looking out, at his pale window behind the screen, upon the dead brick wall; I was quite sure he never visited any refectory or eating house; while his pale face clearly indicated that

4. Ancient city whose ruins are in Jordan, on a slope of Mt. Hor.

5. Gaius Marius, Roman General (157–86 B.C.), who returned to power after exile.

he never drank beer like Turkey, or tea and coffee even, like other men; that he never went any where in particular that I could learn; never went out for a walk, unless indeed that was the case at present; that he had declined telling who he was, or whence he came, or whether he had any relatives in the world; that though so thin and pale, he never complained of ill health. And more than all, I remembered a certain unconscious air of pallid—how shall I call it?—of pallid haughtiness, say, or rather an austere reserve about him, which had positively awed me into my tame compliance with his eccentricities, when I had feared to ask him to do the slightest incidental thing for me, even though I might know, from his long-continued motionlessness, that behind his screen he must be standing in one of those dead-wall reveries of his.

Revolving all these things, and coupling them with the recently discovered fact that he made my office his constant abiding place and home, and not forgetful of his morbid moodiness; revolving all these things, a prudential feeling began to steal over me. My first emotions had been those of pure melancholy and sincerest pity; but just in proportion as the forlornness of Bartleby grew and grew to my imagination, did that same melancholy merge into fear, that pity into repulsion. So true it is, and so terrible too, that up to a certain point the thought or sight of misery enlists our best affections; but, in certain special cases, beyond that point it does not. They err who would assert that invariably this is owing to the inherent selfishness of the human heart. It rather proceeds from a certain hopelessness of remedying excessive and organic ill. To a sensitive being, pity is not seldom pain. And when at last it is perceived that such pity cannot lead to effectual succor, common sense bids the soul be rid of it. What I saw that morning persuaded me that the scrivener was the victim of innate and incurable disorder. I might give alms to his body; but his body did not pain him; it was his soul that suffered, and his soul I could not reach.

I did not accomplish the purpose of going to Trinity Church that morning. Somehow, the things I had seen disqualified me for the time from church-going. I walked homeward, thinking what I would do with Bartleby. Finally, I resolved upon this;—I would put certain calm questions to him the next morning, touching his history, &c., and if he declined to answer them openly and unreservedly (and I supposed he would prefer not), then to give him a twenty dollar bill over and above whatever I might owe him, and tell him his services were no longer required; but that if in any other way I could assist him, I would be happy to do so, especially if he desired to return to his native place, wherever that might be, I would willingly help to defray the expenses. Moreover, if, after reaching home, he found himself at any time in want of aid, a letter from him would be sure of a reply.

The next morning came.

"Bartleby," said I, gently calling to him behind his screen.

No reply.

"Bartleby," said I, in a still gentler tone, "come here; I am not going to ask you to do any thing you would prefer not to do—I simply wish to speak to you."

Upon this he noiselessly slid into view.

"Will you tell me, Bartleby, where you were born?"

"I would prefer not to."

"Will you tell me *any thing* about yourself?"

"I would prefer not to."

"But what reasonable objection can you have to speak to me? I feel friendly towards you."

He did not look at me while I spoke, but kept his glance fixed upon my bust of Cicero, which as I then sat, was directly behind me, some six inches above my head.

"What is your answer, Bartleby?" said I, after waiting a considerable time for a reply, during which his countenance remained immovable, only there was the faintest conceivable tremor of the white attenuated mouth.

"At present I prefer to give no answer," he said, and retired into his hermitage.

It was rather weak in me I confess, but his manner on this occasion nettled me. Not only did there seem to lurk in it a certain calm disdain, but his perverseness seemed ungrateful, considering the undeniable good usage and indulgence he had received from me.

Again I sat ruminating what I should do. Mortified as I was at his behavior, and resolved as I had been to dismiss him when I entered my office, nevertheless I strangely felt something superstitious knocking at my heart, and forbidding me to carry out my purpose, and denouncing me for a villain if I dared to breathe one bitter word against this forlornest of mankind. At last, familiarly drawing my chair behind his screen, I sat down and said: "Bartleby, never mind then about revealing your history; but let me entreat you, as a friend, to comply as far as may be with the usages of this office. Say now you will help to examine papers to-morrow or next day: in short, say now that in a day or two you will begin to be a little reasonable:—say so, Bartleby."

"At present I would prefer not to be a little reasonable," was his mildly cadaverous reply.

Just then the folding-doors opened, and Nippers approached. He seemed suffering from an unusually bad night's rest, induced by severer indigestion than common. He overheard those final words of Bartleby.

"*Prefer not*, eh?" gritted Nippers—"I'd *prefer* him, if I were you, sir," addressing me—"I'd *prefer* him; I'd give him preferences, the stubborn mule! What is it, sir, pray, that he *prefers* not to do now?"

Bartleby moved not a limb.

"Mr. Nippers," said I, "I'd prefer that you would withdraw for the present."

Somehow, of late I had got into the way of involuntarily using this word "prefer" upon all sorts of not exactly suitable occasions. And I trembled to think that my contact with the scrivener had already and seriously affected me in a mental way. And what further and deeper aberration might it not yet produce? This apprehension had not been without efficacy in determining me to summary means.

As Nippers, looking very sour and sulky, was departing, Turkey blandly and deferentially approached.

"With submission, sir," said he, "yesterday I was thinking about Bartleby here, and I think that if he would but prefer to take a quart of good ale every day, it would do much towards mending him, and enabling him to assist in examining his papers."

"So you have got the word too," said I, slightly excited.

"With submission, what word, sir," asked Turkey, respectfully crowding himself into the contracted space behind the screen, and by so doing, making me jostle the scrivener. "What word, sir?"

"I would prefer to be left alone here," said Bartleby, as if offended at being mobbed in his privacy.

"*That's* the word, Turkey," said I—"*that's* it."

"Oh, *prefer*? oh yes—queer word. I never use it myself. But, sir, as I was saying, if he would but prefer—"

"Turkey," interrupted I, "you will please withdraw."

"Oh certainly, sir, if you prefer that I should."

As he opened the folding-door to retire, Nippers at his desk caught a glimpse of me, and asked whether I would prefer to have a certain paper copied on blue paper or white. He did not in the least roguishly accent the word prefer. It was plain that it involuntarily rolled from his tongue. I thought to myself, surely I must get rid of a demented man, who already has in some degree turned the tongues, if not the heads of myself and clerks. But I thought it prudent not to break the dismission at once.

The next day I noticed that Bartleby did nothing but stand at his window in his dead-wall revery. Upon asking him why he did not write, he said that he had decided upon doing no more writing.

"Why, how now? what next?" exclaimed I, "do no more writing?"

"No more."

"And what is the reason?"

"Do you not see the reason for yourself," he indifferently replied.

I looked steadfastly at him, and perceived that his eyes looked dull and glazed. Instantly it occurred to me, that his unexampled diligence in copying by his dim window for the first few weeks of his stay with me might have temporarily impaired his vision.

I was touched. I said something in condolence with him. I hinted that of course he did wisely in abstaining from writing for a while;

and urged him to embrace that opportunity of taking wholesome exercise in the open air. This, however, he did not do. A few days after this, my other clerks being absent, and being in a great hurry to dispatch certain letters by the mail, I thought that, having nothing else earthly to do, Bartleby would surely be less inflexible than usual, and carry these letters to the post-office. But he blankly declined. So, much to my inconvenience, I went myself.

Still added days went by. Whether Bartleby's eyes improved or not, I could not say. To all appearance, I thought they did. But when I asked him if they did, he vouchsafed no answer. At all events, he would do no copying. At last, in reply to my urgings, he informed me that he had permanently given up copying.

"What!" exclaimed I; "suppose your eyes should get entirely well —better than ever before—would you not copy then?"

"I have given up copying," he answered, and slid aside.

He remained as ever, a fixture in my chamber. Nay—if that were possible—he became still more of a fixture than before. What was to be done? He would do nothing in the office: why should he stay there? In plain fact, he had now become a millstone to me, not only useless as a necklace, but afflictive to bear. Yet I was sorry for him. I speak less than truth when I say that, on his own account, he occasioned me uneasiness. If he would but have named a single relative or friend, I would instantly have written, and urged their taking the poor fellow away to some convenient retreat. But he seemed alone, absolutely alone in the universe. A bit of wreck in the mid Atlantic. At length, necessities connected with my business tyrannized over all other considerations. Decently as I could, I told Bartleby that in six days' time he must unconditionally leave the office. I warned him to take measures, in the interval, for procuring some other abode. I offered to assist him in this endeavor, if he himself would but take the first step towards a removal. "And when you finally quit me, Bartleby," added I, "I shall see that you go not away entirely unprovided. Six days from this hour, remember."

At the expiration of that period, I peeped behind the screen, and lo! Bartleby was there.

I buttoned up my coat, balanced myself; advanced slowly towards him, touched his shoulder, and said, "The time has come; you must quit this place; I am sorry for you; here is money; but you must go."

"I would prefer not," he replied, with his back still towards me.

"You *must*."

He remained silent.

Now I had an unbounded confidence in this man's common honesty. He had frequently restored to me sixpences and shillings carelessly dropped upon the floor, for I am apt to be very reckless in such shirt-button affairs. The proceeding then which followed will not be deemed extraordinary.

"Bartleby," said I, "I owe you twelve dollars on account; here are thirty-two; the odd twenty are yours.—Will you take it?" and I handed the bills towards him.

But he made no motion.

"I will leave them here then," putting them under a weight on the table. Then taking my hat and cane and going to the door I tranquilly turned and added—"After you have removed your things from these offices, Bartleby, you will of course lock the door—since every one is now gone for the day but you—and if you please, slip your key underneath the mat, so that I may have it in the morning. I shall not see you again; so good-bye to you. If hereafter in your new place of abode I can be of any service to you, do not fail to advise me by letter. Good-bye, Bartleby, and fare you well."

But he answered not a word; like the last column of some ruined temple, he remained standing mute and solitary in the middle of the otherwise deserted room.

As I walked home in a pensive mood, my vanity got the better of my pity. I could not but highly plume myself on my masterly management in getting rid of Bartleby. Masterly I call it, and such it must appear to any dispassionate thinker. The beauty of my procedure seemed to consist in its perfect quietness. There was no vulgar bullying, no bravado of any sort, no choleric hectoring, and striding to and fro across the apartment, jerking out vehement commands for Bartleby to bundle himself off with his beggarly traps. Nothing of the kind. Without loudly bidding Bartleby depart —as an inferior genius might have done—I *assumed* the ground that depart he must; and upon that assumption built all I had to say. The more I thought over my procedure, the more I was charmed with it. Nevertheless, next morning, upon awakening, I had my doubts,—I had somehow slept off the fumes of vanity. One of the coolest and wisest hours a man has, is just after he awakes in the morning. My procedure seemed as sagacious as ever,—but only in theory. How it would prove in practice—there was the rub. It was truly a beautiful thought to have assumed Bartleby's departure; but, after all, that assumption was simply my own, and none of Bartleby's. The great point was, not whether I had assumed that he would quit me, but whether he would prefer so to do. He was more a man of preferences than assumptions.

After breakfast, I walked down town, arguing the probabilities *pro* and *con*. One moment I thought it would prove a miserable failure, and Bartleby would be found all alive at my office as usual; the next moment it seemed certain that I should see his chair empty. And so I kept veering about. At the corner of Broadway and Canal-street, I saw quite an excited group of people standing in earnest conversation.

"I'll take odds he doesn't," said a voice as I passed.

"Doesn't go?—done!" said I, "put up your money."

I was instinctively putting my hand in my pocket to produce my own, when I remembered that this was an election day. The words I had overheard bore no reference to Bartleby, but to the success or non-success of some candidate for the mayoralty. In my intent frame of mind, I had, as it were, imagined that all Broadway shared in my excitement, and were debating the same question with me. I passed on, very thankful that the uproar of the street screened my momentary absent-mindedness.

As I had intended, I was earlier than usual at my office door. I stood listening for a moment. All was still. He must be gone. I tried the knob. The door was locked. Yes, my procedure had worked to a charm; he indeed must be vanished. Yet a certain melancholy mixed with this: I was almost sorry for my brilliant success. I was fumbling under the door mat for the key, which Bartleby was to have left there for me, when accidentally my knee knocked against a panel, producing a summoning sound, and in response a voice came to me from within—"Not yet; I am occupied."

It was Bartleby.

I was thunderstruck. For an instant I stood like the man who, pipe in mouth, was killed one cloudless afternoon long ago in Virginia, by summer lightning; at his own warm open window he was killed, and remained leaning out there upon the dreamy afternoon, till some one touched him, when he fell.

"Not gone!" I murmured at last. But again obeying that wondrous ascendancy which the inscrutable scrivener had over me, and from which ascendancy, for all my chafing, I could not completely escape, I slowly went down stairs and out into the street, and while walking round the block, considered what I should next do in this unheard-of perplexity. Turn the man out by an actual thrusting I could not; to drive him away by calling him hard names would not do; calling in the police was an unpleasant idea; and yet, permit him to enjoy his cadaverous triumph over me,—this too I could not think of. What was to be done? or, if nothing could be done, was there any thing further that I could *assume* in the matter? Yes, as before I had prospectively assumed that Bartleby would depart, so now I might retrospectively assume that departed he was. In the legitimate carrying out of this assumption, I might enter my office in a great hurry, and pretending not to see Bartleby at all, walk straight against him as if he were air. Such a proceeding would in a singular degree have the appearance of a home-thrust. It was hardly possible that Bartleby could withstand such an application of the doctrine of assumptions. But upon second thoughts the success of the plan seemed rather dubious. I resolved to argue the matter over with him again.

"Bartleby," said I, entering the office, with a quietly severe expression, "I am seriously displeased. I am pained, Bartleby. I had thought better of you. I had imagined you of such a gentlemanly organization, that in any delicate dilemma a slight hint would suffice—in short, an assumption. But it appears I am deceived. Why," I added, unaffectedly starting, "you have not even touched that money yet," pointing to it, just where I had left it the evening previous.

He answered nothing.

"Will you, or will you not, quit me?" I now demanded in a sudden passion, advancing close to him.

"I would prefer *not* to quit you," he replied, gently emphasizing the *not*.

"What earthly right have you to stay here? Do you pay any rent? Do you pay my taxes? Or is this property yours?"

He answered nothing.

"Are you ready to go on and write now? Are your eyes recovered? Could you copy a small paper for me this morning? or help examine a few lines? or step round to the post-office? In a word, will you do any thing at all, to give a coloring to your refusal to depart the premises?"

He silently retired into his hermitage.

I was now in such a state of nervous resentment that I thought it but prudent to check myself at present from further demonstrations. Bartleby and I were alone. I remembered the tragedy of the unfortunate Adams and the still more unfortunate Colt in the solitary office of the latter; and how poor Colt, being dreadfully incensed by Adams,[6] and imprudently permitting himself to get wildly excited, was at unawares hurried into his fatal act—an act which certainly no man could possibly deplore more than the actor himself. Often it had occurred to me in my ponderings upon the subject, that had that altercation taken place in the public street, or at a private residence, it would not have terminated as it did. It was the circumstance of being alone in a solitary office, up stairs, of a building entirely unhallowed by humanizing domestic associations—an uncarpeted office, doubtless, of a dusty, haggard sort of appearance; —this it must have been, which greatly helped to enhance the irritable desperation of the hapless Colt.

But when this old Adam of resentment rose in me and tempted me concerning Bartleby, I grappled him and threw him. How? Why, simply by recalling the divine injunction: "A new commandment give I unto you, that ye love one another." Yes, this it was that saved me. Aside from higher considerations, charity often operates as a vastly wise and prudent principle—a great safeguard to its

6. Notorious murder case which occurred while Melville was in the South Seas. In 1841 Samuel Adams, a printer, called upon John C. Colt (brother of the inventor of the revolver) at Broadway and Chambers Street in lower Manhattan in order to collect a debt. Colt murdered Adams with a hatchet and crated the corpse for shipment to New Orleans. The body was found, and Colt was soon arrested. Despite his pleas of self-defense Colt was convicted the next year, amid continuing newspaper publicity, and stabbed himself to death just before he was to be hanged. The setting of *Bartleby* is not far from the scene of the murder.

possessor. Men have committed murder for jealousy's sake, and anger's sake, and hatred's sake, and selfishness' sake, and spiritual pride's sake; but no man that ever I heard of, ever committed a diabolical murder for sweet charity's sake. Mere self-interest, then, if no better motive can be enlisted, should, especially with high-tempered men, prompt all beings to charity and philanthropy. At any rate, upon the occasion in question, I strove to drown my exasperated feelings towards the scrivener by benevolently construing his conduct. Poor fellow, poor fellow! thought I, he don't mean any thing; and besides, he has seen hard times, and ought to be indulged.

I endeavored also immediately to occupy myself, and at the same time to comfort my despondency. I tried to fancy that in the course of the morning, at such time as might prove agreeable to him, Bartleby, of his own free accord, would emerge from his hermitage, and take up some decided line of march in the direction of the door. But no. Half-past twelve o'clock came; Turkey began to glow in the face, overturn his inkstand, and become generally obstreperous; Nippers abated down into quietude and courtesy; Ginger Nut munched his noon apple; and Bartleby remained standing at his window in one of his profoundest dead-wall reveries. Will it be credited? Ought I to acknowledge it? That afternoon I left the office without saying one further word to him.

Some days now passed, during which, at leisure intervals I looked a little into "Edwards on the Will," and "Priestley on Necessity."[7] Under the circumstances, those books induced a salutary feeling. Gradually I slid into the persuasion that these troubles of mine touching the scrivener, had been all predestinated from eternity, and Bartleby was billeted upon me for some mysterious purpose of an all-wise Providence, which it was not for a mere mortal like me to fathom. Yes, Bartleby, stay there behind your screen, thought I; I shall persecute you no more; you are harmless and noiseless as any of these old chairs; in short, I never feel so private as when I know you are here. At least I see it, I feel it; I penetrate to the predestinated purpose of my life. I am content. Others may have loftier parts to enact; but my mission in this world, Bartleby, is to furnish you with office-room for such period as you may see fit to remain.

I believe that this wise and blessed frame of mind would have continued with me, had it not been for the unsolicited and uncharitable remarks obtruded upon me by my professional friends who visited the rooms. But thus it often is, that the constant friction of illiberal minds wears out at last the best resolves of the more generous. Though to be sure, when I reflected upon it, it was not strange that people entering my office should be struck by the peculiar aspect of the unaccountable Bartleby, and so be tempted to

7. Jonathan Edwards's *Freedom of the Will* (1754) and Joseph Priestley's *Doctrine of Philosophical Necessity Illustrated* (1777). The colonial minister and the English scientist agree that the will is not free.

throw out some sinister observations concerning him. Sometimes an attorney having business with me, and calling at my office, and finding no one but the scrivener there, would undertake to obtain some sort of precise information from him touching my where-abouts; but without heeding his idle talk, Bartleby would remain standing immovable in the middle of the room. So after contemplating him in that position for a time, the attorney would depart, no wiser than he came.

Also, when a Reference[8] was going on, and the room full of lawyers and witnesses and business was driving fast; some deeply occupied legal gentleman present, seeing Bartleby wholly unem-ployed, would request him to run round to his (the legal gentle-man's) office and fetch some papers for him. Thereupon, Bartleby would tranquilly decline, and yet remain idle as before. Then the lawyer would give a great stare, and turn to me. And what could I say? At last I was made aware that all through the circle of my professional acquaintance, a whisper of wonder was running round, having reference to the strange creature I kept at my office. This worried me very much. And as the idea came upon me of his possibly turning out a long-lived man, and keep occupying my chambers, and denying my authority; and perplexing my visitors; and scandalizing my professional reputation; and casting a general gloom over the premises; keeping soul and body together to the last upon his savings (for doubtless he spent but half a dime a day), and in the end perhaps outlive me, and claim possession of my office by right of his perpetual occupancy: as all these dark anticipations crowded upon me more and more, and my friends continually in-truded their relentless remarks upon the apparition in my room; a great change was wrought in me. I resolved to gather all my facul-ties together, and for ever rid me of this intolerable incubus.

Ere revolving any complicated project, however, adapted to this end, I first simply suggested to Bartleby the propriety of his perma-nent departure. In a calm and serious tone, I commended the idea to his careful and mature consideration. But having taken three days to meditate upon it, he apprised me that his original determina-tion remained the same; in short, that he still preferred to abide with me.

What shall I do? I now said to myself, buttoning up my coat to the last button. What shall I do? what ought I to do? what does conscience say I *should* do with this man, or rather ghost. Rid myself of him, I must; go, he shall. But how? You will not thrust him, the poor, pale, passive mortal,—you will not thrust such a helpless creature out of your door? you will not dishonor yourself by such cruelty? No, I will not, I cannot do that. Rather would I let him live and die here, and then mason up his remains in the wall. What then will you do? For all your coaxing, he will not budge.

8. The act of referring a disputed matter to referees.

Bribes he leaves under your own paper-weight on your table; in short, it is quite plain that he prefers to cling to you.

Then something severe, something unusual must be done. What! surely you will not have him collared by a constable, and commit his innocent pallor to the common jail? And upon what ground could you procure such a thing to be done?—a vagrant, is he? What! he a vagrant, a wanderer, who refuses to budge? It is because he will *not* be a vagrant, then, that you seek to count him *as* a vagrant. That is too absurd. No visible means of support: there I have him. Wrong again: for indubitably he *does* support himself, and that is the only unanswerable proof that any man can show of his possessing the means so to do. No more then. Since he will not quit me, I must quit him. I will change my offices; I will move elsewhere; and give him fair notice, that if I find him on my new premises I will then proceed against him as a common trespasser.

Acting accordingly, next day I thus addressed him: "I find these chambers too far from the City Hall; the air is unwholesome. In a word, I propose to remove my offices next week, and shall no longer require your services. I tell you this now, in order that you may seek another place."

He made no reply, and nothing more was said.

On the appointed day I engaged carts and men, proceeded to my chambers, and having but little furniture, every thing was removed in a few hours. Throughout, the scrivener remained standing behind the screen, which I directed to be removed the last thing. It was withdrawn; and being folded up like a huge folio, left him the motionless occupant of a naked room. I stood in the entry watching him a moment, while something from within me upbraided me.

I re-entered, with my hand in my pocket—and—and my heart in my mouth.

"Good-bye, Bartleby; I am going—good-bye, and God some way bless you; and take that," slipping something in his hand. But it dropped upon the floor, and then,—strange to say—I tore myself from him whom I had so longed to be rid of.

Established in my new quarters, for a day or two I kept the door locked, and started at every footfall in the passages. When I returned to my rooms after any little absence, I would pause at the threshold for an instant, and attentively listen, ere applying my key. But these fears were needless. Bartleby never came nigh me.

I thought all was going well, when a perturbed looking stranger visited me, inquiring whether I was the person who had recently occupied rooms at No.—Wall-street.

Full of forebodings, I replied that I was.

"Then sir," said the stranger, who proved a lawyer, "you are responsible for the man you left there. He refuses to do any copying; he refuses to do any thing; he says he prefers not to; and he refuses to quit the premises."

"I am very sorry, sir," said I, with assumed tranquillity, but an inward tremor, "but, really, the man you allude to is nothing to me—he is no relation or apprentice of mine, that you should hold me responsible for him."

"In mercy's name, who is he?"

"I certainly cannot inform you. I know nothing about him. Formerly I employed him as a copyist; but he has done nothing for me now for some time past."

"I shall settle him then,—good morning, sir."

Several days passed, and I heard nothing more; and though I often felt a charitable prompting to call at the place and see poor Bartleby, yet a certain squeamishness of I know not what withheld me.

All is over with him, by this time, thought I at last, when through another week no further intelligence reached me. But coming to my room the day after, I found several persons waiting at my door in a high state of nervous excitement.

"That's the man—here he comes," cried the foremost one, whom I recognized as the lawyer who had previously called upon me alone.

"You must take him away, sir, at once," cried a portly person among them, advancing upon me, and whom I knew to be the landlord of No.—Wall-street. "These gentlemen, my tenants, cannot stand it any longer; Mr. B——" pointing to the lawyer, "has turned him out of his room, and he now persists in haunting the building generally, sitting upon the banisters of the stairs by day, and sleeping in the entry by night. Every body is concerned; clients are leaving the offices; some fears are entertained of a mob; something you must do, and that without delay."

Aghast at this torrent, I fell back before it, and would fain have locked myself in my new quarters. In vain I persisted that Bartleby was nothing to me—no more than to any one else. In vain:—I was the last person known to have any thing to do with him, and they held me to the terrible account. Fearful then of being exposed in the papers (as one person present obscurely threatened) I considered the matter, and at length said, that if the lawyer would give me a confidential interview with the scrivener, in his (the lawyer's) own room, I would that afternoon strive my best to rid them of the nuisance they complained of.

Going up stairs to my old haunt, there was Bartleby silently sitting upon the banister at the landing.

"What are you doing here, Bartleby?" said I.

"Sitting upon the banister," he mildly replied.

I motioned him into the lawyer's room, who then left us.

"Bartleby," said I, "are you aware that you are the cause of great tribulation to me, by persisting in occupying the entry after being dismissed from the office?"

No answer.

"Now one of two things must take place. Either you must do something, or something must be done to you. Now what sort of business would you like to engage in? Would you like to re-engage in copying for some one?"

"No; I would prefer not to make any change."

"Would you like a clerkship in a dry-goods store?"

"There is too much confinement about that. No, I would not like a clerkship; but I am not particular."

"Too much confinement," I cried, "why you keep yourself confined all the time!"

"I would prefer not to take a clerkship," he rejoined, as if to settle that little item at once.

"How would a bar-tender's business suit you? There is no trying of the eyesight in that."

"I would not like it at all; though, as I said before, I am not particular."

His unwonted wordiness inspirited me. I returned to the charge.

"Well then, would you like to travel through the country collecting bills for the merchants? That would improve your health."

"No, I would prefer to be doing something else."

"How then would going as a companion to Europe, to entertain some young gentleman with your conversation,—how would that suit you?"

"Not at all. It does not strike me that there is any thing definite about that. I like to be stationary. But I am not particular."

"Stationary you shall be then," I cried, now losing all patience, and for the first time in all my exasperating connection with him fairly flying into a passion. "If you do not go away from these premises before night, I shall feel bound—indeed I *am* bound—to—to—to quit the premises myself!" I rather absurdly concluded, knowing not with what possible threat to try to frighten his immobility into compliance. Despairing of all further efforts, I was precipitately leaving him, when a final thought occurred to me—one which had not been wholly unindulged before.

"Bartleby," said I, in the kindest tone I could assume under such exciting circumstances, "will you go home with me now—not to my office, but my dwelling—and remain there till we can conclude upon some convenient arrangement for you at our leisure? Come, let us start now, right away."

"No: at present I would prefer not to make any change at all."

I answered nothing; but effectually dodging every one by the suddenness and rapidity of my flight, rushed from the building, ran up Wall-street towards Broadway, and jumping into the first omnibus was soon removed from pursuit. As soon as tranquillity returned I distinctly perceived that I had now done all that I possibly could, both in respect to the demands of the landlord and his tenants, and with regard to my own desire and sense of duty, to benefit Bartleby, and shield him from rude persecution. I now strove to be entirely care-free and quiescent; and my conscience justified me in the attempt; though indeed it was not so successful as I could have wished. So fearful was I of being again hunted out by the incensed landlord and his exasperated tenants, that, surrendering my business to Nippers, for a few days I drove about the upper part of the town and through the suburbs, in my rockaway;[9] crossed over to Jersey City and Hoboken, and paid fugitive visits to Manhattanville and Astoria. In fact I almost lived in my rockaway for the time.

When again I entered my office, lo, a note from the landlord lay upon the desk. I opened it with trembling hands. It informed me that the writer had sent to the police, and had Bartleby removed to the Tombs as a vagrant. Moreover, since I knew more about him than any one else, he wished me to appear at that place, and make a suitable statement of the facts. These tidings had a conflicting effect upon me. At first I was indignant; but at last almost approved. The landlord's energetic, summary disposition, had led him to adopt a procedure which I do not think I would have decided upon myself; and yet as a last resort, under such peculiar circumstances, it seemed the only plan.

As I afterwards learned, the poor scrivener, when told that he must be conducted to the Tombs, offered not the slightest obstacle, but in his pale unmoving way, silently acquiesced.

Some of the compassionate and curious bystanders joined the party; and headed by one of the constables arm in arm with Bartleby, the silent procession filed its way through all the noise, and heat, and joy of the roaring thoroughfares at noon.

The same day I received the note I went to the Tombs, or to speak more properly, the Halls of Justice. Seeking the right officer, I stated the purpose of my call, and was informed that the individual I described was indeed within. I then assured the functionary that Bartleby was a perfectly honest man, and greatly to be compassionated, however unaccountably eccentric. I narrated all I knew, and closed by suggesting the idea of letting him remain in as indulgent confinement as possible till something less harsh might be done—though indeed I hardly knew what. At all events, if nothing else could be decided upon, the alms-house must receive him. I then begged to have an interview.

Being under no disgraceful charge, and quite serene and harmless in all his ways, they had permitted him freely to wander about the prison, and especially in the inclosed grass-platted yards thereof. And so I found him there, standing all alone in the quietest of the yards,

---

9. Light open-sided carriage. The narrator crossed the Hudson River to Jersey City and Hoboken, then drove far up unsettled Manhattan Island to the community of Manhattanville (Grant's Tomb is in what was Manhattanville), and finally crossed the East River to Astoria, on Long Island.

his face towards a high wall, while all around, from the narrow slits of the jail windows, I thought I saw peering out upon him the eyes of murderers and thieves.

"Bartleby!"

"I know you," he said, without looking round,—"and I want nothing to say to you."

"It was not I that brought you here, Bartleby," said I, keenly pained at his implied suspicion. "And to you, this should not be so vile a place. Nothing reproachful attaches to you by being here. And see, it is not so sad a place as one might think. Look, there is the sky, and here is the grass."

"I know where I am," he replied, but would say nothing more, and so I left him.

As I entered the corridor again, a broad meat-like man, in an apron, accosted me, and jerking his thumb over his shoulder said— "Is that your friend?"

"Yes."

"Does he want to starve? If he does, let him live on the prison fare, that's all."

"Who are you?" asked I, not knowing what to make of such an unofficially speaking person in such a place.

"I am the grub-man. Such gentlemen as have friends here, hire me to provide them with something good to eat."

"Is this so?" said I, turning to the turnkey.

He said it was.

"Well then," said I, slipping some silver into the grub-man's hands (for so they called him). "I want you to give particular attention to my friend there; let him have the best dinner you can get. And you must be as polite to him as possible."

"Introduce me, will you?" said the grub-man, looking at me with an expression which seemed to say he was all impatience for an opportunity to give a specimen of his breeding.

Thinking it would prove of benefit to the scrivener, I acquiesced; and asking the grub-man his name, went up with him to Bartleby.

"Bartleby, this is Mr. Cutlets; you will find him very useful to you."

"Your sarvant, sir, your sarvant," said the grub-man, making a low salutation behind his apron. "Hope you find it pleasant here, sir;—spacious grounds—cool apartments, sir—hope you'll stay with us some time—try to make it agreeable. May Mrs. Cutlets and I have the pleasure of your company to dinner, sir, in Mrs. Cutlets' private room?"

"I prefer not to dine to-day," said Bartleby, turning away. "It would disagree with me; I am unused to dinners." So saying he slowly moved to the other side of the inclosure, and took up a position fronting the dead-wall.

"How's this?" said the grub-man, addressing me with a stare of astonishment. "He's odd, aint he?"

"I think he is a little deranged," said I, sadly.

"Deranged? deranged is it? Well now, upon my word, I thought that friend of yourn was a gentleman forger; they are always pale and genteel-like, them forgers. I can't help pity 'em—can't help it, sir. Did you know Monroe Edwards?"[1] he added touchingly, and paused. Then, laying his hand pityingly on my shoulder, sighed, "he died of consumption at Sing-Sing.[2] So you weren't acquainted with Monroe?"

"No, I was never socially acquainted with any forgers. But I cannot stop longer. Look to my friend yonder. You will not lose by it. I will see you again."

Some few days after this, I again obtained admission to the Tombs, and went through the corridors in quest of Bartleby; but without finding him.

"I saw him coming from his cell not long ago," said a turnkey, "may be he's gone to loiter in the yards."

So I went in that direction.

"Are you looking for the silent man?" said another turnkey passing me. "Yonder he lies—sleeping in the yard there. 'Tis not twenty minutes since I saw him lie down."

The yard was entirely quiet. It was not accessible to the common prisoners. The surrounding walls, of amazing thickness, kept off all sounds behind them. The Egyptian character of the masonry weighed upon me with its gloom. But a soft imprisoned turf grew under foot. The heart of the eternal pyramids, it seemed, wherein, by some strange magic, through the clefts, grass-seed, dropped by birds, had sprung.

Strangely huddled at the base of the wall, his knees drawn up, and lying on his side, his head touching the cold stones, I saw the wasted Bartleby. But nothing stirred. I paused; then went close up to him; stooped over, and saw that his dim eyes were open; otherwise he seemed profoundly sleeping. Something prompted me to touch him. I felt his hand, when a tingling shiver ran up my arm and down my spine to my feet.

The round face of the grub-man peered upon me now. "His dinner is ready. Won't he dine to-day, either? Or does he live without dining?"

"Lives without dining," said I, and closed the eyes.

"Eh!—He's asleep, aint he?"

"With kings and counsellors,"[3] murmured I.

---

1. Horace Greeley's *Tribune* called Col. Monroe Edwards (1808–47) "the most distinguished financier since the days of Judas Iscariot"; his trial in New York City (lasting all the second week of June, 1842), caused the greatest public excitement since the trial "of the murderer, Colt." (See note 6, above.) He was convicted of swindling two firms of $25,000 each through forged letters of credit, sending tremors through the "exchange banking and commission business"—like undermining our Security Exchange. Melville was then in the South Seas, but the case was sensational, and his brothers were in New York.

2. Prison at Ossining, New York, not far up the Hudson.

3. Job 3.14.

There would seem little need for proceeding further in this history. Imagination will readily supply the meagre recital of poor Bartleby's interment. But ere parting with the reader, let me say, that if this little narrative has sufficiently interested him, to awaken curiosity as to who Bartleby was, and what manner of life he led prior to the present narrator's making his acquaintance, I can only reply, that in such curiosity I fully share, but am wholly unable to gratify it. Yet here I hardly know whether I should divulge one little item of rumor, which came to my ear a few months after the scrivener's decease. Upon what basis it rested, I could never ascertain; and hence, how true it is I cannot now tell. But inasmuch as this vague report has not been without a certain strange suggestive interest to me, however said, it may prove the same with some others; and so I will briefly mention it. The report was this: that Bartleby had been a subordinate clerk in the Dead Letter Office at Washington, from which he had been suddenly removed by a change in the administration. When I think over this rumor, I cannot adequately express the emotions which seize me. Dead letters! does it not sound like dead men? Conceive a man by nature and misfortune prone to a pallid hopelessness, can any business seem more fitted to heighten it than that of continually handling these dead letters, and assorting them for the flames? For by the cart-load they are annually burned. Sometimes from out the folded paper the pale clerk takes a ring:——the finger it was meant for, perhaps, moulders in the grave; a bank-note sent in swiftest charity: ——he whom it would relieve, nor eats nor hungers any more; pardon for those who died despairing; hope for those who died unhoping; good tidings for those who died stifled by unrelieved calamities. On errands of life, these letters speed to death.

Ah Bartleby! Ah humanity!

1853

Benito Cereno[1]

In the year 1799, Captain Amasa Delano, of Duxbury, in Massachusetts, commanding a large sealer and general trader, lay at anchor, with a valuable cargo, in the harbor of St. Maria—a small, desert, uninhabited island toward the southern extremity of the long coast of Chili. There he had touched for water.

On the second day, not long after dawn, while lying in his berth, his mate came below, informing him that a strange sail was coming into the bay. Ships were then not so plenty in those waters as now. He rose, dressed, and went on deck.

1. This text is based on the first printing, in *Putnam's Monthly* for October, November, and December, 1855, but it also incorporates the many small revisions Melville made for its 1856 republication in *The Piazza Tales*.

Melville based his plot very closely upon a few narrative pages in Chapter 18 of Captain Amasa Delano's *Narra-* tive of Voyages and Travels in the Northern and Southern Hemispheres (Boston, 1817). Melville's "deposition" is roughly half from Delano's much longer section of documents, half his own writing. (A later relative of the real Amasa Delano was Franklin Delano Roosevelt.)

The morning was one peculiar to that coast. Everything was mute and calm; everything gray. The sea, though undulated into long roods of swells, seemed fixed, and was sleeked at the surface like waved lead that has cooled and set in the smelter's mold. The sky seemed a gray surtout.[2] Flights of troubled gray fowl, kith and kin with flights of troubled gray vapors among which they were mixed, skimmed low and fitfully over the waters, as swallows over meadows before storms. Shadows present, foreshadowing deeper shadows to come.

To Captain Delano's surprise, the stranger, viewed through the glass, showed no colors; though to do so upon entering a haven, however uninhabited in its shores, where but a single other ship might be lying, was the custom among peaceful seamen of all nations. Considering the lawlessness and loneliness of the spot, and the sort of stories, at that day, associated with those seas, Captain Delano's surprise might have deepened into some uneasiness had he not been a person of a singularly undistrustful good nature, not liable, except on extraordinary and repeated incentives, and hardly then, to indulge in personal alarms, any way involving the imputation of malign evil in man. Whether, in view of what humanity is capable, such a trait implies, along with a benevolent heart, more than ordinary quickness and accuracy of intellectual perception, may be left to the wise to determine.

But whatever misgivings might have obtruded on first seeing the stranger, would almost, in any seaman's mind, have been dissipated by observing that, the ship, in navigating into the harbor, was drawing too near the land; a sunken reef making out off her bow. This seemed to prove her a stranger, indeed, not only to the sealer, but the island; consequently, she could be no wonted freebooter on that ocean. With no small interest, Captain Delano continued to watch her—a proceeding not much facilitated by the vapors partly mantling the hull, through which the far matin[3] light from her cabin streamed equivocally enough; much like the sun—by this time hemisphered on the rim of the horizon, and apparently, in company with the strange ship, entering the harbor—which, wimpled by the same low, creeping clouds, showed not unlike a Lima intriguante's one sinister eye peering across the Plaza from the Indian loop-hole of her dusk *saya-y-manta*.[4]

It might have been but a deception of the vapors, but, the longer the stranger was watched, the more singular appeared her maneuvers. Ere long it seemed hard to decide whether she meant to come in or no—what she wanted, or what she was about. The wind, which had breezed up a little during the night, was now extremely light and baffling, which the more increased the apparent uncertainty of her movements.

2. Long overcoat.
3. Early morning.
4. Skirt-and-mantle combination, the shawl part of which could be drawn about the face so little more than an eye would show; apt garb for assignations.

Surmising, at last, that it might be a ship in distress, Captain Delano ordered his whale-boat to be dropped, and, much to the wary opposition of his mate, prepared to board her, and, at the least, pilot her in. On the night previous, a fishing-party of the seamen had gone a long distance to some detached rocks out of sight from the sealer, and, an hour or two before day-break, had returned, having met with no small success. Presuming that the stranger might have been long off soundings, the good captain put several baskets of the fish, for presents, into his boat, and so pulled away. From her continuing too near the sunken reef, deeming her in danger, calling to his men, he made all haste to apprise those on board of their situation. But, some time ere the boat came up, the wind, light though it was, having shifted, had headed the vessel off, as well as partly broken the vapors from about her.

Upon gaining a less remote view, the ship, when made signally visible on the verge of the leaden-hued swells, with the shreds of fog here and there raggedly furring her, appeared like a white-washed monastery after a thunder-storm, seen perched upon some dun cliff among the Pyrenees. But it was no purely fanciful resemblance which now, for a moment, almost led Captain Delano to think that nothing less than a ship-load of monks was before him. Peering over the bulwarks were what really seemed, in the hazy distance, throngs of dark cowls; while, fitfully revealed through the open port-holes, other dark moving figures were dimly descried, as of Black Friars[5] pacing the cloisters.

Upon a still nigher approach, this appearance was modified, and the true character of the vessel was plain—a Spanish merchantman of the first class; carrying negro slaves, amongst other valuable freight, from one colonial port to another. A very large, and, in its time, a very fine vessel, such as in those days were at intervals encountered along that main; sometimes superseded Acapulco treasure-ships, or retired frigates of the Spanish king's navy, which, like superannuated Italian palaces, still, under a decline of masters, preserved signs of former state.

As the whale-boat drew more and more nigh, the cause of the peculiar pipe-clayed[6] aspect of the stranger was seen in the slovenly neglect pervading her. The spars, ropes, and great part of the bulwarks, looked woolly, from long unacquaintance with the scraper, tar, and the brush. Her keel seemed laid, her ribs put together, and she launched, from Ezekiel's Valley of Dry Bones.[7]

In the present business in which she was engaged, the ship's general model and rig appeared to have undergone no material change from their original war-like and Froissart pattern.[8] However, no guns were seen.

5. Dominicans, an order of mendicant preaching friars.
6. Whitened.
7. Ezekiel 37.1.
8. Medieval, from Jean Froissart (1337–1410), historian of wars of England and France.

The tops were large, and were railed about with what had once been octagonal net-work, all now in sad disrepair. These tops hung overhead like three ruinous aviaries, in one of which was seen perched, on a ratlin,[9] a white noddy, a strange fowl, so called from its lethargic, somnambulistic character, being frequently caught by hand at sea. Battered and mouldy, the castellated forecastle seemed some ancient turret, long ago taken by assault, and then left to decay. Toward the stern, two high-raised quarter galleries—the balustrades here and there covered with dry, tindery sea-moss—opening out from the unoccupied state-cabin, whose dead lights, for all the mild weather, were hermetically closed and calked—these tenantless balconies hung over the sea as if it were the grand Venetian canal. But the principal relic of faded grandeur was the ample oval of the shield-like stern-piece, intricately carved with the arms of Castile and Leon,[1] medallioned about by groups of mythological or symbolical devices; uppermost and central of which was a dark satyr in a mask, holding his foot on the prostrate neck of a writhing figure, likewise masked.

Whether the ship had a figure-head, or only a plain beak, was not quite certain, owing to canvas wrapped about that part, either to protect it while undergoing a re-furbishing, or else decently to hide its decay. Rudely painted or chalked, as in a sailor freak, along the forward side of a sort of pedestal below the canvas, was the sentence, "*Seguid vuestro jefe,*" (follow your leader); while upon the tarnished head-boards, near by, appeared, in stately capitals, once gilt, the ship's name, "SAN DOMINICK," each letter streakingly corroded with tricklings of copper-spike rust; while, like mourning weeds, dark festoons of sea-grass slimily swept to and fro over the name, with every hearse-like roll of the hull.

As at last the boat was hooked from the bow along toward the gangway amidship, its keel, while yet some inches separated from the hull, harshly grated as on a sunken coral reef. It proved a huge bunch of conglobated barnacles adhering below the water to the side like a wen; a token of baffling airs and long calms passed somewhere in those seas.

Climbing the side, the visitor was at once surrounded by a clamorous throng of whites and blacks, but the latter outnumbering the former more than could have been expected, negro transportation-ship as the stranger in port was. But, in one language, and as with one voice, all poured out a common tale of suffering; in which the negresses, of whom there were not a few, exceeded the others in their dolorous vehemence. The scurvy, together with a fever, had swept off a great part of their number, more especially the Spaniards. Off Cape Horn, they had narrowly escaped ship-

9. Small transverse rope attached to the shrouds and forming a step of a rope ladder; "noddy": a tame, stupid-seeming tern.

1. Old kingdoms of Spain; the arms would include a castle for Castile and a lion for León.

wreck; then, for days together, they had lain tranced without wind; their provisions were low; their water next to none; their lips that moment were baked.

While Captain Delano was thus made the mark of all eager tongues, his one eager glance took in all the faces, with every other object about him.

Always upon first boarding a large and populous ship at sea, especially a foreign one, with a nondescript crew such as Lascars or Manilla men,[2] the impression varies in a peculiar way from that produced by first entering a strange house with strange inmates in a strange land. Both house and ship, the one by its walls and blinds, the other by its high bulwarks like ramparts, hoard from view their interiors till the last moment; but in the case of the ship there is this addition; that the living spectacle it contains, upon its sudden and complete disclosure, has, in contrast with the blank ocean which zones it, something of the effect of enchantment. The ship seems unreal; these strange costumes, gestures, and faces, but a shadowy tableau just emerged from the deep, which directly must receive back what it gave.

Perhaps it was some such influence as above is attempted to be described, which, in Captain Delano's mind, heightened whatever, upon a staid scrutiny, might have seemed unusual; especially the conspicuous figures of four elderly grizzled negroes, their heads like black, doddered willow tops, who, in venerable contrast to the tumult below them, were couched sphynx-like, one on the starboard cat-head,[3] another on the larboard, and the remaining pair face to face on the opposite bulwarks above the main-chains. They each had bits of unstranded old junk[4] in their hands, and, with a sort of stoical self-content, were picking the junk into oakum,[5] a small heap of which lay by their sides. They accompanied the task with a continuous, low, monotonous chant; droning and druling[6] away like so many gray-headed bag-pipers playing a funeral march.

The quarter-deck rose into an ample elevated poop, upon the forward verge of which, lifted, like the oakum-pickers, some eight feet above the general throng, sat along in a row, separated by regular spaces, the cross-legged figures of six other blacks; each with a rusty hatchet in his hand, which, with a bit of brick and a rag, he was engaged like a scullion in scouring; while between each two was a small stack of hatchets, their rusted edges turned forward awaiting a like operation. Though occasionally the four oakum-pickers would briefly address some person or persons in the crowd below, yet the six hatchet-polishers neither spoke to others, nor breathed a whisper among themselves, but sat intent upon their task, except at intervals, when, with the peculiar love in negroes of uniting industry

2. From East India or the Philippines, respectively.
3. Projecting piece of timber near the bow (to which the anchor is hoisted and secured).

4. Worn-out rope.
5. Loose fiber from pieces of rope, used for calking.
6. Driveling.

with pastime, two and two they sideways clashed their hatchets together, like cymbals, with a barbarous din. All six, unlike the generality, had the raw aspect of unsophisticated Africans.

But that first comprehensive glance which took in those ten figures, with scores less conspicuous, rested but an instant upon them, as, impatient of the hubbub of voices, the visitor turned in quest of whomsoever it might be that commanded the ship.

But as if not unwilling to let nature make known her own case among his suffering charge, or else in despair of restraining it for the time, the Spanish captain, a gentlemanly, reserved-looking, and rather young man to a stranger's eye, dressed with singular richness, but bearing plain traces of recent sleepless cares and disquietudes, stood passively by, leaning against the main-mast, at one moment casting a dreary, spiritless look upon his excited people, at the next an unhappy glance toward his visitor. By his side stood a black of small stature, in whose rude face, as occasionally, like a shepherd's dog, he mutely turned it up into the Spaniard's, sorrow and affection were equally blended.

Struggling through the throng, the American advanced to the Spaniard, assuring him of his sympathies, and offering to render whatever assistance might be in his power. To which the Spaniard returned, for the present, but grave and ceremonious acknowledgments, his national formality dusked by the saturnine mood of ill health.

But losing no time in mere compliments, Captain Delano returning to the gangway, had his baskets of fish brought up; and as the wind still continued light, so that some hours at least must elapse ere the ship could be brought to the anchorage, he bade his men return to the sealer, and fetch back as much water as the whale-boat could carry, with whatever soft bread the steward might have, all the remaining pumpkins on board, with a box of sugar, and a dozen of his private bottles of cider.

Not many minutes after the boat's pushing off, to the vexation of all, the wind entirely died away, and the tide turning, began drifting back the ship helplessly seaward. But trusting this would not long last, Captain Delano sought with good hopes to cheer up the strangers, feeling no small satisfaction that, with persons in their condition he could—thanks to his frequent voyages along the Spanish main[7]—converse with some freedom in their native tongue.

While left alone with them, he was not long in observing some things tending to highten his first impressions; but surprise was lost in pity, both for the Spaniards and blacks, alike evidently reduced from scarcity of water and provisions; while long-continued suffering seemed to have brought out the less good-natured qualities of the negroes, besides, at the same time, impairing the Spaniard's

7. Sometimes loosely used for the Caribbean Sea, but here the Atlantic and Pacific coasts of South America (mainland as opposed to islands).

authority over them. But, under the circumstances, precisely this condition of things was to have been anticipated. In armies, navies, cities, or families, in nature herself, nothing more relaxes good order than misery. Still, Captain Delano was not without the idea, that had Benito Cereno been a man of greater energy, misrule would hardly have come to the present pass. But the debility, constitutional or induced by the hardships, bodily and mental, of the Spanish captain, was too obvious to be overlooked. A prey to settled dejection, as if long mocked with hope he would not now indulge it, even when it had ceased to be a mock, the prospect of that day or evening at furthest, lying at anchor, with plenty of water for his people, and a brother captain to counsel and befriend, seemed in no perceptible degree to encourage him. His mind appeared unstrung, if not still more seriously affected. Shut up in these oaken walls, chained to one dull round of command, whose unconditionality cloyed him, like some hypochondriac abbot he moved slowly about, at times suddenly pausing, starting, or staring, biting his lip, biting his finger-nail, flushing, paling, twitching his beard, with other symptoms of an absent or moody mind. This distempered spirit was lodged, as before hinted, in as distempered a frame. He was rather tall, but seemed never to have been robust, and now with nervous suffering was almost worn to a skeleton. A tendency to some pulmonary complaint appeared to have been lately confirmed. His voice was like that of one with lungs half gone, hoarsely suppressed, a husky whisper. No wonder that, as in this state he tottered about, his private servant apprehensively followed him. Sometimes the negro gave his master his arm, or took his handkerchief out of his pocket for him; performing these and similar offices with that affectionate zeal which transmutes into something filial or fraternal acts in themselves but menial; and which has gained for the negro the repute of making the most pleasing body servant in the world; one, too, whom a master need be on no stiffly superior terms with, but may treat with familiar trust; less a servant than a devoted companion.

Marking the noisy indocility of the blacks in general, as well as what seemed the sullen inefficiency of the whites, it was not without humane satisfaction that Captain Delano witnessed the steady good conduct of Babo.

But the good conduct of Babo, hardly more than the ill-behavior of others, seemed to withdraw the half-lunatic Don Benito from his cloudy languor. Not that such precisely was the impression made by the Spaniard on the mind of his visitor. The Spaniard's individual unrest was, for the present, but noted as a conspicuous feature in the ship's general affliction. Still, Captain Delano was not a little concerned at what he could not help taking for the time to be Don Benito's unfriendly indifference towards himself. The Spaniard's manner, too, conveyed a sort of sour and gloomy disdain, which he seemed at no pains to disguise. But this the American in charity ascribed to the harassing effects of sickness, since, in former instances, he had noted that there are peculiar natures on whom prolonged physical suffering seems to cancel every social instinct of kindness; as if forced to black bread themselves, they deemed it but equity that each person coming nigh them should, indirectly, by some slight or affront, be made to partake of their fare.

But ere long Captain Delano bethought him that, indulgent as he was at the first, in judging the Spaniard, he might not, after all, have exercised charity enough. At bottom it was Don Benito's reserve which displeased him; but the same reserve was shown towards all but his faithful personal attendant. Even the formal reports which, according to sea-usage, were, at stated times, made to him by some petty underling, either a white, mulatto or black, he hardly had patience enough to listen to, without betraying contemptuous aversion. His manner upon such occasions was, in its degree, not unlike that which might be supposed to have been his imperial countryman's, Charles V., just previous to the anchoritish retirement of that monarch from the throne.[8]

This splenetic disrelish of his place was evinced in almost every function pertaining to it. Proud as he was moody, he condescended to no personal mandate. Whatever special orders were necessary, their delivery was delegated to his body-servant, who in turn transferred them to their ultimate destination, through runners, alert Spanish boys or slave boys, like pages or pilot-fish[9] within easy call continually hovering round Don Benito. So that to have beheld this undemonstrative invalid gliding about, apathetic and mute, no landsman could have dreamed that in him was lodged a dictatorship beyond which, while at sea, there was no earthly appeal.

Thus, the Spaniard, regarded in his reserve, seemed as the involuntary victim of mental disorder. But, in fact, his reserve might, in some degree, have proceeded from design. If so, then here was evinced the unhealthy climax of that icy though conscientious policy, more or less adopted by all commanders of large ships, which, except in signal emergencies, obliterates alike the manifestation of sway with every trace of sociality; transforming the man into a block, or rather into a loaded cannon, which, until there is call for thunder, has nothing to say.

Viewing him in this light, it seemed but a natural token of the perverse habit induced by a long course of such hard self-restraint, that, notwithstanding the present condition of his ship, the Spaniard should still persist in a demeanor, which, however harmless, or, it may be, appropriate, in a well appointed vessel, such as the San Dominick might have been at the outset of the voyage, was any-

8. Charles V (1500–58), King of Spain and Holy Roman Emperor who spent his last years in a monastery (without, however, relinquishing all political power and material possessions).
9. Fish often swimming in the company of a shark, therefore fancied to pilot him.

thing but judicious now. But the Spaniard perhaps thought that it was with captains as with gods: reserve, under all events, must still be their cue. But more probably this appearance of slumbering dominion might have been but an attempted disguise to conscious imbecility—not deep policy, but shallow device. But be all this as it might, whether Don Benito's manner was designed or not, the more Captain Delano noted its pervading reserve, the less he felt uneasiness at any particular manifestation of that reserve towards himself.

Neither were his thoughts taken up by the captain alone. Wonted to the quiet orderliness of the sealer's comfortable family of a crew, the noisy confusion of the San Dominick's suffering host repeatedly challenged his eye. Some prominent breaches not only of discipline but of decency were observed. These Captain Delano could not but ascribe, in the main, to the absence of those subordinate deck-officers to whom, along with higher duties, is entrusted what may be styled the police department of a populous ship. True, the old oakum-pickers appeared at times to act the part of monitorial constables to their countrymen, the blacks; but though occasionally succeeding in allaying trifling outbreaks now and then between man and man, they could do little or nothing toward establishing general quiet. The San Dominick was in the condition of a transatlantic emigrant ship, among whose multitude of living freight are some individuals, doubtless, as little troublesome as crates and bales; but the friendly remonstrances of such with their ruder companions are of not so much avail as the unfriendly arm of the mate. What the San Dominick wanted was, what the emigrant ship has, stern superior officers. But on these decks not so much as a fourth mate was to be seen.

The visitor's curiosity was roused to learn the particulars of those mishaps which had brought about such absenteeism, with its consequences; because, though deriving some inkling of the voyage from the wails which at the first moment had greeted him, yet of the details no clear understanding had been had. The best account would, doubtless, be given by the captain. Yet at first the visitor was loth to ask it, unwilling to provoke some distant rebuff. But plucking up courage, he at last accosted Don Benito, renewing the expression of his benevolent interest, adding, that did he (Captain Delano) but know the particulars of the ship's misfortunes, he would, perhaps, be better able in the end to relieve them. Would Don Benito favor him with the whole story?

Don Benito faltered; then, like some somnambulist suddenly interfered with, vacantly stared at his visitor, and ended by looking down on the deck. He maintained this posture so long, that Captain Delano, almost equally disconcerted, and involuntarily almost as rude, turned suddenly from him, walking forward to accost one of the Spanish seamen for the desired information. But he had hardly gone five paces, when with a sort of eagerness Don Benito invited

him back, regretting his momentary absence of mind, and professing readiness to gratify him.

While most part of the story was being given, the two captains
stood on the after part of the main-deck, a privileged spot, no one
being near but the servant.

"It is now a hundred and ninety days," began the Spaniard, in his
husky whisper, "that this ship, well officered and well manned, with
several cabin passengers—some fifty Spaniards in all—sailed from
Buenos Ayres bound to Lima, with a general cargo, hardware,
Paraguay tea and the like—and," pointing forward, "that parcel of
negroes, now not more than a hundred and fifty, as you see, but
then numbering over three hundred souls. Off Cape Horn we had
heavy gales. In one moment, by night, three of my best officers,
with fifteen sailors, were lost, with the main-yard; the spar snapping
under them in the slings, as they sought, with heavers,[1] to beat
down the icy sail. To lighten the hull, the heavier sacks of mata[2]
were thrown into the sea, with most of the water-pipes[3] lashed on
deck at the time. And this last necessity it was, combined with the
prolonged detentions afterwards experienced, which eventually
brought about our chief causes of suffering. When——"

Here there was a sudden fainting attack of his cough, brought on,
no doubt, by his mental distress. His servant sustained him, and
drawing a cordial from his pocket placed it to his lips. He a little
revived. But unwilling to leave him unsupported while yet imperfectly restored, the black with one arm still encircled his master, at
the same time keeping his eye fixed on his face, as if to watch for
the first sign of complete restoration, or relapse, as the event might
prove.

The Spaniard proceeded, but brokenly and obscurely, as one in a
dream.

—"Oh, my God! rather than pass through what I have, with joy I
would have hailed the most terrible gales; but——"

His cough returned and with increased violence; this subsiding,
with reddened lips and closed eyes he fell heavily against his supporter.

"His mind wanders. He was thinking of the plague that followed
the gales," plaintively sighed the servant; "my poor, poor master!"
wringing one hand, and with the other wiping the mouth. "But be
patient, Señor," again turning to Captain Delano, "these fits do not
last long; master will soon be himself."

Don Benito reviving, went on; but as this portion of the story was
very brokenly delivered, the substance only will here be set down.

It appeared that after the ship had been many days tossed in
storms off the Cape, the scurvy broke out, carrying off numbers of
the whites and blacks. When at last they had worked round into the
Pacific, their spars and sails were so damaged, and so inadequately

1. Bar, most often used as a lever.   3. Kegs of water.
2. Brazilian cotton.

handled by the surviving mariners, most of whom were become invalids, that, unable to lay her northerly course by the wind, which was powerful, the unmanageable ship for successive days and nights was blown northwestward, where the breeze suddenly deserted her, in unknown waters, to sultry calms. The absence of the water-pipes now proved as fatal to life as before their presence had menaced it. Induced, or at least aggravated, by the less than scanty allowance of water, a malignant fever followed the scurvy; with the excessive heat of the lengthened calm, making such short work of it as to sweep away, as by billows, whole families of the Africans, and a yet larger number, proportionably, of the Spaniards, including, by a luckless fatality, every remaining officer on board. Consequently, in the smart west winds eventually following the calm, the already rent sails having to be simply dropped, not furled, at need, had been gradually reduced to the beggar's rags they were now. To procure substitutes for his lost sailors, as well as supplies of water and sails, the captain at the earliest opportunity had made for Baldivia, the southermost civilized port of Chili and South America; but upon nearing the coast the thick weather had prevented him from so much as sighting that harbor. Since which period, almost without a crew, and almost without canvas and almost without water, and at intervals giving its added dead to the sea, the San Dominick had been battle-dored[4] about by contrary winds, inveigled by currents, or grown weedy in calms. Like a man lost in woods, more than once she had doubled upon her own track.

"But throughout these calamities," huskily continued Don Benito, painfully turning in the half embrace of his servant, "I have to thank those negroes you see, who, though to your inexperienced eyes appearing unruly, have, indeed, conducted themselves with less of restlessness than even their owner could have thought possible under such circumstances."

Here he again fell faintly back. Again his mind wandered: but he rallied, and less obscurely proceeded.

"Yes, their owner was quite right in assuring me that no fetters would be needed with his blacks; so that while, as is wont in this transportation, those negroes have always remained upon deck—not thrust below, as in the Guineamen—they have, also, from the beginning, been freely permitted to range within given bounds at their pleasure."

Once more the faintness returned—his mind roved—but, recovering, he resumed:

"But it is Babo here to whom, under God, I owe not only my own preservation, but likewise to him, chiefly, the merit is due, of pacifying his more ignorant brethren, when at intervals tempted to murmurings."

"Ah, master," sighed the black, bowing his face, "don't speak of

4. Tossed back and forth, as a shuttlecock is hit back and forth by a pair of battle-dores (or paddles).

me; Babo is nothing; what Babo has done was but duty."

"Faithful fellow!" cried Capt. Delano. "Don Benito, I envy you such a friend; slave I cannot call him."

As master and man stood before him, the black upholding the white, Captain Delano could not but bethink him of the beauty of that relationship which could present such a spectacle of fidelity on the one hand and confidence on the other. The scene was hightened by the contrast in dress, denoting their relative positions. The Spaniard wore a loose Chili jacket of dark velvet; white small clothes and stockings, with silver buckles at the knee and instep; a high-crowned sombrero, of fine grass; a slender sword, silver mounted, hung from a knot in his sash; the last being an almost invariable adjunct, more for ornament than utility, of a South American gentleman's dress to this hour. Excepting when his occasional nervous contortions brought about disarray, there was a certain precision in his attire, curiously at variance with the unsightly disorder around; especially in the belittered Ghetto, forward of the main-mast, wholly occupied by the blacks.

The servant wore nothing but wide trowsers, apparently, from their coarseness and patches, made out of some old topsail; they were clean, and confined at the waist by a bit of unstranded rope, which, with his composed, deprecatory air at times, made him look something like a begging friar of St. Francis.

However unsuitable for the time and place, at least in the blunt-thinking American's eyes, and however strangely surviving in the midst of all his afflictions, the toilette of Don Benito might not, in fashion at least, have gone beyond the style of the day among South Americans of his class. Though on the present voyage sailing from Buenos Ayres, he had avowed himself a native and resident of Chili, whose inhabitants had not so generally adopted the plain coat and once plebeian pantaloons; but, with a becoming modification, adhered to their provincial costume, picturesque as any in the world. Still, relatively to the pale history of the voyage, and his own pale face, there seemed something so incongruous in the Spaniard's apparel, as almost to suggest the image of an invalid courtier tottering about London streets in the time of the plague.

The portion of the narrative which, perhaps, most excited interest, as well as some surprise, considering the latitudes in question, was the long calms spoken of, and more particularly the ship's so long drifting about. Without communicating the opinion, of course, the American could not but impute at least part of the detentions both to clumsy seamanship and faulty navigation. Eying Don Benito's small, yellow hands, he easily inferred that the young captain had not got into command at the hawse-hole,[5] but the cabin-

5. Metal-lined hole in the bow of a ship, through which a cable passes; the expression means to begin a career before the mast, not in a position of authority.

window; and if so, why wonder at incompetence, in youth, sickness, and gentility united?

But drowning criticism in compassion, after a fresh repetition of his sympathies, Captain Delano having heard out his story, not only engaged, as in the first place, to see Don Benito and his people supplied in their immediate bodily needs, but, also, now further promised to assist him in procuring a large permanent supply of water, as well as some sails and rigging; and, though it would involve no small embarrassment to himself, yet he would spare three of his best seamen for temporary deck officers; so that without delay the ship might proceed to Conception, there fully to refit for Lima, her destined port.

Such generosity was not without its effect, even upon the invalid. His face lighted up; eager and hectic, he met the honest glance of his visitor. With gratitude he seemed overcome.

"This excitement is bad for master," whispered the servant, taking his arm, and with soothing words gently drawing him aside.

When Don Benito returned, the American was pained to observe that his hopefulness, like the sudden kindling in his cheek, was but febrile and transient.

Ere long, with a joyless mien, looking up towards the poop, the host invited his guest to accompany him there, for the benefit of what little breath of wind might be stirring.

As during the telling of the story, Captain Delano had once or twice started at the occasional cymballing of the hatchet-polishers, wondering why such an interruption should be allowed, especially in that part of the ship, and in the ears of an invalid; and moreover, as the hatchets had anything but an attractive look, and the handlers of them still less so, it was, therefore, to tell the truth, not without some lurking reluctance, or even shrinking, it may be, that Captain Delano, with apparent complaisance, acquiesced in his host's invitation. The more so, since with an untimely caprice of punctilio, rendered distressing by his cadaverous aspect, Don Benito, with Castilian bows, solemnly insisted upon his guest's preceding him up the ladder leading to the elevation; where, one on each side of the last step, sat for armorial supporters and sentries two of the ominous file. Gingerly enough stepped good Captain Delano between them, and in the instant of leaving them behind, like one running the gauntlet, he felt an apprehensive twitch in the calves of his legs.

But when, facing about, he saw the whole file, like so many organ-grinders, still stupidly intent on their work, unmindful of everything beside, he could not but smile at his late fidgety panic.

Presently, while standing with his host, looking forward upon the decks below, he was struck by one of those instances of insubordination previously alluded to. Three black boys, with two Spanish boys, were sitting together on the hatches, scraping a rude wooden

platter, in which some scanty mess had recently been cooked. Suddenly, one of the black boys, enraged at a word dropped by one of
his white companions, seized a knife, and though called to forbear
by one of the oakum-pickers, struck the lad over the head, inflicting
a gash from which blood flowed.

In amazement, Captain Delano inquired what this meant. To
which the pale Don Benito dully muttered, that it was merely the
sport of the lad.

"Pretty serious sport, truly," rejoined Captain Delano. "Had such
a thing happened on board the Bachelor's Delight, instant punishment would have followed."

At these words the Spaniard turned upon the American one of his
sudden, staring, half-lunatic looks; then relapsing into his torpor,
answered, "Doubtless, doubtless, Señor."

Is it, thought Captain Delano, that this hapless man is one of
those paper captains I've known, who by policy wink at what by
power they cannot put down? I know no sadder sight than a commander who has little of command but the name.

"I should think, Don Benito," he now said, glancing towards the
oakum-picker who had sought to interfere with the boys, "that you
would find it advantageous to keep all your blacks employed, especially the younger ones, no matter at what useless task, and no
matter what happens to the ship. Why, even with my little band, I
find such a course indispensable. I once kept a crew on my quarter-
deck thrumming mats[6] for my cabin, when, for three days, I had
given up my ship—mats, men, and all—for a speedy loss, owing to
the violence of a gale, in which we could do nothing but helplessly
drive before it."

"Doubtless, doubtless," muttered Don Benito.

"But," continued Captain Delano, again glancing upon the
oakum-pickers and then at the hatchet-polishers, near by. "I see you
keep some at least of your host employed."

"Yes," was again the vacant response.

"Those old men there, shaking their pows[7] from their pulpits,"
continued Captain Delano, pointing to the oakum-pickers, "seem to
act the part of old dominies to the rest, little heeded as their admonitions are at times. Is this voluntary on their part, Don Benito,
or have you appointed them shepherds to your flock of black
sheep?"

"What posts they fill, I appointed them," rejoined the Spaniard,
in an acrid tone, as if resenting some supposed satiric reflection.

"And these others, these Ashantee[8] conjurors here," continued
Captain Delano, rather uneasily eying the brandished steel of the
hatchet-polishers, where in spots it had been brought to a shine,

6. To thrum is to insert pieces of rope
yarn into canvas, thus making a rough
surface or mat (usually for keeping
ropes from chafing against wood or
metal).
7. Heads.
8. West African race.

"this seems a curious business they are at, Don Benito?"

"In the gales we met," answered the Spaniard, "what of our general cargo was not thrown overboard was much damaged by the brine. Since coming into calm weather, I have had several cases of knives and hatchets daily brought up for overhauling and cleaning."

"A prudent idea, Don Benito. You are part owner of ship and cargo, I presume; but not of the slaves, perhaps?"

"I am owner of all you see," impatiently returned Don Benito, "except the main company of blacks, who belonged to my late friend, Alexandro Aranda."

As he mentioned this name, his air was heart-broken; his knees shook: his servant supported him.

Thinking he divined the cause of such unusual emotion, to confirm his surmise, Captain Delano, after a pause, said, "And may I ask, Don Benito, whether—since awhile ago you spoke of some cabin passengers—the friend, whose loss so afflicts you at the outset of the voyage accompanied his blacks?"

"Yes."

"But died of the fever?"

"Died of the fever.—Oh, could I but——"

Again quivering, the Spaniard paused.

"Pardon me," said Captain Delano lowly, "but I think that, by a sympathetic experience, I conjecture, Don Benito, what it is that gives the keener edge to your grief. It was once my hard fortune to lose at sea a dear friend, my own brother, then supercargo. Assured of the welfare of his spirit, its departure I could have borne like a man; but that honest eye, that honest hand—both of which had so often met mine—and that warm heart; all, all—like scraps to the dogs—to throw all to the sharks! It was then I vowed never to have for fellow-voyager a man I loved, unless, unbeknown to him, I had provided every requisite, in case of a fatality, for embalming his mortal part for interment on shore. Were your friend's remains now on board this ship, Don Benito, not thus strangely would the mention of his name affect you."

"On board this ship?" echoed the Spaniard. Then, with horrified gestures, as directed against some specter, he unconsciously fell into the ready arms of his attendant, who, with a silent appeal toward Captain Delano, seemed beseeching him not again to broach a theme so unspeakably distressing to his master.

This poor fellow now, thought the pained American, is the victim of that sad superstition which associates goblins with the deserted body of man, as ghosts with an abandoned house. How unlike are we made! What to me, in like case, would have been a solemn satisfaction, the bare suggestion, even, terrifies the Spaniard into this trance. Poor Alexandro Aranda! what would you say could you here see your friend—who, on former voyages, when you for months were left behind, has, I dare say, often longed, and longed, for one peep at you—now transported with terror at the least thought of having you anyway nigh him.

At this moment, with a dreary graveyard toll, betokening a flaw, the ship's forecastle bell, smote by one of the grizzled oakum-pickers, proclaimed ten o'clock through the leaden calm; when Captain Delano's attention was caught by the moving figure of a gigantic black, emerging from the general crowd below, and slowly advancing towards the elevated poop. An iron collar was about his neck, from which depended a chain, thrice wound round his body; the terminating links padlocked together at a broad band of iron, his girdle.

"How like a mute Atufal moves," murmured the servant.

The black mounted the steps of the poop, and, like a brave prisoner, brought up to receive sentence, stood in unquailing muteness before Don Benito, now recovered from his attack.

At the first glimpse of his approach, Don Benito had started, a resentful shadow swept over his face; and, as with the sudden memory of bootless rage, his white lips glued together.

This is some mulish mutineer, thought Captain Delano, surveying, not without a mixture of admiration, the colossal form of the negro.

"See, he waits your question, master," said the servant.

Thus reminded, Don Benito, nervously averting his glance, as if shunning, by anticipation, some rebellious response, in a disconcerted voice, thus spoke:—

"Atufal, will you ask my pardon now?"

The black was silent.

"Again, master," murmured the servant, with bitter upbraiding eying his countryman, "Again, master; he will bend to master yet."

"Answer," said Don Benito, still averting his glance, "say but the one word *pardon*, and your chains shall be off."

Upon this, the black, slowly raising both arms, let them lifelessly fall, his links clanking, his head bowed; as much as to say, "no, I am content."

"Go," said Don Benito, with inkept and unknown emotion.

Deliberately as he had come, the black obeyed.

"Excuse me, Don Benito," said Captain Delano, "but this scene surprises me; what means it, pray?"

"It means that that negro alone, of all the band, has given me peculiar cause of offense. I have put him in chains; I——"

Here he paused; his hand to his head, as if there were a swimming there, or a sudden bewilderment of memory had come over him; but meeting his servant's kindly glance seemed reassured, and proceeded:—

"I could not scourge such a form. But I told him he must ask my pardon. As yet he has not. At my command, every two hours he stands before me."

"And how long has this been?"

"Some sixty days."

"And obedient in all else? And respectful?"

"Yes."

"Upon my conscience, then," exclaimed Captain Delano, impulsively, "he has a royal spirit in him, this fellow."

"He may have some right to it," bitterly returned Don Benito, "he says he was king in his own land."

"Yes," said the servant, entering a word, "those slits in Atufal's ears once held wedges of gold; but poor Babo here, in his own land, was only a poor slave; a black man's slave was Babo, who now is the white's."

Somewhat annoyed by these conversational familiarities, Captain Delano turned curiously upon the attendant, then glanced inquiringly at his master; but, as if long wonted to these little informalities, neither master nor man seemed to understand him.

"What, pray, was Atufal's offense, Don Benito?" asked Captain Delano; "if it was not something very serious, take a fool's advice, and, in view of his general docility, as well as in some natural respect for his spirit, remit him his penalty."

"No, no, master never will do that," here murmured the servant to himself, "proud Atufal must first ask master's pardon. The slave there carries the padlock, but master here carries the key."

His attention thus directed, Captain Delano now noticed for the first time that, suspended by a slender silken cord, from Don Benito's neck hung a key. At once, from the servant's muttered syllables divining the key's purpose, he smiled and said:—"So, Don Benito—padlock and key—significant symbols, truly."

Biting his lip, Don Benito faltered.

Though the remark of Captain Delano, a man of such native simplicity as to be incapable of satire or irony, had been dropped in playful allusion to the Spaniard's singularly evidenced lordship over the black; yet the hypochondriac seemed in some way to have taken it as a malicious reflection upon his confessed inability thus far to break down, at least, on a verbal summons, the entrenched will of the slave. Deploring this supposed misconception, yet despairing of correcting it, Captain Delano shifted the subject; but finding his companion more than ever withdrawn, as if still sourly digesting the lees of the presumed affront above-mentioned, by-and-by Captain Delano likewise became less talkative, oppressed, against his own will, by what seemed the secret vindictiveness of the morbidly sensitive Spaniard. But the good sailor himself, of a quite contrary disposition, refrained, on his part, alike from the appearance as from the feeling of resentment, and if silent, was only so from contagion.

Presently the Spaniard, assisted by his servant, somewhat discourteously crossed over from his guest; a procedure which, sensibly enough, might have been allowed to pass for idle caprice of ill-humor, had not master and man, lingering round the corner of the elevated skylight, began whispering together in low voices. This was unpleasing. And more: the moody air of the Spaniard, which at times had not been without a sort of valetudinarian stateliness, now seemed anything but dignified; while the menial familiarity of the servant lost its original charm of simple-hearted attachment.

In his embarrassment, the visitor turned his face to the other side of the ship. By so doing, his glance accidentally fell on a young Spanish sailor, a coil of rope in his hand, just stepped from the deck to the first round of the mizzen-rigging. Perhaps the man would not have been particularly noticed, were it not that, during his ascent to one of the yards, he, with a sort of covert intentness, kept his eye fixed on Captain Delano, from whom, presently, it passed, as if by a natural sequence, to the two whisperers.

His own attention thus redirected to that quarter, Captain Delano gave a slight start. From something in Don Benito's manner just then, it seemed as if the visitor had, at least partly, been the subject of the withdrawn consultation going on—a conjecture as little agreeable to the guest as it was little flattering to the host.

The singular alternations of courtesy and ill-breeding in the Spanish captain were unaccountable, except on one of two suppositions—innocent lunacy, or wicked imposture.

But the first idea, though it might naturally have occurred to an indifferent observer, and, in some respect, had not hitherto been wholly a stranger to Captain Delano's mind, yet, now that, in an incipient way, he began to regard the stranger's conduct something in the light of an intentional affront, of course the idea of lunacy was virtually vacated. But if not a lunatic, what then? Under the circumstances, would a gentleman, nay, any honest boor, act the part now acted by his host? The man was an impostor. Some low-born adventurer, masquerading as an oceanic grandee; yet so ignorant of the first requisites of mere gentlemanhood as to be betrayed into the present remarkable indecorum. That strange ceremoniousness, too, at other times evinced, seemed not uncharacteristic of one playing a part above his real level. Benito Cereno—Don Benito Cereno—a sounding name. One, too, at that period, not unknown, in the surname, to supercargoes and sea captains trading along the Spanish Main, as belonging to one of the most enterprising and extensive mercantile families in all those provinces; several members of it having titles; a sort of Castilian Rothschild,[9] with a noble brother, or cousin, in every great trading town of South America. The alleged Don Benito was in early manhood, about twenty-nine or thirty. To assume a sort of roving cadetship[1] in the maritime affairs of such a house, what more likely scheme for a

9. Great German banking family.
1. Position of on-the-job training for a post of authority, appropriate for a

younger or youngest son of a great house.

young knave of talent and spirit? But the Spaniard was a pale invalid. Never mind. For even to the degree of simulating mortal disease, the craft of some tricksters had been known to attain. To think that, under the aspect of infantile weakness, the most savage energies might be couched—those velvets of the Spaniard but the silky paw to his fangs.

From no train of thought did these fancies come; not from within, but from without; suddenly, too, and in one throng, like hoar frost; yet as soon to vanish as the mild sun of Captain Delano's good-nature regained its meridian.

Glancing over once more towards his host—whose side-face, revealed above the skylight, was now turned towards him—he was struck by the profile, whose clearness of cut was refined by the thinness incident to ill-health, as well as ennobled about the chin by the beard. Away with suspicion. He was a true off-shoot of a true hidalgo Cereno.

Relieved by these and other better thoughts, the visitor, lightly humming a tune, now began indifferently pacing the poop, so as not to betray to Don Benito that he had at all mistrusted incivility, much less duplicity; for such mistrust would yet be proved illusory, and by the event; though, for the present, the circumstance which had provoked that distrust remained unexplained. But when that little mystery should have been cleared up, Captain Delano thought he might extremely regret it, did he allow Don Benito to become aware that he had indulged in ungenerous surmises. In short, to the Spaniard's black-letter text,[2] it was best, for awhile, to leave open margin.

Presently, his pale face twitching and overcast, the Spaniard, still supported by his attendant, moved over towards his guest, when, with even more than his usual embarrassment, and a strange sort of intriguing intonation in his husky whisper, the following conversation began:—

"Señor, may I ask how long you have lain at this isle?"

"Oh, but a day or two, Don Benito."

"And from what port are you last?"

"Canton."

"And there, Señor, you exchanged your seal-skins for teas and silks, I think you said?"

"Yes. Silks, mostly."

"And the balance you took in specie, perhaps?"

Captain Delano, fidgeting a little, answered—

"Yes; some silver; not a very great deal, though."

"Ah—well. May I ask how many men have you, Señor?"

Captain Delano slightly started, but answered—

2. Books printed in early type imitative of medieval script; "open margin": without the elucidatory comments then often printed in margins as a gloss upon the main text rather than printing them at the bottom of the page as footnotes. Delano is deciding to reserve judgment.

"About five-and-twenty, all told."

"And at present, Señor, all on board, I suppose?"

"All on board, Don Benito," replied the Captain, now with satisfaction.

"And will be to-night, Señor?"

At this last question, following so many pertinacious ones, for the soul of him Captain Delano could not but look very earnestly at the questioner, who, instead of meeting the glance, with every token of craven discomposure dropped his eyes to the deck; presenting an unworthy contrast to his servant, who, just then, was kneeling at his feet, adjusting a loose shoe-buckle; his disengaged face meantime, with humble curiosity, turned openly up into his master's downcast one.

The Spaniard, still with a guilty shuffle, repeated his question:—

"And—and will be to-night, Señor?"

"Yes, for aught I know," returned Captain Delano,—"but nay," rallying himself into fearless truth, "some of them talked of going off on another fishing party about midnight."

"Your ships generally go—go more or less armed, I believe, Señor?"

"Oh, a six-pounder or two, in case of emergency," was the intrepidly indifferent reply, "with a small stock of muskets, sealing-spears, and cutlasses, you know."

As he thus responded, Captain Delano again glanced at Don Benito, but the latter's eyes were averted; while abruptly and awkwardly shifting the subject, he made some peevish allusion to the calm, and then, without apology, once more, with his attendant, withdrew to the opposite bulwarks, where the whispering was resumed.

At this moment, and ere Captain Delano could cast a cool thought upon what had just passed, the young Spanish sailor before mentioned was seen descending from the rigging. In act of stooping over to spring inboard to the deck, his voluminous, unconfined frock, or shirt, of coarse woollen, much spotted with tar, opened out far down the chest, revealing a soiled under garment of what seemed the finest linen, edged, about the neck, with a narrow blue ribbon, sadly faded and worn. At this moment the young sailor's eye was again fixed on the whisperers, and Captain Delano thought he observed a lurking significance in it, as if silent signs of some Freemason[3] sort had that instant been interchanged.

This once more impelled his own glance in the direction of Don Benito, and, as before, he could not but infer that himself formed the subject of the conference. He paused. The sound of the hatchet-polishing fell on his ears. He cast another swift side-look at the two.

They had the air of conspirators. In connection with the late questionings and the incident of the young sailor, these things now begat such return of involuntary suspicion, that the singular guilelessness of the American could not endure it. Plucking up a gay and humorous expression, he crossed over to the two rapidly, saying:— "Ha, Don Benito, your black here seems high in your trust; a sort of privy-counselor, in fact."

Upon this, the servant looked up with a good-natured grin, but the master started as from a venomous bite. It was a moment or two before the Spaniard sufficiently recovered himself to reply; which he did, at last, with cold constraint:—"Yes, Señor, I have trust in Babo."

Here Babo, changing his previous grin of mere animal humor into an intelligent smile, not ungratefully eyed his master.

Finding that the Spaniard now stood silent and reserved, as if involuntarily, or purposely giving hint that his guest's proximity was inconvenient just then, Captain Delano, unwilling to appear uncivil even to incivility itself, made some trivial remark and moved off; again and again turning over in his mind the mysterious demeanor of Don Benito Cereno.

He had descended from the poop, and, wrapped in thought, was passing near a dark hatchway, leading down into the steerage, when, perceiving motion there, he looked to see what moved. The same instant there was a sparkle in the shadowy hatchway, and he saw one of the Spanish sailors prowling there hurriedly placing his hand in the bosom of his frock, as if hiding something. Before the man could have been certain who it was that was passing, he slunk below out of sight. But enough was seen of him to make it sure that he was the same young sailor before noticed in the rigging.

What was that which so sparkled? thought Captain Delano. It was no lamp—no match—no live coal. Could it have been a jewel? But how come sailors with jewels?—or with silk-trimmed undershirts either? Has he been robbing the trunks of the dead cabin passengers? But if so, he would hardly wear one of the stolen articles on board ship here. Ah, ah—if now that was, indeed, a secret sign I saw passing between this suspicious fellow and his captain awhile since; if I could only be certain that in my uneasiness my senses did not deceive me, then——

Here, passing from one suspicious thing to another, his mind revolved the strange questions put to him concerning his ship.

By a curious coincidence, as each point was recalled, the black wizards of Ashantee would strike up with their hatchets, as in ominous comment on the white stranger's thoughts. Pressed by such enigmas and portents, it would have been almost against nature, had not, even into the least distrustful heart, some ugly misgivings obtruded.

Observing the ship now helplessly fallen into a current, with enchanted sails, drifting with increased rapidity seaward; and noting that, from a lately intercepted projection of the land, the sealer was hidden, the stout mariner began to quake at thoughts which he barely durst confess to himself. Above all, he began to feel a ghostly dread of Don Benito. And yet when he roused himself, dilated his chest, felt himself strong on his legs, and coolly considered it—what did all these phantoms amount to?

Had the Spaniard any sinister scheme, it must have reference not so much to him (Captain Delano) as to his ship (the Bachelor's Delight). Hence the present drifting away of the one ship from the other, instead of favoring any such possible scheme, was, for the time at least, opposed to it. Clearly any suspicion, combining such contradictions, must need be delusive. Beside, was it not absurd to think of a vessel in distress—a vessel by sickness almost dismanned of her crew—a vessel whose inmates were parched for water—was it not a thousand times absurd that such a craft should, at present, be of a piratical character; or her commander, either for himself or those under him, cherish any desire but for speedy relief and refreshment? But then, might not general distress, and thirst in particular, be affected? And might not that same undiminished Spanish crew, alleged to have perished off to a remnant, be at that very moment lurking in the hold? On heart-broken pretense of entreating a cup of cold water, fiends in human form had got into lonely dwellings, nor retired until a dark deed had been done. And among the Malay pirates, it was no unusual thing to lure ships after them into their treacherous harbors, or entice boarders from a declared enemy at sea, by the spectacle of thinly manned or vacant decks, beneath which prowled a hundred spears with yellow arms ready to upthrust them through the mats. Not that Captain Delano had entirely credited such things. He had heard of them—and now, as stories, they recurred. The present destination of the ship was the anchorage. There she would be near his own vessel. Upon gaining that vicinity, might not the San Dominick, like a slumbering volcano, suddenly let loose energies now hid?

He recalled the Spaniard's manner while telling his story. There was a gloomy hesitancy and subterfuge about it. It was just the manner of one making up his tale for evil purposes, as he goes. But if that story was not true, what was the truth? That the ship had unlawfully come into the Spaniard's possession? But in many of its details, especially in reference to the more calamitous parts, such as the fatalities among the seamen, the consequent prolonged beating about, the past sufferings from obstinate calms, and still continued suffering from thirst; in all these points, as well as others, Don Benito's story had corroborated not only the wailing ejaculations of the indiscriminate multitude, white and black, but likewise—what seemed impossible to be counterfeit—by the very expression and play of every human feature, which Captain Delano saw. If Don

Benito's story was throughout an invention, then every soul on board, down to the youngest negress, was his carefully drilled recruit in the plot: an incredible inference. And yet, if there was ground for mistrusting his veracity, that inference was a legitimate one.

But those questions of the Spaniard. There, indeed, one might pause. Did they not seem put with much the same object with which the burglar or assassin, by day-time, reconnoitres the walls of a house? But, with ill purposes, to solicit such information openly of the chief person endangered, and so, in effect, setting him on his guard; how unlikely a procedure was that? Absurd, then, to suppose that those questions had been prompted by evil designs. Thus, the same conduct, which, in this instance, had raised the alarm, served to dispel it. In short, scarce any suspicion or uneasiness, however apparently reasonable at the time, which was not now, with equal apparent reason, dismissed.

At last he began to laugh at his former forebodings; and laugh at the strange ship for, in its aspect someway siding with them, as it were; and laugh, too, at the odd-looking blacks, particularly those old scissors-grinders, the Ashantees; and those bed-ridden old knitting-women, the oakum-pickers; and almost at the dark Spaniard himself, the central hobgoblin of all.

For the rest, whatever in a serious way seemed enigmatical, was now good-naturedly explained away by the thought that, for the most part, the poor invalid scarcely knew what he was about; either sulking in black vapors, or putting idle questions without sense or object. Evidently, for the present, the man was not fit to be entrusted with the ship. On some benevolent plea withdrawing the command from him, Captain Delano would yet have to send her to Conception, in charge of his second mate, a worthy person and good navigator—a plan not more convenient for the San Dominick than for Don Benito; for, relieved from all anxiety, keeping wholly to his cabin, the sick man, under the good nursing of his servant, would probably, by the end of the passage, be in a measure restored to health, and with that he should also be restored to authority.

Such were the American's thoughts. They were tranquilizing. There was a difference between the idea of Don Benito's darkly pre-ordaining Captain Delano's fate, and Captain Delano's lightly arranging Don Benito's. Nevertheless, it was not without something of relief that the good seaman presently perceived his whale-boat in the distance. Its absence had been prolonged by unexpected detention at the sealer's side, as well as its returning trip lengthened by the continual recession of the goal.

The advancing speck was observed by the blacks. Their shouts attracted the attention of Don Benito, who, with a return of courtesy, approaching Captain Delano, expressed satisfaction at the coming of some supplies, slight and temporary as they must necessarily prove.

Captain Delano responded; but while doing so, his attention was drawn to something passing on the deck below: among the crowd climbing the landward bulwarks, anxiously watching the coming boat, two blacks, to all appearances accidentally incommoded by one of the sailors, violently pushed him aside, which the sailor someway resenting, they dashed him to the deck, despite the earnest cries of the oakum-pickers.

"Don Benito," said Captain Delano quickly, "do you see what is going on there? Look!"

But, seized by his cough, the Spaniard staggered, with both hands to his face, on the point of falling. Captain Delano would have supported him, but the servant was more alert, who, with one hand sustaining his master, with the other applied the cordial. Don Benito restored, the black withdrew his support, slipping aside a little, but dutifully remaining within call of a whisper. Such discretion was here evinced as quite wiped away, in the visitor's eyes, any blemish of impropriety which might have attached to the attendant, from the indecorous conferences before mentioned; showing, too, that if the servant were to blame, it might be more the master's fault than his own, since when left to himself he could conduct thus well.

His glance called away from the spectacle of disorder to the more pleasing one before him, Captain Delano could not avoid again congratulating his host upon possessing such a servant, who, though perhaps a little too forward now and then, must upon the whole be invaluable to one in the invalid's situation.

"Tell me, Don Benito," he added, with a smile—"I should like to have your man here myself—what will you take for him? Would fifty doubloons be any object?"

"Master wouldn't part with Babo for a thousand doubloons," murmured the black, overhearing the offer, and taking it in earnest, and, with the strange vanity of a faithful slave appreciated by his master, scorning to hear so paltry a valuation put upon him by a stranger. But Don Benito, apparently hardly yet completely restored, and again interrupted by his cough, made but some broken reply.

Soon his physical distress became so great, affecting his mind, too, apparently, that, as if to screen the sad spectacle, the servant gently conducted his master below.

Left to himself, the American, to while away the time till his boat should arrive, would have pleasantly accosted some one of the few Spanish seamen he saw; but recalling something that Don Benito had said touching their ill conduct, he refrained, as a ship-master indisposed to countenance cowardice or unfaithfulness in seamen.

While, with these thoughts, standing with eye directed forward towards that handful of sailors, suddenly he thought that one or two of them returned the glance and with a sort of meaning. He rubbed his eyes, and looked again; but again seemed to see the same thing. Under a new form, but more obscure than any previous one, the old suspicions recurred, but, in the absence of Don Benito, with less of panic than before. Despite the bad account given of the sailors, Captain Delano resolved forthwith to accost one of them. Descending the poop, he made his way through the blacks, his movement drawing a queer cry from the oakum-pickers, prompted by whom, the negroes, twitching each other aside, divided before him; but, as if curious to see what was the object of this deliberate visit to their Ghetto, closing in behind, in tolerable order, followed the white stranger up. His progress thus proclaimed as by mounted kings-at-arms, and escorted as by a Caffre[4] guard of honor, Captain Delano, assuming a good humored, off-handed air, continued to advance; now and then saying a blithe word to the negroes, and his eye curiously surveying the white faces, here and there sparsely mixed in with the blacks, like stray white pawns venturously involved in the ranks of the chess-men opposed.

While thinking which of them to select for his purpose, he chanced to observe a sailor seated on the deck engaged in tarring the strap of a large block, with a circle of blacks squatted round him inquisitively eying the process.

The mean employment of the man was in contrast with something superior in his figure. His hand, black with continually thrusting it into the tar-pot held for him by a negro, seemed not naturally allied to his face, a face which would have been a very fine one but for its haggardness. Whether this haggardness had aught to do with criminality, could not be determined; since, as intense heat and cold, though unlike, produce like sensations, so innocence and guilt, when, through casual association with mental pain, stamping any visible impress, use one seal—a hacked one.

Not again that this reflection occurred to Captain Delano at the time, charitable man as he was. Rather another idea. Because observing so singular a haggardness combined with a dark eye, averted as in trouble and shame, and then again recalling Don Benito's confessed ill opinion of his crew, insensibly he was operated upon by certain general notions, which, while disconnecting pain and abashment from virtue, invariably link them with vice.

If, indeed, there be any wickedness on board this ship, thought Captain Delano, be sure that man there has fouled his hand in it, even as now he fouls it in the pitch. I don't like to accost him. I will speak to this other, this old Jack here on the windlass.

He advanced to an old Barcelona tar, in ragged red breeches and dirty night-cap, cheeks trenched and bronzed, whiskers dense as thorn hedges. Seated between two sleepy-looking Africans, this mariner, like his younger shipmate, was employed upon some rigging—splicing a cable—the sleepy-looking blacks performing the

4. Kaffir, very tall Bantu tribe of South Africa.

inferior function of holding the outer parts of the ropes for him.

Upon Captain Delano's approach, the man at once hung his head below its previous level; the one necessary for business. It appeared as if he desired to be thought absorbed, with more than common fidelity, in his task. Being addressed, he glanced up, but with what seemed a furtive, diffident air, which sat strangely enough on his weather-beaten visage, much as if a grizzly bear, instead of growling and biting, should simper and cast sheep's eyes. He was asked several questions concerning the voyage, questions purposely refer-ring to several particulars in Don Benito's narrative, not previously corroborated by those impulsive cries greeting the visitor on first coming on board. The questions were briefly answered, confirming all that remained to be confirmed of the story. The negroes about the windlass joined in with the old sailor, but, as they became talkative, he by degrees became mute, and at length quite glum, seemed morosely unwilling to answer more questions, and yet, all the while, this ursine air was somehow mixed with his sheepish one.

Despairing of getting into unembarrassed talk with such a cen-taur, Captain Delano, after glancing round for a more promising countenance, but seeing none, spoke pleasantly to the blacks to make way for him; and so, amid various grins and grimaces, re-turned to the poop, feeling a little strange at first, he could hardly tell why, but upon the whole with regained confidence in Benito Cereno.

How plainly, thought he, did that old whiskerando yonder betray a consciousness of ill-desert. No doubt, when he saw me coming, he dreaded lest I, apprised by his Captain of the crew's general mis-behavior, came with sharp words for him, and so down with his head. And yet—and yet, now that I think of it, that very old fellow, if I err not, was one of those who seemed so earnestly eying me here awhile since. Ah, these currents spin one's head round almost as much as they do the ship. Ha, there now's a pleasant sort of sunny sight; quite sociable, too.

His attention had been drawn to a slumbering negress, partly disclosed through the lace-work of some rigging, lying, with youth-ful limbs carelessly disposed, under the lee of the bulwarks, like a doe in the shade of a woodland rock. Sprawling at her lapped breasts was her wide-awake fawn, stark naked, its black little body half lifted from the deck, crosswise with its dam's; its hands, like two paws, clambering upon her; its mouth and nose ineffectually rooting to get at the mark; and meantime giving a vexatious half-grunt, blending with the composed snore of the negress.

The uncommon vigor of the child at length roused the mother. She started up, at distance facing Captain Delano. But as if not at all concerned at the attitude in which she had been caught, delight-edly she caught the child up, with maternal transports, covering it with kisses.

There's naked nature, now; pure tenderness and love, thought Captain Delano, well pleased.

This incident prompted him to remark the other negresses more particularly than before. He was gratified with their manners; like most uncivilized women, they seemed at once tender of heart and tough of constitution; equally ready to die for their infants or fight for them. Unsophisticated as leopardesses; loving as doves. Ah! thought Captain Delano, these perhaps are some of the very women whom Ledyard[5] saw in Africa, and gave such a noble account of.

These natural sights somehow insensibly deepened his confidence and ease. At last he looked to see how his boat was getting on; but it was still pretty remote. He turned to see if Don Benito had returned; but he had not.

To change the scene, as well as to please himself with a leisurely observation of the coming boat, stepping over into the mizzen-chains he clambered his way into the starboard quarter-gallery;[6] one of those abandoned Venetian-looking water-balconies previously mentioned; retreats cut off from the deck. As his foot pressed the half-damp, half-dry sea-mosses matting the place, and a chance phantom cats-paw—an islet of breeze, unheralded, unfollowed—as this ghostly cats-paw came fanning his cheek, as his glance fell upon the row of small, round dead-lights, all closed like coppered eyes of the coffined, and the state-cabin door, once connecting with the gallery, even as the dead-lights had once looked out upon it, but now calked fast like a sarcophagus lid, to a purple-black, tarred-over panel, threshold, and post; and he bethought him of the time, when that state-cabin and this state-balcony had heard the voices of the Spanish king's officers, and the forms of the Lima viceroy's daughters had perhaps leaned where he stood—as these and other images flitted through his mind, as the cats-paw through the calm, gradually he felt rising a dreamy inquietude, like that of one who alone on the prairie feels unrest from the repose of the noon.

He leaned against the carved balustrade, again looking off toward his boat; but found his eye falling upon the ribboned grass, trailing along the ship's water-line, straight as a border of green box; and parterres[7] of sea-weed, broad ovals and crescents, floating nigh and far, with what seemed long formal alleys between, crossing the terraces of swells, and sweeping round as if leading to the grottoes below. And overhanging all was the balustrade by his arm, which,

5. John Ledyard (1751–89), American traveler, whose comment appeared in *Proceedings of the Association for Promoting the Discovery of the Interior Parts of Africa* (London, 1790). Melville became confused because the Scottish traveler Mungo Park quoted this passage from Ledyard in his *Travels in the Interior of Africa*. The Putnam's version of *Benito Cereno* miscredits this quotation to Park, but Ledyard's name is properly substituted in *The Piazza Tales*.
6. Balcony projecting from the after part of a vessel's sides.
7. Ornamental arrangement, as of flower beds.

partly stained with pitch and partly embossed with moss, seemed the charred ruin of some summer-house in a grand garden long running to waste.

Trying to break one charm, he was but becharmed anew. Though upon the wide sea, he seemed in some far inland country; prisoner in some deserted château, left to stare at empty grounds, and peer out at vague roads, where never wagon or wayfarer passed.

But these enchantments were a little disenchanted as his eye fell on the corroded main-chains. Of an ancient style, massy and rusty in link, shackle and bolt, they seemed even more fit for the ship's present business than the one for which she had been built.

Presently he thought something moved nigh the chains. He rubbed his eyes, and looked hard. Groves of rigging were about the chains; and there, peering from behind a great stay, like an Indian from behind a hemlock, a Spanish sailor, a marlingspike in his hand, was seen, who made what seemed an imperfect gesture towards the balcony, but immediately, as if alarmed by some advancing step along the deck within, vanished into the recesses of the hempen forest, like a poacher.

What meant this? Something the man had sought to communicate, unbeknown to any one, even to his captain. Did the secret involve aught unfavorable to his captain? Were those previous misgivings of Captain Delano's about to be verified? Or, in his haunted mood at the moment, had some random, unintentional motion of the man, while busy with the stay, as if repairing it, been mistaken for a significant beckoning?

Not unbewildered, again he gazed off for his boat. But it was temporarily hidden by a rocky spur of the isle. As with some eagerness he bent forward, watching for the first shooting view of its beak, the balustrade gave way before him like charcoal. Had he not clutched an outreaching rope he would have fallen into the sea. The crash, though feeble, and the fall, though hollow, of the rotten fragments, must have been overheard. He glanced up. With sober curiosity peering down upon him was one of the old oakum-pickers, slipped from his perch to an outside boom; while below the old negro, and, invisible to him, reconnoitering from a port-hole like a fox from the mouth of its den, crouched the Spanish sailor again. From something suddenly suggested by the man's air, the mad idea now darted into Captain Delano's mind, that Don Benito's plea of indisposition, in withdrawing below, was but a pretense: that he was engaged there maturing his plot, of which the sailor, by some means gaining an inkling, had a mind to warn the stranger against; incited, it may be, by gratitude for a kind word on first boarding the ship. Was it from foreseeing some possible interference like this, that Don Benito had, beforehand, given such a bad character of his sailors, while praising the negroes; though, indeed, the former seemed as docile as the latter the contrary? The whites, too, by nature, were the shrewder race. A man with some evil design, would he not be likely to speak well of that stupidity which was blind to his depravity, and malign that intelligence from which it might not be hidden? Not unlikely, perhaps. But if the whites had dark secrets concerning Don Benito, could then Don Benito be any way in complicity with the blacks? But they were too stupid. Besides, who ever heard of a white so far a renegade as to apostatize[8] from his very species almost, by leaguing in against it with negroes? These difficulties recalled former ones. Lost in their mazes, Captain Delano, who had now regained the deck, was uneasily advancing along it, when he observed a new face; an aged sailor seated cross-legged near the main hatchway. His skin was shrunk up with wrinkles like a pelican's empty pouch; his hair frosted; his countenance grave and composed. His hands were full of ropes, which he was working into a large knot. Some blacks were about him obligingly dipping the strands for him, here and there, as the exigencies of the operation demanded.

Captain Delano crossed over to him, and stood in silence surveying the knot; his mind, by a not uncongenial transition, passing from its own entanglements to those of the hemp. For intricacy such a knot he had never seen in an American ship, or indeed any other. The old man looked like an Egyptian priest, making gordian knots for the temple of Ammon.[9] The knot seemed a combination of double-bowline-knot, treble-crown-knot, back-handed-well-knot, knot-in-and-out-knot, and jamming-knot.

At last, puzzled to comprehend the meaning of such a knot, Captain Delano addressed the knotter:—

"What are you knotting there, my man?"

"The knot," was the brief reply, without looking up.

"So it seems; but what is it for?"

"For some one else to undo," muttered back the old man, plying his fingers harder than ever, the knot being now nearly completed.

While Captain Delano stood watching him, suddenly the old man threw the knot towards him, saying in broken English,—the first heard in the ship,—something to this effect—"Undo it, cut it, quick." It was said lowly, but with such condensation of rapidity, that the long, slow words in Spanish, which had preceded and followed, almost operated as covers to the brief English between.

For a moment, knot in hand, and knot in head, Captain Delano stood mute; while, without further heeding him, the old man was now intent upon other ropes. Presently there was a slight stir behind Captain Delano. Turning, he saw the chained negro, Atufal, stand-

8. To deny or renounce (said of religious faith).
9. In Egypt the oracle of Jupiter Ammon predicted to Alexander the Great that he would conquer the world. Later in Phrygia (in north-central Asia Minor), where the former King Gordius had foretold that whoever would untie his intricate knot would become master of Asia, Alexander cut the knot with his sword.

ing quietly there. The next moment the old sailor rose, muttering, and, followed by his subordinate negroes, removed to the forward part of the ship, where in the crowd he disappeared.

An elderly negro, in a clout like an infant's, and with a pepper and salt head, and a kind of attorney air, now approached Captain Delano. In tolerable Spanish, and with a good-natured, knowing wink, he informed him that the old knotter was simple-witted, but harmless; often playing his odd tricks. The negro concluded by begging the knot, for of course the stranger would not care to be troubled with it. Unconsciously, it was handed to him. With a sort of congé,[1] the negro received it, and turning his back, ferreted into it like a detective Custom House officer after smuggled laces. Soon, with some African word, equivalent to pshaw, he tossed the knot overboard.

All this is very queer now, thought Captain Delano, with a qualmish sort of emotion; but as one feeling incipient sea-sickness, he strove, by ignoring the symptoms, to get rid of the malady. Once more he looked off for his boat. To his delight, it was now again in view, leaving the rocky spur astern.

The sensation here experienced, after at first relieving his uneasiness, with unforeseen efficacy, soon began to remove it. The less distant sight of that well-known boat—showing it, not as before, half blended with the haze, but with outline defined, so that its individuality, like a man's, was manifest; that boat, Rover by name, which, though now in strange seas, had often pressed the beach of Captain Delano's home, and, brought to its threshold for repairs, had familiarly lain there, as a Newfoundland dog; the sight of that household boat evoked a thousand trustful associations, which, contrasted with previous suspicions, filled him not only with lightsome confidence, but somehow with half humorous self-reproaches at his former lack of it.

"What, I, Amasa Delano—Jack of the Beach, as they called me when a lad—I, Amasa; the same that, duck-satchel in hand, used to paddle along the waterside to the school-house made from the old hulk;—I, little Jack of the Beach, that used to go berrying with cousin Nat and the rest; I to be murdered here at the ends of the earth, on board a haunted pirate-ship by a horrible Spaniard?—Too nonsensical to think of! Who would murder Amasa Delano? His conscience is clean. There is some one above. Fie, fie, Jack of the Beach! you are a child indeed; a child of the second childhood, old boy; you are beginning to dote and drule, I'm afraid."

Light of heart and foot, he stepped aft, and there was met by Don Benito's servant, who, with a pleasing expression, responsive to his own present feelings, informed him that his master had recovered from the effects of his coughing fit, and had just ordered him to go present his compliments to his good guest, Don Amasa, and

1. Leavetaking, signal by a low bow.

say that he (Don Benito) would soon have the happiness to rejoin him.

There now, do you mark that? again thought Captain Delano, walking the poop. What a donkey I was. This kind gentleman who here sends me his kind compliments, he, but ten minutes ago, dark-lantern in hand, was dodging round some old grind-stone in the hold, sharpening a hatchet for me, I thought. Well, well; these long calms have a morbid effect on the mind, I've often heard, though I never believed it before. Ha! glancing towards the boat; there's Rover; good dog; a white bone in her mouth. A pretty big bone though, seems to me.—What? Yes, she has fallen afoul of the bubbling tide-rip there. It sets her the other way, too, for the time. Patience.

It was now about noon, though, from the grayness of everything, it seemed to be getting towards dusk.

The calm was confirmed. In the far distance, away from the influence of land, the leaden ocean seemed laid out and leaded up, its course finished, soul gone, defunct. But the current from land-ward, where the ship was, increased; silently sweeping her further and further towards the tranced waters beyond.

Still, from his knowledge of those latitudes, cherishing hopes of a breeze, and a fair and fresh one, at any moment, Captain Delano, despite present prospects, buoyantly counted upon bringing the San Dominick safely to anchor ere night. The distance swept over was nothing; since, with a good wind, ten minutes' sailing would retrace more than sixty minutes' drifting. Meantime, one moment turning to mark "Rover" fighting the tide-rip, and the next to see Don Benito approaching, he continued walking the poop.

Gradually he felt a vexation arising from the delay of his boat; this soon merged into uneasiness; and at last, his eye falling contin-ually, as from a stage-box into the pit, upon the strange crowd before and below him, and by and by recognising there the face—now composed to indifference—of the Spanish sailor who had seemed to beckon from the main chains, something of his old trepidations returned.

Ah, thought he—gravely enough—this is like the ague: because it went off, it follows not that it won't come back.

Though ashamed of the relapse, he could not altogether subdue it; and so, exerting his good nature to the utmost, insensibly he came to a compromise.

Yes, this is a strange craft; a strange history, too, and strange folks on board. But—nothing more.

By way of keeping his mind out of mischief till the boat should arrive, he tried to occupy it with turning over and over, in a purely speculative sort of way, some lesser peculiarities of the captain and crew. Among others, four curious points recurred.

First, the affair of the Spanish lad assailed with a knife by the slave boy; an act winked at by Don Benito. Second, the tyranny in Don Benito's treatment of Atufal, the black; as if a child should lead a bull of the Nile by the ring in his nose. Third, the trampling of the sailor by the two negroes; a piece of insolence passed over without so much as a reprimand. Fourth, the cringing submission to their master of all the ship's underlings, mostly blacks; as if by the least inadvertence they feared to draw down his despotic displeasure.

Coupling these points, they seemed somewhat contradictory. But what then, thought Captain Delano, glancing towards his now nearing boat,—what then? Why, Don Benito is a very capricious commander. But he is not the first of the sort I have seen; though it's true he rather exceeds any other. But as a nation—continued he in his reveries—these Spaniards are all an odd set; the very word Spaniard has a curious, conspirator, Guy-Fawkish[2] twang to it. And yet, I dare say, Spaniards in the main are as good folks as any in Duxbury, Massachusetts. Ah good! At last "Rover" has come.

As, with its welcome freight, the boat touched the side, the oakum-pickers, with venerable gestures, sought to restrain the blacks, who, at the sight of three gurried[3] water-casks in its bottom, and a pile of wilted pumpkins in its bow, hung over the bulwarks in disorderly raptures.

Don Benito with his servant now appeared; his coming, perhaps, hastened by hearing the noise. Of him Captain Delano sought permission to serve out the water, so that all might share alike, and none injure themselves by unfair excess. But sensible, and, on Don Benito's account, kind as this offer was, it was received with what seemed impatience; as if aware that he lacked energy as a commander, Don Benito, with the true jealousy of weakness, resented as an affront any interference. So, at least, Captain Delano inferred.

In another moment the casks were being hoisted in, when some of the eager negroes accidentally jostled Captain Delano, where he stood by the gangway; so that, unmindful of Don Benito, yielding to the impulse of the moment, with good-natured authority he bade the blacks stand back; to enforce his words making use of a half-mirthful, half-menacing gesture. Instantly the blacks paused, just where they were, each negro and negress suspended in his or her posture, exactly as the word had found them—for a few seconds continuing so—while, as between the responsive posts of a telegraph, an unknown syllable ran from man to man among the perched oakum-pickers. While the visitor's attention was fixed by this scene, suddenly the hatchet-polishers half rose, and a rapid cry came from Don Benito.

Thinking that at the signal of the Spaniard he was about to be massacred, Captain Delano would have sprung for his boat, but

2. Guy Fawkes (1570–1606), Catholic conspirator executed for plotting to blow up the House of Lords.

3. Gurry: fish offal; here, "gurried": coated with slime.

paused, as the oakum-pickers, dropping down into the crowd with earnest exclamations, forced every white and every negro back, at the same moment, with gestures friendly and familiar, almost jocose, bidding him, in substance, not be a fool. Simultaneously the hatchet-polishers resumed their seats, quietly as so many tailors, and at once, as if nothing had happened, the work of hoisting in the casks was resumed, whites and blacks singing at the tackle.

Captain Delano glanced towards Don Benito. As he saw his meager form in the act of recovering itself from reclining in the servant's arms, into which the agitated invalid had fallen, he could not but marvel at the panic by which himself had been surprised on the darting supposition that such a commander, who upon a legitimate occasion, so trivial, too, as it now appeared, could lose all self-command, was, with energetic iniquity, going to bring about his murder.

The casks being on deck, Captain Delano was handed a number of jars and cups by one of the steward's aids, who, in the name of his captain, entreated him to do as he had proposed: dole out the water. He complied, with republican impartiality as to this republican element, which always seeks one level, serving the oldest white no better than the youngest black; excepting, indeed, poor Don Benito, whose condition, if not rank, demanded an extra allowance. To him, in the first place, Captain Delano presented a fair pitcher of the fluid; but, thirsting as he was for it, the Spaniard quaffed not a drop until after several grave bows and salutes. A reciprocation of courtesies which the sight-loving Africans hailed with clapping of hands.

Two of the less wilted pumpkins being reserved for the cabin table, the residue were minced up on the spot for the general regalement. But the soft bread, sugar, and bottled cider, Captain Delano would have given the whites alone, and in chief Don Benito; but the latter objected; which disinterestedness not a little pleased the American; and so mouthfuls all around were given alike to whites and blacks; excepting one bottle of cider, which Babo insisted upon setting aside for his master.

Here it may be observed that as, on the first visit of the boat, the American had not permitted his men to board the ship, neither did he now; being unwilling to add to the confusion of the decks.

Not uninfluenced by the peculiar good humor at present prevailing, and for the time oblivious of any but benevolent thoughts, Captain Delano, who from recent indications counted upon a breeze within an hour or two at furthest, dispatched the boat back to the sealer with orders for all the hands that could be spared immediately to set about rafting casks to the watering-place and filling them. Likewise he bade word be carried to his chief officer, that if against present expectation the ship was not brought to anchor by sunset, he need be under no concern, for as there was to be a full moon that night, he (Captain Delano) would remain on board ready to play the pilot, come the wind soon or late.

As the two Captains stood together, observing the departing boat —the servant as it happened having just spied a spot on his master's velvet sleeve, and silently engaged rubbing it out—the American expressed his regrets that the San Dominick had no boats; none, at least, but the unseaworthy old hulk of the long-boat, which, warped as a camel's skeleton in the desert, and almost as bleached, lay pot-wise inverted amidships, one side a little tipped, furnishing a subterraneous sort of den for family groups of the blacks, mostly women and small children; who, squatting on old mats below, or perched above in the dark dome, on the elevated seats, were descried, some distance within, like a social circle of bats, sheltering in some friendly cave; at intervals, ebon flights of naked boys and girls, three or four years old, darting in and out of the den's mouth.

"Had you three or four boats now, Don Benito," said Captain Delano, "I think that, by tugging at the oars, your negroes here might help along matters some.—Did you sail from port without boats, Don Benito?"

"They were stove in the gales, Señor."

"That was bad. Many men, too, you lost then. Boats and men.— Those must have been hard gales, Don Benito."

"Past all speech," cringed the Spaniard.

"Tell me, Don Benito," continued his companion with increased interest, "tell me, were these gales immediately off the pitch of Cape Horn?"

"Cape Horn?—who spoke of Cape Horn?"

"Yourself did, when giving me an account of your voyage," answered Captain Delano with almost equal astonishment at this eating of his own words, even as he ever seemed eating his own heart, on the part of the Spaniard. "You yourself, Don Benito, spoke of Cape Horn," he emphatically repeated.

The Spaniard turned, in a sort of stooping posture, pausing an instant, as one about to make a plunging exchange of elements, as from air to water.

At this moment a messenger-boy, a white, hurried by, in the regular performance of his function carrying the last expired half hour forward to the forecastle, from the cabin time-piece, to have it struck at the ship's large bell.

"Master," said the servant, discontinuing his work on the coat sleeve, and addressing the rapt Spaniard with a sort of timid apprehensiveness, as one charged with a duty, the discharge of which, it was foreseen, would prove irksome to the very person who had imposed it, and for whose benefit it was intended, "master told me never mind where he was, or how engaged, always to remind him, to a minute, when shaving-time comes. Miguel has gone to strike the half-hour afternoon. It is *now*, master. Will master go into the cuddy?"

"Ah—yes," answered the Spaniard, starting, somewhat as from dreams into realities; then turning upon Captain Delano, he said that ere long he would resume the conversation.

"Then if master means to talk more to Don Amasa," said the servant, "why not let Don Amasa sit by master in the cuddy, and master can talk, and Don Amasa can listen, while Babo here lathers and strops."

"Yes," said Captain Delano, not unpleased with this sociable plan, "yes, Don Benito, unless you had rather not, I will go with you."

"Be it so, Señor."

As the three passed aft, the American could not but think it another strange instance of his host's capriciousness, this being shaved with such uncommon punctuality in the middle of the day. But he deemed it more than likely that the servant's anxious fidelity had something to do with the matter; inasmuch as the timely interruption served to rally his master from the mood which had evidently been coming upon him.

The place called the cuddy was a light deck-cabin formed by the poop, a sort of attic to the large cabin below. Part of it had formerly been the quarters of the officers; but since their death all the partitionings had been thrown down, and the whole interior converted into one spacious and airy marine hall; for absence of fine furniture and picturesque disarray, of odd appurtenances, somewhat answering to the wide, cluttered hall of some eccentric bachelor-squire in the country, who hangs his shooting-jacket and tobacco-pouch on deer antlers, and keeps his fishing-rod, tongs, and walking-stick in the same corner.

The similitude was heightened, if not originally suggested, by glimpses of the surrounding sea; since, in one aspect, the country and the ocean seem cousins-german.

The floor of the cuddy was matted. Overhead, four or five old muskets were stuck into horizontal holes along the beams. On one side was a claw-footed old table lashed to the deck; a thumbed missal on it, and over it a small, meager crucifix attached to the bulkhead. Under the table lay a dented cutlass or two, with a hacked harpoon, among some melancholy old rigging, like a heap of poor friars' girdles. There were also two long, sharp-ribbed settees of malacca cane, black with age, and uncomfortable to look at as inquisitors' racks, with a large, misshapen arm-chair, which, furnished with a rude barber's crutch[4] at the back, working with a screw, seemed some grotesque engine of torment. A flag locker was in one corner, open, exposing various colored bunting, some rolled up, others half unrolled, still others tumbled. Opposite was a cumbrous washstand, of black mahogany, all of one block, with a pedestal, like a font, and over it a railed shelf, containing combs, brushes, and other implements of the toilet. A torn hammock of

4. Headrest.

stained grass swung near; the sheets tossed, and the pillow wrinkled up like a brow, as if whoever slept here slept but illy, with alternate visitations of sad thoughts and bad dreams.

The further extremity of the cuddy, overhanging the ship's stern, was pierced with three openings, windows or port holes, according as men or cannon might peer, socially or unsocially, out of them. At present neither men nor cannon were seen, though huge ring-bolts and other rusty iron fixtures of the wood-work hinted of twenty-four-pounders.

Glancing towards the hammock as he entered, Captain Delano said, "You sleep here, Don Benito?"

"Yes, Señor, since we got into mild weather."

"This seems a sort of dormitory, sitting-room, sail-loft, chapel, armory, and private closet all together, Don Benito," added Captain Delano, looking round.

"Yes, Señor; events have not been favorable to much order in my arrangements."

Here the servant, napkin on arm, made a motion as if waiting his master's good pleasure. Don Benito signified his readiness, when, seating him in the malacca arm-chair, and for the guest's convenience drawing opposite it one of the settees, the servant commenced operations by throwing back his master's collar and loosening his cravat.

There is something in the negro which, in a peculiar way, fits him for avocations about one's person. Most negroes are natural valets and hair-dressers; taking to the comb and brush congenially as to the castinets, and flourishing them apparently with almost equal satisfaction. There is, too, a smooth tact about them in this employment, with a marvelous, noiseless, gliding briskness, not ungraceful in its way, singularly pleasing to behold, and still more so to be the manipulated subject of. And above all is the great gift of good humor. Not the mere grin or laugh is here meant. Those were unsuitable. But a certain easy cheerfulness, harmonious in every glance and gesture; as though God had set the whole negro to some pleasant tune.

When to all this is added the docility arising from the unaspiring contentment of a limited mind, and that susceptibility of blind attachment sometimes inhering in indisputable inferiors, one readily perceives why those hypochondriacs, Johnson and Byron[5]—it may be something like the hypochondriac, Benito Cereno—took to their hearts, almost to the exclusion of the entire white race, their serving men, the negroes, Barber and Fletcher. But if there be that in the negro which exempts him from the inflicted sourness of the morbid or cynical mind, how, in his most prepossessing aspects, must he

5. Frank Barber was servant to Samuel Johnson for three decades; at his death in 1784 Johnson left Barber a large annuity. William Fletcher, Lord Byron's valet, was a white Englishman, here confused with an American black servant of Edward Trelawny who accompanied Byron on some journeys.

appear to a benevolent one? When at ease with respect to exterior things, Captain Delano's nature was not only benign, but familiarly and humorously so. At home, he had often taken rare satisfaction in sitting in his door, watching some free man of color at his work or play. If on a voyage he chanced to have a black sailor, invariably he was on chatty, and half-gamesome terms with him. In fact, like most men of a good, blithe heart, Captain Delano took to negroes, not philanthropically, but genially, just as other men to Newfoundland dogs.

Hitherto the circumstances in which he found the San Dominick had repressed the tendency. But in the cuddy, relieved from his former uneasiness, and, for various reasons, more sociably inclined than at any previous period of the day, and seeing the colored servant, napkin on arm, so debonair about his master, in a business so familiar as that of shaving, too, all his old weakness for negroes returned.

Among other things, he was amused with an odd instance of the African love of bright colors and fine shows, in the black's informally taking from the flag-locker a great piece of bunting of all hues, and lavishly tucking it under his master's chin for an apron.

The mode of shaving among the Spaniards is a little different from what it is with other nations. They have a basin, specifically called a barber's basin, which on one side is scooped out, so as accurately to receive the chin, against which it is closely held in lathering; which is done, not with a brush, but with soap dipped in the water of the basin and rubbed on the face.

In the present instance salt-water was used for lack of better; and the parts lathered were only the upper lip, and low down under the throat, all the rest being cultivated beard.

The preliminaries being somewhat novel to Captain Delano, he sat curiously eying them, so that no conversation took place, nor for the present did Don Benito appear disposed to renew any.

Setting down his basin, the negro searched among the razors, as for the sharpest, and having found it, gave it an additional edge by expertly strapping it on the firm, smooth, oily skin of his open palm; he then made a gesture as if to begin, but midway stood suspended for an instant, one hand elevating the razor, the other professionally dabbling among the bubbling suds on the Spaniard's lank neck. Not unaffected by the close sight of the gleaming steel, Don Benito nervously shuddered, his usual ghastliness was hightened by the lather, which lather, again, was intensified in its hue by the contrasting sootiness of the negro's body. Altogether the scene was somewhat peculiar, at least to Captain Delano, nor, as he saw the two thus postured, could he resist the vagary, that in the black he saw a headsman, and in the white, a man at the block. But this was one of those antic conceits, appearing and vanishing in a breath, from which, perhaps, the best regulated mind is not always free.

Meantime the agitation of the Spaniard had a little loosened the bunting from around him, so that one broad fold swept curtain-like over the chair-arm to the floor, revealing, amid a profusion of armorial bars and ground-colors—black, blue, and yellow—a closed castle in a blood-red field diagonal with a lion rampant in a white.

"The castle and the lion," exclaimed Captain Delano—"why, Don Benito, this is the flag of Spain you use here. It's well it's only I, and not the King, that sees this," he added with a smile, "but"— turning towards the black,—"it's all one, I suppose, so the colors be gay;" which playful remark did not fail somewhat to tickle the negro.

"Now, master," he said, readjusting the flag, and pressing the head gently further back into the crotch of the chair; "now master," and the steel glanced nigh the throat.

Again Don Benito faintly shuddered.

"You must not shake so, master.—See, Don Amasa, master always shakes when I shave him. And yet master knows I never yet have drawn blood, though it's true, if master will shake so, I may some of these times. Now master," he continued. "And now, Don Amasa, please go on with your talk about the gale, and all that, master can hear, and between times master can answer."

"Ah yes, these gales," said Captain Delano; "but the more I think of your voyage, Don Benito, the more I wonder, not at the gales, terrible as they must have been, but at the disastrous interval following them. For here, by your account, have you been these two months and more getting from Cape Horn to St. Maria, a distance which I myself, with a good wind, have sailed in a few days. True, you had calms, and long ones, but to be becalmed for two months, that is, at least, unusual. Why, Don Benito, had almost any other gentleman told me such a story, I should have been half disposed to a little incredulity."

Here an involuntary expression came over the Spaniard, similar to that just before on the deck, and whether it was the start he gave, or a sudden gawky roll of the hull in the calm, or a momentary unsteadiness of the servant's hand; however it was, just then the razor drew blood, spots of which stained the creamy lather under the throat; immediately the black barber drew back his steel, and remaining in his professional attitude, back to Captain Delano, and face to Don Benito, held up the trickling razor, saying, with a sort of half humorous sorrow, "See, master,—you shook so—here's Babo's first blood."

No sword drawn before James the First of England, no assassination in that timid King's presence,[6] could have produced a more terrified aspect than was now presented by Don Benito.

6. James I (1566–1625), King of Great Britain and Ireland 1604–25, after having reigned as King of Scotland since 1567, lived in terror of assassination by Catholics, especially after the Gunpowder Plot (1605) and the assassination of King Henry IV of France in 1610.

Poor fellow, thought Captain Delano, so nervous he can't even bear the sight of barber's blood; and this unstrung, sick man, is it credible that I should have imagined he meant to spill all my blood, who can't endure the sight of one little drop of his own? Surely, Amasa Delano, you have been beside yourself this day. Tell it not when you get home, sappy Amasa. Well, well, he looks like a murderer, doesn't he? More like as if himself were to be done for. Well, well, this day's experience shall be a good lesson.

Meantime, while these things were running through the honest seaman's mind, the servant had taken the napkin from his arm, and to Don Benito had said—"But answer Don Amasa, please, master, while I wipe this ugly stuff off the razor, and strop it again."

As he said the words, his face was turned half round, so as to be alike visible to the Spaniard and the American, and seemed by its expression to hint, that he was desirous, by getting his master to go on with the conversation, considerately to withdraw his attention from the recent annoying accident. As if glad to snatch the offered relief, Don Benito resumed, rehearsing to Captain Delano, that not only were the calms of unusual duration, but the ship had fallen in with obstinate currents; and other things he added, some of which were but repetitions of former statements, to explain how it came to pass that the passage from Cape Horn to St. Maria had been so exceedingly long, now and then mingling with his words, incidental praises, less qualified than before, to the blacks, for their general good conduct.

These particulars were not given consecutively, the servant at convenient times using his razor, and so, between the intervals of shaving, the story and panegyric went on with more than usual huskiness.

To Captain Delano's imagination, now again not wholly at rest, there was something so hollow in the Spaniard's manner, with apparently some reciprocal hollowness in the servant's dusky comment of silence, that the idea flashed across him, that possibly master and man, for some unknown purpose, were acting out, both in word and deed, nay, to the very tremor of Don Benito's limbs, some juggling play before him. Neither did the suspicion of collusion lack apparent support, from the fact of those whispered conferences before mentioned. But then, what could be the object of enacting this play of the barber before him? At last, regarding the notion as a whimsy, insensibly suggested, perhaps, by the theatrical aspect of Don Benito in his harlequin ensign, Captain Delano speedily banished it.

The shaving over, the servant bestirred himself with a small bottle of scented waters, pouring a few drops on the head, and then diligently rubbing; the vehemence of the exercise causing the muscles of his face to twitch rather strangely.

His next operation was with comb, scissors and brush; going

round and round, smoothing a curl here, clipping an unruly whisker-hair there, giving a graceful sweep to the temple-lock, with other impromptu touches evincing the hand of a master; while, like any resigned gentleman in barber's hands, Don Benito bore all, much less uneasily, at least, than he had done the razoring; indeed, he sat so pale and rigid now, that the negro seemed a Nubian sculptor finishing off a white statue-head.

All being over at last, the standard of Spain removed, tumbled up, and tossed back into the flag-locker, the negro's warm breath blowing away any stray hair which might have lodged down his master's neck; collar and cravat readjusted; a speck of lint whisked off the velvet lapel; all this being done; backing off a little space, and pausing with an expression of subdued self-complacency, the servant for a moment surveyed his master, as, in toilet at least, the creature of his own tasteful hands.

Captain Delano playfully complimented him upon his achievement; at the same time congratulating Don Benito.

But neither sweet waters, nor shampooing, nor fidelity, nor social-ity, delighted the Spaniard. Seeing him relapsing into forbidding gloom, and still remaining seated, Captain Delano, thinking that his presence was undesired just then, withdrew, on pretense of seeing whether, as he had prophecied, any signs of a breeze were visible.

Walking forward to the mainmast, he stood awhile thinking over the scene, and not without some undefined misgivings, when he heard a noise near the cuddy, and turning, saw the negro, his hand to his cheek. Advancing, Captain Delano perceived that the cheek was bleeding. He was about to ask the cause, when the negro's wailing soliloquy enlightened him.

"Ah, when will master get better from his sickness; only the sour heart that sour sickness breeds made him serve Babo so; cutting Babo with the razor, because, only by accident, Babo had given master one little scratch; and for the first time in so many a day, too. Ah, ah, ah," holding his hand to his face.

Is it possible, thought Captain Delano; was it to wreak in private his Spanish spite against this poor friend of his, that Don Benito, by his sullen manner, impelled me to withdraw? Ah, this slavery breeds ugly passions in man—Poor fellow!

He was about to speak in sympathy to the negro, but with a timid reluctance he now reëntered the cuddy.

Presently master and man came forth; Don Benito leaning on his servant as if nothing had happened.

But a sort of love-quarrel, after all, thought Captain Delano.

He accosted Don Benito, and they slowly walked together. They had gone but a few paces, when the steward—a tall, rajah-looking mulatto, orientally set off with a pagoda turban formed by three or four Madras handkerchiefs wound about his head, tier on tier—approaching with a saalam, announced lunch in the cabin.

On their way thither, the two Captains were preceded by the mulatto, who, turning round as he advanced, with continual smiles and bows, ushered them on, a display of elegance which quite completed the insignificance of the small bare-headed Babo, who, as if not unconscious of inferiority, eyed askance the graceful steward. But in part, Captain Delano imputed his jealous watchfulness to that peculiar feeling which the full-blooded African entertains for the adulterated one. As for the steward, his manner, if not bespeaking much dignity of self-respect, yet evidenced his extreme desire to please; which is doubly meritorious, as at once Christian and Chesterfieldian.[7]

Captain Delano observed with interest that while the complexion of the mulatto was hybrid, his physiognomy was European; classically so.

"Don Benito," whispered he, "I am glad to see this usher-of-the-golden-rod[8] of yours; the sight refutes an ugly remark once made to me by a Barbardoes planter; that when a mulatto has a regular European face, look out for him; he is a devil. But see, your steward here has features more regular than King George's of England; and yet there he nods, and bows, and smiles; a king, indeed —the king of kind hearts and polite fellows. What a pleasant voice he has, too?"

"He has, Señor."

"But, tell me, has he not, so far as you have known him, always proved a good, worthy fellow?" said Captain Delano, pausing, while with a final genuflexion the steward disappeared into the cabin; "come, for the reason just mentioned, I am curious to know."

"Francesco is a good man," a sort of sluggishly responded Don Benito, like a phlegmatic appreciator, who would neither find fault nor flatter.

"Ah, I thought so. For it were strange indeed, and not very creditable to us white-skins, if a little of our blood mixed with the African's, should, far from improving the latter's quality, have the sad effect of pouring vitriolic acid into black broth; improving the hue, perhaps, but not the wholesomeness."

"Doubtless, doubtless, Señor, but"—glancing at Babo—"not to speak of negroes, your planter's remark I have heard applied to the Spanish and Indian intermixtures in our provinces. But I know nothing about the matter," he listlessly added.

And here they entered the cabin.

The lunch was a frugal one. Some of Captain Delano's fresh fish and pumpkins, biscuit and salt beef, the reserved bottle of cider, and the San Dominick's last bottle of Canary.

7. Philip Stanhope, the fourth Earl of Chesterfield (1694–1773), in his letters to his son advocated a worldly code at variance with Jesus' absolute morality.
8. In this English usage, usher means an attendant charged with walking ceremoniously before a person of rank; certain ushers were known by the color of the rod or scepter they traditionally carried.

As they entered, Francesco, with two or three colored aids, was hovering over the table giving the last adjustments. Upon perceiving their master they withdrew, Francesco making a smiling congé, and the Spaniard, without condescending to notice it, fastidiously remarking to his companion that he relished not superfluous attendance.

Without companions, host and guest sat down, like a childless married couple, at opposite ends of the table, Don Benito waving Captain Delano to his place, and, weak as he was, insisting upon that gentleman being seated before himself.

The negro placed a rug under Don Benito's feet, and a cushion behind his back, and then stood behind, not his master's chair, but Captain Delano's. At first, this a little surprised the latter. But it was soon evident that, in taking his position, the black was still true to his master; since by facing him he could the more readily anticipate his slightest want.

"This is an uncommonly intelligent fellow of yours, Don Benito," whispered Captain Delano across the table.

"You say true, Señor."

During the repast, the guest again reverted to parts of Don Benito's story, begging further particulars here and there. He inquired how it was that the scurvy and fever should have committed such wholesale havoc upon the whites, while destroying less than half of the blacks. As if this question reproduced the whole scene of plague before the Spaniard's eyes, miserably reminding him of his solitude in a cabin where before he had had so many friends and officers around him, his hand shook, his face became hueless, broken words escaped; but directly the sane memory of the past seemed replaced by insane terrors of the present. With starting eyes he stared before him at vacancy. For nothing was to be seen but the hand of his servant pushing the Canary over towards him. At length a few sips served partially to restore him. He made random reference to the different constitution of races, enabling one to offer more resistance to certain maladies than another. The thought was new to his companion.

Presently Captain Delano, intending to say something to his host concerning the pecuniary part of the business he had undertaken for him, especially—since he was strictly accountable to his owners—with reference to the new suit of sails, and other things of that sort; and naturally preferring to conduct such affairs in private, was desirous that the servant should withdraw; imagining that Don Benito for a few minutes could dispense with his attendance. He, however, waited awhile; thinking that, as the conversation proceeded, Don Benito, without being prompted, would perceive the propriety of the step.

But it was otherwise. At last catching his host's eye, Captain Delano, with a slight backward gesture of his thumb, whispered,

"Don Benito, pardon me, but there is an interference with the full expression of what I have to say to you."

Upon this the Spaniard changed countenance; which was imputed to his resenting the hint, as in some way a reflection upon his servant. After a moment's pause, he assured his guest that the black's remaining with them could be of no disservice; because since losing his officers he had made Babo (whose original office, it now appeared, had been captain of the slaves) not only his constant attendant and companion, but in all things his confidant.

After this, nothing more could be said; though, indeed, Captain Delano could hardly avoid some little tinge of irritation upon being left ungratified in so inconsiderable a wish, by one, too, for whom he intended such solid services. But it is only his querulousness, thought he; and so filling his glass he proceeded to business.

The price of the sails and other matters was fixed upon. But while this was being done, the American observed that, though his original offer of assistance had been hailed with hectic animation, yet now when it was reduced to a business transaction, indifference and apathy were betrayed. Don Benito, in fact, appeared to submit to hearing the details more out of regard to common propriety, than from any impression that weighty benefit to himself and his voyage was involved.

Soon, his manner became still more reserved. The effort was vain to seek to draw him into social talk. Gnawed by his splenetic mood, he sat twitching his beard, while to little purpose the hand of his servant, mute as that on the wall, slowly pushed over the Canary.

Lunch being over, they sat down on the cushioned transom; the servant placing a pillow behind his master. The long continuance of the calm had now affected the atmosphere. Don Benito sighed heavily, as if for breath.

"Why not adjourn to the cuddy," said Captain Delano; "there is more air there." But the host sat silent and motionless.

Meantime his servant knelt before him, with a large fan of feathers. And Francesco coming in on tiptoes, handed the negro a little cup of aromatic waters, with which at intervals he chafed his master's brow; smoothing the hair along the temples as a nurse does a child's. He spoke no word. He only rested his eye on his master's, as if, amid all Don Benito's distress, a little to refresh his spirit by the silent sight of fidelity.

Presently the ship's bell sounded two o'clock; and through the cabin-windows a slight rippling of the sea was discerned; and from the desired direction.

"There," exclaimed Captain Delano, "I told you so, Don Benito, look!"

He had risen to his feet, speaking in a very animated tone, with a view the more to rouse his companion. But though the crimson curtain of the stern-window near him that moment fluttered against his pale cheek, Don Benito seemed to have even less welcome for the breeze than the calm.

Poor fellow, thought Captain Delano, bitter experience has taught him that one ripple does not make a wind, any more than one swallow a summer. But he is mistaken for once. I will get his ship in for him, and prove it.

Briefly alluding to his weak condition, he urged his host to remain quietly where he was, since he (Captain Delano) would with pleasure take upon himself the responsibility of making the best use of the wind.

Upon gaining the deck, Captain Delano started at the unexpected figure of Atufal, monumentally fixed at the threshold, like one of those sculptured porters of black marble guarding the porches of Egyptian tombs.

But this time the start was, perhaps, purely physical. Atufal's presence, singularly attesting docility even in sullenness, was contrasted with that of the hatchet-polishers, who in patience evinced their industry; while both spectacles showed, that lax as Don Benito's general authority might be, still, whenever he chose to exert it, no man so savage or colossal but must, more or less, bow.

Snatching a trumpet which hung from the bulwarks, with a free step Captain Delano advanced to the forward edge of the poop, issuing his orders in his best Spanish. The few sailors and many negroes, all equally pleased, obediently set about heading the ship towards the harbor.

While giving some directions about setting a lower stu'n'-sail, suddenly Captain Delano heard a voice faithfully repeating his orders. Turning, he saw Babo, now for the time acting, under the pilot, his original part of captain of the slaves. This assistance proved valuable. Tattered sails and warped yards were soon brought into some trim. And no brace or halyard was pulled but to the blithe songs of the inspirited negroes.

Good fellows, thought Captain Delano, a little training would make fine sailors of them. Why see, the very women pull and sing too. These must be some of those Ashantee negresses that make such capital soldiers, I've heard. But who's at the helm. I must have a good hand there.

He went to see.

The San Dominick steered with a cumbrous tiller, with large horizontal pullies attached. At each pully-end stood a subordinate black, and between them, at the tiller-head, the responsible post, a Spanish seaman, whose countenance evinced his due share in the general hopefulness and confidence at the coming of the breeze.

He proved the same man who had behaved with so shame-faced an air on the windlass.

"Ah—it is you, my man," exclaimed Captain Delano—"well, no more sheep's-eyes now;—look straightforward and keep the ship so.

Good hand, I trust? And want to get into the harbor, don't you?"

The man assented with an inward chuckle, grasping the tiller-head firmly. Upon this, unperceived by the American, the two blacks eyed the sailor intently.

Finding all right at the helm, the pilot went forward to the forecastle, to see how matters stood there.

The ship now had way enough to breast the current. With the approach of evening, the breeze would be sure to freshen.

Having done all that was needed for the present, Captain Delano, giving his last orders to the sailors, turned aft to report affairs to Don Benito in the cabin; perhaps additionally incited to rejoin him by the hope of snatching a moment's private chat while the servant was engaged upon deck.

From opposite sides, there were, beneath the poop, two approaches to the cabin; one further forward than the other, and consequently communicating with a longer passage. Marking the servant still above, Captain Delano, taking the nighest entrance—the one last named, and at whose porch Atufal still stood—hurried on his way, till, arrived at the cabin threshold, he paused an instant, a little to recover from his eagerness. Then, with the words of his intended business upon his lips, he entered. As he advanced toward the seated Spaniard, he heard another footstep, keeping time with his. From the opposite door, a salver in hand, the servant was likewise advancing.

"Confound the faithful fellow," thought Captain Delano; "what a vexatious coincidence."

Possibly, the vexation might have been something different, were it not for the brisk confidence inspired by the breeze. But even as it was, he felt a slight twinge, from a sudden indefinite association in his mind of Babo with Atufal.

"Don Benito," said he, "I give you joy; the breeze will hold, and will increase. By the way, your tall man and time-piece, Atufal, stands without. By your order, of course?"

Don Benito recoiled, as if at some bland satirical touch, delivered with such adroit garnish of apparent good-breeding as to present no handle for retort.

He is like one flayed alive, thought Captain Delano; where may one touch him without causing a shrink?

The servant moved before his master, adjusting a cushion; recalled to civility, the Spaniard stiffly replied: "You are right. The slave appears where you saw him, according to my command; which is, that if at the given hour I am below, he must take his stand and abide my coming."

"Ah now, pardon me, but that is treating the poor fellow like an ex-king indeed. Ah, Don Benito," smiling, "for all the license you permit in some things, I fear lest, at bottom, you are a bitter hard master."

Again Don Benito shrank; and this time, as the good sailor thought, from a genuine twinge of his conscience.

Again conversation became constrained. In vain Captain Delano called attention to the now perceptible motion of the keel gently cleaving the sea; with lack-lustre eye, Don Benito returned words few and reserved.

By-and-by, the wind having steadily risen, and still blowing right into the harbor, bore the San Dominick swiftly on. Rounding a point of land, the sealer at distance came into open view.

Meantime Captain Delano had again repaired to the deck, remaining there some time. Having at last altered the ship's course, so as to give the reef a wide berth, he returned for a few moments below.

I will cheer up my poor friend, this time, thought he.

"Better and better, Don Benito," he cried as he blithely reëntered; "there will soon be an end to your cares, at least for awhile. For when, after a long, sad voyage, you know, the anchor drops into the haven, all its vast weight seems lifted from the captain's heart. We are getting on famously, Don Benito. My ship is in sight. Look through this side-light here; there she is; all a-taunt-o! The Bachelor's Delight, my good friend. Ah, how this wind braces one up. Come, you must take a cup of coffee with me this evening. My old steward will give you as fine a cup as ever any sultan tasted. What say you, Don Benito, will you?"

At first, the Spaniard glanced feverishly up, casting a longing look towards the sealer, while with mute concern his servant gazed into his face. Suddenly the old ague of coldness returned, and dropping back to his cushions he was silent.

"You do not answer. Come, all day you have been my host; would you have hospitality all on one side?"

"I cannot go," was the response.

"What? it will not fatigue you. The ships will lie together as near as they can, without swinging foul. It will be little more than stepping from deck to deck; which is but as from room to room. Come, come, you must not refuse me."

"I cannot go," decisively and repulsively repeated Don Benito.

Renouncing all but the last appearance of courtesy, with a sort of cadaverous sullenness, and biting his thin nails to the quick, he glanced, almost glared, at his guest; as if impatient that a stranger's presence should interfere with the full indulgence of his morbid hour. Meantime the sound of the parted waters came more and more gurglingly and merrily in at the windows; as reproaching him for his dark spleen; as telling him that, sulk as he might, and go mad with it, nature cared not a jot; since, whose fault was it, pray?

But the foul mood was now at its depth, as the fair wind at its hight.

There was something in the man so far beyond any mere un-sociality or sourness previously evinced, that even the forbearing good-nature of his guest could no longer endure it. Wholly at a loss to account for such demeanor, and deeming sickness with eccentric-ity, however extreme, no adequate excuse, well satisfied, too, that nothing in his own conduct could justify it, Captain Delano's pride began to be roused. Himself became reserved. But all seemed one to the Spaniard. Quitting him, therefore, Captain Delano once more went to the deck.

The ship was now within less than two miles of the sealer. The whale-boat was seen darting over the interval.

To be brief, the two vessels, thanks to the pilot's skill, ere long in neighborly style lay anchored together.

Before returning to his own vessel, Captain Delano had intended communicating to Don Benito the smaller details of the proposed services to be rendered. But, as it was, unwilling anew to subject himself to rebuffs, he resolved, now that he had seen the San Dominick safely moored, immediately to quit her, without further allusion to hospitality or business. Indefinitely postponing his ul-terior plans, he would regulate his future actions according to future circumstances. His boat was ready to receive him; but his host still tarried below. Well, thought Captain Delano, if he has little breed-ing, the more need to show mine. He descended to the cabin to bid a ceremonious, and, it may be, tacitly rebukeful adieu. But to his great satisfaction, Don Benito, as if he began to feel the weight of that treatment with which his slighted guest had, not indecorously, retaliated upon him, now supported by his servant, rose to his feet, and grasping Captain Delano's hand, stood tremulous; too much agitated to speak. But the good augury hence drawn was suddenly dashed, by his resuming all his previous reserve, with augmented gloom, as, with half-averted eyes, he silently reseated himself on his cushions. With a corresponding return of his own chilled feelings, Captain Delano bowed and withdrew.

He was hardly midway in the narrow corridor, dim as a tunnel, leading from the cabin to the stairs, when a sound, as of the tolling for execution in some jail-yard, fell on his ears. It was the echo of the ship's flawed bell, striking the hour, drearily reverberated in this subterranean vault. Instantly, by a fatality not to be withstood, his mind, responsive to the portent, swarmed with superstitious suspi-cions. He paused. In images far swifter than these sentences, the minutest details of all his former distrusts swept through him.

Hitherto, credulous good-nature had been too ready to furnish excuses for reasonable fears. Why was the Spaniard, so superflu-ously punctilious at times, now heedless of common propriety in not accompanying to the side his departing guest? Did indisposition forbid? Indisposition had not forbidden more irksome exertion that day. His last equivocal demeanor recurred. He had risen to his feet, grasped his guest's hand, motioned toward his hat; then, in an instant, all was eclipsed in sinister muteness and gloom. Did this imply one brief, repentent relenting at the final moment, from some iniquitous plot, followed by remorseless return to it? His last glance seemed to express a calamitous, yet acquiescent farewell to Captain Delano forever. Why decline the invitation to visit the sealer that evening? Or was the Spaniard less hardened than the Jew, who refrained not from supping at the board of him whom the same night he meant to betray?[9] What imported all those day-long enigmas and contradictions, except they were intended to mystify, preliminary to some stealthy blow? Atufal, the pretended rebel, but punctual shadow, that moment lurked by the threshold without. He seemed a sentry, and more. Who, by his own confession, had stationed him there? Was the negro now lying in wait?

The Spaniard behind—his creature before: to rush from darkness to light was the involuntary choice.

The next moment, with clenched jaw and hand, he passed Atufal, and stood unharmed in the light. As he saw his trim ship lying peacefully at anchor, and almost within ordinary call; as he saw his household boat, with familiar faces in it, patiently rising and falling on the short waves by the San Dominick's side; and then, glancing about the decks where he stood, saw the oakum-pickers still gravely plying their fingers; and heard the low, buzzing whistle and industrious hum of the hatchet-polishers, still bestirring themselves over their endless occupation; and more than all, as he saw the benign aspect of nature, taking her innocent repose in the evening; the screened sun in the quiet camp of the west shining out like the mild light from Abraham's tent;[1] as charmed eye and ear took in all these, with the chained figure of the black, clenched jaw and hand relaxed. Once again he smiled at the phantoms which had mocked him, and felt something like a tinge of remorse, that, by harboring them even for a moment, he should, by implication, have betrayed an atheist doubt of the ever-watchful Providence above.

There was a few minutes' delay, while, in obedience to his orders, the boat was being hooked along to the gangway. During this interval, a sort of saddened satisfaction stole over Captain Delano, at thinking of the kindly offices he had that day discharged for a stranger. Ah, thought he, after good actions one's conscience is never ungrateful, however much so the benefited party may be.

Presently, his foot, in the first act of descent into the boat, pressed the first round of the side-ladder, his face presented inward upon the deck. In the same moment, he heard his name courteously sounded; and, to his pleased surprise, saw Don Benito advancing—an unwonted energy in his air, as if, at the last moment, intent upon making amends for his recent discourtesy. With instinctive good

9. The Jew is Judas (Matthew 26).
1. Perhaps a distorted recollection of Genesis 18.1.

feeling, Captain Delano, withdrawing his foot, turned and reciprocally advanced. As he did so, the Spaniard's nervous eagerness increased, but his vital energy failed; so that, the better to support him, the servant, placing his master's hand on his naked shoulder, and gently holding it there, formed himself into a sort of crutch.

When the two captains met, the Spaniard again fervently took the hand of the American, at the same time casting an earnest glance into his eyes, but, as before, too much overcome to speak.

I have done him wrong, self-reproachfully thought Captain Delano; his apparent coldness has deceived me; in no instance has he meant to offend.

Meantime, as if fearful that the continuance of the scene might too much unstring his master, the servant seemed anxious to terminate it. And so, still presenting himself as a crutch, and walking between the two captains, he advanced with them towards the gangway; while still, as if full of kindly contrition, Don Benito would not let go the hand of Captain Delano, but retained it in his, across the black's body.

Soon they were standing by the side, looking over into the boat, whose crew turned up their curious eyes. Waiting a moment for the Spaniard to relinquish his hold, the now embarrassed Captain Delano lifted his foot, to overstep the threshold of the open gangway; but still Don Benito would not let go his hand. And yet, with an agitated tone, he said, "I can go no further; here I must bid you adieu. Adieu, my dear, dear Don Amasa. Go—go!" suddenly tearing his hand loose, "go, and God guard you better than me, my best friend."

Not unaffected, Captain Delano would now have lingered; but catching the meekly admonitory eye of the servant, with a hasty farewell he descended into his boat, followed by the continual adieus of Don Benito, standing rooted in the gangway.

Seating himself in the stern, Captain Delano, making a last salute, ordered the boat shoved off. The crew had their oars on end. The bowsman pushed the boat a sufficient distance for the oars to be lengthwise dropped. The instant that was done, Don Benito sprang over the bulwarks, falling at the feet of Captain Delano; at the same time, calling towards his ship, but in tones so frenzied, that none in the boat could understand him. But, as if not equally obtuse, three sailors, from three different and distant parts of the ship, splashed into the sea, swimming after their captain, as if intent upon his rescue.

The dismayed officer of the boat eagerly asked what this meant. To which, Captain Delano, turning a disdainful smile upon the unaccountable Spaniard, answered that, for his part, he neither knew nor cared; but it seemed as if Don Benito had taken it into his head to produce the impression among his people that the boat wanted to kidnap him. "Or else—give way for your lives," he wildly added, starting at a clattering hubbub in the ship, above which rang the tocsin of the hatchet-polishers; and seizing Don Benito by the throat he added, "this plotting pirate means murder!" Here, in apparent verification of the words, the servant, a dagger in his hand, was seen on the rail overhead, poised, in the act of leaping, as if with desperate fidelity to befriend his master to the last; while, seemingly to aid the black, the three white sailors were trying to clamber into the hampered bow. Meantime, the whole host of negroes, as if inflamed at the sight of their jeopardized captain, impended in one sooty avalanche over the bulwarks.

All this, with what preceded, and what followed, occurred with such involutions of rapidity, that past, present, and future seemed one.

Seeing the negro coming, Captain Delano had flung the Spaniard aside, almost in the very act of clutching him, and, by the unconscious recoil, shifting his place, with arms thrown up, so promptly grappled the servant in his descent, that with dagger presented at Captain Delano's heart, the black seemed of purpose to have leaped there as to his mark. But the weapon was wrenched away, and the assailant dashed down into the bottom of the boat,which now, with disentangled oars, began to speed through the sea.

At this juncture, the left hand of Captain Delano, on one side, again clutched the half-reclined Don Benito, heedless that he was in a speechless faint, while his right foot, on the other side, ground the prostrate negro; and his right arm pressed for added speed on the after oar, his eye bent forward, encouraging his men to their utmost.

But here, the officer of the boat, who had at last succeeded in beating off the towing sailors, and was now, with face turned aft, assisting the bowsman at his oar, suddenly called to Captain Delano, to see what the black was about; while a Portuguese oarsman shouted to him to give heed to what the Spaniard was saying.

Glancing down at his feet, Captain Delano saw the freed hand of the servant aiming with a second dagger—a small one, before concealed in his wool—with this he was snakishly writhing up from the boat's bottom, at the heart of his master, his countenance lividly vindictive, expressing the centred purpose of his soul; while the Spaniard, half-choked, was vainly shrinking away, with husky words, incoherent to all but the Portuguese.

That moment, across the long-benighted mind of Captain Delano, a flash of revelation swept, illuminating in unanticipated clearness, his host's whole mysterious demeanor, with every enigmatic event of the day, as well as the entire past voyage of the San Dominick. He smote Babo's hand down, but his own heart smote him harder. With infinite pity he withdrew his hold from Don Benito. Not Captain Delano, but Don Benito, the black, in leaping into the boat, had intended to stab.

Both the black's hands were held, as, glancing up towards the San Dominick, Captain Delano, now with scales dropped from his eyes, saw the negroes, not in misrule, not in tumult, not as if frantically concerned for Don Benito, but with mask torn away, flourishing hatchets and knives, in ferocious piratical revolt. Like delirious black dervishes, the six Ashantees danced on the poop. Prevented by their foes from springing into the water, the Spanish boys were hurrying up to the topmost spars, while such of the few Spanish sailors, not already in the sea, less alert, were descried, helplessly mixed in, on deck, with the blacks.

Meantime Captain Delano hailed his own vessel, ordering the ports up, and the guns run out. But by this time the cable of the San Dominick had been cut; and the fag-end, in lashing out, whipped away the canvas shroud about the beak, suddenly revealing, as the bleached hull swung round towards the open ocean, death for the figure-head, in a human skeleton; chalky comment on the chalked words below, "*Follow your leader.*"

At the sight, Don Benito, covering his face, wailed out: " 'Tis he, Aranda! my murdered, unburied friend!"

Upon reaching the sealer, calling for ropes, Captain Delano bound the negro, who made no resistance, and had him hoisted to the deck. He would then have assisted the now almost helpless Don Benito up the side; but Don Benito, wan as he was, refused to move, or be moved, until the negro should have been first put below out of view. When, presently assured that it was done, he no more shrank from the ascent.

The boat was immediately dispatched back to pick up the three swimming sailors. Meantime, the guns were in readiness, though, owing to the San Dominick having glided somewhat astern of the sealer, only the aftermost one could be brought to bear. With this, they fired six times; thinking to cripple the fugitive ship by bringing down her spars. But only a few inconsiderable ropes were shot away. Soon the ship was beyond the guns' range, steering broad out of the bay; the blacks thickly clustering round the bowsprit, one moment with taunting cries towards the whites, the next with up-thrown gestures hailing the now dusky moors of ocean—cawing crows escaped from the hand of the fowler.

The first impulse was to slip the cables and give chase. But, upon second thoughts, to pursue with whale-boat and yawl seemed more promising.

Upon inquiring of Don Benito what fire arms they had on board the San Dominick, Captain Delano was answered that they had none that could be used; because, in the earlier stages of the mutiny, a cabin-passenger, since dead, had secretly put out of order the locks of what few muskets there were. But with all his remaining strength, Don Benito entreated the American not to give chase, either with ship or boat; for the negroes had already proved themselves such desperadoes, that, in case of a present assault, nothing but a total massacre of the whites could be looked for. But, regarding this warning as coming from one whose spirit had been crushed by misery, the American did not give up his design.

The boats were got ready and armed. Captain Delano ordered his men into them. He was going himself when Don Benito grasped his arm.

"What! have you saved my life, señor, and are you now going to throw away your own?"

The officers also, for reasons connected with their interests and those of the voyage, and a duty owing to the owners, strongly objected against their commander's going. Weighing their remonstrances a moment, Captain Delano felt bound to remain; appointing his chief mate—an athletic and resolute man, who had been a privateer's-man[2]—to head the party. The more to encourage the sailors, they were told, that the Spanish captain considered his ship good as lost; that she and her cargo, including some gold and silver, were worth more than a thousand doubloons. Take her, and no small part should be theirs. The sailors replied with a shout.

The fugitives had now almost gained an offing. It was nearly night; but the moon was rising. After hard, prolonged pulling, the boats came up on the ship's quarters, at a suitable distance laying upon their oars to discharge their muskets. Having no bullets to return, the negroes sent their yells. But, upon the second volley, Indian-like, they hurtled their hatchets. One took off a sailor's fingers. Another struck the whale-boat's bow, cutting off the rope there, and remaining stuck in the gunwale like a woodman's axe. Snatching it, quivering from its lodgment, the mate hurled it back. The returned gauntlet now stuck in the ship's broken quarter-gallery, and so remained.

The negroes giving too hot a reception, the whites kept a more respectful distance. Hovering now just out of reach of the hurtling hatchets, they, with a view to the close encounter which must soon come, sought to decoy the blacks into entirely disarming themselves of their most murderous weapons in a hand-to-hand fight, by foolishly flinging them, as missiles, short of the mark, into the sea. But ere long perceiving the stratagem, the negroes desisted, though not before many of them had to replace their lost hatchets with handspikes; an exchange which, as counted upon, proved in the end favorable to the assailants.

Meantime, with a strong wind, the ship still clove the water; the boats alternately falling behind, and pulling up, to discharge fresh volleys.

The fire was mostly directed towards the stern, since there, chiefly, the negroes, at present, were clustering. But to kill or maim the negroes was not the object. To take them, with the ship, was the

2. Had served on a privateer—a ship legally commissioned by a government to prey on shipping of other countries.

object. To do it, the ship must be boarded; which could not be done by boats while she was sailing so fast.

A thought now struck the mate. Observing the Spanish boys still aloft, high as they could get, he called to them to descend to the yards, and cut adrift the sails. It was done. About this time, owing to causes hereafter to be shown, two Spaniards, in the dress of sailors and conspicuously showing themselves, were killed; not by volleys, but by deliberate marksman's shots; while, as it afterwards appeared, by one of the general discharges, Atufal, the black, and the Spaniard at the helm likewise were killed. What now, with the loss of the sails, and loss of leaders, the ship became unmanageable to the negroes.

With creaking masts, she came heavily round to the wind; the prow slowly swinging, into view of the boats, its skeleton gleaming in the horizontal moonlight, and casting a gigantic ribbed shadow upon the water. One extended arm of the ghost seemed beckoning the whites to avenge it.

"Follow your leader!" cried the mate; and, one on each bow, the boats boarded. Sealing-spears and cutlasses crossed hatchets and hand-spikes. Huddled upon the long-boat amidships, the negresses raised a wailing chant, whose chorus was the clash of the steel.

For a time, the attack wavered; the negroes wedging themselves to beat it back; the half-repelled sailors, as yet unable to gain a footing, fighting as troopers in the saddle, one leg sideways flung over the bulwarks, and one without, plying their cutlasses like carters' whips. But in vain. They were almost overborne, when, rallying themselves into a squad as one man, with a huzza, they sprang inboard; where, entangled, they involuntarily separated again. For a few breaths' space, there was a vague, muffled, inner sound, as of submerged sword-fish rushing hither and thither through shoals of black-fish. Soon, in a reunited band, and joined by the Spanish seamen, the whites came to the surface, irresistibly driving the negroes toward the stern. But a barricade of casks and sacks, from side to side, had been thrown up by the mainmast. Here the negroes faced about, and though scorning peace or truce, yet fain would have had respite. But, without pause, overleaping the barrier, the unflagging sailors again closed. Exhausted, the blacks now fought in despair. Their red tongues lolled, wolf-like, from their black mouths. But the pale sailors' teeth were set; not a word was spoken; and, in five minutes more, the ship was won.

Nearly a score of the negroes were killed. Exclusive of those by the balls,[3] many were mangled; their wounds—mostly inflicted by the long-edged sealing-spears—resembling those shaven ones of the English at Preston Pans, made by the poled scythes of the Highlanders.[4] On the other side, none were killed, though several were

3. Those killed by musketballs.
4. At the battle of Preston Pans (in East Lothian, Scotland) during 1745, Prince Charles Edward, grandson of James II, led Scottish Highlanders armed with scythes fastened to poles to victory over the royal forces.

wounded; some severely, including the mate. The surviving negroes were temporarily secured, and the ship, towed back into the harbor at midnight, once more lay anchored.

Omitting the incidents and arrangements ensuing, suffice it that, after two days spent in refitting, the ships sailed in company for Conception, in Chili, and thence for Lima, in Peru; where, before the vice-regal courts, the whole affair, from the beginning, underwent investigation.

Though, midway on the passage, the ill-fated Spaniard, relaxed from constraint, showed some signs of regaining health with free-will; yet, agreeably to his own foreboding, shortly before arriving at Lima, he relapsed, finally becoming so reduced as to be carried ashore in arms. Hearing of his story and plight, one of the many religious institutions of the City of Kings opened an hospitable refuge to him, where both physician and priest were his nurses, and a member of the order volunteered to be his one special guardian and consoler, by night and by day.

The following extracts, translated from one of the official Spanish documents, will it is hoped, shed light on the preceding narrative, as well as, in the first place, reveal the true port of departure and true history of the San Dominick's voyage, down to the time of her touching at the island of St. Maria.

But, ere the extracts come, it may be well to preface them with a remark.

The document selected, from among many others, for partial translation, contains the deposition of Benito Cereno; the first taken in the case. Some disclosures therein were, at the time, held dubious for both learned and natural reasons. The tribunal inclined to the opinion that the deponent, not undisturbed in his mind by recent events, raved of some things which could never have happened. But subsequent depositions of the surviving sailors, bearing out the revelations of their captain in several of the strangest particulars, gave credence to the rest. So that the tribunal, in its final decision, rested its capital sentences upon statements which, had they lacked confirmation, it would have deemed it but duty to reject.

———

I, Don Jose de Abos and Padilla, His Majesty's Notary for the Royal Revenue, and Register of this Province, and Notary Public of the Holy Crusade of this Bishopric, etc.

Do certify and declare, as much as is requisite in law, that, in the criminal cause commenced the twenty-fourth of the month of September, in the year seventeen hundred and ninety-nine, against the negroes of the ship San Dominick, the following declaration before me was made.

———

Declaration of the first witness, DON BENITO CERENO

The same day, and month, and year, His Honor, Doctor Juan Martinez de Rozas, Councilor of the Royal Audience of this King-

dom, and learned in the law of this Intendency, ordered the captain of the ship San Dominick, Don Benito Cereno, to appear; which he did in his litter, attended by the monk Infelez; of whom he received the oath, which he took by God, our Lord, and a sign of the Cross; under which he promised to tell the truth of whatever he should know and should be asked;—and being interrogated agreeably to the tenor of the act commencing the process, he said, that on the twentieth of May last, he set sail with his ship from the port of Valparaiso, bound to that of Callao; loaded with the produce of the country beside thirty cases of hardware and one hundred and sixty blacks, of both sexes, mostly belonging to Don Alexandro Aranda, gentleman, of the city of Mendoza; that the crew of the ship consisted of thirty-six men, beside the persons who went as passengers; that the negroes were in part as follows:

[Here, in the original, follows a list of some fifty names, descriptions, and ages, compiled from certain recovered documents of Aranda's, and also from recollections of the deponent, from which portions only are extracted.]

One, from about eighteen to nineteen years, named José, and this was the man that waited upon his master, Don Alexandro, and who speaks well the Spanish, having served him four or five years; · · · a mulatto, named Francisco, the cabin steward, of a good person and voice, having sung in the Valparaiso churches, native of the province of Buenos Ayres, aged about thirty-five years. · · · A smart negro, named Dago, who had been for many years a grave-digger among the Spaniards, aged forty-six years. · · · Four old negroes, born in Africa, from sixty to seventy, but sound, calkers by trade, whose names are as follows:—the first was named Muri, and he was killed (as was also his son named Diamelo); the second, Nacta; the third, Yola, likewise killed; the fourth, Ghofan; and six full-grown negroes, aged from thirty to forty-five, all raw, and born among the Ashantees—Matiluqui, Yan, Lecbe, Mapenda, Yambaio, Akim; four of whom were killed; · · · a powerful negro named Atufal, who, being supposed to have been a chief in Africa, his owners set great store by him. · · · And a small negro of Senegal, but some years among the Spaniards, aged about thirty, which negro's name was Babo; · · · that he does not remember the names of the others, but that still expecting the residue of Don Alexandro's papers will be found, will then take due account of them all, and remit to the court; · · · and thirty-nine women and children of all ages.

[The catalogue over, the deposition goes on:]

· · · That all the negroes slept upon deck, as is customary in this navigation, and none wore fetters, because the owner, his friend Aranda, told him that they were all tractable; · · · that on the seventh day after leaving port, at three o'clock in the morning, all the Spaniards being asleep except the two officers on the watch, who

were the boatswain, Juan Robles, and the carpenter, Juan Bautista
Gayete, and the helmsman and his boy, the negroes revolted suddenly, wounded dangerously the boatswain and the carpenter, and
successively killed eighteen men of those who were sleeping upon
deck, some with hand-spikes and hatchets, and others by throwing
them alive overboard, after tying them; that of the Spaniards upon
deck, they left about seven, as he thinks, alive and tied, to manœuvre
the ship, and three or four more who hid themselves, remained also
alive. Although in the act of revolt the negroes made themselves
masters of the hatchway, six or seven wounded went through it to
the cockpit, without any hindrance on their part; that during the act
of revolt, the mate and another person, whose name he does not
recollect, attempted to come up through the hatchway, but being
quickly wounded, were obliged to return to the cabin; that the deponent resolved at break of day to come up the companionway,
where the negro Babo was, being the ringleader, and Atufal, who
assisted him, and having spoken to them, exhorted them to cease
committing such atrocities, asking them, at the same time, what
they wanted and intended to do, offering, himself, to obey their
commands; that, notwithstanding this, they threw, in his presence,
three men, alive and tied, overboard; that they told the deponent to
come up, and that they would not kill him; which having done, the
negro Babo asked him whether there were in those seas any negro
countries where they might be carried, and he answered them. No;
that the negro Babo afterwards told him to carry them to Senegal,
or to the neighboring islands of St. Nicholas; and he answered, that
this was impossible, on account of the great distance, the necessity
involved of rounding Cape Horn, the bad condition of the vessel,
the want of provisions, sails, and water; but that the negro Babo replied to him he must carry them in any way; that they would do and
conform themselves to everything the deponent should require as to
eating and drinking; that after a long conference, being absolutely
compelled to please them, for they threatened him to kill all the
whites if they were not, at all events, carried to Senegal, he told
them that what was most wanting for the voyage was water; that
they would go near the coast to take it, and thence they would proceed on their course; that the negro Babo agreed to it; and the deponent steered towards the intermediate ports, hoping to meet some
Spanish or foreign vessel that would save them; within ten or eleven
days they saw the land, and continued their course by it in the vicinity of Nasca; that the deponent observed that the negroes were
now restless and mutinous, because he did not effect the taking in of
water, the negro Babo having required, with threats, that it should
be done, without fail, the following day; he told him he saw plainly
that the coast was steep, and the rivers designated in the maps were
not to be found, with other reasons suitable to the circumstances;
that the best way would be to go to the island of Santa Maria, where

they might water easily, it being a solitary island, as the foreigners
did; that the deponent did not go to Pisco, that was near, nor make
any other port of the coast, because the negro Babo had intimated to
him several times, that he would kill all the whites the very moment
he should perceive any city, town, or settlement of any kind on the
shores to which they should be carried: that having determined to
go to the island of Santa Maria, as the deponent had planned, for
the purpose of trying whether, on the passage or near the island itself, they could find any vessel that should favor them, or whether
he could escape from it in a boat to the neighboring coast of Arruco;
to adopt the necessary means he immediately changed his course,
steering for the island; that the negroes Babo and Atufal held daily
conferences, in which they discussed what was necessary for their
design of returning to Senegal, whether they were to kill all the
Spaniards, and particularly the deponent; that eight days after parting from the coast of Nasca, the deponent being on the watch a little
after day-break, and soon after the negroes had their meeting, the
negro Babo came to the place where the deponent was, and told him
that he had determined to kill his master, Don Alexandro Aranda,
both because he and his companions could not otherwise be sure of
their liberty, and that, to keep the seamen in subjection, he wanted
to prepare a warning of what road they should be made to take did
they or any of them oppose him; and that, by means of the death of
Don Alexandro, that warning would best be given; but, that what
this last meant, the deponent did not at the time comprehend, nor
could not, further than that the death of Don Alexandro was intended; and moreover, the negro Babo proposed to the deponent to
call the mate Raneds, who was sleeping in the cabin, before the
thing was done, for fear, as the deponent understood it, that the
mate, who was a good navigator, should be killed with Don Alexandro and the rest; that the deponent, who was the friend, from
youth, of Don Alexandro, prayed and conjured, but all was useless;
for the negro Babo answered him that the thing could not be prevented, and that all the Spaniards risked their death if they should
attempt to frustrate his will in this matter or any other; that, in this
conflict, the deponent called the mate, Raneds, who was forced to
go apart, and immediately the negro Babo commanded the Ashantee
Martinqui and the Ashantee Lecbe to go and commit the murder;
that those two went down with hatchets to the berth of Don Alexandro; that, yet half alive and mangled, they dragged him on deck;
that they were going to throw him overboard in that state, but the
negro Babo stopped them, bidding the murder be completed on
the deck before him, which was done, when, by his orders, the
body was carried below, forward; that nothing more was seen of it
by the deponent for three days; • • • that Don Alonzo Sidonia, an
old man, long resident at Valparaiso, and lately appointed to a civil
office in Peru, whither he had taken passage, was at the time sleeping

in the berth opposite Don Alexandro's; that, awakening at his cries, surprised by them, and at the sight of the negroes with their bloody hatchets in their hands, he threw himself into the sea through a window which was near him, and was drowned, without it being in the power of the deponent to assist or take him up; · · · that, a short time after killing Aranda, they brought upon deck his german-cousin, of middle-age, Don Francisco Masa, of Mendoza, and the young Don Joaquin, Marques de Aramboalaza, then lately from Spain, with his Spanish servant Ponce, and the three young clerks of Aranda, José Morairi, Lorenzo Bargas, and Hermenegildo Gandix, all of Cadiz; that Don Joaquin and Hermenegildo Gandix, the negro Babo for purposes hereafter to appear, preserved alive; but Don Francisco Masa, José Morairi, and Lorenzo Bargas, with Ponce the servant, beside the boatswain, Juan Robles, the boatswain's mates, Manuel Viscaya and Roderigo Hurta, and four of the sailors, the negro Babo ordered to be thrown alive into the sea, although they made no resistance, nor begged for anything else but mercy; that the boatswain, Juan Robles, who knew how to swim, kept the longest above water, making acts of contrition, and, in the last words he uttered, charged this deponent to cause mass to be said for his soul to our Lady of Succor; · · · that, during the three days which followed, the deponent, uncertain what fate had befallen the remains of Don Alexandro, frequently asked the negro Babo where they were, and if, still on board, whether they were to be preserved for interment ashore, entreating him so to order it; that the negro Babo answered nothing till the fourth day, when at sunrise, the deponent coming on deck, the negro Babo showed him a skeleton, which had been substituted for the ship's proper figure-head, the image of Christopher Colon, the discoverer of the New World; that the negro Babo asked him whose skeleton that was, and whether, from its whiteness, he should not think it a white's; that, upon his covering his face, the negro Babo, coming close, said words to this effect: "Keep faith with the blacks from here to Senegal, or you shall in spirit, as now in body, follow your leader," pointing to the prow; · · · that the same morning the negro Babo took up succession each Spaniard forward, and asked him whose skeleton that was, and whether, from its whiteness, he should not think it a white's; that each Spaniard covered his face; that then to each the negro Babo repeated the words in the first place said to the deponent; · · · that they (the Spaniards), being then assembled aft, the negro Babo harangued them, saying that he had now done all; that the deponent (as navigator for the negroes) might pursue his course, warning him and all of them that they should, soul and body, go the way of Don Alexandro if he saw them (the Spaniards) speak or plot anything against them (the negroes)—a threat which was repeated every day; that, before the events last mentioned, they had tied the cook to throw him overboard, for it is not known what thing they heard him speak, but finally the negro Babo spared his life, at the request of the deponent; that a few days after, the deponent, endeavoring not to omit any means to preserve the lives of the remaining whites, spoke to the negroes peace and tranquillity, and agreed to draw up a paper, signed by the deponent and the sailors who could write, as also by the negro Babo, for himself and all the blacks, in which the deponent obliged himself to carry them to Senegal, and they not to kill any more, and he formally to make over to them the ship, with the cargo, with which they were for that time satisfied and quieted. · · · But the next day, the more surely to guard against the sailors' escape, the negro Babo commanded all the boats to be destroyed but the long-boat, which was unseaworthy, and another, a cutter in good condition, which, knowing it would yet be wanted for towing the water casks, he had lowered down into the hold.

· · · · · · · ·

[*Various particulars of the prolonged and perplexed navigation ensuing here follow, with incidents of a calamitous calm, from which portion one passage is extracted, to wit:*]
—That on the fifth day of the calm, all on board suffering much from the heat, and want of water, and five having died in fits, and mad, the negroes became irritable, and for a chance gesture, which they deemed suspicious—though it was harmless—made by the mate, Raneds, to the deponent, in the act of handing a quadrant, they killed him; but that for this they afterwards were sorry, the mate being the only remaining navigator on board, except the deponent.

· · · · · · ·

—That omitting other events, which daily happened, and which can only serve uselessly to recall past misfortunes and conflicts, after seventy-three days' navigation, reckoned from the time they sailed from Nasca, during which they navigated under a scanty allowance of water, and were afflicted with the calms before mentioned, they at last arrived at the island of Santa Maria, on the seventeenth of the month of August, at about six o'clock in the afternoon, at which hour they cast anchor very near the American ship, Bachelor's Delight, which lay in the same bay, commanded by the generous Captain Amasa Delano; but at six o'clock in the morning, they had already descried the port, and the negroes became uneasy, as soon as at distance they saw the ship, not having expected to see one there; that the negro Babo pacified them, assuring them that no fear need be had; that straightway he ordered the figure on the bow to be covered with canvas, as for repairs, and had the decks a little set in order; that for a time the negro Babo and the negro Atufal conferred; that the negro Atufal was for sailing away, but the negro Babo would not, and, by himself, cast about what to do; that at last he came to the deponent, proposing to him to say and do all that the deponent declares to have said and done to the American captain;

• • • • • • that the negro Babo warned him that if he varied in the least, or uttered any word, or gave any look that should give the least intimation of the past events or present state, he would instantly kill him, with all his companions, showing a dagger, which he carried hid, saying something which, as he understood it, meant that that dagger would be alert as his eye; that the negro Babo then announced the plan to all his companions, which pleased them; that he then, the better to disguise the truth, devised many expedients, in some of them uniting deceit and defense; that of this sort was the device of the six Ashantees before named, who were his bravoes;[5] that them he stationed on the break of the poop, as if to clean certain hatchets (in cases, which were part of the cargo), but in reality to use them, and distribute them at need, and at a given word he told them that, among other devices, was the device of presenting Atufal, his right-hand man, as chained, though in a moment the chains could be dropped; that in every particular he informed the deponent what part he was expected to enact in every device, and what story he was to tell on every occasion, always threatening him with instant death if he varied in the least: that, conscious that many of the negroes would be turbulent, the negro Babo appointed the four aged negroes, who were calkers, to keep what domestic order they could on the decks; that again and again he harangued the Spaniards and his companions, informing them of his intent, and of his devices, and of the invented story that this deponent was to tell, charging them lest any of them varied from that story; that these arrangements were made and matured during the interval of two or three hours, between their first sighting the ship and the arrival on board of Captain Amasa Delano; that this happened about half-past seven o'clock in the morning, Captain Amasa Delano coming in his boat, and all gladly receiving him; that the deponent, as well as he could force himself, acting then the part of principal owner, and a free captain of the ship, told Captain Amasa Delano, when called upon, that he came from Buenos Ayres, bound to Lima, with three hundred negroes; that off Cape Horn, and in a subsequent fever, many negroes had died; that also, by similar casualties, all the sea officers and the greatest part of the crew had died.

• • • • • •

[*And so the deposition goes on, circumstantially recounting the fictitious story dictated to the deponent by Babo, and through the deponent imposed upon Captain Delano; and also recounting the friendly offers of Captain Delano, with other things, but all of which is here omitted. After the fictitious story, etc., the deposition proceeds:*]

• • • • • • •

—that the generous Captain Amasa Delano remained on board all the day, till he left the ship anchored at six o'clock in the evening,

5. Savage henchmen.

deponent speaking to him always of his pretended misfortunes, under the fore-mentioned principles, without having had it in his power to tell a single word, or give him the least hint, that he might know the truth and state of things; because the negro Babo, performing the office of an officious servant with all the appearance of submission of the humble slave, did not leave the deponent one moment; that this was in order to observe the deponent's actions and words, for the negro Babo understands well the Spanish; and besides, there were thereabout some others who were constantly on the watch, and likewise understood the Spanish; * * * that upon one occasion, while deponent was standing on the deck conversing with Amasa Delano, by a secret sign the negro Babo drew him (the deponent) aside, the act appearing as if originating with the deponent; that then, he being drawn aside, the negro Babo proposed to him to gain from Amasa Delano full particulars about his ship, and crew, and arms; that the deponent asked "For what?" that the negro Babo answered he might conceive; that, grieved at the prospect of what might overtake the generous Captain Amasa Delano, the deponent at first refused to ask the desired questions, and used every argument to induce the negro Babo to give up this new design; that the negro Babo showed the point of his dagger; that, after the information had been obtained, the negro Babo again drew him aside, telling him that that very night he (the deponent) would be captain of two ships, instead of one, for that, great part of the American's ship's crew being to be absent fishing, the six Ashantees, without any one else, would easily take it; that at this time he said other things to the same purpose; that no entreaties availed; that, before Amasa Delano's coming on board, no hint had been given touching the capture of the American ship: that to prevent this project the deponent was powerless; • • • —that in some things his memory is confused, he cannot distinctly recall every event; • • • —that as soon as they had cast anchor at six of the clock in the evening, as has before been stated, the American Captain took leave to return to his vessel; that upon a sudden impulse, which the deponent believes to have come from God and his angels, he, after the farewell had been said, followed the generous Captain Amasa Delano as far as the gunwale, where he stayed, under pretense of taking leave, until Amasa Delano should have been seated in his boat; that on shoving off, the deponent sprang from the gunwale into the boat, and fell into it, he knows not how, God guarding him; that—

• • • • • •

[Here, in the original, follows the account of what further happened at the escape, and how the San Dominick was retaken, and of the passage to the coast; including in the recital many expressions of "eternal gratitude" to the "generous Captain Amasa Delano." The deposition then proceeds with recapitulatory remarks, and a partial renumeration of the negroes, making record of their indi-

vidual part in the past events, with a view to furnishing, according
to command of the court, the data whereon to found the criminal
sentences to be pronounced. From this portion is the following:]
—That he believes that all the negroes, though not in the first
place knowing to the design of revolt, when it was accomplished,
approved it. · · · That the negro, José, eighteen years old, and in
the personal service of Don Alexandro, was the one who communicated the information to the negro Babo, about the state of things
in the cabin, before the revolt; that this is known, because, in the
preceding midnight, he used to come from his berth, which was
under his master's, in the cabin, to the deck where the ringleader
and his associates were, and had secret conversations with the negro
Babo, in which he was several times seen by the mate; that, one
night, the mate drove him away twice; · · that this same negro
José, was the one who, without being commanded to do so by the
negro Babo, as Lecbe and Martinqui were, stabbed his master, Don
Alexandro, after he had been dragged half-lifeless to the deck; · ·
that the mulatto steward, Francisco, was of the first band of revolt-
ers, that he was, in all things, the creature and tool of the negro
Babo; that, to make his court, he, just before a repast in the cabin,
proposed, to the negro Babo, poisoning a dish for the generous Captain Amasa Delano; this is known and believed, because the negroes have said it; but that the negro Babo, having another design,
forbade Francisco; · · that the Ashantee Lecbe was one of the
worst of them; for that, on the day the ship was retaken, he assisted
in the defense of her, with a hatchet in each hand, one of which
he wounded, in the breast, the chief mate of Amasa Delano, in
the first act of boarding; this all knew; that, in sight of the deponent, Lecbe struck, with a hatchet, Don Francisco Masa when,
by the negro Babo's orders, he was carrying him to throw him overboard, alive; beside participating in the murder, before mentioned,
of Don Alexandro Aranda, and others of the cabin-passengers; that,
owing to the fury with which the Ashantees fought in the engagement with the boats, but this Lecbe and Yau survived; that Yau
was bad as Lecbe; that Yau was the man who, by Babo's command,
willingly prepared the skeleton of Don Alexandro, in a way the
negroes afterwards told the deponent, but which he, so long as reason is left him, can never divulge; that Yau and Lecbe were the
two who, in a calm by night, riveted the skeleton to the bow; this
also the negroes told him; that the negro Babo was he who traced
the inscription below it; that the negro Babo was the plotter from
first to last; he ordered every murder, and was the helm and keel
of the revolt; that Atufal was his lieutenant in all; but Atufal, with
his own hand, committed no murder; nor did the negro Babo; · ·
that Atufal was shot, being killed in the fight with the boats, ere
boarding; · · that the negresses, of age, were knowing to the revolt,
and testified themselves satisfied at the death of their master, Don

Alexandro; that, had the negroes not restrained them, they would have tortured to death, instead of simply killing, the Spaniards slain by command of the negro Babo; that the negresses used their utmost influence to have the deponent made away with; that, in the various acts of murder, they sang songs and danced—not gaily, but solemnly; and before the engagement with the boats, as well as during the action, they sang melancholy songs to the negroes, and that this melancholy tone was more inflaming than a different one would have been, and was so intended; that all this is believed, because the negroes have said it.

—that of the thirty-six men of the crew exclusive of the passengers, (all of whom are now dead), which the deponent had knowledge of, six only remained alive, with four cabin-boys and ship-boys, not included with the crew; • • —that the negroes broke an arm of one of the cabin-boys and gave him strokes with hatchets.

[*Then follow various random disclosures referring to various periods of time. The following are extracted:*]

—That during the presence of Captain Amasa Delano on board, some attempts were made by the sailors, and one by Hermenegildo Gandix, to convey hints to him of the true state of affairs; but that these attempts were ineffectual, owing to fear of incurring death, and furthermore owing to the devices which offered contradictions to the true state of affairs; as well as owing to the generosity and piety of Amasa Delano incapable of sounding such wickedness; • • • that Luys Galgo, a sailor about sixty years of age, and formerly of the king's navy, was one of those who sought to convey tokens to Captain Amasa Delano; but his intent, though undiscovered, being suspected, he was, on a pretense, made to retire out of sight, and at last into the hold, and there was made away with. This the negroes have since said; • • • that one of the ship-boys feeling, from Captain Amasa Delano's presence, some hopes of release, and not having enough prudence, dropped some chance-word respecting his expectations, which being overheard and understood by a slave-boy with whom he was eating at the time, the latter struck him on the head with a knife, inflicting a bad wound, but of which the boy is now healing; that likewise, not long before the ship was brought to anchor, one of the seamen, steering at the time, endangered himself by letting the blacks remark some expression in his countenance, arising from a cause similar to the above; but this sailor, by his heedful after conduct, escape; • • • that these statements are made to show the court that from the beginning to the end of the revolt, it was impossible for the deponent and his men to act otherwise than they did; • • •—that the third clerk, Hermenegildo Gandix, who before had been forced to live among the seamen, wearing a seaman's habit, and in all respects appearing to be one for the time; he, Gandix, was killed by a musket-ball fired through a mistake from the boats before boarding; having in his fright run up the mizzen-rigging, calling to the boats—"don't board," lest upon their boarding the negroes should kill him; that this inducing the Americans to believe he some way favored the cause of the negroes, they fired two balls at him, so that he fell wounded from the rigging, and was drowned in the sea; • • •—that the young Don Joaquin, Marques de Arambaolaza, like Hermenegildo Gandix, the third clerk, was degraded to the office and appearance of a common seaman; that upon one occasion when Don Joaquin shrank, the negro Babo commanded the Ashantee Lecbe to take tar and heat it, and pour it upon Don Joaquin's hands; • • • —that Don Joaquin was killed owing to another mistake of the Americans, but one impossible to be avoided, as upon the approach of the boats, Don Joaquin, with a hatchet tied edge out and upright to his hand, was made by the negroes to appear on the bulwarks; whereupon, seen with arms in his hands and in a questionable attitude, he was shot for a renegade seaman; • • • —that on the person of Don Joaquin was found secreted a jewel, which, by papers that were discovered, proved to have been meant for the shrine of our Lady of Mercy in Lima; a votive offering, beforehand prepared and guarded, to attest his gratitude, when he should have landed in Peru, his last destination, for the safe conclusion of his entire voyage from Spain; • • • —that the jewel, with the other effects of the late Don Joaquin, is in the custody of the brethren of the Hospital de Sacerdotes, awaiting the disposition of the honorable court; • • • —that, owing to the condition of the deponent, as well as the haste in which the boats departed for the attack, the Americans were not forewarned that there were, among the apparent crew, a passenger and one of the clerks disguised by the negro Babo; • • • —that, beside the negroes killed in the action, some were killed after the capture and re-anchoring at night, when shackled to the ring-bolts on deck; that these deaths were committed by the sailors, ere they could be prevented. That so soon as informed of it, Captain Amasa Delano used all his authority, and, in particular with his own hand, struck down Martinez Gola, who, having found a razor in the pocket of an old jacket of his, which one of the shackled negroes had on, was aiming it at the negro's throat; that the noble Captain Amasa Delano also wrenched from the hand of Bartholomew Barlo, a dagger secreted at the time of the massacre of the whites, with which he was in the act of stabbing a shackled negro, who, the same day, with another negro, had thrown him down and jumped upon him; • • • —that, for all the events, befalling through so long a time, during which the ship was in the hands of the negro Babo, he cannot here give account; but that, what he has said is the most substantial of what occurs to him at present, and is the truth under the oath which he has taken; which declaration he affirmed and ratified, after hearing it read to him.

He said that he is twenty-nine years of age, and broken in body and mind; that when finally dismissed by the court, he shall not return home to Chili, but betake himself to the monastery on Mount Agonia without; and signed with his honor, and crossed himself, and, for the time, departed as he came, in his litter, with the monk Infelez, to the Hospital de Sacerdotes. BENITO CERENO DOCTOR ROZAS.

————————————

If the Deposition have served as the key to fit into the lock of the complications which precede it, then, as a vault whose door has been flung back, the San Dominick's hull lies open to-day.

Hitherto the nature of this narrative, besides rendering the intricacies in the beginning unavoidable, has more or less required that many things, instead of being set down in the order of occurrence, should be retrospectively, or irregularly given; this last is the case with the following passages, which will conclude the account:

During the long, mild voyage to Lima, there was, as before hinted, a period during which the sufferer a little recovered his health, or, at least in some degree, his tranquillity. Ere the decided relapse which came, the two captains had many cordial conversations—their fraternal unreserve in singular contrast with former withdrawments.

Again and again, it was repeated, how hard it had been to enact the part forced on the Spaniard by Babo.

"Ah, my dear," Don Benito once said, "at those very times when you thought me so morose and ungrateful, nay, when, as you now admit, you half thought me plotting your murder, at those very times my heart was frozen; I could not look at you, thinking of what, both on board this ship and your own, hung, from other hands, over my kind benefactor. And as God lives, Don Amasa, I know not whether desire for my own safety alone could have nerved me to that leap into your boat, had it not been for the thought that, did you, unenlightened, return to your ship, you, my best friend, with all who might be with you, stolen upon, that night, in your hammocks, would never in this world have wakened again. Do but think how you walked this deck, how you sat in this cabin, every inch of ground mined into honey-combs under you. Had I dropped the least hint, made the least advance towards an understanding between us, death, explosive death—yours as mine—would have ended the scene."

"True, true," cried Captain Delano, starting, "you saved my life, Don Benito, more than I yours; saved it, too, against my knowledge and will."

"Nay, my friend," rejoined the Spaniard, courteous even to the point of religion, "God charmed your life, but you saved mine. To think of some things you did—those smilings and chattings, rash pointings and gesturings. For less than these, they slew my mate, Raneds; but you had the Prince of Heaven's safe conduct through all ambuscades."

"Yes, all is owing to Providence, I know; but the temper of my mind that morning was more than commonly pleasant, while the sight of so much suffering, more apparent than real, added to my good nature, compassion, and charity, happily interweaving the three. Had it been otherwise, doubtless, as you hint, some of my interferences might have ended unhappily enough. Besides, those feelings I spoke of enabled me to get the better of momentary distrust, at times when acuteness might have cost me my life, without saving another's. Only at the end did my suspicions get the better of me, and you know how wide of the mark they then proved."

"Wide, indeed," said Don Benito, sadly; "you were with me all day; stood with me, sat with me, talked with me, looked at me, ate with me, drank with me; and yet, your last act was to clutch for a monster, not only an innocent man, but the most pitiable of all men. To such degree may malign machinations and deceptions impose. So far may even the best man err, in judging the conduct of one with the recesses of whose condition he is not acquainted. But you were forced to it; and you were in time undeceived. Would that, in both respects, it was so ever, and with all men."

"You generalize, Don Benito; and mournfully enough. But the past is passed; why moralize upon it? Forget it. See, yon bright sun has forgotten it all, and the blue sea, and the blue sky; these have turned over new leaves."

"Because they have no memory," he dejectedly replied; "because they are not human."

"But these mild trades[6] that now fan your cheek, do they not come with a human-like healing to you? Warm friends, steadfast friends are the trades."

"With their steadfastness they but waft me to my tomb, señor," was the foreboding response.

"You are saved," cried Captain Delano, more and more astonished and pained; "you are saved; what has cast such a shadow upon you?"

"The negro."

There was silence, while the moody man sat, slowly and unconsciously gathering his mantle about him, as if it were a pall.

There was no more conversation that day.

But if the Spaniard's melancholy sometimes ended in muteness upon topics like the above, there were others upon which he never spoke at all; on which, indeed, all his old reserves were piled. Pass over the worse, and, only to elucidate, let an item or two of these be cited. The dress so precise and costly, worn by him on the day

6. Trade winds; here, dependable winds blowing from southeast to northwest.

whose events have been narrated, had not willingly been put on. And that silver-mounted sword, apparent symbol of despotic command, was not, indeed, a sword, but the ghost of one. The scabbard, artificially stiffened, was empty.

As for the black—whose brain, not body, had schemed and led the revolt, with the plot—his slight frame, inadequate to that which it held, had at once yielded to the superior muscular strength of his captor, in the boat. Seeing all was over, he uttered no sound, and could not be forced to. His aspect seemed to say, since I cannot do deeds, I will not speak words. Put in irons in the hold, with the rest, he was carried to Lima. During the passage Don Benito did not visit him. Nor then, nor at any time after, would he look at him. Before the tribunal he refused. When pressed by the judges he fainted. On the testimony of the sailors alone rested the legal identity of Babo.

Some months after, dragged to the gibbet at the tail of a mule, the black met his voiceless end. The body was burned to ashes; but for many days, the head, that hive of subtlety, fixed on a pole in the Plaza, met, unabashed, the gaze of the whites; and across the Plaza looked towards St. Bartholomew's church, in whose vaults slept then, as now, the recovered bones of Aranda; and across the Rimac bridge looked towards the monastery, on Mount Agonia without; where, three months after being dismissed by the court, Benito Cereno, borne on the bier, did, indeed, follow his leader.

1855, 1856

From BATTLE-PIECES[1]
The Portent
(1859)[2]

Hanging from the beam,
   Slowly swaying (such the law),

Gaunt the shadow on your green,
   Shenandoah!
The cut is on the crown   5
   (Lo, John Brown),
And the stabs shall heal no more.

Hidden in the cap
   Is the anguish none can draw;
So your future veils its face,   10
   Shenandoah!
But the streaming beard is shown
   (Weird John Brown),
The meteor of the war.

1866

1. *Battle-Pieces and Aspects of the War* was published by the Harpers in 1866, the source of the present texts. 2. John Brown's raid at Harpers Ferry, Virginia, in October, 1859, is naturally seen here as a portent of the war.

Misgivings

(1860)[3]

When ocean-clouds over inland hills
    Sweep storming in late autumn brown,
And horror the sodden valley fills,
    And the spire falls crashing in the town,
I muse upon my country's ills—   5
The tempest bursting from the waste of Time
On the world's fairest hope linked with man's foulest crime.

Nature's dark side is heeded now—
    (Ah! optimist-cheer disheartened flown)—
A child may read the moody brow   10
    Of yon black mountain lone.
With shouts the torrents down the gorges go,
And storms are formed behind the storm we feel:
The hemlock shakes in the rafter, the oak in the driving keel.
                                  1866

The March into Virginia,

Ending in the First Manassas
(July, 1861)[4]

Did all the lets and bars appear
    To every just or larger end,
Whence should come the trust and cheer?
    Youth must its ignorant impulse lend—
Age finds place in the rear.   5
    All wars are boyish, and are fought by boys,
The champions and enthusiasts of the state:
    Turbid ardors and vain joys
        Not barrenly abate—
Stimulants to the power mature,   10
    Preparatives of fate.

Who here forecasteth the event?
What heart but spurns at precedent
    And warnings of the wise,
Contemned foreclosures of surprise?   15
The banners play, the bugles call,
The air is blue and prodigal.
    No berrying party, pleasure-wooed,
No picnic party in the May,
Ever went less loth than they   20
    Into that leafy neighborhood.
In Bacchic glee they file toward Fate,

---

3. The "dolorous winter" of 1860–61, between the election of Abraham Lincoln and the first shots fired at Fort Sumter, South Carolina, seemed to Melville a "Sad arch between contrasted eras" (*Clarel* IV.v.80).

4. In July, 1861, at Manassas, Virginia (a railroad junction some 30 miles from Washington), Confederate forces routed the Union Army.

Moloch's uninitiate;[5]
Expectancy, and glad surmise
Of battle's unknown mysteries.   25
All they feel is this: 'tis glory,
A rapture sharp, though transitory,
Yet lasting in belaureled story.
So they gayly go to fight,
Chatting left and laughing right.   30

But some who this blithe mood present,
    As on in lightsome files they fare,
Shall die experienced ere three days are spent—
    Perish, enlightened by the vollied glare;
Or shame survive, and, like to adamant,   35
    The throe of Second Manassas[6] share.

1866

A Utilitarian View of the Monitor's Fight[7]

Plain be the phrase, yet apt the verse,
    More ponderous than nimble;
For since grimed War here laid aside
His Orient pomp, 'twould ill befit
        Overmuch to ply   5
    The rhyme's barbaric cymbal.

Hail to victory without the gaud
    Of glory; zeal that needs no fans
Of banners; plain mechanic power
Plied cogently in War now placed—   10
    Where War belongs—
    Among the trades and artisans.

Yet this was battle, and intense—
    Beyond the strife of fleets heroic;
Deadlier, closer, calm 'mid storm;   15
No passion; all went on by crank,
        Pivot, and screw,
    And calculations of caloric.

Needless to dwell; the story's known.
    The ringing of those plates on plates   20

5. Old Testament heathen god to whom children were burnt in sacrifice (see Leviticus 20.2–5). Melville probably had in mind Milton's *Paradise Lost*, I.392 ff., the description of "*Moloch*, horrid King besmear'd with blood/Of human sacrifice, and parents tears."
6. In August, 1862, Robert E. Lee and "Stonewall" Jackson defeated the Union army once again.
7. The battle between the Confederate ironclad *Merrimack* and the Union iron-clad *Monitor* on May 9, 1862, at Hampton Roads, Virginia, was inconclusive. Later the same year Confederates scuttled the *Merrimack* during an evacuation and the *Monitor* sank in a gale. The point of view adopted in the poem is that of a follower of Jeremy Bentham (1748–1832), English philosopher who advocated that behavior be judged by its utility, not its correspondence to any ideal.

Still ringeth round the world—
The clangor of that blacksmiths' fray.
The anvil-din
Resounds this message from the Fates:

War shall yet be, and to the end;   25
    But war-paint shows the streaks of weather;
War yet shall be, but warriors
Are now but operatives;[8] War's made
    Less grand than Peace,
And a singe runs through lace and feather.   30

1866

The House-top

A Night Piece
(July, 1863)[9]

No sleep. The sultriness pervades the air
And binds the brain—a dense oppression, such
As tawny tigers feel in matted shades,
Vexing their blood and making apt for ravage.
Beneath the stars the roofy desert spreads   5
Vacant as Libya. All is hushed near by.
Yet fitfully from far breaks a mixed surf
Of muffled sound, the Atheist roar of riot.
Yonder, where parching Sirius[1] set in drought,
Balefully glares red Arson—there—and there.   10
The Town is taken by its rats—ship-rats
And rats of the wharves. All civil charms
And priestly spells which late held hearts in awe—
Fear-bound, subjected to a better sway
Than sway of self; these like a dream dissolve,   15
And man rebounds whole æons back in nature.[2]
Hail to the low dull rumble, dull and dead,
And ponderous drag that shakes the wall.
Wise Draco[3] comes, deep in the midnight roll
Of black artillery; he comes, though late;   20
In code corroborating Calvin's[4] creed
And cynic tyrannies of honest kings;
He comes, nor parlies; and the Town, redeemed,

8. Factory workers.
9. In July, 1863, mobs in New York City, composed largely of Irish immigrants, rioted against the new draft laws, destroying property and attacking free Negroes (their economic competitors) in their rage at being required to fight a war for the abolition of slavery.
1. The dog star associated with the miserable "dog days" in which the riots occurred.
2. " 'I dare not write the horrible and inconceivable atrocities committed,' says Froissart, in alluding to the remarkable sedition in France during his time. The like may be hinted of some proceedings of the draft-rioters" [Melville's note]. There were accounts, for instance, of Irish women cutting the genitals off Negro men hanged on lampposts.
3. The Athenian Draco in 621 B.C. set forth laws that became proverbial for their harshness.
4. John Calvin (1509–64), French theologian, believed every human being was born guilty of Original Sin, the consequence of God's curse when Adam and Eve disobeyed Him in the Garden of Eden.

Gives thanks devout; nor, being thankful, heeds
The grimy slur on the Republic's faith implied,   25
Which holds that Man is naturally good,
And—more—is Nature's Roman, never to be scourged.[5]

1866

From JOHN MARR AND OTHER SAILORS[1]

The Maldive Shark

About the Shark, phlegmatical one,
Pale sot of the Maldive sea,[2]
The sleek little pilot-fish, azure and slim,
How alert in attendance be.
From his saw-pit of mouth, from his charnel of maw   5
They have nothing of harm to dread,
But liquidly glide on his ghastly flank
Or before his Gorgonian head;[3]
Or lurk in the port of serrated teeth
In white triple tiers of glittering gates,   10
And there find a haven when peril's abroad,
An asylum in jaws of the Fates!

They are friends; and friendly they guide him to prey,
Yet never partake of the treat—
Eyes and brains to the dotard lethargic and dull,   15
Pale ravener of horrible meat.

1888

To Ned[1]

Where is the world we roved, Ned Bunn?
    Hollows thereof lay rich in shade
By voyagers old inviolate thrown
    Ere Paul Pry cruised with Pelf and Trade.[2]
To us old lads some thoughts come home   5
Who roamed a world young lads no more shall roam.

Nor less the satiate year impends
    When, wearying of routine-resorts,
The pleasure-hunter shall break loose,
    Ned, for our Pantheistic ports:—   10

5. Acts 22.25: "And as they bound him with thongs, Paul said unto the centurion that stood by, Is it lawful for you to scourge a man that is a Roman, and uncondemned?"

1. In 1888 Melville paid to have *John Marr and Other Sailors* printed in an edition of 25 copies; the texts here are those of the 1888 edition.

2. Part of the Indian Ocean around the Maldive Islands, southwest of the southern tip of India.

3. Capable, like the snake-haired Gorgon's head in Greek mythology, of turning the beholder to stone.

1. Involved in the makeup of "Ned Bunn" are Melville's shipmate Richard (Toby) Greene, with whom he deserted ship in the Marquesas, and his own part-fictional character of Toby in *Typee*.

2. *Paul Pry* is a play by the English dramatist John Poole (1786–1872); for Hawthorne, Melville, and many of their time Paul Pry was a type name for an unscrupulous pryer into others' secrets. "Pelf and Trade": outright robbery and commerce.

Marquesas and glenned isles that be
Authentic Edens in a Pagan sea.

The charm of scenes untried shall lure,
    And, Ned, a legend urge the flight—
The Typee-truants under stars   15
    Unknown to Shakespeare's *Midsummer-Night;*[3]
And man, if lost to Saturn's Age,
Yet feeling life no Syrian pilgrimage.[4]

But, tell, shall he the tourist find
    Our isles the same in voilet-glow   20
Enamoring us what years and years—
    Ah, Ned, what years and years ago!
Well, Adam advances, smart in pace,
But scarce by violets that advance you trace.

But we, in anchor-watches calm,   25
    The Indian Psyche's languor won,
And, musing, breathed primeval balm
    From Edens ere yet over-run;
Marvelling mild if mortal twice,
Here and hereafter, touch a Paradise.   30

                                                    1888

From TIMOLEON, ETC.[1]
After the Pleasure Party[2]
LINES TRACED UNDER AN IMAGE OF AMOR THREATENING[3]

> *Fear me, virgin whosoever*
> *Taking pride from love exempt,*
> *Fear me, slighted. Never, never*
> *Brave me, nor my fury tempt:*
> *Downy wings, but wroth they beat*   5
> *Tempest even in reason's seat.*

Behind the house the upland falls
With many an odorous tree—

3. Unknown because different constellations are visible in the Marquesas, below the equator.
4. Melville's own *Typee* was in fact, even before his death, beginning to lure tourists to the South Seas, travelers who are not of the fabled Golden Age of Saturn's rule (before the triumph of the Olympian gods) but who also are post-Christian in not seeing life as a painful progress or pilgrimage through the wastelands of this world.
1. Melville had this small volume privately printed in 1891 by the Caxton Press in an edition limited to 25 copies. It was dedicated to the American artist Elihu Vedder, whom Melville had not met.

2. The punctuation of this poem has proved confusing. After the italicized epigraph, the narrator speaks lines 7–17; then the main speaker takes up, an unmarried woman who is an astronomer. Apparently her long monologue ends with line 110, but she also speaks lines 131–47. The treatment of sexual denial and its consequences is franker and more complex than in any of Melville's American contemporaries except Whitman and Dickinson.
3. Probably the image of Amor (or Cupid) is threatening with the conventional weapons of bow and arrow; as the god warns in the epigraph, his wings are soft but powerful.

White marbles gleaming through green halls,
Terrace by terrace, down and down,   10
And meets the starlit Mediterranean Sea.

  'Tis Paradise. In such an hour
Some pangs that rend might take release.
Nor less perturbed who keeps this bower
Of balm, nor finds balsamic peace?   15
From whom the passionate words in vent
After long revery's discontent?

  Tired of the homeless deep,
Look how their flight yon hurrying billows urge,
Hitherward but to reap   20
Passive repulse from the iron-bound verge!
Insensate, can they never know
'Tis mad to wreck the impulsion so?

  An art of memory is, they tell:
But to forget! forget the glade   25
Wherein Fate sprung Love's ambuscade,
To flout pale years of cloistral life
And flush me in this sensuous strife.
'Tis Vesta struck with Sappho's smart.[4]
No fable her delirious leap: 30
With more of cause in desperate heart,
Myself could take it—but to sleep!

  Now first I feel, what all may ween,
That soon or late, if faded e'en,
One's sex asserts itself. Desire,   35
The dear desire through love to sway,
Is like the Geysers that aspire—
Through cold obstruction win their fervid way.
But baffled here—to take disdain,
To feel rule's instinct, yet not reign;   40
To dote, to come to this drear shame—
Hence the winged blaze that sweeps my soul
Like prairie fires that spurn control,
Where withering weeds incense the flame.

  And kept I long heaven's watch for this,   45
Contemning love, for this, even this?
O terrace chill in Northern air,
O reaching ranging tube I placed
Against yon skies, and fable chased
Till, fool, I hailed for sister there   50

---

4. Vesta is the Roman goddess of the hearth and household, but here Melville has in mind the requirement that her priestesses, the vestal virgins, be chaste; the virgin goddess is struck with sexual desires such as ruled Sappho, Greek lyric poet of the seventh century B.C. A legend has it that she committed suicide by leaping from a cliff into the sea.

Starred Cassiopea in Golden Chair.[5]
In dream I throned me, nor I saw
In cell the idiot crowned with straw.

And yet, ah yet scarce ill I reigned,
Through self-illusion self-sustained,   55
When now—enlightened, undeceived—
What gain I barrenly bereaved!
Than this can be yet lower decline—
Envy and spleen, can these be mine?

The peasant girl demure that trod   60
Beside our wheels that climbed the way,
And bore along a blossoming rod
That looked the sceptre of May-Day—
On her—to fire this petty hell,
His softened glance how moistly fell!   65
The cheat! on briars her buds were strung;
And wiles peeped forth from mien how meek.
The innocent bare-foot! young, so young!
To girls, strong man's a novice weak.
To tell such beads! And more remain,   70
Sad rosary of belittling pain.

When after lunch and sallies gay
Like the Decameron folk[6] we lay
In sylvan groups; and I——let be!
O, dreams he, can he dream that one   75
Because not roseate feels no sun?
The plain lone bramble thrills with Spring
As much as vines that grapes shall bring.

Me now fair studies charm no more.
Shall great thoughts writ, or high themes sung   80
Damask wan cheeks—unlock his arm
About some radiant ninny flung?
How glad with all my starry lore,
I'd buy the veriest wanton's rose
Would but my bee therein repose.   85

Could I remake me! or set free
This sexless bound in sex, then plunge
Deeper than Sappho, in a lunge
Piercing Pan's[7] paramount mystery!

5. In Greek mythology Cassiopeia, Queen of Ethiopia, boasted that she was more beautiful than the sea nymphs; her being made into a constellation was punishment, since the basket she sat in ludicrously turned upside down at some seasons. Melville follows a version of the story which stresses the beauty of her starry chair.

6. Like the bantering sophisticates who take refuge together from the plague and amuse themselves by telling stories (many of them indelicate) in the *Decameron* of the Italian writer Giovanni Boccaccio (1313–75).
7. In Greek mythology, god of woods, fields, and flocks; he had a human torso but goat's legs, ears, and horns.

For, Nature, in no shallow surge   90
Against thee either sex may urge,
Why hast thou made us but in halves—
Co-relatives?[8] This makes us slaves.
If these co-relatives never meet
Self-hood itself seems incomplete.   95
And such the dicing of blind fate
Few matching halves here meet and mate.
What Cosmic jest or Anarch blunder
The human integral clove asunder
And shied the fractions through life's gate?   100

      Ye stars that long your votary knew
Rapt in her vigil, see me here!
Whither is gone the spell ye threw
When rose before me Cassiopea?
Usurped on by love's stronger reign—   105
But lo, your very selves do wane:
Light breaks—truth breaks! Silvered no more,
But chilled by dawn that brings the gale
Shivers yon bramble above the vale,
And disillusion opens all the shore.   110

      One knows not if Urania[9] yet
The pleasure-party may forget;
Or whether she lived down the strain
Of turbulent heart and rebel brain;
For Amor so resents a slight,   115
And her's had been such haught disdain,
He long may wreak his boyish spite,
And boy-like, little reck the pain.

      One knows not, no. But late in Rome
(For queens discrowned a congruous home)   120
Entering Albani's porch she stood
Fixed by an antique pagan stone
Colossal carved.[1] No anchorite seer,
Not Thomas a Kempis,[2] monk austere,
Religious more are in their tone;   125
Yet far, how far from Christian heart
That form august of heathen Art.
Swayed by its influence, long she stood,
Till surged emotion seething down,
She rallied and this mood she won: 130

8. Legend in Plato's *Symposium* that
man was originally a round, four-legged
being which split in two and forever
after is seeking to be reunited through
sex.
9. In Greek mythology, the muse of astronomy; here used of the modern
woman who speaks the central section
of the poem.
1. The Villa Albani, which Melville visited in February, 1857.
2. German religious writer (1380–1471).

Languid in frame for me,
To-day by Mary's convent shrine,
Touched by her picture's moving plea
In that poor nerveless hour of mine,
I mused—A wanderer still must grieve.   135
Half I resolved to kneel and believe,
Believe and submit, the veil take on.
But thee, armed Virgin!³ less benign,
Thee now I invoke, thou mightier one.
Helmeted woman—if such term   140
Befit thee, far from strife
Of that which makes the sexual feud
And clogs the aspirant life—
O self-reliant, strong and free,
Thou in whom power and peace unite,   145
Transcender! raise me up to thee,
Raise me and arm me!

                              Fond appeal.
For never passion peace shall bring,
Nor Art inanimate for long
Inspire. Nothing may help or heal   150
While Amor incensed remembers wrong.
Vindictive, not himself he'll spare;
For scope to give his vengeance play
Himself he'll blaspheme and betray.

        Then for Urania, virgins everywhere,   155
O pray! Example take too, and have care.

                                             1891

Art

In placid hours well-pleased we dream
Of many a brave unbodied scheme.
But form to lend, pulsed life create,
What unlike things must meet and mate:
A flame to melt—a wind to freeze;   5
Sad patience—joyous energies;
Humility—yet pride and scorn;
Instinct and study; love and hate;
Audacity—reverence. These must mate,
And fuse with Jacob's mystic heart,   10
To wrestle with the angel⁴—Art.

                                             1891

3. Not the Virgin Mary, but Athena, the Greek goddess of wisdom.   4. Jacob's wrestling with the angel is in Genesis 32.

EMILY DICKINSON
1830–1886

Emily Elizabeth Dickinson, one of the greatest American poets, was born in Amherst, Massachusetts, a prototypical New England village, on December 10, 1830, the second child of Emily and Edward Dickinson; she died in the same house some fifty-six years later. In between, she left her native land never, her home state apparently only once, her village a small handful of times, and, particularly after 1872, her house and yard scarcely at all. Reasonably gregarious in her youth, in the last years of her life she retreated to a tighter and tighter circle of family and friends. In those later years she dressed in white, avoided strangers, and communicated even with intimates chiefly through cryptic notes and fragments of poems. The doctor who attended her fatal illness was permitted to "examine" her only by observing his patient in the next room walk by a partially opened door. A quasi-mythical figure in Amherst when she died from Bright's disease on May 15, 1886, Emily Dickinson was essentially unknown to the rest of the world. Only seven of her poems had appeared in print, all of them anonymously.

But to think of Emily Dickinson only as an eccentric recluse is a serious mistake. Like Thoreau, she lived simply and deliberately; she fronted the essential facts of life. In Henry James's phrase, she was one of those on whom nothing was lost. Only by thus living austerely and intensely could Dickinson manage both to fulfill what for her were the strenuous physical and emotional obligations of a daughter, a sister, a sister-in-law, citizen, and housekeeper, and write (as she did during the early 1860s) on the average a poem a day.

The welter of facts that constitute the visible life of even this resolutely exclusive and inward person are recorded in Jay Leyda's *The Years and Hours of Emily Dickinson.* Biographers before and since have typically fastened on a few of these facts, particularly those involving the relations of the poet with various men: her severe, beloved father, Edward; her devoted brother, Austin; the law student Benjamin Newton, who apprenticed with her father; the bright, sensitive Amherst College undergraduate Henry Vaughan Emmons; the powerful and attractive Reverend Charles Wadsworth; the sympathetic but baffled critic Thomas Wentworth Higginson; and the loyal, intimate family friend Judge Otis P. Lord. There is no question that in each case there was a genuine emotional bond. But if Emily Dickinson at different points in her life was capable of emotional commitment of various kinds and levels of intensity, it must also be understood that for her every positive attraction carried with it an equal and opposite charge. Her insistence on keeping whole the fragile membrane of her inviolate self made complete submission to God or man, nature or society impossible for Emily Dickinson. As she observed in a letter to her old friend Judge Lord, who about 1880 seems to have wished to marry her: "you ask the divine Crust and that would doom the bread." Emily Dickinson, then, selected her society scrupulously, making few do for many.

Her relationship to books, to literary precedent and example, was similar.

She was no ransacker and devourer of libraries. Like Lincoln, she knew relatively few volumes but knew them deeply. As a girl she attended Amherst Academy and also Mount Holyoke Female Seminary, a few miles distant, during her seventeenth year, but school gave her neither intellectual nor social satisfactions to compensate for the reassuring intimacy of home and family she keenly missed. The standard works she knew best and drew on most commonly for allusions and references in her poetry and vivid letters were the classic myths, the Bible, and Shakespeare. Among the English Romantics, she valued John Keats especially; among her English contemporaries she was particularly attracted by the Brontës, the Brownings, Alfred, Lord Tennyson, and George Eliot. None of these, however, can be said to have influenced her literary practice significantly. Indeed, not the least notable quality of her poetry is its dazzling originality. Thoreau and Emerson, especially the latter, as we know from her letters, were perhaps her most important contemporary American intellectual resources, though their liberal influence seems always to have been tempered by the legacy of a conservative Puritanism best expressed in the writings of Jonathan Edwards. Her chief prosodic and formal model was the commonly used hymnals of the times with their simple patterns of meter and rhyme.

Though her intellectual, literary, and formal resources were comparatively few, Emily Dickinson had a clear and self-conscious conception of the poetic vocation and the poetic process. Her poetry, moreover, has thematic heft and a surprisingly dense verbal texture. She sometimes doubted her power to "write her letter to the world," and asked the famous critic and journalist Thomas Wentworth Higginson to tell her if her poetry was alive; yet many of her poems suggest an unquestionable conviction that she possessed enormous poetic power. The proof of her conviction is the 1,775 poems she left behind at her death.

The poetry itself, despite its ostensible formal simplicity, is remarkable for its variety, subtlety, and richness, and Emily Dickinson from the beginning has attracted both popular and specialized audiences: those who find satisfaction in the sometimes sachet-quaint, aphoristic, generalizing tendency in her work, as well as those who take their pleasure in the experimental, intellectually knotty, often darker awareness that marks her most sophisticated poems. There is very little dispute regarding Dickinson's claim to a very high place among America's poets.

Recognition of Dickinson's true stature has come slowly, partly because early editions of her work frequently "corrected" and "smoothed" out her eccentric, often startling grammatical and poetic inventiveness. These early editions (and much subsequent criticism) also tended to organize her poems (and thus our sense of her mind) into conventionally "poetic" categories—Love, Nature, Friendship, Death, Immortality, and the like. At various times and in various moods Dickinson did return repeatedly to certain subjects, most of them at heart concerned with the paradoxes and dilemmas of the self that is conscious of being trapped in time. But to present scrubbed-up little poems in prepackaged assortments is to falsify her oblique, witty, mercurial human and poetic character.

It was really not until 1955—when Thomas Johnson and Theodora Ward made her complete work available—that the full sweep of her accomplishment could be assessed. Since that time her affinities with the

English metaphysical poets John Donne and George Herbert, with the prophetic William Blake, and with such difficult modern poets as Gerard Manley Hopkins have been observed and celebrated. Understood now as a distinctively modern poet, she has been claimed by William Carlos Williams as his "patron saint" and acclaimed in verse and prose by such fellow poets as Hart Crane, Allen Tate, Archibald MacLeish, Richard Wilbur, and Adrienne Rich.

In very recent years, Emily Dickinson has been understood as a poet of genuinely philosophic and tragic dimensions—one who was acutely responsive to the enduring questions of the nature and meaning of human consciousness in its ceaseless attempt to understand its origins, conditions, relations, and fate. Typically, Dickinson's poems begin with assertion and affirmation only to end in qualification and question, if not outright denial. Few other poets have given us such direct access to the riddling tensions of human consciousness, to the terrible slipperiness of reality. She understood what at times we all understand—that the "sweetness of life" consists in our knowing "that it will never come again," and that this same awareness is the source of the bitterness of life. Emily Dickinson lived a life of ecstatic and agonized instants; her most vital poems record those "nows" made "forever."

The numbers identifying poems follow T. H. Johnson's numbering in his three-volume edition of *The Poems of Emily Dickinson* (1955). Johnson assigned numbers to all surviving manuscripts to reflect the approximate order of their composition, the dates being inferred largely from the internal evidence of differences in the handwriting. Dickinson's irregular and often idiosyncratic punctuation and capitalization were preserved so far as typographical resources permitted in a clear text. The most obvious typographical feature of the poems is the abundance of dashes; in manuscript, these vary in length and angle up or down from horizontal and may have been used systematically for rhetorical emphasis or musical pointing.

Though seven of her poems were published during Dickinson's lifetime, none appeared with her full knowledge and consent. Her public career as a poet did not begin until 1890, when Mabel Loomis Todd, a family friend, and Thomas Wentworth Higginson selected and edited 115 poems for publication. Six additional volumes of selections (edited with varying degrees of faithfulness) were published between 1890 and the appearance of the Johnson edition.

49

I never lost as much but twice,
And that was in the sod.
Twice have I stood a beggar
Before the door of God!

Angels—twice descending   5
Reimbursed my store—
Burglar! Banker—Father!
I am poor once more!

1858   1890

130

These are the days when Birds come back—
A very few—a Bird or two—
To take a backward look.

These are the days when skies resume
The old—old sophistries[1] of June—   5
A blue and gold mistake.

Oh fraud that cannot cheat the Bee—
Almost thy plausibility
Induces my belief.

Till ranks of seeds their witness bear—   10
And softly thro' the altered air
Hurries a timid leaf.

Oh Sacrament of summer days,
Oh Last Communion[2] in the Haze—
Permit a child to join.

Thy sacred emblems to partake—
Thy consecrated bread to take
And thine immortal wine!

1859   1890

185

"Faith" is a fine invention
When Gentlemen can *see*—
But *Microscopes* are prudent
In an Emergency.

1860   1891

187

How many times these low feet staggered—
Only the soldered mouth can tell—
Try—can you stir the awful rivet—
Try—can you lift the hasps of steel!

Stroke the cool forehead—hot so often—   5
Lift—if you care—the listless hair—
Handle the adamantine[3] fingers
Never a thimble—more—shall wear—

Buzz the dull flies—on the chamber window—
Brave—shines the sun through the freckled pane—   10

1. Deceptively subtle reasonings.
2. I.e., the symbolic death of nature in late fall is compared to the death of Christ commemorated by the Christian sacrament of communion.
3. Rigidly firm; literally, made of adamant, a stone believed to be of impenetrable hardness.

Fearless—the cobweb swings from the ceiling—
Indolent Housewife—in Daisies—lain!

1860   1890

214

I taste a liquor never brewed—
From Tankards scooped in Pearl—
Not all the Frankfort Berries[1]
Yield such an Alcohol!

Inebriate of Air—am I—   5
And Debauchee of Dew—
Reeling—thro endless summer days—
From inns of Molten Blue—

When "Landlords" turn the drunken Bee
Out of the Foxglove's door—   10
When Butterflies—renounce their "drams"—
I shall but drink the more!

Till Seraphs[2] swing their snowy Hats—
And Saints—to windows run—
To see the little Tippler   15
From Manzanilla[3] come!

1860   1890

216

Safe in their Alabaster[4] Chambers—
Untouched by Morning
And untouched by Noon—
Sleep the meek members of the Resurrection—
Rafter of satin,   5
And Roof of stone.

Light laughs the breeze
In her Castle above them—
Babbles the Bee in a stolid Ear,
Pipe the Sweet Birds in ignorant cadence—   10
Ah, what sagacity perished here!

version of 1859   1890

216

Safe in their Alabaster Chambers—
Untouched by Morning—
And untouched by Noon—
Lie the meek members of the Resurrection—
Rafter of Satin—and Roof of Stone!   5

1. Grapes from the region around Frankfurt am Main, Germany, used to make a popular wine.
2. Six winged angels believed to guard God's throne.
3. Manzanillo is a busy port in Cuba.
4. Translucent, white, chalky material used in making plaster of Paris.

Grand go the Years—in the Crescent—above them—
Worlds scoop their Arcs—
And Firmaments—row—
Diadems—drop—and Doges[1]—surrender—

Soundless as dots—on a Disc of Snow—   10
version of 1861   1890

241

I like a look of Agony,
Because I know it's true—
Men do not sham Convulsion,
Nor simulate, a Throe—

The Eyes glaze once—and that is Death—   5
Impossible to feign
The Beads upon the Forehead
By homely Anguish strung.

1861   1890

249

Wild Nights—Wild Nights!
Were I with thee
Wild Nights should be
Our luxury!

Futile—the Winds—   5
To a Heart in port—
Done with the Compass—
Done with the Chart!

Rowing in Eden—
Ah, the Sea!   10
Might I but moor—Tonight—
In Thee!

1861   1891

258

There's a certain Slant of light,
Winter Afternoons—
That oppresses, like the Heft
Of Cathedral Tunes—

Heavenly Hurt, it gives us—   5
We can find no scar,
But internal difference,
Where the Meanings, are—

1. Chief magistrates in the republics of Venice and Genoa from the 11th through the
16th centuries.

None may teach it—Any—
'Tis the Seal[1] Despair—   10
An imperial affliction
Sent us of the Air—

When it comes, the Landscape listens—
Shadows—hold their breath—
When it goes, 'tis like the Distance   15
On the look of Death—

1861   1890

280

I felt a Funeral, in my Brain,
And Mourners to and fro
Kept treading—treading—till it seemed
That Sense was breaking through—

And when they all were seated,   5
A Service, like a Drum—
Kept beating—beating—till I thought
My Mind was going numb—

And then I heard them lift a Box
And creak across my Soul   10
With those same Boots of Lead, again,
The Space—began to toll,

As all the Heavens were a Bell,
And Being, but an Ear,
And I, and Silence, some strange Race   15
Wrecked, solitary, here—

And then a Plank in Reason, broke,
And I dropped down, and down—
And hit a World, at every plunge,
And Finished knowing—then—   20

1861   1896

281

'Tis so appalling—it exhilirates—
So over Horror, it half Captivates—
The Soul stares after it, secure—
To know the worst, leaves no dread more—

To scan a Ghost, is faint—   5
But grappling, conquers it—

---

1. In the double sense of a device used to imprint an official mark and an official sign of confirmation.

How easy, Torment, now—
Suspense kept sawing so—

The Truth, is Bald, and Cold—
But that will hold—   10
If any are not sure—
We show them—prayer—
But we, who know,
Stop hoping, now—

Looking at Death, is Dying—   15
Just let go the Breath—
And not the pillow at your Cheek
So Slumbereth—

Others, Can wrestle—
Yours, is done—   20
And so of Wo, bleak dreaded—come,
It sets the Fright at liberty—
And Terror's free—
Gay, Ghastly, Holiday!

1861   1935

287

A Clock stopped—
Not the Mantel's—
Geneva's[1] farthest skill
Cant put the puppet bowing—
That just now dangled still—   5

An awe came on the Trinket!
The Figures hunched, with pain—
Then quivered out of Decimals—
Into Degreeless Noon—

It will not stir for Doctor's—   10
This Pendulum of snow—
The Shopman importunes it—
While cool—concernless No—

Nods from the Gilded pointers—
Nods from the Seconds slim—   15
Decades of Arrogance between
The Dial life—
And Him—

1861   1896

1. Geneva, Switzerland, famous for its clockmakers.

303

The Soul selects her own Society—
Then—shuts the Door—
To her divine Majority—
Present no more—

Unmoved—she notes the Chariots—pausing—   5
At her low Gate—
Unmoved—an Emperor be kneeling
Upon her Mat—

I've known her—from an ample nation—
Choose One—   10
Then—close the Valves of her attention—
Like Stone—

1862   1890

304

The Day came slow—till Five o'clock—
Then sprang before the Hills
Like Hindered Rubies—or the Light
A Sudden Musket—spills—

The Purple could not keep the East—   5
The Sunrise shook abroad
Like Breadths of Topaz—[1]packed a Night—
The Lady just unrolled—

The Happy Winds—their Timbrels[2] took—
The Birds—in docile Rows   10
Arranged themselves around their Prince
The Wind—is Prince of Those—

The Orchard sparkled like a Jew—[3]
How mighty 'twas—to be
A Guest in this stupendous place—   15
The Parlor—of the Day —

1862   1891

311

It sifts from Leaden Sieves—
It powders all the Wood.
It fills with Alabaster[1] Wool
The Wrinkles of the Road—

1.. Yellow, reddish, or pink transparent crystal used as a gem.
2. Small hand drum or tambourine.
3. Charles Anderson suggests that with this phrase Dickinson evokes "nature as magnificent artifice" and recalls the Eastern richness suggested in Shakespeare's *The Merchant of Venice.*

1. Translucent, white, chalky material used in making plaster of Paris.

It makes an Even Face   5
Of Mountain, and of Plain—
Unbroken Forehead from the East
Unto the East again—

It reaches to the Fence—
It wraps it Rail by Rail   10
Till it is lost in Fleeces—
It deals Celestial Vail[2]

To Stump, and Stack—and Stem—
A Summer's empty Room—
Acres of Joints, where Harvests were,   15
Recordless, but for them—

It Ruffles Wrists of Posts
As Ankles of a Queen—
Then stills it's Artisans—like Ghosts—
Denying they have been—   20

1862   1891

326

I cannot dance upon my Toes—
No Man instructed me—
But oftentimes, among my mind,
A Glee possesseth me,

That had I Ballet knowledge—   5
Would put itself abroad
In Pirouette[1] to blanch a Troupe—
Or lay a Prima,[2] mad,

And though I had no Gown of Gauze—
No Ringlet, to my Hair,   10
Nor hopped for Audiences—like Birds,
One Claw upon the Air,

Nor tossed my shape in Eider[3] Balls,
Nor rolled on wheels of snow
Till I was out of sight, in sound,   15
The House encore me so—

Nor any know I know the Art
I mention—easy—Here—
Nor any Placard boast me—
It's full as Opera—   20

1862   1929

2. I.e., veil.
1. A full turn on the toe or ball of one foot.
2. The prima ballerina is the leading dancer in a ballet company.
3. Balls of fine, soft down from an eider duck.

328

A Bird came down the Walk—
He did not know I saw—
He bit an Angleworm in halves
And ate the fellow, raw,

And then he drank a Dew   5
From a convenient Grass—
And then hopped sidewise to the Wall
To let a Beetle pass—

He glanced with rapid eyes
That hurried all around—   10
They looked like frightened Beads, I thought—
He stirred his Velvet Head

Like one in danger, Cautious,
I offered him a Crumb
And he unrolled his feathers   15
And rowed him softer home—

Than Oars divide the Ocean,
Too silver for a seam—
Or Butterflies, off Banks of Noon
Leap, plashless[1] as they swim.   20

1862   1891

341

After great pain, a formal feeling comes—
The Nerves sit ceremonious, like Tombs—
The stiff Heart questions was it He, that bore,
And Yesterday, or Centuries before?

The Feet, mechanical, go round—   5
Of Ground, or Air, or Ought—
A Wooden way
Regardless grown,
A Quartz contentment, like a stone—

This is the Hour of Lead—   10
Remembered, if outlived,
As Freezing persons, recollect the Snow—
First—Chill—then Stupor—then the letting go—

1862   1929

348

I dreaded that first Robin, so,
But He is mastered, now,
I'm some accustomed to Him grown,
He hurts a little, though—

1. I.e., splashless.

I thought if I could only live   5
Till that first Shout got by—
Not all Pianos in the Woods
Had power to mangle me—

I dared not meet the Daffodils—
For fear their Yellow Gown   10
Would pierce me with a fashion
So foreign to my own—

I wished the Grass would hurry—
So—when 'twas time to see—
He'd be too tall, the tallest one   15
Could stretch—to look at me—

I could not bear the Bees should come,
I wished they'd stay away
In those dim countries where they go,
What word had they, for me?   20

They're here, though; not a creature failed—
No Blossom stayed away
In gentle deference to me—
The Queen of Calvary—[1]

Each one salutes me, as he goes,   25
And I, my childish Plumes,
Lift, in bereaved acknowledgement
Of their unthinking Drums—

1862   1891

412

I read my sentence—steadily—
Reviewed it with my eyes,
To see that I made no mistake
In it's extremest clause—
The Date, and manner, of the shame—   5
And then the Pious Form
That "God have mercy" on the Soul
The Jury voted Him—
I made my soul familiar—with her extremity—
That at the last, it should not be a novel Agony—   10
But she, and Death, acquainted—
Meet tranquilly, as friends—
Salute, and pass, without a Hint—
And there, the Matter ends—

1862   1891

1. A woman who has experienced in-   who was crucified on Mount Calvary
tense suffering; Mary, mother of Jesus,   near Jerusalem.

435

Much Madness is divinest Sense—
To a discerning Eye—
Much Sense—the starkest Madness—
'Tis the Majority
In this, as All, prevail— 5
Assent—and you are sane—
Demur—you're straightway dangerous—
And handled with a Chain—

1862   1890

449

I died for Beauty—but was scarce
Adjusted in the Tomb
When One who died for Truth, was lain
In an adjoining Room—

He questioned softly "Why I failed"? 5
"For Beauty", I replied—
"And I—for Truth—Themself are One—
We Bretheren, are", He said—

And so, as Kinsmen, met a Night—
We talked between the Rooms— 10
Until the Moss had reached our lips—
And covered up—our names—

1862   1890

465

I heard a Fly buzz—when I died—
The Stillness in the Room
Was like the Stillness in the Air—
Between the Heaves of Storm—

The Eyes around—had wrung them dry— 5
And Breaths were gathering firm
For that last Onset—when the King
Be witnessed—in the Room—

I willed my Keepsakes—Signed away
What portion of me be 10
Assignable—and then it was
There interposed a Fly—

With Blue—uncertain stumbling Buzz—
Between the light—and me—
And then the Windows failed—and then 15
I could not see to see—

1862   1896

500

Within my Garden, rides a Bird
Upon a single Wheel—
Whose spokes a dizzy Music make
As 'twere a travelling Mill—

He never stops, but slackens   5
Above the Ripest Rose—
Partakes without alighting
And praises as he goes,

Till every spice is tasted—
And then his Fairy Gig[1]   10
Reels in remoter atmospheres—
And I rejoin my Dog,

And He and I, perplex us
If positive, 'twere we—
Or bore the Garden in the Brain   15
This Curiosity—

But He, the best Logician,
Refers my clumsy eye—
To just vibrating Blossoms!
An Exquisite Reply!   20

1862   1929

501

This World is not Conclusion.
A Species stands beyond—
Invisible, as Music—
But positive, as Sound—
It beckons, and it baffles—   5
Philosophy—dont know—
And through a Riddle, at the last—
Sagacity, must go—
To guess it, puzzles scholars—
To gain it, Men have borne   10
Contempt of Generations
And Crucifixion, shown—
Faith slips—and laughs, and rallies—
Blushes, if any see—
Plucks at a twig of Evidence—   15
And asks a Vane, the way—
Much Gesture, from the Pulpit—
Strong Hallelujahs roll—
Narcotics cannot still the Tooth
That nibbles at the soul—   20

1862   1896

---

1. Light, two-wheeled carriage.

512

The Soul has Bandaged moments—
When too appalled to stir—
She feels some ghastly Fright come up
And stop to look at her—

Salute her—with long fingers—   5
Caress her freezing hair—
Sip, Goblin, from the very lips
The Lover—hovered—o'er—
Unworthy, that a thought so mean
Accost a Theme—so—fair—   10

The soul has moments of Escape—
When bursting all the doors—
She dances like a Bomb, abroad,
And swings upon the Hours,

As do the Bee—delirious borne—   15
Long Dungeoned from his Rose—
Touch Liberty—then know no more,
But Noon, and Paradise—

The Soul's retaken moments—
When, Felon led along,   20
With shackles on the plumed feet,
And staples,[1] in the Song,

The Horror welcomes her, again,
These, are not brayed of Tongue—

1862   1945

520

I started Early—Took my Dog—
And visited the Sea—
The Mermaids in the Basement
Came out to look at me—

And Frigates[2]—in the Upper Floor   5
Extended Hempen Hands—
Presuming Me to be a Mouse—
Aground—upon the Sands—

But no Man moved Me—till the Tide
Went past my simple Shoe—   10
And past my Apron—and my Belt
And past my Boddice—too—

1. U-shaped metal loops attached to shackles; in moments of ecstasy these would be pulled out of the surface into which they were driven; when the "felon" soul is "retaken," the staples would be used once again to restrict freedom.
2. Light boats originally propelled by oars, later by sails.

And made as He would eat me up—
As wholly as a Dew
Upon a Dandelion's Sleeve—   15
And then—I started—too—

And He—He followed—close behind—
Ĩ felt His Silver Heel
Upon my Ancle—Then my Shoes
Would overflow with Pearl—   20

Until We met the Solid Town—
No One He seemed to know—
And bowing—with a Mighty look—
At me—The Sea withdrew—

1862   1891

528

Mine—by the Right of the White Election!
Mine—by the Royal Seal!
Mine—by the Sign in the Scarlet prison—
Bars—cannot conceal!

Mine—here—in Vision—and in Veto!   5
Mine—by the Grave's Repeal—
Titled—Confirmed—
Delirious Charter!
Mine—long as Ages steal!

1890

547

I've seen a Dying Eye
Run round and round a Room—
In search of Something—as it seemed—
Then Cloudier become—
And then—obscure with Fog—   5
And then—be soldered down
Without disclosing what it be
'Twere blessed to have seen—

1862   1890

569

I reckon—when I count at all—
First—Poets—Then the Sun—
Then Summer—Then the Heaven of God—
And then—the List is done—

But, looking back—the First so seems   5
To Comprehend the Whole—
The Others look a needless Show—
So I write—Poets—All—

Their Summer—lasts a Solid Year—
They can afford a Sun   10
The East—would deem extravagant—
And if the Further Heaven—

Be Beautiful as they prepare
For Those who worship Them—
It is too difficult a Grace—   15
To justify the Dream—

1862   1929

585

I like to see it lap the Miles—
And lick the Valleys up—
And stop to feed itself at Tanks—
And then—prodigious step

Around a Pile of Mountains—   5
And supercilious peer
In Shanties—by the sides of Roads—
And then a Quarry pare

To fit it's sides
And crawl between   10
Complaining all the while
In horrid—hooting stanza—
Then chase itself down Hill—

And neigh like Boanerges[1]—
Then—prompter than a Star   15
Stop—docile and omnipotent
At it's own stable door—

1862   1891

599

There is a pain—so utter—
It swallows substance up—
Then covers the Abyss with Trance—
So Memory can step
Around—across—upon it—
As one within a Swoon—
Goes safely—where an open eye—
Would drop Him—Bone by Bone.

1862   1929

632

The Brain—is wider than the Sky—
For—put them side by side—
The one the other will contain
With ease—and You—beside—

---

1. "Sons of thunder" (Mark 3.17); originally applied to the zealous Apostles John and James; by extension, any vociferous orator.

The Brain is deeper than the sea—   5
For—hold them—Blue to Blue—
The one the other will absorb—
As Sponges—Buckets—do—

The Brain is just the weight of God—
For—Heft them—Pound for Pound—   10
And they will differ—if they do—
As Syllable from Sound—

1862   1896

664

Of all the Souls that stand create—
I have elected—One—
When Sense from Spirit—files away—
And Subterfuge—is done—
When that which is—and that which was—   5
Apart—intrinsic—stand—
And this brief Tragedy of Flesh—
Is shifted—like a Sand—
When Figures show their royal Front—
And Mists—are carved away,   10
Behold the Atom—I preferred—
To all the lists of Clay!

1862   1891

668

"Nature" is what we see—
The Hill—the Afternoon—
Squirrel—Eclipse—the Bumble bee—
Nay—Nature is Heaven—
Nature is what we hear—   5
The Bobolink—the Sea—
Thunder—the Cricket—
Nay—Nature is Harmony—
Nature is what we know—
Yet have no art to say—   10
So impotent Our Wisdom is
To her Simplicity

1863   1914

675

Essential Oils—are wrung—
The Attar[1] from the Rose
Be not expressed by Suns—alone—
It is the gift of Screws—

The General Rose—decay—   5
But this—in Lady's Drawer

---

1. Fragrant oil (as from rose petals).

Make Summer—When the Lady lie
In Ceaseless Rosemary—[2]

1863   *for remembrance*   1891

709

Publication—is the Auction
Of the Mind of Man—
Poverty—be justifying
For so foul a thing

Possibly—but We—would rather   5
From Our Garret go
White—Unto the White Creator—
Than invest—Our Snow—

Thought belong to Him who gave it—
Then—to Him Who bear   10
It's Corporeal illustration—Sell
The Royal Air—

In the Parcel—Be the Merchant
Of the Heavenly Grace—
But reduce no Human Spirit   15
To Disgrace of Price—

1863   1929

712

Because I could not stop for Death—
He kindly stopped for me—
The Carriage held but just Ourselves—
And Immortality.

We slowly drove—He knew no haste   5
And I had put away
My labor and my leisure too,
For His Civility—

We passed the School, where Children strove
At Recess—in the Ring—   10
We passed the Fields of Gazing Grain—
We passed the Setting Sun—

Or rather—He passed Us—
The Dews drew quivering and chill—
For only Gossamer, my Gown—   15
My Tippet[1]—only Tulle—

We paused before a House that seemed
A Swelling of the Ground—

2. Common fragrant herb used in cooking   1. Shoulder cape.
and perfume making.

The Roof was scarcely visible—
The Cornice—in the Ground—   20

Since then—'tis Centuries—and yet
Feels shorter than the Day
I first surmised the Horses Heads
Were toward Eternity—

1863   1890

721

Behind Me—dips Eternity—
Before Me—Immortality—
Myself—the Term between—
Death but the Drift of Eastern Gray,
Dissolving into Dawn away,   5
Before the West begin—

'Tis Kingdoms—afterward—they say—
In perfect—pauseless Monarchy—
Whose Prince—is Son of None—
Himself—His Dateless Dynasty—   10
Himself—Himself diversify—
In Duplicate divine—

'Tis Miracle before Me—then—
'Tis Miracle behind—between—
A Crescent in the Sea—   15
With Midnight to the North of Her—
And Midnight to the South of Her—
And Maelstrom[1]—in the Sky—

1863   1929

745

Renunciation—is a piercing Virtue—
The letting go
A Presence—for an Expectation—
Not now—
The putting out of Eyes—   5
Just Sunrise—
Lest Day—
Day's Great Progenitor—
Outvie
Renunciation—is the Choosing   10
Against itself—
Itself to justify
Unto itself—
When larger function—
Make that appear—   15
Smaller—that Covered Vision—Here—

1863   1929

1. A powerful whirlpool; turbulence.

764

Presentiment—is that long Shadow—on the Lawn—
Indicative that Suns go down—

The Notice to the startled Grass
That Darkness—is about to pass—

1863

824

The Wind begun to knead the Grass—
As Women do a Dough—
He flung a Hand full at the Plain—
A Hand full at the Sky—
The Leaves unhooked themselves from Trees—   5
And started all abroad—
The Dust did scoop itself like Hands—
And throw away the Road—
The Wagons quickened on the Street—
The Thunders gossiped low—   10
The Lightning showed a Yellow Head—
And then a livid Toe—
The Birds put up the Bars to Nests—
The Cattle flung to Barns—
Then came one drop of Giant Rain—   15
And then, as if the Hands
That held the Dams—had parted hold—
The Waters Wrecked the Sky—
But overlooked my Father's House—
Just Quartering a Tree—   20

[first version, c. 1864]   1891

824

The Wind begun to rock the Grass
With threatening Tunes and low—
He threw a Menace at the Earth—
A Menace at the Sky.

The Leaves unhooked themselves from Trees—   5
And started all abroad
The Dust did scoop itself like Hands
And threw away the Road.

The Wagons quickened on the Streets
The Thunder hurried slow—   10
The Lightning showed a Yellow Beak
And then a livid Claw.

The Birds put up the Bars to Nests—
The Cattle fled to Barns—

There came one drop of Giant Rain   15
And then as if the Hands

That held the Dams had parted hold
The Waters Wrecked the Sky,
But overlooked my Father's House—
Just quartering a Tree—   20
[second version, c. 1864]   1891

861

Split the Lark—and you'll find the Music—
Bulb after Bulb, in Silver rolled—
Scantily dealt to the Summer Morning
Saved for your Ear when Lutes be old.

Loose the Flood—you shall find it patent—[1]   5
Gush after Gush, reserved for you—
Scarlet Experiment! Sceptic Thomas![2]
Now, do you doubt that your Bird was true?

1864   1896

986

A narrow Fellow in the Grass
Occasionally rides—
You may have met Him—did you not
His notice sudden is—

The Grass divides as with a Comb—   5
A spotted shaft is seen—
And then it closes at your feet
And opens further on—

He likes a Boggy Acre
A floor too cool for Corn—   10
Yet when a Boy, and Barefoot—
I more than once at Noon
Have passed, I thought, a Whip lash
Unbraiding in the Sun
When stooping to secure it   15
It wrinkled, and was gone—

Several of Nature's People
I know, and they know me—
I feel for them a transport
Of cordiality—   20

1. In the paradoxical double sense of open to all and/or accessible and ex-clusively reserved by law.
2. Doubting disciple of Jesus.

But never met this Fellow
Attended, or alone
Without a tighter breathing
And Zero at the Bone—

1865   1866

1068

Further in Summer than the Birds
Pathetic from the Grass
A minor Nation[1] celebrates
It's unobtrusive Mass.

No Ordinance be seen   5
So gradual the Grace
A pensive Custom it becomes
Enlarging Loneliness.

Antiquest felt at Noon
When August burning low   10
Arise this spectral Canticle[2]
Repose to typify

Remit as yet no Grace
No Furrow on the Glow
Yet a Druidic[3] Difference   15
Enhances Nature now

1866   1891

1072

Title divine—is mine!
The Wife—without the Sign!
Acute Degree—conferred on me—
Empress of Calvary![1]
Royal—all but the Crown!   5
Betrothed—without the swoon
God sends us Women—
When you—hold—Garnet to Garnet—
Gold—to Gold—
Born—Bridalled—Shrouded—   10
In a Day—
"My Husband"—women say—
Stroking the Melody—
Is *this*—the way?

1862   1924

---

1. I.e., insects.
2. One of several liturgical songs taken from the Bible.
3. Druids were a pre-Christian Celtic religious order of priests, soothsayers, judges, and poets.

1. A woman who has experienced intense suffering; Mary, mother of Jesus, who was crucified on Mount Calvary near Jerusalem.

1078

The Bustle in a House
The Morning after Death
Is solemnest of industries
Enacted upon Earth—

The Sweeping up the Heart   5
And putting Love away
We shall not want to use again
Until Eternity.

1866   1890

1100

The last Night that She lived
It was a Common Night
Except the Dying—this to Us
Made Nature different

We noticed smallest things—   5
Things overlooked before
By this great light upon our Minds
Italicized—as 'twere.

As We went out and in
Between Her final Room   10
And Rooms where Those to be alive
Tomorrow were, a Blame

That Others could exist
While She must finish quite
A Jealousy for Her arose
So nearly infinite—

We waited while She passed—
It was a narrow time—
Too jostled were Our Souls to speak
At length the notice came.   20

She mentioned, and forgot—
Then lightly as a Reed
Bent to the Water, struggled scarce—
Consented, and was dead—

And We—We placed the Hair—   25
And drew the Head erect—
And then an awful leisure was
Belief to regulate—

1866   1890

1125

Oh Sumptuous moment
Slower go
That I may gloat on thee—
'Twill never be the same to starve
Now I abundance see—   5

Which was to famish, then or now—
The difference of Day
Ask him unto the Gallows led—
With morning in the sky

1868   1945

1126

Shall I take thee, the Poet said
To the propounded word?
Be stationed with the Candidates
Till I have finer tried—

The Poet searched Philology[1]   5
And when about to ring
for the Suspended Candidate
There came unsummoned in—

That portion of the Vision
The Word applied to fill   10
Not unto nomination
The Cherubim reveal—

1868   1945

1129

Tell all the Truth but tell it slant—
Success in Circuit lies
Too bright for our infirm Delight
The Truth's superb surprise

As Lightning to the Children eased   5
With explanation kind
The Truth must dazzle gradually
Or every man be blind—

c. 1868   1945

1333

A little Madness in the Spring
Is wholesome even for the King,
But God be with the Clown—

1. I.e., literature and the disciplines that concern themselves with words.

Who ponders this tremendous scene—
This whole Experiment of Green—   5
As if it were his own!

1875   1914

1356

The Rat is the concisest Tenant.
He pays no Rent.
Repudiates the Obligation—
On Schemes intent

Balking our Wit   5
To sound or circumvent—
Hate cannot harm
A Foe so reticent—
Neither Decree prohibit him—
Lawful as Equilibrium.   10

1876   1891

1400

What mystery pervades a well!
The water lives so far—
A neighbor from another world
Residing in a jar

Whose limit none have ever seen,   5
But just his lid of glass—
Like looking every time you please
In an abyss's face!

The grass does not appear afraid,
I often wonder he   10
Can stand so close and look so bold
At what is awe to me.

Related somehow they may be,
The sedge stands next the sea—
Where he is floorless   15
And does no timidity betray

But nature is a stranger yet;
The ones that cite her most
Have never passed her haunted house,
Nor simplified her ghost.   20

To pity those that know her not
Is helped by the regret
That those who know her, know her less
The nearer her they get.

No MS (c. 1877)   1896

1405

Bees are Black, with Gilt Surcingles—[2]
Bucaneers of Buzz.
Ride abroad in ostentation
And subsist on Fuzz.

Fuzz ordained—not Fuzz contingent—   5
Marrows of the Hill.
Jugs—a Universe's fracture
Could not jar or spill.

1877   1945

1445

Death is the supple Suitor
That wins at last—
It is a stealthy Wooing
Conducted first
By pallid innuendos   5
And dim approach
But brave at last with Bugles
And a bisected Coach
It bears away in triumph
To Troth unknown   10
And Kinsmen as divulgeless
As throngs of Down—

1878   1945

1461

"Heavenly Father"—take to thee
The supreme iniquity
Fashioned by thy candid Hand
In a moment contraband—
Though to trust us—seem to us   5
More respectful—"We are Dust"—
We apologize to thee
For thine own Duplicity—

1879   1914

1463

A Route of Evanescence
With a revolving Wheel—
A Resonance of Emerald—
A Rush of Cochineal[1]—
And every Blossom on the Bush   5
Adjust it's tumbled Head—
The mail from Tunis,[2] probably,
An easy Morning's Ride—

1879   1891

2. Belts or girth straps (as for a horse).   2. City on the north coast of Africa.
1. Red dye.

1501

It's little Ether[3] Hood
Doth sit upon it's Head—
The millinery supple
Of the sagacious God—

Till when it slip away   5
A nothing at a time—
And Dandelion's Drama
Expires in a stem.

1880   1945

1540

As imperceptibly as Grief
The Summer lapsed away—
Too imperceptible at last
To seem like Perfidy—
A Quietness distilled   5
As Twilight long begun,
Or Nature spending with herself
Sequestered Afternoon—
The Dusk drew earlier in—
The Morning foreign shone—   10
A courteous, yet harrowing Grace,
As Guest, that would be gone—
And thus, without a Wing
Or service of a Keel
Our Summer made her light escape   15
Into the Beautiful.

1882   1891

1545

The Bible is an antique Volume—
Written by faded Men
At the suggestion of Holy Spectres—
Subjects—Bethlehem—
Eden—the ancient Homestead—   5
Satan—the Brigadier—
Judas—the Great Defaulter—
David—the Troubadour—
Sin—a distinguished Precipice
Others must resist—
Boys that "believe" are very lonesome—
Other Boys are "lost"—
Had but the Tale a warbling Teller—
All the Boys would come—
Orpheus' Sermon[1] captivated—   15
It did not condemn—

1882   1924

---

3. Ethereal, light.
1. Orpheus, a musician of Greek legend, reputedly had the power to soothe wild beasts and make trees dance and rivers stand still with his lyre playing.

1593

There came a Wind like a Bugle—
It quivered through the Grass
And a Green Chill upon the Heat
So ominous did pass
We barred the Windows and the Doors   5
As from an Emerald Ghost—
The Doom's electric Moccasin
That very instant passed—
On a strange Mob of panting Trees
And Fences fled away   10
And Rivers where the Houses ran
Those looked that lived—that Day—
The Bell within the steeple wild
The flying tidings told—
How much can come   15
And much can go,
And yet abide the World!

1883   1891

1601

Of God we ask one favor,
That we may be forgiven—
For what, he is presumed to know—
The Crime, from us, is hidden—
Immured the whole of Life   5
Within a magic Prison
We reprimand the Happiness
That too competes with Heaven.

1884   1894

1624

Apparently with no surprise
To any happy Flower
The Frost beheads it at it's play—
In accidental power—
The blonde Assassin passes on—   5
The Sun proceeds unmoved
To measure off another Day
For an Approving God.

1884   1890

1670

In Winter in my Room
I came upon a Worm
Pink lank and warm
But as he was a worm
And worms presume   5
Not quite with him at home
Secured him by a string

To something neighboring
And went along.

A Trifle afterward   10
A thing occurred
I'd not believe it if I heard
But state with creeping blood
A snake with mottles rare
Surveyed my chamber floor   15
In feature as the worm before
But ringed with power
The very string with which
I tied him—too
When he was mean and new   20
That string was there—

I shrank—"How fair you are"!
Propitiation's claw—
"Afraid he hissed
Of me"?   25
"No cordiality"—
He fathomed me—
Then to a Rhythm *Slim*
Secreted in his Form
As Patterns swim   30
Projected him.

That time I flew
Both eyes his way
Lest he pursue
Nor ever ceased to run   35
Till in a distant Town
Towns on from mine
I set me down
This was a dream—

No MS   1914

American Literature
1865–1914

THE TRANSFORMATION OF A NATION

IN THE SECOND half of the nineteenth century, the fertile, mineral-rich American continent west of the Appalachians and Alleghenies was peopled and exploited. Americans, their numbers doubled by a continuous flow of immigrants, pushed westward to the Pacific coast, displacing Indians and the Spanish settlements where they stood in the way. Vast stands of timber were consumed; numberless herds of buffalo and other game gave way to cattle, sheep, farms, villages, and cities; various technologies converted the country's immense natural resources into industrial products both for its own burgeoning population and for foreign markets.

The result was that, between the end of the Civil War and the beginning of the First World War, the country was wholly transformed. Before the Civil War, America had been essentially a rural, agrarian, isolated republic whose idealistic, confident, and self-reliant inhabitants for the most part believed in God; by the time the United States entered World War I as a world power, it was an industrialized, urbanized, continental nation whose people had been forced to come to terms with the implications of Darwin's theory of evolution as well as with profound changes in its own social institutions and cultural values.

The Civil War cost some $8 billion and claimed 600,000 lives. It seems also to have left the country morally exhausted. Nonetheless, the country prospered materially over the five following decades in part because the war had stimulated technological development and had served as an occasion to test new methods of organization and management that were required to move efficiently large numbers of men and material, and which were then adapted to industrial modernization on a massive scale. The first transcontinental railroad was completed in 1869; industrial output grew at a geometric rate, and agricultural productivity increased dramatically; electricity was introduced on a large scale; new means of communication such as the telephone revolutionized many aspects of daily life; coal, oil, iron, gold, silver, and other kinds of mineral wealth were discovered and extracted to make large numbers of vast individual fortunes and to make the nation as a whole rich enough to capitalize for the first time on its own further development. By the end of the century, no longer a colony politically or economically, the United States could begin its own imperialist expansion (of which the Spanish-American War in 1898 was only one sign).

The central material fact of the period was industrialization, on a scale unprecedented in the earlier experiences of England and Europe. Between 1850 and 1880 capital invested in manufacturing industries more than quadrupled, while factory employment nearly doubled. By 1885 four transcontinental railroad lines were completed, using in their own construction and carrying to manufacturing centers in Pittsburgh, Cleveland, Detroit, and Chicago the nation's quintupled output of steel. This extensive railway system—and the invention of the refrigerated railway car—in turn made possible such economic developments as the centralization of the meat-packing industry in Chicago. Control over this enterprise as well as other industries passed to fewer and larger companies as time went on. In the two decades following the 1870s, a very small number of men controlled without significant competition the enormously profitable steel, railroad, oil, and meat-packing industries.

This group of men, known variously as buccaneers, captains of industry, self-made men, or robber barons, included Jay Gould, Jim Hill, Leland Stanford, Jim Fisk, Andrew Carnegie, J. P. Morgan, and John D. Rockefeller. However different in temperament and public behavior, all of these men successfully squeezed out their competitors and accumulated vast wealth and power. All were good examples of what the English novelist D. H. Lawrence described as "the lone hand and the huge success." These were the men who served as examplars of Mark Twain's Colonel Beriah Sellers, a character who in turn epitomizes much of the spirit of acquisitiveness excoriated by Twain in *The Gilded Age*, 1873 (a novel written in collaboration with Charles Dudley Warner).

In this half century, as industry flourished, America's cities grew. When the Civil War broke out, America, except for the northeastern seaboard, was a country of farms, villages, and small towns. Most of its citizens were involved in agricultural pursuits and small family businesses. By the turn of the century only about one third of the population lived on farms. New York had grown from a city of 500,000 in 1850 to a metropolis of nearly 3,500,000 persons by 1900, many of them recent immigrants from central, eastern, and southern Europe. Chicago, at mid-century a raw town of 20,000, had over 2,000,000 inhabitants by 1910. By the end of the First World War one half of the American population was concentrated in a dozen or so cities; the vast majority of all wage earners were employed by corporations and large enterprises, 8.5 million as factory workers. Millions of people participated in the prosperity that accompanied this explosive industrial expansion, but the social costs were immense.

The transformation of an entire continent, outlined above, was not accomplished, that is, without incalculable suffering. In the countryside increasing numbers of farmers, dependent for transportation of their crops on the monopolistic railroads, were squeezed off the land by what novelist Frank Norris characterized as the giant "octopus" that crisscrossed the continent. Everywhere independent farmers were placed "under the lion's paw" of land speculators and absentee landlords that Hamlin Garland's story made famous. For many, the great cities were also, as the radical novelist Upton Sinclair sensed, jungles where only the strongest, the most ruthless, and the luckiest survived. An oversupply of labor kept wages down and allowed the industrialists to maintain working conditions of notorious danger and discomfort for men, women, and children who competed for the scarce jobs.

Neither farmers nor urban laborers were effectively organized to pursue their own interests, and neither group had any significant political leverage until the 1880s. Legislators essentially served the interests of business and industry, and the scandals of President Grant's administration, the looting of the New York City Treasury by William Marcy ("Boss") Tweed in the 1870s, as well as the later horrors of municipal corruption exposed by journalist Lincoln Steffens and other "muckrakers" were symptomatic of what many writers of the time took to be the age of the "Great Barbecue." Early attempts by labor to organize were crude and often violent, and such groups as the notorious "Molly Maguires," which performed acts of terrorism in Pennsylvania, seemed to confirm the sense of the public and of the courts that labor organizations were "illegal conspiracies" and thus public enemies. Direct violence was probably, as young Emma Goldman believed, a necessary step toward establishing collective bargaining as a means of negotiating disputes between industrial workers and their employers; it was, in any event, not until such an alternative developed—really not until the 1930s—that labor acquired the unquestioned right to strike.

MARK TWAIN, W. D. HOWELLS, AND HENRY JAMES

This rapid transcontinental settlement and these new urban industrial circumstances were accompanied by the development of a national literature of great abundance and variety. New themes, new forms, new subjects, new regions, new authors, new audiences all emerged in the literature of this half century. As a result, at the onset of World War I, the spirit and substance of American literature had evolved remarkably, just as its center of production had shifted from Boston to New York in the late 1880s and the sources of its energy to Chicago and the Midwest. No longer was it produced, at least in its popular forms, in the main by solemn, typically moralistic men from New England and the Old South; no longer were polite, well-dressed, grammatically correct, middle-class young people the only central characters in its narratives; no longer were these narratives to be set in exotic places and remote times; no longer, indeed, were fiction, poetry, drama, and formal history the chief acceptable forms of literary expression; no longer, finally, was literature read primarily by young, middle-class women. In sum, American literature in these years fulfilled in considerable measure the condition Whitman called for in 1867 in describing *Leaves of Grass:* it treats, he said of his own major work, each state and region as peers "and expands from them, and includes the world . . . connecting an American citizen with the citizens of all nations." Self-educated men from the frontier, adventurers, and journalists introduced industrial workers and the rural poor, ambitious businessmen and vagrants, prostitutes and unheroic soldiers as major characters in fiction. At the same time, these years saw the emergence of what the critic Warner Berthoff aptly designates "the literature of argument," powerful works in sociology, philosophy, psychology, many of them impelled by the spirit of exposure and reform. Just as America learned to play a role in this half century as an autonomous international political, economic, and military power, so did its literature establish itself as a producer of major works. In its new security, moreover, it welcomed (in translation) the leading European figures of the time—Tolstoy, Ibsen, Chekhov, Hardy, Zola, Galdós, Verga—often in the columns of Henry James and William Dean Howells, who reviewed their works enthusiastically in *Harper's Weekly* and *Harper's Monthly,* the *North*

American Review, and other leading journals of the era. American writers in this period, like most writers of other times and places, wrote to earn money, earn fame, change the world, and—out of that mysterious compulsion to find the best order for the best words—express themselves in a permanent form and thus exorcise the demon that drove them.

The three figures who dominated prose fiction in the last quarter of the nineteenth century were Mark Twain, William Dean Howells, and Henry James. For half a century Howells was friend, editor, correspondent, and champion of both Twain and James. These latter two, however, knew each other little, and liked each other's work even less.

Twain was without doubt the most popular of the three, in part because of his unusual gift as a humorous public speaker. There is, indeed, much truth in the shrewd observation that Twain's art was in essence the art of the performer. Unlike many of his contemporaries who were successful on the platform, however, Twain had the even rarer ability to convert the humor of stage performance into written language. Twain is one of the few writers of any nationality who makes nearly all of his readers laugh out loud. He had flair; he shared the belief of one of his characters, that "you can't throw too much style into a miracle." To say this is not to deny Twain's skills as a craftsman, his consummate care with words. He was constitutionally incapable of writing—or speaking—a dull sentence. He was a master of style, and there is wide agreement among his fellow writers and literary historians alike that his masterpiece, *Adventures of Huckleberry Finn* (1885), is the fountainhead of American colloquial prose. Nor is there much dispute about Twain's ability to capture the enduring, archetypal, mythic images of America before the writer and the country came of age, or to create some of the most memorable characters in all of American fiction—Colonel Sellers, Tom Sawyer, Huck Finn and his Pap, the king and the duke, and Nigger Jim. Because of his "river" books—*The Adventures of Tom Sawyer* (1876), *Life on the Mississippi* (1883), and *Adventures of Huckleberry Finn* (1885)—we as Americans have a clearer (because it is mythic rather than historical) collective sense of life in the prewar Mississippi valley than we do of any other region of America at any other time.

Huck Finn might, at the end of his adventures, imagine lighting out "for the territory," the still uncivilized frontier, but for Twain there was to be no escape; *Huck Finn* was his last "evasion." Obliged to confront the moral, mental, and material squalor of postwar industrialized America (and to deal with a succession of personal catastrophes), Twain's work became more cerebral, contrived, and embittered. Only when he began in the late 1890s to dictate his autobiography, as the critic Jay Martin observes, did he once again become invisible in his words. Freed of the need to be respectable or to make large coherent narrative structures, Twain once more could function as an artist.

If Twain was the most popular of these three major figures, his friend and adviser Howells was unquestionably the most influential American writer in the last quarter of the century. Relentlessly productive, Howells wrote and published the equivalent of one hundred books during his sixty-year professional career. He wrote novels, travel books, biographies, plays, criticism, essays, autobiography—and made them pay. He was the first American writer self-consciously to conceive, to cite the title of one of his essays, of *The Man of Letters As a Man of Business*. Howells was, however, no mere acquisitive hack. He wrote always with a sense of his genteel, largely female audience; but if he generally observed the proprieties, he often took real risks and opened new territories for fiction. In Marcia Gaylord of *A Modern Instance* (1881) he traces with great subtlety the moral decline of an overindulged country girl as she turns into a vindictively jealous bitch. Nor can those who complain that Howells only wrote of the "smiling aspects" of life have read *The Landlord at Lion's Head* (1897), with its brutal Jeff Durgin, or *An Imperative Duty* (1892), a curious novel of miscegenation.

Starting with *The Rise of Silas Lapham* (1885), Howells addressed with deepening seriousness (and risk to his reputation) the relationship between the economic transformation of America and its moral condition. In *A Hazard of New Fortunes* (1890), written after his "conversion" to socialism by his reading of Tolstoy and what he described as the "civic murder" of the Haymarket anarchists in 1887, Howells offers his most extended interpretation of the decay of American life under the rule of competitive capitalism. Of the physical squalor and human misery he observes in the Bowery, his character Basil March (the central consciousness of the narrative) concludes:

> Accident and then exigency seemed the forces at work to this extraordinary effect; the play of energies as free and planless as those that force the forest from the soil to the sky; and then the fierce struggle for survival, with the stronger life persisting over the deformity, the mutilation, the destruction, the decay of the weaker. The whole at moments seemed to him lawless, Godless; the absence of intelligent, comprehensive purpose in the huge disorder, and the violent struggle to subordinate the result to the greater good, penetrated with its dumb appeal the consciousness of a man who had always been too self-inwrapt to perceive the chaos to which the individual selfishness must always lead.

Howells could see even more clearly in New York than he had in Boston that for the mass of their populations the cities had become infernos, the social environment degraded by the chasm that had opened between rich and poor, leaving the middle class trapped in their attitudes toward the poor between appalled sympathy and fear.

In his criticism, Howells had called for a literary realism that would treat commonplace Americans truthfully; "this truth given, the book *cannot* be wicked and cannot be weak; and without it all graces of style and feats of invention and cunning of construction are so many superfluities of naughtiness," he observed in "The Editor's Study" of *Harper's Monthly* for April, 1887. Critics are now pretty much in agreement that Howells's novels, especially those of the 1880s and 1890s, succeed in providing such a "truthful treatment." His lifelong friend Henry James observed in a letter on the occasion of Howells's seventy-fifth birthday: "Stroke by stroke and book by book your work was to become, for this exquisite notation of our whole democratic light and shade and give and take, in the highest degree *documentary*, so that none other . . . could approach it in value and amplitude." Whether Howells also had a truly "grasping imagination," the imagination to penetrate photographic surfaces, is the kind of question that finally each reader must decide; that it has become an active question at all is a sign of how far Howells's reputation has come since H. L. Mencken characterized him in 1919 as "an urbane and highly respectable old gentleman, a sitter on committees, an intimate of professors . . . , a placid conformist."

On his deathbed Howells was writing an essay on *The American James,* an essay which apparently would have defended James against charges that in moving to England permanently in the mid-1870s he had cut himself off from the sources of his imaginative power as well as sacrificed any hope of having a large audience for his work. In view of the critical praise lavished on James since World War II, and in the light of the highly successful film and television adaptations of his fiction, the need for any such defense may seem puzzling. Yet it is true that during his lifetime few of James's books had much popular success and that he was not a favorite of the American people.

There is more of American life and spirit represented in his literary production than is often thought, but James early came to believe the literary artist should not simply hold a mirror to the surface of social life in particular times and places. Instead, the writer should use language to probe the deepest reaches of the psychological and moral nature of human beings. At a time when novels were widely conceived as mere popular entertainment, James believed that the best fiction illuminates life by revealing it as an immensely complex process; and he demonstrated in work after work what a superb literary sensibility can do to dignify both life and art. He is a realist of the inner life; a dramatizer, typically, to put it crudely, of the tensions that develop between the young, innocent, selfless, and free woman and the older, sophisticated, and convention-bound man. That he is always on the side of freedom, Ezra Pound was to be one of the first to note.

Twain, James, and Howells together brought to fulfillment native trends in the realistic portrayal of the landscape and social surfaces, brought to perfection the vernacular style, and explored and exploited the literary possibilities of the interior life. Among them they recorded and made permanent the essential life of the eastern third of the continent as it was lived in the last half of the nineteenth century on the vanishing frontier, in the village, small town, or turbulent metropolis, and in European watering places and capitals. Among them they established the literary identity of distinctively American protagonists, specifically the vernacular hero and the "American Girl," the baffled and strained middle-class family, the businessman, the psychologically complicated citizens of a new international culture. Together, in short, they set the example and charted the future course for the subjects, themes, techniques, and styles of fiction we still call modern.

OTHER REALISTS AND NATURALISTS

Terms like *realism, naturalism, local color,* while useful shorthand for professors of literature trying to "cover" great numbers of books and long periods of time, probably do as much harm as they do good, especially for readers who are beginning their study of literature. The chief disservice these generalizing terms do to readers and authors is to divert attention away from the distinctive quality of an author's sense of life to a general body of ideas. In a letter turning down one of the many professorships he was offered, Howells observed that the study of literature should begin and end in pleasure, and it is far more rewarding to establish, in Emerson's phrase, "an original relationship" to particular texts and authors than it is an attempt to fit them into movements. However, since these generalizations are still in currency, we need to examine some of them.

One of the most far-reaching intellectual events of the last half of the nineteenth century was the publication in 1859 of Charles Darwin's *The Origin of Species*. This book, together with Darwin's *Descent of Man* (1870), hypothesized that over the millennia, man had evolved from lower forms of life. Humans were special, not—as the Bible taught—because God had created them in His image, but because they had successfully adapted to changing environmental conditions and had passed on their survival-making characteristics genetically. Though few American authors wrote treatises in reaction to Darwinism, nearly every writer had to come to terms somehow with this challenge to traditional conceptions of man, nature, and the social order.

One response was to accept the more negative implications of evolutionary theory and to use it to account for the behavior of characters in literary works. That is, characters were conceived as more or less complex combinations of inherited attributes and habits conditioned by social and economic forces. As Émile Zola, the influential French theorist and novelist, put the matter in his essay *The Experimental Novel*:

> In short, we must operate with characters, passions, human and social data as the chemist and the physicist work on inert bodies, as the physiologist works on living bodies. Determinism governs everything. It is scientific investigation; it is experimental reasoning that combats one by one the hypotheses of the idealists and will replace novels of pure imagination by novels of observation and experiment.

Many American writers adopted this pessimistic form of realism, this so-called "naturalistic" view of man, though each writer, of course, incorporated this assumption, and many others, into his or her work in highly individual ways.

Stephen Crane is a case in point. Crane believed, as he said of *Maggie*, that environment counts for a great deal in determining human fate. Nature is not hostile, he observes in *The Open Boat*, only "indifferent, flatly indifferent." Indeed, the earth, in *The Blue Hotel*, is described in one of the most famous passages in naturalistic fiction, as a "whirling, fire-smitten, ice-locked, disease-stricken, space-lost bulb." In Crane's *The Red Badge of Courage* Henry Fleming responds to the very end to the world of chaos and violence that surrounds him with alternating surges of panic and self-congratulation, not as a man who has understood himself and his place in a world which reveals order below its confused surface.

But after we have granted this ostensibly "naturalistic" perspective to Crane, we are still left with his distinctiveness as a writer, with his "personal honesty" in reporting what he saw (and concomitant rejection of accepted literary conventions), and with his use of impressionistic literary techniques to present incomplete characters and a broken world—a world more random than scientifically predictable. We are also left with the hardly pessimistic lesson of *The Open Boat*—that precisely because human beings are exposed to a savage world of chance where death is always imminent, they must learn the art of sympathetic identification with others and how to practice solidarity, often learned at the price of death. Without this deeply felt human connection, human experience is as meaningless as wind, sharks, and waves. It is Crane's power with words and his ability to live with paradox, then, that make him interesting, not his allegiance to philosophic or scientific theories.

Theodore Dreiser certainly did not share Crane's tendency to use words and images as if he were a composer or a painter. But he did share, at least early in his career, Crane's view that by and large human beings were more like moths drawn to flame than lords of creation. But, again, it is not Dreiser's beliefs that make him an enduring major figure in American letters: it is what his imagination and literary technique do with an extremely rich set of ideas, experiences, and emotions to create the "color of life" in his fictions that earns him an honored place. If Crane gave us through the personal honesty of his vision a new sense of the human consciousness under conditions of extreme pressure, Dreiser gave us for the first time in his unwieldy novels a sense of the fumbling, yearning, confused response to the simultaneously enchanting, exciting, ugly, and dangerous metropolis that had become the familiar residence for such large numbers of Americans by the turn of the century.

Regional writing, another expression of the realistic impulse, resulted from the desire both to preserve distinctive ways of life before industrialization dispersed or homogenized them and to avoid the harsh realities that seemed to replace these early times. At a more practical level, much of the writing was a response to the opportunities presented by the rapid growth of magazines, which created a new market for short fiction. By the end of the century, in any case, virtually every region of the country, from Maine to California, from the northern plains to the Louisiana bayous, had its "local colorist" (the implied comparison is to painters of so-called "genre" scenes) to immortalize its distinctive natural, social, and linguistic features. Though often suffused with sentimentalism and nostalgia, the best work of these regionalists renders both a convincing surface of a particular time and location and penetrates below that surface to the depths that transform the local into the universal.

During these fifty years a vast body of nonfictional prose was devoted to the description, analysis, and critique of social, economic, and political institutions and to the unsolved social problems that were one consequence of the rapid growth and change of the time. Women's rights, political corruption, economic inequity, business deceptions, the exploitation of labor—these became the subjects of articles and books by a long list of journalists, historians, social critics, and economists. A surprising amount of this writing survives as literature, and much of it has genuine power that is often attributed only to the older, "purer" forms. Certainly in that most ambitious of all American works of moral instruction, *The Education of Henry Adams* (1918), Adams registers through a literary sensibility a sophisticated historian's sense of what we now recognize as the disorientation that accompanies rapid and continuous change. The result is one of the most essential books of and about the whole period, and it seems fitting that Adams should have the last—though surely not the conclusive—word about his own problematic times.

In this half century, material, intellectual, social, and psychological changes in America went forward at such extreme speed and on such a massive scale that the enormously diverse writing of the time registers, at its core, degrees of shocked recognition of the human consequences of these radical transformations. Sometimes the shock is expressed in recoil and denial—thus the persistence, in the face of the ostensible triumph of realism, of the literature of diversion: nostalgic poetry, sentimental and melodramatic drama, and swashbuckling historical novels. The more enduring fictional and nonfictional prose forms of the era, however, come to terms imaginatively with the individual and collective dislocations and discontinuities associated with the closing out of the frontier, urbanization, intensified secularism, unprecedented immigration, the surge of national wealth unequally distributed, revised conceptions of human nature and destiny, the reordering of family and civil life, and the pervasive spread of mechanical and organizational technologies. The examples of courage, sympathy, and critical understanding on the part of our writers were a legacy to be drawn on, often unconsciously, often rebelliously, as America entered her next round of triumphs and tragedies, as the country self-consciously began its quest for a usable past.

SAMUEL CLEMENS
1835–1910

Samuel Langhorne Clemens, the third of five children, was born on November 30, 1835, in the village of Florida, Missouri, and grew up in the larger river town of Hannibal, that mixture of idyll and nightmare in and around which his two most famous characters, Tom Sawyer and Huck Finn, live out their adventure-filled summers. Sam's father, an ambitious and respected but unsuccessful country lawyer and storekeeper, died when Sam was twelve, and from that time on Sam worked to support himself and the rest of the family. Perhaps, as more than one critic has remarked, the shortness of his boyhood made him value it the more.

Sam was apprenticed to a printer after his father's death, and in 1851, when his brother Orion became a publisher in Hannibal, Sam went to work for him. In 1853 he began a three-year period of restless travel which took him to St. Louis, New York, Philadelphia, Keokuk (Iowa), and Cincinnati, in each of which he earned his living as a printer hired by the day. In 1856 he set out by steamboat for New Orleans, intending to go to the Amazon, where he expected to find adventure and perhaps wealth and fame besides. This scheme fell through, and instead, he apprenticed himself to Horace Bixby, the pilot of the Mississippi riverboat. After a training period of eighteen months, Clemens satisfied a boyhood ambition when he became a pilot himself. Clemens practiced this lucrative and prestigious trade until the Civil War virtually ended commercial river traffic in 1861. During this period he began to write humorous accounts of his activities for the *Keokuk Saturday Post*; though only three of these articles were published (under the pseudonym Thomas Jefferson Snodgrass), they established the pattern of peripatetic journalism—the pattern for much of the next ten years of his life.

After brief and rather inglorious service in the Confederate militia, Clemens made the first of the trips that would take him farther West and toward his ultimate careers as humorist, lecturer, and writer. In 1861, he accompanied Orion to the Nevada Territory, to which the latter had been appointed Secretary (chief record keeper for the territorial government) by President Lincoln. In *Roughing It*, written a decade later, Clemens told of the brothers' adventures on the way to Carson City and of the various schemes that, once there, Sam devised for getting rich quick on timber and silver. All of these schemes failed, however, and as usual Clemens was to get much more out of refining the ore of his experience into books than he was to earn from the actual experience of prospecting and claim staking. Soon enough Clemens was once again writing for newspapers, first for the *Territorial Enterprise* in Virginia City and then, after 1864, for the *Californian*. The fashion of the time called for a *nom de plume*, and Clemens used "Mark Twain," a term from his piloting days signifying "two fathoms deep" or "safe water." Twain's writing during these early years, while often distinctive and amusing, was largely imitative of the humorous journalism of the time and is important chiefly because it provided him with an opportunity to master the techniques of the short narrative and to try out a variety of subjects in a wide range of tones and modes. No less important than his writing during these formative Western years were his friendships with three master storytellers: the writer Bret Harte, the famous professional lecturer Artemus Ward, and the obscure amateur raconteur Jim Gillis. Twain owed his earliest national audience and critical recognition to his performances as lecturer and to his skillful retelling of a well-known tall tale, *The Jumping Frog of Calaveras County*, first published in 1865.

In this same year Twain signed up with the *Sacramento Union* to cover in a series of amusing letters the newly opened passenger service between San Francisco and Honolulu. In these letters Twain used a fictitious character, Mr. Brown, to present inelegant ideas, attitudes, and information, sometimes in impolite language. In this series Twain discovered that he could say almost anything he wanted—provided he could convincingly claim that he was simply reporting what others said and did. The refinement of this technique—a written equivalent of "dead-pan" lecturing—which allowed his fantasy a long leash and yet required him to anchor it in the circumstantial details of time and place, was to be his major technical accomplishment of the next two decades, the period of his best work.

The first book of this period—and still one of Twain's most popular—was *Innocents Abroad* (1869). It consists of a revised form of letters that Twain wrote for the *Alta California* and the *New York Tribune* during his 1867 excursion on the *Quaker City* to the Mediterranean and the Holy Land. The letters as they appeared in the newspaper—and later the book —were enormously popular, not only because they were exuberantly funny, but also because the satire they leveled against a pretentious, decadent, and undemocratic Old World was especially relished by a young country about to enter a period of explosive economic growth and political consolidation.

Twain still had no literary aspirations, nor any clear plans for a career. Nor did he have a permanent base of operations. To record the second half of his life is to record his acquisition of status, the mature development of his literary powers, and his transformation into a living public legend. A wife came first. Twain courted and finally won the hand of Olivia Langdon, the physically delicate daughter of a wealthy industrialist in Elmira, New York. This unlikely marriage brought both husband and wife the special pleasure that the union of apparent opposites sometimes yields. She was his "Angel"; he was her "Youth." The charge leveled in the 1920s by the critic Van Wyck Brooks that Livy's gentility emasculated Twain as a writer is no longer given serious credence. The complex comfort that this ebullient, whiskey-drinking, cigar-smoking, wild humorist took in his wife is suggested in the letter reproduced below that he wrote to his childhood friend Will Bowen.

The letter also makes clear just how deep the rich material of his Mississippi boyhood ran in his memory and imagination. To get to it, Twain had, in effect, to work chronologically backward and psychically inward. He made a tentative probe of this material as early as 1870 in an early version of *The Adventures of Tom Sawyer* called *A Boy's Manuscript*. But it was not until 1875, when he wrote *Old Times on the Mississippi* in seven installments of the *Atlantic Monthly* (edited by his lifelong friend W. D. Howells), that Twain arrived at the place his deepest imagination called home. For in this work, an account of Twain's apprenticeship to the pilot Horace Bixby, he evokes not only "the great Mississippi, the majestic, the magnificent Mississippi, rolling its mile-wide tide along, shining in the sun," but also his most intimate ties to the life on its surface and shores. With *Old Times* Twain is no longer a writer exploiting material; rather, he is a man expressing imaginatively a period of his life that he had deeply absorbed. The material he added to these installments to make the book *Life on the Mississippi* (1883) is often excellent, but it is clearly grafted on, not an organic development of the original story.

For all its charm and lasting appeal, *The Adventures of Tom Sawyer* (1876) is in certain respects a backward step. There is no mistaking its failure to integrate its self-consciously fine writing, addressed to adults, with its account in plain diction of thrilling adventures designed to appeal to young people. Twain had returned imaginatively to the Hannibal of his youth; before he could realize the deepest potential of this material he would have to put aside the psychological impediments of his civilized adulthood. *Tom Sawyer* creates a compelling myth of the endless summer of childhood pleasures mixed with terror, but as Twain hints in the opening sentence of *Huck Finn*, the earlier novel is important primarily as the place of origin of Huckleberry Finn—often described by critics and such fellow writers as Hemingway as the greatest American character in the greatest American book.

Adventures of Huckleberry Finn took Twain eight years to write. He began it in 1876, and completed it, after several stops and starts, in 1883. The fact that he devoted seven months to subsequent revision suggests that Twain was aware of the novel's sometimes discordant tones and illogical shifts in narrative intention. Any real or imagined flaws in the novel, however, have not bothered most readers; *Huck Finn* has enjoyed extraordinary popularity since its publication nearly one hundred years ago. Its unpretentious, colloquial, yet poetic style, its wide-ranging humor, its embodiment of the enduring and universally shared dream of perfect innocence and freedom, its recording of a vanished way of life in the pre–Civil War Mississippi valley has moved millions of people of all ages and conditions, and all over the world. It is one of those rare works that reveals to us the discrepancy between appearance and reality without leading us to despair of ourselves or others.

Though Twain made a number of attempts to return to the characters, themes, settings, and points of view of Tom and Huck, *Old Times*, *Tom Sawyer*, and *Huck Finn* had exhausted the rich themes of river and boyhood. Twain would live to write successful—even memorable—books, but the critical consensus is that his creative time had passed when he turned fifty. *A Connecticut Yankee at King Arthur's Court* (1889), for instance, is vividly imagined and very entertaining, but the satire, burlesque, moral outrage, and comic invention remain unintegrated.

A similar mood of despair informs and flaws *The Tragedy of Pudd'nhead Wilson* (1894). The book shows the disastrous effects of slavery on victimizer and victim alike—the unearned pride of whites and the undeserved self-hate of blacks. Satire turns to scorn, and beneath the drollery and fun one senses angry contempt for what Twain would soon refer to regularly as the "damned human race."

The decline of Twain's achievement as a writer between *Huck Finn* in 1884 and *Pudd'nhead Wilson* ten years later, however, is not nearly so precipitous as it turned out to be in the next decade, a decade which saw Twain's physical, economic, familial, and psychological supports collapse in a series of calamitous events. For forty years Twain had been fortune's favorite. Now, suddenly, his health was broken, his speculative investments in such enterprises as the Paige typesetting machine bankrupted him in the panic of 1893, his youngest daughter Jean was diagnosed as an epileptic, his oldest daughter Susy died of meningitis while he and Livy were in Europe, his wife began her decline into permanent invalidism, and Twain's grief for a time threatened his own sanity. For several years, as the critic Bernard DeVoto has shown, writing became both agonized labor and necessary therapy. The result of these circumstances was a dull book, *Following the Equator* (1897), which records Twain's round-the-world lecture tour undertaken to pay off debts; a sardonically preachy story, *The Man That Corrupted Hadleyburg* (1900); an embittered treatise on man's foibles, follies, and venality, *What Is Man?* (1906); and the bleakly despairing *The Mysterious Stranger*, first published in an "edited" version by Albert Bigelow Paine in 1916. Though the continuing study of the large bulk of Twain's unfinished (and, until recently, unpublished) work reveals much that is of interest to students of Twain, no one has come forward to claim that his final fifteen years represent a Henry Jamesian "major phase."

Unlike James, however, Twain in his last years became a revered public institution; his opinions were sought by the press on every subject of general interest. Though his opinions on many of these subjects—political, military, and social—were often tinged with vitriol, it was only to his best friends—who understood the complex roots of his despair and anger at the human race in general—that he vented his blackest rages. Much of this bitterness nonetheless informs such works, unpublished in his lifetime, as *The United States of Lyncherdom* (composed in 1901).

But early and late Twain maintained his magical power with language. What he said of one of his characters is a large part of his permanent appeal: "he could curl his tongue around the bulliest words in the language when he was a mind to, and lay them before you without a jint started, anywheres." This love of words and command over their arrangement, his mastery at distilling the rhythms and metaphors of oral speech into written prose, his vivid personality, his identification with the deepest centers of his fellow man's emotional and moral condition—all of these made Twain unique. As his friend Howells observed, he was unlike any of his contemporaries in American letters: "Emerson, Longfellow, Lowell, Holmes—I knew them all and all the rest of our sages, poets, seers, critics, humorists; they were like one another and like other literary men; but Clemens was sole, incomparable, the Lincoln of our literature." Like Lincoln, Clemens spoke to and for the common man of the American heartland that had nourished them both. And like Lincoln, who transcended the local political traditions in which he was trained to become the first great President of a continental nation, Clemens made out of the frontier humor and storytelling conventions of his journalistic influences a body of work of enduring value to the world of letters.

The editor is indebted to Frederick Anderson, late editor of the Mark Twain Papers at the Bancroft Library, University of California, Berkeley, for textual advice and general counsel in preparing the Clemens materials.

Jim Smiley and His Jumping Frog[1]

Mr. A. Ward,

Dear Sir:—Well, I called on good-natured, garrulous old Simon Wheeler, and I inquired after your friend Leonidas W. Smiley, as you requested me to do, and I hereunto append the result. If you can get any information out of it you are cordially welcome to it. I have a lurking suspicion that your Leonidas W. Smiley is a myth— that you never knew such a personage, and that you only conjectured that if I asked old Wheeler about him it would remind him of his infamous *Jim* Smiley, and he would go to work and bore me nearly to death with some infernal reminiscence of him as long and tedious as it should be useless to me. If that was your design, Mr. Ward, it will gratify you to know that it succeeded.

I found Simon Wheeler dozing comfortably by the bar-room stove of the little old dilapidated tavern in the ancient mining camp of Boomerang,[2] and I noticed that he was fat and bald-headed, and had an expression of winning gentleness and simplicity upon his tranquil countenance. He roused up and gave me good-day. I told him a friend of mine had commissioned me to make some inquiries about a cherished companion of his boyhood named Leonidas W. Smiley—Rev. Leonidas W. Smiley—a young minister of the gospel, who he had heard was at one time a resident of this village of Boomerang. I added that if Mr. Wheeler could tell me anything about this Rev. Leonidas W. Smiley, I would feel under many obligations to him.

Simon Wheeler backed me into a corner and blockaded me there with his chair—and then sat down and reeled off the monotonous narrative which follows this paragraph. He never smiled, he never frowned, he never changed his voice from the quiet, gently-flowing key to which he turned the initial sentence, he never betrayed the slightest suspicion of enthusiasm—but all through the interminable narrative there ran a vein of impressive earnestness and sincerity, which showed me plainly that so far from his imagining that there was anything ridiculous or funny about his story, he regarded it as a really important matter, and admired its two heroes as men of transcendent genius in finesse. To me, the spectacle of a man drifting serenely along through such a queer yarn without ever smiling was exquisitely absurd. As I said before, I asked him to tell me what he knew of Rev. Leonidas W. Smiley, and he replied as follows. I let him go on in his own way, and never interrupted him once:

---

1. We reprint the original text as it appeared in *The New York Saturday Press* for November 18, 1865. It was later published as a short book entitled *The Celebrated Jumping Frog of Calaveras County* in 1867. Artemus ("A.") Ward was the nom de plume of Charles Farrar Browne (1834–67), an American humorist and friend of Twain's.

2. A booming camp in Calaveras County, northern California, during the gold rush. Calaveras County is now famous for its annual frog-jumping contest.

There was a feller here once by the name of *Jim* Smiley, in the winter of '49—or maybe it was the spring of '50—I don't recollect exactly, some how, though what makes me think it was one or the other is because I remember the big flume³ wasn't finished when he first come to the camp; but anyway, he was the curiosest man about always betting on anything that turned up you ever see, if he could get anybody to bet on the other side, and if he couldn't he'd change sides—any way that suited the other man would suit *him*— any way just so's he got a bet, *he* was satisfied. But still, he was lucky—uncommon lucky; he most always come out winner. He was always ready and laying for a chance; there couldn't be no solitry thing mentioned but what that feller'd offer to bet on it—and take any side you please, as I was just telling you: if there was a horse race, you'd find him flush or you find him busted at the end of it; if there was a dog-fight, he'd bet on it; if there was a cat-fight, he'd bet on it; if there was a chicken-fight, he'd bet on it; why if there was two birds setting on a fence, he would bet you which one would fly first—or if there was a camp-meeting he would be there regular to bet on parson Walker, which he judged to be the best exhorter about here, and so he was, too, and a good man; if he even see a straddle-bug⁴ start to go any wheres, he would bet you how long it would take him to get wherever he was going to, and if you took him up he would foller that straddle-bug to Mexico but what he would find out where he was bound for and how long he was on the road. Lots of the boys here has seen that Smiley and can tell you about him. Why, it never made no difference to *him*—he would bet on *anything*—the dangdest feller. Parson Walker's wife laid very sick, once, for a good while, and it seemed as if they warn't going to save her; but one morning he come in and Smiley asked him how she was, and he said she was considerable better— thank the Lord for his inf'nit mercy—and coming on so smart that with the blessing of Providence she'd get well yet—and Smiley, before he thought, says, "Well, I'll resk two-and-a-half that she don't, anyway."

Thish-yer Smiley had a mare—the boys called her the fifteen-minute nag, but that was only in fun, you know, because, of course, she was faster than that—and he used to win money on that horse, for all she was so slow and always had the asthma, or the distemper, or the consumption, or something of that kind. They used to give her two or three hundred yards' start, and then pass her under way; but always at the fag-end of the race she'd get excited and desperate-like, and come cavorting and spraddling up, and scattering her legs around limber, sometimes in the air, and sometimes out to one side amongst the fences, and kicking up m-o-r-e dust, and raising m-o-r-e racket with her coughing and sneezing and blowing her nose—and

3. An inclined channel which conveys water from a distance.
4. Long-legged beetle.

always fetch up at the stand just about a neck ahead, as near as you could cipher it down.

And he had a little small bull-pup, that to look at him you'd think he warn't worth a cent, but to set around and look ornery, and lay for a chance to steal something. But as soon as money was up on him he was a different dog—his under-jaw'd begin to stick out like the for'castle of a steamboat, and his teeth would uncover, and shine savage like the furnaces. And a dog might tackle him, and bully-rag him, and bite him, and throw him over his shoulder two or three times, and Andrew Jackson—which was the name of the pup—Andrew Jackson would never let on but what he was satisfied, and hadn't expected nothing else—and the bets being doubled and doubled on the other side all the time, till the money was all up— and then all of a sudden he would grab that other dog just by the joint of his hind legs and freeze to it—not chaw, you understand, but only just grip and hang on till they throwed up the sponge, if it was a year. Smiley always came out winner on that pup till he harnessed a dog once that didn't have no hind legs, because they'd been sawed off in a circular saw, and when the thing had gone along far enough, and the money was all up, and he came to make a snatch for his pet holt, he saw in a minute how he'd been imposed on, and how the other dog had him in the door, so to speak, and he 'peared surprised, and then he looked sorter discouraged like, and didn't try no more to win the fight, and so he got shucked out bad. He gave Smiley a look as much as to say his heart was broke, and it was *his* fault, for putting up a dog that hadn't no hind legs for him to take holt of, which was his main dependence in a fight, and then he limped off a piece, and laid down and died. It was a good pup, was that Andrew Jackson, and would have made a name for hisself if he'd lived, for the stuff was in him, and he had genius—I know it, because he hadn't had no opportunities to speak of, and it don't stand to reason that a dog could make such a fight as he could under them circumstances, if he hadn't no talent. It always makes me feel sorry when I think of that last fight of his'on, and the way it turned out.

Well, thish-yer Smiley had rat-terriers and chicken cocks, and tom-cats, and all them kind of things, till you couldn't rest, and you couldn't fetch nothing for him to bet on but he'd match you. He ketched a frog one day and took him home and said he cal'lated to educate him; and so he never done nothing for three months but set in his back yard and learn that frog to jump. And you bet you he *did* learn him, too. He'd give him a little hunch behind, and the next minute you'd see that frog whirling in the air like a doughnut —see him turn one summerset,[5] or maybe a couple, if he got a good start, and come down flat-footed and all right, like a cat. He

5. Somersault.

got him up so in the matter of ketching flies, and kept him in practice so constant, that he'd nail a fly every time as far as he could see
him. Smiley said all a frog wanted was education, and he could do
most anything—and I believe him. Why, I've seen him set Dan'l
Webster down here on this floor—Dan'l Webster was the name of
the frog—and sing out, "Flies! Dan'l, flies," and quicker'n you
could wink, he'd spring straight up, and snake a fly off'n the counter there, and flop down on the floor again as solid as a gob of mud,
and fall to scratching the side of his head with his hind foot as
indifferent as if he hadn't no idea he'd done any more'n any frog
might do. You never see a frog so modest and straightfor'ard as he
was, for all he was so gifted. And when it come to fair-and-square
jumping on a dead level, he could get over more ground at one
straddle than any animal of his breed you ever see. Jumping on a
dead level was his strong suit, you understand, and when it come to
that, Smiley would ante up money on him as long as he had a red.[6]
Smiley was monstrous proud of his frog, and well he might be, for
fellers that had travelled and ben everywheres all said he laid over
any frog that ever *they* see.

Well, Smiley kept the beast in a little lattice box, and he used to
fetch him down town sometimes and lay for a bet. One day a feller
—a stranger in the camp, he was—come across him with his box,
and says:

"What might it be that you've got in the box?"

And Smiley says, sorter indifferent like, "It might be a parrot, or
it might be a canary, maybe, but it ain't—it's only just a frog."

And the feller took it, and looked at it careful, and turned it
round this way and that, and says, "H'm—so 'tis. Well, what's *he*
good for?"

"Well," Smiley says, easy and careless, "He's good enough for
one thing I should judge—he can out-jump any frog in Calaveras
county."

The feller took the box again, and took another long, particular
look, and give it back to Smiley and says, very deliberate, "Well—I
don't see no points about that frog that's any better'n any other
frog."

"Maybe you don't," Smiley says. "Maybe you understand frogs,
and maybe you don't understand 'em; maybe you've had experience,
and maybe you ain't only a amature, as it were. Anyways, I've got
my opinion, and I'll resk forty dollars that he can outjump ary frog
in Calaveras county."

And the feller studied a minute, and then says, kinder sad, like,
"Well—I'm only a stranger here, and I ain't got no frog—but if I
had a frog I'd bet you."

And then Smiley says, "That's all right—that's all right—if you'll

6. I.e., a red cent.

hold my box a minute I'll go and get you a frog;" and so the feller took the box, and put up his forty dollars along with Smiley's, and set down to wait.

So he set there a good while thinking and thinking to hisself, and then he got the frog out and prized his mouth open and took a tea-spoon and filled him full of quail-shot—filled him pretty near up to his chin—and set him on the floor. Smiley he went out to the swamp and slopped around in the mud for a long time, and finally he ketched a frog and fetched him in and give him to this feller and says:

"Now if you're ready, set him alongside of Dan'l, with his fore-paws just even with Dan'l's, and I'll give the word. Then he says, "one—two—three—jump!" and him and the feller touched up the frogs from behind, and the new frog hopped off lively, but Dan'l give a heave, and hysted up his shoulders—so—like a Frenchman, but it wasn't no use—he couldn't budge; he was planted as solid as a anvil, and he couldn't no more stir than if he was anchored out. Smiley was a good deal surprised, and he was disgusted too, but he didn't have no idea what the matter was, of course.

The feller took the money and started away, and when he was going out at the door he sorter jerked his thumb over his shoulder —this way—at Dan'l, and says again, very deliberate, "Well—I don't see no points about that frog that's any better'n any other frog."

Smiley he stood scratching his head and looking down at Dan'l a long time, and at last he says, "I do wonder what in the nation that frog throwed off for—I wonder if there ain't something the matter with him—he 'pears to look mighty baggy, somehow—and he ketched Dan'l by the nap of the neck, and lifted him up and says, "Why blame my cats if he don't weigh five pound"—and turned him upside down, and he belched out about a double-handful of shot. And then he see how it was, and he was the maddest man— he set the frog down and took out after that feller, but he never ketched him. And——

(Here Simon Wheeler heard his name called from the front-yard, and got up to go and see what was wanted.) And turning to me as he moved away, he said: "Just sit where you are, stranger, and rest easy—I ain't going to be gone a second."

But by your leave, I did not think that a continuation of the his-tory of the enterprising vagabond Jim Smiley would be likely to afford me much information concerning the Rev. Leonidas W. Smiley, and so I started away.

At the door I met the sociable Wheeler returning, and he but-tonholed me and recommenced:

"Well, thish-yer Smiley had a yaller one-eyed cow that didn't have no tail only just a short stump like a bannanner, and——"

"O, curse Smiley and his afflicted cow!" I muttered, good-naturedly, and bidding the old gentleman good-day, I departed.

Yours, truly,

Mark Twain

1865, 1867

Cruelty to Animals: The Histrionic Pig[1]

One of the most praiseworthy institutions in New York, and one which must plead eloquently for it when its wickedness shall call down the anger of the gods, is the Society for the Prevention of Cruelty to Animals. Its office is located on the corner of Twelfth street and Broadway, and its affairs are conducted by humane men who take a genuine interest in their work, and who have got worldly wealth enough to make it unnecessary for them to busy themselves about anything else. They have already put a potent check upon the brutality of draymen[2] and others to their horses, and in future will draw a still tighter rein upon such abuses, a late law of the Legislature having quadrupled their powers, and distinctly marked and specified them. You seldom see a horse beaten or otherwise cruelly used in New York now, so much has the society made itself feared and respected. Its members promptly secure the arrest of guilty parties and relentlessly prosecute them.

The new law gives the Society power to designate an adequate number of agents in every county, and these are appointed by the Sheriff, but work independently of all other branches of the civil organization. They can make arrests of guilty persons on the spot, without calling upon the regular police, and what is better, they can compel a man to stop abusing his horse, his dog, or any other animal, at a moment's warning. The object of the Society, as its name implies, is to prevent cruelty to animals, rather than punish men for being guilty of it.

They are going to put up hydrants and water tanks at convenient distances all over the city, for drinking places for men, horses and dogs.

Mr. Bergh,[3] the President of the Society, is a sort of enthusiast on the subject of cruelty to animals—or perhaps it would do him better justice to say he is full of honest earnestness upon the subject. Nothing that concerns the happiness of a brute is a trifling matter with him—no brute of whatever position or standing, however plebeian or insignificant, is beneath the range of his merciful interest. I have in my mind an example of his kindly solicitude for his dumb and helpless friends.

---

1. First published under the dateline New York, April 30, 1867, in the San Francisco *Alta California*, the source of the present text.

2. Carters.

3. Henry Bergh (1811–88), first president of the ASPCA.

He went to see the dramatic version of "Griffith Gaunt"[4] at Wallack's Theatre. The next morning he entered the manager's office and the following conversation took place:

Mr. Bergh—"Are you the manager of this theatre?"

Manager—"I am, sir. What can I do for you?"

Mr. B.—"I am President of the Society for the Prevention of Cruelty to Animals, and I have come to remonstrate against your treatment of that pig in the last act of the play last night. It is cruel and wrong, and I beg that you will leave the pig out in future."

"That is impossible! The pig is necessary to the play."

"But it is cruel, and you could alter the play in some way so as to leave the pig out."

"It cannot possibly be done, and besides I do not see anything wrong about it at all. What is it you complain of?"

"Why, it is plain enough. They punch the pig with sticks, and chase him and harass him, and contrive all manner of means to make him unhappy. The poor thing runs about in its distress, and tries to escape, but is met at every turn by its tormentors and its hopes blighted. The pig does not understand it. If the pig understood it, it might be well enough, but the pig does not know it is a play, but takes it all as reality, and is frightened and bewildered by the crowd of people and the glare of the lights, and yet no time is given it for reflection—no time is given it to arrive at a just appreciation of its circumstances—but its persecutors constantly assail it and keep its mind in such a chaotic state that it can form no opinion upon any point in the case. And besides, the pig is cast in the play without its consent, is forced to conduct itself in a manner which cannot but be humiliating to it, and leaves that stage every night with a conviction that it would rather die than take a character in a theatrical performance again. Pigs are not fitted for the stage; they have no dramatic talent; all their inclinations are toward a retired and unostentatious career in the humblest walks of life, and——"

Manager—"Say no more, sir. The pig is yours. I meant to have educated him for tragedy and made him a blessing to mankind and an ornament to his species, but I am convinced, now, that I ought not to do this in the face of his marked opposition to the stage, and so I present him to you, who will treat him well, I am amply satisfied. I am the more willing to part with him, since the play he performs in was taken off the stage last night, and I could not conveniently arrange a part for him in the one we shall run for the next three weeks, which is Richard III."[5]

1867

---

4. Dramatization by the producer Augustin Daly of a novel by Charles Reade; it was first produced on November 7, 1866. Wallack's Theatre was located on Broad-way at 13th Street.
5. The popular history play by Shakespeare.

Letter to Will Bowen (February 6, 1870)[1]
[*The Matter of Hannibal*]

Sunday Afternoon,
At Home, 472 Delaware Avenue,
My First, & Oldest & Dearest Friend,   Buffalo Feb. 6. 1870

My heart goes out to you just the same as ever. Your letter has stirred me to the bottom. The fountains of my great deep are broken up & I have rained reminiscences for four & twenty hours. The old life has swept before me like a panorama; the old days have trooped by in their old glory, again; the old faces have looked out of the mists of the past; old footsteps have sounded in my listening ears; old hands have clasped mine, old voices have greeted me, & the songs I loved ages & ages ago have come wailing down the centuries! Heavens what eternities have swung their hoary cycles about us since those days were new!—Since we tore down Dick Hardy's stable; since you had the measles & I went to your house purposely to catch them; since Henry Beebe kept that envied slaughter-house, & Joe Craig sold him cats to kill in it; since old General Gaines used to say, "Whoop! Bow your neck & spread!"; since Jimmy Finn was town drunkard & we stole his dinner while he slept in the vat & fed it to the hogs in order to keep them still till we could mount them & have a ride; since Clint Levering was drowned; since we taught that one-legged nigger, Higgins, to offend Bill League's dignity by hailing him in public with his exasperating "Hello, League!"—since we used to undress & play Robin Hood in our shirt-tails, with lath swords, in the woods on Halliday's Hill on those long summer days; since we used to go in swimming above the still-house branch—& at mighty intervals wandered on vagrant fishing excursions clear up to "the Bay," & wondered what was curtained away in the great world beyond that remote point; since I jumped overboard from the ferry boat in the middle of the river that stormy day to get my hat, & swam two or three miles after it (& *got* it,) while all the town collected on the wharf & for an hour or so looked out across the angry waste of "white-caps" toward where people said Sam. Clemens was last seen before he went down; since we got up a rebellion against Miss Newcomb, under Ed. Stevens' leadership, (to force her to let us all go over to Miss Torry's side of the schoolroom,) & gallantly "sassed" Laura Hawkins when she came out the third time to call us in, & then afterward marched in in threatening & bloodthirsty array,—& meekly yielded, & took each his little thrashing, & resumed his old seat entirely

1. This letter to one of Clemens's best boyhood friends was originally published in 1938; in 1941 it appeared in *Mark* *Twain's Letters to Will Bowen,* the source of the present text.

"reconstructed;" since we used to indulge in that very peculiar performance on that old bench outside the school-house to drive good old Bill Brown crazy while he was eating his dinner; since we used to remain at school at noon & go hungry, in order to persecute Bill Brown in all possible ways—poor old Bill, who *could* be driven to such extremity of vindictiveness as to call us "You *infernal* fools!" & chase us round & round the school-house—& yet who never had the heart to hurt us when he caught us, & who always loved us & always took our part when the big boys wanted to thrash us; since we used to lay in wait for Bill Pitts at the pump & whale him; (I saw him two or three years ago, & was awful polite to his six feet two, & mentioned no reminiscences); since we used to be in Dave Garth's class in Sunday school & on week-days stole his leaf tobacco to run our miniature tobacco presses with; since Owsley shot Smar; since Ben Hawkins shot off his finger; since we accidentally burned up that poor fellow in the calaboose;[2] since we used to shoot spool cannons; & cannons made of keys, while that envied & hated Henry Beebe drowned out our poor little pop-guns with his booming brazen little artillery on wheels; since Laura Hawkins was my sweetheart——

Hold! *That* rouses me out of my dream, & brings me violently back unto this day and this generation. For behold I have at this moment the only sweetheart I ever *loved*, and bless her old heart she is lying asleep upstairs in a bed that I sleep in every night, and for four whole days she has been *Mrs. Samuel L. Clemens!*

I am 34 and she is 24; I am young and very handsome (I make the statement with the fullest confidence, for I got it from her) and she is much the most beautiful girl I ever saw (I said that before she was anything to me,[3] and so it is worthy of all belief) and she is the *best* girl, and the sweetest, and the gentlest, and the daintiest and the most modest and unpretentious, and the wisest in all things she should be wise in, and the most ignorant in all matters it would not grace her to know, and she is sensible and quick, and loving and faithful, forgiving, full of charity—and her beautiful life is ordered by a religion that is all kindliness and unselfishness. Before the gentle majesty of her purity all evil things and evil ways and evil deeds stand abashed—then surrender. Wherefore, without effort, or struggle, or spoken exorcism, all the old vices and shameful habits that have possessed me these many many years, are falling away, one by one, and departing into the darkness.

Bill, I know whereof I speak. I am too old and have moved about

2. Small building that served as a jail; Clemens continued to blame this death on himself, though he had only supplied matches to the prisoner so that the latter could smoke.

3. Clemens fell in love with Olivia Langdon when her brother showed him her picture during a trip to the Holy Land.

too much, and rubbed against too many people not to know human beings as well as we used to know "boils" from "breaks."⁴

She is the very most perfect gem of womankind that ever I saw in my life—and I will stand by that remark till I die.

William, old boy, her father surprised us a little, the other night. We all arrived here in a night train (my little wife and I were going to board) and under pretense of taking us to the private boarding house that had been selected for me while I was absent lecturing in New England, my new father-in-law and some old friends drove us in sleighs to the daintiest, darlingest, lovliest little palace in America—and when I said "Oh, this wont do—people who can afford to live in this sort of style wont take boarders," that same blessed father-in-law let out the secret that this was all *our* property—a present from himself. House & furniture cost $40,000 in cash, (including stable, horse & carriage), & is a most exquisite little palace (I saw no apartment in Europe so lovely as our drawing-room.)

Come along, you & Mollie, just whenever you can, & pay us a visit, (giving us a little notice beforehand,) & if we don't make you comfortable nobody in the world can.

And now my princess has come down for dinner (bless me, isn't it cosy, nobody but just us two, & three servants to wait on us & respectfully call us "Mr." and "Mrs. Clemens" instead of "Sam." & "Livy!") It took me many a year to work up to where I can put on style, but now I'll do it.—My book gives me an income like a small lord, & my paper is a good profitable concern.⁵

Dinner's ready. Good bye & God bless you, old friend, & keep your heart fresh & your memory green for the old days that will never come again.

> Yrs always
> Sam. Clemens

From *Roughing It*¹
[*The Story of the Old Ram*]

Every now and then, in these days, the boys used to tell me I ought to get one Jim Blaine to tell me the stirring story of his grandfather's old ram—but they always added that I must not mention the matter unless Jim was drunk at the time—just comfortably and sociably drunk. They kept this up until my curiosity was on the rack to hear the story. I got to haunting Blaine; but it was of no

4. Eddylike disturbances on the surface of the water; a "break" is caused by a hidden solid object and is thus dangerous.
5. Clemens was part owner of the Buffalo *Express*; "My book": *Innocents Abroad* (1869).
1. The source of the present text is that of the 1872 edition (Hartford, Conn.: American Publishing Co.). The story of the composition and publication of this episodic work is told fully by Franklin R. Rogers in the introduction to Volume 2 of *The Works of Mark Twain* (1972); the basis for the present text is Chapter 53. The time is 1863 in Virginia City; the "boys" are the local roughs. Twain was 28 at the time of this "incident."

use, the boys always found fault with his condition; he was often moderately but never satisfactorily drunk. I never watched a man's condition with such absorbing interest, such anxious solicitude; I never so pined to see a man uncompromisingly drunk before. At last, one evening I hurried to his cabin, for I learned that this time his situation was such that even the most fastidious could find no fault with it—he was tranquilly, serenely, symmetrically drunk—not a hiccup to mar his voice, not a cloud upon his brain thick enough to obscure his memory. As I entered, he was sitting upon an empty powder-keg, with a clay pipe in one hand and the other raised to command silence. His face was round, red, and very serious; his throat was bare and his hair tumbled; in general appearance and costume he was a stalwart miner of the period. On the pine table stood a candle, and its dim light revealed "the boys" sitting here and there on bunks, candle-boxes, powder-kegs, etc. They said:

"Sh—! Don't speak—he's going to commence."

THE STORY OF THE OLD RAM

I found a seat at once, and Blaine said:

"I don't reckon them times will ever come again. There never was a more bullier old ram than what he was. Grandfather fetched him from Illinois—got him of a man by the name of Yates—Bill Yates—maybe you might have heard of him; his father was a deacon—Baptist—and he was a rustler, too; a man had to get up ruther early to get the start of old Thankful Yates; it was him that put the Greens up to jining teams with my grandfather when he moved West. Seth Green was prob'ly the pick of the flock; he married a Wilkerson—Sarah Wilkerson—good cretur, she was—one of the likeliest heifers that was ever raised in old Stoddard, everybody said that knowed her. She could heft a bar'l of flour as easy as I can flirt a flapjack. And spin? Don't mention it! Independent? Humph! When Sile Hawkins come a-browsing around her, she let him know that for all his tin he couldn't trot in harness alongside of *her*. You see, Sile Hawkins was—no, it warn't Sile Hawkins, after all—it was a galoot by the name of Filkins—I disremember his first name; but he *was* a stump—come into pra'r meeting drunk, one night, hooraying for Nixon, becuz he thought it was a primary; and old deacon Ferguson up and scooted him through the window and he lit on old Miss Jefferson's head, poor old filly. She was a good soul—had a glass eye and used to lend it to old Miss Wagner, that hadn't any, to receive company in; it warn't big enough, and when Miss Wagner warn't noticing, it would get twisted around in the socket, and look up, maybe, or out to one side, and every which way, while t'other one was looking as straight ahead as a spy-glass. Grown people didn't mind it, but it most always made the children cry, it was so sort of scary. She tried packing it in raw cotton, but it wouldn't work, somehow—the cotton would get loose and stick out and look so kind of awful that the children couldn't stand it no way. She was always dropping it out, and turned up her old dead-light on the company empty, and making them oncomfortable, becuz *she* never could tell when it hopped out, being blind on that side, you see. So somebody would have to hunch her and say, 'Your game eye has fetched loose, Miss Wagner dear'—and then all of them would have to sit and wait till she jammed it in again—wrong side before, as a general thing, and green as a bird's egg, being a bashful cretur and easy sot back before company. But being wrong side before warn't much difference, anyway, becuz her own eye was sky-blue and the glass one was yaller on the front side, so whichever way she turned it it didn't match nohow. Old Miss Wagner was considerable on the borrow, she was. When she had a quilting, or Dorcas S'iety[2] at her house she gen'ally borrowed Miss Higgins's wooden leg to stump around on; it was considerable shorter than her other pin, but much *she* minded that. She said she couldn't abide crutches when she had company, becuz they were so slow; said when she had company and things had to be done, she wanted to get up and hump herself. She was as bald as a jug, and so she used to borrow Miss Jacops's wig—Miss Jacops was the coffin-peddler's wife—a ratty old buzzard, he was, that used to go roosting around where people was sick, waiting for 'em; and there that old rip would sit all day, in the shade, on a coffin that he judged would fit the can'idate; and if it was a slow customer and kind of uncertain, he'd fetch his rations and a blanket along and sleep in the coffin nights. He was anchored out that way, in frosty weather, for about three weeks, once, before old Robbins's place, waiting for him; and after that, for as much as two years, Jacops was not on speaking terms with the old man, on account of his disapp'inting him. He got one of his feet froze, and lost money, too, becuz old Robbins took a favorable turn and got well. The next time Robbins got sick, Jacops tried to make up with him, and varnished up the same old coffin and fetched it along; but old Robbins was too many for him; he had him in, and 'peared to be powerful weak; he bought the coffin for ten dollars and Jacops was to pay it back and twenty-five more besides if Robbins didn't like the coffin after he'd tried it. And then Robbins died, and at the funeral he bursted off the lid and riz up in his shroud and told the parson to let up on the performances, becuz he could *not* stand such a coffin as that. You see he had been in a trance once before, when he was young, and he took the chances on another, cal'lating that if he made the trip it was money in his pocket, and if he missed fire he couldn't lose a cent. And by George he sued Jacops for the rhino[3] and got jedgment; and he set up the coffin in his back parlor and said he

2. Dorcas Society, common name for charitable church societies. From the Biblical Dorcas (Acts 9.36–42), who per-formed charitable deeds, particularly sewing clothes for the poor.
3. Slang for cash, money.

'lowed to take his time, now. It was always an aggravation to
Jacops, the way that miserable old thing acted. He moved back to
Indiany pretty soon—went to Wellsville—Wellsville was the place
the Hogadorns was from. Mighty fine family. Old Maryland stock.
Old Squire Hogadorn could carry around more mixed licker, and
cuss better than most any man I ever see. His second wife was the
widder Billings—she that was Becky Martin; her dam was deacon
Dunlap's first wife. Her oldest child, Maria, married a missionary
and died in grace—et up by the savages. They et *him*, too, poor feller—biled him. It warn't the custom, so they say, but they explained
to friends of his'n that went down there to bring away his things,
that they'd tried missionaries every other way and never could get
any good out of 'em—and so it annoyed all his relations to find out
that that man's life was fooled away just out of a dern'd experiment, so to speak. But mind you, there ain't anything ever reely
lost; everything that people can't understand and don't see the
reason of does good if you only hold on and give it a fair shake;
Prov'dence don't fire no blank ca'tridges, boys. That there missionary's substance, unbeknowns to himself, actu'ly converted every last
one of them heathens that took a chance at the barbecue. Nothing
ever fetched them but that. Don't tell *me* it was an accident that
he was biled. There ain't no such a thing as an accident. When my
uncle Lem was leaning up agin a scaffolding once, sick, or drunk, or
suthin, an Irishman with a hod full of bricks fell on him out of the
third story and broke the old man's back in two places. People said
it was an accident. Much accident there was about that. He didn't
know what he was there for, but he was there for a good object. If
he hadn't been there the Irishman would have been killed. Nobody
can ever make me believe anything different from that. Uncle
Lem's dog was there. Why didn't the Irishman fall on the dog?
Becuz the dog would a seen him a-coming and stood from under.
That's the reason the dog warn't appinted. A dog can't be
depended on to carry out a special providence. Mark my words it
was a put-up thing. Accidents don't happen, boys. Uncle Lem's dog
—I wish you could a seen that dog. He was a reglar shepherd—or
ruther he was part bull and part shepherd—splendid animal;
belonged to parson Hagar before Uncle Lem got him. Parson Hagar
belonged to the Western Reserve Hagars; prime family; his mother
was a Watson; one of his sisters married a Wheeler; they settled in
Morgan County, and he got nipped by the machinery in a carpet
factory and went through in less than a quarter of a minute; his
widder bought the piece of carpet that had his remains wove in,
and people come a hundred mile to 'tend the funeral. There was
fourteen yards in the piece. She wouldn't let them roll him up, but
planted him just so—full length. The church was middling small
where they preached the funeral, and they had to let one end of the
coffin stick out of the window. They didn't bury him—they planted

one end, and let him stand up, same as a monument. And they nailed a sign on it and put—put on—put on it—sacred to—the m-e-m-o-r-y—of fourteen y-a-r-d-s—of three-ply—car - - - pet—containing all that was—m-o-r-t-a-l—of—of—W-i-l-l-i-a-m—W-h-e—"

Jim Blaine had been growing gradually drowsy and drowsier—his head nodded, once, twice, three times—dropped peacefully upon his breast, and he fell tranquilly asleep. The tears were running down the boys' cheeks—they were suffocating with suppressed laughter—and had been from the start, though I had never noticed it. I perceived that I was "sold." I learned then that Jim Blaine's peculiarity was that whenever he reached a certain stage of intoxication, no human power could keep him from setting out, with impressive unction, to tell about a wonderful adventure which he had once had with his grandfather's old ram—and the mention of the ram in the first sentence was as far as any man had ever heard him get, concerning it. He always maundered off, interminably, from one thing to another, till his whisky got the best of him and he fell asleep. What the thing was that happened to him and his grandfather's old ram is a dark mystery to this day, for nobody has ever yet found out.

1872

From Old Times on the Mississippi[1]

I

When I was a boy, there was but one permanent ambition among my comrades in our village on the west bank of the Mississippi River. That was, to be a steamboatman. We had transient ambitions of other sorts, but they were only transient. When a circus came and went, it left us all burning to become clowns; the first negro minstrel show that came to our section left us all suffering to try that kind of life; now and then we had a hope that if we lived and were good, God would permit us to be pirates. These ambitions faded out, each in its turn; but the ambition to be a steamboatman always remained.

Once a day a cheap, gaudy packet arrived upward from St. Louis, and another downward from Keokuk. Before these events had transpired, the day was glorious with expectancy; after they had transpired, the day was a dead and empty thing. Not only the boys, but the whole village, felt this. After all these years I can picture that old time to myself now, just as it was then: the white town drows-

1. Twain had considered writing an account of his apprenticeship and piloting days a few years after they came to an end, but he did not get around to writing about them until 1875, when he contributed a series of seven papers to the *Atlantic Monthly*, the source of the present text. The first essay appeared in the *Atlantic* for January, 1875, the second in the February issue. Together they constitute the story of his initiation into riverboat piloting. After a trip down the Mississippi in 1882 Twain more than doubled the bulk of the *Old Times* series to make *Life on the Mississippi* (1883).

ing in the sunshine of a summer's morning; the streets empty, or
pretty nearly so; one or two clerks sitting in front of the Water
Street stores, with their splint-bottomed chairs tilted back against
the wall, chins on breasts, hats slouched over their faces, asleep—
with shingle-shavings enough around to show what broke them
down; a sow and a litter of pigs loafing along the sidewalk, doing a
good business in water-melon rinds and seeds; two or three lonely
little freight piles scattered about the "levee;"[2] a pile of "skids" on
the slope of the stone-paved wharf, and the fragrant town drunkard
asleep in the shadow of them; two or three wood flats at the head
of the wharf, but nobody to listen to the peaceful lapping of the
wavelets against them; the great Mississippi, the majestic, the magnificent Mississippi, rolling its mile-wide tide along, shining in the
sun; the dense forest away on the other side; the "point" above the
town, and the "point" below, bounding the river-glimpse and turning it into a sort of sea, and withal a very still and brilliant and
lonely one. Presently a film of dark smoke appears above one of
those remote "points;" instantly a negro drayman,[3] famous for his
quick eye and prodigious voice, lifts up the cry, "S-t-e-a-m-boat a-
comin'!" and the scene changes! The town drunkard stirs, the clerks
wake up, a furious clatter of drays follows, every house and store
pours out a human contribution, and all in a twinkling the dead
town is alive and moving. Drays, carts, men, boys, all go hurrying
from many quarters to a common centre, the wharf. Assembled
there, the people fasten their eyes upon the coming boat as upon a
wonder they are seeing for the first time. And the boat *is* rather a
handsome sight, too. She is long and sharp and trim and pretty; she
has two tall, fancy-topped chimneys, with a gilded device of some
kind swung between them; a fanciful pilot-house, all glass and
"gingerbread," perched on top of the "texas"[4] deck behind them;
the paddle-boxes are gorgeous with a picture or with gilded rays
above the boat's name; the boiler deck, the hurricane deck, and the
texas deck are fenced and ornamented with clean white railings;
there is a flag gallantly flying from the jack-staff; the furnace doors
are open and the fires glaring bravely; the upper decks are black
with passengers; the captain stands by the big bell, calm, imposing,
the envy of all; great volumes of the blackest smoke are rolling and
tumbling out of the chimneys—a husbanded grandeur created with
a bit of pitch pine just before arriving at a town; the crew are
grouped on the forecastle; the broad stage[5] is run far out over the
port bow, and an envied deck-hand stands picturesquely on the end
of it with a coil of rope in his hand; the pent steam is screaming
through the gauge-cocks; the captain lifts his hand, a bell rings, the

2. A river landing place; "skids": low
wooden platforms.
3. A "dray" is a strong, low cart or
wagon without sides.
4. The deck of a Mississippi steamer just

over the largest cabins, those of the
officers; "paddle-boxes": covering of the
paddlewheels on a sidewheel steamboat.
5. Plank for the landing and embarking
of passengers and freight.

wheels stop; then they turn back, churning the water to foam, and the steamer is at rest. Then such a scramble as there is to get aboard, and to get ashore, and to take in freight and to discharge freight, all at one and the same time; and such a yelling and cursing as the mates facilitate it all with! Ten minutes later the steamer is under way again, with no flag on the jack-staff and no black smoke issuing from the chimneys. After ten more minutes the town is dead again, and the town drunkard asleep by the skids once more.

My father was a justice of the peace, and I supposed he possessed the power of life and death over all men and could hang anybody that offended him. This was distinction enough for me as a general thing; but the desire to be a steamboatman kept intruding, nevertheless. I first wanted to be a cabin-boy, so that I could come out with a white apron on and shake a table-cloth over the side, where all my old comrades could see me; later I thought I would rather be the deck-hand who stood on the end of the stage-plank with the coil of rope in his hand, because he was particularly conspicuous. But these were only daydreams—they were too heavenly to be contemplated as real possibilities. By and by one of our boys went away. He was not heard of for a long time. At last he turned up as apprentice engineer or "striker" on a steamboat. This thing shook the bottom out of all my Sunday-school teachings. That boy had been notoriously worldly, and I just the reverse; yet he was exalted to this eminence, and I left in obscurity and misery. There was nothing generous about this fellow in his greatness. He would always manage to have a rusty bolt to scrub while his boat tarried at our town, and he would sit on the inside guard and scrub it, where we could all see him and envy him and loathe him. And whenever his boat was laid up he would come home and swell around the town in his blackest and greasiest clothes, so that nobody could help remembering that he was a steamboatman; and he used all sorts of steamboat technicalities in his talk, as if he were so used to them that he forgot common people could not understand them. He would speak of the "labboard"[6] side of a horse in an easy, natural way that would make one wish he was dead. And he was always talking about "St. Looy" like an old citizen; he would refer casually to occasions when he "was coming down Fourth Street," or when he was "passing by the Planter's House," or when there was a fire and he took a turn on the brakes of "the old Big Missouri;" and then he would go on and lie about how many towns the size of ours were burned down there that day. Two or three of the boys had long been persons of consideration among us because they had been to St. Louis once and had a vague general knowledge of its wonders, but the day of their glory was over now. They lapsed into a humble silence, and learned to disappear when the ruthless "cub"-engineer approached. This fellow had money, too, and hair oil. Also an igno-

6. I.e., larboard or port—the left-hand side of a ship.

rant silver watch and a showy brass watch chain. He wore a leather belt and used no suspenders. If ever a youth was cordially admired and hated by his comrades, this one was. No girl could withstand his charms. He "cut out" every boy in the village. When his boat blew up at last, it diffused a tranquil contentment among us such as we had not known for months. But when he came home the next week, alive, renowned, and appeared in church all battered up and bandaged, a shining hero, stared at and wondered over by everybody, it seemed to us that the partiality of Providence for an undeserving reptile had reached a point where it was open to criticism.

This creature's career could produce but one result, and it speedily followed. Boy after boy managed to get on the river. The minister's son became an engineer. The doctor's and the postmaster's sons became "mud clerks;"[7] the wholesale liquor dealer's son became a bar-keeper on a boat; four sons of the chief merchant, and two sons of the county judge, became pilots. Pilot was the grandest position of all. The pilot, even in those days of trivial wages, had a princely salary—from a hundred and fifty to two hundred and fifty dollars a month, and no board to pay. Two months of his wages would pay a preacher's salary for a year. Now some of us were left disconsolate. We could not get on the river—at least our parents would not let us.

So by and by I ran away. I said I never would come home again till I was a pilot and could come in glory. But somehow I could not manage it. I went meekly aboard a few of the boats that lay packed together like sardines at the long St. Louis wharf, and very humbly inquired for the pilots, but got only a cold shoulder and short words from mates and clerks. I had to make the best of this sort of treatment for the time being, but I had comforting day-dreams of a future when I should be a great and honored pilot, with plenty of money, and could kill some of these mates and clerks and pay for them.

Months afterward the hope within me struggled to a reluctant death, and I found myself without an ambition. But I was ashamed to go home. I was in Cincinnati, and I set to work to map out a new career. I had been reading about the recent exploration of the river Amazon by an expedition sent out by our government. It was said that the expedition, owing to difficulties, had not thoroughly explored a part of the country lying about the head-waters, some four thousand miles from the mouth of the river. It was only about fifteen hundred miles from Cincinnati to New Orleans, where I could doubtless get a ship. I had thirty dollars left; I would go and complete the exploration of the Amazon. This was all the thought I gave to the subject. I never was great in matters of detail. I packed

---

7. Assistants to steamboat clerks, so called because they had the task of re-ceiving and delivering the freight on open wharves in all kinds of weather.

my valise, and took passage on an ancient tub called the Paul Jones, for New Orleans. For the sum of sixteen dollars I had the scarred and tarnished splendors of "her" main saloon principally to myself, for she was not a creature to attract the eye of wiser travelers.

When we presently got under way and went poking down the broad Ohio, I became a new being, and the subject of my own admiration. I was a traveler! A word never had tasted so good in my mouth before. I had an exultant sense of being bound for mysterious lands and distant climes which I never have felt in so uplifting a degree since. I was in such a glorified condition that all ignoble feelings departed out of me, and I was able to look down and pity the untraveled with a compassion that had hardly a trace of contempt in it. Still, when we stopped at villages and wood-yards, I could not help lolling carelessly upon the railings of the boiler deck to enjoy the envy of the country boys on the bank. If they did not seem to discover me, I presently sneezed to attract their attention, or moved to a position where they could not help seeing me. And as soon as I knew they saw me I gaped and stretched, and gave other signs of being mightily bored with traveling.

I kept my hat off all the time, and stayed where the wind and the sun could strike me, because I wanted to get the bronzed and weather-beaten look of an old traveler. Before the second day was half gone, I experienced a joy which filled me with the purest gratitude; for I saw that the skin had begun to blister and peel off my face and neck. I wished that the boys and girls at home could see me now.

We reached Louisville in time—at least the neighborhood of it. We stuck hard and fast on the rocks in the middle of river and lay there four days. I was now beginning to feel a strong sense of being a part of the boat's family, a sort of infant son to the captain and younger brother to the officers. There is no estimating the pride I took in this grandeur, or the affection that began to swell and grow in me for those people. I could not know how the lordly steamboatman scorns that sort of presumption in a mere landsman. I particularly longed to acquire the least trifle of notice from the big stormy mate, and I was on the alert for an opportunity to do him a service to that end. It came at last. The riotous powwow of setting a spar was going on down on the forecastle, and I went down there and stood around in the way—or mostly skipping out of it—till the mate suddenly roared a general order for somebody to bring him a capstan bar.[8] I sprang to his side and said: "Tell me where it is— I'll fetch it!"

If a rag-picker had offered to do a diplomatic service for the Emperor of Russia, the monarch could not have been more astounded than the mate was. He even stopped swearing. He stood

8. Bar used to turn a capstan, a vertical spindle-mounted drum that is rotated to raise heavy weights by winding a heavy cable about it.

and stared down at me. It took him ten seconds to scrape his disjointed remains together again. Then he said impressively: "Well, if
this don't beat hell!" and turned to his work with the air of a man
who had been confronted with a problem too abstruse for solution.

I crept away, and courted solitude for the rest of the day. I did
not go to dinner; I stayed away from supper until everybody else
had finished. I did not feel so much like a member of the boat's
family now as before. However, my spirits returned, in installments,
as we pursued our way down the river. I was sorry I hated the mate
so, because it was not in (young) human nature not to admire him.
He was huge and muscular, his face was bearded and whiskered all
over; he had a red woman and a blue woman tattooed on his right
arm,—one on each side of a blue anchor with a red rope to it; and
in the matter of profanity he was perfect. When he was getting out
cargo at a landing, I was always where I could see and hear. He felt
all the sublimity of his great position, and made the world feel it,
too. When he gave even the simplest order, he discharged it like a
blast of lightning, and sent a long, reverberating peal of profanity
thundering after it. I could not help contrasting the way in which
the average landsman would give an order, with the mate's way of
doing it. If the landsman should wish the gangplank moved a foot
farther forward, he would probably say: "James, or Williams, one of
you, push that plank forward, please;" but put the mate in his
place, and he would roar out: "Here, now, start that gang-plank
for'ard! Lively, now! *What*'re you about! Snatch it! *snatch* it!
There! there! Aft again! aft again! Don't you hear me? Dash it to
dash! are you going to *sleep* over it! 'Vast heaving. 'Vast heaving, I
tell you! Going to heave it clear astern? WHERE're you going with
that barrel! *for'ard* with it 'fore I make you swallow it, you dash-
dash-dash-*dashed* split between a tired mud-turtle and a crippled
hearse-horse!"

I wished I could talk like that.

When the soreness of my adventure with the mate had somewhat worn off, I began timidly to make up to the humblest official
connected with the boat—the night watchman. He snubbed my
advances at first, but I presently ventured to offer him a new chalk
pipe, and that softened him. So he allowed me to sit with him by
the big bell on the hurricane deck, and in time he melted into conversation. He could not well have helped it, I hung with such
homage on his words and so plainly showed that I felt honored by
his notice. He told me the names of dim capes and shadowy islands
as we glided by them in the solemnity of the night, under the winking stars, and by and by got to talking about himself. He seemed
over-sentimental for a man whose salary was six dollars a week—or
rather he might have seemed so to an older person than I. But I
drank in his words hungrily, and with a faith that might have
moved mountains if it had been applied judiciously. What was it to

me that he was soiled and seedy and fragrant with gin? What was it to me that his grammar was bad, his construction worse, and his profanity so void of art that it was an element of weakness rather than strength in his conversation? He was a wronged man, a man who had seen trouble, and that was enough for me. As he mellowed into his plaintive history his tears dripped upon the lantern in his lap, and I cried, too, from sympathy. He said he was the son of an English nobleman—either an earl or an alderman, he could not remember which, but believed he was both; his father, the noble- man, loved him, but his mother hated him from the cradle; and so while he was still a little boy he was sent to "one of them old, ancient colleges"—he couldn't remember which; and by and by his father died and his mother seized the property and "shook" him, as he phrased it. After his mother shook him, members of the nobility with whom he was acquainted used their influence to get him the position of "lob-lolly-boy in a ship;" and from that point my watch- man threw off all trammels of date and locality and branched out into a narrative that bristled all along with incredible adventures; a narrative that was so reeking with blood-shed and so crammed with hair-breadth escapes and the most engaging and unconscious per- sonal villainies, that I sat speechless, enjoying, shuddering, wonder- ing, worshiping.

It was a sore blight to find out afterwards that he was a low, vulgar, ignorant, sentimental, half-witted humbug, an untraveled native of the wilds of Illinois, who had absorbed wildcat literature and appropriated its marvels, until in time he had woven odds and ends of the mess into this yarn, and then gone on telling it to fledg- lings like me, until he had come to believe it himself.

II

A "CUB" PILOT'S EXPERIENCE; OR, LEARNING THE RIVER

What with lying on the rocks four days at Louisville, and some other delays, the poor old Paul Jones fooled away about two weeks in making the voyage from Cincinnati to New Orleans. This gave me a chance to get acquainted with one of the pilots, and he taught me how to steer the boat, and thus made the fascination of river life more potent than ever for me.

It also gave me a chance to get acquainted with a youth who had taken deck passage—more's the pity; for he easily borrowed six dol- lars of me on a promise to return to the boat and pay it back to me the day after we should arrive. But he probably died or forgot, for he never came. It was doubtless the former, since he had said his parents were wealthy, and he only traveled deck passage[9] because it was cooler.

I soon discovered two things. One was that a vessel would not be likely to sail for the mouth of the Amazon under ten or twelve

9. " 'Deck' passage—i.e., steerage passage" [Clemens's note].

years; and the other was that the nine or ten dollars still left in my pocket would not suffice for so imposing an exploration as I had planned, even if I could afford to wait for a ship. Therefore it followed that I must contrive a new career. The Paul Jones was now bound for St. Louis. I planned a siege against my pilot, and at the end of three hard days he surrendered. He agreed to teach me the Mississippi River from New Orleans to St. Louis for five hundred dollars, payable out of the first wages I should receive after graduating. I entered upon the small enterprise of "learning" twelve or thirteen hundred miles of the great Mississippi River with the easy confidence of my time of life. If I had really known what I was about to require of my faculties, I should not have had the courage to begin. I supposed that all a pilot had to do was to keep his boat in the river, and I did not consider that that could be much of a trick, since it was so wide.

The boat backed out from New Orleans at four in the afternoon, and it was "our watch" until eight. Mr. B——, my chief, "straightened her up," plowed her along past the sterns of the other boats that lay at the Levee, and then said, "Here, take her; shave those steamships as close as you'd peel an apple." I took the wheel, and my heart went down into my boots; for it seemed to me that we were about to scrape the side off every ship in the line, we were so close. I held my breath and began to claw the boat away from the danger; and I had my own opinion of the pilot who had known no better than to get us into such peril, but I was too wise to express it. In half a minute I had a wide margin of safety intervening between the Paul Jones and the ships; and within ten seconds more I was set aside in disgrace, and Mr. B—— was going into danger again and flaying me alive with abuse of my cowardice. I was strung, but I was obliged to admire the easy confidence with which my chief loafed from side to side of his wheel, and trimmed the ships so closely that disaster seemed ceaselessly imminent. When he had cooled a little he told me that the easy water was close ashore and the current outside, and therefore we must hug the bank, upstream, to get the benefit of the former, and stay well out, downstream, to take advantage of the latter. In my own mind I resolved to be a down-stream pilot and leave the up-streaming to people dead to prudence.

Now and then Mr. B—— called my attention to certain things. Said he, "This is Six-Mile Point." I assented. It was pleasant enough information, but I could not see the bearing of it. I was not conscious that it was a matter of any interest to me. Another time he said, "This is Nine-Mile Point." Later he said, "This is Twelve-Mile Point." They were all about level with the water's edge; they all looked about alike to me; they were monotonously unpicturesque. I hoped Mr. B—— would change the subject. But no; he would crowd up around a point, hugging the shore with affection, and then say: "The slack water ends here, abreast this bunch of China-trees; now we cross over." So he crossed over. He gave me the wheel once or twice, but I had no luck. I either came near chipping off the edge of a sugar plantation, or else I yawed[1] too far from shore, and so I dropped back into disgrace again and got abused.

The watch was ended at last, and we took supper and went to bed. At midnight the glare of a lantern shone in my eyes, and the night watchman said:—

"Come! turn out!"

And then he left. I could not understand this extraordinary procedure; so I presently gave up trying to, and dozed off to sleep. Pretty soon the watchman was back again, and this time he was gruff. I was annoyed. I said:—

"What do you want to come bothering around here in the middle of the night for? Now as like as not I'll not get to sleep again to-night."

The watchman said;—

"Well, if this ain't good, I'm blest."

The "off-watch" was just turning in, and I heard some brutal laughter from them, and such remarks as "Hello, watchman! an't the new cub turned out yet? He's delicate, likely. Give him some sugar in a rag and send for the chambermaid to sing rock-a-by-baby to him."

About this time Mr. B—— appeared on the scene. Something like a minute later I was climbing the pilot-house steps with some of my clothes on and the rest in my arms. Mr. B—— was close behind, commenting. Here was something fresh—this thing of getting up in the middle of the night to go to work. It was a detail in piloting that had never occurred to me at all. I knew that boats ran all night, but somehow I had never happened to reflect that somebody had to get up out of a warm bed to run them. I began to fear that piloting was not quite so romantic as I had imagined it was; there was something very real and work-like about this new phase of it.

It was a rather dingy night, although a fair number of stars were out. The big mate was at the wheel, and he had the old tub pointed at a star and was holding her straight up the middle of the river. The shores on either hand were not much more than a mile apart, but they seemed wonderfully far away and ever so vague and indistinct. The mate said:—

"We've got to land at Jones's plantation, sir."

The vengeful spirit in me exulted. I said to myself, I wish you joy of your job, Mr. B——; you'll have a good time finding Mr. Jones's plantation such a night as this; and I hope you never *will* find it as long as you live.

1. Swerved.

Mr. B—— said to the mate:—

"Upper end of the plantation, or the lower?"

"Upper."

"I can't do it. The stumps there are out of water at this stage. It's no great distance to the lower, and you'll have to get along with that."

"All right, sir. If Jones don't like it he'll have to lump it, I reckon."

And then the mate left. My exultation began to cool and my wonder to come up. Here was a man who not only proposed to find this plantation on such a night, but to find either end of it you preferred. I dreadfully wanted to ask a question, but I was carrying about as many short answers as my cargo-room would admit of, so I held my peace. All I desired to ask Mr. B—— was the simple question whether he was ass enough to really imagine he was going to find that plantation on a night when all plantations were exactly alike and all the same color. But I held in. I used to have fine inspirations of prudence in those days.

Mr. B—— made for the shore and soon was scraping it, just the same as if it had been daylight. And not only that, but singing—

"Father in heaven the day is declining," etc.

It seemed to me that I had put my life in the keeping of a peculiarly reckless outcast. Presently he turned on me and said:—

"What's the name of the first point above New Orleans?"

I was gratified to be able to answer promptly, and I did. I said I didn't know.

"Don't *know*?"

This manner jolted me. I was down at the foot again, in a moment. But I had to say just what I had said before.

"Well, you're a smart one," said Mr. B——. "What's the name of the *next* point?"

Once more I didn't know.

"Well this beats anything. Tell me the name of *any* point or place I told you."

I studied a while and decided that I couldn't.

"Look-a-here! What do you start out from, above Twelve-Mile Point, to cross over?"

"I—I—don't know."

"You—you—don't know?" mimicking my drawling manner of speech. "What *do* you know?"

"I—I—nothing, for certain."

"By the great Caesar's ghost I believe you! You're the stupidest dunderhead I ever saw or ever heard of, so help me Moses! The idea of *you* being a pilot—*you!* Why, you don't know enough to pilot a cow down a lane."

Oh, but his wrath was up! He was a nervous man, and he shuffled from one side of his wheel to the other as if the floor was hot. He would boil a while to himself, and then overflow and scald me again.

"Look-a-here! What do you suppose I told you the names of those points for?"

I tremblingly considered a moment, and then the devil of temptation provoked me to say:—

"Well—to—to—be entertaining, I thought."

This was a red rag to the bull. He raged and stormed so (he was crossing the river at the time) that I judge it made him blind, because he ran over the steering-oar of a trading-scow. Of course the traders sent up a volley of red-hot profanity. Never was a man so grateful as Mr. B—— was: because he was brim full, and here were subjects who would *talk back*. He threw open a window, thrust his head out, and such an irruption followed as I never had heard before. The fainter and farther away the scowmen's curses drifted, the higher Mr. B—— lifted his voice and the weightier his adjectives grew. When he closed the window he was empty. You could have drawn a seine through his system and not caught curses enough to disturb your mother with. Presently he said to me in the gentlest way:—

"My boy, you must get a little memorandum-book, and every time I tell you a thing, put it down right away. There's only one way to be a pilot, and that is to get this entire river by heart. You have to know it just like A B C."

That was a dismal revelation to me; for my memory was never loaded with anything but blank cartridges. However, I did not feel discouraged long. I judged that it was best to make some allowances, for doubtless Mr. B—— was "stretching." Presently he pulled a rope and struck a few strokes on the big bell. The stars were all gone, now, and the night was as black as ink. I could hear the wheels churn along the bank, but I was not entirely certain that I could see the shore. The voice of the invisible watchman called up from the hurricane deck:—

"What's this, sir?"

"Jones's plantation."

I said to myself, I wish I might venture to offer a small bet that it isn't. But I did not chirp. I only waited to see. Mr. B—— handled the engine bells, and in due time the boat's nose came to the land, a torch glowed from the forecastle, a man skipped ashore, a darky's voice on the bank said, "Gimme de carpet-bag, Mars' Jones," and the next moment we were standing up the river again, all serene. I reflected deeply a while, and then said,—but not aloud, —Well, the finding of that plantation was the luckiest accident that ever happened; but it couldn't happen again in a hundred years. And I fully believed it *was* an accident, too.

By the time we had gone seven or eight hundred miles up the

river, I had learned to be a tolerably plucky up-stream steersman, in daylight, and before we reached St. Louis I had made a trifle of progress in night-work, but only a trifle. I had a note-book that fairly bristled with the names of towns, "points," bars, islands, bends, reaches, etc.; but the information was to be found only in the note-book—none of it was in my head. It made my heart ache to think I had only got half of the river set down; for as our watch was four hours off and four hours on, day and night, there was a long four-hour gap in my book for every time I had slept since the voyage began.

My chief was presently hired to go on a big New Orleans boat, and I packed my satchel and went with him. She was a grand affair. When I stood in her pilot-house I was so far above the water that I seemed perched on a mountain; and her decks stretched so far away, fore and aft, below me, that I wondered how I could ever have considered the little Paul Jones a large craft. There were other differences, too. The Paul Jones's pilot-house was a cheap, dingy, battered rattle-trap, cramped for room: but here was a sumptuous glass temple; room enough to have a dance in; showy red and gold window-curtains; an imposing sofa; leather cushions and a back to the high bench where visiting pilots sit, to spin yarns and "look at the river;" bright, fanciful "cuspadores" instead of a broad wooden box filled with sawdust; nice new oil-cloth on the floor; a hospitable big stove for winter; a wheel as high as my head, costly with inlaid work; a wire tiller-rope; bright brass knobs for the bells; and a tidy, white-aproned, black "texas-tender," to bring up tarts and ices and coffee during mid-watch, day and night. Now this was "something like;" and so I began to take heart once more to believe that piloting was a romantic sort of occupation after all. The moment we were under way I began to prowl about the great steamer and fill myself with joy. She was as clean and as dainty as a drawing-room; when I looked down her long, gilded saloon, it was like gazing through a splendid tunnel; she had an oil-picture, by some gifted sign-painter, on every state-room door; she glittered with no end of prism-fringed chandeliers; the clerk's office was elegant, the bar was marvelous, and the bar-keeper had been barbered and upholstered at incredible cost. The boiler deck (*i.e.*, the second story of the boat, so to speak) was as spacious as a church, it seemed to me; so with the forecastle; and there was no pitiful handful of deckhands, firemen, and roust-abouts down there, but a whole battalion of men. The fires were fiercely glaring from a long row of furnaces, and over them were eight huge boilers! This was unutterable pomp. The mighty engines—but enough of this. I had never felt so fine before. And when I found that the regiment of natty servants respectfully "sir'd" me, my satisfaction was complete.

When I returned to the pilot-house St. Louis was gone and I was lost. Here was a piece of river which was all down in my book, but I could make neither head nor tail of it: you understand, it was turned around. I had seen it, when coming up-stream, but I had never faced about to see how it looked when it was behind me. My heart broke again, for it was plain that I had got to learn this troublesome river *both ways*.

The pilot-house was full of pilots, going down to "look at the river." What is called the "upper river" (the two hundred miles between St. Louis and Cairo, where the Ohio comes in) was low; and the Mississippi changes its channel so constantly that the pilots used to always find it necessary to run down to Cairo to take a fresh look, when their boats were to lie in port a week, that is, when the water was at a low stage. A deal of this "looking at the river" was done by poor fellows who seldom had a berth, and whose only hope of getting one lay in their being always freshly posted and therefore ready to drop into the shoes of some reputable pilot, for a single trip, on account of such pilot's sudden illness, or some other necessity. And a good many of them constantly ran up and down inspecting the river, not because they ever really hoped to get a berth, but because (they being guests of the boat) it was cheaper to "look at the river" than stay ashore and pay board. In time these fellows grew dainty in their tastes, and only infested boats that had an established reputation for setting good tables. All visiting pilots were useful, for they were always ready and willing, winter or summer, night or day, to go out in the yawl and help buoy the channel or assist the boat's pilots in any way they could. They were likewise welcome because all pilots are tireless talkers, when gathered together, and as they talk only about the river they are always understood and are always interesting. Your true pilot cares nothing about anything on earth but the river, and his pride in his occupation surpasses the pride of kings.

We had a fine company of these river-inspectors along, this trip. There were eight or ten; and there was abundance of room for them in our great pilot-house. Two or three of them wore polished silk hats, elaborate shirt-fronts, diamond breastpins, kid gloves, and patent-leather boots. They were choice in their English, and bore themselves with a dignity proper to men of solid means and prodigious reputation as pilots. The others were more or less loosely clad, and wore upon their heads tall felt cones that were suggestive of the days of the Commonwealth.[2]

I was a cipher in this august company, and felt subdued, not to say torpid. I was not even of sufficient consequence to assist at the wheel when it was necessary to put the tiller hard down in a hurry; the guest that stood nearest did that when occasion required—and this was pretty much all the time, because of the crookedness of the channel and the scant water. I stood in a corner; and the talk I lis-

2. I.e., the government established in England by Oliver Cromwell from the execution of Charles I in 1649 to the Restoration of Charles II in 1660.

tened to took the hope all out of me. One visitor said to another:—

"Jim, how did you run Plum Point, coming up?"

"It was in the night, there, and I ran it the way one of the boys on the Diana told me; started out about fifty yards above the wood pile on the false point, and held on the cabin under Plum Point till I raised the reef—quarter less twain[3]—then straightened up for the middle bar[4] till I got well abreast the old one-limbed cotton-wood in the bend, then got my stern on the cotton-wood and head on the low place above the point, and came through a-booming—nine and a half."[5]

"Pretty square crossing, an't it?"

"Yes, but the upper bar's working down fast."

Another pilot spoke up and said:—

"I had better water than that, and ran it lower down; started out from the false point—mark twain[6]—raised the second reef abreast the big snag in the bend, and had quarter less twain."

One of the gorgeous ones remarked: "I don't want to find fault with your leadsmen,[7] but that's a good deal of water for Plum Point, it seems to me."

There was an approving nod all around as this quiet snub dropped on the boaster and "settled" him. And so they went on talk - talk - talking. Meantime, the thing that was running in my mind was, "Now if my ears hear aright, I have not only to get the names of all the towns and islands and bends, and so on, by heart, but I must even get up a warm personal acquaintanceship with every old snag and one-limbed cotton-wood and obscure wood pile that ornaments the banks of this river for twelve hundred miles; and more than that, I must actually know where these things are in the dark, unless these guests are gifted with eyes that can pierce through two miles of solid blackness; I wish the piloting business was in Jericho and I had never thought of it."

At dusk Mr. B—— tapped the big bell three times (the signal to land), and the captain emerged from his drawing-room in the forward end of the texts, and looked up inquiringly. Mr. B—— said:—

"We will lay up here all night, captain."

"Very well, sir."

That was all. The boat came to shore and was tied up for the night. It seemed to me a fine thing that the pilot could do as he pleased without asking so grand a captain's permission. I took my supper and went immediately to bed, discouraged by my day's observations and experiences. My late voyage's note-booking was but a confusion of meaningless names. It had tangled me all up in a

---

3. A quarter of a fathom less than two fathoms; i.e., 10½ feet of water.
4. I.e., submerged dirt bar in the middle of the river.
5. Nine and a half feet of water.

6. A two-fathom (12-foot) sounding; i.e., safe water.
7. Men who use sounding leads to determine the depth of water.

knot every time I had looked at it in the daytime. I now hoped for respite in sleep; but no, it reveled all through my head till sunrise again, a frantic and tireless nightmare.

Next morning I felt pretty rusty and low-spirited. We went booming along, taking a good many chances, for we were anxious to "get out of the river" (as getting out to Cairo was called) before night should overtake us. But Mr. B——'s partner, the other pilot, presently grounded the boat, and we lost so much time getting her off that it was plain the darkness would overtake us a good long way above the mouth. This was a great misfortune, especially to certain of our visiting pilots, whose boats would have to wait for their return, no matter how long that might be. It sobered the pilot-house talk a good deal. Coming up-stream, pilots did not mind low water or any kind of darkness; nothing stopped them but fog. But down-stream work was different; a boat was too nearly helpless, with a stiff current pushing behind her; so it was not customary to run down-stream at night in low water.

There seemed to be one small hope, however: if we could get through the intricate and dangerous Hat Island crossing before night, we could venture the rest, for we would have plainer sailing and better water. But it would be insanity to attempt Hat Island at night. So there was a deal of looking at watches all the rest of the day, and a constant ciphering upon the speed we were making; Hat Island was the eternal subject; sometimes hope was high and sometimes we were delayed in a bad crossing, and down it went again. For hours all hands lay under the burden of this suppressed excitement; it was even communicated to me, and I got to feeling so solicitous about Hat Island, and under such an awful pressure of responsibility, that I wished I might have five minutes on shore to draw a good, full, relieving breath, and start over again. We were standing no regular watches. Each of our pilots ran such portions of the river as he had run when coming up-stream, because of his greater familiarity with it; but both remained in the pilot-house constantly.

An hour before sunset, Mr. B—— took the wheel and Mr. W—— stepped aside. For the next thirty minutes every man held his watch in his hand and was restless, silent, and uneasy. At last somebody said, with a doomful sign.

"Well, yonder's Hat Island—and we can't make it."

All the watches closed with a snap, everybody sighed and muttered something about its being "too bad, too bad—ah, if we could *only* have got here half an hour sooner!" and the place was thick with the atmosphere of disappointment. Some started to go out, but loitered, hearing no bell-tap to land. The sun dipped behind the horizon, the boat went on. Inquiring looks passed from one guest to another; and one who had his hand on the doorknob, and had turned it, waited, then presently took away his hand and let the

knob turn back again. We bore steadily down the bend. More looks
were exchanged, and nods of surprised admiration—but no words.
Insensibly the men drew together behind Mr. B—— as the sky darkened and one or two dim stars came out. The dead silence and
sense of waiting became oppressive. Mr. B—— pulled the cord, and
two deep, mellow notes from the big bell floated off on the night.
Then a pause, and one more note was struck. The watchman's voice
followed, from the hurricane deck:—

"Labboard lead, there! Stabboard lead!"

The cries of the leadsmen began to rise out of the distance, and
were gruffly repeated by the word-passers on the hurricane deck.

"M-a-r-k three! M-a-r-k three! Quarter-less-three! Half twain!
Quarter twain! M-a-r-k twain! Quarter-less"—

Mr. B—— pulled two bell-ropes, and was answered by faint jin-
glings far below in the engine-room, and our speed slackened. The
steam began to whistle through the gauge-cocks. The cries of the
leadsmen went on—and it is a weird sound, always, in the night.
Every pilot in the lot was watching, now, with fixed eyes, and talking under his breath. Nobody was calm and easy but Mr. B——.
He would put his wheel down and stand on a spoke, and as the
steamer swung into her (to me) utterly invisible marks—for we
seemed to be in the midst of a wide and gloomy sea—he would
meet and fasten her there. Talk was going on, now, in low voices:-

"There; she's over the first reef all right!"

After a pause, another subdued voice:—

"Her stern's coming down just *exactly* right, by *George!* Now
she's in the marks;[8] over she goes!"

Somebody else muttered:—

"Oh, it was done beautiful—*beautiful!*"

Now the engines were stopped altogether, and we drifted with
the current. Not that I could see the boat drift, for I could not, the
stars being all gone by this time. This drifting was the dismalest
work; it held one's heart still. Presently I discovered a blacker gloom
than that which surrounded us. It was the head of the island. We
were closing right down upon it. We entered its deeper shadow,
and so imminent seemed the peril that I was likely to suffocate; and
I had the strongest impulse to do *something*, anything, to save the
vessel. But still Mr. B—— stood by his wheel, silent, intent as a
cat, and all the pilots stood shoulder to shoulder at his back.

"She'll not make it!" somebody whispered.

The water grew shoaler and shoaler by the leadsmen's cries, till it
was down to—

"Eight-and-a-half! E-i-g-h-t feet! E-i-g-h-t feet! Seven-and"—

Mr. B—— said warningly through his speaking tube to the engineer:—

"Stand by, now!"

8. Safe water, measurable water on a sounding lead.

"Aye-aye, sir."

"Seven-and-a-half! Seven feet! *Six*-and"——

We touched bottom! Instantly Mr. B—— set a lot of bells ringing, shouted through the tube, "Now let her have it—every ounce
you've got!" then to his partner, "Put her hard down! snatch her!
snatch her!" The boat rasped and ground her way through the sand,
hung upon the apex of disaster a single tremendous instant, and
then over she went! And such a shout as went up at Mr. B——'s
back never loosened the roof of a pilot-house before!

There was no more trouble after that. Mr. B—— was a hero that
night; and it was some little time, too, before his exploit ceased to
be talked about by river men.

Fully to realize the marvelous precision required in laying the
great steamer in her marks in that murky waste of water, one should
know that not only must she pick her intricate way through snags
and blind reefs, and then shave the head of the island so closely as
to brush the overhanging foliage with her stern, but at one place
she must pass almost within arm's reach of a sunken and invisible
wreck that would snatch the hull timbers from under her if she
should strike it, and destroy a quarter of a million dollars' worth of
steamboat and cargo in five minutes, and maybe a hundred and fifty
human lives into the bargain.

The last remark I heard that night was a compliment to Mr.
B——, uttered in soliloquy and with unction by one of our guests.
He said:—

"By the Shadow of Death, but he's a lightning pilot!"

1875, 1883

Whittier Birthday Dinner Speech[1]

Mr. Chairman—This is an occasion peculiarly meet for the digging up of pleasant reminiscences concerning literary folk; therefore
I will drop lightly into history myself. Standing here on the shore of
the Atlantic & contemplating certain of its biggest literary billows, I
am reminded of a thing which happened to me fifteen years ago,
when I had just succeeded in stirring up a little Nevadian literary
ocean-puddle myself, whose spume-flakes were beginning to blow
thinly California-wards. I started on an inspection-tramp through
the Southern mines of California. I was callow & conceited, & I
resolved to try the virtue of my nom de plume. I very soon had an
opportunity. I knocked at a miner's lonely log cabin in the foot-hills

1. This speech, which Twain delivered
on December 17, 1877, was first printed
in this form in the *Harvard Library Bulletin* (Spring, 1955), the source of the
present text. The occasion was a dinner
given by H. O. Houghton & Co., publishers of the *Atlantic Monthly*, in honor
of John Greenleaf Whittier's 70th birthday; among the guests were Ralph Waldo
Emerson, Oliver Wendell Holmes, and
Henry Wadsworth Longfellow. Although
the speech was intended to introduce a
note of humor into the solemn proceedings, some of the listeners thought
it in bad taste, and Twain worried about
the incident for a number of years afterward.

of the Sierras just at nightfall. It was snowing at the time. A jaded, melancholy man of fifty, barefooted, opened to me. When he heard my nom de plume, he looked more dejected than before. He let me in—pretty reluctantly, I thought—& after the customary bacon & beans, black coffee & a hot whisky, I took a pipe. This sorrowful man had not said three words up to this time. Now he spoke up & said in the voice of one who is secretly suffering, "You're the fourth —I'm a-going to move." "The fourth what?" said I. "The fourth littery man that's been here in twenty-four hours—I'm a-going to move." "You don't tell me!" said I; "Who were the others?" "Mr. Longfellow, Mr. Emerson, & Mr. Oliver Wendell Holmes—dad fetch the lot!"

The Miner's Story

You can easily believe I was interested.—I supplicated—three hot whiskies did the rest—& finally the melancholy miner began. Said he—

They came here just at dark yesterday evening, & I let them in, of course. Said they were going to Yo Semite.[2] They were a rough lot —but that's nothing—everybody looks rough that travels afoot. Mr. Emerson was a seedy little bit of a chap—red headed. Mr. Holmes was as fat as a balloon—he weighed as much as three hundred, & had double chins all the way down to his stomach. Mr. Longfellow was built like a prize fighter. His head was cropped & bristly—like as if he had a wig made of hair-brushes. His nose lay straight down his face, like a finger, with the end-joint tilted up. They had been drinking—I could see that. And what queer talk they used! Mr. Holmes inspected this cabin, then he took me by the button-hole, & says he—

"Through the deep caves of thought
I hear a voice that sings:
Build thee more stately mansions,
O my Soul!"[3]

Says I, "I can't afford it, Mr. Holmes, & moreover I don't want to." Blamed if I liked it pretty well, either, coming from a stranger, that way! However, I started to get out my bacon & beans, when Mr. Emerson came & looked on a while, & then *he* takes me aside by the button-hole & says—

"Give me agates for my meat;
Give me cantharids to eat;
From air & ocean bring me foods,
From all zones & altitudes."[4]

2. I.e., the Yosemite Valley, in California.
3. Cf. Holmes's *Chambered Nautilus*. Twain alters some of the following quotations, in order to heighten the joke.
4. Emerson's *Mithridates*; "cantharids": a species of beetle, the "Spanish fly," which in powdered form acts as a stimulant.

Says I, "Mr. Emerson, if you'll excuse me, this ain't no hotel." You see it sort of riled me—I warn't used to the ways of littery swells. But I went on a-sweating over my work, & next comes Mr. Longfellow & button-holes me, & interrupts me. Says he—

> "Honor be to Mudjekeewis!
> You shall hear how Pau-Puk-Kee-wis—"[5]

But I broke in, & says I, "Begging your pardon, Mr. Longfellow, if you'll be so kind as to hold your yawp for about five minutes, & let me get this grub ready, you'll do me proud." Well, sir, after they'd filled up, I set out the jug. Mr. Holmes looks at it, & then he fires up all of a sudden & yells—

> "Flash out a stream of blood-red wine!—
> For I would drink to other days."[6]

By George, I was getting kind of worked up. I don't deny it, I was getting kind of worked up. I turns to Mr. Holmes, & says I, "Looky-here, my fat friend, I'm a-running this shanty, & if the court knows herself, you'll take whisky-straight or you'll go dry!" Them's the very words I said to him. Now I didn't want to sass such famous littery people, but you see they kind of forced me. There ain't nothing onreasonable 'bout me; I don't mind a passel of guests a-tread'n on my tail three or four times, but when it comes to *standing* on it, it's different, & if the court knows herself, you'll take whisky-straight or you'll go dry!" Well, between drinks they'd swell around the cabin & strike attitudes & spout. Says Mr. Longfellow—

> "This is the forest primeval."[7]

Says Mr. Emerson—

> "Here once the embattled farmers stood,
> And fired the shot heard round the world."[8]

Says I, "O, blackguard the premises as much as you want to—it don't cost you a cent." Well, they went on drinking, & pretty soon they got out a greasy old deck & went to playing cut-throat euchre at ten cents a corner—on trust. I begun to notice some pretty suspicious things. Mr. Emerson dealt, looked at his hand, shook his head, says—

> "I am the doubter & the doubt—"

—& calmly bunched the hands & went to shuffling for a new lay-out. Says he—

> "They reckon ill who leave me out;
> They know not well the subtle ways
> I keep. [*pause*] I pass, & deal *again!*"[9]

5. Longfellow's *Hiawatha*.
6. Holmes's *Mare Rubrum*.
7. Longfellow's *Evangeline*.

8. Emerson's *Hymn Sung at the Completion of the Concord Monument*.
9. Emerson's *Brahma*.

Hang'd if he didn't go ahead & do it, too! O, he was a cool one! Well, in about a minute, things were running pretty tight, but all of a sudden I see by Mr. Emerson's eye that he judged he had 'em. He had already coralled two tricks, & each of the others one. So now he kinds of lifts a little, in his chair, & says—

"I tire of globes & aces!—
Too long the game is played!"[1]

—and down he fetches a right bower.[2] Mr. Longfellow smiles as sweet as pie, & says—:

"Thanks, thanks to thee, my worthy friend,
For the lesson thou hast taught!"[3]

—and dog my cats if he didn't down with *another* right bower! Well, sir, up jumps Holmes, a-war-whooping, as usual, & says—

"God help them if the tempest swings
The pine against the palm!"[4]

—and I wish I may go to grass if he didn't swoop down with *another* right bower! Emerson claps his hand on his bowie,[5] Longfellow claps his on his revolver, & I went under a bunk. There was going to be trouble; but that monstrous Holmes rose up, wobbling his double chins, & says he, "Order, gentlemen; the first man that draws, I'll lay down on him & smother him!" All quiet on the Potomac, you bet you! They were pretty how-come-you-so, now, & they begun to blow. Emerson says, "The bulliest thing I ever wrote, was Barbara Frietchie." Says Longfellow, "It don't begin with my Biglow Papers." Says Holmes, "My Thanatopsis lays over 'em both."[6] They mighty near ended in a fight. Then they wished they had some more company—& Mr. Emerson pointed at me & says—

"Is yonder squalid peasant all
That this proud nursery could breed?"[7]

He was a-whetting his bowie on his boot—so I let it pass. Well, sir, next they took it into their heads that they would like some music; so they made me stand up & sing "When Johnny Comes Marching Home" till I dropped—at thirteen minutes past four this morning. That's what I've been through, my friend. When I woke at seven, they were leaving, thank goodness, & Mr. Longfellow had my only boots on, & his own under his arm. Says I, "Hold on, there, Evangeline, what you going to do with *them*?"—He says: "Going to make tracks with 'em; because—

1. Emerson's *Song of Nature*.
2. The jack of trumps in euchre.
3. Longfellow's *Village Blacksmith*.
4. Holmes's *Voice of the Loyal North*.
5. Bowie knife.
6. Twain purposely misattributes these works: *Barbara Frietchie* is by John Greenleaf Whittier, *The Biglow Papers* by James Russell Lowell, and *Thanatopsis* by William Cullen Bryant.
7. Emerson's *Monadnoc*.

"Lives of great men all remind us
We can make our lives sublime;
And departing, leave behind us
Footprints on the sands of Time."[8]

[As I said, Mr. Twain, you are the fourth in twenty-four hours—and I'm going to move; I ain't suited to a'] littery atmosphere."[9]

I said to the miner, "Why my dear sir, *these* were not the gracious singers to whom we & the world pay loving reverence & homage: these were impostors." The miner investigated me with a calm eye for a while, then said he, "Ah—impostors, were they? are *you?*" I did not pursue the subject; and since then I haven't traveled on my nom de plume enough to hurt. Such is the reminiscence I was moved to contribute, Mr. Chairman. In my enthusiasm I may have exaggerated the details a little, but you will easily forgive me that fault since I believe it is the first time I have ever deflected from perpendicular fact on an occasion like this.

1877   1955

Fenimore Cooper's Literary Offences[1]

The *Pathfinder* and *The Deerslayer* stand at the head of Cooper's novels as artistic creations. There are others of his works which contain parts as perfect as are to be found in these, and scenes even more thrilling. Not one can be compared with either of them as a finished whole.

The defects in both of these tales are comparatively slight. They were pure works of art.—*Prof. Lounsbury.*

The five tales reveal an extraordinary fulness of invention.

. . . One of the very greatest characters in fiction, "Natty Bumppo." . . .

The craft of the woodsman, the tricks of the trapper, all the delicate art of the forest, were familiar to Cooper from his youth up.—*Prof. Brander Matthews.*

Cooper is the greatest artist in the domain of romantic fiction yet produced by America.—*Wilkie Collins.*[2]

It seems to me that it was far from right for the Professor of English Literature in Yale, the Professor of English Literature in Columbia, and Wilkie Collins, to deliver opinions on Cooper's literature without having read some of it. It would have been much

8. Longfellow's *Psalm of Life.*
9. A piece of the manuscript is missing here; the omitted words are supplied from the version Twain included in his Autobiographical Dictation.
1. James Fenimore Cooper (1789–1851) is best known for his series of historical novels in which the hero is variously called Leatherstocking, Natty Bumppo, Hawkeye, and Deerslayer. The novels are *The Pioneers* (1823), *The Last of the Mohicans* (1826), *The Prairie* (1827), *The Pathfinder* (1840), and *The*

Deerslayer (1841). This essay was first published in July, 1895, in *North American Review*, the source for the present text, and later in the collection *How to Tell a Story and Other Essays* in 1897.
2. Thomas Raynesford Lounsbury (1838–1915), American scholar and editor; professor at Yale University; James Brander Matthews (1852–1929), American educator and author, professor at Columbia University; William Wilkie Collins (1824–89), English novelist.

more decorous to keep silent and let persons talk who have read Cooper.

Cooper's art has some defects. In one place in *Deerslayer*, and in the restricted space of two-thirds of a page, Cooper has scored 114 offences against literary art out of a possible 115. It breaks the record.

There are nineteen rules governing literary art in the domain of romantic fiction—some say twenty-two. In *Deerslayer* Cooper violated eighteen of them. These eighteen require:

1. That a tale shall accomplish something and arrive somewhere. But the *Deerslayer* tale accomplishes nothing and arrives in the air.

2. They require that the episodes of a tale shall be necessary parts of the tale, and shall help to develop it. But as the *Deerslayer* tale is not a tale, and accomplishes nothing and arrives nowhere, the episodes have no rightful place in the work, since there was nothing for them to develop.

3. They require that the personages in a tale shall be alive, except in the case of corpses, and that always the reader shall be able to tell the corpses from the others. But this detail has often been overlooked in the *Deerslayer* tale.

4. They require that the personages in a tale, both dead and alive, shall exhibit a sufficient excuse for being there. But this detail also has been overlooked in the *Deerslayer* tale.

5. They require that when the personages of a tale deal in conversation, the talk shall sound like human talk, and be talk such as human beings would be likely to talk in the given circumstances, and have a discoverable meaning, also a discoverable purpose, and a show of relevancy, and remain in the neighborhood of the subject in hand, and be interesting to the reader, and help out the tale, and stop when the people cannot think of anything more to say. But this requirement has been ignored from the beginning of the *Deerslayer* tale to the end of it.

6. They require that when the author describes the character of a personage in his tale, the conduct and conversation of that personage shall justify said description. But this law gets little or no attention in the *Deerslayer* tale, as "Natty Bumppo's" case will amply prove.

7. They require that when a personage talks like an illustrated, gilt-edged, tree-calf, hand-tooled, seven-dollar Friendship's Offering[3] in the beginning of a paragraph, he shall not talk like a negro minstrel in the end of it. But this rule is flung down and danced upon in the *Deerslayer* tale.

8. They require that crass stupidities shall not be played upon the reader as "the craft of the woodsman, the delicate art of the

---

3. I.e., like one of the then-popular expensive, illustrated literary miscellanies ("tree-calf" was leather chemically treated to produce a treelike design).

forest," by either the author or the people in the tale. But this rule is persistently violated in the *Deerslayer* tale.

9. They require that the personages of a tale shall confine themselves to possibilities and let miracles alone; or, if they venture a miracle, the author must so plausibly set it forth as to make it look possible and reasonable. But these rules are not respected in the *Deerslayer* tale.

10. They require that the author shall make the reader feel a deep interest in the personages of his tale and in their fate; and that he shall make the reader love the good people in the tale and hate the bad ones. But the reader of the *Deerslayer* tale dislikes the good people in it, is indifferent to the others, and wishes they would all get drowned together.

11. They require that the characters in a tale shall be so clearly defined that the reader can tell beforehand what each will do in a given emergency. But in the *Deerslayer* tale this rule is vacated.

In addition to these large rules there are some little ones. These require that the author shall

12. *Say* what he is proposing to say, not merely come near it.

13. Use the right word, not its second cousin.

14. Eschew surplusage.

15. Not omit necessary details.

16. Avoid slovenliness of form.

17. Use good grammar.

18. Employ a simple and straightforward style.

Even these seven are coldly and persistently violated in the *Deerslayer* tale.

Cooper's gift in the way of invention was not a rich endowment; but such as it was he liked to work it, he was pleased with the effects, and indeed he did some quite sweet things with it. In his little box of stage properties he kept six or eight cunning devices, tricks, artifices for his savages and woodsmen to deceive and circumvent each other with, and he was never so happy as when he was working these innocent things and seeing them go. A favorite one was to make a moccasined person tread in the tracks of the moccasined enemy, and thus hide his own trail. Cooper wore out barrels and barrels of moccasins in working that trick. Another stage-property that he pulled out of his box pretty frequently was his broken twig. He prized his broken twig above all the rest of his effects, and worked it the hardest. It is a restful chapter in any book of his when somebody doesn't step on a dry twig and alarm all the reds and whites for two hundred yards around. Every time a Cooper person is in peril, and absolute silence is worth four dollars a minute, he is sure to step on a dry twig. There may be a hundred handier things to step on, but that wouldn't satisfy Cooper. Cooper requires him to turn out and find a dry twig; and if he can't do it, go and borrow one. In fact the Leather Stocking Series ought to have been called the Broken Twig Series.

I am sorry there is not room to put in a few dozen instances of the delicate art of the forest, as practiced by Natty Bumppo and some of the other Cooperian experts. Perhaps we may venture two or three samples. Cooper was a sailor—a naval officer; yet he gravely tells us how a vessel, driving toward a lee shore[4] in a gale, is steered for a particular spot by her skipper because he knows of an *undertow* there which will hold her back against the gale and save her. For just pure woodcraft, or sailor-craft, or whatever it is, isn't that neat? For several years Cooper was daily in the society of artillery, and he ought to have noticed that when a cannon ball strikes the ground it either buries itself or skips a hundred feet or so; skips again a hundred feet or so—and so on, till it finally gets tired and rolls. Now in one place he loses some "females"—as he always calls women—in the edge of a wood near a plain at night in a fog, on purpose to give Bumppo a chance to show off the delicate art of the forest before the reader. These mislaid people are hunting for a fort. They hear a cannon-blast, and a cannon-ball presently comes rolling into the wood and stops at their feet. To the females this suggests nothing. The case is very different with the admirable Bumppo. I wish I may never know peace again if he doesn't strike out promptly and *follow the track* of that cannon-ball across the plain through the dense fog and find the fort. Isn't it a daisy? If Cooper had any real knowledge of Nature's ways of doing things, he had a most delicate art in concealing the fact. For instance: one of his acute Indian experts, Chingachgook[5] (pronounced Chicago, I think), has lost the trail of a person he is tracking through the forest. Apparently that trail is hopelessly lost. Neither you nor I could ever have guessed out the way to find it. It was very different with Chicago. Chicago was not stumped for long. He turned a running stream out of its course, and there, in the slush in its old bed, were that person's moccasin-tracks. The current did not wash them away, as it would have done in all other like cases—no, even the eternal laws of Nature have to vacate when Cooper wants to put up a delicate job of woodcraft on the reader.

We must be a little wary when Brander Matthews tells us that Cooper's books "reveal an extraordinary fulness of invention." As a rule, I am quite willing to accept Brander Matthews's literary judgments and applaud his lucid and graceful phrasing of them; but that particular statement needs to be taken with a few tons of salt. Bless your heart, Cooper hadn't any more invention than a horse; and I don't mean a high-class horse, either; I mean a clothes-horse. It would be very difficult to find a really clever "situation" in Cooper's books; and still more difficult to find one of any kind which he has failed to render absurd by his handling of it. Look at the epi-

4. Shore that is protected from the wind.   5. Natty Bumppo's Indian friend.

sodes of "the caves;" and at the celebrated scuffle between Maqua and those others on the table-land a few days later; and at Hurry Harry's queer water-transit from the castle to the ark; and at Deerslayer's half hour with his first corpse; and at the quarrel between Hurry Harry and Deerslayer later; and at—but choose for yourself; you can't go amiss.

If Cooper had been an observer, his inventive faculty would have worked better, not more interestingly, but more rationally, more plausibly. Cooper's proudest creations in the way of "situations" suffer noticeably from the absence of the observer's protecting gift. Cooper's eye was splendidly inaccurate. Cooper seldom saw anything correctly. He saw nearly all things as through a glass eye, darkly.[6] Of course a man who cannot see the commonest little everyday matters accurately is working at a disadvantage when he is constructing a "situation." In the *Deerslayer* tale Cooper has a stream which is fifty feet wide, where it flows out of a lake; it presently narrows to twenty as it meanders along for no given reason, and yet, when a stream acts like that it ought to be required to explain itself. Fourteen pages later the width of the brook's outlet from the lake has suddenly shrunk thirty feet, and become "the narrowest part of the stream." This shrinkage is not accounted for. The stream has bends in it, a sure indication that it has alluvial banks, and cuts them; yet these bends are only thirty and fifty feet long. If Cooper had been a nice[7] and punctilious observer he would have noticed that the bends were oftener nine hundred feet long than short of it.

Cooper made the exit of that stream fifty feet wide in the first place, for no particular reason; in the second place, he narrowed it to less than twenty to accommodate some Indians. He bends a "sapling" to the form of an arch over this narrow passage, and conceals six Indians in its foliage. They are "laying" for a settler's scow or ark which is coming up the stream on its way to the lake; it is being hauled against the stiff current by a rope whose stationary end is anchored in the lake; its rate of progress cannot be more than a mile an hour. Cooper describes the ark, but pretty obscurely. In the matter of dimensions "it was little more than a modern canal boat." Let us guess, then, that it was about 140 feet long. It was of "greater breadth than common." Let us guess, then, that it was about sixteen feet wide. This leviathan had been prowling down bends which were but a third as long as itself, and scraping between banks where it had only two feet of space to spare on each side. We cannot too much admire this miracle. A low-roofed log dwelling occupies "two-third's of the ark's length"—a dwelling ninety feet long and sixteen feet wide, let us say—a kind of vestibule train. The dwelling has two rooms—each forty-five feet long and sixteen feet

6. Humorous turn of the Biblical "through a glass, darkly" (1 Corinthians 13.12).

7. I.e., meticulous.

wide, let us guess. One of them is the bed-room of the Hutter girls, Judith and Hetty; the other is the parlor, in the day time, at night it is papa's bed chamber. The ark is arriving at the stream's exit, now, whose width has been reduced to less than twenty feet to accommodate the Indians—say to eighteen. There is a foot to spare on each side of the boat. Did the Indians notice that there was going to be a tight squeeze there? Did they notice that they could make money by climbing down out of that arched sapling and just stepping aboard when the ark scraped by? No; other Indians would have noticed these things, but Cooper's Indians never notice anything. Cooper thinks they are marvellous creatures for noticing, but he was almost always in error about his Indians. There was seldom a sane one among them.

The ark is 140 feet long; the dwelling is 90 feet long. The idea of the Indians is to drop softly and secretly from the arched sapling to the dwelling as the ark creeps along under it at the rate of a mile an hour, and butcher the family. It will take the ark a minute and a half to pass under. It will take the 90-foot dwelling a minute to pass under. Now, then, what did the six Indians do? It would take you thirty years to guess, and even then you would have to give it up, I believe. Therefore, I will tell you what the Indians did. Their chief, a person of quite extraordinary intellect for a Cooper Indian, warily watched the canal boat as it squeezed along under him, and when he had got his calculations fined down to exactly the right shade, as he judged, he let go and dropped. And *missed the house!* That is actually what he did. He missed the house, and landed in the stern of the scow. It was not much of a fall, yet it knocked him silly. He lay there unconscious. If the house had been 97 feet long, he would have made the trip. The fault was Cooper's, not his. The error lay in the construction of the house. Cooper was no architect.

There still remained in the roost five Indians. The boat has passed under and is now out of their reach. Let me explain what the five did—you would not be able to reason it out for yourself. No. 1 jumped for the boat, but fell in the water astern of it. Then No. 2 jumped for the boat, but fell in the water still further astern of it. Then No. 3 jumped for the boat, and fell a good way astern of it. Then No. 4 jumped for the boat, and fell in the water *away* astern. Then even No. 5 made a jump for the boat—for he was a Cooper Indian. In the matter of intellect, the difference between a Cooper Indian and the Indian that stands in front of the cigar shop is not spacious. The scow episode is really a sublime burst of invention; but it does not thrill, because the inaccuracy of the details throws a sort of air of fictitiousness and general improbability over it. This comes of Cooper's inadequacy as an observer.

The reader will find some examples of Cooper's high talent for inaccurate observation in the account of the shooting match in *The*

Pathfinder. "A common wrought nail was driven lightly into the target, its head having been first touched with paint." The color of the paint is not stated—an important omission, but Cooper deals freely in important omissions. No, after all, it was not an important omission; for this nail head is *a hundred yards* from the marksman and could not be seen by them at that distance no matter what its color might be. How far can the best eyes see a common house fly? A hundred yards? It is quite impossible. Very well, eyes that cannot see a house fly that is a hundred yards away cannot see an ordinary nail head at that distance, for the size of the two objects is the same. It takes a keen eye to see a fly or a nail head at fifty yards— one hundred and fifty feet. Can the reader do it?

The nail was lightly driven, its head painted, and game called. Then the Cooper miracles began. The bullet of the first marksman chipped an edge of the nail head; the next man's bullet drove the nail a little way into the target—and removed all the paint. Haven't the miracles gone far enough now? Not to suit Cooper; for the purpose of this whole scheme is to show off his prodigy, Deerslayer-Hawkeye-Long-Rifle-Leather-Stocking-Pathfinder-Bumppo before the ladies.

> "Be all ready to clench it, boys!" cried out Pathfinder, step-ping into his friend's tracks the instant they were vacant. "Never mind a new nail; I can see that, though the paint is gone, and what I can see, I can hit at a hundred yards, though it were only a mosquito's eye. Be ready to clench!"
>
> The rifle cracked, the bullet sped its way and the head of the nail was buried in the wood, covered by the piece of flat-tened lead.

There, you see, is a man who could hunt flies with a rifle, and command a ducal salary in a Wild West show to-day, if we had him back with us.

The recorded feat is certainly surprising, just as it stands; but it is not surprising enough for Cooper. Cooper adds a touch. He has made Pathfinder do this miracle with another man's rifle, and not only that, but Pathfinder did not have even the advantage of load-ing it himself. He had everything against him, and yet he made that impossible shot, and not only made it, but did it with absolute con-fidence, saying, "Be ready to clench." Now a person like that would have undertaken that same feat with a brickbat, and with Cooper to help he would have achieved it, too.

Pathfinder showed off handsomely that day before the ladies. His very first feat was a thing which no Wild West show can touch. He was standing with the group of marksmen, observing—a hundred yards from the target, mind: one Jasper raised his rifle and drove the centre of the bull's-eye. Then the quartermaster fired. The target exhibited no result this time. There was a laugh. "It's a dead miss," said Major Lundie. Pathfinder waited an impressive moment or two, then said in that calm, indifferent, know-it-all way of his, "No, Major—he has covered Jasper's bullet, as will be seen if any one will take the trouble to examine the target."

Wasn't it remarkable! How *could* he see that little pellet fly through the air and enter that distant bullet-hole? Yet that is what he did; for nothing is impossible to a Cooper person. Did any of those people have any deep-seated doubts about this thing? No; for that would imply sanity, and these were all Cooper people.

The respect for Pathfinder's skill and for his *quickness and accuracy of sight* (the italics are mine) was so profound and general, that the instant he made this declaration the spectators began to distrust their own opinions, and a dozen rushed to the target in order to ascertain the fact. There, sure enough, it was found that the quartermaster's bullet had gone through the hole made by Jasper's, and that, too, so accurately as to require a minute examination to be certain of the circumstance, which, however, was soon clearly established by discovering one bullet over the other in the stump against which the target was placed.

They made a "minute" examination; but never mind, how could they know that there were two bullets in that hole without digging the latest one out? for neither probe nor eyesight could prove the presence of any more than one bullet. Did they dig? No; as we shall see. It is the Pathfinder's turn now; he steps out before the ladies, takes aim, and fires.

But alas! here is a disappointment; an incredible, an unimaginable disappointment—for the target's aspect is unchanged; there is nothing there but that same old bullet hole!

"If one dared to hint at such a thing," cried Major Duncan, "I should say that the Pathfinder has also missed the target."

As nobody had missed it yet, the "also" was not necessary; but never mind about that, for the Pathfinder is going to speak.

"No, no, Major," said he, confidently, "that *would* be a risky declaration. I didn't load the piece, and can't say what was in it, but if it was lead, you will find the bullet driving down those of the Quartermaster and Jasper, else is not my name Pathfinder."

A shout from the target announced the truth of this assertion.

Is the miracle sufficient as it stands? Not for Cooper. The Pathfinder speaks again, as he "now slowly advances towards the stage occupied by the females:"

"That's not all, boys, that's not all; if you find the target touched at all, I'll own to a miss. The Quartermaster cut the wood, but you'll find no wood cut by that last messenger."

The miracle is at last complete. He knew—doubtless *saw*—at the distance of a hundred yards—that his bullet had passed into the hole *without fraying the edges*. There were now three bullets in that one hole—three bullets imbedded processionally in the body of the stump back of the target. Everybody knew this—somehow or other—and yet nobody had dug any of them out to make sure. Cooper is not a close observer, but he is interesting. He is certainly always that, no matter what happens. And he is more interesting when he is not noticing what he is about than when he is. This is a considerable merit.

The conversations in the Cooper books have a curious sound in our modern ears. To believe that such talk really ever came out of people's mouths would be to believe that there was a time when time was of no value to a person who thought he had something to say; when it was the custom to spread a two-minute remark out to ten; when a man's mouth was a rolling-mill, and busied itself all day long in turning four-foot pigs[8] of thought into thirty-foot bars of conversational railroad iron by attenuation; when subjects were seldom faithfully stuck to, but the talk wandered all around and arrived nowhere; when conversations consisted mainly of irrelevances, with here and there a relevancy, a relevancy with an embarrassed look, as not being able to explain how it got there.

Cooper was certainly not a master in the construction of dialogue. Inaccurate observation defeated him here as it defeated him in so many other enterprises of his. He even failed to notice that the man who talks corrupt English six days in the week must and will talk it on the seventh, and can't help himself. In the *Deerslayer* story he lets Deerslayer talk the showiest kind of book talk sometimes, and at other times the basest of base dialects. For instance, when some one asks him if he has a sweetheart, and if so, where she abides, this is his majestic answer:

"She's in the forest—hanging from the boughs of the trees, in a soft rain—in the dew on the open grass—the clouds that float about in the blue heavens—the birds that sing in the woods—the sweet springs where I slake my thirst—and in all the other glorious gifts that come from God's Providence!"

And he preceded that, a little before, with this:

8. Crude castings of iron.

"It consarns me as all things that touches a fri'nd consarns a fri'nd."

And this is another of his remarks:

"If I was Injun born, now, I might tell of this, or carry in the scalp and boast of the expl'ite afore the whole tribe; or if my inimy had only been a bear"—and so on.

We cannot imagine such a thing as a veteran Scotch Commander-in-Chief comporting himself in the field like a windy melodramatic actor, but Cooper could. On one occasion Alice and Cora were being chased by the French through a fog in the neighborhood of their father's fort:

"*Point de quartier aux coquins!*"[9] cried an eager pursuer, who seemed to direct the operations of the enemy.
"Stand firm and be ready, my gallant 60ths!" suddenly exclaimed a voice above them; "wait to see the enemy; fire low, and sweep the glacis."[1]
"Father! father!" exclaimed a piercing cry from out the mist; "it is I! Alice! thy own Elsie! spare, O! save your daughters!"
"Hold!" shouted the former speaker, in the awful tones of parental agony, the sound reaching even to the woods, and rolling back in solemn echo. "'Tis she! God has restored me my children! Throw open the sally-port;[2] to the field, 60ths, to the field; pull not a trigger, lest ye kill my lambs! Drive off these dogs of France with your steel."

Cooper's word-sense was singularly dull. When a person has a poor ear for music he will flat and sharp right along without knowing it. He keeps near the tune, but it is *not* the tune. When a person has a poor ear for words, the result is a literary flatting and sharping; you perceive what he is intending to say, but you also perceive that he doesn't *say* it. This is Cooper. He was not a word-musician. His ear was satisfied with the *approximate* word. I will furnish some circumstantial evidence in support of this charge. My instances are gathered from half a dozen pages of the tale called *Deerslayer*. He uses "verbal," for "oral"; "precision," for "facility"; "phenomena," for "marvels"; "necessary," for "predetermined"; "unsophisticated," for "primitive"; "preparation," for "expectancy"; "rebuked," for "subdued"; "dependent on," for "resulting from"; "fact," for "condition"; "fact," for "conjecture"; "precaution," for "caution"; "explain," for "determine"; "mortified," for "disappointed"; "meretricious," for "factitious"; "materially," for "considerably"; "decreasing," for "deepening"; "increasing," for "disappear-

---

9. "No quarter for the rascals!"
1. Slope that runs downward from a fortification.

2. Gate or passage in a fortified place for use of troops making a sortie.

ing"; "embedded," for "enclosed"; "treacherous," for "hostile"; "stood," for "stooped"; "softened," for "replaced"; "rejoined," for "remarked"; "situation," for "condition"; "different," for "differing"; "insensible," for "unsentient"; "brevity," for "celerity"; "distrusted," for "suspicious"; "mental imbecility," for "imbecility"; "eyes," for "sight"; "counteracting," for "opposing"; "funeral obsequies," for "obsequies."

There have been daring people in the world who claimed that Cooper could write English, but they are all dead now—all dead but Lounsbury. I don't remember that Lounsbury makes the claim in so many words, still he makes it, for he says that *Deerslayer* is a "pure work of art." Pure, in that connection, means faultless—faultless in all details—and language is a detail. If Mr. Lounsbury had only compared Cooper's English with the English which he writes himself—but it is plain that he didn't; and so it is likely that he imagines until this day that Cooper's is as clean and compact as his own. Now I feel sure, deep down in my heart, that Cooper wrote about the poorest English that exists in our language, and that the English of *Deerslayer* is the very worst than even Cooper ever wrote.

I may be mistaken, but it does seem to me that *Deerslayer* is not a work of art in any sense; it does seem to me that it is destitute of every detail that goes to the making of a work of art; in truth, it seems to me that *Deerslayer* is just simply a literary *delirium tremens*.

A work of art? It has no invention; it has no order, system, sequence, or result; it has no lifelikeness, no thrill, no stir, no seeming of reality; its characters are confusedly drawn, and by their acts and words they prove that they are not the sort of people the author claims that they are; its humor is pathetic; its pathos is funny; its conversations are—oh! indescribable; its love-scenes odious; its English a crime against the language.

Counting these out, what is left is Art. I think we must all admit that.

1895, 1897

The United States of Lyncherdom[1]

I

And so Missouri has fallen, that great state! Certain of her children have joined the lynchers, and the smirch is upon the rest of us.

1. This essay was not published during Mark Twain's lifetime. When Albert Bigelow Paine edited the essay for inclusion in *Europe and Elsewhere* (1923), he took considerable liberties with the manuscript that Twain had left behind. The present text is freshly edited from the manuscript with the cooperation of Frederick Anderson, late editor of the Mark Twain Papers, University of California, Berkeley. It will appear in a forthcoming volume of the California edition of Twain's works.

That handful of her children have given us a character and labeled us with a name; and to the dwellers in the four quarters of the earth we are "lynchers," now, and ever shall be. For the world will not stop and think—it never does, it is not its way; its way is to generalize from a single sample. It will not say "Those Missourians have been busy eighty years in building an honorable good name for themselves; these hundred lynchers down in the corner of the State are not real Missourians, they are bastards." No, that truth will not enter its mind; it will generalize from the one or two misleading samples and say "The Missourians are lynchers." It has no reflection, no logic, no sense of proportion. With it, figures go for nothing; to it, figures reveal nothing, it cannot reason upon them rationally; it is Brother J.-J. infinitely multiplied; it would say, with him, that China is being swiftly and surely Christianized, since 9 Chinese Christians are being made every day; and it would fail, with him, to notice that the fact that 33,000 pagans are *born* there every day, damages the argument. It would J-J Missouri, and say "There are a hundred lynchers there, therefore the Missourians are lynchers;" the considerable fact that there are two and a half million Missourians who are *not* lynchers would not affect their verdict any more than it would affect Bro. J.-J.'s.

What, then, results from this curious fashion of the world? A chief and dismal result is, that the reputations of States and nations are made by the conduct of a few of their noisiest and worst representatives, and not by the conduct of the clean and sober multitude of their populations. In the eyes of the world are all Englishmen Alfreds the Great? no, they are all Chamberlains; are all Frenchmen Saints Louis? no, they are all Melines; are all New Yorkers Odells? no, they are all Crokers;[2] are all but three of our States honest and honorable? no, in foreign eyes we are all repudiators of our bonds; are we Southerners all Dr. Lazears,[3] General Lees, Stonewall Jacksons? no, we are all lynchers.

Since, then, a nation's foreign reputation must depend upon the conduct of its handful of Chamberlains, bond-repudiators, Quays,[4] Crokers, and such, it ought to be a good idea to suppress these people—is it not so? How is it to be done? It is an easy question—to ask.

2. Alfred (849–899), King of the West Saxons (871–899), known as Alfred the Great; Joseph Chamberlain (1836–1914), British statesman who finally resigned from office in 1903 after controversy surrounded his political affairs; Louis IX (1214–70), King of France (1226–70), went on the Sixth Crusade and planned another; Félix Jules Méline (1838–1925), French statesman forced to resign because he refused to rely solely on the left-wing republicans; Benjamin Barker Odell (1854–1926), Governor of New York, much lauded for reduced direct taxation; Richard Croker (1841–1922), New York City machine politician.
3. Jesse William Lazear (1866–1900), American physician and member of the commission which proved that yellow fever was transmitted through mosquitoes. As proof of the commission's thesis, he allowed a mosquito to bite him and died a week later.
4. Matthew Stanley Quay (1833–1904), Pennsylvania politician who controlled the state's Republican machine.

II

Oh, Missouri!

The tragedy occurred near Pierce City, down in the southwestern corner of the State. On a Sunday afternoon a young white woman who had started homeward alone from church, was found murdered. From church. For there are churches there; in my time religion was more general, more pervasive, in the South than it was in the North, and more virile and earnest, too, I think; I have some reason to believe that this is still the case. The young woman was found murdered. Although it was a region of churches and schools, the people rose, lynched three negroes—two of them very aged ones—burned out five negro households, and drove thirty negro families to the woods.

I do not dwell upon the provocation which moved the people to these crimes, for that has nothing to do with the matter. We know by the recent Keller case in a Jersey court, that evidence to show that an assassin was moved by great provocation is not admissible and cannot be listened to; the only question is, did the assassin *take the law into his own hands?* It is very simple, and very just. If the assassin be proved to have usurped the law's prerogative in righting his wrongs, that ends the matter; a thousand provocations are no defence. The Pierce City people had bitter provocation—indeed, as revealed by certain of the particulars, the bitterest of all provocations—but no matter, they took the law into their own hands, and by the terms of their own statutes they are assassins and should hang—in Jersey, let us believe, that is what would happen to them. Also, in Jersey the assassin of that helpless poor white woman would hang—and we know that Pierce City's courts would have hanged him if the law had been allowed to take its course, for there are but few negroes in that region and they are without authority and without influence in overawing juries.

III

Why has lynching, with various barbaric accompaniments, become a favorite regulator in cases of "the usual crime" in several parts of the country? Is it because men think a lurid and terrible punishment a more forcible object-lesson and a more effective deterrent than a sober and colorless hanging, done privately in a jail, would be? Surely sane men do not think that. Even the average child should know better. It should know that any strange and much-talked-of event is always followed by imitations, the world being so well supplied with excitable people who only need a little stirring up to make them lose what is left of their heads and do mad things which they would not have thought of ordinarily. It should know that if a man jump off Brooklyn Bridge another will imitate him; that if a person venture down Niagara Whirlpool in a barrel another will imitate him; that if a Jack the Ripper make notoriety by slaughtering women in dark alleys he will be imitated; that if a man attempt a king's life and the newspapers carry the noise of it around the globe, regicides will crop up all around. The child should know that one much-talked-of outrage and murder committed by a negro will upset the disturbed intellects of several other negroes and produce a series of the very tragedies the community would so strenuously wish to prevent; that each of these crimes will produce another series, and year by year steadily increase the tale of these disasters instead of diminishing it; that, in a word, the lynchers are themselves the worst enemies of their women. The child should also know that by a law of our make, communities, as well as individuals, are imitators; and that a much-talked-of lynching will infallibly produce other lynchings here and there and yonder, and that in time these will breed a mania, a fashion; a fashion which will spread wide and wider, year by year, covering State after State, as with an advancing disease. Lynching has reached Colorado, it has reached California, it has reached Indiana—and now Missouri! I shall live to see a negro burned in Union Square, New York, with fifty thousand people present, and not a sheriff visible, not a governor, not a constable, not a colonel, not a clergyman, not a law-and-order representative of any sort.

Increase in Lynching.—In 1900 there were eight more cases than in 1899, and probably this year there will be more than there were last year. The year is little more than half gone, and yet there are eighty-eight cases as compared with one hundred and fifteen for all of last year. Four Southern states, Alabama, Georgia, Louisiana, and Mississippi are the worst offenders. Last year there were eight cases in Alabama, sixteen in Georgia, twenty in Louisiana, and twenty in Mississippi—over one-half the total. This year to date there have been nine in Alabama, twelve in Georgia, eleven in Louisiana, and thirteen in Mississippi—again more than one-half the total number in the whole United States.—*Chicago Tribune.*

It must be that the increase comes of the inborn human instinct to imitate—that and man's commonest weakness, his aversion to being unpleasantly conspicuous, pointed at, shunned, as being on the unpopular side. Its other name is Moral Cowardice, and is the supreme feature of the make-up of 9,999 men in the 10,000. I am not offering this as a discovery; privately the dullest of us knows it to be true. History will not allow us to forget or ignore this commanding trait of our character. It persistently and sardonically reminds us that from the beginning of the world no revolt against a public infamy or oppression has ever been begun but by the one daring man in the 10,000, the rest timidly waiting, and slowly and reluctantly joining, under the influence of that man and his fellows from the other ten thousands. The abolitionists remember. Privately the public feeling was with them early, but each man was afraid to speak out until he got some hint that his neighbor was privately feeling as he privately felt himself. Then the boom followed. It always does. It will occur in New York, some day; and even in Pennsylvania.

IV

It has been supposed—and said—that the people at a lynching enjoy the spectacle, and are glad of a chance to see it. It cannot be true; all experience is against it. The people in the South are made like the people in the North—the vast majority of whom are right-hearted and compassionate, and would be cruelly pained by such a spectacle—and *would attend it*, and let on to be pleased with it if the public approval seemed to require it. We are made like that, and we cannot help it. The other animals are not so, but we cannot help that, either. They lack the Moral Sense; we have no way of trading ours off, for a nickel or some other thing above its value. The Moral Sense teaches us what is right and how to avoid it. When unpopular.

It is thought, as I have said, that a lynching-crowd enjoys a lynching. It certainly is not true; it is impossible of belief. It is freely asserted—you have seen it in print many times of late—that the lynching impulse has been misinterpreted: that it is *not* the outcome of a spirit of revenge, but of a "mere atrocious hunger *to look upon human suffering.*" If that were so, the crowds that saw the Windsor Hotel[5] burn down would have enjoyed the horrors that fell under their eyes. Did they? No one will think that of them, no one will make that charge. Many risked their lives to save the men and women who were in peril. Why did they do that? Because *none would disapprove.* There was no restraint; they could follow their natural impulse. Why does a crowd of the same kind of people in Texas, Colorado, Indiana, stand by, smitten to the heart and miserable, and by ostentatious outward signs pretend to enjoy a lynching? Why does it lift no hand nor voice in protest? Only because it would be unpopular to do it, I think; each man is afraid of his neighbor's disapproval—a thing which to the general run of the race is more dreaded than wounds and death. When there is to be a lynching the people hitch up and come miles to see it, bringing their wives and children. Really to see it? No—they come only because they are afraid to stay at home, lest it be noticed and offensively commented upon. We may believe this, for we all know how *we* feel about such spectacles—also, how we would act under the like pressure. We are not any better nor any braver than anybody else, and we must not try to creep out of it.

---

5. The grand Windsor Hotel in New York City was reduced to ashes on March 17, 1899.

V

A Savonarola[6] can quell and scatter a mob of lynchers with a mere glance of his eye; so can a Merrill[7] or a Beloat. For no mob has any sand in the presence of a man known to be splendidly brave. Besides, a lynching-mob would *like* to be scattered, for of a certainty there are never ten men in it who would not prefer to be somewhere else—and would be, if they but had the courage to go. When I was a boy I saw a brave gentleman deride and insult a mob and drive it away; and afterward, in Nevada, I saw a noted desperado make two hundred men sit still, with the house burning under them, until he gave them permission to retire. A plucky man can rob a whole passenger train by himself; and the half of a brave man can hold up a stagecoach and strip its occupants.

Then perhaps the remedy for lynchings comes to this: station a brave man in each affected community to encourage, support, and bring to light the deep disapproval of lynching hidden in the secret places of its heart—for it is there, beyond question. Then those communities will find something better to imitate—of course, being human they must imitate something. Where shall these brave men be found? That is indeed a difficulty; there are not three hundred of them in the earth. If merely *physically* brave men would do, then it were easy; they could be furnished by the cargo. When Hobson[8] called for seven volunteers to go with him to what promised to be certain death, four thousand men responded—the whole fleet, in fact. Because *all the world would approve.* They knew that; but if Hobson's project had been charged with the scoffs and jeers of the friends and associates, whose good opinion and approval the sailors valued, he could not have got his seven.

No, upon reflection, the scheme will not work. There are not enough morally brave men in stock. We are out of moral-courage material; we are in a condition of profound poverty. We have those two sheriffs down South who—but never mind, it is not enough to go around; they have to stay and take care of their own communities.

But if we only *could* have three or four more sheriffs of that great breed! Would it help? I think so. For we are all imitators: other brave sheriffs would follow; to be a dauntless sheriff would come to be recognized as the correct and only thing, and the dreaded disapproval would fall to the share of the other kind; courage in this office would become custom, the absence of it a dishonor, just as

6. Girolamo Savonarola (1452–98), Florentine religious reformer known for his eloquence and rousing speeches.
7. "Merrill": "sheriff of Carroll County, Georgia" [Clemens's note]. "Beloat": "sheriff, Princeton, Indiana. By that formidable power which lies in an established reputation for cold pluck they

faced lynching mobs and securely held the field against them" [Clemens's note].
8. Edward Henry Hobson (1825–1901), Union soldier who recruited the 13th Union Infantry from Kentucky and was promoted to Brigadier-General of the Volunteers.

courage presently replaces the timidity of the new soldier; then the mobs and the lynchings would disappear, and——

However. It can never be done without some starters, and where are we to get the starters? Advertise? Very well, then, let us advertise.

VI

In the meantime, there is another plan. Let us import American missionaries from China, and send them into the lynching field. With 1511 of them out there converting 2 Chinamen apiece per annum against an uphill birthrate of 33,000 pagans per day,[9] it will take upwards of a million years to make the conversions balance the output and bring the Christianizing of the country in sight with the naked eye; therefore, if we can offer our missionaries as rich a field at home, at lighter expense and quite satisfactory in the matter of danger, why shouldn't they find it fair and right to come back and give us a trial? The Chinese are universally conceded to be excellent people, honest, honorable, industrious, trustworthy, kind-hearted, and all that—leave them alone, they are plenty good enough just as they are; and besides, almost every convert runs a risk of catching our Civilization. We ought to be careful. We ought to think twice before we encourage a risk like that; for, *once civilized, China can never be uncivilized again.* We have not been thinking of that. Very well, we ought to think of it now. Our missionaries will find that we have a field for them—and not only for the 1511, but for 15,011. Let them look at the following telegram, and see if they have anything in China that is more appetizing. It is from Texas:

> The negro was taken to a tree, and swung in the air. Wood and fodder were piled beneath his body and a hot fire was made. *Then it was suggested that the man ought not to die too quickly, and he was let down to the ground, while a party went to Dexter about two miles distant, to procure coal oil. This was thrown on the flames and the work completed.*

We implore them to come back and help us in our need. Patriotism imposes this duty upon them. Our country is worse off than China; they are our countrymen, their motherland supplicates their aid in this her hour of deep distress. They are competent, our people are not; they are used to scoffs, sneers, revilings, danger; our people are not; they have the martyr spirit, nothing but the martyr spirit can brave a lynching-mob and cow it and scatter it; they can save their country, we beseech them to come home and do it. We ask them to read that telegram again, and yet again, and picture the scene in their minds, and soberly ponder it; then multiply it by 115; add 88; place the 203 in a row, allowing 600 feet of space for each

9. "These figures are not fanciful; all of them are genuine and authentic. They are from official missionary records in China. See Dr. Morrison's book on his pedestrian journey across China; he quotes them and gives his authorities. For several years he has been the London *Times*'s representative in Peking, and was there through the siege" [Clemens's note].

human torch, so that there may be viewing-room around it for 5,000 Christian American men, women and children, youths and maidens; make it night, for grim effect; have the show in a gradually rising plain, and let the course of the stakes be up-hill; the eye can then take in the whole line of twenty-four miles of blood-and-flesh bonfires unbroken, whereas if it occupied level ground the ends of the line would bend down and be hidden from view by the curvature of the earth; all being ready, now, and the darkness opaque, the stillness impressive—for there should be no sound but the soft moaning of the night-wind and the muffled sobbing of the sacrifices—let all the far stretch of kerosened pyres be touched off simultaneously and the glare and the shrieks and the agonies burst heavenward to the Throne.

There are more than a million persons present; the light from the fires flushes into vague outline against the night the spires of five thousand churches. Oh kind missionary, oh compassionate missionary, leave China! come home and convert these Christians!

VII

I believe that if anything can stop this epidemic of bloody insanities it is martial personalities that can face mobs without flinching; and as such personalities are developed only by familiarity with danger and by the training and seasoning which come of resisting it, the likeliest place to find them must be among the missionaries who have been under tuition in China during the past year or two. We have abundance of work for them, and for hundreds and thousands more, and the field is daily growing and spreading. We shall add 60 lynchings to our 88 before the year is out; we shall reach the 300-mark next year, the 500-mark in 1903, we shall lynch a thousand negroes in 1904; in 1905 there will be lynchings in every State and Territory and Possession, for our morals follow the flag, let the Constitution do as it may; and by 1910 my prediction will mature and I shall see a negro burned in Union Square. For murder? No, for being in the country against the country's will—like the Chinaman in America and the missionary in China. By that time we shall be known abroad as the United States of Lyncherdom, and be no more respected than a Chamberlain-war[1] for South African swag.

This must all happen. That is, unless we find those seasoned great braves—in the China missions or elsewhere—with the Savonarola glance which withers mobs with its stern rebuke and disperses them in shame and fear. Shall we find them? We can try. In 75,000,000 there must be other Merrills and Beloats; and it is the law of our make that each example shall wake up drowsing chevaliers of the same great knighthood and bring them to the front.

1901   1923

1. I.e., the Boer War (1899–1902) be-   sion of South Africa. England eventually
tween the Boers, descendants of the   won.
Dutch settlers, and England over posses-

BRET HARTE
1836–1902

Francis Brett Harte was born August 25, 1836, in Albany, New York, of Jewish, English, and Dutch descent. His schoolteacher father died in 1845, and four years later Harte left school to work in a lawyer's office and then in the counting room of a merchant. In 1854 he followed his mother and elder sister and brother to Oakland, California, and for a year he lived with his mother and stepfather, Colonel Andrew Williams, the first Mayor of Oakland. During this time he contributed some stories and poems to eastern magazines, but he had by no means settled on a career as a writer. When he turned twenty-one he left Williams's house to wander in northern California, where he rode shotgun for Wells Fargo and held less romantic jobs as miner, teacher, apothecary's clerk, and in the printing room of the *Humbolt Times*.

In San Francisco in 1860, Harte set type for the newspaper the *Golden Era* and began contributing to it. Shortly after the Civil War started, he quit the *Golden Era* and became a clerk in the Surveyor-General's office. Soon after he married Anna Griswold in 1862, Harte secured the undemanding position of Secretary to the Superintendent in the U.S. Branch Mint in San Francisco and began to establish himself as a journalist, contributing together with Mark Twain and others to the first issue of the weekly *Californian*. Harte's career as a writer was confirmed in 1868 when he became the first editor of the newly established magazine *Overland Monthly*, for which he wrote the works that made him—and the magazine —famous. Two of his poems were printed in the first issue, followed in the second by *The Luck of Roaring Camp*, the story that made him a celebrity from coast to coast, and then by *The Outcasts of Poker Flat* and his best-known poem, *Plain Language from Truthful James*, commonly referred to as *The Heathen Chinee*.

Harte left San Francisco with his family early in 1871 at the height of his fame. After a journey cross-country that was covered by the daily press, the Hartes were entertained in Boston for a week as guests of the new editor of the *Atlantic Monthly*, William Dean Howells. James T. Fields, who owned the prestigious *Atlantic*, had offered Harte the unheard of sum of ten thousand dollars to write twelve or more poems and sketches within a year to be published in the *Atlantic* and *Every Saturday*. Unfortunately, Harte could not meet his deadline, and it was not until several months later that he satisfied Fields. The contract was not renewed.

The last three decades of Harte's life constitute a decline in his personal and literary fortunes. Though he lectured and wrote feverishly for the next few years, his extravagant life style kept him continually in debt. His novel *Gabriel Conroy* (1876) earned a "small fortune," but he soon spent it. For a time he turned to the theater; though his plays *Two Men of Sandy Bar* (1876) and *Ah Sin* (1877) were produced (the latter after an unhappy attempt to collaborate with Mark Twain), neither was successful. Appointed Consul to Crefield, Prussia, by President Rutherford B. Hayes, Harte, by this time on bad terms with his wife, went alone to this post in

1878. He never saw his wife again. He was transferred to Glasgow, Scotland, in 1880 at his request. On visits to London and Bournemouthe, he often stayed with a Belgian couple, the Van de Veldes. From the beginning, Harte's relationship with Mrs. Van de Velde caused much gossip, and rumors of a *ménage à trois* floated through society. When Grover Cleveland became President in 1885, Harte lost his consulship and became a permanent guest of the Van de Veldes.

Harte's English readers continued to receive his work favorably until his death, although the American audience had long since tired of the sentimental depictions of stock frontier types that brought him initial fame. By the turn of the century, what in 1870 had seemed like a bold, realistic treatment of sex and love seemed rather tame, and the appeal of the Wild West that Harte had pioneered in implanting in the American imagination had become the stuff of myth. When Harte died on May 5, 1902, he had outlived his distinctive contribution to American letters by a generation.

The Outcasts of Poker Flat[1]

As Mr. John Oakhurst, gambler, stepped into the main street of Poker Flat on the morning of the twenty-third of November, 1850, he was conscious of a change in its moral atmosphere from the preceding night. Two or three men, conversing earnestly together, ceased as he approached, and exchanged significant glances. There was a Sabbath lull in the air, which, in a settlement unused to Sabbath influences, looked ominous.

Mr. Oakhurst's calm, handsome face betrayed small concern of these indications. Whether he was conscious of any predisposing cause, was another question. "I reckon they're after somebody," he reflected; "likely it's me." He returned to his pocket the handkerchief with which he had been whipping away the red dust of Poker Flat from his neat boots, and quietly discharged his mind of any further conjecture.

In point of fact, Poker Flat was "after somebody." It had lately suffered the loss of several thousand dollars, two valuable horses, and a prominent citizen. It was experiencing a spasm of virtuous reaction, quite as lawless and ungovernable as any of the acts that had provoked it. A secret committee[2] had determined to rid the town of all improper persons. This was done permanently in regard of two men who were then hanging from the boughs of a sycamore in the gulch, and temporarily in the banishment of certain other objectionable characters. I regret to say that some of these were ladies. It is but due to the sex, however, to state that their impropriety was professional, and it was only in such easily established standards of evil that Poker Flat ventured to sit in judgment.

1. First published in *Overland Monthly,* January, 1869, the source of the present text, and collected in the *Luck of Roaring Camp and Other Sketches* (Boston: Fields, Osgood, 1870).

2. I.e., vigilance committee—a volunteer committee of citizens organized to suppress and punish crime summarily when the processes of law appear inadequate.

Mr. Oakhurst was right in supposing that he was included in this category. A few of the committee had urged hanging him as a possible example, and a sure method of reimbursing themselves from his pockets of the sums he had won from them. "It's agin justice," said Jim Wheeler, "to let this yer young man from Roaring Camp—an entire stranger—carry away our money." But a crude sentiment of equity residing in the breasts of those who had been fortunate enough to win from Mr. Oakhurst, overruled this narrower local prejudice.

Mr. Oakhurst received his sentence with philosophic calmness, none the less coolly, that he was aware of the hesitation of his judges. He was too much of a gambler not to accept Fate. With him life was at best an uncertain game, and he recognized the usual percentage in favor of the dealer.

A body of armed men accompanied the deported wickedness of Poker Flat to the outskirts of the settlement. Besides Mr. Oakhurst, who was known to be a coolly desperate man, and for whose intimidation the armed escort was intended, the expatriated party consisted of a young woman familiarly known as "The Duchess;" another, who had gained the infelicitous title of "Mother Shipton,"[3] and "Uncle Billy," a suspected sluice-robber[4] and confirmed drunkard. The cavalcade provoked no comments from the spectators, nor was any word uttered by the escort. Only when the gulch which marked the uttermost limit of Poker Flat was reached, the leader spoke briefly and to the point. The exiles were forbidden to return at the peril of their lives.

As the escort disappeared, their pent-up feelings found vent in a few hysterical tears from "The Duchess," some bad language from Mother Shipton, and a Partheian[5] volley of expletives from Uncle Billy. The philosophic Oakhurst alone remained silent. He listened calmly to Mother Shipton's desire to cut somebody's heart out, to the repeated statements of "The Duchess" that she would die in the road, and to the alarming oaths that seemed to be bumped out of Uncle Billy as he rode forward. With the easy good-humor characteristic of his class, he insisted upon exchanging his own riding-horse, "Five Spot," for the sorry mule which the Duchess rode. But even this act did not draw the party into any closer sympathy. The young woman reädjusted her somewhat draggled plumes with a feeble, faded coquetry; Mother Shipton eyed the possessor of "Five Spot" with malevolence, and Uncle Billy included the whole party in one sweeping anathema.

The road to Sandy Bar—a camp that not having as yet experienced the regenerating influences of Poker Flat, consequently

3. Mother Shipton (1488–1560), an English witch who was carried off by the devil and bore him an imp. Her prophecies were edited by S. Baker in 1797.
4. A sluice is an inclined trough or flume for washing or separating gold from earth.
5. An ancient people of southwest Asia noted for firing shots while in real or feigned retreat.

seemed to offer some invitation to the emigrants—lay over a steep mountain range. It was distant a day's severe journey. In that advanced season, the party soon passed out of the moist, temperate regions of the foot-hills, into the dry, cold, bracing air of the Sierras. The trail was narrow and difficult. At noon the Duchess, rolling out of her saddle upon the ground, declared her intention of going no further, and the party halted.

The spot was singularly wild and impressive. A wooded amphitheatre, surrounded on three sides by precipitous cliffs of naked granite, sloped gently toward the crest of another precipice that overlooked the valley. It was undoubtedly the most suitable spot for a camp, had camping been advisable. But Mr. Oakhurst knew that scarcely half the journey to Sandy Bar was accomplished, and the party were not equipped or provisioned for delay. This fact he pointed out to his companions curtly, with a philosophic commentary on the folly of "throwing up their hand before the game was played out." But they were furnished with liquor, which in this emergency stood them in place of food, fuel, rest and prescience. In spite of his remonstrances, it was not long before they were more or less under its influence. Uncle Billy passed rapidly from a bellicose state into one of stupor, the Duchess became maudlin, and Mother Shipton snored. Mr. Oakhurst alone remained erect, leaning against a rock, calmly surveying them.

Mr. Oakhurst did not drink. It interfered with a profession which required coolness, impassiveness and presence of mind, and, in his own language, he "couldn't afford it." As he gazed at his recumbent fellow-exiles, the loneliness begotten of his pariah-trade, his habits of life, his very vices, for the first time seriously oppressed him. He bestirred himself in dusting his black clothes, washing his hands and face, and other acts characteristic of his studiously neat habits, and for a moment forgot his annoyance. The thought of deserting his weaker and more pitiable companions never perhaps occurred to him. Yet he could not help feeling the want of that excitement, which singularly enough was most conducive to that calm equanimity for which he was notorious. He looked at the gloomy walls that rose a thousand feet sheer above the circling pines around him; at the sky, ominously clouded; at the valley below, already deepening into shadow. And doing so, suddenly he heard his own name called.

A horseman slowly ascended the trail. In the fresh, open face of the new-comer, Mr. Oakhurst recognized Tom Simson, otherwise known as "The Innocent" of Sandy Bar. He had met him some months before over a "little game," and had, with perfect equanimity, won the entire fortune—amounting to some forty dollars—of that guileless youth. After the game was finished. Mr. Oakhurst drew the youthful speculator behind the door and thus addressed him: "Tommy, you're a good little man, but you can't gamble worth a cent. Don't try it over again." He then handed him his money back, pushed him gently from the room, and so made a devoted slave of Tom Simson.

There was a remembrance of this in his boyish and enthusiastic greeting of Mr. Oakhurst. He had started, he said, to go to Poker Flat to seek his fortune. "Alone?" No, not exactly alone; in fact—a giggle—he had run away with Piney Woods. Didn't Mr. Oakhurst remember Piney? She that used to wait on the table at the Temperance House? They had been engaged a long time, but old Jake Woods had objected, and so they had run away, and were going to Poker Flat to be married, and here they were. And they were tired out, and how lucky it was they had found a place to camp and company. All this The Innocent delivered rapidly, while Piney—a stout, comely damsel of fifteen—emerged from behind the pine tree, where she had been blushing unseen, and rode to the side of her lover.

Mr. Oakhurst seldom troubled himself with sentiment. Still less with propriety. But he had a vague idea that the situation was not felicitous. He retained, however, his presence of mind sufficiently to kick Uncle Billy, who was about to say something, and Uncle Billy was sober enough to recognize in Mr. Oakhurst's kick a superior power that would not bear trifling. He then endeavored to dissuade Tom Simson from delaying further, but in vain. He even pointed out the fact that there was no provision, nor means of making a camp. But, unluckily, "The Innocent" met this objection by assuring the party that he was provided with an extra mule loaded with provisions, and by the discovery of a rude attempt at a log-house near the trail. "Piney can stay with Mrs. Oakhurst," said The Innocent, pointing to the Duchess, "and I can shift for myself."

Nothing but Mr. Oakhurst's admonishing foot saved Uncle Billy from bursting into a roar of laughter. As it was, he felt compelled to retire up the cañon until he could recover his gravity. There he confided the joke to the tall pine trees, with many slaps of his leg, contortions of his face, and the usual profanity. But when he returned to the party, he found them seated by a fire—for the air had grown strangely chill and the sky overcast—in apparently amicable conversation. Piney was actually talking in an impulsive, girlish fashion to the Duchess, who was listening with an interest and animation she had not shown for many days. The Innocent was holding forth, apparently with equal effect, to Mr. Oakhurst and Mother Shipton, who was actually relaxing into amiability. "Is this yer a d——d picnic?" said Uncle Billy, with inward scorn, as he surveyed the sylvan group, the glancing fire-light and the tethered animals in the foreground. Suddenly an idea mingled with the alcoholic fumes that disturbed his brain. It was apparently of a jocular nature, for he felt impelled to slap his leg again and cram his fist into his mouth.

As the shadows crept slowly up the mountain, a slight breeze rocked the tops of the pine trees, and moaned through their long and gloomy aisles. The ruined cabin, patched and covered with pine boughs, was set apart for the ladies. As the lovers parted, they unaffectedly exchanged a parting kiss, so honest and sincere that it might have been heard above the swaying pines. The frail Duchess and the malevolent Mother Shipton were probably too stunned to remark upon this last evidence of simplicity, and so turned without a word to the hut. The fire was replenished, the men lay down before the door, and in a few minutes were asleep.

Mr. Oakhurst was a light sleeper. Toward morning he awoke benumbed and cold. As he stirred the dying fire, the wind, which was now blowing strongly, brought to his cheek that which caused the blood to leave it—snow!

He started to his feet with the intention of awakening the sleepers, for there was no time to lose. But turning to where Uncle Billy had been lying he found him gone. A suspicion leaped to his brain and a curse to his lips. He ran to the spot where the mules had been tethered; they were no longer there. The tracks were already rapidly disappearing in the snow.

The momentary excitement brought Mr. Oakhurst back to the fire with his usual calm. He did not waken the sleepers. The Innocent slumbered peacefully, with a smile on his good-humored, freckled face; the virgin Piney slept beside her frailer sisters as sweetly as though attended by celestial guardians, and Mr. Oakhurst, drawing his blanket over his shoulders, stroked his mustachios and waited for the dawn. It came slowly in a whirling mist of snowflakes, that dazzled and confused the eye. What could be seen of the landscape appeared magically changed. He looked over the valley, and summed up the present and future in two words—"Snowed in!"

A careful inventory of the provisions, which, fortunately for the party, had been stored within the hut, and so escaped the felonious fingers of Uncle Billy, disclosed the fact that with care and prudence they might last ten days longer. "That is," said Mr. Oakhurst, *sotto voce*[6] to The Innocent, "if you're willing to board us. If you aint—and perhaps you'd better not—you can wait till Uncle Billy gets back with provisions." For some occult reason, Mr. Oakhurst could not bring himself to disclose Uncle Billy's rascality, and so offered the hypothesis that he had wandered from the camp and had accidentally stampeded the animals. He dropped a warning to the Duchess and Mother Shipton, who of course knew the facts of their associate's defection. "They'll find out the truth about us *all*, when they find out anything," he added, significantly, "and there's no good frightening them now."

Tom Simson not only put all his worldly store at the disposal of Mr. Oakhurst, but seemed to enjoy the prospect of their enforced seclusion. "We'll have a good camp for a week, and then the

6. In an undertone.

snow'll melt, and we'll all go back together." The cheerful gayety of the young man and Mr. Oakhurst's calm infected the others. The Innocent, with the aid of pine boughs, extemporized a thatch for the roofless cabin, and the Duchess directed Piney in the reärrangement of the interior with a taste and tact that opened the blue eyes of that provincial maiden to their fullest extent. "I reckon now you're used to fine things at Poker Flat," said Piney. The Duchess turned away sharply to conceal something that reddened her cheeks through its professional tint, and Mother Shipton requested Piney not to "chatter." But when Mr. Oakhurst returned from a weary search for the trail, he heard the sound of happy laughter echoed from the rocks. He stopped in some alarm, and his thoughts first naturally reverted to the whiskey—which he had prudently *cachéd*. "And yet it don't somehow sound like whiskey," said the gambler. It was not until he caught sight of the blazing fire through the still blinding storm, and the group around it, that he settled to the conviction that it was "square fun."

Whether Mr. Oakhurst had *cachéd* his cards with the whiskey as something debarred the free access of the community, I cannot say. It was certain that, in Mother Shipton's words, he "didn't say cards once" during that evening. Haply the time was beguiled by an accordeon, produced somewhat ostentatiously by Tom Simson, from his pack. Notwithstanding some difficulties attending the manipulation of this instrument, Piney Woods managed to pluck several reluctant melodies from its keys, to an accompaniment by The Innocent on a pair of bone castinets. But the crowning festivity of the evening was reached in a rude camp-meeting hymn, which the lovers, joining hands, sang with great earnestness and vociferation. I fear that a certain defiant tone and Covenanter's swing[7] to its chorus, rather than any devotional quality, caused it to speedily infect the others, who at last joined in the refrain:

"I'm proud to live in the service of the Lord,
And I'm bound to die in His army."[8]

The pines rocked, the storm eddied and whirled above the miserable group, and the flames of their altar leaped heavenward, as if in token of the vow.

At midnight the storm abated, the rolling clouds parted, and the stars glittered keenly above the sleeping camp. Mr. Oakhurst, whose professional habits had enabled him to live on the smallest possible amount of sleep, in dividing the watch with Tom Simson, somehow managed to take upon himself the greater part of that duty. He excused himself to The Innocent, by saying that he had "often

7. Covenanters were Scottish Presbyterians of the 16th and 17th centuries who bound themselves by a series of oaths or covenants to the Presbyterian doctrine and demanded separation from the Church of England. "Covenanter's swing" would indicate that the hymn would be sung with a vigorous rhythm with a martial beat.
8. Refrain of the early American spiritual *Service of the Lord*.

been a week without sleep." "Doing what?" asked Tom. "Poker!" replied Oakhurst, sententiously; "when a man gets a streak of luck —nigger-luck[9]—he don't get tired. The luck gives in first. Luck," continued the gambler, reflectively, "is a mighty queer thing. All you know about it for certain is that it's bound to change. And it's finding out when it's going to change that makes you. We've had a streak of bad luck since we left Poker Flat—you come along and slap you get into it, too. If you can hold your cards right along you're all right. For," added the gambler, with cheerful irrelevance,

> "I'm proud to live in the service of the Lord,
> And I'm bound to die in His army."

The third day came, and the sun, looking through the white-curtained valley, saw the outcasts divide their slowly decreasing store of provisions for the morning meal. It was one of the peculiarities of that mountain climate that its rays diffused a kindly warmth over the wintry landscape, as if in regretful commiseration of the past. But it revealed drift on drift of snow piled high around the hut; a hopeless, uncharted, trackless sea of white lying below the rocky shores to which the castaways still clung. Through the marvellously clear air, the smoke of the pastoral village of Poker Flat rose miles away. Mother Shipton saw it, and from a remote pinnacle of her rocky fastness, hurled in that direction a final malediction. It was her last vituperative attempt, and perhaps for that reason was invested with a certain degree of sublimity. It did her good, she privately informed the Duchess. "Just you go out there and cuss, and see." She then set herself to the task of amusing 'the child," as she and the Duchess were pleased to call Piney. Piney was no chicken, but it was a soothing and ingenious theory of the pair to thus account for the fact that she didn't swear and wasn't improper.

When night crept up again through the gorges, the reedy notes of the accordeon rose and fell in fitful spasms and long-drawn gasps by the flickering camp-fire. But music failed to fill entirely the aching void left by insufficient food, and a new diversion was proposed by Piney—story-telling. Neither Mr. Oakhurst nor his female companions caring to relate their personal experiences, this plan would have failed, too, but for The Innocent. Some months before he had chanced upon a stray copy of Mr. Pope's[1] ingenious translation of the Iliad. He now proposed to narrate the principal incidents of that poem—having thoroughly mastered the argument and fairly forgotten the words—in the current vernacular of Sandy Bar. And so for the rest of that night the Homeric demi-gods again walked the earth. Trojan bully and wily Greek wrestled in the winds, and the great pines in the cañon seemed to bow to the wrath of the son of Peleus.[2] Mr. Oakhurst listened with quiet satisfaction.

9. I.e., very good fortune.
1. Alexander Pope (1688–1744), English poet who translated Homer's *Iliad* (and *Odyssey*) into the Heroic couplets for which Pope is famous.
2. Achilles, chief hero on the Greek side of the Trojan War.

Most especially was he interested in the fate of "Ash-heels,"[3] as The Innocent persisted in denominating the "swift-footed Achilles."

So with small food and much of Homer and the accordeon, a week passed over the heads of the outcasts. The sun again forsook them, and again from leaden skies the snow-flakes were sifted over the land. Day by day closer around them drew the snowy circle, until at last they looked from their prison over drifted walls of dazzling white, that towered twenty feet above their heads. It became more and more difficult to replenish their fires, even from the fallen trees beside them, now half-hidden in the drifts. And yet no one complained. The lovers turned from the dreary prospect and looked into each other's eyes, and were happy. Mr. Oakhurst settled himself coolly to the losing game before him. The Duchess, more cheerful than she had been, assumed the care of Piney. Only Mother Shipton—once the strongest of the party—seemed to sicken and fade. At midnight on the tenth day she called Oakhurst to her side. "I'm going," she said, in a voice of querulous weakness, "but don't say anything about it. Don't waken the kids. Take the bundle from under my head and open it." Mr. Oakhurst did so. It contained Mother Shipton's rations for the last week, untouched. "Give 'em to the child," she said, pointing to the sleeping Piney. "You've starved yourself," said the gambler. "That's what they call it," said the woman querulously, as she lay down again, and turning her face to the wall, passed quietly away.

The accordeon and the bones were put aside that day, and Homer was forgotten. When the body of Mother Shipton had been committed to the snow, Mr. Oakhurst took The Innocent aside, and showed him a pair of snow-shoes, which he had fashioned from the old pack-saddle. "There's one chance in a hundred to save her yet," he said, pointing to Piney; "but it's there," he added, pointing toward Poker Flat. "If you can reach there in two days she's safe." "And you?" asked Tom Simson. "I'll stay here," was the curt reply.

The lovers parted with a long embrace. "You are not going, too," said the Duchess, as she saw Mr. Oakhurst apparently waiting to accompany him. "As far as the cañon," he replied. He turned suddenly, and kissed the Duchess, leaving her pallid face aflame, and her trembling limbs rigid with amazement.

Night came, but not Mr. Oakhurst. It brought the storm again and the whirling snow. Then the Duchess, feeding the fire, found that some one had quietly piled beside the hut enough fuel to last a few days longer. The tears rose to her eyes, but she hid them from Piney.

The women slept but little. In the morning, looking into each

---

3. The mispronunciation emphasizes Achilles' one vulnerable spot, his heel, by which his mother Thetis held him when she dipped him in the River Styx to make him invulnerable.

other's faces, they read their fate. Neither spoke; but Piney, accepting the position of the stronger, drew near and placed her arm around the Duchess's waist. They kept this attitude for the rest of the day. That night the storm reached its greatest fury, and rending asunder the protecting pines, invaded the very hut.

Toward morning they found themselves unable to feed the fire, which gradually died away. As the embers slowly blackened, the Duchess crept closer to Piney, and broke the silence of many hours: "Piney, can you pray?" "No, dear," said Piney, simply. The Duchess, without knowing exactly why, felt relieved, and putting her head upon Piney's shoulder, spoke no more. And so reclining, the younger and purer pillowing the head of her soiled sister upon her virgin breast, they fell asleep.

The wind lulled as if it feared to waken them. Feathery drifts of snow, shaken from the long pine boughs, flew like white-winged birds, and settled about them as they slept. The moon through the rifted clouds looked down upon what had been the camp. But all human stain, all trace of earthly travail, was hidden beneath the spotless mantle mercifully flung from above.

They slept all that day and the next, nor did they waken when voices and footsteps broke the silence of the camp. And when pitying fingers brushed the snow from their wan faces, you could scarcely have told from the equal peace that dwelt upon them, which was she that had sinned. Even the Law of Poker Flat recognized this, and turned away, leaving them still locked in each other's arms.

But at the head of the gulch, on one of the largest pine trees, they found the deuce of clubs pinned to the bark with a bowie knife. It bore the following, written in pencil, in a firm hand:

†

BENEATH THIS TREE

LIES THE BODY

OF

JOHN OAKHURST,

WHO STRUCK A STREAK OF BAD LUCK

ON THE 23D OF NOVEMBER, 1850,

AND

HANDED IN HIS CHECKS

ON THE 7TH DECEMBER, 1850

‡

And pulseless and cold, with a Derringer[4] by his side and a bullet in his heart, though still calm as in life, beneath the snow, lay he who was at once the strongest and yet the weakest of the outcasts of Poker Flat.   1869, 1870

---

4. Short-barreled pocket pistol, named after its 19th-century American inventor, Henry Deringer.

W. D. HOWELLS
1837–1920

No other American writer has dominated the literary scene the way William Dean Howells did in his prime. As a steadily productive novelist, playwright, critic, essayist, and editor, Howells was always in the public eye, and his influence during the 1880s and 1890s on a growing, serious middle-class readership was incalculable. He made that emerging middle class aware of itself: in his writings an entire generation discovered through his faithful description of familiar places, his dramatizations of ordinary lives, and his shrewd analyses of shared moral issues, its tastes, its social behavior, its values, and its problems. Howells was by temperament genial and modest, but he was also forthright and tough-minded. He was, as the critic Lionel Trilling has observed, a deeply civil man with a balanced sense of life. Perhaps that is why, when he died in 1920, the spontaneous outpouring of sorrow and admiration was the kind reserved for national heroes.

Howells's eminence had its roots in humble beginnings. He was born, one of eight children, in the postfrontier village of Martin's Ferry, Ohio, on March 1, 1837, to a poor, respectable, proud, and culturally informed family. Like his contemporary Mark Twain and their predecessor Ben Franklin, Howells went to school at the printer's office, setting type for the series of unsuccessful newspapers that his good-natured, somewhat impractical father owned. Though the family moved around a good deal in Ohio, Howells's youth was secure and, on the whole, happy. His mother, he observed, had the gift of making each child feel that he or she was the center of the world.

From his earliest years Howells had both literary passions and literary ambitions. When he was not setting type or reading Goldsmith, Irving, Shakespeare, Dickens, Thackeray, or other favorites, he was teaching himself several foreign languages. Howells tried his hand at a number of literary forms in his teens, but his first regular jobs involved writing for newspapers in Columbus and Cincinnati. It was as a journalist that he made his first pilgrimage to New England in 1860, where he was treated with remarkable generosity by such literary leaders as Lowell, Holmes, Emerson, and Hawthorne, who must have recognized that he possessed talent and the will to succeed, as well as courtesy and deference.

A campaign biography of Lincoln, his first significant book, won for Howells the consulship at Venice in 1861. There he wrote the series of travel letters that eventually became *Venetian Life* (1866); more importantly, they made his name known in eastern literary circles. When he returned to America in 1866, he went to work in New York briefly for the *Nation* until James T. Fields offered him the assistant editorship of the enormously prestigious and influential *Atlantic Monthly*, to which he had contributed some of his earliest verse before the war. In effect Howells assumed active control of the magazine from the very beginning, and he succeeded officially to the editorship in 1871, a position he held until he resigned in 1881 in order to have more time to write fiction. Because the *Atlantic* was the pre-eminent literary magazine of the day, Howells had, as a young man, the power to make or break careers, a power he exercised tactfully and responsibly.

Howells had been finding his way as a novelist during his ten years as editor, publishing seven novels in this period, beginning with *Their Wedding Journey* (1872) and concluding with *The Undiscovered Country* (1880). These first novels are short, uncomplicated linear narratives which deliberately eschew the passionate, heroic, action-packed, and exciting adventures that were the staple of American fiction of the time, a fiction read chiefly by middle-class women.

In the 1880s Howells came into his own as novelist and critic. A *Modern Instance* (1882) examines psychic, familial, and social disintegration under the pressure of the secularization and urbanization of post–Civil War America, the disintegration that is Howells's central and deepest subject. Three years later Howells published his most famous novel, *The Rise of Silas Lapham* (1885). Within a year of its publication Howells, who was ostensibly successful and financially secure, suddenly felt that "the bottom had dropped out" of his life, had been profoundly affected by Tolstoy's Christian socialism, and publicly defended the "Haymarket Anarchists," a group of Chicago workers, several of whom were executed without clear proof of their complicity in a dynamiting at a public demonstration. After *Lapham*, Howells offered more direct, ethical criticism of social and economic anomalies and inequities in such succeeding novels as the popular, large-canvassed *A Hazard of New Fortunes* (1890) and the utopian romance *A Traveler from Altruria* (1894). In these same years Howells also penetrated more deeply into individual consciousness, particularly in two short novels, *The Shadow of a Dream* (1890) and *An Imperative Duty* (1892).

In his later years Howells sustained and deepened his varied literary output. Among his novels of consequence in this period are the naturalistic *The Landlord at Lion's Head* (1897) and the elegiac *The Vacation of the Kelwyns* (published posthumously, 1920). He also wrote charming and vivid autobiography and reminiscence in *A Boy's Town* (1890) and *Years of My Youth* (1916), and as he had since the 1870s, Howells continued to produce plays and farces which served, as one critic has remarked, as "finger exercises for his novels."

In the mid 1880s Howells had aggressively argued the case for realism and against the "romanticistic," promoting Henry James in particular at the expense of such English novelists as Scott, Dickens, and Thackeray. In *The Editor's Study* essays he wrote for *Harper's Monthly* starting in 1886 (and some of which in 1891 he made into *Criticism and Fiction*), Howells attacked sentimentality of thought and feeling and the falsification of moral nature and ethical options wherever he found them in fiction. He believed that realism "was nothing more or less than the truthful treatment of material," especially the motives and actions of *ordinary* men and women. He insisted, sooner and more vigorously than any other American critic, that the novel be objective or dramatic in point of view; solidly based in convincingly motivated characters speaking the language of actual men and women; free of contrived events or melodramatic effects; true to the particulars of a recent time and specific place; ethically and aesthetically a seamless piece. Indeed, perhaps the polemical nature of his critical stance in the 1880s did as much as anything to obscure until recently the flexibility and range of his sensibility. Certainly *Novel-Writing and Novel-Reading*, first delivered as a lecture in 1899, suggests more accurately than *Criticism and*

Fiction (1891) the shrewdness, common sense, and penetrating thought-fulness that characterize his criticism at its best.

In the course of his life-long career as literary arbiter, Howells was remarkably international in outlook, and promoted in his diverse critical writings such non-American contemporaries as Ivan Turgenev, Benito Pérez Galdós, Björnstjerne Björnson, Leo Tolstoy, Henrik Ibsen, Émile Zola, George Eliot, and Thomas Hardy. Howells also championed many younger American writers and early recognized many talented women writers in the relentless stream of reviews he wrote over six decades—among them Sarah Orne Jewett, Mary E. Wilkins Freeman, Edith Wharton, and Emily Dickinson. He is even better known for actively promoting the careers of such emerging realists and naturalists as Stephen Crane, Hamlin Garland, and Frank Norris. His chief fault as critic—if it is one—was excessive generosity, though he never falsely flattered or encouraged anyone.

The two contemporaries in whom Howells had the greatest critical confidence were Henry James and Mark Twain, both of whom he served since the 1860s as editor and with both of whom he also sustained a personal friendship of forty years and more. In the late 1860s, Howells had walked the Cambridge streets with James, discussing the present state and future prospects for the substance and techniques of fiction. As editor of *Atlantic Monthly*, Howells had accepted a number of James's early tales. Throughout his career he wrote essays and reviews in praise of James's work; on his deathbed Howells was working on an essay, *The American James*. He genuinely admired and was friendly with the patrician James, but he clearly loved and was more intimate with the rough-textured Twain. *My Mark Twain*, written immediately after his friend's death in 1910, records that affection in one of the enduring memoirs of our literary history.

By the time Howells died, he had served for thirteen years as first president of the American Academy of Arts and Letters, the organization which seeks to identify and honor the most distinguished work in these fields, and was himself a national institution. For rebels and iconoclasts like H. L. Mencken, Sinclair Lewis, and a young Van Wyck Brooks, he epitomized the dead hand of the past, the genteel, Victorian enemy. Since the 1930s, however, Howells's reputation has slowly recovered from these charges. The courage of his liberal—at times radical—perspective in his own time has been acknowledged; his steady, masterful style has been given its due; and his intelligent civility has been commended.

Novel-Writing and Novel-Reading:
An Impersonal Explanation[1]

It was Thackeray[2] who noted how actors, when they had a holiday, always went and saw a play. I fancy that they form the kindest

1. This essay is a corrected transcription of a draft manuscript of one of two lectures Howells gave during a tour of the East and Midwest he made in 1899 under the auspices of the lecture agent James B. Pond. The manuscript survives in the Rutherford B. Hayes Library, Fremont, Ohio; while Howells himself never prepared it for publication, a slightly different version was edited by William M. Gibson and published, first, in the *Bulletin* of the New York Public Library in January, 1958, and later as a monograph, also by the New York Public Library. The present text, copyright 1978 by the Indiana University Press and the Howells Edition Editorial Board, was prepared for *A Selected Edition of W. D. Howells* and is used with permission. The notes were prepared by the present editor.

2. William Makepeace Thackeray (1811–63), English novelist.

and best part of the house at such times. They know how hard it is to do what the people on the stage are doing; if they are quick to what is ill-done, they are quick to what is well done, too; and from what I have seen of their behavior at actors' matinées,[3] as they are called, when the profession pretty much fills the house, I am ready to say, that they are the most lenient, the most generous of all the spectators.

It is much the same, I believe, with novelists, whom I will assume for the purposes of illustration, to be so largely of my own mind and make, that I need not consider those who are otherwise. In fact, I will assume, as a working hypothesis that I am exactly like every other novelist, and I will speak for the whole body of fiction-mongers in saying that when I get a day off from a novel of my own, there is nothing I like so much as to lose myself in the novel of some one else. When I have not a whole day, I am very glad of a half day, or even such hours and halfhours as I can steal from sleep after going to bed at night, and before getting up in the morning. I do not despise other kinds of reading. I like history, I like biography, I like travels, I like poetry, I like drama, I like metaphysics; but I suspect that if I could once be got to tell the whole truth, it would appear that I liked all these in the measure they reminded me of the supreme literary form, the fine flower of the human story, the novel; and if I have anywhere said anything else to the contrary, I take it back, at least for the time being.

You would have thought perhaps that having written so many novels myself,—the procession has now been some twenty-five years in passing a given point,—I would not care to read any; but we novelists, like the actors, are so in love with our art that we cannot get enough of it; and rather than read no novels at all, I would read my own, over and over again. In fact I often do this, and I have probably read them more times than any person present, not because I admire them so very much, but because when I find myself in a difficult place in some new one, I can learn from the old ones how I once behaved in another difficult place. If I go to some other novelist's book to take a leaf from it, I am apt to become so interested in the story, that I forget what I went to it for, and rise from it as honest as I sat down. But I know the story in my own books so thoroughly that I can give myself without hindrance to the study of the method, which is what I want.

That is what we go to one another's novels for. We read them for pleasure, of course, but for a pleasure quite different from that which other readers find in them. The pleasure they yield is probably greater for us than for any other kind of reader; but again we are like the actors at the play: we are all the time, consciously or unconsciously, taking note how the thing is done. We may forget the shop, as I have just now pretended, but the shop does not forget

3. I.e., afternoon benefit performances.

us; sooner or later we find that we have had it with us; and here appears that chasmal difference between the author and the reader, which Goethe[4] says can never be bridged. The reader who is not an author considers what the book is; the author who is a reader, considers, will he, nill he, how the book has been done. It is so in every art. The painter, sculptor, architect, musician feels to his inmost soul the beauty of the picture, statue, edifice, symphony, but he feels still more thoroughly the skill which manifests that beauty. This difference is from everlasting to everlasting, and it disposes instantly of the grotesque pretension that the artist is not the best critic of his art. He is the best of all possible critics. Others may learn to enjoy, to reason and to infer in the presence of a work of art; but he alone who has wrought in the same kind can feel and know concerning it from instinct and from experience. Construction and criticism go hand in hand. No man ever yet imagined beauty without imagining more beauty and less; he *senses*, as the good common phrase has it, the limitations to the expression of beauty; and if he is an artist he puts himself in the place of the man who made the thing of beauty before him, clothes himself in his possibilities, and lives the failure and the success which it records. His word, if honest, is the supreme criticism.

By beauty of course I mean truth, for the one involves the other; it is only the false in art which is ugly, and it is only the false which is immoral. The truth may be indecent, but it cannot be vicious, it can never corrupt or deprave; and I should say this in defence of the grossest material honestly treated in modern novels as against the painted and perfumed meretriciousness of the novels that went before them. I conceive that apart from all the clamor about schools of fiction is the question of truth, how to get it in, so that it may get itself out again as beauty, the divinely living thing, which all men love and worship. So I make truth the prime test of a novel. If I do not find that it is like life, then it does not exist for me as art; it is ugly, it is ludicrous, it is impossible. I do not expect a novel to be wholly true; I have never read one that seemed to me so except Tolstoy's[5] novels; but I expect it to be a constant endeavor for the truth, and I perceive beauty in it so far as it fulfills this endeavor. I am quite willing to recognize and enjoy whatever measure of truth I find in a novel that is partly or mainly false; only, if I come upon the falsehood at the outset I am apt not to read that novel. But I do not bear such a grudge against it as I do against the novel which lures me on with a fair face of truth, and drops the mask midway. If you ask me for illustrations, I am somewhat at a loss, but if you ask me for examples, they are manifold. In English I should say the truthful novelists or those working with an ideal of truth were Jane

4. Johann Wolfgang von Goethe (1749–1832), German poet, dramatist, novelist, and scientist.

5. Leo Tolstoy (1828–1910), Russian novelist and religious philosopher.

Austen, George Eliot, Anthony Trollope, Thomas Hardy, Mrs. Humphrey Ward, George Moore; in French, Flaubert, Maupassant, the Goncourts, Daudet and Zola; in Russian, Tourguenief and Tolstoy; in Spanish, Valdés, Galdós and Pardo-Bazan; in Norwegian, Björnson, Lie, and Kielland. In English, some untruthful novelists, or those working from an ideal of effect, are Thackeray, Dickens, Bulwer, Reade, and all their living followers; in French, Dumas, Feuillet, Ohnet; in Spanish Valera; in Russian, measurably Doystoyevsky;[6] in Norwegian, none that I know of. It is right to say, however that of some of the untruthful novelists, and notably of Thackeray, that they were the victims of their period. If Thackeray had been writing in our time, I have no question but he would have been one of its most truthful artists.

The truth which I mean, the truth which is the only beauty, is truth to human experience, and human experience is so manifold and so recondite, that no scheme can be too remote, too airy for the test. It is a well ascertained fact concerning the imagination that it can work only with the stuff of experience. It can absolutely create nothing; it can only compose. The most fantastic extravagance comes under the same law that exacts likeness to the known as well as the closest and severest study of life. Once for all, then, obedience to this law is the creed of the realist, and rebellion is the creed of the romanticist. Both necessarily work under it, but one willingly, to beautiful effect, and the other unwillingly to ugly effect.

For the reader, whether he is an author too, or not, the only test of a novel's truth is his own knowledge of life. Is it like what he has seen or felt? Then it is true, and for him it cannot otherwise be true, that is to say beautiful. It will not avail that it has style, learning, thinking, feeling; it is no more beautiful without truth than the pretty statue which cannot stand on its feet. It is very astonishing to me that any sort of people can find pleasure in such a thing; but I know that there are many who do; and I should not think of consigning them to the police for their bad taste so long as their taste alone is bad. At the same time I confess that I should suspect an unreality, an insincerity in a mature and educated person whom I found liking an unreal, an insincere novel. You see, I take novels rather seriously, and I would hold them to a much stricter account

6. Jane Austen (1775–1817), George Eliot (1819–80), Anthony Trollope (1815–82), Thomas Hardy (1840–1928), Mrs. Humphrey Ward (1851–1920), and George Moore (1852–1933), English novelists; Gustave Flaubert (1821–80), Guy de Maupassant (1850–93), Edmond de Goncourt (1822–96) and Jules de Goncourt (1830–70), Alphonse Daudet (1840–97), Émile Zola (1840–1902), French novelists; Ivan Turgenev (1818–83) and Leo Tolstoy (1828–1910), Russian novelists; Armando Palacio Valdés (1853–1938), Benito Pérez Galdós (1843–1920), Emilia Pardo Bazán (1852–1921), Spanish novelists; Björnstjerne Martinius Björnson (1832–1910), Jonas Lauritz Idemil Lie (1833–1909), Alexander Lange Kielland (1849–1906), Norwegian novelists; William Makepeace Thackeray (1811–63), Charles Dickens (1812–70), Edward George Earle Lytton Bulwer-Lytton (1803–73), Charles Reade (1814–84), English novelists; Alexandre Dumas (1802–70), Octave Feuillet (1821–90), Georges Ohnet (1848–1918), French novelists; Juan Valera y Alcalá Galiano (1824–1905), Spanish novelist; Feodor Dostoevsky (1821–81), Russian novelist.

than they are commonly held to. If I could, I would have them all subject to the principles that govern an honest man, and do not suffer him to tell lies of any sort. I think the novelist is rarely the victim of such a possession, or obsession, that he does not know when he is representing and when he is misrepresenting life. If he does not know it fully at the time, he cannot fail to be aware of it upon review of his work. In the frenzy of inspiration, he may not know that he has been lying; but a time will quickly come to him, if he is at all an artist, when he will know it, and will see that the work he has done is ugly because of it. That is the time for him to tear up his work, and to begin anew.

Of course, there are several ways of regarding life in fiction, and in order to do justice to the different kinds we ought to distinguish very clearly between them. There are three forms, which I think of, and which I will name in the order of their greatness: the novel, the romance, and the romanticistic novel.

The novel I take to be the sincere and conscientious endeavor to picture life just as it is, to deal with character as we witness it in living people, and to record the incidents that grow out of character. This is the supreme form of fiction, and I offer as supreme examples of it, Pride and Prejudice, Middlemarch, Anna Karenina, Fathers and Sons, Doña Perfecta & Marta y María,[7] sufficiently varied in their origin and material and method, but all of the same absolute honesty in their intention. They all rely for their moral effect simply and solely upon their truth to nature.

The romance is of as great purity of intention as the novel, but it deals with life allegorically and not representatively; it employs types rather than characters, and studies them in the ideal rather than the real; it handles the passions broadly. Altogether the greatest in this kind are The Scarlet Letter and The Marble Faun of Hawthorne, which partake of the nature of poems, and which, as they frankly place themselves outside of familiar experience and circumstance, are not to be judged by the rules of criticism that apply to the novel. In this sort, Judd's Margaret is another eminent example that occurs to me; and some of you will think of Mrs. Shelley's Frankenstein, & of Stevenson's Jekyll and Hyde. I suggest also Chamisso's Peter Schlemihl.[8]

The romanticistic novel professes like the real novel to portray actual life, but it does this with an excess of drawing and coloring

7. *Pride and Prejudice* (1813) by Jane Austen; *Middlemarch* (1871–72) by George Eliot; *Anna Karenina* (1875–77) by Leo Tolstoy; *Fathers and Sons* (1861) by Ivan Turgenev; *Doña Perfecta* (1876) by Benito Pérez Galdós; and *Marta y María* (1883) by Armando Palacio Valdés.
8. *The Scarlet Letter* (1859), *The Marble Faun* (1860). Sylvester Judd III (1813–53), American novelist, published *Margaret, a Tale of the Real and Ideal,* *Including Sketches of a Place Not Before Described, Called Mons Christi* in 1845; Mary Wollstonecraft Godwin Shelley (1797–1851), English novelist, published *Frankenstein* in 1818; Robert Louis Stevenson (1850–94), English novelist, published *The Strange Case of Dr. Jeckyll and Mr. Hyde* in 1886; Adelbert von Chamisso (1781–1838), German writer, published *Peter Schlemihls wunderbare Geschichte* in 1814.

which are false to nature. It attributes motives to people which do not govern real people, and its characters are of the quality of types; they are heroic, for good or for bad. It seeks effect rather than truth; and endeavors to hide in a cloud of incident the deformity and artificiality of its creations. It revels in the extravagant, the unusual and the bizarre. The worst examples of it are to be found in the fictions of two very great men: Charles Dickens and Victor Hugo;[9] but it prevailed in all languages, except the Russian, from the rise of Bulwer and Balzac to the death of Dickens, in spite of the influence of George Eliot and Thackeray. Both these writers contemned it, but not effectively; the one was too much a moralist, the other too much a sentimentalist and caricaturist; I am speaking broadly. In all that time the most artistic, that is to say the most truthful, English novelist was Anthony Trollope, and he was so unconscious of his excellence, that at times he strove hard for the most inartistic, the most untruthful attitudes of Thackeray. Now, all is changed: not one *great* novelist, not a single one in any European language, in any country, has for the last twenty five years been a romanticistic novelist; while literature swarms with second-rate, third-rate romanticistic novelists. The great novelists of China, of Abyssinia, of Polynesia, may still be romanticists, but they are not so in any Western civilization. If you wish to darken council by asking how it is that these inferior romanticists are still incomparably the most popular novelists, I can only whisper, in strict confidence, that by far the greatest number of people in the world, even the civilized world, are people of weak and childish imagination, pleased with gross fables, fond of prodigies, heroes, heroines, portents and improbabilities, without self-knowledge, and without the wish for it. Only in some such exceptional assemblage as the present, do they even prefer truth to lies in art, and it is a great advance for them to prefer the half-lies which they get in romanticistic novels.

I believe, nevertheless, that the novelist has a grave duty to his reader; and I wish his reader realized that he has a grave duty to the novelist, and ought to exact the truth of him. But most readers think that they ought only to exact amusement of him. They are satisfied if they can get that, and often they have to be satisfied without it. In spite of the fact that the novelist is usually so great a novel-reader himself, I doubt if he is fully conscious of the mind the novel reader commonly brings to the work he has taken so much pains with. Once, a great while ago, when a story of mine was appearing from month to month, a young lady wrote me that she was reading it, with nine other serials, besides novels out of the circulating library,[1] and she liked mine the best of all. I thought it was very kind of her, and I could not help wondering what the

9. Victor Hugo (1802–85) and Honoré de Balzac (1779–1850), French novelists.

1. A circulating library rented popular fiction at a low daily rate.

inside of her mind could be like. But the mind of youth, before the world has yet filled it, is hospitable to many guests, and perhaps with all the people of all those stories in it, the mind of this young lady was still tolerably empty. I dare say there is not a person here present but has at some time or other read a novel; it is possible that several may have read two or three serials at the same time; and I would like these to understand that I do not at all object to that way of reading novels. It is much better than not to read novels at all, and I do not know that I felt any reproach for another young lady whose teacher evolved from her the fact that she knew all of what she called the love-parts of my novels, but supposed that I was an Englishman, and that I was dead. It would be no bad thing, I suppose, to be an Englishman if one were dead; and perhaps this may be my palingenesis;[2] but if I were to rise an English novelist I should like to be allowed a choice which.

It is not of novel writers however, that I now wish to speak, but a little more of novel readers, as I have known them. When another story of mine was appearing I had a bill for $30 sent me from a tailor in Chicago for a spring overcoat, against a certain Mr. Ferris. The hero of my novel—he was, as usual,—very unheroic—was Henry Ferris,[3] and the tailor naturally thought that from my intimacy with one member of the family, I would very likely know the address of another.

Only last summer a lady said to me that she wondered I could remember so exactly all that was said and done in a current story of mine; and it appeared that she thought it had all really happened as I had set it down. This was very gratifying in a way, but it was a little dismaying, too; and I fancy it would not be well to peer too earnestly into that chasm which parts authors and readers. Many people read your book without ever looking at the title page, or knowing who wrote it, or caring. This is the wholly unliterary sort, who do not know apparently how books come to be, or how they differ in origin from products of the loom or plough. There is another sort who amiably confuse you with some brother author, and praise you for novels that you have never written. I remember that one night at the White House one of the ladies who was receiving had the goodness to say that she was reading my story of The Bostonians with so much interest. I was forced to disclaim the honor done me; I could only thank her and add that I liked The Bostonians too, as I did everything that Henry James wrote.[4] Upon this we both fell into some embarrassment, I do know why; she excused herself for her blunder; of course, the story was Mr. James's; she knew that; and she asked me if I would be introduced to the Secretary of the Periphery (that was not really the office)

2. I.e., rebirth.
3. Henry Ferris was the sensitive young observer of Howells's *A Foregone Con-* *clusion* (1875).
4. *The Bostonians* was published in 1886.

who liked my books, and greatly wished to see me. The secretary was very cordial, and told me that he always kept my Stillwater Tragedy lying on his desk, he liked so much to take it up and read it at leisure moments. What could I do? I answered that I should be glad to tell Mr. Aldrich[5] when I went back to Boston, what a favorite his book was with the Secretary of the Periphery; and I really forget how we got rid of each other.

But anything so disastrous as this does not often happen to a novelist, I fancy. Those crushing blows, which fell within ten minutes of each other, were probably meant to cure me of vanity, and I can confidently say that they did so. I have not felt since the slightest motion of pride or conceit when the reader has failed to confound me with some one else, or even when he knows distinctly who I am and what I have written, and seizes with exquisite intelligence my lightest and slightest intention in a book. There *are* such readers; and I feel sure that nothing good that the author puts into a novel is ever lost. Some one sees it, feels it, loves it, and loves him for it. This is the sweet compensation for much negligence, much coldness, much dullness. Readers are not so bad, I should like to say to my brother novelists; they are really very good, and at any rate we could not get on without them. I myself think they are better in the small towns, where the excitements and the distractions are few, than in the cities where there are many. I have said before, somewhere, that in the cities people do not read books, they read about them; and I believe that it is far from these nervous centres, that the author finds his closest, truest, loveliest appreciation. For my part I like best to think of my stories, if they are so blest, as befriending the loneliness of outlying farms, dull villages, distant exile. I cannot express the joy it gave me to have General Greeley[6] say that he had read me amidst the frozen blackness of the arctic night; and the other day I had a letter from a man who had followed the fortunes of some imaginary people of mine through a long cruise in the South Seas. I answered him that it was the knowledge of such things which made it sweet to be a novelist; and I may add, to you, that to have a letter like that I would willingly disown all the books that Mr. James ever wrote, or Mr. Aldrich either.

While I am about these confidences, which I make very frank because they are typical rather than personal, and deal with things that happen to all authors more or less, I will confess that I have never yet seen one novel of mine sold. Once, in a book store, I saw a lady take up my latest, and look into it; I waited breathless; but she laid it down again, and went out directly, as if it had perhaps been too much for her. Yet, unless the publishers have abused my fondness,[7] some of my novels have had a pretty sale enough; and I

5. Thomas Bailey Aldrich (1836–1907), American writer and editor, published *The Stillwater Tragedy* in 1880.
6. General Adolphus W. Greely (1844– 1935), American army officer and explorer, led an expedition to the arctic in 1881.
7. I.e., my willingness to believe.

have at least overheard them talked about. In a railway train once, I listened to a gentleman in the seat before me commending them to a young lady for their blameless morality; another day, at the table next mine, in a restaurant, a young man went critically through most of them to his commensal.[8]

I believe he was rather lenient to them; but you cannot always depend upon the flattering quality of such eavesdroppings, and I think it is best to get away from them. At a table d'hôte[9] in Florence a charming young English lady, who knew me for an American by my speech began the talk by saying that she was just from Venice, where she had read a book which a countryman of mine had written about that city, and named a book of my own. I did not think it would be fair to let her go on if she had any censure to pronounce, and I knew it would not be pleasant; so I made haste to say that I was myself the countryman of mine whom she meant; after that she had nothing but praise for my book. The trouble is you cannot be sure what people will say, and it is best to forego anything surreptitious in the collection of opinion. I once wrote a novel in which I thought I had been very deeply, and was perhaps only too subtly, serious; but a young gentleman in a waltz, when his partner asked him if he were reading it, said, "No; too trivial." I did not overhear this, and so I do not feel justly punished by it. Neither do I consider that I quite merit the blame bestowed upon another novel of mine,[1] but I will report the fact because it shows that there may be two views of my morality. This story was of a young girl who, by a series of misunderstandings, finds herself the only woman on board a vessel going to Italy, with three young men for her fellow passengers. They do everything they can to keep her from embarrassment or even consciousness, and one of them marries her when they get to Venice. I thought this a very harmless scheme, and so did a friend of mine,[2] who was in France when the book came out, and who recommended it, perhaps too confidently, to a French mother of daughters anxious for some novel in English proper for a young girl to read. He lent it her, but when she had read it herself, she brought it back, and said the situation imagined in it was immoral and altogether unfit to be presented to the mind of a *jeune fille*.[3]

To tell the truth, I do not think it would be well for the author to aim at the good opinion of the reader in this or anything else. That cannot be trusted to keep the author's literary or moral conscience clean and that is the main thing with him. His affair is to do the best he can with the material he has chosen, to make the truest possible picture of life, and this is what I believe he always does, if he is worthy of the name of artist. He had better not aim to

8. I.e., those who shared the table with him.
9. Common dining table in a hotel.

1. *Lady of the Aroostook* (1879).
2. Henry James.
3. Young lady.

please, and he had still better not aim to instruct; the pleasure and
the instruction will follow from such measure of truth as the author
has in him to such measure of truth as the reader has in him. You
will sometimes find it said by the critics that such and such a novel
has evidently been written with such and such an object; but unless
it is the work of a mere artizan, and no artist at all, I believe this is
never the fact. If it is a work of art, it promptly takes itself out of
the order of polemics or of ethics, and primarily consents to be
nothing if not aesthetical. Its story is the thing that tells, first of all,
and if that does not tell, nothing in it tells. It is said that one
reason why Tolstoy, when he felt the sorrow of the world laid upon
him, decided to write no more novels, because no matter how full
he filled these with the desire of his soul to help them that have no
helper, he found that what went into the minds of most readers was
merely the story.

Then shall the novel have no purpose? Shall it not try to do good?
Shall this unrivalled, this inapproachable form, beside which epic
and drama dwindle to puny dwarfishness, and are so little that
they can both be lost in its vast room, shall this do nothing to
better men and uplift them? Shall it only amuse them? No, and a
thousand times, no! But it shall be a mission to their higher selves
only so far as it shall charm their minds and win their hearts. It
shall do no good directly. It shall not be the bread, but the grain of
wheat which must sprout and grow in the reader's soul, and be har-
vested in his experience, and in the mills of the gods ground slowly
perhaps many years before it shall duly nourish him. I do not mean
that there can never be any immediate good from novels. I do not see
how any one can read The Scarlet Letter, or Middlemarch, or Romola,[4] without being instantly seized with the dread of falsehood. This
is in the way to the love of truth. It is the first step, the indispensable
first step towards that love, but it is by no means arrival at it. The
novel can teach, and for shame's sake, it must teach, but only by
painting life truly. This is what it must above all things strive to do.
If it succeeds, every good effect shall come from it: delight, use, wisdom. If it does not succeed in this, no good can come of it. Let no
reader, and let no intending novelist suppose that this fidelity to life
can be carried too far. After all, and when the artist has given his
whole might to the realization of his ideal, he will have only an
effect of life. I think the effect is like that in those cycloramas
where up to a certain point there is real ground and real grass, and
then carried indivisibly on to the canvas the best that the painter
can do to imitate real ground and real grass. We start in our novels
with something we have known of life, that is, with life itself; and
then we go on and imitate what we have known of life. If we are
very skilful and very patient we can *hide the joint*. But the joint is
always there, and on one side of it are real ground and real grass,

4. *Romola* (1862–63), by George Eliot.

and on the other are the painted images of ground and grass. I do not believe that there was ever any one who longed more strenuously or endeavored more constantly to make the painted ground and grass exactly like the real, than I have done in my cycloramas. But I have to own that I have never yet succeeded to my own satisfaction. Some touch of color, some tone of texture is always wanting; the light is different; it is all in another region. At the same time I have the immense, the sufficient consolation, of knowing that I have not denied such truth as was in me by imitating unreal ground and unreal grass, or even copying the effect of some other's effort to represent real ground and real grass.

Early in the practice of my art I perceived that what I must do in fiction, if I were to do anything worth while, was to get into it from life the things that had not been got into fiction before. At the very first, of course I tried to do the things that I found done already, or the kind of things, especially as I found them in English novels. These had been approved as fit for literature, and they alone were imaginably fit for it. But I tried some other things, and found them fit too. Then I said to myself that I would throw away my English glasses, and look at American life with my own American eyes, and report the things I saw there, whether they were like the things in English fiction or not. In a modest measure this plan succeeded, and I could not commend any other to the American novelist.

I do not mean to say, however, that one's work is always of this intentional, this voluntary sort. On the contrary, there is so much which is unintentional and involuntary, that one might very well believe one's self inspired if one did not know better. For instance, each novel has a law of its own, which it seems to create for itself. Almost from the beginning it has its peculiar temperament and quality, and if you happen to be writing that novel you feel that you must respect its law. You, who are master of the whole affair, cannot violate its law without taking its life. It may grow again, but it will be of another generation and another allegiance. No more can you change the nature of any character in it without spoiling it. You cannot even change the name of a character without running great risk of affecting its vital principle; and by the way where do one's characters get their names? They mostly appear with their names on, an integral part of themselves. This is very curious; but it does not evince inspiration. It merely suggests that the materials which the imagination deals with are not fluid, not flexible, not ductile; but when they have once taken form have a plaster of paris fixity, which is scarcely more subject to the author's will than the reader's. Either one of these may shatter the form, but one is almost as able to reconstitute it as the other. I hope this is not very mystical for I hate anything of that sort, and would have all in plain day if I could. The most that I will allow is that the mind fathers creatures which are apparently as self-regulated as any other

offspring. They are the children of a given mind; they bear a likeness to it; they are qualified by it; but they seem to have their own
life and their own being apart from it. Perhaps this is allowing a
good deal.

Another, and much simpler, fact of my experience has been that
you never master your art as a whole. I used vainly to suppose,
when I began to write fiction, that after I had struck my gait I had
merely to keep on at that pace forever. But I discovered to my vast
surprise that I had struck my gait for this or that book only, and
that the pace would not serve for another novel. I must strike a new
gait, I must get a new pace for every new story. I could issue master
from the last, but I must begin prentice with the next; and I suppose this is the great difference between an art and a trade, or even a
science. The art is always both a teaching and a learning. In virtue
of never being twice the same, it is a perpetual delight, a perpetual
ordeal to the artist. He enjoys and he suffers in it, as no other man
enjoys or suffers in his work.

In fiction you cannot, if you would, strike twice in the same
place, and you certainly had better not, if you could. It is interesting to note how, if you carry a character from one story to another,
it can scarcely be important in both. If you have first given it a
leading part, you have exhausted its possibilities, but if it has been at
first subordinate, then you may develop it into something important
in the second handling. Still less can you twice treat the same
theme twice. For the novelist there is no replica; and I would ask
those readers who sometimes complain of sameness in an author's
books to consider whether it is anything more than that family likeness which they must inevitably have. All Mr. James's book are like
Mr. James; all Tourguenieff's books are like Tourguenieff; all Hawthorne's books are like Hawthorne. You cannot read a page in any
of them without knowing them for this author or that; but the
books of no author resemble one another than through this sort of
blood-relationship.

Indefinite patience is requisite to a fine or true effect in this art
which I am speaking of. In my beginning, I sometimes imagined
that a novel might be blocked out by writing all the vital scenes
first from the earliest to the latest, and then going back, and supplying the spaces of dead color between them. But this is so obviously
impossible that I never even tried it. The events of a real novel
grow slowly and necessarily out of the development of its characters,
and the author cannot fully forecast these. He creates them, but he
has to get acquainted with them in great measure afterwards. He
knows the nature of each, but he does not know how they will
affect one another till he tries. Sometimes, I have hurried forward
to an effect, impatient of intervening detail, but when I have got
the effect by this haste I find that it is weak and false because the
detail was wanting. That is the soil which it must grow out of; with-

out that, and the slow, careful thinking which supplies it, the effect is a sickly and spindling growth.

The novel reader, who is on the outside of all these processes, cannot consider them in liking or disliking a novel. Yet it is the readers and not the writers of novels who decide their fate, and whom novels must first appeal to upon some broad principle common to all men, and especially to that kind of men who are called women. The favor of all the novel writers in the world could not solely make a novel successful; and yet if the novelists liked it I should say it was surely a good novel. I do not say, on the other hand that readers choose falsely, although they often choose foolishly. One could bring up a terrible array of foolish choices against them; novels that sold by the hundred thousand, and yet were disgracefully bad, and are now wholly forgotten. They met a momentary want, they caught a passing fancy; perhaps they touched with artless fortune a chord of real feeling. They pleased vastly, if not mightily, and till they blew over, as Douglass Jerrold[5] used to say of such books, the few who knew better had to hang their heads in shame for the rest.

But there are also novels that please mightily as well as vastly, and then cease to please at all and are as if they had never been. Who is it that now speaks of —— I was going to speak of it, but I will not; everyone knows what I mean and is sick of it. Yet it was a charming book, full of fun and airy fancy, and of a certain truth, generous, spirited, gay and heartbreaking. Why should not it please forever? It must simply be that the principles which in their peculiar combination it appealed to were worn out, as the capacity for being amused by a certain joke is exhausted by familiarity. The joke is as droll as ever: why do not you laugh still? That air, that song, which ravished your sense ninety times was torture the hundredth. Your beloved who died ten years ago, is more lost to you now than then: where are your tears?

> "All things are taken from us and become
> Portions and parcels of the dreadful past."[6]

Laughter passes; grief gluts itself and can no more; laughter dies, spent with its own joy.

That book *was* charming; I say it again; but nothing could make me read it again. Yet there are stories that I can read again and again, and not tire of. They are not such as appeal so much to the passions, or else they appeal to them in a different way. In them, the elements are more fortunately mixed and more skilfully; but it would be hard to say what makes a work of art lastingly please, and what makes a work of art please transiently. If you ask me, I will own frankly I do not know. I can only offer some such makeshift of

5. Douglas William Jerrold (1803–57), English dramatist and journalist.
6. Apparently Howells's parody of a clichéd verse.

an explanation as that it is repose which causes the enduring charm; but who can say just what repose is?

It is taken for granted that one thing which always pleases in a novel is the love-making; but I doubt it. A good deal of the lovemaking in novels is vulgar and offensive. Love is a passion which must be delicately handled by a novelist, or else his lovers will be as disgusting as those who betray their fondness in society, and make the spectator sick; they will be as bad as those poor things who sit with their arms round each other on the benches in the park. Really some of the lovescenes even in so great a novelist as George Eliot, stomach one. But there is nothing better than a love scene when it is well done, though there can be other things quite as good. I think that to make it very acceptable, there should be a little humorous consciousness, a little self-irony in the lovers; though when I think of such noble tragedy as the love-passages in Tourguenieff, I am not sure of my position. Still, still, I think I prefer the love-making of Jane Austen's people; but what do not I prefer of Jane Austen's?

As for my own modest attempts in that direction, I should be far too shame-faced to allege them, if I had not once received a singular proof of their success. I do not mean in the favor of that young lady who had read all the love parts in my books, and supposed I was a dead Englishman, or that other young lady who liked my story best of all out of the nine serials and novels from the library which she was reading. It was such testimony as the boys and the blackbirds bear to the flavor of fruit, and it came about through the printers' leaving the copy of one of my love scenes out overnight where the mice could get at it. The mice ate the delicious morsel all up but a few tattered fragments. It was excessively gratifying to my vanity as author; more, for me, mice could not do; but I did not find it so agreeable when the printers sent me these remnants, and asked me to supply the paragraphs which had been devoured with such eager interest. If you have never had a like experience you can not have any notion how difficult it is to reproduce a love-passage which the mice have eaten.

When I began to write fiction we were under the romantic superstition that the hero must do something to *win* the heroine; perform some valorous or generous act; save her from danger, as a burning building or a breaking bridge, or the like, or at least be nursed by her through a long and dangerous sickness. In compliance with this burdensome tradition, I had my hero rescue my heroine from a ferocious bulldog, which I remember was thought rather *infra dig.*[7] by some of the critics; but I had no other mortal peril handy, and a bulldog is really a very dangerous animal. This was in my first novel; but after that I began to look about me and consider. I observed that none of the loved husbands of the happy wives I knew

7. Abbreviation for *infra dignitatem*, beneath one's dignity.

had done anything to "win" them except pay a certain number of visits, send them flowers, dance or sit out dances with them at parties, and then muster courage to ask if they would have them. Amongst the young people of my acquaintance, I noticed that this simple and convenient sort of conquest was still going on; and I asked myself why it should be different in books. It was certainly very delightful as I saw it in nature, and why try to paint the lily or tint the rose? After that I let my heroes win my heroines by being as nice fellows as I could make them. But even then I felt that they both expected too much of me; and it was about this time that I had many long and serious talks with my friend, Mr. Henry James, as to how we might eliminate the everlasting young man and young woman, as we called them. We imagined a great many intrigues in which they should *not* be the principal personages; I remember he had one very notable scheme for a novel whose interest should centre about a mother and a son. Still, however, he is writing stories, as I still am, about the everlasting man and young woman; though I do think we have managed somewhat to moderate them a little as to their importance in fiction. I suppose we must always have them there, as we must always have them in life, if the race is to go on; but I think the modern novel is more clearly ascertaining their place. Their dominance of course was owing to the belief that young people were the chief readers of fiction. I dare say this is true yet; but I doubt if it is the young people who make the fortune of a novel. Rather, I fancy, its prosperity lies in the favor of women of all ages—and (I was going to say) sexes. These are the most devoted novel-readers, the most intelligent (after the novelists themselves) and the most influential, by far. It is the man of feminine refinement and of feminine culture, with us so much greater than masculine culture, who loves fiction, but amongst other sorts of men I have observed that lawyers are the greatest novel-readers. They read, however, for the story, the distraction, the relief; and after them come physicians, who read novels for much the same reasons, but more for the psychological interest than lawyers. The more liberal sorts of ministers read novels, with an eye to the ethical problems treated; but none of these read so nearly from the novelists' own standpoint as the women. Like the novelists, these read with sympathy for the way the thing is done, with an eye for the shades of character, the distribution of motive, the management of the intrigue, and not merely for the story, or so much for the psychological and ethical aspects of it. Business men, I fancy, seldom read novels at all; they read newspapers.

Fiction is the chief intellectual stimulus of our time, whether we like the fact or not, and taking it in the broad sense if not the deep sense, it is the chief intellectual influence. I should say moral influence, too; but it is often a moral stimulus without being a moral influence; it reaches the mind, and stops short of the conduct. As to the prime fact involved, I think we have but to recall the books of any last year of modern times, and we cannot question it. It is nine-ty-nine chances out of a hundred that the book which at any given moment is making the world talk, and making the world think is a novel. Within the last generation, I can remember only one book making the impression that a dozen of novels have each made, and against Renan's Life of Jesus,[8] I will set Les Miserable, Romola and Middlemarch and Daniel Deronda, Le'Assomoir and Nana, Tess of the D'Urbervilles, Anna Karénina and the Kreuzer Sonata, Robert Elsmere, Trilby, Ben Hur, not all, or at all, of the same artistic value, but all somehow, of a mighty human interest. We must leave Uncle Tom's Cabin out of the count because it was of an earlier period; if we counted it, the proof of my assertion would be overwhelming.

The novel is easily first among books that people read willingly, and it is rightfully first. It has known how to keep the charm of the story, and to add to it the attraction of almost every interest. It still beguiles, as in the hands of the Byzantine romancers,[9] not to go unnumbered centuries back to the Greek novel of Homer, the Odyssey; and it has learnt how to warn, to question, to teach in every concern of life. Scarcely any predicament, moral or psychologi-cal has escaped its study, and it has so refined and perfected its methods that antiseptic surgery itself has hardly made a more benef-icent advance. It began with the merest fable, excluding from the reader's interest all but the fortunes of princes and the other digni-fied personages, for whose entertainment it existed until now it includes all sorts and conditions of men, who turn to it for instruc-tion, inspiration, consolation. It has broadened and deepened down and out till it compasses the whole of human nature; and no cause important to the race has been unfriended of it. Sometimes I have been vexed at its vicious pandering to passion, but I cannot think, after all, of any great modern novel which has not been distinctly moral in effect. I am not sorry to have had it go into the dark places of the soul, the filthy and squalid places of society, high and low, and shed there its great light. Let us know with its help what we are, and where we are. Let all the hidden things be brought into the sun, and let every day be the day of judgment. If the sermon cannot any longer serve this end, let the novel do it.

8. Ernest Renan's *Life of Jesus* (1863) was one of the first such books to take a (what was for the time scandalous) sci-entific-historical approach to its subject. Victor Hugo published *Les Miserables* in 1862; George Eliot published *Daniel De-ronda* in 1876; Émile Zola published *L'Assommoir* in 1877 and *Nana* in 1880; Thomas Hardy published *Tess of the D'Urbervilles* in 1891; Leo Tolstoy pub-lished *Kreuzer Sonata* in 1889; Mary Augusta Ward (1851–1920), American novelist, published *Robert Elsmere* in 1888; George du Maurier (1834–96), English artist and writer, published *Trilby* in 1894; General Lew Wallace (1827–1905), American soldier and nov-elist, published *Ben Hur; A Tale of the Christ* in 1880; Harriett Beecher Stowe (1811–96), American writer, published *Uncle Tom's Cabin* in 1851–52.
9. Perhaps Howells has in mind works such as *The Arabian Nights* or *Thou-sand and One Nights*, though these were not published until the 18th century.

But in doing this it will have to render a stricter account than it has yet been held to. The old superstition of a dramatic situation as the supreme representation of life must be discarded, and the novelist must endeavor to give exactly the effect of life. I believe he will yet come to do this. I can never do it, for I was bred in a false school whose trammels I have never been quite able to burst; but the novelist who begins where I leave off, will yet write the novel which has been my ideal. He will not reject anything because he cannot make it picturesque or dramatic; but he will feel the beauty of truth so intimately, and will value it so supremely that he will seek the effect of that solely. He cannot transport life really into his story, any more than the cycloramist could carry the real ground and the real grass into his picture. But he will not rest till he has made his story as like life as he can, with the same mixed motives, the same voluntary and involuntary actions, the same unaccountable advances and perplexing pauses, the same moments of rapture, the same days and weeks of horrible dullness, the same conflict of the higher and lower purposes, the same vices and virtues, inspirations and propensities. He will not shun any aspect of life because its image will be stupid and gross, still less because its image will be incredibly noble and glorious. He will try to give that general resemblance which can come only from the most devoted fidelity to particulars. As it is now the representation of life in novels, even the most conscientious in its details, is warped and distorted by the novelist's anxiety to produce an image that is startling and impressive, as well as true. But if he can once conceive the notion of letting the reader's imagination care for these things; if he can convince himself that his own affair is to arrange a correct perspective, in which all things shall appear in their very proportion and relation, he will have mastered the secret of repose, which is the soul of beauty in all its forms.

The hope of this may be the vainest of dreams, but I do not think so. Already I see the promise, the prophesy of such a novel in the work of some of the younger men. That work often seems to me crude and faulty, but I feel that it is in the right direction, and I value it for that reason with a faith which only work in the right direction can inspire. Good work in the wrong direction fills me with despair, and my heart sinks lower the better the work is. In a picture of life which is fundamentally or structurally false, I cannot value coloring or drawing, composition or sentiment; the lie at the thing's heart taints and blights every part of it. This happens to me from my own work when I have made a false start, and then I keep trying to hark back to the truth as I know it, and start afresh. A hundred times in the course of a story I have to retrace my steps, and efface them. Often the whole process is a series of arduous experiments, trying it this way, trying it that; testing it by my knowledge of myself and my acquaintance with others; asking if it would be true of me, or true of my friend or my enemy; and not possibly resting content with anything I thought gracious or pleasing in my performance till I have got the setting of truth for it. This sort of scrutiny goes on perpetually in the novelist's mind. His story is never out of it. He lies down with it in his last waking thought and rises up with it in his first. Throughout the day, in crowds or in solitude, it is dimly or distinctly in his thought, a joy, a torment. He shakes hands with a friend and asks after his sick wife, but he is really wondering whether his hero would probably marry his heroine. In his talk at dinner he brings covertly to the test of his neighbor's experience the question of the situation he is developing. He escapes with his life from a cable-car, and at the same instant the solution of a difficult problem flashes upon him. Till he has written finis at the end of his book, it literally obsesses him. He cannot dismiss it; consciously or unconsciously it pervades his being.

Is this a normal, a healthful state for a man to be in? I suppose it is measurably the state of every manner of artist, and I am not describing a condition that will seem strange to any artist. I am not at all sure that it is morbid or unwholesome. The best thing that can fill man's mind is his work, for if his work does not fill it, his self will fill it, and it can have no worse tenant. Of one thing I am certain, and that is that the preoccupation with work that constantly exacts reference to life, makes life incessantly interesting. In my quality of novelist I defy the deadliest bore to afflict me. I have but to test some bore in my story by him, and he becomes a boon, a favor of heaven, an invaluable and exquisitely interesting opportunity.

As to the outward shape of the inward life of the novel, which must invariably be truth, there is some choice, but mainly between three sorts: the autobiographical, the biographical and the historical. The first of these I have always considered the most perfect literary form after the drama. If you tell the story as apparently your own, you are completely master of the situation, and you can report everything as if it were a real incident. What goes on in your own mind concerning persons and events you can give with absolute authority, and you are not tempted to say what goes on in the minds of others, except in the way of conjecture, as one does in life. But the conditions are that you must not go outside of your own observation and experience; you cannot tell what you have not yourself seen and known to happen. If you do, you at once break the illusion; and you cannot even repeat things that you have at second hand, without some danger of this. Within its narrow range the autobiographical story[1] operates itself as much as the play does. Perhaps because of its limitations none of the greatest novels have been written in that form perfect as it is, and delightful as it is to

---

1. I.e., novels with a first-person narrator.

the reader, except Gil Blas[2] only. But Thackeray was always fond of it, and he wrote his best book, The Luck of Barry Lyndon, in it, and his next best, Henry Esmond. In many other novels of his, it is employed; in the very last, The Adventures of Philip, Pendennis tells the story of Firman as if he were knowing to it. Hawthorne chose the autobiographic form for what I think his greatest novel, The Blithedale Romance, and many others have used it. The old fashioned novels in letters, like Pamela and Evalina, were modifications of it; and some next-to-modern novelists, like Wilkie Collins have used the narratives, or statements, of several persons concerning the same fact to much the effect of the autobiographic novel. Gil Blas is possibly the most famous story in this form, and David Copperfield next.

The biographical novel is that in which the author chooses a central figure and refers to it and reports from it all the facts and feelings involved. The central figure must be of very paramount importance to justify this form, which is nearly as cramping as the autobiographical, and has not its intimate charm. Mr. James used it in his Roderick Hudson,[3] but to immeasurably less beautiful effect than he has used the autobiographical, in some of his incomparable short stories. He seems of late to prefer it to any other, and he has cast in it work of really unimpeachable perfection.

After all, however, the historical is the great form, impure and imperfect as it is. But here I wish you to note that I am talking of the historical form in novel writing, and not at all of the historical novel. The historical novel may be written in either the autobiographical, the biographical, or the historical form; but it is not now specifically under discussion. What is under discussion is any sort of novel whose material is treated as if it were real history. In this the novelist supposes himself to be narrating a series of events, indefinite in compass, and known to him from the original documents, as a certain passage in the real life of the race is known to the historian. If, then, he could work entirely in the historian's spirit, and content himself and his reader with conjecture as to his people's motives and with report of them from hearsay, I should not call this form impure or imperfect. But he cannot do this, apparently, or at least he never has done it. He enters into the minds and hearts of his characters; he gives long passages of dialogue among them, and invents speeches for them, as the real historians used to do for their real personages; and he not only does this, but he makes his reader privy to their most secret thoughts, feelings and desires. At times his work is dramatic, and at times narrative; he makes it either at

2. A picaresque novel by Alain René Lesage (1668–1747), French dramatist and novelist. The Luck of Barry Lyndon (1844), Henry Esmond (1852), The Adventures of Philip Pendennis (1848) by William Makepeace Thackeray; The Blithedale Romance (1852) by Nathaniel Hawthorne; Pamela, or Virtue Rewarded (1740–41) by Samuel Richardson; and Evelina (1778) by Frances Burney (1752–1840). Wilkie Collins (1824–89), English novelist. David Copperfield by Charles Dickens was published in 1850. 3. Published in 1876.

will. He dwells in a world of his own creating, where he is a universal intelligence, comprehending and interpreting everything, not indirectly or with any artistic conditions, but frankly and straightforwardly, without accounting in any way for his knowledge of the facts. The form involves a thousand contradictions, impossibilities. There is no point where it cannot be convicted of the most grotesque absurdity. The historian has got the facts from some one who witnessed them; but the novelist employing the historic form has no proof of them; he gives his word alone for them. He visits this situation and that and reports what no one but himself could have seen. He has the intimate confidence of his character in the hour of passion, the hour of remorse, the hour of death itself. Tourguenief and Tolstoy came back from following theirs to the verge of the other world. They tell what they thought and felt as this world faded from them, and nothing in fiction is more impressive, more convincing of its truth.

The historical form, though it involves every contradiction, every impossibility, is the only form which can fully represent any passage of life in its inner and outer entirety. It alone leaves nothing untouched, nothing unsearched. It is the primal form of fiction; it is epic. The first great novels, the Illiad and the Odyssey were cast in it; and the last, if there is ever any last novel while the human race endures, will probably wear it. The subtlest, the greatest achievements of fiction in other forms are nothing beside it. Think of Don Quixote, of Wilhelm Meister, of the Bride of Lammermoor, of I Promessi Sposi, of War and Peace, of Fathers and Sons, of Middlemarch, of Pendennis, of Bleak House, of Uncle Tom's Cabin, of The Scarlet Letter, of L'Assomoir, of The Grandissimes, of Princess Casamassima, of Far from the Madding Crowd:[4] the list of masterpieces in this form is interminable.

When Homer wrote his novels, he feigned that he had his facts from the Muse, and that saved appearances; but hardly any novelist since has seriously done so. The later novelist boldly asks you to believe, as a premise, that he knows all about things that no one man can imaginably know all about, and you are forced to grant it because he has the power of convincing you against your reason. The form which is the least artistic, is the least artificial; the novel of historic form is the novel par excellence; all other forms are clever feats in fiction, literary, conscious. This supreme form is almost shapeless, as it is with the greatest difficulty, with serious limitations of its effects, that you can give it symmetry. Left to

4. *Don Quixote* (1605, 1615), by Cervantes; *Wilhelm Meister* (1796), by Johann Wolfgang von Goethe; *Bride of Lammermoor* (1819), by Sir Walter Scott (1771–1832); *I Promessi Sposi* (1825–26), by Alessandro Manzoni (1785–1873); *War and Peace* (1862–69) by Leo Tolstoy; *The History of Pendennis* (1848–50), by William Makepeace Thackeray; *Bleak House* (1853), by Charles Dickens; *L'Assommoir* (1877), by Émile Zola; *The Grandissimes* (1880), by George Washington Cable (1844–1925); *Princess Casamassima* (1886), by Henry James; *Far from the Madding Crowd* (1874), by Thomas Hardy.

itself, it is sprawling, splay-footed, gangling, proportionless and inchoate; but if it is true to the life which it can give no authority for seeming to know, it is full of beauty and symmetry.

In fine, at the end of the ends, as the Italians say, truth to life is the supreme office of the novel, in whatever form. I am always saying this, and I can say no other. If you like to have it in different words, the business of the novelist is to make you understand the real world through his faithful effigy of it; or, as I have said before, to arrange a perspective for you with everything in its proper relation and proportion to everything else, and this so manifest that you cannot err in it however myopic or astigmatic you may be. It is his function to help you to be kinder to your fellows, juster to yourself, truer to all.

Mostly, I should say, he has failed. I can think of no one, except Tolstoy alone, who has met the high requirements of his gift, though I am tempted to add Björnson in some of his later books. But in spite of his long and almost invariable failure, I have great hopes of the novelist. His art, which is as old as the world, is yet the newest in it, and still very imperfect. But no novelist can think of it without feeling its immeasurable possibilities, without owning that in every instance the weakness, the wrong is in himself, and not in his art.

1899   1958, 1979

HENRY JAMES
1843–1916

Henry James was the first American writer to conceive his career in international terms; he set out, that is, to be a "literary master" in the European sense. Partly because of this grandiose self-conception, partly because he spent most of his adult life in England, partly because his intricate style and choices of cultivated characters ran counter to the dominant vernacular tradition initiated by Mark Twain, James attracted, in his own lifetime, only a select company of admirers. The recognition of his intrinsic importance, as well as his wide influence as novelist and critic, did not emerge until the years between the world wars, when American literary taste reached a new level of sophistication. Only quite recently has his playful prediction that "some day all my buried prose will kick off its various tombstones at once" largely come to pass. James is now firmly established as one of America's major novelists and critics and as a psychological realist of unsurpassed subtlety.

James was born in New York City on April 15, 1843. His father was an eccentric, independently wealthy philosopher and religious visionary; his slightly older brother William was the first notable American psychologist and perhaps our country's most influential philosopher; two younger brothers and a sister completed one of the most remarkable of American families.

First taken to Europe as an infant, James spent his boyhood in a still almost bucolic New York City before the family once again left for the Continent when he was twelve. His father wanted the children to have a rich, "sensuous education," and during the next four years, with stays in England, Switzerland, and France, they were endlessly exposed to galleries, libraries, museums, and (of special interest to Henry) theaters. Henry's formal schooling was unsystematic, but he mastered French well enough to begin his lifelong study of its literature, and he thoroughly absorbed the ambiance of the Old World. From childhood on he was aware of the intricate network of institutions and traditions that he later lamented (in his study of Hawthorne and elsewhere) American novelists had to do without.

James early developed what he described in *A Small Boy and Others* (1913) as the "practice of wondering and dawdling and gaping." In that same memoir he also relates how he suffered the "obscure hurt" to his back which disqualified him from service in the Civil War and which must have helped to reinforce his inclination to be an observer rather than a participator. In his late teens his interest in literature and in writing intensified, and by the time he reached his majority he was publishing reviews and stories in some of the leading American journals—*Atlantic Monthly, North American Review, Galaxy,* and *Nation.* Though the crucial decision to establish his base of operations in England in 1876 remained to be made (after much shuttling back and forth between America and Europe, and after trial residence in France and Italy), the direction of James's single-minded career as man of letters was clearly marked in his early manhood. James never married. He maintained close ties with his family, kept up a large correspondence, was extremely sociable and a famous diner-out, knew most of his great contemporaries in the arts, many intimately—but he lived and worked alone. His emotional life and prodigious creative energy were invested for fifty years in what he called the "sacred rage" of his art.

Leon Edel, James's biographer, divides the writer's mature career into three parts. In the first, which culminated with *The Portrait of a Lady* (1881), he felt his way toward and appropriated the so-called international theme—the drama, comic and tragic, of Americans in Europe and Europeans in America. In the tripartite second period, he experimented with diverse themes and forms—first with novels dealing explicitly with strong social and political currents of the 1870s and 1880s, then with writing for the theater, and finally with shorter fictions that explore the relationship of artists to society and the troubled psychology of oppressed children and haunted or obsessed men and women. In James's last period—the so-called "major phase"—he returned to international or cosmopolitan subjects in an extraordinary series of elaborately developed novels, shorter fiction, and criticism.

Three of his earliest books—*A Passionate Pilgrim,* a collection of stories; *Transatlantic Sketches,* a collection of travel pieces; and *Roderick Hudson,* a novel—were all published in 1875. *The American* (1877) was his first successful and extended treatment of the naïve young American (Christopher Newman) from the New World in tension with the traditions, customs, and values of the Old. *Daisy Miller* (1878) was the work with which he first achieved widespread popularity. In this "sketch," as it was originally subtitled, the dangerously naïve young American girl (a subject to be treated often by James and his friend W. D. Howells) pays for her innocence of European social mores—and her willfulness—with her life. These stories make it clear that James was neither a chauvinist nor a resentful émigré, but a cosmopolitan whose concern was to explore the moral qualities of men and women forced to deal with the dilemmas of cultural displacement.

Despite their appeal, the characters of Daisy Miller and Christopher Newman are reductively simple and typecast, and this makes romance, melodrama, and pathos (whatever their charms) more likely than psychological complexity and genuine tragedy, which require, especially for James, a broad canvas. In the character and career of Isabel Archer—for which he drew on the tragically blighted life of his beloved cousin, Mary Temple, who died of tuberculosis in 1870 at the age of twenty-four—he found the focus for his first masterpiece on the international theme, *The Portrait of a Lady* (1881). Here, for the first time, the complex inner life of his characters—compounded of desire, will, thought, impulse—is fully and realistically projected. All the same, even in a relatively short work like *Daisy Miller*, James's essential themes and procedures are available.

From 1885 to 1890 James was largely occupied writing three novels in the naturalistic mode—*The Bostonians* (1886), *The Princess Casamassima* (1886), and *The Tragic Muse* (1889). James may have been right to put aside temporarily the "American-European legend" as subject, but he could not finally accept philosophic determinism of his characters' behavior or render his materials in documentary detail or depend for interest and effect on violent, physical action. These three novels have their virtues, but they are not the virtues of the mode as practiced by Zola, Norris, or Dreiser. For better *and* for worse, the English novelist Joseph Conrad observed, James was the "historian of fine consciences."

These stories of reformers, radicals, and revolutionaries, better appreciated in our own time than in his, alienated James's hard-won audience. Out of a sense of artistic challenge as well as financial need (he was never rich), James attempted to regain popularity and earn money by turning dramatist. Between 1890 and 1895 he wrote seven plays; two were produced, neither was a success. Humiliated by the boos and hooting of a hostile first-night crowd for *Guy Domville* (1895), James gave up the attempt to master this new form.

Between 1895 and 1900 James returned to fiction, especially to experiment in shorter works with three dominant subjects which he often combined: misunderstood or troubled writers and artists, ghosts and apparitions, and doomed or threatened children and adolescents. *The Real Thing* is an excellent example of a special kind of artistic dilemma that fascinated James, while *The Turn of the Screw* (1898), in which a whole household, including two young children, is terrorized by "ghosts," is the most powerful and famous of those stories in which, as James put it, "the strange and sinister is embroidered on the very type of the normal and the easy." Almost as well known, *The Beast in the Jungle* (1903) projects the pathetic career of a man who allows his obsessive imagination of personal disaster in the future to destroy his chances for love and life in the present. This

theme of the wasted life is played out once again in the tantalizingly autobiographical *The Jolly Corner* (1907).

Following his own advice to other novelists to "dramatize, dramatize,
dramatize," James increasingly removed himself as controlling narrator—became "invisible," in T. S. Eliot's phrase—from the reader's awareness. The
benefits of this heightened emphasis on showing rather than telling were
compression or intensification and enhanced opportunity for ambiguity. The
more the author withdrew, the more the reader was forced to enter the
process of creating meaning. We are accustomed now to having our fiction
thus "objectified," but it is James who is largely responsible for this development in narrative technique.

The Wings of the Dove (1902), *The Ambassadors* (1903), which James
thought was "the best 'all round'" of his productions, and *The Golden
Bowl* (1904) are demanding novels. The first two deal with subjects he had
treated earlier in *The American* and *The Portrait of a Lady*, and all three
concern themselves with James's grand theme of freedom through perception: only awareness of one's own character and others' provides the
wisdom to live well. The treatment of this theme in these books, however,
is characterized by richness of syntax, characterization, point of view, symbolic resonance, metaphoric texture, and organizing rhythms. The world of
these novels is, as a critic has remarked, like the very atmosphere of the
mind. These dramas of perception are widely considered to be James's
most influential contribution to the craft of fiction.

When James was not writing fiction, he was most often writing about it
—either his own or others. He was, as he noted in one of his letters, "a critical, a *non-naif*, a questioning, worrying reader." If he was somewhat narrow
in his reading—restricting himself chiefly to nineteenth-century fiction—he
made his limited experience count for as much in criticism as it did in
fiction. His inquiries into the achievement of other writers—preserved in
such volumes as *French Poets and Novelists* (1878), *Partial Portraits*
(1888), and *Notes on Novelists* (1914)—are remarkable for their breadth,
balance, and acuteness. His taste and judgments have been largely confirmed
by time.

More broadly philosophic than the reviews or even the essays on individual writers, *The Art of the Novel* (1884) fairly represents James's central
aesthetic conceptions. Calling attention to the unparalleled opportunities
open to the artist of fiction and the beauty of the novel which creates a
new form, James also insists that "the deepest quality of a work of art will
always be the quality of the mind of the producer" and that "no good
novel will ever proceed from a superficial mind." James left no better
record than this essay of his always twinned concerns over the moral and
formal qualities of fiction, of the relationship between aesthetic and moral
perception.

James was an extremely self-conscious writer, and his *Notebooks* (published in 1947) reveal a subtle, intense mind in the act of discovering subjects, methods, and principles. The prefaces he wrote for the definitive New
York edition of his extensively revised novels and tales (gathered and published in 1934 as *The Art of the Novel*) contain James's final study of the
works that he considered best represented his achievement. The culmination of an entire lifetime of reflection on the craft of fiction, they provide

extraordinary accounts of the origins and growth of his major writings and exquisite analyses of the fictional problems that each work posed. Despite their occasional opacity, they also serve, as he wrote Howells he hoped they would, as a "sort of comprehensive manual or *vade-mecum* for aspirants in our arduous profession." These prefaces provided both vocabulary and example for the close textual analysis of prose fiction in the New Criticism that was dominant in America in the generation following World War II.

While James was in the United States arranging for Scribner's New York edition of his novels and tales, he also took the occasion to travel extensively and to lecture in his native land and Canada. The chief fruit of this experience was *The American Scene* (1907), which carried the art of travel writing to the same sophisticated level he had carried the art of fiction in *The Ambassadors*, *The Wings of the Dove*, and *The Golden Bowl*. This "absolutely personal" book is perhaps the most vividly particular account we have of the vast and profound changes that occurred in America between the Civil War and World War I, the period James later characterized as the "Age of the Mistake."

The same intricate, ruminative richness marks the three autobiographical reminiscences he wrote late in life: *A Small Boy and Others* (1913), *Notes of a Son and Brother* (1914), and the fragmentary and posthumously published *The Middle Years* (1917). Henry James died in 1916, a year after he became a naturalized British subject out of impatience with America's reluctance to enter World War I.

At one time or another James has been characterized as a snob, a deserter from his native land, an old maid, a mere aesthete; his fiction has been deprecated as narrowly concerned with the rich, the bloodless, and the sexless, as needlessly elaborate and long-winded, and as excessively introspective and autobiographical. But James, who always believed in his own genius, has been vindicated because he understood the mixed nature of men and women profoundly, because he judged them humanely, and because he gave enduring and compelling shape to his sense of life.

The Real Thing[1]

I

When the porter's wife, who used to answer the house-bell, announced "A gentleman and a lady, sir" I had, as I often had in those days—the wish being father to the thought—an immediate vision of sitters. Sitters my visitors in this case proved to be; but not in the sense I should have preferred. There was nothing at first however to indicate that they mightn't have come for a portrait.

---

1. This "little gem of bright, quick, vivid form," as James called it, first appeared in *Black and White* on April 16, 1892, in *The Real Thing and Other Tales* (1893), and finally in Vol. XVIII (1909) of the New York edition, the source of the present text.

The gentleman, a man of fifty, very high and very straight, with a moustache slightly grizzled and a dark grey walking-coat admirably fitted, both of which I noted professionally—I don't mean as a barber or yet as a tailor—would have struck me as a celebrity if celebrities often were striking. It was a truth of which I had for some time been conscious that a figure with a good deal of frontage was, as one might say, almost never a public institution. A glance at the lady helped to remind me of this paradoxical law: she also looked too distinguished to be a "personality." Moreover one would scarcely come across two variations together.

Neither of the pair immediately spoke—they only prolonged the preliminary gaze suggesting that each wished to give the other a chance. They were visibly shy; they stood there letting me take them in—which, as I afterwards perceived, was the most practical thing they could have done. In this way their embarrassment served their cause. I had seen people painfully reluctant to mention that they desired anything so gross as to be represented on canvas; but the scruples of my new friends appeared almost insurmountable. Yet the gentleman might have said "I should like a portrait of my wife," and the lady might have said "I should like a portrait of my husband." Perhaps they weren't husband and wife—this naturally would make the matter more delicate. Perhaps they wished to be done together—in which case they ought to have brought a third person to break the news.

"We come from Mr. Rivet," the lady finally said with a dim smile that had the effect of a moist sponge passed over a "sunk"[2] piece of painting, as well as of a vague allusion to vanished beauty. She was as tall and straight, in her degree, as her companion, and with ten years less to carry. She looked as sad as a woman could look whose face was not charged with expression; that is her tinted oval mask showed waste as an exposed surface shows friction. The hand of time had played over her freely, but to an effect of elimination. She was slim and stiff, and so well-dressed, in dark blue cloth, with lappets and pockets and buttons, that it was clear she employed the same tailor as her husband. The couple had an indefinable air of prosperous thrift—they evidently got a good deal of luxury for their money. If I was to be one of their luxuries it would behove me to consider my terms.

"Ah Claude Rivet recommended me?" I echoed; and I added that it was very kind of him, though I could reflect that, as he only painted landscape, this wasn't a sacrifice.

The lady looked very hard at the gentleman, and the gentleman looked round the room. Then staring at the floor a moment and

2. When colors lose their brilliance after they have dried on the canvas, they are said to have "sunk in."

stroking his moustache, he rested his pleasant eyes on me with the remark: "He said you were the right one."

"I try to be, when people want to sit.

"Yes, we should like to," said the lady anxiously.

"Do you mean together?"

My visitors exchanged a glance. "If you could do anything with *me* I suppose it would be double," the gentleman stammered.

"Oh yes, there's naturally a higher charge for two figures than for one."

"We should like to make it pay," the husband confessed.

"That's very good of you," I returned, appreciating so unwonted a sympathy—for I supposed he meant pay the artist.

A sense of strangeness seemed to dawn on the lady.

"We mean for the illustrations—Mr. Rivet said you might put one in."

"Put in—an illustration?" I was equally confused.

"Sketch her off, you know," said the gentleman, colouring.

It was only then that I understood the service Claude Rivet had rendered me; he had told them how I worked in black-and-white, for magazines, for storybooks, for sketches of contemporary life, and consequently had copious employment for models. These things were true, but it was not less true—I may confess it now; whether because the aspiration was to lead to everything or to nothing I leave the reader to guess—that I couldn't get the honours, to say nothing of the emoluments, of a great painter of portraits out of my head. My "illustrations" were my pot-boilers; I looked to a different branch of art—far and away the most interesting it had always seemed to me—to perpetuate my fame. There was no shame in looking to it also to make my fortune; but that fortune was by so much further from being made from the moment my visitors wished to be "done" for nothing. I was disappointed; for in the pictorial sense I had immediately *seen* them. I had seized their type —I had already settled what I would do with it. Something that wouldn't absolutely have pleased them, I afterwards reflected.

"Ah you're—you're—a—?" I began as soon as I had mastered my surprise. I couldn't bring out the dingy word "models": it seemed so little to fit the case.

"We haven't had much practice," said the lady.

"We've got to *do* something, and we've thought that an artist in your line might perhaps make something of us," her husband threw off. He further mentioned that they didn't know many artists and that they had gone first, on the off-chance—he painted views of course, but sometimes put in figures; perhaps I remembered—to Mr. Rivet, whom they had met a few years before at a place in Norfolk where he was sketching.

"We used to sketch a little ourselves," the lady hinted.

"It's very awkward, but we absolutely *must* do something," her husband went on.

"Of course we're not so *very* young," she admitted with a wan smile.

With the remark that I might as well know something more about them the husband had handed me a card extracted from a neat new pocket-book—their appurtenances were all of the freshest —and inscribed with the words "Major Monarch." Impressive as these words were they didn't carry my knowledge much further; but my visitor presently added: "I've left the army and we've had the misfortune to lose our money. In fact our means are dreadfully small."

"It's awfully trying—a regular strain," said Mrs. Monarch.

They evidently wished to be discreet—to take care not to swagger because they were gentlefolk. I felt them willing to recognise this as something of a drawback, at the same time that I guessed at an underlying sense—their consolation in adversity—that they *had* their points. They certainly had; but these advantages struck me as preponderantly social; such for instance as would help to make a drawing-room look well. However, a drawing-room was always, or ought to be, a picture.

In consequence of his wife's allusion to their age Major Monarch observed: "Naturally it's more for the figure that we thought of going in. We can still hold ourselves up." On the instant I saw that the figure was indeed their strong point. His "naturally" didn't sound vain, but it lighted up the question. "*She* has the best one," he continued, nodding at his wife with a pleasant after-dinner absence of circumlocution. I could only reply, as if we were in fact sitting over our wine, that this didn't prevent his own from being very good; which led him in turn to make answer: "We thought that if you ever have to do people like us we might be something like it. *She* particularly—for a lady in a book, you know."

I was so amused by them that, to get more of it, I did my best to take their point of view; and though it was an embarrassment to find myself appraising physically, as if they were animals on hire or useful blacks, a pair whom I should have expected to meet only in one of the relations in which criticism is tacit, I looked at Mrs. Monarch judicially enough to be able to exclaim after a moment with conviction: "Oh yes, a lady in a book!" She was singularly like a bad illustration.

"We'll stand up, if you like," said the Major; and he raised himself before me with a really grand air.

I could take his measure at a glance—he was six feet two and a perfect gentleman. It would have paid any club in process of formation and in want of a stamp to engage him at a salary to stand in the principal window. What struck me at once was that in coming to me they had rather missed their vocation; they could surely have been turned to better account for advertising purposes. I couldn't of course see the thing in detail, but I could see them make somebody's fortune—I don't mean their own. There was something in them for a waistcoat-maker, an hotel-keeper or a soap-vendor. I could imagine "We always use it" pinned on their bosoms with the greatest effect; I had a vision of the brilliancy with which they would launch a table d'hôte.[3]

Mrs. Monarch sat still, not from pride but from shyness, and presently her husband said to her: "Get up, my dear, and show how smart you are." She obeyed, but she had no need to get up to show it. She walked to the end of the studio and then came back blushing, her fluttered eyes on the partner of her appeal. I was reminded of an incident I had accidentally had a glimpse of in Paris being with a friend there, a dramatist about to produce a play, when an actress came to him to ask to be entrusted with a part. She went through her paces before him, walked up and down as Mrs. Monarch was doing. Mrs. Monarch did it quite as well, but I abstained from applauding. It was very odd to see such people apply for such poor pay. She looked as if she had ten thousand a year. Her husband had used the word that described her: she was in the London current jargon essentially and typically "smart." Her figure was, in the same order of ideas, conspicuously and irreproachably "good." For a woman of her age her waist was surprisingly small; her elbow moreover had the orthodox crook. She held her head at the conventional angle, but why did she come to *me*? She ought to have tried on jackets at a big shop. I feared my visitors were not only destitute but "artistic"—which would be a great complication. When she sat down again I thanked her, observing that what a draughtsman most valued in his model was the faculty of keeping quiet.

"Oh *she* can keep quiet," said Major Monarch. Then he added jocosely: "I've always kept her quiet."

"I'm not a nasty fidget, am I?" It was going to wring tears from me, I felt, the way she hid her head, ostrich-like, in the other broad bosom.

The owner of this expanse addressed his answer to me. "Perhaps it isn't out of place to mention—because we ought to be quite business-like, oughtn't we?—that when I married her she was known as the Beautiful Statue."

"Oh dear!" said Mrs. Monarch ruefully.

"Of course I should want a certain amount of expression," I rejoined.

"Of *course*!"—and I had never heard such unanimity.

"And then I suppose you know that you'll get awfully tired."

"Oh we *never* get tired!" they eagerly cried.

"Have you had any kind of practice?"

3. A common table for guests at a hotel.

They hesitated—they looked at each other. "We've been photographed—*immensely*," said Mrs. Monarch.

"She means the fellows have asked us themselves," added the
Major.

"I see—because you're so good-looking."

"I don't know what they thought, but they were always after us."

"We always got our photographs for nothing," smiled Mrs. Monarch.

"We might have brought some, my dear," her husband
remarked.

"I'm not sure we have any left. We've given quantities away,"
she explained to me.

"With our autographs and that sort of thing," said the Major.

"Are they to be got in the shops?" I enquired as a harmless pleasantry.

"Oh yes, *hers*—they used to be."

"Not now," said Mrs. Monarch with her eyes on the floor.

II

I could fancy the "sort of thing" they put on the presentation
copies of their photographs, and I was sure they wrote a beautiful
hand. It was odd how quickly I was sure of everything that concerned them. If they were now so poor as to have to earn shillings
and pence they could never have had much of a margin. Their good
looks had been their capital, and they had good-humouredly made
the most of the career that this resource marked out for them. It
was in their faces, the blankness, the deep intellectual repose of the
twenty years of country-house visiting that had given them pleasant
intonations. I could see the sunny drawing-rooms, sprinkled with
periodicals she didn't read, in which Mrs. Monarch had continuously sat; I could see the wet shrubberies in which she had walked,
equipped to admiration for either exercise. I could see the rich
covers[4] the Major had helped to shoot and the wonderful garments
in which, late at night, he repaired to the smoking-room to talk
about them. I could imagine their leggings and waterproofs, their
knowing tweeds and rugs, their rolls of sticks and cases of tackle
and neat umbrellas; and I could evoke the exact appearance of their
servants and the compact variety of their luggage on the platforms of
country stations.

They gave small tips, but they were liked; they didn't do anything themselves, but they were welcome. They looked so well
everywhere; they gratified the general relish for stature, complexion
and "form." They knew it without fatuity or vulgarity, and they
respected themselves in consequence. They weren't superficial; they
were thorough and kept themselves up—it had been their line.
People with such a taste for activity had to have some line. I could

4. Magazine covers.

feel how even in a dull house they could have been counted on for
the joy of life. At present something had happened—it didn't
matter what, their little income had grown less, it had grown least
—and they had to do something for pocket-money. Their friends
could like them, I made out, without liking to support them. There
was something about them that represented credit—their clothes,
their manners, their type; but if credit is a large empty pocket in
which an occasional chink reverberates, the chink at least must be
audible. What they wanted of me was to help to make it so. Fortunately they had no children—I soon divined that. They would also
perhaps wish our relations to be kept secret: this was why it was
"for the figure"—the reproduction of the face would betray them.

I liked them—I felt, quite as their friends must have done—they
were so simple; and I had no objection to them if they would suit.
But somehow with all their perfections I didn't easily believe in
them. After all they were amateurs, and the ruling passion of my
life was the detestation of the amateur. Combined with this was
another perversity—an innate preference for the represented subject
over the real one: the defect of the real one was so apt to be a lack
of representation. I liked things that appeared; then one was sure.
Whether they *were* or not was a subordinate and almost always a
profitless question. There were other considerations, the first of
which was that I already had two or three recruits in use, notably a
young person with big feet, in alpaca, from Kilburn, who for a
couple of years had come to me regularly for my illustrations and
with whom I was still—perhaps ignobly—satisfied. I frankly
explained to my visitors how the case stood, but they had taken
more precautions than I supposed. They had reasoned out their
opportunity, for Claude Rivet had told them of the projected
édition de luxe of one of the writers of our day—the rarest of the
novelists—who, long neglected by the multitudinous vulgar and
dearly prized by the attentive (need I mention Philip Vincent?)[5]
had had the happy fortune of seeing, late in life, the dawn and then
the full light of a higher criticism; an estimate in which on the part
of the public there was something really of expiation. The edition
preparing, planned by a publisher of taste, was practically an act of
high reparation; the wood-cuts with which it was to be enriched
were the homage of English art to one of the most independent
representatives of English letters. Major and Mrs. Monarch confessed to me they had hoped I might be able to work *them* into my
branch of the enterprise. They knew I was to do the first of the
books, "Rutland Ramsay," but I had to make clear to them that my
participation in the rest of the affair—this first book was to be a
test—must depend on the satisfaction I should give. If this should
be limited my employers would drop me with scarce common

5. Obviously James, here indulging in some good-natured complaining and fan-
tasizing.

forms. It was therefore a crisis for me, and naturally I was making special preparations, looking about for new people, should they be necessary, and securing the best types. I admitted however that I should like to settle down to two or three good models who would do for everything.

"Should we have often to—a—put on special clothes?" Mrs. Monarch timidly demanded.

"Dear yes—that's half the business."

"And should we be expected to supply our own costumes?"

"Oh no; I've got a lot of things. A painter's models put on—or put off—anything he likes."

"And you mean—a—the same?"

"The same?"

Mrs. Monarch looked at her husband again.

"Oh she was just wondering," he explained, "if the costumes are in *general* use." I had to confess that they were, and I mentioned further that some of them—I had a lot of genuine greasy last-century things—had served their time, a hundred years ago, on living world-stained men and women; on figures not perhaps so far removed, in that vanished world, from *their* type, the Monarchs', *quoi!*[6] of a breeched and bewigged age. "We'll put on anything that *fits*," said the Major.

"Oh I arrange that—they fit in the pictures."

"I'm afraid I should do better for the modern books. I'd come as you like," said Mrs. Monarch.

"She has got a lot of clothes at home: they might do for contemporary life," her husband continued.

"Oh I can fancy scenes in which you'd be quite natural." And indeed I could see the slipshod rearrangements of stale properties— the stories I tried to produce pictures for without the exasperation of reading them—whose sandy tracts the good lady might help to people. But I had to return to the fact that for this sort of work— the daily mechanical grind—I was already equipped: the people I was working with were fully adequate.

"We only thought we might be more like *some* characters," said Mrs. Monarch mildly, getting up.

Her husband also rose; he stood looking at me with a dim wistfulness that was touching in so fine a man.

"Wouldn't it be rather a pull sometimes to have—a—to have—?" He hung fire; he wanted me to help him by phrasing what he meant. But I couldn't—I didn't know. So he brought it out awkwardly: "The *real* thing; a gentleman, you know, or a lady." I was quite ready to give a general assent—I admitted that there was a great deal in that. This encouraged Major Monarch to say, following up his appeal with an unacted gulp: "It's awfully hard—we've tried everything." The gulp was communicative; it proved too much

6. What!

for his wife. Before I knew it Mrs. Monarch had dropped again upon a divan and burst into tears. Her husband sat down beside her, holding one of her hands; whereupon she quickly dried her eyes with the other, while I felt embarrassed as she looked up at me. "There isn't a confounded job I haven't applied for—waited for— prayed for. You can fancy we'd be pretty bad first. Secretaryships and that sort of thing? You might as well ask for a peerage. I'd be *anything*—I'm strong; a messenger or a coalheaver. I'd put on a gold-laced cap and open carriage-doors in front of the haberdasher's; I'd hang about a station to carry portmanteaux; I'd be a postman. But they won't *look* at you; there are thousands as good as yourself already on the ground. *Gentlemen*, poor beggars, who've drunk their wine, who've kept their hunters!"

I was as reassuring as I knew how to be, and my visitors were presently on their feet again while, for the experiment, we agreed on an hour. We were discussing it when the door opened and Miss Churm came in with a wet umbrella. Miss Churm had to take the omnibus to Maida Vale and then walk half a mile. She looked a trifle blowsy and slightly splashed. I scarcely ever saw her come in without thinking fresh how odd it was that, being so little in her- self, she should yet be so much in others. She was a meagre little Miss Churm, but was such an ample heroine of romance. She was only a freckled cockney,[7] but she could represent everything, from a fine lady to a shepherdess; she had the faculty as she might have had a fine voice or long hair. She couldn't spell and she loved beer, but she had two or three "points," and practice, and a knack, and mother-wit, and a whimsical sensibility, and a love of the theatre, and seven sisters, and not an ounce of respect, especially for the *h*. The first thing my visitors saw was that her umbrella was wet, and in their spotless perfection they visibly winced at it. The rain had come on since their arrival.

"I'm all in a soak; there *was* a mess of people in the 'bus. I wish you lived near a stytion," said Miss Churm. I requested her to get ready as quickly as possible, and she passed into the room in which she always changed her dress. But before going out she asked me what she was to get into this time.

"It's the Russian princess, don't you know?" I answered; "the one with the 'golden eyes,' in black velvet, for the long thing in the *Cheapside*."[8]

"Golden eyes? I *say!*" cried Miss Churm, while my companions watched her with intensity as she withdrew. She always arranged herself, when she was late, before I could turn around; and I kept my visitors a little on purpose, so that they might get an idea, from seeing her, what would be expected of themselves. I mentioned that

7. Native of London, especially the East End. The cockney dialect is known for dropping *h*'s; for example, *hair* would be pronounced *air*.
8. Imaginary magazine named after a main business street in London.

she was quite my notion of an excellent model—she was really very clever.

"Do you think she looks like a Russian princess?" Major Monarch asked with lurking alarm.

"When I make her, yes."

"Oh if you have to *make* her—!" he reasoned, not without point.

"That's the most you can ask. There are so many who are not makeable."

"Well now, *here's* a lady"—and with a persuasive smile he passed his arm into his wife's—"who's already made!"

"Oh I'm not a Russian princess," Mrs. Monarch protested a little coldly. I could see she had known some and didn't like them. There at once was a complication of a kind I never had to fear with Miss Churm.

This young lady came back in black velvet—the gown was rather rusty and very low on her lean shoulders—and with a Japanese fan in her red hands. I reminded her that in the scene I was doing she had to look over some one's head. "I forget whose it is; but it doesn't matter. Just look over a head."

"I'd rather look over a stove," said Miss Churm; and she took her station near the fire. She fell into position, settled herself into a tall attitude, gave a certain backward inclination to her head and a certain forward droop to her fan, and looked, at least to my prejudiced sense, distinguished and charming, foreign and dangerous. We left her looking so while I went downstairs with Major and Mrs. Monarch.

"I believe I could come about as near it as that," said Mrs. Monarch.

"Oh, you think she's shabby, but you must allow for the alchemy of art."

However, they went off with an evident increase of comfort founded on their demonstrable advantage in being the real thing. I could fancy them shuddering over Miss Churm. She was very droll about them when I went back, for I told her what they wanted.

"Well, if *she* can sit I'll tyke to bookkeeping," said my model.

"She's very ladylike," I replied as an innocent form of aggravation.

"So much the worse for *you*. That means she can't turn round."

"She'll do for the fashionable novels."

"Oh yes, she'll *do* for them!" my model humorously declared. "Ain't they bad enough without her?" I had often sociably denounced them to Miss Churm.

III

It was for the elucidation of a mystery in one of these works that I first tried Mrs. Monarch. Her husband came with her, to be useful if necessary—it was sufficiently clear that as a general thing he would prefer to come with her. At first I wondered if this were for "propriety's" sake—if he were going to be jealous and meddling. The idea was too tiresome, and if it had been confirmed it would speedily have brought our acquaintance to a close. But I soon saw there was nothing in it and that if he accompanied Mrs. Monarch it was—in addition to the chance of being wanted—simply because he had nothing else to do. When they were separate his occupation was gone and they never *had* been separate. I judged rightly that in their awkward situation their close union was their main comfort and that this union had no weak spot. It was a real marriage, an encouragement to the hesitating, a nut for pessimists to crack. Their address was humble—I remember afterwards thinking it had been the only thing about them that was really professional—and I could fancy the lamentable lodgings in which the Major would have been left alone. He could sit there more or less grimly with his wife—he couldn't sit there anyhow without her.

He had too much tact to try and make himself agreeable when he couldn't be useful; so when I was too absorbed in my work to talk he simply sat and waited. But I liked to hear him talk—it made my work, when not interrupting it, less mechanical, less special. To listen to him was to combine the excitement of going out with the economy of staying at home. There was only one hindrance—that I seemed not to know any of the people this brilliant couple had known. I think he wondered extremely, during the term of our intercourse, whom the deuce I *did* know. He hadn't a stray sixpence of an idea to fumble for, so we didn't spin it very fine; we confined ourselves to questions of leather and even of liquor—saddlers and breeches-makers and how to get excellent claret cheap—and matters like "good trains" and the habits of small game. His lore on these last subjects was astonishing—he managed to interweave the station-master with the ornithologist. When he couldn't talk about greater things he could talk cheerfully about smaller, and since I couldn't accompany him into reminiscences of the fashionable world he could lower the conversation without a visible effort to my level.

So earnest a desire to please was touching in a man who could so easily have knocked one down. He looked after the fire and had an opinion on the draught of the stove without my asking him, and I could see that he thought many of my arrangements not half knowing. I remember telling him that if I were only rich I'd offer him a salary to come and teach me how to live. Sometimes he gave a random sigh of which the essence might have been: "Give me even such a bare old barrack as *this*, and I'd do something with it!" When I wanted to use him he came alone; which was an illustration of the superior courage of women. His wife could bear her solitary second floor, and she was in general more discreet; showing by various small reserves that she was alive to the propriety of keeping

our relations markedly professional—not letting them slide into
sociability. She wished it to remain clear that she and the Major
were employed, not cultivated, and if she approved of me as a superior, who could be kept in his place, she never thought me quite
good enough for an equal.

She sat with great intensity, giving the whole of her mind to it,
and was capable of remaining for an hour almost as motionless as
before a photographer's lens. I could see she had been photographed often, but somehow the very habit that made her good for
that purpose unfitted her for mine. At first I was extremely pleased
with her ladylike air, and it was a satisfaction, on coming to follow
her lines, to see how good they were and how far they could lead
the pencil. But after a little skirmishing I began to find her too
insurmountably stiff; do what I would with it my drawing looked
like a photograph or a copy of a photograph. Her figure had no variety of expression—she herself had no sense of variety. You may say
that this was my business and was only a question of placing her.
Yet I placed her in every conceivable position and she managed to
obliterate their differences. She was always a lady certainly, and into
the bargain was always the same lady. She was the real thing, but
always the same thing. There were moments when I rather writhed
under the serenity of her confidence that she *was* the real thing. All
her dealings with me and all her husband's were an implication that
this was lucky for *me*. Meanwhile I found myself trying to invent
types that approached her own, instead of making her own transform itself—in the clever way that was not impossible for instance
to poor Miss Churm. Arrange as I would and take the precautions I
would, she always came out, in my pictures, too tall—landing me in
the dilemma of having represented a fascinating woman as seven
feet high, which (out of respect perhaps to my own very much
scantier inches) was far from my idea of such personage.

The case was worse with the Major—nothing I could do would
keep *him* down, so that he became useful only for the representation of brawny giants. I adored variety and range, I cherished
human accidents, the illustrative note; I wanted to characterise
closely, and the thing in the world I most hated was the danger of
being ridden by a type. I had quarrelled with some of my friends
about it; I had parted company with them for maintaining that one
had to be, and that if the type was beautiful—witness Raphael[9]
and Leonardo—the servitude was only a gain. I was neither Leonardo nor Raphael—I might only be a presumptuous young modern
searcher; but I held that everything was to be sacrificed sooner than
character. When they claimed that the obsessional form could
easily *be* character I retorted, perhaps superficially, "Whose?" It

9. Raffaello Santi or Sanzio (1483–1520), Italian painter; Leonardo da Vinci (1452–
1519), Florentine painter, sculptor, architect, and engineer.

couldn't be everybody's—it might end in being nobody's.

After I had drawn Mrs. Monarch a dozen times I felt surer even than before that the value of such a model as Miss Churm resided precisely in the fact that she had no positive stamp, combined of course with the other fact that what she did have was a curious and inexplicable talent for imitation. Her usual appearance was like a curtain which she could draw up at request for a capital perform-ance. This performance was simply suggestive; but it was a word to the wise—it was vivid and pretty. Sometimes even I thought it, though she was plain herself, too insipidly pretty; I made it a reproach to her that the figures drawn from her were monotonously (*bêtement*,[1] as we used to say) graceful. Nothing made her more angry: it was so much her pride to feel she could sit for characters that had nothing in common with each other. She would accuse me at such moments of taking away her "reputytion."

It suffered a certain shrinkage, this queer quantity, from the repeated visits of my new friends. Miss Churm was greatly in demand, never in want of employment, so I had no scruple in put-ting her off occasionally, to try them more at my ease. It was cer-tainly amusing at first to do the real thing—it was amusing to do Major Monarch's trousers. They *were* the real thing, even if he did come out colossal. It was amusing to do his wife's back hair—it was so mathematically neat—and the particular "smart" tension of her tight stays. She lent herself especially to positions in which the face was somewhat averted or blurred; she abounded in ladylike back views and *profils perdus*.[2] When she stood erect she took naturally one of the attitudes in which court-painters represent queens and princesses; so that I found myself wondering whether, to draw out this accomplishment, I couldn't get the editor of the *Cheapside* to publish a really royal romance, "A Tale of Buckingham Palace." Sometimes however the real thing and the make-believe came into contact; by which I mean that Miss Churm, keeping an appoint-ment or coming to make one on days when I had much work in hand, encountered her invidious rivals. The encounter was not on their part, for they noticed her no more than if she had been the housemaid; not from intentional loftiness, but simply because as yet, professionally, they didn't know how to fraternise, as I could imagine they would have liked—or at least that the Major would. They couldn't talk about the omnibus—they always walked; and they didn't know what else to try—she wasn't interested in good trains or cheap claret. Besides, they must have felt—in the air—that she was amused at them, secretly derisive of their ever knowing how. She wasn't a person to conceal the limits of her faith if she had had a chance to show them. On the other hand Mrs. Monarch didn't think her tidy; for why else did she take pains to say to me

1. Foolishly.   2. Averted glances.

—it was going out of the way, for Mrs. Monarch—that she didn't like dirty women?

One day when my young lady happened to be present with my other sitters—she even dropped in, when it was convenient, for a chat—I asked her to be so good as to lend a hand in getting tea, a service with which she was familiar and which was one of a class that, living as I did in a small way, with slender domestic resources, I often appealed to my models to render. They liked to lay hands on my property, to break the sitting, and sometimes the china—it made them feel Bohemian. The next time I saw Miss Churm after this incident she surprised me greatly by making a scene about it— she accused me of having wished to humiliate her. She hadn't resented the outrage at the time, but had seemed obliging and amused, enjoying the comedy of asking Mrs. Monarch, who sat vague and silent, whether she would have cream and sugar, and putting an exaggerated simper into the question. She had tried intonations—as if she too wished to pass for the real thing—till I was afraid my other visitors would take offence.

Oh they were determined not to do this, and their touching patience was the measure of their great need. They would sit by the hour, uncomplaining, till I was ready to use them; they would come back on the chance of being wanted and would walk away cheerfully if it failed. I used to go to the door with them to see in what magnificent order they retreated. I tried to find other employment for them—I introduced them to several artists. But they didn't "take," for reasons I could appreciate, and I became rather anxiously aware that after such disappointments they fell back upon me with a heavier weight. They did me the honor to think me most *their* form. They weren't romantic enough for the painters, and in those days there were few serious workers in black-and-white. Besides, they had an eye to the great job I had mentioned to them—they had secretly set their hearts on supplying the right essence for my pictorial vindication of our fine novelist. They knew that for this undertaking I should want no costume-effects, none of the frippery of past ages—that it was a case in which everything would be contemporary and satirical and presumably genteel. If I could work them into it their future would be assured, for the labour would of course be long and the occupation steady.

One day Mrs. Monarch came without her husband—she explained his absence by his having had to go to the City.[3] While she sat there in her usual relaxed majesty there came at the door a knock which I immediately recognised as the subdued appeal of a model out of work. It was followed by the entrance of a young man whom I at once saw to be a foreigner and who proved in fact an Italian acquainted with no English word but my name, which he uttered in a way that made it seem to include all others. I hadn't

3. London.

then visited his country, nor was I proficient in his tongue; but as he was not so meanly constituted—what Italian is?—as to depend only on that member for expression he conveyed to me, in familiar but graceful mimicry, that he was in search of exactly the employment in which the lady before me was engaged. I was not struck with him at first, and while I continued to draw I dropped few signs of interest or encouragement. He stood his ground however—not importunately, but with a dumb dog-like fidelity in his eyes that amounted to innocent impudence, the manner of a devoted servant —he might have been in the house for years—unjustly suspected. Suddenly it struck me that this very attitude and expression made a picture; whereupon I told him to sit down and wait till I should be free. There was another picture in the way he obeyed me, and I observed as I worked that there were others still in the way he looked wonderingly, with his head thrown back, about the high studio. He might have been crossing himself in Saint Peter's. Before I finished I said to myself "The fellow's a bankrupt orange-monger, but a treasure."

When Mrs. Monarch withdrew he passed across the room like a flash to open the door for her, standing there with the rapt pure gaze of the young Dante spellbound by the young Beatrice.[4] As I never insisted, in such situations, on the blankness of the British domestic, I reflected that he had the making of a servant—and I needed one, but couldn't pay him to be only that—as well as of a model; in short I resolved to adopt my bright adventurer if he would agree to officiate in the double capacity. He jumped at my offer, and in the event my rashness—for I had really known nothing about him—wasn't brought home to me. He proved a sympathetic though a desultory ministrant, and had in a wonderful degree the *sentiment de la pose*.[5] It was uncultivated, instinctive, a part of the happy instinct that had guided him to my door and helped him to spell out my name on the card nailed to it. He had had no other introduction to me than a guess, from the shape of my high north window, seen outside, that my place was a studio and that as a studio it would contain an artist. He had wandered to England in search of fortune, like other itinerants, and had embarked, with a partner and a small green hand-cart, on the sale of penny ices. The ices had melted away and the partner had dissolved in their train. My young man wore tight yellow trousers with reddish stripes and his name was Oronte. He was sallow but fair, and when I put him into some old clothes of my own he looked like an Englishman. He was as good as Miss Churm, who could look, when requested, like an Italian.

4. Dante Alighieri (1265–1321), Italian poet, first saw Beatrice Portinari when they were both nine. Though he only saw her a few times, she made a lasting impression on him and became his ideal, his life's inspiration, and direct agent of his salvation (as his greatest work, *The Divine Comedy,* makes clear).
5. Instinct for striking poses.

IV

I thought Mrs. Monarch's face slightly convulsed when, on her coming back with her husband, she found Oronte installed. It was strange to have to recognise in a scrap of a lazzarone[6] a competitor to her magnificent Major. It was she who scented danger first, for the Major was anecdotically unconscious. But Oronte gave us tea, with a hundred eager confusions—he had never been concerned in so queer a process—and I think she thought better of me for having at last an "establishment." They saw a couple of drawings that I had made of the establishment, and Mrs. Monarch hinted that it never would have struck her he had sat for them. "Now the drawings you make from *us*, they look exactly like us," she reminded me, smiling in triumph; and I recognized that this was indeed just their defect. When I drew the Monarchs I couldn't anyhow get away from them—get into the character I wanted to represent; and I hadn't the least desire my model should be discoverable in my picture. Miss Churm never was, and Mrs. Monarch thought I hid her, very properly, because she was vulgar; whereas if she was lost it was only as the dead who go to heaven are lost—in the gain of an angel the more.

By this time I had got a certain start with "Rutland Ramsay," the first novel in the great projected series; that is I had produced a dozen drawings, several with the help of the Major and his wife, and I had sent them in for approval. My understanding with the publishers, as I have already hinted, had been that I was to be left to do my work, in this particular case, as I liked, with the whole book committed to me; but my connexion with the rest of the series was only contingent. There were moments when, frankly, it *was* a comfort to have the real thing under one's hand; for there were characters in "Rutland Ramsay" that were very much like it. There were people presumably as erect as the Major and women of as good a fashion as Mrs. Monarch. There was a great deal of country-house life—treated, it is true, in a fine fanciful ironical generalised way—and there was a considerable implication of knickerbockers and kilts.[7] There were certain things I had to settle at the outset; such things for instance as the exact appearance of the hero and the particular bloom and figure of the heroine. The author of course gave me a lead, but there was a margin for interpretation. I took the Monarchs into my confidence, I told them frankly what I was about, I mentioned my embarrassments and alternatives. "Oh take *him!*" Mrs. Monarch murmured sweetly, looking at her husband; and "What could you want better than my wife?" the Major enquired with the comfortable candour that now prevailed between us.

6. Beggar.
7. Knickerbockers (close-fitting short pants gathered at the knee) and kilts (a knee-length pleated skirt, usually of tartan, worn by Scottish men) are suggestive of rural, outdoors attire.

I wasn't obliged to answer these remarks—I was only obliged to place my sitters. I wasn't easy in mind, and I postponed a little timidly perhaps the solving of my question. The book was a large canvas, the other figures were numerous, and I worked off at first some of the episodes in which the hero and the heroine were not concerned. When once I had set *them* up I should have to stick to them—I couldn't make my young man seven feet high in one place and five feet nine in another. I inclined on the whole to the latter measurement, though the Major more than once reminded me that *he* looked about as young as any one. It was indeed quite possible to arrange him, for the figure, so that it would have been difficult to detect his age. After the spontaneous Oronte had been with me a month, and after I had given him to understand several times over that his native exuberance would presently constitute an insurmountable barrier to our further intercourse, I waked to a sense of his heroic capacity. He was only five feet seven, but the remaining inches were latent. I tried him almost secretly at first, for I was really rather afraid of the judgment my other models would pass on such a choice. If they regarded Miss Churm as little better than a snare what would they think of the representation by a person so little the real thing as an Italian street-vendor of a protagonist formed by a public school?

If I went a little in fear of them it wasn't because they bullied me, because they had got an oppressive foothold, but because in their really pathetic decorum and mysteriously permanent newness they counted on me so intensely. I was therefore very glad when Jack Hawley came home: he was always of such good counsel. He painted badly himself, but there was no one like him for putting his finger on the place. He had been absent from England for a year; he had been somewhere—I don't remember where—to get a fresh eye. I was in a good deal of dread of any such organ, but we were old friends; he had been away for months and a sense of emptiness was creeping into my life. I hadn't dodged a missile for a year.

He came back with a fresh eye, but with the same old black velvet blouse, and the first evening he spent in my studio we smoked cigarettes till the small hours. He had done no work himself, he had only got the eye; so the field was clear for the production of my little things. He wanted to see what I had produced for the *Cheapside,* but he was disappointed in the exhibition. That at least seemed the meaning of two or three comprehensive groans which, as he lounged on my big divan, his leg folded under him, looking at my latest drawings, issued from his lips with the smoke of the cigarette.

"What's the matter with you?" I asked.

"What's the matter with *you?*"

"Nothing save that I'm mystified."

"You are indeed. You're quite off the hinge. What's the meaning

of this new fad?" And he tossed me, with visible irreverence, a
drawing in which I happened to have depicted both my elegant
models. I asked if he didn't think it good, and he replied that it
struck him as execrable, given the sort of thing I had always represented myself to him as wishing to arrive at; but I let that pass—I
was so anxious to see exactly what he meant. The two figures in the
picture looked colossal, but I supposed this was *not* what he meant,
inasmuch as, for aught he knew to the contrary, I might have been
trying for some such effect. I maintained that I was working exactly
in the same way as when he last had done me the honour to tell me
I might do something some day. "Well, there's a screw loose somewhere," he answered; "wait a bit and I'll discover it." I depended
upon him to do so: where else was the fresh eye? But he produced
at last nothing more luminous than "I don't know—I don't like
your types." This was lame for a critic who had never consented to
discuss with me anything but the question of execution, the direction of strokes and the mystery of values.

"In the drawings you've been looking at I think my types are very
handsome."

"Oh they won't do!"

"I've been working with new models."

"I see you have. *They* won't do."

"Are you very sure of that?"

"Absolutely—they're stupid."

"You mean *I* am—for I ought to get round that."

"You *can't*—with such people. Who are they?"

I told him, so far as was necessary, and he concluded heartlessly:
"Ce sont des gens qu'il faut mettre à la porte."[8]

"You've never seen them; they're awfully good"—I flew to their
defence.

"Not seen them? Why all this recent work of yours drops to
pieces with them. It's all I want to see of them."

"No one else has said anything against it—the *Cheapside* people
are pleased."

"Every one else is an ass, and the *Cheapside* people the biggest
asses of all. Come, don't pretend at this time of day to have pretty
illusions about the public, especially about publishers and editors.
It's not for *such* animals you work—it's for those who know, *coloro
che sanno*;[9] so keep straight for *me* if you can't keep straight for
yourself. There was a certain sort of thing you used to try for—and
a very good thing it was. But this twaddle isn't *in* it." When I
talked with Hawley later about "Rutland Ramsay" and its possible
successors he declared that I must get back into my boat again or I

8. "Such people should be shown the door."
9. A misquotation of a phrase Dante applied to Aristotle in *The Divine Comedy*,
"Inferno," IV.131, which reads, "*el mae-stro di color che sanno*," literally "the master of those who know."

should go to the bottom. His voice in short was the voice of warning.

I noted the warning, but I didn't turn my friends out of doors. They bored me a good deal; but the very fact that they bored me admonished me not to sacrifice them—if there was anything to be done with them—simply to irritation. As I look back at this phase they seem to me to have pervaded my life not a little. I have a vision of them as most of the time in my studio, seated against the wall on an old velvet bench to be out of the way, and resembling the while a pair of patient courtiers in a royal ante-chamber. I'm convinced that during the coldest weeks of the winter they held their ground because it saved them fire. Their newness was losing its gloss, and it was impossible not to feel them objects of charity. Whenever Miss Churm arrived they went away, and after I was fairly launched in "Rutland Ramsay" Miss Churm arrived pretty often. They managed to express to me tacitly that they supposed I wanted her for the low life of the book, and I let them suppose it, since they had attempted to study the work—it was lying about the studio—without discovering that it dealt only with the highest circles. They had dipped into the most brilliant of our novelists without deciphering many passages. I still took an hour from them, now and again, in spite of Jack Hawley's warning: it would be time enough to dismiss them, if dismissal should be necessary, when the rigour of the season was over. Hawley had made their acquaintance —he had met them at my fireside—and thought them a ridiculous pair. Learning that he was a painter they tried to approach him, to show him too that they were the real thing; but he looked at them, across the big room, as if they were miles away: they were a compendium of everything he most objected to in the social system of his country. Such people as that, all convention and patent-leather, with ejaculations that stopped conversation, had no business in a studio. A studio was a place to learn to see, and how could you see through a pair of feather-beds?

The main inconvenience I suffered at their hands was that at first I was shy of letting it break upon them that my artful little servant had begun to sit to me for "Rutland Ramsay." They knew I had been odd enough—they were prepared by this time to allow oddity to artists—to pick a foreign vagabond out of the streets when I might have had a person with whiskers and credentials; but it was some time before they learned how high I rated his accomplishments. They found him in an attitude more than once, but they never doubted I was doing him as an organ-grinder. There were several things they never guessed, and one of them was that for a striking scene in the novel, in which a footman briefly figured, it occurred to me to make use of Major Monarch as the menial. I kept putting this off, I didn't like to ask him to don the livery—besides the difficulty of finding a livery to fit him. At last, one day late in the

winter, when I was at work on the despised Oronte, who caught
one's idea on the wing, and was in the glow of feeling myself go
very straight, they came in, the Major and his wife, with their
society laugh about nothing (there was less and less to laugh at);
came on like country-callers—they always reminded me of that—who
have walked across the park after church and are presently persuaded to stay to luncheon. Luncheon was over, but they could stay
to tea—I knew they wanted it. The fit was on me, however, and I
couldn't let my ardour cool and my work wait, with the fading daylight, while my model prepared it. So I asked Mrs. Monarch if she
would mind laying it out—a request which for an instant brought
all the blood to her face. Her eyes were on her husband's for a
second, and some mute telegraphy passed between them. Their folly
was over the next instant; his cheerful shrewdness put an end to it.
So far from pitying their wounded pride, I must add, I was moved
to give it as complete a lesson as I could. They bustled about
together and got out the cups and saucers and made the kettle boil.
I know they felt as if they were waiting on my servant, and when
the tea was prepared I said: "He'll have a cup, please—he's tired."
Mrs. Monarch brought him one where he stood, and he took it
from her, as if he had been a gentleman at a party squeezing a
crush-hat with an elbow.

Then it came over me that she had made a great effort for me—
made it with a kind of nobleness—and that I owed her a compensation. Each time I saw her after this I wondered what the compensation could be. I couldn't go on doing the wrong thing to oblige
them. Oh it *was* the wrong thing, the stamp of the work for which
they sat—Hawley was not the only person to say it now. I sent in a
large number of the drawings I had made for "Rutland Ramsay,"
and I received a warning that was more to the point than Hawley's.
The artistic adviser of the house for which I was working was of
opinion that many of my illustrations were not what had been
looked for. Most of these illustrations were the subjects in which
the Monarchs had figured. Without going into the question of what
had been looked for, I had to face the fact that at this rate I
shouldn't get the other books to do. I hurled myself in despair on
Miss Churm—I put her through all her paces. I not only adopted
Oronte publicly as my hero, but one morning when the Major
looked in to see if I didn't require him to finish a *Cheapside* figure
for which he had begun to sit the week before, I told him I had
changed my mind—I'd do the drawing from my man. At this my
visitor turned pale and stood looking at me. "Is *he* your idea of an
English gentleman?" he asked.

I was disappointed, I was nervous, I wanted to get on with my
work; so I replied with irritation: "Oh my dear Major—I can't be
ruined for *you!*". .

It was a horrid speech, but he stood another moment—after

which, without a word, he quitted the studio. I drew a long breath, for I said to myself that I shouldn't see him again. I hadn't told him definitely that I was in danger of having my work rejected, but I was vexed at his not having felt the catastrophe in the air, read with me the moral of our fruitless collaboration, the lesson that in the deceptive atmosphere of art even the highest respectability may fail of being plastic.

I didn't owe my friends money, but I did see them again. They reappeared together three days later, and, given all the other facts, there was something tragic in that one. It was a clear proof they could find nothing else in life to do. They had threshed the matter out in a dismal conference—they had digested the bad news that they were not in for the series. If they weren't useful to me even for the *Cheapside* their function seemed difficult to determine, and I could only judge at first that they had come, forgivingly, decorously, to take a last leave. This made me rejoice in secret that I had little leisure for a scene; for I had placed both my other models in position together and I was pegging away at a drawing from which I hoped to derive glory. It had been suggested by the passage in which Rutland Ramsay, drawing up a chair to Artemisia's piano-stool, says extraordinary things to her while she ostensibly fingers out a difficult piece of music. I had done Miss Churm at the piano before—it was an attitude in which she knew how to take on an absolutely poetic grace. I wished the two figures to "compose" together with intensity, and my little Italian had entered perfectly into my conception. The pair were vividly before me, the piano had been pulled out; it was a charming show of blended youth and murmured love, which I had only to catch and keep. My visitors stood and looked at it, and I was friendly to them over my shoulder.

They made no response, but I was used to silent company and went on with my work, only a little disconcerted—even though exhilarated by the sense that *this* was at least the ideal thing—at not having got rid of them after all. Presently I heard Mrs. Monarch's sweet voice beside or rather above me: "I wish her hair were a little better done." I looked up and she was staring with a strange fixedness at Miss Churm, whose back was turned to her. "Do you mind my just touching it?" she went on—a question which made me spring up for an instant as with the instinctive fear that she might do the young lady a harm. But she quieted me with a glance I shall never forget—I confess I should like to have been able to paint *that* —and went for a moment to my model. She spoke to her softly, laying a hand on her shoulder and bending over her; and as the girl, understanding, gratefully assented, she disposed her rough curls, with a few quick passes, in such a way as to make Miss Churm's head twice as charming. It was one of the most heroic personal services I've ever seen rendered. Then Mrs. Monarch turned away with a low sigh and, looking about her as if for something to do, stooped to the floor with a noble humility and picked up a dirty rag that had dropped out of my paint-box.

The Major meanwhile had also been looking for something to do, and, wandering to the other end of the studio, saw before him my breakfast-things neglected, unremoved. "I say, can't I be useful *here*?" he called out to me with an irrepressible quaver. I assented with a laugh that I fear was awkward, and for the next ten minutes, while I worked, I heard the light clatter of china and the tinkle of spoons and glass. Mrs. Monarch assisted her husband—they washed up my crockery, they put it away. They wandered off into my little scullery, and I afterwards found that they had cleaned my knives and that my slender stock of plate had an unprecedented surface. When it came over me, the latent eloquence of what they were doing, I confess that my drawing was blurred for a moment—the picture swam. They had accepted their failure, but they couldn't accept their fate. They had bowed their heads in bewilderment to the perverse and cruel law in virtue of which the real thing could be so much less precious than the unreal; but they didn't want to starve. If my servants were my models; then my models might be my servants. They would reverse the parts—the others would sit for the ladies and gentlemen and *they* would do the work. They would still be in the studio—it was an intense dumb appeal to me not to turn them out. "Take us on," they wanted to say—"we'll do *anything*."

My pencil dropped from my hand; my sitting was spoiled and I got rid of my sitters, who were also evidently rather mystified and awestruck. Then, alone with the Major and his wife I had a most uncomfortable moment. He put their prayer into a single sentence: "I say, you know—just let *us* do for you, can't you?" I couldn't—it was dreadful to see them emptying my slops; but I pretended I could, to oblige them, for about a week. Then I gave them a sum of money to go away, and I never saw them again, I obtained the remaining books, but my friend Hawley repeats that Major and Mrs. Monarch did me a permanent harm, got me into false ways. If it be true I'm content to have paid the price—for the memory. 1892, 1909

The Beast in the Jungle[1]

I

What determined the speech that startled him in the course of their encounter scarcely matters, being probably but some words spoken by himself quite without intention—spoken as they lingered

---

1. James initially recorded the "germ" for this story in 1895, but it first appeared in the collection *The Better Sort* (1903). It was reprinted, with minor re- visions, in the *Altar of the Dead* volume of the New York edition, Vol. XVII (1909), the source of the present text.

and slowly moved together after their renewal of acquaintance. He had been conveyed by friends an hour or two before to the house at which she was staying; the party of visitors at the other house, of whom he was one, and thanks to whom it was his theory, as always, that he was lost in the crowd, had been invited over to luncheon. There had been after luncheon much dispersal, all in the interest of the original motive, a view of Weatherend itself and the fine things, intrinsic features, pictures, heirlooms, treasures of all the arts, that made the place almost famous; and the great rooms were so numerous that guests could wander at their will, hang back from the principal group and in cases where they took such matters with the last seriousness give themselves up to mysterious appreciations and measurements. There were persons to be observed, singly or in couples, bending toward objects in out-of-the-way corners with their hands on their knees and their heads nodding quite as with the emphasis of an excited sense of smell. When they were two they either mingled their sounds of ecstasy or melted into silences of even deeper import, so that there were aspects of the occasion that gave it for Marcher much the air of the "look round," previous to a sale highly advertised, that excites or quenches, as may be, the dream of acquisition. The dream of acquisition at Weatherend would have had to be wild indeed, and John Marcher found himself, among such suggestions, disconcerted almost equally by the presence of those who knew too much and by that of those who knew nothing. The great rooms caused so much poetry and history to press upon him that he needed some straying apart to feel in a proper relation with them, though this impulse was not, as happened, like the gloating of some of his companions, to be compared to the movements of a dog sniffing a cupboard. It had an issue promptly enough in a direction that was not to have been calculated.

It led, briefly, in the course of the October afternoon, to his closer meeting with May Bartram, whose face, a reminder, yet not quite a remembrance, as they sat much separated at a very long table, had begun merely by troubling him rather pleasantly. It affected him as the sequel of something of which he had lost the beginning. He knew it, and for the time quite welcomed it, as a continuation, but didn't know what it continued, which was an interest or an amusement the greater as he was also somehow aware —yet without a direct sign from her—that the young woman herself hadn't lost the thread. She hadn't lost it, but she wouldn't give it back to him, he saw, without some putting forth of his hand for it; and he not only saw that, but saw several things more, things odd enough in the light of the fact that at the moment some accident of grouping brought them face to face he was still merely fumbling with the idea that any contact between them in the past

would have had no importance. If it had had no importance he
scarcely knew why his actual impression of her should so seem to
have so much; the answer to which, however, was that in such a life
as they all appeared to be leading for the moment one could but
take things as they came. He was satisfied, without in the least
being able to say why, that this young lady might roughly have
ranked in the house as a poor relation; satisfied also that she was not
there on a brief visit, but was more or less a part of the establishment—almost a working, a remunerated part. Didn't she enjoy at
periods a protection that she paid for by helping, among other services, to show the place and explain it, deal with the tiresome
people, answer questions about the dates of the building, the styles
of the furniture, the authorship of the pictures, the favourite haunts
of the ghost? It wasn't that she looked as if you could have given
her shillings—it was impossible to look less so. Yet when she finally
drifted toward him, distinctly handsome, though ever so much older
—older than when he had seen her before— it might have been as
an effect of her guessing that he had, within the couple of hours,
devoted more imagination to her than to all the others put
together, and had thereby penetrated to a kind of truth that the
others were too stupid for. She *was* there on harder terms than any
one; she was there as a consequence of things suffered, one way and
another, in the interval of years; and she remembered him very
much as she was remembered—only a good deal better.

By the time they at last thus came to speech they were alone in
one of the rooms—remarkable for a fine portrait over the chimney-
place—out of which their friends had passed, and the charm of it
was that even before they had spoken they had practically arranged
with each other to stay behind for talk. The charm, happily, was in
other things too—partly in there being scarce a spot at Weatherend
without something to stay behind for. It was in the way the autumn
day looked into the high windows as it waned; the way the red
light, breaking at the close from under a low sombre sky, reached
out in a long shaft and played over old wainscots, old tapestry, old
gold, old colour. It was most of all perhaps in the way she came to
him as if, since she had been turned on to deal with the simpler
sort, he might, should he choose to keep the whole thing down, just
take her mild attention for a part of her general business. As soon
as he heard her voice, however, the gap was filled up and the missing link supplied; the slight irony he divined in her attitude lost its
advantage. He almost jumped at it to get there before her. "I met
you years and years ago in Rome. I remember all about it." She
confessed to disappointment—she had been so sure he didn't; and
to prove how well he did he began to pour forth the particular recollections that popped up as he called for them. Her face and her
voice, all at his service now, worked the miracle—the impression
operating like the torch of a lamplighter who touches into flame,

one by one, a long row of gas-jets. Marcher flattered himself the illumination was brilliant, yet he was really still more pleased on her showing him, with amusement, that in his haste to make everything right he had got most things rather wrong. It hadn't been at Rome —it had been at Naples; and it hadn't been eight years before—it had been more nearly ten. She hadn't been, either, with her uncle and aunt, but with her mother and her brother; in addition to which it was not with the Pembles *he* had been, but with the Boyers, coming down in their company from Rome—a point on which she insisted, a little to his confusion, and as to which she had her evidence in hand. The Boyers she had known, but didn't know the Pembles, though she had heard of them, and it was the people he was with who had made them acquainted. The incident of the thunderstorm that had raged round them with such violence as to drive them for refuge into an excavation—this incident had not occurred at the Palace of the Cæsars, but at Pompeii,[2] on an occasion when they had been present there at an important find.

He accepted her amendments, he enjoyed her corrections, though the moral of them was, she pointed out, that he *really* didn't remember the least thing about her; and he only felt it as a drawback that when all was made strictly historic there didn't appear much of anything left. They lingered together still, she neglecting her office—for from the moment he was so clever she had no proper right to him—and both neglecting the house, just waiting as to see if a memory or two more wouldn't again breathe on them. It hadn't taken them many minutes, after all, to put down on the table, like the cards of a pack, those that constituted their respective hands; only what came out was that the pack was unfortunately not perfect —that the past, invoked, invited, encouraged, could give them, naturally, no more than it had. It had made them anciently meet—her at twenty, him at twenty-five; but nothing was so strange, they seemed to say to each other, as that, while so occupied, it hadn't done a little more for them. They looked at each other as with the feeling of an occasion missed; the present would have been so much better if the other, in the far distance, in the foreign land, hadn't been so stupidly meagre. There weren't apparently, all counted, more than a dozen little old things that had succeeded in coming to pass between them; trivialities of youth, simplicities of freshness, stupidities of ignorance, small possible germs, but too deeply buried —too deeply (didn't it seem?) to sprout after so many years. Marcher could only feel he ought to have rendered her some service —saved her from a capsized boat in the Bay or at least recovered her dressing-bag, filched from her cab in the streets of Naples by a lazzarone[3] with a stiletto. Or it would have been nice if he could have been taken with fever all alone at his hotel, and she could have

2. Pompeii is near Naples, not Rome.   3. Beggar.

come to look after him, to write to his people, to drive him out in convalescence. *Then* they would be in possession of the something or other that their actual show seemed to lack. It yet somehow presented itself, this show, as too good to be spoiled; so that they were reduced for a few minutes more to wondering a little helplessly why —since they seemed to know a certain number of the same people —their reunion had been so long averted. They didn't use that name for it, but their delay from minute to minute to join the others was a kind of confession that they didn't quite want it to be a failure. Their attempted supposition of reasons for their not having met but showed how little they knew of each other. There came in fact a moment when Marcher felt a positive pang. It was vain to pretend she was an old friend, for all the communities were wanting, in spite of which it was as an old friend that he saw she would have suited him. He had new ones enough—was surrounded with them for instance on the stage of the other house; as a new one he probably wouldn't have so much as noticed her. He would have liked to invent something, get her to make-believe with him that some passage of a romantic or critical kind *had* originally occurred. He was really almost reaching out in imagination—as against time—for something that would do, and saying to himself that if it didn't come this sketch of a fresh start would show for quite awkwardly bungled. They would separate, and now for no second or no third chance. They would have tried and not succeeded. Then it was, just at the turn, as he afterwards made it out to himself, that, everything else failing, she herself decided to take up the case and, as it were, save the situation. He felt as soon as she spoke that she had been consciously keeping back what she said and hoping to get on without it; a scruple in her that immensely touched him when, by the end of three or four minutes more, he was able to measure it. What she brought out, at any rate, quite cleared the air and supplied the link—the link it was so odd he should frivolously have managed to lose.

"You know you told me something I've never forgotten and that again and again has made me think of you since; it was that tremendously hot day when we went to Sorrento,[4] across the bay, for the breeze. What I allude to was what you said to me, on the way back, as we sat under the awning of the boat enjoying the cool. Have you forgotten?"

He had forgotten and was even more surprised than ashamed. But the great thing was that he saw in this no vulgar reminder of any "sweet" speech. The vanity of women had long memories, but she was making no claim on him of a compliment or a mistake. With another woman, a totally different one, he might have feared the recall possibly even some imbecile "offer." So, in having to say that he had indeed forgotten, he was conscious rather of a loss than of a gain; he already saw an interest in the matter of her mention. "I try to think—but I give it up. Yet I remember the Sorrento day."

"I'm not very sure you do," May Bartram after a moment said; "and I'm not very sure I ought to want you to. It's dreadful to bring a person back at any time to what he was ten years before. If you've lived away from it," she smiled, "so much the better."

"Ah if *you* haven't why should I?" he asked.

"Lived away, you mean, from what I myself was?"

"From what *I* was. I was of course an ass," Marcher went on; "but I would rather know from you just the sort of ass I was than—from the moment you have something in your mind—not know anything."

Still, however, she hesitated. "But if you've completely ceased to be that sort—?"

"Why I can then all the more bear to know. Besides, perhaps I haven't."

"Perhaps. Yet if you haven't," she added, "I should suppose you'd remember. Not indeed that I in the least connect with my impression the invidious name you use. If I had only thought you foolish," she explained, "the thing I speak of wouldn't so have remained with me. It was about yourself." She waited as if it might come to him; but as, only meeting her eyes in wonder, he gave no sign, she burnt her ships. "Has it ever happened?"

Then it was that, while he continued to stare, a light broke for him and the blood slowly came to his face, which began to burn with recognition. "Do you mean I told you—?" But he faltered, lest what came to him shouldn't be right, lest he should only give himself away.

"It was something about yourself that it was natural one shouldn't forget—that is if one remembered you at all. That's why I ask you," she smiled, "if the thing you then spoke of has ever come to pass?"

Oh then he saw, but he was lost in wonder and found himself embarrassed. This, he also saw, made her sorry for him, as if her allusion had been a mistake. It took him but a moment, however, to feel it hadn't been, much as it had been a surprise. After the first little shock of it her knowledge on the contrary began, even if rather strangely, to taste sweet to him. She was the only other person in the world then who would have it, and she had had it all these years, while the fact of his having so breathed his secret had unaccountably faded from him. No wonder they couldn't have met as if nothing had happened. "I judge," he finally said, "that I know what you mean. Only I had strangely enough lost any sense of having taken you so far into my confidence."

"Is it because you've taken so many others as well?"

"I've taken nobody. Not a creature since then."

"So that I'm the only person who knows?"

"The only person in the world."

"Well," she quickly replied, "I myself have never spoken. I've never, never repeated of you what you told me." She looked at him so that he perfectly believed her. Their eyes met over it in such a way that he was without a doubt. "And I never will."

She spoke with an earnestness that, as if almost excessive, put him at ease about her possible derision. Somehow the whole question was a new luxury to him—that is from the moment she was in possession. If she didn't take the sarcastic view she clearly took the sympathetic, and that was what he had had, in all the long time, from no one whomsoever. What he felt was that he couldn't at present have begun to tell her, and yet could profit perhaps exquisitely by the accident of having done so of old. "Please don't then. We're just right as it is."

"Oh I am," she laughed, "if you are!" To which she added: "Then you do still feel in the same way?"

It was impossible he shouldn't take to himself that she was really interested, though it all kept coming as perfect surprise. He had thought of himself so long as abominably alone, and lo he wasn't alone a bit. He hadn't been, it appeared, for an hour—since those moments on the Sorrento boat. It was *she* who had been, he seemed to see as he looked at her—she who had been made so by the graceless fact of his lapse of fidelity. To tell her what he had told her—what had it been but to ask something of her? something that she had given, in her charity, without his having, by a remembrance, by a return of the spirit, failing another encounter, so much as thanked her. What he had asked of her had been simply at first not to laugh at him. She had beautifully not done so for ten years, and she was not doing so now. So he had endless gratitude to make up. Only for that he must see just how he had figured to her. "What, exactly, was the account I gave—?"

"Of the way you did feel? Well, it was very simple. You said you had had from your earliest time, as the deepest thing within you, the sense of being kept for something rare and strange, possibly prodigious and terrible, that was sooner or later to happen to you, that you had in your bones the foreboding and the conviction of, and that would perhaps overwhelm you."

"Do you call that very simple?" John Marcher asked.

She thought a moment. "It was perhaps because I seemed, as you spoke, to understand it."

"You do understand it?" he eagerly asked.

Again she kept her kind eyes on him. "You still have the belief?"

"Oh!" he exclaimed helplessly. There was too much to say.

"Whatever it's to be," she clearly made out, "it hasn't yet come."

He shook his head in complete surrender now. "It hasn't yet come. Only, you know, it isn't anything I'm to *do* to achieve in the world, to be distinguished or admired for. I'm not such an ass as *that*. It would be much better, no doubt, if I were."

"It's to be something you're merely to suffer?"

"Well, say to wait for—to have to meet, to face, to see suddenly break out in my life; possibly destroying all further consciousness, possibly annihilating me; possibly, on the other hand, only altering everything, striking at the root of all my world and leaving me to the consequences, however they shape themselves."

She took this in, but the light in her eyes continued for him not to be that of mockery. "Isn't what you describe perhaps but the expectation—or at any rate the sense of danger, familiar to so many people—of falling in love?"

John Marcher wondered. "Did you ask me that before?"

"No—I wasn't so free-and-easy then. But it's what strikes me now."

"Of course," he said after a moment, "it strikes you. Of course it strikes *me*. Of course what's in store for me may be no more than that. The only thing is," he went on, "that I think if it had been that I should by this time know."

"Do you mean because you've *been* in love?" And then as he but looked at her in silence: "You've been in love, and it hasn't meant such a cataclysm, hasn't proved the great affair?"

"Here I am, you see. It hasn't been overwhelming."

"Then it hasn't been love," said May Bartram.

"Well, I at least thought it was. I took it for that—I've taken it till now. It was agreeable, it was delightful, it was miserable," he explained. "But it wasn't strange. It wasn't what *my* affair's to be."

"You want something all to yourself—something that nobody else knows or *has* known?"

It isn't a question of what I 'want'—God knows I don't want anything. It's only a question of the apprehension that haunts me —that I live with day by day."

He said this so lucidly and consistently that he could see it further impose itself. If she hadn't been interested before she'd have been interested now. "Is it a sense of coming violence?"

Evidently now too again he liked to talk of it. "I don't think of it as—when it does come—necessarily violent. I only think of it as natural and as of course above all unmistakeable. I think of it simply as *the* thing. *The* thing will of itself appear natural."

"Then how will it appear strange?"

Marcher bethought himself. "It won't—to *me*."

"To whom then?"

"Well," he replied, smiling at last, "say to you."

"Oh then I'm to be present?"

"Why you *are* present—since you know."

"I see." She turned it over. "But I mean at the catastrophe."

At this, for a minute, their lightness gave way to their gravity; it was as if the long look they exchanged held them together. "It will only depend on yourself—if you'll watch with me."

"Are you afraid?" she asked.

"Don't leave me *now*," he went on.

"Are you afraid?" she repeated.

"Do you think me simply out of my mind?" he pursued instead of answering. "Do I merely strike you as a harmless lunatic?"

"No," said May Bartram. "I understand you. I believe you."

"You mean you feel how my obsession—poor old thing!—may correspond to some possible reality?"

"To some possible reality."

"Then you *will* watch with me?"

She hesitated, then for the third time put her question. "Are you afraid?"

"Did I tell you I was—at Naples?"

"No, you said nothing about it."

"Then I don't know. And I should *like* to know," said John Marcher. "You'll tell me yourself whether you think so. If you'll watch with me you'll see."

"Very good then." They had been moving by this time across the room, and at the door, before passing out, they paused as for the full wind-up of their understanding. "I'll watch with you," said May Bartram.

II

The fact that she "knew"—knew and yet neither chaffed him nor betrayed him—had in a short time begun to constitute between them a goodly bond, which became more marked when, within the year that followed their afternoon at Weatherend, the opportunities for meeting multiplied. The event that thus promoted these occasions was the death of the ancient lady her great-aunt, under whose wing, since losing her mother, she had to such an extent found shelter, and who, though but the widowed mother of the new successor to the property, had succeeded—thanks to a high tone and a high temper—in not forfeiting the supreme position at the great house. The deposition of this personage arrived but with her death, which, followed by many changes, made in particular a difference for the young woman in whom Marcher's expert attention had recognized from the first a dependent with a pride that might ache though it didn't bristle. Nothing for a long time had made him easier than the thought that the aching must have been much soothed by Miss Bartram's now finding herself able to set up a small home in London. She had acquired property, to an amount that made that luxury just possible, under her aunt's extremely complicated will, and when the whole matter began to be straightened out, which indeed took time, she let him know that the happy issue was at last in view. He had seen her again before that day, because she had more than once accompanied the ancient lady to town and because he had paid another visit to the friends who so conveniently made of Weatherend one of the charms of their own hospitality. These friends had taken him back there; he had achieved there again with Miss Bartram some quiet detachment; and he had in London succeeded in persuading her to more than one brief absence from her aunt. They went together, on these latter occasions, to the National Gallery and the South Kensington Museum, where, among vivid reminders, they talked of Italy at large—not now attempting to recover, as at first, the taste of their youth and their ignorance. That recovery, the first day at Weatherend, had served its purpose well, had given them quite enough; so that they were, to Marcher's sense, no longer hovering about the headwaters of their stream, but had felt their boat pushed sharply off and down the current.

They were literally afloat together; for our gentleman this was marked, quite as marked as that the fortunate cause of it was just the buried treasure of her knowledge. He had with his own hands dug up this little hoard, brought to light—that is to within reach of the dim day constituted by their discretions and privacies—the object of value the hiding-place of which he had, after putting it into the ground himself, so strangely, so long forgotten. The rare luck of his having again just stumbled on the spot made him indifferent to any other question; he would doubtless have devoted more time to the odd accident of his lapse of memory if he hadn't been moved to devote so much to the sweetness, the comfort, as he felt, for the future, that this accident itself had helped to keep fresh. It had never entered into his plan that any one should "know," and mainly for the reason that it wasn't in him to tell any one. That would have been impossible, for nothing but the amusement of a cold world would have waited on it. Since, however, a mysterious fate had opened his mouth betimes, in spite of him, he would count that a compensation and profit by it to the utmost. That the right person *should* know tempered the asperity of his secret more even than his shyness had permitted him to imagine; and May Bartram was clearly right, because—well, because there she was. Her knowledge simply settled it; he would have been sure enough by this time had she been wrong. There was that in his situation, no doubt, that disposed him too much to see her as a mere confidant, taking all her light for him from the fact—the fact only—of her interest in his predicament; from her mercy, sympathy, seriousness, her consent not to regard him as the funniest of the funny. Aware, in fine, that her price for him was just in her giving him this constant sense of his being admirably spared, he was careful to remember that she had also a life of her own, with things that might happen to *her*, things that in friendship one should likewise take account of. Something fairly remarkable came to pass with him, for that matter, in this connexion—something represented by a certain passage of his consciousness, in the suddenest way, from one extreme to the other.

He had thought himself, so long as nobody knew, the most disinterested person in the world, carrying his concentrated burden, his perpetual suspense, ever so quietly, holding his tongue about it, giving others no glimpse of it nor of its effect upon his life, asking of them no allowance and only making on his side all those that were asked. He hadn't disturbed people with the queerness of their having to know a haunted man, though he had had moments of rather special temptation on hearing them say they were forsooth "unsettled." If they were as unsettled as he was—he who had never been settled for an hour in his life—they would know what it meant. Yet it wasn't, all the same, for him to make them, and he listened to them civilly enough. This was why he had such good—though possibly such rather colourless—manners; this was why, above all, he could regard himself, in a greedy world, as decently—as in fact perhaps even a little sublimely—unselfish. Our point is accordingly that he valued this character quite sufficiently to measure his present danger of letting it lapse, against which he promised himself to be much on his guard. He was quite ready, none the less, to be selfish just a little, since surely no more charming occasion for it had come to him. "Just a little," in a word, was just as much as Miss Bartram, taking one day with another, would let him. He never would be in the least coercive, and would keep well before him the lines on which consideration for her—the very highest—ought to proceed. He would thoroughly establish the heads under which her affairs, her requirements, her peculiarities—he went so far as to give them the latitude of that name—would come into their intercourse. All this naturally was a sign of how much he took the intercourse itself for granted. There was nothing more to be done about *that*. It simply existed; had sprung into being with her first penetrating question to him in the autumn light there at Weatherend. The real form it should have taken on the basis that stood out large was the form of their marrying. But the devil in this was that the very basis itself put marrying out of the question. His conviction, his apprehension, his obsession, in short, wasn't a privilege he could invite a woman to share; and that consequence of it was precisely what was the matter with him. Something or other lay in wait for him, amid the twists and the turns of the months and the years, like a crouching beast in the jungle. It signified little whether the crouching beast were destined to slay him or to be slain. The definite point was the inevitable spring of the creature;

and the definite lesson from that was that a man of feeling didn't cause himself to be accompanied by a lady on a tiger-hunt. Such was the image under which he had ended by figuring his life.

They had at first, none the less, in the scattered hours spent together, made no allusion to that view of it; which was a sign he was handsomely alert to give that he didn't expect, that he in fact didn't care, always to be talking about it. Such a feature in one's outlook was really like a hump on one's back. The difference it made every minute of the day existed quite independently of discussion. One discussed of course *like* a hunchback, for there was always, if nothing else, the hunchback face. That remained, and she was watching him; but people watched best, as a general thing, in silence, so that such would be predominantly the manner of their vigil. Yet he didn't want, at the same time, to be tense and solemn; tense and solemn was what he imagined he too much showed for with other people. The thing to be, with the one person who knew, was easy and natural—to make the reference rather than be seeming to avoid it, to avoid it rather than be seeming to make it, and to keep it, in any case, familiar, facetious even, rather than pedantic and portentous. Some such consideration as the latter was doubtless in his mind for instance when he wrote pleasantly to Miss Bartram that perhaps the great thing he had so long felt as in the lap of the gods was no more than this circumstance, which touched him so nearly, of her acquiring a house in London. It was the first allusion they had yet again made, needing any other hitherto so little; but when she replied, after having given him the news, that she was by no means satisfied with such a trifle as the climax to so special a suspense, she almost set him wondering if she hadn't even a larger conception of singularity for him than he had for himself. He was at all events destined to become aware little by little, as time went by, that she was all the while looking at his life, judging it, measuring it, in the light of the thing she knew, which grew to be at last, with the consecration of the years, never mentioned between them save as "the real truth" about him. That had always been his own form of reference to it, but she adopted the form so quietly that, looking back at the end of a period, he knew there was no moment at which it was traceable that she had, as he might say, got inside his idea, or exchanged the attitude of beautifully indulging for that of still more beautifully believing him.

It was always open to him to accuse her of seeing him but as the most harmless of maniacs, and this, in the long run—since it covered so much ground—was his easiest description of their friendship. He had a screw loose for her, but she liked him in spite of it and was practically, against the rest of the world, his kind wise keeper, unremunerated but fairly amused and, in the absence of other near ties, not disreputably occupied. The rest of the world of course thought him queer, but she, she only, knew how, and above all why, queer; which was precisely what enabled her to dispose the concealing veil in the right folds. She took his gaiety from him—since it had to pass with them for gaiety—as she took everything else; but she certainly so far justified by her unerring touch his finer sense of the degree to which he had ended by convincing her. *She* at least never spoke of the secret of his life except as "the real truth about you," and she had in fact a wonderful way of making it seem, as such, the secret of her own life too. That was in fine how he so constantly felt her as allowing for him; he couldn't on the whole call it anything else. He allowed for himself, but she, exactly, allowed still more; partly because, better placed for a sight of the matter, she traced his unhappy perversion through reaches of its course into which he could scarce follow it. He knew how he felt, but, besides knowing that, she knew how he *looked* as well; he knew each of the things of importance he was insidiously kept from doing, but she could add up the amount they made, understand how much, with a lighter weight on his spirit, he might have done, and thereby establish how, clever as he was, he fell short. Above all she was in the secret of the difference between the forms he went through—those of his little office under Government, those of caring for his modest patrimony, for his library, for his garden in the country, for the people in London whose invitations he accepted and repaid—and the detachment that reigned beneath them and that made of all behaviour, all that could in the least be called behaviour, a long act of dissimulation. What it had come to was that he wore a mask painted with the social simper, out of the eye-holes of which there looked eyes of an expression not in the least matching the other features. This the stupid world, even after years, had never more than half-discovered. It was only May Bartram who had, and she achieved, by an art indescribable, the feat of at once —or perhaps it was only alternately—meeting the eyes from in front and mingling her own vision, as from over his shoulder, with their peep through the apertures.

So while they grew older together she did watch with him, and so she let this association give shape and colour to her own existence. Beneath *her* forms as well detachment had learned to sit, and behaviour had become for her, in the social sense, a false account of herself. There was but one account of her that would have been true all the while and that she could give straight to nobody, least of all to John Marcher. Her whole attitude was a virtual statement, but the perception of that only seemed called to take its place for him as one of the many things necessarily crowded out of his consciousness. If she had moreover, like himself, to make sacrifices to their real truth, it was to be granted that her compensation might have affected her as more prompt and more natural. They had long periods, in this London time, during which, when they were together, a stranger might have listened to them without in the least pricking up his ears; on the other hand the real truth was equally liable at any moment to rise to the surface, and the auditor would then have wondered indeed what they were talking about. They had from an early hour made up their mind that society was, luckily, unintelligent, and the margin allowed them by this had fairly become one of their commonplaces. Yet there were still moments when the situation turned almost fresh—usually under the effect of some expression drawn from herself. Her expressions doubtless repeated themselves, but her intervals were generous. "What saves us, you know, is that we answer so completely to so usual an appearance: that of the man and woman whose friendship has become such a daily habit—or almost—as to be at last indispensable." That for instance was a remark she had frequently enough had occasion to make, though she had given it at different times different developments. What we are especially concerned with is the turn it happened to take from her one afternoon when he had come to see her in honour of her birthday. This anniversary had fallen on a Sunday, at a season of thick fog and general outward gloom; but he had brought her his customary offering, having known her now long enough to have established a hundred small traditions. It was one of his proofs to himself, the present he made her on her birthday, that he hadn't sunk into real selfishness. It was mostly nothing more than a small trinket, but it was always fine of its kind, and he was regularly careful to pay for it more than he thought he could afford. "Our habit saves you at least, don't you see? because it makes you, after all, for the vulgar, indistinguishable from other men. What's the most inveterate mark of men in general? Why the capacity to spend endless time with dull women—to spend it I won't say without being bored, but without minding that they are, without being driven off at a tangent by it; which comes to the same thing. I'm your dull woman, a part of the daily bread for which you pray at church. That covers your tracks more than anything."

"And what covers yours?" asked Marcher, whom his dull woman could mostly to this extent amuse. "I see of course what you mean by your saving me, in this way and that, so far as other people are concerned—I've seen it all along. Only what is it that saves *you*? I often think, you know, of that."

She looked as if she sometimes thought of that too, but rather in a different way. "Where other people, you mean, are concerned?"

"Well, you're really so in with me, you know—as a sort of result of my being so in with yourself. I mean of my having such an immense regard for you, being so tremendously mindful of all you've done for me. I sometimes ask myself if it's quite fair. Fair I mean to have so involved and—since one may say it—interested you. I almost feel as if you hadn't really had time to do anything else."

"Anything else but be interested?" she asked. "Ah what else does one ever want to be? If I've been 'watching' with you, as we long ago agreed I was to do, watching's always in itself an absorption."

"Oh certainly," John Marcher said, "if you hadn't had your curiosity—! Only doesn't it sometimes come to you as time goes on that your curiosity isn't being particularly repaid?"

May Bartram had a pause. "Do you ask that, by any chance, because you feel at all that yours isn't? I mean because you have to wait so long."

Oh he understood what she meant! "For the thing to happen that never does happen? For the beast to jump out? No, I'm just where I was about it. It isn't a matter as to which I can *choose*, I can decide for a change. It isn't one as to which there *can* be a change. It's in the lap of the gods. One's in the hands of one's law —there one is. As to the form the law will take, the way it will operate, that's its own affair."

"Yes," Miss Bartram replied; "of course one's fate's coming, of course it *has* come in its own form and its own way, all the while. Only, you know, the form and the way in your case were to have been—well, something so exceptional and, as one may say, so particularly *your* own."

Something in this made him look at her with suspicion. "You say 'were to *have* been,' as if in your heart you had begun to doubt."

"Oh!" she vaguely protested.

"As if you believed," he went on, "that nothing will now take place."

She shook her head slowly but rather inscrutably. "You're far from my thought."

He continued to look at her. "What then is the matter with you?"

"Well," she said after another wait, "the matter with me is simply that I'm more sure than ever my curiosity, as you call it, will be but too well repaid."

They were frankly grave now; he had got up from his seat, had turned once more about the little drawing-room to which, year after year, he brought his inevitable topic; in which he had, as he might have said, tasted their intimate community with every sauce, where every object was as familiar to him as the things of his own house and the very carpets were worn with his fitful walk very much as the desks in old counting-houses are worn by the elbows of generations of clerks. The generations of his nervous moods had been at work there, and the place was the written history of his whole middle life. Under the impression of what his friend had just said he knew himself, for some reason, more aware of these things; which made him, after a moment, stop again before her. "Is it possibly that you've grown afraid?"

"Afraid?" He thought, as she repeated the word, that his question had made her, a little, change colour; so that, lest he should have touched on a truth, he explained very kindly: "You remember that that was what you asked *me* long ago—that first day at Weatherend."

"Oh yes, and you told me you didn't know—that I was to see for myself. We've said little about it since, even in so long a time."

"Precisely," Marcher interposed—"quite as if it were too delicate a matter for us to make free with. Quite as if we might find, on pressure, that I *am* afraid. For then," he said, "we shouldn't, should we? quite know what to do."

She had for the time no answer to his question. "There have been days when I thought you were. Only, of course," she added, "there have been days when we have thought almost anything."

"Everything. Oh!" Marcher softly groaned as with a gasp, half-spent, at the face, more uncovered just then than it had been for a long while, of the imagination always with them. It had always had its incalculable moments of glaring out, quite as with the very eyes of the very Beast, and, used as he was to them, they could still draw from him the tribute of a sigh that rose from the depths of his being. All they had thought, first and last, rolled over him; the past seemed to have been reduced to mere barren speculation. This in fact was what the place had just struck him as so full of—the simplification of everything but the state of suspense. That remained only by seeming to hang in the void surrounding it. Even his original fear, if fear it had been, had lost itself in the desert. "I judge, however," he continued, "that you see I'm not afraid now."

"What I see, as I make it out, is that you've achieved something almost unprecedented in the way of getting used to danger. Living with it so long and so closely you've lost your sense of it; you know it's there, but you're indifferent, and you cease even, as of old, to have to whistle in the dark. Considering what the danger is," May Bartram wound up, "I'm bound to say I don't think your attitude could well be surpassed."

John Marcher faintly smiled. "It's heroic?"

"Certainly—call it that."

It was what he would have liked indeed to call it. "I *am* then a man of courage?"

"That's what you were to show me."

He still, however, wondered. "But doesn't the man of courage know what he's afraid of—or *not* afraid of? I don't know *that*, you see. I don't focus it. I can't name it. I only know I'm exposed."

"Yes, but exposed—how shall I say?—so directly. So intimately. That's surely enough."

"Enough to make you feel then—as what we may call the end and the upshot of our watch—that I'm not afraid?"

"You're not afraid. But it isn't," she said, "the end of our watch. That is it isn't the end of yours. You've everything still to see."

"Then why haven't *you?*" he asked. He had had, all along, to-day, the sense of her keeping something back, and he still had it. As this was his first impression of that it quite made a date. The case was the more marked as she didn't at first answer; which in turn made him go on. "You know something I don't." Then his voice, for that of a man of courage, trembled a little. "You know what's to happen." Her silence, with the face she showed, was almost a confession—it made him sure. "You know, and you're afraid to tell me. It's so bad that you're afraid I'll find out."

All this might be true, for she did look as if, unexpectedly to her, he had crossed some mystic line that she had secretly drawn around her. Yet she might, after all, not have worried; and the real climax was that he himself at all events, needn't. "You'll never find out."

III

It was all to have made, none the less, as I have said, a date; which came out in the fact that again and again, even after long intervals, other things that passed between them wore in relation to this hour but the character of recalls and results. Its immediate effect had been indeed rather to lighten insistence—almost to provoke a reaction; as if their topic had dropped by its own weight and as if moreover, for that matter, Marcher had been visited by one of his occasional warnings against egotism. He had kept up, he felt, and very decently on the whole, his consciousness of the importance of not being selfish, and it was true that he had never sinned in that direction without promptly enough trying to press the scales the other way. He often repaired his fault, the season permitting, by inviting his friend to accompany him to the opera; and it not infrequently thus happened that, to show he didn't wish her to have but one sort of food for her mind, he was the cause of her appearing there with him a dozen nights in the month. It even happened that, seeing her home at such times, he occasionally went in with her to finish, as he called it, the evening, and, the better to make his point, sat down to the frugal but always careful little supper that awaited his pleasure. His point was made, he thought, by his not eternally insisting with her on himself; made for instance, at such hours, when it befell that, her piano at hand and each of them familiar with it, they went over passages of the opera together. It chanced to be on one of these occasions, however, that he reminded her of not having answered a certain question he had put to her during the talk that had taken place between them on her last birthday. "What is it that saves *you?*"—saved her, he meant, from that appearance of variation from the usual human type. If he had practically escaped remark, as she pretended, by doing, in the most important particular, what most men do—find the answer to life in patching up an alliance of a sort with a woman no better than himself—how had she escaped it, and how could the alliance, such as it was, since they must suppose it had been more or less noticed, have failed to make her rather positively talked about?

"I never said," May Bartram replied, "that it hadn't made me a good deal talked about."

"Ah well then you're not 'saved.'"

"It hasn't been a question for me. If you've had your woman I've had," she said, "my man."

"And you mean that makes you all right?"

Oh it was always as if there were so much to say! "I don't know why it shouldn't make me—humanly, which is what we're speaking of—as right as it makes you."

"I see," Marcher returned. "'Humanly,' no doubt, as showing that you're living for something. Not, that is, just for me and my secret."

May Bartram smiled. "I don't pretend it exactly shows that I'm not living for you. It's my intimacy with you that's in question."

He laughed as he saw what she meant. "Yes, but since, as you say, I'm only, so far as people make out, ordinary, you're—aren't you?—no more than ordinary either. You help me to pass for a man like another. So if I *am*, as I understand you, you're not compromised. Is that it?"

She had another of her waits, but she spoke clearly enough. "That's it. It's all that concerns me—to help you to pass for a man like another."

He was careful to acknowledge the remark handsomely. "How kind, how beautiful, you are to me! How shall I ever repay you?"

She had her last grave pause, as if there might be a choice of ways. But she chose. "By going on as you are."

It was into this going on as he was that they relapsed, and really for so long a time that the day inevitably came for a further sounding of their depths. These depths, constantly bridged over by a structure firm enough in spite of its lightness and of its occasional oscillation in the somewhat vertiginous air, invited on occasion, in the interest of their nerves, a dropping of the plummet and a measurement of the abyss. A difference had been made moreover, once for all, by the fact that she had all the while not appeared to feel the need of rebutting his charge of an idea within her that she didn't dare to express—a charge uttered just before one of the fullest of their later discussions ended. It had come up for him then that she "knew" something and that what she knew was bad—too bad to tell him. When he had spoken of it as visibly so bad that she was afraid he might find it out, her reply had left the matter too equivocal to be let alone and yet, for Marcher's special sensibility, almost too formidable again to touch. He circled about it at a distance that alternately narrowed and widened and that still wasn't much affected by the consciousness in him that there was nothing she could "know," after all, any better than he did. She had no source of knowledge he hadn't equally—except of course that she might have finer nerves. That was what women had where they were interested; they made out things, where people were concerned, that the people often couldn't have made out for themselves. Their nerves, their sensibility, their imagination, were conductors and revealers, and the beauty of May Bartram was in particular that she had given herself so to his case. He felt in these days what, oddly enough, he had never felt before, the growth of a dread of losing her by some catastrophe—some catastrophe that yet wouldn't at all be *the* catastrophe: partly because she had almost of a sudden begun to strike him as more useful to him than ever yet, and partly by reason of an appearance of uncertainty in her health, coincident and equally new. It was characteristic of the inner detachment he had hitherto so successfully cultivated and to which our whole account of him is a reference, it was characteristic that his complications, such as they were, had never yet seemed so as at this crisis to thicken about him, even to the point of making him ask himself if he were, by any chance, of a truth, within sight or sound, within touch or reach, within the immediate jurisdiction, of the thing that waited.

When the day came, as come it had to, that his friend confessed to him her fear of a deep disorder in her blood, he felt somehow the shadow of a change and the chill of a shock. He immediately began to imagine aggravations and disasters, and above all to think of her peril as the direct menace for himself of personal privation. This indeed gave him one of those partial recoveries of equanimity that were agreeable to him—it showed him that what was still first in his mind was the loss she herself might suffer. "What if she should have to die before knowing, before seeing—?" It would have been brutal, in the early stages of her trouble, to put that question to her; but it had immediately sounded for him to his own concern, and the possibility was what most made him sorry for her. If she did "know," moreover, in the sense of her having had some—what should he think?—mystical irresistible light, this would make the matter not better, but worse, inasmuch as her original adoption of his own curiosity had quite become the basis of her life. She had been living to see what would *be* to be seen, and it would quite lacerate her to have to give up before the accomplishment of the vision. These reflexions, as I say, quickened his generosity; yet, make them as he might, he saw himself, with the lapse of the period, more and more disconcerted. It lapsed for him with a strange steady sweep, and the oddest oddity was that it gave him, independently of the threat of much inconvenience, almost the only positive surprise his career, if career it could be called, had yet offered him. She kept the house as she had never done; he had to go to her to see her—she could meet him nowhere now, though there was scarce a corner of their loved old London in which she hadn't in the past, at one time or another, done so; and he found her always seated by her fire in the deep old-fashioned chair she was less and less able to leave. He had been struck one day, after an absence exceeding his usual measure, with her suddenly looking much older to him than he had ever thought of her being; then he recognised that the suddenness was all on his side—he had just simply and suddenly noticed. She looked older because inevitably, after so many years, she *was* old, or almost; which was of course true in still greater measure of her companion. If she was old, or almost, John Marcher assuredly was, and yet it was her showing of the lesson, not his own, that brought the truth home to him. His surprises began here; when once they had begun they multiplied; they came rather with a rush: it was as if, in the oddest way in the world, they had all been kept back, sown in a thick cluster, for the late afternoon of life, the time at which for people in general the unexpected has died out.

One of them was that he should have caught himself—for he *had* so done—*really* wondering if the great accident would take form now as nothing more than his being condemned to see this charming woman, this admirable friend, pass away from him. He had never so unreservedly qualified her as while confronted in thought with such a possibility; in spite of which there was small doubt for him that as an answer to his long riddle the mere effacement of even so fine a feature of his situation would be an abject anti-climax. It would represent, as connected with his past attitude, a drop of dignity under the shadow of which his existence could only become the most grotesque of failures. He had been far from holding it a failure—long as he had waited for the appearance that was to make it a success. He had waited for quite another thing, not for such a thing as that. The breath of his good faith came short, however, as he recognised how long he had waited, or how long at least his companion had. That she, at all events, might be recorded as having waited in vain—this affected him sharply, and all the more because of his at first having done little more than

amuse himself with the idea. It grew more grave as the gravity of
her condition grew, and the state of mind it produced in him,
which he himself ended by watching as if it had been some definite
disfigurement of his outer person, may pass for another of his surprises. This conjoined itself still with another, the really stupefying
consciousness of a question that he would have allowed to shape
itself had he dared. What did everything mean—what, that is, did
she mean, she and her vain waiting and her probable death and the
soundless admonition of it all—unless that, at this time of day, it
was simply, it was overwhelmingly too late? He had never at any
stage of his queer consciousness admitted the whisper of such a
correction; he had never till within these last few months been so
false to his conviction as not to hold that what was to come to him
had time, whether *he* struck himself as having it or not. That at
last, at last, he certainly hadn't it, to speak of, or had it but in the
scantiest measure—such, soon enough, as things went with him,
became the inference with which his old obsession had to reckon:
and this it was not helped to do by the more and more confirmed
appearance that the great vagueness casting the long shadow in
which he had lived had, to attest itself, almost no margin left. Since
it was in Time that he was to have met his fate, so it was in Time
that his fate was to have acted; and as he waked up to the sense of
no longer being young, which was exactly the sense of being stale,
just as that, in turn, was the sense of being weak, he waked up to
another matter beside. It all hung together; they were subject, he
and the great vagueness, to an equal and indivisible law. When the
possibilities themselves had accordingly turned stale, when the
secret of the gods had grown faint, had perhaps even quite evaporated, that, and that only, was failure. It wouldn't have been failure
to be bankrupt, dishonoured, pilloried, hanged; it was failure not to
be anything. And so, in the dark valley into which his path had
taken its unlooked-for twist, he wondered not a little as he groped.
He didn't care what awful crash might overtake him, with what
ignominy or what monstrosity he might yet be associated—since he
wasn't after all too utterly old to suffer—if it would only be
decently proportionate to the posture he had kept, all his life, in the
threatened presence of it. He had but one desire left—that he
shouldn't have been "sold."

IV

Then it was that, one afternoon, while the spring of the year was
young and new she met all in her own way his frankest betrayal of
these alarms. He had gone in late to see her, but evening hadn't set-

tled and she was presented to him in that long fresh light of
waning April days which affects us often with a sadness sharper
than the greyest hours of autumn. The week had been warm, the
spring was supposed to have begun early, and May Bartram sat, for
the first time in the year, without a fire; a fact that, to Marcher's
sense, gave the scene of which she formed part a smooth and ultimate look, an air of knowing, in its immaculate order and cold
meaningless cheer, that it would never see a fire again. Her own
aspect—he could scarce have said why—intensified this note.
Almost as white as wax, with the marks and signs in her face as
numerous and as fine as if they had been etched by a needle, with
soft white draperies relieved by a faded green scarf on the delicate
tone of which the years had further refined, she was the picture of a
serene and exquisite but impenetrable sphinx, whose head, or
indeed all whose person, might have been powdered with silver. She
was a sphinx, yet with her white petals and green fronds she might
have been a lily too—only an artificial lily, wonderfully imitated
and constantly kept, without dust or stain, though not exempt from
a slight droop and a complexity of faint creases, under some clear
glass bell. The perfection of household care, of high polish and
finish, always reigned in her rooms, but they now looked most as if
everything had been wound up, tucked in, put away, so that she
might sit with folded hands and with nothing more to do. She was
"out of it," to Marcher's vision; her work was over; she communicated with him as across some gulf or from some island of rest that
she had already reached, and it made him feel strangely abandoned.
Was it—or rather wasn't it—that if for so long she had been watching with him the answer to their question must have swum into her
ken and taken on its name, so that her occupation was verily gone?
He had as much as charged her with this in saying to her, many
months before, that she even then knew something she was keeping
from him. It was a point he had never since ventured to press,
vaguely fearing as he did that it might become a difference, perhaps
a disagreement, between them. He had in this later time turned
nervous, which was what he in all the other years had never been;
and the oddity was that his nervousness should have waited till he
had begun to doubt, should have held off so long as he was sure.
There was something, it seemed to him, that the wrong word would
bring down on his head, something that would so at least ease off
his tension. But he wanted not to speak the wrong word; that
would make everything ugly. He wanted the knowledge he lacked to
drop on him, if drop it could, by its own august weight. If she was
to forsake him it was surely for her to take leave. This was why he
didn't directly ask her again what she knew; but it was also why,

approaching the matter from another side, he said to her in the course of his visit: "What do you regard as the very worst that at this time of day *can* happen to me?"

He had asked her that in the past often enough; they had, with the odd irregular rhythm of their intensities and avoidances, exchanged ideas about it and then had seen the ideas washed away by cool intervals, washed like figures traced in sea-sand. It had ever been the mark of their talk that the oldest allusions in it required but a little dismissal and reaction to come out again, sounding for the hour as new. She could thus at present meet his enquiry quite freshly and patiently. "Oh yes, I've repeatedly thought, only it always seemed to me of old that I couldn't quite make up my mind. I thought of dreadful things, between which it was difficult to choose; and so must you have done."

"Rather! I feel now as if I had scarce done anything else. I appear to myself to have spent my life in thinking of nothing *but* dreadful things. A great many of them I've at different times named to you, but there were others I couldn't name."

"They were too, too dreadful?"

"Too, too dreadful—some of them."

She looked at him a minute, and there came to him as he met it an inconsequent sense that her eyes, when one got their full clearness, were still as beautiful as they had been in youth, only beautiful with a strange cold light—a light that somehow was a part of the effect, if it wasn't rather a part of the cause, of the pale hard sweetness of the season and the hour. "And yet," she said at last, "there are horrors we've mentioned."

It deepened the strangeness to see her, as such a figure in such a picture, talk of "horrors," but she was to do in a few minutes something stranger yet—though even of this he was to take the full measure but afterwards—and the note of it already trembled. It was, for the matter of that, one of the signs that her eyes were having again the high flicker of their prime. He had to admit, however, what she said. "Oh yes, there were times when we did go far." He caught himself in the act of speaking as if it all were over. Well, he wished it were; and the consummation depended for him clearly more and more on his friend.

But she had now a soft smile. "Oh far—!"

It was oddly ironic. "Do you mean you're prepared to go further?"

She was frail and ancient and charming as she continued to look at him, yet it was rather as if she had lost the thread. "Do you consider that we went far?"

"Why I thought it the point you were just making—that we *had* looked most things in the face."

"Including each other?" She still smiled. "But you're quite right. We've had together great imaginations, often great fears; but some of them have been unspoken."

"Then the worst—we haven't faced that. I *could* face it, I believe, if I knew what you think it. I feel," he explained, "as if I had lost my power to conceive such things." And he wondered if he looked as blank as he sounded. "It's spent."

"Then why do you assume," she asked, "that mine isn't?"

"Because you've given me signs to the contrary. It isn't a question for you of conceiving, imagining, comparing. It isn't a question now of choosing." At last he came out with it. "You know something I don't. You've shown me that before."

These last words had affected her, he made out in a moment, exceedingly, and she spoke with firmness. "I've shown you, my dear, nothing."

He shook his head. "You can't hide it."

"Oh, oh!" May Bartram sounded over what she couldn't hide. It was almost a smothered groan.

"You admitted it months ago, when I spoke of it to you as of something you were afraid I should find out. Your answer was that I couldn't, that I wouldn't, and I don't pretend I have. But you had something therefore in mind, and I now see how it must have been, how it still is, the possibility that, of all possibilities, has settled itself for you as the worst. This," he went on, "is why I appeal to you. I'm only afraid of ignorance to-day—I'm not afraid of knowledge." And then as for a while she said nothing: "What makes me sure is that I see in your face and feel here, in this air and amid these appearances, that you're out of it. You've done. You've had your experience. You leave me to my fate."

Well, she listened, motionless and white in her chair, as on a decision to be made, so that her manner was fairly an avowal, though still, with a small fine inner stiffness, an imperfect surrender. "It *would* be the worst," she finally let herself say. "I mean the thing I've never said."

It hushed him a moment. "More monstrous than all the monstrosities we've named?"

"More monstrous. Isn't that what you sufficiently express," she asked, "in calling it the worst?"

Marcher thought. "Assuredly—if you mean, as I do, something that includes all the loss and all the shame that are thinkable."

"It would if it *should* happen," said May Bartram. "What we're speaking of, remember, is only my idea."

"It's your belief," Marcher returned. "That's enough for me. I feel your beliefs are right. Therefore if, having this one, you give me no more light on it, you abandon me."

"No, no!" she repeated. "I'm with you—don't you see?—still."
And as to make it more vivid to him she rose from her chair—a
movement she seldom risked in these days—and showed herself, all
draped and all soft, in her fairness and slimness. "I haven't forsaken
you."

It was really, in its effort against weakness, a generous assurance,
and had the success of the impulse not, happily, been great, it
would have touched him to pain more than to pleasure. But the
cold charm in her eyes had spread, as she hovered before him, to all
the rest of her person, so that it was for the minute almost a recovery of youth. He couldn't pity her for that; he could only take her
as she showed—as capable even yet of helping him. It was as if, at
the same time, her light might at any instant go out; wherefore he
must make the most of it. There passed before him with intensity
the three or four things he wanted most to know; but the question
that came of itself to his lips really covered the others. "Then tell
me if I shall consciously suffer."

She promptly shook her head. "Never!"

It confirmed the authority he imputed to her, and it produced on
him an extraordinary effect. "Well, what's better than that? Do you
call that the worst?"

"You think nothing is better?" she asked.

She seemed to mean something so special that he again sharply
wondered, though still with the dawn of a prospect of relief. "Why
not, if one doesn't *know*?" After which, as their eyes, over his question, met in a silence, the dawn deepened and something to his purpose came prodigiously out of her very face. His own, as he took it
in, suddenly flushed to the forehead, and he gasped with the force
of a perception to which, on the instant, everything fitted. The
sound of his gasp filled the air; then he became articulate. "I see—
if I don't suffer!"

In her own look, however, was doubt. "You see what?"

"Why what you mean—what you've always meant."

She again shook her head. "What I mean isn't what I've always
meant. It's different."

"It's something new?"

She hung back from it a little. "Something new. It's not what you
think. I see what you think."

His divination drew breath then; only her correction might be
wrong. "It isn't that I *am* a blockhead?" he asked between faintness
and grimness. "It isn't that it's all a mistake?"

"A mistake?" she pityingly echoed. *That* possibility, for her, he
saw, would be monstrous; and if she guaranteed him the immunity
from pain it would accordingly not be what she had in mind. "Oh
no," she declared; "it's nothing of that sort. You've been right."

Yet he couldn't help asking himself if she weren't, thus pressed,

speaking but to save him. It seemed to him he should be most in a
hole if his history should prove all a platitude. "Are you telling me
the truth, so that I shan't have been a bigger idiot than I can bear
to know? I *haven't* lived with a vain imagination, in the most
besotted illusion? I haven't waited but to see the door shut in my
face?"

She shook her head again. "However the case stands *that* isn't
the truth. Whatever the reality, it *is* a reality. The door isn't shut.
The door's open," said May Bartram.

"Then something's to come?"

She waited once again, always with her cold sweet eyes on him.
"It's never too late." She had, with her gliding step, diminished the
distance between them, and she stood nearer to him, close to him, a
minute, as if still charged with the unspoken. Her movement might
have been for some finer emphasis of what she was at once hesitating and deciding to say. He had been standing by the chimney-piece,
fireless and sparely adorned, a small perfect old French clock and
two morsels of rosy Dresden constituting all its furniture; and her
hand grasped the shelf while she kept him waiting, grasped it a
little as for support and encouragement. She only kept him waiting,
however; that is he only waited. It had become suddenly, from her
movement and attitude, beautiful and vivid to him that she had
something more to give him; her wasted face delicately shone with
it—it glittered almost as with the white lustre of silver in her
expression. She was right, incontestably, for what he saw in her face
was the truth, and strangely, without consequence, while their talk
of it as dreadful was still in the air, she appeared to present it as
inordinately soft. This, prompting bewilderment, made him but gape
the more gratefully for her revelation, so that they continued for
some minutes silent, her face shining at him, her contact imponder-
ably pressing, and his stare all kind but all expectant. The end,
none the less, was that what he had expected failed to come to him.
Something else took place instead, which seemed to consist at first
in the mere closing of her eyes. She gave way at the same instant to
a slow fine shudder, and though he remained staring—though he
stared in fact but the harder—turned off and regained her chair. It
was the end of what she had been intending, but it left him thinking only of that.

"Well, you don't say—?"

She had touched in her passage a bell near the chimney and had
sunk back strangely pale. "I'm afraid I'm too ill."

"Too ill to tell me?" It sprang up sharp to him, and almost to his
lips, the fear she might die without giving him light. He checked
himself in time from so expressing his question, but she answered
as if she had heard the words.

"Don't you know—now?"

" 'Now'—?" She had spoken as if some difference had been made within the moment. But her maid, quickly obedient to her bell, was already with them. "I know nothing." And he was afterwards to say to himself that he must have spoken with odious impatience, such an impatience as to show that, supremely disconcerted, he washed his hands of the whole question.

"Oh!" said May Bartram.

"Are you in pain?" he asked as the woman went to her.

"No," said May Bartram.

Her maid, who had put an arm round her as if to take her to her room, fixed on him eyes that appealingly contradicted her; in spite of which, however, he showed once more his mystification. "What then has happened?"

She was once more, with her companion's help, on her feet, and, feeling withdrawal imposed on him, he had blankly found his hat and gloves and had reached the door. Yet he waited for her answer. "What *was* to," she said.

V

He came back the next day, but she was then unable to see him, and as it was literally the first time this had occurred in the long stretch of their acquaintance he turned away, defeated and sore, almost angry—or feeling at least that such a break in their custom was really the beginning of the end—and wandered alone with his thoughts, especially with the one he was least able to keep down. She was dying and he would lose her; she was dying and his life would end. He stopped in the Park, into which he had passed, and stared before him at his recurrent doubt. Away from her the doubt pressed again; in her presence he had believed her, but as he felt his forlornness he threw himself into the explanation that, nearest at hand, had most of a miserable warmth for him and least of a cold torment. She had deceived him to save him—to put him off with something in which he should be able to rest. What could the thing that was to happen to him be, after all, but just this thing that had begun to happen? Her dying, her death, his consequent solitude—*that* was what he had figured as the Beast in the Jungle, that was what had been in the lap of the gods. He had had her word for it as he left her—what else on earth could she have meant? It wasn't a thing of a monstrous order; not a fate rare and distinguished; not a stroke of fortune that overwhelmed and immortalised; it had only the stamp of the common doom. But poor Marcher at this hour judged the common doom sufficient. It would serve his turn, and even as the consummation of infinite waiting he would bend his pride to accept it. He sat down on a bench in the twilight. He hadn't been a fool. Something had *been*, as she had said, to come. Before he rose indeed it had quite struck him that the final fact really matched with the long avenue through which he had had to reach it. As sharing his suspense and as giving herself all, giving her life, to bring it to an end, she had come with him every step of the way. He had lived by her aid, and to leave her behind would be cruelly, damnably to miss her. What could be more overwhelming than that?

Well, he was to know within the week, for though she kept him a while at bay, left him restless and wretched during a series of days on each of which he asked about her only again to have to turn away, she ended his trial by receiving him where she had always received him. Yet she had been brought out at some hazard into the presence of so many of the things that were, consciously, vainly, half their past, and there was scant service left in the gentleness of her mere desire, all too visible, to check his obsession and wind up his long trouble. That was clearly what she wanted, the one thing more for her own peace while she could still put out her hand. He was so affected by her state that, once seated by her chair, he was moved to let everything go; it was she herself therefore who brought him back, took up again, before she dismissed him, her last word of the other time. She showed how she wished to leave their business in order. "I'm not sure you understood. You've nothing to wait for more. It *has* come."

Oh how he looked at her! "Really?"

"Really."

"The thing that, as you said, *was* to?"

"The thing that we began in our youth to watch for."

Face to face with her once more he believed her; it was a claim to which he had so abjectly little to oppose. "You mean that it has come as a positive definite occurrence, with a name and a date?"

"Positive. Definite. I don't know about the 'name,' but oh with a date!"

He found himself again too helplessly at sea. "But come in the night—come and passed me by?"

May Bartram had her strange faint smile. "Oh no, it hasn't passed you by!"

"But if I haven't been aware of it and it hasn't touched me—?"

"Ah your not being aware of it"—and she seemed to hesitate an instant to deal with this—"your not being aware of it is the strangeness *in* the strangeness. It's the wonder *of* the wonder." She spoke as with the softness almost of a sick child, yet now at last, at the end of all, with the perfect straightness of a sibyl. She visibly knew that she knew, and the effect on him was of something co-ordinate, in its high character, with the law that had ruled him. It was the true voice of the law; so on her lips would the law itself have sounded. "It *has* touched you," she went on. "It has done its office. It has made you all its own."

"So utterly without my knowing it?"

"So utterly without your knowing it." His hand, as he leaned to her, was on the arm of her chair, and, dimly smiling always now, she placed her own on it. "It's enough if *I* know it."

"Oh!" he confusedly breathed, as she herself of late so often had done.

"What I long ago said is true. You'll never know now, and I think you ought to be content. You've *had* it," said May Bartram.

"But had what?"

"Why what was to have marked you out. The proof of your law. It has acted. I'm too glad," she then bravely added, "to have been able to see what it's *not*."

He continued to attach his eyes to her, and with the sense that it was all beyond him, and that *she* was too, he would still have sharply challenged her hadn't he so felt it an abuse of her weakness to do more than take devoutly what she gave him, take it hushed as to a revelation. If he did speak, it was out of the foreknowledge of his loneliness to come. "If you're glad of what it's 'not' it might then have been worse?"

She turned her eyes away, she looked straight before her; with which after a moment: "Well, you know our fears."

He wondered. "It's something then we never feared?"

On this slowly she turned to him. "Did we ever dream, with all our dreams, that we should sit and talk of it thus?"

He tried for a little to make out that they had; but it was as if their dreams, numberless enough, were in solution in some thick cold mist through which thought lost itself. "It might have been that we couldn't talk?"

"Well"—she did her best for him—"not from this side. This, you see," she said, "is the *other* side."

"I think," poor Marcher returned, "that all sides are the same to me." Then, however, as she gently shook her head in correction: "We mightn't, as it were, have got across—?"

"To where we are—no. We're *here*"—she made her weak emphasis.

"And much good does it do us!" was her friend's frank comment.

"It does us the good it can. It does us the good that *it* isn't here. It's past. It's behind," said May Bartram. "Before—" but her voice dropped.

He had got up, not to tire her, but it was hard to combat his yearning. She after all told him nothing but that his light had failed —which he knew well enough without her. "Before—?" he blankly echoed.

"Before, you see, it was always to *come*. That kept it present."

"Oh I don't care what comes now! Besides," Marcher added, "it seems to me I liked it better present, as you say, than I can like it absent with *your* absence."

"Oh mine!"—and her pale hands made light of it.

"With the absence of everything." He had a dreadful sense of standing there before her for—so far as anything but this proved, this bottomless drop was concerned—the last time of their life. It rested on him with a weight he felt he could scarce bear, and this weight it apparently was that still pressed out what remained in him of speakable protest. "I believe you; but I can't begin to pretend I understand. *Nothing*, for me, is past; nothing *will* pass till I pass myself, which I pray my stars may be as soon as possible. Say, however," he added, "that I've eaten my cake, as you contend, to the last crumb—how can the thing I've never felt at all be the thing I was marked out to feel?"

She met him perhaps less directly, but she met him unperturbed. "You take your 'feelings' for granted. You were to suffer your fate. That was not necessarily to know it."

"How in the world—when what is such knowledge but suffering?"

She looked up at him a while in silence. "No— you don't understand."

"I suffer," said John Marcher.

"Don't, don't!"

"How can I help at least *that?*"

"*Don't!*" May Bartram repeated.

She spoke it in a tone so special, in spite of her weakness, that he stared an instant—stared as if some light, hitherto hidden, had shimmered across his vision. Darkness again closed over it, but the gleam had already become for him an idea. "Because I haven't the right—?"

"Don't *know*—when you needn't," she mercifully urged. "You needn't—for we shouldn't."

"Shouldn't?" If he could but know what she meant!

"No—it's too much."

"Too much?" he still asked but, with a mystification that was the next moment of a sudden to give way. Her words, if they meant something, affected him in this light—the light also of her wasted face—as meaning *all*, and the sense of what knowledge had been for herself came over him with a rush which broke through into a question. "Is it of that then you're dying?"

She but watched him, gravely at first, as to see, with this, where he was, and she might have seen something or feared something that moved her sympathy. "I would live for you still—if I could." Her eyes closed for a little, as if, withdrawn into herself, she were for a last time trying. "But I can't!" she said as she raised them again to take leave of him.

She couldn't indeed, as but too promptly and sharply appeared, and he had no vision of her after this that was anything but darkness and doom. They had parted for ever in that strange talk; access to her chamber of pain, rigidly guarded, was almost wholly forbidden him; he was feeling now moreover, in the face of doctors, nurses, the two or three relatives attracted doubtless by the pre-sumption of what she had to "leave," how few were the rights, as they were called in such cases, that he had to put forward, and how odd it might even seem that their intimacy shouldn't have given him more of them. The stupidest fourth cousin had more, even though she had been nothing in such a person's life. She had been a feature of features in *his*, for what else was it to have been so indispensa-ble? Strange beyond saying were the ways of existence, baffling for him the anomaly of his lack, as he felt it to be, of producible claim. A woman might have been, as it were, everything to him, and it might yet present him in no connexion that any one seemed held to recognise. If this was the case in these closing weeks it was the case more sharply on the occasion of the last offices rendered, in the great grey London cemetery, to what had been mortal, to what had been precious, in his freind. The concourse at her grave was not numerous, but he saw himself treated as scarce more nearly con-cerned with it than if there had been a thousand others. He was in short from this moment face to face with the fact that he was to profit extraordinarily little by the interest May Bartram had taken in him. He couldn't quite have said what he expected, but he hadn't surely expected this approach to a double privation. Not only had her interest failed him, but he seemed to feel himself unattended—and for a reason he couldn't seize—by the distinction, the dignity, the propriety, if nothing else, of the man markedly bereaved. It was as if in the view of society he had not *been* mark-edly bereaved, as if there still failed some sign or proof of it, and as if none the less his character could never be affirmed nor the deficiency ever made up. There were moments as the weeks went by when he would have liked, by some almost aggressive act, to take his stand on the intimacy of his loss, in order that it *might* be ques-tioned and his retort, to the relief of his spirit, so recorded; but the moments of an irritation more helpless followed fast on these, the moments during which, turning things over with a good conscience but with a bare horizon, he found himself wondering if he oughtn't to have begun, so to speak, further back.

He found himself wondering at many things, and this last speculation had others to keep it company. What could he have done, after all, in her lifetime, without giving them both, as it were, away? He couldn't have made known she was watching him, for that would have published the superstition of the Beast. This was what closed his mouth now—now that the Jungle had been threshed to vacancy and that the Beast had stolen away. It sounded too foolish and too flat; the difference for him in this particular, the extinction in his life of the element of suspense, was such as in fact to surprise him. He could scarce have said what the effect resem-bled; the abrupt cessation, the positive prohibition, of music per-haps, more than anything else, in some place all adjusted and all accustomed to sonority and to attention. If he could at any rate have conceived lifting the veil from his image at some moment of the past (what had he done, after all, if not lift it to *her*?) so to do this to-day, to talk to people at large of the Jungle cleared and confide to them that he now felt it as safe, would have been not only to see them listen as to a goodwife's tale, but really ro hear himself tell one. What it presently came to in truth was that poor Marcher waded through his beaten grass, where no life stirred, where no breath sounded, where no evil eye seemed to gleam from a possible lair, very much as if vaguely looking for the Beast, and still more as if acutely missing it. He walked about in an existence that had grown strangely more spacious, and, stopping fitfully in places where the undergrowth of life struck him as closer, asked himself yearningly, wondered secretly and sorely, if it would have lurked here or there. It would have at all events *sprung*; what was at least complete was his belief in the truth of the assurance given him. The change from his old sense to his new was absolute and final: what was to happen *had* so absolutely and finally happened that he was as little able to know a fear for his future as to know a hope; so absent in short was any question of anything still to come. He was to live entirely with the other question, that of his unidentified past, that of his having to see his fortune impenetrably muffled and masked.

The torment of this vision became then his occupation; he couldn't perhaps have consented to live but for the possibility of guessing. She had told him, his friend, not to guess; she had forbidden him, so far as he might, to know, and she had even in a sort denied the power in him to learn: which were so many things, precisely, to deprive him of rest. It wasn't that he wanted, he argued for fairness, that anything past and done should repeat itself; it was only that he shouldn't, as an anticlimax, have been taken sleeping so sound as not to be able to win back by an effort of thought the lost stuff of consciousness. He declared to himself at moments that he would either win it back or have done with consciousness for ever; he made this idea his one motive in fine, made it so much his passion that none other, to compare with it, seemed ever to have touched him. The lost stuff of consciousness became thus for him as a strayed or stolen child to an unappeasable father; he hunted it up and down very much as if he were knocking at doors and enquiring of the police. This was the spirit in which, inevitably, he set himself to travel; he started on a journey that was to be as long as he could make it; it danced before him that, as the other side of the globe couldn't possibly have less to say to him, it might, by a possibility of suggestion, have more. Before he quitted London, however, he made a pilgrimage to May Bartram's grave, took his way to it through the endless avenues of the grim suburban metropolis, sought it out in the wilderness of tombs, and, though he had come but for the renewal of the act of farewell, found himself, when he had at last stood by it, beguiled into long intensities. He stood for an hour, powerless to turn away and yet powerless to penetrate the darkness of death; fixing with his eyes her inscribed name and date, beating his forehead against the fact of the secret they kept, drawing his breath, while he waited, as if some sense would in pity of him rise from the stones. He kneeled on the stones, however, in vain; they kept what they concealed; and if the face of the tomb did become a face for him it was because her two names became a pair of eyes that didn't know him. He gave them a last long look, but no palest light broke.

VI

He stayed away, after this, for a year; he visited the depths of Asia, spending himself on scenes of romantic interest, of superlative sanctity; but what was present to him everywhere was that for a man who had known what *he* had known the world was vulgar and vain. The state of mind in which he had lived for so many years shone out to him, in reflexion, as a light that coloured and refined, a light beside which the glow of the East was garish cheap and thin. The terrible truth was that he had lost—with everything else—a distinction as well; the things he saw couldn't help being common when he had become common to look at them. He was simply now one of them himself—he was in the dust, without a peg for the sense of difference; and there were hours when, before the temples of gods and the sepulchres of kings, his spirit turned for nobleness of association to the barely discriminated slab in the London suburb. That had become for him, and more intensely with time and distance, his one witness of a past glory. It was all that was left to him for proof or pride, yet the past glories of Pharaohs were nothing to him as he thought of it. Small wonder then that he came back to it on the morrow of his return. He was drawn there this time as irresistibly as the other, yet with a confidence, almost, that was doubtless the effect of the many months that had elapsed. He had lived, in spite of himself, into his change of feeling, and in wandering over the earth had wandered, as might be said, from the circumference to the centre of his desert. He had settled to his safety and accepted perforce his extinction; figuring to himself, with some colour, in the likeness of certain little old men he remembered to have seen, of whom, all meagre and wizened as they might look, it was related that they had in their time fought twenty duels or been loved by ten princesses. They indeed had been wondrous for others while he was but wondrous for himself; which, however, was exactly the cause of his haste to renew the wonder by getting back, as he might put it, into his own presence. That had quickened his steps and checked his delay. If his visit was prompt it was because he had been separated so long from the part of himself that alone he now valued.

It's accordingly not false to say that he reached his goal with a certain elation and stood there again with a certain assurance. The creature beneath the sod *knew* of his rare experience, so that, strangely now, the place had lost for him its mere blankness of expression. It met him in mildness—not, as before, in mockery; it wore for him the air of conscious greeting that we find, after absence, in things that have closely belonged to us and which seem to confess of themselves to the connexion. The plot of ground, the graven tablet, the tended flowers affected him so as belonging to him that he resembled for the hour a contented landlord reviewing a piece of property. Whatever had happened—well, had happened. He had not come back this time with the vanity of that question, his former worrying "What, *what?*" now practically so spent. Yet he would none the less never again so cut himself off from the spot; he would come back to it every month, for if he did nothing else by its aid he at least held up his head. It thus grew for him, in the oddest way, a positive resource; he carried out his idea of periodical returns, which took their place at last among the most inveterate of his habits. What it all amounted to, oddly enough, was that in his finally so simplified world this garden of death gave him the few square feet of earth on which he could still most live. It was as if, being nothing anywhere else for any one, nothing even for himself, he were just everything here, and if not for a crowd of witnesses or indeed for any witness but John Marcher, then by clear right of the register that he could scan like an open page. The open page was the tomb of his friend, and *there* were the facts of the past, there the truth of his life, there the backward reaches in which he could lose himself. He did this from time to time with such effect that he seemed to wander through the old years with his hand in the arm of a companion who was, in the most extraordinary manner, his other, his younger self; and to wander, which was more extraordinary yet, round and round a third presence—not wandering she, but stationary, still, whose eyes, turning with his revolution, never ceased to follow him, and whose seat was his point, so to speak, of orientation. Thus in short he settled to live—feeding all on the sense that he once *had* lived, and dependent on it not alone for a support but for an identity.

It sufficed him in its way for months and the year elapsed; it would doubtless even have carried him further but for an accident, superficially slight, which moved him, quite in another direction, with a force beyond any of his impressions of Egypt or of India. It was a thing of the merest chance—the turn, as he afterwards felt, of a hair, though he was indeed to live to believe that if light hadn't come to him in this particular fashion it would still have come in

another. He was to live to believe this, I say, though he was not to
live, I may not less definitely mention, to do much else. We allow
him at any rate the benefit of the conviction, struggling up for him
at the end, that, whatever might have happened or not happened,
he would have come round of himself to the light. The incident of
an autumn day had put the match to the train laid from of old by
his misery. With the light before him he knew that even of late his
ache had only been smothered. It was strangely drugged, but it
throbbed; at the touch it began to bleed. And the touch, in the
event, was the face of a fellow mortal. This face, one grey afternoon
when the leaves were thick in the alleys, looked into Marcher's own,
at the cemetery, with an expression like the cut of a blade. He felt
it, that is, so deep down that he winced at the steady thrust. The
person who so mutely assaulted him was a figure he had noticed, on
reaching his own goal, absorbed by a grave a short distance away, a
grave apparently fresh, so that the emotion of the visitor would
probably match it for frankness. This fact alone forbade further
attention, though during the time he stayed he remained vaguely
conscious of his neighbour, a middle-aged man apparently, in
mourning, whose bowed back, among the clustered monuments and
mortuary yews, was constantly presented. Marcher's theory that
these were elements in contact with which he himself revived, had
suffered, on this occasion, it may be granted, a marked, an excessive
check. The autumn day was dire for him as none had recently been,
and he rested with a heaviness he had not yet known on the low
stone table that bore May Bartram's name. He rested without
power to move, as if some spring in him, some spell vouchsafed, had
suddenly been broken for ever. If he could have done that moment
as he wanted he would simply have stretched himself on the slab
that was ready to take him, treating it as a place prepared to receive
his last sleep. What in all the wide world had he now to keep
awake for? He stared before him with the question, and it was then
that, as one of the cemetery walks passed near him, he caught the
shock of the face.

His neighbour at the other grave had withdrawn, as he himself,
with force enough in him, would have done by now, and was advancing along the path on his way to one of the gates. This brought
him close, and his pace was slow, so that—and all the more as there
was a kind of hunger in his look—the two men were for a minute
directly confronted. Marcher knew him at once for one of the
deeply stricken—a perception so sharp that nothing else in the picture comparatively lived, neither his dress, his age, nor his presumable character and class; nothing lived but the deep ravage of the features he showed. He *showed* them—that was was the point; he was
moved, as he passed, by some impulse that was either a signal for
sympathy or, more possibly, a challenge to an opposed sorrow. He
might already have been aware of our friend, might at some pre-

vious hour have noticed in him the smooth habit of the scene, with which the state of his own senses so scantly consorted, and might thereby have been stirred as by an overt discord. What Marcher was at all events conscious of was in the first place that the image of scarred passion presented to him was conscious too—of something that profaned the air; and in the second that, roused, startled, shocked, he was yet the next moment looking after it, as it went, with envy. The most extraordinary thing that had happened to him —though he had given that name to other matters as well—took place, after his immediate vague stare, as a consequence of this impression. The stranger passed, but the raw glare of his grief remained, making our friend wonder in pity what wrong, what wound it expressed, what injury not to be healed. What had the man *had*, to make him by the loss of it so bleed and yet live?

Something—and this reached him with a pang—that *he*, John Marcher, hadn't; the proof of which was precisely John Marcher's arid end. No passion had ever touched him, for this was what passion meant; he had survived and maundered and pined, but where had been *his* deep ravage? The extraordinary thing we speak of was the sudden rush of the result of this question. The sight that had just met his eyes named to him, as in letters of quick flame, something he had utterly, insanely missed, and what he had missed made these things a train of fire, made them mark themselves in an anguish of inward throbs. He had seen *outside* of his life, not learned it within, the way a woman was mourned when she had been loved for herself: such was the force of his conviction of the meaning of the stranger's face, which still flared for him as a smoky torch. It hadn't come to him, the knowledge, on the wings of experience; it had brushed him, jostled him, upset him, with the disrespect of chance, the insolence of accident. Now that the illumination had begun, however, it blazed to the zenith, and what he presently stood there gazing at was the sounded void of his life. He gazed, he drew breath, in pain; he turned in his dismay, and, turning, he had before him in sharper incision than ever the open page of his story. The name on the table smote him as the passage of his neighbour had done, and what it said to him, full in the face, was that *she* was what he had missed. This was the awful thought, the answer to all the past, the vision at the dread clearness of which he grew as cold as the stone beneath him. Everything fell together, confessed, explained, overwhelmed; leaving him most of all stupefied at the blindness he had cherished. The fate he had been marked for he had met with a vengeance—he had emptied the cup to the lees; he had been the man of his time, *the* man, to whom nothing on earth was to have happened. That was the rare stroke—that was his visitation. So he saw it, as we say, in pale horror, while the pieces fitted and fitted. So *she* had seen it while he didn't, and so she served at this hour to drive the truth home. It was the truth, vivid and monstrous, that all the while he had waited the wait was itself his portion. This the companion of his vigil had at a given moment made out, and she had then offered him the chance to baffle his doom. One's doom, however, was never baffled, and on the day she told him his own had come down she had seen him but stupidly stare at the escape she offered him.

The escape would have been to love her; then, *then* he would have lived. *She* had lived—who could say now with what passion? —since she had loved him for himself; whereas he had never thought of her (ah how it hugely glared at him!) but in the chill of his egotism and the light of her use. Her spoken words came back to him—the chain stretched and stretched. The Beast had lurked indeed, and the Beast, at its hour, had sprung; it had sprung in that twilight of the cold April when, pale, ill, wasted, but all beautiful, and perhaps even then recoverable, she had risen from her chair to stand before him and let him imaginably guess. It had sprung as he didn't guess; it had sprung as she hopelessly turned from him, and the mark, by the time he left her, had fallen where it *was* to fall. He had justified his fear and achieved his fate; he had failed, with the last exactitude, of all he was to fail of; and a moan now rose to his lips as he remembered she had prayed he mightn't know. This horror of waking—*this* was knowledge, knowledge under the breath of which the very tears in his eyes seemed to freeze. Through them, none the less, he tried to fix it and hold it; he kept it there before him so that he might feel the pain. That at least, belated and bitter, had something of the taste of life. But the bitterness suddenly sickened him, and it was as if, horribly, he saw in the truth, in the cruelty of his image, what had been appointed and done. He saw the Jungle of his life and saw the lurking Beast; then, while he looked, perceived it, as by a stir of the air, rise, huge and hideous, for the leap that was to settle him. His eyes darkened—it was close; and, instinctively turning, in his hallucination, to avoid it, he flung himself, face down, on the tomb.

1901   1903, 1909

The Jolly Corner[1]
I

"Every one asks me what I 'think' of everything," said Spencer Brydon; "and I make answer as I can—begging or dodging the question, putting them off with any nonsense. It wouldn't matter to any of them really," he went on, "for even were it possible to meet in that stand-and-deliver way so silly a demand on so big a subject, my 'thoughts' would still be almost altogether about something that concerns only myself." He was talking to Miss Staverton, with whom for a couple of months now he had availed himself of

---

1. Shortly after its appearance in the *English Review* for December, 1908, *The Jolly Corner* was thoroughly revised for inclusion as part of Volume XVII of the New York edition (1909), the source of the present text.

every possible occasion to talk; this disposition and this resource, this comfort and support, as the situation in fact presented itself, having promptly enough taken the first place in the considerable array of rather unattenuated surprises attending his so strangely belated return to America. Everything was somehow a surprise; and that might be natural when one had so long and so consistently neglected everything, taken pains to give surprises so much margin for play. He had given them more than thirty years—thirty-three, to be exact; and they now seemed to him to have organised their performance quite on the scale of that licence. He had been twenty-three on leaving New York—he was fifty-six to-day: unless indeed he were to reckon as he had sometimes, since his repatriation, found himself feeling; in which case he would have lived longer than is often allotted to man. It would have taken a century, he repeatedly said to himself, and said also to Alice Staverton, it would have taken a longer absence and a more averted mind than those even of which he had been guilty, to pile up the differences, the newnesses, the queernesses, above all the bignesses, for the better or the worse, that at present assaulted his vision wherever he looked.

The great fact all the while however had been the incalculability; since he *had* supposed himself, from decade to decade, to be allowing, and in the most liberal and intelligent manner, for brilliancy of change. He actually saw that he had allowed for nothing; he missed what he would have been sure of finding, he found what he would never have imagined. Proportions and values were upside-down; the ugly things he had expected, the ugly things of his far-away youth, when he had too promptly waked up to a sense of the ugly—these uncanny phenomena placed him rather, as it happened, under the charm; whereas the "swagger" things, the modern, the monstrous, the famous things, those he had more particularly, like thousands of ingenuous enquirers every year, come over to see, were exactly his sources of dismay. They were as so many set traps for displeasure, above all for reaction, of which his restless tread was constantly pressing the spring. It was interesting, doubtless, the whole show, but it would have been too disconcerting hadn't a certain finer truth saved the situation. He had distinctly not, in this steadier light, come over *all* for the monstrosities; he had come, not only in the last analysis but quite on the face of the act, under an impulse with which they had nothing to do. He had come —putting the thing pompously—to look at his "property," which he had thus for a third of a century not been within four thousand miles of; or, expressing it less sordidly, he had yielded to the humour of seeing again his house on the jolly corner, as he usually, and quite fondly, described it—the one in which he had first seen the light, in which various members of his family had lived and had died, in which the holidays of his overschooled boyhood had been passed and the few social flowers of his chilled adolescence gathered,

and which, alienated then for so long a period, had, through the
successive deaths of his two brothers and the termination of old
arrangements, come wholly into his hands. He was the owner of
another, not quite so "good"—the jolly corner having been, from far
back, superlatively extended and consecrated; and the value of the
pair represented his main capital, with an income consisting, in
these later years, of their respective rents which (thanks precisely to
their original excellent type) had never been depressingly low. He
could live in "Europe," as he had been in the habit of living, on
the product of these flourishing New York leases, and all the better
since, that of the second structure, the mere number in its long
row, having within a twelvemonth fallen in, renovation at a high
advance had proved beautifully possible.

These were items of property indeed, but he had found himself
since his arrival distinguishing more than ever between them. The
house within the street, two bristling blocks westward, was already
in course of reconstruction as a tall mass of flats; he had acceded,
some time before, to overtures for this conversion—in which, now
that it was going forward, it had been not the least of his astonish-
ments to find himself able, on the spot, and though without a previous ounce of such experience, to participate with a certain intelligence, almost with a certain authority. He had lived his life with his
back so turned to such concerns and his face addressed to those of
so different an order that he scarce knew what to make of this lively
stir, in a compartment of his mind never yet penetrated, of a capacity for business and a sense for construction. These virtues, so
common all round him now, had been dormant in his own organism—where it might be said of them perhaps that they had slept
the sleep of the just. At present, in the splendid autumn weather—
the autumn at least was a pure boon in the terrible place—he
loafed about his "work" undeterred, secretly agitated; not in the
least "minding" that the whole proposition, as they said, was vulgar
and sordid, and ready to climb ladders, to walk the plank, to handle
materials and look wise about them, to ask questions, in fine, and
challenge explanations and really "go into" figures.

It amused, it verily quite charmed him; and, by the same stroke,
it amused, and even more, Alice Staverton, though perhaps charming her perceptibly less. She wasn't however going to be better-off
for it, as *he* was—and so astonishingly much: nothing was now
likely, he knew, ever to make her better-off than she found herself,
in the afternoon of life, as the delicately frugal possessor and tenant
of the small house in Irving Place to which she had subtly managed
to cling through her almost unbroken New York career. If he knew
the way to it now better than to any other address among the
dreadful multiplied numberings which seemed to him to reduce the
whole place to some vast ledger-page, overgrown, fantastic, of ruled
and criss-crossed lines and figures—if he had formed, for his consolation, that habit, it was really not a little because of the charm of

his having encountered and recognised, in the vast wilderness of the
wholesale, breaking through the mere gross generalisation of wealth
and force and success, a small still scene where items and shades, all
delicate things, kept the sharpness of the notes of a high voice perfectly trained, and where economy hung about like the scent of a
garden. His old friend lived with one maid and herself dusted her
relics and trimmed her lamps and polished her silver; she stood off,
in the awful modern crush, when she could, but she sallied forth
and did battle when the challenge was really to "spirit," the spirit
she after all confessed to, proudly and a little shyly, as to that of
the better time, that of *their* common, their quite far-away and
antediluvian social period and order. She made use of the street-cars
when need be, the terrible things that people scrambled for as the
panic-stricken at sea scramble for the boats; she affronted, inscrutably, under stress, all the public concussions and ordeals; and yet,
with that slim mystifying grace of her appearance, which defied you
to say if she were a fair young woman who looked older through
trouble, or a fine smooth older one who looked young through successful indifference; with her precious reference, above all, to memories and histories into which he could enter, she was as exquisite for
him as some pale pressed flower (a rarity to begin with), and, failing other sweetnesses, she was a sufficient reward of his effort. They
had communities of knowledge, "their" knowledge (this discriminating possessive was always on her lips) of presences of the other
age, presences all overlaid, in his case, by the experience of a man
and the freedom of a wanderer, overlaid by pleasure, by infidelity,
by passages of life that were strange and dim to her, just by
"Europe" in short, but still unobscured, still exposed and cherished,
under that pious visitation of the spirit from which she had never
been diverted.

She had come with him one day to see how his "apartment-
house" was rising; he had helped her over gaps and explained to her
plans, and while they were there had happened to have, before her,
a brief but lively discussion with the man in charge, the representative of the building-firm that had undertaken his work. He had
found himself quite "standing-up" to this personage over a failure
on the latter's part to observe some detail of one of their noted conditions, and had so lucidly argued his case that, besides ever so prettily flushing, at the time, for sympathy in his triumph, she had afterwards said to him (though to a slightly greater effect of irony) that
he had clearly for too many years neglected a real gift. If he had but
stayed at home he would have anticipated the inventor of the sky-
scraper. If he had but stayed at home he would have discovered
his genius in time really to start some new variety of awful architectural hare and run it till it burrowed in a gold mine. He was to
remember these words, while the weeks elapsed, for the small silver
ring they had sounded over the queerest and deepest of his own
lately most disguised and most muffled vibrations.

It had begun to be present to him after the first fortnight, it had broken out with the oddest abruptness, this particular wanton wonderment: it met him there—and this was the image under which he himself judged the matter, or at least, not a little, thrilled and flushed with it—very much as he might have been met by some strange figure, some unexpected occupant, at a turn of one of the dim passages of an empty house. The quaint analogy quite hauntingly remained with him, when he didn't indeed rather improve it by a still intenser form: that of his opening a door behind which he would have made sure of finding nothing, a door into a room shuttered and void, and yet so coming, with a great suppressed start, on some quite erect confronting presence, something planted in the middle of the place and facing him through the dusk. After that visit to the house in construction he walked with his companion to see the other and always so much the better one, which in the eastward direction formed one of the corners, the "jolly" one precisely, of the street now so generally dishonoured and disfigured in its westward reaches, and of the comparatively conservative Avenue.[2] The Avenue still had pretensions, as Miss Staverton said, to decency; the old people had mostly gone, the old names were unknown, and here and there an old association seemed to stray, all vaguely, like some very aged person, out too late, whom you might meet and feel the impulse to watch or follow, in kindness, for safe restoration to shelter.

They went in together, our friends; he admitted himself with his key, as he kept no one there, he explained, preferring, for his reasons, to leave the place empty, under a simple arrangement with a good woman living in the neighbourhood and who came for a daily hour to open windows and dust and sweep. Spencer Brydon had his reasons and was growingly aware of them; they seemed to him better each time he was there, though he didn't name them all to his companion, any more than he told her as yet how often, how quite absurdly often, he himself came. He only let her see for the present, while they walked through the great blank rooms, that absolute vacancy reigned and that, from top to bottom, there was nothing but Mrs. Muldoon's broomstick, in a corner, to tempt the burglar. Mrs. Muldoon was then on the premises, and she loquaciously attended the visitors, preceding them from room to room and pushing back shutters and throwing up sashes—all to show them, as she remarked, how little there was to see. There was little indeed to see in the great gaunt shell where the main dispositions and the general apportionment of space, the style of an age of ampler allowances, had nevertheless for its master their honest pleading message, affecting him as some good old servant's, some lifelong retainer's appeal for a character, or even for a retiring-pension; yet it was

2. The story is set in the area of Fifth Avenue and 14th Street in Manhattan, where James spent his childhood years.

This detail is one among many which make the story tantalizingly autobiographical.

also a remark of Mrs. Muldoon's that, glad as she was to oblige him by her noonday round, there was a request she greatly hoped he would never make of her. If he should wish her for any reason to come in after dark she would just tell him, if he "plased," that he must ask it of somebody else.

The fact that there was nothing to see didn't militate for the worthy woman against what one *might* see, and she put it frankly to Miss Staverton that no lady could be expected to like, could she? "craping up to thim top storeys in the ayvil hours." The gas and the electric light were off the house, and she fairly evoked a gruesome vision of her march through the great grey rooms—so many of them as there were too!—with her glimmering taper. Miss Staverton met her honest glare with a smile and the profession that she herself certainly would recoil from such an adventure. Spencer Brydon meanwhile held his peace—for the moment; the question of the "evil" hours in his old home had already become too grave for him. He had begun some time since to "crape," and he knew just why a packet of candles addressed to that pursuit had been stowed by his own hand, three weeks before, at the back of a drawer of the fine old sideboard that occupied, as a "fixture," the deep recess in the dining-room. Just now he laughed at his companions—quickly however changing the subject; for the reason that, in the first place, his laugh struck him even at that moment as starting the odd echo, the conscious human resonance (he scarce knew how to qualify it) that sounds made while he was there alone sent back to his ear or his fancy; and that, in the second, he imagined Alice Staverton for the instant on the point of asking him, with a divination, if he ever so prowled. There were divinations he was unprepared for, and he had at all events averted enquiry by the time Mrs. Muldoon had left them, passing on to other parts.

There was happily enough to say, on so consecrated a spot, that could be said freely and fairly; so that a whole train of declarations was precipitated by his friend's having herself broken out, after a yearning look round: "But I hope you don't mean they want you to pull *this* to pieces!" His answer came, promptly, with his re-awakened wrath: it was of course exactly what they wanted, and what they were "at" him for, daily, with the iteration of people who couldn't for their life understand a man's liability to decent feelings. He had found the place, just as it stood and beyond what he could express, an interest and a joy. There were values other than the beastly rent-values, and in short, in short—! But it was thus Miss Staverton took him up. "In short you're to make so good a thing of your sky-scraper that, living in luxury on *those* ill-gotten gains, you can afford for a while to be sentimental here!" Her smile had for him, with the words, the particular mild irony with which he found half her talk suffused; an irony without bitterness and that came, exactly, from her having so much imagination—not, like the cheap sarcasms with which one heard most people, about the world of "society," bid for the reputation of cleverness, from nobody's really having any. It was agreeable to him at this very moment to be sure that when he had answered, after a brief demur, "Well, yes: so, precisely, you may put it!" her imagination would still do him justice. He explained that even if never a dollar were to come to him from the other house he would nevertheless cherish this one; and he dwelt, further, while they lingered and wandered, on the fact of the stupefaction he was already exciting, the positive mystification he felt himself create.

He spoke of the value of all he read into it, the mere sight of the walls, mere shapes of the rooms, mere sound of the floors, mere feel, in his hand, of the old silver-plate knobs of the several mahogany doors, which suggested the pressure of the palms of the dead; the seventy years of the past in fine that these things represented, the annals of nearly three generations, counting his grandfather's, the one that had ended there, and the impalpable ashes of his long-extinct youth, afloat in the very air like microscopic motes. She listened to everything; she was a woman who answered intimately but who utterly didn't chatter. She scattered abroad therefore no cloud of words; she could assent, she could agree, above all she could encourage, without doing that. Only at the last she went a little further than he had done himself. "And then how do you know? You may still, after all, want to live here." It rather indeed pulled him up, for it wasn't what he had been thinking, at least in her sense of the words. "You mean I may decide to stay on for the sake of it?"

"Well, *with* such a home—!" But, quite beautifully, she had too much tact to dot so monstrous an *i*, and it was precisely an illustration of the way she didn't rattle. How could any one—of any wit—insist on any one else's "wanting" to live in New York?

"Oh," he said, "I *might* have lived here (since I had my opportunity early in life); I might have put in here all these years. Then everything would have been different enough—and, I dare say, 'funny' enough. But that's another matter. And then the beauty of it—I mean of my perversity, of my refusal to agree to a 'deal'—is just in the total absence of a reason. Don't you see that if I had a reason about the matter at all it would *have* to be the other way, and would then be inevitably a reason of dollars? There are no reasons here *but* of dollars. Let us therefore have none whatever—not the ghost of one."

They were back in the hall then for departure, but from where they stood the vista was large, through an open door, into the great square main saloon, with its almost antique felicity of brave spaces between windows. Her eyes came back from that reach and met his own a moment. "Are you very sure the 'ghost' of one doesn't, much rather, serve—?"

He had a positive sense of turning pale. But it was as near as they were then to come. For he made answer, he believed, between a glare and a grin: "Oh ghosts—of course the place must swarm with them! I should be ashamed of it if it didn't. Poor Mrs. Muldoon's right, and it's why I haven't asked her to do more than look in."

Miss Staverton's gaze again lost itself, and things she didn't utter, it was clear, came and went in her mind. She might even for the minute, off there in the fine room, have imagined some element dimly gathering. Simplified like the death-mask of a handsome face, it perhaps produced for her just then an effect akin to the stir of an expression in the "set" commemorative plaster. Yet whatever her impression may have been she produced instead a vague platitude. "Well, if it were only furnished and lived in—!"

She appeared to imply that in case of its being still furnished he might have been a little less opposed to the idea of a return. But she passed straight into the vestibule, as if to leave her words behind her, and the next moment he had opened the house-door and was standing with her on the steps. He closed the door and, while he re-pocketed his key, looking up and down, they took in the comparatively harsh actuality of the Avenue, which reminded him of the assault of the outer light of the Desert on the traveller emerging from an Egyptian tomb. But he risked before they stepped into the street his gathered answer to her speech. "For me it *is* lived in. For me it *is* furnished." At which it was easy for her to sigh "Ah yes—!" all vaguely and discreetly; since his parents and his favourite sister, to say nothing of other kin, in numbers, had run their course and met their end there. That represented, within the walls, ineffaceable life.

It was a few days after this that, during an hour passed with her again, he had expressed his impatience of the too flattering curiosity —among the people he met—about his appreciation of New York. He had arrived at none at all that was socially producible, and as for that matter of his "thinking" (thinking the better or the worse of anything there) he was wholly taken up with one subject of thought. It was mere vain egoism, and it was moreover, if she liked, a morbid obsession. He found all things come back to the question of what he personally might have been, how he might have led his life and "turned out," if he had not so, at the outset, given it up. And confessing for the first time to the intensity within him of this absurd speculation—which but proved also, no doubt, the habit of too selfishly thinking—he affirmed the impotence there of any other source of interest, any other native appeal. "What would it have made of me, what would it have made of me? I keep for ever wondering, all idiotically; as if I could possibly know! I see what it has made of dozens of others, those I meet, and it positively aches within me, to the point of exasperation, that it would have made something of me as well. Only I can't make out *what*, and the worry of it, the small rage of curiosity never to be satisfied, brings back what I remember to have felt, once or twice, after judging best, for reasons, to burn some important letter unopened. I've been sorry, I've hated it—I've never known what was in the letter. You may of course say it's a trifle—!"

"I don't say it's a trifle," Miss Staverton gravely interrupted.

She was seated by her fire, and before her, on his feet and restless, he turned to and fro between this intensity of his idea and a
fitful and unseeing inspection, through his single eye-glass, of the
dear little old objects on her chimney-piece. Her interruption made
him for an instant look at her harder. "I shouldn't care if you did!"
he laughed, however; "and it's only a figure, at any rate, for the way
I now feel. *Not* to have followed my perverse young course—and
almost in the teeth of my father's curse, as I may say; not to have
kept it up, so, 'over there,' from that day to this, without a doubt or
a pang; not, above all, to have liked it, to have loved it, so much,
loved it, no doubt, with such an abysmal conceit of my own preference: some variation from *that*, I say, must have produced some different effect for my life and for my 'form.' I should have stuck here
—if it had been possible; and I was too young, at twenty-three, to
judge, *pour deux sous*,[3] whether it *were* possible. If I had waited I
might have seen it was, and then I might have been, by staying
here, something nearer to one of these types who have been hammered so hard and made so keen by their conditions. It isn't that I
admire them so much—the question of any charm in them, or of
any charm, beyond that of the rank money-passion, exerted by their
conditions *for* them, has nothing to do with the matter: it's only a
question of what fantastic, yet perfectly possible, development of
my own nature I mayn't have missed. It comes over me that I had
then a strange *alter ego* deep down somewhere within me, as the
full-blown flower is in the small tight bud, and that I took the
course, I just transferred him to the climate, that blighted him for
once and for ever."

"And you wonder about the flower," Miss Staverton said. "So do
I, if you want to know; and so I've been wondering these several
weeks. I believe in the flower," she continued, "I feel it would
have been quite splendid, quite huge and monstrous."

"Monstrous above all!" her visitor echoed; "and I imagine, by the
same stroke, quite hideous and offensive."

"You don't believe that," she returned; "if you did you wouldn't
wonder. You'd know, and that would be enough for you. What you
feel—and what I feel *for* you—is that you'd have had power."

"You'd have liked me that way?" he asked.

She barely hung fire. "How should I not have liked you?"

"I see. You'd have liked me, have preferred me, a billionaire!"

"How should I not have liked you?" she simply again asked.

He stood before her still—her question kept him motionless. He
took it in, so much there was of it; and indeed his not otherwise
meeting it testified to that. "I know at least what I am," he simply
went on; "the other side of the medal's clear enough. I've not been

---

3. "For two cents," i.e., with my limited experience.

edifying—I believe I'm thought in a hundred quarters to have been barely decent. I've followed strange paths and worshipped strange gods; it must have come to you again and again—in fact you've admitted to me as much—that I was leading, at any time these thirty years, a selfish frivolous scandalous life. And you see what it has made of me."

She just waited, smiling at him. "You see what it has made of *me*."

"Oh you're a person whom nothing can have altered. You were born to be what you are, anywhere, anyway: you've the perfection nothing else could have blighted. And don't you see how, without my exile, I shouldn't have been waiting till now—?" But he pulled up for the strange pang.

"The great thing to see," she presently said, "seems to me to be that it has spoiled nothing. It hasn't spoiled your being here at last. It hasn't spoiled this. It hasn't spoiled your speaking—" She also however faltered.

He wondered at everything her controlled emotion might mean. "Do you believe then—too dreadfully!—that I *am* as good as I might ever have been?"

"Oh no! Far from it!" With which she got up from her chair and was nearer to him. "But I don't care," she smiled.

"You mean I'm good enough?"

She considered a little. "Will you believe it if I say so? I mean will you let that settle your question for you?" And then as if making out in his face that he drew back from this, that he had some idea which, however absurd, he couldn't yet bargain away: "Oh you don't care either—but very differently: you don't care for anything but yourself."

Spencer Brydon recognised it—it was in fact what he had absolutely professed. Yet he importantly qualified. "*He* isn't myself. He's the just so totally other person. But I do want to see him," he added. "And I can. And I shall."

Their eyes met for a minute while he guessed from something in hers that she divined his strange sense. But neither of them otherwise expressed it, and her apparent understanding, with no protesting shock, no easy derision, touched him more deeply than anything yet, constituting for his stifled perversity, on the spot, an element that was like breatheable air. What she said however was unexpected. "Well, *I've* seen him."

"You—?"

"I've seen him in a dream."

"Oh a 'dream'—!" It let him down.

"But twice over," she continued. "I saw him as I see you now."

"You've dreamed the same dream—?"

"Twice over," she repeated. "The very same."

This did somehow a little speak to him, as it also gratified him. "You dream about me at that rate?"

"Ah about *him!*" she smiled.

His eyes again sounded her. "Then you know all about him."
And as she said nothing more: "What's the wretch like?"

She hesitated, and it was as if he were pressing her so hard that,
resisting for reasons of her own, she had to turn away. "I'll tell you
some other time!"

II

It was after this that there was most of a virtue for him, most of
a cultivated charm, most of a preposterous secret thrill, in the particular form of surrender to his obsession and of address to what he
more and more believed to be his privilege. It was what in these
weeks he was living for—since he really felt life to begin but after
Mrs. Muldoon had retired from the scene and, visiting the ample
house from attic to cellar, making sure he was alone, he knew himself in safe possession and, as he tacitly expressed it, let himself go.
He sometimes came twice in the twenty-four hours; the moments
he liked best were those of gathering dusk, of the short autumn twilight; this was the time of which, again and again, he found himself
hoping most. Then he could, as seemed to him, most intimately
wander and wait, linger and listen, feel his fine attention, never in
his life before so fine, on the pulse of the great vague place: he preferred the lampless hour and only wished he might have prolonged
each day the deep crepuscular spell. Later—rarely much before midnight, but then for a considerable vigil—he watched with his glimmering light; moving slowly, holding it high, playing it far, rejoicing
above all, as much as he might, in open vistas, reaches of communication between rooms and by passages; the long straight chance or
show, as he would have called it, for the revelation he pretended to
invite. It was a practice he found he could perfectly "work" without
exciting remark; no one was in the least the wiser for it; even Alice
Staverton, who was moreover a well of discretion, didn't quite fully
imagine.

He let himself in and let himself out with the assurance of calm
proprietorship; and accident so far favoured him that, if a fat
Avenue "officer" had happened on occasion to see him entering at
eleven-thirty, he had never yet, to the best of his belief, been
noticed as emerging at two. He walked there on the crisp November
nights, arrived regularly at the evening's end; it was as easy to do
this after dining out as to take his way to a club or to his hotel.
When he left his club, if he hadn't been dining out, it was ostensibly to go to his hotel; and when he left his hotel, if he had spent a
part of the evening there, it was ostensibly to go to his club. Everything was easy in fine; everything conspired and promoted: there was
truly even in the strain of his experience something that glossed
over, something that salved and simplified, all the rest of consciousness. He circulated, talked, renewed, loosely and pleasantly, old relations—met indeed, so far as he could, new expectations and seemed

to make out on the whole that in spite of the career, of such different contacts, which he had spoken of to Miss Staverton as ministering so little, for those who might have watched it, to edification, he was positively rather liked than not. He was a dim secondary social success—and all with people who had truly not an idea of him. It was all mere surface sound, this murmur of their welcome, this popping of their corks—just as his gestures of response were the extravagant shadows, emphatic in proportion as they meant little, of some game of *ombres chinoises*.[4] He projected himself all day, in thought, straight over the bristling line of hard unconscious heads and into the other, the real, the waiting life; the life that, as soon as he had heard behind him the click of his great house-door, began for him, on the jolly corner, as beguilingly as the slow opening bars of some rich music follows the tap of the conductor's wand.

He always caught the first effect of the steel point of his stick on the old marble of the hall pavement, large black-and-white squares that he remembered as the admiration of his childhood and that had then made in him, as he now saw, for the growth of an early conception of style. The effect was the dim reverberating tinkle as of some far-off bell hung who should say where?— in the depths of the house, of the past, of that mystical other world that might have flourished for him had he not, for weal or woe, abandoned it. On this impression he did ever the same thing; he put his stick noiselessly away in a corner—feeling the place once more in the likeness of some great glass bowl, all precious concave crystal, set delicately humming by the play of a moist finger round its edge. The concave crystal held, as it were, this mystical other world, and the indescribably fine murmur of its rim was the sigh there, the scarce audible pathetic wail to his strained ear, of all the old baffled forsworn possibilities. What he did therefore by this appeal of his hushed presence was to wake them into such measure of ghostly life as they might still enjoy. They were shy, all but unappeasably shy, but they weren't really sinister; at least they weren't as he had hitherto felt them—before they had taken the Form he so yearned to make them take, the Form he at moments saw himself in the light of fairly hunting on tiptoe, the points of his evening-shoes, from room to room and from storey to storey.

That was the essence of his vision—which was all rank folly, if one would, while he was out of the house and otherwise occupied, but which took on the last verisimilitude as soon as he was placed and posted. He knew what he meant and what he wanted; it was as clear as the figure on a cheque presented in demand for cash. His *alter ego* "walked"—that was the note of his image of him, while his image of his motive for his own odd pastime was the desire to waylay him and meet him. He roamed, slowly, warily, but all rest-

4. I.e., Oriental shadow plays.

lessly, he himself did—Mrs. Muldoon had been right, absolutely, with her figure of their "craping"; and the presence he watched for would roam restlessly too. But it would be as cautious and as shifty; the conviction of its probable, in fact its already quite sensible, quite audible evasion of pursuit grew for him from night to night, laying on him finally a rigour to which nothing in his life had been comparable. It had been the theory of many superficially-judging persons, he knew, that he was wasting that life in a surrender to sensations, but he had tasted of no pleasure so fine as his actual tension, had been introduced to no sport that demanded at once the patience and the nerve of this stalking of a creature more subtle, yet at bay perhaps more formidable, than any beast of the forest. The terms, the comparisons, the very practices of the chase positively came again into play; there were even moments when passages of his occasional experience as a sportsman, stirred memories, from his younger time, of moor and mountain and desert, revived for him— and to the increase of his keenness—by the tremendous force of analogy. He found himself at moments—once he had placed his single light on some mantel-shelf or in some recess—stepping back into shelter or shade, effacing himself behind a door or in an embrasure, as he had sought of old the vantage of rock and tree; he found himself holding his breath and living in the joy of the instant, the supreme suspense created by big game alone.

He wasn't afraid (though putting himself the question as he believed gentlemen on Bengal tiger-shoots or in close quarters with the great bear of the Rockies had been known to confess to having put it); and this indeed—since here at least he might be frank!— because of the impression, so intimate and so strange, that he himself produced as yet a dread, produced certainly a strain, beyond the liveliest he was likely to feel. They fell for him into categories, they fairly became familiar, the signs, for his own perception, of the alarm his presence and his vigilance created; though leaving him always to remark, portentously, on his probably having formed a relation, his probably enjoying a consciousness, unique in the experience of man. People enough, first and last, had been in terror of apparitions, but who had ever before so turned the tables and become himself, in the apparitional world, an incalculable terror? He might have found this sublime had he quite dared to think of it; but he didn't too much insist, truly, on that side of his privilege. With habit and repetition he gained to an extraordinary degree the power to penetrate the dusk of distances and the darkness of corners, to resolve back into their innocence the treacheries of uncertain light, the evil-looking forms taken in the gloom by mere shadows, by accidents of the air, by shifting effects of perspective; putting down his dim luminary he could still wander on without it, pass into other rooms and, only knowing it was there behind him in case of need, see his way about, visually project for his purpose a comparative clearness. It made him feel, this acquired faculty, like some monstrous stealthy cat; he wondered if he would have glared at these moments with large shining yellow eyes, and what it mightn't verily be, for the poor hard-pressed *alter ego*, to be confronted with such a type.

He liked however the open shutters; he opened everywhere those Mrs. Muldoon had closed, closing them as carefully afterwards, so that she shouldn't notice: he liked—oh this he did like, and above all in the upper rooms!—the sense of the hard silver of the autumn stars through the window-panes, and scarcely less the flare of the street-lamps below, the white electric lustre which it would have taken curtains to keep out. This was human actual social; this was of the world he had lived in, and he was more at his ease certainly for the countenance, coldly general and impersonal, that all the while and in spite of his detachment it seemed to give him. He had support of course mostly in the rooms at the wide front and the prolonged side; it failed him considerably in the central shades and the parts of the back. But if he sometimes, on his rounds, was glad of his optical reach, so none the less often the rear of the house affected him as the very jungle of his prey. The place was there more subdivided; a large "extension" in particular, where small rooms for servants had been multiplied, abounded in nooks and corners, in closets and passages, in the ramifications especially of an ample back staircase over which he leaned, many a time, to look far down—not deterred from his gravity even while aware that he might, for a spectator, have figured some solemn simpleton playing at hide-and-seek. Outside in fact he might himself make that ironic *rapprochement*;[5] but within the walls, and in spite of the clear windows, his consistency was proof against the cynical light of New York.

It had belonged to that idea of the exasperated consciousness of his victim to become a real test for him; since he had quite put it to himself from the first that, oh distinctly! he could "cultivate" his whole perception. He had felt it as above all open to cultivation— which indeed was but another name for his manner of spending his time. He was bringing it on, bringing it to perfection, by practice; in consequence of which it had grown so fine that he was now aware of impressions, attestations of his general postulate, that couldn't have broken upon him at once. This was the case more specifically with a phenomenon at last quite frequent for him in the upper rooms, the recognition—absolutely unmistakeable, and by a turn dating from a particular hour, his resumption of his campaign after a diplomatic drop, a calculated absence of three nights—of his being definitely followed, tracked at a distance carefully taken and to the express end that he should the less confidently, less arrogantly, appear to himself merely to pursue. It worried, it finally quite broke him up, for it proved, of all the conceivable impressions, the

5. Reconciliation.

one least suited to his book. He was kept in sight while remaining himself—as regards the essence of his position—sightless, and his only recourse then was in abrupt turns, rapid recoveries of ground. He wheeled about, retracing his steps, as if he might so catch in his face at least the stirred air of some other quick revolution. It was indeed true that his fully dislocalised thought of these manœuvres recalled to him Pantaloon, at the Christmas farce, buffeted and tricked from behind by ubiquitous Harlequin;[6] but it left intact the influence of the conditions themselves each time he was re-exposed to them, so that in fact this association, had he suffered it to become constant, would on a certain side have but ministered to his inten-ser gravity. He had made, as I have said, to create on the premises the baseless sense of a reprieve, his three absences; and the result of the third was to confirm the after-effect of the second.

On his return, that night—the night succeeding his last intermis-sion—he stood in the hall and looked up the staircase with a cer-tainty more intimate than any he had yet known. "He's *there*, at the top, and waiting—not, as in general, falling back for disappearance. He's holding his ground, and it's the first time—which is a proof, isn't it? that something has happened for him." So Brydon argued with his hand on the banister and his foot on the lowest stair; in which position he felt as never before the air chilled by his logic. He himself turned cold in it, for he seemed of a sudden to know what now was involved. "Harder pressed?—yes, he takes it in, with its thus making clear to him that I've come, as they say, 'to stay.' He finally doesn't like and can't bear it, in the sense, I mean, that his wrath, his menaced interest, now balances with his dread. I've hunted him till he has 'turned': that, up there, is what has hap-pened—he's the fanged or the antlered animals brought at last to bay." There came to him, as I say—but determined by an influence beyond my notation!—the acuteness of this certainty; under which however the next moment he had broken into a sweat that he would as little have consented to attribute to fear as he would have dared immediately to act upon it for enterprise. It marked none the less a prodigious thrill, a thrill that represented sudden dismay, no doubt, but also represented, and with the selfsame throb, the strangest, the most joyous, possibly the next minute almost the proudest, duplica-tion of consciousness.

"He has been dodging, retreating, hiding, but now, worked up to anger, he'll fight!"—this intense impression made a single mouth-ful, as it were, of terror and applause. But what was wondrous was that the applause, for the felt fact, was so eager, since, if it was his other self he was running to earth, this ineffable identity was thus in the last resort not unworthy of him. It bristled there—some-where near at hand, however unseen still—as the hunted thing, even as the trodden worm of the adage *must* at last bristle; and

6. Harlequin is a character in comedy and pantomime who plays tricks on Panta-loon, a lean, old dotard.

Brydon at this instant tasted probably of a sensation more complex than had ever before found itself consistent with sanity. It was as if it would have shamed him that a character so associated with his own should triumphantly succeed in just skulking, should to the end not risk the open; so that the drop of his danger was, on the spot, a great lift of the whole situation. Yet with another rare shift of the same subtlety he was already trying to measure by how much more he himself might now be in peril of fear; so rejoicing that he could, in another form, actively inspire that fear, and simultaneously quaking for the form in which he might passively know it.

The apprehension of knowing it must after a little have grown in him, and the strangest moment of his adventure perhaps, the most memorable or really most interesting, afterwards, of his crisis, was the lapse of certain instants of concentrated conscious *combat*, the sense of a need to hold on to something, even after the manner of a man slipping and slipping on some awful incline; the vivid impulse, above all, to move, to act, to charge, somehow and upon something —to show himself, in a word, that he wasn't afraid. The state of "holding-on" was thus the state to which he was momentarily reduced; if there had been anything, in the great vacancy, to seize, he would presently have been aware of having clutched it as he might under a shock at home have clutched the nearest chair-back. He had been surprised at any rate—of this he *was* aware—into something unprecedented since his original appropriation of the place; he had closed his eyes, held them tight, for a long minute, as with that instinct of dismay and that terror of vision. When he opened them the room, the other contiguous rooms, extraordinarily, seemed lighter—so light, almost, that at first he took the change for day. He stood firm, however that might be, just where he had paused; his resistance had helped him—it was as if there were something he had tided over. He knew after a little what this was—it had been in the imminent danger of flight. He had stiffened his will against going; without this he would have made for the stairs, and it seemed to him that, still with his eyes closed, he would have descended them, would have known how, straight and swiftly, to the bottom.

Well, as he had held out, here he was—still at the top, among the more intricate upper rooms and with the gauntlet of the others, of all the rest of the house, still to run when it should be his time to go. He would go at his time—only at his time: didn't he go every night very much at the same hour? He took out his watch— there was light for that: it was scarcely a quarter past one, and he had never withdrawn so soon. He reached his lodgings for the most part at two—with his walk of a quarter of an hour. He would wait for the last quarter—he wouldn't stir till then; and he kept his watch there with his eyes on it, reflecting while he held it that this deliberate wait, a wait with an effort, which he recognised, would serve perfectly for the attestation he desired to make. It would prove his courage—unless indeed the latter might most be proved by his budging at last from his place. What he mainly felt now was that, since he hadn't originally scuttled, he had his dignities—which had never in his life seemed so many—all to preserve and to carry aloft. This was before him in truth as a physical image, an image almost worthy of an age of greater romance. That remark indeed glimmered for him only to glow the next instant with a finer light; since what age of romance, after all, could have matched either the state of his mind or, "objectively," as they said, the wonder of his situation? The only difference would have been that, brandishing his dignities over his head as in a parchment scroll, he might then— that is in the heroic time—have proceeded downstairs with a drawn sword in his other grasp.

At present, really, the light he had set down on the mantel of the next room would have to figure his sword; which utensil, in the course of a minute, he had taken the requisite number of steps to possess himself of. The door between the rooms was open, and from the second another door opened to a third. These rooms, as he remembered, gave all three upon a common corridor as well, but there was a fourth, beyond them, without issue save through the preceding. To have moved, to have heard his step again, was appreciably a help; though even in recognising this he lingered once more a little by the chimney-piece on which his light had rested. When he next moved, just hesitating where to turn, he found himself considering a circumstance that, after his first and comparatively vague apprehension of it, produced in him the start that often attends some pang of recollection, the violent shock of having ceased happily to forget. He had come into sight of the door in which the brief chain of communication ended and which he now surveyed from the nearer threshold, the one not directly facing it. Placed at some distance to the left of this point, it would have admitted him to the last room of the four, the room without other approach or egress, had it not, to his intimate conviction, been closed *since* his former visitation, the matter probably of a quarter of an hour before. He stared with all his eyes at the wonder of the fact, arrested again where he stood and again holding his breath while he sounded its sense. Surely it had been *subsequently* closed—that is it had been on his previous passage indubitably open!

He took it full in the face that something had happened between —that he couldn't not have noticed before (by which he meant on his original tour of all the rooms that evening) that such a barrier had exceptionally presented itself. He had indeed since that moment undergone an agitation so extraordinary that it might have muddled for him any earlier view; and he tried to convince himself that he might perhaps then have gone into the room and, inadvertently, automatically, on coming out, have drawn the door after him. The difficulty was that this exactly was what he never did; it was against his whole policy, as he might have said, the essence of which was to keep vistas clear. He had them from the first, as he was well aware, quite on the brain: the strange apparition, at the far end of one of them, of his baffled "prey" (which had become by so sharp an irony so little the term now to apply!) was the form of success his imagination had most cherished, projecting into it always a refinement of beauty. He had known fifty times the start of perception that had afterwards dropped; had fifty times gasped to himself "There!" under some fond brief hallucination. The house, as the case stood, admirably lent itself; he might wonder at the taste, the native architecture of the particular time, which could rejoice so in the multiplication of doors—the opposite extreme to the modern, the actual almost complete proscription of them; but it had fairly contributed to provoke this obsession of the presence encountered telescopically, as he might say, focussed and studied in diminishing perspective and as by a rest for the elbow.

It was with these considerations that his present attention was charged—they perfectly availed to make what he saw portentous. He *couldn't*, by any lapse, have blocked that aperture; and if he hadn't, if it was unthinkable, why what else was clear but that there had been another agent? Another agent?—he had been catching, as he felt, a moment back, the very breath of him; but when had he been so close as in this simple, this logical, this completely personal act? It was so logical, that is, that one might have *taken* it for personal; yet for what did Brydon take it, he asked himself, while, softly panting, he felt his eyes almost leave their sockets. Ah this time at last they *were*, the two, the opposed projections of him, in presence; and this time, as much as one would, the question of danger loomed. With it rose, as not before, the question of courage—for what he knew the blank face of the door to say to him was "Show us how much you have!" It stared, it glared back at him with that challenge; it put to him the two alternatives: should he just push it open or not? Oh to have this consicousness was to *think*—and to think, Brydon knew, as he stood there, was, with the lapsing moments, not to have acted! Not to have acted—that was the misery and the pang—was even still not to act; was in fact *all* to feel the thing in another, in a new and terrible way. How long did he pause and how long did he debate? There was presently nothing to measure it; for his vibration had already changed—as just by the effect of its intensity. Shut up there, at bay, defiant, and with the prodigy of the thing palpably proveably *done*, thus giving notice like some stark signboard—under that accession of accent the situation itself had turned; and Brydon at last remarkably made up his mind on what it had turned to.

It had turned altogether to a different admonition; to a supreme hint, for him, of the value of Discretion! This slowly dawned, no doubt—for it could take its time; so perfectly, on his threshold, had he been stayed, so little as yet had he either advanced or retreated. It was the strangest of all things that now when, by his taking ten steps and applying his hand to a latch, or even his shoulder and his knee, if necessary, to a panel, all the hunger of his prime need might have been met, his high curiosity crowned, his unrest assuaged—it was amazing, but it was also exquisite and rare, that insistence should have, at a touch, quite dropped from him. Discretion—he jumped at that; and yet not, verily, at such a pitch, because it saved his nerves or his skin, but because, much more valuably, it saved the situation. When I say he "jumped" at it I feel the consonance of this term with the fact that—at the end indeed of I know not how long—he did move again, he crossed straight to the door. He wouldn't touch it—it seemed now that he might *if* he would: he would only just wait there a little, to show, to prove, that he wouldn't. He had thus another station, close to the thin partition by which revelation was denied him; but with his eyes bent and his hands held off in a mere intensity of stillness. He listened as if there had been something to hear, but this attitude, while it lasted, was his own communication. "If you won't then—good: I spare you and I give up. You affect me as by the appeal positively for pity: you convince me that for reasons rigid and sublime—what do I know?—we both of us should have suffered. I respect them then, and, though moved and privileged as, I believe, it has never been given to man, I retire, I renounce—never, on my honour, to try again. So rest for ever—and let *me!*"

That, for Brydon was the deep sense of this last demonstration—solemn, measured, directed, as he felt it to be. He brought it to a close, he turned away; and now verily he knew how deeply he had been stirred. He retraced his steps, taking up his candle, burnt, he observed, well-nigh to the socket, and marking again, lighten it as he would, the distinctness of his footfall; after which in a moment, he knew himself at the other side of the house. He did here what he had not yet done at these hours—he opened half a casement, one of those in the front, and let in the air of the night; a thing he would have taken at any time previous for a sharp rupture of his spell. His spell was broken now, and it didn't matter —broken by his concession and his surrender, which made it idle henceforth that he should ever come back. The empty street—its other life so marked even by the great lamplit vacancy—was within call, within touch; he stayed there as to be in it again, high above it though he was still perched; he watched as for some comforting common fact, some vulgar human note, the passage of a scavenger or a thief, some night-bird however base. He would have blessed that sign of life; he would have welcomed positively the slow approach of his friend the policeman, whom he had hitherto only sought to avoid, and was not sure that if the patrol had come into sight he mightn't have felt the impulse to get into relation with it, to hail it, on some pretext, from his fourth floor.

The pretext that wouldn't have been too silly or too compromising, the explanation that would have saved his dignity and kept his name, in such a case, out of the papers, was not definite to him: he was so occupied with the thought of recording his Discretion—as an effect of the vow he had just uttered to his intimate adversary—that the importance of this loomed large and something had overtaken all ironically his sense of proportion. If there had been a ladder applied to the front of the house, even one of the vertiginous perpendiculars employed by painters and roofers and sometimes left standing overnight, he would have managed somehow, astride of the windowsill, to compass by outstretched leg and arm that mode of descent. If there had been some such uncanny thing as he had found in his room at hotels, a workable fire-escape in the form of notched cable or a canvas shoot, he would have availed himself of it as a proof—well, of his present delicacy. He nursed that sentiment, as the question stood, a little in vain, and even—at the end of he scarce knew, once more, how long—found it, as by the action on his mind of the failure of response of the outer world, sinking back to vague anguish. It seemed to him he had waited an age for some stir of the great grim hush; the life of the town was itself under a spell —so unnaturally, up and down the whole prospect of known and rather ugly objects, the blankness and the silence lasted. Had they ever, he asked himself, the hard-faced houses, which had begun to look livid in the dim dawn, had they ever spoken so little to any need of his spirit? Great built voids, great crowded stillnesses put on, often, in the heart of cities, for the small hours, a sort of sinister mask, and it was of this large collective negation that Brydon presently became conscious—all the more that the break of day was, almost incredibly, now at hand, proving to him what a night he had made of it.

He looked again at his watch, saw what had become of his time-values (he had taken hours for minutes—not, as in other tense situations, minutes for hours) and the strange air of the streets was but the weak, the sullen flush of a dawn in which everything was still locked up. His choked appeal from his own open window had been the sole note of life, and he could but break off at last as for a worse despair. Yet while so deeply demoralised he was capable again of an impulse denoting—at least by his present measure—extraordinary resolution; of retracing his steps to the spot where he had turned cold with the extinction of his last pulse of doubt as to there being in the place another presence than his own. This required an effort strong enough to sicken him; but he had his reason, which overmastered for the moment everything else. There was the whole of the rest of the house to traverse, and how should he screw himself to that if the door he had seen closed were at present open? He could hold to the idea that the closing had practically been for him an act of mercy, a chance offered him to descend, depart, get off the ground and never again profane it. This conception held together, it worked; but what it meant for him depended now clearly on the amount of forbearance his recent action, or rather his recent inaction, had engendered. The image of the "presence," whatever it was, waiting there for him to go—this image had not yet been so concrete for his nerves as when he stopped short of the point at which certainty would have come to him. For, with all his resolution, or more exactly with all his dread, he did stop short—he hung back from really seeing. The risk was too great and his fear too definite: it took at this moment an awful specific form.

He knew—yes, as he had never known anything—that, *should* he see the door open, it would all too abjectly be the end of him. It would mean that the agent of his shame—for his shame was the deep abjection—was once more at large and in general possession; and what glared him thus in the face was the act that this would determine for him. It would send him straight about to the window he had left open, and by that window, be long ladder and dangling rope as absent as they would, he saw himself uncontrollably insanely fatally take his way to the street. The hideous chance of this he at least could avert; but he could only avert it by recoiling in time from assurance. He had the whole house to deal with, this fact was still there; only he now knew that uncertainty alone could start him. He stole back from where he had checked himself—merely to do so was suddenly like safety—and, making blindly for the greater staircase, left gaping rooms and sounding passages behind. Here was the top of the stairs, with a fine large dim descent and three spacious landings to mark off. His instinct was all for mildness, but his feet were harsh on the floors, and, strangely, when he had in a couple of minutes become aware of this, it counted somehow for help. He couldn't have spoken, the tone of his voice would have scared him, and the common conceit or resource of "whistling in the dark" (whether literally or figuratively) have appeared basely vulgar; yet he liked none the less to hear himself go, and when he had reached his first landing—taking it all with no rush, but quite steadily—that stage of success drew from him a gasp of relief.

The house, withal, seemed immense, the scale of space again inordinate; the open rooms, to no one of which his eyes deflected, gloomed in their shuttered state like mouths of caverns; only the high skylight that formed the crown of the deep well created for him a medium in which he could advance, but which might have been, for queerness of colour, some watery under-world. He tried to think of something noble, as that his property was really grand, a splendid possession; but this nobleness took the form too of the clear delight with which he was finally to sacrifice it. They might come in now, the builders, the destroyers—they might come as soon as they would. At the end of two flights he had dropped to another zone, and from the middle of the third, with only one more left, he recognised the influence of the lower windows, of half-drawn blinds, of the occasional gleam of street-lamps, of the glazed spaces of the vestibule. This was the bottom of the sea, which showed an illumination of its own and which he even saw paved—when at a given moment he drew up to sink a long look over the banisters—with the marble squares of his childhood. By that time indubitably he felt, as he might have said in a commoner cause, better; it had allowed him to stop and draw breath; and the ease increased with the sight of the old black-and-white slabs. But what he most felt was that now surely, with the element of impunity pulling him as by hard firm hands, the case was settled for what he might have seen above had he dared that last look. The closed door, blessedly remote now, was still closed—and he had only in short to reach that of the house.

He came down further, he crossed the passage forming the access to the last flight; and if here again he stopped an instant it was almost for the sharpness of the thrill of assured escape. It made him shut his eyes—which opened again to the straight slope of the remainder of the stairs. Here was impunity still, but impunity almost excessive; inasmuch as the sidelights and the high fan-tracery of the entrance were glimmering straight into the hall; an appearance produced, he the next instant saw, by the fact that the vestibule gaped wide, that the hinged halves of the inner door had been thrown far back. Out of that again the *question* sprang at him, making his eyes, as he felt, half-start from his head, as they had done, at the top of the house, before the sign of the other door. If he had left that one open, hadn't he left this one closed, and wasn't he now in *most* immediate presence of some inconceivable occult activity? It was as sharp, the question, as a knife in his side, but the answer hung fire still and seemed to lose itself in the vague darkness to which the thin admitted dawn, glimmering archwise over the whole outer door, made a semicircular margin, a cold silvery nimbus that seemed to play a little as he looked—to shift and expand and contract.

It was as if there had been something within it, protected by indistinctness and corresponding in extent with the opaque surface behind, the painted panels of the last barrier to his escape, of which the key was in his pocket. The indistinctness mocked him even while he stared, affected him as somehow shrouding or challenging certitude, so that after faltering an instant on his step he let himself go with the sense that here *was* at last something to meet, to touch, to take, to know—something all unnatural and dreadful, but to advance upon which was the condition for him either of liberation or of supreme defeat. The penumbra, dense and dark, was the virtual screen of a figure which stood in it as still as some image erect in a niche or as some black-vizored sentinel guarding a treasure. Brydon was to know afterwards, was to recall and make out, the particular thing he had believed during the rest of his descent. He saw, in its great grey glimmering margin, the central vagueness diminish, and he felt it to be taking the very form toward which, for so many days, the passion of his curiosity had yearned. It gloomed, it loomed, it was something, it was somebody, the prodigy of a personal presence.

Rigid and conscious, spectral yet human, a man of his own substance and stature waited there to measure himself with his power to dismay. This only could it be—this only till he recognised, with his advance, that what made the face dim was the pair of raised hands that covered it and in which, so far from being offered in defiance, it was buried as for dark deprecation. So Brydon, before him, took him in; with every fact of him now, in the higher light, hard and acute—his planted stillness, his vivid truth, his grizzled bent head and white masking hands, his queer actuality of evening-dress, of dangling double eye-glass, of gleaming silk lappet and white linen, of pearl button and gold watch-guard and polished shoe. No portrait by a great modern master could have presented him with more intensity, thrust him out of his frame with more art, as if there had been "treatment," of the consummate sort, in his every shade and salience. The revulsion, for our friend, had become, before he knew it, immense—this drop, in the act of apprehension, to the sense of his adversary's inscrutable manœuvre. That meaning at least, while he gaped, it offered him; for he could but gape at his other self in this other anguish, gape as a proof that *he*, standing there for the achieved, the enjoyed, the triumphant life, couldn't be faced in his triumph. Wasn't the proof in the splendid covering hands, strong and completely spread?—so spread and so intentional that, in spite of a special verity that surpassed every other, the fact that one of these hands had lost two fingers, which were reduced to stumps, as if accidentally shot away, the face was effectually guarded and saved.

"Saved," though, *would* it be?—Brydon breathed his wonder till the very impunity of his attitude and the very insistence of his eyes produced, as he felt, a sudden stir which showed the next instant as a deeper portent, while the head raised itself, the betrayal of a braver purpose. The hands, as he looked, began to move, to open; then, as if deciding in a flash, dropped from the face and left it uncovered and presented. Horror, with the sight, had leaped into Brydon's throat, gasping there in a sound he couldn't utter; for the bared identity was too hideous as *his*, and his glare was the passion of his protest. The face, *that* face, Spencer Brydon's?—he searched it still, but looking away from it in dismay and denial, falling straight from his height of sublimity. It was unknown, inconceivable, awful, disconnected from any possibility—! He had been "sold," he inwardly moaned, stalking such game as this: the presence before him was a presence, the horror within him a horror, but the waste of his nights had been only grotesque and the success of his adventure an irony. Such an identity fitted his at *no* point, made its alternative monstrous. A thousand times yes, as it came upon him nearer now—the face was the face of a stranger. It came upon him nearer now, quite as one of those expanding fantastic images projected by the magic lantern of childhood; for the stranger, whoever he might be, evil, odious, blatant, vulgar, had advanced as for aggression, and he knew himself give ground. Then harder pressed still, sick with the force of his shock, and falling back as under the hot breath and the roused passion of a life larger than his own, a rage of personality before which his own collapsed, he felt the whole vision turn to darkness and his very feet give way. His head went round; he was going; he had gone.

III

What had next brought him back, clearly—though after how long?—was Mrs. Muldoon's voice, coming to him from quite near, from so near that he seemed presently to see her as kneeling on the ground before him while he lay looking up at her; himself not wholly on the ground, but half-raised and upheld—conscious, yes, of tenderness of support and, more particularly, of a head pillowed in extraordinary softness and faintly refreshing fragrance. He considered, he wondered, his wit but half at his service; then another face intervened, bending more directly over him, and he finally knew that Alice Staverton had made her lap an ample and perfect cushion to him, and that she had to this end seated herself on the lowest degree of the staircase, the rest of his long person remaining stretched on his old black-and-white slabs. They were cold, these marble squares of his youth; but *he* somehow was not, in this rich return of consciousness—the most wonderful hour, little by little, that he had ever known, leaving him, as it did, so gratefully, so abysmally passive, and yet as with a treasure of intelligence waiting all round him for quiet appropriation; dissolved, he might call it, in the air of the place and producing the golden glow of a late autumn afternoon. He had come back, yes—come back from further away than any man but himself had ever travelled; but it was strange how with this sense what he had come back *to* seemed really the great thing, and as if his prodigious journey had been all for the sake of it. Slowly but surely his consciousness grew, his vision of his state thus completing itself: he had been miraculously *carried* back— lifted and carefully borne as from where he had been picked up, the uttermost end of an interminable grey passage. Even with this he was suffered to rest, and what had now brought him to knowledge was the break in the long mild motion.

It had brought him to knowledge, to knowledge—yes, this was the beauty of his state; which came to resemble more and more that of a man who has gone to sleep on some news of a great inherit-

ance, and then, after dreaming it away, after profaning it with matters strange to it, has waked up again to serenity of certitude and
has only to lie and watch it grow. This was the drift of his patience
—that he had only to let it shine on him. He must moreover, with
intermissions, still have been lifted and borne; since why and how
else should he have known himself, later on, with the afternoon
glow intenser, no longer at the foot of his stairs—situated as these
now seemed at that dark other end of his tunnel—but on a deep
window-bench of his high saloon, over which had been spread,
couch-fashion, a mantle of soft stuff lined with grey fur that was
familiar to his eyes and that one of his hands kept fondly feeling as
for its pledge of truth. Mrs. Muldoon's face had gone, but the
other, the second he had recognised, hung over him in a way that
showed how he was still propped and pillowed. He took it all in,
and the more he took it the more it seemed to suffice: he was as
much at peace as if he had had food and drink. It was the two
women who had found him, on Mrs. Muldoon's having plied, at
her usual hour, her latch-key—and on her having above all arrived
while Miss Staverton still lingered near the house. She had been
turning away, all anxiety, from worrying the vain bell-handle—her
calculation having been of the hour of the good woman's visit; but
the latter, blessedly, had come up while she was still there, and they
had entered together. He had then lain, beyond the vestibule, very
much as he was lying now—quite, that is, as he appeared to have
fallen, but all so wondrously without bruise or gash; only in a depth
of stupor. What he most took in, however, at present, with the
steadier clearance, was that Alice Staverton had for a long unspeakable moment not doubted he was dead.

"It must have been that I *was*." He made it out as she held him.
"Yes—I can only have died. You brought me literally to life. Only,"
he wondered, his eyes rising to her, "only, in the name of all the
benedictions, how?"

It took her but an instant to bend her face and kiss him, and
something in the manner of it, and in the way her hands clasped
and locked his head while he felt the cool charity and virtue of her
lips, something in all this beatitude somehow answered everything.
"And now I keep you," she said.

"Oh keep me, keep me!" he pleaded while her face still hung
over him: in response to which it dropped again and stayed close,
clingingly close. It was the seal of their situation—of which he
tasted the impress for a long blissful moment in silence. But he
came back. "Yet how did you know—?"

"I was uneasy. You were to have come, you remember—and you
had sent no word."

"Yes, I remember—I was to have gone to you at one to-day." It
caught on to their "old" life and relation—which were so near and
so far. "I was still out there in my strange darkness—where was it,

what was it? I must have stayed there so long." He could but wonder at depth and the duration of his swoon.

"Since last night?" she asked with a shade of fear for her possible indiscretion.

"Since this morning—it must have been: the cold dim dawn of to-day. Where have I been," he vaguely wailed, "where have I been?" He felt her hold him close, and it was as if this helped him now to make in all security his mild moan. "What a long dark day!"

All in her tenderness she had waited a moment. "In the cold dim dawn?" she quavered.

But he had already gone on piecing together the parts of the whole prodigy. "As I didn't turn up you came straight—?"

She barely cast about. "I went first to your hotel—where they told me of your absence. You had dined out last evening and hadn't been back since. But they appeared to know you had been at your club."

"So you had the idea of *this*—?"

"Of what?" she asked in a moment.

"Well—of what has happened."

"I believed at least you'd have been here. I've known, all along," she said, "that you've been coming."

" 'Known' it—?"

"Well, I've believed it. I said nothing to you after that talk we had a month ago—but I felt sure. I knew you *would*," she declared.

"That I'd persist, you mean?"

"That you'd see him."

"Ah but I didn't!" cried Brydon with his long wail. "There's somebody—an awful beast; whom I brought, too horribly, to bay. But it's not me."

At this she bent over him again, and her eyes were in his eyes. "No—it's not you." And it was as if, while her face hovered, he might have made out in it, hadn't it been so near, some particular meaning blurred by a smile. "No, thank heaven," she repeated— "it's not you! Of course it wasn't to have been."

"Ah but it *was*," he gently insisted. And he stared before him now as he had been staring for so many weeks. "I was to have known myself."

"You couldn't!" she returned consolingly. And then reverting, and as if to account further for what she had herself done, "But it wasn't only *that*, that you hadn't been at home," she went on. "I waited till the hour at which we had found Mrs. Muldoon that day of my going with you; and she arrived, as I've told you, while, failing to bring any one to the door, I lingered in my despair on the steps. After a little, if she hadn't come, by such a mercy, I should have found means to hunt her up. But it wasn't," said Alice Staverton, as if once more with her fine intention—"it wasn't only that."

His eyes, as he lay, turned back to her. "What more then?"

She met it, the wonder she had stirred. "In the cold dim dawn, you say? Well, in the cold dim dawn of this morning I too saw you."

"Saw *me*—?"

"Saw *him*," said Alice Staverton. "It must have been at the same moment."

He lay an instant taking it in—as if he wished to be quite reasonable. "At the same moment?"

"Yes—in my dream again, the same one I've named to you. He came back to me. Then I knew it for a sign. He had come to you."

At this Brydon raised himself; he had to see her better. She helped him when she understood his movements, and he sat up, steadying himself beside her there on the window-bench and with his right hand grasping her left. "*He* didn't come to me."

"You came to yourself," she beautifully smiled.

"Ah I've come to myself now—thanks to you, dearest. But this brute, with his awful face—this brute's a black stranger. He's none of *me*, even as I *might* have been," Brydon sturdily declared.

But she kept the clearness that was like the breath of infallibility. "Isn't the whole point that you'd have been different?"

He almost scowled for it. "As different as *that*—?"

Her look again was more beautiful to him than the things of this world. "Haven't you exactly wanted to know *how* different? So this morning," she said, "you appeared to me."

"Like *him*?"

"A black stranger!"

"Then how did you know it was I?"

"Because, as I told you weeks ago, my mind, my imagination, had worked so over what you might, what you mightn't have been —to show you, you see, how I've thought of you. In the midst of that you came to me—that my wonder might be answered. So I knew," she went on; "and believed that, since the question held you too so fast, as you told me that day, you too would see for yourself. And when this morning I again saw I knew it would be because you had—and also then, from the first moment, because you somehow wanted me. *He* seemed to tell me of that. So why," she strangely smiled, "shouldn't I like him?"

It brought Spencer Brydon to his feet. "You 'like' that horror—?"

"I *could* have liked him. And to me," she said, "he was no horror. I had accepted him."

" 'Accepted'—?" Brydon oddly sounded.

"Before, for the interest of his difference—yes. And as I didn't disown him, as I knew him—which you at last, confronted with him in his difference, so cruelly didn't, my dear—well, he must have been, you see, less dreadful to me. And it may have pleased him that I pitied him."

She was beside him on her feet, but still holding his hand—still with her arm supporting him. But though it all brought for him thus a dim light, "You 'pitied' him?" he grudgingly, resentfully asked.

"He has been unhappy, he has been ravaged," she said.

"And haven't I been unhappy? Am not I—you've only to look at me!— ravaged?"

"Ah I don't say I like him *better*," she granted after a thought. "But he's grim, he's worn—and things have happened to him. He doesn't make shift, for sight, with your charming monocle."

"No"—it struck Brydon: "I couldn't have sported mine 'down-town.' They'd have guyed[7] me there."

"His great convex pince-nez—I saw it, I recognised the kind—is for his poor ruined sight. And his poor right hand—!"

"Ah!" Brydon winced—whether for his proved identity or for his lost fingers. Then, "He has a million a year," he lucidly added. "But he hasn't you."

"And he isn't—no, he isn't—*you!*" she murmured as he drew her to his breast.

1908, 1909

The Art of Fiction[1]

I should not have affixed so comprehensive a title to these few remarks, necessarily wanting in any completeness upon a subject the full consideration of which would carry us far, did I not seem to discover a pretext for my temerity in the interesting pamphlet lately published under this name by Mr. Walter Besant. Mr. Besant's lecture at the Royal Institution—the original form of his pamphlet—appears to indicate that many persons are interested in the art of fiction, and are not indifferent to such remarks, as those who practise it may attempt to make about it. I am therefore anxious not to lose the benefit of this favourable association, and to edge in a few words under cover of the attention which Mr. Besant is sure to have excited. There is something very encouraging in his having put into form certain of his ideas on the mystery of story-telling.

It is a proof of life and curiosity—curiosity on the part of the brotherhood of novelists as well as on the part of their readers. Only a short time ago it might have been supposed that the English novel was not what the French call *discutable*.[2] It had no air of having a theory, a conviction, a consciousness of itself behind it— of being the expression of an artistic faith, the result of choice and comparison. I do not say it was necessarily the worse for that: it would take much more courage than I possess to intimate that the

---

7. Ridiculed.
1. James's most famous (and influential) critical essay was written in response to a lecture on fiction delivered by the English novelist and historian Walter Besant (1836–1901) at the Royal Institution (London) on April 25, 1884. First pub-

lished in *Longman's Magazine* for September, 1884, the essay was reprinted in book form the next year and in *Partial Portraits* (1888), the source of the present text.
2. Debatable.

form of the novel as Dickens and Thackeray (for instance) saw it had any taint of incompleteness. It was, however, *naïf* (if I may help myself out with another French word); and evidently if it be destined to suffer in any way for having lost its *naïveté* it has now an idea of making sure of the corresponding advantages. During the period I have alluded to there was a comfortable, good-humoured feeling abroad that a novel is a novel, as a pudding is a pudding, and that our only business with it could be to swallow it. But within a year or two, for some reason or other, there have been signs of returning animation—the era of discussion·would appear to have been to a certain extent opened. Art lives upon discussion, upon experiment, upon curiosity, upon variety of attempt, upon the exchange of views and the comparison of standpoints; and there is a presumption that those times when no one has anything particular to say about it, and has no reason to give for practice or preference, though they may be times of honour, are not times of development —are times, possibly even, a little of dulness. The successful application of any art is a delightful spectacle, but the theory too is interesting; and though there is a great deal of the latter without the former I suspect there has never been a genuine success that has not had a latent core of conviction. Discussion, suggestion, formulation, these things are fertilising when they are frank and sincere. Mr. Besant has set an excellent example in saying what he thinks, for his part, about the way in which fiction should be written, as well as about the way in which it should be published; for his view of the "art," carried on into an appendix, covers that too. Other labourers in the same field will doubtless take up the argument, they will give it the light of their experience, and the effect will surely be to make our interest in the novel a little more what it had for some time threatened to fail to be—a serious, active, inquiring interest, under protection of which this delightful study may, in moments of confidence, venture to say a little more what it thinks of itself.

It must take itself seriously for the public to take it so. The old superstition about fiction being "wicked" has doubtless died out in England; but the spirit of it lingers in a certain oblique regard directed toward any story which does not more or less admit that it is only a joke. Even the most jocular novel feels in some degree the weight of the proscription that was formerly directed against literary levity: the jocularity does not always succeed in passing for orthodoxy. It is still expected, though perhaps people are ashamed to say it, that a production which is after all only a "make-believe" (for what else is a "story"?) shall be in some degree apologetic—shall renounce the pretension of attempting really to represent life. This, of course, any sensible, wide-awake story declines to do, for it quickly perceives that the tolerance granted to it on such a condition is only an attempt to stifle it disguised in the form of generosity. The old evangelical hostility to the novel, which was as explicit as it was narrow, and which regarded it as little less favourable to our immortal part than a stage-play, was in reality far less insulting. The only reason for the existence of a novel is that it does attempt to represent life. When it relinquishes this attempt, the same attempt that we see on the canvas of the painter, it will have arrived at a very strange pass. It is not expected of the picture that it will make itself humble in order to be forgiven; and the analogy between the art of the painter and the art of the novelist is, so far as I am able to see, complete. Their inspiration is the same, their process (allowing for the different quality of the vehicle), is the same, their success is the same. They may learn from each other, they may explain and sustain each other. Their cause is the same, and the honour of one is the honour of another. The Mahometans think a picture an unholy thing, but it is a long time since any Christian did, and it is therefore the more odd that in the Christian mind the traces (dissimulated though they may be) of a suspicion of the sister art should linger to this day. The only effectual way to lay it to rest is to emphasise the analogy to which I just alluded—to insist on the fact that as the picture is reality, so the novel is history. That is the only general description (which does it justice) that we may give of the novel. But history also is allowed to represent life; it is not, any more than painting, expected to apologise. The subject-matter of fiction is stored up likewise in documents and records, and if it will not give itself away, as they say in California, it must speak with assurance, with the tone of the historian. Certain accomplished novelists have a habit of giving themselves away which must often bring tears to the eyes of people who take their fiction seriously. I was lately struck, in reading over many pages of Anthony Trollope,[3] with his want of discretion in this particular. In a digression, a parenthesis or an aside, he concedes to the reader that he and this trusting friend are only "making believe." He admits that the events he narrates have not really happened, and that he can give his narrative any turn the reader may like best. Such a betrayal of a sacred office seems to me, I confess, a terrible crime; it is what I mean by the attitude of apology, and it shocks me every whit as much in Trollope as it would have shocked me in Gibbon or Macaulay.[4] It implies that the novelist is less occupied in looking for the truth (the truth, of course I mean, that he assumes, the premises that we must grant him, whatever they may be), than the historian, and in doing so it deprives him at a stroke of all his standing-room. To represent and illustrate the past, the actions of men, is the task of either writer, and the only difference that I can see is, in proportion as he succeeds, to the honour of the novelist, consisting as it does in his having more difficulty in collecting his evidence, which is so far from being purely literary. It seems to me to give

3. English novelist (1815–82).
4. Edward Gibbon (1737–94) and
Thomas Babington Macaulay (1800–59), English historians.

him a great character, the fact that he has at once so much in common with the philosopher and the painter; this double analogy is a magnificent heritage.

It is of all this evidently that Mr. Besant is full when he insists upon the fact that fiction is one of the *fine* arts, deserving in its turn of all the honours and emoluments that have hitherto been reserved for the successful profession of music, poetry, painting, architecture. It is impossible to insist too much on so important a truth, and the place that Mr. Besant demands for the work of the novelist may be represented, a trifle less abstractly, by saying that he demands not only that it shall be reputed artistic, but that it shall be reputed very artistic indeed. It is excellent that he should have struck this note, for his doing so indicates that there was need of it, that his proposition may be to many people a novelty. One rubs one's eyes at the thought; but the rest of Mr. Besant's essay confirms the revelation. I suspect in truth that it would be possible to confirm it still further, and that one would not be far wrong in saying that in addition to the people to whom it has never occurred that a novel ought to be artistic, there are a great many others who, if this principle were urged upon them, would be filled with an indefinable mistrust. They would find it difficult to explain their repugnance, but it would operate strongly to put them on their guard. "Art," in our Protestant communities, where so many things have got so strangely twisted about, is supposed in certain circles to have some vaguely injurious effect upon those who make it an important consideration, who let it weigh in the balance. It is assumed to be opposed in some mysterous manner to morality, to amusement, to instruction. When it is embodied in the work of the painter (the sculptor is another affair!) you know what it is: it stands there before you, in the honesty of pink and green and a gilt frame; you can see the worst of it at a glance, and you can be on your guard. But when it is introduced into literature it becomes more insidious—there is danger of its hurting you before you know it. Literature should be either instructive or amusing, and there is in many minds an impression that these artistic preoccupations, the search for form, contribute to neither end, interfere indeed with both. They are too frivolous to be edifying, and too serious to be diverting; and they are moreover priggish and paradoxical and superfluous. That, I think, represents the manner in which the latent thought of many people who read novels as an exercise in skipping would explain itself if it were to become articulate. They would argue, of course, that a novel ought to be "good," but they would interpret this term in a fashion of their own, which indeed would vary considerably from one critic to another. One would say that being good means representing virtuous and aspiring characters, placed in prominent positions; another would say that it depends on a "happy ending," on a distribution at the last of prizes, pensions, husbands, wives, babies, millions, appended paragraphs, and cheerful remarks. Another still would say that it means being full of incident and movement, so that we shall wish to jump ahead, to see who was the mysterious stranger, and if the stolen will was ever found, and shall not be distracted from this pleasure by any tiresome analysis or "description." But they would all agree that the "artistic" idea would spoil some of their fun. One would hold it accountable for all the description, another would see it revealed in the absence of sympathy. Its hostility to a happy ending would be evident, and it might even in some cases render any ending at all impossible. The "ending" of a novel is, for many persons, like that of a good dinner, a course of dessert and ices, and the artist in fiction is regarded as a sort of meddlesome doctor who forbids agreeable aftertastes. It is therefore true that this conception of Mr. Besant's of the novel as a superior form encounters not only a negative but a positive indifference. It matters little that as a work of art it should really be as little or as much of its essence to supply happy endings, sympathetic characters, and an objective tone, as if it were a work of mechanics: the association of ideas, however incongruous, might easily be too much for it if an eloquent voice were not sometimes raised to call attention to the fact that it is at once as free and as serious a branch of literature as any other.

Certainly this might sometimes be doubted in presence of the enormous number of works of fiction that appeal to the credulity of our generation, for it might easily seem that there could be no great character in a commodity so quickly and easily produced. It must be admitted that good novels are much compromised by bad ones, and that the field at large suffers discredit from overcrowding. I think, however, that this injury is only superficial, and that the superabundance of written fiction proves nothing against the principle itself. It has been vulgarised, like all other kinds of literature, like everything else to-day, and it has proved more than some kinds accessible to vulgarisation. But there is as much difference as there ever was between a good novel and a bad one: the bad is swept with all the daubed canvases and spoiled marble into some unvisited limbo, or infinite rubbish-yard beneath the back-windows of the world, and the good subsists and emits its light and stimulates our desire for perfection. As I shall take the liberty of making but a single criticism of Mr. Besant, whose tone is so full of the love of his art, I may as well have done with it at once. He seems to me to mistake in attempting to say so definitely beforehand what sort of an affair the good novel will be. To indicate the danger of such an error as that has been the purpose of these few pages; to suggest that certain traditions on the subject, applied *a priori*, have already had much to answer for, and that the good health of an art which undertakes so immediately to reproduce life must demand that it be perfectly free. It lives upon exercise, and the very meaning of exercise is freedom. The only obligation to which in advance we may hold a novel, without incurring the accusation of being arbitrary, is that it be interesting. That general responsibility rests upon it, but it is the only one I can think of. The ways in which it is at liberty to accomplish this result (of interesting us) strike me as innumerable, and such as can only suffer from being marked out or fenced in by prescription. They are as various as the temperament of man, and they are successful in proportion as they reveal a particular mind, different from others. A novel is in its broadest definition a personal, a direct impression of life: that, to begin with, constitutes its value, which is greater or less according to the intensity of the impression. But there will be no intensity at all, and therefore no value, unless there is freedom to feel and say. The tracing of a line to be followed, of a tone to be taken, of a form to be filled out, is a limitation of that freedom and a suppression of the very thing that we are most curious about. The form, it seems to me, is to be appreciated after the fact: then the author's choice has been made, his standard has been indicated; then we can follow lines and directions and compare tones and resemblances. Then in a word we can enjoy one of the most charming of pleasures, we can estimate quality, we can apply the test of execution. The execution belongs to the author alone; it is what is most personal to him, and we measure him by that. The advantage, the luxury, as well as the torment and responsibility of the novelist, is that there is no limit to what he may attempt as an executant—no limit to his possible experiments, efforts, discoveries, successes. Here it is especially that he works, step by step, like his brother of the brush, of whom we may always say that he has painted his picture in a manner best known to himself. His manner is his secret, not necessarily a jealous one. He cannot disclose it as a general thing if he would; he would be at a loss to teach it to others. I say this with a due recollection of having insisted on the community of method of the artist who paints a picture and the artist who writes a novel. The painter *is* able to teach the rudiments of his practice, and it is possible, from the study of good work (granted the aptitude), both to learn how to paint and to learn how to write. Yet it remains true, without injury to the *rapprochement*, that the literary artist would be obliged to say to his pupil much more than the other, "Ah, well, you must do it as you can!" It is a question of degree, a matter of delicacy. If there are exact sciences, there are also exact arts, and the grammar of painting is so much more definite that it makes the difference.

I ought to add, however, that if Mr. Besant says at the beginning of his essay that the "laws of fiction may be laid down and taught with as much precision and exactness as the laws of harmony, perspective, and proportion," he mitigates what might appear to be an extravagance by applying his remark to "general" laws, and by expressing most of these rules in a manner with which it would certainly be unaccommodating to disagree. That the novelist must write from his experience, that his "characters must be real and such as might be met with in actual life;" that "a young lady brought up in a quiet country village should avoid descriptions of garrison life," and "a writer whose friends and personal experiences belong to the lower middle-class should carefully avoid introducing his characters into society;" that one should enter one's notes in a common-place book; that one's figures should be clear in outline; that making them clear by some trick of speech or of carriage is a bad method, and "describing them at length" is a worse one; that English Fiction should have a "conscious moral purpose;" that "it is almost impossible to estimate too highly the value of careful workmanship—that is, of style;" that "the most important point of all is the story," that "the story is everything": these are principles with most of which it is surely impossible not to sympathise. That remark about the lower middle-class writer and his knowing his place is perhaps rather chilling; but for the rest I should find it difficult to dissent from any one of these recommendations. At the same time, I should find it difficult positively to assent to them, with the exception, perhaps, of the injunction as to entering one's notes in a common-place book. They scarcely seem to me to have the quality that Mr. Besant attributes to the rules of the novelist— the "precision and exactness" of "the laws of harmony, perspective, and proportion." They are suggestive, they are even inspiring, but they are not exact, though they are doubtless as much so as the case admits of: which is a proof of that liberty of interpretation for which I just contended. For the value of these different injunctions —so beautiful and so vague—is wholly in the meaning one attaches to them. The characters, the situation, which strike one as real will be those that touch and interest one most, but the measure of reality is very difficult to fix. The reality of Don Quixote or of Mr. Micawber[5] is a very delicate shade; it is a reality so coloured by the author's vision that, vivid as it may be, one would hesitate to propose it as a model: one would expose one's self to some very embarrassing questions on the part of a pupil. It goes without saying that you will not write a good novel unless you possess the sense of reality; but it will be difficult to give you a recipe for calling that sense into being. Humanity is immense, and reality has a myriad forms; the most one can affirm is that some of the flowers of fiction have the odour of it, and others have not; as for telling you in advance how your nosegay should be composed, that is another affair. It is equally excellent and inconclusive to say that one must write from experience; to our supposititious aspirant such a declaration might savour of mockery. What kind of experience is intended, and where does it begin and end? Experience is never limited, and it is never

5. Main characters in the novel by Cervantes and in Dickens's *David Copperfield*.

complete; it is an immense sensibility, a kind of huge spiderweb of the finest silken threads suspended in the chamber of consciousness, and catching every airborne particle in its tissue. It is the very atmosphere of the mind; and when the mind is imaginative—much more when it happens to be that of a man of genius—it takes to itself the faintest hints of life, it converts the very pulses of the air into revelations. The young lady living in a village has only to be a damsel upon whom nothing is lost to make it quite unfair (as it seems to me) to declare to her that she shall have nothing to say about the military. Greater miracles have been seen than that, imagination assisting, she should speak the truth about some of these gentlemen. I remember an English novelist, a woman of genius,[6] telling me that she was much commended for the impression she had managed to give in one of her tales of the nature and way of life of the French Protestant youth. She had been asked where she learned so much about this recondite being, she had been congratulated on her peculiar opportunities. These opportunities consisted in her having once, in Paris, as she ascended a staircase, passed an open door where, in the household of a *pasteur*,[7] some of the young Protestants were seated at table round a finished meal. The glimpse made a picture; it lasted only a moment, but that moment was experience. She had got her direct personal impression, and she turned out her type. She knew what youth was, and what Protestantism; she also had the advantage of having seen what it was to be French, so that she converted these ideas into a concrete image and produced a reality. Above all, however, she was blessed with the faculty which when you give it an inch takes an ell, and which for the artist is a much greater source of strength than any accident of residence or of place in the social scale. The power to guess the unseen from the seen, to trace the implication of things, to judge the whole piece by the pattern, the condition of feeling life in general so completely that you are well on your way to knowing any particular corner of it—this cluster of gifts may almost be said to constitute experience, and they occur in country and in town, and in the most differing stages of education. If experience consists of impressions, it may be said that impressions *are* experience, just as (have we not seen it?) they are the very air we breathe. Therefore, if I should certainly say to a novice, "Write from experience and experience only," I should feel that this was rather a tantalising monition if I were not careful immediately to add, "Try to be one of the people on whom nothing is lost!"

I am far from intending by this to minimise the importance of exactness—of truth of detail. One can speak best from one's own taste, and I may therefore venture to say that the air of reality (sol-

6. Leon Edel identifies her as novelist Thackeray's daughter Anne, whose first novel contains many of the elements of James's description.
7. Pastor, minister.

idity of specification) seems to me to be the supreme virtue of a novel—the merit on which all its other merits (including that conscious moral purpose of which Mr. Besant speaks) helplessly and submissively depend. If it be not there they are all as nothing, and if these be there, they owe their effect to the success with which the author has produced the illusion of life. The cultivation of this success, the study of this exquisite process, form, to my taste, the beginning and the end of the art of the novelist. They are his inspiration, his despair, his reward, his torment, his delight. It is here in very truth that he competes with life; it is here that he competes with his brother the painter in *his* attempt to render the look of things, the look that conveys their meaning, to catch the colour, the relief, the expression, the surface, the substance of the human spectacle. It is in regard to this that Mr. Besant is well inspired when he bids him take notes. He cannot possibly take too many, he cannot possibly take enough. All life solicits him, and to "render" the simplest surface, to produce the most momentary illusion, is a very complicated business. His case would be easier, and the rule would be more exact, if Mr. Besant had been able to tell him what notes to take. But this, I fear, he can never learn in any manual; it is the business of his life. He has to take a great many in order to select a few, he has to work them up as he can, and even the guides and philosophers who might have most to say to him must leave him alone when it comes to the application of precepts, as we leave the painter in communion with his palette. That his characters "must be clear in outline," as Mr. Besant says —he feels that down to his boots; but how he shall make them so is a secret between his good angel and himself. It would be absurdly simple if he could be taught that a great deal of "description" would make them so, or that on the contrary the absence of description and the cultivation of dialogue, or the absence of dialogue and the multiplication of "incident," would rescue him from his difficulties. Nothing, for instance, is more possible than that he be of a turn of mind for which this odd, literal opposition of description and dialogue, incident and description, has little meaning and light. People often talk of these things as if they had a kind of internecine distinctness, instead of melting into each other at every breath, and being intimately associated parts of one general effort of expression. I cannot imagine composition existing in a series of blocks, nor conceive, in any novel worth discussing at all, of a passage of description that is not in its intention narrative, a passage of dialogue that is not in its intention descriptive, a touch of truth of any sort that does not partake of the nature of incident, or an incident that derives its interest from any other source than the general and only source of the success of a work of art—that of being illustrative. A novel is a living thing, all one and continuous, like any other organism, and in proportion as it lives will it be found, I

think, that in each of the parts there is something of each of the
other parts. The critic who over the close texture of a finished work
shall pretend to trace a geography of items will mark some frontiers
as artificial, I fear, as any that have been known to history. There is
an old-fashioned distinction between the novel of character and the
novel of incident which must have cost many a smile to the intending fabulist who was keen about his work. It apppears to me as little
to the point as the equally celebrated distinction between the novel
and the romance—to answer as little to any reality. There are bad
novels and good novels, as there are bad pictures and good pictures;
but that is the only distinction in which I see any meaning, and I
can as little imagine speaking of a novel of character as I can imagine speaking of a picture of character. When one says picture one
says of character, when one says novel one says of incident, and the
terms may be transposed at will. What is character but the determination of incident? What is incident but the illustration of character? What is either a picture or a novel that is *not* of character?
What else do we seek in it and find in it? It is an incident for a
woman to stand up with her hand resting on a table and look out at
you in a certain way; or if it be not an incident I think it will be
hard to say what it is. At the same time it is an expression of character. If you say you don't see it (character in *that—allons donc!*[8]),
this is exactly what the artist who has reasons of his own for thinking he *does* see it undertakes to show you. When a young man
makes up his mind that he has not faith enough after all to enter
the church as he intended, that is an incident, though you may not
hurry to the end of the chapter to see whether perhaps he doesn't
change once more. I do not say that these are extraordinary or startling incidents. I do not pretend to estimate the degree of interest
proceeding from them, for this will depend upon the skill of the
painter. It sounds almost puerile to say that some incidents are
intrinsically much more important than others, and I need not take
this precaution after having professed my sympathy for the major
ones in remarking that the only classification of the novel that I can
understand is into that which has life and that which has it not.

The novel and the romance, the novel of incident and that of
character—these clumsy separations appear to me to have been
made by critics and readers for their own convenience, and to help
them out of some of their occasional queer predicaments, but to
have little reality or interest for the producer, from whose point of
view it is of course that we are attempting to consider the art of
fiction. The case is the same with another shadowy category which
Mr. Besant apparently is disposed to set up—that of the "modern
English novel"; unless indeed it be that in this matter he has fallen
into an accidental confusion of standpoints. It is not quite clear
whether he intends the remarks in which he alludes to it to be

8. Come now!—i.e., nonsense!

didactic or historical. It is as difficult to suppose a person intending to write a modern English as to suppose him writing an ancient English novel: that is a label which begs the question. One writes the novel, one paints the picture, of one's language and of one's time, and calling it modern English will not, alas! make the difficult task any easier. No more, unfortunately, will calling this or that work of one's fellow-artist a romance—unless it be, of course, simply for the pleasantness of the thing, as for instance when Hawthorne gave this heading to his story of *Blithedale*.⁹ The French, who have brought the theory of fiction to remarkable completeness, have but one name for the novel, and have not attempted smaller things in it, that I can see, for that. I can think of no obligation to which the "romancer" would not be held equally with the novelist; the standard of execution is equally high for each. Of course it is of execution that we are talking—that being the only point of a novel that is open to contention. This is perhaps too often lost sight of, only to produce interminable confusions and cross-purposes. We must grant the artist his subject, his idea, his *donnée*:¹ our criticism is applied only to what he makes of it. Naturally I do not mean that we are bound to like it or find it interesting: in case we do not our course is perfectly simple—to let it alone. We may believe that of a certain idea even the most sincere novelist can make nothing at all, and the event may perfectly justify our belief; but the failure will have been a failure to execute, and it is in the execution that the fatal weakness is recorded. If we pretend to respect the artist at all, we must allow him his freedom of choice, in the face, in particular cases, of innumerable presumptions that the choice will not fructify. Art derives a considerable part of its beneficial exercise from flying in the face of presumptions, and some of the most interesting experiments of which it is capable are hidden in the bosom of common things. Gustave Flaubert has written a story about the devotion of a servant-girl to a parrot,² and the production, highly finished as it is, cannot on the whole be called a success. We are perfectly free to find it flat, but I think it might have been interesting; and I, for my part, am extremely glad he should have written it; it is a contribution to our knowledge of what can be done—or what cannot. Ivan Turgénieff has written a tale about a deaf and dumb serf and a lap-dog,³ and the thing is touching, loving, a little masterpiece. He struck the note of life where Gustave Flaubert missed it—he flew in the face of a presumption and achieved a victory.

Nothing, of course, will ever take the place of the good old fashion of "liking" a work of art or not liking it: the most improved criticism will not abolish that primitive, that ultimate test. I mention this to guard myself from the accusation of intimating that the

---

9. *The Blitheddle Romance* was published in 1852.
1. That which is given.

2. *A Simple Soul*, by Gustave Flaubert (1821–80).
3. *Mumu*, by Ivan Turgenev (1818–83).

idea, the subject, of a novel or a picture, does not matter. It matters,
to my sense, in the highest degree, and if I might put up a prayer it
would be that artists should select none but the richest. Some, as I
have already hastened to admit, are much more remunerative than
others, and it would be a world happily arranged in which persons
intending to treat them should be exempt from confusions and mistakes. This fortunate condition will arrive only, I fear, on the same
day that critics become purged from error. Meanwhile, I repeat, we
do not judge the artist with fairness unless we say to him, "Oh, I
grant you your starting-point, because if I did not I should seem to
prescribe to you, and heaven forbid I should take that responsibility. If I pretend to tell you what you must not take, you will call
upon me to tell you then what you must take; in which case I shall
be prettily caught. Moreover, it isn't till I have accepted your data
that I can begin to measure you. I have the standard, the pitch; I
have no right to tamper with your flute and then criticise your
music. Of course I may not care for your idea at all; I may think it
silly, or stale, or unclean; in which case I wash my hands of you
altogether. I may content myself with believing that you will not
have succeeded in being interesting, but I shall, of course, not
attempt to demonstrate it, and you will be as indifferent to me as I
am to you. I needn't remind you that there are all sorts of tastes:
who can know it better? Some people, for excellent reasons, don't
like to read about carpenters; others, for reasons even better, don't
like to read about courtesans. Many object to Americans. Others (I
believe they are mainly editors and publishers) won't look at Italians. Some readers don't like quiet subjects; others don't like bustling ones. Some enjoy a complete illusion, others the consciousness
of large concessions. They choose their novels accordingly, and if
they don't care about your idea they won't, *a fortiori*,[4] care about
your treatment."

So that it comes back very quickly, as I have said, to the liking:
in spite of M. Zola,[5] who reasons less powerfully than he represents,
and who will not reconcile himself to this absoluteness of taste,
thinking that there are certain things that people ought to like, and
that they can be made to like. I am quite at a loss to imagine anything (at any rate in this matter of fiction) that people *ought* to
like or to dislike. Selection will be sure to take care of itself, for it
has a constant motive behind it. That motive is simply experience.
As people feel life, so they will feel the art that is most closely
related to it. This closeness of relation is what we should never
forget in talking of the effort of the novel. Many people speak of it
as a factitious, artificial form, a product of ingenuity, the business of
which is to alter and arrange the things that surround us, to trans-

4. All the more reason.
5. Émile Zola (1840–1902), French naturalistic novelist and literary theorist

who argued that people should like best
what was represented with the greatest
measure of scientific objectivity.

late them into conventional, traditional moulds. This however, is a view of the matter which carries us but a very short way, condemns the art to an eternal repetition of a few familiar *clichés*, cuts short its development, and leads us straight up to a dead wall. Catching the very note and trick, the strange irregular rhythm of life, that is the attempt whose strenuous force keeps Fiction upon her feet. In proportion as in what she offers us we see life *without* rearrangement do we feel that we are touching the truth; in proportion as we see it *with* rearrangement do we feel that we are being put off with a substitute, a compromise and convention. It is not uncommon to hear an extraordinary assurance of remark in regard to this matter of rearranging, which is often spoken of as if it were the last word of art. Mr. Besant seems to me in danger of falling into the great error with his rather unguarded talk about "selection." Art is essentially selection, but it is a selection whose main care is to be typical, to be inclusive. For many people art means rose-coloured window-panes, and selection means picking a bouquet for Mrs. Grundy.[6] They will tell you glibly that artistic considerations have nothing to do with the disagreeable, with the ugly; they will rattle off shallow commonplaces about the province of art and the limits of art till you are moved to some wonder in return as to the province and the limits of ignorance. It appears to me that no one can ever have made a seriously artistic attempt without becoming conscious of an immense increase—a kind of revelation—of freedom. One perceives in that case—by the light of a heavenly ray—that the province of art is all life, all feeling, all observation, all vision. As Mr. Besant so justly intimates, it is all experience. That is a sufficient answer to those who maintain that it must not touch the sad things of life, who stick into its divine unconscious bosom little prohibitory inscriptions on the end of sticks, such as we see in public gardens—"It is forbidden to walk on the grass; it is forbidden to touch the flowers; it is not allowed to introduce dogs or to remain after dark; it is requested to keep to the right." The young aspirant in the line of fiction whom we continue to imagine will do nothing without taste, for in that case his freedom would be of little use to him; but the first advantage of his taste will be to reveal to him the absurdity of the little sticks and tickets. If he have taste, I must add, of course he will have ingenuity, and my disrespectful reference to that quality just now was not meant to imply that it is useless in fiction. But it is only a secondary aid; the first is a capacity for receiving straight impressions.

Mr. Besant has some remarks on the question of "the story" which I shall not attempt to criticise, though they seem to me to contain a singular ambiguity, because I do not think I understand

---

6. A stock character marked by prudishness who is alluded to in Thomas Mor- ton's *Speed the Plough* (1798).

them. I cannot see what is meant by talking as if there were a part of a novel which is the story and part of it which for mystical reasons is not—unless indeed the distinction be made in a sense in which it is difficult to suppose that any one should attempt to convey anything. "The story," if it represents anything, represents the subject, the idea, the *donnée* of the novel; and there is surely no "school"—Mr. Besant speaks of a school—which urges that a novel should be all treatment and no subject. There must assuredly be something to treat; every school is intimately conscious of that. This sense of the story being the idea, the starting-point, of the novel, is the only one that I see in which it can be spoken of as something different from its organic whole; and since in proportion as the work is successful the idea permeates and penetrates it, informs and animates it, so that every word and every punctuation-point contribute directly to the expression, in that proportion do we lose our sense of the story being a blade which may be drawn more or less out of its sheath. The story and the novel, the idea and the form, are the needle and thread, and I never heard of a guild of tailors who recommended the use of the thread without the needle, or the needle without the thread. Mr. Besant is not the only critic who may be observed to have spoken as if there were certain things in life which constitute stories, and certain others which do not. I find the same odd implication in an entertaining article in the *Pall Mall Gazette*, devoted, as it happens, to Mr. Besant's lecture. "The story is the thing!" says this graceful writer, as if with a tone of opposition to some other idea. I should think it was, as every painter who, as the time for "sending in"[7] his picture looms in the distance, finds himself still in quest of a subject—as every belated artist not fixed about his theme will heartily agree. There are some subjects which speak to us and others which do not, but he would be a clever man who should undertake to give a rule—an index expurgatorius[8]—by which the story and the no-story should be known apart. It is impossible (to me at least) to imagine any such rule which shall not be altogether arbitrary. The writer in the *Pall Mall* opposes the delightful (as I suppose) novel of *Margot la Balafrée*[9] to certain tales in which "Bostonian nymphs" appear to have "rejected English dukes for psychological reasons." I am not acquainted with the romance just designated, and can scarcely forgive the *Pall Mall* critic for not mentioning the name of the author, but the title appears to refer to a lady who may have received a scar in some heroic adventure. I am inconsolable at not being acquainted with this episode,[1] but am utterly at a loss to see why it is a story when the rejection (or acceptance) of a duke is not, and why a reason, psychological or other, is

7. I.e., at the time of shipping his painting to a Royal Academy exhibit.
8. The list of books from which condemned passages must be removed before the book could be read by Catholics.
9. *Margot the Scarred Woman* (1884) by Fortuné du Boisgobey.
1. In truth James's *An International Episode* (1879).

not a subject when a cicatrix[2] is. They are all particles of the multitudinous life with which the novel deals, and surely no dogma which pretends to make it lawful to touch the one and unlawful to touch the other will stand for a moment on its feet. It is the special picture that must stand or fall, according as it seem to possess truth or to lack it. Mr. Besant does not, to my sense, light up the subject by intimating that a story must, under penalty of not being a story, consist of "adventures." Why of adventures more than of green spectacles? He mentions a category of impossible things, and among them he places "fiction without adventure." Why without adventure, more than without matrimony, or celibacy, or parturition, or cholera, or hydropathy,[3] or Jansenism? This seems to me to bring the novel back to the hapless little *rôle* of being an artificial, ingenious thing—bring it down from its large, free character of an immense and exquisite correspondence with life. And what *is* adventure, when it comes to that, and by what sign is the listening pupil to recognise it? It is an adventure—an immense one—for me to write this little article; and for a Bostonian nymph to reject an English duke is an adventure only less stirring, I should say, than for an English duke to be rejected by a Bostonian nymph. I see dramas within dramas in that, and innumerable points of view. A psychological reason is, to my imagination, an object adorably pictorial; to catch the tint of its complexion—I feel as if that idea might inspire one to Titianesque[4] efforts. There are few things more exciting to me, in short, than a psychological reason, and yet, I protest, the novel seems to me the most magnificent form of art. I have just been reading, at the same time, the delightful story of *Treasure Island*,[5] by Mr. Robert Louis Stevenson and, in a manner less consecutive, the last tale from M. Edmond de Goncourt, which is entitled *Chérie*. One of these works treats of murders, mysteries, islands of dreadful renown, hairbreadth escapes, miraculous coincidences and buried doubloons. The other treats of a little French girl who lived in a fine house in Paris, and died of wounded sensibility because no one would marry her. I call *Treasure Island* delightful, because it appears to me to have succeeded wonderfully in what it attempts; and I venture to bestow no epithet upon *Chérie*, which strikes me as having failed deplorably in what it attempts—that is in tracing the development of the moral consciousness of a child. But one of these productions strikes me as exactly as much of a novel as the other, and as having a "story" quite as much. The moral consciousness of a child is as much a part of life as the islands of the Spanish Main, and the one sort of geography seems to me to have those "surprises"—of which Mr. Besant speaks quite as much as the other. For myself (since it comes back

2. Scar.
3. Hydropathy involves the use of water in the treatment of disease; Jansenism is a theological doctrine named after Cornelis Jansen (1585–1638), Dutch theologian who emphasized predestination and the necessity of belonging to the Catholic Church for salvation.
4. Titian (1477–1576), Italian painter highly regarded for his use of color.
5. *Treasure Island* was published in 1883, *Chérie* in 1884.

in the last resort, as I say, to the preference of the individual), the picture of the child's experience has the advantage that I can at successive steps (an immense luxury, near to the "sensual pleasure" of which Mr. Besant's critic in the *Pall Mall* speaks) say Yes or No, as it may be, to what the artist puts before me. I have been a child in fact, but I have been on a quest for a buried treasure only in supposition, and it is a simple accident that with M. de Goncourt I should have for the most part to say No. With George Eliot,[6] when she painted that country with a far other intelligence, I always said Yes.

The most interesting part of Mr. Besant's lecture is unfortunately the briefest passage—his very cursory allusion to the "conscious moral purpose" of the novel. Here again it is not very clear whether he be recording a fact or laying down a principle; it is a great pity that in the latter case he should not have developed his idea. This branch of the subject is of immense importance, and Mr. Besant's few words point to considerations of the widest reach, not to be lightly disposed of. He will have treated the art of fiction but superficially who is not prepared to go every inch of the way that these considerations will carry him. It is for this reason that at the beginning of these remarks I was careful to notify the reader that my reflections on so large a theme have no pretension to be exhaustive. Like Mr. Besant, I have left the question of the morality of the novel till the last, and at the last I find I have used up my space. It is a question surrounded with difficulties, as witness the very first that meets us, in the form of a definite question, on the threshold. Vagueness, in such a discussion, is fatal, and what is the meaning of your morality and your conscious moral purpose? Will you not define your terms and explain how (a novel being a picture) a picture can be either moral or immoral? You wish to paint a moral picture or carve a moral statue: will you not tell us how you would set about it? We are discussing the Art of Fiction; questions of art are questions (in the widest sense) of execution; questions of morality are quite another affair, and will you not let us see how it is that you find it so easy to mix them up? These things are so clear to Mr. Besant that he has deduced from them a law which he sees embodied in English Fiction, and which is "a truly admirable thing and a great cause for congratulation." It is a great cause for congratulation indeed when such thorny problems become as smooth as silk. I may add that in so far as Mr. Besant perceives that in point of fact English Fiction has addressed itself preponderantly to these delicate questions he will appear to many people to have made a vain discovery. They will have been positively struck, on the contrary, with the moral timidity of the usual English novelist; with his (or with her) aversion to face the difficulties with which on every side the treatment of reality bristles. He is apt to be extremely shy (whereas the picture that Mr. Besant draws is a picture of boldness), and the sign of his work, for the most part, is a cautious silence on certain subjects. In the English novel (by which of course I mean the American as well), more than in any other, there is a traditional difference between that which people know and that which they agree to admit that they know, that which they see and that which they speak of, that which they feel to be a part of life and that which they allow to enter into literature. There is the great difference, in short, between what they talk of in conversation and what they talk of in print. The essence of moral energy is to survey the whole field, and I should directly reverse Mr. Besant's remark and say not that the English novel has a purpose, but that it has a diffidence. To what degree a purpose in a work of art is a source of corruption I shall not attempt to inquire; the one that seems to me least dangerous is the purpose of making a perfect work. As for our novel, I may say lastly on this score that as we find it in England to-day it strikes me as addressed in a large degree to "young people," and that this in itself constitutes a presumption that it will be rather shy. There are certain things which it is generally agreed not to discuss, not even to mention, before young people. That is very well, but the absence of discussion is not a symptom of the moral passion. The purpose of the English novel—"a truly admirable thing, and a great cause for congratulation"—strikes me therefore as rather negative.

There is one point at which the moral sense and the artistic sense lie very near together; that is in the light of the very obvious truth that the deepest quality of a work of art will always be the quality of the mind of the producer. In proportion as that intelligence is fine will the novel, the picture, the statue partake of the substance of beauty and truth. To be constituted of such elements is, to my vision, to have purpose enough. No good novel will ever proceed from a superficial mind; that seems to me an axiom which, for the artist in fiction, will cover all needful moral ground: if the youthful aspirant take it to heart it will illuminate for him many of the mysteries of "purpose." There are many other useful things that might be said to him, but I have come to the end of my article, and can only touch them as I pass. The critic in the *Pall Mall Gazette*, whom I have already quoted, draws attention to the danger, in speaking of the art of fiction, of generalising. The danger that he has in mind is rather, I imagine, that of particularising, for there are some comprehensive remarks which, in addition to those embodied in Mr. Besant's suggestive lecture, might without fear of misleading him be addressed to the ingenuous student. I should remind him first of the magnificence of the form that is open to him, which offers to sight so few restrictions and such innumerable opportunities. The other arts, in comparison, appear confined and hampered; the various conditions under which they are exercised are so rigid and definite. But the only condition that I can think of attaching to the composition of the novel is, as I have already said, that it be sincere. This freedom is a splendid privilege, and the first lesson of the young novelist is to learn to be worthy of it. "Enjoy it as it deserves," I should say to him; "take possession of it, explore it to its utmost extent, publish it, rejoice in it. All life belongs to you, and do not listen either to those who would shut you up into corners of it and tell you that it is only here and there that art inhabits, or to those who would persuade you that this heavenly messenger wings her way outside of life altogether, breathing a superfine air, and turning away her head from the truth of things. There is no impression of life, no manner of seeing it and feeling it, to which the plan of the novelist may not offer a place; you have only to remember that talents so dissimilar as those of Alexandre Dumas and Jane Austen, Charles Dickens and Gustave Flaubert have worked in this field with equal glory. Do not think too much about optimism and pessimism; try and catch the colour of life itself. In France to-day we see a prodigious effort (that of Emile Zola, to whose solid and serious work no explorer of the capacity of the novel can allude without respect), we see an extraordinary effort vitiated by a spirit of pessimism on a narrow basis. M. Zola is magnificent, but he strikes an English reader as ignorant; he has an air of working in the dark; if he had as much light as energy, his results would be of the highest value. As for the aberrations of a shallow optimism, the ground (of English fiction especially) is strewn with their brittle particles as with broken glass. If you must indulge in conclusions, let them have the taste of a wide knowledge. Remember that your first duty is to be as complete as possible—to make as perfect a work. Be generous and delicate and pursue the prize."

1884, 1888

KATE CHOPIN
1851–1904

Katherine O'Flaherty did not appear to be destined for a literary career, certainly not one which would end with the overtones of scandal. Her father was a successful businessman, the family enjoyed a high place in St. Louis society, and her mother, grandmother, and great-grandmother were active, pious Catholics. But in part under the influence of her strong-willed great-grandmother (who was also a compelling and tireless storyteller), and long before she began to compose and submit stories in the 1880s for publication, the young Katherine asserted her independence by smoking in company and going about the streets without a companion of either sex—both rather daring acts for the time.

At the age of twenty she married Oscar Chopin (who pronounced his name Show-pan) and spent the next decade in New Orleans, where her husband first prospered, then failed, in the cotton business. After spending a few years in Cloutierville in northwest Louisiana, where her husband had opened a general store and taken over the management of a family cotton plantation, she returned to St. Louis in 1884, a year after her husband's sudden death from swamp fever. A year later her mother died, and at the age of thirty-five Kate Chopin was left essentially alone to raise her children and to fashion a literary career out of her experience of Louisiana life, and her reading of such French contemporary realists as Émile Zola and Guy de Maupassant.

Chopin wrote on a lapboard in the center of a swirl of children. She claimed, moreover, that she wrote on impulse, that she was "completely at the mercy of unconscious selection" of subject, and that "the polishing up process * * * always proved disastrous." This method, while it insured freshness and sincerity, made some of her stories seem anecdotal, and sometimes too loose or thin. In the relatively few years of her writing career— scarcely more than a decade—she completed two novels, over 150 stories and sketches, and a substantial body of poetry, reviews, and criticism. A first novel, *At Fault*, was published in 1890; but it was her early stories of Louisiana rural life, especially the collection *Bayou Folk* (1894), which won her national recognition as a leading practitioner of local-color fiction. She made the Catholic Creoles with their old-fashioned European customs, their polyglot, witty speech, and the lush, semitropical landscape of picturesque Natchitoches Parish as familiar to Americans as Sarah Orne Jewett (whom she admired) was making the townspeople of the coastal region of Maine. A second collection of her stories, *A Night in Acadie*, was published three years later and enhanced that reputation.

Chopin's major work, *The Awakening*, was published in 1899. The novel, which traces the psychological and sexual coming to consciousness of a young woman, predictably aroused hostility among contemporary reviewers, just as Whitman's *Leaves of Grass*, which she admired, had done half a century earlier. The "new woman" of the time, demanding social, economic, and political equality, was already a common topic of public discussion and subject for fiction. But the depiction of such an unrepentant sensualist as Edna Pontellier was more than the critics of the time could allow to pass. The book was described as "trite and sordid," "essentially vulgar," and "unhealthily introspective and morbid in feeling." Chopin was hurt by this criticism and the social ostracism that accompanied it. Her only published response, however, was to claim, tongue in cheek: "I never dreamed of Mrs. Pontellier making such a mess of things and working out her own damnation as she did. If I had had the slightest intimation of such a thing I would have excluded her from the company." Though the book and its author fell into obscurity for half a century, they have now secured an honored place in American literary history.

The Storm

A SEQUEL TO "THE 'CADIAN BALL"[1]

I

The leaves were so still that even Bibi thought it was going to rain. Bobinôt, who was accustomed to converse on terms of perfect equality with his little son, called the child's attention to certain sombre clouds that were rolling with sinister intention from the west, accompanied by a sullen, threatening roar. They were at Fried-heimer's store and decided to remain there till the storm had passed. They sat within the door on two empty kegs. Bibi was four years old and looked very wise.

"Mama'll be 'fraid, yes," he suggested with blinking eyes.

"She'll shut the house. Maybe she got Sylvie helpin' her this evenin'," Bobinôt responded reassuringly.

"No; she ent got Sylvie. Sylvie was helpin' her yistiday," piped Bibi.

Bobinôt arose and going across to the counter purchased a can of shrimps, of which Calixta was very fond. Then he returned to his perch on the keg and sat stolidly holding the can of shrimps while the storm burst. It shook the wooden store and seemed to be rip-ping furrows in the distant field. Bibi laid his little hand on his father's knee and was not afraid.

II

Calixta, at home, felt no uneasiness for their safety. She sat at a side window sewing furiously on a sewing machine. She was greatly occupied and did not notice the approaching storm. But she felt very warm and often stopped to mop her face on which the perspi-ration gathered in beads. She unfastened her white sacque at the throat. It began to grow dark, and suddenly realizing the situation she got up hurriedly and went about closing windows and doors.

Out on the small front gallery she had hung Bobinôt's Sunday clothes to air and she hastened out to gather them before the rain fell. As she stepped outside, Alcée Laballière rode in at the gate. She had not seen him very often since her marriage, and never alone. She stood there with Bobinôt's coat in her hands, and the big rain drops began to fall. Alcée rode his horse under the shelter of a side projection where the chickens had huddled and there were plows and a harrow piled up in the corner.

"May I come and wait on your gallery till the storm is over, Calixta?" he asked.

1. Chopin's notebooks show that this story was written on July 18, 1898, just six months after she had submitted her novel *The Awakening* to a publisher. As its subtitle indicates, it was intended as a sequel to a story published six years earlier in which the planter Alcée Laballière is prevented from running off with the "Spanish vixen" Calixta by Clarisse, who claims him for herself. *The Storm* was apparently never submitted for publication and did not see print until the publication of Per Seyerstad's edition of *The Complete Works of Kate Chopin* in 1969, the basis for the present text.

"Come 'long in, M'sieur Alcée."

His voice and her own startled her as if from a trance, and she seized Bobinôt's vest. Alcée, mounting to the porch, grabbed the trousers and snatched Bibi's braided jacket that was about to be carried away by a sudden gust of wind. He expressed an intention to remain outside, but it was soon apparent that he might as well have been out in the open: the water beat in upon the boards in driving sheets, and he went inside, closing the door after him. It was even necessary to put something beneath the door to keep the water out.

"My! what a rain! It's good two years sence it rain' like that," exclaimed Calixta as she rolled up a piece of bagging and Alcée helped her to thrust it beneath the crack.

She was a little fuller of figure than five years before when she married; but she had lost nothing of her vivacity. Her blue eyes still retained their melting quality; and her yellow hair, dishevelled by the wind and rain, kinked more stubbornly than ever about her ears and temples.

The rain beat upon the low, shingled roof with a force and clatter that threatened to break an entrance and deluge them there. They were in the dining room—the sitting room—the general utility room. Adjoining was her bed room, with Bibi's couch along side her own. The door stood open, and the room with its white, monumental bed, its closed shutters, looked dim and mysterious.

Alcée flung himself into a rocker and Calixta nervously began to gather up from the floor the lengths of a cotton sheet which she had been sewing.

"If this keeps up, *Dieu sait* if the levees[2] goin' to stan' it!" she exclaimed.

"What have you got to do with the levees?"

"I got enough to do! An' there's Bobinôt with Bibi out in that storm—if he only didn' left Friedheimer's!"

"Let us hope, Calixta, that Bobinôt's got sense enough to come in out of a cyclone."

She went and stood at the window with a greatly disturbed look on her face. She wiped the frame that was clouded with moisture. It was stiflingly hot. Alcée got up and joined her at the window, looking over her shoulder. The rain was coming down in sheets obscuring the view of far-off cabins and enveloping the distant wood in a gray mist. The playing of the lightning was incessant. A bolt struck a tall chinaberry tree at the edge of the field. It filled all visible space with a blinding glare and the crash seemed to invade the very boards they stood upon.

Calixta put her hands to her eyes, and with a cry, staggered backward. Alcée's arm encircled her, and for an instant he drew her close and spasmodically to him.

"*Bonté!*"[3] she cried, releasing herself from his encircling arm and

2. Built-up earth banks designed to keep the river (in this case the Red River) from flooding the surrounding land.   *"Dieu sait"*: God only knows.
3. Goodness!

retreating from the window, "the house'll go next! If I only knew w'ere Bibi was!" She would not compose herself; she would not be seated. Alcée clasped her shoulders and looked into her face. The contact of her warm, palpitating body when he had unthinkingly drawn her into his arms, had aroused all the old-time infatuation and desire for her flesh.

"Calixta," he said, "don't be frightened. Nothing can happen. The house is too low to be struck, with so many tall trees standing about. There! aren't you going to be quiet? say, aren't you?" He pushed her hair back from her face that was warm and steaming. Her lips were as red and moist as pomegranate seed. Her white neck and a glimpse of her full, firm bosom disturbed him powerfully. As she glanced up at him the fear in her liquid blue eyes had given place to a drowsy gleam that unconsciously betrayed a sensuous desire. He looked down into her eyes and there was nothing for him to do but to gather her lips in a kiss. It reminded him of Assumption.[4]

"Do you remember—in Assumption, Calixta?" he asked in a low voice broken by passion. Oh! she remembered; for in Assumption he had kissed her and kissed and kissed her; until his senses would well nigh fail, and to save her he would resort to a desperate flight. If she was not an immaculate dove in those days, she was still inviolate; a passionate creature whose very defenselessness had made her defense, against which his honor forbade him to prevail. Now —well, now—her lips seemed in a manner free to be tasted, as well as her round, white throat and her whiter breasts.

They did not heed the crashing torrents, and the roar of the elements made her laugh as she lay in his arms. She was a revelation in that dim, mysterious chamber; as white as the couch she lay upon. Her firm, elastic flesh that was knowing for the first time its birthright, was like a creamy lily that the sun invites to contribute its breath and perfume to the undying life of the world.

The generous abundance of her passion, without guile or trickery, was like a white flame which penetrated and found response in depths of his own sensuous nature that had never yet been reached.

When he touched her breasts they gave themselves up in quivering ecstasy, inviting his lips. Her mouth was a fountain of delight. And when he possessed her, they seemed to swoon together at the very borderland of life's mystery.

He stayed cushioned upon her, breathless, dazed, enervated, with his heart beating like a hammer upon her. With one hand she clasped his head, her lips lightly touching his forehead. The other hand stroked with a soothing rhythm his muscular shoulders.

The growl of the thunder was distant and passing away. The rain beat softly upon the shingles, inviting them to drowsiness and sleep.

4. I.e., Assumption Parish, Louisiana, the setting for "The 'Cadian Ball."

But they dared not yield.

III

The rain was over; and the sun was turning the glistening green world into a palace of gems. Calixta, on the gallery, watched Alcée ride away. He turned and smiled at her with a beaming face; and she lifted her pretty chin in the air and laughed aloud.

Bobinôt and Bibi, trudging home, stopped without at the cistern to make themselves presentable.

"My! Bibi, w'at will yo' mama say! You ought to be ashame'. You oughtn' put on those good pants. Look at 'em! An' that mud on yo' collar! How you got that mud on yo' collar, Bibi? I never saw such a boy!" Bibi was the picture of pathetic resignation. Bobinôt was the embodiment of serious solicitude as he strove to remove from his own person and his son's the signs of their tramp over heavy roads and through wet fields. He scraped the mud off Bibi's bare legs and feet with a stick and carefully removed all traces from his heavy brogans. Then, prepared for the worst—the meeting with an over-scrupulous housewife, they entered cautiously at the back door.

Calixta was preparing supper. She had set the table and was dripping coffee at the hearth. She sprang up as they came in.

"Oh, Bobinôt! You back! My! but I was uneasy. W'ere you been during the rain? An' Bibi? he ain't wet? he ain't hurt?" She had clasped Bibi and was kissing him effusively. Bobinôt's explanations and apologies which he had been composing all along the way, died on his lips as Calixta felt him to see if he were dry, and seemed to express nothing but satisfaction at their safe return.

"I brought you some shrimps, Calixta," offered Bobinôt, hauling the can from his ample side pocket and laying it on the table.

"Shrimps! Oh, Bobinôt! you too good fo' anything!" and she gave him a smacking kiss on the cheek that resounded, "*J'vous réponds,*[5] we'll have a feas' to night! umph-umph!"

Bobinôt and Bibi began to relax and enjoy themselves, and when the three seated themselves at table they laughed much and so loud that anyone might have heard them as far away as Laballière's.

IV

Alcée Laballière wrote to his wife, Clarisse, that night. It was a loving letter, full of tender solicitude. He told her not to hurry back, but if she and the babies liked it at Biloxi, to stay a month longer. He was getting on nicely; and though he missed them, he was willing to bear the separation a while longer—realizing that their health and pleasure were the first things to be considered.

5. I promise you.

V

As for Clarisse, she was charmed upon receiving her husband's letter. She and the babies were doing well. The society was agreeable; many of her old friends and acquaintances were at the bay. And the first free breath since her marriage seemed to restore the pleasant liberty of her maiden days. Devoted as she was to her husband, their intimate conjugal life was something which she was more than willing to forego for a while.

So the storm passed and every one was happy.

1898   1969

STEPHEN CRANE
1871–1900

Stephen Crane was born November 1, 1871, and died on June 5, 1900. By the age of twenty-eight he had published enough material to fill a dozen volumes of a collected edition and had lived a legendary life that has grown in complexity and interest to scholars and readers the more the facts have come to light. His family settled in America in the mid-seventeenth century. He himself was the son of a Methodist minister, but he systematically rejected religious and social traditions, identified with the urban poor, and "married" the mistress of one "of the better houses of ill-fame" in Jacksonville, Florida. Although he was temperamentally a gentle man, Crane was attracted to—even obsessed by—war and other forms of physical and psychic violence. He frequently lived the down-and-out life of a penniless artist; he was also ambitious and something of a snob; he was a poet and an impressionist: a journalist, a social critic and realist. In short, there is much about Crane's life and writing that is paradoxical; he is an original and not easy to be right about.

Crane once explained to an editor: "After all, I cannot help vanishing and disappearing and dissolving. It is my foremost trait." This distinctively restless, peripatetic quality of Crane's life began early. The last of fourteen children, Crane moved with his family at least three times before he entered school at age seven in the small town of Port Jervis, New York. His father died in 1880, and after a tentative return to Paterson the family settled in the coastal resort town of Asbury Park, New Jersey, three years later. The next five years, as his biographer and critic Edwin H. Cady has observed, "confirmed in the sensitive, vulnerable, fatherless preacher's kid his fate as isolato." Crane never took to schooling. At Syracuse University he distinguished himself as a baseball player but was unable to accept the routine of academic life and after one semester left with no intention of returning. He was not sure what he would do with himself, but by 1891 he had begun to write for newspapers, and he hungered for immersion in life of the kind that his early journalistic assignments caused him to witness at close hand.

New York City was the inevitable destination for a young man with literary ambitions and a desire to experience the fullness of life. A couple of jobs with New York newspapers proved abortive, and Crane spent much of the next two years shuttling between the seedy apartments of his artist friends and his brother Edmund's house in nearby Lake View, New Jersey. In these years of extreme privation, Crane developed his powers as an observer of psychological and social reality. Encouraged by the realist credo of Hamlin Garland, whom he had heard lecture in 1891, Crane wrote and then—after it had been rejected by several New York editors—published in 1893, at his own expense, a work he had begun while at Syracuse: *Maggie, A Girl of the Streets*. Though Garland admired the short novel, and the powerful literary arbiter William Dean Howells promoted Crane and his book (and continued to think it his best work), the book did not sell. The mass audience of the time sought from literature what *Maggie* denied: escape, distraction, and easy pleasure in romances which falsified and obscured the social, emotional, and moral nature of life.

Just why Crane turned next to the Civil War as a subject for *The Red Badge of Courage* is not clear. He may have wished to appeal to a popular audience and make some money; he later described the book as a "pot-boiler." Or he may have been compelled by a half-recognized need to test himself imaginatively by placing his young protagonist-counterpart in a psychological and philosophical pressure cooker. In any case, in 1893 Crane began his best-known work, *The Red Badge of Courage. An Episode of the American Civil War*. The narrative, which depicts the education of a young man in the context of struggle, is as old as Homer's *Odyssey* and is a dominant story-type in American literature from Benjamin Franklin through Melville, Hemingway, Malcolm X, and Saul Bellow.

When *Red Badge* was first published as a syndicated newspaper story in December, 1894, Crane's fortunes began to improve. The same syndicate that took *Red Badge* assigned him early in 1895 as a roving reporter in the American West and Mexico, experience which would give him the material for several of his finest tales—*The Bride Comes to Yellow Sky* and *The Blue Hotel* among them. Early in 1895 an established New York publisher agreed to issue *Red Badge* in book form. In the spring of that year, *The Black Riders and Other Lines*, his first volume of poetry, was published. Predictably, Garland and Howells responded intelligently to Crane's spare, original, unflinchingly honest poetry, but it was too experimental in form and too unconventional in philosophic outlook to win wide acceptance. *The Red Badge of Courage* appeared in the fall as a book and became the first widely successful American work to be realistic in the modern way. It won Crane international acclaim at the age of twenty-four.

In all of his poetry, journalism, and fiction Crane clearly demonstrated his religious, social, and literary rebelliousness; his alienated, unconventional stance also led him to direct action. After challenging the New York police force on behalf of a prostitute who claimed harassment at its hands, Crane left the city in the winter of 1896–97 to cover the insurrection against Spain in Cuba. On his way to Cuba he met Cora Howorth Taylor, the proprietress of the aptly named Hotel de Dream in Jacksonville, Florida, with whom he lived for the last three years of his life. On January 2, Crane's ship *The Commodore* sank off the coast of Florida. His report of this harrowing adventure was published a few days later in the New York *Press*. He promptly converted this event into *The Open Boat*. This story, like *Red Badge*, reveals Crane's characteristic subject matter—the physical, emotional, and intellectual responses of men under extreme pressure—and the dominant themes of nature's indifference to humanity's fate and the consequent need for compassionate collective action. In the late stories *The Open Boat* and *The Blue Hotel*, Crane achieved his mature style. In both of these works we can observe his tough-minded irony and his essential vision: a sympathetic but unflinching demand for courage, integrity, grace, and generosity in the face of a universe in which human beings, to quote from *The Blue Hotel*, are so many lice clinging "to a whirling, fire-smote, ice-locked, disease-stricken, space-lost bulb."

In the summer of 1897 Crane covered the Greco-Turkish War and later that year settled in England, where he made friends, most notably with the English writers Joseph Conrad, H. G. Wells, and Ford Maddox Hueffer (later Ford) and the American writers Henry James and Harold Frederic. The following year he covered the Spanish-American War for Joseph Pulitzer's New York *World*. When he wrote *The Red Badge of Courage*, he had never observed battle at firsthand, but his experiences during the Greco-Turkish and Spanish-American Wars confirmed his sense that he had recorded, with more than literal accuracy, the realities of battle.

In the last months of his life Crane's situation became desperate; he was suffering from tuberculosis and was hopelessly in debt. He wrote furiously in a doomed attempt to earn money, but the effort only worsened his health. In 1899 he drafted thirteen stories set in the fictional town Whilomville for *Harper's Magazine*, published his second volume of poetry, *War Is Kind*, the weak novel *Active Service*, and the American edition of *The Monster and Other Stories*. During a Christmas party that year Crane nearly died of a lung hemorrhage. Surviving only a few months, he somehow summoned the strength to write a series of nine articles on great battles and complete the first twenty-five chapters of the novel *The O'Ruddy*. In spite of Cora's hopes for a miraculous cure, and the generous assistance of Henry James and others, Crane died at Badenweiler, Germany, on June 5, 1900.

The Open Boat[1]

A TALE INTENDED TO BE AFTER THE FACT. BEING THE EXPERIENCE OF FOUR MEN FROM THE SUNK STEAMER COMMODORE

I

None of them knew the color of the sky. Their eyes glanced level, and were fastened upon the waves that swept toward them.

---

1. Crane sailed as a correspondent on the steamer *Commodore*, which on January 1, 1897, left Jacksonville, Florida, with munitions for the Cuban insurrectionists. Early on the morning of January 2, the steamer sank. With four others, Crane reached Daytona Beach in a 10-foot dinghy on the following morning. Under the title "Stephen Crane's Own Story" the New York *Press* carried the details of his nearly fatal experience on January 7. In June, 1897, he published his fictional account, *The Open Boat*, in *Scribner's Magazine*. The story gave the title to *The Open Boat and Other Tales of Adventure* (1898). The present text reprints that established for the University of Virginia edition of *The Works of Stephen Crane*, Vol. V, *Tales of Adventure* (1970).

These waves were of the hue of slate, save for the tops, which were of foaming white, and all of the men knew the colors of the sea. The horizon narrowed and widened, and dipped and rose, and at all times its edge was jagged with waves that seemed thrust up in points like rocks.

Many a man ought to have a bath-tub larger than the boat which here rode upon the sea. These waves were most wrongfully and barbarously abrupt and tall, and each froth-top was a problem in small boat navigation.

The cook squatted in the bottom and looked with both eyes at the six inches of gunwale which separated him from the ocean. His sleeves were rolled over his fat forearms, and the two flaps of his unbuttoned vest dangled as he bent to bail out the boat. Often he said: "Gawd! That was a narrow clip." As he remarked it he invariably gazed eastward over the broken sea.

The oiler,[2] steering with one of the two oars in the boat, sometimes raised himself suddenly to keep clear of water that swirled in over the stern. It was a thin little oar and it seemed often ready to snap.

The correspondent, pulling at the other oar, watched the waves and wondered why he was there.

The injured captain, lying in the bow, was at this time buried in that profound dejection and indifference which comes, temporarily at least, to even the bravest and most enduring when, willy nilly, the firm fails, the army loses, the ship goes down. The mind of the master of a vessel is rooted deep in the timbers of her, though he command for a day or a decade, and this captain had on him the stern impression of a scene in the grays of dawn of seven turned faces, and later a stump of a top-mast with a white ball on it that slashed to and fro at the waves, went low and lower, and down. Thereafter there was something strange in his voice. Although steady, it was deep with mourning, and of a quality beyond oration or tears.

"Keep'er a little more south, Billie," said he.

" 'A little more south,' sir," said the oiler in the stern.

A seat in this boat was not unlike a seat upon a bucking broncho, and, by the same token, a broncho is not much smaller. The craft pranced and reared, and plunged like an animal. As each wave came, and she rose for it, she seemed like a horse making at a fence outrageously high. The manner of her scramble over these walls of water is a mystic thing, and, moreover, at the top of them were ordinarily these problems in white water, the foam racing down from the summit of each wave, requiring a new leap, and a leap

---

2. One who oils machinery in the engine room of a ship.

from the air. Then, after scornfully bumping a crest, she would slide, and race, and splash down a long incline and arrive bobbing and nodding in front of the next menace.

A singular disadvantage of the sea lies in the fact that after successfully surmounting one wave you discover that there is another behind it just as important and just as nervously anxious to do something effective in the way of swamping boats. In a ten-foot dingey one can get an idea of the resources of the sea in the line of waves that is not probable to the average experience, which is never at sea in a dingey. As each slaty wall of water approached, it shut all else from the view of the men in the boat, and it was not difficult to imagine that this particular wave was the final outburst of the ocean, the last effort of the grim water. There was a terrible grace in the move of the waves, and they came in silence, save for the snarling of the crests.

In the wan light, the faces of the men must have been gray. Their eyes must have glinted in strange ways as they gazed steadily astern. Viewed from a balcony, the whole thing would doubtlessly have been weirdly picturesque. But the men in the boat had no time to see it, and if they had had leisure there were other things to occupy their minds. The sun swung steadily up the sky, and they knew it was broad day because the color of the sea changed from slate to emerald-green, streaked with amber lights, and the foam was like tumbling snow. The process of the breaking day was unknown to them. They were aware only of this effect upon the color of the waves that rolled toward them.

In disjointed sentences the cook and the correspondent argued as to the difference between a life-saving station and a house of refuge. The cook had said: "There's a house of refuge just north of the Mosquito Inlet Light, and as soon as they see us, they'll come off in their boat and pick us up."

"As soon as who see us?" said the correspondent.

"The crew," said the cook.

"Houses of refuge don't have crews," said the correspondent. "As I understand them, they are only places where clothes and grub are stored for the benefit of shipwrecked people. They don't carry crews."

"Oh, yes, they do," said the cook.

"No, they don't," said the correspondent.

"Well, we're not there yet, anyhow," said the oiler, in the stern.

"Well," said the cook, "perhaps it's not a house of refuge that I'm thinking of as being near Mosquito Inlet Light. Perhaps it's a life-saving station."

"We're not there yet," said the oiler, in the stern.

II

As the boat bounced from the top of each wave, the wind tore through the hair of the hatless men, and as the craft plopped her stern down again the spray slashed past them. The crest of each of these waves was a hill, from the top of which the men surveyed, for a moment, a broad tumultuous expanse, shining and wind-riven. It was probably splendid. It was probably glorious, this play of the free sea, wild with lights of emerald and white and amber.

"Bully good thing it's an on-shore wind," said the cook. "If not, where would we be? Wouldn't have a show."

"That's right," said the correspondent.

The busy oiler nodded his assent.

Then the captain, in the bow, chuckled in a way that expressed humor, contempt, tragedy, all in one. "Do you think we've got much of a show, now, boys?" said he.

Whereupon the three were silent, save for a trifle of hemming and hawing. To express any particular optimism at this time they felt to be childish and stupid, but they all doubtless possessed this sense of the situation in their mind. A young man thinks doggedly at such times. On the other hand, the ethics of their condition was decidedly against any open suggestion of hopelessness. So they were silent.

"Oh, well," said the captain, soothing his children, "we'll get ashore all right."

But there was that in his tone which made them think, so the oiler quoth: "Yes! If this wind holds!"

The cook was bailing. "Yes! If we don't catch hell in the surf."

Canton flannel[3] gulls flew near and far. Sometimes they sat down on the sea, near patches of brown sea-weed that rolled over the waves with a movement like carpets on a line in a gale. The birds sat comfortably in groups, and they were envied by some in the dingey, for the wrath of the sea was no more to them than it was to a covey of prairie chickens a thousand miles inland. Often they came very close and stared at the men with black bead-like eyes. At these times they were uncanny and sinister in their unblinking scrutiny, and the men hooted angrily at them, telling them to be gone. One came, and evidently decided to alight on the top of the captain's head. The bird flew parallel to the boat and did not circle, but made short sidelong jumps in the air in chicken-fashion. His black eyes were wistfully fixed upon the captain's head. "Ugly brute," said the oiler to the bird. "You look as if you were made with a jack-knife." The cook and the correspondent swore darkly at

3. Stout cotton fabric.

the creature. The captain naturally wished to knock it away with the end of the heavy painter, but he did not dare do it, because anything resembling an emphatic gesture would have capsized this freighted boat, and so with his open hand, the captain gently and carefully waved the gull away. After it had been discouraged from the pursuit the captain breathed easier on account of his hair, and others breathed easier because the bird struck their minds at this time as being somehow grewsome and ominous.

In the meantime the oiler and the correspondent rowed. And also they rowed.

They sat together in the same seat, and each rowed an oar. Then the oiler took both oars; then the correspondent took both oars; then the oiler; then the correspondent. They rowed and they rowed. The very ticklish part of the business was when the time came for the reclining one in the stern to take his turn at the oars. By the very last star of truth, it is easier to steal eggs from under a hen than it was to change seats in the dingey. First the man in the stern slid his hand along the thwart and moved with care, as if he were of Sèvres.[4] Then the man in the rowing seat slid his hand along the other thwart. It was all done with the most extraordinary care. As the two sidled past each other, the whole party kept watchful eyes on the coming wave, and the captain cried: "Look out now! Steady there!"

The brown mats of sea-weed that appeared from time to time were like islands, bits of earth. They were travelling, apparently, neither one way nor the other. They were, to all intents, stationary. They informed the men in the boat that it was making progress slowly toward the land.

The captain, rearing cautiously in the bow, after the dingey soared on a great swell, said that he had seen the light-house at Mosquito Inlet. Presently the cook remarked that he had seen it. The correspondent was at the oars, then, and for some reason he too wished to look at the light-house, but his back was toward the far shore and the waves were important, and for some time he could not seize an opportunity to turn his head. But at last there came a wave more gentle than the others, and when at the crest of it he swiftly scoured the western horizon.

"See it?" said the captain.

"No," said the correspondent, slowly, "I didn't see anything."

"Look again," said the captain. He pointed. "It's exactly in that direction."

At the top of another wave, the correspondent did as he was bid, and this time his eyes chanced on a small still thing on the edge of the swaying horizon. It was precisely like the point of a pin. It took

---

4. Fine, often ornately decorated, French porcelain.

an anxious eye to find a light-house so tiny.

"Think we'll make it, Captain?"

"If this wind holds and the boat don't swamp, we can't do much else," said the captain.

The little boat, lifted by each towering sea, and splashed viciously by the crests, made progress that in the absence of sea-weed was not apparent to those in her. She seemed just a wee thing wallowing, miraculously, top-up, at the mercy of five oceans. Occasionally, a great spread of water, like white flames, swarmed into her.

"Bail her, cook," said the captain, serenely.

"All right, Captain," said the cheerful cook.

III

It would be difficult to describe the subtle brotherhood of men that was here established on the seas. No one said that it was so. No one mentioned it. But it dwelt in the boat, and each man felt it warm him. They were a captain, an oiler, a cook, and a correspondent, and they were friends, friends in a more curiously iron-bound degree than may be common. The hurt captain, lying against the water-jar in the bow, spoke always in a low voice and calmly, but he could never command a more ready and swiftly obedient crew than the motley three of the dingey. It was more than a mere recognition of what was best for the common safety. There was surely in it a quality that was personal and heartfelt. And after this devotion to the commander of the boat there was this comradeship that the correspondent, for instance, who had been taught to be cynical of men, knew even at the time was the best experience of his life. But no one said that it was so. No one mentioned it.

"I wish we had a sail," remarked the captain. "We might try my overcoat on the end of an oar and give you two boys a chance to rest." So the cook and the correspondent held the mast and spread wide the overcoat. The oiler steered, and the little boat made good way with her new rig. Sometimes the oiler had to scull sharply to keep a sea from breaking into the boat, but otherwise sailing was a success.

Meanwhile the light-house had been growing slowly larger. It had now almost assumed color, and appeared like a little gray shadow on the sky. The man at the oars could not be prevented from turning his head rather often to try for a glimpse of this little gray shadow.

At last, from the top of each wave the men in the tossing boat could see land. Even as the light-house was an upright shadow on the sky, this land seemed but a long black shadow on the sea. It certainly was thinner than paper. "We must be about opposite New Smyrna," said the cook, who had coasted this shore often in schooners. "Captain, by the way, I believe they abandoned that life-saving station there about a year ago."

"Did they?" said the captain.

The wind slowly died away. The cook and the correspondent were not now obliged to slave in order to hold high the oar. But the waves continued their old impetuous swooping at the dingey, and the little craft, no longer under way, struggled woundily over them. The oiler or the correspondent took the oars again.

Shipwrecks are *apropos* of nothing. If men could only train for them and have them occur when the men had reached pink condition, there would be less drowning at sea. Of the four in the dingey none had slept any time worth mentioning for two days and two nights previous to embarking in the dingey, and in the excitement of clambering about the deck of a foundering ship they had also forgotten to eat heartily.

For these reasons, and for others, neither the oiler nor the correspondent was fond of rowing at this time. The correspondent wondered ingenuously how in the name of all that was sane could there be people who thought it amusing to row a boat. It was not an amusement; it was a diabolical punishment, and even a genius of mental aberrations could never conclude that it was anything but a horror to the muscles and a crime against the back. He mentioned to the boat in general how the amusement of rowing struck him, and the weary-faced oiler smiled in full sympathy. Previously to the foundering, by the way, the oiler had worked double-watch in the engine-room of the ship.

"Take her easy, now, boys," said the captain. "Don't spend yourselves. If we have to run a surf you'll need all your strength, because we'll sure have to swim for it. Take your time."

Slowly the land arose from the sea. From a black line it became a line of black and a line of white—trees and sand. Finally, the captain said that he could make out a house on the shore. "That's the house of refuge, sure," said the cook. "They'll see us before long, and come out after us."

The distant light-house reared high. "The keeper ought to be able to make us out now, if he's looking through a glass," said the captain. "He'll notify the life-saving people."

"None of those other boats could have got ashore to give word of the wreck," said the oiler, in a low voice. "Else the life-boat would be out hunting us."

Slowly and beautifully the land loomed out of the sea. The wind came again. It had veered from the northeast to the southeast. Finally, a new sound struck the ears of the men in the boat. It was the low thunder of the surf on the shore. "We'll never be able to make the light-house now," said the captain. "Swing her head a little more north, Billie."

" 'A little more north,' sir," said the oiler.

Whereupon the little boat turned her nose once more down the wind, and all but the oarsman watched the shore grow. Under the influence of this expansion doubt and direful apprehension was leaving the minds of the men. The management of the boat was still most absorbing, but it could not prevent a quiet cheerfulness. In an hour, perhaps, they would be ashore.

Their back-bones had become thoroughly used to balancing in the boat and they now rode this wild colt of a dingey like circus men. The correspondent thought that he had been drenched to the skin, but happening to feel in the top pocket of his coat, he found therein eight cigars. Four of them were soaked with seawater; four were perfectly scatheless. After a search, somebody produced three dry matches, and thereupon the four waifs rode impudently in their little boat, and with an assurance of an impending rescue shining in their eyes, puffed at the big cigars and judged well and ill of all men. Everybody took a drink of water.

IV

"Cook," remarked the captain, "there don't seem to be any signs of life about your house of refuge."

"No," replied the cook. "Funny they don't see us!"

A broad stretch of lowly coast lay before the eyes of the men. It was of dunes topped with dark vegetation. The roar of the surf was plain, and sometimes they could see the white lip of a wave as it spun up the beach. A tiny house was blocked out black upon the sky. Southward, the slim light-house lifted its little gray length.

Tide, wind, and waves were swinging the dingey northward. "Funny they don't see us," said the men.

The surf's roar was here dulled, but its tone was, nevertheless, thunderous and mighty. As the boat swam over the great rollers, the men sat listening to this roar. "We'll swamp sure," said everybody.

It is fair to say here that there was not a life-saving station within twenty miles in either direction, but the men did not know this fact and in consequence they made dark and opprobrious remarks concerning the eyesight of the nation's life-savers. Four scowling men sat in the dingey and surpassed records in the invention of epithets.

"Funny they don't see us."

The light-heartedness of a former time had completely faded. To their sharpened minds it was easy to conjure pictures of all kinds of incompetency and blindness and, indeed, cowardice. There was the shore of the populous land, and it was bitter and bitter to them that from it came no sign.

"Well," said the captain, ultimately, "I suppose we'll have to make a try for ourselves. If we stay out here too long, we'll none of us have strength left to swim after the boat swamps."

And so the oiler, who was at the oars, turned the boat straight for the shore. There was a sudden tightening of muscles. There was some thinking.

"If we don't all get ashore—" said the captain. "If we don't all get ashore, I suppose you fellows know where to send news of my finish?"

They then briefly exchanged some addresses and admonitions. As for the reflections of the men, there was a great deal of rage in them. Perchance they might be formulated thus: "If I am going to be drowned—if I am going to be drowned—if I am going to be drowned, why, in the name of the seven mad gods who rule the sea, was I allowed to come thus far and contemplate sand and trees? Was I brought here merely to have my nose dragged away as I was about to nibble the sacred cheese of life? It is preposterous. If this old ninny-woman, Fate, cannot do better than this, she should be deprived of the management of men's fortunes. She is an old hen who knows not her intention. If she has decided to drown me, why did she not do it in the beginning and save me all this trouble. The whole affair is absurd. . . . But, no, she cannot mean to drown me. She dare not drown me. She cannot drown me. Not after all this work." Afterward the man might have had an impulse to shake his fist at the clouds. "Just you drown me, now, and then hear what I call you!"

The billows that came at this time were more formidable. They seemed always just about to break and roll over the little boat in a turmoil of foam. There was a preparatory and long growl in the speech of them. No mind unused to the sea would have concluded that the dingey could ascend these sheer heights in time. The shore was still afar. The oiler was a wily surfman. "Boys," he said, swiftly, "she won't live three minutes more and we're too far out to swim. Shall I take her to sea again, Captain?"

"Yes! Go ahead!" said the captain.

This oiler, by a series of quick miracles, and fast and steady oars-manship, turned the boat in the middle of the surf and took her safely to sea again.

There was a considerable silence as the boat bumped over the furrowed sea to deeper water. Then somebody in gloom spoke. "Well, anyhow, they must have seen us from the shore by now."

The gulls went in slanting flight up the wind toward the gray desolate east. A squall, marked by dingy clouds, and clouds brick-red, like smoke from a burning building, appeared from the southeast.

"What do you think of those life-saving people? Ain't they peaches?"

"Funny they haven't seen us."

"Maybe they think we're out here for sport! Maybe they think we're fishin'. Maybe they think we're damned fools."

It was a long afternoon. A changed tide tried to force them southward, but wind and wave said northward. Far ahead, where coast-line, sea, and sky formed their mighty angle, there were little dots which seemed to indicate a city on the shore.

"St. Augustine?"

The captain shook his head. "Too near Mosquito Inlet."

And the oiler rowed, and then the correspondent rowed. Then the oiler rowed. It was a weary business. The human back can become the seat of more aches and pains than are registered in books for the composite anatomy of a regiment. It is a limited area, but it can become the theatre of innumerable muscular conflicts, tangles, wrenches, knots, and other comforts.

"Did you ever like to row, Billie?" asked the correspondent.

"No," said the oiler. "Hang it."

When one exchanged the rowing-seat for a place in the bottom of the boat, he suffered a bodily depression that caused him to be careless of everything save an obligation to wiggle one finger. There was cold sea-water swashing to and fro in the boat, and he lay in it. His head, pillowed on a thwart, was within an inch of the swirl of a wave crest, and sometimes a particularly obstreperous sea came inboard and drenched him once more. But these matters did not annoy him. It is almost certain that if the boat had capsized he would have tumbled comfortably out upon the ocean as if he felt sure that it was a great soft mattress.

"Look! There's a man on the shore!"

"Where?"

"There? See'im? See'im?"

"Yes, sure! He's walking along."

"Now he's stopped. Look! He's facing us!"

"He's waving at us!"

"So he is! By thunder!"

"Ah, now, we're all right! Now we're all right! There'll be a boat out here for us in half an hour."

"He's going on. He's running. He's going up to that house there."

The remote beach seemed lower than the sea, and it required a searching glance to discern the little black figure. The captain saw a floating stick and they rowed to it. A bath-towel was by some weird chance in the boat, and, tying this on the stick, the captain waved it. The oarsman did not dare turn his head, so he was obliged to ask questions.

"What's he doing now?"

"He's standing still again. He's looking, I think. ... There he goes again. Toward the house. ... Now he's stopped again."

"Is he waving at us?"

"No, not now! he was, though."

"Look! There comes another man!"

"He's running."

"Look at him go, would you."

"Why, he's on a bicycle. Now he's met the other man. They're both waving at us. Look!"

"There comes something up the beach."

"What the devil is that thing?"

"Why, it looks like a boat."

"Why, certainly it's a boat."

"No, it's on wheels."

"Yes, so it is. Well, that must be the life-boat. They drag them along shore on a wagon."

"That's the life-boat, sure."

"No, by——, it's—it's an omnibus."

"I tell you it's a life-boat."

"It is not! It's an omnibus. I can see it plain. See? One of those big hotel omnibuses."

"By thunder, you're right. It's an omnibus, sure as fate. What do you suppose they are doing with an omnibus? Maybe they are going around collecting the life-crew, hey?"

"That's it, likely. Look! There's a fellow waving a little black flag. He's standing on the steps of the omnibus. There come those other two fellows. Now they're all talking together. Look at the fellow with the flag. Maybe he ain't waving it!"

"That ain't a flag, is it? That's his coat. Why, certainly, that's his coat."

"So it is. It's his coat. He's taken it off and is waving it around his head. But would you look at him swing it!"

"Oh, say, there isn't any life-saving station there. That's just a winter resort hotel omnibus that has brought over some of the boarders to see us drown."

"What's that idiot with the coat mean? What's he signaling, anyhow?"

"It looks as if he were trying to tell us to go north. There must be a life-saving station up there."

"No! He thinks we're fishing. Just giving us a merry hand. See? Ah, there, Willie."

"Well, I wish I could make something out of those signals. What do you suppose he means?"

"He don't mean anything. He's just playing."

"Well, if he'd just signal us to try the surf again, or to go to sea and wait, or go north, or go south, or go to hell—there would be some reason in it. But look at him. He just stands there and keeps his coat revolving like a wheel. The ass!"

"There come more people."

"Now there's quite a mob. Look! Isn't that a boat?"

"Where? Oh, I see where you mean. No, that's no boat."

"That fellow is still waving his coat."

"He must think we like to see him do that. Why don't he quit it. It don't mean anything."

"I don't know. I think he is trying to make us go north. It must be that there's a life-saving station there somewhere."

"Say, he ain't tired yet. Look at 'im wave."

"Wonder how long he can keep that up. He's been revolving his coat ever since he caught sight of us. He's an idiot. Why aren't they getting men to bring a boat out. A fishing boat—one of those big yawls—could come out here all right. Why don't he do something?"

"Oh, it's all right, now."

"They'll have a boat out here for us in less than no time, now that they've seen us."

A faint yellow tone came into the sky over the low land. The shadows on the sea slowly deepened. The wind bore coldness with it, and the men began to shiver.

"Holy smoke!" said one, allowing his voice to express his impious mood, "if we keep on monkeying out here! If we've got to flounder out here all night!"

"Oh, we'll never have to stay here all night! Don't you worry. They've seen us now, and it won't be long before they'll come chasing out after us."

The shore grew dusky. The man waving a coat blended gradually into this gloom, and it swallowed in the same manner the omnibus and the group of people. The spray, when it dashed uproariously over the side, made the voyagers shrink and swear like men who were being branded.

"I'd like to catch the chump who waved the coat. I feel like soaking him one, just for luck."

"Why? What did he do?"

"Oh, nothing, but then he seemed so damned cheerful."

In the meantime the oiler rowed, and then the correspondent rowed, and then the oiler rowed. Gray-faced and bowed forward, they mechanically, turn by turn, plied the leaden oars. The form of the light-house had vanished from the southern horizon, but finally a pale star appeared, just lifting from the sea. The streaked saffron in the west passed before the all-merging darkness, and the sea to the east was black. The land had vanished, and was expressed only by the low and drear thunder of the surf.

"If I am going to be drowned—if I am going to be drowned—if I am going to be drowned, why, in the name of the seven mad gods who rule the sea, was I allowed to come thus far and contemplate sand and trees? Was I brought here merely to have my nose dragged away as I was about to nibble the sacred cheese of life?"

The patient captain, drooped over the water-jar, was sometimes obliged to speak to the oarsman.

"Keep her head up! Keep her head up!"

" 'Keep her head up,' sir." The voices were weary and low.

This was surely a quiet evening. All save the oarsman lay heavily and listlessly in the boat's bottom. As for him, his eyes were just capable of noting the tall black waves that swept forward in a most sinister silence, save for an occasional subdued growl of a crest.

The cook's head was on a thwart, and he looked without interest at the water under his nose. He was deep in other scenes. Finally he spoke. "Billie," he murmured, dreamfully, "what kind of pie do you like best?"

V

"Pie," said the oiler and the correspondent, agitatedly. "Don't talk about those things, blast you!"

"Well," said the cook, "I was just thinking about ham sandwiches, and——"

A night on the sea in an open boat is a long night. As darkness settled finally, the shine of the light, lifting from the sea in the south, changed to full gold. On the northern horizon a new light appeared, a small bluish gleam on the edge of the waters. These two lights were the furniture of the world. Otherwise there was nothing but waves.

Two men huddled in the stern, and distances were so magnificent in the dingey that the rower was enabled to keep his feet partly warmed by thrusting them under his companions. Their legs indeed extended far under the rowing-seat until they touched the feet of the captain forward. Sometimes, despite the efforts of the tired oarsman, a wave came piling into the boat, an icy wave of the night, and the chilling water soaked them anew. They would twist their bodies for a moment and groan, and sleep the dead sleep once more, while the water in the boat gurgled about them as the craft rocked.

The plan of the oiler and the correspondent was for one to row until he lost the ability, and then arouse the other from his sea-water couch in the bottom of the boat.

The oiler plied the oars until his head drooped forward, and the overpowering sleep blinded him. And he rowed yet afterward. Then he touched a man in the bottom of the boat, and called his name. "Will you spell me for a little while?" he said, meekly.

"Sure, Billie," said the correspondent, awakening and dragging himself to a sitting position. They exchanged places carefully, and the oiler, cuddling down in the sea-water at the cook's side, seemed to go to sleep instantly.

The particular violence of the sea had ceased. The waves came without snarling. The obligation of the man at the oars was to keep the boat headed so that the tilt of the rollers would not capsize her, and to preserve her from filling when the crests rushed past. The black waves were silent and hard to be seen in the darkness. Often one was almost upon the boat before the oarsman was aware.

In a low voice the correspondent addressed the captain. He was not sure that the captain was awake, although this iron man seemed to be always awake. "Captain, shall I keep her making for that light north, sir?"

The same steady voice answered him. "Yes. Keep it about two points off the port bow."

The cook had tied a life-belt around himself in order to get even the warmth which this clumsy cork contrivance could donate, and he seemed almost stove-like when a rower, whose teeth invariably chattered wildly as soon as he ceased his labor, dropped down to sleep.

The correspondent, as he rowed, looked down at the two men sleeping under foot. The cook's arm was around the oiler's shoulders, and, with their fragmentary clothing and haggard faces, they were the babes of the sea, a grotesque rendering of the old babes in the wood.

Later he must have grown stupid at his work, for suddenly there was a growling of water, and a crest came with a roar and a swash into the boat, and it was a wonder that it did not set the cook afloat in his life-belt. The cook continued to sleep, but the oiler sat up, blinking his eyes and shaking with the new cold.

"Oh, I'm awful sorry, Billie," said the correspondent, contritely.

"That's all right, old boy," said the oiler, and lay down again and was asleep.

Presently it seemed that even the captain dozed, and the correspondent thought that he was the one man afloat on all the oceans. The wind had a voice as it came over the waves, and it was sadder than the end.

There was a long, loud swishing astern of the boat, and a gleaming trail of phosphorescence, like blue flame, was furrowed on the black waters. It might have been made by a monstrous knife.

Then there came a stillness, while the correspondent breathed with the open mouth and looked at the sea.

Suddenly there was another swish and another long flash of bluish light, and this time it was alongside the boat, and might almost have been reached with an oar. The correspondent saw an enormous fin speed like a shadow through the water, hurling the crystalline spray and leaving the long glowing trail.

The correspondent looked over his shoulder at the captain. His face was hidden, and he seemed to be asleep. He looked at the babes of the sea. They certainly were asleep. So, being bereft of sympathy, he leaned a little way to one side and swore softly into the sea.

But the thing did not then leave the vicinity of the boat. Ahead or astern, on one side or the other, at intervals long or short, fled the long sparkling streak, and there was to be heard the whiroo of the dark fin. The speed and power of the thing was greatly to be admired. It cut the water like a gigantic and keen projectile.

The presence of this biding thing did not affect the man with the same horror that it would if he had been a picnicker. He simply looked at the sea dully and swore in an undertone.

Nevertheless, it is true that he did not wish to be alone with the thing. He wished one of his companions to awaken by chance and keep him company with it. But the captain hung motionless over the water-jar and the oiler and the cook in the bottom of the boat were plunged in slumber.

VI

"If I am going to be drowned—if I am going to be drowned—if I am going to be drowned, why, in the name of the seven mad gods who rule the sea, was I allowed to come thus far and contemplate sand and trees?"

During this dismal night, it may be remarked that a man would conclude that it was really the intention of the seven mad gods to drown him, despite the abominable injustice of it. For it was certainly an abominable injustice to drown a man who had worked so hard, so hard. The man felt it would be a crime most unnatural. Other people had drowned at sea since galleys swarmed with painted sails, but still——

When it occurs to a man that nature does not regard him as important, and that she feels she would not maim the universe by disposing of him, he at first wishes to throw bricks at the temple, and he hates deeply the fact that there are no bricks and no temples. Any visible expression of nature would surely be pelleted with his jeers.

Then, if there be no tangible thing to hoot he feels, perhaps, the desire to confront a personification and indulge in pleas, bowed to one knee, and with hands supplicant, saying: "Yes, but I love myself."

A high cold star on a winter's night is the word he feels that she says to him. Thereafter he knows the pathos of his situation.

The men in the dingey had not discussed these matters, but each had, no doubt, reflected upon them in silence and according to his mind. There was seldom any expression upon their faces save the general one of complete weariness. Speech was devoted to the business of the boat.

To chime the notes of his emotion, a verse mysteriously entered the correspondent's head. He had even forgotten that he had forgotten this verse, but it suddenly was in his mind.

> A soldier of the Legion lay dying in Algiers,
> There was lack of woman's nursing, there was dearth
>   of woman's tears;
> But a comrade stood beside him, and he took that
>   comrade's hand,
> And he said: "I never more shall see my own, my
>   native land."[5]

In his childhood, the correspondent had been made acquainted with the fact that a soldier of the Legion lay dying in Algiers, but he had never regarded it as important. Myriads of his school-fellows had informed him of the soldier's plight, but the dinning had naturally ended by making him perfectly indifferent. He had never considered it his affair that a soldier of the Legion lay dying in Algiers, nor had it appeared to him as a matter for sorrow. It was less to him than the breaking of a pencil's point.

Now, however, it quaintly came to him as a human, living thing. It was no longer merely a picture of a few throes in the breast of a poet, meanwhile drinking tea and warming his feet at the grate; it was an actuality—stern, mournful, and fine.

The correspondent plainly saw the soldier. He lay on the sand with his feet out straight and still. While his pale left hand was upon his chest in an attempt to thwart the going of his life, the blood came between his fingers. In the far Algerian distance, a city of low square forms was set against a sky that was faint with the last sunset hues. The correspondent, plying the oars and dreaming of the slow and slower movements of the lips of the soldier, was moved by a profound and perfectly impersonal comprehension. He was sorry for the soldier of the Legion who lay dying in Algiers.

The thing which had followed the boat and waited had evidently grown bored at the delay. There was no longer to be heard the slash of the cut-water, and there was no longer the flame of the long trail. The light in the north still glimmered, but it was apparently no nearer to the boat. Sometimes the boom of the surf rang in the correspondent's ears, and he turned the craft seaward then and rowed harder. Southward, some one had evidently built a watch-fire on the beach. It was too low and too far to be seen, but it made a shimmering, roseate reflection upon the bluff back of it, and this could be discerned from the boat. The wind came stronger, and sometimes a wave suddenly raged out like a mountain-cat and there was to be seen the sheen and sparkle of a broken crest.

The captain, in the bow, moved on his water-jar and sat erect.

5. The lines are incorrectly quoted from Caroline E. S. Norton's poem *Bingen on the Rhine* (1883).

"Pretty long night," he observed to the correspondent. He looked at the shore. "Those life-saving people take their time."

"Did you see that shark playing around?"

"Yes, I saw him. He was a big fellow, all right."

"Wish I had known you were awake."

Later the correspondent spoke into the bottom of the boat.

"Billie!" There was a slow and gradual disentanglement. "Billie, will you spell me?"

"Sure," said the oiler.

As soon as the correspondent touched the cold comfortable sea-water in the bottom of the boat, and had huddled close to the cook's life-belt he was deep in sleep, despite the fact that his teeth played all the popular airs. This sleep was so good to him that it was but a moment before he heard a voice call his name in a tone that demonstrated the last stages of exhaustion. "Will you spell me?"

"Sure, Billie."

The light in the north had mysteriously vanished, but the correspondent took his course from the wide-awake captain.

Later in the night they took the boat farther out to sea, and the captain directed the cook to take one oar at the stern and keep the boat facing the seas. He was to call out if he should hear the thunder of the surf. This plan enabled the oiler and the correspondent to get respite together. "We'll give those boys a chance to get into shape again," said the captain. They curled down and, after a few preliminary chatterings and trembles, slept once more the dead sleep. Neither knew they had bequeathed to the cook the company of another shark, or perhaps the same shark.

As the boat caroused on the waves, spray occasionally bumped over the side and gave them a fresh soaking, but this had no power to break their repose. The ominous slash of the wind and the water affected them as it would have affected mummies.

"Boys," said the cook, with the notes of every reluctance in his voice, "she's drifted in pretty close. I guess one of you had better take her to sea again." The correspondent, aroused, heard the crash of the toppled crests.

As he was rowing, the captain gave him some whiskey and water, and this steadied the chills out of him. "If I ever get ashore and anybody shows me even a photograph of an oar——"

At last there was a short conversation.

"Billie. . . . Billie, will you spell me?"

"Sure," said the oiler.

VII

When the correspondent again opened his eyes, the sea and the sky were each of the gray hue of the dawning. Later, carmine and gold was painted upon the waters. The morning appeared finally, in its splendor, with a sky of pure blue, and the sunlight flamed on the tips of the waves.

On the distant dunes were set many little black cottages, and a tall white wind-mill reared above them. No man, nor dog, nor bicycle appeared on the beach. The cottages might have formed a deserted village.

The voyagers scanned the shore. A conference was held in the boat. "Well," said the captain, "if no help is coming, we might better try a run through the surf right away. If we stay out here much longer we will be too weak to do anything for ourselves at all." The others silently acquiesced in this reasoning. The boat was headed for the beach. The correspondent wondered if none ever ascended the tall wind-tower, and if then they never looked seaward. This tower was a giant, standing with its back to the plight of the ants. It represented in a degree, to the correspondent, the serenity of nature amid the struggles of the individual—nature in the wind, and nature in the vision of men. She did not seem cruel to him then, nor beneficent, nor treacherous, nor wise. But she was indifferent, flatly indifferent. It is, perhaps, plausible that a man in this situation, impressed with the unconcern of the universe, should see the innumerable flaws of his life and have them taste wickedly in his mind and wish for another chance. A distinction between right and wrong seems absurdly clear to him, then, in this new ignorance of the grave-edge, and he understands that if he were given another opportunity he would mend his conduct and his words, and be better and brighter during an introduction, or at a tea.

"Now, boys," said the captain, "she is going to swamp sure. All we can do is to work her in as far as possible, and then when she swamps, pile out and scramble for the beach. Keep cool now, and don't jump until she swamps sure."

The oiler took the oars. Over his shoulders he scanned the surf. "Captain," he said, "I think I'd better bring her about, and keep her head-on to the seas and back her in."

"All right, Billie," said the captain. "Back her in." The oiler swung the boat then and, seated in the stern, the cook and the correspondent were obliged to look over their shoulders to contemplate the lonely and indifferent shore.

The monstrous inshore rollers heaved the boat high until the men were again enabled to see the white sheets of water scudding up the slanted beach. "We won't get in very close," said the captain. Each time a man could wrest his attention from the rollers, he turned his glance toward the shore, and in the expression of the eyes during this contemplation there was a singular quality. The correspondent, observing the others, knew that they were not afraid, but the full meaning of their glances was shrouded.

As for himself, he was too tired to grapple fundamentally with the fact. He tried to coerce his mind into thinking of it, but the mind was dominated at this time by the muscles, and the muscles said they did not care. It merely occurred to him that if he should drown it would be a shame.

There were no hurried words, no pallor, no plain agitation. The men simply looked at the shore. "Now, remember to get well clear of the boat when you jump," said the captain.

Seaward the crest of a roller suddenly fell with a thunderous crash, and the long white comber came roaring down upon the boat.

"Steady now," said the captain. The men were silent. They turned their eyes from the shore to the comber and waited. The boat slid up the incline, leaped at the furious top, bounced over it, and swung down the long back of the wave. Some water had been shipped and the cook bailed it out.

But the next crest crashed also. The tumbling boiling flood of white water caught the boat and whirled it almost perpendicular. Water swarmed in from all sides. The correspondent had his hands on the gunwale at this time, and when the water entered at that place he swiftly withdrew his fingers, as if he objected to wetting them.

The little boat, drunken with this weight of water, reeled and snuggled deeper into the sea.

"Bail her out, cook! Bail her out," said the captain.

"All right, Captain," said the cook.

"Now, boys, the next one will do for us, sure," said the oiler. "Mind to jump clear of the boat."

The third wave moved forward, huge, furious, implacable. It fairly swallowed the dingey, and almost simultaneously the men tumbled into the sea. A piece of life-belt had lain in the bottom of the boat, and as the correspondent went overboard he held this to his chest with his left hand.

The January water was icy, and he reflected immediately that it was colder than he had expected to find it off the coast of Florida. This appeared to his dazed mind as a fact important enough to be noted at the time. The coldness of the water was sad; it was tragic. This fact was somehow so mixed and confused with his opinion of his own situation that it seemed almost a proper reason for tears. The water was cold.

When he came to the surface he was conscious of little but the noisy water. Afterward he saw his companions in the sea. The oiler was ahead in the race. He was swimming strongly and rapidly. Off to the correspondent's left, the cook's great white and corked back bulged out of the water, and in the rear the captain was hanging with his one good hand to the keel of the overturned dingey.

There is a certain immovable quality to a shore, and the correspondent wondered at it amid the confusion of the sea.

It seemed also very attractive, but the correspondent knew that it

was a long journey, and he paddled leisurely. The piece of life preserver lay under him, and sometimes he whirled down the incline of
a wave as if he were on a hand-sled.

But finally he arrived at a place in the sea where travel was beset
with difficulty. He did not pause swimming to inquire what manner
of current had caught him, but there his progress ceased. The shore
was set before him like a bit of scenery on a stage, and he looked at
it and understood with his eyes each detail of it.

As the cook passed, much farther to the left, the captain was calling to him, "Turn over on your back, cook! Turn over on your back
and use the oar."

"All right, sir." The cook turned on his back, and, paddling with
an oar, went ahead as if he were a canoe.

Presently the boat also passed to the left of the correspondent
with the captain clinging with one hand to the keel. He would have
appeared like a man raising himself to look over a board fence, if it
were not for the extraordinary gymnastics of the boat. The correspondent marvelled that the captain could still hold to it.

They passed on, nearer to shore—the oiler, the cook, the captain
—and following them went the water-jar, bouncing gayly over the
seas.

The correspondent remained in the grip of this strange new
enemy—a current. The shore, with its white slope of sand and its
green bluff, topped with little silent cottages, was spread like a picture before him. It was very near to him then, but he was impressed
as one who in a gallery looks at a scene from Brittany or Holland.

He thought: "I am going to drown? Can it be possible? Can it
be possible? Can it be possible?" Perhaps an individual must consider his own death to be the final phenomenon of nature.

But later a wave perhaps whirled him out of this small deadly
current, for he found suddenly that he could again make progress
toward the shore. Later still, he was aware that the captain, clinging
with one hand to the keel of the dingey, had his face turned away
from the shore and toward him, and was calling his name. "Come
to the boat! Come to the boat!"

In his struggle to reach the captain and the boat, he reflected
that when one gets properly wearied, drowning must really be a
comfortable arrangement, a cessation of hostilities accompanied by
a large degree of relief, and he was glad of it, for the main thing in
his mind for some moments had been horror of the temporary
agony. He did not wish to be hurt.

Presently he saw a man running along the shore. He was undressing with most remarkable speed. Coat, trousers, shirt, everything
flew magically off him.

"Come to the boat," called the captain.

"All right, Captain." As the correspondent paddled, he saw the
captain let himself down to bottom and leave the boat. Then the

correspondent performed his one little marvel of the voyage. A large
wave caught him and flung him with ease and supreme speed completely over the boat and far beyond it. It struck him even then as
an event in gymnastics, and a true miracle of the sea. An overturned
boat in the surf is not a plaything to a swimming man.

The correspondent arrived in water that reached only to his
waist, but his condition did not enable him to stand for more than
a moment. Each wave knocked him into a heap, and the under-tow
pulled at him.

Then he saw the man who had been running and undressing, and
undressing and running, come bounding into the water. He dragged
ashore the cook, and then waded toward the captain, but the captain waved him away, and sent him to the correspondent. He was
naked, naked as a tree in winter, but a halo was about his head, and
he shone like a saint. He gave a strong pull, and a long drag, and a
bully heave at the correspondent's hand. The correspondent,
schooled in the minor formulæ, said: "Thanks, old man." But suddenly the man cried: "What's that?" He pointed a swift finger.
The correspondent said: "Go."

In the shallows, face downward, lay the oiler. His forehead
touched sand that was periodically, between each wave, clear of the
sea.

The correspondent did not know all that transpired afterward.
When he achieved safe ground he fell, striking the sand with each
particular part of his body. It was as if he had dropped from a roof,
but the thud was grateful to him.

It seems that instantly the beach was populated with men with
blankets, clothes, and flasks, and women with coffee-pots and all the
remedies sacred to their minds. The welcome of the land to the
men from the sea was warm and generous, but a still and dripping
shape was carried slowly up the beach, and the land's welcome for
it could only be the different and sinister hospitality of the grave.

When it came night, the white waves paced to and fro in the
moonlight, and the wind brought the sound of the great sea's voice
to the men on shore, and they felt that they could then be interpreters.

1897, 1898

From THE BLACK RIDERS AND OTHER LINES[1]

I

Black riders came from the sea.
There was clang and clang of spear and shield,

1. These poems from *The Black Riders
and Other Lines* (1895), together with
their numbering, are reprinted from Joseph Katz's *The Poems of Stephen
Crane* (1966).

And clash and clash of hoof and heel,
Wild shouts and the wave of hair
In the rush upon the wind: 5
Thus the ride of Sin.

27

A youth in apparel that glittered
Went to walk in a grim forest.
There he met an assassin
Attired all in garb of old days;
He, scowling through the thickets,   5
And dagger poised quivering,
Rushed upon the youth.
"Sir," said this latter,
"I am enchanted, believe me,
To die, thus,   10
In this medieval fashion,
According to the best legends;
Ah, what joy!"
Then took he the wound, smiling,
And died, content.   15

42

I walked in a desert.
And I cried:
"Ah, God, take me from this place!"
A voice said: "It is no desert."
I cried: "Well, but—   5
The sand, the heat, the vacant horizon."
A voice said: "It is no desert."

56

A man feared that he might find an assassin;
Another that he might find a victim.
One was more wise than the other.

From WAR IS KIND[1]

76

Do not weep, maiden, for war is kind.
Because your lover threw wild hands toward the sky
And the affrighted steed ran on alone,
Do not weep.
War is kind.   5

---

1. These poems from *War Is Kind* (1899) are reprinted from Joseph Katz's critical
edition of *The Poems of Stephen Crane* (1966).

Hoarse, booming drums of the regiment,
Little souls who thirst for fight,
These men were born to drill and die.
The unexplained glory flies above them,
Great is the Battle-God, great, and his Kingdom—   10
A field where a thousand corpses lie.

Do not weep, babe, for war is kind.
Because your father tumbled in the yellow trenches,
Raged at his breast, gulped and died,
Do not weep.   15
War is kind.

Swift blazing flag of the regiment,
Eagle with crest of red and gold,
These men were born to drill and die.
Point for them the virtue of slaughter,   20
Make plain to them the excellence of killing
And a field where a thousand corpses lie.

Mother whose heart hung humble as a button
On the bright splendid shroud of your son,
Do not weep.   25
War is kind.

1896   1899

83

Fast rode the knight
With spurs, hot and reeking
Ever waving an eager sword.
    "To save my lady!"
Fast rode the knight   5
And leaped from saddle to war.
Men of steel flickered and gleamed
Like riot of silver lights
And the gold of the knight's good banner
Still waved on a castle wall.   10

A horse
Blowing, staggering, bloody thing
Forgotten at foot of castle wall.
A horse
Dead at foot of castle wall.   15

1896   1899

87

A newspaper is a collection of half-injustices
Which, bawled by boys from mile to mile,
Spreads its curious opinion
To a million merciful and sneering men,

While families cuddle the joys of the fireside 5
When spurred by tale of dire lone agony.
A newspaper is a court
Where every one is kindly and unfairly tried
By a squalor of honest men.
A newspaper is a market 10
Where wisdom sells its freedom
And melons are crowned by the crowd.
A newspaper is a game
Where his error scores the player victory
While another's skill wins death. 15
A newspaper is a symbol;
It is fetless life's chronicle,
A collection of loud tales
Concentrating eternal stupidities,
That in remote ages lived unhaltered, 20
Roaming through a fenceless world.

1899

89

A slant of sun on dull brown walls
A forgotten sky of bashful blue.
Toward God a mighty hymn
A song of collisions and cries
Rumbling wheels, hoof-beats, bells, 5
Welcomes, farewells, love-calls, final moans,
Voices of joy, idiocy, warning, despair,
The unknown appeals of brutes,
The chanting of flowers
The screams of cut trees, 10
The senseless babble of hens and wise men—
A cluttered incoherency that says at the stars:
"Oh, God, save us."

1895   1899

96

A man said to the universe:
"Sir, I exist!"
"However," replied the universe,
"The fact has not created in me
A sense of obligation."

1899

From POSTHUMOUSLY PUBLISHED POEMS[1]

113

A man adrift on a slim spar
A horizon smaller than the rim of a bottle

1. The poem, first published in *The Bookman* for April, 1929, is reprinted from Joseph Katz's critical edition of *The Poems of Stephen Crane* (1966).

Tented waves rearing lashy dark points
The near whine of froth in circles.
                    God is cold.   5

The incessant raise and swing of the sea
And growl after growl of crest
The sinkings, green, seething, endless
The upheaval half-completed.
                    God is cold.   10

The seas are in the hollow of The Hand;
Oceans may be turned to a spray
Raining down through the stars
Because of a gesture of pity toward a babe.
Oceans may become grey ashes,   15
Die with a long moan and a roar
Amid the tumult of the fishes
And the cries of the ships,
Because The Hand beckons the mice.

A horizon smaller than a doomed assassin's cap,   20
Inky, surging tumults
A reeling, drunken sky and no sky
A pale hand sliding from a polished spar.
                    God is cold.

The puff of a coat imprisoning air: 25
A face kissing the water-death
A weary slow sway of a lost hand
And the sea, the moving sea, the sea.
                    God is cold.

c. 1898   1929

THEODORE DREISER
1871–1945

Theodore Herman Albert Dreiser was born in Terre Haute, Indiana, on August 27, 1871, the twelfth of thirteen children. His gentle and devoted mother was illiterate; his German immigrant father was severe and distant. From the former he seems to have absorbed a quality of compassionate wonder; from the latter he seems to have inherited moral earnestness and the capacity to persist in the face of failure, disappointment, and despair.

Dreiser's childhood was decidedly unhappy. The large family moved from house to house in Indiana dogged by poverty, insecurity, and internal division. One of his brothers became a famous popular songwriter under the name of Paul Dresser, but other brothers and sisters drifted into drunkenness, promiscuity, and squalor. Dreiser as a youth was as ungainly, confused, shy, and full of vague yearnings as most of his fictional protagonists, male and female. In this as in many other ways, Dreiser's novels are direct projections of his inner life as well as careful transcriptions of his experiences.

From the age of fifteen Dreiser was essentially on his own, earning meager support from a variety of menial jobs. A high-school teacher staked him to a year at Indiana University in 1889, but Dreiser's education was to come from experience and from independent reading and thinking. This education began in 1892 when he wangled his first newspaper job with the Chicago *Globe*. Over the next decade as an itinerant journalist Dreiser slowly groped his way to authorship, testing what he knew from direct experience against what he was learning from reading Charles Darwin, Ernst Haeckel, Thomas Huxley, and Herbert Spencer, those late-nineteenth-century scientists and social scientists who lent support to the view that nature and society had no divine sanction.

Sister Carrie (1900), which traces the material rise of Carrie Meeber and the tragic decline of G. W. Hurstwood, was Dreiser's first novel. Because it depicted social transgressions by characters who felt no remorse and largely escaped punishment, and because it used "strong" language and used names of living persons, it was virtually suppressed by its publisher, who printed but refused to promote the book. Since its reissue in 1907 it has steadily risen in popularity and scholarly acceptance as one of the key works in the Dreiser canon. Indeed, though turn-of-the-century readers found Dreiser's point of view crude and immoral, his influence on the fiction of the first quarter of the century is perhaps greater than any other writer's. In this early period some of his best short fictions were written, among them *Nigger Jeff* and *Butcher Rogaum's Daughter*. The best of his short stories —like all of Dreiser's fiction—have the unusual power to compel our sympathy for and wonder over characters whose minds and inner life we never really enter. The lynching of Jeff Ingalls unfolds according to "some axiomatic, mathematical law" and against the incongruous backdrop of a "stillness of purest, summery-est, country-est quality." The events have the air of necessity of a Greek tragedy, a horrifying and moving spectacle.

In the first years of the century Dreiser suffered a breakdown. With the help of his brother Paul, however, he eventually recovered and by 1904 was on the way to several successful years as an editor, the last of them as editorial director of the Butterick Publishing Company. In 1910 he resigned to write *Jennie Gerhardt* (1911), the first of a long succession of books that marked his turn to writing as a full-time career.

In *The Financier* (1912). *The Titan* (1914), and *The Stoic* (not published until 1947) Dreiser shifted from the pathos of helpless protagonists to the power of those unusual individuals who assume dominant roles in business and society. The protagonist of this "Trilogy of Desire" (as Dreiser described it), Frank Cowperwood, is modeled after the Chicago speculator Charles T. Yerkes. These novels of the businessman as buccaneer introduced, even more explicitly than had *Sister Carrie*, the notion that men of high sexual energy were financially successful, a theme that is carried over into the rather weak autobiographical novel *The "Genius"* (1915).

The identification of potency with money is at the heart of Dreiser's greatest and most successful novel, *An American Tragedy* (1925). The center of this immense novel's thick texture of biographical circumstance, social fact, and industrial detail is a young man who acts as if the only way he can be truly fulfilled is by acquiring wealth—through marriage if necessary.

During the last two decades of his life Dreiser turned entirely away from fiction and toward political activism and polemical writing. He visited the Soviet Union in 1927 and published *Dreiser Looks at Russia* the following year. In the 1930s, like many other American intellectuals and writers, Dreiser was increasingly attracted by the philosophical program of the Communist party. Unable to believe in traditional religious credos, yet unable to give up his strong sense of justice, he continued to seek a way to reconcile his determinism with his compassionate sense of the mystery of life.

In his lifetime Dreiser was controversial as a man and as a writer. He was accused, with some justice by conventional standards, of being immoral in his personal behavior, a poor thinker, and a dangerous political radical; his style was said (by critics more than by fellow authors) to be ponderous and his narrative sense weak. As time has passed, however, Dreiser has become recognized as a profound and prescient critic of debased American values and as a powerful novelist.

Nigger Jeff[1]

The city editor was waiting for his good reporter, Eugene Davies. He had cut an item from one of the afternoon papers and laid it aside to give to Mr. Davies. Presently the reporter appeared.

It was one o'clock of a sunny, spring afternoon. Davies wore a new spring suit, a new hat and new shoes. In the lapel of his coat was a small bunch of violets. He was feeling exceedingly well and good-natured. The world seemed worth singing about.

"Read that, Davies," said the city editor, handing him the clipping. "I'll tell you what I want you to do afterward."

The reporter stood by the editorial chair and read:

> "Pleasant Valley, Mo., April 16.
> "A most dastardly crime has just been reported here. Jeff Ingalls, a negro, this morning assaulted Ada Whittier, the nineteen-year-old daughter of Morgan Whittier, a well-to-do farmer, whose home is four miles south of this place. A posse, headed by Sheriff Mathews, has started in pursuit. If he is caught, it is thought he will be lynched."

The reporter raised his eyes as he finished.

"You had better go out there, Davies," said the city editor. "It looks as if something might come of that. A lynching up here would be a big thing."

Davies smiled. He was always pleased to be sent out of town. It was a mark of appreciation. The city editor never sent any of the other boys on these big stories. What a nice ride he would have.

He found Pleasant Valley to be a small town, nestling between green slopes of low hills, with one small business corner and a rambling array of lanes. One or two merchants of St. Louis lived out here, but otherwise it was exceedingly rural. He took note of the whiteness of the little houses, the shimmering beauty of the little creek you had to cross in going from the depot. At the one main

1. Reprinted from *Ainslee's Magazine* (1901).

corner a few men were gathered about a typical village barroom. Davies headed for this as being the most apparent source of information.

In mingling with the company, he said nothing about his errand. He was very shy about mentioning that he was a newspaper man.

The whole company was craving excitement and wanted to see something come of the matter. They hadn't had such a chance to work up wrath and satisfy their animal propensities in years. It was a fine opportunity and such a righteous one.

He went away thinking that he had best find out for himself how the girl was. Accordingly he sought the old man that kept a stable in the village and procured a horse. No carriage was to be had. Davies was not an excellent rider, but he made a shift of it. The farm was not so very far away, and before long he knocked at the front door of the house, set back a hundred feet from the rough country road.

"I'm from the *Republic*," he said, with dignity. His position took very well with farmers. "How is Miss Whittier?"

"She's doing very well," said a tall, raw-boned woman. "Won't you come in? She's rather feverish, but the doctor says she'll be all right."

Davies acknowledged the invitation by entering. He was anxious to see the girl, but she was sleeping, and under the influence of an opiate.

"When did this happen?" he asked.

"About eight o'clock this morning," said the woman. "She started off to go over to our next neighbor here, Mr. Edmonds, and this negro met her. I didn't know anything about it until she came crying through the gate and dropped down in here."

"Were you the first one to meet her?" asked Davies.

"Yes, I was the only one," said Mrs. Whittier. "The men had gone out in the fields."

Davies listened to more of the details, and then rose to go. He was allowed to have a look at the girl, who was rather pretty. In the yard he met a country chap who had come over to hear the news. This man imparted more information.

"They're lookin' all around south of here," said the man, speaking of the crowd supposed to be in search. "I expect they'll make short work of him if they get him."

"Where does this negro live?" asked Davies.

"Oh, right down here a little way. You follow this road to the next crossing and turn to the right. It's a little log house that sits back off the road—something like this, only it's got a lot of chips scattered about."

Davies decided to go there, but changed his mind. It was getting late. He had better return to the village, he thought.

Accordingly, he rode back and put the horse in the hands of its owner. Then he went over to the principal corner. Much the same company was still present. He wondered what these people had been doing all the time. He decided to ingratiate himself by imparting a little information.

Just then a young fellow came galloping up.

"They've got him," he shouted, excitedly, "they've got him."

A chorus of "whos" and "wheres," with sundry other queries, greeted this information as the crowd gathered about the rider.

"Why, Mathews caught him up here at his own house. Says he'll shoot the first man that dares to try to take him away. He's taking him over to Clayton."

"Which way'd he go?" exclaimed the men.

" 'Cross Sellers' Lane," said the rider. "The boys think he's going to Baldwin."

"Whoopee," yelled one of the listeners. "Are you going, Sam?"

"You bet," said the latter. "Wait'll I get my horse."

Davies waited no longer. He saw the crowd would be off in a minute to catch up with the sheriff. There would be information in that quarter. He hastened after his horse.

"He's eating," said the man.

"I don't care," exclaimed Davies. "Turn him out. I'll give you a dollar more."

The man led the horse out, and the reporter mounted.

When he got back to the corner several of the men were already there. The young man who had brought the news had dashed off again.

Davies waited to see which road they would take. Then he did the riding of his life.

In an hour the company had come in sight of the sheriff, who, with two other men, was driving a wagon he had borrowed. He had a revolver in each hand and was sitting with his face toward the group, that trailed after at a respectful distance. Excited as every one was, there was no disposition to halt the progress of the law.

"He's in that wagon," Davies heard one man say. "Don't you see they've got him tied and laid down in there?"

Davies looked.

"We ought to take him away and hang him," said one of the young fellows who rode nearest the front.

"Where's old man Whittier?" asked one of the crowd, who felt that they needed a leader.

"He's out with the other crowd," was the reply.

"Somebody ought to go and tell him."

"Clark's gone," assured another, who hoped for the worst.

Davies rode among the company very much excited. He was astonished at the character of the crowd. It was largely impelled to its excited jaunt by curiosity and a desire to see what would happen.

There was not much daring in it. The men were afraid of the determined sheriff. They thought something ought to be done, but they did not feel like getting into trouble.

The sheriff, a sage, lusty, solemn man, contemplated the recent addition to these trailers with considerable feeling. He was determined to protect his man and avoid injustice. A mob should not have him if he had to shoot, and if he shot, he was going to empty both revolvers, and those of his companions. Finally, since the company thus added to did not dash upon him, he decided to scare them off. He thought he could do it since they trailed like calves.

"Stop a minute," he said to his driver.

The latter pulled up. So did the crowd behind. Then the sheriff stood over the prostrate body of the negro, who lay trembling in the jolting wagon bed and called back to the men.

"Go on away from here, you people," he said. "Go on, now. I won't have you foller after me."

"Give us the nigger," yelled one in a half-bantering, half-derisive tone of voice.

"I'll give you five minutes to go on back out of this road," returned the sheriff grimly. They were about a hundred feet apart. "If you don't, I'll clear you out."

"Give us the nigger!"

"I know you, Scott," answered the sheriff, recognizing the voice. "I'll arrest every last one of you to-morrow. Mark my word!"

The company listened in silence, the horses champing and twisting.

"We've got a right to follow," answered one of the men.

"I give you fair warning," said the sheriff, jumping from his wagon and leveling his pistols as he approached. "When I count five, I'll begin to shoot."

He was a serious and stalwart figure as he approached, and the crowd retreated.

"Get out o' this now," he yelled. "One, two——"

The company turned completely and retreated.

"We'll follow him when he gets farther on," said one of the men in explanation.

"He's got to do it," said another. "Let him get a little ahead."

The sheriff returned to his wagon and drove on. He knew that he would not be obeyed, and that safety lay in haste alone. If he could only make them lose track of him and get a good start it might be possible to get to Clayton and the strong county jail by morning.

Accordingly he whipped up his horses while keeping his grim lookout.

"He's going to Baldwin," said one of the company of which Davies was a member.

"Where is that?" asked Davies.

"Over west of here, about four miles."

The men lagged, hesitating what to do. They did not want to lose sight of him, and yet cowardice controlled them. They did not want to get into direct altercation with the law. It wasn't their place to hang the man, although he ought to be hanged and it would be a stirring and exciting thing if he were. Consequently, they desired to watch and be on hand—to get old Whittier and his son Jake if they could, who were out looking elsewhere. They wanted to see what the father and brother would do.

The quandary was solved by Dick Hewlitt, who suggested that they could get to Baldwin by going back to Pleasant Valley and taking the Sand River pike. It was a shorter cut than this. Maybe they could beat the sheriff there. Accordingly, while one or two remained to track the sheriff, the rest set off at a gallop to Pleasant Valley. It was nearly dusk when they got there and stopped for a few minutes at the corner store. Here they talked, and somehow the zest to follow departed; they were not certain now of going on. It was supper time. The fires of evening meals were marked by upcurling smoke. Evidently the sheriff had them worsted for to-night. Morg Whittier had not been found. Neither had Jake. Perhaps they had better eat. Two or three had already secretly fallen away.

They were telling the news to the one or two storekeepers, when Jake Whittier, the girl's brother, and several companions came riding up. They had been scouring the territory to the north of the town.

"The sheriff's got him," said one of the company. "He's taking him over to Baldwin in a wagon."

"Which way did he go?" asked young Jake, whose hardy figure, worn, hand-me-down clothes and rakish hat showed up picturesquely as he turned on his horse.

" 'Cross Sellers' Lane. You won't get him that way. Better take the short cut."

A babble of voices was making the little corner interesting. One told how he had been caught, another that the sheriff was defiant, a third that men were tracking him, until the chief points of the drama had been spoken, if not heard.

"Come on, boys," said Jake, jerking at the reins and heading up the pike. "I'll get the damn nigger."

Instantly suppers were forgotten. The whole customary order of the evening was neglected. The company started off on another exciting jaunt, up hill and down dale, through the lovely country that lay between Baldwin and Pleasant Valley.

Davies was very weary of his saddle. He wondered when he was to write his story. The night was exceedingly beautiful. Stars were already beginning to shine. Distant lamps twinkled like yellow eyes from the cottages in the valleys on the hillsides. The air was fresh

and tender. Some pea fowls were crying afar off and the east promised a golden moon.

Silently the assembled company trotted on—no more than a
score in all. It was too grim a pilgrimage for joking. Young Jake,
riding silently toward the front, looked as if he meant business. His
friends did not like to say anything to him, seeing that he was the
aggrieved. He was left alone.

After an hour's riding Baldwin came into view, lying in a sheltering cup of low hills. Its lights were twinkling softly, and there was
an air of honest firesides and cheery suppers about it which
appealed to Davies in his hungry state. Still, he had no thought but
of carrying out his mission.

Once in the village they were greeted by calls of recognition.
Everybody knew what they had come for. The local storekeepers
and loungers followed the cavalcade up the street to the sheriff's
house, for the riders had now fallen into a solemn walk.

"You won't get him, boys," said Seavey, the young postmaster
and telegraph operator, as they passed his door. "Mathews says he's
sent him to Clayton."

At the first street corner they were joined by several men who
had followed the sheriff.

"He tried to give us the slip," they said, excitedly, "but he's got
the nigger in the house there, down in the cellar."

"How do you know?"

"I saw him bring him in this way. I think he is, anyhow."

A block from the sheriff's little white cottage the men parleyed.
They decided to go up and demand the negro.

"If he don't turn him out, we'll break in the door and take him,"
said Jake.

"That's right. We'll stand by you, Whittier."

A throng had gathered. The whole village was up in arms. The
one street was alive and running with people. Riders pranced up
and down, hallooing. A few shot off revolvers. Presently the mob
gathered about the sheriff's gate, and Jake stepped forward as
leader.

Their coming was not unexpected. Sheriff Mathews was ready for
them with a double-barreled Winchester. He had bolted the doors
and put the negro in the cellar, pending the arrival of the aid he
had telegraphed for to Clayton. The latter was cowering and chattering in the darkest corner of his dungeon against the cold, damp
earth, as he hearkened to the voices and the firing of the revolvers.
With wide, bulging eyes, he stared into the gloom.

Jake, the son and brother, took the precautionary method of calling to the sheriff.

"Hello, Mathews!"

"Eh, eh, eh," bellowed the crowd.

Suddenly the door flew open, and appearing first in the glow of the lamp came the double barrel of a Winchester, followed by the form of the sheriff, who held his gun ready for a quick throw to the shoulder. All except Jake fell back.

"We want that nigger," said Jake, deliberately.

"He isn't here," said the sheriff.

"Then what you got that gun for?" yelled a voice.

The sheriff made no answer.

"Better give him up, Mathews," called another, who was safe in the crowd, "or we'll come in and take him."

"Lookee here, gentlemen," said the sheriff, "I said the man wasn't here. I say it again. You couldn't have him if he was and you can't come in my house. Now, if you people don't want trouble, you'd better go on away."

"He's down in the cellar," yelled another.

The sheriff waved his gun slightly.

"Why don't you let us see?" said another.

"You'd better go away from here now," cautioned the sheriff.

The crowd continued to simmer and stew, while Jake stood out before. He was very pale and determined, but lacked initiative.

"He won't shoot. Why don't you go in, boys, and get him?"

"He won't, eh?" thought the sheriff. Then he said aloud: "The first man that comes inside that gate takes the consequences."

No one ventured near the gate. It seemed as if the planned assault must come to nothing.

"You'd better go away from here," cautioned the sheriff again. "You can't come in, it'll only mean bloodshed."

There was more chattering and jesting while the sheriff stood on guard. He said no more. Nor did he allow the banter, turmoil and lust for tragedy to disturb him. Only he kept his eye on Jake, on whose movements the crowd hung.

"I'll get him," said Jake, "before morning."

The truth was that he felt the weakness of the crowd. He was, to all intents and purposes, alone, for he did not inspire confidence.

Thus the minutes passed. It became a half hour and then an hour. With the extending time pedestrians dropped out and then horsemen. Some went up the street, several back to Pleasant Valley, more galloped about until there were very few left at the gate. It was plain that organization was lost. Finally Davies smiled and came away. He was sure he had a splendid story.

He began to look for something to eat, and hunted for the telegraph operator.

He found the operator first and told him he wanted to write a story and file it. The latter said there was a table in the little post-office and telegraph station which he could use. He got very much interested in Davies, and when he asked where he could get something to eat, said he would run across the street and tell the proprietor of the only boarding-house to fix him something which he could eat as he wrote.

"You start your story," he said, "and I'll come back and see if I can get the *Republic*."

Davies sat down and started the account.

"Very obliging postmaster," he thought, but he had so often encountered pleasant and obliging people on his rounds, that he soon dropped that thought.

The food was brought and Davies wrote. By eight-thirty the *Republic* answered an often-repeated call.

"Davies at Baldwin," ticked the postmaster, "get ready for quite a story."

"Let 'er go," answered the operator in the *Republic*, who had been expecting this dispatch.

Davies turned over page after page as the events of the day formulated themselves in his mind. He ate a little between whiles, looking out through the small window before him, where afar off he could see a lonely light twinkling in a hillside cottage. Not infrequently he stopped work to see if anything new was happening. The operator also wandered about, waiting for an accumulation of pages upon which he could work, but making sure to catch up with the writer. The two became quite friendly.

Davies finished his dispatch with the caution that more might follow, and was told by the city editor to watch it. Then he and the postmaster sat down to talk.

About twelve o'clock the lights in all the village houses had vanished and the inhabitants had gone to bed. The man-hunters had retired, and the night was left to its own sounds and murmurs, when suddenly the faint beating of hoofs sounded out on the Sand River Pike, which led away toward Pleasant Valley, back of the post-office. The sheriff had not relaxed any of his vigilance. He was not sleeping. There was no sleep for him until the county authorities should come to his aid.

"Here they come back again," exclaimed the postmaster.

"By George, you're right," said Davies.

There was a clattering of hoofs and grunting of saddle girths as a large company of men dashed up the road and turned into the narrow street of the village.

Instantly the place was astir again. Lights appeared in doorways, and windows were thrown open. People were gazing out to see what new movement was afoot. Davies saw that there was none of the hip and hurrah business about this company such as had characterized the previous descent. There was grimness everywhere, and he began to feel that this was the beginning of the end. He ran down the street toward the sheriff's house, arriving a few moments after the crowd, which was in part dismounted.

With the clear moon shining straight overhead, it was nearly as bright as day. Davies made out several of his companions of the afternoon and Jake, the son. There were many more, though, whom he did not know, and foremost among them an old man. He was strong, iron-gray and wore a full beard. He looked very much like a blacksmith.

While he was still looking, the old man went boldly forward to the little front porch of the house and knocked at the door. Some one lifted a curtain at the window and peeped out.

"Hello, in there," the old man cried, knocking again and much louder.

"What do you want?" said a voice.

"We want that nigger."

"Well, you can't have him. I've told you people once."

"Bring him out or we'll break down the door," said the old man.

"If you do, it's at your own risk. I'll give you three minutes to get off that porch."

"We want that nigger."

"If you don't get off that porch I'll fire through the door," said the voice, solemnly. "One, two——"

The old man backed cautiously away.

"Come out, Mathews," yelled the crowd. "You've got to give him up. We ain't going back without him."

Slowly the door opened, as if the individual within was very well satisfied as to his power to handle the mob. It revealed the tall form of Sheriff Mathews, armed with his Winchester. He looked around very stolidly and then addressed the old man as one would a friend.

"You can't have him, Morgan," he said, "it's against the law."

"Law or no law," said the old man, "I want that nigger."

"I can't let you have him, Morgan. It's against the law. You oughtn't to be coming around here at this time of night acting so."

"Well, we'll take him, then," said the old man, making a move.

The sheriff leveled his gun on the instant.

"Stand back, there," he shouted, noticing a movement on the part of the crowd. "I'll blow ye into kingdom come, sure as hell."

The crowd halted at this assurance.

The sheriff lowered his weapon as if he thought the danger were over.

"You all ought to be ashamed of yourselves," he said, softly, his voice sinking to a gentle, neighborly reproof, "tryin' to upset the law this way."

"The nigger didn't upset the law, did he?" asked one, derisively.

The sheriff made no answer.

"Give us that scoundrel, Mathews, you'd better do it," said the old man. "It'll save a heap of trouble."

"I'll not argue with you, Morgan. I said you couldn't have him, and you can't. If you want bloodshed, all right. But don't blame me. I'll kill the first man that tries to make a move this way."

He shifted his gun handily and waited. The crowd stood outside his little fence murmuring.

Presently the old man retired and spoke to several leaders.

There was more murmuring, and then he came back to the dead line.

"We don't want to cause trouble, Mathews," he began, explanatively, moving his hand oratorically, "but we think you ought to see that it won't do you any good to stand out. We think that——"

Davies was watching young Jake, the son, whose peculiar attitude attracted his attention. The latter was standing poised at the edge of the crowd, evidently seeking to remain unobserved. His eyes were on the sheriff, who was hearkening to the old man. Suddenly, when the sheriff seemed for a moment mollified and unsuspecting, he made a quick run for the porch. There was an intense movement all along the line, as the life and death of the deed became apparent. Quickly the sheriff drew his gun to his shoulder. He pressed both triggers at the same time, but not before Jake reached him. The latter knocked the gun barrel upward and fell upon his man. Both shots blazed out over the heads of the crowd in red puffs, and then followed a general onslaught. Men leaped the fence by tens, and crowded upon the little cottage. They swarmed on every side of the house, and crowded about the porch and the door, where four men were scuffling with the sheriff. The latter soon gave up, vowing vengeance. Torches were brought and a rope. A wagon drove up and was backed into the yard. Then began the calls for the negro.

The negro had been crouching in his corner in the cellar, trembling for his fate ever since the first attack. He had not dozed or lost consciousness during the intervening hours, but cowered there, wondering and praying. He was terrified lest the sheriff might not get him away in time. He was afraid that every sound meant a new assault. Now, however, he had begun to have the faintest glimmerings of hope when the new murmurs of contention arose. He heard the gallop of the horses' feet, voices of the men parleying, the ominous knock on the door.

At this sound, his body quaked and his teeth chattered. He began to quiver in each separate muscle and run cold. Already he saw the men at him, beating and kicking him.

"Before God, boss, I didn't mean to," he chattered, contemplating the chimera of his brain with startling eyes. "Oh, my God! boss, no, no. Oh, no, no."

He crowded closer to the wall. Another sound greeted his ears. It was the roar of a shotgun. He fell, groveling upon the floor, his nails digging in the earth.

"Oh, my Lawd, boss," he moaned, "oh, my Lawd, boss, don't kill me. I won't do it no mo'. I didn't go to do it. I didn't." His teeth were in the wet earth.

It was but now that the men were calling each other to the search. Five jumped to the outside entrance way of the low cellar, carrying a rope. Three others followed with their torches. They descended into the dark hole and looked cautiously about.

Suddenly, in the farthest corner, they espied him. In his agony, he had worked himself into a crouching position, as if he were about to spring. His hands were still in the earth. His eyes were rolling, his mouth foaming.

"Oh, my Lawd!" he was repeating monotonously, "oh, my Lawd!"

"Here he is. Pull him out, boys," cried several together.

The negro gave one yell of horror. He quite bounded as he did so, coming down with a dead chug on the earthen floor. Reason had forsaken him. He was a groveling, foaming brute. The last gleam of intelligence was that which notified him of the set eyes of his pursuers.

Davies was standing ten feet back when they began to reappear. He noted the heads of the torches, the disheveled appearance of the men, the scuffling and pulling. Then he clapped his hands over his mouth and worked his fingers convulsively, almost unconscious of what he was doing.

"Oh, my God," he whispered, his voice losing power.

The sickening sight was that of negro Jeff, foaming at the mouth, bloodshot in the eyes, his hands working convulsively, being dragged up the cellar steps, feet foremost. They had tied a rope about his waist and feet, and had hauled him out, leaving his head to hang and drag. The black face was distorted beyond all human semblance.

"Oh, my God!" said Davies again, biting his fingers unconsciously.

The crowd gathered about, more horror-stricken than gleeful at their own work. The negro was rudely bound and thrown like a sack of wheat into the wagon bed. Father and son mounted to drive, and the crowd took their horses. Wide-eyed and brain-racked, Davies ran for his own. He was so excited, he scarcely knew what he was doing.

Slowly the gloomy cavalcade took its way up the Sand River Pike. The moon was pouring down a wash of silvery light. The shadowy trees were stirring with a cool night wind. Davies hurried after and joined the silent, tramping throng.

"Are they going to hang him?" he asked.

"That's what they got him for," answered the man nearest him.

Davies dropped again into silence and tried to recover his nerves. The gloomy company seemed a terrible thing. He drew near the wagon and looked at the negro.

The latter seemed out of his senses. He was breathing heavily and groaning. His eyes were fixed and staring, his face and hands bleeding as if they had been scratched or trampled on. He was bundled up like limp wheat.

Davies could not stand it longer. He fell back, sick at heart, It seemed a ghastly, unmerciful way to do. Still, the company moved on and he followed, past fields lit white by the moon, under dark, silent groups of trees, through which the moonlight fell in patches, up hilltops and down into valleys, until at last the little stream came into view, sparkling like a molten flood of silver in the night. After a time the road drew close to the water and made for a wagon bridge, which could be seen a little way ahead. The company rode up to this and halted. Davies dismounted with the others. The wagon was driven up to the bridge and father and son got out.

Fully a score of men gathered about, and the negro was lifted from the wagon. Davies thought he could not stand it, and went down by the waterside slightly above the bridge. He could see long beams of iron sticking out over the water, where the bridge was braced.

The men fastened a rope to a beam and then he could see that they were fixing the other end around the negro's neck.

Finally the curious company stood back.

"Have you anything to say?" a voice demanded.

The negro only lolled and groaned, slobbering at the mouth. He was out of his mind.

Then came the concerted action of four men, a lifting of a black mass in the air, and then Davies saw the limp form plunge down and pull up with a creaking sound of rope. In the weak moonlight it seemed as if the body were struggling, but he could not tell. He watched, wide-mouthed and silent, and then the body ceased moving. He heard the company depart, but that did not seem important. Only the black mass swaying in the pale light, over the shiny water of the stream seemed wonderful.

He sat down upon the bank and gazed in silence. He was not afraid. Everything was summery and beautiful. The whole cavalcade disappeared, the moon sank. The light of morning began to show as tender lavender and gray in the east. Still he sat. Then came the roseate hue of day, to which the waters of the stream responded, the white pebbles shining beautifully at the bottom. Still the body hung black and limp, and now a light breeze sprang up and stirred it visibly. At last he arose and made his way back to Pleasant Valley.

Since his duties called him to another day's work here, he idled about, getting the details of what was to be done. He talked with citizens and officials, rode out to the injured girl's home, rode to Baldwin to see the sheriff. There was singular silence and placidity in that corner. The sheriff took his defeat as he did his danger, philosophically.

It was evening again before he remembered that he had not discovered whether the body had been removed. He had not heard why the negro came back or how he was caught. The little cabin was two miles away, but he decided to walk, the night was so springlike. Before he had traveled half way, the moon arose and stretched long shadows of budding trees across his path. It was not long before he came upon the cabin, set well back from the road and surrounded with a few scattered trees. The ground between the door and the road was open, and strewn with the scattered chips of a woodpile. The roof was sagged and the windows patched in places, but, for all that, it had the glow of a home. Through the front door, which stood open, the blaze of a fire shone, its yellow light filling the interior with golden fancies.

Davies stopped at the door and knocked, but received no answer. He looked in on the battered cane chairs and aged furniture with considerable interest.

A door in the rear room opened, and a little negro girl entered, carrying a battered tin lamp, without any chimney. She had not heard his knock, and started perceptibly at the sight of his figure in the doorway. Then she raised her smoking lamp above her head in order to see better and approached.

There was something comical about her unformed figure and loose gingham dress. Her black head was strongly emphasized by little pigtails of hair done up in white twine, which stood out all over her head. Her dark skin was made apparently more so by contrast with her white teeth and the whites of her eyes.

Davies looked at her for a moment and asked, "Is this where Ingalls lives?"

The girl nodded her head. She was exceedingly subdued, and looked as if she had been crying.

"Has the body been brought here?" he asked.

"Yes, suh," she answered, with a soft negro accent.

"When did they bring it home?"

"This moanin'."

"Are you his sister?"

"Yes, suh."

"Well, can you tell me how they caught him?" asked Davies, feeling slightly ashamed to intrude thus. "What did he come back for?"

"To see us," said the girl.

"Well, did he want anything? He didn't come just to see you, did he?"

"Yes, suh," said the girl, "he come to say good-by."

Her voice wavered.

"Didn't he know he might get caught?" asked Davies.

"Yes, suh, I think he did."

She still stood very quietly holding the poor battered lamp up, and looking down.

"Well, what did he have to say?" asked Davies.

"He said he wanted tuh see motha'. He was a-goin' away."

The girl seemed to regard Davies as an official of some sort, and he knew it.

"Can I have a look at the body?" he asked.

The girl did not answer, but started as if to lead the way.

"When is the funeral?" he asked.

"To-morrow.".

The girl led him through several bare sheds of rooms to the furthermost one of the line. This last seemed a sort of storage shed for odds and ends. It had several windows, but they were bare of glass, and open to the moonlight, save for a few wooden boards nailed across from the outside. Davies had been wondering all the while at the lonely and forsaken air of the place. No one seemed about but this little girl. If they had colored neighbors, none thought it worth while to call.

Now, as he stepped into this cool, dark, exposed outer room, the desolation seemed complete. The body was there in the middle of the bare room, stretched upon an ironing board, which rested on a box and a chair, and covered with a white sheet. All the corners of the room were quite dark, and only in the middle were shining splotches of moonlight.

Davies came forward, but the girl left him, carrying her lamp. She did not seem able to remain. He lifted the sheet, for he could see well enough, and looked at the stiff, black form. The face was extremely distorted, even in death, and he could see where the rope had tightened. A bar of cool moonlight lay across the face and breast. He was still looking, thinking soon to restore the covering, when a sound, half sigh, half groan, reached his ears.

He started as if a ghost had touched him. His muscles tightened. Instantly his heart was hammering like mad in his chest. His first impression was that it came from the dead.

"Oo-o-ohh," came the sound again, this time whimpering, as if someone were crying.

He turned quickly, for now it seemed to come from the corner. Greatly disturbed, he hesitated, and then as his eyes strained he caught the shadow of something. It was in the extreme corner, huddled up, dark, almost indistinguishable—crouching against the cold walls.

"Oh, oh, oh," was repeated, even more plaintively than before.

Davies began to understand. He approached lightly. Then he made out an old black mammy, doubled up and weeping. She was in the very niche of the corner, her head sunk on her knees, her tears falling, her body rocking to and fro.

Davies drew silently back. Before such grief, his intrusion seemed cold and unwarranted. The sensation of tears came to his eyes. He covered the dead, and withdrew.

Out in the moonlight, he struck a pace, but soon stopped and looked back. The whole dreary cabin, with its one golden door, where the light was, seemed a pitiful thing. He swelled with feeling and pathos as he looked. The night, the tragedy, the grief, he saw it all.

"I'll get that in," he exclaimed, feelingly, "I'll get it all in."

1901

HENRY ADAMS
1838–1918

Henry Adams's great-grandfather John was second President of the United States, his grandfather John Quincy was sixth President, and his father Charles Francis was a distinguished political leader and diplomat. Despite Henry's lifelong penchant for self-deprecation, his own achievements as scholar, teacher, novelist, editor, and cultural historian are worthy of his eminent forebears, for they earn him an important place in American intellectual and literary history.

Adams was born in Boston, on Beacon Hill, and spent his happy childhood in the constant presence of renowned politicians, artists, and intellectuals; the index to his autobiography, *The Education of Henry Adams*, is virtually an *International Who Was Who* of the time. He was to claim that his studies at Harvard College of the 1850s prepared him neither for a career nor for the extraordinary intellectual, technological, and social transformations of the last half of the century, but the records reveal that he was a good student and his disparagement of his college experience should be understood as an example of Adams's persistent self-irony. After graduation, he studied and traveled in Europe before he returned to America in 1860 to become private secretary to his father, a position he held for nine years, first in Washington when the elder Adams was elected to Congress, then in London, where his father served as foreign minister during the Civil War years.

In 1870, after a brief career as a freelance journalist, he rather reluctantly accepted a position as assistant professor of medieval history at Harvard, doubting his own fitness for the position; he observed later that the entire educational system of the college was "fallacious from the beginning to the end," and that "the lecture-room was futile enough, but the faculty-room was worse." In the same period he undertook the editorship of the prestigious *North American Review*, using this position to criticize both of the major political parties in this period of uncontrolled expansion and widespread government corruption. He sardonically remarked of his two vocations that "a professor commonly became a pedagogue or a pedant; an editor an authority on advertising." By all accounts other than his own, however, he served the college and the journal well before he gave up both positions in the late 1870s in order to settle in Washington to devote full time to historical research.

Adams's new career as historian did not begin auspiciously. His biography (1879) of Albert Gallatin, Thomas Jefferson's Secretary of the Treasury, and his subsequent biography of the flamboyant Congressman John Randolph were successful neither with his professional colleagues nor with the public. The research in original documents Adams had undertaken in order to write these biographies, however, brought him to the grander subject of which they were a part: *The History of the United States of America during the Administrations of Thomas Jefferson and James Madison* (1889–91). The nine-volume work (which took almost nine years to complete) is still a standard account of this crucial period of transition from Old to New World dominance. During these years Adams also published two novels rich in social detail: *Democracy* in 1880, and *Esther* in 1884. Both novels stirred considerable interest when they were published and continue to attract readers who seek, in the first novel, a contemporary account of the corruption of government by business interests and, in the second, a representation of the effect of scientific thought on traditional religious belief.

Adams's life was profoundly changed by the suicide, in 1885, of his charming and seemingly contented wife, Marian Hooper. The publication of *The History* also left Adams physically drained and intellectually depleted. For a time it seemed that he would occupy himself with nothing more serious than travel to familiar places in Europe and to more exotic places in the Pacific and Middle East. But a trip to northern France in the summer of 1895 with Mrs. Henry Cabot Lodge, the wife of his friend the Senator from Massachusetts, stirred his intellectual curiosity and ignited his creative energies once again.

What most struck the world-weary Adams that summer was the severe and majestic harmony of the twelfth- and thirteenth-century Norman and Gothic cathedrals they visited, particularly the one at Chartres. *Mont-Saint-Michel and Chartres* (privately printed in 1904; published 1913) not only provided illuminating discussions of the literary and architectural monuments of the period but penetrated to the powerful spiritual forces that lay behind those achievements. Paradoxically, the insights which allowed Adams to represent what he considered a straight-line historical decline from this medieval unity to his own period of confusion and immanent disaster served to renew his own spirits and brought his mind and work to the attention of a wider public.

The Education of Henry Adams, begun as a kind of sequel to *Mont-Saint-Michel*, was privately printed early in 1907 and distributed to nearly one hundred friends and prominent personages mentioned in the text. In it, Adams rehearses his own miseducation by his family, by schools, and by his experience of various social institutions, ironically offers his own "failures" as a negative example to young men of the time, and describes the increasingly rapid collapse of Western culture in the era after the twelfth century, during which the Virgin Mary (the spiritually unifying) gave way to the Industrial Dynamo (the scientifically disunifying) as the object of man's worship.

The Virgin and the Dynamo were not presented simply as historical facts but to serve as symbols. In fact, *The Education* is primarily a literary, not a biographical or historical, work. This book, since it shows imaginatively how all the major intellectual, social, political, military, and economic issues and developments of Adams's day are interrelated, is now considered by many critics to be the one indispensable text for students seeking to understand the period between the Civil War and the First World War. The perspective taken by Adams in *The Education* used to seem excessively pessimistic; its final chapter (dated 1905) prophesied that the disintegrative forces unleashed by science threatened to effect the destruction of the world within a generation. This prophecy has not been borne out literally, but the sense of immanent global disaster has come to be widely shared. "Naturally," as Adams went on to observe in this chapter, "such an attitude of an umpire is apt to infuriate the spectators. Above all, it was profoundly unmoral, and tended to discourage effort. On the other hand, it tended to encourage foresight and to economize waste of mind. If this was not itself education, it pointed out the economies necessary for the education of the new American. There, the duty stopped."

The Education has grown in interest in recent years for two major reasons: for what it tells us about its complex, elusive, and paradoxical author, and for what it tells us about the technology-dominated, dehumanized world whose major power alignments, political and material, he foresaw so clearly. As a prophet, he was, as social analyst Daniel Bell observed, the first American writer who "caught a sense of the quickening change of pace that drives all our lives." As a historian of his own time, he was also, as historian Richard Hofstadter noted, "singular not only for the quality of his prose and the sophistication of his mind but also for the unparalleled mixture of his detachment and involvement." Adams, like many of us, was fascinated with the past, horrified by the present, and skeptical about the future; thus, more than half a century after his death, Adams and his most complex book speak with renewed pertinence to his dilemmas and ours.

From The Education of Henry Adams[1]

Chapter XXV. The Dynamo and the Virgin (1900)

Until the Great Exposition of 1900[2] closed its doors in November, Adams haunted it, aching to absorb knowledge, and helpless to find it. He would have liked to know how much of it could have been grasped by the best-informed man in the world. While he was thus meditating chaos, Langley[3] came by, and showed it to him. At Langley's behest, the Exhibition dropped its superfluous rags and stripped itself to the skin, for Langley knew what to study, and why, and how; while Adams might as well have stood outside in the night, staring at the Milky Way. Yet Langley said nothing new, and taught nothing that one might not have learned from Lord Bacon,[4] three hundred years before; but though one should have

1. The selection from *The Education* reprints Houghton Mifflin Company's Riverside Edition text established by Ernest Samuels. The notes are indebted to Samuels's annotation for that volume.
2. The World's Fair held in Paris from April through November.
3. Samuel P. Langley (1834–1906), American astronomer and inventor, in 1896, of the first airplane to fly successfully.
4. Sir Francis Bacon (1561–1626), English natural philosopher, author of *The Advancement of Learning* (1605).

known the "Advancement of Science" as well as one knew the "Comedy of Errors,"[5] the literary knowledge counted for nothing until some teacher should show how to apply it. Bacon took a vast deal of trouble in teaching King James I[6] and his subjects, American or other, towards the year 1620, that true science was the development or economy of forces; yet an elderly American in 1900 knew neither the formula nor the forces; or even so much as to say to himself that his historical business in the Exposition concerned only the economies or developments of force since 1893, when he began the study at Chicago.[7]

Nothing in education is so astonishing as the amount of ignorance it accumulates in the form of inert facts. Adams had looked at most of the accumulations of art in the storehouses called Art Museums; yet he did not know how to look at the art exhibits of 1900. He had studied Karl Marx[8] and his doctrines of history with profound attention, yet he could not apply them at Paris. Langley, with the ease of a great master of experiment, threw out of the field every exhibit that did not reveal a new application of force, and naturally threw out, to begin with, almost the whole art exhibit. Equally, he ignored almost the whole industrial exhibit. He led his pupil directly to the forces. His chief interest was in new motors to make his airship feasible, and he taught Adams the astonishing complexities of the new Daimler[9] motor, and of the automobile, which, since 1893, had become a nightmare at a hundred kilometres an hour, almost as destructive as the electric tram which was only ten years older; and threatening to become as terrible as the locomotive steam-engine itself, which was almost exactly Adams's own age.

Then he showed his scholar the great hall of dynamos, and explained how little he knew about electricity or force of any kind, even of his own special sun, which spouted heat in inconceivable volume, but which, as far as he knew, might spout less or more, at any time, for all the certainty he felt in it. To him, the dynamo itself was but an ingenious channel for conveying somewhere the heat latent in a few tons of poor coal hidden in a dirty engine-house carefully kept out of sight; but to Adams the dynamo became a symbol of infinity. As he grew accustomed to the great gallery of machines, he began to feel the forty-foot dynamos as a moral force, much as the early Christians felt the Cross. The planet itself seemed less impressive, in its old-fashioned, deliberate, annual or daily revolution, than this huge wheel, revolving within arm's-length

5. Early Shakespearean comedy (1594).
6. King of England 1603–25.
7. The subject of Chapter XXII of the *Education*. The Chicago Exposition of 1893 first stimulated Adams's interest in the "economy of forces."
8. Karl Marx (1818–83), German social philosopher, architect of modern social-ism and communism; formulated his theories on the principle of "dialectical materialism."
9. Gottlieb Daimler (1834–1900), German engineer, inventor of the high-speed internal combusion engine and an early developer of the automobile.

at some vertiginous speed, and barely murmuring—scarcely humming an audible warning to stand a hair's-breadth further for respect of power—while it would not wake the baby lying close against its frame. Before the end, one began to pray to it; inherited instinct taught the natural expression of man before silent and infinite force. Among the thousand symbols of ultimate energy, the dynamo was not so human as some, but it was the most expressive.

Yet the dynamo, next to the steam-engine, was the most familiar of exhibits. For Adams's objects its value lay chiefly in its occult mechanism. Between the dynamo in the gallery of machines and the engine-house outside, the break of continuity amounted to abysmal fracture for a historian's objects. No more relation could he discover between the steam and the electric current than between the Cross and the cathedral. The forces were interchangeable if not reversible, but he could see only an absolute *fiat* in electricity as in faith. Langley could not help him. Indeed, Langley seemed to be worried by the same trouble, for he constantly repeated that the new forces were anarchical, and especially that he was not responsible for the new rays, that were little short of parricidal in their wicked spirit towards science. His own rays,[10] with which he had doubled the solar spectrum, were altogether harmless and beneficent; but Radium denied its God[1]—or, what was to Langley the same thing, denied the truths of his Science. The force was wholly new.

A historian who asked only to learn enough to be as futile as Langley or Kelvin,[2] made rapid progress under this teaching, and mixed himself up in the tangle of ideas until he achieved a sort of Paradise of ignorance vastly consoling to his fatigued senses. He wrapped himself in vibrations and rays which were new, and he would have hugged Marconi[3] and Branly had he met them, as he hugged the dynamo; while he lost his arithmetic in trying to figure out the equation between the discoveries and the economies of force. The economies, like the discoveries, were absolute, supersensual, occult; incapable of expression in horse-power. What mathematical equivalent could he suggest as the value of a Branly coherer? Frozen air, or the electric furnace, had some scale of measurement, no doubt, if somebody could invent a thermometer adequate to the purpose; but X-rays[4] had played no part whatever in man's consciousness, and the atom itself had figured only as a fiction of

10. Langley had invented the bolometer, with which he was able to measure intensities of invisible heat rays in the infrared spectrum.
1. Since radium, first isolated by the Curies in 1898, underwent spontaneous transformation through radioactive emission, it did not fit prevailing scientific distinctions between matter and energy.
2. William Thomson, Baron Kelvin (1824–1907), English mathematician and

physicist known especially for his work in thermodynamics and electrodynamics.
3. Guglielmo Marconi (1874–1937), Italian inventor of radio telegraphy in 1895; Édouard Branly (1844–1940), French physicist and inventor in 1890 of the Branly "coherer" for detecting radio waves.
4. Wilhelm Roentgen (1845–1923) discovered X-rays in 1895.

thought. In these seven years man had translated himself into a new universe which had no common scale of measurement with the old. He had entered a supersensual world, in which he could measure nothing except by chance collisions of movements imperceptible to his senses, perhaps even imperceptible to his instruments, but perceptible to each other, and so to some known ray at the end of the scale. Langley seemed prepared for anything, even for an indeterminable number of universes interfused—physics stark mad in metaphysics.

Historians undertake to arrange sequences,—called stories, or histories—assuming in silence a relation of cause and effect. These assumptions, hidden in the depths of dusty libraries, have been astounding, but commonly unconscious and childlike; so much so, that if any captious critic were to drag them to light, historians would probably reply, with one voice, that they had never supposed themselves required to know what they were talking about. Adams, for one, had toiled in vain to find out what he meant. He had even published a dozen volumes of American history for no other purpose than to satisfy himself whether, by the severest process of stating, with the least possible comment, such facts as seemed sure, in such order as seemed rigorously consequent, he could fix for a familiar moment a necessary sequence of human movement. The result had satisfied him as little as at Harvard College. Where he saw sequence, other men saw something quite different, and no one saw the same unit of measure. He cared little about his experiments and less about his statesmen, who seemed to him quite as ignorant as himself and, as a rule, no more honest; but he insisted on a relation of sequence, and if he could not reach it by one method, he would try as many methods as science knew. Satisfied that the sequence of men led to nothing and that the sequence of their society could lead no further, while the mere sequence of time was artificial, and the sequence of thought was chaos, he turned at last to the sequence of force; and thus it happened that, after ten years' pursuit, he found himself lying in the Gallery of Machines at the Great Exposition of 1900, with his historical neck broken by the sudden irruption of forces totally new.

Since no one else showed much concern, an elderly person without other cares had no need to betray alarm. The year 1900 was not the first to upset schoolmasters. Copernicus and Galileo had broken many professorial necks about 1600;[5] Columbus had stood the world on its head towards 1500; but the nearest approach to the revolution of 1900 was that of 310, when Constantine set up the

---

5. Copernicus (1473–1543), Polish astronomer, proved that the earth rotated around the sun and not vice versa; Galileo (1564–1642), Italian astronomer and developer of the refracting telescope, was condemned by the Inquisition for espousing Copernican heliocentric theory.

Cross.[6] The rays that Langley disowned, as well as those which he fathered, were occult, supersensual, irrational; they were a revelation of mysterious energy like that of the Cross; they were what, in terms of mediæval science, were called immediate modes of the divine substance.

The historian was thus reduced to his last resources. Clearly if he was bound to reduce all these forces to a common value, this common value could have no measure but that of their attraction on his own mind. He must treat them as they had been felt; as convertible, reversible, interchangeable attractions on thought. He made up his mind to venture it; he would risk translating rays into faith. Such a reversible process would vastly amuse a chemist, but the chemist could not deny that he, or some of his fellow physicists, could feel the force of both. When Adams was a boy in Boston, the best chemist in the place had probably never heard of Venus except by way of scandal, or of the Virgin except as idolatry;[7] neither had he heard of dynamos or automobiles or radium; yet his mind was ready to feel the force of all, though the rays were unborn and the women were dead.

Here opened another totally new education, which promised to be by far the most hazardous of all. The knife-edge along which he must crawl, like Sir Lancelot in the twelfth century,[8] divided two kingdoms of force which had nothing in common but attraction. They were as different as a magnet is from gravitation, supposing one knew what a magnet was, or gravitation, or love. The force of the Virgin was still felt at Lourdes,[9] and seemed to be as potent as X-rays; but in America neither Venus nor Virgin ever had value as force—at most as sentiment. No American had ever been truly afraid of either.

This problem in dynamics gravely perplexed an American historian. The Woman had once been supreme; in France she still seemed potent, not merely as a sentiment, but as a force. Why was she unknown in America? For evidently America was ashamed of her, and she was ashamed of herself, otherwise they would not have strewn fig-leaves so profusely all over her.[1] When she was a true force, she was ignorant of fig-leaves, but the monthly-magazine-made American female had not a feature that would have been recognized by Adam. The trait was notorious, and often humorous,

6. Constantine the Great (288?–337), Roman Emperor, issued the Edict of Milan in 313 proclaiming toleration of Christians, which paved the way for the ascendancy of Christianity.
7. That is, the druggist knew of Venus only through selling medication for venereal disease; and since Boston was largely Protestant, he would only have heard of the Virgin as the object of idolatrous worship.

8. In Chrétien de Troyes's *Lancelot*, the hero was obliged to crawl across a bridge composed of a knife in order to enter a castle and rescue Guinevere.
9. A famous shrine in France known for its miraculous cures; the Virgin Mary was said to have appeared to a peasant girl there in 1858.
1. I.e., to conceal sexual organs and sensuality.

but anyone brought up among Puritans knew that sex was sin. In any previous age, sex was strength. Neither art nor beauty was needed. Everyone, even among Puritans, knew that neither Diana of the Ephesians[2] nor any of the Oriental goddesses was worshipped for her beauty. She was goddess because of her force; she was the animated dynamo; she was reproduction—the greatest and most mysterious of all energies; all she needed was to be fecund. Singularly enough, not one of Adams's many schools of education had ever drawn his attention to the opening lines of Lucretius, though they were perhaps the finest in all Latin literature, where the poet invoked Venus exactly as Dante invoked the Virgin:—

"Quae quoniam rerum naturam *sola* gubernas."

The Venus of Epicurean philosophy survived in the Virgin of the Schools:—

"Donna, sei tanto grande, e tanto vali,
Che qual vuol grazia, e a te non ricorre,
Sua disianza vuol volar senz' ali."[4]

All this was to American thought as though it had never existed. The true American knew something of the facts, but nothing of the feelings; he read the letter, but he never felt the law. Before this historical chasm, a mind like that of Adams felt itself helpless; he turned from the Virgin to the Dynamo as though he were a Branly coherer. On one side, at the Louvre and at Chartres, as he knew by the record of work actually done and still before his eyes, was the highest energy ever known to man, the creator of four-fifths of his noblest art, exercising vastly more attraction over the human mind than all the steam-engines and dynamos ever dreamed of; and yet this energy was unknown to the American mind. An American Virgin would never dare command; an American Venus would never dare exist.

The question, which to any plain American of the nineteenth century seemed as remote as it did to Adams, drew him almost violently to study, once it was posed; and on this point Langleys were as useless as though they were Herbert Spencers[5] or dynamos. The idea survived only as art. There one turned as naturally as though

2. The shrine at Ephesus on the West Coast of Asia Minor was dedicated to Artemis, a virgin-goddess mother-figure.
3. Lucretius (c. 99–55 B.C.), in *On the Nature of Things*, I.21: "And since 'tis *thou* / Venus alone / Guidest the Cosmos." Trans. William Ellery Leonard.
4. The Virgin of the Schools is a reference to medieval scholastic philosophers. The lines from Dante translate:
Lady, thou art so great and hath such

worth, that if there be who would have grace yet betaketh not himself to thee, his longing seeketh to fly without wings. (Dante, *Paradiso*, XXXIII, trans. Carlyle-Wicksteed)
5. Herbert Spencer (1820–93), English philosopher and popularizer of Darwinian evolutionary principles. Adams ironically suggests that Spencer's explanations were too general and abstract to explain this primal force.

the artist were himself a woman. Adams began to ponder, asking himself whether he knew of any American artist who had ever insisted on the power of sex, as every classic had always done; but he could think only of Walt Whitman; Bret Harte,[6] as far as the magazines would let him venture; and one or two painters, for the flesh-tones. All the rest had used sex for sentiment, never for force; to them, Eve was a tender flower, and Herodias[7] an unfeminine horror. American art, like the American language and American education, was as far as possible sexless.[8] Society regarded this victory over sex as its greatest triumph, and the historian readily admitted it, since the moral issue, for the moment, did not concern one who was studying the relations of unmoral force. He cared nothing for the sex of the dynamo until he could measure its energy.

Vaguely seeking a clue, he wandered through the art exhibit, and, in his stroll, stopped almost every day before St. Gaudens's[9] General Sherman, which had been given the central post of honor. St. Gaudens himself was in Paris, putting on the work his usual interminable last touches, and listening to the usual contradictory suggestions of brother sculptors. Of all the American artists who gave to American art whatever life it breathed in the seventies, St. Gaudens was perhaps the most sympathetic, but certainly the most inarticulate. General Grant or Don Cameron[1] had scarcely less instinct of rhetoric than he. All the others—the Hunts, Richardson, John La Farge, Stanford White[2]—were exuberant; only St. Gaudens could never discuss or dilate on an emotion, or suggest artistic arguments for giving to his work the forms that he felt. He never laid down the law, or affected the despot, or became brutalized like Whistler[3] by the brutalities of his world. He required no incense; he was no egoist; his simplicity of thought was excessive; he could not imitate, or give any form but his own to the creations of his hand. No one felt more strongly than he the strength of other men, but the idea that they could affect him never stirred an image in his mind.

This summer his health was poor and his spirits were low. For such a temper, Adams was not the best companion, since his own

6. Whitman's *Leaves of Grass* (1855) treated sex boldly and was much criticized on that account; Bret Harte (1836–1902) sympathetically portrayed prostitutes in such stories as *The Outcasts of Poker Flat*.
7. The wife of King Herod who collaborated with her daughter Salome in arranging the beheading of John the Baptist.
8. Just as the American language had no genders, American education was coeducational and in Adams's day entirely excluded sex as a subject from the curriculum.
9. Augustus Saint-Gaudens (1848–1907),

American sculptor.
1. Senator James Donald Cameron (1833–1918), Secretary of War under President Grant, 1876.
2. William Morris Hunt (1824–79), noted painter, and his younger brother Richard Morris Hunt (1828–95), noted architect; Henry Hobson Richardson (1838–86), architect; John La Farge (1835–1910), muralist and maker of stained-glass windows; Stanford White (1853–1906), architect.
3. James Abbott McNeill Whistler (1834–1903), American painter and lithographer.

gaiety was not *folle*;[4] but he risked going now and then to the studio on Mont Parnasse to draw him out for a stroll in the Bois de Boulogne,[5] or dinner as pleased his moods, and in return St. Gaudens sometimes let Adams go about in his company.

Once St. Gaudens took him down to Amiens, with a party of Frenchmen, to see the cathedral. Not until they found themselves actually studying the sculpture of the western portal, did it dawn on Adams's mind that, for his purposes, St. Gaudens on that spot had more interest to him than the cathedral itself. Great men before great monuments express great truths, provided they are not taken too solemnly. Adams never tired of quoting the supreme phrase of his idol Gibbon, before the Gothic cathedrals: "I darted a contemptuous look on the stately monuments of superstition."[6] Even in the footnotes of his history, Gibbon had never inserted a bit of humor more human than this, and one would have paid largely for a photograph of the fat little historian, on the background of Notre Dame of Amiens, trying to persuade his readers—perhaps himself —that he was darting a contemptuous look on the stately monument, for which he felt in fact the respect which every man of his vast study and active mind always feels before objects worthy of it; but besides the humor, one felt also the relation. Gibbon ignored the Virgin, because in 1789 religious monuments were out of fashion. In 1900 his remark sounded fresh and simple as the green fields to ears that had heard a hundred years of other remarks, mostly no more fresh and certainly less simple. Without malice, one might find it more instructive than a whole lecture of Ruskin.[7] One sees what one brings, and at that moment Gibbon brought the French Revolution. Ruskin brought reaction against the Revolution. St. Gaudens had passed beyond all. He liked the stately monuments much more than he liked Gibbon or Ruskin; he loved their dignity; their unity; their scale; their lines; their lights and shadows; their decorative sculpture; but he was even less conscious than they of the force that created it all—the Virgin, the Woman—by whose genius "the stately monuments of superstition" were built, through which she was expressed. He would have seen more meaning in Isis[8] with the cow's horns, at Edfoo, who expressed the same thought. The art remained, but the energy was lost even upon the artist.

Yet in mind and person St. Gaudens was a survival of the 1500's;

---

4. Excessive.
5. A large wooded park on the outskirts of Paris; "Mont Parnasse": a Paris Left Bank district frequented by artists and writers.
6. Apparently Adams's adaptation of a passage in Gibbon's French journal for February 21, 1763.

7. John Ruskin (1819–1900), English art critic and social reformer, famous for his highly imaginative interpretations of great works of the Italian Renaissance.
8. Egyptian earth-mother goddess; Adams visited Edfu, the site of the best-preserved temple in Egypt, in 1872–73 and in 1893.

he bore the stamp of the Renaissance, and should have carried an image of the Virgin round his neck, or stuck in his hat, like Louis XI.[9] In mere time he was a lost soul that had strayed by chance into the twentieth century, and forgotten where it came from. He writhed and cursed at his ignorance, much as Adams did at his own, but in the opposite sense. St. Gaudens was a child of Benvenuto Cellini,[1] smothered in an American cradle. Adams was a quintessence of Boston, devoured by curiosity to think like Benvenuto. St. Gaudens's art was starved from birth, and Adams's instinct was blighted from babyhood. Each had but half of a nature, and when they came together before the Virgin of Amiens they ought both to have felt in her the force that made them one; but it was not so. To Adams she became more than ever a channel of force; to St. Gaudens she remained as before a channel of taste.

For a symbol of power, St. Gaudens instinctively preferred the horse, as was plain in his horse and Victory of the Sherman monument. Doubtless Sherman also felt it so. The attitude was so American that, for at least forty years, Adams had never realized that any other could be in sound taste. How many years had he taken to admit a notion of what Michael Angelo and Rubens[2] were driving at? He could not say; but he knew that only since 1895 had he begun to feel the Virgin or Venus as force, and not everywhere even so. At Chartres—perhaps at Lourdes—possibly at Cnidos[3] if one could still find there the divinely naked Aphrodite of Praxiteles —but otherwise one must look for force to the goddesses of Indian mythology. The idea died out long ago in the German and English stock. St. Gaudens at Amiens was hardly less sensitive to the force of the female energy than Matthew Arnold at the Grande Chartreuse.[4] Neither of them felt goddesses as power—only as reflected emotion, human expression, beauty, purity, taste, scarcely even as sympathy. They felt a railway train as power; yet they, and all other artists, constantly complained that the power embodied in a railway train could never be embodied in art. All the steam in the world could not, like the Virgin, build Chartres.

Yet in mechanics, whatever the mechanicians might think, both energies acted as interchangeable forces on man, and by action on man all known force may be measured. Indeed, few men of science measured force in any other way. After once admitting that a

9. A pious French King (1423–83) who often disguised himself as a pilgrim and wore an old felt hat decorated with the lead statuette of a saint.
1. Flamboyant Italian sculptor and goldsmith (1500–71), author of a famous autobiography.
2. Michelangelo Buonarroti (1475–1564), Italian sculptor, painter, architect, and poet of the High Renaissance. Peter Paul Rubens (1577–1640), 17th-century Flemish painter. Both are known for their exceptional renderings of the human body.
3. Cnidos or Cnidus is an ancient city in Asia Minor, site of the most famous of the statues of Aphrodite by Praxiteles (c. 370–330 B.C.); only a copy survives in the Vatican.
4. Matthew Arnold's (1822–88) poem *Stanzas from the Grande Chartreuse* invokes the Virgin Mary in mourning the loss of faith formerly held by ascetic Carthusian monks.

straight line was the shortest distance between two points, no serious mathematician cared to deny anything that suited his convenience, and rejected no symbol, unproved or unproveable, that helped him to accomplish work. The symbol was force, as a compass-needle or a triangle was force, as the mechanist might prove by losing it, and nothing could be gained by ignoring their value. Symbol or energy, the Virgin had acted as the greatest force the Western world ever felt, and had drawn man's activities to herself more strongly than any other power, natural or super-natural, had ever done; the historian's business was to follow the track of the energy; to find where it came from and where it went to; its complex source and shifting channels; its values, equivalents, conversions. It could scarcely be more complex than radium; it could hardly be deflected, diverted, polarized, absorbed more perplexingly than other radiant matter. Adams knew nothing about any of them, but as a mathematical problem of influence on human progress, though all were occult, all reacted on his mind, and he rather inclined to think the Virgin easiest to handle.

The pursuit turned out to be long and tortuous, leading at last into the vast forests of scholastic science. From Zeno to Descartes, hand in hand with Thomas Aquinas, Montaigne, and Pascal,[5] one stumbled as stupidly as though one were still a German student of 1860. Only with the instinct of despair could one force one's self into this old thicket of ignorance after having been repulsed at a score of entrances more promising and more popular. Thus far, no path had led anywhere, unless perhaps to an exceedingly modest living. Forty-five years of study had proved to be quite futile for the pursuit of power; one controlled no more force in 1900 than in 1850, although the amount of force controlled by society had enormously increased. The secret of education still hid itself somewhere behind ignorance, and one fumbled over it as feebly as ever. In such labyrinths, the staff is a force almost more necessary than the legs; the pen becomes a sort of blind-man's dog, to keep him from falling into the gutters. The pen works for itself, and acts like a hand, modelling the plastic material over and over again to the form that suits it best. The form is never arbitrary, but is a sort of growth like crystallization, as any artist knows too well; for often the pencil or pen runs into side-paths and shapelessness, loses its relations, stops or is bogged. Then it has to return on its trail, and recover, if it can, its line of force. The result of a year's work depends more on what is struck out than on what is left in; on the sequence of the main

5. Probably Zeno of Citium (c. 366–264 B.C.), Greek philosopher; René Descartes (1596–1650), French philosopher; Thomas Aquinas (1225?–1274), Italian philosopher and theologian who is the subject of the last chapter of Adams's *Mont-Saint-Michel and Chartres*; Michel Eyquem de Montaigne (1533–92), French essayist and skeptical philosopher; Blaise Pascal (1623–62), French philosopher.

lines of thought, than on their play or variety. Compelled once more to lean heavily on this support, Adams covered more thousands of pages with figures as formal as though they were algebra, laboriously striking out, altering, burning, experimenting, until the year had expired, the Exposition had long been closed, and winter drawing to its end, before he sailed from Cherbourg, on January 19, 1901, for home.

1907, 1918

American Literature between the Wars 1914–1945

IN THE EARLY decades of the twentieth century the United States emerged as a major power whose actions in diplomacy, warfare, and political and economic affairs had profound and durable consequences on the international scene. There was a comparable accession of power in American writing which gave new urgency to the view of American culture set forth in 1870 in *Democratic Vistas* by Walt Whitman. There the poet issued a double challenge to the American imagination. On the one hand he celebrated the emergence of artists "commensurate with the people" and burdened them with the mission to remold society and to bring to their fulfillment the new eras in human history that he found incarnate in the American experiment. On the other hand he defined a profound crisis for the arts, itemizing the manifest deficiencies, the diseased corrosions of American civilization, which threatened increasingly to betray the high expectations it had aroused, and he warned that they might bring on a catastrophe comparable to the fate in hell of the "fabled damned." In the period encompassing two world wars and the Great Depression of 1929, ambitious missions for the arts comparable to those defined by Whitman, and comparably dire warnings agitated American writers, defining the confidence and boldness with which the best pursued their careers and sought recognition, but defining also the anxiety, even desperate urgency, with which most scrutinized their traditions and their contemporary environment as they weighed the chances for human fulfillment and sought to shape conditions that would enable them to fulfill their aspirations as artists.

They were the heirs of a provincial culture in the nineteenth century, located on the "circumference of civilization," as James wrote of Hawthorne's America, and, therefore, distant from the centers of power and taste in Europe. It was an America where literary pursuits and institutions were comparatively insecure in the experimental, still new society, and where writers with distinctive talents were partially alienated from the institutions and prevailing tastes of their environment. Those very insecurities and that alienation proved, however, in the late nineteenth and twentieth centuries, to be the very basis of the American tradition, a stimulus as much as a threat, and enabled American writers to play a leading role as innovators in the literary revolution that took shape in English and American letters during World War I. From the beginning Americans had been familiar with rootlessness, accelerated historical change, and cultural dislocations, and they were practiced, as Gertrude Stein noted, at coping with them, struggling to avoid the mere imitation of imported or established literary modes, striving to forge new ratios between form and substance in literary art, and seeking new ways to invigorate the language and establish contact with their audiences. The American tradition sustained the effort to confront and express the discontinuities and dislocations which were to be dominant characteristics of modern writing.

REASSESSMENT OF AMERICAN CULTURAL TRADITIONS

A reassessment of American cultural traditions was one of the major undertakings of creative writers and critics alike. Emerson's awareness that his traditions were in a state of flux and that an age of "transition" was a stimulus to creativity was heightened in the twentieth century. To see the epoch of World War I as "America's Coming of Age" (the title of Van Wyck Brooks's critical volume in 1915) aroused concern not only about the future but about the preceding stages, and the southern poet-critic Allen Tate later insisted that a period of transition was a stimulus to the imagination because diverging forms of behavior, conflicting values, or varied tastes and literary conventions were clearly juxtaposed, and they encouraged writers to compare and sift them. The exploration of their native past that the writers undertook reshaped it into a tradition more supportive of their efforts, more varied in its resources, and more firm in its orientation than had been the case earlier. They discovered the poetry of Emily Dickinson, first published in 1890 and issued piecemeal until the 1940s, and established her as one of the chief figures in American letters. They rediscovered Melville in the 1920s and reassessed Henry James in the 1930s, reargued the merits of Whitman and Mark Twain, and defined a durable importance for these writers that continued to be influential after World War II in American cultural circles and in the new British universities as well.

REGIONAL TRADITIONS

The New Englanders Thoreau, Emerson, and Hawthorne remained highly esteemed, and the poets Robinson and Frost paid tribute to their New England predecessors. But the long-dominant New England tradition more widely came under attack for the evasive gentility and prim idealism that it had encouraged. Aspects of the specifically Puritan tradition came under particularly intense scrutiny. It was attacked in the 1920s for its constraints and moral severity by the critic H. L. Mencken and others and for its reinforcement of the inhibiting complacencies of the middle-class "booboisie." But the subtleties of its contributions to American culture and its importance in defining the national character and America's "mission" were emphasized by Harvard historians and others in the 1930s and 1940s.

Groups and regions outside New England asserted their claims to recognition and importance, possibly because shifts in population and economic power were generating new energies there or because these subcultures—their familiar values and the mores of their communities—seemed threatened by industrialism and the increasing standardization of American life. Southern writers, although they broke away from the pieties of the plantation myth that dominated post–Civil War southern literature, explored their ties to an agrarian past that had been repudiated by the industrial North (as in the manifesto *I'll Take My Stand,* 1930); as critics, fiction writers, and poets they made the South the dominant regional force in national letters through the 1940s. Many midwestern writers were more apt than southerners to acknowledge the lure of the East Coast or foreign lands in the themes of their works. Sinclair Lewis and Sherwood Anderson subjected the small cities of Minnesota and Ohio and their commercial aspirations to scathing scrutiny; but their fiction, like F. Scott Fitzgerald's and Willa Cather's, was informed by a sense that a more wholesome past was slipping away and that regional traditions sanctioning innocence, simplicity, and dreams of enterprise and self-fulfillment were standards against which to measure the corrosions of contemporary society. For midwestern writers, even those who left the region in the course of their careers, Chicago became a center of ferment and stimulus. Asserting its prominence as an economic and cultural center, it was, as Carl Sandburg exclaims in his poem *Chicago,* a "tall bold slugger set vivid against the little soft cities." It provided writers with bohemian residential areas, literary salons, opportunities in journalism, and publishing outlets of international importance like Harriet Monroe's *Poetry, a Magazine of Verse* and Margaret Anderson's *The Little Review.*

New York City, however, remained the publishing center of the nation, and two neighborhoods there became important literary centers. In uptown Manhattan, Harlem, its fine avenues, its night spots featuring jazz music, as well as its congested slums, became the metropolitan center of black culture in the United States beginning in the earliest decades of the century, when rural black people were migrating by the thousands to urban centers and blacks in New York City, responding to the pressures of residential and economic segregation, completed the transformation of white suburban Harlem into a black ghetto. With heightened consciousness of their racial and community identity, black intellectuals launched in the 1920s the first important movement in black American literature, the "Harlem Renaissance," to strengthen the cultural traditions of their people and demonstrate their achievements to the white society that habitually ignored them. During the same decades, Greenwich Village in lower Manhattan, with its inexpensive tenements and relative isolation from the institutions of the native-born white middle-class, became the haven of literary and intellectual bohemians, prominent writers and intellectuals as well as "inglorious Miltons by the score, and Rodins, one to every floor," as the radical John Reed wrote mockingly in *The Day in Bohemia.* In the Village they found the freedom to pursue styles of living at odds with notions of propriety prevailing elsewhere, to circumvent the Prohibition laws against the sale of alcoholic beverages, and to experiment with radical political ideas and new literary forms.

THE CRISIS

The excitement of life in Harlem and in the Village (they became tourist attractions for other New Yorkers and outsiders alike) did much to set the tone for which "The Jazz Age" or "The Twenties" became famous. But the experimentation with radical ideas and new modes in the arts which marked the 1920s was not unique to the decade; it was an authentic response to a crisis in American culture that took shape earlier but was intensified during World War I (1914–18) and its aftermath, even though America's distance from the battlefields and relatively late entry into combat (1917) protected it from the worst consequences of a war which strained European economies, decimated their populations, and cut off the careers of countless young intellectuals. To many, like the novelists Hemingway and Dos Passos, who witnessed it, but to many more, like the pacifist philosopher John Dewey or the poet Ezra Pound, who did not, the war which President Woodrow Wilson proclaimed would "make the world safe for democracy" proved to be a senseless slaughter in the name of a "botched civilization." Postwar diplomacy, establishing the League of Nations on the shifting sands of power politics, seemed a futile and cynical charade.

The intellectuals' aroused concern as well as their imminent disenchantment quickened the efforts begun in the post–Civil War decades to scrutinize and challenge more forcefully the ideals of democracy and the institutions that presumably secured it, the institutions of industrial and finance capitalism which were becoming so dominant in all aspects of social and cultural life, and the institutions of the family and sexual codes that had been central in the structure of American values. Racial violence and industrial strife in the late war years and early 1920s, an aroused socialist movement, and scandalous corruption in league with official complacency during the presidency of Warren G. Harding intensified the challenge.

Late in the 1920s new jolts to American society extended the crisis into the 1930s and sharpened writers' sense of urgency about their mission. Open evasion of the Prohibition laws and illicit traffic in "bootleg" liquor were accompanied by the advent of gang warfare in major cities, marking a decay in private and public morals. Poverty spread through rural areas in the 1920s and 1930s, and the stock market crash of 1929 bankrupted individuals and institutions throughout the nation and sent shock waves through the economies of Europe, where fascist Italy and Nazi Germany were preparing the way for the outbreak of World War II in 1939. Industrial retrenchment and widespread unemployment during the Great Depression of the 1930s and the explosive violence of industrial strife between corporations and the labor force occasioned a major reorientation of American political and economic life under President Franklin D. Roosevelt's "New Deal," launched in 1933 to cope with the conditions which had reduced countless Americans to lives of unquiet desperation. To the crises of the 1920s and 1930s American writers responded by re-examining and redefining their distinctively American traditions and by treating social problems more explicitly in fiction and poetry as well as in literary journalism. They also looked to Europe for ideas, programs, and forms that could guide or implement their efforts.

AMERICA AND EUROPE

Though the poet William Carlos Williams and others, like James Fenimore Cooper in the early nineteenth century, warned against subservience to foreign models, Williams and most others responded nonetheless to major currents of thought and taste abroad. Indeed, Americans responded to European precedent with more confidence than before, rivaling while seeking to appropriate and adjust such models to their own purposes and determined to have a reciprocal impact abroad.

Marxism and Freudianism

Karl Marx's indictment of capitalism and his economic and historical theories, though familiar to such earlier American social critics as Thorstein Veblen, became widely influential in literary and **intellectual** circles after

1915. Interest in European radicalism, including anarchism and communism, was heightened by the outbreak in 1917 of the Russian Revolution, which held forth the hope of a strikingly new solution to social and political problems. Marxist thought infused the radicalism of Max Eastman and John Reed, editors of *The Masses* (at the time of its suppression in 1917 for opposition to the war) and of its successor, *The Liberator*, subsequently named *The New Masses*, which became the official journal of the American Communist party. These journals reinforced the critique of American society sustained by the conservative H. L. Mencken of *The American Mercury* and the liberal journalism of *The New Republic* and *The Nation*. And Marxist theory, along with radical organizations pursuing socialist aims, helped shape the careers of the critic Edmund Wilson and such fiction writers as Theodore Dreiser, John Dos Passos, Herbert Gold, and Richard Wright in the 1930s. Marxists and social critics informed by Marxist thought were among those in the 1930s, following the crash of 1929, who demanded that writers take direct cognizance of social and economic problems and enlist their writing in the cause of advancing social change or revolution. The suppression of dissent in Russia, demands for conformity within communist organizations in this country, and the alignment in 1939 of Stalinist Russia with Hitler's Germany disillusioned many intellectuals with communist-oriented radicalism; but Marxist theory continued to be influential in subsequent decades.

Freudianism, the theory and practice of psychoanalysis as defined by the Viennese doctor Sigmund Freud, was another Continental movement that penetrated deeply into American intellectual life. Psychoanalysis stressed the importance of the unconscious or the irrational in the human psyche; the dramas and dream symbols generated in the mind by mechanisms, psychological and social, of sexual repression; the pervasiveness of "discontents" that underlie all civilization; the ways in which psychic experience is masked or veiled by common language and the conscious operations of the mind; and the capacity of "depth psychology" to diagnose psychic illness and try, at least, to cure it. Freud found a readier acceptance in America (where he first lectured in 1909) than in Europe, and his theories, after becoming a fad in the journalism of the 1920s, inspired in subsequent decades a host of novels and such reinterpretations of American culture as Ludwig Lewisohn's *Expression in America* (1932). More significantly, psychoanalytic theory seemed to sanction quests for sexual liberation in actual behavior and disdain for established conventions which inhibited fulfillment of the self. Rival psychoanalysts, including Alfred Adler and Carl Jung, added new versions to the literature of psychoanalysis. By 1940 "the psychiatrist," the "archetypal symbol," and "the Oedipus complex" had become part of modern mythology, and both the mythology and the theories of psychoanalysis had infused literature as varied as the poems of Conrad Aiken, the plays of Eugene O'Neill, and most poetry and fiction that drew heavily on the subconscious implications of symbolic forms or probed the depths of psychic experience.

European Arts

Contemporary European movements in the arts also inspired new undertakings in American literature after 1914. In fiction the Irish novelist James Joyce perfected techniques of "realism" that rendered the surface textures of ordinary life while penetrating the hidden depths of psychological and social reality and, at the same time, radically altered novelistic conventions so as to accommodate patterns of classical myth, parodic manipulations of language, expressionistic effects, and allusive symbolic forms. Comparably innovative achievements in the music of Stravinsky, Hindemith, and Schoenberg (all of whom later emigrated to the United States) sounded cacophonous to large audiences but provocative to vanguard writers, and in 1913 a painting exhibition, the "Armory Show," which opened in New York City and traveled to Boston and Chicago, had a similar effect in introducing contemporary modes of painting. American painters advancing a bolder kind of "realism" than had prevailed in the nineteenth century (some were dubbed derisively the "Ash Can" school) were displayed alongside an immense selection representing the newest modes in European art. Particularly the explosively colorful Fauves and various forms of abstract painting from Europe, defying earlier conventions of representation and engaging in experiments with the basic elements of line, color, and the sheer medium of paint, shocked while intriguing thousands of viewers and sanctioned for vanguard writers new ways of treating their materials and more daring ways of dealing with their audiences. The emigration of European artists to the United States, which accelerated in the 1930s and 1940s, when totalitarian regimes abroad threatened the lives of Jews and radical dissidents, simply facilitated an exchange that was important in American culture from the 1920s on.

The Expatriates

Americans who chose to live abroad, by the very fact of their expatriation, presented a challenge to the American society they left and welcomed the stimulus of contemporary movements and cultural traditions in the countries where they resided. But expatriation, a common pattern in the cultural history of England and European countries since 1800, did not constitute a simple repudiation of American society, though it admittedly ran the risk of severance from familiar materials and a responsive audience, as the critic Van Wyck Brooks warned in his *Pilgrimage of Henry James* (1925). Instead, expatriation was an extension of America's traditions and part of the effort to redefine and strengthen them. Some expatriates settled abroad for frivolous reasons, and all benefited from the fact that living and publishing expenses were cheaper in other countries and that in France publication in English was not subject to censorship. But for earlier Americans, as for Donatello in Hawthorne's *The Marble Faun*, rootlessness was a condition which could eventually bring freedom and responsibility, and it was a condition with which they were familiar, whether they sought liberation and opportunities for their talent by moving into frontier territories and the West, by moving to cultural centers in the East, or by establishing residence in or near the capitals of England and Europe. American writers as diverse as James Fenimore Cooper, Nathaniel Hawthorne, and Henry James had lived for protracted periods of time outside the United States and had incorporated cosmopolitan perspectives in their writing. In the twentieth century some Americans decided to live in Europe early in their careers and stayed permanently; Gertrude Stein, T. S. Eliot, and Ezra Pound were the most famous examples. Others stayed for a protracted period (the poet Archibald MacLeish planned in advance a five-year stay in France) and then returned. The black novelist Richard Wright lived for the last fifteen years of his life in France. Hemingway spent years intermittently abroad, then returned to the edges of the United States to live on islands off Florida and Cuba. For most, the experience sharpened their sense of American identity and entailed choosing deliberately and selectively, like Emerson before them, those traditions, American or foreign, which would sustain their art. In the case of Pound and Eliot, they attained an importance on the London cultural scene comparable to that of the American expatriate painters Benjamin West and James McNeill Whistler in the nineteenth century, and they played leading roles, boldly and shrewdly, in the revolution of taste which established modern modes in Anglo-American literature.

MODERNISM

That revolution took shape in a convergence of tendencies in modern culture, accidental circumstance, and concerted effort on the part of influential writers, some politically conservative and some radical.

Imagism

It began to emerge after 1908 in the movement Imagism and the reaction of the English writer T. E. Hulme and his followers against amorphous abstraction, clichéd "poetic" diction, indulgent ornamentation, and monotonous regularity of meter in Victorian and Edwardian poetry. Ezra Pound issued an anthology of Imagist poems (*Des Imagistes*, 1914), and the American poet Amy Lowell, a propagandist for the movement, issued subsequent anthologies in 1915, 1916, and 1917. The 1915 Imagist "manifesto" warned against "sonorous" generalizations, "merely decorative" diction, and "old rhythms," and called instead for a poetic of the "exact word" and "common speech," "new rhythms," compression of statement, a "poetry that is hard and clear," and "free verse" as a "principle of liberty." The demand for precision and common speech was perfectly compatible with the experiments being worked out in prose at the same time by Gertrude Stein, and Pound defined the Imagists' aims more precisely. But the movement was soon superseded by others and by the more varied principles and practices of innovative writers in prose and verse that revealed the more ambitious aims and governing principles of Modernist writing.

Discontinuity and Fragmentation

One characteristic feature of Modernism is that it dramatizes discontinuity, the sense of disjunction and imminent severance from the past, while making determined efforts to appropriate the past, its values, and its artistic forms by incorporating them in new acts of creation. Pound's translations, the subtle echoes and quotations of earlier verse in Eliot's poems, William Carlos Williams's essays *In the American Grain*, or the disjunctive structure of Faulkner's fiction display in varied ways the urgency of this undertaking. A second characteristic is fragmentation: the sense that individual experience consists of severed or loosely connected pieces or moments, and that communities are atomized into individuals or groups that are antagonistic or distant from one another. The literature incorporates and expresses this fragmentation in varied ways—exposing it often in stunned repudiation or mordant satire, sometimes with a sense of futility verging on despair—while reaching beyond it toward possibilities of both self-integration and personal and social communion.

Language, Form, and Audience

A third feature of literary Modernism is its concern with language, all aspects of its medium, and the problematic relations between literature and its audiences. American writers had been anxious about these matters since the days of Cooper, Melville, and Whitman, but modern writers felt more intensely that the language prevailing in literary works was debilitated and needed resuscitation, that the conventions comprising their medium had become stale and needed either replacement or reworking, and that audiences had been numbed by the spread of minimal literacy and the impact of standardized journalism and cheapened tastes. Most significant writing in the twentieth century undertook to remake the language of literature, to experiment with new conventions of poetic structure and prose narration, and to establish new connections with audiences who needed to be prepared to respond to unfamiliar modes of writing.

Much poetry and popular fiction continued to cater to outmoded tastes and reading habits, and sensational journalism along with the diversions of movies and radio made it increasingly difficult for serious writers to reach large audiences, whose reading habits and preferences were not receptive to innovations. The most ambitious writers met the challenge by attacking the currently established foundations of writing in the attempt to reconstitute them. The variety of their strategies in this effort was striking, and their effects were of lasting importance. They found ways to capture the pace, tensions, and rhythms of urban life; they shaped the cadences of black dialect, of southern rhetoric, of colloquial speech as actually spoken into a medium for prose narration, for dialogue in fiction, and for lyric poetry. At times they abandoned or mocked conventions of punctuation and stanza form in verse; at other times they converted such conventions or the sonnet form to new uses. They wanted to extend the range of literature, to augment its resources and to perfect the techniques for controlling literary statement and the reactions of readers. For these purposes they exploited at times the irregularities of rhymeless free verse or the inconclusiveness of open form and improvisation, and relied at other times on tightly governed narrative form (as in Fitzgerald and Hemingway) or regular stanzas in verse. Some poets revived ancient or foreign stanza patterns (as did Pound in his early poems) or incorporated the forms of popular songs and ballads (as did Langston Hughes in his "blues"), often displaying the relish of sheer ingenuity which appeared also in the unique new stanza patterns devised by Marianne Moore. In fiction as well as in verse, writers drew on the complexities, and often the subtly implicit or subconscious effects, of symbolic forms so as to render them firmly functional rather than merely ornamental or clumsily imposed, as instruments. They adapted traditional epic conventions and seminal myths, ancient and modern, to enlarge the scale, define the ironies, or reveal and make persuasive the moral significance of their works.

Many of these strategies made for writing that was notably difficult to apprehend and appreciate and for writing that was in certain ways devious in its approach to its audiences because it ignored or violated familiar ways of revealing meaning and avoided certain sure ways of appealing to their expectations (pretty ornament, clichés, formulaic adventure plots, complacent optimism, overt moralizing, ingratiating addresses to "dear reader," and the like). Many modern productions in prose or verse are presented without overt solicitation of the reader's interest, emotions, or understanding and are difficult to apprehend because they are designed instead to have the vividness and discrete independence of a well-made artifact, the reality of an object. That objectivity is the work's claim on the reader's attention, and other appeals remain tacit or covert. Yet both the difficulty and the deviousness testify to the deliberateness with which the important writers of the period undertook to address their society. If modern society consisted not of one single audience but a "checkerboard" of different and separate audiences, as Henry James had discerned, one might hope for a response from some but not all audiences or make a variety of appeals designed to educate the audiences and create the taste for modern writing. Common to the strategies pursued by many vanguard writers was that of articulating or dramatizing the barrier between writer and audience by bluntly and explicitly assaulting the audience's expectations or by proceeding in seeming indifference to its needs, in either case catching and holding readers' attention by shocking them into a new alertness. Virtually new to American writing (though there were precedents in Melville and Twain) is the devious and sportive playfulness, the virtuoso posturing, the indulgent frivolity which in some poetry and fiction replaced the solemnity with which authors customarily had addressed their subjects and their readers.

THE INSTITUTIONALIZATION OF MODERNISM

Thanks to whatever luck and astuteness, the new modes gradually became dominant during the decades between the two wars, in part because leading writers campaigned for the new modes in countless journalistic essays and in critical essays of great subtlety and force, in part because they institutionalized their undertaking. Pound was the chief agent in these affairs, issuing manifestoes, serving on editorial boards, editing manuscripts, and goading publishers. Journals promoting the vanguard arts, the so-called "little magazines," multiplied on both sides of the ocean after 1912. Important writers became their editors or founded their own magazines as outlets for new creative productions and for their own critical essays. Pound's and Wyndham Lewis's *Blast* (1914–15) in London, *The Seven Arts* (1916–17) in New York City, *The Fugitive* (1922–25) in Memphis, Tennessee, and *The Dial* (1880–1929) in New York City were among the most important. *The Criterion* (1922–29), founded and edited by T. S. Eliot, was the most influential critical journal. Presses in London and Paris were pressed into service, and on one occasion two poets, Louis Zukofsky and William Carlos Williams, established a press (the short-lived Objectivist Press) which published Williams's own *Collected Poems*. Eliot became an important editor at Faber & Faber in London and the poet Allen Tate became poetry editor at Henry Holt in New York. In the theater, the Provincetown Players, founded in 1915, produced experimental plays by Edna St. Vincent Millay, E. E. Cummings, Sherwood Anderson, and landmark works by Eugene O'Neill which introduced expressionistic techniques (masks and symbolic forms to define subjective experience) into the American theater. Beginning with the poet-critic John Crowe Ransom at the end of World War I and with increasing frequency after 1930, prominent writers, including Allen Tate and the novelists Robert Penn Warren and Katherine Anne Porter, secured posts in colleges and universities which facilitated the acceptance of the new literature in academia and among younger generations of students. After 1940 established poets like E. E. Cummings and William Carlos Williams, along with Robert Frost and T. S. Eliot and, later, novelists, found new and widening audiences by recording their works on phonograph records for libraries and commercial companies and by giving public readings on countless campuses.

TRADITION AND REVOLUTION

The new modes of writing that became in these ways firmly established in the American literary tradition did not win acceptance without opposition. In the late 1950s Walter Allen and a number of other English critics who had been most appreciative of American achievements began to back away from the writing of Eliot's and Pound's generation, deploring the neglect of other English poets and claiming that the revolution in Anglo-American writing had been chiefly an American imposition on the English tradition. Before that, notable American literary historians, including Van Wyck Brooks, had called into question the validity of the new modes of writing, and the poet Archibald MacLeish, on returning from France in 1930, declared that poetry of his generation (including his own) had surrendered to pessimism, alienation, nostalgia, political conservatism, and the cry of individual isolation voiced so eloquently in Eliot's *The Waste Land*. He called for an extension of the revolution in aesthetics into a social revolution and for an art of "public speech" addressed, more directly than he thought his generation's had been, to social, economic, and moral issues. The new poetry and prose, however, had never ignored these issues, and political liberals, if not radicals, along with such notable conservatives as Pound and Eliot, had helped generate the literary revolution emerging in the early decades of the century. MacLeish's demands projected not so much a repudiation of the new literary modes as a renewed effort to extend them which had been part of the dynamics of American cultural history since the 1920s. The poet Hart Crane recognized astutely in 1930 that the literary revolution, by virtue of its very success, had created the problem of how to keep it from petrifying, how to sustain it as a tradition in new and continually varied creative effort. "Revolution flourishes still," he wrote in *Modern Poetry*, but "rather as a contemporary tradition. * * * It persists as a rapid momentum in certain groups or movements, but often in forms that are more constricting than liberating." A decade before, a poet, critic, and novelist who had helped launch the new movement, William Carlos Williams, charted the strategy that had, along with the efforts of expatriates, helped inaugurate the revolution in the first place and did the most to keep it a vital tradition, a sustaining force, for writers after World War II. Expatriate poets had betrayed the aims of Modernism by their undisguised use of foreign settings and the formal techniques of traditional verse, he declared in 1923, and he called for continually new experimentation that drew on specifically American settings and speech patterns, on the American environment and aspects of its traditions that the cosmopolitan writers had either obscured or ignored. However different his program, the impulses and traditions behind his demands were essentially the same that had impelled his antagonists to revamp tradition, to "make it new," as Pound had declared, and change the course of literary history. It was the full variety of American writers' achievements that transmitted to the post–World War II generation a strengthened and still generative literary tradition.

EZRA POUND
1885–1972

In 1908 Ezra Loomis Pound, a young scholar and poet, left his native America for Venice and London, convinced that the authority of the arts in America was unrecognized and insecure, and determined to reconstitute the very foundations of literature and its sustaining civilization in the West. His reception since 1930 has been complicated by his anti-Semitism, his affiliation with fascist Italy during World War II, the charge of treason brought against him by the United States, and the diagnosis of insanity that resulted in the suspension of his trial and eventual dismissal of the charge. But within six years after Pound's arrival in Europe in 1908, he had become one of the dominant figures on the London literary scene. His numerous critical and scholarly productions, his efforts on behalf of other writers, the twelve volumes of verse that formed the basis for the collection *Personae* issued in 1926, and the one hundred sixteen *Cantos* that he published between 1916 and 1969 made him the most important single figure, with the most continuously sustained influence on other authors, in Anglo-American poetry of the twentieth century.

He was born in Hailey, Idaho, to parents who had roots in New England and upper New York State as well as in the West, and who settled, when Pound was an infant, in a comfortable suburb near Philadelphia where his father was employed by the Mint. His parents always maintained a genuine, if often puzzled, interest in his career. He studied American history, the ancient classics, and English literature at the University of Pennsylvania for two years, then studied Old English and the Romance literatures at Hamilton College, where he graduated in 1905. He returned to Pennsylvania for an M.A. (1906) preparatory to a career as a teacher and scholar of Romance literatures. His dismissal from his first post at Wabash College in Indiana (for innocuously offering shelter to a jobless burlesque queen) precipitated his departure for Europe, but he never lost his ambition to teach nor his sense of identification with his American origins. He taught in London at the Regent Street Polytechnic in 1908–9, offering courses in Renaissance and medieval literature while at the same time he was establishing his reputation as a poet. His translations and scholarly studies of these materials appeared as *The Spirit of Romance* in 1910; *A Lume Spento*, his first volume of verse, was published in Venice in 1908, while *A Quinzaine for This Yule* and *Personae* were published in London in 1908 and 1909. He married Dorothy Shakespeare (who later bore him his only son) in 1914. By then his zeal for teaching, combined, as T. S. Eliot remarked, with the tactics of a "campaigner,"[1] was being channeled into his critical essays and pronouncements advancing the cause of his own and others' poetic talents.

Although he remained an expatriate—lured by the dream of a world in which cultural values would not be divorced from political and economic structures as he found them to be in modern America—he became a conspicuously *American* expatriate, conscious of his American identity and

1. Introduction, *Literary Essays of Ezra Pound* (1954), p. xii.

using it strategically like Benjamin Franklin, Walt Whitman, and Benjamin West, the expatriate painter, before him. His memoir *Indiscretions* (1923) dwelt affectionately, if ironically, on his parents' gentility but underscored the Pounds' affiliations with the rough frontier of his lumberman grandfather in the West and the local notoriety of his mother's family in upper New York. (His father had had an Indian nurse; the Loomises had been known with some justification as horse thieves.) Sharing an obsession with his predecessors Cooper, Hawthorne, and Henry James, he had earlier noted that "Mrs. Columbia has no mysterious & shadowy past to make her interesting,"[2] and his compensatory sense of national identity became a consciously cultivated role, a willed Americanism as carefully defined and strenuously pursued as Whitman's, its characteristic signs showing in the boldness of his critical stance, his use of raw slang in his criticism (and obscenities in his letters), and his infusion of colloquialisms and slang into his poetry. Yet his role as an American expatriate included a willed cosmopolitanism that recalls Edgar Allan Poe: he is like Poe in his allegiance to exacting and extranational standards of taste, the bluntness with which he held writers accountable for lapses from such standards, and his determination to codify those standards, if not to redefine them from scratch, in an effort to reconstitute a literary tradition.

To that task Pound brought a rhetoric of revolutionary radicalism more strident than that of the contemporary Georgian poets who were calling for a literary "revolt" against the writers of the decadence, academicism in the arts, and Victorianism. Behind what T. S. Eliot later called "the revolution in taste and practice which [Pound] has brought about"[3] was an ambition as striking as Milton's in his striving for personal supremacy—Richard Aldington noted that Pound wanted to be "literary dictator of London"[4]—and in the mission he defined for the arts as the instrument for refounding civilization. Then as later in his career his tactics threatened to become mere petulance and bellicosity, and he enjoyed the delights of self-dramatization: for example, he wrote Harriet Monroe that "For one man I strike there are ten to strike back at me. I stand exposed."[5] Nevertheless, Pound's motives were genuine when scorning "the abominable dogbiscuit of Milton's rhetoric."[6] His literary criticism was part of a large-scale program to overhaul Anglo-American cultural traditions. (Characteristically, he came to terms with his traditions and predecessors by wrestling with them, as in the case of Whitman.)

Pound had, as he said, a "plymouth-rock conscience landed on a predilection for the arts,"[7] and his undertakings combined the fervor of a missionary with the enterprise of a colonizer. Essays such as *Patria Mia* (1912), *Prolegomena* (1912), and *Renaissance* (1914) constituted the "propaganda" that Pound thought necessary to launch a renaissance in American culture. Scholarship and the various arts should be partners in the search for models that were not to be imitated or plagiarized, but recreated, emulated, and surpassed. Poets were to break away from the poetics of "emotional slither," using "fewer painted adjectives" which impede "the shock and

2. Quoted by Alan Holder, *Three Voyagers in Search of Europe* (1965), p. 216.
3. *Literary Essays*, p. xii.
4. Quoted by Norman, *Ezra Pound* (1960), p. 273.
5. *Letters*, p. 13.
6. *Pavannes and Divisions* (1918), p. 247.
7. *Letters*, p. 12.

stroke" of authentic verse.[8] Poets should go "against the grain of contemporary taste," and writers were to be as boldly eclectic as Emerson wished, ranging through past and foreign cultures in their search for models or "instigations," he declared in *Renaissance*. They should ransack particular eras of history to uncover lost traditions, as Pound did in finding a "New Greece" in the culture of China. The arts could be subsidized by endowed, self-perpetuating trusts; these would free artists from the need to flatter an "ignorant" public or to meet the demands of commercial bureaucracies, which he scorned.

This design for cultural revolution governed Pound's almost incredible efforts on behalf of other artists, as well as the critical and scholarly activities he pursued in the decades before 1925. He was briefly the secretary and a life-long friend of the Irish poet William Butler Yeats, a friend and sparring partner of William Carlos Williams, and promoted the music of the American composer George Antheil—Pound himself composed music. Pound was the first important reviewer to praise Frost's *A Boy's Will* in 1913 and recognized his "VURRY Amur'k'n" talent[9] before Frost was known on either side of the Atlantic. He was instrumental in getting Joyce's *Portrait of the Artist* published serially in *The Egoist*, and arranged for the publication of Marianne Moore's first book. He secured the appearance of Eliot's first published work in *Poetry*, and later, in one of the important instances of creative editing in the twentieth century, he suggested crucial inclusions and deletions in the manuscript of Eliot's *The Waste Land*. He championed the modernist sculpture of Gaudier-Brzeska, later killed in World War I, but Pound's *Memoir* (1916) of the sculptor was not only a personal tribute but a manifesto in Pound's critical campaign and an exploration of his own aesthetic principles.

While serving as foreign correspondent for Harriet Monroe's *Poetry* and editor of the anthology *Des Imagistes* (1913), Pound had joined Richard Aldington, Hilda Doolittle, and others in formulating the doctrines of Imagism, a movement advancing modernism in the arts which concentrated on reforming the medium of poetry. Pound endorsed the group's three main principles—the "direct treatment" of poetic subjects, elimination of merely ornamental or "superfluous" words, and rhythmical composition in the sequence of the supple "musical phrase" rather than in the "sequence of a metronome"—and Pound provided the title "Imagiste" for the group. But he also developed definitions of their aims that disclose his personal conception of the authority to be achieved by a new poetics. An image, he wrote in *A Few Don'ts by an Imagiste*, is "that which presents an intellectual and emotional complex in an instant of time," and he interpreted the term *complex* as a notably energized verbal experience that yields a "sudden liberation," expresses a "sense of freedom" and "sudden growth."[1] While he wanted *imagism* to mean "hard light, clear edges" in the textures of verse, and warned poets to "Go in fear of abstractions," his emphasis on the kinetic charge generated by verbal forms was the basis for his break with the Imagists and his launching (with the sculptors Jacob Epstein and Gaudier-Brzeska, the painter Wyndham Lewis, and others) of the movement Vorticism, whose theories dominated the short-lived magazine

8. *Literary Essays*, p. 12.   1. *Literary Essays*, pp. 4–5.
9. *Letters*, p. 114.

Blast in 1914–15. The principles of Vorticism remained central to Pound's poetic strategies, although he later used other vocabularies for defining them. Imagism, he felt, tended to produce visual and static patterns, while Vorticism called for forms that were both sculptural and generative of movement and power. Pound conceived the vortex as some generative "primary form"[2] that is prior to the specific forms in any art. In cultural history, a vortex is a period of shared ferment in the arts, or the channeling of effort and purpose that makes a city into a cultural capital. In poetry, an image is a "VORTEX," a "word beyond formulated language," "a radiant node or cluster * * * from which, and through which, and into which, ideas are constantly rushing."[3]

During these same years Pound was engaged in the study of Oriental literature, an interest evident in his own versions of Japanese *haiku* and translations from the Chinese in *Cathay* (1915) and *Lustra* (1916). The *haiku*, Pound declared, using his *In a Station of the Metro* as an example, provided a model of compression in verse; a "one-image poem" which renders "the precise instant when a thing outward and objective transforms itself, or darts into a thing inward and subjective,"[4] combining the graphic with the sense of motion and volatile discovery. Given the opportunity to study Ernest Fenollosa's manuscript of *The Chinese Written Character as a Medium of Poetry* (and to complete some translations Fenollosa left unfinished), Pound inferred that the condensed juxtapositions of Chinese ideograms embodied what in effect were events and actions, transferences of energy among words, with the graphic figures presenting, as Fenollosa said, "something of the character of a motion picture."[5] Pound's interest in Oriental culture, the "New Greece," extended to the Confucian ethics which pervades his *Cantos*, and the songs which he rendered freely in *The Classic Anthology Defined by Confucius* (1954). But, as in his earlier studies of Romance literature, his forays into scholarship (though judged deficient by specialized scholars) were inseparable from his campaigns in the critical arena and had provided by 1920 the basis for his "propaganda" and for the poetics he tirelessly sought to make authoritative in his generation.

In 1920 Pound abandoned England for Paris and the Continent, as scornful of what he thought literary London's backwardness as he had earlier been of New York's. He memorialized his departure in a "farewell to London," his early masterpiece *Hugh Selwyn Mauberley* (1920). By the time he had settled in Rapallo, Italy, in 1925, he had left behind the literary arenas of Europe's capitals and was dependent on publication and on his voluminous correspondence to effect his ends as a man of letters, devoting most of his creative effort to gathering materials and composing his long poem, the *Cantos*. Of the critical essays in *Make It New* (1934) and *Literary Essays* (1954), most date from 1920 and earlier. *The ABC of Reading* (1934) and *Polite Essays* (1937) continued his appraisals of literary traditions and of modern writing. His studies of Confucius confirmed Pound's conception of the poet's mission to renovate the language and bring about social and personal regeneration. Confucius also provided models of social order in the relation of the Chinese "Lord" to his community. In his continuing study of medieval culture he found the term *forma* which defined

2. *Blast*, I (January, 1914), 54.
3. *Gaudier-Brzeska* (1916), p. 92.
4. *Ibid.*, p. 103.

5. Quoted by Hugh Kenner, *The Pound Era* (1971), p. 289.

the basic, emergent "dynamic form," the generative shaping process that artists try to incarnate in the finished forms of particular works. His researches in history convinced him that there were crucial similarities among statesmen and art patrons of the Italian past (notably Sigismondo Malatesta), the Americans Thomas Jefferson and John Adams, and the fascist dictator Mussolini (an "artist" as well as a statesman, Pound insisted).[6] Pound's *Guide to Kulchur* (1938) contained some of his most perceptive notations on history and culture, but it displayed also something of the incoherence and hysteria that characterized much of his social program after 1930. His increasingly vicious denunciations of the Jews, vilification of America under Franklin D. Roosevelt, and corresponding endorsement of Italian fascism were justified in his eyes by the vastly simplified and obsessive notion that the root of all social evil and cultural disorder is "usury," and by his conviction that he was trying to awaken America to its true political and cultural tradition.

When the Italian government in World War II invited Pound to deliver radio broadcasts aimed at American troops, with no prescription as to content, he readily delivered the talks that resulted in his being charged by the United States with treason in 1943 (as well as provoking *Poetry* magazine's repudiation and the temporary deletion of his poetry from an anthology); he read from his *Cantos* over the air, abused President Roosevelt and the Jews, advocated monetary reform, and mocked the motives for America's entering and continuing the war. Taken into custody after the fall of Italy, Pound suffered deprivations and shattering humiliations during his imprisonment for months in an open-air cage, which are echoed in the humility and anguish of the *Pisan Cantos* (1948). He was returned to the United States in 1945 for trial, but his case was suspended when the court, accepting the report of a panel of psychiatrists, found Pound "insane and mentally unfit for trial."[7] He spent the years from 1946 to 1958 as a patient in St. Elizabeth's Hospital in Washington, D.C., receiving visits from his devoted wife and admirers, working on Confucius and the *Cantos*. The charge of treason was dropped in 1958 and Pound was released, to return to Italy, as the result of efforts on his behalf by lawyers, congressmen, and writers led by Archibald MacLeish. Pound spent the rest of his life in Europe, chiefly in Italy, cared for by his wife, the concert violinist Olga Rudge, who had been his long-time companion and bore him his only daughter, Mary, the Countess de Rachewiltz. He traveled occasionally to London and Paris, and he visited literary friends in a number of cities when he returned to the United States in 1969 to receive an honorary degree from Hamilton College. He continued to publish cantos and to receive visits from friends, although in his last years he withdrew into protracted periods of unbroken silence.

The award of the Bollingen Prize for poetry in 1948 by a committee of the Library of Congress to Pound's *Pisan Cantos* had touched off an explosive controversy that continues to color the critical reception of his writings, though his impact on poets continued to be immense in the 1960s and 1970s. Opinion was divided about the merits of Pound's poetry, the effect on it of his anti-Semitism, and the propriety of an award from a governmental agency to an alleged traitor. Pound received the Harriet Monroe

6. *Jefferson and/or Mussolini* (1935), p. 34.

7. Quoted by Noel Stock, *The Life of Ezra Pound* (1970), p. 418.

Award for poetry in 1962 and a prize from the Academy of American Poets in 1963. But just before his death in 1972, when a committee of the American Academy of Arts and Sciences nominated him for its Emerson-Thoreau Award, the Academy's governing board overruled its committee and did not award the prize. A remark Pound made to his biographer before his final departure for Italy captured both the imaginative daring and the tragedy always latent in his career: "When I talk it is like an explosion in an art museum, you have to hunt around for the pieces."[8]

THE POETRY THROUGH MAUBERLEY

Three things distinguish Ezra Pound's poetic productions through 1920: the refinements of the dramatic monologue that he developed from the models of Robert Browning; the translations and adaptations that were the proving ground for his many innovations and are among his finest achievements; and the meters and disciplined nuances of effect that both strengthen and refine the surface texture of his verse, fusing the cadences of spoken speech with those of music.

In Browning, the monologue is presented (in an ostensibly direct, prosaic idiom) as if spoken not by the poet but by a speaker resurrected from the past, whose words or thoughts define dramatically a situation in which he is involved; together the speaker and the context are objectively observed. In Pound's treatment the form becomes more personal, the figure from the past and the poet himself become intimately connected so that the figure becomes a mask for the poet's own voice, an instrument or "persona" through which both the speaker's voice and the poet's voice are presented. It becomes what Pound called a "dramatic lyric * * * the poetic part of a drama the rest of which (to me the prose part) is left to the reader's imagination or implied or set in a short note."[9] It becomes also, as in *Sestina: Altaforte*, a bold and complicated act of historical reconstruction, for Pound not only recaptures details of the speaker's life (usually a poet) but echoes or translates fragments of the documents in which the speaker's life is recorded, including images and lines from the speaker's poems. The "persona" became for Pound an important expressive form, no longer limited to dramatic monologues, and enabled him to express the various voices he assumed.

Pound's strategies of translation were an integral feature of his poetics, and while he has long been criticized for careless lapses and flagrant inaccuracies, his procedures have stimulated many of the best translators in the twentieth century while constituting a major resource for his own poetry. The process which Pound called "casting off * * * complete masks of the self in each poem" was part of his "search for the real" which he said he "continued in a long series of translations, which were but more elaborate masks."[1] With scrupulous attention to the original text, but recognizing that the lines are mere notations of the realities suggested by them, he devised equivalences in English that were neither their literal transcriptions nor irresponsible adaptations but creative works in their own right; they manage to recapture the essentials, formal contours, and effects of the original with astonishing fidelity. Pound's translations become "personae" themselves, and the illusion of a translation is often sustained in poems that are

8. Quoted by Norman, *Ezra Pound*, p. 457.

9. *Letters*, pp. 3–4.

1. "Vorticism," *Fortnightly Review*, 96 (September 1, 1914), 463–64.

not strictly translations at all. One brilliant instance of poetry where adaptive translation is basic to Pound's creative effort is the long poem *Homage to Sextus Propertius* (1917), in which he converted the Roman poet and certain of his texts into a persona through which Pound voices his disenchantment with imperial London and the war. Another is his early success *Sestina: Altaforte*. There Pound recreated the stanzaic form of his favorite Provençal poet, Arnaud Daniel, but attributed it to another troubadour who never used that form, Bertrand de Born, whose *Praise of War* is drawn on freely for attitudes, images, and rhetorical devices.

The "persona" was an enabling form through which Pound's voice found expression and by means of which he explored and appropriated his tradition, but it depended for its effectiveness on the achievement of a fully realized poetic surface, a surface that his disciplined meters and poetic line had achieved by the 1920s. Wary of "the borderline between forceful language and violent language,"[2] he could so control colloquial speech and the placement of alliterative "b" sounds as to etch graphically and powerfully the stark futility of World War I:

> There died a myriad,
> And of the best, among them,
> For an old bitch, gone in the teeth,
> For a botched civilization.

The " 'sculpture' of rhyme" and delicately suspended modulations of rhythm expose the affluent Mr. Nixon "In the cream gilded cabin of his steam yacht." What Pound later defined as his attempt "To break the pentameter" (Canto 81) is apparent in the frequency with which he avoids routine regularity in length of lines by breaking them with a caesura, and avoids the metronome's regularity of "hefty swats on alternate syllables" by subtle variations in cadence, modulations in tone, and nuances of rhyme in counterpoint with residual echoes. The discipline of his verse sustains ironies in *Hugh Selwyn Mauberley* that avoid the melodrama of obvious contrast and reversal, and it represents his extension of the resources of poetry to which T. S. Eliot and other modern poets have paid tribute.

THE CANTOS

Pound began to work intently on his epic poem the *Cantos* in 1915 and gathered the first collection of them (after rejecting and revising some of the originals) in *A Draft of XVI Cantos* (1925). From the start, the question of their merits—indeed of their precise nature—has agitated criticism, which now ranges from George P. Elliott's verdict that the series disintegrates into a disorderly "bundle of poetry and mutter" to Eva Hesse's celebration of them as a splendidly "open form" appropriate to the "pluralistic and relativistic mode of thinking" of his age.[3] The latest published collections, including *Section: Rock-Drill* (1956) and *Thrones* (1959), brought the collected number to 116 and marked the near-completion of one of the major undertakings of twentieth-century poets.

Statements by Pound himself help define the design that he hoped would emerge, including one when he speaks of his friend William Carlos Williams's habit of "yanking and hauling as much [material] as possible

2. *Letters*, p. 182.
3. George P. Elliott, *Ezra Pound: A Collection of Critical Essays* (1963), p.

152; Eva Hesse, *New Approaches to Ezra Pound* (1969), pp. 17, 46.

into some sort of order (or beauty), aware of it as both chaos and as potential";[4] in *Cantos* he said that they were a "rag-bag for the 'modern world' to stuff all its thought in."[5] It was an epic designed to appropriate history, a "poem including history,"[6] he declared, with three centers of attention: men of action (statesmen, warriors, bankers), documents and books, and creative genius, all presented in the idiom of anecdote and conversation:

> And men of unusual genius, books, arms,
> Both of ancient times and our own, in short, the usual subjects
> Of conversation between intelligent men. (Canto XI)

The vast stretches of history encompassed by the poem are presented not as a subject to be studied but as the realm of action and thought to be experienced that Emerson earlier had demanded in his essay *History.*

The historical materials so gathered and talked about are more important than is customary in poetry because they are "included" literally in the text, in fragmented phrases or long excerpts from literary works, biographies, letters, and business transactions, or in the form of the poet's personal memories of incident and information. On the basis of such materials it is possible to distinguish related groups of cantos within the larger sequence: the "Malatesta cantos," centering on the Renaissance soldier and patron Sigismondo Malatesta; the "American cantos," including the "John Adams cantos," featuring the statecraft of Presidents Jefferson, the Adamses, and Van Buren; the "Chinese History cantos;" and the "Pisan cantos," drawing on Pound's experience as a prisoner in the American detention camp. These categories, however, obscure the complicated juxtapositions and interconnections among them that fleetingly or recurringly emerge in the movement of the poem.

Singly and together, the *Cantos* are personae in which the voices of history intermingle with Pound's own in a shifting surface that is fluid and linear rather than agitated by dramatic conflict, and whose shape is an exploratory orientation toward form rather than a completion or fulfillment of a formal design. The various themes and values in the poem, with their often tenuous connections, are obscured by the "deliberate disconnectedness"[7] that R. P. Blackmur discerned as a controlling principle of the poem's movement. The *Cantos* do not display a new form so much as a new species of poetic statement in which the raw materials of historical data remain raw and often obscure while approaching the threshold of order, and the process of digesting history's meaning remains ruminative and partial while the poet seeks the full illumination of imaginative vision. At times there emerge moments of luminous beauty and lyric intensity that match the best of modern poetry. The poem's movement in the direction of these moments is quickened by such recurring symbols as the image of the city and the image of light; by such mythological figures as Aphrodite, the goddess of love, and Persephone, goddess of regeneration; and by the voices and figures of countless artists, from Homer, "blind as a bat," to Dante, Cavalcanti, and the troubadour poets, through Browning and Pound's American predecessors Whitman and Henry James, to such con-

---

4. *Polite Essays* (1966), p. 77.
5. *Lustra* (1917), p. 181.
6. *Selected Essays*, p. 86.

7. R. P. Blackmur, *Language as Gesture* (1952), p. 140.

temporaries as Eliot and Yeats. And the poem is governed by three archetypal quests which displace and dissolve into each other in the long series of metamorphoses that recreate them: Odysseus's journey across the sea and through the underworld in quest of home in Homer's epic; the exiled Dante's journey through hell and purgatory in search of the radiant light and angelic "thrones" which he finally glimpses in the paradise of the *Divine Comedy*; and the tragic search for an ideal social order or an imagined city, for the "stones of foundation" on which "to build the city of Dioce whose terraces are the color of stars" (Canto 74).[8]

To Whistler, American[1]

On the loan exhibit of his paintings at the Tate Gallery.

You also, our first great,
Had tried all ways;
Tested and pried and worked in many fashions,
And this much gives me heart to play the game.

Here is a part that's slight, and part gone wrong,   5
And much of little moment, and some few
Perfect as Dürer![2]
"In the Studio" and these two portraits,[3] if I had my choice!
And then these sketches in the mood of Greece?

You had your searches, your uncertainties,   10
And this is good to know—for us, I mean,
Who bear the brunt of our America
And try to wrench her impulse into art.

You were not always sure, not always set
To hiding night or tuning "symphonies";[4]   15
Had not one style from birth, but tried and pried
And stretched and tampered with the media.

You and Abe Lincoln from that mass of dolts
Show us there's chance at least of winning through.

                              1912, 1949

Portrait d'une Femme[1]

Your mind and you are our Sargasso Sea,[2]
London has swept about you this score years

---

8. Deioces, **legendary King of the Medes**, whose terraces and golden tower, Ectaban, are described by the Greek historian Herodotus (480?–420? B.C.) in his *History* (1.98).
1. James Abbott McNeill Whistler (1834–1903), expatriate American painter.
2. Albrecht Dürer (1471–1528), German painter and engraver.
3. "Brown and Gold—de Race," "Grenat et Or—Le Petit Cardinal" ("Garnet and Gold—The Little Cardinal") [Pound's note]: titles and subjects of the portraits.
4. Whistler painted many night scenes and entitled many paintings "symphonies" to suggest their abstract or non-representational quality.
1. Portrait of a Lady.
2. Sea in the North Atlantic where boats were becalmed; named for its large masses of floating seaweed.

And bright ships left you this or that in fee:
Ideas, old gossip, oddments of all things,
Strange spars of knowledge and dimmed wares of price.   5
Great minds have sought you—lacking someone else.
You have been second always. Tragical?
No. You preferred it to the usual thing:
One dull man, dulling and uxorious,
One average mind—with one thought less, each year.   10
Oh, you are patient, I have seen you sit
Hours, where something might have floated up.
And now you pay one. Yes, you richly pay.
You are a person of some interest, one comes to you
And takes strange gain away: 15
Trophies fished up; some curious suggestion;
Fact that leads nowhere; and a tale or two,
Pregnant with mandrakes,[3] or with something else
That might prove useful and yet never proves,
That never fits a corner or shows use,   20
Or finds its hour upon the loom of days:
The tarnished, gaudy, wonderful old work;
Idols and ambergris and rare inlays,
These are your riches, your great store; and yet
For all this sea-hoard of deciduous things,   25
Strange woods half sodden, and new brighter stuff:
In the slow float of differing light and deep,
No! there is nothing! In the whole and all,
Nothing that's quite your own.
        Yet this is you.   30

            1912

A Pact

I make a pact with you, Walt Whitman—
I have detested you long enough.
I come to you as a grown child
Who has had a pig-headed father;
I am old enough now to make friends.   5
It was you that broke the new wood,
Now is a time for carving.
We have one sap and one root—
Let there be commerce between us.

        1913, 1916

The Rest

O helpless few in my country,
O remnant enslaved!
Artists broken against her,
A-stray, lost in the villages,
Mistrusted, spoken-against,   5

---

3. Herb, used as a cathartic; believed
in legend to have human properties, to
shriek when pinched, and to promote
conception in women.

Lovers of beauty, starved,
Thwarted with systems,
Helpless against the control;

You who can not wear yourselves out
By persisting to successes,   10
You who can only speak,
Who can not steel yourselves into reiteration;

You of the finer sense,
Broken against false knowledge,
You who can know at first hand,   15
Hated, shut in, mistrusted:

Take thought:
I have weathered the storm,
I have beaten out my exile.

                                                                1913, 1916

In a Station of the Metro[4]

The apparition of these faces in the crowd;
Petals on a wet, black bough.

                                                                1913, 1916

The River-Merchant's Wife: A Letter[1]

While my hair was still cut straight across my forehead
I played about the front gate, pulling flowers.
You came by on bamboo stilts, playing horse,
You walked about my seat, playing with blue plums.

And we went on living in the village of Chokan:[2]   5
Two small people, without dislike or suspicion.
At fourteen I married My Lord you.
I never laughed, being bashful.
Lowering my head, I looked at the wall.
Called to, a thousand times, I never looked back.   10

At fifteen I stopped scowling,
I desired my dust to be mingled with yours
Forever and forever and forever.
Why should I climb the look out?

---

4. Paris subway.
1. Adaptation from the Chinese of Li Po (701–762), named Rihaku in Japanese. Pound found the material in the papers of Fenollosa and used Japanese paraphrases in making the translation with the result that the proper names are their Japanese equivalents. Among Fenollosa's divergences from some other translations is the substitution of blue for green (line 4), of August for October (line 23), and of the image of walking on bamboo stilts for that of riding a bamboo horse.
2. Suburb of Nanking.

At sixteen you departed, [15]
You went into far Ku-to-yen,[3] by the river of swirling eddies,
And you have been gone five months.
The monkeys make sorrowful noise overhead.

You dragged your feet when you went out.
By the gate now, the moss is grown, the different mosses [20]
Too deep to clear them away!
The leaves fall early this autumn, in wind.
The paired butterflies are already yellow with August,
Over the grass in the West garden;
They hurt me. I grow older. [25]
If you are coming down through the narrows of the river Kiang,
Please let me know beforehand,
And I will come out to meet you
      As far as Cho-fu-Sa.[4]

By Rihaku
1915

Hugh Selwyn Mauberley[1]
(Life and Contacts)[2]

"Vocat aestus in umbram"[3]
—NEMESIANUS, *Ec. IV*

E. P. Ode pour l'election de Son Sepulchre[4]

For three years, out of key with his time,
He strove to resuscitate the dead art
Of poetry; to maintain "the sublime"
In the old sense. Wrong from the start—

3. An island several hundred miles up the river Kiang.
4. Beach several hundred miles above Nanking.
1. This poem was published in 1920 when Pound was on the verge of leaving London and already at work on his more ambitious and innovative *Cantos*. Deftly ironic, the poem measures both the validity and the limitations of his aesthetic practices. At the same time he diagnosed the ills of the social, economic, and cultural environment of England, which threatened to repress imaginative talent or channel it into trivial, escapist, and ineffectual forms. Pound declared that the poem was modeled partly on the technique of Henry James's prose fiction: it presents its subject through the medium of a character's mind or voice, a "center of consciousness" whose mind and standards are also part of the subject being treated and are exposed themselves to scrutiny and assessment. In *Hugh Selwyn Mauberley*, the first 13 lyrics are presented through "E.P.," a persona through which Pound expresses some of his own ambitions and tastes; cultural conditions in London and repre-

sentative literary figures (some clearly opposed, some more closely linked to Pound's own aims and associations) are surveyed through the mind of "E.P.," and reactions to them are expressed in his voice, which verges on futility until it is quickened into passion in the "Envoi." In the next section, entitled "Mauberley" (I through "Medallion"), the persona of "E.P." is absorbed in his attention to the fictitious poet Mauberley, a second persona through whom Pound subtly mocks while simultaneously pursuing his own attempt to explore experience and resuscitate poetry in sculptured forms. Pound attempts by tender but pointed irony to exorcise the nostalgia, isolation, and cult of durable form that impels his own poetic practice but threatens to ennervate it and reduce it to an aesthete's "overblotted / Series / Of intermittences." By exaggerating these motives, Pound incorporates his own "profane protest" against them and triumphs over them in verse that culminates in the highly charged and durably sculptured, though still miniature, "Medallion."
2. Ironic echo of a conventional subtitle

No, hardly, but seeing he had been born   5
In a half savage country, out of date;
Bent resolutely on wringing lilies from the acorn;
Capaneus;[5] trout for factitious bait;

Ἴδμεν γάρ τοι πάνθ᾽, ὅσ᾽ ἐνὶ Τροίῃ[6]
Caught in the unstopped ear;   10
Giving the rocks small lee-way
The chopped seas held him, therefore, that year.

His true Penelope[7] was Flaubert,[8]
He fished by obstinate isles;
Observed the elegance of Circe's[9] hair   15
Rather than the mottoes on sun-dials.

Unaffected by "the march of events,"
He passed from men's memory in *l'an trentiesme
De son eage,*[1] the case presents
No adjunct to the Muses' diadem.   20

II[2]

The age demanded an image
Of its accelerated grimace,
Something for the modern stage,
Not, at any rate, an Attic grace;

Not, not certainly, the obscure reveries   25
Of the inward gaze;
Better mendacities
Than the classics in paraphrase!

---

of literary biographies, "Life and Letters." In a subsequent edition of 1957, Pound reversed the sequence, claiming that "Contacts and Life" followed the "order of the subject matter." To the American edition of *Personae* in 1926, Pound added the following note: "The sequence is so distinctly a farewell to London that the reader who chooses to regard this as an exclusively American edition may as well omit it and turn at once to page 205."
3. "The heat calls us into the shade," from the *Eclogues* (IV.38) of the third-century Carthaginian poet Nemesianus.
4. Adaptation of the title of an ode by Pierre de Ronsard (1524–85) *On the Selection of His Tomb* (*Odes*, IV.5).
5. One of Seven against Thebes whom Zeus struck down by lightning for his rebellious defiance.
6. "For we know all the toils [endured] in wide Troy," part of the sirens' song to detain Odysseus in Homer's *Odyssey* (Book 12, 189). Odysseus stopped his comrades' ears with wax to prevent their succumbing to the lure.
7. Wife of Odysseus who remained faithful during his long absence but alter-

nately enticed and held off the many suitors who sought to displace him.
8. Gustave Flaubert (1821–80), French novelist who cultivated form and stylistic precision.
9. Enchantress with whom Odysseus dallied for a year before returning home.
1. "The thirtieth year of his age," adapted from *The Testament* by the 15th-century French thief and poet François Villon. Since the turning point Pound had in mind was the publication of his *Lustra* in 1916, he later changed the line to "trentunieme" or "thirty–first" to conform to his age at that time.
2. The subtle ironies of this poem derive from three senses of the word *demanded*: what the age wanted and liked, what by contrast it needed, and what it "asked for" in the sense of deserved to get. The poem defines the expectations that make it difficult for the modern poet to fulfill his public role, and also suggests the means that Pound characteristically tried to use: translations that go beyond mere paraphrase, forms with the force of prose and the movement of cinematic art, and the "'sculpture' of rhyme."

The "age demanded" chiefly a mould in plaster,
Made with no loss of time,   30
A prose kinema,[3] not, not assuredly, alabaster
Or the "sculpture" of rhyme.

III

The tea-rose tea-gown, etc.
Supplants the mousseline of Cos,[4]
The pianola "replaces"   35
Sappho's barbitos.[5]

Christ follows Dionysus,[6]
Phallic and ambrosial
Made way for macerations;[7]
Caliban casts out Ariel.[8]   40

All things are a flowing,
Sage Heracleitus[9] says;
But a tawdry cheapness
Shall outlast our days.

Even the Christian beauty   45
Defects—after Samothrace;[1]
We see τὸ καλόν[2]
Decreed in the market place.

Faun's flesh is not to us,
Nor the saint's vision.   50
We have the press for wafer;
Franchise for circumcision.

All men, in law, are equals.
Free of Pisistratus,[3]
We choose a knave or an eunuch   55
To rule over us.

O bright Apollo,
τίν᾿ ἄνδρα, τίν᾿ ἥρωα, τίνα θεόν,[4]

---

3. Movement (Greek), and early spelling of *cinema*, motion pictures.
4. Gauzelike fabric for which the Aegean island Cos was famous.
5. Lyrelike instrument used by the Greek poetess Sappho (fl. 600 B.C.)
6. Greek god of fertility, regenerative suffering, wine, and poetic inspiration; his worshipers were known for destructive frenzies and consuming ecstasies; his festivals included sexual rites, wine tasting, and dramatic performances.
7. Wasting, fasting. Pound compares Christian asceticism to Dionysian rites.
8. In Shakespeare's *Tempest*, Caliban is the earth-bound and rebellious slave of

Prospero, ruler and magician; Ariel is the imaginative spirit whose assistance Prospero commandeers.
9. Greek philosopher (fl. 500 B.C.) who taught that all reality is flux or a "flowing."
1. North Aegean island, center of religious mystery cults, site of the famous statue *Winged Victory*.
2. The beautiful.
3. Athenian tyrant and art patron (fl. sixth century B.C.).
4. "What man, what hero, what god," Pound's version of Pindar's "What god, what hero, what man shall we loudly praise" (*Olympian Odes*, II.2).

What god, man, or hero
Shall I place a tin wreath upon!   60

IV

These fought in any case,
and some believing,
            pro domo,[5] in any case . . .

Some quick to arm,
some for adventure,   65
some from fear of weakness,
some from fear of censure,
some for love of slaughter, in imagination,
learning later . . .
some in fear, learning love of slaughter;   70

Died some, pro patria,
           non "dulce" non "et decor"[6] . . .
walked eye-deep in hell
believing in old men's lies, then unbelieving
came home, home to a lie,   75
home to many deceits,
home to old lies and new infamy;
usury age-old and age-thick
and liars in public places.

Daring as never before, wastage as never before.   80
Young blood and high blood,
fair cheeks, and fine bodies;

fortitude as never before

frankness as never before,
disillusions as never told in the old days,   85
hysterias, trench confessions,
laughter out of dead bellies.

V

There died a myriad,
And of the best, among them,

For an old bitch gone in the teeth,   90
For a botched civilization,

Charm, smiling at the good mouth,
Quick eyes gone under earth's lid,

For two gross of broken statues,
For a few thousand battered books.   95

---

5. "For the home," adapted from Cicero's *De Domo Sua.*
6. "For one's native land, not sweetly, not gloriously," adapted from Horace, "it is sweet and glorious to die for one's fatherland" (*Odes,* III.ii.13).

Yeux Glauques[7]

Gladstone was still respected,
When John Ruskin produced
"King's Treasures"; Swinburne
And Rossetti still abused.

Foetid Buchanan lifted up his voice　　　　　100
When that faun's head of hers
Became a pastime for
Painters and adulterers.

The Burne-Jones cartons[8]
Have preserved her eyes;　　　　　　　　105
Still, at the Tate, they teach
Cophetua to rhapsodize;

Thin like brook-water,
With a vacant gaze.
The English Rubaiyat was still-born[9]　　　110
In those days.

The thin, clear gaze, the same
Still darts out faun-like from the half-ruin'd face,
Questing and passive. . . .
"Ah, poor Jenny's case"[1] . . .　　　　　　115

Bewildered that a world
Shows no surprise
At her last maquero's[2]
Adulteries.

"Siena mi fe'; Disfecemi Maremma"[3]

Among the pickled foetuses and bottled bones,　　120
Engaged in perfecting the catalogue,
I found the last scion of the
Senatorial families of Strasbourg, Monsieur Verog.[4]

7. The brilliant yellow-green eyes of Elizabeth Siddal, the seamstress who became the favorite model of the Pre-Raphaelite painters and later the wife of the painter and poet Dante Gabriel Rossetti (1828–82). She was the model for the beggar maid in *Cophetua and the Beggar Maid* (now hanging in the Tate Gallery, London), by Sir Edward Burne-Jones (1833–98). The Pre-Raphaelites, including the poet Algernon Swinburne (1837–1909), were attacked as "The Fleshly School of Poetry" by Robert W. Buchanan (1841–1901) in 1871, and were defended by the critic John Ruskin (1819–1900), whose *Sesame and Lilies* (1865) contains a chapter entitled "Kings' Treasuries," calling for the diffusion of literature and the improvement of English tastes in the arts. Wil-

liam E. Gladstone (1809–98) was a politician and three times Prime Minister of Britain.
8. Drawings.
9. Edward Fitzgerald (1809–83) translated *The Rubáiyát of Omar Khayyám* in 1859, but it was not read ("still-born") until discovered later by the Pre-Raphaelites.
1. Prostitute, heroine of a poem by Rossetti.
2. Or *magnereau*: sexual exploiter, pimp.
3. "Siena made me, Maremma unmade me," spoken by a Sienese woman, condemned by her husband to die in Maremma marshes for her infidelity, in Dante's *Purgatory* (V.134).
4. Victor Plarr (1863–1929), French poet (*In the Dorian Mood*, 1896) and raconteur from Strasbourg, later librarian of

For two hours he talked of Gallifet;[5]
Of Dowson; of the Rhymers' Club;   125
Told me how Johnson (Lionel) died
By falling from a high stool in a pub . . .

But showed no trace of alcohol
At the autopsy, privately performed—
Tissue preserved—the pure mind   130
Arose toward Newman[6] as the whiskey warmed.

Dowson found harlots cheaper than hotels;
Headlam for uplift; Image impartially imbued
With raptures for Bacchus, Terpsichore[7] and the Church.
So spoke the author of "The Dorian Mood,"   135

M. Verog, out of step with the decade,
Detached from his contemporaries,
Neglected by the young,
Because of these reveries.

Brennbaum

        The sky-like limpid eyes,   140
        The circular infant's face,
        The stiffness from spats to collar
        Never relaxing into grace;

        The heavy memories of Horeb, Sinai and the forty
              years,[8]
        Showed only when the daylight fell   145
        Level across the face
        Of Brennbaum "The Impeccable."

Mr. Nixon

In the cream gilded cabin of his steam yacht
Mr. Nixon advised me kindly, to advance with fewer
Dangers of delay. "Consider   150
        "Carefully the reviewer.

"I was as poor as you are;
"When I began I got, of course,

---

the Royal College of Surgeons and member of the Rhymer's Club. Other members: two Roman Catholic poets and heavy drinkers, Ernest Dowson (1867–1900), of whom Plarr published a memoir, and Lionel Johnson (1867–1902), whose *Poetical Works* Pound edited in 1915; the Reverend Stewart D. Headlam (1847–1924), forced to resign his curacy for lecturing on the dance to working-men's clubs; and Selwyn Image (1849–1930), founder with Headlam of the Church and Stage Guild.

5. Marquis de Galliffet (1830–1909), French General at the battle of Sedan, which the French lost, in the Franco-Prussian War.
6. John Henry Newman (1801–90), editor and Roman Catholic convert and intellectual, later Cardinal.
7. Greek muse of the dance.
8. The children of Israel wandered in the wilderness for 40 years. Moses saw the burning bush at Horeb (Exodus 3.2); he received the Ten Commandments at Sinai (Exodus 19.20 ff.).

"Advance on royalties, fifty at first," said Mr. Nixon,
"Follow me, and take a column,   155
"Even if you have to work free.

"Butter reviewers. From fifty to three hundred
"I rose in eighteen months;
"The hardest nut I had to crack
"Was Dr. Dundas.   160

"I never mentioned a man but with the view
"Of selling my own works.
"The tip's a good one, as for literature
"It gives no man a sinecure.

"And no one knows, at sight, a masterpiece.   165
"And give up verse, my boy,
"There's nothing in it."

.   .   .   .

Likewise a friend of Bloughram's once advised me:[9]
Don't kick against the pricks,[1]
Accept opinion. The "Nineties" tried your game   170
And died, there's nothing in it.

X

Beneath the sagging roof
The stylist has taken shelter,
Unpaid, uncelebrated,
At last from the world's welter   175

Nature receives him;
With a placid and uneducated mistress
He exercises his talents
And the soil meets his distress.

The haven from sophistications and contentions   180
Leaks through its thatch;
He offers succulent cooking;
The door has a creaking latch.

XI

"Conservatrix of Milésien"[2]
Habits of mind and feeling,   185
Possibly. But in Ealing[3]
With the most bank-clerky of Englishmen?

No, "Milésian" is an exaggeration.
No instinct has survived in her
Older than those her grandmother   190
Told her would fit her station.

---

9. In Browning's *Bishop Bloughram's Apology*, the Bishop rationalized his doctrinal laxity.
1. Ironic echo of Christ's statement to Saul: "it is hard for thee to kick against the pricks" (Acts 9.5).

2. I.e., conservator of the erotic indulgence for which the Ionian city of Miletus and Aristides' *Milesian Tales* (second century B.C.) were known.
3. London suburb.

XII

"Daphne with her thighs in bark
Stretches toward me her leafy hands,"[4]—
Subjectively. In the stuffed-satin drawing-room
I await The Lady Valentine's commands,   195

Knowing my coat has never been
Of precisely the fashion
To stimulate, in her,
A durable passion;

Doubtful, somewhat, of the value   200
Of well-gowned approbation
Of literary effort,
But never of The Lady Valentine's vocation:

Poetry, her border of ideas,
The edge, uncertain, but a means of blending   205
With other strata
Where the lower and higher have ending;

A hook to catch the Lady Jane's attention,
A modulation toward the theatre,
Also, in the case of revolution,   210
A possible friend and comforter.
•   •   •   •   •   •   •   •

Conduct, on the other hand, the soul
"Which the highest cultures have nourished"[5]
To Fleet St. where
Dr. Johnson flourished;[6]   215

Beside this thoroughfare
The sale of half-hose has
Long since superseded the cultivation
Of Pierian roses.[7]

Envoi (1919)[8]

Go, dumb-born book,   220
Tell her that sang me once that song of Lawes:

4. The metamorphosis of the nymph Daphne into a laurel tree to escape the embrace of Apollo; Pound's lines are a translation of Théophile Gautier's version of Ovid's story in *Le Château de Souvenir*.
5. A translation of two lines from *Complainte de Pianos* by the short-lived French poet Jules Laforgue (1860–87).
6. Samuel Johnson (1709–84), journalist, poet, critic, and moral essayist, the reigning man of letters in late 18th-century London. "Fleet St.": newspaper publishing center in London.
7. Roses of Pieria, place near Mount Olympus where the Muses were worshiped.
8. This poem is modeled on the rhetoric and cadences of *Goe, Lovely Rose* by Edmund Waller (1606–87), whose poems were set to music by Henry Lawes (1596–1662). "E.P." first addresses his book and dispatches it with a message for the listener (to the poem) who once inspired him by singing Waller's song, then in line 222 breaks off to confess the limited powers of his poetry. The second stanza celebrates the beauty of the singer and listener, and offers to commemorate her in verse. The third warns of the incompleteness of her response and of the transience of all things but beauty. The concert singer whom Pound heard sing Lawes's and Waller's song was Raymonde Collignon.

Hadst thou but song
As thou hast subjects known.
Then were there cause in thee that should condone
Even my faults that heavy upon me lie,　　　225
And build her glories their longevity.

Tell her that sheds
Such treasure in the air,
Recking naught else but that her graces give
Life to the moment,　　　230
I would bid them live
As roses might, in magic amber laid,
Red overwrought with orange and all made
One substance and one colour
Braving time.　　　235

Tell her that goes
With song upon her lips
But sings not out the song, nor knows
The maker of it, some other mouth,
May be as fair as hers,　　　240
Might, in new ages, gain her worshippers,
When our two dusts with Waller's shall be laid,
Siftings on siftings in oblivion,
Till change hath broken down
All things save Beauty alone.　　　245

Mauberley
1920

"*Vacuos exercet aera morsus.*"[9]

I

Turned from the "eau-forte
Par Jaquemart"[1]
To the strait head
Of Messalina:[2]

"His true Penelope　　　250
Was Flaubert,"[3]
And his tool
The engraver's.

Firmness,
Not the full smile,　　　255
His art, but an art
In profile;

---

9. "He snaps vacuously at the empty air," the vain effort of a dog in Ovid's *Metamorphoses* (VII.786) to bite an elusive monster attacking Thebes; both dog and monster are later turned to stone.

1. "Etching by Jacquemart," Jules Jacquemart, French graphic artist (1837–80)
2. Dissolute wife of the Roman Emperor Claudius (c. A.D. 8).
3. Quoted from line 13.

Colourless
Pier Francesca,[4]
Pisanello[5] lacking the skill   260
To forge Achaia.[6]

II

"Qu'est ce qu'ils savent de l'amour, et qu'est ce qu'ils peuvent en com-
prendre?
  S'ils ne comprennent pas la poésie, s'ils ne sentent pas la musique,
qu'est ce qu'ils peuvent comprendre de cette passion en comparaison
avec laquelle la rose est grossiè et le parfum des violettes un
tonnere?"
                                                    —CAID ALI[7]

For three years, diabolus[8] in the scale,
He drank ambrosia,
All passes, ANANGKE[9] prevails,
Came end, at last, to that Arcadia.[1]   265

He had moved amid her phantasmagoria,
Amid her galaxies,
NUKTIS 'AGALMA[2]
  ·   ·   ·   ·   ·   ·   ·
Drifted . . . drifted precipitate,
Asking time to be rid of . . .   270
Of his bewilderment; to designate
His new found orchid. . . .

To be certain . . . certain . . .
(Amid aerial flowers) . . . time for arrangements—
Drifted on   275
To the final estrangement;

Unable in the supervening blankness
To sift TO AGATHON[3] from the chaff
Until he found his sieve . . .
Ultimately, his seismograph: 280

—Given that is his "fundamental passion,"
This urge to convey the relation
Of eye-lid and cheek-bone
By verbal manifestations;

To present the series   285
Of curious heads in medallion—

---

4. Piero della Francesca, Umbrian painter (1420?–92).
5. Vittore Pisano (1397?–1455?), Veronese painter and medalist who made medallions based on Greek coins.
6. Southern Greece.
7. "What do they know of love, and what can they understand? If they do not understand poetry, if they do not respond to music, what can they understand of this passion in comparison to which the rose is gross and the perfume of violets a clap of thunder?" Caid Ali is a pseudonym of Pound.
8. The devil and, in music, the interval of the augmented fourth.
9. Necessity.
1. Greek mountain region thought of as an ideal rustic paradise.
2. "Night's jewel," a phrase from a celebration of the evening star in a pastoral by the Greek poet Bion (c. 100 B.C.).
3. The good.

He had passed, inconscient, full gaze,
The wide-banded irides[4]
And botticellian sprays[5] implied
In their diastasis;[6]   290

Which anaethesis, noted a year late,
And weighed, revealed his great affect,
(Orchid), mandate
Of Eros,[7] a retrospect.

Mouths biting empty air,   295
The still stone dogs,
Caught in metamorphosis, were
Left him as epilogues.

"The Age Demanded"

VIDE[8] POEM II. PAGE 1251

For this agility chance found
Him of all men, unfit   300
As the red-beaked steeds of
The Cytheraean for a chain bit.[9]

The glow of porcelain
Brought no reforming sense
To his perception   305
Of the social inconsequence.

Thus, if her colour
Came against his gaze,
Tempered as if
It were through a perfect glaze   310

He made no immediate application
Of this to relation of the state
To the individual, the month was more temperate
Because this beauty had been.

            The coral isle, the lion-coloured sand   315
            Burst in upon the porcelain revery:
            Impetuous troubling
            Of his imagery.

Mildness, amid the neo-Nietzschean[1] clatter,
His sense of graduations,   320

4. Plural of *iris*: flowers and eyes.
5. Allusion to the Italian painter Sandro Botticelli (1444?–1510) and women carrying flowers in his painting *Primavera* ("Spring").
6. Separation.
7. Greek god of love, son of Aphrodite (Venus).
8. "Vide": see.
9. Doves harnessed to the chariot of Aphrodite, who first landed on the island of Cythera off the Laconian coast.
1. Ideas derived from Friedrich Wilhelm Nietzsche (1844–1900), German philosopher. His paradoxical ethical and aesthetic theories, widely discussed at the time, were a bold challenge to conventional ideas in England, Europe, and America.

Quite out of place amid
Resistance to current exacerbations,

Invitation, mere invitation to perceptivity
Gradually led him to the isolation
Which these presents place   325
Under a more tolerant, perhaps, examination.

By constant elimination
The manifest universe
Yielded an armour
Against utter consternation,   330

A Minoan undulation,[2]
Seen, we admit, amid ambrosial circumstances
Strengthened him against
The discouraging doctrine of chances,

And his desire for survival,   335
Faint in the most strenuous moods,
Became an Olympian *apathein*[3]
In the presence of selected perceptions.

A pale gold, in the aforesaid pattern,
The unexpected palms   340
Destroying, certainly, the artist's urge,
Left him delighted with the imaginary
Audition of the phantasmal sea-surge,

Incapable of the least utterance or composition,
Emendation, conservation of the "better tradition,"   345
Refinement of medium, elimination of superfluities,
August attraction or concentration.

Nothing, in brief, but maudlin confession,
Irresponse to human aggression,
Amid the precipitation, down-float   350
Of insubstantial manna,
Lifting the faint susurrus
Of his subjective hosannah.

Ultimate affronts to
Human redundancies;   355

Non-esteem of self-styled "his betters"
Leading, as he well knew,
To his final
Exclusion from the world of letters.

2. Stylistic feature of portrait sculptures by the Cretan Scopas during the Minoan period (2000–1500 B.C.), according to the French archaeologist Salomon Reinach (1858–1932) in his *Apollo* (1904).
3. The absence of feeling or indifference of the Olympian gods to human affairs.

IV

Scattered Moluccas[4]   360
Not knowing, day to day,
The first day's end, in the next noon;
The placid water
Unbroken by the Simoon;[5]

Thick foliage   365
Placid beneath warm suns,
Tawn fore-shores
Washed in the cobalt of oblivions;

Or through dawn-mist
The grey and rose   370
Of the juridical
Flamingoes;

A consciousness disjunct,
Being but this overblotted
Series   375
Of intermittences;

Coracle[6] of Pacific voyages,
The unforecasted beach;
Then on an oar
Read this: 380

"I was
And I no more exist;
Here drifted
An hedonist."

Medallion

Luini in porcelain![7]   385
The grand piano
Utters a profane
Protest with her clear soprano.

The sleek head emerges
From the gold-yellow frock   390
As Anadyomene[8] in the opening
Pages of Reinach.

Honey-red, closing the face-oval,
A basket-work of braids which seem as if they were
Spun in King Minos'[9] hall   395
From metal, or intractable amber;

4. Spice Islands in the Malay Archipelago.
5. Violent wind and sand storm of Near
Eastern deserts.
6. Small hide-covered wicker-frame boat
used by ancient Britons.

7. Bernardino Luini (1475?–1532?), Italian painter.
8. Epithet of Aphrodite meaning foam-
born. Reinach's *Apollo* included a reproduction of a head of Aphrodite.
9. Legendary King of Crete.

The face-oval beneath the glaze,
Bright in its suave bounding-line, as,
Beneath half-watt rays,
The eyes turn topaz.   400

                                                 1920

From THE CANTOS
XVII[1]

So that the vines burst from my fingers
And the bees weighted with pollen
More heavily in the vine-shoots:
chirr—chirr—chir-rikk—a purring sound,
And the birds sleepily in the branches.   5
          ZAGREUS! IO ZAGREUS.[2]
With the first pale-clear of the heaven
And the cities set in their hills,
And the goddess of the fair knees[3]
Moving there, with the oak-woods behind her,   10
The green slope, with white hounds
          leaping about her;
And thence down to the creek's mouth, until evening,
Flat water before me,
          and the trees growing in water,   15
Marble trunks out of stillness,[4]
On past the palazzi,
          in the stillness,
The light now, not of the sun.
          Chrysophrase,[5]   20

1. Canto 17 represents Pound's treatment of ancient myth and Renaissance history. Three sequences, as in a motion-picture montage, displace each other and interpenetrate in this canto: Ulysses' (Odysseus's) voyage through the Mediterranean in search of home in Ithaca, Jason's voyage to the island of Colchis in search of the Golden Fleece, and a ship's entrance into Venice. The poem suggests the imminence of dangers and corruption as the scene of beautiful Mediterranean islands is translated into the lush and marble setting of mercantile Venice, but these premonitions are held in abeyance with the enchanting vitality and "Splendour" that dominate the verse. The poet evokes moments of tense excitement in ancient myth which immediately precede acts of violence. The epic quests become an experience of seing, a process of imaginative and sensory vision; the scenes emerge in their radiance without becoming static or fixed, and the poet's quest continues without settling into the finality of conclusion. Among the personae that Pound assumes, besides Ulysses and Jason, are Dionysus (Zagreus) the Greek god of wine, rebirth, and ecstasy; Actaeon, who enjoyed a glimpse of the goddess Diana bathing and was punished by being turned into a stag and torn apart by his hounds; and Hades (Pluto), god of the underworld, who loved Kore (Persephone, goddess of regeneration) and who kidnaped her from a meadow to make her queen of the lower regions but was later forced to allow her to return to earth in the spring for nine months each year.

2. "Zagreus. I am Zagreus!" A sacrificial god often identified in Greek myth with Dionysus (meaning "twice-born"), Zagreus was the offspring of Persephone (Koré) and Zeus; Zeus seduced her before she was kidnaped by Hades.

3. Diana, chaste goddess of the hunt.

4. The first glimpse of Venice. The marble façades and columns of its "palazzi" (palaces) are conceived as emerging from nature into art in accordance with the theories of the landscape painter and art critic Adrian Stokes (1854–1935), a friend of Pound in the late 1920s. Stokes (*The Stones of Rimini*, 1934, p. 19) stressed the affinity of Venetian arts with the salt sea, and the origin of marble from sunken forests and watery limestone: "Amid the sea Venice is built from the essence of the sea."

5. Apple-green precious stone.

And the water green clear, and blue clear;
On, to the great cliffs of amber.
                                    Between them,
Cave of Nerea,[6]
        she like a great shell curved,
And the boat drawn without sound,   25
Without odour of ship-work,
Nor bird-cry, nor any noise of wave moving,
Nor splash of porpoise, nor any noise of wave moving,
Within her cave, Nerea,   30
                        she like a great shell curved
In the suavity of the rock,
                        cliff green-gray in the far,
In the near, the gate-cliffs of amber,
And the wave   35
                        green clear, and blue clear,
And the cave salt-white, and glare-purple,
            cool, porphyry smooth,
            the rock sea-worn.
No gull-cry, no sound of porpoise,   40
Sand as of malachite,[7] and no cold there,
            the light not of the sun.

Zagreus, feeding his panthers,
            the turf clear as on hills under light.
And under the almond-trees, gods,   45
            with them, *choros nympharum*.[8] Gods,
Hermes and Athene,[9]
                        As shaft of compass,
Between them, trembled—
To the left is the place of fauns,   50
            sylva nympharum;[1]
The low wood, moor-scrub,
            the doe, the young spotted deer,
            leap up through the broom-plants,
                        as dry leaf amid yellow.   55
And by one cut of the hills,
            the great alley of Memnons.[2]
Beyond, sea, crests seen over dune
Night sea churning shingle,
To the left, the alley of cypress.   60
                        A boat came,
    One man holding her sail,

6. Possibly the nymph Calypso, the temptress and death goddess who detained Ulysses for seven years in her island cave. She was the daughter of Atlas.
7. Green mineral.
8. Chorus of nymphs.
9. Athene (goddess of wisdom) and Hermes (messenger of the gods, patron of merchants and thieves) were Ulysses' protectors. Hermes freed Ulysses from the bonds of Calypso, and Athene calmed the waves for his final voyage home to Ithaca (Odyssey, Book 5).
1. Wood of the nymphs.
2. Commander of Ethiopian troops in the defense of Troy, Memnon was called "son of dawn." His statue near Thebes, here likened to a row of cypress trees, reputedly issued a sound when dawn's light struck it.

Guiding her with oar caught over gunwale, saying:
"   There, in the forest of marble,
"   the stone trees—out of water—   65
"   the arbours of stone—
"   marble leaf, over leaf,
"   silver, steel over steel,
"   silver beaks rising and crossing,
"   prow set against prow,   70
"   stone, ply over ply,
"   the gilt beams flare of an evening"
Borso, Carmagnola,[3] the men of craft, *i vitrei*,[4]
Thither, at one time, time after time,
And the waters richer than glass,   75
Bronze gold, the blaze over the silver,
Dye-pots in the torch-light,
The flash of wave under prows,
And the silver beaks rising and crossing.
            Stone trees, white and rose-white in the darkness,   80
Cypress there by the towers,
            Drift under hulls in the night.

            "In the gloom the gold
Gathers the light about it." . . .[5]

Now supine in burrow, half over-arched bramble,   85
One eye for the sea, through that peek-hole,
Gray light, with Athene.
Zothar[6] and her elephants, the gold loin-cloth,
The sistrum,[7] shaken, shaken,
            the cohorts of her dancers.   90
And Aletha, by bend of the shore,
            with her eyes seaward,
            and in her hands sea-wrack
Salt-bright with the foam.
Koré[8] through the bright meadow,   95
            with green-gray dust in the grass:
"For this hour, brother of Circe."[9]
Arm laid over my shoulder,
Saw the sun for three days, the sun fulvid,
As a lion lift over sand-plain;   100

3. Borse d'Este (1431–71) of Ferrara, patron of learning and unsuccessful peacemaker whose assassination was attempted in Venice; and Francesco Bussone da Carmagnola (1390?–1432), mercenary soldier, tried for treason in Venice and executed between two columns.
4. Glassmakers, craftsmen for whom Venice is famous.
5. "Quoted from an earlier Canto (#11) where 'about' reads 'against.' It is derived from a distich by Pindar" [Hugh Kenner's note].
6. Zothar and Aletha (below) are proba-

bly invented names.
7. An Egyptian metal rattle.
8. Persephone, goddess of regeneration and bride of Hades, ruler of the underworld of the dead.
9. "Circe": the enchantress who detained Ulysses for a year, then instructed him to consult Tiresias in the underworld to learn his route home, and told him how to enter the world of the dead through the Grove of Persephone. Her brother, Aetes, King of Colchis, maintained a cult of the sun (his father) on his island, and held possession of the Golden Fleece sought by Jason.

and that day,
And for three days, and none after,
Splendour, as the splendour of Hermes,
And shipped thence
           to the stone place,   105
Pale white, over water,
        known water,
And the white forest of marble, bent bough over bough,
The pleached arbour of stone,
Thither Borso, when they shot the barbed arrow at him,   110
And Carmagnola, between the two columns,
Sigismundo, after that wreck in Dalmatia.[1]
    Sunset like the grasshopper flying.

                                       1933

1. Sigismondo Malatesta (1417–68), Renaissance ruler of Rimini whom Pound admired; an art patron, antipapist, and builder of the Tempio Malatestiana in Rimini, he fought for Venice and other cities and in 1464 reached the Dalmatian coast in an unsuccessful crusade.

EDWIN ARLINGTON ROBINSON
1869–1935

The most important poet to launch his career between the generation of Whitman and Dickinson and the ascendency of Frost, Pound, and William Carlos Williams was a New Englander, Edwin Arlington Robinson. His lyrics gave a somber orientation to Emersonian Transcendentalism while probing the isolation of the psyche and the theme of tragic failure with an intensiveness comparable to Hawthorne's and Melville's. Compared to his major contemporaries, he had scarcely any influence on other writers, but he shared in the modern effort to appropriate for poetry the effects of good prose.

Robinson was raised in Gardiner, Maine, where cultivated friends undertook to encourage his vocation as a poet even though its cultural resources were limited and his family's misfortunes ruled out sustained support from his parents. His father's lumber business and land speculations were on the brink of failure even before his death in 1892 and the collapse of the family's holdings in 1893. His brothers' promise of success dissolved before his eyes (one, a physician, became addicted to drugs; the other, a businessman, to liquor), and after four exciting years and one postgraduate year in high school he drifted near despair as he tried to find a way to pursue his vocation as poet. The family managed to send him for two years to Harvard (1891–93) where Professor Charles Eliot Norton, among others, helped to widen his intellectual horizons. But he met with little success when he returned to Gardiner and began writing the miniature dramas and portraits (including *The House on the Hill* and *Richard Cory*) that were to identify "Tilbury Town" in his first published volumes. One source of encouragement was reading and conversation in the home of Mrs. Laura Richards, while another was instruction from the local poet Dr. Alanson T. Schumann, who set him to writing in the intricate traditional French forms popular among Pre-Raphaelite poets in England. It was in the 1890s that he discovered the meditative poetry of Wordsworth and the realistic verse of George Crabbe, along with the more recent verse of Whitman, Rudyard Kipling, and Thomas Hardy, the fiction of Hawthorne and James, the essays of Emerson. With his lifelong friend Harry de Forest Smith he read the classics and translated Sophocles. He published his first two volumes of verse at his own expense before moving from Gardiner to New York City; only when a group of Gardiner friends guaranteed the venture did he find a publisher for *Captain Craig* in 1902.

Robinson's first volume was brought to the attention of President Theodore Roosevelt, who secured him a position as customs inspector in New York from 1905 to 1909, but the poet's position remained desperate even after *The Town down the River* (1910) and *The Man against the Sky* (1916) had attracted a larger audience of readers and critics. Heavy drinking had long become a habit. He wrote fiction (which he destroyed) and plays (two of which were published) in an attempt to earn money and gain popularity. He lived in lonely rooms or in the studios and homes of friends, and was grateful for a small bequest in 1914 from a Gardiner friend, Hays Gardiner, and for the income from a trust established in 1917 by anonymous donors for his support. That support continued until 1922, when a measure of success followed the publication of *Avon's Harvest* and *Collected Poems* and the award of the first of his three Pulitzer Prizes. Robinson spent the rest of his life living in New York during the winter but did most of his writing in New England during the summers, which he spent after 1911 in New Hampshire at the MacDowell Colony, founded by Marian MacDowell, widow of the composer Edward MacDowell, for the support of American musicians, artists, and writers.

The deprivations of Robinson's personal experience, his reading in the Bible and in Eastern mysticism, and the Transcendental idealism of Emerson and idealistic philosophers whom he had studied at Harvard combined to form a secular religion which Robinson once called an "optimistic desperation."[1] This attitude underlies his longer narrative poems as well as his shorter poems; the grim residuum of affirmation is often figured in the images of a "Word" and "light." His poems, including his many brilliant sonnets, are characterized by austere diction, by somber wit in the tradition of Yankee humor, and by irony that is seldom subtle but is unrelieved, sustained, and consequently powerful. In his blank-verse narratives and notably in his short lyrics, traditional but simple verse forms are played unobtrusively against the momentum of long sentences and prolonged rhetorical questions which give his best poems the peculiar force of prose. Indeed, the late poet Conrad Aiken discerned, in Robinson's treatment of his painfully isolated characters, a novelistic focus on "relations and contacts" that are "always extraordinarily *conscious*" and thus comparable to the prose fiction of Henry James.[2]

After 1921 Robinson devoted his talents to long narrative poems, including a trilogy based on the Arthurian romances that he began in 1917 with *Merlin* and completed in 1927 with *Tristram*, a best seller which brought him his third Pulitzer Prize. These poems have not been as favorably received by critics as the earlier poems, even by such admirers of Robinson as the post-World War II poet Louis O. Coxe. *King Jasper*, which recounts

1. Letter to Harry de Forest Smith in Charles T. Davis, ed., *Edwin Arlington Robinson: Selected Early Poems and*   *Letters*, p. 210.
2. *Collected Criticism of Conrad Aiken* (1958), p. 342.

the nemesis of a ruthless industrialist in a world he has repudiated, was published posthumously in 1935 and was prefaced by a tribute from his fellow New England poet Robert Frost.

George Crabbe[1]

Give him the darkest inch your shelf allows,
Hide him in lonely garrets, if you will,—
But his hard, human pulse is throbbing still
With the sure strength that fearless truth endows.
In spite of all fine science disavows,   5
Of his plain excellence and stubborn skill
There yet remains what fashion cannot kill,
Though years have thinned the laurel[2] from his brows.

Whether or not we read him, we can feel
From time to time the vigor of his name   10
Against us like a finger for the shame
And emptiness of what our souls reveal
In books that are as altars where we kneel
To consecrate the flicker, not the flame.

1896

Richard Cory

Whenever Richard Cory went down town,
We people on the pavement looked at him:
He was a gentleman from sole to crown,
Clean favored, and imperially slim.

And he was always quietly arrayed,   5
And he was always human when he talked;
But still he fluttered pulses when he said,
"Good—morning," and he glittered when he walked.

And he was rich—yes, richer than a king—
And admirably schooled in every grace: 10
In fine, we thought that he was everything
To make us wish that we were in his place.

So on we worked, and waited for the light,
And went without the meat, and cursed the bread;
And Richard Cory, one calm summer night,   15
Went home and put a bullet through his head.

1896

Credo[1]

I cannot find my way: there is no star
In all the shrouded heavens anywhere;
And there is not a whisper in the air

---

1. **English** physician, curate, and poet (1754–1832), known for his realistic narrative poems.
2. Evergreen shrub from which the ancient Romans wove wreaths for poetry

prizes. Hence an emblem of poetic merit and public recognition.
1. A statement of belief (from the Latin, meaning "I believe").

Of any living voice but one so far
That I can hear it only as a bar   5
Of lost, imperial music, played when fair
And angel fingers wove, and unaware,
Dead leaves to garlands where no roses are.

No, there is not a glimmer, nor a call,
For one that welcomes, welcomes when he fears,   10
The black and awful chaos of the night;
For through it all—above, beyond it all—
I know the far-sent message of the years,
I feel the coming glory of the Light.

1896

Miniver Cheevy

Miniver Cheevy, child of scorn,
    Grew lean while he assailed the seasons;
He wept that he was ever born,
    And he had reasons.

Miniver loved the days of old   5
    When swords were bright and steeds were prancing;
The vision of a warrior bold
    Would set him dancing.

Miniver sighed for what was not,
    And dreamed, and rested from his labors;   10
He dreamed of Thebes and Camelot,
    And Priam's neighbors.[1]

Miniver mourned the ripe renown
    That made so many a name so fragrant;
He mourned Romance, now on the town,   15
    And Art, a vagrant.

Miniver loved the Medici,[2]
    Albeit he had never seen one;
He would have sinned incessantly
    Could he have been one.   20

Miniver cursed the commonplace
    And eyed a khaki suit with loathing;
He missed the mediaeval grace
    Of iron clothing.

Miniver scorned the gold he sought,   25
    But sore annoyed was he without it;

---

1. Thebes was an ancient city in Boeotia, rival of Athens and Sparta for supremacy in Greece and the setting of Sophocles' tragedies about Oedipus; Camelot is the legendary court of King Arthur and the Knights of the Round Table; the neighbors of King Priam in Homer's *Iliad* are his heroic compatriots in the doomed city of Troy.
2. Family of wealthy merchants, statesmen, and art patrons in Renaissance Florence.

Miniver thought, and thought, and thought,
  And thought about it.

Miniver Cheevy, born too late,
  Scratched his head and kept on thinking;   30
Miniver coughed, and called it fate,
  And kept on drinking.

                                                              1910

Eros Turannos[1]

She fears him, and will always ask
  What fated her to choose him;
She meets in his engaging mask
  All reasons to refuse him;
But what she meets and what she fears   5
Are less than are the downward years,
Drawn slowly to the foamless weirs
  Of age, were she to lose him.

Between a blurred sagacity
  That once had power to sound him,   10
And Love, that will not let him be
  The Judas[2] that she found him,
Her pride assuages her almost,
As if it were alone the cost.—
He sees that he will not be lost,   15
  And waits and looks around him.

A sense of ocean and old trees
  Envelops and allures him;
Tradition, touching all he sees,
  Beguiles and reassures him;   20
And all her doubts of what he says
Are dimmed with what she knows of days—
Till even prejudice delays
  And fades, and she secures him.

The falling leaf inaugurates   25
  The reign of her confusion;
The pounding wave reverberates
  The dirge of her illusion;
And home, where passion lived and died,
Becomes a place where she can hide,   30
While all the town and harbor side
  Vibrate with her seclusion.

We tell you, tapping on our brows,
  The story as it should be,—
As if the story of a house   35

---

1. Greek for "Love, the King," an echo   *Oidipous Turranos* (*Oedipus the King*).
of the solemn title to Sophocles' tragedy   2. The disciple who betrayed Christ.

Were told, or ever could be;
We'll have no kindly veil between
Her visions and those we have seen,—
As if we guessed what hers have been,
    Or what they are or would be.   40

Meanwhile we do no harm; for they
    That with a god have striven,
Not hearing much of what we say,
    Take what the god has given;
Though like waves breaking it may be,   45
Or like a changed familiar tree,
Or like a stairway to the sea
    Where down the blind are driven.

        1913, 1916

The Mill

The miller's wife had waited long,
    The tea was cold, the fire was dead;
And there might yet be nothing wrong
    In how he went and what he said:
"There are no millers any more,"   5
    Was all that she had heard him say;
And he had lingered at the door
    So long that it seemed yesterday.

Sick with fear that had no form
    She knew that she was there at last;   10
And in the mill there was a warm
    And mealy fragrance of the past.
What else there was would only seem
    To say again what he had meant;
And what was hanging from a beam   15
    Would not have heeded where she went.

And if she thought it followed her,
    She may have reasoned in the dark
That one way of the few there were
    Would hide her and would leave no mark: 20
Black water, smooth above the weir
    Like starry velvet in the night,
Though ruffled once, would soon appear
    The same as ever to the sight.

        1920

Mr. Flood's Party

Old Eben Flood, climbing alone one night
Over the hill between the town below
And the forsaken upland hermitage
That held as much as he should ever know

On earth again of home, paused warily.   5
The road was his with not a native near;
And Eben, having leisure, said aloud,
For no man else in Tilbury Town[1] to hear:

"Well, Mr. Flood, we have the harvest moon
Again, and we may not have many more;   10
The bird is on the wing, the poet says,[2]
And you and I have said it here before.
Drink to the bird." He raised up to the light
The jug that he had gone so far to fill,
And answered huskily: "Well, Mr. Flood,   15
Since you propose it, I believe I will."

Alone, as if enduring to the end
A valiant armor of scarred hopes outworn,
He stood there in the middle of the road
Like Roland's ghost winding a silent horn.[3]   20
Below him, in the town among the trees,
Where friends of other days had honored him,
A phantom salutation of the dead
Rang thinly till old Eben's eyes were dim.

Then, as a mother lays her sleeping child   25
Down tenderly, fearing it may awake,
He set the jug down slowly at his feet
With trembling care, knowing that most things break;
And only when assured that on firm earth
It stood, as the uncertain lives of men   30
Assuredly did not, he paced away,
And with his hand extended paused again:

"Well, Mr. Flood, we have not met like this
In a long time; and many a change has come
To both of us, I fear, since last it was   35
We had a drop together. Welcome home!"
Convivially returning with himself,
Again he raised the jug up to the light;
And with an acquiescent quaver said:
"Well, Mr. Flood, if you insist, I might.   40

"Only a very little, Mr. Flood—
For auld lang syne.[4] No more, sir; that will do."

1. The fictive town named in a number of Robinson's poems, modeled on Gardiner, Maine.
2. Flood paraphrases lines 25–28 of the Persian poem *The Rubáiyát of Omar Khayyám* in the translation by the English poet Edward FitzGerald (1809–83): "Come, fill the Cup, and in the fire of Spring/Your Winter-garment of Repentance fling:/The Bird of Time has but a little way/To flutter and the Bird is on the Wing."

3. King Charlemagne's nephew, celebrated in the medieval *Chanson de Roland (Song of Roland*, c. 1000). Just before dying in the futile battle of Roncevalles (A.D. 778), he sounded his horn for help.
4. Scottish meaning "for the days of long ago," the title and refrain of a song of comradeship and parting immortalized by the Scottish poet Robert Burns (1759–96).

So, for the time, apparently it did,
And Eben evidently thought so too;
For soon amid the silver loneliness   45
Of night he lifted up his voice and sang,
Secure, with only two moons listening,
Until the whole harmonious landscape rang—

"For auld lang syne." The weary throat gave out,
The last word wavered, and the song was done.   50
He raised again the jug regretfully
And shook his head, and was again alone.
There was not much that was ahead of him,
And there was nothing in the town below—
Where strangers would have shut the many doors   55
That many friends had opened long ago.

                                                 1921

The Long Race

Up the old hill to the old house again
Where fifty years ago the friend was young
Who should be waiting somewhere there among
Old things that least remembered most remain,
He toiled on with a pleasure that was pain   5
To think how soon asunder would be flung
The curtain half a century had hung
Between the two ambitions they had slain.

They dredged an hour for words, and then were done.
"Good-bye! . . . You have the same old weather-vane—   10
Your little horse that's always on the run."
And all the way down back to the next train,
Down the old hill to the old road again,
It seemed as if the little horse had won.

                                                 1921

ROBERT FROST
1874–1963

In 1915, when the expatriate Ezra Pound began writing his *Cantos* in London and T. S. Eliot took up residence there, the poet Robert Frost, who had sojourned in England with his family for three years, returned to the United States. Discouraged by his failure to get his poems published in this country, he had sold his New Hampshire farm and moved to England; within a month after arriving there, his first manuscript was accepted by a London publisher. Ezra Pound recommended his poems to American editors, and Frost's second volume, *North of Boston*, had received critical praise from Pound and other reviewers by the time Frost returned to America. Within the next two years an American publisher brought out his first two volumes and added a third, and Frost was chosen Phi Beta Kappa poet by two American colleges and elected to the National Institute of Arts and Letters. He was well on his way to becoming the most publicly honored and popular of America's first-rate poetic talents. For the next four decades he devoted his efforts to reaching a large audience that would support his writing, and to fashioning, consciously and deliberately, from literary tradition and from the regional materials of New England life and speech, the poetic idiom that is unmistakably his own.

Robert Lee Frost was born in 1874 in California, where his father, a disgruntled Yankee with sympathies for the Confederacy during the Civil War, went to pursue a career in journalism and nourished political ambitions. His father's occasional violent behavior toward his family scarred Frost's memories and heightened his apprehensions about vindictive and self-destructive impulses that became apparent in his own temperament. Frost's Scottish mother, who wrote poetry herself, introduced Frost to the writings of her fellow Scots, and to the poems of Wordsworth, Bryant, and Emerson; she introduced him also to the tradition of Christian piety that led in her own case from Presbyterianism to Unitarianism and finally to the New Jerusalem church of Emanuel Swedenborg (whose doctrines of symbolic correspondence had earlier fascinated Emerson). When Frost's father died in 1885 the family returned to New England, where Mrs. Frost taught in Massachusetts and New Hampshire schools, and Frost attended high school in Lawrence, Massachusetts, specializing in the classics. At his graduation in 1892 he was class poet and shared the post of valedictorian with Elinor White, whom he married three years later. Frost studied for a term at Dartmouth College, then taught intermittently in elementary schools and held jobs in mills and on local newspapers until enrolling as a special student at Harvard from 1897 to 1899, lured by the writings of the pragmatist William James. He took English courses but emphasized the Greek and Latin classics and philosophy, studying James's *Psychology* in one course and attending the philosopher George Santayana's lectures in another. After leaving Harvard he moved to a farm in Derry, New Hampshire, which his grandfather bought for him. These were years of hardship for the Frosts and their four children, and of personal anxieties for Frost himself that culminated in obsessions with suicide. But Frost also became intimate for the first time with the rural environment and with nature, and he wrote poetry, publishing some in such local papers as the Derry *Enterprise*. Particularly in the years 1906–7 he worked on some of the major poems that appeared later in *North of Boston* and *Mountain Interval* (1916). Frost left the farm in 1909 and taught in New Hampshire until he left for England in 1912.

When Frost chose to live in New England on his return in 1915 (he bought another farm in New Hampshire and began in 1917 the first of three teaching appointments at Amherst College), he committed himself consciously not only to a locale but to a tradition and to a poetic instrument that he was perfecting. Long before H. L. Mencken's *The American Language* had enforced Frost's "sense of national difference" between English and American idioms, the English poet Rudyard Kipling had warned that Frost seemed strange to readers from the "elder earth" of England because of his "alien speech."[1] Yet the differences in language habits, Frost insisted, enable members of one culture to entertain the speech of another

---

1. *Selected Prose of Robert Frost*, p. 75.

with relish, recognizing in it the "freshness of a stranger." Moreover, he pointed out that such strangeness or "estrangement" that derives from the peculiarities of national or regional speech is essentially the same as the strangeness of images and other devices, the intriguing "word-shift by metaphor," that characterizes all poetry.[2] The settings and characteristic idiom of Frost's poems have had precisely that appeal (of refreshing strangeness and suggestiveness) for Americans living outside as well as inside New England, and his regionalism constitutes not a literalistic or photographic strategy for his poetry, but a metaphorical one. His settings, characters, and language are engagingly familiar but also intriguingly strange metaphors for experience and realities that are recognizable elsewhere.

The tradition which sanctioned such a strategy had many remote sources, including Wordsworth, the ancient classics, and the Bible, but the tradition in America led from such near contemporaries as Edwin Arlington Robinson and William James back to Thoreau and Emerson. Frost was as conscious of his predecessors as were Pound and Eliot or Crane and William Carlos Williams of theirs, though unlike them Frost was not troubled by discontinuities in his tradition and felt no impulse to reconstitute radically its foundations. He shaped that tradition cautiously to his own uses. He admired Thoreau's cranky independence (and his suspicion of professional reformers) rather than his belligerence, praising Thoreau's "declaration of independence" from the feverish "modern pace" and particularly the tough fiber of the prose with which Thoreau fronted life directly and articulated his convictions in *Walden*.[3] Chiefly he found his precedents in Emerson's combination of allusive meaning with simple diction and simple forms, his reliance, as Emerson said in *Monadnock*, on the speech actually spoken by the "Rude poets of the tavern hearth" (*On Emerson*, 1959).[4] It was by fashioning the cadences and phrasing of such colloquial New England speech into a poetic idiom that Frost appropriated to his own use the familiar conventions of nature poetry and the conventions of classical pastoral poetry.

In enunciating his principles Frost cautioned against the search always for "new ways to be new" in extreme experimentation, and he insisted that instead "We play the words as we find them" (*The Constant Symbol*, 1946).[5] His touchstones for poetry outside of literature were science and philosophy rather than the music or the modern arts that interested his bolder contemporary writers like Pound and Hart Crane; his derivations were more narrowly literary than theirs. Poets should accept and perfect the habitual meters of iambic verse (units of one accented and one unaccented syllables) since "iambic and loose iambic" meters are the only ones natural to English speech (*The Figure a Poem Makes*, 1939).[6] "All folk speech" is itself "musical," he insisted in 1915, and this resource could be converted into a poetic idiom by developing "the sound of sense."[7] By this he meant that discursive and intelligible meaning affected the total experience of a poem and that the poet should develop interconnections of sound and feeling with "sense" or meaning. Discursive and even didactic statement need not be eliminated but should be played against the sounds and tonal qualities of the idiom. Poetic effects are achieved when the cadences of actual speech and the rhythms of sequential thought are played off "against the

2. *Ibid.*, p. 77.
3. *Interviews with Robert Frost*, p. 146.
4. *Selected Prose*, p. 113.

5. *Ibid.*, p. 28.
6. *Ibid.*, pp. 17–18.
7. *Interviews*, p. 6.

rigidity of a limited meter," and when recognizable subjects and themes are treated so as to "steady us down."[8]

Such a fusion of meter with subject and theme, of sound with sense, should also be dramatic, Frost claimed, for Emerson's prose had shown that "writing is unboring to the extent that it is dramatic." Frost insisted that the dramatic effect must penetrate even the sentences of a poem by rendering "the speaking tone of voice somehow entangled in the words and fastened to the page for the ear of the imagination."[9]

A poem so conceived becomes a symbol or "metaphor," which Frost defined as a way of "saying one thing in terms of another," an imaginative activity that yields the "pleasure of ulteriority" and grounds poetry on indirection, tangential or allusive implication, the "guardedness" of wry or tragic irony, and wily cunning.[1] It also founded poetry on disciplined, cautious restraint, for Frost conceived metaphor as something that "tamed" the "enthusiasm" it captured (*Education by Poetry*, 1931).[2] The poem becomes a careful, loving exploration of reality which is at the same time a defensive shield against its tragedies and perils, a "momentary stay against confusion."[3] In its confrontation with reality, which Frost, like William James, regarded as a hazard or trial, the poem is a "figure of the will braving alien entanglements"; it is an expression, however desperate or tentative, of what James called the will to believe. Frost once catalogued the important affirmations of belief from which poetry springs as "self-belief," "love-belief," "literary belief" (the belief in art), "God belief" (a "relation you enter into with Him to bring about the future"), and "national belief."[4]

Governed by these principles, Frost's idiom enabled him to create works of unmistakable beauty and power. The pointed wit of his animal fables, as well as the intensity of his best lyrics and the impact of his dramatic narratives, are enhanced by the suppleness with which he used a wide range of traditional conventions—rhymed couplets, rhyming quatrains, blank verse, and, brilliantly, the sonnet form. Yet the diction and cadences of colloquial speech are so deftly incorporated, the illusion of gestures and scene is so convincingly represented, the facts of nature and everyday occupations in rural New England are so tangibly recreated in the verse, that his poems are a much closer approximation of actuality than those of Wordsworth and Emerson or those of poets with comparable intentions, like Edward Thomas, whom Frost found congenial during his stay in England. At the same time the indirection or "ulteriority" of metaphor is generated by the "whispering" motion of a scythe in *Mowing*, by such images as the "tongue of bloom" in *The Tuft of Flowers*, or by such religious terms as "fall" in *After Apple-Picking* and "annunciation" in *West-Running Brook*. These poems metaphorically suggest dimensions of relevance beyond what is made explicit; they invite the reader to explore unspoken connections to the whispering of Frost's own poem, to the search for communion with God or nature, to the assessment of an apple picker's moral worth before death, or to the revelation of life's meaning in a New England brook.

Though Frost's allusiveness, like his resort to a rural setting removed from the urban world, ran the risk of evading pressing realities of his time, his best poems are somber, even terrifying, in rendering many anxieties that grip the modern imagination: apprehensions about the dissolution of marriages and families, the conformist standardization of urban civilization, the

8. *Selected Prose*, p. 18.
9. *Ibid.*, pp. 114, 13–14.
1. *Selected Letters*, p. 299.

2. *Selected Prose*, p. 36.
3. *Ibid.*, p. 18.
4. *Ibid.*, pp. 25, 45.

bleak isolation of an aging man, the imminence of death and extinction, the desperation with which abject men or women must "provide, provide" in order to survive at all, the horror of a world in which nature's "design" is predatory if indeed any design governs nature at all.

His major poems (all written before 1930, and many of them appearing in his masterpiece *North of Boston* in 1914) are the dramatic monologues or dialogues, such as *A Servant to Servants* and *West-Running Brook*, and the dramatic narratives, including *Home Burial* and *The Witch of Coös*. Compressed in their impact and in the contraries of temperament and attitude they delineate, fully dramatized in speech and gesture, they are probing illuminations of such subjects as repressed passion and the ritualization of grief, and profoundly suggestive meditations on the human predicament.

Frost always gave first priority to his poetry, but his career as teacher, platform reader of his poems, and occasional lecturer was designed to support his writing and augment his audience as well as to stimulate an interest in poetry generally. His longest affiliation was with Amherst (1917–20, 1923–38, 1949), but he was also Poet in Residence at Michigan in the early 1920s, Norton Professor at Harvard in 1936, and Ticknor Fellow at Dartmouth College for six years in the 1940s. He was particularly attached after 1921 to the summer sessions of the Bread Loaf School of English and the Bread Loaf Writers Conference, which he helped to establish at Middlebury College in Vermont. His effectiveness in teaching—Socratic, anecdotal, folksy—was proverbial, but he sought arrangements that would leave much of the week, and much of the year, free for writing. By the time he had won four Pulitzer Prizes (the first in 1924 for *New Hampshire*, the last in 1943 for *A Witness Tree*) he was a public figure. The publication of two ambitious philosophical poems, *The Masque of Reason* and *The Masque of Mercy*, in 1945 and 1947, and *In the Clearing* in 1962 marked the end of his productive career. He received honorary degrees from both Oxford and Cambridge universities in 1957 when he was a "good-will ambassador" abroad for the Department of State, and he received the Emerson-Thoreau Medal of the American Academy of Arts and Sciences in the same year he was appointed Consultant in Poetry in the Library of Congress (1958). His most extraordinary honor, confirming his earlier hope that he "fitted, not into the nature of the Universe, but * * * into the nature of Americans—into their affections,"[5] was his selection by John F. Kennedy to read a poem at his presidential inauguration in 1961. (Though a long-time antagonist of liberal reform and of President Franklin D. Roosevelt's New Deal, Frost became an enthusiast for Kennedy.) A cold wind and blinding sun rendered his text unreadable, but Frost recited *The Gift Outright* from memory and later sent the President an expanded version of the "dedication" he had written for the occasion, in which he prophesied for America "The glory of a next Augustan age," a "golden age of poetry and power."

The Pasture

I'm going out to clean the pasture spring;
I'll only stop to rake the leaves away
(And wait to watch the water clear, I may):
I sha'n't be gone long.—You come too.

5. *Ibid.*, p. 102.

I'm going out to fetch the little calf   5
That's standing by the mother. It's so young
It totters when she licks it with her tongue.
I sha'n't be gone long.—You come too.

1913

Mowing

There was never a sound beside the wood but one,
And that was my long scythe whispering to the ground.
What was it it whispered? I knew not well myself;
Perhaps it was something about the heat of the sun,
Something, perhaps, about the lack of sound—   5
And that was why it whispered and did not speak.
It was no dream of the gift of idle hours,
Or easy gold at the hand of fay or elf:
Anything more than the truth would have seemed too weak
To the earnest love that laid the swale[1] in rows,   10
Not without feeble-pointed spikes of flowers
(Pale orchises), and scared a bright green snake.
The fact is the sweetest dream that labor knows.
My long scythe whispered and left the hay to make.

1913

Mending Wall

Something there is that doesn't love a wall,
That sends the frozen-ground-swell under it,
And spills the upper boulders in the sun;
And makes gaps even two can pass abreast.
The work of hunters is another thing: 5
I have come after them and made repair
Where they have left not one stone on a stone,
But they would have the rabbit out of hiding,
To please the yelping dogs. The gaps I mean,
No one has seen them made or heard them made,   10
But at spring mending-time we find them there.
I let my neighbor know beyond the hill;
And on a day we meet to walk the line
And set the wall between us once again.
We keep the wall between us as we go.   15
To each the boulders that have fallen to each.
And some are loaves and some so nearly balls
We have to use a spell to make them balance:
'Stay where you are until our backs are turned!'
We wear our fingers rough with handling them.   20
Oh, just another kind of outdoor game,
One on a side. It comes to little more:
There where it is we do not need the wall:
He is all pine and I am apple orchard.
My apple trees will never get across   25

---

1. Grasses in a marshy meadow.

And eat the cones under his pines, I tell him.
He only says, 'Good fences make good neighbors.'
Spring is the mischief in me, and I wonder
If I could put a notion in his head:
'*Why* do they make good neighbors? Isn't it   30
Where there are cows? But here there are no cows.
Before I built a wall I'd ask to know
What I was walling in or walling out,
And to whom I was like to give offense.
Something there is that doesn't love a wall,   35
That wants it down.' I could say 'Elves' to him,
But it's not elves exactly, and I'd rather
He said it for himself. I see him there
Bringing a stone grasped firmly by the top
In each hand, like an old-stone savage armed.   40
He moves in darkness as it seems to me,
Not of woods only and the shade of trees.
He will not go behind his father's saying,
And he likes having thought of it so well
He says again, 'Good fences make good neighbors.'   45

1914

Home Burial[1]

He saw her from the bottom of the stairs
Before she saw him. She was starting down,
Looking back over her shoulder at some fear.
She took a doubtful step and then undid it
To raise herself and look again. He spoke   5
Advancing toward her: 'What is it you see
From up there always—for I want to know.'
She turned and sank upon her skirts at that,
And her face changed from terrified to dull.
He said to gain time: 'What is it you see,'   10
Mounting until she cowered under him.
'I will find out now—you must tell me, dear.'
She, in her place, refused him any help
With the least stiffening of her neck and silence.
She let him look, sure that he wouldn't see,   15
Blind creature; and awhile he didn't see.
But at last he murmured, 'Oh,' and again, 'Oh.'

'What is it—what?' she said.

              'Just that I see.'

'You don't,' she challenged. 'Tell me what it is.'   20

1. The title refers to the custom of bury-
ing members of the family on the home
property. Frost's biographer, Lawrance
Thompson (in *Robert Frost, the Early
Years,* p. 597) reports that the poem
commemorated the loss of a child and
the consequent marital difficulties of two
acquaintances, Mr. and Mrs. Nathaniel
Harvey, in 1895, and that the poem
draws on the grief of the Frosts after
the loss of their first child in 1900. Line
110 echoes Mrs. Frost's statement at the
time that "The world's evil."

'The wonder is I didn't see at once.
I never noticed it from here before.
I must be wonted to it—that's the reason.
The little graveyard where my people are!
So small the window frames the whole of it.   25
Not so much larger than a bedroom, it is?
There are three stones of slate and one of marble,
Broad-shouldered little slabs there in the sunlight
On the sidehill. We haven't to mind *those*.
But I understand: it is not the stones,   30
But the child's mound—'

                'Don't, don't, don't, don't,' she cried.

She withdrew shrinking from beneath his arm
That rested on the bannister, and slid downstairs;
And turned on him with such a daunting look,   35
He said twice over before he knew himself:
'Can't a man speak of his own child he's lost?'

'Not you! Oh, where's my hat? Oh, I don't need it!
I must get out of here. I must get air.
I don't know rightly whether any man can.'   40

'Amy! Don't go to someone else this time.
Listen to me. I won't come down the stairs.'
He sat and fixed his chin between his fists.
'There's something I should like to ask you, dear,'

'You don't know how ask it.'   45
                    'Help me, then.'

Her fingers moved the latch for all reply.

'My words are nearly always an offense.
I don't know how to speak of anything
So as to please you. But I might be taught   50
I should suppose. I can't say I see how.
A man must partly give up being a man
With women-folk. We could have some arrangement
By which I'd bind myself to keep hands off
Anything special you're a-mind to name.   55
Though I don't like such things 'twixt those that love.
Two that don't love can't live together without them.
But two that do can't live together with them.'
She moved the latch a little. 'Don't—don't go.
Don't carry it to someone else this time.   60
Tell me about it if it's something human.
Let me into your grief. I'm not so much
Unlike other folks as your standing there
Apart would make me out. Give me my chance.

I do think, though, you overdo it a little. 65
What was it brought you up to think it the thing
To take your mother-loss of a first child
So inconsolably—in the face of love.
You'd think his memory might be satisfied—'

'There you go sneering now!' 70

       'I'm not, I'm not!
You make me angry. I'll come down to you.
God, what a woman! And it's come to this,
A man can't speak of his own child that's dead.'

'You can't because you don't know how to speak. 75
If you had any feelings, you that dug
With your own hand—how could you?—his little grave;
I saw you from that very window there,
Making the gravel leap and leap in air,
Leap up, like that, like that, and land so lightly 80
And roll back down the mound beside the hole.
I thought, Who is that man? I didn't know you.
And I crept down the stairs and up the stairs
To look again, and still your spade kept lifting.
Then you came in. I heard your rumbling voice 85
Out in the kitchen, and I don't know why,
But I went near to see with my own eyes.
You could sit there with the stains on your shoes
Of the fresh earth from your own baby's grave
And talk about your everyday concerns. 90
You had stood the spade up against the wall
Outside there in the entry, for I saw it.'

'I shall laugh the worst laugh I ever laughed.
I'm cursed. God, if I don't believe I'm cursed.'

'I can repeat the very words you were saying. 95
"Three foggy mornings and one rainy day
Will rot the best birch fence a man can build."
Think of it, talk like that at such a time!
What had how long it takes a birch to rot
To do with what was in the darkened parlor. 100
You couldn't care! The nearest friends can go
With anyone to death, comes so far short
They might as well not try to go at all.
No, from the time when one is sick to death,
One is alone, and he dies more alone. 105
Friends make pretense of following to the grave,
But before one is in it, their minds are turned
And making the best of their way back to life
And living people, and things they understand.
But the world's evil. I won't have grief so 110
If I can change it. Oh, I won't, I won't!'

'There, you have said it all and you feel better.
You won't go now. You're crying. Close the door.
The heart's gone out of it: why keep it up.
Amy! There's someone coming down the road!'   115

'You—oh, you think the talk is all. I must go—
Somewhere out of this house. How can I make you—'

'If—you—do!' She was opening the door wider.
'Where do you mean to go? First tell me that.
I'll follow and bring you back by force. I *will!*—'   120

1912–13   1914

A Servant to Servants

I didn't make you know how glad I was
To have you come and camp here on our land.
I promised myself to get down some day
And see the way you lived, but I don't know!
With a houseful of hungry men to feed   5
I guess you'd find. . . . It seems to me
I can't express my feelings any more
Than I can raise my voice or want to lift
My hand (oh, I can lift it when I have to).
Did ever you feel so? I hope you never.   10
It's got so I don't even know for sure
Whether I *am* glad, sorry, or anything.
There's nothing but a voice-like left inside
That seems to tell me how I ought to feel,
And would feel if I wasn't all gone wrong.   15
You take the lake. I look and look at it.
I see it's a fair, pretty sheet of water.
I stand and make myself repeat out loud
The advantages it has, so long and narrow,
Like a deep piece of some old running river   20
Cut short off at both ends. It lies five miles
Straight away through the mountain notch
From the sink window where I wash the plates,
And all our storms come up toward the house,
Drawing the slow waves whiter and whiter and whiter. 25
It took my mind off doughnuts and soda biscuit
To step outdoors and take the water dazzle
A sunny morning, or take the rising wind
About my face and body and through my wrapper,
When a storm threatened from the Dragon's Den,   30
And a cold chill shivered across the lake.
I see it's a fair, pretty sheet of water,
Our Willoughby! How did you hear of it?
I expect, though, everyone's heard of it.
In a book about ferns? Listen to that!   35
You let things more like feathers regulate

Your going and coming. And you like it here?
I can see how you might. But I don't know!
It would be different if more people came,
For then there would be business. As it is,   40
The cottages Len built, sometimes we rent them,
Sometimes we don't. We've a good piece of shore
That ought to be worth something, and may yet.
But I don't count on it as much as Len.
He looks on the bright side of everything,   45
Including me. He thinks I'll be all right
With doctoring. But it's not medicine—
Lowe is the only doctor's dared to say so—
It's rest I want—there, I have said it out—
From cooking meals for hungry hired men   50
And washing dishes after them—from doing
Things over and over that just won't stay done.
By good rights I ought not to have so much
Put on me, but there seems no other way.
Len says one steady pull more ought to do it.   55
He says the best way out is always through,
And I agree to that, or in so far
As that I can see no way out but through—
Leastways for me—and then they'll be convinced.
It's not that Len don't want the best for me.   60
It was his plan our moving over in
Beside the lake from where that day I showed you
We used to live—ten miles from anywhere.
We didn't change without some sacrifice,
But Len went at it to make up the loss.   65
His work's a man's, of course, from sun to sun,
But he works when he works as hard as I do—
Though there's small profit in comparisons,
(Women and men will make them all the same.)
But work ain't all. Len undertakes too much.   70
He's into everything in town. This year
It's highways, and he's got too many men
Around him to look after that make waste.
They take advantage of him shamefully,
And proud, too, of themselves for doing so.   75
We have four here to board, great good-for-nothings,
Sprawling about the kitchen with their talk
While I fry their bacon. Much they care!
No more put out in what they do or say
Than if I wasn't in the room at all.   80
Coming and going all the time, they are:
I don't learn what their names are, let alone
Their characters, or whether they are safe
To have inside the house with doors unlocked.
I'm not afraid of them, though, if they're not   85
Afraid of me. There's two can play at that.
I have my fancies: it runs in the family.
My father's brother wasn't right. They kept him

Locked up for years back there at the old farm.
I've been away once—yes, I've been away. 90
The State Asylum. I was prejudiced;
I wouldn't have sent anyone of mine there;
You know the old idea—the only asylum
Was the poorhouse, and those who could afford,
Rather than send their folks to such a place, 95
Kept them at home; and it does seem more human.
But it's not so: the place is the asylum.
There they have every means proper to do with,
And you aren't darkening other people's lives—
Worse than no good to them, and they no good 100
To you in your condition; you can't know
Affection or the want of it in that state.
I've heard too much of the old-fashioned way.
My father's brother, he went mad quite young.
Some thought he had been bitten by a dog, 105
Because his violence took on the form
Of carrying his pillow in his teeth;
But it's more likely he was crossed in love,
Or so the story goes. It was some girl.
Anyway all he talked about was love. 110
They soon saw he would do someone a mischief
If he wa'n't kept strict watch of, and it ended
In father's building him a sort of cage,
Or room within a room, of hickory poles,
Like stanchions in the barn, from floor to ceiling,— 115
A narrow passage all the way around.
Anything they put in for furniture
He'd tear to pieces, even a bed to lie on.
So they made the place comfortable with straw,
Like a beast's stall, to ease their consciences. 120
Of course they had to feed him without dishes.
They tried to keep him clothed, but he paraded
With his clothes on his arm—all of his clothes.
Cruel—it sounds. I s'pose they did the best
They knew. And just when he was at the height, 125
Father and mother married, and mother came,
A bride, to help take care of such a creature,
And accommodate her young life to his.
That was what marrying father meant to her.
She had to lie and hear love things made dreadful 130
By his shouts in the night. He'd shout and shout
Until the strength was shouted out of him,
And his voice died down slowly from exhaustion.
He'd pull his bars apart like bow and bowstring,
And let them go and make them twang until 135
His hands had worn them smooth as any oxbow.[1]
And then he'd crow as if he thought that child's play
The only fun he had. I've heard them say, though,

---

1. U-shaped collar harnessing an ox to a yoke.

They found a way to put a stop to it.
He was before my time—I never saw him;   140
But the pen stayed exactly as it was
There in the upper chamber in the ell,
A sort of catch-all full of attic clutter.
I often think of the smooth hickory bars.
It got so I would say—you know, half fooling—   145
'It's time I took my turn upstairs in jail'—
Just as you will till it becomes a habit.
No wonder I was glad to get away.
Mind you, I waited till Len said the word.
I didn't want the blame if things went wrong.   150
I was glad though, no end, when we moved out,
And I looked to be happy, and I was,
As I said, for a while—but I don't know!
Somehow the change wore out like a prescription.
And there's more to it than just window-views   155
And living by a lake. I'm past such help—
Unless Len took the notion, which he won't,
And I won't ask him—it's not sure enough.
I s'pose I've got to go the road I'm going:
Other folks have to, and why shouldn't I?   160
I almost think if I could do like you,
Drop everything and live out on the ground—
But it might be, come night, I shouldn't like it,
Or a long rain. I should soon get enough,
And be glad of a good roof overhead.   165
I've lain awake thinking of you, I'll warrant,
More than you have yourself, some of these nights.
The wonder was the tents weren't snatched away
From over you as you lay in your beds.
I haven't courage for a risk like that.   170
Bless you, of course, you're keeping me from work,
But the thing of it is, I need to *be* kept.
There's work enough to do—there's always that;
But behind's behind. The worst that you can do
Is set me back a little more behind.   175
I sha'n't catch up in this world, anyway.
I'd *rather* you'd not go unless you must.

                                        1914

After Apple-Picking

My long two-pointed ladder's sticking through a tree
Toward heaven still,
And there's a barrel that I didn't fill
Beside it, and there may be two or three
Apples I didn't pick upon some bough.   5
But I am done with apple-picking now.
Essence of winter sleep is on the night,
The scent of apples: I am drowsing off.
I cannot rub the strangeness from my sight

I got from looking through a pane of glass   10
I skimmed this morning from the drinking trough
And held against the world of hoary grass.
It melted, and I let it fall and break.
But I was well
Upon my way to sleep before it fell,   15
And I could tell
What form my dreaming was about to take.
Magnified apples appear and disappear,
Stem end and blossom end,
And every fleck of russet showing clear.   20
My instep arch not only keeps the ache,
It keeps the pressure of a ladder-round.
I feel the ladder sway as the boughs bend.
And I keep hearing from the cellar bin
The rumbling sound   25
Of load on load of apples coming in.
For I have had too much
Of apple-picking: I am overtired
Of the great harvest I myself desired.
There were ten thousand thousand fruit to touch,   30
Cherish in hand, lift down, and not let fall.
For all
That struck the earth,
No matter if not bruised or spiked with stubble,
Went surely to the cider-apple heap   35
As of no worth.
One can see what will trouble
This sleep of mine, whatever sleep it is.
Were he not gone,
The woodchuck could say whether it's like his   40
Long sleep, as I describe its coming on,
Or just some human sleep.

                                                                    1914

The Road Not Taken

Two roads diverged in a yellow wood,
And sorry I could not travel both
And be one traveler, long I stood
And looked down one as far as I could
To where it bent in the undergrowth;   5

Then took the other, as just as fair,
And having perhaps the better claim,
Because it was grassy and wanted wear;
Though as for that the passing there
Had worn them really about the same,   10

And both that morning equally lay
In leaves no step had trodden black.

Oh, I kept the first for another day!
Yet knowing how way leads on to way,
I doubted if I should ever come back.   15

I shall be telling this with a sigh
Somewhere ages and ages hence:
Two roads diverged in a wood, and I—
I took the one less traveled by,
And that has made all the difference.   20

1916

An Old Man's Winter Night

All out-of-doors looked darkly in at him
Through the thin frost, almost in separate stars,
That gathers on the pane in empty rooms.
What kept his eyes from giving back the gaze
Was the lamp tilted near them in his hand.   5
What kept him from remembering what it was
That brought him to that creaking room was age.
He stood with barrels round him—at a loss.
And having scared the cellar under him
In clomping here, he scared it once again   10
In clomping off;—and scared the outer night,
Which has its sounds, familiar, like the roar
Of trees and crack of branches, common things,
But nothing so like beating on a box.
A light he was to no one but himself   15
Where now he sat, concerned with he knew what,
A quiet light, and then not even that.
He consigned to the moon, such as she was,
So late-arising, to the broken moon
As better than the sun in any case   20
For such a charge, his snow upon the roof,
His icicles along the wall to keep;
And slept. The log that shifted with a jolt
Once in the stove, disturbed him and he shifted,
And eased his heavy breathing, but still slept.   25
One aged man—one man—can't keep a house,
A farm, a countryside, or if he can,
It's thus he does it of a winter night.

1906   1916

The Oven Bird

There is a singer everyone has heard,
Loud, a mid-summer and a mid-wood bird,
Who makes the solid tree trunks sound again.
He says that leaves are old and that for flowers
Mid-summer is to spring as one to ten.   5
He says the early petal-fall is past
When pear and cherry bloom went down in showers
On sunny days a moment overcast;

And comes that other fall we name the fall.
He says the highway dust is over all.   10
The bird would cease and be as other birds
But that he knows in singing not to sing.
The question that he frames in all but words
Is what to make of a diminished thing.

1906–7   1916

Birches

When I see birches bend to left and right
Across the lines of straighter darker trees,
I like to think some boy's been swinging them.
But swinging doesn't bend them down to stay
As ice-storms do. Often you must have seen them   5
Loaded with ice a sunny winter morning
After a rain. They click upon themselves
As the breeze rises, and turn many-colored
As the stir cracks and crazes their enamel.
Soon the sun's warmth makes them shed crystal shells   10
Shattering and avalanching on the snow-crust—
Such heaps of broken glass to sweep away
You'd think the inner dome of heaven had fallen.
They are dragged to the withered bracken by the load,
And they seem not to break; though once they are bowed   15
So low for long, they never right themselves:
You may see their trunks arching in the woods
Years afterwards, trailing their leaves on the ground
Like girls on hands and knees that throw their hair
Before them over their heads to dry in the sun.   20
But I was going to say when Truth broke in
With all her matter-of-fact about the ice-storm
I should prefer to have some boy bend them
As he went out and in to fetch the cows—
Some boy too far from town to learn baseball,   25
Whose only play was what he found himself,
Summer or winter, and could play alone.
One by one he subdued his father's trees
By riding them down over and over again
Until he took the stiffness out of them,   30
And not one but hung limp, not one was left
For him to conquer. He learned all there was
To learn about not launching out too soon
And so not carrying the tree away
Clear to the ground. He always kept his poise   35
To the top branches, climbing carefully
With the same pains you use to fill a cup
Up to the brim, and even above the brim.
Then he flung outward, feet first, with a swish,
Kicking his way down through the air to the ground.   40
So was I once myself a swinger of birches.
And so I dream of going back to be.

It's when I'm weary of considerations,
And life is too much like a pathless wood
Where your face burns and tickles with the cobwebs   45
Broken across it, and one eye is weeping
From a twig's having lashed across it open.
I'd like to get away from earth awhile
And then come back to it and begin over.
May no fate willfully misunderstand me   50
And half grant what I wish and snatch me away
Not to return. Earth's the right place for love:
I don't know where it's likely to go better.
I'd like to go by climbing a birch tree,
And climb black branches up a snow-white trunk   55
Toward heaven, till the tree could bear no more,
But dipped its top and set me down again.
That would be good both going and coming back.
One could do worse than be a swinger of birches.

1913–14   1916

Range-Finding

The battle rent a cobweb diamond-strung
And cut a flower beside a ground bird's nest
Before it stained a single human breast.
The stricken flower bent double and so hung.
And still the bird revisited her young.   5
A butterfly its fall had dispossessed
A moment sought in air his flower of rest,
Then lightly stooped to it and fluttering clung.
On the bare upland pasture there had spread
O'ernight 'twixt mullein stalks a wheel of thread   10
And straining cables wet with silver dew.
A sudden passing bullet shook it dry.
The indwelling spider ran to greet the fly,
But finding nothing, sullenly withdrew.

1911   1916

'Out, Out—'[1]

The buzz saw snarled and rattled in the yard
And made dust and dropped stove-length sticks of wood,
Sweet-scented stuff when the breeze drew across it.
And from there those that lifted eyes could count
Five mountain ranges one behind the other   5
Under the sunset far into Vermont.
And the saw snarled and rattled, snarled and rattled,
As it ran light, or had to bear a load.
And nothing happened: day was all but done.
Call it a day, I wish they might have said   10
To please the boy by giving him the half hour

1. A quotation from Shakespeare's Macbeth (5.5.23–24): "Out, out, brief candle! /Life's but a walking shadow."

That a boy counts so much when saved from work.
His sister stood beside them in her apron
To tell them 'Supper.' At the word, the saw,
As if to prove saws knew what supper meant,   15
Leaped out at the boy's hand, or seemed to leap—,
He must have given the hand. However it was,
Neither refused the meeting. But the hand!
The boy's first outcry was a rueful laugh,
As he swung toward them holding up the hand   20
Half in appeal, but half as if to keep
The life from spilling. Then the boy saw all—
Since he was old enough to know, big boy
Doing a man's work, though a child at heart—
He saw all spoiled. 'Don't let him cut my hand off—   25
The doctor, when he comes. Don't let him, sister!'
So. But the hand was gone already.
The doctor put him in the dark of ether.
He lay and puffed his lips out with his breath.
And then—the watcher at his pulse took fright.   30
No one believed. They listened at his heart.
Little—less—nothing!—and that ended it.
No more to build on there. And they, since they
Were not the one dead, turned to their affairs.

                                                      1916

The Witch of Coös[1]

I stayed the night for shelter at a farm
Behind the mountain, with a mother and son,
Two old-believers.[2] They did all the talking.

MOTHER. Folks think a witch who has familiar spirits
She could call up to pass a winter evening,   5
But won't, should be burned at the stake or something.
Summoning spirits isn't 'Button, button,
Who's got the button,' I would have them know.

SON. Mother can make a common table rear
And kick with two legs like an army mule.[3]   10

MOTHER. And when I've done it, what good have I done?
Rather than tip a table for you, let me
Tell you what Ralle the Sioux Control[4] once told me.
He said the dead had souls, but when I asked him
How could that be—I thought the dead were souls,   15
He broke my trance. Don't that make you suspicious

1. A county in northern New Hampshire.
2. Believers in traditional folklore and
superstitions.
3. Mediums, supposedly in contact with
supernatural spirits, claim to move furniture about during seances without exerting physical force.
4. In spiritualistic theory, a control is
the agent who voices the utterances of
the dead. American Indians were favorite controls at the turn of the 19th century.

That there's something the dead are keeping back?
Yes, there's something the dead are keeping back.

SON. You wouldn't want to tell him what we have
Up attic, mother?   20

MOTHER. Bones—a skeleton.

SON. But the headboard of mother's bed is pushed
Against the attic door: the door is nailed.
It's harmless. Mother hears it in the night
Halting perplexed behind the barrier   25
Of door and headboard. Where it wants to get
Is back into the cellar where it came from.

MOTHER. We'll never let them, will we, son! We'll never!

SON. It left the cellar forty years ago
And carried itself like a pile of dishes   30
Up one flight from the cellar to the kitchen,
Another from the kitchen to the bedroom,
Another from the bedroom to the attic,
Right past both father and mother, and neither stopped it.
Father had gone upstairs; mother was downstairs.   35
I was a baby: I don't know where I was.

MOTHER. The only fault my husband found with me—
I went to sleep before I went to bed,
Especially in winter when the bed
Might just as well be ice and the clothes snow.   40
The night the bones came up the cellar-stairs
Toffile had gone to bed alone and left me,
But left an open door to cool the room off
So as to sort of turn me out of it.
I was just coming to myself enough   45
To wonder where the cold was coming from,
When I heard Toffile upstairs in the bedroom
And thought I heard him downstairs in the cellar.
The board we had laid down to walk dry-shod on
When there was water in the cellar in spring   50
Struck the hard cellar bottom. And then someone
Began the stairs, two footsteps for each step,
The way a man with one leg and a crutch,
Or a little child, comes up. It was Toffile:
It wasn't anyone who could be there.   55
The bulkhead double-doors were double-locked
And swollen tight and buried under snow.
The cellar windows were banked up with sawdust
And swollen tight and buried under snow.
It was the bones. I knew them—and good reason.   60
My first impulse was to get to the knob
And hold the door. But the bones didn't try

The door; they halted helpless on the landing,
Waiting for things to happen in their favor.
The faintest restless rustling ran all through them.   65
I never could have done the thing I did
If the wish hadn't been too strong in me
To see how they were mounted for this walk.
I had a vision of them put together
Not like a man, but like a chandelier.   70
So suddenly I flung the door wide on him.
A moment he stood balancing with emotion,
And all but lost himself. (A tongue of fire
Flashed out and licked along his upper teeth.
Smoke rolled inside the sockets of his eyes.)   75
Then he came at me with one hand outstretched,
The way he did in life once; but this time
I struck the hand off brittle on the floor,
And fell back from him on the floor myself.
The finger-pieces slid in all directions.   80
(Where did I see one of those pieces lately?
Hand me my button-box—it must be there.)
I sat up on the floor and shouted, 'Toffile,
It's coming up to you.' It had its choice
Of the door to the cellar or the hall.   85
It took the hall door for the novelty,
And set off briskly for so slow a thing,
Still going every which way in the joints, though,
So that it looked like lightning or a scribble,
From the slap I had just now given its hand.   90
I listened till it almost climbed the stairs
From the hall to the only finished bedroom,
Before I got up to do anything;
Then ran and shouted, 'Shut the bedroom door,
Toffile, for my sake!' 'Company?' he said,   95
'Don't make me get up; I'm too warm in bed.'
So lying forward weakly on the handrail
I pushed myself upstairs, and in the light
(The kitchen had been dark) I had to own
I could see nothing. 'Toffile, I don't see it.   100
It's with us in the room though. It's the bones.'
'What bones?' 'The cellar bones—out of the grave.'
That made him throw his bare legs out of bed
And sit up by me and take hold of me.
I wanted to put out the light and see   105
If I could see it, or else mow the room,
With our arms at the level of our knees,
And bring the chalk-pile down. 'I'll tell you what—
It's looking for another door to try.
The uncommonly deep snow has made him think   110
Of his old song, *The Wild Colonial Boy*,
He always used to sing along the tote⁵ road.

5. Colloquial for hauling or carrying.

He's after an open door to get outdoors.
Let's trap him with an open door up attic.'
Toffile agreed to that, and sure enough,   115
Almost the moment he was given an opening,
The steps began to climb the attic stairs.
I heard them. Toffile didn't seem to hear them.
'Quick!' I slammed to the door and held the knob.
'Toffile, get nails.' I made him nail the door shut   120
And push the headboard of the bed against it.
Then we asked was there anything
Up attic that we'd ever want again.
The attic was less to us than the cellar.
If the bones liked the attic, let them have it.   125
Let them stay in the attic. When they sometimes
Come down the stairs at night and stand perplexed
Behind the door and headboard of the bed,
Brushing their chalky skull with chalky fingers,
With sounds like the dry rattling of a shutter,   130
That's what I sit up in the dark to say—
To no one any more since Toffile died.
Let them stay in the attic since they went there.
I promised Toffile to be cruel to them
For helping them be cruel once to him.   135

SON. We think they had a grave down in the cellar.

MOTHER. We know they had a grave down in the cellar.

SON. We never could find out whose bones they were.

MOTHER. Yes, we could too, son. Tell the truth for once.
They were a man's his father killed for me.   140
I mean a man he killed instead of me.
The least I could do was to help dig their grave.
We were about it one night in the cellar.
Son knows the story: but 'twas not for him
To tell the truth, suppose the time had come.   145
Son looks surprised to see me end a lie
We'd kept all these years between ourselves
So as to have it ready for outsiders.
But tonight I don't care enough to lie—
I don't remember why I ever cared.   150
Toffile, if he were here, I don't believe
Could tell you why he ever cared himself. . . .

She hadn't found the finger-bone she wanted
Among the buttons poured out in her lap.
I verified the name next morning: Toffile.   155
The rural letter box said Toffile Lajway.

                                            1923

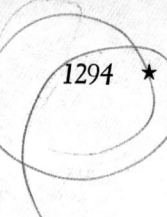

Nothing Gold Can Stay

Nature's first green is gold,
Her hardest hue to hold.
Her early leaf's a flower;
But only so an hour.
Then leaf subsides to leaf.   5
So Eden sank to grief,
So dawn goes down to day.
Nothing gold can stay.

1923

The Runaway

Once when the snow of the year was beginning to fall,
We stopped by a mountain pasture to say, 'Whose colt?
A little Morgan[1] had one forefoot on the wall,
The other curled at his breast. He dipped his head
And snorted at us. And then he had to bolt.   5
We heard the miniature thunder where he fled,
And we saw him, or thought we saw him, dim and gray,
Like a shadow against the curtain of falling flakes.
'I think the little fellow's afraid of the snow.
He isn't winter-broken. It isn't play   10
With the little fellow at all. He's running away.
I doubt if even his mother could tell him, "Sakes,
It's only weather." He'd think she didn't know!
Where is his mother? He can't be out alone.'
And now he comes again with clatter of stone,   15
And mounts the wall again with whited eyes
And all his tail that isn't hair up straight.
He shudders his coat as if to throw off flies.
'Whoever it is that leaves him out so late,
When other creatures have gone to stall and bin,   20
Ought to be told to come and take him in.'

1918, 1923

Stopping by Woods on a Snowy Evening

Whose woods these are I think I know.
His house is in the village though;
He will not see me stopping here
To watch his woods fill up with snow.

My little horse must think it queer   5
To stop without a farmhouse near
Between the woods and frozen lake
The darkest evening of the year.

---

1. Breed of horse developed from the stallion Justin Morgan in Vermont in the 19th century.

He gives his harness bells a shake
To ask if there is some mistake.   10
The only other sound's the sweep
Of easy wind and downy flake.

The woods are lovely, dark and deep,
But I have promises to keep,
And miles to go before I sleep,   15
And miles to go before I sleep.

                                                    1923

Spring Pools

These pools that, though in forests, still reflect
The total sky almost without defect,
And like the flowers beside them, chill and shiver,
Will like the flowers beside them soon be gone,
And yet not out by any brook or river,   5
But up by roots to bring dark foliage on.

The trees that have it in their pent-up buds
To darken nature and be summer woods—
Let them think twice before they use their powers
To blot out and drink up and sweep away   10
These flowery waters and these watery flowers
From snow that melted only yesterday.

                                                    1928

Once by the Pacific

The shattered water made a misty din.
Great waves looked over others coming in,
And thought of doing something to the shore
That water never did to land before.
The clouds were low and hairy in the skies,   5
Like locks blown forward in the gleam of eyes.
You could not tell, and yet it looked as if
The shore was lucky in being backed by cliff,
The cliff in being backed by continent;
It looked as if a night of dark intent   10
Was coming, and not only a night, an age.
Someone had better be prepared for rage.
There would be more than ocean-water broken
Before God's last *Put out the Light* was spoken.[1]

                                                    1928

West-Running Brook

'Fred, where is north?'
                        'North? North is there, my love.
The brook runs west.'
                        'West-running Brook then call it.'

---

1. Ironic reversal of God's creation of the world in Genesis 1.3: "And God said, Let there be light."

(West-running Brook men call it to this day.)
'What does it think it's doing running west
When all the other country brooks flow east   5
To reach the ocean? It must be the brook
Can trust itself to go by contraries
The way I can with you—and you with me—
Because we're—we're—I don't know what we are.
What are we?'
                              'Young or new?'
                                                   'We must be something.   10
We've said we two. Let's change that to we three.
As you and I are married to each other,
We'll both be married to the brook. We'll build
Our bridge across it, and the bridge shall be
Our arm thrown over it asleep beside it.   15
Look, look, it's waving to us with a wave
To let us know it hears me.'
                                        'Why, my dear,
That wave's been standing off this jut of shore—'
(The black stream, catching on a sunken rock,
Flung backward on itself in one white wave,   20
And the white water rode the black forever,
Not gaining but not losing, like a bird
White feathers from the struggle of whose breast
Flecked the dark stream and flecked the darker pool
Below the point, and were at last driven wrinkled 25
In a white scarf against the far shore alders.)
'That wave's been standing off this jut of shore
Ever since rivers, I was going to say,
Were made in heaven. It wasn't waved to us.'

'It wasn't, yet it was. If not to you   30
It was to me—in an annunciation.'[1]

'Oh, if you take it off to lady-land,
As't were the country of the Amazons[2]
We men must see you to the confines of
And leave you there, ourselves forbid to enter,—   35
It is your brook! I have no more to say.'

'Yes, you have, too. Go on. You thought of something.'

'Speaking of contraries, see how the brook
In that white wave runs counter to itself.
It is from that in water we were from   40
Long, long before we were from any creature.
Here we, in our impatience of the steps,
Get back to the beginning of beginnings,

---

1. Act of announcing or proclaiming, usually in reference to the angel Gabriel's revelation to the Virgin Mary that she would give birth to Jesus Christ.
2. Legendary nation of women warriors.

The stream of everything that runs away.
Some say existence like a Pirouot   45
And Pirouette,[3] forever in one place,
Stands still and dances, but it runs away,
It seriously, sadly, runs away
To fill the abyss' void with emptiness.
It flows beside us in this water brook,   50
But it flows over us. It flows between us
To separate us for a panic moment.
It flows between us, over us, and *with* us.
And it is time, strength, tone, light, life, and love—
And even substance lapsing unsubstantial;   55
The universal cataract of death
That spends to nothingness—and unresisted,
Save by some strange resistance in itself,
Not just a swerving,[4] but a throwing back,
As if regret were in it and were sacred.   60
It has this throwing backward on itself
So that the fall of most of it is always
Raising a little, sending up a little.
Our life runs down in sending up the clock.
The brook runs down in sending up our life.   65
The sun runs down in sending up the brook.
And there is something sending up the sun.
It is this backward motion toward the source,
Against the stream, that most we see ourselves in,
The tribute of the current to the source.   70
It is from this in nature we are from.
It is most us.'
         'Today will be the day
You said so.'
            'No, today will be the day
You said the brook was called West-running Brook.'

'Today will be the day of what we both said.'   75

                                            1928

A Drumlin[1] Woodchuck

One thing has a shelving bank,
Another a rotting plank,
To give it cozier skies
And make up for its lack of size.

My own strategic retreat   5
Is where two rocks almost meet,

---

3. Lovers in traditional French pantomime.
4. In the philosophical poem *De Rerum Natura* (*On the Nature of Things*) by the Roman Titus Lucretius Carus (96?–55? B.C.), the universe is explained as an endlessly descending flow of atoms, whose unpredictable swerving from a fixed course is the source of human volition.
1. A mound or hill formed by glacial action.

And still more secure and snug,
A two-door burrow I dug.

With those in mind at my back
I can sit forth exposed to attack   10
As one who shrewdly pretends
That he and the world are friends.

All we who prefer to live
Have a little whistle we give,
And flash, at the least alarm   15
We dive down under the farm.

We allow some time for guile
And don't come out for a while
Either to eat or drink.
We take occasion to think.   20

And if after the hunt goes past
And the double-barreled blast
(Like war and pestilence
And the loss of common sense),

If I can with confidence say   25
That still for another day,
Or even another year,
I will be there for you, my dear,

It will be because, though small
As measured against the All,   30
I have been so instinctively thorough
About my crevice and burrow.

                                          1936

Desert Places

Snow falling and night falling fast, oh, fast
In a field I looked into going past,
And the ground almost covered smooth in snow,
But a few weeds and stubble showing last.

The woods around it have it—it is theirs.   5
All animals are smothered in their lairs.
I am too absent-spirited to count;
The loneliness includes me unawares.

And lonely as it is that loneliness
Will be more lonely ere it will be less—   10
A blanker whiteness of benighted snow
With no expression, nothing to express.

They cannot scare me with their empty spaces
Between stars—on stars where no human race is.

I have it in me so much nearer home 15
To scare myself with my own desert places.

1936

Design

I found a dimpled spider, fat and white,
On a white heal-all,[6] holding up a moth
Like a white piece of rigid satin cloth—
Assorted characters of death and blight
Mixed ready to begin the morning right, 5
Like the ingredients of a witches' broth—
A snow-drop spider, a flower like a froth,
And dead wings carried like a paper kite.

What had that flower to do with being white,
The wayside blue and innocent heal-all? 10
What brought the kindred spider to that height,
Then steered the white moth thither in the night?
What but design of darkness to appall?—
If design govern in a thing so small.

1922, 1936

On a Bird Singing in Its Sleep

A bird half wakened in the lunar noon
Sang halfway through its little inborn tune.
Partly because it sang but once all night
And that from no especial bush's height;
Partly because it sang ventriloquist 5
And had the inspiration to desist
Almost before the prick of hostile ears,
It ventured less in peril than appears.
It could not have come down to us so far
Through the interstices of things ajar 10
On the long bead chain of repeated birth
To be a bird while we are men on earth
If singing out of sleep and dream that way
Had made it much more easily a prey.

1936

Provide, Provide[1]

The witch that came (the withered hag)
To wash the steps with pail and rag,
Was once the beauty Abishag,[2]

6. An albino version of the common field flower *Prunella vulgaris,* whose hooded blossom is normally violet or blue, whose leaves are either toothed or toothless, and which was once widely used as a medicine.
1. Frost's biographer, Lawrance Thompson, asserts that this poem was provoked by a strike of charwomen that some Harvard faculty members helped organize (*Robert Frost, the Years of Triumph,* p. 437).
2. A beautiful maiden brought to comfort King David in his old age (1 Kings 1.2–4).

The picture pride of Hollywood.
Too many fall from great and good   5
For you to doubt the likelihood.

Die early and avoid the fate.
Or if predestined to die late,
Make up your mind to die in state.

Make the whole stock exchange your own!   10
If need be occupy a throne,
Where nobody can call *you* crone.

Some have relied on what they knew;
Others on being simple true.
What worked for them might work for you.   15

No memory of having starred
Atones for later disregard,
Or keeps the end from being hard.

Better to go down dignified
With boughten friendship at your side   20
Than none at all. Provide, provide!

                                           1934, 1936

The Silken Tent

She is as in a field a silken tent
At midday when a sunny summer breeze
Has dried the dew and all its ropes relent,
So that in guys[1] it gently sways at ease,
And its supporting central cedar pole,   5
That is its pinnacle to heavenward
And signifies the sureness of the soul,
Seems to owe naught to any single cord,
But strictly held by none, is loosely bound
By countless silken ties of love and thought   10
To everything on earth the compass round,
And only by one's going slightly taut
In the capriciousness of summer air
Is of the slightest bondage made aware.

                                               1942

The Most of It

He thought he kept the universe alone;
For all the voice in answer he could wake
Was but the mocking echo of his own
From some tree-hidden cliff across the lake.
Some morning from the boulder-broken beach   5

---

1. Ropes holding an upright pole in position.

He would cry out on life, that what it wants
Is not its own love back in copy speech,
But counter-love, original response.
And nothing ever came of what he cried
Unless it was the embodiment that crashed   10
In the cliff's talus[2] on the other side,
And then in the far distant water splashed,
But after a time allowed for it to swim,
Instead of proving human when it neared
And someone else additional to him,   15
As a great buck it powerfully appeared,
Pushing the crumpled water up ahead,
And landed pouring like a waterfall,
And stumbled through the rocks with horny tread,
And forced the underbrush—and that was all.   20

1942

The Gift Outright[1]

The land was ours before we were the land's.
She was our land more than a hundred years
Before we were her people. She was ours
In Massachusetts, In Virginia,
But we were England's, still colonials,   5
Possessing what we still were unpossessed by,
Possessed by what we now no more possessed.
Something we were withholding made us weak
Until we found out that it was ourselves
We were withholding from our land of living,   10
And forthwith found salvation in surrender.
Such as we were we gave ourselves outright
(The deed of gift was many deeds of war)
To the land vaguely realizing westward,
But still unstoried, artless, unenhanced,   15
Such as she was, such as she would become.

1942

2. Sloping bank of rock fragments at the foot of a cliff.
1. The poem recited by Frost at the inauguration of President John F. Kennedy in January, 1963.

CARL SANDBURG
1878–1967

In the three decades following the World's Fair of 1893, the city of Chicago became the center of a midwestern "renaissance" in the arts, a "vortex" (Ezra Pound's term) involving talents as diverse as the architect Frank Lloyd Wright, the novelists Theodore Dreiser, Henry Blake Fuller, and Robert Herrick, the radical journalist and novelist Floyd Dell, and the poet Edgar Lee Masters. Two new literary journals—*Poetry*, founded in 1912 by the poet Harriet Monroe, and *The Little Review*, founded by

Margaret C. Anderson in 1914—soon had an international audience and made Chicago a center for the introduction of new writing on both sides of the Atlantic. One of the important poets discovered by *Poetry* magazine was Carl Sandburg, who became one of the most popularly read poets during the 1920s and 1930s.

Sandburg was the son of Swedish immigrants who settled in Galesburg, Illinois; his father was a machinist's blacksmith. Sandburg had irregular schooling and worked as an itinerant laborer and jack-of-all-trades in the Midwest before enlisting in the army during the Spanish-American War and serving as correspondent for the Galesburg *Evening Mail*. After the war he attended Lombard College, working for the fire department to support himself, but withdrew without a degree in 1902. He worked as advertising writer, roving reporter, and organizer for the Social Democratic party in Wisconsin, and he married the sister of famed photographer Edward Steichen in 1908. He served as secretary to the Socialist mayor of Milwaukee (1910–12) and wrote editorials for the Milwaukee *Leader* before moving to Chicago in 1913. He had published a pamphlet of poems privately in Galesburg in 1904, and *Poetry* magazine published his poem *Chicago* in 1914.

The poems that made Sandburg famous appeared in four volumes: *Chicago Poems* (1914), *Cornhuskers* (1918), *Smoke and Steel* (1920), and *Slabs of the Sunburnt West* (1922). With the precedent of Whitman behind them, they present a sweeping panorama of American life, encompassing prairie, eastern, and western landscapes as well as vignettes of the modern city. They celebrate, from the standpoint of a Populist radical, the lives of outcasts, the contributions of immigrants and common people to urban culture, and the occupations of those who have survived or been sacrificed in the rise of industrial civilization. Sandburg's language draws on the colorful diction of immigrants and the lingo of urban dwellers, but unvarnished directness of statement takes precedence over subtleties of imagery or rhythm in his verse, even in such poems as *Cool Tombs*, where the consistency of tone is impressive, or in *Flash Crimson*, where the techniques of symbolism are used. Sandburg avoided regular stanza patterns and traditional blank verse and wrote an utterly free verse, developing Whitman's long line but moderating its rhetorical impact and intensity, and composing what are often in effect prose paragraphs. As one who undertook as a spokesman for the common people to inscribe "public speech," he was proud late in his career to "favor simple poems for simple people."[1]

His most ambitious attempt to accomplish that aim was *The People, Yes* (1936), consisting of prose vignettes, anecdotes, and verse, which drew on his studies of American folksongs that preoccupied his attention after the publication of *The American Songbag* in 1927. Indeed, from 1918 on, other professional activities took precedence over his verse. He was a columnist, editorial writer, and feature writer on the *Chicago Daily News* from 1918 to 1933, and he published an account of *The Chicago Race Riots* in 1919. He published *The Rootabaga Stories* (for children) and two sequels between 1922 and 1930, and his biographies include *Steichen the Photographer* (1929) and *Mary Lincoln* (1932). His major work in prose was a monumental and celebratory biography of Abraham Lincoln, beginning with the two-volume *Prairie Years* in 1926 and culminating in *The*

---

1. Introduction, *Complete Poems*, p. xxix.

War Years (1939), a four-volume work which won the Pulitzer Prize in 1940.

Chicago

Hog Butcher for the World,
Tool Maker, Stacker of Wheat,
Player with Railroads and the Nation's Freight Handler;
Stormy, husky, brawling,
City of the Big Shoulders: 5
They tell me you are wicked and I believe them, for I have seen
    your painted women under the gas lamps luring the farm boys.
And they tell me you are crooked and I answer: Yes, it is true I
    have seen the gunman kill and go free to kill again.
And they tell me you are brutal and my reply is: On the faces of
    women and children I have seen the marks of wanton hunger.
And having answered so I turn once more to those who sneer at this
    my city, and I give them back the sneer and say to them:
Come and show me another city with lifted head singing so proud
    to be alive and coarse and strong and cunning.   10
Flinging magnetic curses amid the toil of piling job on job, here is a
    tall bold slugger set vivid against the little soft cities;
Fierce as a dog with tongue lapping for action, cunning as a savage
    pitted against the wilderness,
       Bareheaded,
       Shoveling,
       Wrecking,   15
       Planning,
       Building, breaking, rebuilding,
Under the smoke, dust all over his mouth, laughing with white
    teeth,
Under the terrible burden of destiny laughing as a young man
    laughs,
Laughing even as an ignorant fighter laughs who has never lost a
    battle,   20
Bragging and laughing that under his wrist is the pulse, and under
    his ribs the heart of the people,
       Laughing!
Laughing the stormy, husky, brawling laughter of Youth, half-
    naked, sweating, proud to be Hog Butcher, Tool Maker, Stacker
    of Wheat, Player with railroads and Freight Handler to the
    Nation.

                             1914, 1916

Fog

The fog comes
on little cat feet.

It sits looking
over harbor and city
on silent haunches   5
and then moves on.

                                 1916

Cool Tombs

When Abraham Lincoln was shoveled into the tombs, he forgot the copperheads and the assassin . . . in the dust, in the cool tombs.[1]

And Ulysses Grant lost all thought of con men and Wall Street, cash and collateral turned ashes . . . in the dust, in the cool tombs.[2]

Pocahontas' body, lovely as a poplar, sweet as a red haw in November or a pawpaw in May, did she wonder? does she remember? . . . in the dust, in the cool tombs?[3]

Take any streetful of people buying clothes and groceries, cheering a hero or throwing confetti and blowing tin horns . . . tell me if the lovers are losers . . . tell me if any get more than the lovers . . . in the dust . . . in the cool tombs.

1918

Grass[4]

Pile the bodies high at Austerlitz and Waterloo.
Shovel them under and let me work—
I am the grass; I cover all.

And pile them high at Gettysburg
And pile them high at Ypres and Verdun.   5
Shovel them under and let me work.
Two years, ten years, and passengers ask the conductor:
What place is this?
Where are we now?

I am the grass.   10
Let me work.

1918

From The People, Yes

56

The sacred legion of the justborn—
how many thousands born this minute?
how many fallen for soon burial?
what are these deaths and replacements?

1. President Abraham Lincoln (1809–65) was opposed by southern sympathizers in the North, called Copperheads, and assassinated by John Wilkes Booth.
2. The second administration of President Ulysses S. Grant (1822–85) was riddled with bribery and political corruption. After leaving office Grant was exploited in business and underwent bankruptcy.

3. Pocahontas (1595?–1617), daughter of the Indian chief Powhatan, intervened to save the life of Captain John Smith. The red haw is a type of American hawthorn tree; "pawpaw" is colloquial for the fruit of the papaya tree.
4. The proper names in this poem are all famous battlefields in, respectively, the Napoleonic wars, the American Civil War, and World War I.

what is this endless shuttling of shadowlands   5
where the spent and done go marching into one
and from another arrive those crying Mama Mama?

In the people is the eternal child,
the wandering gypsy, the pioneer homeseeker,
the singer of home sweet home.   10

The people say and unsay,
put up and tear down
and put together again—
a builder, wrecker, and builder again—
this is the people.   15

* * *

1936

WALLACE STEVENS

1879–1955

Wallace Stevens stood somewhat apart from the literary controversies that engrossed Ezra Pound, T. S. Eliot, and William Carlos Williams. A lawyer and insurance executive, he pursued a business career along with that of poet until shortly before his death in 1955. By the end of World War II he had produced a body of verse that distinguished him as one of the six or seven major poets writing in English in the twentieth century. He gave renewed strength to the Whitman tradition in American poetry by extending both the sensuous appeal of poetry and its capacity for abstract statement, celebrating the reality of the physical world and the poet's power to shape that reality with his imagination, and affirming that "After the final no there comes a yes / And on that yes the future world depends."

Stevens was raised in Reading, Pennsylvania, the descendant of Pennsylvania farmers to whom he attributed his occasional grumpiness, stubbornness, and taciturnity. His father, a schoolteacher, attorney, and occasional poet, provided an impetus for both of Stevens's careers when his son graduated from public high school and left for Harvard in 1897. He instructed Stevens that the world offers success only to those who put "work and study" above "dreams," but also urged him to perfect his "power of painting pictures in words" and to "paint truth but not always in drab clothes."[1] Stevens later was to enjoy the sharp division between the practical and the imaginative activities, and made of it, along with the interraction between them, a structuring principle of his life. "After writing a poem," Stevens wrote in a comment on *To the One of Fictive Music,* it is "a good thing to walk around the block; after too much midnight, it is pleasant to hear the milkman, and yet, and this is the point of the poem, the imaginative world is the only real world after all."[2] Most of Stevens's associates at the Hartford insurance company where he worked were igno-

1. *Letters,* p. 14.   2. *Ibid.,* pp. 251–52.

rant of his writing career, and for many years even his wife had little idea of his eminence as a poet.

During the three years he attended Harvard as a special student, Stevens published stories and verse in the *Harvard Advocate*, of which he became president, and in the *Harvard Monthly*. His verse then showed the influence of the English poets Keats, Tennyson, and Meredith, but he came to know personally the philosopher and poet George Santayana (for whom he wrote a moving elegy in 1952). Santayana's *Interpretations of Poetry and Religion* (1900), and later *Skepticism and Animal Faith* (1923), helped shape Stevens's belief that poetry, in a skeptical age, fulfills religion's office.

Stevens left Harvard in 1900 to pursue a literary career, but, having determined never to "make a petty struggle for existence—physical or literary"[3] he sought a well-paying job in journalism. He was a reporter on the New York *Tribune* but did not succeed, and he then followed his father's advice to enter the New York Law School. He was admitted to the bar in 1904, and after several unsuccessful years practicing law he joined the New York office of a bonding company in 1908. He married Elsie Moll in 1909. After a few years in New York City Stevens joined The Hartford Accident and Insurance Company and in 1916 moved with his wife to Hartford, Connecticut.

They made Hartford their lifetime home, and Stevens established the routines that enabled him to pursue two separate careers. He wrote his poems at night, on weekends, and during the summers, which his wife spent mostly in the country, where the air was better for her delicate health. Stevens traveled often on business, in the United States and Canada, but he never visited Europe. He enjoyed frequent vacations in Florida and business trips through the South; the contrast between its warm, lush climate and the austerity of chill New England became important in much of his poetry. His poems began to appear in the little magazines in 1914, notably in Harriet Monroe's *Poetry* and Alfred Kreymborg's *Others*, and his verse was published regularly until 1923. He also wrote three experimental one-act plays in 1915–17. During these years he joined occasionally in the activities of a circle of poets and artists in New York City, but he remained a peripheral figure at their gatherings. He became the friend of Monroe and Kreymborg, and a close professional friend of Williams and the poet Marianne Moore, with both of whom he kept up a friendly correspondence throughout his life. Yet the combination of his reticence and his business activities kept him removed from literary circles.

After the publication of his first volume of verse, *Harmonium*, in 1923, and the birth of his daughter Holly in 1924, Stevens's engrossment in family and business affairs kept him from writing for virtually six years. By the middle 1930s his economic situation was comfortable and secure (he became vice-president of his Hartford company in 1934), and his creative energies had revived early in the decade, prompted in part by widening recognition of *Harmonium*, which was reprinted in an expanded version in 1931. New volumes and major works appeared in swift succession: *Ideas of Order* in 1935, *Owl's Clover* in 1936, *The Man with the Blue Guitar* in 1937, *Parts of a World* in 1942, *Transport to Summer* in 1947, and *Auroras of Autumn* in 1950. On rare occasions he consented to give readings of his poems, and he wrote a few appreciations of his friends' poetry and lec-

3. *Ibid.*, p. 34.

tured at Princeton, Mount Holyoke, Harvard, Yale, Columbia, and the English Institute in the Forties. A selection of these essays, *The Necessary Angel* (1951), pursued many tangents in exploring the central problem that had engaged Emerson and Whitman earlier: the role of the imagination in relation to reality, or the reality of poetry in relation to the reality outside it. In the "relentless contact" of their interaction, he believed, the imagination "enables us to perceive the normal in the abnormal, the opposite of chaos in chaos" and thus help persons to survive the conditions of modern life. The poet's role, though he speaks only to an elite, is "to help people to live their lives," Stevens declared. The poet can do this because he "gives to life the supreme fictions without which we are unable to conceive it" and creates "a world of poetry indistinguishable from the world in which we live, or, I ought to say, no doubt, from the world in which we shall come to live."[4]

Stevens's interest in the theory of poetry, though more pronounced in his essays and his later poetry, was never absent from his earlier productions. But what struck readers of the early poems was their sensuous surface and the poet's bravura effects, the flair, and at times the self-mockery of a virtuoso performance. Some, displaying the qualities of "water-colors, little statues,"[5] which he admired in the French symbolist Paul Verlaine or the American David Evans, were unusually refined in their nuances of tone; they led some readers to think him an aesthete without serious concern for the real world. Other poems flaunted dazzling, bold colors, elaborate apostrophes to readers or subjects, comic and at times self-deprecating mockery, or rhetorical acrobatics that led some critics to think him a posturing dandy. Yet the sensuous appeal of his verse displayed the "essential gaudiness" he thought important in all good poetry,[6] and his pose of dandy was a consciously assumed posture in his verse, adopted like Whitman's dandyism as a means of gaining freedom from solemn conventions of the past, so as to liberate the imagination. The comic antics of the speakers in many poems were informed by the French philosopher Henri Bergson's theories of the comic, while the delight in self-dramatization and sheer performance sprang from energies that Stevens thought basic to imaginative activity; in one of his aphorisms he declared that "Authors are actors, books are theatres." His celebration of the senses and the physical world was part of a deliberate attempt to acknowledge the passing of traditional religious beliefs and, like Whitman, to confront a world in which irrationality, process, and change were the inescapable reality. One early masterpiece, *Peter Quince at the Clavier*, celebrates the beauty of the flesh that survives in memory only because it lives and dies in transitory moments. The long mock-epic *The Comedian as the Letter C* presents a comic version of the poet in the protagonist Crispin, who indulges in extravagant flights of the imagination while also undertaking a quasi-allegorical voyage across the Atlantic, to settle colonies in America; his grandiose dreams crumble in the face of the blunt practicalities he faces in the New World, but he finds in the prosaic conditions of everyday life, and in his domestic life with his four daughters, the potential for life and future fulfillments of the imagination.

To relish "gaiety of language," to arouse delight in "flawed words and stubborn sounds" for people living in the "imperfect" world which can be

---

4. *The Necessary Angel*, pp. 153, 30–31.   6. *Ibid.*, p. 263.
5. *Letters*, p. 110.

our only "paradise," was one of the chief aims of Stevens's early verse that remained a principle of his poetics throughout his career. Another aim, however, was to perfect an idiom for rendering astringently, without falsifying, the bareness of the social and natural environments he knew, or the sensations of futility and "nothingness." For either of these purposes he found formal technique, or intricate stanza patterns of the traditional kind that interested Ezra Pound, of little importance; "freedom regardless of form,"[7] he declared, was necessary for the poet. He wrote in blank verse, or devised simple stanzas, often of two unrhymed lines or three, and developed analogies with the arts of painting and music to achieve the effects of sensuous appeal and stark precision that he sought. His poems abound in allusions to singers, instrumental performers, and their musical instruments. Effects of color and precision of line comparable to those of the French painter Henri Matisse, or manipulations of perspective and abstract analytic precision comparable to those of Cubist painters, characterize many of his productions.

In his poems after 1931 Stevens incorporated, more often and directly than before, his theories of poetry and his conception of the heroic role for the poetic imagination. Some critics and poets thought that Stevens's later poetry suffered for becoming more abstractly theoretical and for focusing so intently on poetic theory. Others welcomed what they found to be deeper philosophical significance and Stevens's more direct concern with the aesthetic and moral issues—the relation of art to social realities and the relations of the self to the outer world—which have engrossed poets in the Romantic tradition since Wordsworth, Emerson, and Whitman. In any case he brought a serious and subtle imagination into play to define the "supreme fictions" that can become the ground for belief and commitment in the modern world, and he sustained a high level of achievement that was recognized in the award of the Bollingen Prize in 1950 and both the National Book Award and the Pulitzer prizes in the year of his death. In *The Man with the Blue Guitar* the creative imagination takes the form of a musician, drawn from a painting of Pablo Picasso. His improvisations define a poetics which is constantly changing and an art which alters reality outside it, for "Things as they are / Are changed upon the blue guitar." Yet those changes are part of the transformations occurring in reality itself. In *Notes toward a Supreme Fiction* he plays suggestively with the principal ideas around which his speculation about poetry is organized: that "It Must be Abstract," "It Must Change," and "It Must Give Pleasure." He concludes with a tender tribute to a woman who, in his loving perception of her, becomes the "more than natural figure," the "irrational" or "more than rational distortion," "The Fiction that results from feeling." Even when searching through the "trash" of modernity in *Man on the Dump*, he discovers in the cacophony of grackles the vocation of the poet-priest and the materials for poetic stanzas. In *No Possum, No Sop, No Taters* he could face the "single emptiness of the bad" but find there "The last purity of the knowledge of good." As for Whitman, the prospect of death in Stevens's last poems could not obscure the "scrawny cry" which presaged the dawn of new life.

---

7. Quoted by Samuel F. Morse, *Wallace Stevens* (1970), p. 114.

Sunday Morning[1]

I

Complacencies of the peignoir,[2] and late
Coffee and oranges in a sunny chair,
And the green freedom of a cockatoo
Upon a rug mingle to dissipate
The holy hush of ancient sacrifice.   5
She dreams a little, and she feels the dark
Encroachment of that old catastrophe,
As a calm darkens among water-lights.
The pungent oranges and bright, green wings
Seem things in some procession of the dead,   10
Winding across wide water, without sound.
The day is like wide water, without sound,
Stilled for the passing of her dreaming feet
Over the seas, to silent Palestine,
Dominion of the blood and sepulchre.[3]   15

II

Why should she give her bounty to the dead?
What is divinity if it can come
Only in silent shadows and in dreams?
Shall she not find in comforts of the sun,
In pungent fruit and bright, green wings, or else   20
In any balm or beauty of the earth,
Things to be cherished like the thought of heaven?
Divinity must live within herself:
Passions of rain, or moods in falling snow;
Grievings in loneliness, or unsubdued   25
Elations when the forest blooms; gusty
Emotions on wet roads on autumn nights;
All pleasures and all pains, remembering

1. This, the poem that first brought Stevens to the attention of critics and other poets, is a religious meditation on death, on the conflicting but comparable claims of Christian and pagan rituals, and on the conflicting affirmations of supernatural values. The title suggests a pun on "Sunday": the day named for nature's "sun," and the day consecrated to the Christian "son" of God. Stevens declared in 1928 that it was "an expression of paganism, although, of course, I did not think I was expressing paganism when I wrote it" and that it is not simply a "woman's" but "anybody's meditation on religion and the meaning of life" (*Letters*, p. 250). Yet the poem, in its themes, setting, and structure, draws on Stevens's experience of the death in 1912 of his mother, a devout Christian. Stevens wrote his wife that his preoccupation with his mother's death would be "dissipated" by more pleasant surroundings and urged her to take pleasure in "sweet breaths, sweet fruits, sweet everything, make the most of it."

His journal records his mother's last days in her bedroom, surrounded by the familiar rugs and furnishings she had arranged, enjoying "grape juice, orange juice, lemon and sugar," trying to collect her thoughts, resisting death but reiterating her faith in immortality; her faith could be considered just, Stevens declared, "if there be a God" (*Letters*, pp. 172–74). The poem counterpoints the dying woman's thoughts and the Christian claims voiced through her, and the responses of the poet's voice to her beliefs and the crisis of her death. It was first published in *Poetry* magazine in 1915; the editor Harriet Monroe printed only five of its eight stanzas, but arranged them in the order Stevens suggested when consenting to the deletions (I, VIII, IV, V, and VII); he restored the deletions and the original sequence in subsequent printings.
2. Loose dressing gown.
3. Palestine contains many sacred tombs and was the scene of many blood sacrifices, including those of Jesus.

The bough of summer and the winter branch.
These are the measures destined for her soul. 30

III

Jove[4] in the clouds had his inhuman birth.
No mother suckled him, no sweet land gave
Large-mannered motions to his mythy mind
He moved among us, as a muttering king,
Magnificent, would move among his hinds, 35
Until our blood, commingling, virginal,
With heaven, brought such requital to desire
The very hinds[5] discerned it, in a star.
Shall our blood fail? Or shall it come to be
The blood of paradise? And shall the earth 40
Seem all of paradise that we shall know?
The sky will be much friendlier then than now,
A part of labor and a part of pain,
And next in glory to enduring love,
Not this dividing and indifferent blue. 45

IV

She says, "I am content when wakened birds,
Before they fly, test the reality
Of misty fields, by their sweet questionings;
But when the birds are gone, and their warm fields
Return no more, where, then, is paradise?" 50
There is not any haunt of prophecy,
Nor any old chimera[6] of the grave,
Neither the golden underground, nor isle
Melodious, where spirits gat them home,
Nor visionary south, nor cloudy palm 55
Remote on heaven's hill, that has endured
As April's green endures; or will endure
Like her remembrance of awakened birds,
Or her desire for June and evening, tipped
By the consummation of the swallow's wings. 60

V

She says, "But in contentment I still feel
The need of some imperishable bliss."
Death is the mother of beauty;[7] hence from her,
Alone, shall come fulfilment to our dreams
And our desires. Although she strews the leaves 65
Of sure obliteration on our paths,
The path sick sorrow took, the many paths
Where triumph rang its brassy phrase, or love
Whispered a little out of tenderness,
She makes the willow shiver in the sun 70
For maidens who were wont to sit and gaze

---

4. Roman name of Zeus, supreme god in
Greek and Roman mythology.
5. Farm servants. Possibly an allusion to
the shepherds who saw the star of Bethlehem that signaled the birth of Jesus.
6. Monster with lion's head, here an
image of fanciful beliefs in other worlds

and idealized realms.
7. Stevens refers to "gentle, delicate
death" in a letter about his mother's
dying (*Selected Letters*, p. 174). Cf. the
"delicious word death" in Whitman's
Out of the Cradle Endlessly Rocking,
line 168.

Upon the grass, relinquished to their feet.
She causes boys to pile new plums and pears
On disregarded plate. The maidens taste
And stray impassioned in the littering leaves.[8]   75

VI

Is there no change of death in paradise?
Does ripe fruit never fall? Or do the boughs
Hang always heavy in that perfect sky,
Unchanging, yet so like our perishing earth,
With rivers like our own that seek for seas   80
They never find, the same receding shores
That never touch with inarticulate pang?
Why set the pear upon those river-banks
Or spice the shores with odors of the plum?
Alas, that they should wear our colors there,   85
The silken weavings of our afternoons,
And pick the strings of our insipid lutes!
Death is the mother of beauty, mystical,
Within whose burning bosom we devise
Our earthly mothers waiting, sleeplessly.   90

VII

Supple and turbulent, a ring of men
Shall chant in orgy on a summer morn
Their boisterous devotion to the sun,
Not as a god, but as a god might be,
Naked among them, like a savage source.   95
Their chant shall be a chant of paradise,
Out of their blood, returning to the sky;
And in their chant shall enter, voice by voice,
The windy lake wherein their lord delights,
The trees, like serafin,[9] and echoing hills,   100
That choir among themselves long afterward.
They shall know well the heavenly fellowship
Of men that perish and of summer morn.
And whence they came and whither they shall go
The dew upon their feet shall manifest.[1]   105

VIII

She hears, upon that water without sound,
A voice that cries, "The tomb in Palestine
Is not the porch of spirits lingering.
It is the grave of Jesus, where he lay."
We live in an old chaos[2] of the sun,   110
Or old dependency of day and night,

---

8. Stevens explained: "Plate is used in the sense of so-called family plate [silver-plated dinner service]. Disregarded refers to the disuse into which things fall that have been possessed for a long time. I mean, therefore, that death releases and renews. What the old have come to disregard, the young inherit and make use of" (*Selected Letters*, p. 183).
9. Seraphim, high-ranking angels or celestial beings.
1. Stevens explained: "Life is as fugitive as dew upon the feet of men dancing in dew" (*Selected Letters*, p. 250).
2. Unformed primordial matter; the origin or original god of creation according to the *Theogony* of the Greek poet Hesiod (fl. eighth century B.C.); a traditional term for fiery masses of elements, comparable to those in the sun.

Or island solitude, unsponsored, free,
Of that wide water, inescapable.
Deer walk upon our mountains, and the quail
Whistle about us their spontaneous cries;   115
Sweet berries ripen in the wilderness;
And, in the isolation of the sky,
At evening, casual flocks of pigeons make
Ambiguous undulations as they sink,
Downward to darkness, on extended wings.[3]   120

                                            1915, 1923

Anecdote of the Jar

I placed a jar in Tennessee,
And round it was, upon a hill.
It made the slovenly wilderness
Surround that hill.

The wilderness rose up to it,   5
And sprawled around, no longer wild.
The jar was round upon the ground
And tall and of a port in air.
It took dominion everywhere.
The jar was gray and bare.   10
It did not give of bird or bush,
Like nothing else in Tennessee.

                                            1923

A High-toned Old Christian Woman

Poetry is the supreme fiction, madame.
Take the moral law and make a nave[1] of it
And from the nave build haunted heaven. Thus,
The conscience is converted into palms,
Like windy citherns[2] hankering for hymns.   5
We agree in principle. That's clear. But take
The opposing law and make a peristyle,[3]
And from the peristyle project a masque[4]

---

3. Cf. Francis Higginson's description of the New England wilderness in *New England's Plantation*, 1630 (quoted in Perry Miller and T. H. Johnson, *The Puritans*, 1938, p. 125): "Here are * * * all manner of Berries and Fruits. In the winter time I have seen flocks of pidgeons and have eaten them. They doe fly from tree to tree as other birds doe; they are of all colors as ours are, but their wings and tayles are farr longer and therefore it is likely they fly swifter to escape the terrible Hawkes in this country." Cf. also the closing lines of the ode *To Autumn* by the English poet John Keats (1795–1821): "While barred clouds bloom the soft-dying day, / And touch the stubble-plains with rosey hue; / Then in a wailful choir the small gnats mourn / Among the river sallows, borne aloft / Or sinking as the light wind lives or dies; / And full-grown lambs loud bleat from hilly bourn; / Hedge-crickets sing; and now with treble soft / The red-breast whistles from a garden-croft; / And gathering swallows twitter in the skies."

1. Main body of a church building, especially the vaulted central portion of a Christian Gothic church.

2. Variation of "cittern," a pear-shaped guitar.

3. Colonnade surrounding a building, especially the cells or main chamber of an ancient Greek temple—an "opposing law" to a Christian church.

4. Spectacle or entertainment consisting of music, dancing, mime, and often poetry.

Beyond the planets. Thus, our bawdiness,
Unpurged by epitaph, indulged at last,   10
Is equally converted into palms,
Squiggling like saxophones. And palm for palm,
Madame, we are where we began. Allow,
Therefore, that in the planetary scene
Your disaffected flagellants,[5] well-stuffed,   15
Smacking their muzzy bellies in parade,
Proud of such novelties of the sublime,
Such tink and tank and tunk-a-tunk-tunk,
May, merely may, madame, whip from themselves
A jovial hullabaloo among the spheres.   20
This will make widows wince. But fictive things
Wink as they will. Wink most when widows wince.

1923

The Emperor of Ice-Cream[1]

Call the roller of big cigars,
The muscular one, and bid him whip
In kitchen cups concupiscent curds.[2]
Let the wenches dawdle in such dress
As they are used to wear, and let the boys   5
Bring flowers in last month's newspapers.
Let be be finale[3] of seem.
The only emperor is the emperor of ice-cream.

Take from the dresser of deal,[4]
Lacking the three glass knobs, that sheet   10
On which she embroidered fantails[5] once
And spread it so as to cover her face.
If her horny feet protrude, they come
To show how cold she is, and dumb.

5. Religious devotees who whip themselves to intensify their fervor; "muzzy": sodden with drunkenness.

1. In 1945 Stevens remarked on the "singularity" of this poem: any poem "must have a peculiarity, as if it was the momentarily complete idiom of that which prompts it, even if what prompts it is the vaguest emotion." In 1933 he had written that it was his "favorite" and that he liked its combination of a "deliberately commonplace costume" with "the essential gaudiness of poetry." He said he remembered nothing of what prompted the poem other than his "state of mind": "This poem is an instance of letting myself go * * * . This represented what was in my mind at the moment, with the least possible manipulation" (*Letters*, pp. 500, 263–64). The poem's setting is a funeral or wake, an occasion when the living pay tribute to the dead and when commitments to life are juxtaposed against the face of death.

2. Stevens commented in 1945: "The words 'concupiscent curds' have no genealogy; they are merely expressive * * *.

They express the concupiscence of life, but, by contrast with the things in relation to them in the poem, they express or accentuate life's destitution, and it is this that gives them something more than a cheap lustre" (*Letters*, p. 500).

3. The concluding section of a musical composition, often including striking flourishes and virtuoso performances. Also: the final end, catastrophe. "Let be be finale of seem" suggests contrary readings: let being (actual existence) be consummated in a display of gaudy or brilliant illusions; let being put an end to mere appearances. Stevens claimed in 1939 that "the true sense of Let be be finale of seem is let being become the conclusion or denouement of appearing to be: in short, icecream is an absolute good. The poem is obviously not about icecream, but about being as distinguished from seeming to be" (*Letters*, p. 341).

4. Plain, unfinished wood.

5. Stevens explained that "the word fantails does not mean fans, but fantail pigeons * * *" (*Letters*, p. 340).

Let the lamp[6] affix its beam. 15
The only emperor is the emperor of ice-cream.

1923

To the One of Fictive Music

Sister and mother and diviner love,
And of the sisterhood of the living dead
Most near, most clear, and of the clearest bloom,
And of the fragrant mothers the most dear
And queen, and of diviner love the day 5
And flame and summer and sweet fire, no thread
Of cloudy silver sprinkles in your gown
Its venom of renown, and on your head
No crown is simpler than the simple hair.

Now, of the music summoned by the birth 10
That separates us from the wind and sea,
Yet leaves us in them, until earth becomes,
By being so much of the things we are,
Gross effigy and simulacrum,[1] none
Gives motion to perfection more serene 15
Than yours, out of our imperfections wrought,
Most rare, or ever of more kindred air
In the laborious weaving that you wear.

For so retentive of themselves are men
That music is intensest which proclaims 20
The near, the clear, and vaunts the clearest bloom,
And of all vigils musing the obscure,
That apprehends the most which sees and names,
As in your name, an image that is sure,
Among the arrant[2] spices of the sun, 25
O bough and bush and scented vine, in whom
We give ourselves our likest issuance.

Yet not too like, yet not so like to be
Too near, too clear, saving a little to endow
Our feigning with the strange unlike, whence
  springs 30
The difference that heavenly pity brings.
For this, musician, in your girdle fixed
Bear other perfumes. On your pale head wear
A band entwining, set with fatal stones.
Unreal, give back to us what once you gave: 35
The imagination that we spurned and crave.

1923

---

6. I.e., the light illuminating the corpse; also, the stark illumination of art.
1. Semblance, something having the form without the substance of a material object.
2. Confirmed, thoroughgoing; notorious.

The American Sublime

How does one stand
To behold the sublime,
To confront the mockers,
The mickey mockers
And plated pairs?[1]   5

When General Jackson
Posed for his statue[2]
He knew how one feels.
Shall a man go barefoot
Blinking and blank?   10

But how does one feel?
One grows used to the weather,
The landscape and that;
And the sublime comes down
To the spirit itself,   15

The spirit and space,
The empty spirit
In vacant space.
What wine does one drink?
What bread does one eat?[3]   20

                                                      1935

Dry Loaf[1]

It is equal to living in a tragic land
To live in a tragic time.
Regard now the sloping, mountainous rocks
And the river that batters its way over stones,
Regard the hovels of those that live in this land.   5

That was what I painted behind the loaf,
The rocks not even touched by snow,
The pines along the river and the dry men blown
Brown as the bread, thinking of birds
Flying from burning countries and brown sand
    shores,   10

1. Images opposed to the authentic sublime. "Mickey" is a pejorative slang term for Irishman, while "mickey mockers" evokes the sentimental, tricky Mickey Mouse in animated film cartoons. "Plated pairs": i.e., stereotyped or identical objects, of ordinary material but coated over with some more attractive surface, like silver plate.
2. In *The Noble Rider and the Sound of Words*, Stevens described the statue of General and President Andrew Jackson

(1767–1845) in Lafayette Square, Washington, D.C., by the sculptor Clark Mills (1810–83). He associated it with inflated eloquence and appeals to the complacency of a democratic society and to fixed, stereotyped notions of the ideal.
3. "Wine" and "bread": familiar forms of sustenance. Also the bread and wine of Christian communion.
1. A loaf of bread, but also a mountain or hill in the shape of a loaf of bread.

Birds that came like dirty water in waves
Flowing above the rocks, flowing over the sky,
As if the sky was a current that bore them along,
Spreading them as waves spread flat on the shore,
One after another washing the mountains bare.   15

It was the battering of drums I heard
It was hunger, it was the hungry that cried
And the waves, the waves were soldiers moving,
Marching and marching in a tragic time
Below me, on the asphalt, under the trees.   20

It was soldiers went marching over the rocks
And still the birds came, came in watery flocks,
Because it was spring and the birds had to come.
No doubt that soldiers had to be marching
And that drums had to be rolling, rolling, rolling.   25

1942

Of Modern Poetry

The poem of the mind in the act of finding
What will suffice. It has not always had
To find: the scene was set; it repeated what
Was in the script.
                    Then the theatre was changed   5
To something else. Its past was a souvenir.
It has to be living, to learn the speech of the place.
It has to face the men of the time and to meet
The women of the time. It has to think about war
And it has to find what will suffice. It has   10
To construct a new stage. It has to be on that stage
And, like an insatiable actor, slowly and
With meditation, speak words that in the ear,
In the delicatest ear of the mind, repeat,
Exactly, that which it wants to hear, at the sound   15
Of which, an invisible audience listens,
Not to the play, but to itself, expressed
In an emotion as of two people, as of two
Emotions becoming one. The actor is
A metaphysician in the dark, twanging   20
An instrument, twanging a wiry string that gives
Sounds passing through sudden rightnesses, wholly
Containing the mind, below which it cannot descend,
Beyond which it has no will to rise.
                                It must   25
Be the finding of a satisfaction, and may
Be of a man skating, a woman dancing, a woman
Combing. The poem of the act of the mind.

1942

Asides on the Oboe

The prologues are over. It is a question, now,
Of final belief. So, say that final belief
Must be in a fiction. It is time to choose.

I

That obsolete fiction of the wide river in
An empty land;[1] the gods that Boucher killed;[2]   5
And the metal heroes that time granulates—
The philosophers' man alone still walks in dew,
Still by the sea-side mutters milky lines
Concerning an immaculate imagery.
If you say on the hautboy[3] man is not enough,   10
Can never stand as god, is ever wrong
In the end, however naked, tall, there is still
The impossible possible philosophers' man,
The man who has had the time to think enough,
The central man, the human globe, responsive   15
As a mirror with a voice, the man of glass,
Who in a million diamonds sums us up.[4]

II

He is the transparence of the place in which
He is and in his poems we find peace.
He sets this peddler's pie and cries in summer,   20
The glass man, cold and numbered, dewily cries,
"Thou art not August[5] unless I make thee so."
Clandestine steps upon imagined stairs
Climb through the night, because his cuckoos call.

III

One year, death and war prevented the jasmine scent   25
And the jasmine islands were bloody martyrdoms.
How was it then with the central man? Did we
Find peace? We found the sum of men. We found,
If we found the central evil, the central good.
We buried the fallen without jasmine crowns.   30
There was nothing he did not suffer, no; nor we.

It was not as if the jasmine ever returned.
But we and the diamond globe at last were one.
We had always been partly one. It was as we came
To see him, that we were wholly one, as we heard   35

1. I.e., the Christian Holy Land, with the river Jordan; also America, with the Mississippi River and other wide rivers.
2. Boucher could be Pierre Boucher (1622–1717), French explorer and Canadian settler; and Jonathan Boucher (1738–1804), Anglican clergyman and Loyalist during the American Revolution, author of *A View of the Causes and Consequences of the American Revolution* (1797); and François Boucher (1703–70), French decorative painter in the royal court of Louis XV. They represent, respectively, the displacement of native Indian gods and cultures; the rejection of the political "gods" of revolutionary America; and the trivialization of ancient mythologies in paintings of French court life.
3. Oboe.
4. Cf. Emerson's conception of the ideal poet in *The Poet* (1844): "He is a sovereign, and stands in the center."
5. I.e., the summer month, but also denoting nobility, magnificence.

Him chanting for those buried in their blood,
In the jasmine haunted forests, that we knew
The glass man, without external reference.

1940, 1942

The Motive for Metaphor

You like it under the trees in autumn,
Because everything is half dead.
The wind moves like a cripple among the leaves
And repeats words without meaning.

In the same way, you were happy in spring,   5
With the half colors of quarter-things,
The slightly brighter sky, the melting clouds,
The single bird, the obscure moon—

The obscure moon lighting an obscure world
Of things that would never be quite expressed,   10
Where you yourself were never quite yourself
And did not want nor have to be,

Desiring the exhilarations of changes:
The motive for metaphor, shrinking from
The weight of primary noon,   15
The A B C of being,

The ruddy temper, the hammer
Of red and blue, the hard sound—
Steel against intimation—the sharp flash,
The vital, arrogant, fatal, dominant X.   20

1947

Man Carrying Thing

The poem must resist the intelligence
Almost successfully. Illustration:

A brune[2] figure in winter evening resists
Identity. The thing he carries resists

The most necessitous sense. Accept them, then,   5
As secondary (parts not quite perceived

Of the obvious whole, uncertain particles
Of the certain solid, the primary free from doubt,

Things floating like the first hundred flakes of snow
Out of a storm we must endure all night,   10

---

2. French term intended to resist "intelligent" meaning. The adjective signifies a
feminine figure with brown skin or hair.
The noun signifies dusk.

Out of a storm of secondary things),
A horror of thoughts that suddenly are real.

We must endure our thoughts all night, until
The bright obvious stands motionless in cold.

1947

Questions Are Remarks

In the weed[1] of summer comes this green sprout why.
The sun aches and ails and then returns halloo
Upon the horizon amid adult enfantillages.[2]

Its fire fails to pierce the vision that beholds it,
Fails to destroy the antique acceptances,   5
Except that the grandson[3] sees it as it is,

Peter the voyant,[4] who says "Mother, what is that"—
The object that rises with so much rhetoric,
But not for him. His question is complete.

It is the question of what he is capable.   10
It is the extreme, the expert aetat.[5] 2.
He will never ride the red horse[6] she describes.

His question is complete because it contains
His utmost statement. It is his own array,
His own pageant and procession and display,   15

As far as nothingness permits . . . Hear him.
He does not say, "Mother, my mother, who are you,"
The way the drowsy, infant, old men do.

1950

The Hermitage at the Center

The leaves on the macadam make a noise—
    How soft the grass on which the desired
    Reclines in the temperature of heaven—

Like tales that were told the day before yesterday—
    Sleek in a natural nakedness,   5
    She attends the tintinnabula—[7]

And the wind sways like a great thing tottering—
    Of birds called up by more than the sun,
    Birds of more wit, that substitute—

1. Wild vegetation. Also garment (obs.).
2. Childishnesses.
3. The boy, Peter, though a mere grandson, and his elemental question, though a simple one, are the match for his grandiloquent grandsires and the age-old sun.
4. Clairvoyant, one who can see clearly.
5. Age (Latin abbreviation).
6. In ancient mythology, the god of the sun, Apollo, drives a fiery chariot across the sky.
7. Jingling, tinkling sound as of bells. Cf. "the tintinabulation of the bells" in Edgar Allan Poe's poem *The Bells*.

Which suddenly is all dissolved and gone—   10
    Their intelligible twittering
    For unintelligible thought.

And yet this end and this begining are one,
    And one last look at the ducks is a look
    At lucent children round her in a ring.   15

1954

Not Ideas about the Thing
but the Thing Itself

At the earliest ending of winter,
In March, a scrawny cry from outside
Seemed like a sound in his mind.

He knew that he heard it,
A bird's cry, at daylight or before,   5
In the early March wind

The sun was rising at six,
No longer a battered panache[1] above snow . . .
It would have been outside.

It was not from the vast ventriloquism   10
Of sleep's faded papier-mâché[2] . . .
The sun was coming from outside.

That scrawny cry—it was
A chorister whose c[3] preceded the choir.
It was part of the colossal sun,   15

Surrounded by its choral rings,
Still far away. It was like
A new knowledge of reality.

1954

1. Tuft, plume.
2. Paper pulp used for molding decorative objects and imitations such as stage props and masks.

3. I.e., the note, middle C, sounded by a member of a choir which establishes the correct pitch.

HILDA DOOLITTLE (H. D.)
1886–1961

In January, 1913, Ezra Pound placed in *Poetry* magazine three poems by Hilda Doolittle and ordered that they be signed "H. D., Imagiste." Thus, simultaneously, Hilda Doolittle was "discovered," H. D. was invented, and the Imagist movement was proclaimed. The identification of H. D. with Imagism, persisting long beyond the life of the movement, has perhaps unfairly diminished the amount and limited the scope of critical attention to her work. She was, however, the most faithful adherent to the precepts of that "group of poets who, between 1912 and 1917, joined the reaction against the careless techniques and extra-poetic values of much nineteenth century verse."[1] The essential qualities of her poetry—its commitment to the precise word and to the presentation of hard and clear images, its creation of new rhythms to express new moods, its concentration—fulfill all the dictates of the Imagist Credo outlined by the contributors to the 1915 Imagist anthology.

The prevalence of classical and mythological settings in H. D.'s work and her borrowing of subjects and characters from ancient writers have led some readers unjustly to label her an escapist or "an inspired anachronism."[2] As her literary executor Norman Holmes Pearson comments, "She has been so praised as a kind of Greek publicity girl that people have forgotten that she writes the most intensely personal poems using Greek myth as a metaphor."[3] H. D.'s poems, like the Sapphic fragments of the ancient Greek poet Sappho upon which they sometimes are constructed, examine the problem of being a woman and a poet in a fragmented world. Her classical past is not simply an escapist alternative to modern life; its relation to the present is more intricate and symbiotic, like the relation of the partially erased earlier writing on a palimpsest to the new layer of words being inscribed.

Hilda Doolittle was born in Bethlehem, Pennsylvania, in 1886. Her father, a remote and forbidding figure to her, was a professor of mathematics and astronomy at Lehigh University and later at the University of Pennsylvania. H. D. attended local public and private schools and, in 1904, entered Bryn Mawr where she studied for two years before withdrawing because of poor health. Ezra Pound was a close friend and, according to William Carlos Williams, was in love with her. Williams, in his autobiography, remembers H. D. as a tall, carelessly dressed young woman with "a provocative indifference to rule and order which I liked."[4]

H. D. lived at home, reading mostly Greek and Latin literature and writing poetry, until in 1911 she went to Europe for what she thought would be a summer's visit. In London she renewed her friendship with Pound and was introduced into a circle of young poets who were experimenting in free verse and would soon refer to themselves as Imagists. One of these poets was Richard Aldington, whom H. D. married in 1913. They collaborated on some translations from the Greek, and, when he was called up for military service in 1916, H. D. assumed his position as literary editor of *The Egoist*. By the end of the war, the marriage had collapsed. In 1919, H. D. gave birth to a daughter and separated from Aldington. Both she and the infant were seriously ill, and H. D. credited their recovery to the spiritual and material assistance of Winifred Ellerman, a young admirer of her poetry who had become her friend and benefactress. In the next three years, H. D. and Miss Ellerman, who later wrote historical novels under the name of "Bryher," took trips to Greece, America, and Egypt. In 1923,

1. Stanley K. Coffman, Jr., *Imagism: A Chapter for the History of Modern Poetry* (Norman, Okla., 1951), p. 3.
2. Thomas Burnett Swann, *The Classical World of H. D.* (Lincoln, Neb., 1962), p. 3.

3. "Norman Holmes Pearson on H. D.: An Interview," *Contemporary Literature*, 10 (1969), 441.
4. William Carlos Williams, *The Autobiography of William Carlos Williams* (New York, 1948), p. 68.

H. D. settled in Switzerland where she spent most of the next thirty-eight years.

The settings of H. D.'s works of prose fiction—*Palimpsest* (1926), *Hedylus* (1928), and *Bid Me to Live* (1960)—range from ancient Rome and Greece to modern England and Egypt, but they are similar in theme and technique. Each of *Palimpsest's* three sections, and each of the other books, focuses meditatively on a sensitive woman's attempt to come to terms with herself and with a coarse and violent world. Like her long poem *Helen in Egypt* (1961), a dramatic monologue by Helen of Troy, the prose works are versions of spiritual autobiography.

H. D. chose to live in London during World War II. There she wrote her *Tribute to Freud* (1956), an account of her psychoanalysis by Freud in 1933 and 1934 and a re-creation of the self-portrait that emerged from it. She also wrote a trilogy of war poems—*The Walls Do Not Fall* (1944), *Tribute to the Angels* (1945), and *The Flowering of the Rod* (1946). These poems affirm the survival, even the strengthening, of the human spirit amid chaos and devastation. H. D. finally seeks a personal and universal mystic experience, but it is her language that both preserves her contact with the outside and provides "an inner region of defense."[5]

Oread[1]

Whirl up, sea—
whirl your pointed pines,
splash your great pines
on our rocks,
hurl your green over us,   5
cover us with your pools of fir.

1914, 1924

Heat[2]

O wind, rend open the heat,
cut apart the heat,
rend it to tatters.
Fruit cannot drop
through this thick air—   5
fruit cannot fall into heat
that presses up and blunts
the points of pears
and rounds the grapes.

Cut the heat—   10
plough through it,
turning it on either side
of your path.

1916

---

5. Hilda Doolittle, "A Note on Poetry," in *The Oxford Anthology of American Literature*, ed. William Rose Benét and Norman Holmes Pearson (New York, 1938), p. 1288.

1. A nymph of mountains and hills.
2. In *Sea Garden* (1916) and *Collected Poems* (1925), *Heat* is published as the second part of a two-part poem called *Garden*.

At Baia[3]

I should have thought
in a dream you would have brought
some lovely, perilous thing,
orchids piled in a great sheath,
as who would say (in a dream)   5
I send you this,
who left the blue veins
of your throat unkissed.

Why was it that your hands   10
(that never took mine)
your hands that I could see
drift over the orchid heads
so carefully,
your hands, so fragile, sure to lift
so gently, the fragile flower stuff—   15
ah, ah, how was it

You never sent (in a dream)
the very form, the very scent,
not heavy, not sensuous,
but perilous—perilous—   20
of orchids, piled in a great sheath,
and folded underneath on a bright scroll
some word:

Flower sent to flower;
for white hands, the lesser white,   25
less lovely of flower leaf,

or

Lover to lover, no kiss,
no touch, but forever and ever this.

                                          1921

Fragment 113

"Neither honey nor bee for me."
—SAPPHO[1]

Not honey,
not the plunder of the bee
from meadow or sand-flower
or mountain bush;
from winter-flower or shoot   5
born of the later heat:
not honey, not the sweet
stain on the lips and teeth:

---

3. An ancient Roman resort town.
1. Greek woman lyric poet of Lesbos in the seventh century B.C.

not honey, not the deep
plunge of soft belly   10
and the clinging of the gold-edged
pollen-dusted feet;

though rapture blind my eyes,
and hunger crisp
dark and inert my mouth,   15
not honey, not the south,
not the tall stalk
of red twin-lilies,
nor light branch of fruit tree
caught in flexible light branch;   20
not honey, not the south;
ah flower of purple iris,
flower of white,
or of the iris, withering the grass—
for fleck of the sun's fire,   25
gathers such heat and power,
that shadow-print is light,
cast through the petals
of the yellow iris flower;

not iris—old desire—old passion—   30
old forgetfulness—old pain—
not this, nor any flower,
but if you turn again,
seek strength of arm and throat,
touch as the god;   35
neglect the lyre-note;
knowing that you shall feel,
about the frame,
no trembling of the string
but heat, more passionate   40
of bone and the white shell
and fiery tempered steel.

                                               1922

From Tribute to the Angels

4

Not in our time, O Lord,
the plowshare for the sword,

not in our time, the knife,
sated with life-blood and life,

to trim the barren vine;   5
no grape-leaf for the thorn,

no vine-flower for the crown;
not in our time, O King,

the voice to quell the re-gathering,
thundering storm.   10

6

Never in Rome,
so many martyrs fell;

not in Jerusalem,
never in Thebes,[1]

so many stood and watched   5
chariot-wheels turning,

saw with their very eyes,
the battle of the Titans,[2]

saw Zeus' thunderbolts in action
and how, from giant hands,   10

the lightning shattered earth
and splintered sky, nor fled

to hide in caves,
but with unbroken will,

with unbowed head, watched   15
and though unaware, worshipped

and knew not that they worshipped
and that they were

that which they worshipped;
had they known, the fire   20

of strength, endurance, anger
in their hearts,

was part of that same fire
that in a candle on a candle-stick

or in a star,   25
is known as one of seven,

is named among the seven Angels,
Uriel.[3]

1945

1. Ancient Greek city and battle site.
2. Gods and goddesses of Greek mythology. They were overthrown in a 10-year
battle by Zeus, who thereby became
ruler of the Greek gods.
3. One of the Archangels, the angel of
fire or light.

MARIANNE MOORE
1887–1972

The "poet's poet" among those who rose to prominence toward the end of World War I was a woman, Marianne Craig Moore. She produced a small but unmistakably unique body of poems whose originality was greatly admired by such contemporaries as Pound, Eliot, Stevens, and Williams, and her radically inventive forms had an impact on Robert Lowell, Richard Wilbur, and others of the generation who began writing after World War II. For years an inconspicuous figure in the literary world, she was nevertheless among the finest of this century's American poets. In later years her activities on behalf of the Ford Motor Company and the Brooklyn Dodgers baseball team, along with her many literary honors, made her a charming footnote to the public scene.

She was born in 1887 to a devoutly Presbyterian family in Kirkwood, Missouri, near St. Louis. After her father suffered a business and personal collapse and abandoned the family, she moved with her mother in 1894 to Carlisle, Pennsylvania. She attended the Metzger Institute, where her mother taught, and in 1909 graduated from Bryn Mawr College, where she did better in biology than in literature but tried her hand at writing poetry and discovered the power of seventeenth-century English prose.

Moore took courses at the Carlisle Commercial College and traveled with her mother in England and France in 1911, then from 1911 to 1915 taught commercial subjects at the U.S. Indian School in Carlisle. Her verse was first published in 1915 and 1916 in *The Egoist* (London), *Poetry*, and *Others*, the magazine founded by Alfred Kreymborg to encourage experimental writing, and by the time her poems were appearing regularly in the journals she and her mother moved, first to New Jersey, and later to Brooklyn, to remain near her brother, who had been ordained a Presbyterian minister. (Throughout her life she remained exceptionally close to her mother and brother, whose judgment she often consulted when composing or revising her poems.) She was working as tutor and secretary in a girls' school and as a part-time assistant in a branch library, and attending parties with friends among the vanguard painters and writers associated with *Others*, when her first two books were published. Her jobs, she said in a later interview, had "hardened my muscle considerably, my mental approach to things."[1] Her *Poems* (1921) was brought out in London without her knowledge by two admirers, the writers Bryher (Mrs. Robert McAlmon) and Hilda Doolittle. *Observations* appeared in 1924 and received the Dial Award for that year.

Her tenure as editor of *The Dial* (1925–29) she described as five years of "compacted pleasantness" in an "atmosphere of excited triumph."[2] Her reviews commanded respect and her exacting standards inspired awe; she was criticized, however, for rejecting excerpts from James Joyce's *Finnegans Wake* on the grounds that its language and sexual allusions were offensive, and the alterations she insisted on in one of Hart Crane's poems provoked

1. Marianne Moore, *A Collection of Critical Essays*, p. 23.

2. "The Dial: A Retrospect," *Predilections* (1955), pp. 103, 105.

his exclamation that America's two most important literary journals (*The Dial* and Harriet Monroe's *Poetry*) were edited by "hysterical virgins."[3] She wrote virtually no verse until the magazine was disbanded in 1929 but resumed her poetic career with the publication of *Selected Poems* (1935). T. S. Eliot in his introduction declared that in her poetry "an original sensibility and alert intelligence and deep feeling have been engaged in maintaining the life of the English language."

One of the distinguishing features of her verse is its saturation "in the perfections of prose," as Eliot put it (introduction, 7), the qualities she admired in "prose stylists" from Xenophon, Francis Bacon, and Thomas Browne to Henry James and Ezra Pound.[4] Testifying to the impact of scientific language and procedures on her imagination, she declared that "Precision, economy of statement, logic employed to ends that are disinterested * * * liberate * * * the imagination, it seems to me."[5] Prose sources, as often as poetic ones, provide her materials, which she acknowledges candidly by putting quotation marks around excerpts from history books, reference works, and even travel brochures. The definiteness and solidity of prose, the momentum of its sentences and paragraphs, are incorporated in her verse to attain an "unbearable accuracy," a "precision" that has both "impact and exactitude, as with surgery," she declared.[6]

Her precision is combined with an emotional intensity and compressed drama that reveal Biblical prose and poetry to be one of the governing models for her verse. Precision, she wrote, derives from "diction that is virile because galvanized against inertia," and her stanza forms and versification are designed to achieve this galvanizing effect. She considered the stanza, not the line, to be the unit of poetry. Once words, "clustering like chromosomes," had shaped a stanza, she adjusted it and made subsequent stanzas conform to the first.[7] The results in such elaborately structured poems as *The Fish* or *Virginia Britannia* are numerous. The corresponding lines in all stanzas have the exact same number of syllables; rhymes occur in a strictly regular though unique sequence, even when they depend on unaccented syllables and words broken at the hyphen; concealed or interior and approximate rhymes contribute to a texture that is precise in tactile and auditory as well as visual detail, and also unbroken in the continuity of its syntax. Her poems become at once descriptive and meditative, analytic and emotional. The form provides an instrument for guiding and compressing emotion with exceptional control, and for focusing attention and shifting the focus of attention as the experience dramatized approaches its climax or the meditation on experience proceeds toward its conclusion. As she wrote in *The Past Is Present*, one of her many works which reflect on the nature of poetry, a necessary "expediency determines the form" but "Ecstasy affords the occasion."

Moore attained the height of her powers in *The Pangolin and other Verse* (1936) and in *What Are Years?* (1941) and *Nevertheless* (1944), which include poems written in disturbed response to World War II. She continued to write poems about animals which were none the less accurate in descriptive detail for being wide-ranging reflections on human experience, but the emotional depth and the scale of her meditations became

---

3. Hart Crane, *Letters*, p. 289.
4. "Interview," *Collection of Critical Essays*, pp. 30–31.
5. *Ibid.*, p. 23.

6. *Predilections*, p. 4.
7. "Interview," *Collection of Critical Essays*, pp. 29, 34.

more extensive, and her tone more elegiac. Her treatment of central themes—the search for freedom within restraint or the validity of familial love—becomes more troubled in such poems as *Bird-Witted* and *The Paper Nautilus*. Her most ambitious war poem, *In Distrust of Merits*, attempts to encompass the horrors of World War II and to find that the sacrifice of lives has been redeemed by a humane vision of racial tolerance and amity, but it confesses that the redemptive transformation has not yet been achieved.

Her most ambitious undertaking in later years was her translation of *The Fables of La Fontaine* (1954), but her original verse after 1956 became more perfunctory. *Hometown Piece for Messrs. Alston and Reese* had particular appeal for Brooklyn Dodger and other baseball fans (a fan herself, she once tossed in the ball to open the official season), and her topical poems frequently reflected the public life with which she became more involved after receiving the Bollingen, National Book, and Pulitzer awards for *Collected Poems* in 1951. One poem of 1963 was commissioned by the Steuben Glass Corporation, and in 1955 she advised the Ford Motor Company, in a long, hilarious, and futile series of letters published in *The New Yorker*, on the name for a new car eventually named (by the company) for Edsel Ford. (Among countless suggestions Moore proposed "Utopian Turtletop" and "astranaut" [sic].) Although she disapproved of the explicit sexual elements in some postwar poetry, she appreciated the work of many younger poets.[8]

Her own place in the history of American poetry was well defined by the poet Randall Jarrell, who insisted in 1953 that she had "discovered both a new sort of subject (a queer many-headed one) and a new sort of connection and structure for it, so that she has widened the scope of poetry. * * *"[9]

Except where indicated otherwise, selections are from *The Complete Poems of Marianne Moore* (1967).

The Past Is the Present[1]

If external action is effete
   and rhyme is outmoded,
      I shall revert to you,
Habakkuk,[2] as on a recent occasion I was goaded
      into doing by XY, who was speaking of unrhymed verse. 5
This man said—I think that I repeat
   his identical words:
      'Hebrew poetry is
prose with a sort of heightened consciousness.'[3] Ecstasy affords
      the occasion and expediency determines the form.   10

1915, 1924

To a Snail

   If "compression is the first grace of style,"[4]
   you have it. Contractility is a virtue

---

8. "The Way Our Poets Have Taken Since the War," *Marianne Moore Reader*, pp. 237–43.
9. *Poetry and the Age*, pp. 172–73.
1. Text and format are those of *Selected Poems* (1935).
2. A minor prophet, author of the Book of Habakkuk in the Old Testament.

3. "Dr. E. H. Kellogg in Bible class, Presbyterian Church, Carlisle, Pennsylvania" [Moore's note].
4. " 'The very first grace of style is that which comes from compression.' *Demetrius on Style* translated by W. Hamilton Fyfe. Heinemann, 1932" [Moore's note].

as modesty is a virtue.
It is not the acquisition of any one thing
    that is able to adorn,   5
or the incidental quality that occurs
as a concomitant of something well said,
    that we value in style,
but the principle that is hid:
    in the absence of feet, "a method of conclusions";   10
    "a knowledge of principles,"
in the curious phenomenon of your occipital horn.

                              1924

Peter[1]

Strong and slippery, built for the midnight grass-party con-
    fronted by four cats,
    he sleeps his time away—the detached first claw on the
        foreleg, which corresponds
to the thumb, retracted to its tip; the small tuft of fronds   5
    or katydid legs above each eye, still numbering the units in
        each group;
        the shadbones regularly set about the mouth, to droop
        or rise

in unison like the porcupine's quills—motionless. He lets   10
    himself be flat-
tened out by gravity, as it were a piece of seaweed tamed and
    weakened by
exposure to the sun; compelled when extended, to lie
    stationary. Sleep is the result of his delusion that one   15
    must do as
    well as one can for oneself; sleep—epitome of what is to

him as to the average person, the end of life. Demonstrate on
    him how
the lady caught the dangerous southern snake,[2] placing a   20
    forked stick on either
side of its innocuous neck; one need not try to stir
    him up; his prune-shaped head and alligator eyes are not a
    party to the
    joke. Lifted and handled, he may be dangled like an eel 25
    or set

1. This poem presents a woman's meditation on the masculine character, and a Protestant's meditation on Catholicism. The cat's name "Peter" draws attention to his masculine gender and to religious issues that surface in the poem. Simon Peter, Christ's apostle and author of the First and Second Epistles, denied Christ three times to protect himself (Luke 23.56–62) but later as Bishop held the cathedral chair at Rome and founded the Roman Catholic Church, which requires celibacy of its priests and does not admit women to the priesthood. Miss Moore wryly drew attention to the issues of both personal and religious devotion in her note to the title: "Cat owned by Miss Magdalen Heuber and Miss Maria Weniger." The Virgin Mary and Mary Magdalen (one married, one not) were two women whose devotion to Christ, unlike Peter's, was unwavering. The poem first appeared in *Observations* (1929). The text and format printed here are those of *Selected Poems* (1935).
2. In Christian iconography, the Virgin Mary stamps out evil, which takes the form of a snake.

up on the forearm like a mouse; his eyes bisected by pupils of a
    pin's
    width, are flickeringly exhibited, then covered up. May be?
        I should say   30
    might have been; when he has been got the better of in a
        dream—as in a fight with nature or with cats— we all know
            it. Profound sleep is
        not with him a fixed illusion. Springing about with
            froglike ac-   35

curacy, emitting jerky cries when taken in the hand, he is
        himself
    again; to sit caged by the rungs of a domestic chair would be
        unprofit-
    able—human. What is the good of hypocrisy? It   40
        is permissible to choose one's employment, to abandon the
            wire nail, the
            roly-poly, when it shows signs of being no longer a
                pleas-

ure, to score the adjacent magazine with a double line of
    strokes.   45
            He can
    talk, but insolently says nothing. What of it? When one is
        frank, one's very
    presence is a compliment. It is clear that he can see
        the virtue of naturalness, that he is one of those who do not
            regard   50
        the published fact as a surrender. As for the disposition

invariably to affront, an animal with claws wants to have to use
    them; that eel-like extension of trunk into tail is not an
        accident. To   55
    leap, to lengthen out, divide the air—to purloin, to pursue.
        To tell the hen: fly over the fence, go in the wrong way in
            your perturba-
            tion—this is life; to do less would be nothing but
                dishonesty.   60
c. 1921   1924, 1935

The Fish

    wade
        through black jade.
            Of the crow-blue mussel shells, one keeps
                adjusting the ash heaps;
                    opening and shutting itself like   5

        an
            injured fan.
                The barnacles which encrust the side

of the wave, cannot hide
        there for the submerged shafts of the   10

sun,
split like spun
        glass, move themselves with spotlight swiftness
        into the crevices—
            in and out, illuminating   15

the
turquoise sea
        of bodies. The water drives a wedge
        of iron through the iron edge
        of the cliff; whereupon the stars,   20

pink
rice-grains, ink-
        bespattered jellyfish, crabs like green
        lilies, and submarine
            toadstools, slide each on the other.   25

All
external
        marks of abuse are present on this
        defiant edifice—
            all the physical features of   30

ac-
cident—lack
        of cornice, dynamite grooves, burns, and
        hatchet strokes, these things stand
            out on it; the chasm side is   35

dead.
Repeated
        evidence has proved that it can live
        on what can not revive
            its youth. The sea grows old in it.   40

                                        1918, 1924

The Student[1]

"In America," began
the lecturer, "everyone must have a
degree. The French do not think that
all can have it, they don't say everyone
        must go to college."[2] We   5

1. This poem first appeared in 1932 as the second in a three-part work entitled *Part of a Novel, Part of a Poem, Part of a Play*. Omitted from some later volumes, it was included as a separate poem in *What Are Years?* (1941) and, considerably revised, in *Complete Poems* (1967), the text reprinted here.

2. " 'Les Idéals de l'Education Française,' lecture, December 3, 1931, by M. Auguste Desclos, Director-adjoint, Office National des Universités et Ecoles Françaises de Paris" [Moore's note].

incline to feel
  that although it may be unnecessary

to know fifteen languages,
one degree is not too much. With us, a
school—like the singing tree of which   10
the leaves were mouths singing in concert[3]
  is both a tree of knowledge
and of liberty—
  seen in the unanimity of college

mottoes, *Lux et veritas*,   15
*Christo et ecclesiae, Sapient
felici*.[4] It may be that we
have not knowledge, just opinions, that we
  are undergraduates,
not students; we know   20
  we have been told with smiles, by expatriates

of whom we had asked "When will
your experiment be finished?" "Science
is never finished."[5] Secluded
from domestic strife, Jack Bookworm led a   25
  college life, says Goldsmith;[6]
and here also as
  in France or Oxford, study is beset with

dangers—with bookworms, mildews,
and complaisancies. But someone in New   30
England has known enough to say
the student is patience personified,
  is a variety
of hero, "patient
  of neglect and of reproach"—who can "hold by   35

himself."[7] You can't beat hens to
make them lay. Wolf's wool is the best of wool,
but it cannot be sheared because
the wolf will not comply.[8] With knowledge as
  with the wolf's surliness,   40
the student studies
  voluntarily, refusing to be less

---

3. " 'Each leaf was a mouth, and every
leaf joined in concert.' *Arabian Nights*"
[Moore's note].
4. "Light and Truth," "For Christ and
the Church," "Fortunate Are the
Learned." The first is the motto of Yale
University; the second appeared on Harvard's seal from 1693 to 1935.
5. "Albert Einstein to an American student, *New York Times*" [Moore's note].
6. Moore's note refers readers to the
character in *The Doubled Transforma-*

tion by the British writer Oliver Goldsmith (c. 1730–74).
7. "Emerson in *The American Scholar*:
'There can be no scholar without the heroic mind'; 'Let him hold by himself; . . .
patient of neglect, patient of reproach"
[Moore's note].
8. "Edmund Burke, November, 1781,
in reply to Fox: 'There is excellent wool
on the back of a wolf and therefore he
must be sheared. . . . But will he
comply?' " [Moore's note].

than individual. He
"gives his opinion and then rests on it";[9]
he renders service when there is   45
no reward, and is too reclusive for
   some things to seem to touch
him, not because he
   has no feeling but because he has so much.

                      1932, 1941

Bird-Witted[1]

With innocent wide penguin eyes, three
   large fledgling mockingbirds below
the pussy-willow tree,
   stand in a row,
wings touching, feebly solemn,   5
till they see
      their no longer larger
      mother bringing
something which will partially
feed one of them.   10

Toward the high-keyed intermittent squeak
   of broken carriage springs, made by
the three similar, meek-
   coated bird's-eye
freckled forms she comes; and when   15
from the beak
      of one, the still living
      beetle has dropped
out, she picks it up and puts
it in again.   20

Standing in the shade till they have dressed
   their thickly filamented, pale
pussy-willow-surfaced
   coats, they spread tail
and wings, showing one by one,   25
the modest
      white stripe lengthwise on the
      tail and crosswise
underneath the wing, and the
accordion   30

---

9. Moore's note quotes from the art critic Henry McBride, speaking of the appraisal of art works in the *"New York Sun,* December 12, 1931: 'Dr. Valentiner . . . has the typical reserve of the student. He does not enjoy the active battle of opinion that invariably rages when a decision is announced that can be weighed in great sums of money. He gives his opinion firmly and rests upon that.' "

1. Moore's note quotes the statement "If a boy be bird-witted" from *The Advancement of Learning* (1605), Book II, by Sir Francis Bacon (1561–1626), the English jurist, philosopher, and essayist. The poem, one of four originally published under the covering title of "Old Dominion" in *The Pangolin and Other Verse* (1936), was published separately in subsequent editions.

is closed again. What delightful note
    with rapid unexpected flute
sounds leaping from the throat
    of the astute
grown bird, comes back to one from   35
the remote
        unenergetic sun-
        lit air before
the brood was here? How harsh
the bird's voice has become.   40

A piebald cat observing them,
    is slowly creeping toward the trim
trio on the tree stem.
    Unused to him
the three make room—uneasy   45
new problem.
        A dangling foot that missed
        its grasp, is raised
and finds the twig on which it
planned to perch. The   50

parent darting down, nerved by what chills
    the blood, and by hope rewarded—
of toil—since nothing fills
    squeaking unfed
mouths, wages deadly combat,   55
and half kills
        with bayonet beak and
        cruel wings, the
intellectual cautious-
ly creeping cat.   60

                1936, 1941

The Paper Nautilus[1]

For authorities whose hopes
are shaped by mercenaries?
    Writers entrapped by
    teatime fame and by
commuters' comforts? Not for these   5
    the paper nautilus
    constructs her thin glass shell.

    Giving her perishable
souvenir of hope, a dull
    white outside and smooth-
    edged inner surface   10

---

1. A fish, also named the Argonaut, related to the octopus. The female constructs an extremely fragile shell to serve as an egg case, which she carries on the surface of the water, grasping it with two of her eight arms which were once thought to serve as sails.

glossy as the sea, the watchful
　　maker of it guards it
　　day and night; she scarcely

eats until the eggs are hatched.　　　　　　15
Buried eightfold in her eight
　　arms, for she is in
　　a sense a devil-
fish, her glass ram's-horn-cradled freight
　　is hid but is not crushed;　　　　　　20
　　as Hercules, bitten

by a crab loyal to the hydra,[2]
was hindered to succeed,
　　the intensively
　　watched eggs coming from　　　　　25
the shell free it when they are freed—
　　leaving its wasp-nest flaws
　　of white on white, and close-

laid Ionic chiton-folds[3]
　　like the lines in the mane of　　　　30
　　a Parthenon horse,[4]
round which the arms had
　　wound themselves as if they knew love
　　is the only fortress
　　strong enough to trust to.　　　　　35

　　　　　　　　　　　　　　　　　　1941

2. In Greek mythology, the second of
Hercules' 12 labors required him to
subdue the Hydra, a many-headed monster. Attacked by a crab during the ordeal, he crushed it with his foot.
3. Sleeved linen undergarment worn by
the ancient Greeks, named for the
Greeks in Ionia on the west coast of
Asia Minor.
4. A figure in the frieze of the Parthenon, the temple of Athena in Athens.

T. S. ELIOT
1888–1965

An expatriate American, Thomas Stearns Eliot, was for over two decades
the most influential poet writing in English. Once he settled on a vocation,
the brilliance of his verse as well as the persuasiveness of his criticism
helped effect a major reorientation of the English literary tradition, secured
his supremacy as a man of letters in London, and helped establish modernism as the dominant mode in Anglo-American poetry.

Eliot was raised in St. Louis, Missouri, the son of a successful, cultivated
businessman and grandson of a New England Unitarian minister who
founded Washington University. Eliot later testified to the good fortune of
being born in St. Louis rather than in "Boston, or New York, or London,"
and to the lasting impress of childhood experience—memories of the Mississippi River, of seaside vacations on Cape Ann in Massachusetts, and of

his family's earnest piety, the "Law of Public Service" that familial mores enforced and the respect it inculcated for "symbols of Religion, the Community and Education."[1] These cultural advantages and civic activities—notably his mother's philanthropic endeavors and her writing, including a verse drama about Savonarola which Eliot later published—identify the "Genteel Tradition" which had spread from the East to Midwestern cities during the nineteenth century.

Eliot attended Harvard for both undergraduate and graduate work (1906–10, 1911–14) where he came under the influence of the philosopher George Santayana and the humanist Irving Babbitt, and he began studying philosophy, the poets Donne and Dante, and the Elizabethan and Jacobean dramatists, who figured importantly later in his criticism. He published verse in college periodicals, but his discovery in 1908 of Arthur Symons's *The Symbolist Movement in Literature*, and through it the ironic poetry of Jules LaForgue and other French poets, had an immediate impact on the verse that he wrote later at Harvard and submitted for professional publication between 1915 and 1917. His graduate work, some of it pursued abroad, began to include social anthropology and to center on languages (notably Sanskrit) and philosophy; Eliot was at Oxford studying Greek philosophy at the outset of World War I, and he completed by 1916 a Harvard doctoral dissertation on the philosophy of F. H. Bradley, though wartime circumstances prevented his returning for the oral examination required for the degree.

The year 1915 marked a turning point in his career from the pursuit of philosophy as a profession (which his family hoped for) to the vocation of poet and man of letters. The publication of *The Love Song of J. Alfred Prufrock* in Chicago's *Poetry* magazine and of *Preludes* in London's *Blast* brought him professional attention as a poet, and Ezra Pound, who had facilitated publication of *Prufrock* and judged it "the best poem I have had or seen from an American," began to introduce in literary circles the young American who had "trained himself *and* modernized himself *on his own*."[2] At the same time Eliot decided to settle in England and married Vivian Haigh-Wood after a brief courtship and despite the stern misgivings of his parents. Though some friends thought her uninteresting, she was vivacious, took an interest in Eliot's poetry, and did some creative writing herself, but her mental health was fragile and deteriorated frequently before her death, in confinement, in 1947. The strains imposed on Eliot by her illnesses, and his anxieties about his responsibilities in their marriage, were a continual burden even after he abruptly arranged a separation from her in 1932.

After his marriage Eliot turned for support to teaching in London, then worked (1917–25) in the foreign department of Lloyd's Bank. There the daily routine, though demanding, provided a regimen less distracting than teaching from his writing.

Though most of Eliot's criticism before 1925 was written in the form of book reviews for money, and though virtually all of it was motivated by its bearing on the development of his own poetic techniques, it had a durable impact for several reasons in addition to the unmistakable astuteness of his essays. As an American expatriate Eliot cultivated assiduously an elegant manner and became well acquainted with the élite and vanguard intellec-

---

1. "American Literature and the American Language," *To Criticize the Critic*, 1965, pp. 44–45.

2. *Letters of Ezra Pound*, ed. D. D. Paige (1950), p. 40.

tual circle, the Bloomsbury Group. Moreover, some of his criticism was published in such well-established journals as *The Athenaeum* and the *Times Literary Supplement*. As assistant editor of the *Egoist* (1917–19) and then founding editor of *The Criterion* (1922–39) he controlled subsidized journals that enabled him to circulate his own views regularly and exercise his predilections in selecting the writing of others. His persuasive style combined Olympian judiciousness with a wit more subtle, urbane learning with an earnestness more disciplined, than those of the Victorian Matthew Arnold on whose criticism Eliot to some extent modeled his own. His stance was at once more philosophical and more strictly literary than Pound's—less programmatic, promotional, and scrappy—and Eliot gave the effect of appraising and adjusting traditions rather than assaulting and appropriating them. Indeed Pound, recognizing that Eliot could not effectively operate as a "battering ram," urged him to "try a more oceanic and fluid method of sapping the foundations"[3] of the establishment culture. The essays gathered in *The Sacred Wood* (1920), notably *Tradition and the Individual Talent* and *Rhetoric and Poetic Drama*, and later in *Homage to Dryden* (1924) including *The Metaphysical Poets*, helped gain him an audience in some academic circles that preceded his impact as a poet.

The impact was unmistakable, however, after the publication of *Prufrock and Other Observations* (1917), *Gerontion* (1919), and *Poems* (1920). His urbanity, vivid dramatizations, subtle ironies, and symbolist techniques had already begun to inspire imitators when in 1921 he began work on *The Waste Land*, which he finished during convalescence in a Swiss sanatorium following a nervous collapse. Eliot made some alterations at his wife's suggestion, and major changes on the advice of Pound, before publishing it in 1922 in his own *Criterion*, then in America in the *Dial*. It won the Dial Award for that year, and for publication in book form Eliot added footnotes to expand the volume's size. The poem's learned allusions and unfamiliar structure antagonized some readers. Nevertheless by the time *Poems, 1909–1925* appeared in 1925 *The Waste Land* had presented the exciting challenge of a new poetic structure. Also it bequeathed to the Twenties a myth (recreated from archetypes Christian, Oriental, and primitive) about the deepening decay of Western civilization and the quest for regeneration. Both the poetic structure and the myth had reverberations among writers as diverse as the poet Hart Crane and the novelist William Faulkner.

In 1927 Eliot's assumption of British citizenship and confirmation in the Anglican Church focused professional attention on the point of view he defined in the preface to the collection of essays *For Lancelot Andrews* (1928): he was a "classicist in literature, royalist in politics, and anglo-catholic in religion." (He later regretted the label "classicist," which many critics thought had little application to his criticism and even less to his poetry.) Though Eliot never divorced literary study from social, cultural, or moral concerns, his criticism after 1925 became more emphatically social and cultural in its range, and more specifically Christian in its orientation. His power and influence were buttressed after 1925 by a position as editor and later director of the publishing firm Faber and Faber, which underwrote the expenses of the *Criterion*. Membership in discussion groups (the Chandos Group in the Thirties and the Moot from 1938 to 1947) kept him in touch with a variety of conservative intellectuals who shared Eliot's

---

3. Quoted in Noel Stock, *The Life of Ezra Pound* (1970), p. 206.

concern to diagnose the ills of Western civilization and, in Eliot's words, "to build a new structure in which democracy can live" (1928)[4] and to strengthen the authority of Christian principles. Though liberals and critics from the Left were disturbed by Eliot's sympathy with the authoritarian views of Charles Maurras and the *Action Française*, and by Eliot's emphasis on "order," "hierarchy," and racial homogeneity in his social thought, Eliot saw his position as opposed to secularism in fascism as well as in industrial capitalism and egalitarian democracy. In *After Strange Gods* (1934) he called for a combination of "orthodoxy" and "tradition" (an habitual "way of feeling and acting which characterizes a group through generations")[5] to combat the secularism and rival religions that prevailed among humanist critics and such writers as Pound and Yeats. In *The Idea of a Christian Society* (1940) he extended Coleridge's notion of a "clerisy" to define an "élite" of "superior intellectual and/or spiritual gifts," a "Community of Christians" composed of both Christian clergy and laity, who through example and persuasion would "form the conscious mind and the con-science of the nation."[6] In *Notes toward the Definition of Culture* (1949) he defined a fusion of religion and culture, operative at subcon-scious levels and sustained by family traditions and by "élites" affiliated with a "governing class," but insuring the interconnections among social groups across "a continuous gradation of cultural levels."[7]

To "cut across all the present stratifications of public taste—stratifica-tions which are perhaps a sign of social disintegration," to write "for as large and miscellaneous an audience as possible,"[8] early became Eliot's aim, particularly in writing plays. He had discovered the power of myth to tap subconscious feelings which are the basis of poetic communication, and he had learned from literary anthropology the importance of ritual and rhythm to the communal theater among the ancients. The combination of sheer entertainment with poetry of a high order which he admired in Elizabethan drama, and the infectious rapport between Charlie Chaplin or music hall performers with their audiences, inspired Eliot to recreate the verse drama for the modern theater. In every collection after *Poems, 1909–1935* (1935) he always included the fragments of his first dramatic experiment, *Wanna Go Home, Baby?* 1926–27 (entitled *Sweeney Agonistes* after 1932). Like *The Rock* (1934), for which Eliot wrote the choruses, Eliot's first successful play was a ritual drama commissioned for a religious occasion (*Murder in the Cathedral*, presented in 1935 and later filmed). Later he sought to avoid the obvious trappings of dramatic verse for fear of arousing outdated expectations in a modern audience; he aimed to perfect a verse idiom derived instead from contemporary speech that would not sound like poetry but would operate unobtrusively on several levels at once, convey the play's musical pattern, and reveal the ritual substructure that lies beneath the surface. Myth, ritual, and dramatic conventions derived from ancient drama function more obviously in *The Family Reunion* (1939) than in later plays, but these patterns infuse the manor-house setting and contem-porary idiom to reveal a drama of guilt and purgation. Likewise, in *The Cocktail Party* (1949), his most successful comedy of manners, they dis-close the advent of sainthood and a makeshift search for salvation among

4. Quoted by Roger Kojecky, *T. S. Eliot's Social Criticism* (1971), p. 9.
5. P. 31.
6. P. 43.
7. Pp. 45–47.
8. *The Use of Poetry and the Use of Criticism* (1933), pp. 152–53.

sophisticated patients of a London psychiatrist. *The Confidential Clerk* (1953) presents the search for a Christian vocation. In *The Elder Statesman* (1959), written when Eliot was radiantly happy following his marriage to Valerie Fletcher in 1957, Sophocles' *Oedipus at Colonus* informs an aging statesman's discovery of divine forgiveness through his daughter's love.

Meditative poetry, however, not the verse drama, remained the basis for Eliot's reputation after 1925, and his poems commanded respect even though some readers were disappointed by the less innovative forms of his verse or its explicitly religious themes. While *The Hollow Men* (1925) in effect restated and intensified without resolving the dilemma of *The Waste Land* in its quest for regeneration, the poem was more liturgical in its motives. Yet it was the six devotional lyrics comprising *Ash Wednesday* (1930) that marked Eliot's conversion to Christianity, presenting the agonized incompletion of a doubt-ridden search for penitence and a prayer for forgiveness. *Burnt Norton* (1934) was the first of four meditative poems which Eliot completed in the Forties and comprise his late masterpiece, *Four Quartets* (1943). In 1947 he received an honorary degree from Harvard, and in 1948 he was awarded the Order of Merit by the British crown and received the Nobel Prize for Literature.

ELIOT'S LITERARY CRITICISM

Eliot's essays concentrating on literary matters focused on two questions: the importance of tradition and the perennial need for reshaping it; and the resources of language as a means of objectifying states of feeling and achieving auditory effects in poetry.

Tradition, Eliot declared, gave the artist an "ideal order" which related his own productions to those of the past and served both to liberate and to discipline his talent. With a characteristic American feeling of alienation from cultural traditions, but with an equally typical disdain for automatically inheriting them, he insisted that tradition must be cultivated, indeed obtained "by great labor" (*Tradition and the Individual Talent*, 1919). In the active engagement between the past and present that Eliot defined, earlier and foreign works of art opened up perspectives and enlarged the poetic resources available to the poet, freeing him from dependence on his immediate predecessors. Correspondingly, the artist could reshape the tradition by his contributions to it and by revaluating it. Eliot's essays effect that revaluation, seeking to counteract the "dissociation of sensibility" (the severance of language from feeling) that he thought had occurred in the seventeenth century. He deplored the dominance of Milton (his elaborately baroque idiom) as well as the eccentricities of vision in the romantic poet William Blake and the amorphous sonorities of the Victorian Swinburne. He endorsed instead the conjunction of analytic intellect with passion in John Donne and the metaphysical poets, the "direct sensuous apprehension of thought," the "recreation of thought into feeling," that he aimed for in his own verse (*The Metaphysical Poets*, 1921). Although Eliot conceded, when reinterpreting the English tradition again in 1947, that enslavement to colloquial speech was dangerous and that poets now could learn from Milton's poetic structures, in earlier decades he had joined Pound in disparaging Milton and castigating the exhausted, attenuated language of the Victorian and Georgian periods and had voiced the need to incorporate the cadences of colloquial speech. The colloquial elements of Jacobean blank verse drama, and its rhetorical flexibility, made him assign it a higher priority in dramatic history than it had earlier enjoyed.

The meticulous attention to details of diction and rhetoric in poetry, for which Eliot's essays were famous, reinforced his efforts to reconstitute the language of poetry—the perennial task, as both Emerson and Thoreau had insisted, of the creative writer. Since the poet is both "more civilized" and "more *primitive*" than other men, his idiom must encompass the full range of intellect, learning, and conventionalized forms along with the energies of the subconscious mind. The poet should get "the whole weight of the history of the language behind his word" so that he can give "to the word a new life and to the language a new idiom."[9] Eliot sought to shape an idiom that embraced the full suggestiveness of the Symbolists along with the visual lucidity that he admired in Dante and the precision of the Imagists. Chiefly he wanted a language adequate to what he termed "the auditory imagination," a "feeling for syllable and rhythm, penetrating far below the conscious levels of thought and feeling," reaching "to the most primitive and forgotten" and fusing "the old and the obliterated and the trite, the current, and the new and surprising, the most ancient and the most civilized mentality."[1] Such a language would permit the enlargement and "concentration" of experience, the "fusion" under "pressure," entailed by a creative process in which the artist's mere "personality" is extinguished in a "continual self-sacrifice" to reveal objectively the genuinely intimate apprehensions of the self and create the new feelings enacted in art (*Tradition and the Individual Talent*).

ELIOT'S POETRY

The reorientation of response demanded of readers, once Eliot shifted from the late Victorian modes of his earliest verse under the impact of Jules Laforgue and the Symbolists, was a function chiefly of his ostensibly banal subjects; his subtle irony; his juxtaposition of unexpected images and seemingly discontinuous shifts in feeling without transitional passages or extended comment by the author; the mixture of erudition and common speech in his diction; and the emergence of mythological perspectives through often veiled allusion and sometimes faint echoes of the cadences of earlier poetry. The carefully controlled voice in *Preludes* yields not only the detachment of precise scrutiny but tender feeling as well. The mocking but deft caricature of *Sweeney among the Nightingales*, intensified by the regularity of stanza pattern, gradually evokes not only amusement but concern for the threatened protagonist; the culminating allusions to a convent and the mythological Agamemnon define the contrast in scale between Sweeney's sordid present and the heroic past but suggest also the imminent tragedy of their shared fate. Eliot's refinement of the dramatic monologue in *Prufrock* and *Gerontion* enabled him to project figures in whom objective observation and subjective response coexist; the speakers are symbolic figures in which the observed fragments of an outer world comprise a symbolic landscape but are indistinguishable from a mind's meditations, intimate responses, or fantasies. Eliot is able to define the pathos as well as the ineffectuality of Prufrock's alienation, revealing the aptness of Prufrock's recognitions while in the same allusions deftly mocking the disproportion with which Prufrock dramatizes his aspirations. Eliot's major poems manage to speak simultaneously in the "three voices of poetry" that Eliot defined in a lecture with that title in 1953: "the voice of the poet talking to himself—

---

9. Quoted by F. O. Matthiessen, *The Achievement of T. S. Eliot*, rev. ed.   (1958), pp. 94, 83.   1. *Ibid.*, p. 81.

or nobody," the "voice of the poet addressing an audience," and the voice of "the poet when he attempts to create a dramatic character speaking in verse."[2]

In *The Waste Land,* Eliot's technique enabled him to create a poem that was not only intimate in feeling but virtually epic in scale which encompassed an impending crisis in all civilization and spoke to the disillusionment of readers in the Twenties. A shifting series of modern and ancient prophetic figures displace or dissolve into each other, intertwined with a series of merchants and commercial clerks, seducers, lovers, and married couples drawn from ancient mythology, Elizabethan history, and contemporary London. The poem reveals a landscape of sexual disorder and spiritual desolation and projects a desperate quest for regeneration that remains unfulfilled, though more clearly envisioned, at the end. Various legends, Christian as well as pagan, reinforce each other and are given focus by the legend of the Holy Grail and related Mediterranean fertility myths that Eliot studied in anthropological writings by Jessie Weston and James Frazer. Eliot's poem invokes in fragments the story of the dying god, the Fisher King, whose death or impotence brought desiccation to all vegetation and sterility to men and beasts in his kingdom. Lifebringing rain, sexual and spiritual regeneration, would be assured only if a knight ventured to the Chapel Perilous in the heart of the waste land and, after surviving a nightmarish temptation to despair, asked the proper ritual questions about the grail or cup (female) and the lance (male) harbored there. In Eliot's poem the sequences of bizarre images and estranged voices have the effect at once of cinematic montage and the developmental variations of musical composition. The final version, after Eliot pruned it severely in accordance with Pound's suggestions, is orchestrated in five separate lyrics which build with increasing intensity to an ambiguously portentous crisis (the planting of a corpse, the end of time, purgatorial fire, a drowning, the sound of thunder speaking in an untranslated foreign language), projecting the urgent need and possibility of regeneration but leaving the issue uncertain at the end.

In his later masterpiece, *Four Quartets,* Eliot also used a fivefold structure for each of the four poems that is even more suggestively modeled on musical form (possibly on Beethoven's late quartets), though the materials are openly personal and the themes are philosophical and finally religious. In each Quartet the first and fifth sections combine personal reminiscence (about childhood, his earlier career, or World War I experience) with philosophical meditation on history and "the point of the intersection of the timeless/With time" in the Christian Incarnation (*The Dry Salvages*). The second and fourth sections contain intense and highly formal lyrics. What is new to the *Quartets,* notably in parts of the second and third sections of each, are the passages of relatively discursive poetry of the kind Eliot had deliberately excluded from his earlier verse, providing transitions between more intense passages. Each Quartet contains a meditation on the function of language in art, the strains which "crack and sometimes break" the idiom of the poet (*Burnt Norton*) or conversely (in the concluding Quartet, *Little Gidding*) on the moments when "the old and the new," "the common word" and the "formal word," unite to become "the complete consort dancing together."

2. *On Poetry and Poets* (1957), p. 98.

The Love Song of J. Alfred Prufrock[1]

S'io credessi che mia risposta fosse
a persona che mai tornasse al mondo,
questa fiamma staria senza più scosse.
Ma per ciò che giammai di questo fondo
non tornò vivo alcun, s'i'odo il vero,
senza tema d'infamia ti rispondo.[2]

Let us go then, you and I,
When the evening is spread out against the sky
Like a patient etherised upon a table;
Let us go, through certain half-deserted streets,
The muttering retreats   5
Of restless nights in one-night cheap hotels
And sawdust restaurants with oyster-shells:
Streets that follow like a tedious argument
Of insidious intent
To lead you to an overwhelming question. . .   10
Oh, do not ask, 'What is it?'
Let us go and make our visit.

In the room the women come and go
Talking of Michelangelo.

The yellow fog that rubs its back upon the window-panes,   15
The yellow smoke that rubs its muzzle on the window-panes,
Licked its tongue into the corners of the evening,
Lingered upon the pools that stand in drains,
Let fall upon its back the soot that falls from chimneys,
Slipped by the terrace, made a sudden leap,   20
And seeing that it was a soft October night,
Curled once about the house, and fell asleep.

And indeed there will be time[3]
For the yellow smoke that slides along the street
Rubbing its back upon the window-panes;   25
There will be time, there will be time
To prepare a face to meet the faces that you meet;
There will be time to murder and create,
And time for all the works and days[4] of hands
That lift and drop a question on your plate;   30

1. Composed at Harvard, this dramatic monologue is a symbolist poem written in intermittently rhymed free verse. The title suggests an ironic contrast between a "love song" and a poem that proves to be about the absence of love, while the speaker's name suggests an ironic contrast between his ordinary surname (that of a St. Louis furniture dealer) and the elegant convention of using the first initial and middle name as a form of address, in a world which includes Prufrock but renders him profoundly alien in it.
2. "If I thought that my reply would be to one who would ever return to the world, this flame would stay without further movement; but since none has ever returned alive from this depth, if what I hear is true, I answer you without fear of infamy" (Dante, *Inferno* XXVII, 61–66). The speaker, Guido da Montefeltro, consumed in his flame as punishment for giving false counsel, confesses his shame without fear of its being reported since he believes Dante cannot return to earth.
3. An echo of Andrew Marvell's seductive plea in *To His Coy Mistress* (1681): "Had we but world enough and time * * *"
4. *Works and Days* is a didactic poem about farming by the Greek poet Hesiod (eighth century B.C.).

Time for you and time for me,
And time yet for a hundred indecisions,
And for a hundred visions and revisions,
Before the taking of a toast and tea.

In the room the women come and go   35
Talking of Michelangelo.

And indeed there will be time
To wonder, 'Do I dare?' and, 'Do I dare?'
Time to turn back and descend the stair,
With a bald spot in the middle of my hair—   40
(They will say: 'How his hair is growing thin!')
My morning coat, my collar mounting firmly to the chin,
My necktie rich and modest, but asserted by a simple pin—
(They will say: 'But how his arms and legs are thin!')
Do I dare   45
Disturb the universe?
In a minute there is time
For decisions and revisions which a minute will reverse.

For I have known them all already, known them all—
Have known the evenings, mornings, afternoons,   50
I have measured out my life with coffee spoons;
I know the voices dying with a dying fall[5]
Beneath the music from a farther room.
    So how should I presume?

And I have known the eyes already, known them all—   55
The eyes that fix you in a formulated phrase,
And when I am formulated, sprawling on a pin,
When I am pinned and wriggling on the wall,
Then how should I begin
To spit out all the butt-ends of my days and ways?   60
    And how should I presume?

And I have known the arms already, known them all—
Arms that are braceleted and white and bare
(But in the lamplight, downed with light brown hair!)
Is it perfume from a dress   65
That makes me so digress?
Arms that lie along a table, or wrap about a shawl.
    And should I then presume?
    And how should I begin?

                . . . .

Shall I say, I have gone at dusk through narrow streets   70
And watched the smoke that rises from the pipes
Of lonely men in shirt-sleeves, leaning out of windows? . . .

---

5. Echo of Duke Orsino's self-indulgent invocation of music in Shakespeare's *Twelfth Night* (1.1.4): "If music be the food of love, play on * * * That strain again! It had a dying fall."

I should have been a pair of ragged claws
Scuttling across the floors of silent seas.[6]

        .   .   .   .   .

And the afternoon, the evening, sleeps so peacefully!   75
Smoothed by long fingers,
Asleep . . . tired . . . or it malingers,
Stretched on the floor, here beside you and me.
Should I, after tea and cakes and ices,
Have the strength to force the moment to its crisis?   80
But though I have wept and fasted, wept and prayed,
Though I have seen my head (grown slightly bald) brought in
        upon a platter,[7]
I am no prophet—and here's no great matter;
I have seen the moment of my greatness flicker,
And I have seen the eternal Footman hold my coat, and
        snicker,   85
And in short, I was afraid.

And would it have been worth it, after all,
After the cups, the marmalade, the tea,
Among the porcelain, among some talk of you and me,
Would it have been worth while,   90
To have bitten off the matter with a smile,
To have squeezed the universe into a ball
To roll it towards some overwhelming question,
To say: 'I am Lazarus,[8] come from the dead,
Come back to tell you all, I shall tell you all'—   95
If one, settling a pillow by her head,
        Should say: 'That is not what I meant at all.
        That is not it, at all.'

And would it have been worth it, after all,
Would it have been worth while,   100
After the sunsets and the dooryards and the sprinkled streets,
After the novels, after the teacups, after the skirts that trail along
        the floor—
And this, and so much more?—
It is impossible to say just what I mean!
But as if a magic lantern threw the nerves in patterns on a
        screen: 105
Would it have been worth while
If one, settling a pillow or throwing off a shawl,
And turning toward the window, should say:
        'That is not it at all,
        That is not what I meant, at all.'   110

        .   .   .   .   .

6. Cf. Hamlet's mocking of Polonius in Shakespeare's *Hamlet* (1602), II.ii.205–6: "you yourself, sir, should be old as I am, if like a crab you could go backward."
7. As did John the Baptist, beheaded by the temptress Salome, who presented his head to the spurned queen Herodias, in the story recounted in Mark 6.17–20, Matthew 14.3–11, and Oscar Wilde's play *Salome* (1894).
8. The resurrection of Lazarus is recounted in Luke 16.19–31 and John 11.1–44.

No! I am not Prince Hamlet, nor was meant to be;
Am an attendant lord, one that will do
To swell a progress,[9] start a scene or two,
Advise the prince; no doubt, an easy tool,
Deferential, glad to be of use,   115
Politic, cautious, and meticulous;
Full of high sentence,[1] but a bit obtuse;
At times, indeed, almost ridiculous—
Almost, at times, the Fool.

I grow old . . . I grow old . . .   120
I shall wear the bottoms of my trousers rolled.

Shall I part my hair behind? Do I dare to eat a peach?
I shall wear white flannel trousers, and walk upon the beach.
I have heard the mermaids singing, each to each.

I do not think that they will sing to me.   125

I have seen them riding seaward on the waves
Combing the white hair of the waves blown back
When the wind blows the water white and black.

We have lingered in the chambers of the sea
By sea-girls wreathed with seaweed red and brown   130
Till human voices wake us, and we drown.

1910–11   1915, 1917

From Preludes[2]
IV

His soul stretched tight across the skies
That fade behind a city block,
Or trampled by insistent feet
At four and five and six o'clock;
And short square fingers stuffing pipes,   5
And evening newspapers, and eyes
Assured of certain certainties,
The conscience of a blackened street
Impatient to assume the world.

I am moved by fancies that are curled   10
Around these images, and cling:
The notion of some infinitely gentle,
Infinitely suffering thing.

Wipe your hand across your mouth, and laugh;
The worlds revolve like ancient women   15
Gathering fuel in vacant lots.

1911   1915, 1917

9. A journey or procession made by royal
courts and often portrayed on Elizabethan
stages.
1. Opinions, sententiousness.

2. Written in Paris, this is the last of
four poems treating the cityscape in a
series of disjunctive images. Early drafts
used, as titles, locations in Boston.

Sweeney among the Nightingales[1]

ὤμοι, πέπληγμαι καιρίαν πληγὴν ἔσω.[2]

Apeneck Sweeney spreads his knees
Letting his arms hang down to laugh,
The zebra stripes along his jaw
Swelling to maculate[3] giraffe.

The circles of the stormy moon   5
Slide westward toward the River Plate,[4]
Death and the Raven drift above
And Sweeney guards the hornèd gate.[5]

Gloomy Orion and the Dog[6]
Are veiled; and hushed the shrunken seas;   10
The person in the Spanish cape
Tries to sit on Sweeney's knees

Slips and pulls the table cloth
Overturns a coffee-cup,
Reorganized upon the floor   15
She yawns and draws a stocking up;

The silent man in mocha brown
Sprawls at the window-sill and gapes;
The waiter brings in oranges
Bananas figs and hothouse grapes;   20

The silent vertebrate in brown
Contracts and concentrates, withdraws;
Rachel *née* Rabinovitch
Tears at the grapes with murderous paws;

She and the lady in the cape   25
Are suspect, thought to be in league;
Therefore the man with heavy eyes
Declines the gambit, shows fatigue,

---

1. This poem, dramatizing a tragicomic conspiracy, is one in which Eliot disciplined his verse by employing a regular stanza pattern (rhyming quatrains), inspired by the *Emaux et Camées* (*Enamels and Cameos*, 1852) of the French writer Théophile Gautier (1811–72). The poem generates complex ironies and analogies between the imminent death of Sweeney in a grubby and nonheroic present, the ritual sacrifice of Christ enacted in a convent, the murder (by his wife and her lover) of heroic Agamemnon in Aeschylus's tragedy, and the tragedy of the mythological Philomela. Raped by her sister's brother, who then cut out her tongue to prevent her identifying her attacker, Philomela was transformed into a nightingale whose song springs from the

violation she has suffered but cannot report.
2. Agamemnon's cry from within the palace when he is murdered: "Alas, I am struck a mortal blow within" (Aeschylus, *Agamemnon*, line 1343).
3. Blotched, stained.
4. Estuary between Argentina and Uruguay.
5. The gates of horn, in Hades; true dreams pass through them to the upper world.
6. Orion in mythology was a handsome hunter, slain by Diana, then immortalized as a constellation of stars. The "Dog" is the brilliant Dog Star, Sirius, who in mythology sired the Sphinx and, in Homer, was Orion's dog.

Leaves the room and reappears
Outside the window, leaning in,   30
Branches of wistaria
Circumscribe a golden grin;

The host with someone indistinct
Converses at the door apart,
The nightingales are singing near   35
The Convent of the Sacred Heart,

And sang within the bloody wood
When Agamemnon cried aloud[7]
And let their liquid siftings fall
To stain the stiff dishonoured shroud.   40

1918, 1919

Gerontion[1]

*Thou hast nor youth nor age
But as it were an after dinner sleep
Dreaming of both.*[2]

Here I am, an old man in a dry month,
Being read to by a boy, waiting for rain.
I was neither at the hot gates[3]
Nor fought in the warm rain
Nor knee deep in the salt marsh, heaving a cutlass,   5
Bitten by flies, fought.
My house is a decayed house,
And the Jew squats on the window sill, the owner,
Spawned in some estaminet[4] of Antwerp,
Blistered in Brussels, patched and peeled in London.[5] 10
The goat coughs at night in the field overhead;
Rocks, moss, stonecrop, iron, merds.[6]
The woman keeps the kitchen, makes tea,
Sneezes at evening, poking the peevish gutter.
                              I an old man,   15
A dull head among windy spaces.

Signs are taken for wonders. 'We would see a sign!'[7]

7. The poem transfers Agamemnon's death to the woods where Philomela was ravished and to the "bloody wood" of Nemi where ancient priests were slain by their successors in the fertility rite described by James Frazer in *The Golden Bough*, Chapter 1.

1. This dramatic monologue presents the re-examination of his life and a search for commitment on the part of an old man who, like the world he inhabits, has lost the passion for either sexual and human love or religious devotion. Eliot intended to reprint it as a prelude to *The Waste Land* until persuaded not to by Ezra Pound. The title is a coinage derived from the Greek word meaning "an old man."

2. Lines spoken by the wily Duke to prepare a young lover for death in Shakespeare's *Measure for Measure* (3.1.32–34).

3. A sexual image, and an allusion to Thermopylae, the mountain pass where the Spartans heroically defeated the Persians (480 B.C.).

4. Café.

5. Allusions to symptoms and cures for venereal disease.

6. Dung.

7. Echoes of the "signs and wonders" expected as testimony to Christ's divinity in John 4.48 and the demand "Master, we would see a sign from thee" in Matthew 12.38, the subject of a Nativity sermon by Bishop Lancelot Andrewes (1555–1626).

The word within a word, unable to speak a word,
Swaddled with darkness. In the juvescence of the year
Came Christ the tiger   20

In depraved May,[8] dogwood and chestnut, flowering judas,
To be eaten, to be divided, to be drunk
Among whispers; by Mr. Silvero
With caressing hands, at Limoges
Who walked all night in the next room;   25
By Hakagawa, bowing among the Titians;
By Madame de Tornquist, in the dark room
Shifting the candles; Fräulein von Kulp
Who turned in the hall, one hand on the door.
    Vacant shuttles
Weave the wind. I have no ghosts,   30
An old man in a draughty house
Under a windy knob.[9]

After such knowledge, what forgiveness? Think now
History has many cunning passages, contrived corridors
And issues, deceives with whispering ambitions,   35
Guides us by vanities. Think now
She gives when our attention is distracted
And what she gives, gives with such supple confusions
That the giving famishes the craving. Gives too late
What's not believed in, or is still believed,   40
In memory only, reconsidered passion. Gives too soon
Into weak hands, what's thought can be dispensed with
Till the refusal propagates a fear. Think
Neither fear nor courage saves us. Unnatural vices
Are fathered by our heroism. Virtues   45
Are forced upon us by our impudent crimes.
These tears are shaken from the wrath-bearing tree.

The tiger springs in the new year. Us he devours. Think at
    last
We have not reached conclusion, when I
Stiffen in a rented house. Think at last   50
I have not made this show purposelessly
And it is not by any concitation[1]
Of the backward devils.
I would meet you upon this honestly.
I that was near your heart was removed therefrom   55
To lose beauty in terror, terror in inquisition.

---

8. Cf. *The Education of Henry Adams* (1918), Chapter 18: the "passionate depravity that marked the Maryland May."
9. In lines 18–22, Gerontion imagines Christ as a terrifying though enviably powerful beast, like "the Lion of the tribe of Judah" in Revelation 5.5, but he recalls also that the springtime of Christ's crucifixion was a time of betrayal and he renders the Christian Word (message) as impenetrable and impotent in a world which debases the rites and gestures of Christian devotion. Line 22 alludes to dividing and eating the bread, and drinking the wine, in Christian communion.
1. Stirring up, concerted action.

I have lost my passion: why should I need to keep it
Since what is kept must be adulterated?
I have lost my sight, smell, hearing, taste and touch:[2]
How should I use them for your closer contact?   60

These with a thousand small deliberations
Protract the profit of their chilled delirium,
Excite the membrane, when the sense has cooled,
With pungent sauces, multiply variety
In a wilderness of mirrors. What will the spider do,   65
Suspend its operations, will the weevil
Delay? De Bailhache, Fresca, Mrs. Cammel, whirled[3]
Beyond the circuit of the shuddering Bear
In fractured atoms. Gull against the wind, in the windy straits
Of Belle Isle, or running on the Horn.   70
White feathers in the snow, the Gulf claims,
And an old man driven by the Trades[4]
To a sleepy corner.

                    Tenants of the house,
Thoughts of a dry brain in a dry season.   75

                                      1920

The Waste Land[1]

'Nam Sibyllam quidem Cumis ego ipse oculis meis vidi in ampulla pendere, et cum illi pueri dicerent: Σίβυλλα τί θέλεις: respondebat illa: ἀποθανεῖν θέλω.'[2]

FOR EZRA POUND

il miglior fabbro.

I. The Burial of the Dead[3]

April is the cruellest month, breeding
Lilacs out of the dead land, mixing

2. An echo of Henry Adams's statement that civilization "has reached an age when it can no longer depend, as in childhood, on its taste, or smell, or sight, or hearing, or memory" in *Mont-Saint-Michel and Chartres* (1913), Chapter 8.
3. Unidentified persons whose fate, beyond the polestar in the constellation the Bear, is compared to that of the gull and Gerontion buffeted by the winds.
4. Trade winds.
1. The notes which Eliot provided for the first hard-cover edition of *The Waste Land* opened with his acknowledgment that "not only the title, but the plan and a good deal of the incidental symbolism of the poem were suggested by Miss Jessie L. Weston's book on the Grail Legend: *From Ritual to Romance*" (1920) and that he was indebted also to James G. Frazer's *The Golden Bough* (1890–1915), "especially the two volumes

Adonis, Attis, Osiris," which deal with vegetation myths and fertility rites. Eliot's subsequent notes are incorporated in the present editor's footnotes.
2. A quotation from Petronius's *Satyricon* (first century A.D.) about the Sibyl (prophetess) of Cumae, blessed with eternal life by Apollo but doomed to perpetual old age, who guided Aeneas through Hades in Virgil's *The Aeneid*: "For once I myself saw with my own eyes the Sibyl at Cumae hanging in a cage, and when the boys said to her 'Sibyl, what do you want?' she replied, 'I want to die.'" *"Il miglior fabbro"*: "the better craftsman," the tribute in Dante's *Purgatorio* XXVI.117 to the Provençal poet Arnaut Daniel, here paid by Eliot to his friend Ezra Pound, whose editorial suggestions had contributed to the poem's elliptical structure and compression.
3. A phrase from the Anglican burial service.

Memory and desire, stirring
Dull roots with spring rain.
Winter kept us warm, covering   5
Earth in forgetful snow, feeding
A little life with dried tubers.
Summer surprised us, coming over the Starnbergersee[4]
With a shower of rain; we stopped in the colonnade,
And went on in sunlight, into the Hofgarten,[5]   10
And drank coffee, and talked for an hour.
Bin gar keine Russin, stamm' aus Litauen, echt deutsch.[6]
And when we were children, staying at the arch-duke's,
My cousin's, he took me out on a sled,
And I was frightened. He said, Marie,   15
Marie, hold on tight. And down we went.
In the mountains, there you feel free.
I read, much of the night, and go south in the winter.

What are the roots that clutch, what branches grow
Out of this stony rubbish? Son of man,[7]   20
You cannot say, or guess, for you know only
A heap of broken images, where the sun beats,
And the dead tree gives no shelter, the cricket no relief,[8]
And the dry stone no sound of water. Only
There is shadow under this red rock,[9]   25
(Come in under the shadow of this red rock),
And I will show you something different from either
Your shadow at morning striding behind you
Or your shadow at evening rising to meet you;
I will show you fear in a handful of dust.   30
               Frisch weht der Wind
               Der Heimat zu
               Mein Irisch Kind,
               Wo weilest du?[1]
'You gave me hyacinths first a year ago;   35
'They called me the hyacinth girl.'
—Yet when we came back, late, from the hyacinth garden,
Your arms full, and your hair wet, I could not
Speak, and my eyes failed, I was neither
Living nor dead, and I knew nothing,   40

---

4. A lake near Munich. Lines 8–16 were suggested by the Countess Marie Larisch's memoir, *My Past* (1913).
5. A public park in Munich, with cafés; former grounds of a Bavarian palace.
6. "I am certainly not Russian; I come from Lithuania, a true German."
7. "Cf. Ezekiel II, i" [Eliot's note], where God addresses the prophet Ezekiel as "Son of man" and declares: "stand upon thy feet, and I will speak unto thee."
8. "Cf. Ecclesiastes XII,v" [Eliot's note], where the Preacher describes the bleakness of old age when "the grasshopper shall be a burden, and desire shall fail."

9. Cf. Isaiah 32.1–2 and the prophecy that the reign of the Messiah "shall be * * * as rivers of water in a dry place, as the shadow of a great rock in a weary land."
1. "V. [see] *Tristan und Isolde*, I, verses 5–8" [Eliot's note]. In Wagner's opera, a carefree sailor aboard Tristan's ship recalls his girlfriend in Ireland: "Fresh blows the wind to the homeland; my Irish child, where are you waiting?" The young boy Hyacinth, in Ovid's *Metamorphoses*, X, was beloved by Apollo but slain by a jealous rival. The Greeks celebrated his festival in May.

Looking into the heart of light, the silence.
Oed' und leer das Meer.[2]

Madame Sosostris,[3] famous clairvoyante,
Had a bad cold, nevertheless
Is known to be the wisest woman in Europe,   45
With a wicked pack of cards.[4] Here, said she,
Is your card, the drowned Phoenician Sailor,[5]
(Those are pearls that were his eyes.[6] Look!)
Here is Belladonna, the Lady of the Rocks,[7]
The lady of situations.   50
Here is the man with three staves,[8] and here the Wheel,
And here is the one-eyed merchant,[9] and this card,
Which is blank, is something he carries on his back,
Which I am forbidden to see. I do not find
The Hanged Man. Fear death by water.   55
I see crowds of people, walking round in a ring.
Thank you. If you see dear Mrs. Equitone,
Tell her I bring the horoscope myself:
One must be so careful these days.

Unreal City,[1]   60
Under the brown fog of a winter dawn,

---

2. In *Tristan* "III, verse 24" [Eliot's note] the dying Tristan, awaiting the ship that carries his beloved Isolde, is told that "Empty and barren is the sea."
3. Eliot derived the name from "Sesostris, the Sorceress of Ectabana," the pseudo-Egyptian name assumed by a woman who tells fortunes in Aldous Huxley's novel *Chrome Yellow* (1921). Sesostris was a 12th-dynasty Egyptian King.
4. The tarot deck of cards, once important in Eastern magic but now "fallen somewhat into disrepute, being principally used for purposes of divination," according to Weston's *From Ritual to Romance.* Its four suits are the recurring life symbols of the Grail legend: the cup, lance, sword, and dish. Eliot's note to this passage reads: "I am not familiar with the exact constitution of the Tarot pack of cards, from which I have obviously departed to suit my own convenience. The Hanged Man, a member of the traditional pack, fits my purpose in two ways: because he is associated in my mind with the Hanged God of Frazer, and because I associate him with the hooded figure in the passage of the disciples to Emmaus in Part V. The Phoenician Sailor and the Merchant appear later; also the 'crowds of people,' and Death by Water is executed in Part IV. The Man with Three Staves (an authentic member of the Tarot pack) I associate, quite arbitrarily, with the Fisher King himself."
5. The Phoenician Sailor is a symbolic figure which includes "Mr. Eugenides, the Smyrna merchant" in Part III and "Phlebas the Phoenician" in Part IV. The ancient Phoenicians were sea-going merchants whose crews spread Egyptian fertility cults throughout the Mediterranean.
6. The line is a quotation from Ariel's song in Shakespeare's *The Tempest,* 1.2.398. Prince Ferdinand, disconsolate because he thinks his father had drowned in the storm, is consoled when Ariel sings of a miraculous "sea change" which has transformed disaster into "something rich and strange." Fears attendant on drowning, and the sea as an agent of purification and possible resurrection, are important elements throughout the poem.
7. Belladonna, literally meaning "beautiful lady," is the name of both the poisonous plant "nightshade" and a cosmetic. It suggests also the Christian "madonna" or Virgin Mary, particularly, in context, the painting *Madonna of the Rocks* by Leonardo da Vinci.
8. One tarot card, the Three Scepters, represents a successful merchant surveying his ships with three wands planted in the ground nearby. Another is associated with three phalli and the rebirth of the Egyptian god Osiris. The Wheel on another card is the Wheel of Fortune.
9. Later identified as "Mr. Eugenides," the figure is "one-eyed" because his activities are ominous and because he is shown in profile on the card.
1. "Cf. Baudelaire: 'Fourmillante cité, cité pleine de rêves, / Où le spectre en plein jour raccroche le passant'" [Eliot's note]. The lines are quoted from *Les Sept Viellards* (*The Seven Old Men*),

A crowd flowed over London Bridge, so many,
I had not thought death had undone so many.[2]
Sighs, short and infrequent, were exhaled,[3]
And each man fixed his eyes before his feet.   65
Flowed up the hill and down King William Street,
To where Saint Mary Woolnoth kept the hours
With a dead sound on the final stroke of nine.[4]
There I saw one I knew, and stopped him, crying: 'Stetson!
'You who were with me in the ships at Mylae![5]   70
'That corpse you planted last year in your garden,
'Has it begun to sprout? Will it bloom this year?
'Or has the sudden frost disturbed its bed?
'O keep the Dog far hence, that's friend to men,
'Or with his nails he'll dig it up again![6]   75
'You! hypocrite lecteur!—mon semblable,—mon frère!'[7]

II. A Game of Chess[8]

The Chair she sat in, like a burnished throne,[9]
Glowed on the marble, where the glass
Held up by standards wrought with fruited vines
From which a golden Cupidon peeped out   80
(Another hid his eyes behind his wing)
Doubled the flames of sevenbranched candelabra
Reflecting light upon the table as

---

poem XCIII of *Les Fleurs du Mal* (*The Flowers of Evil*, 1857) by the French Symbolist Charles Baudelaire (1821–67), and may be translated: "Swarming city, city full of dreams, / Where the specter in broad daylight accosts the passerby."
2. "Cf. *Inferno* III, 55–57 * * * " [Eliot's note]. The note continues to quote Dante's lines, which may be translated: "So long a train of people, / That I should never have believed / That death had undone so many."
3. "Cf. *Inferno* IV, 25–27 * * * " [Eliot's note]. Dante describes, in Limbo, the virtuous but pagan dead who, living before Christ, could not achieve the hope of embracing the Christian God. The lines read: "Here, so far as I could tell by listening, / There was no lamentation except sighs, / Which caused the eternal air to tremble."
4. "A phenomenon which I have often noticed" [Eliot's note]. The church named is in the financial district of London, and the allusion is the first to suggest the Chapel Perilous of the Grail legend.
5. "Stetson" is the familiar name of a hat manufacturer; the battle of Mylae (260 B.C.) was a victory for Rome in her commercial war against Carthage.
6. Lines 71–75 present a grotesquely mocking image of the possible resurrection of a fertility god. "Cf. the dirge in Webster's *White Devil*" [Eliot's note]: in the play by John Webster (d. 1625), a crazed Renaissance Roman matron fears that the corpses of her decadent

and murdered relatives might be disinterred: "But keep the wolf far thence, that's foe to men, / For with his nails he'll dig them up again." In echoing the lines Eliot altered "foe" to "friend" and the "wolf" to "Dog," invoking the brilliant Dog Star Sirius, whose rise in the heavens accompanied the flooding of the Nile and promised the return of fertility to Egypt.
7. "V. Baudelaire, Preface to *Fleurs du Mal*" [Eliot's note]. The last line of the introductory poem to *Les Fleurs du Mal*. "Au Lecteur" (To the Reader), may be translated: "Hypocrite reader!—my likeness—my brother!" It dramatizes the reader's and author's mutual involvement in the waste land.
8. Part II juxtaposes two contemporary settings—the first of elegant decadence, the second of lower-class vulgarity—in a bar or pub and interweaves echoes of the historical and literary past. The title suggests two plays by Thomas Middleton: *A Game of Chess* (1627), about a marriage of political expediency, and *Women Beware Women* (1657), containing a scene in which a mother-in-law is engrossed in a chess game while her daughter-in-law is seduced on a balcony onstage. Eliot's note to line 137 below refers readers to this play.
9. "Cf. *Antony and Cleopatra*, II, ii. 1. 190" [Eliot's note]. In Shakespeare's play, Enobarbus's description of Cleopatra's regal sensuous beauty begins: "The barge she sat in, like a burnish'd throne, / Burn'd on the water."

The glitter of her jewels rose to meet it,
From satin cases poured in rich profusion. 85
In vials of ivory and coloured glass
Unstoppered, lurked her strange synthetic perfumes,
Unguent, powdered, or liquid—troubled, confused
And drowned the sense in odours; stirred by the air
That freshened from the window, these ascended 90
In fattening the prolonged candle-flames,
Flung their smoke into the laquearia,[1]
Stirring the pattern on the coffered ceiling.
Huge sea-wood fed with copper
Burned green and orange, framed by the coloured stone, 95
In which sad light a carvèd dolphin swam.
Above the antique mantel was displayed
As though a window gave upon the sylvan scene[2]
The change of Philomel, by the barbarous king
So rudely forced; yet there the nightingale 100
Filled all the desert with inviolable voice
And still she cried, and still the world pursues,
'Jug Jug'[3] to dirty ears.
And other withered stumps of time
Were told upon the walls; staring forms 105
Leaned out, leaning, hushing the room enclosed.
Footsteps shuffled on the stair.
Under the firelight, under the brush, her hair
Spread out in fiery points
Glowed into words, then would be savagely still. 110

'My nerves are bad to-night. Yes, bad. Stay with me.[4]
'Speak to me. Why do you never speak. Speak.
    'What are you thinking of? What thinking? What?
'I never know what you are thinking. Think.'

I think we are in rats' alley[5] 115
Where the dead men lost their bones.

'What is that noise?'

---

1. "Laqueaira. V. *Aeneid*, I, 726 * * * ."
Eliot quotes the passage containing
the term *laquearia* ("paneled ceiling")
and describing the banquet hall where
Queen Dido welcomed Aeneas to Carthage. It reads: "Blazing torches hang
from the gilded paneled ceiling, and
torches conquer the night with flames."
Aeneas became Dido's lover but abandoned her to continue his journey to
found Rome, and she committed suicide.
2. Eliot's notes for lines 98–99 refer the
reader to "Milton, *Paradise Lost*, IV,
140" for the phrase "sylvan scene" and
to "Ovid, *Metamorphoses*, VI, Philomela." The lines ironically splice the setting of Eve's temptation in the Garden
of Eden, first described through Satan's
eyes, with the rape of Philomela by her

sister's husband, King Tereus, and her
transformation into the nightingale.
Eliot's note for line 100 refers the reader
ahead to the nightingale's song as rendered in Part III, line 204, of his own
poem. The myth of Philomela suggests
the transformation of suffering into art.
3. The conventional rendering of the
nightingale's song in Elizabethan poetry.
4. Lines 111–138 may be read as an interior monologue, with the same person
speaking the quoted dialogue outright
while phrasing unspoken thoughts in the
intervening lines; or as a dialogue in
which a woman speaks the quoted remarks while her lover or husband responds silently in the intervening lines.
5. Eliot's note refers readers to "Part
III, 1.195."

        The wind under the door.[6]
'What is that noise now? What is the wind doing?'
        Nothing again nothing.   120
                        'Do
'You know nothing? Do you see nothing? Do you remember
'Nothing?'

       I remember
Those are pearls that were his eyes.   125
'Are you alive, or not? Is there nothing in your head?'
                            But
O O O O that Shakespeherian Rag—
It's so elegant
So intelligent   130
'What shall I do now? What shall I do?'
'I shall rush out as I am, and walk the street
'With my hair down, so. What shall we do tomorrow?
'What shall we ever do?'
                    The hot water at ten.   135
And if it rains, a closed car at four.
And we shall play a game of chess,
Pressing lidless eyes and waiting for a knock upon the door.

When Lil's husband got demobbed,[7] I said—
I didn't mince my words, I said to her myself,   140
HURRY UP PLEASE ITS TIME[8]
Now Albert's coming back, make yourself a bit smart.
He'll want to know what you done with that money he gave
    you
To get yourself some teeth. He did, I was there.
You have them all out, Lil, and get a nice set,   145
He said, I swear, I can't bear to look at you.
And no more can't I, I said, and think of poor Albert,
He's been in the army four years, he wants a good time,
And if you don't give it him, there's others will, I said.
Oh is there, she said. Something o' that, I said.   150
Then I'll know who to thank, she said, and give me a straight
    look.
HURRY UP PLEASE ITS TIME
If you don't like it you can get on with it, I said.
Others can pick and choose if you can't.
But if Albert makes off, it won't be for lack of telling.   155
You ought to be ashamed, I said, to look so antique.
(And her only thirty-one.)
I can't help it, she said, pulling a long face,
It's them pills I took, to bring it off, she said.
(She's had five already, and nearly died of young George.)   160

6. "Cf. Webster: 'Is the wind in that door still?'" [Eliot's note]. In *The Devil's Law Case* (1623), III.ii.162, by John Webster (d. 1625), a Duke is cured of an infection by a wound intended to kill him; a surprised surgeon asks the quoted question meaning "Is he still alive?" 7. British slang for "demobilized," discharged from the army. 8. Routine call of British bartenders to clear the "pub" at closing time.

The chemist[9] said it would be all right, but I've never been
    the same.
You *are* a proper fool, I said.
Well, if Albert won't leave you alone, there it is, I said,
What you get married for if you don't want children?
HURRY UP PLEASE ITS TIME   165
Well, that Sunday Albert was home, they had a hot gammon,[1]
And they asked me in to dinner, to get the beauty of it hot—
HURRY UP PLEASE ITS TIME
HURRY UP PLEASE ITS TIME
Goonight Bill. Goonight Lou. Goonight May. Goonight.   170
Ta ta. Goodnight. Goonight.
Good night, ladies, good night, sweet ladies, good night,
    good night.[2]

III. The Fire Sermon[3]

The river's tent is broken; the last fingers of leaf
Clutch and sink into the wet bank. The wind
Crosses the brown land, unheard. The nymphs are departed. 175
Sweet Thames, run softly, till I end my song.[4]
The river bears no empty bottles, sandwich papers,
Silk handkerchiefs, cardboard boxes, cigarette ends
Or other testimony of summer nights. The nymphs are
    departed.
And their friends, the loitering heirs of City directors;   180
Departed, have left no addresses.
By the waters of Leman I sat down and wept . . .[5]
Sweet Thames, run softly till I end my song,
Sweet Thames, run softly, for I speak not loud or long.
But at my back in a cold blast I hear[6]   185
The rattle of the bones, and chuckle spread from ear to ear.

A rat crept softly through the vegetation
Dragging its slimy belly on the bank

---

9. Druggist.
1. Ham or bacon; in slang, "thigh."
2. A double echo of the popular song *Good night ladies, we're going to leave you now* and of mad Ophelia's pathetic farewell before drowning herself in Shakespeare's *Hamlet*, 4.5.72.
3. Eliot's notes to lines 307–9 of this section identify two governing perspectives: "Buddha's Fire Sermon (which corresponds in importance to the Sermon on the Mount)" and "St. Augustine's *Confessions*." The "collocation of these two representatives of eastern and western asceticism, as the culmination of this part of the poem, is not an accident." Part III unfolds an increasingly drab and ominous series of contemporary seductions and assignations, interwoven with ironic evocations of the past, to define the lusts and sterile parodies of love which Buddha's Fire Sermon denounces as the "fire of passion," "the fire of hatred," and "the fire of infatuation"

which consume the senses.
4. "V. Spenser, *Prothalamion*" [Eliot's note]. The line is the refrain of the marriage song by Edmund Spenser (d. 1599), a pastoral celebration of marriage set along the Thames River near London.
5. The phrasing recalls Psalms 137.1, where the exiled Jews mourn for their homeland: "By the rivers of Babylon, there we sat down, yea, we wept, when we remembered Zion." Lake Leman is another name for Lake Geneva, location of the sanatorium where Eliot wrote the bulk of *The Waste Land*. The archaic term "leman," for illicit mistress, led to the phrase "waters of leman" signifying lusts.
6. This line and line 196 below echo Andrew Marvell (1621–78), *To His Coy Mistress*, lines 21–24: "But at my back I always hear / Time's wingèd chariot hurrying near, / And yonder all before us lie / Deserts of vast eternity."

While I was fishing in the dull canal
On a winter evening round behind the gashouse   190
Musing upon the king my brother's wreck[7]
And on the king my father's death before him.
White bodies naked on the low damp ground
And bones cast in a little low dry garret,
Rattled by the rat's foot only, year to year.   195
But at my back from time to time I hear
The sound of horns and motors, which shall bring
Sweeney to Mrs. Porter in the spring.[8]
O the moon shone bright on Mrs. Porter
And on her daughter   200
They wash their feet in soda water[9]
Et O ces voix d'enfants, chantant dans la coupole![1]

Twit twit twit
Jug jug jug jug jug
So rudely forc'd.   205
Tereu[2]

Unreal City
Under the brown fog of a winter noon
Mr. Eugenides,[3] the Smyrna merchant
Unshaven, with a pocket full of currants   210
C.i.f.[4] London: documents at sight,
Asked me in demotic French
To luncheon at the Cannon Street Hotel
Followed by a weekend at the Metropole.[5]

---

7. "Cf. *The Tempest*, I, ii" [Eliot's
note]. An allusion to Shakespeare's play,
1.2.389–90, where Prince Ferdinand,
thinking his father dead, describes himself as "Sitting on a bank, / Weeping
again the King my father's wreck."
8. "Cf. Day, *Parliament of Bees*: 'When
of the sudden, listening, you shall hear, /
A noise of horns and hunting, which shall
bring / Actaeon to Diana in the spring, /
Where all shall see her naked skin * * *'"
[Eliot's note]. Actaeon was changed
into a stag and hunted to death as punishment for seeing Diana, goddess of
chastity, bathing. Eliot's parody of the
poem by John Day (1574–c. 1640) suggests an analogy between Actaeon's deed
and Sweeney's.
9. "I do not know the origin of the ballad from which these lines are taken: it
was reported to me from Sydney, Australia" [Eliot's note]. The bawdy song,
itself a parody of the American ballad
Little Redwing, was popular among
World War I troops. One decorous version follows: "O the moon shines bright
on Mrs. Porter / And on the daughter /
Of Mrs. Porter. / They wash their feet
in soda water / And so they oughter /
To keep them clean."

1. "V. Verlaine, *Parsifal*" [Eliot's note].
The last line of the sonnet *Parsifal* by
the French Symbolist Paul Verlaine
(1844–96) reads: "And O those children's voices singing in the cupola." In
Wagner's opera, the feet of Parsifal, the
questing knight, are washed before he
enters the sanctuary of the Grail.
2. "Tereu" alludes to Tereus, the viola-
tor of Philomela, and like "jug" is a
conventional Elizabethan term for the
nightingale's songs.
3. The Turkish merchant's name suggests "well born" and the science of improving a species by selective breeding.
4. "The currants were quoted at a price
'carriage and insurance free to London';
and the Bill of Lading, etc. were to be
handed to the buyer upon payment of
the sight draft" [Eliot's note]. His second
wife has corrected the phrase "carriage
and insurance free" to read "cost, insurance and freight."
5. The Canon Street Hotel adjoins a
London station frequented by Continental
travelers. The Metropole, which Mr. Eugenides proposes for a probably lecher-
ous weekend, is a luxury hotel in Brigh-
ton, England.

At the violet hour, when the eyes and back   215
Turn upward from the desk, when the human engine waits
Like a taxi throbbing waiting,
I Tiresias,[6] though blind, throbbing between two lives,
Old man with wrinkled female breasts, can see
At the violet hour, the evening hour that strives   220
Homeward, and brings the sailor home from sea,[7]
The typist home at teatime, clears her breakfast, lights
Her stove, and lays out food in tins.
Out of the window perilously spread
Her drying combinations touched by the sun's last rays,   225
On the divan are piled (at night her bed)
Stockings, slippers, camisoles, and stays.
I Tiresias, old man with wrinkled dugs
Perceived the scene, and foretold the rest—
I too awaited the expected guest.   230
He, the young man carbuncular, arrives,
A small house agent's clerk, with one bold stare,
One of the low on whom assurance sits
As a silk hat on a Bradford[8] millionaire.
The time is now propitious, as he guesses,   235
The meal is ended, she is bored and tired,
Endeavours to engage her in caresses
Which still are unreproved, if undesired.
Flushed and decided, he assaults at once;
Exploring hands encounter no defence;   240
His vanity requires no response,
And makes a welcome of indifference.
(And I Tiresias have foresuffered all
Enacted on this same divan or bed;

6. Eliot's note reads: "Tiresias, although a mere spectator and not indeed a 'character,' is yet the most important personage in the poem, uniting all the rest. Just as the one-eyed merchant, seller of currants, melts into the Phoenician Sailor, and the latter is not wholly distinct from Ferdinand Prince of Naples, so all the women are one woman, and the two sexes meet in Tiresias. What Tiresias *sees*, in fact, is the substance of the poem. The whole passage from Ovid is of great anthropological interest * * * ." The note quotes the Latin passage from Ovid's *Metamorphoses*, II, 421–43 which may be translated: "Jove, [very drunk] said jokingly to Juno: 'You women have greater pleasure in love than that enjoyed by men.' She denied it. So they decided to refer the question to wise Tiresias who knew love from both points of view. For once, with a blow of his staff, he had separated two huge snakes who were copulating in the forest, and miraculously was changed instantly from a man into a woman and remained so for seven years. In the eighth year he saw the snakes again and said: 'If a blow against you is so power-ful that it changes the sex of the author of it, now I shall strike you again.' With these words he struck them, and his former shape and masculinity were restored. As referee in the sportive quarrel, he supported Jove's claim. Juno, overly upset by the decision, condemned the arbitrator to eternal blindness. But the all-powerful father (inasmuch as no god can undo what has been done by another god) gave him the power of prophecy, with this honor compensating him for the loss of sight."
7. "This may not appear as exact as Sappho's lines, but I had in mind the 'longshore' or 'dory' fisherman, who returns at nightfall" [Eliot's note]. Fragment CXLIX, by the Greek woman poet Sappho (fl. 600 B.C.), celebrates the Evening Star who "brings homeward all those / Scattered by the dawn, / The sheep to fold * * * / The children to their mother's side." A more familiar echo is "Home is the sailor, home from sea" in *Requiem* by the Scottish poet Robert Louis Stevenson (1850–94).
8. A Yorkshire, England, manufacturing town where quick fortunes were made during World War I.

I who have sat by Thebes below the wall[9]   245
And walked among the lowest of the dead.)
Bestows one final patronising kiss,
And gropes his way, finding the stairs unlit. .

She turns and looks a moment in the glass,
Hardly aware of her departed lover;   250
Her brain allows one half-formed thought to pass:
'Well now that's done: and I'm glad it's over.'
When lovely woman stoops to folly and
Paces about her room again, alone,
She smoothes her hair with automatic hand,   255
And puts a record on the gramophone.[1]

'This music crept by me upon the waters'[2]
And along the Strand, up Queen Victoria Street.
O City city, I can sometimes hear
Beside a public bar in Lower Thames Street,   260
The pleasant whining of a mandoline
And a clatter and a chatter from within
Where fishmen lounge at noon: where the walls
Of Magnus Martyr hold
Inexplicable splendour of Ionian white and gold.[3]   265

　　　　　The river sweats[4]
　　　　　Oil and tar
　　　　　The barges drift
　　　　　With the turning tide
　　　　　Red sails   270
　　　　　Wide
　　　　　To leeward, swing on the heavy spar.
　　　　　The barges wash
　　　　　Drifting logs
　　　　　Down Greenwich reach   275

9. Tiresias prophesied in the marketplace below the wall of Thebes, witnessed the tragedies of Oedipus and Creon in that city, and retained his prophetic powers in Hades.
1. Eliot's note refers to the novel *The Vicar of Wakefield* (1766) by Oliver Goldsmith (1728–74) and the song sung by Olivia when she revisits the scene of her seduction: "When lovely woman stoops to folly / And finds too late that men betray / What charm can soothe her melancholy, / What art can wash her guilt away? / The only art her guilt to cover, / To hide her shame from every eye, / To give repentance to her lover / And wring his bosom—is to die."
2. Eliot's note refers to Shakespeare's *The Tempest,* the scene where Ferdinand listens in wonder to Ariel's song telling of his father's miraculous "sea-change":

"This music crept by me on the waters, / Allaying both their fury and my passion / With its sweet air * * *" (1.2. 391–93).
3. "The interior of St. Magnus Martyr is to my mind one of the finest among [Christopher] Wren's interiors * * * " [Eliot's note].
4. "The Song of the (three) Thames-daughters begins here. From line 292 to 306 inclusive they speak in turn. V. *Götterdämmerung*, III, i: the Rhine-daughters" [Eliot's note]. A contemporary view of the Thames and the three songs which tell of passionless seductions are interwoven with a refrain (lines 277–78, 290–91) from the lament of the Rhine-maidens for the lost beauty of the Rhine River in the opera by Richard Wagner (1813–83), *Die Götterdämmerung* ("The Twilight of the Gods," 1876).

Past the Isle of Dogs.[5]
Weialala leia
Wallala leialala

Elizabeth and Leicester[6]
Beating oars   280
The stern was formed
A gilded shell
Red and gold
The brisk swell
Rippled both shores   285
Southwest wind
Carried down stream
The peal of bells
White towers
        Weialala leia   290
        Wallala leialala

'Trams and dusty trees.
Highbury bore me. Richmond and Kew
Undid me.[7] By Richmond I raised my knees
Supine on the floor of a narrow canoe.'   295

'My feet are at Moorgate,[8] and my heart
Under my feet. After the event
He wept. He promised "a new start."
I made no comment. What should I resent?'

'On Margate Sands.[9]   300
I can connect
Nothing with nothing.
The broken fingernails of dirty hands.
My people humble people who expect
Nothing.'   305
        la la

To Carthage then I came[1]

5. A peninsula in the Thames opposite Greenwich, a borough of London and the birthplace of Queen Elizabeth.
6. The love affair of Queen Elizabeth and the Earl of Leicester (Robert Dudley), though it transpired in courtly and beautiful settings, was fruitless and conducted with regard to diplomatic expediency. Eliot's note refers to the historian James A. Froude, "*Elizabeth*, Vol. I, ch. iv, letter of [Bishop] De Quadra [the ambassador] to Philip of Spain: 'In the afternoon we were in a barge, watching the games on the river. (The queen) was alone with Lord Robert and myself on the poop, when they began to talk nonsense, and went so far that Lord Robert at last said, as I was on the spot there was no reason why they should not be married if the queen pleased.' "
7. "Cf. *Purgatorio*, V, 133 * * * " [Eliot's note]. Eliot quotes Dante's lines, which he parodied; they may be translated: "Remember me, who am La Pia. / Siena made me, Maremma undid me." Highbury is a London suburb, Richmond a London borough with park and boating facilities, and Kew an adjoining district containing the Kew Botanical Gardens.
8. An East London slum.
9. Resort on the Thames estuary.
1. "V. St. Augustine's *Confessions*: 'to Carthage then I came, where a cauldron of unholy loves sang all about mine ears' " [Eliot's note]. Augustine here recounts his licentious youth.

Burning burning burning burning[2]
O Lord Thou pluckest me out[3]
O Lord Thou pluckest   310

burning

IV. Death by Water[4]

Phlebas the Phoenician, a fortnight dead,
Forgot the cry of gulls, and the deep sea swell
And the profit and loss.
                    A current under sea   315
Picked his bones in whispers. As he rose and fell
He passed the stages of his age and youth
Entering the whirlpool.
                    Gentile or Jew
O you who turn the wheel and look to windward,   320
Consider Phlebas, who was once handsome and tall as you.

V. What the Thunder Said[5]

After the torchlight red on sweaty faces
After the frosty silence in the gardens
After the agony in stony places
The shouting and the crying   325
Prison and palace and reverberation
Of thunder of spring over distant mountains
He who was living is now dead
We who were living are now dying
With a little patience[6]   330

Here is no water but only rock
Rock and no water and the sandy road
The road winding above among the mountains
Which are mountains of rock without water
If there were water we should stop and drink   335

2. Eliot's note to lines 307–9 refers to "Buddha's Fire Sermon (which corresponds in importance to the Sermon on the Mount)" and "St. Augustine's Confessions." The "collocation of these two representatives of eastern and western asceticism, as the culmination of this part of the poem, is not an accident."
3. The line is from Augustine's *Confessions* and echoes also Zechariah 3.2, where Jehovah, rebuking Satan, calls the high priest Joshua "a brand plucked out of the fire."
4. This lyric section presents the fearsome "death by water" that Madame Sosostris warned against in line 55. Critics disagree as to whether Phlebas's death, in the context of the entire poem, is a catastrophe prefiguring annihilation, or a ritual sacrifice prefiguring rebirth.
5. "In the first part of Part V three themes are employed: the journey to Emmaus, the approach to the Chapel Perilous (see Miss Weston's book) and the present decay of eastern Europe" [Eliot's note]. During his disciples' journey to Emmaus, after his crucifixion and resurrection, Jesus walked alongside and conversed with them, but they thought him a stranger until he revealed his identity (Luke 24.13–34). This section interfuses (1) images of personal isolation and disintegration, social upheaval, and makeshift routine with (2) images of redemptive suffering, compassion, sacrificial ritual, and disciplined control.
6. The opening nine lines contain allusions to Christ's imprisonment and trial, to his agony in the garden of Gethsemane, and his crucifixion and burial in the garden of Golgotha; they suggest the imminent despair during the days between the crucifixion and the resurrection on Easter.

Amongst the rock one cannot stop or think
Sweat is dry and feet are in the sand
If there were only water amongst the rock
Dead mountain mouth of carious teeth that cannot spit
Here one can neither stand nor lie nor sit   340
There is not even silence in the mountains
But dry sterile thunder without rain
There is not even solitude in the mountains
But red sullen faces sneer and snarl
From doors of mudcracked houses   345
                              If there were water
     And no rock
     If there were rock
     And also water
     And water   350
     A spring
     A pool among the rock
     If there were the sound of water only
     Not the cicada[7]
     And dry grass singing   355
But sound of water over a rock
Where the hermit-thrush[8] sings in the pine trees
Drip drop drip drop drop drop drop
But there is no water

Who is the third who walks always beside you?[9]   360
When I count, there are only you and I together
But when I look ahead up the white road
There is always another one walking beside you
Gliding wrapt in a brown mantle, hooded
I do not know whether a man or a woman   365
—But who is that on the other side of you?

What is that sound high in the air[1]
Murmur of maternal lamentation
Who are those hooded hordes swarming
Over endless plains, stumbling in cracked earth   370

---

7. Grasshopper. Cf. line 23 and Ecclesiastes 12.5: "the grasshopper shall be a burden, and desire shall fail."
8. "This is * * * the hermit-thrush which I have heard in Quebec Province. Chapman says (*Handbook of Birds of Eastern North America*) 'it is most at home in secluded woodland and thickety retreats. * * * Its 'water dripping song' is justly celebrated" [Eliot's note].
9. "The following lines were stimulated by the account of one of the Antarctic expeditions (I forget which, but I think one of Shackleton's): it was related that the party of explorers, at the extremity of their strength, had the constant delusion that there was *one more member* than could actually be counted" [Eliot's note]. The reminiscence is associated with Christ's unrecognized presence on the journey to Emmaus. (See introductory note to Part V.)
1. Eliot's note quotes a passage in German from *Blick ins Chaos* (1920) by Hermann Hesse (1877–1962), which may be translated: "Already half of Europe, already at least half of Eastern Europe, on the way to Chaos, drives drunk in sacred infatuation along the edge of the precipice, sings drunkenly, as though hymn-singing, as Dimitri Karamazov sang [in the novel *The Brothers Karamazov* (1882) by Feodor Dostoevsky (1821–81)]. The offended bourgeois laughs at the songs; the saint and the seer hear them with tears."

Ringed by the flat horizon only
What is the city over the mountains
Cracks and reforms and bursts in the violet air
Falling towers
Jerusalem Athens Alexandria   375
Vienna London
Unreal

A woman drew her long black hair out tight[2]
And fiddled whisper music on those strings
And bats with baby faces in the violet light   380
Whistled, and beat their wings
And crawled head downward down a blackened wall
And upside down in air were towers
Tolling reminiscent bells, that kept the hours
And voices singing out of empty cisterns and exhausted wells. 385

In this decayed hole among the mountains
In the faint moonlight, the grass is singing
Over the tumbled graves, about the chapel
There is the empty chapel, only the wind's home.
It has no windows, and the door swings,   390
Dry bones can harm no one.
Only a cock stood on the rooftree
Co co rico co co rico[3]
In a flash of lightning. Then a damp gust
Bringing rain   395

Ganga[4] was sunken, and the limp leaves
Waited for rain, while the black clouds
Gathered far distant, over Himavant.[5]
The jungle crouched, humped in silence.
Then spoke the thunder   400
Da[6]
Datta: what have we given?
My friend, blood shaking my heart
The awful daring of a moment's surrender
Which an age of prudence can never retract   405
By this, and this only, we have existed
Which is not to be found in our obituaries
Or in memories draped by the beneficient spider[7]

2. Lines 378–91 suggest the final temptation to despair as the questing knight
encounters the Chapel Perilous in the
Grail legend.
3. A cock's crow in folklore signaled the
departure of ghosts (as in Shakespeare's
Hamlet, 1.1.157 ff); in Matthew 26.34 and
74 a cock crowed, as Christ predicted,
when Peter denied him three times.
4. The Indian river Ganges, sacred to
Hindus and the scene of purgation and
fertility ceremonies.
5. A mountain in the Himalayas.
6. "'Datta, dayadhvam, damyata' (Give,
sympathise, control). The fable of the

meaning of the Thunder is found in the
Brihadaranyaka—Upanishad, 5, 1 * * *"
[Eliot's note]. In the Hindu legend,
the injunction of Prajapati (supreme
diety) is "Da," which is interpreted in
three different ways by gods, men, and
demons, to mean "control ourselves,"
"give alms," and "have compassion."
Prajapati assures them that when "the
divine voice, The Thunder," repeats the
syllable it means all three things and
that therefore "one should practice * * *
Self-Control, Alms-giving, and Compassion."
7. Eliot's note refers to *The White Devil*

Or under seals broken by the lean solicitor
In our empty rooms   410
DA
Dayadhvam: I have heard the key[8]
Turn in the door once and turn once only
We think of the key, each in his prison
Thinking of the key, each confirms a prison   415
Only at nightfall, aethereal rumours
Revive for a moment a broken Coriolanus[9]
DA
Damyata: The boat responded
Gaily, to the hand expert with sail and oar   420
The sea was calm, your heart would have responded
Gaily, when invited, beating obedient
To controlling hands

                    I sat upon the shore
Fishing,[1] with the arid plain behind me   425
Shall I at least set my lands in order?[2]
London Bridge is falling down falling down falling down
Poi s'ascose nel foco che gli affina[3]
Quando fiam uti chelidon[4]—O swallow swallow
Le Prince d'Aquitaine à la tour abolie[5]   430
These fragments I have shored against my ruins[6]
Why then Ile fit you. Hieronymo's mad againe.[7]

(1612) by the English playwright John
Webster (d. 1625), V.vi: " * * * they'll
remarry / Ere the worm pierce your
winding-sheet, ere the spider / Make a
thin curtain for your epitaphs."
8. "Cf. *Inferno*, XXXIII, 46 * * * "
[Eliot's note]. At this point Ugolino recalls his imprisonment with his children,
where they starved to death: "And I
heard below the door of the horrible
tower being locked up." Eliot's note
continues: "Also F. H. Bradley, *Appearance and Reality*, p. 346. 'My external
sensations are no less private to myself
than are my thoughts or my feelings. In
either case my experience falls within my
own circle, a circle closed on the outside; and, with all its elements alike,
every sphere is opaque to the others
which surround it * * * . In brief, regarded as an existence which appears in
a soul, the whole world for each is peculiar and private to that soul.' "
9. The Roman patrician Coriolanus defi-
antly chose self-exile when threatened
with banishment by the leaders of the
populace; he led enemy forces against
Rome, then, persuaded by his family, he
tried too late to reconcile the warring cities. He is the tragic protagonist in
Shakespeare's *Coriolanus* (1608).
1. "V. Weston: *From Ritual to Romance*; chapter on the Fisher King"
[Eliot's note].
2. Cf. Isaiah 38.1: "Thus saith the Lord,
Set thine house in order: for thou shalt

die, and not live."
3. Eliot's note to *Purgatorio*, XXVI,
quotes in Italian the passage (lines 145–
48) where the Provençal poet Arnaut
Daniel, recalling his lusts, addresses
Dante: "I pray you now, by the Goodness that guides you to the summit of this
staircase, reflect in due season on my
suffering." Then, in the line quoted in
The Waste Land, "he hid himself in the
fire that refines them."
4. Eliot's note refers to the *Pervigilium
Veneris* ("The Vigil of Venus"), an
anonymous Latin poem, and suggests a
comparison with "Philomela in Parts II
and III" of *The Waste Land*. The last
stanzas of the *Pervigilium* recreate the
myth of the nightingale in the image of
a swallow, and the poet listening to the
bird speaks the quoted line, "When shall
I be as the swallow," and adds: "that I
may cease to be silent." "O Swallow,
Swallow" are the opening words of one
of the songs interspersed in Tennyson's
narrative poem *The Princess* (1847).
5. "V. Gerard de Nerval, Sonnet *El
Desdichado*" [Eliot's note]. The line
reads: "The Prince of Aquitaine in the
ruined tower."
6. This line, like the invocation of the
swallow in line 428, and the allusion to
mad Hieronymo below, suggests the
poet's attempt to come to terms with the
waste land through his own verse.
7. Eliot's note refers to Thomas Kyd's

Datta. Dayadhvam. Damyata.
Shantih shantih shantih[8]

1921   1922

From FOUR QUARTETS

Burnt Norton[1]

τοῦ λόγου δ' ἐόντος ξυνοῦ ζώουσιν οἱ πολλοί
ὡς ἰδίαν ἔχοντες φρόνησιν.
                              I. p. 77. Fr. 2.
ὁδὸς ἄνω κάτω μία καὶ ὡυτή.
                              I. p. 89. Fr. 60.
—DIELS: *Die Fragmente der Vorsokratiker* (HERAKLEITOS)

I[2]

Time present and time past
Are both perhaps present in time future,
And time future contained in time past.
If all time is eternally present
All time is unredeemable.   5
What might have been is an abstraction
Remaining a perpetual possibility
Only in a world of speculation.
What might have been and what has been
Point to one end,[3] which is always present.   10

---

revenge play, *The Spanish Tragedy*, subtitled *Hieronymo's Mad Againe* (1594). In it Hieronymo is asked to write a court play and he answers "I'll fit you" in the double sense of "oblige" and "get even." He manages, though mad, to kill the murderers of his son by acting in the play and assigning parts appropriately, then commits suicide.

8. "Shantih. Repeated as here, a formal ending to an Upanishad [Vedic treatise, sacred Hindu text]. 'The peace which passeth understanding' is our equivalent to this word" [Eliot's note].

1. Eliot made *Burnt Norton*, though published originally as a separate poem, the basis and formal model for *East Coker* (1940), *The Dry Salvages* (1941), and *Little Gidding* (1942). Together they comprise *Four Quartets* (1943). Exploring the meaning of time in a search for its redemption, the *Quartets* seek to capture those rare moments when eternity "intersects" the temporal continuum, while treating also the relations between those moments and the flux of time. The series and each Quartet in it recapitulate the central theme through contrapuntal variations while moving gradually (through often tentative and enigmatic sequences) toward the explicit naming and more full revelation of its Christian content (the "Word" at the end of *Burnt Norton*, the "purgatorial fires" and "Annunciation" in subsequent Quartets, the coalescence of religious symbols at the end of *Little Gidding*). Central to its structure, and particularly in *Burnt Norton*, is the idea

of the Spanish mystic St. John of the Cross (1542–91) that the ascent of a soul to union with God is facilitated by memory and disciplined meditation but that meditation is superseded by a "dark night of the soul," a passive surrender of the will to God, an emptying of the senses and the self, a descent into darkness that is deepened paradoxically the nearer one approaches the light of God. The Greek epigraphs are from the pre-Socratic philosopher Heraclitus (540?–475 B.C.) and may be translated: "But although the Word is common to all, the majority of people live as though they had each an understanding peculiarly his own" and "The way up and the way down are one and the same."

2. Lines 1–10 introduce the theme of time abstractly and express skepticism about the redemption of time and the significance of a "might have been" in any but theoretical terms. The succeeding lines of Section I turn from abstraction to concrete memories, and the echoing words that accompany them, so as to explore childhood premonitions (glimpsed but not understood) of the sexual awakening and religious illumination that are figured in the rose garden and its empty pool that mysteriously fills with sunlight. Burnt Norton is a manor house in Gloucestershire, England. The opening lines echo Ecclesiastes 3.14–15: "That which hath been is now; and that which is to be hath already been."

3. In the double sense of "termination" and "purpose."

Footfalls echo in the memory
Down the passage which we did not take
Towards the door we never opened
Into the rose-garden.[4] My words echo
Thus, in your mind.
     But to what purpose   15
Disturbing the dust on a bowl of rose-leaves
I do not know.
    Other echoes
Inhabit the garden. Shall we follow?
Quick, said the bird, find them, find them,
Round the corner. Through the first gate, 20
Into our first world, shall we follow
The deception of the thrush? Into our first world.
There they were, dignified, invisible,
Moving without pressure, over the dead leaves,
In the autumn heat, through the vibrant air, 25
And the bird called, in response to
The unheard music hidden in the shrubbery,
And the unseen eyebeam crossed, for the roses
Had the look of flowers that are looked at.
There they were as our guests, accepted and accepting. 30
So we moved, and they, in a formal pattern,
Along the empty alley, into the box circle,[5]
To look down into the drained pool.
Dry the pool, dry concrete, brown edged,
And the pool was filled with water out of sunlight, 35
And the lotos[6] rose, quietly, quietly,
The surface glittered out of heart of light,[7]
And they were behind us, reflected in the pool.
Then a cloud passed, and the pool was empty.
Go, said the bird, for the leaves were full of children, 40
Hidden excitedly, containing laughter.
Go, go, go, said the bird: human kind
Cannot bear very much reality.
Time past and time future
What might have been and what has been 45
Point to one end, which is always present.

II[8]

Garlic and sapphires in the mud[9]
Clot the bedded axle-tree.
The trilling wire in the blood

4. The rose is a symbol of sexual and spiritual love; in Christian traditions it is associated with the harmony of religious truth and with the Virgin Mary, her bower being depicted often as a rose garden.
5. Evergreen boxwood shrubs, planted in a circle.
6. The lotus is a highly erotic symbol, associated particularly with the goddess Lakshmi in Hindu mythology.
7. An echo of Dante's *Paradiso*, XII.28–

29: "from out of the heart of one of the new lights there moved a voice."
8. In this opening lyric, a turbulent and paradoxical experience, compounding the earthly and sensuous with the transcendent, leads to a state of being which is neither static nor agitated but intensely "still" in the double sense of "tranquil" and "enduring." This state is imaged as a dance.
9. Derived from line 10 of the sonnet by the French Symbolist Stéphane Mallarmé

Sings below inveterate scars   50
Appeasing long forgotten wars.
The dance along the artery
The circulation of the lymph
Are figured in the drift of stars
Ascend to summer in the tree   55
We move above the moving tree
In light upon the figured leaf[1]
And hear upon the sodden floor
Below, the boarhound and the boar
Pursue their pattern as before   60
But reconciled among the stars.

At the still point of the turning world. Neither flesh nor
      fleshless;
Neither from nor towards; at the still point, there the dance
      is,
But neither arrest nor movement. And do not call it fixity,
Where past and future are gathered. Neither movement from
      nor towards,   65
Neither ascent nor decline. Except for the point, the still
      point,
There would be no dance, and there is only the dance.
I can only say, *there* we have been: but I cannot say where.
And I cannot say, how long, for that is to place it in time.

The inner freedom from the practical desire,   70
The release from action and suffering, release from the
      inner
And the outer compulsion, yet surrounded
By a grace of sense, a white light still and moving,
Erhebung[2] without motion, concentration
Without elimination, both a new world   75
And the old made explicit, understood
In the completion of its partial ecstasy,
The resolution of its partial horror.
Yet the enchainment of past and future
Woven in the weakness of the changing body,   80
Protects mankind from heaven and damnation
Which flesh cannot endure.
                           Time past and time future
Allow but a little consciousness.
To be conscious is not to be in time   85
But only in time can the moment in the rose-garden,
The moment in the arbour where the rain beat,
The moment in the draughty church at smokefall

(1842–98) "M'introduire dans ton histoire" ("To fit myself into your story").
Line 10 reads: "Tonnère et rubis aux
moyeux," or "Thunder and rubies at the
axles."
1. An echo of the description of death
in Tennyson's *In Memoriam* (1850),
XLIII.10–12: "So that still garden of the
souls / In many a figured leaf enrolls /
The total world since life began."
2. German for "exaltation."

Be remembered; involved with past and future.
Only through time time is conquered.   90

III[3]

Here is a place of disaffection
Time before and time after
In a dim light: neither daylight
Investing form with lucid stillness
Turning shadow into transient beauty   95
With slow rotation suggesting permanence
Nor darkness to purify the soul
Emptying the sensual with deprivation
Cleansing affection from the temporal.
Neither plenitude nor vacancy. Only a flicker   100
Over the strained time-ridden faces
Distracted from distraction by distraction
Filled with fancies and empty of meaning
Tumid apathy with no concentration
Men and bits of paper, whirled by the cold wind   105
That blows before and after time,
Wind in and out of unwholesome lungs
Time before and time after.
Eructation of unhealthy souls
Into the faded air, the torpid   110
Driven on the wind that sweeps the gloomy hills of London,
Hampstead and Clerkenwell, Campden and Putney,
Highgate, Primrose and Ludgate.[4] Not here
Not here the darkness, in this twittering world.

Descend lower, descend only   115
Into the world of perpetual solitude,
World not world, but that which is not world,
Internal darkness, deprivation
And destitution of all property,
Desiccation of the world of sense,   120
Evacuation of the world of fancy,
Inoperancy of the world of spirit;
This is the one way, and the other
Is the same, not in movement
But abstention from movement; while the world moves   125
In appetency, on its metalled ways
Of time past and time future.

IV[5]

Time and the bell have buried the day,
The black cloud carries the sun away.

3. This section presents first a descent into a subway, then a deeper descent into the darkness of the self.
4. Districts and neighborhoods in London.
5. This lyric is poised between anxious anticipation of night and death, and confidence in the eternity of the receding sunlight. The images evoke symbolic associations: the sunflower with light, the clematis (called "virgin's bower") with the Virgin Mary, the yew with death and immortality, the kingfisher with the Fisher King of the Grail legend, the light with Dante's vision of God in *Paradiso*, XXXIII.109–10.

Will the sunflower turn to us, will the clematis 130
Stray down, bend to us; tendril and spray
Clutch and cling?
Chill
Fingers of yew be curled
Down on us? After the kingfisher's wing 135
Has answered light to light, and is silent, the light is still
At the still point of the turning world.

V[6]

Words move, music moves
Only in time; but that which is only living
Can only die. Words, after speech, reach 140
Into the silence. Only by the form, the pattern,
Can words or music reach
The stillness, as a Chinese jar still
Moves perpetually in its stillness.
Not the stillness of the violin, while the note lasts, 145
Not that only, but the co-existence,
Or say that the end precedes the beginning,
And the end and the beginning were always there
Before the beginning and after the end.
And all is always now. Words strain, 150
Crack and sometimes break, under the burden,
Under the tension, slip, slide, perish,
Decay with imprecision, will not stay in place,
Will not stay still. Shrieking voices
Scolding, mocking, or merely chattering, 155
Always assail them. The Word in the desert[7]
Is most attacked by voices of temptation,
The crying shadow in the funeral dance,
The loud lament of the disconsolate chimera.[8]

The detail of the pattern is movement, 160
As in the figure of the ten stairs.[9]
Desire itself is movement
Not in itself desirable;
Love is itself unmoving,
Only the cause and end of movement, 165
Timeless, and undesiring
Except in the aspect of time
Caught in the form of limitation
Between un-being and being.

---

6. This section parallels the world's assaults on actual language with the temptation of the Christian Word in the wilderness, and celebrates the attempt of words and "pattern" in art to reach beyond themselves to the stillness of divine love and so capture the moment of revelation imaged in "the hidden laughter / Of children in the foliage."
7. An allusion to Christ's temptation in the wilderness, Luke 4.1–4.

8. A monster in Greek mythology, and a symbol of fantasies and delusions, slain by Bellerophon with the help of the winged horse Pegasus. Pegasus was a favorite of the Muses of the arts whom Bellerophon had tamed.
9. An allusion to St. John of the Cross's figure for the soul's ascent to God, "The Ten Degrees of the Mystical Ladder of Divine Love."

Sudden in a shaft of sunlight   170
Even while the dust moves
There rises the hidden laughter
Of children in the foliage
Quick, now, here, now always—
Ridiculous the waste sad time   175
Stretching before and after.

1934   1936, 1943

JOHN CROWE RANSOM
1888–1974

More than a decade before the Great Depression of the 1930s, when writers in significant numbers began teaching in American colleges and universities, the Southerner John Crowe Ransom accepted an appointment at
Vanderbilt University, where he began writing poetry and criticism that
soon established him as a leading exponent of southern agrarianism, the
most influential of the "New Critics" writing in America, and the finest
southern poet if his generation.

The son of a Methodist minister, Ransom was born in 1888 in Tennessee, was educated by his father until attending the Bowen School in Nashville, then studied philosophy and the Greek and Latin classics at Vanderbilt and later as a Rhodes Scholar at Oxford University in England. After
teaching Latin in a preparatory school for a year, he began teaching English
literature at Vanderbilt in 1914 and began writing poetry before leaving for
service in the army. He was studying at Grenoble University in France
when he finished his first book, *Poems without God*, which was published
in 1919 just before his return to Vanderbilt and teaching. He remained at
Vanderbilt until 1937 when he joined the faculty of Kenyon College in
Ohio, where he founded in 1939 the influential *Kenyon Review* (which he
continued to edit until his retirement in 1959) and launched in 1948 the
Kenyon School of English (which became the School of Letters of Indiana
University in 1951). Virtually all of his poems had been published by 1927
(in *Chills and Fever*, 1924, and *Two Gentlemen in Bonds*, 1927), and he
added only five new poems to those he consented to reprint (he excluded
Poems without God entirely), in *Selected Poems* (1945). But he revised
his poems continuously, adding stanzas or deleting and altering lines, and
readmitting some poems to revised editions of *Selected Poems* in 1963 and
1969, which were awarded respectively the Bollingen Prize for Poetry and
the National Book Award.

Though literary criticism and cultural commentary took precedence over
his poetry after 1927, the three pursuits were inseparably interwoven earlier
in his career. At Vanderbilt Ransom became the center of a group known as
The Fugitives which met fortnightly to discuss philosophy and to read their
own verse; they included the poet Donald Davidson and later the poet
Allen Tate and the poet and novelist Robert Penn Warren. In the group's
journal, *The Fugitive*, published between 1922 and 1925, much of Ran-

som's best poetry appeared. The poems were conspicuous for their wit and learning but chiefly for the subtle irony which enabled Ransom to affirm traditional southern values without succumbing to sentimentality, and to combine sophisticated perspectives on human experience with the simplicity of ballad forms and children's stories. Ransom's versification and stanza patterns are more traditional than T. S. Eliot's, but approximate rhymes or an occasional rough rhyme and doubled rhyme ("Moo" and "set-to," "rogue" and "prologue," "clack clack") produce bizarre effects, give body to the texture of the verse, and strengthen both the impact and the irony of the poems. He called his irregular metric "accentual meter; that is to say, we just count the number of accented or stressed syllables in the line and let unstressed syllables take care of themselves. So that sometimes * * * you run over two or three, or maybe even four, unstressed syllables at once—which was against the rule of the old poets. * * *"1 Ransom's conspicuously mixed diction—Latinate mixed with Old English terms, learned or abstruse terms fused with childlike colloquialisms, archaic terms or elegant usages joined with commonplaces—enable him to achieve the general aim that he defined in a letter to Allen Tate in 1927: "(1) I want to find the experience that is in the common actuals; (2) I want this experience to carry * * * the dearest possible values to which we have attached ourselves; (3) I want to face the disintegration or nullification of these values as calmly and religiously as possible."2 The result (in *Captain Carpenter,* for instance) is often a parody or mockery of poetic conventions which does not repudiate their power but reinforces it with irony.

When treating his characteristic themes—mortality, the denial of the body by the intellect and codes of decorum, the consecration of the past, and the frustration of moral idealism—Ransom's verse displays many of the characteristics of such southern fiction as the noevls of William Faulkner or the stories of Flannery O'Connor: the repression of energy which breaks through the surface in bizarre gestures and explosive violence, the haunting presence of archetypal figures, redolent of a ghostly chivalric past but still vivid in the present, as in faded tapestries and cracked paintings. Ransom's idiom can also render complex psychological states and densely enigmatic philosophical speculation even when warning against what he regards as the mind's worst enemy, abstraction.

The habit of intellectual abstraction had been instilled in the modern mind, Ransom felt, by science and its attendant industrial technology which dominated culture in the North. Ransom championed instead a regionalism that asserted the primacy of the full sensibility and of a traditionalistic community with ties to the more leisurely pace, the social structures and amenities, and the Protestant religious myths of the South's agrarian past. Myths extolling subsistence farming and a life close to nature informed Ransom's contributions to the collaborative *I'll Take My Stand,* published by "Twelve Southerners" in 1930. The threat posed by science to religious myths in the Scopes trial of 1925 (which examined the right of a teacher to teach Darwinian evolutionary theories in the Tennessee schools) provoked Ransom to write *God without Thunder* (1930), in which he defended not the High Anglicanism of Eliot but the original orthodoxies of the evangelical denominations which comprised the "social solidar-

---

1. Quoted by Robert Buffington, *The Equilibrist,* 88. 2. Quoted by Louise Cowan, *The Fugitive Group,* 178.

ity of my own community."[3] Religious myths did more justice than narrowly rationalistic science, Ransom felt, to the dualities and complexities of the psyche, and the vitalities of the natural world.

Alternatives to science and abstraction were the basis not only of Ransom's conservative agrarianism but of his literary theory as well, which he developed in countless reviews and in two books: *The World's Body* (1938), a collection of essays on poetic strategy in works ranging from Shakespeare's sonnets to Eliot's *Murder in the Cathedral*, and *The New Criticism* (1941), essays on contemporary literary critics and questions of critical method. Ransom warned against not only the abstractions of science but those of "Platonic" idealists (or "monsters") who "worship universals, laws, Platonic ideas, reason, the 'immaterial,' "[4] and he claimed for poetry a different and superior status as a means to knowledge. He claimed that poetry and, paradoxically, abstraction in art encompassed the particulars of actual experience, the concrete body of reality which eluded abstract modes of scientific and rational thought. While poetry aimed for the traditional fusion of a "moral effect with an aesthetic effect,"[5] the knowledge it contained was more extensive than philosophy because of the poem's texture of images, meters, and connotations; its texture could extend so far as to seem irrelevant to the poem's pattern (making for densely enigmatic poems), yet lead the reader into the full increment of knowledge that the poem made possible. Whatever the limitations of Ransom's critical theory, his method encouraged the close attention to details of poetic texts that distinguish them from other forms of discourse, an emphasis on verbal and formal features of poems rather than historical and philosophical contexts. He thereby defined the main aims of what came to be recognized as the "New Criticism" in American literary criticism. Kenyon College and its school of English became important centers for the encouragement of criticism and poetry—attracted by Ransom, the poet Robert Lowell transferred from Harvard to Kenyon—and the *Kenyon Review*, under Ransom's editorial direction, had a lasting impact on the younger ranks in academic circles and on the literary profession at large.

Here Lies a Lady

Here lies a lady of beauty and high degree.
Of chills and fever she died, of fever and chills,
The delight of her husband, her aunt, an infant of three,
And of medicos[1] marveling sweetly on her ills.

For either she burned, and her confident eyes would blaze,   5
And her fingers fly in a manner to puzzle their heads—
What was she making? Why, nothing; she sat in a maze
Of old scraps of laces, snipped into curious shreds—

Or this would pass, and the light of her fire decline
Till she lay discouraged and cold, like a stalk white and blown, 10
And would not open her eyes, to kisses, to wine;
The sixth of these states was her last; the cold settled down.

3. P. 327.
4. *The World's Body*, p. 225.

5. *Ibid.*, p. 57.
1. Doctors (slang).

Sweet ladies, long may ye bloom, and toughly I hope ye may
    thole,[2]
But was she not lucky? In flowers and lace and mourning,
In love and great honor we bade God rest her soul   15
After six little spaces of chill, and six of burning.

1924

Captain Carpenter[1]

Captain Carpenter rose up in his prime
Put on his pistols and went riding out
But had got wellnigh nowhere at that time
Till he fell in with ladies in a rout.

It was a pretty lady and all her train   5
That played with him so sweetly but before
An hour she'd taken a sword with all her main
And twined him of his nose for evermore.

Captain Carpenter mounted up one day
And rode straightway into a stranger rogue   10
That looked unchristian but be that as may
The Captain did not wait upon prologue.

But drew upon him out of his great heart
The other swung against him with a club
And cracked his two legs at the shinny part   15
And let him roll and stick like any tub.

Captain Carpenter rode many a time
From male and female took he sundry harms
He met the wife of Satan crying "I'm
The she-wolf bids you shall bear no more arms."   20

Their strokes and counters whistled in the wind
I wish he had delivered half his blows
But where she should have made off like a hind
The bitch bit off his arms at the elbows.

And Captain Carpenter parted with his ears   25
To a black devil that used him in this wise
O Jesus ere his threescore and ten years
Another had plucked out his sweet blue eyes.

Captain Carpenter got up on his roan
And sallied from the gate in hell's despite   30
I heard him asking in the grimmest tone
If any enemy yet there was to fight?

2. Endure (archaic).
1. When reading from his own works, Ransom declared: "There's an old hero who dies with his boots on, so to speak, and I've been asked if he could not have represented the Old South. Or if he could not have stood for the old-time religion. But those ideas did not enter my mind when I was composing the little ballad." (Quoted by Robert Buffington, *The Equilibrist*, p. 90.)

"To any adversary it is fame
If he risk to be wounded by my tongue
Or burnt in two beneath my red heart's flame   35
Such are the perils he is cast among.

"But if he can he has a pretty choice
From an anatomy with little to lose
Whether he cut my tongue and take my voice
Or whether it be my round red heart he choose."   40

It was the neatest knave that ever was seen
Stepping in perfume from his lady's bower
Who at this word put in his merry mien
And fell on Captain Carpenter like a tower.

I would not knock old fellows in the dust   45
But there lay Captain Carpenter on his back
His weapons were the old heart in his bust
And a blade shook between rotten teeth alack.

The rogue in scarlet and grey soon knew his mind
He wished to get his trophy and depart   50
With gentle apology and touch refined
He pierced him and produced the Captain's heart.

God's mercy rest on Captain Carpenter now
I thought him Sirs an honest gentleman
Citizen husband soldier and scholar enow[2]   55
Let jangling kites[3] eat of him if they can.

But God's deep curses follow after those
That shore him of his goodly nose and ears
His legs and strong arms at the two elbows
And eyes that had not watered seventy years.   60

The curse of hell upon the sleek upstart
That got the Captain finally on his back
And took the red red vitals of his heart
And made the kites to whet their beaks clack clack.

                                                      1924

The Equilibrists[1]

Full of her long white arms and milky skin
He had a thousand times remembered sin.
Alone in the press of people traveled he,
Minding her jacinth, and myrrh,[2] and ivory.

---

2. Enough (archaic).
3. Birds of prey.
1. Tightrope walkers, acrobats.

2. Resin used in perfume, medicine, and incense; "jacinth": reddish-orange gem.

Mouth he remembered: the quaint orifice   5
From which came heat that flamed upon the kiss,
Till cold words came down spiral from the head,
Grey doves from the officious tower illsped.

Body: it was a white field ready for love,
On her body's field, with the gaunt tower above,   10
The lilies grew, beseeching him to take,
If he would pluck and wear them, bruise and break.

Eyes talking: Never mind the cruel words,
Embrace my flowers, but not embrace the swords.
But what they said, the doves came straightway flying   15
And unsaid: Honor, Honor, they came crying.

Importunate her doves. Too pure, too wise,
Clambering on his shoulder, saying, Arise,
Leave me now, and never let us meet,
Eternal distance now command thy feet.   20

Predicament indeed, which thus discovers
Honor among thieves, Honor between lovers.
O such a little word is Honor, they feel!
But the grey word is between them cold as steel.

At length I saw these lovers fully were come   25
Into their torture of equilibrium;
Dreadfully had forsworn each other, and yet
They were bound each to each, and they did not forget.

And rigid as two painful stars, and twirled
About the clustered night their prison world,   30
They burned with fierce love always to come near,
But Honor beat them back and kept them clear.

Ah, the strict lovers, they are ruined now!
I cried in anger. But with puddled brow
Devising for those gibbeted and brave   35
Came I descanting:[3] Man, what would you have?

For spin your period out, and draw your breath,
A kinder saeculum[4] begins with Death.
Would you ascend to Heaven and bodiless dwell?
Or take your bodies honorless to Hell?   40

In Heaven you have heard no marriage is,[5]
No white flesh tinder to your lecheries,

---

3. Singing in full praise; "puddled": stirred or concerned with in an untidy way; "Devising": bequeathing or imagining; "gibbeted": hung on a gallows.
4. Period of time.

5. "For in the resurrection they neither marry nor are given in marriage, but are as the angels of God in heaven" (Matthew 22.30).

Your male and female tissue sweetly shaped
Sublimed away, and furious blood escaped.

Great lovers lie in Hell, the stubborn ones   45
Infatuate of the flesh upon the bones;
Stuprate,[6] they rend each other when they kiss,
The pieces kiss again, no end to this.

But still I watched them spinning, orbited nice.
Their flames were not more radiant than their ice.   50
I dug in the quiet earth and wrought the tomb
And made these lines to memorize their doom:—

Epitaph

Equilibrists lie here; stranger, tread light;
Close, but untouching in each other's sight;
Mouldered the lips and ashy the tall skull,   55
Let them lie perilous and beautiful.

1925, 1927

6. Ravished. The allusion is to Paolo and Francesca, bodiless spirits damned in the torment of their illicit love in Dante's *Divine Comedy* (*Inferno*, V).

E. E. CUMMINGS
1894–1962

A remarkable talent, a flair for self-dramatization, and the accident of being imprisoned in a French detention camp during World War I brought the poet and painter Edward Estlin Cummings into prominence in the early 1920s. In the four ensuing decades he gave a striking redefinition to the tradition of New England individualism, helped invent what has since been termed "pop art," and engaged a wondrously playful imagination in the creation of ingenious poetic forms.

As a child in Cambridge, Massachusetts, where he was born, and later at Harvard, he had the advantages of a favored son in a cultured family, whose respect for the individual and concern for social ethics won Cummings's lasting devotion. (Two of his poems, *if there are any heavens my mother will (all by herself) have* and *my father moved through dooms of love*, are impassioned tributes to his parents.) His father, a Congregational minister, taught English and later social ethics at Harvard. Encouraged by his parents, Cummings painted as a child, and he published poetry in college journals at Harvard while studying languages and the classics, taking sonnets as his models (by Shakespeare, the Pre-Raphaelite Dante Gabriel Rossetti, and the Harvard philosopher-poet George Santayana) along with the intricate stanza forms of earlier English and Continental literature. By the time he graduated from Harvard in 1915 (and received an M.A. in 1916), he had discovered Ezra Pound's poetry and had made life-long friends with the novelist John Dos Passos and with Scofield Thayer and James Sibley Watson, Jr., who were to assume control of *The Dial* and publish Cummings's first poems in 1920. His graduation address ("The New Art") was a bold endorsement of modern art—Stravinsky's music and the Cubists and Futurists in painting.

In the famous Norton Harjes Ambulance Corps in France, he and his friend Slater Brown aroused suspicion by their moustaches, their disdain for the bureaucracy, and Brown's outspoken letters home (which censors thought unpatriotic). Cummings was incarcerated along with his friend in October, 1917 in a French prison camp until letters from Cummings's father to U.S. officials and President Woodrow Wilson brought about his release three months later. Cummings's brilliantly surrealistic treatment of his experience, a celebration of the uniqueness of his fellow prisoners and a savage exposé of the French bureaucracy, was written at his father's urging and published as *The Enormous Room* in 1922. It made Cummings instantly famous.

Cummings lived in Paris from 1921 to 1923, and traveled frequently and regularly throughout his career, never disguising his enthusiasm for what was liberating in European culture but concluding (in pieces he wrote for *Vanity Fair* in 1925 and 1926) that America was more youthful and alive, that "France has happened more than she is happening whereas America is happening more than she has happened."[1] He was nothing less than appalled by Stalinist Russia, which he visited in 1931 and described in *Eimi* (1933), and he rendered the regimentation and deadness he found there in an image of the citizens dutifully lined up to pass by Lenin's tomb: "all toward All budgeshuffle:all Toward standwait. Isn'tish."[2] The America that excited him was divided between the family summer place in New Hampshire, which Cummings eventually inherited, and New York City, where he moved in 1918 and finally established himself in a Greenwich Village studio in 1924. He was still residing there with his third wife, an accomplished photographer, when he died in 1962.

There he pursued his "twin obsessions" of painting and writing,[3] painting during the day and writing at night. By carefully harboring his prizes, meager royalties and commissions, and the small income he received from his mother while she was alive, he managed to avoid the entanglements and routines of regular employment. His first volume of verse, *Tulips and Chimneys* (1923), was followed swiftly by two substantial volumes in 1925, the year he received the Dial Award, and *is 5* in 1926.

Cummings's verse displays a combination of Thoreau's controlled belligerency, the elitism of a privileged Bostonian, and the brash abandon of an uninhibited bohemian. He readily discerned in the modern environment a level of emotional deadness, conformity, barbarous warfare, intellectual abstractions, and clichés which he called a nonworld or "pseudoworld"[4] made up of "mostpeople" playing "impotent nongames of wrongright and rightwrong."[5] With a mixture of sardonic disdain, outright hatred, and earnest moral reproval he castigated that nonworld in mordantly sportive sonnets which caricatured it. Yet in the life around him Cummings also found sources of vitality and authenticity which he celebrated with refreshing candor in the lives of derelicts, unpretentious persons, lovers, his parents, and children, along with folk heroes, popular entertainers, and artists whose finesse as performers Cummings admired. His poems were designed to liberate the emotions and the imaginations of his readers by alternately (or

1. Quoted by Charles Norman, *The Magic-Maker*, p. 197.
2. Quoted by Cummings, *i: six nonlectures*, p. 101.
3. Quoted by Norman, *The Magic-Maker*, p. 251.
4. *i: six nonlectures*, p. 47.
5. Introduction, *Collected Poems* (1938).

simultaneously) shocking and exhilarating them, awakening them to their authentic selfhood as individuals and to a transformed world whose models are the process of birth and growth ("becoming") in nature, and the process of "making" in the arts. He conceived his poems as "competing" with the vitality of nature (its roses and "Niagara Falls"), surpassing "the 4th of July" in their celebration of independence, and rivaling the artifacts ("locomotives") that abound in industrial society (Foreword to *is 5*). His poems became not statements but exuberant games played with the reader and virtuoso performances by the writer.

Accordingly, Cummings's poetic forms, though less ambitious in their attempt to reconstitute poetry than Eliot's and Pound's, Crane's, Stevens's, or Williams's, are radically innovative and amount to more than the "pseudo-experimental poetry" that the critic Yvor Winters judged his work to be.[6] He clung to familiar themes and to the form of the sonnet, but he perfected that form in many tender lyrics and overhauled it in other instances, making it an unexpectedly effective instrument for satire and caricature. Shattering the conventions of syntax, punctuation, and typography, Cummings (with the help of an expert typesetter and friend Samuel A. Jacobs) devised poems that were as open-ended at beginning and conclusion as the processes of nature. These innovations converted print into visual shape, anticipating the "composition by field" of Charles Olson and such "concrete poets" as Aram Saroyan and Richard Kostelanetz. They also converted nouns into verbs and words into motion of the kind he admired as much in burlesque shows and the comic strip "Krazy Kat" as in the cityscapes of the painter John Marin. Cummings could incorporate the inflections of New England speech (in *rain or hail*), but more often he adopted the lingo of the city streets, whether to expose the vulgarity of anti-Semitism and the brutality of war, or to capture the energy and tenderness of urban life. Rivaling the intellectual abstractions he detested in the life around him, he made a poetic vocabulary out of the most abstract components of the language (for instance the syllables "non" and "un," the terms "it," "is," and "anyone") which enabled him to mock the unworld or to create an alternative world of spontaneity and beauty. In many poems Cummings boldly appropriated the clichés of public speech, popular culture, and advertising slogans, mocking their cheapness but capturing their energy and the force of their reality in ways comparable to the paintings of hamburgers, soup cans, and comic strips by "pop artists" of the 1960s.

Too often Cummings's ingenuities remain mere tricks and verbal puzzles, and his career displayed little development in theme, social vision, or technical mastery, but he sustained a very high level of achievement throughout his career in various genres. His first play, *Him*, an expressionistic work in which an artist weighs the demands of art and love, was condemned by critics in 1927 but had an impact on John Dos Passos; he also wrote a scenario for a ballet based on *Uncle Tom's Cabin* entitled *Tom* (1935), and a contemporary morality play entitled *Santa Claus* (1946). He had exhibited his paintings in 1919, but his first major show was in Cleveland in 1931 at a time when he could find no publisher for his poems. Subsequent exhibitions in New York in the 1940s and 1950s coincided with the revival of interest in his poetry occasioned by publication of his *Collected Poems* in 1938. By the time he delivered his idiosyncratic Norton Lectures at Harvard (published as *i: six nonlectures*, 1953) he was one of the most

---

6. *In Defense of Reason* (1947), p. 86.

popular of the established poets among college and high-school students. A special citation by the National Book Award in 1955 and the award of the Bollingen Prize in Poetry in 1957 were further testimony to the widening recognition of this "proudhumble citizen of ecstasies."

✓ in Just-

in Just-
spring   when the world is mud-
luscious the little
lame balloonman

whistles   far   and wee   5

and eddieandbill come
running from marbles and
piracies and it's
spring

when the world is puddle-wonderful   10

the queer
old balloonman whistles
far   and   wee
and bettyandisbel come dancing

from hop-scotch and jump-rope and   15

it's
spring
and
    the

        goat-footed   20

balloonMan   whistles
far
and
wee

                                      1920, 1923

Buffalo Bill 's[1]

Buffalo Bill 's
defunct
        who used to
        ride a watersmooth-silver
                              stallion   5
and break onetwothreefourfive pigeonsjustlikethat
                                        Jesus
he was a handsome man
                    and what i want to know is

---

1. William F. Cody (1846–1917), American scout and Wild West showman.

how do you like your blueeyed boy   10
Mister Death

                                               1920, 1923

the Cambridge ladies who live in furnished souls

the Cambridge ladies who live in furnished souls
are unbeautiful and have comfortable minds
(also,with the church's protestant blessings
daughters,unscented shapeless spirited)
they believe in Christ and Longfellow,[2] both dead,   5
are invariably interested in so many things—
at the present writing one still finds
delighted fingers knitting for the is it Poles?
perhaps.   While permanent faces coyly bandy
scandal of Mrs. N and Professor D   10
. . . . the Cambridge ladies do not care,above
Cambridge if sometimes in its box of
sky lavender and cornerless,the
moon rattles like a fragment of angry candy

                                               1923

it is at moments after i have dreamed

it is at moments after i have dreamed
of the rare entertainment of your eyes,
when(being fool to fancy)i have deemed

with your peculiar mouth my heart made wise;
at moments when the glassy darkness holds   5

the genuine apparition of your smile
(it was through tears always)and silence moulds
such strangeness as was mine a little while;

moments when my once more illustrious arms
are filled with fascination,when my breast   10
wears the intolerant brightness of your charms:

one pierced moment whiter than the rest

—turning from the tremendous lie of sleep
i watch the roses of the day grow deep.

                                               1923

between the breasts

between the breasts
of bestial
Marj lie large
men who praise

---

2. Henry Wadsworth Longfellow (1807–82), American poet and professor of Romance languages at Harvard.

Marj's cleancornered strokable   5
body   these men's
fingers toss trunks
shuffle sacks spin kegs they

curl
loving   10
around
beers

        the world has
these men's hands but their
bodies big and boozing   15
belong to

Marj
the greenslim purse of whose
face opens
on a fatgold   20

grin
hooray
hoorah for the large
men who lie

between the breasts   25
of bestial Marj
for the strong men
who

sleep between the legs of Lil

                                    1925

Spring is like a perhaps hand

Spring is like a perhaps hand
(which comes carefully
out of Nowhere)arranging
a window,into which people look(while
people stare   5
arranging and changing placing
carefully there a strange
thing and a known thing here)and

changing everything carefully

spring is like a perhaps   10
Hand in a window
(carefully to
and fro moving New and
Old things,while
people stare carefully   15

moving a perhaps
fraction of flower here placing
an inch of air there)and

without breaking anything.

1925

Picasso[1]

Picasso
you give us Things
which
bulge:grunting lungs pumped full of sharp thick mind

you make us shrill   5
presents always
shut in the sumptuous screech of
simplicity

(out of the
black unbunged   10
Something gushes vaguely a squeak of planes
or

between squeals of
Nothing grabbed with circular shrieking tightness
solid screams whisper.)   15
Lumberman of The Distinct

your brain's
axe only chops hugest inherent
Trees of Ego,from
whose living and biggest   20

bodies lopped
of every
prettiness

you hew form truly

1925

this is the garden:colours come and go

this is the garden:colours come and go,
frail azures fluttering from night's outer wing
strong silent greens serenely lingering,
absolute lights like baths of golden snow.
This is the garden:pursed lips do blow   5
upon cool flutes within wide glooms,and sing
(of harps celestial to the quivering string)
invisible faces hauntingly and slow.

1. Pablo Picasso (1881–1973), Spanish painter and sculptor.

This is the garden.   Time shall surely reap,
and on Death's blade lie many a flower curled,   10
in other lands where other songs be sung;
yet stand They here enraptured,as among
the slow deep trees perpetual of sleep
some silver-fingered fountain steals the world.

1925

✓Jimmie's got a goil

Jimmie's got a goil
               goil
                   goil,
                       Jimmie
's got a goil and   5
she coitnly can shimmie

when you see her shake
                shake
                    shake,
                        when   10
you see her shake a
shimmie how you wish that you was Jimmie.

Oh for such a gurl
              gurl
                 gurl,   15
                   oh
for such a gurl to
be a fellow's twistandtwirl

talk about your Sal-
              Sal-   20
                 Sal-,
                   talk
about your Salo
-mes[1] but gimmie Jimmie's gal.

1926

"next to of course god america i

"next to of course god america i
love you land of the pilgrims' and so forth oh
say can you see by the dawn's early my
country 'tis of centuries come and go
and are no more what of it we should worry   5
in every language even deafanddumb
thy sons acclaim your glorious name by gorry
by jingo by gee by gosh by gum

---

1. Salome, sensual Biblical princess who
danced before Herod Antipas in return
for his promise to behead John the Baptist (Mark 6.22–27).

why talk of beauty what could be more beautiful than these heroic happy dead   10
who rushed like lions to the roaring slaughter
they did not stop to think they died instead
then shall the voice of liberty be mute?"

He spoke.   And drank rapidly a glass of water

1926

✔look at this)

look at this)
a 75^2 done
this nobody would
have believed
would they no   5
kidding this was my particular

pal
funny aint
it we was
buddies   10
i used to

know
him lift the
poor cuss
tenderly this side up handle   15

with care
fragile
and send him home

to his old mother in
a new nice pine box   20

(collect

1926

if there are any heavens my mother will (all by herself) have

if there are any heavens my mother will (all by herself) have
one.   It will not be a pansy heaven nor
a fragile heaven of lilies-of-the-valley but
it will be a heaven of blackred roses

my father will be (deep like a rose   5
tall like a rose)

standing near my

2. French 75, a piece of artillery used in World War I.

swaying over her
(silent)
with eyes which are really petals and see   10

nothing with the face of a poet really which
is a flower and not a face with
hands
which whisper
This is my beloved my   15

                    (suddenly in sunlight
he will bow,

& the whole garden will bow)

                                                    1931

                    r-p-o-p-h-e-s-s-a-g-r

                              r-p-o-p-h-e-s-s-a-g-r
                    who
        a)s w(e loo)k
        upnowgath
                    PPEGORHRASS   5
                              eringint(o-
        aThe):l
              eA
              !p:
        S   a   10
                    (r
        rIvInG   .gRrEaPsPhOs)
                                        to
        rea(be)rran(com)gi(e)ngly
        ,grasshopper;   15
                              1932, 1935

            kumrads die because they're told)
            kumrads die because they're told)
            kumrads die before they're old
            (kumrads aren't afraid to die
            kumrads don't
            and kumrads won't   5
            believe in life)and death knows whie

            (all good kumrads you can tell
            by their altruistic smell
            moscow pipes good kumrads dance)
            kumrads enjoy   10

s.freud[1] knows whoy
the hope that you may mess your pance

every kumrad is a bit
of quite unmitigated hate
(travelling in a futile groove   15
god knows why)
and so do i
(because they are afraid to love

                                        1935

anyone lived in a pretty how town

anyone lived in a pretty how town
(with up so floating many bells down)
spring summer autumn winter
he sang his didn't he danced his did.

Women and men(both little and small)   5
cared for anyone not at all
they sowed their isn't they reaped their same
sun moon stars rain

children guessed(but only a few
and down they forgot as up they grew   10
autumn winter spring summer)
that noone loved him more by more

when by now and tree by leaf
she laughed his joy she cried his grief
bird by snow and stir by still   15
anyone's any was all to her

someones married their everyones
laughed their cryings and did their dance
(sleep wake hope and then)they
said their nevers they slept their dream 20

stars rain sun moon
(and only the snow can begin to explain
how children are apt to forget to remember
with up so floating many bells down)

one day anyone died i guess   25
(and noone stooped to kiss his face)
busy folk buried them side by side
little by little and was by was

all by all and deep by deep
and more by more they dream their sleep 30

---

1. Sigmund Freud (1856–1939), Austrian psychoanalyst.

noone and anyone earth by april
wish by spirit and if by yes.

Women and men(both dong and ding)
summer autumn winter spring
reaped their sowing and went their came   35
sun moon stars rain

1940

my father moved through dooms of love

my father moved through dooms of love
through sames of am through haves of give,
singing each morning out of each night
my father moved through depths of height

this motionless forgetful where   5
turned at his glance to shining here;
that if(so timid air is firm)
under his eyes would stir and squirm

newly as from unburied which
floats the first who,his april touch   10
drove sleeping selves to swarm their fates
woke dreamers to their ghostly roots

and should some why completely weep
my father's fingers brought her sleep:
vainly no smallest voice might cry   15
for he could feel the mountains grow.

Lifting the valleys of the sea
my father moved through griefs of joy;
praised a forehead called the moon
singing desire into begin   20

joy was his song and joy so pure
a heart of star by him could steer
and pure so now and now so yes
the wrists of twilight would rejoice

keen as midsummer's keen beyond   25
conceiving mind of sun will stand,
so strictly(over utmost him
so hugely)stood my father's dream

his flesh was flesh his blood was blood:
no hungry man but wished him food;   30
no cripple wouldn't creep one mile
uphill to only see him smile.

Scorning the pomp of must and shall
my father moved through dooms of feel;
his anger was as right as rain
his pity was as green as grain

35

septembering arms of year extend
less humbly wealth to foe and friend
than he to foolish and to wise
offered immeasurable is

40

proudly and(by octobering flame
beckoned)as earth will downward climb,
so naked for immortal work
his shoulders marched against the dark

his sorrow was as true as bread:
no liar looked him in the head;
if every friend became his foe
he'd laugh and build a world with snow.

45

My father moved through theys of we,
singing each new leaf out of each tree
(and every child was sure that spring
danced when she heard my father sing)

50

then let men kill which cannot share,
let blood and flesh be mud and mire,
scheming imagine,passion willed,
freedom a drug that's bought and sold

55

giving to steal and cruel kind,
a heart to fear,to doubt a mind,
to differ a disease of same,
conform the pinnacle of am

60

though dull were all we taste as bright,
bitter all utterly things sweet,
maggoty minus and dumb death
all we inherit,all bequeath

and nothing quite so least as truth
—i say though hate were why men breathe—
because my father lived his soul
love is the whole and more than all

65

1940

pity this busy monster,manunkind

pity this busy monster,manunkind,

not.   Progress is a comfortable disease:
your victim(death and life safely beyond)

plays with the bigness of his littleness
—electrons deify one razorblade   5
into a mountainrange;lenses extend

unwish through curving wherewhen till unwish
returns on its unself.   A world of made
is not a world of born—pity poor flesh   10

and trees,poor stars and stones,but never this
fine specimen of hypermagical

ultraomnipotence.   We doctors know

a hopeless case if—listen:there's a hell
of a good universe next door;let's go   15

                                                    1944

✓ rain or hail

rain or hail
sam done
the best he kin
till they digged his hole

:sam was a man
                                                                  5

stout as a bridge
rugged as a bear
slickern a weazel
how be you

(sun or snow)   10

gone into what
like all them kings
you read about
and on him sings

a whippoorwill;   15

heart was big
as the world aint square
with room for the devil
and his angels too

yes,sir   20

what may be better
or what may be worse
and what may be clover
clover clover

(nobody'll know)   25

sam was a man
grinned his grin
done his chores
laid him down.

Sleep well   30

1944

why must itself up every of a park

why must itself up every of a park

anus stick some quote statue unquote to
prove that a hero equals any jerk
who was afraid to dare to answer "no"?

quote citizens unquote might otherwise   5
forget(to err is human;to forgive
divine)that if the quote state unquote says
"kill" killing is an act of christian love.

"Nothing" in 1944 AD

"can stand against the argument of mil   10
itary necessity"(generalissimo e)[1]
and echo answers "there is no appeal

from reason"(freud)[2]—you pays your money and
you doesn't take your choice. Ain't freedom grand

1950

1. General, later President, Dwight D. Eisenhower (1890–1969).
2. Sigmund Freud (1856–1939), Austrian psychologist and founder of psychoanalysis.

HART CRANE
1899–1932

In the course of a brief career that lasted scarcely fifteen years, Harold Hart Crane aroused for his fellow writers some of the highest hopes and some of the most gnawing fears about American culture and modernism in the arts. Without either wide popularity or durable institutional affiliations to sustain him, he devoted a tormented life to the perfection and then the exhaustion of an extraordinary talent, and bequeathed to his successors a body of brilliantly visionary poetry along with an image of tragic failure.

Crane was raised by parents whose marital turmoil and undisguised sexual incompatibility (leading to separations, reconciliations, and finally divorce in 1916) agitated Crane's affections from the age of eight and burdened him

with what he called later, in *The Bridge*, the "curse of sundered paren-
tage." He sought and welcomed the affection of his mother, and time and
again had occasion to be grateful to her for the checks she sent him after
he left home; though they quarreled at one point over an inheritance from
his grandmother, Mrs. Crane in later years was proud of his reputation as a
poet. Yet her needs for companionship and demands for affection were distracting throughout his life. His father, a successful candy manufacturer,
introduced Crane to writers and gave some encouragement to Crane's
poetic career, but he never genuinely understood his son's aspirations, continued to hope that he would follow a business career, and did not provide
the regular support that Crane hoped for.

Crane attended high school in Cleveland, and at his parents' request
made gestures toward preparing for college when he lived in New York
City in 1917–18 and again in 1919 on an allowance from his father. But
Crane, who had been painting since the age of eleven and had published a
poem in *The Pagan* in 1916, quickly decided upon a literary career. He
worked at his poetry and developed his aesthetic theories in conversations
with the painter Carl Schmitt at his Stuyvesant Square studio, published
The Hive in 1917, and became an unsalaried associate editor of *The Pagan*
with the critic Gorham Munson, whose support and friendship were important throughout his early career. When Crane left New York for a job in
his father's business in Cleveland in 1919, it was with the purpose of
finding support for a literary career.

Protracted unemployment, a job in a bookstore, and a position with a
direct-mail advertising house (writing copy for Sieberling Tires and the Fox
Furnace Company) followed a dispute with his father in 1921. Crane
applied himself assiduously to the work, convinced (he declared in 1919)
that even "in an age of violent commercialism" an "artist's creation is
bound to be largely interpretive of his environment" and that he "cannot
remain aloof from the welters without losing the essential, imminent vitality of his vision." While he should avoid "smugness," he should count
patiently on "America's great economic growth, eventually to complete a
structure that will provide a quieter platform * * * for the artist and his
audience."[1] Crane's career was complicated during these same years by
increasingly frequent bouts of heavy drinking and by homosexual affairs—
one an affair of passion in 1919 of the kind that later inspired some of his
best poetry, others passionless encounters in the sexual underworld he
sought out to his peril.

Crane's intellectual horizons widened significantly during his Cleveland
years. Within a larger circle of acquaintances that included the composer
Ernest Bloch, he became a close friend of the Swiss architect and painter
William Lescaze and the painter William Sommer; he sustained a lively
correspondence with associates he knew in New York like Gorham
Munson, Sherwood Anderson, and other literary figures whose views he
sought out and welcomed, including the poet and critic Allen Tate. His
reading ranged from the cultural critics recommended by Anderson, Van
Wyck Brooks, and Waldo Frank (whose *Our America* [1919] envisioned
the transformation of industrial America) to the Continental writers Baudelaire, Dostoevsky, and Proust, who were recommended by Munson or Lescaze. His enthusiasms were diffuse, scattering in all the directions charted

1. Quoted by John Unterecker, *Voyager*, p. 156.

by contemporary criticism but including also the romantic poets; he scorned Milton and Tennyson among others, he wrote in 1921, adding that "I *do* run joyfully toward Messrs. Poe, Shakespeare, Keats, Shelley, Coleridge, John Donne!!!, John Webster!!! Marlowe, Baudelaire, Laforgue, Dante, Cavalcanti, Li Po, and a host of others."[2] He responded early to Pound, William Carlos Williams, Marianne Moore, and notably Wallace Stevens, but he had been steeped in T. S. Eliot since 1919 and Eliot represented the most stimulating challenge. Eliot's achievement, Crane wrote, threatened to become an "absolute impasse" but Crane felt that he had "discovered a safe tangent to strike which, if I can possibly explain the position, goes *through* [Eliot] toward a different goal." After "having absorbed him enough we can trust ourselves as never before," Crane said, adding that "I * * * would like to leave a few of his 'negations' behind me, risk the realm of the obvious more, in quest of new sensations * * * ."[3]

It was in Ohio that Crane produced his first distinctive poems and moreover established the basis for his subsequent career. *My Grandmother's Love Letters,* the first poem for which he was paid, appeared in the *Dial* in 1920, *Chaplinesque* in the *Gargoyle* in 1921, and *Praise for an Urn* in the *Dial* in 1922. By 1923 he had finished his first masterpiece, *For the Marriage of Faustus and Helen,* and composed the first versions of two poems which were to be incorporated in his second masterpiece, *Voyages* (1926). He published one of the latter in Gorham Munson's *Secession* in 1923, entitling it *Poster* because he conceived it as approaching "the 'advertisement' form."[4] He acknowledged to Munson that he was discovering his affinities with Whitman, that he thought himself potentially "the Pindar for the dawn of the machine age," and that he was "ruminating" on the epic and "mystical synthesis of America" that was to be his major undertaking, *The Bridge.*[5]

In 1923 Crane settled on New York City as the most stimulating environment for a writer. Between 1923 and 1927 he wrote the best, indeed most, of his poetry, completing the love poem *Voyages* in 1924, then publishing his first collection, *White Buildings,* and writing ten of the fifteen poems that were to comprise *The Bridge* in the single year 1926. He held intermittent jobs, but he was enabled to pursue his career chiefly by funds sent by his parents, by the generosity of friends with whom he stayed for protracted periods, by sizable grants from the banker and patron Otto Kahn which Crane requested when his needs were most desperate—and by his own strong commitment to an increasingly desperate visionary poetics. He had been inspired by P. D. Ouspensky's *Tertium Organum* (tr. 1922), which had attributed a higher state of consciousness to poets than to ordinary men; but he was influenced more directly by the rhetorical intensity of Melville and the Jacobean dramatists, and the poetry of Whitman and William Blake to undertake a "complete reversal of direction" from the pessimism of Eliot (he wrote in 1923) and to aim for "a more positive, or (if [I] must put it so in a skeptical age) ecstatic goal." The "modern artist," he wrote, in order to get composers like "Strauss, Ravel, Scriabin, and Bloch into *words,*" must "ransack the vocabularies of Shakespeare, Jonson, Webster * * * and add our scientific, street and counter, and psychological terms * * * "; he "needs gigantic assimilative capacities, emotion,—and the

2. *Letters,* p. 67.
3. *Ibid.,* p. 90.
4. *Ibid.,* p. 99.
5. *Ibid.,* pp. 128–29.

greatest of all—*vision*."[6] The "full contemporary function" of modern poetry embraced the imperative to "absorb the machine, i.e., *acclimatize* it as naturally and casually as trees * * * castles and all other human associations of the past" (*Of Modern Poetry*, 1929).[7] This Crane himself undertook to do through images of machines and patterns of motion in the poems derived from technological symbols like the Brooklyn Bridge, aircraft, and subway trains.

In New York as in Cleveland, Crane pursued an "ecstatic goal" in poetry under the stimulus of wine and recorded phonograph music to help generate the production of his first drafts, which he later worked over assiduously. His diction became more abstract, his verse more compressed, his poems consequently more abstruse. He aimed—and this has been of increasing interest to poets after World War II—for a visionary poetry he called "absolute" (*General Aims and Theories*, 1925) in which a poem had "an orbit or predetermined direction of its own" and gave the reader a "single, new *word*, never before spoken and impossible to actually enunciate, but self-evident as an active principle in the reader's consciousness henceforward." Whether employing traditional stanza patterns or free verse, he relied at once on the abstract associations and sensuous impingements of words, drawing on "the implicit emotional dynamics of the materials used" and selecting his terms and images "less for their logical (literal) significance than for their associational meanings." The "entire construction of the poem," he insisted, builds on the " 'logic of metaphor,' which antedates our so-called pure logic, and which is the genetic basis of all speech, hence consciousness and thought-extension."[8]

Plans for *The Bridge* that Crane had been nurturing since 1923 began to be realized early in 1926 when he finished *Atlantis*, its exhilarating concluding lyric. He explained to his patron Otto Kahn in 1927 that his poetic "symphony with an epic theme" was to encompass "the Myth of America" on a scale comparable to Virgil's *Aeneid*, and he had explained to Gorham Munson in 1923 that in it "History and fact, location, etc., all have to be transfigured into abstract form that would almost function independently of its subject matter."[9] The poem's materials included the full panorama of America's history from geological eras to contemporary suburbia and urban New York, and such figures from history and legend as Columbus, Rip Van Winkle, Pocahontas, Whitman, and Poe who define the aspirations, anxieties, and tragedies that have shaped the myth of America. Crane's governing symbol was drawn from Brooklyn Bridge, which epitomized the reconciliation of technological civilization with art, spanned the full range of human possibility, and imaged the communion of love. The "very idea of a bridge," he wrote to the critic Waldo Frank in 1926, depends on "spiritual convictions" and accomplishes in fact an "act of faith" and an act of "communication."[1] One thrust of the poem is to celebrate those values as incarnate in the *Bridge* and to embody in the poem's movement a personal commitment to those values. Nevertheless the poem was animated also by Crane's fears about the validity of the myth and the forces of disintegration in modern life which threatened its fulfillment, to anxieties about his own career and talent that made it difficult for him to complete the poem, and by apprehensions that his willed commitment to the celebration of the

---

6. *Ibid.*, pp. 114–15, 129.
7. *Complete Poems and Selected Letters and Prose*, p. 262.
8. *Ibid.*, pp. 220–21.
9. *Letters*, pp. 305, 309.
1. *Ibid.*, p. 124.

vision was false. The "forms, materials, dynamics" on which his commitment depended "are simply non-existent in the [contemporary] world," he added in the letter to Frank, and "I may amuse and * * * flatter myself as much as I please—but I am only evading a recognition and playing Don Quixote in an immorally conscious way. * * * The bridge as a symbol today has no significance beyond an economical approach to shorter hours, quicker lunches, behaviorism and toothpicks."[2] Impelled by such recognitions, Crane's poem became not only a willed celebration of the myth but a quest for the self-transformation that would validate his commitment to the vision, a quest that is challenged frequently in the poem by imminent disenchantment or disillusion.

Crane did not finish *The Bridge* until 1930, when it was published first in Paris and then in New York, after years troubled by family problems, self-doubts about his capacities as a poet, and by extremities of debauchery which continued when he traveled abroad in 1928 and 1929. Although he received *Poetry* magazine's Levinson Award for *The Bridge*, reviews of the poem were not highly favorable, and Crane was held up by the critics Allen Tate and Yvor Winters as an object lesson in the excesses attributed to romanticism. He received a Guggenheim fellowship in 1930 and went to Mexico in hope of salvaging his talent. He produced some poems, including *The Broken Tower*, and was living happily with Peggy Baird Cowley, the divorced wife of the critic Malcolm Cowley, and talking of marriage. But the violence of his drinking bouts and the sordidness of his homosexual dissipations were conspicuous to the novelist Katherine Anne Porter and other friends in Mexico. Before his death he had completed arranging a volume to be entitled *Key West*. In 1932 he jumped to his death from the deck of the ship carrying him to New York City. In the eyes of the poet Karl Shapiro, Crane died a "wasted death," an object lesson in the folly of total commitment to poetry (*Essay on Rime*, 1945). To the poet Robert Lowell in *Words for Hart Crane* (1958), Crane died in defiance of a world which did not reward his talent.

Legend[1]

As silent as a mirror is believed
Realities plunge in silence by . . .

I am not ready for repentance;
Nor to match regrets. For the moth
Bends no more than the still   5
Imploring flame. And tremorous

In the white falling flakes
Kisses are,—
The only worth all granting.

It is to be learned—   10
This cleaving and this burning,

---

2. *Ibid.*, p. 261.
1. The poems *Legend* through *Voyages VI* are presented in the sequence Crane determined for them in *White Buildings* (1926); the arrangement was designed to make the poems more accessible to the reader. The title *White Buildings* is drawn from a recurring motif in the surrealistic paintings of Giorgio de Chirico (b. 1888).

But only by the one who
Spends out himself again.

Twice and twice
(Again the smoking souvenir,   15
Bleeding eidolon!)[2] and yet again.
Until the bright logic is won
Unwhispering as a mirror
Is believed.

Then, drop by caustic drop, a perfect cry   20
Shall string some constant harmony,—
Relentless caper for all those who step
The legend of their youth into the noon.

1925   1926

Black Tambourine[1]

The interests of a black man in a cellar
Mark tardy judgment on the world's closed door.
Gnats toss in the shadow of a bottle,
And a roach spans a crevice in the floor.

Æsop,[2] driven to pondering, found   5
Heaven with the tortoise and the hare;
Fox brush and sow ear top his grave
And mingling incantations on the air.

The black man, forlorn in the cellar,
Wanders in some mid-kingdom, dark, that lies,   10
Between his tambourine, stuck on the wall,
And, in Africa, a carcass quick with flies.

1921   1921, 1926

Lachrymae Christi[1]

Whitely, while benzine[2]
Rinsings from the moon
Dissolve all but the windows of the mills

2. Specter, phantom.
1. Crane claimed in letters to Sherwood Anderson and Gorham Munson (*Letters*, pp. 77, 58) that his aim in this poem was to create a "form that is so thorough and intense as to dye the words themselves with a peculiarity of meaning, slightly different maybe from the ordinary definition of them separate from the poem," that he had gone back to "rime and rhythm" so as to avoid the disorder of Dada art, and that "the poem is * * * a bundle of insinuations, suggestions, bearing on the Negro's place somewhere between man and beast. That is why Aesop is brought in, etc.—the popular conception of Negro romance,

the tambourine on the wall. The value of the poem is only, to me, in what a painter would call its 'tactile' quality, —an entirely aesthetic feature. A propagandist for either side of the Negro question could find anything he wanted to in it. My only declaration is that I find the Negro (in the popular mind) sentimentally or brutally 'placed' in this midkingdom, etc."
2. Greek author of *Fables* (fl. 620–560 B.C.) and a freed slave.
1. The first version of this poem, begun early in 1924, was enclosed in a letter to Waldo Frank in which Crane wrote of his encouragement at the prospect of a new job in advertising, his excitement at

(Inside the sure machinery
Is still   5
And curdled[3] only where a sill
Sluices its one unyielding smile)

Immaculate venom binds
The fox's teeth, and swart
Thorns freshen on the year's   10
First blood. From flanks unfended,
Twanged red perfidies of spring
Are trillion on the hill.

And the nights opening
Chant pyramids,—   15
Anoint with innocence,—recall
To music and retrieve what perjuries
Had galvanized[4] the eyes.

                      While chime
Beneath and all around   20
Distilling clemencies,—worms'[5]
Inaudible whistle, tunneling
Not penitence
But song, as these
Perpetual fountains, vines,—   25
Thy Nazarene[6] and tinder eyes.

(Let[7] sphinxes from the ripe
Borage[8] of death have cleared my tongue
Once and again; vermin and rod
No longer bind. Some sentient cloud   30
Of tears flocks through the tendoned loam:
Betrayed stones slowly speak.)

Names peeling from Thine eyes
And their undimming lattices of flame,

---

having begun *The Bridge*, and his passionate love for Emil Opffer, Jr., a merchant seaman with whom he soon established residence. The poem celebrates a
tortured process of redemption: the
transfiguration of an industrial factory,
the revitalization of the speaker's sexual
passion and the reawakening of his powers of expression, and the rebirth of nature in springtime. The title ("Tears of
Christ") is the name of an Italian wine,
frequently used as communion wine. The
poem is built on paradoxes: the redemptive process is conceived as being both
natural and supernatural, Christian and
pagan, violently destructive and life-giving, agonizing and ecstatic.
2. A poisonous cleaning fluid.
3. In the sense of "reduced to its richest

essence."
4. Both coated over and charged with
energy.
5. A punning allusion to earthworms and
gears of machinery.
6. From the town of Nazareth: Jesus.
7. A verb in the sense of "may." The
stanza in parenthesis is a silent prayer in
which the speaker begs that his powers
of speech be cleansed while imagining
that they have been revitalized already;
the process of renewal must be repeated
"once and again," but the cleansing
process begins even before he finishes
the prayer.
8. An herb used to flavor salads and
claret punch and believed to cure melancholy and to quiet lunatics.

Spell out in palm and pain   35
Compulsion of the year, O Nazarene.

Lean long from sable, slender boughs,
Unstanched and luminous. And as the nights
Strike from Thee perfect spheres,
Lift up in lilac-emerald breath the grail   40
Of earth again—

Thy face
From charred and riven stakes, O
Dionysus,[9] Thy
Unmangled target smile.   45

1924–25   1926

From Voyages[1]

I

Above the fresh ruffles[2] of the surf
Bright striped urchins flay each other with sand.
They have contrived a conquest for shell shucks,
And their fingers crumble fragments of baked weed
Gaily digging and scattering.   5

And in answer to their treble interjections
The sun beats lightning on the waves,
The waves fold thunder on the sand;
And could they hear me I would tell them:

O brilliant kids, frisk with your dog,   10
Fondle your shells and sticks, bleached
By time and the elements; but there is a line
You must not cross nor ever trust beyond it
Spry cordage[3] of your bodies to caresses
Too lichen-faithful from too wide a breast.   15
The bottom of the sea is cruel.

1921–23   1923, 1926

9. Greek god, inventor of wine, and the center of mystery cults, whose metamorphoses and sacrifice were associated with the rebirth of spring. His rites were characterized by intoxication, religious frenzy, madness, and orgiastic indulgence, while his spring festival in Athens was the occasion for the production of new tragedies and comedies.

1. Crane composed this suite of love poems in 1924 and 1925, revising two earlier poems (*Poster* and *Belle Isle*) which became sections I and VI of *Voyages*. The lover with whom he was living was Emil Opffer, Jr., a merchant seaman; the facts that his knowledge of the sea and exotic islands exceeded Crane's, and that he was separated from Crane while on sea duty, contributed a dynamic to the poem: images of the *consummation* of passion blend with images

of the *anticipation* of its ecstasies and its treacheries, and with fears of its dissolution. Crane presents love as a time-bound passion, tormenting as well as exhilarating, a communion that affords ecstasy and beauty or imaginative vision but also portends the dissolution of passion and death. The poem's central symbol is the sea. Its motion charts the course of the lovers' passion. The sea also images the art which the passion inspired (the "snowy sentences," the "transmemberment of song," Crane's own poem), and the transcendent vision, which lies beyond the poet's experience and expressive power.
2. Cloth gathered in ripples, and also the rhythms of drum beats. Musical images recur in *Voyages*.
3. Rigging of ships, associated here with lovers' bodies.

II

—And yet this great wink of eternity,
Of rimless floods, unfettered leewardings,
Samite[4] sheeted and processioned where
Her undinal[5] vast belly moonward bends,   20
Laughing the wrapt inflections of our love;

Take this Sea, whose diapason[6] knells
On scrolls of silver snowy sentences,
The sceptred terror of whose sessions rends
As her demeanors notion well or ill,   25
All but the pieties of lovers' hands.

And onward, as bells off San Salvador
Salute the crocus lustres of the stars,
In these poinsettia meadows of her tides,—
Adagios of islands,[7] O my Prodigal,   30
Complete the dark confessions her veins spell.

Mark how her turning shoulders wind the hours,
And hasten while her penniless rich palms
Pass superscription of bent foam and wave,—
Hasten, while they are true,—sleep, death, desire,   35
Close round one instant in one floating flower.

Bind us in time, O Seasons clear, and awe.
O minstrel galleons of Carib fire,
Bequeath us to no earthly shore until
Is answered in the vortex of our grave   40
The seal's wide spindrift[8] gaze toward paradise.

1924   1926

VI[1]

Where icy and bright dungeons lift
Of swimmers their lost morning eyes,
And ocean rivers, churning, shift
Green borders under stranger skies,   115

Steadily as a shell secretes
Its beating leagues of monotone,

---

4. Medieval gold-threaded silk cloth.
5. Alluding to Undine, a water nymph capable of becoming human through love.
6. Melody or harmony in music.
7. Crane explained that "adagios of islands" refers "to the motion of a boat through islands clustered thickly, the rhythm of the motion, etc. And it seems a much more direct and creative statement than any more logical employment of words such as 'coasting slowly through the islands.' besides ushering in a whole world of music." ("General Aims and Theories," *Complete Poems and Selected Letters and Prose*, p. 221.) Like the "insular Tahiti" invoked in Melville's *Moby-Dick* (Chapter 58), islands function in Crane's poem as an image of serene joy surrounded by tempestuous, dangerous seas.
8. Sea foam. The suggestions of "drifting" and "spinning" define Crane's conception of the currents and intensity of passion.
1. This section celebrates the endurance of love, transcending death and the wreck of love, as imaged even in the "icy" Belle Isle and in the "imaged Word" of poetry.

Or as many waters trough the sun's
Red kelson[2] past the cape's wet stone;

O rivers mingling toward the sky   120
And harbor of the phoenix' breast—
My eyes pressed black against the prow,
—Thy derelict and blinded guest

Waiting, afire, what name, unspoke,
I cannot claim: let thy waves rear   125
More savage than the death of kings,[3]
Some splintered garland for the seer.

Beyond siroccos harvesting
The solstice thunders, crept away,
Like a cliff swinging or a sail   130
Flung into April's inmost day—

Creation's blithe and petalled word
To the lounged goddess when she rose[4]
Conceding dialogue with eyes
That smile unsearchable repose—   135

Still fervid covenant, Belle Isle,[5]
—Unfolded floating dais before
Which rainbows twine continual hair—
Belle Isle, white echo of the oar!

The imaged Word, it is, that holds   140
Hushed willows anchored in its glow.
It is the unbetrayable reply
Whose accent no farewell can know.

1921–25   1923, 1926

From THE BRIDGE

*From going to and fro in the earth,
and from walking up and down in it.*
—THE BOOK OF JOB[1]

To Brooklyn Bridge

How many dawns, chill from his rippling rest
The seagull's wings shall dip and pivot him,
Shedding white rings of tumult, building high
Over the chained bay waters Liberty—

2. Timber of a ship joining the ribs to the keel.
3. Richard II calls for tales of the "sad death of kings" in Shakespeare's play (3.3.156).
4. Possibly Aphrodite (Venus), goddess of love, who was born of the foam of the sea; possibly Eos, goddess of dawn, the lover of Orion who restored his blinded eyes to sight.

5. An island off Labrador southeast of Cape St. Charles.
1. The formal pattern of the poem (movement up and down on a vertical axis, intersecting movement westward and eastward on a horizontal axis) is suggested by this quotation from *Job*, which gives Satan's answer to Jehovah when asked where he has been (1.7).

Then, with inviolate curve, forsake our eyes   5
As apparitional as sails that cross
Some page of figures to be filed away;
—Till elevators drop us from our day . . .

I think of cinemas, panoramic sleights
With multitudes bent toward some flashing scene   10
Never disclosed, but hastened to again,
Foretold to other eyes on the same screen;

And Thee, across the harbor, silver-paced
As though the sun took step of thee, yet left
Some motion ever unspent in thy stride,—   15
Implicitly thy freedom staying thee!

Out of some subway scuttle, cell or loft
A bedlamite[2] speeds to thy parapets,
Tilting there momently, shrill shirt ballooning,
A jest falls from the speechless caravan.   20

Down Wall, from girder into street noon leaks,
A rip-tooth of the sky's acetylene;
All afternoon the cloud-flown derricks turn . . .
Thy cables breathe the North Atlantic still.

And obscure as that heaven of the Jews,   25
Thy guerdon . . . Accolade thou dost bestow
Of anonymity time cannot raise:
Vibrant reprieve and pardon thou dost show.

O harp and altar, of the fury fused,
(How could mere toil align thy choiring strings!)   30
Terrific threshold of the prophet's pledge,
Prayer of pariah, and the lover's cry,—

Again the traffic lights that skim thy swift
Unfractioned idiom, immaculate sigh of stars,
Beading thy path—condense eternity: 35
And we have seen night lifted in thine arms.

Under thy shadow by the piers I waited;
Only in darkness is thy shadow clear.
The City's fiery parcels all undone,
Already snow submerges an iron year . . .   40

O Sleepless as the river under thee,
Vaulting the sea, the prairies' dreaming sod,
Unto us lowliest sometime sweep, descend
And of the curveship lend a myth to God.

1926   1927, 1930

---

2. Madman, inmate of a hospital for the insane.

From II. Powhatan's Daughter[3]

*"—Pocahuntus, a well-featured but wanton yong girle . . . of the age
of eleven or twelve years, get the boyes forth with her into the market
place, and make them wheele, falling on their hands, turning their heels
upwards, whom she would followe, and wheele so herself, naked as she
was, all the fort over."*

* * *

The River[4]

*. . . and
past the
din and
slogans of
the year—*

Stick your patent name on a signboard
brother—all over—going west—young man
Tintex—Japalac—Certain-teed Overalls ads[5]
and land sakes! under the new playbill ripped
in the guaranteed corner—see Bert Williams[6] what?   5
Minstrels when you steal a chicken just
save me the wing for if it isn't
Erie it ain't for miles around a
Mazda—and the telegraphic night coming on Thomas
a Ediford[1]—and whistling down the tracks   10
a headlight rushing with the sound—can you
imagine—while an EXPRESS makes time like
SCIENCE—COMMERCE and the HOLYGHOST
RADIO ROARS IN EVERY HOME WE HAVE THE NORTHPOLE
WALLSTREET AND VIRGINBIRTH WITHOUT STONES OR   15
WIRES OR EVEN RUNning brooks connecting ears
and no more sermons windows flashing roar
breathtaking—as you like it . . . eh?[2]

So the 20th Century—so
whizzed the Limited—roared by and left   20
three men, still hungry on the tracks, ploddingly
watching the tail lights wizen and converge, slip-

---

3. Pocahontas, the legendary figure whom Crane associated with the American "continent," a "nature symbol" comparable to the "traditional Hertha of ancient Teutonic mythology." In this section the sights and sounds of modernity are displaced first by fleeting glimpses, then persistent memories, of the past, yielding finally a deep communion with the "River of Time" and revealing something vital, tragic, and sacred in the mythic past. The epigraph is from William Strachey, *History of Travaile into Virginia Britannica* (1615).

4. Crane wrote to Mrs. T. W. Simpson (*Letters*, p. 303): "I'm trying in this part of the poem to chart the pioneer experience of our forefathers—and to tell the story backwards * * * on the 'backs' of hoboes. These hoboes are simply 'psychological ponies' to carry the reader across the country and back to the Mississippi, which you will notice is described as a great River of Time. I also unlatch the door to the pure Indian world which opens out in 'The Dance' section, so the reader is gradually led back in time to the pure savage world,

while existing at the same time in the present." The poem probes beneath the surface excitements and tawdry debasements of modern commercial life to reveal deep continuities with folk culture and the pagan past. In the opening lines the image of a subway is translated into an image of a luxury express train, The Twentieth Century Limited, traveling from New York to Chicago. The "marginal gloss" continued here began in the preceding section, and reads, "Like Memory, she is time's truant, shall take you by the hand . . ."

5. Advertising slogans: trade names of a dye, a varnish, and a brand of overalls.

6. Egbert A. Williams (1876–1922), popular, talented black minstrel show entertainer.

1. A mocking reference to Thomas A. Edison (1847–1931), inventor of the electric light bulb (trade name "Mazda"), and Henry Ford (1863–1947), automobile manufacturer.

2. An echo of Shakespeare's *As You Like It* (2.1.16–17): "books in the running brooks / Sermons in stones."

ping gimleted and nearly out of sight.

. . . .

The last bear, shot drinking in the Dakotas
Loped under wires that span the mountain stream. 25
Keen instruments,[3] strung to a vast precision

to those
whose ad-
dresses are
never near

Bind town to town and dream to ticking dream.
But some men take their liquor slow—and count
—Though they'll confess no rosary nor clue—
The river's minute by the far brook's year. 30
Under a world of whistles, wires and steam
Caboose-like they go ruminating through
Ohio, Indiana—blind baggage—
To Cheyenne tagging . . . Maybe Kalamazoo.

Time's rendings, time's blendings they construe 35
As final reckonings of fire and snow;
Strange bird-wit, like the elemental gist
Of unwalled winds they offer, singing low
My Old Kentucky Home and *Casey Jones,*
Some Sunny Day. I heard a road-gang chanting so. 40
And afterwards, who had a colt's eyes—one said,
"Jesus! Oh I remember watermelon days!" And sped
High in a cloud of merriment, recalled
"—And when my Aunt Sally Simpson smiled," he
        drawled—
"It was almost Louisiana, long ago." 45
"There's no place like Booneville though, Buddy,"
One said, excising a last burr from his vest,
"—For early trouting." Then peering in the can,
"—But I kept on the tracks." Possessed, resigned,
He trod the fire down pensively and grinned, 50
Spreading dry shingles of a beard. . . .

                                    Behind
My father's cannery works I used to see
Rail-squatters ranged in nomad raillery,
The ancient men—wifeless or runaway
Hobo-trekkers that forever search 55
An empire wilderness of freight and rails.
Each seemed a child, like me, on a loose perch,
Holding to childhood like some termless play.
John, Jake or Charley, hopping the slow freight
—Memphis to Tallahassee—riding the rods, 60
Blind fists of nothing, humpty-dumpty clods.

Yet they touch something like a key perhaps.

but who
have
touched
her, know-
ing her
without
name

From pole to pole across the hills, the states
—They know a body under the wide rain;
Youngsters with eyes like fjords, old reprobates 65
With racetrack jargon,—dotting immensity
They lurk across her, knowing her yonder breast

---

3. The telephone and telegraph.

Snow-silvered, sumac-stained or smoky blue—
Is past the valley-sleepers, south or west.
—As I have trod the rumorous midnights, too,   70

And past the circuit of the lamp's thin flame
(O Nights that brought me to her body bare!)
Have dreamed beyond the print that bound her name.
Trains sounding the long blizzards out—I heard
Wail into distances I knew were hers.   75
Papooses crying on the wind's long mane
Screamed redskin dynasties that fled the brain,
—Dead echoes! But I knew her body there,
Time like a serpent down her shoulder, dark,
And space, an eaglet's wing, laid on her hair.[4]   80

Under the Ozarks, domed by Iron Mountain,
The old gods of the rain lie wrapped in pools
Where eyeless fish curvet a sunken fountain
And re-descend with corn from querulous crows.
nor the myths of her fathers . . . Such pilferings make up their timeless eatage,   85
Propitiate them for their timber torn
By iron, iron—always the iron dealt cleavage!
They doze now, below axe and powder horn.
And Pullman breakfasters glide glistening steel
From tunnel into field—iron strides the dew—   90
Straddles the hill, a dance of wheel on wheel.
You have a half-hour's wait at Siskiyou,
Or stay the night and take the next train through.
Southward, near Cairo passing, you can see
The Ohio merging,—borne down Tennessee;   95
And if it's summer and the sun's in dusk
Maybe the breeze will lift the River's musk
—As though the waters breathed that you might know
Memphis Johnny, Steamboat Bill, Missouri Joe.
Oh, lean from the window, if the train slows down,   100
As though you touched hands with some ancient
        clown,
—A little while gaze absently below
And hum *Deep River* with them while they go.

Yes, turn again and sniff once more—look see,
O Sheriff, Brakeman and Authority—   105
Hitch up your pants and crunch another quid,[5]
For you, too, feed the River timelessly.
And few evade full measure of their fate;
Always they smile out eerily what they seem.
I could believe he joked at heaven's gate—   110
Dan Midland—jolted from the cold brake-beam.[6]

4. Feathered headdresses and serpents are familiar tokens of Indian culture. William Strachey reported that 17th-century Indians wore small live snakes as earrings.

5. Chunk of chewing tobacco.
6. Structure on a railroad car where hoboes ride; Dan Midland was a legendary hobo who fell to his death from a train.

Down, down—born pioneers in time's despite,
Grimed tributaries to an ancient flow—
They win no frontier by their wayward plight,
But drift in stillness, as from Jordan's[7] brow.   115

You will not hear it as the sea; even stone
Is not more hushed by gravity . . . But slow,
As loth to take more tribute—sliding prone
Like one whose eyes were buried long ago
The River, spreading, flows—and spends your dream.   120
What are you, lost within this tideless spell?
You are your father's father, and the stream—
A liquid theme that floating niggers swell.

Damp tonnage and alluvial march of days—
Nights turbid, vascular with silted shale   125
And roots surrendered down of moraine clays:
The Mississippi drinks the farthest dale.

O quarrying passion, undertowed sunlight!
The basalt surface drags a jungle grace
Ochreous and lynx-barred in lengthening might;   130
Patience! and you shall reach the biding place!

Over De Soto's bones the freighted floors
Throb past the City storied of three thrones.[8]
Down two more turns the Mississippi pours
(Anon tall ironsides up from salt lagoons)   135

And flows within itself, heaps itself free.
All fades but one thin skyline 'round . . . Ahead
No embrace opens but the stinging sea;
The River lifts itself from its long bed,

Poised wholly on its dream, a mustard glow   140
Tortured with history, its one will—flow!
—The Passion spreads in wide tongues, choked and
    slow,
Meeting the Gulf, hosannas silently below.

1926   1930

7. Palestinian river, celebrated in Negro spirituals as the boundary between earth and heaven.
8. New Orleans, at various times under Spanish, French, and English rule. The body of the Spanish explorer Hernando de Soto (1500–42) was secretly committed to the Mississippi River so that hostile Indians would continue to believe in his divinity. In 1862 Admiral David G. Farragut (1801–70) led a Union fleet up the Mississippi from the Gulf and captured New Orleans. "Ironsides" is a term for warships, whether or not ironclad.

The Dance[9]

*Then you
shall see
her truly—
your blood
remember-
ing its first
invasion of
her secrecy,
its first
encounters
with her
kin, her
chieftain
lover . . .
his shade
that haunts
the lakes
and hills*

The swift red flesh, a winter king—
Who squired the glacier woman down the sky?
She ran the neighing canyons all the spring;
She spouted arms; she rose with maize—to die.

And in the autumn drouth, whose burnished hands   5
With mineral wariness found out the stone
Where prayers, forgotten, streamed mesa sands?
He holds the twilight's dim, perpetual throne.

Mythical brows we saw retiring—loth,
Disturbed and destined, into denser green.   10
Greeting they sped us, on the arrow's oath:
Now lie incorrigibly what years between . . .

There was a bed of leaves, and broken play;
There was a veil upon you, Pocahontas, bride—
O Princess whose brown lap was virgin May;   15
And bridal flanks and eyes hid tawny pride.

I left the village for dogwood. By the canoe
Tugging below the mill-race, I could see
Your hair's keen crescent running, and the blue
First moth of evening take wing stealthily.   20

What laughing chains the water wove and threw!
I learned to catch the trout's moon whisper; I
Drifted how many hours I never knew,
But, watching, saw that fleet young crescent die,—

And one star, swinging, take its place, alone,   25
Cupped in the larches of the mountain pass—
Until, immortally, it bled into the dawn.
I left my sleek boat nibbling margin grass . . .

I took the portage climb, then chose
A further valley-shed; I could not stop.   30
Feet nozzled wat'ry webs of upper flows;
One white veil gusted from the very top.

O Appalachian Spring! I gained the ledge;
Steep, inaccessible smile that eastward bends
And northward reaches in that violet wedge   35
Of Adirondacks!—wisped of azure wands,

---

9. This section presents Crane's imaginary consummation of his adventure
westward and quest into the past. Crane
intended to become "identified with the
Indian and his world before it is over.
* * * Pocahontas (the continent) is the
common basis of our meeting, she survives the extinction of the Indian, who
finally, after being assumed into the elements of nature * * * persists only as a
kind of 'eye' in the sky, or a star
* * *" (*Letters*, p. 307).

Over how many bluffs, tarns, streams I sped!
—And knew myself within some boding shade:—
Grey tepees tufting the blue knolls ahead,
Smoke swirling through the yellow chestnut glade . . . 40

A distant cloud, a thunder-bud—it grew,
That blanket of the skies: the padded foot
Within,—I heard it; 'til its rhythm drew,
—Siphoned the black pool from the heart's hot root!

A cyclone threshes in the turbine crest,   45
Swooping in eagle feathers down your back;
Know, Maquokeeta,[1] greeting; know death's best;
—Fall, Sachem, strictly as the tamarack!

A birch kneels. All her whistling fingers fly.
The oak grove circles in a crash of leaves;   50
The long moan of a dance is in the sky.
Dance, Maquokeeta: Pocahontas grieves . . .

And every tendon scurries toward the twangs
Of lightning deltaed down your saber hair.
Now snaps the flint in every tooth; red fangs   55
And splay tongues thinly busy the blue air . . .

Dance, Maquokeeta! snake that lives before,
That casts his pelt, and lives beyond![2] Sprout, horn!
Spark, tooth! Medicine-man, relent, restore—
Lie to us,—dance us back the tribal morn!   60

Spears and assemblies: black drums thrusting on—
O yelling battlements,—I, too, was liege
To rainbows currying each pulsant bone:
Surpassed the circumstance, danced out the siege!

And buzzard-circleted, screamed from the stake;   65
I could not pick the arrows from my side.
Wrapped in that fire, I saw more escorts wake—
Flickering, spring up the hill groins like a tide.

I heard the hush of lava wrestling your arms,
And stag teeth foam about the raven throat;   70
Flame cataracts of heaven in seething swarms
Fed down your anklets to the sunset's moat.

O, like the lizard in the furious noon,
That drops his legs and colors in the sun,
—And laughs, pure serpent, Time itself, and moon   75
Of his own fate, I saw thy change begun!

1. The name of an Indian cab-driver who told Crane it meant "Big River" and signified a god whose rains refreshed the plains. "Sachem" (Algonquin): chief.

2. The snake is an archetypal symbol of time, and shedding its skin is a symbol of time's renewal.

And saw thee dive to kiss that destiny
Like one white meteor, sacrosanct and blent
At last with all that's consummate and free
There, where the first and last gods keep thy tent.   80

. . .

Thewed of the levin,[3] thunder-shod and lean,
Lo, through what infinite seasons dost thou gaze—
Across what bivouacs of thine angered slain,
And see'st thy bride immortal in the maize!

Totem and fire-gall, slumbering pyramid—[4]   85
Though other calendars now stack the sky,
Thy freedom is her largesse, Prince, and hid
On paths thou knewest best to claim her by.

High unto Labrador the sun strikes free
Her speechless dream of snow, and stirred again,   90
She is the torrent and the singing tree;
And she is virgin to the last of men . . .

West, west and south! winds over Cumberland
And winds across the llano[5] grass resume
Her hair's warm sibilance. Her breasts are fanned   95
O stream by slope and vineyard—into bloom!

And when the caribou slant down for salt
Do arrows thirst and leap? Do antlers shine
Alert, star-triggered in the listening vault
Of dusk?—And are her perfect brows to thine?   100

We danced, O Brave, we danced beyond their farms,
In cobalt desert closures made our vows . . .
Now is the strong prayer folded in thine arms,
The serpent with the eagle in the boughs.

1926   1927, 1930

* * *

VII. The Tunnel[1]

To Find the Western path
Right thro' the Gates of Wrath.
—BLAKE[2]

Performances, assortments, résumés—
Up Times Square to Columbus Circle lights

3. Lightning.
4. Suggests the smoldering volcano Popocatepetl near Mexico City, and the huge pyramids of the Mayans, used for sacrificial ceremonies and the astronomical measurement of time. "Fire-gall": charred ashes, like the excrescence made by insects on a tree.
5. Treeless plain.
1. *The Tunnel*, representing ₁ epic convention of a descent into hell, renders

the poet's descent into a subway and the interior of his mind, followed by the resurgence of desperate hope when he emerges to see the Bridge and the East River leading to the distant sea.
2. The opening lines of *Morning*, by the visionary English poet William Blake (1757–1827), which foresees the return of dawn and the triumph of love after facing the crisis which tests it.

Channel the congresses, nightly sessions,
Refractions of the thousand theatres, faces—
Mysterious kitchens. . . . You shall search them all.   5
Someday by heart you'll learn each famous sight
And watch the curtain lift in hell's despite;
You'll find the garden in the third act dead,
Finger your knees—and wish yourself in bed
With tabloid crime-sheets perched in easy sight.   10

> Then let you reach your hat
> and go.
> As usual, let you—also
> walking down—exclaim
> to twelve upward leaving   15
> a subscription praise
> for what time slays.

Or can't you quite make up your mind to ride;
A walk is better underneath the L[3] a brisk
Ten blocks or so before? But you find yourself   20
Preparing penguin flexions of the arms,—
As usual you will meet the scuttle yawn:
The subway yawns the quickest promise home.

Be minimum, then, to swim the hiving swarms
Out of the Square, the Circle burning bright—[4]   25
Avoid the glass doors gyring at your right,
Where boxed alone a second, eyes take fright
—Quite unprepared rush naked back to light:
And down beside the turnstile press the coin
Into the slot. The gongs already rattle.   30

> And so
> of cities you bespeak
> subways, rivered under streets
> and rivers. . . . In the car
> the overtone of motion   35
> underground, the monotone
> of motion is the sound
> of other faces, also underground—

"Let's have a pencil Jimmy—living now
at Floral Park   40
Flatbush—on the fourth of July—
like a pigeon's muddy dream—potatoes
to dig in the field—travlin the town—too—
night after night—the Culver line—the
girls all shaping up—it used to be—"   45

---

3. Abbreviation: elevated railway.
4. An echo of Blake's poem *The Tiger*:
"Tiger! Tiger! burning bright / In the
forests of the night, / What immortal
hand or eye / Could frame thy fearful
symmetry?" Also the lighted sign indicating a subway station.

Our tongues recant like beaten weather vanes.
This answer lives like verdigris,[5] like hair
Beyond extinction, surcease of the bone;
And repetition freezes—"What

"what do you want? getting weak on the links?   50
fandaddle daddy don't ask for change—IS THIS
FOURTEENTH? it's half past six she said—if
you don't like my gate why did you
swing on it, why *didja*
swing on it   55
anyhow—"

      And somehow anyhow swing—

The phonographs of hades in the brain
Are tunnels that re-wind themselves, and love
A burnt match skating in a urinal—   60
Somewhere above Fourteenth TAKE THE EXPRESS
To brush some new presentiment of pain—

"But I want service in this office SERVICE
I said—after
the show she cried a little afterwards but—"   65

Whose head is swinging from the swollen strap?[1]
Whose body smokes along the bitten rails,
Bursts from a smoldering bundle far behind
In back forks of the chasms of the brain,—
Puffs from a riven stump far out behind   70
In interborough fissures of the mind . . . ?

And why do I often meet your visage here,
Your eyes like agate lanterns—on and on
Below the toothpaste and the dandruff ads?
—And did their riding eyes right through your side,   75
And did their eyes like unwashed platters ride?
And Death, aloft,—gigantically down
Probing through you—toward me, O evermore![2]
And when they dragged your retching flesh,
Your trembling hands that night through Baltimore—   80
That last night on the ballot rounds, did you
Shaking, did you deny the ticket, Poe?

For Gravesend Manor change at Chambers Street.
The platform hurries along to a dead stop.

---

5. Green coating or stain on copper.
1. The images of disfigurement and violent death in this section introduce the tortured figure and haunted imagination of Edgar Allan Poe. He died in Baltimore in 1849, reputedly of brutal mistreatment while drunk at the hands of a political gang who wanted him to cast multiple ballots illegally for their "ticket."
2. Echoes of the line "Death looks gigantically down" in *The City in the Sea*, and of the refrain "Nevermore" in *The Raven*, by Poe.

The intent escalator lifts a serenade   85
Stilly
Of shoes, umbrellas, each eye attending its shoe, then
Bolting outright somewhere above where streets
Burst suddenly in rain. . . . The gongs recur:
Elbows and levers, guard and hissing door.   90
Thunder is galvothermic[3] here below. . . . The car
Wheels off. The train rounds, bending to a scream,
Taking the final level for the dive
Under the river—
And somewhat emptier than before,   95
Demented, for a hitching second, humps; then
Lets go. . . . Toward corners of the floor
Newspapers wing, revolve and wing.
Blank windows gargle signals through the roar.

And does the Daemon take you home, also,   100
Wop washerwoman, with the bandaged hair?
After the corridors are swept, the cuspidors—
The gaunt sky-barracks cleanly now, and bare,
O Genoese,[4] do you bring mother eyes and hands
Back home to children and to golden hair?   105

Daemon, demurring and eventful yawn!
Whose hideous laughter is a bellows mirth
—Or the muffled slaughter of a day in birth—
O cruelly to inoculate the brinking dawn
With antennae toward worlds that glow and sink;— 110
To spoon us out more liquid than the dim
Locution of the eldest star, and pack
The conscience navelled in the plunging wind,
Umbilical to call—and straightway die!

O caught like pennies beneath soot and steam,   115
Kiss of our agony thou gatherest;
Condensed, thou takest all—shrill ganglia
Impassioned with some song we fail to keep.
And yet, like Lazarus,[5] to feel the slope,
The sod and billow breaking,—lifting ground,   120
—A sound of waters bending astride the sky
Unceasing with some Word that will not die . . .!

          .   .   .   .   .   .

A tugboat, wheezing wreaths of steam,
Lunged past, with one galvanic blare stove up the
          River.
I counted the echoes assembling, one after one,   125
Searching, thumbing the midnight on the piers.

3. I.e., galvanothermic, producing heat
by electricity.
4. The Italian-American mother is called
Genoese to recall Genoa, Italy, the
birthplace of Columbus.
5. Lazarus was resurrected from the
grave by Jesus in John 11.43–44.

Lights, coasting, left the oily tympanum of waters;
The blackness somewhere gouged glass on a sky.
And this thy harbor, O my City, I have driven under,
Tossed from the coil of ticking towers. . . .   130
    Tomorrow,
And to be. . . . Here by the River that is East—
Here at the waters' edge the hands drop memory;
Shadowless in that abyss they unaccounting lie.
How far away the star has pooled the sea—   135
Or shall the hands be drawn away, to die?

Kiss of our agony Thou gatherest,
              O Hand of Fire
                        gatherest —

1926   1927, 1930

A Name for All

Moonmoth and grasshopper that flee our page
And still wing on, untarnished of the name
We pinion to your bodies to assuage
Our envy of your freedom—we must maim

Because we are usurpers, and chagrined—   5
And take the wing and scar it in the hand.
Names we have, even, to clap on the wind;
But we must die, as you, to understand.

I dreamed that all men dropped their names, and sang
As only they can praise, who build their days   10
With fin and hoof, with wing and sweetened fang
Struck free and holy in one Name always.

1928   1929, 1933

LANGSTON HUGHES
1902–1967

Of the writers who contributed to the "Harlem Renaissance" in the 1920s, the one with the most durable and varied talent was Langston Hughes, who first gained public attention as the "busboy poet" and later was known as the Poet Laureate of Harlem. He introduced the rhythms of jazz and the blues into his lyric verse, developed subjects from Negro life and racial themes, and contributed to the strengthening of black consciousness and racial pride that was the Harlem Renaissance's legacy to the more militant decades of the '50s and '60s.

James Langston Hughes was born in Joplin, Missouri, to parents who soon separated. His principal home before 1914 was with his maternal grandmother. But he lived intermittently with his mother (who wrote poems occasionally, insisted on an integrated school for her son, and took him to the theater) and briefly with his father, who had developed contempt for Negroes who accepted their lot in the United States and had moved to a ranch in Mexico. Hughes learned German and Spanish at the instigation of his father and began to write poetry while living with him in Mexico after graduating from high school in 1920, and his father financed his one unsuccessful year at Columbia University in 1921–22. Yet before leaving Mexico for Columbia, Hughes had been brought to the verge of suicide and to an intense hatred for his father which never healed.

His intellectual awakening had begun earlier at high school in Cleveland while living with his mother and stepfather. The famous Karamu settlement house, social occasions at the home of a high school teacher (Helen M. Chesnutt, daughter of the black novelist Charles W. Chesnutt), the enthusiasm of another teacher, Miss Ethel Weimer, for Shakespeare and Carl Sandburg, and the racial consciousness heightened by personal experience and his reading in the Negro journal *The Crisis*—all helped to stimulate the imagination of the young poet. *The Crisis* published his early poems sporadically during the two years (1923–24) when he shipped out as a merchant seaman, traveled in Africa and the Mediterranean, and worked for a while in a Paris nightclub. It was there that Dr. Alain Locke, chief black proponent of the "Harlem Renaissance," met him and arranged to include eleven poems in his important anthology, *The New Negro* (1925). On returning to the States, Hughes was working as busboy in a Washington, D.C., hotel when he ventured to show some poems to the white poet Vachel Lindsay, who included them in a public reading that night and brought Hughes to the attention of the local press. He was by then familiar with the leaders of the "Harlem Renaissance," and one of its principal white supporters, the novelist Carl Van Vechten, was instrumental in securing publication of Hughes's first volume, *The Weary Blues*, in 1926.

Hughes was enrolled in Lincoln University, supported by grants from the white philanthropist Mrs. Amy Spingarn, when his second volume appeared (*Fine Clothes to the Jew*, 1927) and he had completed an extensive tour of the South, reading his poems, by the time he graduated from Lincoln in 1929. Grants from a white patron, Mrs. Rufus O. Mason, enabled him to live in New York after 1928, but he terminated the arrangement in 1930 shortly after completing his first novel, *Not Without Laughter*, which won the Harmon Gold Award for Literature. From then on he attempted to live on his earnings as a writer. He conducted a long reading tour in the South in 1931, reaching countless communities despite the fact that his protest poems were controversial among Negroes at Hampton Institute and that his reading at the University of North Carolina aroused the animosities of whites and so ruled out later appearances on other white campuses. He toured Russia, writing on Asia and race relations for Moscow journals. His friendship there with the novelist Arthur Koestler and his discovery of the short stories of the English novelist D. H. Lawrence revived his interest in fiction, and shortly after his return in 1933 appeared his first volume of short stories, *The Ways of White Folks* (1934).

Hughes's various literary activities were prodigious during the '30s and later decades. He engaged in radical protest in California and submitted a speech to the leftist First American Writer's Conference in New York City in 1935. In that year an earlier play, *The Mulatto*, appeared on Broadway. He covered the Spanish Civil War for the Baltimore *Afro-American* in

1937. He founded black theaters in Harlem, Los Angeles, and Chicago, published numerous anthologies of Negro writing and histories of eminent Negroes, wrote two autobiographical volumes (*The Big Sea*, 1940, and *I Wonder as I Wander*, 1956), turned his play *Troubled Island* into the libretto for an opera by the black composer William Grant Still (1949), and collaborated on the script for a Hollywood film. In 1943 he began a series of newspaper sketches centering on an engagingly comic urban Negro named Jesse B. Semple which became very popular and were collected in four volumes. The vignettes are a vivid reflection of life in Negro urban communities and present pointed comment on public affairs and the injustices of white racism. Besides the many volumes of his own verse, including *Montage of a Dream Deferred* (1951), he published translations of the Spanish playwright Federico García Lorca's *Blood Wedding* (1951) and the poems of the Chilean poet Gabriela Mistral (1957).

Hughes's poetry, liberated by the example of Sandburg's free verse forms, aimed from the start for utter directness and simplicity, not only in poems like *To a Little Lover-Lass, Dead*, which have no racial content and do not employ the distinctive diction and cadences of Negro idioms, but also in the "90 percent" of his work which he said attempts to "explain and illuminate the Negro condition in America." (Quoted by James Emanuel, *Langston Hughes*, p. 68.) While Hughes never militantly repudiated cooperation with the white community, the poems which protest against white racism are boldly direct.

Simple stanza patterns and strict rhyme schemes derived from blues songs enabled him to capture the ambience of the setting as well as the rhythms of jazz music. At his best, in *Love*, a direct allusion to a Negro folk hero charges a celebration of all human love with paradox and tragic feeling.

The Negro Speaks of Rivers

I've known rivers:
I've known rivers ancient as the world and older than the
    flow of human blood in human veins.

My soul has grown deep like the rivers.

I bathed in the Euphrates[1] when dawns were young.   5
I built my hut near the Congo and it lulled me to sleep.
I looked upon the Nile and raised the pyramids above it.
I heard the singing of the Mississippi when Abe Lincoln
    went down to New Orleans, and I've seen its muddy
    bosom turn all golden in the sunset.   10

I've known rivers:
Ancient, dusky rivers.

My soul has grown deep like the rivers.

1920   1921, 1926

1. The Euphrates River, cradle of ancient Babylonian civilization, flows from Turkey through Syria and Iraq; the Congo flows from the Republic of the Congo in central Africa into the Atlantic Ocean; the Nile, site of ancient Egyptian civilization, empties into the Mediterranean Sea.

Mother to Son

Well, son, I'll tell you:
Life for me ain't been no crystal stair.
It's had tacks in it,
And splinters,
And boards torn up,   5
And places with no carpet on the floor—
Bare.
But all the time
I'se been a-climbin' on,
And reachin' landin's,   10
And turnin' corners,
And sometimes goin' in the dark
Where there ain't been no light.
So boy, don't you turn back.
Don't you set down on the steps   15
'Cause you finds it's kinder hard.
Don't you fall now—
For I'se still goin', honey,
I'se still climbin',
And life for me ain't been no crystal stair.   20

                                                1922, 1926

To a Little Lover-Lass, Dead

She
Who searched for lovers
In the night
Has gone the quiet way
Into the still,   5
Dark land of death
Beyond the rim of day.

Now like a little lonely waif
She walks
An endless street   10
And gives her kiss to nothingness.
Would God his lips were sweet!

                                                        1926

Song for a Dark Girl

Way Down South in Dixie[1]
    (Break the heart of me)
They hung my black young lover
    To a cross roads tree.

Way Down South in Dixie   5
    (Bruised body high in air)

1. This ironic refrain is the last line of *Dixie*, the popular minstrel song, probably composed by Daniel D. Emmett (1815–1904), which became the rallying cry of southern patriotism during and after the Civil War.

I asked the white Lord Jesus
  What was the use of prayer.

Way Down South in Dixie
  (Break the heart of me)   10
Love is a naked shadow
  On a gnarled and naked tree.

                                                  1927

Young Gal's Blues

I'm gonna walk to the graveyard
'Hind ma friend Miss Cora Lee.
Gonna walk to the graveyard
'Hind ma dear friend Cora Lee
Cause when I'm dead some   5
Body'll have to walk behind me.

I'm goin' to the po' house
To see ma old Aunt Clew.
Goin' to the po' house
To see ma old Aunt Clew.   10
When I'm old an' ugly
I'll want to see somebody, too.

The po' house is lonely
An' the grave is cold.
O, the po' house is lonely,   15
The graveyard grave is cold.
But I'd rather be dead than
To be ugly an' old.

When love is gone what
Can a young gal do?   20
When love is gone, O,
What can a young gal do?
Keep on a-lovin' me, daddy,
Cause I don't want to be blue.

                                                  1927

Love

Love is a wild wonder
And stars that sing,
Rocks that burst asunder
And mountains that take wing.

John Henry[1] with his hammer   5
Makes a little spark.

---

1. **Legendary black hero, celebrated in** paced the drill but died from the effort
ballads and tall tales for driving steel in "with the hammer in his hand."
competition with a steam drill; he out-

That little spark is love
Dying in the dark.

1942

Uncle Tom[2]

Within—
The beaten pride.
Without—
The grinning face,
The low, obsequious,   5
Double bow,
The sly and servile grace
Of one the white folks
Long ago
Taught well   10
To know his
Place.

1944

Trumpet Player

The Negro
With the trumpet at his lips
Has dark moons of weariness
Beneath his eyes
Where the smoldering memory   5
Of slave ships
Blazed to the crack of whips
About his thighs.

The Negro
With the trumpet at his lips   10
Has a head of vibrant hair
Tamed down,
Patent-leathered now
Until it gleams
Like jet—   15
Were jet a crown.

The music
From the trumpet at his lips
Is honey
Mixed with liquid fire.   20
The rhythm
From the trumpet at his lips
Is ecstasy
Distilled from old desire—

---

2. The name of the central Negro character in the antislavery novel *Uncle Tom's Cabin*, by Harriet Beecher Stowe (1811–96), which has since come to signify a Negro who accepts domination by whites and masks his resentment or his attempts to circumvent it.

Desire 25
That is longing for the moon
Where the moonlight's but a spotlight
In his eyes,
Desire
That is longing for the sea 30
Where the sea's a bar-glass
Sucker size.

The Negro
With the trumpet at his lips
Whose jacket 35
Has a *fine* one-button roll,
Does not know
Upon what riff the music slips
Its hypodermic needle
To his soul— 40

But softly
As the tune comes from his throat
Trouble
Mellows to a golden note.

1947

WILLIAM CARLOS WILLIAMS
1883–1963

The writer who did most to sponsor younger American poets after World War II was William Carlos Williams, whose own career spanned six decades during which he combined the careers of poet and pediatrician. He became, with Hart Crane, the leading exponent of a Whitman tradition in American letters. To an already long list of poems, critical essays, plays, and fiction, Williams added in the 1940s, 1950s, and 1960s an epic poem, *Paterson*, and a body of love poems that are among the major achievements of twentieth-century American poetry.

Williams was born in 1883 near Paterson in Rutherford, New Jersey, which he made his lifelong home. His immigrant parents—a "too English" father and a Puerto Rican mother who had studied painting in Paris—early exposed him to European culture, but his "mixed ancestry" made him feel the more deeply that America was his only home and that it was his "first business in life to possess it" (*Selected Letters*, pp. 127, 185). He attended schools in Switzerland and Paris as well as local schools, and he went directly from the Horace Mann High School into medical school at the University of Pennsylvania, where he became friends with Ezra Pound, the poet Hilda Doolittle, and the painter Charles Demuth. He graduated in 1906, interned in New York City, and after postgraduate work in pediatrics at Leipzig, Germany, he returned to Rutherford, married Florence Herman, and began his medical practice. Except during European travels in 1924 and 1927, he pursued his medical career full time until the middle 1950s when, following a series of strokes, he turned over his practice to one of his sons. He accomplished his writing at night and between professional appointments, "determined to be a poet," as he wrote later, and convinced that "only medicine, a job I enjoyed, would make it possible for me to live and write as I wanted to * * * " (*Autobiography*, p. 51). His professional career involved him in community activities (mosquito control, medical inspection of schools), brought him directly into the lives of patients with various racial and social backgrounds, and strengthened the respect for sheer physical life, the vitality of the body, that characterizes his poetry.

Williams published his first book, *Poems*, privately in 1909, but it was *The Tempers* (published with Ezra Pound's help in 1913) that first revealed the impact of Pound's verse as well as Williams's struggle to respond to modern modes of poetry while yet developing for himself a distinctly individual voice. At gatherings of vanguard painters and writers in New York City and Grantwood, New Jersey, he became familiar with the painters Francis Picabia and Marcel Duchamp, the poets Marianne Moore and Wallace Stevens, the critic Kenneth Burke. Though he was increasingly suspicious of Pound's expatriation and cultivation of traditional verse forms, Williams endorsed Pound's principles of "imagism"; he kept up an argumentative correspondence with him that remained frequent through the Twenties. He had read Whitman enthusiastically during high school and medical school, and while by 1917 Williams had become critical of Whitman's formlessness, as he saw it, he developed a conception of the poet's role that reaffirmed Whitman's celebration of the body and his views on the liberation of feeling, the authenticity of native materials for poetry, and the poet's crucial mission in a democratic society. By the time he published his manifesto "Prologue to *Kora in Hell*" (1920), essays in the magazine *Contact*, which he edited with Robert McAlmon (1920–23), the poems and prose statements in *Spring and All* (1923), and the historical essays *In the American Grain* (1925), Williams had differentiated his program from that of the expatriates and defined a tradition grounded on his own cranky rebelliousness and the elemental realities of the American scene.

Since new life is always "subversive of life as it was the moment before," Williams declared, poetry should be impelled by the same energy of "disestablishment, something in the nature of an impalpable revolution * * * "; poetry is "a rival government always in opposition to its cruder replicas" (*Selected Letters*, pp. 23–24; *Selected Essays*, p. 180). The expatriates had fled from that "immediate contact with the world" on which original writing depends, with the result that even Eliot's "exquisite work" was a "rehash" and Pound's early poems were "paraphrases" (*Spring and All*)— the work of men "content with the connotations of their masters" ("Prologue to *Kora in Hell*"). *The Waste Land* was a "catastrophe" for American letters that confirmed Williams in his commitment to Whitman (*Autobiography*, p. 146). Williams called for a confrontation with immediate reality, even the "filth" and "ignorance" of the present, and denounced in his antagonists their use of explicit metaphor, "crude symbolism," "strained associations," and "complicated ritualistic forms designed to separate the work from reality" (*Spring and All*). The phrase "no ideas but in things" recurs in Williams's poetry and prose; meanings should be found only in actual things, and universality could be found only in the concrete particulars of the locality. The figures Williams most admired in his study of the American past had, like Poe, a direct sense of their *"new locality"* and the determination to "clear the GROUND" for a new beginning; like the French Jesuit missionary Sebastian Rasles, they had the nerve "to MARRY—to touch" the reality around them (*In the American Grain*, pp. 216, 121).

To make this contact possible in poetry—to confront the "cruder," elemental realities and "the primitive profundity of the personality" (*Selected Letters*, p. 226; *How to Write*, Wagner, *The Poems of William Carlos Williams*, p. 145)—Williams insisted throughout his career that American poets should rely on the diction and cadences of the distinctively American language; they should derive their rhythms and measures from American speech as actually spoken whose very "instability" made it exceptionally receptive to "innovation" (*The Poem as a Field of Action*, 1948). But the poet, Williams insisted, could not simply record that speech or "copy" reality. He must assault conventional ways of perceiving the world and the conventional language patterns for expressing it. To gain access to elemental realities he must destroy these conventions as part of the creative "process of ordering" which produces inventive "imitations," not copies, and distinctly new poetic forms; destruction and creation are simultaneous (*Autobiography*, p. 241; *Selected Essays*, pp. 121, 288). The imagination's exposure to reality in each poem is contingent on the creation of a new, striking, and flexible poetic form which Williams called a "structure" and, later, a "mutation" (*Against the Weather*, 1930, *Selected Essays*, p. 208).

Accordingly Williams's poems intentionally disorient readers by avoiding traditional rhyme or stanza and line patterns, presenting an unsettling mixture of attitudes or tones (as at the opening of Section III, Book II, of *Paterson*), or offering a rough jumble of images and allusions which prepares the reader for the unusual perception or emotion that is created by the poem (as in *Portrait of a Lady*). The objects, figures, and emotions are revealed with exceptional clarity, while the poem is exposed with equal candor as an artifact or structure. The conjunction of "thighs" and "appletrees" in *Portrait of a Lady*, for instance, appears simultaneously as a revealed fact and a mockingly arbitrary, painted arrangement. In Williams's characteristic style, the line breaks do not coincide regularly or easily with the phrasing or syntax of the language. The effect of this ostensible fragmentation is to induce the reader to focus with unusual intentness on concrete objects, phases of emotion and movement, specific aspects of the subject, and particulars of the poem's language and rhythm, while drawing attention at the same time to the discrete lines and the artifice of their sequence which articulate the poem's structure.

The language of painting, as well as American speech, provided Williams with a means of circumventing the traditional conventions of poetry. The transformation of his poetics that culminated in *Sour Grapes* (1921) and *Spring and All* was inspired in part by the excitement of the Armory Show of 1913, which introduced modern modes of European painting to a shocked public in New York City, and Williams not only admired such Europeans as Duchamp, the Dadaist Picabia, and the Cubist Juan Gris, but numbered the Expressionist Marsden Hartley and the "Precisionists" Charles Demuth and Charles Sheeler among his friends. He found models for the clarity and vividness of his own poems in the analytic abstractions and the fragmentation of objects in Cubism, in the precise linear outlines of Demuth's watercolors of flowers, and in the Precisionists' combination of objective realism with geometric abstraction.

The objectification of perception and feeling became more pronounced in Williams's poems of the 1930s, when his poetry became one of the models for the movement "Objectivism" which he helped to sponsor and whose Objectivist Press published his *Collected Poems, 1921–1931* in 1934. Although *The Yachts* and other excellent poems appeared in the 1930s, much of Williams's creative effort went into prose fiction. He had published *The Great American Novel* in 1923 (an autobiographical, Joycean treatment of the writer) and *A Voyage to Pagany* (an autobiographical pilgrimage abroad which explores the contrast between European and American cultures) in 1928. The prose is more spare in his collections of short stories, *The Edge of the Knife* (1932) and *Life along the Passaic River* (1938), which treat the bleakness and heroisms, at times comic, in the lives of immigrants, adolescents, and various professional people during the Depression. The novels *White Mule* (1937) and its sequel *In the Money* (1940) treat the immigrant's search for a community, and then success, in an America dominated by business.

Williams's political activities during the 1930s and 1940s included a commitment to the Democratic party, membership in his local social club (The Fortnightly) and in several societies of the American Social Credit movement, affiliation with such leftist causes as the "Open Letter of the Friends of the Soviet Union" calling for cooperation with Stalinist Russia (1939), and membership in the Cooperative Consumers' Society of Bergen County whose presidency he accepted in 1942. Some of these activities marked him a "fellow traveler" in the eyes of conservative opponents, despite Williams's repudiations of authoritarianism and his commitment to libertarian democracy, and he was deprived of the post of Consultant in Poetry at the Library of Congress which had been offered to him in 1948 and which he planned to take up when his health and medical practice permitted in 1952.

The completion of Williams's epic *Paterson* (Book I in 1946, subsequent volumes in 1948, 1949, 1951, and 1958), signaled a resurgence in his career as poet which was all the more remarkable for the series of heart attacks which began in 1948 and eventually made it difficult physically to write. Born in *"local pride,"* conceived as *"a confession . . . a reply to Greek and Latin with the bare hands"* (Book I), and intended to "lift an environment to expression" (*Selected Letters*, p. 286), the poem celebrated the search for a community and imaginative expression in the modern city. It seemed chaotic to many readers, including such admirers as the poet Randall Jarrell, but was praised as a major undertaking of the twentieth century by other poets; Robert Lowell, for instance, called it "our *Leaves of Grass*" (Miller, *William Carlos Williams: A Collection of Critical Essays*, p. 158). Book IV (originally the final book) concluded with the protagonist, like the epic hero Odysseus, returning to his native land; he returned to the state of New Jersey where Whitman and later Allen Ginsberg (born in Paterson) made their home. A letter from Ginsberg incorporated in the poem testifies that "at least one actual citizen in your community," Ginsberg himself, "has inherited your experience in his struggle to love and know a world-city, through your work. * * *"

Williams's unabashed candor, the directness of his style, and his openness to formal innovation found a larger audience in postwar generations whose poets he inspired, as in the cases of Allen Ginsberg, Denise Levertov, Charles Olson, and Robert Creeley. While he had earlier received the Dial Award (1926) and other honors, his talks and readings on college campuses and such awards as the National Book Award (1950), the Bollingen Award (1953), and the Pulitzer Prize (1962) testified to wider recognition during his later career.

Williams's critical attacks on his antagonists and his defense of his own poetic aims became more moderate, and he devoted renewed effort to exploring the resources of American speech and perfecting the verse form that first appeared in *Paterson* II (Section iii) and which Williams called the "variable foot" and the "triadic line." He believed that he had found in the cadences of the American idiom a new "measure"; it was more regular than free verse (which he thought unstructured or formless), yet more resilient than the "Euclidian" metrics that traditional verse depended on. A "relative order," rather than absolute or fixed order, "is operative elsewhere in our lives," he declared, and the "variable foot," with no fixed number of units, was a *"relatively* stable foot, not a rigid one." This would bring the necessary "discipline" to verse that Whitman neglected, though Whitman had been correct in "breaking [the] bounds" of traditional metrics (*On Measure, Selected Essays*, pp. 337–40). To construct the "triadic line" Williams arranged relatively short feet, of uneven length, in gradually descending steps across the page, establishing a musical pace, as he wrote the poet Richard Eberhart, that gives weight to silent pauses as well as to the words, and accords an even time interval to each of the segments (*Selected Letters*, pp. 326–27). Williams did not always use the triadic line in his later verse, and he frequently allowed himself the license of employing units of two and four feet instead of three. But the triadic form, as used in *The Sparrow*, for instance, achieved the same precision of notation which had characterized his earlier verse, while in such works as *Daphne and Virginia* the form sustained a more meditative tone and resilient line that combined fluency with intense compression: "IN OUR FAMILY we stammer unless, / half mad, / we come to speech at last   ."

Besides *Paterson*, Williams's late verse is distinguished particularly by the long love poems in *Desert Music and Other Poems* (1954) and *Journey to Love* (1955), which were gathered with later poems in *Pictures from Brueghel* (1962). In them and notably in *Asphodel, That Greeny Flower*, he examines with the tender but probing candor of an aging man the deviousness and authenticity of passion, the pained intricacies and redemptive power of marital and familial love, and their integral connection with the dance of life and the reality of poetry to which he consecrated his art.

Overture to a Dance of Locomotives

I[1]

Men with picked voices chant the names
of cities in a huge gallery: promises
that pull through descending stairways
to a deep rumbling.

1. "I" suggests an overture or beginning. There are no other parts to the poem.

                    The rubbing feet   5
of those coming to be carried quicken a
grey pavement into soft light that rocks
to and fro, under the domed ceiling,
across and across from pale
earthcolored walls of bare limestone.   10

Covertly the hands of a great clock
go round and round! Were they to
move quickly and at once the whole
secret would be out and the shuffling
of all ants be done forever.   15

A leaning pyramid of sunlight, narrowing
out at a high window, moves by the clock;
discordant hands straining out from
a center: inevitable postures infinitely
repeated—   20

two—twofour—twoeight!

Porters in red hats run on narrow platforms.

This way ma'am!
                    —important not to take
the wrong train!   25
                    Lights from the concrete
ceiling hang crooked but—
                    Poised horizontal
on glittering parallels the dingy cylinders
packed with a warm glow—inviting entry—   30
pull against the hour. But brakes can
hold a fixed posture till—
                    The whistle!

Not twoeight. Not twofour. Two!

Gliding windows. Colored cooks sweating   35
in a small kitchen. Taillights—
In time: twofour!
In time: twoeight!

—rivers are tunneled: trestles
cross oozy swampland: wheels repeating   40
the same gesture remain relatively
stationary: rails forever parallel
return on themselves infinitely.
                    The dance is sure.

c. 1913   1921

Portrait of a Lady

Your thighs are appletrees
whose blossoms touch the sky.
Which sky? The sky
where Watteau[1] hung a lady's
slipper. Your knees   5
are a southern breeze—or
a gust of snow. Agh! what
sort of man was Fragonard?[2]
—as if that answered
anything. Ah, yes—below   10
the knees, since the tune
drops that way, it is
one of those white summer days,
the tall grass of your ankles
flickers upon the shore—   15
Which shore?—
the sand clings to my lips—
Which shore?
Agh, petals maybe. How
should I know?   20
Which shore? Which shore?
I said petals from an appletree.

1915   1913

Queen-Ann's-Lace[1]

Her body is not so white as
anemone petals nor so smooth—nor
so remote a thing. It is a field
of the wild carrot taking
the field by force; the grass   5
does not raise above it.
Here is no question of whiteness,
white as can be, with a purple mole
at the center of each flower.
Each flower is a hand's span   10
of her whiteness. Wherever
his hand has lain there is
a tiny purple blemish. Each part
is a blossom under his touch
to which the fibres of her being   15
stem one by one, each to its end,
until the whole field is a

1. Jean Antoine Watteau (1684–1721), French Rococo artist. He painted sensuous, refined love scenes, with elegantly dressed lovers in idealized rustic settings.
2. Jean Honoré Fragonard (1732–1806), French painter, depicted fashionable lovers in paintings more wittily and openly erotic than Watteau's. Fragonard's *The Swing* depicts a girl who has kicked her slipper into the air.
1. The common field flower (wild carrot), whose wide white flower is composed of a multitude of minute blossoms, each with a dark spot at the center, joined to the stalk by fibrous stems.

white desire, empty, a single stem,
a cluster, flower by flower,
a pious wish to whiteness gone over—   20
or nothing.

1921

The Widow's Lament in Springtime

Sorrow is my own yard
where the new grass
flames as it has flamed
often before but not
with the cold fire   5
that closes round me this year.
Thirtyfive years
I lived with my husband.
The plumtree is white today
with masses of flowers.   10
Masses of flowers
load the cherry branches
and color some bushes
yellow and some red
but the grief in my heart   15
is stronger than they
for though they were my joy
formerly, today I notice them
and turned away forgetting.
Today my son told me   20
that in the meadows,
at the edge of the heavy woods
in the distance, he saw
trees of white flowers.
I feel that I would like   25
to go there
and fall into those flowers
and sink into the marsh near them.

1921

Spring and All[1]

By the road to the contagious hospital[2]
under the surge of the blue
mottled clouds driven from the
northeast—a cold wind. Beyond, the
waste of broad, muddy fields   5
brown with dried weeds, standing and fallen

patches of standing water
the scattering of tall trees

---

1. In *Spring and All* (as originally published, 1923), prose statements were interspersed through the poems and the
poems were identified by roman numer-
als. Williams added titles later and used
the volume's title for the opening poem.
2. A hospital for treating contagious diseases.

All along the road the reddish
purplish, forked, upstanding, twiggy   10
stuff of bushes and small trees
with dead, brown leaves under them
leafless vines—

Lifeless in appearance, sluggish
dazed spring approaches—   15

They enter the new world naked,
cold, uncertain of all
save that they enter. All about them
the familiar wind—

Now the grass, tomorrow   20
the stiff curl of wildcarrot leaf
One by one objects are defined—
It quickens: clarity, outline of leaf

But now the stark dignity of
entrance—Still, the profound change   25
has come upon them: rooted, they
grip down and begin to awaken

                                          1923

To Elsie

The pure products of America[1]
go crazy—
mountain folk from Kentucky

or the ribbed north end of
Jersey   5
with its isolate lakes and

valleys, its deaf-mutes, thieves
old names
and promiscuity between

devil-may-care men who have taken   10
to railroading
out of sheer lust of adventure—

and young slatterns, bathed
in filth
from Monday to Saturday   15

---

1. What Williams calls "pure products of America" are those whose descent is through generations of inbreeding in isolated communities; they are "pure" because unmixed with later American strains, but also because racial mixture is the essence of the American "melting pot." Elsie herself is descended from the "Jackson Whites" who lived in northern New Jersey. The Whites were offspring of Indians, Hessian deserters during the Revolutionary War, escaped Negro slaves, and white women imported (reputedly by a person named Jackson) as sexual companions for British troops.

to be tricked out that night
with gauds
from imaginations which have no

peasant traditions to give them
character   20
but flutter and flaunt

sheer rags—succumbing without
emotion
save numbed terror

under some hedge of choke-cherry   25
or viburnum—
which they cannot express—

Unless it be that marriage
perhaps
with a dash of Indian blood   30

will throw up a girl so desolate
so hemmed round
with disease or murder

that she'll be rescued by an
agent—   35
reared by the state and

sent out at fifteen to work in
some hard-pressed
house in the suburbs—

some doctor's family, some Elsie—   40
voluptuous water
expressing with broken

brain the truth about us—
her great
ungainly hips and flopping breasts   45

addressed to cheap
jewelry
and rich young men with fine eyes

as if the earth under our feet
were   50
an excrement of some sky

and we degraded prisoners
destined
to hunger until we eat filth

while the imagination strains 55
after deer
going by fields of goldenrod in

the stifling heat of September
Somehow
it seems to destroy us 60

It is only in isolate flecks that
something
is given off

No one
to witness 65
and adjust, no one to drive the car

1923

The Red Wheelbarrow

so much depends
upon

a red wheel
barrow

glazed with rain 5
water

beside the white
chickens.

1923

The Wind Increases

The harried
earth is swept
            The trees
the tulip's bright
        tips 5
            sidle and
toss—
        Loose your love
to flow

Blow! 10

Good Christ what is
a poet—if any
            exists?

a man
whose words will 15

                    bite
                              their way
            home—being actual

            having the form
                         of motion   20

            At each twigtip

            new

            upon the tortured
            body of thought

                         gripping   25
            the ground

            a way
                    to the last leaftip
1928   1934

The Botticellian Trees[1]

            The alphabet of
            the trees

            is fading in the
            song of the leaves

            the crossing   5
            bars of the thin

            letters that spelled
            winter

            and the cold
            have been illumined   10

            with
            pointed green

            by the rain and sun—
            The strict simple

            principles of   15
            straight branches

            are being modified
            by pinched-out

1. An allusion to the painting *Primavera* ("Spring") by the Italian Sandro Botticelli (1444–1510).

ifs of color, devout
conditions   20

the smiles of love—
· · · · · ·

until the stript
sentences

move as a woman's
limbs under cloth   30

and praise from secrecy
quick with desire

love's ascendancy
in summer—

In summer the song   25
sings itself

above the muffled words—

1928   1934

The Yachts

contend in a sea which the land partly encloses
shielding them from the too-heavy blows
of an ungoverned ocean which when it chooses

tortures the biggest hulls, the best man knows
to pit against its beatings, and sinks them pitilessly.   5
Mothlike in mists, scintillant in the minute

brilliance of cloudless days, with broad bellying sails
they glide to the wind tossing green water
from their sharp prows while over them the crew crawls

ant-like, solicitously grooming them, releasing,   10
making fast as they turn, lean far over and having
caught the wind again, side by side, head for the mark.

In a well guarded arena of open water surrounded by
lesser and greater craft which, sycophant, lumbering
and flittering follow them, they appear youthful, rare   15

as the light of a happy eye, live with the grace
of all that in the mind is feckless, free and
naturally to be desired. Now the sea which holds them

is moody, lapping their glossy sides, as if feeling
for some slightest flaw but fails completely.   20
Today no race. Then the wind comes again. The yachts

move, jockeying for a start, the signal is set and they
are off. Now the waves strike at them but they are too
well made, they slip through, though they take in canvas.   25

Arms with hands grasping seek to clutch at the prows.
Bodies thrown recklessly in the way are cut aside.
It is a sea of faces about them in agony, in despair

until the horror of the race dawns staggering the mind,   30
the whole sea become an entanglement of watery bodies
lost to the world bearing what they cannot hold. Broken,

desolate, reaching from the dead to be taken up
failing, failing! their cries rising
skillful yachts pass over.[1]   1935

Burning the Christmas Greens

Their time past, pulled down
cracked and flung to the fire
—go up in a roar

All recognition lost, burnt clean   5
clean in the flame, the green
dispersed, a living red,
flame red, red as blood wakes
on the ash—

and ebbs to a steady burning   10
the rekindled bed become
a landscape of flame

At the winter's midnight
we went to the trees, the coarse
holly, the balsam and   15
the hemlock for their green

At the thick of the dark
the moment of the cold's
deepest plunge we brought branches
cut from the green trees

to fill our need, and over   20
doorways, about paper Christmas
bells covered with tinfoil
and fastened by red ribbons

---

1. An ironic echo of the Jewish feast of Passover, commemorating the occasion when Jewish infants were saved from the plague in Exodus 11–12. Jehovah exempted or "passed over" the children of Jews when he inflicted death on the firstborn Egyptian infants.

we stuck the green prongs
in the windows hung
woven wreaths and above pictures
the living green. On the   25

mantle we built a green forest
and among those hemlock
sprays put a herd of small
white deer as if they

were walking there. All this!
and it seemed gentle and good
to us. Their time past,   35
relief! The room bare

stuffed the dead grate
with th   upon it   hair burnt out
logs smoldering eye, opening
red and closing under them

and we stood there looking down.   40
Green is a solace
a promise of peace, a fort
against the cold (though we

did not say so) a challenge
above the snow's   45
hard shell. Green (we might
have said) that, where

small birds hide and dodge
and lift their plaintive
rallying cries, blocks for them   50
and knocks down

the unseeing bullets of
the storm. Green spruce boughs
pulled down by a weight of
snow—Transformed!   55

Violence leaped and appeared.
Recreant! roared to life
as the flame rose through and
our eyes recoiled from it.

In the jagged flames green   60
to red, instant and alive. Green!
those sure abutments . . . Gone!
lost to mind

and quick in the contracting
tunnel of the grate   65
appeared a world! Black
mountains, black and red—as

yet uncolored—and ash white,
an infant landscape of shimmering
ash and flame and we, in   70
that instant, lost,

breathless to be witnesses,
as if we stood
ourselves refreshed among
the shining fauna of that fire.   75

1944

The Sparrow
TO MY FATHER

This sparrow
    who comes to sit at my window
        is a poetic truth
more than a natural one.
    His voice,   5
        his movements,
his habits—
    how he loves to
        flutter his wings
in the dust—   10
    all attest it;
        granted, he does it
to rid himself of lice
    but the relief he feels
        makes him   15
cry out lustily—
    which is a trait
        more related to music
than otherwise.
    Wherever he finds himself   20
        in early spring,
on back streets
    or beside palaces,
        he carries on
unaffectedly   25
    his amours.
        It begins in the egg,
his sex genders it:
    What is more pretentiously
        useless   30

or about which
        we more pride ourselves?
           It leads as often as not
to our undoing.
      The cockerel, the crow   35
          with their challenging voices
cannot surpass
      the insistence
         of his cheep!
Once   40
    at El Paso
        toward evening,
I saw—and heard!—
      ten thousand sparrows
         who had come in from   45
the desert
      to roost. They filled the trees
        of a small park. Men fled
(with ears ringing!)
      from their droppings,   50
        leaving the premises
to the alligators
     who inhabit
        the fountain. His image
is familiar   55
      as that of the aristocratic
        Unicorn,[1] a pity
there are not more oats eaten
      now-a-days
        to make living easier   60
for him.[2]
     At that,
        his small size,
keen eyes,
      serviceable beak   65
        and general truculence
assure his survival—
     to say nothing
        of his innumerable
brood.   70
     Even the Japanese
        know him
and have painted him
      sympathetically,
        with profound insight   75
into his minor
      characteristics.

1. A mythological creature, horse-like, with a long horn growing from its forehead.
2. Before the introduction of the automobile, when horses were the common mode of transportation, a major source of food for the sparrows was the partially digested oat kernels left in the horse droppings.

Nothing even remotely
subtle
    about his lovemaking.   80
    He crouches
before the female,
    drags his wings,
      waltzing,
throws back his head   85
    and simply—
      yells! The din
is terrific.
    The way he swipes his bill
      across a plank   90
to clean it,
    is decisive.
      So with everything
he does. His coppery
    eyebrows   95
      give him the air
of being always
    a winner—and yet
      I saw once,
the female of his species   100
    clinging determinedly
      to the edge of
a waterpipe,
    catch him
      by his crown-feathers   105
to hold him
    silent,
      subdued,
hanging above the city streets
    until   110
      she was through with him.
What was the use
    of that?
      She hung there
herself,   115
    puzzled at her success.
      I laughed heartily.
Practical to the end,
    it is the poem
      of his existence   120
that trumphed
    finally;
      a wisp of feathers
flattened to the pavement,
    wings spread symmetrically   125
      as if in flight,
the head gone,
    the black escutcheon
      undecipherable,

an effigy of a sparrow,   130
    a dried wafer only,
        left to say
and it says it
    without offense,
        beautifully;   135
This was I,
    a sparrow.
        I did my best;
farewell.

1955

The Lady Speaks

A storm raged among the live oaks
    while my husband and I
        sat in the semi-dark
listening!
    We watched from the windows   5
        the lights off
saw the moss
    whipped upright
        by the wind's force.
Two candles we had lit   10
    side by side
        before us
so solidly had our house been built
    kept their tall flames
        unmoved.   15
May it be so
    when a storm sends the moss
        whipping
back and forth
    upright   20
        above my head
like flames in the final
    fury.

1955

The Dance

When the snow falls the flakes
spin upon the long axis
that concerns them most intimately
two and two to make a dance

the mind dances with itself,   5
taking you by the hand,
your lover follows
there are always two,

yourself and the other,
the point of your shoe setting the pace,   10
if you break away and run
the dance is over

Breathlessly you will take
another partner
better or worse who will keep   15
at your side, at your stops

whirls and glides until he too
leaves off
on his way down as if
there were another direction   20

gayer, more carefree
spinning face to face but always down
with each other secure
only in each other's arms

But only the dance is sure!   25
make it your own.
Who can tell
what is to come of it?

in the woods of your
own nature whatever   30
twig interposes, and bare twigs
have an actuality of their own

this flurry of the storm
that holds us,
plays with us and discards us   35
dancing, dancing as may be credible.

                                                        1962

GERTRUDE STEIN
1874–1946

The first important Jewish writer in America, and one of its most famous
expatriates, was Gertrude Stein, whose witty conversation, continuous experimentation in writing, and widening fame as an eccentric "personality" have
assured her a place in literary history. Although most of her poetry is inde-
cipherable, and much of her later prose is a combination of the patently
obvious and the abstruse, she produced in *Melanctha* an unmistakable masterpiece. She also contributed to the remaking of literary language that was
the joint undertaking of Pound, Eliot, Hemingway, and other major writers of the twentieth century.

The "Mother Goose of Montparnasse," as she was dubbed by an admirer who was amused by her service to younger writers in Paris,[1] was born in Allegheny, Pennsylvania. Her well-to-do German-Jewish family was bourgeois and unintellectual but distinctly unconventional. The Stein grandparents emigrated from Germany to the United States before the Civil War and were well established in business in Baltimore at the time of her birth. She was the youngest child of seven and had an unusually close relationship with Leo, an older brother.

The Stein family lived abroad in Austria and France from 1875 to 1879, so her first spoken language was German, then French, though she later wrote that "there is for me only one language and that is English."[2] Upon the family's return to the United States in 1879 they moved to California, where the five surviving children attended intermittently the Oakland schools. Her mother, for whom Gertrude seemed to have little affection, and her father, whom she portrayed as a capricious tyrant, died during her adolescence. With income from the sale of the family business (a San Francisco railway), the young Steins were free to live where and as they chose. "Thus our life without a father began a very pleasant one."[3]

When Leo transferred from Berkeley to Harvard in 1892, Gertrude followed him, and though she lacked a high-school diploma she was admitted to Harvard's Annex for women, now Radcliffe College, as a special student in 1893. There she studied under the philosophers Josiah Royce and George Santayana and the psychologist and philosopher William James. In James's laboratory she undertook research on the role of repetition in memory. This research led to the publication, with other graduate students, of several articles in 1896. James encouraged her to pursue studies in psychology and philosophy, but her brother Leo decided to study biology at Johns Hopkins and Gertrude followed him there, enrolling in the medical school.

The Steins had a wide circle of friends in Baltimore, including the art patrons the Cone sisters: Dr. Claribel, one of the first women to earn a Hopkins medical degree, and Etta, who became Stein's secretary later in Paris and typed the manuscript of Stein's first important work, *Three Lives* (1909). After two years of creditable work in medical school, Stein seems deliberately to have failed, whether because, as she claimed, she became bored by medical study, because Leo was abandoning Baltimore and biology for aesthetics and Europe, or because she was becoming involved in a triangular love affair with two women (documented in a novel written in 1903 and published posthumously as *Q.E.D.* or *Things As They Are*, 1950). In any case she moved in 1902 to Paris, where the Steins began to accumulate their famous collection of Postimpressionist paintings. Though Leo usually took the lead in determining their purchases of paintings and theorizing about visual aesthetics, Gertrude led in the discovery of Picasso.

In Paris Stein launched her major experiments in writing. She finished *Three Lives* in 1905 and in 1908 finished *The Making of Americans* (1925), a long work that converts the history of her family and friends into a discontinuous and convoluted narration. While working on these manuscripts, as she explained later in a number of lectures, she developed the techniques which distinguish her prose. Impressed by what she had learned from William James and her own studies about the fluidity of the "stream"

1. William Gass, introduction to *The Geographical History of America* (New York: Vintage), p. 20.

2. *Autobiography of Alice B. Toklas*, p. 20.

3. *Everybody's Autobiography*, p. 142.

of consciousness and the importance of memory to perception, sequential thinking, and feeling, she felt that repetitions or recurrences in speech were what expressed the "bottom nature" of an individual's identity and experience.[4] Yet such recurrences were not sheer or exact repetitions, but slight shifts in emphasis or "insistence" as she called it, likening it to what "William James calls the Will to Live."[5] A sequence of recurring phrases or "insistences" defines a process of "beginning again and again" and extends a "prolonged present" into the "continuous present" that characterizes her prose.[6]

Moreover, she felt that the variety of differing perspectives on a person, which are defined by the recurring feelings and reactions of others, creates a process of "composition" in experience as it is lived as well as in writing, so that the "subtle variations" of "insistence" reveal the "complete rhythm of a personality," as it is "coming clear into ordered recognition."[7] The result in the three short novels, *Three Lives*, is that the writing captures without condescension the simplicity and integrity of two white immigrant serving women and black Melanctha Herbert. The tensions among domestic propriety, sensuousness, and self-recognition that agitated Stein's own affairs she located in the experience of a black woman, and *Melanctha* is a landmark in the sensitive treatment by a white writer of a black character, while converting the halting and echoing phrasing of colloquial speech into the cadences of music. The book failed to find a commercial publisher and was published in 1909 at the author's expense.

In 1909 Alice B. Toklas, who was to become Gertrude Stein's lover and life-long companion, joined the household to serve as housekeeper, typist, and editor. Within several years friction in the household and differences of opinion about painting between the two Steins led to Leo's withdrawal, after dividing the painting collection with his sister. Gertrude's home and salon at 22 rue de Fleurus continued to be the mecca of expatriate writers and artists interested in vanguard movements in the arts.

After 1909 the fascination of her experiments with language and writing, and the success with which she made her eccentricities and the force of her personality the center of attention at her salon, made her an important catalyst in literary circles and eventually turned her into a public figure with a wide audience. If she did not have the generative influence on Sherwood Anderson and Ernest Hemingway that she claimed, she nonetheless read and commented on their work; and she encouraged the development of their distinctive styles by her emphasis on precise observation, simplicity of statement, deliberate construction, and the basic elements of sheer language. Her own writing after 1909 undertook to achieve in verbal language the effects of abstract or nonrepresentational painting. *Tender Buttons* (1914) consisted of word pictures or collages, arranged under the headings of "Objects," "Food," and "Rooms," with no coherent syntax or paraphrasable meaning; the nonsensical splicing of discrete words and phrases creates an often witty and teasing sequence in which the effect of spoken speech, and the play of color and visual form, are heightened. Her experiments with the time sense in narration and with the elimination of punctuation and the construction of paragraphs continued in collections of stories and plays,

4. *Ibid.*, p. 231.
5. "Portraits and Repetition," *Lectures in America*, p. 169.
6. "Composition as Explanation," *What*

Are Masterpieces?, pp. 29, 31.
7. "The Gradual Making of Americans," *Lectures in America*, pp. 140, 147.

and she promoted her theories while explaining them in lectures at Oxford and Cambridge in 1926 and a wildly successful lecture tour of the United States in 1934. (*Composition as Explanation*, 1926, *Narration*, 1935, and the *Geographical History of America*, 1936, are the fruits of these lectures, and *Everybody's Autobiography*, 1937, is an account of her American lecture tour.)

Her fame became notoriety with the publication of the best seller *The Autobiography of Alice B. Toklas* (1933), in which Stein, in the guise of her companion, assembled her views of writing and artists in the twentieth century. Her claims to importance and her questionable judgments of painting were attacked by prominent painters in a special supplement of the magazine *Transition* entitled "Testimony against Gertrude Stein" (1935), and the writers Katherine Anne Porter and Hemingway were among those who acknowledged her early experiments but later denigrated the extent and nature of her influence.

Among her popular successes were the opera *Four Saints in Three Acts* (1934) with music by Virgil Thompson, and accounts of her years during World War II in occupied France. Her exhilaration upon becoming acquainted with hundreds of American troops who flocked to see her and enjoyed her hospitality after the liberation of France is recorded in *Wars I Have Seen* (1945) and *Brewsie and Willie* (1946). The war years and her anxieties about the fate of Western civilization had made her acutely conscious of that identity as an American which she had insisted on during her long years of expatriation. She died in Paris after an operation for cancer in 1946.

From Melanctha[1]

[*Conclusion*]

Melanctha Herbert began to feel she must begin again to look and see if she could find what it was she had always wanted. Now Rose Johnson could no longer help her.

And so Melanctha Herbert began once more to wander and with men Rose never thought it was right she should be with.

One day Melanctha had been very busy with the different kinds of ways she wandered. It was a pleasant late afternoon at the end of a long summer. Melanctha was walking along, and she was free and excited. Melanctha had just parted from a white man and she had a bunch of flowers he had left with her. A young buck, a mulatto, passed by and snatched them from her. "It certainly is real sweet in you sister, to be giving me them pretty flowers," he said to her.

"I don't see no way it can make them sweeter to have with you," said Melanctha. "What one man gives, another man had certainly

---

1. *Melanctha* is one of *Three Lives* (1909) in which Stein examines the relation of suffering and dependency in the lives of three serving women. Melanctha Herbert, a promiscuous black woman, has sought security and affection with two friends: Jeff Campbell, her lover and a physician, whose concern for "reg-ular living" has proved to be incompatible with Melanctha's willingness to drift while seeking satisfaction; and Rose Johnson, who, since her recent marriage to Sam, welcomes Melanctha's help as a servant but does not wish to live with her.

just as much good right to be taking." "Keep your old flowers then, I certainly don't never want to have them." Melanctha Herbert laughed at him and took them. "No, I didn't nohow think you really did want to have them. Thank you kindly mister, for them. I certainly always do admire to see a man always so kind of real polite to people." The man laughed, "You ain't nobody's fool I can say for you, but you certainly are a damned pretty kind of girl, now I look at you. Want men to be polite to you? All right, I can love you, that's real polite now, want to see me try it." "I certainly ain't got no time this evening just only left to thank you. I certainly got to be real busy now, but I certainly always will admire to see you." The man tried to catch and stop her, Melanctha Herbert laughed and dodged so that he could not touch her. Melanctha went quickly down a side street near her and so the man for that time lost her.

For some days Melanctha did not see any more of her mulatto. One day Melanctha was with a white man and they saw him. The white man stopped to speak to him. Afterwards Melanctha left the white man and she then soon met him. Melanctha stopped to talk to him. Melanctha Herbert soon began to like him.

Jem Richards, the new man Melanctha had begun to know now, was a dashing kind of fellow, who had to do with fine horses and with racing. Sometimes Jem Richards would be betting and would be good and lucky, and be making lots of money. Sometimes Jem would be betting badly, and then he would not be having any money.

Jem Richards was a straight man. Jem Richards always knew that by and by he would win again and pay it, and so Jem mostly did win again, and then he always paid it.

Jem Richards was a man other men always trusted. Men gave him money when he lost all his, for they all knew Jem Richards would win again, and when he did win they knew, and they were right, that he would pay it.

Melanctha Herbert all her life had always loved to be with horses. Melanctha liked it that Jem knew all about fine horses. He was a reckless man was Jem Richards. He knew how to win out, and always all her life, Melanctha Herbert loved successful power.

Melanctha Herbert always liked Jem Richards better. Things soon began to be very strong between them.

Jem was more game even than Melanctha. Jem always had known what it was to have real wisdom. Jem had always all his life been understanding.

Jem Richards made Melanctha Herbert come fast with him. He never gave her any time with waiting. Soon Melanctha always had Jem with her. Melanctha did not want anything better. Now in Jem Richards, Melanctha found everything she had ever needed to content her.

Melanctha was now less and less with Rose Johnson. Rose did

not think much of the way Melanctha now was going. Jem
Richards was all right, only Melanctha never had no sense of the
right kind of way she should be doing. Rose often was telling Sam
now, she did not like the fast way Melanctha was going. Rose told
it to Sam, and to all the girls and men, when she saw them. But
Rose was nothing just then to Melanctha. Melanctha Herbert now
only needed Jem Richards to be with her.

And things were always getting stronger between Jem Richards
and Melanctha Herbert. Jem Richards began to talk now as if he
wanted to get married to her. Jem was deep in his love now for her.
And as for Melanctha, Jem was all the world now to her. And so
Jem gave her a ring, like white folks, to show he was engaged to
her, and would by and by be married to her. And Melanctha was
filled full with joy to have Jem so good to her.

Melanctha always loved to go with Jem to the races. Jem had
been lucky lately with his betting, and he had a swell turn-out to
drive in, and Melanctha looked very handsome there beside him.

Melanctha was very proud to have Jem Richards want her.
Melanctha loved it the way Jem knew how to do it. Melanctha
loved Jem and loved that he should want her. She loved it too, that
he wanted to be married to her. Jem Richards was a straight decent
man, whom other men always looked up to and trusted. Melanctha
needed badly a man to content her.

Melanctha's joy made her foolish. Melanctha told everybody
about how Jem Richards, that swell man who owned all those fine
horses and was so game, nothing ever scared him, was engaged to be
married to her, and that was the ring he gave her.

Melanctha let out her joy very often to Rose Johnson. Melanctha
had begun again now to go there.

Melanctha's love for Jem made her foolish. Melanctha had to
have some one always now to talk to and so she went often to Rose
Johnson.

Melanctha put all herself into Jem Richards. She was mad and
foolish in the joy she had there.

Rose never liked the way Melanctha did it. "No Sam I don't say
never Melanctha ain't engaged to Jem Richards the way she always
says it, and Jem he is all right for that kind of a man he is, though
he do think himself so smart and like he owns the earth and everything he can get with it, and he sure gave Melanctha a ring like he
really meant he should be married right soon with it, only Sam, I
don't ever like it the way Melanctha is going. When she is engaged
to him Sam, she ain't not right to take on so excited. That ain't no
decent kind of a way a girl ever should be acting. There ain't no
kind of a man going stand that, not like I knows men Sam, and I
sure does know them. I knows them white and I knows them
colored, for I was raised by white folks, and they don't none of

them like a girl to act so. That's all right to be so when you is just only loving, but it ain't no ways right to be acting so when you is engaged to him, and when he says, all right he get really regularly married to you. You see Sam I am right like I am always and I knows it. Jem Richards, he ain't going to the last to get real married, not if I knows it right, the way Melanctha now is acting to him. Rings or anything ain't nothing to them, and they don't never do no good for them, when a girl acts foolish like Melanctha always now is acting. I certainly will be right sorry Sam, if Melanctha has real bad trouble come now to her, but I certainly don't no ways like it Sam the kind of way Melanctha is acting to him. I don't never say nothing to her Sam. I just listens to what she is saying always, and I thinks it out like I am telling to you Sam but I don't never say nothing no more now to Melanctha. Melanctha didn't say nothing to me about that Jem Richards till she was all like finished with him, and I never did like it Sam, much, the way she was acting, not coming here never when she first ran with those men and met him. And I didn't never say nothing to her, Sam, about it, and it ain't nothing ever to me, only I don't never no more want to say nothing to her, so I just listens to what she got to tell like she wants it. No Sam, I don't never want to say nothing to her. Melanctha just got to go her own way, not as I want to see her have bad trouble ever come hard to her, only it ain't in me never Sam, after Melanctha did so, ever to say nothing more to her how she should be acting. You just see Sam like I tell you, what way Jem Richards will act to her, you see Sam I just am right like I always am when I knows it."

Melanctha Herbert never thought she could ever again be in trouble. Melanctha's joy had made her foolish.

And now Jem Richards had some bad trouble with his betting. Melanctha sometimes felt now when she was with him that there was something wrong inside him. Melanctha knew he had had trouble with his betting but Melanctha never felt that that could make any difference to them.

Melanctha once had told Jem, sure he knew she always would love to be with him, if he was in jail or only just a beggar. Now Melanctha said to him, "Sure you know Jem that it don't never make any kind of difference you're having any kind of trouble, you just try me Jem and be game, don't look so worried to me. Jem sure I know you love me like I love you always, and its all I ever could be wanting Jem to me, just your wanting me always to be with you. I get married Jem to you soon ever as you can want me, if you once say it Jem to me. It ain't nothing to me ever, anything like having any money Jem, why you look so worried to me."

Melanctha Herbert's love had surely made her mad and foolish. She thrust it always deep into Jem Richards and now that he had trouble with his betting, Jem had no way that he ever wanted to be made to feel it. Jem Richards never could want to marry any girl while he had trouble. That was no way a man like him should do it. Melanctha's love had made her mad and foolish, she should be silent now and let him do it. Jem Richards was not a kind of man to want a woman to be strong to him, when he was in trouble with his betting. That was not the kind of a time when a man like him needed to have it.

Melanctha needed so badly to have it, this love which she had always wanted, she did not know what she should do to save it. Melanctha saw now, Jem Richards always had something wrong inside him. Melanctha soon dared not ask him. Jem was busy now, he had to sell things and see men to raise money. Jem could not meet Melanctha now so often.

It was lucky for Melanctha Herbert that Rose Johnson was coming now to have her baby. It had always been understood between them, Rose should come and stay then in the house where Melanctha lived with an old colored woman, so that Rose could have the Doctor from the hospital near by to help her, and Melanctha there to take care of her the way Melanctha always used to do it.

Melanctha was very good now to Rose Johnson. Melanctha did everything that any woman could, she tended Rose, and she was patient, submissive, soothing and untiring, while the sullen, childish, cowardly, black Rosie grumbled, and fussed, and howled, and made herself to be an abomination and like a simple beast.

All this time Melanctha was always being every now and then with Jem Richards. Melanctha was beginning to be stronger with Jem Richards. Melanctha was never so strong and sweet and in her nature as when she was deep in trouble, when she was fighting so with all she had, she could not do any foolish thing with her nature.

Always now Melanctha Herbert came back again to be nearer to Rose Johnson. Always now Melanctha would tell all about her troubles to Rose Johnson. Rose had begun now a little again to advise her.

Melanctha always told Rose now about the talks she had with Jem Richards, talks where they neither of them liked very well what the other one was saying. Melanctha did not know what it was Jem Richards wanted. All Melanctha knew was, he did not like it when she wanted to be good friends and get really married, and then when Melanctha would say, "all right, I never wear your ring no more Jem, we ain't not any more to meet ever like we ever going to get really regular married," then Jem did not like it either. What was it Jem Richards really wanted?

Melanctha stopped wearing Jem's ring on her finger. Poor Melanctha, she wore it on a string she tied around her neck so that she could always feel it, but Melanctha was strong now with Jem

Richards, and he never saw it. And sometimes Jem seemed to be awful sorry for it, and sometimes he seemed kind of glad of it. Melanctha never could make out really what it was Jem Richards wanted.

There was no other woman yet to Jem, that Melanctha knew, and so she always trusted that Jem would come back to her, deep in his love, the way once he had had it and had made all the world like she once had never believed anybody could really make it. But Jem Richards was more game than Melanctha Herbert. He knew how to fight to win out, better. Melanctha really had already lost it, in not keeping quiet and waiting for Jem to do it.

Jem Richards was not yet having better luck in his betting. He never before had had such a long time without some good coming to him in his betting. Sometimes Jem talked as if he wanted to go off on a trip somewhere and try some other place for luck with his betting. Jem Richards never talked as if he wanted to take Melanctha with him.

And so Melanctha sometimes was really trusting, and sometimes she was all sick inside her with her doubting. What was it Jem really wanted to do with her? He did not have any other woman, in that Melanctha could be really trusting, and when she said no to him, no she never would come near him, now he did not want to have her, then Jem would change and swear, yes sure he did want her, now and always right here near him, but he never now any more said he wanted to be married soon to her. But then Jem Richards never would marry a girl, he said that very often, when he was in this kind of trouble, and now he did not see any way he could get out of his trouble. But Melanctha ought to wear his ring, sure she knew he never had loved any kind of woman like he loved her. Melanctha would wear the ring a little while, and then they would have some more trouble, and then she would say to him, no she certainly never would any more wear anything he gave her, and then she would wear it on the string so nobody could see it but she could always feel it on her.

Poor Melanctha, surely her love had made her mad and foolish.

And now Melanctha needed always more and more to be with Rose Johnson, and Rose had commenced again to advise her, but Rose could not help her. There was no way now that anybody could advise her. The time when Melanctha could have changed it with Jem Richards was now all past for her. Rose knew it, and Melanctha too, she knew it, and it almost killed her to let herself believe it.

The only comfort Melanctha ever had now was waiting on Rose till she was so tired she could hardly stand it. Always Melanctha did everything Rose ever wanted. Sam Johnson began now to be very gentle and a little tender to Melanctha. She was so good to Rose and Sam was so glad to have her there to help Rose and to do things and to be a comfort to her.

Rose had a hard time to bring her baby to its birth and Melanctha did everything that any woman could.

The baby though it was healthy after it was born did not live long. Rose Johnson was careless and negligent and selfish and when Melanctha had to leave for a few days the baby died. Rose Johnson had liked her baby well enough and perhaps she just forgot it for a while, anyway the child was dead and Rose and Sam were very sorry, but then these things came so often in the negro world in Bridgepoint that they neither of them thought about it very long. When Rose had become strong again she went back to her house with Sam. And Sam Johnson was always now very gentle and kind and good to Melanctha who had been so good to Rose in her bad trouble.

Melanctha Herbert's troubles with Jem Richards were never getting any better. Jem always now had less and less time to be with her. When Jem was with Melanctha now he was good enough to her. Jem Richards was worried with his betting. Never since Jem had first begun to make a living had he ever had so much trouble for such a long time together with his betting. Jem Richards was good enough now to Melanctha but he had not much strength to give her. Melanctha could never any more now make him quarrel with her. Melanctha never now could complain of his treatment of her, for surely, he said it always by his actions to her, surely she must know how a man was when he had trouble on his mind with trying to make things go a little better.

Sometimes Jem and Melanctha had long talks when they neither of them liked very well what the other one was saying, but mostly now Melanctha could not make Jem Richards quarrel with her, and more and more, Melanctha could not find any way to make it right to blame him for the trouble she now always had inside her. Jem was good to her, and she knew, for he told her, that he had trouble all the time now with his betting. Melanctha knew very well that for her it was all wrong inside Jem Richards, but Melanctha had now no way that she could really reach him.

Things between Melanctha and Jem Richards were now never getting any better. Melanctha now more and more needed to be with Rose Johnson. Rose still liked to have Melanctha come to her house and do things for her, and Rose liked to grumble to her and to scold her and to tell Melanctha what was the way Melanctha always should be doing so she could make things come out better and not always be so much in trouble. Sam Johnson in these days was always very good and gentle to Melanctha. Sam was now beginning to be very sorry for her.

Jem Richards never made things any better for Melanctha. Often

Jem would talk so as to make Melanctha almost certain that he never any more wanted to have her. Then Melanctha would get very blue, and she would say to Rose, sure she would kill herself, for that certainly now was the best way she could do.

Rose Johnson never saw it the least bit that way. "I don't see Melanctha why you should talk like you would kill yourself just because you're blue. I'd never kill myself Melanctha cause I was blue. I'd maybe kill somebody else but I'd never kill myself. If I ever killed myself, Melanctha it'd be by accident and if I ever killed myself by accident, Melanctha, I'd be awful sorry. And that certainly is the way you should feel it Melanctha, now you hear me, not just talking foolish like you always do. It certainly is only your way just always being foolish makes you all that trouble to come to you always now, Melanctha, and I certainly right well knows that. You certainly never can learn no way Melanctha ever with all I certainly been telling to you, ever since I know you good, that it ain't never no way like you do always is the right way you be acting ever and talking, the way I certainly always have seen you do so Melanctha always. I certainly am right Melanctha about them ways you have to do it, and I knows it; but you certainly never can noways learn to act right Melanctha, I certainly do know that, I certainly do my best Melanctha to help you with it only you certainly never do act right Melanctha, not to nobody ever, I can see it. You never act right by me Melanctha no more than by everybody. I never say nothing to you Melanctha when you do so, for I certainly never do like it when I just got to say it to you, but you just certainly done with that Jem Richards you always say wanted real bad to be married to you, just like I always said to Sam you certainly was going to do it. And I certainly am real kind of sorry like for you Melanctha, but you certainly had ought to have come to see me to talk to you, when you first was engaged to him so I could show you, and now you got all this trouble come to you Melanctha like I certainly know you always catch it. It certainly ain't never Melanctha I ain't real sorry to see trouble come so hard to you, but I certainly can see Melanctha it all is always just the way you always be having it in you not never to do right. And now you always talk like you just kill yourself because you are so blue, that certainly never is Melanctha, no kind of a way for any decent kind of a girl to do."

Rose had begun to be strong now to scold Melanctha and she was impatient very often with her, but Rose could now never any more to be a help to her. Melanctha Herbert never could know now what it was right she should do. Melanctha always wanted to have Jem Richards with her and now he never seemed to want her, and what could Melanctha do. Surely she was right now when she said she would just kill herself, for that was the only way now she could do.

Sam Johnson always, more and more, was good and gentle to Melanctha. Poor Melanctha, she was so good and sweet to do anything anybody ever wanted, and Melanctha always liked it if she could have peace and quiet, and always she could only find new ways to be in trouble. Sam often said this now to Rose about Melanctha.

"I certainly don't never want Sam to say bad things about Melanctha, for she certainly always do have most awful kind of trouble come hard to her, but I never can say I like it real right Sam the way Melanctha always has to do it. Its now just the same with her like it is always she has got to do it, now the way she is with that Jem Richards. Her certainly now don't never want to have her but Melanctha she ain't got no right kind of spirit. No Sam I don't never like the way any more Melanctha is acting to him, and then Sam, she ain't never real right honest, the way she always should do it. She certainly just don't kind of never Sam tell right what way she is doing with it. I don't never like to say nothing Sam no more to her about the way she always has to be acting. She always say, yes all right Rose, I do the way you say it, and then Sam she don't never noways do it. She certainly is right sweet and good, Sam, is Melanctha, nobody ever can hear me say she ain't always ready to do things for everybody any way she ever can see to do it, only Sam some ways she never does act real right ever, and some ways, Sam, she ain't ever real honest with it. And Sam sometimes I hear awful kind of things she been doing, some girls know about her how she does it, and sometimes they tell me what kind of ways she has to do it, and Sam it certainly do seem to me like more and more I certainly am awful afraid Melanctha never will come to any good. And then Sam, sometimes, you hear it, she always talk like she kill herself all the time she is so blue, and Sam that certainly never is no kind of way any decent girl ever had ought to do. You see Sam, how I am right like I always is when I knows it. You just be careful, Sam, now you hear me, you be careful Sam sure, I tell you, Melanctha more and more I see her I certainly do feel Melanctha no way is really honest. You be careful, Sam now, like I tell you, for I knows it, now you hear to me, Sam, what I tell you, for I certainly always is right, Sam, when I knows it."

At first Sam tried a little to defend Melanctha, and Sam always was good and gentle to her, and Sam liked the ways Melanctha had to be quiet to him, and to always listen as if she was learning, when she was there and heard him talking, and then Sam liked the sweet way she always did everything so nicely for him; but Sam never liked to fight with anybody ever, and surely Rose knew best about Melanctha and anyway Sam never did really care much about Melanctha. Her mystery never had had any interest for him. Sam liked it that she was sweet to him and that she always did everything Rose ever wanted that she should be doing, but Melanctha never could be important to him. All Sam ever wanted was to have a little house and to live regular and to work hard and to come home to his dinner, when he was tired with his working and by and by he wanted to have some children all his own to be good to, and so Sam was real sorry for Melanctha, she was so good and so sweet always to them, and Jem Richards was a bad man to behave so to her, but that was always the way a girl got it when she liked that kind of a fast fellow. Anyhow Melanctha was Rose's friend, and Sam never cared to have anything to do with the kind of trouble always came to women, when they wanted to have men, who never could know how to behave good and steady to their women.

And so Sam never said much to Rose about Melanctha. Sam was always very gentle to her, but now he began less and less to see her. Soon Melanctha never came any more to the house to see Rose and Sam never asked Rose anything about her.

Melanctha Herbert was beginning now to come less and less to the house to be with Rose Johnson. This was because Rose seemed always less and less now to want her, and Rose would not let Melanctha now do things for her. Melanctha was always humble to her and Melanctha always wanted in every way she could to do things for her. Rose said no, she guessed she do that herself like she likes to have it better. Melanctha is real good to stay so long to help her, but Rose guessed perhaps Melanctha better go home now, Rose don't need nobody to help her now, she is feeling real strong, not like just after she had all that trouble with the baby, and then Sam, when he comes home for his dinner he likes it when Rose is all alone there just to give him his dinner. Sam always is so tired now, like he always is in the summer, so many people always on the steamer, and they make so much work so Sam is real tired now, and he likes just to eat his dinner and never have people in the house to be a trouble to him.

Each day Rose treated Melanctha more and more as if she never wanted Melanctha any more to come there to the house to see her. Melanctha dared not ask Rose why she acted in this way to her. Melanctha badly needed to have Rose always there to save her. Melanctha wanted badly to cling to her and Rose had always been so solid for her. Melanctha did not dare to ask Rose if she now no longer wanted her to come and see her.

Melanctha now never any more had Sam to be gentle to her. Rose always sent Melanctha away from her before it was time for Sam to come home to her. One day Melanctha had stayed a little longer, for Rose that day had been good to let Melanctha begin to do things for her. Melanctha then left her and Melanctha met Sam Johnson who stopped a minute to speak kindly to her.

The next day Rose Johnson would not let Melanctha come in to her. Rose stood on the steps, and there she told Melanctha what she thought now of her.

"I guess Melanctha it certainly ain't no ways right for you to come here no more just to see me. I certainly don't Melanctha no ways like to be a trouble to you. I certainly think Melanctha I get along better now when I don't have nobody like you are, always here to help me, and Sam he do so good now with his working, he pay a little girl something to come every day to help me. I certainly do think Melanctha I don't never want you no more to come here just to see me." "Why Rose, what I ever done to you, I certainly don't think you is right Rose to be so bad now to me." "I certainly don't no ways Melanctha Herbert think you got any right ever to be complaining the way I been acting to you. I certainly never do think Melanctha Herbert, you hear to me, nobody ever been more patient to you than I always been to like you, only Melanctha, I hear more things now so awful bad about you, everybody always is telling to me what kind of a way you always have been doing so much, and me always so good to you, and you never no ways, knowing how to be honest to me. No Melanctha it ain't ever in me, not to want you to have good luck come to you, and I like it real well Melanctha when you some time learn how to act the way it is decent and right for a girl to be doing, but I don't no ways ever like it the kind of things everybody tell me now about you. No Melanctha, I can't never any more trust you. I certainly am real sorry to have never any more to see you, but there ain't no other way, I ever can be acting to you. That's all I ever got any more to say to you now Melanctha." "But Rose, deed; I certainly don't know, no more than the dead, nothing I ever done to make you act so to me. Anybody say anything bad about me Rose, to you, they just a pack of liars to you, they certainly is Rose, I tell you true. I certainly never done nothing I ever been ashamed to tell you. Why you act so bad to me Rose. Sam he certainly don't think ever like you do, and Rose I always do everything I can, you ever want me to do for you." "It ain't never no use standing there talking, Melanctha Herbert. I just can tell it to you, and Sam, he don't know nothing about women ever the way they can be acting. I certainly am very sorry Melanctha, to have to act so now to you, but I certainly can't do no other way with you, when you do things always so bad, and everybody is talking so about you. It ain't no use to you to stand there and say it different to me Melanctha. I certainly am always right Melanctha Herbert, the way I certainly always have been when I knows it, to you. No Melanctha, it just is, you never can have no kind of a way to act right, the way a decent girl has to do, and I done my best always to be telling it to you Melanctha Herbert, but it don't never do no good to tell nobody how to act right; they certainly never can learn when they ain't got no sense right to know it, and you never have no sense right Melanctha to be honest, and I ain't never wishing no harm to you ever Melanctha Herbert, only I don't never want any more to see you come here. I just say to you now, like I always been saying to you, you don't know never the right way, any kind of decent girl has to be acting, and so Melanctha Herbert, me and Sam, we don't never any more want you to be setting your foot in my house here Melanctha Herbert, I just tell you. And so you just go along now, Melanctha Herbert, you hear me, and I don't never wish no harm to come to you."

Rose Johnson went into her house and closed the door behind her. Melanctha stood like one dazed, she did not know how to bear this blow that almost killed her. Slowly then Melanctha went away without even turning to look behind her.

Melanctha Herbert was all sore and bruised inside her. Melanctha had needed Rose always to believe her, Melanctha needed Rose always to let her cling to her, Melanctha wanted badly to have somebody who could make her always feel a little safe inside her, and now Rose had sent her from her. Melanctha wanted Rose more than she had ever wanted all the others. Rose always was so simple, solid, decent, for her. And now Rose had cast her from her. Melanctha was lost, and all the world went whirling in a mad weary dance around her.

Melanctha Herbert never had any strength alone ever to feel safe inside her. And now Rose Johnson had cast her from her, and Melanctha could never any more be near her. Melanctha Herbert knew now, way inside her, that she was lost, and nothing any more could ever help her.

Melanctha went that night to meet Jem Richards who had promised to be at the old place to meet her. Jem Richards was absent in his manner to her. By and by he began to talk to her, about the trip he was going to take soon, to see if he could get some luck back in his betting. Melanctha trembled, was Jem too now going to leave her. Jem Richards talked some more then to her, about the bad luck he always had now, and how he needed to go away to see if he could make it come out any better.

Then Jem stopped, and then he looked straight at Melanctha.

"Tell me Melanctha right and true, you don't care really nothing more about me now Melanctha," he said to her.

"Why you ask me that, Jem Richards," said Melanctha.

"Why I ask you that Melanctha, God Almighty, because I just don't give a damn now for you any more Melanctha. That the reason I was asking."

Melanctha never could have for this an answer. Jem Richards waited and then he went away and left her.

Melanctha Herbert never again saw Jem Richards. Melanctha never again saw Rose Johnson, and it was hard to Melanctha never any more to see her. Rose Johnson had worked in to be the deepest of all Melanctha's emotions.

"No, I don't never see Melanctha Herbert no more now," Rose

would say to anybody who asked her about Melanctha. "No, Melanctha she never comes here no more now, after we had all that trouble with her acting so bad with them kind of men she liked so much to be with. She don't never come to no good Melanctha Herbert don't, and me and Sam don't want no more to see her. She didn't do right ever the way I told her. Melanctha just wouldn't, and I always said it to her, if she don't be more kind of careful, the way she always had to be acting, I never did want no more she should come here in my house no more to see me. I ain't no ways ever against any girl having any kind of a way, to have a good time like she wants it, but not that kind of a way Melanctha always had to do it. I expect some day Melanctha kill herself, when she act so bad like she do always, and then she get so awful blue. Melanctha always says that's the only way she ever can think it a easy way for her to do. No, I always am real sorry for Melanctha, she never was no just common kind of nigger, but she don't never know not with all the time I always was telling it to her, no she never no way could learn, what was the right way she should do. I certainly don't never want no kind of harm to come bad to Melanctha, but I certainly do think she will most kill herself some time, the way she always say it would be easy way for her to do. I never see nobody ever could be so awful blue."

But Melanctha Herbert never really killed herself because she was so blue, though often she thought this would be really the best way for her to do. Melanctha never killed herself, she only got a bad fever and went into the hospital where they took good care of her and cured her.

When Melanctha was well again, she took a place and began to work and to live regular. Then Melanctha got very sick again, she began to cough and sweat and be so weak she could not stand to do her work.

Melanctha went back to the hospital, and there the Doctor told her she had the consumption, and before long she would surely die. They sent her where she would be taken care of, a home for poor consumptives, and there Melanctha stayed until she died.

1905   1909

SHERWOOD ANDERSON

1876–1941

Sherwood Anderson brought simultaneously the instincts of a naïf and the earnestness of a devotee to bear on the craft of writing and became one of the important catalysts in the literary world of the 1920s, stimulating other prose writers to achievements beyond his own. He also produced in *Winesburg, Ohio*, one of the masterpieces of American prose fiction.

He was born in southern Ohio near the Kentucky border, the third of seven children in a family that drifted north through small Ohio towns as his father turned from harness-making to odd jobs and sign-painting. In 1894 they settled in Clyde. His memories of that rural town were to provide the setting for *Winesburg*. His memories of his mother's stamina and tenderness, and of his father's wanderlust, his boozing, and facility as a storyteller, were to appear, altered and magnified, in his three volumes of highly imaginative autobiography: *A Story-Teller's Story* (1924), *Tar* (1926), and *Sherwood Anderson's Memoirs* (1942).

His family's migration and his jobs as farm hand, stableboy, and worker in a bicycle factory made for irregular schooling, and he dropped out of high school before graduating. In 1896 he followed his older brother Karl (a painter) to Chicago, and after serving briefly in Cuba during the Spanish-American War he went with his brother to Springfield, Ohio, where he enrolled for a year in Wittenberg Academy and became acquainted with the editors, artists, and advertising men at Crowell Publishing Company, where his brother was employed. The offer of a job as advertising copywriter in 1900 took him to Chicago, where he wrote promotional articles for a house journal extolling the virtues of business enterprise. He married the first of his four wives in 1904, the daughter of a well-to-do Ohio merchant, and in 1906 moved to Ohio, where he managed a mail-order business and two paint firms, living the life of a successful businessman but writing fiction in secret and finding that his business career and his writing craft were increasingly incompatible.

In 1912 he suddenly disappeared for several days, reappeared in Cleveland in a state of nervous collapse, and after a quick recovery made the partial break with the business world that he later dramatized repeatedly in his autobiographical essays and memoirs. He returned temporarily to his advertising job in Chicago but took with him the drafts of two novels and other manuscripts. Advertising had earlier seemed to nourish his facility with words, but he now became acquainted with the serious writers of the Chicago "renaissance," including Floyd Dell, Carl Sandburg, and Theodore Dreiser, who encouraged his serious ambitions. His reading earlier had included such figures as Poe, Whitman, and Twain, the British poets Keats and Browning, the English novelists Arnold Bennett and Thomas Hardy, and the English writer of character sketches George Borrow. Now his reading extended to D. H. Lawrence, Sigmund Freud, the Russian novelist Turgenev, and Gertrude Stein, whose *Three Lives* he had read in 1909 but whose experimental *Tender Buttons* was an exciting discovery in 1914. His first major publication was *Windy McPherson's Son* (1916), the story of a man who flees a small Iowa town to engage in a futile search for life's meaning first as a businessman, then as a vagabond among the common people, finally as father to three adopted children. Anderson's next novel was *Marching Men* (1917), whose protagonist, a charismatic lawyer, combats the dehumanization of an industrial society by organizing its masses into phalanxes which threaten their further dehumanization.

Anderson had started writing the tales that were to comprise *Winesburg*,

Ohio in 1916, and some had appeared in journals between 1916 and 1918 before they were gathered in 1919 to form the loosely but suggestively related group of stories that have the integral coherence of a novel. His models were the comparably related sketches in Turgenev's *Sportsman's Sketches* (1852) and Edgar Lee Masters's collection of elegiac portraits in verse, *Spoon River Anthology* (1915). But Anderson's volume is unique. Each character is a "grotesque" whose conduct and usually fumbling attempts at communication are warped versions of some reality or "truth," whose realization is never achieved in the life of the community. The pathos of the inhabitants' isolation and quest for fulfillment is given focus by the maturing consciousness of the young reporter and future writer, George Willard, who discovers his ties to the community in the process of deciding to leave it. The groping attempts of the townspeople to articulate their feelings and communicate with others—the repressed yearnings and obsessions which surface fleetingly or explosively from the depths of their experience—are surrealistic or expressionistic in their effects. And these effects are captured by the narrative's simple but bold imagery and the stylization of Anderson's prose, which becomes a species of elegiac poetry. The simple vernacular diction and the declarative sentence structure, the repetition of words and phrases, the recurring restatements of ideals and feelings have at once the power of colloquial speech that Anderson admired in Twain and the "pure and beautiful prose" that he found exemplified in Gertrude Stein.[1]

Anderson's best writing in *Winesburg* was matched only by his short stories (collected in *The Triumph of the Egg*, 1921; *Horses and Men*, 1923; and *Death in the Woods and Other Stories*, 1933) and, to a lesser extent, by the novel *Poor White* (1920), the story of an inventor who brings industrialism to a rural town only to find that it blights the community and that his business success is no answer to his alienation from his fellows. Though each contained some impressive sections, the two novels about sexuality, *Many Marriages* (1923) and *Beyond Desire* (1932), and *Kit Brandon* (1936) were neither critical nor popular successes. Anderson published a volume of rhapsodic free verse (*Mid-American Chants*, 1918), a volume of prose poems (*A New Testament*, 1927), four plays (*Winesburg and Others*, 1937), as well as collections of essays on social issues (*Perhaps Women*, 1931, and *Puzzled America*, 1935) and essays on literary figures (*No Swank*, 1934). But his chief importance in the 1920s and afterward, as the reputation of his fiction declined, was the series of autobiographical volumes that centered on his own professional career and his attempt to define the writer's vocation in America, and on his impact on other writers. He had met Hemingway in Europe in 1921 and had helped secure publication of *In Our Time* in the United States. And while Hemingway later mocked Anderson's style and cult of the primitive in *Torrents of Spring*, he had been inspired by the "simplification"[2] of Anderson's prose to make his own, similar break from the more elaborate conventions of earlier fiction. Anderson's treatment of violence and his "grotesques" provided models for Nathanael West in the 1930s. During two periods, in 1922–23 and 1924–25, Anderson had known William Faulkner in New Orleans and had shared an

1. "Gertrude Stein," *No Swank* (1934), p. 82.
2. *Hello, Towns* (1929), p. 325.

apartment with him. Faulkner later testified that Anderson's encouragement, and his example, had enabled him to see that he could rely on the uniqueness of his own talent as a writer and on the milieu of his own native region for material.[3]

Anderson's own search for roots in his native land was never satisfied. He moved with his third wife to Virginia in 1927 and bought two local newspapers to bond his attachment to the region. He published editorials from the papers in *Hello, Towns* (1929). His concern with the dislocations of modern industrialism led to his involvement in leftist political activity in the early 1930s, and his interest in public issues continued until 1941, when he died on a good-will mission to South America for the State Department.

From WINESBURG, OHIO
The Book of the Grotesque

The writer, an old man with a white mustache, had some difficulty in getting into bed. The windows of the house in which he lived were high and he wanted to look at the trees when he awoke in the morning. A carpenter came to fix the bed so that it would be on a level with the window.

Quite a fuss was made about the matter. The carpenter, who had been a soldier in the Civil War, came into the writer's room and sat down to talk of building a platform for the purpose of raising the bed. The writer had cigars lying about and the carpenter smoked.

For a time the two men talked of the raising of the bed and then they talked of other things. The soldier got on the subject of the war. The writer, in fact, led him to that subject. The carpenter had once been a prisoner in Andersonville prison[1] and had lost a brother. The brother had died of starvation, and whenever the carpenter got upon that subject he cried. He, like the old writer, had a white mustache, and when he cried he puckered up his lips and the mustache bobbed up and down. The weeping old man with the cigar in his mouth was ludicrous. The plan the writer had for the raising of his bed was forgotten and later the carpenter did it in his own way and the writer, who was past sixty, had to help himself with a chair when he went to bed at night.

In his bed the writer rolled over on his side and lay quite still. For years he had been beset with notions concerning his heart. He was a hard smoker and his heart fluttered. The idea had got into his mind that he would some time die unexpectedly and always when he got into bed he thought of that. It did not alarm him. The effect in fact was quite a special thing and not easily explained. It made him more alive, there in bed, than at any other time. Perfectly still he lay and his body was old and not of much use

---

3. "Sherwood Anderson: An Appreciation," *Atlantic Monthly*, 1953, p. 29.

1. Notorious Confederate prison in Georgia during the Civil War.

any more, but something inside him was altogether young. He was like a pregnant woman, only that the thing inside him was not a baby but a youth. No, it wasn't a youth, it was a woman, young, and wearing a coat of mail like a knight. It is absurd, you see, to try to tell what was inside the old writer as he lay on his high bed and listened to the fluttering of his heart. The thing to get at is what the writer, or the young thing within the writer, was thinking about.

The old writer, like all of the people in the world, had got, during his long life, a great many notions in his head. He had once been quite handsome and a number of women had been in love with him. And then, of course, he had known people, many people, known them in a peculiarly intimate way that was different from the way in which you and I know people. At least that is what the writer thought and the thought pleased him. Why quarrel with an old man concerning his thoughts?

In the bed the writer had a dream that was not a dream. As he grew somewhat sleepy but was still conscious, figures began to appear before his eyes. He imagined the young indescribable thing within himself was driving a long procession of figures before his eyes.

You see the interest in all this lies in the figures that went before the eyes of the writer. They were all grotesques. All of the men and women the writer had ever known had become grotesques.

The grotesques were not all horrible. Some were amusing, some almost beautiful, and one, a woman all drawn out of shape, hurt the old man by her grotesqueness. When she passed he made a noise like a small dog whimpering. Had you come into the room you might have supposed the old man had unpleasant dreams or perhaps indigestion.

For an hour the procession of grotesques passed before the eyes of the old man, and then, although it was a painful thing to do, he crept out of bed and began to write. Some one of the grotesques had made a deep impression on his mind and he wanted to describe it.

At his desk the writer worked for an hour. In the end he wrote a book which he called "The Book of the Grotesque." It was never published, but I saw it once and it made an indelible impression on my mind. The book had one central thought that is very strange and has always remained with me. By remembering it I have been able to understand many people and things that I was never able to understand before. The thought was involved but a simple statement of it would be something like this:

That in the beginning when the world was young there were a great many thoughts but no such thing as a truth. Man made the truths himself and each truth was a composite of a great many vague thoughts. All about in the world were the truths and they were all beautiful.

The old man had listed hundreds of the truths in his book. I will not try to tell you of all of them. There was the truth of virginity and the truth of passion, the truth of wealth and of poverty, of thrift and of profligacy, of carelessness and abandon. Hundreds and hundreds were the truths and they were all beautiful.

And then the people came along. Each as he appeared snatched up one of the truths and some who were quite strong snatched up a dozen of them.

It was the truths that made the people grotesques. The old man had quite an elaborate theory concerning the matter. It was his notion that the moment one of the people took one of the truths to himself, called it his truth, and tried to live his life by it, he came a grotesque and the truth he embraced became a falsehood.

You can see for yourself how the old man, who had spent all of his life writing and was filled with words, would write hundreds of pages concerning this matter. The subject would become so big in his mind that he himself would be in danger of becoming a grotesque. He didn't, I suppose, for the same reason that he never published the book. It was the young thing inside him that saved the old man.

Concerning the old carpenter who fixed the bed for the writer, I only mentioned him because he, like many of what are called very common people, became the nearest thing to what is understandable and lovable of all the grotesques in the writer's book.

Mother

Elizabeth Willard, the mother of George Willard, was tall and gaunt and her face was marked with smallpox scars. Although she was but forty-five, some obscure disease had taken the fire out of her figure. Listlessly she went about the disorderly old hotel looking at the faded wall-paper and the ragged carpets and, when she was able to be about, doing the work of a chambermaid among beds soiled by the slumbers of fat traveling men. Her husband, Tom Willard, a slender, graceful man with square shoulders, a quick military step, and a black mustache, trained to turn sharply up at the ends, tried to put the wife out of his mind. The presence of the tall ghostly figure, moving slowly through the halls, he took as a reproach to himself. When he thought of her he grew angry and swore. The hotel was unprofitable and forever on the edge of failure and he wished himself out of it. He thought of the old house and the woman who lived there with him as things defeated and done for. The hotel in which he had begun life so hopefully was now a mere ghost of what a hotel should be. As he went spruce and businesslike through the streets of Winesburg, he sometimes stopped and turned quickly about as though fearing that the spirit of the hotel and of the woman would follow him even into the streets. "Damn such a life, damn it!" he sputtered aimlessly.

Tom Willard had a passion for village politics and for years had been the leading Democrat in a strongly Republican community. Some day, he told himself, the tide of things political will turn in my favor and the years of ineffectual service count big in the bestowal of rewards. He dreamed of going to Congress and even of becoming governor. Once when a younger member of the party arose at a political conference and began to boast of his faithful service, Tom Willard grew white with fury. "Shut up, you," he roared, glaring about. "What do you know of service? What are you but a boy? Look at what I've done here! I was a Democrat here in Winesburg when it was a crime to be a Democrat. In the old days they fairly hunted us with guns."

Between Elizabeth and her one son George there was a deep unexpressed bond of sympathy, based on a girlhood dream that had long ago died. In the son's presence she was timid and reserved, but sometimes while he hurried about town intent upon his duties as a reporter, she went into his room and closing the door knelt by a little desk, made of a kitchen table, that sat near a window. In the room by the desk she went through a ceremony that was half a prayer, half a demand, addressed to the skies. In the boyish figure she yearned to see something half forgotten that had once been a part of herself recreated. The prayer concerned that. "Even though I die, I will in some way keep defeat from you," she cried, and so deep was her determination that her whole body shook. Her eyes glowed and she clenched her fists. "If I am dead and see him becoming a meaningless drab figure like myself, I will come back," she declared. "I ask God now to give me that privilege. I demand it. I will pay for it. God may beat me with his fists. I will take any blow that may befall if but this my boy be allowed to express something for us both." Pausing uncertainly, the woman stared about the boy's room. "And do not let him become smart and successful either," she added vaguely.

The communion between George Willard and his mother was outwardly a formal thing without meaning. When she was ill and sat by the window in her room he sometimes went in the evening to make her a visit. They sat by a window that looked over the roof of a small frame building into Main Street. By turning their heads they could see, through another window, along an alleyway that ran behind the Main Street stores and into the back door of Abner Groff's bakery. Sometimes as they sat thus a picture of village life presented itself to them. At the back door of his shop appeared Abner Groff with a stick or an empty milk bottle in his hand. For a long time there was a feud between the baker and a grey cat that belonged to Sylvester West, the druggist. The boy and his mother saw the cat creep into the door of the bakery and presently emerge followed by the baker who swore and waved his arms about. The baker's eyes were small and red and his black hair and beard were filled with flour dust. Sometimes he was so angry that, although the cat had disappeared, he hurled sticks, bits of broken glass, and even some of the tools of his trade about. Once he broke a window at the back of Sinning's Hardware Store. In the alley the grey cat crouched behind barrels filled with torn paper and broken bottles above which flew a black swarm of flies. Once when she was alone, and after watching a prolonged and ineffectual outburst on the part of the baker, Elizabeth Willard put her head down on her long white hands and wept. After that she did not look along the alleyway any more, but tried to forget the contest between the bearded man and the cat. It seemed like a rehearsal of her own life, terrible in its vividness.

In the evening when the son sat in the room with his mother, the silence made them both feel awkward. Darkness came on and the evening train came in at the station. In the street below feet tramped up and down upon a board sidewalk. In the station yard, after the evening train had gone, there was a heavy silence. Perhaps Skinner Leason, the express agent, moved a truck the length of the station platform. Over on Main Street sounded a man's voice, laughing. The door of the express office banged. George Willard arose and crossing the room fumbled for the doorknob. Sometimes he knocked against a chair, making it scrape along the floor. By the window sat the sick woman, perfectly still, listless. Her long hands, white and bloodless, could be seen drooping over the ends of the arms of the chair. "I think you had better be out among the boys. You are too much indoors," she said, striving to relieve the embarrassment of the departure. "I thought I would take a walk," replied George Willard, who felt awkward and confused.

One evening in July, when the transient guests who made the New Willard House their temporary homes had become scarce, and the hallways, lighted only by kerosene lamps turned low, were plunged in gloom, Elizabeth Willard had an adventure. She had been ill in bed for several days and her son had not come to visit her. She was alarmed. The feeble blaze of life that remained in her body was blown into a flame by her anxiety and she crept out of bed, dressed and hurried along the hallway toward her son's room, shaking with exaggerated fears. As she went along she steadied herself with her hand, slipped along the papered walls of the hall and breathed with difficulty. The air whistled through her teeth. As she hurried forward she thought how foolish she was. "He is concerned with boyish affairs," she told herself. "Perhaps he has now begun to walk about in the evening with girls."

Elizabeth Willard had a dread of being seen by guests in the hotel that had once belonged to her father and the ownership of which still stood recorded in her name in the county courthouse. The hotel was continually losing patronage because of its shabbiness and she thought of herself as also shabby. Her own room was in an obscure corner and when she felt able to work she voluntarily worked among the beds, preferring the labor that could be done when the guests were abroad seeking trade among the merchants of Winesburg.

By the door of her son's room the mother knelt upon the floor and listened for some sound from within. When she heard the boy moving about and talking in low tones a smile came to her lips. George Willard had a habit of talking aloud to himself and to hear him doing so had always given his mother a peculiar pleasure. The habit in him, she felt, strengthened the secret bond that existed between them. A thousand times she had whispered to herself of the matter. "He is groping about, trying to find himself," she thought. "He is not a dull clod, all words and smartness. Within him there is a secret something that is striving to grow. It is the thing I let be killed in myself."

In the darkness in the hallway by the door the sick woman arose and started again toward her own room. She was afraid that the door would open and the boy come upon her. When she had reached a safe distance and was about to turn a corner into a second hallway she stopped and bracing herself with her hands waited, thinking to shake off a trembling fit of weakness that had come upon her. The presence of the boy in the room had made her happy. In her bed, during the long hours alone, the little fears that had visited her had become giants. Now they were all gone. "When I get back to my room I shall sleep," she murmured gratefully.

But Elizabeth Willard was not to return to her bed and sleep. As she stood trembling in the darkness the door of her son's room opened and the boy's father, Tom Willard, stepped out. In the light that streamed out at the door he stood with the knob in his hand and talked. What he said infuriated the woman.

Tom Willard was ambitious for his son. He had always thought of himself as a successful man, although nothing he had ever done had turned out successfully. However, when he was out of sight of the New Willard House and had no fear of coming upon his wife, he swaggered and began to dramatize himself as one of the chief men of the town. He wanted his son to succeed. He it was who had secured for the boy the position on the *Winesburg Eagle.* Now, with a ring of earnestness in his voice, he was advising concerning some course of conduct. "I tell you what, George, you've got to wake up," he said sharply. "Will Henderson has spoken to me three times concerning the matter. He says you go along for hours not hearing when you are spoken to and acting like a gawky girl. What ails you?" Tom Willard laughed good-naturedly. "Well, I guess you'll get over it," he said. "I told Will that. You're not a fool and you're not a woman. You're Tom Willard's son and you'll wake up. I'm not afraid. What you say clears things up. If being a newspaper man had put the notion of becoming a writer into your mind that's all right. Only I guess you'll have to wake up to do that too, eh?"

Tom Willard went briskly along the hallway and down a flight of stairs to the office. The woman in the darkness could hear him laughing and talking with a guest who was striving to wear away a dull evening by dozing in a chair by the office door. She returned to the door of her son's room. The weakness had passed from her body as by a miracle and she stepped boldly along. A thousand ideas raced through her head. When she heard the scraping of a chair and the sound of a pen scratching upon paper, she again turned and went back along the hallway to her own room.

A definite determination had come into the mind of the defeated wife of the Winesburg Hotel keeper. The determination was the result of long years of quiet and rather ineffectual thinking. "Now," she told herself, "I will act. There is something threatening my boy and I will ward it off." The fact that the conversation between Tom Willard and his son had been rather quiet and natural, as though an understanding existed between them, maddened her. Although for years she had hated her husband, her hatred had always before been a quite impersonal thing. He had been merely a part of something else that she hated. Now, and by the few words at the door, he had become the thing personified. In the darkness of her own room she clenched her fists and glared about. Going to a cloth bag that hung on a nail by the wall she took out a long pair of sewing scissors and held them in her hand like a dagger. "I will stab him." she said aloud. "He has chosen to be the voice of evil and I will kill him. When I have killed him something will snap within myself and I will die also. It will be a release for all of us."

In her girlhood and before her marriage with Tom Willard, Elizabeth had borne a somewhat shaky reputation in Winesburg. For years she had been what is called "stage-struck" and had paraded through the streets with traveling men guests at her father's hotel, wearing loud clothes and urging them to tell her of life in the cities out of which they had come. Once she startled the town by putting on men's clothes and riding a bicycle down Main Street.

In her own mind the tall girl had been in those days much confused. A great restlessness was in her and it expressed itself in two ways. First there was an uneasy desire for change, for some big definite movement to her life. It was this feeling that had turned her mind to the stage. She dreamed of joining some company and wandering over the world, seeing always new faces and giving something out of herself to all people. Sometimes at night she was quite beside herself with the thought, but when she tried to talk of the matter to the members of the theatrical com-

panies that came to Winesburg and stopped at her father's hotel,
she got nowhere. They did not seem to know what she meant, or
if she did get something of her passion expressed, they only
laughed. "It's not like that," they said. "It's as dull and uninteresting as this here. Nothing comes of it."

With the traveling men when she walked about with them, and
later with Tom Willard, it was quite different. Always they seemed
to understand and sympathize with her. On the side streets of the
village, in the darkness under the trees, they took hold of her hand
and she thought that something unexpressed in herself came forth
and became a part of an unexpressed something in them.

And then there was the second expression of her restlessness.
When that came she felt for a time released and happy. She did
not blame the men who walked with her and later did not blame
Tom Willard. It was always the same, beginning with kisses and
ending, after strange wild emotions, with peace and then sobbing
repentance. When she sobbed she put her hand upon the face of
the man and had always the same thought. Even though he were
large and bearded she thought he had become suddenly a little
boy. She wondered why he did not sob also.

In her room, tucked away in a corner of the old Willard House,
Elizabeth Willard lighted a lamp and put it on a dressing table
that stood by the door. A thought had come into her mind and
she went to a closet and brought out a small square box and set
it on the table. The box contained material for make-up and had
been left with other things by a theatrical company that had once
been stranded in Winesburg. Elizabeth Willard had decided that
she would be beautiful. Her hair was still black and there was a
great mass of it braided and coiled about her head. The scene
that was to take place in the office below began to grow in her
mind. No ghostly worn-out figure should confront Tom Willard,
but something quite unexpected and startling. Tall and with dusky
cheeks and hair that fell in a mass from her shoulders, a figure
should come striding down the stairway before the startled loungers in the hotel office. The figure would be silent— it would be
swift and terrible. As a tigress whose cub had been threatened
would she appear, coming out of the shadows, stealing noiselessly
along and holding the long wicked scissors in her hand.

With a little broken sob in her throat, Elizabeth Willard blew
out the light that stood upon the table and stood weak and trembling in the darkness. The strength that had been as a miracle in
her body left and she half reeled across the floor, clutching at the
back of the chair in which she had spent so many long days staring
out over the tin roofs into the main street of Winesburg. In the
hallway there was the sound of footsteps and George Willard came
in at the door. Sitting in a chair beside his mother he began to
talk. "I'm going to get out of here," he said. "I don't know where
I shall go or what I shall do but I am going away."

The woman in the chair waited and trembled. An impulse came to her. "I suppose you had better wake up," she said. "You think that? You will go to the city and make money, eh? It will be better for you, you think, to be a business man, to be brisk and smart and alive?" She waited and trembled.

The son shook his head. "I suppose I can't make you understand, but oh, I wish I could," he said earnestly. "I can't even talk to father about it. I don't try. There isn't any use. I don't know what I shall do. I just want to go away and look at people and think."

Silence fell upon the room where the boy and woman sat together. Again, as on the other evenings, they were embarrassed. After a time the boy tried again to talk. "I suppose it won't be for a year or two but I've been thinking about it," he said, rising and going toward the door. "Something father said makes it sure that I shall have to go away." He fumbled with the door knob. In the room the silence became unbearable to the woman. She wanted to cry out with joy because of the words that had come from the lips of her son, but the expression of joy had become impossible to her. "I think you had better go out among the boys. You are too much indoors," she said. "I thought I would go for a little walk," replied the son stepping awkwardly out of the room and closing the door.

Departure

Young George Willard got out of bed at four in the morning. It was April and the young tree leaves were just coming out of their buds. The trees along the residence streets in Winesburg are maple and the seeds are winged. When the wind blows they whirl crazily about, filling the air and making a carpet underfoot.

George came down stairs into the hotel office carrying a brown leather bag. His trunk was packed for departure. Since two o'clock he had been awake thinking of the journey he was about to take and wondering what he would find at the end of his journey. The boy who slept in the hotel office lay on a cot by the door. His mouth was open and he snored lustily. George crept past the cot and went out into the silent deserted main street. The east was pink with the dawn and long streaks of light climbed into the sky where a few stars still shone.

Beyond the last house on Trunion Pike in Winesburg there is a great stretch of open fields. The fields are owned by farmers who live in town and drive homeward at evening along Trunion Pike in light creaking wagons. In the fields are planted berries and small fruits. In the late afternoon in the hot summers when the road and the fields are covered with dust, a smoky haze lies over the great flat basin of land. To look across it is like looking out across the sea. In the spring when the land is green the effect is somewhat different. The land becomes a wide green billiard table on which tiny human insects toil up and down.

All through his boyhood and young manhood George Willard had been in the habit of walking on Trunion Pike. He had been in the midst of the great open place on winter nights when it was covered with snow and only the moon looked down at him; he had been there in the fall when bleak winds blew and on summer evenings when the air vibrated with the song of insects. On the April morning he wanted to go there again, to walk again in the silence. He did walk to where the road dipped down by a little stream two miles from town and then turned and walked silently back again. When he got to Main Street clerks were sweeping the sidewalks before the stores. "Hey, you George. How does it feel to be going away?" they asked.

The west bound train leaves Winesburg at seven forty-five in the morning. Tom Little is conductor. His train runs from Cleveland to where it connects with a great trunk line railroad with terminals in Chicago and New York. Tom has what in railroad circles is called an "easy run." Every evening he returned to his family. In the fall and spring he spends his Sundays fishing in Lake Erie. He has a round red face and small blue eyes. He knows the people in the towns along his railroad better than a city man knows the people who live in his apartment building.

George came down the little incline from the New Willard House at seven o'clock. Tom Willard carried his bag. The son had become taller than the father.

On the station platform everyone shook the young man's hand. More than a dozen people waited about. Then they talked of their own affairs. Even Will Henderson, who was lazy and often slept until nine, had got out of bed. George was embarrassed. Gertrude Wilmot, a tall thin woman of fifty who worked in the Winesburg post office, came along the station platform. She had never before paid any attention to George. Now she stopped and put out her hand. In two words she voiced what everyone felt. "Good luck," she said sharply and then turning went on her way.

When the train came into the station George felt relieved. He scampered hurriedly aboard. Helen White[2] came running along Main Street hoping to have a parting word with him, but he had found a seat and did not see her. When the train started Tom Little punched his ticket, grinned and, although he knew George well and knew on what adventure he was just setting out, made no comment. Tom had seen a thousand George Willards go out of their towns to the city. It was a commonplace enough incident with him. In the smoking car there was a man who had just invited Tom to go on a fishing trip to Sandusky Bay. He wanted to accept the invitation and talk over details.

2. George Willard's girl friend, daughter of a Winesburg banker.

George glanced up and down the car to be sure no one was looking then took out his pocketbook and counted his money. His mind was occupied with a desire not to appear green. Almost the last words his father had said to him concerned the matter of his behavior when he got to the city. "Be a sharp one," Tom Willard had said. "Keep your eyes on your money. Be awake. That's the ticket. Don't let any one think you're a greenhorn."

After George counted his money he looked out of the window and was surprised to see that the train was still in Winesburg.

The young man, going out of his town to meet the adventure of life, began to think but he did not think of anything very big or dramatic. Things like his mother's death, his departure from Winesburg, the uncertainty of his future life in the city, the serious and larger aspects of his life did not come into his mind.

He thought of little things—Turk Smallet wheeling boards through the main street of his town in the morning, a tall woman, beautifully gowned, who had once stayed over night at his father's hotel, Butch Wheeler the lamp lighter of Winesburg hurrying through the streets on a summer evening and holding a torch in his hand, Helen White standing by a window in the Winesburg post office and putting a stamp on an envelope.

The young man's mind was carried away by his growing passion for dreams. One looking at him would not have thought him particularly sharp. With the recollection of little things occupying his mind he closed his eyes and leaned back in the car seat. He stayed that way for a long time and when he aroused himself and again looked out of the car window the town of Winesburg had disappeared and his life there had become but a background on which to paint the dreams of his manhood.

1919

KATHERINE ANNE PORTER
1890–

One of the finest stylists among important American writers of fiction is Katherine Anne Porter, who is acutely conscious of her southern origins. She herself has declared that she was "the grandchild of a Lost War," born in Indian Creek, Texas, into a family that was conscious of former wealth and position in the older South of Louisiana and Kentucky.[1]

Little is known about her life, and it is uncertain whether she was raised in early childhood as a Roman Catholic or a Methodist. She was brought up by her paternal grandmother after her mother's death in 1892. Like her fictional surrogate Miranda in *Old Mortality* and other stories, she was educated in convent schools where she received a "frag-

---

1. "Portrait: Old South," *The Days Before*, p. 155.

mentary, but strangely useless and ornamental education."[2] She eloped
from such a school when about sixteen and subsequently was married to
and divorced from Eugene Pressly, a foreign-service officer, and Albert
Erskine, an editor of the *Southern Review*. As she candidly explained,
"They couldn't live with me because I was a writer, and, now as then,
writing took first place."[3] She worked on newspapers in Texas, Denver,
and Chicago before and during World War I and lived between 1918
and 1921 in Mexico, where she studied art and became involved in left-
wing politics. (She has declared recently that "my political thinking was
the lamentable 'political illiteracy' of a liberal idealist—we might say a
species of Jeffersonian.")[4] She returned to Mexico on a Guggenheim
fellowship in 1931, then journeyed to Europe, residing in Switzerland,
France, and Germany until returning permanently to the States in 1937.

Her first short story (*Maria Concepcion*) had been published in 1922,
and her first collections of stories and short novels, *Flowering Judas* and
Noon Wine, in 1930 and 1937. These gained her a secure reputation
among critics that was sustained by *Pale Horse, Pale Rider* (1939) and
The Leaning Tower (1944). But it was not until 1962 that she gained a
large audience with the publication of her first and only novel, *The Ship
of Fools*. Though it received mixed reviews, its account of cosmopolitan
passengers traveling from Mexico to Germany at the onset of Hitler's
Nazi regime was widely read, and it formed the basis of a popular Hollywood film. She did some screen writing for MGM in 1948–50, and after
1950 she lectured frequently and gave public readings from her works.
She was the Fulbright Professor at Liège, Belgium, in 1945 and succeeded
William Faulkner as Writer in Residence at the University of Virginia
in 1958. A selection of her reviews, essays, and translations, *The Days
Before*, appeared in 1952. Her *Collected Stories* (1965) brought both the
National Book Award and the Pulitzer Prize in 1966, as well as election
to the American Academy of Letters. In 1967 she won the Gold Medal
for Fiction of the National Institute of Arts and Letters to which she had
been elected in 1941. She now lives, in delicate health and semiretire-
ment, near Washington, D.C.

The power and subtlety of Porter's fiction were acclaimed by critics
and fellow writers from the 1930s on, despite the fact that her output
consists of only one full-length novel and a mere twenty-seven stories and
short novels. Her prose, with its spare use of figures of speech and its
deceptive simplicity, displays what the critic Edmund Wilson called a
"purity and precision almost unique in contemporary American fiction."[5]
While some tales deal with Irish immigrants, a larger and more powerful
number deal with cultural conflict and revolutionary ferment in Mexico,
exploring the Indian's half-Christian, half-pagan world of primitive emotion (in *Maria Concepcion*) or the reactions of North Americans to the
exotic in Spanish America (in *Flowering Judas and Hacienda*). The most
clearly autobiographical stories center on the stresses of marriage, the
family pieties and social proprieties that have been important in southern
fiction, and the growing up of the young girl Miranda in *Old Mortality*.
The precision with which Porter captures the nuances of anger, fear, and
wounded pathos when people encounter death or the death of cherished

2. *Ibid.*, p. 159.
3. *The Never Ending Wrong*, p. 13.
4. *Baltimore Evening Sun*, Feb. 28,
1969.
5. Edmund Wilson, *Classics and Commercials*, p. 219.

dreams mark *Pale Horse, Pale Rider* and *The Jilting of Granny Weatherall* as masterful pieces of short fiction. *The Jilting of Granny Weatherall* is reprinted from *The Collected Stories of Katherine Anne Porter*.

The Jilting of Granny Weatherall

She flicked her wrist neatly out of Doctor Harry's pudgy careful fingers and pulled the sheet up to her chin. The brat ought to be in knee breeches. Doctoring around the country with spectacles on his nose! "Get along now, take your schoolbooks and go. There's nothing wrong with me."

Doctor Harry spread a warm paw like a cushion on her forehead where the forked green vein danced and made her eyelids twitch. "Now, now, be a good girl, and we'll have you up in no time."

"That's no way to speak to a woman nearly eighty years old just because she's down. I'd have you respect your elders, young man."

"Well, Missy, excuse me." Doctor Harry patted her cheek. "But I've got to warn you, haven't I? You're a marvel, but you must be careful or you're going to be good and sorry."

"Don't tell me what I'm going to be. I'm on my feet now, morally speaking. It's Cornelia. I had to go to bed to get rid of her."

Her bones felt loose, and floated around in her skin, and Doctor Harry floated like a balloon around the foot of the bed. He floated and pulled down his waistcoat and swung his glasses on a cord. "Well, stay where you are, it certainly can't hurt you."

"Get along and doctor your sick," said Granny Weatherall. "Leave a well woman alone. I'll call for you when I want you. . . . Where were you forty years ago when I pulled through milk-leg[1] and double pneumonia? You weren't even born. Don't let Cornelia lead you on," she shouted, because Doctor Harry appeared to float up to the ceiling and out. "I pay my own bills, and I don't throw my money away on nonsense!"

She meant to wave good-bye, but it was too much trouble. Her eyes closed of themselves, it was like a dark curtain drawn around the bed. The pillow rose and floated under her, pleasant as a hammock in a light wind. She listened to the leaves rustling outside the window. No, somebody was swishing newspapers: no, Cornelia and Doctor Harry were whispering together. She leaped broad awake, thinking they whispered in her ear.

"She was never like this, *never* like this!" "Well, what can we expect?" "Yes, eighty years old. . . ."

Well, and what if she was? She still had ears. It was like Cornelia to whisper around doors. She always kept things secret in such a public way. She was always being tactful and kind. Cornelia

1. Swelling of the legs after childbirth.

was dutiful; that was the trouble with her. Dutiful and good: "So good and dutiful," said Granny, "that I'd like to spank her." She saw herself spanking Cornelia and making a fine job of it.

"What'd you say, Mother?"

Granny felt her face tying up in hard knots.

"Can't a body think, I'd like to know?"

"I thought you might want something."

"I do. I want a lot of things. First off, go away and don't whisper."

She lay and drowsed, hoping in her sleep that the children would keep out and let her rest a minute. It had been a long day. Not that she was tired. It was always pleasant to snatch a minute now and then. There was always so much to be done, let me see: tomorrow.

Tomorrow was far away and there was nothing to trouble about. Things were finished somehow when the time came; thank God there was always a little margin over for peace: then a person could spread out the plan of life and tuck in the edges orderly. It was good to have everything clean and folded away, with the hair brushes and tonic bottles sitting straight on the white embroidered linen: the day started without fuss and the pantry shelves laid out with rows of jelly glasses and brown jugs and white stone-china jars with blue whirligigs and words painted on them: coffee, tea, sugar, ginger, cinnamon, allspice: and the bronze clock with the lion on top nicely dusted off. The dust that lion could collect in twenty-four hours! The box in the attic with all those letters tied up, well, she'd have to go through that tomorrow. All those letters —George's letters and John's letters and her letters to them both —lying around for the children to find afterwards made her uneasy. Yes, that would be tomorrow's business. No use to let them know how silly she had been once.

While she was rummaging around she found death in her mind and it felt clammy and unfamiliar. She had spent so much time preparing for death there was no need for bringing it up again. Let it take care of itself now. When she was sixty she had felt very old, finished, and went around making farewell trips to see her children and grandchildren, with a secret in her mind: This is the very last of your mother, children! Then she made her will and came down with a long fever. That was all just a notion like a lot of other things, but it was lucky too, for she had once for all got over the idea of dying for a long time. Now she couldn't be worried. She hoped she had better sense now. Her father had lived to be one hundred and two years old and had drunk a noggin of strong hot toddy on his last birthday. He told the reporters it was his daily habit, and he owed his long life to that. He had made quite a scandal and was very pleased about it. She believed she'd just plague Cornelia a little.

"Cornelia! Cornelia!" No footsteps, but a sudden hand on her cheek. "Bless you, where have you been?"

"Here, Mother."

"Well, Cornelia, I want a noggin of hot toddy."

"Are you cold, darling?"

"I'm chilly, Cornelia. Lying in bed stops the circulation. I must have told you that a thousand times."

Well, she could just hear Cornelia telling her husband that Mother was getting a little childish and they'd have to humor her. The thing that most annoyed her was that Cornelia thought she was deaf, dumb, and blind. Little hasty glances and tiny gestures tossed around her and over her head saying, "Don't cross her, let her have her way, she's eighty years old," and she sitting there as if she lived in a thin glass cage. Sometimes Granny almost made up her mind to pack up and move back to her own house where nobody could remind her every minute that she was old. Wait, wait, Cornelia, till your own children whisper behind your back!

In her day she had kept a better house and had got more work done. She wasn't too old yet for Lydia to be driving eighty miles for advice when one of the children jumped the track, and Jimmy still dropped in and talked things over: "Now, Mammy, you've a good business head, I want to know what you think of this? . . ." Old. Cornelia couldn't change the furniture around without asking. Little things, little things! They had been so sweet when they were little. Granny wished the old days were back again with the children young and everything to be done over. It had been a hard pull, but not too much for her. When she thought of all the food she had cooked, and all the clothes she had cut and sewed, and all the gardens she had made—well, the children showed it. There they were, made out of her, and they couldn't get away from that. Sometimes she wanted to see John again and point to them and say, Well, I didn't do so badly, did I? But that would have to wait. That was for tomorrow. She used to think of him as a man, but now all the children were older than their father, and he would be a child beside her if she saw him now. It seemed strange and there was something wrong in the idea. Why, he couldn't possibly recognize her. She had fenced in a hundred acres once, digging the post holes herself and clamping the wires with just a negro boy to help. That changed a woman. John would be looking for a young woman with the peaked Spanish comb in her hair and the painted fan. Digging post holes changed a woman. Riding country roads in the winter when women had their babies was another thing: sitting up nights with sick horses and sick negroes and sick children and hardly ever losing one. John, I hardly ever lost one of them! John would see that in a minute, that would be something he could understand, she wouldn't have to explain anything!

It made her feel like rolling up her sleeves and putting the whole

place to rights again. No matter if Cornelia was determined to be
everywhere at once, there were a great many things left undone on
this place. She would start tomorrow and do them. It was good to
be strong enough for everything, even if all you made melted and
changed and slipped under your hands, so that by the time you
finished you almost forgot what you were working for. What was it
I set out to do? she asked herself intently, but she could not remember. A fog rose over the valley, she saw it marching across the
creek swallowing the trees and moving up the hill like an army of
ghosts. Soon it would be at the near edge of the orchard, and then
it was time to go in and light the lamps. Come in, children, don't
stay out in the night air.

Lighting the lamps had been beautiful. The children huddled up
to her and breathed like little calves waiting at the bars in the
twilight. Their eyes followed the match and watched the flame rise
and settle in a blue curve, then they moved away from her. The
lamp was lit, they didn't have to be scared and hang on to mother
any more. Never, never, never more. God, for all my life I thank
Thee. Without Thee, my God, I could never have done it. Hail,
Mary, full of grace.

I want you to pick all the fruit this year and see that nothing is
wasted. There's always someone who can use it. Don't let good
things rot for want of using. You waste life when you waste good
food. Don't let things get lost. It's bitter to lose things. Now, don't
let me get to thinking, not when I am tired and taking a little nap
before supper. . . .

The pillow rose about her shoulders and pressed against her
heart and the memory was being squeezed out of it: oh, push
down the pillow, somebody: it would smother her if she tried to
hold it. Such a fresh breeze blowing and such a green day with no
threats in it. But he had not come, just the same. What does a
woman do when she has put on the white veil and set out the white
cake for a man and he doesn't come? She tried to remember. No, I
swear he never harmed me but in that. He never harmed me but in
that . . . and what if he did? There was the day, the day, but a
whirl of dark smoke rose and covered it, crept up and over into
the bright field where everything was planted so carefully in orderly rows. That was hell, she knew hell when she saw it. For sixty
years she had prayed against remembering him and against losing
her soul in the deep pit of hell, and now the two things were
mingled in one and the thought of him was a smoky cloud from
hell that moved and crept in her head when she had just got rid of
Doctor Harry and was trying to rest a minute. Wounded vanity,
Ellen, said a sharp voice in the top of her mind. Don't let your
wounded vanity get the upper hand of you. Plenty of girls get
jilted. You were jilted, weren't you? Then stand up to it. Her
eyelids wavered and let in streamers of blue-gray light like tissue

paper over her eyes. She must get up and pull the shades down or she'd never sleep. She was in bed again and the shades were not down. How could that happen? Better turn over, hide from the light, sleeping in the light gave you nightmares. "Mother, how do you feel now?" and a stinging wetness on her forehead. But I don't like having my face washed in cold water!

Hapsy? George? Lydia? Jimmy? No, Cornelia, and her features were swollen and full of little puddles. "They're coming, darling, they'll all be here soon." Go wash your face, child, you look funny.

Instead of obeying, Cornelia knelt down and put her head on the pillow. She seemed to be talking but there was no sound. "Well, are you tongue-tied? Whose birthday is it? Are you going to give a party?"

Cornelia's mouth moved urgently in strange shapes. "Don't do that, you bother me, daughter."

"Oh, no, Mother. Oh, no. . . ."

Nonsense. It was strange about children. They disputed your every word. "No what, Cornelia?"

"Here's Doctor Harry."

"I won't see that boy again. He just left five minutes ago."

"That was this morning, Mother. It's night now. Here's the nurse."

"This is Doctor Harry, Mrs. Weatherall. I never saw you look so young and happy!"

"Ah, I'll never be young again—but I'd be happy if they'd let me lie in peace and get rested."

She thought she spoke up loudly, but no one answered. A warm weight on her forehead, a warm bracelet on her wrist, and a breeze went on whispering, trying to tell her something. A shuffle of leaves in the everlasting hand of God, He blew on them and they danced and rattled. "Mother, don't mind, we're going to give you a little hypodermic." "Look here, daughter, how do ants get in this bed? I saw sugar ants yesterday." Did you send for Hapsy too?

It was Hapsy she really wanted. She had to go a long way back through a great many rooms to find Hapsy standing with a baby on her arm. She seemed to herself to be Hapsy also, and the baby on Hapsy's arm was Hapsy and himself and herself, all at once, and there was no surprise in the meeting. Then Hapsy melted from within and turned flimsy as gray gauze and the baby was a gauzy shadow, and Hapsy came up close and said, "I thought you'd never come," and looked at her very searchingly and said, "You haven't changed a bit!" They leaned forward to kiss, when Cornelia began whispering from a long way off, "Oh, is there anything you want to tell me? Is there anything I can do for you?"

Yes, she had changed her mind after sixty years and she would like to see George. I want you to find George. Find him and be sure to tell him I forgot him. I want him to know I had my husband just the same and my children and my house like any other woman. A good house too and a good husband that I loved and fine children out of him. Better than I hoped for even. Tell him I was given back everything he took away and more. Oh, no, oh, God, no, there was something else besides the house and the man and the children. Oh, surely they were not all? What was it? Something not given back. . . . Her breath crowded down under her ribs and grew into a monstrous frightening shape with cutting edges; it bored up into her head, and the agony was unbelievable: Yes, John, get the Doctor now, no more talk, my time has come.

When this one was born it should be the last. The last. It should have been born first, for it was the one she had truly wanted. Everything came in good time. Nothing left out, left over. She was strong, in three days she would be as well as ever. Better. A woman needed milk in her to have her full health.

"Mother, do you hear me?"

"I've been telling you—"

"Mother, Father Connolly's here."

"I went to Holy Communion only last week. Tell him I'm not so sinful as all that."

"Father just wants to speak to you."

He could speak as much as he pleased. It was like him to drop in and inquire about her soul as if it were a teething baby, and then stay on for a cup of tea and a round of cards and gossip. He always had a funny story of some sort, usually about an Irishman who made his little mistakes and confessed them, and the point lay in some absurd thing he would blurt out in the confessional showing his struggles between native piety and original sin. Granny felt easy about her soul. Cornelia, where are your manners? Give Father Connolly a chair. She had her secret comfortable understanding with a few favorite saints who cleared a straight road to God for her. All as surely signed and sealed as the papers for the new Forty Acres. Forever . . . heirs and assigns[2] forever. Since the day the wedding cake was not cut, but thrown out and wasted. The whole bottom dropped out of the world, and there she was blind and sweating with nothing under her feet and the walls falling away. His hand had caught her under the breast, she had not fallen, there was the freshly polished floor with the green rug on it, just as before. He had cursed like a sailor's parrot and said, "I'll kill him for you." Don't lay a hand on him, for my sake leave something to God. "Now, Ellen, you must believe what I tell you. . . ."

So there was nothing, nothing to worry about any more, except sometimes in the night one of the children screamed in a nightmare, and they both hustled out shaking and hunting for the

2. Legal term for those persons charged with carrying out the provisions of a will.

matches and calling, "There, wait a minute, here we are!" John, get the doctor now. Hapsy's time has come. But there was Hapsy standing by the bed in a white cap. "Cornelia, tell Hapsy to take off her cap. I can't see her plain."

Her eyes opened very wide and the room stood out like a picture she had seen somewhere. Dark colors with the shadows rising towards the ceiling in long angles. The tall black dresser gleamed with nothing on it but John's picture, enlarged from a little one, with John's eyes very black when they should have been blue. You never saw him, so how do you know how he looked? But the man insisted the copy was perfect, it was very rich and handsome. For a picture, yes, but it's not my husband. The table by the bed had a linen cover and a candle and a crucifix. The light was blue from Cornelia's silk lampshades. No sort of light at all, just frippery. You had to live forty years with kerosene lamps to appreciate honest electricity. She felt very strong and she saw Doctor Harry with a rosy nimbus[3] around him.

"You look like a saint, Doctor Harry, and I vow that's as near as you'll ever come to it."

"She's saying something."

"I heard you, Cornelia. What's all this carrying-on?"

"Father Connolly's saying—"

Cornelia's voice staggered and bumped like a cart in a bad road. It rounded corners and turned back again and arrived nowhere. Granny stepped up in the cart very lightly and reached for the reins, but a man sat beside her and she knew him by his hands, driving the cart. She did not look in his face, for she knew without seeing, but looked instead down the road where the trees leaned over and bowed to each other and a thousand birds were singing a Mass. She felt like singing too, but she put her hand in the bosom of her dress and pulled out a rosary, and Father Connolly murmured Latin in a very solemn voice and tickled her feet. My God, will you stop that nonsense? I'm a married woman. What if he did run away and leave me to face the priest by myself? I found another a whole world better. I wouldn't have exchanged my husband for anybody except St. Michael himself, and you may tell him that for me with a thank you in the bargain.

Light flashed on her closed eyelids, and a deep roaring shook her. Cornelia, is that lightning? I hear thunder. There's going to be a storm. Close all the windows. Call the children in. . . . "Mother, here we are, all of us." "Is that you, Hapsy?" "Oh, no, I'm Lydia. We drove as fast as we could." Their faces drifted above her, drifted away. The rosary fell out of her hands and Lydia put it back. Jimmy tried to help, their hands fumbled together, and Granny closed two fingers around Jimmy's thumb.

---

3. Bright halo around or over the head.

Beads wouldn't do, it must be something alive. She was so amazed her thoughts ran round and round. So, my dear Lord, this is my death and I wasn't even thinking about it. My children have come to see me die. But I can't, it's not time. Oh, I always hated surprises. I wanted to give Cornelia the amethyst set—Cornelia, you're to have the amethyst set, but Hapsy's to wear it when she wants, and, Doctor Harry, do shut up. Nobody sent for you. Oh, my dear Lord, do wait a minute. I meant to do something about the Forty Acres, Jimmy doesn't need it and Lydia will later on, with that worthless husband of hers. I meant to finish the altar cloth and send six bottles of wine to Sister Borgia for her dyspepsia.[4] I want to send six bottles of wine to Sister Borgia, Father Connolly, now don't let me forget.

Cornelia's voice made short turns and tilted over and crashed. "Oh, Mother, oh, Mother, oh, Mother. . . ."

"I'm not going Cornelia. I'm taken by surprise. I can't go."

You'll see Hapsy again. What about her? "I thought you'd never come." Granny made a long journey outward, looking for Hapsy. What if I don't find her? What then? Her heart sank down and down, there was no bottom to death, she couldn't come to the end of it. The blue light from Cornelia's lampshade drew into a tiny point in the center of her brain, it flickered and winked like an eye, quietly it fluttered and dwindled. Granny lay curled down within herself, amazed and watchful, staring at the point of light that was herself; her body was now only a deeper mass of shadow in an endless darkness and this darkness would curl around the light and swallow it up. God, give a sign!

For the second time there was no sign. Again no bridegroom and the priest in the house. She could not remember any other sorrow because this grief wiped them all away. Oh, no, there's nothing more cruel than this—I'll never forgive it. She stretched herself with a deep breath and blew out the light.

1930

4. Indigestion.

JOHN DOS PASSOS
1896–1970

Some of the most ambitious experiments in twentieth-century American fiction were those of John Dos Passos. In decades when many contemporary writers were concentrating on specific regions, perfecting shorter forms of fiction, or fashioning novels designed for the intensive scrutiny of characters and institutions, Dos Passos was impelled by an aroused social conscience to survey modern American society in its larger contours and to create panoramic forms that could accommodate the multiplicity of historical currents, characters, and events on so large a scene.

Born in Chicago, John Roderigo Dos Passos was raised by parents who lived apart but provided him with the advantages of travel in Europe and private schooling in England and at Choate in Connecticut. His father (who refused to acknowledge his son until 1916) was of Portuguese descent, a well-to-do lawyer who wrote a book on America's rise to dominance in the twentieth century. In 1916 Dos Passos graduated with honors from Harvard, where he was introduced to the incisive critiques of capitalistic society by the sociologist Thorstein Veblen and to such progenitors of modernism in literature as Walter Pater and James Joyce. He also wrote verse and published reviews in the *Advocate*, among them commentaries on the new poetry of Pound and Eliot and on the radical John Reed's *Insurgent Mexico*. He followed his father's wishes in beginning the study of architecture in Spain in 1916, but in 1917 he joined the famous Morton-Harjes volunteer ambulance corps, serving in France and Italy, and later became a medical corpsman in the U.S. Army. Much of the next ten years he spent as a free-lance journalist, traveling in Spain and Europe and serving with the Near East relief mission. In 1926 he joined the executive board of the Communist *New Masses*.

By that time Dos Passos's particular kind of radicalism was finding expression both in political action and in a variety of literary productions. In 1922 a volume of poems, *A Pushcart at the Curb*, appeared along with *Rosinante to the Road Again*, in which essays on Spanish culture and a dialogue between two travelers define Dos Passos's growing dissatisfaction with existing institutions in modern America and his attempt to identify with the masses. Two plays, expressionistic in technique, *The Garbage Man* and *Airways, Inc.*, appeared between 1925 and 1928, and a second travel book, *Orient Express* (1927), deplored the encroachment of American technological civilization on the ancient worlds of the Near East. But it was his fiction that secured his reputation. *One Man's Initiation—1917* (1920) consists of vignettes describing the impact of World War I on France and on an imaginative young American who serves in the ambulance corps. The more bluntly powerful *Three Soldiers* (1921) turns World War I and the U.S. Army into images of a repressive, mechanized society that grinds under an optical worker from San Francisco, a farm boy from Indiana, and a Harvard-educated composer named Andrews who rebels, then escapes from a labor battalion with the help of an anarchist and deserts the army so as to write music in Paris. At the end he is working on a composition entitled *The Soul and Body of John Brown*, the fanatical abolitionist who led the attack on slave-owning society in 1859 at Harpers Ferry and whom Andrews calls "the madman who wanted to free people." A comparable combination of explosive protest with a sense of futility governs the brilliant *Manhattan Transfer* (1925) in which Dos Passos devised kaleidoscopic techniques to render the contemporary urban scene: swiftly shifting episodes of crime and violence shatter the illusions of opportunity that entice the reporter Jimmy Herf; a roller coaster, a nickelodeon, and revolving doors provide a phantasmagoric imagery for the energetic but aimless society which threatens to stifle all individuality.

Between 1926 and 1936 Dos Passos's disgust with the institutions of American capitalism and the mechanization of life was intensified to the point of hatred, but it was complicated by the profound distrust of all institutions and all power that he displayed even while seeking to exert power as a writer and activist. He was aroused by industrial strikes and the efforts to suppress them, by the onset of the Great Depression, and notably by the prolonged trial and subsequent execution in 1927 of the two immigrant anarchists Sacco and Vanzetti, which to him and many other intellectuals seemed a gross miscarriage of justice perpetrated by the American establishment. He wrote for Communist journals, served on committees to aid political prisoners and strikers sponsored by the Communist party, and publicly supported Communist candidates in the 1932 elections, but he never joined the Communist party. He insisted that editors of Communist journals allow room for dissent, that the struggle against oppression be conducted as humanely as possible, and that he preferred the looser affiliation of a fellow traveler or "scavenger and campfollower."[1] Soon after Communists broke up a rally of Socialists in Madison Square Garden in 1934, Dos Passos withdrew from Communist front activities and supported Franklin D. Roosevelt for President in 1936.

His masterpiece, gathered as the trilogy *U.S.A.* in 1937, consists of three novels that were published between 1930 and 1936: *The 42nd Parallel*, *1919*, and *The Big Money*. Its span encompasses the decades before World War I to 1936; its geography extends from Seattle, San Francisco, Hollywood, and Mexico to Chicago, New York, Miami, and France; its social settings include the union offices of the International Workers of the World, the public relations office and the affluent circle of the magnate J. Ward Moorehouse, the airplane factory of Charlie Anderson, the tinsel world of the film actress Margo Dowling, and the strike headquarters of radical organizers in Pennsylvania and New Jersey. The cast of characters includes eleven major figures whose private and public lives intersect or diverge, terminating often in obscurity, accidental death, or suicide. Three of their careers are typical. Charlie Anderson, a Dakota farm boy and later a Socialist, joins the ambulance corps and returns from World War I to found an airplane factory, sells out to a large Detroit firm and marries an heiress, then spends out his life in unsuccessful business deals until killed in an auto crash. Ben Compton, a Brooklyn construction worker, union organizer, and Communist party member, is eventually expelled from the party for deviation from party policy. Mary French, a liberal college graduate and Communist supporter though never a member of the party, is shattered when her Communist lover deserts her and when another Communist co-worker is shot by strike breakers in the Pennsylvania mining camps. The trilogy closes with a panoramic view that contrasts a well-fed but airsick executive on a transcontinental plane with a hungry, valueless vagabond bypassed by speeding traffic on the highway below. The effect is to underscore the bleak declarations earlier in the triology that all individuals have been rendered "foreigners" by modern society, that America stands "defeated," sharply divided into "two nations."

The trilogy is sustained by a style as harshly blunt, as driven, as energized and machinelike as the powers and conflicts it depicts, and the vast scale of the fictive action is extended by three structural devices that Dos Passos used in later fiction. One he called the "Newsreel," a collage

---

1. Quoted by Granville Hicks in Andrew Hook, ed., *Dos Passos: A Collection of*   *Critical Essays*, p. 23.

of newspaper excerpts and headlines, snatches of popular songs, and quotations from official oratory and reports, which, scattered through the narrative, establish a public context for the themes and incidents of the fiction. The second he called the "Camera Eye," a sequence of impressionistic fragments, emotional and subjective, in effect lyrical or elegiac, giving an interval of submerged response to the issues and conflicts that appear in the narrative and in the public world. The third is the series of brilliantly concise biographies of historical personages that are interspersed throughout each volume, from the labor leader Eugene Debs, to such figures as the dancer Isadora Duncan and the film star Rudolph Valentino, the politicians Robert La Follette and Woodrow Wilson, the inventors Thomas Edison and the Wright Brothers, the financier J. P. Morgan, the sociologist Thorstein Veblen, the architect Frank Lloyd Wright. The biographies measure the cost of success—or of failure—in American life. And they give particular point to Dos Passos's hope that by "writing straight," instead of either overtly preaching or cranking out prose as mechanical as the machine civilization he detested, a writer could become an "architect of history."[2]

A historical point of view—non-Marxist and increasingly patriotic—proved to be the governing perspective of Dos Passos's career in later decades. His concern for personal liberty, and his fear of encroaching institutions and concentrations of power, shifted its base to the right; his antagonists became the Washington bureaucracy, the labor unions, and the Communists; his sanctions for his position became increasingly the remote past of Jeffersonianism; his castigation of the present and warnings about the future became more shrill. He went to Spain during the Spanish Civil War but returned in protest against Communist interference with the Loyalist cause, and continued a prolific career. Following the death of his first wife in an auto accident, he remarried and lived with his wife and daughter in the vicinity of Washington and Baltimore. He published two volumes of reports on World War II (*State of the Nation*, 1944, and *Tour of Duty*, 1946), a number of historical works (including *The Heart and Head of Thomas Jefferson*, 1954, and *The Men Who Made the Nation*, 1957), and a series of increasingly bitter novels. The three comprising the trilogy *District of Columbia* (1952) deal with the amatory and political entanglements of characters involved in the Spanish war, the Huey Long regime in Louisiana, and the Washington bureaucracy under the New Deal. In *Midcentury* (1961) Dos Passos returned to the format of *U.S.A.*, but it serves only to mask a hopeless denunciation of power struggles among financiers and labor leaders.

From U.S.A.

The Big Money[1]
Newsreel LXVIII

WALL STREET STUNNED

This is not Thirty-eight, but it's old Ninety-seven
You must put her in Center on time[2]

2. *Occasions and Protests*, pp. 7–8.
1. The following selections are the concluding sections of *The Big Money* (1936), the last novel in the trilogy

U.S.A.
2. The engineer of the locomotive Ninety-seven was the legendary Casey Jones, who died in a head-on collision

MARKET SURE TO RECOVER FROM SLUMP
Decline in Contracts
POLICE TURN MACHINE GUNS ON COLORADO MINE STRIKERS
KILL 5 WOUND 40

sympathizers appeared on the scene just as thousands of office workers were pouring out of the buildings at the lunch hour. As they raised their placard high and started an indefinite march from one side to the other, they were jeered and hooted not only by the office workers but also by workmen on a building under construction

NEW METHODS OF SELLING SEEN
Rescue Crews Try To Upend Ill-fated Craft While Waiting For Pontoons

He looked 'round an' said to his black greasy fireman
Jus' shovel in a little more coal
And when we cross that White Oak Mountain
You can watch your Ninety-seven roll

I find your column interesting and need advice. I have saved four thousand dollars which I want to invest for a better income. Do you think I might buy stocks?

POLICE KILLER FLICKS CIGARETTE AS HE GOES TREMBLING TO DOOM
PLAY AGENCIES IN RING OF SLAVE GIRL MARTS
Maker of Love Disbarred as Lawyer

Oh the right wing clothesmakers
And the Socialist fakers
They make by the workers . . .
Double cross

They preach Social-ism
But practice Fasc-ism
To keep capitalism
By the boss[3]

MOSCOW CONGRESS OUSTS OPPOSITION[4]

It's a mighty rough road from Lynchburg to Danville
An' a line on a three mile grade
It was on that grade he lost his average
An' you see what a jump he made

MILL THUGS IN MURDER RAID

here is the most dangerous example of how at the decisive moment the bourgeois ideology liquidates class solidarity and turns a friend of the workingclass of yesterday into a most miserable propagandist for imperialism today

---

when he tried to get his train to San Francisco on time. Excerpts from the ballad celebrating Jones are interspersed in this "Newsreel."
3. Excerpts from this labor-union protest song are also interspersed in this "Newsreel."
4. Probably the Seventh Congress of the Third International held in Moscow in 1935.

RED PICKETS FINED FOR PROTEST HERE

We leave our home in the morning
We kiss our children goodby

OFFICIALS STILL HOPE FOR RESCUE OF MEN

He was goin' downgrade makin' ninety miles an hour
When his whistle broke into a scream
He was found in the wreck with his hand on the throttle
An' was scalded to death with the steam

RADICALS FIGHT WITH CHAIRS AT UNITY MEETING

PATROLMEN PROTECT REDS

U.S. CHAMBER OF COMMERCE URGES CONFIDENCE

REAL VALUES UNHARMED

While we slave for the bosses
Our children scream an' cry
But when we draw our money
Our grocery bills to pay

PRESIDENT SEES PROSPERITY NEAR

Not a cent to spend for clothing
Not a cent to lay away

STEAMROLLER IN ACTION AGAINST MILITANTS

MINERS BATTLE SCABS

But we cannot buy for our children
Our wages are too low
Now listen to me you workers
Both you women and men
Let us win for them the victory
I'm sure it ain't no sin

CARILLON PEALS IN SINGING TOWER[5]

the President declared it was impossible to view the increased advantages for the many without smiling at those who a short time ago expressed so much fear lest our country might come under the control of a few individuals of great wealth

HAPPY CROWDS THRONG CEREMONY

on a tiny island nestling like a green jewel in the lake that mirrors the singing tower, the President today participated in the dedication of a bird sanctuary and its pealing carillon, fulfilling the dream of an immigrant boy

The Camera Eye (51)

at the head of the valley in the dark of the hills on the broken floor of a lurchedover cabin a man halfsits halflies propped up by an old woman two wrinkled girls that might be young   chunks of coal flare in the hearth flicker in his face white and sagging as

5. The Singing Tower was erected in Florida as a monument to Edward W. Bok (1863–1930), Dutch-born editor and philanthropist, who is buried at its base. President Calvin Coolidge spoke at its dedication in 1929.

dough   blacken the cavedin mouth the taut throat the belly
swelled enormous with the wound he got working on the mine-
tipple[6]   the barefoot girl brings him a tincup of water   the
woman wipes sweat off his streaming face with a dirty denim
sleeve   the firelight flares in his eyes stretched big with fever
in the women's scared eyes and in the blanched faces of the
foreigners

without help in the valley hemmed by dark strike-silent hills
the man will die (my father died we know what it is like to see
a man die) the women will lay him out on the rickety cot the
miners will bury him

in the jail it's light too hot the steamheat hisses we talk through
the greenpainted iron bars to a tall white mustachioed old man
some smiling miners in shirtsleeves a boy   faces white from mining have already the tallowy look of jailfaces

foreigners what can we say to the dead?   foreigners what
can we say to the jailed?   the representative of the political
party talks fast through the bars join up with us and no other
union we'll send you tobacco candy solidarity our lawyers will write
briefs speakers will shout your names at meetings they'll carry
your names on cardboard on picketlines   the men in jail shrug
their shoulders smile thinly our eyes look in their eyes through the
bars   what can I say?   (in another continent I have seen the
faces looking out through the barred basement windows behind
the ragged sentry's boots I have seen before day the straggling
footsore prisoners herded through the streets limping between
bayonets   heard the volley

I have seen the dead lying out in those distant deeper valleys)
      what can we say to the jailed?

in the law's office we stand against the wall the law is a big
man with eyes angry in a big pumpkinface   who sits and stares
at us meddling foreigners through the door the deputies crane
with their guns   they stand guard at the mines   they
blockade the miners' soupkitchens they've cut off the road up the
valley   the hiredmen with guns stand ready to shoot (they have
made us foreigners in the land where we were born they are the
conquering army that has filtered into the country unnoticed they
have taken the hilltops by stealth they levy toll they stand at the
minehead   they stand at the polls   they stand by when the
bailiffs carry the furniture of the family evicted from the city tenement out on the sidewalk they are there when the bankers foreclose
on a farm they are ambushed and ready to shoot down the strikers

---

6. The apparatus at a coal mine which tips cars at an angle to unload the coal.

marching behind the flag up the switchback road to the mine
those that the guns spare they jail)

the law stares across the desk out of angry eyes his face reddens
in splotches like a gobbler's neck with the strut of the power of
submachineguns sawedoffshotguns teargas and vomitinggas the
power that can feed you or leave you to starve

sits easy at his desk his back is covered he feels strong behind
him he feels the prosecutingattorney the judge an owner himself
the political boss the minesuperintendent the board of directors
the president of the utility the manipulator of the holding-
company

he lifts his hand towards the telephone
the deputies crowd in the door
we have only words against

Power Superpower

In eighteen eighty when Thomas Edison's[7] agent was hooking
up the first telephone in London, he put an ad in the paper for
a secretary and stenographer. The eager young cockney with sprouting muttonchop whiskers who answered it[8]

had recently lost his job as officeboy. In his spare time he had
been learning shorthand and bookkeeping and taking dictation
from the editor of the English *Vanity Fair* at night and jotting
down the speeches in Parliament for the papers. He came of temperance smallshopkeeper stock; already he was butting his bullet-
head against the harsh structure of caste that doomed boys of his
class to a life of alpaca jackets, penmanship, subordination. To
get a job with an American firm was to put a foot on the rung of
a ladder that led up into the blue.

He did his best to make himself indispensable; they let him
operate the switchboard for the first half-hour when the telephone
service was opened. Edison noticed his weekly reports on the
electrical situation in England

and sent for him to be his personal secretary.

Samuel Insull landed in America on a raw March day in eighty-
one. Immediately he was taken out to Menlo Park, shown about
the little group of laboratories, saw the strings of electriclightbulbs
shining at intervals across the snowy lots, all lit from the world's
first central electric station. Edison put him right to work and he
wasn't through till midnight. Next morning at six he was on the
job; Edison had no use for any nonsense about hours or vacations.
Insull worked from that time on until he was seventy without a

7. Thomas A. Edison (1847–1931), inventor of the phonograph whose engineering and entrepreneurial skills paved
the way for the commercial development of the electric light and the organization of the motion-picture industry. His laboratory was at Menlo Park,
New Jersey.
8. Samuel Insull (1859–1938), British-
born American utilities executive.

break; no nonsense about hours or vacations. Electric power turned the ladder into an elevator.

Young Insull made himself indispensable to Edison and took more and more charge of Edison's business deals. He was tireless, ruthless, reliable as the tides, Edison used to say, and fiercely determined to rise.

In ninetytwo he induced Edison to send him to Chicago and put him in as President of the Chicago Edison Company. Now he was on his own. *My engineering,* he said once in a speech, when he was sufficiently czar of Chicago to allow himself the luxury of plain speaking, *has been largely concerned with engineering all I could out of the dollar.*

He was a stiffly arrogant redfaced man with a closecropped mustache; he lived on Lake Shore Drive and was at the office at 7:10 every morning. It took him fifteen years to merge the five electrical companies into the Commonwealth Edison Company. *Very early I discovered that the first essential, as in other public utility business, was that it should be operated as a monopoly.*

When his power was firm in electricity he captured gas, spread out into the surrounding townships in northern Illinois. When politicians got in his way, he bought them, when laborleaders got in his way he bought them. Incredibly his power grew. He was scornful of bankers, lawyers were his hired men. He put his own lawyer in as corporation counsel and through him ran Chicago. When he found to his amazement that there were men (even a couple of young lawyers, Richberg and Ickes)[9] in Chicago that he couldn't buy, he decided he'd better put on a show for the public;

> Big Bill Thompson, the Builder:[1]
> *punch King George in the nose,*
> the hunt for the treeclimbing fish,
> the Chicago Opera.

It was too easy; the public had money, there was one of them born every minute,[2] with the founding of Middlewest Utilities in nineteen twelve Insull began to use the public's money to spread his empire. His companies began to have open stockholders' meetings, to ballyhoo service, the small investor could sit there all day hearing the bigwigs talk. It's fun to be fooled. Companyunions hypnotized his employees; everybody had to buy stock in his com-

---

9. Donald R. Richberg (1881–1960), Washington bureaucrat and leading architect of President Franklin D. Roosevelt's New Deal; Harold L. Ickes (1874–1952), Progressive party politician, Secretary of Labor and director of the Public Works Administration under President Roosevelt.
1. William H. Thompson (1868–1944), flamboyant politician, self-styled "the Builder," repeatedly Mayor of Chicago between 1915 and 1930, fanatical anglophobe.
2. "There's a sucker born every minute": a proverbial saying of Phineas T. Barnum (1810–91), the American circus impresario and showman.

panies, employees had to go out and sell stock, officeboys, linemen, trolleyconductors. Even Owen D. Young[3] was afraid of him. *My experience is that the greatest aid in the efficiency of labor is a long line of men waiting at the gate.*

War shut up the progressives (no more nonsense about trustbusting, controlling monopoly, the public good) and raised Samuel Insull to the peak.

He was head of the Illinois State Council of Defense. *Now,* he said delightedly, *I can do anything I like.* With it came the perpetual spotlight, the purple taste of empire. If anybody didn't like what Samuel Insull did he was a traitor. Chicago damn well kept its mouth shut.

The Insull companies spread and merged put competitors out of business until Samuel Insull and his stooge brother Martin controlled through the leverage of holdingcompanies and directorates and blocks of minority stock.

light and power, coalmines and tractioncompanies

in Illinois, Michigan, the Dakotas, Nebraska, Arkansas, Oklahoma, Missouri, Maine, Kansas, Wisconsin, Virginia, Ohio, North Carolina, Indiana, New York, New Jersey, Texas, in Canada, in Louisiana, in Georgia, in Florida and Alabama.

(It has been figured out that one dollar in Middle West Utilities controlled seventeen hundred and fifty dollars invested by the public in the subsidiary companies that actually did the work of producing electricity. With the delicate lever of a voting trust controlling the stock of the two top holdingcompanies he controlled a twelfth of the power output of America.)

Samuel Insull began to think he owned all that the way a man owns the roll of bills in his back pocket.

Always he'd been scornful of bankers. He owned quite a few in Chicago. But the New York bankers were laying for him; they felt he was a bounder, whispered that this financial structure was unsound. Fingers itched to grasp the lever that so delicately moved this enormous power over lives,

superpower, Insull liked to call it.

A certain Cyrus S. Eaton[4] of Cleveland, an exBaptistminister, was the David that brought down this Goliath. Whether it was so or not he made Insull believe that Wall Street was behind him.

He started buying stock in the three Chicago utilities. Insull in a panic for fear he'd lose his control went into the market to buy against him. Finally the Reverend Eaton let himself be bought

3. Owen D. Young (1874–1962), lawyer and financier, organizer of R.C.A. in 1919, and head of General Electric Corporation in 1922.

4. Cyrus S. Eaton (b. 1883), Canadian-born financier, utilities magnate, founder of Republic Steel.

out, shaking down the old man for a profit of twenty million dollars.

The stockmarket crash.

Paper values were slipping. Insull's companies were intertwined in a tangle that no bookkeeper has ever been able to unravel.

The gas hissed out of the torn balloon. Insull threw away his imperial pride and went on his knees to the bankers.

The bankers had him where they wanted him. To save the face of the tottering czar he was made a receiver of his own concerns. But the old man couldn't get out of his head the illusion that the money was all his. When it was discovered that he was using the stockholders' funds to pay off his brothers' brokerage accounts it was too thick even for a federal judge. Insull was forced to resign.

He held directorates in eightyfive companies, he was chairman of sixtyfive, president of eleven: it took him three hours to sign his resignations.

As a reward for his services to monopoly his companies chipped in on a pension of eighteen thousand a year. But the public was shouting for criminal prosecution. When the handouts stopped newspapers and politicians turned on him. Revolt against the moneymanipulators was in the air. Samuel Insull got the wind up and ran off to Canada with his wife.

Extradition proceedings. He fled to Paris. When the authorities began to close in on him there he slipped away to Italy, took a plane to Tirana, another to Saloniki and then the train to Athens. There the old fox went to earth. Money talked as sweetly in Athens as it had in Chicago in the old days.

The American ambassador tried to extradite him. Insull hired a chorus of Hellenic lawyers and politicos and sat drinking coffee in the lobby of the Grande Bretagne, while they proceeded to tie up the ambassador in a snarl of chicanery as complicated as the bookkeeping of his holdingcompanies. The successors of Demosthenes[5] were delighted. The ancestral itch in many a Hellenic palm was temporarily assuaged. Samuel Insull settled down cozily in Athens, was stirred by the sight of the Parthenon, watched the goats feeding on the Pentelic slopes, visited the Areopagus, admired marble fragments ascribed to Phidias, talked with the local bankers about reorganizing the public utilities of Greece, was said to be promoting Macedonian lignite. He was the toast of the Athenians; Mme. Kouryoumdjouglou the vivacious wife of a Bagdad datemerchant devoted herself to his comfort. When the first effort

---

5. Famed Athenian orator of the fifth century B.C. References in the succeeding lines are to sites in or near Athens (the Temple of Athens, the slopes of Mount Pentelicus, and the hill Areopagus after which a quasi-judicial elite tribunal was named) and to Phidias, a famous fifth-century-B.C. Athenian sculptor.

at extradition failed, the old gentleman declared in the courtroom, as he struggled out from the embraces of his four lawyers: *Greece is a small but great country.*

The idyll was interrupted when the Roosevelt Administration began to put the heat on the Greek foreign office. Government lawyers in Chicago were accumulating truckloads of evidence and chalking up more and more drastic indictments.

Finally after many a postponement (he had hired physicians as well as lawyers, they cried to high heaven that it would kill him to leave the genial climate of the Attic plain),

he was ordered to leave Greece as an undesirable alien to the great indignation of Balkan society and of Mme. Kouryoumdjouglou.

He hired the *Maiotis* a small and grubby Greek freighter and panicked the foreignnews services by slipping off for an unknown destination.

It was rumored that the new Odysseus was bound for Aden, for the islands of the South Seas, that he'd been invited to Persia. After a few days he turned up rather seasick in the Bosporus on his way, it was said, to Rumania where Madame Kouryoumdjouglou had advised him to put himself under the protection of her friend la Lupescu.[6]

At the request of the American ambassador the Turks were delighted to drag him off the Greek freighter and place him in a not at all comfortable jail. Again money had been mysteriously wafted from England, the healing balm began to flow, lawyers were hired, interpreters expostulated, doctors made diagnoses;

but Angora[7] was boss

and Insull was shipped off to Smyrna to be turned over to the assistant federal districtattorney who had come all that way to arrest him.

The Turks wouldn't even let Mme. Kouryoumdjouglou, on her way back from making arrangements in Bucharest, go ashore to speak to him. In a scuffle with the officials on the steamboat the poor lady was pushed overboard

and with difficulty fished out of the Bosporus.

Once he was cornered the old man let himself tamely be taken home on the *Exilona,* started writing his memoirs, made himself agreeable to his fellow passengers, was taken off at Sandy Hook and rushed to Chicago to be arraigned.

In Chicago the government spitefully kept him a couple of nights in jail; men he'd never known, so the newspapers said, stepped forward to go on his twohundredandfiftythousanddollar bail. He was moved to a hospital that he himself had endowed.

---

6. Madame Magda Lupescu (b. 1902), Rumanian adventuress, the "Cleopatra of the Near East," mistress of King   Carol II of Rumania.
7. City in central Turkey.

Solidarity. The leading businessmen in Chicago were photographed visiting him there. Henry Ford[8] paid a call.

The trial was very beautiful. The prosecution got bogged in finance technicalities. The judge was not unfriendly. The Insulls stole the show.

They were folks, they smiled at reporters, they posed for photographers, they went down to the courtroom by bus. Investors might have been ruined but so, they allowed it to be known, were the Insulls; the captain had gone down with the ship.

Old Samuel Insull rambled amiably on the stand, told his life-story: from officeboy to powermagnate, his struggle to make good, his love for his home and the kiddies. He didn't deny he'd made mistakes; who hadn't, but they were honest errors. Samuel Insull wept. Brother Martin wept. The lawyers wept. With voices choked with emotion headliners of Chicago business told from the witnessstand how much Insull had done for business in Chicago. There wasn't a dry eye in the jury.

Finally driven to the wall by the prosecutingattorney Samuel Insull blurted out that yes, he had made an error of some ten million dollars in accounting but that it had been an honest error.

Verdict: Not Guilty.

Smiling through their tears the happy Insulls went to their towncar amid the cheers of the crowd. Thousands of ruined investors, at least so the newspapers said, who had lost their life savings sat crying over the home editions at the thought of how Mr. Insull had suffered. The bankers were happy, the bankers had moved in on the properties.

In an odor of sanctity the deposed monarch of superpower, the officeboy who made good, enjoys his declining years spending the pension of twenty-one hundred a year that the directors of his old companies dutifully restored to him. *After fifty years of work*, he said, *my job is gone*.

                                                                    1936, 1937

F. SCOTT FITZGERALD
1896–1940

In Fitzgerald's essay *The Crack-Up*, where he traces the process of emotional disintegration that followed his spectacular early success as a writer and moneymaker, he declares that "the test of a first-rate intelligence is the ability to hold two opposed ideas in the mind at the same time, and still retain the ability to function."[1] Fitzgerald's own career may be viewed as a series of struggles to achieve this ideal convergence

1. *The Crack-Up*, p. 69.

of intellectual tension and productive capacity. His work at its best sustains the tensions that his personal life finally buckled under: his dual roles as representative and critic of the decade he labeled "The Jazz Age"; his fascination with wealth, along with his horror at the "vast carelessness" and varieties of dehumanization which afflict its possessors; his romantic idealization of sexual love along with his suspicion of its economic underpinnings and its risks of self-destruction. As the critic Lionel Trilling remarked, Fitzgerald's success lies less in his narrative power than in the delicate and elegant "voice of his prose,"[2] which rarely loses its irony in his involvement and never sacrifices its sympathy in his detachment.

Francis Scott Key Fitzgerald was born in St. Paul, Minnesota, in 1896. His father never attained the social status of his maternal ancestors—southern landowners and legislators from the early colonial period—but he retained their upper-class manners despite numerous commercial failures. His mother, the daughter of a prosperous Irish immigrant, was a dominant and eccentric woman whose devotion to her only son grew more fervent as the family moved from apartment to apartment and city to city in order to keep marginally solvent. At fifteen Fitzgerald was sent to a Catholic boarding school in New Jersey (an aunt paid the tuition in the hope of gaining a more disciplined and pious nephew); two years later he enrolled at Princeton. There, he helped write and produce the Triangle Club's annual musical comedies and participated on or behind stage in campus theatricals, and wrote for the college literary magazines. Friendships with campus intellectuals—Professor Christian Gauss, Edmund Wilson, and John Peale Bishop, who later became prominent as critic and poet—nourished his ambition to become a writer. But neglect of academic study disqualified him from extracurricular activities, and he left college in 1917 to accept an army commission.

Stationed near Montgomery, Alabama, he fell in love with Zelda Sayre, a talented, high-spirited beauty. Enshrined in his idealizing imagination, she seemed entitled to the luxurious and romantic life they both desired; after their engagement her demands for his financial and professional success made for a protracted and tempestuous courtship. After his discharge from the army in 1919 Fitzgerald went to New York—the "land of ambition and success," he called it[3]—to make a fortune. He made less than that at his job with an advertising agency; his long-distance engagement became strained and Zelda soon broke it off. Fitzgerald then quit his job and returned to St. Paul to finish a novel entitled *The Romantic Egoist* which he had begun in college and had worked on in army training camps, the story of a flamboyant young man's coming of age at Princeton. It was published in 1920 as *This Side of Paradise*, and one week later, while the book's sales skyrocketed, Fitzgerald and Zelda Sayre were married.

This Side of Paradise sold forty thousand copies in its first year and Fitzgerald became a hero of popular culture. He looked back in the 1930s on his early success as an illusion-ridden period in his life when success came inexplicably but tempted him to believe in his personal destiny and enticed him to play a role, for which he was not equipped, as "spokesman" for his time and "its typical product."[4] Yet Fitzgerald was

---

2. *The Liberal Imagination*, p. 253.
3. Quoted by Arthur Mizener, *The Far Side of Paradise*, p. 86.
4. "My Lost City," *The Crack-Up*, p. 27.

better equipped than he recognized to represent his age and his profession. The monetary success of his first novel had enabled him to ransom his fairy princess and to re-create his life in the shape of his fantasy, producing in his personal experience a precarious and highly charged union of material success and idealization. This union of the material and the ideal corresponded to the quest for material affluence, excitement, and intellectual liberation that emerged among the privileged young in the 1920s. At the same time the tension between the monetary and the ideal, and Fitzgerald's scrupulously ambivalent treatment of it in such works as *The Rich Boy*, placed his fiction in line with that recurring theme in such earlier writers as Melville, Hawthorne, and Henry James and lent his writing an historical significance beyond its value as a document of the Twenties. The hero of one masterpiece, Jay Gatsby, is a wealthy gangster who longs to win a sweetheart in the glamorous upper-class society of Long Island, but he is also a midwestern boy who is enchanted, like Benjamin Franklin earlier, by the American dream of success and recognition; he is a person whose identity is created by his ideal "Platonic conception of himself" and whose beloved seems an incarnation of his "unutterable visions."[5] His tragic career is not only a phenomenon of the Jazz Age but a haunting version of America's history and Fitzgerald's own life. Gatsby's illusions render him vulnerable to delusion and the disdain of others, but they give him a dignity and pathos that is not matched in the fashionably wealthy society, feckless and deceptively elegant, that excludes him.

The Fitzgeralds lived extravagantly in New York City, in St. Paul, and on Long Island, spending money faster than two collections of short stories (*Flappers and Philosophers*, 1921, and *Tales of the Jazz Age*, 1922) and his second novel (*The Beautiful and Damned*, 1922) could earn it. For two and a half years after 1924 the Fitzgeralds lived in Europe among such expatriated Americans as Hemingway, Gertrude Stein, and Ezra Pound. During this time *The Great Gatsby* appeared (1925) and in 1926 he published *All the Sad Young Men*, which, along with his last volume of stories, *Taps at Reveille* (1935), contains his best short fiction. The market for short fiction was a lucrative one, and throughout the Twenties and Thirties Fitzgerald often allowed the prospect of easy money to seduce him into writing substandard stories. This practice gradually eroded his self-esteem and contributed to his subsequent bouts of despair and profligacy.

In December, 1926, the family returned to the United States. Fitzgerald stopped briefly in Hollywood to fulfill his first assignment as a screen writer. Then he settled with Zelda and their daughter, Scottie, near Wilmington, Delaware, where financial problems, Fitzgerald's alcoholism, and Zelda's frustrated artistic ambitions dominated their lives. (She was a writer whose talent Fitzgerald denigrated and discouraged, and in 1928 she became obsessed with a belated desire to become a ballerina). Back in Europe in 1930, Zelda suffered the first in a series of breakdowns which necessitated her residence in mental institutions for most of the remaining seventeen years of her life. Fitzgerald returned permanently to the United States in 1931, establishing residence eventually in the vicinity of Baltimore to be near Zelda's hospitals. In 1934 he published his

5. *The Great Gatsby*, pp. 99, 112.

second masterpiece, *Tender Is the Night,* which traces the decline of a young American psychiatrist whose personal energies are sapped, and his professional career corroded, by his marriage to a beautiful and wealthy patient. The protagonist's attempt to salvage the rituals of domestic harmony and social affability in a community of expatriates on the French Riviera is an elegy to values that are vanishing in his society.

In 1937, after several alcoholic and infirm years, when his income from writing all but disappeared, Fitzgerald went to work regularly as a screenwriter in Hollywood. His salaries enabled him to recoup his finances and provide for the education of his daughter, and, bolstered by a love affair with the journalist Sheilah Graham, he hoped to revive his career as a writer of fiction. He died of a heart attack at the age of forty-four in Hollywood, leaving unfinished his novel about a self-made mogul of the film industry, *The Last Tycoon.*

The posthumous publication of *The Last Tycoon* in 1941, and of the collection of essays and miscellanies, *The Crack-Up* (edited by his friend Edmund Wilson), in 1945, revived critical interest in Fitzgerald and secured his reputation as a celebrant, and a diagnostician, of twentieth-century manners and morals. As he learned to discipline the self-indulgence and extravagance of his early novels, he achieved a firmness of narrative structure that he modeled on the fiction of the British novelist Joseph Conrad and that of Henry James: intelligent and wary, but sensitive and sympathetic, observers (like the narrator in *The Rich Boy*) enable him to focus on the fascinating behavior of his protagonists and the enchanting surfaces of their worlds, while exposing beneath them the ambiguous motives, the violence that threatens, the affluence that paralyzes while seeming to release affection, the presumption to dominate that makes one incapable of love.

The Rich Boy is printed from *All the Sad Young Men.*

The Rich Boy

Begin with an individual, and before you know it you find that you have created a type; begin with a type, and you find that you have created—nothing. That is because we are all queer fish, queerer behind our faces and voices than we want any one to know or than we know ourselves. When I hear a man proclaiming himself an "average, honest, open fellow," I feel pretty sure that he has some definite and perhaps terrible abnormality which he has agreed to conceal—and his protestation of being average and honest and open is his way of reminding himself of his misprision.

There are no types, no plurals. There is a rich boy, and this is his and not his brothers' story. All my life I have lived among his brothers but this one has been my friend. Besides, if I wrote about his brothers I should have to begin by attacking all the lies that the poor have told about the rich and the rich have told about themselves—such a wild structure they have erected that when we pick up a book about the rich, some instinct prepares us for unreality. Even the intelligent and impassioned reporters of life have made the country of the rich as unreal as fairy-land.

Let me tell you about the very rich. They are different from you and me. They possess and enjoy early, and it does something to them, makes them soft where we are hard, and cynical where we are trustful, in a way that, unless you were born rich, it is very difficult to understand. They think, deep in their hearts, that they are better than we are because we had to discover the compensations and refuges of life for ourselves. Even when they enter deep into our world or sink below us, they still think that they are better than we are. They are different. The only way I can describe young Anson Hunter is to approach him as if he were a foreigner and cling stubbornly to my point of view. If I accept his for a moment I am lost—I have nothing to show but a preposterous movie.

II

Anson was the eldest of six children who would some day divide a fortune of fifteen million dollars, and he reached the age of reason—is it seven?—at the beginning of the century when daring young women were already gliding along Fifth Avenue in electric "mobiles." In those days he and his brother had an English governess who spoke the language very clearly and crisply and well, so that the two boys grew to speak as she did—their words and sentences were all crisp and clear and not run together as ours are. They didn't talk exactly like English children but acquired an accent that is peculiar to fashionable people in the city of New York.

In the summer the six children were moved from the house on 71st Street to a big estate in northern Connecticut. It was not a fashionable locality—Anson's father wanted to delay as long as possible his children's knowledge of that side of life. He was a man somewhat superior to his class, which composed New York society, and to his period, which was the snobbish and formalized vulgarity of the Gilded Age,[1] and he wanted his sons to learn habits of concentration and have sound constitutions and grow up into right-living and successful men. He and his wife kept an eye on them as well as they were able until the two older boys went away to school, but in huge establishments this is difficult—it was much simpler in the series of small and medium-sized houses in which my own youth was spent—I was never far out of the reach of my mother's voice, of the sense of her presence, her approval or disapproval.

Anson's first sense of his superiority came to him when he realized the half-grudging American deference that was paid to him in the Connecticut village. The parents of the boys he played with always inquired after his father and mother, and were vaguely

1. Era of economic expansion and political corruption following the Civil War, named from Mark Twain's and Charles Dudley Warner's novel *The Gilded Age* (1873).

excited when their own children were asked to the Hunters' house. He accepted this as the natural state of things, and a sort of impatience with all groups of which he was not the centre—in money, in position, in authority—remained with him for the rest of his life. He disdained to struggle with other boys for precedence—he expected it to be given him freely, and when it wasn't he withdrew into his family. His family was sufficient, for in the East money is still a somewhat feudal thing, a clan-forming thing. In the snobbish West, money separates families to form "sets."

At eighteen, when he went to New Haven, Anson was tall and thick-set, with a clear complexion and a healthy color from the ordered life he had led in school. His hair was yellow and grew in a funny way on his head, his nose was beaked—these two things kept him from being handsome—but he had a confident charm and a certain brusque style, and the upper-class men who passed him on the street knew without being told that he was a rich boy and had gone to one of the best schools. Nevertheless, his very superiority kept him from being a success in college—the independence was mistaken for egotism, and the refusal to accept Yale standards with the proper awe seemed to belittle all those who had. So, long before he graduated, he began to shift the centre of his life to New York.

He was at home in New York—there was his own house with "the kind of servants you can't get any more"—and his own family, of which, because of his good humor and a certain ability to make things go, he was rapidly becoming the centre, and the débutante parties, and the correct manly world of the men's clubs, and the occasional wild spree with the gallant girls whom New Haven only knew from the fifth row.[2] His aspirations were conventional enough—they included even the irreproachable shadow he would some day marry, but they differed from the aspirations of the majority of young men in that there was no mist over them, none of that quality which is variously known as "idealism" or "illusion." Anson accepted without reservation the world of high finance and high extravagance, of divorce and dissipation, of snobbery and of privilege. Most of our lives end as a compromise—it was as a compromise that his life began.

He and I first met in the late summer of 1917 when he was just out of Yale, and, like the rest of us, was swept up into the systematized hysteria of the war. In the blue-green uniform of the naval aviation he came down to Pensacola, where the hotel orchestras played "I'm sorry, dear," and we young officers danced with the girls. Every one liked him, and though he ran with the drinkers and wasn't an especially good pilot, even the instructors treated him with a certain respect. He was always having

---

2. I.e., chorus girls or performers in burlesque theaters.

long talks with them in his confident, logical voice—talks which ended by his getting himself, or, more frequently, another officer, out of some impending trouble. He was convivial, bawdy, robustly avid for pleasure, and we were all surprised when he fell in love with a conservative and rather proper girl.

Her name was Paula Legendre, a dark, serious beauty from somewhere in California. Her family kept a winter residence just outside of town, and in spite of her primness she was enormously popular; there is a large class of men whose egotism can't endure humor in a woman. But Anson wasn't that sort, and I couldn't understand the attraction of her "sincerity"—that was the thing to say about her—for his keen and somewhat sardonic mind.

Nevertheless, they fell in love—and on her terms. He no longer joined the twilight gathering at the De Sota bar, and whenever they were seen together they were engaged in a long, serious dialogue, which must have gone on several weeks. Long afterward he told me that it was not about anything in particular but was composed on both sides of immature and even meaningless statements—the emotional content that gradually came to fill it grew up not out of the words but out of its enormous seriousness. It was a sort of hypnosis. Often it was interrupted, giving way to that emasculated humor we call fun; when they were alone it was resumed again, solemn, low-keyed, and pitched so as to give each other a sense of unity in feeling and thought. They came to resent any interruptions of it, to be unresponsive to facetiousness about life, even to the mild cynicism of their contemporaries. They were only happy when the dialogue was going on, and its seriousness bathed them like the amber glow of an open fire. Toward the end there came an interruption they did not resent—it began to be interrupted by passion.

Oddly enough, Anson was as engrossed in the dialogue as she was and as profoundly affected by it, yet at the same time aware that on his side much was insincere, and on hers much was merely simple. At first, too, he despised her emotional simplicity as well, but with his love her nature deepened and blossomed, and he could despise it no longer. He felt that if he could enter into Paula's warm safe life he would be happy. The long preparation of the dialogue removed any constraint—he taught her some of what he had learned from more adventurous women, and she responded with a rapt holy intensity. One evening after a dance they agreed to marry, and he wrote a long letter about her to his mother. The next day Paula told him that she was rich, that she had a personal fortune of nearly a million dollars.

III

It was exactly as if they could say "Neither of us has anything: we shall be poor together"—just as delightful that they should be rich instead. It gave them the same communion of adventure. Yet when Anson got leave in April, and Paula and her mother accompanied him North, she was impressed with the standing of his family in New York and with the scale on which they lived. Alone with Anson for the first time in the rooms where he had played as a boy, she was filled with a comfortable emotion, as though she were pre-eminently safe and taken care of. The pictures of Anson in a skull cap at his first school, of Anson on horseback with the sweetheart of a mysterious forgotten summer, of Anson in a gay group of ushers and bridesmaid at a wedding, made her jealous of his life apart from her in the past, and so completely did his authoritative person seem to sum up and typify these possessions of his that she was inspired with the idea of being married immediately and returning to Pensacola as his wife.

But an immediate marriage wasn't discussed—even the engagement was to be secret until after the war. When she realized that only two days of his leave remained, her dissatisfaction crystallized in the intention of making him as unwilling to wait as she was. They were driving to the country for dinner and she determined to force the issue that night.

Now a cousin of Paula's was staying with them at the Ritz,[3] a severe, bitter girl who loved Paula but was somewhat jealous of her impressive engagement, and as Paula was late in dressing, the cousin, who wasn't going to the party, received Anson in the parlor of the suite.

Anson had met friends at five o'clock and drunk freely and indiscreetly with them for an hour. He left the Yale Club at a proper time, and his mother's chauffeur drove him to the Ritz, but his usual capacity was not in evidence, and the impact of the steam-heated sitting-room made him suddenly dizzy. He knew it, and he was both amused and sorry.

Paula's cousin was twenty-five, but she was exceptionally naïve, and at first failed to realize what was up. She had never met Anson before, and she was surprised when he mumbled strange information and nearly fell off his chair, but until Paula appeared it didn't occur to her that what she had taken for the odor of a dry-cleaned uniform was really whiskey. But Paula understood as soon as she appeared; her only thought was to get Anson away before her mother saw him, and at the look in her eyes the cousin understood too.

When Paula and Anson descended to the limousine they found two men inside, both asleep; they were the men with whom he had been drinking at the Yale Club, and they were also going to the party. He had entirely forgotten their presence in the car. On the way to Hempstead[4] they awoke and sang. Some of the songs were rough, and though Paula tried to reconcile herself to the fact

---

3. Swank hotel in New York City.   4. Town on Long Island.

that Anson had few verbal inhibitions, her lips tightened with shame and distaste.

Back at the hotel the cousin, confused and agitated, considered the incident, and then walked into Mrs. Legendre's bedroom, saying: "Isn't he funny?"

"Who is funny?"

"Why—Mr. Hunter. He seemed so funny."

Mrs. Legendre looked at her sharply.

"How is he funny?"

"Why, he said he was French. I didn't know he was French."

"That's absurd. You must have misunderstood." She smiled: "It was a joke."

The cousin shook her head stubbornly.

"No. He said he was brought up in France. He said he couldn't speak any English, and that's why he couldn't talk to me. And he couldn't!"

Mrs. Legendre looked away with impatience just as the cousin added thoughtfully, "Perhaps it was because he was so drunk," and walked out of the room.

This curious report was true. Anson, finding his voice thick and uncontrollable, had taken the unusual refuge of announcing that he spoke no English. Years afterward he used to tell that part of the story, and he invariably communicated the uproarious laughter which the memory aroused in him.

Five times in the next hour Mrs. Legendre tried to get Hempstead on the phone. When she succeeded, there was a ten-minute delay before she heard Paula's voice on the wire.

"Cousin Jo told me Anson was intoxicated."

"Oh, no. . . ."

"Oh, yes. Cousin Jo says he was intoxicated. He told her he was French, and fell off his chair and behaved as if he was very intoxicated. I don't want you to come home with him."

"Mother, he's all right! Please don't worry about ——"

"But I do worry. I think it's dreadful. I want you to promise me not to come home with him."

"I'll take care of it, mother. . . ."

"I don't want you to come home with him."

"All right, mother. Good-by."

"Be sure now, Paula. Ask some one to bring you."

Deliberately Paula took the receiver from her ear and hung it up. Her face was flushed with helpless annoyance. Anson was stretched asleep out in a bedroom up-stairs, while the dinner-party below was proceeding lamely toward conclusion.

The hour's drive had sobered him somewhat—his arrival was merely hilarious—and Paula hoped that the evening was not spoiled, after all, but two imprudent cocktails before dinner completed the disaster. He talked boisterously and somewhat offensively to the party at large for fifteen minutes, and then slid silently under the table; like a man in an old print—but, unlike an old print, it was rather horrible without being at all quaint. None of the young girls present remarked upon the incident—it seemed to merit only silence. His uncle and two other men carried him up-stairs, and it was just after this that Paula was called to the phone.

An hour later Anson awoke in a fog of nervous agony, through which he perceived after a moment the figure of his uncle Robert standing by the door.

". . . I said are you better?"

"What?"

"Do you feel better, old man?"

"Terrible," said Anson.

"I'm going to try you on another bromo-seltzer. If you can hold it down, it'll do you good to sleep."

With an effort Anson slid his legs from the bed and stood up.

"I'm all right," he said dully.

"Take it easy."

"I thin' if you gave me a glassbrandy I could go down-stairs."

"Oh, no——"

"Yes, that's the only thin'. I'm all right now. . . . I suppose I'm in Dutch[5] dow' there."

"They know you're a little under the weather," said his uncle deprecatingly. "But don't worry about it. Schuyler didn't even get here. He passed away in the locker-room over at the Links."[6]

Indifferent to any opinion, except Paula's, Anson was nevertheless determined to save the débris of the evening, but when after a cold bath he made his appearance most of the party had already left. Paula got up immediately to go home.

In the limousine the old serious dialogue began. She had known that he drank, she admitted, but she had never expected anything like this—it seemed to her that perhaps they were not suited to each other, after all. Their ideas about life were too different, and so forth. When she finished speaking, Anson spoke in turn, very soberly. Then Paula said she'd have to think it over; she wouldn't decide to-night; she was not angry but she was terribly sorry. Nor would she let him come into the hotel with her, but just before she got out of the car she leaned and kissed him unhappily on the cheek.

The next afternoon Anson had a long talk with Mrs. Legendre while Paula sat listening in silence. It was agreed that Paula was to brood over the incident for a proper period and then, if mother and daughter thought it best, they would follow Anson to Pensacola. On his part he apologized with sincerity and dignity—that was all; with every card in her hand Mrs. Legendre was unable to

5. Slang: in trouble.　　　　6. Golf course.

establish any advantage over him. He made no promises, showed no humility, only delivered a few serious comments on life which brought him off with rather a moral superiority at the end. When they came South three weeks later, neither Anson in his satisfaction nor Paula in her relief at the reunion realized that the psychological moment had passed forever.

IV

He dominated and attracted her, and at the same time filled her with anxiety. Confused by his mixture of solidity and self-indulgence, of sentiment and cynicism—incongruities which her gentle mind was unable to resolve—Paula grew to think of him as two alternating personalities. When she saw him alone, or at a formal party, or with his casual inferiors, she felt a tremendous pride in his strong, attractive presence, the paternal, understanding stature of his mind. In other company she became uneasy when what had been a fine imperviousness to mere gentility showed its other face. The other face was gross, humorous, reckless of everything but pleasure. It startled her mind temporarily away from him, even led her into a short covert experiment with an old beau, but it was no use—after four months of Anson's enveloping vitality there was an anæmic pallor in all other men.

In July he was ordered abroad, and their tenderness and desire reached a crescendo. Paula considered a last-minute marriage— decided against it only because there were always cocktails on his breath now, but the parting itself made her physically ill with grief. After his departure she wrote him long letters of regret for the days of love they had missed by waiting. In August Anson's plane slipped down into the North Sea. He was pulled onto a destroyer after a night in the water and sent to hospital with pneumonia; the armistice was signed before he was finally sent home.

Then, with every opportunity given back to them, with no material obstacle to overcome, the secret weavings of their temperaments came between them, drying up their kisses and their tears, making their voices less loud to one another, muffling the intimate chatter of their hearts until the old communcation was only possible by letters, from far away. One afternoon a society reporter waited for two hours in the Hunters' house for a confirmation of their engagement. Anson denied it; nevertheless an early issue carried the report as a leading paragraph—they were "constantly seen together at Southampton, Hot Springs, and Tuxedo Park."[7] But the serious dialogue had turned a corner into a long-sustained quarrel, and the affair was almost played out. Anson got drunk flagrantly and missed an engagement with her, whereupon Paula made certain behavioristic demands. His despair was helpless before

7. A wealthy Long Island suburb, an Arkansas resort, and a resort near New York City, respectively.

his pride and his knowledge of himself; the engagement was definitely broken.

"Dearest," said their letters now, "Dearest, Dearest, when I wake up in the middle of the night and realize that after all it was not to be, I feel that I want to die. I can't go on living any more. Perhaps when we meet this summer we may talk things over and decide differently—we were so excited and sad that day, and I don't feel that I can live all my life without you. You speak of other people. Don't you know there are no other people for me, but only you. . . ."

But as Paula drifted here and there around the East she would sometimes mention her gaieties to make him wonder. Anson was too acute to wonder. When he saw a man's name in her letters he felt more sure of her and a little disdainful—he was always superior to such things. But he still hoped that they would some day marry.

Meanwhile he plunged vigorously into all the movement and glitter of post-bellum[8] New York, entering a brokerage house, joining half a dozen clubs, dancing late, and moving in three worlds— his own world, the world of young Yale graduates, and that section of the half-world which rests one end on Broadway. But there was always a thorough and infractible eight hours devoted to his work in Wall Street, where the combination of his influential family connection, his sharp intelligence, and his abundance of sheer physical energy brought him almost immediately forward. He had one of those invaluable minds with partitions in it; sometimes he appeared at his office refreshed by less than an hour's sleep, but such occurrences were rare. So early as 1920 his income in salary and commissions exceeded twelve thousand dollars.

As the Yale tradition slipped into the past he became more and more of a popular figure among his classmates in New York, more popular than he had ever been in college. He lived in a great house, and had the means of introducing young men into other great houses. Moreover, his life already seemed secure, while theirs, for the most part, had arrived again at precarious beginnings. They commenced to turn to him for amusement and escape, and Anson responded readily, taking pleasure in helping people and arranging their affairs.

There were no men in Paula's letters now, but a note of tenderness ran through them that had not been there before. From several sources he heard that she had "a heavy beau,"[9] Lowell Thayer, a Bostonian of wealth and position, and though he was sure she still loved him, it made him uneasy to think that he might lose her, after all. Save for one unsatisfactory day she had not been in New York for almost five months, and as the rumors

---

8. Postwar, after World War I.   9. A very attentive boyfriend.

multiplied he became increasingly anxious to see her. In February he took his vacation and went down to Florida.

Palm Beach sprawled plump and opulent between the sparkling sapphire of Lake Worth, flawed here and there by house-boats at anchor, and the great turquoise bar of the Atlantic Ocean. The huge bulks of the Breakers and the Royal Poinciana[1] rose as twin paunches from the bright level of the sand, and around them clustered the Dancing Glade, Bradley's House of Chance, and a dozen modistes and milliners with goods at triple prices from New York. Upon the trellissed veranda of the Breakers two hundred women stepped right, stepped left, wheeled, and slid in that then celebrated calisthenic known as the double-shuffle, while in half-time to the music two thousand bracelets clicked up and down on two hundred arms.

At the Everglades Club after dark Paula and Lowell Thayer and Anson and a casual fourth played bridge with hot cards. It seemed to Anson that her kind, serious face was wan and tired—she had been around now for four, five, years. He had known her for three.

"Two spades."

"Cigarette? . . . Oh, I beg your pardon. By me."

"By."

"I'll double three spades."

There were a dozen tables of bridge in the room, which was filling up with smoke. Anson's eyes met Paula's, held them persistently even when Thayer's glance fell between them. . . .

"What was bid?" he asked abstractedly.

"Rose of Washington Square"

sang the young people in the corners:

> *"I'm withering there*
> *In basement air——"*

The smoke banked like fog, and the opening of a door filled the room with blown swirls of ectoplasm. Little Bright Eyes streaked past the tables seeking Mr. Conan Doyle among the Englishmen who were posing as Englishmen about the lobby.[2]

"You could cut it with a knife."

". . . cut it with a knife."

". . . a knife."

At the end of the rubber Paula suddenly got up and spoke to Anson in a tense, low voice. With scarcely a glance at Lowell Thayer, they walked out the door and descended a long flight of

---

1. Swank hotels in Palm Beach, Florida.
2. To the narrator, the silly girl (Little Miss Bright Eyes) who seeks out the posturing Englishmen in the smoke-filled room resembles a character surrounded by the effusions (ectoplasm) of a spiritualistic medium in a mystery thriller by A. Conan Doyle (1859–1930), British author of Sherlock Holmes mystery stories.

stone steps—in a moment they were walking hand in hand along the moonlit beach.

"Darling, darling. . . ." They embraced recklessly, passionately, in a shadow. . . . Then Paula drew back her face to let his lips say what she wanted to hear—she could feel the words forming as they kissed again. . . . Again she broke away, listening, but as he pulled her close once more she realized that he had said nothing—only "*Darling! Darling!*" in that deep, sad whisper that always made her cry. Humbly, obediently, her emotions yielded to him and the tears streamed down her face, but her heart kept on crying: "Ask me—oh, Anson, dearest, ask me!"

"Paula. . . . *Paula!*"

The words wrung her heart like hands, and Anson, feeling her tremble, knew that emotion was enough. He need say no more, commit their destinies to no practical enigma. Why should he, when he might hold her so, biding his own time, for another year—forever? He was considering them both, her more than himself. For a moment, when she said suddenly that she must go back to her hotel, he hesitated, thinking, first, "This is the moment, after all," and then: "No, let it wait—she is mine. . . ."

He had forgotten that Paula too was worn away inside with the strain of three years. Her mood passed forever in the night.

He went back to New York next morning filled with a certain restless dissatisfaction. Late in April, without warning, he received a telegram from Bar Harbor in which Paula told him that she was engaged to Lowell Thayer, and that they would be married immediately in Boston. What he never really believed could happen had happened at last.

Anson filled himself with whiskey that morning, and going to the office, carried on his work without a break—rather with a fear if what would happen if he stopped. In the evening he went out as usual, saying nothing of what had occurred; he was cordial, humorous, unabstracted. But one thing he could not help—for three days, in any place, in any company, he would suddenly bend his head into his hands and cry like a child.

V

In 1922 when Anson went abroad with the junior partner to investigate some London loans, the journey intimated that he was to be taken into the firm. He was twenty-seven now, a little heavy without being definitely stout, and with a manner older than his years. Old people and young people liked him and trusted him, and mothers felt safe when their daughters were in his charge, for he had a way, when he came into a room, of putting himself on a footing with the oldest and most conservative people there. "You and I," he seemed to say, "we're solid. We understand."

He had an instinctive and rather charitable knowledge of the weaknesses of men and women, and, like a priest, it made him the more concerned for the maintenance of outward forms. It was typical of him that every Sunday morning he taught in a fashionable Episcopal Sunday-school—even though a cold shower and a quick change into a cutaway coat were all that separated him from the wild night before.

After his father's death he was the practical head of his family, and, in effect, guided the destinies of the younger children. Through a complication his authority did not extend to his father's estate, which was administrated by his Uncle Robert, who was the horsey member of the family, a good-natured, hard-drinking member of that set which centres about Wheatley Hills.

Uncle Robert and his wife, Edna, had been great friends of Anson's youth, and the former was disappointed when his nephew's superiority failed to take a horsey form. He backed him for a city club which was the most difficult in America to enter—one could only join if one's family had "helped to build up New York" (or, in other words, were rich before 1880)—and when Anson, after his election, neglected it for the Yale Club, Uncle Robert gave him a little talk on the subject. But when on top of that Anson declined to enter Robert Hunter's own conservative and somewhat neglected brokerage house, his manner grew cooler. Like a primary teacher who has taught all he knew, he slipped out of Anson's life.

There were so many friends in Anson's life—scarcely one for whom he had not done some unusual kindness and scarcely one whom he did not occasionally embarrass by his bursts of rough conversation or his habit of getting drunk whenever and however he liked. It annoyed him when any one else blundered in that regard—about his own lapses he was always humorous. Odd things happened to him and he told them with infectious laughter.

I was working in New York that spring, and I used to lunch with him at the Yale Club, which my university was sharing until the completion of our own. I had read of Paula's marriage, and one afternoon, when I asked him about her, something moved him to tell me the story. After that he frequently invited me to family dinners at his house and behaved as though there was a special relation between us, as though with his confidence a little of that consuming memory had passed into me.

I found that despite the trusting mothers, his attitude toward girls was not indiscriminately protective. It was up to the girl—if she showed an inclination toward looseness, she must take care of herself, even with him.

"Life," he would explain sometimes, "has made a cynic of me."

By life he meant Paula. Sometimes, especially when he was drinking, it became a little twisted in his mind, and he thought that she had callously thrown him over.

This "cynicism," or rather his realization that naturally fast girls were not worth sparing, led to his affair with Dolly Karger. It wasn't his only affair in those years, but it came nearest to touching him deeply, and it had a profound effect upon his attitude toward life.

Dolly was the daughter of a notorious "publicist"[3] who had married into society. She herself grew up into the Junior League, came out at the Plaza, and went to the Assembly;[4] and only a few old families like the Hunters could question whether or not she "belonged," for her picture was often in the papers, and she had more enviable attention than many girls who undoubtedly did. She was dark-haired, with carmine lips and a high, lovely color, which she concealed under pinkish-gray powder all through the first year out, because high color was unfashionable—Victorian-pale was the thing to be. She wore black, severe suits and stood with her hands in her pockets leaning a little forward, with a humorous restraint on her face. She danced exquisitely—better than anything she liked to dance—better than anything except making love. Since she was ten she had always been in love, and, usually, with some boy who didn't respond to her. Those who did—and there were many— bored her after a brief encounter, but for her failures she reserved the warmest spot in her heart. When she met them she would always try once more—sometimes she succeeded, more often she failed.

It never occurred to this gypsy of the unattainable that there was a certain resemblance in those who refused to love her—they shared a hard intuition that saw through to her weakness, not a weakness of emotion but a weakness of rudder. Anson perceived this when he first met her, less than a month after Paula's marriage. He was drinking rather heavily, and he pretended for a week that he was falling in love with her. Then he dropped her abruptly and forgot—immediately he took up the commanding position in her heart.

Like so many girls of that day Dolly was slackly and indiscreetly wild. The unconventionality of a slightly older generation had been simply one facet of a post-war movement to discredit obsolete manners—Dolly's was both older and shabbier, and she saw in Anson the two extremes which the emotionally shiftless woman seeks, an abandon to indulgence alternating with a protective strength. In his character she felt both the sybarite and the solid rock, and these two satisfied every need of her nature.

She felt that it was going to be difficult, but she mistook the reason—she thought that Anson and his family expected a more spectacular marriage, but she guessed immediately that her advantage lay in his tendency to drink.

They met at the large débutante dances, but as her infatuation

---

3. Public relations agent.
4. Annual dance for debutantes; "Junior League": organization of socialite women for charitable work and social diversion; "Plaza": New York City hotel, location of many socialite "coming out" parties.

increased they managed to be more and more together. Like most mothers, Mrs. Karger believed that Anson was exceptionally reliable, so she allowed Dolly to go with him to distant country clubs and suburban houses without inquiring closely into their activities or questioning her explanations when they came in late. At first these explanations might have been accurate, but Dolly's worldly ideas of capturing Anson were soon engulfed in the rising sweep of her emotion. Kisses in the back of taxis and motor-cars were no longer enough; they did a curious thing:

They dropped out of their world for a while and made another world just beneath it where Anson's tippling and Dolly's irregular hours would be less noticed and commented on. It was composed, this world, of varying elements—several of Anson's Yale friends and their wives, two or three young brokers and bond salesmen and a handful of unattached men, fresh from college, with money and a propensity to dissipation. What this world lacked in spaciousness and scale it made up for by allowing them a liberty that it scarcely permitted itself. Moreover it centred around them and permitted Dolly the pleasure of a faint condescension—a pleasure which Anson, whose whole life was a condescension from the certitudes of his childhood, was unable to share.

He was not in love with her, and in the long feverish winter of their affair he frequently told her so. In the spring he was weary— he wanted to renew his life at some other source—moreover, he saw that either he must break with her now or accept the responsibility of a definite seduction. Her family's encouraging attitude precipitated his decision—one evening when Mr. Karger knocked discreetly at the library door to announce that he had left a bottle of old brandy in the dining-room, Anson felt that life was hemming him in. That night he wrote her a short letter in which he told her that he was going on his vacation, and that in view of all the circumstances they had better meet no more.

It was June. His family had closed up the house and gone to the country, so he was living temporarily at the Yale Club. I had heard about his affair with Dolly as it developed—accounts salted with humor, for he despised unstable women, and granted them no place in the social edifice in which he believed—and when he told me that night that he was definitely breaking with her I was glad. I had seen Dolly here and there, and each time with a feeling of pity at the hopelessness of her struggle, and of shame at knowing so much about her that I had no right to know. She was what is known as "a pretty little thing," but there was a certain recklessness which rather fascinated me. Her dedication to the goddess of waste would have been less obvious had she been less spirited—she would most certainly throw herself away, but I was glad when I heard that the sacrifice would not be consummated in my sight.

Anson was going to leave the letter of farewell at her house next morning. It was one of the few houses left open in the Fifth Avenue district, and he knew that the Kargers, acting upon erroneous information from Dolly, had foregone a trip abroad to give their daughter her chance. As he stepped out the door of the Yale Club into Madison Avenue the postman passed him, and he followed back inside. The first letter that caught his eye was in Dolly's hand.

He knew what it would be—a lonely and tragic monologue, full of the reproaches he knew, the invoked memories, the "I wonder if's"—all the immemorial intimacies that he had communicated to Paula Legendre in what seemed another age. Thumbing over some bills, he brought it on top again and opened it. To his surprise it was a short, somewhat formal note, which said that Dolly would be unable to go to the country with him for the week-end, because Perry Hull from Chicago had unexpectedly come to town. It added that Anson had brought this on himself: "—if I felt that you loved me as I love you I would go with you at any time, any place, but Perry is *so* nice, and he so much wants me to marry him——"

Anson smiled contemptuously—he had had experience with such decoy epistles. Moreover, he knew how Dolly had labored over this plan, probably sent for the faithful Perry and calculated the time of his arrival—even labored over the note so that it would make him jealous without driving him away. Like most compromises, it had neither force nor vitality but only a timorous despair.

Suddenly he was angry. He sat down in the lobby and read it again. Then he went to the phone, called Dolly and told her in his clear, compelling voice that he had received her note and would call for her at five o'clock as they had previously planned. Scarcely waiting for the pretended uncertainty of her "Perhaps I can see you for an hour," he hung up the receiver and went down to his office. On the way he tore his own letter into bits and dropped it in the street.

He was not jealous—she meant nothing to him—but at her pathetic ruse everything stubborn and self-indulgent in him came to the surface. It was a presumption from a mental inferior and it could not be overlooked. If she wanted to know to whom she belonged she would see.

He was on the door-step at quarter past five. Dolly was dressed for the street, and he listened in silence to the paragraph of "I can only see you for an hour," which she had begun on the phone.

"Put on your hat, Dolly," he said, "we'll take a walk."

They strolled up Madison Avenue and over to Fifth while Anson's shirt dampened upon his portly body in the deep heat. He talked little, scolding her, making no love to her, but before they had walked six blocks she was his again, apologizing for the note, offering not to see Perry at all as an atonement, offering anything. She thought that he had come because he was beginning to love her.

"I'm hot," he said when they reached 71st Street. "This is a winter suit. If I stop by the house and change, would you mind waiting for me down-stairs? I'll only be a minute."

She was happy; the intimacy of his being hot, of any physical fact about him, thrilled her. When they came to the iron-grated door and Anson took out his key she experienced a sort of delight.

Down-stairs it was dark, and after he ascended in the lift Dolly raised a curtain and looked out through opaque lace at the houses over the way. She heard the lift machinery stop, and with the notion of teasing him pressed the button that brought it down. Then on what was more than an impulse she got into it and sent it up to what she guessed was his floor.

"Anson," she called, laughing a little.

"Just a minute," he answered from his bedroom . . . then after a brief delay: "Now you can come in."

He had changed and was buttoning his vest. "This is my room," he said lightly. "How do you like it?"

She caught sight of Paula's picture on the wall and stared at it in fascination, just as Paula had stared at the pictures of Anson's childish sweethearts five years before. She knew something about Paula—sometimes she tortured herself with fragments of the story.

Suddenly she came close to Anson, raising her arms. They embraced. Outside the area window a soft artificial twilight already hovered, though the sun was still bright on a back roof across the way. In half an hour the room would be quite dark. The uncalculated opportunity overwhelmed them, made them both breathless, and they clung more closely. It was eminent, inevitable. Still holding one another, they raised their heads—their eyes fell together upon Paula's picture, staring down at them from the wall.

Suddenly Anson dropped his arms, and sitting down at his desk tried the drawer with a bunch of keys.

"Like a drink?" he asked in a gruff voice.

"No, Anson."

He poured himself half a tumbler of whiskey, swallowed it, and then opened the door into the hall.

"Come on," he said.

Dolly hesitated.

"Anson—I'm going to the country with you tonight, after all. You understand that, don't you?"

"Of course," he answered brusquely.

In Dolly's car they rode on to Long Island, closer in their emotions than they had ever been before. They knew what would happen—not with Paula's face to remind them that something was lacking, but when they were alone in the still, hot Long Island night they did not care.

The estate in Port Washington[5] where they were to spend the week-end belonged to a cousin of Anson's who had married a Montana copper operator. An interminable drive began at the lodge and twisted under imported poplar saplings toward a huge, pink, Spanish house. Anson had often visited there before.

After dinner they danced at the Linx Club. About midnight Anson assured himself that his cousins would not leave before two —then he explained that Dolly was tired; he would take her home and return to the dance later. Trembling a little with excitement, they got into a borrowed car together and drove to Port Washington. As they reached the lodge he stopped and spoke to the night-watchman.

"When are you making a round, Carl?"

"Right away."

"Then you'll be here till everybody's in?"

"Yes, sir."

"All right. Listen: if any automobile, no matter whose it is, turns in at this gate, I want you to phone the house immediately." He put a five-dollar bill into Carl's hand. "Is that clear?"

"Yes, Mr. Anson." Being of the Old World, he neither winked nor smiled. Yet Dolly sat with her face turned slightly away.

Anson had a key. Once inside he poured a drink for both of them—Dolly left hers untouched—then he ascertained definitely the location of the phone, and found that it was within easy hearing distance of their rooms, both of which were on the first floor.

Five minutes later he knocked at the door of Dolly's room.

"Anson?" He went in, closing the door behind him. She was in bed, leaning up anxiously with elbows on the pillow; sitting beside her he took her in his arms.

"Anson, darling."

He didn't answer.

"Anson. . . . Anson! I love you. . . . Say you love me. Say it now—can't you say it now? Even if you don't mean it?"

He did not listen. Over her head he perceived that the picture of Paula was hanging here upon this wall.

He got up and went close to it. The frame gleamed faintly with thrice-reflected moonlight—within was a blurred shadow of a face that he saw he did not know. Almost sobbing, he turned around and stared with abomination at the little figure on the bed.

"This is all foolishness," he said thickly. "I don't know what I was thinking about. I don't love you and you'd better wait for somebody that loves you. I don't love you a bit, can't you understand?"

His voice broke, and he went hurriedly out. Back in the salon

5. **Town on Long Island.**

he was pouring himself a drink with uneasy fingers, when the front door opened suddenly, and his cousin came in.

"Why, Anson, I hear Dolly's sick," she began solicitously. "I hear she's sick. . . ."

"It was nothing," he interrupted, raising his voice so that it would carry into Dolly's room. "She was a little tired. She went to bed."

For a long time afterward Anson believed that a protective God sometimes interfered in human affairs. But Dolly Karger, lying awake and staring at the ceiling, never again believed in anything at all.

VI

When Dolly married during the following autumn, Anson was in London on business. Like Paula's marriage, it was sudden, but it affected him in a different way. At first he felt that it was funny, and had an inclination to laugh when he thought of it. Later it depressed him—it made him feel old.

There was something repetitive about it—why, Paula and Dolly had belonged to different generations. He had a foretaste of the sensation of a man of forty who hears that the daughter of an old flame has married. He wired congratulations and, as was not the case with Paula, they were sincere—he had never really hoped that Paula would be happy.

When he returned to New York, he was made a partner in the firm, and, as his responsibilities increased, he had less time on his hands. The refusal of a life-insurance company to issue him a policy made such an impression on him that he stopped drinking for a year, and claimed that he felt better physically, though I think he missed the convivial recounting of those Celliniesque adventures[6] which, in his early twenties, had played such a part of his life. But he never abandoned the Yale Club. He was a figure there, a personality, and the tendency of his class, who were now seven years out of college, to drift away to more sober haunts was checked by his presence.

His day was never too full nor his mind too weary to give any sort of aid to any one who asked it. What had been done at first through pride and superiority had become a habit and a passion. And there was always something—a younger brother in trouble at New Haven, a quarrel to be patched up between a friend and his wife, a position to be found for this man, an investment for that. But his specialty was the solving of problems for young married people. Young married people fascinated him and their apartments were almost sacred to him—he knew the story of their love-affair, advised them where to live and how, and remembered their babies'

6. Amatory adventures like those of   sculptor and a legendary lover.
Benvenuto Cellini (1500–71), Italian

names. Toward young wives his attitude was circumspect: he never abused the trust which their husbands—strangely enough in view of his unconcealed irregularities—invariable reposed in him.

He came to take a vicarious pleasure in happy marriages, and to be inspired to an almost equally pleasant melancholy by those that went astray. Not a season passed that he did not witness the collapse of an affair that perhaps he himself had fathered. When Paula was divorced and almost immediately remarried to another Bostonian, he talked about her to me all one afternoon. He would never love any one as he had loved Paula, but he insisted that he no longer cared.

"I'll never marry," he came to say; "I've seen too much of it, and I know a happy marriage is a very rare thing. Besides, I'm too old."

But he did believe in marriage. Like all men who spring from a happy and successful marriage, he believed in it passionately—nothing he had seen would change his belief, his cynicism dissolved upon it like air. But he did really believe he was too old. At twenty-eight he began to accept with equanimity the prospect of marrying without romantic love; he resolutely chose a New York girl of his own class, pretty, intelligent, congenial, above reproach—and set about falling in love with her. The things he had said to Paula with sincerity, to other girls with grace, he could no longer say at all without smiling, or with the force necessary to convince.

"When I'm forty," he told his friends, "I'll be ripe. I'll fall for some chorus girl like the rest."

Nevertheless, he persisted in his attempt. His mother wanted to see him married, and he could now well afford it—he had a seat on the Stock Exchange, and his earned income came to twenty-five thousand a year. The idea was agreeable: when his friends—he spent most of his time with the set he and Dolly had evolved—closed themselves in behind domestic doors at night, he no longer rejoiced in his freedom. He even wondered if he should have married Dolly. Not even Paula had loved him more, and he was learning the rarity, in a single life, of encountering true emotion.

Just as this mood began to creep over him a disquieting story reached his ear. His aunt Edna, a woman just this side of forty, was carrying on an open intrigue with a dissolute, hard-drinking young man named Cary Sloane. Every one knew of it except Anson's Uncle Robert, who for fifteen years had talked long in clubs and taken his wife for granted.

Anson heard the story again and again with increasing annoyance. Something of his old feeling for his uncle came back to him, a feeling that was more than personal, a reversion toward that family solidarity on which he had based his pride. His intuition singled out the essential point of the affair, which was that his uncle shouldn't be hurt. It was his first experiment in unsolicited meddling, but with his knowledge of Edna's character he felt that he could handle the matter better than a district judge or his uncle.

His uncle was in Hot Springs. Anson traced down the sources of the scandal so that there should be no possibility of mistake and then he called Edna and asked her to lunch with him at the Plaza next day. Something in his tone must have frightened her, for she was reluctant, but he insisted, putting off the date until she had no excuse for refusing.

She met him at the appointed time in the Plaza lobby, a lovely, faded, gray-eyed blonde in a coat of Russian sable. Five great rings, cold with diamonds and emeralds, sparkled on her slender hands. It occurred to Anson that it was his father's intelligence and not his uncle's that had earned the fur and the stones, the rich brilliance that buoyed up her passing beauty.

Though Edna scented his hostility, she was unprepared for the directness of his approach.

"Edna, I'm astonished at the way you've been acting," he said in a strong, frank voice. "At first I couldn't believe it."

"Believe what?" she demanded sharply.

"You needn't pretend with me, Edna. I'm talking about Cary Sloane. Aside from any other consideration, I didn't think you could treat Uncle Robert——"

"Now look here, Anson——" she began angrily, but his peremptory voice broke through hers:

"——and your children in such a way. You've been married eighteen years, and you're old enough to know better."

"You can't talk to me like that! You——"

"Yes, I can. Uncle Robert has always been my best friend." He was tremendously moved. He felt a real distress about his uncle, about his three young cousins.

Edna stood up, leaving her crab-flake cocktail untasted.

"This is the silliest thing——"

"Very well, if you won't listen to me I'll go to Uncle Robert and tell him the whole story—he's bound to hear it sooner or later. And afterward I'll go to old Moses Sloane."

Edna faltered back into her chair.

"Don't talk so loud," she begged him. Her eyes blurred with tears. "You have no idea how your voice carries. You might have chosen a less public place to make all these crazy accusations."

He didn't answer.

"Oh, you never liked me, I know," she went on. "You're just taking advantage of some silly gossip to try and break up the only interesting friendship I've ever had. What did I ever do to make you hate me so?"

Still Anson waited. There would be the appeal to his chivalry, then to his pity, finally to his superior sophistication—when he had shouldered his way through all these there would be admissions, and he could come to grips with her. By being silent, by being impervious, by returning constantly to his main weapon, which was his own true emotion, he bullied her into frantic despair as the luncheon hour slipped away. At two o'clock she took out a mirror and a handkerchief, shined away the marks of her tears and powdered the slight hollows where they had lain. She had agreed to meet him at her own house at five.

When he arrived she was stretched on a *chaise-longue*[7] which was covered with cretonne for the summer, and the tears he had called up at luncheon seemed still to be standing in her eyes. Then he was aware of Cary Sloane's dark anxious presence upon the cold hearth.

"What's this idea of yours?" broke out Sloane immediately. "I understand you invited Edna to lunch and then threatened her on the basis of some cheap scandal."

Anson sat down.

"I have no reason to think it's only scandal."

"I hear you're going to take it to Robert Hunter, and to my father."

Anson nodded.

"Either you break it off—or I will," he said.

"What God damned business is it of yours, Hunter?"

"Don't lose your temper, Cary," said Edna nervously. "It's only a question of showing him how absurd——"

"For one thing, it's my name that's being handed around," interrupted Anson. "That's all that concerns you, Cary."

"Edna isn't a member of your family."

"She most certainly is!" His anger mounted. "Why—she owes this house and the rings on her fingers to my father's brains. When Uncle Robert married her she didn't have a penny."

They all looked at the rings as if they had a significant bearing on the situation. Edna made a gesture to take them from her hand.

"I guess they're not the only rings in the world," said Sloane.

"Oh, this is absurd," cried Edna. "Anson, will you listen to me? I've found out how the silly story started. It was a maid I discharged who went right to the Chilicheffs—all these Russians pump things out of their servants and then put a false meaning on them." She brought down her fist angrily on the table: "And after Tom lent them the limousine for a whole month when we were South last winter——"

"Do you see?" demanded Sloane eagerly. "This maid got hold of the wrong end of the thing. She knew that Edna and I were friends, and she carried it to the Chilicheffs. In Russia they assume that if a man and a woman——"

7. **Reclining lounge chair.**

He enlarged the theme to a disquisition upon social relations in the Caucasus.

"If that's the case it better be explained to Uncle Robert," said Anson dryly, "so that when the rumors do reach him he'll know they're not true."

Adopting the method he had followed with Edna at luncheon he let them explain it all away. He knew that they were guilty and that presently they would cross the line from explanation into justification and convict themselves more definitely than he could ever do. By seven they had taken the desperate step of telling him the truth—Robert Hunter's neglect, Edna's empty life, the casual dalliance that had flamed up into passion—but like so many true stories it had the misfortune of being old, and its enfeebled body beat helplessly against the armor of Anson's will. The threat to go to Sloane's father sealed their helplessness, for the latter, a retired cotton broker out of Alabama, was a notorious fundamentalist who controlled his son by a rigid allowance and the promise that at his next vagary the allowance would stop forever.

They dined at a small French restaurant, and the discussion continued—at one time Sloane resorted to physical threats, a little later they were both imploring him to give them time. But Anson was obdurate. He saw that Edna was breaking up, and that her spirit must not be refreshed by any renewal of their passion.

At two o'clock in a small night-club on 53d Street, Edna's nerves suddenly collapsed, and she cried to go home. Sloane had been drinking heavily all evening, and he was faintly maudlin, leaning on the table and weeping a little with his face in his hands. Quickly Anson gave them his terms. Sloane was to leave town for six months, and he must be gone within forty-eight hours. When he returned there was to be no resumption of the affair, but at the end of a year Edna might, if she wished, tell Robert Hunter that she wanted a divorce and go about it in the usual way.

He paused, gaining confidence from their faces for his final word.

"Or there's another thing you can do," he said slowly, "if Edna wants to leave her children, there's nothing I can do to prevent your running off together."

"I want to go home!" cried Edna again. "Oh, haven't you done enough to us for one day?"

Outside it was dark, save for a blurred glow from Sixth Avenue down the street. In that light those two who had been lovers looked for the last time into each other's tragic faces, realizing that between them there was not enough youth and strength to avert their eternal parting. Sloane walked suddenly off down the street and Anson tapped a dozing taxi-driver on the arm.

It was almost four; there was a patient flow of cleaning water along the ghostly pavement of Fifth Avenue, and the shadows of two night women[8] flitted over the dark façade of St. Thomas's church. Then the desolate shrubbery of Central Park where Anson had often played as a child, and the mounting numbers, significant as names, of the marching streets. This was his city, he thought, where his name had flourished through five generations. No change could alter the permanence of its place here, for change itself was the essential substratum by which he and those of his name identified themselves with the spirit of New York. Resourcefulness and a powerful will—for his threats in weaker hands would have been less than nothing—had beaten the gathering dust from his uncle's name, from the name of his family, from even this shivering figure that sat beside him in the car.

Cary Sloane's body was found next morning on the lower shelf of a pillar of Queensboro Bridge. In the darkness and in his excitement he had thought that it was the water flowing black beneath him, but in less than a second it made no possible difference—unless he had planned to think one last thought of Edna, and call out her name as he struggled feebly in the water.

VII

Anson never blamed himself for his part in this affair—the situation which brought it about had not been of his making. But the just suffer with the unjust, and he found that his oldest and somehow his most precious friendship was over. He never knew what distorted story Edna told, but he was welcome in his uncle's house no longer.

Just before Christmas Mrs. Hunter retired to a select Episcopal heaven, and Anson became the responsible head of his family. An unmarried aunt who had lived with them for years ran the house, and attempted with helpless inefficiency to chaperone the younger girls. All the children were less self-reliant than Anson, more conventional both in their virtues and in their shortcomings. Mrs. Hunter's death had postponed the début of one daughter and the wedding of another. Also it had taken something deeply material from all of them, for with her passing the quiet, expensive superiority of the Hunters came to an end.

For one thing, the estate, considerably diminished by two inheritance taxes and soon to be divided among six children, was not a notable fortune any more. Anson saw a tendency in his youngest sisters to speak rather respectfully of families that hadn't "existed" twenty years ago. His own feeling of precedence was not echoed in them—sometimes they were conventionally snobbish, that was all. For another thing, this was the last summer they would spend on the Connecticut estate; the clamor against it was too loud: "Who wants to waste the best months of the year shut up in that dead old town?" Reluctantly he yielded—the house would go into

8. Prostitutes.

the market in the fall, and next summer they would rent a smaller place in Westchester County. It was a step down from the expensive simplicity of his father's idea, and, while he sympathized with the revolt, it also annoyed him; during his mother's lifetime he had gone up there at least every other week-end—even in the gayest summers.

Yet he himself was part of this change, and his strong instinct for life had turned him in his twenties from the hollow obsequies of that abortive leisure class. He did not see this clearly—he still felt that there was a norm, a standard of society. But there was no norm, it was doubtful if there had ever been a true norm in New York. The few who still paid and fought to enter a particular set succeeded only to find that as a society it scarcely functioned—or, what was more alarming, that the Bohemia from which they fled sat above them at table.[9]

At twenty-nine Anson's chief concern was his own growing loneliness. He was sure now that he would never marry. The number of weddings at which he had officiated as best man or usher was past all counting—there was a drawer at home that bulged with the official neckties of this or that wedding-party, neckties standing for romances that had not endured a year, for couples who had passed completely from his life. Scarf-pins, gold pencils, cuff-buttons, presents from a generation of grooms had passed through his jewel-box and been lost—and with every ceremony he was less and less able to imagine himself in the groom's place. Under his hearty good-will toward all those marriages there was despair about his own.

And as he neared thirty he became not a little depressed at the inroads that marriage, especially lately, had made upon his friendships. Groups of people had a disconcerting tendency to dissolve and disappear. The men from his own college—and it was upon them he had expended the most time and affection—were the most elusive of all. Most of them were drawn deep into domesticity, two were dead, one lived abroad, one was in Hollywood writing continuities for pictures that Anson went faithfully to see.

Most of them, however, were permanent commuters with an intricate family life centring around some suburban country club, and it was from these that he felt his estrangement most keenly.

In the early days of their married life they had all needed him; he gave them advice about their slim finances, he exorcised their doubts about the advisability of bringing a baby into two rooms and a bath, especially he stood for the great world outside. But now their financial troubles were in the past and the fearfully expected child had evolved into an absorbing family. They were always glad to see old Anson, but they dressed up for him and

9. I.e., that the social set they disdained (ironically called an idyllic Bohemia) actually was accorded higher status in seating arrangements at formal dinners.

tried to impress him with their present importance, and kept their troubles to themselves. They needed him no longer.

A few weeks before his thirtieth birthday the last of his early and intimate friends was married. Anson acted in his usual rôle of best man, gave his usual silver tea-service, and went down to the usual *Homeric*[1] to say good-by. It was a hot Friday afternoon in May, and as he walked from the pier he realized that Saturday closing had begun and he was free until Monday morning.

"Go where?" he asked himself.

The Yale Club, of course; bridge until dinner, then four or five raw cocktails in somebody's room and a pleasant confused evening. He regretted that this afternoon's groom wouldn't be along —they had always been able to cram so much into such nights: they knew how to attach women and how to get rid of them, how much consideration any girl deserved from their intelligent hedonism. A party was an adjusted thing—you took certain girls to certain places and spent just so much on their amusement; you drank a little, not much, more than you ought to drink, and at a certain time in the morning you stood up and said you were going home. You avoided college boys, sponges, future engagements, fights, sentiment, and indiscretions. That was the way it was done. All the rest was dissipation.

In the morning you were never violently sorry—you made no resolutions, but if you had overdone it and your heart was slightly out of order, you went on the wagon for a few days without saying anything about it, and waited until an accumulation of nervous boredom projected you into another party.

The lobby of the Yale Club was unpopulated. In the bar three very young alumni looked up at him, momentarily and without curiosity.

"Hello there, Oscar," he said to the bartender. "Mr. Cahill been around this afternoon?"

"Mr. Cahill's gone to New Haven."

"Oh . . . that so?"

"Gone to the ball game. Lot of men gone up."

Anson looked once again into the lobby, considered for a moment, and then walked out and over to Fifth Avenue. From the broad window of one of his clubs—one that he had scarcely visited in five years—a gray man with watery eyes stared down at him. Anson looked quickly away—that figure sitting in vacant resignation, in supercilious solitude, depressed him. He stopped and, retracing his steps, started over 47th Street toward Teak Warden's apartment. Teak and his wife had once been his most familiar friends—it was a household where he and Dolly Karger had been used to go in the days of their affair. But Teak had taken to drink, and his wife had remarked publicly that Anson was a bad influence

---

1. Ocean liner departing on a cruise.

on him. The remark reached Anson in an exaggerated form—when it was finally cleared up, the delicate spell of intimacy was broken, never to be renewed.

"Is Mr. Warden at home?" he inquired.

"They've gone to the country."

The fact unexpectedly cut at him. They were gone to the country and he hadn't known. Two years before he would have known the date, the hour, come up at the last moment for a final drink, and planned his first visit to them. Now they had gone without a word.

Anson looked at his watch and considered a weekend with his family, but the only train was a local that would jolt through the aggressive heat for three hours. And to-morrow in the country, and Sunday—he was in no mood for porch-bridge with polite undergraduates, and dancing after dinner at a rural roadhouse, a diminutive of gaiety which his father had estimated too well.

"Oh, no," he said to himself. . . . "No."

He was a dignified, impressive young man, rather stout now, but otherwise unmarked by dissipation. He could have been cast for a pillar of something—at times you were sure it was not society, at others nothing else—for the law, for the church. He stood for a few minutes motionless on the sidewalk in front of a 47th Street apartment-house; for almost the first time in his life he had nothing whatever to do.

Then he began to walk briskly up Fifth Avenue, as if he had just been reminded of an important engagement there. The necessity of dissimulation is one of the few characteristics that we share with dogs, and I think of Anson on that day as some well-bred specimen who had been disappointed at a familiar back door. He was going to see Nick, once a fashionable bartender in demand at all private dances, and now employed in cooling non-alcoholic champagne among the labyrinthine cellars of the Plaza Hotel.

"Nick," he said, "what's happened to everything?"

"Dead," Nick said.

"Make me a whiskey sour." Anson handed a pint bottle over the counter.[2] "Nick, the girls are different; I had a little girl in Brooklyn and she got married last week without letting me know."

"That a fact? Ha-ha-ha," responded Nick diplomatically. "Slipped it over on you."

"Absolutely," said Anson. "And I was out with her the night before."

"Ha-ha-ha," said Nick, "ha-ha-ha!"

"Do you remember the wedding, Nick, in Hot Springs where I

2. From 1919 to 1933, a federal "Prohibition" law and a constitutional amendment prohibited the manufacture and sale of intoxicating alcoholic beverages. Violation of the law was wide- spread; individuals secured contraband or "bootleg" liquor from illegal distillers or stores and brought it with them to bars and parties.

had the waiters and the musicians singing 'God save the King'?"

"Now where was that, Mr. Hunter?" Nick concentrated doubt-fully. "Seems to me that was——"

"Next time they were back for more, and I began to wonder how much I'd paid them," continued Anson.

"—seems to me that was at Mr. Trenholm's wedding."

"Don't know him," said Anson decisively. He was offended that a strange name should intrude upon his reminiscences; Nick perceived this.

"Naw—aw—" he admitted, "I ought to know that. It was one of *your* crowd—Brakins . . . . Baker——"

"Bicker Baker," said Anson responsively. "They put me in a hearse after it was over and covered me up with flowers and drove me away."

"Ha-ha-ha," said Nick. "Ha-ha-ha."

Nick's simulation of the old family servant paled presently and Anson went up-stairs to the lobby. He looked around—his eyes met the glance of an unfamiliar clerk at the desk, then fell upon a flower from the morning's marriage hesitating in the mouth of a brass cuspidor. He went out and walked slowly toward the blood-red sun over Columbus Circle. Suddenly he turned around and, retracing his steps to the Plaza, immured himself in a telephone-booth.

Later he said that he tried to get me three times that afternoon, that he tried every one who might be in New York—men and girls he had not seen for years, an artist's model of his college days whose faded number was still in his address book—Central[3] told him that even the exchange existed no longer. At length his quest roved into the country, and he held brief disappointing conversations with emphatic butlers and maids. So-and-so was out, riding, swimming, playing golf, sailed to Europe last week. Who shall I say phoned?

It was intolerable that he should pass the evening alone—the private reckonings which one plans for a moment of leisure lose every charm when the solitude is enforced. There were always women of a sort, but the ones he knew had temporarily vanished, and to pass a New York evening in the hired company of a stranger never occurred to him—he would have considered that that was something shameful and secret, the diversion of a travelling salesman in a strange town.

Anson paid the telephone bill—the girl tried unsuccessfully to joke with him about its size—and for the second time that afternoon started to leave the Plaza and go he knew not where. Near the revolving door the figure of a woman, obviously with child, stood sideways to the light—a sheer beige cape fluttered at her shoulders when the door turned and, each time, she looked impatiently toward it as if she were weary of waiting. At the first

3. Telephone operator.

sight of her a strong nervous thrill of familiarity went over him,
but not until he was within five feet of her did he realize that it
was Paula.

"Why, Anson Hunter!"

His heart turned over.

"Why, Paula——"

"Why, this is wonderful. I can't believe it, *Anson!*"

She took both his hands, and he saw in the freedom of the gesture that the memory of him had lost poignancy to her. But not to
him—he felt that old mood that she evoked in him stealing over
his brain, that gentleness with which he had always met her optimism as if afraid to mar its surface.

"We're at Rye for the summer. Pete had to come East on business—you know of course I'm Mrs. Peter Hagerty now—so we
brought the children and took a house. You've got to come out and
see us."

"Can I?" he asked directly. "When?"

"When you like. Here's Pete." The revolving door functioned,
giving up a fine tall man of thirty with a tanned face and a trim
mustache. His immaculate fitness made a sharp contrast with
Anson's increasing bulk, which was obvious under the faintly tight
cut-away coat.

"You oughtn't to be standing," said Hagerty to his wife. "Let's
sit down here." He indicated lobby chairs, but Paula hesitated.

"I've got to go right home," she said. "Anson, why don't you—
why don't you come out and have dinner with us to-night? We're
just getting settled, but if you can stand that——"

Hagerty confirmed the invitation cordially.

"Come out for the night."

Their car waited in front of the hotel, and Paula with a tired
gesture sank back against silk cushions in the corner.

"There's so much I want to talk to you about," she said, "it
seems hopeless."

"I want to hear about you."

"Well"—she smiled at Hagerty—"that would take a long time
too. I have three children—by my first marriage. The oldest is five,
then four, then three." She smiled again. "I didn't waste much
time having them, did I?"

"Boys?"

"A boy and two girls. Then—oh, a lot of things happened, and
I got a divorce in Paris a year ago and married Pete. That's all—
except that I'm awfully happy."

In Rye they drove up to a large house near the Beach Club, from
which there issued presently three dark, slim children who broke
from an English governess and approached them with an esoteric
cry. Abstractedly and with difficulty Paula took each one into her
arms, a caress which they accepted stiffly, as they had evidently

been told not to bump into Mummy. Even against their fresh faces Paula's skin showed scarcely any weariness—for all her physical languor she seemed younger than when he had last seen her at Palm Beach seven years ago.

At dinner she was preoccupied, and afterward, during the homage to the radio, she lay with closed eyes on the sofa, until Anson wondered if his presence at this time were not an intrusion. But at nine o'clock, when Hagerty rose and said pleasantly that he was going to leave them by themselves for a while, she began to talk slowly about herself and the past.

"My first baby," she said—"the one we call Darling, the biggest little girl—I wanted to die when I knew I was going to have her, because Lowell was like a stranger to me. It didn't seem as though she could be my own. I wrote you a letter and tore it up. Oh, you were *so* bad to me, Anson."

It was the dialogue again, rising and falling. Anson felt a sudden quickening of memory.

"Weren't you engaged once?" she asked—"a girl named Dolly something?"

"I wasn't ever engaged. I tried to be engaged, but I never loved anybody but you, Paula."

"Oh," she said. Then after a moment: "This baby is the first one I ever really wanted. You see, I'm in love now—at last."

He didn't answer, shocked at the treachery of her remembrance. She must have seen that the "at last" bruised him, for she continued:

"I was infatuated with you, Anson—you could make me do anything you liked. But we wouldn't have been happy. I'm not smart enough for you. I don't like things to be complicated like you do." She paused. "You'll never settle down," she said.

The phrase struck at him from behind—it was an accusation that of all accusations he had never merited.

"I could settle down if women were different," he said. "If I didn't understand so much about them, if women didn't spoil you for other women, if they had only a little pride. If I could go to sleep for a while and wake up into a home that was really mine— why, that's what I'm made for, Paula, that's what women have seen in me and liked in me. It's only that I can't get through the preliminaries any more."

Hagerty came in a little before eleven; after a whiskey Paula stood up and announced that she was going to bed. She went over and stood by her husband.

"Where did you go, dearest?" she demanded.

"I had a drink with Ed Saunders."

"I was worried. I thought maybe you'd run away."

She rested her head against his coat.

"He's sweet, isn't he, Anson?" she demanded.

"Absolutely," said Anson, laughing.

She raised her face to her husband.

"Well, I'm ready," she said. She turned to Anson: "Do you want to see our family gymnastic stunt?"

"Yes," he said in an interested voice.

"All right. Here we go!"

Hagerty picked her up easily in his arms.

"This is called the family acrobatic stunt," said Paula. "He carries me up-stairs. Isn't it sweet of him?"

"Yes," said Anson.

Hagerty bent his head slightly until his face touched Paula's.

"And I love him," she said. "I've just been telling you, haven't I, Anson?"

"Yes," he said.

"He's the dearest thing that ever lived in this world; aren't you, darling? . . . Well, good night. Here we go. Isn't he strong?"

"Yes," Anson said.

"You'll find a pair of Pete's pajamas laid out for you. Sweet dreams—see you at breakfast."

"Yes," Anson said.

VIII

The older members of the firm insisted that Anson should go abroad for the summer. He had scarcely had a vacation in seven years, they said. He was stale and needed a change. Anson resisted.

"If I go," he declared, "I won't come back any more."

"That's absurd, old man. You'll be back in three months with all this depression gone. Fit as ever."

"No." He shook his head stubbornly. "If I stop, I won't go back to work. If I stop, that means I've given up—I'm through."

"We'll take a chance on that. Stay six months if you like—we're not afraid you'll leave us. Why, you'd be miserable if you didn't work."

They arranged his passage for him. They liked Anson—every one liked Anson—and the change that had been coming over him cast a sort of pall over the office. The enthusiasm that had invariably signalled up business, the consideration toward his equals and his inferiors, the lift of his vital presence—within the past four months his intense nervousness had melted down these qualities into the fussy pessimism of a man of forty. On every transaction in which he was involved he acted as a drag and a strain.

"If I go I'll never come back," he said.

Three days before he sailed Paula Legendre Hagerty died in childbirth. I was with him a great deal then, for we were crossing together, but for the first time in our friendship he told me not a word of how he felt, nor did I see the slightest sign of emotion. His chief preoccupation was with the fact that he was thirty years old—

he would turn the conversation to the point where he could remind you of it and then fall silent, as if he assumed that the statement would start a chain of thought sufficient to itself. Like his partners, I was amazed at the change in him, and I was glad when the *Paris* moved off into the wet space between the worlds, leaving his principality behind.

"How about a drink?" he suggested.

We walked into the bar with that defiant feeling that characterizes the day of departure and ordered four Martinis. After one cocktail a change came over him—he suddenly reached across and slapped my knee with the first joviality I had seen him exhibit for months.

"Did you see that girl in the red tam?" he demanded, "the one with the high color who had the two police dogs down to bid her good-by."

"She's pretty," I agreed.

"I looked her up in the purser's office and found out that she's alone. I'm going down to see the steward in a few minutes. We'll have dinner with her to-night."

After a while he left me, and within an hour he was walking up and down the deck with her, talking to her in his strong, clear voice. Her red tam was a bright spot of color against the steel-green sea, and from time to time she looked up with a flashing bob of her head, and smiled with amusement and interest, and anticipation. At dinner we had champagne, and were very joyous—afterward Anson ran the pool[4] with infectious gusto, and several people who had seen me with him asked me his name. He and the girl were talking and laughing together on a lounge in the bar when I went to bed.

I saw less of him on the trip than I had hoped. He wanted to arrange a foursome, but there was no one available, so I saw him only at meals. Sometimes, though, he would have a cocktail in the bar, and he told me about the girl in the red tam, and his adventures with her, making them all bizarre and amusing, as he had a way of doing, and I was glad that he was himself again, or at least the self that I knew, and with which I felt at home. I don't think he was ever happy unless some one was in love with him, responding to him like filings to a magnet, helping him to explain himself, promising him something. What it was I do not know. Perhaps they promised that there would always be women in the world who would spend their brightest, freshest, rarest hours to nurse and protect that superiority he cherished in his heart.

1926

ERNEST HEMINGWAY
1899–1961

By the time he was twenty-four, when he published his first book in Paris, Ernest Hemingway possessed much of the raw material out of which he would fashion a forty-year literary career and a durable personal myth. He had spent his boyhood shuttling with his family between a comfortable Chicago suburb and a remnant of the earlier frontier in the backwoods of northern Michigan. From his experience as a contestant in the academic and athletic competitions of his high school, and from his exposure as hunter and fisherman to the landscape and primitive society of the Michigan woods, emerged a set of values stressing a cult of elemental experience, courage, and masculine prowess like those of a frontiersman, along with the habit of comparing the social codes of different environments, that were displayed in Hemingway's own later conduct and that of his fictional heroes.

When he served with the ambulance corps in Italy in World War I, both his legs were shattered by Austrian shrapnel while he was rescuing a wounded Italian soldier; he had accepted baptism by a Roman Catholic priest in a battlefield hospital, suffered the pain of the wound and the surgery afterward, and had survived. The reality of war, the imminence of death, and desperate commitments to survival haunted his fiction throughout his career.

Finally, as a newspaperman and then as an expatriate writer in Paris, he had honed his prose style on the principles of the best journalism and on the example of such contemporaries as Sherwood Anderson and Hemingway's fellow expatriates, Gertrude Stein and Ezra Pound. Within a decade Hemingway's style was internationally famous. Moreover, the desperate necessity for style, both in actual conduct and in writing, as a triumph over meaningless death or a shield against the disintegration of one's values and talent, had become one of his central themes in fiction, along with the quest by expatriates for rituals of commitment to vocations and actions that would constitute a stoic heroism.

Ernest Miller Hemingway was born in Oak Park, Illinois. His mother taught music, directed the choir in the Congregational church where Hemingway was confirmed in 1911, and took her six children frequently to painting exhibitions, concerts, and plays. His father was a successful physician with a relish for hunting, fishing, cooking, and taxidermy. The Hemingways spent summers at their cottage in northern Michigan. Alone, with his father, or with some Ottawa Indian children from a nearby settlement, Hemingway explored the surrounding woods. Dr. Hemingway initiated his son into the rituals of the sportsman.

In Oak Park, Hemingway was a sociable and competitive suburban youth. He learned to box, played on several high school athletic teams ("At Oak Park if you could play football you had to play it"),[1] joined civic and debating societies, and played the cello in the high school

---

1. Quoted by Charles Fenton, *The Apprenticeship of Ernest Hemingway*, p. 7.

orchestra ("I played it worse than anyone on earth").[2] He also wrote regularly for the school's newspaper and literary magazine.

Hemingway refused to attend college, and, beginning to break away from the family circle whose pressures and stern piety seemed constricting, he got a job as cub reporter on the Kansas City *Star*. His assignments familiarized him with the violence and the seamy side of the city. The newspaper's style sheet instructed reporters to "Avoid the use of adjectives, especially such extravagant ones as *splendid, gorgeous, grand, magnificent,* etc.," and enjoined "short sentences," "short first paragraphs," and "vigorous English"; Hemingway later called these "the best rules I ever learned for the business of writing."[3] Though an eye defect kept him from enlisting in the army, Hemingway joined the ambulance corps in 1918 and was wounded after three weeks of active duty. He recuperated for six months and shipped home in 1919 to a hero's welcome. In 1921, after two years of loafing and sporadic journalism for *The Toronto Star,* he married Hadley Richardson and took her to live in Paris.

There he set out to bring his writing up to his own high standards of proficiency. He sent articles to Toronto journals, and he covered the Greco-Turkish war in Constantinople in 1922, but he made more exacting demands of his writing than those required for journalism. He said later that mere "timeliness" was a crutch on which newspaper writing could rely to secure emotional effect, but he recalled that in the early 1920s, in his creative writing, he was aiming for "the real thing, the sequence of motion and fact that made the motion and which would be as valid in a year or in ten years, or, with luck * * * always. * * *"[4] He traveled with Hadley in Spain, Italy, and Switzerland and shared the excitement of postwar Paris with artists, adventurers, and intellectuals, many of them British or American expatriates, whose common bond was a bitter disillusionment with traditional institutions, vague ideologies, hypocritical values, and hackneyed, imprecise language. He responded to the writing of his contemporaries, as to those of such predecessors as Turgenev or Conrad, with the canniness and at times the aggressiveness of an equal contestant or rival, and he sought models for his writing also in the paintings of such artists as Cézanne and Goya. A *Moveable Feast* (1964), containing mordant recollections of F. Scott Fitzgerald and Gertrude Stein among others, is Hemingway's nostalgic memoir of his early years in Paris.

Hemingway's first book, *Three Stories and Ten Poems,* was published in Paris in 1923, but it was his next volumes that secured his critical reputation and after 1929 brought him financial security. *In Our Time,* published first in Paris and then, thanks to efforts by Fitzgerald and Sherwood Anderson, published in an enlarged edition in America in 1925, traced in separate, but thematically related, short stories the growth of a youth called Nick Adams from his childhood in the Michigan woods to his return as a war veteran. Interspersed in the sequence are brief prose sketches or vignettes—including a hanging in Kansas City, a massacre in Turkey, the culmination of a bullfight in Spain—which underscore the violence and imminent tragedy that are manifest in Nick's world. Both the stories and the vignettes display the compressed irony and the understatement that characterize Hemingway's style. In 1926 appeared *The Torrents of Spring,*

2. George Plimpton, "Interview with Hemingway," in *Hemingway and His Critics,* ed. Carlos Baker, p. 28.

3. Quoted by Fenton, *The Apprenticeship of Ernest Hemingway,* pp. 31–34.
4. *Death in the Afternoon,* p. 2.

a savage parody of primitivistic attitudes and styles that Hemingway attributed to Sherwood Anderson, and his first and best novel, *The Sun Also Rises*. The novel's protagonist is Jake Barnes, a writer rendered impotent by a war wound, who tries to bring some measure of order to the lives of his expatriate friends in Paris and finds the epitome of courage and honor, finally, in the figure of a young bullfighter in Spain. A collection of stories, *Men without Women*, appeared in 1927. The novel *A Farewell to Arms* (1929) wrote the epitaph to a decade in its story of the tragic love affair of a wounded American lieutenant who falls in love with his British nurse and deserts the army, making "a separate peace," to flee with her to Switzerland. Their brief idyll ends when she dies in childbirth.

Hemingway's cult of athletic prowess and masculinity, his unending quest for adventure abroad and his exploits as a war correspondent in the Spanish Civil War and later in World War II, and his contributions to popular magazines made him a conspicuous public figure whose image threatened to distract readers (and to divert Hemingway himself) from his devotion to the writer's craft. He married four times and lived or traveled abroad for much of his life, even after settling in locations on the edge of America where he enjoyed frequent reunions with his three sons, first at Key West, then on the island of Bimini off Miami, later in Cuba, and finally in the American West at Ketchum, Idaho. Yet Hemingway's writings make clear that his expatriation and various threats to his vocation were precisely the material with which his imagination worked.

His tribute to Spanish culture, *Death in the Afternoon* (1932), centers on the "tragedy and ritual"[5] of the bullfight where Hemingway found an art comparable to the art of writing he wanted to perfect. The co-presence of "life and death" in the bullring, the imminence of "violent death" as "one of the simplest things of all and the most fundamental," is the elemental reality he wanted to test his feelings and his writing against.[6] The feelings aroused were more intense because grounded in bodily sensations. The bullfighter not only confronted mortal danger but, through his stylized ritual performance, "increased" and "created" the danger to his body,[7] and the ritual which thus threatened his survival also permitted him to triumph by achieving the grace and power of the matador Romero whom Hemingway had celebrated in his first novel. Such is the disciplined heroism of the bullfighters and other heroes, even those who are defeated, in the early fiction and in Hemingway's second collection of stories (*Winner Take Nothing*, 1933).

The writer's struggle to survive, against the threats to his talent of genteel traditions in America, success, money, and domestic entanglements, is central in Hemingway's account of his safari in Africa, *The Green Hills of Africa* (1935). There he asserts a connection between big game hunting and writing, and he defines his expatriation as a continuation of the settlement of America and the movement west, claiming that "I would go, now, somewhere else as we had always had the right to go somewhere else and as we had always done."[8] The attempt by a mortally wounded American writer, on safari in Africa, to redeem his imagination from the

---

5. *Ibid.*, p. 9.
6. *Ibid.*, p. 2.
7. *Ibid.*, pp. 16, 21.
8. P. 192.

corrosions of wealth and domestic strife is the subject of the brilliant short story *The Snows of Kilimanjaro* (1936).

Since the early 1930s Hemingway's leftist friends and critics had condemned his abstention from political affiliations and his avoidance of overtly political themes in his fiction; these were deliberate strategies in keeping with Hemingway's feeling that "anyone is cheating who takes politics as a way out" from the writer's responsibility "to write straight honest prose on human beings."[9] His next novels, while they treated more openly the theme of social solidarity, kept to Hemingway's characteristic pattern of a lonely, individualized heroism. *To Have and Have Not* (1937) experiments with narrative structure to encompass the economic hardships and cruel ironies in the life of a desperate Florida boatman who takes to smuggling during the Depression.

It culminates in the dying hero's recognition that "one man alone ain't got no bloody f - - - ing chance,"[1] but it articulates no political or social alternative to his plight. *For Whom the Bell Tolls* (1940) takes Madrid and the mountain areas of Spain during the Spanish Civil War for its canvas, and for its hero a Montana Spanish instructor who fights with a guerrilla band in the Loyalist cause, carrying out his military mission even after it proves to be absurdly futile and, while mortally wounded, mustering the strength to protect the escape of the Spanish girl with whom he has fallen in love. Yet his vaguely Jeffersonian principles have little to do with his commitment to the cause of anti-Fascism, and his experience in Madrid exposes the political machinations of Spanish and Communist politicians who have betrayed the cause. His heroism constitutes a commitment to personal responsibility and intimate love in the face of a lost gamble in a losing game.

"You are all a lost generation," Gertrude Stein had once told the young Hemingway.[2] And the concept of loss informs everything he wrote, though Hemingway's heroes seek ways to minimize the loss and to find the dignity of bravery and disciplined commitment even in defeat. Like Jake Barnes in *The Sun Also Rises*, making a staunch commitment to playing an unwinnable game properly, they care less about understanding the world than about the desperate and pragmatic need to learn "how to live in it."[3]

The requirements for playing properly—a view of the field uncluttered by delusions, and close attention to the game's technicalities—are Hemingway's stylistic requirements for his art in seeking the "discipline" that enables the imagination to "survive."[4] With such models as Biblical prose, the painter Van Gogh, the fiction of Stendhal, and, among Americans, the "good writers" Henry James, Stephen Crane, and Mark Twain,[5] he forged the discipline that amounted to a poetics of prose. "All our words from loose using have lost their edge," he charged.[6] A writer could restore that edge by using words sparingly, savoring each word as one savors good wine, and by making sentences that minutely render physical reality and physical sensation so as to capture their particularity. If a writer's prose were transparently simple and honest, Hemingway felt, he

9. *By-Line: Ernest Hemingway*, ed. William White, p. 183.
1. P. 225.
2. Quoted as an epigraph to *The Sun Also Rises*.
3. *Ibid.*, p. 148.
4. *The Green Hills of Africa*, p. 23.
5. *Ibid.*, p. 19.
6. *Ibid.*, p. 71.

could leave his meaning implicit and the reader would discern it and even help provide it: "I always try to write on the principle of the iceberg. There is seven eighths of it under water for every part that shows."[7] Accordingly Hemingway's seemingly matter-of-fact, laconic statements are highly stylized sentences that can incorporate the rhythms and diction of colloquial speech along with archaic inversions of sentence structure and parodies of Gertrude Stein's or the Bible's prose. Their simplicity, like Twain's in *Huckleberry Finn*, assigns a priority to feeling over thought and to action over comment. The result is that Hemingway can suggest, through oblique implication, compressed irony, or understatement, a sense of moral urgency along with a considerable range and astonishing intensity of feeling, while rendering outward actions and scenes with a vividness seldom matched. His prose style had an immense impact on contemporary and younger writers in France and Italy as well as in England and America.

For Whom the Bell Tolls, though widely popular, marked the beginning of a decline in Hemingway's work, for the language had lost some of its edge and the book indulged in some self-conscious meditation and sentiment that Hemingway earlier kept better controlled. His wartime love idyll, *Across the River and into the Trees* (1950), elicited boredom and even scorn. Only the novella *The Old Man and the Sea* (1952), which reached a mass audience through *Life* magazine, *The Moveable Feast* (1964), and parts of the posthumous novel *Islands in the Stream* (1970) approached the quality of his earlier work. *The Old Man and the Sea* won the Pulitzer Prize in 1953 and occasioned the award of the Nobel Prize a year later. But for several years Hemingway suffered from physical ailments, and his emotional breakdowns that required clinical treatment were the sign of his despair that he could no longer write consistently and well. In July, 1961, in Ketchum, Idaho, following his father's example, he used a gun to take his own life.

Big Two-Hearted River
Part I[1]

The train went on up the track out of sight, around one of the hills of burnt timber. Nick sat down on the bundle of canvas and bedding the baggage man had pitched out of the door of the baggage car. There was no town, nothing but the rails and the burned-over country. The thirteen saloons that had lined the one street

---

7. George Plimpton, "Interview with Hemingway," in *Hemingway and His Critics*, ed. Carlos Baker, p. 34.

1. "Big Two-Hearted River" is the concluding story in *In Our Time*. In it Nick Adams returns from the ravage of World War I to the northern Michigan he had known as a child and younger man. Alone and in virtual silence, he seeks to construct for himself a substitute for home and, through the ritual of fishing, to renew his contact with nature's vi-

tality, discipline his feelings, bring order to the violence his sport entails, and find in the experience an alternative to the desiccation of the "burned over" area around him and the imminent tragedy he may encounter in the future. The story is preceded by a vignette describing the death of a matador. The two parts are separated by a vignette describing the hanging of a criminal in Kansas City. The story is followed by a vignette describing an exiled King.

of Seney had not left a trace. The foundations of the Mansion House hotel stuck up above the ground. The stone was chipped and split by the fire. It was all that was left of the town of Seney. Even the surface had been burned off the ground.

Nick looked at the burned-over stretch of hillside, where he had expected to find the scattered houses of the town and then walked down the railroad track to the bridge over the river. The river was there. It swirled against the log spiles of the bridge. Nick looked down into the clear, brown water, colored from the pebbly bottom, and watched the trout keeping themselves steady in the current with wavering fins. As he watched them they changed their positions by quick angles, only to hold steady in the fast water again. Nick watched them a long time.

He watched them holding themselves with their noses into the current, many trout in deep, fast moving water, slightly distorted as he watched far down through the glassy convex surface of the pool, its surface pushing and swelling smooth against the resistance of the log-driven piles of the bridge. At the bottom of the pool were the big trout. Nick did not see them at first. Then he saw them at the bottom of the pool, big trout looking to hold themselves on the gravel bottom in a varying mist of gravel and sand, raised in spurts by the current.

Nick looked down into the pool from the bridge. It was a hot day. A kingfisher flew up the stream. It was a long time since Nick had looked into a stream and seen trout. They were very satisfactory. As the shadow of the kingfisher moved up the stream, a big trout shot upstream in a long angle, only his shadow marking the angle, then lost his shadow as he came through the surface of the water, caught the sun, and then, as he went back into the stream under the surface, his shadow seemed to float down the stream with the current, unresisting, to his post under the bridge where he tightened facing up into the current.

Nick's heart tightened as the trout moved. He felt all the old feeling.

He turned and looked down the stream. It stretched away, pebbly-bottomed with shallows and big boulders and a deep pool as it curved away around the foot of a bluff.

Nick walked back up the ties to where his pack lay in the cinders beside the railway track. He was happy. He adjusted the pack harness around the bundle, pulling straps tight, slung the pack on his back, got his arms through the shoulder straps and took some of the pull off his shoulders by leaning his forehead against the wide band of the tump-line. Still, it was too heavy. It was much too heavy. He had his leather rod-case in his hand and leaning forward to keep the weight of the pack high on his shoulders he walked along the road that paralleled the railway track, leaving the burned town behind in the heat, and then turned off around a hill with a high, fire-scarred hill on either side onto a road that went back into the country. He walked along the road feeling the ache from the pull of the heavy pack. The road climbed steadily. It was hard work walking up-hill. His muscles ached and the day was hot, but Nick felt happy. He felt he had left everything behind, the need for thinking, the need to write, other needs. It was all back of him.

From the time he had gotten down off the train and the baggage man had thrown his pack out of the open car door things had been different. Seney was burned, the country was burned over and changed, but it did not matter. It could not all be burned. He knew that. He hiked along the road, sweating in the sun, climbing to cross the range of hills that separated the railway from the pine plains.

The road ran on, dipping occasionally, but always climbing. Nick went on up. Finally the road after going parallel to the burnt hillside reached the top. Nick leaned against a stump and slipped out of the pack harness. Ahead of him, as far as he could see, was the pine plain. The burned country stopped off at the left with the range of hills. On ahead islands of dark pine trees rose out of the plain. Far off to the left was the line of the river. Nick followed it with his eye and caught glints of the water in the sun.

There was nothing but the pine plain ahead of him, until the far blue hills that marked the Lake Superior height of land. He could hardly see them, faint and far away in the heat-light over the plain. If he looked too steadily they were gone. But if he only half-looked they were there, the far off hills of the height of land.

Nick sat down against the charred stump and smoked a cigarette. His pack balanced on the top of the stump, harness holding ready, a hollow molded in it from his back. Nick sat smoking, looking out over the country. He did not need to get his map out. He knew where he was from the position of the river.

As he smoked, his legs stretched out in front of him, he noticed a grasshopper walk along the ground and up onto his woolen sock. The grasshopper was black. As he had walked along the road, climbing, he had started many grasshoppers from the dust. They were all black. They were not the big grasshoppers with yellow and black or red and black wings whirring out from their black wing sheathing as they fly up. These were just ordinary hoppers, but all a sooty black in color. Nick had wondered about them as he walked, without really thinking about them. Now, as he watched the black hopper that was nibbling at the wool of his sock with its fourway lip, he realized that they had all turned black from living in the burned-over land. He realized that the fire must have come the year before, but the grasshoppers were all black now. He wondered how long they would stay that way.

Carefully he reached his hand down and took hold of the hopper by the wings. He turned him up, all his legs walking in the air, and looked at his jointed belly. Yes, it was black too, iridescent where the back and head were dusty.

"Go on, hopper," Nick said, speaking out loud for the first time, "Fly away somewhere."

He tossed the grasshopper up into the air and watched him sail away to a charcoal stump across the road.

Nick stood up. He leaned his back against the weight of his pack where it rested upright on the stump and got his arms through the shoulder straps. He stood with the pack on his back on the brow of the hill looking out across the country, toward the distant river and then struck down the hillside away from the road. Underfoot the ground was good walking. Two hundred yards down the hillside the fire line stopped. Then it was sweet fern, growing ankle high, to walk through, and clumps of jack pines; a long undulating country with frequent rises and descents, sandy underfoot and the country alive again.

Nick kept his direction by the sun. He knew where he wanted to strike the river and he kept on through the pine plain, mounting small rises to see other rises ahead of him and sometimes from the top of a rise a great solid island of pines off to his right or his left. He broke off some sprigs of the heathery sweet fern, and put them under his pack straps. The chafing crushed it and he smelled it as he walked.

He was tired and very hot, walking across the uneven, shadeless pine plain. At any time he knew he could strike the river by turning off to his left. It could not be more than a mile away. But he kept on toward the north to hit the river as far upstream as he could go in one day's walking.

For some time as he walked Nick had been in sight of one of the big islands of pine standing out above the rolling high ground he was crossing. He dipped down and then as he came slowly up to the crest of the ridge he turned and made toward the pine trees.

There was no underbrush in the island of pine trees. The trunks of the trees went straight up or slanted toward each other. The trunks were straight and brown without branches. The branches were high above. Some interlocked to make a solid shadow on the brown forest floor. Around the grove of trees was a bare space. It was brown and soft underfoot as Nick walked on it. This was the over-lapping of the pine needle floor, extending out beyond the width of the high branches. The trees had grown tall and the branches moved high, leaving in the sun this bare space they had once covered with shadow. Sharp at the edge of this extension of the forest floor commenced the sweet fern.

Nick slipped off his pack and lay down in the shade. He lay on his back and looked up into the pine trees. His neck and back and the small of his back rested as he stretched. The earth felt good against his back. He looked up at the sky, through the branches, and then shut his eyes. He opened them and looked up again. There was a wind high up in the branches. He shut his eyes again and went to sleep.

Nick woke stiff and cramped. The sun was nearly down. His pack was heavy and the straps painful as he lifted it on. He leaned over with the pack on and picked up the leather rod-case and started out from the pine trees across the sweet fern swale, toward the river. He knew it could not be more than a mile.

He came down a hillside covered with stumps into a meadow. At the edge of the meadow flowed the river. Nick was glad to get to the river. He walked upstream through the meadow. His trousers were soaked with the dew as he walked. After the hot day, the dew had come quickly and heavily. The river made no sound. It was too fast and smooth. At the edge of the meadow, before he mounted to a piece of high ground to make camp, Nick looked down the river at the trout rising. They were rising to insects come from the swamp on the other side of the stream when the sun went down. The trout jumped out of water to take them. While Nick walked through the little stretch of meadow alongside the stream, trout had jumped high out of water. Now as he looked down the river, the insects must be settling on the surface, for the trout were feeding steadily all down the stream. As far down the long stretch as he could see, the trout were rising, making circles all down the surface of the water, as though it were starting to rain.

The ground rose, wooded and sandy, to overlook the meadow, the stretch of river and the swamp. Nick dropped his pack and rod-case and looked for a level piece of ground. He was very hungry and he wanted to make his camp before he cooked. Between two jack pines, the ground was quite level. He took the ax out of the pack and chopped out two projecting roots. That leveled a piece of ground large enough to sleep on. He smoothed out the sandy soil with his hand and pulled all the sweet fern bushes by their roots. His hands smelled good from the sweet fern. He smoothed the uprooted earth. He did not want anything making lumps under the blankets. When he had the ground smooth, he spread his three blankets. One he folded double, next to the ground. The other two he spread on top.

With the ax he slit off a bright slab of pine from one of the stumps and split it into pegs for the tent. He wanted them long and solid to hold in the ground. With the tent unpacked and spread on the ground, the pack, leaning against a jackpine, looked much smaller. Nick tied the rope that served the tent for a ridge-pole to the trunk of one of the pine trees and pulled the tent up off the ground with the other end of the rope and tied it to the other pine. The tent hung on the rope like a canvas blanket on a clothes line. Nick poked a pole he had cut up under the back peak of the canvas and then made it a tent by pegging out the sides. He pegged the sides out taut and drove the pegs deep, hitting them down into the ground with the flat of the ax until the rope loops were buried and the canvas was drum tight.

Across the open mouth of the tent Nick fixed cheese cloth to keep out mosquitoes. He crawled inside under the mosquito bar with various things from the pack to put at the head of the bed under the slant of the canvas. Inside the tent the light came through the brown canvas. It smelled pleasantly of canvas. Already there was something mysterious and homelike. Nick was happy as he crawled inside the tent. He had not been unhappy all day. This was different though. Now things were done. There had been this to do. Now it was done. It had been a hard trip. He was very tired. That was done. He had made his camp. He was settled. Nothing could touch him. It was a good place to camp. He was there, in the good place. He was in his home where he had made it. Now he was hungry.

He came out, crawling under the cheese cloth. It was quite dark outside. It was lighter in the tent.

Nick went over to the pack and found, with his fingers, a long nail in a paper sack of nails, in the bottom of the pack. He drove it into the pine tree, holding it close and hitting it gently with the flat of the ax. He hung the pack up on the nail. All his supplies were in the pack. They were off the ground and sheltered now.

Nick was hungry. He did not believe he had ever been hungrier. He opened and emptied a can of pork and beans and a can of spaghetti into the frying pan.

"I've got a right to eat this kind of stuff, if I'm willing to carry it," Nick said. His voice sounded strange in the darkening woods. He did not speak again.

He started a fire with some chunks of pine he got with the ax from a stump. Over the fire he stuck a wire grill, pushing the four legs down into the ground with his boot. Nick put the frying pan on the grill over the flames. He was hungrier. The beans and spaghetti warmed. Nick stirred them and mixed them together. They began to bubble, making little bubbles that rose with difficulty to the surface. There was a good smell. Nick got out a bottle of tomato catchup and cut four slices of bread. The little bubbles were coming faster now. Nick sat down beside the fire and lifted the frying pan off. He poured about half the contents out into the tin plate. It spread slowly on the plate. Nick knew it was too hot. He poured on some tomato catchup. He knew the beans and spaghetti were still too hot. He looked at the fire, then at the tent, he was not going to spoil it all by burning his tongue. For years he had never enjoyed fried bananas because he had never been able to wait for them to cool. His tongue was very sensitive. He was very hungry. Across the river in the swamp, in the almost dark, he saw a mist rising. He looked at the tent once more. All right. He took a full spoonful from the plate.

"Chrise," Nick said, "Geezus Chrise," he said happily.

He ate the whole plateful before he remembered the bread. Nick finished the second plateful with the bread, mopping the plate shiny. He had not eaten since a cup of coffee and a ham sandwich in the station restaurant at St. Ignace. It had been a very fine experience. He had been that hungry before, but had not been able to satisfy it. He could have made camp hours before if he had wanted to. There were plenty of good places to camp on the river. But this was good.

Nick tucked two big chips of pine under the grill. The fire flared up. He had forgotten to get water for the coffee. Out of the pack he got a folding canvas bucket and walked down the hill, across the edge of the meadow, to the stream. The other bank was in the white mist. The grass was wet and cold as he knelt on the bank and dipped the canvas bucket into the stream. It bellied and pulled hard in the current. The water was ice cold. Nick rinsed the bucket and carried it full up to the camp. Up away from the stream it was not so cold.

Nick drove another big nail and hung up the bucket full of water. He dipped the coffee pot half full, put some more chips under the grill onto the fire and put the pot on. He could not remember which way he made coffee. He could remember an argument about it with Hopkins, but not which side he had taken. He decided to bring it to a boil. He remembered now that was Hopkins's way. He had once argued about everything with Hopkins. While he waited for the coffee to boil, he opened a small can of apricots. He liked to open cans. He emptied the can of apricots out into a tin cup. While he watched the coffee on the fire, he drank the juice syrup of the apricots, carefully at first to keep from spilling, then meditatively, sucking the apricots down. They were better than fresh apricots.

The coffee boiled as he watched. The lid came up and coffee and grounds ran down the side of the pot. Nick took it off the grill. It was a triumph for Hopkins. He put sugar in the empty apricot cup and poured some of the coffee out to cool. It was too hot to pour and he used his hat to hold the handle of the coffee pot. He would not let it steep in the pot at all. Not the first cup. It should be straight Hopkins all the way. Hop deserved that. He was a very serious coffee maker. He was the most serious man Nick had ever known. Not heavy, serious. That was a long time ago. Hopkins spoke without moving his lips. He had played polo. He made made millions of dollars in Texas. He had borrowed car-fare to go to Chicago, when the wire came that his first big well had come in. He could have wired for money. That would have been too slow. They called Hop's girl the Blonde Venus. Hop did not mind because she was not his real girl. Hopkins said very confidently that none of them would make fun of his real girl. He was right. Hopkins went away when the telegram came. That was on the Black River. It took eight days for the telegram to reach him. Hopkins gave away his .22 caliber Colt automatic pistol to Nick. He gave his camera to Bill. It was to remember him always by. They were all going fishing again next summer. The Hop Head was rich. He would get a yacht and they would all cruise along the north shore of Lake Superior. He was excited but serious. They said good-bye and all felt bad. It broke up the trip. They never saw Hopkins again. That was a long time ago on the Black River.

Nick drank the coffee, the coffee according to Hopkins. The coffee was bitter. Nick laughed. It made a good ending to the story. His mind was starting to work. He knew he could choke it because he was tired enough. He spilled the coffee out of the pot and shook the grounds loose into the fire. He lit a cigarette and went inside the tent. He took off his shoes and trousers, sitting on the blankets, rolled the shoes up inside the trousers for a pillow and got in between the blankets.

Out through the front of the tent he watched the glow of the fire, when the night wind blew on it. It was a quiet night. The swamp was perfectly quiet. Nick stretched under the blanket comfortably. A mosquito hummed close to his ear. Nick sat up and lit a match. The mosquito was on the canvas, over his head. Nick moved the match quickly up to it. The mosquito made a satisfactory hiss in the flame. The match went out. Nick lay down again under the blankets. He turned on his side and shut his eyes. He was sleepy. He felt sleep coming. He curled up under the blanket and went to sleep.

Big Two-Hearted River

Part II

In the morning the sun was up and the tent was starting to get hot. Nick crawled out under the mosquito netting stretched across the mouth of the tent, to look at the morning. The grass was wet on his hands as he came out. He held his trousers and his shoes in his hands. The sun was just up over the hill. There was the meadow, the river and the swamp. There were birch trees in the green of the swamp on the other side of the river.

The river was clear and smoothly fast in the early morning. Down about two hundred yards were three logs all the way across the stream. They made the water smooth and deep above them. As Nick watched, a mink crossed the river on the logs and went into the swamp. Nick was excited. He was excited by the early morning and the river. He was really too hurried to eat breakfast, but he knew he must. He built a little fire and put on the coffee pot. While the water was heating in the pot he took an empty bottle and went down over the edge of the high ground to the meadow. The meadow was wet with dew and Nick wanted to catch grasshoppers for bait before the sun dried the grass. He found plenty of good grasshoppers. They were at the base of the grass stems. Sometimes they clung to a grass stem. They were cold and wet with the dew, and could not jump until the sun warmed them. Nick picked them up, taking only the medium sized brown ones, and put them into the bottle. He turned over a log and just under the shelter of the edge were several hundred hoppers. It was a grasshopper lodging house. Nick put about fifty of the medium browns into the bottle. While he was picking up the hoppers the others warmed in the sun and commenced to hop away. They flew when they hopped. At first they made one flight and stayed stiff when they landed, as though they were dead.

Nick knew that by the time he was through with breakfast they would be as lively as ever. Without dew in the grass it would take him all day to catch a bottle full of good grasshoppers and he would have to crush many of them, slamming at them with his hat. He washed his hands at the stream. He was excited to be near it. Then he walked up to the tent. The hoppers were already jumping stiffly in the grass. In the bottle, warmed by the sun, they were jumping in a mass. Nick put in a pine stick as a cork. It plugged the mouth of the bottle enough, so the hoppers could not get out and left plenty of air passage.

He had rolled the log back and knew he could get grasshoppers there every morning.

Nick laid the bottle full of jumping grasshoppers against a pine trunk. Rapidly he mixed some buckwheat flour with water and stirred it smooth, one cup of flour, one cup of water. He put a handful of coffee in the pot and dipped a lump of grease out of a can and slid it sputtering across the hot skillet. On the smoking skillet he poured smoothly the buckwheat batter. It spread like lava, the grease spitting sharply. Around the edges the buckwheat cake began to firm, then brown, then crisp. The surface was bubbling slowly to porousness. Nick pushed under the browned under surface with a fresh pine chip. He shook the skillet sideways and the cake was loose on the surface. I won't try and flop it, he thought. He slid the chip of clean wood all the way under the cake, and flopped it over onto its face. It sputtered in the pan.

When it was cooked Nick greased the skillet. He used all the batter. It made another big flapjack and one smaller one.

Nick ate a big flapjack and a smaller one, covered with apple butter. He put apple butter on the third cake, folded it over twice, wrapped it in oiled paper and put it in his shirt pocket. He put the apple butter jar back in the pack and cut bread for two sandwiches.

In the pack he found a big onion. He sliced it in two and peeled the silky outer skin. Then he cut one half into slices and made onion sandwiches. He wrapped them in oiled paper and buttoned them in the other pocket of his khaki shirt. He turned the skillet upside down on the grill, drank the coffee, sweetened and yellow brown with the condensed milk in it, and tidied up the camp. It was a nice little camp.

Nick took his fly rod out of the leather rod-case, jointed it, and shoved the rod-case back into the tent. He put on the reel and threaded the line through the guides. He had to hold it from hand to hand, as he threaded it, or it would slip back through its own weight. It was a heavy, double tapered fly line. Nick had paid eight dollars for it a long time ago. It was made heavy to lift back in the air and come forward flat and heavy and straight to make it possible to cast a fly which has no weight. Nick opened the aluminum leader box. The leaders were coiled between the damp flannel pads. Nick had wet the pads at the water cooler on the train up to St. Ignace. In the damp pads the gut leaders had softened and Nick unrolled one and tied it by a loop at the end to the heavy fly line. He fastened a hook on the end of the leader. It was a small hook; very thin and springy.

Nick took it from his hook book, sitting with the rod across his lap. He tested the knot and the spring of the rod by pulling the line taut. It was a good feeling. He was careful not to let the hook bite into his finger.

He started down to the stream, holding his rod, the bottle of grasshoppers hung from his neck by a thong tied in half hitches around the neck of the bottle. His landing net hung by a hook from his belt. Over his shoulder was a long flour sack tied at each corner into an ear. The cord went over his shoulder. The sack flapped against his legs.

Nick felt awkward and professionally happy with all his equipment hanging from him. The grasshopper bottle swung against his chest. In his shirt the breast pockets bulged against him with the lunch and his fly book.

He stepped into the stream. It was a shock. His trousers clung tight to his legs. His shoes felt the gravel. The water was a rising cold shock.

Rushing, the current sucked against his legs. Where he stepped in, the water was over his knees. He waded with the current. The gravel slid under his shoes. He looked down at the swirl of water below each leg and tipped up the bottle to get a grasshopper.

The first grasshopper gave a jump in the neck of the bottle and went out into the water. He was sucked under in the whirl by Nick's right leg and came to the surface a little way down stream.

1532 ★ *Ernest Hemingway*

He floated rapidly, kicking. In a quick circle, breaking the smooth surface of the water, he disappeared. A trout had taken him.

Another hopper poked his head out of the bottle. His antennæ wavered. He was getting his front legs out of the bottle to jump. Nick took him by the head and held him while he threaded the slim hook under his chin, down through his thorax and into the last segments of his abdomen. The grasshopper took hold of the hook with his front feet, spitting tobacco juice on it. Nick dropped him into the water.

Holding the rod in his right hand he let out line against the pull of the grasshopper in the current. He stripped off line from the reel with his left hand and let it run free. He could see the hopper in the little waves of the current. It went out of sight.

There was a tug on the line. Nick pulled against the taut line. It was his first strike. Holding the now living rod across the current, he brought in the line with his left hand. The rod bent in jerks, the trout pumping against the current. Nick knew it was a small one. He lifted the rod straight up in the air. It bowed with the pull.

He saw the trout in the water jerking with his head and body against the shifting tangent of the line in the stream.

Nick took the line in his left hand and pulled the trout, thumping tiredly against the current, to the surface. His back was mottled the clear, water-over-gravel color, his side flashing in the sun. The rod under his right arm, Nick stooped, dipping his right hand into the current. He held the trout, never still, with his moist right hand, while he unhooked the barb from his mouth, then dropped him back into the stream.

He hung unsteadily in the current, then settled to the bottom beside a stone. Nick reached down his hand to touch him, his arm to the elbow under water. The trout was steady in the moving stream, resting on the gravel, beside a stone. As Nick's fingers touched him, touched his smooth, cool, underwater feeling he was gone, gone in a shadow across the bottom of the stream.

He's all right, Nick thought. He was only tired.

He had wet his hand before he touched the trout, so he would not disturb the delicate mucus that covered him. If a trout was touched with a dry hand, a white fungus attacked the unprotected spot. Years before when he had fished crowded streams, with fly fishermen ahead of him and behind him, Nick had again and again come on dead trout, furry with white fungus, drifted against a rock, or floating belly up in some pool. Nick did not like to fish with other men on the river. Unless they were of your party, they spoiled it.

He wallowed down the stream, above his knees in the current, through the fifty yards of shallow water above the pile of logs that crossed the stream. He did not rebait his hook and held it in his hand as he waded. He was certain he could catch small trout in the shallows, but he did not want them. There would be no big trout in the shallows this time of day.

Now the water deepened up his thighs sharply and coldly. Ahead was the smooth dammed-back flood of water above the logs. The water was smooth and dark; on the left, the lower edge of the meadow; on the right the swamp.

Nick leaned back against the current and took a hopper from the bottle. He threaded the hopper on the hook and spat on him for good luck. Then he pulled several yards of line from the reel and tossed the hopper out ahead onto the fast, dark water. It floated down towards the logs, then the weight of the line pulled the bait under the surface. Nick held the rod in his right hand, letting the line run out through his fingers.

There was a long tug. Nick struck and the rod came alive and dangerous, bent double, the line tightening, coming out of water, tightening, all in a heavy, dangerous, steady pull. Nick felt the moment when the leader would break if the strain increased and let the line go.

The reel ratcheted into a mechanical shriek as the line went out in a rush. Too fast. Nick could not check it, the line rushing out, the reel note rising as the line ran out.

With the core of the reel showing, his heart feeling stopped with the excitement, leaning back against the current that mounted icily his thighs, Nick thumbed the reel hard with his left hand. It was awkward getting his thumb inside the fly reel frame.

As he put on pressure the line tightened into sudden hardness and beyond the logs a huge trout went high out of water. As he jumped, Nick lowered the tip of the rod. But he felt, as he dropped the tip to ease the strain, the moment when the strain was too great; the hardness too tight. Of course, the leader had broken. There was no mistaking the feeling when all spring left the line and it became dry and hard. Then it went slack.

His mouth dry, his heart down, Nick reeled in. He had never seen so big a trout. There was a heaviness, a power not to be held, and then the bulk of him, as he jumped. He looked as broad as a salmon.

Nick's hand was shaky. He reeled in slowly. The thrill had been too much. He felt, vaguely, a little sick, as though it would be better to sit down.

The leader had broken where the hook was tied to it. Nick took it in his hand. He thought of the trout somewhere on the bottom, holding himself steady over the gravel, far down below the light, under the logs, with the hook in his jaw. Nick knew the trout's teeth would cut through the snell of the hook. The hook would imbed itself in his jaw. He'd bet the trout was angry. Anything that size would be angry. That was a trout. He had been solidly hooked. Solid as a rock. He felt like a rock, too, before he started off. By God, he was a big one. By God, he was the biggest one I ever heard of.

Nick climbed out onto the meadow and stood, water running down his trousers and out of his shoes, his shoes squlchy. He went over and sat on the logs. He did not want to rush his sensations any.

He wriggled his toes in the water, in his shoes, and got out a cigarette from his breast pocket. He lit it and tossed the match into the fast water below the logs. A tiny trout rose at the match, as it swung around in the fast current. Nick laughed. He would finish the cigarette.

He sat on the logs, smoking, drying in the sun, the sun warm on his back, the river shallow ahead entering the woods, curving into the woods, shallows, light glittering, big water-smooth rocks, cedars along the bank and white birches, the logs warm in the sun, smooth to sit on, without bark, gray to the touch; slowly the feeling of disappointment left him. It went away slowly, the feeling of disappointment that came sharply after the thrill that made his shoulders ache. It was all right now. His rod lying out on the logs, Nick tied a new hook on the leader, pulling the gut tight until it grimped into itself in a hard knot.

He baited up, then picked up the rod and walked to the far end of the logs to get into the water, where it was not too deep. Under and beyond the logs was a deep pool. Nick walked around the shallow shelf near the swamp shore until he came out on the shallow bed of the stream.

On the left, where the meadow ended and the woods began, a great elm tree was uprooted. Gone over in a storm, it lay back into the woods, its roots clotted with dirt, grass growing in them, rising a solid bank beside the stream. The river cut to the edge of the uprooted tree. From where Nick stood he could see deep channels, like ruts, cut in the shallow bed of the stream by the flow of the current. Pebbly where he stood and pebbly and full of boulders beyond; where it curved near the tree roots, the bed of the stream was marly and between the ruts of deep water green weed fronds swung in the current.

Nick swung the rod back over his shoulder and forward, and the line, curving forward, laid the grasshopper down on one of the deep channels in the weeds. A trout struck and Nick hooked him.

Holding the rod far out toward the uprooted tree and sloshing backward in the current, Nick worked the trout, plunging, the rod bending alive, out of the danger of the weeds into the open river. Holding the rod, pumping alive against the current, Nick brought the trout in. He rushed, but always came, the spring of the rod yielding to the rushes, sometimes jerking under water, but always bringing him in. Nick eased downstream with the rushes. The rod above his head he led the trout over the net, then lifted.

The trout hung heavy in the net, mottled trout back and silver sides in the meshes. Nick unhooked him; heavy sides, good to hold, big undershot jaw, and slipped him, heaving and big sliding, into the long sack that hung from his shoulders in the water.

Nick spread the mouth of the sack against the current and it filled, heavy with water. He held it up, the bottom in the stream, and the water poured out through the sides. Inside at the bottom was the big trout, alive in the water.

Nick moved downstream. The sack out ahead of him, sunk, heavy in the water, pulling from his shoulders.

It was getting hot, the sun hot on the back of his neck.

Nick had one good trout. He did not care about getting many trout. Now the stream was shallow and wide. There were trees along both banks. The trees of the left bank made short shadows on the current in the forenoon sun. Nick knew there were trout in each shadow. In the afternoon, after the sun had crossed toward the hills, the trout would be in the cool shadows on the other side of the stream.

The very biggest ones would lie up close to the bank. You could always pick them up there on the Black. When the sun was down they all moved out into the current. Just when the sun made the water blinding in the glare before it went down, you were liable to strike a big trout anywhere in the current. It was almost impossible to fish then, the surface of the water was blinding as a mirror in the sun. Of course, you could fish upstream, but in a stream like the Black, or this, you had to wallow against the current and in a deep place, the water piled up on you. It was no fun to fish upstream with this much current.

Nick moved along through the shallow stretch watching the banks for deep holes. A beech tree grew close beside the river, so that the branches hung down into the water. The stream went back in under the leaves. There were always trout in a place like that.

Nick did not care about fishing that hole. He was sure he would get hooked in the branches.

It looked deep though. He dropped the grasshopper so the current took it under water, back in under the overhanging branch. The line pulled hard and Nick struck. The trout threshed heavily, half out of water in the leaves and branches. The line was caught. Nick pulled hard and the trout was off. He reeled in and holding the hook in his hand, walked down the stream.

Ahead, close to the left bank, was a big log. Nick saw it was hollow; pointing up river the current entered it smoothly, only a little ripple spread each side of the log. The water was deepening. The top of the hollow log was gray and dry. It was partly in the shadow.

Nick took the cork out of the grasshopper bottle and a hopper clung to it. He picked him off, hooked him and tossed him out. He held the rod far out so that the hopper on the water moved into the current flowing into the hollow log. Nick lowered the rod and the hopper floated in. There was a heavy strike. Nick swung the rod against the pull. It felt as though he were hooked into the log itself, except for the live feeling.

He tried to force the fish out into the current. It came, heavily. The line went slack and Nick thought the trout was gone. Then he saw him, very near, in the current, shaking his head, trying to get the hook out. His mouth was clamped shut. He was fighting the hook in the clear flowing current.

Looping in the line with his left hand, Nick swung the rod to make the line taut and tried to lead the trout toward the net, but he was gone, out of sight, the line pumping. Nick fought him against the current, letting him thump in the water against the spring of the rod. He shifted the rod to his left hand, worked the trout upstream, holding his weight, fighting on the rod, and then let him down into the net. He lifted him clear of the water, a heavy half circle in the net, the net dripping, unhooked him and slid him into the sack.

He spread the mouth of the sack and looked down in at the two big trout alive in the water.

Through the deepening water, Nick waded over to the hollow log. He took the sack off, over his head, the trout flopping as it came out of water, and hung it so the trout were deep in the water. Then he pulled himself up on the log and sat, the water from his trousers and boots running down into the stream. He laid his rod down, moved along to the shady end of the log and took the sandwiches out of his pocket. He dipped the sandwiches in the cold water. The current carried away the crumbs. He ate the sandwiches and dipped his hat full of water to drink, the water running out through his hat just ahead of his drinking.

It was cool in the shade, sitting on the log. He took a cigarette out and struck a match to light it. The match sunk into the gray wood, making a tiny furrow. Nick leaned over the side of the log, found a hard place and lit the match. He sat smoking and watching the river.

Ahead the river narrowed and went into a swamp. The river became smooth and deep and the swamp looked solid with cedar trees, their trunks close together, their branches solid. It would not be possible to walk through a swamp like that. The branches grew so low. You would have to keep almost level with the ground to move at all. You could not crash through the branches. That must be why the animals that lived in swamps were built the way they were, Nick thought.

He wished he had brought something to read. He felt like

reading. He did not feel like going on into the swamp. He looked down the river. A big cedar slanted all the way across the stream. Beyond that the river went into the swamp.

Nick did not want to go in there now. He felt a reaction against deep wading with the water deepening up under his armpits, to hook big trout in places impossible to land them. In the swamp the banks were bare, the big cedars came together overhead, the sun did not come through, except in patches; in the fast deep water, in the half light, the fishing would be tragic. In the swamp fishing was a tragic adventure. Nick did not want it. He did not want to go down the stream any further today.

He took out his knife, opened it and stuck it in the log. Then he pulled up the sack, reached into it and brought out one of the trout. Holding him near the tail, hard to hold, alive, in his hand, he whacked him against the log. The trout quivered, rigid. Nick laid him on the log in the shade and broke the neck of the other fish the same way. He laid them side by side on the log. They were fine trout.

Nick cleaned them, slitting them from the vent to the tip of the jaw. All the insides and the gills and tongue came out in one piece. They were both males; long gray-white strips of milt, smooth and clean. All the insides clean and compact, coming out all together. Nick tossed the offal ashore for the minks to find.

He washed the trout in the stream. When he held them back up in the water they looked like live fish. Their color was not gone yet. He washed his hands and dried them on the log. Then he laid the trout on the sack spread out on the log, rolled them up in it, tied the bundle and put it in the landing net. His knife was still standing, blade stuck in the log. He cleaned it on the wood and put it in his pocket.

Nick stood up on the log, holding his rod, the landing net hanging heavy, then stepped into the water and splashed ashore. He climbed the bank and cut up into the woods, toward the high ground. He was going back to camp. He looked back. The river just showed through the trees. There were plenty of days coming when he could fish the swamp.

1923, 1925

NATHANAEL WEST

1903–1940

A fatal automobile accident in 1940 took the life of Nathanael West and his bride of eight months, and thus cut short the career, begun less than a decade earlier, of America's first important Jewish novelist and one of its most radical writers. His was one of the most bizarre careers in the history of American letters.

West was born Nathan Weinstein, the son of Jewish immigrants from Russia. In 1925, before beginning his first novel, he changed his name, relishing the Biblical derivation of "Nathanael" (meaning "gift of God") but trying to screen the Jewish identity that, despite his fondness for his family, had embarrassed him since college days at Brown University. He was particularly close to the younger of his two sisters and to his father, a prosperous New York contractor, though he rebelled early against pressures to conform to middle-class educational and career patterns.

He read avidly even in his preteen years, encouraged by his family's devotion to learning and by their gifts of books—Shakespeare, Thomas Hardy, and the Continental classics. But he neglected his academic work in public school and left high school without graduating. He gained admission to Tufts College (from which he quickly flunked out), and then to Brown University, by forging admissions records. At Brown (1922–24) he began to cultivate the dress and manner of the gentile Ivy League elite, but also he took seriously his studies in literature, philosophy, and history, and he edited the campus literary magazine. Moreover, he read the literature of medieval Catholicism and that of the nineteenth-century aesthetic movement, the French Symbolists, contemporary Dadaists, and James Joyce, who nourished his taste for the surreal and the fantastic. He also read the vanguard contemporary American writers Gertrude Stein, Sherwood Anderson, and Hemingway, and the poets Pound, E. E. Cummings, Wallace Stevens, and William Carlos Williams. Though he worked at times on construction jobs and his family hoped he would enter business or a conventional profession after graduating from Brown, he persuaded his father and an uncle to underwrite two years in Paris (1926–27). There he sustained his virtually compulsive devotion to literature while working on his first short novel, *The Dream Life of Balso Snell* (1931), a garish comic fantasy exposing the pretensions of artists and the decadence of modern life.

West's efforts to support a literary career were based on a series of expedients. For intervals between 1927 and 1932 he worked as a night clerk in two New York City hotels, where he could read while on duty, gather material for his fiction by surreptitiously opening the residents' mail, and fraudulently secure rooms at little or no cost for his literary friends. From 1933 until his death he worked in Hollywood on a series of routine scripts for inconsequential films to earn money that would enable him to continue his serious writing. He was co-editor of the literary journal *Contact* with William Carlos Williams in 1932, and he edited the magazine *Americana* briefly in 1933. After 1933 he took part in political controversy as a Communist sympathizer, though he never joined the party or made political ideology programmatic in his fiction.

West's four novels brought the Gothic novel (fiction marked by terror and compulsive fascination with forbidden and decadent themes) to a new pitch of intensity, and brought the comic grotesque to new heights of absurdity and inventiveness. He worked (as in *The Day of the Locust*) by parodying or caricaturing the themes and effects of masterpieces in literature and painting, to portray a world in which violence, utterly repulsive tastes, and shallow fantasies are outrageously grotesque but inescapably real, crying out for redemption but finding none. (*The Dream Life of Balso Snell* takes place inside the Trojan horse of ancient legend, which the poet Snell enters via the rectum.) In one masterpiece, *Miss Lonelyhearts*, the male author of an advice-to-the-lovelorn column in a newspaper is consumed by his sympathy for his correspondents, whose lives are a nightmare of vulgarity and bizarre disasters. He is a "priest of our time who has a religious experience" comparable, West declared, to cases analyzed by William James in *Varieties of Religious Experience*.[1] Yet his imitation of Christ's compassion proves to be as sterile as the lives of the victims he would help. In *A Cool Million* (1934), a jailhouse, a museum, and Wu Fong's whorehouse provide the setting for West's mockery of the Horatio Alger success myth, which he shows to be fraudulent but ineradicable in an America whose vulgarities and fantasies are the seedbed of fascism. West's last novel, *The Day of the Locust* (1939), found the fantasies and the realities of Hollywood to be indistinguishable in their fraudulence, and haunting as a portent of annihilation. West's works are, as he once declared, "lyric novels" constructed "according to Poe's definition of a lyric poem,"[2] and part of their bleak horror, as West recognized, is that "there is nothing to root for in my work and what is even worse, no rooters."[3]

During his brief lifetime West's fiction gained some favorable critical attention in America but earned scarcely a penny and found very few readers. In the aftermath of World War II, however, the publication of his works in Britain and France attracted favorable attention from both writers and critics, and paperback editions gained him hundreds of thousands of readers in the United States. West brought Biblical visions of national destiny and apocalypse, which had earlier shaped the imaginations of New England Puritans and of Hawthorne and Melville, into an explosive confrontation with the facts of modern life, a confrontation which had vivid meaning after the horrors of World War II. His combination of the agonized conscience with the comic had much in common with the fiction of such Jewish writers as Bernard Malamud and Saul Bellow who flourished after the war, and his parodic and grotesque effects proved to be in the vanguard of the later fiction of John Barth, Thomas Pynchon, and Joseph Heller.

From The Day of the Locust[1]

Chapter Eighteen

Faye moved out of the San Berdoo the day after the funeral. Tod didn't know where she had gone and was getting up the courage to call Mrs. Jenning[2] when he saw her from the window of his office.

1. "Some Notes on Miss Lonelyhearts," *Contempo*, 3 (May 15, 1933), 1–2.
2. *Ibid.*, 1.
3. Letter to George Milburn, quoted by Richard B. Lehman, introduction to *The Day of the Locust* (1950), p. xx.
1. The title derives from the accounts in Exodus 10.4–15 of Jehovah's laying waste to Egypt by a plague of locusts, and the prophecy in the apocalyptic Book of Revelation 9.3–10 of locusts sent by God to destroy those not sealed in his faith. The novel (originally entitled, *The Cheated*) depicts an assortment of characters who seek the Promised Land in Los Angeles, California. The cast includes Tod Hackett, a graduate of Yale's school of fine art, now in training as a costume and set designer and plan-

ning a mammoth painting entitled *The Burning of Los Angeles;* Harry Greener, a broken-down vaudevillian who peddles silver polish until he dies; his daughter Faye, a would-be film actress who enters a callhouse for prostitutes to earn the expenses of her father's funeral, then moves into the house of a reclusive eccentric, Homer Simpson; and Simpson, a painfully inarticulate Iowa bookkeeper residing in California to recover from pneumonia. The novel ends when a crowd riots at a Hollywood film premier, enflamed because Homer Simpson, gone berserk, has trampled child actor Adore Loomis to death.
2. Proprietress of the callhouse where Faye works.

She was dressed in the costume of a Napoleonic vivandière.[3] By the time he got the window open, she had almost turned the corner of the building. He shouted for her to wait. She waved, but when he got downstairs she was gone.

From her dress, he was sure that she was working in the picture called "Waterloo."[4] He asked a studio policeman where the company was shooting and was told on the back lot. He started toward it at once. A platoon of cuirassiers,[5] big men mounted on gigantic horses, went by. He knew that they must be headed for the same set and followed them. They broke into a gallop and he was soon outdistanced.

The sun was very hot. His eyes and throat were choked with the dust thrown up by the horses' hooves and his head throbbed. The only bit of shade he could find was under an ocean liner made of painted canvas with real lifeboats hanging from its davits. He stood in its narrow shadow for a while then went on toward a great forty-foot papier mâché[6] sphinx that loomed up in the distance. He had to cross a desert to reach it, a desert that was continually being made larger by a fleet of trucks dumping white sand. He had gone only a few feet when a man with a megaphone ordered him off.

He skirted the desert, making a wide turn to the right, and came to a Western street with a plank sidewalk. On the porch of the "Last Chance Saloon" was a rocking chair. He sat down on it and lit a cigarette.

From there he could see a jungle compound with a water buffalo tethered to the side of a conical grass hut. Every few seconds the animal groaned musically. Suddenly an Arab charged by on a white stallion. He shouted at the man, but got no answer. A little while later he saw a truck with a load of snow and several malamute dogs. He shouted again. The driver shouted something back, but didn't stop.

Throwing away his cigarette, he went through the swinging doors of the saloon. There was no back to the building and he found himself in a Paris street. He followed it to its end, coming out in a Romanesque[7] courtyard. He heard voices a short distance away and went toward them. On a lawn of fiber, a group of men and women in riding costume were picnicking. They were eating cardboard food in front of a cellophane waterfall. He started toward them to ask his way, but was stopped by a man who scowled and held up a sign— "Quiet, Please, We're Shooting." When Tod took another step forward, the man shook his fist threateningly.

Next he came to a small pond with large celluloid swans floating

3. Woman accompanying French troops to sell provisions and liquor.
4. Battle of Waterloo in Belgium where the British, under General Wellington, and their allies defeated the French under Napoleon in 1815.
5. Cavalry soldiers wearing armored breastplates.
6. Substance made by mixing paper pulp and glue, which hardens after being shaped in molds.
7. European architectural style prevailing between the Roman and Gothic periods, c. 775–1200.

on it. Across one end was a bridge with a sign that read, "To Kamp Komfit." He crossed the bridge and followed a little path that ended at a Greek temple dedicated to Eros.[8] The god himself lay face downward in a pile of old newspapers and bottles.

From the steps of the temple, he could see in the distance a road lined with Lombardy poplars. It was the one on which he had lost the cuirassiers. He pushed his way through a tangle of briars, old flats and iron junk, skirting the skeleton of a Zeppelin, a bamboo stockade, an adobe fort, the wooden horse of Troy, a flight of baroque palace stairs that started in a bed of weeds and ended against the branches of an oak, part of the Fourteenth Street elevated station, a Dutch windmill, the bones of a dinosaur, the upper half of the Merrimac,[9] a corner of a Mayan temple, until he finally reached the road.

He was out of breath. He sat down under one of the poplars on a rock made of brown plaster and took off his jacket. There was a cool breeze blowing and he soon felt more comfortable.

He had lately begun to think not only of Goya and Daumier but also of certain Italian artists of the seventeenth and eighteenth centuries,[1] of Salvator Rosa, Francesco Guardi and Monsu Desiderio, the painters of Decay and Mystery. Looking downhill now, he could see compositions that might have actually been arranged from the Calabrian[2] work of Rosa. There were partially demolished buildings and broken monuments, half-hidden by great, tortured trees, whose exposed roots writhed dramatically in the arid ground, and by shrubs that carried, not flowers or berries, but armories of spikes, hooks and swords.

For Guardi and Desiderio there were bridges which bridged nothing, sculpture in trees, palaces that seemed of marble until a whole stone portico began to flap in the light breeze. And there were figures as well. A hundred yards from where Tod was sitting a man in a derby hat leaned drowsily against the gilded poop of a Venetian barque and peeled an apple. Still farther on, a charwoman on a stepladder was scrubbing with soap and water the face of a Buddha thirty feet high.

He left the road and climbed across the spine of the hill to look down on the other side. From there he could see a ten-acre field of cockleburs spotted with clumps of sunflowers and wild gum. In the center of the field was a gigantic pile of sets, flats and props. While

8. Greek god of love.
9. I.e., a model of the *Merrimack*, warship salvaged and armored by the Confederacy during the Civil War, defeated by the Union's ironclad *Monitor* in the battle of Hampton Roads, 1862, later burned and abandoned.
1. Painters of the fantastic and grotesque: Francisco José de Goya y Lucientes (1746–1828), Spanish painter and graphic artist who recorded the horrors of the Napoleonic wars in *The Disasters of War* (1810–14); Honoré Daumier (1808–79), French caricaturist and painter; Salvator Rosa (1615–73), Italian painter of wild landscapes; Francesco Guardi (1712–93), Venetian painter; "Monsu Desiderio," the name used by several minor 17th-century painters of architectural fantasies.
2. Pertaining to Calabria, department in southern Italy.

he watched, a ten-ton truck added another load to it. This was the final dumping ground. He thought of Janvier's "Sargasso Sea."[3] Just as that imaginary body of water was a history of civilization in the form of a marine junkyard, the studio lot was one in the form of a dream dump. A Sargasso of the imagination! And the dump grew continually, for there wasn't a dream afloat somewhere which wouldn't sooner or later turn up on it, having first been made photographic by plaster, canvas, lath and paint. Many boats sink and never reach the Sargasso, but no dream ever entirely disappears. Somewhere it troubles some unfortunate person and some day, when that person has been sufficiently troubled, it will be reproduced on the lot.

When he saw a red glare in the sky and heard the rumble of cannon, he knew it must be Waterloo. From around a bend in the road trotted several cavalry regiments. They wore casques and chest armor of black cardboard and carried long horse pistols in their saddle holsters. They were Victor Hugo's[4] soldiers. He had worked on some of the drawings for their uniforms himself, following carefully the descriptions in "Les Miserables."

He went in the direction they took. Before long he was passed by the men of Lefebvre-Desnouttes,[5] followed by a regiment of gendarmes d'élite, several companies of chasseurs of the guard and a flying detachment of Rimbaud's lancers.

They must be moving up for the disastrous attack on La Haite Santée. He hadn't read the scenario and wondered if it had rained yesterday. Would Grouchy or Bulcher arrive? Grotenstein, the producer, might have changed it.

The sound of cannon was becoming louder all the time and the red fan in the sky more intense. He could smell the sweet, pungent odor of blank powder. It might be over before he could get there. He started to run. When he topped a rise after a sharp bend in the road, he found a great plain below him covered with early nineteenth-century troops, wearing all the gay and elaborate uniforms that used to please him so much when he was a child and spent long hours looking at the soldiers in an old dictionary. At the far end of the field, he could see an enormous hump around which the English and their allies were gathered. It was Mont St. Jean and they were getting ready to defend it gallantly. It wasn't quite finished, however, and swarmed with grips, property men, set dressers, carpenters and painters.

Tod stood near a eucalyptus tree to watch, concealing himself

---

3. The Sargasso Sea is a large region of relatively still water in the middle of the North Atlantic where seaweed and debris accumulate.
4. French poet, dramatist, and novelist (1802–85), whose novel *Les Miserables* (1862) contains a description of the Battle of Waterloo.

5. Names in this and the following paragraphs, except fictitious names, identify military commanders and their units. "Gendarmes d'élite": elite cavalry troops; "chasseurs": fast-moving cavalry or infantry troops; "lancers": cavalrymen armed with lances.

behind a sign that read, " 'Waterloo'—A Charles H. Grotenstein Production." Nearby a youth in a carefully torn horse guard's uniform was being rehearsed in his lines by one of the assistant directors.

"Vive L'Empereur!"[6] the young man shouted, then clutched his breast and fell forward dead. The assistant director was a hard man to please and made him do it over and over again.

In the center of the plain, the battle was going ahead briskly. Things looked tough for the British and their allies. The Prince of Orange commanding the center, Hill the right and Picton the left wing, were being pressed hard by the veteran French. The desperate and intrepid Prince was in an especially bad spot. Tod heard him cry hoarsely above the din of battle, shouting to the Hollande-Belgians, "Nassau! Brunswick! Never retreat!" Nevertheless, the retreat began. Hill, too, fell back. The French killed General Picton with a ball through the head and he returned to his dressing room. Alten was put to the sword and also retired. The colors of the Lunenberg battalion, borne by a prince of the family of Deux-Ponts, were captured by a famous child star in the uniform of a Parisian drummer boy. The Scotch Greys were destroyed and went to change into another uniform. Ponsonby's heavy dragoons were also cut to ribbons. Mr. Grotenstein would have a large bill to pay at the Western Costume Company.

Neither Napoleon nor Wellington was to be seen. In Wellington's absence, one of the assistant directors, a Mr. Crane, was in command of the allies. He reinforced his center with one of Chasse's brigades and one of Wincke's. He supported these with infantry from Brunswick, Welsh foot, Devon yeomanry and Hanoverian light horse with oblong leather caps and flowing plumes of horsehair.

For the French, a man in a checked cap ordered Milhaud's cuirassiers to carry Mont St. Jean. With their sabers in their teeth and their pistols in their hands, they charged. It was a fearful sight.

The man in the checked cap was making a fatal error. Mont St. Jean was unfinished. The paint was not yet dry and all the struts were not in place. Because of the thickness of the cannon smoke, he had failed to see that the hill was still being worked on by property men, grips and carpenters.

It was the classic mistake, Tod realized, the same one Napoleon had made. Then it had been wrong for a different reason. The Emperor had ordered the cuirassiers to charge Mont St. Jean not knowing that a deep ditch was hidden at its foot to trap his heavy cavalry. The result had been disaster for the French; the beginning of the end.

This time the same mistake had a different outcome. Waterloo,

6. "Long live the Emperor."

instead of being the end of the Grand Army, resulted in a draw. Neither side won, and it would have to be fought over again the next day. Big losses, however, were sustained by the insurance company in workmen's compensation. The man in the checked cap was sent to the dog house by Mr. Grotenstein just as Napoleon was sent to St. Helena.[7]

When the front rank of Milhaud's heavy division started up the slope of Mont St. Jean, the hill collapsed. The noise was terrific. Nails screamed with agony as they pulled out of joists. The sound of ripping canvas was like that of little children whimpering. Lath and scantling snapped as though they were brittle bones. The whole hill folded like an enormous umbrella and covered Napoleon's army with painted cloth.

It turned into a rout. The victors of Bersina, Leipsic, Austerlitz,[8] fled like schoolboys who had broken a pane of glass. "Sauve qui peut!"[9] they cried, or, rather, "Scram!"

The armies of England and her allies were too deep in scenery to flee. They had to wait for the carpenters and ambulances to come up. The men of the gallant Seventy-Fifth Highlanders were lifted out of the wreck with block and tackle. They were carted off by the stretcher-bearers, still clinging bravely to their claymores.[10]

1939

7. Island in the Mediterranean to which Napoleon was exiled from 1815 to 1821.
8. Sites of Napoleon's victories.
9. "Every man for himself."
10. Large two-edged Scottish sword.

RICHARD WRIGHT
1908–1960

Richard Wright—a black novelist, for ten years a member of the Communist party, and later an expatriate—was the first Negro writer to reach a large white audience with a militant attack on racism in America. His explosively powerful novel *Native Son* and his autobiographical volume *Black Boy* (1945) achieved immediate success and established him as the most important Negro writer before Ralph Ellison and James Baldwin.

The cruel debasements that Wright suffered as a black child in a dominantly white America find no full equivalents in the biographies of other American writers. Wright's father, an impoverished sharecropper on a Mississippi farm, abandoned his family when Wright was five, and his mother struggled as a housemaid to support the family until she was incapacitated when Wright was twelve. Wright was raised by a series of relatives in Jackson, Mississippi, and Memphis, Tennessee. His schooling was irregular (though he was valedictorian when he graduated from ninth grade in Memphis in 1925), and his grandmother's fervent evangelism ruled out children's games and forbade all but religious reading. His mother demanded his obedience to the codes of white supremacy, enforcing them with exceedingly harsh punishment when he violated them. Left to his own devices during the day when his relatives were working, he was introduced to hard liquor at the age of six and experienced the ruthlessness and desperation of ghetto street life. His autobiography records vividly the shock with which he learned that an uncle had been shot by whites who coveted his thriving liquor business, the vulgar contempt with which Wright was treated by white prostitutes and their customers when he worked as a hall boy in a disreputable hotel, his initiation into deceit and theft. Long before Wright decided to leave the South at the age of seventeen and follow his Aunt Maggie Wilson to Chicago, he had known the vehement feelings that were to impel his fiction: the "dread of white people [that] now came to live permanently in my feelings and imagination," the consciousness that "a white censor was standing over me," the "sustained expectation of violence," the fear that "Negroes had never been allowed to catch the full Western civilization, that they lived somehow in it but not of it."[1] In retrospect he felt so alienated that, unlike some other Negroes who found tenderness and sustaining strength within the black community, he found that his "life at home had cut me off, not only from white people but from Negroes as well" and he wondered if such qualities as "tenderness, love, honor, [and] loyalty," instead of being "native with man," must be "fostered, won, struggled and suffered for, preserved in ritual from one generation to another."[2] The only thing that had bolstered his identity and his hopes was reading, which he managed to do despite the censorious restrictions of his mother and grandmother and the refusal of southern libraries to grant him access to books. While working as an errand boy for an optical company in Memphis, he persuaded a white employee to lend him his library card so that he might surreptitiously charge books and forged the white man's name to a note to the librarian reading "*Will you please let this nigger boy * * * have some books by* [the critic] *H. L. Mencken?*"[3] When Wright left Memphis for Chicago in 1927 he had discovered Mencken's prefaces, Dreiser's *Sister Carrie* and *Jennie Gerhardt*, and the fiction by Sherwood Anderson and Sinclair Lewis that he thought "defensively critical of the straitened American environment." He wrote one story at the age of fifteen and published it in a Memphis Negro publication. He declared later that "it had been only through books—at best, no more than vicarious cultural transfusions—that I had managed to keep myself alive" and that it had been "my accidental reading of fiction and literary criticism that had evoked in me vague glimpses of life's possibilities."[4]

In Chicago Wright worked as a porter, burial-insurance salesman, and postal clerk before going on relief during the Depression and joining the WPA Writers Project, first as a writer of guidebooks, then as a director of the Federal Negro Theater. He steeped himself in the writings of the urban sociologists at the University of Chicago at the time when he was exploring Marxist theory, and Mrs. Mary Wirth, wife of Professor Louis Wirth, who had helped get him a job with the federal Writers Project, encouraged him in his writing. He joined the leftist John Reed Club in 1932 and published poems in its journal *Left Front*. Shortly afterward he joined the Communist party, whose journals published his poems and essays. Wright drew on the experience of fellow black party members for his first work of fiction, *Uncle*

---

1. *Black Boy*, pp. 64, 225, 171, 33.   3. *Ibid.*, p. 216.
2. *Ibid.*, pp. 187, 33.   4. *Ibid.*, pp. 226–27.

Tom's Children (1938), a collection of short novels, including *Long Black Song*. The tales, which won first prize in a WPA competition, treat the southern black's entanglement in the values of a white society that he or she takes for models of achievement, the violence inflicted on those blacks who defy the codes of white supremacy, and, in *Fire and Cloud*, a black preacher's commitment to a protest demonstration organized by Communists.

Wright moved to New York City in 1937 to become Harlem editor of the *Daily Worker*. His "legman" in the office was the future novelist Ralph Ellison, a musician and sculptor whom Wright persuaded to take up writing. While Wright continued to take part in Communist affairs until his break with the party in 1942, he found himself increasingly at odds with party policies that sought to constrain the artist's freedom and that subordinated black protests against racial injustice to the aims of Communist revolution. When Wright published *Native Son* in 1940 it shocked party members as well as non-Marxist middle-class readers, both black and white, because its criminal black hero, Bigger Thomas, is never fully understood by the Communist lawyer who defends him in court, nor by the blacks, nor by the whites who find in his baffled rage the threat of rape and a challenge to their power structure. Bigger's crimes—the accidental killing of the liberal white daughter of a wealthy slum landlord, his attempt to bury the evidence of that offense, the rape and intentional murder later of a black girl friend who might betray him—have the effect of exposing the combination of exploitative capitalism and condescending charity toward Negroes that traps him, but they also enable him to discover his identity and to attain the dignity of one who in puzzling over "why all the killing was" begins "to feel what I wanted, what I am." The book, Wright's first novel, won him the Spingarn Award of the National Association for the Advancement of Colored People and, along with his autobiographical *Black Boy* (1945), brought him financial security and international fame.

Wright's first marriage, to a white dancer in 1940, was short-lived. But in 1941 he married Ellen Poplar, another white woman and a fellow party worker who remained with Wright when he broke with the Communists in 1942 and who bore him two daughters. In 1947 they moved permanently to France, where Wright became an outspoken anti-Communist, an advisor to African diplomats, a frequent lecturer, the familiar friend of Gertrude Stein and of the Marxist Existentialist philosopher Sartre and other French intellectuals who admired Wright's work and introduced him to the writings of contemporary philosophers. The sense of alienation that was pronounced in Bigger Thomas became more central to his vision and governed both the fiction that he resumed writing in the 1950s and the nonfiction that increasingly preoccupied him. But his later writings seldom achieved the impact of his earlier works. In collections of travelogues and political essays (*Black Power*, 1954; *The Color Curtain*, 1956; *Pagan Spain*, 1957; and *White Man, Listen!*, 1957) he tried to affect the policies of the developing nations in Asia and Africa, asserting the advantages of a hybrid, Westernized "Asian-African elite."[5] But none of the developing nations would accept the repudiation of native traditions, their "ancestor-worshiping attitudes,"[6] that Wright's program seemed to call for. Neither *The Outsider* (1953) nor *The Long Dream* (1958), set in Mississippi, has the power of *Black Boy* or Wright's best fiction with a southern setting. Of the stories

5. *White Man, Listen!*, p. 65.   6. *Ibid.*, p. 133.

published posthumously in *Eight Men* (1961), only the brilliant *The Man Who Lived Underground*, written in 1942, has the power of his best writing.

That power derives from Wright's combination of bluntly direct narration with surrealistic or expressionistic gestures and imagery which reveal the explosive pressures beneath the façades of order in his world, and the intensity of obsessive fantasies which motivate his characters and challenge their capacity to articulate their feelings in word or deed. Though both the later black novelists James Baldwin and Ralph Ellison questioned Wright's resort to melodrama and the explicitness of protest in his fiction, Wright's example paved the way for the flourishing of black writing in subsequent decades and, in Ellison's words, "converted the American Negro impulse toward self-annihilation and 'going-underground' into a will to confront the world * * * and throw his findings unashamedly into the guilty conscience of America."[7]

The Man Who Was Almost a Man

Dave struck out across the fields, looking homeward through paling light. Whuts the usa talkin wid em niggers in the field? Anyhow, his mother was putting supper on the table. Them niggers can't understand *nothing*. One of these days he was going to get a gun and practice shooting, then they can't talk to him as though he were a little boy. He slowed, looking at the ground. Shucks, Ah ain scareda them even ef they are biggern me! Aw, Ah know whut Ahma do. . . . Ahm going by ol Joe's sto n git that Sears Roebuck catlog n look at them guns. Mabbe Ma will lemme buy one when she gits mah pay from ol man Hawkins. Ahma beg her t gimme some money. Ahm ol ernough to hava gun. Ahm seventeen. Almos a man. He strode, feeling his long, loose-jointed limbs. Shucks, a man oughta hava little gun aftah he done worked hard all day. . . .

He came in sight of Joe's store. A yellow lantern glowed on the front porch. He mounted steps and went through the screen door, hearing it bang behind him. There was a strong smell of coal oil and mackerel fish. He felt very confident until he saw fat Joe walk in through the rear door, then his courage began to ooze.

'Howdy, Dave! Whutcha want?'

'How yuh, Mistah Joe? Aw, Ah don wanna buy nothing. Ah jus wanted t see ef yuhd lemme look at tha ol catlog erwhile.'

'Sure! You wanna see it here?'

'Nawsuh. Ah wans t take it home wid me. Ahll bring it back termorrow when Ah come in from the fiels.'

'You plannin on buyin something?'

'Yessuh.'

'Your ma letting you have your own money now?'

'Shucks. Mistah Joe, Ahm gittin t be a man like anybody else!'

7. "Richard Wright's Blues," *Shadow and Act*, p. 94.

Joe laughed and wiped his greasy white face with a red bandanna.

'Whut you plannin on buyin?'

Dave looked at the floor, scratched his head, scratched his thigh, and smiled. Then he looked up shyly.

'Ahll tell yuh, Mistah Joe, ef yuh promise yuh won't tell.'

'I promise.'

'Waal, Ahma buy a gun.'

'A gun? Whut you want with a gun?'

'Ah wanna keep it.'

'You ain't nothing but a boy. You don't need a gun.'

'Aw, lemme have the catalog, Mistah Joe. Ahll bring it back.'

Joe walked through the rear door. Dave was elated. He looked around at barrels of sugar and flour. He heard Joe coming back. He craned his neck to see if he were bringing the book. Yeah, he's got it! Gawddog, he's got it!

'Here; but be sure you bring it back. It's the only one I got.'

'Sho, Mistah Joe.'

'Say, if you wanna buy a gun, why don't you buy one from me. I gotta gun to sell.'

'Will it shoot?'

'Sure it'll shoot.'

'Whut kind is it?'

'Oh, it's kinda old. . . . A Lefthand Wheeler. A pistol. A big one.'

'Is it got bullets in it?'

'It's loaded.'

'Kin Ah see it?'

'Where's your money?'

'Whut yuh wan fer it?'

'I'll let you have it for two dollars.'

'Just *two* dollahs? Shucks, Ah could buy tha when Ah git mah pay.'

'I'll have it here when you want it.'

'Awright, suh. Ah be in fer it.'

He went through the door, hearing it slam again behind him. Ahma git some money from Ma n buy me a gun! Only *two* dollahs! He tucked the thick catalogue under his arm and hurried.

'Where yuh been, boy?' His mother held a steaming dish of black-eyed peas.

'Aw, Ma, Ah jus stopped down the road t talk wid th boys.'

'Yuh know bettah than t keep suppah waitin.'

He sat down, resting the catalogue on the edge of the table.

'Yuh git up from there and git to the well n wash yosef! Ah ain feedin no hogs in mah house!'

She grabbed his shoulder and pushed him. He stumbled out of the room, then came back to get the catalogue.

'Whut this?'

'Aw, Ma, it's jusa catlog.'

'Who yuh git it from?'

'From Joe, down at the sto.'

'Waal, thas good. We kin use it around the house.'

'Naw, Ma.' He grabbed for it. 'Gimme mah catlog, Ma.'

She held onto it and glared at him.

'Quit hollerin at me! Whuts wrong wid yuh? Yuh crazy?'

'But Ma, please. It ain mine! It's Joe's! He tol me t bring it back t im termorrow.'

She gave up the book. He stumbled down the back steps, hugging the thick book under his arm. When he had splashed water on his face and hands, he groped back to the kitchen and fumbled in a corner for the towel. He bumped into a chair; it clattered to the floor. The catalogue sprawled at his feet. When he had dried his eyes he snatched up the book and held it again under his arm. His mother stood watching him.

'Now, ef yuh gonna acka fool over that ol book, Ahll take it n burn it up.'

'Naw, Ma, please.'

'Waal, set down n be still!'

He sat and drew the oil lamp close. He thumbed page after page, unaware of the food his mother set on the table. His father came in. Then his small brother.

'Whutcha got there, Dave?' his father asked.

'Jusa catlog,' he answered, not looking up.

'Ywah, here they is!' His eyes glowed at blue and black revolvers. He glanced up, feeling sudden guilt. His father was watching him. He eased the book under the table and rested it on his knees. After the blessing was asked, he ate. He scooped up peas and swallowed fat meat without chewing. Buttermilk helped to wash it down. He did not want to mention money before his father. He would do much better by cornering his mother when she was alone. He looked at his father uneasily out of the edge of his eye.

'Boy, how come yuh don quit foolin wid tha book n eat yo suppah?'

'Yessuh.'

'How yuh n ol man Hawkins gittin erlong?'

'Suh?'

'Can't yuh hear? Why don yuh lissen? Ah ast yuh how wuz yuh n ol man Hawkins gittin erlong?'

'Oh, swell, Pa. Ah plows mo lan than anybody over there.'

'Waal, yuh oughta keep yo min on whut yuh doin.'

'Yessuh.'

He poured his plate full of molasses and sopped at it slowly with a chunk of cornbread. When all but his mother had left the kitchen, he still sat and looked again at the guns in the catalogue. Lawd, ef

Ah only had tha pretty one! He could almost feel the slickness of the weapon with his fingers. If he had a gun like that he would polish it and keep it shining so it would never rust. N Ahd keep it loaded, by Gawd!

'Ma?'

'Hunh?'

'Ol man Hawkins give yuh mah money yit?'

'Yeah, but ain no usa yuh thinkin bout thowin nona it erway. Ahm keepin tha money sos yuh kin have cloes t go to school this winter.'

He rose and went to her side with the open catalogue in his palms. She was washing dishes, her head bent low over a pan. Shyly he raised the open book. When he spoke his voice was husky, faint.

'Ma, Gawd knows Ah wans one of these.'

'One of whut?' she asked, not raising her eyes.

'One of *these*,' he said again, not daring even to point. She glanced up at the page, then at him with wide eyes.

'Nigger is yuh gone plum crazy?'

'Ah, Ma——'

'Git outta here! Don yuh talk t me bout no gun! Yuh a fool!'

'Ma, Ah kin buy one fer *two* dollahs.'

'Not ef Ah knows it yuh ain!'

'But yuh promised me one——'

'Ah don care whut Ah promised! Yuh ain nothing but a boy yit!'

'Ma, ef yuh lemme buy one Ahll *never* ast yuh fer nothing no mo.'

'Ah tol yuh t git outta here! Yuh ain gonna toucha penny of tha money fer no gun! Thas how come Ah has Mistah Hawkins t pay yo wages t me, cause Ah knows yuh ain got no sense.'

'But Ma, we needa gun. Pa ain got no gun. We needa gun in the house. Yuh kin never tell whut might happen.'

'Now don yuh try to maka fool outta me, boy! Ef we did hava gun yuh wouldn't have it!'

He laid the catalogue down and slipped his arm around her waist.

'Aw, Ma, Ah done worked hard alla summer n ain ast yuh fer nothin, is Ah, now?'

'Thas whut yuh spose t do!'

'But Ma, Ah wans a gun. Yuh kin lemme have two dollahs outta mah money. Please, Ma. I kin give it to Pa . . . Please, Ma! Ah loves yuh, Ma.'

When she spoke her voice came soft and low.

'Whut yuh wan wida gun, Dave? Yuh don need no gun. Yuhll git in trouble. N ef yo Pa jus *thought* Ah let yuh have money t buy a gun he'd hava fit.'

'Ahll hide it, Ma, it ain but two dollahs.'

'Lawd, chil, whuts wrong wid yuh?'

'Ain nothing wrong, Ma. Ahm almos a man now. Ah wans a gun.'

'Who gonna sell yuh a gun?'

'Ol Joe at the sto.'

'N it don cos but two dollahs?'

'Thas all, Ma. Just two dollahs. Please, Ma.'

She was stacking the plates away; her hands moved slowly, reflectively. Dave kept an anxious silence. Finally, she turned to him.

'Ahll let yuh git tha gun ef yuh promise me one thing.'

'Whuts tha, Ma?'

'Yuh bring it straight back t *me*, yuh hear? Itll be fer Pa.'

'Yessum! Lemme go now, Ma.'

She stooped, turned slightly to one side, raised the hem of her dress, rolled down the top of her stocking, and came up with a slender wad of bills.

'Here,' she said. 'Lawd knows yuh don need no gun. But yer Pa does. Yuh bring it right back t *me*, yuh hear? Ahma put it up. Now ef yuh don, Ahma have yuh Pa lick yuh so hard yuh won ferget it.'

'Yessum.'

He took the money, ran down the steps, and across the yard.

'Dave! Yuuuuuh Daaaaave!'

He heard, but he was not going to stop now. 'Naw, Lawd!'

The first movement he made the following morning was to reach under his pillow for the gun. In the gray light of dawn he held it loosely, feeling a sense of power. Could killa man wida gun like this. Kill anybody, black er white. And if he were holding his gun in his hand nobody could run over him; they would have to respect him. It was a big gun, with a long barrel and a heavy handle. He raised and lowered it in his hand, marveling at its weight.

He had not come straight home with it as his mother had asked; instead he had stayed out in the fields, holding the weapon in his hand, aiming it now and then at some imaginary foe. But he had not fired it; he had been afraid that his father might hear. Also he was not sure he knew how to fire it.

To avoid surrendering the pistol he had not come into the house until he knew that all were asleep. When his mother had tiptoed to his bedside late that night and demanded the gun, he had first played 'possum; then he had told her that the gun was hidden outdoors, that he would bring it to her in the morning. Now he lay turning it slowly in his hands. He broke it, took out the cartridges, felt them, and then put them back.

He slid out of bed, got a long strip of old flannel from a trunk, wrapped the gun in it, and tied it to his naked thigh while it was still loaded. He did not go in to breakfast. Even though it was not yet daylight, he started for Jim Hawkins' plantation. Just as the sun was rising he reached the barns where the mules and plows were kept.

'Hey! That you, Dave?'

He turned. Jim Hawkins stood eying him suspiciously.

'Whatre yuh doing here so early?'

'Ah didn't know Ah wuz gittin up so early, Mistah Hawkins. Ah wuz fixin t hitch up ol Jenny n take her t the fiels.'

'Good. Since you're here so early, how about plowing that stretch down by the woods?'

'Suits me. Mistah Hawkins.'

'O.K. Go to it!'

He hitched Jenny to a plow and started across the fields. Hot dog! This was just what he wanted. If he could get down by the woods, he could shoot his gun and nobody would hear. He walked behind the plow, hearing the traces creaking, feeling the gun tied tight to his thigh.

When he reached the woods, he plowed two whole rows before he decided to take out the gun. Finally, he stopped, looked in all directions, then untied the gun and held it in his hand. He turned to the mule and smiled.

'Know whut this is, Jenny? Naw, yuh wouldn't know! Yuhs jusa ol mule! Anyhow, this is a gun, n it kin shoot, by Gawd!'

He held the gun at arm's length. Whut t hell, Ahma shoot this thing! He looked at Jenny again.

'Lissen here, Jenny! When Ah pull this ol trigger Ah don wan yuh t run n acka fool now.'

Jenny stood with head down, her short ears pricked straight. Dave walked off about twenty feet, held the gun far out from him, at arm's length, and turned his head. Hell, he told himself, Ah ain afraid. The gun felt loose in his fingers; he waved it wildly for a moment. Then he shut his eyes and tightened his forefinger. *Blooom!* A report half-deafened him and he thought his right hand was torn from his arm. He heard Jenny whinnying and galloping over the field, and he found himself on his knees, squeezing his fingers hard between his legs. His hand was numb; he jammed it into his mouth, trying to warm it, trying to stop the pain. The gun lay at his feet. He did not quite know what had happened. He stood up and stared at the gun as though it were a live thing. He gritted his teeth and kicked the gun. Yuh almos broke mah arm! He turned to look for Jenny; she was far over the fields, tossing her head and kicking wildly.

'Hol on there, ol mule!'

When he caught up with her she stood trembling, walling her big white eyes at him. The plow was far away; the traces had broken. Then Dave stopped short, looking, not believing. Jenny was bleeding. Her left side was red and wet with blood. He went closer. Lawd have mercy! Wondah did Ah shoot this mule? He grabbed for Jenny's mane. She flinched, snorted, whirled, tossing her head.

'Hol on now! Hol on.'

Then he saw the hole in Jenny's side, right between the ribs. It was round, wet, red. A crimson stream streaked down the front leg, flowing fast. Good Gawd! Ah wuznt shootin at tha mule. . . . He felt panic. He knew he had to stop that blood, or Jenny would bleed to death. He had never seen so much blood in all his life. He ran the mule for half a mile, trying to catch her. Finally she stopped, breathing hard, stumpy tail half arched. He caught her mane and led her back to where the plow and gun lay. Then he stopped and grabbed handfuls of damp black earth and tried to plug the bullet hole. Jenny shuddered, whinnied, and broke from him.

'Hol on! Hol on now!'

He tried to plug it again, but blood came anyhow. His fingers were hot and sticky. He rubbed dirt hard into his palms, trying to dry them. Then again he attempted to plug the bullet hole, but Jenny shied away, kicking her heels high. He stood helpless. He had to do something. He ran at Jenny; she dodged him. He watched a red stream of blood flow down Jenny's leg and form a bright pool at her feet.

'Jenny . . . Jenny . . .' he called weakly.

His lips trembled. She's bleeding t death! He looked in the direction of home, wanting to go back, wanting to get help. But he saw the pistol lying in the damp black clay. He had a queer feeling that if he only did something, this would not be; Jenny would not be there bleeding to death.

When he went to her this time, she did not move. She stood with sleepy, dreamy eyes; and when he touched her she gave a low-pitched whinny and knelt to the ground, her front knees slopping in blood.

'Jenny . . . Jenny . . .' he whispered.

For a long time she held her neck erect; then her head sank, slowly. Her ribs swelled with a mighty heave and she went over.

Dave's stomach felt empty, very empty. He picked up the gun and held it gingerly between his thumb and forefinger. He buried it at the foot of a tree. He took a stick and tried to cover the pool of blood with dirt—but what was the use? There was Jenny lying with her mouth open and her eyes walled and glassy. He could not tell Jim Hawkins he had shot his mule. But he had to tell something. Yeah, Ahll tell em Jenny started gittin wil n fell on the joint of the plow. . . . But that would hardly happen to a mule. He walked across the field slowly, head down.

It was sunset. Two of Jim Hawkins' men were over near the edge of the woods digging a hole in which to bury Jenny. Dave was surrounded by a knot of people; all of them were looking down at the dead mule.

'I don't see how in the world it happened,' said Jim Hawkins for the tenth time.

The crowd parted and Dave's mother, father, and small brother pushed into the center.

'Where Dave?' his mother called.

'There he is,' said Jim Hawkins.

His mother grabbed him.

'Whut happened, Dave? Whut yuh done?'

'Nothing.'

'C'mon, boy, talk,' his father said.

Dave took a deep breath and told the story he knew nobody believed.

'Waal,' he drawled. 'Ah brung ol Jenny down here sos Ah could do mah plowin. Ah plowed bout two rows, just like yuh see.' He stopped and pointed at the long rows of upturned earth. 'Then something musta been wrong wid ol Jenny. She wouldn't ack right a-tall. She started snortin n kickin her heels. Ah tried to hol her, but she pulled erway, rearin n goin on. Then when the point of the plow was stickin up in the air, she swung erroun n twisted hersef back on it. . . . She stuck hersef n started t bleed. N fo Ah could do anything, she wuz dead.'

'Did you ever hear of anything like that in all your life?' asked Jim Hawkins.

There were white and black standing in the crowd. They murmured. Dave's mother came close to him and looked hard into his face.

'Tell the truth, Dave,' she said.

'Looks like a bullet hole ter me,' said one man.

'Dave, whut yuh do wid tha gun?' his mother asked.

The crowd surged in, looking at him. He jammed his hands into his pockets, shook his head slowly from left to right, and backed away. His eyes were wide and painful.

'Did he hava gun?' asked Jim Hawkins.

'By Gawd, Ah tol yuh tha wuz a *gun* wound,' said a man, slapping his thigh.

His father caught his shoulders and shook him till his teeth rattled.

'Tell whut happened, yuh rascal! Tell whut . . .'

Dave looked at Jenny's stiff legs and began to cry.

'Whut yuh do wid tha gun?' his mother asked.

'Whut wuz he doin wida gun?' his father asked.

'Come on and tell the truth,' said Hawkins. 'Ain't nobody going to hurt you . . .'

His mother crowded close to him.

'Did yuh shoot tha mule, Dave?'

Dave cried, seeing blurred white and black faces.

"Ahh ddinnt gggo tt sshoooot hher. . . . Ah ssswear off Gawd Ahh ddint. . . . Ah wuz a-tryin' t sssee ef the ol gggun wouid sshoot——'

'Where yuh git the gun from?' his father asked.

'Ah got it from Joe, at the sto.'

'Where yuh git the money?'

'Ma give it t me.'

'He kept worryin me, Bob. . . . Ah had t. . . . Ah tol im t bring the gun right back t me. . . . It was fer yuh, the gun.'

'But how yuh happen to shoot that mule?' asked Jim Hawkins.

'Ah wuznt shootin at the mule, Mistah Hawkins. The gun jumped when Ah pulled the trigger . . . N fo Ah knowed anything Jenny wuz there a-bleedin.'

Somebody in the crowd laughed. Jim Hawkins walked close to Dave and looked into his face.

'Well, looks like you have bought you a mule, Dave.'

'Ah swear fo Gawd, Ah didn't go t kill the mule, Mistah Hawkins!'

'But you killed her!'

All the crowd was laughing now. They stood on tiptoe and poked heads over one another's shoulders.

'Well, boy, looks like yuh done bought a dead mule! Hahaha!'

'Ain tha ershame.'

'Hohohohoho.'

Dave stood head down, twisting his feet in the dirt.

'Well, you needn't worry about it, Bob,' said Jim Hawkins to Dave's father. 'Just let the boy keep on working and pay me two dollars a month.'

'Whut yuh wan fer yo mule, Mistah Hawkins?'

Jim Hawkins screwed up his eyes.

'Fifty dollars.'

'Whut yuh do wid tha gun?' Dave's father demanded.

Dave said nothing.

'Yuh wan me t take a tree lim n beat yuh till yuh talk!'

'Nawsuh!'

'Whut yuh do wid it?'

'Ah thowed it erway.'

'Where?'

'Ah . . . Ah thowed it in the creek.'

'Waal, c mon home. N firs thing in the mawnin git to tha creek n fin tha gun.'

'Yessuh.'

'Whut yuh pay fer it?'

'Two dollahs.'

'Take tha gun n git yo money back n carry it t Mistah Hawkins, yuh hear? N don fergit Ahma lam yo black bottom good fer this! Now march yosef on home, suh!'

Dave turned and walked slowly. He heard people laughing. Dave glared, his eyes welling with tears. Hot anger bubbled in him. Then he swallowed and stumbled on.

That night Dave did not sleep. He was glad that he had gotten out of killing the mule so easily, but he was hurt. Something hot seemed to turn over inside him each time he remembered how they had laughed. He tossed on his bed, feeling his hard pillow. N Pa says he's gonna beat me. . . . He remembered other beatings, and his back quivered. Naw, naw, Ah sho don wan im t beat me tha way no mo. . . . Dam em *all!* Nobody ever gave him anything. All he did was work. They treat me lika mule. . . . N then they beat me. . . . He gritted his teeth. N Ma had t tell on me.

Well, if he had to, he would take old man Hawkins that two dollars. But that meant selling the gun. And he wanted to keep that gun. Fifty dollahs fer a dead mule.

He turned over, thinking of how he had fired the gun. He had an itch to fire it again. Ef other men kin shoota gun, by Gawd, Ah kin! He was still listening. Mebbe they all sleepin now. . . . The house was still. He heard the soft breathing of his brother. Yes, now! He would go down and get that gun and see if he could fire it! He eased out of bed and slipped into overalls.

The moon was bright. He ran almost all the way to the edge of the woods. He stumbled over the ground, looking for the spot where he had buried the gun. Yeah, here it is. Like a hungry dog scratching for a bone he pawed it up. He puffed his black cheeks and blew dirt from the trigger and barrel. He broke it and found four cartridges unshot. He looked around; the fields were filled with silence and moonlight. He clutched the gun stiff and hard in his fingers. But as soon as he wanted to pull the trigger, he shut his eyes and turned his head. Naw, Ah can't shoot wid mah eyes closed n mah head turned. With effort he held his eyes open; then he squeezed. *Blooooom!* He was stiff, not breathing. The gun was still in his hands. Dammit, he'd done it! He fired again. *Blooooom!* He smiled. *Bloooom! Bloooom! Click, click.* There! It was empty. If anybody could shoot a gun, he could. He put the gun into his hip pocket and started across the fields.

When he reached the top of a ridge he stood straight and proud in the moonlight, looking at Jim Hawkins' big white house, feeling the gun sagging in his pocket. Lawd, ef Ah had jus one mo bullet Ahd taka shot at tha house. Ahd like t scare ol man Hawkins jusa little. . . . Jussa enough t let im know Dave Sanders is a man.

To his left the road curved, running to the tracks of the Illinois Central. He jerked his head, listening. From far off came a faint *hoooof-hoooof; hoooof-hoooof; hoooof-hoooof* . . . Tha's number eight. He took a swift look at Jim Hawkins' white house; he thought of pa, of ma, of his little brother, and the boys. He thought of the dead mule and heard *hoooof-hoooof; hoooof-hoooof; hoooof-hoooof* . . . He stood rigid. Two dollahs a mont. Les see now. . . . Tha means itll take bout two years. Shucks! Ahll be dam!

He started down the road, toward the tracks. Yeah, here she comes! He stood beside the track and held himself stiffly. Here she comes, erroun the ben. . . . C mon, yuh slow poke! C mon! He had his hand on his gun; something quivered in his stomach. Then the train thundered past, the gray and brown box cars rumbling and clinking. He gripped the gun tightly; then he jerked his hand out of his pocket. Ah betcha Bill wouldn't do it! Ah betcha. . . . The cars slid past, steel grinding upon steel. Ahm riding yuh ternight so hep me Gawd! He was hot all over. He hesitated just a moment; then he grabbed, pulled atop of a car, and lay flat. He felt his pocket; the gun was still there. Ahead the long rails were glinting in the moonlight, stretching away, away to somewhere, somewhere where he could be a man. . . .

1961

EUDORA WELTY
1909–

Eudora Welty's three volumes of short stories, seven novels, several essays on the art of fiction, and an album of photographs identify not only an important talent among American women but a leading figure in the literature of the South.

Welty was born in Jackson, Mississippi, to parents who had come from the North, and was raised in comfortable circumstances (her father headed an insurance company). She attended Mississippi State College for Women, graduated from the University of Wisconsin in 1929, and then went to New York for a course in advertising at Columbia University's Business School. Returning home to a depression-blighted South, she worked as a radio continuity writer and society editor before traveling throughout the state of Mississippi for the Works Progress Administration, taking photographs and interviewing local residents. Those travels are reflected in her fiction and also in a book of her photographs published in 1971, *One Time and Place*.

In 1941 Katherine Anne Porter wrote a warm introduction to *A Curtain of Green and Other Stories*, hailing the arrival of yet another gifted southern fiction writer. From then on her productions were greeted with comparable acclaim. During the next decades her stories were published in leading national journals, *Southern Review* and *Sewanee Review* as well as *Atlantic Monthly* and *The New Yorker*, and her writing was praised by the critic John Crowe Ransom and the novelist Robert Penn Warren. Even a less favorable judge, Diana Trilling, was struck by the emotional impact of Welty's fiction.

Her first collection included the famous *Why I Live at the P.O.* and *Death of a Traveling Salesman*, her first story which had appeared in 1936, and the book introduced many of the characters and themes that make up her unique fictional world: the fulfillment of marriage; the satisfactions and the constraints of domestic rituals; the involuted, extended southern families; the physically handicapped, mentally retarded, or unstable kinfolk; the more somber undercurrents of death, violence, and degradation that characterize so much southern fiction. They emerge in highly charged descriptions of delta- or hill-country communities, and they are treated with a gentle tolerance of human foibles and with an antic humor that has characterized the literature of the Southwest since Mark Twain.

Her next book, *The Robber Bridegroom* (1942), was an extended fantasy, a folk tale mixing elements of Grimm's fairy tales and W. B. Yeats's Celtic tales with those of ancient classical myths, and treating the materials with the bizarre exaggerations of American frontier humor and the "tall tale." Set in the Natchez Trace country of Mississippi in the eighteenth century, its fantastic trappings (ghosts, talking heads, Indians and outlaws, magic potions) are grounded in the reality of that time and place. At its center is a quietly poignant tale of self-deception, self-discovery, and the ambiguity of love.

Her characters' search for identity, rather than plot or character development, provides the structure for her best work, along with a strong sense of place. Though Welty has refused the label of regionalist, she once wrote that "like a good many other writers, I am myself touched off by place, the place where I am and the place I know." Indeed "location is the ground-conductor of all the currents of emotion and belief and moral conviction that charge out from the story in its course." While "it is through place that we put out roots," yet "where these roots reach toward * * * is the deep running vein * * * that feeds and is fed by human understanding."[1] Her stories are evocative of season and soil, the topography and atmosphere of her Mississippi.

Welty's contention that character is brought out by a strong sense of place is borne out by the striking gallery of amiably eccentric characters who inhabit her fiction: the wizened, club-footed old black forced to play *Keela the Outcast Indian Maiden* in a freak show; hotelkeeper Edna Earle, the garrulous chronicler of *The Ponder Heart* (1954), and her dim-witted Uncle Daniel Ponder, whose weakness is to go on "giving away" sprees. Other memorable characters are the strong, not overly bright, young heroes of *The Wide Net* (1943) and *Losing Battles* (1970), as well as the vulgar yet touchingly vulnerable "common" women of *Going to Naples* and *The Optimist's Daughter* (1972).

An overabundance of metaphor and strained similes, which the novelist Robert Penn Warren complained about in her early fiction, is less evident in her later writing. The profusion of metaphor may have been encouraged by her acknowledged models, Virginia Woolf, Katherine Mansfield, and William Faulkner, who in any case sanctioned her deliberate strategy of baffling or mystifying her readers, a strategy that was conspicuous in *The Burning* and the title story of *The Bride of the Innisfallen* (1955). Welty herself has written that "fine story writers seem to be in a sense obstructionists."[2]

1. "Place in Fiction," *Three Papers on Fiction* (1962).   2. *Three Essays.*

Her narratives unfold on the principle of varied repetitions or reiterations that have, Welty has claimed, the function of a deliberate double exposure in photography. Hers is a distinctly visual imagination; scenes seem composed as by a camera frame, with carefully delineated foregrounds and backgrounds and a strong play of light and shade on surface detail. One of the functions of "place," she wrote, "is to focus the gigantic, voracious eye of genius and bring its gaze to a point."[3]

Since her early success in securing publication and gaining recognition, Welty has had an uninterrupted career and has received many of the major awards for fiction, including most recently the Howells Gold Medal for Fiction of the Academy of Arts and Letters (1955), and a term as Honorary Consultant at the Library of Congress (1958–61). She has held Guggenheim fellowships and lectureships at Cambridge University (1955) and Smith College (1952). But she has avoided the role of a public literary figure, returning always to the Mississippi which she celebrates in her fiction.

Why I Live at the P.O.[1]

I was getting along fine with Mama, Papa-Daddy and Uncle Rondo until my sister Stella-Rondo just separated from her husband and came back home again. Mr. Whitaker! Of course I went with Mr. Whitaker first, when he first appeared here in China Grove, taking "Pose Yourself" photos, and Stella-Rondo broke us up. Told him I was one-sided. Bigger on one side than the other, which is a deliberate, calculated falsehood: I'm the same. Stella-Rondo is exactly twelve months to the day younger than I am and for that reason she's spoiled.

She's always had anything in the world she wanted and then she'd throw it away. Papa-Daddy gave her this gorgeous Add-a-Pearl necklace when she was eight years old and she threw it away playing baseball when she was nine, with only two pearls.

So as soon as she got married and moved away from home the first thing she did was separate! From Mr. Whitaker! This photographer with the popeyes she said she trusted. Came home from one of those towns up in Illinois and to our complete surprise brought this child of two.

Mama said she like to made her drop dead for a second. "Here you had this marvelous blonde child and never so much as wrote your mother a word about it," says Mama. "I'm thoroughly ashamed of you." But of course she wasn't.

Stella-Rondo just calmly takes off this *hat*, I wish you could see it. She says, "Why, Mama, Shirley-T.'s adopted, I can prove it."

"How?" says Mama, but all I says was, "H'm!" There I was over the hot stove, trying to stretch two chickens over five people and a completely unexpected child into the bargain, without one moment's notice.

---

3. "Place in Fiction," *Three Essays.*   1. "P.O.": post office.

"What do you mean—'H'm!'?" says Stella-Rondo, and Mama says, "I heard that, Sister."

I said that oh, I didn't mean a thing, only that whoever Shirley-T. was, she was the spit-image of Papa-Daddy if he'd cut off his beard, which of course he'd never do in the world. Papa-Daddy's Mama's papa and sulks.

Stella-Rondo got furious! She said, "Sister, I don't need to tell you you got a lot of nerve and always did have and I'll thank you to make no future reference to my adopted child whatsoever."

"Very well," I said. "Very well, very well. Of course I noticed at once she looks like Mr. Whitaker's side too. That frown. She looks like a cross between Mr. Whitaker and Papa-Daddy."

"Well, all I can say is she isn't."

"She looks exactly like Shirley Temple[2] to me," says Mama, but Shirley-T. just ran away from her.

So the first thing Stella-Rondo did at the table was turn Papa-Daddy against me.

"Papa-Daddy," she says. He was trying to cut up his meat. "Papa-Daddy!" I was taken completely by surprise. Papa-Daddy is about a million years old and's got this long-long beard. "Papa-Daddy, Sister says she fails to understand why you don't cut off your beard."

So Papa-Daddy l-a-y-s down his knife and fork! He's real rich. Mama says he is, he says he isn't. So he says, "Have I heard correctly? You don't understand why I don't cut off my beard?"

"Why," I says, "Papa-Daddy, of course I understand, I did not say any such of a thing, the idea!"

He says, "Hussy!"

I says, "Papa-Daddy, you know I wouldn't any more want you to cut off your beard than the man in the moon. It was the farthest thing from my mind! Stella-Rondo sat there and made that up while she was eating breast of chicken."

But he says, "So the postmistress fails to understand why I don't cut off my beard. Which job I got you through my influence with the government. 'Bird's nest'—is that what you call it?"

Not that it isn't the next to smallest P.O. in the entire state of Mississippi.

I says, "Oh, Papa-Daddy," I says, "I didn't say any such of a thing, I never dreamed it was a bird's nest, I have always been grateful though this is the next to smallest P.O. in the state of Mississippi, and I do not enjoy being referred to as a hussy by my own grandfather."

But Stella-Rondo says, "Yes, you did say it too. Anybody in the world could of heard you, that had ears."

"Stop right there," says Mama, looking at *me*.

---

2. **Shirley Temple (b. 1928), tap-dancing child star in saccharine motion pictures.**

So I pulled my napkin straight back through the napkin ring and left the table.

As soon as I was out of the room Mama says, "Call her back, or she'll starve to death," but Papa-Daddy says, "This is the beard I started growing on the Coast when I was fifteen years old." He would of gone on till nightfall if Shirley-T. hadn't lost the Milky Way she ate in Cairo.

So Papa-Daddy says, "I am going out and lie in the hammock, and you can all sit here and remember my words: I'll never cut off my beard as long as I live, even one inch, and I don't appreciate it in you at all." Passed right by me in the hall and went straight out and got in the hammock.

It would be a holiday. It wasn't five minutes before Uncle Rondo suddenly appeared in the hall in one of Stella-Rondo's flesh-colored kimonos, all cut on the bias, like something Mr. Whitaker probably thought was gorgeous.

"Uncle Rondo!" I says. "I didn't know who that was! Where are you going?"

"Sister," he says, "get out of my way, I'm poisoned."

"If you're poisoned stay away from Papa-Daddy," I says. "Keep out of the hammock. Papa-Daddy will certainly beat you on the head if you come within forty miles of him. He thinks I deliberately said he ought to cut off his beard after he got me the P.O., and I've told him and told him and told him, and he acts like he just don't hear me. Papa-Daddy must of gone stone deaf."

"He picked a fine day to do it then," says Uncle Rondo, and before you could say "Jack Robinson" flew out in the yard.

What he'd really done, he'd drunk another bottle of that prescription. He does it every single Fourth of July as sure as shooting, and it's horribly expensive. Then he falls over in the hammock and snores. So he insisted on zigzagging right on out to the hammock, looking like a half-wit.

Papa-Daddy woke up with this horrible yell and right there without moving an inch he tried to turn Uncle Rondo against me. I heard every word he said. Oh, he told Uncle Rondo I didn't learn to read till I was eight years old and he didn't see how in the world I ever got the mail put up at the P.O., much less read it all, and he said if Uncle Rondo could only fathom the lengths he had gone to to get me that job! And he said on the other hand he thought Stella-Rondo had a brilliant mind and deserved credit for getting out of town. All the time he was just lying there swinging as pretty as you please and looping out his beard, and poor Uncle Rondo was *pleading* with him to slow down the hammock, it was making him as dizzy as a witch to watch it. But that's what Papa-Daddy likes about a hammock. So Uncle Rondo was too dizzy to get turned against me for the time being. He's Mama's only brother and is a good case of a one-track mind. Ask anybody. A certified pharmacist.

Just then I heard Stella-Rondo raising the upstairs window. While she was married she got this peculiar idea that it's cooler with the windows shut and locked. So she has to raise the window before she can make a soul hear her outdoors.

So she raises the window and says, "*Oh!*" You would have thought she was mortally wounded.

Uncle Rondo and Papa-Daddy didn't even look up, but kept right on with what they were doing. I had to laugh.

I flew up the stairs and threw the door open! I says, "What in the wide world's the matter, Stella-Rondo? You mortally wounded?"

"No," she says, "I am not mortally wounded but I wish you would do me the favor of looking out that window there and telling me what you see."

So I shade my eyes and look out the window.

"I see the front yard," I says.

"Don't you see any human beings?" she says.

"I see Uncle Rondo trying to run Papa-Daddy out of the hammock," I says. "Nothing more. Naturally, it's so suffocating-hot in the house, with all the windows shut and locked, everybody who cares to stay in their right mind will have to go out and get in the hammock before the Fourth of July is over."

"Don't you notice anything different about Uncle Rondo?" asks Stella-Rondo.

"Why, no, except he's got on some terrible-looking flesh-colored contraption I wouldn't be found dead in, is all I can see," I says.

"Never mind, you won't be found dead in it, because it happens to be part of my trousseau, and Mr. Whitaker took several dozen photographs of me in it," says Stella-Rondo. "What on earth could Uncle Rondo *mean* by wearing part of my trousseau out in the broad open daylight without saying so much as 'Kiss my foot,' *knowing* I only got home this morning after my separation and hung my negligee up on the bathroom door, just as nervous as I could be?"

"I'm sure I don't know, and what do you expect me to do about it?" I says. "Jump out the window?"

"No, I expect nothing of the kind. I simply declare that Uncle Rondo looks like a fool in it, that's all," she says. "It makes me sick to my stomach."

"Well, he looks as good as he can," I says. "As good as anybody in reason could." I stood up for Uncle Rondo, please remember. And I said to Stella-Rondo, "I think I would do well not to criticize so freely if I were you and came home with a two-year-old child I had never said a word about, and no explanation whatever about my separation."

"I asked you the instant I entered this house not to refer one more time to my adopted child, and you gave me your word of honor you would not," was all Stella-Rondo would say, and started pulling out every one of her eyebrows with some cheap Kress tweezers.

So I merely slammed the door behind me and went down and made some green-tomato pickle. Somebody had to do it. Of course Mama had turned both the niggers loose; she always said no earthly power could hold one anyway on the Fourth of July, so she wouldn't even try. It turned out that Jaypan fell in the lake and came within a very narrow limit of drowning.

So Mama trots in. Lifts up the lid and says, "H'm! Not very good for your Uncle Rondo in his precarious condition, I must say. Or poor little adopted Shirley-T. Shame on you!"

That made me tired. I says, "Well, Stella-Rondo had better thank her lucky stars it was her instead of me came trotting in with that very peculiar-looking child. Now if it had been me that trotted in from Illinois and brought a peculiar-looking child of two, I shudder to think of the reception I'd of got, much less controlled the diet of an entire family."

"But you must remember, Sister, that you were never married to Mr. Whitaker in the first place and didn't go up to Illinois to live," says Mama, shaking a spoon in my face. "If you had I would of been just as overjoyed to see you and your little adopted girl as I was to see Stella-Rondo, when you wound up with your separation and came on back home."

"You would not," I says.

"Don't contradict me, I would," says Mama.

But I said she couldn't convince me though she talked till she was blue in the face. Then I said, "Besides, you know as well as I do that that child is not adopted."

"She most certainly is adopted," says Mama, stiff as a poker.

I says, "Why Mama, Stella-Rondo had her just as sure as anything in this world, and just too stuck up to admit it."

"Why, Sister," said Mama. "Here I thought we were going to have a pleasant Fourth of July, and you start right out not believing a word your own baby sister tells you!"

"Just like Cousin Annie Flo. Went to her grave denying the facts of life," I remind Mama.

"I told you if you ever mentioned Annie Flo's name I'd slap your face," says Mama, and slaps my face.

"All right, you wait and see," I says.

"I," says Mama, "I prefer to take my children's word for anything when it's humanly possible." You ought to see Mama, she weighs two hundred pounds and has real tiny feet.

Just then something perfectly horrible occurred to me.

"Mama," I says, "can that child talk?" I simply had to whisper! "Mama, I wonder if that child can be—you know—in any way? Do you realize," I says, "that she hasn't spoken one single, solitary word to a human being up to this minute? This is the way she looks," I says, and I looked like this.

Well, Mama and I just stood there and stared at each other. It was horrible!

"I remember well that Joe Whitaker frequently drank like a fish," says Mama. "I believed to my soul he drank *chemicals*." And without another word she marches to the foot of the stairs and calls Stella-Rondo.

"Stella-Rondo? O-o-o-o-o! Stella-Rondo!"

"What?" says Stella-Rondo from upstairs. Not even the grace to get up off the bed.

"Can that child of yours talk?" asks Mama.

Stella-Rondo says, "Can she what?"

"Talk! Talk!" says Mama. "Burdyburdyburdyburdy!"

So Stella-Rondo yells back, "Who says she can't talk?"

"Sister says so," says Mama.

"You didn't have to tell me, I know whose word of honor don't mean a thing in this house," says Stella-Rondo.

And in a minute the loudest Yankee voice I ever heard in my life yells out, "OE'm Pop-OE the Sailor-r-r-r Ma-a-an!"[3] and then somebody jumps up and down in the upstairs hall. In another second the house would of fallen down.

"Not only talks, she can tap-dance!" calls Stella-Rondo. "Which is more than some people I won't name can do."

"Why, the little precious darling thing!" Mama says, so surprised. "Just as smart as she can be!" Starts talking baby talk right there. Then she turns on me. "Sister, you ought to be thoroughly ashamed! Run upstairs this instant and apologize to Stella-Rondo and Shirley-T."

"Apologize for what?" I says. "I merely wondered if the child was normal, that's all. Now that she's proved she is, why, I have nothing further to say."

But Mama just turned on her heel and flew out, furious. She ran right upstairs and hugged the baby. She believed it was adopted. Stella-Rondo hadn't done a thing but turn her against me from upstairs while I stood there helpless over the hot stove. So that made Mama, Papa-Daddy and the baby all on Stella-Rondo's side.

Next, Uncle Rondo.

I must say that Uncle Rondo has been marvelous to me at various times in the past and I was completely unprepared to be made to jump out of my skin, the way it turned out. Once Stella-Rondo did something perfectly horrible to him—broke a chain letter from Flanders Field[4]—and he took the radio back he had given her and gave it to me. Stella-Rondo was furious! For six months we all had to call her Stella instead of Stella-Rondo, or she wouldn't answer. I

---

3. Title song of the popular film cartoon series *Popeye the Sailor.*   4. World War I battle site and military cemetery.

always thought Uncle Rondo had all the brains of the entire family. Another time he sent me to Mammoth Cave, with all expenses paid.

But this would be the day he was drinking that prescription, the Fourth of July.

So at supper Stella-Rondo speaks up and says she thinks Uncle Rondo ought to try to eat a little something. So finally Uncle Rondo said he would try a little cold biscuits and ketchup, but that was all. So *she* brought it to him.

"Do you think it wise to disport with ketchup in Stella-Rondo's flesh-colored kimono?" I says. Trying to be considerate! If Stella-Rondo couldn't watch out for her trousseau, somebody had to.

"Any objections?" asks Uncle Rondo, just about to pour out all the ketchup.

"Don't mind what she says, Uncle Rondo," says Stella-Rondo. "Sister has been devoting this solid afternoon to sneering out my bedroom window at the way you look."

"What's that?" says Uncle Rondo. Uncle Rondo has got the most terrible temper in the world. Anything is liable to make him tear the house down if it comes at the wrong time.

So Stella-Rondo says, "Sister says, 'Uncle Rondo certainly does look like a fool in that pink kimono!'"

Do you remember who it was really said that?

Uncle Rondo spills out all the ketchup and jumps out of his chair and tears off the kimono and throws it down on the dirty floor and puts his foot on it. It had to be sent all the way to Jackson to the cleaners and re-pleated.

"So that's your opinion of your Uncle Rondo, is it?" he says. "I look like a fool, do I? Well, that's the last straw. A whole day in this house with nothing to do, and then to hear you come out with a remark like that behind my back!"

"I didn't say any such of a thing, Uncle Rondo," I says, "and I'm not saying who did, either. Why, I think you look all right. Just try to take care of yourself and not talk and eat at the same time," I says. "I think you better go lie down."

"Lie down my foot," says Uncle Rondo. I ought to of known by that he was fixing to do something perfectly horrible.

So he didn't do anything that night in the precarious state he was in—just played Casino[5] with Mama and Stella-Rondo and Shirley-T. and gave Shirley-T. a nickel with a head on both sides. It tickled her nearly to death, and she called him "Papa." But at 6:30 A.M. the next morning, he threw a whole five-cent package of some unsold one-inch firecrackers from the store as hard as he could into my bedroom and they every one went off. Not one bad one in the string. Anybody else, there'd be one that wouldn't go off.

5. Card game.

Well, I'm just terribly susceptible to noise of any kind, the doctor has always told me I was the most sensitive person he had ever seen in his whole life, and I was simply prostrated. I couldn't eat! People tell me they heard it as far as the cemetery, and old Aunt Jep Peterson, that had been holding her own so good, thought it was Judgment Day and she was going to meet her whole family. It's usually so quiet here.

And I'll tell you it didn't take me any longer than a minute to make up my mind what to do. There I was with the whole entire house on Stella-Rondo's side and turned against me. If I have anything at all I have pride.

So I just decided I'd go straight down to the P.O. There's plenty of room there in the back, I says to myself.

Well! I made no bones about letting the family catch on to what I was up to. I didn't try to conceal it.

The first thing they knew, I marched in where they were all playing Old Maid[6] and pulled the electric oscillating fan out by the plug, and everything got real hot. Next I snatched the pillow I'd done the needlepoint on right off the davenport from behind Papa-Daddy. He went "Ugh!" I beat Stella-Rondo up the stairs and finally found my charm bracelet in her bureau drawer under a picture of Nelson Eddy.[7]

"So that's the way the land lies," says Uncle Rondo. There he was, piecing on the ham. "Well, Sister, I'll be glad to donate my army cot if you got any place to set it up, providing you'll leave right this minute and let me get some peace." Uncle Rondo was in France.

"Thank you kindly for the cot and 'peace' is hardly the word I would select if I had to resort to firecrackers at 6:30 A.M. in a young girl's bedroom," I says back to him. "And as to where I intend to go, you seem to forget my position as postmistress of China Grove, Mississippi," I says. "I've always got the P.O."

Well, that made them all sit up and take notice.

I went out front and started digging up some four-o'clocks to plant around the P.O.

"Ah-ah-ah!" says Mama, raising the window. "Those happen to be my four-o'clocks. Everything planted in that star is mine. I've never known you to make anything grow in your life."

"Very well," I says. "But I take the fern. Even you, Mama, can't stand there and deny that I'm the one watered that fern. And I happen to know where I can send in a box top and get a packet of one thousand mixed seeds, no two the same kind, free."

"Oh, where?" Mama wants to know.

But I says, "Too late. You 'tend to your house, and I'll 'tend to mine. You hear things like that all the time if you know how to

6. Card game.
7. Nelson Eddy (1901–67), singing star in movie operettas.

listen to the radio. Perfectly marvelous offers. Get anything you want free."

So I hope to tell you I marched in and got that radio, and they could of all bit a nail in two, especially Stella-Rondo, that it used to belong to, and she well knew she couldn't get it back, I'd sue for it like a shot. And I very politely took the sewing-machine motor I helped pay the most on to give Mama for Christmas back in 1929, and a good big calendar, with the first-aid remedies on it. The thermometer and the Hawaiian ukulele certainly were rightfully mine, and I stood on the step-ladder and got all my watermelon-rind preserves and every fruit and vegetable I'd put up, every jar. Then I began to pull the tacks out of the bluebird wall vases on the archway to the dining room.

"Who told you you could have those, Miss Priss?" says Mama, fanning as hard as she could.

"I bought 'em and I'll keep track of 'em," I says. "I'll tack 'em up and I'll keep track of 'em," I says. "I'll tack 'em up one on each side the postoffice window, and you can see 'em when you come to ask me for your mail, if you're so dead to see 'em."

"Not I! I'll never darken the door to that post office again if I live to be a hundred," Mama says. "Ungrateful child! After all the money we spent on you at the Normal."[8]

"Me either," says Stella-Rondo. "You just let my mail lie there and *rot*, for all I care. I'll never come and relieve you of a single, solitary piece."

"I should worry," I says. "And who you think's going to sit down and write you all those big fat letters and postcards, by the way? Mr. Whitaker? Just because he was the only man ever dropped down in China Grove and you got him—unfairly—is he going to sit down and write you a lengthy correspondence after you come home giving no rhyme nor reason whatsoever for your separation and no explanation for the presence of that child? I may not have your brilliant mind, but I fail to see it."

So Mama says, "Sister, I've told you a thousand times that Stella-Rondo simply got homesick, and this child is far too big to be hers," and she says, "Now, why don't you all just sit down and play Casino?"

Then Shirley-T. sticks out her tongue at me in this perfectly horrible way. She has no more manners than the man in the moon. I told her she was going to cross her eyes like that some day and they'd stick.

"It's too late to stop me now," I says. "You should have tried that yesterday. I'm going to the P.O. and the only way you can possibly see me is to visit me there."

So Papa-Daddy says, "You'll never catch me setting foot in that post office, even if I should take a notion into my head to write a

8. Normal School, teachers' college.

letter some place." He says, "I won't have you reachin' out of that little old window with a pair of shears and cuttin' off any beard of mine. I'm too smart for you!"

"We all are," says Stella-Rondo.

But I said, "If you're so smart, where's Mr. Whitaker?"

So then Uncle Rondo says, "I'll thank you from now on to stop reading all the orders I get on postcards and telling everybody in China Grove what you think is the matter with them," but I says, "I draw my own conclusions and will continue in the future to draw them." I says, "If people want to write their inmost secrets on penny postcards, there's nothing in the wide world you can do about it, Uncle Rondo."

"And if you think we'll ever *write* another postcard you're sadly mistaken," says Mama.

"Cutting off your nose to spite your face then," I says. "But if you're all determined to have no more to do with the U.S. mail, think of this: What will Stella-Rondo do now, if she wants to tell Mr. Whitaker to come after her?"

"Wah!" says Stella-Rondo. I knew she'd cry. She had a conniption fit right there in the kitchen.

"It will be interesting to see how long she holds out," I says. "And now—I am leaving."

"Good-bye," says Uncle Rondo.

"Oh, I declare," says Mama, "to think a family of mine should quarrel on the Fourth of July, or the day after, over Stella-Rondo leaving old Mr. Whitaker and having the sweetest little adopted child! It looks like we'd all be glad!"

"Wah!" says Stella-Rondo, and has a fresh conniption fit.

"*He* left *her*—you mark my words," I says. "That's Mr. Whitaker. I know Mr. Whitaker. After all, I knew him first. I said from the beginning he'd up and leave her. I foretold every single thing that's happened."

"Where did he go?" asks Mama.

"Probably to the North Pole, if he knows what's good for him," I says.

But Stella-Rondo just bawled and wouldn't say another word. She flew to her room and slammed the door.

"Now look what you've gone and done, Sister," says Mama. "You go apologize."

"I haven't got time, I'm leaving," I says.

"Well, what are you waiting around for?" asks Uncle Rondo.

So I just picked up the kitchen clock and marched off, without saying "Kiss my foot" or anything, and never did tell Stella-Rondo goodbye.

There was a nigger girl going along on a little wagon right in front.

"Nigger girl," I says, "come help me haul these things down the hill, I'm going to live in the post office."

Took her nine trips in her express wagon. Uncle Rondo came out on the porch and threw her a nickel.

And that's the last I've laid eyes on any of my family or my family laid eyes on me for five solid days and nights. Stella-Rondo may be telling the most horrible tales in the world about Mr. Whitaker, but I haven't heard them. As I tell everybody, I draw my own conclusions.

But oh, I like it here. It's ideal, as I've been saying. You see, I've got everything cater-cornered, the way I like it. Hear the radio? All the war news. Radio, sewing machine, book ends, ironing board and that great big piano lamp—peace, that's what I like. Butter-bean vines planted all along the front where the strings are.

Of course, there's not much mail. My family are naturally the main people in China Grove, and if they prefer to vanish from the face of the earth, for all the mail they get or the mail they write, why, I'm not going to open my mouth. Some of the folks here in town are taking up for me and some turned against me. I know which is which. There are always people who will quit buying stamps just to get on the right side of Papa-Daddy.

But here I am, and here I'll stay. I want the world to know I'm happy.

And if Stella-Rondo should come to me this minute, on bended knees, and *attempt* to explain the incidents of her life with Mr. Whitaker, I'd simply put my fingers in both my ears and refuse to listen.

1941

WILLIAM FAULKNER
1897–1962

The most imposing body of fiction written by an American after World War I consisted of the novels and tales of a southerner from Mississippi, William Faulkner. His brilliantly inventive fiction drew admiration from a growing number of writers and critics in America and France after 1929 during decades when wider audiences disdained him. Many readers were offended by his sensationalism; others were baffled by the density and intricacies of his style. In 1944, scarcely any of his works was in print. After World War II, however, larger audiences in Europe and America responded to the power of his prose. They were gripped by Faulkner's devotion to traditional values in a world of crisis which threatened them, and by the anguish and boldness with which he himself challenged those values, recognizing the need to redefine and reaffirm them.

William Cuthbert Falkner was born near Oxford, Mississippi, where he was raised and in which he later made his permanent home. His family had figured prominently in the history of the region, whose habit of violence, codes of honor, cult of white social status and racial solidarity, and entrepreneurial ambitions provided Faulkner with material for his fiction. His great-grandfather had fought as a colonel in the Civil War and returned to pursue a career as lawyer, railroad financier, civic benefactor, politician, and novelist. (Faulkner is thought to have added the "u" to the spelling of his family name "Falkner" so as to avoid confusion with his novel-writing great-grandfather.) The Colonel had written two novels and a play, run for the state legislature, and helped found the thriving Gulf and Chicago Railroad before being shot dead by an envious political rival and business partner in 1889. Faulkner's father worked for the family railroad until it was sold in 1902, then bought and operated a livery stable and a hardware store, and became business manager of the University of Mississippi at the end of World War I.

Faulkner attended grammar school and two years of high school in Oxford. The only attraction for him at high school was football, and the temptations of hunting, drinking parties, and other diversions continued to be strong after he dropped out in 1915; the family saw to it that he took a clerical job in his grandfather's bank. He had begun drawing early in school, and was trying his hand at writing as early as the sixth grade, but his frequent truancy was a clear sign that academic work was neither stimulating nor disciplining his mind. He found Melville's *Moby-Dick*, Shakespeare's plays, and the novels of Fielding and Conrad elsewhere, notably in his grandfather's library.

The most durable influence on his reading was a slightly older friend, Phil Stone, who encouraged his ambition to be a writer. Educated at Mississippi and Yale, Stone shared Faulkner's interest in regional history and made available to him his family's large library. Faulkner read in the ancient classics, began to teach himself French, and discovered Balzac's novels under Stone's tutelage. In 1918 when his sweetheart decided to marry another young man, Faulkner went for a year to New Haven where Stone was in law school, working for a time as a clerk in an arms factory but continuing his writing and his discussions of literature with Stone. When Faulkner became eligible for military service, he enlisted in the Royal Air Force in Canada and was trained as a pilot, but he returned to Oxford at the end of the war and resumed his writing. Stone, by then a practicing lawyer, began to play a more active role in Faulkner's career; he entered into negotiations with publishers, criticized manuscripts, helped to finance his friend's first published volume, and eventually wrote biographical sketches of Faulkner to bring the author to the attention of critics and readers. Though Faulkner studied languages and English literature for over a year as a special student at the University of Mississippi, worked in a bookstore for a short time in New York City in 1920, worked briefly (and with marked inefficiency) as university postmaster in Oxford, and held jobs as a skilled carpenter and housepainter, his chosen vocation after 1920 was writing.

His first published works were poems. Many drew on the conventions of Shakespearean and pastoral poetry, and Victorian or Edwardian modes of Swinburne and A. E. Housman, while others echoed Yeats, T. S. Eliot, and the French Symbolists Mallarmé and Verlaine. He subsequently published two volumes of verse, *The Marble Faun* (1924) and *A Green Bough* (1931). Faulkner also experimented with plays written for the university's dramatic association. His first published short story (an account of the mis-adventures of an RAF pilot) appeared in *The Mississippian* in 1920, and he wrote other short fiction in the early 1920s.

By 1924, when *The Double Dealer*, a new literary magazine, accepted the second of his early poems, Faulkner's literary horizons were widening and he spent much of 1925 living in New Orleans, where *The Double Dealer* and the *Times Picayune* published his verse, articles, and the sketches later collected as *New Orleans Sketches* (1968). A thriving literary colony in New Orleans drew Faulkner into the discussion of Freud's psychology, James Joyce's vanguard fiction, and the anthropology of Sir James Frazer's *Golden Bough*. At the center of the colony was the fiction writer Sherwood Anderson, with whom Faulkner became closely acquainted. Personal relations between them became distant after Faulkner collaborated in a parody of Anderson's prose style (*Sherwood Anderson and Other Famous Creoles*, 1926). But the two writers developed some ideas for fiction together, and it was Anderson, Faulkner later testified, who encouraged him to develop his own style as a writer and to draw on his native region for material. Though he had not read it, Anderson recommended to his own publisher the novel that Faulkner wrote in New Orleans, the story of a dying air force veteran entitled *Soldier's Pay* (1926). Faulkner left New Orleans in 1925 for Europe, where he visited museums and took long hiking tours through Italy, France, and England, and visited New Orleans again briefly before settling in Mississippi in 1926.

During the next four years, with time out for odd jobs, hunting expeditions, escapades in Memphis, vacations on the Gulf beaches, and politicking in the unsuccessful political campaigns of his brother John, he finished the work that brought the first phase of his career to its culmination. In 1927 appeared his second novel, *Mosquitoes*, a mocking and at times surrealistic treatment of New Orleans intellectuals and the "jazz age" mores of their wealthy acquaintances aboard a yacht in the Mississippi River basin. Two years later he published *Sartoris*. Its account of a wounded veteran's return home in 1919, his marriage and death, focuses on the interconnections between a prominent southern family and the local community later named Yoknapatawpha County that became legendary in Faulkner's mature fiction. In the same year he finished two of his most ambitious and radically inventive works, which caught the attention of Jean-Paul Sartre in France and other leading critics and charted the main directions of his later writing. *The Sound and the Fury* (1929), Faulkner's favorite among his novels, is the tragic account of a disintegrating southern "aristocratic" family, its vestiges of starved and warped affection, kept perilously intact by its charade of family honor and the ministrations of loyal Negro servants. Of its four chapters, one is narrated through the mind of a thirty-three-year-old idiot, another through the mind of his brother at Harvard who is on the verge of suicide. *As I Lay Dying* (1930), grotesquely comic and grimly tragic at the same time (and written, Faulkner claimed, without revision during six weeks when he worked a night shift in the university power plant), recounts the desperate struggle of a family of "poor whites" to get the mother's body to her home town for proper burial while pursuing their own mundane objectives in the city: an abortion, the purchase of some false teeth, the acquisition of a new wife. Each chapter (one of them only one sentence long) is the "interior monologue" of one of the characters.

In 1929 Faulkner married Mrs. Estelle Franklin, his former sweetheart, who had divorced her first husband and returned with her two children to live in Oxford. They purchased a pillared mansion, Rowan Oak, in 1930, where their own daughter was born in 1933, and made it their permanent home.

While Faulkner's fiction had begun to attract favorable attention, and he was able to sell some stories to national magazines after 1930, it was often refused publication, and his earnings from writing were not enough to support his immediate family and the relatives who were dependent on him. One expedient he turned to was sensationalism in the novel *Sanctuary* (1931); it was written to sell, he acknowledged, but he revised it extensively to meet the high standards of his best writing. It is the story of an arsonist, gangster, and murderer and the college girl he corrupts, who together betray a bootlegger and his devoted common-law wife. The book brought brief notoriety but not lasting success. A second expedient was to write for the films. He regarded this work as distinctly separate from his writing career, but from 1932 to 1936, 1942 to 1945, and again in 1951 and 1954, he spent some time each year in Hollywood or writing on contract for various movie studios.

During the same years he published the brilliant and inventive novels that were to bring him international recognition. *Light in August* (1932) enclosed an explosive drama, about a white but allegedly Negro murderer and his subsequent lynching, within an amusing pastoral idyl about a pregnant girl journeying from Alabama to Mississippi in search of the father of her child. *Absalom, Absalom!* (1936) was at once a tour de force of historical reconstruction and a lyric tragedy; it probed the incestuous love, the fear of miscegenation, and the dream of founding a dynasty in a Mississippi plantation owner's family, and it celebrated the agony of a Mississippi community during and after the Civil War that could neither achieve, nor purge and purify, its flawed ambitions. *Wild Palms* (1939) ingeniously spliced two stories (one of them *Old Man*, below) in a counterpoint of alternating chapters. At the outbreak of World War II Faulkner published two of his most impressive works, each composed of separate but related stories he had been working on since the early Thirties. *The Hamlet* (1940), riotously comic and horrifying at the same time, describes a southern community's attempt to cope with the rise to prominence of a proliferating family of "poor whites" named Snopes. *Go Down, Moses* (1942). comic at times but elegiac in tone, explores the interconnections over generations between two branches of the McCaslin family in Yoknapatawpha County, the white branch epitomized by Ike McCaslin, who relinquishes his inheritance in an effort to atone for racial injustice, the black branch epitomized by Lucas Beauchamp, who insists on his claim to his rights and the family tradition.

The publication of the anthology *The Portable Faulkner* in 1946, edited by Malcolm Cowley, helped to enlarge the American audience for Faulkner's works and marked a new era in his critical esteem. Elected to the National Institute of Arts and Letters in 1939 and to the American Academy of Arts and Letters in 1948, he received in the next three years the American Academy's Howells Award for American Fiction, the National Book Award, and the Nobel Prize for Literature. Though he consistently kept his distance from literary circles, in his later years he appeared often on college campuses to lecture, read from his works, or confer with students and faculty members about the literary profession. He was writer in residence at the University of Virginia in 1957–58 and joined the faculty in 1960, and he conducted seminars in Japan, Greece, and Venezuela under the auspices of the U.S. State Department.

Faulkner commented publicly on race relations during the struggle to desegregate public schools in the South. He acknowledged the rightful claims, the stamina, and the human dignity of the blacks, and he deplored the brutality and injustice inflicted on them, but he warned against federal interference in southern affairs, begged time for the gradual solution of racial problems, and on occasion belligerently defended the desire of southern whites to work out their own solutions to racial conflict. His fiction, not his public pronouncements, remained the most sensitive register of his anguished recognition of the power and human capacities of the black people. There he had "explored," in the words of the black novelist Ralph Ellison, "perhaps more successfully than anyone else, white or black, certain forms of Negro humanity." In portraying American Negroes he "had been more willing perhaps than any other artist to start with the stereotype, accept it as true, and then seek out the human truth which it hides."[1]

Faulkner's major productions after World War II were *The Fable* (1954), an allegorical story about a World War I general and a corporal who is his illegitimate son, and two chronicles which, with the earlier *The Hamlet*, comprise a trilogy of the Snopes clan: *The Town* (1957) and *The Mansion* (1959). Just before his death from a heart attack in 1962, Faulkner dedicated to his grandsons *The Reivers*, a comic and nostalgic recapitulation of episodes from his own life and from his earlier fiction.

When speaking in public of his literary themes and methods, Faulkner often took the stance of a devoted but amateur craftsman, likening his use of religious symbols to the selection of an ax by a woodsman, speaking as a farmer or carpenter who stumbled upon significant Biblical images while looking "for a tool that will make a better chicken-house."[2] Such ostensibly deprecating remarks reveal both the boldness with which he improvised and the deliberateness with which he "built" or composed his fiction. The sequence of sections or chapters of his stories does not follow the chronology of the events they recount; rather, the narration is deliberately constructed to defy and fracture chronology, intriguing readers by puzzling them until gradually they apprehend, or are suddenly shocked into recognizing, what has been going on and what its significance is. (In *Old Man*, the account of the convict's deeds on the flooded Mississippi is spliced with incidents that take place afterward in the prison where he tells about his seven weeks' adventure, and information about his earlier life is scattered in the opening and closing pages of the story.) The narrative tries to encompass complexities in the course of history, or different rhythms and phases in the flow of time and in the consciousness of time, that lie beneath the surface of chronology.

The violation of chronology in the narrative structure is matched by a violation of everyday language habits in Faulkner's prose style. He assaults outright, or strains to the utmost, the conventions of grammar and syntax, so as to achieve a number of effects: to channel energies emerging in the past and extending into the future in a "long sentence" that gets a character's "past and possibly his future into the instant in which he does some-

1. Ellison, *Shadow and Act* (Vintage),   2. *Faulkner in the University*, p. 68.
pp. 30, 43.

thing * * *";[3] to achieve a level of lyric, even rhapsodic, intensity appropriate to the feelings of rage or longing or astonishment being evoked, and to the sense of urgency he wanted to arouse in readers; and to convert prose into quickened and fluid movement. Though he was fearful that language might not be capable of sustaining the burdens of expression that he assigned it—that, as one character says in *As I Lay Dying*, words could merely "say at" reality and always miss or distort it—that fear simply gave urgency to Faulkner's early conviction that the "earthy strength" of the English language as spoken in America could provide a substitute for the richer literary traditions it lacked.[4] Faulkner's incorporation in his style of the diction and cadences of American colloquial speech, along with his overtly rhetorical intensity and elaboration, provide links to two of the books he thought best in American literature, Twain's *Huckleberry Finn* and (his favorite) Melville's *Moby-Dick*.

Faulkner's literary form is distinctly appropriate to the sense of desperate urgency with which he confronts the world he envisions. The contortions into which he twists his materials (which he defended as the "artist's prerogative * * * to underline, to blow up facts, distort facts in order to state a truth")[5] and the sensationalism of his themes and formal effects (which he admitted was a "betrayal" of a writer's "vocation" unless subordinated to the governing purposes of his story)[6] define Faulkner's world. It is a world in which "despair" and "doom" are recurring motifs because social and moral orders prove to be founded on racial exploitation and violence; civil wars, macabre murders, suicides, labor pains, and flood waters on the rampage are its characteristic events; social and moral traditions are threatened with enervation and perversion, and human destiny is faced with the alternatives of annihilation or apocalypse. In a world threatened late in his life by the portent of the atomic bomb, Faulkner declared at his daughter's college graduation (in the speech printed below) that only a revolution accomplished within the textures of everyday life could save mankind from extinction. Such an immanent and ultimate crisis had been the central challenge in all his fiction, where the remnants of an upper-class family are located perilously on the "ultimate edge" of the earth in *The Sound and the Fury*, or where the frantic members of a poor white family in *As I Lay Dying* struggle to keep standing in a crisis "where the motion of the wasted world accelerates, just before the final precipice," or where a defeated South in *Absalom, Absalom!* tries to salvage new life from the despair of defeat, to "salvage something anyway of the old lost enchantment of the heart." Faulkner's characters fail to cleanse their world of guilt, to escape completely from its paralysis, or to achieve lasting solutions to its problems, but some (like the Tall Convict in *Old Man*) have the desperate courage to try. Their often bizarre heroism signals the urgency of the challenge life presents to human values, and their suffering reveals the need for reconstituting those values and prefigures the transformation their world seeks. The mission for young writers that Faulkner defined in the Nobel Prize speech of 1950 was to remind readers "of the courage and honor and hope and pride and compassion and pity and sacrifice which have been the glory of [mankind's] past" so as to enable men and women not only to "endure" the crisis of life but to "prevail."[7]

---

3. *Ibid.*, p. 84.
4. *Early Prose and Poetry*, pp. 87, 96.
5. *Faulkner in the University*, p. 282.

6. *Ibid.*, p. 49.
7. *Essays, Speeches, & Public Letters*, p. 120.

Old Man[1]

Once (it was in Mississippi, in May, in the flood year 1927) there were two convicts. One of them was about twenty-five, tall, lean, flat-stomached, with a sunburned face and Indian-black hair and pale, china-colored outraged eyes—an outrage directed not at the men who had foiled his crime, not even at the lawyers and judges who had sent him here, but at the writers,[2] the uncorporeal names attached to the stories, the paper novels—the Diamond Dicks and Jesse Jameses[3] and such—whom he believed had led him into his present predicament through their own ignorance and gullibility regarding the medium in which they dealt and took money for, in accepting information on which they placed the stamp of verisimilitude and authenticity (this so much the more criminal since there was no sworn notarised statement attached and hence so much the quicker would the information be accepted by one who expected the same unspoken good faith, demanding, asking, expecting no certification, which he extended along with the dime or fifteen cents to pay for it) and retailed for money and which on actual application proved to be impractical and (to the convict) criminally false; there would be times when he would halt his mule and plow in midfurrow (there is no walled penitentiary in Mississippi; it is a cotton plantation which the convicts work under the rifles and shotguns of guards and trusties) and muse with a kind of enraged impotence, fumbling among the rubbish left him by his one and only experience with courts and law, fumbling until the meaningless and verbose shibboleth took form at last (himself seeking justice at the same blind fount where he had met justice and been hurled back and down): Using the mails to defraud: who felt that he had been defrauded by the third-class mail system not of crass and stupid money which he did not particularly want anyway, but of liberty and honor and pride.

He was in for fifteen years (he had arrived shortly after his nineteenth birthday) for attempted train robbery. He had laid his plans

1. Afro-American nickname for the Mississippi River. This story, dramatizing the heroism of a nameless convict who rescues a pregnant woman during a Mississippi flood and assists at the birth of her child, takes the form of a frontier "tall tale." Faulkner drew attention to the combination of tragedy and comedy in *Old Man* when he declared later that the tale is about people's "aspirations and their struggles and the bizarre, the comic, and the tragic conditions they get themselves into simply coping with themselves and one another and environment" (*Faulkner in the University,* p. 177). It was originally published as part of the novel *The Wild Palms* (1939), where the chapters of *Old Man* alternated with chapters recounting a different story: the tragic love affair of an artist named Charlotte and a doctor named Harry Wilbourne who devote their lives to intense personal passion. Convinced that a child would interfere, Charlotte persuades Harry to perform an abortion and dies because he is not sufficiently skillful; Harry is tried for murder, convicted, and imprisoned in the Mississippi prison farm at Parchman, which is the setting of *Old Man.* Faulkner intended that *Old Man* serve as "counterpoint," as in music, to the other story.

2. Cf. Isaiah 10.1: "Woe unto them that decree unrighteous decrees, And to the writers that write perverseness."

3. I.e., stories of famous desperadoes, like the train robber Jesse James (1847–82).

in advance, he had followed his printed (and false) authority to the letter; he had saved the paperbacks for two years, reading and rereading them, memorising them, comparing and weighing story and method against story and method, taking the good from each and discarding the dross as his workable plan emerged, keeping his mind open to make the subtle last-minute changes, without haste and without impatience, as the newer pamphlets appeared on their appointed days as a conscientious dressmaker makes the subtle alterations in a court presentation costume as the newer bulletins appear. And then when the day came, he did not even have a chance to go through the coaches and collect the watches and the rings, the brooches and the hidden money-belts, because he had been captured as soon as he entered the express car where the safe and the gold would be. He had shot no one because the pistol which they took away from him was not that kind of a pistol although it was loaded; later he admitted to the District Attorney that he had got it, as well as the dark lantern in which a candle burned and the black handkerchief to wear over the face, by peddling among his pinehill neighbors subscriptions to the *Detectives' Gazette*. So now from time to time (he had ample leisure for it) he mused with that raging impotence, because there was something else he could not tell them at the trial, did not know how to tell them. It was not the money he had wanted. It was not riches, not the crass loot; that would have been merely a bangle to wear upon the breast of his pride like the Olympic runner's amateur medal—a symbol, a badge to show that he too was the best at his chosen gambit in the living and fluid world of his time. So that at times as he trod the richly shearing black earth behind his plow or with a hoe thinned the sprouting cotton and corn or lay on his sullen back in his bunk after supper, he cursed in a harsh steady unrepetitive stream, not at the living men who had put him where he was but at what he did not even know were pennames, did not even know were not actual men but merely the designations of shades who had written about shades.

The second convict was short and plump. Almost hairless, he was quite white. He looked like something exposed to light by turning over rotting logs or planks and he too carried (though not in his eyes like the first convict) a sense of burning and impotent outrage. So it did not show on him and hence none knew it was there. But then nobody knew very much about him, including the people who had sent him here. His outrage was directed at no printed word but at the paradoxical fact that he had been forced to come here of his own free choice and will. He had been forced to choose between the Mississippi State penal farm and the Federal Penitentiary at Atlanta, and the fact that he, who resembled a hairless and pallid slug, had chosen the out-of-doors and the sunlight was merely another manifestation of the close-guarded and solitary enigma of his character, as something recognisable roils momentarily into view from beneath stagnant and opaque water, then sinks again. None of his fellow prisoners knew what his crime had been, save that he was in for a hundred and ninety-nine years—this incredible and impossible period of punishment or restraint itself carrying a vicious and fabulous quality which indicated that his reason for being here was such that the very men, the paladins[4] and pillars of justice and equity who had sent him here had during that moment become blind apostles not of mere justice but of all human decency, blind instruments not of equity but of all human outrage and vengeance, acting in a savage personal concert, judge, lawyer and jury, which certainly abrogated justice and possibly even law. Possibly only the Federal and State's Attorneys knew what the crime actually was. There had been a woman in it and a stolen automobile transported across a state line, a filling station robbed and the attendant shot to death. There had been a second man in the car at the time and anyone could have looked once at the convict (as the two attorneys did) and known he would not even have had the synthetic courage of alcohol to pull trigger on anyone. But he and the woman and the stolen car had been captured while the second man, doubtless the actual murderer, had escaped, so that, brought to bay at last in the State's Attorney's office, harried, dishevelled and snarling, the two grimly implacable and viciously gleeful attorneys in his front and the now raging woman held by two policemen in the anteroom in his rear, he was given his choice. He could be tried in Federal Court under the Mann Act[5] and for the automobile, that is, by electing to pass through the anteroom where the woman raged he could take his chances on the lesser crime in Federal Court, or by accepting a sentence for manslaughter in the State Court he would be permitted to quit the room by a back entrance, without having to pass the woman. He had chosen; he stood at the bar and heard a judge (who looked down at him as if the District Attorney actually had turned over a rotten plank with his toe and exposed him) sentence him to a hundred and ninety-nine years at the State Farm. Thus (he had ample leisure too; they had tried to teach him to plow and had failed, they had put him in the blacksmith shop and the foreman trusty[6] himself had asked to have him removed: so that now, in a long apron like a woman, he cooked and swept and dusted in the deputy wardens' barracks) he too mused at times with that sense of impotence and outrage though it did not show on him as on the first convict since he leaned on no halted broom to do it and so none knew it was there.

It was this second convict who, toward the end of April, began to read aloud to the others from the daily newspapers when, chained

---

4. Palace officers, knights.
5. U.S. law forbidding the transportation of women across state lines for immoral purposes.
6. Prisoner trusted with responsibility for supervising other prisoners.

ankle to ankle and herded by armed guards, they had come up from
the fields and had eaten supper and were gathered in the bunkhouse. It was the Memphis newspaper which the deputy wardens
had read at breakfast; the convict read aloud from it to his companions who could have had but little active interest in the outside
world, some of whom could not have read it for themselves at all
and did not even know where the Ohio and Missouri river basins
were, some of whom had never even seen the Mississippi River
although for past periods ranging from a few days to ten and twenty
and thirty years (and for future periods ranging from a few months
to life) they had plowed and planted and eaten and slept beneath
the shadow of the levee[7] itself, knowing only that there was water
beyond it from hearsay and because now and then they heard the
whistles of steamboats from beyond it and during the last week or
so had seen the stacks and pilot houses moving along the sky sixty
feet above their heads.

But they listened, and soon even those who like the taller convict
had probably never before seen more water than a horse pond
would hold knew what thirty feet on a river gauge at Cairo or
Memphis meant and could (and did) talk glibly of sandboils.[8] Perhaps what actually moved them was the accounts of the conscripted
levee gangs, mixed blacks and whites working in double shifts
against the steadily rising water; stories of men, even though they
were Negroes, being forced like themselves to do work for which
they received no other pay than coarse food and a place in a mud-
floored tent to sleep on—stories, pictures, which emerged from the
shorter convict's reading voice: the mudsplashed white men with
the inevitable shotguns, the antlike lines of Negroes carrying sand-
bags, slipping and crawling up the steep face of the revetment to
hurl their futile ammunition into the face of a flood and return for
more. Or perhaps it was more than this. Perhaps they watched the
approach of the disaster with that same amazed and incredulous
hope of the slaves—the lions and bears and elephants, the grooms
and bathmen and pastrycooks—who watched the mounting flames
of Rome from Ahenobarbus'[9] gardens. But listen they did and presently it was May and the wardens' newspaper began to talk in headlines two inches tall—those black staccato slashes of ink which, it
would almost seem, even the illiterate should be able to read: *Crest
Passes Memphis at Midnight 4000 Homeless in White River Basin
Governor Calls out National Guard Martial Law Declared in Following Counties Red Cross Train with Secretary Hoover[1] Leaves
Washington Tonight*; then, three evenings later (It had been raining all day—not the vivid brief thunderous downpours of April and

7. Earth embankment, built to contain
the swollen river so as to prevent flooding of adjacent land.
8. Bubbling springs, bursting out at the
base of levees, owing to the pressure of
flood water on the other side.

9. Aristocratic Roman family whose
scion, the Emperor Nero (A.D. 54–68),
watched Rome burn.
1. Herbert C. Hoover (1874–1964), in
1927 Secretary of Commerce; later President of the United States (1929–33).

May, but the slow steady gray rain of November and December before a cold north wind. The men had not gone to the fields at all during the day, and the very secondhand optimism of the almost twenty-four-hour-old news seemed to contain its own refutation.):
Crest Now Below Memphis 22,000 *Refugees Safe at Vicksburg Army Engineers Say Levees Will Hold.*

"I reckon that means it will bust tonight," one convict said.

"Well, maybe this rain will hold on until the water gets here," a second said. They all agreed to this because what they meant, the living unspoken thought among them, was that if the weather cleared, even though the levees broke and the flood moved in upon the Farm itself, they would have to return to the fields and work, which they would have had to do. There was nothing paradoxical in this, although they could not have expressed the reason for it which they instinctively perceived: that the land they farmed and the substance they produced from it belonged neither to them who worked it nor to those who forced them at guns' point to do so, that as far as either—convicts or guards—were concerned, it could have been pebbles they put into the ground and papier-mâché cotton- and corn-sprouts which they thinned. So it was that, what between the sudden wild hoping and the idle day and the evening's headlines, they were sleeping restlessly beneath the sound of the rain on the tin roof when at midnight the sudden glare of the electric bulbs and the guards' voices waked them and they heard the throbbing of the waiting trucks.

"Turn out of there!" the deputy shouted. He was fully dressed—rubber boots, slicker and shotgun. "The levee went out at Mound's Landing an hour ago. Get up out of it!"

When the belated and streaming dawn broke, the two convicts, along with twenty others, were in a truck. A trusty drove, two armed guards sat in the cab with him. Inside the high, stall-like top-less body the convicts stood, packed like matches in an upright box or like the pencil-shaped ranks of cordite[2] in a shell, shackled by the ankles to a single chain which wove among the motionless feet and swaying legs and a clutter of picks and shovels among which they stood, and was riveted by both ends to the steel body of the truck.

Then and without warning they saw the flood about which the plump convict had been reading and they listening for two weeks or more. The road ran south. It was built on a raised levee, known locally as a dump, about eight feet above the flat surrounding land, bordered on both sides by the barrow pits from which the earth of the levee had been excavated. These barrow pits had held water all winter from the fall rains, not to speak of the rain of yesterday, but now they saw that the pit on either side of the road had vanished

---

2. Explosive powder, shaped into stringlike strands, used in ammunition.

and instead there lay a flat still sheet of brown water which extended into the fields beyond the pits, ravelled out into long motionless shreds in the bottom of the plow furrows and gleaming faintly in the gray light like the bars of a prone and enormous grating. And then (the truck was moving at good speed) as they watched quietly (they had not been talking much anyway but now they were all silent and quite grave, shifting and craning as one to look soberly off to the west side of the road) the crests of the furrows vanished too and they now looked at a single perfectly flat and motionless steel-colored sheet in which the telephone poles and the straight hedgerows which marked section lines seemed to be fixed and rigid as though set in concrete.

It was perfectly motionless, perfectly flat. It looked, not innocent, but bland. It looked almost demure. It looked as if you could walk on it. It looked so still that they did not realise it possessed motion until they came to the first bridge. There was a ditch under the bridge, a small stream, but ditch and stream were both invisible now, indicated only by the rows of cypress and bramble which marked its course. Here they both saw and heard movement—the slow profound eastward and upstream ("It's running backward," one convict said quietly.) set of the still rigid surface, from beneath which came a deep faint subaquean rumble which (though none in the truck could have made the comparison) sounded like a subway train passing far beneath the street and which implied a terrific and secret speed. It was as if the water itself were in three strata, separate and distinct, the bland and unhurried surface bearing a frothy scum and a miniature flotsam of twigs and screening as though by vicious calculation the rush and fury of the food itself, and beneath this in turn the original stream, trickle, murmuring along in the opposite direction, following undisturbed and unaware its appointed course and serving its Lilliputian[3] end, like a thread of ants between the rails on which an express train passes, they (the ants) as unaware of the power and fury as if it were a cyclone crossing Saturn.

Now there was water on both sides of the road and now, as if once they had become aware of movement in the water the water seemed to have given over deception and concealment, they seemed to be able to watch it rising up the flanks of the dump; trees which a few miles back had stood on tall trunks above the water now seemed to burst from the surface at the level of the lower branches like decorative shrubs on barbered lawns. The truck passed a Negro cabin. The water was up to the window ledges. A woman clutching two children squatted on the ridgepole, a man and a half-grown youth, standing waist-deep, were hoisting a squealing pig onto the slanting roof of a barn, on the ridgepole of which sat a row of

3. I.e., very small. Lilliput, in Jonathan Swift's satire, *Gulliver's Travels* (1726), was an island inhabited by miniature people.

chickens and a turkey. Near the barn was a haystack on which a cow stood tied by a rope to the center pole and bawling steadily; a yelling Negro boy on a saddleless mule which he flogged steadily, his legs clutching the mule's barrel and his body leaned to the drag of a rope attached to a second mule, approached the haystack, splashing and floundering. The woman on the housetop began to shriek at the passing truck, her voice carrying faint and melodious across the brown water, becoming fainter and fainter as the truck passed and went on, ceasing at last, whether because of distance or because she had stopped screaming those in the truck did not know.

Then the road vanished. There was no perceptible slant to it yet it had slipped abruptly beneath the brown surface with no ripple, no ridgy demarcation, like a flat thin blade slipped obliquely into flesh by a delicate hand, annealed into the water without disturbance, as if it had existed so for years, had been built that way. The truck stopped. The trusty descended from the cab and came back and dragged two shovels from among their feet, the blades clashing against the serpentining of the chain about their ankles. "What is it?" one said. "What are you fixing to do?" The trusty didn't answer. He returned to the cab, from which one of the guards had descended, without his shotgun. He and the trusty, both in hip boots and each carrying a shovel, advanced into the water, gingerly, probing and feeling ahead with the shovel handles. The same con- vict spoke again. He was a middle-aged man with a wild thatch of iron-gray hair and a slightly mad face. "What the hell are they doing?" he said. Again nobody answered him. The truck moved, on into the water, behind the guard and the trusty, beginning to push ahead of itself a thick slow viscid ridge of chocolate water. Then the gray-haired convict began to scream. "God damn it, unlock the chain!" He began to struggle, thrashing violently about him, strik- ing at the men nearest him until he reached the cab, the roof of which he now hammered on with his fists, screaming. "God damn it, unlock us! Unlock us! Son of a bitch!" he screamed, addressing no one. "They're going to drown us! Unlock the chain!" But for all the answer he got the men within radius of his voice might have been dead. The truck crawled on, the guard and the trusty feeling out the road ahead with the reversed shovels, the second guard at the wheel, the twenty-two convicts packed like sardines into the truck bed and padlocked by the ankles to the body of the truck itself. They crossed another bridge—two delicate and paradoxical iron railings slanting out of the water, travelling parallel to it for a distance, then slanting down into it again with an outrageous qual- ity almost significant yet apparently meaningless like something in a dream not quite nightmare. The truck crawled on.

Along toward noon they came to a town, their destination. The streets were paved; now the wheels of the truck made a sound like tearing silk. Moving faster now, the guard and the trusty in the cab again, the truck even had a slight bone in its teeth, its bow-wave spreading beyond the submerged sidewalks and across the adjacent lawns, lapping against the stoops and porches of houses where people stood among piles of furniture. They passed through the business district; a man in hip boots emerged knee-deep in water from a store, dragging a flat-bottomed skiff containing a steel safe.

At last they reached the railroad. It crossed the street at right angles, cutting the town in two. It was on a dump, a levee, also, eight or ten feet above the town itself; the street ran blankly into it and turned at right angles beside a cotton compress[4] and a loading platform on stilts at the level of a freight-car door. On this platform was a khaki army tent and a uniformed National Guard sentry with a rifle and bandolier.[5]

The truck turned and crawled out of the water and up the ramp which cotton wagons used and where trucks and private cars filled with household goods came and unloaded onto the platform. They were unlocked from the chain in the truck and shackled ankle to ankle in pairs they mounted the platform and into an apparently inextricable jumble of beds and trunks, gas and electric stoves, radios and tables and chairs and framed pictures which a chain of Negroes under the eye of an unshaven white man in muddy corduroy and hip boots carried piece by piece into the compress, at the door of which another guardsman stood with his rifle, they (the convicts) not stopping here but herded on by the two guards with their shotguns, into the dim and cavernous building where among the piled heterogeneous furniture the ends of cotton bales and the mirrors on dressers and sideboards gleamed with an identical mute and unreflecting concentration of pallid light.

They passed on through, onto the loading platform where the army tent and the first sentry were. They waited here. Nobody told them for what nor why. While the two guards talked with the sentry before the tent the convicts sat in a line along the edge of the platform like buzzards on a fence, their shackled feet dangling above the brown motionless flood out of which the railroad embankment rose, pristine and intact, in a kind of paradoxical denial and repudiation of change and portent, not talking, just looking quietly across the track to where the other half of the amputated town seemed to float, house shrub and tree, ordered and pageant-like and without motion, upon the limitless liquid plain beneath the thick gray sky.

After a while the other four trucks from the Farm arrived. They came up, bunched closely, radiator to tail light, with their four separate sounds of tearing silk and vanished beyond the compress. Pres-

---

4. Apparatus for compressing cotton bales.
5. Cartridge belt worn over one shoulder and diagonally across the chest under the other arm.

ently the ones on the platform heard the feet, the mute clashing of the shackles, the first truckload emerged from the compress, the second, the third; there were more than a hundred of them now in their bed-ticking overalls and jumpers and fifteen or twenty guards with rifles and shotguns. The first lot rose and they mingled, paired, twinned by their clanking and clashing umbilicals; then it began to rain, a slow steady gray drizzle like November instead of May. Yet not one of them made any move toward the open door of the compress. They did not even look toward it, with longing or hope or without it. If they thought at all, they doubtless knew that the available space in it would be needed for furniture, even if it were not already filled. Or perhaps they knew that, even if there were room in it, it would not be for them, not that the guards would wish them to get wet but that the guards would not think about getting them out of the rain. So they just stopped talking and with their jumper collars turned up and shackled in braces like dogs at a field trial they stood, immobile, patient, almost ruminant, their backs turned to the rain as sheep and cattle do.

After another while they became aware that the number of soldiers had increased to a dozen or more, warm and dry beneath rubberised ponchos, there was an officer with a pistol at his belt, then and without making any move toward it, they began to smell food and, turning to look, saw an army field kitchen set up just inside the compress door. But they made no move, they waited until they were herded into line, they inched forward, their heads lowered and patient in the rain, and received each a bowl of stew, a mug of coffee, two slices of bread. They ate this in the rain. They did not sit down because the platform was wet, they squatted on their heels as country men do, hunching forward, trying to shield the bowls and mugs into which nevertheless the rain splashed steadily as into miniature ponds and soaked, invisible and soundless, into the bread.

After they had stood on the platform for three hours, a train came for them. Those nearest the edge saw it, watched it—a passenger coach apparently running under its own power and trailing a cloud of smoke from no visible stack, a cloud which did not rise but instead shifted slowly and heavily aside and lay upon the surface of the aqueous earth with a quality at once weightless and completely spent. It came up and stopped, a single old-fashioned open-ended wooden car coupled to the nose of a pushing switch engine considerably smaller. They were herded into it, crowding forward to the other end where there was a small cast-iron stove. There was no fire in it, nevertheless they crowded about it—the cold and voiceless lump of iron stained with fading tobacco and hovered about by the ghosts of a thousand Sunday excursions to Memphis or Moorhead and return—the peanuts, the bananas, the soiled garments of infants—huddling, shoving for places near it. "Come on, come on," one of the guards shouted. "Sit down, now." At last three of the guards, laying aside their guns, came among them and broke up the huddle, driving them back and into seats.

There were not enough seats for all. The others stood in the aisle, they stood braced, they heard the air hiss out of the released brakes, the engine whistled four blasts, the car came into motion with a snapping jerk; the platform, the compress fled violently as the train seemed to transpose from immobility to full speed with that same quality of unreality with which it had appeared, running backward now though with the engine in front where before it had moved forward but with the engine behind.

When the railroad in its turn ran beneath the surface of the water, the convicts did not even know it. They felt the train stop, they heard the engine blow a long blast which wailed away un-echoed across the waste, wild and forlorn, and they were not even curious; they sat or stood behind the rain-streaming windows as the train crawled on again, feeling its way as the truck had while the brown water swirled between the trucks and among the spokes of the driving wheels and lapped in cloudy steam against the dragging fire-filled belly of the engine; again it blew four short harsh blasts filled with the wild triumph and defiance yet also with repudiation and even farewell, as if the articulated steel itself knew it did not dare stop and would not be able to return. Two hours later in the twilight they saw through the streaming windows a burning planta-tion house. Juxtaposed to nowhere and neighbored by nothing it stood, a clear steady pyre-like flame rigidly fleeing its own reflection, burning in the dusk above the watery desolation with a quality para-doxical, outrageous and bizarre.

Some time after dark the train stopped. The convicts did not know where they were. They did not ask. They would no more have thought of asking where they were than they would have asked why and what for. They couldn't even see, since the car was unlighted and the windows fogged on the outside by rain and on the inside by the engendered heat of the packed bodies. All they could see was a milky and sourceless flick and glare of flashlights. They could hear shouts and commands, then the guards inside the car began to shout; they were herded to their feet and toward the exit, the ankle chains clashing and clanking. They descended into a fierce hissing of steam, through ragged wisps of it blowing past the car. Laid-to alongside the train and resembling a train itself was a thick blunt motor launch to which was attached a string of skiffs and flat boots. There were more soldiers; the flashlights played on the rifle barrels and bandolier buckles and flicked and glinted on the ankle chains of the convicts as they stepped gingerly down into knee-deep water and entered the boats; now car and engine both vanished com-pletely in steam as the crew began dumping the fire from the firebox.

After another hour they began to see lights ahead—a faint wavering row of red pin-pricks extending along the horizon and apparently hanging low in the sky. But it took almost another hour to reach them while the convicts squatted in the skiffs, huddled into the soaked garments (they no longer felt the rain any more at all as separate drops) and watched the lights draw nearer and nearer until at last the crest of the levee defined itself; now they could discern a row of army tents stretching along it and people squatting about the fires, the wavering reflections from which, stretching across the water, revealed an involved mass of other skiffs tied against the flank of the levee which now stood high and dark overheard. Flashlights glared and winked along the base, among the tethered skiffs; the launch, silent now, drifted in.

When they reached the top of the levee they could see the long line of khaki tents, interspersed with fires about which people—men, women and children, Negro and white—crouched or stood among shapeless bales of clothing, their heads turning, their eyeballs glinting in the firelight as they looked quietly at the striped garments and the chains; further down the levee, huddled together too though untethered, was a drove of mules and two or three cows. Then the taller convict became conscious of another sound. He did not begin to hear it all at once, he suddenly became aware that he had been hearing it all the time, a sound so much beyond all his experience and his powers of assimilation that up to this point he had been as oblivious of it as an ant or a flea might be of the sound of the avalanche on which it rides; he had been travelling upon water since early afternoon and for seven years now he had run his plow and harrow and planter within the very shadow of the levee on which he now stood, but this profound deep whisper which came from the further side of it he did not at once recognise. He stopped. The line of convicts behind jolted into him like a line of freight cars stopping, with an iron clashing like cars. "Get on!" a guard shouted.

"What's that?" the convict said. A Negro man squatting before the nearest fire answered him:

"Dat's him. Dat's de Ole Man."

"The old man?" the convict said.

"Get on! Get on up there!" the guard shouted. They went on; they passed another huddle of mules, the eyeballs rolling too, the long morose faces turning into and out of the firelight; they passed them and reached a section of empty tents, the light pup tents of a military campaign, made to hold two men. The guards herded the convicts into them, three brace of shackled men to each tent.

They crawled in on all fours, like dogs into cramped kennels, and settled down. Presently the tent became warm from their bodies. Then they became quiet and then all of them could hear it, they lay listening to the bass whisper deep, strong and powerful. "The old man?" the train-robber convict said.

"Yah," another said. "He dont have to brag."[6]

At dawn the guards waked them by kicking the soles of the projecting feet. Opposite the muddy landing and the huddle of skiffs an army field kitchen was set up, already they could smell the coffee. But the taller convict at least, even though he had had but one meal yesterday and that at noon in the rain, did not move at once toward the food. Instead and for the first time he looked at the River within whose shadow he had spent the last seven years of his life but had never seen before; he stood in quiet and amazed surmise and looked at the rigid steel-colored surface not broken into waves but merely slightly undulant. It stretched from the levee on which he stood, further than he could see—a slowly and heavily roiling chocolate-frothy expanse broken only by a thin line a mile away as fragile in appearance as a single hair, which after a moment he recognised. *It's another levee,* he thought quietly. *That's what we look like from there. That's what I am standing on looks like from there.* He was prodded from the rear; a guard's voice carried forward: "Go on! Go on! You'll have plenty of time to look at that!"

They received the same stew and coffee and bread as the day before; they squatted again with their bowls and mugs as yesterday, though it was not raining yet. During the night an intact wooden barn had floated up. It now lay jammed by the current against the levee while a crowd of Negroes swarmed over it, ripping off the shingles and planks and carrying them up the bank; eating steadily and without haste, the taller convict watched the barn dissolve rapidly down to the very water-line exactly as a dead fly vanished beneath the moiling industry of a swarm of ants.

They finished eating. Then it began to rain again, as upon a signal, while they stood or squatted in their harsh garments which had not dried out during the night but had merely become slightly warmer than the air. Presently they were haled to their feet and told off into two groups, one of which was armed from a stack of mud-clogged picks and shovels nearby, and marched away up the levee. A little later the motor launch with its train of skiffs came up across what was, fifteen feet beneath its keel, probably a cotton field, the skiffs loaded to the gunwales with Negroes and a scattering of white people nursing bundles on their laps. When the engine

6. On the awesome force of the Mississippi, Faulkner commented: "[The valley is] rich because the river for hundreds of years has deposited the rich silt on top of the ground, and so the river dominates not only the economy of that country but it dominates the spiritual life. * * * the river is Master, and any time the Old Man wants to he can break the levee and can ruin the cotton crop. * * * the planter is at armistice with him, and the superstitious planter believes that he has got to make libations, make sacrifices to him * * * like the ancients with the dragon, the Minotaur, the symbols of destructiveness which they had to placate, sacrifice" (*Faulkner in the University,* p. 178).

shut off the faint plinking of a guitar came across the water. The skiffs warped in and unloaded; the convicts watched the men and women and children struggle up the muddy slope, carrying heavy towsacks and bundles wrapped in quilts. The sound of the guitar had not ceased and now the convicts saw him—a young, black, lean-hipped man, the guitar slung by a piece of cotton plowline about his neck. He mounted the levee, still picking it. He carried nothing else, no food, no change of clothes, not even a coat.

The taller convict was so busy watching this that he did not hear the guard until the guard stood directly beside him shouting his name. "Wake up!" the guard shouted. "Can you fellows paddle a boat?"

"Paddle a boat where?" the taller convict said.

"In the water," the guard said. "Where in hell do you think?"

"I aint going to paddle no boat nowhere out yonder," the tall convict said, jerking his head toward the invisible river beyond the levee behind him.

"No, it's on this side," the guard said. He stooped swiftly and unlocked the chain which joined the tall convict and the plump hairless one. "It's just down the road a piece." He rose. The two convicts followed him down to the boats. "Follow them telephone poles until you come to a filling station. You can tell it, the roof is still above water. It's on a bayou and you can tell the bayou because the tops of the trees are sticking up. Follow the bayou until you come to a cypress snag with a woman in it. Pick her up and then cut straight back west until you come to a cotton house with a fellow sitting on the ridgepole—" He turned, looking at the two convicts, who stood perfectly still, looking first at the skiff and then at the water with intense sobriety. "Well? What are you waiting for?"

"I cant row a boat," the plump convict said.

"Then it's high time you learned," the guard said. "Get in."

The tall convict shoved the other forward. "Get in," he said. "That water aint going to hurt you. Aint nobody going to make you take a bath."

As, the plump one in the bow and the other in the stern, they shoved away from the levee, they saw other pairs being unshackled and manning the other skiffs. "I wonder how many more of them fellows are seeing this much water for the first time in their lives too," the tall convict said. The other did not answer. He knelt in the bottom of the skiff, pecking gingerly at the water now and then with his paddle. The very shape of his thick soft back seemed to wear that expression of wary and tense concern.

Some time after midnight a rescue boat filled to the guard rail with homeless men and women and children docked at Vicksburg. It was a steamer, shallow of draft; all day long it had poked up and down cypress- and gum-choked bayous and across cotton fields (where at times instead of swimming it waded) gathering its sorry cargo from the tops of houses and barns and even out of trees, and now it warped into that mushroom city of the forlorn and despairing where kerosene flares smoked in the drizzle and hurriedly strung electrics glared upon the bayonets of martial policemen and the Red Cross brassards of doctors and nurses and canteen-workers. The bluff overhead was almost solid with tents, yet still there were more people than shelter for them; they sat or lay, single and by whole families, under what shelter they could find or sometimes under the rain itself, in the little death of profound exhaustion while the doctors and the nurses and the soldiers stepped over and around and among them.

Among the first to disembark was one of the penitentiary deputy wardens, followed closely by the plump convict and another white man—a small man with a gaunt unshaven wan face still wearing an expression of incredulous outrage. The deputy warden seemed to know exactly where he wished to go. Followed closely by his two companions he threaded his way swiftly among the piled furniture and the sleeping bodies and stood presently in a fiercely lighted and hastily established temporary office, almost a military post of command in fact, where the Warden of the Penitentiary sat with two army officers wearing majors' leaves. The deputy warden spoke without preamble. "We lost a man," he said. He called the tall convict's name.

"Lost him?" the Warden said.

"Yah. Drowned." Without turning his head he spoke to the plump convict. "Tell him," he said.

"He was the one that said he could row a boat," the plump convict said. "I never. I told him myself—" he indicated the deputy warden with a jerk of his head "—I couldn't. So when we got to the bayou—"

"What's this?" the Warden said.

"The launch brought word in," the deputy warden said. "Woman in a cypress snag on the bayou, then this fellow—" he indicated the third man; the Warden and the two officers looked at the third man "—on a cotton house. Never had room in the launch to pick them up. Go on."

"So we come to where the bayou was," the plump convict continued in a voice perfectly flat, without any inflection whatever. "Then the boat got away from him. I dont know what happened. I was just sitting there because he was so positive he could row a boat. I never saw any current. Just all of a sudden the boat whirled clean around and begun to run fast backward like it was hitched to a train and it whirled around again and I happened to look up and there was a limb right over my head and I grabbed it just in time and that boat was snatched out from under me like you'd snatch off a sock and I saw it one time more upside down and that fellow that said he knew all about rowing holding to it with one hand and still holding the paddle in the other—" He ceased. There was no dying fall to his voice, it just ceased and the convict stood looking quietly at a half-full quart of whiskey sitting on the table.

"How do you know he's drowned?" the Warden said to the deputy. "How do you know he didn't just see his chance to escape, and took it?"

"Escape where?" the other said. "The whole Delta's flooded. There's fifteen foot of water for fifty miles, clean back to the hills. And that boat was upside down."

"That fellow's drowned," the plump convict said. "You dont need to worry about him. He's got his pardon; it wont cramp nobody's hand signing it, neither."

"And nobody else saw him?" the Warden said. "What about the woman in the tree?"

"I dont know," the deputy said. "I aint found her yet. I reckon some other boat picked her up. But this is the fellow on the cotton house."

Again the Warden and the two officers looked at the third man, at the gaunt, unshaven wild face in which an old terror, an old blending of fear and impotence and rage still lingered. "He never came for you?" the Warden said. "You never saw him?"

"Never nobody came for me," the refugee said. He began to tremble though at first he spoke quietly enough. "I set there on that sonabitching cotton house, expecting hit to go any minute. I saw that launch and them boats come up and they never had no room for me. Full of bastard niggers and one of them setting there playing a guitar but there wasn't no room for me. A guitar!" he cried; now he began to scream, trembling, slavering, his face twitching and jerking. "Room for a bastard nigger guitar but not for me—"

"Steady now," the Warden said. "Steady now."

"Give him a drink," one of the officers said. The Warden poured the drink. The deputy handed it to the refugee, who took the glass in both jerking hands and tried to raise it to his mouth. They watched him for perhaps twenty seconds, then the deputy took the glass from him and held it to his lips while he gulped, though even then a thin trickle ran from each corner of his mouth, into the stubble on his chin.

"So we picked him and—" the deputy called the plump convict's name now "—both up just before dark and come on in. But that other fellow is gone."

"Yes," the Warden said. "Well. Here I haven't lost a prisoner in ten years, and now, like this—I'm sending you back to the Farm tomorrow. Have his family notified, and his discharge papers filled out at once."

"All right," the deputy said. "And listen, chief. He wasn't a bad fellow and maybe he never had no business in that boat. Only he did say he could paddle one. Listen. Suppose I write on his discharge, Drowned while trying to save lives in the great flood of nineteen twenty-seven, and send it down for the Governor to sign it. It will be something nice for his folks to have, to hang on the wall when neighbors come in or something. Maybe they will even give his folks a cash bonus because after all they sent him to the Farm to raise cotton, not to fool around in a boat in a flood."

"All right," the Warden said. "I'll see about it. The main thing is to get his name off the books as dead before some politician tries to collect his food allowance."

"All right," the deputy said. He turned and herded his companions out. In the drizzling darkness again he said to the plump convict: "Well, your partner beat you. He's free. He's done served his time out but you've got a right far piece to go yet."

"Yah," the plump convict said. "Free. He can have it."

As the short convict had testified, the tall one, when he returned to the surface, still retained what the short one called the paddle. He clung to it, not instinctively against the time when he would be back inside the boat and would need it, because for a time he did not believe he would ever regain the skiff or anything else that would support him, but because he did not have time to think about turning it loose. Things had moved too fast for him. He had not been warned, he had felt the first snatching tug of the current, he had seen the skiff begin to spin and his companion vanish violently upward like in a translation out of Isaiah,[7] then he himself was in the water, struggling against the drag of the paddle which he did not know he still held each time he fought back to the surface and grasped at the spinning skiff which at one instant was ten feet away and the next poised above his head as though about to brain him, until at last he grasped the stern, the drag of his body becoming a rudder to the skiff, the two of them, man and boat and with the paddle perpendicular above them like a jackstaff,[8] vanishing from the view of the short convict (who had vanished from that of the tall one with the same celerity though in a vertical direction) like a tableau snatched offstage intact with violent and incredible speed.

He was now in the channel of a slough, a bayou, in which until today no current had run probably since the old subterranean outrage which had created the country. There was plenty of current in it now though; from his trough behind the stern he seemed to see

7. Cf. Isaiah 8.7–8: "* * * and now therefore, behold, the Lord bringeth up upon them the waters of the River * * * and he shall come up over all his channels, and go over all his banks; * * * and he shall overthrow and pass through; he shall reach even to the neck; and the stretching out of his wings shall fill the breadth of thy land, O Immanuel."
8. Flag staff at the bow of a ship.

the trees and sky rushing past with vertiginous speed, looking down at him between the gouts of cold yellow in lugubrious and mournful amazement. But they were fixed and secure in something; he thought of that, he remembered in an instant of despairing rage the firm earth fixed and founded strong and cemented fast and stable forever by the generations of laborious sweat, somewhere beneath him, beyond the reach of his feet, when, and again without warning, the stern of the skiff struck him a stunning blow across the bridge of his nose. The instinct which had caused him to cling to it now caused him to fling the paddle into the boat in order to grasp the gunwale with both hands just as the skiff pivoted and spun away again. With both hands free he now dragged himself over the stern and lay prone on his face, streaming with blood and water and panting, not with exhaustion but with that furious rage which is terror's aftermath.

But he had to get up at once because he believed he had come much faster (and so farther) than he had. So he rose, out of the watery scarlet puddle in which he had lain, streaming, the soaked denim heavy as iron on his limbs, the black hair plastered to his skull, the blood-infused water streaking his jumper, and dragged his forearm gingerly and hurriedly across his lower face and glanced at it then grasped the paddle and began to try to swing the skiff back upstream. It did not even occur to him that he did not know where his companion was, in which tree among all which he had passed or might pass. He did not even speculate on that for the reason that he knew so incontestably that the other was upstream from him, and after his recent experience the mere connotation of the term upstream carried a sense of such violence and force and speed that the conception of it as other than a straight line was something which the intelligence, reason, simply refused to harbor, like the notion of a rifle bullet the width of a cotton field.

The bow began to swing back upstream. It turned readily, it outpaced the aghast and outraged instant in which he realised it was swinging far too easily, it had swung on over the arc and lay broadside to the current and began again that vicious spinning while he sat, his teeth bared in his bloody streaming face while his spent arms flailed the impotent paddle at the water, that innocent-appearing medium which at one time had held him in iron-like and shifting convolutions like an anaconda yet which now seemed to offer no more resistance to the thrust of his urge and need than so much air, like air; the boat which had threatened him and at last actually struck him in the face with the shocking violence of a mule's hoof now seemed to poise weightless upon it like a thistle bloom, spinning like a wind vane while he flailed at the water and thought of, envisioned, his companion safe, inactive and at ease in the tree with nothing to do but wait, musing with impotent and terrified fury upon that arbitrariness of human affairs which had abrogated to the one the secure tree and to the other the hysterical and unmanageable boat for the very reason that it knew that he alone of the two of them would make any attempt to return and rescue his companion.

The skiff had paid off and now ran with the current again. It seemed again to spring from immobility into incredible speed, and he thought he must already be miles away from where his companion had quitted him, though actually he had merely described a big circle since getting back into the skiff, and the object (a clump of cypress trees choked by floating logs and debris) which the skiff was now about to strike was the same one it had careened into before when the stern had struck him. He didn't know this because he had not yet ever looked higher than the bow of the boat. He didn't look higher now, he just saw that he was going to strike; he seemed to feel run through the very insentient fabric of the skiff a current of eager gleeful vicious incorrigible wilfulness; and he who had never ceased to flail at the bland treacherous water with what he had believed to be the limit of his strength now from somewhere, some ultimate absolute reserve, produced a final measure of endurance, will to endure which adumbrated mere muscle and nerves, continuing to flail the paddle right up to the instant of striking, completing one last reach thrust and recover out of pure desperate reflex, as a man slipping on ice reaches for his hat and moneypocket, as the skiff struck and hurled him once more flat on his face in the bottom of it.

This time he did not get up at once. He lay flat on his face, slightly spread-eagled and in an attitude almost peaceful, a kind of abject meditation. He would have to get up sometime, he knew that, just as all life consists of having to get up sooner or later and then having to lie down again sooner or later after a while. And he was not exactly exhausted and he was not particularly without hope and he did not especially dread getting up. It merely seemed to him that he had accidentally been caught in a situation in which time and environment, not himself, was mesmerised; he was being toyed with by a current of water going nowhere, beneath a day which would wane toward no evening; when it was done with him it would spew him back into the comparatively safe world he had been snatched violently out of and in the meantime it did not much matter just what he did or did not do. So he lay on his face, now not only feeling but hearing the strong quiet rustling of the current on the underside of the planks, for a while longer. Then he raised his head and this time touched his palm gingerly to his face and looked at the blood again, then he sat up onto his heels and leaning over the gunwale he pinched his nostrils between thumb and finger and expelled a gout of blood and was in the act of wiping his fingers on his thigh when a voice slightly above his line of sight said quietly, "It's taken you a while," and he who up to this moment had had neither reason nor time to raise his eyes higher than the bows looked up and saw, sitting in a tree and looking at him, a woman. She was not ten feet away. She sat on the lowest limb of one of the trees holding the jam he had grounded on, in a calico wrapper and an army private's tunic and a sunbonnet, a woman whom he did not even bother to examine since that first startled glance had been ample to reveal to him all the generations of her life and background, who could have been his sister if he had a sister, his wife if he had not entered the penitentiary at an age scarcely out of adolescence and some years younger than that at which even his prolific and monogamous kind married—a woman who sat clutching the trunk of the tree, her stockingless feet in a pair of man's unlaced brogans less than a yard from the water, who was very probably somebody's sister and quite certainly (or certainly should have been) somebody's wife, though this too he had entered the penitentiary too young to have had more than mere theoretical female experience to discover yet. "I thought for a minute you wasn't aiming to come back."

"Come back?"

"After the first time. After you run into this brush pile the first time and got into the boat and went on." He looked about, touching his face tenderly again; it could very well be the same place where the boat had hit him in the face.

"Yah," he said. "I'm here now though."

"Could you maybe get the boat a little closer? I taken a right sharp strain getting up here; maybe I better . . ." He was not listening; he had just discovered that the paddle was gone; this time when the skiff hurled him forward he had flung the paddle not into it but beyond it. "It's right there in them brush tops," the woman said. "You can get it. Here. Catch a holt of this." It was a grapevine. It had grown up into the tree and the flood had torn the roots loose. She had taken a turn with it about her upper body; she now loosed it and swung it out until he could grasp it. Holding to the end of the vine he warped the skiff around the end of the jam, picking up the paddle, and warped the skiff on beneath the limb and held it and now he watched her move, gather herself heavily and carefully to descend—that heaviness which was not painful but just excruciatingly careful, that profound and almost lethargic awkwardness which added nothing to the sum of that first aghast amazement which had served already for the catafalque of invincible dream since even in durance he had continued (and even with the old avidity, even though they had caused his downfall) to consume the impossible pulp-printed fables carefully censored and as carefully smuggled into the penitentiary; and who to say what Helen, what living Garbo,[9] he had not dreamed of rescuing from what craggy

---

9. Helen of Troy (whose beauty occa-   Garbo (b. 1905), film star.
sioned the Trojan War) and Greta

pinnacle or dragoned keep when he and his companion embarked in the skiff. He watched her, he made no further effort to help her beyond holding the skiff savagely steady while she lowered herself from the limb—the entire body, the deformed swell of belly bulging the calico, suspended by its arms, thinking, *And this is what I get. This, out of all the female meat that walks, is what I have to be caught in a runaway boat with.*

"Where's that cottonhouse?" he said.

"Cottonhouse?"

"With that fellow on it. The other one."

"I dont know. It's a right smart of cottonhouses around here. With folks on them too, I reckon." She was examining him. "You're bloody as a hog," she said. "You look like a convict."

"Yah," he said, snarled. "I feel like I done already been hung. Well, I got to pick up my pardner and then find that cottonhouse." He cast off. That is, he released his hold on the vine. That was all he had to do, for even while the bow of the skiff hung high on the log jam and even while he held it by the vine in the comparatively dead water behind the jam, he felt steadily and constantly the whisper, the strong purring power of the water just one inch beyond the frail planks on which he squatted and which, as soon as he released the vine, took charge of the skiff not with one powerful clutch but in a series of touches light, tentative, and catlike; he realised now that he had entertained a sort of foundationless hope that the added weight might make the skiff more controllable. During the first moment or two he had a wild (and still foundationless) belief that it had; he had got the head upstream and managed to hold it so by terrific exertion continued even after he discovered that they were travelling straight enough but stern-first and continued somehow even after the bow began to wear away and swing; the old irresistible movement which he knew well by now, too well to fight against it, so that he let the bow swing on downstream with the hope of utilising the skiff's own momentum to bring it through the full circle and so upstream again, the skiff travelling broadside then bow-first then broadside again, diagonally across the channel, toward the other wall of submerged trees; it began to flee beneath him with terrific speed, they were in an eddy but did not know it; he had no time to draw conclusions or even wonder; he crouched, his teeth bared in his blood-caked and swollen face, his lungs bursting, flailing at the water while the trees stooped hugely down at him. The skiff struck, spun, struck again; the woman half lay in the bow, clutching the gunwales, as if she were trying to crouch behind her own pregnancy; he banged now not at the water but at the living sap-blooded wood with the paddle, his desire now not to go anywhere, reach any destination, but just to keep the skiff from beating itself to fragments against the tree trunks. Then something exploded, this time against the back of his head, and stooping trees and dizzy water, the woman's face and all, fled together and vanished in bright soundless flash and glare.

An hour later the skiff came slowly up an old logging road and so out of the bottom, the forest, and into (or onto) a cottonfield—a gray and limitless desolation now free of turmoil, broken only by a thin line of telephone poles like a wading millipede. The woman was now paddling, steadily and deliberately, with that curious lethargic care, while the convict squatted, his head between his knees, trying to stanch the fresh and apparently inexhaustible flow of blood from his nose with handfuls of water. The woman ceased paddling, the skiff drifted on, slowing, while she looked about. "We're done out," she said.

The convict raised his head and also looked about. "Out where?"

"I thought maybe you might know."

"I dont even know where I used to be. Even if I knowed which way was north, I wouldn't know if that was where I wanted to go." He cupped another handful of water to his face and lowered his hand and regarded the resulting crimson marbling on his palm, not with dejection, not with concern, but with a kind of sardonic and vicious bemusement. The woman watched the back of his head.

"We got to get somewhere."

"Dont I know it? A fellow on a cottonhouse. Another in a tree. And now that thing in your lap."

"It wasn't due yet. Maybe it was having to climb that tree quick yesterday, and having to set in it all night. I'm doing the best I can. But we better get somewhere soon."

"Yah," the convict said. "I thought I wanted to get somewhere too and I ain't had no luck at it. You pick out a place to get to now and we'll try yours. Gimme that oar." The woman passed him the paddle. The boat was a double-ender; he had only to turn around.

"Which way you fixing to go?" the woman said.

"Never you mind that. You just keep on holding on." He began to paddle, on across the cottonfield. It began to rain again, though not hard at first. "Yah," he said. "Ask the boat. I been in it since breakfast and I aint never knowed, where I aimed to go or where I was going either."

That was about one oclock. Toward the end of the afternoon the skiff (they were in a channel of some sort again, they had been in it for some time; they had got into it before they knew it and too late to get out again, granted there had been any reason to get out, as, to the convict anyway, there was certainly none and the fact that their speed had increased again was reason enough to stay in it) shot out upon a broad expanse of debris-filled water which the convict recognised as a river and, from its size, the Yazoo River though it was little enough he had seen of this country which he had not quitted for so much as one single day in the last seven years of his life. What he did not know was that it was now running backward.

So as soon as the drift of the skiff indicated the set of the current, he began to paddle in that direction which he believed to be downstream, where he knew there were towns—Yazoo City, and as a last resort, Vicksburg, if his luck was that bad, if not, smaller towns whose names he did not know but where there would be people, houses, something, anything he might reach and surrender his charge to and turn his back on her forever, on all pregnant and female life forever and return to that monastic existence of shotguns and shackles where he would be secure from it. Now, with the imminence of habitations, release from her, he did not even hate her. When he looked upon the swelling and unmanageable body before him it seemed to him that it was not the woman at all but rather a separate demanding threatening inert yet living mass of which both he and she were equally victims; thinking, as he had been for the last three or four hours, of that minute's—nay, second's—aberration of eye or hand which would suffice to precipitate her into the water to be dragged down to death by that senseless millstone which in its turn would not even have to feel agony, he no longer felt any glow of revenge toward her as its custodian, he felt sorry for her as he would for the living timber in a barn which had to be burned to rid itself of vermin.

He paddled on, helping the current, steadily and strongly, with a calculated husbandry of effort, toward what he believed was downstream, towns, people, something to stand upon, while from time to time the woman raised herself to bail the accumulated rain from the skiff. It was raining steadily now though still not hard, still without passion, the sky, the day itself dissolving without grief; the skiff moved in a nimbus, an aura of gray gauze which merged almost without demarcation with the roiling spittle-frothed debris-choked water. Now the day, the light, definitely began to end and the convict permitted himself an extra notch or two of effort because it suddenly seemed to him that the speed of the skiff had lessened. This was actually the case though the convict did not know it. He merely took it as a phenomenon of the increasing obfuscation, or at most as a result of the long day's continuous effort with no food, complicated by the ebbing and fluxing phases of anxiety and impotent rage at his absolutely gratuitous predicament. So he stepped up his stroke a beat or so, not from alarm but on the contrary, since he too had received that lift from the mere presence of a known stream, a river known by its ineradicable name to generations of men who had been drawn to live beside it as man always has been drawn to dwell beside water, even before he had a name for water and fire, drawn to the living water, the course of his destiny and his actual physical appearance rigidly coerced and postulated by it. So he was not alarmed. He paddled on, upstream without knowing it, unaware that all the water which for forty hours now

had been pouring through the levee break to the north was somewhere ahead of him, on its way back to the River.

It was full dark now. That is, night had completely come, the
gray dissolving sky had vanished, yet as though in perverse ratio surface visibility had sharpened, as though the light which the rain of
the afternoon had washed out of the air had gathered upon the
water as the rain itself had done, so that the yellow flood spread on
before him now with a quality almost phosphorescent, right up to
the instant where vision ceased. The darkness in fact had its advantages; he could now stop seeing the rain. He and his garments had
been wet for more than twenty-four hours now so he had long since
stopped feeling it, and now that he could no longer see it either it
had in a certain sense ceased for him. Also, he now had to make no
effort even not to see the swell of his passenger's belly. So he was
paddling on, strongly and steadily, not alarmed and not concerned
but just exasperated because he had not yet begun to see any reflection on the clouds which would indicate the city or cities which he
believed he was approaching but which were actually now miles
behind him, when he heard a sound. He did not know what it was
because he had never heard it before and he would never be
expected to hear such again since it is not given to every man to
hear such at all and to none to hear it more than once in his life.
And he was not alarmed now either because there was not time, for
although the visibility ahead, for all its clarity, did not extend very
far, yet in the next instant to the hearing he was also seeing something such as he had never seen before. This was that the sharp line
where the phosphorescent water met the darkness was now about
ten feet higher than it had been an instant before and that it was
curled forward upon itself like a sheet of dough being rolled out for
a pudding. It reared, stooping; the crest of it swirled like the mane
of a galloping horse and, phosphorescent too, fretted and flickered
like fire. And while the woman huddled in the bows, aware or not
aware the convict did not know which, he (the convict), his swollen and blood-streaked face gaped in an expression of aghast and
incredulous amazement, continued to paddle directly into it. Again
he simply had not had time to order his rhythm-hypnotised muscles
to cease. He continued to paddle though the skiff had ceased to
move forward at all but seemed to be hanging in space while the
paddle still reached thrust recovered and reached again; now instead
of space the skiff became abruptly surrounded by a welter of fleeing
debris—planks, small buildings, the bodies of drowned yet antic
animals, entire trees leaping and diving like porpoises above which
the skiff seemed to hover in weightless and airy indecision like a
bird above a fleeing countryside, undecided where to light or whether to light at all, while the convict squatted in it still going through
the motions of paddling, waiting for an opportunity to scream. He

never found it. For an instant the skiff seemed to stand erect on its stern and then shoot scrabbling and scrambling up the curling wall of water like a cat, and soared on above the licking crest itself and hung cradled into the high actual air in the limbs of a tree, from which bower of new-leafed boughs and branches the convict, like a bird in its nest and still waiting his chance to scream and still going through the motions of paddling though he no longer even had the paddle now, looked down upon a world turned to furious motion and in incredible retrograde.

Some time about midnight, accompanied by a rolling cannonade of thunder and lightning like a battery going into action, as though some forty hours' constipation of the elements, the firmament itself, were discharging in clapping and glaring salute to the ultimate acquiescence to desperate and furious motion, and still leading its charging welter of dead cows and mules and outhouses and cabins and hen-coops, the skiff passed Vicksburg. The convict didn't know it. He wasn't looking high enough above the water; he still squatted, clutching the gunwales and glaring at the yellow turmoil about him out of which entire trees, the sharp gables of houses, the long mournful heads of mules which he fended off with a splintered length of plank snatched from he knew not where in passing (and which seemed to glare reproachfully back at him with sightless eyes, in limber-lipped and incredulous amazement) rolled up and then down again, the skiff now travelling forward now sideways now sternward, sometimes in the water, sometimes riding for yards upon the roofs of houses and trees and even upon the backs of the mules as though even in death they were not to escape that burden-bearing doom with which their eunuch race was cursed. But he didn't see Vicksburg; the skiff, travelling at express speed, was in a seething gut between soaring and dizzy banks with a glare of light above them but he did not see it; he saw the flotsam ahead of him divide violently and begin to climb upon itself, mounting, and he was sucked through the resulting gap too fast to recognise it as the trestling of a railroad bridge; for a horrible moment the skiff seemed to hang in static indecision before the looming flank of a steamboat as though undecided whether to climb over it or dive under it, then a hard icy wind filled with the smell and taste and sense of wet and boundless desolation blew upon him; the skiff made one long bounding lunge as the convict's native state, in a final paroxysm, regurgitated him onto the wild bosom of the Father of Waters.

This is how he told about it seven weeks later, sitting in new bed-ticking garments, shaved and with his hair cut again, on his bunk in the barracks:

During the next three or four hours after the thunder and lightning had spent itself the skiff ran in pitch streaming darkness upon a roiling expanse which, even if he could have seen, apparently had no boundaries. Wild and invisible, it tossed and heaved about and beneath the boat, ridged with dirty phosphorescent foam and filled with a debris of destruction—objects nameless and enormous and invisible which struck and slashed at the skiff and whirled on. He did not know he was now upon the River. At that time he would have refused to believe it, even if he had known. Yesterday he had known he was in a channel by the regularity of the spacing between the bordering trees. Now, since even by daylight he could have seen no boundaries, the last place under the sun (or the streaming sky rather) he would have suspected himself to be would have been a river; if he had pondered at all about his present whereabouts, about the geography beneath him, he would merely have taken himself to be travelling at dizzy and inexplicable speed above the largest cotton field in the world; if he who yesterday had known he was in a river, had accepted that fact in good faith and earnest, then had seen that river turn without warning and rush back upon him with furious and deadly intent like a frenzied stallion in a lane—if he had suspected for one second that the wild and limitless expanse on which he now found himself was a river, consciousness would simply have refused; he would have fainted.

When daylight—a gray and ragged dawn filled with driving scud between icy rain-squalls—came and he could see again, he knew he was in no cottonfield. He knew that the wild water on which the skiff tossed and fled flowed above no soil tamely trod by man, behind the straining and surging buttocks of a mule. That was when it occurred to him that its present condition was no phenomenon of a decade, but that the intervening years during which it consented to bear upon its placid and sleepy bosom the frail mechanicals of man's clumsy contriving was the phenomenon and this the norm and the river was now doing what it liked to do, had waited patiently the ten years in order to do, as a mule will work for you ten years for the privilege of kicking you once. And he also learned something else about fear too, something he had even failed to discover on that other occasion when he was really afraid—that three or four seconds of that night in his youth while he looked down the twice-flashing pistol barrel of the terrified mail clerk before the clerk could be persuaded that his (the convict's) pistol would not shoot: that if you just held on long enough a time would come in fear after which it would no longer be agony at all but merely a kind of horrible outrageous itching, as after you have been burned bad.

He did not have to paddle now, he just steered (who had been without food for twenty-four hours now and without any sleep to speak of for fifty) while the skiff sped on across that boiling desolation where he had long since begun to not dare believe he could possibly be where he could not doubt he was, trying with his fragment of splintered plank merely to keep the skiff intact and afloat among the houses and trees and dead animals (the entire towns, stores, residences, parks and farmyards, which leaped and played about him like fish), not trying to reach any destination, just trying to keep the skiff afloat until he did. He wanted so little. He wanted nothing for himself. He just wanted to get rid of the woman, the belly, and he was trying to do that in the right way, not for himself, but for her. He could have put her back into another tree at any time—

"Or you could have jumped out of the boat and let her and it drown," the plump convict said. "Then they could have given you the ten years for escaping and then hung you for the murder and charged the boat to your folks."

"Yah," the tall convict said.—but he had not done that. He wanted to do it the right way, find somebody, anybody he could surrender her to, something solid he could set her down on and then jump back into the river, if that would please anyone. That was all he wanted—just to come to something, anything. That didn't seem like a great deal to ask. And he couldn't do it. He told how the skiff fled on—

"Didn't you pass nobody?" the plump convict said. "No steamboat, nothing?"

"I dont know," the tall one said.—while he tried merely to keep it afloat, until the darkness thinned and lifted and revealed—

"Darkness?" the plump convict said. "I thought you said it was already daylight."

"Yah," the tall one said. He was rolling a cigarette, pouring the tobacco carefully from a new sack, into the creased paper. "This was another one. They had several while I was gone."—the skiff to be moving still rapidly up a winding corridor bordered by drowned trees which the convict recognised again to be a river running again in the direction that, until two days ago, had been upstream. He was not exactly warned through instinct that this one, like that of two days ago, was in reverse. He would not say that he now believed himself to be in the same river, though he would not have been surprised to find that he did believe this, existing now, as he did and had and apparently was to continue for an unnamed period, in a state in which he was toy and pawn on a vicious and inflammable geography. He merely realised that he was in a river again, with all the subsequent inferences of a comprehensible, even if not familiar, portion of the earth's surface. Now he believed that all he had to do would be to paddle far enough and he would come to something horizontal and above water even if not dry and perhaps even populated; and, if fast enough, in time, and that his only other crying urgency was to refrain from looking at the woman who, as vision, the incontrovertible and apparently inescapable presence of his passenger, returned with dawn, had ceased to be a human being and (you could add twenty-four more hours to the first twenty-four and the first fifty now, even counting the hen. It was dead, drowned, caught by one wing under a shingle on a roof which had rolled momentarily up beside the skiff yesterday and he had eaten some of it raw though the woman would not) had become instead one single inert monstrous sentient womb which, he now believed, if he could only turn his gaze away and keep it away, would disappear, and if he could only keep his gaze from pausing again at the spot it had occupied, would not return. That's what he was doing this time when he discovered the wave was coming.

He didn't know how he discovered it was coming back. He heard no sound, it was nothing felt nor seen. He did not even believe that finding the skiff to be now in slack water—that is, that the motion of the current which, whether right or wrong, had at least been horizontal, had now stopped that and assumed a vertical direction— was sufficient to warn him. Perhaps it was just an invincible and almost fanatic faith in the inventiveness and innate viciousness of that medium on which his destiny was now cast, apparently forever; a sudden conviction far beyond either horror or surprise that now was none too soon for it to prepare to do whatever it was it intended doing. So he whirled the skiff, spun it on its heel like a running horse, whereupon, reversed, he could not even distinguish the very channel he had come up. He did not know whether he simply could not see it or if it had vanished some time ago and he not aware at the time; whether the river had become lost in a drowned world or if the world had become drowned in one limitless river. So now he could not tell if he were running directly before the wave or quartering across its line of charge; all he could do was keep that sense of swiftly accumulating ferocity behind him and paddle as fast as his spent and now numb muscles could be driven, and try not to look at the woman, to wrench his gaze from her and keep it away until he reached something flat and above water. So, gaunt, hollow-eyed, striving and wrenching almost physically at his eyes as if they were two of those suction-tipped rubber arrows shot from the toy gun of a child, his spent muscles obeying not will now but that attenuation beyond mere exhaustion which, mesmeric, can continue easier than cease, he once more drove the skiff full tilt into something it could not pass and, once more hurled violently forward onto his hands and knees, crouching, he glared with his wild swollen face up at the man with the shotgun and said in a harsh, croaking voice: "Vicksburg? Where's Vicksburg?"

Even when he tried to tell it, even after the seven weeks and he safe, secure, riveted warranted and doubly guaranteed by the ten years they had added to his sentence for attempted escape, something of the old hysteric incredulous outrage came back into his face, his voice, his speech. He never did even get on the other boat. He told how he clung to a strake[1] (it was a dirty unpainted shanty

1. Plank on the bottom or side of a ship.

boat with a drunken rake of tin stove pipe, it had been moving when he struck it and apparently it had not even changed course even though the three people on it must have been watching him all the while—a second man, barefoot and with matted hair and beard also at the steering sweep, and then—he did not know how long—a woman leaning in the door, in a filthy assortment of men's garments, watching him too with the same cold speculation) being dragged violently along, trying to state and explain his simple (and to him at least) reasonable desire and need; telling it, trying to tell it, he could feel again the old unforgettable affronting like an ague fit as he watched the abortive tobacco rain steadily and faintly from between his shaking hands and then the paper itself part with a thin dry snapping report:

"Burn my clothes?" the convict cried. "Burn them?"

"How in hell do you expect to escape in them billboards?" the man with the shotgun said. He (the convict) tried to tell it, tried to explain as he had tried to explain not to the three people on the boat alone but to the entire circumambience—desolate water and forlorn trees and sky—not for justification because he needed none and knew that his hearers, the other convicts, required none from him, but rather as, on the point of exhaustion, he might have picked dreamily and incredulously at a suffocation. He told the man with the gun how he and his partner had been given the boat and told to pick up a man and a woman, how he had lost his partner and failed to find the man, and now all in the world he wanted was something flat to leave the woman on until he could find an officer, a sheriff. He thought of home, the place where he had lived almost since childhood, his friends of years whose ways he knew and who knew his ways, the familiar fields where he did work he had learned to do well and to like, the mules with characters he knew and respected as he knew and respected the characters of certain men; he thought of the barracks at night, with screens against the bugs in summer and good stoves in winter and someone to supply the fuel and the food too; the Sunday ball games and the picture shows—things which, with the exception of the ball games, he had never known before. But most of all, his own character (Two years ago they had offered to make a trusty of him. He would no longer need to plow or feed stock, he would only follow those who did with a loaded gun, but he declined. "I reckon I'll stick to plowing," he said, absolutely without humor. "I done already tried to use a gun one time too many."), his good name, his responsibility not only toward those who were responsible toward him but to himself, his own honor in the doing of what was asked of him, his pride in being able to do it, no matter what it was. He thought of this and listened to the man with the gun talking about escape and it seemed to him that, hanging there, being dragged violently along (It was here he said that he first noticed the goats' beards of moss in the trees, though it could have been there for several days so far as he knew. It just happened that he first noticed it here.), he would simply burst.

"Cant you get it into your head that the last thing I want to do is run away?" he cried. "You can set there with that gun and watch me; I give you fair lief. All I want is to put this woman—"

"And I told you she could come aboard," the man with the gun said in his level voice. "But there aint no room on no boat of mine for nobody hunting a sheriff in no kind of clothes, let alone a penitentiary suit."

"When he steps aboard, knock him in the head with the gun barrel," the man at the sweep said. "He's drunk."

"He aint coming aboard," the man with the gun said. "He's crazy."

Then the woman spoke. She didn't move, leaning in the door, in a pair of faded and patched and filthy overalls like the two men: "Give them some grub and tell them to get out of here." She moved, she crossed the deck and looked down at the convict's companion with her cold sullen face. "How much more time have you got?"

"It wasn't due till next month," the woman in the boat said. "But I—" The woman in overalls turned to the man with the gun.

"Give them some grub," she said. But the man with the gun was still looking down at the woman in the boat.

"Come on," he said to the convict. "Put her aboard, and beat it."

"And what'll happen to you," the woman in overalls said, "when you try to turn her over to an officer? When you lay alongside a sheriff and the sheriff asks you who you are?" Still the man with the gun didn't even look at her. He hardly even shifted the gun across his arm as he struck the woman across the face with the back of his other hand, hard. "You son of a bitch," she said. Still the man with the gun did not even look at her.

"Well?" he said to the convict.

"Dont you see I cant?" the convict cried. "Cant you see that?"

Now, he said, he gave up. He was doomed. That is, he knew now that he had been doomed from the very start never to get rid of her, just as the ones who sent him out with the skiff knew that he never would actually give up; when he recognised one of the objects which the woman in overalls was hurling into the skiff to be a can of condensed milk, he believed it to be a presage, gratuitous and irrevocable as a death-notice over the telegraph, that he was not even to find a flat stationary surface in time for the child to be born on it. So he told how he held the skiff alongside the shanty-boat while the first tentative toying of the second wave made up beneath him, while the woman in overalls passed back and forth between house and rail, flinging the food—the hunk of salt meat, the ragged and filthy quilt, the scorched lumps of cold bread which she poured into the skiff from a heaped dishpan like so much garbage—while he clung to the strake against the mounting pull of the current, the new wave which for the moment he had forgotten because he was still trying to state the incredible simplicity of his desire and need until the man with the gun (the only one of the three who wore shoes) began to stamp at his hands, he snatching his hands away one at a time to avoid the heavy shoes, then grasping the rail again until the man with the gun kicked at his face, he flinging himself sideways to avoid the shoe and so breaking his hold on the rail, his weight canting the skiff off at a tangent on the increasing current so that it began to leave the shanty boat behind and he paddling again now, violently, as a man hurries toward the precipice for which he knows at last he is doomed, looking back at the other boat, the three faces sullen derisive and grim and rapidly diminishing across the widening water and at last, apoplectic, suffocating with the intolerable fact not that he had been refused but that he had been refused so little, had wanted so little, asked for so little, yet there had been demanded of him in return the one price out of all breath which (they must have known) if he could have paid it, he would not have been where he was, asking what he asked, raising the paddle and shaking it and screaming curses back at them even after the shotgun flashed and the charge went scuttering past along the water to one side.

So he hung there, he said, shaking the paddle and howling, when suddenly he remembered that other wave, the second wall of water full of houses and dead mules building up behind him back in the swamp. So he quit yelling then and went back to paddling. He was not trying to outrun it. He just knew from experience that when it overtook him, he would have to travel in the same direction it was moving in anyway, whether he wanted to or not, and when it did overtake him, he would begin to move too fast to stop, no matter what places he might come to where he could leave the woman, land her in time. Time: that was his itch now, so his only chance was to stay ahead of it as long as he could and hope to reach something before it struck. So he went on, driving the skiff with muscles which had been too tired so long they had quit feeling it, as when a man has had bad luck for so long that he ceases to believe it is even bad, let alone luck. Even when he ate—the scorched lumps the size of baseballs and the weight and durability of cannel coal even after having lain in the skiff's bilge where the shanty boat woman had thrown them—the iron-like lead-heavy objects which no man would have called bread outside of the crusted and scorched pan in which they had cooked—it was with one hand, begrudging even that from the paddle.

He tried to tell that too—that day while the skiff fled on among the bearded trees while every now and then small quiet tentative ex-

ploratory feelers would come up from the wave behind and toy for a
moment at the skiff, light and curious, then go on with a faint hissing sighing, almost a chuckling, sound, the skiff going on, driving
on with nothing to see but trees and water and solitude: until after a
while it no longer seemed to him that he was trying to put space
and distance behind him or shorten space and distance ahead but
that both he and the wave were now hanging suspended simultaneous and unprogressing in pure time, upon a dreamy desolation in
which he paddled on not from any hope even to reach anything at
all but merely to keep intact what little of distance the length of the
skiff provided between himself and the inert and inescapable mass
of female meat before him; then night and the skiff rushing on, fast
since any speed over anything unknown and invisible is too fast,
with nothing before him and behind him the outrageous idea of a
volume of moving water toppling forward, its crest frothed and
shredded like fangs, and then dawn again (another of those dreamlike alterations day to dark then back to day again with that quality
truncated, anachronic and unreal as the waxing and waning of
lights in a theatre scene) and the skiff emerging now with the
woman no longer supine beneath the shrunken soaked private's coat
but sitting bolt upright, gripping the gunwales with both hands, her
eyes closed and her lower lip caught between her teeth and he driving the splintered board furiously now, glaring at her out of his wild
swollen sleepless face and crying, croaking, "Hold on! For God's
sake hold on!"

"I'm trying to," she said. "But hurry! Hurry!" He told it, the unbelievable: hurry, hasten: the man falling from a cliff being told to
catch onto something and save himself; the very telling of it emerging shadowy and burlesque, ludicrous, comic and mad, from the
ague of unbearable forgetting with a quality more dreamily furious
than any fable behind proscenium lights:

He was in a basin now—"A basin?" the plump convict said.
"That's what you wash in."

"All right," the tall one said, harshly, above his hands. "I did."
With a supreme effort he stilled them long enough to release the
two bits of cigarette paper and watched them waft in light fluttering indecision to the floor between his feet, holding his hands motionless even for a moment longer—a basin, a broad peaceful yellow
sea which had an abruptly and curiously ordered air, giving him,
even at that moment, the impression that it was accustomed to
water even if not total submersion; he even remembered the name
of it, told to him two or three weeks later by someone: Atchafalaya—

"Louisiana?" the plump convict said. "You mean you were clean
out of Mississippi? Hell fire." He stared at the tall one. "Shucks,"
he said. "That aint but just across from Vicksburg."

"They never named any Vicksburg across from where I was," the

tall one said. "It was Baton Rouge they named." And now he
began to talk about a town, a little neat white portrait town nestling among enormous very green trees, appearing suddenly in the
telling as it probably appeared in actuality, abrupt and airy and miragelike and incredibly serene before him behind a scattering of
boats moored to a line of freight cars standing flush to the doors in
water. And now he tried to tell that too: how he stood waist-deep
in water for a moment looking back and down at the skiff in which
the woman half lay, her eyes still closed, her knuckles white on the
gunwales and a tiny thread of blood creeping down her chin from
her chewed lip, and he looking down at her in a kind of furious desperation.

"How far will I have to walk?" she said.

"I dont know, I tell you!" he cried. "But it's land somewhere
yonder! It's land, houses."

"If I try to move, it wont even be born inside a boat," she said.
"You'll have to get closer."

"Yes," he cried, wild, desperate, incredulous. "Wait. I'll go and
surrender, then they will have—" He didn't finish, wait to finish; he
told that too: himself splashing, stumbling, trying to run, sobbing
and gasping; now he saw it—another loading platform standing
above the yellow flood, the khaki figures on it as before, identical,
the same; he said how the intervening days since that first innocent
morning telescoped, vanished as if they had never been, the two
contiguous succeeding instants (succeeding? simultaneous) and he
transported across no intervening space but merely turned in his
own footsteps, plunging, splashing, his arms raised, croaking
harshly. He heard the startled shout, "There's one of them!", the
command, the clash of equipment, the alarmed cry: "There he
goes! There he goes!"

"Yes!" he cried, running, plunging. "Here I am! Here! Here!"
running on, into the first scattered volley, stopping among the bullets, waving his arms, shrieking, "I want to surrender! I want to surrender!" watching not in terror but in amazed and absolutely
unbearable outrage as a squatting clump of the khaki figures
parted and he saw the machine gun, the blunt thick muzzle slant
and drop and probe toward him and he still screaming in his hoarse
crow's voice, "I want to surrender! Cant you hear me?" continuing
to scream even as he whirled and plunged splashing, ducking, went
completely under and heard the bullets going thuck-thuck-thuck on
the water above him and he scrabbling still on the bottom, still
trying to scream even before he regained his feet and still all submerged save his plunging unmistakable buttocks, the outraged
screaming bubbling from his mouth and about his face since he
merely wanted to surrender. Then he was comparatively screened,
out of range, though not for long. That is (he didn't tell how or
where) there was a moment in which he paused, breathed for a sec-

ond before running again, the course back to the skiff open for the
time being though he could still hear the shouts behind him and
now and then a shot, and he panting, sobbing, a long savage tear in
the flesh of one hand, got when and how he did not know, and he
wasting precious breath, speaking to no one now any more than the
scream of the dying rabbit is addressed to any mortal ear but rather
an indictment of all breath and its folly and suffering, its infinite
capacity for folly and pain, which seems to be its only immortality:
"All in the world I want is just to surrender."

He returned to the skiff and got in and took up his splintered
plank. And now when he told this, despite the fury of element
which climaxed it, it (the telling) became quite simple; he now
even creased another cigarette paper between fingers which did not
tremble at all and filled the paper from the tobacco sack without
spilling a flake, as though he had passed from the machine-gun's
barrage into a bourne beyond any more amazement: so that the
subsequent part of his narrative seemed to reach his listeners as
though from beyond a sheet of slightly milky though still transparent glass, as something not heard but seen[2]—a series of shadows,
edgeless yet distinct, and smoothly flowing, logical and unfrantic
and making no sound: They were in the skiff, in the center of the
broad placid trough which had no boundaries and down which the
tiny forlorn skiff flew to the irresistible coercion of a current going
once more he knew not where, the neat small liveoak-bowered
towns unattainable and miragelike and apparently attached to
nothing upon the airy and unchanging horizon. He did not believe
them, they did not matter, he was doomed; they were less than the
figments of smoke or of delirium, and he driving his unceasing paddle without destination or even hope now, looking now and then at
the woman sitting with her knees drawn up and locked and her entire body one terrific clench while the threads of bloody saliva crept
from her teeth-clenched lower lip. He was going nowhere and
fleeing from nothing, he merely continued to paddle because he had
paddled so long now that he believed if he stopped his muscles
would scream in agony. So when it happened he was not surprised.
He heard the sound which he knew well (he had heard it but once
before, true enough, but no man needed hear it but once) and he
had been expecting it; he looked back, still driving the paddle, and
saw it, curled, crested with its strawlike flotsam of trees and debris and
dead beasts, and he glared over his shoulder at it for a full minute
out of that attenuation far beyond the point of outragement where
even suffering, the capability of being further affronted, had ceased,
from which he now contemplated with savage and invulnerable curiosity the further extent to which his now anesthetised nerves

2. Cf. Isaiah 32.1–4: "Behold, a king
shall reign in righteousness. * * * And
the eyes of them that see shall not be
dim, and the ears of them that hear
shall hearken. The heart also of the rash
shall understand knowledge, and the
tongue of the stammerers shall be ready
to speak plainly."

could bear, what next could be invented for them to bear, until the wave actually began to rear above his head into its thunderous climax. Then only did he turn his head. His stroke did not falter, it neither slowed nor increased; still paddling with that spent hypnotic steadiness, he saw the swimming deer. He did not know what it was nor that he had altered the skiff's course to follow it, he just watched the swimming head before him as the wave boiled down and the skiff rose bodily in the old familiar fashion on a welter of tossing trees and houses and bridges and fences, he still paddling even while the paddle found no purchase save air and still paddled even as he and the deer shot forward side by side at arm's length, he watching the deer now, watching the deer begin to rise out of the water bodily until it was actually running along upon the surface, rising still, soaring clear of the water altogether, vanishing upward in a dying crescendo of splashings and snapping branches, its damp scut flashing upward, the entire animal vanishing upward as smoke vanishes. And now the skiff struck and canted and he was out of it too, standing knee-deep, springing out and falling to his knees, scrambling up, glaring after the vanished deer. "Land!" he croaked. "Land! Hold on! Just hold on!" He caught the woman beneath the arms, dragging her out of the boat, plunging and panting after the vanished deer. Now earth actually appeared—an acclivity smooth and swift and steep, bizarre, solid and unbelievable; an Indian mound, and he plunging at the muddy slope, slipping back, the woman struggling in his muddy hands.

"Let me down!" she cried. "Let me down!" But he held her, panting, sobbing, and rushed again at the muddy slope; he had almost reached the flat crest with his now violently unmanageable burden when a stick under his foot gathered itself with thick convulsive speed. *It was a snake,* he thought as his feet fled beneath him and with the indubitable last of his strength he half pushed and half flung the woman up the bank as he shot feet first and face down back into that medium upon which he had lived for more days and nights than he could remember and from which he himself had never completely emerged, as if his own failed and spent flesh were attempting to carry out his furious unflagging will for severance at any price, even that of drowning, from the burden with which, unwitting and without choice, he had been doomed. Later it seemed to him that he had carried back beneath the surface with him the sound of the infant's first mewling cry.

When the woman asked him if he had a knife, standing there in the streaming bed-ticking garments which had got him shot at, the second time by a machine gun, on the two occasions when he had seen any human life after leaving the levee four days ago, the convict felt exactly as he had in the fleeing skiff when the woman suggested that they had better hurry. He felt the same outrageous affronting of a condition purely moral, the same raging impotence to find any answer to it; so that, standing above her, spent suffocating and inarticulate, it was a full minute before he comprehended that she was now crying, "The can! The can in the boat!" He did not anticipate what she could want with it; he did not even wonder nor stop to ask. He turned running; this time he thought, *It's another moccasin* as the thick body truncated in that awkward reflex which had nothing of alarm in it but only alertness, he not even shifting his stride though he knew his running foot would fall within a yard of the flat head. The bow of the skiff was well up the slope now where the wave had set it and there was another snake just crawling over the stern into it and as he stooped for the bailing can he saw something else swimming toward the mound, he didn't know what —a head, a face at the apex of a vee of ripples. He snatched up the can; by pure juxtaposition of it and water he scooped it full, already turning. He saw the deer again, or another one. That is, he saw a deer—a side glance, the light smoke-colored phantom in a cypress vista then gone, vanished, he not pausing to look after it, galloping back to the woman and kneeling with the can to her lips until she told him better.

It had contained a pint of beans or tomatoes, something, hermetically sealed and opened by four blows of an axe heel, the metal flap turned back, the jagged edges razor-sharp. She told him how, and he used this in lieu of a knife, he removed one of his shoelaces, and cut it in two with the sharp tin. Then she wanted warm water—"If I just had a little hot water," she said in a weak serene voice without particular hope; only when he thought of matches it was again a good deal like when she had asked him if he had a knife, until she fumbled in the pocket of the shrunken tunic (it had a darker double vee on one cuff and a darker blotch on the shoulder where service stripes and a divisional emblem had been ripped off but this meant nothing to him) and produced a match-box contrived by telescoping two shotgun shells. So he drew her back a little from the water and went to hunt wood dry enough to burn, thinking this time, *It's just another snake,* only, he said, he should have thought *ten thousand other snakes*: and now he knew it was not the same deer because he saw three at one time, does or bucks he did not know which since they were all anterless in May and besides he had never seen one of any kind anywhere before except on a Christmas card; and then the rabbit, drowned, dead anyway, already torn open, the bird, the hawk, standing upon it—the erected crest, the hard vicious patrician nose, the intolerant omnivorous yellow eye—and he kicking at it, kicking it lurching and broadwinged into the actual air.

When he returned with the wood and the dead rabbit, the baby, wrapped in the tunic, lay wedged between two cypress-knees and the woman was not in sight, though while the convict knelt in the mud, blowing and nursing his meagre flame, she came slowly and weakly from the direction of the water. Then, the water heated at last and there produced from some where he was never to know, she herself perhaps never to know until the need comes, no woman perhaps ever to know, only no woman will even wonder, that square of something somewhere between sackcloth and silk—squatting, his own wet garments steaming in the fire's heat, he watched her bathe the child with a savage curiosity and interest that became amazed unbelief, so that at last he stood above them both, looking down at the tiny terra-cotta-colored creature resembling nothing, and thought, *And this is all. This is what severed me violently from all I ever knew and did not wish to leave and cast me upon a medium I was born to fear, to fetch up at last in a place I never saw before and where I do not even know where I am.*

Then he returned to the water and refilled the bailing can. It was drawing toward sunset now (or what would have been sunset save for the high prevailing overcast) of this day whose beginning he could not even remember; when he returned to where the fire burned in the interlaced gloom of the cypresses, even after this short absence, evening had definitely come, as though darkness too had taken refuge upon that quarter-acre mound, that earthen Ark out of Genesis, that him wet cypress-choked life-teeming constricted desolation in what direction and how far from what and where he had no more idea than of the day of the month, and had now with the setting of the sun crept forth again to spread upon the waters. He stewed the rabbit in sections while the fire burned redder and redder in the darkness where the shy wild eyes of small animals— once the tall mild almost plate-sized stare of one of the deer— glowed and vanished and glowed again, the broth hot and rank after the four days; he seemed to hear the roar of his own saliva as he watched the woman sip the first canful. Then he drank too; they ate the other fragments which had been charring and scorching on willow twigs; it was full night now. "You and him better sleep in the boat," the convict said. "We want to get an early start tomorrow." He shoved the bow of the skiff off the land so it would lie level, he lengthened the painter with a piece of grapevine and returned to the fire and tied the grapevine about his wrist and lay down. It was mud he lay upon, but it was solid underneath, it was earth, it did not move; if you fell upon it you broke your bones against its incontrovertible passivity sometimes but it did not accept you substanceless and enveloping and suffocating, down and down and down; it was hard at times to drive a plow through, it sent you spent, weary, and cursing its light-long insatiable demands back to your bunk at sunset at times but it did not snatch you violently out of all familar knowing and sweep you thrall and impotent for days against any returning. *I dont know where I am and I dont reckon I know the way back to where I want to go,* he thought. *But at least*

the boat has stopped long enough to give me a chance to turn it
around.

He waked at dawn, the light faint, the sky jonquil-colored; the
day would be fine. The fire had burned out; on the opposite side of
the cold ashes lay three snakes motionless and parallel as underscor-
ing, and in the swiftly making light others seemed to materialise:
earth which an instant before had been mere earth broke up into
motionless coils and loops, branches which a moment before had
been mere branches now become immobile ophidian[3] festoons even
as the convict stood thinking about food, about something hot before they started. But he decided against this, against wasting this
much time, since there still remained in the skiff quite a few of the
rocklike objects which the shanty woman had flung into it; besides
(thinking this), no matter how fast nor successfully he hunted, he
would never be able to lay up enough food to get them back to
where they wanted to go. So he returned to the skiff, paying himself
back to it by his vine-spliced painter, back to the water on which a
low mist thick as cotton batting (though apparently not very tall,
deep) lay, into which the stern of the skiff was already beginning to
disappear although it lay with its prow almost touching the mound.
The woman waked, stirred. "We fixing to start now?" she said.

"Yah," the convict said. "You aint aiming to have another one
this morning, are you?" He got in and shoved the skiff clear of the
land, which immediately began to dissolve into the mist. "Hand me
the oar," he said over his shoulder, not turning yet.

"The oar?"

He turned his head. "The oar. You're laying on it." But she was
not, and for an instant during which the mound, the island continued to fade slowly into the mist which seemed to enclose the skiff
in weightless and impalpable wool like a precious or fragile bauble
or jewel, the convict squatted not in dismay but in that frantic and
astonished outrage of a man who, having just escaped a falling safe,
is struck by the following two-ounce paper weight which was sitting
on it: this the more unbearable because he knew that never in his
life had he less time to give way to it. He did not hesitate. Grasping
the grapevine end he sprang into the water, vanishing in the violent
action of climbing, and reappeared still climbing and (who had
never learned to swim) plunged and threshed on toward the al-
most-vanished mound, moving through the water then upon it as
the deer had done yesterday and scrabbled up the muddy slope and
lay gasping and panting, still clutching the grapevine end.

Now the first thing he did was to choose what he believed to be
the most suitable tree (for an instant in which he knew he was insane he thought of trying to saw it down with the flange of the bailing can) and build a fire against the butt of it. Then he went to
seek food. He spent the next six days seeking it while the tree

3. Pertaining to snakes.

burned through and fell and burned through again at the proper
length and he nursing little constant cunning flames along the
flanks of the log to make it paddle-shaped, nursing them at night too
while the woman and baby (it was eating, nursing now, he turning
his back or even returning into the woods each time she prepared to
open the faded tunic) slept in the skiff. He learned to watch for
stooping hawks and so found more rabbits and twice possums; they
ate some drowned fish which gave them both a rash and then a violent flux and one snake which the woman thought was turtle and
which did them no harm, and one night it rained and he got up and
dragged brush, shaking the snakes (he no longer thought, *It aint
nothing but another moccasin,* he just stepped aside for them as
they, when there was time, telescoped sullenly aside for him) out of
it with the old former feeling of personal invulnerability and built a
shelter and the rain stopped at once and did not recommence and
the woman went back to the skiff.

　　Then one night—the slow tedious charring log was almost a paddle now—one night and he was in bed, in his bed in the bunkhouse
and it was cold, he was trying to pull the covers up only his mule[4]
wouldn't let him, prodding and bumping heavily at him, trying to
get into the narrow bed with him and now the bed was cold too
and wet and he was trying to get out of it only the mule would not
let him, holding him by his belt in its teeth, jerking and bumping
him back into the cold wet bed and, leaning, gave him a long swipe
across the face with its cold limber musculated tongue and he
waked to no fire, no coal even beneath where the almost-finished
paddle had been charring and something else prolonged and coldly
limber passed swiftly across his body where he lay in four inches of
water while the nose of the skiff alternately tugged at the grapevine
tied about his waist and bumped and shoved him back into the
water again. Then something else came up and began to nudge at
his ankle (the log, the oar, it was) ever as he groped frantically for
the skiff, hearing the swift rustling going to and fro inside the hull
as the woman began to thrash about and scream. "Rats!" she cried.
"It's full of rats!"

　　"Lay still!" he cried. "It's just snakes. Cant you hold still long
enough for me to find the boat?" Then he found it, he got into it
with the unfinished paddle; again the thick muscular body convulsed under his foot; it did not strike; he would not have cared,
glaring astern where he could see a little—the faint outer luminosity
of the open water. He poled toward it, thrusting aside the snake-
looped branches, the bottom of the skiff resounding faintly to thick
solid plops, the woman shrieking steadily. Then the skiff was clear
of the trees, the mound, and now he could feel the bodies whipping
about his ankles and hear the rasp of them as they went over the

4. A mule, the offspring of a horse and　　tains to the sex life of convicts.
an ass, is usually sterile. The dream per-

gunwale. He drew the log in and scooped it forward along the bottom of the boat and up and out; against the pallid water he could see three more of them in lashing convolutions before they vanished. "Shut up!" he cried. "Hush! I wish I was a snake so I could get out too!"

When once more the pale and heatless wafer disc of the early sun stared down at the skiff (whether they were moving or not the convict did not know) in its nimbus of fine cotton batting, the convict was hearing again that sound which he had heard twice before and would never forget—that sound of deliberate and irresistible and monstrously disturbed water. But this time he could not tell from what direction it came. It seemed to be everywhere, waxing and fading; it was like a phantom behind the mist, at one instant miles away, the next on the point of overwhelming the skiff within the next second; suddenly, in the instant he would believe (his whole weary body would spring and scream) that he was about to drive the skiff point-blank into it and with the unfinished paddle of the color and texture of sooty bricks, like something gnawed out of an old chimney by beavers and weighing twenty-five pounds, he would whirl the skiff frantically and find the sound dead ahead of him again. Then something bellowed tremendously above his head, he heard human voices, a bell jangled and the sound ceased and the mist vanished as when you draw your hand across a frosted pane, and the skiff now lay upon a sunny glitter of brown water flank to flank with, and about thirty yards away from, a steamboat. The decks were crowded and packed with men women and children sitting or standing beside and among a homely conglomeration of hurried furniture, who looked mournfully and silently down into the skiff while the convict and the man with a megaphone in the pilot house talked to each other in alternate puny shouts and roars above the chuffing of the reversed engines:

"What in hell are you trying to do? Commit suicide?"

"Which is the way to Vicksburg?"

"Vicksburg? Vicksburg? Lay alongside and come aboard."

"Will you take the boat too?"

"Boat? Boat?" Now the megaphone cursed, the roaring waves of blasphemy and biological supposition empty cavernous and bodiless in turn, as if the water, the air, the mist had spoken it, roaring the words then taking them back to itself and no harm done, no scar, no insult left anywhere. "If I took aboard every floating sardine can you sonabitchin mushrats want me to I wouldn't even have room forrard for a leadsman.[5] Come aboard! Do you expect me to hang here on stern engines till hell freezes?"

"I aint coming without the boat," the convict said. Now another voice spoke, so calm and mild and sensible that for a moment it

5. Crewman aboard ship who ascertains the depth of the water by lowering a piece of lead on a rope.

sounded more foreign and out of place then even the megaphone's bellowing and bodiless profanity:

"Where is it you are trying to go?"

"I aint trying," the convict said. "I'm going. Parchman." The man who had spoken last turned and appeared to converse with a third man in the pilot house. Then he looked down at the skiff again.

"Carnarvon?"

"What?" the convict said. "Parchman?"

"All right. We're going that way. We'll put you off where you can get home. Come aboard."

"The boat too?"

"Yes, yes. Come along. We're burning coal just to talk to you." So the convict came alongside then and watched them help the woman and baby over the rail and he came aboard himself, though he still held to the end of the vine-spliced painter until the skiff was hoisted onto the boiler deck. "My God," the man, the gentle one, said, "is that what you have been using for a paddle?"

"Yah," the convict said. "I lost the plank."

"The plank," the mild man, (the convict told how he seemed to whisper it), "the plank. Well. Come along and get something to eat. Your boat is all right now."

"I reckon I'll wait here," the convict said. Because now, he told them, he began to notice for the first time that the other people, the other refugees who crowded the deck, who had gathered in a quiet circle about the upturned skiff on which he and the woman sat, the grapevine painter wrapped several times about his wrist and clutched in his hand, staring at him and the woman with queer hot mournful intensity, were not white people—

"You mean niggers?" the plump convict said.

"No. Not Americans."

"Not Americans? You was clean out of *America* even?"

"I dont know," the tall one said. "They called it Atchafalaya."— Because after a while he said, "What?" to the man and the man did it again, gobble-gobble—

"Gobble-gobble?" the plump convict said.

"That's the way they talked," the tall one said. "Gobble-gobble, whang, caw-caw-to-to."—And he sat there and watched them gobbling at one another and then looking at him again, then they fell back and the mild man (he wore a Red Cross brassard) entered, followed by a waiter with a tray of food. The mild man carried two glasses of whiskey.

"Drink this," the mild man said. "This will warm you." The woman took hers and drank it but the convict told how he looked at his and thought, *I aint tasted whiskey in seven years.* He had not tasted it but once before that; it was at the still itself back in a pine hollow; he was seventeen, he had gone there with four companions, two of whom were grown men, one of twenty-two or -three, the other about forty; he remembered it. That is, he remembered perhaps a third of that evening—a fierce turmoil in the hell-colored firelight, the shock and shock of blows about his head (and likewise of his own fists on other hard bone), then the waking to a splitting and blinding sun in a place, a cowshed, he had never seen before and which later turned out to be twenty miles from his home. He said he thought of this and he looked about at the faces watching him and he said,

"I reckon not."

"Come, come," the mild man said. "Drink it."

"I dont want it."

"Nonsense," the mild man said. "I'm a doctor. Here. Then you can eat." So he took the glass and even then he hesitated but again the mild man said, "Come along, down with it; you're still holding us up," in that voice still calm and sensible but a little sharp too—the voice of a man who could keep calm and affable because he wasn't used to being crossed—and he drank the whiskey and even in the second between the sweet full fire in his belly and when it began to happen he was trying to say, "I tried to tell you! I tried to!" But it was too late now in the pallid sun-glare of the tenth day of terror and hopelessness and despair and impotence and rage and outrage and it was himself and the mule, his mule (they had let him name it—John Henry) which no man save he had plowed for five years now and whose ways and habits he knew and respected and who knew his ways and habits so well that each of them could anticipate the other's very movements and intentions; it was himself and the mule, the little gobbling faces flying before them, the familiar hard skull-bones shocking against his fists, his voice shouting, "Come on, John Henry! Plow them down! Gobble them down, boy!" even as the bright hot red wave turned back, meeting it joyously, happily, lifted, poised, then hurling through space, triumphant and yelling, then again the old shocking blow at the back of his head: he lay on the deck, flat on his back and pinned arm and leg and cold sober again, his nostrils gushing again, the mild man stooping over him with behind the thin rimless glasses the coldest eyes the convict had ever seen—eyes which the convict said were not looking at him but at the gushing blood with nothing in the world in them but complete impersonal interest.

"Good man," the mild man said. "Plenty of life in the old carcass yet, eh? Plenty of good red blood too. Anyone ever suggest to you that you were hemophilic?"[6] ("What?" the plump convict said. "Hemophilic? You know what that means?" The tall convict had his cigarette going now, his body jackknifed backward into the coffinlike space between the upper and lower bunks, lean, clean,

6. Subject to a disease that causes uncontrollable bleeding. The convicts mistakenly think the term means bisexuality or sexual ambiguity.

motionless, the blue smoke wreathing across his lean dark aquiline shaven face. "That's a calf that's a bull and a cow at the same time."

"No, it aint," a third convict said. "It's a calf or a colt that aint neither one."

"Hell fire," the plump one said. "He's got to be one or the other to keep from drounding." He had never ceased to look at the tall one in the bunk; now he spoke to him again: "You let him call you that?") The tall one had done so. He did not answer the doctor (this was where he stopped thinking of him as the mild man) at all. He could not move either, though he felt fine, he felt better than he had in ten days. So they helped him to his feet and steadied him over and lowered him onto the upturned skiff beside the woman, where he sat bent forward, elbows on knees in the immemorial attitude, watching his own bright crimson staining the mudtrodden deck, until the doctor's clean clipped hand appeared under his nose with a phial.

"Smell," the doctor said. "Deep." The convict inhaled, the sharp ammoniac sensation burned up his nostrils and into his throat. "Again," the doctor said. The convict inhaled obediently. This time he choked and spat a gout of blood, his nose now had no more feeling than a toenail, other than it felt about the size of a ten-inch shovel, and as cold.

"I ask you to excuse me," he said. "I never meant—"

"Why?" the doctor said. "You put up as pretty a scrap against forty or fifty men as I ever saw. You lasted a good two seconds. Now you can eat something. Or do you think that will send you haywire again?"

They both ate, sitting on the skiff, the gobbling faces no longer watching them now, the convict gnawing slowly and painfully at the thick sandwich, hunched, his face laid sideways to the food and parallel to the earth as a dog chews; the steamboat went on. At noon there were bowls of hot soup and bread and more coffee; they ate this too, sitting side by side on the skiff, the grapevine still wrapped about the convict's wrist. The baby waked and nursed and slept again and they talked quietly:

"Was it Parchman he said he was going to take us?"

"That's where I told him I wanted to go."

"It never sounded exactly like Parchman to me. It sounded like he said something else." The convict had thought that too. He had been thinking about that fairly soberly ever since they boarded the steamboat and soberly indeed ever since he had remarked the nature of the other passengers, those men and women definitely a little shorter than he and with skin a little different in pigmentation from any sunburn, even though the eyes were sometimes blue or gray, who talked to one another in a tongue he had never heard before and who apparently did not understand his own, people the like of whom he had never seen about Parchman nor anywhere else and who he did not believe were going there or beyond there either. But after his hillbilly country fashion and kind he would not ask, because to his raising asking information was asking a favor and you did not ask favors of strangers; if they offered them perhaps you accepted and you expressed gratitude almost tediously recapitulant, but you did not ask. So he would watch and wait, as he had done before, and do or try to do to the best of his ability what the best of his judgment dictated.

So he waited, and in midafternoon the steamboat chuffed and thrust through a willow-choked gorge and emerged from it, and now the convict knew it was the River. He could believe it now—the tremendous reach, yellow and sleepy in the afternoon—("Because it's too big," he told them soberly. "Aint no flood in the world big enough to make it do more than stand a little higher so it can look back and see just where the flea is, just exactly where to scratch. It's the little ones, the little piddling creeks that run backward one day and forward the next and come busting down on a man full of dead mules and hen houses.")—and the steamboat moving up this now (*like a ant crossing a plate,* the convict thought, sitting beside the woman on the upturned skiff, the baby nursing again, apparently looking too out across the water where, a mile away on either hand, the twin lines of levee resembled parallel unbroken floating thread) and then it was nearing sunset and he began to hear, to notice, the voices of the doctor and of the man who had first bawled at him through the megaphone now bawling again from the pilot house overhead:

"Stop? Stop? Am I running a street car?"

"Stop for the novelty then," the doctor's pleasant voice said. "I dont know how many trips back and forth you have made in yonder nor how many of what you call mushrats you have fetched out. But this is the first time you ever had two people—no, three—who not only knew the name of some place they wished to go to but were actually trying to go there." So the convict waited while the sun slanted more and more and the steamboat-ant crawled steadily on across its vacant and gigantic plate turning more and more to copper. But he did not ask, he just waited. *Maybe it was Carrollton he said,* he thought. *It begun with a C.* But he did not believe that either. He did not know where he was, but he did know that this was not anywhere near the Carrollton he remembered from that day seven years ago when, shackled wrist to wrist with the deputy sheriff, he had passed through it on the train—the slow spaced repeated shattering banging of trucks where two railroads crossed, a random scattering of white houses tranquil among trees on green hills lush with summer, a pointing spire, the finger of the hand of God. But there was no river there. *And you aint never close to this river without knowing it,* he thought. *I dont care who you are nor where you have been all your life.* Then the head of the steamboat began to swing across the stream, its shadow swinging too, travelling long before it across the water, toward the vacant ridge of willow-massed earth empty of all life. There was nothing there at all, the convict could not even see either earth or water beyond it; it was as though the steamboat were about to crash slowly through the thin low frail willow barrier and embark into space, or lacking this, slow and back and fill and disembark him into space, granted it was about to disembark him, granted this was that place which was not near Parchman and was not Carrollton either, even though it did begin with C. Then he turned his head and saw the doctor stooping over the woman, pushing the baby's eyelid up with his forefinger, peering at it.

"Who else was there when he came?" the doctor said.

"Nobody," the convict said.

"Did it all yourselves, eh?"

"Yes," the convict said. Now the doctor stood up and looked at the convict.

"This is Carnarvon," he said.

"Carnarvon?" the convict said. "That aint—" Then he stopped, ceased. And now he told about that—the intent eyes as dispassionate as ice behind the rimless glasses, the clipped quick-tempered face that was not accustomed to being crossed or lied to either. ("Yes," the plump convict said. "That's what I was aiming to ask. Them clothes. Anybody would know them. How if this doctor was as smart as you claim he was—"

"I had slept in them for ten nights, mostly in the mud," the tall one said. "I had been rowing since midnight with that sapling oar I had tried to burn out that I never had time to scrape the soot off. But it's being scared and worried and then scared and then worried again in clothes for days and days and days that changes the way they look. I dont mean just your pants." He did not laugh. "Your face too. That doctor knowed."

"All right," the plump one said. "Go on.")

"I know it," the doctor said. "I discovered that while you were lying on the deck yonder sobering up again. Now dont lie to me. I dont like lying. This boat is going to New Orleans."

"No," the convict said immediately, quietly, with absolute finality. He could hear them again—the thuck-thuck-thuck on the water where an instant before he had been. But he was not thinking of the bullets. He had forgotten them, forgiven them. He was thinking of himself crouching, sobbing, panting before running again—the voice, the indictment, the cry of final and irrevocable repudiation of the old primal faithless Manipulator of all the lust and folly and injustice: *All in the world I wanted was just to surrender;* thinking of it, remembering it but without heat now, without passion now and briefer than an epitaph: *No. I tried that once. They shot at me.*

"So you dont want to go to New Orleans. And you didn't exactly plan to go to Carnarvon. But you will take Carnarvon in preference to New Orleans." The convict said nothing. The doctor looked at him, the magnified pupils like the heads of two bridge nails. "What were you in for? Hit him harder than you thought, eh?"

"No. I tried to rob a train."

"Say that again." The convict said it again. "Well? Go on. You dont say that in the year 1927 and just stop, man." So the convict told it, dispassionately too—about the magazines, the pistol which would not shoot, the mask and the dark lantern in which no draft had been arranged to keep the candle burning so that it died almost with the match but even then left the metal too hot to carry, won with subscriptions. *Only it aint my eyes or my mouth either he's watching,* he thought. *It's like he is watching the way my hair grows on my head.* "I see," the doctor said. "But something went wrong. But you've had plenty of time to think about it since. To decide what was wrong, what you failed to do."

"Yes," the convict said. "I've thought about it a right smart since."

"So next time you are not going to make that mistake."

"I dont know," the convict said. "There aint going to be a next time."

"Why? If you know what you did wrong, they wont catch you next time."

The convict looked at the doctor steadily. They looked at each other steadily; the two sets of eyes were not so different after all. "I reckon I see what you mean," the convict said presently. "I was eighteen then. I'm twenty-five now."

"Oh," the doctor said. Now (the convict tried to tell it) the doctor did not move, he just simply quit looking at the convict. He produced a pack of cheap cigarettes from his coat. "Smoke?" he said.

"I wouldn't care for none," the convict said.

"Quite," the doctor said in that affable clipped voice. He put the cigarettes away. "There has been conferred upon my race (the Medical race) also the power to bind and to loose, if not by Jehovah perhaps, certainly by the American Medical Association—on which incidentally, in this day of Our Lord, I would put my money, at any odds, at any amount, at any time. I dont know just how far out of bounds I am on this specific occasion but I think we'll put it to the touch." He cupped his hands to his mouth, toward the pilot house overhead. "Captain!" he shouted. "We'll put these three passengers ashore here." He turned to the convict again. "Yes," he said, "I think I shall let your native state lick its own vomit. Here." Again his hand emerged from his pocket, this time with a bill in it.

"No," the convict said.

"Come, come; I dont like to be disputed either."

"No," the convict said. "I aint got any way to pay it back."

"Did I ask you to pay it back?"

"No," the convict said. "I never asked to borrow it either."

So once more he stood on dry land, who had already been toyed with twice by that risible and concentrated power of water, once more than should have fallen to the lot of any one man, any one lifetime, yet for whom there was reserved still another unbelievable recapitulation, he and the woman standing on the empty levee, the sleeping child wrapped in the faded tunic and the grapevine painter still wrapped about the convict's wrist, watching the steamboat back away and turn and once more crawl onward up the platter-like reach of vacant water burnished more and more to copper, its trailing smoke roiling in slow copper-edged gouts, thinning out along the water, fading, stinking away across the vast serene desolation, the boat growing smaller and smaller until it did not seem to crawl at all but to hang stationary in the airy substanceless sunset, dissolving into nothing like a pellet of floating mud.

Then he turned and for the first time looked about him, behind him, recoiling, not through fear but through pure reflex and not physically but the soul, the spirit, that profound sober alert attentiveness of the hillman who will not ask anything of strangers, not even information, thinking quietly, *No. This aint Carrollton neither.* Because he now looked down the almost perpendicular landward slope of the levee through sixty feet of absolute space, upon a surface, a terrain flat as a waffle and of the color of a waffle or perhaps of the summer coat of a claybank horse and possessing that same piled density of a rug or peltry, spreading away without undulation yet with that curious appearance of imponderable solidity like fluid, broken here and there by thick humps of arsenical green which nevertheless still seemed to possess no height and by writhen veins of the color of ink which he began to suspect to be actual water but with judgment reserved, with judgment still reserved even when presently he was walking in it. That's what he said, told: So they went on. He didn't tell how he got the skiff singlehanded up the revetment and across the crown and down the opposite sixty-foot drop, he just said he went on, in a swirling cloud of mosquitoes like hot cinders, thrusting and plunging through the saw-edged grass which grew taller than his head and which whipped back at his arms and face like limber knives, dragging by the vine-spliced painter the skiff in which the woman sat, slogging and stumbling knee-deep in something less of earth than water, along one of those black winding channels less of water than earth: and then (he was in the skiff too now, paddling with the charred log, what footing there had been having given away beneath him without warning thirty minutes ago, leaving only the air-filled bubble of his jumper-back ballooning lightly on the twilit water until he rose to the surface and scrambled into the skiff) the house, the cabin a little larger than a horse-box, of cypress boards and an iron roof, rising on ten-foot stilts slender as spiders' legs, like a shabby and death-stricken (and probably poisonous) wading creature which had got that far into that flat waste and died with nothing nowhere in reach or sight to lie down upon, a pirogue[7] tied to the foot of a crude ladder, a man standing in the open door holding a lantern (it was that dark now) above his head, gobbling down at them.

He told it—of the next eight or nine or ten days, he did not remember which, while the four of them—himself and the woman and baby and the little wiry man with rotting teeth and soft wild bright eyes like a rat or a chipmunk, whose language neither of them could understand—lived in the room and a half. He did not tell it that way, just as he apparently did not consider it worth the breath to tell how he had got the hundred-and-sixty-pound skiff single-handed up and across and down the sixty-foot levee. He just said, "After a while we come to a house and we stayed there eight or nine days then they blew up the levee with dynamite so we had to leave." That was all. But he remembered it, but quietly now, with the cigar now, the good one the Warden had given him (though not lighted yet) in his peaceful and steadfast hand, remembering that first morning when he waked on the thin pallet beside his host (the woman and baby had the one bed) with the fierce sun already latticed through the warped rough planking of the wall, and stood on the rickety porch looking out upon that flat fecund waste neither earth nor water, where even the senses doubted which was which, which rich and massy air and which mazy and impalpable vegetation, and thought quietly, *He must do something here to eat and live. But I dont know what. And until I can go on again, until I can find where I am and how to pass that town without them seeing me I will have to help him do it so we can eat and live too, and I dont know what.* And he had a change of clothing too, almost at once on that first morning, not telling any more than he had about the skiff and the levee how he had begged borrowed or bought from the man whom he had not laid eyes on twelve hours ago and with whom on the day he saw him for the last time he still could exchange no word, the pair of dungaree pants which even the Cajan[8] had discarded as no longer wearable, filthy, buttonless, the legs slashed and frayed into fringe like that on an 1890 hammock, in which he stood naked from the waist up and holding out to her the mud-caked and soot-stained jumper and overall when the woman waked on that first morning in the crude bunk nailed into one corner and filled with dried grass, saying, "Wash them. Good. I want all them stains out. All of them."

"But the jumper," she said. "Aint he got ere old shirt too? That sun and them mosquitoes—" But he did not even answer, and she said no more either, though when he and the Cajan returned at dark the garments were clean, stained a little still with the old mud and soot, but clean, resembling again what they were supposed to resemble as (his arms and back already a fiery red which would be blisters by tomorrow) he spread the garments out and examined them and then rolled them up carefully in a six-months-old New Orleans paper and thrust the bundle behind a rafter, where it remained while day followed day and the blisters on his back broke and suppurated and he would sit with his face expressionless as a wooden mask beneath the sweat while the Cajan doped his back with something on a filthy rag from a filthy saucer, she still saying nothing since she too doubtless knew what his reason was, not from that rapport of the wedded conferred upon her by the two weeks during which they had jointly suffered all the crises emotional social economic and even moral which do not always occur even in the ordinary fifty married years (the old married: you have seen them, the electroplate reproductions, the thousand identical coupled faces with only a collarless stud or a fichu out of Louisa Alcott[9] to denote the sex, looking in pairs like the winning braces of dogs after a field trial, out from among the packed columns of disaster and alarm and baseless assurance and hope and incredible insensitivity and insulation from tomorrow propped by a thousand morning sugar bowls or coffee urns; or singly, rocking on porches or sitting in the sun beneath the tobacco-stained porticoes of a thousand county courthouses, as though with the death of the other having inherited a sort of rejuvenescence, immortality; relict, they take a new lease on breath and seem to live forever, as though that flesh which the old ceremony or ritual had morally purified and made legally one had actually become so with long tedious habit and he or she who entered the ground first took all of it with him or her, leaving only the old permanent enduring bone, free and tramelless)—not because of this but because she too had stemmed at some point from the same dim hill-bred Abraham.

So the bundle remained behind the rafter and day followed day while he and his partner (he was in partnership now with his host, hunting alligators on shares, on the halvers he called it—"Halvers?" the plump convict said. "How could you make a business agreement with a man you claim you couldn't even talk to?"

"I never had to talk to him," the tall one said. "Money aint got but one language.") departed at dawn each day, at first together in the pirogue but later singly, the one in the pirogue and the other in the skiff, the one with the battered and pitted rifle, the other with the knife and a piece of knotted rope and a lightwood club the

9. Three-cornered cape or neckerchief worn by 19th-century women such as those in the fiction of Louisa May Alcott (1832–88).

size and weight and shape of a Thuringian mace,[1] stalking their pleistocene nightmares up and down the secret inky channels which writhed the flat brass-colored land. He remembered that too: that first morning when turning in the sunrise from the rickety platform he saw the hide nailed drying to the wall and stopped dead, looking at it quietly, thinking quietly and soberly, *So that's it. That's what he does in order to eat and live,* knowing it was a hide, a skin, but from what animal, by association, ratiocination or even memory of any picture out of his dead youth, he did not know but knowing that it was the reason, the explanation, for the little lost spider-legged house (which had already begun to die, to rot from the legs upward almost before the roof was nailed on) set in that teeming and myriad desolation, enclosed and lost within the furious embrace of flowing mare earth and stallion sun, divining through pure rapport of kind for kind, hillbilly and bayou rat, the two one and identical because of the same grudged dispensation and niggard fate of hard and unceasing travail not to gain future security, a balance in the bank or even in a buried soda can for slothful and easy old age, but just permission to endure and endure to buy air to feel and sun to drink for each's little while, thinking (the convict), *Well, anyway I am going to find out what it is sooner than I expected to,* and did so, re-entered the house where the woman was just waking in the one sorry built-in straw-filled bunk which the Cajan had surrendered to her, and ate the breakfast (the rice, a semi-liquid mess violent with pepper and mostly fish considerably high, the chicory-thickened coffee) and, shirtless, followed the little scuttling bobbing bright-eyed rotten-toothed man down the crude ladder and into the pirogue. He had never seen a pirogue either and he believed that it would not remain upright—not that it was light and precariously balanced with its open side upward but that there was inherent in the wood, the very log, some dynamic and unsleeping natural law, almost will, which its present position outraged and violated—yet accepting this too as he had the fact that that hide had belonged to something larger than any calf or hog and that anything which looked like that on the outside would be more than likely to have teeth and claws too, accepting this, squatting in the pirogue, clutching both gunwales, rigidly immobile as though he had an egg filled with nitroglycerin in his mouth and scarcely breathing, thinking, *If that's it, then I can do it too and even if he cant tell me how I reckon I can watch him and find out.* And he did this too, he remembered it, quietly even yet, thinking, *I thought that was how to do it and I reckon I would still think that even if I had it to do again now for the first time*—the brazen day already fierce upon his naked back, the crooked channel like a voluted thread of ink, the

1. Spiked clublike weapon and ceremonial symbol of authority, belonging to the inhabitants of Thuringia in central Germany. "Pleistocene": geologic period in the Cenozoic era.

pirogue moving steadily to the paddle which both entered and left the water without a sound; then the sudden cessation of the paddle behind him and the fierce hissing gobble of the Cajan at his back and he squatting bate-breathed and with that intense immobility of complete sobriety of a blind man listening while the frail wooden shell stole on at the dying apex of its own parted water. Afterward he remembered the rifle too—the rust-pitted single-shot weapon with a clumsily wired stock and a muzzle you could have driven a whiskey cork into, which the Cajan had brought into the boat—but not now; now he just squatted, crouched, immobile, breathing with infinitesimal care, his sober unceasing gaze going here and there constantly as he thought, *What? What? I not only dont know what I am looking for, I dont even know where to look for it.* Then he felt the motion of the pirogue as the Cajan moved and then the tense gobbling hissing actually, hot rapid and repressed, against his neck and ear, and glancing downward saw projecting between his own arm and body from behind, the Cajan's hand holding the knife, and glaring up again saw the flat thick spit of mud which as he looked at it divided and became a thick mud-colored log which in turn seemed, still immobile, to leap suddenly against his retinae in three—no, four—dimensions: volume, solidity, shape, and another: not fear but pure and intense speculation and he looking at the scaled motionless shape, thinking not, *It looks dangerous* but *It looks big*, thinking, *Well, maybe a mule standing in a lot looks big to a man that never walked up to one with a halter before*, thinking, *Only if he could just tell me what to do it would save time*, the pirogue drawing nearer now, creeping now, with no ripple now even and it seemed to him that he could even hear his companion's held breath and he taking the knife from the other's hand now and not even thinking this since it was too fast, a flash; it was not a surrender, not a resignation, it was too calm, it was a part of him, he had drunk it with his mother's milk and lived with it all his life: *After all a man cant only do what he has to do, with what he has to do it with, with what he has learned, to the best of his judgment. And I reckon a hog is still a hog, no matter what it looks like. So here goes*, sitting still for an instant longer until the bow of the pirogue grounded lighter than the falling of a leaf and stepped out of it and paused just for one instant while the words *It does look big* stood for just a second, unemphatic and trivial, somewhere where some fragment of his attention could see them and vanished, and stooped straddling, the knife driving even as he grasped the near foreleg, this all in the same instant when the lashing tail struck him a terrific blow upon the back. But the knife was home, he knew that even on his back in the mud, the weight of the thrashing beast longwise upon him, its ridged back clutched to his stomach, his arm about its throat, the hissing head clamped against his jaw, the furious tail lashing and flailing, the knife in his other hand probing for the life and finding it, the hot fierce gush: and now sitting be-side the profound up-bellied carcass, his head again between his knees in the old attitude while his own blood freshened the other which drenched him, thinking, *It's my durn nose again.*

So he sat there, his head, his streaming face, bowed between his knees in an attitude not of dejection but profoundly bemused, con-templative, while the shrill voice of the Cajan seemed to buzz at him from an enormous distance; after a time he even looked up at the antic wiry figure bouncing hysterically about him, the face wild and grimacing, the voice gobbling and high; while the convict, hold-ing his face carefully slanted so the blood would run free, looked at him with the cold intentness of a curator or custodian paused be-fore one of his own glass cases, the Cajan threw up the rifle, cried "Boom-boom-boom!" flung it down and in pantomime re-enacted the recent scene then whirled his hands again, crying "Magnifique! Magnifique, Cent d'argent! Mille d'argent! Tout l'argent sous le ciel de Dieu!"[2] But the convict was already looking down again, cup-ping the coffee-colored water to his face, watching the constant bright carmine marble it, thinking, *It's a little late to be telling me that now,* and not even thinking this long because presently they were in the pirogue again, the convict squatting again with that unbreathing rigidity as though he were trying by holding his breath to decrease his very weight, the bloody skin in the bows before him and he looking at it, thinking, *And I cant even ask him how much my half will be.*

But this not for long either, because as he was to tell the plump convict later, money has but one language. He remembered that too (they were at home now, the skin spread on the platform, where for the woman's benefit now the Cajan once more went through the pantomime—the gun which was not used, the hand-to-hand battle; for the second time the invisible alligator was slain amid cries, the victor rose and found this time that not even the woman was watch-ing him. She was looking at the once more swollen and inflamed face of the convict. "You mean it kicked you right in the face?" she said.

"Nah," the convict said harshly, savagely. "It never had to. I done seem to got to where if that boy was to shoot me in the tail with a bean blower my nose would bleed.")—remembered that too but he did not try to tell it. Perhaps he could not have—how two people who could not even talk to one another made an agreement which both not only understood but which each knew the other would hold true and protect (perhaps for this reason) better than any written and witnessed contract. They even discussed and agreed somehow that they should hunt separately, each in his own vessel, to double the chances of finding prey. But this was easy: the con-

2. "Magnificent! Magnificent, a hundred in silver! A thousand in silver! All the silver under God's heaven!"

vict could almost understand the words in which the Cajan said, "You do not need me and the rifle; we will only hinder you, be in your way." And more than this, they even agreed about the second rifle: that there was someone, it did not matter who—friend, neighbor, perhaps one in business in that line—from whom they could rent a second rifle; in their two patois, the one bastard English, the other bastard French—the one volatile, with his wild bright eyes and his voluble mouth full of stumps of teeth, the other sober, almost grim, swollen-faced and with his naked back blistered and scoriated like so much beef—they discussed this, squatting on either side of the pegged-out hide like two members of a corporation facing each other across a mahogany board table, and decided against it, the convict deciding: "I reckon not," he said. "I reckon if I had knowed enough to wait to start out with a gun, I still would. But since I done already started out without one, I dont reckon I'll change." Because it was a question of the money in terms of time, days. (Strange to say, that was the one thing which the Cajan could not tell him: how much the half would be. But the convict knew it was half.) He had so little of them. He would have to move on soon, thinking (the convict), *All this durn foolishness will stop soon and I can get on back,* and then suddenly he found that he was thinking, *Will have to get on back,* and he became quite still and looked about at the rich strange desert which surrounded him, in which he was temporarily lost in peace and hope and into which the last seven years had sunk like so many trivial pebbles into a pool, leaving no ripple, and he thought quietly, with a kind of bemused amazement, *Yes. I reckon I had done forgot how good making money was. Being let to make it.*

So he used no gun, his the knotted rope and the Thuringian mace, and each morning he and the Cajan took their separate ways in the two boats to comb and creep the secret channels about the lost land from (or out of) which now and then still other pint-sized dark men appeared gobbling, abruptly and as though by magic from nowhere, in other hollowed logs, to follow quietly and watch him at his single combats—men named Tine and Toto and Theule, who were not much larger than and looked a good deal like the muskrats which the Cajan (the host did this too, supplied the kitchen too, he expressed this too like the rifle business, in his own tongue, the convict comprehending this too as though it had been English: "Do not concern yourself about food, O Hercules. Catch alligators; I will supply the pot.") took now and then from traps as you take a shoat pig at need from a pen, and varied the eternal rice and fish (the convict did tell this: how at night, in the cabin, the door and one sashless window battened against mosquitoes—a form, a ritual, as empty as crossing the fingers or knocking on wood—sitting beside the bug-swirled lantern on the plank table in a temperature close to blood heat he would look down at the swimming segment of meat on his sweating plate and think, *It must be Theule. He was the fat one.*)—day following day, unemphatic and identical, each like the one before and the one which would follow while his theoretical half of a sum to be reckoned in pennies, dollars, or tens of dollars he did not know, mounted—the mornings when he set forth to find waiting for him like the *matador* his *aficionados* the small clump of constant and deferential pirogues, the hard noons when ringed half about by little motionless shells he fought his solitary combats, the evenings, the return, the pirogues departing one by one into inlets and passages which during the first few days he could not even distinguish, then the platform in the twilight where before the static woman and the usually nursing infant and the one or two bloody hides of the day's take the Cajan would perform his ritualistic victorious pantomime before the two growing rows of knife-marks in one of the boards of the wall; then the nights when, the woman and child in the single bunk and the Cajan already snoring on the pallet and the reeking lantern set close, he (the convict) would sit on his naked heels, sweating steadily, his face worn and calm, immersed and indomitable, his bowed back raw and savage as beef beneath the suppurant old blisters and the fierce welts of tails, and scrape and chip at the charred sapling which was almost a paddle now, pausing now and then to raise his head while the cloud of mosquitoes about it whined and whirled, to stare at the wall before him until after a while the crude boards themselves must have dissolved away and let his blank unseeing gaze go on and on unhampered, through the rich oblivious darkness, beyond it even perhaps, even perhaps beyond the seven wasted years during which, so he had just realised, he had been permitted to toil but not to work. Then he would retire himself, he would take a last look at the rolled bundle behind the rafter and blow out the lantern and lie down as he was beside his snoring partner, to lie sweating (on his stomach, he could not bear the touch of anything to his back) in the whining ovenlike darkness filled with the forlorn bellowing of alligators, thinking not, *They never gave me time to learn* but *I had forgot how good it is to work.*

Then on the tenth day it happened. It happened for the third time. At first he refused to believe it, not that he felt that now he had served out and discharged his apprenticeship to mischance, had with the birth of the child reached and crossed the crest of his Golgotha[3] and would now be, possibly not permitted so much as ignored, to descend the opposite slope free-wheeling. That was not his feeling at all. What he declined to accept was the fact that a power, a force such as that which had been consistent enough to concentrate upon him with deadly undeviation for weeks, should with all the wealth of cosmic violence and disaster to draw from, have been so barren of invention and imagination, so lacking in

3. Place outside Jerusalem where Christ was crucified.

pride of artistry and craftmanship, as to repeat itself twice. Once he had accepted, twice he even forgave, but three times he simply declined to believe, particularly when he was at last persuaded to realise that this third time was to be instigated not by the blind potency of volume and motion but by human direction and hands: that now the cosmic joker, foiled twice, had stooped in its vindictive concentration to the employing of dynamite.

He did not tell that. Doubtless he did not know himself how it happened, what was happening. But he doubtless remembered it (but quietly above the thick rich-colored pristine cigar in his clean steady hand), what he knew, divined of it. It would be evening, the ninth evening, he and the woman on either side of their host's empty place at the evening meal, he hearing the voices from without but not ceasing to eat, still chewing steadily, because it would be the same as though he were seeing them anyway—the two or three or four pirogues floating on the dark water beneath the platform on which the host stood, the voices gobbling and jabbering, incomprehensible and filled not with alarm and not exactly with rage or ever perhaps absolute surprise but rather just cacophony like those of disturbed marsh fowl, he (the convict) not ceasing to chew but just looking up quietly and maybe without a great deal of interrogation or surprise too as the Cajan burst in and stood before them, wild-faced, glaring, his blackened teeth gaped against the inky orifice of his distended mouth, watching (the convict) while the Cajan went through his violent pantomime of violent evacuation, ejection, scooping something invisible into his arms and hurling it out and downward and in the instant of completing the gesture changing from instigator to victim of that which he had set into pantomimic motion, clasping his head and, bowed over and not otherwise moving, seeming to be swept on and away before it, crying "Boom! Boom! Boom!", the convict watching him, his jaw not chewing now, though for just that moment, thinking, *What? What is it he is trying to tell me?* thinking (this a flash too, since he could not have expressed this, and hence did not even know that he have ever thought it) that though his life had been cast here, circumscribed by this environment, accepted by this environment and accepting it in turn (and he had done well here—this quietly, soberly indeed, if he had been able to phrase it, think it instead of merely knowing it—better than he had ever done, who had not even known until now how good work, making money, could be), yet it was not his life, he still and would ever be no more than the water bug upon the surface of the pond, the plumbless and lurking depths of which he would never know, his only actual contact with it being the instants when on lonely and glaring mudspits under the pitiless sun and amphitheatred by his motionless and riveted semicircle of watching pirogues, he accepted the gambit which he had not elected, entered the lashing radius of the armed tail and beat at the thrashing and hissing head with his lightwood club, or this failing, embraced without hesitation the armored body itself with the frail web of flesh and bone in which he walked and lived and sought the raging life with an eight-inch knife-blade.

So he and the woman merely watched the Cajan as he acted out the whole charade of eviction—the little wiry man gesticulant and wild, his hysterical shadow leaping and falling upon the rough wall as he went through the pantomime of abandoning the cabin, gathering in pantomime his meagre belongings from the walls and corners—objects which no other man would want and only some power or force like blind water or earthquake or fire would ever dispossess him of, the woman watching too, her mouth slightly open upon a mass of chewed food, on her face an expression of placid astonishment, saying, "What? What's he saying?"

"I dont know," the convict said. "But I reckon if it's something we ought to know we will find it out when it's ready for us to." Because he was not alarmed, though by now he had read the other's meaning plainly enough. *He's fixing to leave*, he thought. *He's telling me to leave too*—this later, after they had quitted the table and the Cajan and the woman had gone to bed and the Cajan had risen from the pallet and approached the convict and once more went through the pantomime of abandoning the cabin, this time as one repeats a speech which may have been misunderstood, tediously, carefully repetitional as to a child, seeming to hold the convict with one hand while he gestured, talked, with the other, gesturing as though in single syllables, the convict (squatting, the knife open and the almost-finished paddle across his lap) watching, nodding his head, even speaking in English: "Yah; sure. You bet. I got you."—trimming again at the paddle but no faster, with no more haste than on any other night, serene in his belief that when the time came for him to know whatever it was, that would take care of itself, having already and without even knowing it, even before the possibility, the question, ever arose, declined, refused to accept even the thought of moving also, thinking about the hides, thinking, *If there was just some way he could tell me where to carry my share to get the money* but thinking this only for an instant between two delicate strokes of the blade because almost at once he thought, *I reckon as long as I can catch them I wont have no big trouble finding whoever it is that will buy them.*

So the next morning he helped the Cajan load his few belongings —the pitted rifle, a small bundle of clothing (again they traded, who could not even converse with one another, this time the few cooking vessels, a few rusty traps by definite allocation, and something embracing and abstractional which included the stove, the crude bunk, the house or its occupancy—something—in exchange for one alligator hide)—into the pirogue, then, squatting and as two children divide sticks they divided the hides, separating them into two piles, one-for-me-and-one-for-you, two-for-me-and-two-for-you, and the Cajan loaded his share and shoved away from the platform and paused again, though this time he only put the paddle down, gathered something invisibly into his two hands and flung it violently upward, crying "Boom? Boom?" on a rising inflection, nodding violently to the half-naked and savagely scoriated man on the platform who stared with a sort of grim equability back at him and said, "Sure. Boom. Boom." Then the Cajan went on. He did not look back. They watched him, already paddling rapidly, or the woman did; the convict had already turned.

"Maybe he was trying to tell us to leave too," she said.

"Yah," the convict said. "I thought of that last night. Hand me the paddle." She fetched it to him—the sapling, the one he had been trimming at nightly, not quite finished yet though one more evening would do it (he had been using a spare one of the Cajan's. The other had offered to let him keep it, to include it perhaps with the stove and the bunk and the cabin's freehold, but the convict had declined. Perhaps he had computed it by volume against so much alligator hide, this weighed against one more evening with the tedious and careful blade.)—and he departed too with his knotted rope and mace, in the opposite direction, as though not only not content with refusing to quit the place he had been warned against, he must establish and affirm the irrevocable finality of his refusal by penetrating even further and deeper into it. And then and without warning the high fierce drowsing of his solitude gathered itself and struck at him.

He could not have told this if he had tried—this not yet mid-morning and he going on, alone for the first time, no pirogue emerging anywhere to fall in behind him, but he had not expected this anyway, he knew that the others would have departed too; it was not this, it was his very solitude, his desolation which was now his alone and in full since he had elected to remain; the sudden cessation of the paddle, the skiff shooting on for a moment yet while he thought, *What? What?* Then, *No. No. No,* as the silence and solitude and emptiness roared down upon him in a jeering bellow: and now reversed, the skiff spun violently on its heel, he the betrayed driving furiously back toward the platform where he knew it was already too late, that citadel where the very crux and dear breath of his life—the being allowed to work and earn money, that right and privilege which he believed he had earned to himself unaided, asking no favor of anyone or anything save the right to be let alone to pit his will and strength against the sauric[4] protagonist of a land, a region, which he had not asked to be projected into—was being threatened, driving the homemade paddle in grim fury, coming in sight of the platform at last and seeing the motor launch lying alongside it with no surprise at all but actually with a kind of pleas-

4. Prehistoric, pertaining to the era of the dinosaurs.

ure as though at a visible justification of his outrage and fear, the privilege of saying *I told you so* to his own affronting, driving on toward it in a dreamlike state in which there seemed to be no progress at all, in which, unimpeded and suffocating, he strove dreamily with a weightless oar, with muscles without strength or resiliency, at a medium without resistance, seeming to watch the skiff creep infinitesimally across the sunny water and up to the platform while a man in the launch (there were five of them in all) gobbled at him in that same tongue he had been hearing constantly now for ten days and still knew no word of, just as a second man, followed by the woman carrying the baby and dressed again for departure in the faded tunic and the sunbonnet, emerged from the house, carrying (the man carried several other things but the convict saw nothing else) the paper-wrapped bundle which the convict had put behind the rafter ten days ago and no other hand had touched since, he (the convict) on the platform too now, holding the skiff's painter in one hand and the bludgeon-like paddle in the other, contriving to speak to the woman at last in a voice dreamy and suffocating and incredibly calm: "Take it away from him and carry it back into the house."

"So you can talk English, can you?" the man in the launch said. "Why didn't you come out like they told you to last night?"

"Out?" the convict said. Again he even looked, glared, at the man in the launch, contriving even again to control his voice: "I aint got time to take trips. I'm busy," already turning to the woman again, his mouth already open to repeat as the dreamy buzzing voice of the man came to him and he turning once more, in a terrific and absolutely unbearable exasperation, crying, "Flood? What flood? Hell a mile, it's done passed me twice months ago! It's gone! What flood?" and then (he did not think this in actual words either but he knew it, suffered that flashing insight into his own character or destiny: how there was a peculiar quality of repetitiveness about his present fate, how not only the almost seminal crises recurred with a certain monotony, but the very physical circumstances followed a stupidly unimaginative pattern) the man in the launch said, "Take him" and he was on his feet for a few minutes yet, lashing and striking in panting fury, then once more on his back on hard unyielding planks while the four men swarmed over him in a fierce wave of hard bones and panting curses and at last the thin dry vicious snapping of handcuffs.

"Damn it, are you mad?" the man in the launch said. "Cant you understand they are going to dynamite that levee at noon today?[5] —Come on," he said to the others. "Get him aboard. Let's get out of here."

"I want my hides and boat," the convict said.

5. Levees are sometimes dynamited to flood one area so as to protect a more seriously threatened region downstream.

"Damn your hides," the man in the launch said. "If they dont get that levee blowed pretty soon you can hunt plenty more of them on the capitol steps at Baton Rouge. And this is all the boat you will need and you can say your prayers about it."

"I aint going without my boat," the convict said. He said it calmly and with complete finality, so calm, so final that for almost a minute nobody answered him, they just stood looking quietly down at him as he lay, half-naked, blistered and scarred, helpless and manacled hand and foot, on his back, delivering his ultimatum in a voice peaceful and quiet as that in which you talk to your bedfellow before going to sleep. Then the man in the launch moved; he spat quietly over the side and said in a voice as calm and quiet as the convict's:

"All right. Bring his boat." They helped the woman, carrying the baby and the paper-wrapped parcel, into the launch. Then they helped the convict to his feet and into the launch too, the shackles on his wrists and ankles clashing. "I'd unlock you if you'd promise to behave yourself," the man said. The convict did not answer this at all.

"I want to hold the rope," he said.

"The rope?"

"Yes," the convict said. "The rope." So they lowered him into the stern and gave him the end of the painter after it had passed the towing cleat, and they went on. The convict did not look back. But then, he did not look forward either, he lay half sprawled, his shackled legs before him, the end if the skiff's painter in one shackled hand. The launch made two other stops; when the hazy wafer of the intolerable sun began to stand once more directly overhead there were fifteen people in the launch; and then the convict, sprawled and motionless, saw the flat brazen land begin to rise and become a greenish-black mass of swamp, bearded and convoluted, this in turn stopping short off and there spread before him an expanse of water embraced by a blue dissolution of shoreline and glittering thinly under the noon, larger than he had ever seen before, the sound of the launch's engine ceasing, the hull sliding on behind its fading bow-wave. "What are you doing?" the leader said.

"It's noon," the helmsman said. "I thought we might hear the dynamite." So they all listened, the launch lost of all forward motion, rocking slightly, the glitter-broken small waves slapping and whispering at the hull, but no sound, no tremble even, came anywhere under the fierce hazy sky; the long moment gathered itself and turned on and noon was past. "All right," the leader said. "Let's go." The engine started again, the hull began to gather speed. The leader came aft and stooped over the convict, key in hand. "I guess you'll have to behave now, whether you want to or not," he said, unlocking the manacles. "Wont you?"

"Yes," the convict said. They went on; after a time the shore vanished completely and a little sea got up. The convict was free now but he lay as before, the end of the skiff's painter in his hand, bent now with three or four turns about his wrist; he turned his head now and then to look back at the towing skiff as it slewed and bounced in the launch's wake; now and then he even looked out over the lake, the eyes alone moving, the face grave and expressionless, thinking, *This is a greater immensity of water, of waste and desolation, than I have ever seen before*; perhaps not; thinking three or four hours later, the shoreline raised again and broken into a clutter of sailing sloops and power cruisers, *These are more boats than I believed existed, a maritime race of which I also had no cognizance* or perhaps not thinking it but just watching as the launch opened the shore gut of the ship canal, the low smoke of the city beyond it, then a wharf, the launch slowing in; a quiet crowd of people watching with that same forlorn passivity he had seen before and whose race he did recognise even though he had not seen Vicksburg when he passed it—the brand, the unmistakable hallmark of the violently homeless, he more so than any, who would have permitted no man to call him one of them.

"All right," the leader said to him. "Here you are."

"The boat," the convict said.

"You've got it. What do you want me to do—give you a receipt for it?"

"No," the convict said. "I just want the boat."

"Take it. Only you ought to have a bookstrap or something to carry it in." ("Carry it in?" the plump convict said. "Carry it where? Where would you have to carry it?")

He (the tall one) told that: how he and the woman disembarked and how one of the men helped him haul the skiff up out of the water and how he stood there with the end of the painter wrapped around his wrist and the man bustled up, saying, "All right. Next load! Next load!" and how he told this man too about the boat and the man cried, "Boat? Boat?" and how he (the convict) went with them when they carried the skiff over and racked, berthed, it with the others and how he lined himself up by a Coca-Cola sign and the arch of a draw bridge so he could find the skiff again quick when he returned, and how he and the woman (he carrying the paper-wrapped parcel) were herded into a truck and after a while the truck began to run in traffic, between close houses, then there was a big building, an armory—

"Armory?" the plump one said. "You mean a jail."

"No. It was a kind of warehouse, with people with bundles laying on the floor." And how he thought maybe his partner might be there and how he even looked about for the Cajan while waiting for a chance to get back to the door again, where the soldier was and how he got back to the door at last, the woman behind him and his chest actually against the dropped rifle.

"Gwan, gwan," the soldier said. "Get back. They'll give you some clothes in a minute. You cant walk around the streets that way. And something to eat too. Maybe your kinfolks will come for you by that time." And he told that too: how the woman said,

"Maybe if you told him you had some kinfolks here he would let us out." And how he did not; he could not have expressed this either, it too deep, too ingrained; he had never yet had to think it into words through all the long generations of himself—his hill-man's sober and jealous respect not for truth but for the power, the strength, of lying—not to be niggared with lying but rather to use it with respect and even care, delicate quick and strong, like a fine and fatal blade. And how they fetched him clothes—a blue jumper and overalls, and then food too (a brisk starched young woman saying, "But the baby must be bathed, cleaned. It will die if you dont," and the woman saying, "Yessum. He might holler some, he aint never been bathed before. But he's a good baby.") and now it was night, the unshaded bulbs harsh and savage and forlorn above the snorers and he rising, gripping the woman awake, and then the window. He told that: how there were doors in plenty, leading he did not know where, but he had a hard time finding a window they could use but he found one at last, he carrying the parcel and the baby too while he climbed through first—"You ought to tore up a sheet and slid down it," the plump convict said. But he needed no sheet, there were cobbles under his feet now, in the rich darkness. The city was there too but he had not seen it yet and would not— the low constant glare; Bienville had stood there too, it had been the figment of an emasculate also calling himself Napoleon but no more, Andrew Jackson had found it one step from Pennsylvania Avenue.[6] But the convict found it considerably further than one step back to the ship canal and the skiff, the Coca-Cola sign dim now, the draw bridge arching spidery against the jonquil sky at dawn: nor did he tell, any more than about the sixty-foot levee, how he got the skiff back into the water. The lake was behind him now; there was but one direction he could go. When he saw the River again he knew it at once. He should have; it was now inerad-icably a part of his past, his life; it would be a part of what he would bequeath, if that were in store for him. But four weeks later it would look different from what it did now, and did: he (the Old Man) had recovered from his debauch, back in banks again, the

6. The narrator compares the convict to three famous heroes associated with New Orleans. Sieur de Bienville (Jean Baptiste Lemoyne, 1680–1768) founded New Orleans in 1718 and was twice Governor of colonial Louisiana. Napoleon Bonaparte (1769–1821), Emperor of France 1804–15, dreamed (his "figment") of ruling a French empire in America until military affairs forced him to sell the Louisiana territory to the United States in 1803. When forced into exile in 1815 he hoped to come to the United States, but was imprisoned on St. Helena Island. General Andrew Jackson (1767–1845) defeated the British at the Battle of Orleans in 1815; his ensuing fame led to his election to the U.S. presidency in 1828 and residence in the White House on Pennsylvania Avenue in Washington, D.C.

Old Man, rippling placidly toward the sea, brown and rich as chocolate between levees whose inner faces were wrinkled as though in a
frozen and aghast amazement, crowned with the rich green of summer in the willows; beyond them, sixty feet below, slick mules
squatted against the broad pull of middle-busters in the richened
soil which would not need to be planted, which would need only to
be shown a cotton seed to sprout and make; there would be the
symmetric miles of strong stalks by July, purple bloom in August, in
September the black fields snowed over, spilled, the middles
dragged smooth by the long sacks, the long black limber hands
plucking, the hot air filled with the whine of gins,[7] the September
air then but now June air heavy with locust and (the towns) the
smell of new paint and the sour smell of the paste which holds wall
paper—the towns, the villages, the little lost wood landings on stilts
on the inner face of the levee, the lower storeys bright and rank
under the new paint and paper and even the marks on spile and
post and tree of May's raging water-height fading beneath each
bright silver gust of summer's loud and inconstant rain; there was a
store at the levee's lip, a few saddled and rope-bridled mules in the
sleepy dust, a few dogs, a handful of Negroes sitting on the steps beneath the chewing tobacco and malaria medicine signs, and three
white men, one of them a deputy sheriff canvassing for votes to
beat his superior (who had given him his job) in the August primary, all pausing to watch the skiff emerge from the glitter-glare of
the afternoon water and approach and land, a woman carrying a
child stepping out, then a man, a tall man who, approaching,
proved to be dressed in a faded but recently washed and quite clean
suit of penitentiary clothing, stopping in the dust where the mules
dozed and watching with pale cold humorless eyes while the deputy
sheriff was still making toward his armpit that gesture which everyone present realised was to have produced a pistol in one flashing
motion for a considerable time while still nothing came of it. It was
apparently enough for the newcomer, however.

"You a officer?" he said.

"You damn right I am," the deputy said. "Just let me get this
damn gun—"

"All right," the other said. "Yonder's your boat, and here's the
woman. But I never did find that bastard on the cottonhouse."

One of the Governor's young men arrived at the Penitentiary the
next morning. That is, he was fairly young (he would not see thirty
again though without doubt he did not want to, there being that
about him which indicated a character which never had and never
would want anything it did not, or was not about to, possess), a Phi
Beta Kappa out of an Eastern university, a colonel on the Gover-

7. Machines for separating cotton from its seeds.

nor's staff who did not buy it with a campaign contribution, who had stood in his negligent Eastern-cut clothes and his arched nose and lazy contemptuous eyes on the galleries of any number of little lost backwoods stores and told his stories and received the guffaws of his overalled and spitting hearers and with the same look in his eyes fondled infants named in memory of the last administration and in honor (or hope) of the next, and (it was said of him and doubtless not true) by lazy accident the behinds of some who were not infants any longer though still not old enough to vote. He was in the Warden's office with a briefcase, and presently the deputy warden of the levee was there too. He would have been sent for presently though not yet, but he came anyhow, without knocking, with his hat on, calling the Governor's young man loudly by a nickname and striking him with a flat hand on the back and lifted one thigh to the Warden's desk, almost between the Warden and the caller, the emissary. Or the vizier with the command, the knotted cord, as began to appear immediately.

"Well," the Governor's young man said, "you've played the devil, haven't you?" The Warden had a cigar. He had offered the caller one. It had been refused, though presently, while the Warden looked at the back of his neck with hard immobility even a little grim, the deputy leaned and reached back and opened the desk drawer and took one.

"Seems straight enough to me," the Warden said. "He got swept away against his will. He came back as soon as he could and surrendered."

"He even brought that damn boat back," the deputy said. "If he'd a throwed the boat away he could a walked back in three days. But no sir. He's got to bring the boat back. 'Here's your boat and here's the woman but I never found no bastard on no cottonhouse.'" He slapped his knee, guffawing. "Them convicts. A mule's got twice as much sense."

"A mule's got twice as much sense as anything except a rat," the emissary said in his pleasant voice. "But that's not the trouble."

"What is the trouble?" the Warden said.

"This man is dead."

"Hell fire, he aint dead," the deputy said. "He's up yonder in that bunkhouse right now, lying his head off probly. I'll take you up there and you can see him." The Warden was looking at the deputy.

"Look," he said. "Bledsoe was trying to tell me something about that Kate mule's leg. You better go up to the stable and—"

"I done tended to it," the deputy said. He didn't even look at the Warden. He was watching, talking to, the emissary. "No sir. He aint—"

"But he has received an official discharge as being dead. Not a pardon nor a parole either: a discharge. He's either dead, or free. In either case he doesn't belong here." Now both the Warden and the deputy looked at the emissary, the deputy's mouth open a little, the cigar poised in his hand to have its tip bitten off. The emissary spoke pleasantly, extremely distinctly: "On a report of death forwarded to the Governor by the Warden of the Penitentiary." The deputy closed his mouth, though otherwise he didn't move. "On the official evidence of the officer delegated at the time to the charge and returning of the body of the prisoner to the Penitentiary." Now the deputy put the cigar into his mouth and got slowly off the desk, the cigar rolling across his lip as he spoke:

"So that's it. I'm to be it, am I?" He laughed shortly, a stage laugh, two notes. "When I done been right three times running through three separate administrations? That's on a book somewhere too. Somebody in Jackson can find that too. And if they cant, I can show—"

"Three administrations?" the emissary said. "Well, well. That's pretty good."

"You damn right it's good," the deputy said. "The woods are full of folks that didn't." The Warden was again watching the back of the deputy's neck.

"Look," he said. "Why dont you step up to my house and get that bottle of whiskey out of the sideboard and bring it down here?"

"All right," the deputy said. "But I think we better settle this first. I'll tell you what we'll do—"

"We can settle it quicker with a drink or two," the Warden said. "You better step on up to your place and get a coat so the bottle—"

"That'll take too long," the deputy said. "I wont need no coat." He moved to the door, where he stopped and turned. "I'll tell you what to do. Just call twelve men in here and tell him it's a jury—he never seen but one before and he wont know no better—and try him over for robbing that train. Hamp can be the judge."

"You cant try a man twice for the same crime," the emissary said. "He might know that even if he doesn't know a jury when he sees one."

"Look," the Warden said.

"All right. Just call it a new train robbery. Tell him it happened yesterday, tell him he robbed another train while he was gone and just forgot it. He couldn't help himself. Besides, he wont care. He'd just as lief be here as out. He wouldn't have nowhere to go if he was out. None of them do. Turn one loose and be damned if he aint right back here by Christmas like it was a reunion or something, for doing the very same thing they caught him at before." He guffawed again. "Them convicts."

"Look," the Warden said. "While you're there, why dont you

open the bottle and see if the liquor's any good. Take a drink or two. Give yourself time to feel it. If it's not good, no use in bringing it."

"O.K.," the deputy said. He went out this time.

"Couldn't you lock the door?" the emissary said. The Warden squirmed faintly. That is, he shifted his position in his chair.

"After all, he's right," he said. "He's guessed right three times now. And he's kin to all the folks in Pittman County except the niggers."

"Maybe we can work fast then." The emissary opened the briefcase and took out a sheaf of papers. "So there you are," he said.

"There what are?"

"He escaped."

"But he came back voluntarily and surrendered."

"But he escaped."

"All right," the Warden said. "He escaped. Then what?" Now the emissary said look. That is, he said,

"Listen. I'm on per diem. That's taxpayers, votes. And if there's any possible chance for it to occur to anyone to hold an investigation about this, there'll be ten senators and twenty-five representatives here on a special train maybe. On per diem. And it will be mighty hard to keep some of them from going back to Jackson by way of Memphis or New Orleans—on per diem."

"All right," the Warden said. "What does he say to do?"

"This. The man left here in charge of one specific officer. But he was delivered back here by a different one."

"But he surren—" This time the Warden stopped of his own accord. He looked, stared almost, at the emissary. "All right. Go on."

"In specific charge of an appointed and delegated officer, who returned here and reported that the body of the prisoner was no longer in his possession; that, in fact, he did not know where the prisoner was. That's correct, isn't it?" The Warden said nothing. "Isn't that correct?" the emissary said, pleasantly, insistently.

"But you cant do that to him. I tell you he's kin to half the—"

"That's taken care of. The Chief has made a place for him on the highway patrol."

"Hell," the Warden said. "He cant ride a motorcycle. I dont even let him try to drive a truck."

"He wont have to. Surely an amazed and grateful state can supply the man who guessed right three times in succession in Mississippi general elections with a car to ride in and somebody to run it if necessary. He wont even have to stay in it all the time. Just so he's near enough so when an inspector sees the car and stops and blows the horn of it he can hear it and come out."

"I still dont like it," the Warden said.

"Neither do I. Your man could have saved all of this if he had just gone on and drowned himself, as he seems to have led everybody to believe he had. But he didn't. And the Chief says do. Can you think of anything better?" The Warden sighed.

"No," he said.

"All right." The emissary opened the papers and uncapped a pen and began to write. "Attempted escape from the Penitentiary, ten years' additional sentence," he said. "Deputy Warden Buckworth transferred to highway patrol. Call it for meritorious service even if you want to. It wont matter now. Done?"

"Done," the Warden said.

"Then suppose you send for him. Get it over with." So the Warden sent for the tall convict and he arrived presently, saturnine and grave, in his new bed-ticking, his jowls blue and close under the sunburn, his hair recently cut and neatly parted and smelling faintly of the prison barber's (the barber was in for life, for murdering his wife, still a barber) pomade. The Warden called him by name.

"You had bad luck, didn't you?" The convict said nothing. "They are going to have to add ten years to your time."

"All right," the convict said.

"It's hard luck. I'm sorry."

"All right," the convict said. "If that's the rule." So they gave him the ten years more and the Warden gave him the cigar and now he sat, jackknifed backward into the space between the upper and lower bunks, the unlighted cigar in his hand while the plump convict and four others listened to him. Or questioned him, that is, since it was all done, finished, now and he was safe again, so maybe it wasn't even worth talking about any more.

"All right," the plump one said. "So you come back into the River. Then what?"

"Nothing. I rowed."

"Wasn't it pretty hard rowing coming back?"

"The water was still high. It was running pretty hard still. I never made much speed for the first week or two. After that it got better." Then, suddenly and quietly, something—the inarticulateness, the innate and inherited reluctance for speech, dissolved and he found himself, listened to himself, telling it quietly, the words coming not fast but easily to the tongue as he required them: how he paddled on (he found out by trying it that he could make better speed, if you could call it speed, next the bank—this after he had been carried suddenly and violently out to midstream before he could prevent it and found himself, the skiff, travelling back toward the region from which he had just escaped and he spent the better part of the morning getting back inshore and up to the canal again from which he had emerged at dawn) until night came and they tied up to the bank and ate some of the food he had secreted in his jumper before leaving the armory in New Orleans and the woman and the

infant slept in the boat as usual and when daylight came they went
on and tied up again that night too and the next day the food gave
out and he came to a landing, a town, he didn't notice the name of
it, and he got a job. It was a cane[8] farm—

"Cane?" one of the other convicts said. "What does anybody
want to raise cane for? You cut cane. You have to fight it where I
come from. You burn it just to get shut of it."

"It was sorghum,"[9] the tall convict said.

"Sorghum?" another said. "A whole farm just raising sorghum?
Sorghum? What did they do with it?" The tall one didn't know.
He didn't ask, he just came up the levee and there was a truck wait-
ing full of niggers and a white man said, "You there. Can you run a
shovel plow?" and the convict said, "Yes," and the man said,
"Jump in then," and the convict said, "Only I've got a—"

"Yes," the plump one said. "That's what I been aiming to ask.
What did—" The tall convict's face was grave, his voice was calm,
just a little short:

"They had tents for the folks to live in. They were behind." The
plump one blinked at him.

"Did they think she was your wife?"

"I dont know. I reckon so." The plump one blinked at him.

"Wasn't she your wife? Just from time to time kind of, you
might say?" The tall one didn't answer this at all. After a moment
he raised the cigar and appeared to examine a loosening of the
wrapper because after another moment he licked the cigar carefully
near the end. "All right," the plump one said. "Then what?" So he
worked there four days. He didn't like it. Maybe that was why: that
he too could not quite put credence in that much of what he believed to be sorghum. So when they told him it was Saturday and
paid him and the white man told him about somebody who was
going to Baton Rouge the next day in a motor boat, he went to see
the man and took the six dollars he had earned and bought food
with it and tied the skiff behind the motor boat and went to Baton
Rouge. It didn't take long and even after they left the motor boat
at Baton Rouge and he was paddling again it seemed to the convict
that the River was lower and the current not so fast, so hard, so
they made fair speed, tying up to the bank at night among the willows, the woman and baby sleeping in the skiff as of old. Then the
food gave out again. This time it was a wood landing, the wood
stacked and waiting, a wagon and team being unladen of another
load. The men with the wagon told him about the sawmill and
helped him drag the skiff up the levee; they wanted to leave it there
but he would not so they loaded it onto the wagon too and he and
the woman got on the wagon too and they went to the sawmill.

8. Hollow-stemmed tall grass.
9. Cane crop raised chiefly for animal feed.

They gave them one room in a house to live in here. They paid two dollars a day and furnish. The work was hard. He liked it. He stayed there eight days.

"If you liked it so well, why did you quit?" the plump one said. The tall convict examined the cigar again, holding it up where the light fell upon the rich chocolate-colored flank.

"I got in trouble," he said.

"What trouble?"

"Woman. It was a fellow's wife."

"You mean you had been toting one piece up and down the country day and night for over a month, and now the first time you have a chance to stop and catch your breath almost you got to get in trouble over another one?" The tall convict had thought of that. He remembered it: how there were times, seconds, at first when if it had not been for the baby he might have, might have tried. But they were just seconds because in the next instant his whole being would seem to flee the very idea in a kind of savage and horrified revulsion; he would find himself looking from a distance at this millstone which the force and power of blind and risible Motion had fastened upon him, thinking, saying aloud actually, with harsh and savage outrage even though it had been two years since he had had a woman and that a nameless and not young Negress, a casual, a straggler whom he had caught more or less by chance on one of the fifth-Sunday visiting days, the man—husband or sweetheart—whom she had come to see having been shot by a trusty a week or so previous and she had not heard about it: "She aint even no good to me for that."

"But you got this one, didn't you?" the plump convict said.

"Yah," the tall one said. The plump one blinked at him.

"Was it good?"

"It's all good," one of the others said. "Well? Go on. How many more did you have on the way back? Sometimes when a fellow starts getting it it looks like he just cant miss even if—" That was all, the convict told them. They left the sawmill fast, he had no time to buy food until they reached the next landing. There he spent the whole sixteen dollars he had earned and they went on. The River was lower now, there was no doubt of it, and sixteen dollars' worth looked like a lot of food and he thought maybe it would do, would be enough. But maybe there was more current in the River still than it looked like. But this time it was Mississippi, it was cotton; the plow handles felt right to his palms again, the strain and squat of the slick buttocks against the middle-buster's blade was what he knew, even though they paid but a dollar a day here. But that did it. He told it: they told him it was Saturday again and paid him and he told about it—night, a smoked lantern in a disc of worn and barren earth as smooth as silver, a circle of crouching figures, the importunate murmurs and ejaculations, the meagre piles of worn bills beneath the crouching knees, the dotted cubes clicking and scuttering in the dust; that did it. "How much did you win?" the second convict said.

"Enough," the tall one said.

"But how much?"

"Enough," the tall one said. It was enough exactly; he gave it all to the man who owned the second motor boat (he would not need food now), he and the woman in the launch now and the skiff towing behind, the woman with the baby and the paper-wrapped parcel beneath his peaceful hand, on his lap; almost at once he recognised, not Vicksburg because he had never seen Vicksburg, but the trestle beneath which on his roaring wave of trees and houses and dead animals he had shot, accompanied by thunder and lightning, a month and three weeks ago; he looked at it once without heat, even without interest as the launch went on. But now he began to watch the bank, the levee. He didn't know how he would know but he knew he would, and then it was early afternoon and sure enough the moment came and he said to the launch owner: "I reckon this will do."

"Here?" the launch owner said. "This dont look like anywhere to me."

"I reckon this is it," the convict said. So the launch put inshore, the engine ceased, it drifted up and lay against the levee and the owner cast the skiff loose.

"You better let me take you on until we come to something," he said. "That was what I promised."

"I reckon this will do," the convict said. So they got out and he stood with the grapevine painter in his hand while the launch purred again and drew away, already curving; he did not watch it. He laid the bundle down and made the painter fast to a willow root and picked up the bundle and turned. He said no word, he mounted the levee, passing the mark, the tide-line of the old raging, dry now and lined, traversed by shallow and empty cracks like foolish and deprecatory senile grins, and entered a willow clump and removed the overalls and shirt they had given him in New Orleans and dropped them without even looking to see where they fell and opened the parcel and took out the other, the known, the desired, faded a little, stained and worn, but clean, recognisable, and put them on and returned to the skiff and took up the paddle. The woman was already in it.

The plump convict stood blinking at him. "So you come back," he said. "Well, well." Now they all watched the tall convict as he bit the end from the cigar neatly and with complete deliberation and spat it out and licked the bite smooth and damp and took a match from his pocket and examined the match for a moment as though to be sure it was a good one, worthy of the cigar perhaps, and raked it up his thigh with the same deliberation—a motion almost too slow to set fire to it, it would seem—and held it until the flame burned clear and free of sulphur, then put it to the cigar. The plump one watched him, blinking rapidly and steadily. "And they give you ten years more for running. That's bad. A fellow can get used to what they give him at first, to start off with, I dont care how much it is, even a hundred and ninety-nine years. But ten more years. Ten years more, on top of that. When you never expected it. Ten more years to have to do without no society, no female companionship—" He blinked steadily at the tall convict. But he (the tall convict) had thought of that too. He had had a sweetheart. That is, he had gone to church singings and picnics with her —a girl a year or so younger than he, short-legged, with ripe breasts and a heavy mouth and dull eyes like ripe muscadines,[1] who owned a baking-powder can almost full of earrings and brooches and rings bought (or presented at suggestion) from ten-cent stores. Presently he had divulged his plan to her, and there were times later when, musing, the thought occurred to him that possibly if it had not been for her he would not actually have attempted it—this a mere feeling, unworded, since he could not have phrased this either: that who to know what Capone's[2] uncandled bridehood she might not have dreamed to be her destiny and fate, what fast car filled with authentic colored glass and machine guns, running traffic lights. But that was all past and done when the notion first occurred to him, and in the third month of his incarceration she came to see him. She wore earrings and a bracelet or so which he had never seen before and it never became quite clear how she had got that far from home, and she cried violently for the first three minutes though presently (and without his ever knowing either exactly how they had got separated or how she had made the acquaintance) he saw her in animated conversation with one of the guards. But she kissed him before she left that evening and said she would return the first chance she got, clinging to him, sweating a little, smelling of scent and soft young female flesh, slightly pneumatic. But she didn't come back though he continued to write to her, and seven months later he got an answer. It was a postcard, a colored lithograph of a Birmingham hotel, a childish X inked heavily across one window, the heavy writing on the reverse slanted and primer-like too: *This is where were honnymonning at. Your friend (Mrs) Vernon Waldrip*

The plump convict stood blinking at the tall one, rapidly and steadily. "Yes, sir," he said. "It's them ten more years that hurt. Ten more years to do without a woman, no woman a tall a fellow wants—" He blinked steadily and rapidly, watching the tall one. The other did not move, jackknifed backward between the two bunks, grave and clean, the cigar burning smoothly and richly in his

---

1. A variety of grapes.
2. Al Capone (1899–1947), notorious American gangster.

clean steady hand, the smoke wreathing upward across his face saturnine, humorless, and calm. "Ten more years—"

"Women, —t," the tall convict said.

1939

Pantaloon in Black[1]

I

He stood in the worn, faded clean overalls which Mannie herself had washed only a week ago, and heard the first clod strike the pine box. Soon he had one of the shovels himself, which in his hands (he was better than six feet and weighed better than two hundred pounds) resembled the toy shovel a child plays with at the shore, its half cubic foot of flung dirt no more than the light gout of sand the child's shovel would have flung. Another member of his sawmill gang touched his arm and said, "Lemme have hit, Rider." He didn't even falter. He released one hand in midstroke and flung it backward, striking the other across the chest, jolting him back a step, and restored the hand to the moving shovel, flinging the dirt with that effortless fury so that the mound seemed to be rising of its own volition, not built up from above but thrusting visibly upward out of the earth itself, until at last the grave, save for its rawness, resembled any other marked off without order about the barren plot by shards of pottery and broken bottles and old brick and other objects insignificant to sight but actually of a profound meaning and fatal to touch, which no white man could have read. Then he straightened up and with one hand flung the shovel quivering upright in the mound like a javelin and turned and began to walk away, walking on even when an old woman came out of the meagre clump of his kin and friends and a few old people who had known him and his dead wife both since they were born, and grasped his forearm. She was his aunt. She had raised him. He could not remember his parents at all.

"Whar you gwine?" she said.

"Ah'm goan home," he said.

"You dont wants ter go back dar by yoself," she said. "You needs to eat. You come on home and eat."

"Ah'm goan home," he repeated, walking out from under her hand, his forearm like iron, as if the weight on it were no more than that of a fly, the other members of the mill gang whose head he was giving way quietly to let him pass. But before he reached the fence one of them overtook him; he did not need to be told it was his aunt's messenger.

1. The Pantaloon was a stock character in Italian comedy, usually the butt of cruel jokes. The story appeared in *Go Down, Moses* (1942), a unified fiction comprised of related stories originally published separately. The story is one of many in that volume that reveal the black people's profound capacity for experience and explore the violence and the tragedy that result from the estrangement of blacks from whites.

"Wait, Rider," the other said. "We gots a jug in de bushes—"
Then the other said what he had not intended to say, what he
had never conceived of himself saying in circumstances like these,
even though everybody knew it—the dead who either will not or
cannot quit the earth yet although the flesh they once lived in has
been returned to it, let the preachers tell and reiterate and affirm
how they left it not only without regret but with joy, mounting toward glory: "You dont wants ter go back dar. She be wawkin yit."

He didn't pause, glancing down at the other, his eyes red at the
inner corners in his high, slightly backtilted head. "Lemme lone,
Acey," he said. "Doan mess wid me now," and went on, stepping
over the three-strand wire fence without even breaking his stride,
and crossed the road and entered the woods. It was middle dusk
when he emerged from them and crossed the last field, stepping
over that fence too in one stride, into the lane. It was empty at this
hour of Sunday evening—no family in wagon, no rider, no walkers
churchward to speak to him and carefully refrain from looking after
him when he had passed—the pale, powder-light, powder-dry dust
of August from which the long week's marks of hoof and wheel had
been blotted by the strolling and unhurried Sunday shoes, with
somewhere beneath them, vanished but not gone, fixed and held in
the annealing dust, the narrow, splay-toed prints of his wife's bare
feet where on Saturday afternoon she would talk to the commissary to buy their next week's supplies while he took his bath; himself, his own prints, setting the period now as he strode on, moving
almost as fast as a smaller man could have trotted, his body breasting the air her body had vacated, his eyes touching the objects—
post and tree and field and house and hill—her eyes had lost.

The house was the last one in the lane, not his but rented from
Carothers Edmonds, the local white landowner. But the rent was
paid promptly in advance, and even in just six months he had re-
floored the porch and rebuilt and roofed the kitchen, doing the work
himself on Saturday afternoon and Sunday with his wife helping
him, and bought the stove. Because he made good money: sawmill-
ing ever since he began to get his growth at fifteen and sixteen and
now, at twenty-four, head of the timber gang itself because the gang
he headed moved a third again as much timber between sunup and
sundown as any other moved, handling himself at times out of the
vanity of his own strength logs which ordinarily two men would
have handled with canthooks; never without work even in the old
days when he had not actually needed the money, when a lot of
what he wanted, needed perhaps, didn't cost money—the women
bright and dark and for all purposes nameless he didn't need to buy
and it didn't matter to him what he wore and there was always food
for him at any hour of day or night in the house of his aunt who
didn't even want to take the two dollars he gave her each Saturday
—so there had been only the Saturday and Sunday dice and whis-

key that had to be paid for until that day six months ago when he saw Mannie, whom he had known all his life, for the first time and said to himself: "Ah'm thu wid all dat," and they married and he rented the cabin from Carothers Edmonds and built a fire on the hearth[2] on their wedding night as the tale told how Uncle Lucas Beauchamp, Edmonds' oldest tenant, had done on his forty-five years ago and which had burned ever since; and he would rise and dress and eat his breakfast by lamplight to walk the four miles to the mill by sunup, and exactly one hour after sundown he would enter the house again, five days a week, until Saturday. Then the first hour would not have passed noon when he would mount the steps and knock, not on post or doorframe but on the underside of the gallery roof itself, and enter and ring the bright cascade of silver dollars onto the scrubbed table in the kitchen where his dinner simmered on the stove and the galvanised tub of hot water and the baking powder can of soft soap and the towel made of scalded flour sacks sewn together and his clean overalls and shirt waited, and Mannie would gather up the money and walk the half-mile to the commissary and buy their next week's supplies and bank the rest of the money in Edmonds' safe and return and they would eat once again without haste or hurry after five days—the sidemeat, the greens, the cornbread, the buttermilk from the well-house, the cake which she baked every Saturday now that she had a stove to bake in.

But when he put his hand on the gate it seemed to him suddenly that there was nothing beyond it. The house had never been his anyway, but now even the new planks and sills and shingles, the hearth and stove and bed, were all a part of the memory of somebody else, so that he stopped in the half-open gate and said aloud, as though he had gone to sleep in one place and then waked suddenly to find himself in another: "Whut's Ah doin hyar?" before he went on. Then he saw the dog. He had forgotten it. He remembered neither seeing nor hearing it since it began to howl just before dawn yesterday—a big dog, a hound with a strain of mastiff from somewhere (he had told Mannie a month after they married: "Ah needs a big dawg. You's de onliest least thing whut ever kep up wid me one day, leff alone fo weeks.") coming out from beneath the gallery and approaching, not running but seeming rather to drift across the dusk until it stood lightly against his leg, its head raised until the tips of his fingers just touched it, facing the house and making no sound; whereupon, as if the animal controlled it, had lain guardian before it during his absence and only this instant relinquished, the shell of planks and shingles facing him solidified, filled, and for the moment he believed that he could not possibly enter it. "But Ah needs to eat," he said. "Us bofe needs to eat," he said, moving on though the dog did not follow until he turned and

2. Symbol of the home, a recurring motif in *Go Down, Moses*.

cursed it. "Come on hyar!" he said. "What you skeered of? She lacked you too, same as me," and they mounted the steps and crossed the porch and entered the house—the dusk-filled single room where all those six months were now crammed and crowded into one instant of time until there was no space left for air to breathe, crammed and crowded about the hearth where the fire which was to have lasted to the end of them, before which in the days before he was able to buy the stove he would enter after his four-mile walk from the mill and find her, the shape of her narrow back and haunches squatting, one narrow spread hand shielding her face from the blaze over which the other hand held the skillet, had already fallen to a dry, light soilure of dead ashes when the sun rose yesterday—and himself standing there while the last of light died about the strong and indomitable beating of his heart and the deep steady arch and collapse of his chest which walking fast over the rough going of woods and fields had not increased and standing still in the quiet and fading room had not slowed down.

Then the dog left him. The light pressure went off his flank; he heard the click and hiss of its claws on the wooden floor as it surged away and he thought at first that it was fleeing. But it stopped just outside the front door, where he could see it now, and the upfling of its head as the howl began, and then he saw her too. She was standing in the kitchen door, looking at him. He didn't move. He didn't breathe nor speak until he knew his voice would be all right, his face fixed too not to alarm her. "Mannie," he said. "Hit's awright. Ah aint afraid." Then he took a step toward her, slow, not even raising his hand yet, and stopped. Then he took another step. But this time as soon as he moved she began to fade. He stopped at once, not breathing again, motionless, willing his eyes to see that she had stopped too. But she had not stopped. She was fading, going. "Wait," he said, talking as sweet as he had ever heard his voice speak to a woman: "Den lemme go wid you, honey." But she was going. She was going fast now, he could actually feel between them the insuperable barrier of that very strength which could handle alone a log which would have taken any two other men to handle, of the blood and bones and flesh too strong, invincible for life, having learned at least once with his own eyes how tough, even in sudden and violent death, not a young man's bones and flesh perhaps but the will of that bone and flesh to remain alive, actually was.

Then she was gone. He walked through the door where she had been standing, and went to the stove. He did not light the lamp. He needed no light. He had set the stove up himself and built the shelves for the dishes, from among which he took two plates by feel and from the pot sitting cold on the cold stove he ladled onto the plates the food which his aunt had brought yesterday and of which he had eaten yesterday though now he did not remember when he had eaten it nor what it was, and carried the plates to the scrubbed bare table beneath the single small fading window and drew two chairs up and sat down, waiting again until he knew his voice would be what he wanted it to be. "Come on hyar, now," he said roughly. "Come on hyar and eat yo supper. Ah aint gonter have no—" and ceased, looking down at his plate, breathing the strong, deep pants, his chest arching and collapsing until he stopped it presently and held himself motionless for perhaps a half minute, and raised a spoonful of the cold and glutinous peas to his mouth. The congealed and lifeless mass seemed to bounce on contact with his lips. Not even warmed from mouth-heat, peas and spoon spattered and rang upon the plate; his chair crashed backward and he was standing, feeling the muscles of his jaw beginning to drag his mouth open, tugging upward the top half of his head. But he stopped that too before it became sound, holding himself again while he rapidly scraped the food from his plate onto the other and took it up and left the kitchen, crossed the other room and the gallery and set the plate on the bottom step and went on toward the gate.

The dog was not there, but it overtook him within the first half mile. There was a moon then, their two shadows flitting broken and intermittent among the trees or slanted long and intact across the slope of pasture or old abandoned fields upon the hills, the man moving almost as fast as a horse could have moved over that ground, altering his course each time a lighted window came in sight, the dog trotting at heel while their shadows shortened to the moon's curve until at last they trod them and the last far lamp had vanished and the shadows began to lengthen on the other hand, keeping to heel even when a rabbit burst from almost beneath the man's foot, then lying in the gray of dawn beside the man's prone body, beside the labored heave and collapse of the chest, the loud harsh snoring which sounded not like groans of pain but like someone engaged without arms in prolonged single combat.

When he reached the mill there was nobody there but the fireman—an older man just turning from the woodpile, watching quietly as he crossed the clearing, striding as if he were going to walk not only through the boiler shed but through (or over) the boiler too, the overalls which had been clean yesterday now draggled and soiled and drenched to the knees with dew, the cloth cap flung onto the side of his head, hanging peak downward over his ear as he always wore it, the whites of his eyes rimmed with red and with something urgent and strained about them. "Whar yo bucket?" he said. But before the fireman could answer he had stepped past him and lifted the polished lard pail down from a nail in a post. "Ah just wants a biscuit," he said.

"Eat hit all," the fireman said. "Ah'll eat outen de yuthers' buckets at dinner. Den you gawn home and go to bed. You dont looks good."

"Ah aint come hyar to look," he said, sitting on the ground, his back against the post, the open pail between his knees, cramming the food into his mouth with his hands, wolfing it—peas again, also gelid and cold, a fragment of yesterday's Sunday fried chicken, a few rough chunks of this morning's fried sidemeat, a biscuit the size of a child's cap—indiscriminate, tasteless. The rest of the crew was gathering now, with voices and sounds of movement outside the boiler shed; presently the white foreman rode into the clearing on a horse. He did not look up, setting the empty pail aside, rising, looking at no one, and went to the branch and lay on his stomach with, or ~~as whi~~ his face to the water, drawing the water into himself with, or ~~as whi~~ ~~strong,~~ troubled inhalations that he had snored day, trying to get air. ~~in~~ the empty house at dusk yester-day, trying to get air.

Then the trucks were rolling. The air pulsed with the rapid beating of the exhaust and the whine ~~and clang of th~~ ~~the trucks~~ rolling one by one up to the skidway, he mounting the trucks in turn, to stand balanced on the load he freed, knocking the chocks out and casting loose the shackle chains and with his cant-hook squaring the sticks of cypress and gum and oak one by one to the incline and holding them until the next two men of his gang were ready to receive and guide them, until the discharge of each truck became one long rumbling roar punctuated by grunting shouts and, as the morning grew and the sweat came, chanted phrases of song tossed back and forth. He did not sing with them. He rarely ever did, and this morning might have been no different from any other —himself man-height again above the heads which carefully refrained from looking at him, stripped to the waist now, the shirt removed and the overalls knotted about his hips by the suspender straps, his upper body bare except for the handkerchief about his neck and the cap clapped and clinging somehow over his right ear, the mounting sun sweat-glinted steel-blue on the midnight-colored bunch and slip of muscles until the whistle blew for noon and he said to the two men at the head of the skidway: "Look out. Git out de way," and rode the log down the incline, balanced erect upon it in short rapid backward-running steps above the headlong thunder.

His aunt's husband was waiting for him—an old man, as tall as he was, but lean, almost frail, carrying a tin pail in one hand and a covered plate in the other; they too sat in the shade beside the branch a short distance from where the others were opening their dinner pails. The bucket contained a fruit jar of buttermilk packed in a clean damp towsack. The covered dish was a peach pie, still warm. "She baked hit fer you dis mawin," the uncle said. "She say fer you to come home." He didn't answer, bent forward a little, his elbows on his knees, holding the pie in both hands, wolfing at it, the syrupy filling smearing and trickling down his chin, blinking rapidly as he chewed, the whites of his eyes covered a little more by the creeping red. "Ah went to yo house last night, but you want dar. She sont me. She wants you to come on home. She kept de lamp burnin all last night fer you."

"Ah'm awright," he said.

"You aint awright. De Lawd guv, and He tuck away. Put yo faith and trust in Him. And she kin help you."

"Whut faith and trust?" he said. "Whut Mannie ever done ter Him? Whut He wanter come messin wid me and ――"

"Hush!" the old man said. "Hush!"

Then the trucks were rolling again. Then he could ~~s~~ ne could to invent to himself reasons for his breathing, ~~u~~ ~~s~~. of the rolling logs; began to believe he had forgot about ~~thunder~~ of the rolling logs; not hear it himself above th whereupon as soon ―― ～ found himself believing he had forgotten it, he knew that he had not, so that instead of tipping the final log onto the skidway he stood up and cast his cant-hook away as if it were a burnt match and in the dying reverberation of the last log's rumbling descent he vaulted down between the two slanted tracks of the skid, facing the log which still lay on the truck. He had done it before—taken a log from the truck onto his hands, balanced, and turned with it and tossed it onto the skidway, but never with a stick of this size, so that in a complete cessation of all sound save the pulse of the exhaust and the light free-running whine of the disengaged saw since every eye there, even that of the white foreman, was upon him, he nudged the log to the edge of the truckframe and squatted and set his palms against the underside of it. For a time there was no movement at all. It was as if the unrational and inanimate wood had invested, mesmerised the man with some of its own primal inertia. Then a voice said quietly: "He got hit. Hit's off de truck," and they saw the crack and gap of air, watching the infinitesimal straightening of the braced legs until the knees locked, the movement mounting infinitesimally through the belly's insuck, the arch of the chest, the neck cords, lifting the lip from the white clench of teeth in passing, drawing the whole head backward and only the bloodshot fixity of the eyes impervious to it, moving on up the arms and the straightening elbows until the balanced log was higher than his head. "Only he aint gonter turn wid dat un," the same voice said. "And when he try to put hit back on de truck, hit gonter kill him." But none of them moved. Then—there was no gathering of supreme effort—the log seemed to leap suddenly backward over his head of its own volition, spinning, crashing and thundering down the incline; he turned and stepped over the slanting track in one stride and walked through them as they gave way and went on across the clearing toward the woods even though the foreman called after him: "Rider!" and again: "You, Rider!"

At sundown he and the dog were in the river swamp four miles away—another clearing, itself not much larger than a room, a hut, a hovel partly of planks and partly of canvas, an unshaven white man standing in the door beside which a shotgun leaned, watching him as he approached, his hand extended with four silver dollars on the palm. "Ah wants a jug," he said.

"A jug?" the white man said. "You mean a pint. This is Monday. Aint you all running this week?"

"Ah laid off," he said. "Whar's my jug?" waiting, looking at nothing apparently, blinking his bloodshot eyes rapidly in his high, slightly back-tilted head, then turning, the jug hanging from his crooked middle finger against his leg, at which moment the white man looked suddenly and sharply at his eyes as though seeing them for the first time—the eyes which had been strained and urgent this morning and which now seemed to be without vision too and in which no white showed at all—and said,

"Here. Gimme that jug. You dont need no gallon. I'm going to give you that pint, give it to you. Then you get out of here and stay out. Dont come back until—" Then the white man reached and grasped the jug, whereupon the other swung it behind him, sweeping his other arm up and out so that it struck the white man across the chest.

"Look out, white folks," he said. "Hit's mine. Ah done paid you."

The white man cursed him. "No you aint. Here's your money. Put that jug down, nigger."

"Hit's mine," he said, his voice quiet, gentle even, his face quiet save for the rapid blinking of the red eyes. "Ah done paid for hit," turning on, turning his back on the man and the gun both, and recrossed the clearing to where the dog waited beside the path to come to heel again. They moved rapidly on between the close walls of impenetrable cane-stalks which gave a sort of blondness to the twilight and possessed something of that oppression, that lack of room to breathe in, which the walls of his house had had. But this time, instead of fleeing it, he stopped and raised the jug and drew the cob stopper from the fierce duskreek of uncured alcohol and drank, gulping the liquid solid and cold as ice water, without either taste or heat until he lowered the jug and the air got in. "Hah," he said. "Dat's right. Try me. Try me, big boy. Ah gots something hyar now dat kin whup you."

And, once free of the bottom's unbreathing blackness, there was the moon again, his long shadow and that of the lifted jug slanting away as he drank and then held the jug poised, gulping the silver air into his throat until he could breathe again, speaking to the jug: "Come on now. You always claim you's a better man den me. Come on now. Prove it." He drank again, swallowing the chill liquid tamed of taste or heat either while the swallowing lasted, feeling it flow solid and cold with fire, past then enveloping the strong steady panting of his lungs until they too ran suddenly free as his moving body ran in the silver solid wall of air he breasted. And he was all right, his striding shadow and the trotting one of the dog travelling swift as those of two clouds along the hill; the long cast of his motionless shadow and that of the lifted jug slanting across the slope as he watched the frail figure of his aunt's husband toiling up the hill.

"Dey tole me at de mill you was gone," the old man said. "Ah knowed whar to look. Come home, son. Dat ar cant help you."

"Hit done awready hope me," he said. "Ah'm awready home. Ah'm snakebit now and pizen cant hawm me."

"Den stop and see her. Leff her look at you. Dat's all she axes: just leff her look at you—" But he was already moving. "Wait!" the old man cried. "Wait!"

"You cant keep up," he said, speaking into the silver air, breasting aside the silver solid air which began to flow past him almost as fast as it would have flowed past a moving horse. The faint frail voice was already lost in the night's infinitude, his shadow and that of the dog scudding the free miles, the deep strong panting of his chest running free as air now because he was all right.

Then, drinking, he discovered suddenly that no more of the liquid was entering his mouth. Swallowing, it was no longer passing down his throat, his throat and mouth filled now with a solid and unmoving column which without reflex or revulsion sprang, columnar and intact and still retaining the mold of his gullet, outward glinting in the moonlight, splintering, vanishing into the myriad murmur of the dewed grass. He drank again. Again his throat merely filled solidly until two icy rills ran from his mouth-corners; again the intact column sprang silvering, glinting, shivering, while he panted the chill of air into his throat, the jug poised before his mouth while he spoke to it: "Awright. Ah'm ghy try you again. Soon as you makes up yo mind to stay whar I puts you, Ah'll leff you alone." He drank, filling his gullet for the third time and lowered the jug one instant ahead of the bright intact repetition, panting, indrawing the cool of air until he could breathe. He stoppered the cob carefully back into the jug and stood, panting, blinking, the long cast of his solitary shadow slanting away across the hill and beyond, across the mazy infinitude of all the night-bound earth. "Awright," he said. "Ah just misread de sign wrong. Hit's done done me all de help Ah needs. Ah'm awright now. Ah doan needs no mo of hit."

He could see the lamp in the window as he crossed the pasture, passing the black-and-silver yawn of the sandy ditch where he had played as a boy with empty snuff-tins and rusted harness-buckles and fragments of trace-chains and now and then an actual wheel, passing the garden patch where he had hoed in the spring days while his aunt stood sentry over him from the kitchen window, crossing the grassless yard in whose dust he had sprawled and crept before he learned to walk. He entered the house, the room, the light itself, and stopped in the door, his head backtilted a little as if he could not see, the jug hanging from his crooked finger, against his leg. "Unc Alec say you wanter see me," he said.

"Not just to see you," his aunt said. "To come home, whar we kin help you."

"Ah'm awright," he said. "Ah doan needs no help."

"No," she said. She rose from the chair and came and grasped his arm as she had grasped it yesterday at the grave. Again, as on yesterday, the forearm was like iron under her hand. "No! When Alec come back and tole me how you had wawked off de mill and de sun not half down, Ah knowed why and whar. And dat cant help you."

"Hit done awready hope me. Ah'm awright now."

"Dont lie to me," she said. "You aint never lied to me. Dont lie to me now."

Then he said it. It was his own voice, without either grief or amazement, speaking quietly out of the tremendous panting of his chest which in a moment now would begin to strain at the walls of this room too. But he would be gone in a moment.

"Nome," he said, "Hit aint done me no good."

"And hit cant! Cant nothing help you but Him! Ax Him! Tole Him about hit! He wants to hyar you and help you!"

"Efn He God, Ah dont needs to tole Him. Efn He God, He awready know hit. Awright. Hyar Ah is. Leff Him come down hyar and do me some good."

"On yo knees!" she cried. "On yo knees and ax Him!" But it was not his knees on the floor, it was his feet. And for a space he could hear her feet too on the planks of the hall behind him and her voice crying after him from the door: "Spoot! Spoot!"—crying after him across the moon-dappled yard the name he had gone by in his childhood and adolescence, before the men he worked with and the bright dark nameless women he had taken in course and forgotten until he saw Mannie that day and said, "Ah'm thu wid all dat," began to call him Rider.

It was just after midnight when he reached the mill. The dog was gone now. This time he could not remember when nor where. At first he seemed to remember hurling the empty jug at it. But later the jug was still in his hand and it was not empty, although each time he drank now the two icy runnels streamed from his mouthcorners, sopping his shirt and overalls until he walked constantly in the fierce chill of the liquid tamed now of flavor and heat and odor too even when the swallowing ceased. "Sides that," he said, "Ah wouldn't thow nothin at him. Ah mout kick him efn he needed hit and was close enough. But Ah wouldn't ruin no dog chunkin hit."

The jug was still in his hand when he entered the clearing and paused among the mute soaring of the moon-blond lumber-stacks. He stood in the middle now of the unimpeded shadow which he was treading again as he had trod it last night, swaying a little, blinking about at the stacked lumber, the skidway, the piled logs waiting for tomorrow, the boiler-shed all quiet and blanched in the moon. And then it was all right. He was moving again. But he was not moving, he was drinking, the liquid cold and swift and tasteless and requiring no swallowing, so that he could not tell if it were going down inside or outside. But it was all right. And now he was moving, the jug gone now and he didn't know the when or where of that either. He crossed the clearing and entered the boiler shed and went on through it, crossing the junctureless backloop of time's trepan, to the door of the tool-room, the faint glow of the lantern beyond the plank-joints, the surge and fall of living shadow, the mutter of voices, the mute click and scutter of the dice, his hand loud on the barred door, his voice loud too: "Open hit. Hit's me. Ah'm snakebit and bound to die."

Then he was through the door and inside the tool-room. They were the same faces—three members of his timber gang, three or four others of the mill crew, the white night-watchman with the heavy pistol in his hip pocket and the small heap of coins and worn bills on the floor before him, one who was called Rider and was Rider standing above the squatting circle, swaying a little, blinking, the dead muscles of his face shaped into smiling while the white man stared up at him. "Make room, gamblers," he said. "Make room. Ah'm snakebit and de pizen cant hawm me."

"You're drunk," the white man said. "Get out of here. One of you niggers open the door and get him out of here."

"Dass awright, boss-man," he said, his voice equable, his face still fixed in the faint rigid smiling beneath the blinking of the red eyes; "Ah aint drunk. Ah just cant wawk straight fer dis yar money weighin me down."

Now he was kneeling too, the other six dollars of his last week's pay on the floor before him, blinking, still smiling at the face of the white man opposite, then, still smiling, he watched the dice pass from hand to hand around the circle as the white man covered the bets, watching the soiled and palm-worn money in front of the white man gradually and steadily increase, watching the white man cast and win two doubled bets in succession then lose one for twenty-five cents, the dice coming to him at last, the cupped snug clicking of them in his fist. He spun a coin into the center.

"Shoots a dollar," he said, and cast, and watched the white man pick up the dice and flip them back to him. "Ah lets hit lay," he said. "Ah'm snakebit. Ah kin pass wid anything," and cast, and this time one of the negroes flipped the dice back. "Ah lets hit lay," he said, and cast, and moved as the white man moved, catching the white man's wrist before his hand reached the dice, the two of them squatting, facing each other above the dice and the money, his left hand grasping the white man's wrist, his face still fixed in the rigid and deadened smiling, his voice equable, almost deferen-

tial: "Ah kin pass even wid missouts. But dese hyar yuther boys—"
until the white man's hand sprang open and the second pair of dice
clattered onto the floor beside the first two and the white man
wrenched free and sprang up and back and reached the hand backward toward the pocket where the pistol was.

The razor hung between his shoulder-blades from a loop of cotton string round his neck inside his shirt. The same motion of the
hand which brought the razor forward over his shoulder flipped the
blade open and freed it from the cord, the blade opening on until
the back edge of it lay across the knuckles of his fist, his thumb
pressing the handle into his closing fingers, so that in the second before the half-drawn pistol exploded he actually struck at the white
man's throat not with the blade but with a sweeping blow of his
fist, following through in the same motion so that not even the first
jet of blood touched his hand or arm.

2

After it was over—it didn't take long; they found the prisoner on
the following day, hanging from the bell-rope in a negro school-
house about two miles from the sawmill, and the coroner had pronounced his verdict of death at the hands of a person or persons unknown and surrendered the body to its next of kin all within five
minutes—the sheriff's deputy who had been officially in charge of
the business was telling his wife about it. They were in the kitchen.
His wife was cooking supper. The deputy had been out of bed and
in motion ever since the jail delivery shortly before midnight of yesterday and had covered considerable ground since, and he was spent
now from lack of sleep and hurried food at hurried and curious
hours and, sitting in a chair beside the stove, a little hysterical too.

"Them damn niggers," he said. "I swear to godfrey, it's a wonder
we have as little trouble with them as we do. Because why? Because
they aint human. They look like a man and they walk on their
hind legs like a man, and they can talk and you can understand
them and you think they are understanding you, at least now and
then. But when it comes to the normal human feelings and sentiments of human beings, they might just as well be a damn herd of
wild buffaloes. Now you take this one today——"

"I wish you would," his wife said harshly. She was a stout
woman, handsome once, graying now and with a neck definitely too
short, who looked not harried at all but composed in fact, only
choleric. Also, she had attended a club rook-party that afternoon
and had won the first, the fifty-cent, prize until another member
had insisted on a recount of the scores and the ultimate throwing
out of one entire game. "Take him out of my kitchen, anyway. You
sheriffs! Sitting around that courthouse all day long, talking. It's no
wonder two or three men can walk in and take prisoners out from
under your very noses. They would take your chairs and desks and

window sills too if you ever got your feet and backsides off of them that long."

"It's more of them Birdsongs than just two or three," the deputy said. "There's forty-two active votes in that connection. Me and Maydew taken the poll-list and counted them one day. But listen —" The wife turned from the stove, carrying a dish. The deputy snatched his feet rapidly out of the way as she passed him, passed almost over him, and went into the dining room. The deputy raised his voice a little to carry the increased distance: "His wife dies on him. All right. But does he grieve? He's the biggest and busiest man at the funeral. Grabs a shovel before they even got the box into the grave they tell me, and starts throwing dirt onto her faster than a slip scraper could have done it. But that's all right—" His wife came back. He moved his feet again and altered his voice again to the altered range: "—maybe that's how he felt about her. There aint any law against a man rushing his wife into the ground, provided he never had nothing to do with rushing her to the cemetery too. But here the next day he's the first man back at work except the fireman, getting back to the mill before the fireman had his fire going, let alone steam up; five minutes earlier and he could even have helped the fireman wake Birdsong up so Birdsong could go home and go back to bed again, or he could even have cut Birdsong's throat then and saved everybody trouble.

"So he comes to work, the first man on the job, when McAndrews and everybody else expected him to take the day off since even a nigger couldn't want no better excuse for a holiday than he had just buried his wife, when a white man would have took the day off out of pure respect no matter how he felt about his wife, when even a little child would have had sense enough to take a day off when he would still get paid for it too. But not him. The first man there, jumping from one log truck to another before the starting whistle quit blowing even, snatching up ten-foot cypress logs by himself and throwing them around like matches. And then, when everybody had finally decided that that's the way to take him, the way he wants to be took, he walks off the job in the middle of the afternoon without by-your-leave or much obliged or goodbye to McAndrews or nobody else, gets himself a whole gallon of bust-skull white-mule whisky, comes straight back to the mill and to the same crap game where Birdsong has been running crooked dice on them mill niggers for fifteen years, goes straight to the same game where he has been peacefully losing a probably steady average ninety-nine percent of his pay ever since he got big enough to read the spots on them miss-out dice, and cuts Birdsong's throat clean to the neckbone five minutes later." The wife passed him again and went to the dining room. Again he drew his feet back and raised his voice:

"So me and Maydew go out there. Not that we expected to

do any good, as he had probably passed Jackson, Tennessee, about daylight; and besides, the simplest way to find him would be just to stay close behind them Birdsong boys. Of course there wouldn't be nothing hardly worth bringing back to town after they did find him, but it would close the case. So it's just by the merest chance that we go by his house; I dont even remember why we went now, but we did; and there he is. Sitting behind the barred front door with a open razor on one knee and a loaded shotgun on the other? No. He was asleep. A big pot of field peas et clean empty on the stove, and him laying in the back yard asleep in the broad sun with just his head under the edge of the porch in the shade and a dog that looked like a cross between a bear and a Polled Angus steer yelling fire and murder from the back door. And we wake him and he sets up and says, 'Awright, white folks. Ah done it. Jest dont lock me up,' and Maydew says, 'Mr Birdsong's kinfolks aint going to lock you up neither. You'll have plenty of fresh air when they get hold of you,' and he says, 'Ah done it. Jest dont lock me up'—advising, instructing the sheriff not to lock him up; he done it all right and it's too bad but it aint convenient for him to be cut off from the fresh air at the moment. So we loaded him into the car, when here come the old woman—his ma or aunt or something—panting up the road at a dog-trot, wanting to come with us too, and Maydew trying to explain to her what would maybe happen to her too if them Birdsong kin catches us before we can get him locked up, only she is coming anyway, and like Maydew says, her being in the car too might be a good thing if the Birdsongs did happen to run into us, because after all interference with the law cant be condoned even if the Birdsong connection did carry that beat for Maydew last summer.

"So we brought her along too and got him to town and into the jail all right and turned him over to Ketcham and Ketcham taken him on up stairs and the old woman coming too, right on up to the cell, telling Ketcham, 'Ah tried to raise him right. He was a good boy. He aint never been in no trouble till now. He will suffer for what he done. But dont let the white folks get him,' until Ketcham says, 'You and him ought to thought of that before he started barbering white men without using no lather first.' So he locked them both up in the cell because he felt like Maydew did, that her being in there with him might be a good influence on the Birdsong boys if anything started if he should happen to be running for sheriff or something when Maydew's term was up. So Ketcham come on back down stairs and pretty soon the chain gang come in and went on up to the bull pen and he thought things had settled down for a while when all of a sudden he begun to hear the yelling, not howling: yelling, though there wasn't no words in it, and he grabbed his pistol and run back up stairs to the bull pen where the chain gang was and Ketcham could see into the cell where the old woman was kind of squinched down in one corner and where that nigger had done tore that iron cot clean out of the floor it was bolted to and was standing in the middle of the cell, holding the cot over his head like it was a baby's cradle, yelling, and says to the old woman, 'Ah aint goan hurt you,' and throws the cot against the wall and comes and grabs holt of that steel barred door and rips it out of the wall, bricks hinges and all, and walks out of the cell toting the door over his head like it was a guaze window-screen, hollering, 'It's awright. It's awright. Ah aint trying to git away.'

"Of course Ketcham could have shot him right there, but like he said, if it wasn't going to be the law, then them Birdsong boys ought to have the first lick at him. So Ketcham dont shoot. Instead, he jumps in behind where them chain gang niggers was kind of backed off from that steel door, hollering, 'Grab him! Throw him down!' except the niggers hung back at first too until Ketcham gets in where he can kick the ones he can reach, batting at the others with the flat of the pistol until they rush him. And Ketcham says that for a full minute that nigger would grab them as they come in and fling them clean across the room like they was rag dolls, saying, 'Ah aint tryin to git out. Ah aint trying to git out,' until at last they pulled him down—a big mass of nigger heads and arms and legs boiling around on the floor and even then Ketcham says every now and then a nigger would come flying out and go sailing through the air across the room, spraddled out like a flying squirrel and with his eyes sticking out like car headlights, until at last they had him down and Ketcham went in and begun peeling away niggers until he could see him laying there under the pile of them, laughing, with tears big as glass marbles running across his face and down past his ears and making a kind of popping sound on the floor like somebody dropping bird eggs, laughing and laughing and saying, 'Hit look lack Ah just cant quit thinking. Look lack Ah just cant quit.' And what do you think of that?"

"I think if you eat any supper in this house you'll do it in the next five minutes," his wife said from the dining room. "I'm going to clear this table then and I'm going to the picture show."

1942

Contemporary
American Prose
1945 –

W HEN IN 1942 a young critic named Alfred Kazin published his first
book, *On Native Grounds*, a survey of American writing from late
nineteenth-century realism up through the literature of the 1930s,
he judged "the greatest single fact about our modern American writing" to
be "our writers' absorption in every last detail of their American world together with their deep and subtle alienation from it." Three years later, on
August 6, 1945, the explosion of an atomic bomb over Hiroshima in Japan
brought about a hasty conclusion to World War II and also introduced into
human life a new reality so unimaginable as to make terms like "crisis" and
"alienation" seem understatements, scarcely adequate to the nature of the
postwar era. What was one to feel, what were American writers to feel
about such an event, and what difference would it make to the kind of work
they were hoping to do?

More than three decades later we can look back upon a number of
cataclysmic upheavals which followed the ones at Hiroshima and Nagasaki.
A short list might include the cold war between America and Russia which,
though never breaking into open violence, reached dangerous proportions in
the 1950s; the attendant fears of nuclear annihilation culminating in the
Cuban missile crisis of 1962; the civil rights movement of the 1960s
with its stark message that there were races in this country who lived
neither on equal nor on amicable terms; the assassination of John F. Kennedy in November of 1963, and the assassinations five years later of his
brother Robert and of Martin Luther King, chief spokesman for civil rights
and the leader of black Americans; the extended, seemingly endless war of
attrition and folly in Vietnam; violence in the urban ghettos; the killing of
four students by the National Guard at Kent State University in 1970,
bringing violence in the universities to a head; the resignation in 1974,
under intense pressure, of President Richard Nixon.

But literature is not life, even though certain episodes from our recent
history wear the aspects of both tragedy and farce; and no list of crucial
public events should be seized upon as responsible for the literature which
is contemporaneous with or which succeeded them. Even though a novel-
fantasy like Robert Coover's *The Public Burning* (1977) could not have
been written without the public events which the novelist uses as his imaginative materials—in this case the execution of the Rosenbergs in 1953

for treason, and the career of Richard Nixon—the directness of relation between literature and contemporary events depends on the books or writers one selects. In the best of postwar American novels and stories the relation has seldom been that direct. To name just three examples from serious writers in the 1950s: the work of Saul Bellow, Flannery O'Connor, and Ralph Ellison surely manifested, in Kazin's terms, "absorption in every last detail of their American world" as well as their "deep and subtle alienation from it." Yet no public event can account for why *The Adventures of Augie March* or *A Good Man Is Hard to Find* or *Invisible Man* should have occurred when they did, or indeed should have occurred at all.

Certain novels did, however, get written as a direct result of World War II, most notably Norman Mailer's *The Naked and the Dead* (1948) and James Jones's *From Here to Eternity* (1951). Mailer had been planning his novel ever since the war began; with hindsight, we can see that he got it out of his system—writing a best seller in the process—then went on to produce many different though not unrelated books. Jones, on the other hand, never found another subject; the final volume of his trilogy about the war was not published until after his death in 1978. World War II novels were long, usually swollen chronicles that looked backward to the naturalism of 1930s writers and were unambitious in their form and style. Perhaps the war novelists were intimidated by Hemingway's *A Farewell to Arms* and persuaded that there was nothing new or profound to say about their subject. Perhaps the novels sank under their conventional story lines, or perhaps it was just too soon to expect highly imaginative writing about the event. At any rate the most compellingly original treatments of that war are to be found in Joseph Heller's *Catch-22* (1961) and Thomas Pynchon's *Gravity's Rainbow* (1973), both written many years after it had ended.

SOUTHERN WRITERS

Taking the years 1945–60 as a unit, we can identify two main groupings of literary energies and principles in the postwar period. The "southern" writers are much less to be thought of as a group than as a number of individually talented novelists and short-story writers, notable for its predominance of women and generally touched by the large shadow of William Faulkner. (Faulkner remained busily at work completing the Snopes trilogy he had begun with *The Hamlet* in 1940, while visiting universities and producing other work until his death in 1962.) The older writers of this group, Katherine Anne Porter and Eudora Welty, remained active—the former occupied with the writing of a lengthy novel *Ship of Fools* (1962), the latter providing a host of distinguished short fiction and novels of which *The Golden Apples* (1947) is perhaps the finest. The younger and, at the time, highly acclaimed writers were Carson McCullers—whose first novel *The Heart Is a Lonely Hunter* (1940) was a critical success and whose short novel *The Member of the Wedding* (1946) is her best work— and Truman Capote, whose *Other Voices, Other Rooms* was published to much fanfare in 1948 when he was twenty-four (Carson McCullers was twenty-three when her first novel appeared).

The absorption of these younger writers in the grotesque, their fascination with extreme and perverse incongruities of character and scene, and their cultivation of verbal effects can be understood as a commitment to "art"— to a use of the creative imagination and language unchecked by any presumed realities of life as it was lived in America in 1948 or 1960. On the other hand, such features of their writing may be defended as the only true and adequate response the artist can make to that bizarre life. Or so Flannery O'Connor, the most talented and humorous of younger southern writers, implied when she remarked wryly, "Of course I have found that any fiction that comes out of the South is going to be called grotesque by the Northern reader—unless it is grotesque, in which case it's going to be called realistic." In any case these artists (to whom one should add the gifted short-story writer Peter Taylor and the novelist Walker Percy) absorbed—often brilliantly created—American speech, manners, habits of eating or praying or loving, while holding back from any topical engagement with the public and social happenings around them.

NEW YORK WRITERS

With proper consciousness of the umbrellalike nature of the label, we can distinguish another main group of writers and the critics who wrote about and publicized them, as "New York." Here the milieu is urban-Jewish, the concerns recognizably more public and political—although less overtly so in the novelists than in the essayist-intellectuals who criticized them. The major periodical for these writers was *Partisan Review* (for a time *Commentary* shared some of the same interests and personnel), a magazine published monthly, bimonthly, or quarterly throughout the 1940s and 1950s and still extant. *Partisan* was remarkable for the way it managed, despite its inception in highly political circumstances and controversies in the 1930s, to maintain an extremely wide range of interests in poetry, fiction, drama, fine arts; and in Continental literature, politics, and sociological thought. Its favorite fiction writers during the postwar years were Bellow (parts of *The Adventures of Augie March* and all of *Seize the Day* appeared there), Bernard Malamud, Mailer (occasionally), and Delmore Schwartz. But *Partisan* also looked beyond urban-Jewish writing, publishing stories by Flannery O'Connor and James Baldwin and the prologue to Ellison's *Invisible Man*. Among the distinguished critics who contributed steadily to the magazine over these years on a variety of literary and political subjects, one should mention Lionel Trilling—whose *The Liberal Imagination* (1948) was perhaps the most widely read and influential of "New York" critical works—as well as Philip Rahv, Irving Howe, Elizabeth Hardwick, and Diana Trilling. Mary McCarthy reviewed plays and published some of her fiction there, and Hannah Arendt's political writing and Clement Greenberg's art criticism were regular features of the journal.

By 1960 regional and ethnic senses of identity became diluted as the various parts of America began more and more to resemble each other. This cultural dilution can be observed at work in some of the "assimilated" Jews depicted in Philip Roth's *Goodbye, Columbus* (1959) or in the southerners who inhabit Walker Percy's novels generally. To be a southern novelist in 1960 was no more exotic than to be a Jewish novelist. At about that time *Partisan Review* began to lose its distinctiveness, especially after 1963, when during a New York City newspaper strike a new magazine was formed, *The New York Review of Books*. This organ, while making use of many of the *Partisan* writers, appeared biweekly rather than quarterly and was able to give essayists and reviewers as much room as they desired. Its striking success has continued to the present day.

THE "TRANQUILLIZED FIFTIES"

The subheading is taken from *Memories of West Street and Lepke*, a poem Robert Lowell published in 1958 during the years of Eisenhower's presidency (1953–61) when

. . . even the man
scavenging filth in the back alley trash cans,
has two children, a beach wagon, a helpmate,
and is a "young Republican."

They were years when attention to serious public matters probably meant attention to the cold war, and what was to be done about the real or mythical communists whom Senator Joseph McCarthy was crusading against, or whether a ban on nuclear testing could be accomplished before the United States and the Soviet Union contrived to blow up the world. At home, although there was a famously named "recession" (a Republican word brought into play so as not to recall the great economic depression of 1929 and beyond), the majority of Americans were employed, paid better than they had ever been paid, and exposed to an ever-increasing number of household "labor-saving devices" and automotive finery like the beach wagon of Lowell's poem. Citizens were generally encouraged to think well of their country, while poor people largely remained, in a phrase coined by the socialist critic Michael Harrington, "Our Invisible Poor." And although there was a flurry of confrontations over the Supreme Court's decision in 1954 that public schools could not be maintained on a segregated, "separate but equal" basis, the dramatic beginning to the civil rights struggles which were to mark the 1960s took place largely in the South and was mainly unremarked by novelists preoccupied with other matters.

For all the abuse the 1950s have received as a success-oriented, socially and ecologically irresponsible, fearfully smug decade, they look in retrospect (though they were surely not felt as such at the time) to have been a good time for serious American writers; and not the less so for the individual writer's assurance that he could not possibly be appreciated nor understood by a philistine and materialistic nation run by businessmen, generals, and golfers. So the "deep and subtle alienation" Kazin spoke of was also seen as a necessary, sometimes even an attractive, condition, which furnished a rich vein for fictional exploration.

OTHER CRITICISMS OF AMERICA

The most stunning fictional success of the 1950s was J. D. Salinger's *The Catcher in the Rye*, a book with a young hero who is out of step with the educational, commercial, and sexual customs of adult society and able to see through and expose its falsities and "phoniness." For a time in the late 1950s and early '60s, one could assume that all college students had read this book, as attested by their "identification" with Holden Caulfield, who summed up and rendered both poignant and funny their own presumed predicaments, enabling them to feel better about being unhappy, even to cherish the illusion that it was virtuous to be so. By 1975 this slippery and astonishing book, though never on the best-seller lists for any one year, had sold nearly six million copies.

Readers of Salinger in the 1950s were probably also reading sociological studies like David Riesman's *The Lonely Crowd* (1950), which classified American behavior as mainly "other-directed," a condition where the accepted standards and criteria for behavior came from outside the self rather than from within. Or they were learning about the "gray-flannel suit" ethos by reading Vance Packard's popular reporting or Sloan Wilson's forgotten best seller, *The Man in the Gray Flannel Suit* (1955). At a more polemical and radical level, a smaller, predominantly university audience read the social criticism of C. Wright Mills in *White Collar* (1951) and *The Power Elite* (1956). There was also much general interest in the popularized psychological, religious, or philosophical movements that surfaced after World War II. The meaning of Existentialism, as imported from France and abstracted from the novels of Albert Camus and Jean-Paul Sartre or from Mailer's novels and essays, was solemnly discussed by those in search of a post-Christian, philosophically respectable attitude toward life. There was much interest in the psychological theories of Erich Fromm or Erik Erikson or, at a lower level, in the "self-help" best sellers like Norman Vincent Peale's *The Power of Positive Thinking* (1952), which sold five million copies. One could even decide that all of Western thinking was at fault and so turn to the East and learn about Zen Buddhism through books by Alan Watts or D. T. Suzuki. The stories of Salinger increasingly featured young people unable to talk to their parents or accept "the American way of life," who became prey to extreme versions of experience, sometimes mystical and sometimes suicidal.

As the 1950s drew to a close, these criticisms of American institutional and social styles over the decade grew more extreme and more specific in the hands of radical critics. Mailer's pamphlet *The White Negro* (1957) defined the American world as essentially "square," stultifying, sickly, infected with pieties and timidities, obedient to authoritarian voices. As a cure for this sickness he proposed a "new breed of adventurer" named "the hipster" who would live according to what Mailer termed the black man's code of "ever-threatening danger" and who would dramatically confront and reject the "square" world. At a no less passionate level, though less violent in its emphasis, was Paul Goodman's *Growing Up Absurd* (1959), the most forceful of many books he wrote in opposition to the technological America he saw around him. Goodman pushed for a return to decentralized, Jeffersonian principles of social organization, to smaller cars, and to meaningful work; his plea especially focused on the "kids" (as he engagingly called them) who needed jobs more worthwhile than making tailfins for America's enormous late-Fifties automobiles if they were to respect either their country or themselves.

But the most publicized literary expression of disaffection with "official" American life was made by the "Beat" writers, of whom Jack Kerouac was the prose laureate. Advertised as "The Beat Generation" and known also, more accurately, as the San Francisco school, the group included most notably the poets Allen Ginsberg and Gregory Corso. Its influence radiated from the City Lights Bookstore run by Lawrence Ferlinghetti, who was himself a poet. Beat poets sometimes read their works to a jazz accompaniment; among their objects of veneration were Whitman, Buddha and Eastern religions generally, and (in Kerouac's case at least) large quantities of Western beer. Their experiments with drugs anticipated the more drastic and often disastrous use of them in the 1960s. The Beats were in favor of "spontaneity" and against constricting forms, poetic or political; indeed, Kerouac proved it was possible to let oneself go and write a novel (*The Subterraneans*, 1958) in three nights. Briefly, the Beat writers constituted a challenge to the many carefully worked-over lyrics or ingeniously worked-out novels which had been characteristic products of the 1950s. They were also good at clowning, and the comic touches that dot their work are probably the parts which will prove most enduring.

LETTING GO IN THE 1960S

The first years of the Sixties decade seemed truly a time when new possibilities and opportunities presented themselves on both public and private levels. The election of John F. Kennedy to the presidency brought to Washington a glamorous and humorous leader who was thought to be committed both to social justice and to culture—perhaps even bringing in the "new Augustan age of poetry and power" Robert Frost wrote about in his inaugural poem. The agreement with Russia to cease nuclear testing in the atmosphere; the increasing concern for changes in the relationship between whites and blacks; the loosening up of sexual codes and of official censorship, coincident with the marketing of an effective new oral contraceptive—these and other events seemed in the minds of some to promise a more life-affirming, less restrictive era than the preceding one. With particular regard to the matter of censorship, it may be noted that for many years Americans who wished to read Henry Miller's novels had to smuggle in from Paris their copies of *Tropic of Cancer* or *Tropic of Capricorn*, while in the mid-1950s Vladimir Nabokov's *Lolita* had to be obtained in a similar way. When in 1956, however, *Lolita* was first published in this country, there was no legal prosecution; and in 1959 the successful publication by Grove Press of D. H. Lawrence's *Lady Chatterley's Lover* cleared the way for novels of more explicit sexual reference. When Mailer published *The Naked and the Dead* in 1948, a well-known four-letter word had to be spelled "fug," evidently to spare the delicate sensibilities of its readers. By 1959, when he published his comic tour de force *The Time of Her Time*, about a sexual warrior's candidly explicit adventures with women in his Greenwich Village loft, hardly anyone raised an eyebrow.

The novelists also seemed, as in the title of Philip Roth's first novel in 1962, to be "letting go" by indulging to the fullest their verbal, storytelling propensities. Between 1961 and 1963 four novels were published (three of them first efforts) by writers whose interest was in the active, exuberant exploration of fantasies, of extremities of experience, and of comic modes which would later on come to be known as "black humor." These books, like many which followed them later in the decade, turned their back on ordinary experience, on the mundane continuities of existence in small town or big city which American writers had so devotedly investigated over the first fifty-odd years of the century. John Barth's *The Sot-Weed Factor* (1961), Joseph Heller's *Catch-22* (1961), Ken Kesey's *One Flew over the Cuckoo's Nest* (1962), and most bizarrely Thomas Pynchon's *V* (1963) all attested to the fact that no verbal resource, no gimmick or extravagance of style need be rejected in the pursuit of putting on a brilliantly entertaining performance. These novels (Kesey's perhaps excluded) were highly self-conscious, parodied other literary styles, and built those parodies into their individually odd designs, at the same time as they were ironic and playful about their own fictional assumptions.

The 1960s were also to see a corresponding "liberation" from official standards of correctness in the realm of the journalistic essay, or—as everyone who wrote in that mode called it—a "piece." Norman Podhoretz noted that everyone he knew was engaged in writing lively essays instead of laboring over novels and poems, and there were many collections of such pieces on subjects ranging from the Beat Generation phenomenon to the trial of Adolf Eichmann. Mailer's *Advertisements for Myself* (1959), the father of the mode, had shown how unconventional and various a book of essays—which also included stories, newspaper columns, interviews—might be, and he produced two more such books in the first half of the new decade. In the hands of Tom Wolfe, style became something to cultivate and exaggerate; the subject of his "New Journalism" might be the doings of a racing-car star or a New York disco celebrity, but it didn't matter since it was merely there for the style to perform upon. A young critic, Susan Sontag, entitled a group of her essays *Against Interpretation* (1966) and made the case for more playful, aesthetically oriented responses to both life and art. That the title essay was originally published in *Partisan Review*, a serious "high culture" periodical, suggested that the winds were changing; as did the fact that literary critics like Richard Poirier or Benjamin DeMott were to be observed writing full-dress "pieces" on listening to the Beatles or on the morals of *Playboy* magazine.

SATIRIC PERFORMANCES

The relation of such expansive and experimental writing to the major public disaster of this time—the assassination of John Kennedy in November 1963—we are not in a position to determine; nor can we determine the effect on literary modes of such events as the following: the increasingly desperate adventure in Vietnam; the riots in the black ghettos of our decaying cities; the turmoil in the universities consequent on the war; the murder of Robert Kennedy and Martin Luther King; the omnipresence of drugs, hard and soft; the rise of pornography as a feature of the sexual "revolution"; the decline of "the family" and the exacerbations in the relations between men and women pointed up at the decade's end by the women's movement and documented in the divorce statistics. Taken together or singly, these severe dislocations proved to be unavailable for writers to deal with in the representational and realistic modes of portrayal handed down to them by novelists such as Howells, Dreiser, or Dos Passos.

Although satire has been a traditional way of dealing with disasters or upheavals, what sort of "satire" could possibly be adequate to matter which seemed beyond the reach of even a gifted writer's words? The middle and late 1960s saw the term "black humor" employed as a tent to cover any literary creation which played fast and loose with ordinary values and standards, frequently employing elements of cruelty and shock to make us see the awful, the ugly, the "sick" in a new way, for what it was. The great humorist in this line—he was a moralist as well—may turn out to have been not a novelist at all but a stand-up comedian, Lenny Bruce, whose violent and obscene rehearsals of clichés in language and in American life gave novelists something to live up to. His was an art of solo performance, dependent (as Frost once said all poetic performance must be) on the prowess and feats of association Bruce's verbal and auditory powers were capable of.

Similar displays by the writer as satiric performer may be viewed in the works of major novelists and prose entertainers from the later 1960s and early '70s: in Mailer's speech to Lyndon Johnson (in 1965) urging him, with as much obscenity and crude familiarity as the lecturer could display, to get us out of Vietnam immediately; in the eloquent pleadings and threatening lashing of James Baldwin's attempts to make whites see and accept blacks; in the daffy brilliance of Pynchon's language throughout *The Crying of Lot 49* (1966); or in the forceful charm with which Roth as comedian came up with one amusing routine after another in *Portnoy's Complaint* (1969). The major performer of them all was undoubtedly Vladimir Nabokov, who in a series of startling novels projected his comic fantasies in a style by turns antic, icy, and weird. When at the end of the decade two of our best, but more conservative, novelists tried to render critically, in vivid detail but in a more traditional narrative, their reservations about American life in the late 1960s, they offended some readers by acting as if such a rendering by an individual imagination could still be made. Bellow's *Mr. Sammler's Planet* (1970) and John Updike's *Rabbit Redux* (1971) remain interestingly combative and tendentious views of the "liberations" of a just-ended decade.

SUMMARY: OLD AND NEW WRITERS IN THE 1970S

In retrospect the major American novelists of the past thirty years, if one agrees to list only five, are arguably Bellow, Mailer, Nabokov, Updike, and Pynchon; opposites in most things except verbal brilliance but likely candidates for positions near the top of any list. But lists are dangerous as well as provocative, and it is more important to note that the dreary cry lamenting the novel's decline is now seldom heard; rather one is impressed by the number of good novels and stories that have been written since the end of the Vietnam war. The women's movement, which became a powerful cultural fact at the end of the 1960s, undoubtedly played an important part in creating a situation where more women than ever in American history are writing fiction, memoir, cultural and social criticism in many distinctive styles and from many points of view. No single female imagination has clearly made its mark as a major one of the stature of older writers like Porter, Welty, O'Connor, or Mary McCarthy; although such a claim could plausibly be made for the work of Susan Sontag, less plausibly for Joyce Carol Oates. But it may be inadvisable to divide up writers according to sex; we can at least observe that new subjects, new materials, for fiction and memoir have been opened up, while new arguments about whether there is or should be a "feminist" criticism of literature have contributed to an enlivening of the literary situation generally.

Much of the vitality in prose writing during the years since the war is found in "nonfiction," as can be seen in this volume by examples from figures so disparate as Mailer, Updike, and Baldwin. Concurrently, many of the best novels from the last thirty or so years are less severe or difficult, more relaxed about their own status as "art," less certain that their only task is to aspire to the condition of a masterpiece—as the great novels by modernist writers like Joyce and Faulkner surely aspired. It is a fact that Pynchon's massive *Gravity's Rainbow*, a very difficult book indeed, has had much impact on college and university readers and teachers; to what extent, however, it will become a book one rereads with pleasure, rather than plows through once as a curiosity, is very much an open question. But generally the jokier, loudly "far-out" or "trip" novels of the late 1960s have given way to more substantial blends of fantasy and realism, as in John Gardner's *The Sunlight Dialogues* (1972) and *October Light* (1976), E. L. Doctorow's *Ragtime* (1975), and Coover's *The Public Burning* (1977). In Gardner's feel for regional speech and behavior, in Robert Stone's ability (in *Dog Soldiers*, 1975) to express post-Vietnam attitudes through dialogue both funny and scarifying, in Paul Theroux's striking comic talents, we see how some of our best new novelists are both original and traditional in their conceptions. These writers and many others must be saluted by a list commemorating older writers, unjustly neglected ones, and ones just beginning to be noticed: William Styron, James Gould Cozzens, J. F. Powers, John Cheever, Evan Connell, Jr., Thomas Savage, Tillie Olsen, Grace Paley, Frederick Buechner, Maureen Howard, Richard Yates, Alison Lurie, Diane Johnson, James Alan McPherson, Cynthia Seton, Henry Bromell, Toni Morrison, Theodore Weesner, Ann Beattie. America is a large country.

VLADIMIR NABOKOV
1899–1977

In Vladimir Nabokov's brilliant memoir *Speak, Memory*, he describes the composition of chess problems as "a beautiful, complex, and sterile art" related to the game itself "only insofar as, say the properties of a sphere are made use of both by a juggler in weaving a new act and by a tennis player in winning a tournament." Nabokov's art of fiction is related to the behavior of ordinary novels no less ambiguously. The best of his many books teaches us how fictional composition may be beautiful, complex, yet avoid sterility because of its fascinated attachment to human life as traced in the actions of an exiled, perverse, and doomed man—inevitably an artist.

Although he did not emigrate to the United States until 1940 (and did not begin writing novels in English until 1938), Nabokov has given us over the past fifty years a rich and ample body of work: novels, memoir, stories, poems, a book on the Russian novelist Gogol, and a translation of the Russian poet Pushkin's *Eugene Onegin*. Like the heroes of his fiction, he has lived in many places. Born in Czarist Russia (St. Petersburg) he moved with his family after the Revolution to London and Berlin, was educated at Trinity College, Cambridge, and lived in Berlin until the Nazis assumed power. After a time in Paris he came to the States, teaching and lecturing during the 1940s and '50s at Wellesley and Cornell. In 1960 he returned to Europe, establishing residence at the Palace Hotel in Montreux, Switzerland, where he eventually died. He rejected numerous invitations to return to America for lectures or residence at a university, preferring instead to pursue, in less academic circumstances, his lifetime avocation as a lepidopterist: "My pleasures are the most intense known to man: writing and butterfly hunting," as he once put it.

In one of his many interviews Nabokov announced that he "was a perfectly normal trilingual child in a family with a large library," a remark which perfectly expresses the imperious charm and sardonically eloquent tenor of his style. Over the past two decades he was partly occupied with turning into English the novels written in Russian during the 1920s and '30s, of which the two most interesting are probably *The Defence*, a poignant account of a doomed chess genius, and *Laughter in the Dark*, a painfully comic fable. Of the novels originally written in English, *Pnin* stands today as most immediately appealing, and is an excellent way to begin reading Nabokov. Its first chapter, printed here, is typical of the way his creator manages to view Timofey Pnin, an extraordinarily passionate and incompetent exile who finds himself teaching in an American college, with fondness and irony. In *Pnin* and in the much more complicated *Pale Fire* five years later, Nabokov gives us marvelously incisive scenes from academic life, as well as the look and feel of lawn, houses, streets in American small towns. Mary McCarthy insisted that *Pale Fire* is "one of the very great works of art of this century"; it is surely one of the most ingenious ones. Cast in the form of a poem, supposedly written by a dead poet named John Shade, then explicated and commented upon by the narrator, Charles Kinbote, until its meanings overwhelm the explicator, the book still manages to remain anchored to a realistic and vibrant sense of place, while giving Nabokov's superb gifts as a parodist their freest reign.

But it is to *Lolita* that appreciation of Nabokov's art most strongly directs itself. Originally published by the Olympia Press in Paris (1955), and until 1958 surreptitiously smuggled into this country as a "dirty book" (Olympia had a large list of pornography), the novel made Nabokov's reputation, indeed his notoriety. Narrated by the most famous of his doomed and duped lovers, Humbert Humbert as he usually calls himself, the book crackles with sustained inventive life, vigorous wordplay, declamation, confession, mock-confession, soliloquy, nonce poems, gibberish, outrageous puns. It is also immensely and entertainingly readable. But its most remarkable achievement is to make us "pity the monsters" (in Robert Lowell's phrase from a poem) by taking to our hearts its repulsive hero—as he himself assures us he is—with eventual sympathy and even love. We can do this only because Nabokov provides Humbert with a language dazzlingly alive to the possibilities of life and to its inevitable fadings, changings, dyings.

Lolita is Nabokov's most eloquent tribute to the power of art, but that power is charmingly saluted in some lines from John Shade's poem in *Pale Fire*, where we hear, clearly, the voice of his author:

> I feel I understand
> Existence, or at least a minute part
> Of my existence, only through my art,
> In terms of combinational delight;
> And if my private universe scans right,
> So does the verse of galaxies divine
> Which I suspect is an iambic line.

Even when, as with the very long novel *Ada* (1969) or with his last fictional efforts, the result seems labored, it is still the "combinational delight" of words brilliantly employed that greets the reader on every page and provides both our first and our deepest pleasure in reading Nabokov.

From Pnin[1]

1

The elderly passenger sitting on the north-window side of that inexorably moving railway coach, next to an empty seat and facing two empty ones, was none other than Professor Timofey Pnin. Ideally bald, sun-tanned, and clean-shaven, he began rather impressively with that great brown dome of his, tortoise-shell glasses (masking an infantile absence of eyebrows), apish upper lip, thick neck, and strong-man torso in a tightish tweed coat, but ended, somewhat disappointingly, in a pair of spindly legs (now flanneled and crossed) and frail-looking, almost feminine feet.

His sloppy socks were of scarlet wool with lilac lozenges; his conservative black oxfords had cost him about as much as all the rest of his clothing (flamboyant goon tie[2] included). Prior to the

1. The first chapter of Nabokov's *Pnin* (1957), first published in *The New Yorker*, November 28, 1953. 2. Large loud necktie in a loutish 1940s style.

nineteen-forties, during the staid European era of his life, he had always worn long underwear, its terminals tucked into the tops of neat silk socks, which were clocked,[3] soberly colored, and held up on his cotton-clad calves by garters. In those days, to reveal a glimpse of that white underwear by pulling up a trouser leg too high would have seemed to Pnin as indecent as showing himself to ladies minus collar and tie; for even when decayed Mme. Roux, the concierge of the squalid apartment house in the Sixteenth Arrondissement of Paris—where Pnin, after escaping from Leninized Russia and completing his college education in Prague, had spent fifteen years—happened to come up for the rent while he was without his *faux col*,[4] prim Pnin would cover his front stud with a chaste hand. All this underwent a change in the heady atmosphere of the New World. Nowadays, at fifty-two, he was crazy about sunbathing, wore sport shirts and slacks, and when crossing his legs would carefully, deliberately, brazenly display a tremendous stretch of bare shin. Thus he might have appeared to a fellow passenger; but except for a soldier asleep at one end and two women absorbed in a baby at the other, Pnin had the coach to himself.

Now a secret must be imparted. Professor Pnin was on the wrong train. He was unaware of it, and so was the conductor, already threading his way through the train to Pnin's coach. As a matter of fact, Pnin at the moment felt very well satisfied with himself. When inviting him to deliver a Friday-evening lecture at Cremona—some two hundred versts[5] west of Waindell, Pnin's academic perch since 1945—the vice-president of the Cremona Women's Club, a Miss Judith Clyde, had advised our friend that the most convenient train left Waindell at 1:52 P.M., reaching Cremona at 4:17; but Pnin—who, like so many Russians, was inordinately fond of everything in the line of timetables, maps, catalogues, collected them, helped himself freely to them with the bracing pleasure of getting something for nothing, and took especial pride in puzzling out schedules for himself—had discovered, after some study, an inconspicuous reference mark against a still more convenient train (Lv. Waindell 2:19 P.M., Ar. Cremona 4:32 P.M.); the mark indicated that Fridays, and Fridays only, the two-nineteen stopped at Cremona on its way to a distant and much larger city, graced likewise with a mellow Italian name. Unfortunately for Pnin, his timetable was five years old and in part obsolete.

He taught Russian at Waindell College, a somewhat provincial institution characterized by an artificial lake in the middle of a landscaped campus, by ivied galleries connecting the various halls, by murals displaying recognizable members of the faculty in the act of passing on the torch of knowledge from Aristotle, Shakespeare, and Pasteur to a lot of monstrously built farm boys and farm girls,

---

3. Ornamented with patterns on the side.
4. Detachable collar. "Stud": ornamental fastener for dress shirt used in place of buttons.
5. An Old Russian unit of measurement, the verst was .66 miles.

and by a huge, active, buoyantly thriving German Department which its Head, Dr. Hagen, smugly called (pronouncing every syllable very distinctly) "a university within a university."

In the Fall Semester of that particular year (1950), the enrollment in the Russian Language courses consisted of one student, plump and earnest Betty Bliss, in the Transitional Group, one, a mere name (Ivan Dub, who never materialized) in the Advanced, and three in the flourishing Elementary: Josephine Malkin, whose grandparents had been born in Minsk; Charles McBeth, whose prodigious memory had already disposed of ten languages and was prepared to entomb ten more; and languid Eileen Lane, whom somebody had told that by the time one had mastered the Russian alphabet one could practically read "Anna Karamazov" in the original. As a teacher, Pnin was far from being able to compete with those stupendous Russian ladies, scattered all over academic America, who, without having had any formal training at all, manage somehow, by dint of intuition, loquacity, and a kind of maternal bounce, to infuse a magic knowledge of their difficult and beautiful tongue into a group of innocent-eyed students in an atmosphere of Mother Volga songs, red caviar, and tea; nor did Pnin, as a teacher, ever presume to approach the lofty halls of modern scientific linguistics, that ascetic fraternity of phonemes,[6] that temple wherein earnest young people are taught not the language itself, but the method of teaching others to teach that method; which method, like a waterfall splashing from rock to rock, ceases to be a medium of rational navigation but perhaps in some fabulous future may become instrumental in evolving esoteric dialects—Basic Basque and so forth—spoken only by certain elaborate machines. No doubt Pnin's approach to his work was amateurish and lighthearted, depending as it did on exercises in a grammar brought out by the Head of a Slavic Department in a far greater college than Waindell—a venerable fraud whose Russian was a joke but who would generously lend his dignified name to the products of anonymous drudgery. Pnin, despite his many shortcomings, had about him a disarming, old-fashioned charm which Dr. Hagen, his staunch protector, insisted before morose trustees was a delicate imported article worth paying for in domestic cash. Whereas the degree in sociology and political economy that Pnin had obtained with some pomp at the University of Prague around 1925 had become by mid-century a doctorate in desuetude, he was not altogether miscast as a teacher of Russian. He was beloved not for any essential ability but for those unforgettable digressions of his, when he would remove his glasses to beam at the past while massaging the lenses of the present. Nostalgic excursions in broken English. Autobiographical tidbits. How Pnin came to the *Soedinyonnïe Shtatï* (the United

6. Smallest unit of speech in the set of sounds that distinguish one utterance from another in a language.

States). "Examination on ship before landing. Very well! 'Nothing to declare?' 'Nothing.' Very well! Then political questions. He asks: 'Are you anarchist?' I answer"—time out on the part of the narrator for a spell of cozy mute mirth—" 'First what do we understand under "Anarchism"? Anarchism practical, metaphysical, theoretical, mystical, abstractical, individual, social? When I was young,' I say, 'all this had for me signification.' So we had a very interesting discussion, in consequence of which I passed two whole weeks on Ellis Island"—abdomen beginning to heave; heaving; narrator convulsed.

But there were still better sessions in the way of humor. With an air of coy secrecy, benevolent Pnin, preparing the children for the marvelous treat he had once had himself, and already revealing, in an uncontrollable smile, an incomplete but formidable set of tawny teeth, would open a dilapidated Russian book at the elegant leatherette marker he had carefully placed there; he would open the book, whereupon as often as not a look of the utmost dismay would alter his plastic features; agape, feverishly, he would flip right and left through the volume, and minutes might pass before he found the right page—or satisfied himself that he had marked it correctly after all. Usually the passage of his choice would come from some old and naïve comedy of merchant-class habitus rigged up by Ostrovski almost a century ago, or from an equally ancient but even more dated piece of trivial Leskovian jollity dependent on verbal contortions.[7] He delivered these stale goods with the rotund gusto of the classical Alexandrinka (a theater in Petersburg), rather than with the crisp simplicity of the Moscow Artists; but since to appreciate whatever fun those passages still retained one had to have not only a sound knowledge of the vernacular but also a good deal of literary insight, and since his poor little class had neither, the performer would be alone in enjoying the associative subtleties of his text. The heaving we have already noted in another connection would become here a veritable earthquake. Directing his memory, with all the lights on and all the masks of the mind a-miming, toward the days of his fervid and receptive youth (in a brilliant cosmos that seemed all the fresher for having been abolished by one blow of history), Pnin would get drunk on his private wines as he produced sample after sample of what his listeners politely surmised was Russian humor. Presently the fun would become too much for him; pear-shaped tears would trickle down his tanned cheeks. Not only his shocking teeth but also an astonishing amount of pink upper-gum tissue would suddenly pop out, as if a jack-in-the-box had been sprung, and his hand would fly to his mouth, while his big shoulders shook and rolled. And although the speech he smothered behind his dancing hand was now doubly unintelligible to the class, his com-

---

7. Alexander Ostrovski (1823–86), Russian playwright. Nikolai Leskov (1831– 95), Russian novelist and short-story writer.

plete surrender to his own merriment would prove irresistible. By the time he was helpless with it he would have his students in stitches, with abrupt barks of clockwork hilarity coming from Charles and a dazzling flow of unsuspected lovely laughter transfiguring Josephine, who was not pretty, while Eileen, who was, dissolved in a jelly of unbecoming giggles.

All of which does not alter the fact that Pnin was on the wrong train.

How should we diagnose his sad case? Pnin, it should be particularly stressed, was anything but the type of that good-natured German platitude of last century, *der zerstreute Professor.*[8] On the contrary, he was perhaps too wary, too persistently on the lookout for diabolical pitfalls, too painfully on the alert lest his erratic surroundings (unpredictable America) inveigle him into some bit of preposterous oversight. It was the world that was absent-minded and it was Pnin whose business it was to set it straight. His life was a constant war with insensate objects that fell apart, or attacked him, or refused to function, or viciously got themselves lost as soon as they entered the sphere of his existence. He was inept with his hands to a rare degree; but because he could manufacture in a twinkle a one-note mouth organ out of a pea pod, make a flat pebble skip ten times on the surface of a pond, shadowgraph with his knuckles a rabbit (complete with blinking eye), and perform a number of other tame tricks that Russians have up their sleeves, he believed himself endowed with considerable manual and mechanical skill. On gadgets he doted with a kind of dazed, superstitious delight. Electric devices enchanted him. Plastics swept him off his feet. He had a deep admiration for the zipper. But the devoutly plugged-in clock would make nonsense of his mornings after a storm in the middle of the night had paralyzed the local power station. The frame of his spectacles would snap in mid-bridge, leaving him with two identical pieces, which he would vaguely attempt to unite, in the hope, perhaps, of some organic marvel of restoration coming to the rescue. The zipper a gentleman depends on most would come loose in his puzzled hand at some nightmare moment of haste and despair.

And he still did not know that he was on the wrong train.

A special danger area in Pnin's case was the English language. Except for such not very helpful odds and ends as "the rest is silence," "nevermore," "weekend," "who's who," and a few ordinary words like "eat," "street," "fountain pen," "gangster," "Charleston," "marginal utility," he had had no English at all at the time he left France for the States. Stubbornly he sat down to the task of learning the language of Fenimore Cooper, Edgar Poe, Edison, and thirty-one Presidents. In 1941, at the end of one year of study, he was proficient enough to use glibly terms like "wishful thinking"

8. The absent-minded professor.

and "okey-dokey." By 1942 he was able to interrupt his narration with the phrase, "To make a long story short." By the time Truman entered his second term, Pnin could handle practically any topic; but otherwise progress seemed to have stopped despite all his efforts, and by 1950 his English was still full of flaws. That autumn he supplemented his Russian courses by delivering a weekly lecture in a so-called symposium ("Wingless Europe: A Survey of Contemporary Continental Culture") directed by Dr. Hagen. All our friend's lectures, including sundry ones he gave out of town, were edited by one of the younger members of the German Department. The procedure was somewhat complicated. Professor Pnin laboriously translated his own Russian verbal flow, teeming with idiomatic proverbs, into patchy English. This was revised by young Miller. Then Dr. Hagen's secretary, a Miss Eisenbohr, typed it out. Then Pnin deleted the passages he could not understand. Then he read it to his weekly audience. He was utterly helpless without the prepared text, nor could he use the ancient system of dissimulating his infirmity by moving his eyes up and down—snapping up an eyeful of words, reeling them off to his audience, and drawing out the end of the sentence while diving for the next. Pnin's worried eye would be bound to lose its bearings. Therefore he preferred reading his lectures, his gaze glued to his text, in a slow, monotonous baritone that seemed to climb one of those interminable flights of stairs used by people who dread elevators.

The conductor, a gray-headed fatherly person with steel spectacles placed rather low on his simple, functional nose and a bit of soiled adhesive tape on his thumb, had now only three coaches to deal with before reaching the last one, where Pnin rode.

Pnin in the meantime had yielded to the satisfaction of a special Pninian craving. He was in a Pninian quandary. Among other articles indispensable for a Pninian overnight stay in a strange town, such as shoe trees, apples, dictionaries, and so on, his Gladstone bag contained a relatively new black suit he planned to wear that night for the lecture ("Are the Russian People Communist?") before the Cremona ladies. It also contained next Monday's symposium lecture ("Don Quixote and Faust"), which he intended to study the next day, on his way back to Waindell, and a paper by the graduate student, Betty Bliss ("Dostoevski and Gestalt Psychology"), that he had to read for Dr. Hagen, who was her main director of cerebration. The quandary was as follows: If he kept the Cremona manuscript—a sheaf of typewriter-size pages, carefully folded down the center—on his person, in the security of his body warmth, the chances were, theoretically, that he would forget to transfer it from the coat he was wearing to the one he would wear. On the other hand, if he placed the lecture in the pocket of the suit in the bag *now*, he would, he knew, be tortured by the possibility of his luggage being stolen. On the third hand (these mental states sprout additional forelimbs all the time), he carried in the inside pocket of his present coat a precious wallet with two ten-dollar bills, the newspaper clipping of a letter he had written, with my help, to the New York Times in 1945 anent the Yalta conference,[9] and his certificate of naturalization; and it was physically possible to pull out the wallet, if needed, in such a way as fatally to dislodge the folded lecture. During the twenty minutes he had been on the train, our friend had already opened his bag twice to play with his various papers. When the conductor reached the car, diligent Pnin was perusing with difficulty Betty's last effort, which began, "When we consider the mental climate wherein we all live, we cannot but notice——"

The conductor entered; did not awake the soldier; promised the women he would let them know when they would be about to arrive; and presently was shaking his head over Pnin's ticket. The Cremona stop had been abolished two years before.

"Important lecture!" cried Pnin. "What to do? It is a cata-stroph!"

Gravely, comfortably, the gray-headed conductor sank into the opposite seat and consulted in silence a tattered book full of dog-eared insertions. In a few minutes, namely at 3:08, Pnin would have to get off at Whitchurch; this would enable him to catch the four-o'clock bus that would deposit him, around six, at Cremona.

"I was thinking I gained twelve minutes, and now I have lost nearly two whole hours," said Pnin bitterly. Upon which, clearing his throat and ignoring the consolation offered by the kind gray-head ("You'll make it."), he took off his reading glasses, collected his stone-heavy bag, and repaired to the vestibule of the car so as to wait there for the confused greenery skimming by to be cancelled and replaced by the definite station he had in mind.

2

Whitchurch materialized as scheduled. A hot, torpid expanse of cement and sun lay beyond the geometrical solids of various clean-cut shadows. The local weather was unbelievably summery for October. Alert, Pnin entered a waiting room of sorts, with a need-less stove in the middle, and looked around. In a solitary recess, one could make out the upper part of a perspiring young man who was filling out forms on the broad wooden counter before him.

"Information, please," said Pnin. "Where stops four-o'clock bus to Cremona?"

"Right across the street," briskly answered the employee without looking up.

"And where possible to leave baggage?"

"That bag? I'll take care of it."

And with the national informality that always nonplused Pnin,

9. Roosevelt, Churchill, and Stalin met in February, 1945, at Yalta in the Crimea to plan how Europe would be admin- istered by the Allies after Germany was defeated in World War II.

the young man shoved the bag into a corner of his nook.

"Quittance?" queried Pnin, Englishing the Russian for "receipt" (*kvitantsiya*).

"What's that?"

"Number?" tried Pnin.

"You don't need a number," said the fellow, and resumed his writing.

Pnin left the station, satisfied himself about the bus stop, and entered a coffee shop. He consumed a ham sandwich, ordered another, and consumed that too. At exactly five minutes to four, having paid for the food but not for an excellent toothpick which he carefully selected from a neat little cup in the shape of a pine cone near the cash register, Pnin walked back to the station for his bag.

A different man was now in charge. The first had been called home to drive his wife in all haste to the maternity hospital. He would be back in a few minutes.

"But I must obtain my valise!" cried Pnin.

The substitute was sorry but could not do a thing.

"It is there!" cried Pnin, leaning over and pointing.

This was unfortunate. He was still in the act of pointing when he realized that he was claiming the wrong bag. His index finger wavered. That hesitation was fatal.

"My bus to Cremona!" cried Pnin.

"There is another at eight," said the man.

What was our poor friend to do? Horrible situation! He glanced streetward. The bus had just come. The engagement meant an extra fifty dollars. His hand flew to his right side. *It* was there, *slava Bogu* (thank God)! Very well! He would not wear his black suit—*vot i vsyo* (that's all). He would retrieve it on his way back. He had lost, dumped, shed many more valuable things in his day. Energetically, almost lightheartedly, Pnin boarded the bus.

He had endured this new stage of his journey only for a few city blocks when an awful suspicion crossed his mind. Ever since he had been separated from his bag, the tip of his left forefinger had been alternating with the proximal edge of his right elbow in checking a precious presence in his inside coat pocket. All of a sudden he brutally yanked it out. It was Betty's paper.

Emitting what he thought were international exclamations of anxiety and entreaty, Pnin lurched out of his seat. Reeling, he reached the exit. With one hand the driver grimly milked out a handful of coins from his little machine, refunded him the price of the ticket, and stopped the bus. Poor Pnin landed in the middle of a strange town.

He was less strong than his powerfully puffed-out chest might imply, and the wave of hopeless fatigue that suddenly submerged his topheavy body, detaching him, as it were, from reality, was a sensation not utterly unknown to him. He found himself in a damp, green, purplish park, of the formal and funereal type, with the stress laid on somber rhododendrons, glossy laurels, sprayed shade trees and closely clipped lawns; and hardly had he turned into an alley of chestnut and oak, which the bus driver had curtly told him led back to the railway station, then that eerie feeling, that tingle of unreality overpowered him completely. Was it something he had eaten? That pickle with the ham? Was it a mysterious disease that none of his doctors had yet detected? My friend wondered, and I wonder, too.

I do not know if it has ever been noted before that one of the main characteristics of life is discreteness. Unless a film of flesh envelops us, we die. Man exists only insofar as he is separated from his surroundings. The cranium is a space-traveler's helmet. Stay inside or you perish. Death is divestment, death is communion. It may be wonderful to mix with the landscape, but to do so is the end of the tender ego. The sensation poor Pnin experienced was something very like that divestment, that communion. He felt porous and pregnable. He was sweating. He was terrified. A stone bench among the laurels saved him from collapsing on the sidewalk. Was his seizure a heart attack? I doubt it. For the nonce I am his physician, and let me repeat, I doubt it. My patient was one of those singular and unfortunate people who regard their heart ("a hollow, muscular organ," according to the gruesome definition in *Webster's New Collegiate Dictionary*, which Pnin's orphaned bag contained) with a queasy dread, a nervous repulsion, a sick hate, as if it were some strong slimy untouchable monster that one had to be parasitized with, alas. Occasionally, when puzzled by his tumbling and tottering pulse, doctors examined him more thoroughly, the cardiograph outlined fabulous mountain ranges and indicated a dozen fatal diseases that excluded one another. He was afraid of touching his own wrist. He never attempted to sleep on his left side, even in those dismal hours of the night when the insomniac longs for a third side after trying the two he has.

And now, in the park of Whitchurch, Pnin felt what he had felt already on August 10, 1942, and February 15 (his birthday), 1937, and May 18, 1929, and July 4, 1920—that the repulsive automaton he lodged had developed a consciousness of its own and not only was grossly alive but was causing him pain and panic. He pressed his poor bald head against the stone back of the bench and recalled all the past occasions of similar discomfort and despair. Could it be pneumonia this time? He had been chilled to the bone a couple of days before in one of those hearty American drafts that a host treats his guests to after the second round of drinks on a windy night. And suddenly Pnin (was he dying?) found himself sliding back into his own childhood. This sensation had the sharpness of retrospective detail that is said to be the dramatic privilege of drowning individuals, especially in the former Russian Navy—a phenomenon of suffocation that a veteran psychoanalyst, whose name escapes me, has explained as being the subconsciously evoked shock of one's baptism which causes an explosion of intervening recollections between the first immersion and the last. It all happened in a flash but there is no way of rendering it in less than so many consecutive words.

Pnin came from a respectable, fairly well-to-do, St. Petersburg family. His father, Dr. Pavel Pnin, an eye specialist of considerable repute, had once had the honor of treating Leo Tolstoy for a case of conjunctivitis. Timofey's mother, a frail, nervous little person with a waspy waist and bobbed hair, was the daughter of the once famous revolutionary Umov (rhymes with "zoom off") and of a German lady from Riga. Through his half swoon, he saw his mother's approaching eyes. It was a Sunday in midwinter. He was eleven. He had been preparing lessons for his Monday classes at the First Gymnasium[1] when a strange chill pervaded his body. His mother took his temperature, looked at her child with a kind of stupefaction, and immediately called her husband's best friend, the pediatrician Belochkin. He was a small, beetle-browed man, with a short beard and cropped hair. Easing the skirts of his frock coat, he sat down on the edge of Timofey's bed. A race was run between the doctor's fat golden watch and Timofey's pulse (an easy winner). Then Timofey's torso was bared, and to it Belochkin pressed the icy nudity of his ear and the sandpapery side of his head. Like the flat sole of some monopode, the ear ambulated all over Timofey's back and chest, gluing itself to this or that patch of skin and stomping on to the next. No sooner had the doctor left than Timofey's mother and a robust servant girl with safety pins between her teeth encased the distressed little patient in a strait-jacket-like compress. It consisted of a layer of soaked linen, a thicker layer of absorbent cotton, and another of tight flannel, with a sticky diabolical oilcloth—the hue of urine and fever—coming between the clammy pang of the linen next to his skin and the excruciating squeak of the cotton around which the outer layer of flannel was wound. A poor cocooned pupa, Timosha (Tim) lay under a mass of additional blankets; they were of no avail against the branching chill that crept up his ribs from both sides of his frozen spine. He could not close his eyes because his eyelids stung so. Vision was but oval pain with oblique stabs of light; familiar shapes became the breeding places of evil delusions. Near his bed was a four-section screen of polished wood, with pyrographic designs representing a bridle path felted with fallen leaves, a lily pond, an old man hunched up on a bench, and a squirrel holding a reddish object in its front paws. Timosha, a methodical child, had often wondered what that object could be (a nut? a pine cone?), and now that he had nothing else to do, he set himself to solve this dreary riddle, but the fever that hummed in his head drowned every effort in pain and panic. Still more oppressive was his tussle with the wallpaper. He had always been able to see

1. European public school.

that in the vertical plane a combination made up of three different clusters of purple flowers and seven different oak leaves was repeated a number of times with soothing exactitude; but now he was bothered by the undismissable fact that he could not find what system of inclusion and circumscription governed the horizontal recurrence of the pattern; that such a recurrence existed was proved by his being able to pick out here and there, all along the wall from bed to wardrobe and from stove to door, the reappearance of this or that element of the series, but when he tried traveling right or left from any chosen set of three inflorescences and seven leaves, he forthwith lost himself in a meaningless tangle of rhododendron and oak. It stood to reason that if the evil designer—the destroyer of minds, the friend of fever—had concealed the key of the pattern with such monstrous care, that key must be as precious as life itself and, when found, would regain for Timofey Pnin his everyday health, his everyday world; and this lucid—alas, too lucid—thought forced him to persevere in the struggle.

A sense of being late for some appointment as odiously exact as school, dinner, or bedtime added the discomfort of awkward haste to the difficulties of a quest that was grading into delirium. The foliage and the flowers, with none of the intricacies of their warp disturbed, appeared to detach themselves in one undulating body from their pale-blue background which, in its turn, lost its papery flatness and dilated in depth till the spectator's heart almost burst in response to the expansion. He could still make out through the autonomous garlands certain parts of the nursery more tenacious of life than the rest, such as the lacquered screen, the gleam of a tumbler, the brass knobs of his bedstead, but these interfered even less with the oak leaves and rich blossoms than would the reflection of an inside object in a windowpane with the outside scenery perceived through the same glass. And although the witness and victim of these phantasms was tucked up in bed, he was, in accordance with the twofold nature of his surroundings, simultaneously seated on a bench in a green and purple park. During one melting moment, he had the sensation of holding at last the key he had sought; but, coming from very far, a rustling wind, its soft volume increasing as it ruffled the rhododendrons—now blossomless, blind—confused whatever rational pattern Timofey Pnin's surroundings had once had. He was alive and that was sufficient. The back of the bench against which he still sprawled felt as real as his clothes, or his wallet, or the date of the Great Moscow Fire—1812.

A gray squirrel sitting on comfortable haunches on the ground before him was sampling a peach stone. The wind paused, and presently stirred the foliage again.

The seizure had left him a little frightened and shaky, but he argued that had it been a real heart attack, he would have surely felt a good deal more unsettled and concerned, and this roundabout piece of reasoning completely dispelled his fear. It was now four-twenty. He blew his nose and trudged to the station.

The initial employee was back. "Here's your bag," he said cheerfully. "Sorry you missed the Cremona bus."

"At least"—and what dignified irony our unfortunate friend tried to inject into that "at least"—"I hope everything is good with your wife?"

"She'll be all right. Have to wait till tomorrow, I guess."

"And now," said Pnin, "where is located the public telephone?"

The man pointed with his pencil as far out and sideways as he could without leaving his lair. Pnin, bag in hand, started to go, but he was called back. The pencil was now directed streetward.

"Say, see those two guys loading that truck? They're going to Cremona right now. Just tell them Bob Horn sent you. They'll take you."

3

Some people—and I am one of them—hate happy ends. We feel cheated. Harm is the norm. Doom should not jam. The avalanche stopping in its tracks a few feet above the cowering village behaves not only unnaturally but unethically. Had I been reading about this mild old man, instead of writing about him, I would have preferred him to discover, upon his arrival to Cremona, that his lecture was not this Friday but the next. Actually, however, he not only arrived safely but was in time for dinner—a fruit cocktail, to begin with, mint jelly with the anonymous meat course, chocolate syrup with the vanilla ice cream. And soon afterwards, surfeited with sweets, wearing his black suit, and juggling three papers, all of which he had stuffed into his coat so as to have the one he wanted among the rest (thus thwarting mischance by mathematical necessity), he sat on a chair near the lectern, while, at the lectern, Judith Clyde, an ageless blonde in aqua rayon, with large, flat cheeks stained a beautiful candy pink and two bright eyes basking in blue lunacy behind a rimless pince-nez, presented the speaker:

"Tonight," she said, "the speaker of the evening——This, by the way, is our third Friday night; last time, as you all remember, we all enjoyed hearing what Professor Moore had to say about agriculture in China. Tonight we have here, I am proud to say, the Russian-born, and citizen of this country, Professor—now comes a difficult one, I am afraid—Professor Pun-neen. I hope I have it right. He hardly needs any introduction, of course, and we are all happy to have him. We have a long evening before us, a long and rewarding evening, and I am sure you would all like to have time to ask him questions afterwards. Incidentally, I am told his father was Dostoevski's family doctor, and he has traveled quite a bit on both sides of the Iron Curtain. Therefore I will not take up your precious time any longer and will only add a few words about our next Friday lecture in this program. I am sure you will all be delighted to know that there is a grand surprise in store for all of us. Our next lecturer is the distinguished poet and prose writer, Miss Linda Lacefield. We all know she has written poetry, prose, and some short stories. Miss Lacefield was born in New York. Her ancestors on both sides fought on both sides in the Revolutionary War. She wrote her first poem before graduation. Many of her poems—three of them, at least—have been published in *Response, A Hundred Love Lyrics by American Women*. In 1922 she received the cash prize offered by——"

But Pnin was not listening. A faint ripple stemming from his recent seizure was holding his fascinated attention. It lasted only a few heartbeats, with an additional systole here and there—last, harmless echoes—and was resolved in demure reality as his distinguished hostess invited him to the lectern; but while it lasted, how limpid the vision was! In the middle of the front row of seats he saw one of his Baltic aunts, wearing the pearls and the lace and the blond wig she had worn at all the performances given by the great ham actor Khodotov, whom she had adored from afar before drifting into insanity. Next to her, shyly smiling, sleek dark head inclined, gentle brown gaze shining up at Pnin from under velvet eyebrows, sat a dead sweetheart of his, fanning herself with a program. Murdered, forgotten, unrevenged, incorrupt, immortal, many old friends were scattered throughout the dim hall among more recent people, such as Miss Clyde, who had modestly regained a front seat. Vanya Bednyashkin, shot by the Reds in 1919 in Odessa because his father had been a Liberal, was gaily signaling to his former schoolmate from the back of the hall. And in an inconspicuous situation Dr. Pavel Pnin and his anxious wife, both a little blurred but on the whole wonderfully recovered from their obscure dissolution, looked at their son with the same life-consuming passion and pride that they had looked at him with that night in 1912 when, at a school festival, commemorating Napoleon's defeat, he had recited (a bespectacled lad all alone on the stage) a poem by Pushkin.

The brief vision was gone. Old Miss Herring, retired Professor of History, author of *Russia Awakes* (1922), was bending across one or two intermediate members of the audience to compliment Miss Clyde on her speech, while from behind that lady another twinkling old party was thrusting into her field of vision a pair of withered, soundlessly clapping hands.

1953, 1957

RALPH ELLISON
1914–

'If the Negro, or any other writer, is going to do what's expected of him, he's lost the battle before he takes the field." This remark of Ralph Ellison's, taken from his *Paris Review* interview of 1953, serves in more than one sense as an appropriate motto for his own career. He has not done what his critics, literary or political ones, suggested that he ought to be doing, but has insisted on being a writer rather than a spokesman for a cause or a representative figure. His importance to American letters is partly due to this independence. It is also true, however, that he has done the unexpected in not following his fine first novel with the others that were predicted; and *Invisible Man*, after a quarter of a century still stands as the sole indicator of Ellison's artistic significance.

He was born in Oklahoma, grew up in Oklahoma City, won a state scholarship and attended Tuskegee Institute where he was a music major, his instrument the trumpet. His musical life was wide enough to embrace both "serious music" and the world of Southwest-Kansas City jazz just reaching its heyday when Ellison was a young man. He became friends with Jimmy Rushing, the blues singer, and was acquainted with other members of what would be the great Count Basie band of the 1930s; this "deep, rowdy stream of jazz" figured for him as an image of the power and control that constituted art. Although he was a serious student of music and composition, his literary inclinations eventually dominated his musical ones; but testimony to his abiding knowledge and love of music may be found in some of the essays from *Shadow and Act* (1964) a collection of his prose.

When Ellison left Tuskegee he went north to make his way in New York City. There, in 1936, he met Richard Wright, who encouraged him as a writer, and Ellison began to publish reviews and short stories. *Invisible Man*, begun in 1945, was published seven years later and won the National Book Award. Ellison subsequently received a number of awards and lectureships, taught at the Salzburg Seminar, at Bard College, and at the University of Chicago, and in 1970 was named Albert Schweitzer Professor of the Humanities at New York University. Yet he has admitted to being troubled by the terms in which *Invisible Man*'s success—and perhaps his own career as well—have been defined. In the *Paris Review* interview he deprecatingly referred to his novel as largely a failure, wished that rather than a "statement" about the American Negro it could be read "simply as a novel," and hoped that in twenty years it would be so read, casually adding, "if it's around that long."

That twenty-year period is now well passed and *Invisible Man* is very much around, though the attempt to view it "simply as a novel" is a complex activity. Near the beginning of the book the young hero dreams that the following message is engraved on a document presented to him at his high school graduation: "To Whom It May Concern: Keep This Nigger-Boy Running." Ellison keeps him running throughout the novel, from his term at the Southern College to his flight North into the dizzying sequence of adventures and the various brands of political and racial rhetoric he undergoes in Harlem. The novel presents a gallery of extraordinary charac-ters and circumstances, expressed through different narrative styles, against which the ordinary hero defines—or fails to define—himself. And we are to remember that the whole book is conceived and understood to be nar-rated from a "hole" into which the Invisible Man has retired and where he thinks about the great black jazz trumpet player, Louis Armstrong: "Perhaps I like Louis Armstrong because he's made poetry out of being invisible."

In the late 1960s some black intellectuals and writers found Ellison's work irrelevant to their more activist designs. And earlier in the decade the white critic Irving Howe published an essay praising Richard Wright for writing "protest" works like *Native Son* while (Howe claimed) Ellison and James Baldwin were evading their tragic responsibilities as black victims by being too sanguine about the possibilities of human freedom. Ellison's answer to Howe's charge is printed in an important essay called *The World and the Jug* (in *Shadow and Act*) and in essence runs like this: "I am a human being, not just the black successor to Richard Wright, and there are ways of celebrating my experience more complex than terms like 'protest' can suggest." In Ellison's own words, "To deny in the interest of revolutionary posture that * * * such possibilities of human richness exist for others" is to impoverish both life and art.

Ellison has surely felt the sting and seriousness of these charges by black and white critics; but one of the things he shares with his invisible-man protagonist is that, as with Armstrong or any good jazz improviser, he never quite stays on the beat, especially when the beat is laid down by somebody else. Although thus far only sections have been published from the long-awaited successor to *Invisible Man*, the depth and humanness of this writer's imagination make it important that readers not stop waiting.

From Invisible Man

Chapter I

[BATTLE ROYAL]

It goes a long way back, some twenty years. All my life I had been looking for something, and everywhere I turned someone tried to tell me what it was. I accepted their answers too, though they were often in contradiction and even self-contradictory. I was naïve. I was looking for myself and asking everyone except myself questions which I, and only I, could answer. It took me a long time and much painful boomeranging of my expectations to achieve a realization everyone else appears to have been born with: That I am nobody but myself. But first I had to discover that I am an invisible man!

And yet I am no freak of nature, nor of history. I was in the cards, other things having been equal (or unequal) eighty-five years ago. I am not ashamed of my grandparents for having been slaves. I am only ashamed of myself for having at one time been ashamed. About eighty-five years ago they were told that they were free, united with others of our country in everything pertaining to the common good, and, in everything social, separate like the fingers of the hand. And they believed it. They exulted in it. They stayed in their place, worked hard, and brought up my father to do the same. But my grandfather is the one. He was an odd old guy, my grandfather, and I am told I take after him. It was he who caused the trouble. On his deathbed he called my father to him and said, "Son, after I'm gone I want you to keep up the good fight. I never told you, but our life is a war and I have been a traitor all my born days, a spy in the enemy's country ever since I give up my gun back in the Reconstruction. Live with your head in the lion's mouth. I want you to overcome 'em with yeses, undermine 'em with grins, agree 'em to death and destruction, let 'em swoller you till they vomit or bust wide open." They thought the old man had gone out of his mind. He had been the meekest of men. The younger children were rushed from the room, the shades drawn and the flame of the lamp turned so low that it sputtered on the wick like the old man's breathing. "Learn it to the younguns," he whispered fiercely; then he died.

But my folks were more alarmed over his last words than over his dying. It was as though he had not died at all, his words caused so much anxiety. I was warned emphatically to forget what he had said and, indeed, this is the first time it has been mentioned outside the family circle. It had a tremendous effect upon me, however. I could never be sure of what he meant. Grandfather had been a quiet old man who never made any trouble, yet on his deathbed he had called himself a traitor and a spy, and he had spoken of his meekness as a dangerous activity. It became a constant puzzle which lay unanswered in the back of my mind. And whenever things went well for me I remembered my grandfather and felt guilty and uncomfortable. It was as though I was carrying out his advice in spite of myself. And to make it worse, everyone loved me for it. I was praised by the most lily-white men of the town. I was considered an example of desirable conduct—just as my grandfather had been. And what puzzled me was that the old man had defined it as *treachery*. When I was praised for my conduct I felt a guilt that in some way I was doing something that was really against the wishes of the white folks, that if they had understood they would have desired me to act just the opposite, that I should have been sulky and mean, and that that really would have been what they wanted, even though they were fooled and thought they wanted me to act as I did. It made me afraid that some day they would look upon me as a traitor and I would be lost. Still I was more afraid to act any other way because they didn't like that at all. The old man's words were like a curse. On my graduation day I delivered an oration in which I showed that humility was the secret, indeed, the very essence of progress. (Not that I believed this—how could I, remembering my grandfather?—I only believed that it worked.) It was a great success. Everyone praised me and I was invited to give the speech at a gathering of the town's leading white citizens. It was a triumph for our whole community.

It was in the main ballroom of the leading hotel. When I got there I discovered that it was on the occasion of a smoker, and I was told that since I was to be there anyway I might as well take part in the battle royal to be fought by some of my schoolmates as part of the entertainment. The battle royal came first.

All of the town's big shots were there in their tuxedoes, wolfing

down ..e buffet foods, drinking beer and whiskey and smoking
bla. cigars. It was a large room with a high ceiling. Chairs were
arranged in neat rows around three sides of a portable boxing ring.
The fourth side was clear, revealing a gleaming space of polished
floor. I had some misgivings over the battle royal, by the way. Not
from a distaste for fighting, but because I didn't care too much for
the other fellows who were to take part. They were tough guys who
seemed to have no grandfather's curse worrying their minds. No
one could mistake their toughness. And besides, I suspected that
fighting a battle royal might detract from the dignity of my speech.
In those pre-invisible days I visualized myself as a potential Booker
T. Washington. But the other fellows didn't care too much for me
either, and there were nine of them. I felt superior to them in my
way, and I didn't like the manner in which we were all crowded
together into the servants' elevator. Nor did they like my being
there. In fact, as the warmly lighted floors flashed past the elevator
we had words over the fact that I, by taking part in the fight, had
knocked one of their friends out of a night's work.

We were led out of the elevator through a rococo hall into an
anteroom and told to get into our fighting togs. Each of us was
issued a pair of boxing gloves and ushered out into the big mirrored
hall, which we entered looking cautiously about us and whispering,
lest we might accidentally be heard above the noise of the room. It
was foggy with cigar smoke. And already the whiskey was taking
effect. I was shocked to see some of the most important men of the
town quite tipsy. They were all there—bankers, lawyers, judges,
doctors, fire chiefs, teachers, merchants. Even one of the more
fashionable pastors. Something we could not see was going on up
front. A clarinet was vibrating sensuously and the men were standing up and moving eagerly forward. We were a small tight group,
clustered together, our bare uppers touching and shining with anticipatory sweat; while up front the big shots were becoming increasingly excited over something we still could not see. Suddenly I heard
the school superintendent, who had told me to come, yell, "Bring up
the shines, gentlemen! Bring up the little shines!"

We were rushed up to the front of the ballroom, where it smelled
even more strongly of tobacco and whiskey. Then we were pushed
into place. I almost wet my pants. A sea of faces, some hostile,
some amused, ringed around us, and in the center, facing us, stood
a magnificent blonde—stark naked. There was dead silence. I felt a
blast of cold air chill me. I tried to back away, but they were behind
me and around me. Some of the boys stood with lowered heads,
trembling. I felt a wave of irrational guilt and fear. My teeth chattered, my skin turned to goose flesh, my knees knocked. Yet I was
strongly attracted and looked in spite of myself. Had the price of
looking been blindness, I would have looked. The hair was yellow
like that of a circus kewpie doll, the face heavily powdered and

rouged, as though to form an abstract mask, the eyes hollow and smeared a cool blue, the color of a baboon's butt. I felt a desire to spit upon her as my eyes brushed slowly over her body. Her breasts were firm and round as the domes of East Indian temples, and I stood so close as to see the fine skin texture and beads of pearly perspiration glistening like dew around the pink and erected buds of her nipples. I wanted at one and the same time to run from the room, to sink through the floor, or go to her and cover her from my eyes and the eyes of the others with my body; to feel the soft thighs, to caress her and destroy her, to love her and murder her, to hide from her, and yet to stroke where below the small American flag tattooed upon her belly her thighs formed a capital V. I had a notion that of all in the room she saw only me with her impersonal eyes.

And then she began to dance, a slow sensuous movement; the smoke of a hundred cigars clinging to her like the thinnest of veils. She seemed like a fair bird-girl girdled in veils calling to me from the angry surface of some gray and threatening sea. I was transported. Then I became aware of the clarinet playing and the big shots yelling at us. Some threatened us if we looked and others if we did not. On my right I saw one boy faint. And now a man grabbed a silver pitcher from a table and stepped close as he dashed ice water upon him and stood him up and forced two of us to support him as his head hung and moans issued from his thick bluish lips. Another boy began to plead to go home. He was the largest of the group, wearing dark red fighting trunks much too small to conceal the erection which projected from him as though in answer to the insinuating low-registered moaning of the clarinet. He tried to hide himself with his boxing gloves.

And all the while the blonde continued dancing, smiling faintly at the big shots who watched her with fascination, and faintly smiling at our fear. I noticed a certain merchant who followed her hungrily, his lips loose and drooling. He was a large man who wore diamond studs in a shirtfront which swelled with the ample paunch underneath, and each time the blonde swayed her undulating hips he ran his hand through the thin hair of his bald head and, with his arms upheld, his posture clumsy like that of an intoxicated panda, wound his belly in a slow and obscene grind. This creature was completely hypnotized. The music had quickened. As the dancer flung herself about with a detached expression on her face, the men began reaching out to touch her. I could see their beefy fingers sink into the soft flesh. Some of the others tried to stop them and she began to move around the floor in graceful circles, as they gave chase, slipping and sliding over the polished floor. It was mad. Chairs went crashing, drinks were spilt, as they ran laughing and howling after her. They caught her just as she reached a door, raised her from the floor, and tossed her as college boys are tossed at a hazing, and above her red, fixed-smiling lips I saw the terror and disgust in her eyes, almost like my own terror and that which I saw in some of the other boys. As I watched, they tossed her twice and her soft breasts seemed to flatten against the air and her legs flung wildly as she spun. Some of the more sober ones helped her to escape. And I started off the floor, heading for the anteroom with the rest of the boys.

Some were still crying and in hysteria. But as we tried to leave we were stopped and ordered to get into the ring. There was nothing to do but what we were told. All ten of us climbed under the ropes and allowed ourselves to be blindfolded with broad bands of white cloth. One of the men seemed to feel a bit sympathetic and tried to cheer us up as we stood with our backs against the ropes. Some of us tried to grin. "See that boy over there?" one of the men said. "I want you to run across at the bell and give it to him right in the belly. If you don't get him, I'm going to get you. I don't like his looks." Each of us was told the same. The blindfolds were put on. Yet even then I had been going over my speech. In my mind each word was as bright as flame. I felt the cloth pressed into place, and frowned so that it would be loosened when I relaxed.

But now I felt a sudden fit of blind terror. I was unused to darkness. It was as though I had suddenly found myself in a dark room filled with poisonous cottonmouths. I could hear the bleary voices yelling insistently for the battle royal to begin.

"Get going in there!"

"Let me at that big nigger!"

I strained to pick up the school superintendent's voice, as though to squeeze some security out of that slightly more familiar sound.

"Let me at those black sonsabitches!" someone yelled.

"No, Jackson, no!" another voice yelled. "Here, somebody, help me hold Jack."

"I want to get at that ginger-colored nigger. Tear him limb from limb," the first voice yelled.

I stood against the ropes trembling. For in those days I was what they called ginger-colored, and he sounded as though he might crunch me between his teeth like a crisp ginger cookie.

Quite a struggle was going on. Chairs were being kicked about and I could hear voices grunting as with a terrific effort. I wanted to see, to see more desperately than ever before. But the blindfold was tight as a thick skin-puckering scab and when I raised my gloved hands to push the layers of white aside a voice yelled, "Oh, no you don't, black bastard! Leave that alone!"

"Ring the bell before Jackson kills him a coon!" someone boomed in the sudden silence. And I heard the bell clang and the sound of the feet scuffling forward.

A glove smacked against my head. I pivoted, striking out stiffly as someone went past, and felt the jar ripple along the length of my arm to my shoulder. Then it seemed as though all nine of the boys had turned upon me at once. Blows pounded me from all sides while I struck out as best I could. So many blows landed upon me that I wondered if I were not the only blindfolded fighter in the ring, or if the man called Jackson hadn't succeeded in getting me after all.

Blindfolded, I could no longer control my motions. I had no dignity. I stumbled about like a baby or a drunken man. The smoke had become thicker and with each new blow it seemed to sear and further restrict my lungs. My saliva became like hot bitter glue. A glove connected with my head, filling my mouth with warm blood. It was everywhere. I could not tell if the moisture I felt upon my body was sweat or blood. A blow landed hard against the nape of my neck. I felt myself going over, my head hitting the floor. Streaks of blue light filled the black world behind the blindfold. I lay prone, pretending that I was knocked out, but felt myself seized by hands and yanked to my feet. "Get going, black boy! Mix it up!" My arms were like lead, my head smarting from blows. I managed to feel my way to the ropes and held on, trying to catch my breath. A glove landed in my mid-section and I went over again, feeling as though the smoke had become a knife jabbed into my guts. Pushed this way and that by the legs milling around me, I finally pulled erect and discovered that I could see the black, sweat-washed forms weaving in the smoky-blue atmosphere like drunken dancers weaving to the rapid drum-like thuds of blows.

Everyone fought hysterically. It was complete anarchy. Everybody fought everybody else. No group fought together for long. Two, three, four, fought one, then turned to fight each other, were themselves attacked. Blows landed below the belt and in the kidney, with the gloves open as well as closed, and with my eye partly opened now there was not so much terror. I moved carefully, avoiding blows, although not too many to attract attention, fighting from group to group. The boys groped about like blind, cautious crabs crouching to protect their mid-sections, their heads pulled in short against their shoulders, their arms stretched nervously before them, with their fists testing the smoke-filled air like the knobbed feelers of hypersensitive snails. In one corner I glimpsed a boy violently punching the air and heard him scream in pain as he smashed his hand against a ring post. For a second I saw him bent over holding his hand, then going down as a blow caught his unprotected head. I played one group against the other, slipping in and throwing a punch then stepping out of range while pushing the others into the melee to take the blows blindly aimed at me. The smoke was agonizing and there were no rounds, no bells at three minute intervals to relieve our exhaustion. The room spun round me, a swirl of lights, smoke, sweating bodies surrounded by tense white faces. I bled from both nose and mouth, the blood spattering upon my chest.

The men kept yelling, "Slug him, black boy! Knock his guts out!"

"Uppercut him! Kill him! Kill that big boy!"

Taking a fake fall, I saw a boy going down heavily beside me as though we were felled by a single blow, saw a sneaker-clad foot shoot into his groin as the two who had knocked him down stumbled upon him. I rolled out of range, feeling a twinge of nausea.

The harder we fought the more threatening the men became. And yet, I had begun to worry about my speech again. How would it go? Would they recognize my ability? What would they give me?

I was fighting automatically when suddenly I noticed that one after another of the boys was leaving the ring. I was surprised, filled with panic, as though I had been left alone with an unknown danger. Then I understood. The boys had arranged it among themselves. It was the custom for the two men left in the ring to slug it out for the winner's prize. I discovered this too late. When the bell sounded two men in tuxedoes leaped into the ring and removed the blindfold. I found myself facing Tatlock, the biggest of the gang. I felt sick at my stomach. Hardly had the bell stopped ringing in my ears than it clanged again and I saw him moving swiftly toward me. Thinking of nothing else to do I hit him smash on the nose. He kept coming, bringing the rank sharp violence of stale sweat. His face was a black blank of a face, only his eyes alive—with hate of me and aglow with a feverish terror from what had happened to us all. I became anxious. I wanted to deliver my speech and he came at me as though he meant to beat it out of me. I smashed him again and again, taking his blows as they came. Then on a sudden impulse I struck him lightly and as we clinched, I whispered, "Fake like I knocked you out, you can have the prize."

"I'll break your behind," he whispered horasely.

"For *them*?"

"For *me*, sonofabitch!"

They were yelling for us to break it up and Tatlock spun me half around with a blow, and as a joggled camera sweeps in a reeling scene, I saw the howling red faces crouching tense beneath the cloud of blue-gray smoke. For a moment the world wavered, unraveled, flowed, then my head cleared and Tatlock bounced before me. That fluttering shadow before my eyes was his jabbing left hand. Then falling forward, my head against his damp shoulder, I whispered,

"I'll make it five dollars more."

"Go to hell!"

But his relaxed a trifle beneath my pressure and I breathed, "Seven?"

"Give it to your ma," he said, ripping me beneath the heart.

And while I still held him I butted him and moved away. I felt myself bombarded with punches. I fought back with hopeless desperation. I wanted to deliver my speech more than anything else in the world, because I felt that only these men could judge truly my ability, and now this stupid clown was ruining my chances. I began fighting carefully now, moving in to punch him and out again with my greater speed. A lucky blow to his chin and I had him going too—until I heard a loud voice yell, "I got my money on the big boy."

Hearing this, I almost dropped my guard. I was confused: Should I try to win against the voice out there? Would not this go against my speech, and was not this a moment for humility, for nonresistance? A blow to my head as I danced about sent my right eye popping like a jack-in-the-box and settled my dilemma. The room went red as I fell. It was a dream fall, my body languid and fastidious as to where to land, until the floor became impatient and smashed up to meet me. A moment later I came to. An hypnotic voice said FIVE emphatically. And I lay there, hazily watching a dark red spot of my own blood shaping itself into a butterfly, glistening and soaking into the soiled gray world of the canvas.

When the voice drawled TEN I was lifted up and dragged to a chair. I sat dazed. My eye pained and swelled with each throb of my pounding heart and I wondered if now I would be allowed to speak. I was wringing wet, my mouth still bleeding. We were grouped along the wall now. The other boys ignored me as they congratulated Tatlock and speculated as to how much they would be paid. One boy whimpered over his smashed hand. Looking up front, I saw attendants in white jackets rolling the portable ring away and placing a small square rug in the vacant space surrounded by chairs. Perhaps, I thought, I will stand on the rug to deliver my speech.

Then the M.C. called to us, "Come on up here boys and get your money."

We ran forward to where the men laughed and talked in their chairs, waiting. Everyone seemed friendly now.

"There it is on the rug," the man said. I saw the rug covered with coins of all dimensions and a few crumpled bills. But what excited me, scattered here and there, were the gold pieces.

"Boys, it's all yours," the man said. "You get all you grab."

"That's right, Sambo," a blond man said, winking at me confidentially.

I trembled with excitement, forgetting my pain. I would get the gold and the bills, I thought. I would use both hands. I would throw my body against the boys nearest me to block them from the gold.

"Get down around the rug now," the man commanded, "and don't anyone touch it until I give the signal."

"This ought to be good," I heard.

As told, we got around the square rug on our knees. Slowly the man raised his freckled hand as we followed it upward with our eyes.

I heard, "These niggers look like they're about to pray!"

Then, "Ready," the man said. "Go!"

I lunged for a yellow coin lying on the blue design of the carpet, touching it and sending a surprised shriek to join those rising around me. I tried frantically to remove my hand but could not let go. A hot, violent force tore through my body, shaking me like a wet rat. The rug was electrified. The hair bristled up on my head as I shook myself free. My muscles jumped, my nerves jangled, writhed. But I saw that this was not stopping the other boys. Laughing in fear and embarrassment, some were holding back and scooping up the coins knocked off by the painful contortions of the others. The men roared above us as we struggled.

"Pick it up, goddamnit, pick it up!" someone called like a bass-voiced parrot. "Go on, get it!"

I crawled rapidly around the floor, picking up the coins, trying to avoid the coppers and to get greenbacks and the gold. Ignoring the shock by laughing, as I brushed the coins off quickly, I discovered that I could contain the electricity—a contradiction, but it works. Then the men began to push us onto the rug. Laughing embarrassedly, we struggled out of their hands and kept after the coins. We were all wet and slippery and hard to hold. Suddenly I saw a boy lifted into the air, glistening with sweat like a circus seal, and dropped, his wet back landing flush upon the charged rug, heard him yell and saw him literally dance upon his back, his elbows beating a frenzied tattoo upon the floor, his muscles twitching like the flesh of a horse stung by many flies. When he finally rolled off, his face was gray and no one stopped him when he ran from the floor amid booming laughter.

"Get the money," the M.C. called. "That's good hard American cash!"

And we snatched and grabbed, snatched and grabbed. I was careful not to come too close to the rug now, and when I felt the hot whiskey breath descend upon me like a cloud of foul air I reached out and grabbed the leg of a chair. It was occupied and I held on desperately.

"Leggo, nigger! Leggo!"

The huge face wavered down to mine as he tried to push me free. But my body was slippery and he was too drunk. It was Mr. Colcord, who owned a chain of movie houses and "entertainment palaces." Each time he grabbed me I slipped out of his hands. It became a real struggle. I feared the rug more than I did the drunk, so I held on, surprising myself for a moment by trying to topple *him* upon the rug. It was such an enormous idea that I found myself actually carrying it out. I tried not to be obvious, yet when I grabbed his leg, trying to tumble him out of the chair, he raised up roaring with laughter, and, looking at me with soberness dead in the eye, kicked me viciously in the chest. The chair leg flew out of my hand and I felt myself going and rolled. It was though I had rolled through a bed of hot coals. It seemed a whole century would pass before I would roll free, a century in which I was seared through the deepest levels of my body to the fearful breath within me and the breath seared and heated to the point of explosion. It'll all be over in a flash, I thought as I rolled clear. It'll all be over in a flash.

But not yet, the men on the other side were waiting, red faces swollen as though from apoplexy as they bent forward in their chairs. Seeing their fingers coming toward me I rolled away as a fumbled football rolls off the receiver's fingertips, back into the coals. That time I luckily sent the rug sliding out of place and heard the coins ringing against the floor and the boys scuffling to pick them up and the M.C. calling, "All right, boys, that's all. Go get dressed and get your money."

I was limp as a dish rag. My back felt as though it had been beaten with wires.

When we had dressed the M.C. came in and gave us each five dollars, except Tatlock, who got ten for being last in the ring. Then he told us to leave. I was not to get a chance to deliver my speech, I thought. I was going out into the dim alley in despair when I was stopped and told to go back. I returned to the ballroom, where the men were pushing back their chairs and gathering in groups to talk.

The M.C. knocked on a table for quiet. "Gentlemen," he said, "we almost forgot an important part of the program. A most serious part, gentlemen. This boy was brought here to deliver a speech which he made at his graduation yesterday . . ."

"Bravo!"

"I'm told that he is the smartest boy we've got out there in Greenwood. I'm told that he knows more big words than a pocket-sized dictionary."

Much applause and laughter.

"So now, gentlemen, I want you to give him your attention."

There was still laughter as I faced them, my mouth dry, my eye throbbing. I began slowly, but evidently my throat was tense, because they began shouting, "Louder! Louder!"

"We of the younger generation extol the wisdom of that great leader and educator," I shouted, "who first spoke these flaming words of wisdom: 'A ship lost at sea for many days suddenly sighted a friendly vessel. From the mast of the unfortunate vessel was seen a signal: "Water, water; we die of thirst!" The answer from the friendly vessel came back: "Cast down your bucket where you are." The captain of the distressed vessel, at last heeding the injunction, cast down his bucket, and it came up full of fresh sparkling water from the mouth of the Amazon River.' And like him I say, and in his words, 'To those of my race who depend upon bettering their condition in a foreign land, or who underestimate the importance of cultivating friendly relations with the Southern white man, who is his next-door neighbor, I would say: "Cast down your bucket where you are"—cast it down in making friends in every manly way of the people of all races by whom we are surrounded . . .'"

I spoke automatically and with such fervor that I did not realize that the men were still talking and laughing until my dry mouth, filling up with blood from the cut, almost strangled me. I coughed, wanting to stop and go to one of the tall brass, sand-filled spittoons to relieve myself, but a few of the men, especially the superintendent, were listening and I was afraid. So I gulped it down, blood, saliva and all, and continued. (What powers of endurance I had during those days! What enthusiasm! What a belief in the rightness of things!) I spoke even louder in spite of the pain. But still they talked and still they laughed, as though deaf with cotton in dirty ears. So I spoke with greater emotional emphasis. I closed my ears and swallowed blood until I was nauseated. The speech seemed a hundred times as long as before, but I could not leave out a single word. All had to be said, each memorized nuance considered, rendered. Nor was that all. Whenever I uttered a word of three or more syllables a group of voices would yell for me to repeat it. I used the phrase "social responsibility" and they yelled:

"What's that word you say, boy?"

"Social responsibility," I said.

"What?"

"Social . . ."

"Louder."

". . . responsibility."

"More!"

"Respon—"

"Repeat!"

"—sibility."

The room filled with the uproar of laughter until, no doubt, distracted by having to gulp down my blood, I made a mistake and yelled a phrase I had often seen denounced in newspaper editorials, heard debated in private.

"Social . . ."

"What?" they yelled.

". . . equality—"

The laughter hung smokelike in the sudden stillness. I opened my eyes, puzzled. Sounds of displeasure filled the room. The M.C. rushed forward. They shouted hostile phrases at me. But I did not understand.

A small dry mustached man in the front row blared out, "Say that slowly, son!"

"What, sir?"

"What you just said!"

"Social responsibility, sir," I said.

"You weren't being smart, were you, boy?" he said, not unkindly.

"No, sir!"

"You sure that about 'equality' was a mistake?"

"Oh, yes, sir," I said. "I was swallowing blood."

"Well, you had better speak more slowly so we can understand. We mean to do right by you, but you've got to know your place at ... All right, now, go on with your speech."

... said. I wanted to leave but I wanted also to speak and I ... atch me down.

"Gentlemen, you see that I did not over... forth with a package thunderous applause. I was ... beginning where I had left off, and ... quiet, address the a good speech and some day he'll lead his people in the proper makes paths. And I don't have to tell you that that is important in these days and times. This is a good, smart boy, and so to encourage him in the right direction, in the name of the Board of Education I wish to present him a prize in the form of this . . ."

He paused, removing the tissue paper and revealing a gleaming calfskin brief case.

". . . in the form of this first-class article from Shad Whitmore's shop."

"Boy," he said, addressing me, "take this prize and keep it well. Consider it a badge of office. Prize it. Keep developing as you are and some day it will be filled with important papers that will help shape the destiny of your people."

I was so moved that I could hardly express my thanks. A rope of bloody saliva forming a shape like an undiscovered continent drooled upon the leather and I wiped it quickly away. I felt an importance that I had never dreamed.

"Open it and see what's inside," I was told.

My fingers a-tremble, I complied, smelling the fresh leather and finding an official-looking document inside. It was a scholarship to the state college for Negroes. My eyes filled with tears and I ran awkwardly off the floor.

I was overjoyed; I did not even mind when I discovered that the gold pieces I had scrambled for were brass pocket tokens advertising a certain make of automobile.

When I reached home everyone was excited. Next day the neighbors came to congratulate me. I even felt safe from grandfather, whose deathbed curse usually spoiled my triumphs. I stood beneath his photograph with my brief case in hand and smiled triumphantly into his stolid black peasant's face. It was a face that fascinated me. The eyes seemed to follow everywhere I went.

That night I dreamed I was at a circus with him and that he

refused to laugh at the clowns no matter what they did. Then
he told me to open my brief case and read what was insi
did, finding an official envelope stamped with the st
inside the envelope I found another and another
thought I would fall of weariness. "Them's
open that one." And I did and in it I four
containing a short message in letters c ream again for many
father said. "Out loud!"

"To Whom It May Concer no insight into its meaning. First
Running."

I awoke with the old

(It was a dream.)

years after Du

I had to att

1952

SAUL BELLOW

1915–

When Saul Bellow was awarded the Nobel Prize, his citation read: "For the human understanding and subtle analysis of contemporary culture that are combined in his work." Except for the poet Robert Lowell, no American writer of the post-Second World War period has a better claim to these virtues than Bellow, who has devoted himself almost exclusively and passionately to the novel and its attempt to imagine life in the United States, particularly in the great cities of Chicago and New York.

He was born in Lachine, Quebec, grew up in the Jewish ghetto of Montreal, and moved to Chicago when he was nine. He attended the University of Chicago, then transferred to Northwestern, where he took a degree in anthropology and sociology, the effects of which study are everywhere evident in the novels he was to write. He taught English for a time, served in the Merchant Marine during World War II, then after the war spent some fifteen years away from Chicago, teaching at N.Y.U. and Princeton, and living in Paris. In 1962 he returned to Chicago and since then has been a lecturer at the University of Chicago. He has married four times.

Bellow's first novel, *Dangling Man*, was not published until he was nearly thirty and is a short series of elegantly morose meditations, told through the journal of a young man waiting to be inducted into the army and with the "freedom" of having nothing to do but wait. Eventually he is drafted: "Long live regimentation," he sardonically exults. His second novel, *The Victim* (1947), continues the investigation of ways people strive to be relieved of self-determination. This book concerns a week in the life of Asa Leventhal, alone in New York City while his wife visits a relative, and suddenly confronted by a figure from the past (Kirby Allbee, a Gentile) who succeeds in implicating Leventhal with the past and its present manifestations. *The Victim* is Bellow's most somberly naturalistic depiction of a man brought up against forces larger than himself; yet from the opening sentence ("On some nights New York is as hot as Bangkok") a poetic dimension makes itself felt and helps create the sense of mystery and disturbance felt by both the main character and the reader.

Dangling Man and *The Victim* are highly wrought, mainly humorless books; in two long novels published in the 1950s Bellow opened up into new ranges of aspiration and situational zaniness which brought him respectful admiration from many critics. *The Adventures of Augie March* (1953) and *Henderson the Rain King* (1959) are each narrated by an "I" who, like his predecessor Huck Finn, is good at lighting out for the territory ahead of whoever means to tie him down. The hero's adventures, whether occurring in Chicago, Mexico, or Africa, are exuberantly delivered in an always stimulating and sometimes overactive prose. *Augie* is filled with sights and sounds, colors and surfaces; its tone is self-involved, affectionate, and affirmative in its ring; *Henderson,* Bellow's most extravagant narrative, has the even more fabulous air of a quest-romance in which the hero returns home from Africa at peace with the world he had been warring against.

The ironic motto for Bellow's novels of the 1950s may well be "Seize the Day," as in the title of perhaps his finest piece of fiction. This short novel is both painful and exhilarating because it so fully exposes its hero (Tommy Wilhelm, an aging out-of-work ex-actor) to the insults of other people who don't understand him, to a city (New York's Upper West Side) impervious to his needs, and to a narrative prose which mixes ridicule and affection so thoroughly as to make them scarcely distinguishable. *Seize the Day* combines, within Tommy's monologues, a wildness and pathos of bitter comedy which was a powerful new element in Bellow's work.

In *Where Do We Go from Here,* an essay published in 1965, Bellow pointed out that nineteenth-century American literature—Emerson, Thoreau, Whitman, Melville—was highly didactic in its efforts to "instruct a young and raw nation." Bellow sees himself in this instructive tradition, and in the international company of "didactic" novelists like Dostoyevsky, D. H. Lawrence, and Joseph Conrad; he believes also that "the imagination is looking for new ways to express virtue * * * we have barely begun to comprehend what a human being is." These concerns animate the novels Bellow has written over the past fifteen years. In *Herzog* (1964) the hero is another down-and-outer, a professor-intellectual, a student of Romanticism and of the glorification of Self which Herzog believes both modern life and modernist literature have been working to undercut. At the same time he is a comic and pathetic victim of marital disorder; like all of Bellow's heroes, Herzog has a terrible time with women, yet cannot live without them. In *Mr. Sammler's Planet* (1970), written out of the disorders of the late 1960s, the atmosphere is grimmer. Sammler, an aging Jew living (again) on New York's West Side, analyzes and judges but cannot understand the young or blacks, or the mass of people gathered at Broadway and 96th Street. He sees about him everywhere "poverty of soul" but admits that he too has "a touch of the same disease—the disease of the single self explaining what was what and who was who."

These novels and the recent *Humboldt's Gift* (1975) have been accused of parading too single-mindedly attitudes to which their author is sympathetic. But what Bellow finds moving in Theodore Dreiser's work, "his balkiness and sullenness, and then his allegiance to life," is found also in his own: complaint and weariness, fault-finding, accusation of self and others—these gestures directed at "life" also make up the stuff of life and the "allegiance" out of which Bellow's heroes are made. We read him for this range of interest, for flexibility and diversity of style and idiom, and for the eloquences of nostalgia, invective, and lamentation which make up his intensely imagined world.

Looking for Mr. Green[1]

Whatsoever thy hand findeth to do, do it with thy might. . . .

Hard work? No, it wasn't really so hard. He wasn't used to walking and stair-climbing, but the physical difficulty of his new job was not what George Grebe felt most. He was delivering relief checks in the Negro district, and although he was a native Chicagoan this was not a part of the city he knew much about—it needed a depression to introduce him to it. No, it wasn't literally hard work, not as reckoned in foot-pounds, but yet he was beginning to feel the strain of it, to grow aware of its peculiar difficulty. He could find the streets and numbers, but the clients were not where they were supposed to be, and he felt like a hunter inexperienced in the camouflage of his game. It was an unfavorable day, too—fall, and cold, dark weather, windy. But, anyway, instead of shells in his deep trenchcoat pocket he had the cardboard of checks, punctured for the spindles of the file, the holes reminding him of the holes in player-piano paper. And he didn't look much like a hunter, either; his was a city figure entirely, belted up in this Irish conspirator's coat.[2] He was slender without being tall, stiff in the back, his legs looking shabby in a pair of old tweed pants gone through and fringy at the cuffs. With this stiffness, he kept his head forward, so that his face was red from the sharpness of the weather; and it was an indoors sort of face with gray eyes that persisted in some kind of thought and yet seemed to avoid definiteness of conclusion. He wore sideburns that surprised you somewhat by the tough curl of the blond hair and the effect of assertion in their length. He was not so mild as he looked, nor so youthful; and nevertheless there was no effort on his part to seem what he was not. He was an educated man; he was a bachelor; he was in some ways simple; without lushing, he liked a drink; his luck had not been good. Nothing was deliberately hidden.

He felt that his luck was better than usual today. When he had reported for work that morning he had expected to be shut up in the relief office at a clerk's job, for he had been hired downtown as a clerk, and he was glad to have, instead, the freedom of the streets

1. From *Mosby's Memoirs and Other Stories* (1968). The epigraph comes from Ecclesiastes 9.10; the verse continues, "for there is no work, nor device, nor knowledge, nor wisdom, in the grave, whither thou goest."
2. Like those worn by members of the anti-British underground in Ireland.

and welcomed, at least at first, the vigor of the cold and even the blowing of the hard wind. But on the other hand he was not getting on with the distribution of the checks. It was true that it was a city job; nobody expected you to push too hard at a city job. His supervisor, that young Mr. Raynor, had practically told him that. Still, he wanted to do well at it. For one thing, when he knew how quickly he could deliver a batch of checks, he would know also how much time he could expect to clip for himself. And then, too, the clients would be waiting for their money. That was not the most important consideration, though it certainly mattered to him. No, but he wanted to do well, simply for doing-well's sake, to acquit himself decently of a job because he so rarely had a job to do that required just this sort of energy. Of this peculiar energy he now had a superabundance; once it had started to flow, it flowed all too heavily. And, for the time being anyway, he was balked. He could not find Mr. Green.

So he stood in his big-skirted trenchcoat with a large envelope in his hand and papers showing from his pocket, wondering why people should be so hard to locate who were too feeble or sick to come to the station to collect their own checks. But Raynor had told him that tracking them down was not easy at first and had offered him some advice on how to proceed. "If you can see the postman, he's your first man to ask, and your best bet. If you can't connect with him, try the stores and tradespeople around. Then the janitor and the neighbors. But you'll find the closer you come to your man the less people will tell you. They don't want to tell you anything."

"Because I'm a stranger."

"Because you're white. We ought to have a Negro doing this, but we don't at the moment, and of course you've got to eat, too, and this is public employment. Jobs have to be made. Oh, that holds for me too. Mind you, I'm not letting myself out. I've got three years of seniority on you, that's all. And a law degree. Otherwise, you might be back of the desk and I might be going out into the field this cold day. The same dough pays us both and for the same, exact, identical reason. What's my law degree got to do with it? But you have to pass out these checks, Mr. Grebe, and it'll help if you're stubborn, so I hope you are."

"Yes, I'm fairly stubborn."

Raynor sketched hard with an eraser in the old dirt of his desk, left-handed, and said, "Sure, what else can you answer to such a question. Anyhow, the trouble you're going to have is that they don't like to give information about anybody. They think you're a plain-clothes dick or an installment collector, or summons-server or something like that. Till you've been seen around the neighborhood for a few months and people know you're only from the relief."

It was dark, ground-freezing, pre-Thanksgiving weather; the wind played hob with the smoke, rushing it down, and Grebe missed his gloves, which he had left in Raynor's office. And no one would admit knowing Green. It was past three o'clock and the postman had made his last delivery. The nearest grocer, himself a Negro, had never heard the name Tulliver Green, or said he hadn't. Grebe was inclined to think that it was true, that he had in the end convinced the man that he wanted only to deliver a check. But he wasn't sure. He needed experience in interpreting looks and signs and, even more, the will not to be put off or denied and even the force to bully if need be. If the grocer did know, he had got rid of him easily. But since most of his trade was with reliefers, why should he prevent the delivery of a check? Maybe Green, or Mrs. Green, if there was a Mrs. Green, patronized another grocer. And was there a Mrs. Green? It was one of Grebe's great handicaps that he hadn't looked at any of the case records. Raynor should have let him read files for a few hours. But he apparently saw no need for that, probably considering the job unimportant. Why prepare systematically to deliver a few checks?

But now it was time to look for the janitor. Grebe took in the building in the wind and gloom of the late November day—trampled, frost-hardened lots on one side; on the other, an automobile junk yard and then the infinite work of Elevated frames, weak-looking, gaping with rubbish fires, two sets of leaning brick porches three stories high and a flight of cement stairs to the cellar. Descending, he entered the underground passage, where he tried the doors until one opened and he found himself in the furnace room. There someone rose toward him and approached, scraping on the coal grit and bending under the canvas-jacketed pipes.

"Are you the janitor?"

"What do you want?"

"I'm looking for a man who's supposed to be living here. Green."

"What Green?"

"Oh, you maybe have more than one Green?" said Grebe with new, pleasant hope. "This is Tulliver Green."

"I don't think I c'n help you, mister. I don't know any."

"A crippled man."

The janitor stood bent before him. Could it be that he was crippled? Oh, God! what if he was. Grebe's gray eyes sought with excited difficulty to see. But no, he was only very short and stooped. A head awakened from meditation, a strong-haired beard, low, wide shoulders. A staleness of sweat and coal rose from his black shirt and the burlap sack he wore as an apron.

"Crippled how?"

Grebe thought and then answered with the light voice of unmixed candor, "I don't know. I've never seen him." This was damaging, but his only other choice was to make a lying guess, and he was not up to it. "I'm delivering checks for the relief to shut-in cases. If he weren't crippled he'd come to collect himself. That's why I said crippled. Bedridden, chair-ridden—is there anybody like that?"

This sort of frankness was one of Grebe's oldest talents, going back to childhood. But it gained him nothing here.

"No suh. I've got four buildin's same as this that I take care of. I don' know all the tenants, leave alone the tenants' tenants. The rooms turn over so fast, people movin' in and out every day. I can't tell you."

The janitor opened his grimy lips but Grebe did not hear him in the piping of the valves and the consuming pull of air to flame in the body of the furnace. He knew, however, what he had said.

"Well, all the same, thanks. Sorry I bothered you. I'll prowl around upstairs again and see if I can turn up someone who knows him."

Once more in the cold air and early darkness he made the short circle from the cellarway to the entrance crowded between the brickwork pillars and began to climb to the third floor. Pieces of plaster ground under his feet; strips of brass tape from which the carpeting had been torn away marked old boundaries at the sides. In the passage, the cold reached him worse than in the street; it touched him to the bone. The hall toilets ran like springs. He thought grimly as he heard the wind burning around the building with a sound like that of the furnace, that this was a great piece of constructed shelter. Then he struck a match in the gloom and searched for names and numbers among the writings and scribbles on the walls. He saw WHOODY-DOODY GO TO JESUS, and zigzags, caricatures, sexual scrawls, and curses. So the sealed rooms of pyramids were also decorated, and the caves of human dawn.

The information on his card was, TULLIVER GREEN—APT 3D. There were no names, however, and no numbers. His shoulders drawn up, tears of cold in his eyes, breathing vapor, he went the length of the corridor and told himself that if he had been lucky enough to have the temperament for it he would bang on one of the doors and bawl out "Tulliver Green!" until he got results. But it wasn't in him to make an uproar and he continued to burn matches, passing the light over the walls. At the rear, in a corner off the hall, he discovered a door he had not seen before and he thought it best to investigate. It sounded empty when he knocked, but a young Negress answered, hardly more than a girl. She opened only a bit, to guard the warmth of the room.

"Yes suh?"

"I'm from the district relief station on Prairie Avenue. I'm looking for a man named Tulliver Green to give him his check. Do you know him?"

No, she didn't; but he thought she had not understood anything of what he had said. She had a dream-bound, dream-blind face, very soft and black, shut off. She wore a man's jacket and pulled the ends together at her throat. Her hair was parted in three directions, at the sides and tranversely, standing up at the front in a dull puff.

"Is there somebody around here who might know?"

"I jus' taken this room las' week."

He observed that she shivered, but even her shiver was somnambulistic and there was no sharp consciousness of cold in the big smooth eyes of her handsome face.

"All right, miss, thank you. Thanks," he said, and went to try another place.

Here he was admitted. He was grateful, for the room was warm. It was full of people, and they were silent as he entered—ten people, or a dozen, perhaps more, sitting on benches like a parliament. There was no light, properly speaking, but a tempered darkness that the window gave, and everyone seemed to him enormous, the men padded out in heavy work clothes and winter coats, and the women huge, too, in their sweaters, hats, and old furs. And, besides, bed and bedding, a black cooking range, a piano piled towering to the ceiling with papers, a dining-room table of the old style of prosperous Chicago. Among these people Grebe, with his cold-heightened fresh color and his smaller stature, entered like a schoolboy. Even though he was met with smiles and good will, he knew, before a single word was spoken, that all the currents ran against him and that he would make no headway. Nevertheless he began. "Does anybody here know how I can deliver a check to Mr. Tulliver Green?"

"Green?" It was the man that had let him in who answered. He was in short sleeves, in a checkered shirt, and had a queer, high head, profusely overgrown and long as a shako;[3] the veins entered it strongly, from his forehead. "I never heard mention of him. Is this where he live?"

"This is the address they gave me at the station. He's a sick man, and he'll need his check. Can't anybody tell me where to find him?"

He stood his ground and waited for a reply, his crimson wool scarf wound about his neck and drooping outside his trenchcoat, pockets weighted with the block of checks and official forms. They must have realized that he was not a college boy employed afternoons by a bill collector, trying foxily to pass for a relief clerk, recognized that he was an older man who knew himself what need was, who had had more than an average seasoning in hardship. It was evident enough if you looked at the marks under his eyes and at the sides of his mouth.

"Anybody know this sick man?"

"No suh." On all sides he saw heads shaken and smiles of denial. No one knew. And maybe it was true, he considered, standing silent in the earthen, musky human gloom of the place as the rumble continued. But he could never really be sure.

"What's the matter with this man?" said shako-head.

3. Stiff military headdress with a high crown and a plume.

"I've never seen him. All I can tell you is that he can't come in person for his money. It's my first day in this district."

"Maybe they given you the wrong number?"

"I don't believe so. But where else can I ask about him?" He felt that this persistence amused them deeply, and in a way he shared their amusement that he should stand up so tenaciously to them. Though smaller, though slight, he was his own man, he retracted nothing about himself, and he looked back at them, gray-eyed, with amusement and also with a sort of courage. On the bench some man spoke in his throat, the words impossible to catch, and a woman answered with a wild, shrieking laugh, which was quickly cut off.

"Well, so nobody will tell me?"

"Ain't nobody who knows."

"At least, if he lives here, he pays rent to someone. Who manages the building?"

"Greatham Company. That's on Thirty-ninth Street."

Grebe wrote it in his pad. But, in the street again, a sheet of wind-driven paper clinging to his leg while he deliberated what direction to take next, it seemed a feeble lead to follow. Probably this Green didn't rent a flat, but a room. Sometimes there were as many as twenty people in an apartment; the real-estate agent would know only the lessee. And not even the agent could tell you who the renters were. In some places the beds were even used in shifts, watchmen or jitney drivers or short-order cooks in night joints turning out after a day's sleep and surrendering their beds to a sister, a nephew, or perhaps a stranger, just off the bus. There were large numbers of newcomers in this terrific, blight-bitten portion of the city between Cottage Grove and Ashland, wandering from house to house and room to room. When you saw them, how could you know them? They didn't carry bundles on their backs or look picturesque. You only saw a man, a Negro, walking in the street or riding in the car, like everyone else, with his thumb closed on a transfer. And therefore how were you supposed to tell? Grebe thought the Greatham agent would only laugh at his question.

But how much it would have simplified the job to be able to say that Green was old, or blind, or consumptive. An hour in the files, taking a few notes, and he needn't have been at such a disadvantage. When Raynor gave him the block of checks he asked, "How much should I know about these people?" Then Raynor had looked as though he were preparing to accuse him of trying to make the job more important than it was. He smiled, because by then they were on fine terms, but nevertheless he had been getting ready to say something like that when the confusion began in the station over Staika and her children.

Grebe had waited a long time for this job. It came to him through the pull of an old schoolmate in the Corporation Counsel's office, never a close friend, but suddenly sympathetic and interested—

pleased to show, moreover, how well he had done, how strongly he
was coming on even in these miserable times. Well, he was coming
through strongly, along with the Democratic administration itself.
Grebe had gone to see him in City Hall, and they had had a counter
lunch or beers at least once a month for a year, and finally it had been
possible to swing the job. He didn't mind being assigned the lowest
clerical grade, nor even being a messenger, though Raynor thought
he did.

This Raynor was an original sort of guy and Grebe had taken to
him immediately. As was proper on the first day, Grebe had come
early, but he waited long, for Raynor was late. At last he darted into
his cubicle of an office as though he had just jumped from one of
those hurtling huge red Indian Avenue cars. His thin, rough face was
wind-stung and he was grinning and saying something breathlessly
to himself. In his hat, a small fedora, and his coat, the velvet collar a
neat fit about his neck, and his silk muffler that set off the nervous
twist of his chin, he swayed and turned himself in his swivel chair,
feet leaving the ground; so that he pranced a little as he sat. Meanwhile he took Grebe's measure out of his eyes, eyes of an unusual
vertical length and slightly sardonic. So the two men sat for a while,
saying nothing, while the supervisor raised his hat from his mis-
combed hair and put it in his lap. His cold-darkened hands were not
clean. A steel beam passed through the little makeshift room, from
which machine belts once had hung. The building was an old factory.

"I'm younger than you; I hope you won't find it hard taking orders
from me," said Raynor. "But I don't make them up, either. You're
how old, about?"

"Thirty-five."

"And you thought you'd be inside doing paper work. But it so
happens I have to send you out."

"I don't mind."

"And it's mostly a Negro load we have in this district."

"So I thought it would be."

"Fine. You'll get along. *C'est un bon boulot.*[4] Do you know
French?"

"Some."

"I thought you'd be a university man."

"Have you been in France?" said Grebe.

"No, that's the French of the Berlitz School. I've been at it for
more than a year, just as I'm sure people have been, all over the
world, office boys in China and braves in Tanganyika. In fact, I damn
well know it. Such is the attractive power of civilization. It's overrated, but what do you want? *Que voulez-vous?*[5] I get *Le Rire* and all
the spicy papers, just like in Tanganyika. It must be mystifying, out
there. But my reason is that I'm aiming at the diplomatic service. I

---

4. "It's a good meal" (French slang).   5. "What do you want?"

have a cousin who's a courier, and the way he describes it is awfully attractive. He rides in the *wagon-lits*[6] and reads books. While we— What did you do before?"

"I sold."

"Where?"

"Canned meat at Stop and Shop. In the basement."

"And before that?"

"Window shades, at Goldblatt's."

"Steady work?"

"No, Thursdays and Saturdays. I also sold shoes."

"You've been a shoe-dog too. Well. And prior to that? Here it is in your folder." He opened the record. "Saint Olaf's College, instructor in classical languages. Fellow, University of Chicago, 1926– 27. I've had Latin, too. Let's trade quotations—'*Dum spiro spero.*' "

" '*Da dextram misero.*' "

" '*Alea jacta est.*' "

" '*Excelsior.*' "[7]

Raynor shouted with laughter, and other workers came to look at him over the partition. Grebe also laughed, feeling pleased and easy. The luxury of fun on a nervous morning.

When they were done and no one was watching or listening, Raynor said rather seriously, "What made you study Latin in the first place? Was it for the priesthood?"

"No."

"Just for the hell of it? For the culture? Oh, the things people think they can pull!" He made his cry hilarious and tragic. "I ran my pants off so I could study for the bar, and I've passed the bar, so I get twelve dollars a week more than you as a bonus for having seen life straight and whole. I'll tell you, as a man of culture, that even though nothing looks to be real, and everything stands for something else, and that thing for another thing, and that thing for a still further one—there ain't any comparison between twenty-five and thirty-seven dollars a week, regardless of the last reality. Don't you think that was clear to your Greeks? They were a thoughtful people, but they didn't part with their slaves."

This was a great deal more than Grebe had looked for in his first interview with his supervisor. He was too shy to show all the astonishment he felt. He laughed a little, aroused, and brushed at the sunbeam that covered his head with its dust. "Do you think my mistake was so terrible?"

"Damn right it was terrible, and you know it now that you've had the whip of hard times laid on your back. You should have been preparing yourself for trouble. Your people must have been well off to send you to the university. Stop me, if I'm stepping on your toes.

6. European railroad sleeping-cars.
7. "Where there's life there's hope" (literally, "while I breathe I hope"); "give the right hand to the wretched"; "the die is cast"; "higher!"

Did your mother pamper you? Did your father give in to you? Were you brought up tenderly, with permission to go and find out what were the last things that everything else stands for while everybody else labored in the fallen world of appearances?"

"Well, no, it wasn't exactly like that." Grebe smiled. *The fallen world of appearances!* no less. But now it was his turn to deliver a surprise. "We weren't rich. My father was the last genuine English butler in Chicago—"

"Are you kidding?"

"Why should I be?"

"In a livery?"

"In livery. Up on the Gold Coast."

"And he wanted you to be educated like a gentleman?"

"He did not. He sent me to the Armour Institute to study chemical engineering. But when he died I changed schools."

He stopped himself, and considered how quickly Raynor had reached him. In no time he had your valise on the table and all your stuff unpacked. And afterward, in the streets, he was still reviewing how far he might have gone, and how much he might have been led to tell if they had not been interrupted by Mrs. Staika's great noise.

But just then a young woman, one of Raynor's workers, ran into the cubicle exclaiming, "Haven't you heard all the fuss?"

"We haven't heard anything."

"It's Staika, giving out with all her might. The reporters are coming. She said she phoned the papers, and you know she did."

"But what is she up to?" said Raynor.

"She brought her wash and she's ironing it here, with our current, because the relief won't pay her electric bill. She has her ironing board set up by the admitting desk, and her kids are with her, all six. They never are in school more than once a week. She's always dragging them around with her because of her reputation."

"I don't want to miss any of this," said Raynor, jumping up. Grebe, as he followed with the secretary, said, "Who is this Staika?"

"They call her the 'Blood Mother of Federal Street.' She's a professional donor at the hospitals. I think they pay ten dollars a pint. Of course it's no joke, but she makes a very big thing out of it and she and the kids are in the papers all the time."

A small crowd, staff and clients divided by a plywood barrier, stood in the narrow space of the entrance, and Staika was shouting in a gruff, mannish voice, plunging the iron on the board and slamming it on the metal rest.

"My father and mother came in a steerage, and I was born in our house, Robey by Huron. I'm no dirty immigrant. I'm a U.S. citizen. My husband is a gassed veteran from France with lungs weaker'n paper, that hardly can he go to the toilet by himself. These six children of mine, I have to buy the shoes for their feet with my own blood. Even a lousy little white Communion necktie, that's a couple of drops of blood; a little piece of mosquito veil for my Vadja so she won't be ashamed in church for the other girls, they take my blood for it by Goldblatt. That's how I keep goin'. A fine thing if I had to depend on the relief. And there's plenty of people on the rolls—fakes! There's nothin' they can't get, that can go and wrap bacon at Swift and Armour any time. They're lookin' for them by the Yards. They never have to be out of work. Only they rather lay in their lousy beds and eat the public's money." She was not afraid, in a predominantly Negro station, to shout this way about Negroes. Grebe and Raynor worked themselves forward to get a closer view

... and with pleasure at her—... woman who wore a cotton cap ... and had on black gym ... not much worked

...Staika's dress on the ironing board. And the children, silent and white, with a kind of locked obstinacy, in sheepskins and lumber-jackets, stood behind her. She had captured the station, and the pleasure this gave her was enormous. Yet her grievances were true grievances. She was telling the truth. But she behaved like a liar. The look of her small eyes was hidden, and while she raged she also seemed to be spinning and planning.

"They send me out college case workers in silk pants to talk me out of what I got comin'. Are they better'n me? Who told them? Fire them. Let 'em go and get married, and then you won't have to cut electric from people's budget."

The chief supervisor, Mr. Ewing, couldn't silence her and he stood with folded arms at the head of his staff, bald, bald-headed, saying to his subordinates like the ex-school principal he was, "Pretty soon she'll be tired and go."

"No she won't," said Raynor to Grebe. "She'll get what she wants She knows more about the relief even then Ewing. She's been on the rolls for years, and she always gets what she wants because she puts on a noisy show. Ewing knows it. He'll give in soon. He's only saving face. If he gets bad publicity, the Commissioner'll have him on the carpet, downtown. She's got him submerged; she'll submerge everybody in time, and that includes nations and governments."

Grebe replied with his characteristic smile, disagreeing complete-ly. Who would take Staika's orders, and what changes could her yelling ever bring about?

No, what Grebe saw in her, the power that made people listen, was that her cry expressed the war of flesh and blood, perhaps turned a little crazy and certainly ugly, on this place and this condi-tion. And at first, when he went out, the spirit of Staika somehow presided over the whole district for him, and it took color from her;

8. Woman's coverall, popular during World War I.

he saw her color, in the spotty curb fires, and the fires under the El,
the straight alley of flamy gloom. Later, too, when he went into a
tavern for a shot of rye, the sweat of beer, association with West Side
Polish streets, made him think of her again.

He wiped the corners of his mouth with his muffler, his handkerchief being inconvenient to reach for, and went out again to get
on with the delivery of his checks. The air bit cold and hard, a
few flakes of snow formed near him. A train struck its wave,
quiver in the frames and a bristling icy hiss o—

Crossing the street, he descended —
basement grocery, setting off g, 'an Italian with a long, hollow face
it caught you with it....g,
and fish. The—
and stubborn bristles. He kept his hands warm under his apron.

No, he didn't know Green. You knew people but not names. The
same man might not have the same name twice. The police didn't
know, either, and mostly didn't care. When somebody was shot or
knifed they took the body away and didn't look for the murderer.
In the first place, nobody would tell them anything. So they made
up a name for the coroner and called it quits. And in the second
place, they didn't give a goddamn anyhow. But they couldn't get to
the bottom of a thing even if they wanted to. Nobody would get to
know even a tenth of what went on among these people. They
stabbed and stole, they did every crime and abomination you ever
heard of, men and men, women and women, parents and children,
worse than the animals. They carried on their own way, and the horrors passed off like a smoke. There was never anything like it in the
history of the whole world.

It was a long speech, deepening with every word in its fantasy
and passion and becoming increasingly senseless and terrible: a
swarm amassed by suggestion and invention, a huge, hugging, despairing knot, a human wheel of heads, legs, bellies, arms, rolling
through his shop.

Grebe felt that he must interrupt him. He said sharply, "What
are you talking about! All I asked was whether you knew this man."

"That isn't even the half of it. I been here six years. You probably
don't want to believe this. But suppose it's true?"

"All the same," said Grebe, "there must be a way to find a
person."

The Italian's close-spaced eyes had been queerly concentrated, as
were his muscles, while he leaned across the counter trying to convince Grebe. Now he gave up the effort and sat down on his stool.
"Oh—I suppose. Once in a while. But I been telling you, even the
cops don't get anywhere."

"They're always after somebody. It's not the same thing."

"Well, keep trying if you want. I can't help you." But he didn't keep trying. He had no more time to spend on Green. He slipped Green's check to the back of the block. The next name on the list was FIELD, WINSTON.

He found the back-yard bungalow without the least trouble; it shared a lot with another house, a few feet of yard between. Grebe knew these two-shack arrangements. They had been built in vast numbers in the days before the swamps were filled and the streets raised, and they were all the same—a boardwalk along the fence, three or four ball-headed posts for clothes-... street level, ...ead shingles, and a long, long flight of stairs ...drew out his checks. "Go bring me my box of papers." He cleared a space on the table.

"Oh, you don't have to go to all that trouble," said Grebe. But Field laid out his papers: Social Security card, relief certification, letters from the state hospital in Manteno, and a naval discharge dated San Diego, 1920.

"That's plenty," Grebe said. "Just sign."

"You got to know who I am," the old man said. "You're from the Government. It's not your check, it's a Government check and you got no business to hand it over till everything is proved."

He loved the ceremony of it, and Grebe made no more objections. Field emptied his box and finished out the circle of cards and letters.

"There's everything I done and been. Just the death certificate and they can close book on me." He said this with a certain happy pride and magnificence. Still he did not sign; he merely held the little pen upright on the golden-green corduroy of his thigh. Grebe did not hurry him. He felt the old man's hunger for conversation.

"I got to get better coal," he said. "I send my little gran'son to the yard with my order and they fill his wagon with screening. The stove ain't made for it. It fall through the grate. The order says Franklin County egg-size coal."

"I'll report it and see what can be done."

"Nothing can be done, I expect. You know and I know. There ain't no little ways to make things better, and the only big thing is money. That's the only sunbeams, money. Nothing is black where it shines, and the only place you see black is where it ain't shining. What we colored have to have is our own rich. There ain't no other way."

Grebe sat, his reddened forehead bridged levelly by his close-cut hair and his cheeks lowered in the wings of his collar—the caked fire shone hard within the isinglass-and-iron frames but the room was not comfortable—sat and listened while the old man unfolded his scheme. This was to create one Negro millionaire a month by subscription. One clever, good-hearted young fellow elected every month would sign a contract to use the money to start a business employing Negroes. This would be advertised by chain letters and word of mouth, and every Negro wage earner would contri... dollar a month. Within five years there would be sixty... "That'll fetch respect," he said with a throa... was reddish. came out like a foreign syllable. "You ... golden sun in this dark room, the money that gets thrown a... race. As long as they can ... for you. Money, th... of mixed bl... And ... shaggy and slab-headed, with the mingled blood of his face and broad lips, the little pen still upright in his hand, like one of the underground kings of mythology, old judge Minos himself.

And now he accepted the check and signed. Not to soil the slip, he held it down with his knuckles. The table budged and creaked, the center of the gloomy, heathen midden of the kitchen covered with bread, meat, and cans, and the scramble of papers.

"Don't you think my scheme'd work?"

"It's worth thinking about. Something ought to be done, I agree."

"It'll work if people will do it. That's all. That's the only thing, any time. When they understand it in the same way, all of them."

"That's true," said Grebe, rising. His glance met the old man's.

"I know you got to go," he said. "Well, God bless you, boy, you ain't been sly with me. I can tell it in a minute."

He went back through the buried yard. Someone nursed a candle in a shed, where a man unloaded kindling wood from a sprawl-wheeled baby buggy and two voices carried on a high conversation. As he came up the sheltered passage he heard the hard boost of the wind in the branches and against the house fronts, and then, reaching the sidewalk, he saw the needle-eye red of cable towers in the open icy height hundreds of feet above the river and the factories—those keen points. From here, his view was obstructed all the way to the South Branch and its timber banks, and the cranes beside the water. Rebuilt after the Great Fire,[9] this part of the city was, not fifty years later, in ruins again, factories boarded up, buildings deserted or fallen, gaps of prairie between. But it wasn't desolation that this made you feel, but rather a faltering of organization that set free a huge energy, an escaped, unattached, unregulated power from the giant raw place. Not only must people feel it but, it seemed to Grebe, they were compelled to match it. In their very bodies. He no less than others, he realized. Say that his parents had been ser-

9. The Chicago Fire of 1871.

vants in their time, whereas he was not supposed to be one. He thought that they had never done any service like this, which no one visible asked for, and probably flesh and blood could not even perform. Nor could anyone show why it should be performed; or see where the performance would lead. That did not mean that he wanted to be released from it, he realized with a grimly pensive face. On the contrary. He had something to do. To be compelled to feel this energy and yet have no task to do—that was horrible; that was suffering; he knew what that was. It was now quitting time. Six o'clock. He could go home if he liked, to his room, that is, to wash in hot water, to pour a drink, lie down on his quilt, read the paper, eat some liver paste on crackers before going out to dinner. But to think of this actually made him feel a little sick as though he had swallowed hard air. He had six checks left, and he was determined to deliver at least one of these: Mr. Green's check.

So he started again. He had four or five dark blocks to go, past open lots, condemned houses, old foundations, closed schools, black churches, mounds, and he reflected that there must be many people alive who had once seen the neighborhood rebuilt and new. Now there was a second layer of ruins; centuries of history accomplished through human massing. Numbers had given the place forced growth; enormous numbers had also broken it down. Objects once so new, so concrete that it could have occurred to anyone they stood for other things, had crumbled. Therefore, reflected Grebe, the secret of them was out. It was that they stood for themselves by agreement, and were natural and not unnatural by agreement, and when the things themselves collapsed the agreement became visible. What was it, otherwise, that kept cities from looking peculiar? Rome, that was almost permanent, did not give rise to thoughts like these. And was it abidingly real? But in Chicago, where the cycles were so fast and the familiar died out, and again rose changed, and died again in thirty years, you saw the common agreement or covenant, and you were forced to think about appearances and realities. (He remembered Raynor and he smiled. Raynor was a clever boy.) Once you had grasped this, a great many things became intelligible. For instance, why Mr. Field should conceive such a scheme. Of course, if people were to agree to create a millionaire, a real millionaire would come into existence. And if you wanted to know how Mr. Field was inspired to think of this, why, he had within sight of his kitchen window the chart, the very bones of a successful scheme— the El with its blue and green confetti of signals. People consented to pay dimes and ride the crash-box cars, and so it was a success. Yet how absurd it looked; how little reality there was to start with. And yet Yerkes,[1] the great financier who built it, had known that he could get people to agree to do it. Viewed as itself, what a scheme

1. Charles Tyson Yerkes (1837–1905), American financier.

of a scheme it seemed, how close to an appearance. Then why wonder at Mr. Field's idea? He had grasped a principle. And then
Grebe remembered, too, that Mr. Yerkes had established the Yerkes
Observatory and endowed it with millions. Now how did the notion
come to him in his New York museum of a palace or his Aegean-
bound yacht to give money to astronomers? Was he awed by the
success of his bizarre enterprise and therefore ready to spend money
to find out where in the universe being and seeming were identical?
Yes, he wanted to know what abides; and whether flesh is Bible
grass; and he offered money to be burned in the fire of suns. Okay,
then, Grebe thought further, these things exist because people consent to exist with them—we have got so far—and also there is a
reality which doesn't depend on consent but within which consent
is a game. But what about need, the need that keeps so many vast
thousands in position? You tell me that, you *private* little gentleman
and *decent* soul—he used these words against himself scornfully.
Why is the consent given to misery? And why so painfully ugly?
Because there is *something* that is dismal and permanently ugly?
Here he sighed and gave it up, and thought it was enough for the
present moment that he had a real check in his pocket for a Mr.
Green who must be real beyond question. If only his neighbors
didn't think they had to conceal him.

This time he stopped at the second floor. He struck a match and
found a door. Presently a man answered his knock and Grebe had
the check ready and showed it even before he began. "Does Tulliver
Green live here? I'm from the relief."

The man narrowed the opening and spoke to someone at his back.
"Does he live here?"

"Uh-uh. No."

"Or anywhere in this building? He's a sick man and he can't come
for his dough." He exhibited the check in the light, which was smoky
—the air smelled of charred lard—and the man held off the brim of
his cap to study it.

"Uh-uh. Never seen the name."

"There's nobody around here that uses crutches?"

He seemed to think, but it was Grebe's impression that he was
simply waiting for a decent interval to pass.

"No, suh. Nobody I ever see."

"I've been looking for this man all afternoon"—Grebe spoke out
with sudden force—"and I'm going to have to carry this check back
to the station. It seems strange not to be able to find a person to
give him something when you're looking for him for a good reason.
I suppose if I had bad news for him I'd find him quick enough."

There was a responsive motion in the other man's face. "That's
right, I reckon."

"It almost doesn't do any good to have a name if you can't be found by it. It doesn't stand for anything. He might as well not have any," he went on, smiling. It was as much of a concession as he could make to his desire to laugh.

"Well, now, there's a little old knot-back man I see once in a while. He might be the one you lookin' for. Downstairs."

"Where? Right side or left? Which door?"

"I don't know which. Thin-face little knot-back with a stick."

But no one answered at any of the doors on the first floor. He went to the end of the corridor, searching by matchlight, and found only a stairless exit to the yard, a drop of about six feet. But there was a bungalow near the alley, an old house like Mr. Field's. To jump was unsafe. He ran from the front door, through the underground passage and into the yard. The place was occupied. There was a light through the curtains, upstairs. The name on the ticket under the broken, scoop-shaped mailbox was Green! He exultantly rang the bell and pressed against the locked door. Then the lock clicked faintly and a long staircase opened before him. Someone was slowly coming down—a woman. He had the impression in the weak light that she was shaping her hair as she came, making herself presentable, for he saw her arms raised. But is was for support that they were raised; she was feeling her way downward, down the wall, stumbling. Next he wondered about the pressure of her feet on the treads; she did not seem to be wearing shoes. And it was a freezing stairway. His ring had got her out of bed, perhaps, and she had forgotten to put them on. And then he saw that she was not only shoeless but naked; she was entirely naked, climbing down while she talked to herself, a heavy woman, naked and drunk. She blundered into him. The contact of her breasts, though they touched only his coat, made him go back against the door with a blind shock. See what he had tracked down, in his hunting game!

The woman was saying to herself, furious with insult, "So I cain't ——k, huh? I'll show that son-of-a-bitch kin I, cain't I."

What should he do now? Grebe asked himself. Why, he should go. He should turn away and go. He couldn't talk to this woman. He couldn't keep her standing naked in the cold. But when he tried he found himself unable to turn away.

He said, "Is this where Mr. Green lives?"

But she was still talking to herself and did not hear him.

"Is this Mr. Green's house?"

At last she turned her furious drunken glance on him. "What do you want?"

Again her eyes wandered from him; there was a dot of blood in their enraged brilliance. He wondered why she didn't feel the cold.

"I'm from the relief."

"Awright, what?"

"I've got a check for Tulliver Green."

This time she heard him and put out her hand.

"No, no, for *Mr.* Green. He's got to sign," he said. How was he going to get Green's signature tonight!

"I'll take it. He cain't."

He desperately shook his head, thinking of Mr. Field's precautions about identification. "I can't let you have it. It's for him. Are you Mrs. Green?"

"Maybe I is, and maybe I ain't. Who want to know?"

"Is he upstairs?"

"Awright. Take it up yourself, you goddamn fool."

Sure, he was a goddamn fool. Of course he could not go up because Green would probably be drunk and naked, too. And perhaps he would appear on the landing soon. He looked eagerly upward. Under the light was a high narrow brown wall. Empty! It remained empty!

"Hell with you, then!" he heard her cry. To deliver a check for coal and clothes, he was keeping her in the cold. She did not feel it, but his face was burning with frost and self-ridicule. He backed away from her.

"I'll come tomorrow, tell him."

"Ah, hell with you. Don' never come. What you doin' here in the nighttime? Don' come back." She yelled so that he saw the breadth of her tongue. She stood astride in the long cold box of the hall and held on to the banister and the wall. The bungalow itself was shaped something like a box, a clumsy, high box pointing into the freezing air with its sharp, wintry lights.

"If you are Mrs. Green, I'll give you the check," he said, changing his mind.

"Give here, then." She took it, took the pen offered with it in her left hand, and tried to sign the receipt on the wall. He looked around, almost as though to see whether his madness was being observed, and came near believing that someone was standing on a mountain of used tires in the auto-junking shop next door.

"But are you Mrs. Green?" he now thought to ask. But she was already climbing the stairs with the check, and it was too late, if he had made an error, if he was now in trouble, to undo the thing. But he wasn't going to worry about it. Though she might not be Mrs. Green, he was convinced that Mr. Green was upstairs. Whoever she was, the woman stood for Green, whom he was not to see this time. Well, you silly bastard, he said to himself, so you think you found him. So what? Maybe you really did find him—what of it? But it was important that there was a real Mr. Green whom they could not keep him from reaching because he seemed to come as an emissary from hostile appearances. And though the self-ridicule was slow to diminish and his face still blazed with it, he had, nevertheless, a feeling of elation, too. "For after all," he said, "he *could* be found!"

1951, 1968

NORMAN MAILER

1923–

"The sour truth is that I am imprisoned with a perception which will settle for nothing less than making a revolution in the consciousness of our time. Whether rightly or wrongly, it is then obvious that I should go so far as to think it is my present and future work which will have the deepest influence of any work being done by an American novelist in these years." Taken from the introductory "advertisement" in Norman Mailer's first collection of his occasional journalism, reviews, and stories, these two sentences lay out the terms in which the writer insists his work be considered and judged. As Richard Poirier has put it, Mailer "has exhibited a literary ambition that can best be called imperialistic. He has wanted to translate his life into a literary career and then to translate that literary career into history." To a remarkable degree this writer has succeeded in realizing his aim.

He was born in New Jersey, grew up in Brooklyn, went to Harvard College, where by the end of his first semester (and just before turning seventeen) he "formed the desire to be a major writer." Mailer took a number of writing courses at Harvard (as well as ones in engineering), won various literary prizes, then upon graduation entered the army with the intention of writing a "great war novel" about either the European or the Pacific conflict, whichever was to be his personal fate. He served in the Pacific and came home to write, over the course of fifteen months, *The Naked and the Dead*. If not obviously a "great war novel," it is a good one, strong when it evokes place and action, lively as a drama of debate, of ideas in conflict. It was also a best-seller and got Mailer off to a success which would take some work to live up to.

His uncertain and turbulent career in the ten years after the war is described in *Advertisements for Myself* (1959): marriages and divorces, disillusionment with leftist politics (Mailer supported former Vice-President Henry Wallace's ill-fated campaign for the presidency in 1948), experiments with drugs and with new fictional styles. The novels from these years, *Barbary Shore* (1951) and *The Deer Park* (1955), were largely treated as fallings-off from his first book. But in *Barbary Shore* Mailer ambitiously attempted and at least partially succeeded in weaving an eerie Cold War fantasy about sex, power, and totalitarianism, all played out within a Brooklyn rooming house. And *The Deer Park* moved across country to Hollywood, continuing to explore relationships among sex, politics, and money, as well as demonstrating Mailer's stylistic debt to F. Scott Fitzgerald.

After *The Deer Park* appeared in 1955, Mailer began to write a column for the *Village Voice*, a Greenwich Village weekly he had helped found. Concurrently he became fascinated with a phenomenon he called "Hip" and with its embodiment in the Hipster, a heroic, cool, but potentially violent figure whose essential being contradicted everything Mailer saw as the dull pieties of social adjustment fashionable in the years of Eisenhower's presidency. "The shits are killing us," was Mailer's war cry, as he set out to confound the "square" mentality by finding suitable antagonists for it, such as the "Existential hero" he saw in John F. Kennedy, or the comic warrior Sergius O'Shaugnessy who appears in Mailer's fiction; or (most luridly and pompously) in the notion of himself as an investigator of "psychic mysteries" like rape, suicide, and orgasm. The fullest development of these theories is found in his essay *The White Negro* (1959); while his culminating fictional portrayal of the hero-as-hipster is Stephen Rojack, the narrator of *An American Dream* (1964) who murders his wife in the novel's second chapter. Adverse critics of the book saw Rojack as no more than a crude projection of his author's destructive fantasies. But Mailer is larger than his character, and the novel an ingenious attempt to create a more human and healthier consciousness.

As his work became increasingly subversive of traditional distinctions among "fiction," "essay," "poetry," or "sociology," Mailer's style grew bolder in metaphor, more daring in its improvisations. And as the country plunged ever deeper into the Vietnam tragedy, his imagination was stung to the composition of a brilliantly obscene improvisation (officially a novel) *Why Are We in Vietnam?* (1967), then to perhaps his finest book, *The Armies of the Night* (1968). Here, for once, the power of the man and of the moment came together in an extended piece of writing that is at once history, fiction, and prophecy. Springing from Mailer's involvement in the 1967 march on the Pentagon in protest of the war, it plays off private, even ignoble concerns against the public significances borne by the event. The "Mailer" who moves through these pages is a less-than-heroic figure, often available for comic and satiric treatment by the author, though not exclusively so; and in that quality *Armies* resembles no book in American letters so much as *The Education of Henry Adams*. Thick with often humorously observed particularity, its tone is eventually graver and more serious than anything he had written before or has written since.

Since 1967 he has been at work on a gigantic novel to end all American novels. Meanwhile he has run for mayor of New York City, and made a number of films—of which perhaps the most interesting is *Maidstone*, to whose published text he wrote a long introduction. He has also, along with a book on spaceflight (*A Fire on the Moon*, 1970), written a polemic against Kate Millett and certain aspects of the women's movement (*The Prisoner of Sex*, 1971), a biography of Marilyn Monroe, and an account of the Foreman-Ali heavyweight struggle in Zaire. Throughout these books, as well as in his personal appearances on late-night talk shows, his comic sense has shown to good advantage the ability in uncertain or unpleasant situations to interpolate, mimic, turn a phrase against the world or himself, sometimes against both at the same time. Whether Mailer's literary production has as yet or will ever live up to the immense promise he made for himself—to make "a revolution in the consciousness of our time"—there is fertility and variety in his performances as an entertainer, diagnostician, and prophet committed to nothing less than what Emerson called the "conversion of the world."

From The Armies of the Night[1]
Part I
5. TOWARD A THEATER OF IDEAS

The guests were beginning to leave the party for the Ambassador, which was two blocks away. Mailer did not know this yet, but the audience there had been waiting almost an hour. They were being entertained by an electronic folk rock guitar group, so presumably the young were more or less happy, and the middle-aged dim. Mailer was feeling the high sense of clarity which accompanies the light show of the aurora borealis when it is projected upon the inner universe of the chest, the lungs, and the heart. He was happy. On leaving, he had appropriated a coffee mug and filled it with bourbon. The fresh air illumined the bourbon, gave it a cerebrative edge; words entered his brain with the agreeable authority of fresh minted coins. Like all good professionals, he was stimulated by the chance to try a new if related line of work. Just as professional football players love sex because it is so close to football, so he was fond of speaking in public because it was thus near to writing. An extravagant analogy? Consider that a good half of writing consists of being sufficiently sensitive to the moment to reach for the next promise which is usually hidden in some word or phrase just a shift to the side of one's conscious intent. (Consciousness, that blunt tool, bucks in the general direction of the truth; instinct plucks the feather. Cheers!) Where public speaking is an exercise from prepared texts to demonstrate how successfully a low order of consciousness can beat upon the back of a collective flesh, public speaking being, therefore, a sullen expression of human possibility metaphorically equal to a bugger on his victim, speaking-in-public (as Mailer liked to describe any speech which was more or less improvised, impromptu, or dangerously written) was an activity like writing; one had to trick or seize or submit to the grace of each moment, which, except for those unexpected and sometimes well-deserved moments when consciousness and grace came together (and one felt on the consequence, heroic) were usually occasions of some mystery. The pleasure of speaking in public was the sensitivity it offered: with every phrase one was better or worse, close or less close to the existential promise of truth, *it feels true*, which hovers on good occasions like a presence between speaker and audience.

---

1. The following selection consists of Chapters 5 and 6 (Part 1) from *The Armies of the Night* (1968). The occasion is the anti-Vietnam war march to the Pentagon in October, 1967. Mailer, despite some reservations, has agreed to participate in this march, and also in a premarch evening rally at a Washington theater. In the first two chapters, which describe that evening, the principal actors are Mailer; the poet Robert Lowell; Dwight Macdonald, a leading American critic and journalist of politics and literature; Paul Goodman, poet and cultural critic, author of *Growing Up Absurd* (1959); and Ed de Grazia, one of the organizers of the march.

Sometimes one was better, and worse, at the same moment; so strategic choices on the continuation of the attack would soon have to be decided, a moment to know the blood of the gambler in oneself.

Intimations of this approaching experience, obviously one of Mailer's preferred pleasures in life, at least when he did it well, were now connected to the professional sense of intrigue at the new task: tonight he would be both speaker and master of ceremonies. The two would conflict, but interestingly. Already he was looking in his mind for kind even celebrative remarks about Paul Goodman which would not violate every reservation he had about Goodman's dank glory. But he had it. It would be possible with no violation of truth to begin by saying that the first speaker looked very much like Nelson Algren,[2] because in fact the first speaker was Paul Goodman, and both Nelson Algren and Paul Goodman looked like old cons. Ladies and Gentlemen, without further ado let me introduce one of young America's favorite old cons, Paul Goodman! (It would not be necessary to add that where Nelson Algren looked like the sort of skinny old con who was in on every make in the joint, and would sign away Grandma's farm to stay in the game, Goodman looked like the sort of old con who had first gotten into trouble in the YMCA, and hadn't spoken to anyone since.)

All this while, Mailer had in clutch *Why Are We in Vietnam?*[3] He had neglected to bring his own copy to Washington and so had borrowed the book from his hostess on the promise he would inscribe it. (Later he was actually to lose it—working apparently on the principle that if you cannot make a hostess happy, the next best charity is to be so evil that the hostess may dine out on tales of your misconduct.) But the copy of the book is now noted because Mailer, holding it in one hand and the mug of whisky in the other, was obliged to notice on entering the Ambassador Theater that he had an overwhelming urge to micturate. The impulse to pass urine, being for some reason more difficult to restrain when both hands are occupied, there was no thought in the Master of Ceremonies' mind about the alternatives—he would have to find The Room before he went on stage.

That was not so immediately simple as one would have thought. The twenty guests from the party, looking a fair piece subdued under the fluorescent lights, had therefore the not unhaggard look of people who have arrived an hour late at the theater. No matter that the theater was by every evidence sleazy (for neighborhood movie houses built on the dream of the owner that some day Garbo or Harlow or Lombard would give a look in, aged immediately they were not used for movies anymore) no matter, the guests had the

2. Chicago-based contemporary American novelist, author of *The Man with the*
Golden Arm (1949) and other works.
3. Mailer's most recent novel (1967).

uneasiness of very late arrivals. Apologetic, they were therefore in haste for the speakers to begin.

Mailer did not know this. He was off already in search of The Room, which, it developed was up on the balcony floor. Imbued with the importance of his first gig as Master of Ceremonies, he felt such incandescence of purpose that he could not quite conceive it necessary to notify de Grazia he would be gone for a minute. Incandescence is the *satori*[4] of the Romantic spirit which spirit would insist—this is the essence of the Romantic—on accelerating time. The greater the power of any subjective state, the more total is a Romantic's assumption that everyone understands exactly what he is about to do, therefore waste not a moment by stopping to tell them.

Flush with his incandescence, happy in all the anticipations of liberty which this Götterdämmerung[5] of a urination was soon to provide, Mailer did not know, but he had already and unwitting to himself metamorphosed into the Beast. Wait and see!

He was met on the stairs by a young man from *Time* magazine, a stringer presumably, for the young man lacked that I-am-damned look in the eye and rep tie of those whose work for *Time* has become a life addiction. The young man had a somewhat ill-dressed look, a map showed on his skin of an old adolescent acne, and he gave off the unhappy furtive presence of a fraternity member on probation for the wrong thing, some grievous mis-deposit of vomit, some hanky panky with frat-house tickets.

But the Beast was in a great good mood. He was soon to speak; that was food for all. So the Beast greeted the *Time* man with the geniality of a surrogate Hemingway unbending for the Luce-ites (Loo-sights was the pun)[6] made some genial cryptic remark or two about finding Herr John, said cheerfully in answer to why he was in Washington that he had come to protest the war in Vietnam, and taking a sip of bourbon from the mug he kept to keep all fires idling right, stepped off into the darkness of the top balcony floor, went through a door into a pitch-black men's room, and was alone with his need. No chance to find the light switch for he had no matches, he did not smoke. It was therefore a matter of locating what's what with the probing of his toes. He found something finally which seemed appropriate, and pleased with the precision of these generally unused senses in his feet, took aim between them at a point twelve inches ahead, and heard in the darkness the sound of his water striking the floor. Some damn mistake had been made, an assault from the side doubtless instead of the front, the bowl was

---

4. Zen Buddhist term for spiritual enlightenment.
5. "Twilight of the Gods"; i.e., on a grand scale.
6. Employees of the publishing empire owned by Henry Luce (*Time, Life*, etc.). "Lucite" is a durable synthetic or plastic, but also "Loo" (English for bathroom) "sights" (sites).

relocated now, and Master of Ceremonies breathed deep of the great reveries of this utterly non-Sisyphian[7] release—at last!!—and thoroughly enjoyed the next forty-five seconds, being left on the aftermath not a note depressed by the condition of the premises. No, he was off on the Romantic's great military dream, which is: seize defeat, convert it to triumph. Of course, pissing on the floor was bad; very bad; the attendant would probably gossip to the police, (if the *Time* man did not sniff it out first) and The Uniformed in turn would report it to The Press who were sure to write about the scandalous condition in which this meeting had left the toilets. And all of this contretemps merely because the management, bitter with their lost dream of Garbo and Harlow and Lombard, were now so pocked and stingy they doused the lights. (Out of such stuff is a novelist's brain.)

Well, he could convert this deficiency to an asset. From gap to gain is very American. He would confess straight out to all aloud that he was the one who wet the floor in the men's room, he alone! While the audience was recovering from the existential anxiety of encountering an orator who confessed to such a crime, he would be able—their attention now riveted—to bring them up to a contemplation of deeper problems, of, indeed, the deepest problems, the most chilling alternatives, and would from there seek to bring them back to a restorative view of man. Man might be a fool who peed in the wrong pot, man was also a scrupulous servant of the self-damaging admission; man was therefore a philosopher who possessed the magic stone; he could turn loss to philosophical gain, and so illumine the deeps, find the poles, and eventually learn to cultivate his most special fool's garden: *satori*, incandescence, and the hard gemlike flame[8] of bourbon burning in the furnaces of metabolism.

Thus composed, illumined by these first stages of Emersonian transcendence, Mailer left the men's room, descended the stairs, entered the back of the orchestra, all opening remarks held close file in his mind like troops ranked in order before the parade, and then suddenly, most suddenly saw, with a cancerous swoop of albatross wings, that de Grazia was on the stage, was acting as M.C., was—no calling it back—launched into the conclusion of a gentle stammering stumbling—small orator, de Grazia!—introduction of Paul Goodman. All lost! The magnificent opening remarks about the forces gathered here to assemble on Saturday before the Pentagon, this historic occasion, let us hold it in our mind and focus on a puddle of passed water on the floor above and see if we assembled here can as leftists and proud dissenters contain within our minds the grandeur of the two—all lost!—no chance to do more than pick

---

7. Sisyphus was condemned to push a huge boulder up a hill with, unlike Mailer, no prospect of relief.
8. The English critic Walter Pater (1839–

94) advocated in *Studies in the History of the Renaissance* that one should cultivate such an intense response to art.

up later—later! after de Grazia and Goodman had finished dead-assing the crowd. Traitor de Grazia! Sicilian de Grazia!

As Mailer picked his way between people sitting on the stone floor (orchestra seats had been removed—the movie house was a dance hall now with a stage) he made a considerable stir in the orchestra. Mailer had been entering theaters for years, mounting stages—now that he had put on weight, it would probably have been fair to say that he came to the rostrum like a poor man's version of Orson Welles,[9] some minor note of the same contemplative presence. A titter and rise of expectation followed him. He could not resist its appeal. As he passed de Grazia, he scowled, threw a look from Lower Shakespearian "Et tu Bruté," and proceeded to slap the back of his hand against de Grazia's solar plexus. It was not a heavy blow, but then de Grazia was not a heavy man; he wilted some hint of an inch. And the audience pinched off a howl, squeaked on their squeal. It was not certain to them what had taken place.

Picture the scene two minutes later from the orchestra floor. Paul Goodman, now up at the microphone with no podium or rostrum, is reading the following lines:

> . . . these days my contempt for the misrulers of my country is icy and my indignation raucous.

It is impossible to tell what he is reading. Off at the wing of the stage where the others are collected—stout Macdonald, noble Lowell, beleaguered de Grazia, and Mailer, Prince of Bourbon, the acoustics are atrocious. One cannot hear a word the speaker is saying. Nor are there enough seats. If de Grazia and Macdonald are sitting in folding chairs, Mailer is squatting on his haunches, or kneeling on one knee like a player about to go back into the ball game. Lowell has the expression on his face of a dues payer who is just about keeping up with the interest on some enormous debt. As he sits on the floor with his long arms clasped mournfully about his long Yankee legs, "I am here," says his expression, "but I do not have to pretend I like what I see." The hollows in his cheeks give a hint of the hanging judge. Lowell is of a good weight, not too heavy, not too light, but the hollows speak of the great Puritan gloom in which the country was founded—man was simply not good enough for God.

At this moment, it is hard not to agree with Lowell. The cavern of the theater seems to resonate behind the glare of the footlights, but this is no resonance of a fine bass voice—it is rather electronics on the march. The public address system hisses, then rings in a random chorus of electronic music, sounds of cerebral mastication from some horror machine of Outer Space (where all that elec-

9. Actor and film director (b. 1915) of large bulk.

tricity doubtless comes from, child!) then a hum like the squeak in the hinges of the gates of Hell—we are in the penumbra of psyche-delic netherworlds, ghost-odysseys from the dead brain cells of adolescent trysts with LSD, some ultrapurple spotlight from the balcony (not ultraviolet—ultrapurple, deepest purple one could conceive) there out in the dark like some neon eye of the night, the media is the message,[1] and the message is purple, speaks of the monarchies of Heaven, madnesses of God, and clam-vaults of peo-ple on a stone floor. Mailer's senses are now tuned to absolute pitch or sheer error—he marks a ballot for absolute pitch—he is certain there is a profound pall in the audience. Yes, they sit there, stricken, inert, in terror of what Saturday will bring, and so are unable to rise to a word the speaker is offering them. It will take dynamite to bring life. The shroud of burned-out psychedelic dreams is in this audience, Cancer Gulch with open maw—and Mailer thinks of the vigor and the light (from marijuana?) in the eyes of those Ameri-can soldiers in Vietnam who have been picked by the newsreel cameras to say their piece, and the happy healthy never unintelli-gent faces of all those professional football players he studies so assiduously on television come Sunday (he has neglected to put his bets in this week) and wonders how they would poll out on senti-ment for the war.

HAWKS 95   DOVES 6

NFL Footballers Approve Vietnam War

Doubtless. All the healthy Marines, state troopers, professional athletes, movie stars, rednecks, sensuous life-loving Mafia, cops, mill workers, city officials, nice healthy-looking easy-grafting politi-cians full of the light (from marijuana?) in their eye of a life they enjoy—yes, they would be for the war in Vietnam. Arrayed against them as hard-core troops: an elite! the Freud-ridden embers of Marxism, good old American anxiety strata—the urban middle-class with their proliferated monumental adenoidal resentments, their secret slavish love for the oncoming hegemony of the com-puter and the suburb, yes, they and their children, by the sheer ironies, the sheer ineptitude, the *kinks* of history, were now being compressed into more and more militant stands, their resistance to the war some hopeless melange, somehow firmed, of Pacifism and closet Communism. And their children—on a freak-out from the suburbs to a love-in on the Pentagon wall.

It was the children in whom Mailer had some hope, a gloomy hope. These mad middle-class children with their lobotomies from sin, their nihilistic embezzlement of all middle-class moral funds,

---

1. Herbert Marshall McLuhan (b. 1911), a Canadian critic and communications theorist, coined this slogan in *Under-standing Media* (1964) to say that each medium of communication (TV, print, etc.) determines its appropriate mode of response, whatever the "subject" of its address.

their innocence, their lust for apocalypse, their unbelievable indifference to waste: twenty generations of buried hopes perhaps
engraved in their chromosomes, and now conceivably burning like
faggots in the secret inquisitional fires of LSD. It was a devil's
drug—designed by the Devil to consume the love of the best, and
leave them liver-wasted, weeds of the big city. If there had been a
player piano, Mailer might have put in a quarter to hear "In the
Heart of the City Which Has No Heart."

Yes, these were the troops: middle-class cancer-pushers and drug-
gutted flower children. And Paul Goodman to lead them. Was he
now reading this?

> Once American faces were beautiful to me but now they look
> cruel and as if they had narrow thoughts.

Not much poetry, but well put prose. And yet there was always
Goodman's damnable tolerance for all the varieties of sex. Did he
know nothing of evil or entropy? Sex was the superhighway to your
own soul's entropy if it was used without a constant sharpening of
the taste. And orgies? What did Goodman know of orgies, real
ones, not lib-lab college orgies to carry out the higher program of
the Great Society, but real ones with murder in the air, and witches
on the shoulder. The collected Tory[2] in Mailer came roaring to the
surface like a cocked hat in a royal coach.

"When Goodman finishes, I'm going to take over as M.C.," he
whispered to de Grazia. (The revery we have just attended took no
more in fact than a second. Mailer's melancholy assessment of the
forces now mounting in America took place between two consecutive lines of Goodman's poem—not because Mailer cerebrated that
instantly, but because he had had the revery many a time before—
he had to do no more than sense the audience, whisper Cancer
Gulch to himself and the revery went by with a mental ch-ch-ch
Click! reviewed again.) In truth, Mailer was now in a state. He had
been prepared to open the evening with apocalyptic salvos to announce the real gravity of the situation, and the intensely peculiar
American aspect of it—which is that the urban and suburban middle class were to be offered on Saturday an opportunity for glory—
what other nation could boast of such option for its middle class?
Instead—lost. The benignity and good humor of his planned opening remarks now subjugated to the electronic hawking and squab-
bling and *hum* of the P.A., the maniacal necessity to *wait* was on
this hiatus transformed into a violent concentration of purpose, all
intentions reversed. He glared at de Grazia. "How could you do
this?" he whispered to his ear.

---

2. Member of the Conservative party in England, generally in support of established
authority.

De Grazia looked somewhat confused at the intensity. Meetings to de Grazia were obviously just meetings, assemblages of people who coughed up for large admissions or kicked in for the pitch; at best, some meetings were less boring than others. De Grazia was much too wise and guilty-spirited to brood on apocalypse. "I couldn't find you," he whispered back.

"You didn't trust me long enough to wait one minute?"

"We were over an hour late," de Grazia whispered again. "We had to begin."

Mailer was all for having the conversation right then on stage: to hell with reciprocal rights and polite incline of the ear to the speaker. The Beast was ready to grapple with the world. "Did you think I wouldn't show up?" he asked de Grazia.

"Well, I was wondering."

In what sort of mumbo-jumbo of promise and betrayal did de Grazia live? How could de Grazia ever suppose he would not show up? He had spent his life showing up at the most boring and onerous places. He gave a blast of his eyes to de Grazia. But Macdonald gave a look at Mailer, as if to say, "You're creating disturbance."

Now Goodman was done.

Mailer walked to the stage. He did not have any idea any longer of what he would say, his mind was empty, but in a fine calm, taking for these five instants a total rest. While there was no danger of Mailer ever becoming a demagogue since if the first idea he offered could appeal to a mob, the second in compensation would be sure to enrage them, he might nonetheless have made a fair country orator, for he loved to speak, he loved in fact to holler, and liked to hear a crowd holler back. (Of how many New York intellectuals may that be said?)

"I'm here as your original M.C., temporarily displaced owing to a contretemps"—which was pronounced purposefully as contretempse —"in the men's room," he said into the microphone for opening, but the gentle high-strung beast of a device pushed into a panic by the electric presence of a real Beast, let loose a squeal which shook the welds in the old foundation of the Ambassador. Mailer immediately decided he had had enough of public address systems, electronic fields of phase, impedance, and spooks in the circuitry. A hex on collaborating with Cancer Gulch. He pushed the microphone away, squared off before the audience. "Can you hear me?" he bellowed.

"Yes."

"Can you hear me in the balcony?"

"Yes."

"Then let's do away with electronics," he called out.

Cries of laughter came back. A very small pattern of applause.

(Not too many on his side for electrocuting the public address system, or so his orator's ear recorded the vote.)

"Now I missed the beginning of this occasion, or I would have been here to introduce Paul Goodman, for which we're all sorry, right?"

Confused titters. Small reaction.

"What are you, dead-heads?" he bellowed at the audience. "Or are you all"—here he put on his false Irish accent—"in the nature of becoming dead ahsses?" Small laughs. A whistle or two. "No," he said, replying to the whistles, "I invoke these dead asses as part of the gravity of the occasion. The middle class plus one hippie sur-realistic symbolic absolutely insane March on the Pentagon, bless us all," beginning of a big applause which offended Mailer for it came on "bless" and that was too cheap a way to win votes, "bless us all—shit!" he shouted, "I'm trying to say the middle class plus shit, I mean plus revolution, is equal to one big collective dead ass." Some yells of approval, but much shocked curious rather stricken silence. He had broken the shank of his oratorical charge. Now he would have to sweep the audience together again. (Perhaps he felt like a surgeon delivering a difficult breech—nothing to do but plunge to the elbows again.)

"To resume our exposition," a good warm titter, then a ripple of laughter, not unsympathetic to his ear; the humor had been unwit-ting, but what was the life of an orator without some bonus? "To resume this orderly marshalling of concepts"—a conscious attempt at humor which worked less well; he was beginning to recognize for the first time that bellowing without a mike demanded a more forthright style—"I shall now *engage* in confession." More Irish accent. (He blessed Brendan Behan[3] for what he had learned from him.) "A public speaker may offer you two opportunities. Instruc-tion or confession." Laughter now. "Well, you're all college heads, so my instruction would be as pearls before—I dare not say it." Laughs. Boos. A voice from the balcony: "Come on, Norman, say something!"

"Is there a black man in the house?" asked Mailer. He strode up and down the stage pretending to peer at the audience. But in fact they were illumined just well enough to emphasize one sad dis-covery—if black faces there were they were certainly not in plenty. "Well ah'll just have to be the *impromptu* Black Power for tonight. Woo-eeeeee! Woo-eeeeee! HMmmmmmm." He grunted with some partial success, showing hints of Cassius Clay.[4] "Get your white butts moving."

"The confession. The confession!" screamed some adolescents from up front.

3. Large, loud Irish playwright and per-former (1925–1964).

4. Muhammad Ali (b. 1942), boxer and world heavyweight champion.

He came to a stop, shifted his voice. Now he spoke in a relaxed tone. "The confession, yeah!" Well, at least the audience was awake. He felt as if he had driven away some sepulchral phantoms of a variety which inhabited the profound middle-class schist. Now to charge the center of vested spookery.

"Say," he called out into the semidarkness with the ultrapurple light coming off the psychedelic lamp on the rail of the balcony, and the spotlights blaring against his eyes, "say," all happiness again, "I think of Saturday, and that March and do you know, fellow carriers of the holy unendurable grail, for the first time in my life I don't know whether I have the piss or the shit scared out of me most." It was an interesting concept, thought Mailer, for there was a difference between the two kinds of fear—pursue the thought, he would, in quieter times—"we are up, face this, all of you, against an existential situation—we do not know how it is going to turn out, and what is even more inspiring of dread is that the government doesn't know either."

Beginning of a real hand, a couple of rebel yells. "We're going to try to stick it up the government's ass," he shouted, "right into the sphincter of the Pentagon." Wild yells and chills of silence from different reaches of the crowd. Yeah, he was cooking now. "Will reporters please get every word accurately," he called out dryly to warm the chill.

But humor may have been too late. *The New Yorker* did not have strictures against the use of sh*t for nothing; nor did Dwight Macdonald love *The New Yorker* for nothing, he also had stricture against sh*t's metaphorical associations. Mailer looked to his right to see Macdonald approaching, a book in his hands, arms at his side, a sorrowing look of concern in his face. "Norman," said Macdonald quietly, "I can't possibly follow you after all this. Please introduce me, and get it over with."

Mailer was near to stricken. On the one hand interrupted on a flight; on the other, he had fulfilled no duty whatsoever as M.C. He threw a look at Macdonald which said: give me this. I'll owe you one.

But de Grazia was there as well. "Norman, let me be M.C. now," he said.

They were being monstrous unfair, thought Mailer. They didn't understand what he had been doing, how good he had been, what he would do next. Fatal to walk off now—the verdict would claim he was unbalanced. Still, he could not hold the stage by force. That was unthinkably worse.

For the virtuous, however, deliverance (like buttercups) pops up everywhere. Mailer now took the microphone and turned to the audience. He was careful to speak in a relaxed voice. "We are having a disagreement about the value of the proceedings. Some think de Grazia should resume his post as Master of Ceremonies. I would like to keep the position. It is an existential moment. We do not know how it will turn out. So let us vote on it." Happy laughter from the audience at these comic effects. Actually Mailer did not believe it was an existential situation any longer. He reckoned the vote would be well in his favor. "Will those," he asked, "who are in favor of Mr. de Grazia succeeding me as Master of Ceremonies please say aye."

A good sound number said aye.

Now for the ovation. "Will those opposed to this, please say no." The no's to Mailer's lack of pleasure were no greater in volume. "It seems the ayes and no's are about equal," said Mailer. (He was thinking to himself that he had posed the issue all wrong—the ayes should have been reserved for those who would keep him in office.) "Under the circumstances," he announced, "I will keep the chair." Laughter at this easy cheek. He stepped into the middle of such laughter. "You have all just learned an invaluable political lesson." He waved the microphone at the audience. "In the absence of a definitive vote, the man who holds the power, keeps it."

"Hey, de Grazia," someone yelled from the audience, "why do you let him have it?"

Mailer extended the microphone to de Grazia who smiled sweetly into it. "Because if I don't," he said in a gentle voice, "he'll beat the shit out of me." The dread word had been used again.

"Please, Norman," said Macdonald retreating.

So Mailer gave his introduction to Macdonald. It was less than he would have attempted if the flight had not been grounded, but it was certainly respectable. Under the military circumstances, it was a decent cleanup operation. For about a minute he proceeded to introduce Macdonald as a man with whom one might seldom agree, but could never disrespect because he always told the truth as he saw the truth, a man therefore of the most incorruptible integrity. "Pray heaven, I am right," said Mailer to himself, and walked past Macdonald who was on his way to the mike. Both men nodded coolly to each other.

In the wing, visible to the audience, Paul Goodman sat on a chair clearly avoiding any contaminatory encounter with The Existentialist. De Grazia gave his "It's tough all over" smile. Lowell sat in a mournful hunch on the floor, his eyes peering over his glasses to scrutinize the metaphysical substance of his boot, now hide? now machine? now, where the joining and to what? foot to boot, boot to earth—cease all speculations as to what was in Lowell's head. "The one mind a novelist cannot enter is the mind of a novelist superior to himself," said once to Mailer by Jean Malaquais.[5] So, by corollary, the one mind a minor poet may not enter . . .

Lowell looked most unhappy. Mailer, minor poet, had often ob-

---

5. French novelist whose *Reflections on Hip* Mailer admired.

served that Lowell had the most disconcerting mixture of strength and weakness in his presence, a blending so dramatic in its visible sign of conflict that one had to assume he would be sensationally attractive to women. He had something untouchable, all insane in its force; one felt immediately there were any number of causes for which the man would be ready to die, and for some he would fight, with an axe in his hand and a Cromwellian[6] light in his eye. It was even possible that physically he was very strong—one couldn't tell at all—he might be fragile, he might have the sort of farm mechanic's strength which could manhandle the rear axle and differential off a car and into the back of a pickup. But physical strength or no, his nerves were all too apparently delicate. Obviously spoiled by everyone for years, he seemed nonetheless to need the spoiling. These nerves—the nerves of a consummate poet—were not tuned to any battering. The squalls of the mike, now riding up a storm on the erratic piping breath of Macdonald's voice, seemed to tear along Lowell's back like a gale. He detested tumult—obviously. And therefore saw everything which was hopeless in a rife situation: the dank middle-class depths of the audience, the strident squalor of the mike, the absurdity of talent gathered to raise money—for what, dear God? who could finally know what this March might convey, or worse, purvey, and worst of all—to be associated now with Mailer's butcher boy attack. Lowell's eyes looked up from the shoe, and passed one withering glance by the novelist, saying much, saying, "Every single bad thing I have ever heard about you is not exaggerated."

Mailer, looking back, thought bitter words he would not say: "You, Lowell, beloved poet of many, what do you know of the dirt and the dark deliveries of the necessary? What do you know of dignity hard-achieved, and dignity lost through innocence, and dignity lost by sacrifice for a cause one cannot name. What do you know about getting fat against your will, and turning into a clown of an arriviste baron when you would rather be an eagle or a count, or rarest of all, some natural aristocrat from these damned democratic states. No, the only subject we share, you and I, is that species of perception which shows that if we are not very loyal to our unendurable and most exigent inner light, then some day we may burn. How dare you condemn me! You know the diseases which inhabit the audience in this accursed psychedelic house. How dare you scorn the explosive I employ?"

And Lowell with a look of the greatest sorrow as if all this *mess* were finally too shapeless for the hard Protestant smith of his own brain, which would indeed burst if it could not forge his experience into the iron edge of the very best words and the most unsinkable relation of words, now threw up his eyes like an epileptic as if

6. Oliver Cromwell (1599–1658) led the Puritan revolt against Charles I.

turned out of orbit by a turn of the vision—and fell backward, his head striking the floor with no last instant hesitation to cushion the blow, but like a baby, downright sudden, savagely to himself, as if from the height of a foot he had taken a pumpkin and dropped it splat on the floor. "There, much-regarded, much-protected brain, you have finally taken a blow," Lowell might have said to himself, for he proceeded to lie there, resting quietly, while Macdonald went on reading from "The White Man's Burden," Lowell seeming as content as if he had just tested the back of his cranium against a policeman's club. What a royal head they had all to lose!

6. A TRANSFER OF POWER

That evening went on. It was in fact far from climax. Lowell, resting in the wing on the floor of the stage, Lowell recuperating from the crack he had given his head, was a dreamy figure of peace in the corner of the proscenium, a reclining shepherd contemplating his flute, although a Washington newspaper was to condemn him on Saturday in company with Mailer for "slobbish behavior" at this unseemly lounging.

Now Macdonald finished. What with the delays, the unmanageable public address system, and the choppy waters of the audience at his commencement, for Mailer had obviously done him no good, Macdonald had been somewhat less impressive than ever. A few people had shown audible boredom with him. (Old-line Communists perhaps. Dwight was by now one of the oldest anti-Communists in America.)

> Take up the White Man's burden—
> Ye dare not stoop to less—
> Nor call too loud on Freedom
> To cloak your weariness;
> By all ye cry or whisper,
> By all ye leave or do,
> The silent, sullen peoples
> Shall weigh your Gods and you.[7]

read Macdonald from Kipling's poem, and the wit was in the selection, never the presentation.

He was done. He walked back to the wings with an air of no great satisfaction in himself, at most the sense of an obligation accomplished. Lowell's turn had arrived. Mailer stood up to introduce him.

The novelist gave a fulsome welcome to the poet. He did not speak of his poetry (with which he was not conspicuously familiar) nor of his prose which he thought excellent—Mailer told instead of

---

7. Penultimate stanza of Rudyard Kipling's poem *The White Man's Burden* (1899), usually considered an imperialist utterance.

why he had respect for Lowell as a man. A couple of years ago, the poet had refused an invitation from President Johnson to attend a garden party for artists and intellectuals, and it had attracted much attention at the time for it was one of the first dramatic acts of protest against the war in Vietnam, and Lowell was the only invited artist of first rank who had refused. Saul Bellow, for example, had attended the garden party. Lowell's refusal could not have been easy, the novelist suggested, because artists were attracted to formal afternoons of such elevated kind since that kind of experience was often stimulating to new perception and new work. So, an honorific occasion in full panoply was not easy for the mature artist to eschew. Capital! Lowell had therefore bypassed the most direct sort of literary capital. Ergo, Mailer respected him—he could not be certain he would have done the same himself, although, of course, he assured the audience he would not probably have ever had the opportunity to refuse. (Hints of merriment in the crowd at the thought of Mailer on the White House lawn.)

If the presentation had been formal up to here, it had also been somewhat graceless. On the consequence, our audience's amusement tipped the slumbering Beast. Mailer now cranked up a vaudeville clown for finale to Lowell's introduction. "Ladies and gentlemen, if novelists come from the middle class, poets tend to derive from the bottom and the top. We all know good poets at the bot'—ladies and gentlemen, here is a poet from the top, Mr. Robert Lowell." A large vigorous hand of applause, genuine enthusiasm for Lowell, some standing ovation.

But Mailer was depressed. He had betrayed himself again. The end of the introduction belonged in a burlesque house—he worked his own worst veins, like a man on the edge of bankruptcy trying to collect hopeless debts. He was fatally vulgar! Lowell passing him on the stage had recovered sufficiently to cast him a nullifying look. At this moment, they were obviously far from friends.

Lowell's shoulders had a slump, his modest stomach was pushed forward a hint, his chin was dropped to his chest as he stood at the microphone, pondering for a moment. One did not achieve the languid grandeurs of that slouch in one generation—the grandsons of the first sons had best go through the best troughs in the best eating clubs at Harvard before anyone in the family could try for such elegant note. It was now apparent to Mailer that Lowell would move by instinct, ability, and certainly by choice, in the direction most opposite from himself.

"Well," said Lowell softly to the audience, his voice dry and gentle as any New England executioner might ever hope to be, "this has been a zany evening." Laughter came back, perhaps a little too much. It was as if Lowell wished to reprove Mailer, not humiliate him. So he shifted, and talked a bit uneasily for perhaps a minute about very little. Perhaps it was too little. Some of the audience, encouraged by earlier examples, now whistled. "We can't hear you," they shouted, "speak louder."

Lowell was annoyed. "I'll bellow," he said, "but it won't do any good." His firmness, his distaste for the occasion, communicated some subtle but impressive sense of his superiority. Audiences are moved by many cues but the most satisfactory to them is probably the voice of their abdomen. There are speakers who give a sense of security to the abdomen, and they always elicit the warmest kind of applause. Mailer was not this sort of speaker; Lowell was. The hand of applause which followed this remark was fortifying. Lowell now proceeded to read some poetry.

He was not a splendid reader, merely decent to his own lines, and he read from that slouch, that personification of ivy climbing a column, he was even diffident, he looked a trifle helpless under the lights. Still, he made no effort to win the audience, seduce them, dominate them, bully them, amuse them, no, they were there for him, to please *him*, a sounding board for the plucked string of his poetic line, and so he endeared himself to them. They adored him—for his talent, his modesty, his superiority, his melancholy, his petulance, his weakness, his painful, almost stammering shyness, his noble strength—*there* was the string behind other strings.

> O to break loose, like the chinook
> salmon jumping and falling back,
> nosing up to the impossible
> stone and bone-crushing waterfall—
> raw-jawed, weak-fleshed there, stopped by ten
> steps of the roaring ladder, and then
> to clear the top on the last try,
> alive enough to spawn and die.[8]

Mailer discovered he was jealous. Not of the talent. Lowell's talent was very large, but then Mailer was a bulldog about the value of his own talent. No, Mailer was jealous because he had worked for this audience, and Lowell without effort seemed to have stolen them: Mailer did not know if he was contemptuous of Lowell for playing *grand maître*,[9] or admiring of his ability to do it. Mailer knew his own version of *grand maître* did not compare. Of course no one would be there there to accept his version either. Then pain of bad reviews was not in the sting, but in the subsequent pressure which, like water on a joint, collected over the decade. People who had not read your books in fifteen years were certain they were missing nothing of merit. A buried sorrow, not very attractive, (for bile was in it and the bitterness of unrequited literary injustice)

8. First stanza of Lowell's *Waking Early Sunday Morning* (from *Near the Ocean*, 1967). After the next paragraph the final stanza is quoted.
9. Grand master.

released itself from some ducts of the heart, and Mailer felt hot anger at how Lowell was loved and he was not, a pure and surprising recognition of how much emotion, how much simple and childlike bitter sorrowing emotion had been concealed from himself for years under the manhole cover of his contempt for bad reviews.

> Pity the planet, all joy gone
> from this sweet volcanic cone;
> peace to our children when they fall
> in small war on the heels of small
> war—until the end of time
> to police the earth, a ghost
> orbiting forever lost
> in our monotonous sublime.

They gave Lowell a good standing ovation, much heartiness in it, much obvious pleasure that they were there on a night in Washington when Robert Lowell had read from his work—it was as nice as that—and then Lowell walked back to the wings, and Mailer walked forward. Lowell did not seem particularly triumphant. He looked still modest, still depressed, as if he had been applauded too much for too little and so the reservoir of guilt was still untapped.

Nonetheless, to Mailer it was now *mano a mano*.[1] Once, on a vastly larger scale of applause, perhaps people had reacted to Manolete[2] not unlike the way they reacted to Lowell, so stirred by the deeps of sorrow in the man, that the smallest move produced the largest emotion. If there was any value to the comparison then Mailer was kin to the young Dominguin, taking raucous chances, spitting in the eye of the bull, an excess of variety in his passes. But probably there was no parallel at all. He may have felt like a matador in the flush of full competition, going out to do his work after the other torero has had a triumph, but for fact he was probably less close in essence now to the bullfighter than the bull. We must not forget the Beast. He had been sipping the last of the bourbon out of the mug. He had been delayed, piqued, twisted from his purpose and without anything to eat in close to ten hours. He was on the hunt. For what, he hardly knew. It is possible the hunt existed long before the victim was ever conceived.

"Now, you may wonder who I am," he said to the audience, or bellowed to them, for again he was not using the mike, "and you may wonder why I'm talking in a Southern accent which is phony" —the Southern accent as it sounded to him in his throat, was actually not too bad at this moment—"and the reason is that I want to make a presentation to you." He did not have a notion of what he would say next, but it never occurred to him something would

---

1. Hand to hand.
2. Manolete and his younger countryman Dominguin were famous matadors.

not come. His impatience, his sorrow, his jealousy were gone, he just wanted to live on the edge of that rhetorical sword he would soon try to run through the heart of the audience. "We are gathered here"—shades of Lincoln in hippieland—"to make a move on Saturday to invest the Pentagon and halt and slow down its workings, and this will be at once a symbolic act and a real act"—he was roaring—"for real heads may possibly get hurt, and soldiers will be there to hold us back, and some of us may be arrested"—how, wondered the wise voice at the rear of this roaring voice, could one ever leave Washington now without going to jail?—"some blood conceivably will be shed. If I were the man in the government responsible for controlling this March, I would not know what to do." Sonorously—"I would not wish to arrest too many or hurt anyone for fear the repercussions in the world would be too large for my bureaucrat's heart to bear—it's so full of shit." Roars and chills from the audience again. He was off into obscenity. It gave a heartiness like the blood of beef tea to his associations. There was no villainy in obscenity for him, just—paradoxically, characteristically—his love for America: he had first come to love America when he served in the U.S. Army, not the America of course of the flag, the patriotic unendurable fix of the television programs and the newspapers, no, long before he was ever aware of the institutional oleo of the most suffocating American ideas he had come to love what editorial writers were fond of calling the democratic principle with its faith in the common man. He found that principle and that man in the Army, but what none of the editorial writers ever mentioned was that that noble common man was obscene as an old goat, and his obscenity was what saved him. The sanity of said common democratic man was in his humor, his humor was in his obscenity. And his philosophy as well—a reductive philosophy which looked to restore the hard edge of proportion to the overblown values overhanging each small military existence—viz: being forced to salute an overconscientious officer with your back stiffened into an exaggerated posture. "That Lieutenant is chickenshit," would be the platoon verdict, and a blow had somehow been struck for democracy and the sanity of good temper. Mailer once heard a private end an argument about the merits of a general by saying, "his spit don't smell like ice cream either," only the private was not speaking of spit. Mailer thought enough of the line to put it into *The Naked and the Dead*, along with a good many other such lines the characters in his mind and his memory of the Army had begun to offer him. The common discovery of America was probably that Americans were the first people on earth to live for their humor; nothing was so important to Americans as humor. In Brooklyn, he had taken this for granted, at Harvard he had thought it was a by-product of being at Harvard, but in the Army he discovered that the humor was probably in the veins and the roots of the local history of every state and county in America—the truth of the way it really felt over the years passed on a river of obscenity from small-town storyteller to storyteller there down below the bankers and the books and the educators and the legislators—so Mailer never felt more like an American than when he was naturally obscene—all the gifts of the American language came out in the happy play of obscenity upon concept, which enabled one to go back to concept again. What was magnificent about the word shit is that it enabled you to use the word noble: a skinny Southern cracker with a beatific smile on his face saying in the dawn in a Filipino rice paddy, "Man, I just managed to take me a noble shit." Yeah, that was Mailer's America. If he was going to love something in the country, he would love that. So after years of keeping obscene language off to one corner of his work, as if to prove after *The Naked and the Dead* that he had many an arrow in his literary quiver, he had come back to obscenity again in the last year—he had kicked goodbye in his novel *Why Are We in Vietnam?* to the old literary corset of good taste, letting his sense of language play on obscenity as freely as it wished, so discovering that everything he knew about the American language (with its incommensurable resources) went flying in and out of the line of his prose with the happiest beating of wings—it was the first time his style seemed at once very American to him and very literary in the best way, at least as he saw the best way. But the reception of the book had been disappointing. Not because many of the reviews were bad (he had learned, despite all sudden discoveries of sorrow, to live with that as one lived with smog) no, what was disappointing was the crankiness across the country. Where fusty conservative old critics had once defended the obscenity in *The Naked and the Dead*, they, or their sons, now condemned it in the new book, and that *was* disappointing. The country was not growing up so much as getting a premature case of arthritis.

At any rate, he had come to the point where he liked to use a little obscenity in his public speaking. Once people got over the shock, they were sometimes able to discover that the humor it provided was not less powerful than the damage of the pain. Of course he did not do it often and he tried not to do it unless he was in good voice—Mailer was under no illusion that public speaking was equal to candid conversation; an obscenity uttered in a voice too weak for its freight was obscene, since obscenity probably resides in the quick conversion of excitement to nausea—which is why Lyndon Johnson's speeches are called obscene by some. The excitement of listening to the American President alters abruptly into the nausea of wandering down the blind alleys of his voice.

This has been a considerable defense of the point, but then the point was at the center of his argument and it could be put thus: the American corporation executive, who was after all the foremost representative of Man in the world today, was perfectly capable of burning unseen women and children in the Vietnamese jungles, yet felt a large displeasure and fairly final disapproval at the generous use of obscenity in literature and in public.

The apology may now be well taken, but what in fact did Mailer say on the stage of the Ambassador before the evening was out? Well, not so very much, just about enough to be the stuff of which footnotes are made, for he did his best to imitate a most high and executive voice.

"I had an experience as I came to the theater to speak to all of you, which is that before appearing on this stage I went upstairs to the men's room as a prelude to beginning this oratory so beneficial to all"—laughs and catcalls—"and it was dark, so—ahem—I missed the bowl—all men will know what I mean. Forgiveness might reign. But tomorrow, they will blame that puddle of water on Communists which is the way we do things here in Amurrica, anyone of you pinko poos want to object, lemme tell ya, the reason nobody was in the men's room, and it so dark, is that if there been a light they'd had to put a CIA man in there and the hippies would grope him silly, see here, you know who I am, why it just came to me, ah'm so phony, I'm as full of shit as Lyndon Johnson. Why, man, I'm nothing but his little old alter ego. That's what you got right here working for you, Lyndon Johnson's little old *dwarf* alter ego. How you like him? How you like him?" (Shades of Cassius Clay again.)

And in the privacy of his brain, quiet in the glare of all that sound and spotlight, Mailer thought quietly, "My God, that is probably exactly what you are at this moment, Lyndon Johnson with all his sores, sorrows, and vanity, squeezed down to five foot eight," and Mailer felt for the instant possessed, as if he had seized some of the President's secret soul, or the President seized some of his—the bourbon was as luminous as moonshine to the spores of insanity in the flesh of his brain, a smoke of menace swished in the air, and something felt real, almost as if he had caught Lyndon Johnson by the toe and now indeed, bugger the rhyme, should never let him go.

"Publicity hound," shouted someone from the upper balcony.

"Fuck you," cried Mailer back with absolute delight, all the force of the Texas presidency in his being. Or was it Lucifer's fire? But let us use asterisks for these obscenities to emphasize how happily he used the words, they went off like fireworks in his orator's heart, and asterisks look like rocket-bursts and the orbs from Roman candles * * *. F*ck you he said to the heckler but with such gusto the vowel was doubled. F*-*ck you! was more like it. So, doubtless, had the President disposed of all opposition in private session. Well, Mailer was here to bring the presidency to the public.

"This yere dwarf alter ego has been telling you about his im-broglio with the p*ssarooney up on the top floor, and will all the reporters please note that I did not talk about defecation commonly known as sheeee-it!"—full imitation of LBJ was attempted there— "but to the contrary, speak of you-rye-nation! I p*ssed on the floor. Hoo-ee! Hoo-ee! How's that for Black Power full of white p*ss? You just know all those reporters are going to say it was sh*t tomorrow. F*ck them. F*ck all of them. Reporters, will you stand up and be counted?"

A wail of delight from the students in the audience. What would the reporters do? Would they stand?

One lone figure arose.

"Where are *you* from?" asked Mailer.

"Washington *Free Press*." A roar of delight from the crowd. It was obviously some student or hippie paper.

"Ah want *The Washington Post*," said Mailer in his best Texas tones, "and the *Star*. Ah know there's a *Time* magazine man here for one, and twenty more like him no doubt." But no one stood. So Mailer went into a diatribe. "Yeah, people," he said, "watch the reporting which follows. Yeah, these reporters will kiss Lyndon Johnson's *ss and Dean Rusk's *ss and Man Mountain Mc-Namara's *ss,[3] they will rush to kiss it, but will they stand up in public? No! Because they are the silent assassins of the Republic. They alone have done more to destroy this nation than any force in it." They will certainly destroy me in the morning, he was thinking. But it was for this moment worth it, as if two very different rivers, one external, one subjective, had come together; the frustrated bile, piss, pus, and poison he had felt at the progressive contamination of all American life in the abscess of Vietnam, all of that, all heaped in lighted coals of brimstone on the press' collective ear, represented one river, and the other was the frustrated actor in Mailer—ever since seeing *All the King's Men* years ago he had wanted to come on in public as a Southern demagogue.

The speech went on, and a few fine things possibly were said next to a few equally obscene words, and then Mailer thought in passing of reading a passage from *Why Are We in Vietnam?* but the passage was full of plays of repetition on the most famous four-letter word of them all, and Mailer thought that was conceivably redundant now and so he ended modestly with a final, "See you on Saturday!"

The applause was fair. Not weak, but empty of large demonstration. No standing ovation for certain. He felt cool, and in a quiet, pleasant, slightly depressed mood. Since there was not much conversation between Macdonald, Lowell, and himself, he turned after a moment, left the stage, and walked along the floor where the

3. Dean Rusk (b. 1909), Secretary of State during the Kennedy and Johnson administrations, 1961–69. Robert Mc-Namara (b. 1916), Secretary of Defense during the Kennedy and Johnson administrations, 1961–68. The two cabinet officials held most responsible for perpetuating the war in Vietnam.

audience had sat. A few people gathered about him, thanked him, shook his hand. He was quiet and reserved now, with genial slightly muted attempts to be cordial. He had noticed this shift in mood before, even after readings or lectures which had been less eventful. There was a mutual embarrassment between speaker and audience once the speaker had left the stage and walked through the crowd. It was due no doubt to the intimacy—that most special intimacy— which can live between a speaker and the people he has addressed, yes they had been so intimate then, that the encounter now, after- ward, was like the eye-to-the-side maneuvers of client and whore once the act is over and dressing is done.

Mailer went on from there to a party of more liberal academics, and drank a good bit more and joked with Macdonald about the superiority of the introduction he had given to Lowell over the introduction Dwight had received.

"Next time don't interrupt me," he teased Macdonald, "and I'll give you a better introduction."

"Goodness, I couldn't hear a word you said," said Macdonald, "you just sounded awful. Do you know, Norman, the acoustics were terrible on the wing. I don't think any of us heard anything anyone else said."

Some time in the early morning, or not so early, Mailer got to bed at the Hay-Adams and fell asleep to dream no doubt of fancy parties in Georgetown when the Federal period in architecture was young. Of course if this were a novel, Mailer would spend the rest of the night with a lady. But it is history, and so the Novelist is for once blissfully removed from any description of the hump-your- backs of sex. Rather he can leave such matters to the happy or unhappy imagination of the reader.

1968

JAMES BALDWIN

1924–

James Baldwin was born in Harlem, the first of nine children. From his novel *Go Tell It on the Mountain* (1951) and his story *The Rockpile,* we learn how extremely painful was the relationship between the father and his eldest son. David Baldwin, son of a slave, was a lay preacher rigidly committed to a vengeful God who would eventually judge white people as they deserved; in the meantime much of the vengeance was taken out on James. His father's "unlimited capacity for introspection and rancor," as the son later put it, must have had a profound effect on the sermonizing style Baldwin was to develop. Just as important was his conversion and re- sulting service as a preacher in his father's church, as we can see from both the rhythm and message of his prose—which is very much a *spoken* prose.

Baldwin did well in school and having received a hardship deferment from military service (his father was dying) began to attach himself to Greenwich Village, where he concentrated on the business of becoming a writer. In 1944 he met Richard Wright, at that time "the greatest black writer in the world for me," in whose early books Baldwin "found expressed, for the first time in my life, the sorrow, the rage, and the murderous bitterness which was eating up my life and the lives of those about me." Wright helped him win a Eugene Saxton Fellowship, and in 1948, when Baldwin went to live in Paris, he was following Wright's footsteps (Wright had become an expatriate to the same city two years earlier). It is perhaps for this reason that in his early essays written for *Partisan Review* and published in 1955 as *Notes of a Native Son* (with the title's explicit reference to Wright's novel) Baldwin dissociated himself from the image of American life found in Wright's "protest work" and, as Ralph Ellison was also doing, went about protesting in his own way.

As far as his novels are concerned, Baldwin's way involved a preoccupation with the intertwining of sexual with racial concerns, particularly in America. His interest in what it means to be black and homosexual in relation to white society is most fully and interestingly expressed in his long and somewhat ragged third novel, *Another Country* (1962). (He had previously written *Go Tell It on the Mountain*, and a second novel, *Giovanni's Room* [1956] about a white expatriate in Paris and his male lover.) *Another Country* contains scenes full of lively detail and intelligent reflection, though it lacks—as do all his novels—a compelling design that draws the book together. Thus far Baldwin has made little fictional use of the talents for irony and sly teasing he is master of in his essays; nor, unlike Ellison or Mailer, has he shown much interest in stylistic experimentation.

Baldwin's imagination is intensely social and reveals itself most passionately and variously in his collections of essays, of which *Notes of a Native Son* is probably the best, and in what many would judge his finest piece of writing, *Letter from a Region of My Mind* (in *The Fire Next Time*, 1963). This essay, the first half of which is printed here, was first published in the *New Yorker* and probably had a greater effect on white liberals than on the blacks who read it. In its firm rejection of separatism between the races as preached by the Black Muslims and their leader, Elijah Muhammad, it spoke out for love as the difficult and necessary way out of slavery and race hatred. Today, with all that has happened since its appearance, it is still a fresh and moving utterance, directed as the best of Baldwin's essays are by a beautifully controlled speaking voice, alternately impressing upon us its accents of polite directness, sardonic irony, or barely controlled fury.

Like Mailer, Baldwin risks advertising himself too strenuously, and sometimes falls into stridency and sentimentality. Like Ellison he has experienced many pressures to be something more than just a novelist; but he has produced a respectable series of novels, stories, and a play, even if no single fiction of his is comparable in breadth and daring to *Invisible Man*. There is surely no black writer better able to imagine white experience, to speak in various tones of different kinds and behaviors of people or places other than his own. In its sensitivity to shades of discrimination and moral shape, and in its commitment—despite everything—to America, his voice is comparable in importance to that of any person of letters writing today.

From The Fire Next Time[1]

[*Part I*]

I underwent, during the summer that I became fourteen, a prolonged religious crisis. I use the word "religious" in the common, and arbitrary, sense, meaning that I then discovered God, His saints and angels, and His blazing Hell. And since I had been born in a Christian nation, I accepted this Deity as the only one. I supposed Him to exist only within the walls of a church—in fact, of *our* church—and I also supposed that God and safety were synonymous. The word "safety" brings us to the real meaning of the word "religious" as we use it. Therefore, to state it in another, more accurate way, I became, during my fourteenth year, for the first time in my life, afraid—afraid of the evil within me and afraid of the evil without. What I saw around me that summer in Harlem was what I had always seen; nothing had changed. But now, without any warning, the whores and pimps and racketeers on the Avenue[2] had become a personal menace. It had not before occurred to me that I could become one of them, but now I realized that we had been produced by the same circumstances. Many of my comrades were clearly headed for the Avenue, and my father said that I was headed that way, too. My friends began to drink and smoke, and embarked—at first avid, then groaning—on their sexual careers. Girls, only slightly older than I was, who sang in the choir or taught Sunday school, the children of holy parents, underwent, before my eyes, their incredible metamorphosis, of which the most bewildering aspect was not their budding breasts or their rounding behinds but something deeper and more subtle, in their eyes, their heat, their odor, and the inflection of their voices. Like the strangers on the Avenue, they became, in the twinkling of an eye, unutterably different and fantastically *present*. Owing to the way I had been raised, the abrupt discomfort that all this aroused in me and the fact that I had no idea what my voice or my mind or my body was likely to do next caused me to consider myself one of the most depraved people on earth. Matters were not helped by the fact that these holy girls seemed rather to enjoy my terrified lapses, our grim, guilty, tormented experiments, which were at once as chill and joyless as the Russian steppes and hotter, by far, than all the fires of Hell.

Yet there was something deeper than these changes, and less definable, that frightened me. It was real in both the boys and the

---

1. *Letter from a Region of My Mind*, of which the first half is printed here, was originally published in the *New Yorker* in 1962, then combined with a shorter essay to make up *The Fire Next Time* (1963). In the second half of the *Letter* Baldwin describes in detail his impressions of the Black Muslims and their leader, Elijah Muhammad. The book's epigraph—"God gave Noah the rainbow sign,/No more water, the fire next time!" —alludes to what Baldwin terms at the book's conclusion "the fulfilment of that prophecy, recreated from the Bible in song by a slave," which we will suffer if we do not "end the racial nightmare, and achieve our country."

2. Lenox Avenue, the main street running through Harlem.

girls, but it was, somehow, more vivid in the boys. In the case of the girls, one watched them turning into matrons before they had become women. They began to manifest a curious and really rather terrifying single-mindedness. It is hard to say exactly how this was conveyed: something implacable in the set of the lips, something farseeing (seeing what?) in the eyes, some new and crushing determination in the walk, something peremptory in the voice. They did not tease us, the boys, any more; they reprimanded us sharply, saying, "You better be thinking about your soul!" For the girls also saw the evidence on the Avenue, knew what the price would be, for them, of one misstep, knew that they had to be protected and that we were the only protection there was. They understood that they must act as God's decoys, saving the souls of the boys for Jesus and binding the bodies of the boys in marriage. For this was the beginning of our burning time, and "It is better," said St. Paul—who elsewhere, with a most unusual and stunning exactness, described himself as a "wretched man"—"to marry than to burn."[3] And I began to feel in the boys a curious, wary, bewildered despair, as though they were now settling in for the long, hard winter of life. I did not know then what it was that I was reacting to; I put it to myself that they were letting themselves go. In the same way that the girls were destined to gain as much weight as their mothers, the boys, it was clear, would rise no higher than their fathers. School began to reveal itself, therefore, as a child's game that one could not win, and boys dropped out of school and went to work. My father wanted me to do the same. I refused, even though I no longer had any illusions about what an education could do for me; I had already encountered too many college-graduate handymen. My friends were now "downtown," busy, as they put it, "fighting the man." They began to care less about the way they looked, the way they dressed, the things they did; presently, one found them in twos and threes and fours, in a hallway, sharing a jug of wine or a bottle of whiskey, talking, cursing, fighting, sometimes weeping: lost, and unable to say what it was that oppressed them, except that they knew it was "the man"—the white man. And there seemed to be no way whatever to remove this cloud that stood between them and the sun, between them and love and life and power, between them and whatever it was that they wanted. One did not have to be very bright to realize how little one could do to change one's situation; one did not have to be abnormally sensitive to be worn down to a cutting edge by the incessant and gratuitous humiliation and danger one encountered every working day, all day long. The humiliation did not apply merely to working days, or workers; I was thirteen and was crossing Fifth Avenue on my way to the Forty-second Street library, and the cop in the middle of the street muttered as I passed him, "Why don't you niggers stay uptown where you be-

3. 1 Corinthians 7.8–9.

long?" When I was ten, and didn't look, certainly, any older, two policemen amused themselves with me by frisking me, making comic (and terrifying) speculations concerning my ancestry and probable sexual prowess, and for good measure, leaving me flat on my back in one of Harlem's empty lots. Just before and then during the Second World War, many of my friends fled into the service, all to be changed there, and rarely for the better, many to be ruined, and many to die. Others fled to other states and cities—that is, to other ghettos. Some went on wine or whiskey or the needle, and are still on it. And others, like me, fled into the church.

For the wages of sin were visible everywhere, in every wine-stained and urine-splashed hallway, in every clanging ambulance bell, in every scar on the faces of the pimps and their whores, in every helpless, newborn baby being brought into this danger, in every knife and pistol fight on the Avenue, and in every disastrous bulletin: a cousin, mother of six, suddenly gone mad, the children parcelled out here and there; an indestructible aunt rewarded for years of hard labor by a slow, agonizing death in a terrible small room; someone's bright son blown into eternity by his own hand; another turned robber and carried off to jail. It was a summer of dreadful speculations and discoveries, of which these were not the worst. Crime became real, for example—for the first time—not as *a* possibility but as *the* possibility. One would never defeat one's circumstances by working and saving one's pennies; one would never, by working, acquire that many pennies, and, besides, the social treatment accorded even the most successful Negroes proved that one needed, in order to be free, something more than a bank account. One needed a handle, a lever, a means of inspiring fear. It was absolutely clear that the police would whip you and take you in as long as they could get away with it, and that everyone else— housewives, taxi-drivers, elevator boys, dishwashers, bartenders, lawyers, judges, doctors, and grocers—would never, by the operation of any generous human feeling, cease to use you as an outlet for his frustrations and hostilities. Neither civilized reason nor Christian love would cause any of those people to treat you as they presumably wanted to be treated; only the fear of your power to retaliate would cause them to do that, or to seem to do it, which was (and is) good enough. There appears to be a vast amount of confusion on this point, but I do not know many Negroes who are eager to be "accepted" by white people, still less to be loved by them; they, the blacks, simply don't wish to be beaten over the head by the whites every instant of our brief passage on this planet. White people in this country will have quite enough to do in learning how to accept and love themselves and each other, and when they have achieved this—which will not be tomorrow and may very well be never—the Negro problem will no longer exist, for it will no longer be needed.

People more advantageously placed than we in Harlem were, and are, will no doubt find the psychology and the view of human nature sketched above dismal and shocking in the extreme. But the Negro's experience of the white world cannot possibly create in him any respect for the standards by which the white world claims to live. His own condition is overwhelming proof that white people do not live by these standards. Negro servants have been smuggling odds and ends out of white homes for generations, and white people have been delighted to have them do it, because it has assuaged a dim guilt and testified to the intrinsic superiority of white people. Even the most doltish and servile Negro could scarcely fail to be impressed by the disparity between his situation and that of the people for whom he worked; Negroes who were neither doltish nor servile did not feel that they were doing anything wrong when they robbed white people. In spite of the Puritan-Yankee equation of virtue with well-being, Negroes had excellent reasons for doubting that money was made or kept by any very striking adherence to the Christian virtues; it certainly did not work that way for black Christians. In any case, white people, who had robbed black people of their liberty and who profited by this theft every hour that they lived, had no moral ground on which to stand. They had the judges, the juries, the shotguns, the law—in a word, power. But it was a criminal power, to be feared but not respected, and to be outwitted in any way whatever. And those virtues preached but not practiced by the white world were merely another means of holding Negroes in subjection.

It turned out, then, that summer, that the moral barriers that I had supposed to exist between me and the dangers of a criminal career were so tenuous as to be nearly nonexistent. I certainly could not discover any principled reason for not becoming a criminal, and it is not my poor, God-fearing parents who are to be indicted for the lack but this society. I was icily determined—more determined, really, than I then knew—never to make my peace with the ghetto but to die and go to Hell before I would let any white man spit on me, before I would accept my "place" in this republic. I did not intend to allow the white people of this country to tell me who I was, and limit me that way, and polish me off that way. And yet, of course, at the same time, I *was* being spat on and defined and described and limited, and could have been polished off with no effort whatever. Every Negro boy—in my situation during those years, at least—who reaches this point realizes, at once, profoundly, because he wants to live, that he stands in great peril and must find, with speed, a "thing," a gimmick, to lift him out, to start him on his way. *And it does not matter what the gimmick is.* It was this last realization that terrified me and—since it revealed that the door opened on so many dangers—helped to hurl me into the church. And, by an unforeseeable paradox, it was my career in the church that turned out, precisely, to be my gimmick.

For when I tried to assess my capabilities, I realized that I had almost none. In order to achieve the life I wanted, I had been dealt, it seemed to me, the worst possible hand. I could not become a prizefighter—many of us tried but very few succeeded. I could not sing. I could not dance. I had been well conditioned by the world in which I grew up, so I did not yet dare take the idea of becoming a writer seriously. The only other possibility seemed to involve my becoming one of the sordid people on the Avenue, who were not really as sordid as I then imagined but who frightened me terribly, both because I did not want to live that life and because of what they made me feel. Everything inflamed me, and that was bad enough, but I myself had also become a source of fire and temptation. I had been far too well raised, alas, to suppose that any of the extremely explicit overtures made to me that summer, sometimes by boys and girls but also, more alarmingly, by older men and women, had anything to do with my attractiveness. On the contrary, since the Harlem idea of seduction is, to put it mildly, blunt, whatever these people saw in me merely confirmed my sense of my depravity.

It is certainly sad that the awakening of one's senses should lead to such a merciless judgment of oneself—to say nothing of the time and anguish one spends in the effort to arrive at any other—but it is also inevitable that a literal attempt to mortify the flesh should be made among black people like those with whom I grew up. Negroes in this country—and Negroes do not, strictly or legally speaking, exist in any other—are taught really to despise themselves from the moment their eyes open on the world. This world is white and they are black. White people hold the power, which means that they are superior to blacks (intrinsically, that is: God decreed it so), and the world has innumerable ways of making this difference known and felt and feared. Long before the Negro child perceives this difference, and even longer before he understands it, he has begun to react to it, he has begun to be controlled by it. Every effort made by the child's elders to prepare him for a fate from which they cannot protect him causes him secretly, in terror, to begin to await, without knowing that he is doing so, his mysterious and inexorable punishment. He must be "good" not only in order to please his parents and not only to avoid being punished by them; behind their authority stands another, nameless and impersonal, infinitely harder to please, and bottomlessly cruel. And this filters into the child's consciousness through his parents' tone of voice as he is being exhorted, punished, or loved; in the sudden, uncontrollable note of fear heard in his mother's or his father's voice when he has strayed beyond some particular boundary. He does not know what the boundary is, and he can get no explanation of it, which is frightening enough, but the fear he hears in the voices of his elders is more frightening still. The fear that I heard in my father's voice, for example, when he realized that I really *believed* I could do anything a white boy could do, and had every intention of proving it, was not at all like the fear I heard when one of us was ill or had fallen down the stairs or strayed too far from the house. It was another fear, a fear that the child, in challenging the white world's assumptions, was putting himself in the path of destruction. A child cannot, thank Heaven, know how vast and how merciless is the nature of power, with what unbelievable cruelty people treat each other. He reacts to the fear in his parents' voices because his parents hold up the world for him and he has no protection without them. I defended myself, as I imagined, against the fear my father made me feel by remembering that he was very old-fashioned. Also, I prided myself on the fact that I already knew how to outwit him. To defend oneself against a fear is simply to insure that one will, one day, be conquered by it; fears must be faced. As for one's wits, it is just not true that one can live by them—not, that is, if one wishes really to live. That summer, in any case, all the fears with which I had grown up, and which were now a part of me and controlled my vision of the world, rose up like a wall between the world and me, and drove me into the church.

As I look back, everything I did seems curiously deliberate, though it certainly did not seem deliberate then. For example, I did not join the church of which my father was a member and in which he preached. My best friend in school, who attended a different church, had already "surrendered his life to the Lord," and he was very anxious about my soul's salvation. (I wasn't, but any human attention was better than none.) One Saturday afternoon, he took me to his church. There were no services that day, and the church was empty, except for some women cleaning and some other women praying. My friend took me into the back room to meet his pastor—a woman. There she sat, in her robes, smiling, an extremely proud and handsome woman, with Africa, Europe, and the America of the American Indian blended in her face. She was perhaps forty-five or fifty at this time, and in our world she was a very celebrated woman. My friend was about to introduce me when she looked at me and smiled and said, "Whose little boy are you?" Now this, unbelievably, was precisely the phrase used by pimps and racketeers on the Avenue when they suggested, both humorously and intensely, that I "hang out" with them. Perhaps part of the terror they had caused me to feel came from the fact that I unquestionably wanted to be *somebody's* little boy. I was so frightened, and at the mercy of so many conundrums, that inevitably, that summer, *someone* would have taken me over; one doesn't, in Harlem, long remain standing on any auction block. It was my good luck—perhaps—that I found myself in the church racket instead of some other, and surrendered to a spiritual seduction long before I came to any carnal knowledge. For when the pastor asked me, with that marvellous smile, "Whose little boy are you?" my heart replied at once, "Why, yours."

The summer wore on, and things got worse. I became more guilty and more frightened, and kept all this bottled up inside me, and naturally, inescapably, one night, when this woman had finished preaching, everything came roaring, screaming, crying out, and I fell to the ground before the altar. It was the strangest sensation I have ever had in my life—up to that time, or since. I had not known that it was going to happen, or that it could happen. One moment I was on my feet, singing and clapping and, at the same time, working out in my head the plot of a play I was working on then; the next moment, with no transition, no sensation of falling, I was on my back, with the lights beating down into my face and all the vertical saints above me. I did not know what I was doing down so low, or how I had got there. And the anguish that filled me cannot be described. It moved in me like one of those floods that devastate counties, tearing everything down, tearing children from their parents and lovers from each other, and making everything an unrecognizable waste. All I really remember is the pain, the unspeakable pain; it was as though I were yelling up to Heaven and Heaven would not hear me. And if Heaven would not hear me, if love could not descend from Heaven—to wash me, to make me clean—then utter disaster was my portion. Yes, it does indeed mean something—something unspeakable—to be born, in a white country, an Anglo-Teutonic, antisexual country, black. You very soon, without knowing it, give up all hope of communion. Black people, mainly, look down or look up but do not look at each other, not at you, and white people, mainly, look away. And the universe is simply a sounding drum; there is no way, no way whatever, so it seemed then and has sometimes seemed since, to get through a life, to love your wife and children, or your friends, or your mother and father, or to be loved. The universe, which is not merely the stars and the moon and the planets, flowers, grass, and trees, but *other people*, has evolved no terms for your existence, has made no room for you, and if love will not swing wide the gates, no other power will or can. And if one despairs—as who has not?—of human love, God's love alone is left. But God—and I felt this even then, so long ago, on that tremendous floor, unwillingly—is white. And if His love was so great, and if He loved all His children, why were we, the blacks, cast down so far? Why? In spite of all I said thereafter, I found no answer on the floor—not *that* answer, anyway—and I was on the floor all night. Over me, to bring me "through," the saints sang and rejoiced and prayed. And in the morning, when they raised me, they told me that I was "saved."

Well, indeed I was, in a way, for I was utterly drained and exhausted, and released, for the first time, from all my guilty torment. I was aware then only of my relief. For many years, I could not ask myself why human relief had to be achieved in a fashion at once so pagan and so desperate—in a fashion at once so unspeakably old and so unutterably new. And by the time I was able to ask myself this question, I was also able to see that the principles governing the rites and customs of the churches in which I grew up did not differ from the principles governing the rites and customs of other churches, white. The principles were Blindness, Loneliness, and Terror, the first principle necessarily and actively cultivated in order to deny the two others. I would love to believe that the principles were Faith, Hope, and Charity, but this is clearly not so for most Christians, or for what we call the Christian world.

I was saved. But at the same time, out of a deep, adolescent cunning I do not pretend to understand, I realized immediately that I could not remain in the church merely as another worshipper. I would have to give myself something to do, in order not to be too bored and find myself among all the wretched unsaved of the Avenue. And I don't doubt that I also intended to best my father on his own ground. Anyway, very shortly after I joined the church, I became a preacher—a Young Minister—and I remained in the pulpit for more than three years. My youth quickly made me a much bigger drawing card than my father. I pushed this advantage ruthlessly, for it was the most effective means I had found of breaking his hold over me. That was the most frightening time of my life, and quite the most dishonest, and the resulting hysteria lent great passion to my sermons—for a while. I relished the attention and the relative immunity from punishment that my new status gave me, and I relished, above all, the sudden right to privacy. It had to be recognized, after all, that I was still a schoolboy, with my schoolwork to do, and I was also expected to prepare at least one sermon a week. During what we may call my heyday, I preached much more often than that. This meant that there were hours and even whole days when I could not be interrupted—not even by my father. I had immobilized him. It took rather more time for me to realize that I had also immobilized myself, and had escaped from nothing whatever.

The church was very exciting. It took a long time for me to disengage myself from this excitement, and on the blindest, most visceral level, I never really have, and never will. There is no music like that music, no drama like the drama of the saints rejoicing, the sinners moaning, the tambourines racing, and all those voices coming together and crying holy unto the Lord. There is still, for me, no pathos quite like the pathos of those multicolored, worn, somehow triumphant and transfigured faces, speaking from the depths of a visible, tangible, continuing despair of the goodness of the Lord. I have never seen anything to equal the fire and excitement that sometimes, without warning, fill a church, causing the church, as

Leadbelly[4] and so many others have testified, to "rock." Nothing that has happened to me since equals the power and the glory that I sometimes felt when, in the middle of a sermon, I knew that I was somehow, by some miracle, really carrying, as they said, "the Word"—when the church and I were one. Their pain and their joy were mine, and mine were theirs—they surrendered their pain and joy to me, I surendered mine to them—and their cries of "Amen!" and "Hallelujah!" and "Yes, Lord!" and "Praise His name!" and "Preach it, brother!" sustained and whipped on my solos until we all became equal, wringing wet, singing and dancing, in anguish and rejoicing, at the foot of the altar. It was, for a long time, in spite of—or, not inconceivably, because of—the shabbiness of my motives, my only sustenance, my meat and drink. I rushed home from school, to the church, to the altar, to be alone there, to commune with Jesus, my dearest Friend, who would never fail me, who knew all the secrets of my heart. Perhaps He did, but I didn't, and the bargain we struck, actually, down there at the foot of the cross, was that He would never let me find out.

He failed His bargain. He was a much better Man than I took Him for. It happened, as things do, imperceptibly, in many ways at once. I date it—the slow crumbling of my faith, the pulverization of my fortress—from the time, about a year after I had begun to preach, when I began to read again. I justified this desire by the fact that I was still in school, and I began, fatally, with Dostoevski. By this time, I was in a high school that was predominantly Jewish. This meant that I was surrounded by people who were, by definition, beyond any hope of salvation, who laughed at the tracts and leaflets I brought to school, and who pointed out that the Gospels had been written long after the death of Christ. This might not have been so distressing if it had not forced me to read the tracts and leaflets myself, for they were indeed, unless one believed their message already, impossible to believe. I remember feeling dimly that there was a kind of blackmail in it. People, I felt, ought to love the Lord *because* they loved Him, and not because they were afraid of going to Hell. I was forced, reluctantly, to realize that the Bible itself had been written by men, and translated by men out of languages I could not read, and I was already, without quite admitting it to myself, terribly involved with the effort of putting words on paper. Of course, I had the rebuttal ready: These men had all been operating under divine inspiration. *Had* they? *All* of them? And I also knew by now, alas, far more about divine inspiration than I dared admit, for I knew how I worked myself up into my own visions, and how frequently—indeed, incessantly—the visions God granted to me differed from the visions He granted to my father. I

4. Huddie Ledbetter (1888–1949) or Leadbelly: folk and blues singer who had enormous influence on other singers.

did not understand the dreams I had at night, but I knew that they were not holy. For that matter, I knew that my waking hours were far from holy. I spent most of my time in a state of repentance for things I had vividly desired to do but had not done. The fact that I was dealing with Jews brought the whole question of color, which I had been desperately avoiding, into the terrified center of my mind. I realized that the Bible had been written by white men. I knew that, according to many Christians, I was a descendant of Ham,[5] who had been cursed, and that I was therefore predestined to be a slave. This had nothing to do with anything I was, or contained, or could become; my fate had been sealed forever, from the beginning of time. And it seemed, indeed, when one looked out over Christendom, that this was what Christendom effectively believed. It was certainly the way it behaved. I remembered the Italian priests and bishops blessing Italian boys who were on their way to Ethiopia.

Again, the Jewish boys in high school were troubling because I could find no point of connection between them and the Jewish pawnbrokers and landlords and grocery-store owners in Harlem. I knew that these people were Jews—God knows I was told it often enough—but I thought of them only as white. Jews, as such, until I got to high school, were all incarcerated in the Old Testament, and their names were Abraham, Moses, Daniel, Ezekiel, and Job, and Shadrach, Meshach, and Abednego. It was bewildering to find them so many miles and centuries out of Egypt, and so far from the fiery furnace.[6] My best friend in high school was a Jew. He came to our house once, and afterward my father asked, as he asked about everyone, "Is he a Christian?"—by which he meant "Is he saved?" I really do not know whether my answer came out of innocence or venom, but I said coldly, "No. He's Jewish." My father slammed me across the face with his great palm, and in that moment everything flooded back—all the hatred and all the fear, and the depth of a merciless resolve to kill my father rather than allow my father to kill me—and I knew that all those sermons and tears and all that repentance and rejoicing had changed nothing. I wondered if I was expected to be glad that a friend of mine, or anyone, was to be tormented forever in Hell, and I also thought, suddenly, of the Jews in another Christian nation, Germany. They were not so far from the fiery furnace after all, and my best friend might have been one of them. I told my father, "He's a better Christian than you are," and walked out of the house. The battle between us was in the open, but that was all right; it was almost a relief. A more deadly struggle had begun.

---

5. One of the sons of Noah, Ham was cursed with slavery for seeing his drunken father's nakedness (Genesis 9. 18–27). He was also regarded as the progenitor of the African peoples.

6. Shadrach, Meshach, and Abednego ap-pear in the Old Testament (Book of Daniel 3.19–24), where they are cast into the fiery furnace for refusing to worship the divinities of Babylon and disobeying King Nebuchadnezzar.

Being in the pulpit was like being in the theatre; I was behind the scenes and knew how the illusion was worked. I knew the other ministers and knew the quality of their lives. And I don't mean to suggest by this the "Elmer Gantry"[7] sort of hypocrisy concerning sensuality; it was a deeper, deadlier, and more subtle hypocrisy than that, and a little honest sensuality, or a lot, would have been like water in an extremely bitter desert. I knew how to work on a congregation until the last dime was surrendered—it was not very hard to do—and I knew where the money for "the Lord's work" went. I knew, though I did not wish to know it, that I had no respect for the people with whom I worked. I could not have said it then, but I also knew that if I continued I would soon have no respect for myself. And the fact that I was "the young Brother Baldwin" increased my value with those same pimps and racketeers who had helped to stampede me into the church in the first place. They still saw the little boy they intended to take over. They were waiting for me to come to my senses and realize that I was in a very lucrative business. They knew that I did not yet realize this, and also that I had not yet begun to suspect where my own needs, *coming up* (they were very patient), could drive me. They themselves did know the score, and they knew that the odds were in their favor. And, really, I knew it, too. I was even lonelier and more vulnerable than I had been before. And the blood of the Lamb had not cleansed me in any way whatever. I was just as black as I had been the day that I was born. Therefore, when I faced a congregation, it began to take all the strength I had not to stammer, not to curse, not to tell them to throw away their Bibles and get off their knees and go home and organize, for example, a rent strike. When I watched all the children, their copper, brown, and beige faces staring up at me as I taught Sunday school, I felt that I was committing a crime in talking about the gentle Jesus, in telling them to reconcile themselves to their misery on earth in order to gain the crown of eternal life. Were only Negroes to gain this crown? Was Heaven, then, to be merely another ghetto? Perhaps I might have been able to reconcile myself even to this if I had been able to believe that there was any loving-kindness to be found in the haven I represented. But I had been in the pulpit too long and I had seen too many monstrous things. I don't refer merely to the glaring fact that the minister eventually acquires houses and Cadillacs while the faithful continue to scrub floors and drop their dimes and quarters and dollars into the plate. I really mean that there was no love in the church. It was a mask for hatred and self-hatred and despair. The transfiguring power of the Holy Ghost ended when the service ended, and salvation stopped at the church door. When we were

7. Hero of Sinclair Lewis's novel of that title whose preaching is at odds with his lechery.

told to love everybody, I had thought that that meant *everybody*. But no. It applied only to those who believed as we did, and it did not apply to white people at all. I was told by a minister, for example, that I should never, on any public conveyance, under any circumstances, rise and give my seat to a white woman. White men never rose for Negro women. Well, that was true enough, in the main—I saw his point. But what was the point, the purpose, of *my* salvation if it did not permit me to behave with love toward others, no matter how they behaved toward me? What others did was their responsibility, for which they would answer when the judgment trumpet sounded. But what *I* did was *my* responsibility, and I would have to answer, too—unless, of course, there was also in Heaven a special dispensation for the benighted black, who was not to be judged in the same way as other human beings, or angels. It probably occurred to me around this time that the vision people hold of the world to come is but a reflection, with predictable wishful distortions, of the world in which they live. And this did not apply only to Negroes, who were no more "simple" or "spontaneous" or "Christian" than anybody else—who were merely more oppressed. In the same way that we, for white people, were the descendants of Ham, and were cursed forever, white people were, for us, the descendants of Cain. And the passion with which we loved the Lord was a measure of how deeply we feared and distrusted and, in the end, hated almost all strangers, always, and avoided and despised ourselves.

But I cannot leave it at that; there is more to it than that. In spite of everything, there was in the life I fled a zest and a joy and a capacity for facing and surviving disaster that are very moving and very rare. Perhaps we were, all of us—pimps, whores, racketeers, church members, and children—bound together by the nature of our oppression, the specific and peculiar complex of risks we had to run; if so, within these limits we sometimes achieved with each other a freedom that was close to love. I remember, anyway, church suppers and outings, and, later, after I left the church, rent and waistline parties[8] where rage and sorrow sat in the darkness and did not stir, and we ate and drank and talked and laughed and danced and forgot all about "the man." We had the liquor, the chicken, the music, and each other, and had no need to pretend to be what we were not. This is the freedom that one hears in some gospel songs, for example, and in jazz. In all jazz, and especially in the blues, there is something tart and ironic, authoritative and double-edged. White Americans seem to feel that happy songs are *happy* and sad songs are *sad*, and that, God help us, is exactly the way most white Americans sing them—sounding, in both cases, so helplessly, defenselessly fatuous that one dare not speculate on the temperature of the deep freeze from which issue their brave and sexless little voices. Only people who have been "down the line," as the song puts it, know what this music is about. I think it was Big Bill Broonzy[9] who used to sing "I Feel So Good," a really joyful song about a man who is on his way to the railroad station to meet his girl. She's coming home. It is the singer's incredibly moving exuberance that makes one realize how leaden the time must have been while she was gone. There is no guarantee that she will stay this time, either, as the singer clearly knows, and, in fact, she has not yet actually arrived. Tonight, or tomorrow, or within the next five minutes, he may very well be singing "Lonesome in My Bedroom," or insisting, "Ain't we, ain't we, going to make it all right? Well, if we don't today, we will tomorrow night." White Americans do not understand the depths out of which such an ironic tenacity comes, but they suspect that the force is sensual, and they are terrified of sensuality and do not any longer understand it. The word "sensual" is not intended to bring to mind quivering dusky maidens or priapic black studs. I am referring to something much simpler and much less fanciful. To be sensual, I think, is to respect and rejoice in the force of life, of life itself, and to be *present* in all that one does, from the effort of loving to the breaking of bread. It will be a great day for America, incidentally, when we begin to eat bread again, instead of the blasphemous and tasteless foam rubber that we have substituted for it. And I am not being frivolous now, either. Something very sinister happens to the people of a country when they begin to distrust their own reactions as deeply as they do here, and become as joyless as they have become. It is this individual uncertainty on the part of white American men and women, this inability to renew themselves at the fountain of their own lives, that makes the discussion, let alone elucidation, of any conundrum—that is, any reality—so supremely difficult. The person who distrusts himself has no touchstone for reality—for this touchstone can be only oneself. Such a person interposes between himself and reality nothing less than a labyrinth of attitudes. And these attitudes, furthermore, though the person is usually unaware of it (is unaware of so much!), are historical and public attitudes. They do not relate to the present any more than they relate to the person. Therefore, whatever white people do not know about Negroes reveals, precisely and inexorably, what they do not know about themselves.

White Christians have also forgotten several elementary historical details. They have forgotten that the religion that is now identified with their virtue and their power—"God is on our side," says Dr. Verwoerd[1]—came out of a rocky piece of ground in what is now

---

8. Gatherings held in houses or apartments in order to raise money to pay the rent.
9. Blues singer and guitarist (1893–1958).

1. Dr. Henrik Verwoerd was a dedicated proponent of *apartheid* (separation of the races) and Prime Minister of South Africa in the late 1950s.

known as the Middle East before color was invented, and that in order for the Christian church to be established, Christ had to be put to death, by Rome, and that the real architect of the Christian church was not the disreputable, sun-baked Hebrew who gave it his name but the mercilessly fanatical and self-righteous St. Paul. The energy that was buried with the rise of the Christian nations must come back into the world; nothing can prevent it. Many of us, I think, both long to see this happen and are terrified of it, for though this transformation contains the hope of liberation, it also imposes a necessity for great change. But in order to deal with the untapped and dormant force of the previously subjugated, in order to survive as a human, moving, moral weight in the world, America and all the Western nations will be forced to reexamine themselves and release themselves from many things that are now taken to be sacred, and to discard nearly all the assumptions that have been used to justify their lives and their anguish and their crimes so long.

"The white man's Heaven," sings a Black Muslim minister, "is the black man's Hell." One may object—possibly—that this puts the matter somewhat too simply, but the song is true, and it has been true for as long as white men have ruled the world. The Africans put it another way: When the white man came to Africa, the white man had the Bible and the African had the land, but now it is the white man who is being, reluctantly and bloodily, separated from the land, and the African who is still attempting to digest or to vomit up the Bible. The struggle, therefore, that now begins in the world is extremely complex, involving the historical role of Christianity in the realm of power—that is, politics—and in the realm of morals. In the realm of power, Christianity has operated with an unmitigated arrogance and cruelty—necessarily, since a religion ordinarily imposes on those who have discovered the true faith the spiritual duty of liberating the infidels. This particular true faith, moreover, is more deeply concerned about the soul than it is about the body, to which fact the flesh (and the corpses) of countless infidels bears witness. It goes without saying, then, that whoever questions the authority of the true faith also contests the right of the nations that hold this faith to rule over him—contests, in short, their title to his land. The spreading of the Gospel, regardless of the motives or the integrity or the heroism of some of the missionaries, was an absolutely indispensable justification for the planting of the flag. Priests and nuns and schoolteachers helped to protect and sanctify the power that was so ruthlessly being used by people who were indeed seeking a city, but not one in the heavens, and one to be made, very definitely, by captive hands. The Christian church itself—again, as distinguished from some of its ministers—sanctified and rejoiced in the conquests of the flag, and encouraged, if it did not formulate, the belief that conquest, with the resulting relative well-being of the Western populations, was proof of the favor of God. God had come a long way from the desert—but then so had Allah, though in a very different direction. God, going north, and rising on the wings of power, had become white, and Allah, out of power, and on the dark side of Heaven, had become—for all practical purposes, anyway—black. Thus, in the realm of morals the role of Christianity has been, at best, ambivalent. Even leaving out of account the remarkable arrogance that assumed that the ways and morals of others were inferior to those of Christians, and that they therefore had every right, and could use any means, to change them, the collision between cultures—and the schizophrenia in the mind of Christendom—had rendered the domain of morals as chartless as the sea once was, and as treacherous as the sea still is. It is not too much to say that whoever wishes to become a truly moral human being (and let us not ask whether or not this is possible; I think we must *believe* that it is possible) must first divorce himself from all the prohibitions, crimes, and hypocrisies of the Christian church. If the concept of God has any validity or any use, it can only be to make us larger, freer, and more loving. If God cannot do this, then it is time we got rid of Him.

1962, 1963

FLANNERY O'CONNOR
1925–1964

Flannery O'Connor, one of this century's finest writers of short stories, was born in Savannah, lived with her mother in Milledgeville, Georgia, for much of her life, and died before her fortieth birthday—victim like her father of disseminated lupus, a rare and incurable blood disease. She was stricken with the disease in 1950 while at work on her first novel, but injections of a cortisone derivative managed to arrest it, though the cortisone weakened her bones to the extent that from 1955 on she got around on aluminum crutches. She was able to write, travel, and lecture until 1964 when the lupus reactivated itself and killed her. A Roman Catholic throughout her life, she is quoted as having remarked, apropos a trip to Lourdes, "I had the best-looking crutches in Europe." This remark suggests the kind of hair-raising jokes which centrally inform her writing, as well as a refusal to indulge in self-pity over her fate.

She published two novels, *Wise Blood* (1952) and *The Violent Bear It Away* (1960), both weighty with symbolic and religious concerns, and ingeniously contrived in the black-humored manner of Nathanael West, her American predecessor in this mode. But her really memorable creations of characters and actions take place in the stories, which are extremely funny, sometimes unbearably so, and finally we may wonder just what it is we are laughing at. Another American "regionalist," the poet Robert Frost, whose own work contains its share of "dreadful jokes," once confessed to being more interested in people's speech than in the people themselves. A typical Flannery O'Connor story consists at its most vital level in people talking, clucking their endless reiterations of clichés about life death, and the universe. These clichés are captured with beautiful accuracy by an artist who had spent her life listening to them, lovingly and maliciously keeping track until she could put them to use. Early in her life she hoped to be a cartoonist, and there is cartoonlike mastery in her vivid renderings of character through speech and other gesture. Critics have called her a maker of grotesques, a label which like other ones—Regionalist, Southern Lady, or Roman Catholic Novelist—might have annoyed if it didn't obviously amuse her too. She once remarked tartly that "anything that comes out of the South is going to be called grotesque by the Northern reader, unless it is grotesque, in which case it is going to be called realistic."

Of course this capacity for mockery, along with a facility in portraying perverse behavior, may work against other demands we make of the fiction writer; and it is true that Flannery O'Connor seldom suggests that her characters have inner lives that are imaginable, let alone worth respect. Instead the emphasis is on the sharp eye and the ability to tell a tale and keep it moving inevitably toward completion. These completions are usually violent, occurring when the character—in many cases a woman—must confront an experience which she cannot handle by the old trustworthy language and habit-hardened responses. O'Connor's art lies partly in making it impossible for us merely to scorn the banalities of expression and behavior by which these people get through their lives. However dark the comedy, it keeps in touch with the things of this world, even when some force from another world threatens to annihilate the embattled protagonist. And although the stories are filled with religious allusions and parodies, they do not try to inculcate a doctrine. One of her best ones is titled *Revelation*, but a reader often finishes a story with no simple, unambiguous sense of what has been revealed. Instead we must trust the internal fun and richness of each tale to reveal what it has to reveal. We can agree also, in sadness, with the critic Irving Howe's conclusion to his review of her posthumous collection of stories that it is intolerable for such a writer to have died at the age of thirty-nine.

The Life You Save May Be Your Own[1]

The old woman and her daughter were sitting on their porch when Mr. Shiftlet came up their road for the first time. The old woman slid to the edge of her chair and leaned forward, shading her eyes from the piercing sunset with her hand. The daughter could not see far in front of her and continued to play with her fingers. Although the old woman lived in this desolate spot with only her daughter and she had never seen Mr. Shiftlet before, she could tell, even from a distance, that he was a tramp and no one to be afraid of. His left coat sleeve was folded up to show there was only half an arm in it

---

1. From *A Good Man Is Hard to Find and Other Stories* (1955).

and his gaunt figure listed slightly to the side as if the breeze were pushing him. He had on a black town suit and a brown felt hat that was turned up in the front and down in the back and he carried a tin tool box by a handle. He came on, at an amble, up her road, his face turned toward the sun which appeared to be balancing itself on the peak of a small mountain.

The old woman didn't change her position until he was almost into her yard; then she rose with one hand fisted on her hip. The daughter, a large girl in a short blue organdy dress, saw him all at once and jumped up and began to stamp and point and make excited speechless sounds.

Mr. Shiftlet stopped just inside the yard and set his box on the ground and tipped his hat at her as if she were not in the least afflicted; then he turned toward the old woman and swung the hat all the way off. He had long black slick hair that hung flat from a part in the middle to beyond the tips of his ears on either side. His face descended in forehead for more than half its length and ended suddenly with his features just balanced over a jutting steel-trap jaw. He seemed to be a young man but he had a look of composed dissatisfaction as if he understood life thoroughly.

"Good evening," the old woman said. She was about the size of a cedar fence post and she had a man's gray hat pulled down low over her head.

The tramp stood looking at her and didn't answer. He turned his back and faced the sunset. He swung both his whole and his short arm up slowly so that they indicated an expanse of sky and his figure formed a crooked cross. The old woman watched him with her arms folded across her chest as if she were the owner of the sun, and the daughter watched, her head thrust forward and her fat helpless hands hanging at the wrists. She had long pink-gold hair and eyes as blue as a peacock's neck.

He held the pose for almost fifty seconds and then he picked up his box and came on to the porch and dropped down on the bottom step. "Lady," he said in a firm nasal voice, "I'd give a fortune to live where I could see me a sun do that every evening."

"Does it every evening," the old woman said and sat back down. The daughter sat down too and watched him with a cautious sly look as if he were a bird that had come up very close. He leaned to one side, rooting in his pants pocket, and in a second he brought out a package of chewing gum and offered her a piece. She took it and unpeeled it and began to chew without taking her eyes off him. He offered the old woman a piece but she only raised her upper lip to indicate she had no teeth.

Mr. Shiftlet's pale sharp glance had already passed over everything in the yard—the pump near the corner of the house and the big fig tree that three or four chickens were preparing to roost in—and had moved to a shed where he saw the square rusted back of an automobile. "You ladies drive?" he asked.

"That car ain't run in fifteen year," the old woman said. "The day my husband died, it quit running."

"Nothing is like it used to be, lady," he said. "The world is almost rotten."

"That's right," the old woman said. "You from around here?"

"Name Tom T. Shiftlet," he murmured, looking at the tires.

"I'm pleased to meet you," the old woman said. "Name Lucynell Crater and daughter Lucynell Crater. What you doing around here, Mr. Shiftlet?"

He judged the car to be about a 1928 or '29 Ford. "Lady," he said, and turned and gave her his full attention, "lemme tell you something. There's one of these doctors in Atlanta that's taken a knife and cut the human heart—the human heart," he repeated, leaning forward, "out of a man's chest and held it in his hand," and he held his hand out, palm up, as if it were slightly weighted with the human heart, "and studied it like it was a day-old chicken, and lady," he said, allowing a long significant pause in which his head slid forward and his clay-colored eyes brightened, "he don't know no more about it than you or me."

"That's right," the old woman said.

"Why, if he was to take that knife and cut into every corner of it, he still wouldn't know no more than you or me. What you want to bet?"

"Nothing," the old woman said wisely. "Where you come from, Mr. Shiftlet?"

He didn't answer. He reached into his pocket and brought out a sack of tobacco and a package of cigarette papers and rolled himself a cigarette, expertly with one hand, and attached it in a hanging position to his upper lip. Then he took a box of wooden matches from his pocket and struck one on his shoe. He held the burning match as if he were studying the mystery of flame while it traveled dangerously toward his skin. The daughter began to make loud noises and to point to his hand and shake her finger at him, but when the flame was just before touching him, he leaned down with his hand cupped over it as if he were going to set fire to his nose and lit the cigarette.

He flipped away the dead match and blew a stream of gray into the evening. A sly look came over his face. "Lady," he said, "nowadays, people'll do anything anyways. I can tell you my name is Tom T. Shiftlet and I come from Tarwater, Tennessee, but you never have seen me before: how you know I ain't lying? How you know my name ain't Aaron Sparks, lady, and I come from Singleberry, Georgia, or how you know it's not George Speeds and I come from Lucy, Alabama, or how you know I ain't Thompson Bright from Toolafalls, Mississippi?"

"I don't know nothing about you," the old woman muttered, irked.

"Lady," he said, "people don't care how they lie. Maybe the best I can tell you is, I'm a man; but listen lady," he said and paused and made his tone more ominous still, "what is a man?"

The old woman began to gum a seed. "What you carry in that tin box, Mr. Shiftlet?" she asked.

"Tools," he said, put back. "I'm a carpenter."

"Well, if you come out here to work, I'll be able to feed you and give you a place to sleep but I can't pay, I'll tell you that before you begin," she said.

There was no answer at once and no particular expression on his face. He leaned back against the two-by-four that helped support the porch roof. "Lady," he said slowly, "there's some men that some things mean more to them than money." The old woman rocked without comment and the daughter watched the trigger that moved up and down in his neck. He told the old woman then that all most people were interested in was money, but he asked what a man was made for. He asked her if a man was made for money, or what. He asked her what she thought she was made for but she didn't answer, she only sat rocking and wondered if a one-armed man could put a new roof on her garden house. He asked a lot of questions that she didn't answer. He told her that he was twenty-eight years old and had lived a varied life. He had been a gospel singer, a foreman on the railroad, an assistant in an undertaking parlor, and he had come over the radio for three months with Uncle Roy and his Red Creek Wranglers. He said he had fought and bled in the Arm Service of his country and visited every foreign land and that everywhere he had seen people that didn't care if they did a thing one way or another. He said he hadn't been raised thataway.

A fat yellow moon appeared in the branches of the fig tree as if it were going to roost there with the chickens. He said that a man had to escape to the country to see the world whole and that he wished he lived in a desolate place like this where he could see the sun go down every evening like God made it to do.

"Are you married or are you single?" the old woman asked.

There was a long silence. "Lady," he asked finally, "where would you find you an innocent woman today? I wouldn't have any of this trash I could just pick up."

The daughter was leaning very far down, hanging her head almost between her knees, watching him through a triangular door she had made in her overturned hair; and she suddenly fell in a heap on the floor and began to whimper. Mr. Shiftlet straightened her out and helped her get back in the chair.

"Is she your baby girl?" he asked.

"My only," the old woman said, "and she's the sweetest girl in the world. I wouldn't give her up for nothing on earth. She's smart too. She can sweep the floor, cook, wash, feed the chickens, and hoe. I wouldn't give her up for a casket of jewels."

"No," he said kindly, "don't ever let any man take her away from you."

"Any man come after her," the old woman said, "'ll have to stay around the place."

Mr. Shiftlet's eye in the darkness was focused on a part of the automobile bumper that glittered in the distance. "Lady," he said, jerking his short arm up as if he could point with it to her house and yard and pump, "there ain't a broken thing on this plantation that I couldn't fix for you, one-arm jackleg or not. I'm a man," he said with a sullen dignity, "even if I ain't a whole one. I got," he said, tapping his knuckles on the floor to emphasize the immensity of what he was going to say, "a moral intelligence!" and his face pierced out of the darkness into a shaft of doorlight and he stared at her as if he were astonished himself at this impossible truth.

The old woman was not impressed with the phrase. "I told you you could hang around and work for food," she said, "if you don't mind sleeping in that car yonder."

"Why listen, Lady," he said with a grin of delight, "the monks of old slept in their coffins!"

"They wasn't as advanced as we are," the old woman said.

The next morning he began on the roof of the garden house while Lucynell, the daughter, sat on a rock and watched him work. He had not been around a week before the change he had made in the place was apparent. He had patched the front and back steps, built a new hog pen, restored a fence, and taught Lucynell, who was completely deaf and had never said a word in her life, to say the word "bird." The big rosy-faced girl followed him everywhere, saying "Burrttddt ddbirrrttdt," and clapping her hands. The old woman watched from a distance, secretly pleased. She was ravenous for a son-in-law.

Mr. Shiftlet slept on the hard narrow back seat of the car with his feet out the side window. He had his razor and a can of water on a crate that served him as a bedside table and he put up a piece of mirror against the back glass and kept his coat neatly on a hanger that he hung over one of the windows.

In the evenings he sat on the steps and talked while the old woman and Lucynell rocked violently in their chairs on either side of him. The old woman's three mountains were black against the dark blue sky and were visited off and on by various planets and by the moon after it had left the chickens. Mr. Shiftlet pointed out that the reason he had improved this plantation was because he had taken a personal interest in it. He said he was even going to make the automobile run.

He had raised the hood and studied the mechanism and he said he could tell that the car had been built in the days when cars were really built. You take now, he said, one man puts in one bolt and another man puts in another bolt and another man puts in another bolt so that it's a man for a bolt. That's why you have to pay so much for a car: you're paying all those men. Now if you didn't have to pay but one man, you could get you a cheaper car and one that had had a personal interest taken in it, and it would be a better car. The old woman agreed with him that this was so.

Mr. Shiftlet said that the trouble with the world was that nobody cared, or stopped and took any trouble. He said he never would have been able to teach Lucynell to say a word if he hadn't cared and stopped long enough.

"Teach her to say something else," the old woman said.

"What you want her to say next?" Mr. Shiftlet asked.

The old woman's smile was broad and toothless and suggestive. "Teach her to say 'sugarpie,'" she said.

Mr. Shiftlet already knew what was on her mind.

The next day he began to tinker with the automobile and that evening he told her that if she would buy a fan belt, he would be able to make the car run.

The old woman said she would give him the money. "You see that girl yonder?" she asked, pointing to Lucynell who was sitting on the floor a foot away, watching him, her eyes blue even in the dark. "If it was ever a man wanted to take her away, I would say, 'No man on earth is going to take that sweet girl of mine away from me!' but if he was to say, 'Lady, I don't want to take her away, I want her right here,' I would say, 'Mister, I don't blame you nine, I wouldn't pass up a chance to live in a permanent place and get the sweetest girl in the world myself. You ain't no fool,' I would say."

"How old is she?" Mr. Shiftlet asked casually.

"Fifteen, sixteen," the old woman said. The girl was nearly thirty but because of her innocence it was impossible to guess.

"It would be a good idea to paint it too," Mr. Shiftlet remarked. "You don't want it to rust out."

"We'll see about that later," the old woman said.

The next day he walked into town and returned with the parts he needed and a can of gasoline. Late in the afternoon, terrible noises issued from the shed and the old woman rushed out of the house, thinking Lucynell was somewhere having a fit. Lucynell was sitting on a chicken crate, stamping her feet and screaming, "Burrddttt! bddurrddtttt!" but her fuss was drowned out by the car. With a volley of blasts it emerged from the shed, moving in a fierce and stately way. Mr. Shiftlet was in the driver's seat, sitting very erect. He had an expression of serious modesty on his face as if he had just raised the dead.

That night, rocking on the porch, the old woman began her business at once. "You want you an innocent woman, don't you?" she asked sympathetically. "You don't want none of this trash."

"No'm, I don't," Mr. Shiftlet said.

"One that can't talk," she continued, "can't sass you back or use foul language. That's the kind for you to have. Right there," and she pointed to Lucynell sitting cross-legged in her chair, holding both feet in her hands.

"That's right," he admitted. "She wouldn't give me any trouble."

"Saturday," the old woman said, "you and her and me can drive into town and get married."

Mr. Shiftlet eased his position on the steps.

"I can't get married right now," he said. "Everything you want to do takes money and I ain't got any."

"What you need with money?" she asked.

"It takes money," he said. "Some people'll do anything anyhow these days, but the way I think, I wouldn't marry no woman that I couldn't take on a trip like she was somebody. I mean take her to a hotel and treat her. I wouldn't marry the Duchesser Windsor," he said firmly, "unless I could take her to a hotel and give her something good to eat.

"I was raised thataway and there ain't a thing I can do about it. My old mother taught me how to do."

"Lucynell don't even know what a hotel is," the old woman muttered. "Listen here, Mr. Shiftlet," she said, sliding forward in her chair, "you'd be getting a permanent house and a deep well and the most innocent girl in the world. You don't need no money. Lemme tell you something: there ain't any place in the world for a poor disabled friendless drifting man."

The ugly words settled in Mr. Shiftlet's head like a group of buzzards in the top of a tree. He didn't answer at once. He rolled himself a cigarette and lit it and then he said in an even voice, "Lady, a man is divided into two parts, body and spirit."

The old woman clamped her gums together.

"A body and a spirit," he repeated. "The body, lady, is like a house: it don't go anywhere; but the spirit, lady, is like a automobile: always on the move, always . . ."

"Listen, Mr. Shiftlet," she said, "my well never goes dry and my house is always warm in the winter and there's no mortgage on a thing about this place. You can go to the courthouse and see for yourself. And yonder under that shed is a fine automobile." She laid the bait carefully. "You can have it painted by Saturday. I'll pay for the paint."

In the darkness, Mr. Shiftlet's smile stretched like a weary snake waking up by a fire. After a second he recalled himself and said, "I'm only saying a man's spirit means more to him than anything else. I would have to take my wife off for the week end without no regards at all for cost. I got to follow where my spirit says to go."

"I'll give you fifteen dollars for a week-end trip," the old woman said in a crabbed voice. "That's the best I can do."

"That wouldn't hardly pay for more than the gas and the hotel," he said. "It wouldn't feed her."

"Seventeen-fifty," the old woman said. "That's all I got so it isn't any use you trying to milk me. You can take a lunch."

Mr. Shiftlet was deeply hurt by the word "milk." He didn't doubt that she had more money sewed up in her mattress but he had already told her he was not interested in her money. "I'll make that do," he said and rose and walked off without treating with her further.

On Saturday the three of them drove into town in the car that the paint had barely dried on and Mr. Shiftlet and Lucynell were married in the Ordinary's office while the old woman witnessed. As they came out of the courthouse, Mr. Shiftlet began twisting his neck in his collar. He looked morose and bitter as if he had been insulted while someone held him. "That didn't satisfy me none," he said. "That was just something a woman in an office did, nothing but paper work and blood tests. What do they know about my blood? If they was to take my heart and cut it out," he said, "they wouldn't know a thing about me. It didn't satisfy me at all."

"It satisfied the law," the old woman said sharply.

"The law," Mr. Shiftlet said and spit. "It's the law that don't satisfy me."

He had painted the car dark green with a yellow band around it just under the windows. The three of them climbed in the front seat and the old woman said, "Don't Lucynell look pretty? Looks like a baby doll." Lucynell was dressed up in a white dress that her mother had uprooted from a trunk and there was a Panama hat on her head with a bunch of red wooden cherries on the brim. Every now and then her placid expression was changed by a sly isolated little thought like a shoot of green in the desert. "You got a prize!" the old woman said.

Mr. Shiftlet didn't even look at her.

They drove back to the house to let the old woman off and pick up the lunch. When they were ready to leave, she stood staring in the window of the car, with her fingers clenched around the glass. Tears began to seep sideways out of her eyes and run along the dirty creases in her face. "I ain't ever been parted with her for two days before," she said.

Mr. Shiftlet started the motor.

"And I wouldn't let no man have her but you because I seen you would do right. Good-by, Sugarbaby," she said, clutching at the sleeve of the white dress. Lucynell looked straight at her and didn't seem to see her there at all. Mr. Shiftlet eased the car forward so that she had to move her hands.

The early afternoon was clear and open and surrounded by pale blue sky. Although the car would go only thirty miles an hour, Mr. Shiftlet imagined a terrific climb and dip and swerve that went entirely to his head so that he forgot his morning bitterness. He had always wanted an automobile but he had never been able to afford one before. He drove very fast because he wanted to make Mobile by nightfall.

Occasionally he stopped his thoughts long enough to look at Lucynell in the seat beside him. She had eaten the lunch as soon as they were out of the yard and now she was pulling the cherries off the hat one by one and throwing them out the window. He became depressed in spite of the car. He had driven about a hundred miles when he decided that she must be hungry again and at the next small town they came to, he stopped in front of an aluminum-painted eating place called The Hot Spot and took her in and ordered her a plate of ham and grits. The ride had made her sleepy and as soon as she got up on the stool, she rested her head on the counter and shut her eyes. There was no one in The Hot Spot but Mr. Shiftlet and the boy behind the counter, a pale youth with a greasy rag hung over his shoulder. Before he could dish up the food, she was snoring gently.

"Give it to her when she wakes up," Mr. Shiftlet said. "I'll pay for it now."

The boy bent over her and stared at the long pink-gold hair and the half-shut sleeping eyes. Then he looked up and stared at Mr. Shiftlet. "She looks like an angel of Gawd," he murmured.

"Hitch-hiker," Mr. Shiftlet explained. "I can't wait. I got to make Tuscaloosa."

The boy bent over again and very carefully touched his finger to a strand of the golden hair and Mr. Shiftlet left.

He was more depressed than ever as he drove on by himself. The late afternoon had grown hot and sultry and the country had flattened out. Deep in the sky a storm was preparing very slowly and without thunder as if it meant to drain every drop of air from the earth before it broke. There were times when Mr. Shiftlet preferred not to be alone. He felt too that a man with a car had a responsibility to others and he kept his eye out for a hitch-hiker. Occasionally he saw a sign that warned: "Drive carefully. The life you save may be your own."

The narrow road dropped off on either side into dry fields and here and there a shack or a filling station stood in a clearing. The sun began to set directly in front of the automobile. It was a reddening ball that through his windshield was slightly flat on the bottom and top. He saw a boy in overalls and a gray hat standing on the edge of the road and he slowed the car down and stopped in front of him. The boy didn't have his hand raised to thumb the ride, he was only standing there, but he had a small cardboard suitcase and his hat was set on his head in a way to indicate that he had left somewhere for good. "Son," Mr. Shiftlet said, "I see you want a ride."

The boy didn't say he did or he didn't but he opened the door of the car and got in, and Mr. Shiftlet started driving again. The child held the suitcase on his lap and folded his arms on top of it. He turned his head and looked out the window away from Mr. Shiftlet. Mr. Shiftlet felt oppressed. "Son," he said after a minute, "I got the best old mother in the world so I reckon you only got the second best."

The boy gave him a quick dark glance and then turned his face back out the window.

"It's nothing so sweet," Mr. Shiftlet continued, "as a boy's mother. She taught him his first prayers at her knee, she give him love when no other would, she told him what was right and what wasn't, and she seen that he done the right thing. Son," he said, "I never rued a day in my life like the one I rued when I left that old mother of mine."

The boy shifted in his seat but he didn't look at Mr. Shiftlet. He unfolded his arms and put one hand on the door handle.

"My mother was a angel of Gawd," Mr. Shiftlet said in a very strained voice. "He took her from heaven and giver to me and I left her." His eyes were instantly clouded over with a mist of tears. The car was barely moving.

The boy turned angrily in the seat. "You go to the devil!" he cried. "My old woman is a flea bag and yours is a stinking pole cat!" and with that he flung the door open and jumped out with his suitcase into the ditch.

Mr. Shiftlet was so shocked that for about a hundred feet he drove along slowly with the door still open. A cloud, the exact color of the boy's hat and shaped like a turnip, had descended over the sun, and another, worse looking, crouched behind the car. Mr. Shiftlet felt that the rottenness of the world was about to engulf him. He raised his arm and let it fall again to his breast. "Oh Lord!" he prayed. "Break forth and wash the slime from this earth!"

The turnip continued slowly to descend. After a few minutes there was a guffawing peal of thunder from behind and fantastic raindrops, like tin-can tops, crashed over the rear of Mr. Shiftlet's car. Very quickly he stepped on the gas and with his stump sticking out the window he raced the galloping shower into Mobile.

1955

JOHN BARTH

1930–

During a prolonged student strike at the State University of New York in Buffalo, John Barth, teaching there at the time, when asked his opinion of the strike answered that it was important but boring. To judge from his fiction, this response also characterizes Barth's attitude toward everyday human affairs. Although gifted with strong representational powers, as demonstrated in his first two novels, he is uninterested in providing carefully rendered imitations of life but does care passionately about the activity of storytelling, of narrative itself—to the extent that a reviewer of one of his most elaborate pieces of fiction (*Chimera*, 1972) called him a "narrative chauvinist pig."

He was born in Cambridge, Maryland, and after a brief time spent at the Juilliard School of Music, entered Johns Hopkins University in Baltimore, from which he graduated and where he currently teaches writing. In the first three and the last three months of 1955, Barth accomplished the noteworthy feat of writing his first two novels, books preoccupied with "ultimate" philosophical and moral questions a gifted young man might have asked at Johns Hopkins. The hero of *The Floating Opera* (1956) is such a young man named Todd Andrews who plans to kill himself on the day of which the novel treats, and who describes the title thus: "It's a floating opera, friend, chock-full of curiosities, melodrama, spectacle, instruction and entertainment, but it floats willy-nilly on the tide of my vagrant prose." No better description has been made of Barth's work as a whole.

The Floating Opera was nominated for the National Book Award and remains compelling for its playful, colloquial speech, and for Barth's ability to spin marvelously funny yarns fringed with metaphysical speculation. It is also notable for its intimate feeling for the Maryland coastal region, at moments reminding us of Melville's similarly genial treatment of New England in *Moby-Dick*, our literature's grandest floating opera. Barth's second novel, *The End of the Road* (1958) was to be his last "realistic" one, though it is also filled with narrative games. In the manic monologues and dialectical skirmishes with others, engaged in by its hero Jacob Horner, Barth practices an inventive and amusing questioning of moral and philosophical values. This extremely funny book turns suddenly grim at its conclusion, when Horner is left with nothing but words (he is an evasive teacher of grammar) to deal with the horror of an event beyond them and for which he is in part responsible.

After the rapid composition of these books, Barth ceased to be interested in writing even as minimally tied to conventionally realistic fiction as they were. In the immensely long and sometimes labored books which followed—*The Sot-Weed Factor* (1960) and *Giles Goat-Boy* (1966)—he constructed gigantic parody-histories filled with gags, sexual bawdy, and lore of all sorts in order to dislocate and entertain the reader. Barth is saying in these books, among other things, that fiction is stranger than "fiction," that the reader must understand he is not reading about Life— events which really take place in a world out there—but is instead reading words. So *Giles* is as much "about" itself as a book can be: its comically pedantic and self-footnoting procedure and its strategy of pretending that the Universe is really the University are only two of the devices by which Barth attempts to dazzle and ensnare us.

The selection printed here is taken from a volume (*Lost in the Funhouse*, 1968) subtitled *Fiction for Print, Tape, Live Voice*, and at one point a voice addresses us as follows: "The reader! You, dogged, uninsultable, print-oriented bastard, it's you I'm addressing, who else, from inside this monstrous fiction." *Life-Story* is one of Barth's most extreme attempts to confuse the realms of art and life, imagination and reality. How much further he can go along the disruptively playful line taken by this story, or by the recent retelling and complicating of classical legends in *Chimera*, is anybody's guess. But Barth has never yet lacked the resourcefulness to create something beyond what we had hitherto imagined as possible or proper narrative behavior, and he will probably do it again.

Life-Story[1]

I

Without discarding what he'd already written he began his story afresh in a somewhat different manner. Whereas his earlier version had opened in a straight-forward documentary fashion and then degenerated or at least modulated intentionally into irrealism and dissonance he decided this time to tell his tale from start to finish in a conservative, "realistic," unself-conscious way. He being by vocation an author of novels and stories it was perhaps inevitable that one afternoon the possibility would occur to the writer of these lines that his own life might be a fiction, in which he was the leading or an accessory character. He happened at the time[2] to be in his study attempting to draft the opening pages of a new short story; its general idea had preoccupied him for some months along with other general ideas, but certain elements of the conceit, without which he could scarcely proceed, remained unclear. More specifically: narrative plots may be imagined as consisting of a "ground-situation" (Scheherazade desires not to die) focused and dramatized by a "vehicle-situation" (Scheherazade beguiles the King with endless stories), the several incidents of which have their final value in terms of their bearing upon the "ground-situation." In our author's case it was the "vehicle" that had vouchsafed itself, first as a germinal proposition in his commonplace book—D comes to suspect that the world is a novel, himself a fictional personage—subsequently as an articulated conceit explored over several pages of the workbook in which he elaborated more systematically his casual inspirations: since D is writing a fictional account of this conviction he has indisputably a fictional existence in his account, replicating what he suspects to be his own situation. Moreover E, hero of D's account,

---

1. From *Lost in the Funhouse: Fiction for Print, Tape, Live Voice* (1968).   2. "9:00 A.M., Monday, June 20, 1966" [Barth's note].

is said to be writing a similar account, and so the replication is in both ontological directions, et cetera. But the "ground-situation"— some state of affairs on D's part which would give dramatic resonance to his attempts to prove himself factual, assuming he made such attempts obstinately withheld itself from his imagination. As is commonly the case the question reduced to one of stakes: what were to be the consequences of D's—and finally E's—disproving or verifying his suspicion, and why should a reader be interested?

What a dreary way to begin a story he said to himself upon reviewing his long introduction. Not only is there no "ground-situation," but the prose style is heavy and somewhat old-fashioned, like an English translation of Thomas Mann,[3] and the so-called "vehicle" itself is at least questionable: self-conscious, vertiginously arch, fashionably solipsistic, unoriginal—in fact a convention of twentieth-century literature. Another story about a writer writing a story! Another regressus in infinitum! Who doesn't prefer art that at least overtly imitates something other than its own processes? That doesn't continually proclaim "Don't forget I'm an artifice!"? That takes for granted its mimetic nature instead of asserting it in order (not so slyly after all) to deny it, or vice-versa? Though his critics sympathetic and otherwise described his own work as avant-garde, in his heart of hearts he disliked literature of an experimental, self-despising, or overtly metaphysical character, like Samuel Beckett's, Marian Cutler's, Jorge Borges's.[4] The logical fantasies of Lewis Carroll pleased him less than straight-forward tales of adventure, subtly sentimental romances, even densely circumstantial realisms like Tolstoy's. His favorite contemporary authors were John Updike, Georges Simenon, Nicole Riboud.[5] He had no use for the theater of absurdity, for "black humor," for allegory in any form, for apocalyptic preachments meretriciously tricked out in dramatic garb.

Neither had his wife and adolescent daughters, who for that matter preferred life to literature and read fiction when at all for entertainment. Their kind of story (his too, finally) would begin if not once upon a time at least with arresting circumstance, bold character, trenchant action. C flung away the whining manuscript and pushed impatiently through the french doors leading to the terrace from his oak-wainscoted study. Pausing at the stone balustrade to light his briar he remarked through a lavender cascade of wisteria that lithe-limbed Gloria, Gloria of timorous eye and militant breast, had once again chosen his boat-wharf as her basking-place.

---

3. German novelist (1875–1955), author of *The Magic Mountain* and others.
4. Samuel Beckett (b. 1906), Irish writer, dramatist, poet. Marian Cutler, one of Barth's invented authors. Jorge Luis Borges (b. 1899), Argentinian short-story writer.
5. Georges Simenon (b. 1903), Belgian-Swiss writer of detective novels. Nicole Riboud, another of Barth's invented authors.

By Jove he exclaimed to himself. It's particularly disquieting to suspect not only that one is a fictional character but that the fiction one's in—the fiction one is—is quite the sort one least prefers. His wife entered the study with coffee and an apple-pastry, set them at his elbow on his work table, returned to the living room. Ed' pelut' kondo nedode; nyoing nyang.[6] One manifestation of schizophrenia as everyone knows is the movement from reality toward fantasy, a progress which not infrequently takes the form of distorted and fragmented representation, abstract formalism, an increasing preoccupation, even obsession, with pattern and design for their own sakes—especially patterns of a baroque, enormously detailed character—to the (virtual) exclusion of representative "content." There are other manifestations. Ironically, in the case of graphic and plastic artists for example the work produced in the advanced stages of their affliction may be more powerful and interesting than the realistic productions of their earlier "sanity." Whether the artists themselves are gratified by this possibility is not reported.

B called upon a literary acquaintance, B_____, summering with Mrs. B and children on the Eastern Shore of Maryland. "You say you lack a ground-situation. Has it occurred to you that that circumstance may be your ground-situation? What occurs to me is that if it is it isn't. And conversely. The case being thus, what's really wanting after all is a well-articulated vehicle, a foreground or upstage situation to dramatize the narrator's or author's grundlage. His what. To write merely C comes to suspect that the world is a novel, himself a fictional personage is but to introduce the vehicle; the next step must be to initiate its uphill motion by establishing and complicating some conflict. I would advise in addition the eschewal of overt and self-conscious discussion of the narrative process. I would advise in addition the eschewal of overt and self-conscious discussion of the narrative process. The via negativa and its positive counterpart are it is to be remembered poles after all of the same cell. Returning to his study.

If I'm going to be a fictional character G declared to himself I want to be in a rousing good yarn as they say, not some piece of avant-garde preciousness. I want passion and bravura action in my plot, heroes I can admire, heroines I can love, memorable speeches, colorful accessory characters, poetical language. It doesn't matter to me how naively linear the anecdote is; never mind modernity! How reactionary J appears to be. How will such nonsense sound thirty-six years from now?[7] As if. If he can only get K through his story I reflected grimly; if he can only retain his self-possession to the end of this sentence; not go mad; not destroy himself and/or others. Then what I wondered grimly. Another sentence fast, another story. Scheherazade my only love! All those nights you kept your secret

6. Nonsense words.
7. "10:00 A.M., Monday, June 20, 1966" [Barth's note].

from the King my rival, that after your defloration he was unnecessary, you'd have killed yourself in any case when your invention failed.

Why could he not begin his story afresh X wondered, for example with the words why could he not begin his story afresh et cetera? Y's wife came into the study as he was about to throw out the baby with the bathwater. "Not for an instant to throw out the baby while every instant discarding the bathwater is perhaps a chief task of civilized people at this hour of the world.[8] I used to tell B_____ that without success. What makes you so sure it's not a film he's in or a theater-piece?

Because U responded while he certainly felt rather often that he was merely acting his own role or roles he had no idea who the actor was, whereas even the most Stanislavsky-methodist would presumably if questioned closely recollect his offstage identity even onstage in mid-act. Moreover a great part of T's "drama," most of his life in fact, was non-visual, consisting entirely in introspection, which the visual dramatic media couldn't manage easily. He had for example mentioned to no one his growing conviction that he was a fictional character, and since he was not given to audible soliloquizing a "spectator" would take him for a cheerful, conventional fellow, little suspecting that et cetera. It was of course imaginable that much goes on in the mind of King Oedipus in addition to his spoken sentiments; any number of interior dramas might be being played out in the actors' or characters' minds, dramas of which the audience is as unaware as are V's wife and friends of his growing conviction that he's a fictional character. But everything suggested that the medium of his life was prose fiction—moreover a fiction narrated from either the first-person or the third-person-omniscient point of view.

Why is it L wondered with mild disgust that both K and M for example choose to write such stuff when life is so sweet and painful and full of such a variety of people, places, situations, and activities other than self-conscious and after all rather blank introspection? Why is it N wondered et cetera that both M and O et cetera when the world is in such parlous explosive case? Why et cetera et cetera et cetera when the word, which was in the beginning, is now evidently nearing the end of its road? Am I being strung out in this ad libitum fashion I wondered merely to keep my author from the pistol? What sort of story is it whose drama lies always in the next frame out? If Sinbad sinks it's Scheherazade who drowns; whose neck one wonders is on her line?

2

Discarding what he'd already written as he could wish to discard the mumbling pages of his life he began his story afresh, resolved

---

8. "11:00 A.M., Monday, June 20, 1966" [Barth's note].

this time to eschew overt and self-conscious discussion of his narrative process and to recount instead in the straight-forwardest manner possible the several complications of his character's conviction that he was a character in a work of fiction, arranging them into dramatically ascending stages if he could for his readers' sake and leading them (the stages) to an exciting climax and dénouement if he could.

He rather suspected that the medium and genre in which he worked—the only ones for which he felt any vocation—were moribund if not already dead. The idea pleased him. One of the successfullest men he knew was a blacksmith of the old school who et cetera. He meditated upon the grandest sailing-vessel ever built, the *France II*, constructed in Bordeaux in 1911 not only when but because the age of sail had passed. Other phenomena that consoled and inspired him were the great flying-boat *Hercules*, the zeppelin *Hindenburg*, the *Tsar Pushka* cannon, the then-record Dow-Jones industrial average of 381.17 attained on September 3, 1929.

He rather suspected that the society in which he persisted—the only one with which he felt any degree of identification—was moribund if not et cetera. He knew beyond any doubt that the body which he inhabited—the only one et cetera—was et cetera. The idea et cetera. He had for thirty-six years lacking a few hours been one of our dustmote's three billion tenants give or take five hundred . million, and happening to be as well a white male citizen of the United States of America he had thirty-six years plus a few hours more to cope with one way or another unless the actuarial tables were mistaken, not bloody likely, or his term was unexpectedly reduced.

Had he written for his readers' sake? The phrase implied a thitherto-unappreciated metaphysical dimension. Suspense. If his life was a fictional narrative it consisted of three terms—teller, tale, told—each dependent on the other two but not in the same ways. His author could as well tell some other character's tale or some other tale of the same character as the one being told as he himself could in his own character as author; his "reader" could as easily read some other story, would be well advised to; but his own "life" depended absolutely on a particular author's original persistence, thereafter upon some reader's. From this consideration any number of things followed, some less tiresome than others. No use appealing to his author, of whom he'd come to dislike even to think. The idea of his playing with his characters' and his own self-consciousness! He himself tended in that direction and despised the tendency. The idea of his or her smiling smugly to himself as the "words" flowed from his "pen" in which his the protagonist's unhappy inner life was exposed! Ah he had mistaken the nature of his narrative; he had thought it very long, longer than Proust's, longer than any German's, longer than *The Thousand Nights and a Night* in ten quarto volumes. Moreover he'd thought it the most prolix and pedestrian *tranche-de-vie* realism, unredeemed by even the limited virtues of colorful squalor, solid specification, an engaging variety of scenes and characters—in a word a bore, of the sort he himself not only would not write but would not read either. Now he understood that his author might as probably resemble himself and the protagonist of his own story-in-progress. Like himself, like his character afore-mentioned, his author not impossibly deplored the obsolescence of humanism, the passing of *savoir-vivre*, et cetera; admired the out-moded values of fidelity, courage, tact, restraint, amiability, self-discipline, et cetera; preferred fictions in which were to be found stirring actions, characters to love as well as ditto to despise, speeches and deeds to affect us strongly, et cetera. He too might wish to make some final effort to put by his fictional character and achieve factuality or at least to figure in if not be hero of a more attractive fiction, but be caught like the writer of these lines in some more or less desperate tour de force. For him to attempt to come to an understanding with such an author were as futile as for one of his own creations to et cetera.

But the reader! Even if his author were his only reader as was he himself of his work-in-progress as of the sentence-in-progress and his protagonist of his, et cetera, his character as reader was not the same as his character as author, a fact which might be turned to account. What suspense.

As he prepared to explore this possibility one of his mistresses whereof he had none entered his brown study unannounced. "The passion of love," she announced, "which I regard as no less essential to a satisfying life than those values itemized above and which I infer from my presence here that you too esteem highly, does not in fact play in your life a role of sufficient importance to sustain my presence here. It plays in fact little role at all outside your imagina-tive and/or ary life. I tell you this not in a criticizing spirit, for I judge you to be as capable of the sentiment aforementioned as any other imagin[ative], deep-feeling man in good physical health more or less precisely in the middle of the road of our life. What hampers, even cripples you in this regard is your final preference, which I refrain from analyzing, for the sedater, more responsible pleasures of monogamous fidelity and the serener affections of domesticity, notwithstanding the fact that your enjoyment of these is correspond-ingly inhibited though not altogether spoiled by an essentially romantical, unstable, irresponsible, death-wishing fancy. V. S. Pritchett,[9] English critic and author, will put the matter succinctly in a soon-to-be-written essay on Flaubert, whose work he'll say depicts the course of ardent longings and violent desires that rise from the horrible, the sensual, and the sadistic. They turn into the virginal and mystical, only to become numb by satiety. At this point

9. English literary critic and autobiographer (b. 1900).

pathological boredom leads to a final desire for death and nothingness—the Romantic syndrome. If, not to be unfair, we qualify
somewhat the terms horrible and sadistic and understand satiety to
include a large measure of vicariousness, this description undeniably
applies to one aspect of yourself and your work; and while your
ditto has other, even contrary aspects, the net fact is that you have
elected familial responsibilities and rewards—indeed, straight-laced
middle-classness in general—over the higher expenses of spirit and
wastes of shame attendant upon a less regular, more glamorous
style of life. So to elect is surely admirable for the layman, even
essential if the social fabric, without which there can be no culture,
is to be preserved. For the artist, however, and in particular the
writer, whose traditional material has been the passions of men and
women, the choice is fatal. You having made it I bid you goodnight
probably forever."

Even as she left he reached for the sleeping pills cached conveniently in his writing desk and was restrained from their administration only by his being in the process of completing a sentence,
which he cravenly strung out at some sacrifice of rhetorical effect
upon realizing that he was et cetera. Moreover he added hastily he
had not described the intruder for his readers' vicarious satiety: a
lovely woman she was, whom he did not after all describe for his
readers' et cetera inasmuch as her appearance and character were
inconstant. Her interruption of his work inspired a few sentences
about the extent to which his fiction inevitably made public his
private life, though the trespasses in this particular were as nothing
beside those of most of his profession. That is to say, while he did
not draw his characters and situations directly from life nor permit
his author-protagonist to do so, any moderately attentive reader of
his oeuvre, his what, could infer for example that its author feared
for example schizophrenia, impotence creative and sexual, suicide—
in short living and dying. His fictions were preoccupied with these
fears among their other, more serious preoccupations. Hot dog. As
of the sentence-in-progress he was not in fact unmanageably
schizophrenic, impotent in either respect, or dead by his own hand,
but there was always the next sentence to worry about. But there
was always the next sentence to worry about. In sum he concluded
hastily such limited self-exposure did not constitute a misdemeanor,
representing or mis as it did so small an aspect of his total self,
negligible a portion of his total life—even which totalities were they
made public would be found remarkable only for their being so
unremarkable. Well shall he continue.

Bearing in mind that he had not developed what he'd mentioned
earlier about turning to advantage his situation vis-à-vis his "reader"
(in fact he deliberately now postponed his return to that subject,
sensing that it might well constitute the climax of his story) he
elaborated one or two ancillary questions, perfectly aware that he

was trying, even exhausting, whatever patience might remain to whatever readers might remain to whoever elaborated yet another ancillary question. Was the novel of his life for example a *roman à clef*.[1]? Of that genre he was as contemptuous as of the others aforementioned; but while in the introductory adverbial clause it seemed obvious to him that he didn't "stand for" anyone else, any more than he was an actor playing the role of himself, by the time he reached the main clause he had to admit that the question was unanswerable, since the "real" man to whom he'd correspond in a *roman à clef* would not be also in the *roman à clef* and the characters in such works were not themselves aware of their irritating correspondences.

Similarly unanswerable were such questions as when "his" story (so he regarded it for convenience and consolement though for all he knew he might be not the central character; it might be his wife's story, one of his daughters's, his imaginary mistress's, the man-who-once-cleaned-his-chimney's) began. Not impossibly at his birth or even generations earlier: a *Bildungsroman*, an *Erziehungsroman*, a *roman fleuve*.[2]! More likely at the moment he became convinced of his fictional nature: that's where he'd have begun it, as he'd begun the piece currently under his pen. If so it followed that the years of his childhood and younger manhood weren't "real," he'd suspected as much, in the first-order sense, but a mere "background" consisting of a few well-placed expository insinuations, perhaps misleading, or inferences, perhaps unwarranted, from strategic hints in his present reflections. God so to speak spare his readers from heavy-footed forced expositions of the sort that begin in the countryside near ——— in May of the year ——— it occurred to the novelist ——— that his own life might be a ———, in which he was the leading or an accessory character. He happened at the time to be in the oak-wainscoted study of the old family summer residence; through a lavender cascade of hysteria he observed that his wife had once again chosen to be the subject of this clause, itself the direct object of his observation. A lovely woman she was, whom he did not describe in keeping with his policy against drawing characters from life as who should draw a condemnee to the gallows. Begging his pardon. Flinging his tiresome tale away he pushed impatiently through the french windows leading from his study to a sheer drop from the then-record high into a nearly fatal depression.

He clung onto his narrative depressed by the disproportion of its ratiocination to its dramatization, reflection to action. One had heard *Hamlet* criticized as a collection of soliloquies for which the implausible plot was a mere excuse; witnessed Italian operas whose dramatic portions were no more than interstitial relief and arbitrary

---

1. Novel about "real" persons and "actual" events.
2. "*Bildungsroman*": novel about education and development of its hero.

"*Erziehungsroman*": novel with a thesis.
"*Roman fleuve*": extended chronicle novel about related people.

continuity between the arias. If it was true that he didn't take his "real" life seriously enough even when it had him by the throat, the fact didn't lead him to consider whether the fact was a cause or a consequence of his tale's tedium or both.

Concluding these reflections he concluded these reflections: that there was at this advancèd page still apparently no ground-situation suggested that his story was dramatically meaningless. If one regarded the absence of a ground-situation, more accurately the protagonist's anguish at that absence and his vain endeavors to supply the defect, as itself a sort of ground-situation, did his life-story thereby take on a kind of meaning? A "dramatic" sort he supposed, though of so sophistical a character as more likely to annoy than to engage.

3

The reader! You, dogged, uninsultable, print-oriented bastard, it's you I'm addressing, who else, from inside this monstrous fiction. You've read me this far, then? Even this far? For what discreditable motive? How is it you don't go to a movie, watch TV, stare at a wall, play tennis with a friend, make amorous advances to the person who comes to your mind when I speak of amorous advances? Can nothing surfeit, saturate you, turn you off? Where's your shame?

Having let go this barrage of rhetorical or at least unanswered questions and observing himself nevertheless in midst of yet another sentence he concluded and caused the "hero" of his story to conclude that one or more of three things must be true: 1) his author was his sole and indefatigable reader; 2) he was in a sense his own author, telling his story to himself, in which case in which case; and/or 3) his reader was not only tireless and shameless but sadistic, masochistic if he was himself.

For why do you suppose—you! you!—he's gone on so, so relentlessly refusing to entertain you as he might have at a less desperate than this hour of the world[3] with felicitous language, exciting situation, unforgettable character and image? Why has he as it were ruthlessly set about not to win you over but to turn you away? Because your own author bless and damn you his life is in your hands! He writes and reads himself; don't you think he knows who gives his creatures their lives and deaths? Do they exist except as he or others read their words? Age except we turn their pages? And can he die until you have no more of him? Time was obviously when his author could have turned the trick; his pen had once to left-to-right it through these words as does your kindless eye and might have ceased at any one. This. This. And did not as you see but went on like an Oriental torturemaster to the end.

But you needn't! He exclaimed to you. In vain. Had he petitioned

3. "11:00 P.M., Monday, June 20, 1966" [Barth's note].

you instead to read slowly in the happy parts, what happy parts, swiftly in the painful no doubt you'd have done the contrary or cut him off entirely. But as he longs to die and can't without your help you force him on, force him on. Will you deny you've read this sentence? This? To get away with murder doesn't appeal to you, is that it? As if your hands weren't inky with other dyings! As if he'd know you'd killed him! Come on. He dares you.

In vain. You haven't: the burden of his knowledge. That he continues means that he continues, a fortiori you too. Suicide's impossible: he can't kill himself without your help. Those petitions aforementioned, even his silly plea for death—don't you think he understands their sophistry, having authored their like for the wretches he's authored? Read him fast or slow, intermittently, continuously, repeatedly, backward, not at all, he won't know it; he only guesses someone's reading or composing his sentences, such as this one, because he's reading or composing sentences such as this one; the net effect is that there's a net effect, of continuity and an apparently consistent flow of time, though his pages do seem to pass more swiftly as they near his end.

To what conclusion will he come? He'd been about to append to his own tale inasmuch as the old analogy between Author and God, novel and world, can no longer be employed unless deliberately as a false analogy, certain things follow: 1) fiction must acknowledge its fictiousness and metaphoric invalidity or 2) choose to ignore the question or deny its relevance or 3) establish some other, acceptable relation between itself, its author, its reader. Just as he finished doing so however his real wife and imaginary mistresses entered his study; "It's a little past midnight" she announced with a smile; "do you know what that means?"

Though she'd come into his story unannounced at a critical moment he did not describe her, for even as he recollected that he'd seen his first light just thirty-six years before the night incumbent he saw his last: that he could not after all be a character in a work of fiction inasmuch as such a fiction would be of an entirely different character from what he thought of as fiction. Fiction consisted of such monuments of the imagination as Cutler's *Morganfield*, Riboud's *Tales Within Tales*, his own creations; fact of such as for example read those fictions. More, he could demonstrate by syllogism that the story of his life was a work of fact: though assaults upon the boundary between life and art, reality and dream, were undeniably a staple of his own and his century's literature as they'd been of Shakespeare's and Cervantes's, yet it was a fact that in the corpus of fiction as far as he knew no fictional character had become convinced as had he that he was a character in a work of fiction. This being the case and he having in fact become thus convinced it followed that his conviction was false. "Happy birthday," said his wife et cetera, kissing him et cetera to obstruct his view of the end of the sentence he was nearing the end of, playfully refusing to be nay-said so that in fact he did at last as did his fictional character end his ending story endless by interruption, cap his pen.

1968

JOHN UPDIKE

1932–

"To transcribe middleness with all its grits, bumps and anonymities, in its fullness of satisfaction and mystery: is it possible * * * or worth doing?" John Updike's novels and stories give a positive answer to the question he asks in his memoir, *The Dogwood Tree: A Boyhood*; for he is arguably the most significant transcriber, or creator rather, of "middleness" in American writing since William Dean Howells. Falling in love in high school, meeting a college roommate, going to the eye-doctor or dentist, eating supper on Sunday night, visiting your mother with your wife and son—these activities are made to yield up their possibilities to a writer as responsively curious in imagination and delicately precise in his literary expression as Updike has shown himself to be.

Born in Reading, Pennsylvania, an only child, he grew up in the small town of Shillington. He was gifted at drawing and caricaturing, and after graduating summa cum laude from Harvard in 1954 he spent a year studying art in England, then returned to America and went to work for the *New Yorker*, where his first stories appeared and to which he is still a regular contributor. When later in the 1950s he left the magazine, he also left New York City and with his wife and children settled in Ipswich, Massachusetts. There he pursued "his solitary trade as methodically as the dentist practiced his" (*The Dogwood Tree*), resisting the temptations of university teaching as successfully as the blandishments of media talk-shows. Like Howells, his ample production has been achieved through dedicated, steady work; his books are the fruit of patience, leisure, and craft.

Since 1958 when his first novel, *The Poorhouse Fair*, appeared, Updike has published not only many novels and stories but also four books of poems, a play, and much occasional journalism. He is most admired by some readers as the author of the "Olinger" stories, about life in an imaginary Pennsylvania town which takes on its colors from the real Shillington of his youth. The heroes of these stories are adolescents straining to break out of their fast-perishing environments, as they grow up and as their small town turns into something else. Updike treats them with a blend of affection and ironic humor that is wonderfully assured in its touch, while his sense of place, of growing up in the Depression and the years of World War II, is always vividly present. Like Howells (whose fine memoir of his youthful days in Ohio, *A Boy's Town*, is an ancestor of Updike's *The Dogwood Tree*) he shows how one's spirit takes on its coloration from the material circumstances—houses, clothes, landscape, food, parents—one is bounded by.

This sense of place, which is also a sense of life, is found in the stories and in the novels too, although Updike has found it harder to invent convincing forms in which to tell longer tales. His most ambitious novel is probably *The Centaur* (1964), memorable for its portrayal of three days of confusion and error in the life of an American high-school teacher seen through his son's eyes; but the book is also burdened with an elaborate set of mythical trappings that seem less than inevitable. *Couples* (1968), a novel which gained him a good deal of notoriety as a chronicler of sexual relationships, marital and adulterous, is jammed with much interesting early-1960s lore about suburban life but seems uncertain whether it is an exercise in realism or a creative fantasy, as does his recent *Marry Me* (1976).

It is in the two "Rabbit" novels that Updike found his most congenial and engaging subject for longer fiction. In each book he has managed to render the sense of an era—the 1950s in *Rabbit, Run*; the late sixties in *Rabbit Redux*—through the eyes of a hero who both is and is not like his creator. Harry "Rabbit" Angstrom, ex-high school basketball star, a prey to nostalgia and in love with his own past, perpetually lives in a present he can't abide. *Rabbit, Run* shows him trying to escape from his town, his job, his wife and child, by a series of disastrously sentimental and humanly irresponsible actions; yet Updike makes us feel Rabbit's yearnings even as we judge the painful consequences of yielding to them. Ten years later, the fading basketball star has become a fortyish, dispirited printer with a way-ward wife and a country (America in summer, 1970) which is both landing on the moon and falling to pieces. *Rabbit Redux* is masterly in presenting a Pennsylvania small town rotting away from its past certainties; it also attempts to deal with the Vietnam war and the black revolution. With respect to these large public events there is sometimes an identity, sometimes a divergence, at other times a confusion between Rabbit's consciousness and the larger one of his author, resulting in a book one must argue with as well as about. Still, his best work is in his stories and in his distinguished short novel, *Of the Farm*.

Near the end of his memoir he summarized his boyish dream of becoming an artist:

> He saw art—between drawing and writing he ignorantly made no distinction—as a method of riding a thin pencil line out of Shillington, out of time altogether, into an infinity of unseen and even unborn hearts. He pictured this infinity as radiant. How innocent!

Most writers would name that innocence only to deplore it. Updike maintains instead that, as with the Christian faith he still professes, succeeding years have given him no better assumptions with which to replace it. In any case his fine sense of fact has protected him from fashionable extravagances in black humor and experimental narratives, while enabling him to be both a satirist and a celebrator of our social and domestic conditions.

The Dogwood Tree[1]

A BOYHOOD

When I was born, my parents and my mother's parents planted a dogwood tree in the side yard of the large white house in which we lived throughout my boyhood. This tree, I learned quite early, was

1. From *Assorted Prose* (1965).

exactly my age, was, in a sense, me. But I never observed it closely, am not now sure what color its petals were; its presence was no more distinct than that of my shadow. The tree was my shadow, and had it died, had it ceased to occupy, each year with increasing volume and brilliance, its place in the side yard, I would have felt that a blessing like the blessing of light had been withdrawn from my life.

Though I cannot ask you to see it more clearly than I myself saw it, yet mentioning it seems to open the possibility of my boyhood home coming again to life. With a sweet damp rush the grass of our yard seems to breathe again on me. It is just cut. My mother is pushing the mower, to which a canvas catch is attached. My grandmother is raking up the loose grass in thick heaps, small green haystacks impregnated with dew, and my grandfather stands off to one side, smoking a cigar, elegantly holding the elbow of his right arm in the palm of his left hand while the blue smoke twists from under his mustache and dissolves in the heavy evening air—that misted, too-rich Pennsylvania air. My father is off, doing some duty in the town; he is a conscientious man, a schoolteacher and deacon, and also, somehow, a man of the streets.

In remembering the dogwood tree I remember the faintly speckled asbestos shingles of the chicken house at the bottom of our yard, fronting on the alley. We had a barn as well, which we rented as a garage, having no car of our own, and between the chicken house and the barn there was a narrow space where my grandfather, with his sly country ways, would urinate. I, a child, did also, passing through this narrow, hidden-feeling passage to the school grounds beyond our property; the fibrous tan-gray of the shingles would leap up dark, silky and almost black, when wetted.

The ground in this little passsage seems a mysterious trough of pebbles of all colors and bits of paper and broken glass. A few weeds managed to grow in the perpetual shadow. All the ground at the lower end of the yard had an ungrateful quality; we had an ash heap on which we used to burn, in an extravagant ceremony that the war's thrift ended, the preceding day's newspaper. The earth for yards around the ashpile was colored gray. Chickens clucked in their wire pen. My grandmother tended them, and when the time came, beheaded them with an archaic efficiency that I don't recall ever witnessing, though I often studied the heavy log whose butt was ornamented with fine white neck-feathers pasted to the wood with blood.

A cat crosses our lawn, treading hastily on the damp grass, crouching low with distaste. Tommy is the cat's name; he lives in our chicken house but is not a pet. He is perfectly black; a rarity, he has no white dab on his chest. The birds scold out of the walnut tree and the apple and cherry trees. We have a large grape arbor,

and a stone birdbath, and a brick walk, and a privet hedge the
height of a child and many bushes behind which my playmates
hide. There is a pansy bed that in winter we cover with straw. The
air is green, and heavy, and flavored with the smell of turned earth;
in our garden grows, among other vegetables, a bland, turniplike
cabbage called kohlrabi, which I have never seen, or eaten, since the
days when, for a snack, I would tear one from its row with my
hands.

History

My boyhood was spent in a world made tranquil by two invisible
catastrophes: the Depression and World War II. Between 1932,
when I was born, and 1945, when we moved away, the town of
Shillington changed, as far as I could see, very little. The vacant lot
beside our home on Philadelphia Avenue remained vacant. The
houses along the street were neither altered nor replaced. The high-
school grounds, season after season, continued to make a placid
plain visible from our rear windows. The softball field, with its
triptych backstop, was nearest us. A little beyond, on the left, were
the school and its boilerhouse, built in the late 1920s of the same
ochre brick. In the middle distance a cinder track circumscribed the
football field. At a greater distance there were the tennis courts and
the poor farm fields and the tall double rows of trees marking the
Poorhouse Lane. The horizon was the blue cloud, scarred by a
gravel pit's orange slash, of Mount Penn, which overlooked the city
of Reading.

A little gravel alley, too small to be marked with a street sign but
known in the neighborhood as Shilling Alley, wound hazardously
around our property and on down, past an untidy sequence of back
buildings (chicken houses, barns out of plumb, a gunshop, a small
lumber mill, a shack where a blind man lived, and the enchanted
grotto of a garage whose cement floors had been waxed to the lustre
of ebony by oil drippings and in whose greasy-black depths a silver
drinking fountain spurted the coldest water in the world, silver
water so cold it made your front teeth throb) on down to Lancaster
Avenue, the main street, where the trolley cars ran. All through
those years, the trolley cars ran. All through those years Pappy
Shilling, the surviving son of the landowner after whom the town
was named, walked up and down Philadelphia Avenue with his thin
black cane and his snow-white bangs; a vibrating chain of perfect-
Sunday-school-attendance pins dangled from his lapel. Each autumn the horse-chestnut trees dropped their useless, treasurable
nuts; each spring the dogwood tree put forth a slightly larger spread
of blossoms; always the leaning walnut tree in our back yard fretted
with the same tracery of branches the view we had.

Within our house, too, there was little change. My grandparents
did not die, though they seemed very old. My father continued to

teach at the high school; he had secured the job shortly after I was born. No one else was born. I was an only child. A great many only children were born in 1932; I make no apologies. I do not remember ever feeling the space for a competitor within the house. The five of us already there locked into a star that would have shattered like crystal at the admission of a sixth. We had no pets. We fed Tommy on the porch, but he was too wild to set foot in the kitchen, and only my grandmother, in a way wild herself, could touch him. Tommy came to us increasingly battered and once did not come at all. As if he had never existed: that was death. And then there was a squirrel, Tilly, that we fed peanuts to; she became very tame, and under the grape arbor would take them from our hands. The excitement of those tiny brown teeth shivering against my fingertips: that was life. But she, too, came from the outside, and returned to her tree, and did not dare intrude in our house.

The arrangement inside, which seemed to me so absolute, had been achieved, beyond the peripheries of my vision, drastically and accidentally. It may, at first, have been meant to be temporary. My father and grandfather were casualties of the early thirties. My father lost his job as a cable splicer with the telephone company; he and my mother had been living—for how long I have never understood—in boarding-houses and hotels throughout western Pennsylvania, in towns whose names (Hazleton, Altoona) even now make their faces light up with youth, a glow flowing out of the darkness preceding my birth. They lived through this darkness, and the details of the adventure that my mother recalls—her lonely closeted days, the games of solitaire, the novels by Turgenev, the prostitutes downstairs, the men sleeping and starving in the parks of Pittsburgh —seem to waken in her an unjust and unreasonable happiness that used to rouse jealousy in my childish heart. I remember waiting with her by a window for my father to return from weeks on the road. It is in the Shillington living room. My hands are on the radiator ridges, I can see my father striding through the hedge toward the grape arbor, I feel my mother's excitement beside me mingle with mine. But she says this cannot be; he had lost his job before I was born.

My grandfather came from farming people in the south of the county. He prospered, and prematurely retired; the large suburban house he bought to house his good fortune became his fortune's shell, the one fragment of it left him. The two men pooled their diminished resources of strength and property and, with their women, came to live together. I do not believe they expected this arrangement to last long. For all of them—for all four of my adult guardians—Shillington was a snag, a halt in a journey that had begun elsewhere. Only I belonged to the town. The accidents that had planted me here made uneasy echoes in the house, but, like Tilly and Tommy, their source was beyond my vision.

Geography

As in time, so it was in space. The town was fringed with things
that appeared awesome and ominous and fantastic to a boy. At the
end of our street there was the County Home—an immense yellow
poorhouse, set among the wide orchards and lawns, surrounded by
a sandstone wall that was low enough on one side for a child to
climb easily, but that on the other side offered a drop of twenty or
thirty feet, enough to kill you if you fell. Why this should have
been, why the poorhouse grounds should have been so deeply re-
cessed on the Philadelphia Avenue side, puzzles me now. What
machinery, then, could have executed such a massive job of grading
I don't know. But at the time it seemed perfectly natural, a dreadful
pit of space congruent with the pit of time into which the old people
(who could be seen circling silently in the shade of the trees whose
very tops were below my feet) had been plunged by some mystery
that would never touch me. That I too would come to their condition was as unbelievable as that I would really fall and break my
neck. Even so, I never acquired the daring that some boys had in
racing along the top of the wall. In fact—let it be said now—I was
not a very daring boy.

The poorhouse impinged on us in many ways. For one thing, my
father, whose favorite nightmare was poverty, often said that he
liked living so close to a poorhouse; if worse came to worse, he
could walk there. For another, the stench of the poorhouse pigs,
when the wind was from the east, drifted well down Philadelphia
Avenue. Indeed, early in my life the poorhouse livestock were still
herded down the street on their way to the slaughterhouse on the
other side of town. Twice, in my childhood, the poorhouse barn
burnt, and I remember my father (he loved crowds) rushing out of
the house in the middle of one night, and my begging to go and my
mother keeping me with her, and the luckier, less sheltered children
the next day telling me horrific tales of cooked cows and screaming
horses. All I saw were the charred ruins, still smoldering, settling
here and there with an unexpected crackle, like the underbrush the
morning after an ice storm.

Most whiffs of tragedy came, strangely, from the east end of the
street. I remember two, both of them associated with the early
morning and with the same house a few doors away in the neighborhood. When I was a baby, a man was run over and crushed by a
milk wagon—a horse-drawn milk wagon. It had happened within
sight of our windows, and I grew to know the exact path of asphalt,
but could never picture it. I believed all horse-drawn milk wagons
were as light as the toy one I had; by the time I understood about
this accident they had vanished from the streets. No matter how
many times I visited the patch of asphalt, I could not understand
how it had happened. And then, the family that succeeded the

widow in her house—or it may have been in the other side; it was a brick semi-detached, set back and beclouded by several tall fir or cedar trees—contained a young man who, while being a counsellor at a summer camp, had had one of his boys dive into shallow water and, neck broken, die in his arms. The young counsellor at dawn one day many months later put a bullet through his head. I seem to remember hearing the shot; certainly I remember hearing my parents bumping around in their bedroom, trying to locate what had wakened them.

Beyond the poorhouse, where Philadelphia Avenue became a country lane, and crossed a little brook where water cress grew, there was a path on the right that led to the poorhouse dam. It was a sizable lake, where people fished and swam. Its mud bottom bristled with broken bottles and jagged cans. A little up from one of its shores, the yellow walls and rotten floor of the old pesthouse survived. Beyond the lake was a woods that extended along the south of the town. Here my parents often took me on walks. Every Sunday afternoon that was fair, we would set out. Sun, birds, and treetops rotated above us as we made our way. There were many routes. Farther down the road, toward Grille, another road led off, and went past a gravel cliff and sad little composition-shingled farmhouses whose invisible inhabitants I imagined as gravel-colored skeletons. By way of a lane we could leave this road and walk down toward the dam. Or we could walk up by the dam until we struck this road, and walk on until we came to a road that took us back into the town by way of the cemetery. I disliked these walks. I would lag farther and farther behind, until my father would retrace his steps and mount me on his shoulders. Upon this giddy, swaying perch—I hesitated to grip his ears and hair as tightly as I needed to—I felt as frightened as exultant, and soon confusedly struggled to be put down. In the woods I would hurl myself against dead branches for the pleasure of feeling them shatter and of disturbing whatever peace and solace my parents were managing to gather. If, at moments, I felt what they wanted me to feel—the sweet moist breath of mulching leaves, the delicate scratch of some bird in the living silence, the benevolent intricacy of moss and rocks and roots and ferns all interlocked on some bank torn by an old logging trail—I did not tell them. I was a small-town child. Cracked pavements and packed dirt were my ground.

This broad crescent of woods is threaded with our walks and suffused with images of love. For it was here, on the beds of needles under the canopies of low pine boughs, that our girls—and this is later, not boyhood at all, but the two have become entangled—were rumored to give themselves. Indeed, I was told that one of the girls in our class, when we were in the ninth grade, had boasted that she liked nothing so much as skinny-dipping in the dam and then making love under the pines. As for myself, this was beyond me, and may be myth, but I do remember, when I was seventeen, taking a girl on one of those walks from my childhood so (then) long ago. We had moved from town, but only ten miles, and my father and I drove in to the high school every day. We walked, the girl and I, down the path where I had smashed so many branches, and sat down on a damp broad log—it was early spring, chilly, a timid froth of leaves overhead—and I dared lightly embrace her from behind and cup my hands over her breasts, small and shallow within the stiffness of her coat, and she closed her eyes and tipped her head back, and an adequate apology seemed delivered for the irritable innocence of these almost forgotten hikes with my parents.

The road that came down into Shillington by way of the cemetery led past the Dives estate, another ominous place. It was guarded by a wall topped with spiky stones. The wall must have been a half-mile long. It was so high my father had to hold me up so I could look in. There were so many buildings and greenhouses I couldn't identify the house. All the buildings were locked and boarded up; there was never anybody there. But in the summer the lawns were mowed; it seemed by ghosts. There were tennis courts, and even—can it be?—a few golf flags. In any case there was a great deal of cut lawn, and gray driveway, and ordered bushes; I got the impression of wealth as a vast brooding absence, like God Himself. The road here was especially overshadowed by trees, so a humid, stale, cloistered smell flavored my glimpses over the wall.

The cemetery was on the side of a hill, bare to the sun, which quickly faded the little American flags and killed the potted geraniums. It was a holiday place; on Memorial Day the parade, in which the boys participated mounted on bicycles whose wheels were threaded with tricolor crêpe paper, ended here, in a picnic of speeches and bugle music and leapfrog over the tombstones. From here you had a perfect view of the town, spread out in the valley between this hill and Slate Hill, the chimneys smoking like just-snuffed cigarettes, the cars twinkling down on Lancaster Avenue, the trolleys moving with the dreamlike slow motion distance imposes.

A little to one side of the cemetery, just below the last trees of the love-making woods, was a small gravel pit where, during the war, we played at being guerrillas. Our leader was a sickly and shy boy whose mother made him wear rubbers whenever there was dew on the grass. His parents bought him a helmet and khaki jacket and even leggings, and he brought great enthusiasm to the imitation of war. G.I.'s and Japs, shouting "Geronimo!" and "Banzai!," we leaped and scrambled over boulders and cliffs in one of whose clefts I always imagined, for some reason, I was going to discover a great deal of money, in a tan cloth bag tied with a leather thong. Though I visualized the bag very clearly, I never found it.

Between this pit and the great quarry on the far edge of town, I lose track of Shillington's boundaries. I believe it was just fields, in which a few things were still farmed. The great quarry was immense, and had a cave, and an unused construction covered with fine gray dust and filled with mysterious gears, levers, scoops, and stairs. The quarry was a mile from my home; I seldom went there. It wears in memory a gritty film. The tougher section of town was nearby. Older boys with .22s used the quarry as a rifle range, and occasionally wounded each other, or smaller children. To scale its sides was even more dangerous than walking along the top of the poorhouse wall. The legends of love that scattered condums along its grassy edges seemed to be of a coarser love than that which perfumed the woods. Its cave was short, and stumpy, yet long enough to let you envision a collapse blocking the mouth of light and sealing you in; the walls were of a greasy golden clay that seemed likely to collapse. The one pure, lovely thing about the quarry, beside its grand size, was the frozen water that appeared on its floor in the winter, and where you could skate sheltered from the wind, and without the fear of drowning that haunted the other skating place, the deep dam. Though the area of ice was smaller, the skaters seemed more skillful down at the quarry: girls in red tights and bouncy short skirts that gave their fannies the effect of a pompon turned and swirled and braked backward to a stop on their points, sparkling ice chips sprinkling in twin fans of spray.

Near the quarry was the Shillington Friday Market, where the sight of so many naked vegetables depressed me, and the Wyomissing Creek, a muddy little thing to skip pebbles in, and the hilly terrain where, in those unbuilding years, a few new houses were built. The best section of town, called Lynoak, was farther on, more toward Reading, around the base of Slate Hill, where I sometimes sledded. It was a long walk back through the streets, under the cold street lights, the sled runners rattling on the frozen ruts, my calves aching with the snow that always filtered through my galoshes. This hill in summer was another place my parents hiked to. The homes of the well-off (including an amazingly modern house, of white brick, with a flat roof, and blue trim, like something assembled from the two dimensions of a Hollywood movie) could be seen climbing its sides, but there was still plenty of room for, during the war, Victory gardens,[2] and above that, steep wilderness. The top was a bare, windy, primeval place. Looking north, you saw the roofs of Shillington merge with the roofs of other suburbs in a torn carpet that went right into the bristling center of Reading, under the blue silhouette of Mount Penn. As Shillington on the south faced the country, northward it faced the city.

2. Vegetable plots planted and used by civilians during World War II.

Reading: a very powerful and fragrant and obscure city—who has ever heard of it? Wallace Stevens was born there, and John Philip Sousa died there.[3] For a generation it had a Socialist mayor. Its railroad is on the Monopoly Board. It is rumored to be endeared to gangsters, for its citizens make the most tolerant juries in the East. Unexpectedly, a pagoda overlooks it from the top of Mount Penn. This is the meagre list of its singularities as I know them. Larger than Harrisburg and Wilkes-Barre and Lancaster, it is less well known. Yet to me Reading is the master of cities, the one at the center that all others echo. How rich it smelled! Kresge's swimming in milk chocolate and violet-scented toilet water, Grant's barricaded with coconut cookies, the vast velveted movie theatres dusted with popcorn and a cold whiff of leather, the bakeshops exhaling hearty brown drafts of molasses and damp dough and crisp grease and hot sugar, the beauty parlors with their gingerly stink of singeing, the bookstores glistening with fresh paper and bubbles of hardened glue, the shoe-repair nooks blackened by Kiwi Wax and aromatic shavings, the public lavatory with its emphatic veil of soap, the hushed, brick-red side streets spiced with grit and the moist seeds of maples and ginkgos, the goblin stench of the trolley car that made each return to Shillington a race with nausea—Reading's smells were most of what my boyhood knew of the Great World that was suspended, at a small but sufficient distance, beyond my world.

For the city and the woods and the ominous places were peripheral; their glamour and menace did not intrude into the sunny area where I lived, where the seasons arrived like issues of a magazine and I moved upward from grade to grade and birthday to birthday on a notched stick that itself was held perfectly steady. There was the movie house, and the playground, and the schools, and the grocery stores, and our yard, and my friends, and the horse-chestnut trees. My geography went like this: in the center of the world lay our neighborhood of Shillington. Around it there was greater Shillington, and around that, Berks County. Around Berks County there was the State of Pennsylvania, the best, the least eccentric, state in the Union. Around Pennsylvania, there was the United States, with a greater weight of people on the right and a greater weight of land on the left. For clear geometrical reasons, not all children could be born, like me, at the center of the nation. But that some children chose to be born in other countries and even continents seemed sad and fantastic. There was only one possible nation: mine. Above this vast, rectangular, slightly (the schoolteachers insisted) curved field of the blessed, there was the sky, and the flag, and, mixed up with both, Roosevelt.

3. Wallace Stevens (1879–1955), American poet. John Philip Sousa (1854–1932), American bandmaster, composer of band marches.

Democrats

We were Democrats. My grandfather lived for ninety years, and always voted, and always voted straight Democrat. A marvellous chain of votes, as marvellous as the chain of Sunday-school-attendance pins that vibrated from Pappy Shilling's lapel. The political tradition that shaped his so incorruptible prejudice I am not historian enough to understand; it had something to do with Lincoln's determination to drive all the cattle out of this section of Pennsylvania if Lee won the Battle of Gettysburg.

My parents are closer to me. The events that shaped their views are in my bones. At the time when I was conceived and born, they felt in themselves a whole nation stunned, frightened, despairing. With Roosevelt, hope returned. This simple impression of salvation is my political inheritance. That this impression is not universally shared amazes me. It is as if there existed a class of people who deny that the sun is bright. To me as a child Republicans seemed blind dragons; their prototype was my barber—an artist, a charmer, the only man, my mother insists, who ever cut my hair properly. Nimble and bald, he used to execute little tap-dance figures on the linoleum floor of his shop, and with engaging loyalty he always had the games of Philadelphia's two eighth-place teams tuned in on the radio. But on one subject he was rabid; the last time he cut my hair he positively asserted that our President had died of syphilis. I cannot to this day hear a Republican put forth his philosophy without hearing the snip of scissors above my ears and feeling the little ends of hair crawling across my hot face, reddened by shame and the choking pressure of the paper collar.

Now

Roosevelt was for me the cap on a steadfast world, its emblem and crown. He was always there. Now he is a weakening memory, a semi-legend; it has begun to seem fabulous—like an episode in a medieval chronicle—that the greatest nation in the world was led through the world's greatest war by a man who could not walk. Now, my barber has retired, my hair is a wretched thatch grizzled with gray, and, of the two Philadelphia ball clubs, one has left Philadelphia and one is not always in last place. Now the brick home of my boyhood is owned by a doctor, who has added an annex to the front, to contain his offices. The house was too narrow for its lot and its height; it had a pinched look from the front that used to annoy my mother. But that thin white front with its eyes of green window sash and its mouth of striped awning had been a face to me; it has vanished. My dogwood tree still stands in the side yard, taller than ever, but the walnut tree out back has been cut down. My grandparents are dead. Pappy Shilling is dead. Shilling

Alley has been straightened, and hardtopped, and rechristened Brobst Street. The trolley cars no longer run. The vacant lots across the town have been filled with new houses and stores. New homes have been built far out Philadelphia Avenue and all over the poorhouse property. The poorhouse has been demolished. The poorhouse dam and its aphrodisiac groves have been trimmed into a town park and a chlorinated pool where all females must sheathe their hair in prophylactic bathing caps. If I could go again into 117 Philadelphia Avenue and look out the rear windows, I would see, beyond the football field and the cinder track, a new, two-million-dollar high school, and beyond it, where still stands one row of the double line of trees that marked the Poorhouse Lane, a gaudy depth of postwar housing and a Food Fair like a hideous ark breasting an ocean of parked cars. Here, where wheat grew, loudspeakers unremittingly vomit commercials. It has taken me the shocks of many returnings, more and more widely spaced now, to learn, what seems simple enough, that change is the order of things. The immutability, the steadfastness, of the site of my boyhood was an exceptional effect, purchased for me at unimaginable cost by the paralyzing calamity of the Depression and the heroic external effort of the Second World War.

Environment

The difference between a childhood and a boyhood must be this: our childhood is what we alone have had; our boyhood is what any boy in our environment would have had. My environment was a straight street about three city blocks long, with a slight slope that was most noticeable when you were on a bicycle. Though many of its residents commuted to Reading factories and offices, the neighborhood retained a rural flavor. Corn grew in the strip of land between the alley and the school grounds. We ourselves had a large vegetable garden, which we tended not as a hobby but in earnest, to get food to eat. We sold asparagus and eggs to our neighbors. Our peddling things humiliated me, but then I was a new generation. The bulk of the people in the neighborhood were not long off the farm. One old lady down the street, with an immense throat goiter, still wore a bonnet. The most aristocratic people in the block were the full-fashioned knitters; Reading's textile industry prospered in the Depression. I felt neither prosperous nor poor. We kept the food money in a little recipe box on top of the icebox, and there were nearly always a few bills and coins in it. My father's job paid him poorly but me well; it gave me a sense of, not prestige, but *place*. As a schoolteacher's son, I was assigned a role; people knew me. When I walked down the street to school, the houses called, "Chonny." I had a place to be.

Schools

The elementary school was a big brick cube set in a square of black surfacing chalked and painted with the diagrams and runes of children's games. Wire fences guarded the neighboring homes from the playground. Whoever, at soccer, kicked the ball over the fence into Snitzy's yard had to bring it back. It was very terrible to have to go into Snitzy's yard, but there was only one ball for each grade. Snitzy was a large dark old German who might give you the ball or lock you up in his garage, depending upon his mood. He did not move like other men; suddenly the air near your head condensed, and his heavy hands were on you.

On the way to school, walking down Lancaster Avenue, we passed Henry's, a variety store where we bought punch-out licorice belts and tablets with Edward G. Robinson and Hedy Lamarr smiling on the cover. In October, Halloween masks appeared, hung on wire clotheslines. Hanging limp, these faces of Chinamen and pirates and witches were distorted, and, thickly clustered and rustling against each other, they seemed more frightening masking empty air than they did mounted on the heads of my friends—which was frightening enough. It is strange how fear resists the attacks of reason, how you can know with absolute certainty that it is only Mark Wenrich or Jimmy Trexler whose eyes are moving so weirdly in those almond-shaped holes, and yet still be frightened. I abhorred that effect of double eyes a mask gives; it was as bad as seeing a person's mouth move upside down.

I was a Crow. That is my chief memory of what went on inside the elementary school. In music class the singers were divided into three groups: Nightingales, Robins, and Crows. From year to year the names changed. Sometimes the Crows were Parrots. When visitors from the high school, or elsewhere "outside," came to hear us sing, the Crows were taken out of the room and sent upstairs to watch with the fifth grade an educational film about salmon fishing in the Columbia River. Usually there were only two of us, me and a girl from Philadelphia Avenue whose voice was in truth very husky. I never understood why I was a Crow, though it gave me a certain derisive distinction. As I heard it, I sang rather well.

The other Crow was the first girl I kissed. I just did it, one day, walking back from school along the gutter where the water from the ice plant ran down, because somebody dared me to. And I continued to do it every day, when we reached that spot on the pavement, until a neighbor told my mother, and she, with a solemn weight that seemed unrelated to the airy act, forbade it.

I walked to school mostly with girls. It happened that the mothers of Philadelphia Avenue and, a block up, of Second Street had borne female babies in 1932. These babies now teased me, the lone boy in their pack, by singing the new song, "Oh, Johnny, oh Johnny, how you can love!" and stealing my precious rubber-lined bookbag. The queen of these girls later became the May Queen of our senior class. She had freckles and thick pigtails and green eyes and her mother made her wear high-top shoes long after the rest of us had switched to low ones. She had so much vitality that on the way back from school her nose would start bleeding for no reason. We would be walking along over the wings of the maple seeds and suddenly she would tip her head back and rest it on a wall while someone ran and soaked a handkerchief in the ice-plant water and applied it to her streaming, narrow, crimson-shining nostrils. She was a Nightingale. I loved her deeply, and ineffectually.

My love for that girl carries through all those elementary-school cloakrooms; they always smelled of wet raincoats and rubbers. That tangy, thinly resonant, lonely smell: can love have a better envelope? Everything I did in grammar school was meant to catch her attention. I had a daydream wherein the stars of the music class were asked to pick partners and she, a Nightingale, picked me, a Crow. The teacher was shocked; the class buzzed. To their amazement I sang superbly; my voice, thought to be so ugly, in duet with hers was beautiful. Still singing, we led some sort of parade.

In the world of reality, my triumph was getting her to slap me once, in the third grade. She was always slapping boys in those years; I could not quite figure out what they did. Pull her pigtails, untie her shoes, snatch at her dress, tease her (they called her "Pug")—this much I could see. But somehow there seemed to be under these offensive acts a current running the opposite way; for it was precisely the boys who behaved worst to her that she talked to solemnly at recess, and walked with after school, and whose names she wrote on the sides of her books. Without seeing this current, but deducing its presence, I tried to jump in; I entered a tussle she was having with a boy in homeroom before the bell. I pulled the bow at the back of her dress, and was slapped so hard that children at the other end of the hall heard the crack. I was overjoyed; the stain and pain on my face seemed a badge of initiation. But it was not. The distance between us remained as it was. I did not really want to tease her, I wanted to rescue her, and to be rescued by her. I lacked—and perhaps here the only child suffers a certain deprivation—that kink in the instincts on which childish courtship turns. He lacks a certain easy roughness with other children.

All the years I was at the elementary school the high school loomed large in my mind. Its students—tall, hairy, smoke-breathing—paced the streets seemingly equal with adults. I could see part of its immensity from our rear windows. It was there that my father performed his mysteries every day, striding off from breakfast, down through the grape arbor, his coat pocket bristling with defective pens. He now and then took me over there; the incorruptible smell of varnish and red sweeping wax, the size of the desks, the height of the drinking fountains, the fantastic dimensions of the combination gymnasium-auditorium made me feel that these were halls in which a race of giants had ages ago labored through lives of colossal bliss. At the end of each summer, usually on Labor Day Monday, he and I went into his classroom, Room 201, and unpacked the books and arranged the tablets and the pencils on the desks of his homeroom pupils. Sharpening forty pencils was a chore, sharing it with him a solemn pleasure. To this day I look up at my father through the cedar smell of pencil shavings. To see his key open the front portals of oak, to share alone with him for an hour the pirate hoard of uncracked books and golden pencils, to switch off the lights and leave the room and walk down the darkly lustrous perspective of the corridor and perhaps halt for a few words by an open door that revealed another teacher, like a sorcerer in his sanctum, inscribing forms beside a huge polished globe of the Earth—such territories of wonder boyhood alone can acquire.

The Playground

The periphery I have traced; the center of my boyhood held a calm collection of kind places that are almost impossible to describe, because they are so fundamental to me, they enclosed so many of my hours, that they have the neutral color of my own soul, which I have always imagined as a pale oblong just under my ribs. In the town where I now live, and where I am writing this, seagulls weep overhead on a rainy day. No seagulls found their way inland to Shillington; there were sparrows, and starlings, and cowbirds, and robins, and occasionally a buzzard floating high overhead on immobile wings like a kite on a string too high to be seen.

The playground: up from the hardball diamond, on a plateau bounded on three sides by cornfields, a pavilion contained some tables and a shed for equipment. I spent my summer weekdays there from the age when I was so small that the dust stirred by the feet of roof-ball players got into my eyes. Roof ball was the favorite game. It was played with a red rubber ball smaller than a basketball. The object was to hit it back up on the roof of the pavilion, the whole line of children in succession. Those who failed dropped out. When there was just one person left, a new game began with the cry "Noo-oo gay-ame," and we lined up in the order in which we had gone out, so that the lines began with the strongest and tallest and ended with the weakest and youngest. But there was never any doubt that everybody could play; it was perfect democracy. Often the line contained as many as thirty pairs of legs, arranged chronologically. By the time we moved away, I had become a regular front-runner; I knew how to flick the ball to give it spin, how to leap up and send the ball skimming the length of the roof edge, how to plump it with my knuckles when there was a high bounce. Somehow the game never palled. The sight of the ball bouncing along the tarpaper of the foreshortened roof was always important. Many days I was at the playground from nine o'clock, when they ran up the American flag, until four, when they called the equipment in, and played nothing else.

If you hit the ball too hard, and it went over the peak of the roof, you were out, and you had to retrieve the ball, going down a steep bank into a field where the poorhouse men had stopped planting corn because it all got mashed down. If the person ahead of you hit the ball into the air without touching the roof, or missed it entirely, you had the option of "saving," by hitting the ball onto the roof before it struck the ground; this created complex opportunities for strategy and gallantry. I would always try to save the Nightingale, for instance, and there was a girl who came from Louisiana with a French name whom everybody wanted to save. At twelve, she seemed already mature, and I can remember standing with a pack of other boys under the swings looking up at the undersides of her long tense dark-skinned legs as she kicked into the air to give herself more height, the tendons on the underside of her smooth knees jumping, her sneakered feet pointing like a ballerina's shoes.

The walls of the pavilion shed were scribbled all over with dirty drawings and words and detailed slanders on the prettier girls. After hours, when the supervisors were gone, if you were tall enough you could grab hold of a cross beam and get on top of the shed, where there was an intimate wedge of space under the slanting roof; here no adult ever bothered to scrub away the pencillings, and the wood fairly breathed of the forbidden. The very silence of the pavilion, after the day-long click of checkers and *pokabok* of ping-pong, was like a love-choked hush.

Reality seemed more intense at the playground. There was a dust, a daring. It was a children's world; nowhere else did we gather in such numbers with so few adults over us. The playground occupied a platform of earth; we were exposed, it seems now, to the sun and sky. Looking up, one might see a buzzard or witness a portent.

The Enormous Cloud

Strange, that I remember it. One day, playing roof ball—and I could be six, nine, or twelve when it happened—my head was tipped back and there was an enormous cloud. Someone, maybe I, even called, "Look at the cloud!" It was a bright day; out of nowhere had materialized a cloud, roughly circular in shape, as big as a continent, leaden-blue in the mass, radiant silver along the edges. Its size seemed overwhelming; it was more than a portent, it was the fulfillment of one. I had never seen, and never saw again, such a big cloud.

For of course what is strange is that clouds have no size. Moving in an immaterial medium at an indeterminate distance, they offer no hold for measurement, and we do not even judge them relative to each other. Even, as on a rainy day, when the sky is filled from horizon to horizon, we do not think, "What an enormous cloud." It is as if the soul is a camera shutter customarily set at "ordinary"; but now and then, through some inadvertence, it is tripped wide open and the film is flooded with an enigmatic image.

Another time the sky spoke at the playground, telling me that treachery can come from above. It was our Field Day. One of the events was a race in which we put our shoes in a heap, lined up at a distance, ran to the heap, found our shoes, put them on, and raced back. The winner got a ticket to the Shillington movie theatre. I was the first to find my shoes, and was tying my laces when, out of the ring of adults and older children who had collected to watch, a voice urged, "Hurry! Don't tie the laces." I didn't, and ran back, and was disqualified. My world reeled at the treachery of that unseen high voice: I loved the movies.

The Movie House

It was two blocks from my home; I began to go alone from the age of six. My mother, so strict about my kissing girls, was strangely indulgent about this. The theatre ran three shows a week, for two days each, and was closed on Sundays. Many weeks I went three times. I remember a summer evening in our yard. Supper is over, the walnut tree throws a heavy shadow. The fireflies are not out yet. My father is off, my mother and her parents are turning the earth in our garden. Some burning sticks and paper on our ash heap fill the damp air with low smoke; I express a wish to go to the movies, expecting to be told No. Instead, my mother tells me to go into the house and clean up; I come into the yard again in clean shorts, the shadows slightly heavier, the dew a little wetter; I am given eleven cents and run down Philadelphia Avenue in my ironed shorts and fresh shirt, down past the running ice-plant water, the dime and the penny in my hand. I always ran to the movies. If it was not a movie with Adolphe Menjou, it was a horror picture. People turning into cats—fingers going stubby into paws and hair being blurred in with double exposure—and Egyptian tombs and English houses where doors creak and wind disturbs the curtains and dogs refuse to go into certain rooms because they sense something supersensory. I used to crouch down into the seat and hold my coat in front of my face when I sensed a frightening scene coming, peeking through the buttonhole to find out when it was over. Through the buttonhole Frankenstein's monster glowered; lightning flash; sweat poured over the bolts that held his face together. On the way home, I ran again, in terror now. Darkness had come, the first show was from seven to nine, by nine even the longest summer day was ending. Each porch along the street seemed to be a tomb crammed with shadows, each shrub seemed to shelter a grasping arm. I ran with a frantic high step, trying to keep my ankles away from the reaching hand. The last and worst terror was our own porch; low brick walls on either side concealed possible cat people. Leaping high, I launched myself at the door and, if no one was in the front of the house, fled through suffocating halls past gaping doorways to the kitchen, where there was always someone working, and a light bulb burning. The icebox. The rickety worn table, oilcloth-covered, where we ate. The windows painted solid black by the interior brightness. But even then I kept my legs away from the furry space beneath the table.

These were Hollywood's comfortable years. The theatre, a shallowly sloped hall too narrow to have a central aisle, was usually crowded. I liked it most on Monday nights, when it was emptiest. It seemed most mine then. I had a favorite seat—rear row, extreme left—and my favorite moment was the instant when the orange side lights, Babylonian in design, were still lit, and the curtain was closed but there was obviously somebody up in the projection room, for the camera had started to whir. In the next instant, I knew, a broad dusty beam of light would fill the air above me, and the titles of the travelogue would appear on the curtains, their projected steadiness undulating as with an unhurried, composed screech the curtains were drawn back, revealing the screen alive with images that then would pass through a few focal adjustments. In that delicate, promissory whir was my favorite moment.

On Saturday afternoons the owner gave us all Hershey bars as we came out of the matinee. On Christmas morning he showed a free hour of cartoons and the superintendent of the Lutheran Sunday school led us in singing carols, gesticulating in front of the high blank screen, no bigger than the shadow of the moth that sometimes landed on the lens. His booming voice would echo curiously on the bare walls, usually so dark and muffling but that on this one morning, containing a loud sea of Christmas children, had a bare, clean, morning quality that echoed. After this special show we all went down to the Town Hall, where the plumpest borough employee, disguised as Santa Claus, gave us each a green box of chocolates. Shillington was small enough to support such traditions.

Three Boys

A, B, and C, I'll say, in case they care. A lived next door; he *loomed* next door, rather. He seemed immense—a great wallowing fatso stuffed with possessions; he was the son of a full-fashioned knitter. He seemed to have a beer-belly; after several generations beer-bellies may become congenital. Also his face had no features. It was just a blank ball on his shoulders. He used to call me "Ostrich," after Disney's Ollie Ostrich. My neck was not very long; the name seemed horribly unfair; it was its injustice that made me cry. But nothing I could say, or scream, would make him stop. And I still, now and then—in reading, say, a book review by one of the

apple-cheeked savants of the quarterlies or one of the pious grem-
lins who manufacture puns for *Time*—get the old sensations: my
ears close up, my eyes go warm, my chest feels thin as an eggshell,
my voice churns silently in my stomach. From A I received my first
impression of the smug, chinkless, irresistible *power* of stupidity; it
is the most powerful force on earth. It says "Ostrich" often enough,
and the universe crumbles.

A was more than a boy, he was a force-field that could manifest
itself in many forms, that could take the wiry, disconsolate shape of
wide-mouthed, tiny-eared boys who would now and then beat me
up on the way back from school. I did not greatly mind being
beaten up, though I resisted it. For one thing, it firmly involved me,
at least during the beating, with the circumambient humanity that
so often seemed evasive. Also, the boys who applied the beating
were misfits, periodic flunkers, who wore corduroy knickers with
threadbare knees and men's shirts with the top button buttoned—
this last an infallible sign of deep poverty. So that I felt there was
some justice, some condonable revenge, being applied with their
fists to this little teacher's son. And then there was the delicious
alarm of my mother and grandmother when I returned home
bloody, bruised, and torn. My father took the attitude that it was
making a boy of me, an attitude I dimly shared. He and I both were
afraid of me becoming a sissy—he perhaps more afraid than I.

When I was eleven or so I met B. It was summer and I was down
at the playground. He was pushing a little tank with moving rubber
treads up and down the hills in the sandbox. It was a fine little toy,
mottled with camouflage green; patriotic manufacturers produced
throughout the war millions of such authentic miniatures which we
maneuvered with authentic, if miniature, militance. Attracted by the
toy, I spoke to him; though taller and a little older than I, he had
my dull straight brown hair and a look of being also alone. We
became fast friends. He lived just up the street—toward the poorhouse, the east part of the street, from which the little winds of
tragedy blew. He had just moved from the Midwest, and his mother
was a widow. Beside wage war, we did many things together. We
played marbles for days at a time, until one of us had won the
other's entire coffee-canful. With jigsaws we cut out of plywood
animals copied from comic books. We made movies by tearing the
pages from Big Little Books[4] and coloring the drawings and pasting
them in a strip, and winding them on toilet-paper spools, and making a cardboard carton a theatre. We rigged up telephones, and
racing wagons, and cities of the future, using orange crates and
cigar boxes and peanut-butter jars and such potent debris. We loved
Smokey Stover[5] and were always saying "Foo." We had an intense
spell of Monopoly. He called me "Uppy"—the only person who

4. Small, squat children's books, usually
featuring comic-strip heroes. Popular be-

fore and during World War II.
5. Comic-strip character.

ever did. I remember once, knowing he was coming down that afternoon to my house to play Monopoly, in order to show my joy I set up the board elaborately, with the Chance and Community Chest cards fanned painstakingly, like spiral staircases. He came into the room, groaned, "Uppy, what are you doing?" and impatiently scrabbled the cards together in a sensible pile. The older we got, the more the year between us told, and the more my friendship embarrassed him. We fought. Once, to my horror, I heard myself taunting him with the fact that he had no father. The unmentionable, the unforgivable. I suppose we patched things up, children do, but the fabric had been torn. He had a long, pale, serious face, with buck-teeth, and is probably an electronics engineer somewhere now, doing secret government work.

So through *B* I first experienced the pattern of friendship. There are three stages. First, acquaintance: we are new to each other, making each other laugh in surprise, and demand nothing beyond politeness. The death of the one would startle the other, no more. It is a pleasant stage, a stable stage; on austere rations of exposure it can live a lifetime, and the two parties to it always feel a slight gratification upon meeting, will feel vaguely confirmed in their human state. Then comes intimacy: now we laugh before two words of the joke are out of the other's mouth, because we know what we will say. Our two beings seem marvellously joined, from our toes to our heads, along tingling points of agreement; everything we venture is right, everything we put forth lodges in a corresponding socket in the frame of the other. The death of one would grieve the other. To be together is to enjoy a mounting excitement, a constant echo and amplification. It is an ecstatic and unstable stage, bound of its own agitation to tip into the third: revulsion. One or the other makes a misjudgment; presumes; puts forth that which does not meet agreement. Sometimes there is an explosion; more often the moment is swallowed in silence, and months pass before its nature dawns. Instead of dissolving, it grows. The mind, the throat, are clogged; forgiveness, forgetfulness, that have arrived so often, fail. Now everything jars and is distasteful. The betrayal, perhaps a tiny fraction in itself, has inverted the tingling column of agreement, made all pluses minuses. Everything about the other is hateful, despicable; yet he cannot be dismissed. We have confided in him too many minutes, too many words; he has those minutes and words as hostages, and his confidences are embedded in us where they cannot be scraped away, and even rivers of time cannot erode them completely, for there are indelible stains. Now—though the friends may continue to meet, and smile, as if they had never trespassed beyond acquaintance—the death of the one would please the other.

An unhappy pattern to which *C* is an exception. He was my friend before kindergarten, he is my friend still. I go to his home now, and he and his wife serve me and my wife with alcoholic drinks and slices of excellent cheese on crisp crackers, just as twenty years ago he served me with treats from his mother's refrigerator. He was a born host, and I a born guest. Also he was intelligent. If my childhood's brain, when I look back at it, seems a primitive mammal, a lemur or shrew, his brain was an angel whose visitation was widely hailed as wonderful. When in school he stood to recite, his cool rectangular forehead glowed. He tucked his right hand into his left armpit and with his left hand mechanically tapped a pencil against his thigh. His answers were always correct. He beat me at spelling bees and, in another sort of competition, when we both collected Big Little Books, he outbid me for my supreme find (in the attic of a third boy), the first Mickey Mouse. I can still see that book, I wanted it so badly, its paper tan with age and its drawings done in Disney's primitive style, when Mickey's black chest is naked like a child's and his eyes are two nicked oblongs. Losing it was perhaps a lucky blow; it helped wean me away from hope of ever having possessions.

C was fearless. He deliberately set fields on fire. He engaged in rock-throwing duels with tough boys. One afternoon he persisted in playing quoits with me although—as the hospital discovered that night—his appendix was nearly bursting. He was enterprising. He peddled magazine subscriptions door-to-door; he mowed neighbors' lawns; he struck financial bargains with his father. He collected stamps so well his collection blossomed into a stamp company that filled his room with steel cabinets and mimeograph machinery. He collected money—every time I went over to his house he would get out a little tin box and count the money in it for me: $27.50 one week, $29.95 the next, $30.90 the next—all changed into new bills nicely folded together. It was a strange ritual, whose meaning for me was: since he was doing it, I didn't have to. His money made me richer. We read Ellery Queen[6] and played chess and invented board games and discussed infinity together. In later adolescence, he collected records. He liked the Goodman quintets but loved Fats Waller.[7] Sitting there in that room so familiar to me, where the machinery of the Shilco Stamp Company still crowded the walls and for that matter the tin box of money might still be stashed, while my thin friend grunted softly along with that dead dark angel on "You're Not the Only Oyster in the Stew," I felt, in the best sense, patronized: the perfect guest of the perfect host. What made it perfect was that we had both spent our entire lives in Shillington.

Concerning the Three Great Secret Things: (I) Sex

In crucial matters, the town was evasive. Sex was an unlikely, though persistent, rumor. My father slapped my mother's bottom and made a throaty noise and I thought it was a petty form of

6. Fictional crime investigator-hero of many novels and a weekly radio program.
7. Benny Goodman (b. 1909), swing clarinetist and orchestra leader. Thomas

("Fats") Waller (1904–1943), jazz pianist, singer, composer whose *You're Not the Only Oyster in the Stew* is one of his best recordings.

sadism. The major sexual experience of my boyhood was a section of a newsreel showing some women wrestling in a pit of mud. The mud covered their bathing suits so they seemed naked. Thick, interlocking, faceless bodies, they strove and fell. The sight was so disturbingly resonant that afterward, in any movie, two women pulling each other's hair or slapping each other—there was a good deal of this in movies of the early forties; Ida Lupino was usually one of the women—gave me a tense, watery, drawn-out feeling below the belt. Thenceforth my imaginings about girls moved through mud. In one recurrent scene I staged in my bed, the girl and I, dressed in our underpants and wrapped around with ropes, had been plunged from an immense cliff into a secret pond of mud, by a villain who resembled Peg-Leg Pete.[8] I usually got my hands free and rescued her; sometimes she rescued me; in any case there hovered over our spattered, elastic-clad bodies the idea that these were the last minutes of our lives, and all our shames and reservations were put behind us. It turned out that she had loved me all along. We climbed out, into the light. The ropes had hurt our wrists; yet the sweet kernel of the fantasy lay somehow in the sensations of being tightly bound, before we rescued each other.

(2) Religion

Pragmatically, I have become a Congregationalist, but in the translucent and tactful church of my adoption my eyes sting, my throat goes grave, when we sing—what we rarely sang in the Lutheran church of my childhood—Luther's mighty hymn:

> For still our ancient foe
> Doth seek to work us woe;
> His craft and power are great,
> And arm'd with cruel hate,
> On earth is not his equal.[9]

This immense dirge of praise for the Devil and the world, thunderous, slow, opaquely proud, nourishes a seed in me I never knew was planted. How did the patently vapid and drearily businesslike teachings to which I was lightly exposed succeed in branding me with a Cross? And a brand so specifically Lutheran, so distinctly Nordic; an obdurate insistence that at the core of the core there is a right-angled clash to which, of all verbal combinations we can invent, the Apostles' Creed[1] offers the most adequate correspondence and response.

Of my family, only my father attended the church regularly, returning every Sunday with the Sunday Reading *Eagle* and the complaint that the minister prayed too long. My own relations with

---

8. Villainous tormentor of Mickey Mouse in comic strips.
9. *"Ein Feste Burg"* ("A mighty fortress is our God") is a famous hymn composed by Martin Luther (1483–1546),

German leader of the Protestant Reformation.
1. Christian statement of belief ascribed to the Twelve Apostles ("I believe in God the Father Almighty * * *").

the church were unsuccessful. In Sunday school, I rarely received the perfect attendance pin, though my attendance seemed to me and my parents as perfect as anybody's. Instead, I was given a pencil stamped KINDT'S FUNERAL HOME. Once, knowing that a lot of racy social activity was going on under its aegis, I tried to join the Luther League; but I had the misfortune to arrive on the night of their Halloween party, and was refused admittance because I was not wearing a costume. And, the worst rebuff, I was once struck by a car on the way to Sunday school. I had the collection nickel in my hand, and held on to it even as I was being dragged fifteen feet on the car's bumper. For this heroic churchmanship I received no palpable credit; the Lutheran Church seemed positively to dislike me.

Yet the crustiness, the inhospitality of the container enhanced the oddly lucid thing contained. I do not recall my first doubts; I doubted perhaps abnormally little. And when they came, they never roosted on the branches of the tree, but attacked the roots; if the first article of the Creed stands, the rest follows as water flows downhill. That God, at a remote place and time, took upon Himself the form of a Syrian carpenter and walked the earth willfully healing and abusing and affirming and grieving, appeared to me quite in the character of the Author of the grass. The mystery that more puzzled me as a child was the incarnation of my ego—that omnivorous and somehow preëxistent "I"—in a speck so specifically situated amid the billions of history. Why was I I? The arbitrariness of it astounded me; in comparison, nothing was too marvellous.

Shillington bred a receptivity to the supernatural unrelated to orthodox religion. This is the land of the hex signs, and in the neighboring town of Grille a "witch doctor" hung out a shingle like a qualified M.D. I was struck recently, on reading Frazer's contemptuous list of superstitions in *The Golden Bough*,[2] by how many of them seemed good sense. My grandmother was always muttering little things; she came from a country world of spilled salt and cracked mirrors and new moons and omens. She convinced me, by contagion, that our house was haunted. I punished her by making her stand guard outside the bathroom when I was in it. If I found she had fallen asleep in the shadowy hallway crawling with ghosts, I would leap up on her back and pummel her with a fury that troubles me now.

Imagine my old neighborhood as an African village; under the pointed roofs tom-toms beat, premonitions prowl, and in the darkness naked superstition in all her plausibility blooms:

The Night-blooming Cereus

It was during the war; early in the war, 1942. *Collier's* had printed a cover showing Hirohito, splendidly costumed and fanged,

---

2. *The Golden Bough*, by Sir James Frazer (1854–1941), a Scottish anthro- pologist who traced the primitive origin of myths.

standing malevolently in front of a bedraggled, bewildered Hitler and an even more decrepit and craven Mussolini. Our troops in the Pacific reeled from island to island; the Japanese seemed a race of demons capable of anything. The night-blooming cereus was the property of a family who lived down the street in a stucco house that on the inside was narrow and dark in the way that houses in the middle of the country sometimes are. The parlor was crowded with obscure furniture decked out with antimacassars and porcelain doodads. At Christmas a splendiferous tree appeared in that parlor, hung with pounds of tinsel and strung popcorn and paper chains and pretzels and balls and intricate, figurative ornaments that must have been rescued from the previous century.

The blooming of the cereus was to be an event in the neighborhood; for days we had been waiting. This night—a clear warm night, in August or September—word came, somehow. My mother and grandmother and I rushed down Philadelphia Avenue in the dark. It was late; I should have been in bed. I remembered the night I was refused permission to go to the poorhouse fire. The plant stood at the side of the house, in a narrow space between the stucco wall and the hedge. A knot of neighborhood women had already gathered; heavy shoulders and hair buns were silhouetted in an indeterminate light. On its twisted, unreal stem the flower had opened its unnaturally brilliant petals. But no one was looking at it. For overhead, in the north of a black sky strewn with stars like thrown salt, the wandering fingers of an aurora borealis gestured, now lengthening and brightening so that shades of blue and green could be distinguished, now ebbing until it seemed there was nothing there at all. It was a rare sight this far south. The women muttered, sighed, and, as if involuntarily, out of the friction of their bodies, moaned. Standing among their legs and skirts, I was slapped by a sudden cold wave of fear. "Is it the end of the world?" one of the women asked. There was no answer. And then a plane went over, its red lights blinking, its motors no louder than the drone of a wasp. Japanese. The Japanese were going to bomb Shillington, the center of the nation. I waited for the bomb, and without words prayed, expecting a miracle, for the appearance of angels and Japanese in the sky was restrained by the same impossibility, an impossibility that the swollen waxy brilliant white of the flower by my knees had sucked from the night.

The plane of course passed; it was one of ours; my prayer was answered with the usual appearance of absence. We went home, and the world reconstituted its veneer of reason, but the moans of the women had rubbed something in me permanently bare.

(3) Art

Leafing through a scrapbook my mother long ago made of my childhood drawings, I was greeted by one I had titled "Mr. Sun talking to Old Man Winter in his Office." Old Man Winter, a cloud with stick legs, and his host, a radiant ball with similar legs, sit at ease, both smiling, on two chairs that are the only furniture of the solar office. That the source of all light should have, somewhere, an office, suited my conception of an artist, who was someone who lived in a small town like Shillington, and who, equipped with pencils and paper, practiced his solitary trade as methodically as the dentist practiced his. And indeed, that is how it is at present with me.

Goethe[3]—probably among others—says to be wary of our youthful wishes, for in maturity we are apt to get them. I go back, now, to Pennsylvania, and on one of the walls of the house in which my parents now live there hangs a photograph of myself as a boy. I am smiling, and staring with clear eyes at something in the corner of the room. I stand before that photograph, and am disappointed to receive no flicker, not the shadow of a flicker, of approval, of gratitude. The boy continues to smile at the corner of the room, beyond me. That boy is not a ghost to me, he is real to me; it is I who am a ghost to him. I, in my present state, was one of the ghosts that haunted his childhood. Like some phantom conjured by this child from a glue bottle, I have executed his commands; acquired pencils, paper, and an office. Now I wait apprehensively for his next command, or at least a nod of appreciation, and he smiles through me, as if I am already transparent with failure.

He saw art—between drawing and writing he ignorantly made no distinction—as a method of riding a thin pencil line out of Shillington, out of time altogether, into an infinity of unseen and even unborn hearts. He pictured this infinity as radiant. How innocent! But his assumption here, like his assumptions on religion and politics, is one for which I have found no certain substitute. He loved blank paper and obedience to this love led me to a difficult artistic attempt. I reasoned thus: just as the paper is the basis for the marks upon it, might not events be contingent upon a never-expressed (because featureless) ground? Is the true marvel of Sunday skaters the pattern of their pirouettes or the fact that they are silently upheld? Blankness is not emptiness; we may skate upon an intense radiance we do not see because we see nothing else. And in fact there is a color, a quiet but tireless goodness that things at rest, like a brick wall or a small stone, seem to affirm. A wordless reassurance these things are pressing to give. An hallucination? To transcribe middleness with all its grits, bumps, and anonymities, in its fullness of satisfaction and mystery: is it possible or, in view of the suffering that violently colors the periphery and that at all moments threatens to move into the center, worth doing? Possibly not; but the horse-chestnut trees, the telephone poles, the porches, the green hedges

3. Johann Wolfgang von Goethe (1749–1832), German poet and dramatist.

recede to a calm point that in my subjective geography is still the center of the world.

End of Boyhood

I was walking down this Philadelphia Avenue one April and was just stepping over the shallow little rain gutter in the pavement that could throw you if you were on roller skates—though it had been years since I had been on roller skates—when from the semidetached house across the street a boy broke and ran. He was the youngest of six sons. All of his brothers were in the armed services, and five blue stars hung in his home's front window.[4] He was several years older than I, and used to annoy my grandparents by walking through our yard, down past the grape arbor, on his way to high school. Long-legged, he was now running diagonally across the high-crowned street. I was the only other person out in the air. "Chonny!" he called. I was flattered to have him, so tall and grown, speak to me. "Did you hear?"

"No. What?"

"On the radio. The President is dead."[5]

That summer the war ended, and that fall, suddenly mobile, we moved away from the big white house. We moved on Halloween night. As the movers were fitting the last pieces of furniture, furniture that had not moved since I was born, into their truck, little figures dressed as ghosts and cats flitted in and out of the shadows of the street. A few rang our bell, and when my mother opened the door they were frightened by the empty rooms they saw behind her, and went away without begging. When the last things had been packed, and the kitchen light turned off, and the doors locked, the three of us—my grandparents were already at the new house—got into the old Buick my father had bought—in Shillington we had never had a car, for we could walk everywhere—and drove up the street, east, toward the poorhouse and beyond. Somewhat self-consciously and cruelly dramatizing my grief, for I was thirteen and beginning to be cunning, I twisted and watched our house recede through the rear window. Moonlight momentarily caught in an upper pane; then the reflection passed, and the brightest thing was the white brick wall itself. Against the broad blank part where I used to bat a tennis ball for hours at a time, the silhouette of the dogwood tree stood confused with the shapes of the other bushes in our side yard, but taller. I turned away before it would have disappeared from sight; and so it is that my shadow has always remained in one place.

1962, 1965

4. During World War II some American families displayed "service flags" in their windows, each star denoting a son in one of the branches of the armed forces. When a serviceman was killed, a gold star was substituted for the blue.

5. Franklin Delano Roosevelt (1882–1945) served as President for 12 years (1933–45). His death, after reelection to a fourth term, came as a grievous shock to many Americans, as though a long period of stability had ended.

Contemporary
American Poetry
1945 –

MORE THAN a decade after the end of World War II, two important
and transforming shocks were administered to American poetry:
Allen Ginsberg's *Howl* (1956) and Robert Lowell's *Life Studies*
(1959). Ginsberg, in a single stroke and with the energy of a reborn
Whitman, made poetry one of the rallying points for underground protest
and prophetic denunciation of the prosperous, complacent, gray-spirited
Eisenhower years. Lowell, a more "difficult," less popular poet, brought a
new autobiographical intensity into American poetry by exposing the
neuroses and bouts with insanity suffered by an inbred New Englander.
The connection between such volumes and the times in which they were
written is direct and apparent. The poems anticipated and explored strains
in American social relationships which were to issue in the open conflicts of
the 1960s and 1970s: public unrest about the uses of government and industrial power; the institutions of marriage and the family; the rights and
powers of racial minorities, of women, and of homosexuals; the use of
drugs; and the causes and consequences of mental disorders.

But social pressures cannot fully explain why American poets such as
Lowell and Ginsberg felt ready to claim new authority and new areas of
experience in their writing. However radical the changes in the style and
content of American poetry of the 1950s and 1960s, its assurance was
rooted in subtle, far-reaching developments of the decade before. American
poetry had flourished in the late 1940s because of a new light in which
American poets perceived themselves and because of the confidence they
drew from the achievements of their predecessors in the first half of the
century. Some paradoxes about what was "American," of course, remained.
A prominent critic, referring to T. S. Eliot and Ezra Pound, could still
assert: "During the past generation we have seen our two chief poets make
their escape from America, the one to become an English subject and the
other a partisan of Mussolini and, ultimately, a prisoner of our government." Arguably the best "American" poem of the 1940s proved to be
Eliot's *Four Quartets* (1943), which had been written in England.

Nevertheless the accomplishments of other members of the generation
of Pound and Eliot—poets who had remained in America and whose
early reputations these two expatriates had overshadowed—were becoming

solidly recognized and evaluated. Robert Frost, whose first *Collected Poems* (1930) received the Pulitzer Prize, was already well known. But William Carlos Williams and Wallace Stevens became influential for younger poets only through the major works they published after the war. The first two books of Williams's *Paterson* appeared in 1946 and 1948; Stevens published *Transport to Summer* in 1947, *The Auroras of Autumn* in 1950, and his *Collected Poems* in 1954. Younger poets who were to prove themselves the strongest of their generation began publishing notable books just after 1945: Elizabeth Bishop's first volume, *North & South* (1946); Robert Lowell's *Lord Weary's Castle* (1946); Richard Wilbur's *The Beautiful Changes* (1947); John Berryman's *The Dispossessed* (1948); and the important second volume by Theodore Roethke, *The Lost Son* (1948). After the death of the great Irish poet W. B. Yeats in 1939 and the immigration to the United States of his most notable English successor, W. H. Auden, it became clear that the center of poetic activity in the English language had shifted from Britain to America.

If there was a new confidence expressed by American poetry after World War II, there was also a new visibility for American poets. Earlier poets had been relatively isolated from the American public: Pound and Eliot lived in Europe; Wallace Stevens was a businessman in Hartford; William Carlos Williams was a small-town doctor in New Jersey. Poetry readings had been relatively rare performances by the few famous poets of the familiar poems the audience already knew but wanted to hear from the illustrious presence himself. After the war, writers' conferences and workshops, recordings, published and broadcast interviews became more common. A network of poets traveling to give readings and of poets-in-residence at universities began to form. In the 1960s and 1970s, readings became less formal, more numerous and accessible, held not only in auditoriums but in coffee houses, bars, and lofts. The purpose of these more casual readings was often to introduce new poets or, perhaps, new poems by an already recognized writer. The poet coming of age after the war was, as one of them, Richard Wilbur, put it, more a "poet-citizen" than an alienated artist. He often made his living by putting together a combination of teaching positions, readings, and foundation grants.

A poet's education in the 1950s probably differed from that of poets in an earlier generation. Writing poetry in the postwar years became firmly linked to the English literature curriculum in ways that it had not been in the past. Many of the young poets were taught to read verse and sometimes to write it by influential literary critics who were often poets themselves: John Crowe Ransom, Yvor Winters, Robert Penn Warren, R. P. Blackmur, Allen Tate. There was, for better or worse, the college major in English literature, as there had not been in so narrow and disciplined a sense for the poetic giants of a generation earlier. Eliot, for example, had done graduate work in philosophy, Pound in Romance philology, Williams had gone to medical school; and each had forged his own literary criticism. A younger poet, on the other hand, *studied* Eliot's essays, or learned critical approaches to literature in English courses such as the ones Allen Ginsberg took from Mark Van Doren and Lionel Trilling at Columbia or James Merrill from Reuben Brower at Amherst. A popular critical text, *Understanding Poetry* (1938) by Cleanth Brooks and Robert Penn Warren, taught students to be close readers of English metaphysical poems of the seventeenth century, such as those by John Donne and Andrew Marvell. As the poet W. D. Snodgrass testifies, "In school we had been taught to write a very difficult and very intellectual poem. We tried to achieve the obscure and dense texture of the French symbolists (very intuitive and often deranged poets), but by using methods similar to those of the very intellectual and conscious poets of the English Renaissance, especially the Metaphysical poets."

A young writer, thus trained to read intricate traditional lyrics, did not expect to encounter much, if any, contemporary verse in the classroom. The student had to seek out modern poems in the literary quarterlies or come upon them through the chance recommendations of informed friends and teachers. And whether a beginning poet fell, in this private, accidental way, under the influence of Eliot's ironic elegies or Stevens's high rhapsodies or William Carlos Williams's homemade documentaries, he was prepared to think of a poem as something "other" than the poet himself, objective, free from the quirks of the personal.

In the 1950s, while there was no dominant prescription for a poem, the short lyric meditation was held in high regard. Avoiding the first person, poets would find an object, a landscape, or an observed encounter which epitomized and clarified their feelings. A poem was the product of retrospection, a gesture of composure following the initial shock or stimulus which provided the occasion for writing. Often composed in intricate stanzas and skillfully rhymed, such a poem deployed its mastery of verse form as one sign of the civilized mind's power to explore, tame, and distill raw experience. Richard Wilbur's verse was especially valued for its speculative neatness, a poise which was often associated with the awareness of the historical values of European culture. It was a time of renewed travel in Europe; there were Fulbright fellowships for American students to study abroad, prizes for writers who wanted to travel and write in Europe. Wilbur and others wrote poems about European art and artifacts and landscapes as a way of testing American experience against the alternative ways of life, for example, American Puritanism or notions of virtue, viewed against such complicated pleasures as those embodied in the seventeenth-century sculpted fountain described in Wilbur's *A Baroque Wall-Fountain in the Villa Sciarra*. Unlike the pessimistic Eliot of *The Waste Land*, such poets found the treasures of the past—its art and literature—nourishing in poems whose chief pleasure was that of evaluation and balancing, of weighing such alternatives as spirituality and worldliness.

That was one side of the picture. The other side, equally important, was the way many of these same young poets reacted to (*to*, rather than *against*) their training. Richard Howard, in a happy phrase, calls this postwar generation of poets "the children of Midas." He is thinking of the last phases of the classical myth, when King Midas, having discovered that everything he touches ("his food, his women, his words") inconveniently turns to gold, prays to lose the gift of the golden touch. "What seems to me especially proper to these poets," Howard says, "* * * is the last development, the longing to *lose* the gift of order, despoiling the self of all that had been, merely, *propriety*." In the 1950s and 1960s there were some very extreme examples of poets transforming themselves: Allen Ginsberg, who began by writing formal quatrains, became the free and rambunctious poet of *Howl*; Sylvia Plath, who began as a well-mannered imitator of Eliot and Dylan Thomas, turned into the intense suicidal protagonist of *Ariel* (1966). It is a special mark of this period that a poet as bookish, as literary, as academic as John Berryman, who started out writing like Auden and Yeats, should also have written the wildest and most disquieting lyrics of his time, *The Dream Songs* (1964, 1968).

The new confidence and technical sophistication of American poetry in the 1940s fostered the more exploratory styles of the 1950s and 1960s. Some changes were more noticeable and notorious than others. For one thing, poetry seemed to be extending its subject matter to more explicit and extreme areas of autobiography: insanity, sex, divorce, alcoholism. The convenient but not very precise label "confessional" came to be attached to certain books: Robert Lowell's *Life Studies*, which explored the disorders of several generations of his New England family; Anne Sexton's *To Bedlam and Part Way Back* (1960) and *All My Pretty Ones* (1962), which dealt openly with abortion, the sex life of women, the poet's own life in mental hospitals; W. D. Snodgrass's *Heart's Needle* (1969), whose central lyric sequence chronicled the stages of divorce from the point of view of a husband separated from his wife and child; John Berryman's *Dream Songs*, which exposed his alcoholism and struggle with insanity. Allen Ginsberg's *Howl* celebrated his homosexuality. Sylvia Plath's *Ariel* explored the frantic energies of a woman on the edge of suicide.

Some of the poetry of this period was avowedly political, tending in the 1950s to general protest and in the 1960s to more specifically focused critiques. The Beats of the 1950s—with *Howl* as their manifesto—had no one particular object of protest. Their work envisioned freer life styles and explored underground alternatives to life in a standardized or mechanized society. The pun on the word "Beat" linked them on the one side to a downtrodden drifting underground community—drugs, homosexuality, political radicalism—and, on the other, to a new "beatitude," made available by Eastern religious cults that many members of this generation espoused. Gary Snyder, who in the 1950s was with the Beats in San Francisco, is one example of how their protests were extended and focused in the next decades. In his books of the Sixties and Seventies such as *Earth House Hold* and *Turtle Island* he dramatizes a very specific alternative to American suburban and urban sprawl: he describes and advocates a life of almost Thoreauvian simplicity in a commune in the Sierras.

Many poets in the 1960s identified themselves with specific reform and protest movements. Denise Levertov, Adrienne Rich, and Robert Lowell, among others, directed poems against American participation in the Vietnam war and our government's support of the corrupt South Vietnam regime. Robert Lowell publicly refused President Johnson's invitation to a White House dinner and was a participant in the 1967 march against the Pentagon which Norman Mailer describes in *Armies of the Night*. Robert Bly and others used the occasion of receiving poetry prizes to make antiwar statements. The important freedom movements of the 1960s—advocating black power, women's liberation, and gay rights—had spokesmen among committed poets. Black poets such as Gwendolyn Brooks and LeRoi Jones (later Imamu Baraka) who had already had considerable success with white audiences, turned to address exclusively black constituencies. Small presses, notably the Broadside Press, were founded for the publication of black poets, and others devoted themselves to feminist writing. Some poetry of the late 1960s had the insistence, urgency, and singlemindedness of political tracts. But the more enduring effect of political protest on poetry was to make a broader, more insistent range of voices available to verse; poems dramatized individual predicaments, stressing the under-

lying angers and desires that also issue in political action. Black poets experimented in bringing out the distinctive speech rhythms of Black English;
they stressed the oral values of verse—its openness to song, to angry chant,
and to the cadenced complaint of the "blues." Among feminists, Adrienne
Rich, in books such as *Diving into the Wreck* and *The Dream of a Common Language*, stressed the power of poetic language to explore the fantasies, buried rage, and special awareness of women.

In an indirect but vital sense the heightened energies of almost all poetry in the 1960s and 1970s had political implications. With the increasing
standardization of speech, a documented decline in reading skills in the
schools, the dominance of nationwide television, poems provided a special
resource for individual expression, a resistance to the leveling force of official
language, and access to profoundly individual areas of consciousness. In
that context poets as superficially apolitical as James Wright, W. S. Merwin, James Merrill, and John Ashberry were by their very cultivation of
private vision making distinctly political choices.

In response to the pressures, inner and outer, of the Fifties and Sixties,
new kinds of poems took their place alongside the favored "objective"
poems of the late Forties. As some poets aimed more at exposing than at
composing the self, they demanded more open forms to suggest vagaries,
twists, and confusions of mind or else its potential directness and spontaneity. Their poems depended on less rhyme, sparer use of regular stanzas
and metrics, even new ways of spacing a poem on the page. Critics talked
of organic form, using free verse which took its length of line or its visual
form on the page from the poet's provisional or intense feelings at the
moment of composition. The most insistent formulations of this attitude
are to be found in the manifestoes of the so-called Black Mountain School,
a group of poets gathered at Black Mountain College in North Carolina
and very much influenced by its rector, Charles Olson. Olson's famous pronouncements on "Projective Verse" were first published in 1950. Ordinary
lineation, straight left-hand margins, regular meters and verse forms were
to be discarded in favor of a free placement of lines and phrases over the
page. The unit of poetic expression was to be not a predetermined metrical
foot but the poet's "breath"; breaks in the line corresponded to the mental
and physical energy enlisted to get the words on the page. Olson's purpose
was to put the poem in touch—as in certain forms of meditation or yoga—
with an individual poet's natural rhythms, often buried by acquired verbal
skills. The poet was not to revise his poems to any great extent; he might
make considerable mental preparation or store up intense feeling before
writing, but the poem itself was to represent feeling at the moment of
composition. Another corollary of Olson's theories, not part of the Black
Mountain credo but growing logically from it, was the notion that a poem
was in some way provisional. In contrast to the notion of the 1940s that a
poem was a completed and permanent object, some poets saw their work
as transitory, incomplete, an instrument of passage. Allen Ginsberg and
Adrienne Rich carefully date each of their poems as if to suggest that the
feelings involved are peculiarly subject to revision by later experience.

A parallel development—only very loosely related to the San Francisco
Beat explosion and Black Mountain manifestoes—took place among a
group of poets involved with and inspired by the work of nonrepresentational or Abstract Expressionist painters in New York. The so-called New
York School included John Ashbery, Kenneth Koch, James Schuyler, and

the figure whose friendship and enthusiasm held them all together, Frank O'Hara. It was O'Hara with his breezy diary poems, almost throwaways, who most typified their belief in the poem as a chronicle of its occasion and of the act of composing it. As O'Hara said in his offhand parody of sober poetic credos, *Personism: A Manifesto* (1959); "The poem is at last between two persons instead of two pages. * * * In all modesty, I confess that it may be the death of literature as we know it."

American poetry emerged from the 1960s enriched, its technical and thematic range enormously expanded. The only single common denominator among these various directions of change was the desire to make poetry less distant—in both sound and subject matter—from its potential readers. The poems of the 1970s were not to be, indeed could not be, as technically innovative as those of the previous decade. The political impulse behind much radical poetry had waned; the main exception was women's poetry, which, especially in response to the work of Adrienne Rich, continued to explore areas special to feminism.

The 1960s had changed the face of American poetry, and it seemed the task of poets in the 1970s not so much to innovate as to consolidate and perhaps reinterpret the achievement of the previous three decades. What characterized the world of American poetry was its pluralism and the power of its best poets to absorb a variety of influences. Traditional verse and metrical forms had not been left behind, but they took their place among a number of poetic resources rather than as obligatory models of poetic decorum. The single perfect and self-contained lyric ceased to define the boundary of poetic ambitions, and the high ambitions of Pound in his *Cantos* or William Carlos Williams in *Paterson* (reaching back to Whitman and *Leaves of Grass*) stood out in bold relief as extensions of poetic possibility. Robert Lowell in *Notebook* showed one of the ways the poetic virtues of the three decades since the war might be combined: it consisted of a series of unrhymed sonnets, which served as a journal. "If I saw something one day, I wrote it that day, or the next, or the next. Things I felt or saw or read were drift to the whirlpool, the squeeze of the sonnet and the loose ravel of blank verse." Berryman's *Dream Songs*, however strict their stanzaic form and their use of rhyme, was a sequence with enormous flexibility of voice and richly varied exposures of the self. Sequences of poems provided one way of countering neat closure and of suggesting both the fluidity of external life and the complexity of consciousness. Long poems provided another alternative. In a bold stroke James Merrill's series of three book-length "Divine Comedies" was a sign that poetic ambitions were taking a new direction. Many American poets did not want to think of their poems as the shards or fragments of modern literature. They perhaps remembered that Wallace Stevens, whose first published volume was *Harmonium*, wanted to call his collected poems *The Whole of Harmonium*. For some writers the poem was once again to fulfill the Romantic poet Wordsworth's ambition to record "the growth of a poet's mind" in a modern epic of consciousness. As poetry moved away from the single lyrics of the immediate postwar period, ambitions once again focused on allowing shorter poetic efforts to grow into larger construction; the strength of longer poems or related series of poems lay not in their paraphrasable message or their power to project an "objective correlative" for a small poetic truth but in their extended power to present particularized models of the mind's continuing struggle for insight.

THEODORE ROETHKE
1908–1963

Theodore Roethke was born in Saginaw, Michigan, where both his German grandfather and his father kept greenhouses for a living. "I was born under a glass heel and have always lived there," he was to remark. He was haunted not only by the ordered, protected world of the paternal greenhouse and its cultivated flowers, but also by the desolate landscape of that part of Michigan. "The marsh, the mire, the Void, is always there, immediate and terrifying. It is a splendid place for schooling the spirit. It is America."

Roethke's poetry often re-enacted this "schooling" of the spirit by revisiting the landscapes of his childhood: the nature poems which make up the largest part of his early work try to bridge the distance between a child's consciousness and the adult mysteries presided over by his father. Roethke arranged and rearranged these poems to give the sense of a spiritual autobiography, especially in preparing the volumes *The Lost Son* (1948), *Praise to the End!* (1951), and *The Waking* (1953). The greenhouse world emerged as a "reality harsher than reality," the cultivator's activity pulsating and threatening. Its overseers, like "Frau Bauman, Frau Schmidt, and Frau Schwartze," emerge as gods, fates, and witches all in one. It was by focusing on the minute processes of botanical growth—the rooting, the budding—that the poet found a way of participating in the mysteries of this once alien world, "alive in a slippery grave."

Another of Roethke's ways of regenerating himself was to explore prerational speech, like children's riddles, to recapture a nonlogical state of being. One of the sections of *The Lost Son* is called "The Gibber," a pun referring both to meaningless utterance and to the technical name for the pouch at the base of the calyx of a flower. The pun identifies a principle of growth with prehuman and childish speech. *The Lost Son* suggests that recapturing a childlike merger with nature might revive the spirit of an adult life: "A lively understandable spirit / Once entertained you. / It will come again. / Be still. / Wait."

If the nature poems of Roethke's first four books explore the anxieties with him since childhood, his later love poems show him in periods of release and momentary pleasure.

> And I dance round and round,
> A fond and foolish man,
> And see and suffer myself
> In another being at last.

The love poems, many of them included in *Words for the Wind* (1958) and *The Far Field* (1964), are among the most appealing in recent American verse. They stand in sharp relief to the suffering Roethke experienced in other areas of his personal life—several mental breakdowns and periods of alcoholism—which led to a premature death. *The Far Field*, a posthumous volume, includes fierce, strongly rhymed lyrics in which Roethke tried "bare, even terrible statement." In these last great poems he moves beyond nature and physical sensation, pressing the senses toward the threshold of spiritual insight:

> I teach my eyes to hear, my ears to see
> How body from spirit slowly does unwind
> Until we are pure spirit at the end.

Even the nature poems of this last volume, gathered as "The North American Sequence," are "journeys to the interior." They use extended landscape to find natural analogies for the human passage toward death, hoping "in their rhythms to catch the very movement of the mind itself."

Roethke is remembered as one of the great teachers of poetry, especially by those young poets and critics who studied with him at the University of Washington from 1948 until the time of his death in 1963. James Wright, David Wagoner, and Richard Hugo, among others, attended his classes. He was noted for his mastery of sound and metrics. Though his own poetry was intensely personal, his starting advice to students always de-emphasized undisciplined self-expression. "Write like someone else," was his instruction to beginners. His own apprenticeship to Blake and to Yeats had shown him how their lyric voices allowed him to discover related but unsuspected voices within himself.

Roethke was much honored later in his career: a Pulitzer Prize for *The Waking* (1953); a National Book Award and Bollingen Prize for the collected poems, *Words for the Wind* (1958), and a posthumous National Book Award for *The Far Field* (1964).

Cuttings

> Sticks-in-a-drowse droop over sugary loam,
> Their intricate stem-fur dries;
> But still the delicate slips keep coaxing up water;
> The small cells bulge;
>
> One nub of growth   5
> Nudges a sand-crumb loose,
> Pokes through a musty sheath
> Its pale tendrilous horn.

1948

Weed Puller

> Under the concrete benches,
> Hacking at black hairy roots,—
> Those lewd monkey-tails hanging from drainholes,—
> Digging into the soft rubble underneath,
> Webs and weeds,   5
> Grubs and snails and sharp sticks,
> Or yanking tough fern-shapes,
> Coiled green and thick, like dripping smilax,[1]

1. A type of fern.

Tugging all day at perverse life:
The indignity of it!—   10
With everything blooming above me,
Lilies, pale-pink cyclamen, roses,
Whole fields lovely and inviolate,—
Me down in that fetor[2] of weeds,
Crawling on all fours,   15
Alive, in a slippery grave.

1948

Frau Bauman, Frau Schmidt, and Frau Schwartze[1]

Gone the three ancient ladies
Who creaked on the greenhouse ladders,
Reaching up white strings
To wind, to wind
The sweet-pea tendrils, the smilax,   5
Nasturtiums, the climbing
Roses, to straighten
Carnations, red
Chrysanthemums; the stiff
Stems, jointed like corn,   10
They tied and tucked,—
These nurses of nobody else.
Quicker than birds, they dipped
Up and sifted the dirt;
They sprinkled and shook;   15
They stood astride pipes,
Their skirts billowing out wide into tents,
Their hands twinkling with wet;
Like witches they flew along rows
Keeping creation at ease;   20
With a tendril for needle
They sewed up the air with a stem;
They teased out the seed that the cold kept asleep,—
All the coils, loops, and whorls.
They trellised the sun; they plotted for more than themselves.   25

I remember how they picked me up, a spindly kid,
Pinching and poking my thin ribs
Till I lay in their laps, laughing,
Weak as a whiffet;[2]
Now, when I'm alone and cold in my bed,   30
They still hover over me,
These ancient leathery crones,
With their bandannas stiffened with sweat,
And their thorn-bitten wrists,
And their snuff-laden breath blowing lightly over me in my first
    sleep.   35

1948

---

2. Stench.
1. Women who worked in the greenhouse owned by Roethke's father.

2. A small, young or unimportant person (probably from *whippet*, a small dog).

My Papa's Waltz

The whiskey on your breath
Could make a small boy dizzy;
But I hung on like death:
Such waltzing was not easy.

We romped until the pans   5
Slid from the kitchen shelf;
My mother's countenance
Could not unfrown itself.

The hand that held my wrist
Was battered on one knuckle;   10
At every step you missed
My right ear scraped a buckle.

You beat time on my head
With a palm caked hard by dirt,
Then waltzed me off to bed   15
Still clinging to your shirt.

1948

The Waking[1]

I wake to sleep, and take my waking slow.
I feel my fate in what I cannot fear.
I learn by going where I have to go.

We think by feeling. What is there to know?
I hear my being dance from ear to ear.   5
I wake to sleep, and take my waking slow.

Of those so close beside me, which are you?
God bless the Ground! I shall walk softly there,
And learn by going where I have to go.

Light takes the Tree; but who can tell us how?   10
The lowly worm climbs up a winding stair;
I wake to sleep, and take my waking slow.

Great Nature has another thing to do
To you and me; so take the lively air,
And, lovely, learn by going where to go.   15

This shaking keeps me steady. I should know.
What falls away is always. And is near.
I wake to sleep, and take my waking slow.
I learn by going where I have to go.

1953

1. The poem's form is that of a villanelle.

I Knew a Woman

I knew a woman, lovely in her bones,
When small birds sighed, she would sigh back at them;
Ah, when she moved, she moved more ways than one:
The shapes a bright container can contain!
Of her choice virtues only gods should speak,⁣   5
Or English poets who grew up on Greek
(I'd have them sing in chorus, cheek to cheek).

How well her wishes went! She stroked my chin,
She taught me Turn, and Counter-turn, and Stand;[1]
She taught me Touch, that undulant white skin;   10
I nibbled meekly from her proffered hand;
She was the sickle; I, poor I, the rake,
Coming behind her for her pretty sake
(But what prodigious mowing we did make).

Love likes a gander, and adores a goose: 15
Her full lips pursed, the errant note to seize;
She played it quick, she played it light and loose;
My eyes, they dazzled at her flowing knees;
Her several parts could keep a pure repose,
Or one hip quiver with a mobile nose   20
(She moved in circles, and those circles moved).

Let seed be grass, and grass turn into hay:
I'm martyr to a motion not my own;
What's freedom for? To know eternity.
I swear she cast a shadow white as stone.   25
But who would count eternity in days?
These old bones live to learn her wanton ways:
(I measure time by how a body sways).

     1958

Her Longing

Before this longing,
I lived serene as a fish,
At one with the plants in the pond,
The mare's tail, the floating frogbit,
Among my eight-legged friends,   5
Open like a pool, a lesser parsnip,
Like a leech, looping myself along,
A bug-eyed edible one,
A mouth like a stickleback,—
A thing quiescent!   10

But now—
The wild stream, the sea itself cannot contain me:

---

1. Parts of a Pindaric ode.

I dive with the black hag, the cormorant,[1]
Or walk the pebbly shore with the humpbacked heron,
Shaking out my catch in the morning sunlight,   15
Or rise with the gar-eagle, the great-winged condor.[2]
Floating over the mountains,
Pitting my breast against the rushing air,
A phoenix, sure of my body,
Perpetually rising out of myself,   20
My wings hovering over the shorebirds,
Or beating against the black clouds of the storm,
Protecting the sea-cliffs.

1964

Her Time

When all
My waterfall
Fancies sway away
From me, in the sea's silence;
In the time   5
When the tide moves
Neither forward nor back,
And the small waves
Begin rising whitely,
And the quick winds   10
Flick over the close whitecaps,
And two scoters fly low,
Their four wings beating together,
And my salt-laden hair
Flies away from my face   15
Before the almost invisible
Spray, and the small shapes
Of light on the far
Cliff disappear in a last
Glint of the sun, before   20
The long surf of the storm booms
Down on the near shore,
When everything—birds, men, dogs—
Runs to cover:
I'm one to follow,   25
To follow.

1964

Wish for a Young Wife

My lizard, my lively writher,
May your limbs never wither,
May the eyes in your face
Survive the green ice
Of envy's mean gaze;   5

---

1. A large, black, voracious seabird.   2. Mountain birds of prey.

May you live out your life
Without hate, without grief,
And your hair ever blaze,
In the sun, in the sun,
When I am undone,   10
When I am no one.

1964

In a Dark Time

In a dark time, the eye begins to see,
I meet my shadow in the deepening shade;
I hear my echo in the echoing wood—
A lord of nature weeping to a tree.
I live between the heron and the wren,   5
Beasts of the hill and serpents of the den.

What's madness but nobility of soul
At odds with circumstance? The day's on fire!
I know the purity of pure despair,
My shadow pinned against a sweating wall.   10
That place among the rocks—is it a cave,
Or winding path? The edge is what I have.

A steady storm of correspondences!
A night flowing with birds, a ragged moon,
And in broad day the midnight come again!   15
A man goes far to find out what he is—
Death of the self in a long, tearless night,
All natural shapes blazing unnatural light.

Dark, dark my light, and darker my desire.
My soul, like some heat-maddened summer fly,   20
Keeps buzzing at the sill. Which I is I?
A fallen man, I climb out of my fear.
The mind enters itself, and God the mind,
And one is One, free in the tearing[1] wind.

1964

Infirmity

In purest song one plays the constant fool
As changes shimmer in the inner eye.
I stare and stare into a deepening pool
And tell myself my image cannot die.
I love myself: that's my one constancy.   5
Oh, to be something else, yet still to be!

Sweet Christ, rejoice in my infirmity;
There's little left I care to call my own.

1. With a pun, referring to "tearless" in line 17 (according to Roethke's note).

Today they drained the fluid from a knee
And pumped a shoulder full of cortisone;   10
Thus I conform to my divinity
By dying inward, like an aging tree.

The instant ages on the living eye;
Light on its rounds, a pure extreme of light
Breaks on me as my meager flesh breaks down—   15
The soul delights in that extremity.
Blessed the meek; they shall inherit wrath;[1]
I'm son and father of my only death.

A mind too active is no mind at all;
The deep eye sees the shimmer on the stone;   20
The eternal seeks, and finds, the temporal,
The change from dark to light of the slow moon,
Dead to myself, and all I hold most dear,
I move beyond the reach of wind and fire.

Deep in the greens of summer sing the lives   25
I've come to love. A vireo whets its bill.
The great day balances upon the leaves;
My ears still hear the bird when all is still;
My soul is still my soul, and still the Son,
And knowing this, I am not yet undone.   30

Things without hands take hands: there is no choice,—
Eternity's not easily come by.
When opposites come suddenly in place,
I teach my eyes to hear, my ears to see
How body from spirit slowly does unwind   35
Until we are pure spirit at the end.

1964

1. This is a paraphrase of Jesus' words in the Sermon on the Mount, "Blessed are the meek, for they shall inherit the earth" (Matthew 5.4).

ELIZABETH BISHOP

1911–1979

Elizabeth Bishop has often been praised for her "famous eye," her direct and unflinching descriptive powers, as if her accuracy were an end in itself. Yet, however detailed her immersion in a particular scene, her poems often end by reminding us, as Wallace Stevens did, that "we live in a place / that is not our own and much more, not ourselves, / and hard it is in spite of blazoned days." Practicing shifts of physical scale, opening long perspectives of time which dwarf the merely human, Bishop does more than merely observe; she emphasizes the dignified frailty of a human observer and the pervasive mysteries which surround her. Examining her own case, she traces the observer's instinct to early childhood. *In the*

Waiting Room, a poem written in the early 1970s, probes the sources and motives behind her interest in detail. Using an incident from 1918 when she was seven—a little girl waits for her aunt in the dentist's ante-room—Bishop shows how in the course of the episode she became aware, as if wounded, of the utter strangeness and engulfing power of the world. The spectator in that poem hangs on to details as a kind of life-jacket; she observes because she has to. In a related autobiographical story, *In the Village*, she tells of a little girl haunted by the screams of her mother, just back from a sanitarium. Like *In the Waiting Room,* the story is told through the child's eyes. It links childhood losses to a maturing need to recapture the physicality of the world and to reclaim vanished primal pleasures, "the elements speaking: earth, air, fire, water."

Elizabeth Bishop was born in 1911 in Worcester, Massachusetts. Her father died when she was eight months old. When she was five, her mother was taken to a sanitarium and Bishop never saw her again. Thereafter she lived with relatives in New England and in Nova Scotia. She graduated from Vassar College in 1934, where while a student she had been intro-duced to Marianne Moore. Moore's meticulous taste for fact was to in-fluence Bishop's poetry, but more immediately Moore's independent life as a poet made that life seem an alternative to Bishop's vaguer intentions to attend medical school. Bishop lived in New York City and in Key West, Florida; a traveling fellowship took her to Brazil, which so appealed to her that she stayed there for over sixteen years.

Exile and travel were at the heart of Bishop's poems from the very start. The title of her very first book, *North & South* (1946), looks forward to the tropical worlds she was to choose so long as home—Florida and, later, Brazil—and backward to her childhood and the northern seas of Nova Scotia. Her poems are set among these landscapes where she can stress the sweep and violence of encircling and eroding geological powers or, in the case of Brazil, a bewildering botanical plenty. *Questions of Travel* (1965), her third volume, constitutes a sequence of poems initiating her, with her botanist-geologist-anthropologist's curiosity, into the life of Brazil and the mysteries of what questions a traveler-exile should ask. In this series with its increasing penetration of a new country, a process is at work similar to one Bishop identifies in the great English zoologist and naturalist Charles Darwin, of whom Bishop once said: "One admires the beautiful solid case being built up out of his endless, heroic observations, almost unconscious or automatic—and then comes a sudden relaxation, a forgetful phrase, and one feels the strangeness of his undertaking, sees the lonely young man, his eyes fixed on facts and minute details, sinking or sliding giddily off into the unknown."

In 1969 Bishop's *Complete Poems* appeared, an ironic title in light of the fact that she continued to write and publish new poetry. *Geog-raphy III* (1976) contains some of her very best work, poems which, from the settled perspective of her return to America, look back and evaluate the appetite for exploration apparent in her earlier verse. Having left Brazil, Bishop lived in Boston since 1970 and taught at Harvard Uni-versity until 1977. She received the Pulitzer Prize for the combined volume *North & South and A Cold Spring* (1955); the National Book Award for *The Complete Poems*; and in 1976 was the first woman and the first American to receive the Books Abroad Neustadt International for Lit-erature Prize.

The Fish

I caught a tremendous fish
and held him beside the boat
half out of water, with my hook
fast in a corner of his mouth.
He didn't fight.   5
He hadn't fought at all.
He hung a grunting weight,
battered and venerable
and homely. Here and there
his brown skin hung in strips   10
like ancient wallpaper,
and its pattern of darker brown
was like wallpaper:
shapes like full-blown roses
stained and lost through age.   15
He was speckled with barnacles,
fine rosettes of lime,
and infested
with tiny white sea-lice,
and underneath two or three   20
rags of green weed hung down.
While his gills were breathing in
the terrible oxygen
—the frightening gills,
fresh and crisp with blood,   25
that can cut so badly—
I thought of the coarse white flesh
packed in like feathers,
the big bones and the little bones,
the dramatic reds and blacks   30
of his shiny entrails,
and the pink swim-bladder
like a big peony.
I looked into his eyes
which were far larger than mine   35
but shallower, and yellowed,
the irises backed and packed
with tarnished tinfoil
seen through the lenses
of old scratched isinglass.[1]   40
They shifted a little, but not
to return my stare.
—It was more like the tipping
of an object toward the light.
I admired his sullen face,   45
the mechanism of his jaw,

1. A whitish, semitransparent substance, originally obtained from the swim-bladders of some fresh-water fishes and occasionally used for windows.

and then I saw
that from his lower lip
—if you could call it a lip—
grim, wet, and weaponlike,   50
hung five old pieces of fish-line,
or four and a wire leader
with the swivel still attached,
with all their five big hooks
grown firmly in his mouth.   55
A green line, frayed at the end
where he broke it, two heavier lines,
and a fine black thread
still crimped from the strain and snap
when it broke and he got away.   60
Like medals with their ribbons
frayed and wavering,
a five-haired beard of wisdom
trailing from his aching jaw.
I stared and stared   65
and victory filled up
the little rented boat,
from the pool of bilge
where oil had spread a rainbow
around the rusted engine   70
to the bailer rusted orange,
the sun-cracked thwarts,
the oarlocks on their strings,
the gunnels—until everything
was rainbow, rainbow, rainbow!   75
And I let the fish go.

                                              1946

At the Fishhouses

Although it is a cold evening,
down by one of the fishhouses
an old man sits netting,
his net, in the gloaming almost invisible
a dark purple-brown,   5
and his shuttle worn and polished.
The air smells so strong of codfish
it makes one's nose run and one's eyes water.
The five fishhouses have steeply peaked roofs
and narrow, cleated gangplanks slant up   10
to storerooms in the gables
for the wheelbarrows to be pushed up and down on.
All is silver: the heavy surface of the sea,
swelling slowly as if considering spilling over,
is opaque, but the silver of the benches,   15
the lobster pots, and masts, scattered
among the wild jagged rocks,

is of an apparent translucence
like the small old buildings with an emerald moss
growing on their shoreward walls.   20
The big fish tubs are completely lined
with layers of beautiful herring scales
and the wheelbarrows are similarly plastered
with creamy iridescent coats of mail,
with small iridescent flies crawling on them.   25
Up on the little slope behind the houses,
set in the sparse bright sprinkle of grass,
is an ancient wooden capstan,[1]
cracked, with two long bleached handles
and some melancholy stains, like dried blood,   30
where the ironwork has rusted.
The old man accepts a Lucky Strike.[2]
He was a friend of my grandfather.
We talk of the decline in the population
and of codfish and herring   35
while he waits for a herring boat to come in.
There are sequins on his vest and on his thumb.
He has scraped the scales, the principal beauty,
from unnumbered fish with that black old knife,
the blade of which is almost worn away.   40

Down at the water's edge, at the place
where they haul up the boats, up the long ramp
descending into the water, thin silver
tree trunks are laid horizontally
across the gray stones, down and down   45
at intervals of four or five feet.

Cold dark deep and absolutely clear,
element bearable to no mortal,
to fish and to seals . . . One seal particularly
I have seen here evening after evening.   50
He was curious about me. He was interested in music;
like me a believer in total immersion,[3]
so I used to sing him Baptist hymns.
I also sang "A Mighty Fortress Is Our God."
He stood up in the water and regarded me   55
steadily, moving his head a little.
Then he would disappear, then suddenly emerge
almost in the same spot, with a sort of shrug
as if it were against his better judgment.
Cold dark deep and absolutely clear,   60
the clear gray icy water . . . Back, behind us,
the dignified tall firs begin.

---

1. Cylindrical drum around which rope is wound, used for hauling.
2. Brand of cigarettes.
3. Form of baptism used by Baptists.

Bluish, associating with their shadows,
a million Christmas trees stand
waiting for Christmas. The water seems suspended   65
above the rounded gray and blue-gray stones.
I have seen it over and over, the same sea, the same,
slightly, indifferently swinging above the stones,
icily free above the stones,
above the stones and then the world.   70
If you should dip your hand in,
your wrist would ache immediately,
your bones would begin to ache and your hand would burn
as if the water were a transmutation of fire
that feeds on stones and burns with a dark gray flame.   75
If you tasted it, it would first taste bitter,
then briny, then surely burn your tongue.
It is like what we imagine knowledge to be:
dark, salt, clear, moving, utterly free,
drawn from the cold hard mouth   80
of the world, derived from the rocky breasts
forever, flowing and drawn, and since
our knowledge is historical, flowing, and flown.

                                           1955

The Armadillo

FOR ROBERT LOWELL

This is the time of year
when almost every night
the frail, illegal fire balloons appear.
Climbing the mountain height,

rising toward a saint   5
still honored in these parts,
the paper chambers flush and fill with light
that comes and goes, like hearts.

Once up against the sky it's hard
to tell them from the stars—   10
planets, that is—the tinted ones:
Venus going down, or Mars,

or the pale green one. With a wind,
they flare and falter, wobble and toss;
but if it's still they steer between   15
the kite sticks of the Southern Cross,

receding, dwindling, solemnly
and steadily forsaking us,
or, in the downdraft from a peak,
suddenly turning dangerous.   20

Last night another big one fell.
It splattered like an egg of fire
against the cliff behind the house.
The flame ran down. We saw the pair

of owls who nest there flying up   25
and up, their whirling black-and-white
stained bright pink underneath, until
they shrieked up out of sight.

The ancient owls' nest must have burned.
Hastily, all alone,   30
a glistening armadillo left the scene,
rose-flecked, head down, tail down,

and then a baby rabbit jumped out,
short-eared, to our surprise.
So soft!—a handful of intangible ash   35
with fixed, ignited eyes.

Too pretty, dreamlike mimicry!
O falling fire and piercing cry
and panic, and a weak mailed fist[1]
clenched ignorant against the sky!   40

                                                1965

In the Waiting Room

In Worcester, Massachusetts,
I went with Aunt Consuelo
to keep her dentist's appointment
and sat and waited for her
in the dentist's waiting room.   5
It was winter. It got dark
early. The waiting room
was full of grown-up people,
arctics and overcoats,
lamps and magazines.   10
My aunt was inside
what seemed like a long time
and while I waited I read
the *National Geographic*
(I could read) and carefully   15
studied the photographs:
the inside of a volcano,
black, and full of ashes;
then it was spilling over
in rivulets of fire.   20

---

1. The armadillo, curled tight. It is protected against everything but fire.

Osa and Martin Johnson[1]
dressed in riding breeches,
laced boots, and pith helmets.
A dead man slung on a pole
—"Long Pig,"[2] the caption said.　　　　　25
Babies with pointed heads
wound round and round with string;
black, naked women with necks
wound round and round with wire
like the necks of light bulbs.　　　　　30
Their breasts were horrifying.
I read it right straight through.
I was too shy to stop.
And then I looked at the cover:
the yellow margins, the date.　　　　　35

Suddenly, from inside,
came an *oh!* of pain
—Aunt Consuelo's voice—
not very loud or long.
I wasn't at all surprised;　　　　　40
even then I knew she was
a foolish, timid woman.
I might have been embarrassed,
but wasn't. What took me
completely by surprise　　　　　45
was that it was *me:*
my voice, in my mouth.
Without thinking at all
I was my foolish aunt,
I—we—were falling, falling,　　　　　50
our eyes glued to the cover
of the *National Geographic,*
February, 1918.
I said to myself: three days
and you'll be seven years old.　　　　　55
I was saying it to stop
the sensation of falling off
the round, turning world
into cold, blue-black space.
But I felt: you are an *I,*　　　　　60
you are an *Elizabeth,*
you are one of *them.*
Why should you be one, too?
I scarcely dared to look
to see what it was I was.　　　　　65
I gave a sidelong glance
—I couldn't look any higher—

1. Famous explorers and travel writers.
2. Polynesian cannibals' name for the human carcass.

at shadowy gray knees,
trousers and skirts and boots
and different pairs of hands   70
lying under the lamps.
I knew that nothing stranger
had ever happened, that nothing
stranger could ever happen.
Why should I be my aunt,   75
or me, or anyone?
What similarities—
boots, hands, the family voice
I felt in my throat, or even
the *National Geographic*   80
and those awful hanging breasts—
held us all together
or made us all just one?
How—I didn't know any
word for it—how "unlikely" . . .   85
How had I come to be here,
like them, and overhear
a cry of pain that could have
got loud and worse but hadn't?

The waiting room was bright   90
and too hot. It was sliding
beneath a big black wave,
another, and another.

Then I was back in it.
The War[3] was on. Outside,   95
in Worcester, Massachusetts,
were night and slush and cold,
and it was still the fifth
of February, 1918.

                                                  1976

One Art

The art of losing isn't hard to master;
so many things seem filled with the intent
to be lost that their loss is no disaster.

Lose something every day. Accept the fluster
of lost door keys, the hour badly spent.   5
The art of losing isn't hard to master.

Then practice losing farther, losing faster:
places, and names, and where it was you meant
to travel. None of these will bring disaster.

---

3. World War I.

I lost my mother's watch. And look! my last, or   10
next-to-last, of three loved houses went.
The art of losing isn't hard to master.

I lost two cities, lovely ones. And, vaster,
some realms I owned, two rivers, a continent.
I miss them, but it wasn't a disaster.   15

—Even losing you (the joking voice, a gesture
I love) I shan't have lied. It's evident
the art of losing's not too hard to master
though it may look like (*Write* it!) like disaster.

                                                                   1976

RANDALL JARRELL
1914–1965

"The gods who had taken away the poet's audience had given him students," said Randall Jarrell. He was describing the state of American poetry after World War II, and from any other poet it would have been an entirely bitter statement. In fact, among the American poets and critics emerging after 1945, Jarrell enjoyed being *the* teacher, both in and out of the classroom. The novelist Peter Taylor recalls that when he came to Vanderbilt University as a freshman in the mid-1930s, Jarrell, then a graduate student, had already turned the literary students into disciples; he held court even on the sidelines of touch football games, where Taylor heard him deliver brilliant analyses of Chekhov short stories. Later in life, he was to give unrelenting attention, as poet and friend, to works-in-progress by his contemporaries—such writers as Robert Lowell, Delmore Schwartz, and John Berryman.

Jarrell was born in Nashville, Tennessee, but spent much of his childhood in Long Beach, California. When his parents divorced, the child, then eleven, remained for a year with his grandparents in Hollywood, then returned to live with his mother in Nashville. He majored in psychology at Vanderbilt and stayed on there to do graduate work in English. In 1937 he left Nashville to teach at Kenyon College (Gambier, Ohio) on the invitation of his old Vanderbilt professor, John Crowe Ransom, the New Critic and Fugitive poet. From that time on, Jarrell almost always had some connection with a university: after Kenyon, the University of Texas, Sarah Lawrence College, and from 1947 until his death, the Women's College of the University of North Carolina at Greensboro. But, as his novel *Pictures from an Institution* (1954) with its mixed satiric and tender views of academic life suggests, he was never satisfied with a cloistered education. His literary criticism and teaching were aimed at recapturing and re-educating an audience either lost to poetry or upon whom poetry was lost. As poetry editor of *The Nation* (1946), and then in a series of essays and reviews collected as *Poetry and the Age* (1953), he introduced readers to the work of their contemporaries—Elizabeth Bishop, Robert Lowell, John Berryman, the William Carlos Williams of *Paterson*—and influentially reassessed the reputations of Whitman and Robert Frost.

Jarrell was not only the ablest critic of verse but also one of the best poets to emerge after World War II. While others—Richard Wilbur and Robert Lowell among them—were writing highly structured poems with complicated imagery, Jarrell was writing with a rare colloquial plainness. Many of his poems are dramatic monologues whose speakers express disappointment with the "dailiness of life." These figures of loneliness and aging are to be found especially in the collection *The Woman at the Washington Zoo* (1960), the heroine of whose title poem sees her frustrations mirrored in the caged animals around her. Jarrell often adopts the points of view of women and children or of the soldiers in his extraordinary war poems, the strongest to come out of the 1941–45 conflict. He had been trained as an Army Air Force pilot, and after that as a control tower operator, hence his sense of the war's special casualties—what one critic calls "the soldier-technician * * * the clumsy mechanical animals and their child-guardians," so often portrayed in Jarrell's volumes *Little Friend, Little Friend* (1945) and *Losses* (1948). In *The Death of the Ball Turret Gunner,* nothing seems to have happened to the soldier-protagonist between conception and the moment he is shot down, which is presented as if it were an abortion.

Against the blasted or unrealized possibilities of adult life Jarrell often poised the rich mysteries of childhood. Many of the poems, like *Cinderella,* reinterpret fairy tales. Jarrell himself translated Grimms' fairy tales and wrote several children's books, among them *The Bat-Poet.* The title poem of his last book, *The Lost World* (1965), looks back to his Los Angeles playtime, the movie sets and plaster dinosaurs and pterodactyls against whose eternal gay presence he measures his own aging: "I hold in my own hands in happiness, / Nothing: the nothing for which there's no reward."

Jarrell suffered a nervous breakdown in February, 1965, but returned to teaching that fall. In October he was struck down by a car and died. His *Complete Poems* were published posthumously (1969) as was a translation of Goethe's *Faust,* Part I, in preparation at his death.

The Death of the Ball Turret Gunner[1]

From my mother's sleep I fell into the State,
And I hunched in its belly till my wet fur froze.
Six miles from earth, loosed from its dream of life,
I woke to black flak[2] and the nightmare fighters.
When I died they washed me out of the turret with a hose.   5

                                                      1945

1. "A ball turret was a plexiglass sphere set into the belly of a B-17 or B-24 [bomber], and inhabited by two .50 caliber machine-guns and one man, a short, small man. When this gunner tracked with his machine-guns a fighter attacking his bomber from below, he revolved with the turret; hunched upside-down in his little sphere, he looked like the foetus in the womb. The fighters which attacked him were armed with cannon firing explosive shells. The hose was a steam hose" [Jarrell's note].

2. Antiaircraft fire.

Next Day

Moving from Cheer to Joy, from Joy to All,
I take a box
And add it to my wild rice, my Cornish game hens.
The slacked or shorted, basketed, identical
Food-gathering flocks
Are selves I overlook. Wisdom, said William James,[1]   5

Is learning what to overlook. And I am wise
If that is wisdom.
Yet somehow, as I buy All from these shelves
And the boy takes it to my station wagon,   10
What I've become
Troubles me even if I shut my eyes.

When I was young and miserable and pretty
And poor, I'd wish
What all girls wish: to have a husband,   15
A house and children. Now that I'm old, my wish
Is womanish:
That the boy putting groceries in my car

See me. It bewilders me he doesn't see me.
For so many years   20
I was good enough to eat: the world looked at me
And its mouth watered. How often they have undressed me,
The eyes of strangers!
And, holding their flesh within my flesh, their vile

Imaginings within my imagining,   25
I too have taken
The chance of life. Now the boy pats my dog
And we start home. Now I am good.
The last mistaken,
Ecstatic, accidental bliss, the blind   30

Happiness that, bursting, leaves upon the palm
Some soap and water—
It was so long ago, back in some Gay
Twenties, Nineties, I don't know . . . Today I miss
My lovely daughter   35
Away at school, my sons away at school,

My husband away at work—I wish for them.
The dog, the maid,
And I go through the sure unvarying days

1. From *Principles of Psychology*, by the American philosopher William James
(1842–1910).

At home in them. As I look at my life,   40
I am afraid
Only that it will change, as I am changing:

I am afraid, this morning, of my face.
It looks at me
From the rear-view mirror, with the eyes I hate,   45
The smile I hate. Its plain, lined look
Of gray discovery
Repeats to me: "You're old." That's all, I'm old.

And yet I'm afraid, as I was at the funeral
I went to yesterday.   50
My friend's cold made-up face, granite among its flowers,
Her undressed, operated-on, dressed body
Were my face and body.
As I think of her I hear her telling me

How young I seem; I *am* exceptional;   55
I think of all I have.
But really no one is exceptional,
No one has anything, I'm anybody,
I stand beside my grave
Confused with my life, that is commonplace and solitary.   60

1965

Well Water

What a girl called "the dailiness of life"
(Adding an errand to your errand. Saying,
"Since you're up . . ." Making you a means to
A means to a means to) is well water.
Pumped from an old well at the bottom of the world.   5
The pump you pump the water from is rusty
And hard to move and absurd, a squirrel-wheel
A sick squirrel turns slowly, through the sunny
Inexorable hours. And yet sometimes
The wheel turns of its own weight, the rusty   10
Pump pumps over your sweating face the clear
Water, cold, so cold! you cup your hands
And gulp from them the dailiness of life.

1965

The Player Piano[1]

I ate pancakes one night in a Pancake House
Run by a lady my age. She was gay.
When I told her that I came from Pasadena
She laughed and said, "I lived in Pasadena
When Fatty Arbuckle[2] drove the El Molino bus."   5

1. A piano which, when fitted with perforated rolls, plays automatically.
2. A corpulent actor of the silent film
era, whose career was destroyed when
he was involved in a sex-murder scandal.

I felt that I had met someone from home.
No, not Pasadena, Fatty Arbuckle.
Who's that? Oh, something that we had in common
Like—like—the false armistice.[3] Piano rolls.
She told me her house was the first Pancake House   10

East of the Mississippi, and I showed her
A picture of my grandson. Going home—
Home to the hotel—I began to hum,
"Smile a while, I bid you sad adieu,
When the clouds roll back I'll come to you."[4]   15

Let's brush our hair before we go to bed,
I say to the old friend who lives in my mirror.
I remember how I'd brush my mother's hair
Before she bobbed[5] it. How long has it been
Since I hit my funnybone? had a scab on my knee?   20

Here are Mother and Father in a photograph,
Father's holding me. . . . They both look so *young*.
I'm so much older than they are. Look at them,
Two babies with their baby. I don't blame you,
You weren't old enough to know any better;   25

If I could I'd go back, sit down by you both,
And sign our true armistice: you weren't to blame.
I shut my eyes and there's our living room.
The piano's playing something by Chopin,
And Mother and Father and their little girl   30

Listen. Look, the keys go down by themselves!
I go over, hold my hands out, play I play—
If only, somehow, I had learned to live!
The three of us sit watching, as my waltz
Plays itself out a half-inch from my fingers.   35

1969

3. Truce.   5. Popular short hairstyle of the period.
4. Lines from a popular song of the 1920s.

JOHN BERRYMAN
1914–1972

From a generation whose ideal poem was short, self-contained, and ironic,
John Berryman emerged as the author of two extended and passionate
works: *Homage to Mistress Bradstreet* and the lyric sequence called *The
Dream Songs*. It was as if Berryman needed more space than the single
lyric provided—a larger theater to play out an unrelenting psychic drama.
He had written shorter poems—songs and sonnets—but it was his dis-

covery of large-scale dramatic situations and strange new voices that astonished his contemporaries.

Berryman seemed fated to intense suffering and self-preoccupation. His father, a banker, shot himself outside his son's window when the boy was twelve. The suicide haunted Berryman to the end of his own life, which also came by suicide. He turns to the subject obsessively in *The Dream Songs.*

> I spit upon this dreadful banker's grave
> who shot his heart out in a Florida dawn
> O ho alas alas
> When will indifference come, I moan and rave

Berryman, who was born John Smith, took a new name from his stepfather, also a banker. His childhood was a series of displacements: ten years near McAlester, Oklahoma, then Tampa, Florida, and, after his father's suicide, Gloucester, Massachusetts, and New York City. His mother's second marriage ended in divorce, but his stepfather sent him to private school in Connecticut. Berryman graduated from Columbia College in 1936 and won a fellowship to Clare College, Cambridge, England.

He was later to say of himself, "I masquerade as a writer. Actually I am a scholar." However misleading this may be about his poetry, it reminds us that all his life Berryman drew nourishment from teaching—at Wayne State, at Harvard (1940–43), then off and on at Princeton, and, from 1955 until his death, at the University of Minnesota. He chose to teach, not creative writing, but literature and the "history of civilization," and claimed that such teaching forced him into areas in which he wouldn't otherwise have done detailed work. A mixture of bookishness and wildness characterizes all his writing: five years of research lay behind the intensities of *Homage to Mistress Bradstreet,* while an important constituent of "huffy Henry's" personality in *The Dream Songs* is his professorial awkwardness and exhibitionism.

Berryman seemed drawn to borrowing identities in his poetry. In his first important volume, *The Dispossessed* (1948), he had experimented with various dramatic voices in short *Nervous Songs: The Song of the Demented Priest, A Professor's Song, The Song of the Tortured Girl, The Song of the Man Forsaken and Obsessed.* The "dispossession" of the book's title had two opposite and urgent meanings for him: "the miserable, *put out of one's own,* and the relieved, saved, undevilled, de-spelled." Taking on such roles was for Berryman both a revelation of his cast-out, fatherless state, and an exorcism of it. It was perhaps in that spirit that he entered into an imaginary dialogue with what he felt as the kindred nature of the Puritan poet Anne Bradstreet. "Both of our worlds unhanded us." What started out to be a poem of fifty lines emerged as the fifty-seven stanzas of *Homage to Mistress Bradstreet* (1956), a work so absorbing that after completing it Berryman claimed to be "a ruin for two years." It was not Bradstreet's poetry which engaged him. Quite the contrary: he was fascinated by the contrast between her "bald abstract rime" and her life of passionate suffering. The poem explores the kinship between Bradstreet and Berryman as figures of turbulence and rebellion.

Berryman took literary encouragement from another American poet of the past, Stephen Crane, about whom he wrote a book-length critical study in 1950. Crane's poems, he said, have "the character of a 'dream,' something seen naively in a new relation. * * * His poetry has the inimitable sincerity of a frightened savage anxious to learn what his dream means." Berryman's attraction to a poetry which accommodated the nightmare antics of the dream world became apparent in his own long work, *The Dream Songs*. It was modeled, he claimed, on "the greatest American poem," Whitman's *Song of Myself*, a less tortured long poem but one in which the speaker assumes, as Berryman's protagonist does, a number of roles and a fluid, ever-changing persona. 77 *Dream Songs* was published in 1964. Additional poems, to a total of 385, appeared in *His Toy, His Dream, His Rest* (1968). (Some uncollected "Dream Songs" were published posthumously in *Henry's Fate*, 1977, and drafts of others remain in manuscript.) There are obvious links between Berryman and other so-called confessional writers such as Robert Lowell, Sylvia Plath, and Anne Sexton. But the special autobiographical flavor of *The Dream Songs* is that of a psychic vaudeville; as in dreams, the poet represents himself through a fluid series of *alter egos*, whose voices often flow into one another in single poems. Despite the suffering which these poems enact, Berryman seemed to find a secret strength through the staginess, variety, resourcefulness, and renewals of these almost 400 poems.

The Dream Songs brought Berryman a success which was not entirely beneficial. The collection *Love and Fame* (1970) shows him beguiled by his own celebrity and wrestling with some of its temptations. In an unfinished, posthumously published novel, *Recovery*, he portrays himself as increasingly prey to alcoholism. Berryman had been married twice before and his hospitalization for drinking and for periods of insanity had put a strain on his third marriage. He came to distrust his poetry as a form of exhibitionism and was clearly, in his use of the discipline of prose and in the prayers which crowd his last two volumes of poetry (*Delusions, Etc.* appeared posthumously) in search of some new and humbling style. Having been raised a strict Catholic and fallen away from the church, he tried to return to it in his last years, speaking of his need for a "God of rescue." On January 7, 1972, Berryman committed suicide by leaping from a Minneapolis bridge onto the frozen Mississippi River.

From DREAM SONGS[1]

I

Huffy Henry hid   the day,
unappeasable Henry sulked.
I see his point,—a trying to put things over.

---

1. These poems were written over a period of 13 years. (77 *Dream Songs* was published in 1964, and the remaining poems appeared in *His Toy, His Dream, His Rest* in 1968. Some uncollected Dream Songs were included in the volume *Henry's Fate*, which appeared five years after Berryman committed suicide in 1972.) Berryman placed an introductory note at the head of *His Toy, His Dream, His Rest*: "The poem then, whatever its wide cast of characters, is essentially about an imaginary character (not the poet, not me) named Henry, a white American in early middle age sometimes in blackface, who has suffered an irreversible loss and talks about himself sometimes in the first person, sometimes in the third, sometimes even in the second; he has a friend, never named, who addresses him as Mr. Bones and variants thereof. Requiescant in pace."

It was the thought that they thought
they could *do* it made Henry wicked & away.   5
But he should have come out and talked.

All the world like a woolen lover
once did seem on Henry's side.
Then came a departure.
Thereafter nothing fell out as it might or ought.   10
I don't see how Henry, pried
open for all the world to see, survived.

What he has now to say is a long
wonder the world can bear & be.
Once in a sycamore I was glad   15
all at the top, and I sang.
Hard on the land wears the strong sea
and empty grows every bed.

14

Life, friends, is boring. We must not say so.
After all, the sky flashes, the great sea yearns,
we ourselves flash and yearn,
and moreover my mother told me as a boy
(repeatingly) 'Ever to confess you're bored   5
means you have no

Inner Resources.' I conclude now I have no
inner resources, because I am heavy bored.
Peoples bore me,
literature bores me, especially great literature,   10
Henry bores me, with his plights & gripes
as bad as achilles,[1]

who loves people and valiant art, which bores me.
And the tranquil hills, & gin, look like a drag
and somehow a dog   15
has taken itself & its tail considerably away
into mountains or sea or sky, leaving
behind: me, wag.

29

There sat down, once, a thing on Henry's heart
so heavy, if he had a hundred years
& more, & weeping, sleepless, in all them time
Henry could not make good.
Starts again always in Henry's ears   5
the little cough somewhere, an odour, a chime.

1. Greek hero of Homer's *The Iliad*, who, angry at slights against his honor, sulked in his tent and refused to fight against the Trojans.

And there is another thing he has in mind
like a grave Sienese face[2] a thousand years
would fail to blur the still profiled reproach of. Ghastly,
with open eyes, he attends, blind.   10
All the bells say: too late. This is not for tears;
thinking.

But never did Henry, as he thought he did,
end anyone and hacks her body up
and hide the pieces, where they may be found.   15
He knows: he went over everyone, & nobody's missing.
Often he reckons, in the dawn, them up.
Nobody is ever missing.

40

I'm scared a lonely. Never see my son,
easy be not to see anyone,
combers[1] out to sea
know they're goin somewhere but not me.
Got a little poison, got a little gun,   5
I'm scared a lonely.

I'm scared a only one thing, which is me,
from othering I don't take nothin, see,
for any hound dog's sake.
But this is where I livin, where I rake   10
my leaves and cop my promise,[2] this' where we
cry oursel's awake.

Wishin was dyin but I gotta make
it all this way to that bed on these feet
where peoples said to meet.   15
Maybe but even if I see my son
forever never, get back on the take,
free, black & forty-one.[3]

45

He stared at ruin. Ruin stared straight back.
He thought they was old friends. He felt on the stair
where her papa found them bare
they became familiar. When the papers were lost
rich with pals' secrets, he thought he had the knack   5
of ruin. Their paths crossed

2. Alluding to the somber, austere mo-
saic-like religious portraits by the Italian
painters who worked in Siena during the
13th and 14th centuries.
1. Waves which roll over and break with
a foamy crest.
2. Slang for "pile up potential."
3. Playing on "free, white, and 21," colloquial for legally independent.

and once they crossed in jail; they crossed in bed;
and over an unsigned letter their eyes met,
and in an Asian city
directionless & lurchy at two & three,   10
or trembling to a telephone's fresh threat,
and when some wired his head

to reach a wrong opinion, 'Epileptic'.
But he noted now that they were not old friends.
He did not know this one.   15
This one was a stranger, come to make amends
for all the imposters, and to make it stick.
Henry nodded, un-.

76. Henry's Confession

Nothin very bad happen to me lately.
How you explain that? —I explain that, Mr Bones,
terms o' your bafflin odd sobriety,
Sober as man can get, no girls, no telephones,
what could happen bad to Mr Bones?   5
—*If* life is a handkerchief sandwich,

in a modesty of death I join my father
who dared so long agone leave me.
A bullet on a concrete stoop
close by a smothering southern sea   10
spreadeagled on an island, by my knee.
—You is from hunger, Mr Bones,

I offers you this handkerchief, now set
your left foot by my right foot,
shoulder to shoulder, all that jazz,   15
arm in arm, by the beautiful sea,[1]
hum a little, Mr Bones.
—I saw nobody coming, so I went instead.

143

—That's enough of that, Mr Bones. *Some* lady you make.
Honour the burnt cork,[2] be a vaudeville man,
I'll sing you now a song
the like of which may bring your heart to break:
he's gone! and we don't know where. When he began   5
taking the pistol out & along,

you was just a little; but gross fears
accompanied us along the beaches, pal.

---

1. **Phrase from a popular song.**   2. **Used by actors to blacken the face.**

My mother was scared almost to death.
He was going to swim out, with me, forevers,   10
and a swimmer strong he was in the phosphorescent Gulf,
but he decided on lead.

That mad drive wiped out my childhood. I put him down
while all the same on forty years I love him
stashed in Oklahoma   15
besides his brother Will. Bite the nerve of the town
for anyone so desperate. I repeat: I love him
until I fall into coma.

384

The marker slants, flowerless, day's almost done,
I stand above my father's grave with rage,
often, often before
I've made this awful pilgrimage to one
who cannot visit me, who tore his page   5
out: I come back for more,

I spit upon this dreadful banker's grave
who shot his heart out in a Florida dawn
O ho alas alas
When will indifference come, I moan & rave   10
I'd like to scrabble till I got right down
away down under the grass

and ax the casket open ha to see
just how he's taking it, which he sought so hard
we'll tear apart   15
the mouldering grave clothes ha & then Henry
will heft the ax once more, his final card,
and fell it on the start.

385

My daughter's heavier. Light leaves are flying.
Everywhere in enormous numbers turkeys will be dying
and other birds, all their wings.
They never greatly flew. Did they wish to?
I should know. Off away somewhere once I knew   5
such things.

Or good Ralph Hodgson[1] back then did, or does.
The man is dead whom Eliot[2] praised. My praise
follows and flows too late.

1. Ralph Hodgson (1871–1962), an English poet who wrote balladlike lyrics.
Berryman may be alluding to Hodgson's
Hymn to Moloch, in which the poet protests the slaughter of birds for commercial uses.
2. T. S. Eliot (1888–1965), American poet
and critic.

Fall is grievy, brisk. Tears behind the eyes 10
almost fall. Fall comes to us as a prize
to rouse us toward our fate.

My house is made of wood and it's made well,
unlike us. My house is older than Henry;
that's fairly old. 15
If there were a middle ground between things and the soul
or if the sky resembled more the sea,
I wouldn't have to scold

my heavy daughter.
1968

ROBERT LOWELL

1917–1977

In 1964 a literary critic called Robert Lowell "our truest historian." Lowell
had earned that emblematic position not simply because his poetry was
often concerned with historical and political issues but because the poet
himself was a figure of historical interest. Descendant of a long line of
New Englanders, he bore powerful witness to their traditions, memories,
eccentricities, and neuroses. He was born into a secure world, but his
poetry records instability and change: personal rebellion, mental illness,
and political stress. Writing under those pressures, he helped give American poetry a new autobiographical cast, a new urgency and directness of
speech, and a new openness to public subjects.

Lowell came from patrician families closely associated with Boston. His
grandfather, another R. T. S. Lowell, was a well-known Episcopal minister
and head of the fashionable St. Mark's School which the poet was later to
attend. His great-granduncle, James Russell Lowell, had been a poet and
ambassador to England. Another relation, A. Lawrence, was president of
Harvard just before the young poet came there as a freshman. Perhaps the
only light note in the family was struck by the poet Amy Lowell, "big
and a scandal, as if Mae West were a cousin." Robert Lowell's father was
a naval officer who, after his retirement, did badly in business. Much of
the poet's life was a reaction against his background of Bostonian eminence and public service. At first he turned angrily against it, but later in
life he tried to draw upon its strength without being crippled by it.

Lowell's first act of separation was to leave the East after two years at
Harvard (1935–37) in order to study at Kenyon College with John Crowe
Ransom, the poet and critic. The move brought him in closer touch with
the New Criticism and its predilections for "formal difficult poems," the
wit and irony of English metaphysical writers such as John Donne. He
also, through Ransom and the poet Allen Tate, came into contact with
(though never formally joined) the Fugitive movement, whose members
were southern agrarians opposed to what they regarded as the corrupting
values of northern industrialism.

Even more decisive was his conversion in 1940 to Roman Catholicism and his resistance to American policies in World War II. Although he did try to enlist in the navy, he refused to be drafted into the army. He opposed the saturation bombing of Hamburg and the Allied policy of unconditional surrender and was as a result sentenced to a year's confinement in New York's West Street jail and admonished by the presiding judge for "marring" his family traditions. In his first book, *Lord Weary's Castle* (1946), Lowell's Catholicism and antagonism to Protestant mercantile Boston were most sharply expressed. Catholicism, he was later to say, gave him not so much a subject as "some kind of form, and I could begin a poem and build it to a climax." It provided a set of ritual symbols and a distanced platform from which to attack "our sublunary secular sprawl," the commercial-minded Boston of *Children of Light*. The climaxes to which Lowell refers are the apocalyptic conclusions of these early poems—devastating sentences such as: "The Lord survives the rainbow of His will"; "The blue kingfisher dives on you in fire."

Alongside those poems drawing upon Old Testament anger, there was another strain in *Lord Weary's Castle*. In poems such as *After the Surprising Conversions* and *Jonathan Edwards and the Spider*, he explored from within the nervous intensity which underlay Puritan revivalism. Similarly in later dramatic narratives with modern settings such as *The Mills of the Kavanaghs* and *Falling Asleep over the Aeneid* he exposed his morbid fascination with and psychological interest in ruined New England families.

In *Life Studies* (1959) his subjects became explicitly autobiographical. In 1957 he had been giving poetry readings on the West Coast. Allen Ginsberg and the other Beats had just made their strongest impact in San Francisco, and Lowell felt that by contrast to their candid, breezy writing, his own style seemed "distant, symbol-ridden, and willfully difficult. * * * I felt my old poems hid what they were really about, and many times offered a stiff, humorless and even impenetrable surface." His own new style was to be more controlled and severe than "Beat" writing, but he was stimulated by Ginsberg's self-revelations to write more openly than he had about his parents and grandparents, about the mental breakdowns he suffered in the 1950s, and about the difficulties of marriage. (Lowell had divorced his first wife, the novelist Jean Stafford, and had married the critic Elizabeth Hardwick in 1949.)

Life Studies opens with a poem, *Beyond the Alps*, which recalls Lowell's departure from the Catholic Church (1950). The book touches on his prison experience during World War II. But by and large it records his ambivalence toward the New England where he resettled after the war, on Boston's "hardly passionate Marlborough Street." He no longer denounces the city of his fathers as if he were a privileged outsider. In complicated psychological portraits of his childhood, his relation to his parents and his wives, he assumes a portion of the weakness and vulnerability for himself.

In 1960 Lowell left Boston to live in New York City. *For the Union Dead* (1964), the book that followed, continued the autobiographical vein of *Life Studies*. Lowell called it a book about "witheredness * * * lemony, soured and dry, the drouth I had touched with my own hands." These poems seem more carefully controlled than his earlier "life studies." Poems like *Neo-Classical Urn* and *For the Union Dead* consist much less in the miscellaneous memories of dreams and jagged experience; instead they present data as a skilled psychoanalyst might organize it, finding key images from the past which explain the present. The book includes a number of political or public poems, such as *July in Washington*, which uses Lowell's scarred personal past and his sense of the nourishing traditions of the Republic as a way of criticizing the bland or evasive politics of the Eisenhower years.

Lowell's *History* (1973) is again a new departure. He had begun these poems as a kind of poetic journal and had originally published them in two editions of *Notebooks*. These unrhymed fourteen-line poems were reactions to his reading, his personal life, and the Vietnam war, of which he was an outspoken critic. "Things I felt or saw, or read were drift in the whirlpool." The poems were more committed to the present than to the recollections which had been the subjects of *Life Studies* and *For the Union Dead*: "always the instant, something changing to the lost. A flash of haiku to lighten the distant." In 1973, in a characteristic act of self-assessment, Lowell rearranged and revised these poems which had begun as a diary. Those dealing with public subjects, present and past, were published, along with some new poems, as *History*. The more personal poems, recording the breakup of his second marriage and separation from his wife and daughter, were published under the title *For Lizzie and Harriet*. At the same time a new collection of sonnets, *The Dolphin*, records his remarriage to Lady Caroline Blackwood. He divided his time between her home in England and periods of teaching writing and literature at Harvard—a familiar pattern for him in which the old tensions between New England and "elsewhere" were being constantly explored and renewed. His last book, *Day by Day* (1977), records those stresses as well as new marital difficulties. It also contains some of his most powerful poems about his childhood.

Lowell had been active in the theater with a version of *Prometheus Bound* and a translation of Racine's *Phaedra*, as well as adaptations of Melville and Hawthorne stories gathered as *The Old Glory*. He translated from modern European poetry and the classics, often freely as "imitations" which brought important poetic voices into English currency. His *Selected Poems* (his own choices) appeared in 1976. When he died suddenly at the age of sixty, he was the dominant and most honored poet of his generation—not only for his ten volumes of verse but for his broad activity as a man of letters. He stood out as one who kept alive—sometimes at great psychological cost—the notion of the poet's public responsibility. He was with the group of writers who led Vietnam war protesters against the Pentagon in 1967, where Norman Mailer, a fellow protester, observed that "Lowell gave off at times the unwilling haunted saintliness of a man who was repaying the moral debts of ten generations of ancestors."

Children of Light[1]

Our fathers wrung their bread from stocks and stones
And fenced their gardens with the Redman's bones;
Embarking from the Nether Land of Holland,
Pilgrims unhouseled[2] by Geneva's night,
They planted here the Serpent's seeds[3] of light;   5
And here the pivoting searchlights probe to shock
The riotous glass houses[4] built on rock,
And candles gutter by an empty altar,
And light is where the landless blood of Cain[5]
Is burning, burning the unburied grain.   10

1946

After the Surprising Conversions[1]

September twenty-second, Sir: today
I answer. In the latter part of May,
Hard on our Lord's Ascension,[2] it began
To be more sensible.[3] A gentleman
Of more than common understanding, strict   5
In morals, pious in behavior, kicked
Against our goad. A man of some renown,
An useful, honored person in the town,
He came of melancholy parents; prone
To secret spells, for years they kept alone—   10
His uncle, I believe, was killed of it:
Good people, but of too much or little wit.
I preached one Sabbath on a text from Kings;
He showed concernment for his soul. Some things
In his experience were hopeful. He   15
Would sit and watch the wind knocking a tree
And praise this countryside our Lord has made.
Once when a poor man's heifer died, he laid
A shilling on the doorsill; though a thirst
For loving shook him like a snake, he durst   20

1. In Luke 16, the children of this world are distinguished from the children of light, the Lord's chosen.
2. I.e., not having received Holy Communion. John Calvin's theology was strongly established in Geneva. He denied that in Holy Communion the Eucharistic bread and wine actually became the body and blood of Christ.
3. Alluding to the channeling of Protestant energies into business. The serpent, associated with Satan the tempter or Eve, seems to pervert the "seeds," the elect or "children of light."
4. Modern houses by the shore.
5. Likening the modern descendants of the Puritans to the race of Cain, presumably because, in dispossessing the Indians, they had committed a crime like Cain's against his brother.

1. This poem draws upon letters and accounts by the Puritan preacher and theologian Jonathan Edwards (1703–58). Edwards describes the effects of an amazing religious revival, the "surprising conversions" of 1734–35. On June 1, 1735, his uncle, Joseph Hawley, slit his own throat; the suicide channeled and countered the effects of the revival in ways that Lowell describes.
2. Christ's ascension into Heaven, celebrated 40 days after Easter.
3. I.e., apparent to the senses. In a letter of May 30, 1735, Edwards wrote, "It began to be very sensible that the spirit of God was gradually withdrawing from us."

Not entertain much hope of his estate
In heaven. Once we saw him sitting late
Behind his attic window by a light
That guttered on his Bible; through that night
He meditated terror, and he seemed   25
Beyond advice or reason, for he dreamed
That he was called to trumpet Judgment Day
To Concord. In the latter part of May
He cut his throat. And though the coroner
Judged him delirious, soon a noisome stir   30
Palsied our village. At Jehovah's nod
Satan seemed more let loose amongst us: God
Abandoned us to Satan, and he pressed
Us hard, until we thought we could not rest
Till we had done with life. Content was gone.   35
All the good work was quashed. We were undone.
The breath of God had carried out a planned
And sensible withdrawal from this land;
The multitude, once unconcerned with doubt,
Once neither callous, curious nor devout,   40
Jumped at broad noon, as though some peddler groaned
At it in its familiar twang: "My friend,
Cut your own throat. Cut your own throat. Now! Now!"
September twenty-second, Sir, the bough
Cracks with the unpicked apples, and at dawn   45
The small-mouth bass breaks water, gorged with spawn.

                                                      1946

Memories of West Street and Lepke[1]

Only teaching on Tuesdays, book-worming
in pajamas fresh from the washer each morning,
I hog a whole house on Boston's
"hardly passionate Marlborough Street,"[2]
where even the man   5
scavenging filth in the back alley trash cans,
has two children, a beach wagon, a helpmate,
and is a "young Republican."
I have a nine months' daughter,
young enough to be my granddaughter.   10
Like the sun she rises in her flame-flamingo infants' wear.

These are the tranquillized *Fifties*,
and I am forty. Ought I to regret my seedtime?
I was a fire-breathing Catholic C.O.,[3]
and made my manic statement,   15

---

1. In 1943 Lowell was sentenced to a year in New York's West Street jail for his refusal to serve in the armed forces. Among the prisoners was Lepke Buchalter, head of Murder Incorporated, an organized crime syndicate, who had been convicted of murder.
2. William James's phrase for a street in the elegant Back Bay section of Boston, where Lowell lived in the 1950s.
3. Conscientious objector (to war).

telling off the state and president, and then
sat waiting sentence in the bull pen
beside a Negro boy with curlicues
of marijuana in his hair.

Given a year,   20
I walked on the roof ot the West Street Jail, a short
enclosure like my school soccer court,
and saw the Hudson River once a day
through sooty clothesline entanglements
and bleaching khaki tenements.   25
Strolling, I yammered metaphysics with Abramowitz,
a jaundice-yellow ("it's really tan")
and fly-weight pacifist,
so vegetarian,
he wore rope shoes and preferred fallen fruit.   30
He tried to convert Bioff and Brown,
the Hollywood pimps, to his diet.
Hairy, muscular, suburban,
wearing chocolate double-breasted suits,
they blew their tops and beat him black and blue.   35

I was so out of things, I'd never heard
of the Jehovah's Witnesses.[4]
"Are you a C.O.?" I asked a fellow jailbird.
"No," he answered, "I'm a J.W."
He taught me the "hospital tuck,"[5]   40
and pointed out the T-shirted back
of *Murder Incorporated's* Czar Lepke,
there piling towels on a rack,
or dawdling off to his little segregated cell full
of things forbidden the common man: 45
a portable radio, a dresser, two toy American
flags tied together with a ribbon of Easter palm.
Flabby, bald, lobotomized,
he drifted in a sheepish calm,
where no agonizing reappraisal   50
jarred his concentration on the electric chair—
hanging like an oasis in his air
of lost connections. . . .

                                                                         1959

Man and Wife

Tamed by *Miltown*,[1] we lie on Mother's bed;
the rising sun in war paint dyes us red;
in broad daylight her gilded bed-posts shine,
abandoned, almost Dionysian.[2]

4. A Christian revivalist sect strongly
opposed to war and denying the power
of the state in matters of conscience.
5. The authorized, efficient way of making beds in a hospital.

1. Brand name of a barbiturate tran-
quilizer.
2. Orgiastic, as in the Greek celebration
honoring Dionysus, the god of wine.

At last the trees are green on Marlborough Street,[3]   5
blossoms on our magnolia ignite
the morning with their murderous five days' white.
All night I've held your hand,
as if you had
a fourth time faced the kingdom of the mad—   10
its hackneyed speech, its homicidal eye—
and dragged me home alive. . . . Oh my *Petite*,
clearest of all God's creatures, still all air and nerve:
you were in your twenties, and I,
once hand on glass   15
and heart in mouth,
outdrank the Rahvs[4] in the heat
of Greenwich Village, fainting at your feet—
too boiled and shy
and poker-faced to make a pass,   20
while the shrill verve
of your invective scorched the traditional South.

Now twelve years later, you turn your back.
Sleepless, you hold
your pillow to your hollows like a child;   25
your old-fashioned tirade—
loving, rapid, merciless—
breaks like the Atlantic Ocean on my head.

                                                        1959

Skunk Hour

[FOR ELIZABETH BISHOP][1]

Nautilus Island's[2] hermit
heiress still lives through winter in her Spartan cottage;
her sheep still graze above the sea.
Her son's a bishop. Her farmer
is first selectman in our village;   5
she's in her dotage.

Thirsting for
the hierarchic privacy
of Queen Victoria's century,
she buys up all   10
the eyesores facing her shore,
and lets them fall.

The season's ill—
we've lost our summer millionaire,

3. In Boston's once elegant Back Bay section.
4. Philip Rahv (1908–74), a literary critic and one of the founders of *Partisan Review*, and a friend from the period when Lowell was courting his second wife, Elizabeth Hardwick.

1. Lowell's poem is a response to Elizabeth Bishop's *The Armadillo*.
2. The poem is set in Castine, Maine, where Lowell had a summer house.

who seemed to leap from an L. L. Bean [15]
catalogue.[3] His nine-knot yawl
was auctioned off to lobstermen.
A red fox stain covers Blue Hill.

And now our fairy
decorator brightens his shop for fall; [20]
his fishnet's filled with orange cork,
orange, his cobbler's bench and awl;
there is no money in his work,
he'd rather marry.

One dark night, [25]
my Tudor Ford climbed the hill's skull;
I watched for love-cars. Lights turned down,
they lay together, hull to hull,
where the graveyard shelves on the town. . . .
My mind's not right. [30]

A car radio bleats,
"Love, O careless Love. . . ." I hear
my ill-spirit sob in each blood cell,
as if my hand were at its throat. . . .
I myself am hell;[4] [35]
nobody's here—

only skunks, that search
in the moonlight for a bite to eat.
They march on their soles up Main Street:
white stripes, moonstruck eyes' red fire [40]
under the chalk-dry and spar spire
of the Trinitarian Church.

I stand on top
of our back steps and breathe the rich air—
a mother skunk with her column of kittens swills the garbage pail. [45]
She jabs her wedge-head in a cup
of sour cream, drops her ostrich tail,
and will not scare.

1959

July in Washington

The stiff spokes of this wheel[1]
touch the sore spots of the earth.

---

3. A mail-order house in Maine, which deals primarily with sporting and camping goods.
4. "Which way I fly is Hell, myself am Hell" (Satan in Milton's *Paradise Lost*, IV.75).
1. I.e., Washington, D.C., which is laid out in a formal plan of concentric circles. Some of the intersecting circles contain equestrian statues (line 7).

On the Potomac, swan-white
power launches keep breasting the sulphurous wave.

Otters slide and dive and slick back their hair,   5
raccoons clean their meat in the creek.

On the circles, green statues ride like South American
liberators above the breeding vegetation—

prongs and spearheads of some equatorial
backland that will inherit the globe.   10

The elect, the elected . . . they come here bright as dimes,
and die disheveled and soft.

We cannot name their names, or number their dates—
circle on circle, like rings on a tree—

but we wish the river had another shore,   15
some farther range of delectable mountains,²

distant hills powdered blue as a girl's eyelid.
It seems the least little shove would land us there,

that only the slightest repugnance of our bodies
we no longer control could drag us back.   20

1964

Night Sweat

Work-table, litter, books and standing lamp,
plain things, my stalled equipment, the old broom—
but I am living in a tidied room,
for ten nights now I've felt the creeping damp
float over my pajamas' wilted white . . .   5
Sweet salt embalms me and my head is wet,
everything streams and tells me this is right;
my life's fever is soaking in night sweat—
one life, one writing! But the downward glide
and bias of existing wrings us dry—   10
always inside me is the child who died,
always inside me is his will to die—
one universe, one body . . . in this urn
the animal night sweats of the spirit burn.
Behind me! You! Again I feel the light   15
lighten my leaded eyelids, while the gray
skulled horses whinny for the soot of night.
I dabble in the dapple of the day,

2. In *Inferno*, Canto I, Dante is separated by his sins from the Delectable
Mountain, representing Virtue. Perhaps
this is also an allusion to the passage
from Hell to the mountain of Purgatory.

a heap of wet clothes, seamy, shivering,
I see my flesh and bedding washed with light,   20
my child exploding into dynamite,
my wife . . . your lightness alters everything,
and tears the black web from the spider's sack,
as your heart hops and flutters like a hare.
Poor turtle, tortoise, if I cannot clear   25
the surface of these troubled waters here,
absolve me, help me, Dear Heart, as you bear
this world's dead weight and cycle on your back.

1964

For the Union Dead[1]

"Relinquunt Omnia Servare Rem Publicam."[2]

The old South Boston Aquarium stands
in a Sahara of snow now. Its broken windows are boarded.
The bronze weathervane cod has lost half its scales.
The airy tanks are dry.

Once my nose crawled like a snail on the glass;   5
my hand tingled
to burst the bubbles
drifting from the noses of the cowed, compliant fish.

My hand draws back. I often sigh still
for the dark downward and vegetating kingdom   10
of the fish and reptile. One morning last March,
I pressed against the new barbed and galvanized

fence on the Boston Common. Behind their cage,
yellow dinosaur steamshovels were grunting
as they cropped up tons of mush and grass   15
to gouge their underworld garage.

Parking spaces luxuriate like civic
sandpiles in the heart of Boston.
A girdle of orange, Puritan-pumpkin colored girders
braces the tingling Statehouse,   20

1. First published under the title *Colonel Shaw and the Massachusetts' 54th* in a paperback edition of *Life Studies* (1960). With a change of title, it became the title poem of *For the Union Dead* (1964).
2. Robert Gould Shaw (1837–63) led the first all-Negro regiment in the North during the Civil War. He was killed in the attack against Fort Wagner, South Carolina. A bronze relief by the sculptor Augustus Saint-Gaudens (1848–97), dedicated in 1897, standing opposite the Massachusetts State House on Boston Common, commemorates the deaths. A Latin inscription on the monument reads *"Omnia Reliquit Servare Rem Publicam"* —"He leaves all behind to serve the Republic." Lowell's epigraph alters the inscription slightly, changing the third person singular ("he") to the third person plural: *"They* give up everything to serve the Republic."

shaking over the excavations, as it faces Colonel Shaw
and his bell-cheeked Negro infantry
on St. Gaudens' shaking Civil War relief,
propped by a plank splint against the garage's earthquake.

Two months after marching through Boston, 25
half the regiment was dead;
at the dedication,
William James[3] could almost hear the bronze Negroes breathe.

Their monument sticks like a fishbone
in the city's throat. 30
Its Colonel is as lean
as a compass-needle.

He has an angry wrenlike vigilance,
a greyhound's gentle tautness;
he seems to wince at pleasure, 35
and suffocate for privacy.

He is out of bounds now. He rejoices in man's lovely,
peculiar power to choose life and die—
when he leads his black soldiers to death,
he cannot bend his back. 40

On a thousand small town New England greens,
the old white churches hold their air
of sparse, sincere rebellion; frayed flags
quilt the graveyards of the Grand Army of the Republic.

The stone statues of the abstract Union Soldier 45
grow slimmer and younger each year—
wasp-waisted, they doze over muskets
and muse through their sideburns . . .

Shaw's father wanted no monument
except the ditch, 50
where his son's body was thrown[4]
and lost with his "niggers."

The ditch is nearer.
There are no statues for the last war[5] here;
on Boylston Street,[6] a commercial photograph 55
shows Hiroshima boiling

over a Mosler Safe, the "Rock of Ages"
that survived the blast. Space is nearer.

---

3. Philosopher and psychologist (1842–
1910) who taught at Harvard.
4. By the Confederate soldiers at Fort

Wagner.
5. World War II.
6. In Boston, where the poem is set.

When I crouch to my television set,
the drained faces of Negro school-children rise like balloons.[7] 60

Colonel Shaw
is riding on his bubble,
he waits
for the blessed break.

The Aquarium is gone. Everywhere, 65
giant finned cars nose forward like fish;
a savage servility
slides by on grease.

1960, 1964

Robert Frost[1]

Robert Frost at midnight,[2] the audience gone
to vapor, the great act laid on the shelf in mothballs,
his voice is musical and raw—he writes in the flyleaf:
For Robert from Robert, his friend in the art.
"Sometimes I feel too full of myself," I say. 5
And he, misunderstanding, "When I am low,
I stray away. My son[3] wasn't your kind. The night
we told him Merrill Moore[4] would come to treat him,
he said, 'I'll kill him first.' One of my daughters thought things,
thought every male she met was out to make her; 10
the way she dressed, she couldn't make a whorehouse."
And I, "Sometimes I'm so happy I can't stand myself."
And he, "When I am too full of joy, I think
how little good my health did anyone near me."

1969, 1973

Ulysses[1]

Shakespeare stand-ins,[2] same string hair, gay, dirty ...
there's a new poetry in the air, it's youth's
patent, lust coolly led on by innocence—
late-flowering Garden, far from Eden fallen,
and still fair! None chooses as his model 5
Ulysses landhugging from port to port for girls ...
his marriage a cover for the underworld,

7. Probably news photographs connected with contemporary civil rights demonstrations to secure desegregation of schools in the South.
1. American poet (1874–1963).
2. Alludes to the poem *Frost at Midnight* by Samuel Taylor Coleridge (1772–1834), in which the English poet celebrates a peaceful moment with his infant son.
3. Frost's son committed suicide.
4. Poet and psychiatrist (b. 1903).
1. The protagonist of *The Odyssey*, by the Greek poet Homer (8th century B.C.); the epic poem recounts the adventures of its hero on his return to Ithaca and his wife, Penelope, 20 years after the fall of Troy. Lowell imagines Ulysses as a Don Juan in search of sexual adventures (as with the nymph Nausicaa, line 9) before he returns to his patient wife.
2. The so-called "flower children" of the 1960s, whose characteristic long hair and colorful outfits remind Lowell of minor actors in Shakespearean plays.

dark harbor of suctions and the second chance.
He won Nausicaa twenty years too late. . . .
Scarred husband and wife sit naked, one Greek smile,   10
thinking *we were bound to fall in love*
if only we stayed married long enough—
because our ships are burned and all friends lost.
How we wish we were friends with half our friends!

1969, 1973

Reading Myself

Like thousands, I took just pride and more than just,
struck matches that brought my blood to a boil;
I memorized the tricks to set the river on fire—
somehow never wrote something to go back to.
Can I suppose I am finished with wax flowers   5
and have earned my grass on the minor slopes of Parnassus.[3] . . .
No honeycomb is built without a bee
adding circle to circle, cell to cell,
the wax and honey of a mausoleum—
this round dome proves its maker is alive;   10
the corpse of the insect lives embalmed in honey,
prays that its perishable work live long
enough for the sweet-tooth bear to desecrate—
this open book . . . my open coffin.

1969, 1973

Epilogue[1]

Those blessèd structures, plot and rhyme—
why are they no help to me now
I want to make
something imagined, not recalled?
I hear the noise of my own voice: 5
The painter's vision is not a lens,
it trembles to caress the light.
But sometimes everything I write
with the threadbare art of my eye
seems a snapshot,   10
lurid, rapid, garish, grouped,
heightened from life,
yet paralyzed by fact.
All's misalliance.
Yet why not say what happened?   15
Pray for the grace of accuracy

---

3. In mythology, the mountain home of
Pegasus, the winged horse of poetry, and
traditionally symbol of the summits of
poetic achievement.

1. The poem is printed as the last piece
(excluding a few translations) in Lowell's
last book, *Day by Day* (1977).

Vermeer[2] gave to the sun's illumination
stealing like the tide across a map
to his girl solid with yearning.
We are poor passing facts,   20
warned by that to give
each figure in the photograph
his living name.

1977

2. Jan Vermeer (1632–75), Dutch painter noted for his subtle handling of the effects of light.

RICHARD WILBUR

1921–

Richard Wilbur was born in New York City, the son of a painter, and grew up in the country in New Jersey. He was educated at Amherst College, where Robert Frost was a frequent guest and teacher. Whether through the direct influence of the older poet or not, Wilbur is recognized, along with Frost, as probably the most accomplished prosodist in recent American poetry. Of the effects of those years, Wilbur says: "Most American poets of my generation were taught to admire the English metaphysical poets of the seventeenth century and such contemporary masters of irony as John Crowe Ransom. We were led by our teachers and by the critics whom we read to feel that the most adequate and convincing poetry is that which accommodates mixed feelings, clashing ideas, and incongruous images." Wilbur was to remain true to this preference for the ironic meditative lyric, the single perfect poem, rather than longer narratives or dramatic sequences.

After graduation and service in the infantry in Italy and France (1943–45), Wilbur returned to study for an M.A. at Harvard, with a firm notion of what he expected to get out of poetry. "My first poems were written in answer to the inner and outer disorders of the second World War and they helped me * * * to take ahold of raw events and convert them, provisionally, into experience." He reasserted the balance of mind against instinct and violence: "The praiseful, graceful soldier / Shouldn't be fired by his gun." The poised lyrics in *The Beautiful Changes* (1947) and *Ceremony* (1950) also reclaimed a sense of Europe obscured by the war: the value of pleasure, defined as an interplay of intelligence with sensuous enjoyment. Whether looking at a real French landscape, as in *Grasse: The Olive Trees*, or a French landscape painting, as in *Ceremony*, the point was to show the witty shaping power of the mind in nature. Hence the importance of "emblem" poems: poems which develop the symbolic significance of a natural object (the "unearthly" pallor of olive trees which "teach the South it is not paradise") or of works of art that incorporate nature.

Wilbur prefers strict stanzaic forms and meters; "limitation makes for power: the strength of the genie comes of his being confined in a bottle."

In individual lines and the structure of an entire poem, his emphasis is on a civilized balancing of perceptions. *A World without Objects Is a Sensible Emptiness* begins with the "tall camels of the spirit," but qualifies our views of lonely spiritual impulses. The poem summons us back to find visionary truth grasped through sensual experience. "All shining things need to be shaped and borne." Wilbur favors "a spirituality which is not abstracted, not dissociated and world-renouncing. A good part of my work could, I suppose, be understood as a public quarrel with the aesthetics of Edgar Allan Poe"—presumably with Poe's notion that poetry provided *indefinite* sensations and aspired to the abstract condition of music.

Wilbur was among the first of the younger postwar poets to adopt a style of living and working different from the masters of an earlier generation—from Eliot, ironic priestlike modernist who lived as a publisher-poet in England, or William Carlos Williams, a doctor in Paterson, or Wallace Stevens, a remote insurance executive in Hartford. Wilbur was a teacher-poet and gave frequent readings. Instead of thinking of himself as an alienated artist, he came to characterize himself as a "poet-citizen," part of what he judged a widening community of poets addressing themselves to an audience increasingly responsive to poetry. Wilbur's taste for civilized wit and his metrical skill made him an ideal translator of the seventeenth-century satirical comedies of Molière, *Tartuffe* (1963) and *The Misanthrope* (1955). They are frequently played, as is the musical version of Voltaire's *Candide* for which Wilbur was one of the collaborating lyricists. Wilbur, who teaches at Smith, received the Pulitzer Prize for his volume *Things of This World* (1956) and has since published *Advice to a Prophet* (1961) and *Walking to Sleep* (1969).

Grasse:[1] The Olive Trees

FOR MARCELLE AND FERDINAND SPRINGER

Here luxury's the common lot. The light
Lies on the rain-pocked rocks like yellow wool
And around the rocks the soil is rusty bright
From too much wealth of water, so that the grass
Mashes under the foot, and all is full   5
Of heat and juice and a heavy jammed excess.

Whatever moves moves with the slow complete
Gestures of statuary. Flower smells
Are set in the golden day, and shelled in heat,
Pine and columnar cypress stand. The palm   10
Sinks its combs in the sky. This whole South swells
To a soft rigor, a rich and crowded calm.

Only the olive contradicts. My eye,
Traveling slopes of rust and green, arrests
And rests from plenitude where olives lie   15
Like clouds of doubt against the earth's array.
Their faint disheveled foliage divests
The sunlight of its color and its sway.

1. A city in southern France.

Not that the olive spurns the sun; its leaves
Scatter and point to every part of the sky, 20
Like famished fingers waving. Brilliance weaves
And sombers down among them, and among
The anxious sliver branches, down to the dry
And twisted trunk, by rooted hunger wrung.

Even when seen from near, the olive shows 25
A hue of far away. Perhaps for this
The dove brought olive back, a tree which grows
Unearthly pale, which ever dims and dries,
And whose great thirst, exceeding all excess, 30
Teaches the South it is not paradise.

1950

The Death of a Toad

A toad the power mower caught,
Chewed and clipped of a leg, with a hobbling hop has got
To the garden verge, and sanctuaried him
Under the cineraria leaves, in the shade
Of the ashen heartshaped leaves, in a dim, 5
Low, and a final glade.

The rare original heartsblood goes,
Spends on the earthen hide, in the folds and wizenings, flows
In the gutters of the banked and staring eyes. He lies
As still as if he would return to stone, 10
And soundlessly attending, dies
Toward some deep monotone,

Toward misted and ebullient seas
And cooling shores, toward lost Amphibia's emperies.[1]
Day dwindles, drowning, and at length is gone 15
In the wide and antique eyes, which still appear
To watch, across the castrate lawn,
The haggard daylight steer.

1950

Ceremony

A striped blouse in a clearing by Bazille[2]
Is, you may say, a patroness of boughs
Too queenly kind[3] toward nature to be kin.
But ceremony never did conceal,
Save to the silly[4] eye, which all allows, 5
How much we are the woods we wander in.

---

1. Archaic for *empires*. Amphibia is imagined to be the spiritual ruler of the toad's universe.
2. Jean Frédéric Bazille (1841–71), French painter noted for painting figures in forest landscapes; he was associated with the Impressionists.
3. An original meaning of "Nature" was "kind."
4. Innocent, homely.

Let her be some Sabrina[5] fresh from stream,
Lucent as shallows slowed by wading sun,
Bedded on fern, the flowers' cynosure:[6]
Then nymph and wood must nod and strive to dream   10
That she is airy earth, the trees, undone,
Must ape her languor natural and pure.

Ho-hum. I am for wit and wakefulness,
And love this feigning lady by Bazille.
What's lightly hid is deepest understood,   15
And when with social smile and formal dress
She teaches leaves to curtsey and quadrille,[7]
I think there are most tigers in the wood.

1950

"A World without Objects Is a Sensible Emptiness"[1]

The tall camels of the spirit
  Steer for their deserts, passing the last groves loud
With the sawmill shrill of the locust, to the whole honey
      of the arid
Sun. They are slow, proud,
  And move with a stilted stride   5
To the land of sheer horizon, hunting Traherne's
Sensible emptiness, there where the brain's lantern-slide
  Revels in vast returns.

  O connoisseurs of thirst,
  Beasts of my soul who long to learn to drink   10
Of pure mirage, those prosperous islands are accurst
    That shimmer on the brink

  Of absence; auras, lustres,
  And all shinings need to be shaped and borne.
Think of those painted saints, capped by the early masters   15
    With bright, jauntily-worn

  Aureate plates, or even
  Merry-go-round rings. Turn, O turn
From the fine sleights of the sand, from the long empty oven
    Where flames in flamings burn   20

---

5. A nymph, the presiding deity of the river Severn in Milton's masque *Comus* (1634).
6. Center of attraction; also from the constellation Ursa Minor, whose center is the Pole Star.
7. A square dance, of French origin, performed by four couples.
1. From *Second Century*, Meditation 65, by Thomas Traherne (c. 1638–74): "Life without objects is sensible emptiness, and that is a greater misery than death or nothing." "Sensible" is used to mean "palpable to the senses."

Back to the trees arrayed
In bursts of glare, to the halo-dialing run
Of the country creeks, and the hills' bracken tiaras made
Gold in the sunken sun,

Wisely watch for the sight   25
Of the supernova[2] burgeoning over the barn,
Lampshine blurred in the steam of beasts, the spirit's right
Oasis, light incarnate.

1950

Year's End

Now winter downs the dying of the year,
And night is all a settlement of snow;
From the soft street the rooms of houses show
A gathered light, a shapen atmosphere,
Like frozen-over lakes whose ice is thin   5
And still allows some stirring down within.

I've known the wind by water banks to shake
The late leaves down, which frozen where they fell
And held in ice as dancers in a spell
Fluttered all winter long into a lake;   10
Graved on the dark in gestures of descent,
They seemed their own most perfect monument.

There was perfection in the death of ferns
Which laid their fragile cheeks against the stone
A million years. Great mammoths overthrown   15
Composedly have made their long sojourns,
Like palaces of patience, in the gray
And changeless lands of ice. And at Pompeii[3]

The little dog lay curled and did not rise
But slept the deeper as the ashes rose   20
And found the people incomplete, and froze
The random hands, the loose unready eyes
Of men expecting yet another sun
To do the shapely thing they had not done.

These sudden ends of time must give us pause.   25
We fray into the future, rarely wrought
Save in the tapestries of afterthought.
More time, more time. Barrages of applause
Come muffled from a buried radio.
The New-year bells are wrangling with the snow.   30

1950

2. A scientific term for an exploding star, here associated with the Star of Bethlehem.

3. Roman city buried by and partly preserved in lava after the eruption of Mount Vesuvius (A.D. 79).

ALLEN GINSBERG
1926–

"Hold back the edges of your gowns, Ladies, we are going through hell." William Carlos Williams's introduction to Allen Ginsberg's *Howl* (1956) was probably the most auspicious public welcome from one poet to another since Emerson had hailed the unknown Whitman in a letter that Whitman prefaced to the second edition of *Leaves of Grass* one hundred years before. *Howl* combined apocalyptic criticism of the dull, prosperous Eisenhower years with exuberant celebration of an emerging counterculture. It was the best known and most widely circulated book of poems of its time, and with its appearance Ginsberg became part of the history of publicity as well as the history of poetry. *Howl* and Jack Kerouac's novel *On the Road* were the pocket Bibles of the generation whose name Kerouac had coined— "Beat," with its punning overtones of "beaten down" and "beatified."

Allen Ginsberg was born in 1926, son of Louis Ginsberg, a schoolteacher in New Jersey, himself a poet, and of Naomi Ginsberg, a Russian émigrée, whose madness and eventual death her son memorialized in *Kaddish* (1959). His official education took place at Columbia University, but for him as for Jack Kerouac the presence of William Burroughs in New York was equally influential. Burroughs (b. 1914), later the author of *Naked Lunch*, one of the most inventive experiments in American prose, was at that time a drug addict about to embark on an expatriate life in Mexico and Tangier. He helped Ginsberg discover modern writers: Kafka, Yeats, Céline, Rimbaud. Ginsberg responded to Burroughs's liberated kind of life, to his comic-apocalyptic view of American society, and to his bold literary use of autobiography, as when writing about his own experience with addicts and addiction in *Junkie*, whose chapters Ginsberg was reading in manuscript form in 1950.

Ginsberg's New York career has passed into mythology. In 1945, his sophomore year, he was expelled from Columbia: he had sketched some obscene drawings and phrases in the dust of his dormitory window to draw the attention of a neglectful cleaning woman to the grimy state of his room. Then, living periodically with Burroughs and Kerouac, he shipped out for short trips as a messman on merchant tankers and worked in addition as a welder, a night-porter, and a dishwasher.

One summer, in a Harlem apartment, Ginsberg underwent what he was always to represent as the central conversion experience of his life. He had an "auditory vision" of the English poet William Blake reciting his poems: first *Ah! Sunflower*, and then a few minutes later the same oracular voice intoning *The Sick Rose*. It was "like hearing the doom of the whole universe, and at the same time the inevitable beauty of that doom." Ginsberg was convinced of the presence of "this big god over all * * * and that the whole purpose of being born was to wake up to Him."

Ginsberg eventually finished Columbia in 1948 with high grades but under a legal cloud. Herbert Huncke, a colorful but irresponsible addict friend, had been using Ginsberg's apartment as a storage depot for the goods he stole to support his "habit." To avoid prosecution as an accomplice, Ginsberg had to plead insanity and spent eight months in Columbia Psychiatric Institute.

After more odd jobs—a ribbon factory in New Jersey, a local AFL newspaper—and a considerable success as a market researcher in San Francisco, Ginsberg left the "straight," nine-to-five world for good. He was drawn to San Francisco, he said, by its "long honorable * * * tradition of Bohemian—Buddhist—Wobbly [the I.W.W., an early radical labor movement]—mystical—anarchist social involvement." In the years after 1954 he met San Francisco poets such as Robert Duncan, Kenneth Rexroth, Gary Snyder (who was studying Chinese and Japanese at Berkeley) and Lawrence Ferlinghetti, whose City Lights Bookshop became the publisher of *Howl.* The night Ginsberg read the new poem aloud at the Six Gallery has been called "the birth trauma of the Beat Generation"; it focused and released an angry call for attention to the spirited "misfits" and downtrodden talents of American life.

Howl's spontaneity of surface conceals but grows out of Ginsberg's care and self-consciousness about rhythm and meter. Under the influence of William Carlos Williams, who had befriended him in Paterson after he left the mental hospital, Ginsberg had started carrying around a notebook to record the rhythms of voices around him. Kerouac's *On the Road* gave him further examples of "frank talk," and in addition of an "oceanic" prose "sometimes as sublime as epic line." Under Kerouac's influence Ginsberg began the long tumbling lines which were to become his trademark. He carefully explained that all of *Howl and Other Poems* was an experiment in what could be done with the long line, the longer unit of breath which seemed natural for him. "My feeling is for a big long clanky statement," one which accommodates "not the way you would *say* it, a thought, but the way you would think it—i.e., we think rapidly, in visual images as well as words, and if each successive thought were transcribed in its confusion * * * you get a slightly different prosody than if you were talking slowly."

The long line is something Ginsberg learned as well from Biblical rhetoric, from the 18th-century English poet Christopher Smart, and, above all, from Whitman and Blake. His first book pays tribute to both these latter poets. *A Supermarket in California,* with its movement from exclamations to sad questioning, is Ginsberg's melancholy reminder of what has become, after a century, of Whitman's vision of American plenty. In *Sunflower Sutra* he celebrates the battered nobility beneath our industrial "skin of grime." Ginsberg at his best gives a sense of both doom and beauty, whether in the denunciatory impatient prophecies of *Howl* or in the catalogue of suffering in *Kaddish.* His disconnected phrases can accumulate as narrative shrieks or, at other moments, can build as a litany of praise.

By the end of the 1960s Ginsberg was widely known and widely traveled. For him it was a decade in which he conducted publicly his own pursuit of inner peace during a long stay with Buddhist instructors in India and at home served as a kind of guru himself for many young people disoriented by the Vietnam war. Ginsberg read his poetry and held "office hours" in universities all over America, a presence at everything from "be-ins"—mass outdoor festivals of chanting, costumes, and music—to antiwar protests. He was a gentle and persuasive presence at hearings for many kinds of reform: revision of severe drug laws and laws against homosexuality. Ginsberg himself had lived for years with the poet Peter Orlovsky and wrote frankly about their relationship. His poems record his drug experiences as well, and *The Change*, written in Japan in 1963, marks his decision to keep away from what he considered the nonhuman domination of drugs and to lay new stress on "living in and inhabiting the human form."

In *The Fall of America* (1972) Ginsberg turned to "epic," a poem including history and registering the ups and downs of his travels across the United States. These "transit" poems sometimes seem like tape-recorded random lists of sights, sounds, and names, but at their best they give a sense of how far America has fallen, by measuring the provisional and changing world of nuclear America against the traces of nature still visible in our landscape and place names. Ginsberg now lives on a farm near Woodstock, New York, and has added ecology to the causes for which he is a patient and attractive spokesman.

From Howl[1]

FOR CARL SOLOMON

I

I saw the best minds of my generation destroyed by madness, starving hysterical naked,

dragging themselves through the negro streets at dawn looking for an
angry fix,

angelheaded hipsters[2] burning for the ancient heavenly connection[3]
to the starry dynamo in the machinery of night,

who poverty and tatters and hollow-eyed and high sat up smoking in
the supernatural darkness of cold-water flats floating across the
tops of cities contemplating jazz,

who bared their brains to Heaven under the El[4] and saw Mohammedan angels staggering on tenement roofs illuminated,   5

who passed through universities with radiant cool eyes hallucinating
Arkansas and Blake-light[5] tragedy among the scholars of war,

who were expelled from the academies for crazy & publishing
obscene odes on the windows of the skull,[6]

who cowered in unshaven rooms in underwear, burning their money
in wastebaskets and listening to the Terror through the wall,

1. With Jack Kerouac's *On the Road* (1957), *Howl* is a central document of the Beat Movement of the 1950s, drawing on incidents in the lives of Ginsberg, Kerouac, and their friends. Ginsberg met Carl Solomon (b. 1928) while both were patients in Columbia Psychiatric Institute in 1949, and called him "an intuitive Bronx Dadaist and prose-poet." Many details in *Howl* come from the "apocryphal history" which Solomon told Ginsberg in 1949. In *More Mishaps* (1968), Solomon admits that these adventures were "compounded partly of truth, but for the most [of] raving self-justification, crypto-bohemian boasting * * * effeminate prancing and esoteric aphorisms."

2. The Beats who went "on the road" in the 1950s; "hip": slang for "aware."
3. Person who supplies dope.
4. Elevated railway in New York City.
5. Referring to Ginsberg's apocalyptic vision of the English poet William Blake (1757–1827).
6. In 1945, Ginsberg was expelled from Columbia for drawing obscene pictures and phrases in the grime of his dormitory windows, in an effort to attract the cleaning woman's attention to the filth in the room. In 1948, in danger of prosecution as an accessory to burglary, Ginsberg volunteered to undergo psychiatric treatment.

who got busted in their pubic beards returning through Laredo with
    a belt of marijuana for New York,

who ate fire in paint hotels or drank turpentine in Paradise Alley,[7]
    death, or purgatoried their torsos night after night   10

with dreams, with drugs, with waking nightmares, alcohol and cock
    and endless balls,

incomparable blind streets of shuddering cloud and lightning in the
    mind leaping toward poles of Canada & Paterson,[8] illuminating
    all the motionless world of Time between,

Peyote solidities of halls, backyard green tree cemetery dawns, wine
    drunkenness over the rooftops, storefront boroughs of teahead
    joyride neon blinking traffic light, sun and moon and tree
    vibrations in the roaring winter dusks of Brooklyn, ashcan rant-
    ings and kind king light of mind,

who chained themselves to subways for the endless ride from Battery
    to holy Bronx[9] on benzedrine until the noise of wheels and
    children brought them down shuddering mouth-wracked and
    battered bleak of brain all drained of brilliance in the drear
    light of Zoo,

who sank all night in submarine light of Bickford's[1] floated out and
    sat through the stale beer afternoon in desolate Fugazzi's,[2]
    listening to the crack of doom on the hydrogen jukebox,   15

who talked continuously seventy hours from park to pad to bar to
    Bellevue[3] to museum to the Brooklyn Bridge,

a lost battalion of platonic conversationalists jumping down the
    stoops off fire escapes off windowsills off Empire State out of
    the moon,

yacketayakking screaming vomiting whispering facts and memories
    and anecdotes and eyeball kicks and shocks of hospitals and
    jails and wars,

whole intellects disgorged in total recall for seven days and nights
    with brilliant eyes, meat for the Synagogue cast on the pave-
    ment,

who vanished into nowhere Zen New Jersey leaving a trail of am-
    biguous picture postcards of Atlantic City Hall,   20

suffering Eastern sweats and Tangerian bone-grindings and migraines
    of China[4] under junk-withdrawal in Newark's bleak furnished
    room,

who wandered around and around at midnight in the railroad yard
    wondering where to go, and went, leaving no broken hearts,

who lit cigarettes in boxcars boxcars boxcars racketing through snow
    toward lonesome farms in grandfather night,

---

7. A tenement courtyard in New York's East Village; setting of Kerouac's *The Subterraneans* (1958).

8. Ginsberg's hometown, also the town celebrated by William Carlos Williams in his long poem *Paterson*.

9. Opposite ends of a New York subway line; the Bronx Zoo was the northern terminus.

1. All-night cafeteria where Ginsberg worked during his college years.

2. Bar near Greenwich Village frequented by the Beats.

3. New York public hospital to which mental patients are generally committed.

4. African and Asian sources of drugs.

who studied Plotinus Poe St. John of the Cross[5] telepathy and bop
kaballa[6] because the cosmos instinctively vibrated at their feet
in Kansas,

who loned it through the streets of Idaho seeking visionary indian
angels who were visionary indian angels, 25

who thought they were only mad when Baltimore gleamed in supernatural ecstasy,

who jumped in limousines with the Chinaman of Oklahoma on the
impulse of winter midnight streetlight smalltown rain,

who lounged hungry and lonesome through Houston seeking jazz
or sex or soup, and followed the brilliant Spaniard to converse
about America and Eternity, a hopeless task, and so took ship
to Africa,

who disappeared into the volcanoes of Mexico leaving behind nothing but the shadow of dungarees and the lava and ash of
poetry scattered in fireplace Chicago,

who reappeared on the West Coast investigating the F.B.I. in beards
and shorts with big pacifist eyes sexy in their dark skin passing
out incomprehensible leaflets, 30

who burned cigarette holes in their arms protesting the narcotic
tobacco haze of Capitalism,

who distributed Supercommunist pamphlets in Union Square[7]
weeping and undressing while the sirens of Los Alamos[8]
wailed them down, and wailed down Wall,[9] and the Staten
Island ferry also wailed,

who broke down crying in white gymnasiums naked and trembling
before the machinery of other skeletons,

who bit detectives in the neck and shrieked with delight in police-
cars for committing no crime but their own wild cooking
pederasty and intoxication,

who howled on their knees in the subway and were dragged off the
roof waving genitals and manuscripts, 35

who let themselves be fucked in the ass by saintly motorcyclists, and
screamed with joy,

who blew and were blown by those human seraphim, the sailors,
caresses of Atlantic and Caribbean love,

who balled in the morning in the evenings in rosegardens and the
grass of public parks and cemeteries scattering their semen
freely to whomever come who may,

who hiccupped endlessly trying to giggle but wound up with a sob
behind a partition in a Turkish Bath when the blonde & naked
angel came to pierce them with a sword,

who lost their loveboys to the three old shrews of fate the one eyed
shrew of the heterosexual dollar the one eyed shrew that winks

---

5. Spanish visionary and poet (1542–91),
author of *The Dark Night of the Soul*;
Edgar Allan Poe (1809–49), American
poet and author of supernatural tales;
Plotinus (205–70), visionary philosopher.
Ginsberg was interested in these figures
because of their mysticism.
6. Bop was the jazz style of the 1940's;
the kaballa is a mystical tradition of interpretation of Hebrew scripture.
7. A gathering spot for radical speakers
in New York in the 1930's.
8. In New Mexico, a center for the development of the atomic bomb.
9. Wall Street, but also alludes to the
Wailing Wall, a place of public lamentation in Jerusalem.

out of the womb and the one eyed shrew that does nothing
 but sit on her ass and snip the intellectual golden threads of the
 craftsman's loom,   40

who copulated ecstatic and insatiate with a bottle of beer a sweetheart a package of cigarettes a candle and fell off the bed, and
 continued along the floor and down the hall and ended fainting
 on the wall with a vision of ultimate cunt and come eluding
 the last gyzym of consciousness,

who sweetened the snatches of a million girls trembling in the
 sunset, and were red eyed in the morning but prepared to
 sweeten the snatch of the sunrise, flashing buttocks under barns
 and naked in the lake,

who went out whoring through Colorado in myriad stolen night-
 cars, N.C.,[1] secret hero of these poems, cocksman and Adonis
 of Denver—joy to the memory of his innumerable lays of
 girls in empty lots & diner backyards, moviehouses' rickety rows,
 on mountaintops in caves or with gaunt waitresses in familiar
 roadside lonely petticoat upliftings & especially secret gas-
 station solipsisms[2] of johns, & hometown alleys too,

who faded out in vast sordid movies, were shifted in dreams, woke
 on a sudden Manhattan, and picked themselves up out of
 basements hungover with heartless Tokay and horrors of Third
 Avenue iron dreams & stumbled to unemployment offices,

who walked all night with their shoes full of blood on the snowbank
 docks waiting for a door in the East River to open to a room
 full of steamheat and opium,   45

who created great suicidal dramas on the apartment cliff-banks of
 the Hudson[3] under the wartime blue floodlight of the moon &
 their heads shall be crowned with laurel in oblivion,

who ate the lamb stew of the imagination or digested the crab at
 the muddy bottom of the rivers of Bowery,[4]

who wept at the romance of the streets with their pushcarts full of
 onions and bad music,

who sat in boxes breathing in the darkness under the bridge, and
 rose up to build harpsichords in their lofts,

who coughed on the sixth floor of Harlem crowned with flame under
 the tubercular sky surrounded by orange crates of theology, 50

who scribbled all night rocking and rolling over lofty incantations
 which in the yellow morning were stanzas of gibberish,

who cooked rotten animals lung heart feet tail borsht & tortillas
 dreaming of the pure vegetable kingdom,

who plunged themselves under meat trucks looking for an egg,

who threw their watches off the roof to cast their ballot for
 Eternity outside of Time, & alarm clocks fell on their heads
 every day for the next decade,

1. Neal Cassady, "hip" companion of
Jack Kerouac and the original Dean
Moriarty, one of the leading figures in
On the Road.
2. A solipsism is the philosophical notion
that all truth is subjective.

3. The steep apartment-lined embank-
ments of the Hudson River along Riverside Drive in New York City.
4. Southern extension of Third Avenue in
New York City; traditional haunt of
derelicts and alcoholics.

who cut their wrists three times successively unsuccessfully, gave up
and were forced to open antique stores where they thought
they were growing old and cried,   55

who were burned alive in their innocent flannel suits on Madison
Avenue[5] amid blasts of leaden verse & the tanked-up clatter
of the iron regiments of fashion & the nitroglycerine shrieks of
the fairies of advertising & the mustard gas of sinister intelligent editors, or were run down by the drunken taxicabs of
Absolute Reality,

who jumped off the Brooklyn Bridge this actually happened and
walked away unknown and forgotten into the ghostly daze of
Chinatown soup alleyways & firetrucks, not even one free beer,

who sang out of their windows in despair, fell out of the subway
window, jumped in the filthy Passaic,[6] leaped on negroes, cried
all over the street, danced on broken wineglasses barefoot
smashed phonograph records of nostalgic European 1930's
German jazz finished the whiskey and threw up groaning into
the bloody toilet, moans in their ears and the blast of colossal
steamwhistles,

who barreled down the highways of the past journeying to each
other's hotrod-Golgotha[7] jail-solitude watch or Birmingham
jazz incarnation,

who drove crosscountry seventytwo hours to find out if I had a
vision or you had a vision or he had a vision to find out
Eternity,   60

who journeyed to Denver, who died in Denver, who came back to
Denver & waited in vain, who watched over Denver & brooded
& loned in Denver and finally went away to find out the
Time, & now Denver is lonesome for her heroes,

who fell on their knees in hopeless cathedrals praying for each
other's salvation and light and breasts, until the soul illumi-
ated its hair for a second,

who crashed through their minds in jail waiting for impossible criminals with golden heads and the charm of reality in their
hearts who sang sweet blues to Alcatraz,

who retired to Mexico to cultivate a habit, or Rocky Mount to
tender Buddha or Tangiers[8] to boys or Southern Pacific to the
black locomotive or Harvard to Narcissus to Woodlawn[9] to the
daisychain or grave,

who demanded sanity trials accusing the radio of hypnotism &
were left with their insanity & their hands & a hung jury,   65

---

5. Center of New York advertising agen-
cies, and of the stereotype "Man in the
Gray Flannel Suit."
6. River flowing past Paterson, New
Jersey.
7. Site of Christ's crucifixion.
8. William Burroughs (b. 1914), homosexual author, drug addict, and Ginsberg's fellow Beat, lived in Mexico and
Tangier, Morocco; Kerouac lived for a
while in Rocky Mount, North Carolina;
Neal Cassady worked for the Southern
Pacific Railroad.
9. A cemetery in the Bronx; Narcissus,
in Greek mythology, fell in love with his
own reflection in a pool.

who threw potato salad at CCNY lecturers[1] on Dadaism[2] and subsequently presented themselves on the granite steps of the madhouse with shaven heads and harlequin speech of suicide, demanding instantaneous lobotomy,

and who were given instead the concrete void of insulin metrasol electricity hydrotherapy psychotherapy occupational therapy pingpong & amnesia,

who in humorless protest overturned only one symbolic pingpong table, resting briefly in catatonia,

returning years later truly bald except for a wig of blood, and tears and fingers, to the visible madman doom of the wards of the madtowns of the East,

Pilgrim State's Rockland's and Greystone's[3] foetid halls, bickering with the echoes of the soul, rocking and rolling in the midnight solitude-bench dolmen-realms of love, dream of life a nightmare, bodies turned to stone as heavy as the moon, 70

with mother finally ******, and the last fantastic book flung out of the tenement window, and the last door closed at 4 AM and the last telephone slammed at the wall in reply and the last furnished room emptied down to the last piece of mental furniture, a yellow paper rose twisted on a wire hanger in the closet, and even that imaginary, nothing but a hopeful little bit of hallucination—

ah, Carl,[4] while you are not safe I am not safe, and now you're really in the total animal soup of time—

and who therefore ran through the icy streets obsessed with a sudden flash of the alchemy of the use of the ellipse the catalog the meter & the vibrating plane,

who dreamt and made incarnate gaps in Time & Space through images juxtaposed, and trapped the archangel of the soul between 2 visual images and joined the elemental verbs and set the noun and dash of consciousness together jumping with sensation of Pater Omnipotens Aeterna Deus[5]

to recreate the syntax and measure of poor human prose and stand before you speechless and intelligent and shaking with shame, rejected yet confessing out the soul to conform to the rhythm of thought in his naked and endless head, 75

the madman bum and angel beat in Time, unknown, yet putting down here what might be left to say in time come after death,

1. This and the following incidents probably derived from the "apocryphal history of my adventures" related by Solomon to Ginsberg.

2. Artistic cult of absurdity (c. 1916–20); "CCNY": City College of New York.

3. Three mental hospitals near New York. Solomon was institutionalized at Pilgrim State and Rockland; Ginsberg's mother, Naomi, was permanently institutionalized at Greystone after years of suffering hallucinations and paranoid attacks. She died there in 1956, the year after *Howl* was written.

4. Solomon.

5. "All Powerful Father, Eternal God." An allusion to a phrase used by the French painter Paul Cézanne (1839–1906), in a letter describing the effects of nature (1904). Ginsberg, in an interview, compared his own method of sharply juxtaposed images to Cézanne's foreshortening of perspective in landscape painting.

and rose reincarnate in the ghostly clothes of jazz in the goldhorn
    shadow of the band and blew the suffering of America's
    naked mind for love into an eli eli lamma lamma sabacthani[6]
    saxophone cry that shivered the cities down to the last radio
with the absolute heart of the poem of life butchered out of their
    own bodies good to eat a thousand years.

                                                 1956

A Supermarket in California

What thoughts I have of you tonight, Walt Whitman,[1] for I
walked down the sidestreets under the trees with a headache self-
conscious looking at the full moon.

In my hungry fatigue, and shopping for images, I went into the
neon fruit supermarket, dreaming of your enumerations!

What peaches and what penumbras![2] Whole families shopping at
night! Aisles full of husbands! Wives in the avocados, babies in the
tomatoes!—and you, Garcia Lorca,[3] what were you doing down by
the watermelons?

I saw you, Walt Whitman, childless, lonely old grubber, poking
among the meats in the refrigerator and eyeing the grocery boys.

I heard you asking questions of each: Who killed the pork
chops? What price bananas? Are you my Angel?   5

I wandered in and out of the brilliant stacks of cans following
you, and followed in my imagination by the store detective.

We strode down the open corridors together in our solitary
fancy tasting artichokes, possessing every frozen delicacy, and never
passing the cashier.

Where are we going, Walt Whitman? The doors close in an
hour. Which way does your beard point tonight?

(I touch your book and dream of our odyssey in the supermarket
and feel absurd.)

Will we walk all night through solitary streets? The trees add
shade to shade, lights out in the houses, we'll both be lonely.   10

Will we stroll dreaming of the lost America of love past blue
automobiles in driveways, home to our silent cottage?

Ah, dear father, graybeard, lonely old courage-teacher, what
America did you have when Charon quit poling his ferry and you
got out on a smoking bank and stood watching the boat disappear on
the black waters of Lethe?[4]

Berkeley 1955   1956

6. Christ's last words on the Cross: "My God, my God, why have you forsaken me?"

1. American poet (1819–92), author of *Leaves of Grass*, against whose homosexuality and vision of American plenty Ginsberg measures himself.

2. Partial shadows.

3. Spanish poet and dramatist (1899–1936), author of *A Poet in New York*, whose work is characterized by surrealist and homoerotic inspiration.

4. Forgetfulness. In Greek mythology, one of the rivers of Hades; Charon was the boatman who ferried the dead to Hell.

Sunflower Sutra[1]

I walked on the banks of the tincan banana dock and sat down
    under the huge shade of a Southern Pacific locomotive to look
    at the sunset over the box house hills and cry.
Jack Kerouac[2] sat beside me on a busted rusty iron pole, companion,
    we thought the same thoughts of the soul, bleak and blue
    and sad-eyed, surrounded by the gnarled steel roots of trees of
    machinery.
The oily water on the river mirrored the red sky, sun sank on top of
    final Frisco peaks, no fish in that stream, no hermit in those
    mounts, just ourselves rheumy-eyed and hungover like old
    bums on the riverbank, tired and wily.
Look at the Sunflower, he said, there was a dead gray shadow
    against the sky, big as a man, sitting dry on top of a pile of
    ancient sawdust—
—I rushed up enchanted—it was my first sunflower, memories of
    Blake[3]—my visions—Harlem   5
and Hells of the Eastern rivers, bridges clanking Joes Greasy Sand-
    wiches, dead baby carriages, black treadless tires forgotten and
    unretreaded, the poem of the riverbank, condoms & pots, steel
    knives, nothing stainless, only the dank muck and the razor
    sharp artifacts passing into the past—
and the gray Sunflower poised against the sunset, crackly bleak and
    dusty with the smut and smog and smoke of olden locomotives
    in its eye—
corolla[4] of bleary spikes pushed down and broken like a battered
    crown, seeds fallen out of its face, soon-to-be-toothless mouth
    of sunny air, sunrays obliterated on its hairy head like a dried
    wire spiderweb,
leaves stuck out like arms out of the stem, gestures from the sawdust
    root, broke pieces of plaster fallen out of the black twigs, a
    dead fly in its ear,
Unholy battered old thing you were, my sunflower O my soul, I
    loved you then!   10
The grime was no man's grime but death and human locomotives,
all that dress of dust, that veil of darkened railroad skin, that smog
    of cheek, that eyelid of black mis'ry, that sooty hand or
    phallus or protuberance of artificial worse-than-dirt—industrial
    —modern—all that civilization spotting your crazy golden
    crown—
and those blear thoughts of death and dusty loveless eyes and ends
    and withered roots below, in the home-pile of sand and saw-
    dust, rubber dollar bills, skin of machinery, the guts and
    innards of the weeping coughing car, the empty lonely tincans

---

1. Sanskrit for "thread"; the word refers to Brahmin or Buddhist religious
texts of ritual instruction.
2. Fellow Beat (1922–69), author of *On
the Road* (1957).
3. In Harlem in 1948, Ginsberg had had

a hallucinatory revelation in which he
heard the English poet William Blake
(1757–1827) reciting his poem *Ah! Sunflower.*
4. Petals forming the inner envelope of
a flower.

with their rusty tongues alack, what more could I name, the smoked ashes of some cock cigar, the cunts of wheelbarrows and the milky breasts of cars, wornout asses out of chairs & sphincters of dynamos—all these

entangled in your mummied roots—and you there standing before me in the sunset, all your glory in your form!

A perfect beauty of a sunflower! a perfect excellent lovely sunflower existence! a sweet natural eye to the new hip moon, woke up alive and excited grasping in the sunset shadow sunrise golden monthly breeze!   15

How many flies buzzed round you innocent of your grime, while you cursed the heavens of the railroad and your flower soul?

Poor dead flower? when did you forget you were a flower? when did you look at your skin and decide you were an impotent dirty old locomotive? the ghost of a locomotive? the specter and shade of a once powerful mad American locomotive?

You were never no locomotive, Sunflower, you were a sunflower!

And you Locomotive, you are a locomotive, forget me not!

So I grabbed up the skeleton thick sunflower and stuck it at my side like a scepter,   20

and deliver my sermon to my soul, and Jack's soul too, and anyone who'll listen,

—We're not our skin of grime, we're not our dread bleak dusty imageless locomotive, we're all beautiful golden sunflowers inside, we're blessed by our own seed & golden hairy naked accomplishment-bodies growing into mad black formal sunflowers in the sunset, spied on by our eyes under the shadow of the mad locomotive riverbank sunset Frisco hilly tincan evening sitdown vision.

Berkeley 1955   1956

To Aunt Rose

Aunt Rose—now—might I see you
with your thin face and buck tooth smile and pain
        of rheumatism—and a long black heavy shoe
                for your bony left leg
        limping down the long hall in Newark on the running carpet 5
                past the black grand piano
                        in the day room
                                where the parties were
                and I sang Spanish loyalist[1] songs
                        in a high squeaky voice   10
                                (hysterical) the committee listening
                        while you limped around the room
                                collected the money—

---

1. During the Spanish Civil War (1936–39), many left-wing Americans—among them Ginsberg's relatives in Newark—sympathized with the Spanish loyalists who were resisting Francisco Franco's (1892–1975) efforts to become dictator of Spain.

Aunt Honey, Uncle Sam, a stranger with a cloth arm
  in his pocket   15
    and huge young bald head
      of Abraham Lincoln Brigade[2]

—your long sad face
  your tears of sexual frustration
    (what smothered sobs and bony hips   20
      under the pillows of Osborne Terrace)
—the time I stood on the toilet seat naked
  and you powdered my thighs with Calomine
    against the poison ivy—my tender
      and shamed first black curled hairs 25
what were you thinking in secret heart then
    knowing me a man already—
and I an ignorant girl of family silence on the thin pedestal
    of my legs in the bathroom—Museum of Newark.
        Aunt Rose   30
Hitler is dead, Hitler is in Eternity; Hitler is with
    Tamburlane and Emily Brontë[3]

Though I see you walking still, a ghost on Osborne Terrace
  down the long dark hall to the front door
    limping a little with a pinched smile   35
    in what must have been a silken
        flower dress
welcoming my father, the Poet, on his visit to Newark
  —see you arriving in the living room
    dancing on your crippled leg   40
    and clapping hands his book
      had been accepted by Liveright[4]

Hitler is dead and Liveright's gone out of business
The Attic of the Past and *Everlasting Minute* are out of print
    Uncle Harry sold his last silk stocking 45
    Claire quit interpretive dancing school
    Buba sits a wrinkled monument in Old
      Ladies Home blinking at new babies

last time I saw you was the hospital
    pale skull protruding under ashen skin 50
    blue veined unconscious girl

---

2. American volunteers who fought
against the Fascists in the Spanish Civil
War.
3. English poet and novelist (1818–48),
author of *Wuthering Heights*; Tamburlane was the Mideastern "scourge" and
conqueror (hero of Christopher Marlowe's *Tamburlane*, 1588).

4. Leading American publisher of the
1920s and 1930s (now a subsidiary of
W. W. Norton & Company), published
The Everlasting Minute (1937), poems
by Allen Ginsberg's father Louis (b.
1895), whose first book was *The Attic
of the Past* (Boston, 1920).

in an oxygen tent
the war in Spain has ended long ago
Aunt Rose

Paris 1958   1961

A. R. AMMONS
1926–

A. R. Ammons writes that he "was born big and jaundiced (and ugly) on February 18, 1926, in a farmhouse 4 miles southwest of Whiteville, N.C., and 2 miles northwest of New Hope Elementary School and New Hope Baptist Church." It is characteristic of Ammons to be laconic, self-deprecating, unfailingly local, and unfailingly exact. He belongs to the "homemade" strain of American writers rather than the Europeanized or cosmopolitan breed. His poems are filled with the landscapes in which he has lived: North Carolina, the South Jersey coast, and the surroundings of Ithaca, New York, where he now lives and teaches in the English department of Cornell University.

Ammons's career did not start out with a traditional literary education. At Wake Forest College in North Carolina he studied mostly scientific subjects, especially biology and chemistry, and that scientific training has strongly colored his poems. Only later (1951–52) did he study English literature for three semesters at the University of California in Berkeley. He had worked briefly as a high school principal in North Carolina. When he returned from Berkeley he spent twelve years as an executive for a firm which made biological glass in southern New Jersey.

In 1955, his thirtieth year, Ammons published his first book of poems, *Ommateum*. The title refers to the compound structure of an insect's eye and foreshadows a twofold impulse in Ammons's work. On one hand he is involved in the minute observation of natural phenomena; on the other hand he is frustrated by the physical limitations analogous to those of the insects' vision. We see the world, as insects do, in small portions and in impulses which take in but do not totally resolve the many images we receive. "Overall is beyond me," says Ammons in *Corsons Inlet*, a important poem in which the shifting details of shoreline and dunes represent a severe challenge to the poet-observer. There are no straight lines. The contours differ every day, every hour, and they teach the poet the endless adjustments he must make to nature's fluidity.

"A poem is a walk," Ammons has said, and in even the most casual of such encounters he looks closely at things: vegetation, small animals, the minute shifts of wind and weather and light. Yet over and over he seems drawn to Emerson's visionary aspirations for poetry. "Poetry," Emerson remarked, "was all written before time was, and whenever we are so finely organized that we can penetrate into that region where the air is music, we hear those primal warnings and attempt to write them down." Much of Ammons's poetry is given over to testing such transcendental promise for a glimpse of supernatural order and calm.

A typical Ammons poem may move, like *Gravelly Run*, from visionary promptings to sober rebukes from a nature more complicated than the poet dare imagine. At the outset Ammons relaxes and gives himself over: "I don't know somehow it seems sufficient / to see and hear whatever coming and going is, / losing the self to the victory / of stones and trees." By the end of the poem he is in a more embattled position: "so I look and reflect, but the air's glass / jail seals each thing in its entity." The word *reflect*, offered as a gesture of understanding, is drained of its meditative meaning before our eyes. Human gesture becomes nothing but a reflection, a mirror. The poem turns everything brutally physical, a world of unconnected particles; the poet's visionary effort becomes merely that of the "surrendered self among unwelcoming forms."

Ammons began his career writing short lyrics, almost journal entries in an unending career of observation. But the laconic notations—of a landslide, a shift in the shoreline from one day to the next—often bore abstract titles—*Clarity, Saliences*—as if to suggest his appetite for more extended meanings. Ammons has been driven to many kinds of poetic experiments in his effort to make his verse fully responsive to the engaging but evasive particularity of natural process. "Stop on any word and language gives way: / the blades of reason, unlightened by motion, sink in," he remarks in his *Essay on Poetics*. Preparing *Tape for the Turn of the Year* (1965) he typed a book-length day-to-day verse diary along an adding machine tape. The poem ended when the tape did. This was his first and most flamboyant attempt to turn his verse into something beyond mere gatherings. Since then he has discovered that the long poem is the form best adapted to his continuing, indeed endless, dialogue between the specific and the general. As punctuation, the poems tend to use the colon instead of the period so as to keep the poem from grinding to a complete halt or stopping the flow in which the mind feverishly suggests analogies among its minutely perceived experiences. Ammons published several notable examples (*Hibernaculum, Extremes and Moderations*) in his *Collected Poems 1951–1971*. But his intention has been most fully realized in *Sphere: the Form of a Motion*, a single poem, book-length, with no full stops, 155 sections of four tercets each, whose very title suggests its ambition to be what Wallace Stevens would call "the poem of the act of the mind." *Sphere* is committed, as Ammons himself is, to the provisional, the self-revising, its only unity the mind's power to make analogies between the world's constant "diversifications."

Gravelly Run

I don't know somehow it seems sufficient
to see and hear whatever coming and going is,
losing the self to the victory
    of stones and trees,
of bending sandpit lakes, crescent   5
round groves of dwarf pine:

for it is not so much to know the self
as to know it as it is known
    by galaxy and cedar cone,

as if birth had never found it   10
and death could never end it:

the swamp's slow water comes
down Gravelly Run fanning the long
    stone-held algal[1]
hair and narrowing roils[2] between   15
the shoulders of the highway bridge:

holly grows on the banks in the woods there,
and the cedars' gothic-clustered
    spires could make
    green religion in winter bones: 20

so I look and reflect, but the air's glass
jail seals each thing in its entity:

no use to make any philosophies here:
    I see no
god in the holly, hear no song from   25
    the snowbroken weeds: Hegel[3] is not the winter
yellow in the pines: the sunlight has never
heard of trees: surrendered self among
    unwelcoming forms: stranger,
hoist your burdens, get on down the road.   30

1965

Corsons Inlet

I went for a walk over the dunes again this morning
to the sea,
then turned right along
    the surf

        rounded a naked headland   5
        and returned

    along the inlet shore:

it was muggy sunny, the wind from the sea steady and high,
crisp in the running sand,
    some breakthroughs of sun   10
    but after a bit

continuous overcast:

the walk liberating, I was released from forms,
from the perpendiculars,

---

1. From algae: aquatic weeds such as kelp.
2. Grows agitated.

3. Georg Wilhelm Friedrich Hegel (1770–1831), German philosopher who believed in a unifying spirit immanent in Nature.

straight lines, blocks, boxes, binds   15
of thought
into the hues, shadings, rises, flowing bends and blends
            of sight:

                    I allow myself eddies of meaning:
yield to a direction of significance   20
running
like a stream through the geography of my work:
    you can find
in my sayings
                swerves of action   25
                like the inlet's cutting edge:
        there are dunes of motion,
organizations of grass, white sandy paths of remembrance
in the overall wandering of mirroring mind:
but Overall is beyond me: is the sum of these events   30
I cannot draw, the ledger I cannot keep, the accounting
beyond the account:

in nature there are few sharp lines: there are areas of
primrose
        more or less dispersed;   35
disorderly orders of bayberry; between the rows
of dunes,
irregular swamps of reeds,
though not reeds alone, but grass, bayberry, yarrow, all . . .
predominantly reeds: 40

I have reached no conclusions, have erected no boundaries,
shutting out and shutting in, separating inside
        from outside: I have
        drawn no lines:
        as   45

manifold events of sand
change the dune's shape that will not be the same shape
tomorrow,

so I am willing to go along, to accept
the becoming   50
thought, to stake off no beginnings or ends, establish
        no walls:

by transitions the land falls from grassy dunes to creek
to undercreek: but there are no lines, though
        change in that transition is clear   55
        as any sharpness: but "sharpness" spread out,
allowed to occur over a wider range
than mental lines can keep:

the moon was full last night: today, low tide was low:
black shoals of mussels exposed to the risk   60
of air
and, earlier, of sun,
waved in and out with the waterline, waterline inexact,
caught always in the event of change:
      a young mottled gull stood free on the shoals   65
      and ate
to vomiting: another gull, squawking possession, cracked a crab,
picked out the entrails, swallowed the soft-shelled legs, a ruddy
turnstone[1] running in to snatch leftover bits:

risk is full: every living thing in   70
siege: the demand is life, to keep life: the small
white blacklegged egret, how beautiful, quietly stalks and spears
        the shallows, darts to shore
           to stab—what? I couldn't
    see against the black mudflats—a frightened   75
    fiddler crab?

        the news to my left over the dunes and
reeds and bayberry clumps was
        fall: thousands of tree swallows
        gathering for flight: 80
        an order held
        in constant change: a congregation
rich with entropy: nevertheless, separable, noticeable
      as one event,
           not chaos: preparations for   85
flight from winter,
cheet, cheet, cheet, cheet, wings rifling the green clumps,
beaks
at the bayberries
    a perception full of wind, flight, curve,   90
    sound:
    the possibility of rule as the sum of rulelessness:
the "field" of action
with moving, incalculable center:

in the smaller view, order tight with shape: 95
blue tiny flowers on a leafless weed: carapace of crab:
snail shell:
      pulsations of order
      in the bellies of minnows: orders swallowed,
broken down, transferred through membranes   100
to strengthen larger orders: but in the large view, no
lines or changeless shapes: the working in and out, together
      and against, of millions of events: this,
         so that I make

---

1. A plover-like migratory bird.

no form of 105
formlessness:

orders as summaries, as outcomes of actions override
or in some way result, not predictably (seeing me gain
the top of a dune,
the swallows 110
could take flight—some other fields of bayberry
could enter fall
berryless) and there is serenity:

no arranged terror: no forcing of image, plan,
or thought: 115
no propaganda, no humbling of reality to precept:

terror pervades but is not arranged, all possibilities
of escape open: no route shut, except in
the sudden loss of all routes:

I see narrow orders, limited tightness, but will 120
not run to that easy victory:
still around the looser, wider forces work:
I will try
to fasten into order enlarging grasps of disorder, widening
scope, but enjoying the freedom that 125
Scope eludes my grasp, that there is no finality of vision,
that I have perceived nothing completely,
that tomorrow a new walk is a new walk.

1965

The City Limits

When you consider the radiance, that it does not withhold
itself but pours its abundance without selection into every
nook and cranny not overhung or hidden; when you consider

that birds' bones make no awful noise against the light but
lie low in the light as in a high testimony; when you consider 5
the radiance, that it will look into the guiltiest

swervings of the weaving heart and bear itself upon them,
not flinching into disguise or darkening; when you consider
the abundance of such resource as illuminates the glow-blue

bodies and gold-skeined wings of flies swarming the dumped 10
guts of a natural slaughter or the coil of shit and in no
way winces from its storms of generosity; when you consider

that air or vacuum, snow or shale, squid or wolf, rose or lichen,
each is accepted into as much light as it will take, then
the heart moves roomier, the man stands and looks about, the 15

leaf does not increase itself above the grass, and the dark
work of the deepest cells is of a tune with May bushes
and fear lit by the breadth of such calmly turns to praise.

1971

JAMES MERRILL
1926–

When James Merrill's *First Poems* were published in 1950, he was immediately recognized as one of the most gifted and polished poets of his
generation. But it was not until *Water Street* (1962), his third volume
of poems, that Merrill began to enlist his brilliant technique and sophisticated tone in developing a poetic autobiography. The book takes its title
from the street where he lives in the seaside village of Stonington, Connecticut. The opening poem, *An Urban Convalescence*, explores his decision to leave New York, which he sees as a distracting city that destroys
its past. He portrays his move as a rededication to his personal past, and
an attempt through poetry "to make some kind of house / Out of the
life lived, out of the love spent."

The metaphor of "home" is an emotional center to which Merrill's
writing often returns, as in *Lost in Translation*, where the narrator recalls
a childhood summer in a home mysteriously without parents. *The Broken
Home* similarly recalls elements of Merrill's own experience as the son of
parents who divorced when he was young. He had been born to the second marriage of Charles E. Merrill, financier and founder of the best
known brokerage firm in America. In the short narrative poems of *Water
Street* and the book that followed he returns again and again to inner
dramas connected to key figures of his childhood and to key childhood
scenes. *The Broken Home* and *Lost in Translation* were to show how
memory and the act of writing have the power to reshape boyhood pain
and conflict so as to achieve "the unstiflement of the entire story." Such
an attitude distinguishes Merrill from his contemporaries (Robert Lowell,
Anne Sexton, Sylvia Plath), whose autobiographical impulse expresses itself
primarily in the present tense and the use of poems as an urgent journal
true to the moment.

As an undergraduate at Amherst College, Merrill had written an honors
thesis on the French novelist Marcel Proust (1871–1922). His poetry was
clearly affected by Proust's notion that the literary exercise of memory
slowly discloses the patterns of childhood experience that we are destined
to relive. Proust showed in his novel how such power over chaotic material of the past is often triggered involuntarily by an object or an episode
in the present whose associations reach back into formative childhood encounters. The questions he asked were asked by Freud as well: What
animates certain scenes—and not others—for us? It is to answer such
questions that some of Merrill's poems are told from the viewpoint of an
observant child. In other poems the poet is explicitly present, at his desk,
trying to incorporate into his adult understanding of the contours of his
life the pain and freshness of childhood memories. The poems are narra-

tive (one of his early books was called *Short Stories*) as often as lyric, in the hope that dramatic *action* will reveal the meanings with which certain objects have become charged. As Merrill sees it, "You hardly ever need to *state* your feelings. The point is to feel and keep the eyes open. Then what you feel is expressed, is mimed back at you by the scene. A room, a landscape. I'd go a step further. We don't *know* what we feel until we see it distanced by this kind of translation."

Merrill has traveled extensively and has presented landscapes from his travels not as external descriptions but rather as ways of exploring alternative or buried states of his own mind, the "translations" of which he speaks above. Poems such as *Days of 1964* and *After the Fire* reflect his experience of Greece, his home for a portion of each year. They respectively anticipate and comment on *The Fire Screen* (1969), a sequence of poems describing the rising and falling curve of a love affair partly in terms of an initiation into Greece with its power to strip away urban sophistication. The books that followed served as initiations into other psychic territories. Problems of family relationships and erotic entanglements previously seen on an intimate scale were in *Braving the Elements* (1972) acted out against a wider backdrop: the long landscapes, primitive geological perspectives, and erosions of the American Far West. Here human experience, examined in his earlier work in close-up, is seen as part of a longer process of evolution comprehensible in terms of enduring nonhuman patterns.

In *Divine Comedies* (which received the Pulitzer Prize in 1977) Merrill began his most ambitious work: two thirds of it is devoted to *The Book of Ephraim*, a long narrative, the first of a trilogy of book-length poems. It is not only a recapitulation of his career but also an attempt to locate individual psychic energies as part of a larger series of nourishing influences: friends living and dead, literary predecessors, scientific theories of the growth of the universe and the mind, the life of other periods and even other universes—all conducted through a set of encounters with the "other world" in seances at the Ouija board. It is a witty and original and assured attempt to take the intimate material of the short lyric which has characterized his earlier work and cast it onto an epic scale.

The Broken Home

Crossing the street,
I saw the parents and the child
At their window, gleaming like fruit
With evening's mild gold leaf.

In a room on the floor below,   5
Sunless, cooler—a brimming
Saucer of wax, marbly and dim—
I have lit what's left of my life.

I have thrown out yesterday's milk
And opened a book of maxims.   10
The flame quickens. The word stirs.

Tell me, tongue of fire,
That you and I are as real
At least as the people upstairs.

     •

My father, who had flown in World War I,   15
Might have continued to invest his life
In cloud banks well above Wall Street and wife.
But the race was run below, and the point was to win.

Too late now, I make out in his blue gaze
(Through the smoked glass of being thirty-six)   20
The soul eclipsed by twin black pupils, sex
And business; time was money in those days.

Each thirteenth year he married. When he died
There were already several chilled wives
In sable orbit—rings, cars, permanent waves.   25
We'd felt him warming up for a green bride.

He could afford it. He was "in his prime"
At three score ten. But money was not time.

     •

When my parents were younger this was a popular act:
A veiled woman would leap from an electric, wine-dark car 30
To the steps of no matter what—the Senate or the Ritz Bar—
And bodily, at newsreel speed, attack

No matter whom—Al Smith[1] or José Maria Sert[2]
Or Clemenceau[3]—veins standing out on her throat
As she yelled *War mongerer! Pig! Give us the vote!*,   35
And would have to be hauled away in her hobble skirt.

What had the man done? Oh, made history.
Her business (he had implied) was giving birth,
Tending the house, mending the socks.

Always that same old story—   40
Father Time and Mother Earth,[4]
A marriage on the rocks.

     •

One afternoon, red, satyr-thighed
Michael, the Irish setter, head
Passionately lowered, led   45
The child I was to a shut door. Inside,

---

1. Alfred E. Smith (1873–1944), a Governor of New York and in 1928 candidate for the presidency.
2. Spanish painter (1876–1945) who decorated the lobby of the Waldorf Astoria hotel in New York (1930).
3. Georges Clemenceau (1841–1929), Premier of France during World War I, visited the U.S. in 1922.
4. In one sense a reference to Cronus (Greek for Time) and Rhea, Mother of the Gods, the parents of Zeus, who dethroned his father.

Blinds beat sun from the bed.
The green-gold room throbbed like a bruise.
Under a sheet, clad in taboos
Lay whom we sought, her hair undone, outspread,   50

And of a blackness found, if ever now, in old
Engravings where the acid bit.
I must have needed to touch it
Or the whiteness—was she dead?
Her eyes flew open, startled strange and cold.   55
The dog slumped to the floor. She reached for me. I fled.

·

Tonight they have stepped out onto the gravel.
The party is over. It's the fall
Of 1931. They love each other still.

She: Charlie, I can't stand the pace.   60
He: Come on, honey—why, you'll bury us all!

A lead soldier guards my windowsill:
Khaki rifle, uniform, and face.
Something in me grows heavy, silvery, pliable.

How intensely people used to feel!   65
Like metal poured at the close of a proletarian novel,[5]
Refined and glowing from the crucible,
I see those two hearts, I'm afraid,
Still. Cool here in the graveyard of good and evil,
They are even so to be honored and obeyed.   70

·

. . . Obeyed, at least, inversely. Thus
I rarely buy a newspaper, or vote.
To do so, I have learned, is to invite
The tread of a stone guest[6] within my house.

Shooting this rusted bolt, though, against him,   75
I trust I am no less time's child than some
Who on the heath impersonate Poor Tom[7]
Or on the barricades risk life and limb.

Nor do I try to keep a garden, only
An avocado in a glass of water—   80
Roots pallid, gemmed with air. And later,

---

5. I.e., a novel dealing with labor, here
with making iron and steel.
6. The commendatore in Mozart's *Don
Giovanni* (1787) returns as a statue to
get his revenge.
7. Edgar, in Shakespeare's *King Lear,*
disowned by his father, wanders the
heath as a madman.

When the small gilt leaves have grown
Fleshy and green, I let them die, yes, yes,
And start another. I am earth's no less.

•

A child, a red dog roam the corridors,　　85
Still, of the broken home. No sound. The brilliant
Rag runners halt before wide-open doors.
My old room! Its wallpaper—cream, medallioned
With pink and brown—brings back the first nightmares,
Long summer colds, and Emma, sepia-faced,　　90
Perspiring over broth carried upstairs
Aswim with golden fats I could not taste.

The real house became a boarding-school.
Under the ballroom ceiling's allegory
Someone at last may actually be allowed　　95
To learn something; or, from my window, cool
With the unstiflement of the entire story,
Watch a red setter stretch and sink in cloud.

　　　　　　　　　　　　　　　　1966

Lost in Translation

FOR RICHARD HOWARD[1]

Diese Tage, die leer dir scheinen
und wertlos für das All,
haben Wurzeln zwischen den Steinen
und trinken dort überall.[2]

A card table in the library stands ready
To receive the puzzle which keeps never coming.
Daylight shines in or lamplight down
Upon the tense oasis of green felt.
Full of unfulfillment, life goes on,　　5
Mirage arisen from time's trickling sands
Or fallen piecemeal into place:
German lesson, picnic, see-saw, walk
With the collie who "did everything but talk"—
Sour windfalls of the orchard back of us.　　10
A summer without parents is the puzzle,
Or should be. But the boy, day after day,
Writes in his Line-a-Day No *puzzle*.

He's in love, at least. His French Mademoiselle,[3]
In real life a widow since Verdun,　　15
Is stout, plain, carrot-haired, devout.

1. American poet and translator from the French (b. 1929).
2. Part of a translation by the Austrian poet Rainer Maria Rilke (1875–1926) of *Palme* by the French poet Paul Valéry (1871–1945). See lines 32, 33. "These days which seem empty and entirely fruitless to you have roots between the stones and drink from everywhere."
3. A French-speaking governess; "Verdun": site of a battle in World War I; "curé": a French priest.

She prays for him, as does a curé in Alsace,
Sews costumes for his marionettes,
Helps him to keep behind the scene
Whose sidelit goosegirl, speaking with his voice,   20
Plays Guinevere as well as Gunmoll Jean.
Or else at bedtime in his tight embrace
Tells him her own French hopes, her German fears,
Her—but what more is there to tell?
Having known grief and hardship, Mademoiselle   25
Knows little more. Her languages. Her place.
Noon coffee. Mail. The watch that also waited
Pinned to her heart, poor gold, throws up its hands—
No puzzle! Steaming bitterness
Her sugars draw pops back into his mouth, translated: 30
"Patience, chéri. Geduld, mein Schatz."[4]
(Thus, reading Valéry the other evening
And seeming to recall a Rilke version of "Palme,"
That sunlit paradigm whereby the tree
Taps a sweet wellspring of authority,   35
The hour came back. Patience dans l'azur.
Geduld im . . . Himmelblau? Mademoiselle.)

Out of the blue, as promised, of a New York
Puzzle-rental shop the puzzle comes—
A superior one, containing a thousand hand-sawn,   40
Sandal-scented pieces. Many take
Shapes known already—the craftsman's repertoire
Nice in its limitation—from other puzzles:
Witch on broomstick, ostrich, hourglass,
Even (surely not just in retrospect)   45
An inchling, innocently branching palm.
These can be put aside, made stories of
While Mademoiselle spreads out the rest face-up,
Herself excited as a child; or questioned
Like incoherent faces in a crowd,   50
Each with its scrap of highly colored
Evidence the Law must piece together.
Sky-blue ostrich? Likely story.
Mauve of the witch's cloak white, severed fingers
Pluck? Detain her. The plot thickens   55
As all at once two pieces interlock.

Mademoiselle does borders—(Not so fast.
A London dusk, December last.
Chatter silenced in the library
This grown man reenters, wearing grey.   60

---

4. French and German phrases for "Have patience, my dear." The recalled phrases remind him of a line in Valéry's *Palme* and Rilke's translation; "Patience in the blue": a way of characterizing the slow nurture of the palm tree.

A medium. All except him have seen
Panel slid back, recess explored,
An object at once unique and common
Displayed, planted in a plain tole
Casket the subject now considers   65
Through shut eyes, saying in effect:
"Even as voices reach me vaguely
A dry saw-shriek drowns them out,
Some loud machinery—a lumber mill?
Far uphill in the fir forest   70
Trees tower, tense with shock,
Groaning and cracking as they crash groundward.
But hidden here is a freak fragment
Of a pattern complex in appearance only.
What it seems to show is superficial   75
Next to that long-term lamination
Of hazard and craft, the karma that has
Made it matter in the first place.
Plywood. Piece of a puzzle." Applause
Acknowledged by an opening of lids   80
Upon the thing itself. A sudden dread—
But to go back. All this lay years ahead.)

Mademoiselle does borders. Straight-edge pieces
Align themselves with earth or sky
In twos and threes, naive cosmogonists   85
Whose views clash. Nomad inlanders meanwhile
Begin to cluster where the totem
Of a certain vibrant egg-yolk yellow
Or pelt of what emerging animal
Acts on the straggler like a trumpet call   90
To form a more sophisticated unit.
By suppertime two ragged wooden clouds
Have formed. In one, a Sheik with beard
And flashing sword hilt (he is all but finished)
Steps forward on a tiger skin. A piece   95
Snaps shut, and fangs gnash out at us!
In the second cloud—they gaze from cloud to cloud
With marked if undecipherable feeling—
Most of a dark-eyed woman veiled in mauve
Is being helped down from her camel (kneeling)   100
By a small backward-looking slave or page-boy
(Her son, thinks Mademoiselle mistakenly)
Whose feet have not been found. But lucky finds
In the last minutes before bed
Anchor both factions to the scene's limits   105
And, by so doing, orient
Them eye to eye across the green abyss.
The yellow promises, oh bliss,
To be in time a sumptuous tent.

Puzzle begun I write in the day's space,   110
Then, while she bathes, peek at Mademoiselle's
Page to the curé: ". . . cette innocente mère,
Ce pauvre enfant, que deviendront-ils?"[5]
Her azure script is curlicued like pieces
Of the puzzle she will be telling him about.   115
(Fearful incuriosity of childhood!
"Tu as l'accent allemand,"[6] said Dominique.
Indeed. Mademoiselle was only French by marriage.
Child of an English mother, a remote
Descendant of the great explorer Speke,   120
And Prussian father. No one knew. I heard it
Long afterwards from her nephew, a UN
Interpreter. His matter-of-fact account
Touched old strings. My poor Mademoiselle,
With 1939 about to shake   125
This world where "each was the enemy, each the friend"
To its foundations, kept, though signed in blood,
Her peace a shameful secret to the end.)
"Schlaf wohl, chéri." Her kiss. Her thumb
Crossing my brow against the dreams to come.   130

This World that shifts like sand, its unforeseen
Consolidations and elate routine,
Whose Potentate had lacked a retinue?
Lo! it assembles on the shrinking Green.

Gunmetal-skinned or pale, all plumes and scars,   135
Of Vassalage the noblest avatars—
The very coffee-bearer in his vair
Vest is a swart Highness, next to ours.

Kef[7] easing Boredom, and iced syrups, thirst,
In guessed-at glooms old wives who know the worst   140
Outsweat that virile fiction of the New:
"Insh'Allah, he will tire—" "—or kill her first!"

(Hardly a proper subject for the Home,
Work of—dear Richard, I shall let *you* comb
Archives and learned journals for his name—   145
A minor lion attending on Gérôme.)[8]

While, thick as Thebes[9] whose presently complete
Gates close behind them, Houri and Afreet[1]
Both claim the Page. He wonders whom to serve,
And what his duties are, and where his feet,   150

5. "This innocent mother, this poor child, what will become of them?"
6. "You have a German accent."
7. A narcotic made from Indian hemp.
8. French painter (1824–1904), noted for his historical paintings, often of Near Eastern scenes.
9. The ancient capital of Upper Egypt.
1. Near Eastern mythological figures: "Houri": a virgin awarded to those who attain paradise; "Afreet": an evil genie.

And if we'll find, as some before us did,
That piece of Distance deep in which lies hid
Your tiny apex sugary with sun,
Eternal Triangle, Great Pyramid!

Then Sky alone is left, a hundred blue   155
Fragments in revolution, with no clue
To where a Niche will open. Quite a task,
Putting together Heaven, yet we do.

It's done. Here under the table all along
Were those missing feet. It's done.   160

The dog's tail thumping. Mademoiselle sketching
Costumes for a coming harem drama
To star the goosegirl. All too soon the swift
Dismantling. Lifted by two corners,
The puzzle hung together—and did not.   165
Irresistibly a populace
Unstitched of its attachments, rattled down.
Power went to pieces as the witch
Slithered easily from Virtue's gown.
The blue held out for time, but crumbled, too.   170
The city had long fallen, and the tent,
A separating sauce mousseline,
Been swept away. Remained the green
On which the grown-ups gambled. A green dusk.
First lightning bugs. Last glow of west   175
Green in the false eyes of (coincidence)
Our mangy tiger safe on his bared hearth.

Before the puzzle was boxed and readdressed
To the puzzle shop in the mid-Sixties,
Something tells me that one piece contrived   180
To stay in the boy's pocket. How do I know?
I know because so many later puzzles
had missing pieces—Maggie Teyte's[2] high notes
Gone at the war's end, end of the vogue for collies,
A house torn down; and hadn't Mademoiselle   185
Kept back her pitiful bit of truth as well?
I've spent the last days, furthermore,
Ransacking Athens for that translation of "Palme."
Neither the Goethehaus nor the National Library
Seems able to unearth it. Yet I can't   190
Just be imagining. I've seen it. Know
How much of the sun-ripe original
Felicity Rilke made himself forego
(Who loved French words—verger, mûr, parfumer)[3]

2. English soprano (1888–1976), famous for her singing of French opera and songs.

3. "Orchard, ripe, to scent."

In order to render its underlying sense.   195
Know already in that tongue of his
What Pains, what monolithic Truths
Shadow stanza to stanza's symmetrical
Rhyme-rutted pavement. Know that ground plan left
Sublime and barren, where the warm Romance   200
Stone by stone faded, cooled; the fluted nouns
Made taller, lonelier than life
By leaf-carved capitals in the afterglow.
The owlet umlaut[4] peeps and hoots
Above the open vowel. And after rain   205
A deep reverberation fills with stars.

Lost, is it, buried? One more missing piece?

But nothing's lost. Or else: all is translation
And every bit of us is lost in it
(Or found—I wander through the ruin of S[5]   210
Now and then, wondering at the peacefulness)
And in that loss a self-effacing tree,
Color of context, imperceptibly
Rustling with its angel, turns the waste
To shade and fiber, milk and memory.   215

                                                   1976

---

4. A German accent mark (¨).   5. Initial of former lover.

FRANK O'HARA

1926–1966

After Frank O'Hara's death, when the critic Donald Allen gathered O'Hara's
Collected Poems, he was surprised to discover that there were more than
500, many not published before. Some had to be retrieved from letters
or from scraps of paper in boxes and trunks. O'Hara's poems were often
spontaneous acts, revised minimally or not at all, then scattered generously,
half forgotten. His work was published, not by large commercial presses
but by art galleries such as Tibor de Nagy and by small presses. These influential but fugitive paperbacks—*A City Winter* (1952), *Meditations
in an Emergency* (1956), *Lunch Poems* (1964), and *Second Avenue*
(1960)—included love poems, "letter" poems, "post-cards," and odes,
each bearing the mark of its occasion: a birthday, a thank you, memories
of a lunch hour, or simply *Having a Coke with You*. They are filled, like
diaries, with the names of Manhattan streets, writers, artists, restaurants,
cafés, and films. O'Hara practiced what he once called, in mockery of
sober poetic manifestoes, "personism." The term came to him one day at
the office when he was writing a poem for someone he loved. "While I
was writing it I was realizing that if I wanted to I could use the telephone
instead of writing the poem, and so Personism was born. * * * It puts

the poem squarely between the poet and the person, Lucky Pierre style, and the poem is correspondingly gratified. The poem is at last between two persons instead of two pages."

O'Hara came to live in New York in 1951. He was born in Baltimore and grew up in Worcester, Massachusetts. He was in the navy for two years (with service in the South Pacific and Japan), then at Harvard, where he majored in music and English. In New York he became involved in the art world, working at different times as an editor and critic for *Art News* and a curator for the Museum of Modern Art. But this was more than a way of making a living; it was also making a life. These were the years in which Abstract Expressionism—nonrepresentational painting— flourished, and New York replaced Paris as the art capital of the world. O'Hara met and wrote about painters such as Willem de Kooning, Franz Kline, and Jackson Pollock, then producing their most brilliant work. After 1955, as a special assistant in the International Program of the Museum of Modern Art, O'Hara helped organize important traveling exhibitions which introduced and impressed the new American painting upon the art world abroad.

As friends, many of these painters and sculptors were the occasion for and recipients of O'Hara's poems. Even more important, their way of working served as a model for his own style of writing. As the poet John Ashbery puts it, "The poem [is] the chronicle of the creative act that produces it." At the simplest level this means including the random jumps, distractions, and loose associations involved in writing about a particular moment, and sometimes recording the pauses in the writing of the poem. ("And now that I have finished dinner I can continue.") In O'Hara's work the casual is often, unexpectedly, the launching point for the visionary. The offhand chronicle of a lunch-hour walk can suddenly crystallize around a thunderclap memory of three friends, artists who died young: "First / Bunny died, then John Latouche, / then Jackson Pollock. But is the / earth as full as life was full, of them?"

Frank O'Hara was indisputably, for his generation, *the* poet of New York. The very title of one of his poems, *To the Mountains in New York,* suggests that the city was for him what pastoral or rural worlds were for other writers, a source of refreshment and fantasy.

> I love this hairy city.
> It's wrinkled like a detective story
> and noisy and getting fat and smudged
> lids hood the sharp hard black eyes.

Behind the exultation of O'Hara's cityscapes, a reader can often sense the melancholy which is made explicit in poems such as *A Step Away from Them.* Part of the city's allure was that it answered O'Hara's driving need to reach out for friends, events, animation. His eagerness is balanced on "the wilderness wish / of wanting to be everything to everybody everywhere." There is also in O'Hara's poetry an understanding of how urban life and the world of machines can devour the spirit; he was fascinated with, and wrote several poems about, the young actor James Dean, whose addiction to racing culminated in a fatal automobile accident.

O'Hara's example encouraged other poets—John Ashbery, Kenneth Koch, and James Schuyler. Loosely known as the New York School of

Poets, they occasionally collaborated on poems, plays, and happenings. O'Hara's bravado was a rallying point for these writers outside the more traditional and historically conscious modernism of Pound and Eliot. As John Ashbery remembers, "He was more influenced by contemporary music and art than by what had been going on in American poetry." His poems were like "inspired rambling," open to all levels and areas of experience, expressed in a colloquial tone which could easily shade into Surrealistic dream. "I'm too blue, / An elephant takes up his trumpet, / money flutters from the windows of cries."

A few days after his fortieth birthday in 1966, O'Hara was struck down at night by a beach-buggy on Fire Island, New York. He died a few hours later. With his death, the nourishing interaction of painting and writing in New York, as well as their fertilizing effect on dance and theater, seemed over. Without his central communicating figure, the fields tended to seal themselves off once more.

To the Harbormaster

I wanted to be sure to reach you;
though my ship was on the way it got caught
in some moorings. I am always tying up
and then deciding to depart. In storms and
at sunset, with the metallic coils of the tide   5
around my fathomless arms, I am unable
to understand the forms of my vanity
or I am hard alee with my Polish rudder[1]
in my hand and the sun sinking. To
you I offer my hull and the tattered cordage   10
of my will. The terrible channels where
the wind drives me against the brown lips
of the reeds are not all behind me. Yet
I trust the sanity of my vessel; and
if it sinks, it may well be in answer   15
to the reasoning of the eternal voices,
the waves which have kept me from reaching you.

1954?   1957

Sleeping on the Wing

Perhaps it is to avoid some great sadness,
as in a Restoration tragedy[2] the hero cries "Sleep!
O for a long sound sleep and so forget it!"
that one flies, soaring above the shoreless city,
veering upward from the pavement as a pigeon   5

1. Probably a submerged comic reference to "The Polish Rider" by Rembrandt. O'Hara said this poem was about his friend, the contemporary painter Larry Rivers, who expressed a continuing fascination with Rembrandt's painting of a knight on horseback. "Hard alee": a movement toward the lee or sheltered side of a sailboat; i.e., away from the wind.

2. An especially melodramatic or rhetorical brand of tragedy from the period of English history after the restoration of Charles II to the throne (1660).

does when a car honks or a door slams, the door
of dreams, life perpetuated in parti-colored loves
and beautiful lies all in different languages.

Fear drops away too, like the cement, and you
are over the Atlantic. Where is Spain? where is   10
who? The Civil War was fought to free the slaves,
was it? A sudden down-draught reminds you of gravity
and your position in respect to human love. But
here is where the gods are, speculating, bemused.
Once you are helpless, you are free, can you believe   15
that? Never to waken to the sad struggle of a face?
to travel always over some impersonal vastness,
to be out of, forever, neither in nor for!
The eyes roll asleep as if turned by the wind
and the lids flutter open slightly like a wing.   20
The world is an iceberg, so much is invisible!
and was and is, and yet the form, it may be sleeping
too. Those features etched in the ice of someone
loved who died, you are a sculptor dreaming of space
and speed, your hand alone could have done this.   25
Curiosity, the passionate hand of desire. Dead,
or sleeping? Is there speed enough? And, swooping,
you relinquish all that you have made your own,
the kingdom of your self sailing, for you must awake
and breathe your warmth in this beloved image   30
whether it's dead or merely disappearing,
as space is disappearing and your singularity.

1955   1957

A Step Away from Them

It's my lunch hour, so I go
for a walk among the hum-colored
cabs. First, down the sidewalk
where laborers feed their dirty
glistening torsos sandwiches   5
and Coca-Cola, with yellow helmets
on. They protect them from falling
bricks, I guess. Then onto the
avenue where skirts are flipping
above heels and blow up over   10
grates. The sun is hot, but the
cabs stir up the air. I look
at bargains in wristwatches. There
are cats playing in sawdust.
                    On   15
to Times Square, where the sign[1]
blows smoke over my head, and higher

---

1. Famous steam-puffing billboard advertising cigarettes.

the waterfall pours lightly. A
Negro stands in a doorway with a
toothpick, languorously agitating.   20
A blonde chorus girl clicks: he
smiles and rubs his chin. Everything
suddenly honks: it is 12:40 of
a Thursday.

       Neon in daylight is a   25
great pleasure, as Edwin Denby[2] would
write, as are light bulbs in daylight.
I stop for a cheeseburger at JULIET'S
CORNER. Giulietta Masina, wife of
Federico Fellini, *è bell' attrice*.[3]   30
And chocolate malted. A lady in
foxes on such a day puts her poodle
in a cab.

       There are several Puerto
Ricans on the avenue today, which   35
makes it beautiful and warm. First
Bunny died, then John Latouche,
then Jackson Pollock.[4] But is the
earth as full as life was full, of them?
And one has eaten and one walks,   40
past the magazines with nudes
and the posters for BULLFIGHT and
the Manhattan Storage Warehouse,
which they'll soon tear down. I
used to think they had the Armory[5]   45
Show there.

       A glass of papaya juice
and back to work. My heart is in my
pocket, it is Poems by Pierre Reverdy.[6]

1956   1964

The Day Lady[1] Died

It is 12:20 in New York a Friday
three days after Bastille day,[2] yes
it is 1959 and I go get a shoeshine
because I will get off the 4:19 in Easthampton[3]

---

2. Fellow poet (b. 1923) and influential ballet critic.
3. "A beautiful actress." Masina starred in many of director Fellini's best-known films, such as *La Strada* (1954) and *Nights of Cabiria* (1956).
4. V. R. Lang (1924–56), poet and director of The Poets' Theater in Cambridge, Massachusetts, where she produced several of O'Hara's plays; Latouche (1917–56), lyricist for several New York musicals, such as *The Golden Apple*; Pollock (1912–1956), Abstract Expressionist painter, considered the originator of "action" painting. All three were gifted friends of the poet who met tragic deaths.
5. Site of the influential and controversial first American showing of European Post-Impressionist painters in 1913.
6. French poet (1899–1960), whose work strongly influenced O'Hara's writing.
1. Billie Holiday (1915–59), also known as Lady Day, the immortal black singer of jazz and the blues.
2. July 14, the French national holiday.
3. Town in eastern Long Island, popular summer resort among New York artists.

at 7:15 and then go straight to dinner   5
and I don't know the people who will feed me

I walk up the muggy street beginning to sun
and have a hamburger and a malted and buy
an ugly NEW WORLD WRITING to see what the poets
in Ghana are doing these days   10
                              I go on to the bank
and Miss Stillwagon (first name Linda I once heard)
doesn't even look up my balance for once in her life
and in the GOLDEN GRIFFIN[4] I get a little Verlaine
for Patsy with drawings by Bonnard although I do   15
think of Hesiod, trans. Richmond Lattimore or
Brendan Behan's new play or *Le Balcon* or *Les Nègres*
of Genet, but I don't, I stick with Verlaine
after practically going to sleep with quandariness

and for Mike I just stroll into the PARK LANE   20
Liquor Store and ask for a bottle of Strega and
then I go back where I came from to 6th Avenue
and the tobacconist in the Ziegfeld Theatre and
casually ask for a carton of Gauloises and a carton
of Picayunes, and a NEW YORK POST with her face on it   25

and I am sweating a lot by now and thinking of
leaning on the john door in the 5 SPOT
while she whispered a song along the keyboard
to Mal Waldron[5] and everyone and I stopped breathing

1959   1960

Ave Maria[1]

Mothers of America
                    let your kids go to the movies!
get them out of the house so they won't know what you're up to
it's true that fresh air is good for the body
                              but what about the soul   5
that grows in darkness, embossed by silvery images
and when you grow old as grow old you must
                              they won't hate you
they won't criticize you they won't know
                    they'll be in some glamorous country   10
they first saw on a Saturday afternoon or playing hookey

---

they may even be grateful to you
                              for their first sexual experience
which only cost you a quarter
                              and didn't upset the peaceful home   15
they will know where candy bars come from
                              and gratuitous bags of popcorn
as gratuitous as leaving the movie before it's over
with a pleasant stranger whose apartment is in the Heaven on Earth
       Bldg
near the Williamsburg Bridge[2]   20
                         oh mothers you will have made the little tykes
so happy because if nobody does pick them up in the movies
they won't know the difference
                         and if somebody does it'll be sheer gravy
and they'll have been truly entertained either way   25
instead of hanging around the yard
                         or up in their room
                                        hating you
prematurely since you won't have done anything horribly mean yet
except keeping them from the darker joys   30
                              it's unforgivable the latter
so don't blame me if you won't take this advice
                              and the family breaks up
and your children grow old and blind in front of a TV set
                                        seeing   35
movies you wouldn't let them see when they were young
1960   1964

2. Bridge connecting lower Manhattan with the Williamsburg section of Brooklyn.

JOHN ASHBERY
1927–

John Ashbery has described his writing this way: "I think that any one of my poems might be considered to be a snapshot of whatever is going on in my mind at the time—first of all the desire to write a poem, after that wondering if I've left the oven on or thinking about where I must be in the next hour." Ashbery has developed a style hospitable to quicksilver changes in tone and attention, to the awkward comedy of the unrelated thoughts and things at any given moment pressing in on what we say or do. So, for example, *Grand Galop* (1965) begins:

All things seem mention of themselves
And the names which stem from them branch out to
                         other referents.
Hugely, spring exits again. The weigela does its
                         dusty thing
In fire-hammered air. And garbage cans are heaved against
The railing as the tulips yawn and crack open and fall apart.

The poem goes on to give some lunch menus which include "sloppy joe on bun." The proximity of "fire-hammered air" and "sloppy joe on bun" suggest a tension central to Ashbery's work: intuitions of vision and poetic ecstasy side by side with the commonplace. Ashbery's work implies that the two kinds of experience are inseparable in the mind.

Ashbery's poetry was not always so open to contradictory notions and impulses. His early books rejected the mere surfaces of realism and the momentary in order to get at "remoter areas of consciousness." The protagonist of *Illustration* (from his first book, *Some Trees*) is a cheerful nun about to leave behind the irrelevancies of the world by leaping from a skyscraper. Her act implies that "Much that is beautiful must be discarded / So that we may resemble a taller / impression of ourselves." To reach the "remoter areas of consciousness," Ashbery tried various technical experiments. He used highly patterned forms such as the sestina in *Some Trees* and *The Tennis Court Oath* (1962) not with any show of mechanical brilliance, but to explore: "I once told somebody that writing a sestina was rather like riding downhill on a bicycle and having the pedals push your feet. I wanted my feet to be pushed into places they wouldn't normally have taken. * * *"

Ashbery was born in Rochester, N.Y., in 1927. He attended Deerfield Academy and Harvard, graduating in 1949. He received an M.A. in English from Columbia in 1951. As a Fulbright scholar in French literature, Ashbery lived in Montpellier and Paris (1955–57) and wrote about the French avant-garde novelist Raymond Roussel. Again in France (1958–65), he was art critic for the European edition of the New York *Herald Tribune* and reported the European shows and exhibitions for *Art News and Arts International*. He returned to New York in 1965 to be executive editor of *Art News*, a position he held until 1972. He is currently a professor of English in the creative writing program of Brooklyn College.

Ashbery's interest in art played a formative role in his poetry. He is often associated with Frank O'Hara, James Schuyler, and Kenneth Koch as part of the "New York School" of poets. The name refers to their common interest in the New York school of abstract painters of the 1940s and 1950s, whose energies and techniques they wished to adapt in poetry. These painters avoided realism in order to stress the work of art as a representation of the creative act which produced it—as in the "action paintings" of Jackson Pollock. Ashbery's long poem *Self-Portrait in a Convex Mirror* gives as much attention to the rapidly changing feelings of the poet in the act of writing his poem as it does to the Renaissance painting which inspired him. The poem moves back and forth between the distracted energies which feed a work of art and the completed composition, which the artist feels as both a triumph and as a falsification of complex feelings. Ashbery shares with O'Hara a sense of the colloquial brilliance of daily life in New York and sets this in tension with the concentration and stasis of art.

The book *Self-Portrait in a Convex Mirror* (1975) won the three major poetry prizes of its year, a tribute to the fact that it had perfected the play of contrasting voices—visionary and colloquial—which had long characterized Ashbery's work. He has since published *Houseboat Days* (1977).

Some Trees

These are amazing: each
Joining a neighbor, as though speech
Were a still performance.
Arranging by chance

To meet as far this morning   5
From the world as agreeing
With it, you and I
Are suddenly what the trees try

To tell us we are:
That their merely being there   10
Means something; that soon
We may touch, love, explain.

And glad not to have invented
Such comeliness, we are surrounded:
A silence already filled with noises,   15
A canvas on which emerges

A chorus of smiles, a winter morning.
Placed in a puzzling light, and moving,
Our days put on such reticence
These accents seem their own defense.   20

1956

Soonest Mended

Barely tolerated, living on the margin
In our technological society, we were always having to be rescued
On the brink of destruction, like heroines in *Orlando Furioso*[1]
Before it was time to start all over again.
There would be thunder in the bushes, a rustling of coils,   5
And Angelica, in the Ingres painting,[2] was considering
The colorful but small monster near her toe, as though wondering
     whether forgetting
The whole thing might not, in the end, be the only solution.
And then there always came a time when
Happy Hooligan[3] in his rusted green automobile   10
Came plowing down the course, just to make sure everything was
     O.K.

1. Fantastic epic poem by Ludovico Ariosto (1474–1533), whose romantic heroine Angelica is constantly being rescued from imminent perils such as monsters and ogres.
2. *Roger Delivering Angelica* (1819), a painting based on a scene from Ariosto, by the French artist Jean Auguste Dominique Ingres (1780–1867).
3. The good-natured, simple title character of a popular comic strip of the 1920s and 1930s.

Only by that time we were in another chapter and confused
About how to receive this latest piece of information.
Was it information? Weren't we rather acting this out
For someone else's benefit, thoughts in a mind   15
With room enough to spare for our little problems (so they began
        to seem),
Our daily quandary about food and the rent and bills to be paid?
To reduce all this to a small variant,
To step free at last, minuscule on the gigantic plateau—
This was our ambition: to be small and clear and free.   20
Alas, the summer's energy wanes quickly,
A moment and it is gone. And no longer
May we make the necessary arrangements, simple as they are.
Our star was brighter perhaps when it had water in it.
Now there is no question even of that, but only   25
Of holding on to the hard earth so as not to get thrown off,
With an occasional dream, a vision: a robin flies across
The upper corner of the window, you brush your hair away
And cannot quite see, or a wound will flash
Against the sweet faces of the others, something like: 30
This is what you wanted to hear, so why
Did you think of listening to something else? We are all talkers
It is true, but underneath the talk lies
The moving and not wanting to be moved, the loose
Meaning, untidy and simple like a threshing floor.[4]   35
These then were some hazards of the course,
Yet though we knew the course *was* hazards and nothing else
It was still a shock when, almost a quarter of a century later,
The clarity of the rules dawned on you for the first time.
They were the players, and we who had struggled at the game   40
Were merely spectators, though subject to its vicissitudes
And moving with it out of the tearful stadium, borne on shoulders,
        at last.
Night after night this message returns, repeated
In the flickering bulbs of the sky, raised past us, taken away from us,
Yet ours over and over until the end that is past truth,   45
The being of our sentences, in the climate that fostered them,
Not ours to own, like a book, but to be with, and sometimes
To be without, alone and desperate.
But the fantasy makes it ours, a kind of fence-sitting
Raised to the level of an esthetic ideal. These were moments,
        years,
                                                                            50
Solid with reality, faces, namable events, kisses, heroic acts,
But like the friendly beginning of a geometrical progression
Not too reassuring, as though meaning could be cast aside some day
When it had been outgrown. Better, you said, to stay cowering
Like this in the early lessons, since the promise of learning   55
Is a delusion, and I agreed, adding that

4. Used at harvest time to separate the wheat from the chaff, which is to be discarded.

Tomorrow would alter the sense of what had already been learned,
That the learning process is extended in this way, so that from this
    standpoint
None of us ever graduates from college,
For time is an emulsion,[5] and probably thinking not to grow up   60
Is the brightest kind of maturity for us, right now at any rate.
And you see, both of us were right, though nothing
Has somehow come to nothing; the avatars[6]
Of our conforming to the rules and living
Around the home have made—well, in a sense, "good citizens"
    of us,   65
Brushing the teeth and all that, and learning to accept
The charity of the hard moments as they are doled out,
For this is action, this not being sure, this careless
Preparing, sowing the seeds crooked in the furrow,
Making ready to forget, and always coming back   70
To the mooring of starting out, that day so long ago.

                1970

Definition of Blue

The rise of capitalism parallels the advance of romanticism
And the individual is dominant until the close of the nineteenth
    century.
In our own time, mass practices have sought to submerge the per-
    sonality
By ignoring it, which has caused it instead to branch out in all di-
    rections
Far from the permanent tug that used to be its notion of "home."   5
These different impetuses are received from everywhere
And are as instantly snapped back, hitting through the cold atmo-
    sphere
In one steady, intense line.

There is no remedy for this "packaging" which has supplanted the
    old sensations.
Formerly there would have been architectural screens at the point
    where the action became most difficult   10
As a path trails off into shrubbery—confusing, forgotten, yet contin-
    uing to exist.
But today there is no point in looking to imaginative new methods
Since all of them are in constant use. The most that can be said for
    them further
Is that erosion produces a kind of dust or exaggerated pumice[1]
Which fills space and transforms it, becoming a medium   15
In which it is possible to recognize oneself.

---

5. A chemical solution in which the particles of one liquid are suspended in another.

6. Incarnations.
1. The cooled, spongy residue of volcanic lava used as an abrasive to polish surfaces.

Each new diversion adds its accurate touch to the ensemble, and so
A portrait, smooth as glass, is built up out of multiple corrections
And it has no relation to the space or time in which it was lived.
Only its existence is a part of all being, and is therefore, I suppose,
    to be prized   20
Beyond chasms of night that fight us
By being hidden and present.

And yet it results in a downward motion, or rather a floating one
In which the blue surroundings drift slowly up and past you
To realize themselves some day, while, you, in this nether world that
    could not be better,
Waken each morning to the exact value of what you did and said,
    which remains.

                                                 1970

Summer

There is that sound like the wind
Forgetting in the branches that means something
Nobody can translate. And there is the sobering "later on,"
When you consider what a thing meant, and put it down.

For the time being the shadow is ample   5
And hardly seen, divided among the twigs of a tree,
The trees of a forest, just as life is divided up
Between you and me, and among all the others out there.

And the thinning-out phase follows
the period of reflection. And suddenly, to be dying   10
Is not a little or mean or cheap thing,
Only wearying, the heat unbearable,

And also the little mindless constructions put upon
Our fantasies of what we did: summer, the ball of pine needles,
The loose fates serving our acts, with token smiles,   15
Carrying out their instructions too accurately—

Too late to cancel them now—and winter, the twitter
Of cold stars at the pane, that describes with broad gestures
This state of being that is not so big after all.
Summer involves going down as a steep flight of steps   20

To a narrow ledge over the water. Is this it, then,
This iron comfort, these reasonable taboos,
Or did you mean it when you stopped? And the face
Resembles yours, the one reflected in the water.

                                               1970

Wet Casements

When Eduard Raban, coming along the passage,
walked into the open doorway, he saw that it
was raining. It was not raining much.
  —KAFKA,[1] *Wedding Preparations in the Country*

The conception is interesting: to see, as though reflected
In streaming windowpanes, the look of others through
Their own eyes. A digest of their correct impressions of
Their self-analytical attitudes overlaid by your
Ghostly transparent face. You in falbalas[2]   5
Of some distant but not too distant era, the cosmetics,
The shoes perfectly pointed, drifting (how long you
Have been drifting; how long I have too for that matter)
Like a bottle-imp[3] toward a surface which can never be approached,
Never pierced through into the timeless energy of a present   10
Which would have its own opinions on these matters,
Are an epistemological[4] snapshot of the processes
That first mentioned your name at some crowded cocktail
Party long ago, and someone (not the person addressed)
Overheard it and carried that name around in his wallet   15
For years as the wallet crumbled and bills slid in
And out of it. I want that information very much today,

Can't have it, and this makes me angry.
I shall use my anger to build a bridge like that
Of Avignon, on which people may dance for the feeling   20
Of dancing on a bridge.[5] I shall at last see my complete face
Reflected not in the water but in the worn stone floor of my bridge.

I shall keep to myself.
I shall not repeat others' comments about me.

                                                    1977

1. Franz Kafka (1883–1924), Czech writer known for his surreal and bleak vision of life.
2. Furbelows, ruffles.
3. A figure suspended in liquid. When pressure is exerted above, the figure sinks to the bottom and is then sent back to the top by counterpressure.
4. Having to do with the nature and limits of knowledge.
5. A reference to the French folksong: *"Sur le pont d'Avignon, l'on y danse.* * * * (On the bridge of Avignon, one dances there)."

ANNE SEXTON
1928–1975

Anne Sexton's first book of poems, *To Bedlam and Part Way Back* (1960), was published at a time when the label "confessional" came to be attached to poems more frankly autobiographical than had been usual in American verse. For Sexton the term "confessional" is particularly apt. Though she had abandoned the Roman Catholicism into which she was born, her poems enact something analogous to preparing for and receiving religious absolution. She dedicates a poem to a friend who "urges me to make an appointment for the Sacrament of Confession,"

> My friend, my friend, I was born
> doing reference work in sin, and born
> confessing it. This is what poems are;
> with mercy
> for the greedy,
> they are the tongue's wrangle,
> the world's pottage, the rat's star.

Sexton's own confessions were to be made in terms more startling than the traditional Catholic images of her childhood. The purpose of her poems was not to analyze or explain behavior but to make it palpable in all its ferocity of feeling. Poetry "should be a shock to the senses. It should also hurt." This is apparent both in the themes she chooses and the particular ways in which she chooses to exhibit her subjects. Sexton writes about sex, illegitimacy, guilt, madness, suicide. Her first book portrays her own mental breakdown, her time in a mental hospital, her efforts at reconciliation with her young daughter and husband when she returns. Her second book, *All My Pretty Ones* (1962) takes its title from *Macbeth* and refers to the death of both her parents within three months of one another. Later books act out a continuing debate about suicide: *Live or Die* (1967), *The Death Notebooks* (1974), and *The Awful Rowing toward God* (1975—posthumous), titles that prefigure the time when she took her own life (1975).

Sexton spoke of images as "the heart of poetry. Images come from the unconscious. Imagination and the unconscious are one and the same." In the mental hospital she sees herself as "the laughing bee on a stalk / of death"; after an operation, as a little girl sent out to play "my stomach laced up like a football / for the game." The poems are very often addressed to other people or entities with whom she is trying to reconnect herself: her parents, her husband, a lover, her child, even her own public self or her body. Powerful images substantiate the strangeness of her own feelings, and attempt to redefine experiences so as to gain understanding, absolution, or revenge. These poems poised between, as her titles suggest, life and death, or "bedlam and part way back" are efforts at establishing a middle ground of self-assertion, substituting surreal images for the reductive versions of life visible to the exterior eye.

Anne Sexton was born in 1928 in Newton, Massachusetts, and attended Garland Junior College. She came to poetry fairly late—when she was twenty-eight, after seeing the critic I. A. Richards lecturing about the sonnet on television. In the late 1950s she attended poetry workshops in the Boston area, including Robert Lowell's poetry seminars at Boston University. One of her fellow students was Sylvia Plath, whose suicide she commemorated in a poem and whose fate she later followed. Sexton claimed that she was less influenced by Lowell's *Life Studies* than by W. D. Snodgrass's autobiographical *Heart's Needle* (1959), but certainly Lowell's support and the association with Plath left their mark upon her and made it possible for her to publish. Though her career was relatively brief, she received several major literary prizes, including the Pulitzer Prize for *Live or Die* and an American Academy of Arts and Letters traveling fellowship. Her suicide came after a series of mental breakdowns.

You, Doctor Martin[1]

You, Doctor Martin, walk
    from breakfast to madness. Late August,
I speed through the antiseptic tunnel
    where the moving dead still talk
of pushing their bones against the thrust   5
of cure. And I am queen of this summer hotel
    or the laughing bee on a stalk

    of death. We stand in broken
lines and wait while they unlock
the door and count us at the frozen gates   10
    of dinner. The shibboleth[2] is spoken
and we move to gravy in our smock
of smiles. We chew in rows, our plates
    scratch and whine like chalk

    in school. There are no knives   15
for cutting your throat. I make
moccasins all morning. At first my hands
    kept empty, unraveled for the lives
they used to work. Now I learn to take
them back, each angry finger that demands   20
    I mend what another will break

    tomorrow. Of course, I love you;
you lean above the plastic sky,
god of our block, prince of all the foxes.
    The breaking crowns are new   25
that Jack wore.[3] Your third eye[4]
moves among us and lights the separate boxes
    where we sleep or cry.

    What large children we are
here. All over I grow most tall   30
in the best ward. Your business is people,
    you call at the madhouse, an oracular
eye in our nest. Out in the hall

1. Sexton's doctor in a mental hospital, the setting of the poem.
2. Password (as contrasted to saying grace before meals).
3. As in the nursery rhyme *Jack and Jill*, where "Jack fell down and broke his crown."
4. In Buddhism, the doorway of the soul, located between and above the two eyes, between the prefrontal lobes of the brain.

the intercom pages you. You twist in the pull
of the foxy children who fall   35

like floods of life in frost.
And we are magic talking to itself,
noisy and alone. I am queen of all my sins
forgotten. Am I still lost?
Once I was beautiful. Now I am myself,   40
counting this row and that row of moccasins
waiting on the silent shelf.

1960

All My Pretty Ones[1]

Father, this year's jinx rides us apart
where you followed our mother to her cold slumber;
a second shock boiling its stone to your heart,
leaving me here to shuffle and disencumber
you from the residence you could not afford: 5
a gold key, your half of a woolen mill,
twenty suits from Dunne's, an English Ford,
the love and legal verbiage of another will,
boxes of pictures of people I do not know.
I touch their cardboard faces. They must go.   10

But the eyes, as thick as wood in this album,
hold me. I stop here, where a small boy
waits in a ruffled dress for someone to come . . .
for this soldier who holds his bugle like a toy
or for this velvet lady who cannot smile.   15
Is this your father's father, this commodore
in a mailman suit? My father, time meanwhile
has made it unimportant who you are looking for.
I'll never know what these faces are all about.
I lock them into their book and throw them out.   20

This is the yellow scrapbook that you began
the year I was born; as crackling now and wrinkly
as tobacco leaves: clippings where Hoover outran
the Democrats, wiggling his dry finger at me
and Prohibition;[2] news where the *Hindenburg* went   25
down[3] and recent years where you went flush

1. From the epigraph to the volume of
which this is the title poem. In Shakespeare's *Macbeth* 4.3.216 ff. Macduff reacts to the news that Macbeth has had
his wife and children slaughtered: "All
my pretty ones? / Did you say all? O
hell-kite! All? / What! all my pretty
chickens and their dam / At one fell
swoop? * * * / I cannot but remember
such things were, / That were most precious to me."
Macduff later kills Macbeth. Sexton's
poem is an elegy for her father, who died
in June, 1959, four months after the death
of his wife.
2. Herbert Hoover (1874–1964) won the
presidential election in 1928; "Prohibition": the period (1920–33) when the
sale of alcoholic beverages was outlawed.
3. The *Hindenburg* was a German passenger zeppelin, which exploded and
crashed in flames at Lakehurst, New Jersey, in 1937.

on war. This year, solvent but sick, you meant
to marry that pretty widow in a one-month rush.
But before you had that second chance, I cried
on your fat shoulder. Three days later you died.   30

These are the snapshots of marriage, stopped in places.
Side by side at the rail toward Nassau⁴ now;
here, with the winner's cup at the speedboat races,
here, in tails at the Cotillion⁵, you take a bow,
here, by our kennel of dogs with their pink eyes,   35
running like show-bred pigs in their chain-link pen;
here, at the horseshow where my sister wins a prize;
and here, standing like a duke among groups of men.
Now I fold you down, my drunkard, my navigator,
my first lost keeper, to love or look at later.   40

I hold a five-year diary that my mother kept
for three years, telling all she does not say
of your alcoholic tendency. You overslept,
she writes. My God, father, each Christmas Day
with your blood, will I drink down your glass   45
of wine? The diary of your hurly-burly years
goes to my shelf to wait for my age to pass.
Only in this hoarded span will love persevere.
Whether you are pretty or not, I outlive you,
bend down my strange face to yours and forgive you.   50

1962

Sylvia's Death

FOR SYLVIA PLATH¹

O Sylvia, Sylvia,
with a dead box of stones and spoons,

with two children, two meteors
wandering loose in the tiny playroom,

with your mouth into the sheet,   5
into the roofbeam, into the dumb prayer,

(Sylvia, Sylvia,
where did you go
after you wrote me   10
from Devonshire
about raising potatoes
and keeping bees?)

4. In the Bahamas.
5. Usually a ball at which debutantes (age 18) are presented to society.
1. American poet (1932–63), friend of Sexton's, and a suicide. Plath was living in Devonshire, England, with her husband, the poet Ted Hughes, at the time of her death.

what did you stand by,
just how did you lie down into?

Thief!— 15
how did you crawl into,

crawl down alone
into the death I wanted so badly and for so long,

the death we said we both outgrew,
the one we wore on our skinny breasts, 20

the one we talked of so often each time
we downed three extra dry martinis in Boston,

the death that talked of analysts and cures,
the death that talked like brides with plots,

the death we drank to, 25
the motives and then the quiet deed?

(In Boston
the dying
ride in cabs,
yes death again, 30
that ride home
with *our* boy.)

O Sylvia, I remember the sleepy drummer
who beat on our eyes with an old story,

how we wanted to let him come 35
like a sadist or a New York fairy

to do his job,
a necessity, a window in a wall or a crib,

and since that time he waited
under our heart, our cupboard, 40

and I see now that we store him up
year after year, old suicides

and I know at the news of your death,
a terrible taste for it, like salt.

(And me, 45
me too.
And now, Sylvia,
you again

with death again,
that ride home   50
with *our* boy.)

And I say only
with my arms stretched out into that stone place,

what is your death
but an old belonging,   55

a mole that fell out
of one of your poems?

(O friend,
while the moon's bad,
and the king's gone,   60
and the queen's at her wit's end
the bar fly ought to sing!)

O tiny mother,
you too!
O funny duchess!   65
O blonde thing!
February 17, 1963   1966

ADRIENNE RICH

1929–

Adrienne Rich began her career with a prize particularly important for
poets of her generation. She was chosen, as were James Wright, John Ashbery, and W. S. Merwin in other years, to have her first book of poems
published in the Yale Series of Younger Poets. W. H. Auden, judge for
the series in the 1950s, said of Rich's volume, *A Change of World* (1951),
that her poems "were neatly and modestly dressed * * * respect their
elders but are not cowed by them, and do not tell fibs." Rich, looking
back at that period from the vantage of 1972, gives a more complicated
sense of it. In an influential essay on women writers, *When We Dead
Awaken*, she remembers this period during and just after her years at Radcliffe College as one in which the chief models for young women who
wanted to write poetry were the admired male poets of the time: Robert
Frost, Dylan Thomas, W. H. Auden, Wallace Stevens, W. B. Yeats. From
their work she learned her craft. Even in looking at the poetry of older
women writers she found herself "looking * * * for the same things I found
in the poetry of men, because I wanted women poets to be the equals of
men, and to be equal was still confused with sounding the same." She felt
then—and was to feel even more sharply in the years following her marriage (in 1953 to the economist Alfred Conrad)—"the split * * * between

the girl * * * who defined herself in writing poems, and the girl who was to define herself by her relationships with men."

Rich has worked, first tentatively and then more securely, to describe and explore this conflict. Twenty years and five volumes after *A Change of World* she published a collection called *The Will to Change*. Those titles accurately reflect the important turn in Rich's work: from acceptance of change as the way of the world to a resolute relocation of power and decision within the self. Speaking twenty years later of *Storm Warnings* from *A Change of World*, Rich remarked, "I'm amazed at the number of images of glass breaking—as if you're the one on the inside and the glass is being broken from without. You're somehow menaced. Now I guess I think of it in reverse. I think of the whole necessity of smashing panes if you're going to save yourself from a burning building."

In her twenties, Rich gave birth to three children within four years; "a radicalizing experience" she said and the chief transforming event in both her personal and writing life. Trying to be the ideal faculty wife and hostess, undergoing difficult pregnancies, and taking care of three small sons, she had little time or energy for writing. In 1955 her second book, *The Diamond Cutters*, appeared; but eight years were to pass before she published another. It was during this time that Rich experienced most severely that gap between what she calls the "energy of creation" and the "energy of relation. * * * In those early years I always felt the conflict as a failure of love in myself." The identification of the writer's imagination with feelings of guilt and cold egotism is voiced in the poem *Orion*, in which a young housewife leaves the demands of home, husband, and children behind and twins herself in fantasy with a male persona, her childhood hero, the giant constellation Orion.

With her third and fourth books, *Snapshots of a Daughter-in-Law* (1963) and *Necessities of Life* (1966), Rich began explicitly to treat problems which have engaged her ever since. The title poem of *Snapshots* exposes the gap between literary versions of women's experience and the day-to-day truths of their lives. Fragments of familiar poems praising and glamorizing women are juxtaposed with scenes from ordinary home life—in a way which adapts the technique of Eliot's *The Waste Land* to more overtly social protest.

When Rich and her husband moved to New York City in 1966 they became increasingly involved in radical politics, especially in the opposition to the Vietnam war. These new concerns are reflected in the poems of *Leaflets* (1969) and *The Will to Change* (1971). But along with new subject matter came equally important changes in style. Rich's poems throughout the 1960s moved away from formal verse patterns to more jagged utterance. Sentence fragments, lines of varying length, irregular spacing to mark off phrases—all these devices emphasized a voice of greater urgency. Ever since *Snapshots of a Daughter-in-Law* Rich had been dating each poem, as if to mark them as provisional, true to the moment but instruments of passage, like entries in a journal where feelings are subject to continual revision. Rich experimented with forms which would give greater prominence to the images in her poetry. Hence, in *The Will to Change*, she refers to and tries to find verbal equivalents for the devices of films, such as quick cuts, close-ups. (One of the poems is called *Shoot-*

ing Script.) In speeded-up and more concentrated attention to images, she saw a more intense way to expose the contradictions of consciousness, the simultaneity of opposing emotions.

Rich sees a strong relation between poetry and feminism. Poetry becomes a means for discovering the inner world of women, just as the telescope was for exploring the heavens. As she says in *Planetarium*,

> I am an instrument in the shape
> of a woman trying to translate pulsations
> into images   for the relief of the body
> and the reconstruction of the mind.

The kind of communication Rich most values is envisioned in *Face to Face*. The speaker, a nineteenth-century American pioneer woman, waits in stillness on the frontier for a reunion with her absent husband. Her winter's isolation has allowed her privacy and a concentration to find "plain words." The two will meet

> each with his God-given secret,
> spelled out through months of snow and silence,
> burning under the bleached scalp; behind dry lips
> a loaded gun.

The last line, quoted from Emily Dickinson, suggests how such meetings with loved ones are both dreaded and desired, how they are potentially explosive but also reach out for understanding, "a poetry of dialogue and of the furious effort to break through to dialogue," as one critic remarks. Rich's poems aim at self-definition, at establishing boundaries of the self, but they also fight off the notion that insights remain solitary and unshared. Many of her poems proceed in a tone of intimate argument, as if understanding, political as well as personal, is only manifest in the terms with which we explain ourselves to lovers, friends, and our closest selves. Others give full and powerful expression to the stored-up angers and resentments in women's experience.

In the 1970s Rich's increasing commitment to feminism and to the study of the writings of women resulted in a book of prose, *Of Woman Born: Motherhood as Experience and Institution* (1976). Partly autobiographical and partly historical and anthropological, the book weighs the actual feelings involved in bearing and rearing children against the myths and expectations fostered by our medical, social, and political institutions. Like Rich's poetry, *Of Woman Born* refuses to separate the immediacy of private feelings from political commentary and decision; it suggests that all change originates in changes of individual consciousness.

The Roofwalker

FOR DENISE LEVERTOV

Over the half-finished houses
night comes. The builders
stand on the roof. It is

quiet after the hammers,
the pulleys hang slack.   5
Giants, the roofwalkers,
on a listing deck, the wave
of darkness about to break
on their heads. The sky
is a torn sail where figures   10
pass magnified, shadows
on a burning deck.

I feel like them up there:
exposed, larger than life,
and due to break my neck.   15

Was it worth while to lay—
with infinite exertion—
a roof I can't live under?
—All those blueprints,
closings of gaps,   20
measurings, calculations?
A life I didn't choose
chose me: even
my tools are the wrong ones
for what I have to do.   25
I'm naked, ignorant,
a naked man fleeing
across the roofs
who could with a shade of difference
be sitting in the lamplight   30
against the cream wallpaper
reading—not with indifference—
about a naked man
fleeing across the roofs.

1961   1962

Face to Face

Never to be lonely like that—
the Early American figure on the beach
in black coat and knee-breeches
scanning the didactic storm in privacy,

never to hear the prairie wolves   5
in their lunar hilarity
circling one's little all, one's claim
to be Law and Prophets

for all that lawlessness,
never to whet the appetite   10
weeks early, for a face, a hand
longed-for and dreaded—

How people used to meet!
starved, intense, the old
Christmas gifts saved up till spring,   15
and the old plain words,

and each with his God-given secret,
spelled out through months of snow and silence,
burning under the bleached scalp; behind dry lips
a loaded gun.[2]   20

1965   1966

Orion[1]

Far back when I went zig-zagging
through tamarack pastures
you were my genius, you
my cast-iron Viking, my helmed
lion-heart king[2] in prison.   5
Years later now you're young

my fierce half-brother, staring
down from that simplified west
your breast open, your belt dragged down
by an oldfashioned thing, a sword   10
the last bravado you won't give over
though it weighs you down as you stride

and the stars in it are dim
and maybe have stopped burning.
But you burn, and I know it;   15
as I throw back my head to take you in
an old transfusion happens again:
divine astronomy is nothing to it.

Indoors I bruise and blunder,
break faith, leave ill enough   20
alone, a dead child born in the dark.
Night cracks up over the chimney,
pieces of time, frozen geodes[3]
come showering down in the grate.

A man reaches behind my eyes   25
and finds them empty
a woman's head turns away
from my head in the mirror
children are dying my death
and eating crumbs of my life.   30

2. Allusion to Emily Dickinson's poem, *My Life had stood—a Loaded Gun.*
1. A constellation of the winter sky, popularly known as the Hunter, which appears as a giant with a belt and sword.
2. Allusion to King Richard the Lion-Heart of England (1157–99), imprisoned in Austria on his return from the Crusades.
3. Small, spheroid stones, with a cavity usually lined with crystals.

Pity is not your forte.
Calmly you ache up there
pinned aloft in your crow's nest,[4]
my speechless pirate!
You take it all for granted   35
and when I look you back

it's with a starlike eye
shooting its cold and egotistical[5] spear
where it can do least damage.
Breathe deep! No hurt, no pardon   40
out here in the cold with you
you with your back to the wall.

1965   1969

Planetarium

THINKING OF CAROLINE HERSCHEL (1750–1848)
ASTRONOMER, SISTER OF WILLIAM,[1] AND OTHERS.

A woman in the shape of a monster
a monster in the shape of a woman
the skies are full of them

a woman   'in the snow
among the Clocks and instruments   5
or measuring the ground with poles'

in her 98 years to discover
8 comets
she whom the moon ruled
like us   10
levitating into the night sky
riding the polished lenses

Galaxies of women, there
doing penance for impetuousness
ribs chilled   15
in those spaces   of the mind

An eye,

            'virile, precise and absolutely certain'
            from the mad webs of Uranusborg[2]

---

4. Lookout post on the mast of old ships.
5. "One of two phrases suggested by Gottfried Benn's essay, *Artists and Old Age* in *Primal Vision*, edited by E. B. Ashton, New Directions" [Rich's note]. Benn's advice to the modern artist is: "Don't lose sight of the cold and egotistical element in your mission. * * * With your back to the wall, care-worn and weary, in the gray light of the void, read Job and Jeremiah and keep going" (pp. 206–7).

1. William Herschel (1738–1822), discoverer of the planet Uranus.
2. Uranienborg, an observatory built by Danish astronomer Tycho Brahe (1546–1601).

encountering the NOVA[3]   20

every impulse of light exploding
from the core
as life flies out of us

        Tycho whispering at last
        'Let me not seem to have lived in vain'   25

What we see, we see
and seeing is changing

the light that shrivels a mountain
and leaves a man alive

Heartbeat of the pulsar[4]   30
heart sweating through my body

The radio impulse
pouring in from Taurus[5]

        I am bombarded yet   I stand

I have been standing all my life in the   35
direct path of a battery of signals
the most accurately transmitted most
untranslatable language in the universe
I am a galactic[6] cloud so deep   so invo-   40
luted that a light wave could take 15
years to travel through me   And has
taken   I am an instrument in the shape
of a woman trying to translate pulsations
into images   for the relief of the body
and the reconstruction of the mind.   45
1968   1971

Diving into the Wreck

First having read the book of myths,
and loaded the camera,
and checked the edge of the knife-blade,
I put on
the body-armor of black rubber   5
the absurd flippers
the grave and awkward mask.
I am having to do this
not like Cousteau[1] with his

---

3. A new star discovered by Brahe in
1573 in the constellation Cassiopeia.
4. Interstellar short period radio signal.
5. The constellation in the northern
hemisphere.
6. Pertaining to the Milky Way.
1. Jacques-Yves Cousteau (b. 1910), a
French underwater explorer and author.

assiduous team   10
aboard the sun-flooded schooner
but here alone.

There is a ladder.
The ladder is always there
hanging innocently   15
close to the side of the schooner.
We know what it is for,
we who have used it.
Otherwise
it's a piece of maritime floss   20
some sundry equipment.

I go down.
Rung after rung and still
the oxygen immerses me
the blue light   25
the clear atoms
of our human air.
I go down.
My flippers cripple me,
I crawl like an insect down the ladder   30
and there is no one
to tell me when the ocean
will begin.

First the air is blue and then
it is bluer and then green and then   35
black I am blacking out and yet
my mask is powerful
it pumps my blood with power
the sea is another story
the sea is not a question of power   40
I have to learn alone
to turn my body without force
in the deep element.

And now: it is easy to forget
what I came for   45
among so many who have always
lived here
swaying their crenellated fans
between the reefs
and besides   50
you breathe differently down here.

I came to explore the wreck.
The words are purposes.
The words are maps.
I came to see the damage that was done   55
and the treasures that prevail.

I stroke the beam of my lamp
slowly along the flank
of something more permanent
than fish or weed   60

the thing I came for:
the wreck and not the story of the wreck
the thing itself and not the myth
the drowned face² always staring
toward the sun   65
the evidence of damage
worn by salt and sway into this threadbare beauty
the ribs of the disaster
curving their assertion
among the tentative haunters.   70

This is the place.
And I am here, the mermaid whose dark hair
streams black, the merman in his armored body
We circle silently
about the wreck   75
we dive into the hold.
I am she: I am he

whose drowned face sleeps with open eyes
whose breasts still bear the stress
whose silver, copper, vermeil³ cargo lies   80
obscurely inside barrels
half-wedged and left to rot
we are the half-destroyed instruments
that once held to a course
the water-eaten log   85
the fouled compass

We are, I am, you are
by cowardice or courage
the one who find our way
back to this scene   90
carrying a knife, a camera
a book of myths
in which
our names do not appear.

1972   1973

For the Dead

I dreamed I called you on the telephone
to say: *Be kinder to yourself*
but you were sick and would not answer

2. Referring to the ornamental female   many old sailing ships.
figurehead which formed the prow of   3. Gilded silver or bronze.

The waste of my love goes on this way
trying to save you from yourself   5

I have always wondered about the leftover
energy, water rushing down a hill
long after the rains have stopped

or the fire you want to go to bed from
but cannot leave, burning-down but not burnt-down   10
the red coals more extreme, more curious
in their flashing and dying
than you wish they were
sitting there long after midnight

1972   1973

Blood-Sister

FOR CYNTHIA

Shoring up the ocean. A railroad track
ran close to the coast for miles
through the potato-fields, bringing us
to summer. Weeds blur the ties,
sludge clots the beaches.   5

During the war, the shells we found—
salmon-and-silver coins
sand dollars dripping sand
like dust. We were dressed
in navy dotted-swiss dresses in the train   10
not to show the soot. Like dolls
we sat with our dolls in the station.

When did we begin to dress ourselves?

Now I'm wearing jeans spider-webbed
with creases, a black sweater bought years ago   15
worn almost daily since
the ocean has undergone a tracheotomy
and lost its resonance
you wear a jersey the color of
Navaho turquoise and sand   20
you are holding a naked baby girl
she laughs into your eyes
we sit at your table drinking coffee
light flashes off unwashed sheetglass
you are more beautiful than you have ever been   25

we talk of destruction and creation
ice fits itself around each twig of the lilac

like a fist of law and order
your imagination burns like a bulb in the frozen soil
the fierce shoots knock   30
at the roof of waiting

when summer comes the ocean may be closed for good
we will turn
to the desert
where survival   35
takes naked and fiery forms

1973   1975

Toward the Solstice

The thirtieth of November.
Snow is starting to fall.
A peculiar silence is spreading
over the fields, the maple grove.
It is the thirtieth of May,   5
rain pours on ancient bushes, runs
down the youngest blade of grass.
I am trying to hold in one steady glance
all the parts of my life.
A spring torrent races   10
on this old slanting roof,
the slanted field below
thickens with winter's first whiteness.
Thistles dried to sticks in last year's wind
stand nakedly in the green,   15
stand sullenly in the slowly whitening,
field.

        My brain glows
more violently, more avidly
the quieter, the thicker   20
the quilt of crystals settles,
the louder, more relentlessly
the torrent beats itself out
on the old boards and shingles.
It is the thirtieth of May,   25
the thirtieth of November,
a beginning or an end,
we are moving into the solstice
and there is so much here
I still do not understand.   30
If I could make sense of how
my life is still tangled
with dead weeds, thistles,
enormous burdocks, burdens
slowly shifting under   35
this first fall of snow,

beaten by this early, racking rain
calling all new life to declare itself strong
or die,
            if I could know   40
in what language to address
the spirits that claim a place
beneath these low and simple ceilings,
tenants that neither speak nor stir
yet dwell in mute insistence   45
till I can feel utterly ghosted in this house.

If history is a spider-thread
spun over and over though brushed away
it seems I might some twilight
or dawn in the hushed country light   50
discern its greyness stretching
from molding or doorframe, out
into the empty dooryard
and following it climb
the path into the pinewoods,   55
tracing from tree to tree
in the failing light, in the slowly
lucidifying day
its constant, purposive trail,
till I reach whatever cellar hole   60
filling with snowflakes or lichen,
whatever fallen shack
or unremembered clearing
I am meant to have found
and there, under the first or last   65
star, trusting to instinct
the words would come to mind
I have failed or forgotten to say
year after year, winter
after summer, the right rune[1]   70
to ease the hold of the past
upon the rest of my life
and ease my hold on the past.
If some rite of separation
is still unaccomplished   75
between myself and the long-gone
tenants of this house,
between myself and my childhood,
and the childhood of my children,
it is I who have neglected   80
to perform the needed acts,
set water in corners, light and eucalyptus
in front of mirrors,

---

1. I.e., spell, incantation.

or merely pause and listen
to my own pulse vibrating   85
lightly as falling snow,
relentlessly as the rainstorm,
and hear what it has been saying.
It seems I am still waiting
for them to make some clear demand   90
some articulate sound or gesture,
for release to come from anywhere
but from inside myself.

A decade of cutting away
dead flesh, cauterizing   95
old scars ripped open over and over
and still it is not enough.
A decade of performing
the loving humdrum acts
of attention to this house   100
transplanting lilac suckers,
washing panes, scrubbing
wood-smoke from splitting paint,
sweeping stairs, brushing the thread
of the spider aside,   105
and so much yet undone,
a woman's work, the solstice nearing,
and my hand still suspended
as if above a letter
I long and dread to close.   110

1977   1978

GARY SNYDER

1930–

Gary Snyder, who has called himself "a hobo and a worker," was loosely
associated with Allen Ginsberg, Jack Kerouac, and others in forming what
came to be called the Beat movement of the 1950s (in *The Dharma Bums*,
Kerouac models a central character after Snyder). But what marked Snyder
off from the others of his generation who went "on the road" was the rigor
with which he pursued alternatives to city life and Western systems of
value. He combined his wanderings with periods of studying linguistics,
mythology, and Oriental languages.

   Snyder grew up on a farm in Washington and studied anthropology at
Reed College in Oregon. Living in the Pacific Northwest, he developed a
serious interest in the folklore and mythology of the American Indians. He
did graduate work in linguistics at Indiana University and classical Chinese
at the University of California in Berkeley. In 1956–57 and again in the
mid-1960s, Snyder took formal instruction in Buddhism under Zen masters
in Japan.

Like Thoreau at Walden Pond, Snyder writes artful journals of a life in nature—"an ecological survival technique," he prefers to call it. "The rhythms of my poems follow the rhythms of the physical work I'm doing and life I'm leading at any given time." His experience as a logger and forest ranger in the Pacific Northwest are reflected in the *Myths and Texts* he was writing from 1952 to 1956 and in poems such as *Piute Creek* and *Milton by Firelight. The Back Country* includes poems from the Far West and from his years at a monastery in Japan. Snyder's service on a tanker in the South Pacific is treated in some of the prose journals of *Earth House Hold* (1969). *Turtle Island* (1974) grows out of his life as a hunter and gatherer of food, a member—along with his wife and children—of a commune in the foothills of the Sierras.

Though the material of his poems is what the naturalist, hunter, and logger would observe, the experiences Snyder tries to capture are those of ritual feelings and natural power. He describes poetry as "the skilled use of the voice and language to embody rare and powerful states of mind that are in immediate origin personal to the singer, but at deep levels common to all who listen." A quenching of a forest fire in the American West—one of his "Texts"—will lead in the accompanying "Myth" to parallels with the burning of Troy and allusions to Buddhist ritual silence and incense. He more characteristically refers to poetry as a "riprap." This is a forester's term, the title of his first book (1959). A riprap, he explains, is "a cobble of stone laid on steep slick rock to make a trail for horses in the mountains." The combination of know-how and ascent appeals to him, and symbolizes his belief that awakened spiritual insight is inseparable from a strong documentary impulse; "Poetry a riprap on the slick rock of metaphysics." It is a statement Thoreau would have understood and a use of poetry he would have endorsed.

Milton[1] by Firelight

PIUTE CREEK, AUGUST 1955

"O hell, what do mine eyes
     with grief behold?"[2]
Working with an old
Singlejack miner, who can sense
The vein and cleavage   5
In the very guts of rock, can
Blast granite, build
Switchbacks[3] that last for years
Under the beat of snow, thaw, mule-hooves.
What use, Milton, a silly story   10
Of our lost general[4] parents,
     eaters of fruit?

The Indian, the chainsaw boy,
And a string of six mules

---

1. John Milton (1608–74), major English poet and author of *Paradise Lost*, which retells the Biblical story of the Fall of Man.
2. Satan's words when he first sees Adam and Eve in the Garden of Eden (*Paradise Lost*, IV.358).
3. Roads ascending a steep incline in a zigzag pattern.
4. I.e., of our race.

Came riding down to camp   15
Hungry for tomatoes and green apples.
Sleeping in saddle-blankets
Under a bright night-sky
Han River slantwise by morning.
Jays squall   20
Coffee boils

In ten thousand years the Sierras
Will be dry and dead, home of the scorpion.
Ice-scratched slabs and bent trees.
No paradise, no fall,   25
Only the weathering land
The wheeling sky,
Man, with his Satan
Scouring the chaos of the mind.
Oh Hell!   30

Fire down
Too dark to read, miles from a road
The bell-mare clangs in the meadow
That packed dirt for a fill-in
Scrambling through loose rocks   35
On an old trail
All of a summer's day.[5]

1959

Riprap[1]

Lay down these words
Before your mind like rocks.
        placed solid, by hands
In choice of place, set
Before the body of the mind   5
        in space and time:
Solidity of bark, leaf, or wall
        riprap of things:
Cobble of milky way,
        straying planets,   10
These poems, people,
        lost ponies with
Dragging saddles—
        and rocky sure-foot trails.
The worlds like an endless   15
        four-dimensional
Game of Go.[2]
        ants and pebbles

---

5. Alludes to an epic simile describing Satan's fall: "From morn / to noon he fell, from noon to dewy eve, / A summer's day" (*Paradise Lost*, I.742–44).
1. "A cobble of stone laid on steep slick rock to make a trail for horses in the mountains" [Snyder's note].
2. An ancient Japanese game played with black and white stones, placed one after the other on a checkered board.

In the thin loam, each rock a word
           a creek-washed stone   20
Granite: ingrained
           with torment of fire and weight
Crystal and sediment linked hot
           all change, in thoughts,
As well as things.   25

1959

From Myths and Texts: Burning, Section 17

THE TEXT

Sourdough mountain called a fire in:
Up Thunder Creek, high on a ridge.
Hiked eighteen hours, finally found
A snag and a hundred feet around on fire:
All afternoon and into night   5
Digging the fire line
Falling the burning snag
It fanned sparks down like shooting stars
Over the dry woods, starting spot-fires
Flaring in wind up Skagit valley   10
From the Sound.
Toward morning it rained.
We slept in mud and ashes,
Woke at dawn, the fire was out,
The sky was clear, we saw   15
The last glimmer of the morning star.

THE MYTH

Fire up Thunder Creek and the mountain—troy's burning!
The cloud mutters
The mountains are your mind.
The woods bristle there,
Dogs barking and children shrieking   5
Rise from below.

Rain falls for centuries
Soaking the loose rocks in space
Sweet rain, the fire's out
The black snag glistens in the rain   10
& the last wisp of smoke floats up
Into the absolute cold
Into the spiral whorls of fire
The storms of the Milky Way
"Buddha incense in an empty world"   15
Black pit cold and light-year
Flame tongue of the dragon
Licks the sun

The sun is but a morning star[1]

1960

1. The last words of *Walden* by the American Transcendentalist author Henry David Thoreau (1817–62).

Vapor Trails

Twin streaks twice higher than cumulus,
Precise plane icetracks in the vertical blue
Cloud-flaked light-shot shadow-arcing
Field of all future war, edging off to space.

Young expert U.S. pilots waiting   5
The day of criss-cross rockets
And white blossoming smoke of bomb,
The air world torn and staggered for these
Specks of brushy land and ant-hill towns—

    I stumble on the cobble rockpath,   10
Passing through temples,
Watching for two-leaf pine
    —spotting that design.

in Daitoku-ji[1]   1968

I Went into the Maverick Bar

    I went into the Maverick Bar
    In Farmington, New Mexico.
And drank double shots of bourbon
        backed with beer.
My long hair was tucked up under a cap   5
I'd left the earring in the car.

Two cowboys did horseplay
        by the pool tables,
A waitress asked us
        where are you from?   10
a country-and-western band began to play
"We don't smoke Marijuana in Muskokie"
And with the next song,
        a couple began to dance.

They held each other like in High School dances   15
        in the fifties;
I recalled when I worked in the woods
        and the bars of Madras, Oregon.
That short-haired joy and roughness—
        America—your stupidity.   20
I could almost love you again.

We left—onto the freeway shoulders—
        under the tough old stars—
In the shadow of bluffs

1. In Japan (where Snyder studied Buddhism).

I came back to myself,   25
To the real work, to
"What is to be done."[1]

1974

1. Title of a revolutionary pamphlet by Lenin (1902).

SYLVIA PLATH

1932–1963

Sylvia Plath is best known for those poems she wrote feverishly, two or three a day, in the last months of her life. Most of them were collected in a volume called *Ariel*, published in 1965, two years after her suicide in London. The book had an eloquent preface by Robert Lowell, who commented on the astonishing transformation which had overtaken Plath's work in a relatively brief period before her death. Lowell had known the young women in 1958 when she and another poet, Anne Sexton, made regular visits to his poetry seminar at Boston University. He remembered her "air of maddening docility * * * I sensed her abashment and distinction, and never guessed her later appalling and triumphant fulfillment."

Lowell's impression was characteristic of most of the witnesses of Plath's life. Even when she retold her story in the autobiographical novel *The Bell Jar* (1963), there seemed to be a terrible gap between the tone and behavior of a gifted, modest suburban girl of the 1950s and the hidden turmoil which burst out in a nervous breakdown and a first suicide attempt when she was nineteen. Like Plath, the heroine of *The Bell Jar* won the chance to spend one summer in New York as a guest editor of a young women's magazine (in real life *Mademoiselle*). *The Bell Jar* details the trivial stereotyping of women's roles in both the business and social worlds. Then, with little psychological preparation for the reader, Esther Greenwood, the heroine, attempts suicide and has to be hospitalized (as was Plath herself between her junior and senior years in college). What is astonishing about *The Bell Jar* is that although it was written just before the *Ariel* poems, it betrays very little of the strong feeling, the ferocity, that Plath was able to discover and express through her last poetry.

As the poem *Daddy* suggests, Plath's final outburst had a great deal to do with the breakup of her marriage to the English poet Ted Hughes and its connection to buried feelings about her childhood and her parents' marriage. Otto Plath died in 1940, when his daughter was eight years old. A Prussian, Sylvia's father was twenty-one years older than her mother and had been his wife's German professor. He was a biologist, author of a treatise on bumblebees, and, in his household, an autocrat. His death from diabetes might have been prevented if he had consented to see a doctor when signs of physical deterioration set in. Instead, diagnosing his own condition as cancer, he virtually turned their home into a hospital, demanded constant family attendance, and for several years kept his wife and children under the strain of living with what he insisted was an incurable disease.

After his death the strain became financial. Sylvia's mother, left penniless with two small children, worked hard as a teacher and reinforced her daughter's ambitions to be literary, successful, and well rounded. On scholarship Sylvia went to Smith College, from which she graduated *summa cum laude*. On a Fulbright grant she studied in England at Cambridge University, where she met and married Ted Hughes. For the first years the marriage seemed to fulfill a romantic dream—a handsome English husband, two poets beginning careers together, encouraging one another's work. (Plath's first book, *The Colossus*, appeared in 1960.) They were known in literary circles in London and spent time in New England when Plath returned for a year to teach at Smith.

In 1960, back in England, their daughter Frieda was born and, in 1962, a son Nicholas. Hughes began seeing another woman; Plath spent two harsh winters in their country house in Devon and finally asked Hughes for a legal separation. It was under the stress of repeating her mother's experience, left alone with two small children, that Plath's deep anger against and attachment to her father surfaced in poetry. In *Daddy* he is identified with her husband and made partially responsible for her blasted marriage.

From Robert Lowell and Anne Sexton, Plath adopted the license to write about "private and taboo subjects," such as their experiences of breakdowns and mental hospitals. In her case the poems touch deep suspicions of family life, voiced resentments of her young children as well as of her husband. But the *Ariel* poems, unlike Lowell's melancholy, often speak through figures of grotesque gaiety. In *Lady Lazarus* Plath was able to refer to her suicide attempts through the mask of a "sort of walking miracle." "Out of the ash / I rise with my red hair / And I eat men like air."

The other side of Plath's disappointment with failed human relationships was to imagine herself and others absorbed by nature. Human contacts become small and unreal while natural processes go on, palpable, unruffled by human pain. Like death, this was another way to escape living with a flawed self. Through writing Plath faced that problem very clearly: "I cannot sympathize with those cries from the heart that are informed by nothing except a needle or a knife. * * * I believe that one should be able to control and manipulate experiences, even the most terrifying * * * with an informed and intelligent mind. * * * Certaintly it [experience] shouldn't be a kind of shut-box and mirror-looking, narcissistic experience." Her lucid stanzas, clear diction, and resolute dramatic speakers embodied those beliefs. Before she died, Plath had managed through poetry to discover, express, and control the hidden violence of her inner experience. Whether she would have been able to sustain those truths in her day-to-day living is an unanswerable question. She took her life on February 11, 1963.

Morning Song

Love set you going like a fat gold watch.
The midwife slapped your footsoles, and your bald cry
Took its place among the elements.

Our voices echo, magnifying your arrival. New statue.
In a drafty museum, your nakedness   5
Shadows our safety. We stand round blankly as walls.

I'm no more your mother
Than the cloud that distils a mirror to reflect its own slow
Effacement at the wind's hand.

All night your moth-breath   10
Flickers among the flat pink roses. I wake to listen:
A far sea moves in my ear.

One cry, and I stumble from bed, cow-heavy and floral
In my Victorian nightgown.
Your mouth opens clean as a cat's. The window square   15

Whitens and swallows its dull stars. And now you try
Your handful of notes;
The clear vowels rise like balloons.

1961   1966

Lady Lazarus[1]

I have done it again.
One year in every ten
I manage it——

A sort of walking miracle, my skin
Bright as a Nazi lampshade,[2]   5
My right foot

A paperweight,
My face a featureless, fine
Jew linen.

Peel off the napkin   10
O my enemy.
Do I terrify?——

The nose, the eye pits, the full set of teeth?
The sour breath
Will vanish in a day.   15

Soon, soon the flesh
The grave cave ate will be
At home on me

1. Lazarus was raised from the dead by   of victims were sometimes used to make
Jesus (John 11.1–45).   lampshades.
2. In the Nazi death camps, the skins

And I a smiling woman.
I am only thirty.   20
And like the cat I have nine times to die.

This is Number Three.
What a trash
To annihilate each decade.

What a million filaments.   25
The peanut-crunching crowd
Shoves in to see

Them unwrap me hand and foot——
The big strip tease.
Gentleman, ladies,   30

These are my hands,
My knees.
I may be skin and bone,

Nevertheless, I am the same, identical woman.
The first time it happened I was ten.   35
It was an accident.

The second time I meant
To last it out and not come back at all.
I rocked shut

As a seashell.   40
They had to call and call
And pick the worms off me like sticky pearls.

Dying
Is an art, like everything else.
I do it exceptionally well.   45

I do it so it feels like hell.
I do it so it feels real.
I guess you could say I've a call.

It's easy enough to do it in a cell.
It's easy enough to do it and stay put.   50
It's the theatrical

Comeback in broad day
To the same place, the same face, the same brute
Amused shout:

"A miracle!"   55
That knocks me out.
There is a charge

For the eyeing of my scars, there is a charge
For the hearing of my heart——
It really goes.   60

And there is a charge, a very large charge,
For a word or a touch
Or a bit of blood

Or a piece of my hair or my clothes.
So, so, Herr[3] Doktor.   65
So, Herr Enemy.

I am your opus,
I am your valuable,
The pure gold baby

That melts to a shriek.   70
I turn and burn.
Do not think I underestimate your great concern.

Ash, ash——
You poke and stir.
Flesh, bone, there is nothing there——   75

A cake of soap,
A wedding ring,
A gold filling.

Herr God, Herr Lucifer,
Beware
Beware.   80

Out of the ash[4]
I rise with my red hair
And I eat men like air.

1962   1966

Daddy

You do not do, you do not do
Any more, black shoe
In which I have lived like a foot
For thirty years, poor and white,
Barely daring to breathe or Achoo.   5

Daddy, I have had to kill you.
You died before I had time——

3. Mr.
4. An allusion to the phoenix, a mythical bird (the only one of its kind) said to live in the Arabian desert; after 500–600 years, the bird dies by fire and is reborn out of its own ashes.

Marble-heavy, a bag full of God,
Ghastly statue with one grey toe[1]
Big as a Frisco seal   10

And a head in the freakish Atlantic
Where it pours bean green over blue
In the waters off beautiful Nauset.[2]
I used to pray to recover you.
Ach, du.[3]   15

In the German tongue, in the Polish town[4]
Scraped flat by the roller
Of wars, wars, wars.
But the name of the town is common.
My Polack friend   20

Says there are a dozen or two.
So I never could tell where you
Put your foot, your root,
I never could talk to you.
The tongue stuck in my jaw.   25

It stuck in a barb wire snare.
Ich, ich, ich, ich,[5]
I could hardly speak.
I thought every German was you.
And the language obscene   30

An engine, an engine
Chuffing me off like a Jew.
A Jew to Dachau, Auschwitz, Belsen.[6]
I began to talk like a Jew.
I think I may well be a Jew.   35

The snows of the Tyrol,[7] the clear beer of Vienna
Are not very pure or true.
With my gypsy ancestress and my weird luck
And my Taroc[8] pack and my Taroc pack
I may be a bit of a Jew.   40

I have always been scared of *you*,
With your Luftwaffe,[9] your gobbledygoo.
And your neat moustache

1. Plath's father's toe turned black from diabetes.
2. Beach north of Boston.
3. Ah, you: the first of a series of references to her father's German origins.
4. Grabow, the birthplace of Otto Plath, the poet's father.
5. I.

6. German concentration camps, where millions of Jews were murdered during World War II.
7. Austrian Alpine region.
8. Variant of Tarot, ancient fortune-telling cards.
9. The German air force.

And your Aryan eye, bright blue.
Panzer-man, panzer-man,[1] O You——   45

Not God but a swastika
So black no sky could squeak through.
Every woman adores a Fascist,
The boot in the face, the brute
Brute heart of a brute like you.   50

You stand at the blackboard, daddy,
In the picture I have of you,
A cleft in your chin instead of your foot
But no less a devil for that, no not
Any less the black man who   55

Bit my pretty red heart in two.
I was ten when they buried you.
At twenty I tried to die
And get back, back, back to you.
I thought even the bones would do.   60

But they pulled me out of the sack,
And they stuck me together with glue.[2]
And then I knew what to do.
I made a model of you,
A man in black with a Meinkampf[3] look   65

And a love of the rack and the screw.
And I said I do, I do.
So daddy, I'm finally through.
The black telephone's off at the root,
The voices just can't worm through.   70

If I've killed one man, I've killed two——
The vampire who said he was you
And drank my blood for a year,
Seven years, if you want to know.
Daddy, you can lie back now.   75

There's a stake in your fat black heart
And the villagers never liked you.
They are dancing and stamping on you.
They always *knew* it was you.
Daddy, daddy, you bastard, I'm through.   80

1962   1966

---

1. Hitler, who had a neat moustache, preached the superiority of the Aryan race—persons with blond hair and blue eyes, as opposed to those with darker, Semitic characteristics. "Panzer" means "armor" in German.
2. An allusion to Plath's first suicide at-tempt.
3. A reference to Hitler's political auto-biography, *Mein Kampf* (*My Struggle*), written and published before his rise to power, in which the future dictator out-lined his plans for world conquest.

Words

Axes
After whose stroke the wood rings,
And the echoes!
Echoes travelling
Off from the centre like horses.   5

The sap
Wells like tears, like the
Water striving
To re-establish its mirror
Over the rock   10

That drops and turns,
A white skull,
Eaten by weedy greens.
Years later I
Encounter them on the road——   15

Words dry and riderless,
The indefatigable hoof-taps.
While
From the bottom of the pool, fixed stars
Govern a life.   20

1963   1966

Blackberrying

Nobody in the lane, and nothing, nothing but blackberries,
Blackberries on either side, though on the right mainly,
A blackberry alley, going down in hooks, and a sea
Somewhere at the end of it, heaving. Blackberries
Big as the ball of my thumb, and dumb as eyes   5
Ebon in the hedges, fat
With blue-red juices. These they squander on my fingers.
I had not asked for such a blood sisterhood; they must love me.
They accommodate themselves to my milkbottle, flattening their
    sides.

Overhead go the choughs[1] in black, cacophonous flocks—   10
Bits of burnt paper wheeling in a blown sky.
Theirs is the only voice, protesting, protesting.
I do not think the sea will appear at all.
The high, green meadows are glowing, as if lit from within.
I come to one bush of berries so ripe it is a bush of flies,   15
Hanging their bluegreen bellies and their wing panes in a Chinese
    screen.

1. Small, chattering birds of the crow family.

The honey-feast of the berries has stunned them; they believe in
heaven.
One more hook, and the berries and bushes end.

The only thing to come now is the sea.
From between two hills a sudden wind funnels at me,   20
Slapping its phantom laundry in my face.
These hills are too green and sweet to have tasted salt.
I follow the sheep path between them. A last hook brings me
To the hills' northern face, and the face is orange rock
That looks out on nothing, nothing but a great space   25
Of white and pewter lights, and a din like silversmiths
Beating and beating at an intractable metal.

1971

IMAMU AMIRI BARAKA (LeRoi Jones)

1934–

The poet who in 1966 returned to the slums of Newark, New Jersey, as a
black activist named Imamu Baraka had been born in that city thirty-two
years before as LeRoi Jones. Baraka came to say that "the Black Artists's
role in America is to aid in the destruction of America as he knows it."
But he arrived at that attitude after almost fifteen years of attempting to
make a go of the idea of an integrated society, or at least of a multiracial
literary world.

Jones was a precocious child who graduated from high school two
years early. He attended Howard University, spent two and a half years
as an aerial-climatographer in the Air Force, then studied at Columbia,
taking an M.A. in German literature. He associated himself with Beat
poets such as Allen Ginsberg and with poets of the New York School such
as Frank O'Hara. "For me, Lorca, Williams, Pound, and Charles Olson
have had the greatest influence," Jones said in 1959. At that time his
desire was to find a poetic style that would fully and freely express "whatever I think I am." In more particular terms, this meant new metrics and
ways of notation on the page so that the poet could stress the way he
sounds, his individual cadence. It was natural that he should turn for
tonal models on the one side to the "projective verse" of Olson, and on
the other to the rhythms he knew from black music like the blues.

Jones's early poems tried to describe or convey a personal anguish for
which the poet projected no political solution. "I am inside someone— /
who hates me, I look / out from his eyes." He returns over and over to
the predicament of a divided self, part of whose pain stems from being a
black intellectual in a white world. For the most part, the speakers in these
poems as in *I Substitute for the Dead Lecturer*, are paralyzed by their
guilt and fears:

And I am frightened
that the flame of my sickness
will burn off my face. And leave
the bones, my stewed black skull,
an empty cage of failure.

Jones's first two volumes of poetry were *Preface to a Twenty Volume Suicide Note* (1961) and *The Dead Lecturer* (1964). Later in his career, he was to look back at those earlier poems as symptomatic not of personal disorders but of a sick society. "You notice the preoccupation with death, with suicide. * * * Always my own, caught up in the deathurge of the twisted society. The work a cloud of abstraction and disjointedness, that was just whiteness. * * * There is a spirituality always trying to get through, to triumph, to walk across those dead bodies like stuntin for disciples, walking the water of dead bodies europeans call their minds."

Looking back at his university and army experience, Jones was to say: "The Howard thing let me understand the Negro sickness. They teach you how to pretend to be white. But the Air Force made me understand the white sickness."

Feeling these pressures, Jones became increasingly outspoken and separate from the white literary circles of which he was an important part. In 1958 with his first wife, who was white, he had started *Yugen* magazine and then the Totem Press, printing work by poets such as Charles Olson, Robert Duncan, Gary Snyder, and Frank O'Hara. But in a series of influential plays in the mid-1960s, Jones turned to explore the violent bases of relationships between blacks and whites. In *Dutchman* (1964), an encounter involving a young black and a white woman in a subway ends in the gratuitous murder of the black.

In his book reviews, Jones became strongly critical of what he called "The Myth of a Negro Literature." He claimed that most Negro literature was mediocre because "most of the Negroes who've found themselves in a position to become writers were middle-class Negroes who thought of literature as a way of proving they were not 'inferior' (and quite a few who wanted to prove they were)."

The answer was to move out of the range of white literary institutions. He first went to Harlem and began the Black Arts Repertory Theater. Then the focus of his activities became avowedly social and political. In 1966 he set up Spirit House in Newark. Jones was involved in the Newark riots in the summer of 1967 and, after appealing a stiff sentence, acquitted of carrying a concealed weapon. His book of poems *Black Magic* (1969) marked a real point of departure in his writings. The titles of its three sections tell a great deal about its concerns: Sabotage, Target Study, and Black Art. "We want a black poem. And a / Black World. / Let all the world be a Black Poem" he says at the conclusion of "Black Art," "And Let All Black People Speak This Poem / Silently / or LOUD." The capital letters suggest how strongly the poems have moved toward chant, oral presentation, and political rallying. Taking a Muslim name, Imamu Amiri Baraka, Jones helped found in 1968 the Black Community Development and Defense Organization, an enclave in Newark which has remained strongly separatist and politically and economically radical.

An Agony. As Now.

I am inside someone—
who hates me. I look
out from his eyes. Smell
what fouled tunes come in
to his breath. Love his   5
wretched women.

Slits in the metal, for sun. Where
my eyes sit turning, at the cool air
the glance of light, or hard flesh
rubbed against me, a woman, a man,   10
without shadow, or voice, or meaning.

This is the enclosure (flesh,
where innocence is a weapon. An
abstraction. Touch. (Not mine.
Or yours, if you are the soul I had   15
and abandoned when I was blind and had
my enemies carry me as a dead man
(if he is beautiful, or pitied.

It can be pain. (As now, as all his
flesh hurts me.) It can be that. Or   20
pain. As when she ran from me into
that forest.
            Or pain, the mind
silver spiraled whirled against the
sun, higher than even old men thought   25
God would be. Or pain. And the other. The
yes. (Inside his books, his fingers. They
are withered yellow flowers and were never
beautiful.) The yes. You will, lost soul, say
'beauty.' Beauty, practiced, as the tree. The   30
slow river. A white sun in its wet sentences.

Or, the cold men in their gale. Ecstasy. Flesh
or soul. The yes. (Their robes blown. Their bowls
empty. They chant at my heels, not at yours.) Flesh
or soul, as corrupt. Where the answer moves too quickly.   35
Where the God is a self, after all.)

Cold air blown through narrow blind eyes. Flesh,
white hot metal. Glows as the day with its sun.
It is a human love, I live inside. A bony skeleton
you recognize as words or simple feeling.   40

But it has no feeling. As the metal, is hot, it is not,
given to love.

It burns the thing
inside it. And that thing
screams.   45

                                               1964

I Substitute for the Dead Lecturer

What is most precious, because it is lost. What is lost, because it is most precious.

They have turned, and say that I am dying. That
I have thrown
my life
away. They
have left me alone, where   5
there is no one, nothing
save who I am. Not a note
nor a word.

                    Cold air batters
the poor (and their minds   10
turn open
like sores). What kindness
What wealth
can I offer? Except
what is, for me,   15
ugliest. What is
for me, shadows, shrieking
phantoms. Except
they have need
of life. Flesh   20
at least,
            should be theirs.

The Lord has saved me
to do this. The Lord
has made me strong. I   25
am as I must have
myself. Against all
thought, all music, all
my soft loves.

                    For all these wan roads   30
I am pushed to follow, are
my own conceit. A simple muttering
elegance, slipped in my head
pressed on my soul, is my heart's
worth. And I am frightened   35
that the flame of my sickness

will burn off my face. And leave
the bones, my stewed black skull,
an empty cage of failure.

1964

After the Ball

The magic dance

of the second ave ladies,

        in the artificial glare
        of the world, silver-green curls sparkle
        and the ladies' arms jingle   5
        with new Fall pesos, sewn on grim bracelets
        the poet's mother-in-law thinks are swell.

        So much for America, let it sweep in grand style
up the avenues of its failure. Let it promenade smartly
beneath the marquees of its despair.   10
        Bells swing lazily in New Mexico
ghost towns. Where the wind celebrates
afternoon, and leftover haunts stir a little
out of vague instinct,

        hanging their messy sheets   15
        in slow motion against the intrepid dust
        or the silence
            which they cannot scare.

1969

Three Modes of History and Culture

Chalk mark sex of the nation, on walls we drummers
know
as cathedrals. Cathedra,[1] in a churning meat milk.

Women glide through looking for telephones. Maps
weep   5
and are mothers and their daughters listening to

music teachers. From heavy beginnings. Plantations,
learning
America, as speech, and a common emptiness. Songs knocking

inside old women's faces. Knocking through cardboard trunks.   10
Trains
leaning north, catching hellfire in windows, passing through

---

1. Literally, chairs of authority, here referring to graffiti as official pronouncements.

the first ignoble cities of missouri, to illinois, and the panting
Chicago.
And then all ways, we go where flesh is cheap. Where factories   15

sit open, burning the chiefs. Make your way! Up through fog and
history
Make your way, and swing the general, that it come flash open

and spill the innards of that sweet thing we heard, and gave theory
to.   20
Breech, bridge, and reach, to where all talk is energy. And there's

enough, for anything singular. All our lean prophets and rhythms.
Entire
we arrive and set up shacks, hole cards,[2] Western hearts at the edge

of saying. Thriving to balance the meanness of particular skies.   25
Race
of madmen and giants.

Brick songs. Shoe songs. Chants of open weariness.
Knife wiggle early evenings of the wet mouth. Tongue
dance midnight, any season shakes our house. Don't   30
tear my clothes! To doubt the balance of misery
ripping meat hug shuffle fuck. The Party of Insane
Hope. I've come from there too. Where the dead told lies
about clever social justice. Burning coffins voted
and staggered through cold white streets listening   35
to Willkie or Wallace or Dewey[3] through the dead face
of Lincoln. Come from there, and belched it out.

I think about a time when I will be relaxed.
When flames and non-specific passion wear themselves
away. And my eyes and hands and mind can turn   40
and soften, and my songs will be softer
and lightly weight the air.

                                                                  1969

2. Probably punched computer cards.
3. Wendell Willkie (1892–1944), Henry
Wallace (1888–1965), and Thomas Dewey
(1902–71), presidential candidates defeated in the elections of the 1940s. Politicians appear to be capitalizing on the
image of strong leadership offered by
Abraham Lincoln.

Selected Bibliographies

The best general handbook for research is Clarence Gohdes's *Bibliographical Guide to the Study of the Literature of the U.S.A.*, 4th ed. (1976). Lists of significant works are given under such general headings as "American studies or American civilization," "American history: general tools," "American history: some special studies," "Selected histories of ideas in the U.S.," "Philosophy and psychology in the U.S.," "Transcendentalism in the U.S.," "Religion in the U.S.," "Arts other than literature," "Chief general bibliographies of American literature," "Chief general histories and selected critical discussions of American literature," "Selected studies of regional literature," and "Literature on or by racial and other minorities," and under the various period headings.

Other useful, critically annotated guides are Lewis Leary's *American Literature: A Study and Research Guide* (1976) and Robert C. Schweik and Dieter Riesner's *Reference Sources in English and American Literature: An Annotated Bibliography* (1977). Two valuable encyclopedic reference volumes are James D. Hart's *Oxford Companion to American Literature*, 4th ed. (1965) and Max J. Herzberg's *The Reader's Encyclopedia of American Literature* (1962).

The most ambitious comprehensive history is *Literary History of the United States*, ed. Robert E. Spiller *et al.* (1948, 1974). The discursive bibliographical supplements to this volume may also be recommended.

EARLY AMERICAN LITERATURE 1620–1820

The classic study of the Puritans remains Perry Miller's *The New England Mind: The Seventeenth Century* (1939) and *The New England Mind: From Colony to Province* (1953). Useful essays by Miller are included in *Errand into the Wilderness* ("The Marrow of Puritan Divinity" and "From Edwards to Emerson" are especially important) (1956). Miller and Thomas H. Johnson edited an anthology of *The Puritans: A Sourcebook of Their Writings* in 1938 (rev. ed., 1963), which contains useful introductions. A valuable collection, *Seventeenth Century American Poetry* (1968), was compiled by Harrison T. Meserole. Other books of general interest are Edmund S. Morgan's *Visible Saints: The History of a Puritan Idea* (1963) and Kenneth B. Murdock's *Literature and Theology in Colonial New England* (1949). On Puritan historians, see Peter Gay's *Loss of Mastery* (1966) and David Levin's *In Defense of Historical Literature* (1967). Sacvan Bercovich has edited a useful collection of essays on the Puritans in *The American Puritan Imagination: Essays in Revaluation* (1974); it contains a good selected bibliography. Another useful revaluation of Puritans which brings together some hard-to-come-by essays is Michael McGiffert's *Puritanism and the American Experience* (1969). Some of the best accounts of the intellectual and cultural life of the 17th century are to be found in the biographies of writers included in this anthology. One important study of a writer not included is Edmund Morgan's life of Ezra Stiles, *The Gentle Puritan* (1962).

Although the bicentennial year produced a number of useful studies of American culture in the 18th century, nothing has replaced Moses Coit Tyler's monumental *A History of American Literature during the Colonial Period, 1607–1765*, 2 vols. (1878; rev. ed., 1897; 1-vol. reprint, 1949) and *The Literary History of the American Revolution, 1763–1783*, 2 vols. (1897; 1-vol. reprint, 1941). Readers should be warned, however, to check biographical details with more recent studies. Russel B. Nye's *The Cultural Life of the New Nation* (1960), Kenneth Silverman's *A Cultural History of the American Revolution* (1976), and Michael Kammer's *A Season of Youth* (1978) are useful surveys of the arts of this period. On the American Revolution, see Edmund S. Morgan's *The American Revolution: Two Centuries of Interpretation* (1965). *Fifteen American Authors before 1900: Bibliographic Essays on Research and Criticism*, ed. Robert A. Rees and Earl N. Harbert (1971), contains discursive chapters on the scholarship and criticism that had appeared up to that date on three authors in this period: Edwards, Franklin, and Taylor.

William Bradford

Samuel Eliot Morison's edition of *Of Plymouth Plantation* (1952) is the standard edition of Bradford's history, and his life of Bradford in the *Dictionary of American Biography* (1933) is useful. For a discussion of Puritan historians, see Kenneth B. Murdock's *Literature and Theology in Colonial New England* (1949) and David Levin's "William Bradford: The Value of Puritan Historiography," in E. H. Emerson's *Major Writers of Early American Literature* (1972), pp. 11–31.

Anne Bradstreet

The standard edition is now Jeannine Hensley's *The Works of Anne Bradstreet* (1967), with an introduction by Adrienne Rich. Useful critical discussions may be found in Josephine K. Piercy's *Anne Bradstreet* (1965) and Elizabeth W. White's *Anne Bradstreet: The Tenth Muse* (1971).

William Byrd

Byrd's prose works are available in an edition edited by Louis B. Wright (1966). *The Secret Diary of William Byrd of Westover, 1709–1712* was edited by Louis B. Wright and Marion Tinling (1941). *Another Secret Diary of William Byrd of Westover, 1739–1741*, ed. M. H. Woodfin and M. Tinling, appeared in 1942. Byrd's *London Diary* (1958) was edited by Louis B. Wright and Marion Tinling. A useful critical essay is Lewis P. Simpson's "William Byrd and the South," *Early American Literature*, Vol. 7 (1972), pp. 187–195.

St. Jean de Crèvecoeur

Sketches of Eighteenth Century America, More Letters of St. John de Crèvecoeur (1925) was edited by H. L. Bourdin *et al. Eighteenth Century Travels in Pennsylvania and New York* (1962) was edited by P. G. Adams. The best critical biography is by Thomas Philbrick (1970).

Jonathan Edwards

Yale University Press is in the process of publishing a complete edition of the works of Jonathan Edwards. A selection from Edwards's writings edited by Clarence H. Faust and Thomas H. Johnson (1935) contains a useful introduction and bibliography. David Levin's *Jonathan Edwards: A Profile* (1969) includes Samuel Hopkins's *Life and Character of the Late Rev. Mr. Jonathan Edwards* (1765). Perry Miller's *Jonathan Edwards* (1949) and essays included in *Errand into the Wilderness* (1956) are indispensable. Daniel B. Shea, Jr., includes a discussion of the *Personal Narrative* in his *Spiritual Autobiography in Early America* (1968). Richard Bushman writes of "Jonathan Edwards as a Great Man" in *Soundings*, Vol. 52 (1969), pp. 15–46.

Benjamin Franklin

The Papers of Benjamin Franklin are being published by Yale University Press under the general editorship of Leonard W. Labaree. A good selection of Franklin's writings may be found in *Benjamin Franklin: Representative Selections* (1936), ed. F. L. Mott and C. L.

Jorgensen. The standard biography is by Carl Van Doren (1938). Useful critical studies include Bruce I. Granger's *Benjamin Franklin, an American Man of Letters* (1964) and Alfred O. Aldridge's *Benjamin Franklin and Nature's God* (1967). See also David Levin, "The Autobiography of Benjamin Franklin: The Puritan Experimenter in Life and Art," *Yale Review*, Vol. 53 (1964), pp. 258–275. A famous unfavorable response to Franklin can be found in D. H. Lawrence's *Studies in Classic American Literature* (1923).

Philip Freneau

The standard edition of *The Poems of Philip Freneau* is that of F. L. Pattee, 3 vols. (1902–1907). Lewis Leary, who edited *The Last Poems of Philip Freneau* (1945), wrote the best biography, *That Rascal Freneau: A Study in Literary Failure* (1941). A very useful introduction to Freneau can be found in Harry H. Clark's edition of *Philip Freneau: Representative Selections* (1929).

Thomas Jefferson

Princeton University Press, under the general editorship of Julian P. Boyd, is currently engaged in publishing the *Papers of Thomas Jefferson*. The best biography is Dumas Malone's five-volume *Jefferson and His Time* (1948–1974). Mr. Malone's article on Jefferson in the *Dictionary of American Biography* is very helpful. Recent critical studies include those of M. D. Peterson, *Thomas Jefferson and the New Nation* (1970), and Garry Wills, *Inventing America: Jefferson's Declaration of Independence* (1978). Fawn Brodie published a controversial biography in 1974.

Sarah Kemble Knight

A useful biographical sketch can be found in the *Dictionary of American Biography* by Sidney Gunn (1933). G. P. Winship's edition of the *Private Journal* (1920) contains Theodore Dwight's original introduction to the first edition (1825).

Cotton Mather

Kenneth B. Murdock's *Selections from Cotton Mather* (1926) contains a useful biographical introduction. The standard biography remains Barrett Wendell's *Cotton Mather: The Puritan Priest* (1891; reprinted, 1963). See also David Levin's *Cotton Mather: The Young Life of the Lord's Remembrancer* (1978). Useful critical essays may be found in Sacvan Bercovich's *The Puritan Origins of the American Self* (1975) and Robert Middlekauff's *The Mathers: Three Generations of Puritan Intellectuals* (1971).

Thomas Paine

Moncure D. Conway, who edited *The Writings of Thomas Paine*, 4 vols. (1894–1896), wrote the best life of Paine (1892). H. H. Clark's *Thomas Paine: Representative Selec-*

tions (1944) contains a very helpful introduction. Cecil Kenyon's "Where Paine Went Wrong," *American Political Science Review*, Vol. 45 (1951), pp. 1086–1099, is a challenging critical assessment. A more recent biography is that by David F. Hawke (1974).

Edward Taylor

The standard edition of *The Poems of Edward Taylor* is by Donald E. Stanford (1960). Taylor's *Diary* was edited by Francis Murphy (1964) and his *Christographia* sermons (1962) and *Treatise Concerning the Lord's Supper* (1966) by Norman S. Grabo, who also did a useful critical biography (1961). Useful critical essays may be found in Thomas H. Johnson's edition of *The Poetical Works of Edward Taylor* (1939) and in Austin Warren's *Rage for Order* (1948).

Phillis Wheatley

The best edition of the poems is that edited by Julian D. Mason (1966). Shirley Graham's *The Story of Phillis Wheatley* (1969) and William G. Allen's *Wheatley, Banneker and Horton* (1970) are useful critical biographies.

John Winthrop

The standard edition of Winthrop's work is *The Winthrop Papers*, ed. A. Forbes, 5 vols. (1929–1945). For critical biographies see D. B. Rutman's *Winthrop's Boston* (1965) and Edmund Morgan's *The Puritan Dilemma: The Story of John Winthrop* (1965).

AMERICAN LITERATURE 1820–1865

Eight American Authors: A Review of Research and Criticism, rev. ed., ed. James Woodress (1971), contains chapters on the scholarship and criticism that had appeared on Poe, Emerson, Hawthorne, Thoreau, Whitman, and Melville up to 1969 (the other two of the eight authors are Clemens and James). *Fifteen American Authors before 1900: Bibliographic Essays on Research and Criticism*, ed. Robert A. Rees and Earl N. Harbert (1971), contains chapters on Bryant, Cooper, Dickinson, Holmes, Irving, Longfellow, Lowell, and Whittier, as well as a section on the southwestern humorists. The essays in these two volumes may be supplemented (especially for recent scholarship and criticism) by the various volumes of *American Literary Scholarship: An Annual*, published each year since 1965 under the editorship of James Woodress or, for some of the recent volumes, J. Albert Robbins. Separate chapters in *American Literary Scholarship* are now devoted to Hawthorne, Poe, and Melville, while one chapter is devoted to "Emerson, Thoreau, and Transcendentalism" and another to "Whitman and Dickinson." Work on other writers of this period is covered in the chapter on "19th-Century Literature." (The volumes appear two years after the works they discuss.)

Documentary discoveries in the last two or three decades have rendered all histories of American literature badly out of date, although the chapters on this period in the most ambitious work, the mammoth *Literary History of the United States*, ed. Robert E. Spiller *et al.* (1948, 1974) are still useful; the second volume of the 1974 edition is an updated bibliography of studies on American writers. The most influential critical study of this period remains F. O. Matthiessen's *American Renaissance: Art and Expression in the Age of Emerson and Whitman* (1941, 1968), although it, too, is outdated, especially in its reliance on now-superseded biographies. The best study of the poetry of the period is in Roy Harvey Pearce's *The Continuity of American Poetry* (1961), now supplemented by Hyatt H. Waggoner's *American Poets from the Puritans to the Present* (1968). Major new studies in literary history, biography, and criticism will surely appear in the 1980s as scholars and critics assimilate new information about the period, most notably that being published in volumes sponsored by the Center for Editions of American Authors and, subsequently, by the Center for Scholarly Editions.

William Cullen Bryant

Bryant's poetry has not been re-edited according to modern standards, and his newspaper writing is mostly buried in the files of the New York *Evening Post*. William Cullen Bryant II and Thomas G. Voss are editing *The Letters of William Cullen Bryant* (1975–). Earlier biographies are superseded by Charles H. Brown's *William Cullen Bryant* (1971). Judith Turner Phair has compiled *A Bibliography of William Cullen Bryant and His Critics, 1808–1972* (1975). The best guide to work on Bryant is the chapter by James E. Rocks in *Fifteen American Authors before 1900*, ed. Robert A. Rees and Earl N. Harbert (1971), supplemented by any new work in the annual *American Literary Scholarship* (1963–).

James Fenimore Cooper

James Franklin Beard is editor-in-chief of a long-planned collected edition of Cooper, but for now almost all available editions are inferior except for some facsimiles of early texts. Beard edited the six-volume *Letters and Journals of James Fenimore Cooper* (1960–1968), the most important work of Cooper scholarship. Cooper prohibited his family to authorize a biography, with the result that early biographers were less accurate and less sympathetic than they might have been. Biographical studies based on fresh examination of documentary evidence include William B. Shubrick Clymer's *James Fenimore Cooper* (1900; reprinted, 1968); Ethel R. Outland's *The "Effingham" Libels on Cooper* (1929); Robert E. Spiller's *Fenimore Cooper: Critic of His Times* (1931); Henry W. Boynton's *James Fenimore Cooper* (1931); Dorothy Waples's *The Whig Myth of James Fenimore Cooper* (1938); and Marcel Clavel's *Fenimore Cooper: sa vie et son oeuvre* (1938). A brisk recent biography is James Grossman's *James Fenimore Cooper* (1949). Beard plans a full-length biography.

Important reputation studies are Marcel Clavel's *Fenimore Cooper and His Critics: American, British and French Criticisms of the Novelist's Early Work* (1938) and George Dekker and John P. McWilliams's *Fenimore Cooper: The Critical Heritage* (1973). The early collection *James Fenimore Cooper: A Re-Appraisal*, ed. Mary E. Cunningham (1954), lived up to its title, consolidating gains in the understanding of Cooper and pointing toward needed work. A few recent studies are Thomas L. Philbrick's *James Fenimore Cooper and the Development of American Sea Fiction* (1961); George Dekker's *James Fenimore Cooper: The American Scott* (1967); Blake Nevius's *Cooper's Landscapes: An Essay on the Picturesque Vision* (1976); and H. Daniel Peck's *A World by Itself: The Pastoral Moment in Cooper's Fiction* (1977). The best guide to work on Cooper is the chapter by James Franklin Beard in *Fifteen American Authors before 1900*, ed. Robert A. Rees and Earl N. Harbert (1971), supplemented by the discussions in the annual *American Literary Scholarship* (1963–). The indefatigable Beard is preparing for publication *James Fenimore Cooper: An Annotated Checklist of Criticism*.

Emily Dickinson

Three volumes of *The Poems of Emily Dickinson* (1955) were edited by Thomas H. Johnson, and he and Theodora Ward edited three companion volumes of *The Letters of Emily Dickinson* (1958). Jay Leyda's *The Years and Hours of Emily Dickinson* (1960) provides two volumes of documents crucial to the understanding of her life and time. Richard B. Sewall's *The Life of Emily Dickinson* (1974) is the most ambitious and detailed life, but Thomas H. Johnson's *Emily Dickinson: An Interpretive Biography* (1955) remains the standard one-volume treatment; John Cody's *After Great Pain: The Inner Life of Emily Dickinson* (1971) is a provocative psychoanalytic study.

Since the rediscovery of Dickinson in the 1930s and 1940s, the quantity of critical studies has been immense. George Frisbie Whicher's *This Was a Poet: A Critical Biography of Emily Dickinson* (1939) remains useful in no small part because it is plainly and engagingly written. John B. Pickard's *Emily Dickinson: An Introduction and Appreciation* (1967) and Richard Chase's *Emily Dickinson* (1951) are two good general critical studies. Among the best of the other book-length studies are Albert J. Gelpi's incisive *Emily Dickinson: The Mind of the Poet* (1965), Charles R. Anderson's indispensable *Emily Dickinson's Poetry* (1960), Clark Griffith's *The Long Shadow: Emily Dickinson's Tragic Poetry* (1964), Brita Lindberg-Seyersted's *The Voice of the Poet: Aspects of Style in the Poetry of Emily Dickinson* (1968), Ruth Miller's *The Poetry of Emily Dickinson* (1968), and Robert Weisbuck's *Emily Dickinson's Poetry* (1975).

Frederick Douglass

No uniform edition of Douglass's writings is available. Philip S. Foner edited *The Life and Writings of Frederick Douglass* in four volumes (1950–1955). Foner also wrote the most comprehensive biography, *Frederick Douglass: A Biography* (1964); other biographies of interest include Charles W. Chesnutt's *Frederick Douglass* (1899), Benjamin Quarles's *Frederick Douglass* (1948), and Arna W. Bontemp's *Free at Last: The Life of Frederick Douglass* (1971).

Ralph Waldo Emerson

The outstanding achievement in Emerson scholarship is the 14-volume *Journals and Miscellaneous Notebooks*, ed. George P. Clark, Merrell R. Davis, Alfred R. Ferguson, Harrison Hayford, and Merton M. Sealts, Jr., later joined by Linda Allardt, Ralph H. Orth, J. E. Parsons, A. W. Plumstead, and Susan Sutton Smith under the editor-in-chief William H. Gilman (1960–1977). The *Collected Works of Ralph Waldo Emerson* has made less progress, but Robert E. Spiller and Alfred R. Ferguson edited the 1849 *Nature, Addresses, and Lectures* (1971). Stephen E. Whicher, Robert E. Spiller, and Wallace E. Williams edited *Early Lectures*, 3 vols. (1959–1972). Whicher's *Selections from Ralph Waldo Emerson: An Organic Anthology* (1957) has been influential, as has William H. Gilman's *Selected Writings of Ralph Waldo Emerson* (1965), notable for the new texts of the journal passages. Merton M. Sealts, Jr., and Alfred R. Ferguson edited *Emerson's "Nature": Origin, Growth, Meaning* (1969), of which Sealts has prepared a revision (1979). Emerson's poems are being edited by Carl Strauch, who has written widely about them. Ralph L. Rusk's six-volume *Letters* (1939) must be supplemented by *The Correspondence of Emerson and Carlyle*, ed. Joseph Slater (1964), and other compilations.

The most detailed biography is Ralph L. Rusk's *The Life of Ralph Waldo Emerson* (1949). Kenneth W. Cameron prepared *Emerson's Workshop: An Analysis of His Reading in Periodicals through 1836* (1964), which should be used along with Walter Harding's *Emerson's Library* (1967). Cameron's *Emerson among His Contemporaries* (1967) includes reviews of Emerson's books and reminiscences from those who knew him. William J. Sowder edited *Emerson's Reviewers and Commentators: A Biographical and Bibliographical Analysis of Nineteenth-Century Periodical Criticism* (1968). Milton R. Konvitz and Stephen E. Whicher edited *Emerson: A Collection of Critical Essays* (1962), and Jackson R. Bryer and Robert A. Rees compiled *A Checklist of Emerson Criticism, 1951–1961* (1964). Buford Jones is preparing a full bibliography of writings on Emerson for G. K. Hall's Reference Guide series. Among the critical books most frequently cited are Sherman Paul's *Emerson's Angle of Vision* (1952); Stephen E. Whicher's *Freedom and Fate: An Inner Life of Ralph Waldo Emerson* (1953; reprinted, 1971); Joel Porte's *Emerson and Thoreau: Transcendentalists in Conflict* (1966); Hyatt H. Waggoner's *Emerson as Poet* (1974); and Sacvan Bercovitch's *The Puritan Origins of the American Self* (1975).

Since Frederic Ives Carpenter's *Emerson Handbook* (1953) is out of print and out of date, the best guide to scholarship and criticism on Emerson is Floyd Stovall's chapter in *Eight American Authors*, rev. ed., ed. James Woodress (1971), supplemented by the annual chapter on the Transcendentalists in *American Literary Scholarship* (1963–).

Margaret Fuller

The guarded, sanitized *Memoirs of Margaret Fuller Ossoli* (1852) was prepared by W. H. Channing, James Freeman Clarke, and Ralph Waldo Emerson, friends who casually vandalized the manuscripts they worked with. There are two substantial recent collections: Bell Gale Chevigny's *The Woman and the Myth: Margaret Fuller's Life and Writings* (1976), a selection of Fuller's writings interspersed with contemporary comments on her; and Joel Myerson's *Margaret Fuller: Essays on American Life and Letters* (1978). Wilma R. Ebbitt and Robert N. Hudspeth are editing Fuller's New York *Tribune* pieces, and Hudspeth is editing her letters. A good brief life is Arthur W. Brown's *Margaret Fuller* (1964). More detailed are Madeleine B. Stern's *The Life of Margaret Fuller* (1942) and Joseph J. Deiss's *The Roman Years of Margaret Fuller* (1969). In *Margaret Fuller: From Transcendentalism to Revolution* (1978), Paula Blanchard examines Fuller's feminism not as an aberration, but as a positive force in her life. The best discursive guide to work by and about Fuller is Myerson's "Biographical Note" in his *Margaret Fuller: Essays on American Life and Letters*.

Nathaniel Hawthorne

The Centenary Edition of Hawthorne published by Ohio State University has been under way since 1963. The texts of the Centenary volumes of short stories and romances, all edited by Fredson Bowers, have proved unreliable, but in all these volumes there is much information about the history of composition and publication, including quotations from previously unpublished letters. Three volumes of the Centenary Edition are genuinely distinguished: Claude M. Simpson's edition of *The American Notebooks* (1972) and the editions of the romances Hawthorne left unfinished at his death, *The American Claimant Manuscripts* (1977) and *The Elixir of Life Manuscripts* (1977), both edited by Edward H. Davidson, Claude M. Simpson, and L. Neal Smith. Randall Stewart edited *The English Notebooks* (1941), and L. Neal Smith and Thomas Woodson have edited *The French and Italian Notebooks* (1979), always known in the incomplete *Passages* published in 1872. Smith and Woodson are remedying the darkest shame of Hawthorne scholarship, the lack of a collected edition of the letters; their first volume of the letters should appear in 1981. The *Love Letters of Nathaniel Hawthorne, 1839–1863* (1907) and the *Letters of Hawthorne to William D. Ticknor, 1851–1864* (1910) were both reprinted in facsimile in 1972, but merely as a stopgap measure; both volumes are incomplete, and Smith has found that one reason the love letters sound disconnected is that they were published out of chronological order.

Lacking access to many important letters, scholars have been unable to produce a solid, comprehensive biography. Students of Hawthorne must still rely on documents cited in works such as Julian Hawthorne's two-volume *Hawthorne and His Wife* (1884; reprinted, 1968); Lawrence S. Hall, *Hawthorne: Critic of Society* (1944); Robert Cantwell's *Nathaniel Hawthorne: The American Years* (1948); Louise H. Tharp, *The Peabody Sisters of Salem* (1950); and Vernon Loggins, *The Hawthornes: The Story of Seven Generations* (1951). Randall Stewart's *Nathaniel Hawthorne: A Biography* (1948) is often cited as standard, but it lacks richness of detail and sophistication of critical judgment.

These books survey Hawthorne's reputation: Kenneth W. Cameron's *Hawthorne among His Contemporaries* (1968); B. Barnard Cohen's *The Recognition of Nathaniel Hawthorne* (1969); and J. Donald Crowley's *Hawthorne: The Critical Heritage* (1970). Other useful collections are *Hawthorne Centenary Essays*, ed. Roy Harvey Pearce (1964); *Hawthorne: A Collection of Critical Essays*, ed. A. N. Kaul (1966); and *Nathaniel Hawthorne: A Collection of Criticism*, ed. J. Donald Crowley (1975). Of the dozens of books on Hawthorne, these are among the most often cited and debated: Roy R. Male's *Hawthorne's Tragic Vision* (1957); Hyatt H. Waggoner's *Hawthorne: A Critical Study*, rev. ed. (1963); Frederick C. Crews's *The Sins of the Fathers: Hawthorne's Psychological Themes* (1966); John Caldwell Stubbs's *The Pursuit of Form: A Study of Hawthorne and the Romance* (1970); Neal Frank Doubleday's *Hawthorne's Early Tales: A Critical Study* (1972); and Nina Baym's *The Shape of Hawthorne's Career* (1976). The Hawthorne chapter in Martin Green's *Re-appraisals: Some Commonsense Readings in American Literature* (1965) hilariously puts down Hawthorne and his admirers.

The best overview of scholarship and criticism on Hawthorne is Walter Blair's chapter in *Eight American Authors*, rev. ed., ed. James Woodress (1971), but it must be supplemented by Buford Jones's *A Checklist of Hawthorne Criticism: 1951–1966* (1967) and Jones's "Hawthorne Studies: The Seventies," *Studies in the Novel*, Vol. 2 (Winter, 1970). Also essential is the annual Hawthorne chapter in *American Literary Scholarship* (1963–) as well as the new trash-and-treasures annual *Nathaniel Hawthorne Journal* (1971–).

Oliver Wendell Holmes

There is no modern scholarly edition of Holmes's works. There has never been a collected edition of his letters, but many are quoted in John T. Morse, Jr.'s *Life and Letters of Oliver Wendell Holmes* (1896). A balanced biography is M. A. de Wolfe Howe's *Holmes of the Breakfast-Table* (1939; reprinted, 1972). Eleanor M. Tilton's *Amiable Aristocrat: A Biography of Dr. Oliver Wendell Holmes* (1947) is standard. The best guide to work on Holmes is Barry Menikoff's chapter in *Fifteen American Authors before 1900*, ed. Robert A. Rees and Earl N. Harbert (1971), supplemented by any mentions in the annual *American Literary Scholarship* (1963–).

Washington Irving

In good progress is the *Complete Works of Washington Irving*, organized under the chief editorship of Henry A. Pochmann. It includes the *Journals and Notebooks*, edited in three volumes by Nathalia Wright and Walter A. Reichart (1969–1970); it is to include all the works Irving published in his lifetime and, for the first time, a comprehensive collection of his letters. In the various introductory essays to this edition the history of Irving's career is being written in greater detail than ever before. The standard biography is Stanley T. Williams's two-volume *Life of Washington Irving* (1935), a brilliant and learned study marred only by Williams's unremitting disparagement of Irving. William L. Hedges's *Washington Irving: An American Study, 1802–1832* (1965) reacts too strongly against Williams in exalting Irving's "relevance." Important documentary studies are Walter A. Reichart's *Washington Irving in Germany* (1957) and Ben Harris McClary's *Washington Irving and the House of Murray* (1969). The best discursive guide to work on Irving is Pochmann's chapter in *Fifteen American Authors before 1900*, ed. Robert A. Rees and Earl N. Harbert (1971), supplemented by the annual review of current work in *American Literary Scholarship* (1963–   ). Two valuable recent works are Haskell Springer's comprehensive *Washington Irving: A Reference Guide* (1976) and Andrew B. Myers's *A Century of Commentary on the Works of Washington Irving* (1976).

Abraham Lincoln

The most comprehensive collection is *The Collected Works of Abraham Lincoln*, ed. Roy P. Basler *et al.* (1953). Basler also edited in one volume *Abraham Lincoln: His Speeches and Writings* (1946); his introduction is useful both for the critical reflections on Lincoln as a man and as a writer and for information on the history of the texts. Writings on Lincoln constitute a library in themselves. For beginners, Carl Sandburg's detailed, passionate, and adulatory two-volume biography, *Abraham Lincoln: The Prairie Years* (1926) and his four-volume *Abraham Lincoln: The War Years* (1939) were abridged in one volume in 1954. More recent, more dispassionate accounts may be found in Benjamin Thomas's *Abraham Lincoln* (1952), perhaps the best short account, and David Donald's *Lincoln Reconsidered* (1956). A good specialized examination of *Lincoln the Writer: The Development of His Style* was published by Herbert Joseph Edwards and John Erskine Hankins in 1962.

Henry Wadsworth Longfellow

Longfellow's novels and poems have not been edited by modern scholars. The outstanding instance of modern work on Longfellow is Andrew Hilen's *The Letters of Henry Wadsworth Longfellow* (1966–   ). The fullest biography is still Samuel Longfellow's two-volume *Life of Henry Wadsworth Longfellow* (1886–1887) and his *Final Memorials of Henry Wadsworth Longfellow* (1887), but Hilen's edition of the letters reveals much that Samuel Longfellow—the poet's brother—suppressed. Recent well-documented biographies are Lawrance R. Thompson's *Young Longfellow (1807–1843)* (1938; reprinted, 1969) and Edward Wagenknecht's *Longfellow: A Full-Length Portrait* (1955). Useful are Andrew Hilen's edition of *Clara Crowninshield's Diary: A European Tour with Longfellow, 1835–1836* (1956) and Wagenknecht's edition of *Mrs. Longfellow: Selected Letters and Journals of Fanny Appleton Longfellow (1817–1861)* (1956). There are important discussions of Longfellow's earnings in William Charvat's *The Profession of Authorship in Amer-*

ica, *1800–1870* (1968). The best guide to work on Longfellow is Richard Dilworth Rust's chapter in *Fifteen American Authors before 1900*, ed. Robert A. Rees and Earl N. Harbert (1971), supplemented by the annual *American Literary Scholarship* (1963–   ).

James Russell Lowell

The only recent scholarly edition of any of Lowell's works is that of *The Biglow Papers, First Series*, ed. Thomas Wortham (1977). Charles Eliot Norton edited a bowdlerized two-volume collection, *Letters of James Russell Lowell* (1894), and M. A. de Wolfe Howe edited *New Letters of James Russell Lowell* (1932). The fullest biography, based on manuscript sources, is Martin Duberman's *James Russell Lowell* (1966), but a better account of Lowell as worker and thinker is in Leon Howard's *Victorian Knight-Errant: The Early Literary Career of James Russell Lowell* (1952). The best guide to work on Lowell is Robert A. Rees's chapter in *Fifteen American Authors before 1900*, ed. Robert A. Rees and Earl N. Harbert (1971), supplemented by any mention in the annual volumes of *American Literary Scholarship* (1963–   ).

Herman Melville

The Northwestern-Newberry Edition of *The Writings of Herman Melville*, ed. Harrison Hayford, Hershel Parker, and G. Thomas Tanselle (1968–   ) will be standard. It will include revised texts of certain works already in print: the lectures reconstructed in Merton M. Sealts, Jr.'s *Melville as Lecturer* (1957); *The Letters of Herman Melville*, ed. Merrell R. Davis and William H. Gilman (1960); *Billy Budd, Sailor*, ed. Harrison Hayford and Merton M. Sealts, Jr. (1962); and the Norton Critical Edition of *Moby-Dick*, ed. Harrison Hayford and Hershel Parker (1967). Some volumes of the incomplete Hendricks House Edition of Melville (1947–1969) remain valuable for their interpretive introductions or explanatory notes; two in effect opened up books for the ordinary reader, or at least for the ordinary academic reader: Elizabeth S. Foster's edition of *The Confidence-Man* (1954) and Walter S. Bezanson's edition of *Clarel* (1960). Until Robert C. Ryan's volume appears in the Northwestern-Newberry Edition, there will be no satisfactory texts of Melville's poems, but otherwise good scholarly editions or facsimiles of most of Melville's works are now available.

The 19th-century biographical accounts are reprinted and analyzed in Merton M. Sealts, Jr.'s *The Early Lives of Melville* (1974). The first full-length biography, Raymond Weaver's *Herman Melville: Mariner and Mystic* (1921), has been superseded by Leon Howard's *Herman Melville: A Biography* (1951) and by Jay Leyda's monumental compilation of documents, *The Melville Log* (1951; reprinted with supplement, 1969). (A new edition of the *Log* is planned with an expanded supplement edited by Leyda and Hershel Parker.)

Merton M. Sealts, Jr.'s *Melville's Reading* (1966) is a compilation of books Melville owned or used. Melville's annotations in his surviving books are recorded in W. Walker Cowen's Harvard dissertation, *Melville's Marginalia* (1965).

Steven Mailloux and Hershel Parker's *Checklist of Melville Reviews* (1975) is complete to that date, and all known reviews are included in a volume edited by Brian Higgins and Mailloux for future publication. Some reviews and later essays are reprinted in *The Recognition of Herman Melville*, ed. Hershel Parker (1967). *"Moby-Dick" as Doubloon*, ed. Parker and Hayford (1970), includes all the then-known reviews of *Moby-Dick* and much later criticism. Watson G. Branch's *Melville: The Critical Heritage* (1974) prints a lavish sampling of reviews. Of the compilations of modern criticism on Melville's masterpiece, the *Doubloon* is fullest, but *"Moby-Dick" Centennial Essays*, ed. Tyrus Hillway and Luther S. Mansfield (1953), is important, and some criticism is reprinted in the Norton Critical Edition (1967). A bibliography of criticism on *Moby-Dick* is in the *Doubloon*, and the Norton Critical Edition of *The Confidence-Man*, ed. Hershel Parker (1971), ends with Watson G. Branch's annotated bibliography of criticism on that book. The best bibliography of Melville criticism is Brian Higgins's *Herman Melville: An Annotated Bibliography, 1846–1930* (with Vol. 2 planned for 1980).

The best guide to work on Melville is Nathalia Wright's chapter in *Eight American Authors*, rev. ed., ed. James Woodress (1971), supplemented by the annual chapter in *American Literary Scholarship* (1963–   ) as well as news in *Melville Society Extracts*.

Edgar Allan Poe

There is no modern scholarly edition of Poe, although there is a number of facsimile reprints of early editions as well as two modern editions of the poems, one by Floyd Stovall (1965), the other by Thomas O. Mabbott (1969). Of recent popular editions, *The Short Fiction of Edgar Allan Poe*, ed. Stuart and Susan Levine (1976), is especially well annotated. John W. Ostrom edited *The Letters of Edgar Allan Poe* (1948; reprinted with additional letters, 1966). Partly because of forgeries and calumnies by Poe's literary executor Rufus Griswold, Poe biography became and has remained enmeshed in legends. John Carl Miller's *Building Poe Biography* (1977) lucidly traces the gradual emergence of documents concerning Poe. Although weak in its literary judgments and given to excessive argument in favor of certain dubious biographical points, Arthur Hobson Quinn's *Edgar Allan Poe: A Critical Biography* (1941) remains the best single factual account. Particular episodes are better told in Sidney P. Moss's *Poe's Literary Battles: The Critic in the Context of His Literary Milieu* (1963) and Moss's *Poe's Major Crisis: His Libel Suit and New York's Literary World* (1970). Robert D. Jacobs's *Poe: Journalist and Critic* (1969) is highly detailed and discriminating.

Poe's early reputation is traced in Eric Carlson's *The Recognition of Edgar Allan Poe* (1966) and in Jean Alexander's *Affidavits of Genius: Edgar Allan Poe and the French Critics, 1847–1924* (1971). Modern criticism is sampled in *Poe: A Collection of Critical Essays*, ed. Robert Regan (1967); *Twentieth Century Interpretations of Poe's Tales*, ed. William L. Howarth (1971); and *Papers on Poe*, ed. Richard P. Veler (1972). Among the most discussed recent books are David Halliburton's *Edgar Allan Poe: A Phenomenological View* (1973) and G. R. Thompson's *Poe's Fiction: Romantic Irony in the Gothic Tales* (1973). A comprehensive listing of works on Poe is Esther F. Hyneman's *Edgar Allan Poe: An Annotated Bibliography of Books and Articles in English, 1827–1973* (1974). The best guide to work on Poe is Jay B. Hubbell's chapter in *Eight American Authors*, rev. ed., ed. James Woodress (1971), supplemented—as in mentions of newly published letters—by the annual discussion in *American Literary Scholarship* (1963–   ), where Poe has been given a chapter to himself since 1973, and by *Poe Studies* (1968–   ).

Henry David Thoreau

The Princeton Edition (1971–   ) will be standard. Of the volumes published so far, *Walden*, ed. J. Lyndon Shanley, must be supplemented by Shanley's own *The Making of "Walden"* (1957; reprinted, 1966) and by further study of the next-to-final form of the manuscript. The Princeton Edition will include the journals, now available in an imperfectly transcribed edition (1906; reprinted 1962) as well as Walter Harding's re-edited *Correspondence*. For now, Thoreau's letters are mostly available in Carl Bode and Walter Harding's *Correspondence of Henry David Thoreau* (1958), supplemented by Kenneth W. Cameron's *Companion to Thoreau's Correspondence* (1964). *Collected Poems*, ed. Carl Bode (1964) is being re-edited for Princeton by Elizabeth Wetherell. Among the separate editions of *Walden* the miscalled "Variorum" edited by Harding is very valuable for its lavish annotations, as is Philip Van Doren Stern's *Annotated Walden* (1970).

The most reliable biography is Walter Harding's *The Days of Henry Thoreau: A Biography* (1965). Also valuable is Harding's *Thoreau: Man of Concord* (1960), recollections by dozens of people who knew Thoreau, and Harding's *Thoreau's Library* (1957). William L. Howarth edited Robert F. Stowell's *A Thoreau Gazetteer* (1970), which includes many maps and photographs of places Thoreau knew. Thoreau's reputation is surveyed in *Thoreau: A Century of Criticism*, ed. Walter Harding (1954), and in Wendell Glick's *The Recognition of Henry David Thoreau* (1969). Among the more frequently cited recent books about Thoreau are Sherman Paul's *The Shores of America: Thoreau's Inward Exploration* (1958); Joel Porte's *Emerson and Thoreau: Transcendentalists in Conflict* (1966); Michael Meyer's *Several More Lives to Live; Thoreau's Political Reputation in America* (1977); and Robert F. Sayre's *Thoreau and the American Indians* (1977). Sherman Paul edited *Thoreau: A Collection of Critical Essays* (1962). In preparation is a two-volume G. K. Hall Reference Guide to writings on Thoreau, the first volume by Steven J. Mailloux, the second by Mailloux and Michael Meyer. Walter Harding's *Thoreau Handbook* (1959), long out of print, is being revised by Harding and Meyer. Until the new *Handbook* appears, the best guide to work on Thoreau is Lewis Leary's chapter in *Eight American Authors*, rev. ed., ed. James Woodress (1971), supplemented by the annual chapter on the Transcendentalists in *American Literary Scholarship* (1963–   ) and issues of the *Thoreau Society Bulletin* (1941–   ).

T. B. Thorpe

None of Thorpe's books is in print. Meager selections are found in anthologies such as Hennig Cohen and William B. Dillingham's *Humor of the Old Southwest*, rev. ed. (1975). The only biography is Milton Rickles's *Thomas Bangs Thorpe: Humorist of the Old Southwest* (1962), which contains useful bibliographies. Essential background is in Norris W. Yates, *William T. Porter and the "Spirit of the Times": A Study of the "Big Bear" School of Humor* (1957). Beginning in 1963, the annual volume of *American Literary Scholarship* lists any new work on Thorpe.

Walt Whitman

The *Collected Writings of Walt Whitman* is in progress under the general editorship of Gay Wilson Allen; part of this edition is the five-volume *Walt Whitman: The Correspondence*, ed. Edwin H. Miller (1961–1969). The growth of *Leaves of Grass* may be studied conveniently in the *Comprehensive Reader's Edition*, ed. Harold W. Blodgett and Sculley Bradley (1965; reprinted, 1968); the same two edited the Norton Critical Edition of *Leaves of Grass* (1973). There are a variety of useful facsimiles of early editions, especially of the 1855 *Leaves of Grass*, and Arthur Golden edited *Walt Whitman's Blue Book* (1968), a facsimile of Whitman's marked-up copy of the 1860 edition.

The standard biography, essential for its week-by-week story of Whitman's life, is Gay Wilson Allen's *The Solitary Singer*, rev. ed. (1967). Allen also wrote *A Reader's Guide to Walt Whitman* (1970) and the extremely valuable *New Walt Whitman Handbook* (1975). Some other recent works based on study of documentary evidence are Thomas L. Brasher's *Whitman as Editor of the Brooklyn Daily Eagle* (1970); Joseph Jay Rubin's *The Historic Whitman* (1973); Floyd Stovall's *The Foreground of "Leaves of Grass"* (1974); and Jerome M. Loving's *Civil War Letters of George Washington Whitman* (1975), the writer of the letters being Whitman's brother. Of recent collections of writing on Whitman, these are especially useful: *The Presence of Walt Whitman*, ed. R. W. B. Lewis (1962); *Whitman: A Collection of Critical Essays*, ed. Roy Harvey Pearce (1962); *The Poet and the President: Whitman's Lincoln Poems*, ed. William Coyle (1962); *Whitman the Poet*, ed. John C. Broderick (1962); *Whitman's "Song of Myself": Origin, Growth, Meaning*, ed. James E. Miller, Jr. (1964); *A Century of Whitman Criticism*, ed. Edwin H. Miller (1969); *Walt Whitman*, ed. Francis Murphy (1969); and *Whitman: The Critical Heritage*, ed. Milton Hindus (1971). A notable recent critical book is Edwin H. Miller's *Walt Whitman's Poetry: A Psychological Journey* (1968).

The best guide to scholarship and criticism on Whitman is Roger Asselineau's chapter in *Eight American Authors*, rev. ed., ed. James Woodress (1971), supplemented by the discussions in *American Literary Scholarship* (1963–   ) and by the *Walt Whitman Review* (1955–   ).

John Greenleaf Whittier

Whittier's poetry has been steadily available, but much of his prose, especially his newspaper writings, has never been collected or else was collected but is out of print. Samuel T. Pickard's *Life and Letters of John Greenleaf Whittier* (1894), long standard, is being superseded by recent works such as John A. Pollard's *John Greenleaf Whittier; Friend of Man* (1949), John B. Pickard's *John Greenleaf Whittier; An Introduction and Interpretation* (1961), and John B. Pickard's three-volume *Letters of John Greenleaf Whittier* (1975). John B. Pickard also edited an important collection of criticism, *Memorabilia of John Greenleaf Whittier* (1968), and Donald C. Freeman, John B. Pickard, and Roland H. Woodwell assembled the illustration-filled *Whittier and Whittierland: Portrait of a Poet and His World* (1976). The best guide to work on Whittier is Karl Keller's chapter in *Fifteen American Authors before 1900*, ed. Robert A. Rees and Earl N. Harbert (1971), supplemented by the annual *American Literary Scholarship* (1963–   ).

AMERICAN LITERATURE 1865–1914

The titles of the following highly selective list of general historical and critical studies suggest their coverage within this period: Daniel Aaron, *The Unwritten War: American Writers and the Civil War* (1973); Louis Auchincloss, *Pioneers and Caretakers: A Study of Nine American Woman Novelists* (1965); Warner Berthoff, *The Ferment of Realism: American Literature, 1884–1919* (1965); Richard Bridgman, *The Colloquial Style in America* (1966); Edwin H. Cady, *The Light of Common Day* (1971); Richard Chase, *The American Novel and Its Tradition* (1957); Henry Steele Commager, *The American Mind: An Interpretation of American Thought and Character since the 1880's* (1950); Marcus Cunliffe, *The Literature of the United States* (1961); Robert Falk, *The Victorian Mode in American Fiction, 1865–1885* (1965); Leslie A. Fiedler, *Love and Death in the American Novel* (1960); Seymour L. Gross and John Edward Hardy, eds., *Images of the Negro in American Literature* (1966); Daniel Hoffman, *Form and Fable in American Fiction* (1961); Leon Howard, *Literature and the American Tradition* (1960); Howard Mumford Jones, *The Age of Energy: Varieties of American Experience, 1865–1915* (1971); Alfred Kazin, *On Native Grounds: An Interpretation of Modern American Prose Literature* (1942); Jay Martin, *Harvests of Change: American Literature, 1865–1914* (1967); Leo Marx, *The Machine in the Garden: Technology and the Pastoral Ideal in America* (1964); Ruth Miller, *Backgrounds to Black American Literature* (1971); Vernon L. Parrington, *Main Currents in American Thought: An Interpretation of American Literature from the Beginnings to 1920*, 3 vols. (1927–1930); Roy Harvey Pearce, *The Continuity of American Poetry* (1961); Donald Pizer, *Realism and Naturalism in Nineteenth-Century American Literature* (1966); Richard Poirier, *A World Elsewhere: The Place of Style in American Literature* (1966); Constance Rourke, *American Humor: A Study of the National*

Character (1931); Henry Nash Smith, *Virgin Land: The American West as Symbol and Myth* (1950) and *Democracy and the Novel: Popular Resistance to Classic American Writers* (1978); Tony Tanner, *The Reign of Wonder* (1965); Hyatt H. Waggoner, *American Poets from the Puritans to the Present* (1968); Charles C. Walcutt, *American Literary Naturalism: A Divided Stream* (1956); Edmund Wilson, *Patriotic Gore: Studies in the Literature of the American Civil War* (1962); Larzer Ziff, *The American 1890's* (1966).

Eight American Authors: A Review of Research and Criticism, rev. ed., ed. James Woodress (1971), contains discursive chapters that had appeared, up to 1969, on two authors in this period–Clemens and James; *Fifteen American Authors before 1900: Bibliographic Essays on Research and Criticism*, ed. Robert A. Rees and Earl N. Harbert (1971), contains chapters on Adams, Stephen Crane, and Howells.

Henry Adams

No uniform, authoritatively edited collection of Henry Adams's writings exists. Ernest Samuels's annotated edition of *The Education* sets a high standard for those who would undertake such a project in the future. In 1920, 1930, and 1938, Worthington C. Ford edited three volumes of Adams's excellent letters. The standard biography is a trilogy by Ernest Samuels: *The Young Henry Adams* (1948); *Henry Adams, the Middle Years* (1958); and *Henry Adams, the Major Phase* (1964). A sound one-volume life is Elizabeth Stevenson's *Henry Adams, a Biography* (1956).

George Hochfield's *Henry Adams: An Introduction and Interpretation* (1962) delivers well what its title promises. Different aspects of Adams are dealt with in Max I. Baym's *The French Education of Henry Adams* (1951); John J. Conder's *A Formula of His Own: Henry Adams's Literary Experiment* (1970); William H. Jordy's *Henry Adams: Scientific Historian* (1952); J. C. Levenson's *The Mind and Art of Henry Adams* (1957), a difficult and distinguished work; Ernest Scheyer's *The Circle of Henry Adams: Art and Artists* (1970); and Vern Wagner's *The Suspension of Henry Adams: A Study of Manner and Matter* (1969), which effectively treats Adams's humor.

Kate Chopin

The Complete Works of Kate Chopin (1969) was edited in two volumes by Per Seyersted, who also wrote the excellent *Kate Chopin: A Critical Biography* (1969), the most important full-length study of this rediscovered author. An early critical biography is Daniel Rankin's *Kate Chopin and Her Creole Stories* (1932).

Samuel Clemens

The present standard edition of *The Writings of Mark Twain* (1922–1925), 37 vols., ed. Albert Bigelow Paine, is being superseded by the ongoing editions of *The Mark Twain Papers* (1969–), supervised by the late Frederick Anderson until his death in 1979, and *The Works of Mark Twain* (1972–), under the supervision of John Gerber. Two volumes of letters have appeared so far in the *Papers*, and other Twain letters are available in several scattered volumes, the most interesting of which is *The Correspondence of Samuel L. Clemens and William Dean Howells, 1872–1910*, ed. Henry Nash Smith and William M. Gibson (1960).

Albert Bigelow Paine's *Mark Twain, a Biography* (1912) is vivid, unreliable, and still indispensable. Justin Kaplan's *Mr. Clemens and Mark Twain, a Biography* (1966) is a lively, popular account. Van Wyck Brooks's polemical and controversial *The Ordeal of Mark Twain* (1920) should be read together with Bernard De Voto's corrective *Mark Twain's America* (1932). De Lancey Ferguson's *Mark Twain: Man and Legend* (1943) is still the best full-length critical biography.

The criticism of Mark Twain is enormous in quantity, variety, and quality. Two excellent general studies are Henry Nash Smith's *Mark Twain: The Development of a Writer* (1962) and William M. Gibson's *The Art of Mark Twain*. James Cox's perceptive *Mark Twain: The Fate of Humor* (1966) and Gladys Bellamy's *Mark Twain as a Literary Artist* (1950), Edgar M. Branch's *The Literary Apprenticeship of Mark Twain* (1950) and Albert E. Stone's *The Innocent Eye: Childhood in Mark Twain's Imagination* (1961) are all distinguished contributions to Twain scholarship. More specialized studies of distinction include Henry Seidel Canby's *Turn West, Turn East* (1951), which concentrates on Twain and Henry James; Louis J. Budd's *Mark Twain: Social Philosopher* (1962); Bernard De Voto's *Mark Twain at Work* (1942); Paul Fatout's *Mark Twain on the Lecture Circuit* (1960); Sydney J. Krause's *Mark Twain as Critic* (1967); Paul Baetzhold's *Mark Twain and John Bull* (1970); and Walter Blair's *Mark Twain and Huckleberry Finn* (1960).

Stephen Crane

Fredson Bowers is the textual editor of *The Works of Stephen Crane* (1969–1976), a complete edition which has come in for some criticism despite its endorsement by the Center for Editions of American Authors. *Stephen Crane: Letters* (1960) was edited by R. W. Stallman and Lillian Gilkes. Joseph Katz edited both *The Portable Stephen Crane* (1969) and *The Poems of Stephen Crane: A Critical Edition* (1966).

Thomas Beer's early biography *Stephen Crane: A Study in American Letters* (1923) is still important, though it lacks documentation. The poet John Berryman prepared a stimulating and more factually reliable biography in 1950, and R. W. Stallman's massive, polemical *Stephen Crane: A Critical Biography* (1968) incorporates the substantial new material discovered in the interim.

Edwin H. Cady's *Stephen Crane* (1962) is an excellent critical introduction to Crane's life and work. Eric Solomon's *Stephen Crane: From Parody to Realism* (1966), Donald B. Gibson's *The Fiction of Stephen Crane* (1968), Daniel G. Hoffman's *The Poetry of Stephen Crane* (1957), Marston La France's *A Reading of Stephen Crane* (1971), and Frank Bergon's *Stephen Crane's Artistry* are among the best of the surprisingly few significant book-length studies.

Theodore Dreiser

Most of Dreiser's novels and short stories are in print, but his poetry, plays, and other writings generally are not, and a well-edited, uniform edition of Dreiser's writings is badly needed. Robert H. Elias has edited three volumes of *Letters of Theodore Dreiser* (1959).

The most complete biography is W. A. Swanberg's lively *Dreiser* (1965), but Robert H. Elias's *Theodore Dreiser: Apostle of Nature* (1949) is still valuable. Marguerite Tjader's *Theodore Dreiser: A New Dimension* (1965) sheds light on his later life especially. Ellen Moers's *Two Dreisers* (1969) examines the biographical circumstances surrounding (and the compositional histories of) *Sister Carrie* and *An American Tragedy*.

Good introductions to Dreiser's life and work may be found in Philip L. Gerber's *Theodore Dreiser* (1964) and John J. McAleer's *Theodore Dreiser: An Introduction and Interpretation*. F. O. Matthiessen's *Theodore Dreiser* (1951) is still illuminating, but more recent critical studies by Charles Shapiro, *Theodore Dreiser: Our Bitter Patriot* (1962), Richard Lehan, *Theodore Dreiser: His World and His Novels* (1969), R. N. Mookerjee, *Theodore Dreiser: His Thought and Social Criticism* (1974), and Donald Pizer, *The Novels of Theodore Dreiser* (1976), have advanced our appreciation and understanding considerably.

A bibliographical study may be found in Jackson R. Bryer's *Fifteen Modern American Authors* (1969), later reissued as *Sixteen Modern American Authors* (1973).

Bret Harte

The most complete collection is *The Works of Bret Harte* (1914) in 25 volumes. Harte's poetry may be found in *The Complete Poetical Works of Bret Harte* (1899). Geoffrey Bret Harte edited *The Letters of Bret Harte* (1926).

The standard life is George R. Stewart, Jr.'s *Bret Harte: Argonaut and Exile* (1931). Among the most important studies are Franklin Walker's in his *San Francisco's Literary Frontier* (1939) and Margaret Duckett's *Mark Twain and Bret Harte* (1964).

W. D. Howells

A Selected Edition of W. D. Howells (1968–) will provide the first authoritatively edited collection (in 30 or more volumes) of this major figure's enormous literary production. In the meantime, individual reprints must be sought out and may be supplemented by such collections of his letters as the two volumes edited by Henry Nash Smith and William M. Gibson, *The Correspondence of Samuel L. Clemens and William Dean Howells, 1872–1910* (1960) and his daughter Mildred Howells's edition in two volumes of *Life in Letters of William Dean Howells* (1928) and by Walter J. Meserve's *The Complete Plays of W. D. Howells* (1960).

The standard critical biography is Edwin H. Cady's two-volume work, *The Road to Realism: The Early Years, 1837–1885* (1956) and *The Realist at War: The Mature Years, 1885–1920* (1958). Van Wyck Brooks's *Howells: His Life and Work* (1959) reveals Howells in relation to his contemporaries. A more recent biography is Kenneth S. Lynn's *William Dean Howells: An American Life* (1971).

In the last three decades critical studies of Howells have appeared regularly. The introduction to Clara M. and Rudolph Kirk's *William Dean Howells: Representative Selections*** (1950) contains much useful information. An excellent brief introduction is William M. Gibson's *William Dean Howells* (1967). Everett Carter's *Howells and the Age of Realism* (1954) was one of the first to argue Howells's centrality to his age and its chief literary development. Other general studies of interest are George N. Bennett's *William Dean Howells: The Development of a Novelist* (1959); George C. Carrington's *The Immense Complex Drama* (1966); William McMurray's *The Literary Realism of William Dean Howells* (1967); and Robert L. Hough's *The Quiet Rebel: William Dean Howells as a Social Commentator* (1959). Kermit Vanderbilt's fresh and illuminating *The Achievement of William Dean Howells: A Reinterpretation* (1968) concentrates on five of Howells's novels. Three important, more specialized studies are James Woodress's *Howells and Italy* (1952); Olov W. Fryckstedt's *In Quest of America: A Study of Howells' Early Development as a Novelist* (1958); and James L. Dean's *Howells' Travels toward Art* (1970).

Henry James

The Novels and Tales of Henry James (The New York Edition) in 26 volumes, originally published 1907–1917, was reissued 1962–1965. Other collections of James's diverse writings are available in Leon Edel's edition of *The Complete Plays of Henry James* (1949); R. P. Blackmur's *The Art of the Novel* (1935); Morris Shapiro's edition of *Selected Literary Criticism* (1963); Morton D. Zabel's edition of James's travel writings, *The Art of Travel* (1958); F. O. Matthiessen and Kenneth B. Murdoch's *The Notebooks of Henry James* (1947); F. W. Dupee's edition of the three volumes of *Henry James: Autobiography* (1956); and in various collections of essays and letters edited by Leon Edel.

Edel is also author of the definitive, "Freudian" biography of James in five volumes, published between 1953 and 1972. F. O. Matthiessen's *The James Family* (1947) reveals Henry's relationships to this extraordinary American family, which included his brother William, the celebrated psychologist and philosopher.

F. W. Dupee's *Henry James* (1951) and Bruce McElderry's *Henry James* (1965) are among the best of the introductory books that survey this life and career. Since the 1940s, critical studies of James have abounded. Among the best of these, in chronological order, are J. W. Beach's pioneering *The Method of Henry James* (1918); Morris Roberts's *Henry James's Criticism* (1929); F. O. Matthiessen's *Henry James, the Major Phase* (1944); Quentin Anderson's *The American Henry James* (1957); Richard Poirier's *The Comic Sense of Henry James* (1960); Oscar Cargill's thorough *The Novels of Henry James* (1961); Dorothea Krook's *The Ordeal of Consciousness in Henry James* (1962); Laurence B. Holland's *The Expense of Vision* (1964); J. A. Ward's *Search for Form: Studies in the Structure of James's Fiction* (1967); Sallie Sears's *The Negative Imagination: Form and Perspective in the Novels of Henry James* (1968); James Kraft's *The Early Tales of Henry James* (1969); and Viola Hopkins Winner's *Henry James and the Visual Arts* (1970).

AMERICAN LITERATURE BETWEEN THE WARS 1914–1945

The best treatments of American literature during the period are Frederick J. Hoffman, *The Twenties: American Writing in the Postwar Decade* (1955); and two books by Hugh Kenner: *The Pound Era* (1971) and *A Homemade World: The American Modernist Writers* (1975). Graham Hough, in *Image and Experience: Studies in a Literary Revolution* (1960), explores modernism in the context of British literature. Two brief surveys are Louise Brogan's *Achievement in American Poetry, 1900–1950* (1951) and Alan S. Downer's *Fifty Years of American Drama, 1900–1950* (1951). Extensive treatments of separate areas are in Alfred Kazin, *On Native Grounds: An Interpretation of Modern American Prose Literature* (1942); Maxwell Geismar, *Writers in Crisis: The American Novel between Two Wars* (1942) and *The Last of the Provincials: The American Novel between Two Wars* (1948); Roy Harvey Pearce, *The Continuity of American Poetry* (1961); *Literary History of the United States*, ed. Robert E. Spiller et al., vol. III (1948); and Walter Sutton, *Modern American Criticism* (1963).

Interpretations of special aspects of American writing against the backgrounds of their traditions are Daniel Aaron's *Writers on the Left* (1961); Richard Bridgman's *The Colloquial Style in America* (1966); Nathan I. Huggins's *Harlem Renaissance* (1971); and Wright Morris's *The Territory Ahead* (1958).

The critic Malcolm Cowley, in *After the Genteel Tradition: American Writers, 1910–1930* (1937), gathered essays by important writers and critics commenting on contemporary literary issues, and wrote *Exile's Return: A Narrative of Ideas* (1934; rev. ed., 1951), an analysis of the 1920s from the standpoint of a participant. Edmund Wilson, in *The Shock of Recognition* (1943), presents writings by American writers about each other. His collections of his own essays and book reviews provide a lively survey of the period: *Classics and Commercials: A Literary Chronicle of the Thirties* (1951) and *The Shores of Light: A Literary Chronicle of the Twenties and Thirties* (1952). In *Earthly Delights, Unearthly Adornments: American Writers as Image-Makers* (1978), Wright Morris comments on his contemporaries and predecessors from the standpoint of a novelist writing after World War II.

Full bibliographical material on several authors in this period may be found in *Fifteen Modern American Authors: A Survey of Research and Criticism*, ed. Jackson R. Bryer (1969): Anderson, Hart Crane, Eliot, Faulkner, Fitzgerald, Frost, Hemingway, Pound, Robinson, and Stevens; a revised and updated edition was issued in 1973 as *Sixteen Modern American Authors* (Williams being added). See also *Bibliography*, ed. Richard M. Ludwig, Supplement to *Literary History of the United States* (1963), and *American Literary Scholarship: An Annual*, ed. James Woodress (1965–).

Sherwood Anderson

There is no collected edition of Anderson's writing. In addition to works mentioned in the author's introduction, Anderson published *The Modern Writer* (1925); *Sherwood Anderson's Notebook* (1925); *Alice and the Lost Novel* (1929); *Nearer the Grass Roots* (1929); *The American Country Fair* (1930); and *Home Town*

(1940). *The Letters of Sherwood Anderson* (1953) was edited by Howard Mumford Jones and Walter Rideout. Eugene P. Seehy and Kenneth A. Lohf compiled *Sherwood Anderson: A Bibliography* (1960). For commentary see David D. Anderson, *Sherwood Anderson: An Introduction and Interpretation* (1967); *Homage to Sherwood Anderson*, ed. Paul P. Appel (1970); Rex Burbank, *Sherwood Anderson* (1964); Irving Howe, *Sherwood Anderson* (1951); James Schevill, *Sherwood Anderson* (1951); Brom Weber, *Sherwood Anderson* (1964); and *The Achievement of Sherwood Anderson: Essays in Criticism*, ed. R. L. White (1966).

Hart Crane

Brom Weber edited *The Complete Poems and Selected Letters and Prose of Hart Crane* (1966), the best current text. Crane's *Collected Poems*, edited by his friend Waldo Frank, appeared in 1933. Other sources of primary and bibliographical material are *The Letters of Hart Crane*, ed. Brom Weber (1965); *The Letters of Hart Crane and His Family*, ed. Thomas S. W. Lewis (1974); *Hart Crane: An Annotated Critical Bibliography*, ed. Joseph Schwartz (1970); and *Hart Crane: A Descriptive Bibliography*, ed. Joseph Schwartz and Robert C. Schweik (1972). Gary Lane compiled *A Concordance to the Poems of Hart Crane* (1972).

Philip Horton's moving biography, *Hart Crane: The Life of an American Poet* (1937; reprint, 1957), was followed by Brom Weber's *Hart Crane: A Biographical and Critical Study* (1948) and John Unterecker's detailed *Voyager: A Life of Hart Crane* (1969). Susan Jenkins Brown, *Robber Rocks: Letters and Memories of Hart Crane, 1923–1932* (1969), provides an unusual portrait of Crane from the perspective of three close friends.

Three critics, themselves poets, contributed important early essays on Crane: Richard P. Blackmur in *The Double Agent* (1935); Allen Tate in *Reactionary Essays on Poetry and Ideas* (1936); and Yvor Winters in *In Defense of Reason* (1947). For more recent studies, see L. S. Dembo's *Hart Crane's Sanscrit Charge: A Study of "The Bridge"* (1960); Richard W. B. Lewis's *The Poetry of Hart Crane* (1967); and Helge N. Nilsen's *Hart Crane's "The Bridge": A Study in Sources and Interpretations* (1976). Brief introductions are Vincent Quinn's *Hart Crane* (1963); Samuel Hazo's *Hart Crane: An Introduction and Interpretation* (1963); Monroe K. Spears's *Hart Crane* (1965); and Herbert A. Leibowitz's *Hart Crane: An Introduction to the Poetry* (1968). R. W. Butterfield's *The Broken Arc: A Study of Hart Crane* (1969) and M. D. Uroff's *Hart Crane: The Patterns of His Poetry* (1974) analyze the often-neglected Key West poems. Hunce Voelcker's *The Hart Crane Voyages* (1967) is an impressionistic re-creation of Crane's process of composing *Voyages*.

E. E. Cummings

Complete Poems, 1913–1962 appeared in a two-volume edition in 1968 and in a single volume in 1972. George J. Firmage edited *Three Plays and a Ballet* (1967). Cummings's prose fiction includes *The Enormous Room* (1922) and *EIMI* (1933). *E. E. Cummings: A Miscellany Revised*, ed. George J. Firmage (1965), contains previously uncollected short prose pieces. *Six Nonlectures* (1953) consists of the talks that Cummings delivered at Harvard in the same year. *Selected Letters of E. E. Cummings* (1969) was edited by F. W. Dupee and George Stade. George J. Firmage edited *E. E. Cummings: A Bibliography* (1960).

Several of Cummings's earlier individual works have been re-edited by George J. Firmage in the Cummings Typescript Editions. These include *Tulips & Chimneys* (1976), *No Thanks* (1978), *The Enormous Room*, including illustrations by Cummings (1978), *ViVa* (1979), and *XAIPE* (1979).

The Magic Maker, a biography by Cummings's friend Charles Norman, was published in 1958 and reissued after the poet's death. A new biography is *Dreams in the Mirror*, by Richard S. Kennedy (1979). Norman Friedman's *E. E. Cummings: The Art of His Poetry* (1960) focuses on Cummings's language and the techniques of his poetry. Friedman's second book, *E. E. Cummings: The Growth of a Writer* (1964), discusses Cummings's texts in roughly chronological order. Barry A. Marks's *E. E. Cummings* (1964) analyzes individual poems and considers the relation between Cummings's graphic art and the principles of his poetry. Robert E. Wegner's *The Poetry and Prose of E. E. Cummings* (1965) and Bethany K. Dumas's *E. E. Cummings: A Remembrance of Miracles* (1974) provide introductions to Cummings's work.

E. E. Cummings and the Critics, ed. S. V. Baum (1962), traces critical response to Cummings over the course of his career. *Harvard Wake*, Vol. I (1946), is devoted to Cummings and contains commentaries by prominent literary figures. Norman Friedman edited *E. E. Cummings: A Collection of Critical Essays* (1972).

Hilda Doolittle (H. D.)

Collected Poems of H. D. was published in 1925. Subsequent volumes of poetry include *Red Rose for Bronze* (1931); the trilogy of war poems *The Walls Do Not Fall* (1944), *Tribute to the Angels* (1945), and *The Flowering of the Rod* (1946); and the dramatic monologue *Helen in Egypt* (1961). H. D.'s other book-length verse dramas are *Hippolytus Temporizes* (1927) and the translation of Euripides' *Ion* (1937). *Palimpsest* (1926), *Hedylus* (1928), and *Bid Me to Live* (1961) comprise her major prose fiction. *By Avon River* (1949) celebrates Shakespeare in prose and verse. *Tribute to Freud* (1956) is her account of her psychoanalysis by Freud.

Thomas Burnett Swann's *The Classical World of H. D.* (1962) and Vincent Quinn's *Hilda Doolittle* (1967) are the only full-length studies of H. D. H. D. figures prominently in *The Heart to Artemis: A Writer's Memoirs* (1962), the autobiography of her closest friend and traveling companion, Bryher (Winifred Ellerman). A chapter entitled "The Perfect Imagist" in Glenn Hughes's *Imagism and the Imagists* (1931) considers H. D.'s early poetry. In "H. D. and the Age of Myth," *The Sewanee Review*, Vol. LVI (1948), Harold H. Watts discusses the war poems and H. D.'s use of history. The autumn, 1969, issue of *Contemporary Literature* is devoted exclusively to H. D. and contains an interview with her literary executor Norman Holmes Pearson, articles by various

critics, and a selection of her poetry, prose, and letters. In the same issue, Jackson R. Bryer and Pamela Roblyer present the best bibliography of material by and about H. D.

John Dos Passos

In addition to works mentioned in the author's introduction, Dos Passos's works include the novels *Streets of Night* (1923) and *Chosen Country* (1951) and the collections of political and historical commentary *In All Countries* (1934), *Journeys between Two Wars* (1938), *The Living Thoughts of Tom Paine* (1940), *The Ground We Stand On* (1941), *The Theme Is Freedom* (1956), and *Occasions and Protests* (1964). Jack Potter's *A Bibliography of John Dos Passos* appeared in 1950, and *Dos Passos: A Collection of Critical Essays*, ed. Andrew Hook, in 1974. Full-length studies include George-Albert Astre, *Thèmes et structures dans l'oeuvre de John Dos Passos* (1956); John D. Brantley, *The Fiction of John Dos Passos* (1968); John H. Wrenn, *John Dos Passos* (1961); and Melvile Landsberg, *Dos Passos' Path to U.S.A.* (1970).

T. S. Eliot

Eliot's poetic works have been collected in *Collected Poems, 1909–1962* (1963) and *The Complete Poems and Plays of T. S. Eliot* which includes *Poems Written in Early Youth* (1969). The indispensable manuscript to *The Waste Land* is in *The Waste Land: A Facsimile and Transcript of the Original Drafts Including the Annotations of Ezra Pound*, ed. Valerie Eliot (1971). Collections of Eliot's criticism include *Selected Essays*, 3rd ed. (1951), and *Selected Prose of T. S. Eliot*, ed. Frank Kermode (1975). Other important critical writings include *The Use of Poetry and the Use of Criticism* (1933); *Poetry and Drama* (1951); *The Three Voices of Poetry* (1953); *On Poetry and Poets* (1957); and *To Criticize the Critic and Other Writings* (1965). Three volumes of social commentary are *After Strange Gods* (1934); *The Idea of a Christian Society* (1939); and *Notes toward the Definition of Culture* (1948). Eliot's Ph.D. dissertation was published as *Knowledge and Experience in the Philosophy of F. H. Bradley* in 1964.

There are biographical materials in *T. S. Eliot: A Symposium*, ed. Richard Marsh and Thurairajah Tambimuttu (1948); Herbart Howarth's *Notes on Some Figures behind T. S. Eliot* (1964); *T. S. Eliot: The Man and His Work*, ed. Allen Tate (1967); and *Affectionately, T. S. Eliot*, ed. William Turner Levy and Victor Scherle (1968). Full-length studies include Russell Kirk's *Eliot and His Age* (1971); Robert Sencourt's *T. S. Eliot: A Memoir* (1971); Lyndall Gordon's *Eliot's Early Years* (1977); and James E. Miller's *T. S. Eliot's Personal Waste Land: Exorcism of the Demons* (1977), a psychosexual study.

Of the countless critical studies of Eliot, the best are F. O. Matthiessen's *The Achievement of T. S. Eliot*, rev. ed. (1947); Helen Gardner's *The Art of T. S. Eliot* (1950); Hugh Kenner's *The Invisible Poet* (1959); Northrup Frye's *T. S. Eliot* (1963); Bernard Bergonzi's *T. S. Eliot* (1972); and Stephen Spender's *T. S. Eliot* (1976). Useful introductions include Elizabeth

Drew's *T. S. Eliot: The Design of His Poetry* (1949); George Williamson's *A Reader's Guide to T. S. Eliot* (1953); and Grover Smith's *T. S. Eliot's Poetry and Plays* (1956). Special studies include Carol Smith's *T. S. Eliot's Dramatic Theory and Practice* (1963); David E. Jones's *The Plays of T. S. Eliot* (1960); John Margolies's *T. S. Eliot's Intellectual Development, 1922–1939* (1972); and Helen Gardner's *The Composition of Four Quartets* (1978). A recent study is Derek Traversi's *T. S. Eliot: The Longer Poems* (1976). For collections of critical essays about Eliot, see *T. S. Eliot: A Study of His Writings by Various Hands*, ed. B. Rajan (1947); *T. S. Eliot, a Selected Critique*, ed. Leonard Unger (1948); Hugh Kenner's *T. S. Eliot: A Collection of Critical Essays* (1962); *Twentieth Century Interpretations of The Waste Land*, ed. Jay Martin (1968); and *T. S. Eliot: A Collection of Criticism*, ed. Linda W. Wagner (1974).

Donald C. Gallup compiled the standard bibliography, *T. S. Eliot: A Bibliography*, rev. ed. (1969).

William Faulkner

There is no collected edition of Faulkner's works. In addition to volumes mentioned in the author's introduction are *Early Prose and Poetry*, ed. Carvel Collins (1962); *Dr. Martino and Other Stories* (1934); *Pylon* (1935); *The Unvanquished* (1938); *Intruder in the Dust* (1948); *Knight's Gambit* (1949); *Collected Stories of William Faulkner* (1950); *Requiem for a Nun* (1951); *Notes on a Horsethief* (1951); *Big Woods* (1955); and *Essays, Speeches, and Public Letters*, ed. James B. Meriwether (1965). Transcripts of discussion sessions and interviews with Faulkner include *Faulkner at Nagano*, ed. Robert A. Jelliffe (1956); *Faulkner in the University*, ed. Frederick L. Gwynn and Joseph L. Blotner (1959); *Faulkner at West Point*, ed. Joseph L. Fant and Robert Ashley (1964); and *The Lion in the Garden*, ed. James B. Meriwether and Michael Millgate (1968). Letters are collected in *The Faulkner–Cowley File*, ed. Malcolm Cowley (1961), and *Selected Letters of William Faulkner*, ed. Joseph L. Blotner (1977).

The full-length *Faulkner: A Biography*, 2 vols. (1974), by Joseph L. Blotner, may be supplemented by biographical material and memoirs in John Cullen and Floyd C. Watkins's *Old Times in the Faulkner Country* (1961); John Faulkner's *My Brother Bill* (1963); *William Faulkner of Oxford*, ed. James B. Webb and A. Wigfall Green (1965); and Murry C. Falkner's *The Falkners of Mississippi* (1967).

Full-length critical studies include Michael Millgate's *The Achievement of William Faulkner* (1963); Cleanth Brooks's *The Yoknapatawpha Country* (1963), and *Toward Yoknapatawpha and Beyond* (1978); Olga Vickery's *The Novels of William Faulkner* (1959; rev. ed., 1964); and Richard P. Adams's *Faulkner: Myth and Motion* (1968). Others are Irving Howe's *William Faulkner* (1952; rev. ed., 1962); Hyatt H. Waggoner's *William Faulkner* (1959); Walter Slatoff's *Quest for Failure* (1960); Frederick J. Hoffman's *William Faulkner* (1961); Lawrance Thompson's *William Faulkner, an Introduction and Interpretation* (1963; rev. ed., 1967); and Edward L. Volpe's *A Reader's Guide to William Faulkner* (1964). More specialized studies include Warren Beck's *Man in Motion: Faulkner's*

Trilogy (1961); John Longley, Jr.'s *The Tragic Mask: A Study of Faulkner's Heroes* (1963); John Hunt's *William Faulkner: Art in Theological Tension* (1965); Peter Swiggert's *The Art of Faulkner's Novels* (1962); Charles H. Nilon's *Faulkner and the Negro* (1965); Melvin Backman's *Faulkner: The Major Years* (1966); Walter Brylowski's *Faulkner's Olympian Laugh* (1968); H. Richardson's *William Faulkner, the Journey to Self-Discovery* (1969); and John T. Irwin's *Doubling and Incest, Repetition and Revenge: A Speculative Reading of Faulkner* (1975).

Collections of critical essays include *William Faulkner: Two Decades of Criticism*, ed. Frederick J. Hoffman and Olga Vickery (1951); Hoffman and Vickery's *William Faulkner: Three Decades of Criticism* (1960); Linda Wagner's *William Faulkner: Four Decades of Criticism* (1973); *Faulkner: A Collection of Critical Essays*, ed. Robert Penn Warren (1966); and *William Faulkner: A Collection of Cricitism*, ed. Dean M. Schmitter (1973). Useful reference books include Robert W. Kirk and Marvin Klots's *Faulkner's People: A Complete Guide and Index* (1963); Harry Runyan's *A Faulkner Glossary* (1964); Dorothy Tuck's *Apollo Handbook of Faulkner* (1964); I. L. Sleeth's *William Faulkner: A Bibliography of Criticism* (1962); and the authoritative bibliography by James B. Meriwether, *The Literary Career of William Faulkner: A Bibliographic Study* (1961).

F. Scott Fitzgerald

Fitzgerald's novels are *This Side of Paradise* (1920); *The Beautiful and Damned* (1922); *The Great Gatsby* (1925); *Tender Is the Night* (1934; rev. ed., 1939; reprint, 1953); and the unfinished *The Last Tycoon* (1941). His collections of stories are *Flappers and Philosophers* (1921); *Tales of the Jazz Age* (1922); *All the Sad Young Men* (1926); and *Taps at Reveille* (1935). A satirical play, *The Vegetable, or From President to Postman*, was published in 1923. Among the valuable collections of Fitzgerald's writings are *The Apprentice Fiction of F. Scott Fitzgerald*, ed. John Kuehl (1965); *F. Scott Fitzgerald in His Own Time: A Miscellany*, ed. Jackson R. Bryer and Matthew J. Bruccoli (1971); and *Afternoon of an Author*, ed. Arthur Mizener (1957). *The Crack-Up*, ed. Edmund Wilson (1945), collects essays, notebook entries, and letters from the 1930s which provide an indispensable firsthand impression of Fitzgerald's early success and subsequent emotional, economic, and marital collapse.

The Letters of F. Scott Fitzgerald (1963) was edited by Andrew Turnbull. Other volumes of letters are *Dear Scott/Dear Max: The Fitzgerald–Perkins Correspondence*, ed. John Kuehl and Jackson R. Bryer (1971); and *As Ever, Scott Fitz*—(letters between Fitzgerald and his literary agent, Harold Ober), ed. Matthew J. Bruccoli and Jennifer McCabe Atkinson (1972); Fitzgerald's biographers are Arthur Mizener, *The Far Side of Paradise* (1951; rev. ed., 1965), and Andrew Turnbull, *Scott Fitzgerald* (1962). Other sources of biographical information include Sheila Graham's reminiscences (*Beloved Infidel* [1958] and *College of One* [1967]) of her affair with Fitzgerald; Nancy Milford's *Zelda* (1970); and Aaron Latham's *Crazy Sundays: F. Scott Fitzgerald in Hollywood* (1971).

Critical introductions to Fitzgerald's work include Kenneth Eble's *F. Scott Fitzgerald* (1963); Sergio Perosa's *The Art of F. Scott Fitzgerald* (1965); Henry Dan Piper's *F. Scott Fitzgerald: A Critical Portrait* (1965) and Richard D. Lehan's *F. Scott Fitzgerald and the Craft of Fiction* (1966). *The Fictional Technique of F. Scott Fitzgerald* (1957) by James E. Miller offers close readings of the first three novels. Robert Sklar's *F. Scott Fitzgerald: The Last Laocoön* (1967) sees the effort to master the genteel tradition as an impetus for complex artistry, while Milton R. Stern's *The Golden Moment: The Novels of F. Scott Fitzgerald* (1970) focuses on autobiography and personality in the fiction. Other useful works include William Goldhurst's *F. Scott Fitzgerald and His Contemporaries* (1963), which studies the relations between Fitzgerald and four influential contemporaries—Wilson, Mencken, Lardner, and Hemingway; Matthew J. Bruccoli's *The Composition of Tender Is the Night* (1963); and John F. Callahan's *The Illusions of a Nation: Myth and History in the Novels of F. Scott Fitzgerald* (1972). *F. Scott Fitzgerald: The Man and His Work*, ed. Alfred Kazin (1951), and *F. Scott Fitzgerald*, ed. Arthur Mizener (1963), are collections of critical essays. Jackson R. Bryer has edited *The Critical Reputation of F. Scott Fitzgerald: A Bibliographical Study* (1967).

Robert Frost

The Poetry of Robert Frost (1969) incorporated *Complete Poems* (1949) and Frost's last volume, *In the Clearing* (1962). Eleven early essays on farming, edited by Edward C. Lathem and Lawrance Thompson, appear in *Robert Frost: Farm Poultryman* (1963). Important collections of letters appear in *Selected Letters of Robert Frost*, ed. Lawrance Thompson (1964); *The Letters of Robert Frost to Louis Untermeyer* (1963); and Margaret Anderson, *Robert Frost and John Bartlett: The Record of a Friendship* (1963). His critical essays have been collected in *Selected Prose*, ed. Hyde Cox and Edward C. Lathem (1966). Collections of conversations and interviews with Frost include Edward C. Lathem's *Interviews with Robert Frost* (1966); Reginald L. Cook's *The Dimensions of Robert Frost* (1964); Daniel Smythe's *Robert Frost Speaks* (1954); and Louis Mertin's *Robert Frost: Life and Talks — Walking* (1965).

Lawrance Thompson finished two volumes of the authorized biography before his death: *Robert Frost: The Early Years, 1874–1915* (1966) and *Robert Frost: The Years of Triumph, 1915–1938* (1970). The early *Robert Frost: A Bibliography*, ed. W. B. Shubrick Clymer and Charles Green (1937), must be supplemented by Una Parameswaran, "Robert Frost: A Bibliography of Articles and Books, 1958–1964," *Bulletin of Bibliography* (January/Spring, May/August, 1967), and by W. B. S. Clymer, *Robert Frost: A Bibliography* (1972).

A brief critical introduction to Frost's poetry is Lawrance Thompson's *Robert Frost* (1964). The best full-length critical studies are Reuben Brower's *The Poetry of Robert Frost: Constellations of Intention* (1963) and John F. Lynen's *The Pastoral Art of Robert Frost* (1964). Other useful studies are Reginald L. Cook's *The Dimensions of Robert Frost* (1958); Radcliffe Squires's *The Major Themes of Robert Frost* (1963); Philip L. Gerber's *Robert Frost*

(1966); George W. Nitchie's adversely critical *Human Values in the Poetry of Robert Frost* (1960); and John R. Doyle, Jr.'s *The Poetry of Robert Frost: An Analysis* (1962). Collections of critical essays include *Robert Frost: An Introduction*, ed. Robert A. Greenberg and James G. Hepburn (1961); *Robert Frost: A Collection of Critical Essays*, ed. James M. Cox (1962); and *Recognition of Robert Frost*, ed. Richard Thornton (1970).

Ernest Hemingway

There is no collected edition of Hemingway's writings. In addition to works mentioned in the author's introduction, his fiction includes *Today Is Friday* (1926) and *God Rest You Merry Gentlemen* (1933). His tribute to Spain, *The Spanish Earth*, and his play *The Fifth Column* appeared in 1938, the latter in a collection *The Fifth Column and the First Forty-nine Stories*. His journalism has been collected in *The Wild Years*, ed. Gene Z. Hanrahan (1962), and *By-Line: Ernest Hemingway, Selected Articles and Dispatches of Four Decades*, ed. William White (1967). Audre Hanneman's *Ernest Hemingway: A Comprehensive Bibliography* (1967) is exceptionally thorough.

The authorized biography is Carlos Baker's *Ernest Hemingway: A Life Story* (1969). Of numerous memoirs by relatives and acquaintances, the best are Leicester Hemingway's *My Brother, Ernest Hemingway* (1962); Marcelline Hemingway Sanford's *At the Hemingways: A Family Portrait* (1962); and Gregory H. Hemingway's *Papa* (1976).

Philip Young, *Ernest Hemingway* (1959), Earl Rovit, *Ernest Hemingway* (1963), and Sheridan Baker, *Ernest Hemingway* (1967) are introductory critical studies. Full-length critical works include Carlos Baker's *Ernest Hemingway: The Writer As Artist* (1952; rev. ed., 1956) and Philip Young's *Ernest Hemingway* (1952; rev. ed., 1966). For collections of critical essays on Hemingway, see *Ernest Hemingway: The Man and His Work*, ed. John K. M. McCaffrey (1950); *Hemingway and His Critics*, ed. Carlos Baker (1961); *Hemingway: A Collection of Critical Essays*, ed. Robert P. Weeks (1962); and *The Literary Reputation of Hemingway in Europe*, ed. Roger Asselineau (1965). Specialized studies include Charles A. Fenton's *The Apprenticeship of Ernest Hemingway: The Early Years* (1954); Robert W. Lewis's *Hemingway on Love* (1965); Robert O. Stephen's *Hemingway's Nonfiction: The Public Voice* (1968); and Emily A. Watts's *Ernest Hemingway and the Arts* (1971).

Langston Hughes

The Selected Poems of Langston Hughes (1959 and 1965) draws on his earlier volumes, the most important of which were *The Weary Blues* (1926); *Fine Clothes to the Jew* (1927); *The Dream Keeper and Other Poems* (1932); *Scottsboro Limited: Four Poems and a Play in Verse* (1932); and *Montage of a Dream Deferred* (1951). More recent is *The Panther and the Lash: Poems of Our Times* (1967). Faith Berry has edited *Good Morning Revolution: Uncollected Social Protest Writings by Langston Hughes* (1973). Hughes's autobiographical volumes include *The Big Sea* (1940) and *I Wonder as I Wander* (1956).

Biographies of Hughes include Milton Meltzer's *Langston Hughes: A Biography* (1968); C. Rollins's *Black Troubador* (1970); and James S. Haskins's *Always Movin' On: The Life of Langston Hughes* (1976). James A. Emmanuel's *Langston Hughes* (1967) is an excellent introduction to Hughes's writing. Full-length critical studies are Onwuchekwa Jemie's *Langston Hughes: An Introduction to the Poetry* (1976) and Richard K. Barksdale's *Langston Hughes: The Poet and His Critics* (1977). Therman B. O'Daniel has edited *Langston Hughes, Black Genius: A Critical Evaluation* (1972). Useful reference works include Donald C. Dickinson's *A Bio-Bibliography of Langston Hughes, 1902–1967* (1967) and the compilation by Peter Mandelik and Stanley Schatt, *A Concordance to the Poetry of Langston Hughes* (1975).

Marianne Moore

No collection of Moore's verse is authoritative because each excluded some earlier poems while subjecting others to extensive revisions and changes in format. Her separate volumes include *Poems* (1921); *Observations* (1924); *The Pangolin and Other Verse* (1936); *What Are Years?* (1941); *Nevertheless* (1944); *Like a Bulwark* (1956); *O to Be a Dragon* (1959); and *Tell Me, Tell Me* (1966). Collections include *Selected Poems* (1935); *Collected Poems* (1951); and *The Complete Poems of Marianne Moore* (1967), which included selections from her translation of *The Fables of La Fontaine* (1954). She also translated Adalbert Stifter's *Rock Crystal: A Christmas Tale* (1945). A collection of her critical essays, *Predilections*, appeared in 1955.

Four important and perceptive essays on Moore are by poet-critics: T. S. Eliot's introduction to Moore's *Selected Poems*; Richard P. Blackmur's "The Method of Marianne Moore," in *The Double Agent* (1935); Kenneth Burke's "Motives and Motifs in the Poetry of Marianne Moore," reprinted in *A Grammar of Motives* (1945); and Randall Jarrell's "Thoughts about Marianne Moore," reprinted in *Poetry and the Age* (1953). Hugh Kenner's incisive essay "Disliking It" is in *A Homemade World: The American Modernist Writers* (1975). Book-length studies include Bernard F. Engel's *Marianne Moore* (1963); Jean Garrigue's *Marianne Moore* (1965); George W. Nitchie's *Marianne Moore: An Introduction to the Poetry* (1969); Pamela W. Hadas's *Marianne Moore, Poet of Affection* (1977); and Donald Hall's *Marianne Moore: The Cage and the Animal* (1970). The most recent is Laurence Stapleton's *Marianne Moore: The Poet's Advance* (1978). M. J. Tambimuttu edited *Festschrift for Marianne Moore's Seventy-fifth Birthday–by Various Hands* in 1964, and Charles Tomlinson edited *Marianne Moore: A Collection of Critical Essays* in 1969. Eugene P. Sheehy and Kenneth A. Lohf compiled *The Achievement of Marianne Moore: A Bibliography, 1907–1957* in 1958, and Craig S. Abbott compiled *Marianne Moore: A Descriptive Bibliography* in 1977. Gary Lane's *A Concordance to the Poems of Marianne Moore* appeared in 1972.

Katherine Anne Porter

There is no standard collection of Porter's works, nor a biography. In addition to the works mentioned in the author's introduction, she published an expanded edition of *Flowering Judas* in 1935, edited a translation from the Spanish by her second husband, Eugene Pressly, of J. F. Lizardi's *The Itching Parrot* (1942), and published *The Collected Essays and Occasional Writings* (1970) and essays and reviews in *The Never Ending Wrong* (1977).

The bibliographies are Edward Schwartz's *Katherine Anne Porter, a Critical Bibliography: Bulletin of the New York Public Library* (1957), and *A Bibliography of the Works of Katherine Anne Porter*, ed. Louise Waldrip and Shirley Ann Bauer (1969). Two important essays on Porter are in Glenway Wescott's *Images of Truth* (1962) and Robert P. Warren's *Selected Essays* (1958). *Katherine Anne Porter: A Critical Symposium* was edited by Lodwicj Hartley and George Gore (1969). An interview with Porter by George Plimpton appeared in the *Paris Review Interviews* (1963). For surveys of her career and writings, see Ray West's *Katherine Anne Porter* (1963); George Hendrick's *Katherine Anne Porter* (1965); M. M. Liberman's *Katherine Anne Porter's Fiction* (1971); Harry J. Mooney's *The Fiction and Criticism of Katherine Anne Porter* (1962); William L. Nance's *Katherine Anne Porter and the Art of Rejection* (1964); W. S. Emmons's *Katherine Anne Porter: The Regional Stories* (1967); and Paul R. Baumgarter's *Katherine Anne Porter* (1969).

Ezra Pound

Pound's early poetry, from *A Lume Spento* (1908) through *Ripostes* (1912), has been collected in the authoritative *Collected Early Poems of Ezra Pound*, ed. Michael King (1976). The same volumes, along with later volumes *Hugh Selwyn Mauberley* and *Homage to Sextus Propertius*, are collected in *Personae: The Collected Poems*, rev. ed., with appendices (1949). Earlier editions of *The Cantos* have been superseded by *The Cantos*, I through CXX with the exception of LXXII and LXXIII, which have never been published (1976). The most important of Pound's critical writings are *The Spirit of Romance*, rev. ed. (1952); *Gaudier-Brzeska: A Memoir* (1916); *Instigations* (1920); *Make It New* (1934); *Guide to Kulchur* (1938); and *Patria Mia* (1950). His criticism has been collected in *Literary Essays of Ezra Pound*, ed. T. S. Eliot (1954), and *Selected Prose, 1909–1965*, ed. William Cookson (1973). *Ezra Pound Speaking: Radio Speeches of World War II*, ed. Leonard W. Doob (1978), is a collection of wartime radio addresses. D. D. Paige edited *The Letters of Ezra Pound, 1907–1941* (1950). A revised edition of *Selected Poems of Ezra Pound* (1949) and *Ezra Pound: Selected Cantos* appeared in 1957 and 1967 respectively.

There is biographical material in Pound's autobiographical pamphlet *Indiscretions* (1923); Noel Stock's *Ezra Pound's Pennsylvania* (1976); Michael Reck's *Ezra Pound: A Close-Up* (1967); Harry M. Meachman's *The Caged Panther: Ezra Pound at St. Elizabeth's* (1968); *The Trial of Ezra Pound* (1966) by Pound's lawyer, Julien Cornell; and the memoir by Pound's daughter, Mary de Rachewiltz, *Discretions* (1972). Full-length biographies are Charles Norman's *Ezra Pound* (1960); Noel Stock's *The Life of Ezra Pound* (1970); and C. David Haymann's *Ezra Pound, the Last Rower: A Political Profile* (1976).

The best critical volumes on Pound are Hugh Kenner's *The Pound Era* (1972); Donald Davie's brief *Pound* (1975); and the latter's *Ezra Pound: Poet as Sculptor* (1964). M. L. Rosenthal's *A Primer of Ezra Pound* (1960) is a good introduction, and Christine Brooke-Rose's *A ZBC of Ezra Pound* (1971) is a perceptive study. Collections of essays on Pound include *Ezra Pound: A Collection of Critical Essays*, ed. Walter Sutton (1963); *New Approaches to Ezra Pound*, ed. Eva Hesse (1969); and *Ezra Pound: The Critical Heritage*, ed. Eric Homberger (1972). On Pound's early verse see N. Christophe de Nagy's *The Poetry of Ezra Pound: The Pre-Imagist Stage*, rev. ed. (1968); Thomas H. Jackson's *The Early Poetry of Ezra Pound* (1968); Hugh Witemeyer's *The Poetry of Ezra Pound: Forms and Renewal, 1908–1920* (1969); and K. K. Ruthven's *A Guide to Ezra Pound's Personae* (1969). Studies of particular works include John Espey's *Ezra Pound's Mauberley: A Study in Composition* (1955); J. P. Sullivan's *Ezra Pound and Sextus Propertius: A Study in Creative Translation* (1964); L. S. Dembo's *The Confucian Odes of Ezra Pound* (1963); the indispensable *Annotated Index to the Cantos of Ezra Pound* (through Canto LXXXIV), ed. John H. Edwards and William W. Vasse (1957); and the following books on *The Cantos: Motive and Method in the Cantos of Ezra Pound*, ed. Lewis Leary (1954); Clark Emery's *Ideas into Action* (1958); George Dekker's *The Cantos of Ezra Pound* (1963); Noel Stock's *Reading the Cantos* (1967); Walter Baumann's *The Rose in the Steel Dust: An Examination of the Cantos of Ezra Pound* (1967); Daniel D. Pearlman's *The Barb of Time: The Unity of Pound's Cantos* (1969); Eugene P. Nasser's *The Cantos of Ezra Pound* (1975); and Ronald Bush's *The Genesis of Pound's Cantos* (1976).

Donald C. Gallup's *A Bibliography of Ezra Pound* appeared in 1963, and Gary Lane issued *A Concordance to Personae: The Shorter Poems of Ezra Pound* in 1974.

John Crowe Ransom

There is no collected edition of Ransom's writings. A selection from his first two volumes, *Poems about God* (1919) and *Chills and Fever* (1924), appeared as *Grace after Meat* in 1924. *Selected Poems* (1945) excluded many of his previously published poems. *Selected Poems* (1963), a more generous selection, included many revisions of his earlier works. His critical volumes are *God without Thunder: An Unorthodox Defense of Orthodoxy* (1931); *The World's Body* (1938); and *Beating the Bushes: Selected Essays, 1941–1970* (1972).

Writings about Ransom include *John Crowe Ransom: Critical Essays and a Bibliography*, ed. Thomas D. Young (1968); John L. Stewart's *John Crowe Ransom* (1962); Thornton H. Parsons's *John Crowe Ransom* (1969); Robert Buffington's *The Equilibrist: A Study of John Crowe Ransom's Poems, 1916–1963* (1976); Karl F. Knight's *The Poetry of John Crowe Ransom: A Study of Diction, Metaphor, and Symbol* (1971); Miller Williams's *The Poetry of John Crowe Ransom* (1972); and Thomas D. Young's *Gentleman in a Dustcoat: A Biography of John Crowe Ransom* (1976).

Edwin Arlington Robinson

Collected Poems of Edwin Arlington Robinson (1921) was enlarged periodically through 1937. In addition to a considerable body of shorter verse, Robinson wrote a number of long narrative poems. *Merlin* (1917), *Lancelot* (1920), and *Tristram* (1927) evoke an Arthurian world embroiled in "modern" moral and political crises. Robinson's other explorations of human character in modern life include *Roman Bartholomew* (1923); *The Man Who Died Twice* (1924); *Cavender's House* (1929); *The Glory of the Nightingales* (1930); *Matthias at the Door* (1931); *Talifer* (1933); *Amaranth* (1934); and *King Jasper* (1935). Ridgely Torrence compiled *Selected Letters* (1940). Denham Sutcliffe edited *Untriangulated Stars: Letters of Edwin Arlington Robinson to Harry de Forest Smith, 1890–1905* (1947). Richard Cary edited *Edwin Arlington Robinson's Letters to Edith Brower* (1968). *A Bibliography of Edwin Arlington Robinson*, ed. Charles Bescher Hogan, appeared in 1936. William White compiled *Edwin Arlington Robinson: A Supplementary Bibliography* (1971).

Two biographies, both entitled *Edwin Arlington Robinson*, are by Hermann Hagedorn (1938) and Emery Neff (1948). Three early critical studies, all entitled *Edwin Arlington Robinson*, were written by Mark Van Doren (1927), Charles Cestre (1930), and Yvor Winters (1946). Ellsworth Barnard's *Edwin Arlington Robinson: A Critical Study* (1952) is the best introduction to Robinson's work. Others include Chard Powers Smith's *Where the Light Falls: A Portrait of Edwin Arlington Robinson* (1965); Wallace L. Anderson's *Edwin Arlington Robinson: A Critical Introduction* (1967); Hoyt C. Franchere's *Edwin Arlington Robinson* (1968); and Louis O. Coxe's *Edwin Arlington Robinson: The Life of Poetry* (1969). Edwin S. Fussell examines the cultural and intellectual influences on Robinson in *Edwin Arlington Robinson: The Literary Background of a Traditional Poet* (1954). W. R. Robinson's *Edwin Arlington Robinson: A Poetry of the Act* (1967) focuses on the poet's aesthetics and on his methods of constructing and exploring philosophical themes and motifs. Ellsworth Barnard edited *Edwin Arlington Robinson: Centenary Essays* (1969). Other collections are *Appreciation of Edwin Arlington Robinson: Twenty-eight Interpretive Essays*, ed. Richard Cary (1969), and *Edwin Arlington Robinson: A Collection of Critical Essays*, ed. Francis Murphy (1970).

Carl Sandburg

The Complete Poems of Carl Sandburg (1950) was revised and expanded in 1969. In addition to seven full-length volumes of poetry, Sandburg wrote a Pulitzer Prize–winning study of Abraham Lincoln (*Abraham Lincoln: The Prairie Years*, 2 vols. [1926], and *Abraham Lincoln: The War Years*, 4 vols. [1939]), two collections of journalism and social commentary, a novel, and several books for children. *The Letters of Carl Sandburg* (1968) was edited by Herbert Mitgang. Mark Van Doren published *Carl Sandburg: With a Bibliography of Sandburg Materials in the Collections of the Library of Congress* (1969).

Always the Young Strangers (1952) is Sandburg's autobiography. Other sources of biographical information are Karl Detzer's *Carl*

Sandburg: A Study in Personality and Background (1941); Harry Golden's *Carl Sandburg* (1961); and North Callahan's *Carl Sandburg: Lincoln of Our Literature* (1970). Critical studies include Richard Crowder's *Carl Sandburg* (1964); Hazell Durnell's *The America of Carl Sandburg* (1965); and Gay Wilson Allen's *Carl Sandburg* (1972). Sandburg's work has occasioned numerous reviews, journal articles, and anthology pieces.

Gertrude Stein

There is no complete edition of Gertrude Stein's profuse and eclectic writings. *The Yale Edition of the Unpublished Writings of Gertrude Stein*, ed. Carl Van Vechten (1951–1958), contains eight volumes of poetry, prose fiction, portraits, essays, and miscellany. Among the works published during Stein's lifetime are *Tender Buttons* (1914), cycles of prose poems on objects, food, and rooms; and the children's book *The World Is Round* (1926). Her prose fiction includes *Three Lives* (1909); *The Making of Americans* (1925); *Lucy Church Amiably* (1930); *Ida: A Novel* (1941); and *Brewsie and Willie* (1946). *Geography and Plays* (1922), *Operas and Plays* (1932), and *Last Operas and Plays*, ed. Carl Van Vechten (1949), contain dramatic pieces. Biographical material and portraits of Stein's contemporaries may be found in *The Autobiography of Alice B. Toklas* (1933); *Portraits and Prayers* (1934); *Everybody's Autobiography* (1937); and *Wars I Have Seen* (1945). Other works, including meditations, essays, sketches, and sociolinguistic treatises, are *Useful Knowledge* (1928); *How to Write* (1931); *Lectures in America* (1935); *Narration: Four Lectures by Gertrude Stein* (1935); *What Are Masterpieces* (1940); and *Four in America* (1947). Donald C. Gallup has edited *Fernhurst, Q.E.D. and Other Early Writings by Gertrude Stein* (1971), and also a collection of letters to Gertrude Stein, *The Flowers of Friendship* (1953). Julian Sawyer's *Gertrude Stein: A Bibliography* (1940) was published before the end of Stein's career.

Biographical studies of Gertrude Stein are Elizabeth Sprigge's *Gertrude Stein: Her Life and Work* (1957); John Malcolm Brinnin's *The Third Rose: Gertrude Stein and Her World* (1959); Howard Greenfield's *Gertrude Stein: A Biography* (1973); and J. Michael Hoffman's *The Development of Abstractionism in the Writings of Gertrude Stein* (1965). W. G. Rogers's *When This You See Remember Me: Gertrude Stein in Person* (1948) and Alice B. Toklas's *What Is Remembered* (1963) are reminiscences of Stein by friends. Critical studies include Donald Sutherland's *Gertrude Stein: A Biography of Her Work* (1951); Allegra Stewart's *Gertrude Stein and the Present* (1967); Norman Weinstein's *Gertrude Stein and the Literature of the Modern Consciousness* (1970); and Robert B. Haas's *A Primer for the Gradual Understanding of Gertrude Stein* (1971). A number of recent essays on Stein's work are collected in a special issue of *Widening Circle*, Vol. I, No. 4 (1973). The most recent biographical and critical studies are Richard Bridgman's *Gertrude Stein in Pieces* (1970); Janet Hobhouse's *Everybody Who Was Anybody* (1975); and James R. Mellow's *Charmed Circle: Gertrude Stein and Company* (1974).

Wallace Stevens

The Collected Poems of Wallace Stevens was published in 1954. *Opus Posthumous*, ed. Samuel French Morse (1957), includes previously uncollected poems, plays, and essays primarily found among Stevens's numerous contributions to magazines and anthologies. Holly Stevens arranges a large selection of her father's poems in their probable order of composition and makes various textual corrections in editing *The Palm at the End of the Mind* (1971). *The Necessary Angel: Essays on Reality and the Imagination* (1951) is Stevens's prose statement on poetry. Other sources of primary and bibliographical material are *Letters of Wallace Stevens*, ed. Holly Stevens (1966); *Concordance to the Poetry of Wallace Stevens*, ed. Thomas Walsh (1963); and *Wallace Stevens: A Descriptive Bibliography*, compiled by J. M. Edelstein (1973).

Samuel French Morse's *Wallace Stevens: Poetry as Life* (1970) is a critical biography. Two early studies, William Van O'Connor's *The Shaping Spirit* (1950) and Robert Pack's *Wallace Stevens* (1958), provide introductions to Stevens's work. Robert Buttel's *Wallace Stevens: The Making of Harmonium* (1967) and Walton A. Litz's *Introspective Voyager: The Poetic Development of Wallace Stevens* (1972) trace the maturation of Stevens's thought and style through his early and middle years. Frank Doggett's *Stevens' Poetry of Thought* (1966), Michel Benamou's *Wallace Stevens and the Symbolist Imagination* (1972), and Harold Bloom's *Wallace Stevens: The Poems of Our Climate* (1977) consider some intellectual influences on Stevens. John J. Enck's *Wallace Stevens: Images and Judgments* (1964), Joseph N. Riddle's *The Clairvoyant Eye: The Poetry and Poetics of Wallace Stevens* (1965), Ronald Sukenick's *Wallace Stevens: Musing the Obscure* (1967), and Merle E. Brown's *Wallace Stevens: The Poem as Act* (1971) offer readings of many individual poems. Other useful full-length studies include Daniel Fuchs's *The Comic Spirit of Wallace Stevens* (1963); Eugene Paul Nassar's *Wallace Stevens: An Anatomy of Figuration* (1965); James Baird's *The Dome and the Rock: Structure in the Poetry of Wallace Stevens* (1968); and Helen H. Vendler's *On Extended Wings: Wallace Stevens' Longer Poems* (1969).

Eudora Welty

There is no collected edition of Eudora Welty's fiction. In addition to the volumes mentioned in the author's introduction, she has written *Delta Wedding* (1946) and *The Golden Apples* (1949) and one book for children, *The Shoe Bird* (1964).

The best introduction to her writing is Joseph A. Bryant, Jr.'s *Eudora Welty* (1968). Ruth M. Vande Kieft's *Eudora Welty* (1962) and Alfred Appel's *A Season of Dreams: The Fiction of Eudora Welty* (1965) are more thorough but less up to date. Valuable essays are Katherine Anne Porter's introduction to *Selected Stories of Eudora Welty* (1954); Robert Penn Warren's "Love and Separateness in Eudora Welty," in *Selected Essays* (1958); Louis D. Rubin's "The Golden Apples of the Sun," in *Writers of the Modern South* (1963); and Burnice Glenn's "Fantasy in the Fiction of Eudora Welty," in *A Southern Vanguard*, ed. Allen Tate (1947). Seymour L.

Gross compiled "Eudora Welty: A Bibliography of Criticism and Comment," *Secretary's News Sheet*, Bibliographical Society, University of Virginia, No. 45 (April, 1960).

Nathanael West

West's novels have been collected in *The Complete Works of Nathanael West*, ed. Alan Ross (1957). William White has compiled two bibliographies: "Nathanael West: A Bibliography," *Studies in Bibliography*, Vol. 11 (1958), pp. 207–224; and "Nathanael West: Further Bibliographical Notes," *Serif*, Vol. 2 (1965), pp. 28–31. The best study of West's work is Randall Reid's *The Fiction of Nathanael West: No Redeemer, No Promised Land* (1971). Other full-length studies are James F. Light's *Nathanael West: An Interpretive Study* (1961); Stanley E. Hyman's *Nathanael West* (1962); and Victor Comerchero's *Nathanael West: The Ironic Prophet* (1964). Jay Martin has written a biography, *Nathanael West: The Art of His Life* (1970), and has edited *Nathanael West: A Collection of Critical Essays* (1971).

William Carlos Williams

Williams's poems have been collected in three volumes: *Collected Earlier Poems of William Carlos Williams* (1951); *Collected Later Poems of William Carlos Williams* (1950); and *Pictures from Brueghel* (1962). The five books of Williams's long poem *Paterson*, published individually between 1946 and 1958, were issued in one volume in 1963. Williams's short stories are collected in *Make Light of It* (1950) and again, with several additions, in *The Farmers' Daughters* (1961). His novels are *A Voyage to Pagany* (1928); *White Mule* (1937); *In the Money* (1940); and *The Build Up* (1946). His dramatic pieces were published together in *Many Loves and Other Plays* (1961). His most important essays are contained in *In the American Grain* (1925; reissued, 1940); *Selected Essays of William Carlos Williams* (1954); and *Imaginations*, ed. Webster Schott (1970). Williams's prose also includes *The Autobiography of William Carlos Williams* (1951); a book of recollections dictated to Edith Heal, *I Wanted to Write a Poem* (1958); and *Yes, Mrs. Williams* (1959), a portrait of the poet's mother. J. C. Thirlwall edited *The Selected Letters of William Carlos Williams* (1957); Ron Loewinsohn edited *William Carlos Williams: The Embodiment of Knowledge* (1974); and Emily Mitchell Wallace compiled *A Bibliography of William Carlos Williams* (1968).

James E. Breslin's *William Carlos Williams: An American Artist* (1970) is the best overview of Williams's work. Other general introductions are Linda Wagner's *The Poems of William Carlos Williams* (1964); Alan Ostrom's *The Poetic World of William Carlos Williams* (1966); Thomas Whitaker's *William Carlos Williams* (1968); and James Guimond's *The Art of William Carlos Williams: A Discovery and Possession of America* (1968). Bram Dijkstra's *The Hieroglyphics of a New Speech: Cubism, Stieglitz, and the Early Poetry of William Carlos Williams* (1969) and Mike Weaver's *William Carlos Williams: The American Background* (1971) focus on the social, intellectual, and aesthetic influences on Williams and on the milieu in which he wrote. Joseph N. Riddel's *The Inverted Bell: Modernism and the Counter-Poetics of William Carlos Williams* (1974) considers the problematic of language for Williams the theorist and experimenter. Reed Whittemore's *William Carlos Williams: Poet from Jersey* (1975) is a critical biography.

Linda Wagner, *The Prose of William Carlos Williams* (1970), and Robert Coles, *William Carlos Williams: The Knack of Survival in America* (1975), discuss Williams's prose. Other full-length studies are W. S. Peterson's *An Approach to Paterson* (1967); Joel Conarroe's *William Carlos Williams' Paterson: Language and Landscape* (1970); Benjamin Sankey's *A Companion to William Carlos Williams's Paterson* (1971); Sherman Paul's *The Music of Survival: A Biography of a Poem by William Carlos Williams* (1968); and Jerome Mazzaro's *William Carlos Williams: The Later Poems* (1973). J. Hillis Miller's *Poets of Reality* (1966) and Hugh Kenner's *A Homemade World: The American Modernist Writers* (1975) contain valuable chapters on Williams. Paul Mariani, *William Carlos Williams: The Poet and His Critics* (1975), considers Williams's critical reception.

Richard Wright

No standard edition of Wright's works exists. His fiction includes *Uncle Tom's Children* (1938); *Native Son* (1940); *The Outsider* (1953); *Savage Holiday* (1954); *The Long Dream* (1958); *Eight Men* (1961); and *Lawd Today* (written in 1936; published in 1963). His full-length works of nonfiction are *Twelve Million Black Voices* (1941); *Black Boy* (1945); *Black Power* (1954); *The Color Curtain* (1956); *Pagan Spain* (1957); and *White Man, Listen!* (1957). Among his important essays are "The Ethics of Living Jim Crow," in *American Stuff: A WPA Writers' Anthology* (1937), reprinted in the Harper Perennial paperback edition of *Uncle Tom's Children* (1965); "Blueprint for Negro Literature," in *New Challenge* (1937), reprinted in *Armistad 2*, ed. John A. Williams and Charles F. Harris (1971); *How Bigger Was Born*, a pamphlet (1940), reprinted in the Harper Perennial paperback edition of *Native Son* (1966); and the chapter "Black Boy," Wright's account of his membership in the Communist party, in Richard Crossman's *The God That Failed* (1949). Full-length critical and biographical studies are Dan McCall's *The Example of Richard Wright* (1969); Edward Margolies's *The Art of Richard Wright* (1969); Michele Febre's *The Unfinished Quest of Richard Wright*, trans. Isabel Benson (1973); Constance Webb's *Richard Wright: A Biography* (1968). Important essays on Wright include Irving Howe's "Black Boys and Native Sons," *Dissent* (1963), reprinted in *A World More Attractive* (1963); Ralph Ellison's reply to Howe in "The World and the Jug," *New Leader* (1963), reprinted in *Shadow and Act* (1964); James Baldwin's "Everybody's Protest Novel" and "Many Thousands Gone," *Partisan Review* (1949 and 1951), reprinted in *Notes of a Native Son* (1955), and "Alas, Poor Richard," in *Nobody Knows My Name* (1961); and Eldridge Cleaver's "Notes on a Native Son," in *Soul on Ice* (1968). *American Hunger* (1977) is an autobiographical fragment, originally intended to be part of *Black Boy*, which covers Wright's experience in the Communist party.

CONTEMPORARY AMERICAN PROSE 1945–

There are no indispensable surveys or histories of the period, and the most useful discussions of books and writers are found in collections of essays by several critics. Warner Berthoff's *Fictions and Events* (1971) has an excellent study of Edmund Wilson as well as an essay on books by Mailer and Malcolm X. *A Literature without Qualities: American Writing since 1945* (1979), also by Warner Berthoff, is a brilliantly succinct consideration of, among other things, postwar novelists. Richard Poirier's *The Performing Self* (1971) has material on Barth, Pynchon, and others of recent reputation. In *Bright Book of Life* (1973), Alfred Kazin discusses American novelists and storytellers from Hemingway to Mailer. Tony Tanner's *City of Words: American Fiction 1935–1970* (1971) provides a full treatment of many figures from this period, while Frank MacConnell's *Four Postwar American Novelists* (1977) contains analyses of the literary careers of Bellow, Mailer, Barth, and Pynchon. Irving Howe's essay on Ralph Ellison and on "The New York Intellectuals" may be found in his *Decline of the New* (1971). Leslie Fiedler's *Collected Essays* (1971) and Norman Podhoretz's *Doings and Undoings* (1964) include essays on and reviews of most of the writers anthologized here. Arthur Mizener's *The Sense of Life in the Modern Novel* (1964) is useful on Salinger and Updike. Roger Sale's *On Not Being Good Enough* (1979) has lively writing about many recent novelists, some of them not very well known. *Black Fiction* (1974), by Roger Rosenblatt, is thoughtful and incisive. Morris Dickstein's *Gates of Eden: American Culture in the Sixties* (1977) is a wide-ranging, personal view of the period. Finally, the 26 issues of *The New American Review*, edited by Theodore Solotaroff, is an excellent place to encounter much interesting fiction and other prose published between 1968 and 1978.

James Baldwin

Baldwin has written six novels: *Go Tell It on the Mountain* (1953); *Giovanni's Room* (1956); *Another Country* (1962); *Tell Me How Long the Train's Been Gone* (1968); *If Beale Street Could Talk* (1974); and *Just Above My Head* (1979). A collection of short stories, *Going to Meet the Man*, was published in 1965.

In addition to *The Fire Next Time* (1963), Baldwin has published two collections of essays: *Notes of a Native Son* (1955) and *Nobody Knows My Name* (1961). *No Name in the Street* (1972) is a later and grimmer look at the racial situation. *The Devil Finds Work* (1976) is a book about the movies.

The Furious Passage of James Baldwin by Fern Eckman (1966) contains biographical information. Louis H. Pratt's *James Baldwin* (1978) is a more recent study. Kenneth Kinnamon has edited the Prentice-Hall collection of critical essays on Baldwin (1974). *Squaring Off: Mailer vs. Baldwin* by W. J. Weatherby (1976) is a lively account of their relationship.

John Barth

Barth is the author of five novels, the first three of which were altered when republished: *The Floating Opera* (1956, 1967); *The End of the Road* (1958, 1967); *The Sot-Weed Factor* (1960, 1967); *Giles Goat-Boy; or, The Revised New Syllabus* (1966); and *Letters* (1979). *Lost in the Funhouse: Fiction for Print, Tape, Live Voice* (1968) is a collection of stories. *Chimera* (1972) consists of three novellas.

John Barth, by Gerhard Joseph (1970) is a biographical and critical pamphlet in the University of Minnesota series. *John Barth: An Introduction* by David Morell (1976) is an intelligent survey of Barth's work, containing much lively information and manuscript material plus an excellent bibliography. The first two chapters of Richard Poirier's *The Performing Self* (1971) are also relevant.

Saul Bellow

Among Bellow's works are the following novels: *Dangling Man* (1944); *The Victim* (1947); *The Adventures of Augie March* (1953); *Henderson the Rain King* (1959); *Herzog* (1964); *Mr. Sammler's Planet* (1970); *Humboldt's Gift* (1975). His play, *The Last Analysis*, produced in New York in 1964, was published in 1965. *To Jerusalem and Back* (1976) is an account of a visit to Israel. Also of interest is a lecture, "Some Notes on Recent American Fiction," *Encounter* (November, 1963).

A useful collection of critical essays on Bellow's work which also includes his essay "Where Do We Go from Here: The Future of Fiction" is *Saul Bellow and the Critics*, ed. Irving Malin (1967). Books about him include *Saul Bellow* by Robert Dutton (1971), which takes him up through *Mr. Sammler's Planet* and provides a selective bibliography of further studies. *Saul Bellow: A Comprehensive Bibliography* has been compiled by B. A. Sokoloff and Mark E. Posner (1973).

Ralph Ellison

Ellison's two books are *Invisible Man* (1952) and *Shadow and Act* (1964). There are also a number of uncollected short stories and excerpts from a novel in progress. (See Bernard Benoit and Michel Fabre, "A Bibliography of Ralph Ellison's Published Writings," in *Studies in Black Literature* [1971].)

The Prentice-Hall collection of critical essays (1974), edited by John Hersey, has a number of useful items, including interviews with Ellison by Hersey and James McPherson and a further selected bibliography. *Studies in Invisible Man*, ed. Ronald Gottesman (1971), is a shorter collection. Irving Howe's essay contrasting Ellison with Richard Wright appears in *A World More Attractive* (1963).

Norman Mailer

Mailer's novels are *The Naked and the Dead* (1948); *Barbary Shore* (1951); *The Deer Park* (1955); *An American Dream* (1965); and *Why Are We in Vietnam?* (1967). Among collections of essays and miscellaneous materials are *Advertisements for Myself* (1959); *The Presidential Papers* (1963); and *Cannibals and Christians* (1966). *The Armies of the Night: History as a Novel, the Novel as History,* was published in 1968. His writings about political conventions are collected in *Some Honorable Men* (1975). Some further titles are *Of a Fire on the Moon* (1971); *The Prisoner of Sex* (1972); *Marilyn* (1973); and *The Executioner's Song* (1979). His most ambitious film is *Maidstone* (1968), for which he later wrote a preface (in *Existential Essays* [1972]).

The best study is Richard Poirier's *Norman Mailer* in the Modern Masters series (1973). Also good is *Down Mailer's Way* (1974) by Robert Solotaroff. *Squaring Off* (1976) by W. J. Weatherby has some interesting personal glimpses.

Vladimir Nabokov

Among Nabokov's many novels are *The Defense* (1930; translated into English, 1964); *Laughter in the Dark* (1933; translated into English, 1961); *The Real Life of Sebastian Knight* (1941); *Lolita* (1955, 1958); *Pnin* (1975); *Pale Fire* (1962); *Ada* (1969); *Look at the Harlequins!* (1974). He published a number of volumes of short stories, of which the most recent was *Tyrants Destroyed and Other Stories* (1975). Among his other works are *Poems* (1959); *Speak Memory: An Autobiography Revisited* (1966, originally published 1951); and *Strong Opinions,* essays and interviews (1973). He wrote a critical book, *Nikolai Gogol* (1944), and edited and translated Pushkin's *Eugene Onegin* (1964).

Andrew Field has compiled *Nabokov: A Bibliography* (1974) and has written a partial biography, *Nabokov: His Life in Part* (1977). Critical studies include Page Stegner's *The Art of Vladimir Nabokov: Escape into Aesthetics* (1966); Andrew Field's *Nabokov: His Life in Art* (1967); *Nabokov: The Man and His Work,* ed. L. S. Dembo (1967); and *For Vladimir Nabokov on His Seventieth Birthday,* ed. Charles Newman and Alfred Appel, Jr. (1970). Julian Moynahan's *Vladimir Nabokov* (1971) is a short introduction. *The Viking Portable Nabokov,* ed. Page Stegner (1968, 1971), is a good selection from the work. Martin Green's *Yeats's Blessings on Von Hügel* (1967) has a fine essay on *Lolita*.

Flannery O'Connor

Flannery O'Connor's novels are *Wise Blood* (1952) and *The Violent Bear It Away* (1960). Two collections of stories, *A Good Man Is Hard to Find* (1955) and *Everything That Rises Must Converge* (1965), are included—plus some earlier previously uncollected stories—in *Complete Stories* (1971). Her letters are collected in *The Habit of Being: The Letters of Flannery O'Connor* (1979).

A collection of her essays and other occasional prose is *Mystery and Manners,* ed. Sally and Robert Fitzgerald (1969). There are a number of books about her fiction, the most interesting of which is probably *The Added Dimension: The Art and Mind of Flannery O'Connor,* ed. M. J. Freedman and L. A. Lawson (1966). Robert Fitzgerald's memoir, prefaced to *Everything That Rises Must Converge,* is an excellent portrait. Stanley Edgar Hyman's pamphlet in the University of Minnesota series (1966) is a good short introduction.

John Updike

Updike's novels are *The Poorhouse Fair* (1958); *Rabbit, Run* (1961); *The Centaur* (1963); *Of the Farm* (1965); *Couples* (1968); *Rabbit Redux* (1971); *A Month of Sundays* (1975); *Marry Me* (1976); and *Coup* (1978). Collections of short fiction are *The Same Door* (1959); *Pigeon Feathers* (1962); *The Music School* (1966); *Bech: A Book* (1970); *Museums and Women* (1973); and *Problems* (1979). *Midpoint* (1969) and *Tossing and Turning* (1977) are collections of poems. His occasional articles, interviews, and reviews are found in *Assorted Prose* (1965) and *Picked-Up Pieces* (1975).

Charles Thomas Samuels's pamphlet in the University of Minnesota series (1969) is usefully critical. Alfred Kazin's discussion of Updike in *Bright Book of Life* (1973) is also of interest.

CONTEMPORARY AMERICAN POETRY 1945–

Randall Jarrell's *Poetry and the Age* (1953) is still the most engaging general introduction for the student of contemporary American poetry. Valuable studies of the period and of particular poets are to be found in Harold Bloom's *The Ringers in the Tower* (1971) and *Figures of Capable Imagination* (1976); *The Survival of Poetry,* ed. Martin Dodsworth (1970); Richard Howard's *Alone with America* (1969); David Kalstone's *Five Temperaments* (1977); Howard Nemerov's *Figures of Thought* (1978); and Robert Pinsky's *The Situation of Poetry* (1976). Useful interview material is to be found in Anthony Ostroff's *The Contemporary Poet as Artist and Critic: Eight Symposia* (1964); *Preferences,* ed. Richard Howard and Thomas Victor (1974); and the various volumes of *Writers at Work: The Paris Review Interviews.*

A. R. Ammons

Ammons's *Collected Poems: 1951–1971* (1972) were followed by several shorter volumes: *Sphere: The Form of a Motion* (1974); *Diversifications* (1975); and *The Snow Poems* (1977). There are useful critical essays on his work in Richard Howard, *Alone with America* (1969) and in Harold Bloom, *The Ringers in the Tower* (1971). An interview with Ammons appeared in a special issue of *Diacritics,* Vol. III, No. 5 (Winter, 1973), devoted to studies of his work.

John Ashbery

John Ashbery's volumes of poetry include *Some Trees* (1956); *The Tennis Court Oath* (1962); *Rivers and Mountains* (1966); *The Double Dream of Spring* (1970); *Three Poems* (1972); *Self-Portrait in a Convex Mirror* (1975); and *Houseboat Days* (1977). With James Schuyler he is co-author of the comic novel *A Nest of Ninnies* (1969). Among useful critical essays are the chapters on Ashbery in Richard Howard's *Alone with America* (1969) and David Kalstone's *Five Temperaments* (1977), and a study by Harold Bloom in *Contemporary Poetry in America: Essays and Introductions*, ed. Robert Boyers (1973). David K. Kermani's *John Ashbery: A Comprehensive Bibliography* (1975) is indispensable and amusing.

Imamu Amiri Baraka (LeRoi Jones)

Baraka's volumes of poetry include *Preface to a Twenty Volume Suicide Note* (1962); *The Dead Lecturer* (1965); *Black Art* (1966); *Black Magic: Poetry 1961–1967* (1969). He is the author of several important plays including *Dutchman* and *Slave* (1964), *The Baptism* and *The Toilet* (1967) and *Four Black Revolutionary Plays* (1969). His prose writings include *Home: Social Essays* (1966) and *Raise Race Rays Raze* (1972).

John Berryman

Homage to Mistress Bradstreet (1956) and a selection of Berryman's *Short Poems* (1948) were issued together in paperback format in 1968. His other volumes of poetry include *77 Dream Songs* (1964); *Berryman's Sonnets* (1967); *His Toy, His Dream, His Rest* (1968); *Love and Fame* (1970; rev. ed., 1972); *Delusions, Etc.* (1972); and a posthumous volume, *Henry's Fate* (1977). Berryman's critical biography, *Stephen Crane*, appeared in 1950, and a collection of his short fiction and literary essays was issued under the title *The Freedom of the Poet* (1976). Berryman's unfinished novel about his alcoholism, *Recovery*, appeared in 1973. A valuable critical introduction to Berryman's poetry is to be found in Martin Dodsworth's essay on him in *The Survival of Poetry*, ed. Martin Dodsworth (1970). Joel Conarroe's *John Berryman* (1977) is a book-length biographical and critical study.

Elizabeth Bishop

The Complete Poems of Elizabeth Bishop was published in 1969. Since then she wrote another book of poems, *Geography III* (1976). The moving and important short story "In the Village" was included in her *Questions of Travel* (1965), and another story, "In Prison," has been reprinted in *The Poet's Story*, ed. Howard Moss (1973). Bishop translated from the Portuguese *The Diary of Helena Morley* (1957, 1977), an enchanting memoir of provincial life in Brazil. Anne Stevenson's biographical and critical study *Elizabeth Bishop* appeared in 1966. Useful critical essays on Bishop's poetry appear in the issue of *World Literature Today* (1976) devoted to her and in David Kalstone's *Five Temperaments* (1977).

Allen Ginsberg

Ginsberg's poems have, with the exception of *Empty Mirror* (early poems collected in 1961), been issued in the now unmistakable City Lights paperbacks. They include *Howl* (1956); *Kaddish and Other Poems, 1958–1960* (1961); *Reality Sandwiches* (1963); *Planet News, 1961–1967* (1968); *The Fall of America, Poems of These States, 1965–1971* (1973); and *Mind Breaths, Poems 1972–1977* (1977). Ginsberg has published a great number of pages from his journals, dealing with his travels, such as the *Indian Journals* (1970). *Allen Verbatim* (1974) includes transcripts of some of his "lectures on poetry." In 1977 he published *Letters Early Fifties Early Sixties* (1977). Among the many interviews, the Paris Review dialogue conducted by Tom Clarkin, *Writers at Work: Third Series* (1967), is extremely valuable. Jane Kramer's *Allen Ginsberg in America* (1969) is a brilliant documentary piece on Ginsberg in the 1960s.

Randall Jarrell

Randall Jarrell's *Complete Poems* were published in 1969. His novel satirizing American academic life, *Pictures from an Institution*, appeared in 1954. Jarrell's critical essays have been collected in *Poetry and the Age* (1953), *A Sad Heart at the Supermarket* (1962), and *The Third Book of Criticism* (1969). Some of the most valuable commentary on Jarrell's life and work are to be found in a memorial volume of essays edited by Robert Lowell, Peter Taylor, and Robert Penn Warren, *Randall Jarrell, 1914–1965* (published in 1967). Suzanne Ferguson's *The Poetry of Randall Jarrell* is a useful book-length study.

Robert Lowell

In 1976, the year before his death, Lowell prepared his *Selected Poems*, which drew upon all his volumes of poetry except *Day by Day* which appeared in 1977. The interested reader will still have to look back to individual volumes since Lowell frequently revised his poems and could only print a small proportion of his work in *Selected Poems*. His earlier books of verse include *Lord Weary's Castle* (1946); *The Mills of the Kavanaughs* (1951); *Life Studies* (1959); *Imitations* (1961); *For the Union Dead* (1964); *Near the Ocean* (1967); *Notebook* (1969; rev. and expanded, 1970); *The Dolphin* (1973); *For Lizzie and Harriet* (1973); and *History* (1973). For the stage Lowell prepared adaptations of Racine's *Phaedra* (1961) and Aeschylus's *Prometheus Bound* (1969) as well as versions of Hawthorne and Melville stories grouped under the title *The Old Glory* (1965). Among the many useful critical works on Lowell are Philip Cooper's *The Autobiographical Myth of Robert Lowell* (1970); Jerome Mazzaro's *The Poetic Themes of Robert Lowell* (1965); Hugh B. Staples's *Robert Lowell, the First Twenty Years* (1962); Stephen Yenser's *Circle to Circle* (1975); and Steven Axelrod's *Robert Lowell: Life and Art* (1978). *Robert Lowell: A Collection of Critical Essays*, ed. Thomas Parkinson (1968) includes an important interview with the poet. A partial bibliography of Lowell's work can be found in Jerome Mazzaro, *The Achievement of Robert Lowell, 1939–1959* (1960).

James Merrill

Merrill has published the following volumes of poetry: *First Poems* (1951); *The Country of a Thousand Years of Peace* (1959; rev. ed., 1970); *Water Street* (1962); *Nights and Days* (1966); *The Fire Screen* (1969); *Braving the Elements* (1972); *Divine Comedies* (1976); and *Mirabell: Books of Number* (1978). He has also written two novels: *The Seraglio* (1957) and *The (Diblos) Notebook* (1965). Among useful critical essays are the chapters on Merrill in Richard Howard's *Alone with America* (1969) and in David Kalstone's *Five Temperaments* (1977) as well as "James Merrill's Oedipal Fire," an article by Richard Saez in *Parnassus: Poetry in Review* (Fall–Winter, 1974), pp. 159–184.

Frank O'Hara

The Collected Poems of Frank O'Hara was published in 1971. It has since been supplemented by two volumes: *Early Poems* (1977) and *Poems Retrieved* (1977). O'Hara's critical articles and other prose can be found in *Art Chronicles 1954–1966* (1975) and *Standing Still and Walking in New York* (1975). A volume of anecdotes and reminiscences, *Homage to Frank O'Hara*, ed. Bill Berkson and Joe LeSueur, appeared in 1978. An informative critical study is Marjorie Perloff's *Frank O'Hara: Poet among Painters* (1977).

Sylvia Plath

Plath's books of poetry include *The Colossus and Other Poems* (1962); *Ariel* (1966); *Crossing the Water* (1971); and *Winter Trees* (1972). Her novel *The Bell Jar* was first published in England in 1963. Plath's mother Aurelia Schober Plath selected and edited the useful *Letters Home: Correspondence 1950–63* (1975). A number of informative essays appeared in *The Art of Sylvia Plath*, ed. Charles Newman (1970). Among the book-length studies, *Chapters in a Mythology* by Judith Kroll (1976) is especially helpful.

Adrienne Rich

A convenient starting point is Rich's *Poems Selected and New, 1950–1974*, which draws poems from the poet's earlier books: *A Change of World* (1951); *The Diamond Cutters* (1955); *Snapshots of a Daughter-in-Law* (1963); *Necessities of Life* (1966); *Leaflets* (1969); *The Will to Change* (1971); and *Diving into the Wreck* (1973). Since then, Rich has published a further book of poems, *The Dream of a Common Language* (1978). There are valuable critical essays and an interview in *Adrienne Rich's Poetry*, ed. Barbara Charlesworth Gelpi and Albert Gelpi (1975). Rich is also the author of an important study, *Of Woman Born: Motherhood as Experience and Institution* (1976). She has collected her essays, lectures, and speeches in *On Lies, Secrets, and Silence: Selected Prose 1966–1978* (1979).

Theodore Roethke

The Collected Poems of Theodore Roethke was published in 1966. Useful comments on poetic tradition and his own poetic practice are to be found in several collections of Roethke's prose: *On the Poet and His Craft: Selected Prose of Theodore Roethke*, ed. Ralph J. Mills, Jr. (1965); *Straw for the Fire: From the Notebooks of Theodore Roethke, 1948–63*, ed. David Wagoner (1972); and *The Selected Letters of Theodore Roethke*, ed. Ralph J. Mills, Jr. (1970). Among the useful studies of Roethke's work are *Theodore Roethke: An Introduction to the Poetry* by Karl Malkoff (1966); Rosemary Sullivan's *Theodore Roethke: The Garden Master* (1975); and *Theodore Roethke: Essays on the Poetry*, ed. Arnold Stein (1965). Allan Seager's *The Glass House* (1968) contains useful biographical material.

Anne Sexton

Sexton's books of poems include *To Bedlam and Part Way Back* (1960); *All My Pretty Ones* (1962); *Live or Die* (1966); *Love Poems* (1969); *Transformations* (1971); *The Book of Folly* (1973); *The Death Notebooks* (1974); and *The Awful Rowing toward God* (1975). A *Self-Portrait in Letters* (1977) gives useful biographical material. Important interviews and critical essays are to be found in J. D. McClatchy's *Anne Sexton: The Artist and Her Critics* (1978).

Gary Snyder

Snyder's poems have been published by several different presses, often with duplication of poems. *Riprap*, originally published in 1959, is most easily obtained in *Riprap, and Cold Mountain Poems* (1965). *Myths and Texts* first appeared in 1960. Other volumes include *The Back Country* (1968); *Regarding Wave* (1970); *Six Sections from Mountains and Rivers without End plus One* (1970); and *Turtle Island* (1974). Snyder's book of prose dealing with ecology, *Earth House Hold*, was published in 1969.

Richard Wilbur

The most convenient single volume is *The Poems of Richard Wilbur* (1963), which draws from *The Beautiful Changes* (1947), *Ceremony and Other Poems* (1950), *Things of This World* (1956), and *Advice to a Prophet* (1961). Since then, he has published *Walking to Sleep* (1969) and *The Mind-Reader* (1976). His translations from Molière include *The Misanthrope* (1955) and *Tartuffe* (1963). A selection of prose is available in *Responses: Prose Pieces, 1948–1976* (1976).

Index